Volkswagen Passat

Service Manual

1998, 1999, 2000, 2001, 2002, 2003, 2004, 2005
1.8L turbo, 2.8L V6, 4.0L W8
including wagon and 4MOTION™

Volkswagen platform code: B5

[B]

Bentley Publishers
Cambridge, Massachusetts

B BENTLEY PUBLISHERS | Automotive Books & Manuals

Bentley Publishers, a division of Robert Bentley, Inc.
1734 Massachusetts Avenue
Cambridge, MA 02138 USA
800-423-4595 / 617-547-4170

Information that makes the difference®

www.BentleyPublishers.com

Technical contact information

We welcome your feedback. Please submit corrections and additions to our Volkswagen technical discussion forum at:

http://www.BentleyPublishers.com

Errata information

We will evaluate submissions and post appropriate editorial changes online as text errata or tech discussion. Appropriate errata will be incorporated with the book text in future printings. Read errata information for this book before beginning work on your vehicle. See the following web address for additional information:

http://www.BentleyPublishers.com/errata/

WARNING—Important Safety Notice

Do not use this manual unless you are familiar with basic automotive repair procedures and safe workshop practices. This manual illustrates the workshop procedures required for most service work; it is not a substitute for full and up-to-date information from the vehicle manufacturer or for proper training as an automotive technician. Note that it is not possible for us to anticipate all of the ways or conditions under which vehicles may be serviced or to provide cautions as to all of the possible hazards that may result.

The vehicle manufacturer will continue to issue service information updates and parts retrofits after the editorial closing of this manual. Some of these updates and retrofits will apply to procedures and specifications in this manual. We regret that we cannot supply updates to purchasers of this manual.

We have endeavored to ensure the accuracy of the information in this manual. Please note, however, that considering the vast quantity and the complexity of the service information involved, we cannot warrant the accuracy or completeness of the information contained in this manual.

FOR THESE REASONS, NEITHER THE PUBLISHER NOR THE AUTHOR MAKES ANY WARRANTIES, EXPRESS OR IMPLIED, THAT THE INFORMATION IN THIS BOOK IS FREE OF ERRORS OR OMISSIONS AND WE EXPRESSLY DISCLAIM THE IMPLIED WARRANTIES OF MERCHANTABILITY AND OF FITNESS FOR A PARTICULAR PURPOSE, EVEN IF THE PUBLISHER OR AUTHOR HAVE BEEN ADVISED OF A PARTICULAR PURPOSE, AND EVEN IF A PARTICULAR PURPOSE IS INDICATED IN THE MANUAL. THE PUBLISHER AND AUTHOR ALSO DISCLAIM ALL LIABILITY FOR DIRECT, INDIRECT, INCIDENTAL OR CONSEQUENTIAL DAMAGES THAT RESULT FROM ANY USE OF THE EXAMPLES, INSTRUCTIONS OR OTHER INFORMATION IN THIS BOOK. IN NO EVENT SHALL OUR LIABILITY WHETHER IN TORT, CONTRACT OR OTHERWISE EXCEED THE COST OF THIS MANUAL.

Your common sense and good judgment are crucial to safe and successful service work. Read procedures through before starting them. Think about whether the condition of your car, your level of mechanical skill, or your level of reading comprehension might result in or contribute in some way to an occurrence which might cause you injury, damage your car, or result in an unsafe repair. If you have doubts for these or other reasons about your ability to perform safe repair work on your car, have the work done at an authorized Volkswagen dealer or other qualified shop.

Part numbers listed in this manual are for identification purposes only, not for ordering. Always check with your authorized Volkswagen dealer to verify part numbers and availability before beginning service work that may require new parts.

Before attempting any work on your Volkswagen, read the warnings and cautions on page vi, and any warning or caution that accompanies a procedure in the service manual. Review the warnings and cautions on page vi each time you prepare to work on your Volkswagen.

Special tools required to perform certain service operations are identified in the manual and are recommended for use. Use of tools other than those recommended in this service manual may be detrimental to the car's safe operation as well as the safety of the person servicing the car.

Copies of this manual may be purchased from most automotive accessories and parts dealers specializing in Volkswagen automobiles, from selected booksellers, or directly from the publisher.

This manual is published by Robert Bentley, Inc., Publishers, Cambridge, Massachusetts. Volkswagen has not reviewed and does not vouch for the accuracy of the technical specifications and procedures described in this manual.

Library of Congress Cataloging-in-Publication Data

Volkswagen Passat service manual : 1998, 1999, 2000, 2001, 2002, 2003, 2004, 2005 1.8L turbo, 2.8L V6, 4.0L W8, including Wagon and 4MOTION.
p. cm.
"Volkswagen platform code: B5."
Includes index.
ISBN-13: 978-0-8376-1483-0 (alk. paper)
ISBN-10: 0-8376-1483-X (alk. paper)
1. Passat automobile--Maintenance and repair--Handbooks, manuals, etc. I. Robert Bentley, inc.

TL215.P37V65 2006

629.28'722--dc22

2006034008

VWoA, Inc. Part No. LPV 800 209
Bentley Stock No. VP05

Editorial closing 10/03

11 10 09 08 07 6 5 4 3 2 1

The paper used in this publication is acid free and meets the requirements of the National Standard for Information Sciences-Permanence of Paper for Printed Library Materials. ∞

Manufactured in the United States of America

Foreword and Disclaimer

For the Volkswagen owner with basic mechanical skills and for independent auto service professionals, this book includes many of the specifications and procedures that were available in an authorized Volkswagen dealer service department as this manual went to press. The Volkswagen owner with no intention of working on his or her car will find that owning and referring to this manual makes it possible to be better informed and to discuss repairs with a professional automotive technician more knowledgeably.

If you are a Volkswagen owner intending to do maintenance and repair work, make sure you have screwdrivers, a set of metric wrenches and sockets, and metric Allen and Torx wrenches, since these basic hand tools are needed for most of the work described in this manual. Many procedures also require a torque wrench to ensure that fasteners are tightened properly and in accordance with specifications. In some cases, the text refers to special tools that are recommended or required to accomplish adjustments or repairs. These tools are often identified by their Volkswagen special tool number and illustrated.

Disclaimer

We have endeavored to ensure the accuracy of the information in this manual. When the vast array of data presented in the manual is taken into account, however, no claim to infallibility can be made. We therefore cannot be responsible for the result of any errors that may have crept into the text. Please also read **WARNING–Important Safety Notice** on the copyright page at the beginning of this book.

Prior to starting a repair procedure, read the procedure, **Warnings and Cautions** on page vi and the warnings and cautions that accompany the procedure. Reading a procedure before beginning work helps you determine in advance the need for specific skills, identify hazards, prepare for appropriate capture and handling of hazardous materials, and the need for particular tools and replacement parts such as gaskets.

Bentley Publishers encourages comments from the readers of this manual with regard to errors, and suggestions for improvement of our product. These communications have been and will be carefully considered in the preparation of this and other manuals. If you identify inconsistencies in the manual, you may have found an error. Please contact the publisher and we will endeavor to post applicable corrections on our website. Review corrections (errata) that we have posted before beginning work. Please see the following web address:

`http://www.BentleyPublishers.com/errata/`

Volkswagen offers extensive warranties, especially on components of the fuel delivery and emission control systems. Therefore, before deciding to repair a Volkswagen that may be covered wholly or in part by any warranties issued by Volkswagen of America, Inc., consult your authorized Volkswagen dealer. You may find that the dealer can make the repair either free or at minimum cost. Regardless of its age, or whether it is under warranty, your Volkswagen is both an easy car to service and an easy car to get serviced. So if at any time a repair is needed that you feel is too difficult to do yourself, a trained Volkswagen technician is ready to do the job for you.

Bentley Publishers

Vehicle Identification Number (VIN), decoding

Some of the information in this manual applies only to cars of a particular model year or range of years. For example, 2002 refers to the 2002 model year but does not necessarily match the calendar year in which the car was manufactured or sold. To be sure of the model year of a particular car, check the Vehicle Identification Number (VIN) on the car.

The VIN is a unique sequence of 17 characters assigned by Volkswagen to identify each individual car. When decoded, the VIN tells the country and year of manufacture; make, model and serial number; assembly plant and even some equipment specifications.

The Volkswagen VIN is on a plate mounted on the top of the dashboard, on the driver's side where the number can be seen through the windshield. The 10th character is the model year code. The letters I, O, Q and U are not used for model year designation. Examples: V for 1997, W for 1998, X for 1999, Y for 2000, 1 for 2001, 2 for 2002, etc. The table below explains some of the codes in the VIN for 1998-2005 Volkswagen Passats covered by this manual.

Sample VIN: WVWBH63B91E150381

position: 1 2 3 4 5 6 7 8 9 10 11 12 - 17

VIN position	Description	Decoding information	
1	Manufacturing Country	W	Germany
2	Manufacturer	V	Volkswagen
3	Vehicle Type	W	Passenger vehicle
4	Series	B	Varies with model year
5	Engine code.	H	Varies with model year
6	Restraint system	6	Varies with model year
7-8	Model type	3B	Passat
9	Check digit		0 - 9 or X, calculated by NHTSA
10	Model year	W X Y 1 2 3 4 5	1998 1999 2000 2001 2002 2003 2004 2005
11	Assembly plant	E W M	Emden Wolfsburg Mosel
12-17	Serial number		Sequential production number for specific vehicle

Please read these warnings and cautions before proceeding with maintenance and repair work.

WARNINGS—
*See also **CAUTIONS** on next page.*

- Read the important safety notice on the copyright page at the beginning of the book.
- Some repairs may be beyond your capability. If you lack the skills, tools and equipment, or a suitable workplace for any procedure described in this manual, we suggest you leave such repairs to an authorized Volkswagen dealer service department or other qualified shop.
- Thoroughly read each procedure and the **WARNINGS** and **CAUTIONS** that accompany the procedure. Also review posted corrections at `www.BentleyPublishers.com/errata/` before beginning work.
- If any procedure, tightening torque, wear limit, specification or data presented in this manual does not appear to be appropriate for a specific application, contact the publisher or the vehicle manufacturer for clarification before using the information in question.
- Volkswagen is constantly improving its cars. Sometimes these changes, both in parts and specifications, are made applicable to earlier models. Therefore, before starting any major jobs or repairs to components on which passenger safety may depend, consult your authorized Volkswagen dealer about technical bulletins that may have been issued.
- Do not reuse any fasteners that are worn or deformed in normal use. Many fasteners are designed to be used only once and become unreliable and may fail when used a second time. This includes, but is not limited to, nuts, bolts, washers, self-locking nuts or bolts, circlips and cotter pins. Replace these fasteners with new parts.
- Do not work under a lifted car unless it is solidly supported on stands designed for the purpose. Do not support a car on cinder blocks, hollow tiles or other props that may crumble under continuous load. Do not work under a car that is supported solely by a jack. Do not work under the car while the engine is running.
- If you are going to work under a car on the ground, make sure that the ground is level. Block the wheels to keep the car from rolling. Disconnect the battery negative (–) terminal (ground strap) to prevent others from starting the car while you are under it.
- Do not run the engine unless the work area is well ventilated. Carbon monoxide kills.
- Remove finger rings, bracelets and other jewelry so that they cannot cause electrical shorts, get caught in running machinery or be crushed by heavy parts.
- Tie long hair behind your head. Do not wear a necktie, a scarf, loose clothing, or a necklace when you work near machine tools or running engines. If your hair, clothing, or jewelry were to get caught in the machinery, severe injury could result.
- Do not attempt to work on your car if you do not feel well. You increase the danger of injury to yourself and others if you are tired, upset or have taken medication or any other substance that may keep you from being fully alert.
- Illuminate your work area adequately but safely. Use a portable safety light for working inside or under the car. Make sure the bulb is enclosed by a wire cage. The hot filament of an accidentally broken bulb can ignite spilled fuel, vapors or oil.
- Catch draining fuel, oil or brake fluid in suitable containers. Do not use food or beverage containers that might mislead someone into drinking from them. Store flammable fluids away from fire hazards. Wipe up spills at once but do not store the oily rags, which can ignite and burn spontaneously.
- Observe good workshop practices. Wear goggles when you operate machine tools or work with battery acid. Wear gloves or other protective clothing whenever the job requires working with harmful substances.
- Greases, lubricants and other automotive chemicals contain toxic substances, many of which are absorbed directly through the skin. Read the manufacturer's instructions and warnings carefully. Use hand and eye protection. Avoid direct skin contact.
- Disconnect the battery negative (–) terminal (ground strap) whenever you work on the fuel system or the electrical system. Do not smoke or work near heaters or other fire hazards. Keep an approved fire extinguisher handy.
- Friction components (such as brake pads or shoes or clutch discs) contain asbestos fibers or other friction materials. Do not create dust by grinding, sanding, or by cleaning with compressed air. Avoid breathing dust. Breathing any friction material dust can lead to serious diseases and may result in death.
- Batteries give off explosive hydrogen gas during charging. Keep sparks, lighted matches and open flame away from the top of the battery. If hydrogen gas escaping from the cap vents is ignited, it will ignite gas trapped in the cells and cause the battery to explode.
- Battery acid (electrolyte) can cause severe burns. Flush contact area with water and seek medical attention.
- Do not quick-charge the battery (for boost starting) for longer than one minute. Wait at least one minute before boosting the battery a second time.
- Connect and disconnect a battery charger only with the battery charger switched OFF.
- Connect and disconnect battery cables, jumper cables or a battery charger only with the ignition switched OFF. Do not disconnect the battery while the engine is running.
- Do not allow the battery charging voltage to exceed 16.5 volts. If the battery begins producing gas or boiling violently, reduce the charging rate. Boosting a sulfated battery at a high rate can cause an explosion.

continued on next page

Please read these warnings and cautions before proceeding with maintenance and repair work.

WARNINGS— (continued)

- The air-conditioning system is filled with a hazardous refrigerant. Make sure the system is serviced only by a trained technician using approved refrigerant recovery/recycling equipment, trained in related safety precautions, and familiar with regulations governing the discharging and disposal of automotive chemical refrigerants.
- Do not expose any part of the air-conditioning system to high temperatures such as open flame. Excessive heat increases system pressure and may cause the system to burst.
- Some aerosol tire inflators are highly flammable. Be extremely cautious when repairing a tire that may have been inflated using an aerosol tire inflator. Keep sparks, open flame or other sources of ignition away from the tire repair area. Inflate and deflate the tire at least four times before breaking the bead from the rim. Completely remove the tire from the rim before attempting any repair.
- Cars covered by this manual are equipped with a supplemental restraint system, that automatically deploys airbags and pyrotechnic seat belt reels in case of a frontal or side impact. These are explosive devices. Handled improperly or without adequate safeguards, they can be accidently activated and cause serious injury.
- The ignition system produces high voltages that can be fatal. Avoid contact with exposed terminals and use extreme care when working on a car with the engine running or the ignition switched ON.
- Place jack stands only at locations specified by the manufacturer. The vehicle lifting jack supplied with the vehicle is intended for tire changes only. Use a heavy duty floor jack to lift the vehicle before installing jack stands. See **03 Maintenance**.
- Aerosol cleaners and solvents may contain hazardous or deadly vapors and are highly flammable. Use only in a well ventilated area. Do not use on hot surfaces (engines, brakes, etc.).
- Due to risk of personal injury, be sure the engine is cold before beginning work on the cooling system.
- Use extreme care when draining and disposing of engine coolant. Coolant is poisonous and lethal to humans and pets. Pets are attracted to coolant because of its sweet smell and taste. Seek medical attention immediately if coolant is ingested.

CAUTIONS—
See also ***WARNINGS*** *on previous page.*

- If you lack the skills, tools and equipment, or a suitable workshop for any procedure described in this manual, we suggest you leave such repairs to an authorized Volkswagen dealer or other qualified shop.
- Volkswagen is constantly improving its cars and sometimes these changes, both in parts and specifications, are made applicable to earlier models. Therefore, part numbers listed in this manual are for reference only. Always check with your authorized Volkswagen dealer parts department for the latest information.
- Before starting a job, make certain that you have the necessary tools and parts on hand. Read the instructions thoroughly and do not attempt shortcuts. Use tools appropriate to the work and use replacement parts meeting Volkswagen specifications. Makeshift tools, parts and procedures do not make good repairs
- Do not use pneumatic and electric tools to tighten fasteners, especially on light alloy parts. Use a torque wrench to tighten fasteners to the tightening torque specification listed.
- Be mindful of the environment. Before you drain the crankcase, find out the proper way to dispose of the oil. Do not pour oil onto the ground, down a drain, or into a stream, pond or lake. Dispose of in accordance with Federal, State and Local laws.
- Do not expose electronic control modules to temperatures from a paint-drying booth or a heat lamp in excess of 203°F (95°C). Do not expose control modules to temperatures in excess of 185°F (85°C) for more than two hours.
- Before doing any electrical welding on cars equipped with ABS, disconnect the battery negative (–) terminal (ground strap) and the ABS control module connector.
- Make sure ignition is OFF before disconnecting the battery.
- Label battery cables before disconnecting. On some models, battery cables are not color coded.
- Disconnecting the battery may erase fault code(s) stored in control module memory. Use Volkswagen diagnostic equipment or equivalent to check for fault codes prior to disconnecting the battery cables. If the Check Engine light (malfunction indicator light or MIL) is illuminated or any other system faults are detected (indicated by an illuminated warning light), see an authorized Volkswagen dealer.
- If a normal or rapid charger is used to charge battery, disconnect the battery and remove from the vehicle in order to avoid damaging paint and upholstery.
- Do not quick-charge the battery (for boost starting) for longer than one minute. Wait at least one minute before boosting the battery a second time.
- Slow-charge a sealed or “maintenance free” battery at an amperage rate that is approximately 10% of the battery’s ampere-hour (Ah) rating.

continued on next page

Product Familiarization

General

The information included in this section is based on introductory and preliminary information for the 1998 through 2005 Volkswagen Passat sold in the USA and Canada. The content provided here is intended to serve as a product familiarization guide and is subject to change.

> ***WARNING—***
> *• Check the Volkswagen factory repair information or the publisher's website at www.bentleypublishers.com for information that may supersede any information included in this section.*

Product Overview

The fifth generation of the Volkswagen Passat, sometimes referred to as the B5 platform, was totally redesigned for 1998 and established new standards for quality, comfort and convenience in a passenger car. Beginning in 2001, the B5 Passat received a face lift. The most noticeable changes were in the front end (including hood, fenders and grille).

The B5 Passat is slightly larger than its predecessor, but boasts improved fuel economy, largely due to its aerodynamic contour. The shape of the Passat is extremely aerodynamic, with an impressive drag coefficient of 0.27.

Under the hood the powertrain is installed in a north/south (longitudinal) orientation. Two engine choices were available during its introduction; the 1.8 liter 4-cylinder turbocharged gasoline engine and the 2.8 liter 90° V6 engine. Both of these engines use dual overhead camshafts and five valves per cylinder for excellent high-rpm breathing. In addition, the naturally-aspirated V6 featured variable valve timing.

In 2001 the Passat received a facelift involving some styling changes and mechanical modifications. Although some of the body panels remained unchanged, new projector-optic headlights and bumpers gave the car a fresh look. The restyled Passat was unofficially known as the B5.5 platform.

A 2.0 liter TDI diesel engine introduced to the model line in 2004 uses advanced "Pumpe Düse" technology, as it is called in Germany, or "pump injectors" as translated in the U.S.

A 4.0 liter, 32-valve eight cylinder version of the Passat went on sale in April 2001 as a 2002 model. The W8 model was named after its unique eight-cylinder engine. The W8 engine has a "double-V" layout which provides extremely compact packaging. Unlike a V8, which has two parallel banks of four cylinders on each side, the W8 has four banks of two cylinders.

More detailed engine information can be found later in this section under **1 Engine and Engine Management**.

While the Passat is categorized as a midsize car, the ample dimensions and quality of its interior defy such limitations. Keeping this spacious environment at just the right temperature is easily accomplished through the Passat's Climatronic system (initially available only on the GLX model); the first electronic temperature control system ever available on a North American Volkswagen.

The interior offers state-of-the-art occupant protection. Along with driver and passenger passive restraints, there are side airbags located within the front seats. In addition, pyrotechnic seat belt tensioners with belt force limiters combine to make the new Passat a *tour de force* in vehicle safety.

General specifications (model year 1998)	
Engine code	AEB: 1.8 liter turbochargerd 4-cyl. dual overhead camshafts
	AHA: 2.8 liter, 90° V6, dual overhead camshafts, 5 valves per cylinder
Horsepower (hp)	AEB: 150 @ 5700 rpm
	AHA: 190 @ 5800
Torque (lb-ft)	AEB: 155 @ 1750 - 4600
	AHA: 206 @ 3200
Fuel capacity	16.4 gallons (62 liters)
Fuel requirement	unleaded premium (91 octane)
Brakes, front/rear	vented disc / solid disc, antilock standard
Suspension, front	Four-link front suspension system, independent MacPherson strut
Suspension, rear	FWD: torsion beam axle with trailing arms and coil springs
Drivetrain	front engine, front-wheel drive

Exterior dimensions and weights

The basic body dimensions of the 1998 Passat sedan are detailed in the table below.

General exterior dimensions and weights (model year 1998)	
Length	184.1 in (4675 mm)
Width	68.5 in (1740 mm)
Height	57.6 in (1462 mm)
Wheelbase	106.4 in (2702 mm)
Front track width	59.0 in (1499 mm)
Rear track width	59.1 in (1501 mm)
Max. permissible gross weight	sedan: 4400 lb (1996 kg)
Curb weight	3195 lbs (1449 kg)
Front seat headroom	39.7 in (1008 mm)
Rear seat headroom	37.8 in (960 mm)
Cargo volume	14.2 cu ft

2001- 2005 Passat sedan

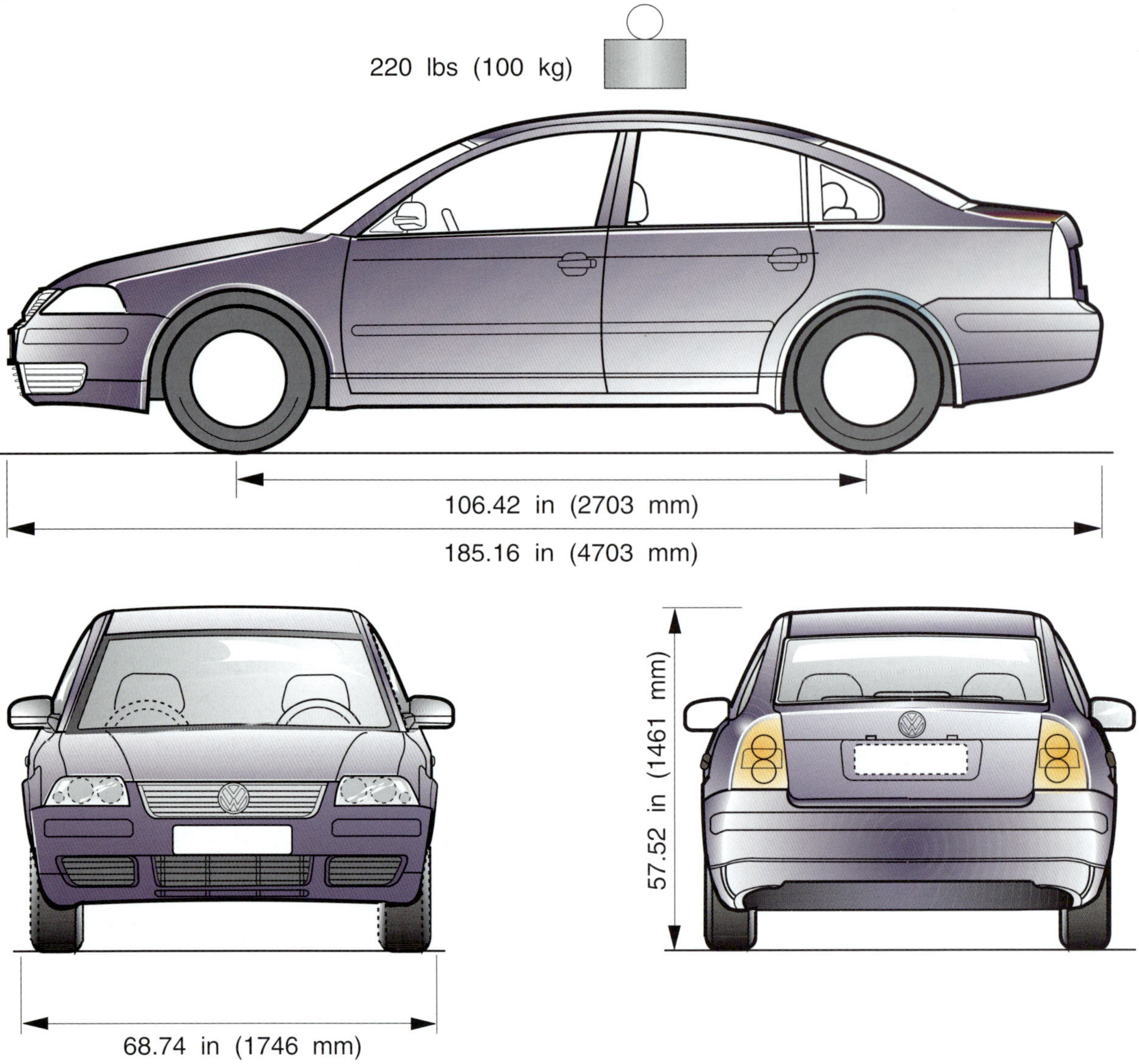

- Length
  185.16 inches (4703 mm)
- Width
  68.74 inches (1746 mm)
- Height
  57.52 inches (1461 mm)
- Wheelbase
  106.42 inches (2703 mm)
- Turning circle
  37.7 feet (11.5 m)
- Fuel tank capacity
  Ex. with W8 engine: 16.4 gallons (62 liters)
  With W8 engine: 21.1 gallons (80 liters)
- Front track width
  59.49 inches (1511 mm)
- Rear track width
  59.45 inches (1510 mm)
- Permitted gross vehicle weight
  4,894 pounds (2220 kg)
- Curb weight
  3,671 pounds (1665 kg)
- Permitted roof load
  220 pounds (100 kg)
- Drag coefficient
  0.27 C_D

2001 - 2005 Passat wagon

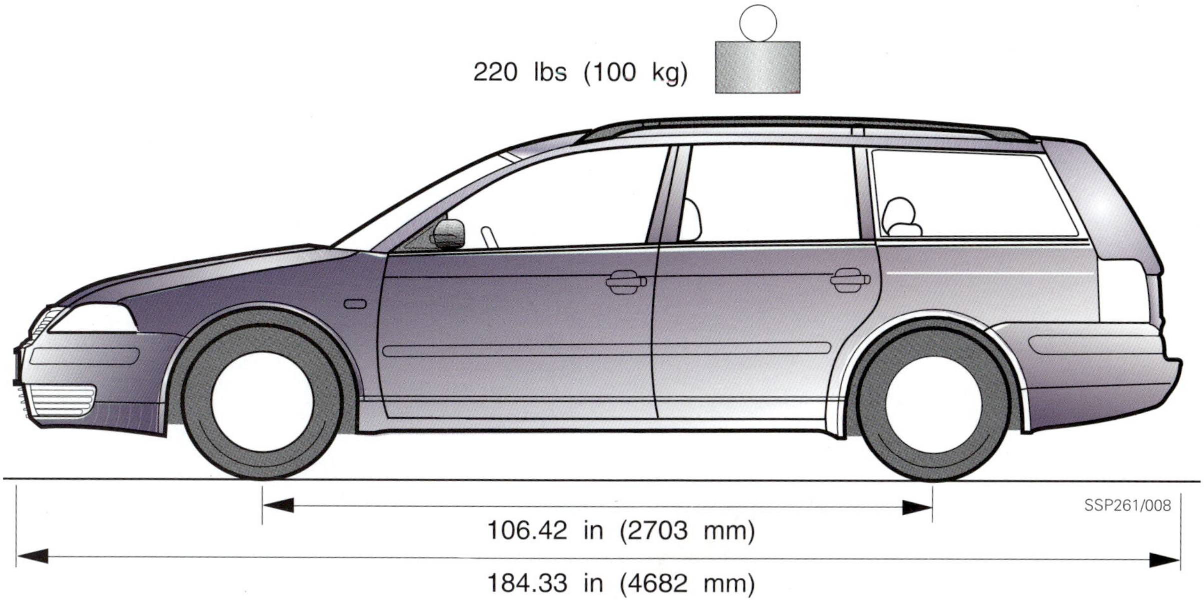

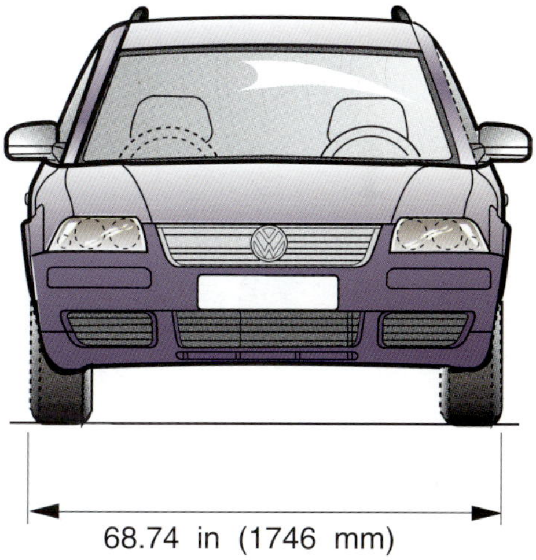

- Length
  184.33 inches (4682 mm)
- Width
  68.74 inches (1746 mm)
- Height
  58.94 inches (1497 mm)
- Wheelbase
  106.42 inches (2703 mm)
- Turning circle
  37.7 feet (11.5 m)
- Fuel tank capacity
  Ex. with W8 engine: 16.4 (62 liters)
  With W8 engine: 21.1 gallons (80 liters)

- Front track width
  59.49 inches (1511 mm)
- Rear track width
  59.45 inches (1510 mm)
- Permitted gross vehicle weight
  5,049 pounds (2290 kg)
- Curb weight
  3,803 pounds (1725 kg)
- Permitted roof load
  220 pounds (100 kg)
- Drag coefficient
  0.27 C_D

1 Engine
2 Engine Management

Multiple engine and transaxles were available for 1998 through 2005 Volkswagen Passat models. For 1998, engine choices were limited to the 150 hp 1.8 liter engine and a 190 hp V6 engine. Additional engines were introduced in the model line up during the vehicle's production.

Engine applications	Engine code
1.8 liter 4-cylinder turbo gasoline	AEB. AUG, ATW, AWM
2.0 liter 4-cylinder TDI PD diesel	BHW
2.8 liter V6 5-valve gasoline (BHW)	AHA, ATQ
4.0 liter W8 (BDP)	BDP

1.8 liter 4-cylinder gasoline engine (engine codes AEB, AUG, ATW, AWM)

The 1.8-liter engine combines the advantages of a long stroke engine in terms of weight, space savings, and exhaust emission performance with the benefits of multi-valve technology. Each of the four combustion chambers employs a centrally-located spark plug surrounded by 3 intake and 2 exhaust valves.

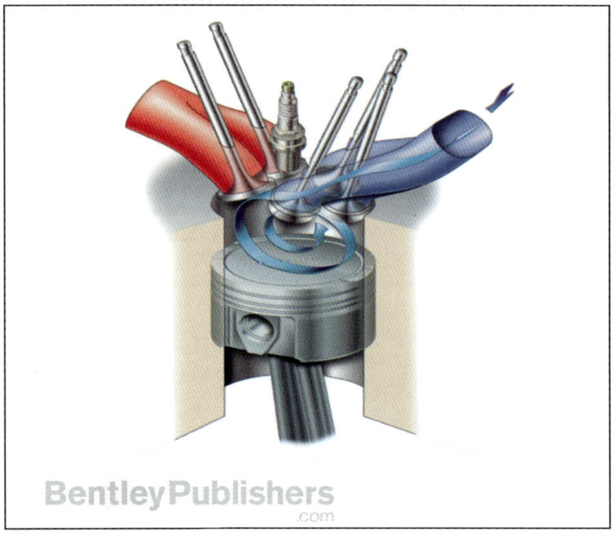

Of the three intake camshaft lobes per cylinder, the center lobe is slightly offset. This offset is necessary to accommodate the angle of the center intake valve, which is different than the outer two. The position of the intake valves allows the intake air to adopt a swirling motion which produces the required turbulence for optimum combustion.

Valve train mass is reduced by using hollow camshafts, smaller diameter valve stems, and only one spring per valve. This weight reduction allows the engine to rev higher, which complements the use of five valves per cylinder.

To improve serviceability, the cylinder head can be removed with the camshafts in place. Head bolts on the 1.8-liter engine are not reusable.

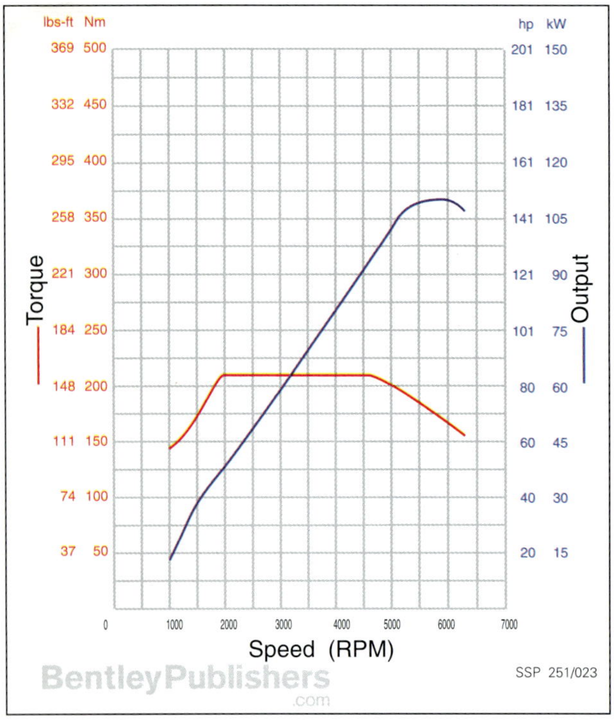

1.8 liter (150 hp) engine performance graph

1.8 liter engine, general specifications	
Engine codes	AEB (1998 - 1999) AUG (2000) ATW (2000 - 2001) AWM (2001 - 2005)
Engine type	dual overhead belt and chain-driven camshafts, 20-valve in-line 4-cylinder
Engine size	1.8 liters / 108.6 cu. in.
Horsepower (hp)	
AEB, AUG, ATW	150 @ 5700 rpm
AWM	170 @ 5900 rpm
Torque (lb-ft)	
AEB, AUG, ATW	155 @ 1750 - 4600 rpm
AWM	166 @ 1950 - 5000 rpm

To achieve maximum torque at such a low rpm, a low-inertia, water-cooled turbocharger is used. Boost pressure is determined by the engine control module via a charge pressure solenoid valve. This valve modifies the vacuum signal to the wastegate actuator. In the event of a system malfunction, the engine control module is programmed with a minimum boost default strategy.

To minimize turbo lag, a bypass valve has been installed on the compressor side of the turbocharger. The valve is operated by manifold vacuum and allows intake air to recirculate during idle and deceleration conditions, when the throttle plate is closed and manifold vacuum is high. This setup allows the turbo to spin freely without developing any boost pressure and helps the compressor maintain a higher speed and quicker response time.

Another feature of the 1.8 liter engine includes a hot-film mass air flow (MAF) sensor, which offers better contamination resistance over the earlier-generation hot-wire type sensor.

The throttle valve control module is a single assembly that incorporates throttle position sensing and idle speed management.

1.8 liter turbocharger with bypass control

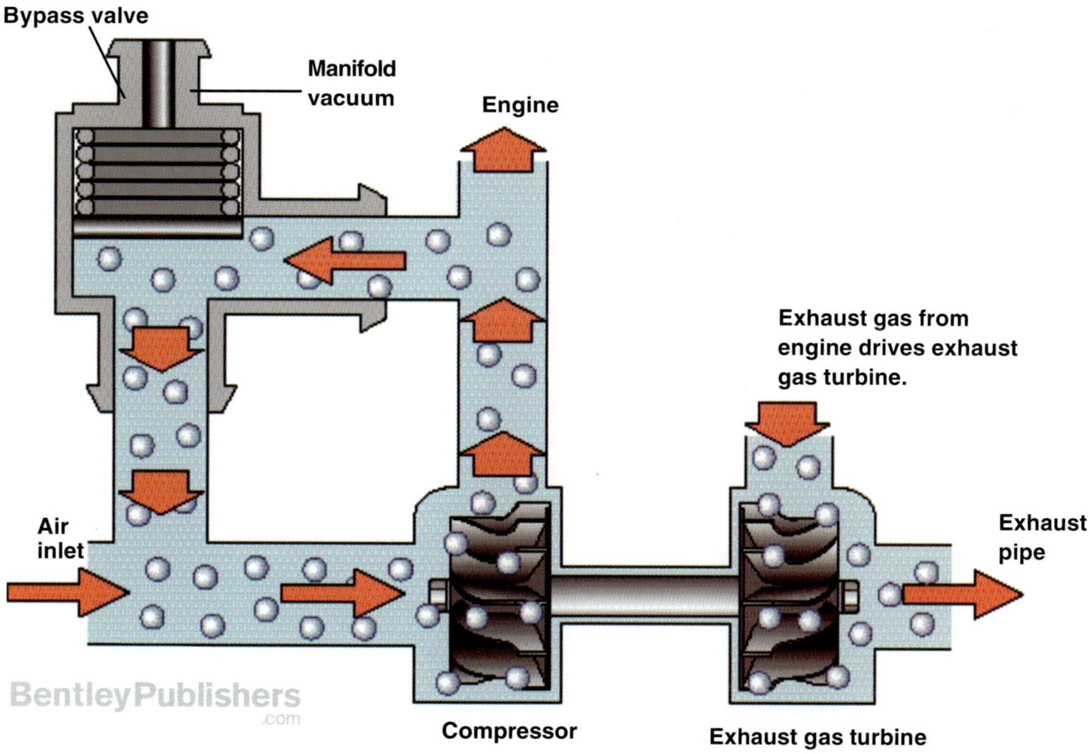

2.8 liter V6 engine (engine codes AHA, ATQ)

The V6 engine is a 2.8 liter naturally-aspirated engine with cylinder banks positioned 90° apart.

A redesigned combustion chamber and centrally-located spark plug are responsible for improving combustion efficiency, reducing fuel consumption and lowering exhaust emissions.

Other features include 5 valves per cylinder, distributorless ignition and variable valve timing. Two variants of this engine were available.

2.8 liter engine, general specifications	
Engine codes	AHA (1998 - 1999) ATQ (2000 - 2005)
Engine type	dual overhead belt and chain-driven camshafts, 30-valve V6
Engine size	2.8 liters / 169 cu. in.
Horsepower (hp)	
AHA	190 @ 6000 rpm
ATQ	190 @ 6000 rpm
Torque (lb-ft)	
AHA	206 @ 3200 rpm
ATQ	206 @ 3200 rpm

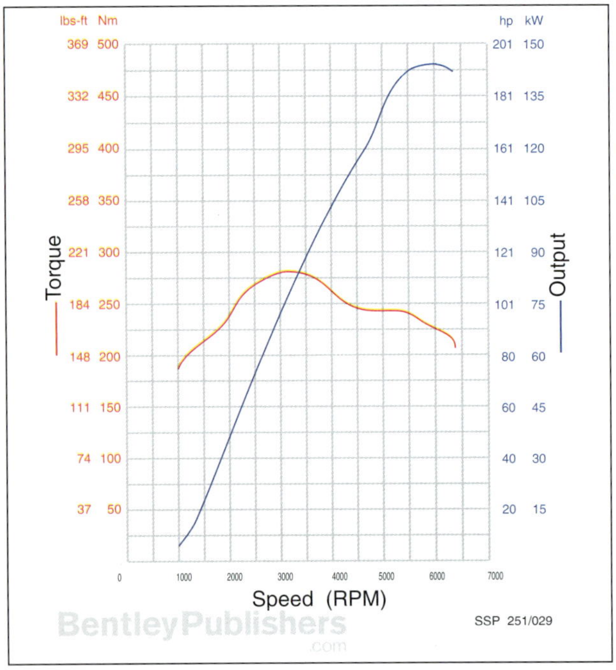

2.8 liter (190 hp) engine performance graph.

As of model year 2001, the 2.8L V6 'ATQ' gasoline engine conforms to the low emissions vehicle (LEV) exhaust emissions standard for all of the U.S. and Canada.

Variable valve timing

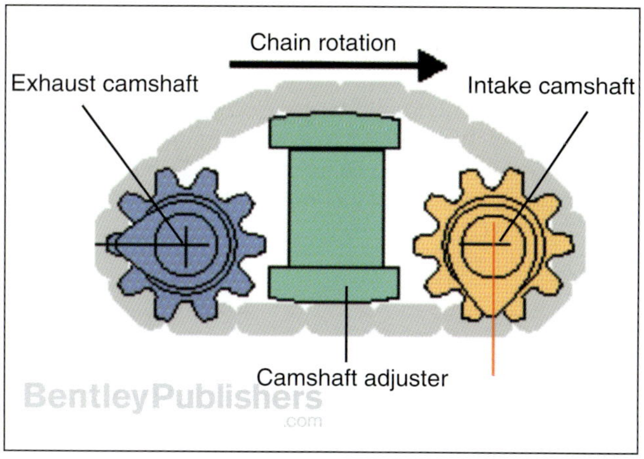

The 2.8 liter V6 engine features a variable valve timing system that enables the intake and exhaust camshafts to be controlled independently. Adjustable valve timing systems overcome the limitations of static valve timing by altering the points in the four-stroke cycle when the valves open and close. This allows for greater torque throughout a wider rpm range.

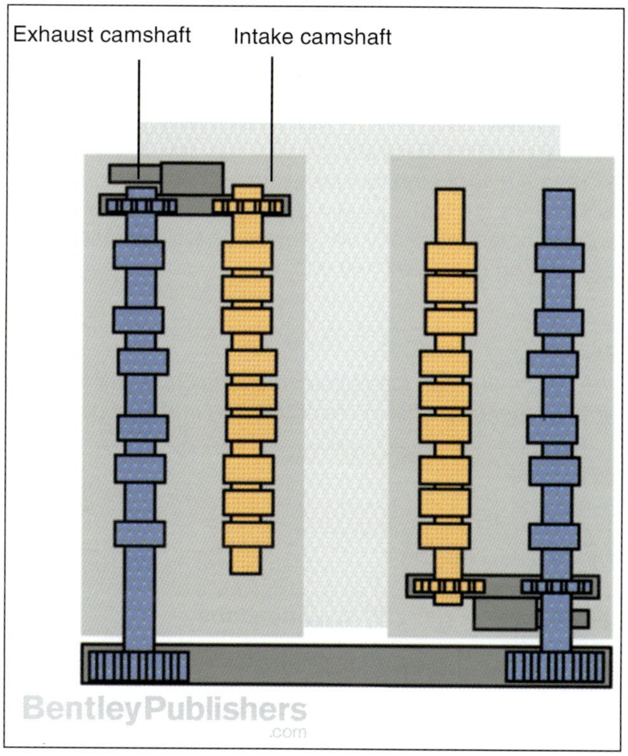

The exhaust camshafts are positioned on the outboard side of the cylinder head, and the intake camshafts are on the inboard side. Because of this arrangement, the camshaft adjuster of the left and right cylinder banks must operate in opposite directions.

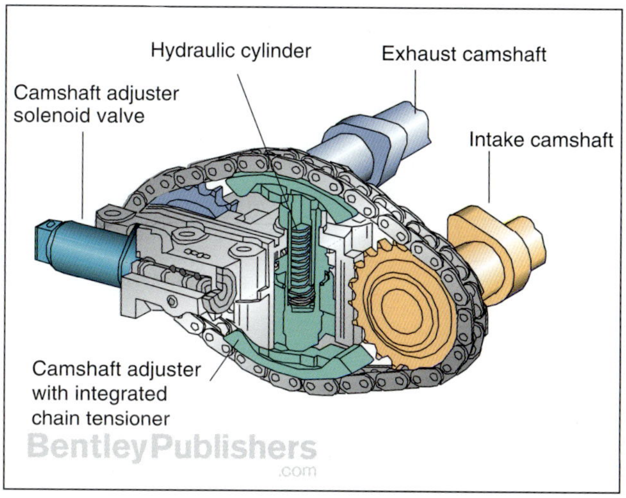

An internal hydraulic cylinder fed by the engine's lubrication system operates the camshaft adjusters. A camshaft adjustment valve regulates oil pressure to the hydraulic cylinder based on commands it receives from the engine control module (ECM). This device is mounted directly to the camshaft adjuster housing. The action of each camshaft adjuster is monitored by the camshaft position sensors. A malfunction in either camshaft adjuster assembly will set a diagnostic trouble code.

2.0 liter TDI PD diesel engine, (engine code BHW)

The TDI Passat was offered for model years 2004 and 2005. The acronym TDI stands for Turbo Direct Injection. The acronym PD is derived from the German "Pumpe Düse" meaning Pump Injection and refers to the pump function of the individual injectors.

Developed in combination with Bosch®, pump injectors are located at each cylinder. Each injector is a small injection pump by itself. Because of exacting electronic control, the new Pumpe Düse technology creates a very high fuel pressure and better atomizes and precisely meters the fuel injection into the cylinders, resulting in increased engine efficiency and power, as well as quieter engine operation and optimized fuel economy.

> ***CAUTION—***
> *• Due to the extremely high loads placed on the camshaft lobes by this arrangement, special oil must be used in this engine.*

2.0 liter TDI PD diesel engine

2.0 liter diesel engine, general specifications	
Engine code	BHW (2004 - 2005)
Type	4-cylinder inline, 2 valves/cylinder
Engine size	2.0 liter (1968 cc), 120.1 cubic inches
Ignition sequence	1-3-4-2
Bore	81.0 mm
Stroke	95.5 mm
Compression ratio	18.5 : 1
Horsepower (hp)	134 @ 4000 rpm
Torque	247 lb-ft (335 Nm) @1900 - 3500 rpm
Fuel requirement	diesel min. 49 CN

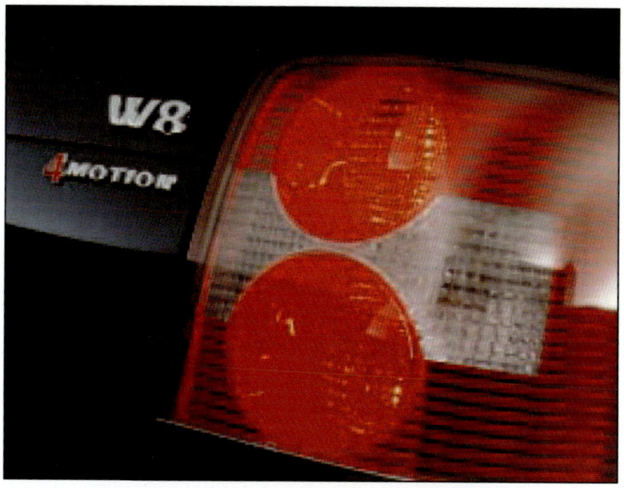

W8 8-cylinder engine (engine code BDP)

Starting with model year 2002, the top model of the Passat range was equipped with an 8-cylinder engine. The Passat W8 engine was offered in sedan and wagon versions.

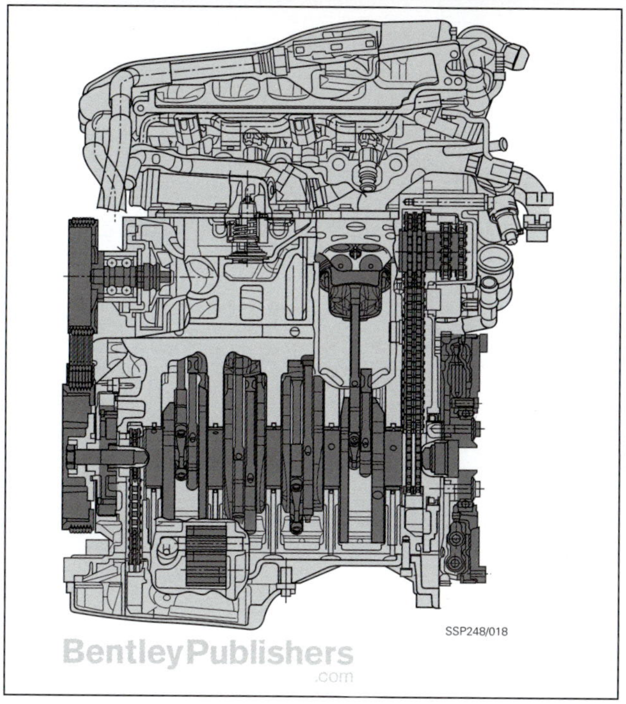

One of the most important technical features of the W8 engine is its compact size.

Whether in the wagon or sedan, the new Passat W8 is distinguishable from other Passat models on the outside by:

- Dual exhaust with two chrome tailpipes on each side.
- Bi-xenon headlights.
- Additional chrome body trim.
- 7J x 16 alloy wheels.
- P205/55 R16H all-season tires.

4.0 liter W8 engine, general specifications	
Engine code	BDP (2002 - 2004)
Engine type	dual overhead belt-driven camshafts, 32-valve W8
Engine size	4.0 liters / 244 cu. in.
Horsepower	270 @ 6000 rpm
Torque (lb-ft)	272 (370 Nm) @ 2750 rpm

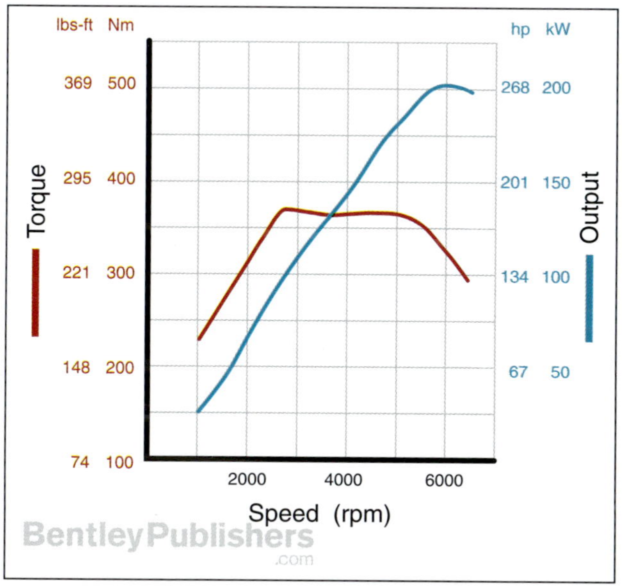

4.0 liter (270 hp) engine performance graph.

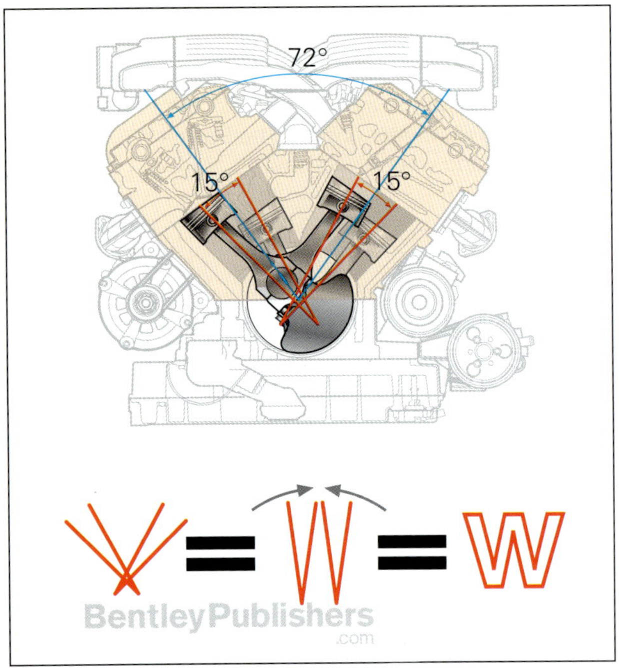

The W engine design incorporates features of a V engine. As with V engines, the cylinders are arranged in two banks. In the W8 engine, these banks of cylinders are aligned at a V-angle of 72 degrees in relation to one another. This layout achieves a very compact eight-cylinder engine with relatively small dimensions.

When the W8 engine is viewed from the front, the cylinder arrangement looks like a double-V. Put the two Vs together, and you get a W. Hence the designation "W engine."

Compared with other eight-cylinder engines, the W8 is relatively lightweight due to its size and use of aluminum alloy castings, slim connecting rods, and short pistons. Four valves per cylinder, low-friction roller rocker fingers, and chain drive for the four overhead camshafts with camshaft timing adjusters are also features of this modern engine.

W8 engine

The Passat W8 has a combined fuel tank capacity of approximately 21.1 gallons (80 liters). The fuel delivery system includes the 15.85-gallon (60-liter) fuel tank as used on the 2001 and later Passat 4MOTION and an auxiliary tank with a capacity of 5.28 gallons (20 liters).

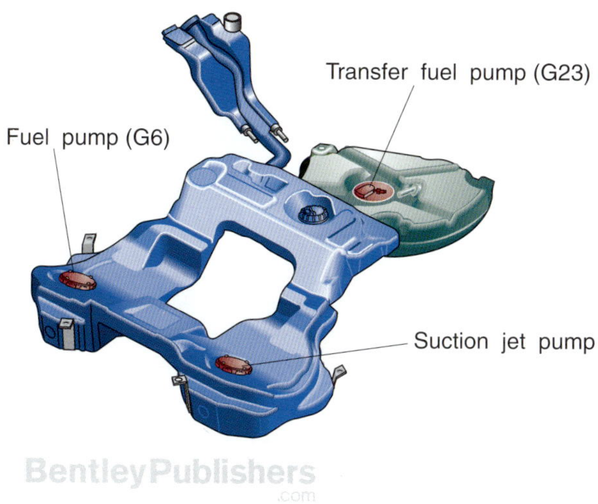

The two fuel tanks are connected by a tube. The auxiliary fuel tank is located under the spare-wheel well, which is reduced in size to accommodate the tank. There is still enough room to store a temporary spare in the reduced spare-wheel well.

The main fuel tank contains an electric fuel pump (G6) and a suction jet pump. There is also an electric transfer fuel pump (G23) in the 5.28-gallon (20-liter) auxiliary fuel tank. The different heights of the main tank and the auxiliary tank require a controlled evacuation of the two tanks.

When both tanks are full, fuel is first pumped from the main tank. When the level in the main tank drops to 10.57 gallons (40 liters), the pump in the auxiliary fuel tank is activated and pumps the fuel from the auxiliary tank to the main tank. The suction jet pump delivers fuel from the left chamber of the main tank to the fuel pump in the right chamber.

Fuel delivery system, W8 engine

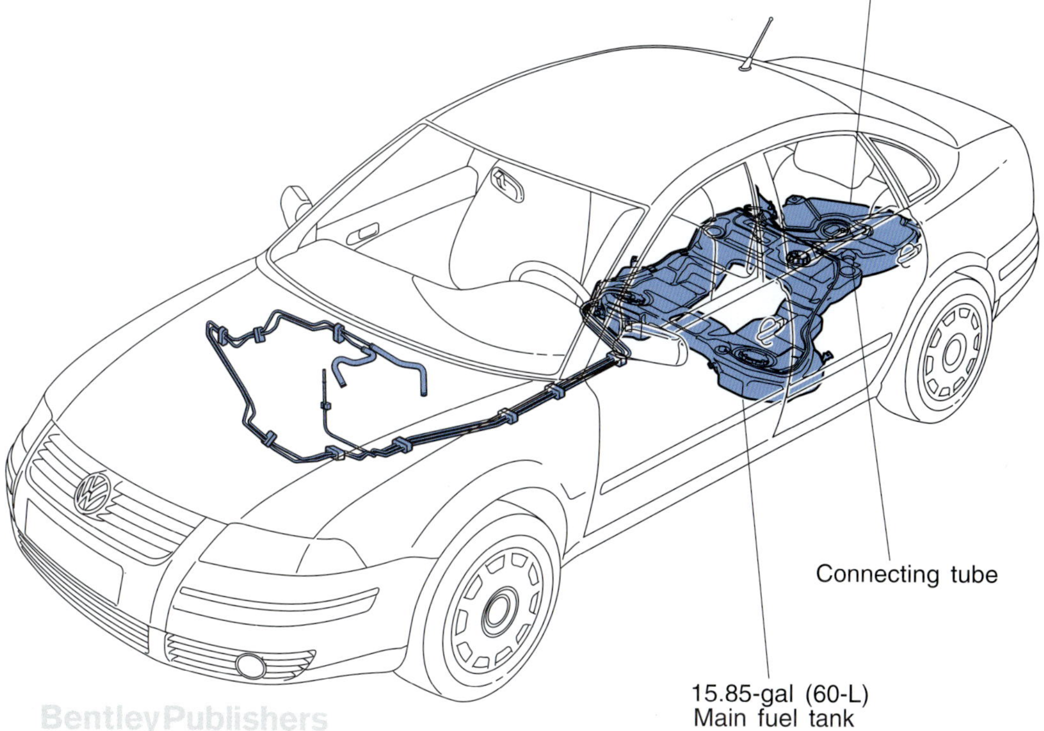

Engine control module (ECM)

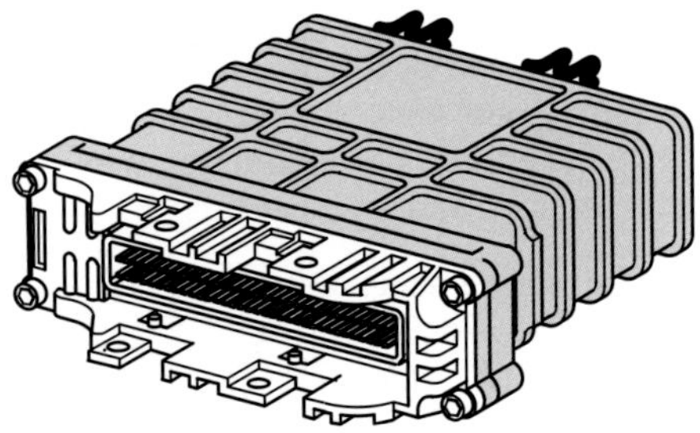

Fuel injection

Multiple variants of Bosch engine management systems were used on the 1998 - 2005 Volkswagen Passat engines.

Engine (engine code)	
1.8 liter AEB, AWP, AMU, BEA)	Bosch Motronic M5.9.2 Bosch Motronic ME 7.5
2.0 liter PD diesel	Bosch EDC
2.8 liter AHA ATQ	Bosch Motronic M5.9.2 Bosch Motronic ME 7.5
4.0 liter W8	Bosch ME 7.1.1

Motronic engine management systems fully comply with federal and state government mandated On-Board Diagnostic (OBD) II standards. There are no mechanical engine settings or adjustments to be made.

The OBD II diagnostic system is integrated into the engine management system, and continuously monitors its own emissions related components, as well as other vehicle systems that affect emissions. Component monitors allow the Motronic engine control module (ECM) to check for plausible signals by cross checking against another component.

Example:

The Motronic ECM compares the Engine Coolant Temperature (ECT) sensor signal against a map based on the Intake Air Temperature (IAT) sensor signal at the time the engine was started. If after a specified time the ECT signal is more than 20° C from the mapped point, the ECM believes this to be implausible for the running condition, and sets an appropriate diagnostic trouble code (DTC).

OBD-II standards include:

- Standardized diagnostic connection located in driver's area
- Standardized DTCs for all manufacturers
- Retrieval of DTCs by commercially available diagnostic equipment (generic scan tools)
- Retention of operating conditions present during a monitored malfunction
- Standards governing when and how a monitored malfunction must be displayed
- Standardized names for components and systems.

OBD II monitors and diagnosis can include:

- Monitoring of catalytic converter function
- Diagnosis of heated oxygen sensor response, voltage, aging, and heater circuit
- Diagnosis of fuel tank and venting system integrity
- Diagnosis of evaporative emissions flow
- Engine misfire detection.

- Enhanced diagnostics for input and output components by checking function and signal plausibility in addition to open and short circuits
- Expanded and standardized DTC capabilities
- Status of emissions-related diagnostic routines (readiness code)

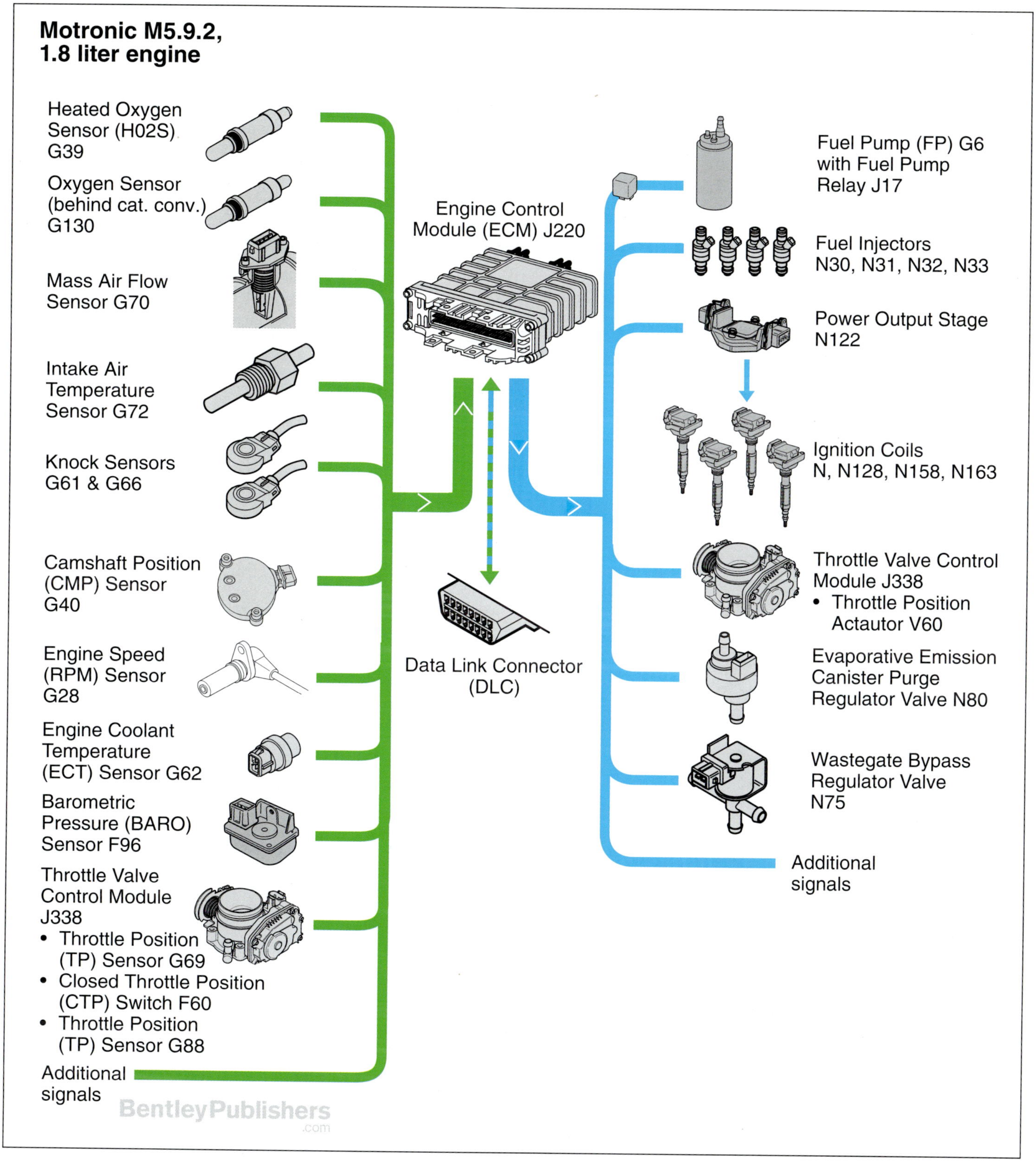

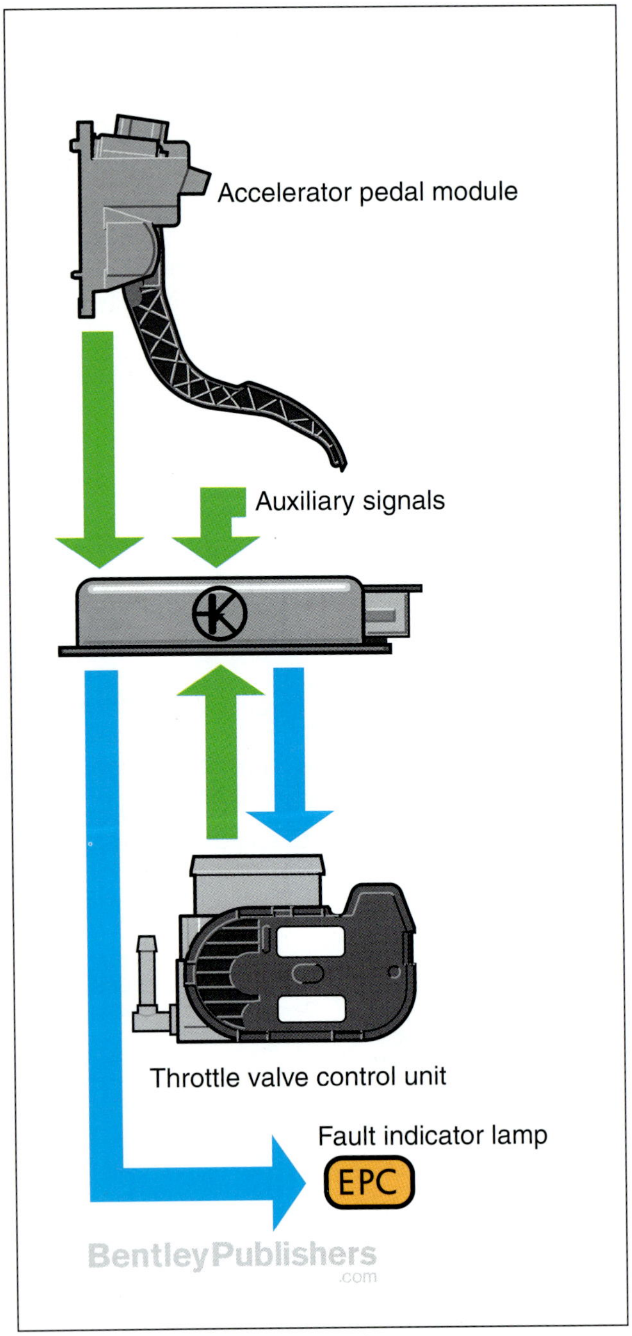

The Bosch Motronic ME 7 family of engine management systems is a further development of the M5.9.2 system. The ME 7 systems are referred to as torque-based engine management systems.

The Bosch Motronic ME 7 engine management system uses a new internal circuit design to calculate engine torque. The engine control module (ECM) receives all of the relevant signal data from input sensors and computes the required amount of engine torque needed based on pre-established priorities.

With ME 7, there is no longer a throttle cable between the accelerator pedal and the throttle valve. This has been replaced by an electronic control system.

The system has the following components:

- Throttle valve control unit
- Accelerator pedal module

The electronic throttle valve control system is used to reduce and increase torque without adversely affecting the exhaust emissions.

The ideal combination of throttle opening and charge pressure can be controlled for greatest efficiency. Under acceleration, throttle movement is programmed to best increase intake charge velocity. On deceleration, the throttle valve is held open to reduce emissions.

3 Clutch and Transmission

Manual transaxles were available with 5 speeds or 6 speeds. Automatic transaxles weres available with 5 speeds.

Transmission applications	Transmission code
5-speed manual, front-wheel drive	012 / 01W
5-speed manual, 4Motion	01A
6-speed manual, 4Motion (8-cylinder)	01E
5-speed automatic Tiptronic or Tiptronic with 4Motion	01V

Manual transmission

Multiple manual gearbox variants are used for power transmission. All feature cable operated shift linkage to help reduce the transfer of noise and vibration to the interior of the vehicle. The clutch is hydraulically operated.

The 6-speed 01E manual transmission with 4Motion (all-wheel drive).

Automatic transmission

Transmission control module (TCM)

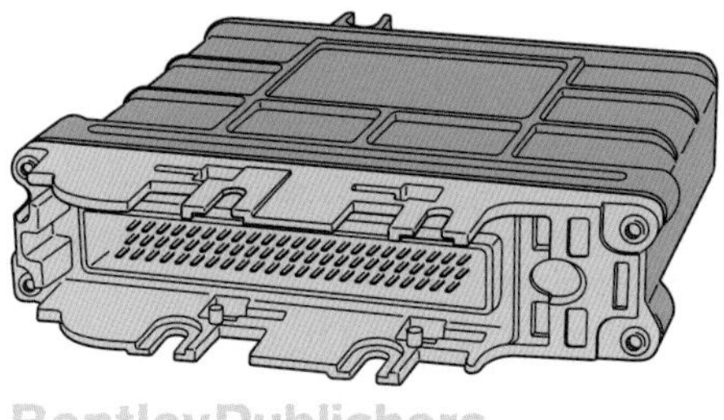

All of the automatic transmissions feature electronic controls. The transmission control module (TCM) controls all the electrical, and thus all the hydraulic operations of the transmission. The TCM also establishes driving programs based on an individual's driving style and driving situations. Determined by "fuzzy logic", shift points are moved to higher or lower engine rpm based on driving style.

Various types of driving resistances such as climbing a hill, towing a trailer, driving into a stiff wind, as well as descending a hill are recognized and accounted for by the TCM.

The computer system includes self-diagnostic capabilities. These self diagnostics use the computer to examine many of the parameters in the system and identify problems. The TCM can check the electrical system during normal operation and store codes in memory to identify any faults it encounters.

Because failures in certain sensors and/or actuators can result in damage to the transmission, the TCM recognizes this and default to an emergency running mode.

Operating the selector lever when the transmission is in the emergency running mode engages the following gears mechanically in the valve body:

- 1st gear hydraulic
- 3rd gear hydraulic
- Reverse gear

When the selector lever is in DRIVE, the vehicle starts off in 3rd gear hydraulic.

The 5-speed 01V automatic transmission with 4Motion (all-wheel drive).

4Motion (all-wheel drive)

The 4Motion system transforms the Passat into an all-wheel drive, confidence inspiring vehicle. The Torsen center differential, electronic differential lock (EDL), a redesigned rear suspension system and fuel tank are the key components that make this happen.

All front to rear power transfer is done automatically and no driver input is necessary. This results in:

- Superior all-weather performance
- Excellent stability on the road
- Excellent traction

Unlike many traditional four-wheel drive systems, 4Motion is a full-time traction enhancing system that distributes power to all four wheels, all the time, at any speed.

The system uses the automatic-locking TORSEN (as in TORque SENsing) center differential to distribute engine torque to the front and rear axles. The Torsen differential detects wheel slip in one axle and distributes the drive power to the wheels of an axle with better wheel grip.

The Torsen differential permits limited differences in speed between the front and rear axles to allow for ABS control processes. It operates automatically and reacts independent of driver input.

There are no electrical connections or computer controls. The Torsen differential is a sealed unit and is located in the rear of the transmission case.

Passat with 4Motion

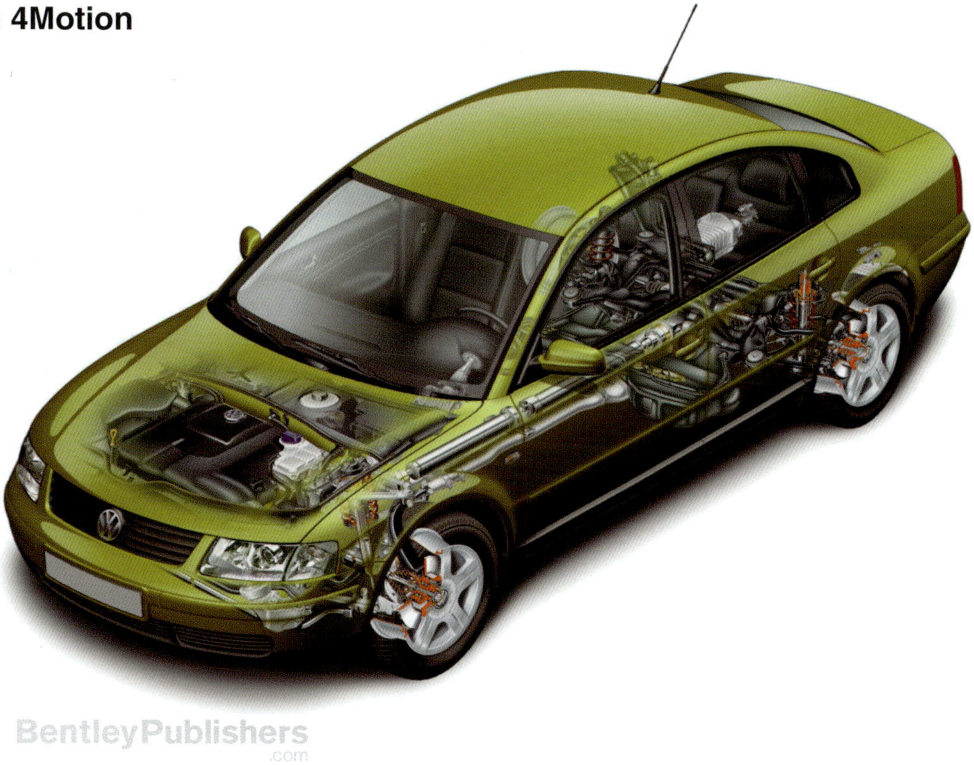

4Motion driveline

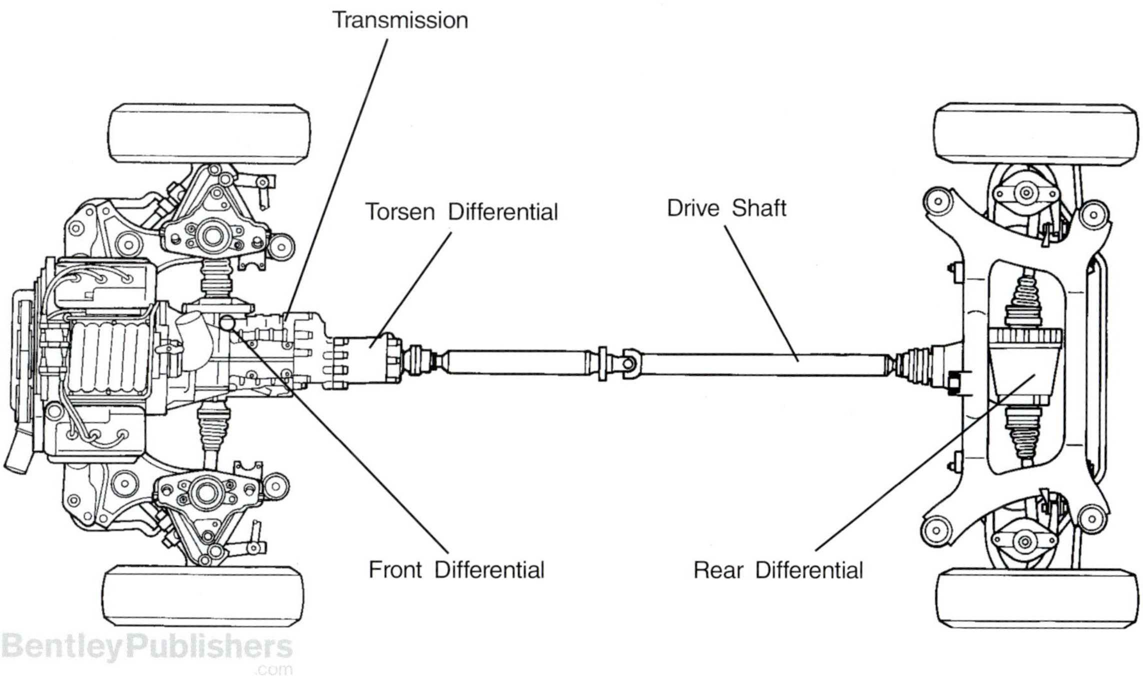

The Torsen differential is composed of the following components:

- Differential case
- Planet gears (with spur gears)
- Front axle side gear
- Rear axle side gear
- Front axle drive
- Rear axle drive

Inside the Torsen housing, there are pairs of helical planet gears. The planet gears are held in tight-fitting pockets inside the housing, and are splined together through spur gears at their ends. These spur gears do not allow the planet gears to rotate in the same direction.

The teeth on each of the planet gears mesh with the teeth of one side gear. When the vehicle is moving in a straight line with no slip, the transmission drives the Torsen unit. The Torsen unit in turn drives the planet gears, which drive the side gears.

Torsen differential

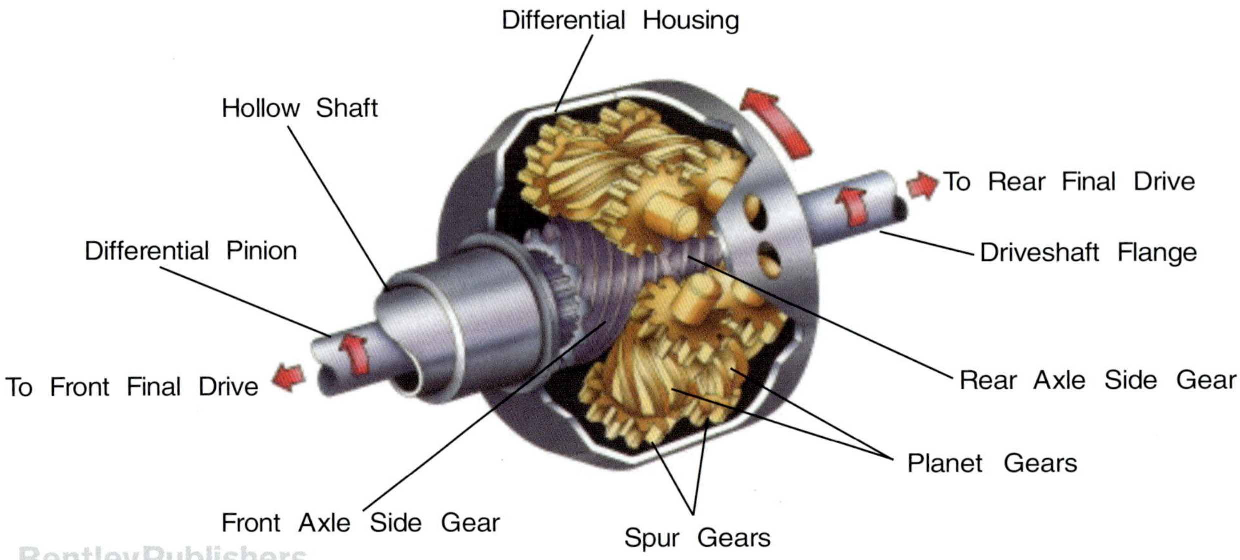

Electronic differential lock (EDL) controls any wheels that are spinning by applying the brake to that wheel to slow it down. Slowing the spinning wheel transfers torque to the wheel that has traction. EDL and the Torsen differential work together to maintain the most traction in any situation.

EDL uses the same sensors and components that the anti-lock brake system (ABS) uses to control traction. EDL does not operate when the brakes are applied or when the vehicle is traveling over 25 mph.

EDL/ABS

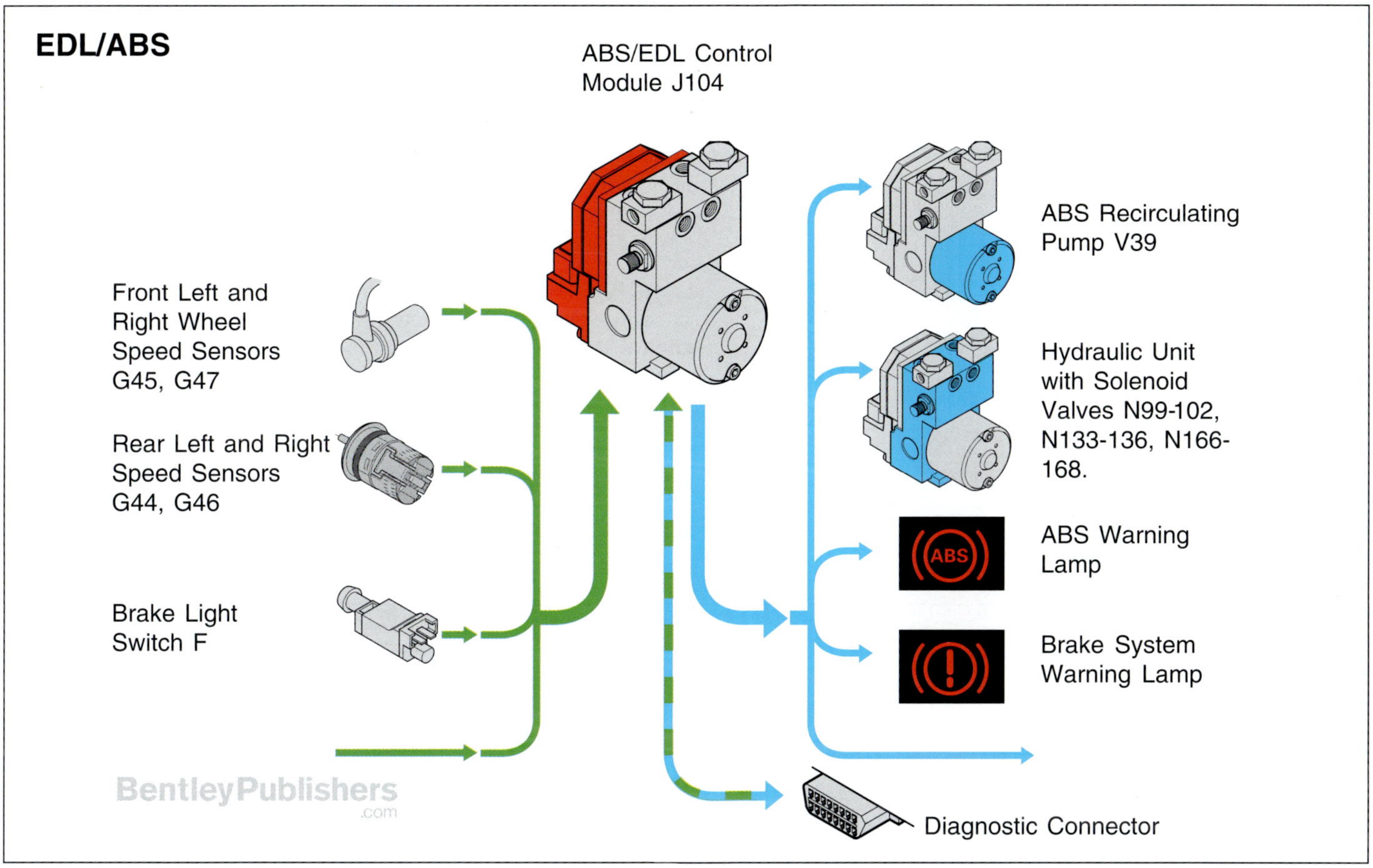

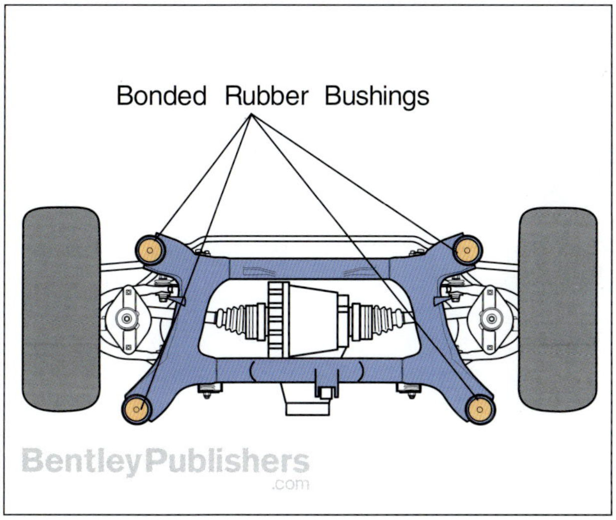

The rear suspension is fully independent and features a double wishbone design. Four large bonded rubber bushings connect the subframe to the body. These bushings inhibit noise transfer to the vehicle. A coil-over-shock assembly is used to maximize space around the rear axle.

The rear shock absorbers are gas pressurized and have a single tube design. Their smaller diameter helps to preserve trunk space despite a physically larger rear suspension. The relatively low position of the upper wishbone helps to preserve the low, flat floor of the trunk area.

29

4 Suspension, Steering and Brakes

Front axle

A four-link front suspension system is used on the Passat B5 platform. This new arrangement virtually eliminates torque steer, even when the vehicle is accelerated hard through a curve. In addition, bump steer is minimal and directional control is excellent.

The suspension components are attached to a subframe using large rubber mountings. This configuration suppresses the road noise and powertrain vibrations that would normally be transmitted to the passenger compartment.

The front wheels are located using four individually pivoted aluminum control arms. The upper control arms are mounted to the body through rubber bushings, and to the wheel bearing housing via low-friction ball joints. The lower control arms are attached to the subframe and are divided into one weight-bearing member and one locating member.

Large fluid-filled bushings are used on the two inside lower control arms to isolate the body from road vibrations. The outer two lower control arms use narrower bushings that allow the arms to move laterally as the front wheels are turning.

The control arms are arranged to compensate for lift and dive, which occurs as a result of vehicle acceleration and deceleration. A tubular anti-roll bar is attached to the subframe and connected by a short link to each of the load-bearing lower control arms. The struts are fixed at both the bottom and top and do not rotate as the front wheels are turned.

Front axle

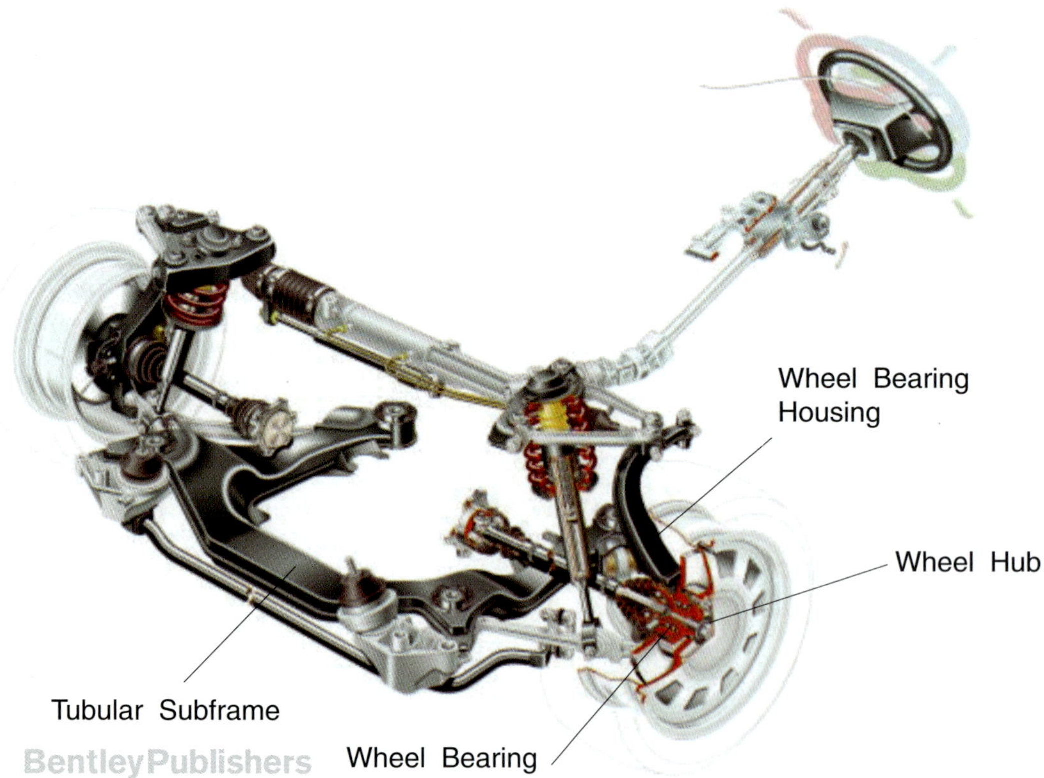

Rear axle (front-wheel drive)

Front-wheel drive models use a rear torsion beam axle with internal stabilizer bar.

Advantages of torsion beam rear suspension:

- Greater cargo area (wider) created by using separate coil springs and shocks
- Cooler operating single-tube shock absorbers
- Maintains consistent wheel camber angles during suspension travel
- Downward-facing V-section of axle beam produces more neutral handling characteristics.

When cornering, lateral forces acting on the suspension alter the track and cause a self-steering effect. The shock absorbers are supported in the wheel housings and the coil springs are supported below the side member.

The axle is stabilized by a tubular stabilizer bar located within the cross section of the transverse bar. Due to the separate layout of the springs and shock absorbers, the vehicle has a large luggage compartment and driving noise in the passenger compartment is reduced.

Rear axle (front-drive)

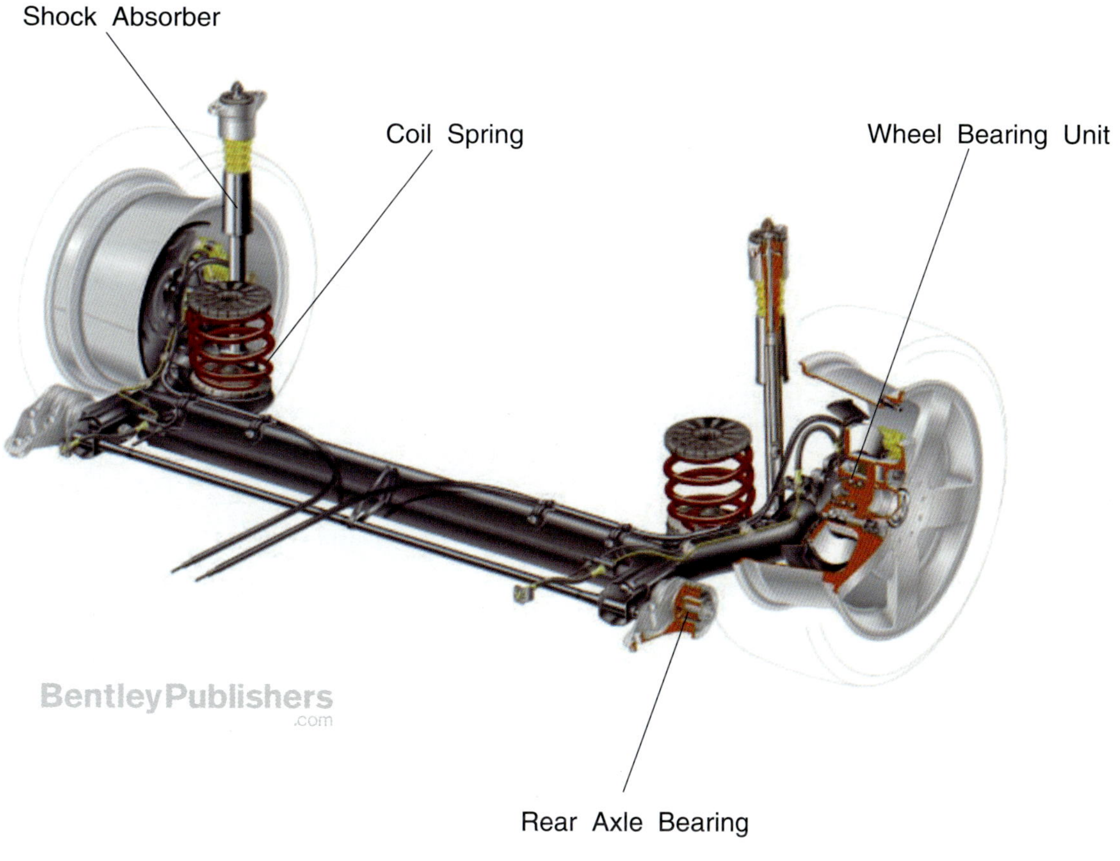

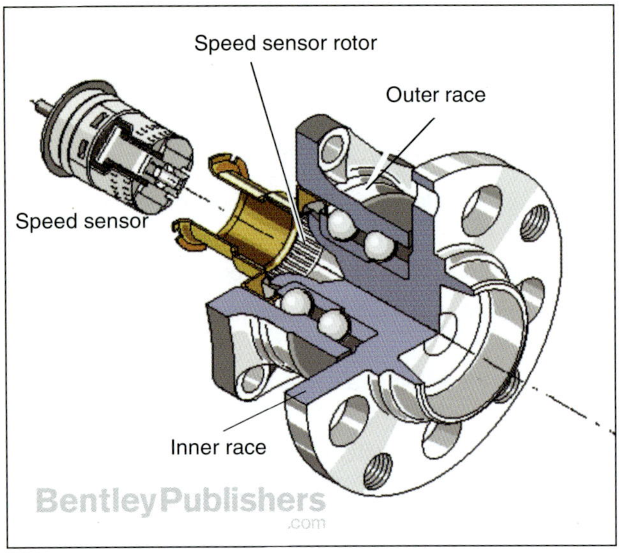

The twin-tracked rear wheel bearing has a stationary outer race that bolts to the rear axle mounting plate. The bearing inner race assembly serves as the wheel hub, which eliminates the need for an axle pivot. In addition, the ABS speed sensor rotor is integrated into the wheel bearing assembly.

The advantages of the new wheel bearing design include:

- Minimal wear due to improved sealing
- Speed sensor rotor is protected from road debris
- No necessary adjustments due to designed-in bearing preload

Rear axle (4Motion)

The 4MOTION all-wheel drive version of the Passat has a double wishbone rear suspension with a closed tubular subframe. The track control arms for wheel location and the differential are attached to the subframe. Insulating bonded rubber bushings provide good acoustic isolation from the vehicle body.

The low-profile construction of this rear axle provides a spacious interior. The dual link trailing arm axle on on cars with 4Motion is fixed via the subframe (4-point attachment) and each of the track correcting bushings are attached to the trailing arm.

Rear axle (4Motion)

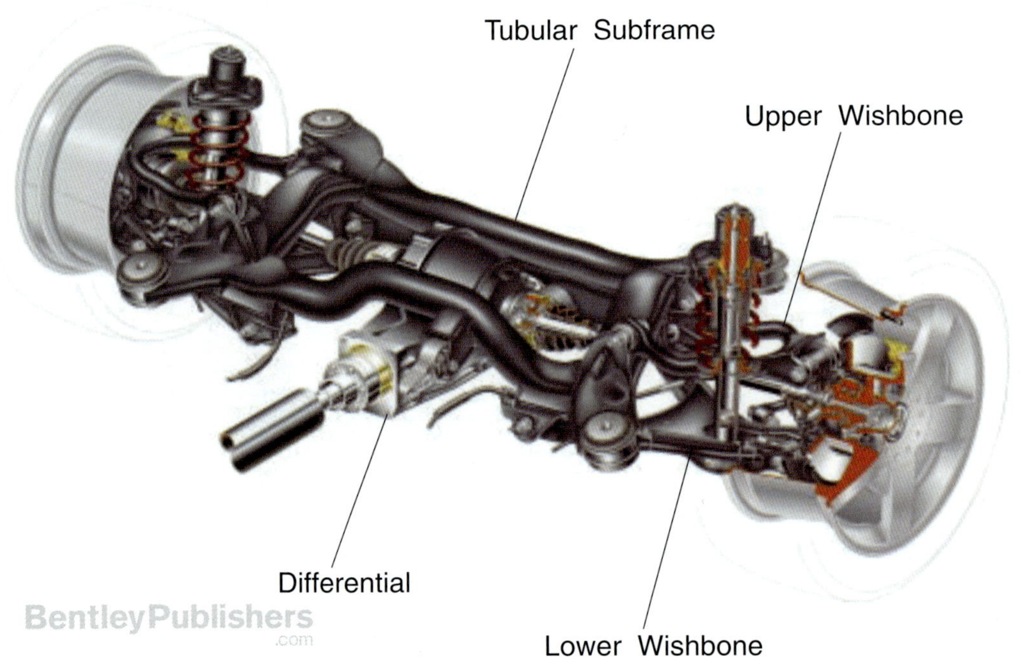

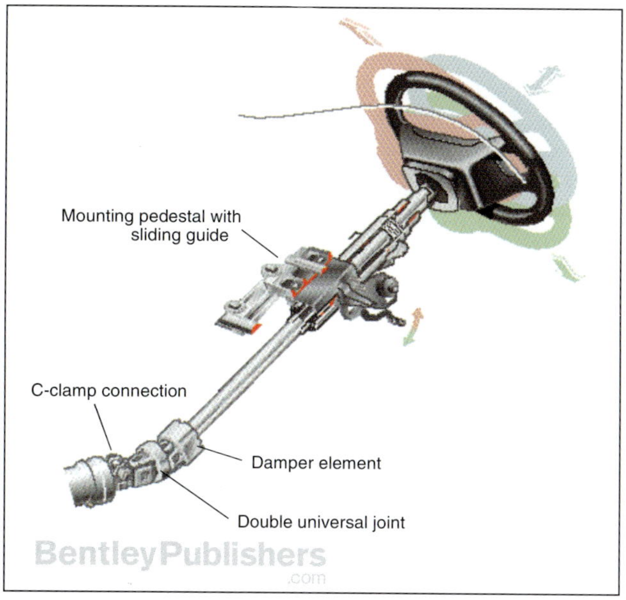

Steering

All B5 Passat models include power steering. The steering column is manually adjustable for both height and reach. The wheel can be moved toward the driver almost 2 inches (50 mm), and can be raised or lowered over one inch (28 mm) from the center position.

The steering column attaches to the body using a mounting pedestal with a sliding guide. To prevent vibration or noise from being transmitted through the column, a damper element is installed above a double universal joint. The column is secured to the steering rack using a C-clamp and bolt.

Brakes

The brake system is a dual-diagonal circuit with anti-lock braking (ABS) All models use discs at all 4 wheels (ventilated front discs).

The ABS hydraulic control module is integrated with the ABS electronic control module.

Power-assisted dual circuit brake system with ABS

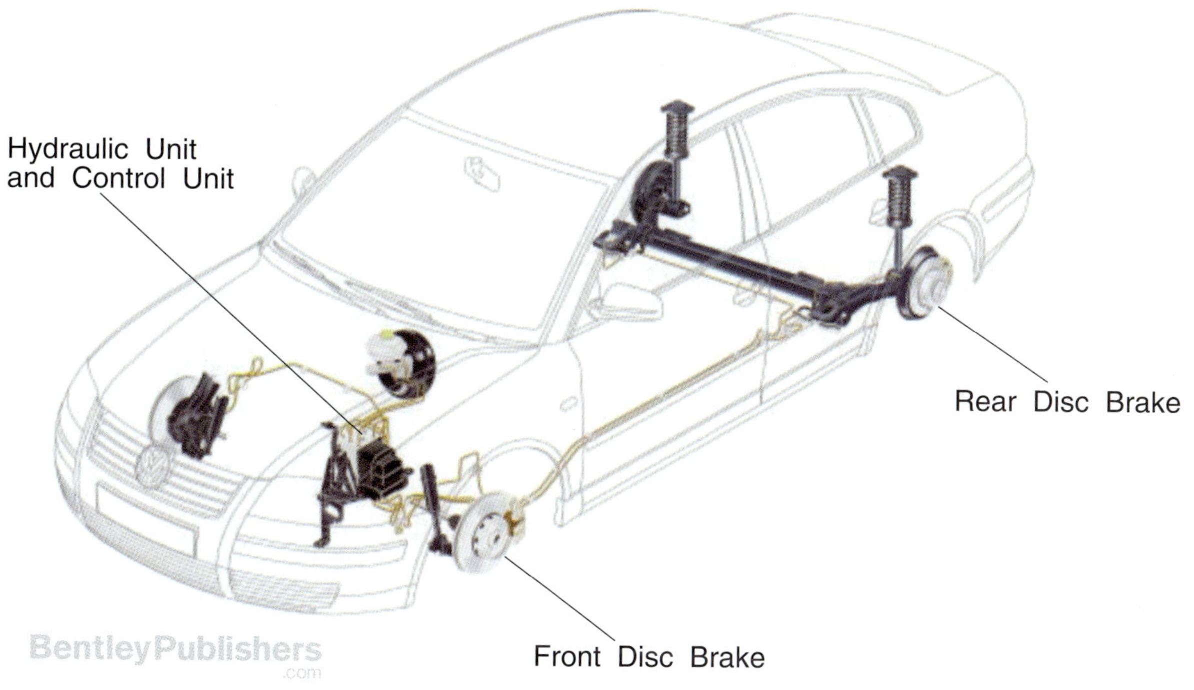

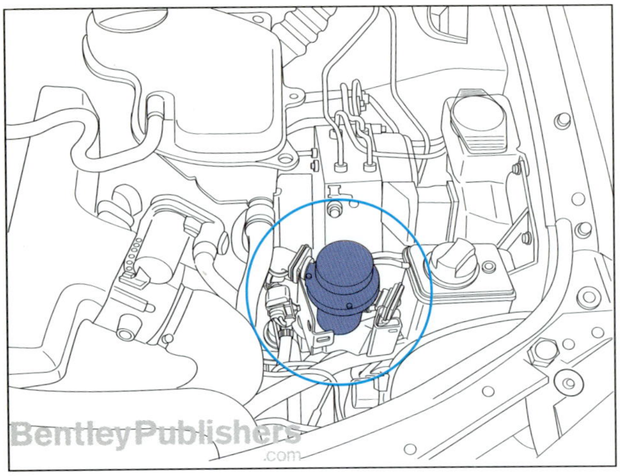

To maintain maximum brake function, an electric vacuum pump for brake boost is used on 2001 - 2005 Passats with automatic transmissions. The pump is used to maintain vacuum boost mainly during the cold start phase of the engine.

The system has a pressure sensor located in the vacuum line to the brake booster. This sensor informs the Motronic engine control module when vacuum pressure falls below a predetermined value. When the pressure drops, the Motronic ECM activates the relay to run the electric vacuum pump.

5 Body – Assembly
6 Body – Components and Accessories

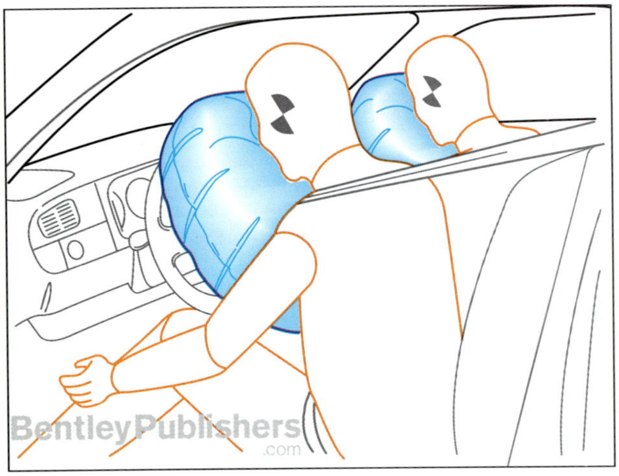

The Volkswagen Passat is equipped with advanced safety features. Occupants are surrounded by a sturdy safety cage and further protected by front and side airbags. Four-wheel antilock disc brakes stop it quickly and effectively. Safety features include:

- ABS (anti-lock braking system) with traction control
- Front and side impact airbags.
- 2001 - 2005 models: head airbags
- Safety belts with added features. Pyrotechnic belt tensioners with force limiters to help moderate chest forces in serious accidents. 3-point rear seat belts.
- Central locking system revisions. In an accident serious enough to deploy an airbag, the central locking system unlocks the doors and turns on the emergency flashers and interior lights.

Head airbags used on 2001 and later models

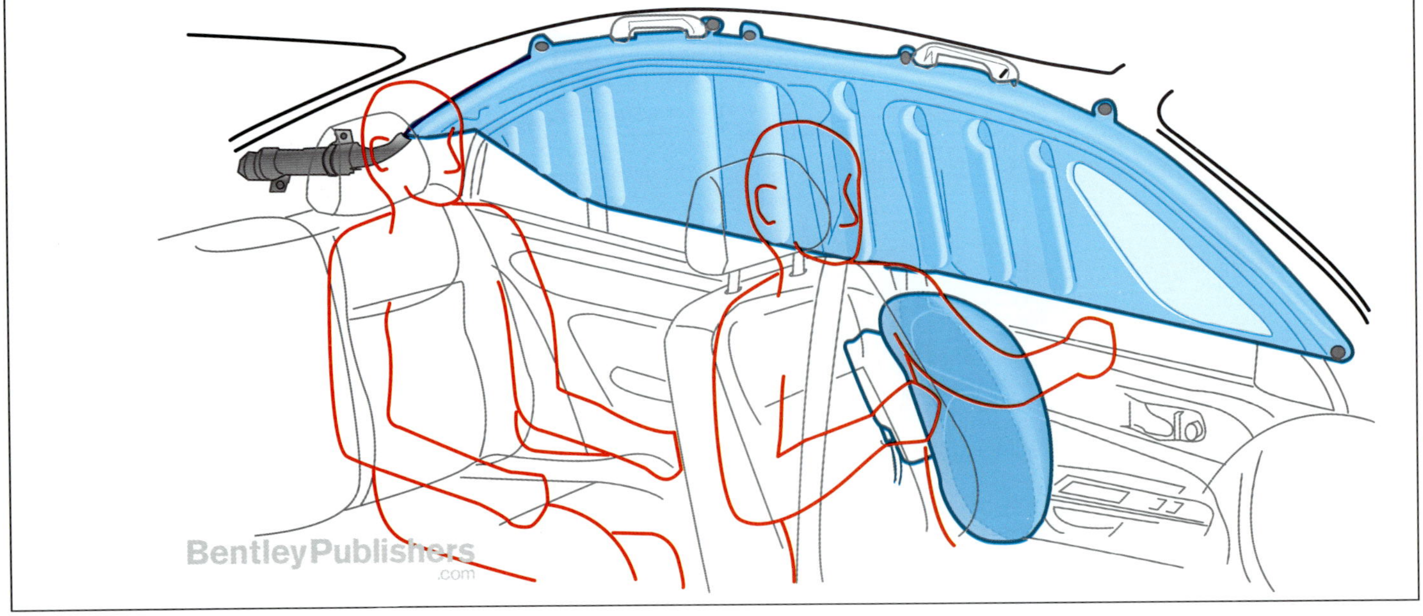

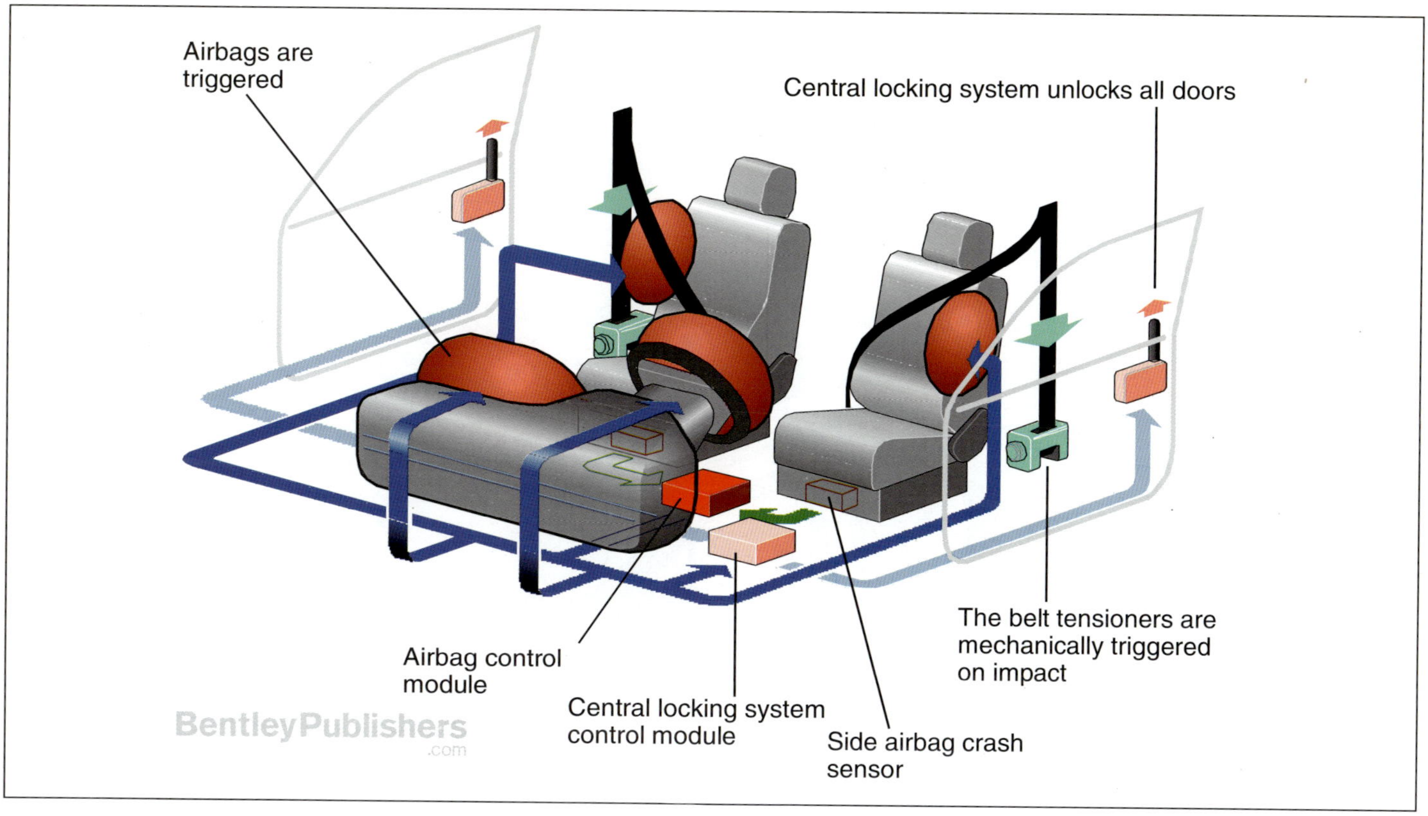

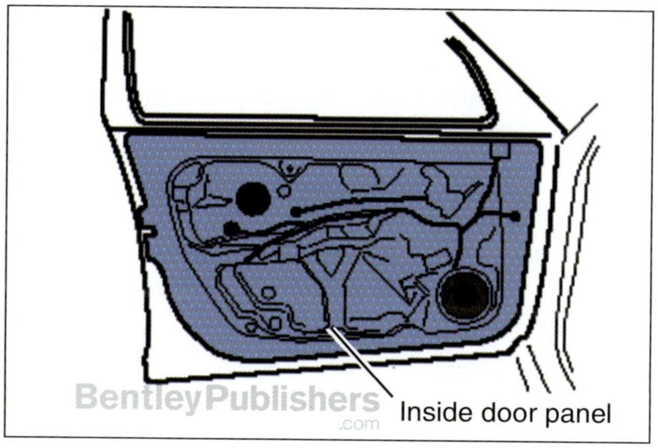

The door is reinforced using an inside door panel. This panel is a structural member that bolts to the inside of the door. The panel contains no opening or holes that would diminish its rigidity or allow the intrusion of door hardware in the event of a side impact.

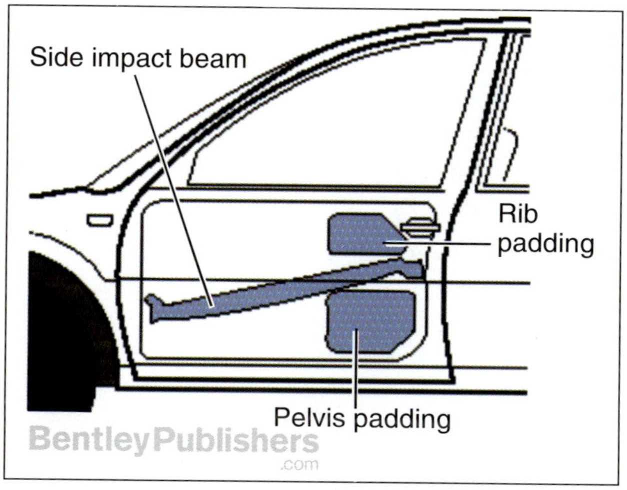

The door also contains a side impact beam made from pressed sheet metal. The beam is positioned diagonally inside the door, and is glued to the outer door panel to increase the door's strength. The side impact beam allows impact energy to be better absorbed, distributed and converted. Added protection is provided by the installation of pelvis and rib padding.

For the first time in its history, the B5 Passat used a completely galvanized body. The illustration below indicates the parts that are either hot-dip or electrolytically galvanized. Since the hot-dip galvanizing process leaves surface patterns on the metal which are visible even after painting, the exterior body panels are electrolytically galvanized to improve their appearance.

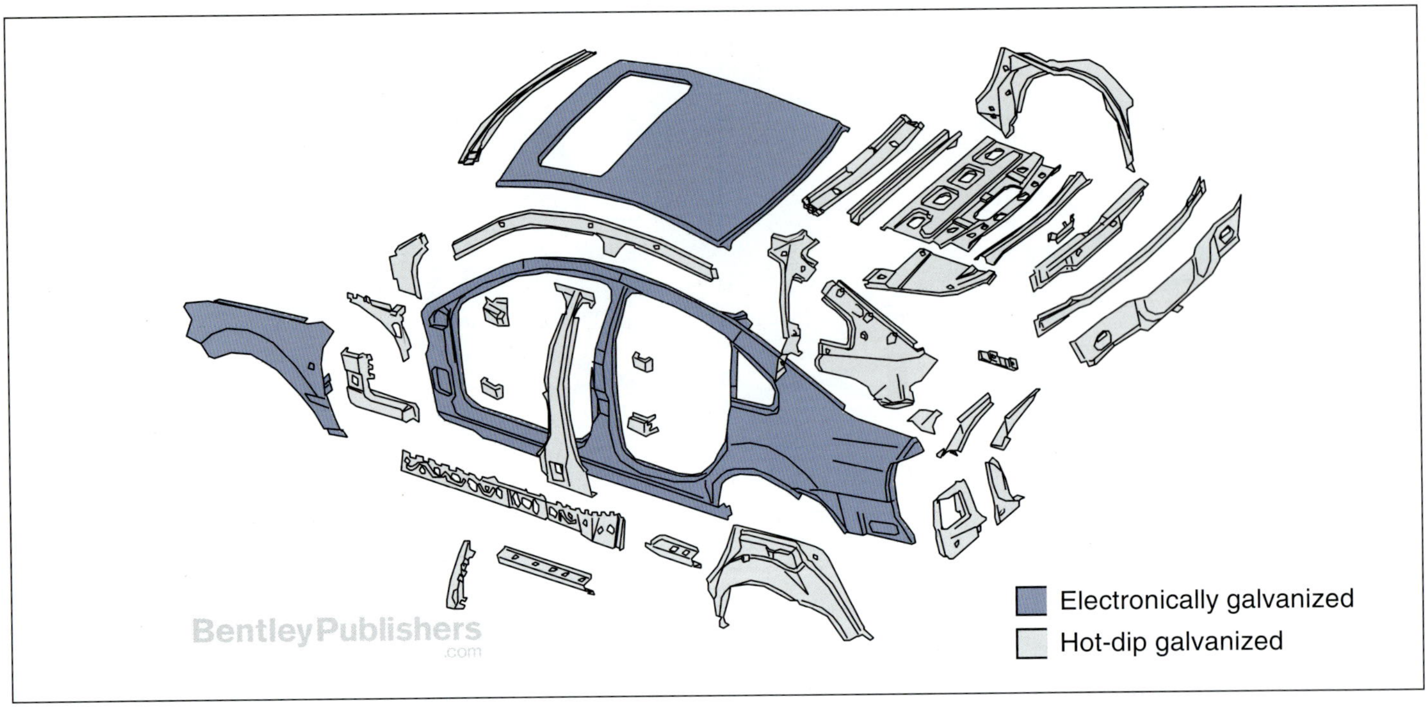

The body of the 2001 - 2005 Passat was reinforced in the zones described below. Panels were thickened and spot welds were supplemented with laser-welded seams. Two impact beams were added to the front doors, and the door-window channels were reinforced. Occupant safety was improved and natural body vibration was reduced 40 Hz as a result of these modifications.

Body reinforcements on 2001 and later models

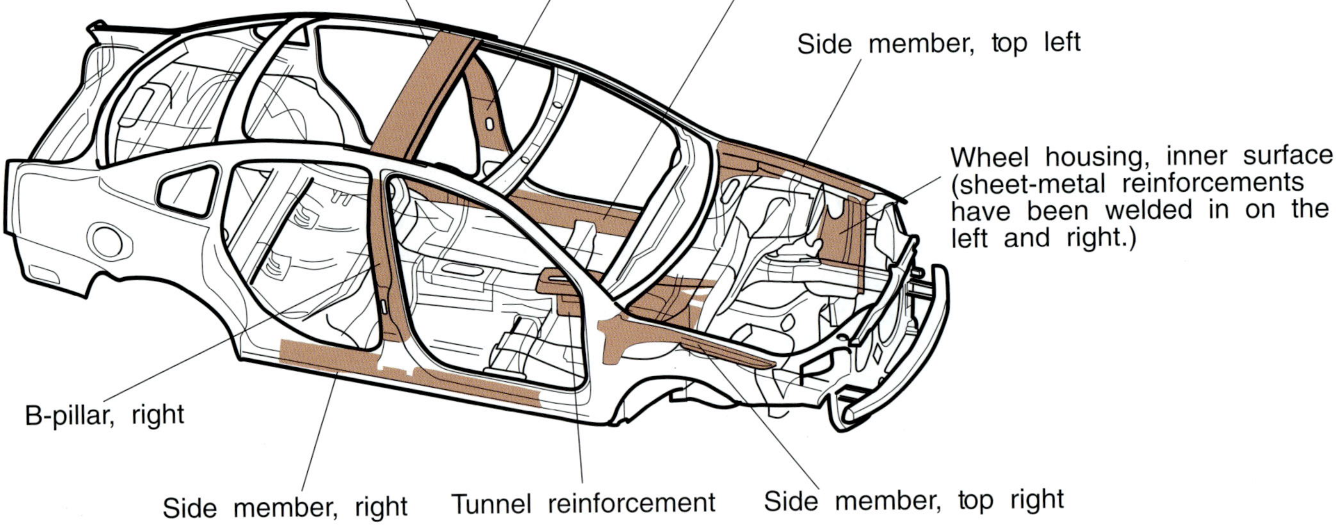

8 Heating and Air-conditioning

CLIMAtronic (introduced on the GLX model in 1998) was the first electronic control system ever used on a North American Volkswagen. The CLIMAtronic control module combines the microprocessor, display head, interior temperature sensor and temperature sensor fan into a single unit. The small fan is used to draw air past the interior sensor for obtaining accurate readings of the vehicle's interior temperature.

CLIMAtronic control module

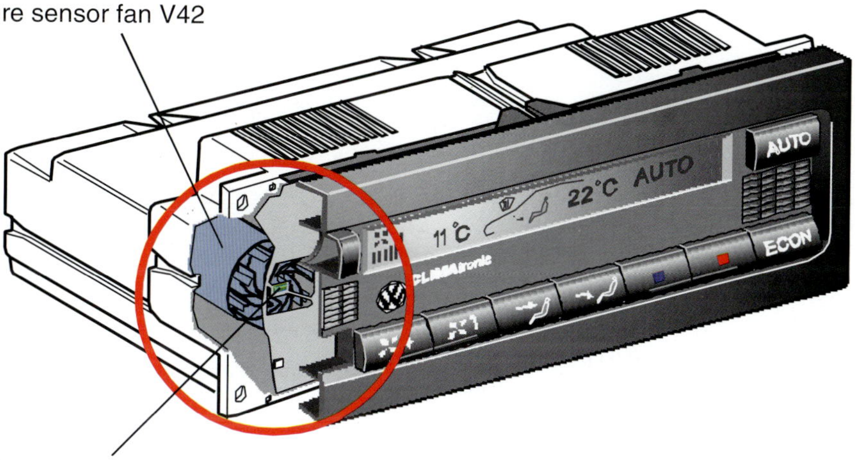

To prevent the vapor from a hot drink being drawn past the interior temperature sensor and affecting its accuracy, the fan and sensor are positioned on the left side of the CLIMAtronic panel away from the cup holder.

When the cup holder symbol on the dashboard is depressed, the cup holder swings out toward the passenger seat. This position keeps drinks far enough away from the interior temperature sensor.

The CLIMAtronic system contains self diagnostic capability and a long-term fault memory. The factory scan tool can be used to perform diagnostics. The scan tool works the same way for testing the CLIMAtronic system as it does for checking other systems such as engine management.

9 Electrical System

The decentralized electrical system is a major advance in vehicle wiring configuration. The primary feature of this arrangement is that the electrical system is subdivided into separate connector stations, relay and fuse carriers. These sub-modules are positioned locally, meaning they are located in close proximity to the assemblies and functional units that they serve.

The main fuse carrier is located on the far left side of the dashboard against the driver's door. The fuse carrier contains both standard size fuses and mini-fuses. The minifuses are used so that more electrical circuits can be individually protected. Connector stations are provided on the B-pillars to make diagnosing door circuits more convenient.

Advantages:

- Shorter wiring harnesses make cable connections easier to locate and identify. Shorter cables reduce vehicle weight
- Test points can be assigned more easily
- Better moisture protection for vehicle electrical components
- Better serviceability.

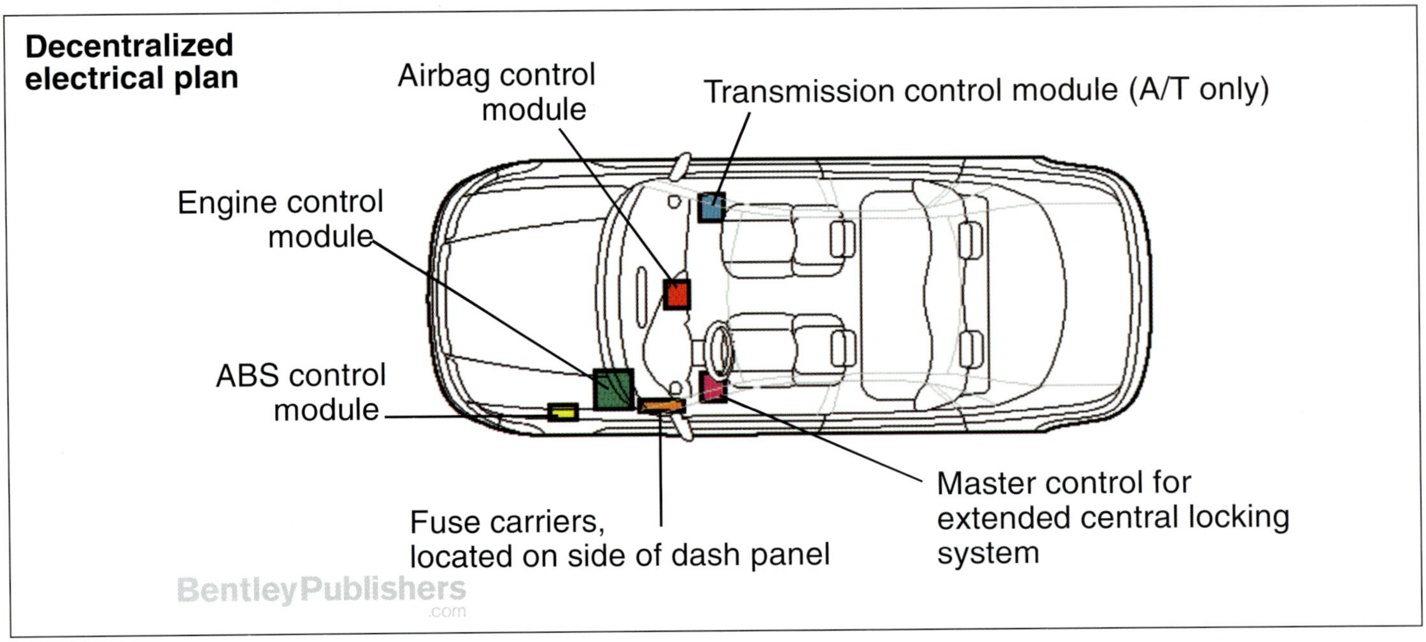

Controller Area Network (CAN)

To reduce the number of wires used on the vehicle and to increase reliability, many electrical functions are accomplished using serial data transmission over a controller area network (CAN). Similar to a telegraph system, this process known as multiplexing, is used to convert information into a series of pulses which represent coded data messages understood by other components.

For example, the extended central locking system uses an individual control module inside each door. These modules send and receive information over the CAN bus. The 16-pin data link connector is tied into the CAN bus to allow the diagnostic scan tool to interface with the electronics.

New features, 2001 and later models

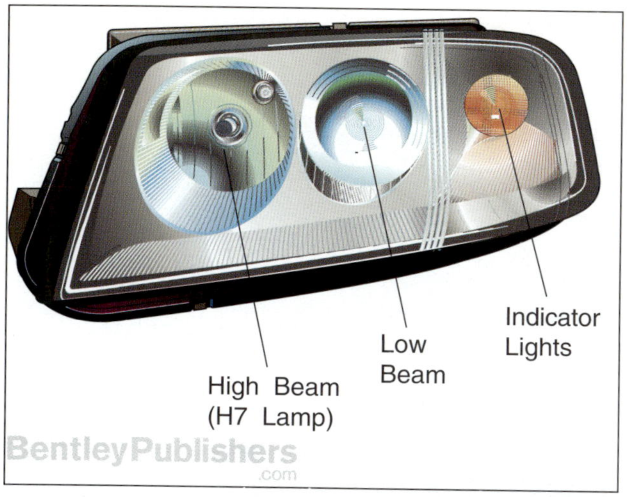

Headlights for the 2001-2005 Passat consist of a clear glass triaxial ellipsoid headlight with an H7 bulb used for high beam. Indicator lights are built into the headlight assembly.

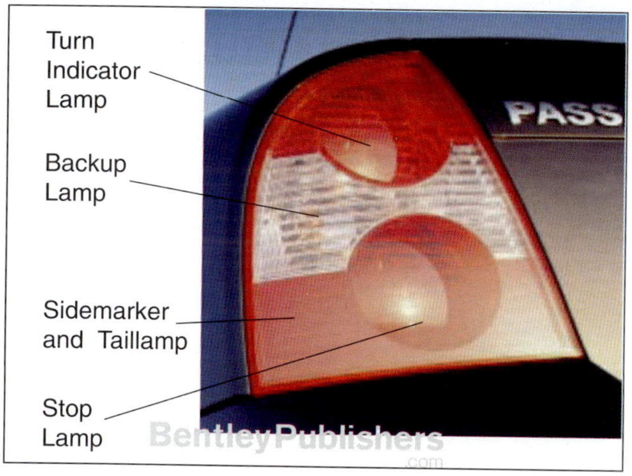

The rear end redesigns for the 2001 and later Passat sedan and wagon include new taillights. Although the wagon and sedan taillights are different shapes, both new designs include stop, turn, tail, sidemarker, and backup lamps in a single assembly.

The instrument cluster has blue backlighting and translucent red dials. The modified design and chrome-bezeled instruments improve appearance and legibility. The instruments are now operated by stepper motors. The instrument cluster is available in two versions, diffeing in the display options in the central display.

2001 - 2005 Instrument cluster

Instrument cluster, dashboard and controls

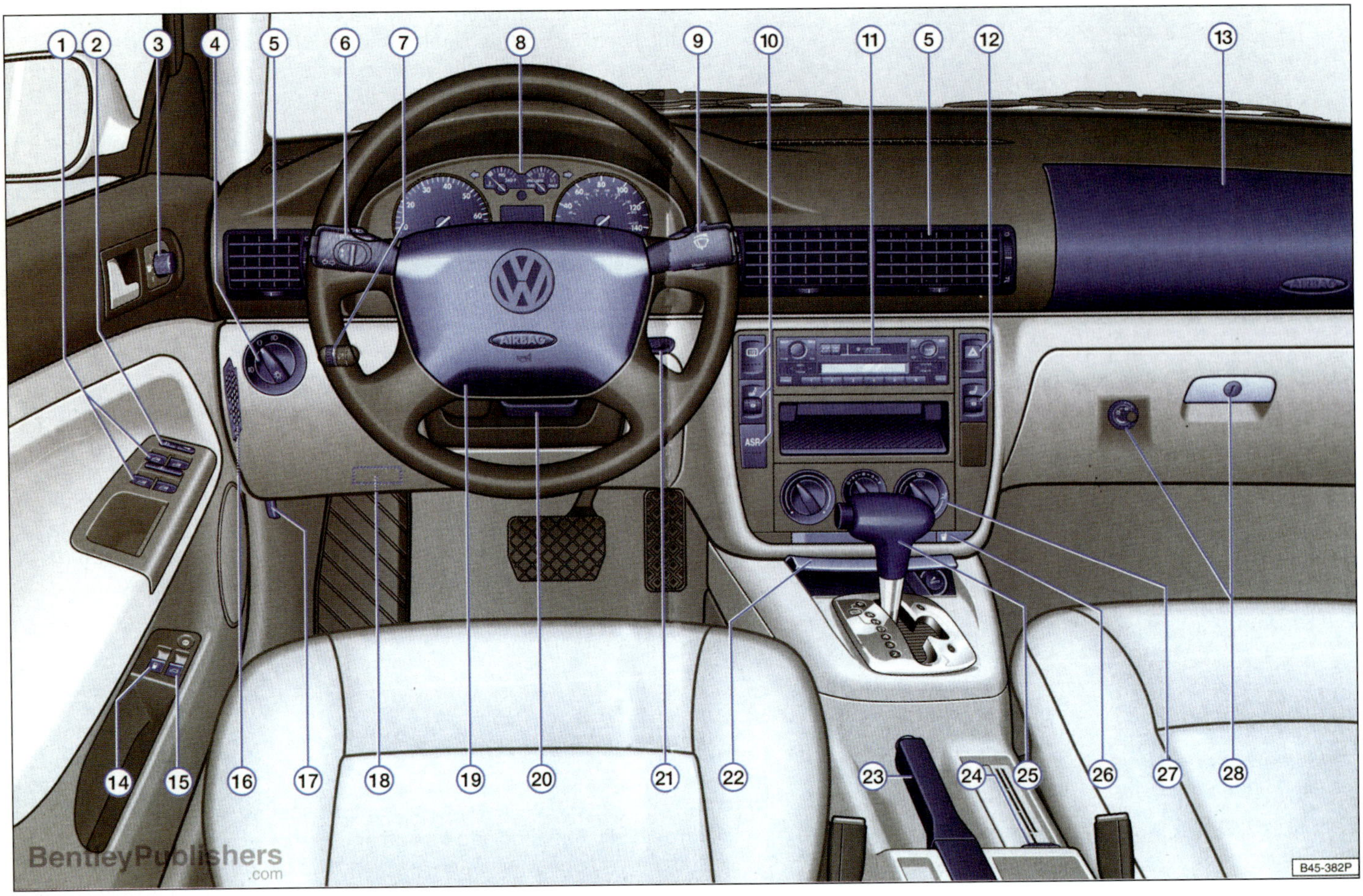

1. Power window switches
2. Power locking switch
3. Outside mirrors adjuster
4. Light switch
5. Air vents
6. Turn signals / high beam lever with buttons for cruise control
7. Instrument cluster illumination
8. Instrument cluster
9. Windshield wiper / washer lever with buttons for multifunction indicator
10. Switches for rear window defogger, heated front driver's seat, anti-slip regulation
11. Radio
12. Switches for emergency flasher and heated front passenger seat
13. Front passenger airbag
14. Fuel filler flap release switch (sedan models)
15. Switch and lock for rear lid
16. Fuse panel
17. Release lever for engine hood
18. Data link connector (DLC) for on-board diagnostics (OBD)
19. Horn (powered only when ignition switch is on), driver's airbag
20. Lever for adjustable steering wheel
21. Steering lock / ignition / starter switch
22. Ashtray, cigarette lighter
23. Parking brake lever
24. Fuel filler flap release switch (wagon models)
25. Selector lever
26. Cup holder
27. Climate controls for air conditioning, climatronic
28. Glove compartment, glove compartment air vent

Maintenance

Section

A

Maintenance

Page

Maintenance

Section

A

General Information

All of the maintenance work described in this section is important and should be carried out at the specified time or mileage interval. The Owner's Manual, the Maintenance Record, and the Warranty Booklet originally supplied with the car contain the maintenance schedules that apply to your Volkswagen. Following these schedules will ensure safe and dependable operation. In addition, many of the maintenance procedures are necessary to maintain warranty protection.

NOTE —
Volkswagen is constantly updating their recommended maintenance procedures and requirements. The information contained here may not include updates or revisions made by Volkswagen since publication of the Owner's Manual, the Maintenance Record, and the Warranty Booklet supplied with the car. If there is any doubt about what procedures apply to a specific model or model year, or what intervals should be followed, remember that an authorized Volkswagen dealer has the latest maintenance information.

CAUTION —
Disconnecting the negative (–) battery cable may erase fault codes and basic settings in the engine management and automatic transmission control modules. Some driveability problems may be noticed until the system re-adapts to operating conditions. OBD II readiness codes, which may be required for emissions testing, may also be erased. Convenience electronics (alarm system, interior light control, power locks, mirrors, and windows) may need to be re-set using a VAS 5051, VAG 1551/1552 or equivalent aftermarket scan tool.

How to use this manual

The manual is divided into several main sections:

Maintenance
Engine Mechanical
Fuel Supply System
Fuel Injection and Ignition
Transmission
Brake System
Suspension, Wheels, Steering
Body–Exterior
Body–Interior
Heating and Air Conditioning
Electrical Equipment
Radio
Wiring Diagrams and Component Locations
Scan Tool

Section **A Maintenance** covers the recommended schedules and service procedures needed to do the Volkswagen-specified scheduled maintenance work. This section has been prepared using the official factory service information as a guideline and by adding any necessary procedures that the editors of this manual have deemed relevant based on hands-on experience.

The remaining sections are repair oriented and are broken down into individual repair groups. Most sections begin with a general information repair group, e.g. **00 General Information.** These general groups frequently have technical data tables and information on how to identify code numbers of various components. The remainder of the repair groups in each section contain the more involved and detailed system repair information.

A listing of the various sections and the corresponding individual repair groups can be found at the front of the manual. At the beginning of each section is a comprehensive index of all procedures covered in that section. There is also a large black tab on the right-hand side of the first index page for each section to easily identify the beginning of the section.

At the end of the manual is a section (**Scan Tool**) that describes many of the numerous Diagnostic Trouble Codes (DTCs). These can be retrieved using a Volkswagen VAG 1551/1552 scan tool, VAS 5051/5052 diagnostic tool, or other aftermarket scan tool or scan tool program. This table lists the various scan tool codes in proprietary Volkswagen and generic P-code formats.

The information contained in this manual is based on official factory repair manual information. Due to the size and scope of the complete factory repair manual information set we have had to do some editorial content selection in an effort to keep this manual to a manageable size.

We have also tried to add additional information throughout the manual in an effort to clarify some of the official factory information. This information is easily identifiable as an ***EDITOR'S NOTE*** —.

Warnings, Cautions and Notes

Throughout this manual are many passages with the headings **WARNING**, **CAUTION**, or **NOTE**. These very important headings have different meanings.

> ***WARNING —***
> *The text under this heading warns of unsafe practices that are very likely to cause injury, either by direct threat to the person(s) doing the work or by increased risk of accident or mechanical failure while driving.*

> ***CAUTION —***
> *A caution calls attention to important precautions to be observed during the repair work that will help prevent accidentally damaging the car or its parts.*

NOTE —
A note contains helpful information, tips that will help in doing a better job and completing it more easily.

Please read every **WARNING**, **CAUTION**, and **NOTE** at the front of the manual and as they appear in repair procedures. They are very important. Read them before you begin any maintenance or repair job.

Some **WARNING**s and **CAUTION**s are repeated wherever they apply. Read them all. Do not skip any. They are important, even to the owner who never intends to work on the car.

Work safety

Although an automobile presents many hazards, common sense and good equipment can help ensure safety. Many accidents happen because of carelessness. Pay attention and stick to these few important safety rules.

> ***WARNING—***
>
> - *Never run the engine in the work area unless it is well-ventilated. The exhaust should be vented to the outside. Carbon Monoxide (CO) is an odorless and colorless gas. Carbon Monoxide (CO) in the exhaust kills.*
>
> - *Remove all neckties, scarfs, loose clothing, and jewelry when working near running engines or power tools. Tuck in shirts. Tie long hair and secure it under a cap. Severe injury can result from these items being caught in rotating parts.*
>
> - *Remove rings, watches, and bracelets. Aside from the dangers of moving parts, metallic jewelry conducts electricity and may cause shorts, sparks, burns, or damage to electrical system when accidentally contacting battery or other electrical terminals.*
>
> - *Disconnect the battery negative (–) (GND) cable whenever working on or near the fuel system, airbag system or anything that is electrically powered. Accidental electrical contact may damage the electrical system, cause fire, or result in serious personal injury.*
>
> - *Never work under a lifted car unless it is solidly supported on jack stands that are intended for that purpose. Do not support a car on cinder blocks, bricks, or other objects that may shift or crumble under continuous load. Never work under a car that is supported only by a jack.*
>
> - *The fuel system retains pressure even when ignition is off. Loosen fuel lines very slowly to allow residual pressure to dissipate. Cover fittings with a suitable shop cloth to avoid spraying fuel.*
>
> - *Fuel is highly flammable. When working around fuel, do not smoke or work near fire hazards. Keep an approved fire extinguisher handy. Be aware that water and building heaters may have standing pilot lights.*
>
> - *Illuminate the work area adequately and safely. Use a portable safety light for working inside or under the car. A fluorescent type light is best because it gives off less heat. If using a light with a normal incandescent bulb, use rough service bulbs to avoid breakage. The hot filament of an accidentally broken bulb can ignite spilled fuel or oil.*
>
> - *Keep sparks, lighted matches, and open flame away from top of battery. Hydrogen gas emitted by the battery is highly flammable. Any nearby source of ignition may cause battery to explode.*
>
> - *Never lay tools or parts in the engine compartment or on top of the battery. They may fall and be difficult to retrieve, become caught in belts or other rotating parts, or cause electrical shorts and damage to the electrical system.*

Identification Plates and Labels

Vehicle Identification Number (VIN)

The vehicle year, model, equipment level, and serial number can be determined from the 17 digit Vehicle Identification Number (VIN). The VIN is located on plates or stickers attached to the body in several locations. Additionally, a partial or production VIN is stamped into the body on the rear cross panel at the rear of the engine compartment and can be seen with the hood raised.

◄ The full 17-digit VIN is stamped into a plate attached to the dashboard (**arrow**). It is visible from outside the vehicle as required by Federal law.

◄ The full VIN or a partial (production) VIN is also stamped into the body on the rear cross panel or bulkhead (**arrow**) in the engine compartment at the beginning of the assembly process when the body is built. If this is a partial VIN, it will have the letter "Z" in several positions as place holders because the specific vehicle equipment level may not have been finalized at the time the body was produced.

Throughout this manual an abbreviated VIN (last 11 digits) are used to identify various running production changes. For example:

(➤ VIN 3B 1E050000) means "up to" vehicles with a production number (last six digits of VIN) of 050000

(VIN 3B 1E050001 ➤) means "from" vehicles with a production number of 050001.

The "1" (next to the "E") in the above VIN indicates the 2001 model year (m.y.). There is digit between the "B" and the "1" (denoted by a space in the above example) that is a mathematically calculated check digit, indicated with blank spaces.

NOTE—

Additional information on decoding the full 17-digit VIN is given in the **Foreword** *section at the front of this manual.*

◄ A vehicle data sticker (**arrow**) is located in the luggage compartment on the left side of the spare tire recess.

The data sticker is computer printed and lists the following vehicle information:

1. Production control number
2. Vehicle identification number (17 digits)
3. Model identification number (6 digits)
4. Model description/market area/engine output/emissions/ Transmission
5. Engine code (3 digits) and transmission code (3 digits)
6. Paint number (4 digits) and interior trim identification code (2 digits)
7. Optional equipment identification numbers (3 digits)

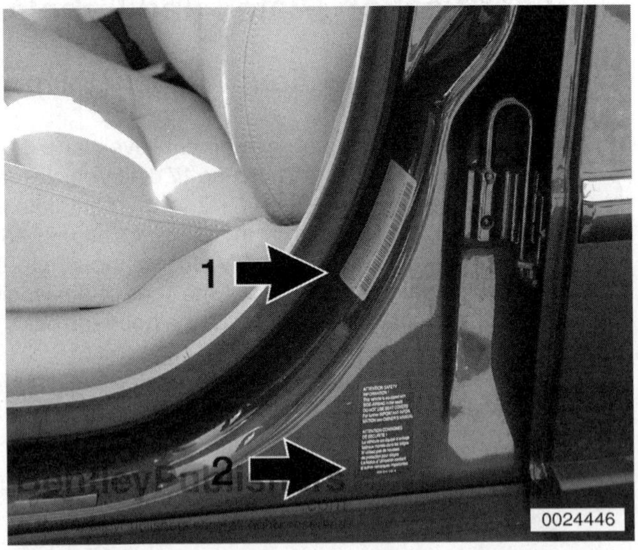

The information printed on this sticker is duplicated on the sticker attached to the factory-supplied maintenance book.

Additional vehicle identification information can also found on the certification sticker attached to the driver's side door post ("B" pillar).

◄ Door pillar certification sticker (**1**) lists the full VIN, vehicle weight ratings and the production date. Emission compliance certification is also listed on the sticker. Also note airbag warning label (**2**).

Engine identification

The engines used in vehicles covered by this manual are identified by a three letter code followed by a six digit number such as: AHA 000 307. Location of the engine code varies between engine types.

Engine Codes

- AEB. . . . . 1.8L 4-cylinder turbo, M5.9.2
- ATW, AUG, AWM . . . . 1.8L 4-cylinder turbo, ME7.5
- AHA. 2.8L 6-cylinder 5V, M5.9.2
- ATQ. 2.8L 6-cylinder 5V, ME7.1
- BDP. 4.0L 8-cylinder 4V W8, ME7.1.1

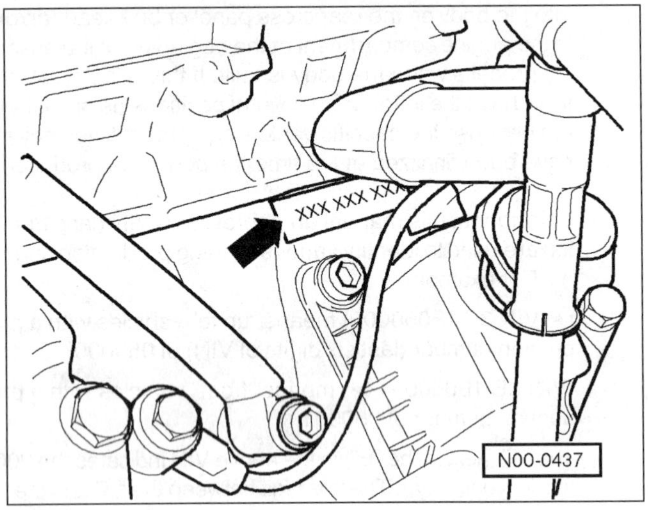

◄ **1.8 Liter:** Engines are stamped with the code letters and serial number on the left side of the cylinder block above the oil filter flange. Engine identification is also stamped on the cylinder head at the front lifting eye. A sticker with the code letters and serial number is usually attached to the cylinder head cover.

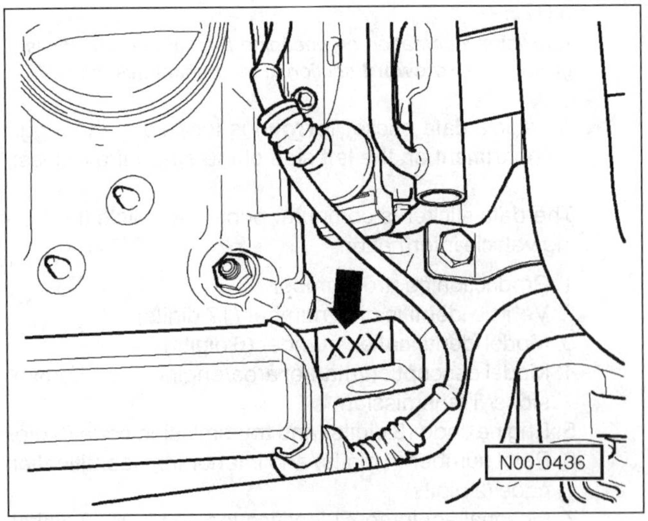

◄ **1.8 Liter:** Engine identification is also stamped on the cylinder head at the front lifting eye (**arrow**). A sticker with the code letters and serial number is usually attached to the cylinder head cover.

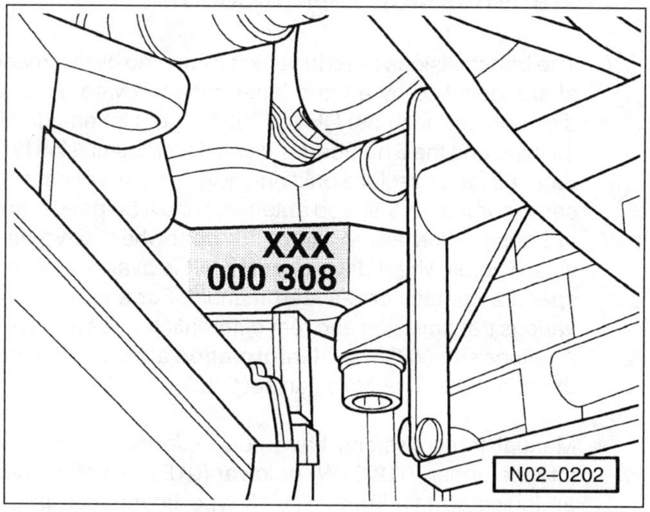

2.8 Liter: engines (engine code: AHA, ATQ) are stamped with the code letters and serial number on the engine block slightly to the right of the center line between the cylinder head and the hydraulic pump. Most engines will also have a sticker with the code letters and serial number attached to the drive belt cover.

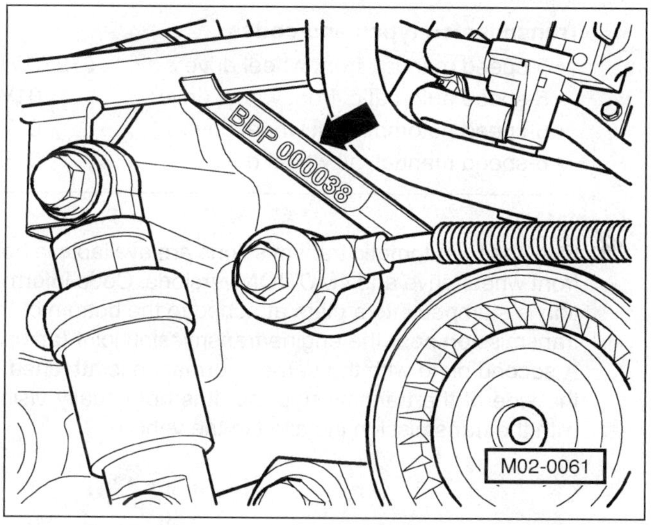

4.0 Liter: The engine code and number (**arrow**) can be found on the cylinder block on the front left.

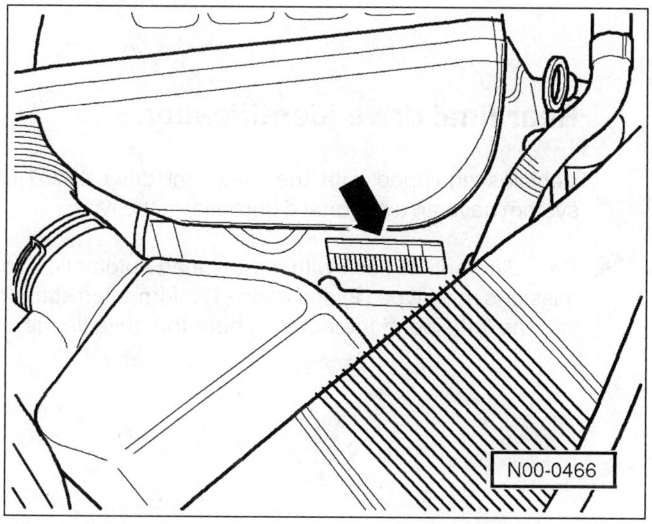

4.0 Liter: A sticker with the code letters and serial number is usually attached to the right side cylinder head cover trim.

Transmission identification

The transmissions used in vehicles covered by this manual are identified by a three letter code followed by a five digit number such as: DHW 17077. The 3 letters identify the type and the 5 numbers indicate the transmission build date. All transmissions built on any given day will have the same number. A 5-speed automatic and a 5-speed manual transmission are available for front-wheel drive vehicles. The all wheel drive (4MOTION) is available as a 5-speed automatic or 6-speed manual. For a listing of the various transmission and rear differential codes and specifications see **00 General Information** at the beginning of the applicable Transmission section.

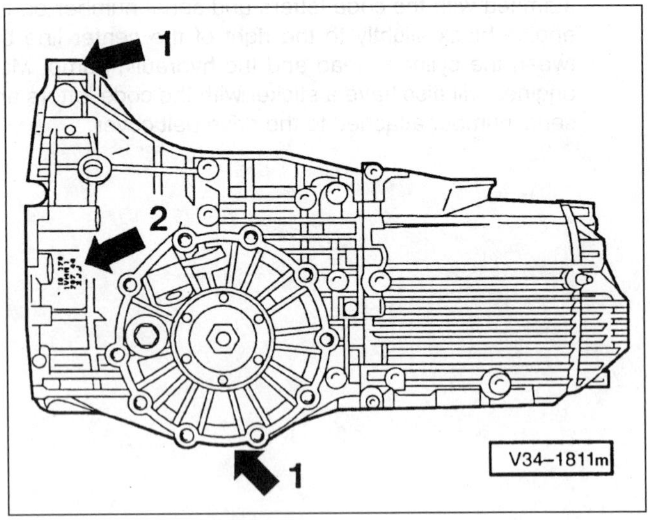

Manual transmissions have code information stamped into the upper (012/01W) or lower (01E) part of the case as indicated (**1**). Transmission type information is cast into the left side of the housing (**2**).

Transmission types and codes

- 5-speed manual, front wheel drive 012/01W
- 5-speed automatic, front wheel drive 01V
- 5-speed automatic, all wheel drive 01V
- 6-speed manual, all wheel drive 01E

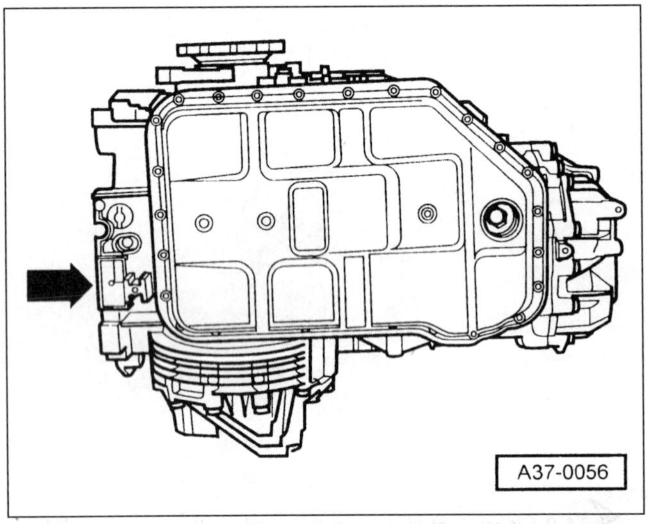

Five speed automatic transmissions are available in both front wheel drive and 4MOTION versions. Code information is stamped into a plate attached to the bottom of the transmission near the engine/transmission joint (**arrow**). A second plate with the same information is attached to the side of the transmission, but it is not usually visible with the transmission installed in the vehicle.

Rear final drive identification

Vehicles equipped with the all-wheel drive 4MOTION system have an additional differential in the rear.

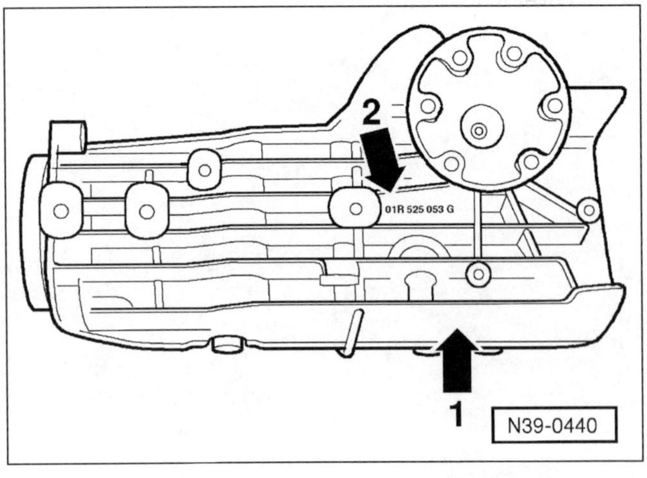

Rear differentials used with the 5-speed automatic transmissions have type (**2**) and code (**1**) information stamped into the left side of the housing near the drive flange.

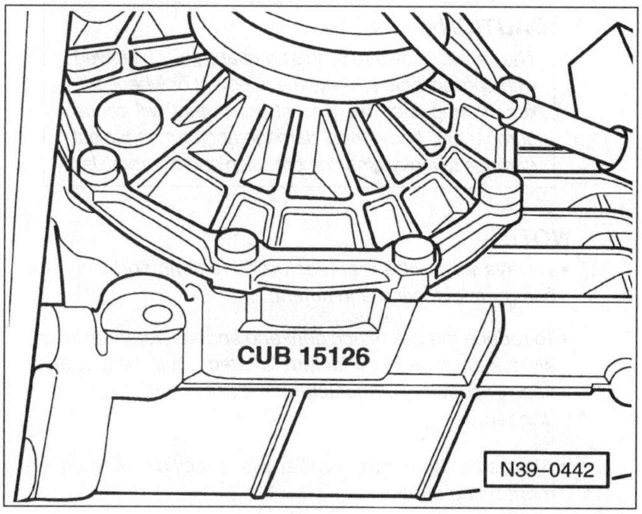

If the rear final drive is to be replaced, it is absolutely necessary to note the code letters of the rear final drive (In this illustration, CUB 15126). The transmission and engine codes must also be noted. This is the only way to insure correct application.

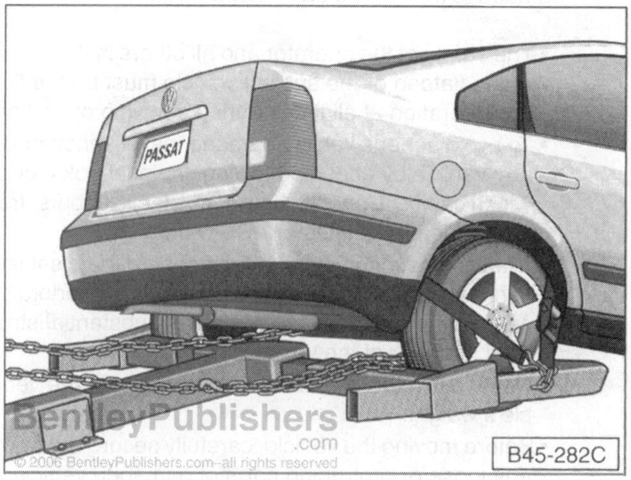

Towing

Emergency towing with commercial equipment, all vehicles

If it becomes necessary to tow a Passat, particular attention must be given to the transmission type.

- Front wheel drive manual transmission vehicles can be towed with a **FRONT** hook-up (rear wheels on the ground) using commercial wheel lift equipment. There is no distance limitation, but maximum speed should not exceed 50 MPH (80 KPH).
- Front wheel drive automatic transmission vehicles can be towed with a **FRONT** hook-up (rear wheels on the ground) using commercial wheel lift equipment. There is no distance limitation, but maximum speed should not exceed 50 MPH (80 KPH).
- Front wheel drive manual transmission vehicles can be towed with a **REAR** hook-up (front wheels on the ground) using commercial wheel lift equipment. Maximum distance should not exceed 50 miles (80 KM), and maximum speed should not exceed 50 MPH (80 KPH).
- Front wheel drive automatic transmission vehicles cannot be towed with the front wheels on the ground. If circumstances do not permit a front hook-up, then the vehicle must be transported with flatbed equipment. Both front and rear loading is allowable.
- 4MOTION vehicles can only be towed with flatbed equipment. Both front and rear loading is allowable.

If the vehicle must be moved over distances greater than noted above, commercial flatbed equipment must be used for transport.

The following information is for use by commercial flatbed truck operators who know how to operate their equipment safely. To prevent damage to the vehicle, read and understand all of the following information before proceeding.

> ***WARNING —***
> *Never allow passengers to ride in a towed vehicle for any reason.*

0024430

0024431

CAUTION—
The vehicles covered by this manual cannot be towed with conventional sling-type equipment or wheel dollies. Towing with this type of equipment will cause bumper and body panel damage and/or mechanical damage to the engine, transmission and differentials.

NOTE—
- *Always follow the manufacturer's recommendations for commercial towing equipment.*
- *To reduce the risk of accident and serious personal injury, always observe manufacturer's rated equipment capacities. Exceeding manufacturer's design specifications is dangerous.*
- *Always be thoroughly familiar with specific towing equipment being used.*
- *Always observe all Federal, State, Provincial and Local laws regarding such items as warning signals, night illumination, speed, licensing, etc.*
- *To load the vehicle onto flatbed equipment, attach brindle to the oblong holes in the front or rear long member.*
- *Attach safety chains to unused oblong hole in longmember at other end of vehicle.*
- *Attach safety straps through appropriate vehicle wheels.*
- *Reinstall rubber sealing plugs at completion.*
- *Allow approximately 12 - 18 inches (30 - 46 cm) between the tires and the ground when towing with wheel lift equipment.*

The following additional points should be observed to insure the safety of all personnel involved and to prevent damage to the vehicle and flatbed truck.

- The safety of the operator and all others in the vicinity of the flatbed or the subject vehicle must be the first consideration at all times during a towing operation.
- To avoid serious vehicle damage, never winch or pull any vehicle by attaching towing chains, hooks, or winch cables to shock absorbers, stabilizer bars, front strut rods, brake lines or ABS wiring.
- Always inspect points of attachment to the disabled vehicle. If they appear to be damaged or deteriorated, select other attachment points at a substantial structural member of the vehicle.
- Never allow the fuel tank to support any of the vehicle's weight.
- Before moving the vehicle, carefully secure or remove any loose or protruding parts of damaged vehicles; such as hoods, fenders, trim, etc.
- Never go under the vehicle while it is being lifted by flatbed equipment, unless the vehicle is adequately supported by safety stands.
- Safe operations always depend on weather, road, traffic, and visibility conditions in addition to the condition of the disabled vehicle.

Towing eyes

Attachment points for vehicle recovery in the form of towing eyes or anchorages are provided at the front and rear of the vehicle on the right side.

When winching a disabled vehicle for recovery purposes, use the towing eyes (also known as loops or anchorages) and observe the following points:

- Attach the winch cable hook to the appropriate towing eye (front or rear).
- Check carefully to make sure that the hookup is secure before moving the vehicle.

> ***WARNING —***
> - *To prevent damage and the risk of personal injury and because rope towing requires considerable skills for the driver of the pulling car and the driver of the car to be pulled, it is strongly recommend that the services of a commercial tow truck be obtained instead of attempting to tow the vehicle by the towing eyes.*
> - *The towing eyes must only be used in exceptional cases and only if the pulling forces are light to moderate, for example, when the car is slightly stuck in mud or snow.*

The front towing eye is accessed by removing the intake grill on the lower right side. The rear towing eye is below the right side of the rear bumper.

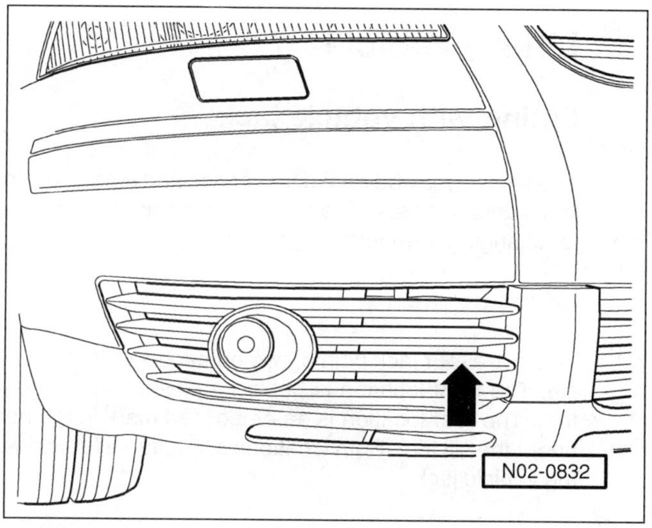

Removal of front intake grills vary slightly depending on model year and equipment. In some cases, the screwdriver from the vehicle tool kit can be used. In other cases, the grill can be simply pulled out.

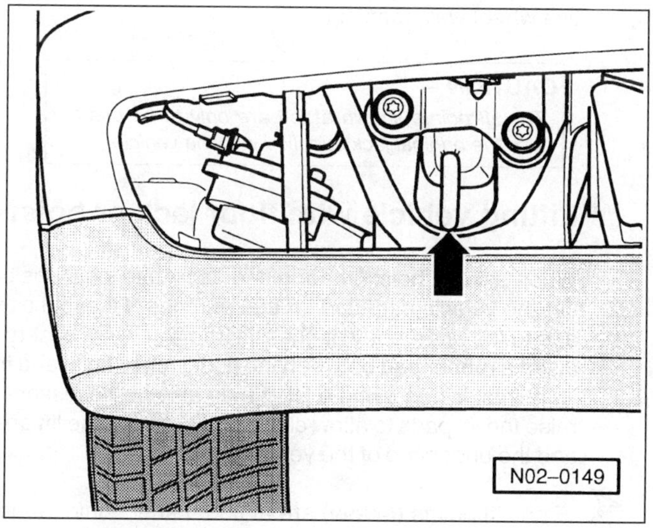

Front towing eye (**arrow**) is welded to body structure.

Rear towing eye (**arrow**) is welded horizontally to body structure.

Lifting Vehicle

Lifting with vehicle jack

The vehicle may be raised for service and repairs by floor jacks and/or hoists. The vehicle jack must only be used for changing a wheel.

Lifting points

There are four jack points, two on each side of the vehicle. The front location is behind a plastic flap in the sill trim. The rear location is an embossed mark in the body under the sill trim. Both locations are suitable for use with the vehicle jack.

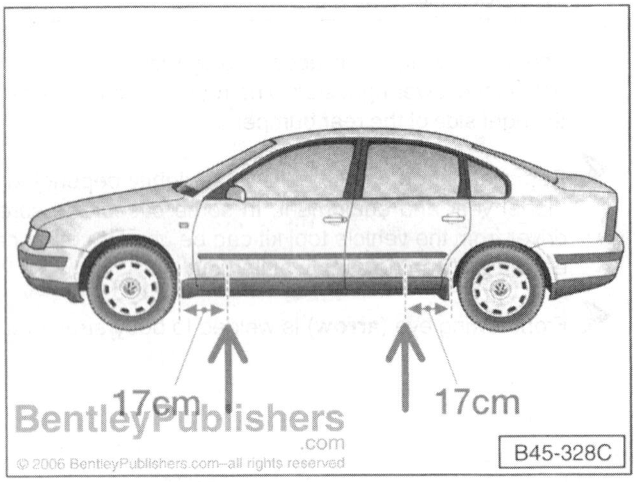

Lift points for the vehicle jack (large **arrows**) are approximately 7 inches (17 cm) from the edges of their respective wheel well openings.

> ***CAUTION—***
> *The lift points shown at left are only to be used with the original jack supplied with the vehicle.*

Lifting vehicle with floor jack or hoist

When using a hoist (lift) or a floor jack, use locations under the vehicle intended for that purpose. To avoid damage to the underbody or chassis frame, it is necessary to insert a rubber pad between the floor jack or hoist and the lift points. When using a lift, it may also be necessary to raise the lift pads to allow clearance between the lift arms and the underside of the vehicle

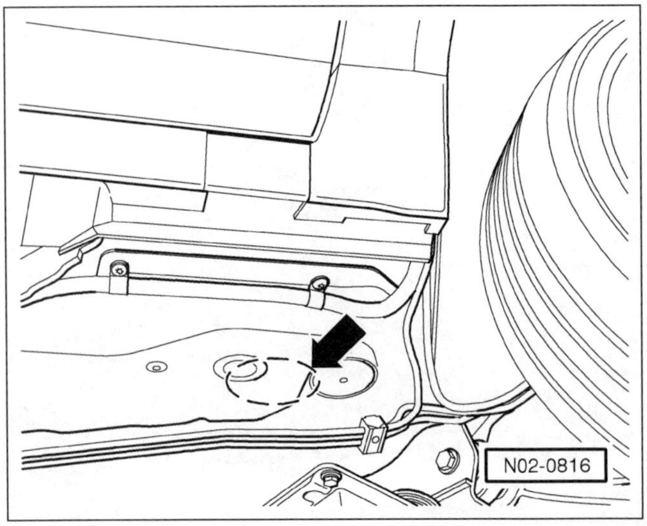

Front lift points (**arrow**) are at floor panel reinforcement on each side of the vehicle.

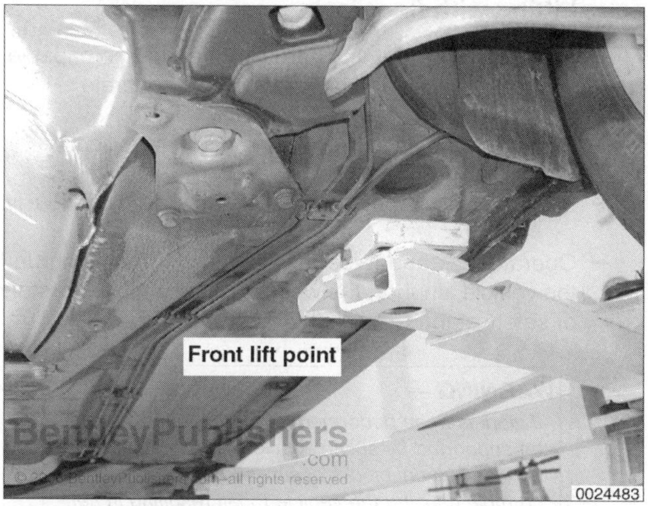

Front lift point shown with lift arm in position.

> ***CAUTION —***
> - *Vehicle may only be lifted at points indicated in order to avoid damaging vehicle floor pan and to prevent the vehicle from tipping.*
> - *Do not lift vehicle at engine oil pan, transmission, rear differential, or on front or rear axle as serious damage may result.*
> - *Never start engine and engage a gear with vehicle lifted if any drive wheel has contact with the floor! There is severe danger of an accident if this is not observed, especially with all-wheel drive (4MOTION) vehicles.*

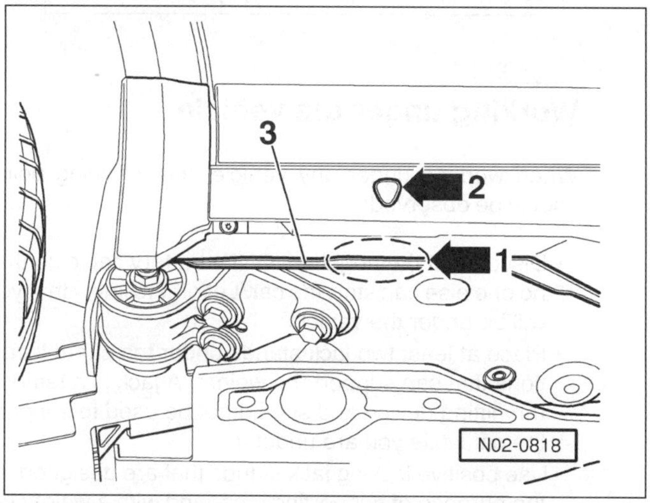

Rear lift point is on sill panel vertical reinforcement in the indicated area (**1**). The center of the jack or hoist pad must be aligned with marking (**2**). Insure that the center vertical reinforcement (**3**) makes contact with the jack or hoist pad.

Lifting with hoist (lift)

– Position the vehicle on the hoist being sure that there is sufficient clearance between the hoist mechanism and lifting arms and the underside of the vehicle.

Position the lift arms at the appropriate points on the vehicle body according to the hoist manufacturers instructions and lock in place.

– Raise the hoist until the lift pads just contact the lift points and verify correct placement.

– Continue raising the hoist to the desired height and secure according to hoist manufacturer's instructions.

> ***WARNING —***
> *When removing heavy components such as the engine or transmission, anchor the vehicle to the hoist or support the vehicle from the opposite end with suitable safety stands. Failure to support the vehicle will cause it to slip off the hoist due to the change in the center of gravity resulting in serious damage and/or personal injury. Observe the same precautions when re-installing heavy components.*

Lifting with floor jack

– Place car on a flat level area with a surface capable of supporting jacks and jackstands, e.g. concrete.

– Block the wheels to prevent the car from rolling.

– Place jack at proper lifting points as shown previously.

– Operate jack and raise the car slowly. If working under the vehicle, support the weight using suitable jackstands or safety stands.

> ***WARNING —***
> *If work is to be performed under the vehicle it must be supported by suitable jack or safety stands. Do not use cement blocks or similar materials; they will crumble causing the vehicle to fall resulting in serious damage and/or personal injury.*

Working under the vehicle

When working under any vehicle, the following points should be observed:

- Disconnect the negative (GND) battery cable so that no one else can start the car. Let others know that you will be under the vehicle.
- Place at least two jack stands under the car at locations that can support the weight. A jack is a temporary lifting device and should not be used to support the car while you are under it.
- Use positive locking jack stands that are designed for the purpose of supporting a car and with a weight rating sufficient for the vehicle.
- Do not place jack stands under sheet metal panels or engine, transmission, rear differential or drive axles.
- If you are using a hoist or lift, be sure that the safety locks are engaged and that the weight of the vehicle is resting on the locks, not the hydraulic system.

> ***WARNING —***
> *When working on a hoist or lift, be sure to remove all safety stands before lowering vehicle!*

- Lower car slowly until its weight is fully supported by the jack stands or hoist. Watch to make sure that jack stands do not tip or lean as the car settles on them and that jack stands and hoist arms are placed solidly and will not move.

> ***WARNING —***
> *Check to make sure vehicle is stable before working under car.*

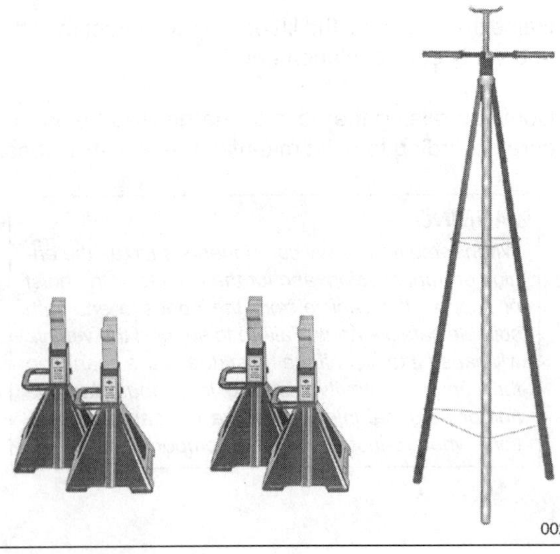
0024408

When working under the vehicle, it must be properly supported with the appropriate jack stands or safety stands.

- Observe all jacking precautions again when raising the car to remove the jack stands.

Engine Compartment Maintenance

Engine compartment, visual inspection

Visually inspect engine and engine compartment for leaks and damage. Also inspect lines, hoses, and connections for fuel system, brake system, cooling system, heating and air conditioning system for leaks, worn areas, porosity, and deterioration.

Fluid levels must also be between specified MIN and max marks on the appropriate reservoirs and tanks.

The following illustrations show representative engine compartment layouts for the three major engine types. There may be minor differences between model years, such as a cover over the battery.

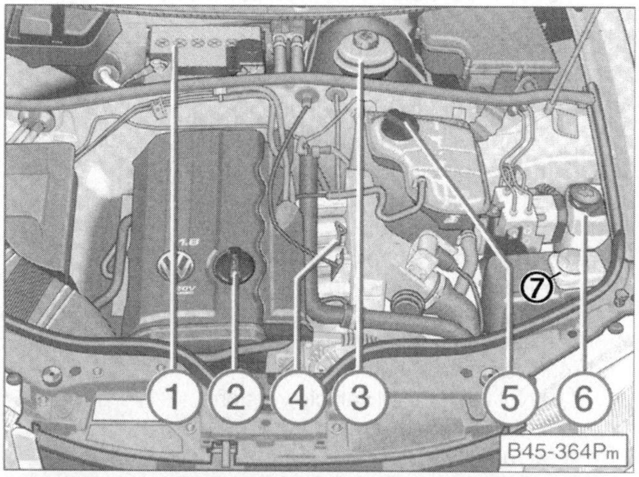

◄ Engine compartment, 1.8 liter 4-cylinder turbo engine

1. Battery
2. Engine oil filler cap
3. Brake fluid reservoir
4. Engine oil dipstick
5. Coolant expansion tank
6. Windshield washer tank
7. Power steering fluid reservoir

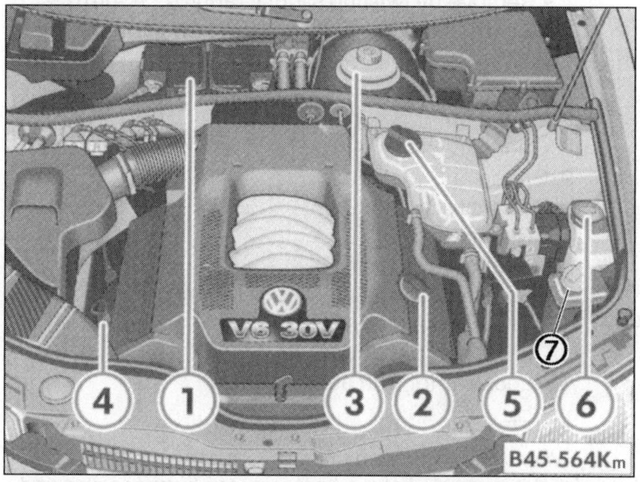

◄ Engine compartment, 2.8 liter V6 engine

1. Battery (some versions under trim cover)
2. Engine oil filler cap
3. Brake fluid reservoir
4. Engine oil dipstick
5. Coolant expansion tank
6. Windshield washer tank
7. Power steering fluid reservoir

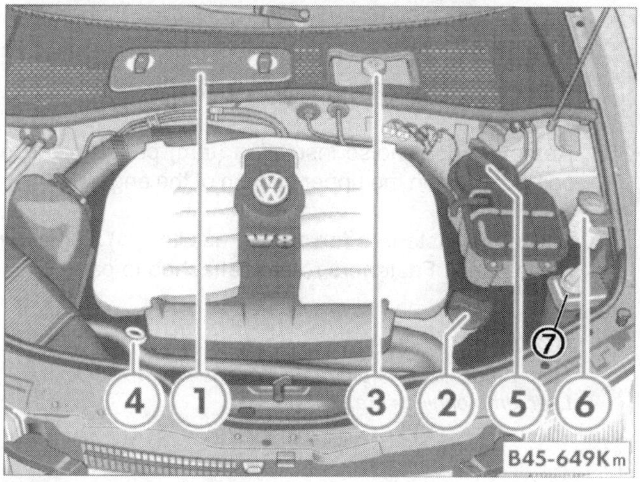

◄ Engine compartment, 4.0 liter W8 engine

1. Battery (under trim cover)
2. Engine oil filler cap
3. Brake fluid reservoir
4. Engine oil dipstick
5. Coolant expansion tank
6. Windshield and headlight washer tank
7. Power steering fluid reservoir

Engine covers (noise insulation panels), lower, removing and installing

Remove lower noise insulation panels (belly pans) from below the engine to access lower engine and transmission components.

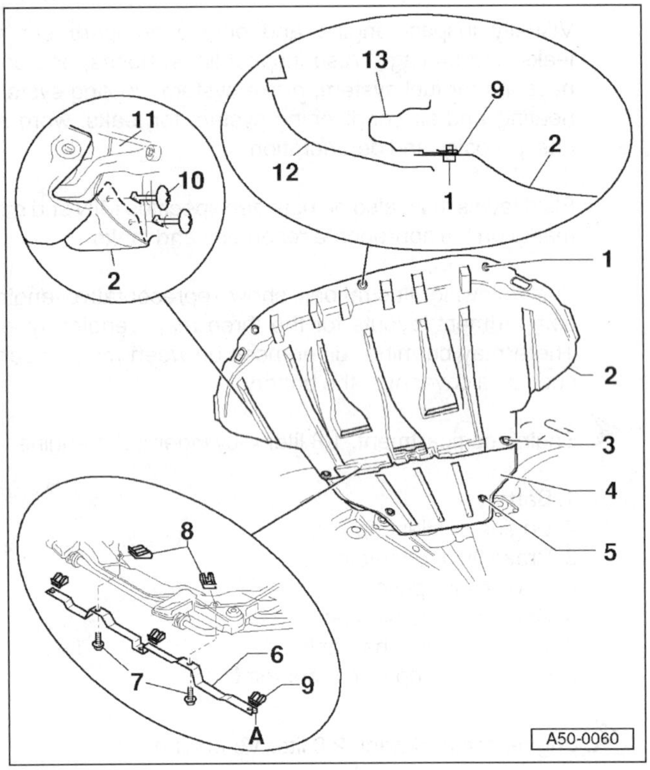

Remove and/or loosen fasteners as noted. Slide front panel to the rear, off of bumper cover to remove. All engine and vehicle types are similar, but some versions do not use rear noise insulation panel (**4**).

1. Quick release screw
   - Quantity 3
   - Tighten to 2 Nm (18 in-lb)
   - Remove by twisting slightly (approx. ¼ turn)
2. Front noise insulation panel
3. Quick release screw
   - Quantity 3
   - Tighten to 2 Nm (18 in-lb)
   - Remove by twisting slightly (approx. ¼ turn)
   - If rear noise insulation panel (**4**) is installed, use longer screws
4. Rear noise insulation panel
   - Not used on all vehicle versions
5. Quick release screw
   - Quantity 2
   - Remove by twisting slightly (approx. ¼ turn)
6. Noise insulation mounting reinforcement
7. Bolt
   - Quantity 2
8. Speed nut
   - Quantity 2
9. Speed nut
   - Quantity 3
10. Clamping pin
    - Quantity 4
    - Remove by twisting slightly (approx. ¼ turn)
11. Wheel housing liner
12. Front bumper cover
13. Lock carrier with attachments

***NOTE*—**

When installing lower front noise insulation panel, be sure that it is inserted at the front, over the bumper cover, and that all fasteners are properly installed.

Engine covers (noise insulation panels), upper, removing and installing

Remove upper noise insulation (trim) panels to access components on the upper section of the engine.

1.8T: Turn fasteners (**arrows**) a quarter (¼) turn and remove panel. Fasteners remain attached to panel.

***NOTE*—**

Later production 1.8 Liter turbo engine shown, earlier 1.8 Liter turbo engines are similar.

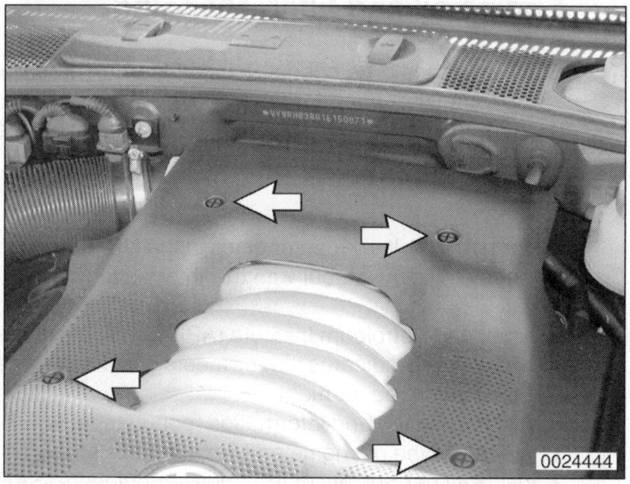

V6: Turn fasteners (**arrows**) a quarter (¼) turn and remove panel. Fasteners remain attached to panel.

NOTE—

Later production 2.8 Liter engine shown, earlier 2.8 Liter engines are similar.

Engine oil level, checking

It is normal for your engine to consume a small amount of oil. The rate of consumption will depend on the quality and viscosity of the oil and operating conditions. Make it a habit to check oil level at every fuel filling.

– After switching ignition OFF, wait at least 3 minutes to allow the oil to flow back into the oil pan.

– Remove dipstick and wipe with a clean cloth. Fully re-insert dipstick into tube.

– Remove dipstick again and read oil level.

– Two different types of dipsticks are used.

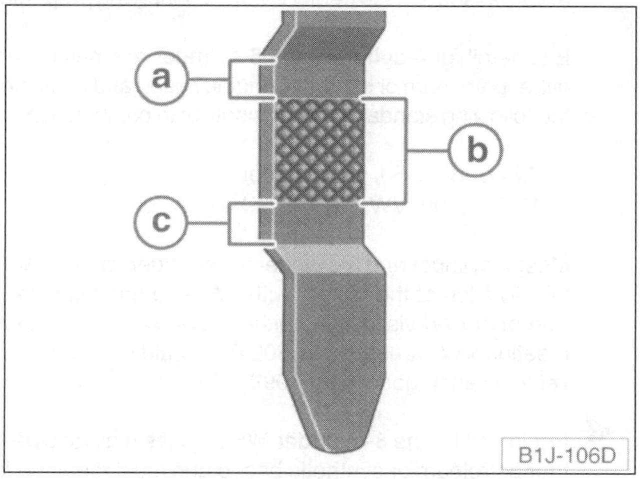

Dipstick is divided into 3 areas:

A. Maximum level, **DO NOT** add additional oil.
B. Oil **CAN** be added and can rise to the **A** level, but must not be above the **A** range.
C. You **MUST** add oil so that level is within the **B** range, but not above the **A** range.

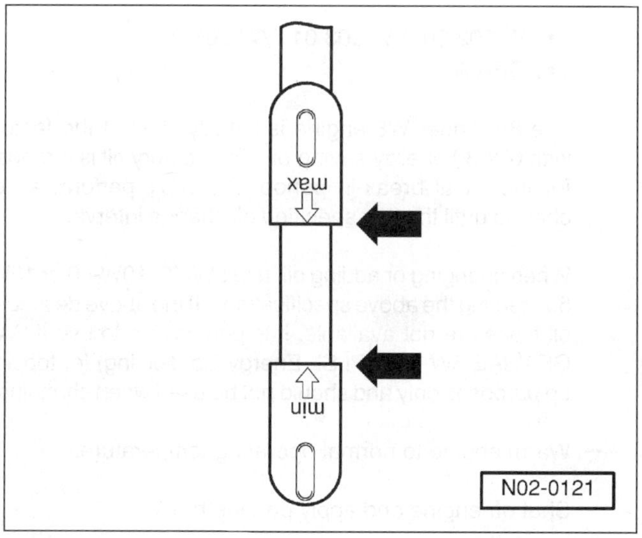

The oil level **MUST** be between the MIN and the MAX marks on the dipstick. Ensure that oil level is not above the MAX mark.

> ***WARNING—***
> *DO NOT OVERFILL! If the oil level is above the **A** range or the **MAX** mark as shown on the above examples, serious damage to the catalytic converter(s) can result.*

If oil must be added, be sure to observe specifications regarding type and viscosity. Always re-check level after adding. Ensure dipstick is fully re-inserted dip into tube upon completion.

Engine oil and filter, changing

Engine oil should be changed according to manufacturers recommendations for time and/or mileage. Consult the appropriate maintenance table at the end of this section for intervals.

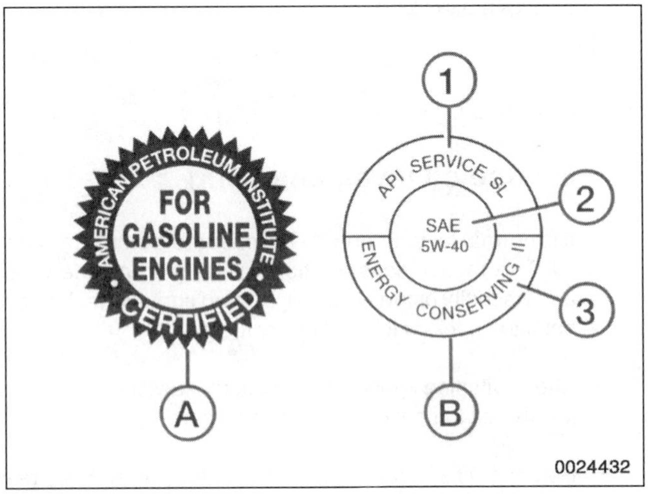

◄ Oil for use in Volkswagen engines in the USA and Canada should display a label with this information:

A. American Petroleum Institute (API) Certification "starburst" label, indicates that oil meets gasoline engine certification requirements.
B. API application and information "donut" label.
1. Indicates type of engine and service classification.
2. Viscosity per Society of Automotive Engineers (SAE) standards.
3. Oil passes tests to certify energy conserving properties.

Engine oil for **4-cylinder** and **6-cylinder** engines may be either petroleum or synthetic engine based and must meet the following standards either singly or in combination:

- API Service SJ, SL or better.
- VW 502 00, VW 501 01, or VW 500 00

Most 4-cylinder and 6-cylinder Volkswagen engines were initially filled at the factory with 5W-40 energy saving oil. The preferred viscosity is either 5W-40 or 10W-40. Oils meeting VW standard VW 503 00 should not be used on vehicles after model year 1999.

◄ Engine oil for the **8-cylinder W8** engines may also be either petroleum or synthetic based and must meet the following standards either singly or in combination:

- VW 502 00, VW 503 01, VW 505 01
- ACEA A3

The 8-cylinder W8 engine is initially filled at the factory with 0W-30 energy saving oil. This factory fill is intended for the initial break-in period. DO NOT perform an oil change until the first specified oil change interval.

When changing or adding oil, use 5W-40, 10W-40 or 15W-50 meeting the above specifications. If the above described oil types are not available, it is permissible to use ILSAC GF3 SAE 5W-30 (API SL Energy Conserving) for topping up purposes only and should not be used when changing.

– Warm engine to normal operating temperature.

– Shut off engine and apply parking brake.

– Raise car with floor jack or hoist and properly support. Use jack stands and/or safety stands where appropriate.

> ***WARNING —***
> *Jack stands should be placed on hard level surface (e.g. concrete).*

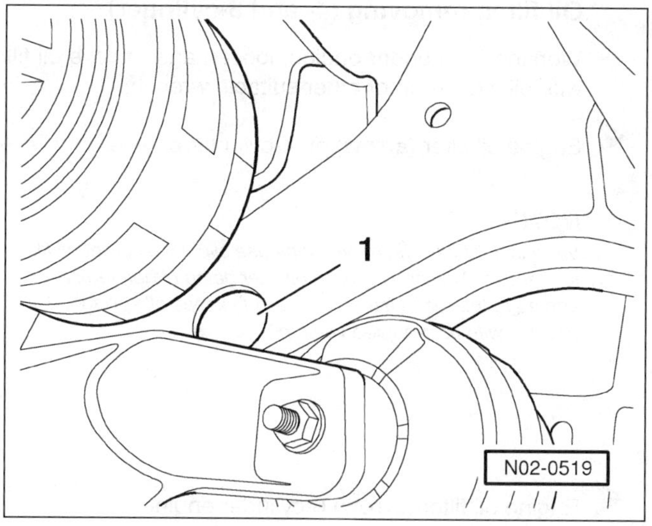

– Remove lower noise insulation panel (belly pan) from beneath engine compartment.

< On 4-cylinder and 6-cylinder vehicles, cover hole (**1**) in engine/transmission carrier with suitable shop rag. This prevents old oil from entering carrier.

– Position a suitable container under vehicle and remove drain plug from oil pan. Allow oil to drain completely into the container.

> ***WARNING —***
> *Hot oil can scald. Wear protective clothing, gloves and eye protection.*

> ***CAUTION —***
> - *Always replace oil drain plug and/or sealing ring (or drain plug as required) anytime the drain plug is removed.*
> - *Dispose of oil properly at a facility equipped for recycling or storage.*

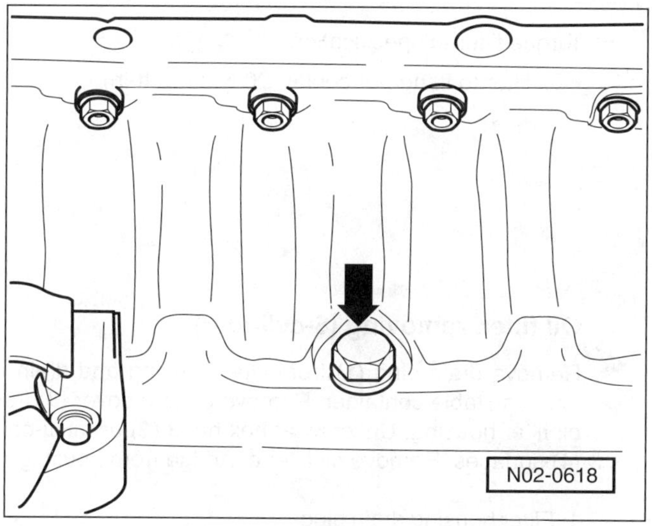

< Oil drain plug (**arrow**) for 4- and 6-cylinder engines.

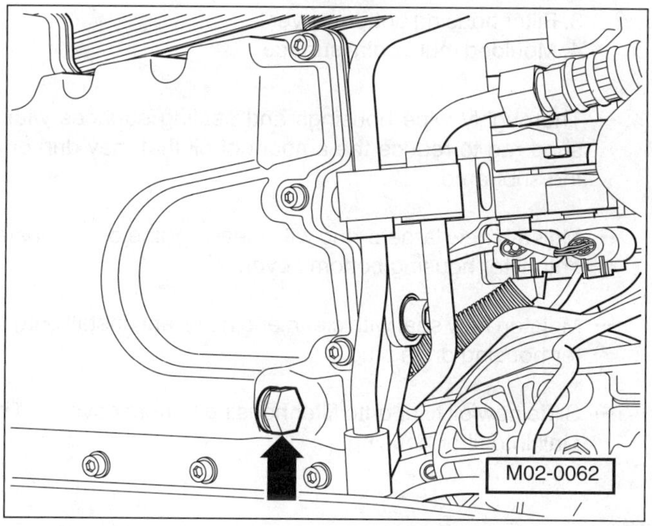

< Oil drain plug (**arrow**) for 8-cylinder engine.

Oil filter, removing (4- and 6-cylinder)

- Working from under engine, loosen and remove oil filter with oil filter strap or other suitable wrench.

◄ Engine oil filter (**arrow**) on 4-cylinder engine.

NOTE—

Various 1.8 Liter engine versions use slightly different oil filter flanges. This can result in oil filter being either vertical or angled to the rear. Use care to avoid excess oil spillage, especially with rear-angled version.

◄ Engine oil filter (**arrow**) 6-cylinder engine.

- Clean sealing surface on oil filter flange and/or oil cooler.
- Lightly lubricate rubber seal with fresh engine oil.
- Thread on new filter and tighten by hand.
- Torque filter to specification.
  - Oil filter to flange or cooler: 20 Nm (15 ft-lb)

Oil filter, removing (8-cylinder)

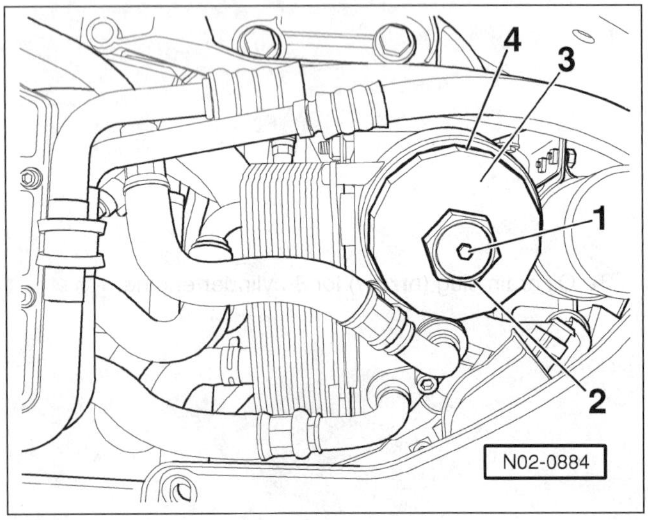

◄ Remove drain plug (**1**) from filter housing and drain oil into a suitable container. Remove bottom cover (**3**) from oil filter housing. Unscrew at hex head (**2**) or multi-point (**4**) surfaces. Remove oil filter cartridge from housing.

1. Filter housing drain plug
2. Moulded hex head in cover
3. Filter housing bottom cover
4. Moulded multi-point surface

- Thoroughly wipe housings and sealing surfaces with a shop rag to reduce the amount of oil that may drip onto the subframe.
- Moisten new large o-ring with clean engine oil and install onto filter housing bottom cover.
- Moisten new seal with clean engine oil and install onto filter housing drain plug.
- Slide new oil filter onto filter housing bottom cover and install into filter housing.

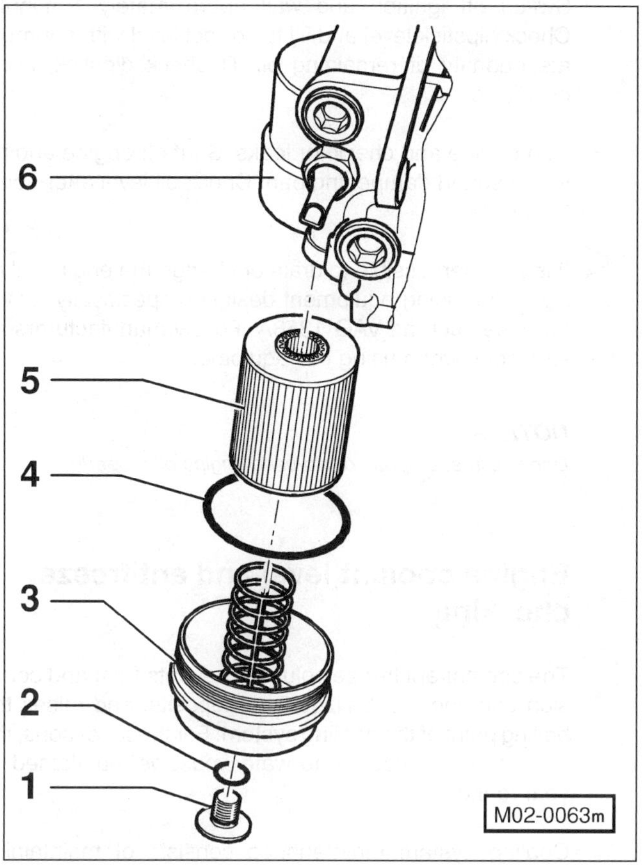

Component parts of 8-cylinder oil filter assembly.

1. Filter housing drain plug
   - Tighten to 10 Nm (7 ft-lb)
2. Filter housing drain plug seal
3. Filter housing bottom cover
   - Tighten to 25 Nm (18 ft-lb)
4. Filter housing bottom cover o-ring seal
5. Oil filter cartridge
6. Filter housing

Continued for all vehicles

– Install new oil drain plug and/or new seal. Tighten drain plug to proper torque.

EDITOR'S NOTE —
Most engine codes will require replacement of the drain plug. Volkswagen no longer supplies the seal; it is now integral with the drain plug.

Oil drain plug to oil pan tightening torque
- 1.8L . . . . .30 Nm (22 ft-lb)
- 2.8L . . . . .30 Nm (22 ft-lb)
- 4.0L . . . . .30 Nm (22 ft-lb)

– Add oil of the proper type to engine except for approximately the last ½ liter (½ quart).

Engine oil capacity (with filter)
- 1.8L (1998 - on) . . . . .3.5 liters (3.7 qt.)
- 1.8L (with long-life filter "06A") . . .3.9 liters (4.1 qt.)
- 2.8L (1998) . . . . .5.7 liters (6.0 qt.)
- 2.8L (1999) . . . . .6.0 liters (6.3 qt.)
- 2.8L (2000 - on) . . . . .6.2 liters (6.5 qt.)
- 4.0L (2002 - on) . . . . .8.5 liters (9.0 qt.)

NOTE —
Quantities listed above are approximate. DO NOT OVERFILL! Overfilling engine oil can cause misfire DTCs to be stored in Engine Control Module (ECM).

– Start engine and allow to run at idle until operating temperature is approximately 140°F (60°C). Do not rev engine until oil light goes out.

CAUTION —
- *After changing the engine oil and the oil filter, observe the following at the first engine start:*
- *The engine must only run idle speed as long as oil pressure warning light in the instrument cluster is on. Do not rev the engine! If the engine is revved with the warning light on, the turbocharger on 1.8L engines could be damaged or fail completely.*
- *Full oil pressure is not reached until the warning light has gone out. Only then can the engine be revved.*

V.A.G 1358 A

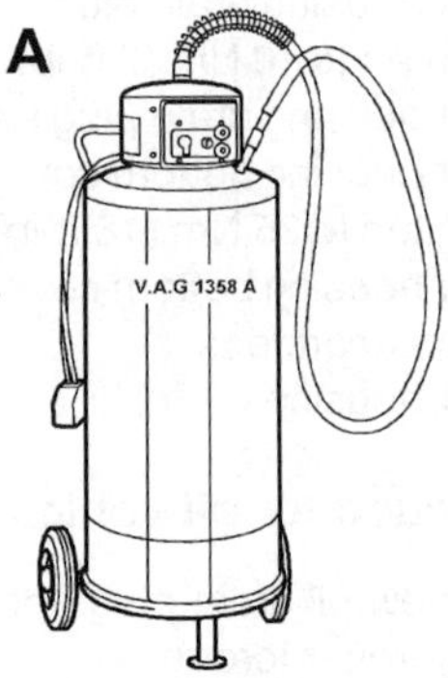

W00-0438

- Switch off ignition and wait approximately 1 minute. Check dipstick level and fill to correct level with appropriate quantity of remaining oil. Recheck dipstick; avoid overfilling.

- Run engine and check for leaks. Shut off engine and re-install sound dampening pan. Check oil level after about 3 minutes.

◄ It is also permissible to drain or change the engine oil by siphoning using equipment designed specifically for the purpose such as VAG 1358A. Follow manufacturers instructions when using this equipment.

NOTE—
Don't pollute! Dispose of old/used engine oil properly.

Engine coolant level and antifreeze, checking

The coolant/antifreeze solution prevents frost and corrosion damage, neutralizes lime deposits, and raises the boiling point of the cooling system. For these reasons, the proper ratio of coolant to water must be maintained all year round.

Cooling system maintenance consists of maintaining coolant level, checking coolant freezing point, and inspecting hoses. Coolant flushing is not part of Volkswagen's scheduled maintenance.

◄ Freeze protection chart for Volkswagen-supplied G12 coolant/antifreeze in quarts and degrees Fahrenheit.

Antifreeze Protection Levels (Quarts and °F)

		Quarts of antifreeze required for protection to temperature shown (°F)						
		2	**3**	**4**	**5**	**6**	**7**	**8**
Total cooling system capacity (quarts)	**5**	-12	-62					
	6	0	-34					
	7	6	-17	-54				
	8		-7	-34	-69			
	9		0	-21	-50			
	10		4	-12	-34	-62		
	11		8	-6	-23	-47		
	12		10	0	-15	-34	-57	
	13			3	-9	-25	-45	
	14			6	-5	-18	-34	-54

Notes:

For best overall protection, antifreeze and water mixture should be within the shaded areas. This represents an antifreeze concentration of between 40% and 60% which is preferred.

When new, vehicles meant for sale in the USA are shipped with a 50/50 coolant/water mixture. Canadian vehicles get a 60/40 coolant/water mixture.

Under no circumstances should the engine be run with 100% antifreeze (or other high concentrations). Engine will overheat and heater output will be minimal. Maximum heat transfer only occurs when antifreeze is mixed with pure water.

Under no circumstances should engine be run with 100% water. Freeze, boil over, and corrosion protection will be eliminated and severe engine damage will result.

Volkswagen G12 coolant as originally supplied is pink in color. Beginning with m.y. 2003, G12 as supplied by VW is purple in color. References to "G12" apply to both versions. Pink and purple G12 can be safely mixed with each other. Do not mix either version of G12 with G11 (blue/green coolant).

- Part number G 012 A8F A4: purple G12
- Part number ZVW 237 G12: pink G12

NOTE—
Volkswagen does not require replacing the coolant as part of routine maintenance. The cooling system has been filled at the factory with a permanent coolant.

WARNING—
- *Hot coolant can scald. Do not work on the cooling system until it has fully cooled.*
- *Use extreme care when draining and disposing of coolant. Coolant is poisonous and lethal. Children and pets are attracted to coolant because of its sweet smell and taste. See a doctor or veterinarian immediately if any amount is ingested or suspected of being ingested.*

Antifreeze Protection Levels (Liters and °C)								
		Liters of antifreeze required for protection to temperature shown (°C)						
		2	3	4	5	6	7	8
Total cooling system capacity in liters	5	-24.4	-52.2					
	6	-17.8	-36.7					
	7	-14.4	-27.2	-47.8				
	8		-21.7	-36.7	-56.1			
	9		-17.8	-29.4	-45.6			
	10		-15.6	-24.4	-36.7	-52.2		
	11		-13.3	-21.1	-30.6	-43.9		
	12		-12.2	-17.8	-26.1	-36.7	-49.4	
	13			-16.1	-22.8	-31.7	-42.8	
	14			-14.4	-20.6	-27.8	-36.7	-47.8

Note:
See notes in Quarts/Fahrenheit table given previously.

> ***CAUTION—***
> - *Use only Volkswagen original anti-freeze when filling the cooling system. Use of any other anti-freeze may be harmful to the cooling system. Do not use an anti-freeze containing phosphates.*
> - *Never mix green coolant (G11) with red coolant (G12). Mixing can cause serious engine damage.*
> - *Do not use tap water in cooling system. Use distilled water only to mix anti-freeze.*

Freeze protection chart for Volkswagen-supplied G12 antifreeze in liters and degrees Celsius.

Cooling system capacity
- 1.8L (1998 - 1999, m/t) . . . . 5.6 liters (5.9 qt.)
- 1.8L (1998- 1999, a/t) . . . . 6.2 liters (6.5 qt.)
- 1.8L (2000 - on) . . . . 7.0 liters (7.4 qt.)
- 2.8L (1998-1999) . . . . 10.0 liters (10.6 qt.)
- 2.8L (2000 - on) . . . . 8.0 liters (8.5 qt.)
- 4.0L (2002 - on) . . . . 8.0 liters (8.5 qt.)

NOTE—
- *Quantities listed above are approximate and can vary slightly depending on model year and equipment level.*
- *Volkswagen uses either pink or purple coolant in vehicles covered by this manual. This coolant/antifreeze is phosphate free and silicate free and is installed in all Passat engine versions in production. When supplied by Volkswagen, this coolant is known as G12.*

On no account should any other type of coolant be used, particularly the older blue/green colored versions. The advantages of G12 over earlier types (blue-green G11) include improvements in corrosion protection, thermal stability, heat transfer/control, hard water tolerance and environmental protection.

Freeze and Boil Protection Chart (°F and °C)		
Percentage of cooling system capacity	Protects from:	
	Freezing down to:	Boiling up to:
33.3%	0°F (-17.8°C)	220°F (104.4°C)
40%	-12°F (-24.4°C)	222°F (105.6°C)
50%	-34°F (-36.7°C)	226°F (107.8°C)
60%	-62°F (-52.2°C)	231°F (110.6°C)

Note:
Boiling points at sea level with no system pressure. Boiling points increase by 3° F (1.7°C) for each pound of pressure (PSI) (0.07 bar) maintained by the cooling system pressure cap. A 21¾ psi (1.5 bar) pressure cap raises boiling point by approx. 65° F (37° C)

> ***CAUTION—***
> - *G12 (pink/purple coolant) must NEVER be mixed with ANY other coolant. Engine damage will result!*
> - *Contamination of G12 with other colored coolants is identifiable by discoloration (brown, gray, etc.). This mixture causes a foamy deposit to appear in the expansion tank/radiator and MUST be drained immediately.*
> - *The cooling system must be completely free of this mixture before refilling with the correct type of coolant/antifreeze.*

Antifreeze also raises the boiling point as noted. Because of the dual role, the mixture of antifreeze and water is more commonly known as coolant.

A translucent expansion tank, or overflow reservoir, on the left side of the engine compartment provides easy monitoring of coolant level without opening the system. The coolant level should always be checked when engine is cold. The coolant level should be between the maximum (MAX) and minimum (MIN) mark on the expansion tank.

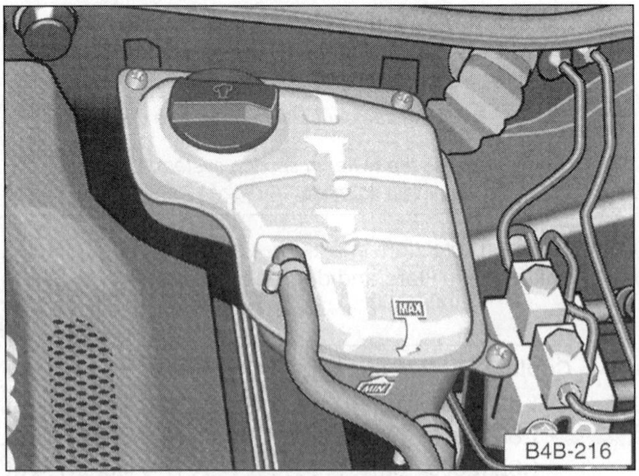

B4B-216

Cooling system level must be between the **MIN** and **MAX** marks on the expansion tank when the engine is cold.

Inspect hoses by first checking that all connections are tight and dry. Coolant seepage indicates that either the hose clamp is loose, that the hose is damaged, or that the connection is dirty or corroded. Dried coolant has a chalky appearance. Check hose condition by pinching them. Hoses should be firm and springy. Replace any hose that is cracked, dry rotted, that has become soft and limp, or has been continuously exposed to oil or fuel.

To check freezing point of the coolant, (and also boiling point), use a refractometer. With the engine cold, remove cap from expansion tank and draw up some coolant with refractometer siphon. Volkswagen recommends using either a 50/50 mixture of anti-freeze and water, which has a freezing point of -34°F (-36.7°C) or a 60/40 mixture which has a freezing point of -62°F (-52.2°C). If the freezing point is too high, drain a small amount of coolant from the system and add new anti-freeze until proper protection level is obtained.

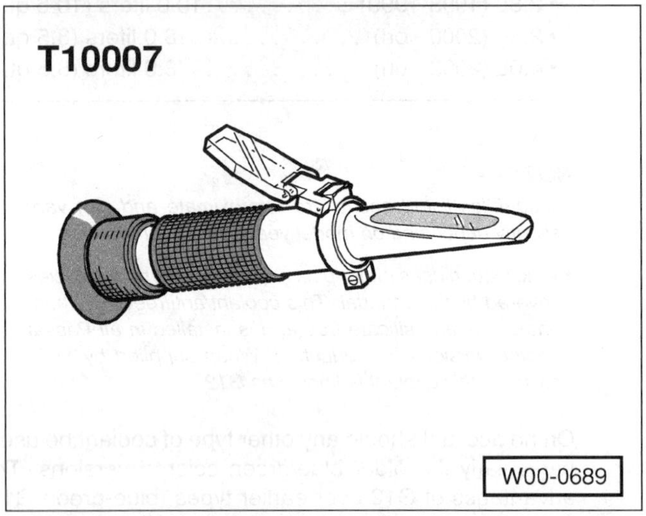

W00-0689

Accurate testing of coolant strength is possible with a refractometer.

Brake fluid level, checking

Routine maintenance of the brake system includes maintaining an adequate level of brake fluid in the reservoir and replacement of the brake fluid every 2 years regardless of the mileage. The system should also be inspected for fluid leaks or other damage.

B8D-071

Brake fluid level must be between the **MIN** and **MAX** marks on the reservoir.

Additional maintenance to the brake system includes checking brake pads and rotors (discs) for condition and wear, and checking hand brake function.

During normal operation, the brake fluid level will drop slightly as the brake pads wear.

WARNING —

- *Use only new, previously unopened brake fluid conforming to the DOT 4 Super standard.*
- *Brake fluid is poisonous. DO NOT ingest brake fluid. Wash thoroughly with soap and water if brake fluid comes into contact with skin.*

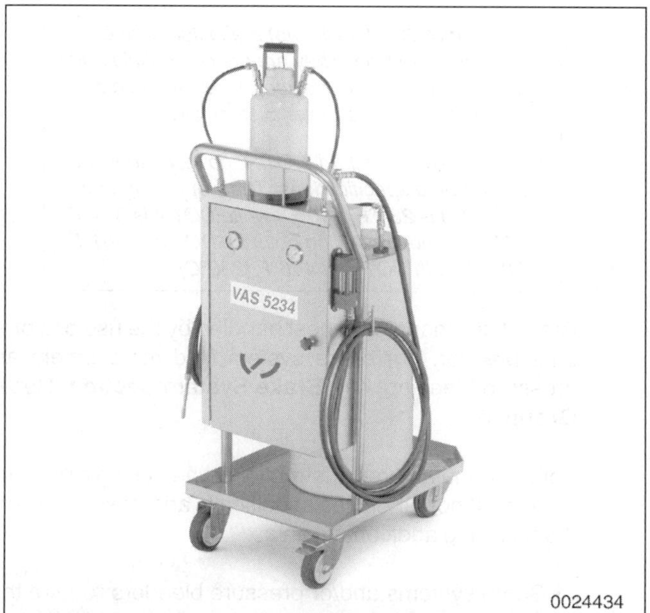

0024434

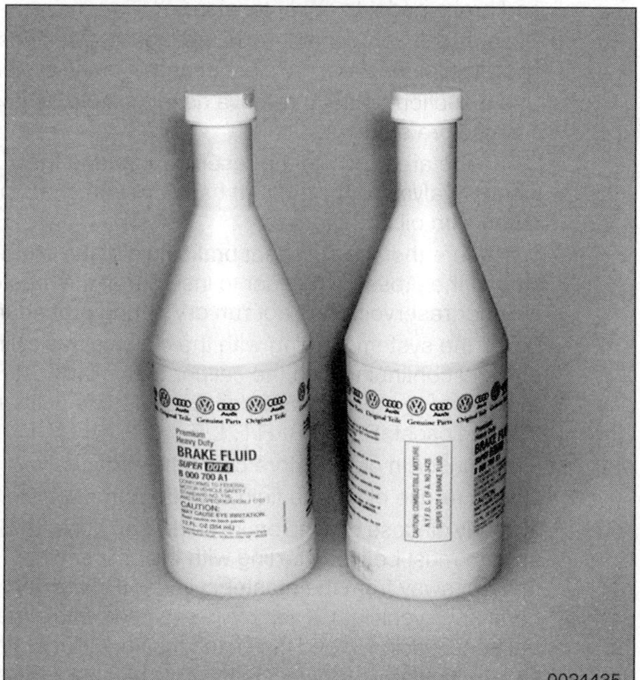

0024435

CAUTION—
- *DO NOT let brake fluid come in contact with painted surfaces. It will remove most types of paint due to its caustic nature. Wash off immediately with soap and water.*
- *Brake fluid absorbs moisture (hygroscopic) from the air and must be stored in an airtight container.*
- *DO NOT allow brake fluid to exceed the maximum level in the fluid reservoir.*

Brake fluid, changing

Brake fluid absorbs moisture from the atmosphere and so is said to be hygroscopic. If the moisture content in the brake fluid is too high, corrosion in the brake system may result. The boiling point of the brake fluid will also be considerably lowered, resulting in the possible vaporization of the brake fluid and reduction in braking force under some conditions.

Brake fluid is also used as the operating fluid for the hydraulic clutch on vehicles with a manual transmission. Fluid is drawn into the clutch master cylinder via a small fitting in the upper part of the reservoir. In the event of a hydraulic clutch leak, fluid loss will only be to the level of the fitting, which retains an adequate reserve for the brake system. Because the hydraulic clutch and the brake system share a common fluid and reservoir, the clutch hydraulic system must be considered when replacing the brake fluid.

EDITOR'S NOTE—
Volkswagen currently specifies that brake system bleeding and brake fluid replacement be performed with the VAS 5234 Brake Systems Filling & Bleeding Machine. While earlier Passats can use older pressure bleeders, W8 Passats MUST use the VAS 5234 due to higher charge pressure settings.

Volkswagen specifies use of brake fluid that meets the DOT 4 Super specification in addition to Federal Motor Vehicle Safety Standard (FMVSS) 116 and Society of Automotive Engineers (SAE) standard J1703. DOT 4 Super is also known as DOT 4+. Brake fluid to this specification is also available in 30 and 50 liter containers for use in pressure bleeders.

Brake fluid specifications
- DOT 4 Super (Volkswagen B 000 700 A1, B 000 700 A7, or B 000 700 A8)
  Minimum wet boiling point. . . . . 329°F (165°C)
  Dry boiling point (exceeds) . . . . 500°F (260°C)

Brake Bleeding Sequence and Bleed Quantity		
	m.y. 1998 - 1999	m.y. 2000➤
Cylinder sequence order:	Brake fluid quantity to be bled:	
Clutch slave	0.1 liters (0.1 qt.)	0.1 liters (0.1 qt.)
Right rear	0.4 to 0.5 liters (0.4 to 0.5 qt.)	0.2 liters (0.21 qt.)
Left rear	0.4 to 0.5 liters (0.4 to 0.5 qt.)	0.2 liters (0.21 qt.)
Right front	0.4 to 0.5 liters (0.4 to 0.5 qt.)	0.2 liters (0.21 qt.)
Left front	0.4 to 0.5 liters (0.4 to 0.5 qt.)	0.2 liters (0.21 qt.)
Approximate total quantity to be bled: (includes fluid siphoned from reservoir)	2.0 liters (2.10 qt.)	1.0 liters (1.05 qt.)

WARNING —

- *Do not use DOT 5 (silicone) brake fluid in any Volkswagen brake system under any circumstances. Brake system performance is designed around the specific characteristics of DOT 4 Super brake fluid only. Safety and performance may be compromised if DOT 5 brake fluid is used instead.*
- *Do not mix DOT 5 (silicone) brake fluid with DOT 4 Super brake fluid. They are not compatible and severe component corrosion will result. Such corrosion could lead to brake system failure.*
- *DOT 3 and standard DOT 4 do not meet the above noted specifications. The wet boiling point for DOT 3 is 284°F (140°C) while DOT 4 is 311°F (155°C). The dry boiling point for DOT 3 is 401°F (205°C) with DOT 4 at 446°F (230°C).*

Brake fluid changing is best handled by the use of a pressure bleeder. For brake system fluid replacement and pressure bleeding, see **Brake System Section, Repair Group 47.**

Follow manufacturers instructions when using a pressure bleeder. If not otherwise specifically addressed, observe the following additional points:

- Some systems and/or pressure bleeders require the use of a brake pedal depressor between driver's seat and brake pedal to apply tension.
- Observe the appropriate minimum pressure settings specified for the pressure bleeder and/or brake system.
- Use a suction bottle to remove as much fluid as possible before starting.
- Clean the area around the reservoir cap and the bleeder valves with cloth that has not been contaminated with oil and grease.
- Make sure there is sufficient brake fluid of the correct type in the pressure bleeder to insure that the master cylinder reservoir does not run dry during procedure.
- Bleed the system starting with the clutch slave cylinder and continuing with the caliper that is farthest away from the master cylinder. End with the caliper that is closest; R/R, L/R, R/F, L/F.
- Fluid must run clear before closing bleeder valve.
- Do not re-use old brake fluid bled from the system.
- Properly dispose of used brake fluid when done. Brake system must be bled starting with the caliper which is farthest away from the master cylinder (R/R on left hand drive vehicles). If vehicle is equipped with a manual transmission, bleed the clutch master cylinder first.
- Be sure fluid level is between the MIN and MAX marks on the reservoir when done.

Power steering fluid level, checking

Maintaining the power steering system generally only requires fluid level checking. The correct fluid level in the reservoir is important for the proper functioning of the power steering system.

Since hydraulic fluid also functions as a lubricant, serious damage to the pump and/or other system components can result if there is insufficient or the wrong fluid in the system.

NOTE—
The power steering system uses Volkswagen-specific fluid and not ATF.

> ***WARNING—***
> *Do not confuse the brake fluid reservoir with the power steering fluid reservoir. Do not mis-identify brake and hydraulic fluids. Incorrect fluids will cause failure.*

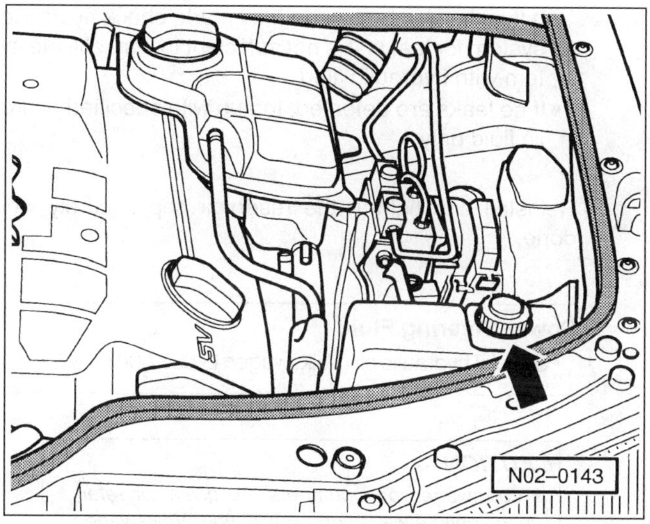

The power steering fluid reservoir is located under the hood on the left side of the engine compartment near the windshield and headlight washer tank.

Power steering fluid level can be checked either cold or at operating temperature (±50°C or 122°F). Observe the correct procedure.

COLD: To check the power steering fluid level on a cold system, proceed as follows:

– With the engine NOT running, use the steering wheel to move the front wheels to the straight ahead position.

– Unscrew the reservoir cap and wipe off the attached dipstick with a clean shop rag. Screw cap on fully and remove again to read.

NOTE—
Screw cap on fully to get an accurate fluid level reading.

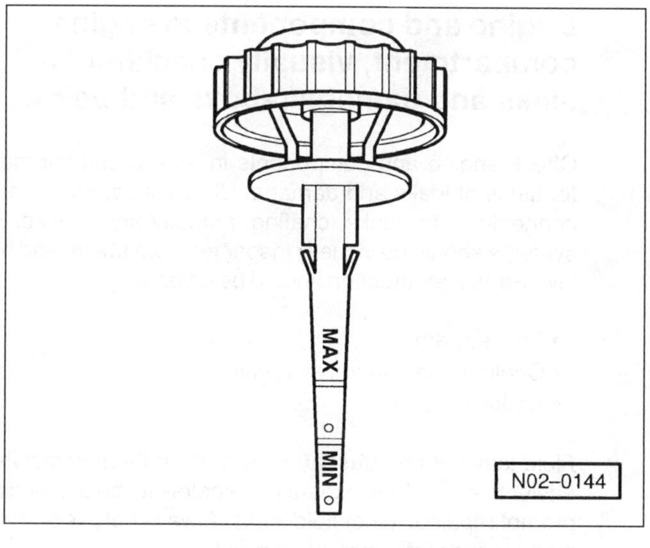

Fluid level must be in area of MIN mark (within 2 mm above or below mark).

- If fluid level is above the MIN mark, some fluid must be siphoned off.
- If fluid level is below the specified level, check hydraulic system for leaks. It is not sufficient to just add fluid.
- If no leaks are detected, top up with specified hydraulic fluid only.

– Reinstall cap and tighten by hand.

Operating temperature: To check the power steering fluid level on a system that is at operating temperature, proceed as follows:

– With the engine running, use the steering wheel to move the front wheels to the straight ahead position.

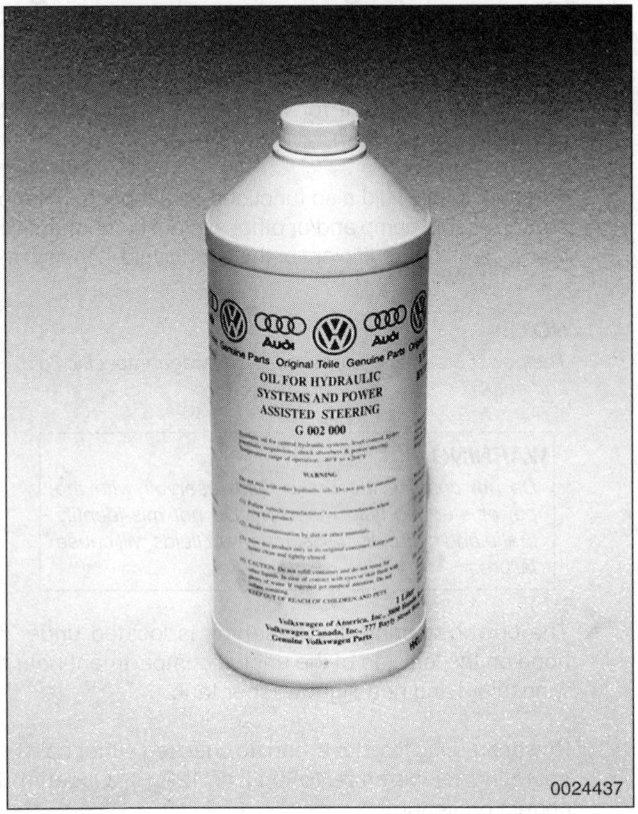

0024437

– Switch off engine.

– Unscrew the reservoir cap and wipe off the attached dipstick with a clean shop rag. Screw cap on fully and remove again to read.

NOTE—
Screw cap on fully to get an accurate fluid level reading.

– Fluid level must be between the MIN and MAX marks on the dipstick.

- If fluid level is above the MAX mark, some fluid must be siphoned off.
- If fluid level is below the MIN mark, check hydraulic system for leaks. It is not sufficient to just refill the system with hydraulic fluid.
- If no leaks are detected, top up with specified hydraulic fluid only.

– Reinstall and tighten the reservoir cap securely when done.

Power Steering Fluid

- Special hydraulic oil (Volkswagen G 002 000 or G 002 012)

CAUTION—
- *Volkswagen part numbers are given for reference only! Always consult with your Volkswagen Parts Department or aftermarket parts specialist for the latest parts information.*
- *Use only Volkswagen hydraulic oil G 002 000 or G 002 012 in the power steering system. Do not use ATF or other non-approved types of power steering fluid. If the wrong fluid is used, power steering components may fail as a result.*

Engine and components in engine compartment, visually checking for leaks and damage (above and below)

Check engine and components in engine compartment for signs of leaks and damage. Check lines, hoses, and connections for leaks, chafing, porosity and breaks. All systems should be visually inspected from above and below. Particular attention should be given to:

- Fuel system
- Cooling and heating systems
- Brake System.

Fluid leaks attract dust and dirt making them somewhat easier to spot. Many expensive repairs can be avoided by prompt repair of minor fluid leaks. Always note/report any malfunctions for subsequent repair.

Fuel filter

Volkswagen specifies that the fuel filter used on all engines is designed to last for the life of the vehicle. There is no routine replacement required. Replacement is only necessary if the filter becomes clogged due to contaminated fuel or failure to keep the fuel tank sealed. If dirt in the fuel filter is suspected, also check the strainer on the bottom of the fuel supply unit in the tank for contamination. The fuel filter is located near the fuel tank, see **Section F, Fuel Supply System, Fuel supply system, servicing**.

Battery service

The battery is one of the most important electrical components in the vehicle. To ensure a long service life, the battery must be checked regularly, serviced and maintained.

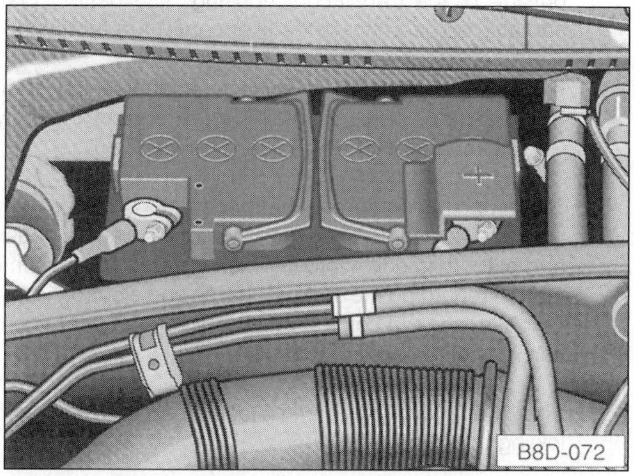

◄ Battery on early production vehicle without trim cover. Note plastic protective cover over positive terminal.

In addition to starting the engine, the battery has additional functions as a buffer and supplier of electrical power for the entire vehicle electrical system.

The battery is located in the air plenum chamber at the rear of the engine compartment in all Passats. Vehicles produced after October 2000 have a trim panel covering the entire plenum area, including the battery. A small section of the cover over the battery is easily removed for partial (service) access. The entire trim cover must be removed for complete access to the battery. The positive (+) battery terminal on some vehicles is also protected with a plastic cover which must be moved for access to the positive (+) terminal.

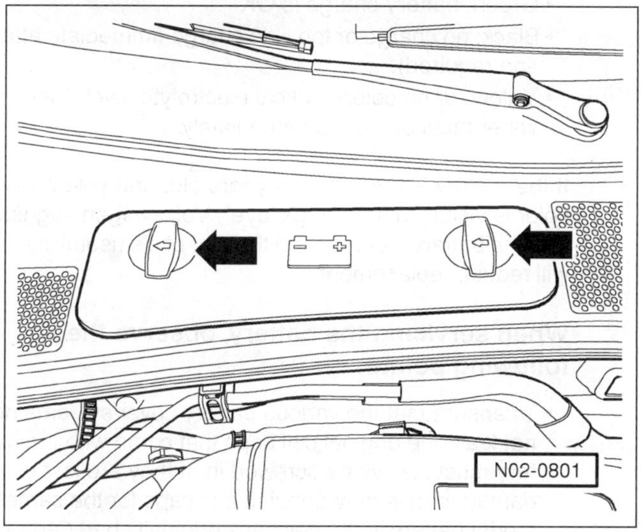

◄ Battery cover in air plenum (note battery symbol) is removed by pushing the handles in the direction of the arrows. Handles do not turn.

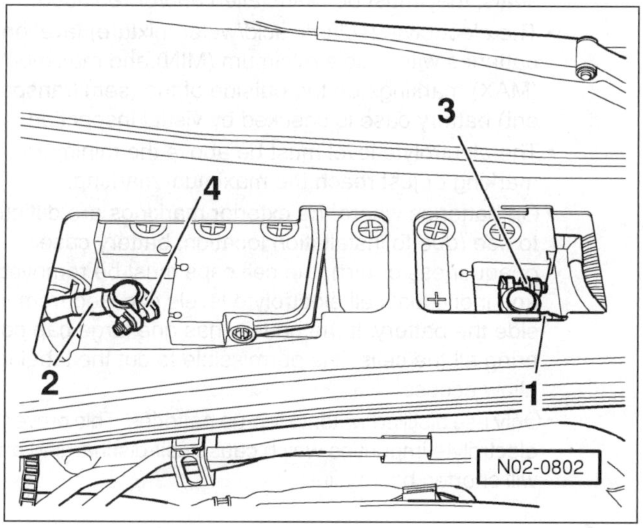

◄ Battery access available when cover is removed.

1. Battery positive (+) cable (partially obscured).
2. Battery negative (-) terminal.
3. Positive (+) terminal cable end.
4. Negative terminal cable end.

Two different types of original equipment batteries are installed depending upon the production date of the vehicle. Both types have translucent outer cases that allow visual inspection of the electrolyte level. Later production vehicles are equipped with batteries with a charge indicator "magic eye" which provides information about the electrolyte level and the state of charge.

Under most normal operating conditions, the battery is generally maintenance free. At high outside temperatures, however, it is advisable to check the electrolyte level through the translucent battery housing.

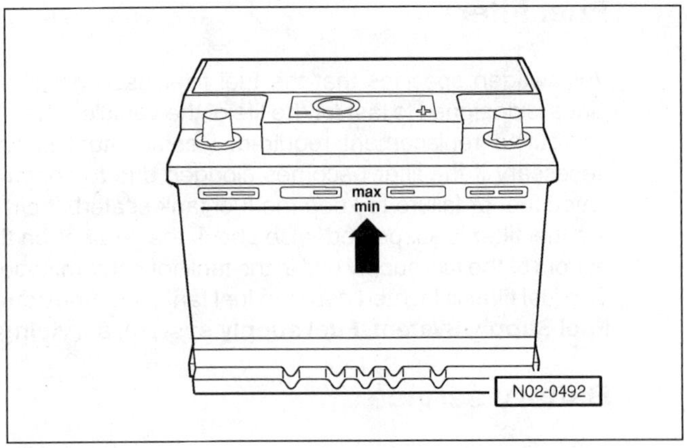

The electrolyte level should be just above the battery plates and their separators. This corresponds to a fluid level between the MIN and MAX marks (**arrow**) on the outside of the translucent housing. The battery plates can also be seen when the filler cap(s) are removed.

If the electrolyte (acid) level is low, replenish it by adding distilled water only. Maintaining the correct electrolyte level is important for long battery service life.

Some original equipment batteries have a device known as a "magic eye" located in the top of the battery housing and extending down into it. This indicator provides information regarding the state of charge and electrolyte level. By observing the color of the "magic eye", battery condition may be determined.

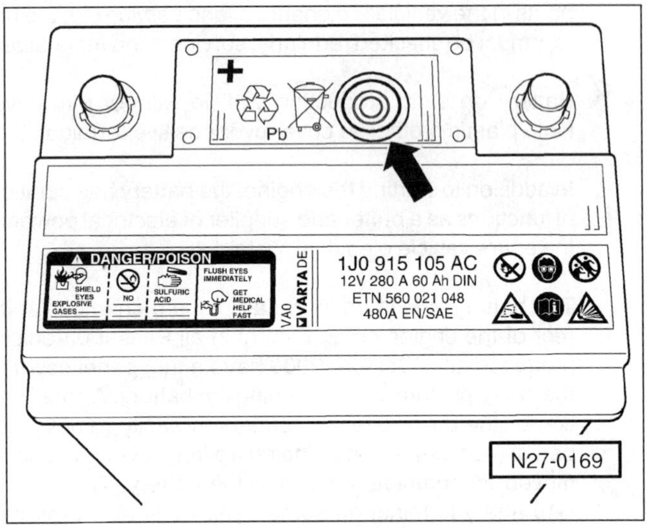

The "magic eye" (**arrow**), by color, can indicate the following conditions:

- Green: battery charge is OK.
- Black: no charge or too little charge (immediate attention required).
- Yellow (or no color): critical electrolyte level. Distilled water must be added immediately.

If the battery is more than 4 years old, and yellow or no color is visible in the "magic eye", Volkswagen suggests that the battery has reached the end of it's useful life and will require replacement.

When servicing the battery, observe the following points:

- To ensure that the various battery cover systems do not leak, the original cell caps that come with the battery must always be screwed in. If they are lost or damaged, use only genuine cell caps for the same model battery. If the cell caps originally had O-ring seals, they must be re-installed and/or replaced.
- The electrolyte (sulfuric acid/water mixture) level on batteries with visible minimum (MIN) and maximum (MAX) markings on the outside of the (semitransparent) battery case is checked by visual inspection.
- The electrolyte level must be above the minimum marking or just reach the maximum marking.
- On batteries where the exterior markings are difficult to see (due to installation location, battery case opaqueness or dirt), the cell caps must be removed and individual cell electrolyte levels checked from inside the battery. If the battery has one large cap covering all the cells, it is permissible to cut the label to allow access.
- Only use distilled water to top-up batteries. This prevents electrolyte impurities which cause self-discharging and will shorten battery life.

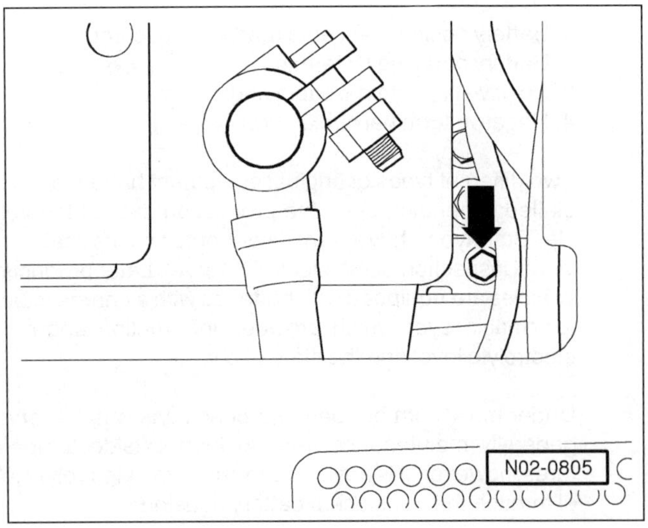

Check that battery is securely mounted and clamping bolt (**arrow**) is torqued to 22 Nm (16 ft-lb).

- When electrolyte level is too low and cell plates are exposed, drying of the cell plates will result in a loss of battery capacity (power). If cell plates are not fully immersed in electrolyte, corrosion will occur on plates, bridges and cell connectors. Optimum battery function is not possible under these conditions.
- When electrolyte level is too high, electrolyte may leak out and damage the surrounding areas such as the battery tray and air plenum.
- DO NOT overfill the battery. Overfilled batteries tend to boil over easily.

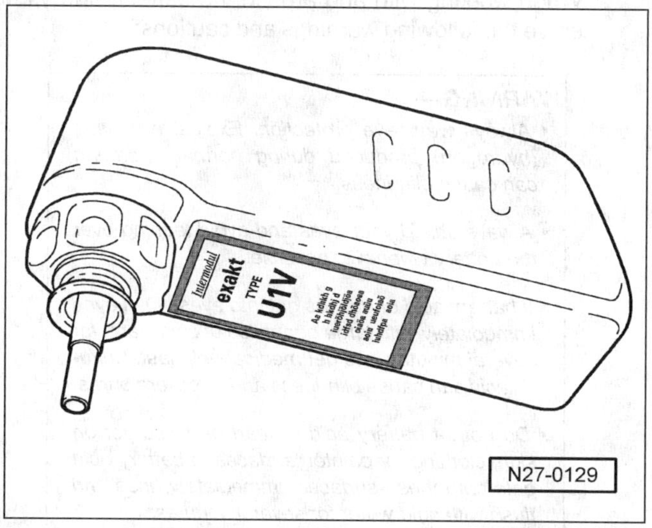

A battery filling bottle such as the VAS 5045 is specifically designed for the purpose and will prevent overfilling of the battery.

- Too little electrolyte greatly reduces the service life of the battery.
- Excess electrolyte MUST be extracted using a hydrometer.
- Dispose of removed electrolyte (sulfuric acid) and old batteries properly! Waste electrolyte must only be disposed of in appropriate waste disposal sites. Refer to local regulations pertaining to electrolyte and used battery disposal.

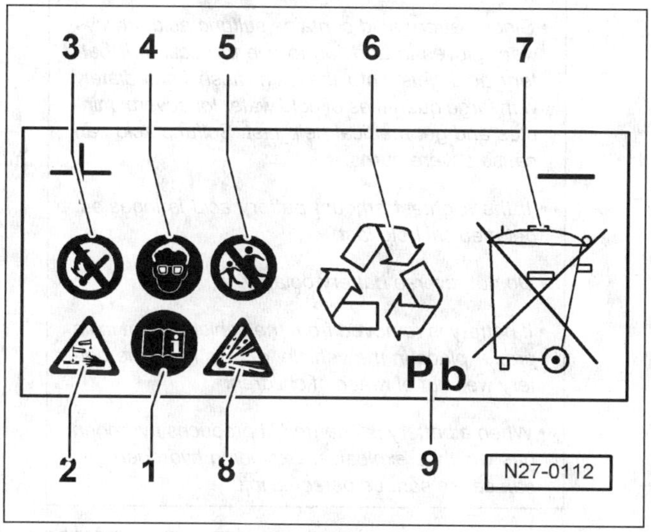

A warning and safety label of this type is found on most original equipment Volkswagen batteries.

1. Follow the recommendations on the battery, in the vehicle Owners's Manual, and in this Manual.
2. Danger of acid burns. Follow warnings listed for working with acid.
3. Keep open flames, sparks, unenclosed light bulbs and lit smoking materials away from battery. Avoid electrical short circuits.
4. Wear eye protection.
5. Keep children away.
6. Dispose of old batteries properly. Recycle when possible. Comply with all applicable laws regarding hazardous waste disposal due to lead and acid content.
7. Do not discard in household trash.
8. Explosion hazard.
9. Contains LEAD (Pb).

- Some batteries are equipped with a central gas venting system and integrated flame trap. This system allows gases produced during charging to escape through a central opening on the side of the upper cover. The integrated flame trap prevents flammable gases in the battery from igniting. Insure that vent line is secure upon completion of repairs.

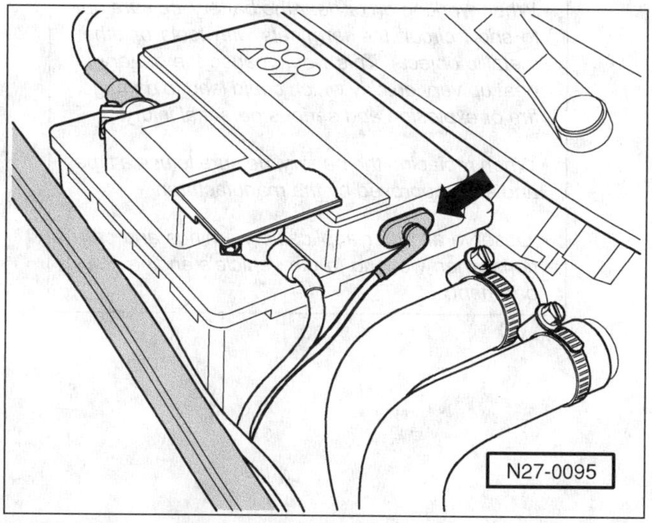

Check battery gas vent and hose (**arrow**) for proper installation upon completion of work around the battery.

NOTE—
For battery load testing, removal and installation and specific gravity test, see **Electrical Equipment section, Repair Group 27**.

When working with and around the battery, always observe the following warnings and cautions:

> ***WARNING—***
> - *Always wear eye protection. Explosive gasses (hydrogen) produced during normal operation can cause blindness.*
> - *Always shield your eyes and avoid leaning over the battery whenever possible.*
> - *If battery acid or gasses get into eyes, flush eyes immediately with large quantities of cold water for several minutes and get medical help fast. Sulfuric acid can cause blindness and/or severe burns.*
> - *Do not let battery acid or lead particles get on skin, clothing, or painted surfaces. If battery acid gets onto these surfaces, immediately rinse and flush with cold water for several minutes.*
> - *Do not smoke when working with or around batteries. Do not expose the battery to open flames, electric sparks or open lights.*
> - *Since battery acid contains sulfuric acid, always wear gloves in addition to eye protection. If battery acid gets onto the skin, flush immediately with large quantities of cold water for several minutes and get medical help fast. Sulfuric acid can cause severe burns.*
> - *If the slightest amount battery acid is ingested, get medical help fast.*
> - *Do not reverse battery polarity.*
> - *If battery is removed from the vehicle or not in its proper place in the vehicle, always keep the battery well out of reach of children.*
> - *When a battery is charged, it produces hydrogen gas which is explosive. Exploding hydrogen gas can cause serious personal injury.*

> ***CAUTION—***
> - *Before working on the electrical system, disconnect the negative (ground) battery cable.*
> - *When working on or near the battery, be sure not to short circuit the terminals with tools or other metallic objects. This would cause the battery to heat up very quickly which could lead to damage, fire or explosion and serious personal injury.*
> - *When replacing the battery, be sure to use a type and style approved by the manufacturer.*
> - *Observe all other applicable warnings and cautions when working in the vehicle's engine compartment.*

Guidelines following interruption of vehicle voltage supply

After disconnecting and re-connecting the battery, some vehicle systems may require initialization (reset of basic settings). The number and type of system will vary considerably by model year and equipment level. Systems that could be affected include any or all of the following:

- Engine management system, including throttle adaptation, mixture adaptation, DTC memory, and readiness codes.
- Automatic transmission.
- Comfort system.
- Climate control.
- Instrument cluster, including clock reset.
- Radio, including unlocking and station presets.
- Memory seats and mirrors.

Refer to the individual Repair Groups for additional information and procedures. Also see **Electrical Equipment, Repair Group 27.**

EDITOR'S NOTE —

Always check and activate all of the vehicle's electrical equipment (radio, clock, comfort and convenience features, etc.) after repairs to confirm proper operation.

V-belt and ribbed V-belt, checking

All Volkswagen engines use a ribbed V-belt to drive engine accessories. Some also use a more standard V-belt to drive the coolant pump. Although belts of these designs are known for their smooth running, high efficiency, and long life; periodic inspection is required.

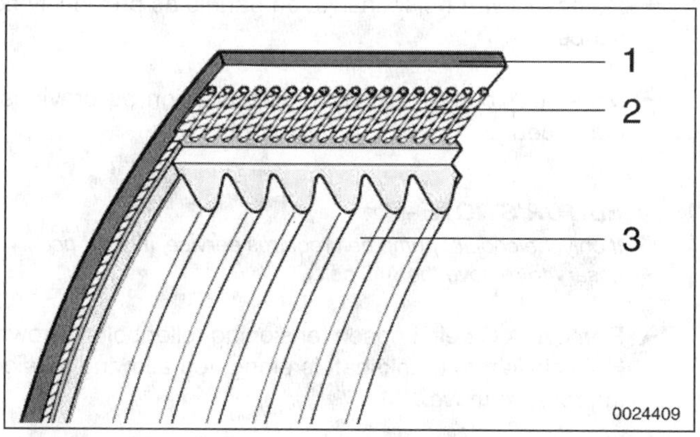

Cross section of Volkswagen ribbed V-belt.

1. Top layer, cover
2. Tension cords
3. Base material

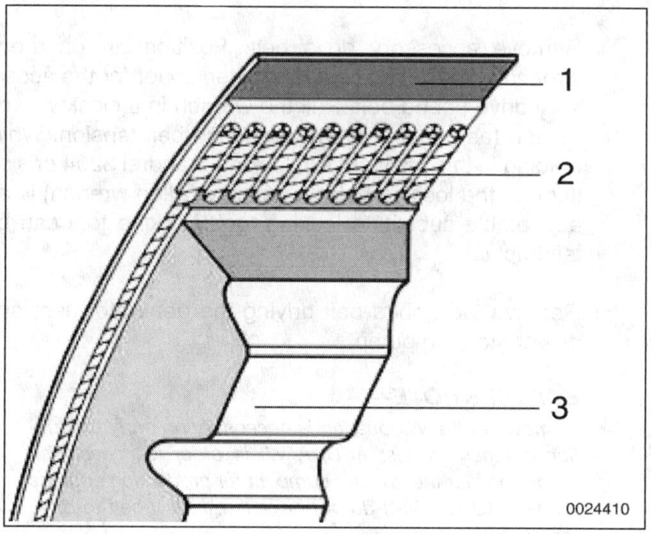

Cross section of Volkswagen V-belt. Use callout numbers above.

Ribbed V-belts and standard V-belts should be periodically checked. Manually turn engine at vibration damper or pulley with a suitable wrench. Inspect belt(s) from above or below for the following conditions:

- Splits in the base material (substrata) including cracks, core fractures and cross section fractures.
- Separation between the layers, especially between the top cover and the tension cord layers.
- Chunking or breaking of the base or lower layer material.
- Fraying of the tension cords.
- Flank (side) wear including flaking, fraying, chunking, glazing, hardening, and surface cracking.
- Oil and grease contamination.

If any of the above conditions exist, the belt(s) must be replaced. Consult the maintenance tables at the end of this section for appropriate routine replacement intervals.

V-belt and ribbed V-belts, replacing, (4-cylinder)

Early production (engine codes: AEB, ATW) 4-cylinder engines use a coolant pump driven by a short V-belt from the power steering pump pulley. Later production engines (engine codes: AUG, AWM) use a coolant pump driven by the toothed belt for the camshaft. Differences for removing the V-belt on later versions are noted below.

This section will cover removal and reinstallation of these belts in this order:

- Air Conditioning (A/C) compressor ribbed belt
- Accessory drive ribbed belt
- Coolant pump V-belt

NOTE—
If used belts will be reinstalled, mark the running direction and reinstall in the same direction. Failure to do so will usually result in premature belt failure.

V-belts, removing (4-cylinder)

– Remove lower noise insulation panels as previously described.

– Move lock carrier to the service position as previously described.

EDITOR'S NOTE—
If only the coolant pump belt requires service, it is not necessary to remove the A/C belt.

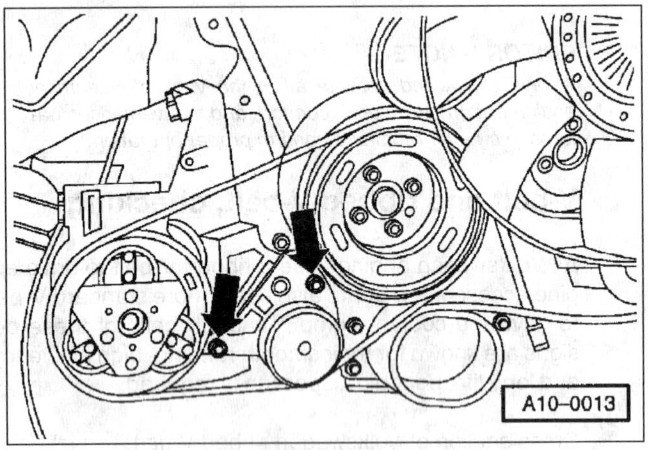

◄ Remove A/C belt. Loosen tensioning roller bolts (arrows). Hole closest to crankshaft is elongated allowing tensioning roller to move.

– Move tensioning roller out of the way and remove ribbed belt for A/C compressor.

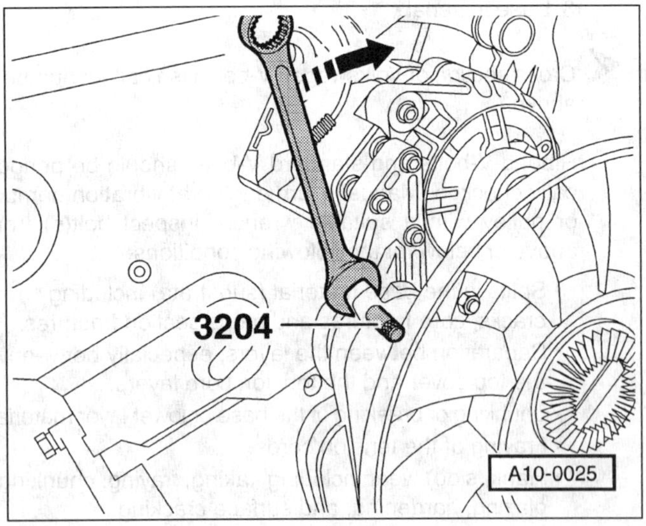

◄ Remove accessory drive belt. Position an open-end wrench onto the tab cast on the tensioner for the accessory drive ribbed belt. Pull the wrench in a clockwise direction (as you look at it) to release belt tension. While holding in this position, slip locking mandrel 3204 or similar into the locking hole. A hex key (Allen wrench) is an acceptable substitute locking tool. Locking tool can be left in place.

– Remove the ribbed belt driving the generator, fan, and power steering pump.

EDITOR'S NOTE—
Removal of the viscous fan is necessary on early production engines (codes: AEB, ATW) in order to access the third V-belt for the coolant pump. Later production engines (codes: AUG, AWM) do not have the third V-belt to drive the coolant pump.

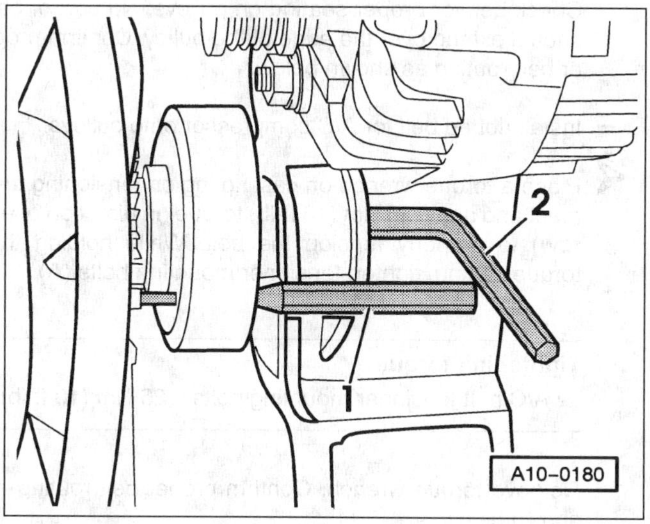

Engine codes: AEB, ATW

Remove viscous fan unit. Lock viscous fan pulley with a drift (**1**), or M5 x 60 bolt, through both the hole in the pulley and the slot in the casting.

– Remove the viscous fan securing bolt with an 8 mm hex wrench (**2**).

- Securing bolt is located in center of the fan and pulley, removal is from the rear.

– Remove viscous fan.

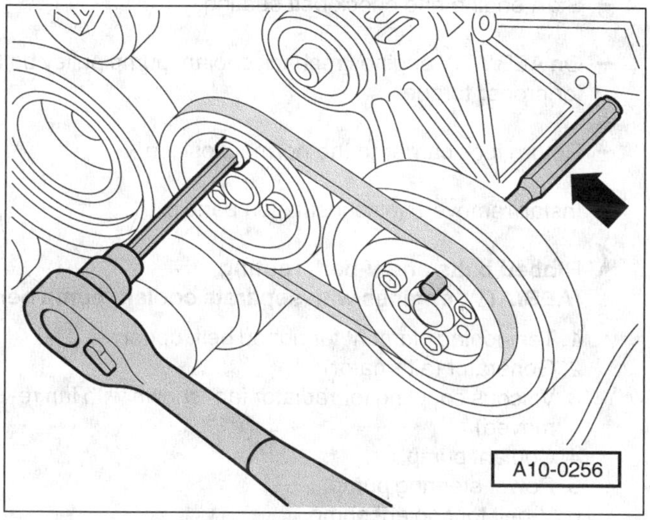

Remove coolant pump belt. Counterhold coolant pump pulley by inserting a drift (**arrow**) or other suitable tool into the hole provided in drive pulley.

– Remove 3 bolts holding V-belt pulley to coolant pump.

– Remove coolant pump pulley and V-belt.

V-belts, installing (4-cylinder)

Installation is in the reverse order of removal. If reinstalling used belts, be sure to note the running direction.

Engine codes: AEB, ATW

– Install coolant pump belt. Place V-belt on both pulleys and assemble two sections of pulley on coolant pump without tightening bolts.

– Tighten the 3 bolts securing the two sections of the coolant pump pulley in stages while turning the coolant pump. As the bolts are tightened, the V-belt should gradually move outward until seated on the running surface.

Tightening torques

- Coolant pump pulley to pump . . . . .25 Nm (18 ft-lb)
- Viscous fan to hub (center bolt) . . .45 Nm (33 ft-lb)

NOTE—
Coolant pump belt tension is not adjustable. The correct belt, properly installed should be tensioned correctly.

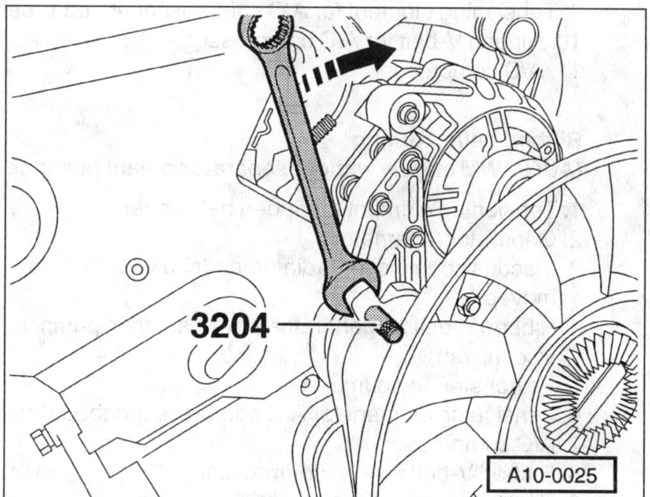

All engines

Install accessory drive belt. Place belt driving the generator, fan, and power steering pump onto pulleys. Slide an open-end wrench onto the tab cast into the tension element. Pull the wrench in a clockwise direction (**arrow**) against spring tension and remove the locking hex wrench or mandrel. Release tensioning element.

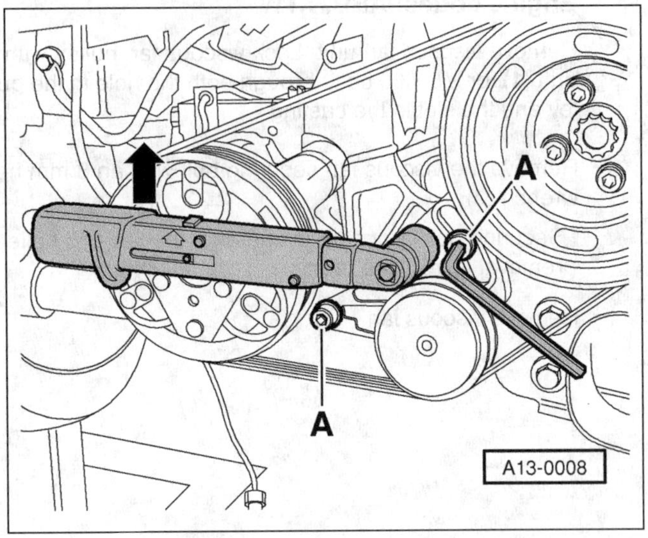

- Check belt for proper seating on pulleys. No part of belt should extend over the edge of the pulley. Confirm proper belt routing as shown below.

- Install ribbed belt for A/C compressor onto pulleys.

- Place a torque wrench on casting lug on tensioning element and apply 25 Nm (18 ft-lb) torque (in direction of **arrow**) to properly tension the belt. While holding this torque setting, tighten tensioner mounting bolts (**A**).

Tightening torque

- A/C belt tensioner mounting bolts . . 25 Nm (18 ft-lb)

- Remove torque wrench. Confirm proper belt routing as shown below.

- Start engine and check belt seating.

- On AEB/ATW engines recheck coolant pump pulley bolts for proper torque.

- Return lock carrier to the normal position.

- Install removed noise insulation panels.

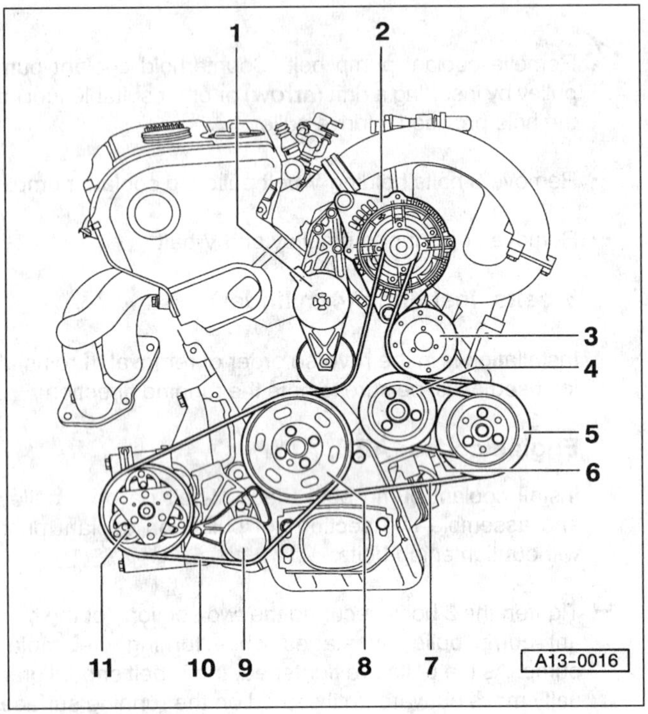

Ribbed belts and V-belt, routing (AEB/ATW, engines with separate coolant pump belt)

1. Tensioning element for ribbed belt, upper
2. Generator (alternator)
3. Viscous coupling for radiator fan (shown with fan removed)
4. Coolant pump
5. Power steering pump
6. V-belt for coolant pump
7. Ribbed V-belt for generator, power steering pump, and viscous fan
8. Vibration damper (front pulley)
9. Tensioning element for A/C compressor ribbed V-belt
10. Ribbed V-belt for A/C compressor
11. A/C compressor

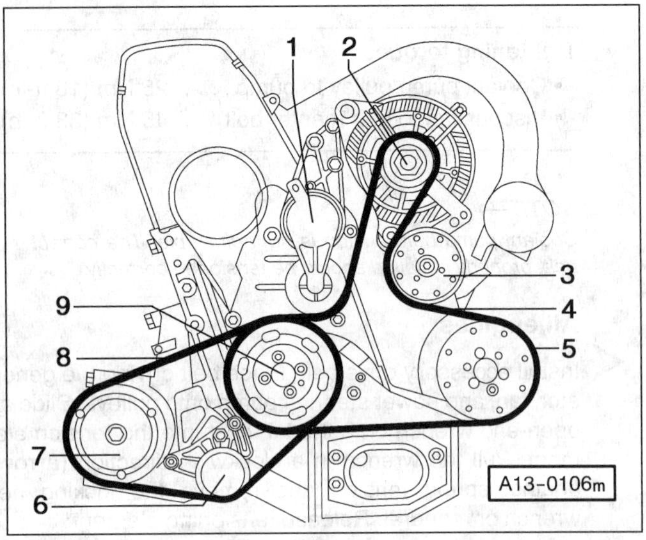

Ribbed belts, routing (AUG/AWM engine without separate coolant pump belt)

1. Tensioning element for ribbed belt, upper
2. Generator (alternator)
3. Viscous coupling for radiator fan (shown with fan removed)
4. Ribbed V-belt for generator, power steering pump, and viscous fan
5. Power steering pump
6. Tensioning element for A/C compressor ribbed V-belt
7. A/C compressor
8. Ribbed V-belt for A/C compressor
9. Vibration damper (front pulley)

Ribbed V-belt, replacing, (6-cylinder)

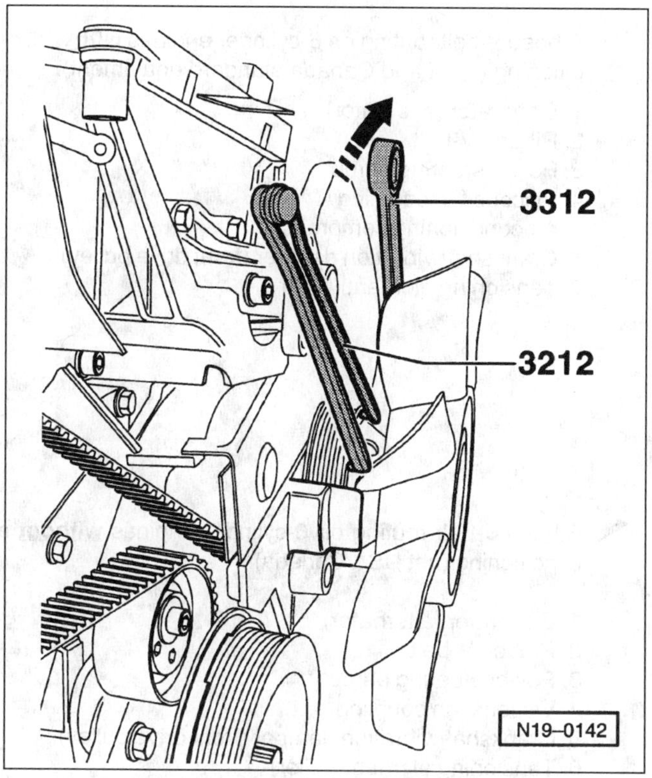

Ribbed V-belt, removing (6-cylinder)

NOTE—

If used belt will be reinstalled, mark the running direction and reinstall in the same direction. Failure to do so will usually result in premature belt failure.

– Remove lower noise insulation panels as previously described.

– Move lock carrier to the service position as previously described.

EDITOR'S NOTE—

It may be possible to perform this procedure without moving the lock carrier to the service position, but ease and accuracy are significantly enhanced.

– Remove intake manifold cover(s) and ribbed V-belt cover.

◄ Remove viscous fan.

- Hold pulley with spanner 3212 and loosen (direction of **arrow**) left-hand threaded bolt with open end wrench 3312 (32 mm).

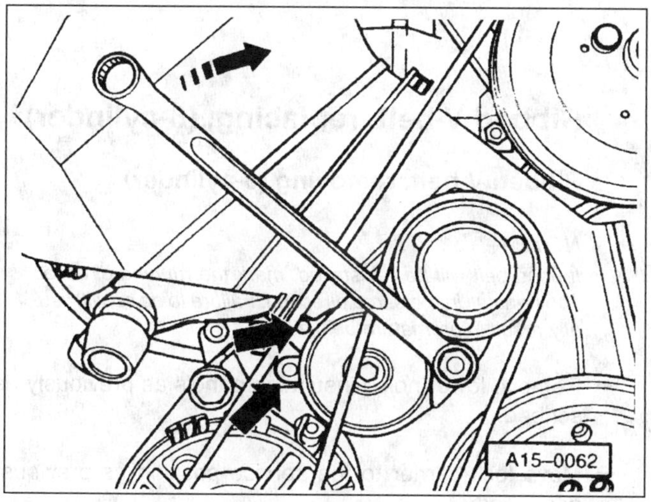

◄ Remove the single ribbed belt by placing a 17 mm wrench or socket onto the tension element. Pull wrench against spring tension in direction shown (**dashed arrow**). Lock tensioning element in position with mandrel 3204 or a suitable hex (Allen) key through front hole and rear hole (**solid arrows**) in tensioning element.

Ribbed V-belt, installing (6-cylinder)

Installation is in the reverse order of removal.

– Install ribbed V-belt over the pulleys in the following order: vibration damper (front drive pulley), A/C compressor, power steering, generator, and tensioning element.

– Pull wrench against spring tension as in removing, to allow removal of locking hex wrench or mandrel. Remove hex key or mandrel 3204 while holding wrench and then carefully ease wrench in opposite direction to tension the belt. Confirm proper belt routing.

– Install viscous fan onto pulley to torque specification.

Tightening torque
- Viscous fan to pulley (using spanner 3312)
  without VW torque wrench 1331 . . . . 70 Nm (52 ft-lb)
  with VW torque wrench 1331 . . . . . . 37 Nm (27 ft-lb)

– Start engine and check for correct belt seating on all pulleys. Switch off ignition.

– Install intake manifold cover(s) and ribbed V-belt cover.

– Return lock carrier to the normal position.

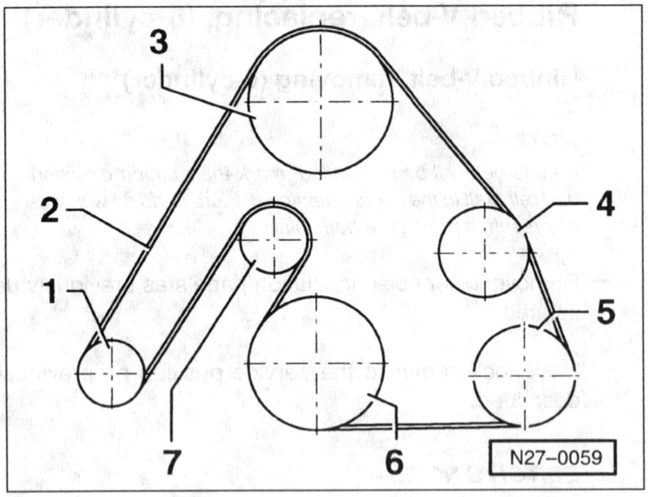

Ribbed V-belt routing on 6-cylinder engines **with air** conditioning (USA and Canada standard equipment).

1. Generator (alternator)
2. Ribbed V-belt
3. Power steering pump
4. Viscous fan coupling
5. Air conditioning compressor
6. Crankshaft vibration damper (front drive pulley)
7. Tensioning element (roller)

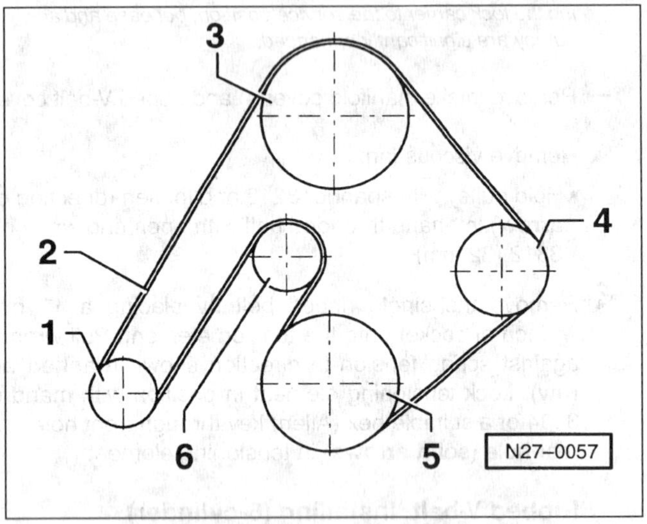

Ribbed V-belt routing on 6-cylinder engines **without** air conditioning (not USA/Canada).

1. Generator (alternator)
2. Ribbed V-belt
3. Power steering pump
4. Viscous fan coupling
5. Crankshaft vibration damper (front drive pulley)
6. Tensioning element (roller)

Ribbed V-belt, replacing, (8-cylinder)

Ribbed V-belt, removing (8-cylinder)

NOTE —

If used belt will be reinstalled, mark the running direction and reinstall in the same direction. Failure to do so will usually result in premature belt failure.

– Remove lower noise insulation panels as previously described.

– Move lock carrier to the service position as previously described.

Remove the single ribbed belt by placing a 30 mm wrench or socket onto the outer hex of the tension element. Pull wrench against spring tension in direction shown (**dashed arrow**).

– Slide belt off of tension element while holding wrench against spring tension.

Ribbed V-belt, installing (8-cylinder)

Installation is the reverse of removal, noting the following:

- Before installing ribbed belt, be sure that all belt-driven components are properly mounted and torqued.
- Observe marked running direction of belt and check for proper seating into pulley grooves.

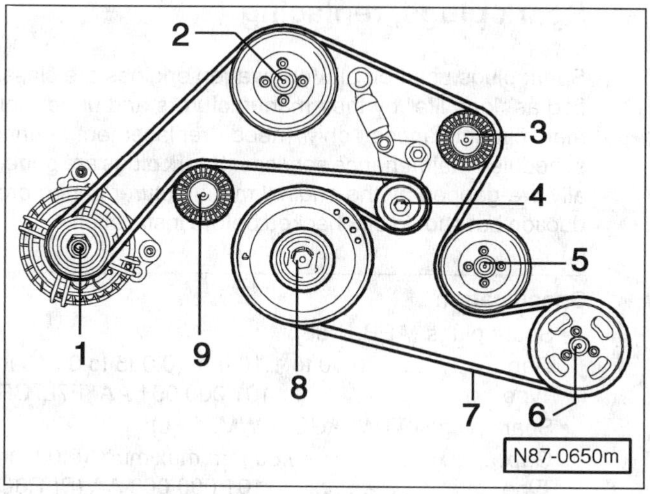

- Start engine and check for proper routing and running before moving lock carrier from the service position.
- Check and adjust headlight aim as required when lock carrier is moved from service position to normal position.

Ribbed V-belt routing on 8-cylinder W8 engines.

1. Generator (alternator)
2. Coolant pump
3. Deflection pulley, bank 2
4. Hydraulic belt tensioner with deflection pulley
5. Air conditioning compressor
6. Power steering pump
7. Ribbed V-belt
8. Crankshaft vibration damper (front drive pulley)
9. Deflection pulley, bank 1

Camshaft belt, checking (4- and 6-cylinder)

The camshaft belt, also called the spur, toothed, or timing belt, should be inspected periodically.

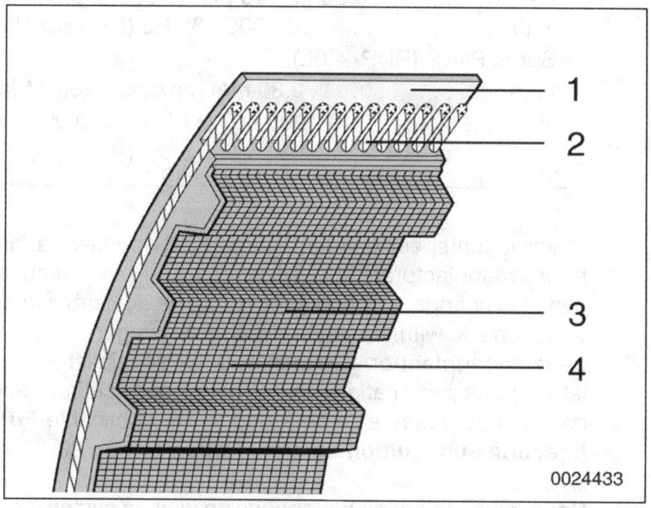

Cross section of Volkswagen camshaft belt.

1. Top layer, backing, cover
2. Tension cords
3. Base material and drive teeth
4. Base material fabric

– Unsnap or open the clips of the upper belt cover and inspect belt from above for the following general conditions:

- Separation between the layers, especially between the top cover and the tension cord layers
- Chunking or breaking of base or lower layer material
- Oil and grease contamination
- Missing and/or damaged drive cogs
- Insufficient tension as evidenced by contact with belt covers

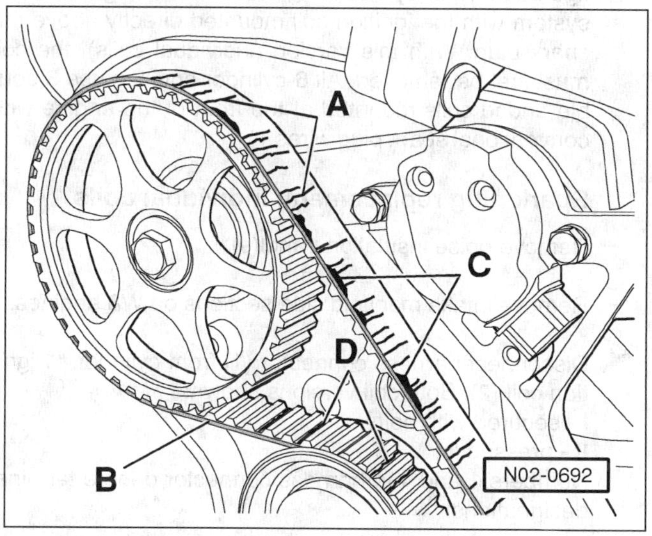

Specific conditions and areas to inspect include:

A. Cracks in the cover material, including splits and fractures. Also glazing from belt cover contact.
B. Chafing and wear on the edges.
C. Fraying of the cord material.
D. Cracks on the drive (gear) side.

In addition to the overall condition of the belt, inspect the tensioner for wear, damage, and excessive bearing noise. If any of the above conditions exist, the belt and tensioner should both be replaced. Consult the maintenance tables at the end of this section for appropriate routine replacement intervals. Refer to the specific **Engine Mechanical** sections for additional information and procedures.

> ***WARNING —***
> *If damage is found, the belt must be replaced! Failure of camshaft belt will result in serious engine damage.*

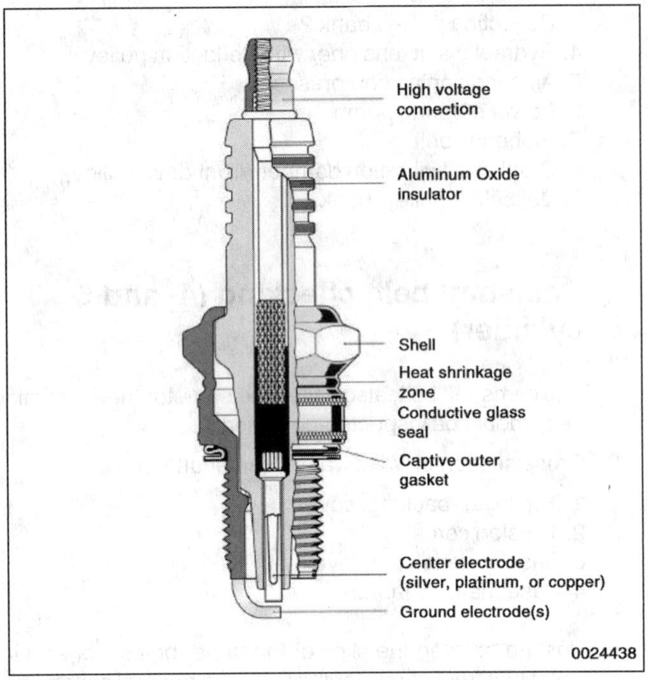

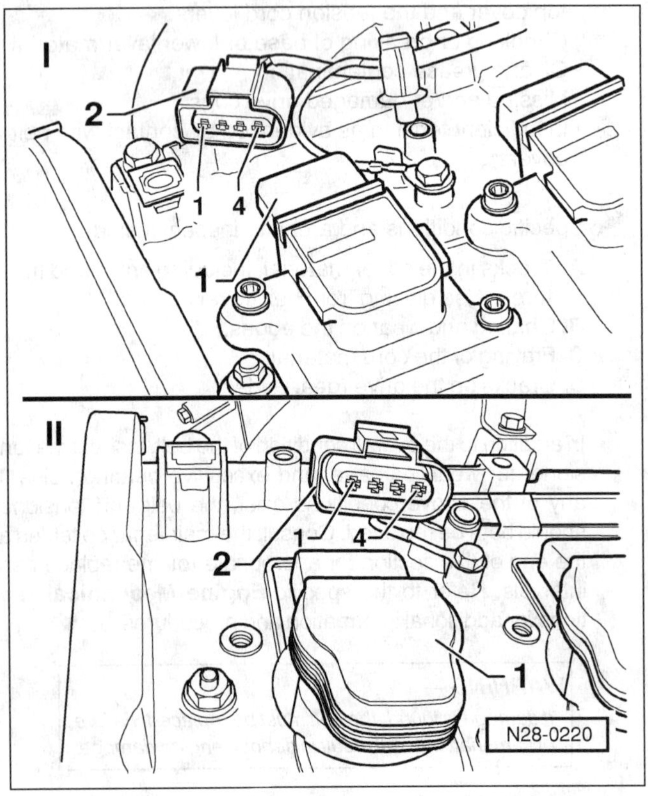

Spark plugs, replacing

Spark plugs used on all Volkswagen engines are classified as "long life" by their manufacturers and under normal circumstances only need replacement during scheduled maintenance services. Spark plugs are generally pre-gapped by the original manufacturer during production but should be checked before installation.

Specification

- Spark plugs (AEB, 1.8L):
  Gap.......... 0.90 to 1.10 mm (0.035 to 0.043 in)
  Type .... 101 000 051 AA (F7LTCR)
- Spark plugs (ATW, AUG, AWM, 1.8L):
  Gap............... 0.80 mm maximum (0.031 in)
  Type ....101 000 063 AA (PFR6Q)
- Spark Plugs (AHA, 2.8L):
  Gap........................ 1.6 mm (0.063 in)
  Type .... 101 000 035 HJ (BKR6EKUB)
- Spark Plugs (ATQ, 2.8L):
  Gap.......... 0.90 to 1.10 mm (0.035 to 0.043 in)
  Type .... 101 000 035 HJ (BKR6EKUB)
- Spark Plugs (BDP, 4.0L):
  Gap............... 0.80 mm maximum (0.031 in)
  Type .... 101 905 607 HJ / 101 000 062 AA (BKR6EKUE)

Volkswagen specifies spark plugs made by several different manufacturers. Specifications and part numbers can also change. For these reasons, it is advisable to always check with an authorized Volkswagen Dealer's parts department or aftermarket parts specialist for the latest parts information. Spark plug specifications and part numbers may also be found in the applicable **Fuel Injection and Ignition section, Repair Group 28**.

Spark plug replacement depends on ignition system configuration. Four cylinder engines and W8 engines use a system with the ignition coil mounted directly above the spark plug. With this version (individual coils), the coil must first be removed. All 6-cylinder engines use 3 double ended coils mounted at the front of the engine with conventional spark plug wires.

Spark plug replacement, individual coils

– Remove noise insulation panel(s).

– Remove intake manifold side sections on W8 engines.

◄ Disconnect harness connector (**1**) from cylinder #1 ignition coil (**2**). Both coil versions shown:
I - secured with bolts,
II - pressed in.
Numbers 1 and 4 in harness connector denote terminal designations.

– Remove coil: On coils secured with bolts, remove. Pull coil up and off. If coil is stuck, use a twisting movement to remove.

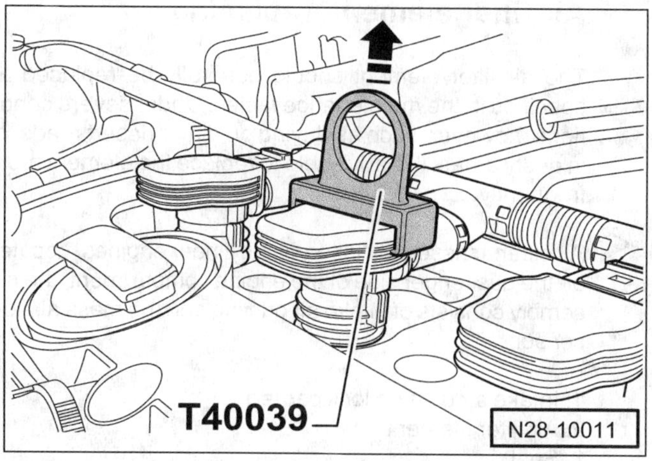

◄ Remove coil: On press-in coils, carefully pull coil up and off with special tool.

- 1.8L engine (code: AUG/AWM): special tool T40039
- 4.0L W8 (code: BDP): special tool T10166

EDITOR'S NOTE —
It is advisable to replace spark plugs one at a time to avoid confusion when connecting harness connectors to ignition coils, particularly on W8 engines.

– Remove spark plug with a 5/8 inch (16 mm) spark plug socket and extension or with special tool 3122B.

– Replace spark plug.

Tightening torque

- Spark plug to cylinder head (4 and 8-cylinder engines) . . . . 30 Nm (22 ft-lb)

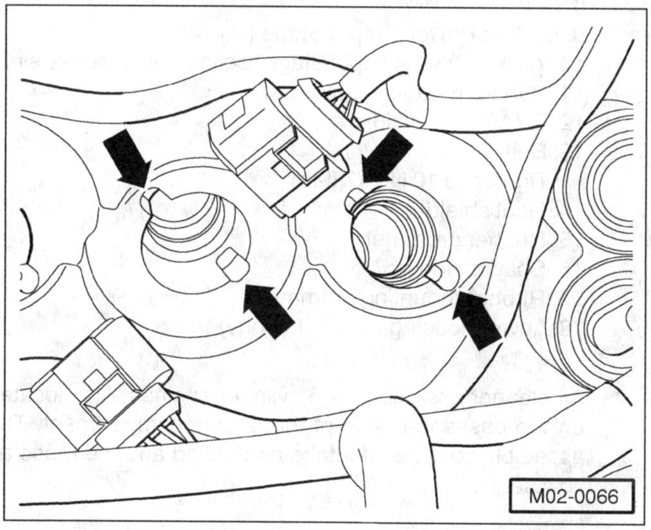

◄ Align coil to guide grooves (**arrows**), where equipped, when installing.

– Reinstall ignition coil and harness connector for cylinder #1.

– Repeat procedure for remaining cylinders.

– Replace intake manifold side sections and gaskets (if damaged) on W8 engines.

– Reinstall sound absorber(s) as required.

Spark plug replacement, double ended coils

– Remove sound absorber panel(s) as previously described.

– Remove first spark plug wire with a suitable puller. Do not pull directly on wires.

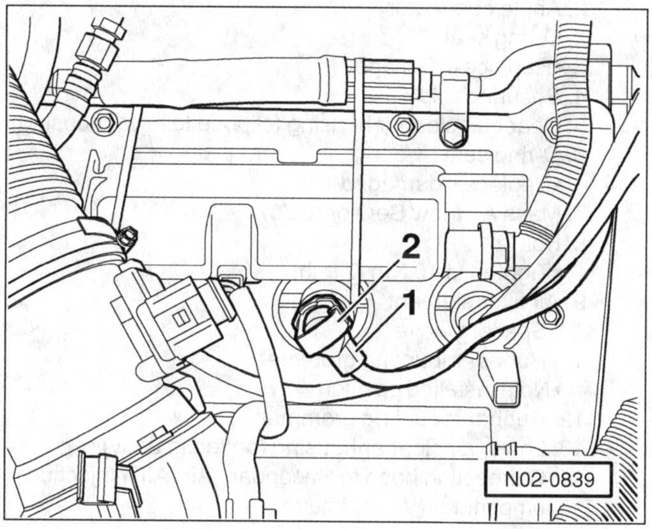

◄ Remove spark plug connector (**1**) with attached tool (**2**). If tool is not present, remove connectors with removal/installation tool T10029.

– Remove spark plug with a 5/8 inch (16 mm) spark plug socket and extension or with special tool 3122B.

– Replace spark plug and reinstall wire.

– Repeat procedure for remaining cylinders.

– Reinstall sound absorber(s) as required.

– If ignition was switched on while components were disconnected, clear any DTCs and set readiness codes.

Tightening torque

- Spark plug to cylinder head (6-cylinder engines) . . . . 30 Nm (22 ft-lb)

Air filter element, replacing

The air filter element should generally be replaced as specified in the maintenance tables. Under severe conditions, however, such as driving on dirt or desert roads, or in dusty areas, it is advisable to replace the element more frequently.

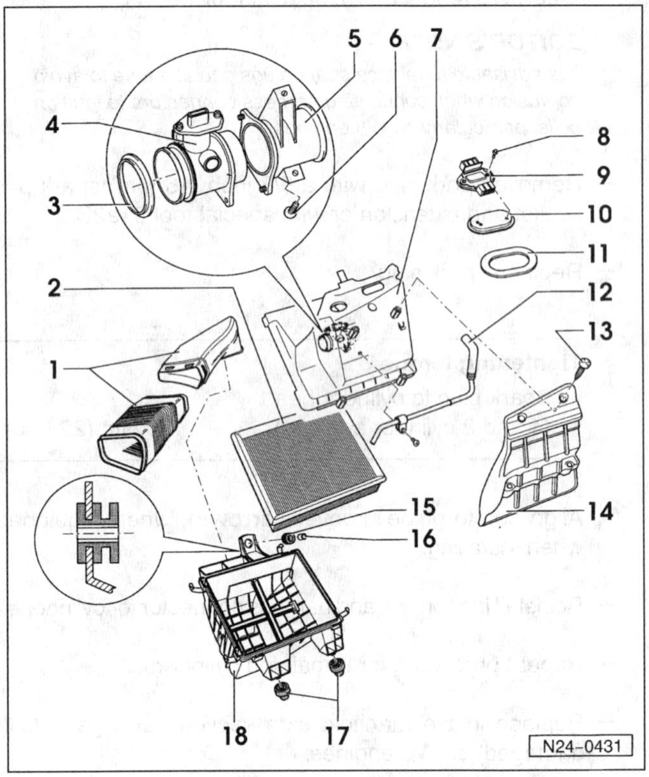

Air cleaner assembly for early 4-cylinder engines. Located on the passenger side of the engine compartment, the assembly consists of intake air ducting and the mass air flow sensor.

1. Intake air duct (to lock carrier)
2. Air filter element
3. Seal
4. Mass Air Flow Sensor (G70)
5. Connecting air duct
6. Bolt
   - Tighten to 6 Nm (53 in-lb)
7. Upper housing (clipped to lower housing)
8. Bolt
9. Ignition Power Output Stage (N122)
10. Ignition Power Output Stage cooling element (heat sink)
11. Rubber mounting grommet
12. EVAP connection
13. Bolt
    - Tighten to 10 Nm (7 ft-lb)
14. Heat shield
15. Rubber grommet
16. Spacer sleeve
17. Rubber mounting grommet
18. Lower housing (bolted to body)

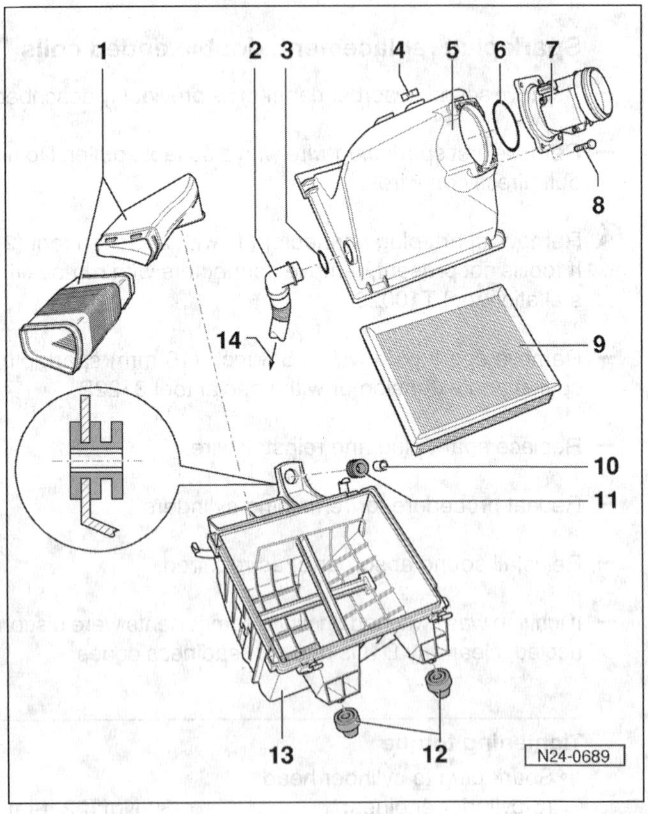

Air cleaner assembly for 6-cylinder engines. Also located on the passenger side of the engine compartment. The assembly consists of intake air ducting and the mass air flow sensor.

1. Intake air duct (to lock carrier)
2. Angle connector
3. O-ring seal
   - Replace if damaged
4. Securing Clip
5. Upper air cleaner housing (clipped to lower housing)
6. O-ring seal
   - Replace if damaged
7. Mass Air Flow Sensor (G70)
8. Bolt
   - Tighten to 10 Nm (7 ft-lb)
9. Air filter element
10. Spacer sleeve
11. Rubber mounting grommet
    - Note installed position
12. Rubber mounting grommet
13. Lower air cleaner housing (bolted to body)
14. Connecting hose to Secondary Air (AIR) Injection pump motor (V101) intake.

NOTE—

For 8-cylinder air filter assembly, see **4.0 Liter Fuel Injection and Ignition section, Repair Group 24**.

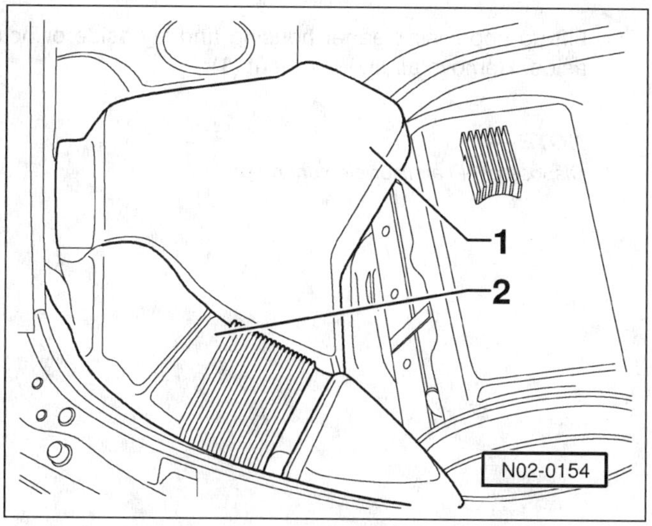

Lift up or unbolt cover (**1**) and remove. Remove Intake air duct (**2**) to lock carrier.

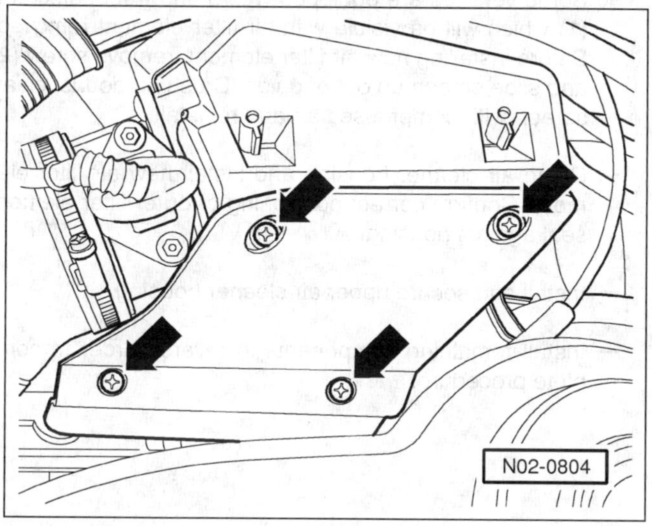

Remove heat shield attaching screws (**arrows**), where applicable, and remove heat shield.

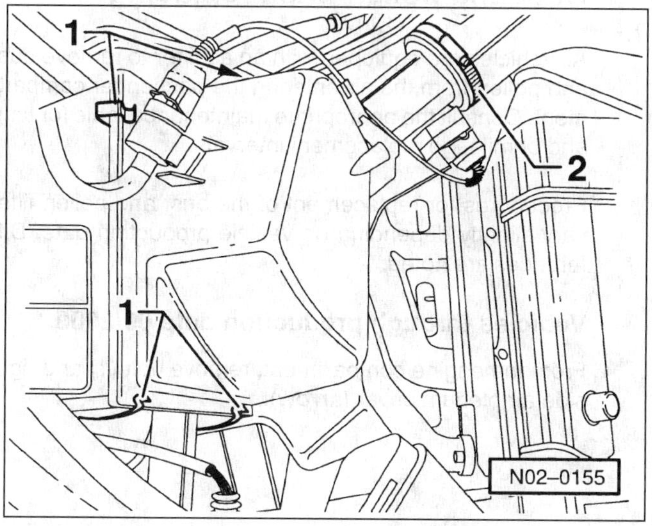

Un-snap securing clips (**1**). Loosen hose clamps (**2**) from Mass Air Flow sensor or connecting air duct as applicable and remove or lay aside intake air hose.

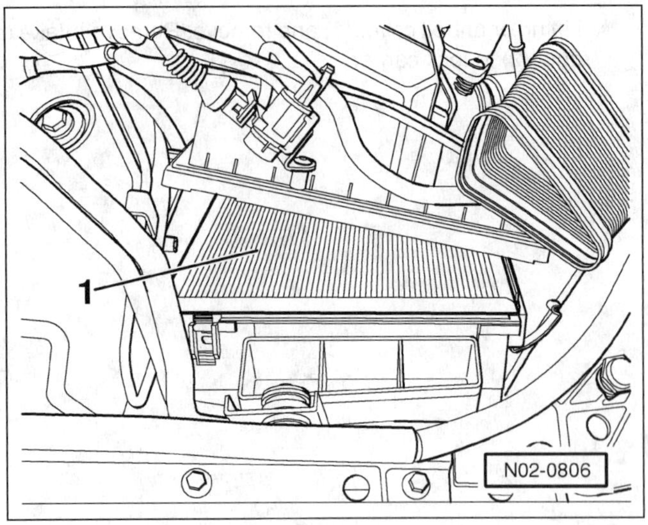

Lift up upper air cleaner housing and lay aside or hold aside. Remove air filter element (**1**).

NOTE —
Dispose of old air filter element properly.

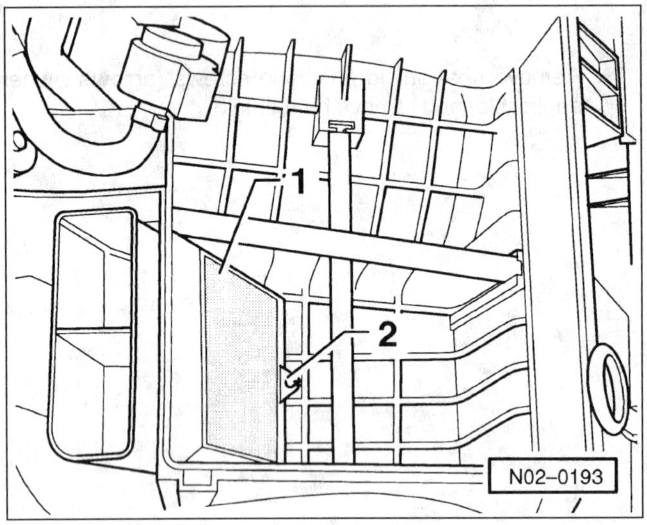

Some vehicles are equipped with a snow/debris screen (**1**) which will be visible with air filter element removed. Before installing new air filter element, remove screw (**2**) and slide screen up out of guide. Brush off debris, clean screen with compressed air and reinstall.

– Clean air cleaner housing and install new air filter element. Confirm correct positioning of outer filter element seal against housing.

– Install and secure upper air cleaner housing.

– Install remaining components in reverse order to complete procedure.

Dust and pollen filter, replacing

All vehicles are equipped with an air filter to remove dust and pollen from the air entering the passenger compartment. Consult the appropriate maintenance table for time and/or mileage replacement intervals.

Procedures for replacement of the dust and pollen filter vary slightly depending on vehicle production date. Differences are noted.

Vehicles through production date 09/2000

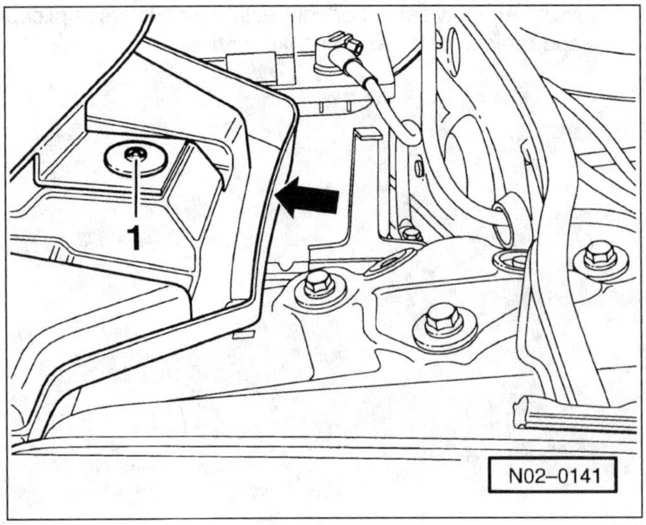

From the engine compartment, remove bolt (**1**) and right side air plenum cover (**arrow**).

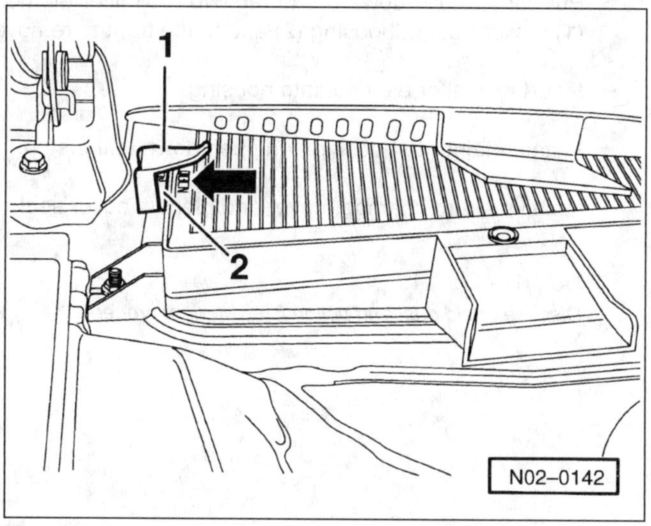

Pull retainer (**1**) up out of spring clip (**2**). Push up filter and pull forward to remove.

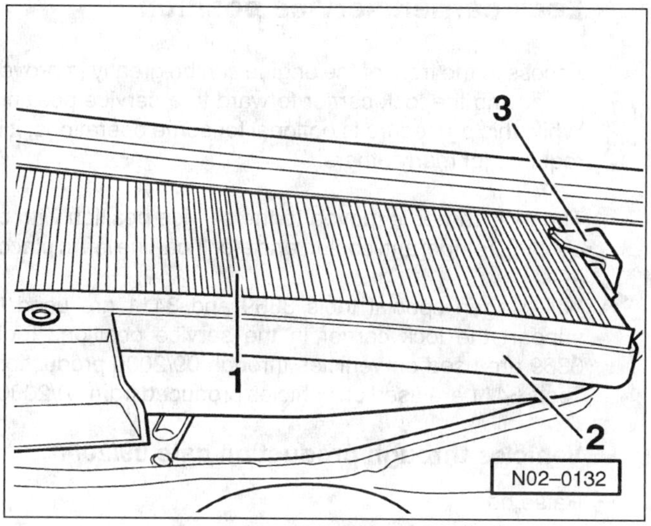

Install new dust and pollen filter (**1**) into housing (**2**) and push filter element under retainer (**3**).

NOTE—
When installing a new filter element, be sure the word "TOP" or "OBEN" on the filter is pointing up.

– Push retainer into spring clip.

NOTE—
Both retainers hold filter element into housing.

– Reinstall air plenum cover and bolt.

NOTE—
Dispose of old dust and pollen filter element properly.

Vehicles from production date 10/2000

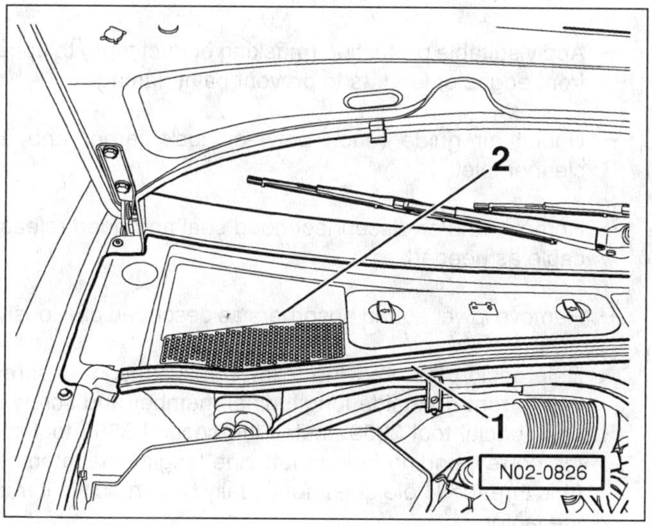

From the engine compartment, carefully remove plenum cover seal (**1**) by pulling toward the front of the vehicle. Gently remove plenum chamber cover (**2**) upward.

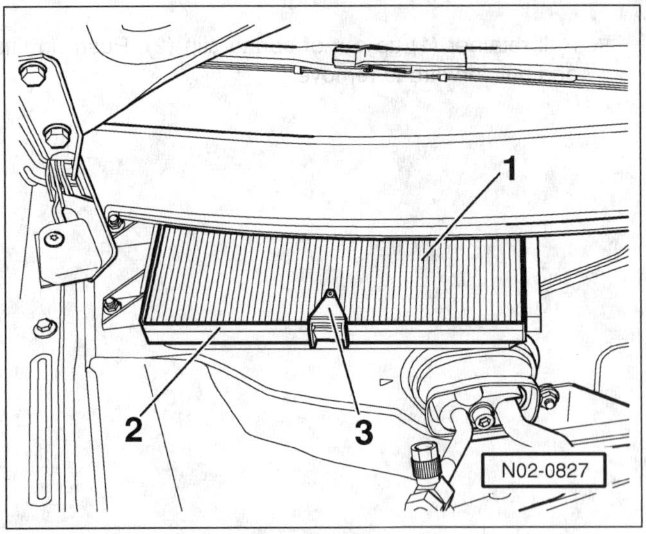

Pull retainer (**3**) upward and remove. Pull filter element (**1**) upward out of housing (**2**) and to the front to remove.

- Insert new filter element into housing.
- Install retainer to housing and push down to the stop.
- Install plenum chamber cover and plenum cover seal.

NOTE—
Dispose of old dust and pollen filter element properly.

Lock carrier, service position

Access to the front of the engine can be greatly improved by moving the lock carrier forward to a service position. While this procedure is optional for some operations, it is required for many others.

The lock carrier is moved to the service position with all coolant, refrigerant, cooler, and electrical line left in place.

Volkswagen special tools 3369 and 3411 are used to support the lock carrier in the service position. Tools 3369 are used on vehicles through 09/2000 production. Tools 3411 are used on vehicles produced from 10/2000.

Vehicles through production date 09/2000

- Raise hood.
- Remove front bumper cover and inner support. Handle carefully and move to a secure location so as to prevent damage.
- Apply suitable protection (masking or duct tape) to upper front edges of fenders to prevent paint damage.
- Unbolt air guide (duct) between lock carrier and air cleaner inlet.
- Remove and/or disconnect hood seal and hood release cable as needed.
- Remove lower sound absorbers as described previously.

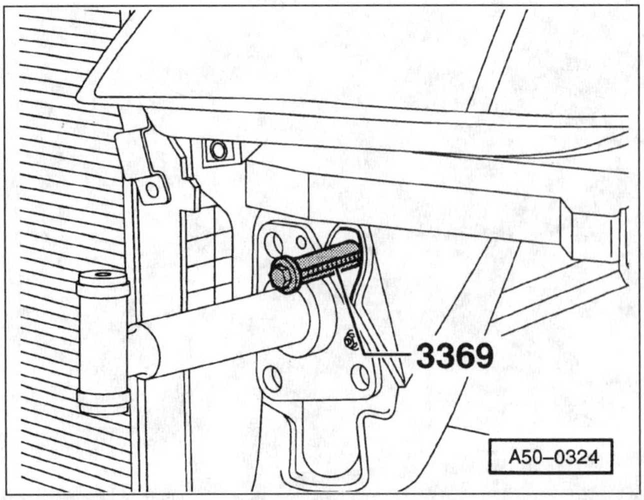

Referencing this and the following illustration, unscrew bolt **3** from right side longitudinal member and screw in one special tool 3369. Install the second 3369 tool into the corresponding hole in left side longitudinal member (this threaded hole does not usually have a bolt in it from the factory).

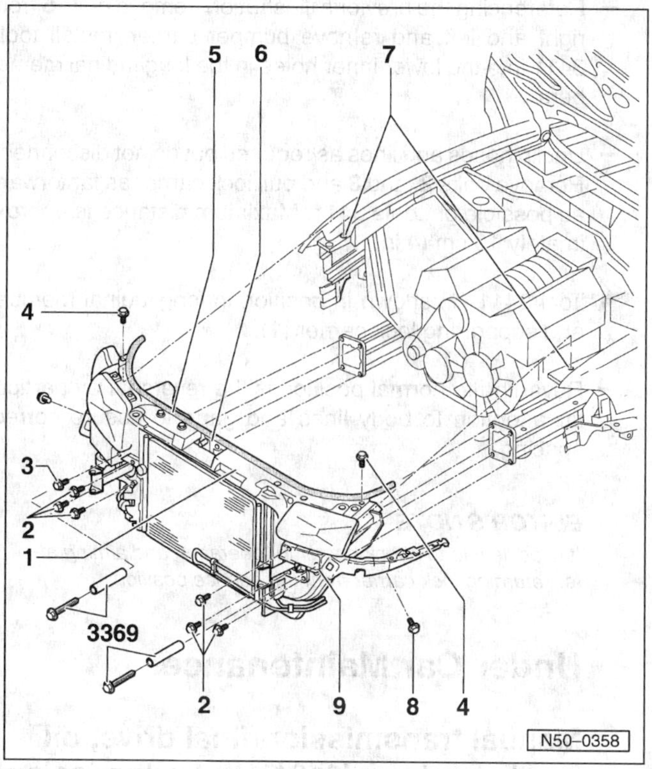

- Unclip hoses and lines as required but do not disconnect. Remove bolts **2**, **4** and **8** and pull lock carrier as far forward as possible on tools 3369.

◀ Lock carrier assembly for Passats produced through 10/2000.

1. Lock carrier and front bumper reinforcement.
2. Bolt
   - Quantity 6
   - Tighten to 50 Nm (37 ft-lb)
3. Bolt
   - Quantity 1
   - Tighten to 50 Nm (37 ft-lb)
4. Bolt
   - Quantity 2
   - Tighten to 8 Nm (6 ft-lb)
5. Seal
6. Bowden cable for hood release
7. Hole in inner fender
8. Bolt
   - Quantity 2
9. Hole in longitudinal member for special tool 3369

- Reverting to normal position is the reverse. Pay particular attention to body lines and gaps to assure correct spacing.

Vehicles from production date 10/2000

- Raise hood.
- Remove front bumper cover and inner support. Handle carefully and move to a secure location so as to prevent damage.
- Apply suitable protection (masking or duct tape) to upper front edges of fenders to prevent paint damage.
- Unbolt air guide (duct) between lock carrier and air cleaner inlet.
- Remove and/or disconnect hood seal and hood release cable as needed.
- Remove lower sound absorbers as described previously.

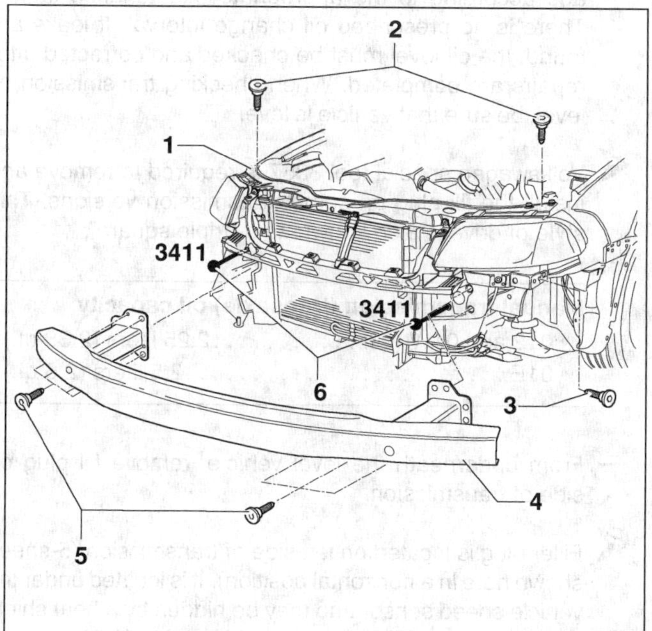

◀ Lock carrier assembly for Passats produced from 10/2000.

1. Lock carrier assembly.
2. Bolt
   - Quantity 2
   - Tighten to 8 Nm (6 ft-lb)
3. Bolt
   - Quantity 3
4. Bumper carrier
5. Bolt
   - Quantity 7
   - Tighten to 50 Nm (37 ft-lb)
6. Special tools 3411 (guide rods)

– Referencing the previous illustration, remove bolts **5** from right and left and remove bumper carrier. Install tools 3411 into the lower inner holes in the longitudinal meme-bers.

– Unclip hoses and lines as required but do not disconnect. Remove bolts **2**, and **3** and pull lock carrier as far forward as possible on tools 3411. Maximum distance is approx-imately 10 cm (4 in.).

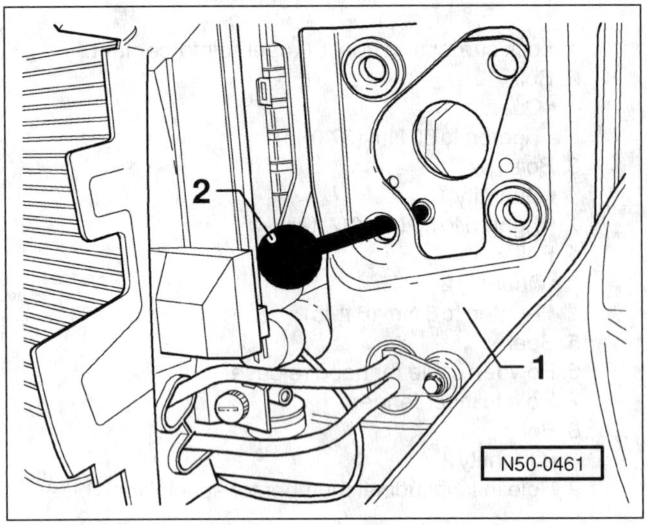

◄ Tool 3411 (**2**) shown in position in longitudinal member and supporting lock carrier (**1**).

– Reverting to normal position is the reverse. Pay particu-lar attention to body lines and gaps to assure correct spacing.

EDITOR'S NOTE —
It is advisable to check headlight operation and aiming af-ter returning lock carrier from the service position.

Under Car Maintenance

Manual transmission/final drive, oil level checking (012/01W transmission)

NOTE —
To check transmission oil level in the 01E 6-speed manual transmission, see **6-speed Manual Transmission 01E section, Repair Group 34**.

The manual transmission should be checked for oil leak-age according to the appropriate maintenance tables. There is no prescribed oil change interval. If leaks are found, the oil level must be checked and corrected after repairs are completed. When checking transmission oil level, be sure that vehicle is level.

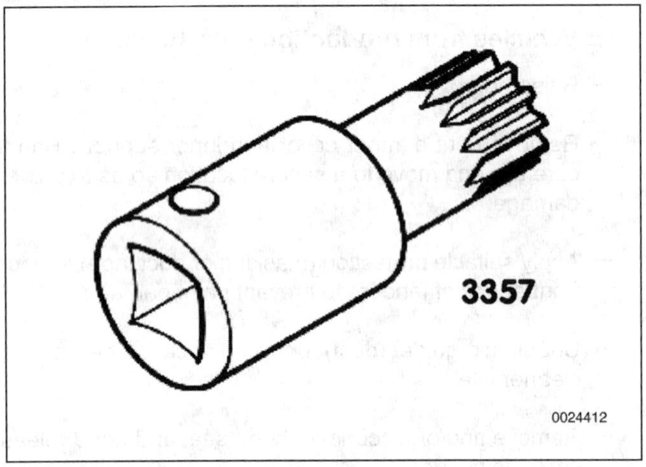

◄ Volkswagen special tool 3357 is required to remove and install the fill plug on some transmission versions. This style of drive is also known as a "triple square".

Manual transmission (transaxle) oil capacity
- 012 and 01W (all) . . . . .2.25 liters (2.38 qt.)
- 01E . . . . .2.5 liters (2.6 qt.)

– From underneath the level vehicle, remove fill plug on side of transmission.

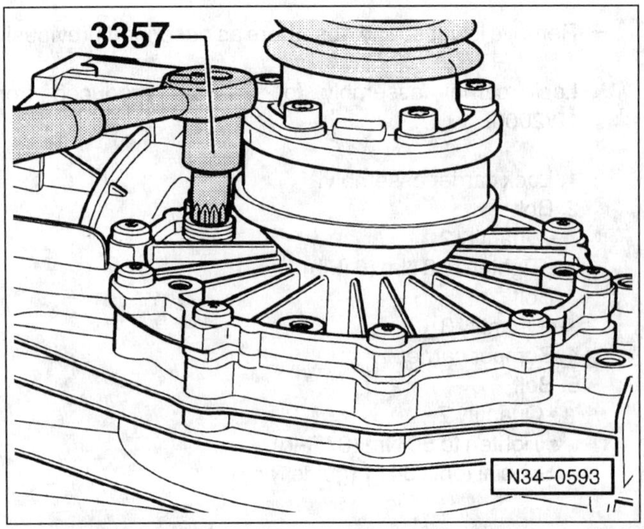

◄ Filler plug is located on left side of transmission (5-speed shown here in a horizontal position). It is located under the vehicle speed sensor and may be hidden by a heat shield over the drive axle. Version requiring tool 3357 is shown.

– Oil level is correct when it is full to the lower edge of filler hole.

– If level is below specification, fill oil to the lower edge of the filler hole only. Do not overfill. Correct transmission oil level is critical to proper operation.

– Install oil filler plug.

Specification

- Manual transmission lubricant, 012 and 01W (all) G 052 911 A1 . . . . SAE 75 W90 (Synthetic oil).

> ***CAUTION—***
> *Volkswagen part numbers are given for reference only! Always consult with your Volkswagen Parts Department or aftermarket parts specialist for the latest parts information.*

Tightening torque

- Oil filler plug, 012 and 01W (all) 5-speed manual transmissions . . . . 25 Nm (18 ft-lb)

NOTE—
Transmission cases are manufactured from aluminum or magnesium. As a result, all housings, covers and hardware, including the drain plug, MUST be matched to the case material to prevent contact corrosion and damage. Always consult with your Volkswagen Parts Department or aftermarket parts specialist for the latest parts information

Automatic transmission, fluid level (ATF) checking and adjusting

The automatic transmission should be checked for oil leakage according to the appropriate maintenance tables. There is no prescribed fluid change interval. If leaks are found, the fluid level must be checked and corrected after repairs are completed.

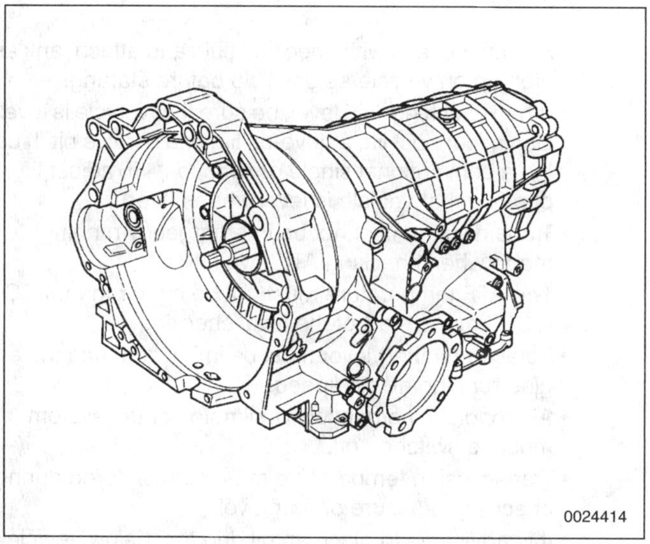

◄ 5-speed automatic transmissions are available in front wheel (4-cylinder and 6-cylinder) and all wheel drive, 4MOTION, (6-cylinder) versions. The transmission designation is 01V.

The automatic transmission (ATF) fluid is checked from under the vehicle by a combination filler/level well. There is no provision for a dip stick.

If transmission needs to be filled or topped up, an appliance such as the VAG 1924 must be used.

NOTE—
Use only Automatic Transmission Fluid (ATF) as specified by Volkswagen. Use of any other type of fluid may adversely affect the automatic transmission's operation and may result in damage.

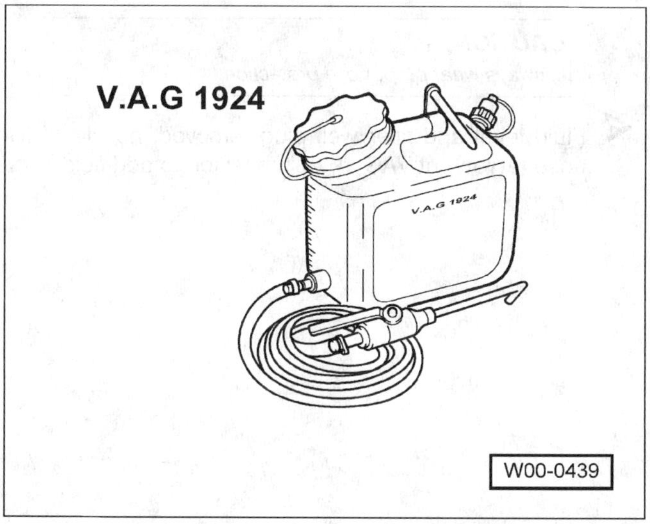

◄ ATF service tank VAG 1924 as supplied by Volkswagen.

Transmission fluid level is extremely dependent on fluid temperature due to fluid and component expansion. A suitable scan tool or computer scan tool program must be used when checking level.

Because there are numerous types of scan tools and computer scan tool programs available, always consult the manufacturers' instructions for information on accessing the Transmission Electronics and/or the Transmission Control Module (TCM). ATF temperature values will need to be continuously displayed.

If using Volkswagen scan tools VAG 1551 or 1552, connect the scan tool to the Data Link Connector (DLC) as described elsewhere in the manual and use the following to access the TCM for ATF temperature information:

- Function 1 - Rapid Data Transfer
- Address word 02 - Transmission electronics
- Function 08 - Read measuring value blocks
- Display group - 004
- Display field - 1 (ATF temperature in degrees C)

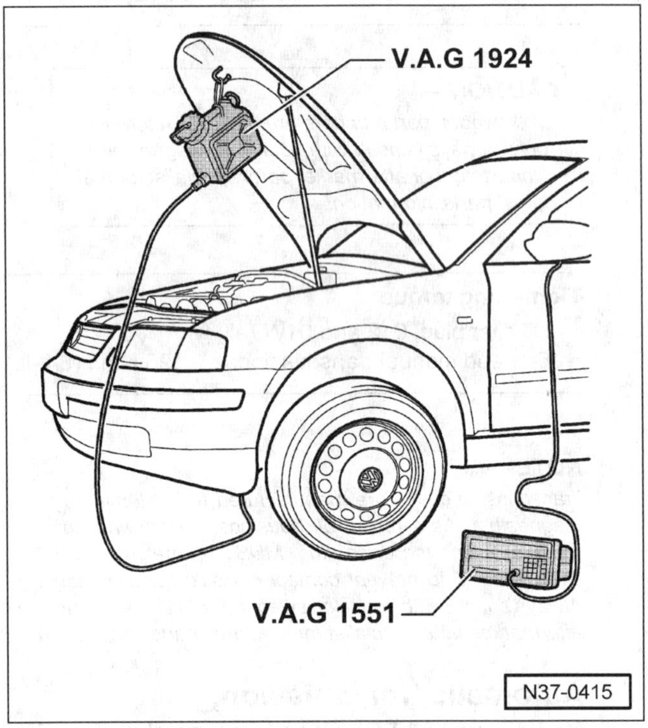

Before checking the ATF level, observe the following notes, conditions and cautions:

- Fill service tank with specified fluid and attach tank as high up on vehicle as possible before starting.
- When checking ATF level, be sure that vehicle is level. Use a four post lift, a drive-on lift, or a service pit. If using a conventional single or twin post lift, adjust lift pads so that vehicle is level when raised.
- Transmission must not be in emergency running mode when checking ATF level.
- The ATF temperature must not be higher than 30°C (86°F) at the start of the level checking.
- Selector (shifter) lever must be in P (Park) and the engine running at idle speed.
- Air conditioning/heating or climate control system must be switched off.
- Transmission temperature must be monitored during checking to assure proper level.
- It is advisable to place an oil drip tray below vehicle.

> ***CAUTION—***
> *Always wear proper eye protection.*

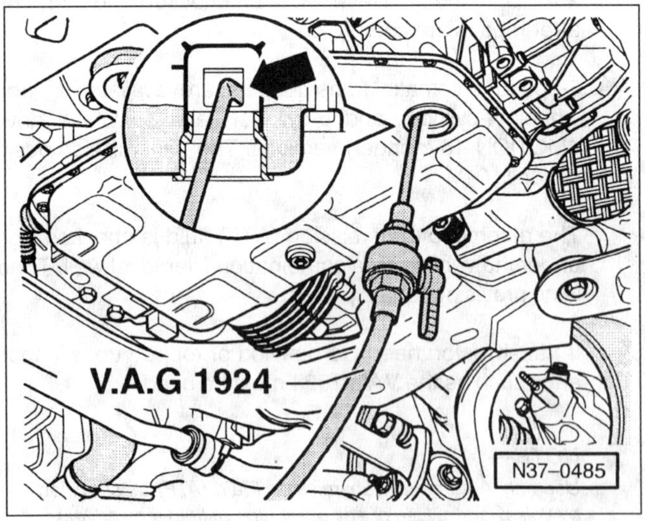

Fluid level and filler well plug removed; nozzle of filler hose (**arrow**) of VAG 1924 in position to add fluid.

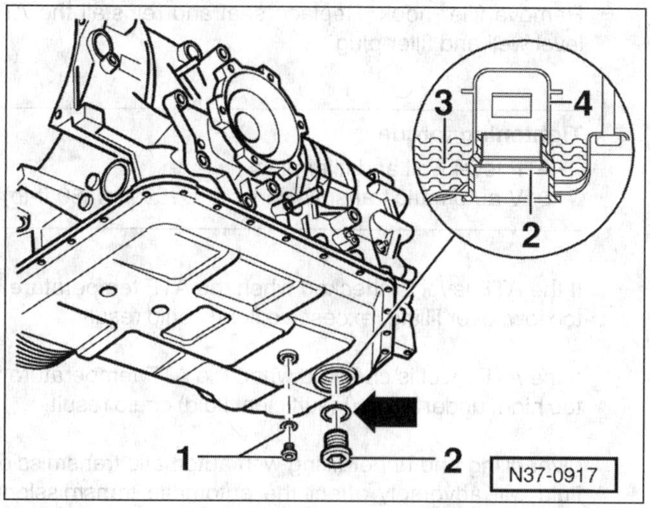

01V Automatic transmission oil pan configuration.

1. ATF drain plug and seal
   - Tighten to 40 Nm (30 ft-lb)
   - Always replace seal
2. ATF level well and filler plug and seal
   - Tighten to 80 Nm (59 ft-lb)
   - Always replace seal (**arrow**)
3. Automatic transmission fluid
   - Shown at correct level
   - Use only Volkswagen specified ATF
4. Fluid level and filler well plug
   - With holes on side for filling and deflector cap on top

– With all notes and conditions observed, No fluid should spill out except for several drops that may have been in the well.

– At a fluid temperature of between 35°C and 45°C (95°F and 113°F), remove ATF level well and filler plug (**2** from above illustration) from the transmission.

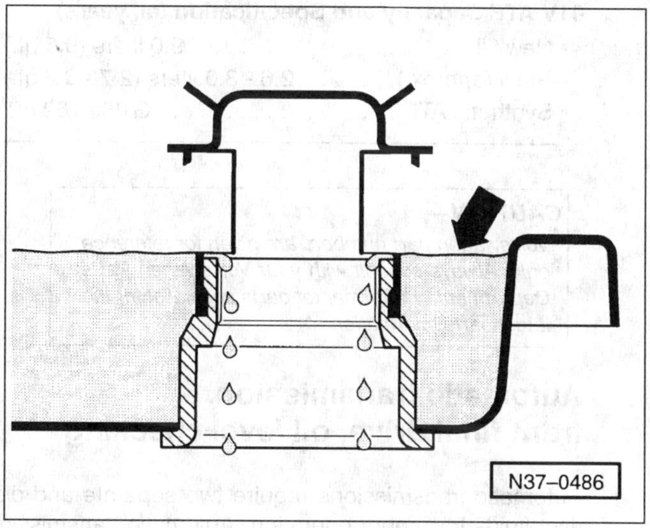

If ATF DRIPS out of plug hole:

- No ATF needs to be added, fluid level is correct (**arrow**).
- Replace seal and reinstall the ATF level well and filler plug (**2** from above illustration).

– **If ATF RUNS out of plug hole:**

- Transmission was overfilled.
- Allow to run out until flow becomes a drip.
- Make sure temperature is still within range.
- Replace seal and reinstall the ATF level well and filler plug (**2** from above illustration).

– **If NO ATF drips out of plug hole:**

- Transmission was under filled.
- Top up fluid with VAG 1924 until it drips out.
- Make sure temperature is still within range,
- Replace seal and reinstall the ATF level well and filler plug (**2** from above illustration.)

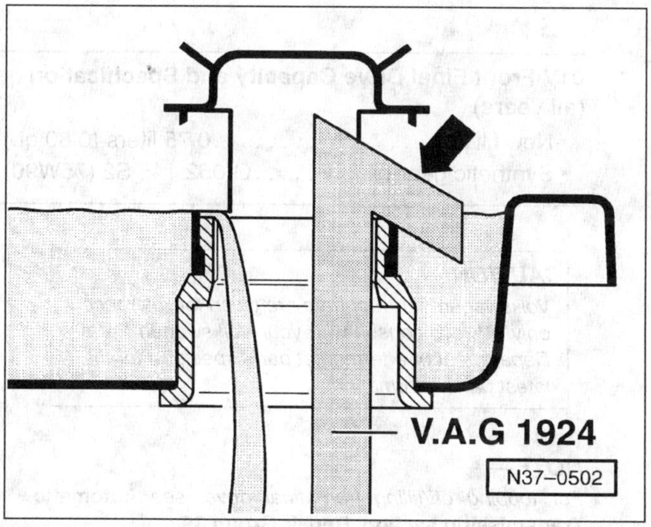

If no fluid flowed out of ATF level well with the ATF temperature between 35°C and 45°C (95°F and 113°F) insert filler hook into level well and into one of the openings in the well. Open filler hook valve and allow fluid to flow until it begins to drip out of hole. Close filler hook valve. If temperature is within the appropriate range, the level is correct at this point.

NOTE—
Do not press the filler hook upwards more than needed to insert into an opening in the well. It is possible to push the deflector cap off of the top of the well. Use care.

– Remove filler hook. Replace seal and reinstall the ATF level well and filler plug.

Tightening torque

- ATF level well and filler plug
  01V automatic transmission . . . . . .59 Nm (80 ft-lb)

If the ATF level is checked when the ATF temperature is too low, over filling (excessive fluid) could result.

If the ATF level is checked when the ATF temperature is too high, under filling (insufficient fluid) could result.

Over filling and under filling with automatic transmission fluid will adversely affect the automatic transmission's operation and may ultimately result in damage.

01V ATF Capacity and Specification (all years)

- New fill . . . . .9.0 liters (9.5 qt.)
- Refill (approx.). . . . . . . 2.6 - 3.0 liters (2.7 - 3.2 qt.)
- Synthetic ATF . . . . G 052 162 A2

> ***CAUTION—***
> *Volkswagen part numbers are given for reference only! Always consult with your Volkswagen Parts Department or aftermarket parts specialist for the latest parts information.*

Automatic transmission, front final drive, oil level checking

Automatic transmissions require two separate and distinct fluids for proper operation. Automatic transmission fluid is used in the automatic section and special gear oil is used in the differential. There is no prescribed oil change interval. If leaks are found, the fluid level must be checked and corrected after repairs are completed.

01V Front Final Drive Capacity and Specification (all years)

- New fill .0.75 liters (0.80 qt.)
- Synthetic gear oil . . . . .G 052 145 S2 (75W90)

> ***CAUTION—***
> *Volkswagen part numbers are given for reference only! Always consult with your Volkswagen Parts Department or aftermarket parts specialist for the latest parts information.*

NOTE—
For topping off/filling front final drive, see **Automatic Transmission section, Repair Group 39**.

When checking the automatic transmission front differential oil level, note the following:

- Be sure that vehicle is level. Use a four post lift, a drive-on lift, or a service pit. If using a conventional single or twin post lift, adjust lift pads so that vehicle is level when raised.
- Use only Volkswagen specified final drive oil.
- It is advisable to place an oil drip tray below vehicle.
- Always replace drain plug seal.

NOTE —
Always wear proper eye protection.

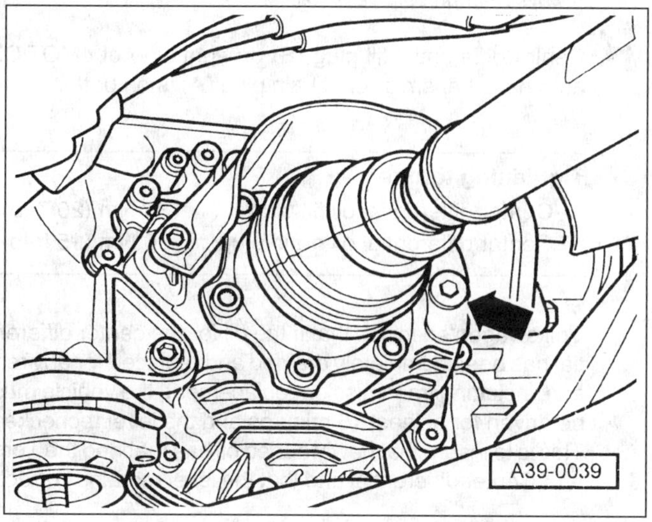

Front final drive fill plug (**arrow**) on left side of 01V automatic transmission. Use 8 mm hex key for removal.

Tightening torque

- Automatic transmission front final drive oil fill plug . . . . .30 Nm (22 ft-lb)

Automatic transmission, 4MOTION center differential, oil level checking

All wheel drive vehicles with automatic transmission have a separate center differential. Some Volkswagen service literature also refers to this as a transfer gearbox.

There is no prescribed oil change interval for the special gear oil used in this differential, Volkswagen specifies that it is filled for life. If leaks are found, the fluid level must be checked and corrected after repairs are completed.

4MOTION Center Differential Capacity and Specification (all years)

- New fill .0.8 liters (0.85 qt.)
- Synthetic gear oil . . . . .G 052 145 S2 (75W90)

CAUTION —
Volkswagen part numbers are given for reference only! Always consult with your Volkswagen Parts Department or aftermarket parts specialist for the latest parts information.

When checking the center differential oil level, note the following;

- Be sure that vehicle is level. Use a four post lift, a drive-on lift, or a service pit. If using a conventional single or twin post lift, adjust lift pads so that vehicle is level when raised.
- It will be necessary to separate right side exhaust pipes on some vehicle configurations to allow access to the filler/level plug.
- Use only Volkswagen specified center differential oil.
- It is advisable to place an oil drip tray below vehicle.

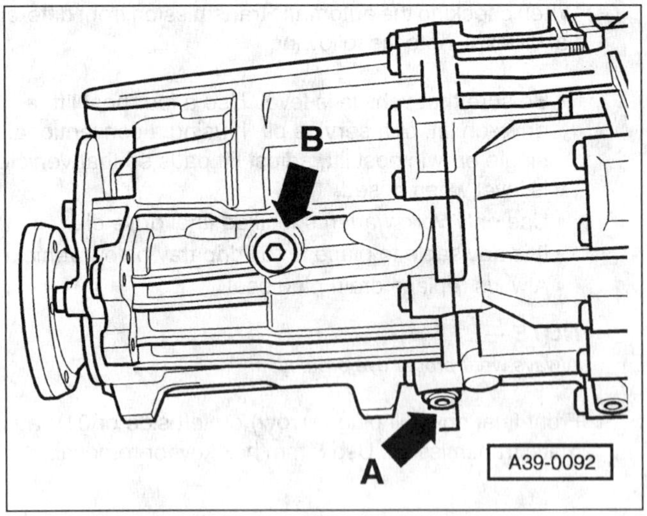

- The oil level is correct when the level is full to the lower edge of the filler hole.
- Always replace filler plug and/or drain plug.

NOTE—
Always wear proper eye protection.

Center differential fill plug (**B**) on right side of 4MOTION automatic transmission. Drain plug (**A**) is on bottom.

Tightening torque
- Center differential oil fill plug. . . . . .35 Nm (26 ft-lb)
- Center differential oil drain plug . . .20 Nm (15 ft-lb)

Volkswagen specifies that if the oil for the center differential has been completely drained and replaced for any reason, including transmission replacement, the vehicle must be driven for at least 15 minutes and the level rechecked. This is to insure that newly added oil has reached the Torsen® Center differential and that the level remains correct.

4MOTION rear differential (final drive) oil level, checking

All wheel drive (4MOTION) vehicles require a separate rear differential or final drive. There is no prescribed oil change interval for the special gear oil used in this differential, Volkswagen specifies that it is filled for life. If leaks are found, the fluid level must be checked and corrected after repairs are completed.

4MOTION rear differential capacity and specification (manual and automatic transmissions)
- New fill . . . . .1.5 liters (1.6 qt.)
- Synthetic gear oil . . . . G 052 145 S2 (75W90)

CAUTION—
Volkswagen part numbers are given for reference only! Always consult with your Volkswagen Parts Department or aftermarket parts specialist for the latest parts information.

When checking the rear differential oil level, note the following;

- When checking rear differential oil level, be sure that vehicle is level. Use a four post lift, a drive-on lift, or a service pit. If using a conventional single or twin post lift, adjust lift pads so that vehicle is level when raised.
- Use only Volkswagen specified final drive oil.
- It is advisable to place an oil drip tray below vehicle.

NOTE—
Always wear proper eye protection.

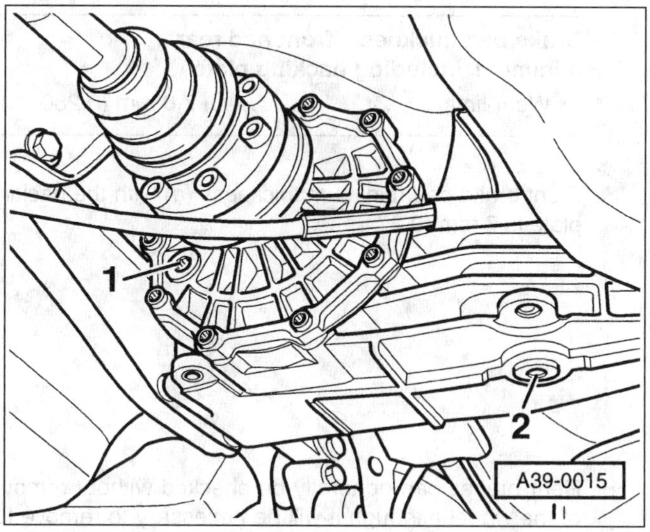

Rear final drive fill plug (**1**) on right side of rear differential. Drain plug (**2**) is on bottom.

Tightening torque

- Center differential oil fill plug. . . . . .35 Nm (26 ft-lb)
- Center differential oil drain plug . . .35 Nm (26 ft-lb)

Brake system, visual inspection

Aside from fluid changes, periodic brake system maintenance includes checks of the brake master cylinder, vacuum brake booster, hydraulic unit, brake pressure regulator, and brake calipers. Components should be checked for leaks, wear and damage.

- Check brake master cylinder, vacuum brake booster, hydraulic unit, brake pressure regulator, brake fluid reservoir, cap and level sensors, and brake calipers for leaks and damage.
- Brake hoses must not be twisted.
- Brake hoses must not touch any part of the vehicle when steering is at full lock.
- Check brake hoses for porosity and deterioration, brake hoses and brake lines for chafing points.
- Check brake connections and attachments for correct seating, leaks and corrosion.

Any faults found must be repaired.

Brake pad thickness (front and rear), checking

Inspect condition and measure the thickness of the front and rear brake pads. In the front, this is best accomplished by removal of a wheel.

EDITOR'S NOTE —

Most vehicles have at least one special anti-theft lug bolt per wheel installed from the factory. This lug requires a unique adapter for removal. Before starting repairs, consult with vehicle owner/driver for it's location. Be sure to return the adapter to this location upon completion of repairs.

– **Full wheel cover:** obtain removal hook from vehicle tool kit or use other suitable tool.

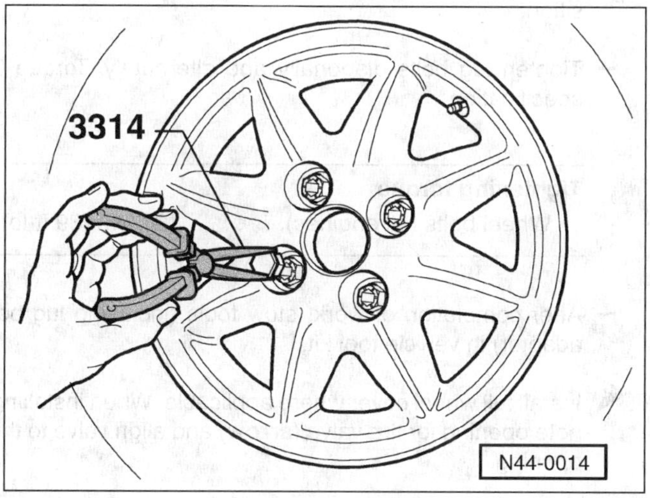

Alloy wheels: remove wheel lug covers with appropriate tool. VW 3314 shown.

– Mark installed position of wheel in relation to brake rotor so that it can be reinstalled in that position. This avoids possible re-installation mismatch conditions.

– Remove lug bolts and wheel.

– Measure thickness of inner and outer brake pads (including backing plate).

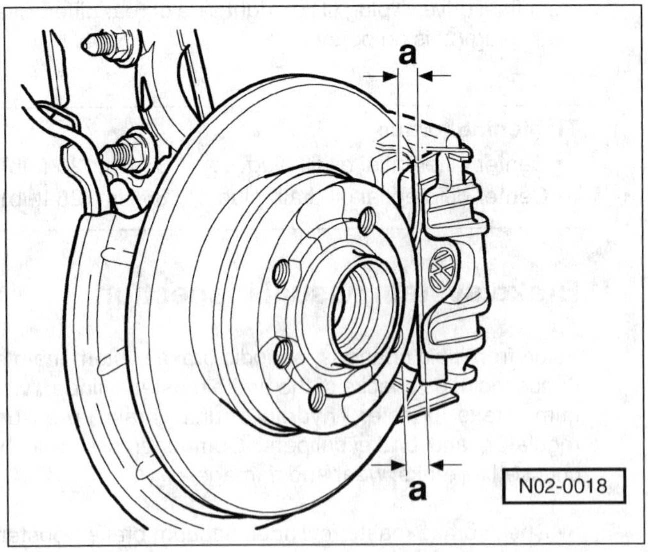

Brake pad thickness, front and rear minimum, including backing plate

- Wear limit . . . . .7.0 mm (0.280 in.)

Front brake pad minimum thickness (**a**) with the backing plate is 7 mm (0.27 in.).

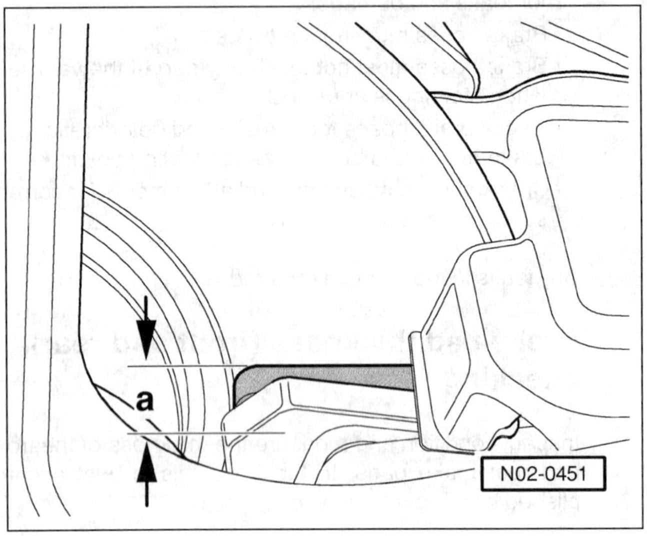

– Rear brakes can generally be checked without removal of the wheel although it will be necessary to remove full wheel covers from steel wheels.

– Check inner and outer rear pad thickness through wheel opening using a flashlight and mirror.

Rear brake pad minimum thickness (**a**) with the backing plate is also 7 mm (0.27 in.).

– If pad thickness (including backing plate) is 7.0 mm (0.28 in.) or less, the brake pads have reached or exceeded their wear limit and must be replaced.

EDITOR'S NOTE —

- *Whenever brakes are inspected, it is advisable to record the pad thickness along with the date and mileage in the vehicle documents for future reference.*
- *If a person other than the Technician performing this check is responsible for the vehicle, always notify this individual before incurring an expense such as replacing brake pads.*

– If wheel was removed, re-install in previously marked position.

– Tighten lug bolts diagonally and alternately. Torque to specification.

Tightening torque

- Wheel bolts (all engines). . . . . . . .120 Nm (89 ft-lb)

– After completion of work, stow tools (including lug bolt adapter) in vehicle tool kit.

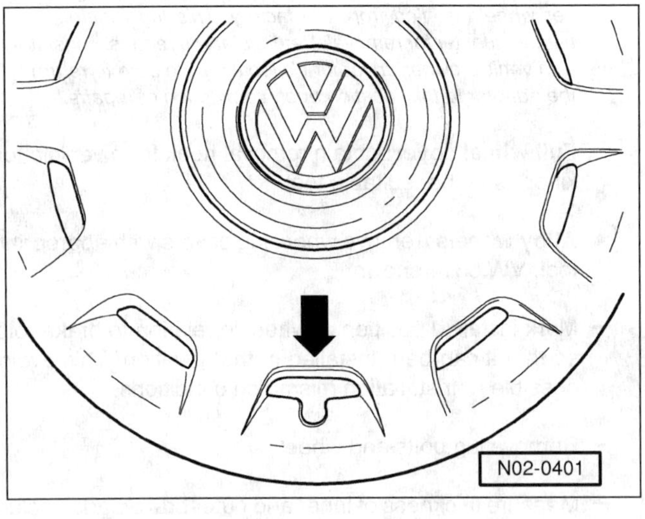

Install full wheel cover where applicable. When installing, note opening for tire valve (**arrow**) and align valve to this opening.

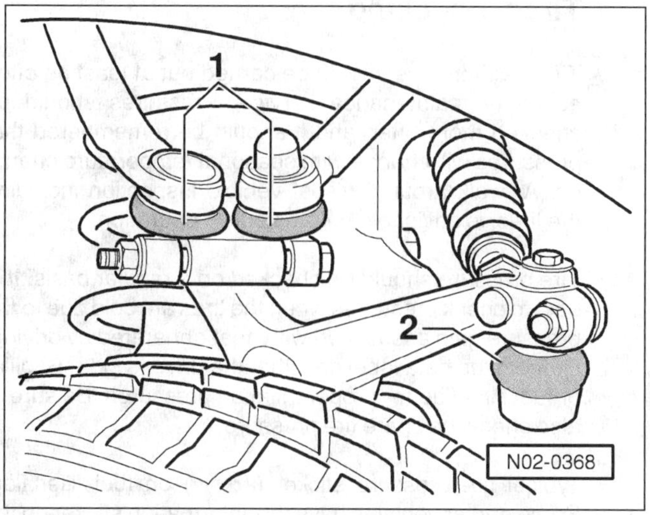

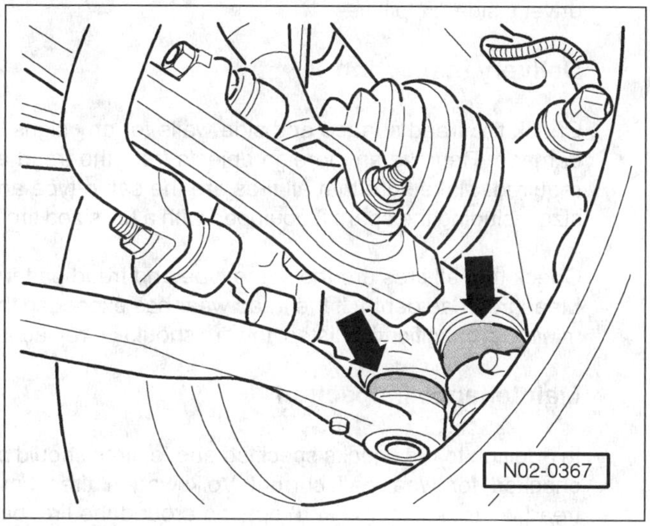

Tie rod ends, control arm joints, checking

When the vehicle is raised with the wheels off the ground, inspect left and right tie rods at both ends. Also inspect control arm joints and boots.

– Move tie rods and wheels. There must be no play at inner or outer ends.

– Check attaching points for looseness and damage,

◄ Check dust boots (**2**) of each tie rod at both ends for damage and correct installation (not twisted). Check upper control arm boots (**1**) for damage and proper seating.

– If vehicle is equipped with a steering dampener on the right side tie rod, inspect for fluid leakage and mounting.

◄ Check lower control arm boots (**arrows**) for damage, proper seating, and leaks.

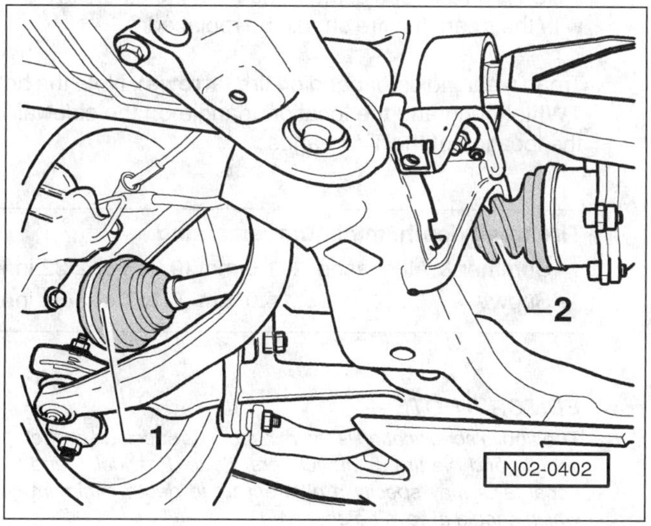

Axle boots, checking

Axle boots protect the constant velocity and tripod joints connecting the driveline to the wheels. Periodic inspection and replacement of damaged boots will greatly prolong service life of joints.

◄ Visually inspect outer (**1**) and inner (**2**) axle boots for damage and leaks. Pay particular attention to stones and debris in the folds. Also inspect rear axle boots on 4MOTION all wheel drive vehicles.

EDITOR'S NOTE—

The drive shaft (propeller shaft) connecting the front and the rear differentials on 4MOTION vehicles also has constant velocity and universal joints. These joints are protected by internal seals and are not serviceable.

Tires, checking

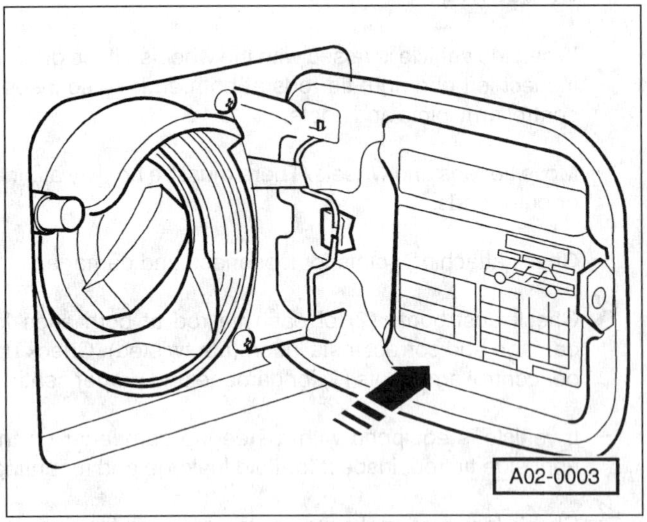

Tire maintenance should be carried out at least as often as every maintenance service. Pressures should be checked more often and it should be remembered that pressures will change with seasonal temperature changes. A "walk around" visual vehicle inspection including the tires, is advisable before vehicle operation.

Tire pressure should be checked on a regular basis. It is best to check pressures when the tires are cold due to the pressure gain associated with heat generated by driving. Refer to the data label on either the driver's side "B" pillar or fuel filler flap for proper inflation pressures. Be sure to also check the spare tire pressure.

< Typical tire pressure sticker (**arrow**) on fuel filler flap. Some earlier vehicles have this information sticker on the driver's side "B" pillar.

Delivery inspection

Check tire tread surface and side walls for any signs of damage. Remove any foreign objects from the tread as required. Make sure that all tires are the same type and size including the spare if equipped with a full sized tire.

Check that all tires are the same type and tread pattern. Measure tread depth. If the tread wear has exceeded the minimum specification listed, the tire should be replaced.

Maintenance inspection

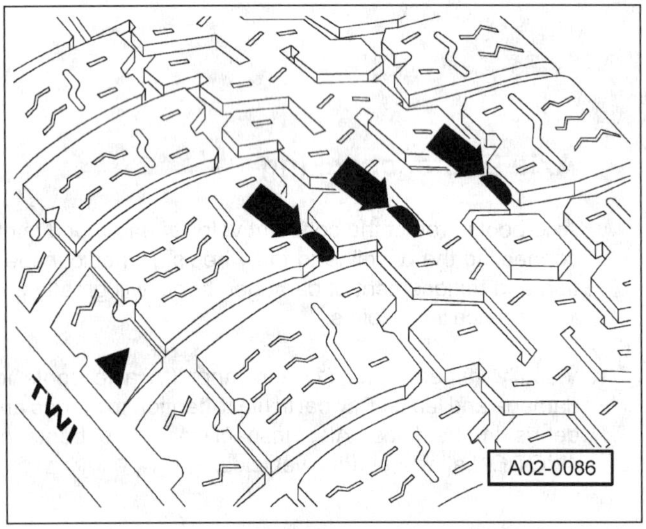

In addition to the checks specified above, tires should be checked for wear. All current Volkswagen tires have tread wear indicators (TWI) spaced around the tire, perpendicular to the tread, at marked intervals. When the TWIs start to become visible, the tire is reaching the wear limit. When the tops of the tread wear indicators are even with the tread, the tire should be replaced.

< Tread wear indicator band on tire (**arrows**). Note the bold TWI lettering and the location triangle on the sidewall at the location of the TWI bands.

Tire tread depth, minimum remaining

- Summer & All-season . . 1.6 mm (0.063 or 2/32 in.)
- Snow . . . . 4.0 mm (0.157 or 5/32 in.)

EDITOR'S NOTE —

The above specification is the minimum as specified by the vehicle and the tire manufacturers. State, Provincial, and Local laws may specify a different tread depth minimum which should always be observed.

EDITOR'S NOTE —
Snow tires (or winter tires) have a deep lug tread pattern and MUST be used on all 4 wheels. They are not generally suitable for warm weather use. They require a 3 psi increase in cold inflation pressure and have different speed ratings. Consult the tire manufacturer for additional information and applications.

Tires should also be checked for scuffing, cupping, feathering, irregular tread wear, sidewall checking, dry rot, cuts and fractures. Rims (both steel and alloy) should also be checked for damage.

Irregular tread wear on tires, especially front tires, can be an indication of an out-of-alignment condition which should be checked and corrected.

- Feather edging or "feathering" on the tread indicates incorrect toe settings.
- Wear on the inside or the outside of the tread indicates incorrect camber settings.

If irregular wear patterns are identified, determine the causes by checking the wheel alignment (front and rear).

Vibrations felt through the steering wheel can be an indication of an out of balance condition which should be checked and corrected. An out of balance condition can shorten the tire life, cause wear on the steering and suspension components, and may result in an uncomfortable ride.

For safety reasons, any faults or defects noted with the tires should be recorded on the appropriate vehicle repair documents and reported to the customer.

Tires should be rotated periodically for maximum service life and quiet running. Volkswagen specifies tire rotation at the first 5000 mile service and then defers to the tire manufacturers recommendations from then on. These recommendations can vary slightly between manufacturers, but experience has shown that rotation at 5000 to 7500 mile intervals results in optimal wear.

Volkswagen specifies that under most conditions, tires should remain on the same side of the vehicle (front to back) when they are rotated. Tires should only be rotated diagonally if they show unusual wear. Do not rotate tires diagonally if they are directional tires as noted by directional arrows on the sidewalls.

Old tires (6 years or more) may not function or perform as originally intended. These tires should only be used with caution. Volkswagen specifies replacement after this time period regardless of tread wear.

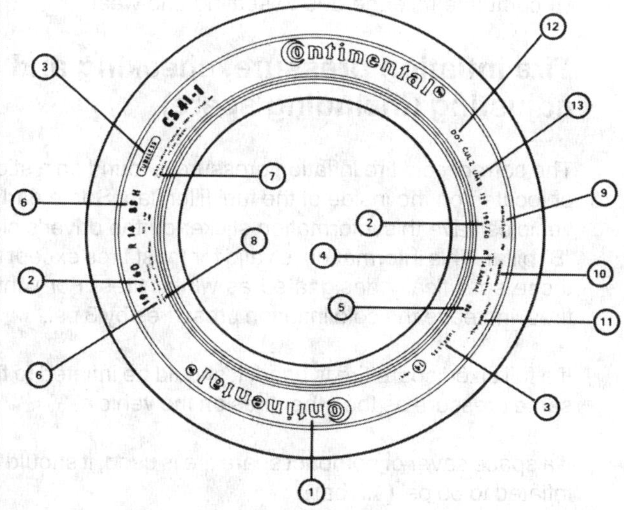

1. Manufacturers brand
2. Tire dimensions (size)
3. Type (tubeless)
4. Country of origin
5. Load index and speed rating
6. Load range (A, B, C)
7. Tread & sidewall ply and construction
8. Maximum cold inflation pressure
9. Treadwear index number
10. Traction index number (AA, A, B, C)
11. Temperature index number (A, B, C)
12. USA DOT compliant
13. Manufacturers production coding

0024416

◄ Overview of typical information found on tire sidewall that complies with United States Federal Motor Vehicle Safety Standard (FMVSS) 109.

> ***CAUTION—***
> *Tires of the same type, size and tread pattern MUST be used on 4MOTION all wheel drive to prevent damage to the center differential.*

When fitting new tires, the following points should be observed:

- Volkswagen specifies that only radial tires be used.
- For safety reasons, it is generally advisable to replace tires in pairs and on the same axle.
- Tires with the best (deepest) tread should be mounted to the front of the vehicle.
- New tires should also be balanced and their rims (wheels) fitted with new valve stems.
- New tires tend to be slippery and require a break-in period of at least 100 miles (160 Km).
- New tires should be of the same size as originally installed. If you change sizes, be aware that gear ratios and speedometer readings will change.
- Do not use tires of a lesser speed rating and/or load range than originally installed.
- Wheel and tires may require re-balancing as they accumulate mileage due to settling and wear.

Tire inflation pressure, checking and adjusting (including spare)

The correct cold tire inflation pressure is found on a sticker located on the inside of the fuel filler flap. Some earlier vehicles have this information sticker on the driver's side "B" pillar. This information is valid for most tires except for those specifically designated as winter tires. For winter tires, increase the cold inflation pressures by 3 psi.

If a full sized spare tire is used, it should be inflated to the same pressure as the other tires on the vehicle.

If a space saver or compact spare tire is used, it should be inflated to 60 psi (4.2 bar).

Always replace dust caps on tire valves when checking tire pressures. If valve extensions are used (as on steel wheels and wheel covers), inspect extensions for damage and replace as required. Make sure that wheel covers are installed correctly so as to not cut valve stems and/or extensions.

Wheel bolts, tightening to specified torque

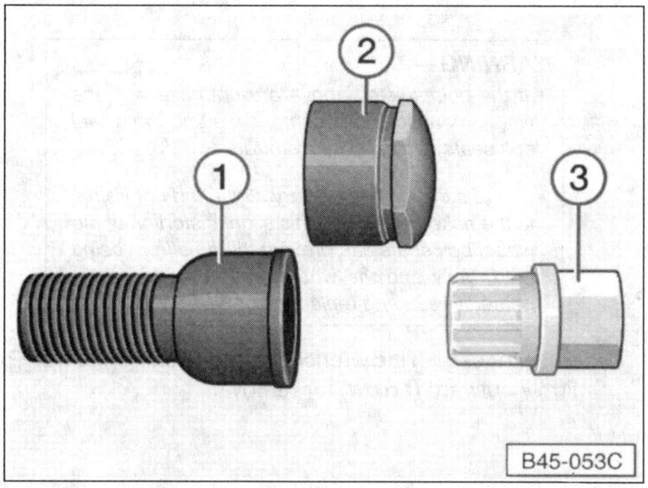

When removing and installing the road wheels (alloy and steel), note the following points to be sure that the wheels remain secure at all times:

Configuration of factory-installed anti-theft wheel bolts.

1. Wheel (lug) bolt
   - Tighten to 120 Nm (89 ft-lb)
2. Wheel bolt cover
   - Removal hook from vehicle tool kit fits into hole in center
   - May not be installed on vehicles with full wheel covers
3. Wheel bolt adapter
   - Many different versions
   - Stowed in small innner pocket of vehicle tool kit

Wheel bolts should be snugged down in the standard diagonal pattern as shown and then tightened to final torque with the same pattern.

Tightening torque (to pattern)

- Wheel bolt to wheel hub . . . . .120 Nm (89 ft-lb)

- Always tighten the wheel bolts in the specified pattern and to the specified torque. Excessive torque can distort the wheel bolt seat in the wheel.
- When installing a wheel, do not torque down the first wheel bolt immediately (for instance, with an impact wrench). This would prevent the other bolts from centering the wheel properly when they are threaded in. The rim could then come loose, even though the bolts have been tightened to the specified torque.
- Wheel bolts must be clean and easy to turn by hand. Do not apply grease or oil.
- Before installing the wheels, examine the seats of the wheel bolts and the contact surfaces between the rims and the hubs for rust, corrosion and paint. Clean off these parts if necessary, otherwise this can affect the clamping force of the bolts after installation, and the rims can come loose.
- A puller is available in the vehicle tool kit for removing the wheel cover where applicable.
- A puller is available in the vehicle tool kit for removing the plastic lug bolt covers where applicable. Lug wrench will not fit over these covers.
- An adapter for removing and installing wheel locks is available in the vehicle tool kit and is vehicle specific.
- A plastic mounting pin is available in the vehicle tool kit that threads into one of the lug bolt holes and facilitates attaching the wheel to the rotor.

> ***WARNING —***
> - *If the points listed above are not observed, the wheel mountings can come loose and the wheel bolt seats can become distorted.*
> - *This distortion may be very slight and not visible to the naked eye. Even this slight distortion of the wheel bolt seats can prevent the rims from being held firmly and the wheels will come loose repeatedly requiring replacement.*

◀ Puller hook with lug wrench fits into opening in cover. Pull firmly outward (**arrow**) for removal.

Transport locks, removing

EDITOR'S NOTE —
Removal of the transport locks is typically done by the Volkswagen dealer before vehicle is sold or delivered.

New Passats are often chained down on the carrier during transportation to prevent suspension and body damage from the motion of the transport vehicle. Plastic spacers are installed in the front suspension to limit travel during this period. These spacers must be removed for proper suspension movement. Failure to remove them will also cause suspension clattering noises when the vehicle is driven.

Remove transport locks by raising vehicle high enough to allow front coil spring to be fully extended. Reach into coil spring and remove plastic lock(s) through spring coils.

NOTE —
It is not necessary to remove wheels for access.

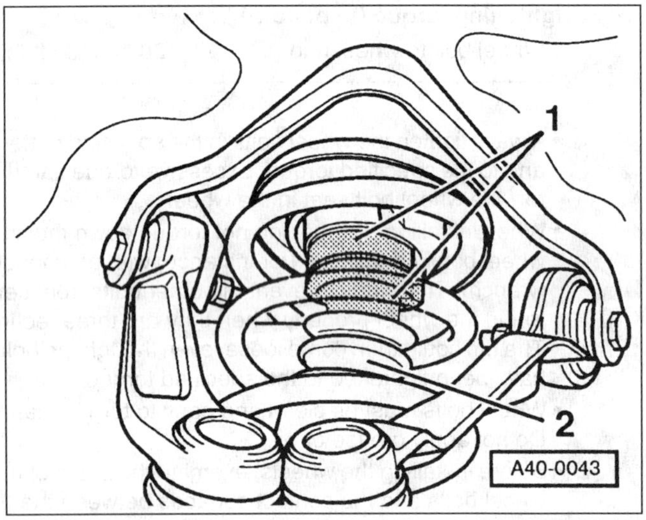

◀ Transport lock(s) (**1**) are installed at top of front shock absorber. Pull bump stop (**2**) down for easier access. Locks are usually red or yellow plastic and are slotted to allow for removal.

Underbody sealants and coatings, visual inspection for damage

The factory recommends that the engine compartment and underside of the vehicle be inspected twice a year for any damage to the protective coatings. The best times for inspections are just before and just after the winter driving seasons, particularly in areas of heavy road salt use. Damage can be caused by normal wear and tear or by driving over road debris.

Whenever the vehicle is raised on a lift or a hoist, it is advisable to visually inspect the underbody, wheel wells, and sill/rocker panels for damage to the underbody sealants and coatings. Inspection should also be made after major repairs to vehicle systems.

Any damage or defects found must be promptly repaired. Only use wax-based or tar-based anti-corrosion compounds. Do not use oil-based anti-corrosion sprays due to incompatibility with factory applied protection. Consult with an authorized Volkswagen Dealer's Parts Department for the appropriate authorized sealing compounds.

Body and Interior Maintenance

Lock cylinders, lubricating

EDITOR'S NOTE —

Volkswagen no longer has a published interval for lock cylinder lubrication. The manufacturer recommends periodic lubrication with Volkswagen G 052 778 A2 lock lubricant to insure smooth operation, particularly in areas where winter temperatures are very low. Exterior door and trunk/tailgate lock cylinders should be included in this operation.

Door hinge and check mechanism, lubricating

Door hinges and check mechanisms should be lubricated to insure smooth operation and also to prevent creaking and groaning noises associated with normal chassis flexing over irregular road surfaces.

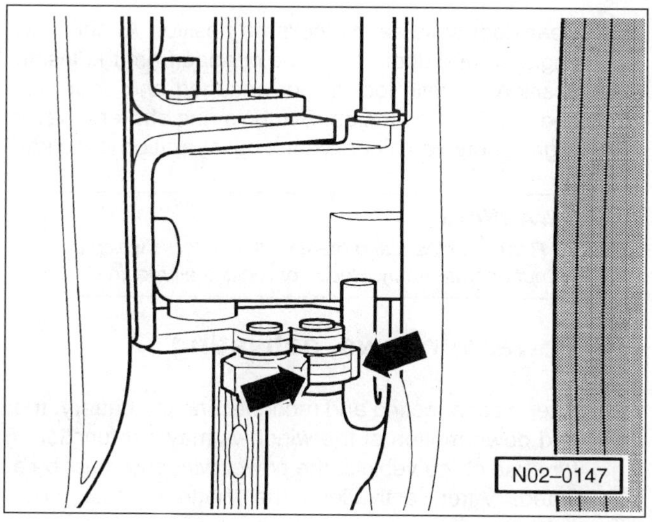

◄ Lubricate both front and rear door checks at points indicated (**arrows**).

Door locks, locking buttons, child safety locks, checking operation

EDITOR'S NOTE —
Volkswagen no longer has a published interval for checking door lock operation. It is good practice, however to routinely check and confirm proper operation. The following points are general guidelines.

All vehicle locking mechanisms must operate smoothly.

Door locks and locking buttons

– Open and close driver and front passenger door from inside. Check whether locking buttons move up and down, respectively, when this is done.

– Press locking button on front passenger door and rear doors down, close doors. Doors must be locked.

– It must not be possible to press down locking button on drivers door when as door is open or improperly closed. This minimizes the chances of unintentionally locking keys in car.

Child safety locks (rear doors)

– Using a key, move turn the slot on rear door lock down in direction of arrow (see marking on door lock). Door handle on inside must be blocked and door will only work from the outside if the locking button is up.

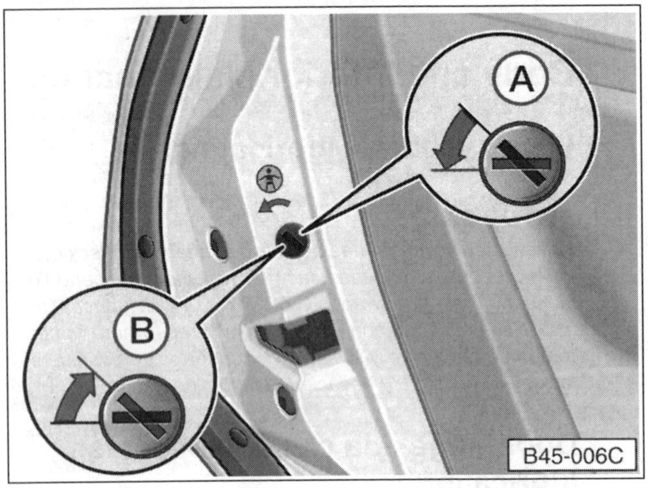

◄ Rear door child safety locks. In position **A**, child lock is engaged and door cannot be opened from the inside. In position **B**, child lock is not engaged and door can be opened from the inside. Symbols and stickers may vary slightly between model years, but operation is the same.

> ***WARNING —***
> *Remove, clean, and grease all door locks which do not operate freely. Repair or replace as required.*

Power windows, activating

After disconnecting and reconnecting the battery, the up and down motion of the windows may not function. Before use of the vehicle, the power windows must be activated. After activation, the battery must not be disconnected.

> ***WARNING —***
> *After disconnecting and reconnecting the battery, the "pinch protection" feature is deactivated. Damage and personal injury can result!*

To activate the power windows:

– Completely close all windows and doors.

– Lock vehicle with the key from the outside either from the driver's or passenger's doors.

– Unlock vehicle with the key.

– Again lock vehicle either from the driver's or passenger's door while holding key in the locked position for at least 1 second.

– Confirm operation with interior power window switches.

Power sunroof, checking and lubricating

The power sunroof should be checked for smooth operation in all modes and in all directions. Air deflector should also move without binding. Fixed aftermarket tinted acrylic Plexiglas air deflectors must not interfere.

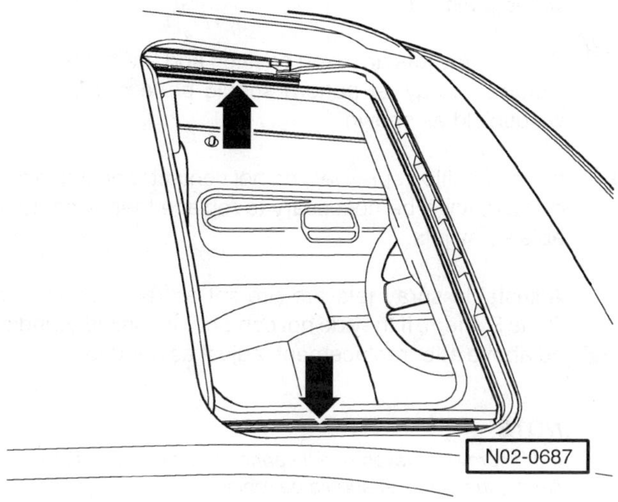

N02-0687

< Clean and lubricate sunroof guide rails (**arrows**) as required. Use only specified lubricant to minimize dirt buildup and caking.

Sunroof lubricant (all years)

- Special silicone grease . . . . G 000 450 02

> ***CAUTION—***
>
> *Volkswagen part numbers are given for reference only! Always consult with your Volkswagen Parts Department or aftermarket parts specialist for the latest parts information.*

Windshield Wiper and Washer System

Washer fluid, checking and filling

The windshield washer reservoir should be filled with a mixture of washer fluid and water for year round protection from freezing and for the strong cleaning properties.

Windshield Washer Mixing Ratio Table ZVW 177 901 / ZVW 177 903 Concentrate ZVW 177 904 / ZVW 177 906 Premium Concentrate		
Freeze protection to:	Fluid concentrate	Water
0°F (-18°C)	1 part	3 parts
-9°F (-23°C)	1 part	2 parts
-36°F (-38°C)	1 part	1 part

Notes:

1. California and Texas users should consult local regulations before mixing; different ratios are required by law.
2. Other States and Provinces may also have local regulations: always check before mixing.
3. Fluid reservoir should be filled to the top.
4. Part numbers valid for USA only.
5. Protection temperatures are approximate.

< Windshield washer fluid provides freeze protection when mixed with the appropriate quantity of water. Table applies to Volkswagen supplied concentrated fluid. When using other brands, defer to their directions.

Windshield washer fluid freeze protection level can be measured with the same refractometer (T10007) used for checking anti-freeze protection levels by referencing the appropriately marked scales.

- All vehicles equipped with fan spray jets must be filled with factory approved fluid or equivalent. These fluids have a low viscosity when below freezing which keeps the complex sprayer system from becoming clogged with crystallized washer fluid. This would interfere with the fan-effect of the spray jets.
- During warm seasons, fill the reservoir with factory approved fluid or equivalent. The strong cleaning action effectively cleans the oily film and residue from the windshield.
- Suggested anti-freeze protection levels for most areas should be maintained at approximately 9°F (-15C°), with arctic climates at approximately -31F° (-35°C).

Windshield washer spray jets, front checking and adjusting

Several different types of windshield washer spray jets were installed in product and different types are available as replacements.

NOTE—
Do not attempt to clean spray jets by blowing back into them from the discharge end. Internal valving mechanism will be damaged.

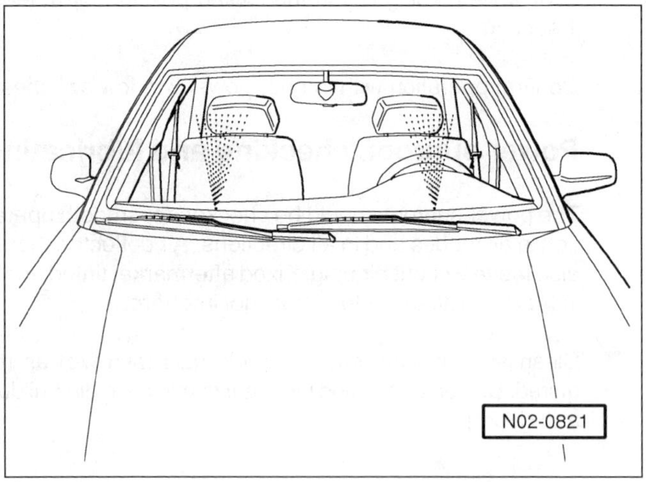

◀ Some spray jets are fixed position and are not height adjustable. Spray jet pattern should be level and contact windshield as shown.

If fixed position spray jets do not contact windshield as indicated, it will be necessary to replace them with adjustable spray jets.

Adjustable spray jets are pre-set as received from the Parts System. If they do not contact windshield as indicated above after replacement, adjust as needed.

NOTE—
If the spray is uneven or if it cannot be adjusted to specification, the spray jet should be replaced.

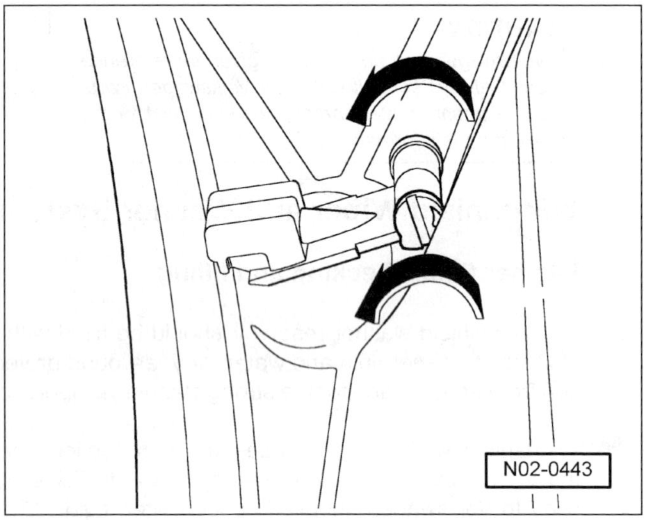

◀ Turn eccentric with a screwdriver in direction of **arrows** to adjust spray pattern upward. Turn in reverse direction to lower spray pattern. Repeat adjustment as required for the other side.

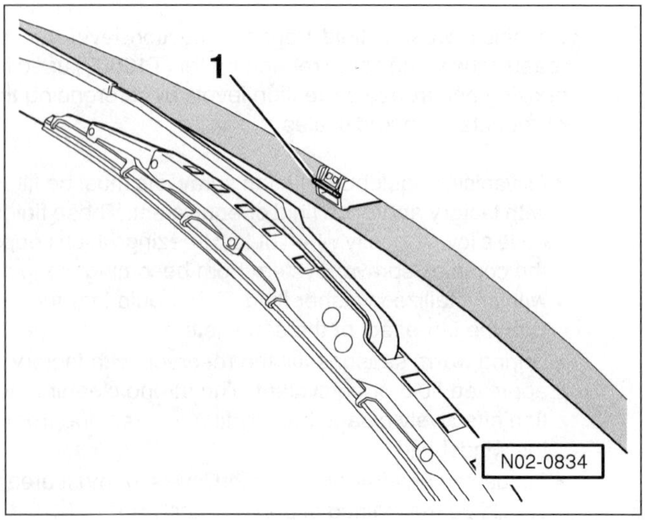

◀ Turn adjuster (**1**) by hand either up or down. Repeat adjustment as required for the other side.

Windshield Washer Spray Jets, Rear (Wagon) Checking and adjusting

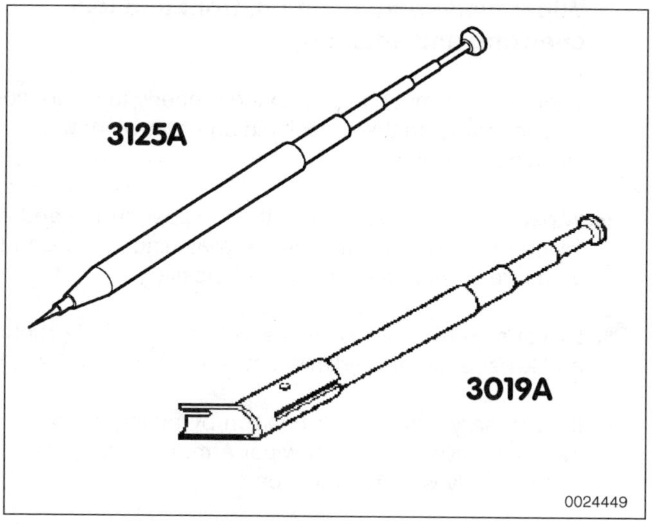

Tools required for rear window and headlight washer aiming. Tool 3125A is used for rear window aiming while tool 3019A is used for headlight washers.

– Adjust spray to contact rear window at center of wiped area using windshield washer aiming tool 3125A.

If spray pattern is uneven or if jet is unable to be adjusted as specified, replace washer jet assembly.

NOTE—
- *Do not attempt to clean spray jets by blowing back into them from the discharge end. Internal valving mechanism will be damaged.*
- *Do not use a needle or similar tool for adjustment. Internal valving mechanism may be damaged.*

Headlight Washer Spray Jets, Checking and adjusting

Headlight washers became available on some Passat models for the North American markets beginning with October 2000 production.

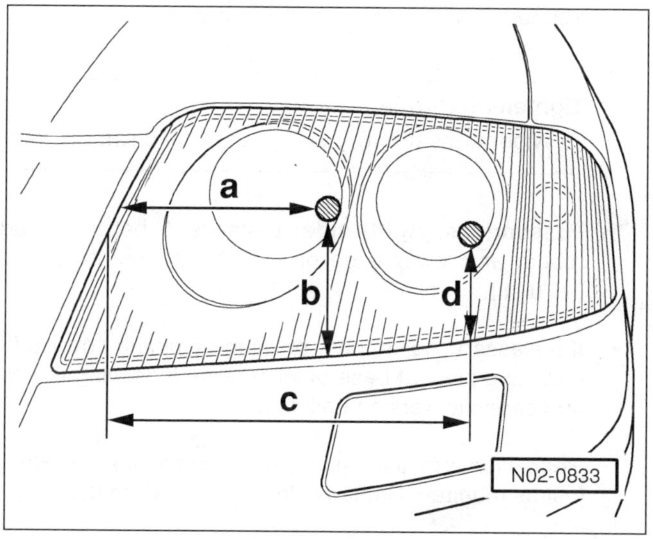

Headlight washers should contact lens in areas shown. Left side headlight illustrated, right side headlight is symmetrically opposite.

a. 110 mm (4.33 in.)
b. 75 mm (2.95 in.)
c. 230 mm (9.05 in.)
d. 50 mm (1.97 in.)

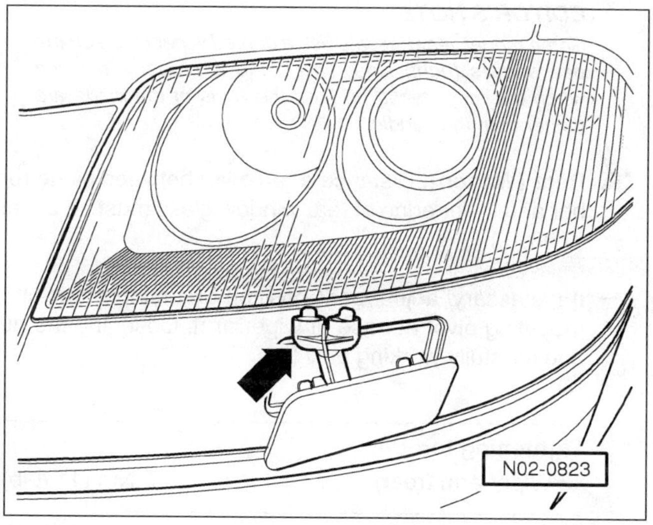

Carefully pull headlight washer cover all the way out by hand, against spring tension. Washer spray jets live in housings (**arrow**). Aim with tool 3019A. Left side headlight illustrated, procedure is the same for right side.

NOTE—
If the spray is uneven or if it cannot be adjusted to specification, the spray jet should be replaced.

Wiper blade park position, front and rear, checking and adjusting

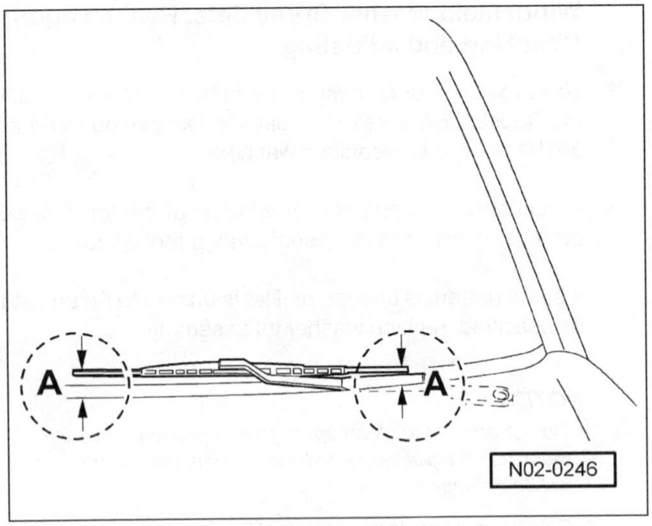

Wiper blades must be positioned properly to insure correct coverage on the windshield and rear window and to avoid body damage.

– Wipers should be shut off with the wiper switch and allowed to stop completely before switching off ignition to measure. Wipers must be in park position.

◄ **Driver's (left) side:** Distance **A** between blade rubber and lower edge of windshield must be 20 mm (0.79 in.).

– If necessary, adjust park position by prying cover off of mounting pivot at base of wiper arm, loosening the nut, and carefully working until free.

– Reposition arm, tighten nut, and recheck position. Readjust as required. Replace plastic cover when done.

EDITOR'S NOTE —
Some model year Passats use a wiper blade with a deflector or airfoil on the driver's (left) side. This blade is properly installed when the airfoil is on the lower side of the blade.

Tightening torque:

- Wiper arms (front) . . . . .20 Nm (15 ft-lb)

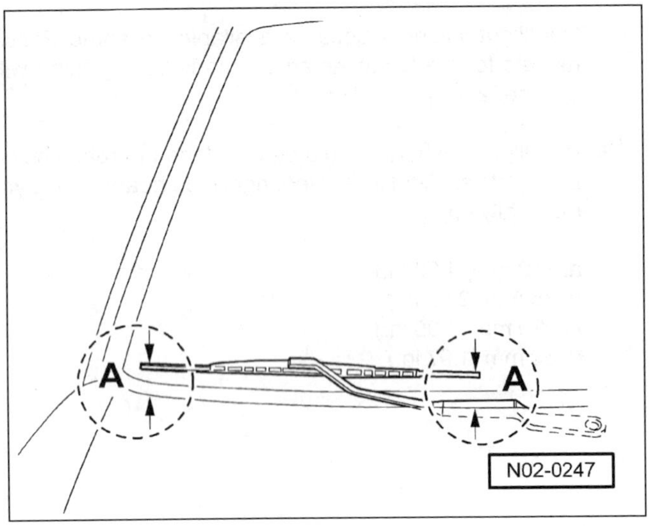

◄ **Passenger's (right) side:** Distance **A** between blade rubber and lower edge of windshield must be 20 mm (0.79 in.).

– If necessary, adjust park position by prying cover off of mounting pivot at base of wiper arm, loosening the nut, and carefully working until free.

– Reposition arm, tighten nut, and recheck position. Readjust as required. Replace plastic cover when done.

EDITOR'S NOTE —
Some model year Passats use a curved wiper blade on the passenger's (right) side. This blade is properly installed when the "cup" faces up (i.e.: the ends of the blade are higher than the middle.

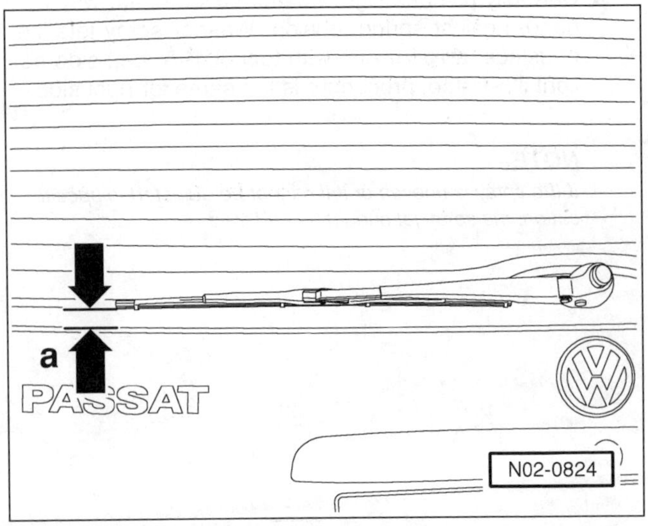

◄ **Rear (Wagon):** Distance **a** (**arrows**) between blade rubber and lower edge of rear window glass must be 25 mm (0.98 in.).

– If necessary, adjust park position by prying cover off of mounting pivot at base of wiper arm, loosening the nut, and carefully working until free.

Tightening torque:

- Wiper arm (rear) . . . . .15 Nm (11 ft-lb)

Wiper blade contact angle, front and rear, checking and adjusting

If wipers blades become noisy or skip, and a thorough cleaning of the blades and windshield has not corrected the problem, check the contact angle of the wiper arm. Adjust as wiper arm as required.

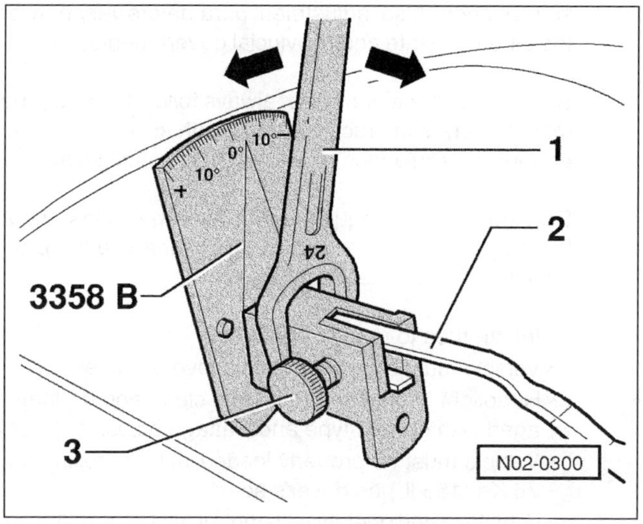

– Shut off wipers with wiper switch and allowed to stop completely before switching off ignition to measure. Wipers must be in park position.

– Remove appropriate wiper blade and set aside.

– Insert wiper arm (**2**) in adjustment tool 3358B as shown and lock with thumbscrew (**3**).

– Check contact angle and compare to table.

– If adjustment is needed, apply 24 mm open ended wrench (**1** on above illustration) to adjustment tool and twist until correct angle is reached.

– Remove wrench. Loosen and retighten thumbscrew to allow tool to centralized. Recheck angle. Repeat as needed until correct angle is reached.

Wiper Arm Contact Angles		
	Left hand drive vehicles	Right hand drive vehicles (**NOT** for North America)
Driver's side	- 3°	+ 3°
Passenger's side	- 5°	+ 5°
Rear (Wagon)	0°	0°
Tolerance	± 2°	± 2°

Passat wiper arm contact table. Note that table includes specifications for non-North American markets.

– Remove adjustment tool.

– Reinstall wiper blade.

– Check wiper operation.

Wiper blades, temporary

Passats are usually shipped to the Dealer with short, temporary plastic wiper blades in all positions. These blades are not intended for regular street operation and should be removed and discarded at the time of delivery. These are usually blue or yellow plastic and should be removed carefully to avoid wiper arm/blade misalignment.

NOTE—

- *Wiper arm/blade alignment has been set at the factory, and then checked and adjusted as required.*
- *Do not twist or bend the wiper arm. If wiper arm/blade alignment is incorrect, chattering and/or streaking will occur.*
- *To replace wiper blades, see* **Electrical Equipment section, Repair Group 92**.

Fuel Tank Filling Compensation Table (for adjusting headlights)	
Fuel level as indicated by fuel gauge	Additional weight (ballast) to add in luggage or cargo area
¼	45 Kg (99 lb)
½	30 Kg (66 lb)
¾	15 Kg (33 lb)
full	none (0)
Note: For additional weight use containers filled with water. A container filled with 5 liters (1.3 gal) of water weighs approx. 5kg (11 lb).	

Headlights, adjusting

Proper adjustment (aim) of the headlights requires specialized equipment which can vary from service shop to service shop. Also, adjustment parameters vary between the different State and Provincial governments.

Because of these variables, always follow the equipment manufacturers instructions along with all applicable governmental regulations when adjusting headlight aim.

Regardless of the equipment or the regulations, the following points are common and preliminary to the proper adjustment of the headlights:

- Inflate tires to correct pressures.
- Vehicle must be standing on a level surface.
- Headlight lenses and reflectors clean and undamaged with correct type and wattage bulbs.
- Vehicle must be properly loaded with approximately 75 Kg (165 lb) on driver's seat.
- Vehicle at curb weight with tool kit, jack, spare tire, etc. and 90% fuel load. If fuel tank is not at least 90% full, add weight to the trunk for compensation according to the table.
- Springs settled by rolling vehicle several meters (yards) or bouncing suspension.
- Disregard inclination % molded into headlight housing. Information is not intended for North American market Passats due to equipment and legal differences. Use will result in headlight mis-alignment.

Vehicles up to through 09/2000 production

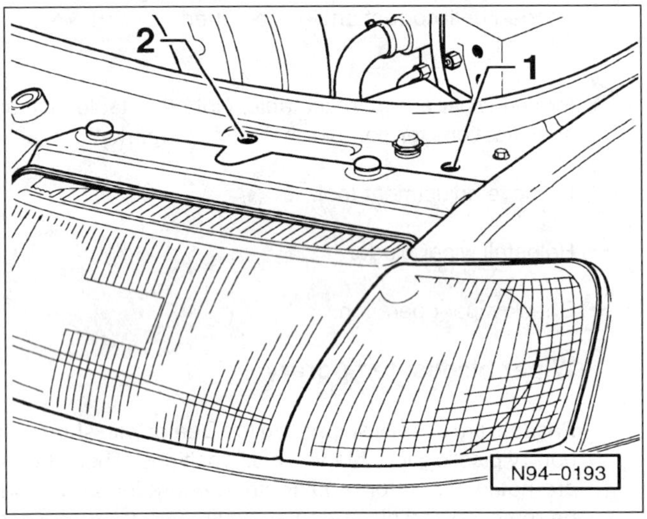

Headlights are adjusted from above with a screwdriver or hex key wrench at **1** and **2**. Left side is shown, right side is symmetrically opposite.

1 = Lateral (side - side) adjuster
2 = Vertical (up - down) adjuster

Adjustment is made with headlights on low beam; when low beam is correctly adjusted, high beam will also be correctly adjusted. Integral factory fog lights are automatically adjusted correctly when headlights are correctly adjusted.

Vehicles from 10/2000 production

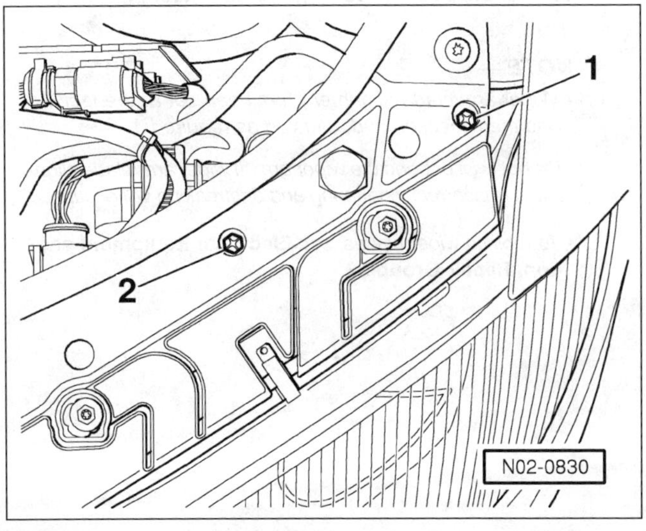

Headlights are adjusted from above with a screwdriver or hex key wrench at **1** and **2**. Left side is shown, right side is symmetrically opposite.

1 = Lateral (side - side) adjuster
2 = Vertical (up - down) adjuster

Adjustment is made with headlights on low beam; when low beam is correctly adjusted, high beam will also be correctly adjusted. Integral factory fog lights are automatically adjusted correctly when headlights are correctly adjusted.

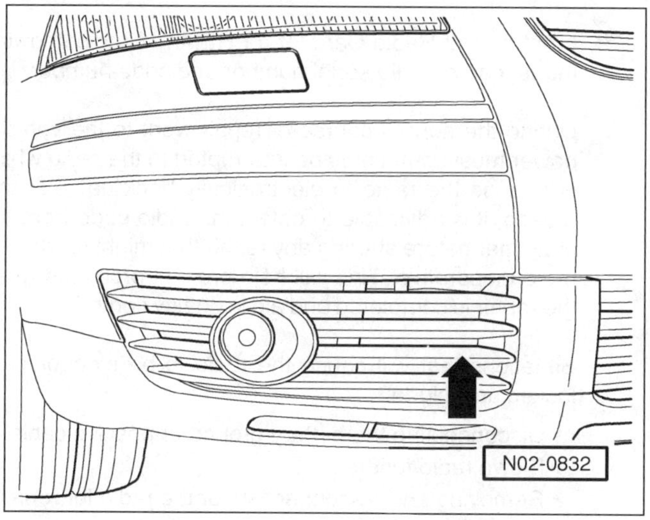

Vehicles with bumper mounted foglights

Bumper mounted foglights are adjusted from the front after removal of the small grill section surrounding the mechanism.

- Remove outboard grill section by unclipping (**arrow**) from lower part of bumper.

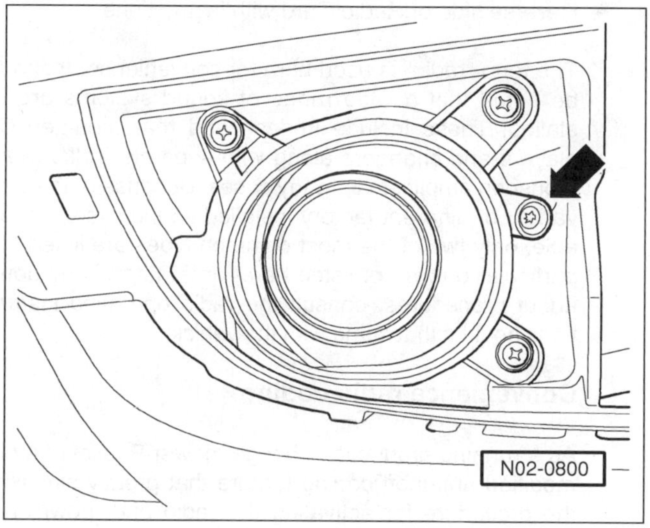

- With grill removed, switch on foglight and make vertical (up - down) adjustments by turning screw (**arrow**). Right side shown, left side is symmetrically opposite. There is only provision for vertical adjustment; there is no lateral (side - side) adjustment.

NOTE —
*See **Electrical Equipment** Section, Repair Group 94 for headlight and foglight bulb replacement.*

Radio anti-theft coding

All factory installed Volkswagen radios have an anti-theft feature that disables the radio whenever the power is removed. This feature deters theft because the radio will not work when power is reconnected unless a numeric code is entered via the radio pushbuttons. The code is assigned to the radio during manufacture.

Radios that have had the power interrupted will indicate that the anti-theft coding is active by showing only the word "SAFE" in the display area and by being inoperative.

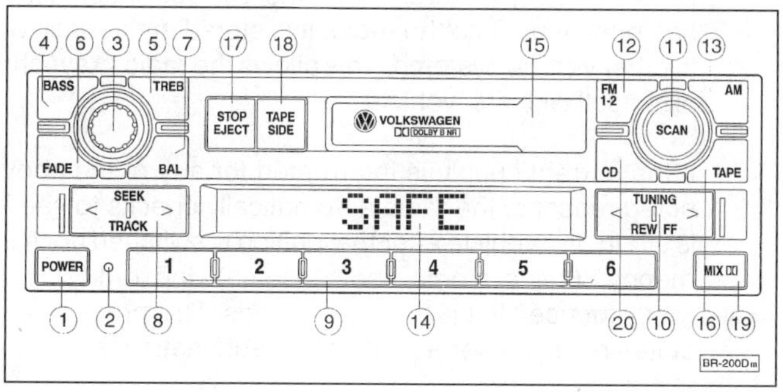

- Locked-up radio showing "SAFE" in the display area (14).

Radio code information is usually attached to a sticker in the luggage compartment near the spare tire during assembly. During the pre-delivery inspection or at the time of vehicle delivery, this information sticker is attached to the Radio Card in the Sound System Manual and placed with the vehicle documents. If, through oversight, the sticker remains attached to the vehicle, the customer should remove the code sticker from the vehicle and attach it to the Radio Card, In all cases, the card should be placed the with the vehicle documents in a safe place away from the vehicle.

Radio Card

Radio: PREMIUM

Serial Number: VWZ5Z7T0758474

Code Number: 1892

Place the self-adhesive labels for the serial number and code number in the appropriate boxes.

0024429

Front side of Radio Card for a PREMIUM radio showing model name, radio serial number and code number.

During the normal course of repair work to the vehicle, power must sometimes be interrupted to the radio which will cause the radio to electronically "lock up". For this reason, it is advisable to obtain the radio code from the customer before starting any repair that might render the radio inoperative. This will help prevent problems when the repair is completed and aid in quality control.

Situations that will cause the radio to be electronically locked up include:

- Disconnecting the battery ground or positive cable.
- Blown radio fuse(s).
- Removing and disconnecting for the radio for repairs to nearby systems.

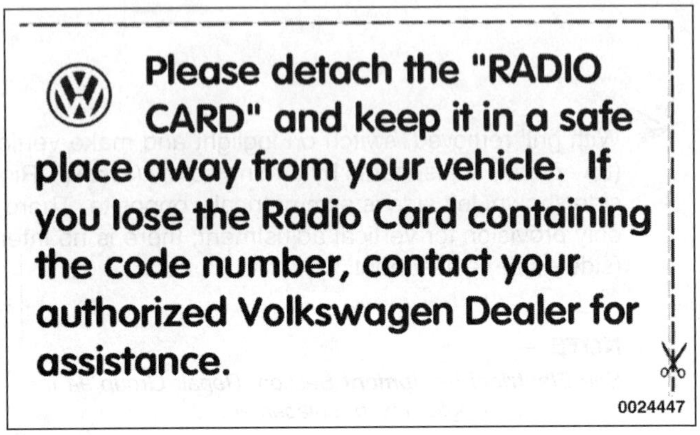
Please detach the "RADIO CARD" and keep it in a safe place away from your vehicle. If you lose the Radio Card containing the code number, contact your authorized Volkswagen Dealer for assistance.

0024447

Reverse side of Radio Card with instructions.

The term "radio" is used here for convenience. It should be noted that a wide range of sound systems are installed. These include in-dash and rear mounted CD players and changers along with separate Volkswagen Monsoon amplifiers in many cases. Because of the wide variety of different factory installed sound system available, only two of the most common types are listed. If a particular unit is not listed here, or if there is any doubt about procedures, consult the radio operating instructions booklet that came with the vehicle.

Convenience radio coding

As a running production change, newer Passats have a modified anti-theft coding feature that greatly simplifies the procedure for activating the radio after power has been interrupted.

In previous versions, the radio would have to be activated after a power interruption by manually entering the correct code. This is still the procedure in older Passats and in some circumstances where the radio is replaced.

That has all changed with the introduction of the convenience coding feature. Once the correct code number has been input into the radio, it is stored in the radio and in other vehicle systems. This allows the radio to synchronize itself with "its" vehicle.

If the current supply is interrupted for any of the above-noted reasons, the radio automatically checks to see if it is still in "its" vehicle after the ignition is switched on. If it is "happy" (the systems codes match), the radio will be ready for use after just a few seconds. This procedure requires no user interface; it occurs automatically.

If a radio with this system is installed in a different vehicle, the code will have to be entered manually to cancel the lock-up and enable radio operation.

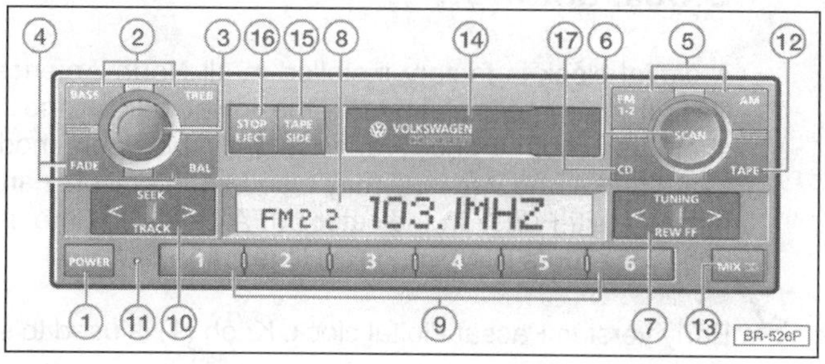

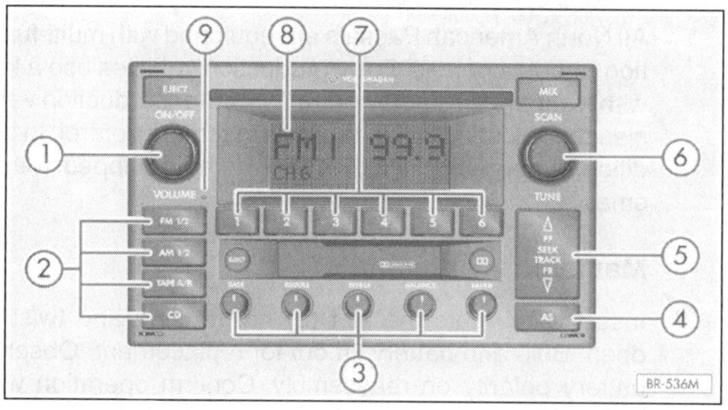

Canceling electronic lock-up

In situations where the convenience radio coding feature is not available or not active, the radio must be unlocked manually. The procedure for cancelling or unlocking a locked up radio is the same for all of the versions listed. The only differences are in which buttons to use.

For the Volkswagen ***Premium*** or ***Premium Monsoon*** Sound System, the word "SAFE" will appear in the display window. After 3 seconds, the display will read "1000".

If the word "SAFE" appears in the display area (**8**) when the radio is turned on, it is locked. Use the accompanying illustration, and proceed with unlocking as follows:

– Switch radio on.

– Use the first 4 station preset buttons (**9**) to enter the appropriate code number. For example, if the code was 1892 as in the Radio Card above, push preset button 1 as many times as needed until the number 1 appears in the window. Repeat for preset button 2 for the 8, preset button 3 for the 9, and preset button 4 for the 2. Preset buttons 5 and 6 are not used for this function.

– Once the correct code is displayed, depress the SEEK/TRACK > button (**10**) and hold for longer than 2 seconds until you hear an audible signal.

– If you have entered the correct code, the display will read "LSM". Release buttons and a station frequency should be displayed. The unit is now ready for use.

For the Volkswagen ***Premium CD*** or ***Premium Monsoon CD*** Sound System; the word "SAFE" will appear in the display window. After 3 seconds, the display will read "1000".

If the word "SAFE" appears in the display area (**8**) when the radio is turned on, it is locked. Use the accompanying illustration, and proceed with unlocking as follows:

– Use the first 4 station preset buttons (**3**) to enter the appropriate code number. For example, if the code was 1892 as in the Radio Card above, push preset button 1 as many times as needed until the number 1 appears in the window. Repeat for preset button 2 for the 8, preset button 3 for the 9, and preset button 4 for the 2. Preset button 5 is not used for this function.

– Once the code is displayed, depress the upper side (△) of the SEEK/TRACK button **(5)** or the SCAN button (**6**) and hold for longer than 2 seconds until you hear an audible signal. Release buttons and station frequency should be displayed.

– f you have entered the correct code, the display will read "LSM". Release buttons and a station frequency should be displayed. The unit is now ready for use.

Clock, setting

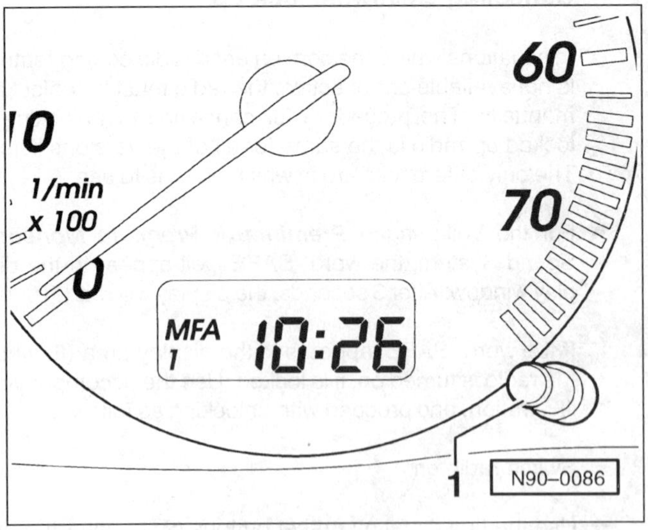

A digital clock is factory installed in all North American Passat instrument clusters in the lower part of the tachometer. Appearance varies slightly between model years and some versions may also use the display area for the Multi-Function Indicator (MFA or MFI). Time setting procedures however, are all the same.

◄ Early version Passat digital clock. Knob (**1**) is used to set the time.

- To set the hour:
  Slightly turning the knob counter-clockwise advances the clock by one hour. Turning and holding the knob advances the hour continuously. Turn knob until desired hour is reached.
- To set the minutes:
  Slightly turning the knob clockwise advances the clock by one minute. Turning and holding the knob advances the minute continuously. Turn knob until desired minute is reached.
- To set the time exactly (synchronize with a known accurate audible time source):
  Set the hour as noted above. Set the minute as noted above but set it 1 minute less than desired. When tone is heard (such as an hourly radio station time check), immediately turn the knob clockwise to advance clock to the exact time.

Keys and remote controls, replacing batteries

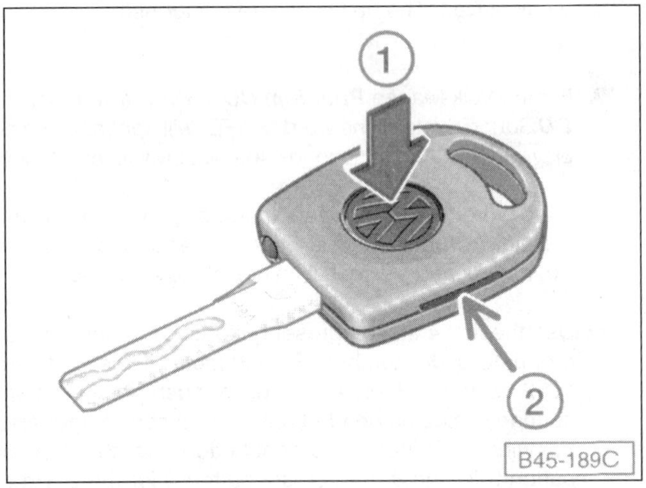

All North American Passats are equipped with multi-function remote controls. Early production vehicles use a key with a separate remote control fob. Later production vehicles use a folding key with a built in remote control. In addition, some early vehicles use a key equipped with a small light.

Master key with light

◄ Insert a coin into the slot (**2**) on the side and twist to open. Bulb and battery lift out for replacement. Observe battery polarity on reassembly. Confirm operation with button (**1**).

- Battery quantity and type: one 1.5 Volt alkaline V625U
- Bulb: one 1.5 Volt, 90 mA (0.090 A) pre-focus

EDITOR'S NOTE —

The master key with light is usually paired with the remote control key fob described below.

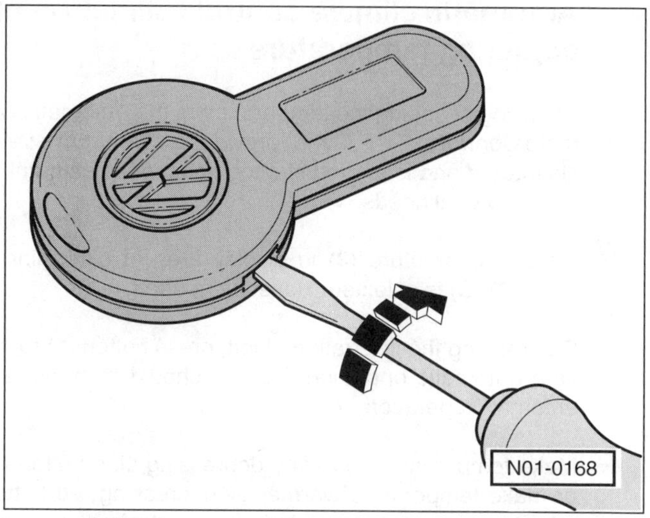

Remote control key fob

Insert small screwdriver into the slot on the side and twist (**arrow**) to open. Batteries lift out for replacement. Observe battery polarity and note spacer plate between batteries on reassembly.

- Battery quantity and type: two 3.0 Volt lithium CR2016

EDITOR'S NOTE —

Passat remote control key fobs look and operate the same as those used on some models of Golf, Jetta, Cabrio and New Beetle. However, they are not interchangeable due to operation on different radio frequencies.

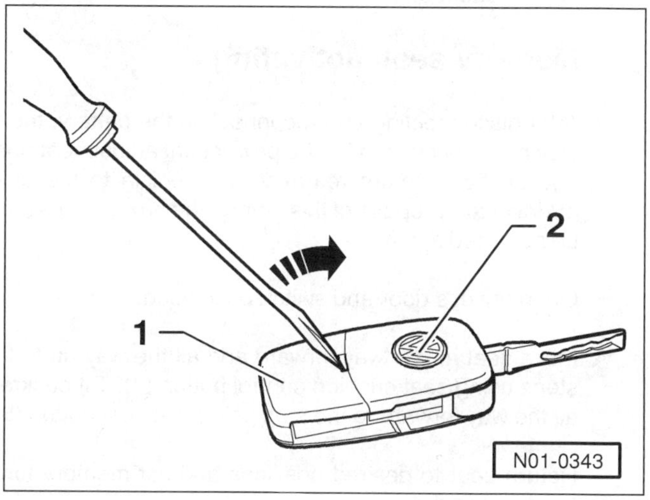

Remote control folding key

The folding key is supplied with two slightly different transmitter types. Procedures are the same for both types. Insert small screwdriver into the slot between the transmitter unit (**1**) and the main folding key body (**2**). Move screwdriver in direction of arrow to unclip and separate the two sections. Sections slide apart.

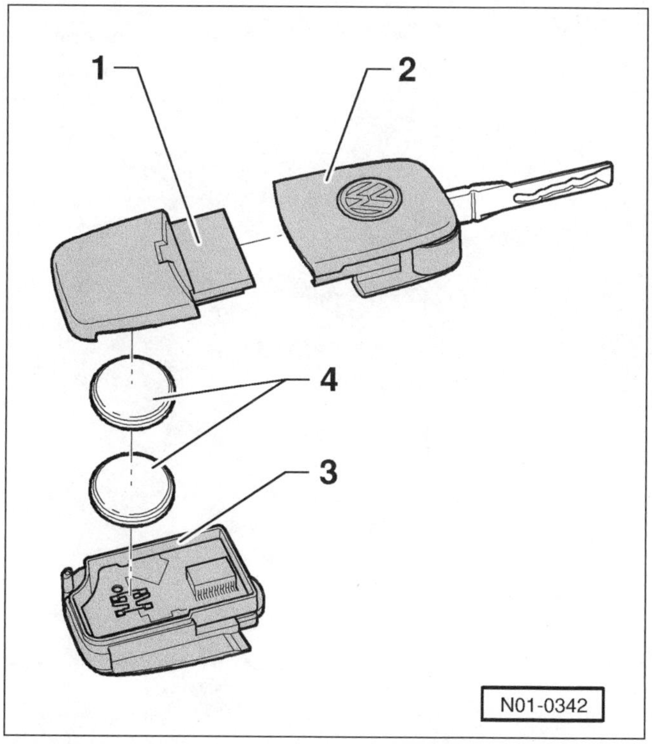

After separating the transmitter unit (**1**) and the main folding key body (**2**), remove the lower section of the transmitter (**3**) by carefully prying apart. Batteries lift out for replacement. Observe battery polarity on reassembly. Battery package (**4**) configuration depends on version.

- Battery quantity and type: two 3.0 Volt lithium CR1620
  or
- Battery quantity and type: one 3.0 Volt lithium CR2032

NOTE —

Dispose of all dead batteries properly. Recycle when possible. Comply with all applicable laws regarding disposal so as not to pollute the environment.

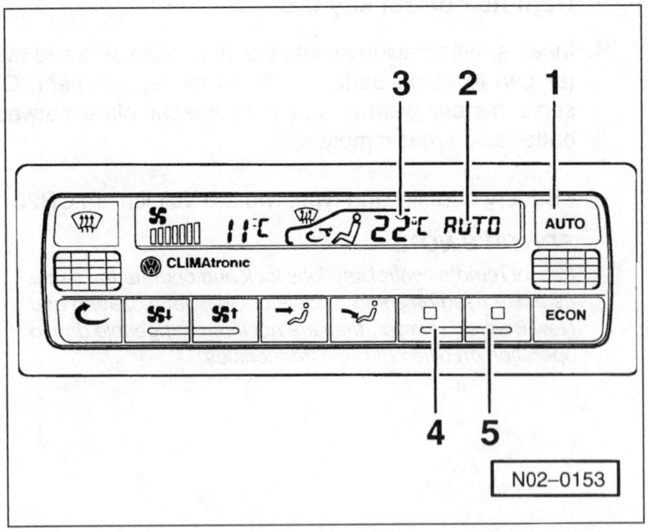

N02–0153

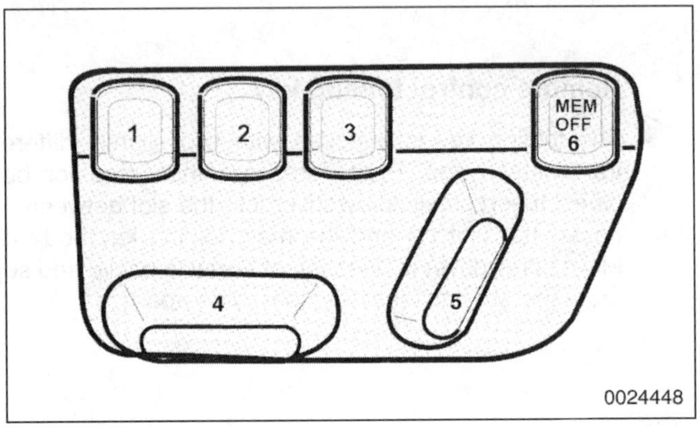

0024448

Automatic climate control (Climatronic), adjusting temperature

Adjusting or setting the temperature of the Automatic Climate Control to 72°F (22°C) provides a comfortable vehicle interior and is a good starting point for adjustment to suit individual needs.

Check temperature (**3**) in display area to determine if 72°F (22°C) is selected. Reset as needed.

- Referencing the illustration at left, press button (**1**) to initiate automatic operation. Display should then indicate automatic operation (**2**).

- Make temperature cooler by depressing blue button (**4**) or make temperature warmer by depressing red button (**5**). Depress appropriate button until desired temperature is reached.

Memory seat, activating

After disconnecting and reconnecting the battery, the 8-way memory system for the power seat adjustment does not function. Before returning the vehicle to the owner/driver after repairs of this nature, the power seats must be activated again.

- Open driver's door and switch on ignition.

Move seat all the way forward and all the way up to the stops using seat cushion control button (**4**). Tilt backrest all the way forward to the stop using backrest button (**5**).

- Return seat to desired positions and set memory functions as required.

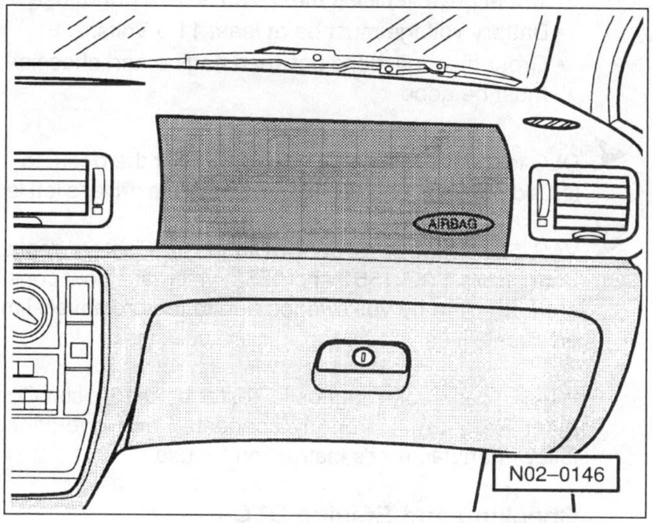

Airbag units, driver and passenger, visually checking

Driver's and passenger's front airbags should periodically be given a visual inspection.

◄ Visually check airbag cover on steering wheel for signs of exterior damage. Observe warning noted below.

> ***WARNING —***
> - *The covers of the airbag units on the steering wheel, instrument panel and side airbags must not be covered or have any objects affixed to them such as stickers, etc.*
> - *DO NOT apply any chemical treatment to air bag covers. Clean with dry or water moistened cloth only.*

◄ Visually check airbag cover on instrument panel for signs of exterior damage. Observe warning noted above.

EDITOR'S NOTE —
It is also advisable to observe the airbag warning light when starting the car. It should come on for a few seconds when the engine is started, as an electronic systems self-test, and then immediately go off and stay off. It should only be illuminated during this self-test or if there is a system malfunction.

Diagnostic Trouble Code (DTC) memory, data link connector and fault checking

All Volkswagen Passats have On Board Diagnostics that allow information to be exchanged between the system and a scan tool. Methods for reading the data stream and working with it will vary between the scan tool or computer program manufacturers.

In all cases, it is necessary to connect the scan tool or computer to the vehicle via the Data Link Connector (DLC).

Before connecting and using any scan tool, the following requirements must be met.

- Fuses 28, 29, 30, 32, and 34 must be good. Front wheel drive vehicles must also have a good fuse 31.
- Battery voltage must be at least 11.5 Volts.
- Ground connections between engine and chassis must be good.

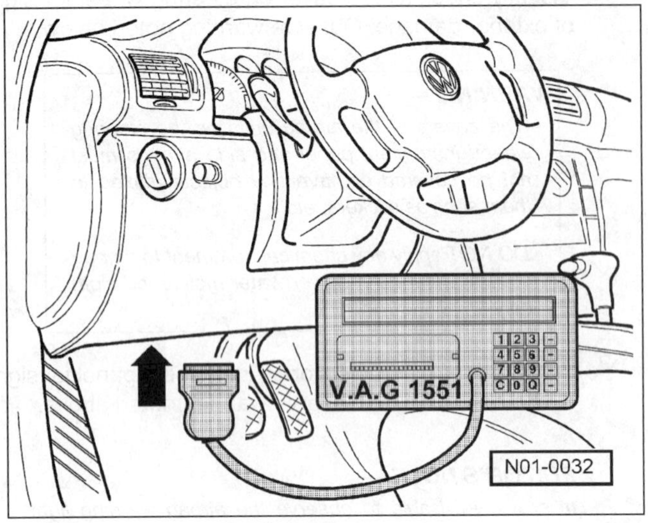

◀ DLC for all model years is located under the dash, to the left side of the steering column, above the driver's left leg.

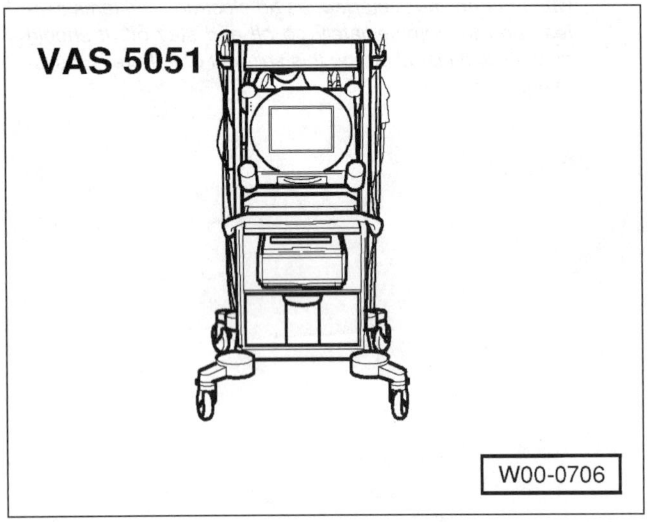

◀ VAS 5051 (and VAS 5052) Diagnostic Testers replace Scan Tools VAG 1551 and 1552 which are now considered obsolete by Volkswagen due to their relatively limited capabilities.

– When Scan Tool, Diagnostic Tester or Diagnostic Computer Program is properly connected and operational, follow manufacturer's instruction for use.

Checking and Erasing DTCs

General instructions for reading and erasing DTCs are outlined below. Procedures vary slightly with the type of equipment used, but follow similar formats. In all cases, DTCs must be read first and than erased. System safeguards prevent erasing without first reading. If your scan tool has a printer or provisions for saving data, it is advisable to make a copy of the data for later reference before erasing to avoid confusion and/or questions.

When using a VAG 1551 or VAG 1552, proceed as follows:

– Connect scan tool as described above.

– From the initial Menu, select:
1 - Rapid Data.

– From the Address Word Menu, select the appropriate vehicle system from **Scan Tool section**
(e.g. 01 - Engine Electronics).

– Advance to the Function Menu when the ECM responds with version and coding information.

- From the Function Menu, select:
  02 - Check DTC memory.
  Read DTCs and return and return to the Function Menu.
  If no DTCs are present, scan tool will respond with a message in the display to indicate that no DTCs are recognized.

- From the Function Menu, select:
  05 - Erase DTC memory.
  Scan tool will respond with a message in the display to indicate that memory is erased.

- Advance to the Function Menu.

- From the Function Menu, select:
  05 - End output.

- Disconnect scan tool.

EDITOR'S NOTE—

After erasing DTCs, it is usually a good idea to go back into the system via Function 02, and re-check the fault memory to make sure that the codes have, in fact, have been erased.

Rapguard®, cleaning residue

After final assembly at the factory, a thin plastic cling-type sheeting material is applied to many of the exposed body panels and interior. This white or blue colored substance is applied to both painted and plated surfaces for protection during transportation and storage and is called Rapguard®. In most cases, the end user (driver/owner) will never see this protection on the vehicle because it is removed during the course of the pre-delivery inspections.

◄ New Passat W8 prior to pre-delivery inspection showing partially removed Rapguard®. Remaining Rapguard® (doors, fenders, lids, and roof) will simply pull off.

Occasionally, the attaching adhesive may leave a residue on painted surfaces. This can be removed with a non-abrasive polishing product such as 3M #5995 for light colors and #5996 for dark colors.

If thick lines of adhesive are visible, these can be removed with a general purpose adhesive cleaner such as 3M #08987 (spray) or #08984 (quart can).

If it becomes necessary for the owner/driver to remove remnants of Rapguard® that were inadvertently left in place either on the exterior or interior, it can be simply pulled off. Dispose of Rapguard® in the same way as normal paper waste. If unsure about proper disposal procedures in your area, check local regulations.

Quality Review

When you have finished working under the hood and around other areas of the vehicle, it is advisable to take a moment to quality check or review your work. This helps to insure that the operation or repair has been completed properly with all affected systems functioning within normal parameters. The following are general guidelines and depending on equipment, may include:

Road test

Upon completion of the maintenance work, the vehicle should be road tested. Check the following systems during the road test:

Automatic transmission shift lock

– Turn ignition on, but do not start engine. Apply parking brake.

– With gearshift selector in park attempt to shift into a drive or reverse. Selector lever should not move.

– Press brake pedal. Selector lever should move into gear.

– Place gear selector lever in neutral and release brake pedal.

– Attempt to shift into gear. Selector lever should not move.

– Press the brake pedal. Selector lever should move into gear.

Automatic park/neutral position switch

– Apply the parking brake and step on the brake pedal. Place the gear shift selector in park and attempt to start car. Repeat for all gears. The engine should only start in park or neutral.

Brake pedal and parking brake

– Check pedal and handbrake travel (free play) and operation.

- Travel of brake pedal: max. 1/3 of overall pedal travel
- Travel of parking brake lever: 2 notches (clicks)

Manual gearshift

– Check for smooth, easy operation of the transmission and its gearshift.

Steering

– Check steering play with vehicle weight on the wheels by turning steering wheel back and forth (wheels in straight ahead position).

- Steering play: zero play with engine running

Kickdown operation

– Depress accelerator pedal fully to floor. Depending on vehicle speed and engine speed, upshift is either delayed or the transmission shifts down into the next lower gear.

Air conditioner and heater

– Check the full range of the heater and the air conditioner for proper function. Cold or warm air must flow out of the vents.

Additional checks

Perform the following checks in accordance with the vehicle equipment and the road conditions available during the road test:

- Engine performance, idle speed, acceleration, cold and hot starting.
- Clutch operation, smoothness and pedal pressure.
- Brake operation, noise and function of ABS.
- Cruise control system operation.
- Radio and CD operation, and reception.
- General vehicle handling and dynamics such as pulling to the side, cornering, vibrations in steering wheel and unusual noises.

Maintenance Schedules

NOTE—
- *Aside from keeping your Volkswagen in the best possible condition, scheduled maintenance plays a role in maintaining full coverage under Volkswagen's extensive warranties. If in doubt about the terms and conditions of your car's warranty, consult an authorized Volkswagen dealer.*
- *Volkswagen continually updates maintenance schedules to suit changing conditions through the issuance of Maintenance Schedule Service Circulars. If in doubt about any of the requirements for your vehicle, consult an authorized Volkswagen dealer.*

EDITOR'S NOTE—
- *Single page maintenance tables cover all engine types from model years 1998 to 2001. Beginning with model year 2002, each engine type has its own individual maintenance table.*
- *The maintenance tables on the following pages contain information for all Volkswagen models available at the time of publication.*
- *The maintenance tables are based on Volkswagen recommended maintenance intervals. However, field experience has shown that replacement intervals for timing belts and tensioners on the 1.8T and V6 engines should be more frequent. These engines are known as "interference engines" and if the tensioner or timing belt fails the possibility exists for significant valve and/or piston damage. The publisher's additional recommended intervals can be found listed under **Editor's Note** at the bottom of each maintenance table.*

1998 Maintenance Schedule

Miles Kilometers	5 8	10 16	15 24	20 32	25 40	30 48	35 56	40 64	45 72	50 80	55 88	60 96	65 104	70 112	75 120	80 128	85 136	90 144	95 152	100 160	105 168
TDI																					
Engine Oil – change	•	•		•		•		•		•		•		•		•		•		•	
Engine Filter – change	•	•		•		•		•		•		•		•		•		•		•	
Water Sep. – drain	•	•		•		•		•		•		•		•		•		•		•	
Timing Belt – check condition	•	•		•		•		•		•		•		•		•		•		•	
Fuel Filter – replace				•				•				•				•				•	
Timing Belt – replace†												•									
Timing Belt Tensioner – replace***												•									
1.8T																					
Engine Oil – change	•	•	•	•	•	•	•	•	•	•	•	•	•	•	•	•	•	•	•	•	•
Engine Filter – change	•	•	•	•	•	•	•	•	•	•	•	•	•	•	•	•	•	•	•	•	•
V-Belt – replace																•					
Spark Plugs – replace								•								•					
Timing Belt – check condition								•								•					
Timing Belt – replace																					•
Timing Belt Tensioner – replace																					•
2.0																					
Engine Oil – change	•	•		•		•		•		•		•		•		•		•		•	
Engine Filter – change	•	•		•		•		•		•		•		•		•		•		•	
Spark Plugs – replace				•				•				•				•				•	
Timing Belt – check condition								•								•					
VR6																					
Engine Oil – change	•	•		•		•		•		•		•		•		•		•		•	
Engine Filter – change	•	•		•		•		•		•		•		•		•		•		•	
Spark Plugs – replace								•								•					
V6																					
Engine Oil – change	•	•	•	•	•	•	•	•	•	•	•	•	•	•	•	•	•	•	•	•	•
Engine Filter – change	•	•	•	•	•	•	•	•	•	•	•	•	•	•	•	•	•	•	•	•	•
Spark Plugs – replace								•								•					
Timing Belt – check condition								•								•					
Timing Belt – replace																					•
Timing Belt Tensioner – replace																					•
All Cars																					
W/W Fluid – check level		•		•		•		•		•		•		•		•		•		•	
Dust Pollen Filter* – replace		•		•		•		•		•		•		•		•		•		•	
Auto-Shift Lock – check		•		•		•		•		•		•		•		•		•		•	
Brake System – check		•		•		•		•		•		•		•		•		•		•	
Wheels – rotate front to rear		•		•		•		•		•		•		•		•		•		•	
Battery – check electrolyte level				•				•				•				•				•	
Manual Trans. – check for leaks				•				•				•				•				•	
Automatic Trans. – check for leaks				•				•				•				•				•	
Tires/Spare – check condition				•				•				•				•				•	
Drive Shafts – check boots				•				•				•				•				•	
OBD – check DTC memory				•				•				•				•				•	
Door Check Straps – lubricate				•				•				•				•				•	
During Road Test				•				•				•				•				•	
After Road Test				•				•				•				•				•	
Air Cleaner** – replace filter								•								•					
Cooling System – check level				•				•				•				•				•	
V-Belt – check tension/condition								•								•					
Ribbed Belt – check condition								•								•					
Front Axle – check								•								•					
Check ATF and diff. level								•								•					
Automatic Trans. AFT – replace												•									

*Where applicable **Two years or 40,000 miles ***Recommended; replace once only †Recommended
Brake Fluid – change every 2 years regardless of mileage
Air Bag System – check for function and damage after 4 years, 8 years and every 2 years thereafter

EDITOR'S NOTE —

Publisher's additional recommended replacement interval:

- *1.8T timing belt and tensioner, inspect every 20K miles, replace every 60k miles*
- *V6 timing belt and tensioner, inspect every 20K miles, replace every 70k miles*

1999 Maintenance Schedule

Miles Kilometers	5 8	10 16	15 24	20 32	25 40	30 48	35 56	40 64	45 72	50 80	55 88	60 96	65 104	70 112	75 120	80 128	85 136	90 144	95 152	100 160	105 168
TDI																					
Engine Oil – change	•	•		•		•		•		•		•		•		•		•		•	
Engine Filter – change	•	•		•		•		•		•		•		•		•		•		•	
Water Sep. – drain	•	•		•		•		•		•		•		•		•		•		•	
Timing Belt – check condition	•	•		•		•		•		•		•		•		•		•		•	
Fuel Filter – replace				•				•				•				•				•	
Timing Belt – replace**												•									
Timing Belt Tensioner – replace***												•									
1.8T																					
Engine Oil – change	•	•	•	•	•	•	•	•	•	•	•	•	•	•	•	•	•	•	•	•	•
Engine Filter – change	•	•	•	•	•	•	•	•	•	•	•	•	•	•	•	•	•	•	•	•	•
V-Belt – replace																•					
Spark Plugs – replace								•								•					
Timing Belt – check condition								•								•					
Timing Belt – replace																					•
Timing Belt Tensioner – replace																					•
2.0																					
Engine Oil – change	•	•		•		•		•		•		•		•		•		•		•	
Engine Filter – change	•	•		•		•		•		•		•		•		•		•		•	
Spark Plugs – replace (all but New Beetle)				•				•				•				•				•	
Spark Plugs – replace (NB)								•								•					
Timing Belt – check condition								•								•					
VR6																					
Engine Oil – change	•	•		•		•		•		•		•		•		•		•		•	
Engine Filter – change	•	•		•		•		•		•		•		•		•		•		•	
Spark Plugs – replace								•								•					
V6																					
Engine Oil – change	•	•	•	•	•	•	•	•	•	•	•	•	•	•	•	•	•	•	•	•	•
Engine Filter – change	•	•	•	•	•	•	•	•	•	•	•	•	•	•	•	•	•	•	•	•	•
Spark Plugs – replace								•								•					
Timing Belt – check condition								•								•					
Timing Belt – replace																					•
Timing Belt Tensioner – replace																					•
All Cars																					
W/W Fluid – check level		•		•		•		•		•		•		•		•		•		•	
Auto-Shift Lock – check		•		•		•		•		•		•		•		•		•		•	
Brake System – check		•		•		•		•		•		•		•		•		•		•	
Wheels – rotate front to rear		•		•		•		•		•		•		•		•		•		•	
Battery – check electrolyte level				•				•				•				•				•	
Dust Pollen Filter – replace				•				•				•				•				•	
Manual Trans. – check for leaks				•				•				•				•				•	
Automatic Trans. – check for leaks				•				•				•				•				•	
Tires/Spare – check condition				•				•				•				•				•	
Drive Shafts – check boots				•				•				•				•				•	
OBD – check DTC memory				•				•				•				•				•	
Door Check Straps – lubricate				•				•				•				•				•	
During Road Test				•				•				•				•				•	
After Road Test				•				•				•				•				•	
Air Cleaner* – replace filter								•								•					
Cooling System – check level				•				•				•				•				•	
V-Belt – check tension/condition								•								•					
Ribbed Belt – check condition								•								•					
Front Axle – check								•								•					
Check ATF and diff. level								•								•					
Headlights – adjust				•				•				•				•				•	

*Every 40,000 miles and 4 years; except Passat and EuroVan, every 2 years **Recommended ***Recommended; replace once only
Brake Fluid – change every 2 years regardless of mileage
Air Bag System – check for function and damage after 4 years, 8 years and every 2 years thereafter

EDITOR'S NOTE—

Publisher's additional recommended maintenance interval:

- *1.8T timing belt and tensioner, inspect every 20K miles, replace every 60k miles*
- *V6 timing belt and tensioner, inspect every 20K miles, replace every 70k miles*

2000 Maintenance Schedule

Miles	5	10	15	20	25	30	35	40	45	50	55	60	65	70	75	80	85	90	95	100	105
Kilometers	8	16	24	32	40	48	56	64	72	80	88	96	104	112	120	128	136	144	152	160	168
TDI																					
Engine Oil – change	•	•		•		•		•		•		•		•		•		•		•	
Engine Filter – change	•	•		•		•		•		•		•		•		•		•		•	
Water Sep. – drain	•	•		•		•		•		•		•		•		•		•		•	
Timing Belt – check condition	•	•		•		•		•		•		•		•		•		•		•	
Fuel Filter – replace				•				•				•				•				•	
Timing Belt – replace*								A				M				A					
1.8T																					
Engine Oil – change	•	•	•	•	•	•	•	•	•	•	•	•	•	•	•	•	•	•	•	•	•
Engine Filter – change	•	•	•	•	•	•	•	•	•	•	•	•	•	•	•	•	•	•	•	•	•
Spark Plugs – replace								•								•					
V-Belt – replace																•					
Timing Belt – check condition								•								•					
Timing Belt – replace																					•
Timing Belt Tensioner – replace																					•
2.0																					
Engine Oil – change	•	•		•		•		•		•		•		•		•		•		•	
Engine Filter – change	•	•		•		•		•		•		•		•		•		•		•	
Spark Plugs – replace (Cabrio)				•				•				•				•				•	
Spark Plugs – replace (except Cabrio)								•								•					
Timing Belt – check condition								•								•					
VR6																					
Engine Oil – change	•	•		•		•		•		•		•		•		•		•		•	
Engine Filter – change	•	•		•		•		•		•		•		•		•		•		•	
Spark Plugs – replace								•								•					
V6																					
Engine Oil – change	•	•	•	•	•	•	•	•	•	•	•	•	•	•	•	•	•	•	•	•	•
Engine Filter – change	•	•	•	•	•	•	•	•	•	•	•	•	•	•	•	•	•	•	•	•	•
Spark Plugs – replace								•								•					
Timing Belt – check condition								•								•					
Timing Belt – replace																					•
Timing Belt Tensioner – replace																					•
All Cars																					
W/W Fluid – check level		•		•		•		•		•		•		•		•		•		•	
Auto-Shift Lock – check		•		•		•		•		•		•		•		•		•		•	
Brake System – check		•		•		•		•		•		•		•		•		•		•	
Wheels – rotate front to rear		•		•		•		•		•		•		•		•		•		•	
Air Bag System **		•		•		•		•		•		•		•		•		•		•	
Battery – check electrolyte level				•				•				•				•				•	
Dust Pollen Filter – replace				•				•				•				•				•	
Cooling System – check level				•				•				•				•				•	
Manual Trans. – check for leaks				•				•				•				•				•	
Automatic Trans. – check for leaks				•				•				•				•				•	
Tires/Spare – check condition				•				•				•				•				•	
Drive Shafts – check boots				•				•				•				•				•	
OBD – check DTC memory				•				•				•				•				•	
Door Hinge				•				•				•				•				•	
Headlights – adjust				•				•				•				•				•	
During Road Test				•				•				•				•				•	
After Road Test				•				•				•				•				•	
Air Cleaner*** – replace filter								•								•					
V-Belt – check tension/condition								•								•					
Ribbed Belt – check condition								•								•					
Brake Fluid****								•								•					
Check ATF and diff. level								•								•					
Front Axle (tie rod, etc.)								•								•					

*Recommended **Every 12 months regardless of mileage ***Every 2 years: Cabrio, Passat, EuroVan. Every 4 years: Golf, Jetta, New Beetle
****Replace every 2 years regardless of mileage

EDITOR'S NOTE—

Publisher's additional recommended maintenance interval:

- *1.8T timing belt and tensioner, inspect every 20K miles, replace every 60k miles*
- *V6 timing belt and tensioner, inspect every 20K miles, replace every 70k miles*

2001 Maintenance Schedule

Miles Kilometers	5 8	10 16	15 24	20 32	25 40	30 48	35 56	40 64	45 72	50 80	55 88	60 96	65 104	70 112	75 120	80 128	85 136	90 144	95 152	100 160	105 168
TDI																					
Engine Oil – change	•	•		•		•		•		•		•		•		•		•		•	
Engine Filter – change	•	•		•		•		•		•		•		•		•		•		•	
Water Sep. – drain	•	•		•		•		•		•		•		•		•		•		•	
Timing Belt – check condition	•	•		•		•		•		•		•		•		•		•		•	
Fuel Filter – replace				•				•				•				•				•	
Timing Belt – replace								A				M				A					
1.8T																					
Engine Oil – change	•	•	•	•	•	•	•	•	•	•	•	•	•	•	•	•	•	•	•	•	•
Engine Filter – change	•	•	•	•	•	•	•	•	•	•	•	•	•	•	•	•	•	•	•	•	•
Spark Plugs – replace								•								•					
V-Belt – replace																•					
Timing Belt – check condition								•								•					
Timing Belt – replace																					•
Timing Belt Tensioner – replace																					•
2.0																					
Engine Oil – change	•	•		•		•		•		•		•		•		•		•		•	
Engine Filter – change	•	•		•		•		•		•		•		•		•		•		•	
Spark Plugs – replace (Cabrio)				•				•				•				•				•	
Spark Plugs – replace (except Cabrio)								•								•					
Timing Belt – check condition								•								•					
VR6																					
Engine Oil – change	•	•		•		•		•		•		•		•		•		•		•	
Engine Filter – change	•	•		•		•		•		•		•		•		•		•		•	
Spark Plugs – replace								•								•					
V6																					
Engine Oil – change	•	•	•	•	•	•	•	•	•	•	•	•	•	•	•	•	•	•	•	•	•
Engine Filter – change	•	•	•	•	•	•	•	•	•	•	•	•	•	•	•	•	•	•	•	•	•
Spark Plugs – replace								•								•					
Timing Belt – check condition								•								•					
Timing Belt Tensioner – replace																					•
Timing Belt – replace																					•
All Cars																					
W/W Fluid – check level		•		•		•		•		•		•		•		•		•		•	
Auto-Shift Lock – check operation		•		•		•		•		•		•		•		•		•		•	
Brake System – check		•		•		•		•		•		•		•		•		•		•	
Wheels – rotate from front to rear		•		•		•		•		•		•		•		•		•		•	
Air Bag System – check function/ damage*		•		•		•		•		•		•		•		•		•		•	
Battery – check electrolyte level		•		•		•		•		•		•		•		•		•		•	
Dust Pollen Filter – replace				•				•				•				•				•	
Cooling System – check level				•				•				•				•				•	
Manual Transmission – for leaks				•				•				•				•				•	
Automatic Transmission – for leaks				•				•				•				•				•	
Tire/Spare Tire – check condition				•				•				•				•				•	
Drive Shafts – check boots				•				•				•				•				•	
OBD – check DTC memory				•				•				•				•				•	
Door Hinge				•				•				•				•				•	
Headlights – check and adjust if necessary				•				•				•				•				•	
Road Test – during				•				•				•				•				•	
Road Test – after				•				•				•				•				•	
Air Cleaner – replace filter**								•								•					
V-Belt – check tension/condition								•								•					
Ribbed Belt – check condition								•								•					
Brake Fluid***								•								•					
Check ATF and diff. level								•								•					
Front Axle (tie rods, etc.)								•								•					

*Check the air bag system every 12 months, regardless of mileage. Check the air bag system at 4 years, 8 years, and then every 2 years (10, 12, 14, etc.)

**Replace the air filter every 2 years or 40,000 miles, whichever occurs first: Cabrio, Passat, EuroVan. Replace the air filter every 4 years or 40,000 miles, whichever occurs first: Golf, Jetta, New Beetle ***Replace the brake fluid every 2 years, regardless of mileage

EDITOR'S NOTE —

Publisher's additional recommended maintenance interval:

- *1.8T timing belt and tensioner, inspect every 20K miles, replace every 60k miles*
- *V6 timing belt and tensioner, inspect every 20K miles, replace every 70k miles*

2002-2005 Maintenance Schedule, 1.8 liter

Kilometers	8	16	24	32	40	48	56	64	72	80	88	96	104	112	120	128	136	144	152	160	168
Miles	**5**	**10**	**15**	**20**	**25**	**30**	**35**	**40**	**45**	**50**	**55**	**60**	**65**	**70**	**75**	**80**	**85**	**90**	**95**	**100**	**105**
Engine Oil – change	•	•	•	•	•	•	•	•	•	•	•	•	•	•	•	•	•	•	•	•	•
Engine Filter – change	•	•	•	•	•	•	•	•	•	•	•	•	•	•	•	•	•	•	•	•	•
Timing Belt – check								•								•					
Timing Belt – replace																					•
Timing Belt Tensioner – replace																					•
V-belt/Ribbed Belt – check								•								•					
V-belt – replace																•					
Spark Plugs – replace								•								•					
Air Cleaner – replace filter								•								•					
W/W Fluid – check level		•		•		•		•		•		•		•		•		•		•	
Auto-shift Lock – check operation		•		•		•		•		•		•		•		•		•		•	
Brake System – check damage/ leaks, pad thickness, fluid level		•		•		•		•		•		•		•		•		•		•	
Wheels – rotate front to rear		•		•		•		•		•		•		•		•		•		•	
Airbag System – check function/ damage*		•		•		•		•		•		•		•		•		•		•	
Battery – check electrolyte level		•		•		•		•		•		•		•		•		•		•	
Engine – check for leaks				•				•				•				•				•	
Pollen Filter – replace				•				•				•				•				•	
Cooling System – check level				•				•				•				•				•	
Transmission – check for leaks**				•				•				•				•				•	
Tires/Spare – check condition				•				•				•				•				•	
Driveshaft – check boots				•				•				•				•				•	
OBD – check DTC memory				•				•				•				•				•	
Door Hinge – lube				•				•				•				•				•	
Headlights – check and adjust if necessary				•				•				•				•				•	
Exhaust System – check for damage and leaks				•				•				•				•				•	
Sunroof Frame – clean and lubricate				•				•				•				•				•	
Road Test				•				•				•				•				•	
Power Steering – check fluid level								•								•					
Brake Fluid***								•								•					
Check ATF and Differential for leaks and level; add if necessary								•								•					
Front Axle – check dust seals, ball joints, tie rods								•								•					

*Air Bag System – visual check every 12 months regardless of mileage **Manual Transmission – check level; add if necessary
***Brake Fluid – change every 2 years regardless of mileage

EDITOR'S NOTE—
Publisher's additional recommended maintenance interval:

- *1.8T timing belt and tensioner, inspect every 20K miles, replace every 60k miles*

2002-2005 Maintenance Schedule, 2.8 liter

Kilometers	8	16	24	32	40	48	56	64	72	80	88	96	104	112	120	128	136	144	152	160	168
Miles	**5**	**10**	**15**	**20**	**25**	**30**	**35**	**40**	**45**	**50**	**55**	**60**	**65**	**70**	**75**	**80**	**85**	**90**	**95**	**100**	**105**
Engine Oil – change	•	•	•	•	•	•	•	•	•	•	•	•	•	•	•	•	•	•	•	•	•
Engine Filter – change	•	•	•	•	•	•	•	•	•	•	•	•	•	•	•	•	•	•	•	•	•
Timing Belt – check								•								•					
Timing Belt – replace																					•
Timing Belt Tensioner – replace																					•
V-belt/Ribbed Belt – check								•								•					
Spark Plugs – replace								•								•					
Air Cleaner – replace filter								•								•					
W/W Fluid – check level		•		•		•		•		•		•		•		•		•		•	
Auto-shift Lock – check operation		•		•		•		•		•		•		•		•		•		•	
Brake System – check damage/ leaks, pad thickness, fluid level		•		•		•		•		•		•		•		•		•		•	
Wheels – rotate front to rear		•		•		•		•		•		•		•		•		•		•	
Airbag System – check function/ damage*		•		•		•		•		•		•		•		•		•		•	
Battery – check electrolyte level		•		•		•		•		•		•		•		•		•		•	
Engine – check for leaks				•				•				•				•				•	
Pollen Filter – replace				•				•				•				•				•	
Cooling System – check level				•				•				•				•				•	
Transmission – check for leaks**				•				•				•				•				•	
Tires/Spare – check condition				•				•				•				•				•	
Driveshaft – check boots				•				•				•				•				•	
OBD – check DTC memory				•				•				•				•				•	
Door Hinge – lube				•				•				•				•				•	
Headlights – check and adjust if necessary				•				•				•				•				•	
Exhaust System – check for damage and leaks				•				•				•				•				•	
Sunroof Frame – clean and lubricate				•				•				•				•				•	
Road Test				•				•				•				•				•	
Power Steering – check fluid level								•								•					
Brake Fluid***								•								•					
Check ATF and Differential for leaks and level; add if necessary								•								•					
Front Axle – check dust seals, ball joints, tie rods								•								•					

*Air Bag System – visual check every 12 months regardless of mileage **Manual Transmission – check level; add if necessary
***Brake Fluid – change every 2 years regardless of mileage ****Change ATF EuroVan only

EDITOR'S NOTE—

Publisher's additional recommended maintenance interval:

- *V6 timing belt and tensioner, inspect every 20K miles, replace every 70k miles*

2002-2005 Maintenance Schedule, 4.0 liter

Kilometers	8	16	24	32	40	48	56	64	72	80	88	96	104	112	120	128	136	144	152	160	168
Miles	**5**	**10**	**15**	**20**	**25**	**30**	**35**	**40**	**45**	**50**	**55**	**60**	**65**	**70**	**75**	**80**	**85**	**90**	**95**	**100**	**105**
Engine Oil – change	•	•		•		•		•		•		•		•		•		•		•	
Engine Filter – change	•	•		•		•		•		•		•		•		•		•		•	
V-belt/Ribbed Belt – check								•								•					
Spark Plugs – replace								•								•					
Air Cleaner – replace filter								•								•					
W/W Fluid – check level		•		•		•		•		•		•		•		•		•		•	
Auto-Shift Lock – check operation		•		•		•		•		•		•		•		•		•		•	
Brake System – check damage/ leaks, pad thickness, fluid level		•		•		•		•		•		•		•		•		•		•	
Wheels – rotate front to rear		•		•		•		•		•		•		•		•		•		•	
Airbag System – check function/ damage*		•		•		•		•		•		•		•		•		•		•	
Battery – check electrolyte level		•		•		•		•		•		•		•		•		•		•	
Engine – check for leaks				•				•				•				•				•	
Pollen Filter – replace				•				•				•				•				•	
Cooling System – check level				•				•				•				•				•	
Transmission – check for leaks**				•				•				•				•				•	
Tires/Spare – check condition				•				•				•				•				•	
Driveshaft – check boots				•				•				•				•				•	
OBD – check DTC memory				•				•				•				•				•	
Door Hinge – lube				•				•				•				•				•	
Headlights – check and adjust if necessary				•				•				•				•				•	
Exhaust System – check for damage and leaks				•				•				•				•				•	
Sunroof Frame – clean and lubricate				•				•				•				•				•	
Road Test				•				•				•				•				•	
Power Steering – check fluid level								•								•					
Brake Fluid***								•								•					
Check ATF and Differential for leaks and level; add if necessary								•								•					
Front Axle – check dust seals, ball joints, tie rods								•								•					

*Airbag System – visual check every 12 months regardless of mileage **Manual Transmission – check level; add if necessary
***Brake Fluid – change every 2 years regardless of mileage

Notes

Notes

Notes

Notes

1.8 Liter 4-Cylinder 5V Turbo Engine Mechanical (code: AEB, ATW)

Section **B**

Section B – 1.8 Liter 4-Cylinder 5V Trubo Engine Mechanical (code: AEB, ATW)

00 General, Technical Data — Page

13 Engine–Crankshaft, Cylinder Block — Page

15 Engine–Cylinder Head, Valvetrain — Page

17 Engine–Lubrication System — Page

19 Engine–Cooling System — Page

20 Fuel Supply

Page

21 Turbocharger

Page

26 Exhaust System

Page

00 General, Technical Data

Repair Group

B 00

Technical data

Engine number

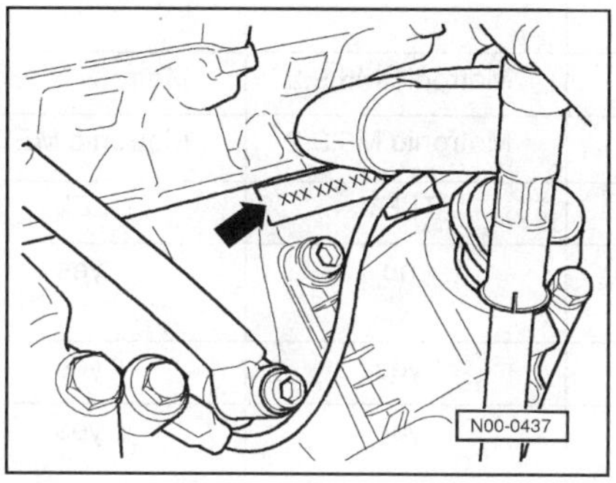

The engine number (engine code letters and serial number) is located on the left side of the cylinder block (arrow).

Additionally there is a sticker on the cylinder head cover with the engine number.

The engine number consists of up to nine characters (alphanumeric). The first part (maximal 3 characters) makes up the "engine code", and the second part (6 characters), the "serial number". If more than 999,999 engines with the same engine code are produced, the first of the six characters is replaced with a letter.

The engine code is also included on the vehicle data plate.

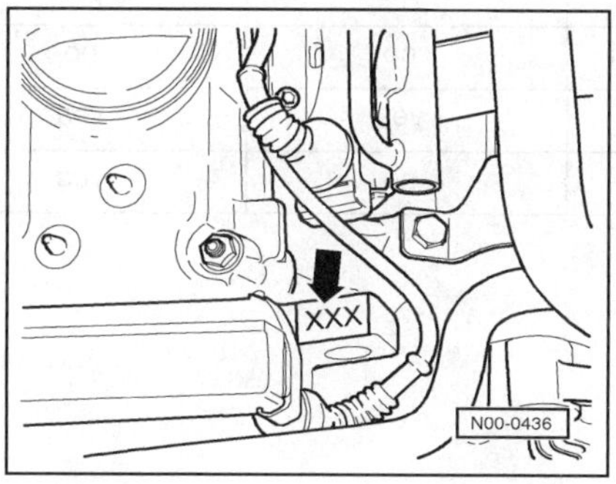

The engine code (arrow) is also stamped in at the front of the engine lifting eye (visible after removing cover above fuel injectors).

Engine data

Code letters		AEB	ATW
Manufacturing dates		08.97➤08.99	09.99➤
Displacement	liters	1.8	1.8
Output	hp (kw) @ rpm	150 (110) @ 5700	150 (110) @ 5700
Torque	ft-lb (Nm) @ rpm	155 (210) @1750 - 4600	155 (210) @1750 - 4600
Bore	diameter (mm)	81.0	81.0
Stroke	mm	86.4	86.4
Compression ratio		9.5 : 1	9.5 : 1
RON	Minimum	95 unleaded	95 unleaded
Fuel injection system		Motronic M5.9.2	Motronic ME7.5
Ignition system		Motronic M5.9.2	Motronic ME7.5
Emissions standard type		TLEV	LEV
Electronic Power Control (EPC) System		no	yes
Knock sensor control		yes	yes
On Board Diagnostic (OBD)		yes	yes
Oxygen sensor control		yes	yes
Three-way catalyst		yes	yes
Exhaust gas recirculation		no	no
Turbocharging		yes	yes
Secondary air system		no	yes

13 Engine–Crankshaft, Cylinder Block

Repair Group
B 13

Engine Block Assembly

NOTE—
Metal shavings or large quantities of small metal particles found in the engine oil during repairs can be caused, for example, by the partial seizure of crankshaft and conrod bearings. If such particles are found, apart from thoroughly cleaning the oil passages, it is necessary to replace the oil cooler.

For timing belt and accessory drive belt information, see **Engine belts.**

For cylinder block, engine bracket, and seal information, see **Engine block, components.**

EDITOR'S NOTE—
Volkswagen identifies electrical components by a letter and/or a number in the electrical schematics. See **Wiring Diagrams, Fuses and Relays**. *These electrical identifiers are listed -between dashes- as an aid to electrical troubleshooting.*

Engine belts

1. 25 Nm (18 ft-lb)
2. **Tensioning device for ribbed belt**
   - Turn with open-end wrench to loosen ribbed belt, see **Maintenance.**
3. 20 Nm (14 ft-lb)
4. **Toothed belt guard - upper part**
5. 10 Nm (7 ft-lb)
6. 10 Nm (7 ft-lb)
7. **Tensioning roller**
8. **Toothed belt**
   - Mark direction of rotation before removing
   - Check for wear
   - Do not kink
   - see **Toothed belt, removing and installing**
9. 20 Nm (14 ft-lb)
10. **Rear toothed belt guard**
11. **Intermediate shaft sprocket**
12. 80 Nm (60 ft-lb)
13. **Idler wheel**
14. **Tensioner for toothed belt**
15. **Crankshaft toothed belt sprocket**
    - To remove, counterhold with 3099 Counter support.
16. 10 Nm (7 ft-lb)
17. 90 Nm (67 ft-lb)+ $^{1}/_{4}$ turn (90°) further
    - Always replace
    - Oil threads
    - Counter-hold with 3099 to loosen and tighten
    - The quarter turn further can be done in several stages.
18. 25 Nm (18 ft-lb)
19. **Toothed belt guard - lower part**
20. 10 Nm (7 ft-lb)
21. **Pulley**
    - For viscous fan coupling
22. **Pulley**
    - For power assisted steering pump
23. **V-belt**
    - See **Maintenance**
24. **Pulley**
    - V-belt version

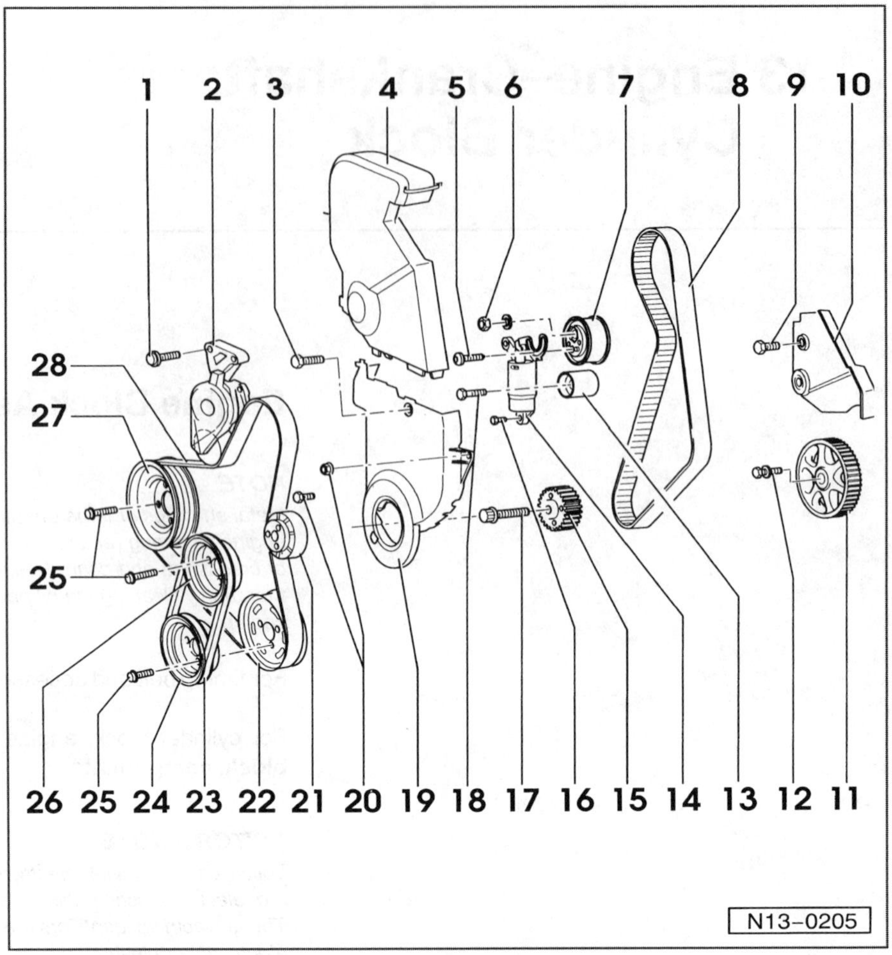

25. 25 Nm (18 ft-lb)
26. **Pulley**
    - For coolant pump
    - V-belt version
27. **Belt pulley/vibration damper**
    - Can only be installed in one position. Holes are offset
    - Note position when installing toothed belt, see **Toothed belt, removing and installing**
28. **Ribbed belt**
    - Mark direction of rotation before removing
    - Removing and installing, see **Maintenance.**

Engine block, components

1. **Cylinder block**
2. **Woodruff key**
   - Check for tight fit
3. **Support**
   - Between turbocharger and cylinder block
4. **30 Nm (22 ft-lb)**
5. **O-Ring**
   - Replace if damaged
6. **Knock sensor 1 -G61-**
   - Checking resistance, see **1.8 Liter Fuel Injection & Ignition, Repair Group 28**
7. **20 Nm (14 ft-lb)**
   - Tightening torque influences the function of the knock sensor
8. **Knock sensor 2 -G66-**
   - Checking resistance, see **1.8 Liter Fuel Injection & Ignition, Repair Group 28**
9. **Input gear**
   - For oil pump
10. **Cap**
    - With play compensation
11. **25 Nm (18 ft-lb)**
12. **Clamp**
13. **Gasket**
    - Always replace
14. **Oil filter bracket**
15. **25 Nm (18 ft-lb)**
16. **Sealing ring**
    - Always replace
17. **25 Nm (18 ft-lb)**
18. **Crankcase breather**
19. **Oil cooler**
    - Coat contact area to flange, outside the seal, with AMV 188 100 02
    - Ensure clearance to adjacent components
20. **25 Nm (18 ft-lb)**
21. **Oil filter**
    - Loosen with strap wrench
    - Tighten by hand
    - Note instructions on filter when installing
22. **10 Nm (7 ft-lb)**

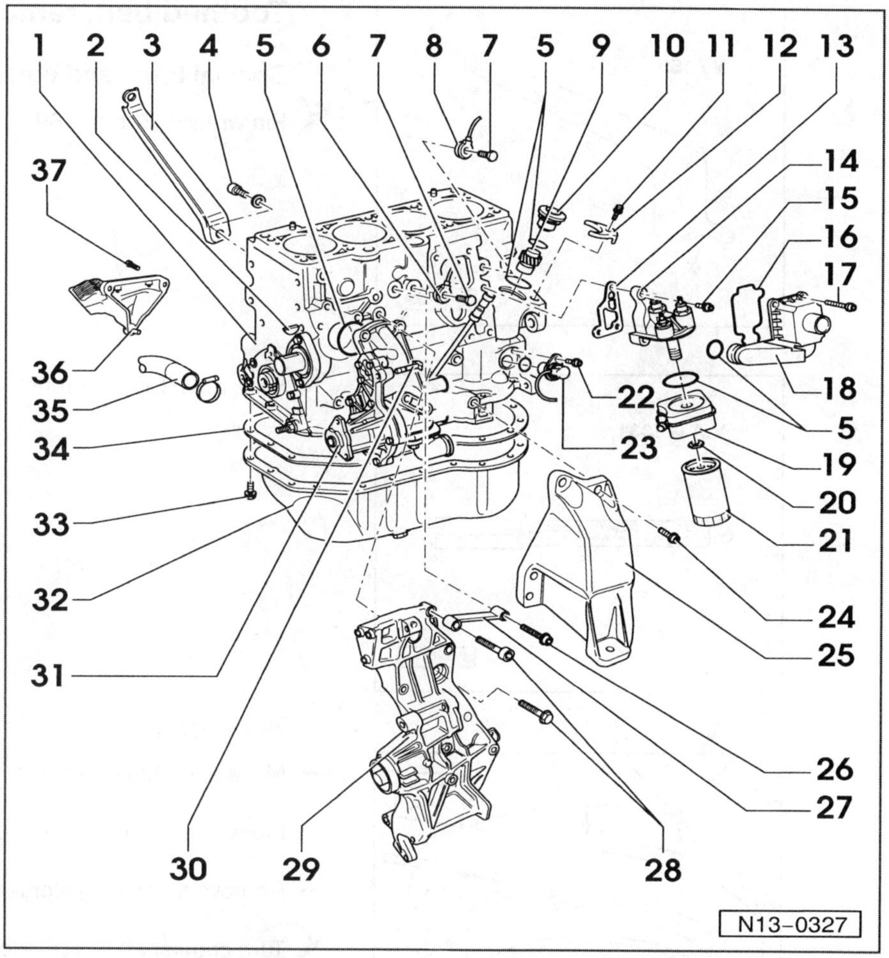

23. **Engine speed (RPM) sensor**
24. **30 Nm (22 ft-lb)**
25. **Left engine bracket**
26. **25 Nm**
27. **Support**
    - Between compact bracket and cylinder block
28. **30 Nm (22 ft-lb)**
29. **Compact bracket**
    - For coolant pump, coolant thermostat, ribbed belt tensioning roller, Generator and power steering pump
30. **Hammer head bolt**
31. **Coolant pump**
    - See **Repair Group 19, Coolant pump, disassembling and assembling**
32. **Oil pan**
    - Clean sealing surface before installing
33. **20 Nm**
    - Remove and install both rear bolts transmission end with jointed wrench 3185
34. **Gasket**
    - Always replace
    - Before installing gasket coat oil pan flange/cylinder block flange with D 454 300 A2
35. **Coolant hose**
    - Between turbocharger and cylinder block
36. **Right engine bracket**
37. **30 Nm (22 ft-lb)**

Toothed belt, removing and installing

Special tools and equipment

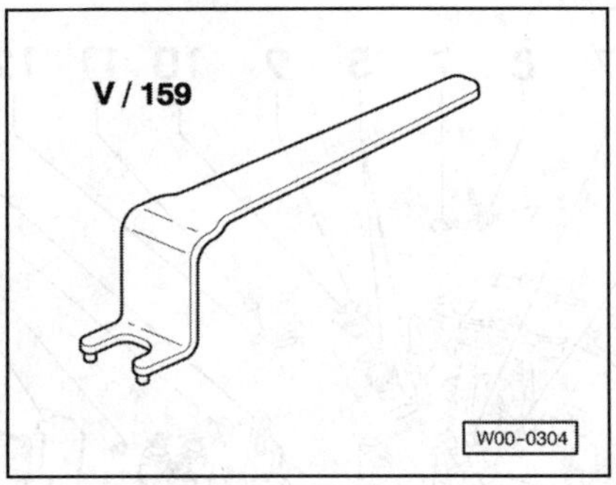

➤ Pin wrench Matra V159

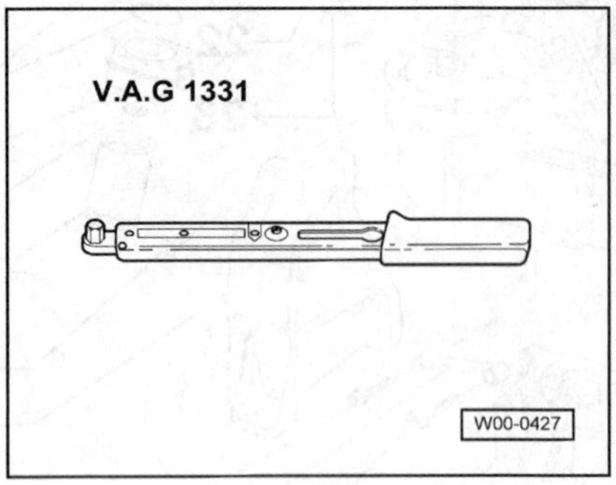

➤ VAG 1331 Torque wrench (5 to 50 Nm)

Removing

– Make sure that coolant temperature at least 30°C

– Move lock carrier into service position, see **Maintenance**

– Remove tensioning element for ribbed belt.

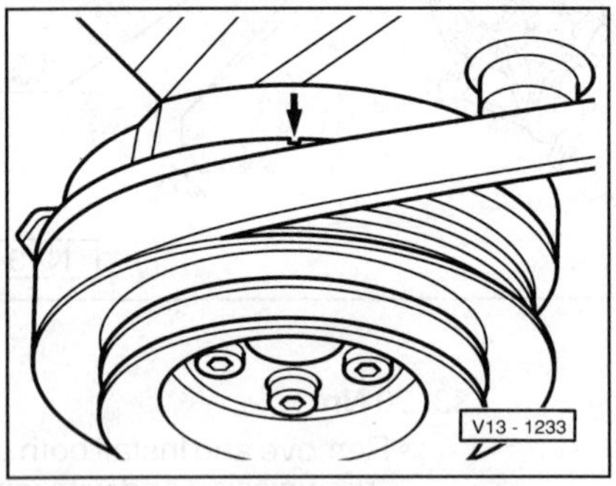

➤ Turn crankshaft to TDC No. 1 cylinder.

– Remove toothed belt guard - upper part.

– Mark toothed belt direction of rotation.

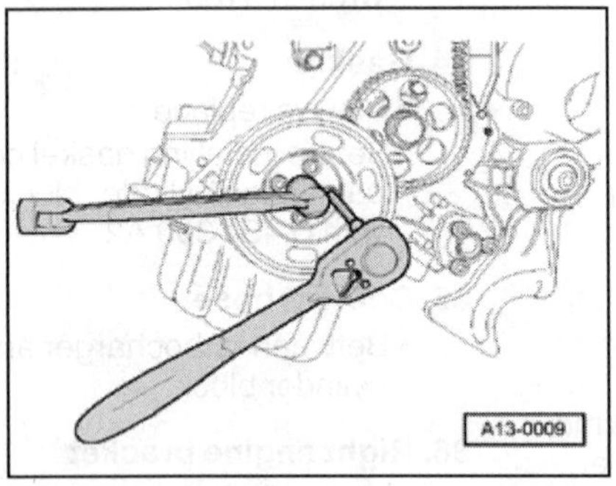

➤ Remove vibration damper/belt pulley.

– Remove lower toothed belt guard.

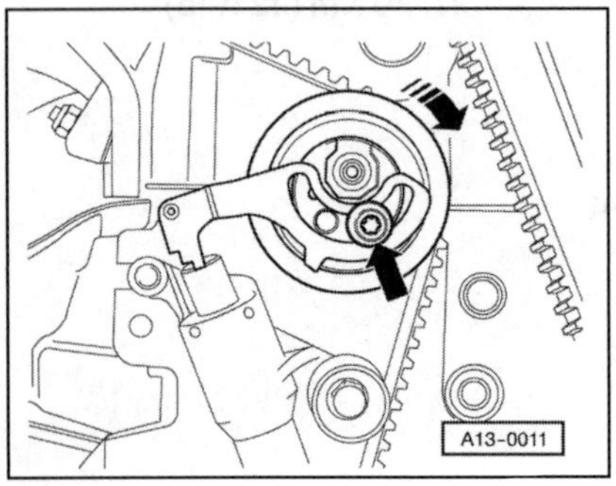

➤ Loosen toothed belt tensioner (arrow) and remove toothed belt.

– Turn crankshaft back slightly.

Installing

NOTE —

When turning the camshaft the crankshaft must not be at TDC or there is a potential for damage to the valves/piston crown.

A - Toothed belt completely removed

If toothed belt has only been removed from camshaft, see **B - Toothed belt only taken off camshaft sprocket**

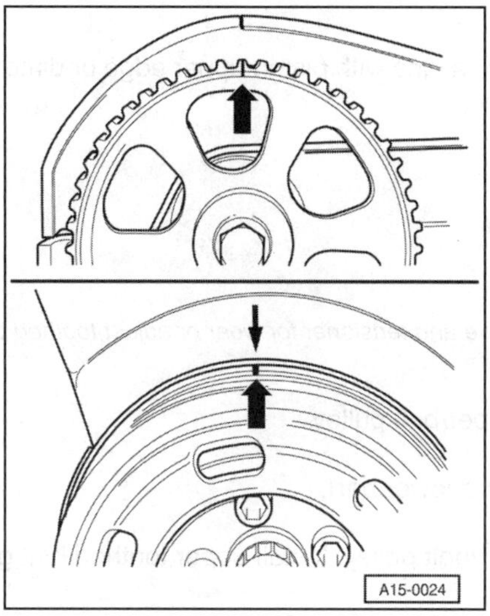

> ***WARNING —***
> *The pistons must not be positioned at TDC.*

Work sequence

< Bring mark on camshaft in line with mark on cylinder head cover.

– Place toothed belt on crankshaft sprocket (observe direction of rotation).

– Secure vibration damper/belt pulley with one bolt (note installation).

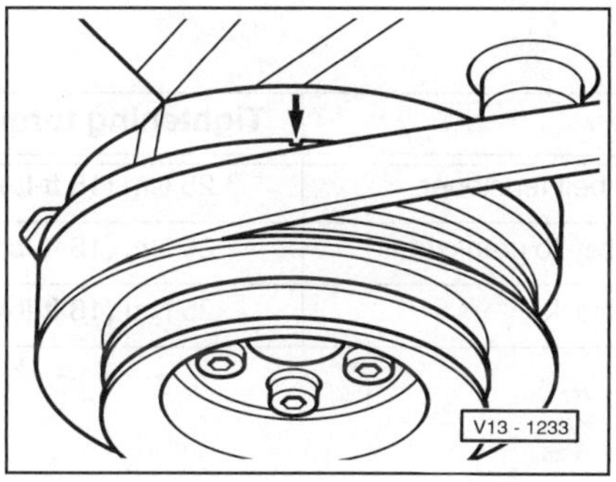

< Turn crankshaft to TDC No. 1 cylinder.

– Place toothed belt on tensioning roller and camshaft sprocket and tension, see **Tensioning toothed belt**.

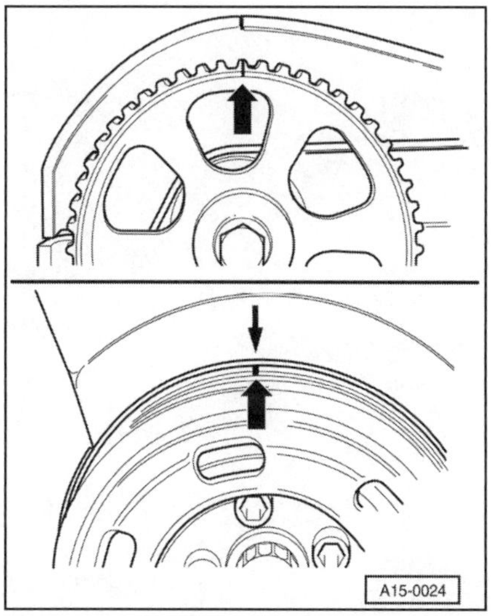

B - Toothed belt only taken off camshaft sprocket

< Bring mark on camshaft in line with mark on cylinder head cover.

– Turn crankshaft to TDC No. 1 cylinder.

– Place toothed belt on the camshaft sprocket and tension, see **Tensioning toothed belt**.

Tensioning toothed belt

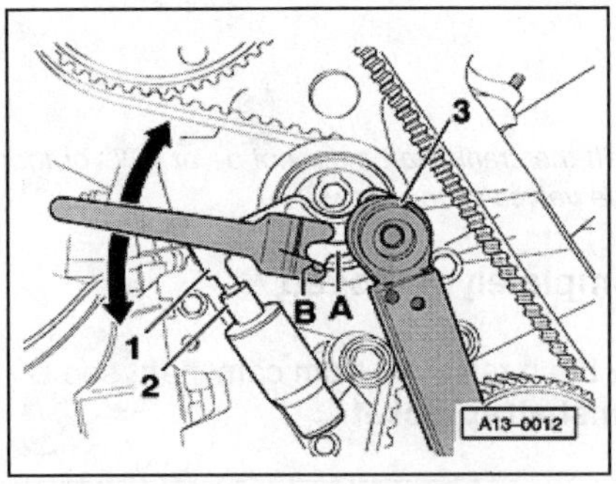

◄ Tension toothed belt by turning tensioner -A- with pin wrench -B- (e.g. Matra V159) to right until piston -1- is fully extended and tensioner piston -2- lifts approx. 1 mm.

– Tighten securing screw -3- on tensioner.

– Turn crankshaft two full turns and check whether camshaft and crankshaft marks align with their reference points.

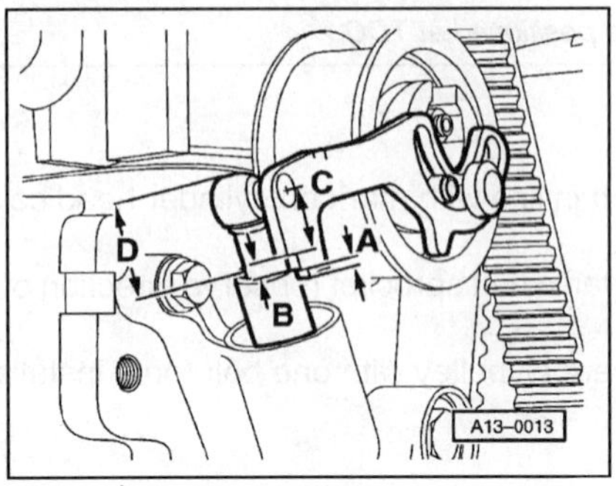

◄ Check whether area -A- aligns with piston upper edge or dimension -D- is: 25 to 29 mm.

NOTE—

• Area -A-: Adjustment OK.

• Area -B-: Wear area

• Area -C-: Check belt drive and tensioner for wear or adjust toothed belt tension.

– Remove vibration damper/belt pulley.

– Install toothed belt guard lower part.

– Install vibration damper/belt pulley. Install upper toothed belt guard.

– Install ribbed belt and tensioner, see **Maintenance**.

Tightening torques

Component	Tightening torques
Securing screw to timing belt tensioner	25 Nm (18 ft-lb)
Vibration damper/belt pulley to crankshaft	25 Nm (18 ft-lb)
Ribbed belt tensioner to block	25 Nm (18 ft-lb)

Sealing flanges and flywheel/drive plate, removing and installing

NOTE—
For repairs to the clutch, see **Repair group 30, 5 Spd. Manual Transmission 012/01W**

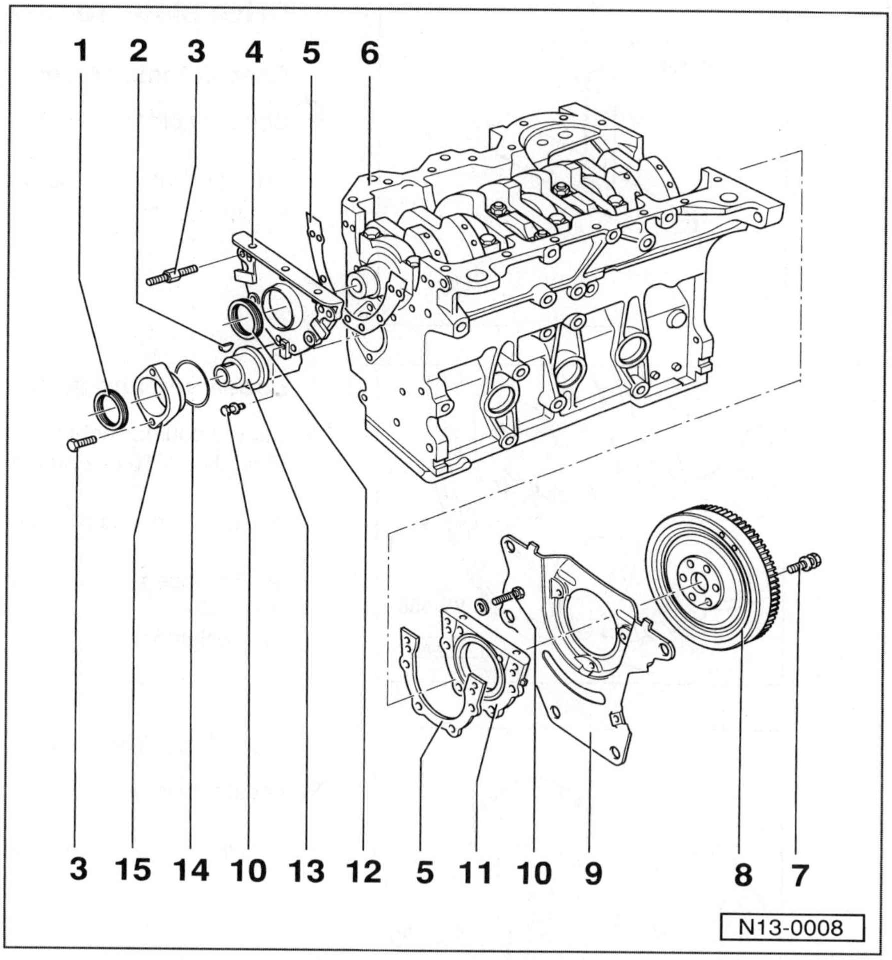

1. **Oil seal**
   - To remove oil seal, remove sealing flange
   - Install with 10-203
   - Lightly oil sealing lip
2. **Woodruff key**
   - Ensure tight fit
3. **25 Nm (18 ft-lb)**
4. **Sealing flange**
5. **Gasket**
   - Always replace
6. **Cylinder block**
7. **60 Nm (44 ft-lb) + $^{1}/_{4}$ turn (90°) further**
   - Always replace
8. **Dual-mass flywheel/drive plate**
   - Lock with 3067 to remove and install flywheel
   - See **Drive plate, removing and installing**
9. **Intermediate plate**
   - Must be located on dowel sleeves
   - Do not damage/bend when assembling
10. **10 Nm (7 ft-lb)**
11. **Sealing flange with oil seal**
    - Only replace complete
    - Lightly oil sealing lip of oil seal
12. **Oil seal**
    - Remove and install with 3083 puller and 2085 oil seal extractor
13. **Intermediate shaft**
    - Axial clearance max. 0.25 mm
14. **O-ring**
    - Replace if damaged
15. **Sealing flange intermediate shaft**

Drive plate, removing and installing

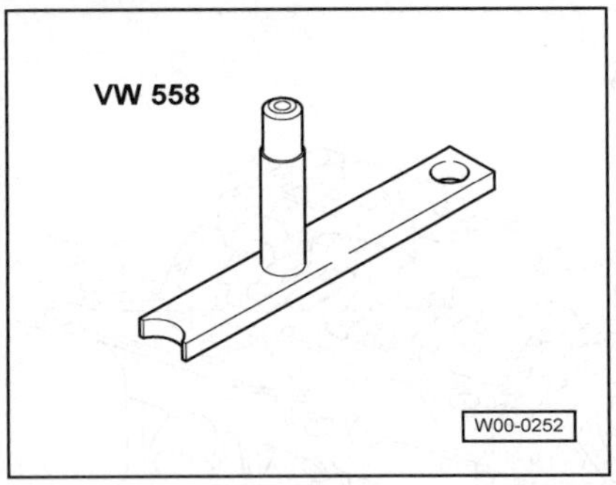

Special tools and equipment

Counter-hold tool VW 558

- Hex bolt M8 x 45 and two M10 hex nuts
- Depth gauge

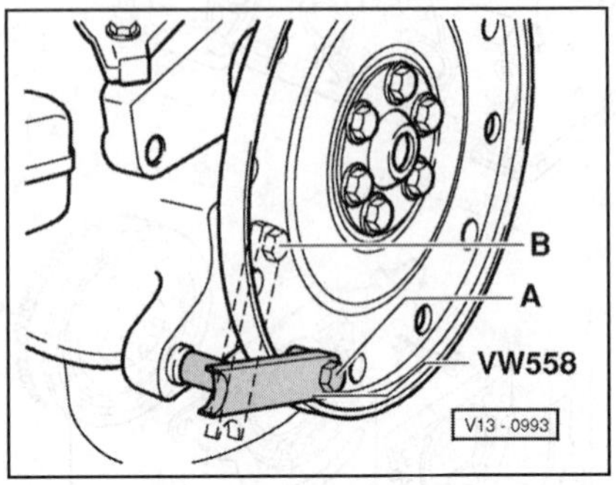

Loosening and tightening drive plate

Secure counter-hold tool VW 558 to drive plate with hex bolt M8 x 45. Place two M10 hex nuts between counter-hold tool and drive plate.

Position of counter-hold tool:

A - To loosen

B - To tighten

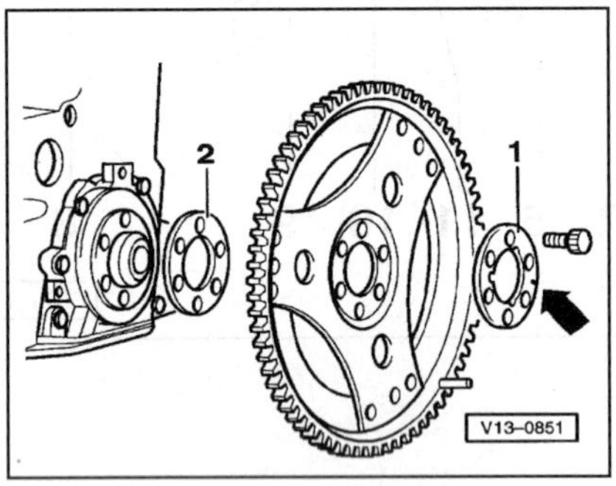

Installing drive plate

Locate drive plate and packing plate with recesses -1-.

– Install new bolts and tighten to 30 Nm.

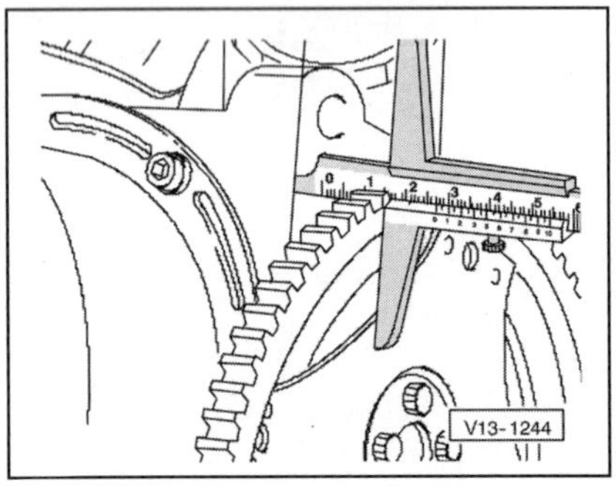

Calculate dimension -a- at three points and calculate average.

- Specification: 26 to 28 mm

If the specification is not attained:

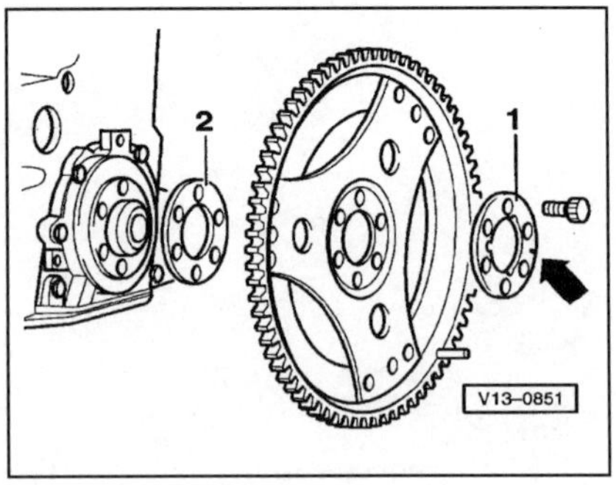

Remove drive plate again and install shim -2-. Tighten bolts again to 30 Nm.

– Tighten bolts to 60 Nm and turn 90° (1/4 turn) further (additional turn can be done in several stages).

Tightening torque

Component	Tightening torque
Drive plate bolts to crankshaft	60 Nm (44 ft-lb) + 90°

15 Engine–Cylinder Head, Valvetrain

Repair Group

B 15

Cylinder Head, Servicing

Checking compression, see **Compression, checking**

NOTE —

- *When installing an exchange cylinder head (with camshaft installed), the contact surfaces between the hydraulic lifters and the cam must be oiled before installing the cylinder head cover.*
- *The plastic packing pieces for protecting the open valves must not be removed until immediately before installing the cylinder head.*
- *If the cylinder head is replaced, all the coolant in the system must also be replaced.*

Cylinder head, components

1. **Cap**
   - Replace seal if damaged
2. **10 Nm (7 ft-lb)**
3. **Cylinder head cover**
4. **Cylinder head cover gasket**
   - Replace if damaged
   - Before installing gasket, coat transition between bearing cap/cylinder head with D 454 300 A2
   - Coat half-moon shaped surface with D 454 300 A2
5. **Oil deflector**
   - Note installation position: Above intake camshaft between No. 1 and No. 2 cylinders
6. **Cylinder head**
   - Check for distortion see **Fig. 1**
   - See **Cylinder head removing and installing**
   - After replacing, completely replace engine coolant
7. **Intake manifold gasket**
   - Always replace
8. **Cylinder head gasket**
   - Always replace
   - Metal gasket
   - If replaced, completely replace engine coolant
   - Note installation position: Identification: Part No. must be readable from intake side
9. **25 Nm (18 ft-lb)**
   - For tensioner
10. **Exhaust manifold gasket**
    - Always replace
    - Note installation position
11. **Cylinder head bolt**
    - Always replace
    - Sequence when loosening and tightening, see **Cylinder head, removing**
    - Remove and install Polydrive cylinder head bolt with 3452
12. **Heat shield**
13. **25 Nm (18 ft-lb)**
14. **10 Nm (7 ft-lb)**

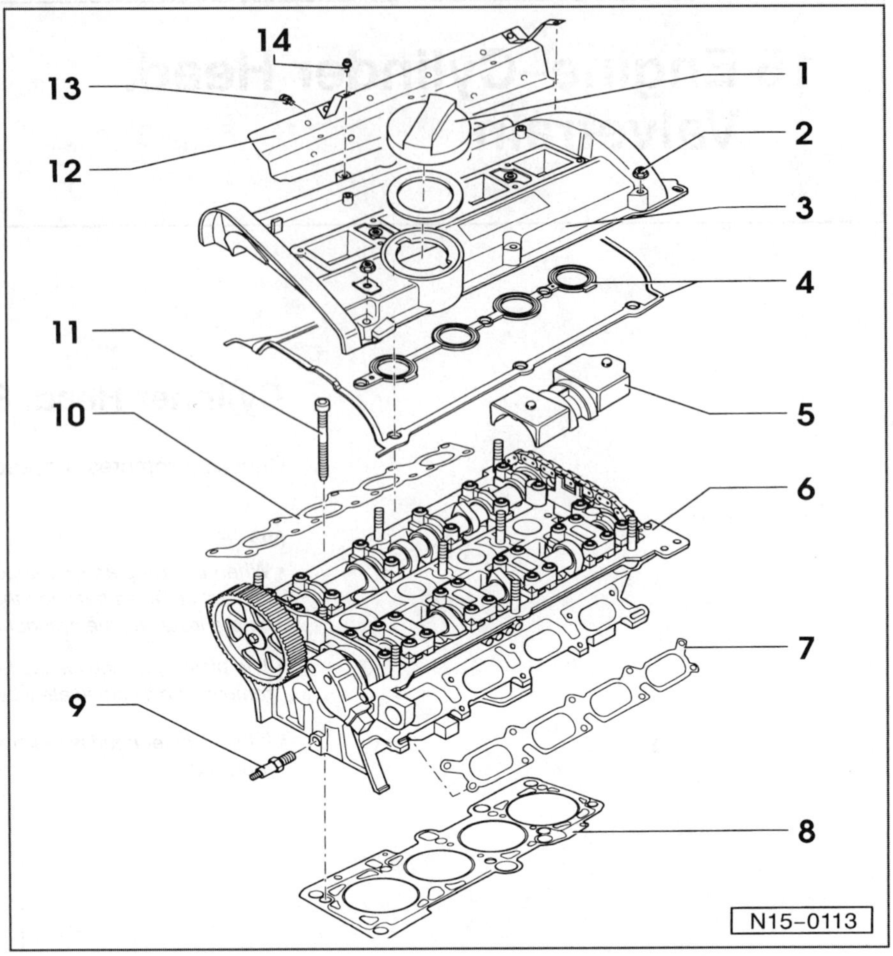

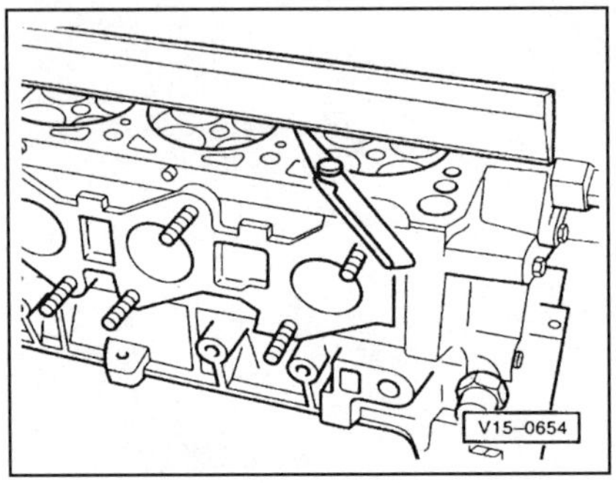

Fig. 1 Checking cylinder head for distortion

- Max. permissible distortion: 0.1 mm

Cylinder head, removing and installing

Special tools and equipment

- VAG 1306 Drip tray
- VAS 5024 Spring clip pliers
- 3070 Guide pins & handle
- VAG 1332 Torque wrench (40 to 200 Nm)

Engines with Polydrive cylinder head bolts:

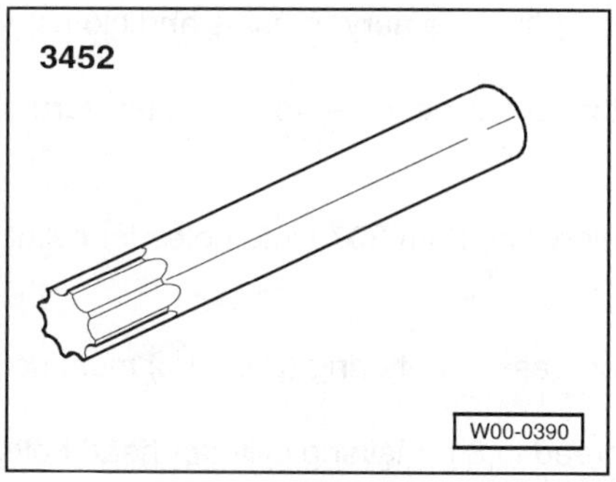

3452 Key

Requirements

- The engine must be no more than warm to touch.
- The pistons must not be positioned at TDC.

Work sequence

– Check whether a coded radio is installed. If so, obtain anti-theft coding.

– With ignition switched off disconnect battery Ground strap.

– Drain coolant, see **Repair Group 19, Cooling system, draining and filling**.

– Separate fuel supply and fuel return lines at connection on fuel rail.

NOTE—

The fuel system is under pressure. Place a cloth over the connection before the hoses are pulled off.

– Seal lines so that fuel system is not contaminated by dirt etc.

– Disconnect front exhaust pipe or catalyst from exhaust manifold.

– Remove toothed belt guard - upper part.

– Release tensioner and remove toothed belt from camshaft sprocket.

– Remove cylinder head cover.

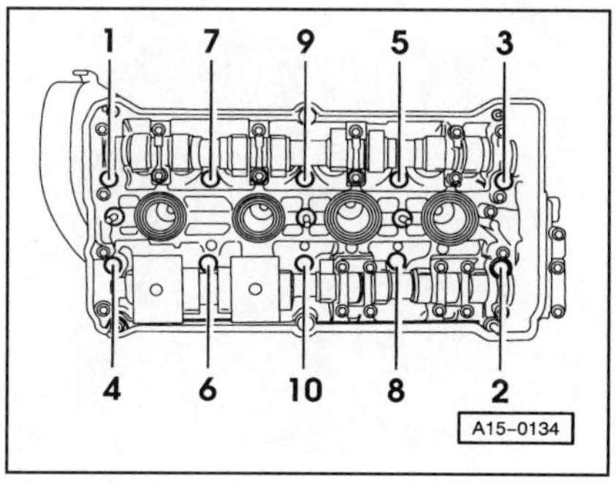

– Loosen socket head bolts in sequence given and then remove completely.

NOTE—

Use tool 3452 for Polydrive cylinder head bolts.

– Carefully lift off cylinder head.

Installing

NOTE—

- *There must not be any oil or coolant in the cylinder head bolt pockets in the cylinder head.*
- *Only remove the new cylinder head gasket from its packing immediately before installing.*
- *Handle the new gasket with extreme care. Damage will lead to leaks.*

– Place a clean cloth in cylinders so that no dirt or emery cloth particles can get between cylinder wall and piston.

NOTE—
Ensure that dirt and emery cloth particles do not get into the coolant.

– Carefully clean cylinder head and cylinder block sealing surfaces to ensure that no scores or scratches are formed (when using abrasive paper the grade must not be less than 100).

– Carefully remove metal particles, emery remains and cloths.

– Set No. 1 cylinder piston to top dead center and then turn crankshaft back slightly.

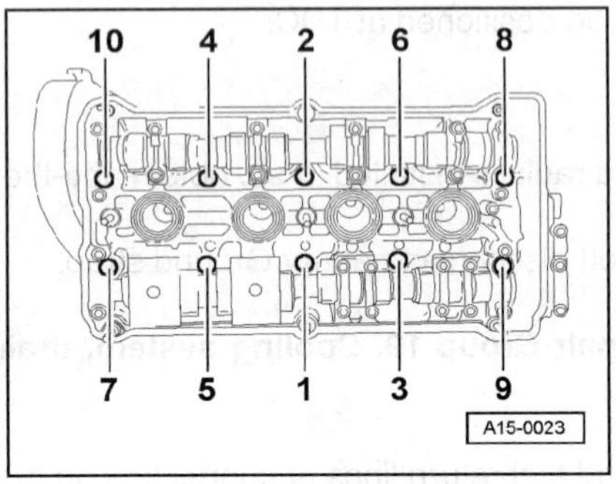

◄ To centralize, thread guide pins from 3070 into holes for cylinder head bolts 8 and 10.

– Install new cylinder head gasket. Lettering (Part No.) must be readable.

– Install cylinder head, thread in 8 remaining cylinder head bolts and tighten by hand.

– Remove guide pins with removal tool from 3070 by turning removal tool counter-clockwise until pins are free.

– Install remaining cylinder head bolts and tighten hand-tight.

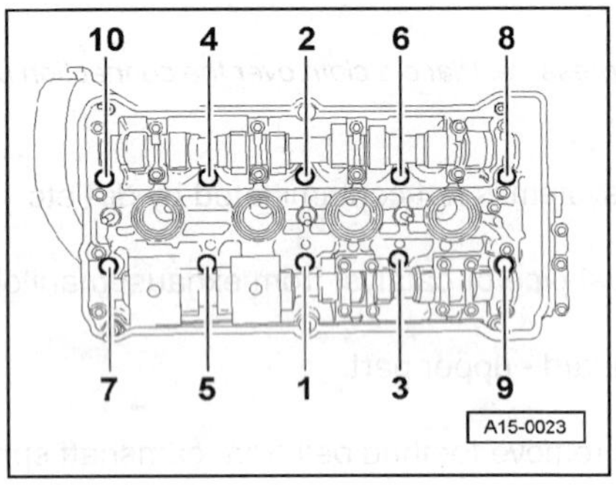

◄ Tighten cylinder head in following tightening sequence:

- Pretighten all bolts to 40 Nm.
- Then tighten all bolts to 60 Nm.
- Then tighten all bolts $^1/_4$ turn (90°) further using a rigid wrench.
- Then tighten all bolts again $^1/_4$ turn (90°) further.

NOTE—

- *The rest of the assembly is basically a reverse of the dismantling sequence.*
- *Installing toothed belt, see* **Repair Group 13, Toothed belt, removing and installing.**
- *Filling with new coolant, see* **Repair Group 19, Cooling system, draining and filling.**

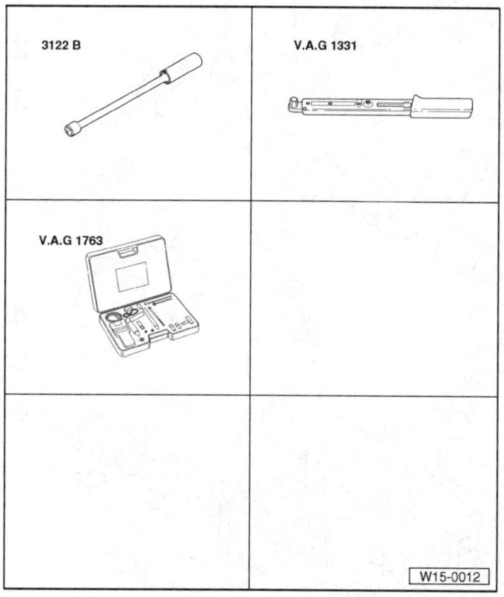

Compression, checking

Special tools and equipment

- 3122 B Spark plug wrench
- VAG 1331 Torque wrench (5 to 50 Nm)
- VAG 1763 Compression tester

Conditions

- Engine oil temperature must be at least 30° C (86° F).

Work sequence

– Disconnect ignition coil connector(s).

– Remove ignition coils.

– Remove spark plugs with spark plug 3122B.

– Have a second technician fully depress accelerator pedal.

– Check compression using compression tester VAG 1381/VAG 1763.

NOTE —
see compression tester operating instructions.

– Operate starter until tester shows no further pressure increase.

Compression pressure

Engine code	new	Wear limit
AEB, ATW	10 to 13 bar (145-190 psi)	7.0 bar (100 psi)

Permissible difference between all cylinders: 3 bar (45 psi)

– Install ignition coils and tighten to 10 Nm (7 ft-lb).

Valve train, components

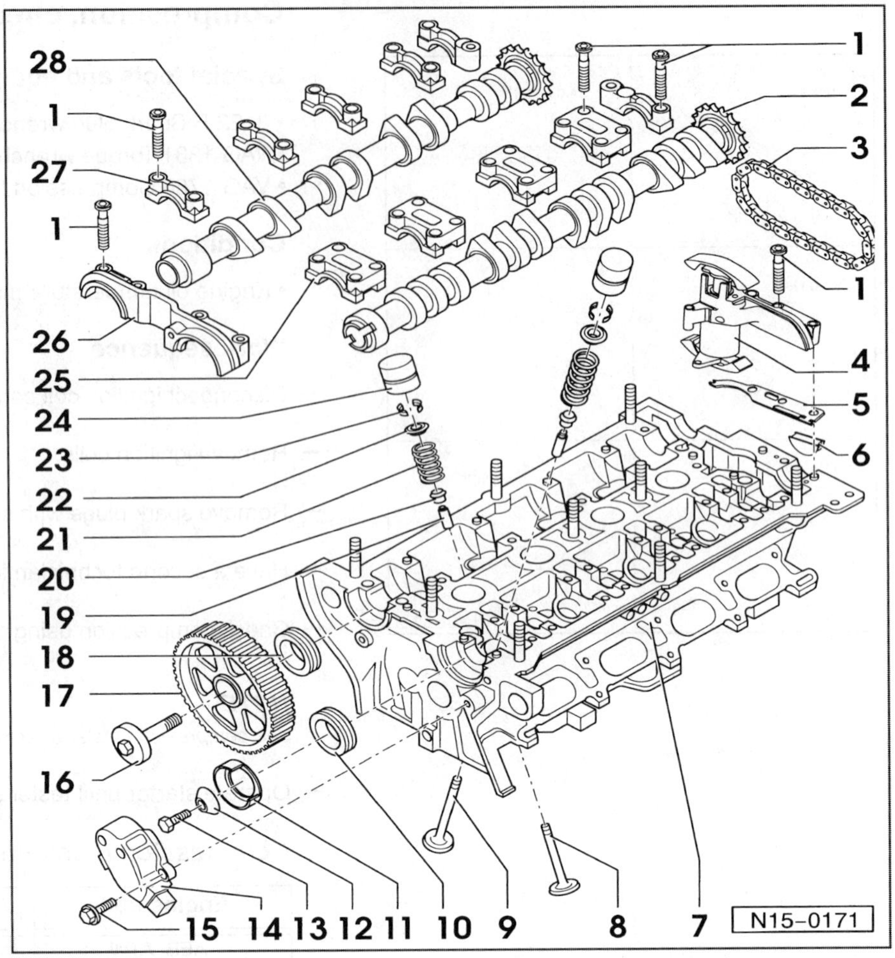

1. **10 Nm (7 ft-lb)**
2. **Intake camshaft**
   - Checking axial clearance, see **Fig. 2**
   - See **Camshafts, removing and installing**
   - Checking radial clearance with plastigage, wear limit: 0.1 mm
   - Run-out: max. 0.01 mm
3. **Drive chain**
   - Before removing mark direction of rotation (installation position), see **Camshafts, removing and installing**
4. **Chain tensioner**
5. **Seal**
   - Rubber/metal seal
   - Always replace
6. **Seal**
7. **Cylinder head**
   - Reworking sealing surface, see **Fig. 1**
   - Sealing transition points, see **Fig. 3** and **Fig. 4**
8. **Exhaust valves**
   - Do not rework; only lapping-in is permitted
9. **Intake valves**
   - Do not rework; only lapping-in is permitted
10. **Oil seal**
    - Lightly oil sealing lip of oil seal
    - See **Camshaft position sensor seal, removing**
11. **Shutter wheel (hood)**
    - For Camshaft Position sensor
    - When installing note component arrangement
12. **Washer**
    - Conical
13. **25 Nm (18 ft-lb)**
14. **Camshaft Position sensor -G40-**
    - Checking, see **1.8 Liter Fuel Injection & Ignition, Repair Group 28**
15. **10 Nm (7 ft-lb)**
16. **65 Nm (48 ft-lb)**
    - Use counter-hold tool 3036 to loosen and tighten
17. **Camshaft sprocket**
    - Note installation position: the small web of the camshaft sprocket points outward and the No. 1 Cyl. TDC marking is visible
18. **Oil seal**
    - See **Camshaft oil seals, replacing**
19. **Valve guide**
    - See **Valve guides, checking**
    - Service version with collar
20. **Valve stem seal**
21. **Valve spring**
22. **Upper valve spring plate**
23. **Valve keepers**
24. **Hydraulic lifter**
    - Do not interchange
    - With hydraulic valve clearance compensation
    - See **Hydraulic lifters, checking**
    - Store with cam contact surface downward
    - Before installing check camshaft axial clearance, see **Fig. 2**
25. **Intake camshaft bearing cap**
    - Installation position and sequence, see **Camshafts, removing and installing**
26. **Double bearing cap**
    - Lightly coat bearing cap contact surface with sealant AMV 174 004 01
27. **Exhaust camshaft bearing cap**
    - Installation position and sequence, see **Camshafts, removing and installing**
    - Lightly coat last (outer) bearing cap contact surface with sealant AMV 174 004 01
28. **Exhaust camshaft**
    - Checking axial clearance, see **Fig. 2**
    - See **Camshafts, removing and installing**
    - Checking radial clearance with plastigage, Wear limit: 0.1 mm
    - Run-out: max. 0.01 mm

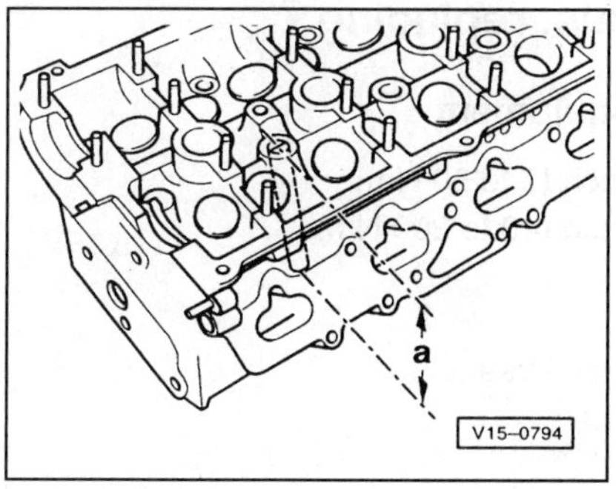

Fig. 1 Reworking cylinder head sealing surface

The dimension for reworking is measured through the holes for the cylinder head bolts

a = minimum 139.2 mm

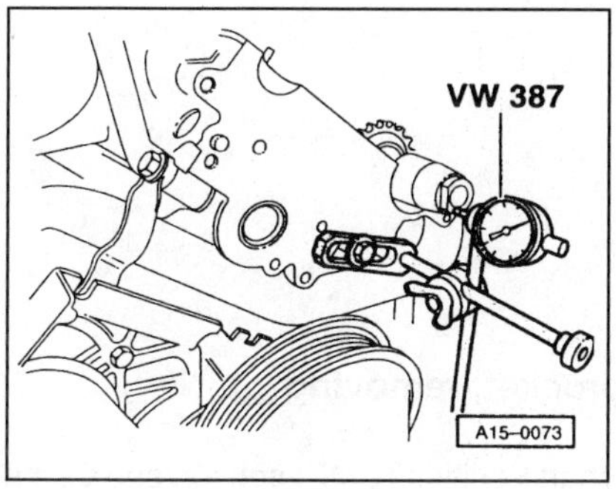

Fig. 2 Checking camshaft axial clearance

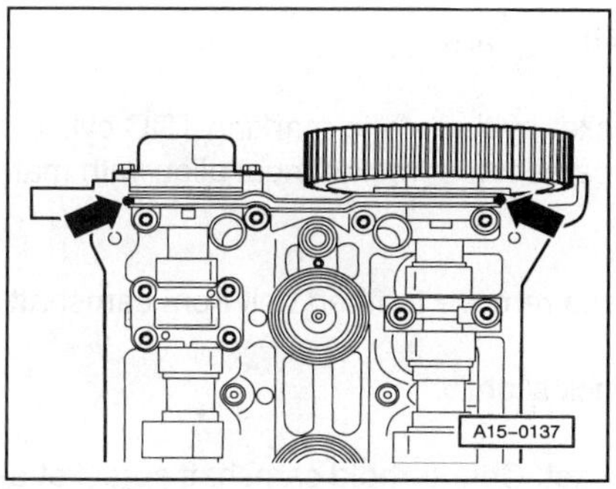

Fig. 3 Sealing transition points between double bearing cap and cylinder head

– Carefully apply a thin coat of sealant D 454 300 A2 to both edges of sealing surfaces between double bearing cap and cylinder head (arrows) using a small screwdriver.

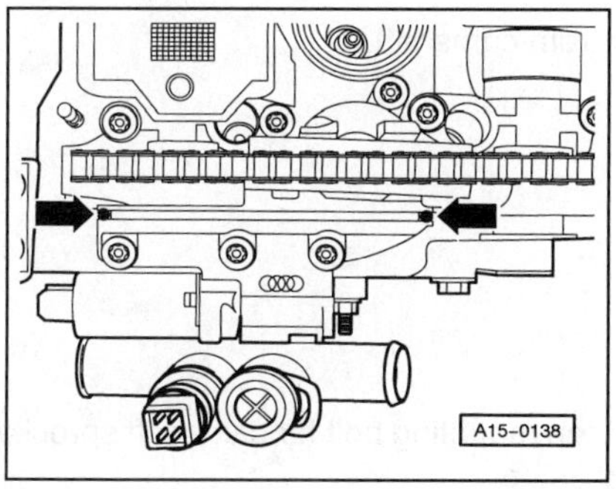

Fig. 4 Sealing transition points between camshaft adjuster and cylinder head

– Carefully apply a thin coat of sealant D 454 300 A2 to half-moon shaped surface and both edges of sealing surfaces between camshaft adjuster and cylinder head (arrows) using a small screwdriver.

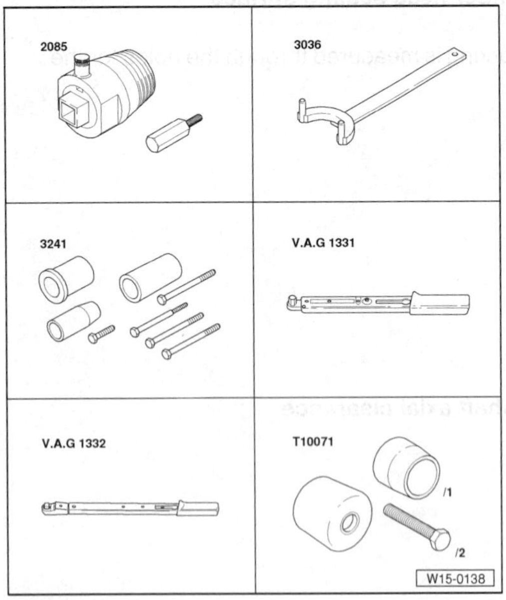

Camshaft oil seals, replacing

Special tools and equipment

- VAG1331 torque wrench (5 to 50 Nm)
- VAG1332 torque wrench (40 to 200 Nm)
- 3036 counterhold
- 3241 seal installer
- 2085 seal puller - general usage
- T10071 installation tool

Seal at camshaft sprocket, removing

Removing seal at Camshaft Position (CMP) sensor, see **Camshaft position sensor seal, replacing**.

– Remove upper toothed belt guard.

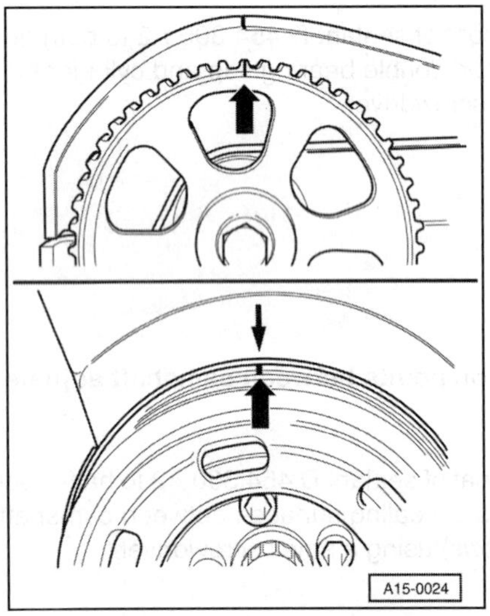

◄ Position camshaft sprocket by turning to marking TDC cyl. 1 at crankshaft. Marking at camshaft sprocket must align with marking at cylinder head cover.

– Loosen belt tensioner and remove toothed belt from camshaft sprocket.

– Then turn crankshaft back slightly.

– Remove camshaft sprocket. Counterhold camshaft sprocket using 3036 counterhold to loosen bolt.

– Remove spring washer from camshaft.

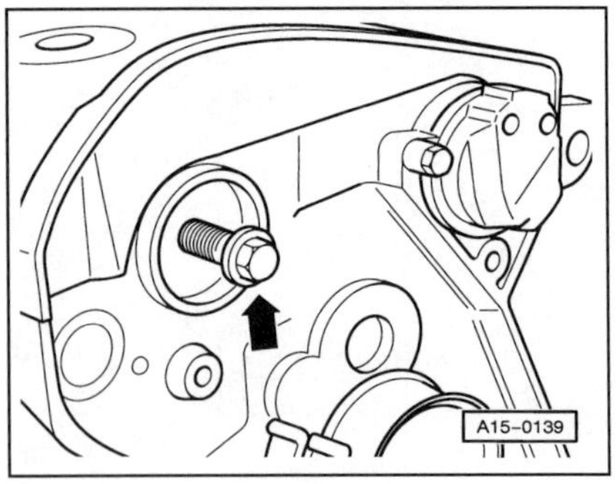

◄ To guide seal puller, tighten mounting bolt for camshaft sprocket (arrow) in camshaft by hand until stop.

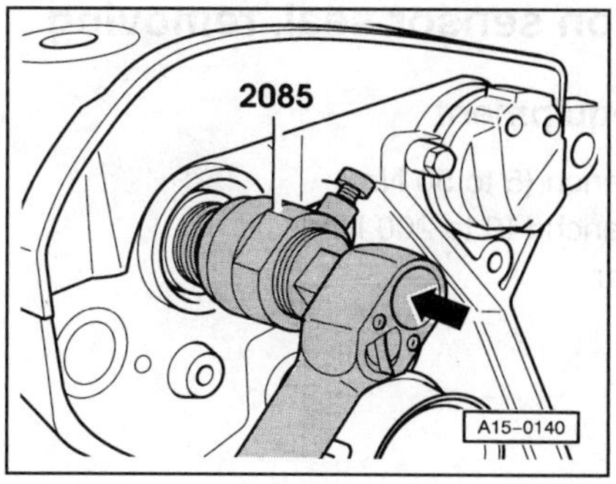

- Turn inner part of 2085 seal puller two rotations (approx. 3 mm) out of outer part and secure using knurled-head bolt.
- Lubricate threaded head of seal puller, position and thread into seal as far as possible under significant pressure.
- Loosen knurled-head bolt and turn inner part against camshaft until seal has been pulled out.

Installing

NOTE —

Implementation of PTFE seals (identifying characteristics: without spring, sealing lip extends further). The sealing lip of these seals must not be oiled or lubricated.

- Remove residual oil at camshaft pins using a clean rag.

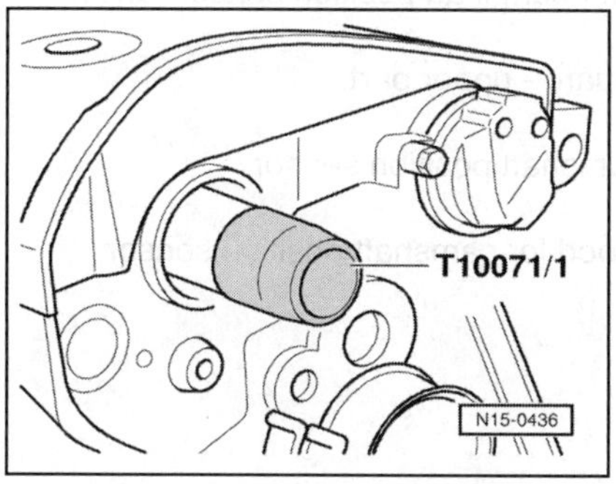

- Position guide sleeve T10071/1 on camshaft pins.
- Slide seal over guide sleeve.

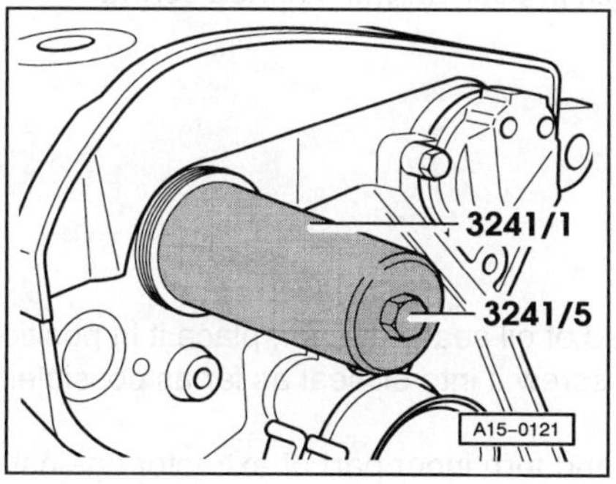

- Press in seal up to stop using 3241/1 seal installer. Use 3241/5 hex bolt for this.
- The remainder of the installation is in the reverse order of removal.
  - Install toothed belt, see **Repair Group 13, Toothed belt, removing and installing**.

Camshaft position sensor seal, removing

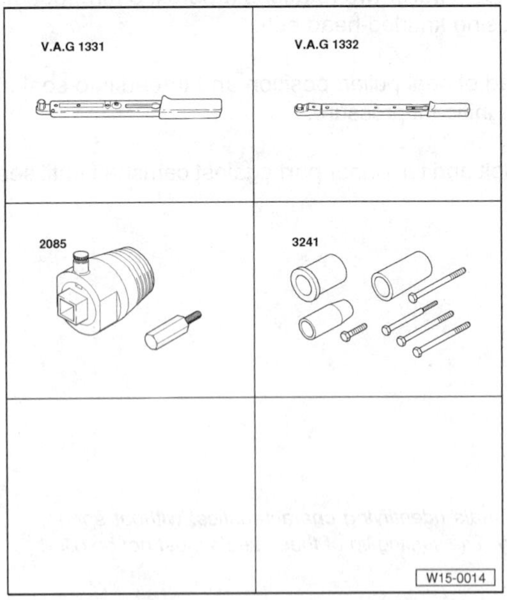

Special tools and equipment

- VAG 1331 Torque wrench (5 to 50 Nm)
- VAG 1332 Torque wrench (40 to 200 Nm)
- 2085 Oil seal extractor
- 3241 Installation tool

Removing

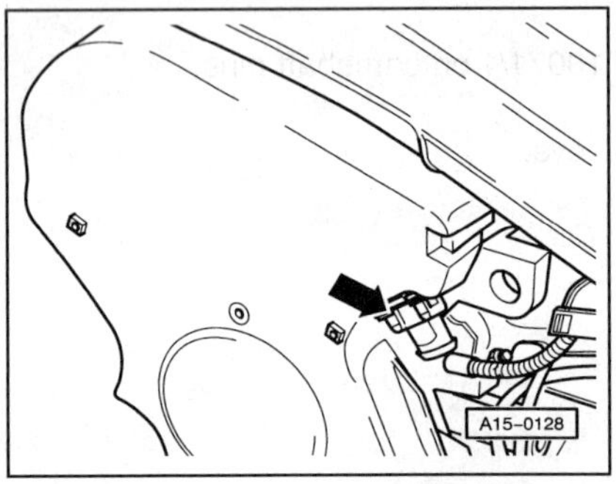

◄ Disconnect connector for camshaft position sensor (arrow).

– Remove toothed belt guard - upper part.

– Remove housing for camshaft position sensor.

– Remove washer and hood for camshaft position sensor.

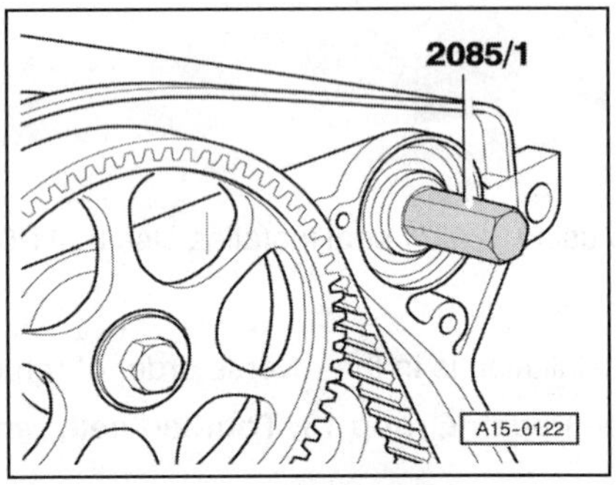

◄ Screw bolt 2085/1 into camshaft onto stop by hand, to guide seal extractor.

– Unscrew inner part of oil seal extractor 2085 two turns (approx. 3 mm) out of outer part and lock in position with knurled screw.

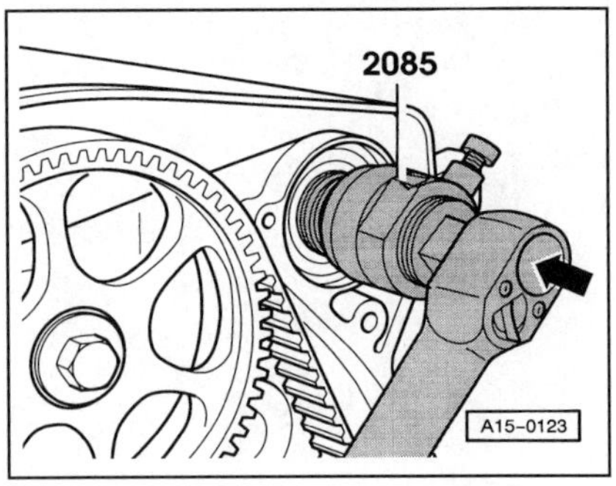

◄ Lubricate threaded head of oil seal extractor, place it in position and exerting firm pressure, screw it into oil seal as far as possible.

– Loosen knurled screw and turn inner part of extractor against camshaft until oil seal has been extracted.

Installing

EDITOR'S NOTE—

When installing Teflon (PTFE) seals (without spring and with extented sealing lips), do not lubricate or oil during installation. If seals are not Teflon (PTFE), lightly oil sealing lip prior to installation.

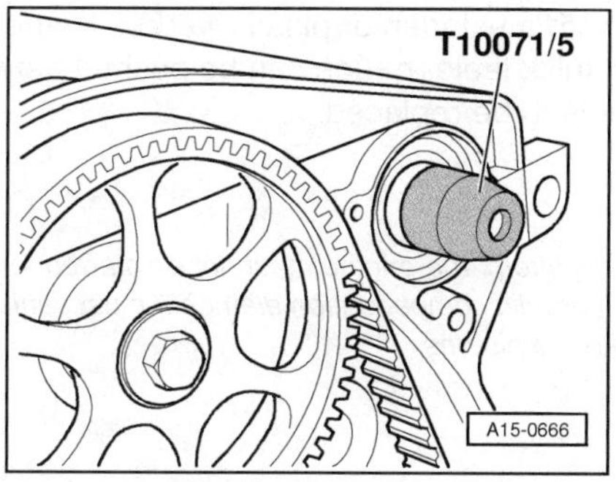

- Install guide sleeveT10071/5 over end of camshaft.

EDITOR'S NOTE—

Special tool 3241/2 can also be used if available.

- Slide oil seal over guide sleeve.
- Remove guide sleeve.

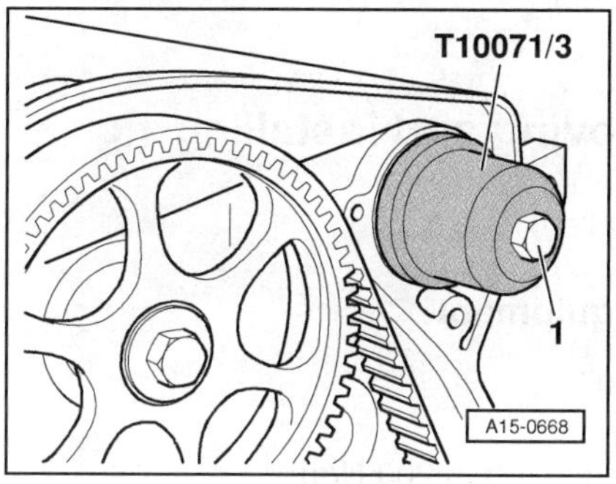

- Press seal to bottom of recess with press sleeve T10071/3 and an M8x60 bolt.

EDITOR'S NOTE—

Special tool 3241/1 press sleeve and 3241/3 bolt can also be used if available.

- Install camshaft position sensor shutter wheel (hood) onto notch in intake camshaft.
- Install washer (taper facing outward) and tighten securing bolt.
- Install housing for camshaft position sensor.
- Install upper toothed belt guard.
- Reconnect harness connector to camshaft position sensor.

Tightening torques

Component	Tightening torques
Securing bolt	25 Nm (18 ft-lb)
Camshaft position sensor housing to head	10 Nm (7 ft-lb)

Hydraulic lifters, checking

Special tools and equipment

- Feeler gauge
- Wood or plastic wedge

NOTE—

- *Replace faulty lifters complete (cannot be adjusted or repaired).*
- *Irregular valve noises when starting engine are normal.*

- Start engine and run until radiator fan has switched on once.
- Increase engine speed to about 2500 rpm for approx. 2 minutes.
  - If the hydraulic lifters are still noisy, locate faulty lifters as follows:
- Remove cylinder head cover.
- Rotate crankshaft clockwise, until cam lobe of lifter to be checked is pointing upward.

– Determine play between cam and lifter contact surface.

- If play is in excess of 0.2 mm, replace lifter.
- If play is less than 0.1 mm or no play, proceed with check as follows:

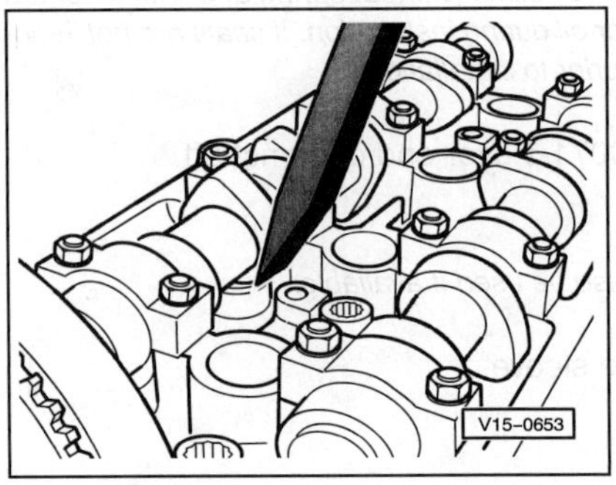

< Push lifter down lightly with a wooden or plastic wedge. If when lifter is pushed down a 0.2 mm thick feeler gauge can be pushed between camshaft and lifter, lifter is to be replaced.

NOTE —
After installing new hydraulic lifters, the engine must not be started for approx. 30 minutes. The hydraulic compensation elements must settle, otherwise valves may strike the pistons.

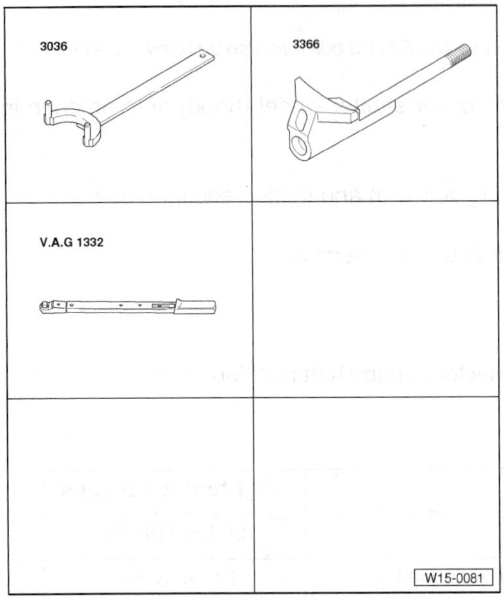

Camshafts, removing and installing

(with cylinder head installed)

Special tools and equipment

- 3036 Counter-hold tool
- 3366 Retainer
- VAG 1332 Torque wrench (40 to 200 Nm)
- D 454 300 A2 Sealant

Removing

– Remove toothed belt guard - upper part.

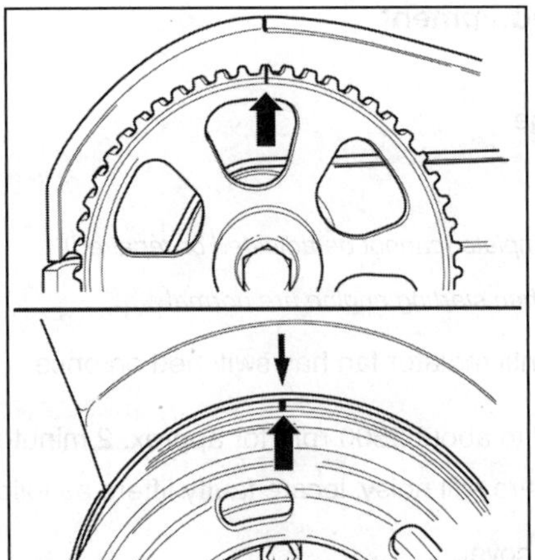

< Set camshaft sprocket to TDC No. 1 cylinder by turning crankshaft.

- Mark on camshaft sprocket must align with mark on cylinder head cover.

– Remove cylinder head cover.

– Release tensioner and remove toothed belt from camshaft sprocket.

– Turn crankshaft back slightly.

– Remove camshaft sprocket. When loosening camshaft sprocket bolt counter-hold with tool 3036.

– Remove woodruff key from camshaft.

– Remove camshaft position sensor housing.

– Remove washer and hood for camshaft position sensor.

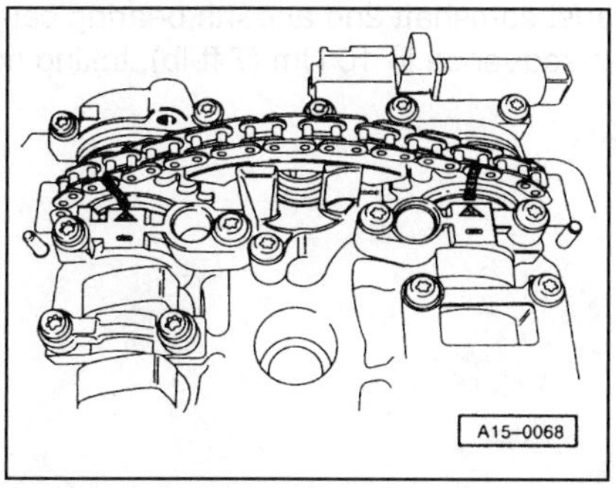

Clean drive chain and camshaft chain sprocket in area of arrows on bearing caps and mark installation position with a colored mark.

NOTE—

- *Do not mark chain with a punched mark, notch or anything that could damage the chain.*
- *The distance between both arrows or the colored marks equals 16 rollers on the drive chain.*

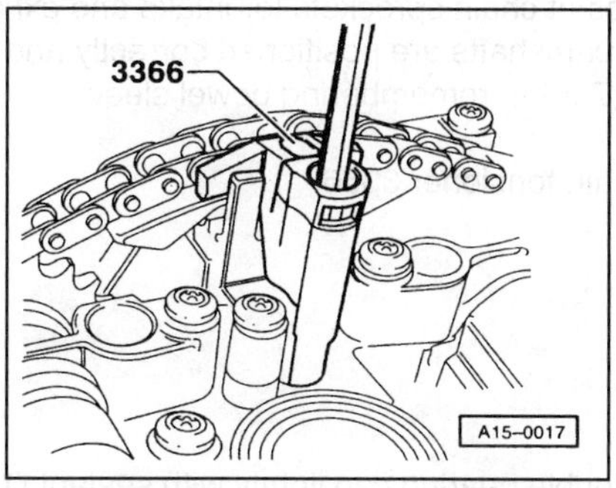

Secure camshaft adjuster or chain tensioner with retainer for chain tensioner 3366.

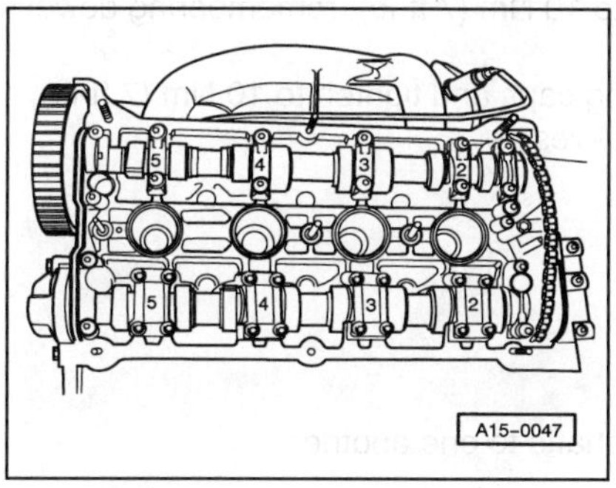

First remove intake and exhaust camshafts 3rd and 5th bearing caps.

– Remove double bearing cap.

– Remove both bearing caps at chain sprockets on intake and exhaust camshafts.

– Remove camshaft adjuster or chain tensioner securing bolts.

– Remove intake and exhaust camshaft 2nd and 4th bearing caps using alternate and cross-over sequence.

– Remove intake and exhaust camshafts with camshaft adjuster or chain tensioner and retainer for chain tensioner 3366.

Installing

NOTE—

- *When installing the camshafts the cams for No. 1 Cyl. must point upward.*
- *When installing the bearing caps, ensure that the cap markings can be read from the intake side of the cylinder head.*

– Install drive chain onto both camshafts relevant to colored markings.

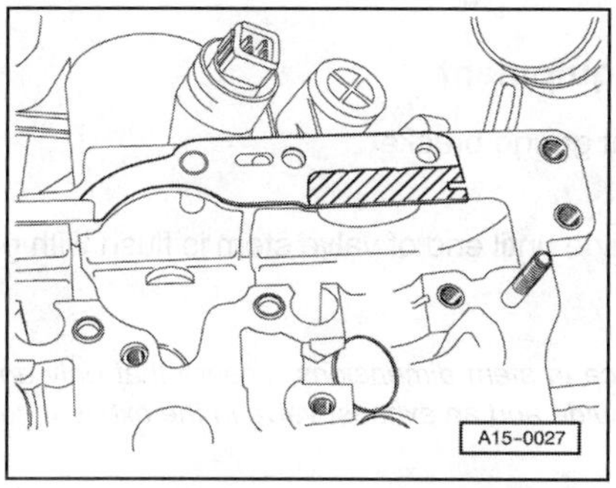

Replace rubber/metal seal for camshaft adjuster or chain tensioner and lightly coat hatched area with sealant D 454 300 A2.

– Slide camshaft adjuster or chain tensioner between drive chain.

– Oil running surfaces of both camshafts.

– Place camshafts with drive chain and camshaft adjuster or chain tensioner in cylinder head.

– Tighten camshaft adjuster or chain tensioner to 10 Nm (7 ft-lb) remembering dowel sleeves.

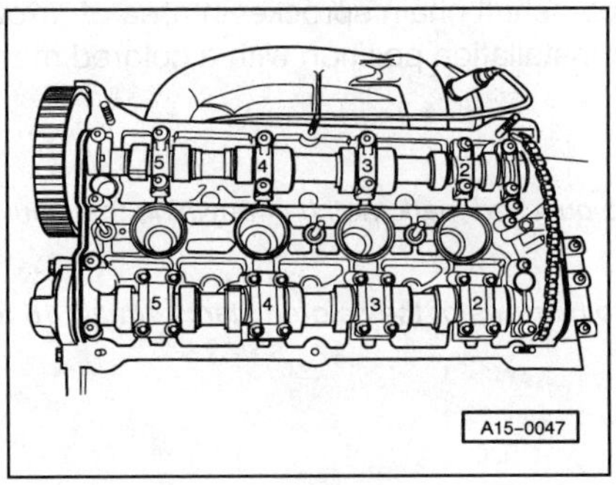

Tighten intake and exhaust camshaft 2nd and 4th bearing caps using alternate and cross-over sequence to 10 Nm (7 ft-lb), taking note of dowel sleeves.

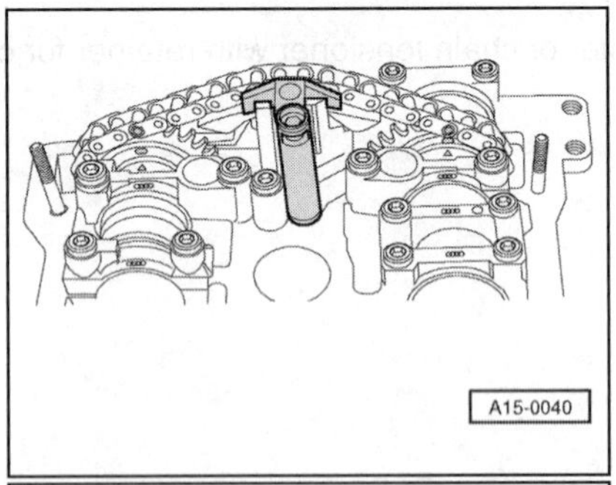

Install both bearing caps at chain sprockets for intake and exhaust camshafts. Check that camshafts are positioned correctly and tighten bearing cap to 10 Nm (7 ft-lb), remembering dowel sleeves.

– Remove retainer for chain tensioner 3366.

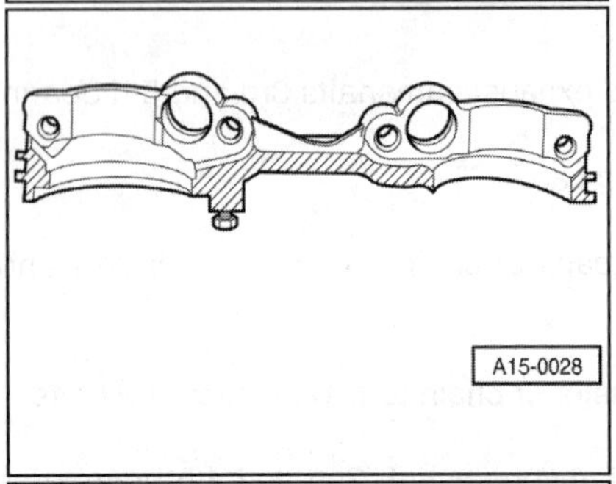

Coat hatched area of double bearing cap lightly with sealant D 454 300 A2, install and tighten to 10 Nm (7 ft-lb), remembering dowel sleeves.

– Install remaining bearing caps and tighten to 10 Nm (7 ft-lb), remembering dowel sleeves.

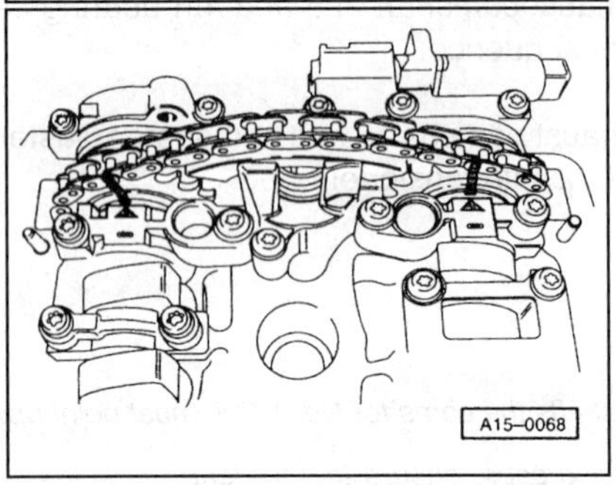

Check position of camshafts to one another.

NOTE—
- *The remaining assembly is basically a reverse of the disassembly sequence.*
- *Installing toothed belt, see* **Repair Group 13, Toothed belt, removing and installing**.
- *Wait about 30 minutes after installing the camshafts before starting the engine. The hydraulic valve compensating elements must settle, otherwise the valves may strike the pistons.*

Valve guides, checking

Special tools and equipment

- VW 387 Universal dial gauge bracket

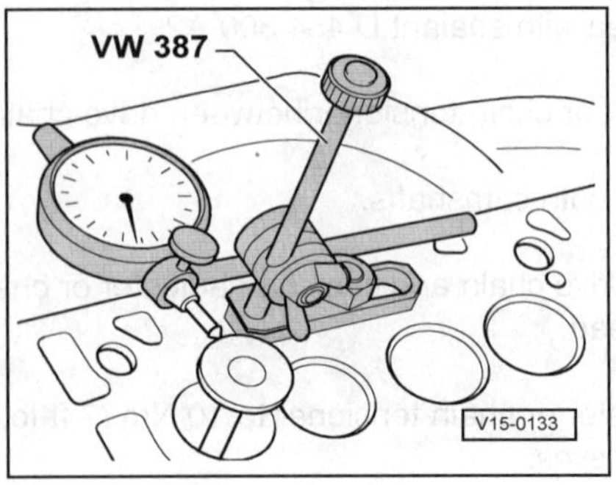

Insert new valve into guide until end of valve stem is flush with guide end.

NOTE—
Due to the slight difference in stem dimensions, ensure that only an inlet valve is used in the inlet guide and an exhaust valve in the exhaust guide.

– Determine rock.

- Wear limit: 0.8 mm

17 Engine–Lubrication System

Repair Group

B 17

Lubrication System Components

***NOTE*—**

- *If large quantities of metal particles or other deposits (caused, for example, by partial seizure of the crankshaft or conrod bearings) are found in the engine oil when performing repairs, clean the oil passages thoroughly and replace the oil cooler in order to prevent further damage from occurring later.*
- *The oil level must not be above the Max. mark - danger of damage to catalyst! See* **Maintenance**

Checking oil pressure, see **Oil pressure and pressure switch, checking.**

Oil system capacity

- With oil filter 3.7 liters (3.9 quarts)

Engine oil specifications

The following items must appear on the oil container singly or in combination with other designations: VW 502 00, VW 500 00 or VW 501 01, ACEA A2 or ACEA A3, API-SJ or API-SL.

Engine oils are graded according to their viscosity. The proper grade to be used in your engine depends on existing climatic or seasonal conditions.

The engine oil which was first put into your engine has a viscosity grade of SAE 5W-40. You can use this oil over all temperature ranges for normal driving. If engine oil viscosity grade SAE 5W-40 is not available, you can also use SAE 5W-30.

***NOTE*—**

Volkswagen identifies electrical components by a letter and/or a number in the electrical schematics. See **Wiring Diagrams, Fuses and Relays**. *These electrical identifiers are listed -between dashes- as an aid to electrical troubleshooting.*

Vehicles without oil level / temperature sensor

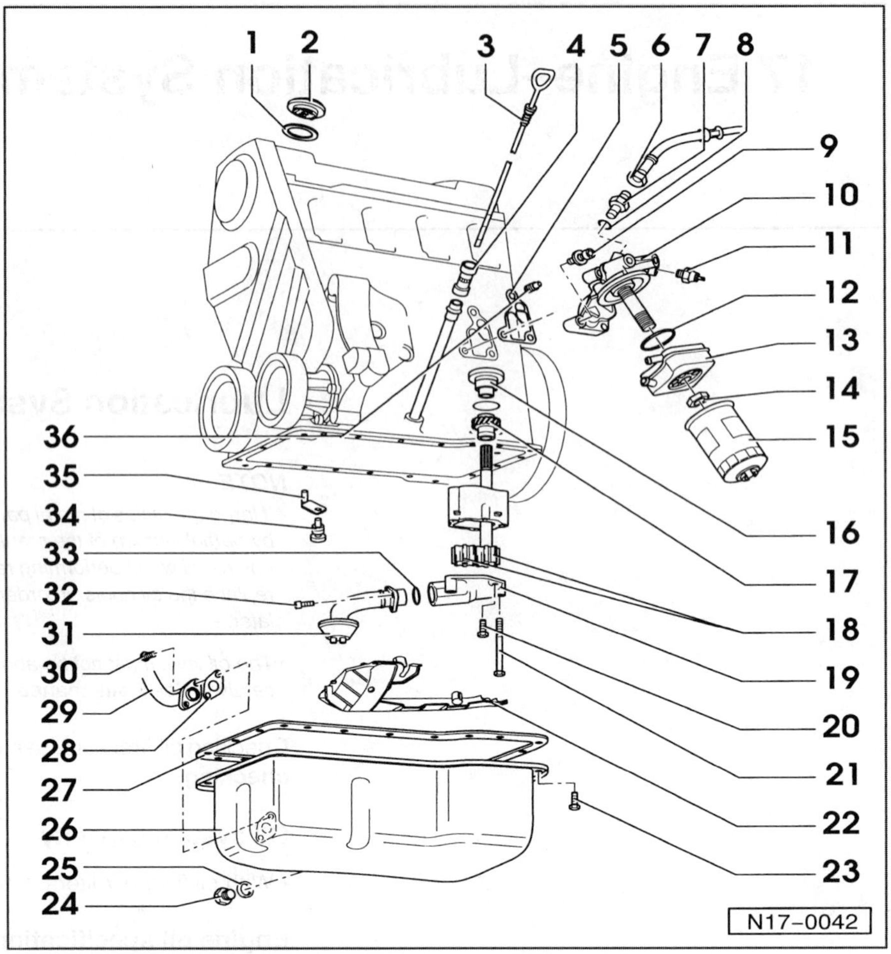

1. **Gasket**
   - Replace if damaged
2. **Cap**
3. **Dipstick**
   - Add oil so that oil level does not exceed the max. mark!
4. **Guide**
   - Pull off to extract oil
5. **Gasket**
   - Always replace
6. **Oil supply line,**
   - Tighten to 20 Nm (14 ft-lb)
   - To turbocharger
7. **30 Nm (22 ft-lb)**
8. **Seal**
   - Always replace
9. **25 Nm (18 ft-lb)**
10. **Oil filter bracket**
11. **1.4 bar oil pressure switch -F1-**
    - Tighten to 25 Nm (18 ft-lb)
    - Black
    - See **Oil pressure and pressure switch, checking**
    - If sealing ring is leaking cut open and replace.
12. **O-ring**
    - Always replace
13. **Oil cooler**
    - Coat contact area to flange, outside the seal, with AMV 188 100 02
    - Ensure clearance to adjacent components
14. **25 Nm (18 ft-lb)**
15. **Oil filter**
    - Loosen with strap wrench
    - Tighten by hand
    - Observe installation instructions on oil filter
16. **Cap**
    - With play compensation
17. **Drive gear**
    - For oil pump
18. **Gears**
    - Checking backlash, see **Fig. 1**
    - Checking axial clearance, see **Fig. 2**
19. **Oil pump cover with pressure relief valve**
    - Opening pressure: 5.7 to 6.7 bar (60-97 psi)
20. **10 Nm (7 ft-lb)**
21. **25 Nm (18 ft-lb)**
22. **Baffle plate**
23. **10 Nm (7 ft-lb)**
    - Remove and install both rear bolts (transmission end) with jointed wrench 3185
24. **Oil drain plug**
    - Tighten to 30 Nm (22 ft-lb)
    - Replace if leaking. In some cases, seal and drain plug are combined; do not interchange with separate seal and drain plug.
25. **Seal**
    - Always replace
26. **Oil pan**
    - Clean sealing surface before installing
    - Aluminium
27. **Gasket**
    - Before installing gasket, coat oil pan flange/cylinder block flange with D 454 300 A2
28. **Gasket**
    - Always replace
29. **Oil return line**
    - From turbocharger
30. **10 Nm (7 ft-lb)**
31. **Suction line**
    - Clean strainer if soiled
32. **10 Nm (7 ft-lb)**
33. **O-ring**
    - Always replace
34. **Pressure relief valve**
    - Tighten to 27 Nm (20 ft-lb)
    - Opening pressure: 2.5 to 3.2 bar (36-46 psi)
35. **Oil spray jet**
    - For piston cooling
36. **Check valve**
    - Tighten to 5 Nm (4 ft-lb)

Vehicles with oil level / temperature sensor

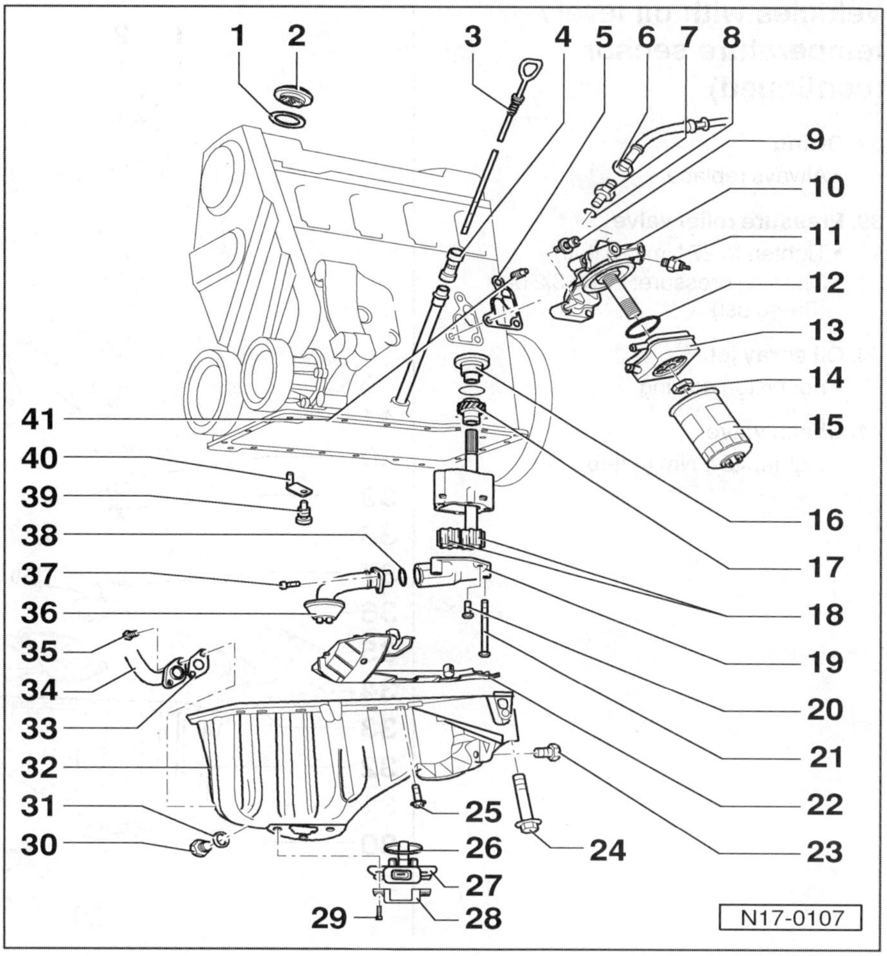

1. **Gasket**
   - Replace if damaged
2. **Cap**
3. **Dipstick**
   - Add oil so that oil level does not exceed the max. mark!
4. **Guide**
   - Pull off to extract oil
5. **Gasket**
   - Always replace
6. **Oil supply line**
   - Tighten to 20 Nm (14 ft-lb)
   - To turbocharger
7. **30 Nm (22 ft-lb)**
8. **Seal**
   - Always replace
9. **25 Nm (18 ft-lb)**
10. **Oil filter bracket**
11. **1.4 bar oil pressure switch -F1-**
    - Tighten to 25 Nm (18 ft-lb)
    - Black
    - See **Oil pressure and pressure switch, checking**
    - If sealing ring is leaking cut open and replace.
12. **O-ring**
    - Always replace
13. **Oil cooler**
    - Coat contact area to flange, outside the seal, with AMV 188 100 02
    - Ensure clearance to adjacent components
14. **25 Nm (18 ft-lb)**
15. **Oil filter**
    - Loosen with strap wrench
    - Tighten by hand
    - Observe installation instructions on oil filter
16. **Cap**
    - With play compensation
17. **Drive gear**
    - For oil pump
18. **Gears**
    - Checking backlash, see **Fig. 1**
    - Checking axial clearance, see **Fig. 2**
19. **Oil pump cover with pressure relief valve**
    - Opening pressure: 5.7 to 6.7 bar (60-97 psi)
20. **10 Nm (7 ft-lb)**
21. **25 Nm (18 ft-lb)**
22. **Baffle plate**
23. **45 Nm (33 ft-lb)**
    - Oil pan/transmission connecting bolt
24. **45 Nm (33 ft-lb)**
25. **10 Nm (7 ft-lb)**
26. **Seal**
    - Always replace
27. **Oil level / temperature sensor -G 266-**
    - Checking, see **Electrical Wiring Diagrams, Troubleshooting & Component Locations**
28. **Cover plate**
29. **10 Nm (7 ft-lb)**
30. **Oil drain plug**
    - Tighten to 30 Nm (22 ft-lb)
    - Replace if leaking. In some cases, seal and drain plug are combined; do not interchange with separate seal and drain plug.
31. **Seal**
    - Always replace
32. **Oil pan**
    - Clean sealing surface before installing
    - Install with silicone sealant D 176 404 A2
    - Aluminium
33. **Gasket**
    - Always replace
34. **Oil return line**
    - From turbocharger
35. **10 Nm (7 ft-lb)**
36. **Suction line**
    - Clean strainer if soiled
37. **10 Nm (7 ft-lb)**

(continued on next page)

Vehicles with oil level / temperature sensor (continued)

38. O-ring
- Always replace

39. Pressure relief valve
- Tighten to 27 Nm (20 ft-lb)
- Opening pressure: 2.5 to 3.2 bar (36-46 psi)

40. Oil spray jet
- For piston cooling

41. Check valve
- Tighten to 5 Nm (4 ft-lb)

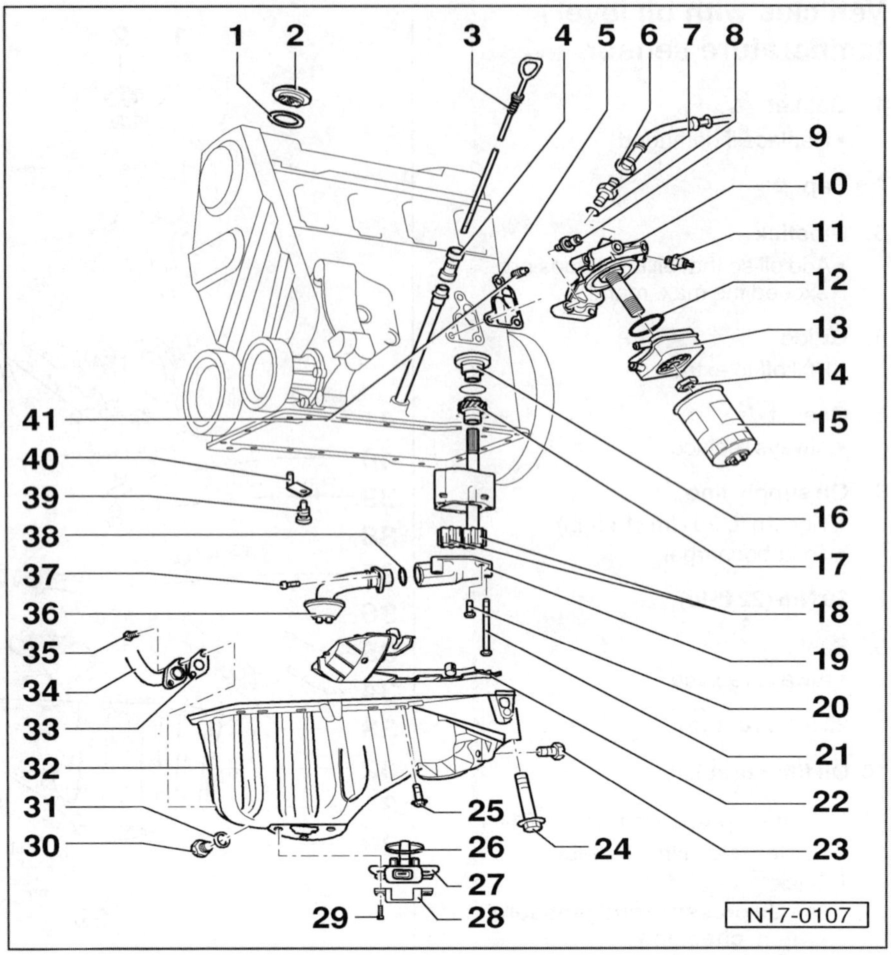

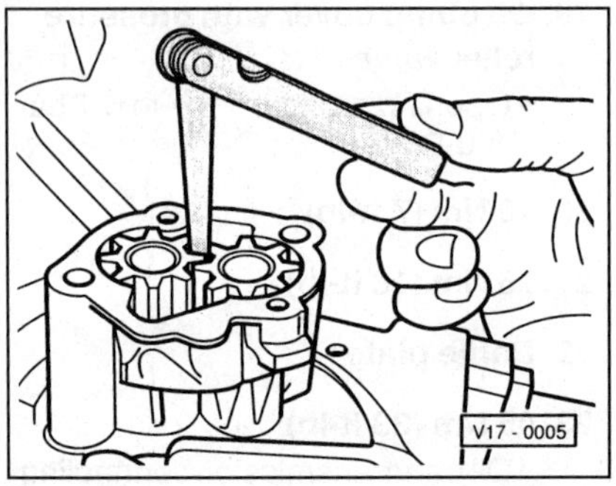

Fig. 1 Checking oil pump backlash

New: 0.05 mm

Wear limit: 0.20 mm

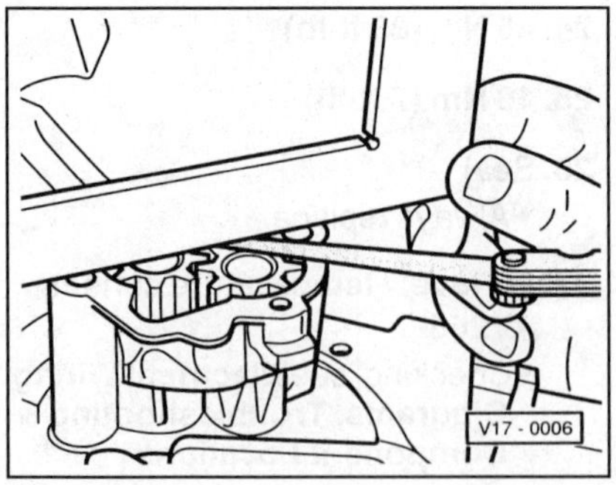

Fig. 2 Checking oil pump axial clearance

Wear limit: 0.15 mm

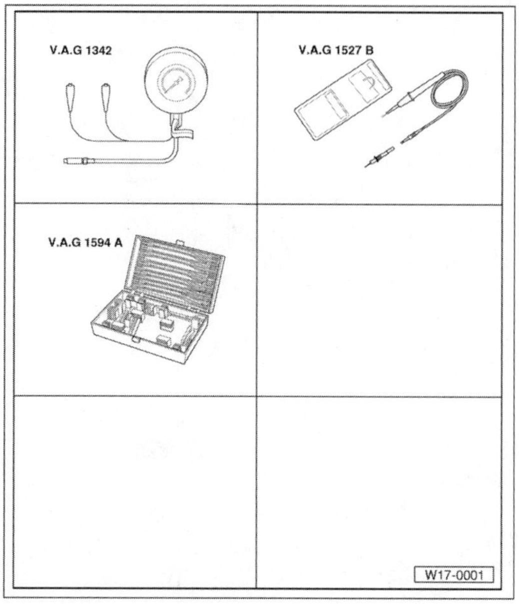

Oil pressure and oil pressure switch, checking

Special tools and equipment

• VAG 1342 Oil pressure tester
• VAG 1527 B Voltage tester
• VAG 1594 A Adapter set

NOTE —
Functional check and servicing the optical and acoustic oil pressure warning:

NOTE —
Electrical Wiring Diagrams, Troubleshooting & Component Locations

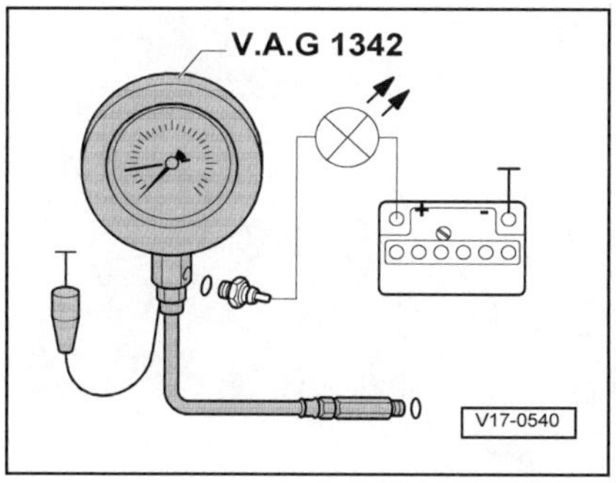

Checking sequence

◄ Remove oil pressure switch -F1- and screw into tester.

– Screw tester into oil filter bracket in place of oil pressure switch.

– Connect brown wire of tester to Ground (-).

– Connect voltage tester VAG 1527 B using auxiliary cables from VAG 1594 A to battery positive (+) and oil pressure switch.

  • LED must not light up.

– If LED lights up, replace 1.4 bar oil pressure switch -F1-.

– If the LED does not light up:

  • Start engine and run at idling speed:
  • At 1.2 to 1.6 bar (17.5-26.0 psi) LED must light up

– If the LED does not light up:

  • Replace oil pressure switch.
  • Oil pressure at idling: min. 2 bar (30 psi)

– Increase engine speed further.

  • At 2000 rpm and an oil temperature of 80°C (176°F), oil pressure should be 3.0 to 4.5 bar (44-65 psi).

– If the specifications are not obtained:

  • Correct possible mechanical damage, e.g. bearing damage.
  • Replace oil filter bracket with pressure relief valve, or oil pump.

NOTE —
At higher engine speeds the oil pressure must not exceed 7.0 bar (101 psi).

– If specified value is exceeded:

  • Check oil passage.
  • Replace oil filter bracket together with pressure release valve.

19 Engine–Cooling System

Repair Group

B 19

Cooling System Components

NOTE —

- *When the engine is warm the cooling system is under pressure. If necessary release pressure before starting repair work.*
- *Hoses are secured with spring-type clips. In cases of repair only use spring-type clips.*
- *VAS 5024 pliers are recommended when installing spring-type clips.*
- *When installing coolant hoses, route stress-free so that they do not come into contact with other components (observe markings on coolant connection and hose).*
- *Perform cooling system leakage test with cooling system tester VAG 1274 and adapters VAG 1274/8 and VAG 1274/9.*

EDITOR'S NOTE —

Volkswagen identifies electrical components by a letter and/or a number in the electrical schematics. See **Wiring Diagrams, Fuses and Relays**. *These electrical identifiers are listed -between dashes- as an aid to electrical troubleshooting.*

Cooling system components, body side

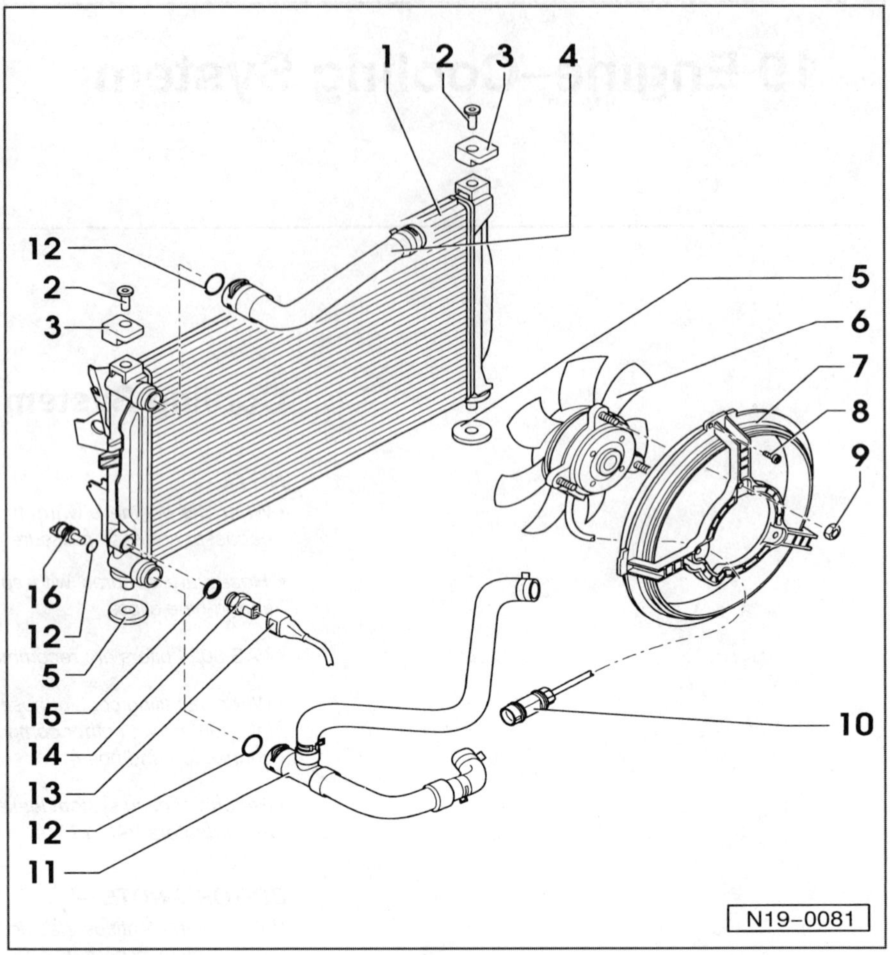

1. **Radiator**
   - See **Radiator, removing and installing**
   - After replacing radiator, replace engine coolant
   - With ATF cooler on vehicles with automatic transmission, see **Repair group 37, 5 Spd. Automatic Transmission 01V**
2. **Securing clip**
3. **Securing rubber**
4. **Upper coolant hose**
   - Secured to radiator with retaining clip
   - Ensure securely seated
   - See **Coolant hose connection diagram**
5. **Rubber washer**
6. **Radiator fan -V7-**
   - Electric fan as optional extra
7. **Fan ring**
   - Clipped into air ducting and secured with securing bolt (**8**)
8. **Locking bolt**
9. **10 Nm (7 ft-lb)**
10. **2-pin connector**
11. **Lower coolant hose**
    - Secured to radiator with retaining clip
    - Ensure securely seated
    - See **Coolant hose connection diagram**
12. **O-ring**
    - Replace if damaged
13. **Connector**
    - Black, 3-pin
    - For radiator fan thermo switch
14. **Radiator fan thermo switch -F18-**
    - Tighten to 35 Nm (26 ft-lb)
    - For electric fan
    - Switching temperatures:
      Stage 1 on:
      92 to 97°C (198 to 207°F
      Stage 1 off:
      84 to 91°C (183 to 196°F)
      Stage 2 on:
      99 to 105°C (210 to 221°F)
      Stage 2 off:
      91 to 98 °C (196 to 208°F)
15. **Seal**
    - Always replace
16. **Drain screw**
    - Tighten to 10 Nm (7 ft-lb)

Cooling system components, engine side

1. **20 Nm (15 ft-lb)**
2. **Toothed belt guard - lower part**
3. **O-ring**
   - Always replace
4. **Coolant pump**
   - Check for ease of movement
   - See **Coolant pump and compact bracket, removing and installing**
   - See **Coolant pump, disassembling and assembling**
5. **Oil cooler**
   - See **1.8 Liter 5V Turbo Engine Mechanical, Repair Group 17**
6. **Retaining clip**
   - Ensure seated securely
7. **Connector**
   - 4-pin
   - Wiring chamber 1 and wiring chamber 3 for -G62-
8. **Engine Coolant Temperature sensor -G62-**
   - With Engine Coolant Temperature gauge sensor -G2-
   - For engine control module
   - If necessary, release pressure in cooling system before removing
   - Electrical testing, see **1.8 Liter Fuel Injection & Ignition section**, **Repair group 24.**
9. **Plug**
10. **To heat exchanger**
    - See **Coolant hose connection diagram**
11. **Connection**
12. **10 Nm (7 ft-lb)**
13. **Upper coolant pipe**
    - Secured to intake manifold
    - **See Coolant hose connection diagram**
14. **From heat exchanger**
    - See **Coolant hose connection diagram**
15. **Lower coolant line**
    - Secured to oil filter bracket
    - See **Coolant hose connection diagram**

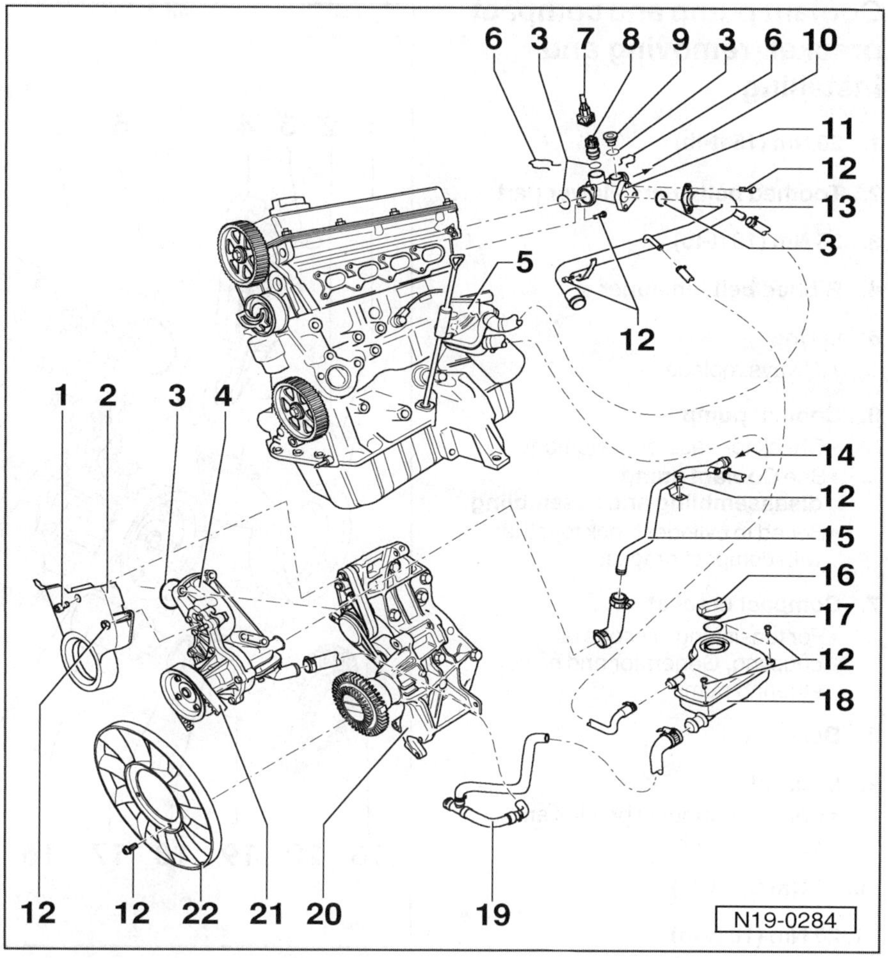

16. **Cap**
    - Check with cooling system tester VAG 1274 and adapter VAG 1274/9
    - Test pressure: 1.4 to 1.6 bar (20-23 psi)
17. **O-ring**
    - Replace if damaged
18. **Expansion tank**
    - Leak test cooling system with cooling system tester VAG 1274 and adapter VAG1274/8
19. **Lower coolant hose**
    - From bottom of radiator
    - See **Coolant hose connection diagram**
    - Secured to radiator with retaining clip
    - Make sure securely seated
20. **Compact bracket**
    - For tensioning viscous fan coupling, Generator and power steering pump
    - See **Coolant pump and compact bracket, removing and installing**
21. **V-belt**
    - Removal and installation, see **Maintenance**
    - See **Coolant pump and compact bracket, removing and installing**
22. **Radiator fan**

Coolant pump and compact bracket, removing and installing

1. **20 Nm (15 ft-lb)**
2. **Toothed belt guard - lower part**
3. **25 Nm (18 ft-lb)**
4. **Ribbed belt tensioner**
5. **O-ring**
   - Always replace
6. **Coolant pump**
   - Check for ease of movement
   - See **Coolant pump, disassembling and assembling**
   - Bolted to cylinder block together with compact bracket
7. **Compact bracket**
   - For tensioning, viscous fan coupling, Generator and power steering pump
8. **Bushing**
9. **Support**
   - Between compact bracket and cylinder block
10. **30 Nm (22 ft-lb)**
11. **25 Nm (18 ft-lb)**
12. **30 Nm (22 ft-lb)**
    - Note tightening sequence, see **Fig. 1**
13. **45 Nm (33 ft-lb)**
    - For viscous fan coupling
14. **Support**
    - Between intake manifold and compact bracket
15. **V-belt**
    - Removal and installation, see **Maintenance**
16. **Bushing**
    - For viscous fan coupling
    - See **Viscous fan coupling bushing, removing and installing**
17. **Circlip**
18. **30 Nm (22 ft-lb)**
19. **Belt pulley**
    - See **Viscous fan coupling, removing and installing**
20. **Viscous fan coupling**
    - See **Viscous fan coupling, removing and installing**

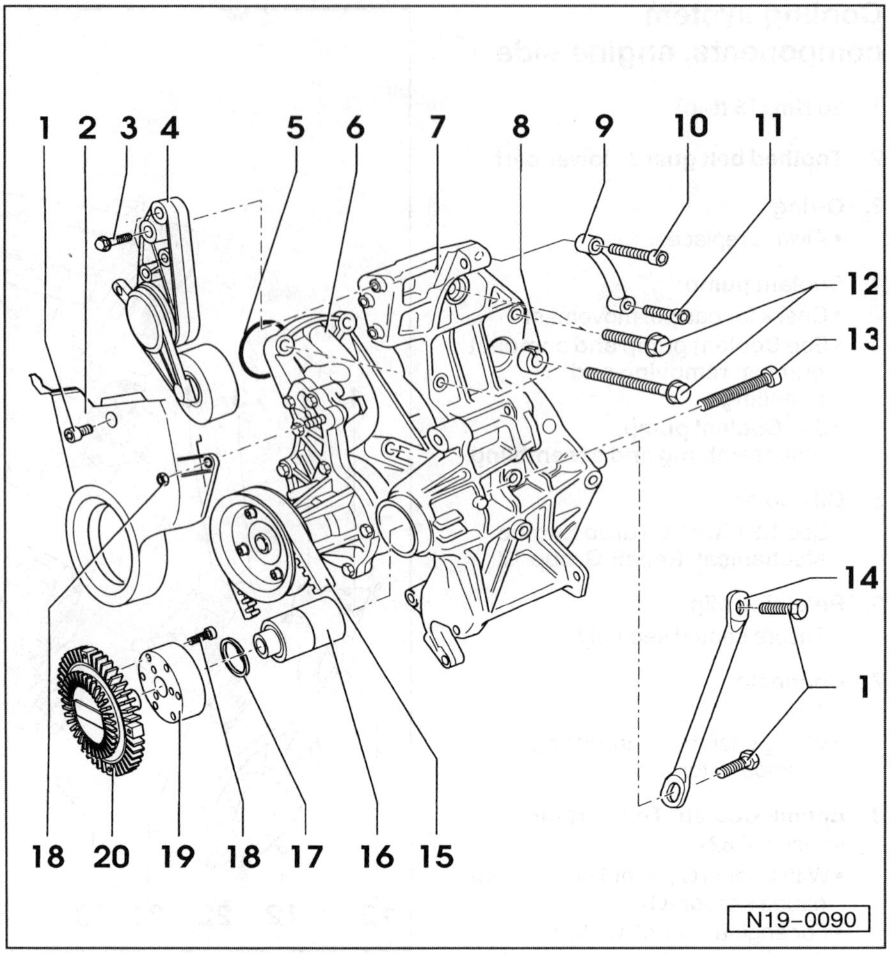

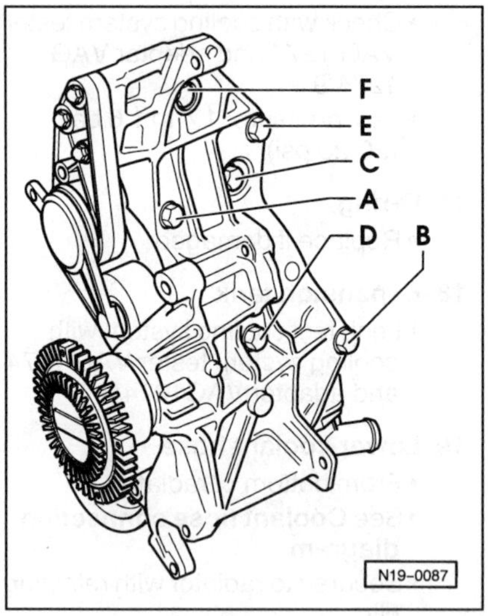

Fig. 1 Tightening sequence for compact bracket to cylinder block

Special tools and equipment

- VAG 1331 Torque wrench (5 to 50 Nm)

Work sequence

– Fit compact bracket with coolant pump to cylinder block.

– Tighten compact bracket to 30 Nm (22 ft-lb) in tightening sequence as follows:

1 - Tighten bolt -A-
2 - Tighten bolt -B-
3 - Tighten bolt -C-
4 - Tighten bolt -D-
5 - Tighten bolt -E-
6 - Tighten bolt -F-

Coolant pump, disassembling and assembling

1. **20 Nm (15 ft-lb)**
2. **Toothed belt guard - lower part**
3. **10 Nm (7 ft-lb)**
4. **Coolant pump**
   - Check for ease of movement
5. **Gasket**
   - Always replace
6. **O-ring**
   - Always replace
7. **Hammer head bolt**
   - Insert before installing
8. **Coolant thermostat**
   - Checking: heat-up thermostat in water
   - Opening: approx. 85°C (185°F
   - Ends: approx. 105°C (221°C)
   - Opening lift: min. 7 mm (0.27")
9. **Connection**
10. **Drain plug**
    - Tighten to 30 Nm (22 ft-lb)
11. **Seal**
    - Always replace
12. **Belt pulley**
    - Before removing mark installation position of both parts
13. **25 Nm (18 ft-lb)**

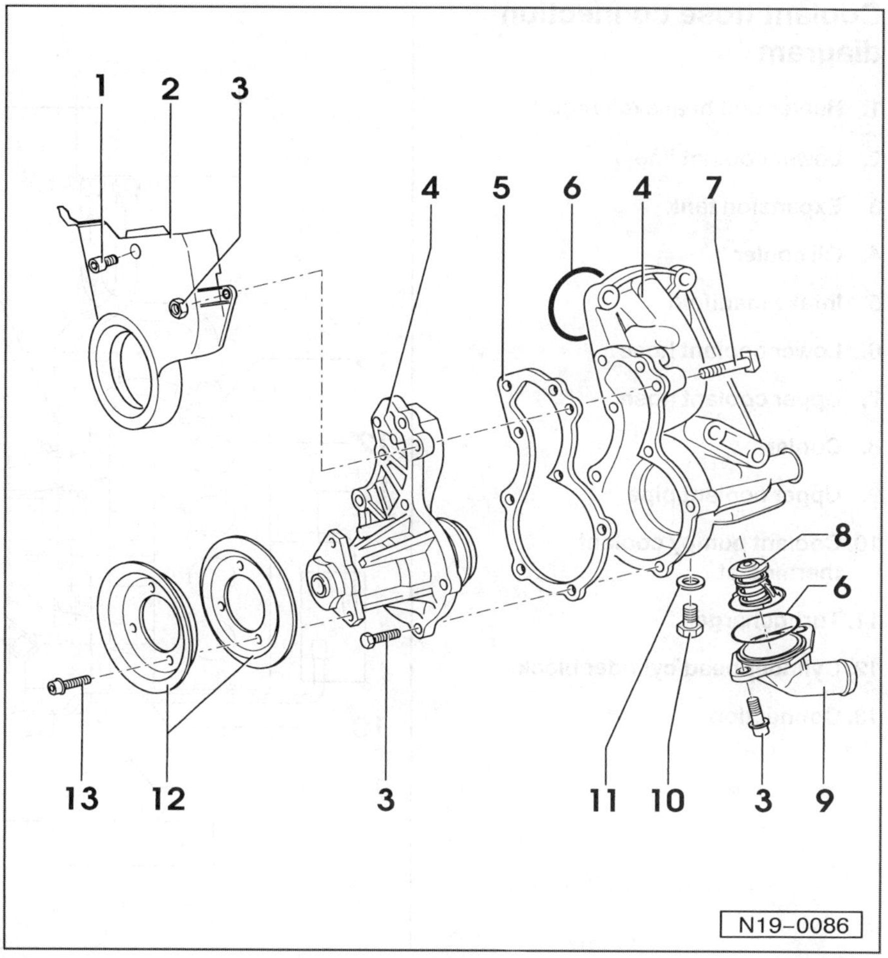

Coolant hose connection diagram

1. Heater unit heat exchanger
2. Lower coolant line
3. Expansion tank
4. Oil cooler
5. Intake manifold
6. Lower coolant hose
7. Upper coolant hose
8. Cooler
9. Upper coolant pipe
10. Coolant pump / coolant thermostat
11. Turbocharger
12. Cylinder head/cylinder block
13. Connection

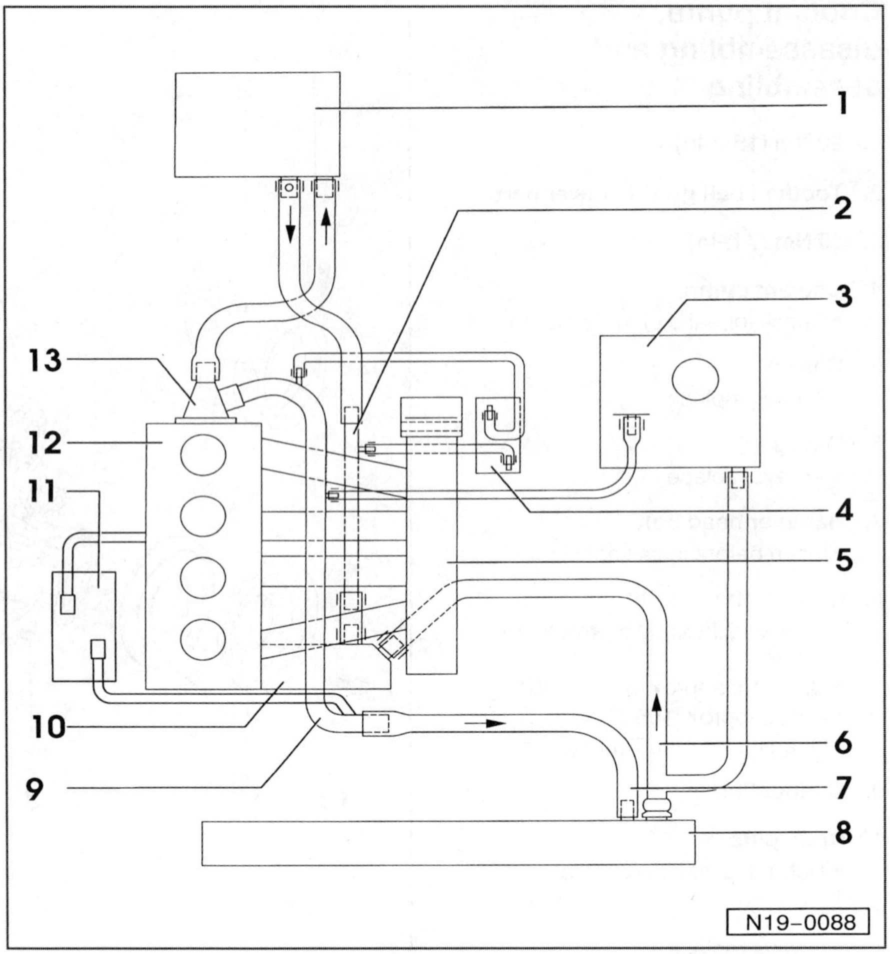

Cooling system, draining and filling

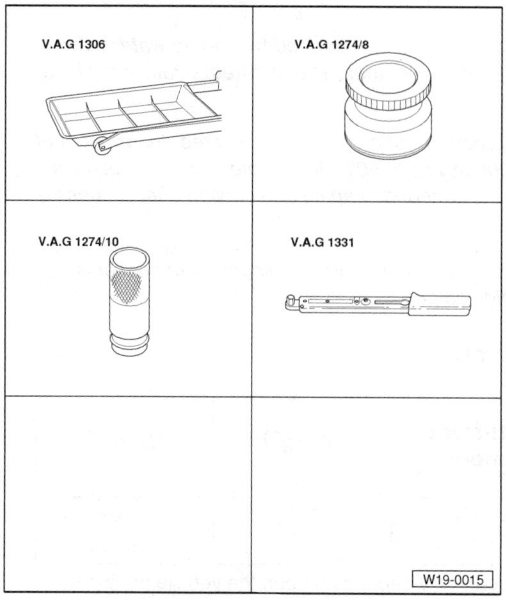

Special tools and equipment

- VAG 1306 Drip tray
- VAG1274/8 Adapter
- VAG1274/10 Adapter
- VAG 1331 Torque wrench (5 to 50 Nm)
- Refractometer

Draining

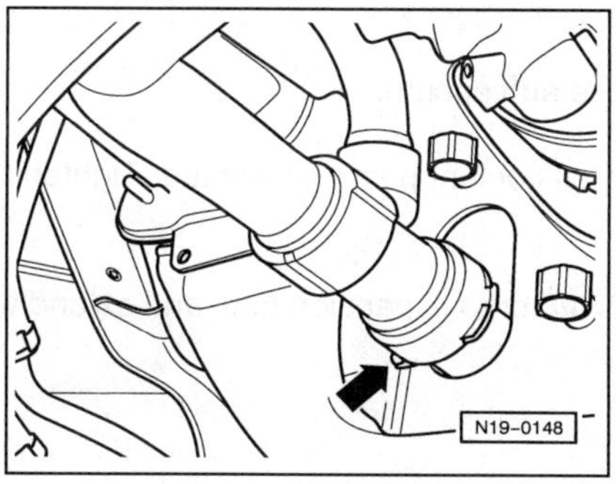

– Open cap on coolant expansion tank.

– Remove insulation tray, see **Maintenance.**

< Pull off bottom coolant hose retaining clip (arrow) and remove coolant hose from radiator.

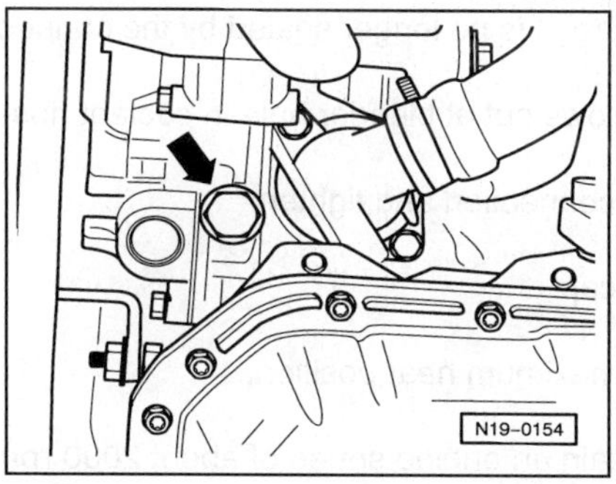

< Unscrew coolant pump drain plug (arrow).

NOTE —
Observe disposal regulations for coolant!

Filling

NOTE —
- *Only use coolant additive G 12. Identification: colored red*
- *On no account must G 12 be mixed with other coolant additives!*
- *If the fluid in expansion tank is brown, G 12 has been mixed with another coolant. In this case the coolant must be changed.*

NOTE —
- *G 12 and coolant additives marked "In accordance with TL VW 774 D" prevent frost and corrosion damage, scaling and also raise boiling point of coolant. For this reason the system must be filled all year round with frost and corrosion protection additives.*
- *Because of its high boiling point, the coolant improves engine reliability under heavy loads, particularly in countries with tropical climates.*
- *Protection against frost must be assured to about -25°C (in arctic climatic countries to about -35°C).*

NOTE —

- *The coolant concentration must not be reduced by adding water even in warmer seasons and in warmer countries. The anti-freeze ratio must be at least 40%.*
- *If for climatic reasons a greater frost protection is required, the amount of G 12 can be increased, but only up to 60% (frost protection to about -40°C) as otherwise frost protection is reduced again and cooling effectiveness is also reduced.*
- *If a radiator, heat exchanger, cylinder head or cylinder head gasket is replaced, do not reuse old coolant.*

Recommended mixture ratios:

Frost protection to	Anti-freeze amount	G 12[1)]	Water[1)]
-25°C -35°C	40% 50%	3.0 l 3.5 l	4.0 l 3.5 l

1) The quantity of coolant can vary depending upon the vehicle equipment.

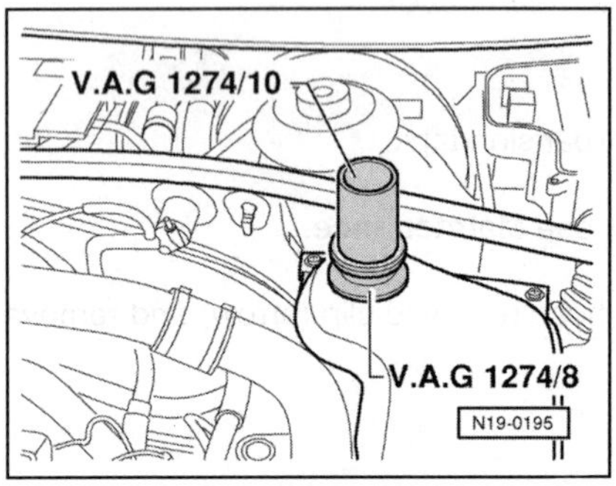

Work sequence

– Install lower coolant hose and secure.

– Replace O-ring and install coolant pump drain plug. Tightening torque: 30 Nm (22 ft-lb)

◀ Screw adapter VAG 1274/8 onto expansion tank and extend with VAG 1274/10.

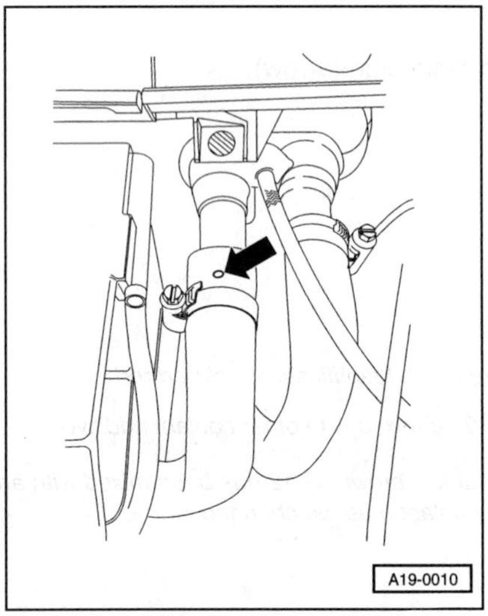

◀ Loosen coolant hose on heat exchanger and pull back hose sufficiently so that bleeder hole (arrow) is no longer sealed by the connection.

– Top up coolant until it flows out at bleeder hole in coolant hose.

– Push coolant hose on connection and tighten.

– Install expansion tank cap.

– Set heater controls to maximum heat position.

– Start engine and maintain an engine speed of about 2000 rpm for approx. 3 minutes.

– Allow engine to run at idling speed until lower hose on radiator becomes hot.

– Check coolant level and top-up if necessary.

- When engine is at normal operating temperature, coolant level must be on the Max. mark.
- When engine is cold, coolant level must be between Min. and Max. marks.

Radiator, removing and installing

Special tools and equipment

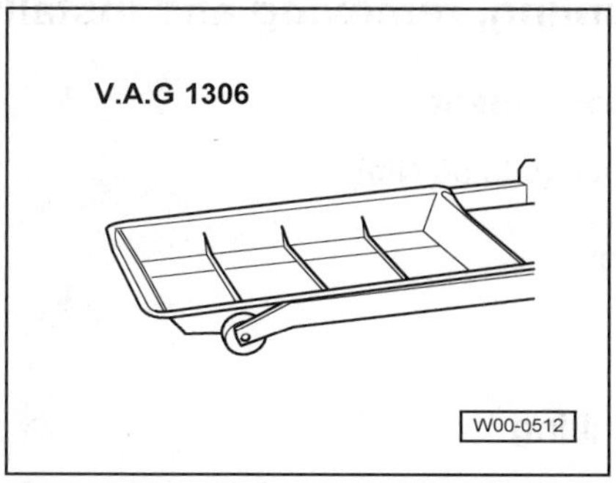

- Drip tray VAG 1306

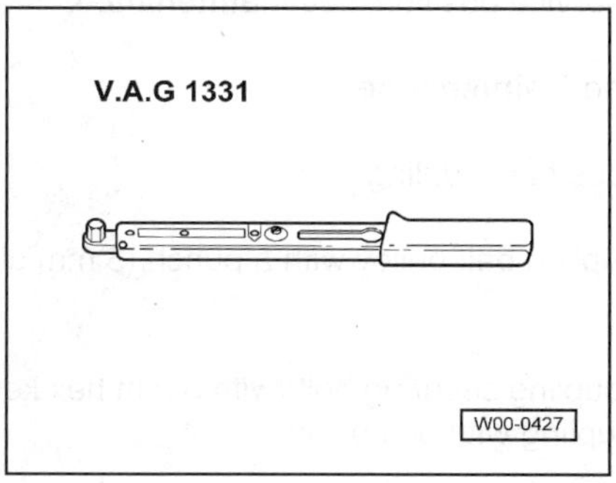

- VAG 1331 Torque wrench (5 to 50 Nm)

Removing and installing

– Drain coolant, see **Cooling system, draining and filling**.

– Remove front bumper, see **Body-Exterior section, Repair Group 63.**

– Pull coolant hoses off radiator.

– Disconnect connectors from thermo-switch.

– On vehicles equipped with automatic transmissions, remove ATF lines/hoses from radiator, see **5-Speed Automatic Transmission 01V section**, **Repair Group 37.**

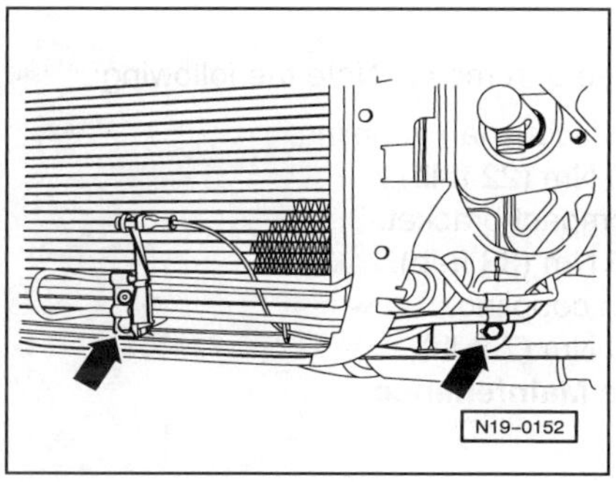

– Unbolt power steering hydraulic oil cooling line (arrows) and ensure it is free. Do not open hydraulic oil circuit.

– Remove retaining clamp(s) from air conditioning refrigerant lines.

> ***WARNING —***
> *The air conditioner refrigerant circuit must not be opened.*
> - *To prevent damage to the condenser and to the refrigerant lines/hoses, ensure that the lines and hoses are not stretched, kinked or bent.*

– Remove condenser from radiator and pull forward as far as possible.

– Secure condenser to body so that refrigerant lines/hoses are not stressed.

– Remove upper radiator securing clips and take radiator out forward. Pull radiator out between condenser and lock carrier.

– Installation is carried out in the reverse order, when doing this note the following:

- Vehicles with automatic transmission, check ATF level, see **Maintenance**.
- Fill with coolant, see **Cooling system, draining and filling**.
- For electrical connections and routing, see **Wiring Diagrams & Component Locations**
- Install front bumper, see **Body-Exterior section, Repair Group 63**
- Check headlight alignment and adjust if necessary, see **Maintenance.**

Viscous fan coupling, removing and installing

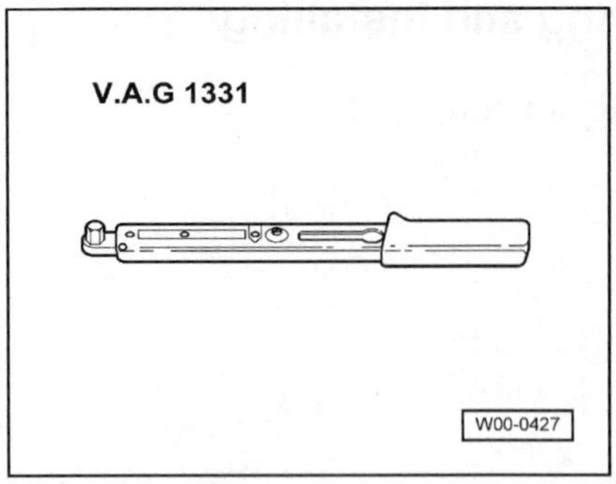

Special tools and equipment

◀ VAG 1331 Torque wrench (5 to 50 Nm)

• Punch, 5 mm diameter

Removing and installing

– Bring lock carrier into service position, see **Maintenance.**

– Remove ribbed belt, see **Maintenance**.

– Remove fan from viscous fan coupling.

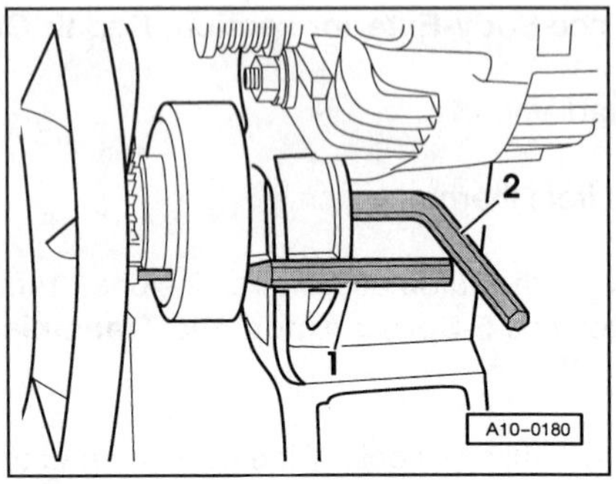

◀ Secure viscous fan coupling belt pulley with a punch (5 mm diameter) -1-.

– Remove viscous fan coupling securing bolt (with 8 mm hex key -2-) and remove viscous fan coupling with belt pulley.

– Remove belt pulley from viscous fan coupling.

Installing

– Installation is the reverse of removal. Note the following:

• Tighten fan pulley to viscous fan coupling.
Tightening torque: 30 Nm (22 ft-lb).
• Bolt viscous fan to compact bracket.
Tightening torque: 45 Nm (33 ft-lb).
• Bolt fan to viscous fan coupling.
Tightening torque: 10 Nm (7 ft-lb).
• Install ribbed belt, see **Maintenance.**

Viscous fan coupling bushing, removing and installing

Special tools and equipment

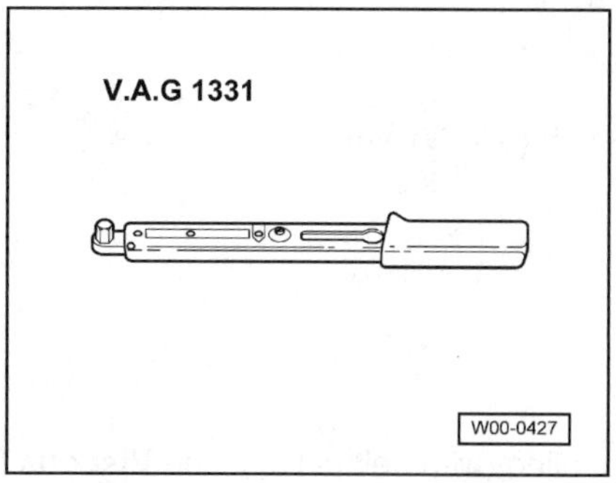

- Hex nut from assembly device 3301
- Extractor tube from assembly device 3350
- Press tool for viscous fan coupling bushing 3367

Removing

– Remove viscous fan coupling (with belt pulley), see **Viscous fan coupling, removing and installing**.

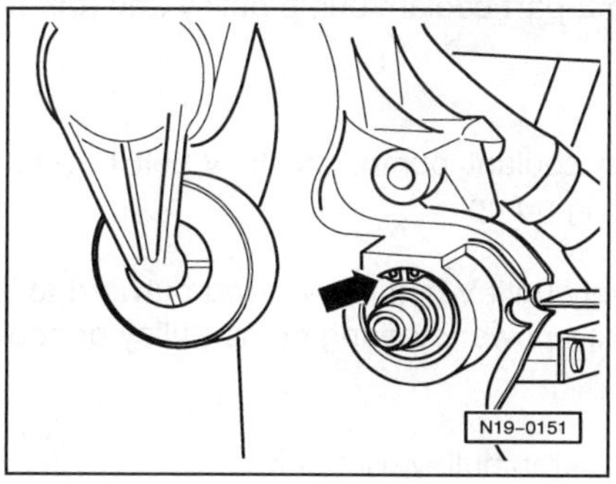

◀ Remove bushing circlip (arrow).

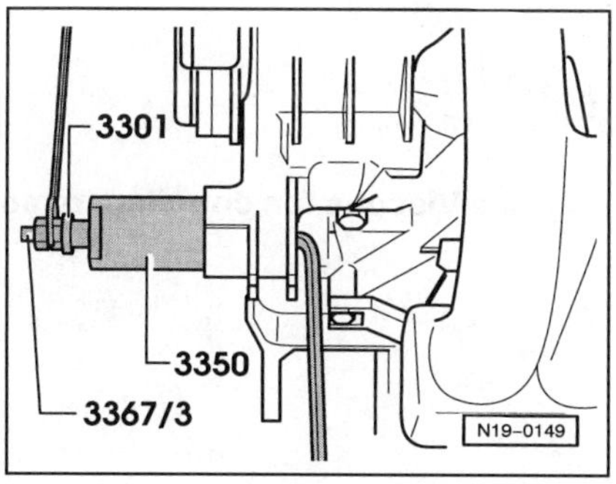

◀ Press bushing out of compact bracket with aid of socket head bolt 3367/3, extractor tube from assembly device 3350 and hex nut from assembly device 3301.

Installing

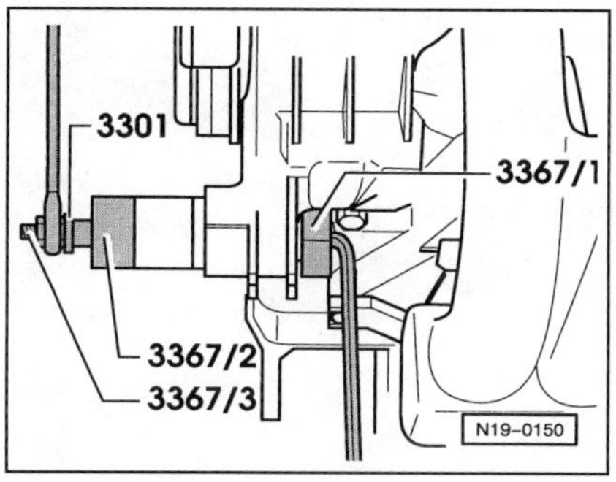

◀ Press bushing into compact bracket onto stop with aid of compensating pieces 3367/1, press piece 3367/2, hex socket head bolt 3367/3 and hex nut from assembly device 3301.

– Secure bushing with circlip after installation.

– Install viscous fan coupling, see **Viscous fan coupling, removing and installing**.

Engine coolant pump V-belt, removing and installing

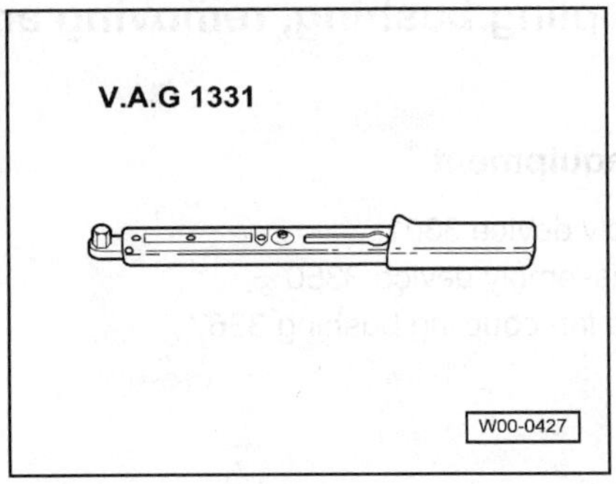

Special tools and equipment

◄ Punch

- VAG 1331 Torque wrench (5 to 50 Nm)

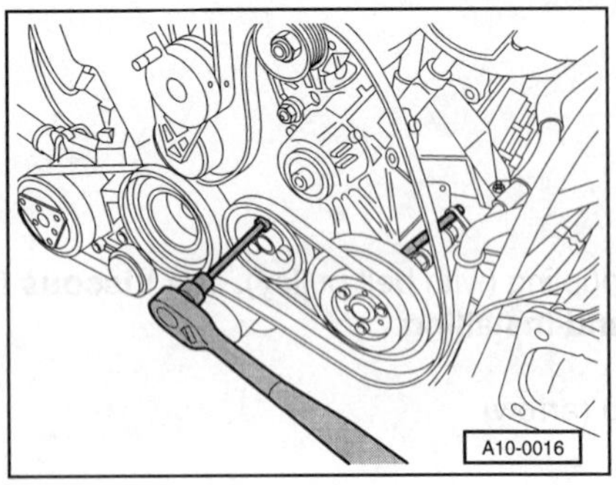

Removing

– Remove viscous fan coupling (with belt pulley), see **Viscous fan coupling, removing and installing**.

– Mark installation position of 2-part coolant pump pulley.

◄ Secure power steering pump belt pulley with a drift.

– Unscrew coolant pump 2-part coolant pump pulley and remove V-belt.

Installing

– Install 2-part pulley on coolant pump and fit V-belt between coolant pump and power steering pump.

– Tighten belt pulley evenly until V-belt has moved outward to its running surface. If necessary turn power steering pump pulley or coolant pump slightly.

– Secure power steering pump pulley with drift.

– Tighten coolant pump pulley to 25 Nm (18 ft-lb).

NOTE—
It is not necessary to adjust V-belt tension.

– Install viscous fan coupling, see **Viscous fan coupling, removing and installing**.

20 Fuel Supply

Repair Group

B 20

Fuel Supply System

Fuel supply system, safety precautions

> ***WARNING —***
> *Fuel supply lines are under pressure! Before removing from hose connection wrap a cloth around the connection. Then release pressure by carefully pulling hose off connection.*

When removing and installing the fuel gauge sender or fuel pump (fuel delivery unit) from a full or partly full fuel tank observe the following:

- Before starting work, switch on exhaust extraction system and place an extraction hose close to the sender opening in the fuel tank to extract escaping gasoline fumes. If no exhaust extraction system is available, a radial fan (as long as motor is not in air flow) with a displacement greater than 15 m^3/h may be used.
- Prevent skin contact with fuel! Wear fuel-resistant gloves!

Rules for cleanliness

When working on the fuel supply/injection system, pay careful attention to the following "5 rules":

1. Thoroughly clean all unions and the adjacent areas before disconnecting fuel lines.
2. Place parts that have been removed on a clean surface and cover. Do not use fluffy cloths!
3. Carefully cover opened components or seal, if the repair cannot be carried out immediately.
4. Only install clean components: Only unpack replacement parts immediately prior to installation. Do not use parts that have been stored loose (e.g. in tool boxes etc.).
5. When the system is open: Do not work with compressed air if this can be avoided. Do not move vehicle unless absolutely necessary.

EDITOR'S NOTE —
Volkswagen identifies electrical components by a letter and/or a number in the electrical schematics. See **Wiring Diagrams, Fuses and Relays***. These electrical identifiers are listed -between dashes- as an aid to electrical troubleshooting.*

Fuel tank/attachments/ fuel filter, components

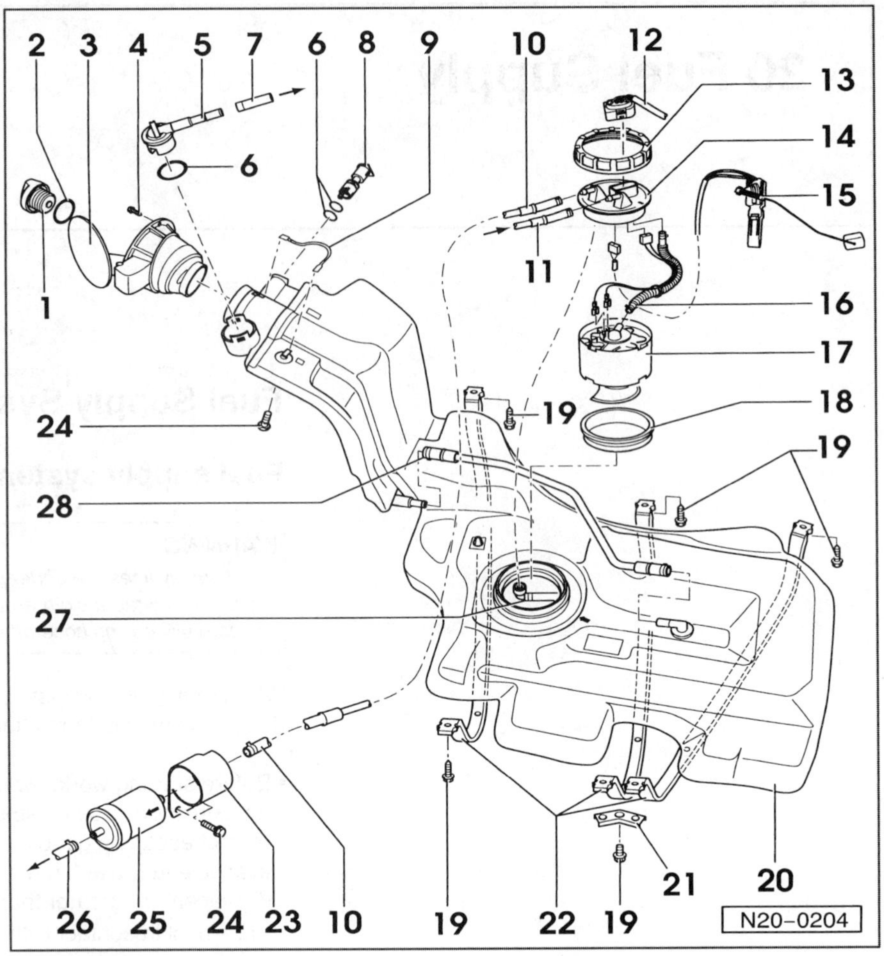

NOTE—

- *Hose connections are secured with either spring-type or clamp-type clips.*
- *Always replace clamp-type clips with spring-type clips.*
- *Fuel hoses in engine compartment must only be secured with spring-type clips. The use of clamp or screw-type clips is not permitted.*
- *VAG 1921 pliers are recommended when installing spring-type clips.*

1. **Sealing cap**
2. **Seal**
   - Replace if damaged
3. **Tank flap unit**
   - With rubber cup
4. **Securing screw**
5. **Gravity valve**
   - To remove valve, unclip upward out of support
   - Check valve for flow: vertical: open valve tilted 45°: closed
6. **O-ring**
   - Replace if damaged
7. **Vent pipe**
   - Black
   - Clipped onto side of fuel tank
   - Check securely seated
   - Secure with spring-type clips
   - To EVAP canister
8. **Vent valve**
   - To remove valve unclip sideways out of support
   - Before installing remove sealing cap **-1-**
   - Checking, see **Fig. 2**
9. **Ground connection**
   - Ensure securely seated
10. **Supply line**
    - Black
    - Clipped onto side of fuel tank
    - Ensure securely seated
    - Secure with spring-type clips
    - Between flange **-14-** and fuel filter **-25-**
11. **Return line**
    - Blue or with blue marking
    - Clipped onto side of fuel tank
    - Ensure securely seated
    - Secure with spring-type clips
    - From fuel return line on fuel rail
12. **Connector**
    - Black, 4-pin
    - For fuel gauge sender -G- and fuel pump -G6-
13. **Union nut**
    - Tighten to 60 Nm (44 ft-lb)
    - Remove and install with spanner 3217
14. **Flange**
    - Note installation position on fuel tank, see **Fig. 1**
15. **Sender for fuel gauge -G-**
    - Clipped onto reservoir housing in fuel tank
16. **Supply line**
    - Ensure securely seated
    - Between flange **-14-** and fuel delivery unit **-17-**
17. **Fuel delivery unit**
    - See **Fuel delivery unit, removing and installing**
    - See **Fuel pump, checking**
    - Clean strainer if soiled
    - Only replace complete with flange and supply line
18. **Sealing ring**
    - Replace if damaged
    - Coat with fuel when installing
19. **25 Nm (18 ft-lb)**
20. **Fuel tank**
21. **Bracket**
    - For securing straps
    - Note installation position

Continued on following page

Fuel tank/attachments/ fuel filter, components (continued)

22. Securing strap
- Note differing length
- Installation position: Attaching point (holes) point in direction of travel

23. Retaining ring
- Secured to body

24. 10 Nm (7 ft-lb)

25. Fuel filter
- Installation position: Arrow points in direction of flow

26. Supply line
- Black
- Ensure securely seated
- Secure with spring-type clips
- To fuel supply line on fuel rail

27. Return line
- Ensure securely seated

28. Breather
- Ensure securely seated
- Clipped onto top of fuel tank

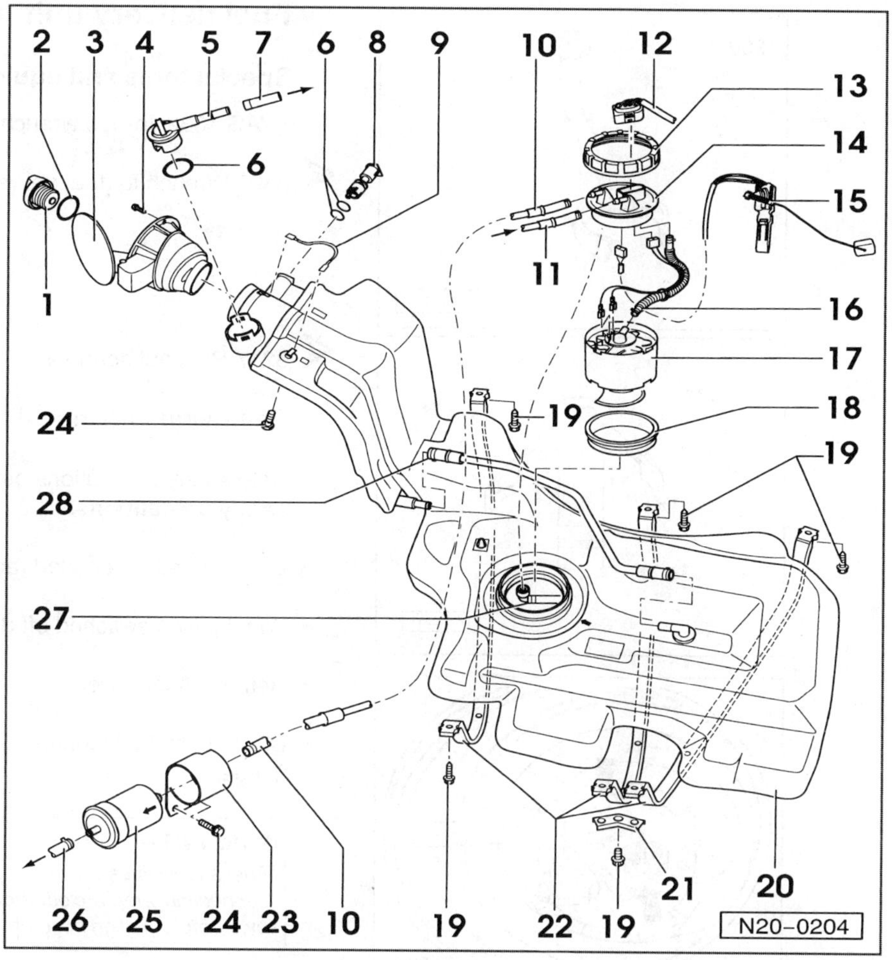

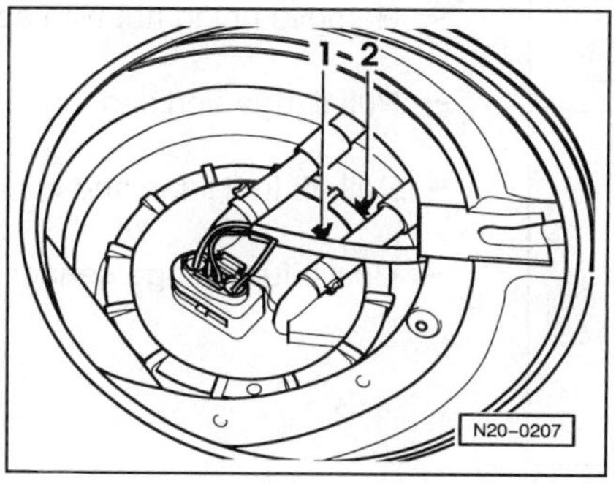

Fig. 1 Installed position of fuel delivery unit flange

Mark on flange -1- must align with mark on fuel tank -2-.

NOTE —
After installing fuel delivery unit flange, check that the supply, return and venting lines are still clipped onto the fuel tank.

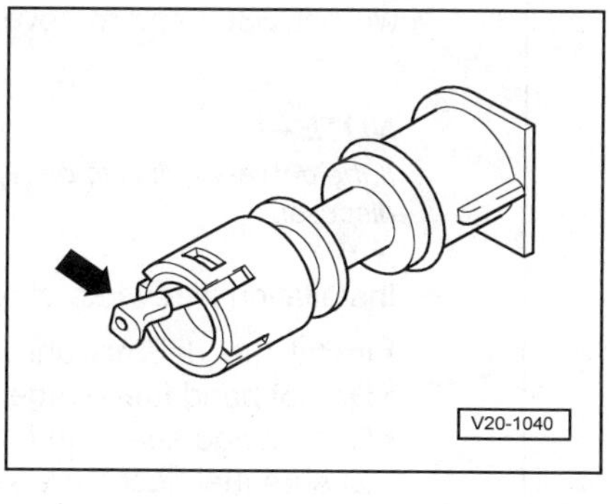

Fig. 2 Checking vent valve

Lever in rest position: Closed

Lever pushed in direction of arrow: Open

NOTE —
Before installing vent valve remove fuel tank cap.

Fuel delivery unit, removing and installing

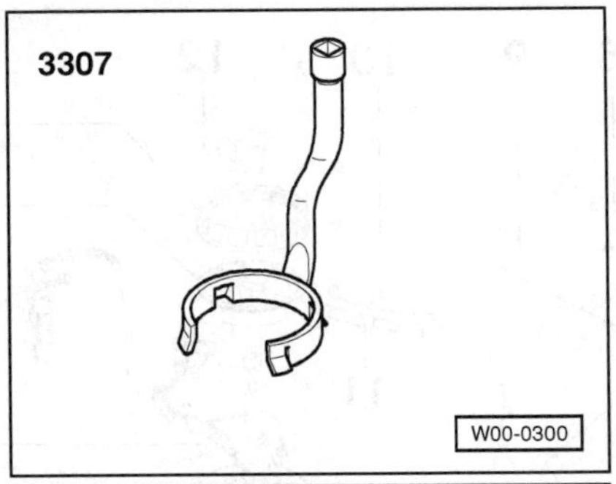

Special tools and equipment

• VAG 1332 Torque wrench (40 to 200 Nm) (30-150 ft-lb)

◀ 3307 Wrench for fuel delivery unit (fuel pump)

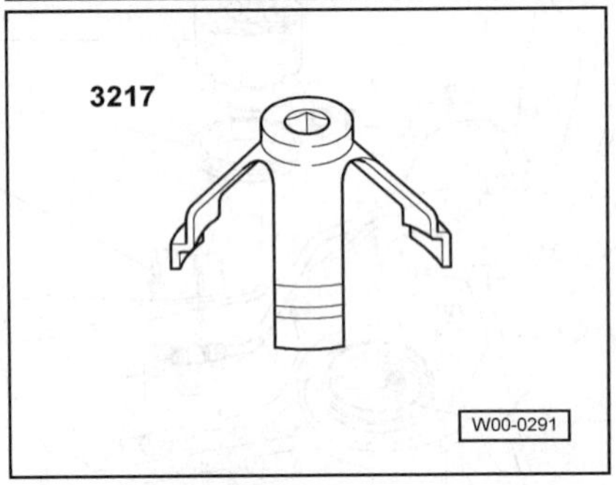

◀ 3217 Ring nut spanner

Removing and installing

Note safety precautions before starting work, see **Fuel supply system, safety precautions.**

– Check whether a coded radio is installed. If so, obtain anti-theft coding.

– With ignition switched off disconnect battery Ground strap.

– Remove floor cover.

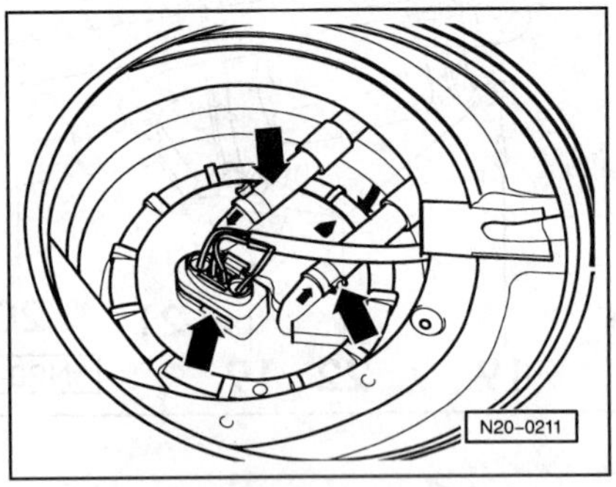

◀ Disconnect 4-pin connector and pull supply and return lines (arrows) off flange.

> ***WARNING —***
> *Fuel supply lines are under pressure! Before removing from hose connection wrap a cloth around the connection. Then release pressure by carefully pulling hose off connection.*

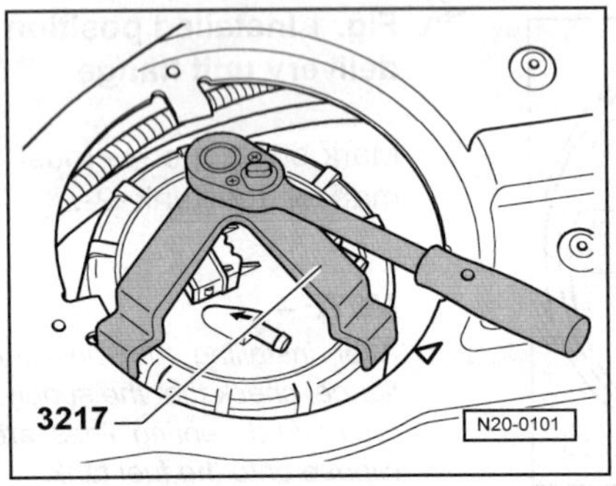

◀ Remove union nut with spanner 3217.

– Pull flange and seal out of opening in fuel tank.

– Pull off fuel lines and connector on lower side of flange.

– Unclip fuel gauge sender -G- on reservoir housing and remove.

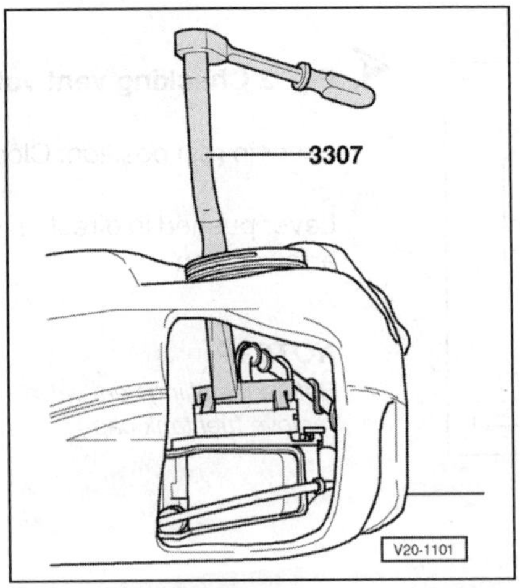

◀ Turn fuel delivery unit approx. 15° counter-clockwise onto stop with wrench 3307 and remove.

> ***NOTE —***
> *If the delivery unit is to be replaced then drain old delivery unit before disposal.*

– Installation is reverse of removal, noting the following:
  - Install fuel delivery unit in reverse order of removal.
  - Do not bend fuel gauge sender when installing.
  - Coat flange seal with fuel when installing.
  - Ensure fuel hoses are seated securely.
  - Do not interchange supply and return hoses (return hose -blue-).
  - Secure fuel hoses with spring-type clamps.

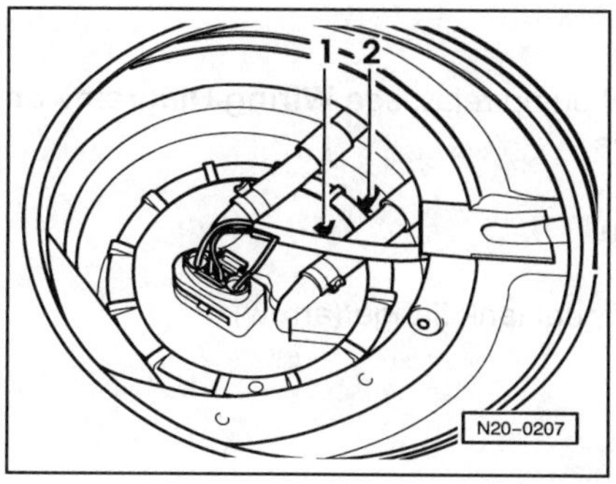

◀ Note installation position of fuel delivery unit flange: Mark on flange -1- must align with mark -2- on fuel tank.

- After installing fuel delivery unit flange, check that the supply, return and breather lines are still clipped onto the fuel tank.

Fuel pump, checking

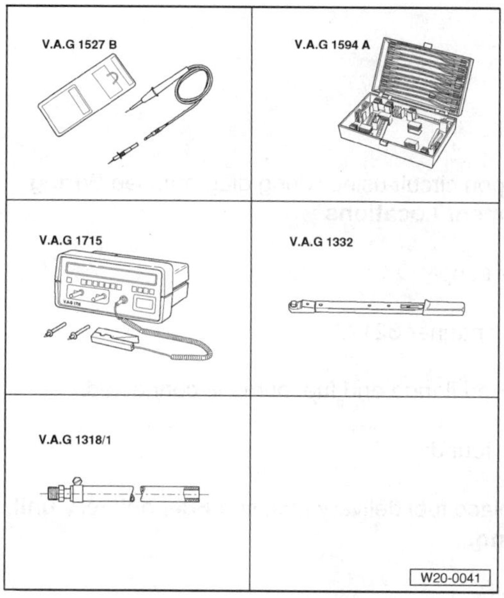

Special tools and equipment

- VAG 1318 Pressure gauge
- VAG 1318/10 Adapter
- VAG 1318/11 Adapter
- VAG 1318/12 Adapter
- VAG 1348 Remote control
- VAG 1466 A Test system
- VAG 1527B voltage tester
- VAG 1594 Adapter set
- VAG 1715 Multimeter
- VAG 1332 Torque wrench (40 to 200 Nm)
- VAG 1318/1 Hose
- VAG 3217 Ring nut spanner
- Measuring container
- Wiring diagram

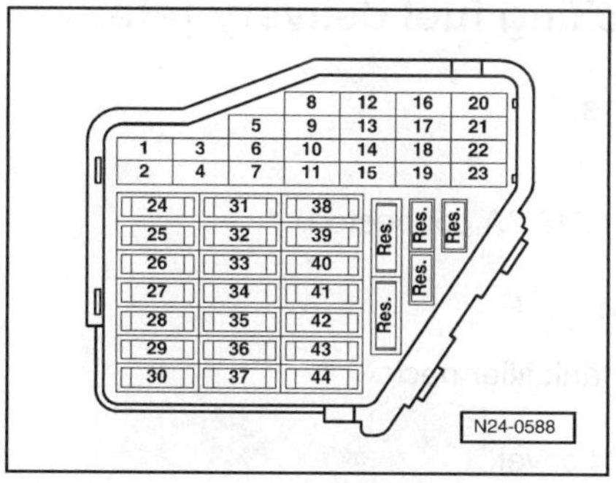

Fuel pump, checking conditions

– Battery voltage at least 11.5 V

◀ Fuse No. 28 OK.

Checking function and voltage supply

– Remove floor cover.

– Briefly operate starter.

- Fuel pump must be heard to run.

– Switch ignition off.

If the fuel pump does not run:

– Remove cover in front of fuse holder.

– Pull fuse No. 28 out of fuse holder.

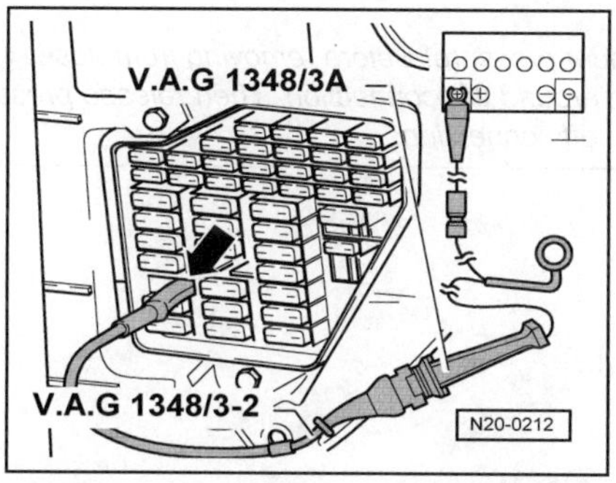

◀ Connect remote control VAG 1348/3A with adapter cable VAG 1348/3-2 to contact 28a to fuel pump and battery positive (+).

– Operate remote control.

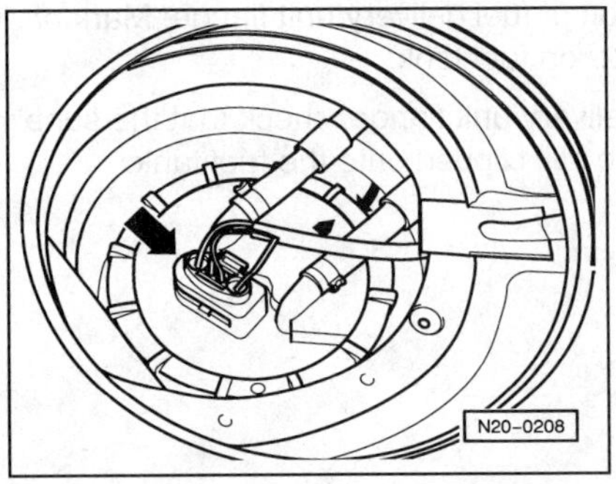

Fuel pumps run:

– Check activation of fuel pump relay, see **Wiring Diagrams and Component Locations**

Fuel pump does not run:

◄ Pull 4-pin connector off fuel tank flange (arrow).

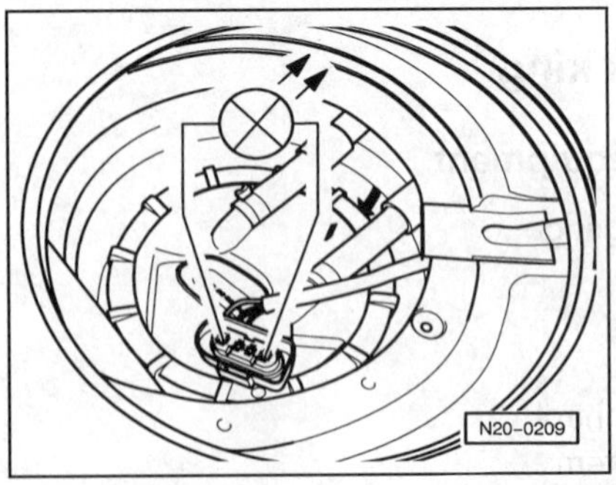

◄ Connect voltage tester VAG 1527B with auxiliary cables from VAG 1594 to outer contacts of connector.

– Operate remote control.

• LED must light up

If LED does not light up:

– Locate and eliminate open circuit using wiring diagram, see **Wiring Diagrams and Component Locations**

If LED lights up (voltage supply OK):

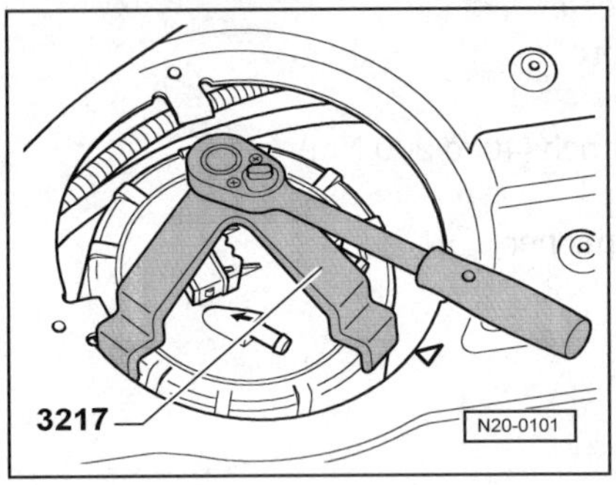

◄ Remove union nut with spanner 3217.

– Check that wiring between flange and fuel pump is connected.

If no open circuit can be found:

– Fuel pump is faulty, replace fuel delivery unit, see **Fuel delivery unit, removing and installing.**

Fuel pump, checking fuel delivery rate

Checking conditions

• Voltage supply OK.

• Remote control VAG 1348/3A connected.

Checking sequence

– Remove cap from fuel tank filler neck.

– Remove intake manifold cover.

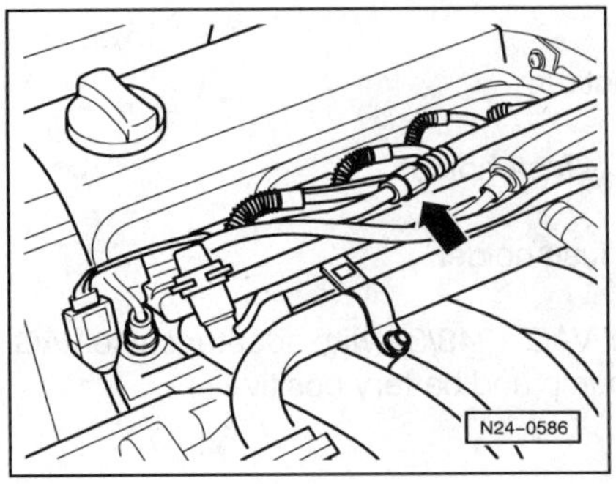

◄ Open threaded union (arrow) and catch escaping fuel with a cloth.

> ***WARNING —***
> *Fuel supply lines are under pressure! Before removing from hose connection wrap a cloth around the connection. Then release pressure by carefully pulling hose off connection.*

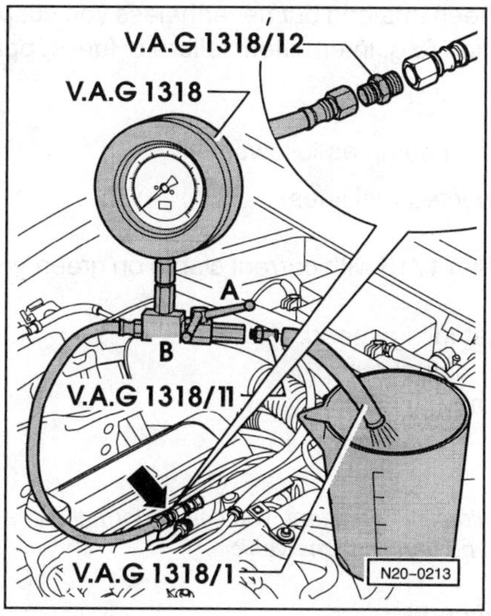

Connect pressure gauge VAG 1318 with adapter 1318/12 to fuel supply line.

– Push hose VAG 1318/1 onto adapter VAG 1318/11 on pressure gauge and hold hose in a measuring container.

– Open pressure gauge shut-off tap. Lever then points in direction of flow -A-.

– Operate remote control VAG 1348/3A and slowly close shut-off tap until pressure gauge shows 3 bar (43.5 psi). From this point on do not move position of shut-off tap.

– Empty measuring container.

– Connect multimeter with adapter cables from VAG 1594 to vehicle battery (fuel pump delivery range is dependent upon battery voltage).

– Operate remote control for 30 seconds and measure battery voltage.

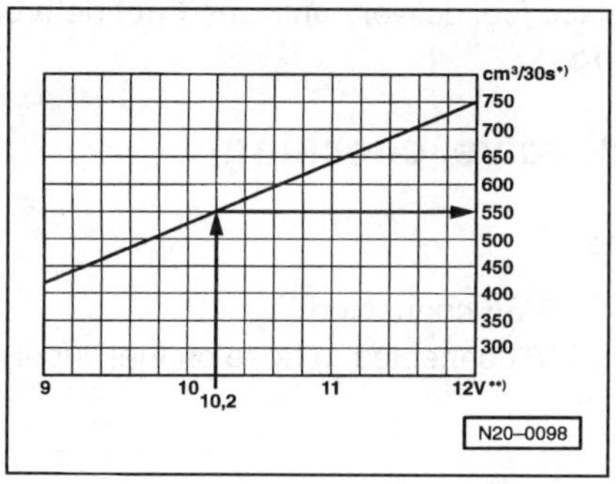

Compare quantity of fuel delivered with specification.

*) Minimum delivery cm^3/30 seconds

**) Voltage at fuel pump with engine stationary and pump running (approx. 2 volts less than battery voltage).

Example:

During the test a voltage of 12.2 volts is measured at the battery. As the voltage at the pump is approx. 2 volts less than the battery voltage, this will equate to a minimum delivery of 550 cm^3/30 s.

If the minimum delivery rate is not attained:

– Check fuel lines for possible restrictions (kinks) or blockages.

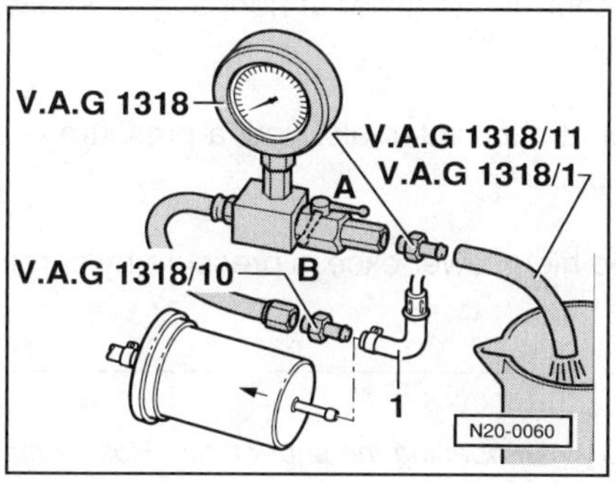

Pull supply hose -1- off inlet to fuel filter.

– Connect pressure gauge VAG 1318 to hose with adapter 1318/10.

– Repeat delivery rate check.

If the minimum delivery rate is now attained:

– Replace fuel filter.

If the minimum delivery rate is again not attained:

– Remove fuel delivery unit (see **Fuel delivery unit, removing and installing**) and check strainer for blockages.

Only if up to now no faulty components have been detected:

– Fuel pump is faulty, replace fuel delivery unit, see **Fuel delivery unit, removing and installing**.

If the delivery rate has been attained but nevertheless you suspect a fuel supply system malfunction (e.g. intermittent failure of fuel supply system):

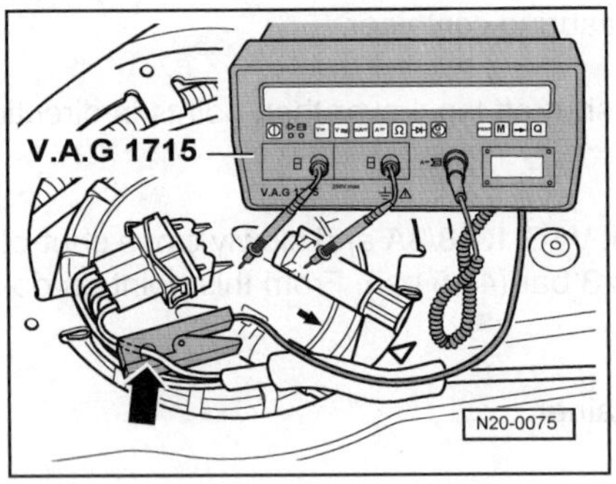

Check current draw of fuel pump as follows:

- Reconnect all disconnected fuel lines.
- Remove floor cover.
- Connect multimeter VAG 1715 with current clamp on green/yellow wire in wiring harness.
- Start engine and run at idling speed.
- Measure current draw of fuel pump.
- Specification: max. 8 amps.

NOTE—
If the fuel system malfunction is intermittent, you can perform the check during a test drive with a 2nd person in the vehicle.

If the current draw is exceeded:

– Fuel pump is faulty, replace fuel delivery unit, see **Fuel delivery unit, removing and installing**.

Fuel pump check valve, checking

Checking conditions

- Remote control VAG 1348/3A connected
- Pressure gauge VAG 1318 connected to hose on fuel supply line.

Checking sequence

NOTE—
Using this procedure the fuel supply line connections from the fuel delivery unit to the point at which the pressure gauge VAG 1318 is connected will also be checked for leaks.

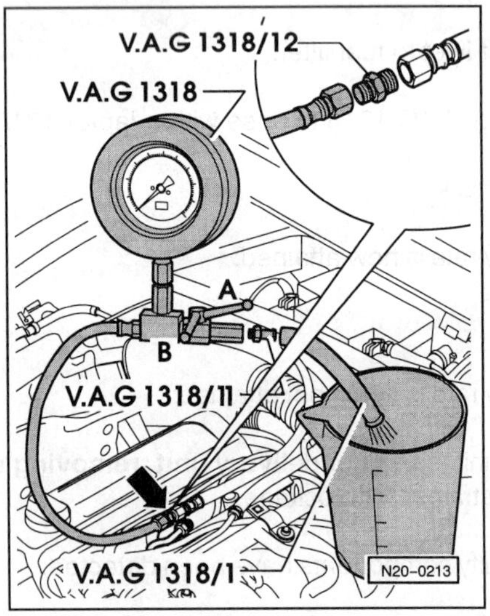

Close pressure gauge shut-off tap (lever at right angles to direction of flow, position -B-.

– Operate remote control at short intervals, until a pressure of approx. 3 bar (43.5 psi) has built-up.

– If pressure builds up too high, lower excess pressure by carefully opening shut-off tap.

> ***WARNING—***
> *A danger of spray exists when opening the shut-off tap. Hold container under the pressure gauge discharge hose.*

– Watch pressure drop on gauge. After 10 minutes pressure must not drop below 2.5 bar (36 psi).

If the pressure drops further:

– Check line connections for leaks.

If no faulty connections are detected:

– Fuel pump is faulty, replace fuel delivery unit, see **Fuel delivery unit, removing and installing**.

Throttle Control

Cable operated accelerator components (engine code: AEB)

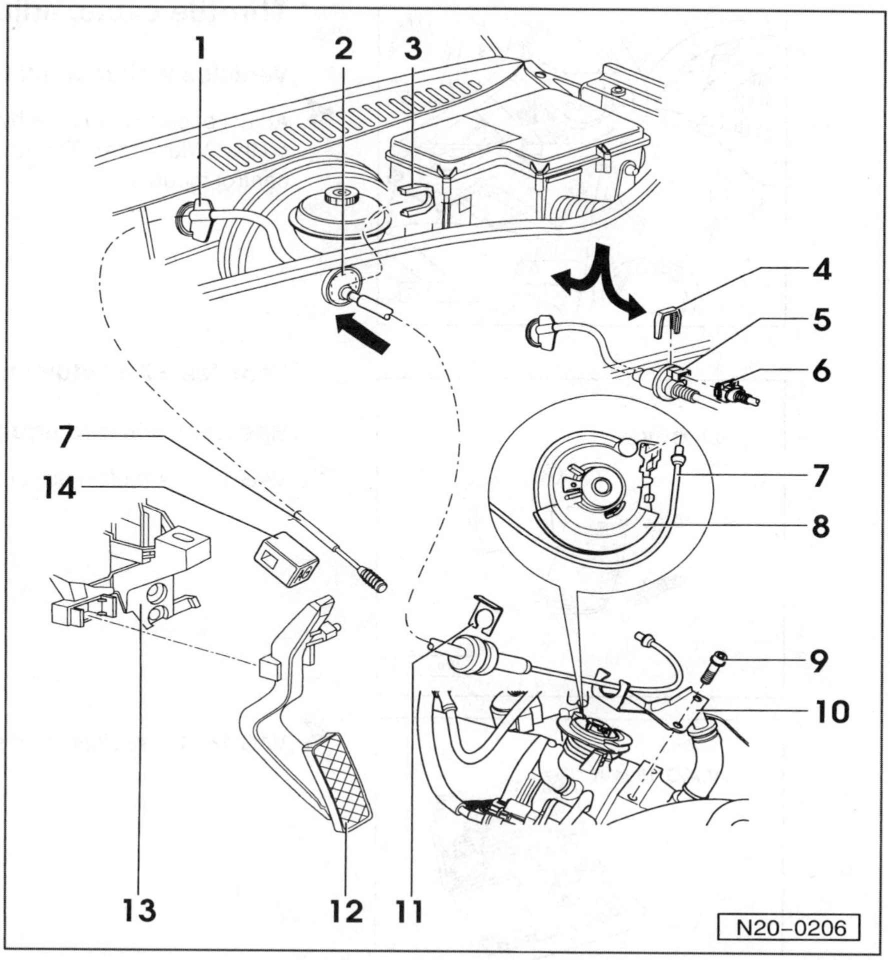

1. **Throttle cable mounting**
   - On bulkhead in plenum chamber
   - Removing/installing, see **Fig. 1**
2. **Throttle cable mounting**
   - Only vehicles with manual transmission
3. **Retaining clip**
   - Only vehicles with manual transmission
   - Ensure seated securely
   - Secured behind bulkhead
4. **Retaining clip**
   - Only vehicles with automatic transmission
   - Ensure seated securely
   - Secured before bulkhead
5. **Kickdown switch -F8-**
   - On bulkhead
   - See **Throttle cable, adjusting, (vehicles with automatic transmission)**
   - Only on vehicles with automatic transmission
6. **Connector**
   - Black, 2-pin
   - Only on vehicles with automatic transmission
7. **Accelerator cable**
   - See **Throttle cable, adjusting**
   - Do not reuse damaged accelerator cable, replace
8. **Cable cam**
   - On throttle valve control part
9. **10 Nm (7 ft-lb)**
10. **Support bracket**
    - For accelerator cable
11. **Retainer**
    - See **Throttle cable, adjusting**
12. **Accelerator pedal**
13. **Mounting bracket**
    - Removing and installing, see **Brake System section, Repair Group 46**
14. **Accelerator pedal stop**
    - Installation position of accelerator pedal stop on mounting bracket, see **Fig. 2**

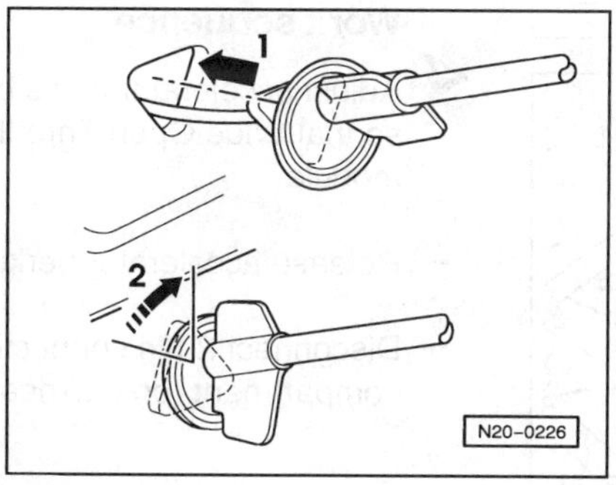

Fig. 1 Removing and installing throttle cable mounting

– Turn throttle cable mounting approx. 90° counter-clockwise and remove from bulkhead in plenum chamber.

– To install, push throttle cable through bulkhead in plenum chamber (arrow 1).

– Push throttle cable mounting into bulkhead and turn approx. 90°clockwise (arrow 2).

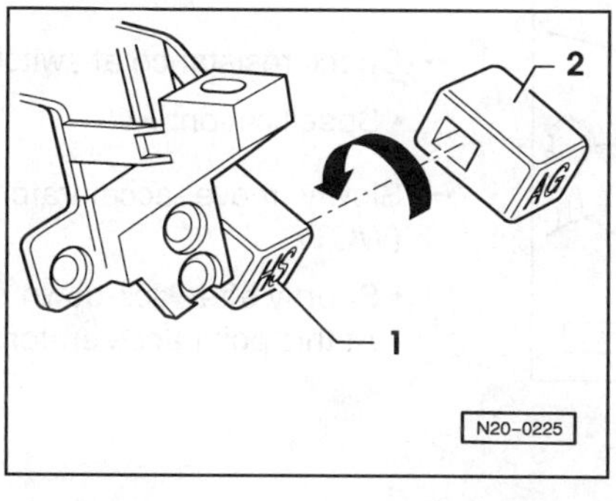

Fig. 2 Installation position of accelerator pedal stop on mounting bracket

- Letters "HS" -1- are visible when installed in vehicles with a manual transmission.
- Letters "AG" -2- are visible when installed in vehicles with automatic transmission.

Throttle cable, adjusting

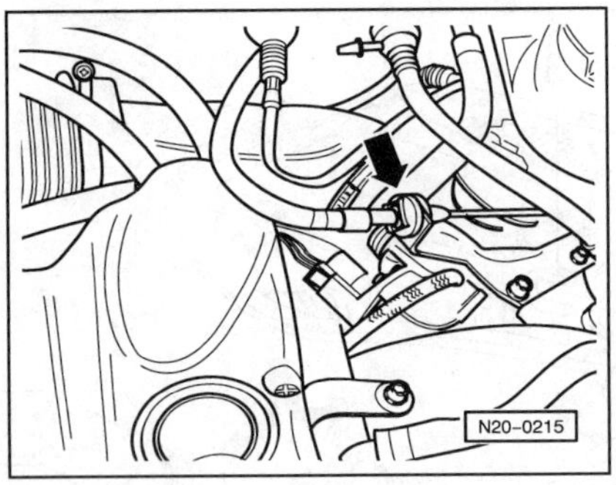

Vehicles with manual transmission

< Adjust accelerator cable by moving retainer (arrow) at support bracket, so that Wide Open Throttle (WOT) position is attained at throttle valve control module.

Vehicles with automatic transmission

Special tools and equipment

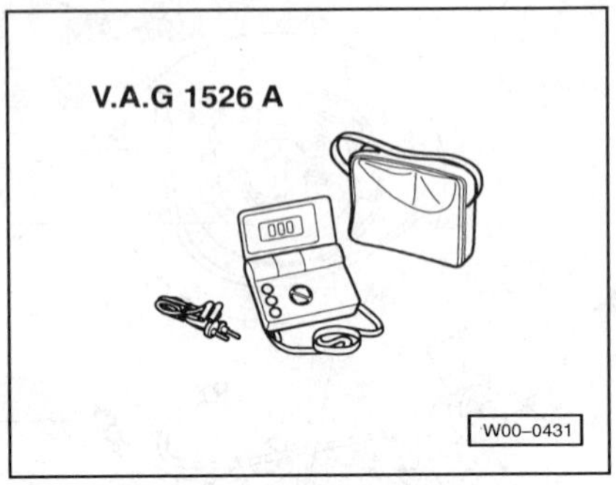

< VAG 1526 A multimeter or Fluke 83 mulitmeter

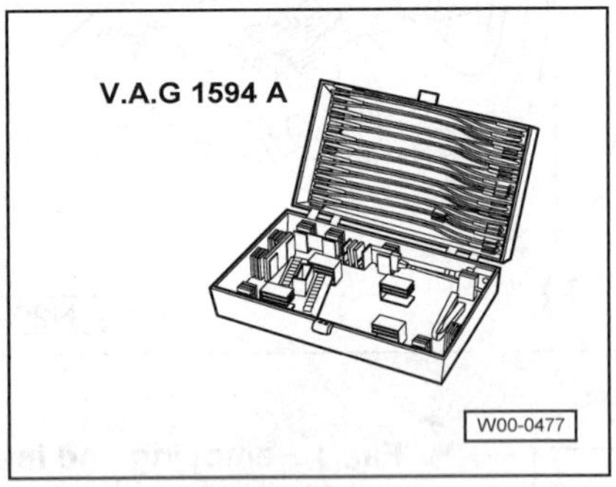

< VAG 1594 A Auxiliary test set

Work sequence

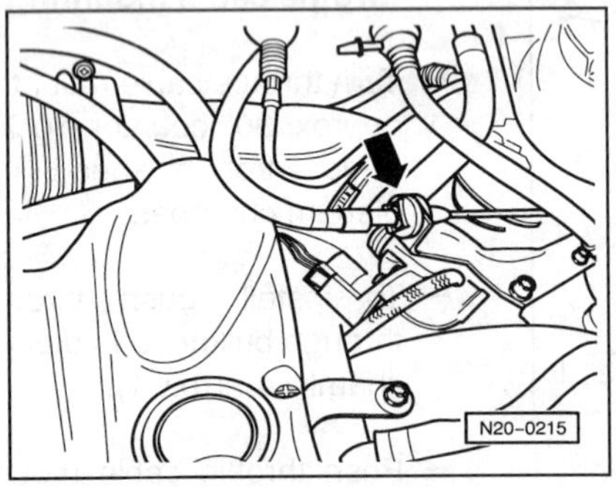

< Adjust accelerator cable by moving retainer (arrow) at support bracket, so that Wide Open Throttle (WOT) position is attained at throttle valve module.

– Release accelerator pedal.

– Disconnect 2-pin connector from kick-down switch -F8- in engine compartment (on bulkhead).

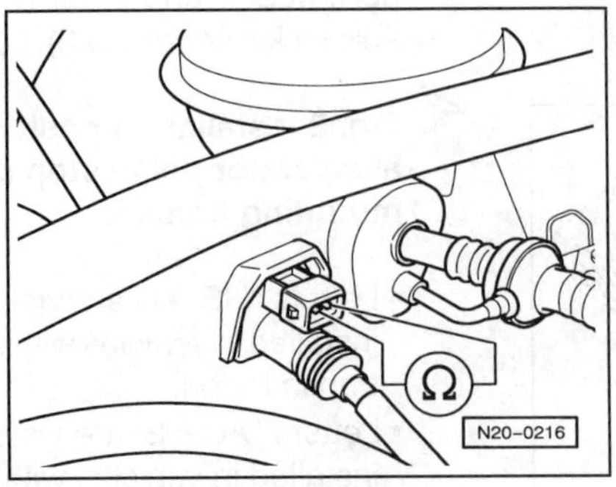

< Connect multimeter VAG 1526 or Fluke 83 with auxiliary cables from VAG 1594 to kick-down switch.

– Check resistance at switch.

• Specification: ∞Ω

– Slowly move accelerator pedal in direction of Wide Open Throttle (WOT).

• Shortly after kick-down pressure point resistance must drop to 0Ω. At this point accelerator pedal must be just off stop.

Electronic accelerator pedal, assembly (engine code: ATW)

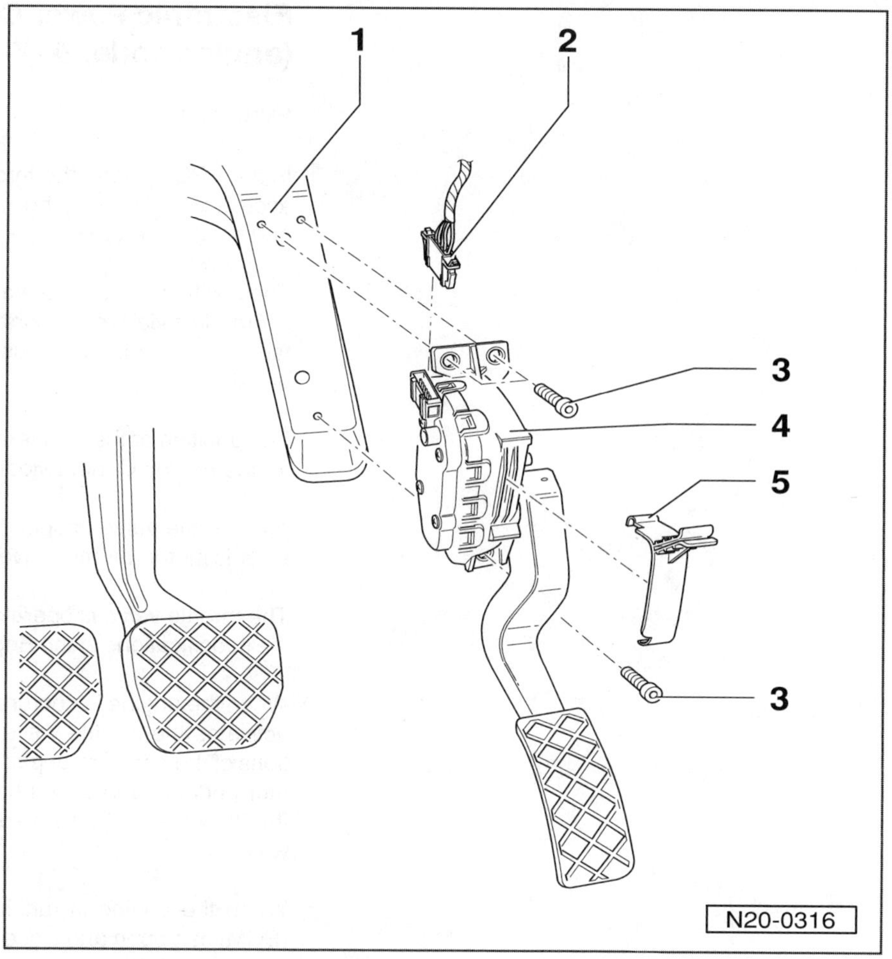

1. **Pedal Bracket**
   - Removing and installing, see **Brake System section, Repair Group 46**
2. **Connector**
   - Black, 6-pin
3. **10 Nm (7 ft-lb)**
4. **Electronic accelerator pedal**
   - With Throttle Position (TP) sensor -G79-
   - With Sender 2 for accelerator pedal position -G185-
   - See **Electronic accelerator pedal, checking**
   - When removing, remove footwell cover
5. **Retainer**
   - For footwell cover
   - Clipped onto accelerator pedal

Electronic Power Control (EPC) system (engine code: ATW)

Function

In the EPC system, the throttle valve is not operated via cable from the accelerator. There is no mechanical connection between accelerator and the throttle valve.

The position of the accelerator pedal is communicated to the engine control module via two sensors for accelerator pedal position (variable resistances, stored in one housing) that are connected with the accelerator pedal.

The position of the accelerator (drivers intention) is the main input value for the Engine Control Module (ECM).

The throttle valve is operated over the entire RPM and load range via electric motor (throttle valve actuator) in the throttle valve control module.

The throttle valve is operated by the throttle valve actuator according to the specifications of the Engine Control Module (ECM).

When the engine is not running and the ignition is switched on, the ECM activates the throttle valve actuator exactly according to the specifications of the accelerator pedal position sensor. This means, if the accelerator pedal is depressed halfway, the throttle valve actuator opens the throttle valve to the same degree; the throttle valve is then opened halfway.

When the engine is running (under load) the Engine Control Module (ECM) can open and close the throttle valve independently of the accelerator pedal position sensor.

In this way, the throttle valve can already be completely open even though the accelerator pedal has only been depressed half way.

This also gives vastly improved fuel consumption and exhaust emission levels under certain load conditions.

The torque required by the engine can be produced by the ECM by optimizing the combination of throttle valve cross-section and charge pressure.

It is incorrect to believe that the EPC system is comprised of only one or two components. EPC is a system containing all components which have an influence, control or monitor the position of the throttle valve, e.g. the accelerator pedal position sensor, the throttle valve control module, the Electronic Power Control (EPC) warning light, the Engine Control Module (ECM).

NOTE —
Volkswagen identifies electrical components by a letter and/or a number in the electrical schematics. See **Wiring Diagrams, Fuses and Relays**. *These electrical identifiers are listed -between dashes- as an aid to electrical troubleshooting.*

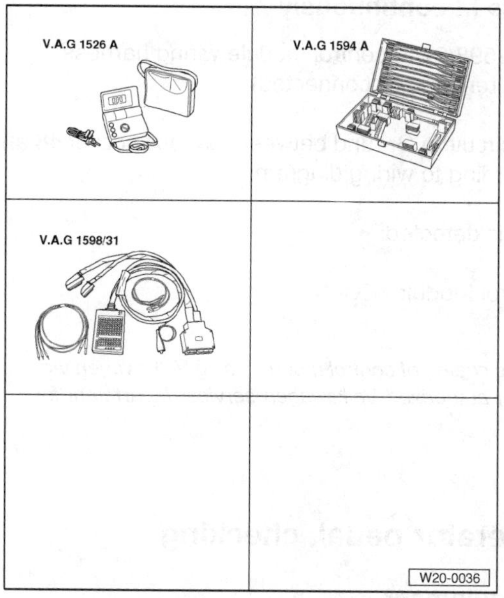

EPC warning lamp -K132-, checking

Special tools and equipment

- VAG 1526 or Fluke 83 multimeter
- VAG 1594 A Adapter set
- VAG 1598/31 Test box
- Wiring diagram

Checking function of EPC warning lamp

– Switch ignition on.

- The EPC warning lamp must light up.

– Start engine and run at idling speed.

- The EPC warning lamp must go out when no EPC malfunction has been stored.

EPC warning lamp does not light up

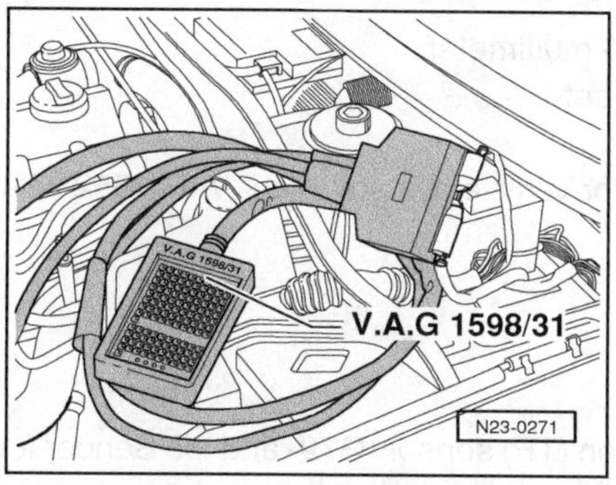

< Connect test box VAG 1598/31 to control module wiring harness (The engine control module remains disconnected).

– Bridge test box sockets 1 + 48 using cable from VAG 1594.

– Switch ignition on.

- The EPC warning lamp must light up.

If the EPC warning lamp does not light up:

– Switch ignition off.

– Remove instrument cluster and check bulb and its voltage supply, see **Electrical Equipment section, Repair Group 90**

If no malfunction is found:

– Check wiring for open circuit between test box socket 48 and instrument cluster according to wiring diagram.

If no wiring malfunction is detected:

- Replace Engine Control Module (ECM).

EDITOR'S NOTE—

ECM replacement requires coding of control module using Volkswagen factory scan tool. Consult an authorized Volkswagen service department for additional information.

EPC warning lamp is lit continuously

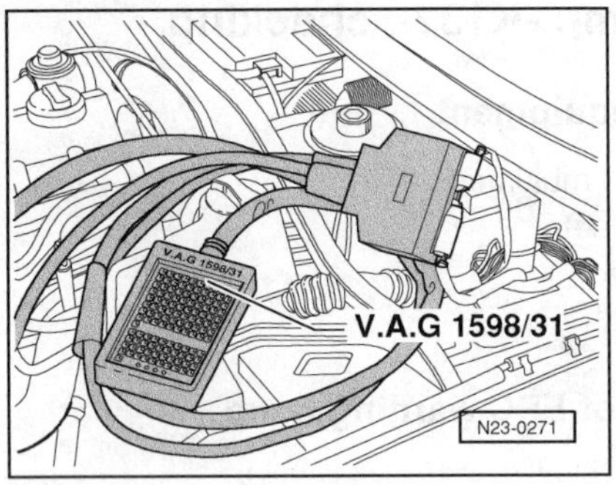

◀ Connect test box VAG 1598/31 to control module wiring harness (Engine control module remains disconnected).

– Check wiring for short circuit to Ground between test box socket 48 and instrument cluster according to wiring diagram.

If no wiring malfunction is detected:

• Replace Engine Control Module (ECM).

EDITOR'S NOTE —
ECM replacement requires coding of control module using Volkswagen factory scan tool. Consult an authorized Volkswagen service department for additional information.

Electronic accelerator pedal, checking

Special tools and equipment

• VAG 1526 or Fluke 83 multimeter
• VAG 1594 A Adapter set
• VAG 1598/31 Test box
• VAG 1551 Scan Tool or vehicle system tester VAG 1552 with cable VAG 1551/3
• Wiring diagram

Function

Both the Throttle Position (TP) sensor -G79- and the Sender for Accelerator Pedal position -G185- are located on the electronic accelerator pedal and communicate the driver's intention to the Engine Control Module (ECM) independently of one another. Both components are located in one housing.

Checking sequence

– Connect VAG 1551 (VAG 1552) and with ignition switched on select engine electronics control module with "Address word" 01.

– Press -0- and -8- buttons to select "Read measured value block" and confirm entry with -Q- button.

– Press -0-, -6- and -2- buttons to select "Display group number 62" and confirm entry with -Q- button.

– Check specifications of TP sensor -G79- at idling stop in display field 3.

• Specification: 12 to 97%

– Check specifications of sender -G185- at idling stop in display field 4.

• Specification: 4 to 49%

NOTE —
The ECM calculates the voltage values of the angle sensor in percent relative to 5 Volts and displays these as a percentage figure (5 Volt supply voltage equates to 100%).

– Operate accelerator pedal slowly until Wide Open Throttle and observe angle indication in display fields 3 and 4:

- The percent figures in display field 3 must rise evenly. The tolerance range of 12 to 97% is not covered completely.
- The percent figures in display field 4 must rise evenly. The tolerance range of 4 to 49% is not covered completely.

NOTE —
The figure displayed in display field 3 must always be approximately double that displayed in display field 4.

– If the display does not indicate as described:

- Check voltage supply and wiring connections to -G79- and -G185-.

Checking voltage supply and wiring to control module

– Remove driver's side stowage compartment, see **Body-Interior section, Repair Group 68**

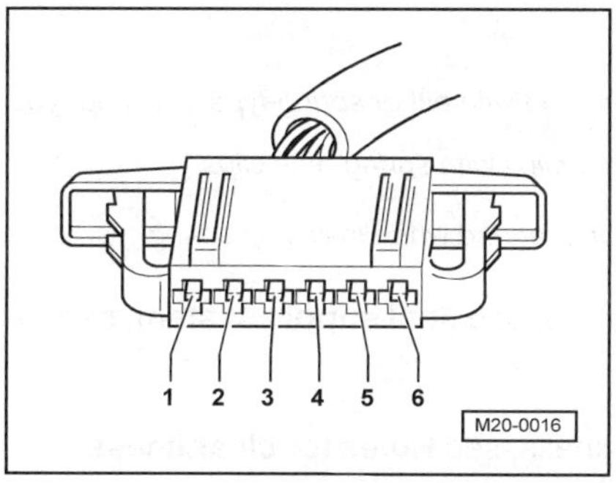

– Disconnect 6-pin connector for sensor.

– Switch ignition on.

Connect multimeter to measure voltage at following contacts of connector:

- Contact 1 + Ground
- Contact 1 + 5
- Contact 2 + Ground
- Contact 2 + 3
- Specification: minimum 4.5 V

– Switch ignition off.

If the specifications are not obtained:

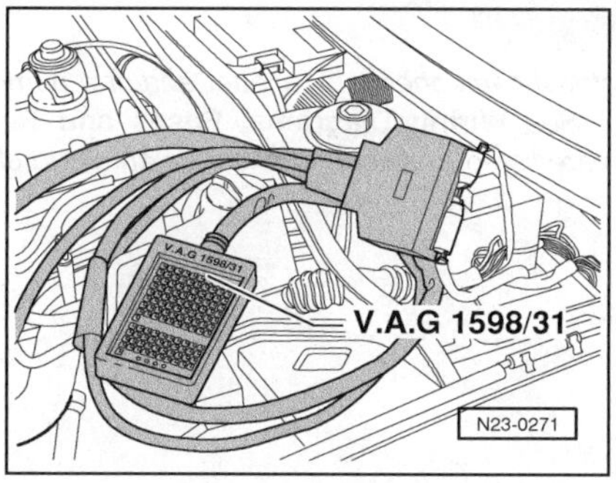

Connect test box VAG 1598/31 to control module wiring harness (The ECM remains disconnected).

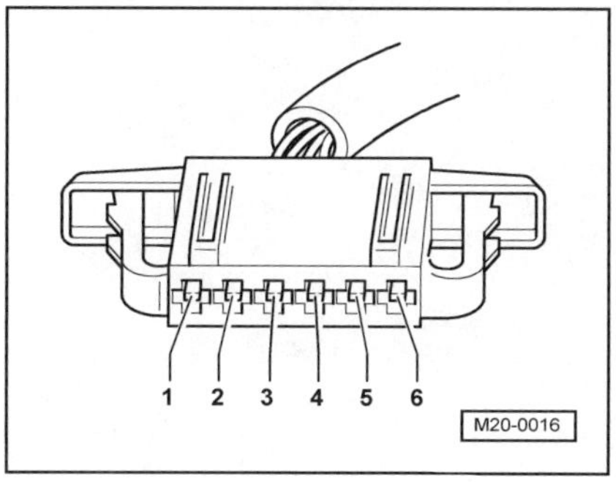

Check wiring between test box and connector for open circuit according to wiring diagram.

- Contact 1+ socket 72
- Contact 2+ socket 73
- Contact 3+ socket 36
- Contact 4+ socket 35
- Contact 5+ socket 33
- Contact 6+ socket 34
- Wire resistance: max. 1.5Ω

– Check wiring for short to battery positive or Ground.

– Check wires for short to one another.

– If no wiring malfunction is detected, replace electronic accelerator pedal.

Evaporative Emissions (EVAP) System

EVAP functional description

The evaporative emissions system (including fuel tank) is equipped with a self-checking leak diagnosis capability.

The diagnosis operates by first pressurizing the system and then measuring the leakdown rate between tank pressurizations. A Diagnostic Trouble Code (DTC) is generated when the rate is equivalent to a leak in excess of a 1 mm diameter hole.

During the diagnosis, the Leak Detection Pump (LDP) -V144- builds up a pressure of approx. 30 mbar (.453 psi) in the system. The pump will switch off when the preset pressure is attained. When the pressure falls below a certain value, the pump will switch on again. The On Board Diagnostic monitors and compares the time between the switched on periods and, if too short, will store a malfunction in the DTC memory.

NOTE —
• Hose connections are secured with either spring-type or clamp-type clips.

• Always replace clamp-type clips with spring-type clips.

• VAG 1921 pliers are recommended when installing spring-type clips.

Observe safety precautions, see **Fuel supply system, safety precautions.**

Observe rules for cleanliness, see **Rules for cleanliness.**

NOTE —
Volkswagen identifies electrical components by a letter and/or a number in the electrical schematics. See **Wiring Diagrams, Fuses and Relays***. These electrical identifiers are listed -between dashes- as an aid to electrical troubleshooting.*

EVAP system components

1. **Leak Detection Pump (LDP) -V144-**
   - In rear right wheel housing under the wheel housing liner
   - See **Leak Detection Pump (LDP), checking**
2. **10 Nm (7 ft-lb)**
3. **Bracket for LDP**
4. **Connector**
   - 3-pin
5. **Coupling**
   - Press together at front to release
6. **Connecting hose**
   - Pressure side
7. **Connecting hose**
   - Suction side
8. **Air filter for LDP**
9. **Connector**
10. **Connecting pipe**
11. **EVAP canister**
    - In spare wheel well on floor
12. **10 Nm (7 ft-lb)**
13. **EVAP canister purge regulator valve -N80-**
    - See **Canister purge regulator valve, checking**
14. **Vent line**
15. **Vacuum line**
16. **Connecting hose**
    - Vacuum side
17. **Elbow**

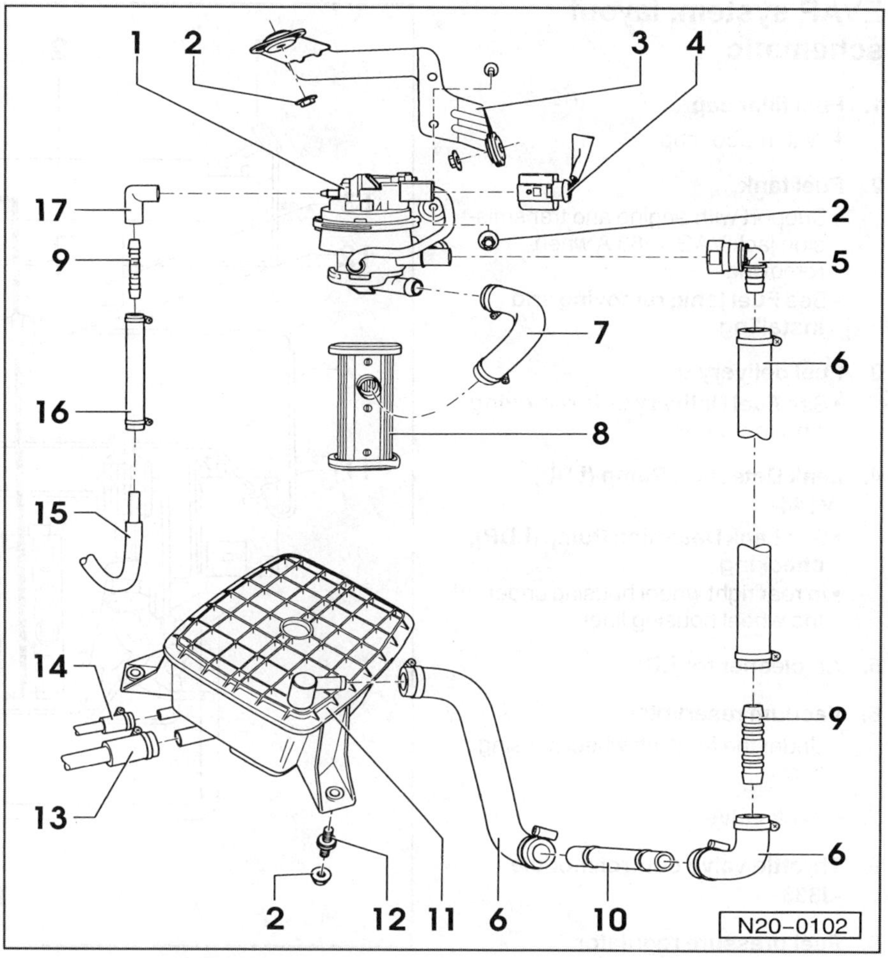

EVAP system, layout schematic

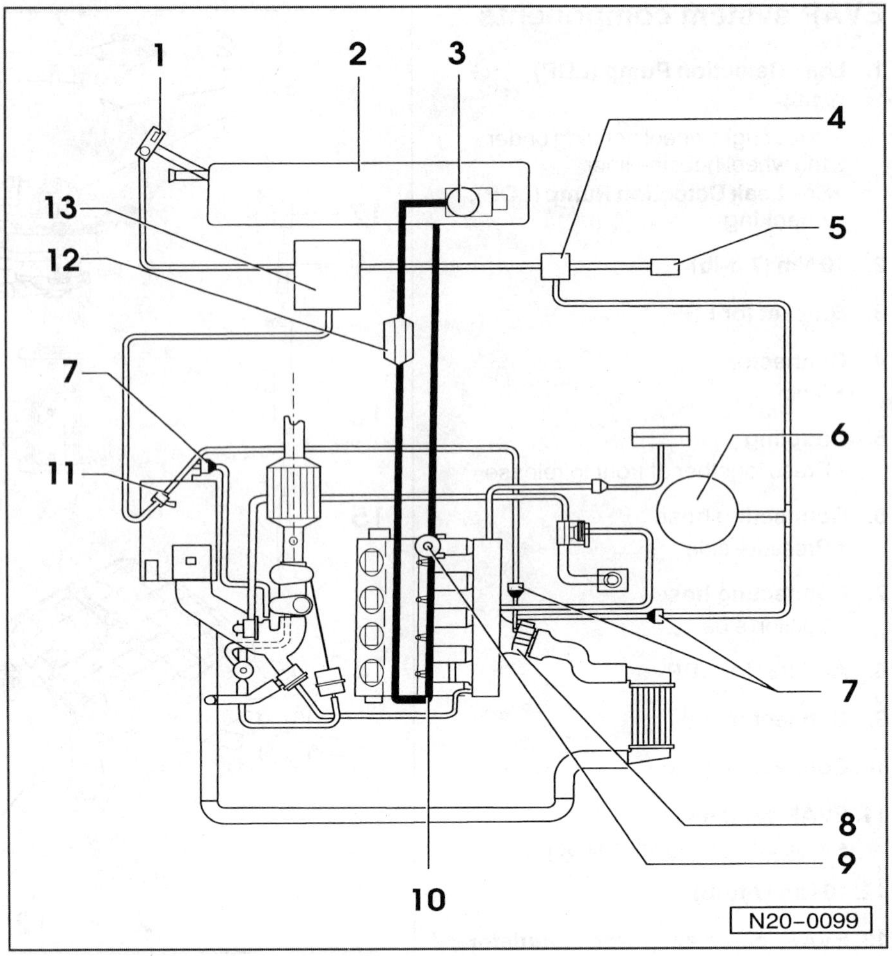

1. **Fuel filler cap**
   - With rubber cup
2. **Fuel tank**
   - Support with engine and transmission jack VAG 1383 A when removing
   - See **Fuel tank, removing and installing**
3. **Fuel delivery unit**
   - See **Fuel Delivery unit, removing and installing**
4. **Leak Detection Pump (LDP) -V144-**
   - See **Leak Detection Pump (LDP), checking**
   - In rear right wheel housing under the wheel housing liner
5. **Air cleaner for LDP**
6. **Vacuum reservoir**
   - Under the front left wheel housing liner
7. **Check valve**
8. **Throttle valve control module -J338-**
9. **Fuel pressure regulator**
10. **Fuel rail with injectors**
11. **EVAP canister purge regulator valve -N80-**
    - See **EVAP canister purge regulator valve, checking**
12. **Fuel filter**
    - Installed position: Arrow points in direction of flow
13. **EVAP Canister**
    - In spare wheel well on vehicle floor

EVAP canister purge regulator valve, checking

In a no-current situation the solenoid valve is closed.

Special tools and equipment

• V.A.G 1551 Scan Tool or vehicle system tester V.A.G 1552 with cable V.A.G 1551/3

Checking conditions

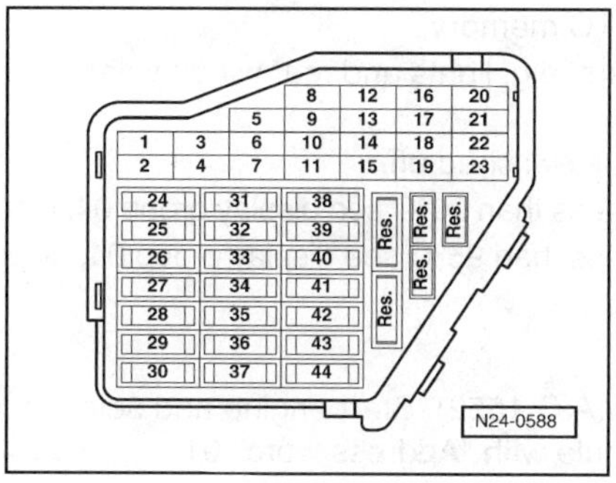

Fuse 29 must be OK.

• Coolant temperature at least 80°C

Functional check

– Connect V.A.G 1551 (V.A.G 1552). Start engine and select engine control module with the "Address word" 01.

– Press -0- and -8- buttons to select "Read measured value block" function and confirm entry with -Q- button.

– Press -0-, -7- and -0- buttons to select "Display group number 70" and confirm entry with -Q- buttons.

– Press button -4- to select "Basic setting" function ("Tank venting valve diagnosis" will be initiated).

NOTE—
The engine must not be loaded during this diagnosis because the diagnosis will be interrupted and will not be started again until the engine is revved-up.

When the diagnosis is initiated by the engine control module, the display in display field 4 jumps from "Test OFF" to "Test ON".

– Allow engine to run at idling speed until specification "TVV OK" appears in display field 4.

– If the display jumps from "Test ON" to "Test OFF" during the test, rev up briefly and the test will be repeated.

If "TVV not OK" appears in display field 4:

– Initiate Output Diagnostic Test Mode (DTM) and activate EVAP canister purge regulator valve -N80-.

EDITOR'S NOTE—
Component activation is accomplished using Volkswagen diagnostic scan tool, output diagnostic test mode (function 03).

– If the specification "TVV OK" is obtained:

• Press → button.
• Press -0- and -6- buttons to select "End data transfer" function and confirm input with the -Q- button.

EVAP canister, checking

Special tools and equipment

- V.A.G 1551 Scan Tool or vehicle system tester V.A.G 1552 with cable V.A.G 1551/3

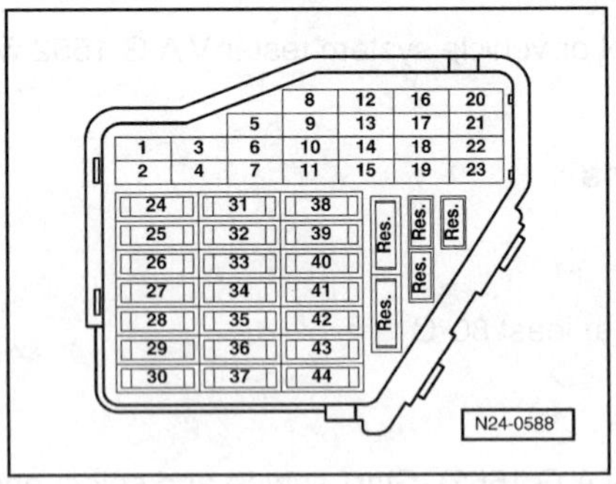

Checking conditions

◄ Fuse 29 must be OK.

- No codes stored in DTC memory
- All electrical consumers, e.g. lights and rear window defogger must be switched off
- Air conditioner must be switched off.
- Intake air temperature less than 90°C see display group 04, display field 4
- Coolant temperature less than 95°C see display group 04, display field 3

Functional check

– Connect V.A.G 1551 (V.A.G 1552). Start engine and select engine electronics control module with "Address word" 01.

– Press -0- and -8- buttons to select "Read measured value block" and confirm entry with -Q- button.

– Press -0-, -7- and -1- to select "Display group number 71" and confirm entry with -Q- button.

– Press button -4- to select "Introduction of basic setting" function.

- The test of the EVAP system will be initiated. The function of the Reed contact in the LDP will be checked by the control module.
- After approx. 30 seconds the LDP will be activated and build up pressure in the fuel system.
- The test will be initiated after approx. 45 seconds (when there is sufficient pressure available in the fuel system).
- The LDP switches off and the ECM measures the time until the Reed contact closes.

– Check display in display fields 3 and 4.

If the specifications are not attained:

– Check system for leaks, see **Checking for leaks.**

– If the specifications are obtained:

- Press →button.
- Press -0- and -6- to select "End data transfer" function and confirm input with the -Q- button.

Checking for leaks

– If a small or large leak is detected during the check of the EVAP system, proceed as follows.

- Switch ignition off.
- Pull hose from LDP off EVAP canister and seal it.
- Repeat functional check.

– If the malfunction is still present:

• Remove rear right wheel housing liner, see **Body-Exterior section, Repair Group 66**
• Check connecting lines and hoses between LDP and EVAP canister, see **EVAP system**.

If the malfunction is still present:

– Check whether fuel filler cap is fully closed or if seal is faulty, replace if necessary.

– Check whether locking ring for fuel delivery unit is tight (found under luggage compartment floor trim). Replace seal if faulty.

– Tighten locking ring with special tool 3217.

– Check supply and return lines on fuel delivery unit for leaks.

– Remove EVAP from spare wheel well and check for cracks and leaks, replace if necessary.

– Check connecting lines and hoses between EVAP canister and fuel tank flap unit/EVAP canister purge regulator valve, see **EVAP system**.

– Check that hoses and hose clamps are tight.

– Repeat functional check.

Leak Detection Pump (LDP), checking

Special tools and equipment

• Multimeter VAG 1526 or Fluke 83
• Voltage tester VAG 1527B
• Adapter set VAG 1594
• Wiring diagram

Electrically checking LDP

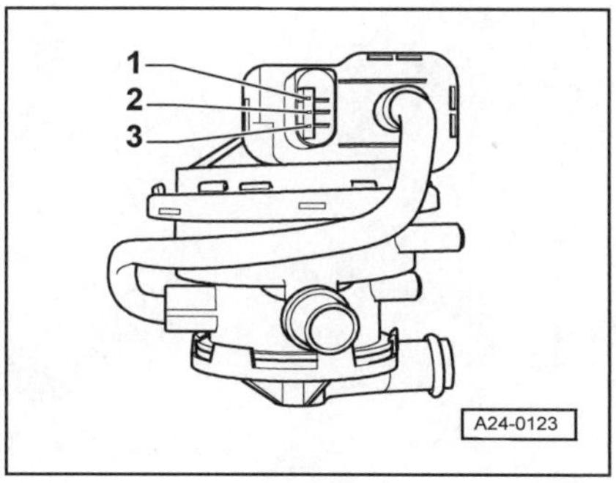

◄ Disconnect 3-pin connector from LDP.

– Measure resistance between terminals 1 and 3 of LDP.

• Specification: 640 to 720 Ω

– Measure resistance between terminals 2 and 3 of LDP.

• Specification for engine code AEB: 17.5 to 27.5 Ω
• Specification for engine code ATW: 15.0 to 17.0 Ω

– If the specifications are not attained:

• Replace Leak Detection Pump.

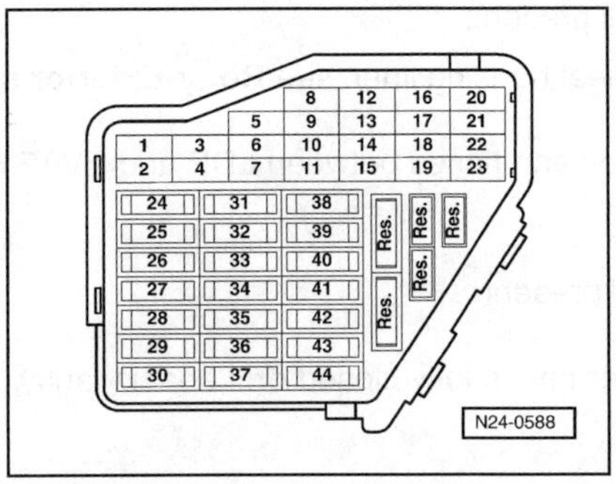

LDP, checking voltage supply

Conditions

Engine code AEB

Fuse 29 must be OK

Engine code ATW

- Fuse 34 must be OK

Checking sequence

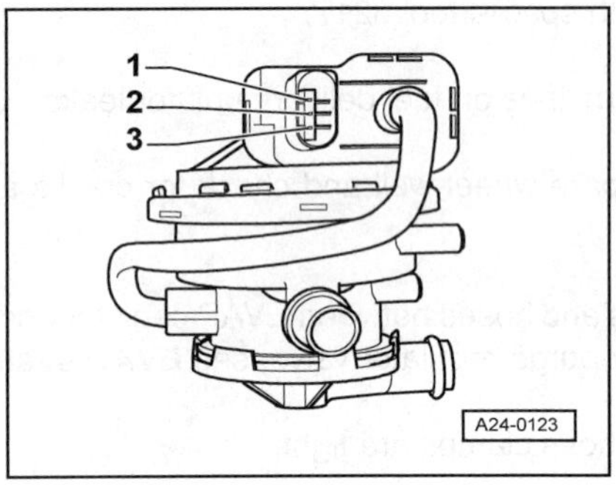

Disconnect 3-pin connector from Leak Detection Pump.

- Connect LED voltage tester VAG 1527B between LDP harness connector terminal 3 (positive) and engine Ground.
- Start engine and run at idling speed: LED must light up.

If the LED does not light up:

- Check wire between LDP harness connector terminal 3 and central electric for open circuit using wiring diagram, repair as necessary.

21 Turbocharger

Repair Group

B 21

Turbocharger System

NOTE —

- *Charge pressure regulation connection diagram, see* **Charge air system diagram**.
- *All hose connections are secured with hose clamps:*
- *Charge pressure system must be free of leaks.*
- *Replace self-locking nuts.*
- *Always replace gaskets.*

Safety precautions

NOTE —

If special testing equipment is required during test drive, note the following:

- *Do not operate test and measuring equipment from the passenger seat.*
- *In the event of an accident involving deployment of the passenger-side airbag, the person seated there could be injured.*
- *When driving or riding in an airbag-equipped vehicle, NEVER hold the scan tool or other test equipment in your hands or lap while in motion. Objects between you and the airbag increase the risk of injury in an accident.*
- *During a road test in an airbag-equipped vehicle, test equipment must always be fastened to and operated from the rear seat by a second technician.*

EDITOR'S NOTE —

Volkswagen identifies electrical components by a letter and/or a number in the electrical schematics. See **Wiring Diagrams, Fuses and Relays**. *These electrical identifiers are listed -between dashes- as an aid to electrical troubleshooting.*

Turbocharger, components

When working on the turbocharger and system components, observe rules for cleanliness, see **Rules for cleanliness.**

> *NOTE —*
> - *All hose connections are secured with clips.*
> - *Charge air system must be free of leaks.*
> - *Replace self-locking nuts*

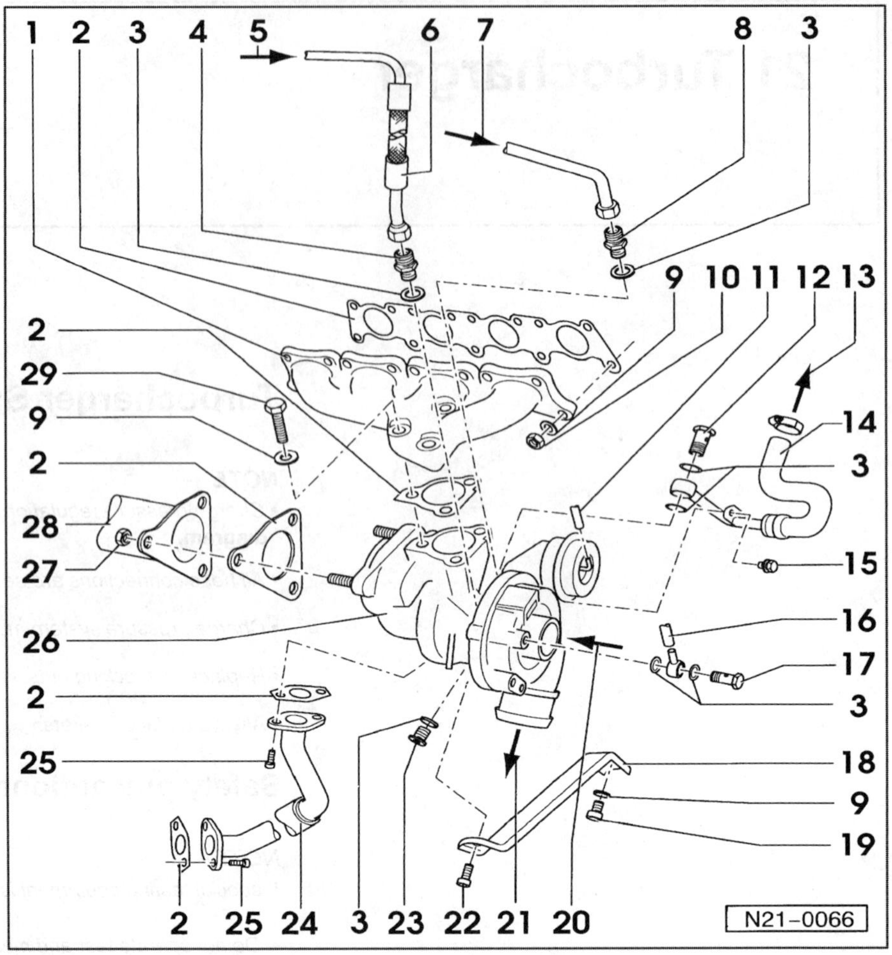

1. **Exhaust manifold**
2. **Gasket**
   - Note installation position
3. **Seal**
   - Always replace
4. **Connection**
   - Tighten to 30 Nm (22 ft-lb)
5. **From oil filter bracket**
6. **Oil supply line**
   - Tighten to 25 Nm (18 ft-lb)
7. **From coolant pipe on intake manifold**
   - Tighten to 30 Nm (22 ft-lb)
8. **Connection**
   - Tighten to 35 Nm (26 ft-lb)
9. **Washer**
10. **25 Nm (18 ft-lb)**
    - Coat threads with G 052 112 A3
11. **Hose**
    - To wastegate bypass regulator valve
12. **Banjo bolt**
    - Tighten to 25 Nm (17 ft-lb)
13. **To cylinder block**
14. **Coolant supply pipe**
15. **10 Nm (7 ft-lb)**
16. **Hose**
    - To wastegate bypass regulator valve
17. **Banjo bolt**
    - Tighten to 15 Nm (11 ft-lb)
18. **Retainer**
    - Between turbocharger and cylinder block
19. **40 Nm (30 ft-lb)**
20. **From air cleaner**
21. **To charge air cooler**
22. **10 Nm (7 ft-lb)**
23. **Plug**
    - Tighten to 15 Nm (11 ft-lb)
24. **Oil return line**
    - To oil pan
25. **10 Nm (7 ft-lb)**
26. **Turbocharger**
    - Charge pressure regulator valve and pressure unit are components of the turbocharger and cannot be replaced individually
27. **25 Nm (18 ft-lb)**
    - Coat threads with G 052 112 A3
28. **Three-Way Catalytic Converter**
29. **35 Nm (26 ft-lb)**
    - Always replace
    - Coat threads and bolt head seating surface with G 052 112 A3

Charge air system, cooling components

NOTE—
- *All hose connections secured by clips.*
- *Charge air system must be free of leaks.*
- *When installing, note the assembly markings on the hoses and components.*

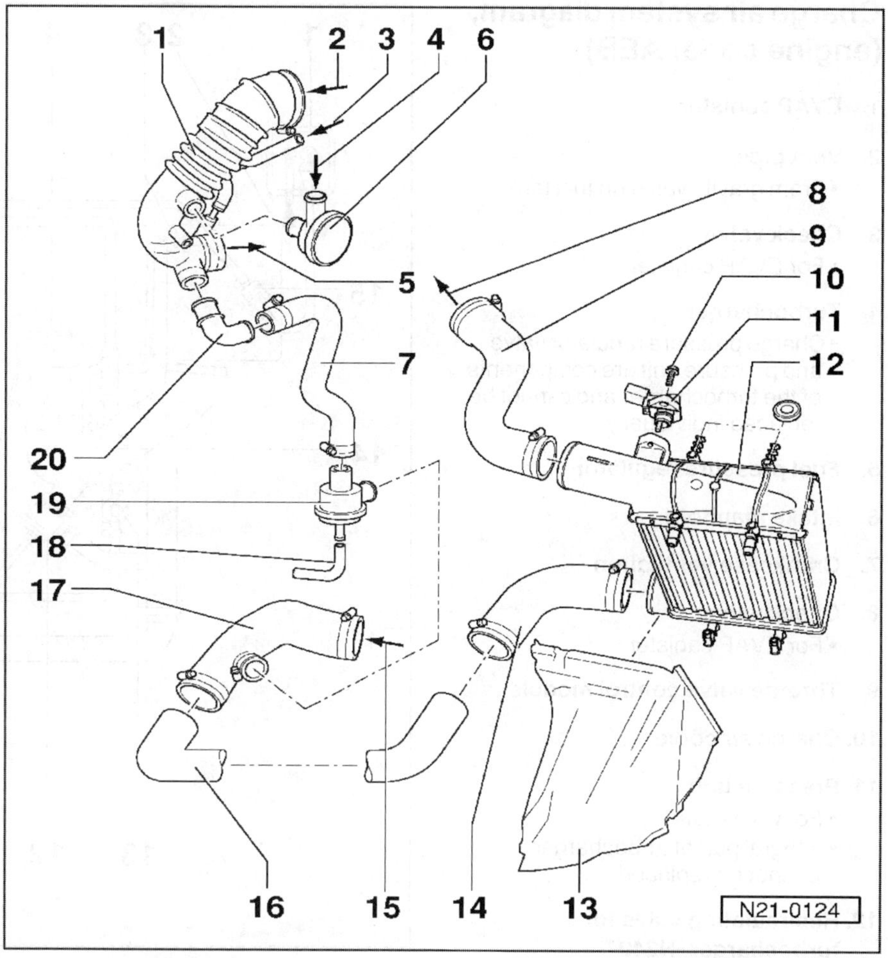

1. **Intake hose**
2. **From air cleaner**
3. **From EVAP canister**
4. **From crankcase breather**
5. **To turbocharger**
6. **Pressure regulator valve**
   - For Positive Crankcase Ventilation valve
7. **Connecting line/hose**
   - Note assembly markings
8. **To throttle valve control module**
9. **Connecting pipe/hose**
   - Charge air cooler/throttle valve control module
   - Note assembly markings
10. **10 Nm (7 ft-lb)**
11. **Charge air pressure sensor -G31-**
    - Only for engine code ATW
    - Replace O-ring if damaged
    - See **Charge air pressure sensor -G31-, checking**
12. **Charge air cooler**
13. **Air duct**
14. **Connecting pipe/hose**
    - Cross tube/charge air cooler
    - Note assembly markings
15. **From turbocharger**
16. **Cross tube**
17. **Connecting pipe/hose**
    - Turbocharger/cross tube
    - Note assembly markings
18. **Vacuum hose/line**
19. **Recirculating valve for turbocharger -N249-**
    - See **Recirculating valve for turbocharger -N249-, checking**
20. **Angle piece**

Charge air system diagram, (engine code: AEB)

1. **EVAP canister**
2. **Vent pipe**
   - From gravity valve on fuel tank
3. **Check valve**
   - For EVAP canister
4. **Turbocharger**
   - Charge pressure regulator valve and pressure unit are components of the turbocharger and cannot be replaced individually
5. **Fuel pressure regulator**
6. **Intake manifold**
7. **Crankcase ventilation**
8. **Check valve**
   - For EVAP canister
9. **Throttle valve control module**
10. **Charge air cooler**
11. **Pressure unit**
    - For wastegate
    - Integral part of turbocharger; cannot be replaced
12. **Recirculating valve for turbocharger -N249-**
13. **Wastegate bypass regulator valve -N75-**
    - The valve will be activated from the Engine Control Module (pulsed)
    - Closed with no current, charge pressure limited
14. **Pressure regulator valve**
    - For Positive Crankcase Ventilation
15. **EVAP canister purge regulator valve -N80-**
    - Checking, see **Repair Group 20**

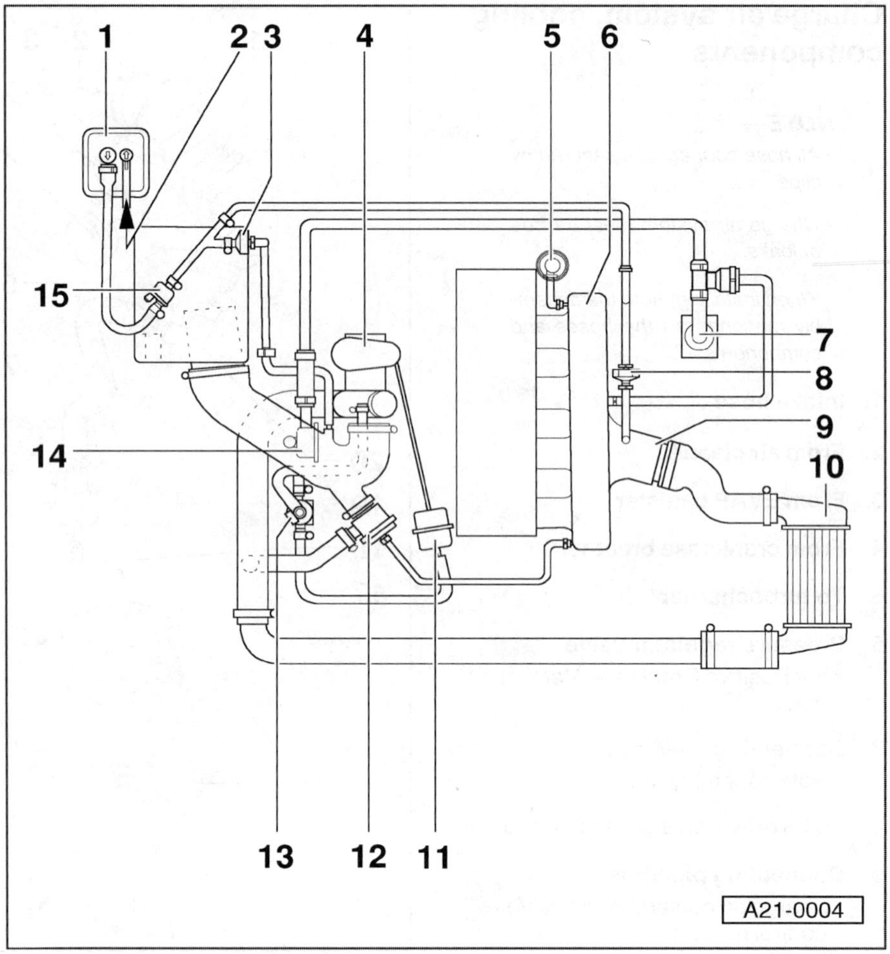

Charge air system diagram, (engine code: ATW)

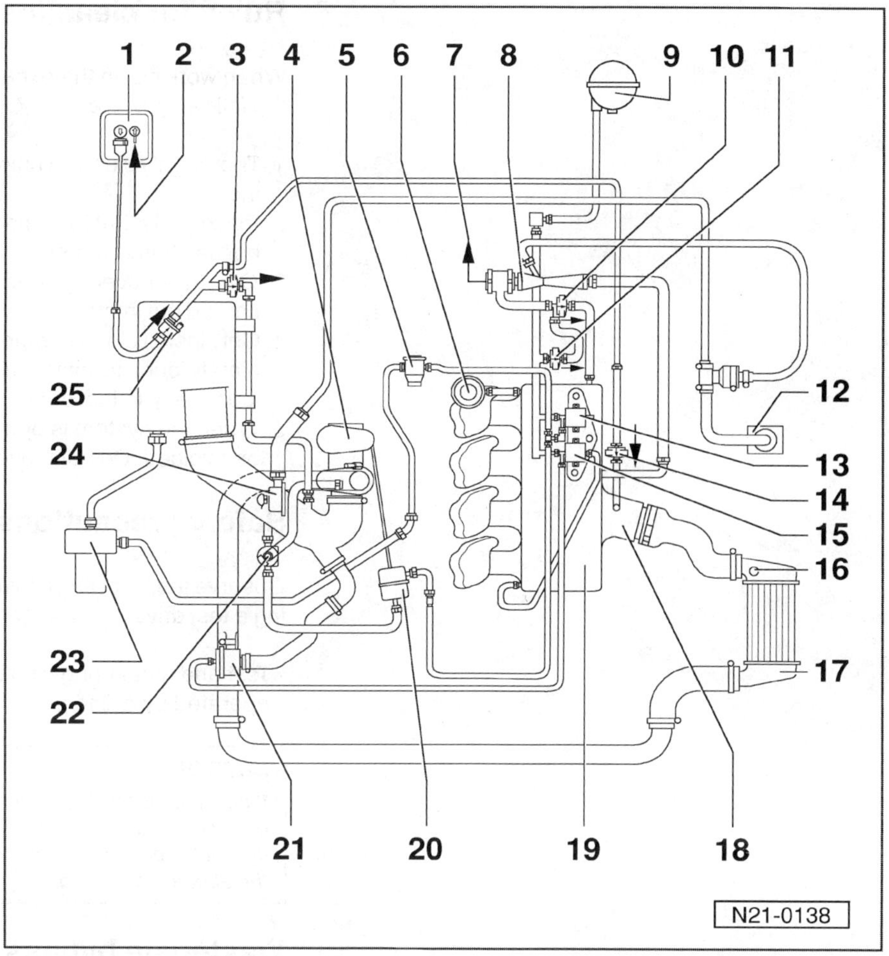

1. **EVAP canister**
2. **Vent pipe**
   - From gravity valve on fuel tank, see **Repair Group 20**
3. **Check valve**
   - For EVAP canister
4. **Turbocharger**
   - Charge pressure regulator valve and pressure unit are components of the turbocharger and cannot be replaced individually
5. **Combination valve**
   - For secondary air system
   - See **Repair Group 26**
6. **Fuel pressure regulator**
7. **To brake servo**
8. **Vacuum booster**
9. **Vacuum reservoir**
   - Secured in wheel housing, front left
10. **Check valve**
11. **Check valve**
12. **Crankcase ventilation**
13. **Secondary Air Injection (AIR) solenoid valve -N112-**
    - Checking, see **Repair Group 26**
14. **Check valve**
    - For EVAP canister
15. **Recirculating valve for turbocharger -N249-**
    - See **Recirculating valve for turbocharger -N249-, checking.**
16. **Charge air pressure sensor -G31-**
    - See **Charge air pressure sensor -G31-, checking**
17. **Charge air cooler**
18. **Throttle valve control module**
19. **Intake manifold**
20. **Pressure unit**
    - For wastegate
    - Integral part of turbocharger; cannot be replaced
21. **Recirculating valve for turbocharger -N249-**
22. **Wastegate bypass regulator valve -N75-**
    - The valve will be activated from the engine control module (pulsed)
    - Closed with no current, charge pressure limited

    EDITOR'S NOTE —

    Component activation is accomplished using Volkswagen diagnostic scan tool, output diagnostic test mode (function 03).
23. **Secondary Air Injection (AIR) pump motor -V101-**
    - Checking see **Repair Group 26**
24. **Pressure regulator valve**
    - For Positive Crankcase Ventilation
25. **EVAP canister purge regulator valve -N80-**
    - Checking, see **Repair Group 20**

Rules for cleanliness

When working on the exhaust gas turbocharger, pay careful attention to the following 5 rules:

1. Thoroughly clean all unions and the adjacent areas before disconnecting.
2. Place parts that have been removed on a clean surface and cover. Do not use fluffy cloths!
3. Carefully cover opened components or seal, if the repair cannot be carried out immediately.
4. Only install clean components: Only unpack replacement parts immediately prior to installation. Do not use parts that have been stored loose (e.g. in tool boxes etc.).
5. When the system is open: Do not work with compressed air if this can be avoided. Do not move the vehicle unless absolutely necessary.

Safety precautions

Observe the following if test and measuring instruments are required during a test drive:

- Test and measuring instruments must be secured to the rear seat and operated by a 2nd person from this location.

CAUTION—
If test and measuring instruments are operated from the front passenger's seat and the vehicle is involved in an accident, there is a possibility that the person sitting in this seat may receive serious injuries when the airbag is triggered.

Wastegate bypass regulator valve -N75-, checking

Special tools and equipment

- Multimeter V.A.G 1526, V.A.G 1715 or Fluke 83
- Adapter set V.A.G 1594

Checking conditions

- Output Diagnostic Test Mode (DTM) must be performed.

EDITOR'S NOTE—
Component activation is accomplished using Volkswagen diagnostic scan tool, output diagnostic test mode (function 03).

Checking sequence

– Switch ignition off.

– Disconnect connector from wastegate bypass regulator valve.

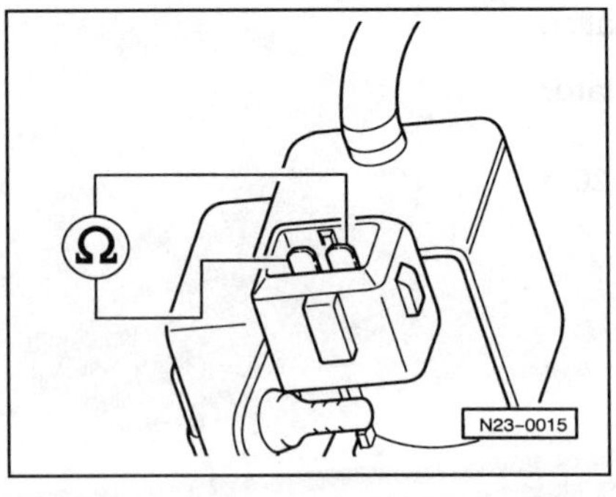

◀ Measure resistance between valve contacts.

- Specification: 25 to 35 Ω

If the specification is not obtained:

– Replace Wastegate Bypass Regulator valve -N75-.

Pressure unit for charge pressure regulator valve, checking

Special tools and equipment

- V.A.G 1551 Scan Tool or vehicle system tester V.A.G 1552 with cable V.A.G 1551/3

Checking conditions

- Engine oil temperature min. 80° C (176°F)

Checking sequence

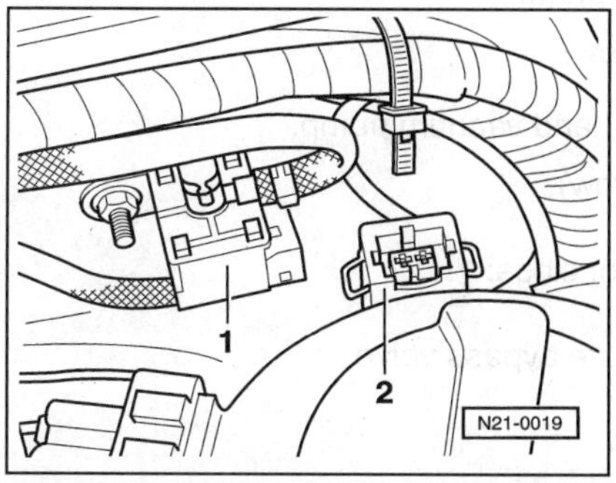

Disconnect connector -2- from Wastegate Bypass Regulator valve -1-.

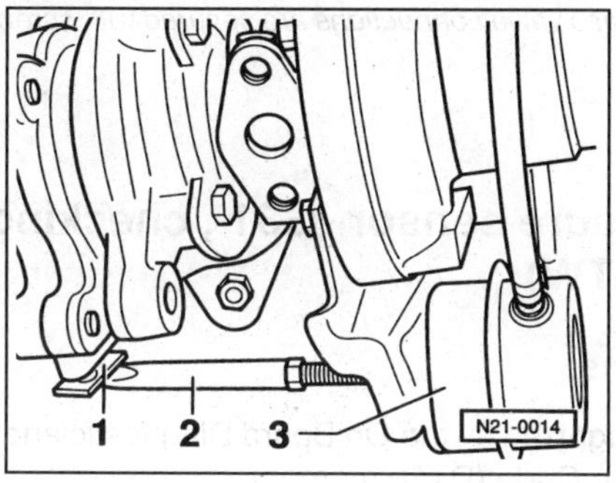

Start engine and raise to maximum rpm briefly using throttle.

- Operating rod -2- must move.

If the operating rod does not move:

– Check valve lever -1- for ease of movement.

If corroded together:

– Replace turbocharger.

If the operating rod does not move, even though the lever is free to move:

– Test wastegate bypass regulator valve.

- Hose from turbocharger via valve to pressure unit must not be blocked when connector is pulled off

If the wastegate bypass regulator valve is OK:

– Replace turbocharger, see **Turbocharger, components**.

Charge pressure bypass valve, checking (engine code: AEB)

NOTE—
Charge pressure bypass valve (engine code: ATW), see **Recirculating valve for turbocharger -N249-, checking.**

Special tools and equipment

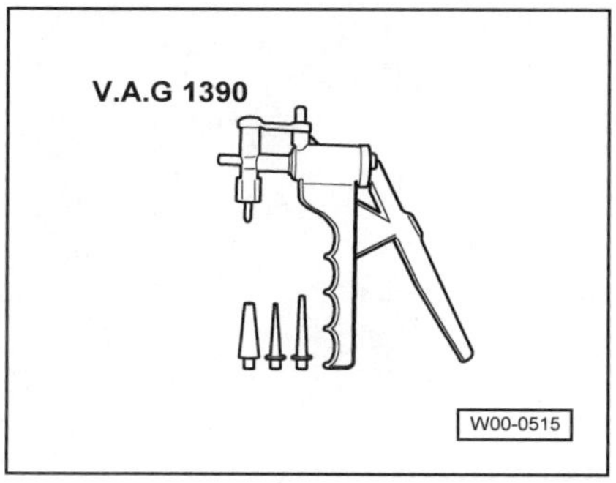

V.A.G 1390 Hand vacuum pump

Checking conditions

- Noticeable reduced performance or load change judders

Checking sequence

NOTE—
The charge pressure bypass valve is located in front of the turbocharger. It is opened by vacuum pressure during the overrun phase and at idling speed.

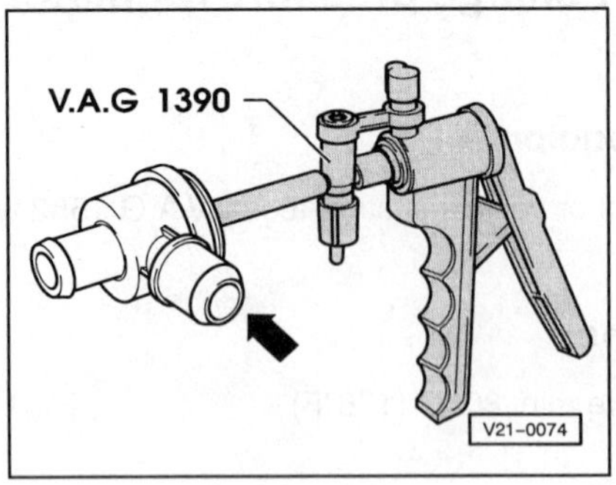

- Connect hand vacuum pump V.A.G 1390 to vacuum connection on valve.

– Operate hand vacuum pump.

  • Valve must open (arrow).

– Operate vent valve on hand vacuum pump.

  • Valve must close (arrow).

If valve does not open or close:

– Replace charge pressure bypass valve.

NOTE—
The charge pressure bypass valve connections are secured with screw-type clips.

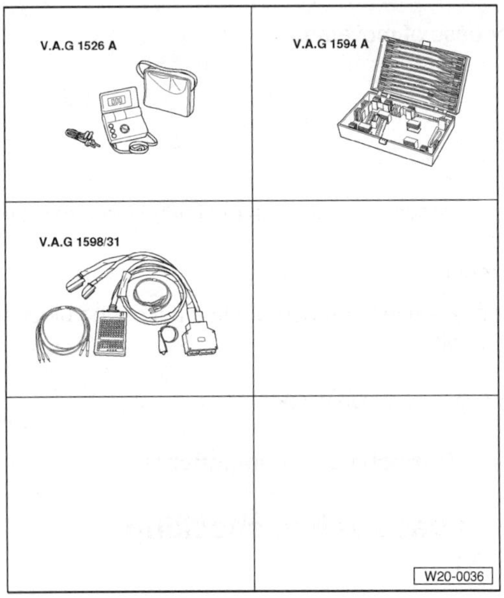

Charge air pressure sensor -G31-, checking (engine code: ATW)

Checking conditions

• A code has been recognized by the On Board Diagnostic and stored in the Diagnostic Trouble Code (DTC) memory

Special tools and equipment

• V.A.G 1526 or Fluke 83 multimeter
• V.A.G 1594 A Adapter set
• V.A.G 1598/31 Test box
• Wiring diagram

Checking voltage supply and wiring to control module

– Switch ignition off.

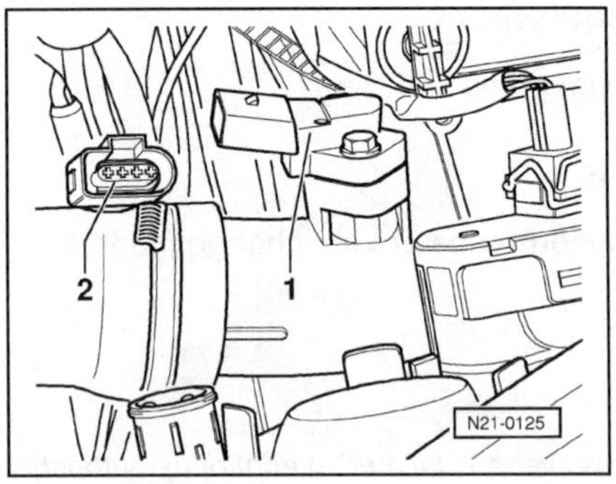

- Disconnect 4-pin connector -2- from charge air pressure sensor -1-.

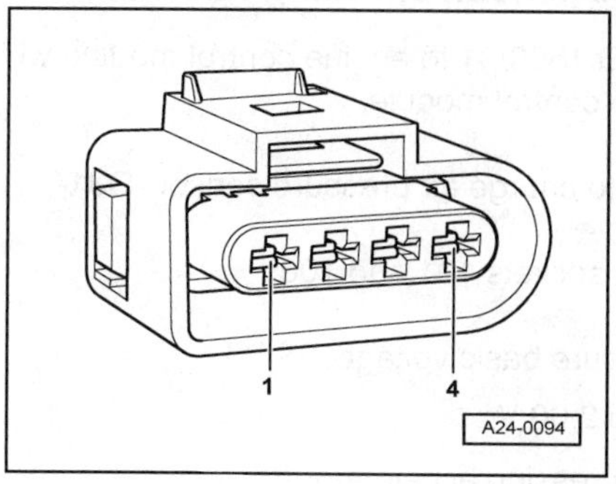

◄ Connect multimeter to measure voltage using auxiliary cables from V.A.G 1594 to connector terminals 1 + 3.

– Switch ignition on.
  - Specification: min. 4.5 V

– Switch ignition off.

If the specification is not obtained:

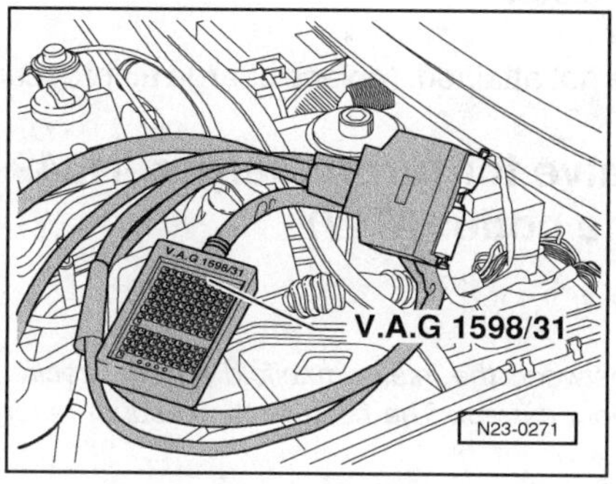

◄ Connect test box V.A.G 1598/31 to control module wiring harness (Engine control module remains disconnected).

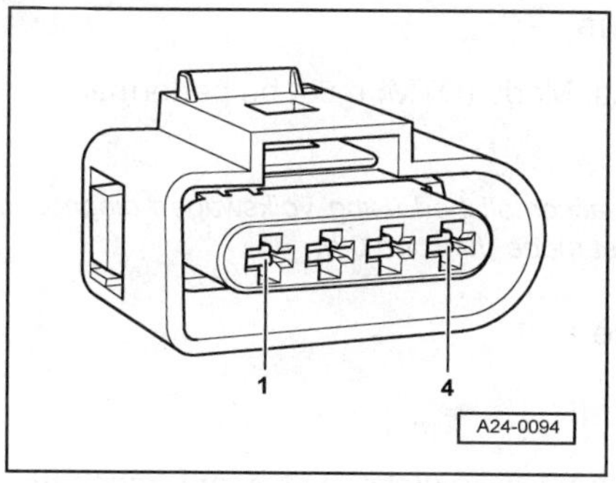

◄ Check wiring between test box and connector for open circuit according to wiring diagram.
  - Terminal 1 + socket 108
  - Terminal 2 + socket 98
  - Terminal 3 + socket 101
  - Wire resistance: Max. 1.5 Ω

– Additionally check wires for short to one another.

If no faulty wiring is detected and voltage was present between terminals1 + 3:

– Check control module function, see **Checking function**.

If no faulty wiring is detected and no voltage was present between terminals 1 + 3:

– Replace engine control module.

EDITOR'S NOTE—

ECM replacement requires coding of control module using Volkswagen factory scan tool. Consult an authorized Volkswagen service department for additional information.

Charge air pressure sensor, checking function

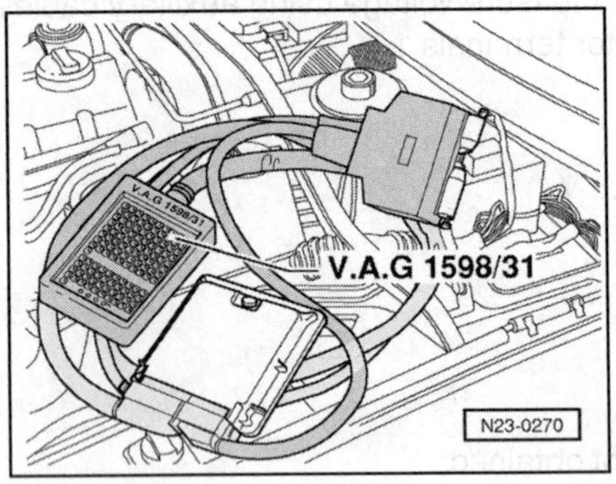

< Connect test box V.A.G 1598/31 to engine control module wiring harness and to engine control module.

– Reconnect connector to charge air pressure sensor -G31-.

– Connect multimeter to sockets 101 and 108.

– Start engine and measure basic voltage.

• Specification: 1.80 to 2.00 V

– Raise engine revs by pressing accelerator to:

• Specification: 2.00 to 3.00 V

If the specifications are not attained, replace charge air pressure sensor.

Recirculating valve for turbocharger -N249-, checking (engine code: ATW)

NOTE —

The valve in the hose between the intake manifold and the recirculating valve is open if it is without current. The hose to the vacuum reservoir is closed.

Checking conditions

• Output Diagnostic Test Mode (DTM) must be performed.

EDITOR'S NOTE —

Component activation is accomplished using Volkswagen diagnostic scan tool, output diagnostic test mode (function 03).

Checking sequence

– Switch off ignition.

– Disconnect connector from recirculating valve for turbocharger.

< Measure resistance between valve contacts.

• Specification: 25 to 35 Ω

If the specification is not obtained, replace Recirculating valve for turbocharger -N249-.

26 Exhaust System and Emission controls

Repair Group

B 26

Exhaust System

NOTE —

- *After work on the exhaust system ensure that the system is not under stress and that it has sufficient clearance from the bodywork. If necessary, loosen the double and single clamps and align the muffler and exhaust pipe so that sufficient clearance is maintained to the bodywork and the support rings are evenly loaded.*
- *Always replace self-locking nuts.*

EDITOR'S NOTE —

Volkswagen identifies electrical components by a letter and/or a number in the electrical schematics. See **Wiring Diagrams, Fuses and Relays***. These electrical identifiers are listed -between dashes- as an aid to electrical troubleshooting.*

Exhaust system components

1. **25 Nm (18 ft-lb)**
2. **Exhaust manifold**
   - See **Repair Group 20, Turbocharger, components**
3. **Gasket**
   - Always replace
4. **Heated Oxygen sensor (HO2S) -G39-, 50 Nm (37 ft-lb)**
   - Grease only the threads of sensor with G 052 112 A3
   - G 052 112 A3 must not get into the slots on the sensor body
   - Remove and install with 3337
   - Checking, see **Fuel injection and ignition section, Repair Group 24**
5. **Three-Way Catalytic Converter (TWC)**
   - See **Three-Way Catalytic (TWC) Converter, checking**
6. **Oxygen sensor (O2S) behind Three-Way Catalytic Converter (TWC) -G130-**
   - Grease only the threads of sensor with G 052 112 A3
   - G 052 112 A3 must not get into the slots on the sensor body
   - Remove and install with 3337
   - Checking, see **Fuel injection and ignition section, Repair Group 24**
7. **Mounting**
   - With retaining ring
8. **Mounting**
9. **25 Nm (18 ft-lb)**
10. **Rear muffler**
    - As standard, center and rear mufflers are installed as a single component. For repairs, the mufflers are supplied individually and with a double clamp for connecting.
    - Removing and installing: cut through connecting pipe separating point, -**16**- at right angles with body saw, e.g. V.A.G 1523.
11. **Dimension -a- = approx. 7 to 9 mm**
    - Exhaust system must be cold
    - Arrow points in direction of travel
12. **Bolt head must face fuel tank**
13. **Angle -α- = 30 ±5°**

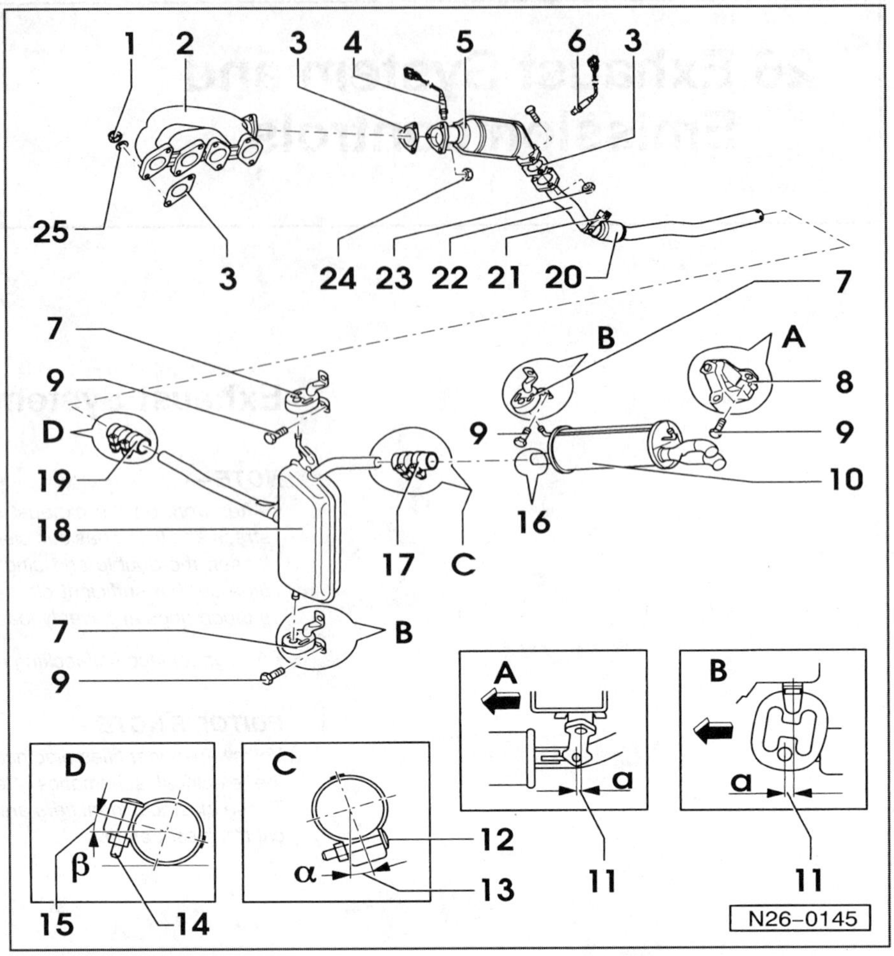

14. **End of bolt must not be below double clamp**
    - Double clamp as seen from normal direction of travel
15. **Angle -β- = 10 +5°**
16. **Separating point**
    - Approx. 160 to 180 mm before the rear muffler, marked by a crimp in pipe.
17. **Double clamp**
    - Tighten evenly to 40 Nm (30 ft-lb)
18. **Center muffler**
19. **Double clamp**
    - Tighten evenly to 40 Nm (30 ft-lb)
20. **Decoupling element**
    - Max. misalignment = 10°
21. **Mounting**
    - Individual components, see **Fig. 1**
22. **Front exhaust pipe**
23. **25 Nm (18 ft-lb)**
24. **30 Nm (22 ft-lb)**
25. **Washer**

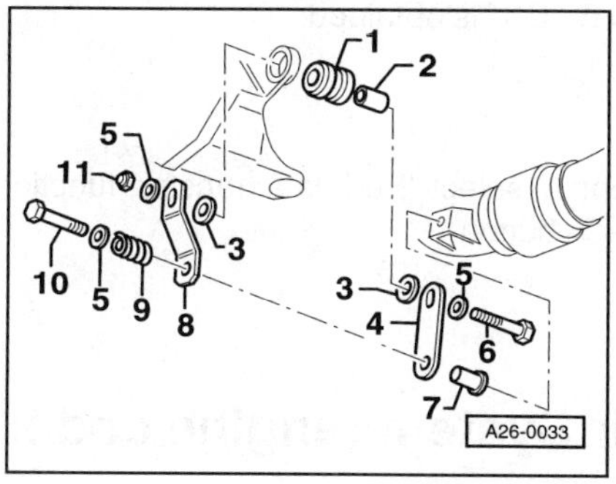

Fig. 1 Individual mounting components

1. Buffer
2. Spacer pipe
3. Washer
4. Right-hand plate
5. Washer
6. Hex bolt
7. Spacer sleeve
8. Left-hand plate
9. Spring
10. Hex bolt
11. Self-locking nut, 25 Nm (18 ft-lb)

Three-Way Catalytic Converter (TWC), checking

Special tools and equipment

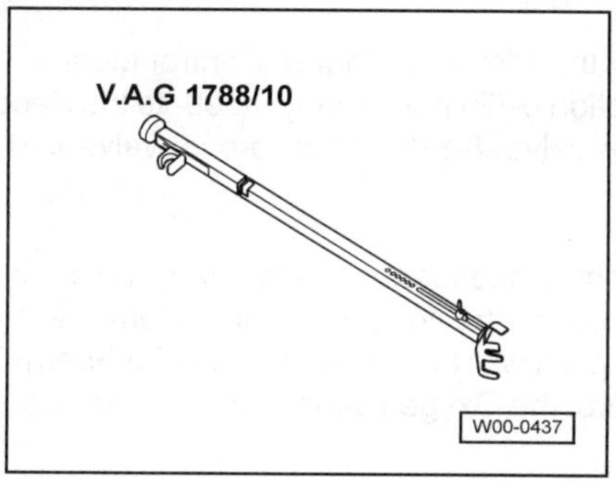

V.A.G 1551 Scan Tool or vehicle system tester V.A.G 1552 with cable V.A.G 1551/3

• V.A.G 1788/10 Engine speed adjuster

Checking sequence

– Connect V.A.G 1551 (V.A.G 1552). Start engine and select engine control module with "Address word" 01.

– Press -0- and -4- buttons to select "Introduction of basic setting" and confirm entry with -Q- button.

– Set engine speed to 1800 to 2200 rpm using speed regulator V.A.G 1788/10.

– Press -0-, -3- and -4- buttons to select "Display group number 34" and confirm entry with -Q- button.

– Maintain engine speed at 1800 to 2200 rpm until display in display field 4 jumps from "Test OFF" to "Test ON".

• Catalyst temperature in display field 2 must be between 350 and 500°C.

– Continue to maintain speed at 1800 to 2200 rpm until "B1-P1 OK" appears in display field 4.

– Press -C- button.

– Press -0-, -4- and -6- buttons to select "Display group number 46" and confirm entry with -Q- button.

– Maintain engine speed at 1800 to 2200 rpm until display in display field 4 jumps from "Test OFF" to "Test ON".

• Catalyst temperature in display field 2 must be at least 350°C.

– Check amplitude ratio in display field 3.

• Specification: max. 0.4

– Continue to maintain engine speed at 1800 to 2200 rpm until "Cat B1 OK" appears in display field 4.

– Check Diagnostic Trouble Code (DTC) memory.

If the specification "Cat B1 OK" is obtained:

- Press →button.
- Press -0- and -6- buttons to select "End data transfer" function and confirm input with the -Q- button.

Secondary Air System (engine code: ATW)

Function

The secondary air system blows air in behind the exhaust valve for 100 seconds during a cold start (5°C (41°F) to 33°C (91°F) coolant temperature). This produces an oxygen rich exhaust gas, causes afterburning and reduces the heating-up phase of the catalyst.

Activation occurs from the Motronic Engine Control Module -J220- via the Secondary Air Injection (AIR) pump relay -J299- to the Secondary Air Injection (AIR) solenoid valve -N112-, change-over valve and combination valve.

Additionally, after each subsequent engine start (up to max. 96°C (205°F) engine temperature) the secondary air system will switch in for 10 seconds during idling and will be checked by the On Board Diagnostic (OBD). When this occurs, the Oxygen sensor control must be active.

Secondary air system components, removing and installing

NOTE—

- *Components marked with an asterisk (*) are checked by On Board Diagnostic (OBD).*
- *Components marked with a double asterisk (**) are checked by the Output Diagnostic Test Mode (DTM).*

EDITOR'S NOTE—

Component activation is accomplished using Volkswagen diagnostic scan tool, output diagnostic test mode (function 03).

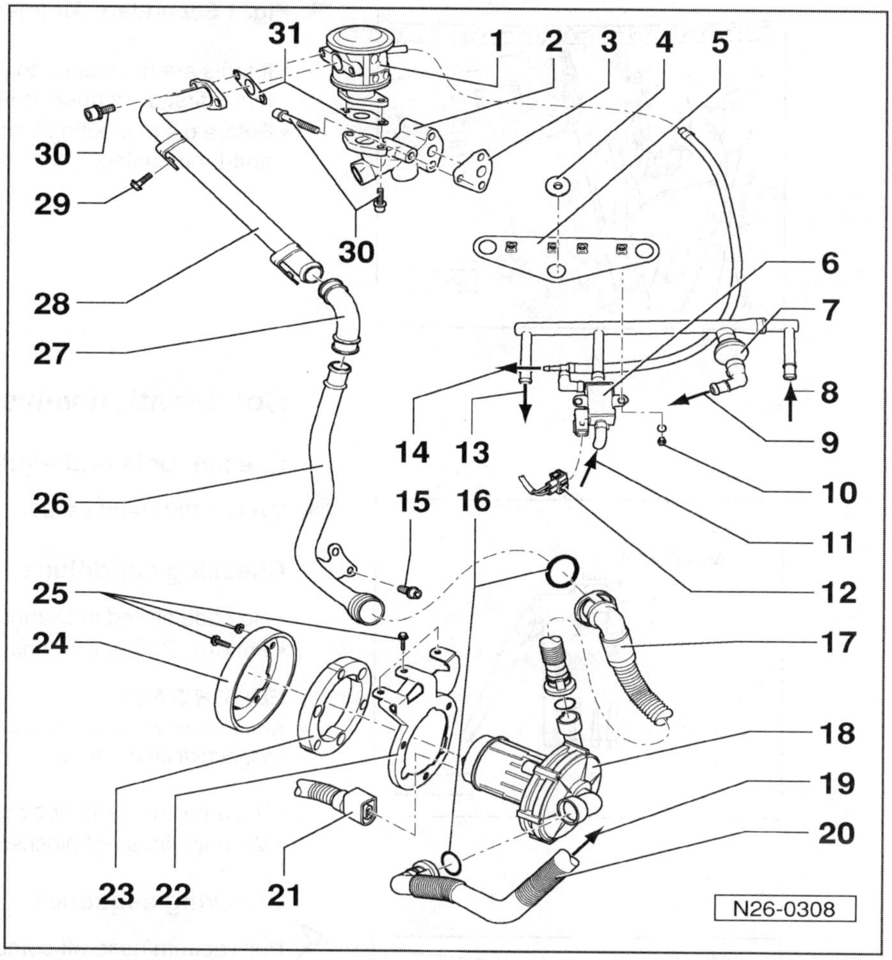

1. **Combination valve**
   - See **Combination valve, checking**
2. **Bracket**
   - Secured to cylinder head
3. **Gasket**
   - Always replace
4. **Rubber mounting**
5. **Bracket**
   - Secured to intake manifold
6. **Secondary Air Injection (AIR) solenoid valve -N112-*/****
   - Resistance: 25 to 35 Ω
7. **Check valve**
8. **From vacuum reservoir**
9. **To brake servo**
10. **6 Nm (53 in-lb)**
11. **Breather connection**
12. **Connector**
    - 2-pin
13. **To Recirculating valve for turbocharger -N249-**
14. **To pressure unit on turbocharger**
15. **10 Nm**
16. **O-Ring**
    - Replace if damaged
17. **Pressure hose**
    - Ensure seated tightly
    - Press together at front to release
18. **Secondary Air Injection pump motor -V101-****
    - See **Secondary Air Injection pump motor, checking**
19. **To air cleaner**
20. **Intake hose**
    - For secondary air pump
21. **Connector**
    - 2-pin
    - For secondary air injection pump motor
22. **Bracket**
    - For secondary air injection pump motor
23. **Insulation ring**
24. **Protective cap**
25. **10 Nm (7 ft-lb)**
26. **Connecting pipe**
27. **Angle piece**
    - Secure with screw-type or spring-type clips
28. **Pressure pipe**
29. **6 Nm (53 in-lb)**
30. **10 Nm (7 ft-lb)**
31. **Gasket**
    - Always replace

Fig. 1 Secondary Air Injection pump relay -J299- (arrow)

- If tools are necessary to pull relays or control modules out of the relay plate, first disconnect battery Ground strap.
- Before disconnecting battery Ground strap obtain code for radios with anti-theft coding.

Combination valve, checking

Special tools and equipment

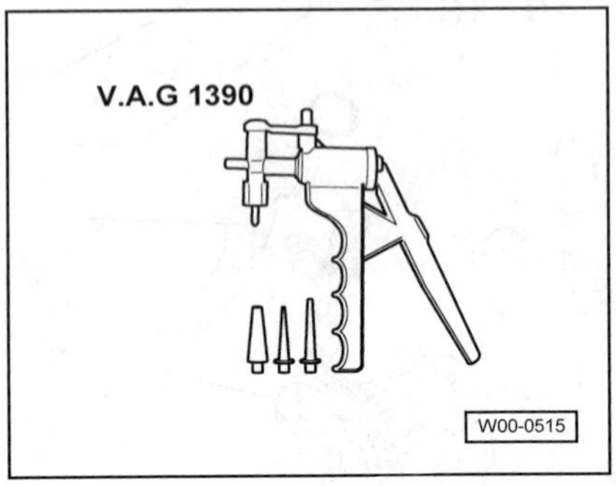

V.A.G 1390 Hand vacuum pump

Checking conditions

- No code stored in Diagnostic Trouble Code (DTC) memory.
- Perform Output Diagnostic Test Mode (DTM).

EDITOR'S NOTE —

Component activation is accomplished using Volkswagen diagnostic scan tool, output diagnostic test mode (function 03).

- Vacuum lines and hose connections free of leaks.
- Vacuum lines not blocked/kinked.

Checking sequence

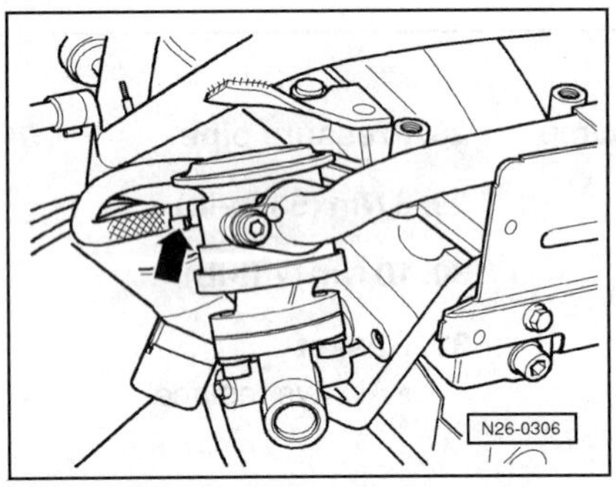

Pull vacuum hose off combination valve (arrow) and connect hand vacuum pump V.A.G 1390.

NOTE —

Do not use compressed air during the following check!

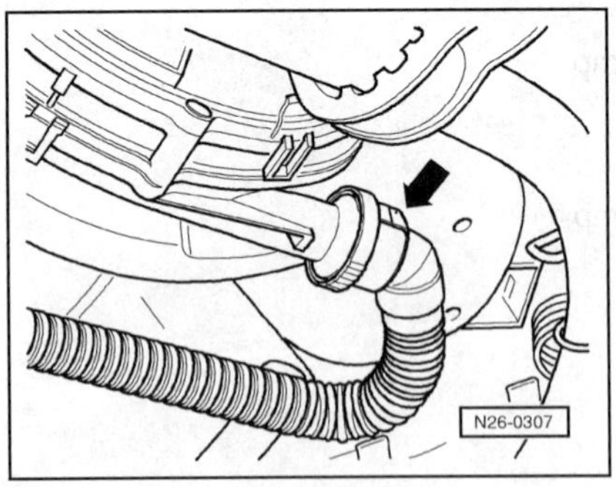

Remove pressure hose from secondary air injection pump motor and blow into hose lightly.

- Combination valve must be closed.

– Operate hand vacuum pump.

- Combination valve must open.

– If the combination valve does not open, replace combination valve

Secondary air injection pump motor, checking

Special tools and equipment

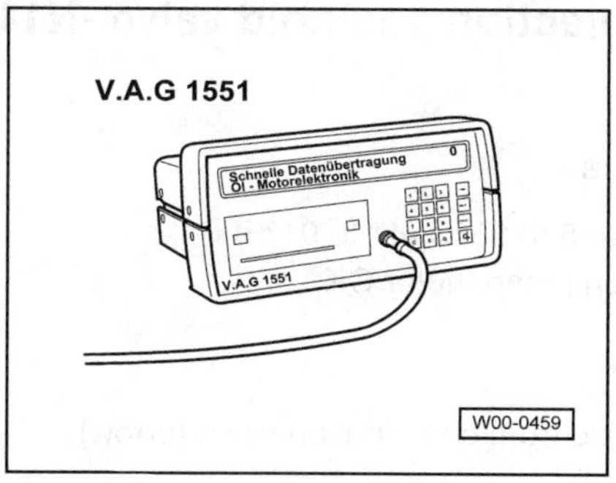

V.A.G 1551 Scan Tool

NOTE—

The vehicle system tester V.A.G 1552 can be used instead of the V.A.G 1551 Scan Tool, however a print-out is not possible.

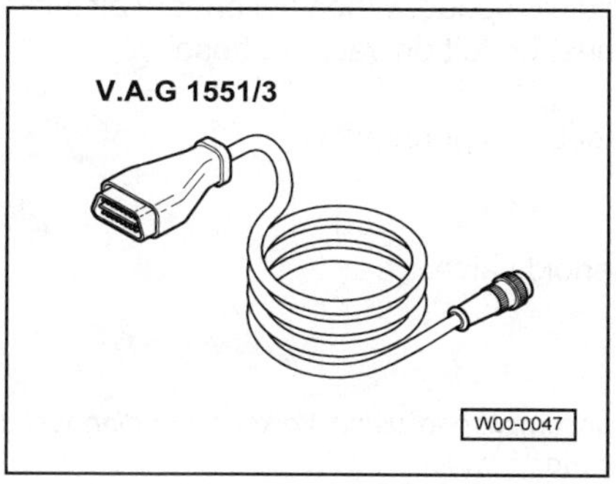

V.A.G 1551/3 Cable

Checking conditions

- Fuse for secondary air injection pump relay OK.
- Suction hose for secondary air injection pump is not blocked or kinked

Checking sequence

– Remove insulation tray, see **Body-Exterior section, Repair Group 50**

– Disconnect pressure hose from the secondary air injection pump motor.

– Activate secondary air injection pump relay via Output DTM.

EDITOR'S NOTE—

Component activation is accomplished using Volkswagen diagnostic scan tool, output diagnostic test mode (function 03).

- Secondary air injection pump motor must run at intervals, and air must exit outlet connections.

If the motor runs, but no air exits:

– Replace secondary air injection pump motor.

If the secondary air injection pump motor does not run at intervals:

– Check activation of secondary air injection pump motor.

EDITOR'S NOTE—

Component activation is accomplished using Volkswagen diagnostic scan tool, output diagnostic test mode (function 03).

Secondary air injection solenoid valve -N112-, checking

Checking conditions

• Engine oil temperature 5 to 33°C (41 to 91°F)

• Secondary air injection pump motor OK.

Checking sequence

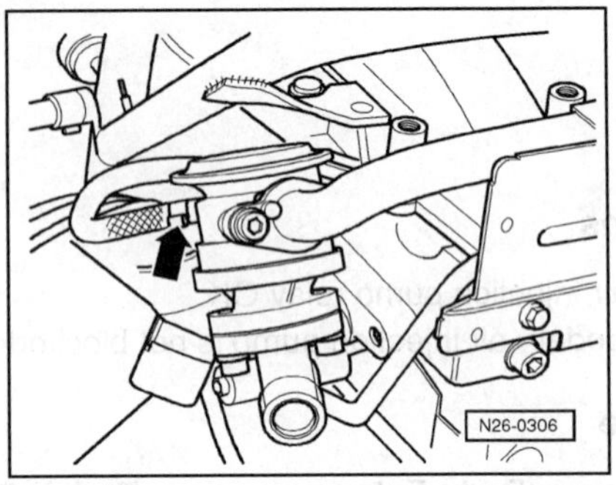

◄ Disconnect vacuum hose from combination valve (arrow).

– Start engine and run at idling speed. When secondary air injection pump motor starts, vacuum must be felt on vacuum hose.

– If no intake manifold vacuum is perceptible:

• Switch engine off.

– Check activation of solenoid valve.

EDITOR'S NOTE —

Component activation is accomplished using Volkswagen diagnostic scan tool, output diagnostic test mode (function 03).

1.8 Liter 4-Cylinder 5V Turbo Engine Mechanical (code: AUG, AWM)

Section
C

20 Fuel Supply

Page

21 Turbocharger

Page

26 Exhaust System and Emission Controls

Page

00 General, Technical Data

Repair Group

c 00

Technical data

Engine identification

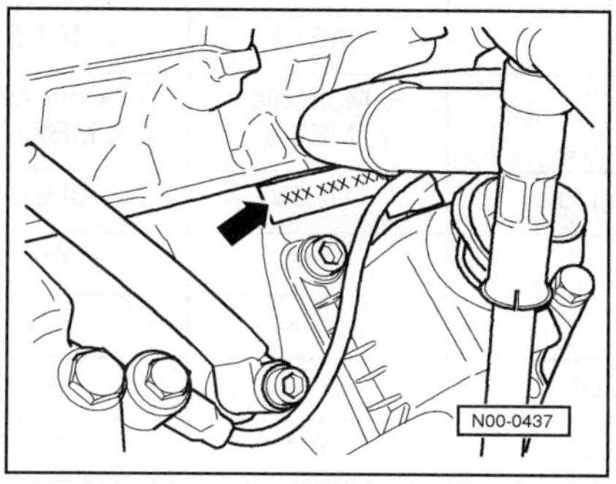

The engine code -arrow- ("engine code" and "production number") are located on the left side, on the cylinder head.

In addition, a sticker containing "engine code" and "production number" is placed on the cylinder head.

The engine identification consists of up to nine characters (alphanumeric). The first portion (maximum 3 characters) represents the "engine code", the second portion (six characters) represents the serial number. If more than 999 999 engines were produced under the same code, the first of the six characters is replaced with a letter.

The engine code is also included on the vehicle data plate.

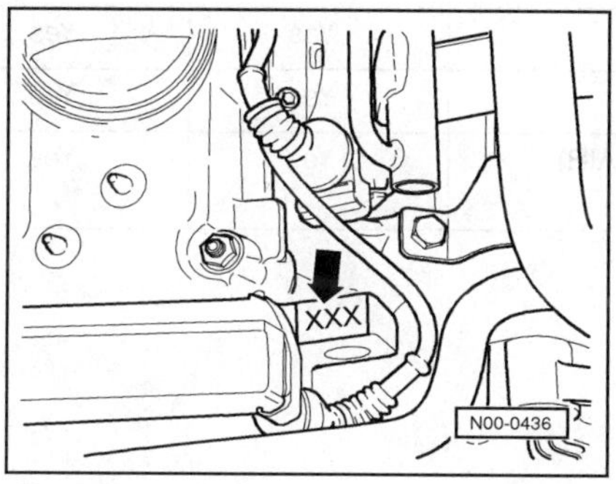

The engine code (arrow) is also stamped in at the front of the engine lifting eye (visible after removing cover above fuel injectors).

Engine data

Code letters	AUG	AWM
Manufactured	05.00➤10.00	11.00➤
Displacement Liters	1.8	1.8
Output hp (kw) @ RPM	150 (110) @ 5700	170 (125) @ 5900
Torque ft-lb (Nm) @ RPM	155 (210) @1750 - 4600	166 (225) @1950 - 5000
Bore diameter mm	81.0	81.0
Stroke mm	86.4	86.4
Compression ratio	9.3	9.3
RON	95 unleaded	95 unleaded
Fuel injection	Motronic ME7.5	Motronic ME7.5
Ignition	Motronic ME7.5	Motronic ME7.5
Emissions limits according to	ULEV	ULEV
E-gas system	Yes	Yes
Knock sensor control	Yes	Yes
On Board Diagnostic (OBD)	Yes	Yes
Oxygen sensor control	Yes	Yes
Catalytic converter	Yes	Yes
Exhaust Gas Recirculation (EGR)	No	No
Camshaft adjustment	Yes	Yes
Turbocharger	Yes	Yes
Secondary Air Injection (AIR) system	Yes	Yes

13 Engine-Crankshaft, Cylinder Block

Repair Group

C 13

Engine Block Assembly

NOTE—

If large quantities of metal particles or other deposits (caused, for example, by partial seizure of the crankshaft or connecting rod bearings) are found in the engine oil, the oil passages must be cleaned thoroughly and the oil cooler replaced in order to prevent further damage from re-occurring.

For timing belt and accessory drive belt information, see **Engine Belts**.

For cylinder block, engine bracket, and seal information, see **Engine block, components**.

Engine belts

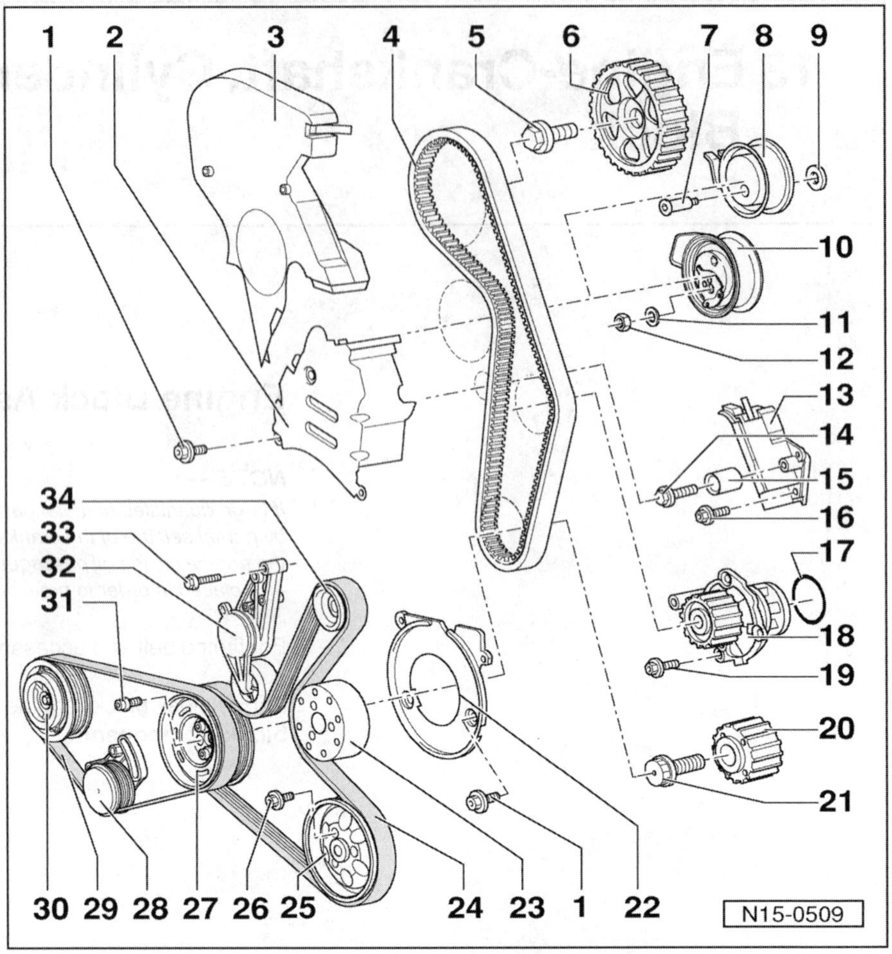

NOTE—

From 08.00 a new tensioning roller and toothed belt gear are being installed for toothed belt drive. Tensioning roller must be adjusted in order to tension toothed belt.

1. **10 Nm (7 ft-lb)**
   - Replace
2. **Toothed belt guard, center piece**
3. **Toothed belt guard, upper part**
4. **Toothed belt**
   - Mark direction of rotation before removing
   - Check for wear
   - Do not kink
   - See **Repair Group 15 , Toothed belt, removing and installing**
5. **65 Nm (48 ft-lb)**
   - Use 3036 retainer for loosening and tightening
6. **Camshaft gear**
   - Note installation position: The narrow rib of the camshaft gear points outward and the TDC mark for cyl. 1 is visible
7. **27 Nm (20 ft-lb)**
8. **Tensioning roller**
   - Up to 07.00, not adjustable
9. **Washer**
10. **Tensioning roller**
    - From 08.00, must be adjusted using 3387 pin wrench
11. **Washer**
12. **27 Nm (20 ft-lb)**
13. **Tensioning device for toothed belt**
    - Up to 07.00 secure using T40011
    - From 08.00 secure using T10008
14. **25 Nm (18 ft-lb)**
15. **Idler roller**
16. **15 Nm (11 ft-lb)**
17. **O-ring**
    - Replace if damaged
18. **Coolant pump**
    - Removing and installing, see **Repair Group 19, Coolant pump, removing and installing**
19. **15 Nm (11 ft-lb)**
20. **Crankshaft toothed belt gear**
21. **90 Nm (66 ft-lb) + $^1/_4$ turn (90 °) additional**
    - Replace
    - Thread and collar must be free of oil and grease
    - Use 3415 counter support to loosen and tighten
22. **Toothed belt guard, lower part**
23. **Belt pulley**
    - for viscous-fan clutch
24. **Ribbed belt**
    - Mark direction of rotation before removing
    - Removing and installing, see **Maintenance**
25. **Belt pulley**
    - for power steering vane pump
26. **25 Nm (18 ft-lb)**
27. **Belt pulley/harmonic balancer**
    - for 2 belt drives
    - Ribbed belt, removing and installing, see **Maintenance**
28. **Ribbed belt tensioner**
29. **Ribbed belt**
    - for A/C compressor
    - Mark direction of rotation before removing
    - Removing and installing, see **Maintenance**
30. **Belt pulley**
    - For A/C compressor
    - Securing A/C compressor to bracket /cylinder block, see **Heating & Air Conditioning section, Repair Group 87**
31. **10 Nm (7 ft-lb) + $^1/_4$ turns (90 °) additional**
    - Replace
32. **Tensioning device for ribbed belt**
    - See **Maintenance**
33. **25 Nm (18 ft-lb)**
34. **Belt pulley**
    - For generator

Engine block, components

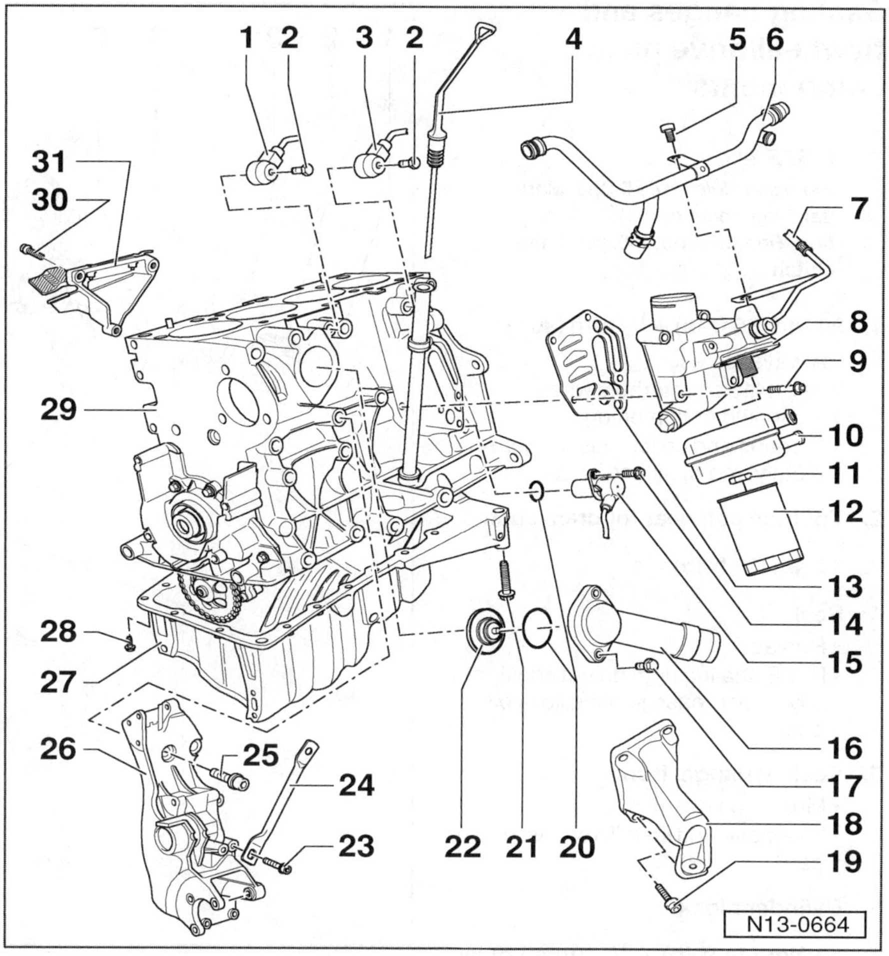

1. **Knock Sensor (KS) 1 -G61-**
   - Checking, see **Fuel injection and ignition section, 1.8 Liter 4-Cyl. 5V Turbo Fuel Injection & Ignition (code: ATW, AUG, AWM), Repair Group 28**
2. **20 Nm (15 ft-lb)**
   - Tightening torque affects function of Knock Sensor (KS)
3. **Knock Sensor (KS) 2 -G66-**
   - Checking, see **Fuel injection and ignition section, 1.8 Liter 4-Cyl. 5V Turbo Fuel Injection & Ignition (code: ATW, AUG, AWM), Repair Group 28**
4. **Dipstick**
   - Oil level must not exceed max. marking!
5. **20 Nm (15 ft-lb)**
6. **Lower coolant line**
7. **Oil supply line**
   - To turbocharger
8. **Oil filter bracket**
   - Disassembling and assembling, see **Repair Group 17,Oil filter bracket, components**
9. **15 Nm + 1/4 turn (90 °)**
   - Always replace
10. **Oil cooler**
    - Coat contact surfaces to oil filter bracket outside seal with AMV 188 100 02.
    - ensure sufficient clearance to surrounding components
    - Observe note, see **Engine, disassembling and assembling**
11. **25 Nm (18 ft-lb)**
12. **Oil filter housing**
    - Remove with tension strap
    - Tighten by hand
    - Follow installation notes found on oil filter
13. **Gasket**
    - Always replace
14. **10 Nm (7 ft-lb)**
15. **Engine speed (RPM) sensor**
16. **Connecting piece**
17. **15 Nm (11 ft-lb)**
18. **Left engine mount**
19. **40 Nm (30 ft-lb)**
20. **O-ring**
    - Replace if damaged
21. **45 Nm (33 ft-lb)**
22. **Coolant thermostat**
    - Checking: Warm thermostat in hot water bath.
    - Starts opening at approx. 86 °C (187°F)
    - Opens at least 7 mm
23. **20 Nm (15 ft-lb)**
24. **Brace**
    - Between bracket and intake manifold
25. **50 Nm (37 ft-lb)**
    - Observe tightening sequence, see **Repair Group 19, Bracket, removing and installing, Fig. 1**
26. **Bracket**
    - See **Repair Group 19, Bracket, removing and installing**
    - Viscous fan bushing, removing and installing, see **Repair Group 19**
27. **Oil pan**
28. **15 Nm (11 ft-lb)**
29. **Cylinder block**
    - Sealing flange and flywheel /drive plate, removing and installing, see **Sealing flanges and flywheel /drive plate, removing and installing**
30. **40 Nm (16 ft-lb)**
31. **Engine bracket, right**

Sealing flanges and flywheel/drive plate, components

NOTE—

Servicing clutch, see **5 Spd. Manual Transmission 012/01W section, Repair Group 30, servicing clutch**

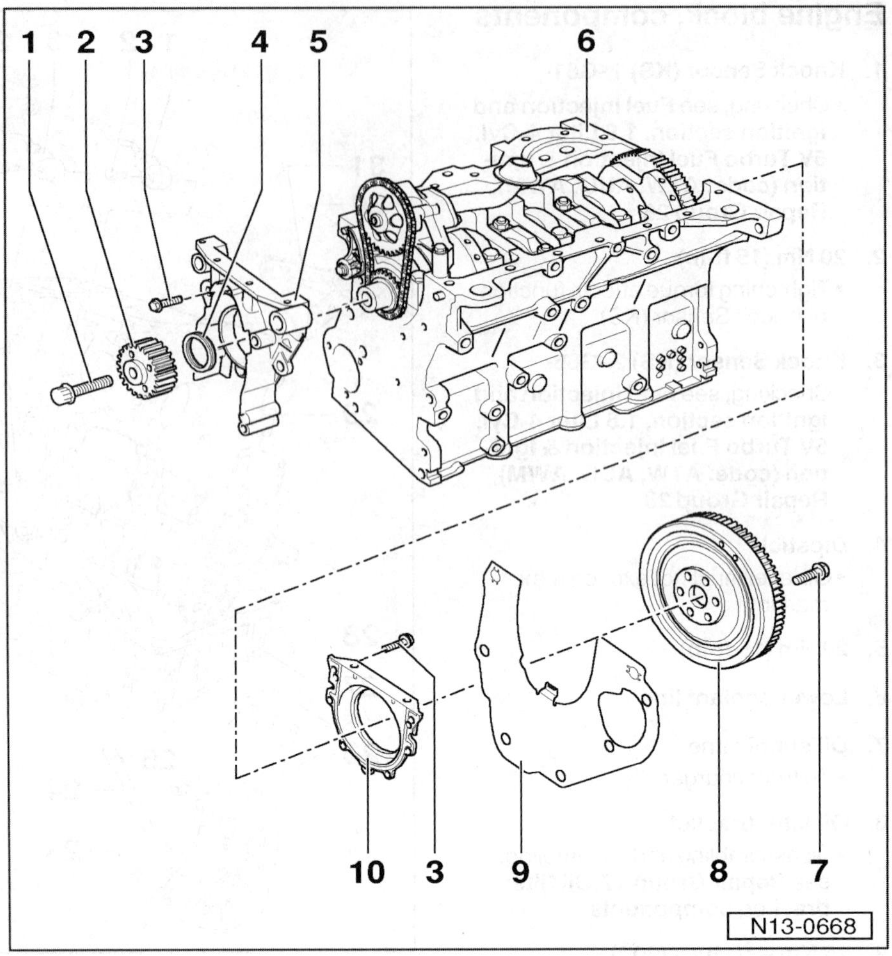

1. **90 Nm (66 ft-lb) + 1/4 turn (90 °)**
   - Always replace
   - Use 3415 counterhold when loosening or tightening
   - Threads and collar must be completely free of oil and grease.
2. **Toothed belt gear for crankshaft**
3. **15 Nm (11 ft-lb)**
4. **Seal**
   - Replace
   - PTFE sealing ring: do not additionally oil or grease sealing lip of oil seal
5. **Sealing flange, front**
   - Must fit on bushing
   - to remove and install, remove oil pan
6. **Cylinder block**
7. **60 Nm (44 ft-lb) + 1/4 turns (90 °) turn additionally**
   - Always replace
8. **Flywheel/drive plate**
   - Secure with 3067 flywheel retainer when removing and installing flywheel
   - See **Drive plate, removing and installing**
9. **Intermediate plate**
   - Must fit on bushing
   - Do not damage or bend while performing repairs
10. **Sealing flange with oil seal**
    - Replace only as complete unit
    - To install, use provided supporting sleeve
    - To remove and install, remove oil pan
    - Do not additionally oil or grease sealing lip of oil seal
    - Before installing, remove oil residue at crankshaft pin using a clean rag.
    - Supporting sleeve may only be removed after the sealing flange has been slid onto the crankshaft pin.

Drive plate, removing and installing

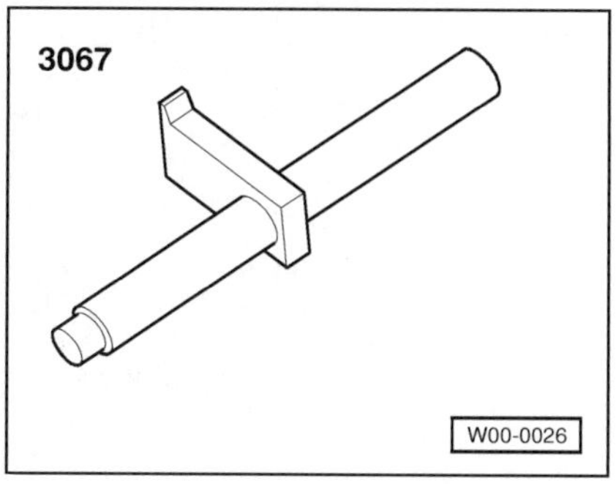

Special tools and equipment

3067 Counterhold

- Depth gauge

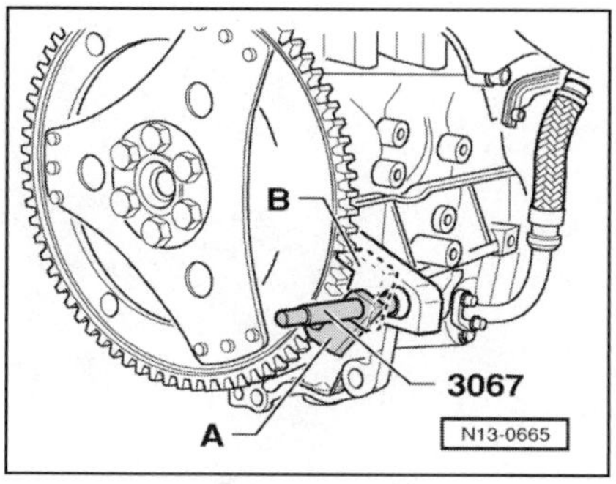

Loosening and tightening drive plate

Re-position 3067 retainer to loosen and tighten bolts.

Installation position of flywheel lock adapter.

A - to loosen

B - to tighten

Installing drive plate

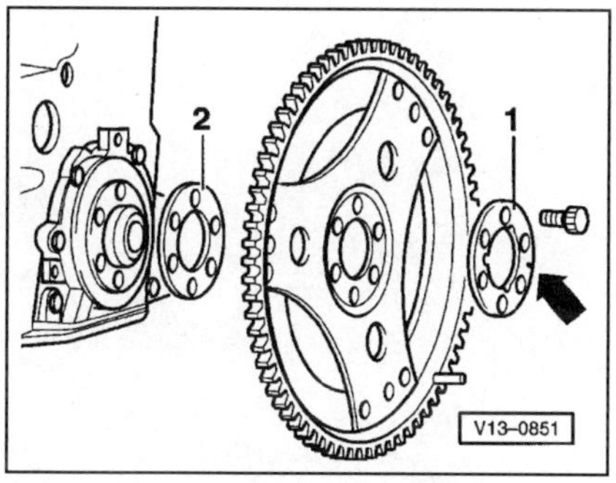

Position drive plate using grooved washer -1-.

– Insert new bolts and tighten to 30 Nm (22 ft-lb).

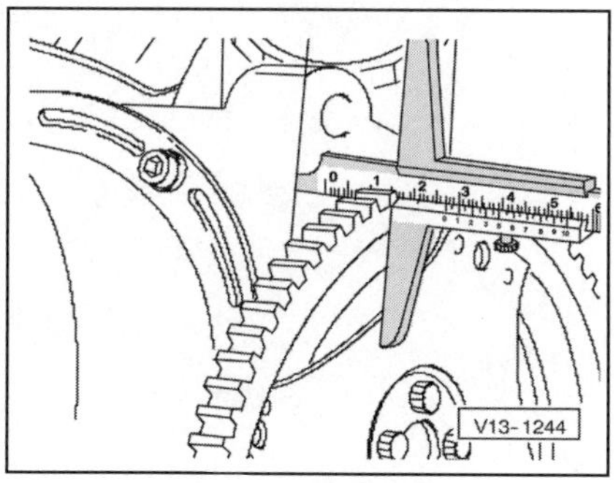

Measure distance -a- at three places and calculate mean.

Specification: 26 to 28 mm

NOTE—
The measurement takes place on the milled surface of the cylinder block.

If measured value is less than specified value:

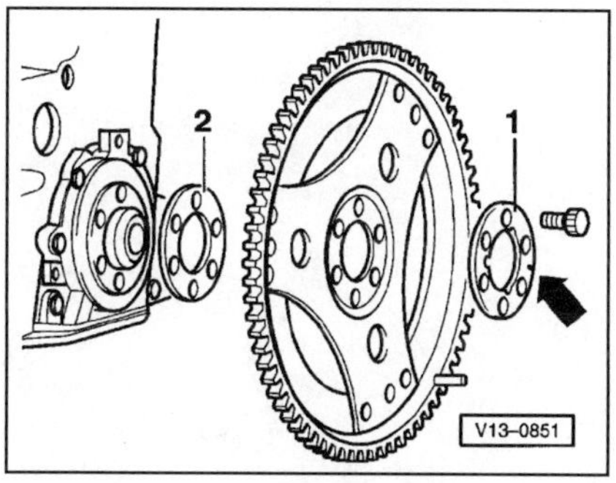

Remove drive plate again and use spacer washer -2-. Tighten bolts to 30 Nm (22 ft-lb) again.

– Tighten bolts to 60 Nm (44 ft-lb) and turn an additional 90° (1/4 rotation, additional rotation may occur in several stages).

- For mounting bracket, bearing 3
- Check locating point

15 Engine–Cylinder Head, Valvetrain

Repair Group

C 15

Cylinder Head, Servicing

From 08.00 a new tightening roller and tightening device for the toothed belt is being installed. Tightening roller must be adjusted in order to tension toothed belt.

Checking compression pressure, see **Compression, checking**

NOTE—

- *When installing a replacement cylinder head with camshafts attached, grease contact surfaces between valve lifters and cam lubricating surfaces before attaching cylinder head cover.*
- *Do not remove plastic covers protecting exposed valves until immediately before installing cylinder head.*
- *When replacing cylinder head or cylinder head gasket, coolant must be replaced.*

EDITOR'S NOTE—

Volkswagen identifies electrical components by a letter and/or a number in the electrical schematics. See **Wiring Diagrams, Fuses and Relays**. *These electrical identifiers are listed -between dashes- as an aid to electrical troubleshooting.*

Cylinder head, components

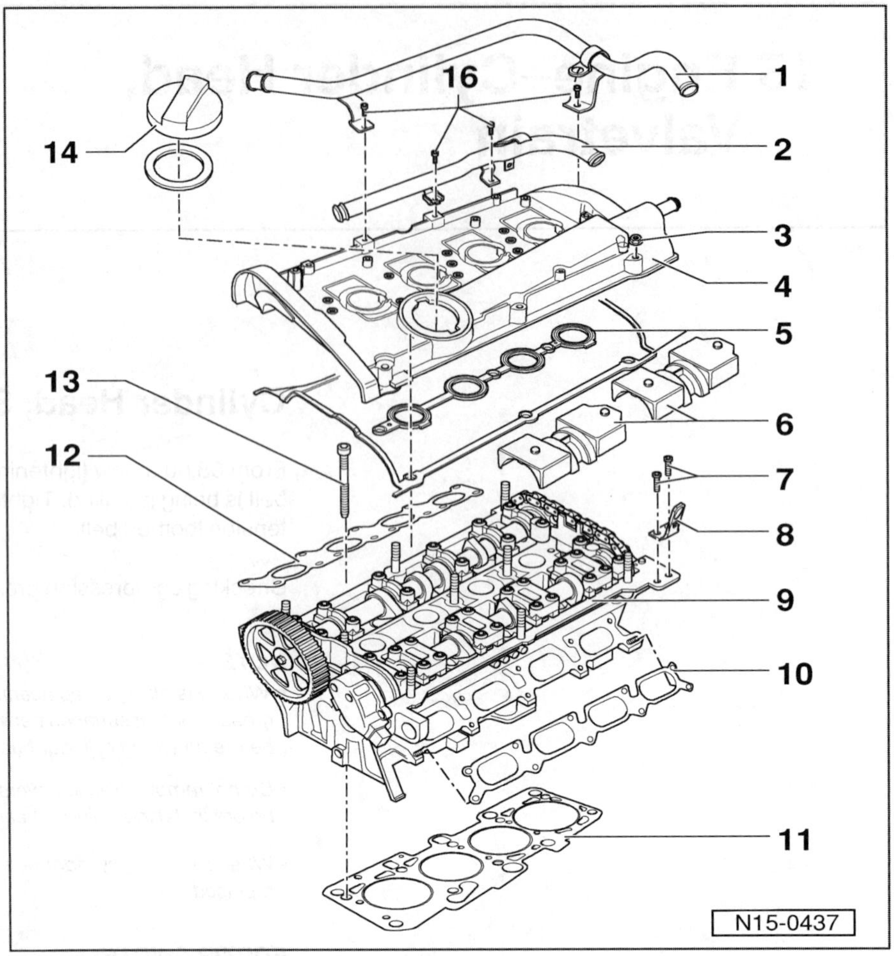

1. **Line**
   - For crankcase ventilation
2. **Pressure line**
   - For combination valve
3. **10 Nm (7 ft-lb)**
4. **Cylinder head cover**
5. **Cylinder head cover gasket**
   - Replace if damaged
   - Before installing, coat sealing contact surfaces between bearing cap and cylinder head with D 454 300 A2
6. **Oil deflector**
   - Note position: above intake camshaft
7. **25 Nm (18 ft-lb)**
8. **Lifting eye**
9. **Cylinder head**
   - Check for warpage, see **Fig. 1**
   - Reface sealing surface, see **Valve train, components**, **Fig. 1**
   - Removing and installing, see **Cylinder head, removing and installing**
   - After replacing, completely replace coolant
10. **Seal for intake line**
    - Always replace
11. **Cylinder head gasket**
    - Always replace
    - Metal gasket
    - After replacing, replace coolant
    - Note position: Identification: Part number must be visible from intake side.
12. **Seal for exhaust manifold**
    - Always replace
    - Note installed position
13. **Cylinder head bolt**
    - Always replace
    - Sequence for loosening and tightening, see **Cylinder head, removing and installing**
14. **Cap**
    - Replace seal if damaged
15. **10 Nm (7 ft-lb)**

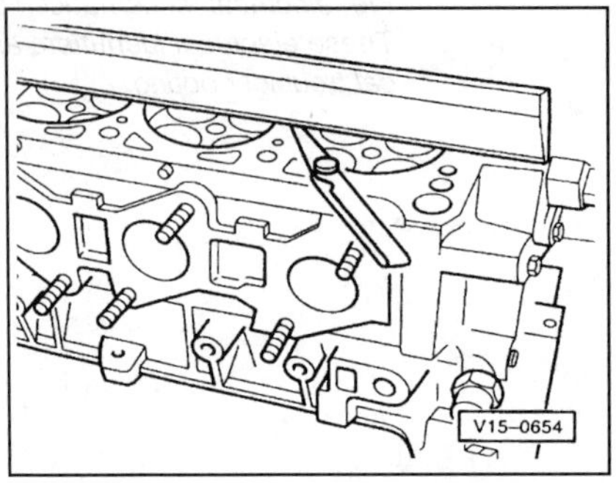

Fig. 1 Check cylinder head for warpage

Cylinder head warpage must not exceed: 0.1 mm (0.0004")

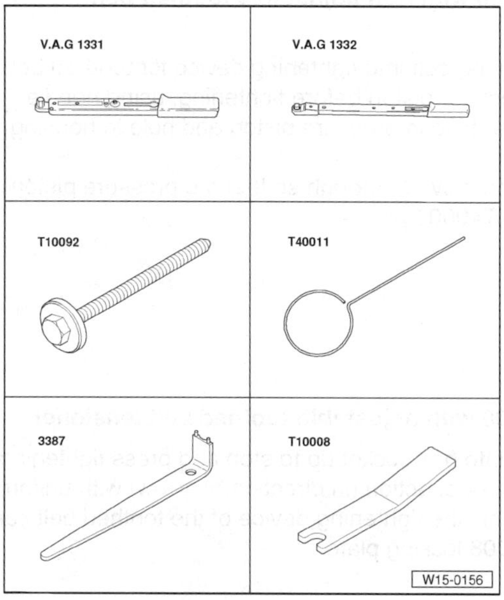

Toothed belt, removing and installing

Special tools and equipment

- VAG 1331 Torque wrench (5 to 50 Nm)
- VAG 1332 Torque wrench (40 to 200 Nm)

Tensioner up to 07.00 (not adjustable)

- T10092 Tightening bolt
- T40011 Pin

Tensioner from 08.00 (must be adjusted)

- 3387 Pin wrench
- T10008 Locking plate

Removing

– Move lock carrier into service position, see **Maintenance**.

– Remove ribbed belt, see **Maintenance**.

– Remove ribbed belt tensioner.

– Remove upper toothed belt guard.

– Mark direction of rotation of toothed belt, if reinstalling.

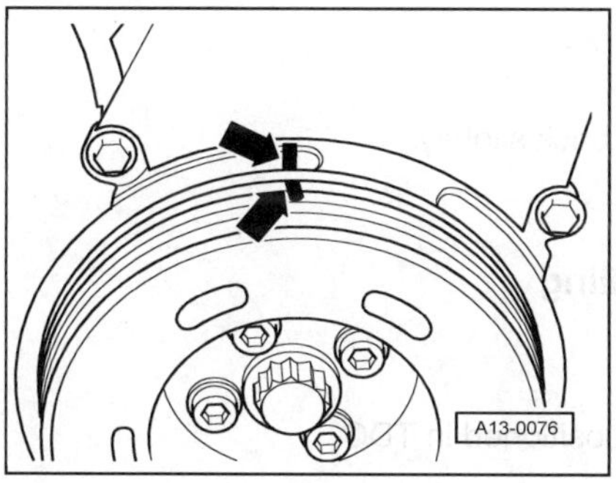

◄ Adjust camshaft to TDC cylinder 1.

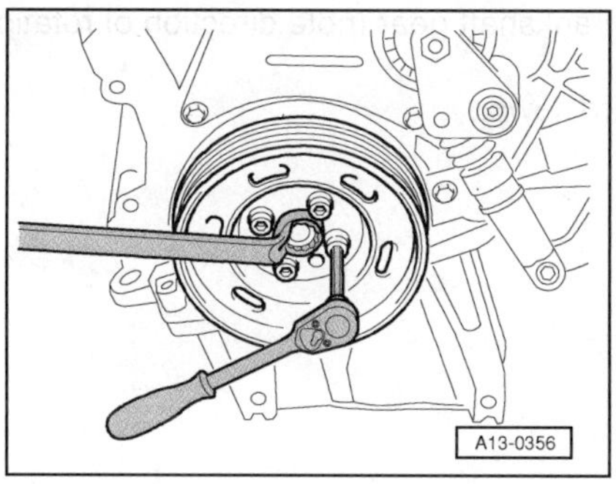

◄ Remove harmonic balancer/belt pulley.

– Remove lower and center toothed belt guard.

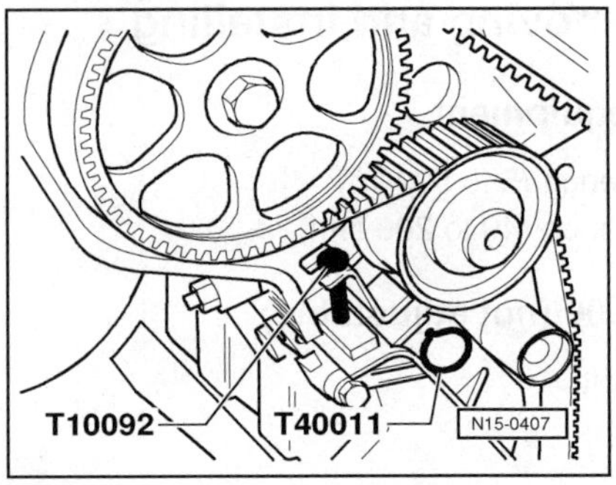

For vehicles up to 07.00 with non-adjustable toothed belt tensioner:

- Thread T10092 tightening bolt into tightening device for toothed belt.
- If necessary, align pressure piston before tightening, using needle nose pliers or thin wire (hole in pressure piston and hole in housing must overlap.
- Tighten pressure piston only far enough so that the pressure piston can be secured using T40001 pin.

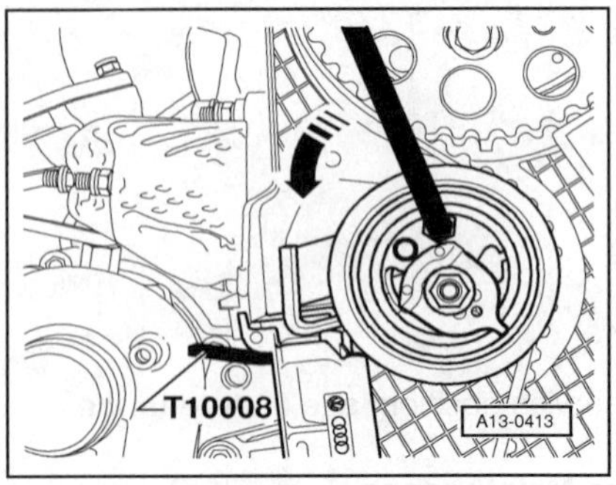

For vehicles from 08.00 with adjustable toothed belt tensioner:

- Insert an hex wrench into hex socket up to stop and press tightening roller in counterclockwise direction (in direction of arrow) with uniform non-excessive force until the tightening device of the toothed belt can be aligned using T10008 locking plate.

NOTE —
Toothed belt tightening device is an oil damper device and can only be compressed slowly and with equal force. Excessive force during compression can damage tightening roller.

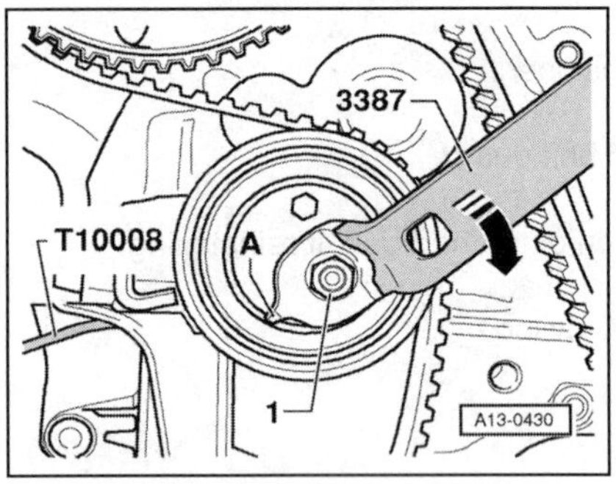

To loosen toothed belt, loosen nut -1- of tightening roller and turn pin wrench in direction of arrow using 3387 pin wrench.

NOTE —
Impact tab -A- of pin wrench must not be bent.

– **All vehicles:**
- Remove toothed belt.
- Then turn crankshaft back slightly.

Toothed belt, installing

Requirement

- Pistons must not be positioned in TDC.

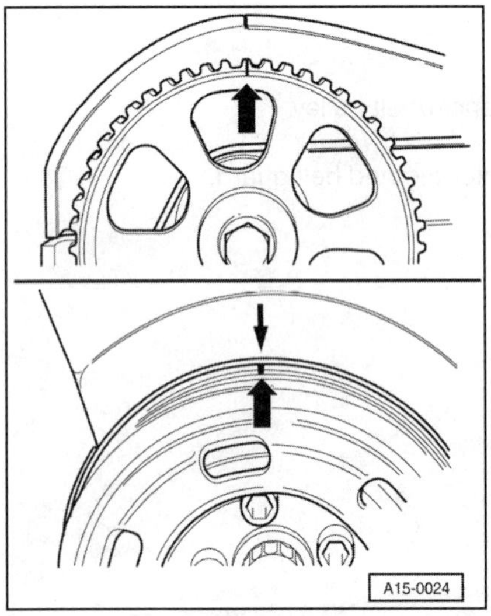

Align mark on camshaft gear and mark at cylinder head cover.

– Install toothed belt on crankshaft gear (note direction of rotation)

– Install toothed belt guard, lower.

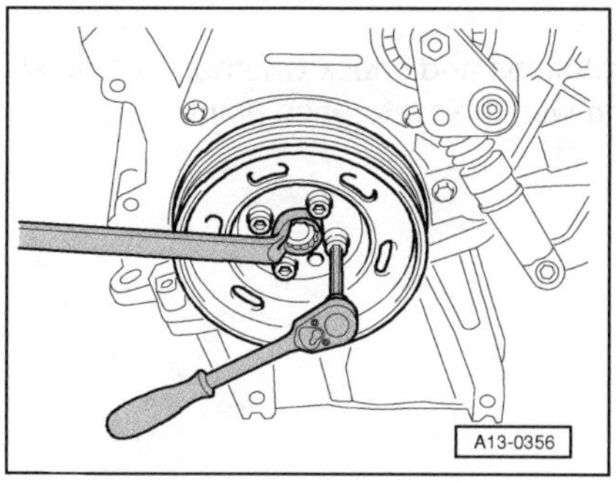

- Install harmonic balancer/belt pulley using new bolts.
  - Tightening torque: 10 Nm + $^1/_4$ turn (90°) additional

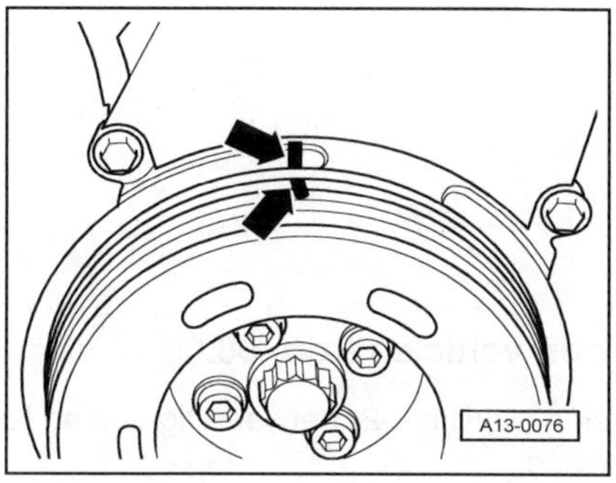

- Adjust crankshaft to TDC cylinder 1.

– Install toothed belt in the following order:

- Coolant pump
- Tightening roller
- Camshaft gear.

> ***WARNING —***
> - *Engine must not be more than hand-warm.*
> - *Engine must be at TDC cylinder 1.*

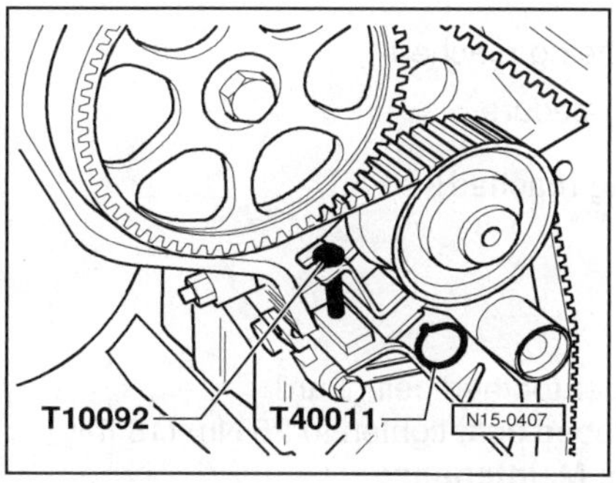

Vehicles up to 07.00 with non-adjustable toothed belt tensioner:

- Tighten toothed belt by pulling out T40001 pin and unbolting T10092 tightening bolt.

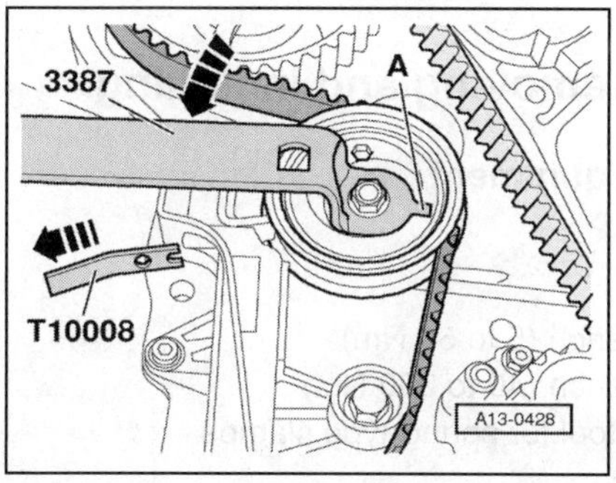

Vehicles from 08.00 with adjustable toothed belt tensioner:

NOTE —

If the toothed belt tightening device is completely extended, it must be pressed back in installed position with the tightening roller. This procedure can take up to 5 minutes. Excessive force during compression can damage tightening roller.

- Turn pin wrench counterclockwise (tab -A- of pin wrench must not be bent) using 3387 pin wrench until T10008 locking plate can be removed without tension.

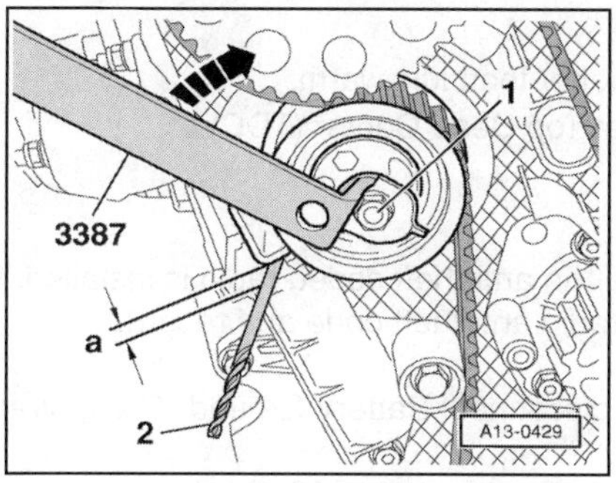

- Turn pin wrench to the right in clockwise direction (direction of arrow) until a drill bit -2- with dimension -a- can be pulled through tightening lever and housing of tightening device.
  - Specified value dimension -a-: 8 mm.

– Hold tightening roller in this position and tighten securing nut of tightening roller as follows:

- Specified value: 27 Nm (20 ft-lb)

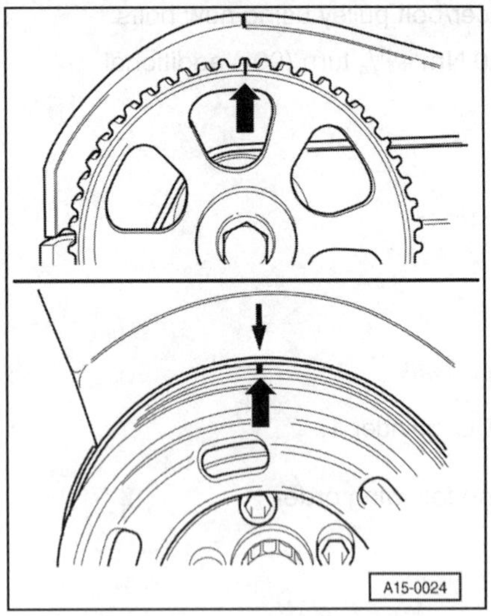

All vehicles:

- Turn crankshaft two rotations and check whether the camshaft and crankshaft marks align with their reference points.

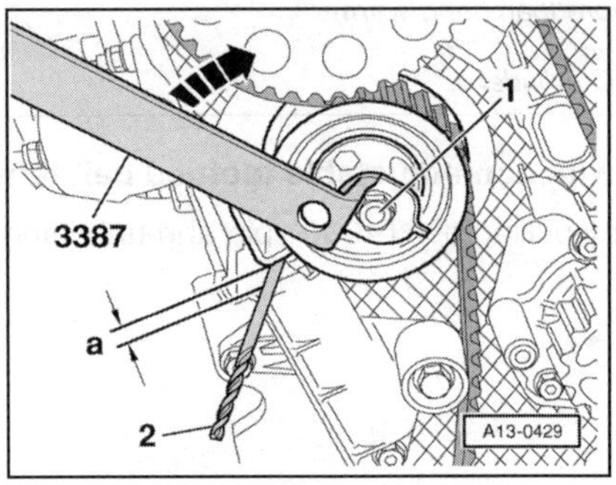

Adjusting tensioner on vehicles from 08.00

Check dimension -a- using a drill bit -2- between tightening lever and housing of tightening device.

- Specified value: 6 to10 mm (0.23" to 0.39")

– If dimension -a- is not being reached
  - Repeat installation procedure.

– If dimension -a- is being reached

– For all vehicles:
  - Install center and upper toothed belt guard.
  - Install tensioner for ribbed belt, tighten to 25 Nm (18 ft-lb)
  - Install ribbed belt, see **Maintenance**.

Cylinder head, removing and installing

Special tools and equipment

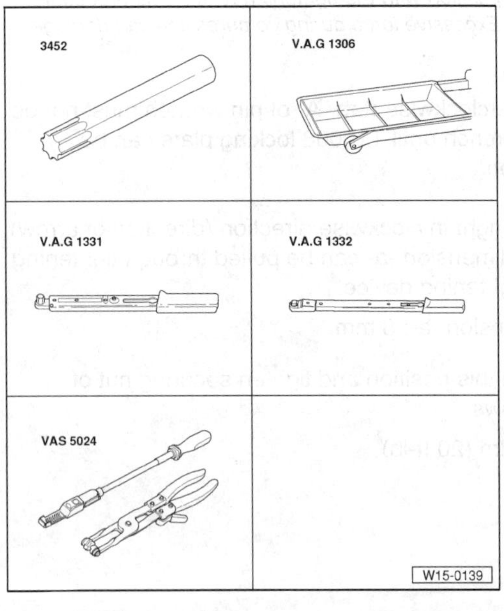

- 3452 polydrive key
- VAG 1306 drip tray
- VAG 1331 torque wrench (5 to 50 Nm)
- VAG 1332 torque wrench (40 to 200 Nm)
- VAS 5024 assembly tool for spring-type clamps

Requirements

- Engine must not be hotter than lukewarm.
- Pistons must not be at Top Dead Center (TDC).

Procedure

– First determine whether an anti-theft coded radio is installed. If so, determine the correct radio anti-theft code and record.

– Switch ignition off and disconnect Battery Ground (GND) strap.

– Drain engine coolant, see **Repair Group 19, Cooling system, draining and filling**.

– Disconnect fuel supply line and return line at connection point at fuel rail. Fuel system is under pressure. Therefore place rags around the connection before disconnecting hoses.

– Seal off lines so that no dirt will get into fuel system.

– Unbolt catalytic converter from turbocharger.

– Disconnect all necessary connections:
- Cooling system
- Crankcase breather valve
- Secondary air system
- Exhaust turbocharger/regulation
- Fuel injection and ignition system

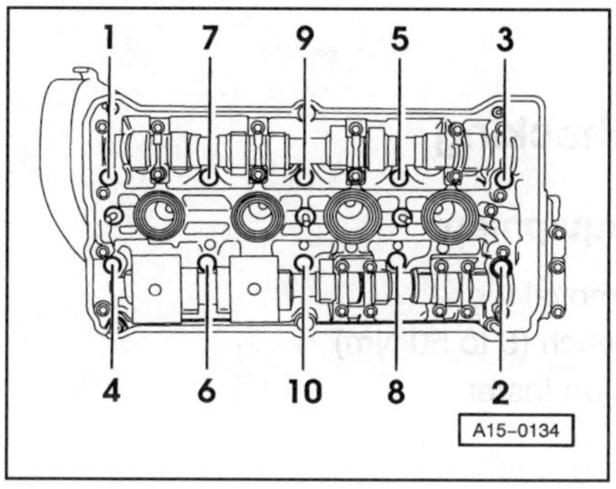

– Remove upper portion of toothed belt guard.

– Loosen tensioner and lift toothed belt from camshaft gear.

– Remove cylinder head cover.

◄ Loosen and unbolt cylinder head bolts in specified sequence using 3452 polydrive key.

– Carefully remove cylinder head.

Installing

NOTE —

- *Pocket holes in engine block for cylinder head bolts must be free of oil and coolant.*

- *Do not remove new cylinder head gasket from package until immediately before installing.*

- *Handle new gasket extremely carefully. Damage will cause leaking.*

– Stuff clean cloths into cylinders so that no dirt or abrasive material can get between cylinder wall and piston.

– Do not allow dirt or abrasive powder to get into coolant either.

– Carefully clean sealing surfaces of cylinder head and cylinder block. Avoid introducing scratches or scoring (do not use sandpaper with grit below 100).

– Thoroughly remove all sanding and grinding residue and remove cloths.

– Set piston of 1. Cylinder in TDC position and slightly turn crankshaft backward.

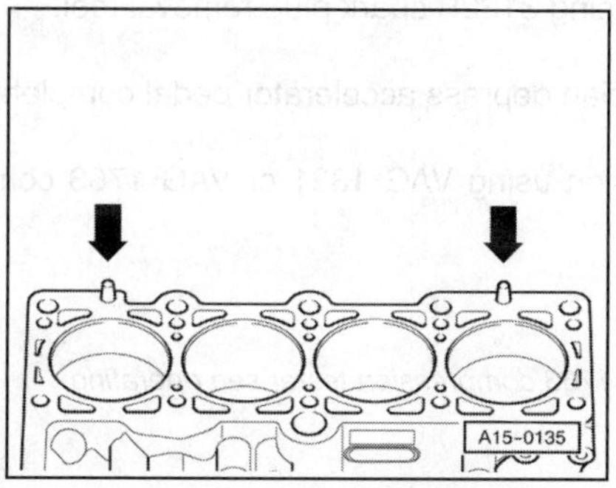

◄ Position new cylinder head gasket.
- Pay close attention to pilot pins in cylinder block (arrows).
- Pay close attention to cylinder head seal; replacement part number must be legible.

– Install cylinder head.

– Insert new cylinder head bolts and hand-tighten.

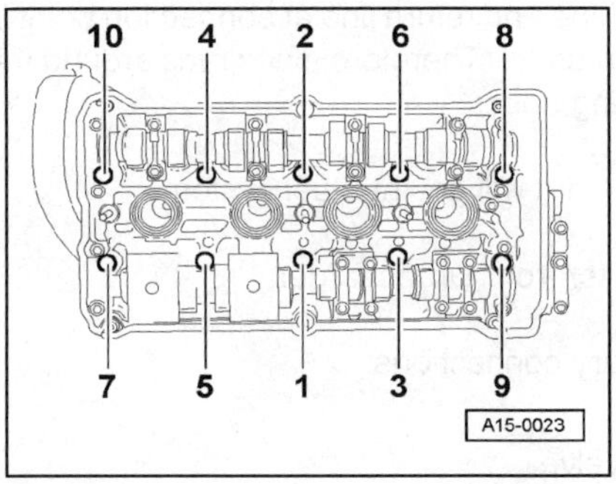

◄ Tighten cylinder head bolts in specified sequence as follows:

- Tighten all bolts to 40 Nm (30 ft-lb).
- Then tighten all bolts to 60 Nm (44 ft-lb).
- Then turn all bolts $^1/_4$ rotation (90 °) using a rigid wrench.
- Then turn all bolts an additional $^1/_4$ rotation (90 °) turn.

– Installation is the reverse of the removal, noting the following:

- Install toothed belt and adjust valve timing, see **Toothed belt, removing and installing**
- Fill cooling system with new coolant, see **Repair Group 19, Cooling system, draining and filling**.
- See **Electrical Equipment section**, **Repair Group 27, Procedure following interruption of voltage supply**
- Perform road test and check DTC memory, see **Maintenance.**

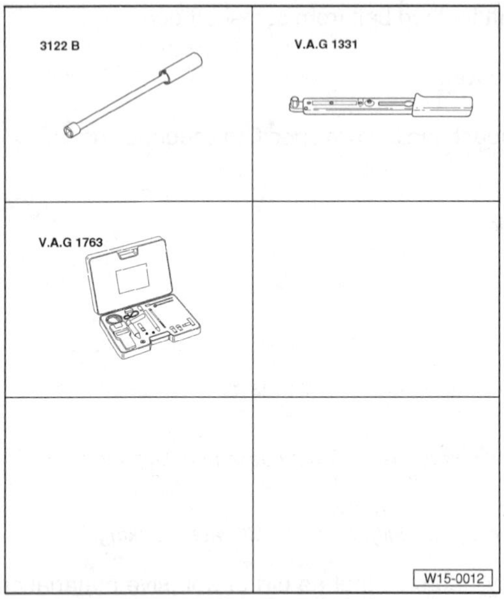

Compression, checking

Special tools and equipment

- 3122B Spark plug removal tool
- VAG 1331 Torque wrench (5 to 50 Nm)
- VAG 1763 Compression tester

Test requirements

- Engine oil temperature must be at least 30 °C.
- Voltage supply OK

Procedure

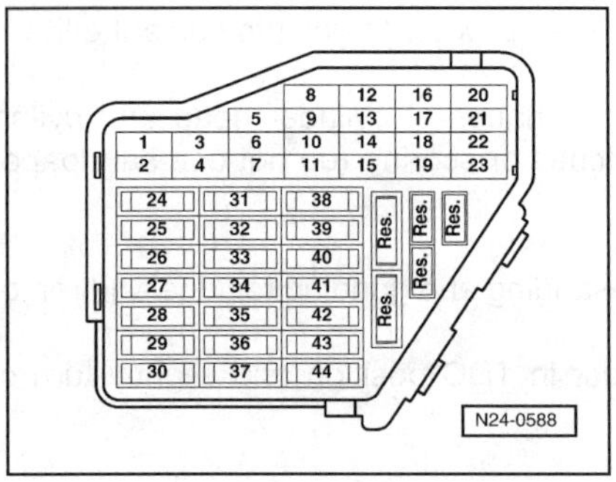

◄ Remove fuse 32 from fuse holder.

NOTE —
By removing fuse 32, voltage supply to the fuel injectors is interrupted.

– Disconnect connectors for ignition coils.

– Remove ignition coils, see **1.8 Liter Fuel Injection & Ignition (code: ATW, AUG, AWM) section, Repair Group 28**

– Remove spark plugs using 3122B spark plug removal tool.

– Have a second technician depress accelerator pedal completely.

– Test engine compression using VAG 1381 or VAG 1763 compression tester.

NOTE —
Using VAG 1381 or VAG 1763 compression tester see operating instructions.

– Have a second technician operate the starter.

– Activate starter until no further pressure increases are indicated by VAG 1381 or VAG 1763 compression tester.

Compression values:

- New: 10 to 14 bar (145-203 psi) positive pressure
- Wear limit: 7.0 bar (102 psi) positive pressure
- Tolerated deviation between cylinders: max. 3 bar (44 psi)

– Install spark plugs and ignition coils.

– Check DTC memory, see **Maintenance.**

Valve train, components

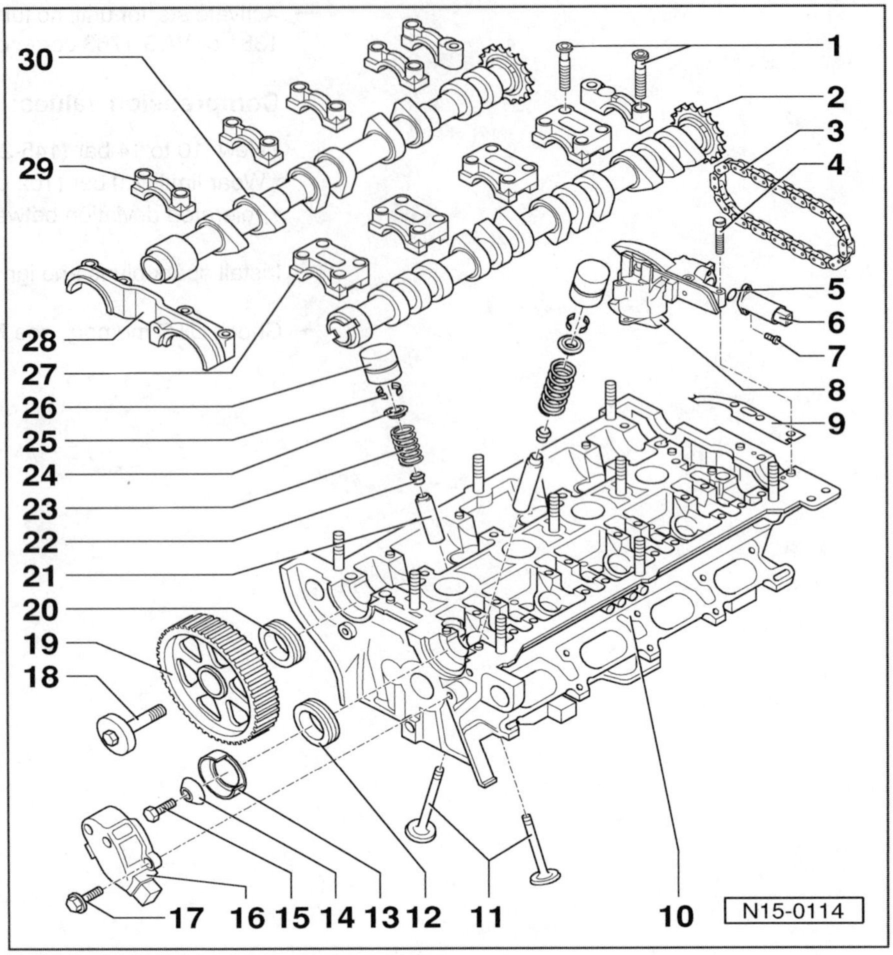

1. **10 Nm (7 ft-lb)**
2. **Intake camshaft**
   - Axial play, checking, see **Fig. 2**
   - Removing and installing, see **Camshaft, removing and installing**
   - Check radial play using Plastigage, Wear limit: 0.1 mm
   - Max. run-out: max. 0.035 mm
3. **Drive chain**
   - Mark direction of travel (installation position) before removing, see **Camshaft, removing and installing**
4. **10 Nm (7 ft-lb)**
5. **O-ring**
   - Always replace
6. **Solenoid valve 1 for camshaft adjustment -N205-**
   - Checking, see **Camshaft adjustment, checking**
7. **3 Nm (27 in-lb)**
8. **Camshaft adjuster**
   - Secure using 3366 bracket-tensioner before removing, see **Camshaft, removing and installing**
   - Checking, see **Camshaft adjustment, checking**
9. **Gasket**
   - Rubber/metal seal
   - Always replace
10. **Cylinder head**
    - Reface sealing surface, see **Fig. 1**
    - Seal off transition points, see **Fig. 3** and **Fig. 4**
11. **Valves**
    - Do not reface. Only hand-lapping is permitted.
12. **Seal**
    - Replace, see **Camshaft oil seals, replacing**
    - PTFE sealing ring, do not oil sealing lip of sealing ring.
13. **Face plate**
    - For Camshaft Position (CMP) sensor
    - Note position when installing
14. **Washer**
    - Conical
15. **25 Nm (18 ft-lb)**
16. **Camshaft Position (CMP) sensor/sensor 2 -G40-/-G163-**
    - See **1.8 Liter Fuel Injection & Ignition (code: ATW, AUH, AWM) section**, **Repair Group 28**
17. **10 Nm (7 ft-lb)**
18. **65 Nm (48 ft-lb)**
    - Use 3036 Retainer when loosening or tightening
19. **Camshaft gear**
    - Note position: Thin rib of camshaft gear points outward and TDC marking cylinder 1 is visible
20. **Seal**
    - Replace, see **Camshaft oil seals, replacing**
    - PTFE sealing ring, do not oil sealing lip of sealing ring.
21. **Valve guide**
    - Checking, see **Valve guides, checking**
22. **Valve stem oil seal**
23. **Valve spring**
24. **Valve spring retainer**
25. **Valve keepers**
26. **Valve lifters**
    - Do not interchange
    - With hydraulic play adjuster
    - Checking, see **Hydraulic valve lifters, checking**
    - Install with journal surface facing down
    - Before installing, check axial play of camshaft, see **Fig. 2**
    - Oil contact surface
27. **Intake camshaft bearing cap**
    - Installation position and installation sequence, see **Camshaft, removing and installing**
28. **Double bearing cap**
    - Coat contact surfaces lightly with ABV 174 004 01 sealant

(continued on following page)

Valve train, components (continued)

29. Exhaust camshaft bearing cap
- Installation position and sequence, see **Camshaft, removing and installing**

30. Exhaust camshaft
- Axial play, checking, see **Fig. 2**
- Removing and installing, see **Camshaft, removing and installing**
- Check radial play using Plastigage, Wear limit: 0.1 mm
- Max. run-out: max. 0.035 mm

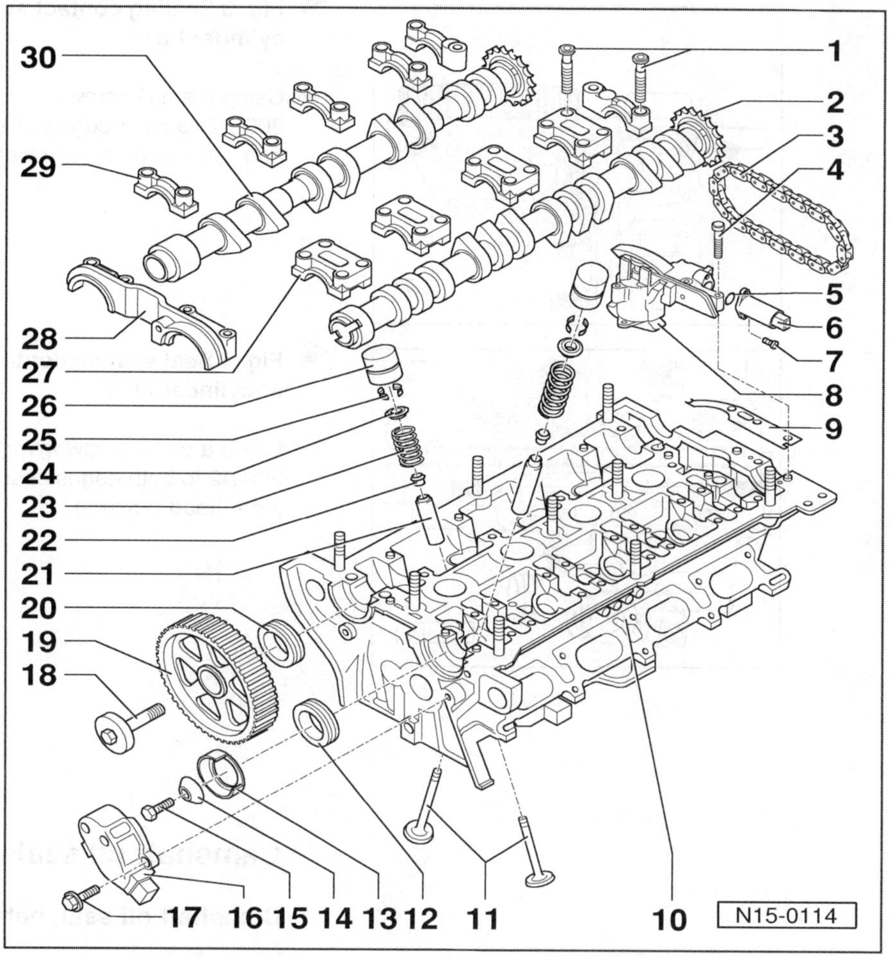

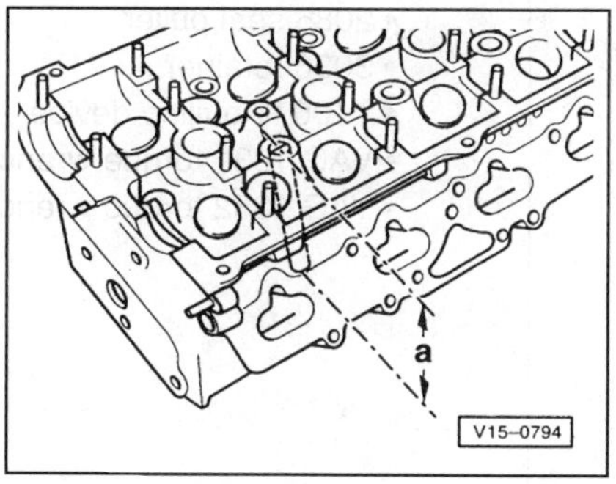

Fig. 1 Cylinder head - sealing surface, resurfacing

Resurfacing measurement is made via the holes for cylinder head bolts

a = at least 139.2 mm

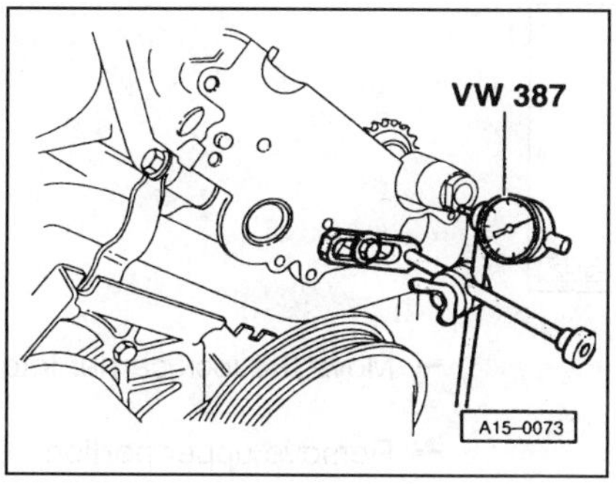

Fig. 2 Camshaft, checking axial play

Special tools and equipment

- VW387 Dial indicator holder
- Dial indicator

Perform measurement with valve lifters removed, chain removed and bearing caps 2 and 4 installed.

Wear limit: max. 0.2 mm

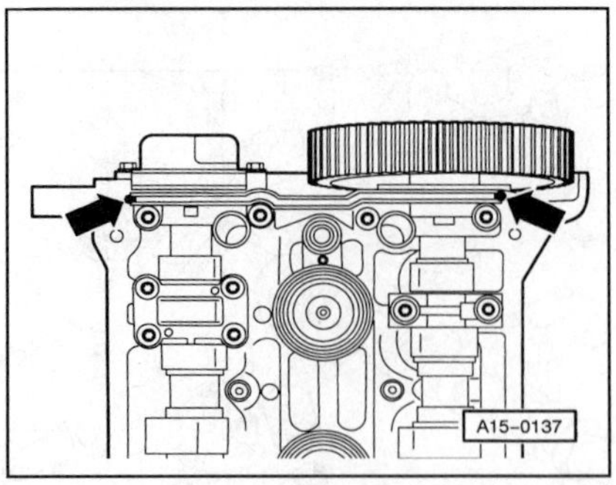

Fig. 3 Sealing contact surfaces between double bearing cap and cylinder head

– Using a small screwdriver, carefully apply a thin layer of sealant D 454 300 A2 to both edges of sealing surface between double bearing cap and cylinder head (arrows).

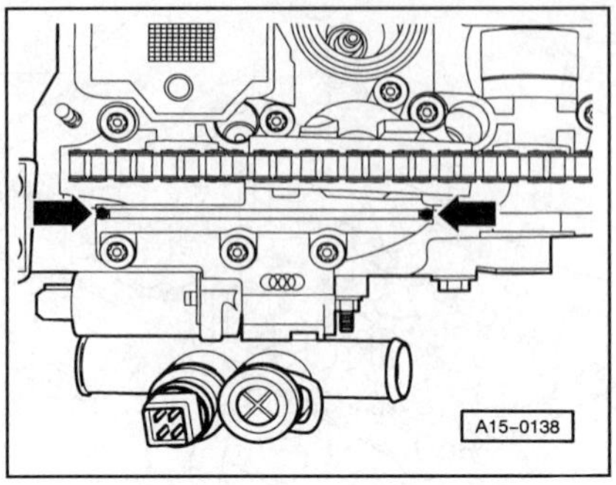

Fig. 4 Seal sealing contact surfaces between camshaft adjuster/cylinder head

– Using a small screwdriver, carefully apply a thin layer of sealant D 454 300 A2 to both edges of sealing surface between camshaft adjuster/cylinder head (arrows).

Camshaft oil seals, replacing

Camshaft oil seal, behind camshaft gear, removing

Special tools and equipment

- 2085 seal puller
- 3036 retainer
- T10071 pulling device
- VAG 1331 torque wrench (5 to 50 Nm)
- VAG 1332 torque wrench (40 to 200 Nm)

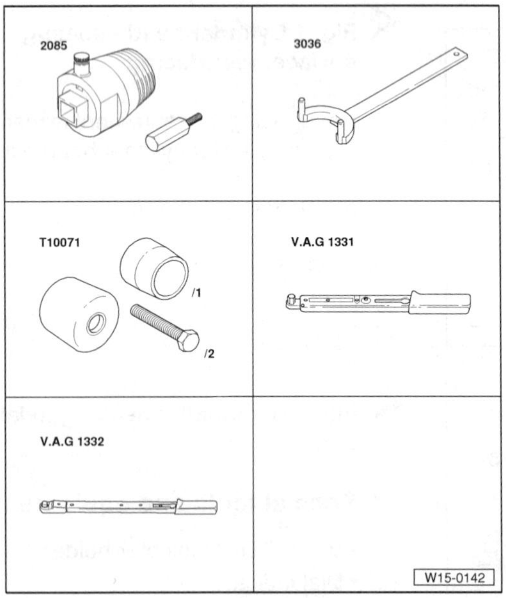

– Move the lock carrier into service position, see **Maintenance.**

– Remove upper portion of toothed belt guard.

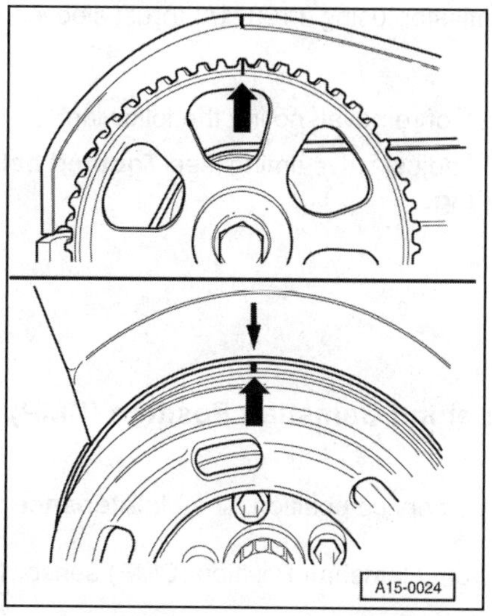

Bring camshaft gear to marking for TDC cylinder 1 by turning crankshaft. Marking on camshaft gear must be flush with marking on cylinder head cover.

– Loosen tensioner and lift toothed belt from camshaft gear.

– Turn crankshaft back slightly.

– Remove camshaft gear. To loosen bolt, counterhold camshaft gear using 3036 retainer.

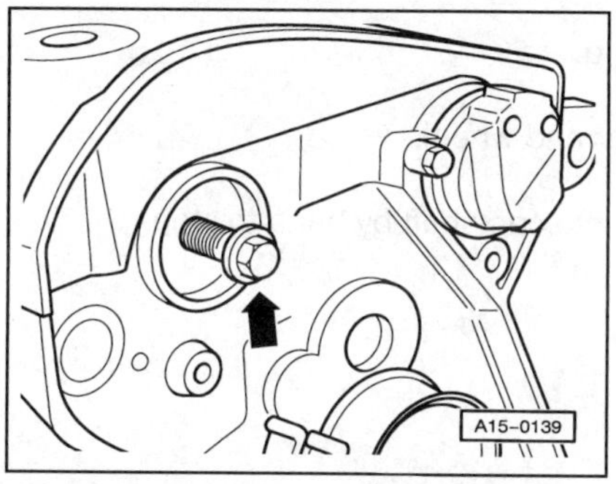

Insert camshaft gear retaining bolt (arrow) into camshaft by hand to stop to guide 2085 Seal Puller.

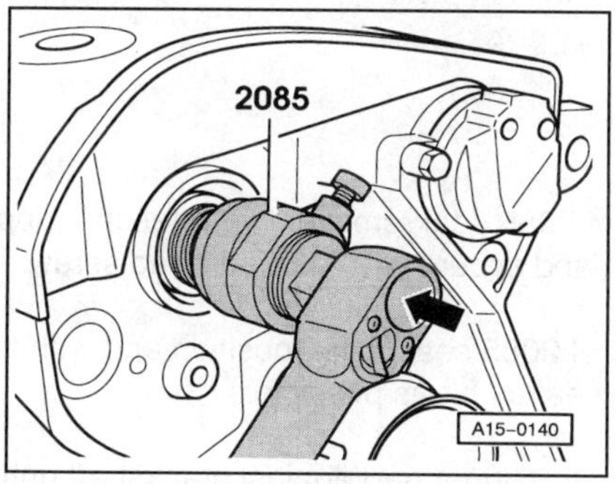

Remove inner portion of 2085 seal remover two rotations (approx. 3 mm) from outer portion and secure with knurled-head screw.

– Grease threaded head of 2085 seal puller, position and with forced pressure screw into oil seal as far as possible.

– Loosen knurled bolt and turn inner part against crankshaft until O-ring is removed.

Camshaft oil seal, installing

NOTE —

Introduction of PTFE-oil seals (identification: without annular spring, sealing lip has a wider design). The sealing lip of these oil seals may not be oiled or greased.

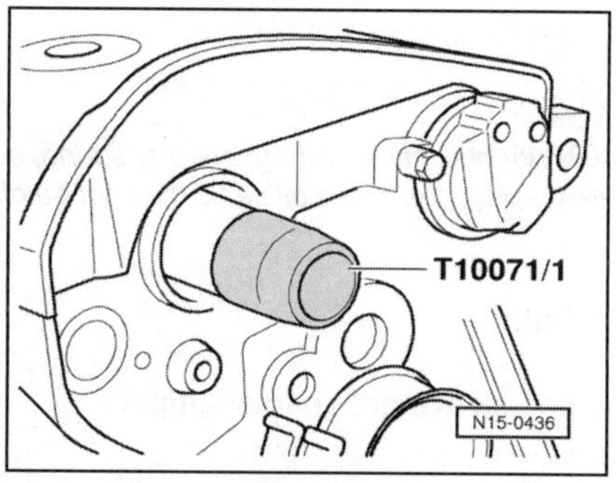

– Remove oil residue at camshaft pin using a clean rag.

Set T10071/1 on camshaft pin.

– Slide sealing ring over guide sleeve onto camshaft pin.

– Remove guide sleeve.

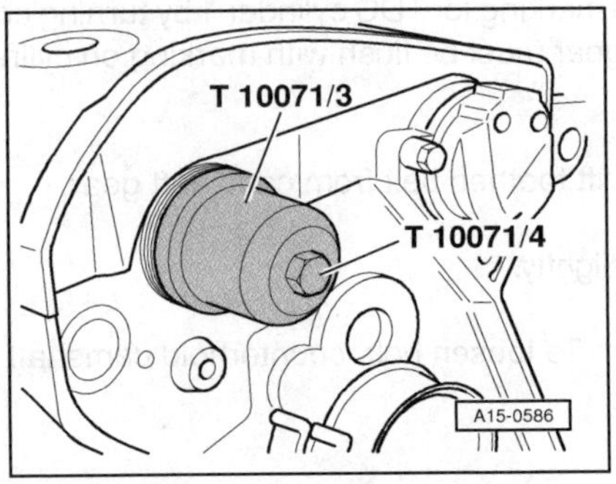

◀ Press in sealing ring until stop using T10071/3 thrust sleeve. Use T10071/4 bolt for this.

– Installation is the reverse of removal, noting the following:
  - Install toothed belt and adjust valve timing, see **Toothed belt, removing and installing.**

Camshaft oil seal, behind Camshaft Position (CMP) sensor, removing

– Bring the lock carrier into service position, see **Maintenance**

◀ Disconnect connector from Camshaft Position (CMP) sensor (arrow).

– Remove upper portion of toothed belt guard.

– Remove sensor housing.

– Remove sensor washer and wheel.

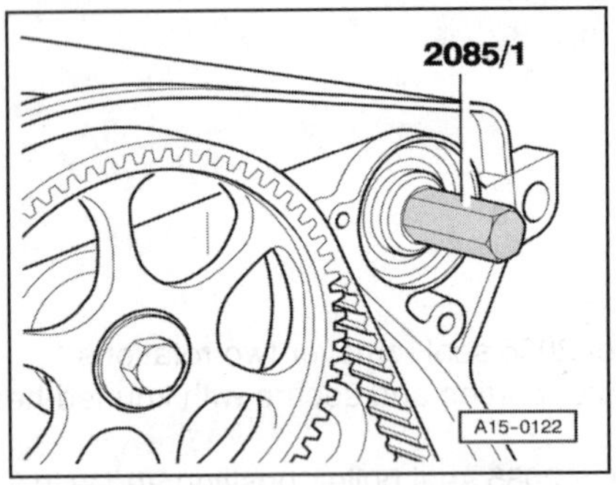

◀ Insert 2085/1 Adapter into camshaft by hand to stop.

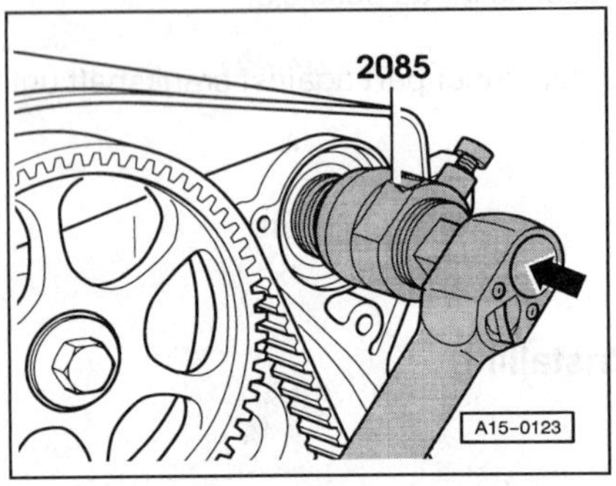

◀ Remove inner portion of 2085 seal remover two rotations (approx. 3 mm) from outer portion and secure with knurled-head screw.

– Grease threaded head of 2085 seal puller, position and with forced pressure screw into oil seal as far as possible.

– Loosen knurled bolt and turn inner part against crankshaft until O-ring is removed.

Installing

NOTE—
Introduction of PTFE-oil seals (identification: without annular spring, sealing lip has a wider design). The sealing lip of these oil seals may not be oiled or greased.

– Place T10071/5 on camshaft pin.

– Slide sealing ring over guide sleeve onto camshaft pin.

– Remove guide sleeve.

– Press in sealing ring until stop using T10071/3 thrust sleeve. Use M8 x 60 bolt for this.

– Insert notched shutter wheel for Camshaft Position (CMP) sensor into intake camshaft.

– Install gear (with cone facing outward) and tighten mounting bolt to 25 Nm (18 ft-lb).

– Install camshaft housing and tighten to 10 Nm (7 ft-lb).

– Install upper portion of toothed belt guard.

– Connect Camshaft Position (CMP) sensor harness connector.

– Installation is the reverse of removal.

Hydraulic valve lifters, checking

Special tools and equipment

- Feeler gauge
- Wooden or plastic wedge

NOTE —

- *Only replace hydraulic valve lifters in complete sets. Hydraulic valve lifters cannot be adjusted or repaired.*
- *Irregular valve noise when starting engine is normal.*

Test sequence

– Start engine and let run until radiator fan has switched on once.

– Increase engine speed to approx. 2500 RPM for 2 minutes.

– If hydraulic valve lifters are still loud, locate faulty valve lifter(s) as follows:

- Remove cylinder head cover.
- Turn crankshaft clockwise until cams of valve lifter to be tested point up.
- Now calculate play between cams and valve lifters.

– Replace valve lifter if play is more than 0.2 mm. If a play of less than 0.1 mm is calculated or if no play is calculated, continue test as follows:

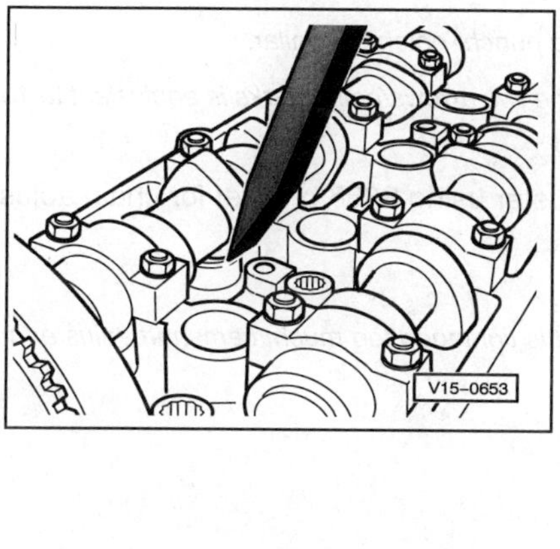

◄ Press valve lifter downward slightly using wooden or plastic wedge. If 0.20 mm thick feeler gauge can be inserted between camshaft and valve lifter, replace valve lifter.

NOTE —

After installing new hydraulic valve lifters, do not start engine for about 30 minutes. The hydraulic valve lifters have to settle (otherwise valves will strike the pistons).

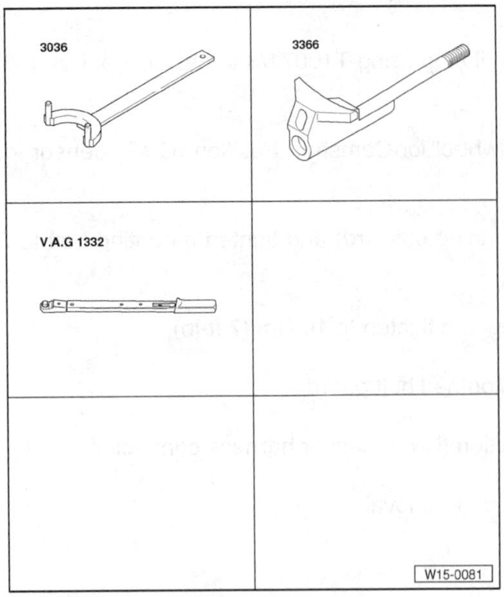

Camshaft, removing and installing

(with cylinder head installed)

Special tools and equipment

- 3036 Retainer
- 3366 Bracket for chain adjustment
- VAG 1332 Torque wrench (40 to 200 Nm)
- Sealant D 454 300 A2

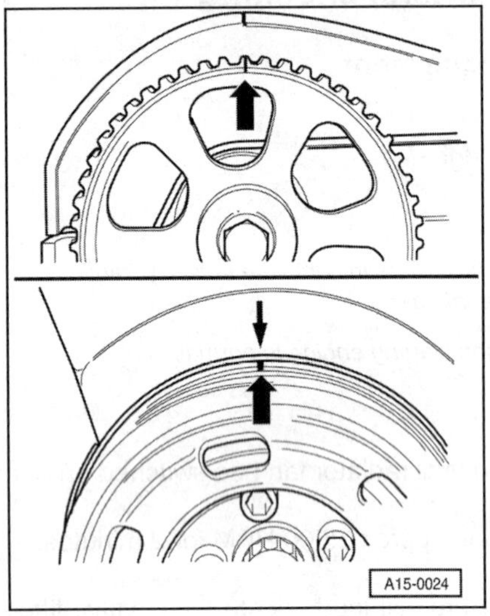

Removing

– Bring the lock carrier into service position, see **Maintenance.**

– Remove upper portion of toothed belt guard.

◄ Bring camshaft gear to marking for TDC cylinder 1 by turning crankshaft. Marking on camshaft gear must be flush with marking on cylinder head cover.

– Remove cylinder head cover.

– Loosen tensioner and lift toothed belt from camshaft gear.

– Turn crankshaft back slightly.

– Remove camshaft gear. To loosen bolt, counterhold camshaft gear using 3036 retainer.

– Remove Camshaft Position (CMP) sensor housing.

– Remove sensor washer and wheel.

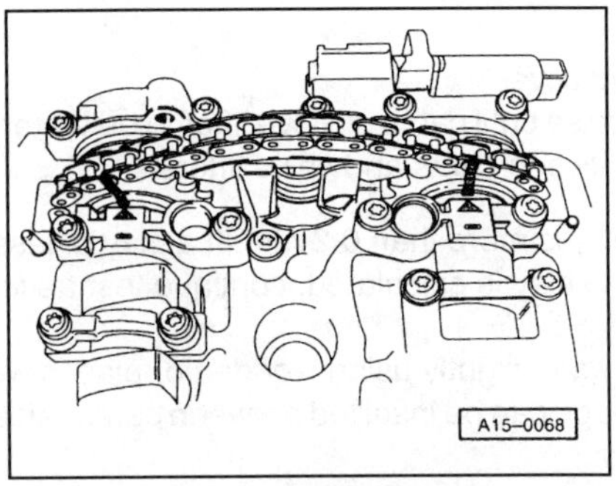

◄ Clean camshaft drive chain and camshaft chain gears opposite both arrows on bearing caps. Mark installed position with paint.

NOTE—

- *Do not mark chain using punch, notch or similar.*
- *The distance between the two arrows/paint marks is equivalent to 16 drive chain rollers.*

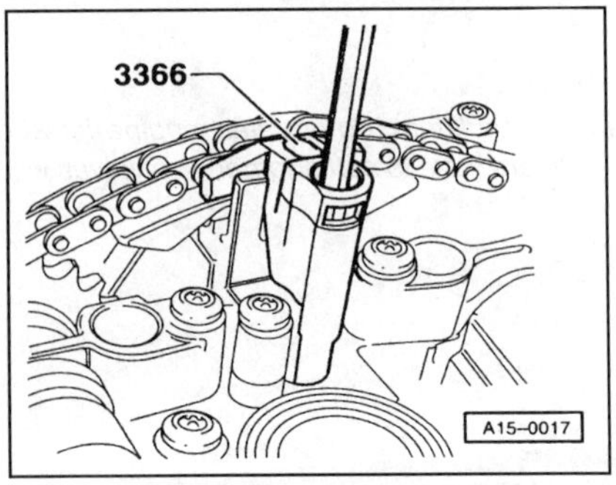

◄ Secure crankshaft adjuster using 3366 bracket for chain adjustment.

NOTE—

If chain tensioner bracket is tightened too much, camshaft adjuster may be damaged.

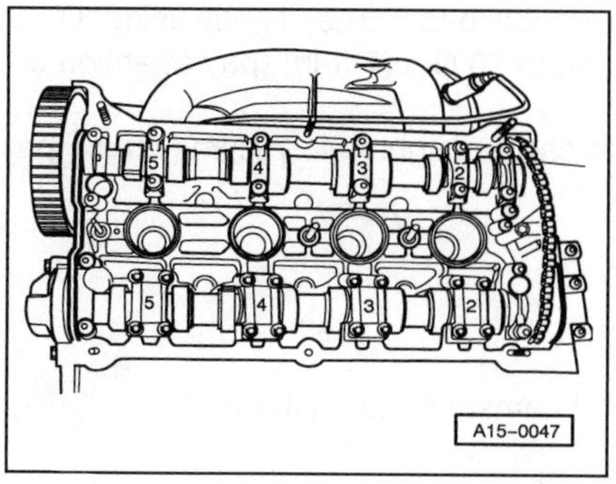

First remove bearing caps 3 and 5 from intake and exhaust camshafts.

– Remove double bearing cap.

– Remove both bearing caps from chain gears on intake and exhaust camshafts.

– Remove mounting bolts for camshaft adjuster.

– Alternating in diagonal sequence, loosen bearing caps 2 and 4 of intake and exhaust camshafts and remove.

– Remove intake and exhaust camshafts along with camshaft position adjuster and 3366 bracket for chain adjustment.

Camshaft, installing

NOTE —

- *When installing camshafts, cams for cylinder 1 must point upward.*
- *When installing bearing caps, verify marking on cap is readable from intake side of cylinder head.*

– Place drive chain on both camshafts according to colored markings.

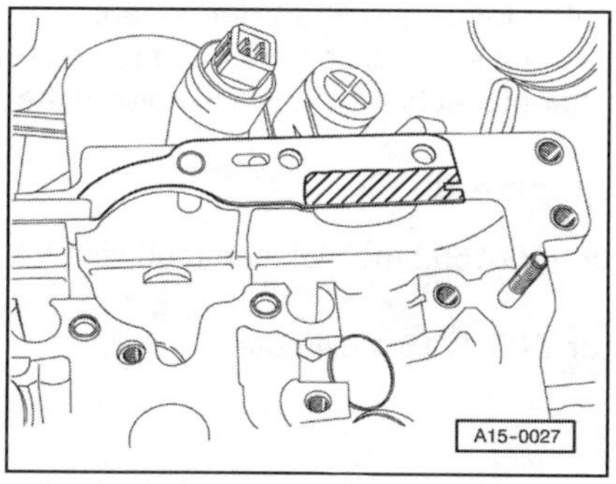

Replace rubber/metal seal for camshaft adjuster or chain tensioner and coat hatched surface with thin coat of "D 454 300 A2" sealant.

– Slide in camshaft adjuster between drive chain.

– Oil camshaft journal surface.

– Insert camshafts with drive chain and camshaft adjuster into cylinder head.

– Tighten camshaft adjuster to 10 Nm (7 ft-lb) (pay attention to bushing).

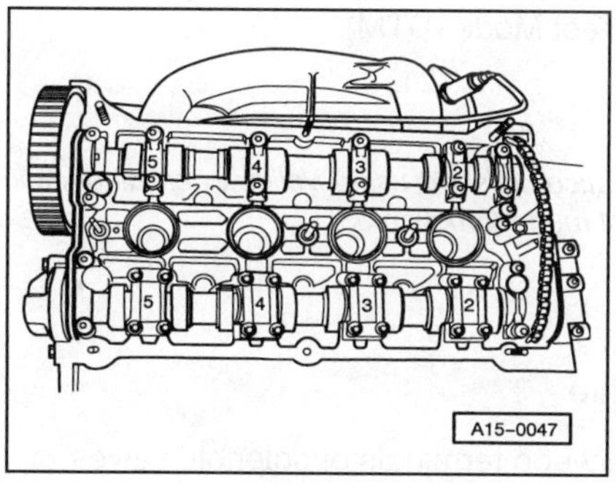

Alternating in diagonal sequence, tighten bearing caps 2 and 4 of intake and exhaust camshafts slightly and tighten to 10 Nm (7 ft-lb) (pay attention to bushing).

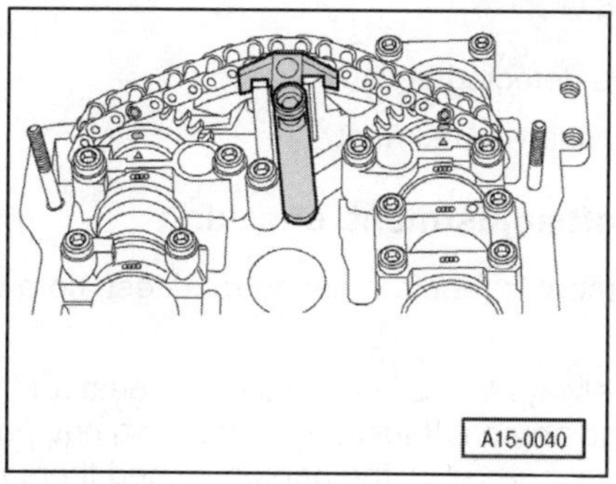

Install both bearing caps from chain gears on intake and exhaust camshafts. Check camshaft for proper adjustment and tighten bearing cap to 10 Nm (7 ft-lb) (pay attention to bushing).

– Remove 3366 bracket-tensioner.

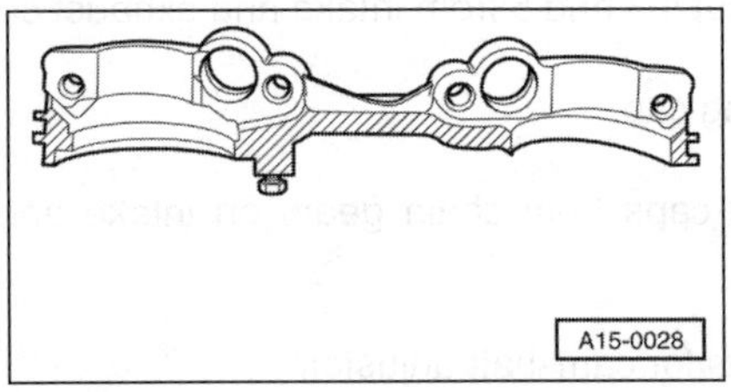

◄ Coat hatched surface of double bearing cap lightly using "D 454 300 A2" sealant, install and tighten to 10 Nm (7 ft-lb) (pay attention to bushing).

– Install remaining bearing cap and tighten to 10 Nm (7 ft-lb) (pay attention to bushing).

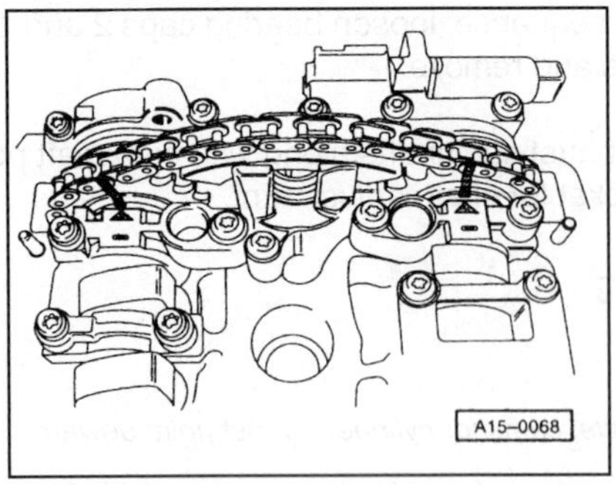

◄ Check relative position of camshafts to each other.

– Installation is the reverse of removal, noting the following:

- Install toothed belt and adjust valve timing, see **Toothed belt, removing and installing.**

NOTE —

- *After installing camshafts, do not start engine for at least 30 minutes. The hydraulic valve lifters have to settle (otherwise valves will strike the pistons).*

Camshaft adjustment, checking

The camshaft adjustment is load- and RPM dependant. The electrical valve for camshaft adjustment switches oil pressure onto camshaft adjuster (mechanical adjustment mechanism), which adjusts the camshaft.

Special tools and equipment

- VAG 1551 scan tool (or VAG 1552 mobile scan tool) with VAG 1551/3 cable
- VAG 1526 multimeter or VAG 1715 multimeter

Checking activation

– First check activation of the solenoid valve for camshaft adjustment via the Output Diagnostic Test Mode (DTM).

EDITOR'S NOTE —

Component activation is accomplished using Volkswagen diagnostic scan tool, output diagnostic test mode (function 03).

If activation is OK:

Check solenoid valve

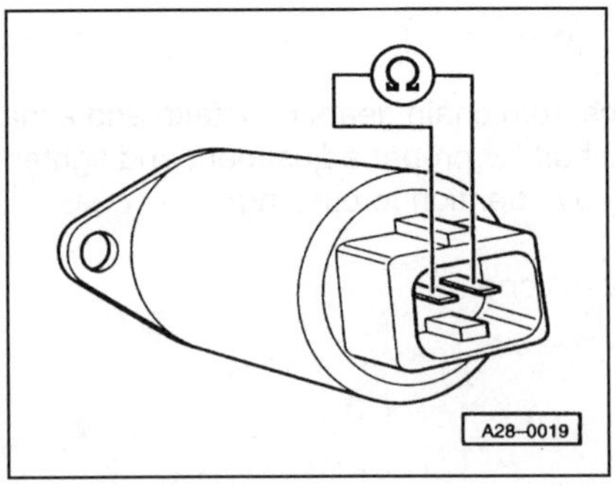

◄ Measure resistance between terminals of solenoid valves.

- Specification: 10 to 18 Ω

– If no malfunction can be detected at valve:

- Check function of camshaft adjustment.

Function of camshaft adjustment, checking

– If special testing equipment is required during road test, note the following:

- Test equipment must always be secured to the rear seat and operated from there by a second person. If test and measuring equipment is operated from the passenger seat, the person seated there could be injured in the event of an accident involving deployment of the passenger-side airbag.

Test requirement

- Coolant Temperature at least 80°C (176°F)

NOTE —

To test camshaft adjustment, vehicle must be driven. Doing this requires a second person.

> ***WARNING —***
>
> *Secure Scan Tool (ST) on the rear seat and have operated from there by second technician.*

Test sequence

– Connect VAG 1551 scan tool (VAG 1552). Start engine and select the Engine Control Module (ECM) using address word 01, see **Maintenance.**

– Press buttons -0- and -8- to select "Read measuring value block" function 08. Press -Q- button to confirm input.

– Press -0-, -9-, and -4- for "Display group number 94". Press -Q- button to confirm input.

– Observe indication in display field 2.

- AUG Engine: Cam.pos.OFF
- AWM engine: CAM-ADJ.ON

– Accelerate vehicle from stand-still in 1. gear or driving mode 2. Camshaft adjuster must be active between 1800 to 3200 RPM.

– Press → button.

– Press 0 and 6 for the function "End Output", and then press Q to exit from input mode.

– Switch ignition off.

– Check DTC memory, see **Maintenance**.

– If a malfunction is indicated, the following malfunction can occur:

- The solenoid valve for camshaft adjuster properly switches the oil pressure to the mechanical camshaft adjuster, however the adjuster cannot reach its end position (e.g. due to difficulty of movement).

– Replace camshaft adjuster.

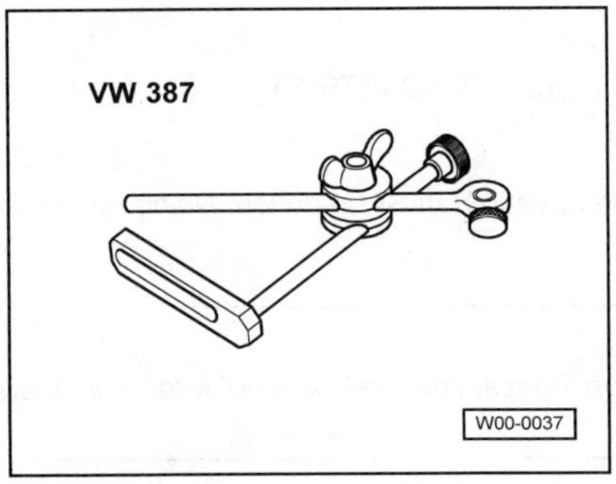

Valve guides, checking

Special tools and equipment

◄ VW387 Dial indicator holder

• Dial indicator

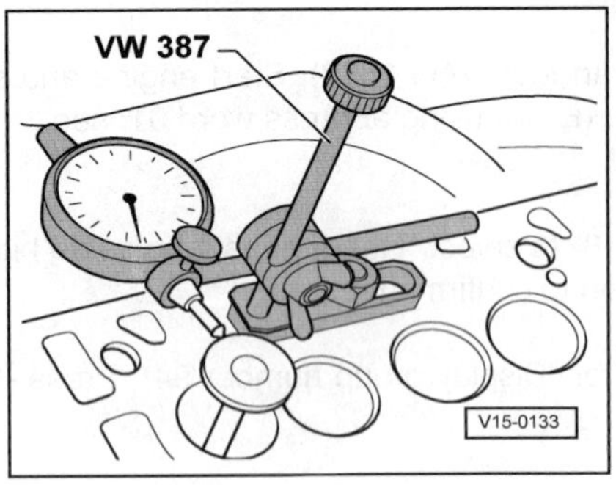

Test sequence

◄ Insert a new valve into guide. Valve stem tip must seal with guide. Due to differences in valve stem diameter, make sure that only intake valves are used to check intake guides, and only exhaust valves are used to check exhaust guides.

– Measure side play. Wear limit: 0.8 mm

If the side play is greater than 0.8 mm:

– Replace cylinder head.

17 Engine–Lubrication System

Repair Group

C 17

Lubrication System

NOTE—

- *If large quantities of metal particles or other deposits (caused, for example, by partial seizure of the crankshaft or connecting rod bearings) are found in the engine oil, the oil passages must be cleaned thoroughly and the oil cooler replaced in order to prevent further damage from re-occurring.*
- *Oil level must not exceed max. marking - danger of catalytic converter damage!*

EDITOR'S NOTE—

For oil system capacities and oil change information, see **Maintenance**.

Engine oils are graded according to their viscosity. The proper grade to be used depends on existing climatic or seasonal conditions.

The original oil in the engine has a viscosity of SAE 5W-40. Use this oil over all temperature ranges for normal driving. If the engine oil viscosity grade SAE 5W-40 is not available, use SAE 5W-30.

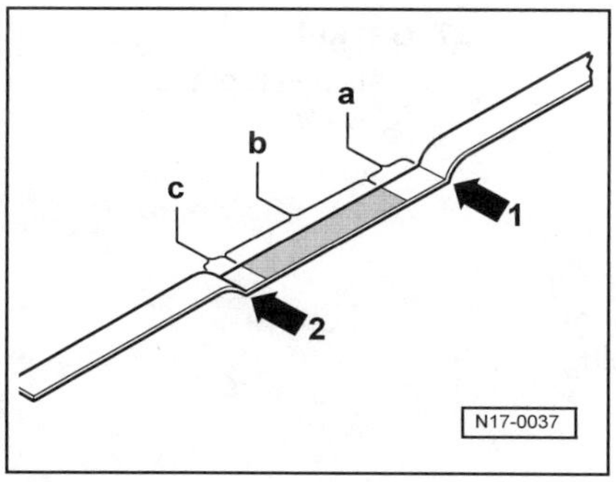

Check oil level using markings on dipstick:

1. Max. marking
2. Min. marking

a - Area above hatched area up to max. marking: Do not add engine oil!

b - Oil level in grid: Oil may be topped off

c - Range from MIN. marking to grid: Do not add more than 0.5 liter (1/2 qt.) oil!

EDITOR'S NOTE—

Volkswagen identifies electrical components by a letter and/or a number in the electrical schematics. See **Wiring Diagrams, Fuses and Relays**. *These electrical identifiers are listed -between dashes- as an aid to electrical troubleshooting.*

Oil pump and oil pan, components

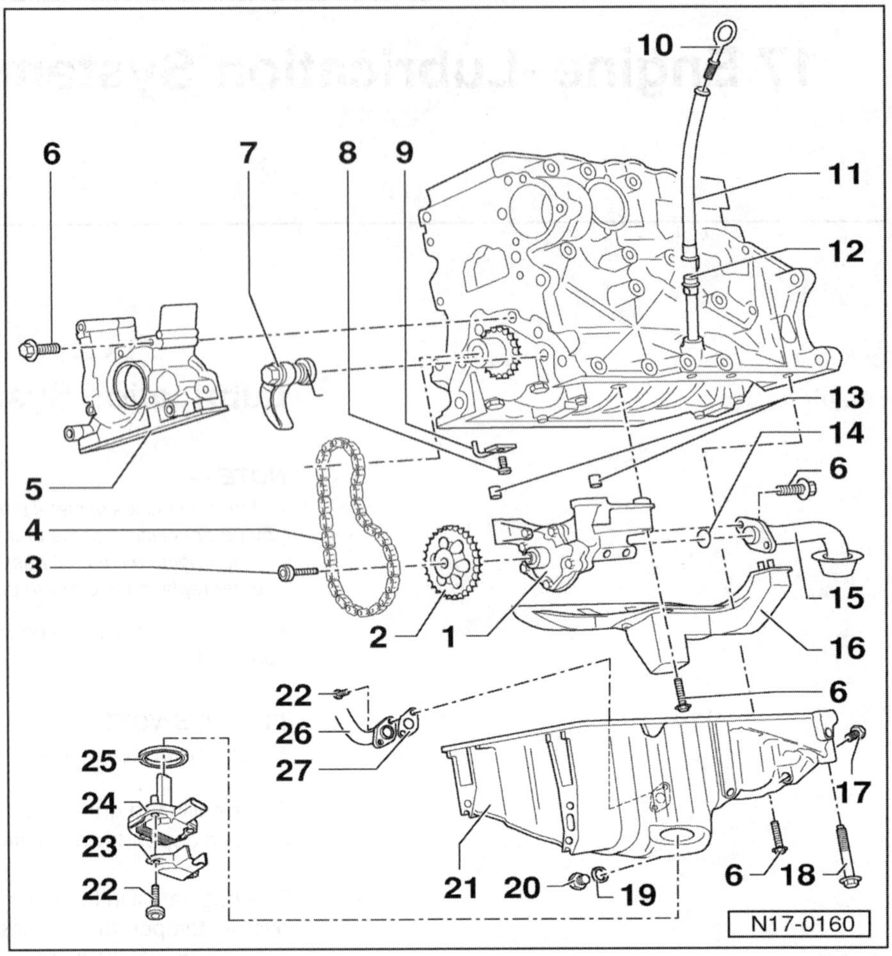

1. **Oil pump**
   - With positive pressure valve 12 bar (174 psi)
   - Before installing, check to be sure both alignment bushings are present (for centering oil pump/cylinder block)
   - Replace if there is striation on contact surfaces and gears
2. **Chain sprocket**
3. **20 Nm (15 ft-lb)**
4. **Chain**
5. **Sealing flange**
   - With seal
   - Must fit on bushing
   - Removing and installing, see **Repair Group 13, Sealing flanges and flywheel/drive plate, components**
   - Insert with silicon sealant D 176 404 A2
6. **15 Nm (11 ft-lb)**
7. **Chain tensioner with tensioning track**
   - Tighten to 15 Nm (11 ft-lb)
   - When installing, pre-tighten spring and engage
8. **25 Nm (18 ft-lb)**
   - Insert without sealant
9. **Oil injection jet**
   - To cool pistons
10. **Dipstick**
    - Oil level must not exceed max. marking!
11. **Insertion funnel**
    - Remove when siphoning oil
12. **Guide tube**
13. **Bushings**
14. **O-ring**
    - Always replace
15. **Intake line**
    - Clean strainer if dirty
16. **Baffle**
17. **45 Nm (33 ft-lb)**
    - Connecting bolt oil pan/transmission
18. **45 Nm (33 ft-lb)**
19. **Seal**
    - Always replace
20. **Oil drain plug**
    - Tighten to 30 Nm (22 ft-lb)
    - Replace if leaking.
    - In some cases, seal and drain plug are combined; do not interchange with separate seal and drain plug.
21. **Oil pan**
    - Clean sealing surface before installing
    - Install using silicon sealant D 176 404 A2
22. **10 Nm (7 ft-lb)**
23. **Cover**
24. **Oil Level Thermal Sensor -G266-**
    - Checking, see **Wiring Diagrams and Component Locations**
25. **Seal**
    - Always replace
    - Oil before installing
26. **Oil return line**
    - from turbocharger
27. **Gasket**
    - Always replace

Oil filter bracket, components

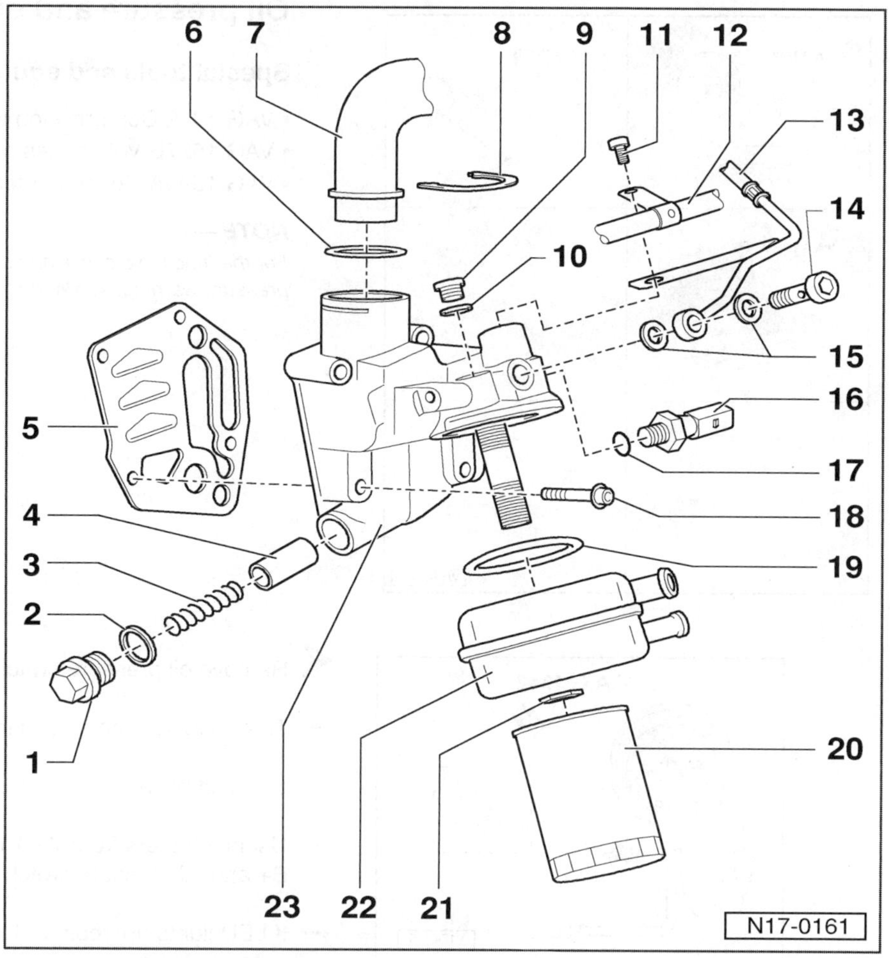

1. **Locking screw**
   - Tighten to 40 Nm (30 ft-lb)
2. **Seal**
   - If seal is leaking, cut open seal and replace seal
3. **Spring**
   - For positive pressure valve, approx. 4 bar (58 psi)
4. **Pistons**
   - For positive pressure valve,
   - approx. 4 bar (58 psi)
5. **Gasket**
   - Always replace
6. **Seal**
   - Slide on up to collar of tube **-7-**
7. **Line**
8. **Retaining clip**
9. **Locking screw**
   - Tighten to 15 Nm (11 ft-lb)
10. **Seal**
    - If seal is leaking, cut open seal and replace seal
11. **20 Nm (15 ft-lb)**
12. **Lower coolant line**
13. **Oil supply line**
    - To turbocharger
14. **Banjo fitting**
    - Tighten to 30 Nm (22 ft-lb)
15. **Gasket**
    - Always replace
16. **1.4 bar oil pressure switch -F1-**
    - Tighten to 25 Nm (18 ft-lb)
    - Checking, see **Oil pressure and oil pressure switch, checking**
17. **Seal**
    - If seal is leaking, cut open seal and replace seal
18. **15 Nm (11 ft-lb) + 1/4 turn (90 °)**
    - Always replace
19. **Gasket**
    - Always replace
    - Snap into grooves on oil cooler
20. **Oil filter housing**
    - Remove with tension strap
    - Tighten by hand
    - Follow installation notes found on oil filter
21. **25 Nm (18 ft-lb)**
22. **Oil cooler**
    - ensure sufficient clearance to surrounding components
    - Coat contact surfaces to oil filter bracket outside seal with AMV 188 100 02.
23. **Oil filter bracket**
    - With positive pressure valve, approx. 4 bar (58 psi)
    - With pressed in check-valve

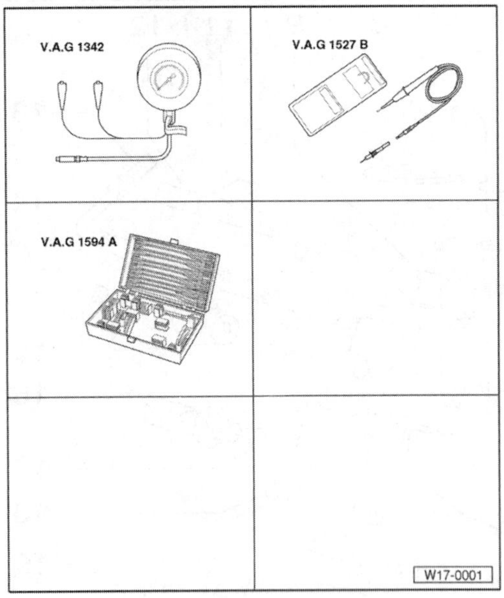

Oil pressure and oil pressure switch, checking

Special tools and equipment

- VAG 1342 Compression tester
- VAG 1527B Voltage tester
- VAG 1594A Connector test kit

NOTE—
For the functional check and servicing of the optical and acoustic oil pressure warning, see **Wiring Diagrams and Component Locations**

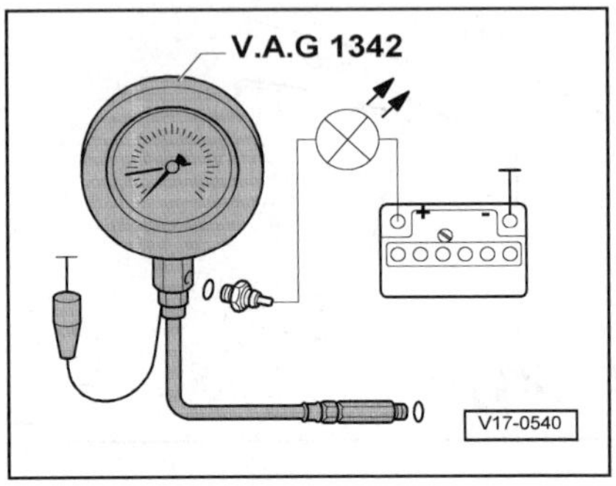

◀ Remove oil pressure switch -F1- and install in tester.

– Thread tester into oil pan in place of oil pressure switch.

– Connect brown wire of tester to Ground (GND).

– Using adapters from VAG 1594A, connect VAG 1527B Voltage Tester to B+ and oil pressure switch. LED must not light up.

– If LED lights up, replace 1.4 bar oil pressure switch -F1-.

– If LED does not light up:
  - Start engine and let run at idle: LED must light up with 1.2 to 1.6 bar (17-23 psi) positive pressure, otherwise replace oil pressure switch.
  - Oil pressure at idle: min. 2 bar (29 psi)

– Increase engine speed (RPM) further. At 2000 RPM and an oil temperature of 80 °C (176°F).
  - Oil pressure must be between 3.0 and 4.5 bar (43-65 psi).

– If specified values are not obtained:
  - Repair mechanical damage, e.g. bearing damage.
  - Check oil filter bracket with pressure valve, see **Oil filter bracket, components**
  - If necessary, replace oil pump.

Oil pressure must not exceed 7.0 bar (101 psi) at higher engine speeds.

19 Engine–Cooling System

Repair Group

C 19

Cooling System Components

> ***WARNING —***
> *Observe the following for all installations, especially in engine compartment due to lack of room:*
>
> - *Route lines of all types (e.g. for fuel, hydraulic, EVAP canister system, coolant and refrigerant, brake fluid, vacuum) and electrical wiring so that the original path is followed.*
> - *Watch for sufficient clearance to all moving or hot components.*

NOTE —

- *The cooling system is under pressure when engine is warm. If necessary relieve pressure before repair.*
- *All hose connections are secured with spring clips. If replacing, only use spring clips.*
- *VAS 5024 spring clip installation tool is recommended for installing spring clips.*
- *When installing coolant hoses, make sure they are free of stress and do not come into contact with other components (observer markings on coolant connections and hoses).*

Use VAG 1274 cooling system tester with VAG 1274/8 and VAG 1274/9 adapters to test coolant system for proper seal

NOTE —

Volkswagen identifies electrical components by a letter and/or a number in the electrical schematics. See **Wiring Diagrams, Fuses and Relays***. These electrical identifiers are listed -between dashes- as an aid to electrical troubleshooting.*

Cooling system components (on body)

> ***EDITOR'S NOTE —***
> *For coolant hose connections, see* **Coolant hose connection diagram**

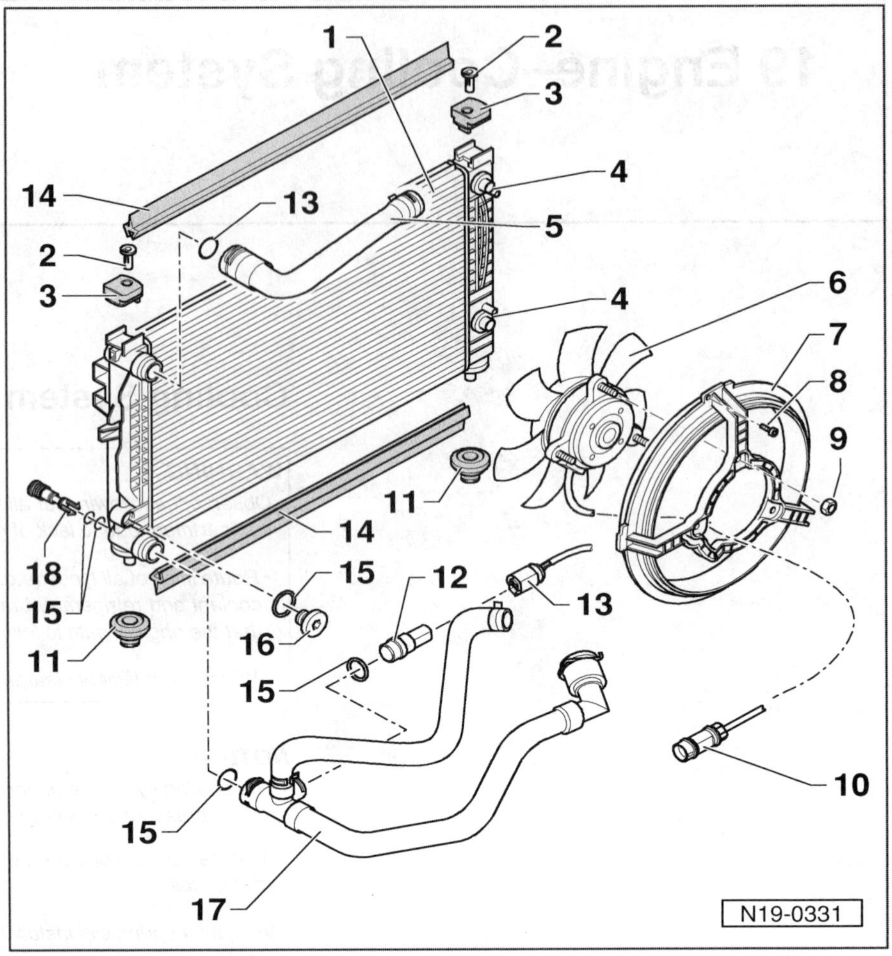

1. **Radiator**
   - See **Radiator, removing and installing**
   - After replacing, completely replace coolant
   - With ATF cooler on vehicles with automatic transmission, see **5-Speed Automatic Transmission 01V section, Repair Group 37**
   - Clean ATF-cooler, see **5-Speed Automatic Transmission 01V section, Repair Group 37**
2. **Securing pin**
3. **Securing rubber**
4. **Connecting piece for ATF line**
   - Only for vehicles with automatic transmission
5. **Upper coolant hose**
   - Secured to radiator with retaining clip
   - Check for secure fitting
6. **Coolant fan -V7-**
7. **Fan ring**
   - Clipped into the intake air elbow and secured with securing bolt **-8-**
8. **Mounting bolt**
9. **10 Nm (7 ft-lb)**
10. **2-way connector**
11. **Rubber bearing**
12. **Coolant Fan Control (FC) thermal switch -F18-**
    - For electric fan.
    - Switching temperatures:
      1st level on: 92 to 97 °C (197 to 206°F), off: 84 to 91 °C (183 to 196°F).
      2nd level on: 99 to 105 °C (210 to 221°F), off: 91 to 98 °C (196 to 208°F)
    - Secured with retaining clip
13. **Connector**
    - Black, 4-pin
    - For Coolant Fan Control (FC) thermal switch -F18-
14. **Sealing strip**
    - Installed on radiator at top and bottom
15. **O-ring**
    - Replace if damaged
16. **Locking screw**
    - Tighten to 10 Nm (7 ft-lb)
17. **Lower coolant hose**
    - Secured with retaining clip at radiator and connecting piece
    - Check for secure fitting
18. **Drain plug**
    - Tighten to 10 Nm (7 ft-lb)

Cooling system components (on engine)

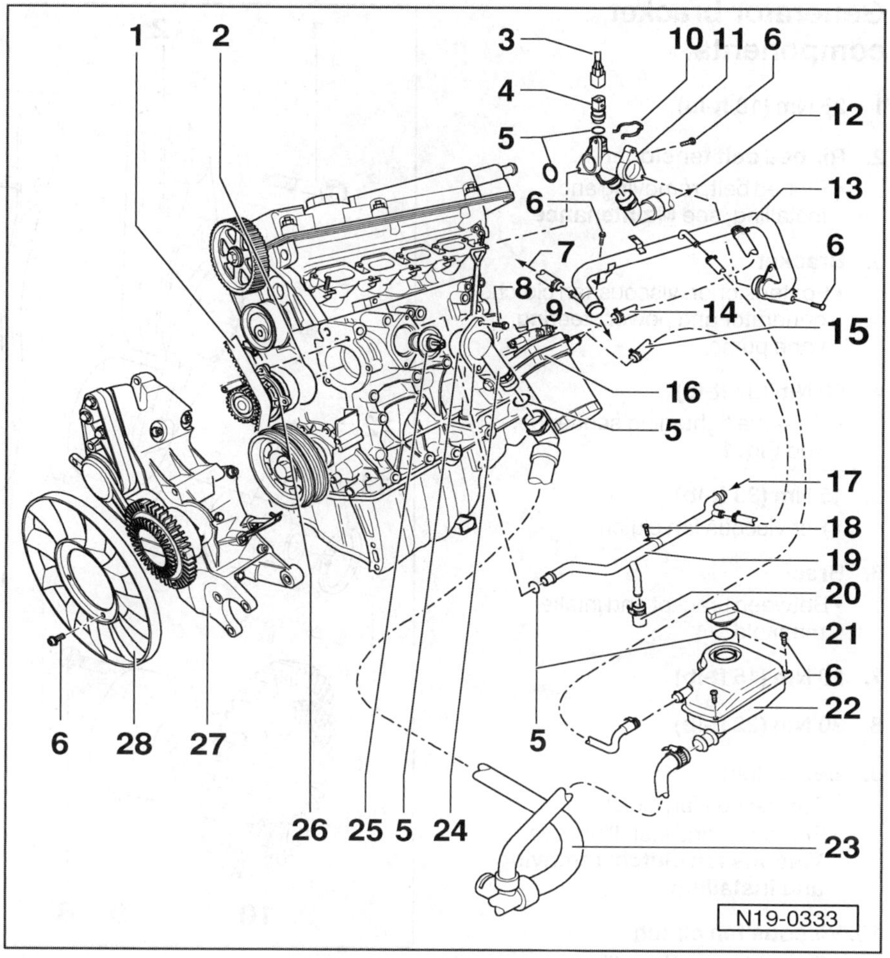

EDITOR'S NOTE —

For coolant hose connections, see **Coolant hose connection diagram**

1. **Toothed belt**
   - Mark direction of travel before removing
   - Check for wear
   - Do not bend
   - Removing and installing, see **Repair Group 15**
2. **O-ring**
   - Always replace
3. **Connector**
   - 4 pin
4. **Engine Coolant Temperature (ECT) sensor -G62-**
   - For Engine Control Module (ECM)
   - If necessary, release pressure in cooling system before removing
   - Checking, see **1.8 Liter Fuel Injection & Ignition, (code: ATW, AUG, AWM) section, Repair Group 24**
5. **O-ring**
   - Replace if damaged
6. **10 Nm (7 ft-lb)**
7. **To turbocharger**
8. **15 Nm (11 ft-lb)**
9. **To top of radiator**
10. **Retaining clip**
    - Check for secure fitting
11. **Connecting piece**
12. **To heater core**
13. **Upper coolant line**
    - Secured at cylinder head
14. **Coolant hose**
15. **Coolant hose**
16. **Oil cooler**
17. **From heater core**
18. **20 Nm (15 ft-lb)**
19. **Lower coolant line**
    - Secured at oil filter bracket
20. **Sealing plug**
    - Check for secure fitting
21. **Cap**
    - Check using VAG 1274 cooling system tester with VAG 1274/9 adapter
    - Test pressure: 1.4 to 1.6 bar (20-23 psi)
22. **Expansion tank**
    - Perform cooling system leakage test using VAG 1274 cooling system tester with VAG 1274/8 adapter
23. **Lower coolant hose**
    - From bottom of radiator
    - secured with retaining clip at radiator and connecting piece
    - Check for secure fitting
24. **Connecting piece**
25. **Coolant thermostat**
    - Installation position: Bracket must be vertical
    - Checking: Warm thermostat in hot water bath.
    - Starts opening at approx. 86 °C (176°F)
    - Opens at least 7 mm (0.27")
26. **Coolant pump**
    - Check for ease of movement
    - Completely replace if damaged or leaking
    - See **Coolant pump, removing and installing**
27. **Bracket**
    - For tensioner, viscous-fan clutch, generator and power steering vane pump
    - Removing and installing, see following page
28. **Fan**

Generator bracket, components

1. **25 Nm (18 ft-lb)**
2. **Ribbed belt tensioner**
   - Ribbed belt, removing and installing, see **Maintenance**
3. **Bracket**
   - For tensioner, viscous-fan clutch, generator and power steering vane pump
4. **50 Nm (37 ft-lb)**
   - Observe tightening sequence, see **Fig. 1**
5. **45 Nm (33 ft-lb)**
   - For viscous fan clutch
6. **Brace**
   - Between bracket and intake manifold
7. **20 Nm (15 ft-lb)**
8. **30 Nm (22 ft-lb)**
9. **Belt pulley**
   - For viscous fan clutch
   - Removing and installing, see **Viscous fan clutch, removing and installing**
10. **Viscous fan clutch**
    - Removing and installing, see **Viscous fan clutch, removing and installing**
11. **Snap ring**
12. **Bearing bushing**
    - For viscous fan clutch
    - Removing and installing, see **Viscous fan clutch bushing, removing and installing**

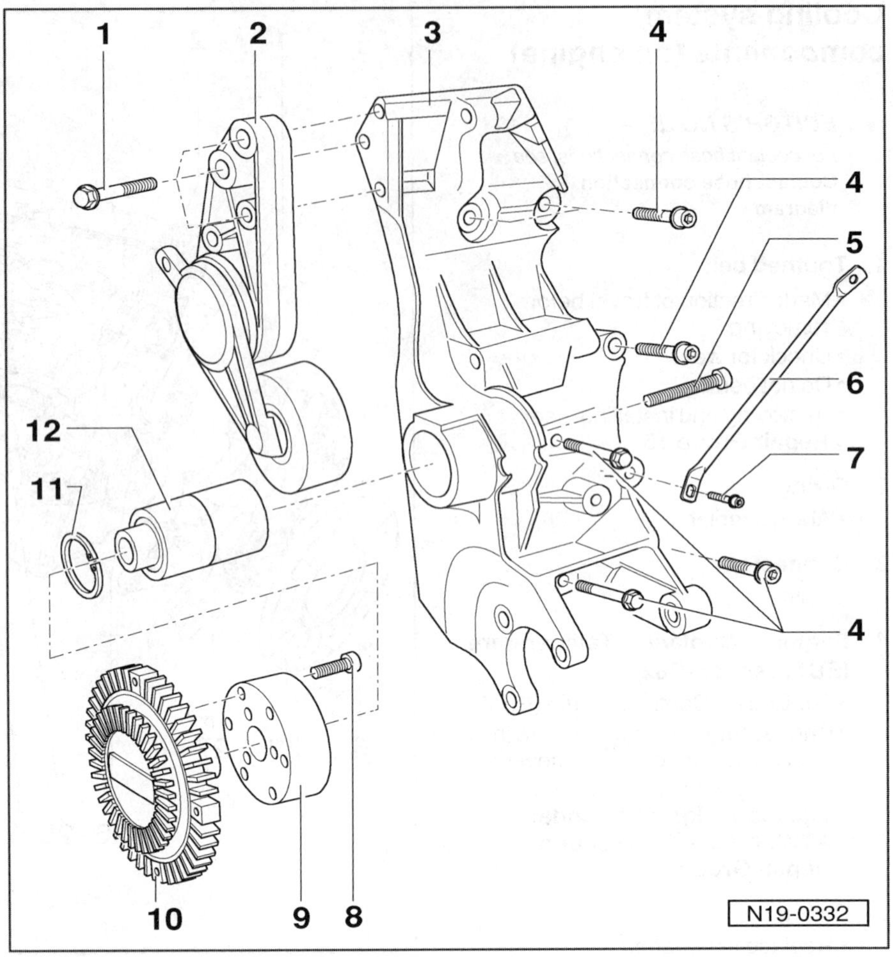

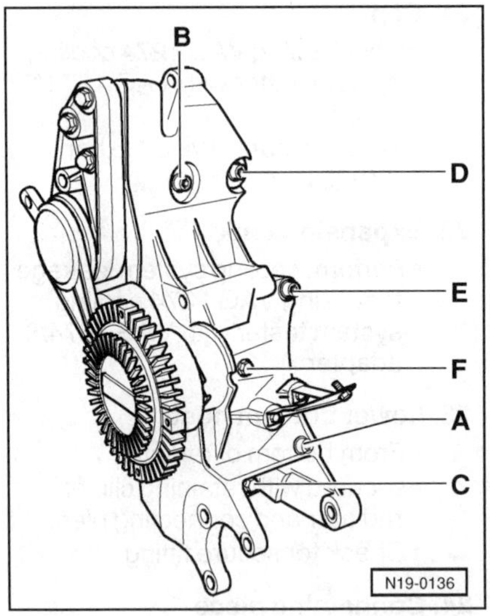

Fig. 1 Tightening sequence, bracket to cylinder block

Special tools and equipment

- VAG 1331 torque wrench (5 to 50 Nm)

Procedure

– Place bracket on cylinder block.

– Tighten bracket to 50 Nm (37 ft-lb) in shown tightening sequence below:

1. Tighten bolt -A-
2. Tighten bolt -B-
3. Tighten bolt -C-
4. Tighten bolt -D-
5. Tighten bolt -E-
6. Tighten bolt -F-

Coolant hose connection diagram

1. Heater core
2. Lower coolant pipe
3. Expansion tank
4. Oil cooler
5. Intake manifold
6. Lower coolant hose
7. Upper coolant hose
8. Radiator
9. Upper coolant line
10. Coolant pump and coolant thermostat
11. Turbocharger
12. Cylinder head/Cylinder block
13. Connecting piece

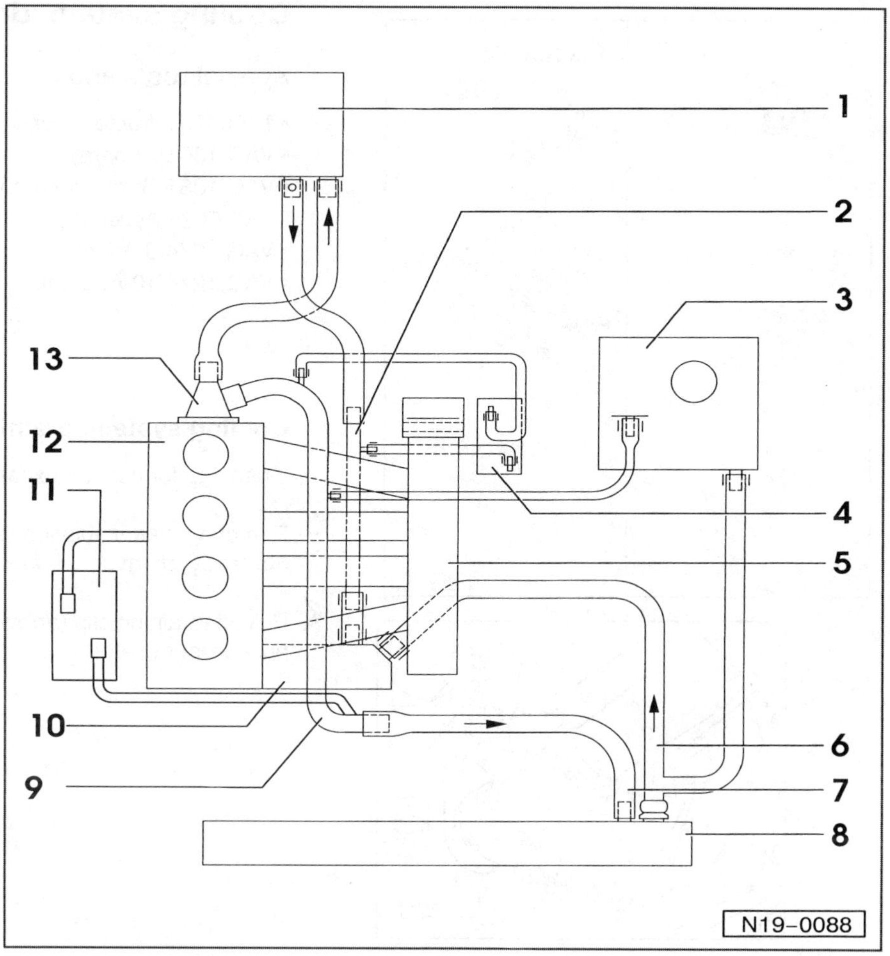

Cooling system, draining and filling

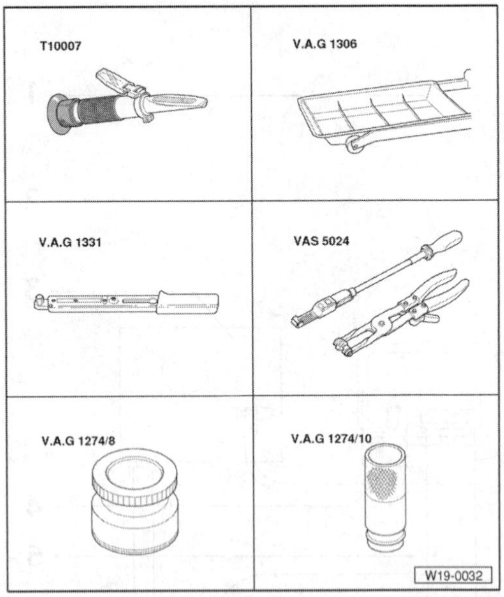

Special tools and equipment

- T10007 Refractometer
- VAG 1306 Drip tray
- VAG 1331 Torque wrench (5 to 50 Nm)
- VAS 5024 Assembly tool for spring-type clamps
- VAG 1274/8 Adapter
- VAG 1274/10 Adapter

Cooling system, draining

– Open cap for coolant expansion tank.

– Remove noise insulation, see **Body–Exterior section, Repair Group 50, Body, front**

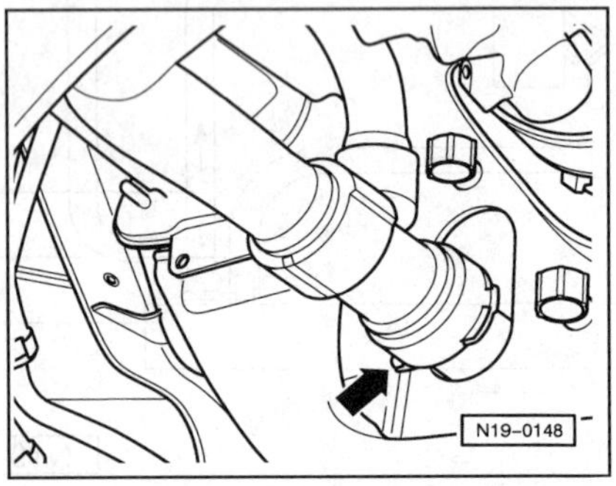

◄ Pull off retaining clip (arrow) for lower coolant hose and remove coolant hose from radiator.

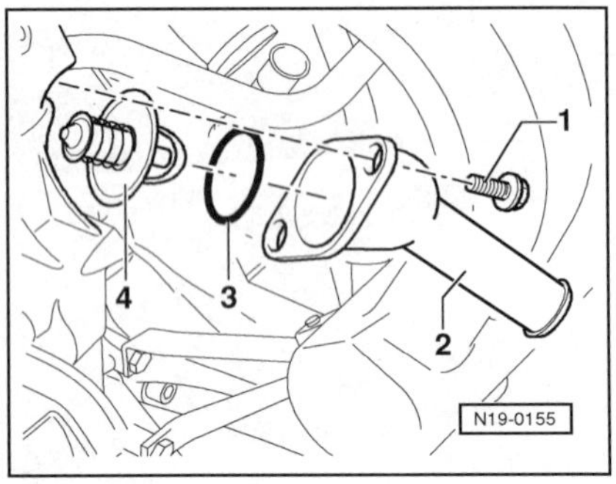

◄ Disconnect coolant hose from connecting piece -2-.

– Remove bolts -1-, connecting piece -2-, O-ring -3- and coolant thermostat -4-.

NOTE—
Follow coolant disposal regulations!

Cooling system, filling

NOTE—
- *Only use coolant additive G12 (according to TL VW 774 D). Characteristic: red color*
- *Do not mix G12 with any other coolant additive!*
- *If the fluid in expansion tank is brown, G12 has been mixed with other coolant. In this is the case, coolant must be replaced.*
- *"G12" and coolant additives marked "In accordance with TL VW 774 D" prevent frost and corrosion damage, scaling and also raise the boiling point of the coolant. Therefore, cooling system must be filled year round with frost and corrosion protection additives.*
- *A higher boiling point improves engine reliability under heavy load particularly in countries with tropical climates.*

Cooling system, filling (continued)

NOTE —

- *Frost protection additives protect against frost to about -25 °C (-13°F) (about -35 °C (-31°F) in arctic climates).*
- *The coolant concentration must not be reduced by adding water even in warmer seasons and in warmer countries. The anti-freeze ratio must be at least 40%.*
- *If for climatic reasons greater frost protection is required, the amount of G 12 can be increased, but only up to 60% (frost protection to about -40 °C) (-40°F), otherwise frost protection and cooling effectiveness will be reduced.*
- *If the radiator, heater core, cylinder head or cylinder head gasket is replaced, completely replace the engine coolant.*

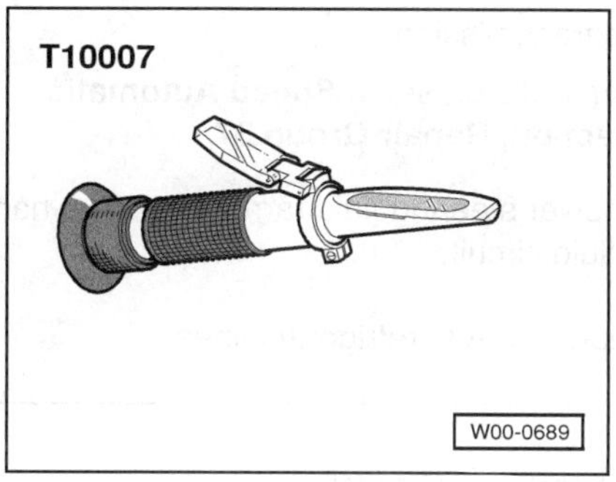

The T10007 refractometer is recommended for determining frost protection density.

Recommended mixture ratios:

Frost protection to	Anti-freeze amount	-G12-[1]	Water[1]
-25°C (-13°F)	40%	3.0 liters	4.0 liters
-35°C (-31°F)	50%	3.5 liters	3.5 liters

[1] The quantity of coolant can vary depending upon vehicle options.

– Install lower radiator hose and secure.

– Replace O-ring and remove coolant thermostat and connecting piece. Bracket of coolant thermostat must be vertical.

- Tightening torque: 15 Nm (11 ft-lb)

– Attach coolant hose to connecting piece.

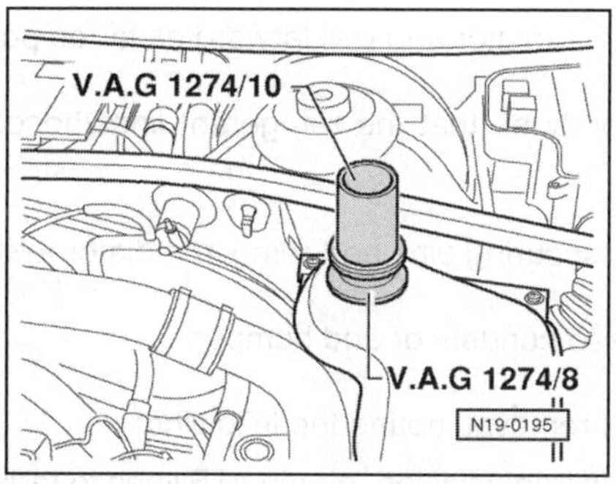

Thread VAG 1274/8 onto coolant reservoir and lengthen using 1274/10 tube

– Move back protective cover for coolant hose at connection of heater core.

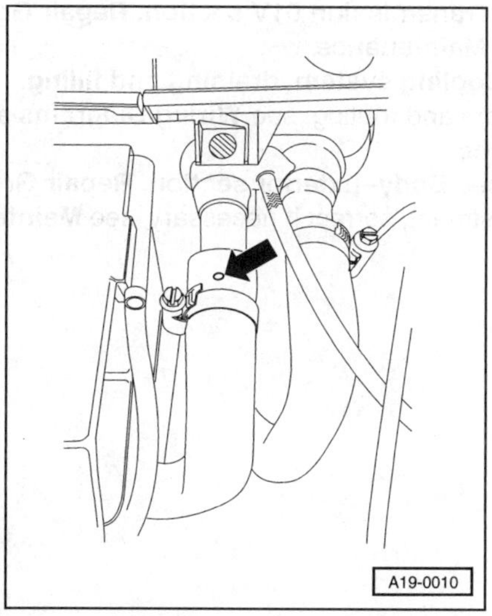

Loosen coolant hose on heater core and pull back hose sufficiently so that the bleed hole in coolant hose (arrow) is no longer covered by the connecting piece.

– Fill up coolant until it flows out of the coolant hose bleed hole.

– Slide coolant hose onto connecting piece and secure it again.

– Install coolant expansion tank cap.

– Start engine and maintain an engine speed of about 2000 RPM for approx. 3 minutes.

– Allow engine to idle until lower coolant hose at radiator is hot.

– Check coolant level and top off if necessary. With engine at operating temperature, coolant level must be at max. marking, with engine cold, it must be between min. and max. marking.

Radiator, removing and installing

Special tools and equipment

- VAG 1306 drip tray
- VAG 1331 torque wrench (5 to 50 Nm)

Removing

— Drain coolant, see **Cooling system, draining and filling**.

— Remove front bumper, see **Body–Exterior section, Repair Group 63, Bumper, front**

— Remove coolant hoses from radiator.

— Vehicles with automatic transmission

- Remove ATF lines from radiator, see **5-Speed Automatic Transmission 01V section, Repair Group 37.**

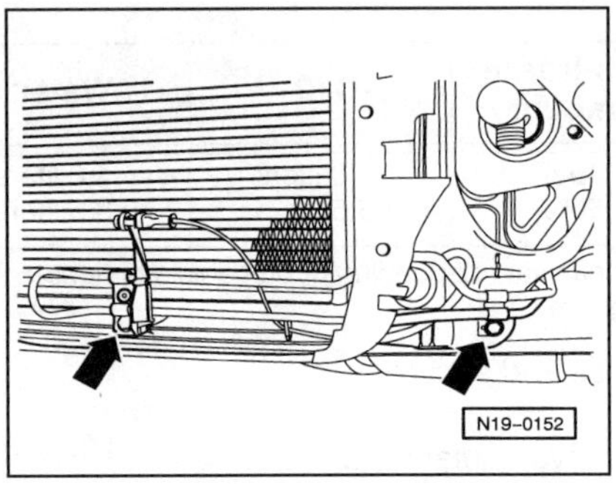

< Unbolt cooling coil for power steering fluid (arrows) and let hang free. Do not open hydraulic fluid circuit.

— Remove retaining clamp(s) for A/C refrigerant lines.

> ***WARNING —***
> *Do not open refrigerant circuit for A/C system.*

NOTE —
Do not bend or stretch lines or hoses as condenser and/or refrigerant lines/hoses may be damaged.

— Remove condenser from radiator and pull forward as far as possible.

— Secure condenser to body so that the refrigerant lines/hoses are not burdened.

— Remove upper radiator securing pins and remove radiator toward front.

— Remove radiator between condenser and bumper.

— Installation is reverse of removal, noting the following:

- Vehicles with automatic transmission, attach ATF lines to radiator, see **5-Speed Automatic Transmission 01V section, Repair Group 37.**
- Check ATF level, see **Maintenance**.
- Top up coolant, see **Cooling system, draining and filling.**
- For harness connectors and routing, see **Wiring Diagrams and Component Locations**
- Install front bumper, see **Body–Exterior section, Repair Group 63.**
- Check headlight adjustment, correct if necessary, see **Maintenance.**

Coolant pump, removing and installing

Special tools and equipment

- T10007 Refractometer
- VAG 1306 Drip tray
- VAG 1331 Torque wrench (5 to 50 Nm)
- VAS 5024 Assembly tool for spring-type clamps
- VAG 1274/8 Adapter
- VAG 1274/10 Adapter

NOTE —

- *Always replace seals and gaskets.*
- *Lower portion of toothed belt guard can remain installed.*
- *Toothed belt stays attached to crankshaft toothed belt gear.*
- *Cover the toothed belt with a cloth to protect it from coolant before removing the coolant pump.*

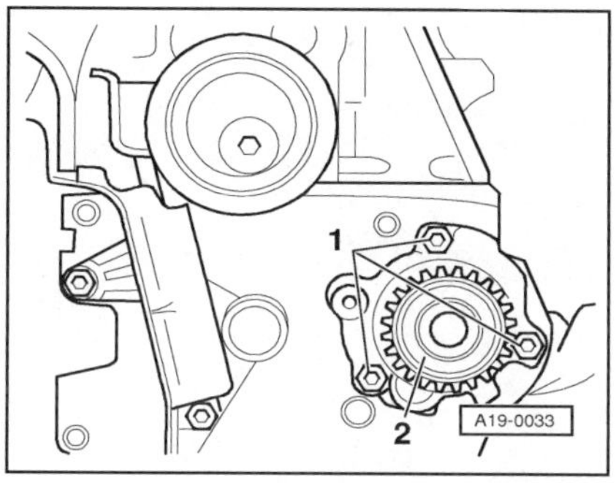

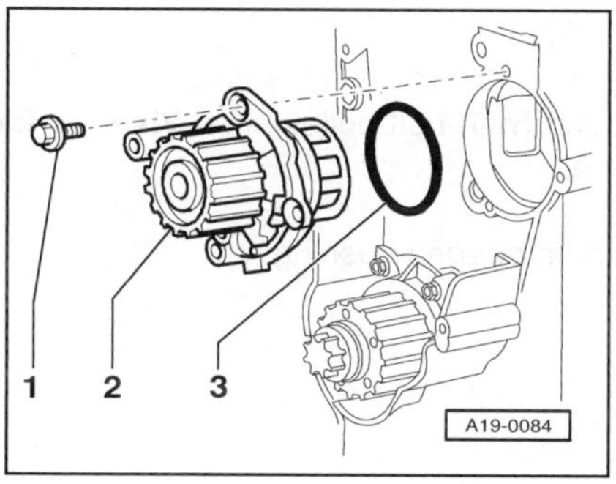

– Drain coolant, see **Cooling system, draining and filling.**

– Remove ribbed belt, see **Maintenance.**

– Remove toothed belt from toothed belt gear of coolant pump, see **Repair Group 15, Toothed belt, removing and installing**.

◄ Remove coolant pump mounting bolts -1- and remove coolant pump -2-.

◄ Moisten new O-Ring -3- with coolant.

– Insert coolant pump -2- so that sealing plug in housing points downward.

– Tighten mounting bolts -1-. Tightening torque: 15 Nm (11 ft-lb)

– Install toothed belt and adjust valve timing, see **Repair Group 15.**

– Install ribbed belt, see **Maintenance.**

– Top up coolant, see **Cooling system, draining and filling**.

Viscous fan clutch, removing and installing

Special tools and equipment

- VAG 1331 torque wrench (5 to 50 Nm)
- 5 mm drift

– Move the lock carrier into service position, see **Maintenance.**

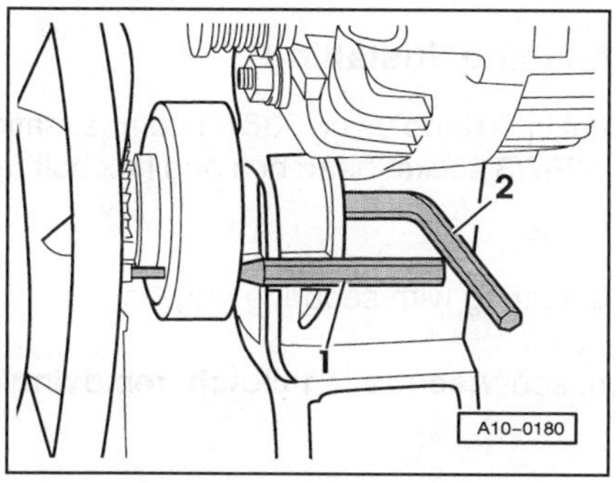

– Remove ribbed belt, see **Maintenance**.

– Remove fan wheel from viscous fan clutch.

◄ Secure belt pulley of viscous-fan clutch using a drift (5 mm) -1-.

– Unbolt securing bolt for viscous fan clutch (with 8 mm wrench insert hex socket -2-) and remove viscous fan clutch with belt pulley.

– Disconnect belt pulley from viscous fan clutch.

– Installation is reverse of removal, noting the following:

- Tighten belt pulley at viscous fan clutch to 30 Nm (22 ft-lb).
- Tighten viscous fan clutch at bracket to 45 Nm (33 ft-lb).
- Tighten fan wheel at viscous fan clutch to 10 Nm (7 ft-lb).
- Install ribbed belt, see **Maintenance**.

Viscous fan clutch bushing, removing and installing

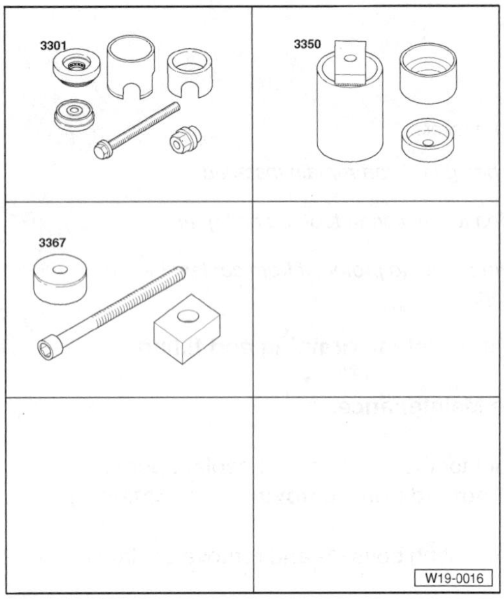

Special tools and equipment

- Hex nut from 3301 Assembly Tool
- Extractor line from 3350 Installation Device
- 3367 Press tool for viscous fan

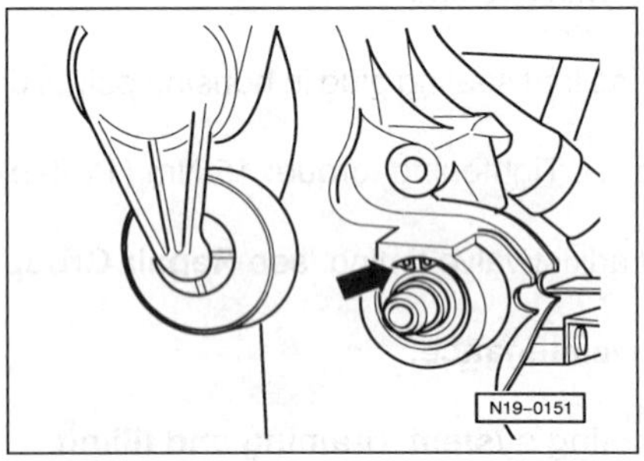

Removing

– Remove viscous fan clutch (with belt pulley), see **Viscous fan clutch, removing and installing**.

◄ Remove circlip -arrow- from bearing bushing.

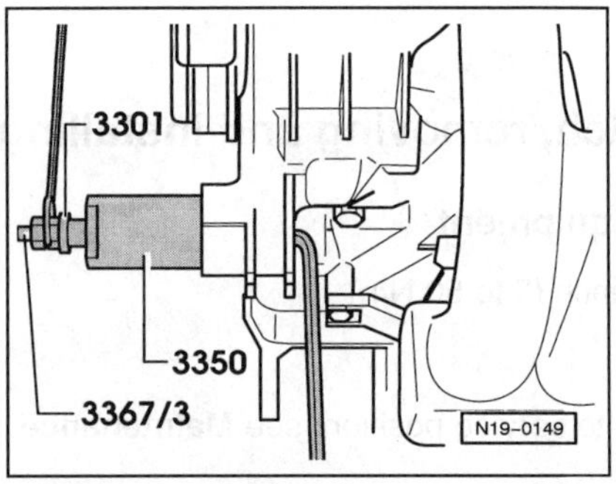

◄ Press out bushing from bracket using 3367/3 socket head bolt, tube from 3350 carrier bearing inst. Tool and hex nut from 3301 assembly tool.

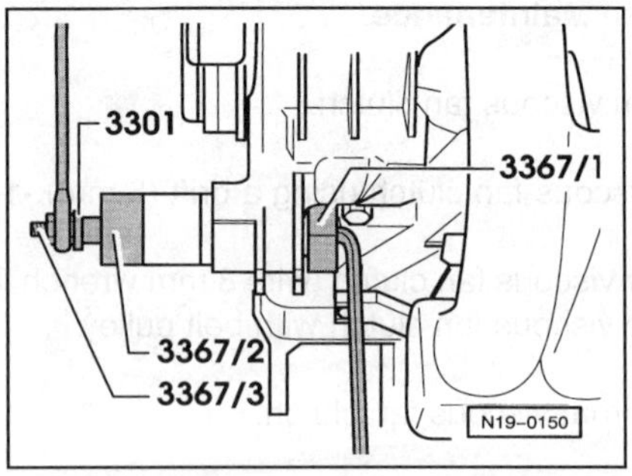

Viscous fan clutch bushing, installing

◄ Press bushing into bracket up to stop using 3367/1 compensation piece, 3367/2 pressure piece, 3367/3 socket head bolt and hex bolt from 3301 assembly tool.

– After installation, secure bearing with securing ring.

– Install viscous fan clutch, see **Viscous fan clutch, removing and installing**.

20 Fuel Supply

Repair Group

c 20

Fuel Supply System

NOTE —

- *All hose connections are secured with spring-type clamps or tensioning clamps.*
- *Always replace tensioning clamps with spring-type clamps.*
- *Fuel hoses at engine must only be secured with spring-type clamps. The use of clamping collars or screw clamps is not permitted.*
- *VAS 5024 Assembly tool or VAG 1921 Pliers are recommended for installing spring clips.*

Safety precautions when working on fuel supply

> ***WARNING —***
> *Fuel supply line is under pressure! Before loosening hose connections, place rags around the connection point. Then release pressure by carefully pulling on hose.*

Always observe the following when removing and installing the sensor for fuel gauge or the Fuel Pump (FP) from partially filled fuel tanks.

- Before starting work, switch on exhaust extraction system and place an extraction hose close to the sensor opening in fuel tank to extract escaping fuel fumes. If no exhaust extraction system is available, a radial fan (as long as motor is not in air flow) with a displacement greater than 15 m^3/h can be used.
- Prevent gasoline from contacting skin! Wear fuel-resistant gloves!

Rules of cleanliness

When working on the fuel supply or fuel injection systems, the following 5 rules should be carefully observed:

1. Thoroughly clean all connections and the surrounding area before disconnecting.
2. Place parts that have been removed on a clean surface and cover. Use lint-free cloths only!
3. Carefully cover over opened components or seal, if repairs are not carried out immediately.
4. Only install clean components: Only unpack replacement parts immediately prior to installation. Do not use parts that have been stored loose (in tool boxes etc.)
5. When the system is open: Do not work with compressed air if it can be avoided. Do not move vehicle unless absolutely necessary.

EDITOR'S NOTE —

Volkswagen identifies electrical components by a letter and/or a number in the electrical schematics. See **Wiring Diagrams, Fuses and Relays**. *These electrical identifiers are listed -between dashes- as an aid to electrical troubleshooting.*

Fuel tank with attachments and fuel filter

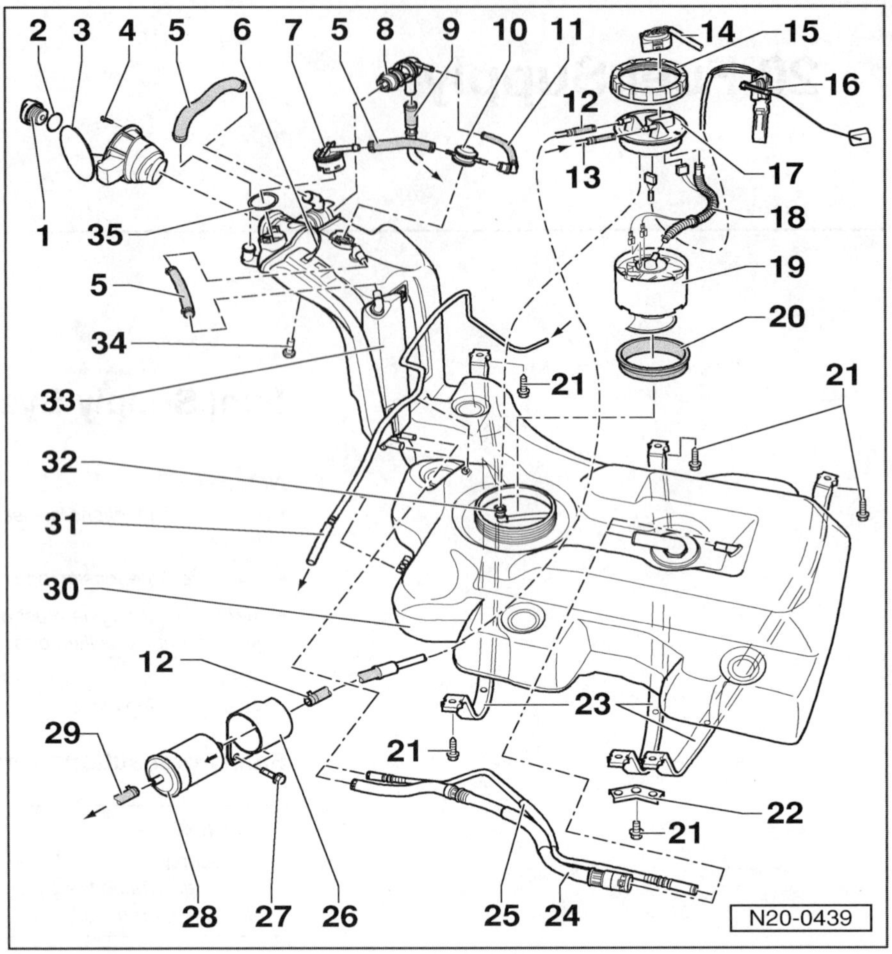

1. **Cap**
2. **Seal**
   - Replace if damaged
3. **Tank lid unit**
   - With rubber cap
4. **Mounting screw**
5. **Connection hose**
6. **Ground (GND) connection**
   - Check for secure fitting
7. **Gravity valve**
   - To remove, unclip valve upward and lift out of shaft.
   - Check valve for throughput. Valve vertical: open, Valve tilted 45 °: closed
8. **Change-over valve**
   - To remove valve, clip out sideways and remove from delivery unit.
   - Before installing, remove cap, **-1-**
   - Checking, see **Fig. 2**
9. **Vent line**
   - Black
   - Clipped in on side of fuel tank
   - Check for secure fitting
   - Secure with spring type clamps
   - To EVAP canister, see **Repair Group 21, Turbocharger components**, **-1-**
10. **Pressure retention valve**
    - For fuel tank ventilation
    - To remove, unclip valve upward and lift out of shaft.
    - Checking, see **Fig. 3**
11. **Vent line**
    - with quick-release coupling
12. **Supply line**
    - Black
    - Clipped in on side of fuel tank
    - Check for secure fitting
    - Mount with spring type clamps
    - Between flange -**17**- and fuel filter -**28**-
13. **Return line**
    - Blue or with blue marking
    - Clipped in on side of fuel tank
    - Check for secure fitting
    - Mount with spring type clamps
    - From fuel return line at fuel rail, see **1.8 Liter Fuel Injection & Ignition (code: ATW, AUG, AWM) section, Repair Group 24**
14. **Connector**
    - Black, 4-pin
    - For sensor for fuel gauge -G- and Fuel Pump (FP) -G6-
15. **Union nut**
    - Tighten to 60 Nm (44 ft-lb)
    - Use ring nut spanner 3217 to remove and install
16. **Sensor for fuel gauge -G-**
    - Clipped in at baffle housing in fuel tank
    - Removing and installing, see **Sensor for fuel gauge, removing and installing**
17. **Flange**
    - Note installation position at fuel tank, see **Fig. 1**
18. **Supply line**
    - Check for secure fitting
    - Between flange -**17**- and Fuel Pump (FP) -**19**-
19. **Fuel pump**
    - Removing and installing, see **Fuel pump, removing and installing**
    - Checking Fuel Pump (FP), see **Fuel pump, checking**
    - Clean strainer if dirty
20. **Seal**
    - to install, place into opening of fuel tank when dry
    - Replace if damaged
    - Only coat flange seal with fuel when installing.
21. **25 Nm (18 ft-lb)**
22. **Bracket**
    - For tension straps
    - Note installed position
23. **Tension strap**
    - Note varying lengths
    - Installation position: Locating point (hole) faces direction of travel
24. **Vent line**
    - Check for secure fitting

(continued on following page)

Fuel tank with attachments and fuel filter (continued)

25. Vent line
- From drip reservoir

26. Retaining ring
- Secured to body

27. 10 Nm (7 ft-lb)

28. Fuel filter
- Installation position: Arrow points in direction of flow

29. Supply line
- Black
- Check for secure fitting
- secure with spring type clamps
- To fuel supply line at fuel distributor, see **1.8 Liter Fuel Injection & Ignition (code: ATW, AUG, AWM) section, Repair Group 24**

30. Fuel tank
- When removing, support using VAG 1383A engine/transmission hoist

31. Vent line
- From EVAP canister to Evaporative Emission (EVAP) canister purge regulator valve -N80-
- Clipped onto fuel tank

32. Return line
- Check for secure fitting
- Between flange -**17**- and baffle housing in fuel tank

33. Drip reservoir
- Clipped in at filler tube

34. 25 Nm

35. O-ring
- Replace if damaged

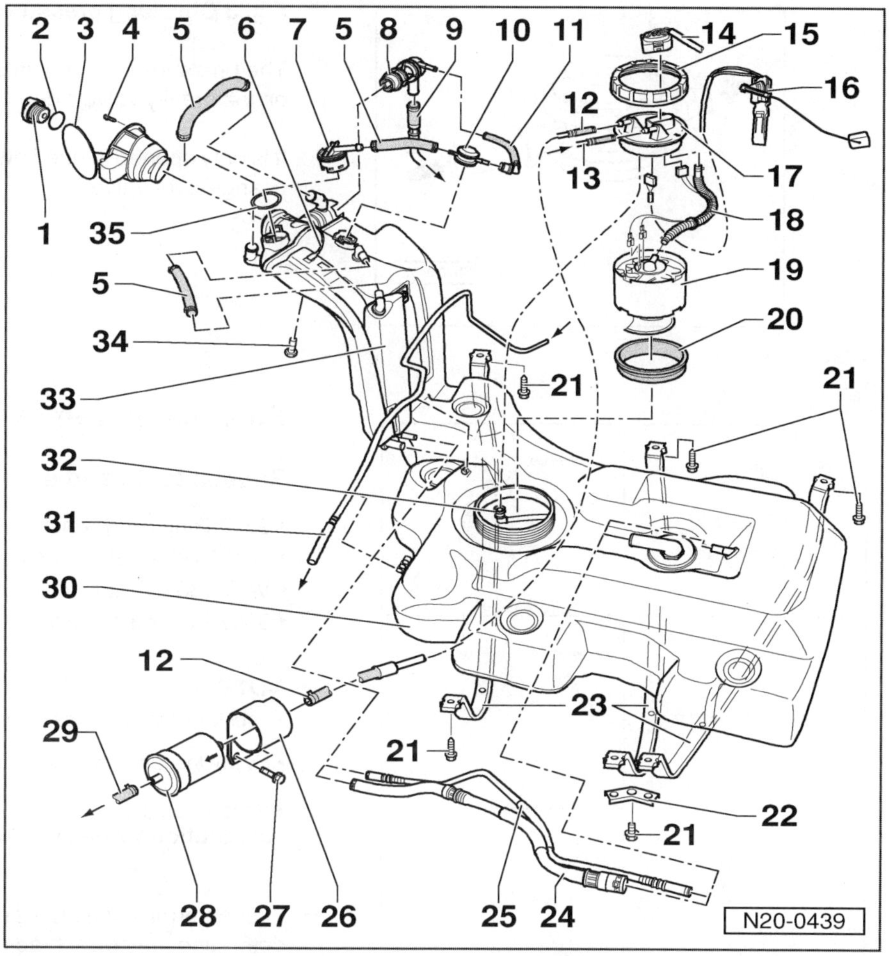

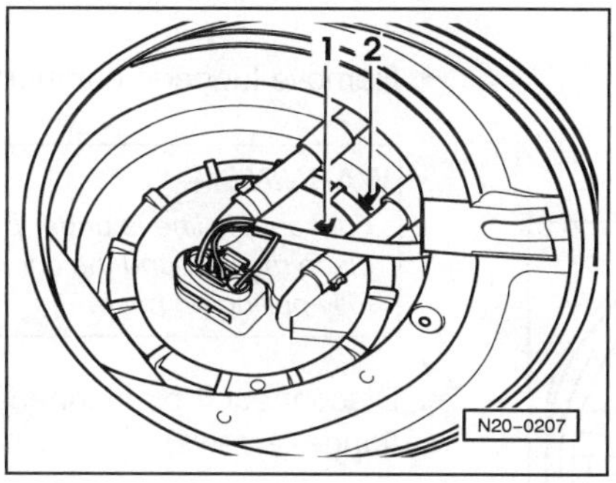

Fig. 1 Installation position of fuel pump flange

Marking on flange -1- must align with marking on fuel tank -2-.

NOTE —
After installing fuel pump flange, check to be sure supply, return and vent lines are still clipped in at fuel tank.

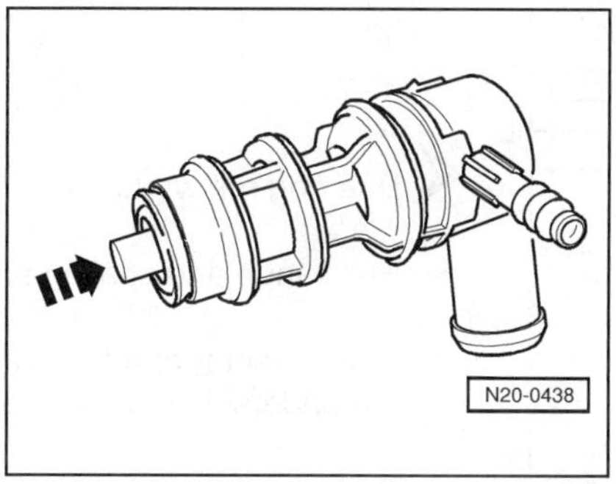

Fig. 2 Checking change-over valve

Lever at rest: closed

Lever pressed in direction of arrow: open

NOTE —
Before installing change-over valve, remove cover from fuel tank.

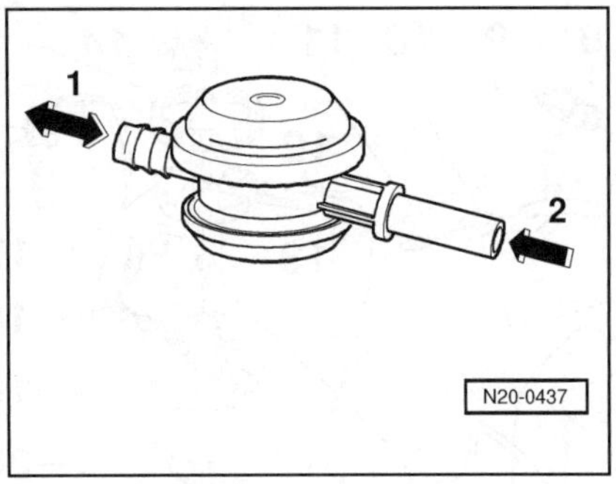

Fig. 3 Checking pressure retaining valve

The pressure retaining valve is opened in both flow directions -arrow 1- on the gravity valve side.

It is only opened in one flow direction -arrow 2- on the side of the change-over valve.

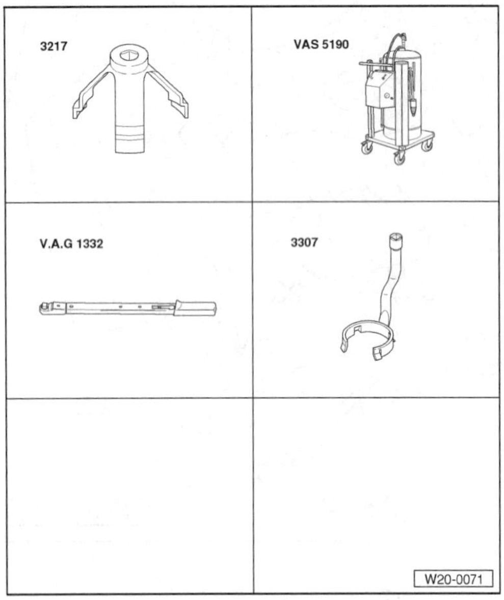

Fuel pump, removing and installing

Special tools and equipment

- 3217 Ring nut wrench
- VAS 5190 Fuel siphoning device
- VAG 1332 Torque wrench (40 to 200 Nm)
- 3307 Fuel pump wrench

NOTE —

- *Fuel tank may be a maximum of 1/3 full.*
- *Empty the fuel tank if necessary using VAG 5190 Fuel siphoning device.*
- *Read safety precautions before beginning work, see* **Safety precautions when working on fuel supply**.

– First determine whether an anti-theft coded radio is installed. If so, determine the correct radio anti-theft code and record.

– Switch ignition off and disconnect battery Ground (GND) strap.

– Remove luggage compartment floor cover.

> ***WARNING —***
> *Fuel supply line is under pressure! Before loosening hose connections, place rags around the connection point. Then release pressure by carefully pulling on hose.*

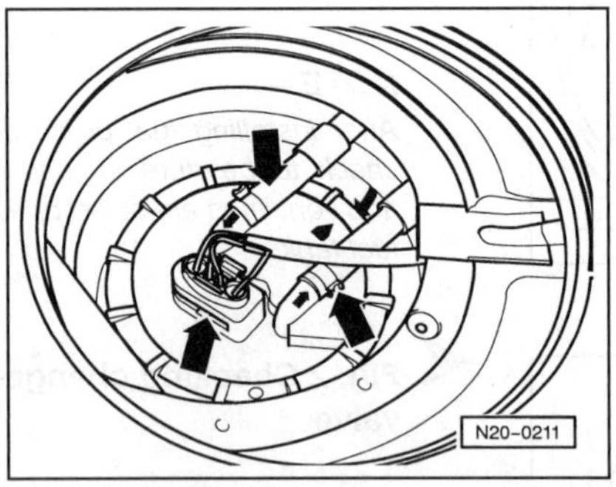

Disconnect 4-pin connector as well as supply and return lines from flange (arrows).

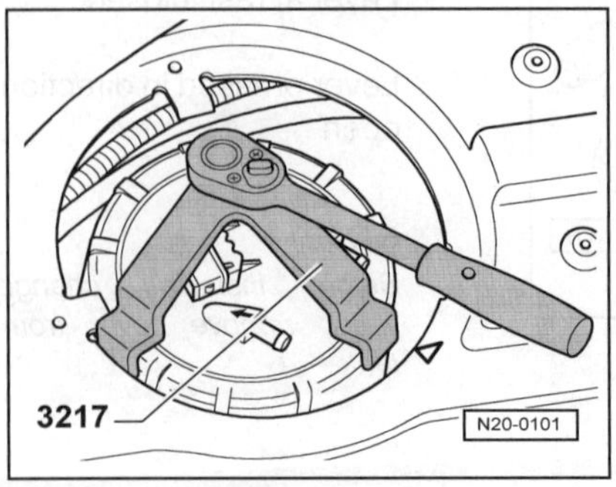

Use ring nut wrench 3217 to remove union nut.

– Pull flange and sealing ring from opening of fuel tank

– Disconnect fuel return line and connector for sensor for fuel gauge on the underside of flange.

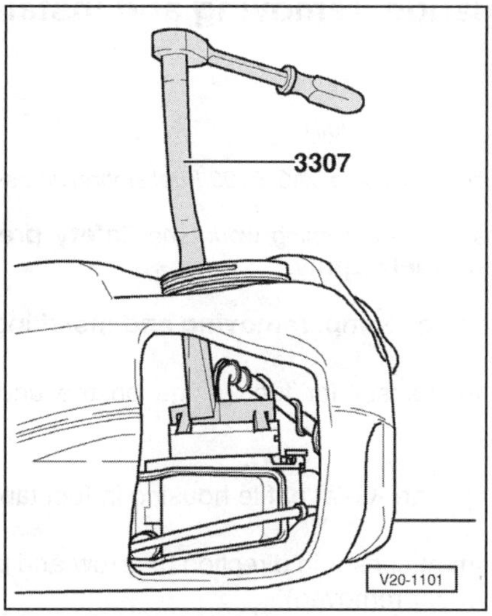

Turn Fuel Pump (FP) approx. 15° toward left until impact using 3307 fuel pump wrench and remove with flange.

NOTE —
When replacing Fuel Pump, empty old pump before disposing.

Fuel pump, installing

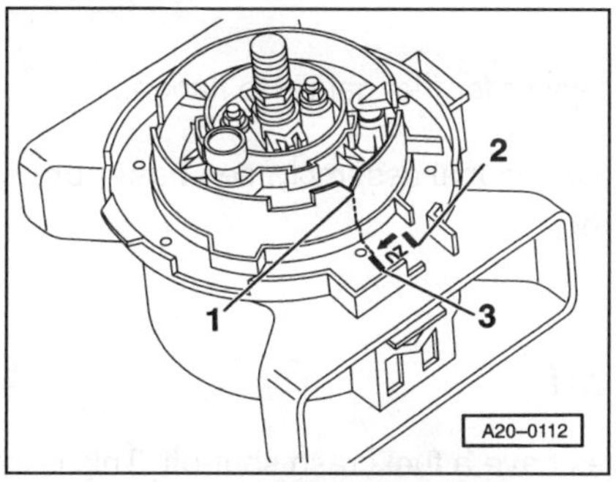

Insert Fuel Pump (FP) into baffle housing in tank so that the notch -1- at the Fuel Pump (FP) aligns with the marking -2- at the baffle housing.

NOTE —
Baffle housing is shown without tank for the sake of illustration.

– Turn Fuel Pump (FP) approx. 15° toward right until impact using 3307 fuel pump wrench.
  - Notch -1- aligns with marking -3- (bayonet connection), see illustration A20-0112.

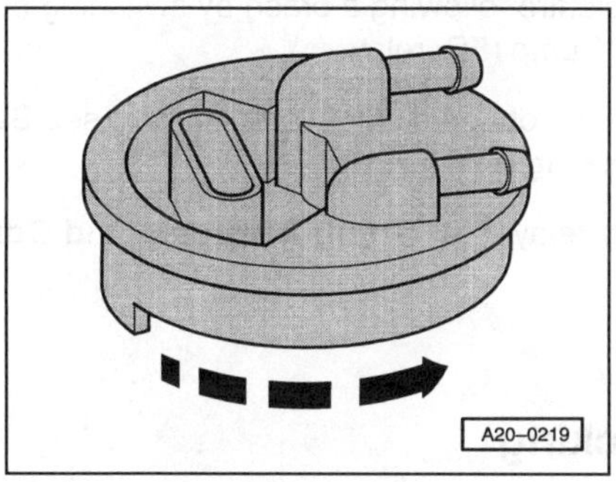

Turn flange toward left in direction of arrow.

– Connect connector for sensor for fuel gauge and fuel return line to flange.

– Coat sealing ring with fuel and install.

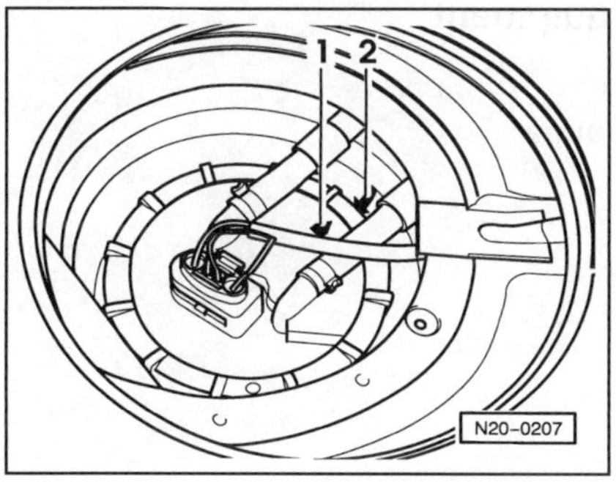

Turn flange toward left into installation position and insert in fuel tank.
  - Marking on flange -1- must align with marking on fuel tank -2-.

– Tighten union nut. Tightening torque: 60 Nm (44 ft-lb)

– Installation is the reverse of the removal noting the following:
  - Secure fuel lines with spring type clamps
  - Do not mix-up supply and return lines (return line is blue or has blue marking).
  - Make sure fuel lines are securely fastened.
  - After installing fuel pump flange, check to be sure supply, return and vent lines are still clipped in at fuel tank.
  - Perform work step **Procedure following interruption of voltage supply**, see **Electrical Equipment section, Repair Group 27**

Sensor for fuel gauge, removing and installing

NOTE —

- *Fuel tank may be a maximum of 1/3 full.*
- *Empty the fuel tank if necessary using VAG 5190 Fuel siphoning device.*
- *Read safety precautions before beginning work, see* **Safety precautions when working on fuel supply**.

– Remove fuel pump, see **Fuel pump, removing and installing**

– Disconnect connector for sensor for fuel gauge on the underside of flange.

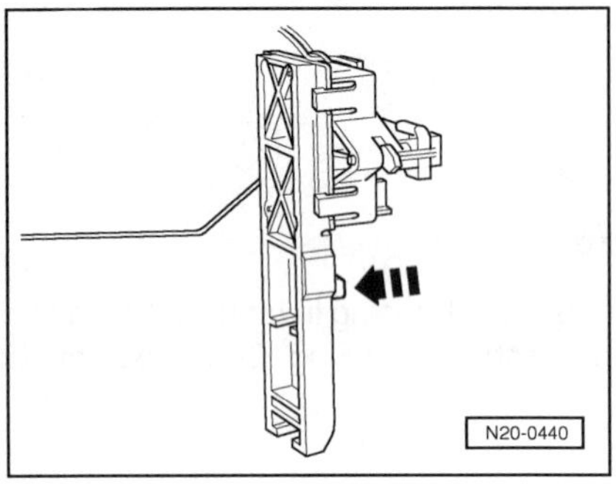

< Remove sensor for fuel gauge -G- at baffle housing in fuel tank.

– Press locking mechanism at sensor in direction of arrow and pull upward (figure shows sensor removed).

– Insert sensor for fuel gauge into guide at baffle housing and press down until engagement.

– Connect connector for sensor for fuel gauge to flange.

– Installing fuel pump and additional assembly, see **Fuel pump, removing and installing**

Crash fuel shut-off

Airbag-equipped vehicles have a fuel crash shut-off. This is intended to reduce the risk of vehicle fire following a crash by switching off the Fuel Pump (FP) via the Fuel Pump (FP) relay.

When opening fuel system, observe safety precautions, see **Safety precautions when working on fuel supply**.

– Check Fuel Pump (FP) relay, see **Wiring Diagrams and Component Locations**

Fuel pump, checking

Special tools and equipment

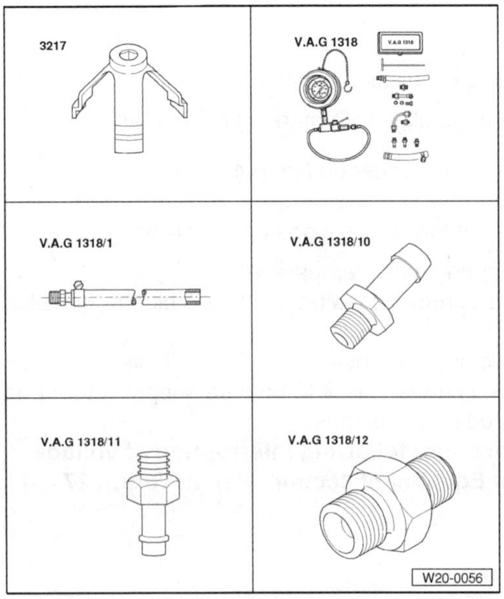

- 3217 Ring nut wrench
- VAG 1318 Pressure gauge
- VAG 1318/1 Adapter
- VAG 1318/10 Adapter
- VAG 1318/11 Adapter
- VAG 1318/12 Adapter

Fuel pump, checking (continued)

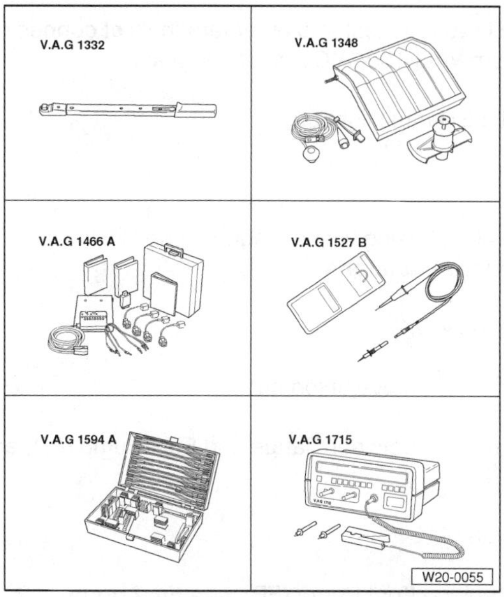

- VAG 1332 Torque wrench (40 to 200 Nm)
- VAG 1348 Quantity comparison tool
- VAG 1466 Electrical circuit tester
- VAG 1527B Voltage tester
- VAG 1594A Connector test kit
- VAG 1715 Multimeter

Test requirements

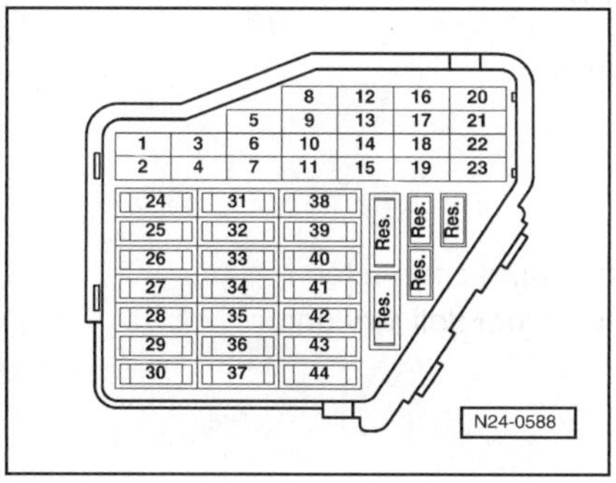

- Battery voltage at least 11.5 V

Fuse No. 28 OK

Fuel pump, checking function and voltage supply

– Remove luggage compartment floor cover.

– Operate starter briefly. Fuel Pump (FP) must run audibly.

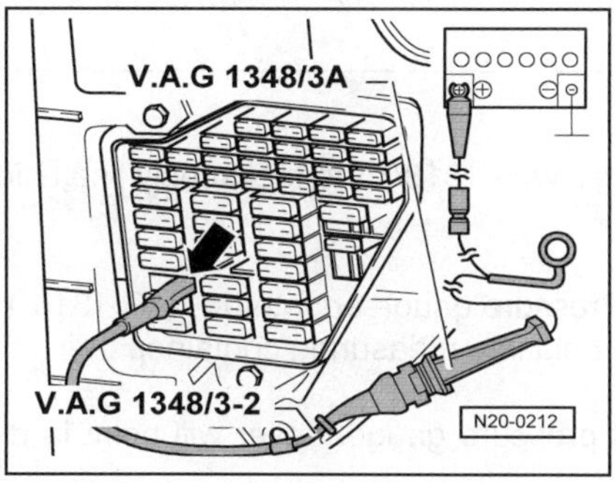

– Switch off ignition.

If Fuel Pump (FP) does not start up:

– Remove cover in front of fuse holder.

Remove fuse 28 from fuse holder.

– Connect VAG 1348/3A Remote control to Fuel Pump (FP) and B+ at terminal 28a using VAG 1348/3-2 Remote control connection.

– Operate remote control.

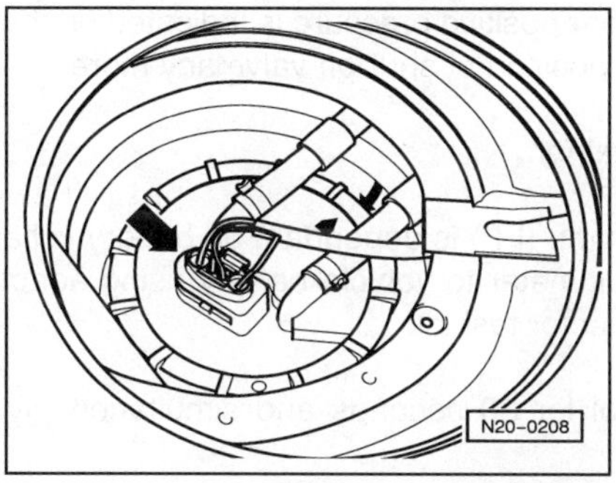

If Fuel Pump (FP) runs:

– Use VAG 1466A to check activation of the Fuel Pump (FP) relay according to wiring diagram, see **Wiring Diagrams and Component Locations**

If Fuel Pump (FP) does not run:

Disconnect 4-pin connector from flange at fuel tank (arrow).

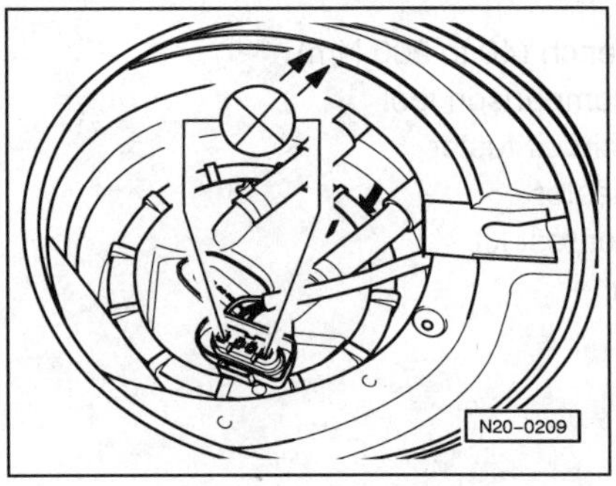

Connect VAG 1527B LED voltage tester to outer terminals of connector using adapter cables from VAG 1594A Connector test kit.

– Operate remote control. LED must light up.

LED does not light up:

– Locate and repair open circuit using wiring diagram, see **Wiring Diagrams and Component Locations**

LED lights up (voltage supply OK):

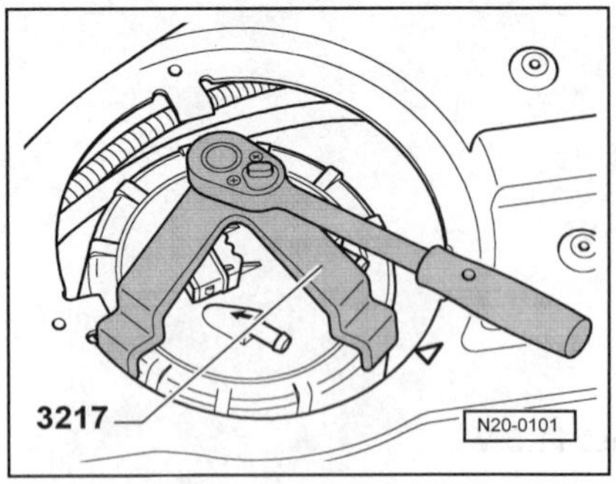

Use ring nut wrench 3217 to remove union nut.

– Check whether electrical wires between flange and Fuel Pump (FP) are connected.

If there are no open circuits:

– Fuel Pump (FP) faulty, replace Fuel Pump (FP), see **Fuel pump, removing and installing**

Checking fuel delivery

Test requirements

- Voltage supply OK
- VAG 1348/3A Remote control connected

– Remove fuel filler cap from fuel delivery unit.

– Remove engine cover.

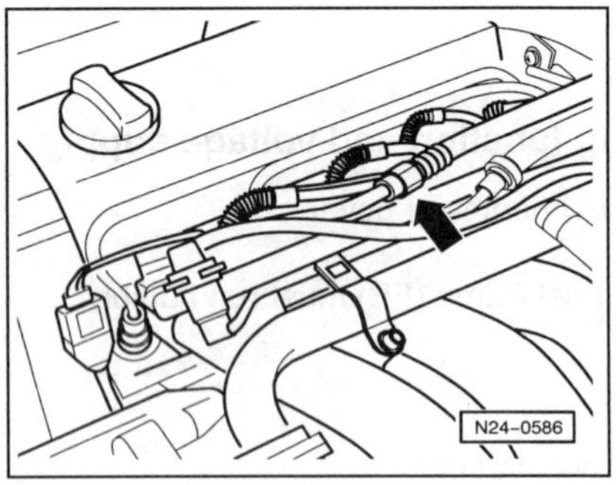

Open threaded connection (arrow) and catch exiting fuel using a rag.

> ***WARNING*** —
> *Fuel supply line is under pressure! Before loosening hose connections, place rags around the connection point. Then release pressure by carefully pulling on hose.*

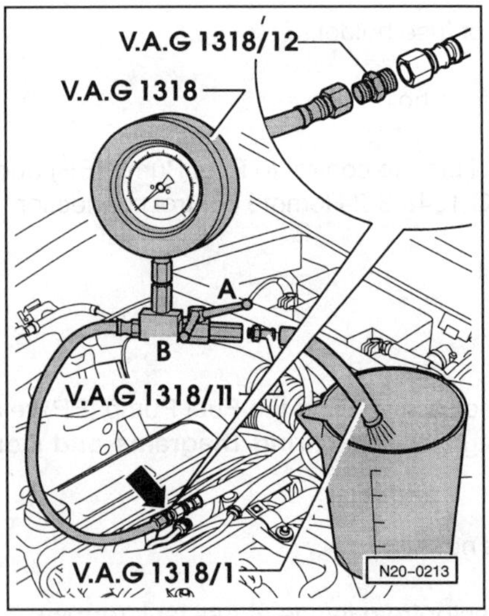

Connect Pressure gauge VAG 1318/11 to fuel supply line using adapters VAG1318/12.

– Connect VAG 1318/1 Pressure gauge adapter to VAG 1318/11 adapter of pressure gauge and hold into measuring container.

– Open shut-off valve of pressure gauge. Lever will point in direction of flow -A-.

– Operate VAG 1348/3A Remote control. Simultaneously, close the shut-off valve slowly until 3 bar positive pressure is indicated on the pressure gauge. Do not change position of shut-off valve any more.

– Empty measuring container.

– Delivery rate of Fuel Pump (FP) is dependant on battery voltage. Therefore, connect multimeter to vehicle battery using adapter cables from VAG 1594A Connector test kit.

– Operate remote control for 30 seconds and simultaneously measure battery voltage.

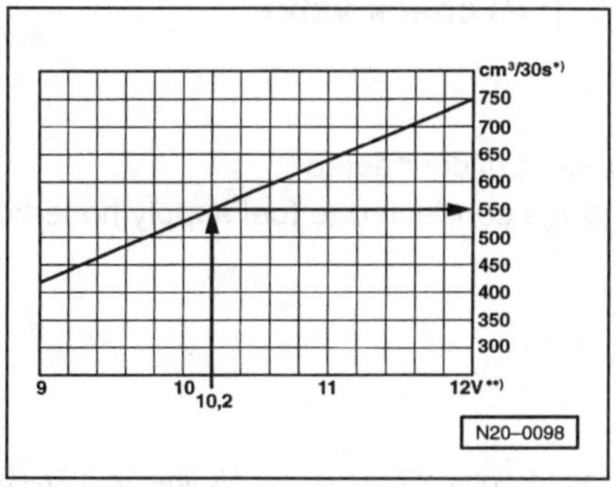

◄ Compare fuel quantity pumped with specified value.

- (*) Minimum delivery rate cm^3/30 s
- (**) Voltage at fuel pump with engine off and pump running (approx. 2 volts less than battery voltage).

Read-out example:

During test, a voltage of 12.2 volts is measured at battery. Since the voltage at the pump is approx. 2 volts less than battery voltage, there is a minimum delivery rate of 550 cm^3/30 s.

If minimum delivery quantity is not obtained:

– Check fuel lines for kinks or clogging.

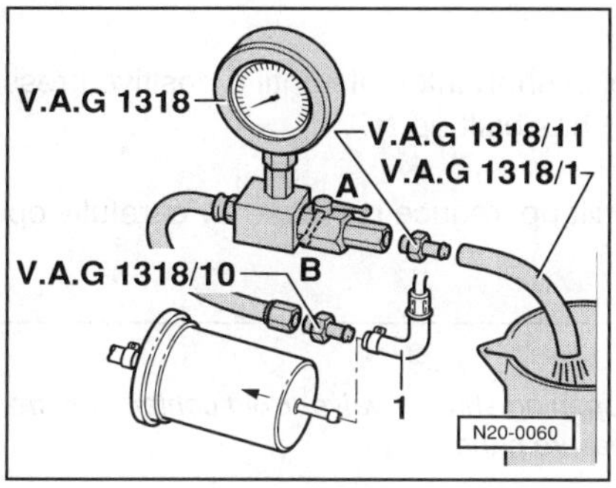

◄ Disconnect supply line -1- from fuel filter input.

– Connect Pressure gauge VAG 1318 to hose using adapters VAG 1318/10.

– Repeat test of delivery quantity.

– If minimum delivery quantity is obtained this time:
  - Replace fuel filter.

– If minimum delivery quantity is still not obtained:
  - Remove Fuel Pump (FP), see **Fuel pump, removing and installing**)and check filter strainer for dirt.

– Only if no malfunctions have been found yet:
  - Fuel Pump (FP) faulty, replace Fuel Pump (FP), see **Fuel pump, removing and installing**.

Delivery rate has been obtained, but malfunctions are still suspected in the fuel supply (e.g. sporadic loss of fuel supply):

– Check Fuel Pump (FP) current draw as follows:

– Connect all disconnected fuel lines.

– Remove luggage compartment floor cover.

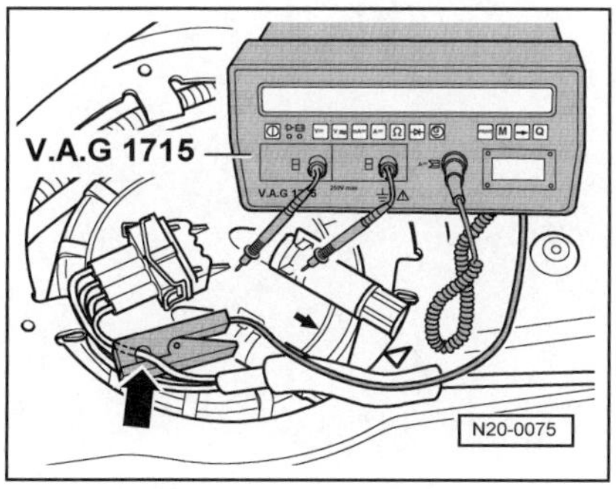

◄ Connect VAG 1715 Multimeter with current probe to green/yellow wire from wiring harness.

– Start engine and let run at idle.

– Measure Fuel Pump (FP) current draw.
  - Specification: max: 8 amps

NOTE—
If disturbance in fuel system is sporadic, test can also be performed during a road test, but a 2nd person is required.

– If current draw is exceeded:
  - Fuel Pump (FP) faulty, replace Fuel Pump (FP), see **Fuel pump, removing and installing**.

Checking Fuel Pump (FP) check-valve

Test requirements

- VAG 1348/3A Remote control connected
- VAG 1318 Pressure gauge connected to fuel supply hose from fuel supply line.

Test sequence

NOTE —

This test also checks the connections of the fuel supply line from Fuel Pump (FP) to connection point of VAG 1318 pressure gauge for proper seal.

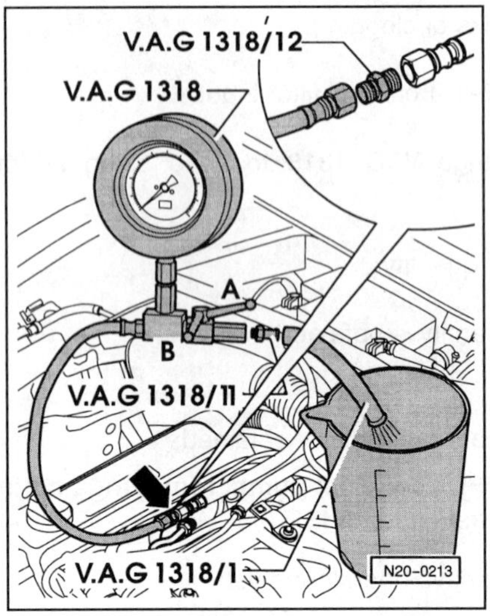

◀ Close pressure gauge shut-off valve (lever perpendicular to direction of flow, -position B-).

– Operate remote control in short intervals until a positive pressure of approx. 3 bar (43.5 psi) has built up.

– If excess pressure is built up, reduce pressure by carefully opening shut-off valve.

> ***WARNING —***
> *Danger of spray when opening shut-off valve. Hold container in front of open connection on pressure gauge.*

– Observe pressure drop on pressure gauge. Pressure must not fall below 2.5 bar (36 psi) within 10 minutes.

If pressure drops further:

– Check line connections for proper seal.

If no malfunctions are found in wires:

– Fuel Pump (FP) faulty, replace Fuel Pump (FP), see **Fuel pump, removing and installing**.

Throttle Control

EPC system, function

For EPC, the throttle valve is not operated by a cable from the accelerator pedal. There is no mechanical connection between the accelerator pedal and the throttle valve.

The position of the accelerator pedal is communicated to the Engine Control Module (ECM) via two sensors for accelerator pedal position (variable resistances; stored in one housing) that are connected with the accelerator pedal.

The accelerator pedal position (driver's intention) is a main input unit for the ECM.

Operation of the throttle valve occurs via an electric motor (throttle valve actuator) in the throttle valve control module. This is true across the entire engine speed and engine load spectrum.

The throttle valve is operated by the throttle valve actuator according to the instructions of the ECM.

With ignition switched on and engine off, the ECM activates the throttle valve actuator precisely according to the sensor for accelerator pedal position. This means, if the accelerator pedal is depressed half way, the throttle valve actuator opens the throttle valve to the same degree; i.e. throttle valve would be opened half way.

With engine running (under load), the ECM can open or close the throttle valve independently of the sensor for accelerator pedal position.

This means, for example, that the throttle valve could be fully opened even though the accelerator pedal has only been depressed half way. This has the advantage of preventing torque losses at the throttle valve.

In addition to this, it results in significantly better values relative to emissions and consumption under certain load conditions.

The required engine torque can be achieved by the Engine Control Module (ECM) via the optimal combination of throttle valve diameter and boost pressure.

It would be incorrect to think that "EPC" consists of only one or two component. EPC is much more of a system containing all components that contribute to recognizing, controlling and monitoring the position of the throttle valve, e.g. sensor for accelerator pedal position, throttle valve control module, EPC warning lamp, Engine Control Module (ECM).

Electronic accelerator pedal assembly, checking

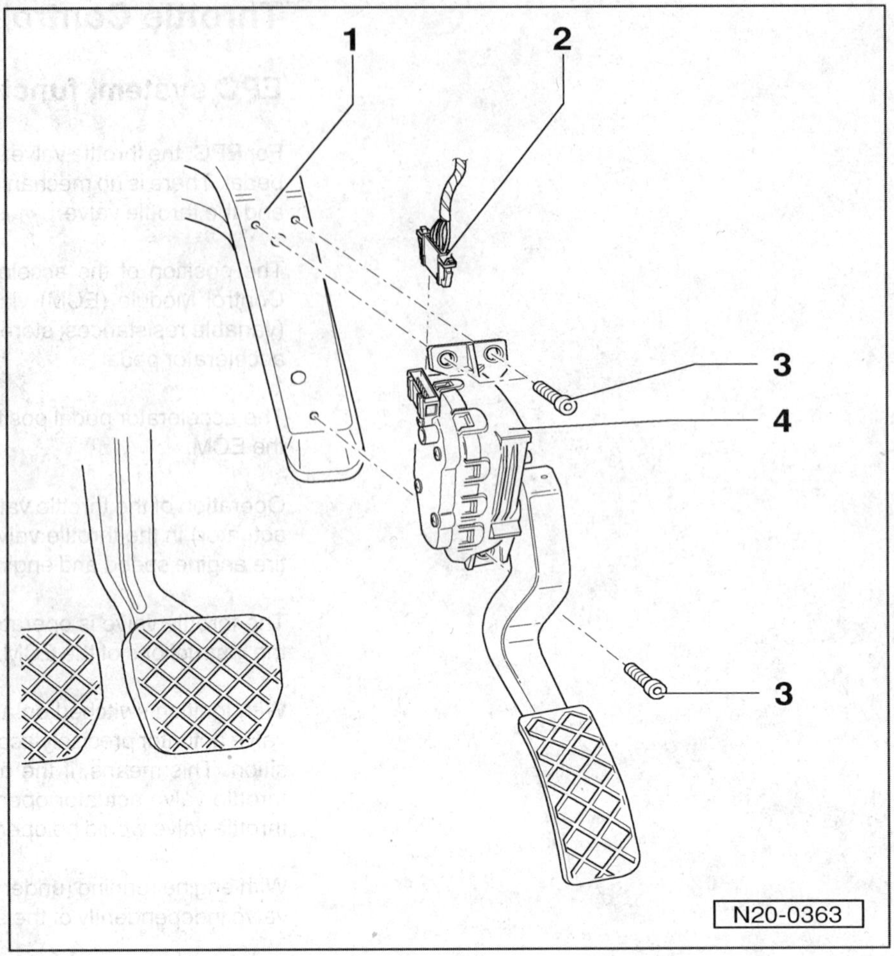

1. **Mounting bracket**
   - Removing and installing, see **Brake System section, Repair Group 46**
2. **Connector**
   - Black, 6-pin
3. **10 Nm (7 ft-lb)**
4. **Throttle Position (TP) sensor -G79- and sensor -2- for accelerator pedal position -G185-**
   - Not adjustable.
   - Checking, see **Throttle Position (TP) sensor, checking**
   - To remove sensor, remove cover in footwell
   - When replacing in vehicles with automatic transmission, adapt kick-down shift point, see **1.8 Liter Fuel Injection & Ignition (code: ATW, AUG, AWM) section, Repair Group 24**

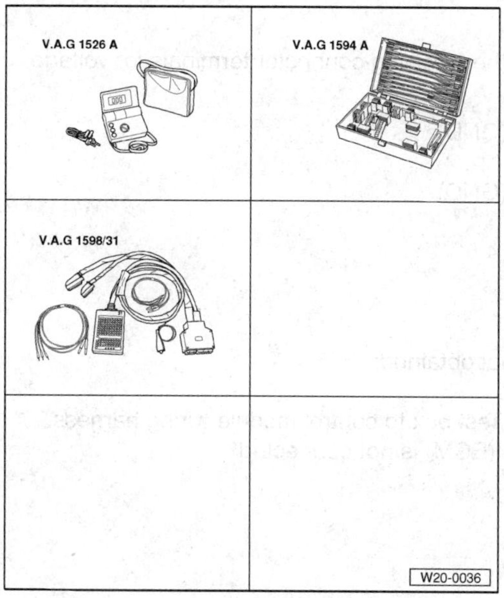

Throttle Position (TP) sensor, checking

Special tools and equipment

- VAG 1526A Multimeter
- VAG 1594A Connector test kit
- VAG 1598/31 Test box
- VAG 1551 Scan tool (or VAG1552 Mobile scan tool) with VAG 1551/3 Cable
- Wiring diagram

Both the Throttle Position (TP) sensor -G79- and the Sensor -2- for accelerator pedal position -G185- are located at the accelerator pedal and communicate the driver's intentions to the ECM completely independently of each other. Both sensors are stored in one housing.

– Connect VAG 1551 Scan tool (or VAG 1552) and select the control module for engine electronics via address word 01. Ignition must be switched on for this, see **Maintenance.**

– Press buttons -0- and -8- to select "Read measuring value block" function 08. Press -Q- button to confirm input.

– Press buttons -0-, -6- and -2- to select "Display group number 62". Press -Q- button to confirm input.

– Check specified value for Throttle Position (TP) sensor -G79- at idle stop in display field 3.

- Specification: 12 to 97%

– Check specified value for sensor -2- for accelerator pedal position -G185- at idle stop in display field 4.

- Specification: 4 to 49%

NOTE —
Engine Control Module (ECM) calculates the voltage values of the angle sensors in percent relative to 5 volts and displays these percentage values. (5 volts supply voltage represents 100%).

– Depress accelerator pedal slowly until Wide Open Throttle (WOT), and observe percent indication in display fields 3 and 4:

- Percent indication in display field 3 must rise uniformly. The tolerance range 12 to 97% is not used completely during this.
- Percent indication in display field 4 must rise uniformly. In the process, the tolerance range of 4 to 49% is not completely utilized.

NOTE —
The indicated value in display field 3 must always be about twice as large as the value in display field 4.

– If indications do not resemble description:

- Check voltage supply and wire connections of sensor for accelerator pedal position, see **Throttle Position (TP) sensor, checking**.

Checking voltage supply and wiring to control module

– Remove driver-side storage compartment, see **Body–Interior section, Repair Group 68.**

– Disconnect 6-pin harness connector for accelerator pedal position sensor.

– Switch ignition on.

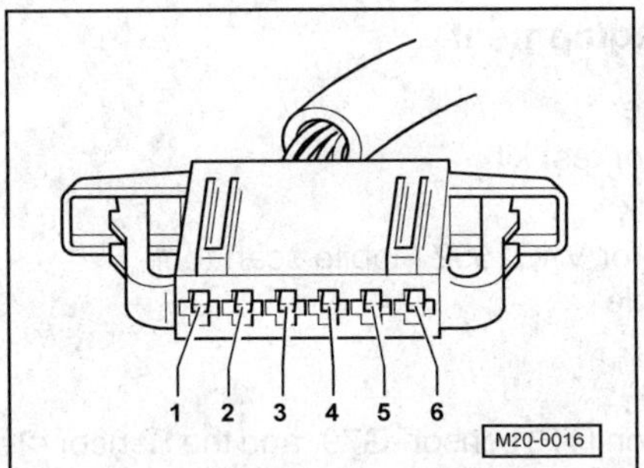

◄ Connect multimeter to the following connector terminals for voltage measurement.

- Terminal 1 + Ground (GND)
- Terminal 1 + 5
- Terminal 2 + Ground (GND)
- Terminal 2 + 3
- Specification: at least 4.5 V

– Switch ignition off.

If specified values are not obtained:

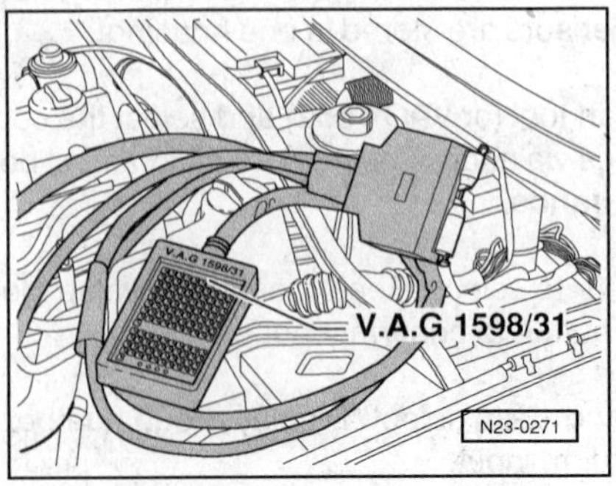

◄ Connect VAG 1598/31 Test box to control module wiring harness. Engine Control Module (ECM) is not connected.

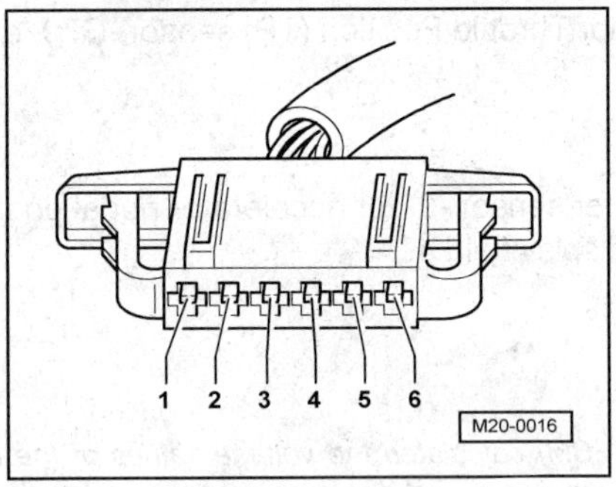

◄ Check for open circuits in wires between test box and connector according to wiring diagram.

- Terminal 1 + socket 72
- Terminal 2 + socket 73
- Terminal 3 + socket 36
- Terminal 4 + socket 35
- Terminal 5 + socket 33
- Terminal 6 + socket 34
- Wire resistance: max: 1.5 Ω

– Also check wires for short circuits to B+ or Ground (GND).

– Also check wires for short circuit to each other.

– If no malfunctions are found in wires:

- Replace accelerator pedal position sensor, see **Electronic engine power control (EPC), checking, -4-**

Evaporative Emissions (EVAP) system

Function description of leak diagnostic

The EVAP canister system (incl. fuel tank) is equipped with a leak diagnostic that detects leaks (improper seal) in the system.

The diagnostic is based on pressure characteristics and is intended to detect leaks with diameters greater than 1mm.

During the diagnostic, the Leak Detection Pump (LDP) -V144- generates a positive pressure of approx. 30 mbar in the EVAP canister system. The pump is switched off once the pressure is reached. If pressure falls below a specified value, the pump is switched on again. On Board Diagnostic monitors the switch condition and stores a DTC in DTC memory if the intervals are too short.

EVAP system components

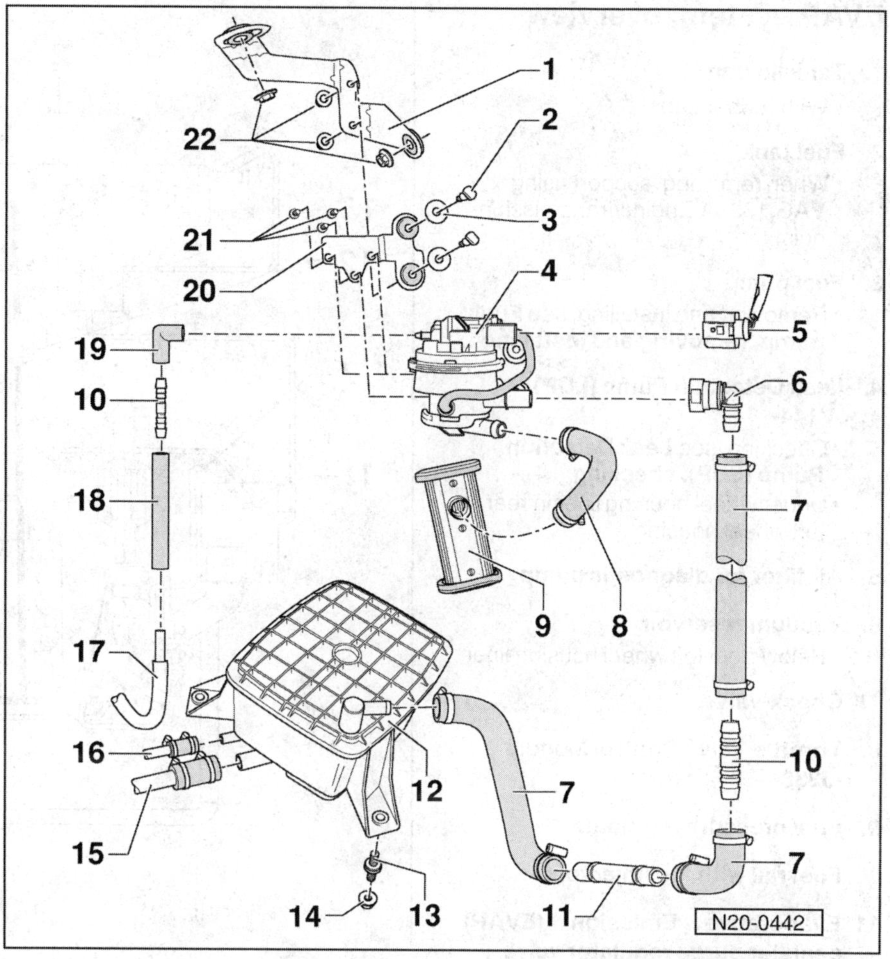

1. **Bracket for diagnostic pump**
   - Bolted to chassis
2. **Bolt**
3. **Washer**
4. **Leak Detection Pump (LDP) -V144-**
   - Under wheelhousing liner in rear left wheelhousing
   - Checking, see **Leak Detection Pump (LDP), checking**
5. **Connector**
   - 3-pin
6. **Clutch**
   - To disengage, squeeze together at front
7. **Connection hose**
   - Pressure side
8. **Connection hose**
   - Intake side
9. **Air filter for diagnostic pump**
10. **Connecting piece**
11. **Connection tube**
12. **EVAP canister**
    - In spare tire well at vehicle floor
13. **15 Nm (11 ft-lb)**
14. **10 Nm (7 ft-lb)**
15. **Vent line**
    - From change-over valve at fuel tank, see **Fuel tank with attachments and fuel filter,** -7-
16. **To Evaporative Emission (EVAP) canister purge regulator valve -N80-**
    - In engine compartment at air filter, see **EVAP system, overview,** -item 11-
17. **Vacuum line**
    - To check-valve at intake manifold, see **EVAP system, overview,** -item 7-
18. **Connection hose**
    - Vacuum side
19. **Angled connector**
20. **Bracket for diagnostic pump**
21. **2 Nm (18 in-lb)**
22. **8 Nm (71 in-lb)**

EVAP system, overview

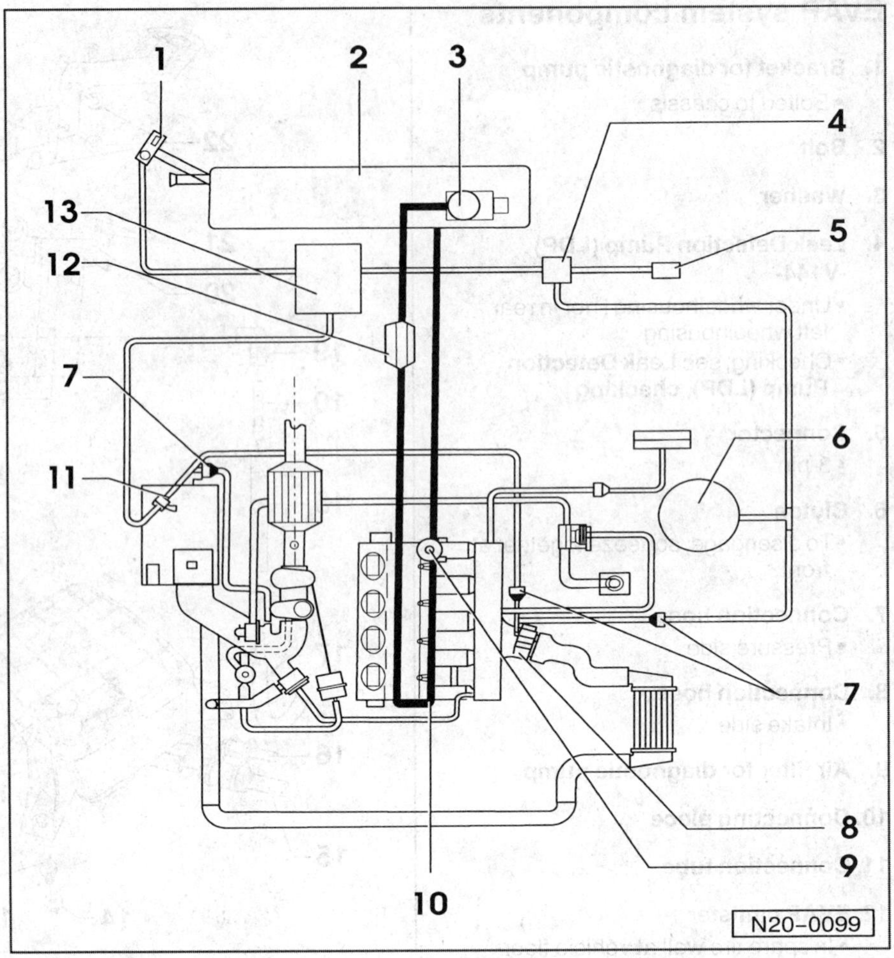

1. **Tank lid unit**
   - With rubber cap
2. **Fuel tank**
   - When removing, support using VAG 1383A Engine/transmission hoist
3. **Fuel pump**
   - Removing and installing, see **Fuel pump, removing and installing**
4. **Leak Detection Pump (LDP) -V144-**
   - Checking, see **Leak Detection Pump (LDP), checking**
   - Under wheel housing liner in rear left wheel housing
5. **Air filter for diagnostic pump**
6. **Vacuum reservoir**
   - Below front left wheel housing liner
7. **Check-valve**
8. **Throttle Valve Control Module -J338-**
9. **Fuel pressure regulator**
10. **Fuel rail with fuel injectors**
11. **Evaporative Emission (EVAP) canister purge regulator valve -N80-**
    - Checking activation with scat tool, see **Maintenance**

    EDITOR'S NOTE —
    Component activation is accomplished using Volkswagen diagnostic scan tool, output diagnostic test mode (function 03).

    - Checking function, see **Evaporative Emission (EVAP) canister purge regulator valve, checking**
12. **Fuel filter**
    - Installation position: Arrow points in direction of flow
13. **EVAP canister**
    - In spare tire well at vehicle floor

Evaporative Emission (EVAP) canister purge regulator valve, checking

Solenoid valve is closed when no voltage is present.

Checking activation, see **Maintenance.**

EDITOR'S NOTE —
Component activation is accomplished using Volkswagen diagnostic scan tool, output diagnostic test mode (function 03).

Special tools and equipment

- VAG 1551 scan tool (or VAG 1552 mobile scan tool) with VAG 1551/3 Cable

Test requirements

- Coolant Temperature at least 80°C (176°F)

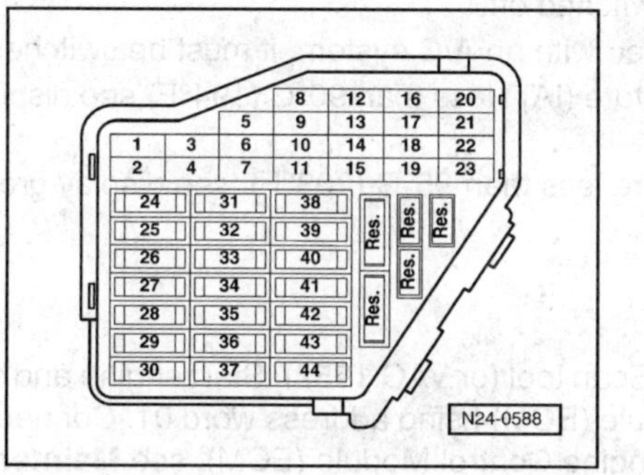

Fuse no. 29 must be OK.

Function test

– Connect VAG 1551 Scan tool (or VAG 1552). Start engine and select the Engine Control Module (ECM) using address word 01. Connecting scan tool and selecting Engine Control Module (ECM), see **Maintenance.**

– Press buttons -0- and -8- to select "Read measuring value block" function 08. Press -Q- button to confirm input.

– Press buttons -0-, -7- and -0- to select "Display group number 70". Press -Q- button to confirm input.

– Press button 4 to select "basic setting" function ("diagnostic of tank vent valve" will be initiated)

NOTE —
Do not generate engine load during diagnostic or else it will be terminated and cannot be initiated again until after accelerator pedal is depressed.

If diagnostic is initiated by the Engine Control Module (ECM), the indication in display field 4 will jump from "Test OFF" to "Test ON".

– Continue to let engine run at idle until specified value "TVV OK" is indicated in display field 4.

NOTE —
If indication changes from "Test ON" to "Test OFF" during the test, depress accelerator pedal briefly; test will be repeated.

If display field 4 indicates "TVV not OK":

– Initiate output Diagnostic Test Mode (DTM) and activate the Evaporative Emission (EVAP) canister purge regulator valve -N80- using scan tool , see **Maintenance.**

EDITOR'S NOTE —
Component activation is accomplished using Volkswagen diagnostic scan tool, output diagnostic test mode (function 03).

If specified value "TVV OK" is obtained:

– Press →button.

– Press buttons -0- and -6- to select "End data transfer" function 06. Press -Q- button to confirm input.

EVAP system, checking for proper seal

Special tools and equipment

• VAG 1551 Scan tool (or VAG 1552 Mobile scan tool) with VAG 1551/3 Cable

Test requirements

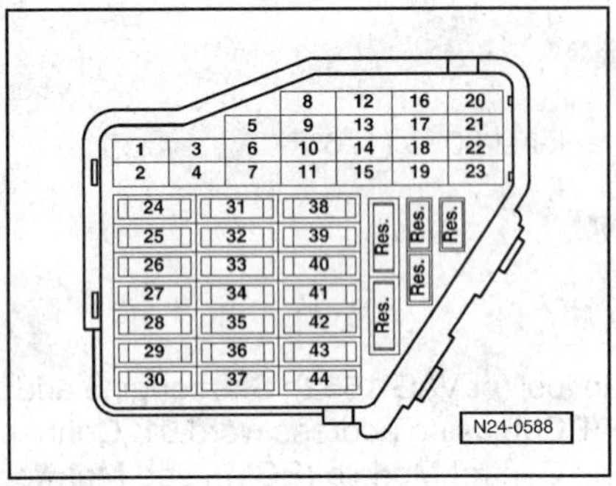

◀ Fuse no. 29 must be OK.

• No DTCs in DTC memory
• All electrical consumers like, for example, lights and rear window heating must be switched off.
• If vehicle is equipped with an A/C system, it must be switched off.
• Intake Air Temperature (IAT) less than 90°C (194°F),see display group 04, display field 4
• Coolant temperature less than 95°C (203°F), see display group 04, display field 3

Function test

– Connect VAG 1551 Scan tool (or VAG 1552). Start engine and select the Engine Control Module (ECM) using address word 01. Connecting scan tool and selecting Engine Control Module (ECM), see **Maintenance**.

– Press buttons -0- and -8- to select "Read measuring value block" function 08. Press -Q- button to confirm input.

– Press buttons -0-, -7- and -1- to select "Display group number 71". Press -Q- button to confirm input.

– Press button -4- to select function "Initiate Basic Setting"

Test of EVAP canister system will be initiated. First, the Reed contact (in leak detection pump -V144-) will be checked for function by the control module.

After approx. 30 seconds, the Leak Detection Pump (LDP) -V144- will be activated and will generate pressure in the fuel system.

Test will be initiated after approx. 45 seconds.

The Leak Detection Pump (LDP) -V144- switches off and the control module measures the time that elapses until the reed contact closes.

– Check indication in display fields 3 and 4:

If specified values are not obtained:

– Check for proper seal, see **EVAP system, checking for proper seal**

If specified values are obtained:

– Press →button.

- Press buttons -0- and -6- to select "End data transfer" function 06. Press -Q- button to confirm input.

EVAP system, checking for proper seal

If a small or large leak was recognized during the test of the EVAP canister system, proceed as follows.

- Switch ignition off.
- Remove hose from diagnostic pump at EVAP canister and seal it, see **EVAP system components, removing and installing**.
- Repeat function test.

If malfunction persists:

- Removing rear right wheelhousing liner, see **Body–Exterior section, Repair Group 66**
- Check the connection lines and hoses between diagnostic pump and EVAP canister, see **EVAP system components**.
- Repeat function test.

If malfunction persists:

- Close pressure side of pump.
- Repeat function test.

If malfunction persists:

- Check seal of vacuum line from connection at intake manifold to connection at diagnostic pump, see **EVAP system, overview**.
- Repeat function test.

If malfunction persists:

- Replace Leak Detection Pump (LDP).

If malfunction is no longer present after first test:

- Re-connect hose from diagnostic pump to EVAP canister.
- Repeat function test after each of the following test steps:

Check whether filler cap for fuel tank is completely closed and whether seal is faulty; replace if necessary.

Check under lining in luggage compartment to see whether lock ring for Fuel Pump (FP) is tightened or if seal is faulty, replace if necessary. Use 3217 ring nut wrench to tighten lock ring. Check supply and return lines at Fuel Pump (FP) for proper seal.

Check connecting lines and hoses from EVAP canister to fuel flap unit and EVAP valve, see **EVAP system, overview**.

Remove EVAP canister from spare tire well at vehicle floor and check for tears and leaks, replace if necessary.

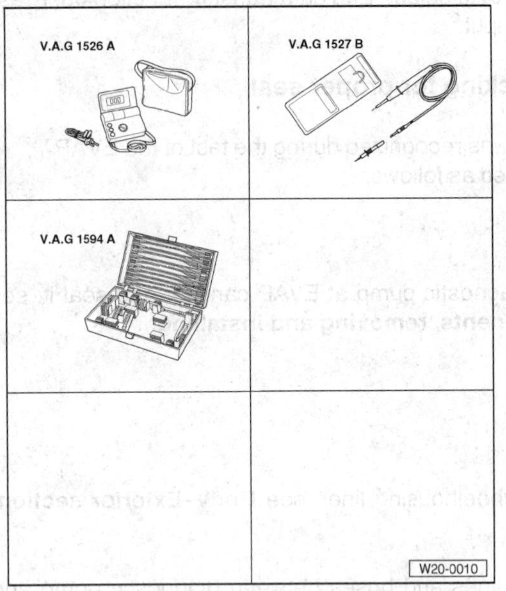

Leak Detection Pump (LDP), checking

Special tools and equipment

- VAG 1526A Multimeter
- VAG 1527B Voltage tester
- VAG 1594A Connector test kit
- Wiring diagram

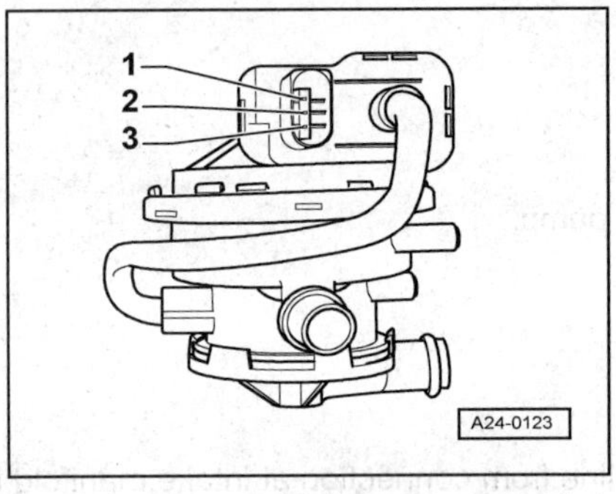

Electrically checking Leak Detection Pump (LDP)

◀ Remove 3-pin connector for Leak Detection Pump (LDP) -V144-.

– Connect VAG 1526 Multimeter between terminals 1 and 3 of Leak Detection Pump (LDP) -V144- for resistance measurement.

Specification: 640 to 720 Ω

– Connect VAG 1526 Multimeter between terminals 2 and 3 of Leak Detection Pump (LDP) -V144- for resistance measurement.

Specification: 15.0 to17.0 Ω

If specified values are not obtained:

– Replace Leak Detection Pump (LDP) -V144-

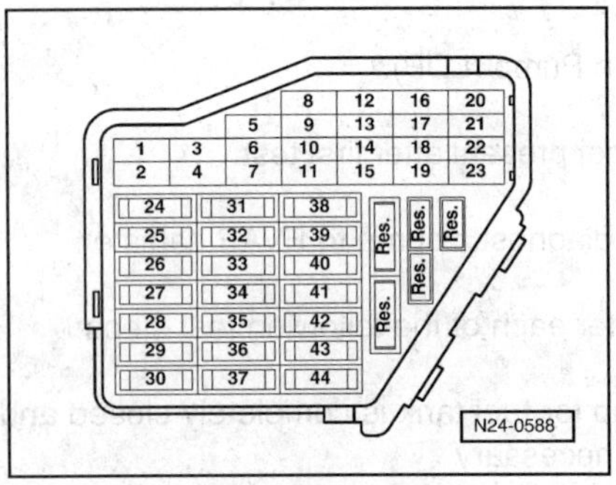

Checking voltage supply

Test requirement

◀ Fuse no. 34 must be OK.

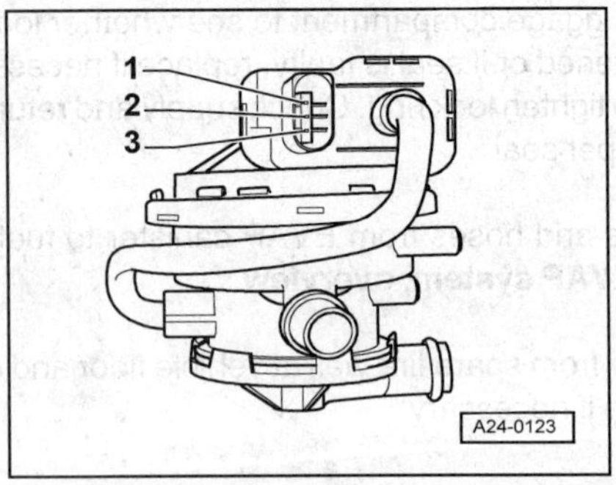

Test sequence

◀ Remove 3-pin connector for Leak Detection Pump (LDP) -V144-.

– Connect VAG 1527 Voltage tester between terminal 3 (B+) of connector and engine Ground (GND).

– Start engine, and let run at idle LED must light up.

– If LED does not light up:
  - Check for open circuits in wire between connector terminal 3 and central electronics according to wiring diagram.

21 Turbocharger

Repair Group
C 21

Charge Air Pressure System with Turbocharger

> ***WARNING —***
> *Observe the following for all installations, especially in engine compartment due to lack of room:*
>
> - *Route lines of all types (e.g. for fuel, hydraulic, EVAP canister system, coolant and refrigerant, brake fluid, vacuum) and electrical wiring so that the original path is followed.*
> - *Watch for sufficient clearance to all moving or hot components.*

NOTE —

- *All hose connections are secured with clamps.*
- *Charge air pressure system must be properly sealed.*
- *Always replace self-locking nuts.*
- *Charge air pressure hoses and their connections must be free of grease before re-installation.*

Rules of cleanliness

When working on turbocharger, carefully observe the following 5 rules:

1. Thoroughly clean all connections and the surrounding area before disconnecting.
2. Place parts that have been removed on a clean surface and cover. Use lint-free cloths only!
3. Carefully cover over opened components or seal, if repairs are not carried out immediately.
4. Only install clean components: Only unpack replacement parts immediately prior to installation. Do not use parts that have been stored loose (in tool boxes etc.)
5. When the system is open: Do not work with compressed air if it can be avoided. Do not move vehicle unless absolutely necessary.

Safety precautions

If special testing equipment is required during road test, note the following:

- Test equipment must always be secured to the rear seat and operated from there by a second person.

If test and measuring equipment is operated from the passenger seat, the person seated there could be injured in the event of an accident involving deployment of the passenger-side airbag.

Turbocharger with attachments

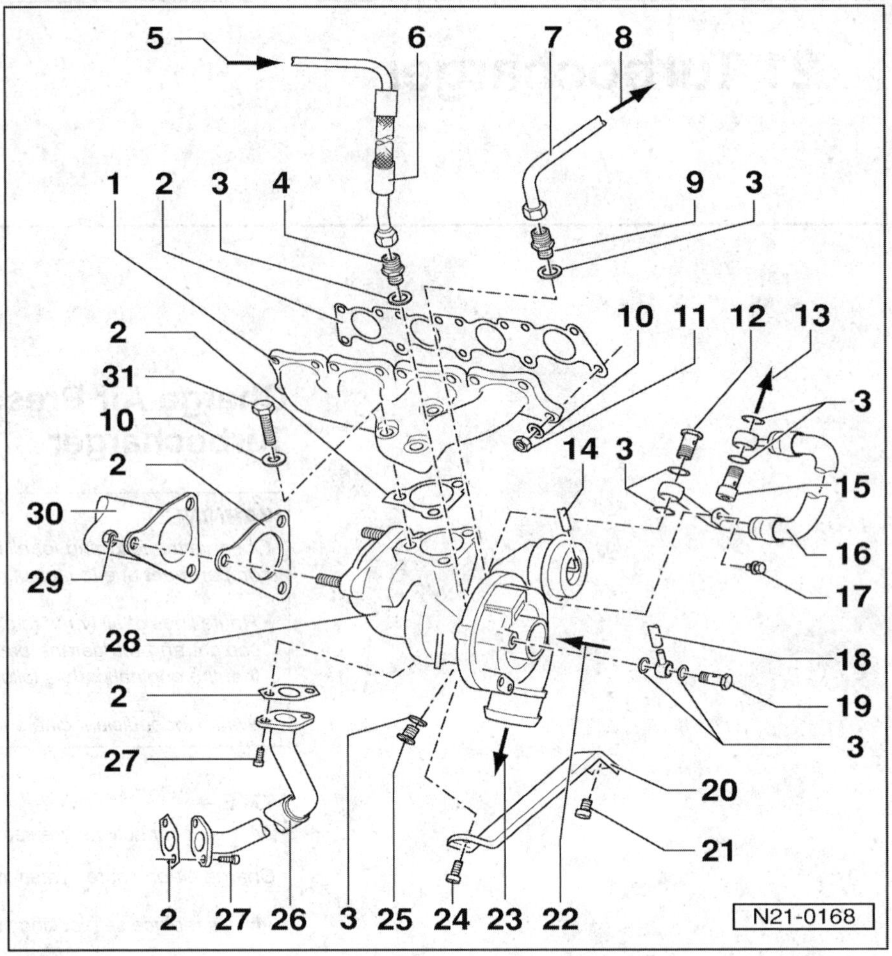

1. **Exhaust manifold**
2. **Gasket**
   - Always replace
   - Note installed position
3. **Seal**
   - Always replace
4. **Connecting piece**
   - Tighten to 30 Nm (22 ft-lb)
5. **From oil filter bracket**
6. **Oil supply line**
   - Tighten to 25 Nm (18 ft-lb)
7. **Coolant return line**
   - Tighten to 30 Nm (22 ft-lb)
8. **To upper coolant line**
9. **Connecting piece**
   - Tighten to 35 Nm (26 ft-lb)
10. **Washer**
11. **30 Nm (22 ft-lb)**
    - Always replace
    - Coat threads with G 052 112 A3
12. **Banjo fitting**
    - Tighten to 35 Nm (26 ft-lb)
13. **To engine block**
14. **Hose**
    - To wastegate bypass regulator valve -N75-, see **Turbocharging system, overview**, -item 24-
15. **Banjo fitting**
    - Tighten to 35 Nm (26 ft-lb)
16. **Coolant supply line**
17. **10 Nm (7 ft-lb)**
18. **Hose**
    - To wastegate bypass regulator valve -N75-, see **Turbocharging system, overview**, -item 24-
19. **Banjo fitting**
    - Tighten to 15 Nm (11 ft-lb)
20. **Bracket**
    - Between turbocharger and cylinder block
21. **25 Nm (18 ft-lb)**
22. **From air filter**
23. **To charge air pressure cooler**
24. **10 Nm (7 ft-lb)**
25. **Locking screw**
    - Tighten to15 Nm (11 ft-lb)
26. **Oil return line**
    - To oil pan
27. **10 Nm (7 ft-lb)**
28. **Turbocharger**
    - Checking charge air pressure regulation, see **Charge air pressure regulation, checking**
    - The charge air pressure regulation valve and vacuum diaphragm for charge air pressure regulation are part of the exhaust turbocharger and cannot be replaced individually.
    - Before connecting oil supply line, fill turbocharger with oil at filler tube
    - After installing turbocharger, let engine idle for approx. 1 minute without increasing engine speed. This ensures adequate oil supply to the turbocharger.
29. **30 Nm (22 ft-lb)**
    - Always replace
    - Coat threads with G 052 112 A3
30. **Catalytic converter**
31. **35 Nm (26 ft-lb)**
    - Always replace
    - Coat threads and bolt head contact surface with G 052 112 A3

Charge air pressure cooler components, removing and installing

NOTE —

- *All hose connections are secured with clamps.*
- *Charge air pressure system must be properly sealed.*
- *Note assembly markings at hoses and components during installation.*

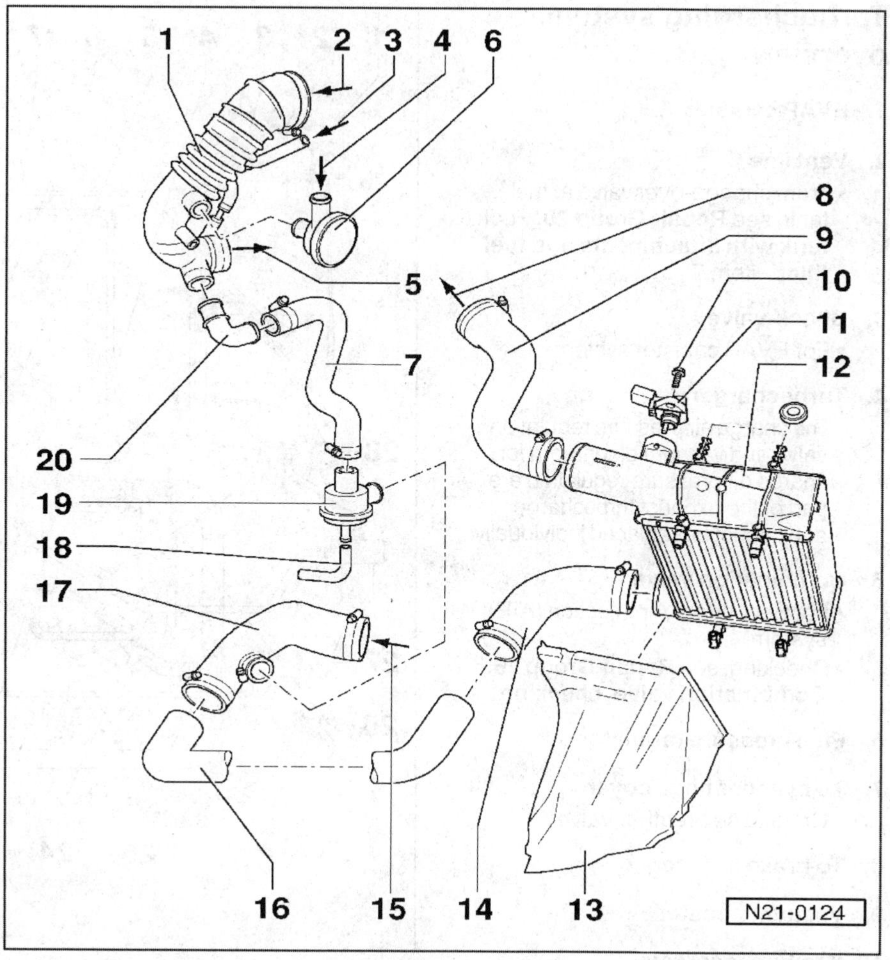

1. **Intake hose**
2. **From air filter**
3. **From EVAP canister**
4. **From crankcase ventilation**
5. **To turbocharger**
6. **Pressure regulation valve**
   - For crankcase ventilation
7. **Connection hose**
   - Note assembly markings
8. **To throttle valve control module**
9. **Connection hose**
   - Charge air pressure cooler/throttle valve control module
   - Note assembly markings
10. **10 Nm (7 ft-lb)**
11. **Charge air pressure sensor -G31-**
    - Replace O-ring when damaged
    - Checking, see **Charge air pressure sensor, checking**
12. **Charge air pressure cooler**
13. **Air duct**
14. **Connection hose**
    - Transverse line/charge air pressure cooler
    - Note assembly markings
15. **From turbocharger**
16. **Transverse line**
17. **Connection hose**
    - Exhaust turbocharger/transverse line
    - Note assembly markings
18. **Vacuum line**
19. **Deceleration shut-off valve**
    - Checking, see **Deceleration shut-off valve, checking**
20. **Elbow**

Turbocharging system, overview

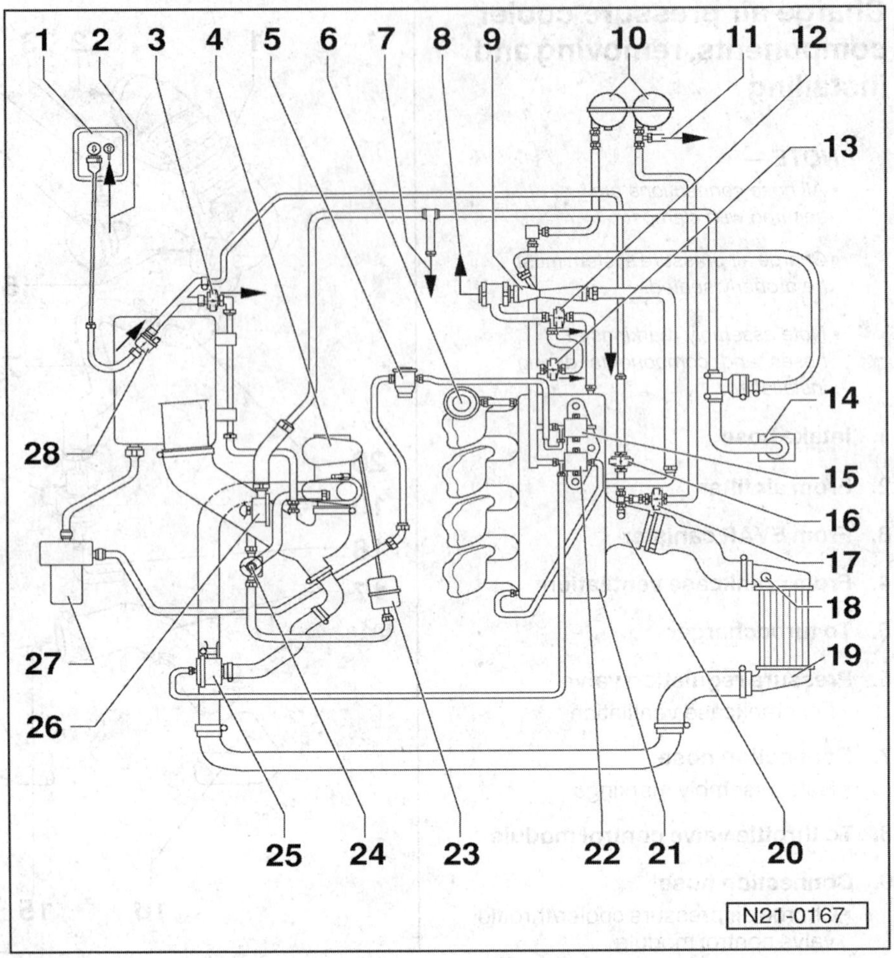

1. **EVAP canister**
2. **Vent line**
   - From change-over valve at fuel tank, see **Repair Group 20, Fuel tank with attachments and fuel filter**, -item 7-
3. **Check-valve**
   - For EVAP canister system
4. **Turbocharger**
   - The charge air pressure regulation valve and vacuum diaphragm for charge air pressure regulation are part of the exhaust turbocharger and cannot be replaced individually.
5. **Combination valve**
   - For Secondary Air Injection (AIR) system
   - Checking, see **Repair Group 26, Combination valve, checking**
6. **Fuel pressure regulator**
7. **To cylinder head cover**
   - Crankcase breather valve
8. **To brake booster**
9. **Vacuum booster**
10. **Vacuum reservoir**
    - To separate chambers
11. **To diagnostic pump**
12. **Check-valve**
    - Bypass to vacuum booster
13. **Check-valve**
    - For engine vacuum control
14. **Crankcase ventilation**
15. **Secondary Air Injection (AIR) solenoid valve -N112-**
    - Checking, see **Repair Group 26, Secondary Air Injection (AIR) valve, checking**
16. **Check-valve**
    - For EVAP canister system
17. **Check-valve**
    - For diagnostic pump
18. **Charge air pressure sensor -G31-**
    - Checking, see **Charge air pressure sensor, checking**
19. **Charge air pressure cooler**
20. **Throttle valve control module**
21. **Intake manifold**
22. **Recirculating valve for turbocharger -N249-**
    - Checking, see **Recirculating valve for turbocharger, checking**
23. **Pressure unit**
    - For charge air pressure regulation valve
    - Part of turbocharger, cannot be replaced
24. **Wastegate bypass regulator valve -N75-**
    - Valve is activated (pulsed) by the Engine Control Module (ECM)
    - Closed when no voltage present, limited by charge air pressure
    - Checking activation:

    EDITOR'S NOTE —

    Component activation is accomplished using Volkswagen diagnostic scan tool, output diagnostic test mode (function 03).
25. **Deceleration shut-off valve**
26. **Pressure regulation valve**
    - For crankcase ventilation
27. **Secondary Air Injection (AIR) pump motor -V101-**
    - Checking function, see **Repair Group 26**
28. **Evaporative Emission (EVAP) canister purge regulator valve -N80-**
    - Checking:

    EDITOR'S NOTE —

    Component activation is accomplished using Volkswagen diagnostic scan tool, output diagnostic test mode (function 03).

Charge air pressure system, checking

Charge air pressure regulation, checking

Special tools and equipment

• VAG 1551 Scan tool (or VAG 1552 Mobile scan tool) with VAG 1551/3 Cable

Test requirements

• No leaks on intake or exhaust side.
• Engine oil temperature at least 80 °C

Test sequence

– Connect VAG 1551 Scan tool (or VAG 1552). Start engine and select the Engine Control Module (ECM) using address word 01. Connecting scan tool and selecting Engine Control Module (ECM), see **Maintenance.**

– Press buttons -1-, -1- and -4- to select "display group number 114" and then press -Q- button to confirm input.

– Check duty cycle of wastegate bypass regulator valve during a road test or on a chassis dynamometer (at Wide Open Throttle between 1800 and 2300 RPM) in display field 4.

– Press -C- button.

– Press buttons -1-, -1- and -5- to select "display group number 115" and then press -Q- button to confirm input.

– Check actual charge pressure value in display field 4 at Wide Open Throttle (WOT):

On dynamometer

In 3rd gear or in 2nd drive mode

While driving

Vehicles with manual transmission

– Accelerate vehicle at Wide Open Throttle (WOT) in 2nd gear

Vehicles with automatic transmission

– Manually (Tiptronic) place vehicle in 4th gear and accelerate vehicle from low speed at Wide Open Throttle (without kick down, transmission does not downshift any more).

– Press print button between 1800 and 2300 RPM and check actual charge air pressure value in display field 4.

Specification: 1350 to 1750 mbar

– Compare actual charge air pressure with specification in display field 3.

Difference: max. 100 mbar

NOTE —
Repeat measurement if charge air pressure was not fully charged or if the difference between actual and specified values was too large.

If charge air pressure is exceeded:

– Check charge air pressure sensor -G31-, see **Charge air pressure sensor, checking**

– Check wastegate bypass regulator valve -N75-. (Throughput in hose from turbocharger via valve to vacuum diaphragm with connector disconnected)

– Check for proper seating of vacuum diaphragm for wastegate bypass regulator valve -N75- at turbocharger.

– Check vacuum diaphragm, see **Vacuum diaphragm for wastegate bypass regulator valve -N75-, checking**

– Check bearing for shaft of wastegate bypass regulator valve in turbocharger for ease of movement. If rusted solid, replace turbocharger.

If charge air pressure is not reached:

– Check charge air pressure sensor -G31-, see **Charge air pressure sensor, checking**

– Check wastegate bypass regulator valve -N75-.

– Check bearing for shaft of wastegate bypass regulator valve -N75- in turbocharger for ease of movement. If rusted solid, replace turbocharger.

– Turbocharger faulty, replace.

Wastegate bypass regulator valve -N75-, checking

Special tools and equipment

- VAG 1526 Multimeter or VAG 1715 Multimeter
- VAG 1594 Connector test kit

Test requirement

- Output Diagnostic Test Mode (DTM) performed, see **Maintenance.**

Test sequence

– Switch ignition off.

– Disconnect connector at solenoid valve for charge air pressure regulation.

◄ Measure resistance between terminals of valve.

Specification: 25 to 35Ω

If specified value is not obtained:

– Replace wastegate bypass regulator valve -N75-.

Vacuum diaphragm for wastegate bypass regulator valve -N75-, checking

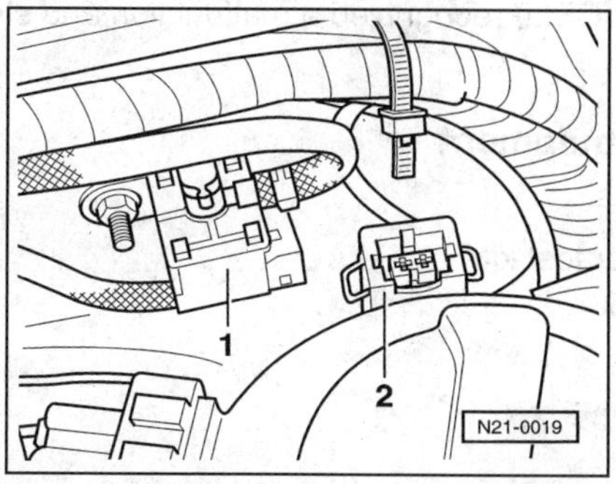

Special tools and equipment

- VAG 1551 Scan tool (or VAG 1552 Mobile scan tool) with VAG1551/3 Cable

Test requirement

- Engine oil temperature at least 80°C (176°F)

◀ Disconnect connector -2- at wastegate bypass regulator valve -N75-.

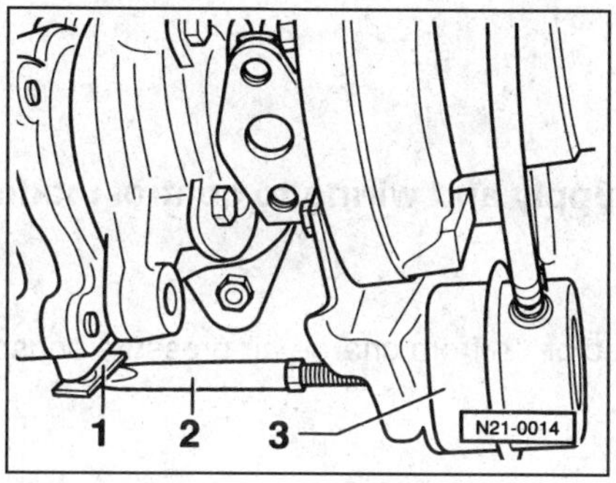

◀ Bring engine to maximum RPM by depressing accelerator pedal suddenly. Operating rod -2- for wastegate bypass regulator valve -N75- must move.

– If operating rod does not move:

- Check lever for wastegate bypass regulator valve -N75- -1- for ease of movement. If rusted solid, replace turbocharger.

– If operating rod does not move despite freely moving lever:

- Check wastegate bypass regulator valve -N75-. (Throughput in hose from turbocharger via valve to vacuum diaphragm with connector disconnected)

– If solenoid valve is OK:

- Replace turbocharger, see **Turbocharger with attachments.**

Deceleration shut-off valve, checking

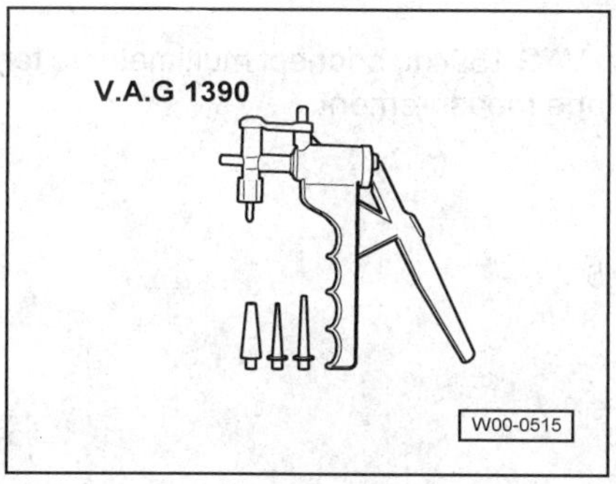

Special tools and equipment

◀ VAG 1390 Hand vacuum pump

Test requirement

- Performance deficiency or tip-in shocks

NOTE —
The deceleration shut-off valve is located in front of the turbocharger. It is opened by vacuum during deceleration and at idle.

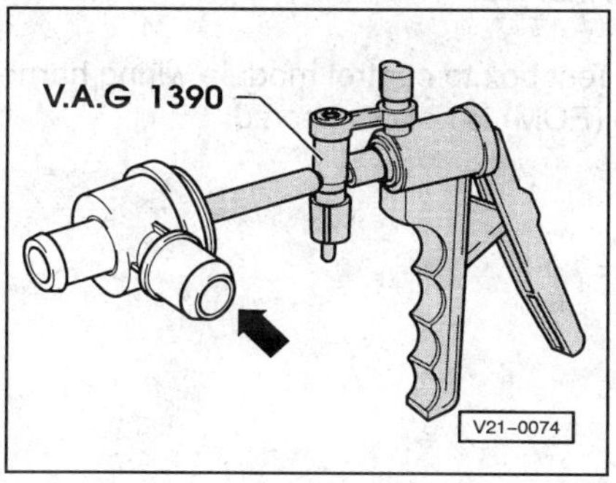

◀ Connect VAG 1390 Hand vacuum pump to vacuum connection for deceleration shut-off valve.

– Operate hand vacuum pump. Deceleration shut-off valve must open (arrow).

– Operate ventilation valve at vacuum pump. Deceleration shut-off valve must close (arrow)

If deceleration shut-off valve does not open or does not close, replace valve.

NOTE —
Secure connections of deceleration valve using screw-type clamps.

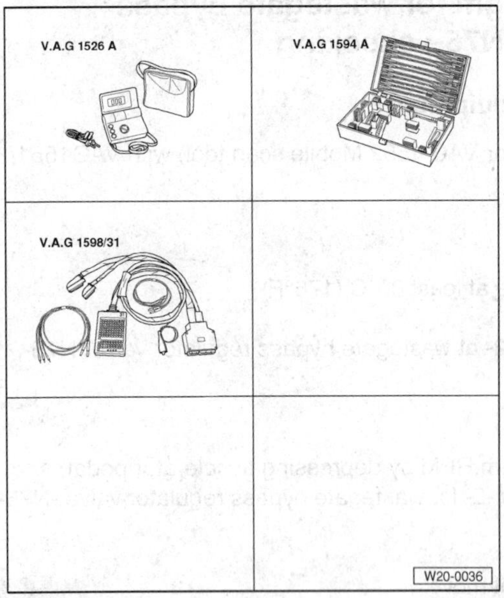

Charge air pressure sensor, checking

Test requirement

- On Board Diagnostic (OBD) recognized a malfunction and stored a DTC in DTC memory.

Special tools and equipment

- VAG 1526A Multimeter
- VAG 1594A Connector test kit
- VAG 1598/31 Test box
- Wiring diagram

Checking voltage supply and wiring to control module

– Switch ignition off.

Disconnect 4-pin connector -2- from charge air pressure sensor -G31- -2-.

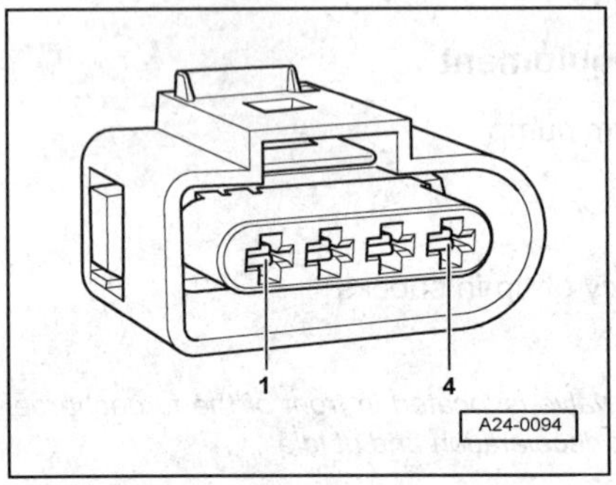

Use adapter cables from VAG 1594 to connect multimeter to terminals 1 + 3 of connector for voltage measurement.

– Switch ignition on.

Specification: at least 4.5 V

– Switch ignition off.

If specified value is not obtained:

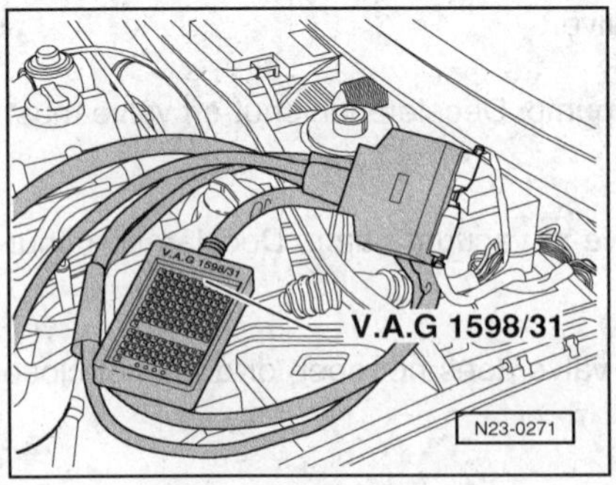

Connect VAG 1598/31 Test box to control module wiring harness. Engine Control Module (ECM) is not connected.

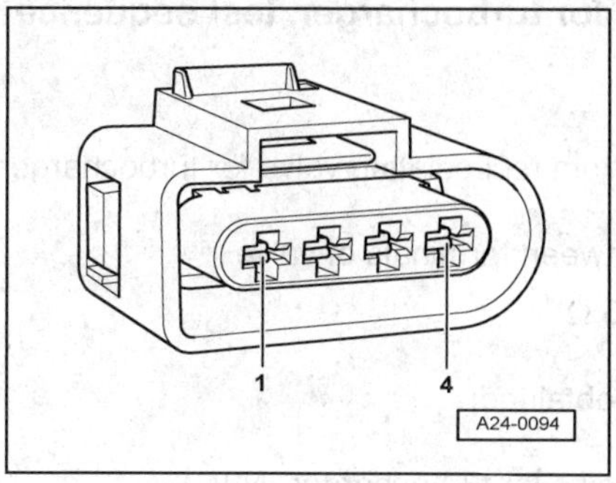

- Check for open circuits in wires between test box and connector according to wiring diagram.
  - Terminal 1 + socket 108
  - Terminal 2 + socket 98
  - Terminal 4 + socket 101
  - Wire resistance: max: 1.5Ω

– Check wires for shorts among one another.

If no malfunctions are found in the wires and there was voltage between terminals 1 + 3:

– Check function, see **Charge air pressure sensor, checking**

If no errors are found in wires and there was no voltage between terminals 1 + 3:

– Replace Engine Control Module (ECM).

EDITOR'S NOTE —
ECM replacement requires coding of control module using Volkswagen factory scan tool. Consult an authorized Volkswagen service department for additional information.

Checking function

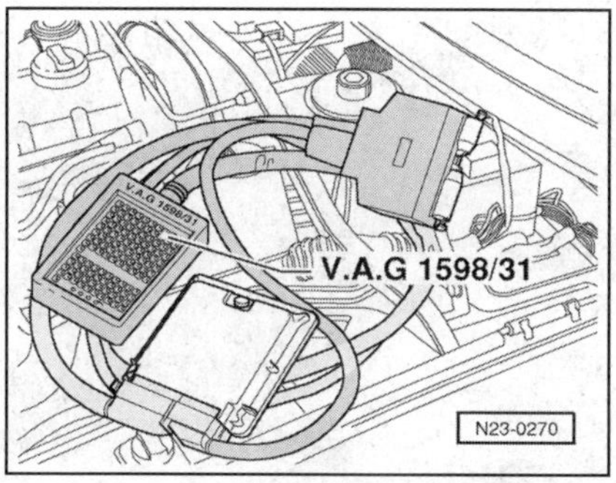

- Connect VAG 1598/31 Test box to control module wiring harness and to control module.

– Re-connect connector at charge air pressure sensor -G31-.

– Connect multimeter to socket 101 and socket 108 for voltage measurement.

– Start engine and measure voltage.

Specification: 1.80 to 2.00 volts

– Bring engine to RPM by depressing accelerator pedal suddenly.

Specification: 2.00 to 3.00 volts

If specified values are not obtained:

– Replace the charge air pressure sensor -G31-.

Recirculating valve for turbocharger, checking

Valve opens between connection to intake manifold and connection to deceleration shut-off valve with no voltage Connection to vacuum reservoir is closed.

Test requirement

Not for engine code AUG

- Output Diagnostic Test Mode (DTM) performed.)

EDITOR'S NOTE —
Component activation is accomplished using Volkswagen diagnostic scan tool, output diagnostic test mode (function 03).

Recirculating valve for turbocharger, test sequence

– Switch ignition off.

– Disconnect connector from recirculating valve for turbocharger -N249-.

◀ Measure resistance between terminals of valve.

• Specification: 25 to 35 Ω

If specified value is not obtained:

– Replace recirculating valve for turbocharger -N249-.

26 Exhaust System and Emission Controls

Repair Group

C 26

Exhaust System

WARNING —

Observe the following for all installations, especially in engine compartment due to lack of room:

- *Route lines of all types (e.g. for fuel, hydraulic, EVAP canister system, coolant and refrigerant, brake fluid, vacuum) and electrical wiring so that the original path is followed.*
- *Watch for sufficient clearance to all moving or hot components.*

NOTE —

- *After exhaust system repairs, make sure exhaust system is not under stress and that it has sufficient clearance from the bodywork. If necessary, loosen double clamps and clamp and align exhaust pipe so that sufficient clearance is maintained to the bodywork and support rings carry uniform loads.*
- *Always replace self-locking nuts.*

EDITOR'S NOTE —

Volkswagen identifies electrical components by a letter and/or a number in the electrical schematics. See **Wiring Diagrams, Fuses and Relays**. *These electrical identifiers are listed -between dashes- as an aid to electrical troubleshooting.*

Exhaust system components

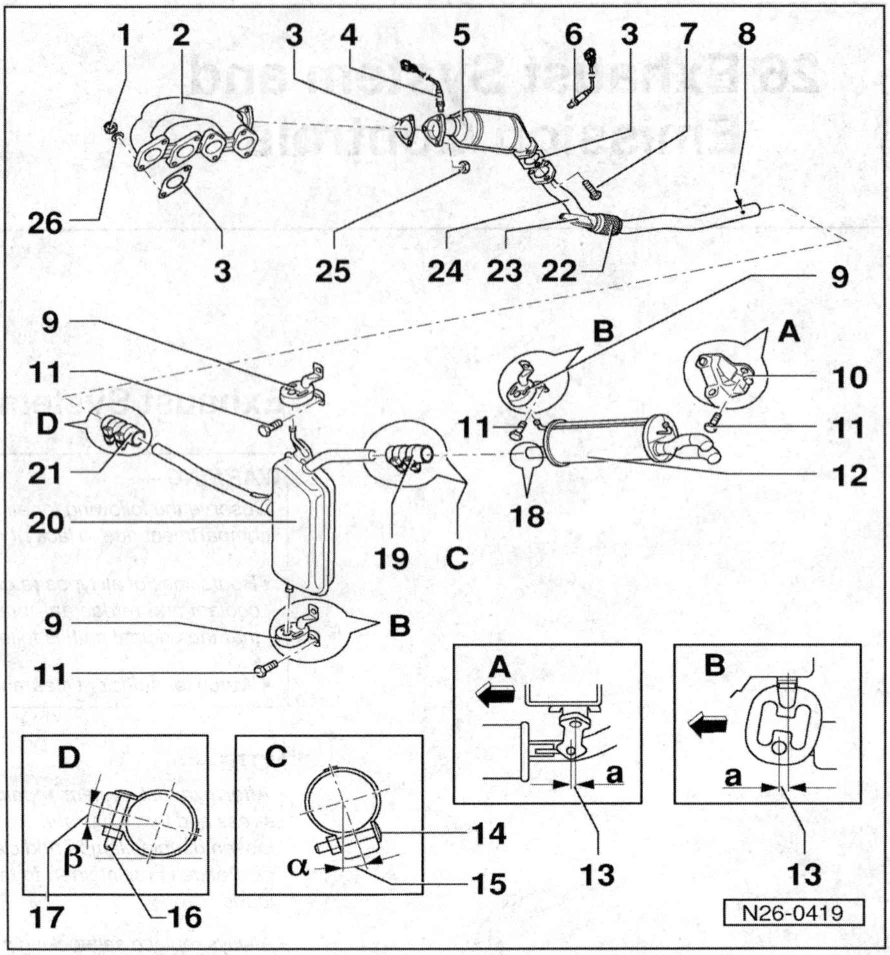

1. **30 Nm (22 ft-lb)**
   - Always replace
   - Coat threads with G 052 112 A3
2. **Exhaust manifold**
   - With turbocharger, see **Repair Group 21, Turbocharger with attachments**
3. **Gasket**
   - Always replace
4. **Heated Oxygen Sensor (HO2S) -G39- before catalytic converter**
   - Tighten to 50 Nm (37 ft-lb)
   - Only use "G 052 112 A3" to grease thread; do not let "G 052 112 A3" get onto slit of oxygen sensor
   - Use 3337 to remove and install
   - Checking, see **1.8 Liter Fuel Injection & Ignition (code: ATW, AUG, AWM) section, Repair Group 24**
5. **Catalytic converter**
   - Checking, see **Three Way Catalytic Converter (TWC), checking**
6. **Oxygen Sensor (O2S) Behind Three Way Catalytic Converter (TWC) -G130-**
   - Only use "G 052 112 A3" to grease thread; do not let "G 052 112 A3" get onto slit of oxygen sensor
   - Use 3337 to remove and install
   - Checking, see **1.8 Liter Fuel Injection & Ignition (code: ATW, AUG, AWM) section, Repair Group 24**
7. **25 Nm (18 ft-lb)**
   - Always replace
   - Coat threads with G 052 112 A3
8. **Marking**
   - 3x at perimeter
   - Slide double clamp **-21-** up to marking on tube
9. **Support ring**
   - With retainer ring
10. **Support ring**
11. **25 Nm (18 ft-lb)**
12. **Rear muffler**
    - Removing and installing: Using VAG 1523 Chassis saw, cut at right angle through connecting pipe at separation point **-18-**.
13. **Dimension -a- = approx. 7 to 9 mm (0.275 to 0.355")**
    - Exhaust system must be cold
    - Arrow shows direction of travel
    - To adjust exhaust system, slide double clamp **-21-** onto exhaust pipe up to marking **-8-** and tighten slightly Slide center and rear mufflers far enough forward that dimension -a- is obtained. Tighten double clamp.
14. **Bolt head must face fuel tank**
15. **Angle = 30 ± 5°**
16. **Bolt end must not be below double clamp**
    - Double clamp viewed in direction of travel
17. **Angle = 10 + 5°**
18. **Separating point**
    - Approx. 160 to 180 mm (6.3 to 7.0") in front of rear muffler, marked by a recess in the pipe.
19. **Double clamp**
    - Tighten evenly to 40 Nm (30 ft-lb)
20. **Center muffler**
21. **Double clamp**
    - Tighten evenly to 40 Nm (30 ft-lb)
22. **Decoupling element**
    - Max. flex 10°
23. **Support ring**
    - Individual components, see **Fig. 1**
24. **Front exhaust pipe**
25. **30 Nm (22 ft-lb)**
    - Always replace
    - Coat threads with G 052 112 A3
26. **Washer**

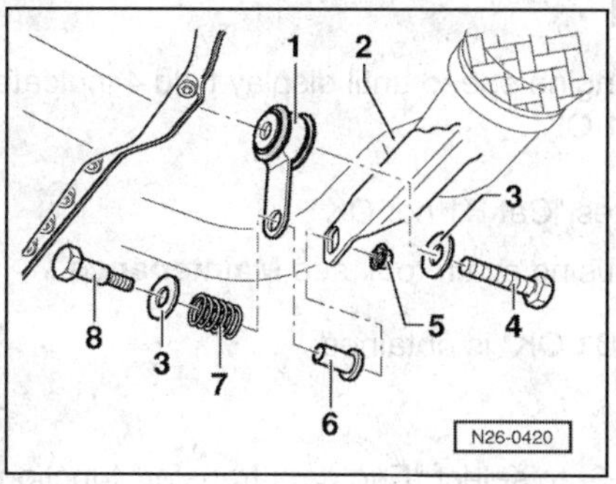

Fig. 1 Individual components of suspension

1. Strap with buffer
2. Exhaust pipe
3. Washer
4. Hex bolt, 25 Nm (18 ft-lb)
5. Self-locking nut, 25 Nm (18 ft-lb)
6. Spacer sleeve
7. Compression spring
8. Hex bolt

Three Way Catalytic Converter (TWC), checking

Special tools and equipment

- VAG 1551 Scan tool (or VAG 1552 Mobile scan tool) with VAG 1551/3 Cable

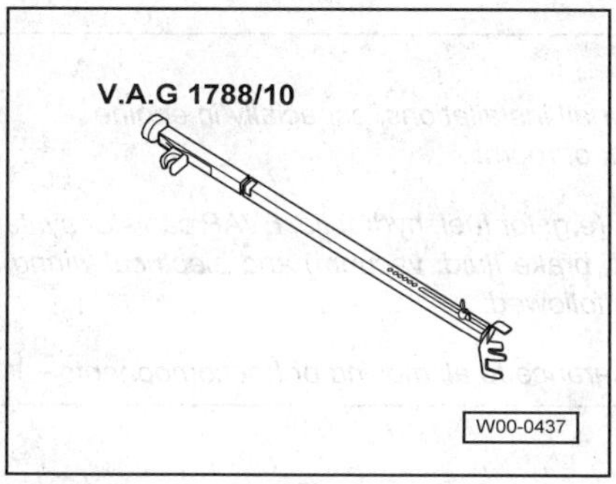

VAG 1788/10 RPM adjuster

Test sequence

– Connect VAG 1551 Scan tool (or VAG 1552). Start engine and select the Engine Control Module (ECM) using address word 01. Connecting scan tool and selecting Engine Control Module (ECM), see **Maintenance.**

– Press buttons -0- and -4- to select "Basic Setting" function 04. Press -Q- button to confirm input.

– Use VAG 1788/10 RPM Adjuster to adjust engine RPM to between 2300 and 2500 RPM:

– Press buttons -0-, -3- and -4- to select "Display group number 34". Press -Q- button to confirm input.

– Maintain engine speed until indication in display field 4 jumps from "Test OFF" to "Test ON". The catalytic converter temperature in display field 2 must be at least 350 °C (662°F).

– Continue to maintain engine speed until display field 4 indicates specified value "B1-S1 OK".

– Press -C- button.

– Press buttons -0-, -4- and -6- to select "Display group number 46". Press -Q- button to confirm input.

– Maintain engine speed until indication in display field 4 jumps from "Test OFF" to "Test ON".

- The catalytic converter temperature in display field 2 must be at least:
  Engine code AUG: 430 °C (806°F)
  Engine code AWM: 550 °C (1022°F)

Engine code AUG

– Check amplitude ratio in display field 3:

- Specification: max: 0.4

Engine code AWM

– Check diagnostic value in display field 3:

- Specification: MIN: 1.0

Continuation for all

– Continue to maintain engine speed until display field 4 indicates specified value "Cat-B1 OK".

– If display field 4 indicates "Cat B1 not OK":
  - Check DTC memory using scatn tool, see **Maintenance**.

– If specified value "Cat B1 OK" is obtained:
  - Press →button.

– Press buttons -0- and -6- to select "End data transfer" function 06. Press -Q- button to confirm input.

Secondary Air Injection (AIR) system

> ***WARNING —***
> *Observe the following for all installations, especially in engine compartment due to lack of room:*
>
> - *Route lines of all types (e.g. for fuel, hydraulic, EVAP canister system, coolant and refrigerant, brake fluid, vacuum) and electrical wiring so that the original path is followed.*
> - *Watch for sufficient clearance to all moving or hot components.*

Function

Air is blown behind the exhaust valves by the Secondary Air Injection (AIR) system for a maximum of 100 seconds following cold start (coolant temperature of 5 to 33 °C (41 to 91°F)). This adds oxygen to the exhaust, initiating a secondary combustion and thereby shortening the warm-up phase of the catalytic converter. Activation is triggered by the motronic Engine Control Module (ECM) -J220- and sent via the Secondary Air Injection (AIR) pump relay -J299- to the Secondary Air Injection (AIR) solenoid valve -N112- (change-over valve) and the combination valve. In addition, the Secondary Air Injection (AIR) system is switched on for 10 seconds during idle after every subsequent engine start (up to max. 96°C (205°F) engine temperature) and is checked via On Board Diagnostic (OBD). Oxygen sensor control must be active for this.

Secondary Air Injection (AIR) system components, removing and installing

NOTE—

- *Components marked with * are checked via On Board Diagnostic (OBD), see* ***Maintenance***
- *Components marked with ** are tested via output Diagnostic Test Mode (DTM).*

EDITOR'S NOTE—

Component activation is accomplished using Volkswagen diagnostic scan tool, output diagnostic test mode (function 03).

NOTE—

*Secondary Air Injection (AIR) pump relay -J299- **, see* **Fig. 1**

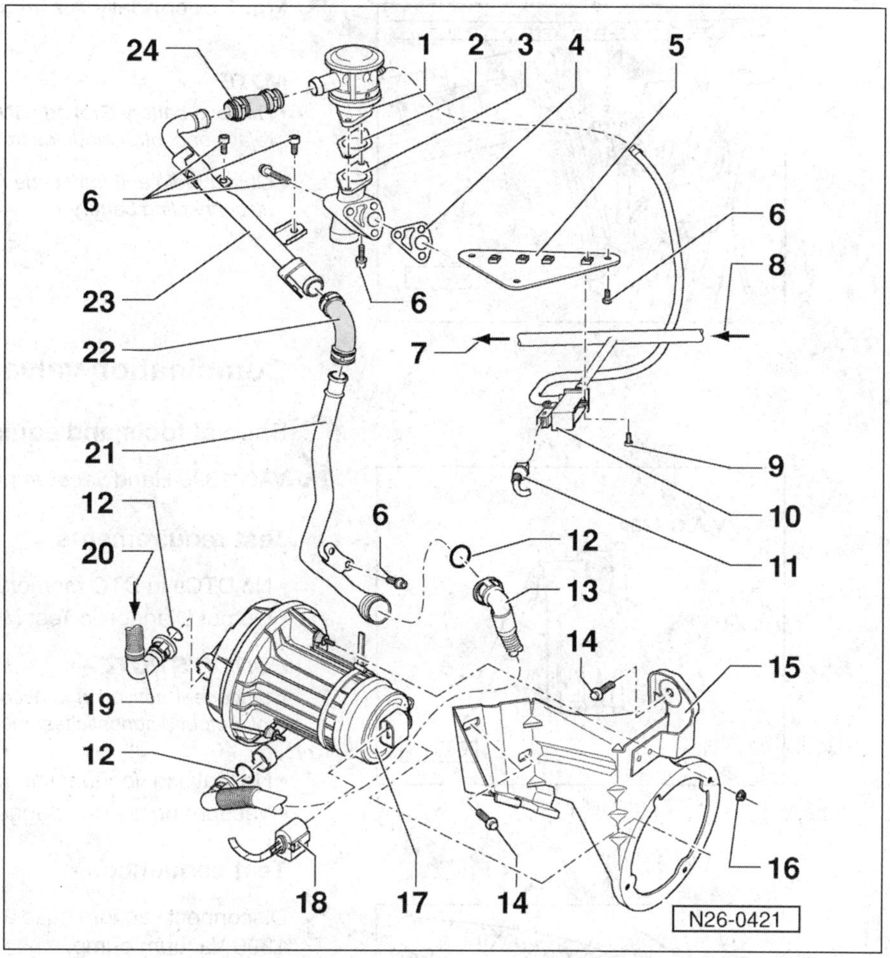

1. **Combination valve**
   - Checking, see **Combination valve, checking**
2. **Gasket**
   - Always replace
3. **Bracket**
   - Bolted at cylinder head
4. **Gasket**
   - Always replace
5. **Bracket**
   - Bolted at intake manifold
6. **10 Nm (7 ft-lb)**
7. **To recirculating valve for turbocharger -N249-**
8. **From vacuum reservoir/intake manifold connection**
9. **6 Nm (5 in-lb)**
10. **Secondary Air Injection (AIR) solenoid valve -N112-*/****
    - Resistance: 25 to 35 Ω
11. **Connector**
    - 2-pin
    - green
12. **O-ring**
    - Replace if damaged
13. **Pressure hose**
    - Check for secure fitting
    - To disengage, squeeze together at front
14. **8 Nm (71 in-lb)**
15. **Bracket**
    - For Secondary Air Injection (AIR) pump motor
16. **8 Nm (71 in-lb)**
17. **Secondary Air Injection (AIR) pump motor -V101-****
    - Checking function, see **Secondary Air Injection (AIR) pump motor, checking**
18. **Connector**
    - 2-pin
    - For Secondary Air Injection (AIR) pump motor
19. **Intake hose**
    - For Secondary Air Injection (AIR) pump
20. **From air filter**
21. **Connection tube**
22. **Elbow**
    - Secure with screw-type clamps or spring clamps
23. **Pressure line**
24. **Connecting piece**
    - Secure with screw-type clamps or spring clamps

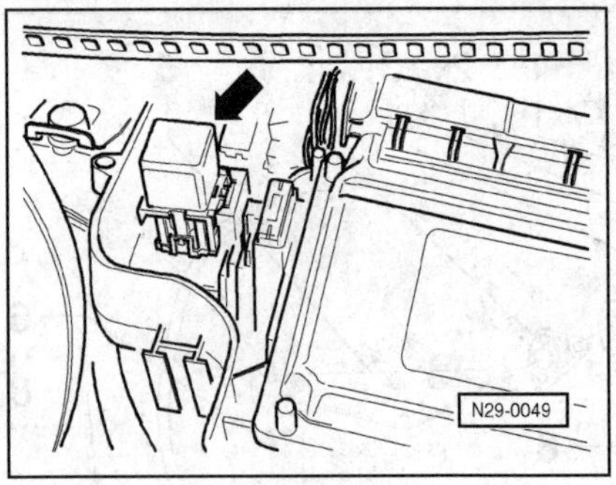

Fig. 1 Secondary Air Injection (AIR) pump relay -J299- (arrow)

NOTE—

- *Remove battery Ground (GND) strap if tools are necessary for removal of relays or control modules from the relay carrier.*
- *Obtain radio anti-theft code for radios with anti-theft coding before disconnecting battery.*

Combination valve, checking

Special tools and equipment

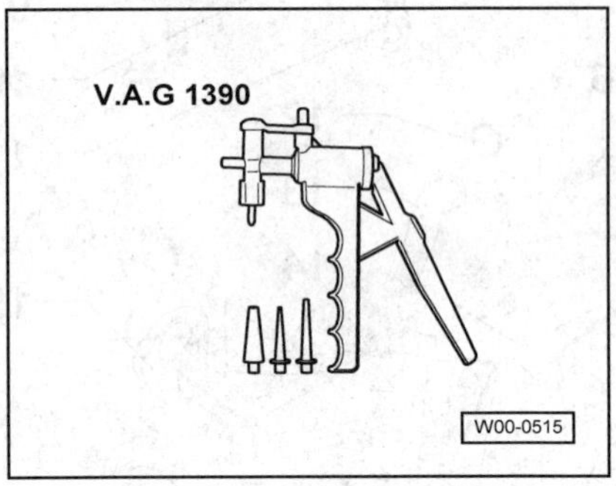

VAG 1390 Hand vacuum pump

Test requirements

- No DTCs in DTC memory, see **Maintenance**
- Output Diagnostic Test Mode (DTM) performed.

EDITOR'S NOTE—

Component activation is accomplished using Volkswagen diagnostic scan tool, output diagnostic test mode (function 03).

- No leaks in vacuum lines or hose connections
- Vacuum hoses not clogged or kinked

Test sequence

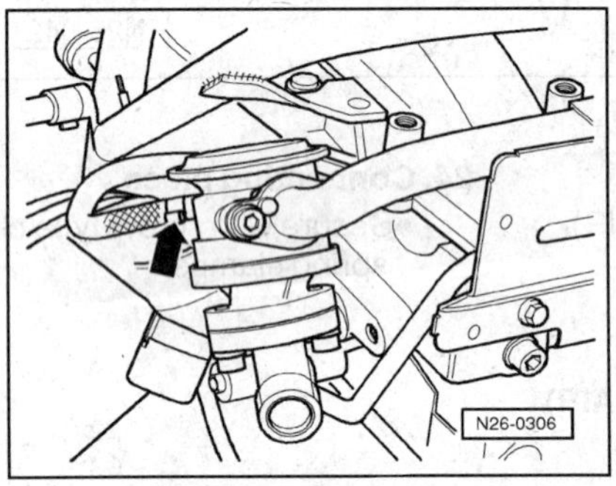

Disconnect vacuum hose at combination valve (arrow) and connect VAG 1390 Vacuum pump.

NOTE—

Do not use pressurized air for the following test!

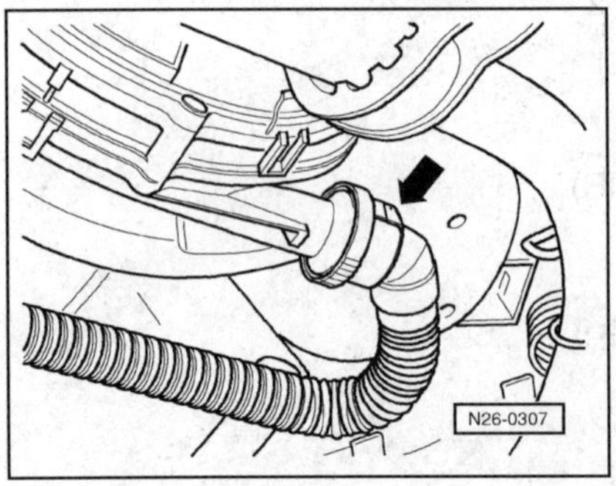

Disconnect pressure hose (arrow) at Secondary Air Injection (AIR) pump motor and blow air in with light pressure. Combination valve must close.

– Operate vacuum pump. Valve must open.

– If combination valve does not open:

- Replace combination valve

Secondary Air Injection (AIR) pump motor, checking

Special tools and equipment

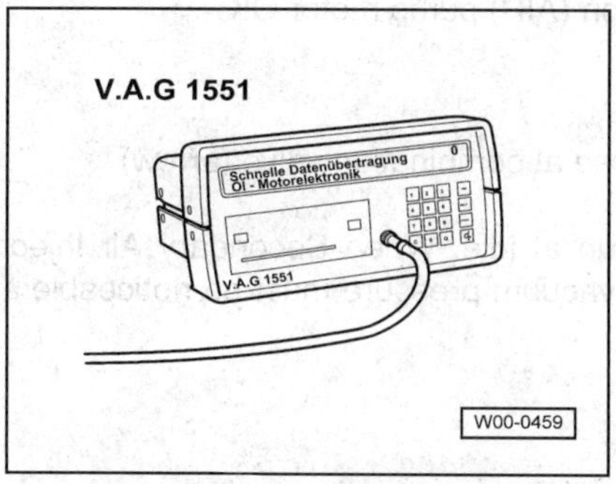

VAG 1551 Scan Tool (ST)

NOTE—

If VAG 1551 Scan Tool (ST) is not available, use VAG 1552 Mobile Scan Tool (ST) instead, although this tool cannot produce a print-out.

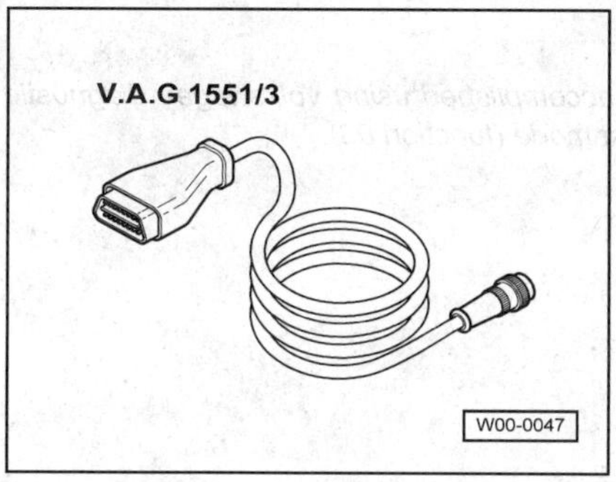

VAG 1551/3 Cable

Test requirements

- Fuse for Secondary Air Injection (AIR) pump relay -J299- OK
- Intake hose for Secondary Air Injection (AIR) pump not clogged or kinked

Test sequence

– Remove noise insulation, see **Body–Exterior section, Repair Group 50.**

– Disconnect pressure hose at Secondary Air Injection (AIR) pump motor.

– Activate Secondary Air Injection (AIR) pump relay via output Diagnostic Test Mode (DTM).

EDITOR'S NOTE—

Component activation is accomplished using Volkswagen diagnostic scan tool, output diagnostic test mode (function 03).

– The Secondary Air Injection (AIR) pump motor must start up in intervals and air must exit from exhaust tube.

– If motor starts, but no air exits:

- Replace Secondary Air Injection (AIR) pump motor.

– If the Secondary Air Injection (AIR) pump motor does not start up in intervals:

- Check activation of the Secondary Air Injection (AIR) pump motor.

EDITOR'S NOTE—

Component activation is accomplished using Volkswagen diagnostic scan tool, output diagnostic test mode (function 03).

Secondary Air Injection (AIR) valve, checking

Test requirement

- Coolant temperature 5 to 33°C (41 to 91°F)
- Secondary Air Injection (AIR) pump motor OK

Test sequence

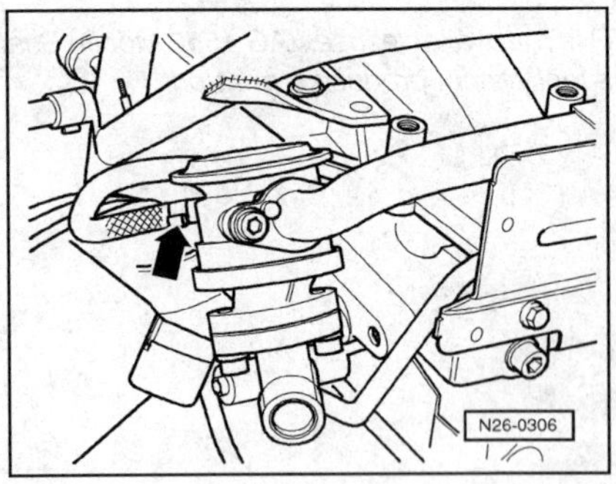

< Disconnect vacuum hose at combination valve (arrow).

– Start engine and let run at idle. When Secondary Air Injection (AIR) pump motor starts up vacuum pressure must be noticeable at vacuum hose.

– If no vacuum is felt:

- Switch off engine.
- Check activation of valve.

EDITOR'S NOTE —
Component activation is accomplished using Volkswagen diagnostic scan tool, output diagnostic test mode (function 03).

2.8 Liter V6 5V Engine Mechanical (code: AHA, ATQ)

Section

D

19 Engine–Cooling System — Page

20 Fuel Supply — Page

26 Exhaust System and Emission Controls — Page

00 General, Technical Data

Repair Group

D 00

Technical data

Engine number

The engine number consists of up to nine characters (alphanumeric). The first part (maximum 3 characters) makes up the "engine code," and the second part (6 characters), the "serial number." If more than 999,999 engines with the same engine code are produced, the first of the six characters is replaced with a letter.

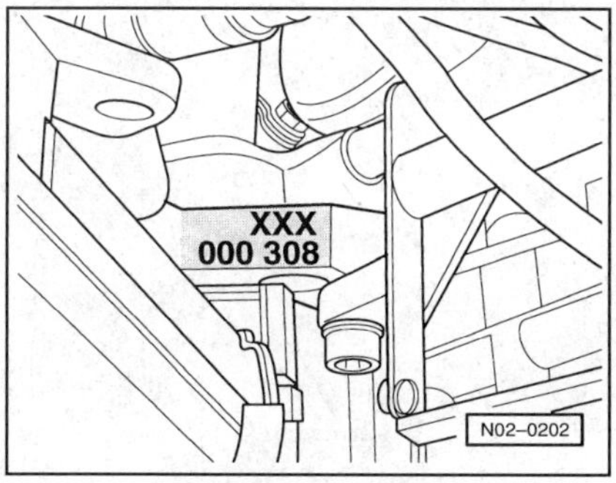

The engine number ("Engine code" and "Serial number") can be found on the flat surface of the cylinder block, on front of right cylinder head.

The engine code is also included on the vehicle data plate.

Engine data

Engine code		AHA	ATQ
Manufactured		08.97 ➤	10.99 ➤
Displacement	cm^3	2771	2771
Output	hp (kW) @ RPM	190 (142) @ 6000	190 (142) @ 6000
Torque	ft-lb (Nm) @ RPM	207 (280) @ 3200	207 (280) @ 3200
Bore	diameter mm	82.5	82.5
Stroke	mm	86.4	86.4
Compression ratio		10.6	10.6
RON	min.	98 unleaded[1)]	98 unleaded[1)]
Injection, ignition		Motronic	Motronic
Knock control		Yes	Yes
On Board Diagnostic (OBD)		OBD II	OBD II
Oxygen sensor control		Yes	Yes
Exhaust Gas Recirculation (EGR)		No	No
Three Way Catalytic Converter (TWC)		Yes	Yes
Secondary Air Injection (AIR) system		Yes	Yes

[1)] In exceptional circumstances min. 95 RON, however with reduced performance

13 Engine–Crankshaft, Cylinder Block

Repair Group

D 13

Engine Block Assembly

NOTE —
If during engine repairs, metal shavings or large quantities of small metal particles are found in the engine oil, the oil passages in the cylinder block must be thoroughly cleaned out and the oil cooler replaced.

EDITOR'S NOTE —
Audi part numbers are given for reference only! Always consult with your Audi Parts Department or aftermarket parts specialist for the latest parts information.

EDITOR'S NOTE —
Audi identifies electrical components by a letter and/or a number in the electrical schematics. See **Wiring Diagrams, Fuses and Relays**. *These electrical identifiers are listed -between dashes- as an aid to electrical troubleshooting.*

Engine belts

NOTE—
Servicing the clutch, see **Repair Manual, 5 Speed Manual Transmission section, Repair Group 30.**

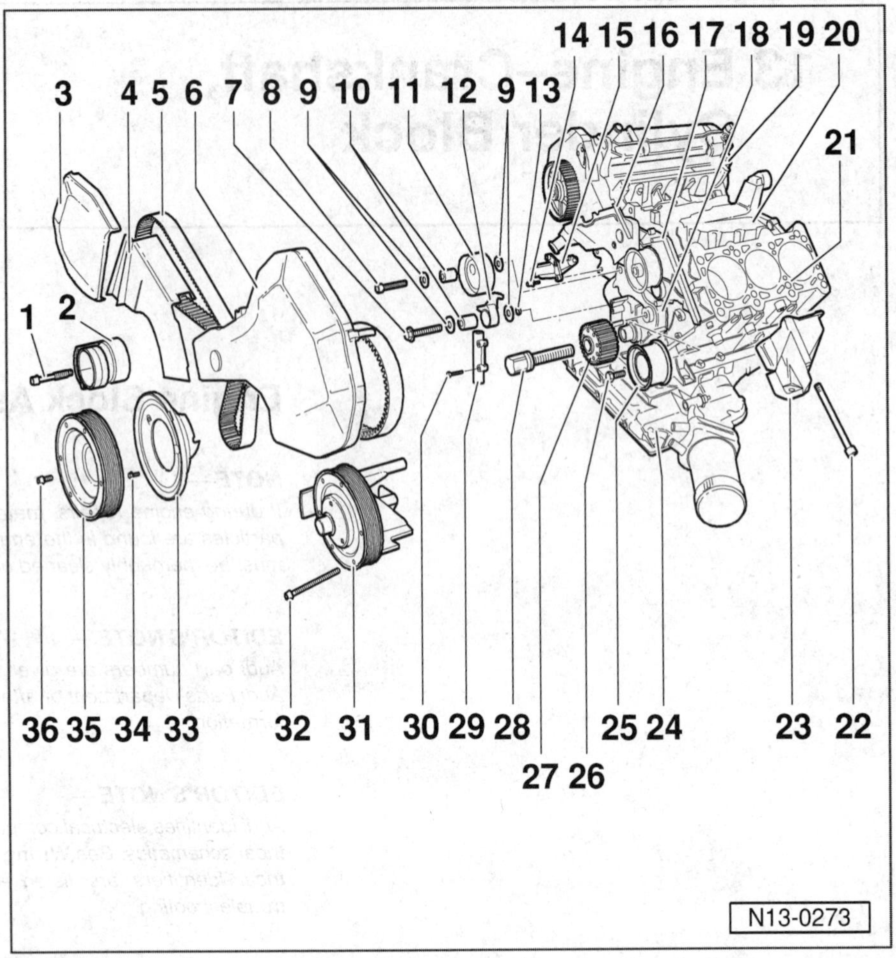

1. **55 Nm (41 ft-lb)**
2. **Tensioner**
   - Secure with 3204 before removing
3. **Upper toothed belt guard, right**
4. **Toothed belt guard, center**
5. **Toothed belt**
   - Mark direction of rotation before removing
   - Check for wear
   - Do not kink
   - Removing, installing and tensioning, see **Toothed belt, removing, installing and tensioning**
6. **Upper toothed belt guard, left**
7. **25 Nm (18 ft-lb)**
8. **25 Nm (18 ft-lb)**
9. **Washer**
10. **Bearing bushing**
11. **Tensioner**
    - See **Toothed belt, removing, installing and tensioning**
12. **Tensioning lever**
13. **O-Ring**
14. **25 Nm (18 ft-lb)**
15. **Bracket**
16. **Connection**
    - Removing and installing thermostat, see **Repair Group 19**
17. **Coolant pump**
    - Removing and installing, see **Repair Group 19**
18. **Oil pump**
    - Removing and installing, see **Repair Group 17**
19. **Cylinder head**
    - Removing and installing, see **Repair Group 15**
20. **Breather housing**
    - Removing and installing, see **Repair Group 17**
21. **Cylinder block**
    - See **Cylinder block, crankshaft and flywheel/drive plate**
22. **45 Nm (33 ft-lb)**
23. **Console**
24. **Sump**
    - Removing and installing, see **Repair Group 17**
25. **Idler roller**
26. **45 Nm**
27. **Crankshaft toothed belt sprocket**
28. **200 Nm + 1/2 turn (180°) further**
    - Always replace
    - Oil threads
    - The turning further can be done in several stages
    - The turning further angle can be checked with a commercial protractor e.g. Hazet 6690
29. **Tensioning element**
30. **10 Nm (7 ft-lb)**
    - Install with D6
31. **Fan retainer**
32. **Bolt**
    - M8: 25 Nm (18 ft-lb)
    - M6: 10 Nm (7 ft-lb)
33. **Toothed belt pulley guard**
34. **10 Nm (7 ft-lb)**
35. **Vibration damper**
    - When installing note fixing arrangement
    - See **Toothed belt, removing, installing and tensioning**
    - Removing and installing ribbed belt, see **Maintenance**
36. **25 Nm (18 ft-lb)**

Toothed belt, removing, installing and tensioning

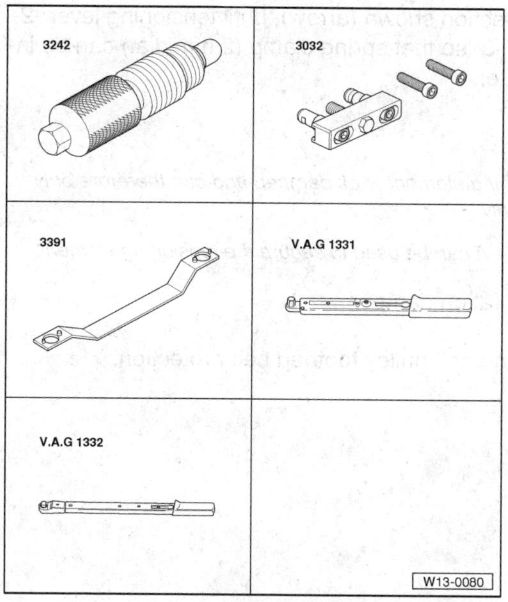

Special tools and equipment

- 3242 Clamping bolt
- 3032 Puller
- 3391 Camshaft clamp
- V.A.G 1331 Torque wrench (5-50 Nm)
- V.A.G 1332 Torque wrench (40-200 Nm)

Removing

– Remove ribbed V-belt first, see **Maintenance.**

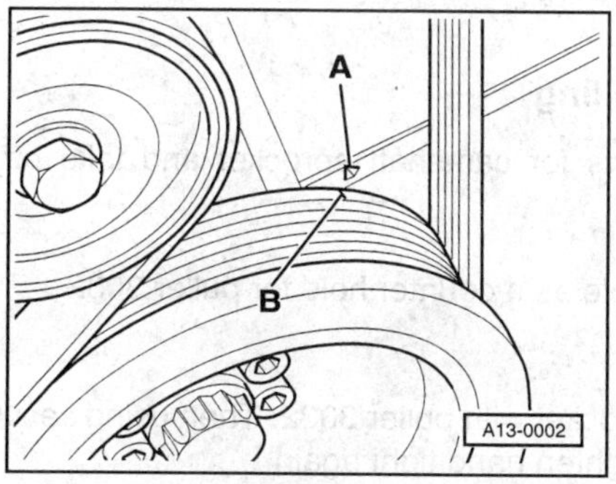

◄ Position crankshaft to top dead center No. 3 cylinder. Mark on toothed belt guard -A- must align with notch on belt pulley -B-.

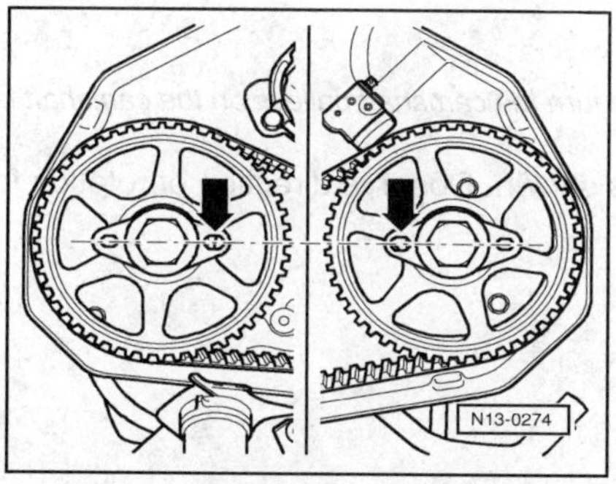

◄ Check position of camshaft sprockets. Larger holes in securing plate -arrows- must align opposite one another on inside.

NOTE—

If the larger holes are positioned to the outside, the crankshaft must be turned one revolution further.

– Remove sealing plugs from cylinder block, left. Crankshaft TDC drilling must be visible (or be able to be felt) in sealing plug hole.

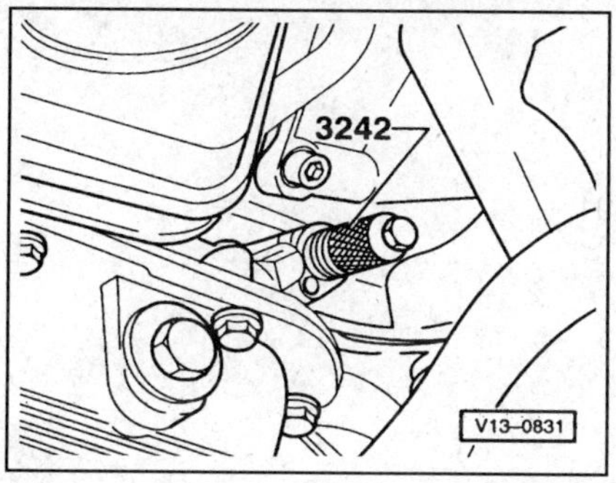

◄ Screw clamping bolt 3242 into hole so that crankshaft is secured against turning.

– Remove tensioner for ribbed V-belt.

– Remove center and right-hand toothed belt guards.

– Mark direction of rotation of toothed belt.

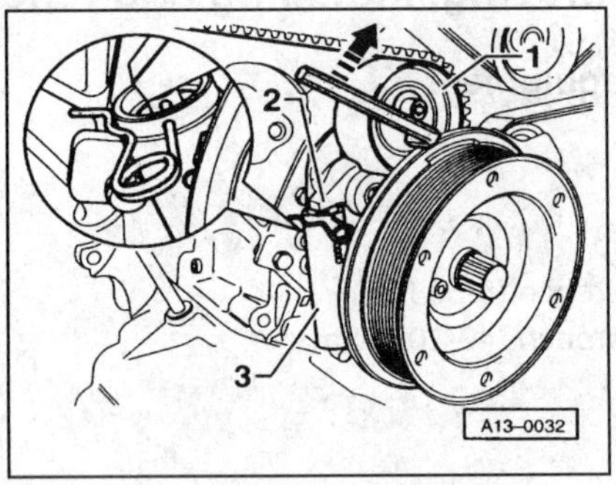

Using 8 mm hex wrench, turn toothed belt tensioner roller -1- counter-clockwise, opposite direction shown (arrow), until tensioning lever -2- compresses tensioner -3- so that spring clamp (2 mm dia.) can be inserted into bore and lifter.

NOTE—
- *The toothed belt tensioning element is oil damped and can therefore only be pushed together slowly.*
- *The spring pin from 2024 A can be used to secure the tensioning element.*

– Remove crankshaft vibration damper.

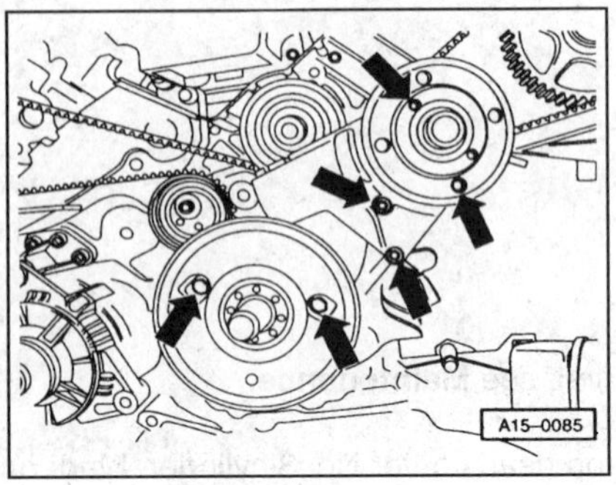

Remove fan retainer and belt pulley toothed belt protection.

– Take off toothed belt.

Toothed belt, installing

– Remove securing bolts for camshaft sprocket and take off securing plate.

– Fit an M10 bolt into hole as a counter hold for puller 3032.

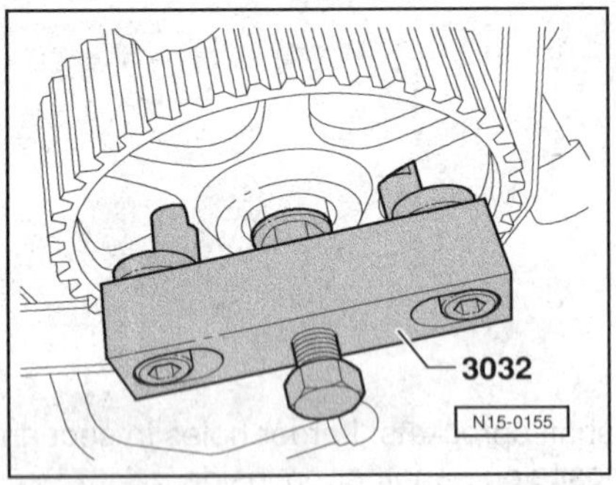

Release camshaft sprocket with puller 3032. Then using securing bolts and securing plates tighten hand tight again.

NOTE—
It must be possible to just turn the camshaft pulleys on the camshaft tapers.

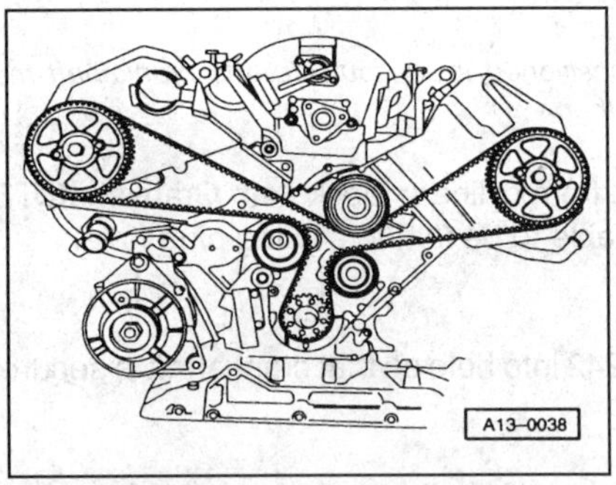

Route toothed belts as shown. Observe direction of rotation for used toothed belts.

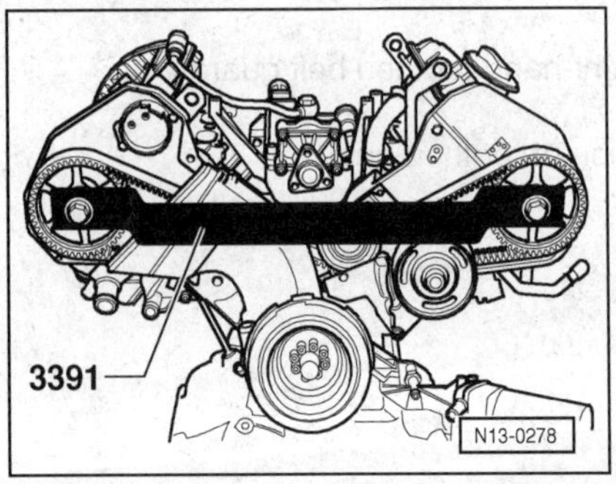

Fit camshaft clamp 3391.

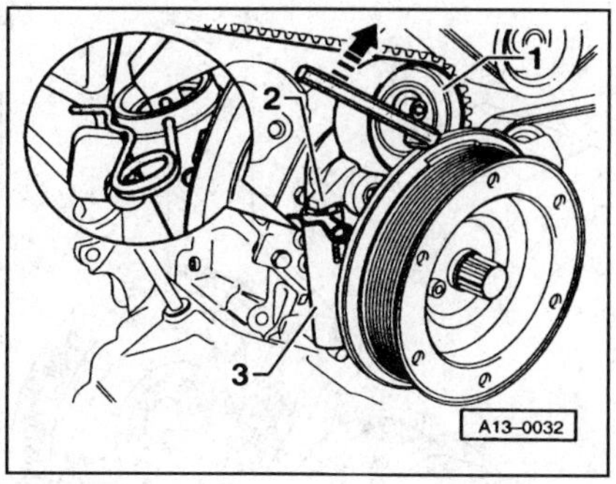

◄ Using 8 mm hex wrench, turn toothed belt tensioner roller -1 counter-clockwise, opposite direction shown (arrow), until tensioning lever -2- compresses tensioner -3- so that spring clamp (2 mm dia.) can be removed and then remove clamp.

– Before starting engine, pre-load timing tensioner roller as follows:

Using a torque wrench with 8 mm hex bit, apply tension to toothed belt clockwise in direction shown (arrow) until a torque of 15 Nm (11 ft-lb) is attained. This ensures tensioner has expanded completely and is properly engaging toothed belt.

– Tighten camshaft sprocket to 55 Nm (41 ft-lb) after pretensioning toothed belt.

– Remove clamping bolt 3242.

The rest of the assembly is basically a reverse of the disassembly sequence.

Cylinder block, crankshaft and flywheel/drive plate

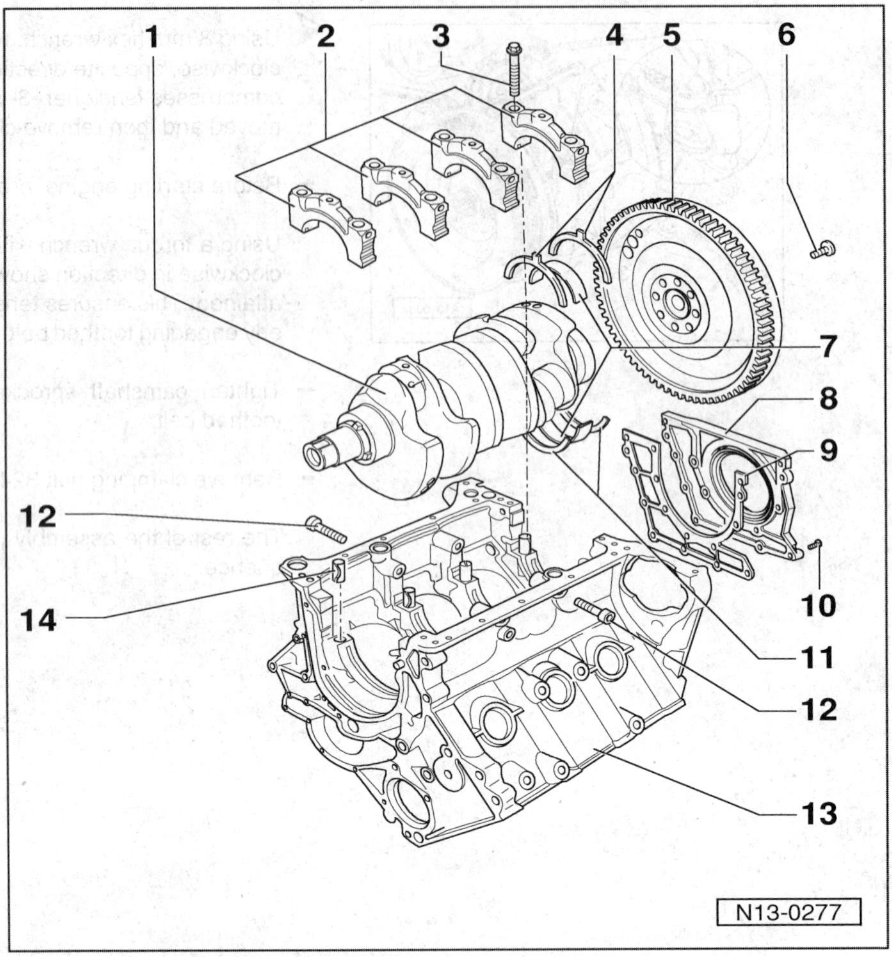

1. **Crankshaft**
   - Axial clearance
     New: 0.07-0.23 mm
     Wear limit: 0.25 mm
   - Measure radial clearance with Plastigage
     New: 0.018-0.045 mm
     Wear limit: 0.10 mm
   - Do not rotate the crankshaft when checking the radial clearance
   - For vehicles with automatic transmission with bearing bushings for torque converter, see **Fig. 3**
2. **Bearing cap**
   - Installation position see **Fig. 1**
   - Bearing cap 4 with recesses for thrust washers
3. **Connecting rod bolt**
   - Tighten to 60 Nm (44 ft-lb) + $^1/_4$ turn (90°)
   - Always replace
   - To measure radial clearance tighten to 60 Nm (44 ft-lb) but do not turn further
4. **Thrust washer**
   - For bearing cap 4
   - Note fixing arrangement
5. **Flywheel/drive plate**
   - Removing and installing needle bearing in flywheel, see **Fig. 2**
   - Removing and installing drive plate, see **Drive plate, removing and installing**
6. **Bolt**
   - Tightening torque for flywheel: 60 Nm (44 ft-lb) + $^1/_2$ turn (180°)
   - Tightening torque for drive plate: 60 Nm + $^1/_4$ turn (90°)
   - Always replace
   - Lock with clamping bolt 3242 when loosening and tightening, see **Fig. 4**
7. **Bearing caps**
   - Do not interchange used bearing shells (mark)
8. **Sealing flange with oil seal**
   - Only replace complete
   - Drain coolant before removing
   - Fit spare part sealing flange with supplied sleeve
9. **Gasket**
   - Always replace
10. **10 Nm (7 ft-lb)**
11. **Thrust washer**
    - For cylinder block, bearing 4
12. **25 Nm (18 ft-lb)**
    - Tighten by hand before tightening the bearing cap bolts
13. **Cylinder block**
14. **Guide sleeve**

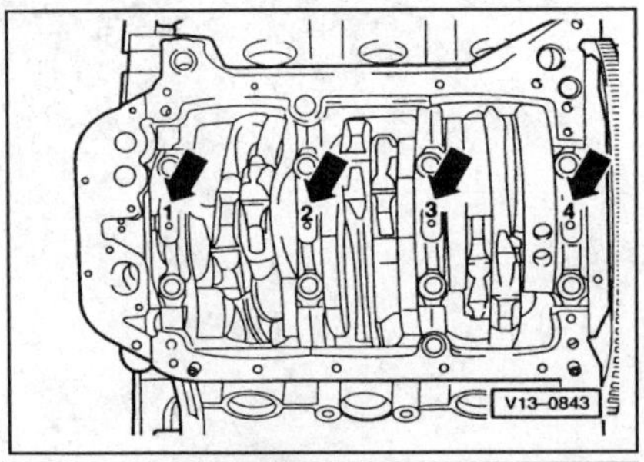

Fig. 1 Bearing cap installation position

- Bearing cap 1 - Belt pulley side
- Bearing cap 4 - Flywheel side

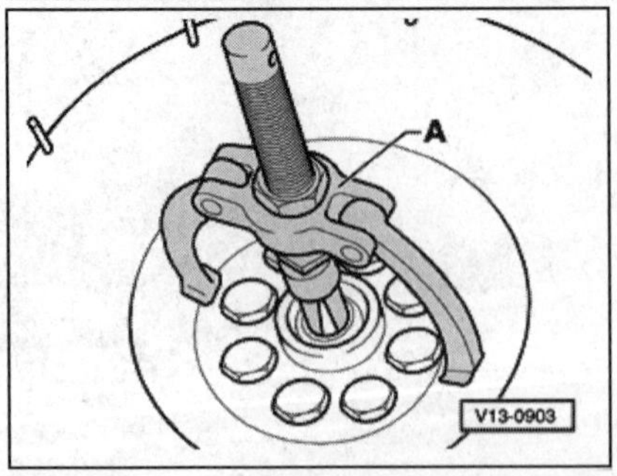

Fig. 2 Removing and installing needle bearing

– Pull out needle bearing with commercially available puller e.g. KUKKO 21/2 or KUKKO 22-1.

– Drive in needle bearing with fitting drift 3264.

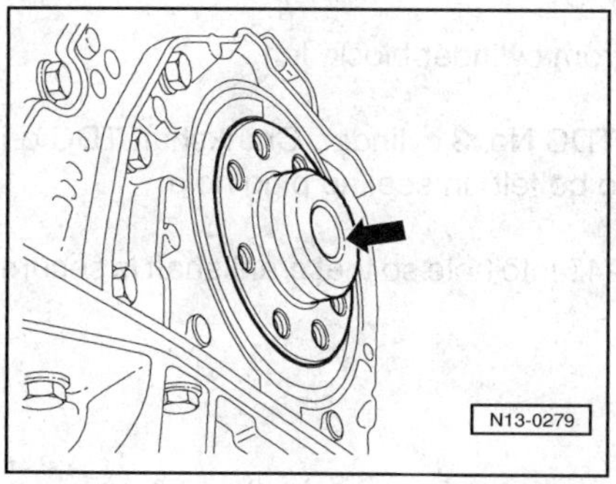

Fig. 3 Bearing bushing for torque converter

NOTE —
Short and exchange engines as well as new and exchange crankshafts are supplied without bushings. The bushings must be driven in before installing the drive plate.

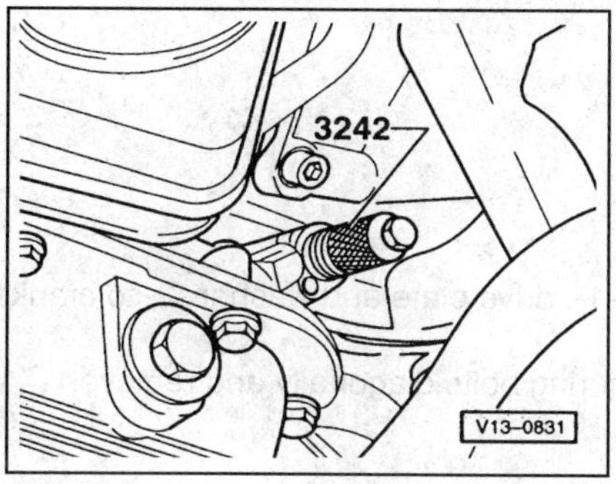

Fig. 4 Locking crankshaft with clamping bolt 3242

Remove sealing plugs and bring crankshaft to TDC No. 3 cylinder, then screw clamping bolt into TDC hole.

Drive plate, removing and installing

Special tools and equipment

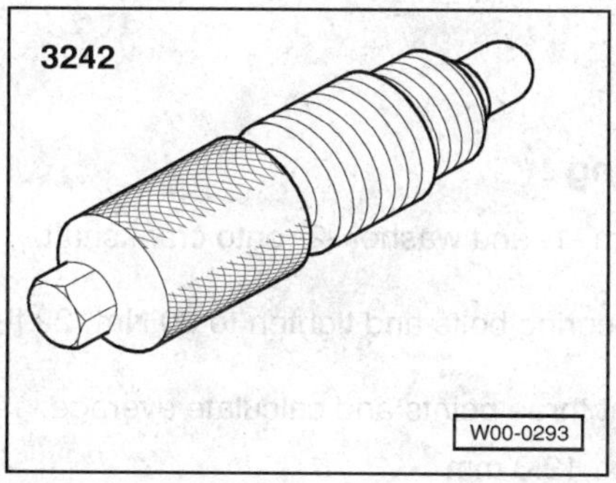

3242 Clamping bolt

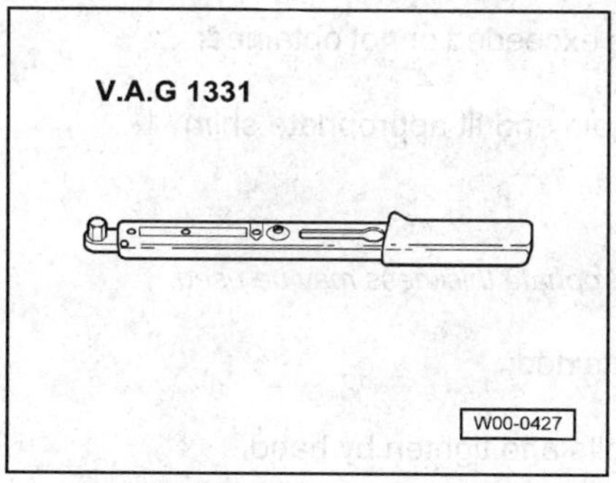

V.A.G 1331 Torque wrench (5-50 Nm)

- V.A.G 1332 Torque wrench (40-200 Nm)
- Depth gauge

Drive plate, removing

– Remove sealing plug from cylinder block, left.

– Position crankshaft to TDC No. 3 cylinder. Crankshaft TDC drilling must be visible (or be able to be felt) in sealing plug hole.

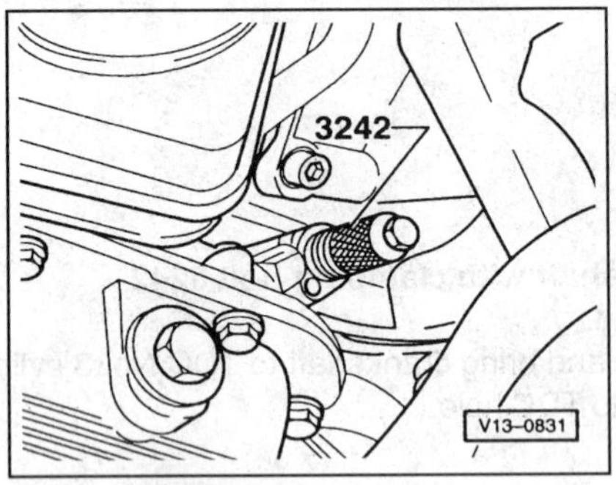

◄ Screw clamping bolt 3242 into hole so that crankshaft is secured against turning.

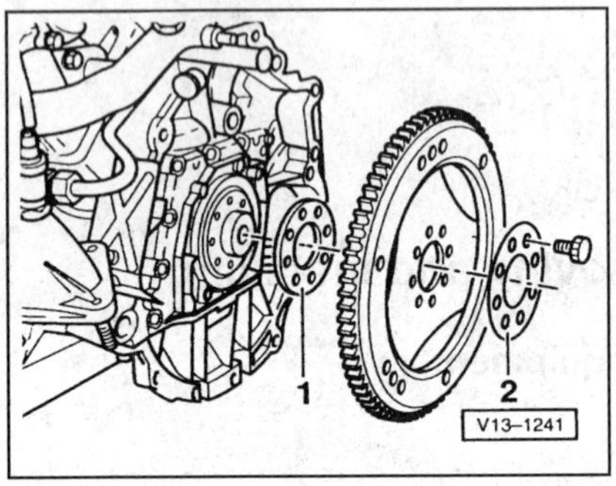

◄ Mark position of shim -1-, drive plate and washer -2- to crankshaft.

– Loosen drive plate securing bolts diagonally and remove.

– Remove drive plate.

Drive plate, installing

– Fit drive plate with shim -1- and washer -2- onto crankshaft.

– Insert at least 3 old securing bolts and tighten to 30 Nm (22 ft-lb).

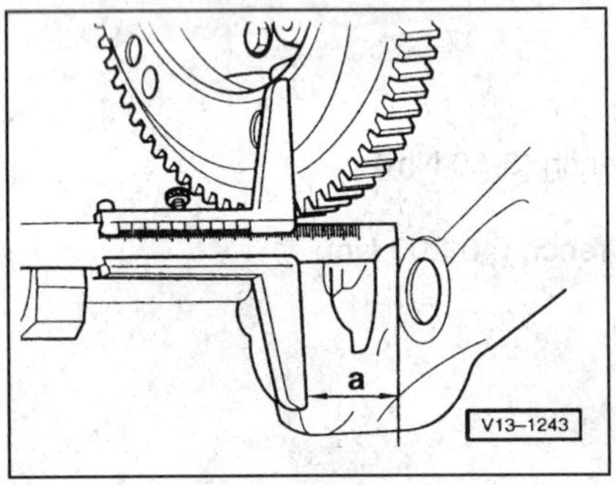

◄ Check dimension -a- at three points and calculate average.

- Specification: approx. 12.3 mm

If the specifications are exceeded or not obtained:

– Remove drive plate again and fit appropriate shim -1-.

NOTE—
Only one shim of the appropriate thickness may be used.

If the specification is attained:

– Insert new securing bolts and tighten by hand.

– Tighten securing bolts to 60 Nm (44 ft-lb) and then turn 90° (1/4 turn) further (turning further can be done in several stages).

15 Engine–Cylinder Head, Valvetrain

Repair Group

D 15

Cylinder Head, Servicing

NOTE—

- *When installing an exchange cylinder head with fitted camshaft, the contact surfaces between valve lifter and cam must be oiled after installation.*
- *The plastic protectors fitted to protect the open valves must only be removed immediately before fitting the cylinder head.*
- *When replacing the cylinder head the complete coolant must be replaced.*
- *If a cylinder head is replaced, drive a sealing cap (core plug) into the front of the respective head as described later.*

CAUTION—

Care must be taken when disposing of old sodium filled valves. They must be cut through in the center of the stem with a hacksaw. When doing this they must be perfectly dry. The valves prepared in this way can then be thrown into a bucket of water a maximum of ten at a time. When doing this step back quickly because a sudden chemical reaction causes the sodium filling to burn. After this treatment, the valves can be put in with the normal scrap.

Cylinder head, assembly

1. **Upper toothed belt guard, right**
2. **Toothed belt guard, center**
3. **Upper toothed belt guard, left**
4. **10 Nm (7 ft-lb)**
5. **Rear toothed belt guard**
6. **Lifting eye**
7. **25 Nm (18 ft-lb)**
8. **Cylinder head cover gasket**
   - Replace if damaged
   - Before fitting gasket, coat transitions between bearing cap and cylinder head with D 454 300 A2
9. **Cylinder head cover**
10. **Seal**
11. **Cap**
    - Replace seal if damaged
12. **Retainer**
    - For engine cover
13. **10 Nm (7 ft-lb)**
    - Tighten from center outward
14. **Cylinder head bolt**
    - Always replace
    - Note installation instructions and sequence when loosening and tightening
15. **Cylinder head**
    - See cylinder head, removing and installing
    - Check for distortion see **Fig. 1**
    - After replacing replace the complete coolant
16. **Oil deflector**
17. **Cylinder head gasket**
    - Always replace
    - Metal gasket
    - After replacing replace the complete coolant
18. **Camshaft sprocket**
    - Note position when installing the toothed belt
19. **Securing plate**
20. **55 Nm (41 ft-lb)**

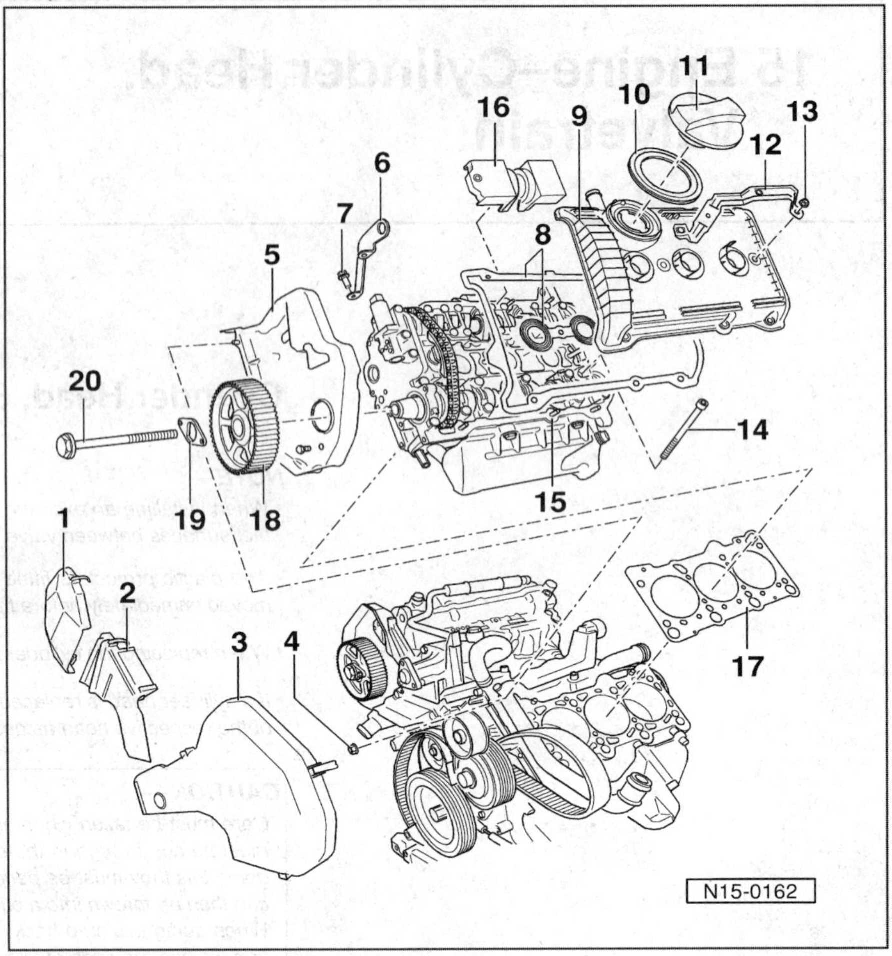

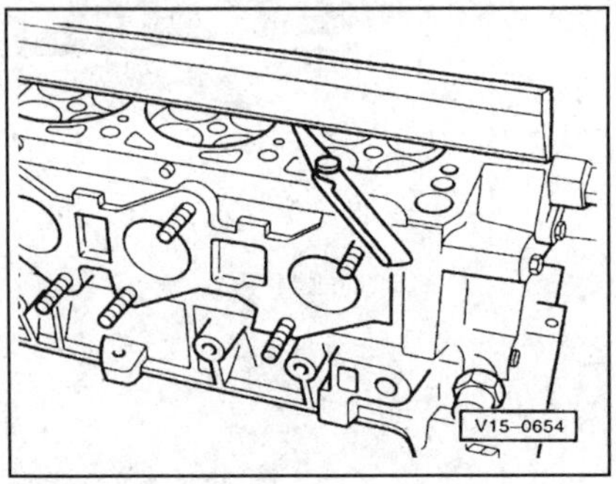

Fig. 1 Checking cylinder head for distortion

Max. permissible distortion: 0.05 mm

Cylinder head, removing and installing

Special tools and equipment

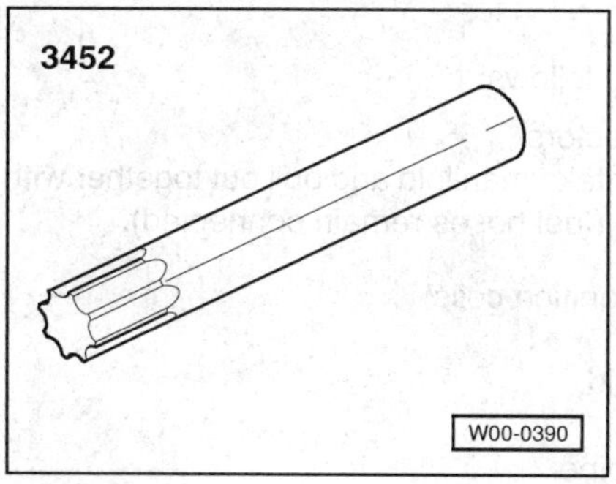

◄ 3452 Socket

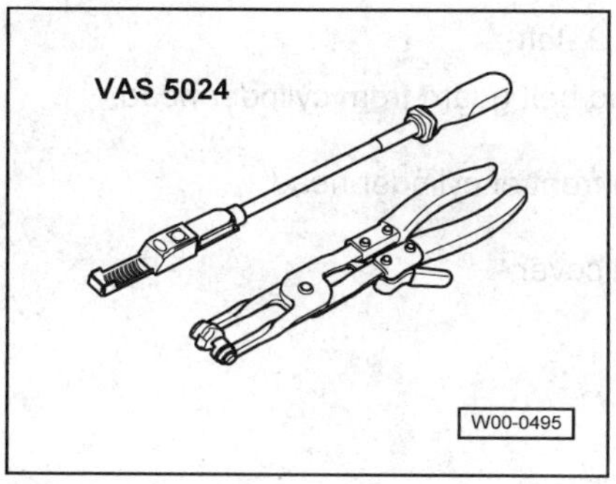

◄ VAS 5024 Assembly tool for spring-type clips

- VAG 1306 Drip tray
- VAG 1331 Torque wrench (5-50 Nm)
- VAG 1332 Torque wrench (40-200 Nm)

Prerequisites

- The engine must be no more than warm to touch.
- The pistons must not be positioned at TDC.

Removing

– First check whether a coded radio is fitted. If so, obtain anti-theft coding.

– Remove engine cover first.

– With ignition switched off disconnect battery Ground strap.

– Removing front noise insulation, see **Body–Exterior section, Repair Group 50**

– Drain coolant, see **Repair Group 19.**

– Remove air duct pipe, air duct elbow between air mass meter and throttle valve control module.

– Loosen or separate the following components:

- accelerator cable
- wiring/hoses from EVAP canister purge regulator solenoid valve 1
- hoses for crankcase breather on cylinder head cover
- connecting pipe between secondary air pump motor and combi valves at the combi valves (engine code AHA only)
- connector from twin path intake manifold change-over valve and from intake air temperature sender
- connector from secondary air inlet valve (engine code AHA only)
- connector from throttle valve control module
- connector from ignition coil output stage
- connector from valves for camshaft adjustment and camshaft position sensors
- connector from sensor for coolant temperature
- the vacuum hose from fuel pressure regulator

– Remove retaining plate for twin path intake manifold change-over valve.

– On AHA engines, remove retaining plate for twin path intake manifold change-over valve and secondary air inlet valve.

– Remove toothed belt, see **Repair Group 13**.

– Remove front exhaust pipes from exhaust manifold, see **Repair Group 26**.

– Remove the fuel rail as follows:

- Pull connector off injectors.
- Unbolt fuel rail from intake manifold and pull out together with injectors.
- Place fuel rail to side (fuel hoses remain connected).

– Remove bracket with ignition coils.

– Remove intake manifold.

– Remove rear coolant pipe.

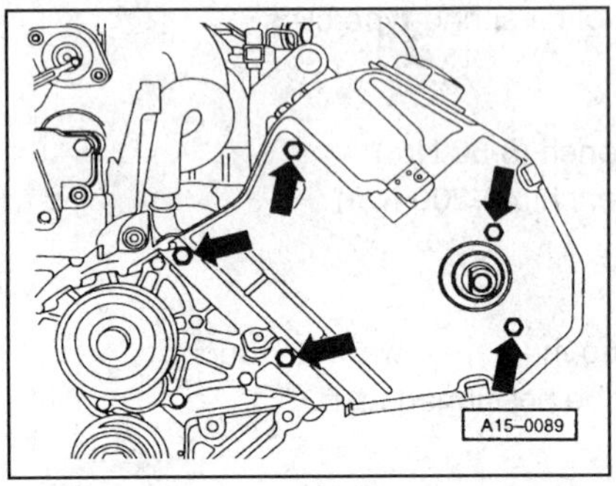

◀ By cylinder head bank 2 -left-:

– Disconnect rear toothed belt guard from cylinder head.

– Unbolt coolant pipe on front of cylinder head.

– Remove cylinder head cover.

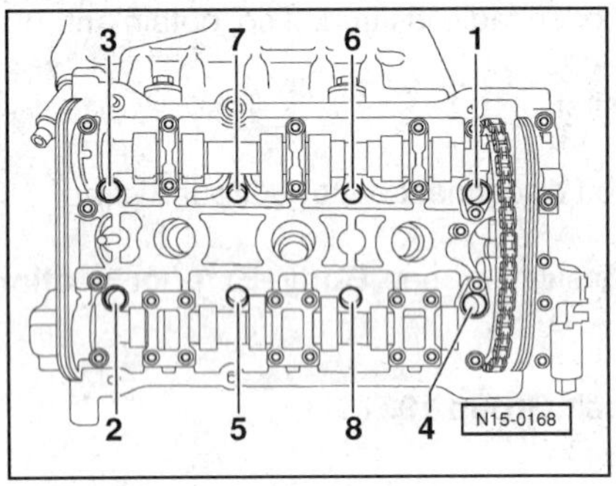

◀ Loosen socket head bolts in sequence given and remove completely.

NOTE—
Use key 3452 for Polydrive cylinder head bolts.

– Carefully lift cylinder head off.

Cylinder head, installing

NOTE—
- *Only remove the new cylinder head gasket from its packing immediately before installing.*
- *Handle the new gasket with extreme care. Damaging will lead to leaks.*

– Place a clean cloth in cylinders so that no dirt or emery cloth particles can get in between cylinder wall and piston.

– Also prevent dirt and emery cloth particles from getting into coolant.

– Carefully clean cylinder head and cylinder block sealing surfaces, thereby ensuring that no scoring or scratches are formed (when using abrasive paper, grade must not be less than 100).

– Clean pockets for cylinder head bolts of oil or coolant (blow out carefully with compressed air).

– Carefully remove metal particles, emery remains and cloths.

– Set No. 3 cylinder piston to top dead center and then turn crankshaft back slightly.

– Fit new cylinder head gasket. Lettering (Part No.) must be readable.

– Place cylinder head on block. When doing this observe centralizing pins in cylinder block.

– Fit new cylinder head bolts and tighten hand tight.

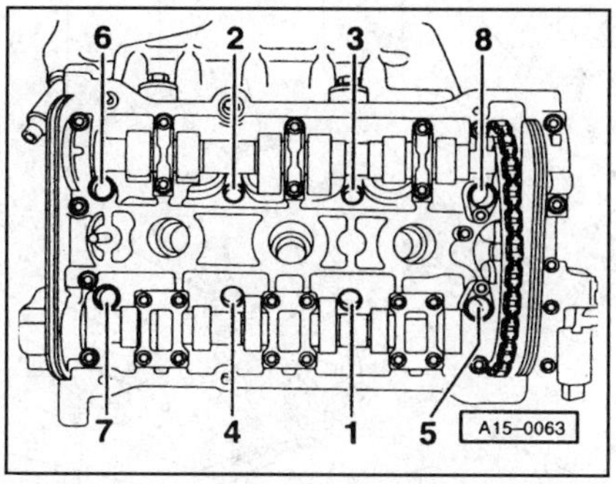

◄ Tighten cylinder head in tightening sequence as follows:

- Tighten all bolts to 60 Nm (44 ft-lb).
- Then tighten all bolts $^{1}/_{4}$ turn (90°) further using a rigid wrench.
- Then tighten all bolts again $^{1}/_{4}$ turn (90°) further.

The rest of the assembly is basically a reverse of the disassembly sequence.

- Tightening torque for front exhaust pipe to exhaust manifold: 25 Nm (18 ft-lb)
- Tightening torque for intake manifold securing bolts: 10 Nm (7 ft-lb)

Installing toothed belt and adjusting timing, see **Repair Group 13.**

– Fill with fresh coolant, see **Repair Group 19.**

Compression pressures, checking

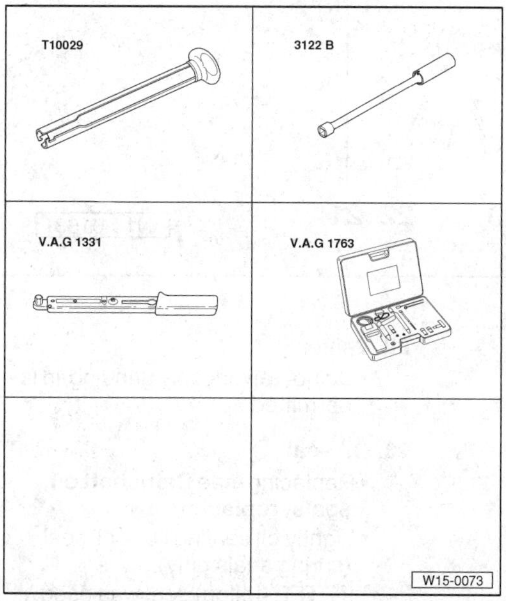

Special tools and equipment

- 3122 B Spark plug spanner
- VAG 1331 Torque wrench (5-50 Nm)
- VAG 1763 Compression tester

Test conditions

- Engine oil temperature must be at least 30°C

Work sequence

– Remove engine cover.

– Disconnect spark plug connector and remove spark plugs with spark plug spanner 3122B.

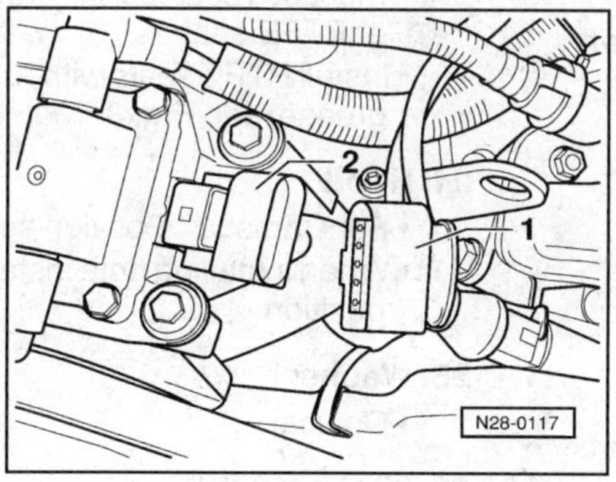

◄ Pull 5 pin connector -1- off ignition coil final stage -2-.

– Remove fuse 28 (fuel pump) or pull connectors off all injectors.

– Have a second technician fully depress accelerator pedal.

– Check compression with compression tester VAG1381/ VAG1763, or equivalent. Operate starter until tester shows no further pressure increase.

Compression pressure:

- New: 9-14 bar (131-203 psi)
- Wear limit: 7.5 bar (109 psi)

Permissible difference between all cylinders: 3 bar (45 psi)

Valve gear, servicing

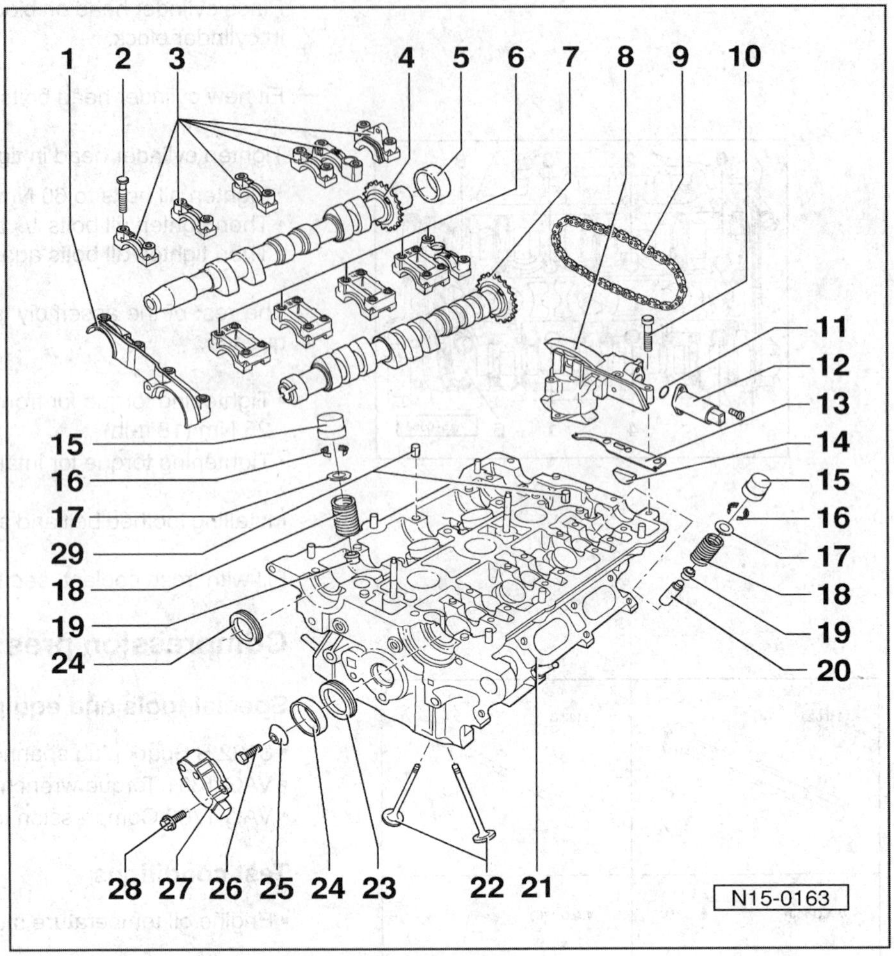

1. **Double bearing cap**
   - Lightly coat bearing cap contact surface with AMV 174 004 01 sealant
2. **10 Nm (7 ft-lb)**
3. **Exhaust camshaft bearing cap**
   - Installation position and installation sequence, see **Camshaft, removing and installing**
   - Lightly coat last (outer) bearing cap contact surface with sealant AMV 174 004 01
4. **Exhaust camshaft**
   - Checking axial clearance, see **Fig. 2**
   - See **Camshaft, removing and installing**
   - Check radial clearance with plastigage:
     Wear limit: 0.1 mm
     Run-out: max. 0.01 mm
5. **Cap**
   - Always replace
   - Take off bearing cap to remove
   - With bearing cap installed drive in carefully using 3202 and synthetic head hammer
6. **Inlet camshaft bearing cap**
   - Installation position and sequence, see **Camshaft, removing and installing**
7. **Inlet camshaft**
   - Check axial clearance, see **Fig. 2**
   - See **Camshaft, removing and installing**
   - Check radial clearance with plastigage
     Wear limit: 0.1 mm
     Run-out: max. 0.01 mm
8. **Camshaft adjuster**
   - Secure with 3366 before removing
9. **Drive chain**
   - Before removing mark direction of rotation (installation position)
10. **10 Nm (7 ft-lb)**
11. **O-ring**
    - Always replace
12. **Camshaft adjuster valve**
    - See **Camshaft adjustment, checking**
    - Resistance between terminals: 10-18 Ω
13. **Gasket**
    - Rubber/metal seal
    - Always replace
14. **Gasket**
    - Always replace
15. **Valve lifter**
    - Do not interchange
    - With hydraulic valve clearance compensation
    - See **Hydraulic valve lifters, checking**
    - Store with cam contact surface downward
    - Before installing check camshaft axial clearance, see **Fig. 2**
    - Oil contact surface
16. **Valve keepers**
17. **Upper valve spring plate**
18. **Valve spring**
19. **Valve stem seal**
20. **Valve guide**
    - See **Valve guides, checking**
    - Service version with collar
21. **Cylinder head**
    - Reworking sealing surface, see **Fig. 1**
22. **Valves**
    - Do not rework only grinding-in is permitted
23. **Oil seal**
    - Replacing, see **Camshaft oil seals, replacing**
    - Lightly oil sealing lip of oil seal (spring seals only)
    - If PTFE (teflon) seal was used, it must be replaced only with another PTFE seal
    - Install PTFE seals without oil or grease, see **Fig. 3**
24. **Hood**
    - For Camshaft Position sensor
    - When installing note installation position
25. **Washer**
    - Conical
26. **25 Nm (18 ft-lb)**
27. **Camshaft position sensor**
28. **10 Nm (7 ft-lb)**
29. **Guide sleeve**

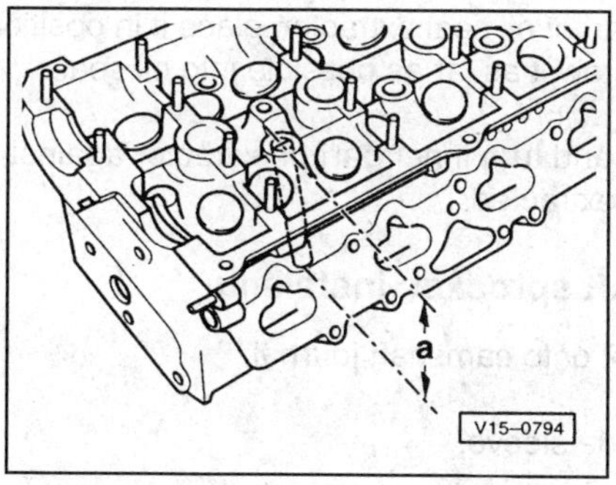

Fig. 1 Reworking cylinder head sealing surface

The dimension for reworking is measured through the holes for the cylinder head bolts

- a = minimum 139.2 mm

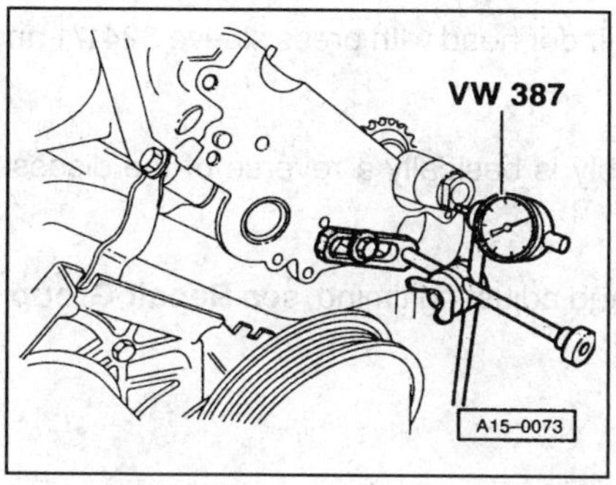

Fig. 2 Checking camshaft axial clearance

Special tools and equipment

- VW 387 Universal dial gauge bracket
- Dial gauge

Check with valve lifters and chain removed and No. 2 and 4 bearing caps fitted.

- Wear limit: max. 0.2 mm

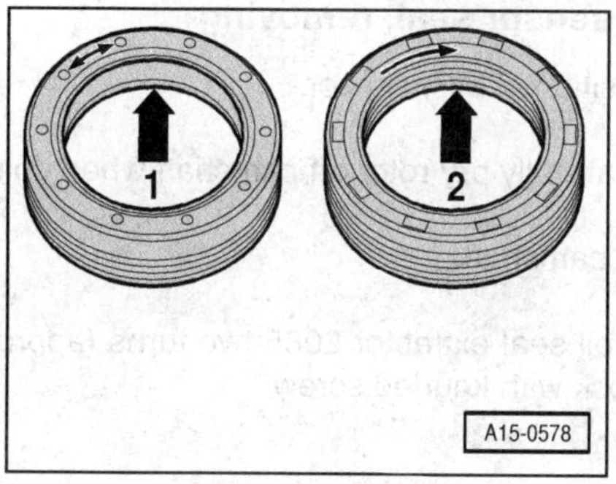

Fig. 3 Removing and installing PTFE seals for camshafts

Instead of the spring seal -1-, a PTFE (teflon) seal -2- is used on some engines. The PTFE seal has a wider sealing lip.

If a PTFE seal was used, it must be replaced only with another PTFE seal. Install PTFE seals without oil or grease.

Camshaft oil seals, replacing

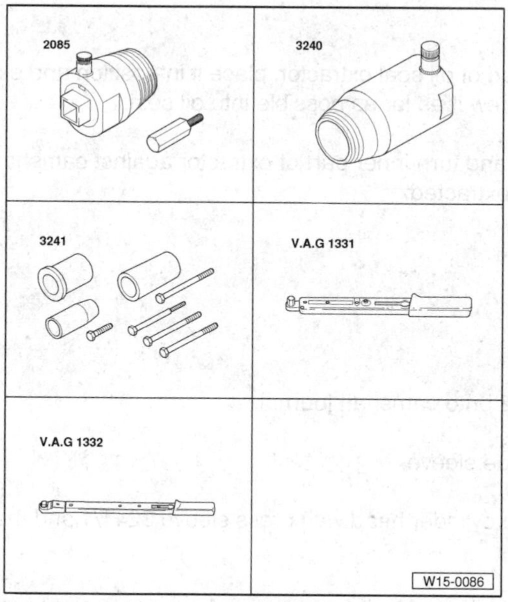

Special tools and equipment

- 2085 Oil seal extractor
- 3240 Oil seal extractor
- 3241 Fitting sleeves
- VAG 1331 Torque wrench (5-50 Nm)
- VAG 1332 Torque wrench (40-200 Nm)

Oil seal on camshaft sprocket, removing

– Remove toothed belts, see **Repair Group 13.**

– Remove camshaft sprockets.

– Unscrew inner part of oil seal extractor 3240 two turns (approx. 3 mm) out of outer part and lock in position with knurled screw.

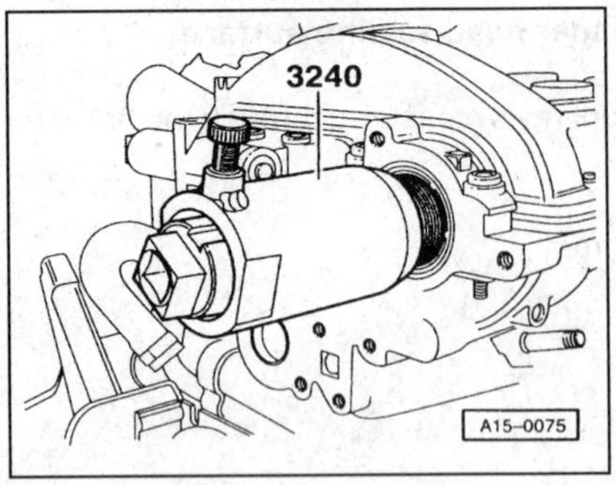

- Lubricate threaded head of oil seal extractor, place it in position and exerting firm pressure screw it as far as possible into oil seal.
- Loosen knurled screw and turn inner part of extractor against camshaft until oil seal has been extracted.

Oil seal on camshaft sprocket, installing

- Fit guide sleeve 3241/2 onto camshaft journal.
- Slide dry seal over guide sleeve.

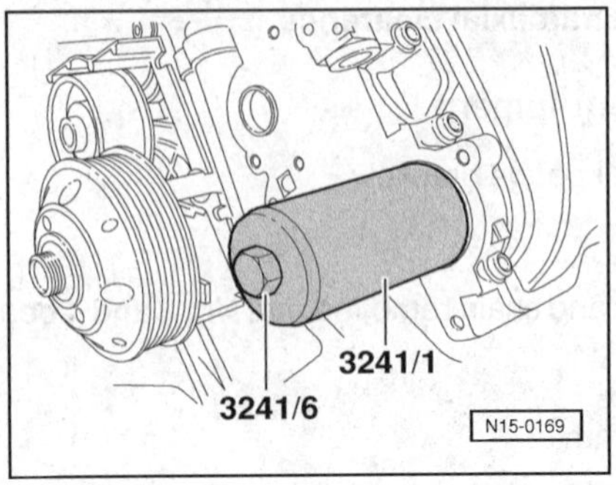

- Press seal flush into cylinder head with press sleeve 3241/1 and the bolt 3241/6.

The rest of the assembly is basically a reverse of the disassembly sequence.

Installing toothed belt and adjusting timing, see **Repair Group 13**.

Camshaft position sensor seal, removing

- Unscrew Camshaft Position (CMP) sensor.
- Unscrew CMP rotor. Carefully pry rotor off camshaft when doing this.

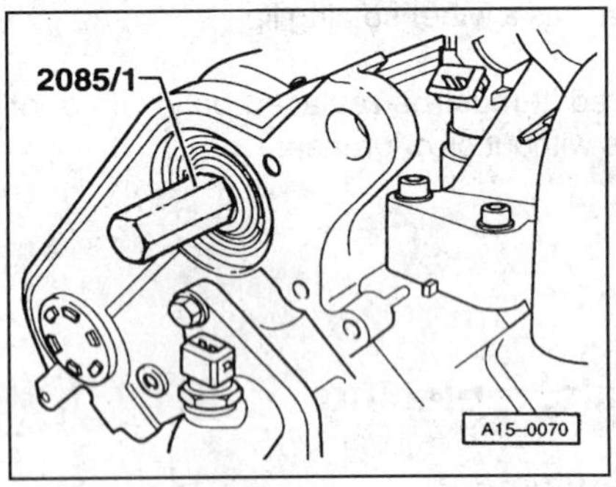

- Screw bolt 2085/1 into camshaft.
- Unscrew inner part of oil seal extractor 2085 two turns (approx. 3 mm) out of outer part and lock with knurled screw.

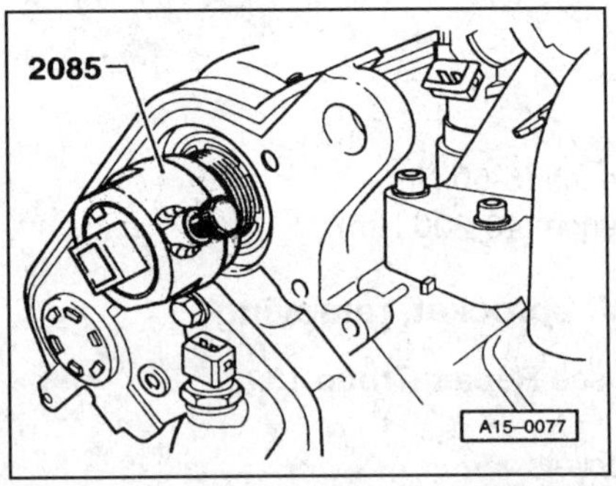

- Lubricate threaded head of oil seal extractor, place it in position and exerting firm pressure screw it as far as possible into oil seal.
- Loosen knurled screw and turn inner part of extractor against camshaft until oil seal has been extracted.

Installing

- Fit guide sleeve 3241/2 onto camshaft journal.
- Slide dry seal over guide sleeve.

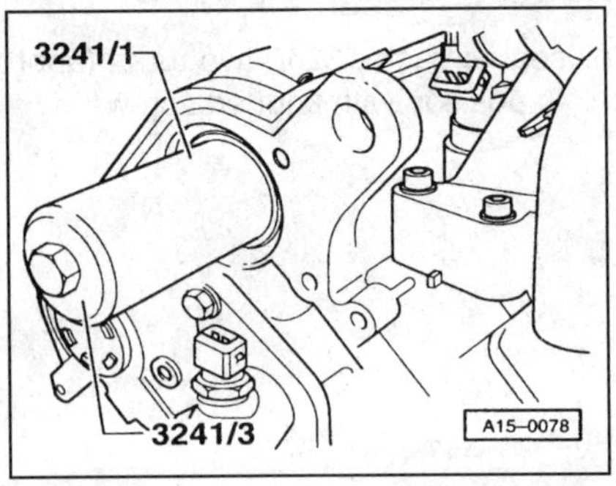

- Press oil seal flush into cylinder head with press sleeve 3241/1 and the bolt 3241/3.

The rest of the assembly is basically a reverse of the disassembly sequence.

Hydraulic valve lifters, checking

Special tools and equipment

- Feeler gauge
- Wood or plastic wedge

NOTE —

- *Replace malfunctioning lifter complete (cannot be adjusted or repaired).*
- *Irregular valve noises when starting engine are normal.*

Test sequence

– Start engine and run until radiator fan has switched on once.

– Increase engine speed to about 2500 rpm for approx. 2 minutes.

If the hydraulic lifters are still noisy, locate malfunctioning lifters as follows:

– Remove cylinder head cover.

– Rotate crankshaft clockwise, until cam of lifter to be checked is pointing upward.

– Determine play between cam and valve lifter.

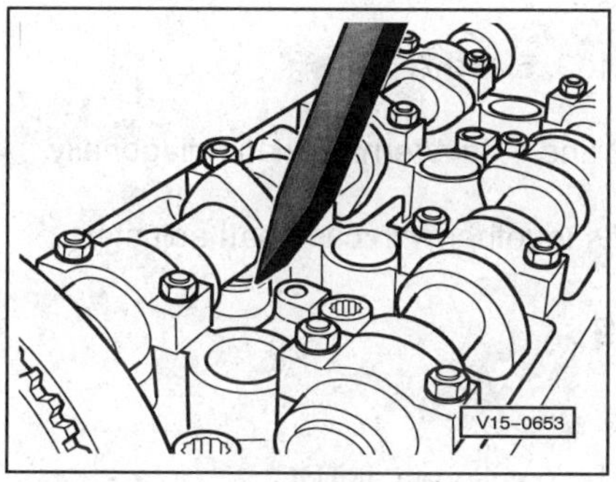

◄ Push valve lifter down lightly with a wooden or plastic wedge. If when valve lifter is pushed down a 0.2 mm thick feeler gauge can be pushed between camshaft and valve lifter, valve lifter is to be replaced.

NOTE —

- *After installing new lifters the engine must not be started for approx. 30 minutes. Hydraulic compensation elements must settle (otherwise valves will strike pistons).*
- *The oil check valve must be replaced if irregular valve noises disappear after longer journeys but reappear after short journeys, see* **Repair Group 17**.

Camshaft, removing and installing

Special tools and equipment

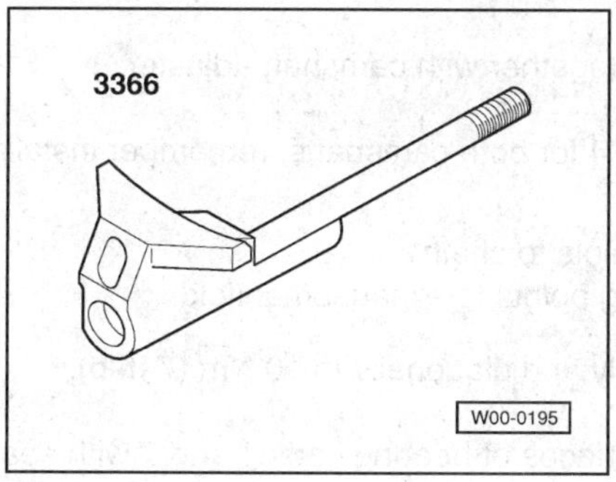

- 3366 Retainer
- ◄ VAG 1331 Torque wrench (5-50 Nm)
- VAG 1332 Torque wrench (40-200 Nm)
- Sealing compound AMV 174 004 01

Removing

– Remove toothed belts, see **Repair Group 13.**

– Turn crankshaft back slightly.

– Take off camshaft sprocket.

– Remove rear toothed belt guard.

– Remove cylinder head cover.

– Remove CMP sensor and rotor.

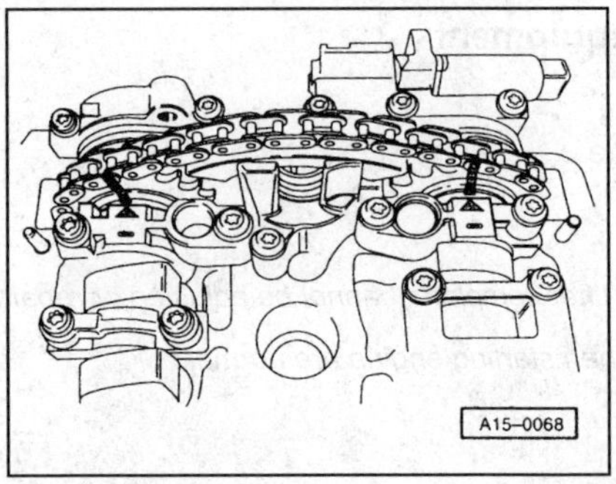

Clean drive chain and chain sprocket before removing and mark (paint etc.) opposite arrows on bearing caps. Distance equates to 16 rollers on chain.

NOTE—
Do not mark chain with a center punch or any similar means!

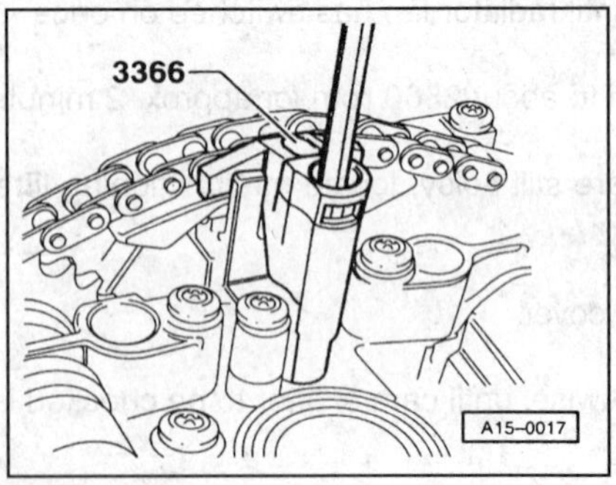

Secure camshaft adjuster now with retainer 3366.

NOTE—
Do not overtighten retainer 3366 otherwise the camshaft adjuster will be damaged.

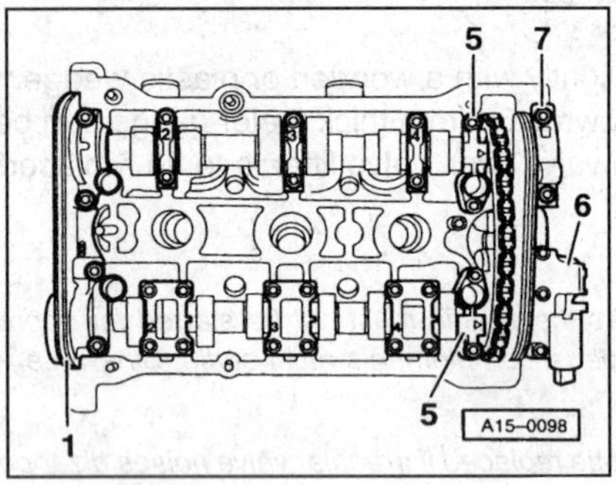

Remove camshaft adjuster -6-.

– Remove bearing caps 1, 3, 5 and 7.

– Loosen bearing caps 2 and 4 off alternately and diagonally.

– Take out both camshafts together with camshaft adjusters.

Camshaft, installing

Requirements

• The pistons must not be positioned at TDC.

– Place chain onto chain sprockets and camshaft as per markings (distance between marks equates to 16 rollers on chain).

– Slide camshaft adjuster between.

– Oil camshaft contact surfaces.

– Insert both camshafts together with camshaft adjuster.

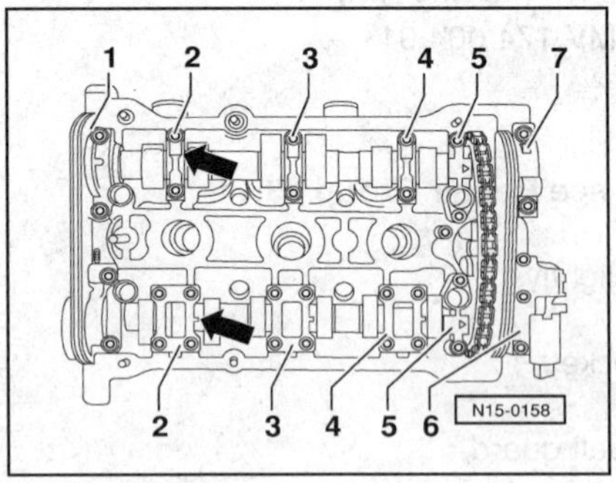

Fit bearing caps 2 and 4 for both camshafts, remember installation position -arrows-.

• Inlet shaft: casting points to chain
• Exhaust shaft: casting points to exhaust manifold

– Tighten caps alternately and diagonally to 10 Nm (7 ft-lb).

– Lightly coat contact surfaces of bearing caps 1 and 7 with sealant AMV 174 004 01.

– Install bearing caps 1, 3, 5 and 7 and tighten to 10 Nm (7 ft-lb).

– Remove retainer for camshaft adjuster 3366.

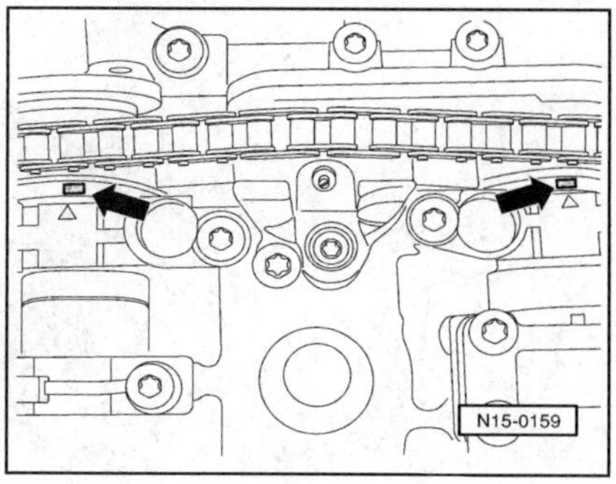

Check position of camshafts to one another. Arrows on bearing caps must align with cut-outs on camshafts -arrows-. To do this it may be necessary to move a camshaft onto markings.

The rest of the assembly is basically a reverse of the disassembly sequence.

Installing toothed belt and adjusting timing, see **Repair Group 13.**

NOTE —

When new lifters have been installed the engine must not be started for about 30 minutes. Hydraulic compensation elements must settle (otherwise valves will strike pistons).

Sealing cap (core plug), driving into replacement cylinder head

Special tools and equipment

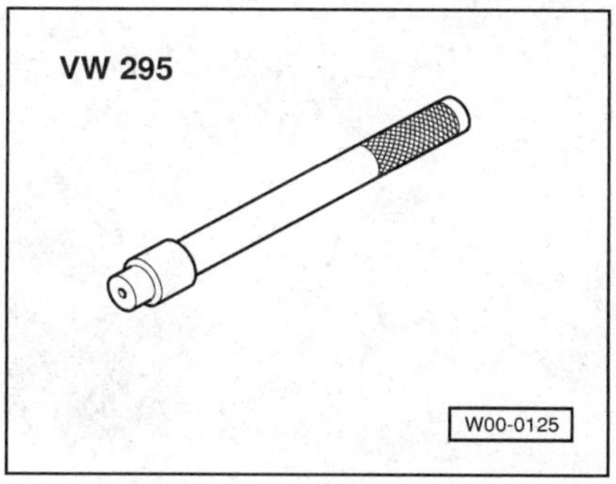

VW 295 Drift

The cylinder head supplied as a spare part can be used on the left or right-hand sides. But a sealing cap (core plug) must be fitted to the relevant front side.

– Coat edge of sealing cap (core plug) with sealant AMV 188 001 02.

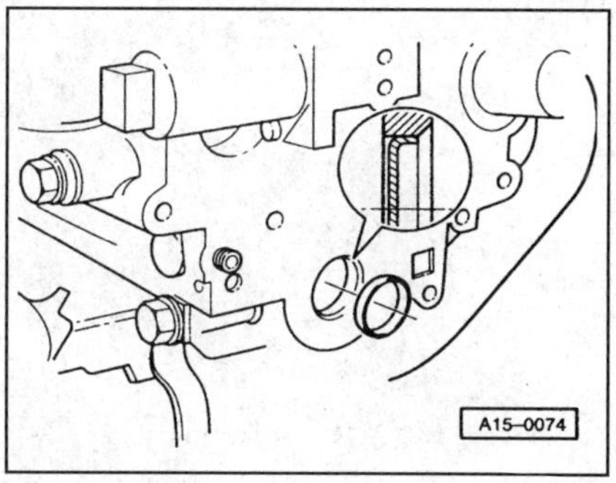

Drive sealing cap (core plug) in with drift VW 295 until outer edge of cap is flush with chamfer on cylinder head.

Valve guides, checking

Special tools and equipment

- VW 387 Universal dial gauge bracket
- Dial gauge

Test sequence

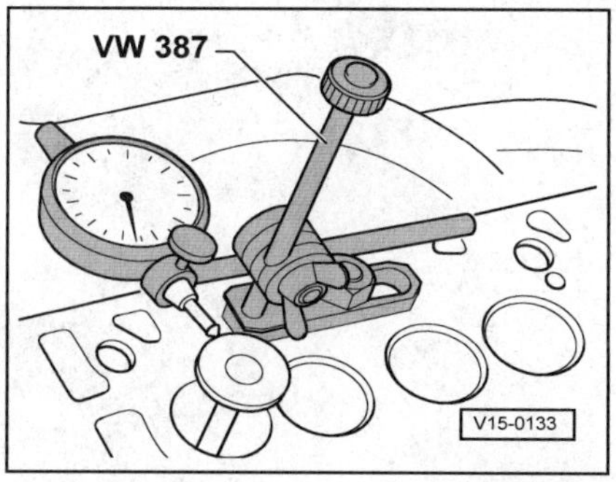

Insert new valve into guide until end of valve stem is flush with end of guide. Due to slight difference in stem dimensions, ensure that only an inlet valve is used in inlet guide and an exhaust valve in exhaust guide.

– Determine rock.

Wear limit: 0.8 mm

17 Engine–Lubrication System

Repair Group

D 17

Lubrication System

NOTE—

- *If large quantities of small metal particles (caused, for example, by partial seizure of crankshaft and connecting rod bearings) are found in the engine oil, the oil passages must be thoroughly cleaned and the oil cooler replaced to prevent further damage.*
- *The oil level must not be above the max. mark - danger of damaging catalyst!*

EDITOR'S NOTE—

For oil system capacities and oil change information, see **Maintenance**.

Engine oils are graded according to their viscosity. The proper grade to be used depends on existing climatic or seasonal conditions.

The original oil in the engine has a viscosity grade of SAE 5W-40. Use this oil over all temperature ranges for normal driving. If engine oil viscosity grade SAE 5W-40 is not available, use SAE 5W-30.

Oil pressure and oil pressure switch, checking

Special tools and equipment

- VAG 1342 Oil pressure tester
- VAG 1527 B Diode test lamp
- VAG 1594 A Adapter set

EDITOR'S NOTE—

See **Electical Wiring Diagrams and Component Locations** *for schematics to check the oil pressure warning light and buzzer.*

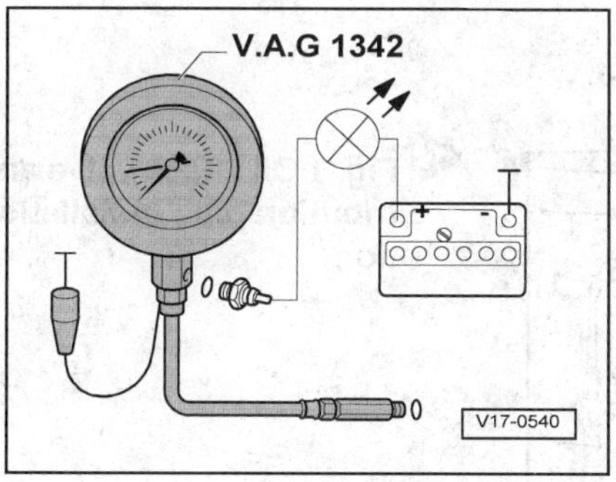

◀ Remove oil pressure switch (F1) and screw into tester.

– Screw tester into oil pan in place of oil pressure switch.

– Connect brown wire of tester to Ground (-).

– Connect diode test lamp VAG 1527 using auxiliary cables from VAG 1594 to battery positive (+) and oil pressure switch.

– Start engine and increase speed slowly. At 1.2-1.6 bar (17-23 psi) LED must light up, otherwise replace oil pressure switch.

– Increase engine speed further. At 2000 rpm and at an oil temperature of 80°C (176°F) oil pressure should be at least 2.0 bar (29 psi).

Oil pump, crankcase breather and oil spray jets, removing and installing

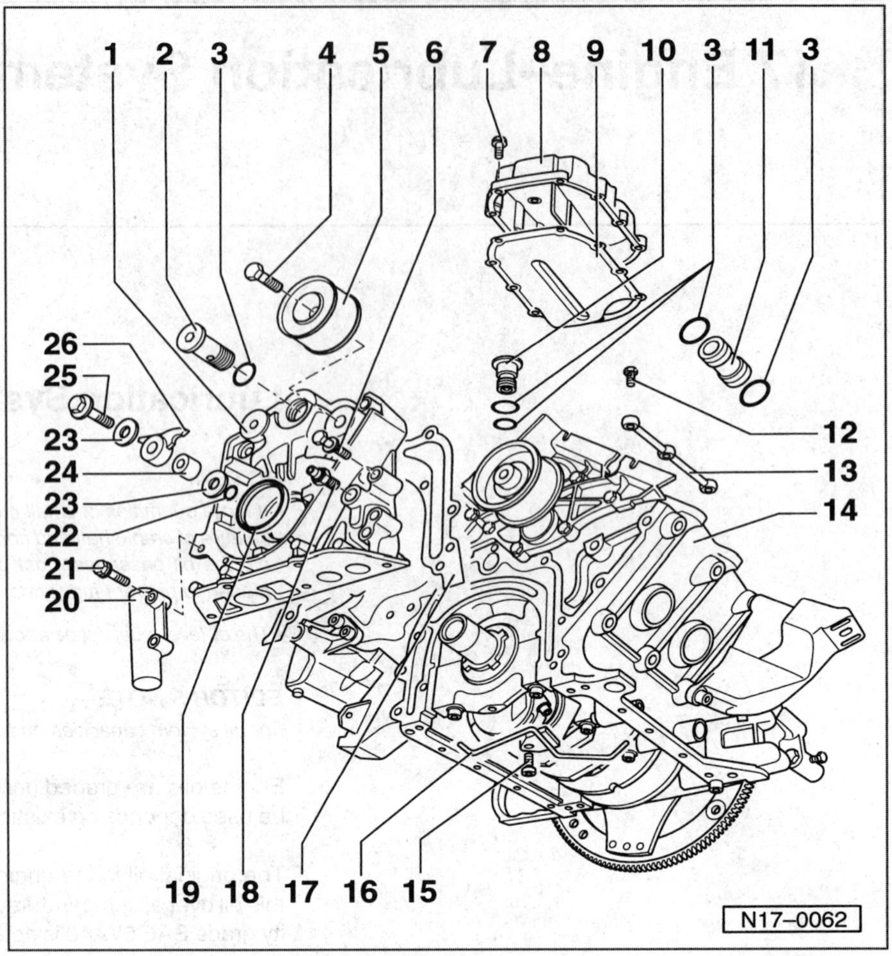

1. **Oil pump**
   - Replace as complete assembly
2. **Spray jet valve**
   - 40 Nm (30 ft-lb)
3. **O-ring**
   - Always replace
4. **45 Nm (33 ft-lb)**
5. **Idler roller**
6. **10 Nm (7 ft-lb)**
7. **10 Nm (7 ft-lb)**
8. **Breather housing**
9. **Gasket/breather plate**
   - Always replace
10. **Oil check valve,**
    - 25 Nm (18 ft-lb)
    - Location, see **Fig. 1**
11. **Connector**
12. **15 Nm (11 ft-lb)**
13. **Oil distribution pipe**
    - For oil spray jets for piston cooling
    - Location, see **Fig. 1**
14. **Cylinder block**
15. **10 Nm (7 ft-lb)**
    - Insert with locking fluid "D6"
16. **Oil spray jet**
    - For piston cooling
17. **Gasket**
    - Metal
    - Always replace
18. **Oil temperature sender (G8)**
    - 10 Nm (7 ft-lb)
    - For oil temperature gauge
    - White
    - If sealing ring is leaking nip open and replace
19. **Oil seal**
    - If a PTFE (teflon) seal was used, it must be replaced only with another PTFE seal
    - Install PTFE seals without oil or grease
20. **Tensioning element**
    - Secure before removing toothed belt, see **Repair Group 13**
21. **10 Nm (7 ft-lb)**
    - Insert with locking fluid "D6"
22. **O-ring**
23. **Washer**
24. **Bearing bushing**
25. **25 Nm (18 ft-lb)**
26. **Tensioning lever**

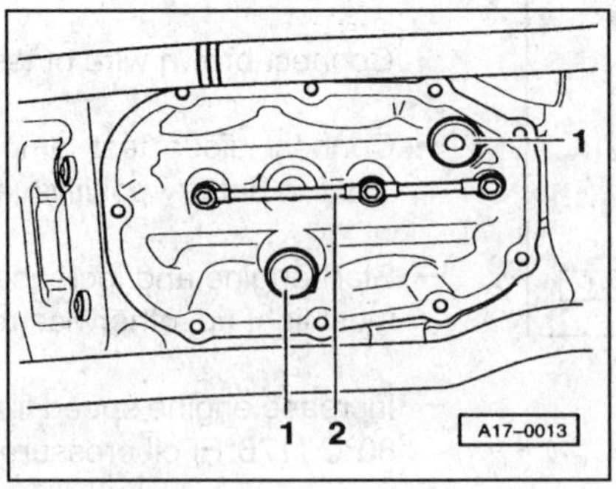

Fig. 1 Oil check valve and oil distribution pipe installation position

Oil pan, oil filter and oil cooler, removing and installing

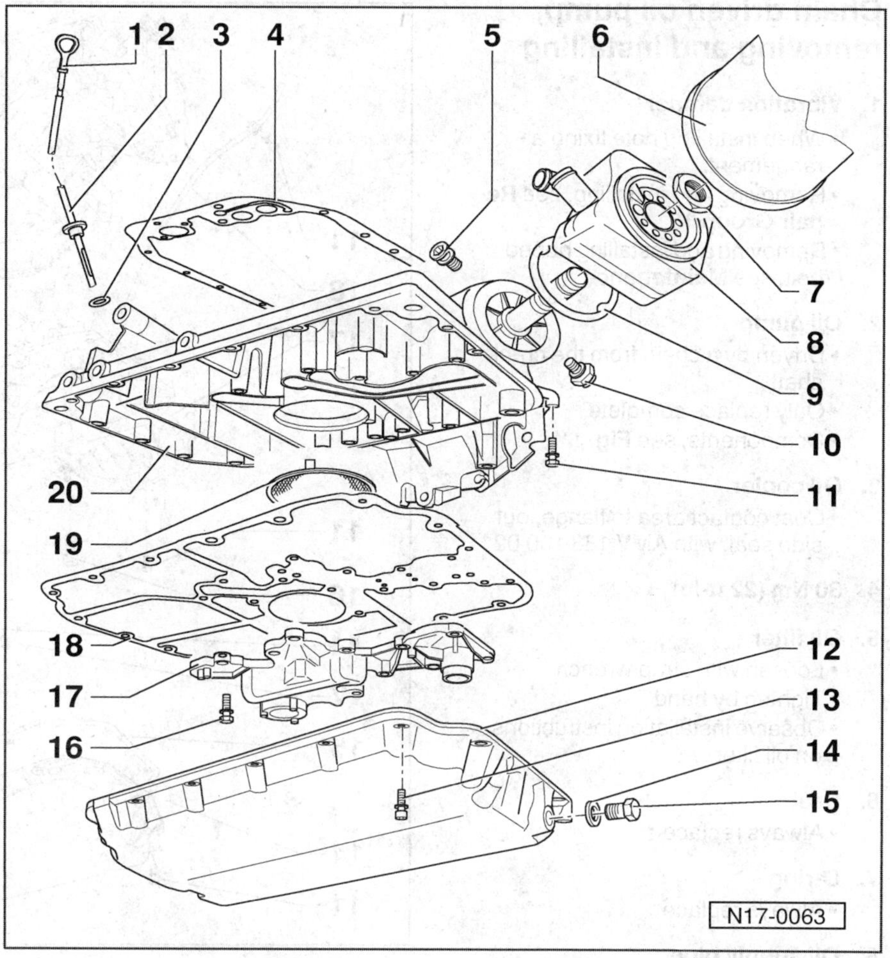

1. **Dipstick**
   - The oil level must not be above the max. mark!
   - Marking, see **Fig. 1**
2. **Guide tube**
3. **O-ring**
   - Always replace
4. **Gasket**
   - Always replace
   - Before fitting coat cylinder block/sealing flange or oil pump/cylinder block with "D2"
5. **Screw plug**
   - 25 Nm (18 ft-lb)
   - If sealing ring is leaking nip open and replace.
6. **Oil filter**
   - Loosen with strap wrench
   - Tighten by hand
   - Observe installation instructions on oil filter
7. **30 Nm (22 ft-lb)**
8. **Oil cooler**
   - Coat contact area to flange, outside seal, with AMV 188 100 02
9. **O-ring**
   - Always replace
10. **1.4 bar oil pressure switch (F1)**
    - 25 Nm (18 ft-lb)
    - White paint mark on housing
    - Checking, see **Oil pressure and oil pressure switch, checking**
    - If sealing ring is leaking nip open and replace
11. **10 Nm (7 ft-lb)**
12. **Oil pan - lower part**
13. **10 Nm (7 ft-lb)**
14. **Seal**
    - Always replace
15. **Oil drain plug**
    - 30 Nm (22 ft-lb)
    - Replace if leaking. See factory parts information for correct application. In some cases, seal and drain plug are combined; do not interchange with separate seal and drain plug.
16. **10 Nm (7 ft-lb)**
17. **Cap**
    - With excess pressure valves, opening pressure 5 and 11 bar (72 and 160 psi)
18. **Gasket**
    - Always replace
19. **Oil strainer**
    - Clean strainer if soiled
20. **Oil pan - upper part**

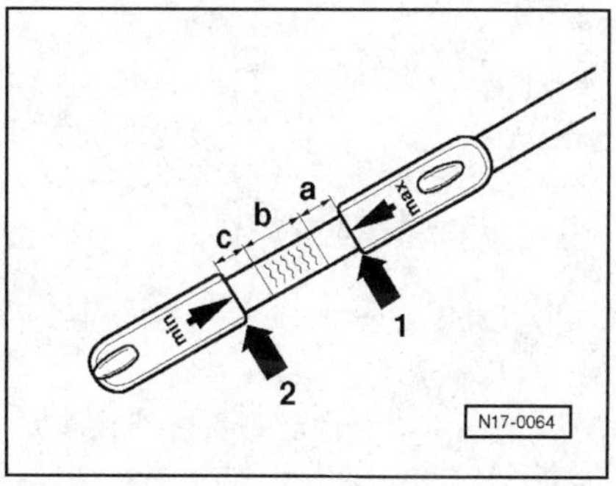

Fig. 1 Marking on oil dipstick

1. max. mark
2. min. mark

- a - Area above hatched field up to max. mark: Do not top-up with engine oil!
- b - Oil level within hatched field: can be topped-up with engine oil
- c - Area from min. mark up to hatched field: fill with max. 0.5 ltr. of engine oil!

Chain driven oil pump, removing and installing

1. **Vibration damper**
   - When installing note fixing arrangement
   - Removing and installing, see **Repair Group 13**
   - Removing and installing ribbed belt, see **Maintenance**
2. **Oil pump**
   - Driven by a chain from the crankshaft
   - Only replace complete
   - Components, see **Fig. 1**
3. **Oil cooler**
   - Coat contact area to flange, outside seal, with AMV 188 100 02
4. **30 Nm (22 ft-lb)**
5. **Oil filter**
   - Loosen with strap wrench
   - Tighten by hand
   - Observe installation instructions on oil filter
6. **Seal**
   - Always replace
7. **O-ring**
   - Always replace
8. **Oil supply pipe**
   - From oil filter to engine oil circuit
9. **Oil supply pipe**
   - From oil pump to oil filter
10. **Oil supply pipe bracket**
11. **10 Nm (7 ft-lb)**
12. **Oil drain plug,**
    - 30 Nm (22 ft-lb)
    - Replace if leaking. See Parts Catalog for correct application. In some cases, seal and drain plug are combined; do not interchange with separate seal and drain plug.
13. **Seal**
    - Always replace
14. **Oil pan - lower part**
15. **Gasket**
    - If fitted, replace
16. **25 Nm**
17. **Oil pan - upper part**
18. **Guide tube**

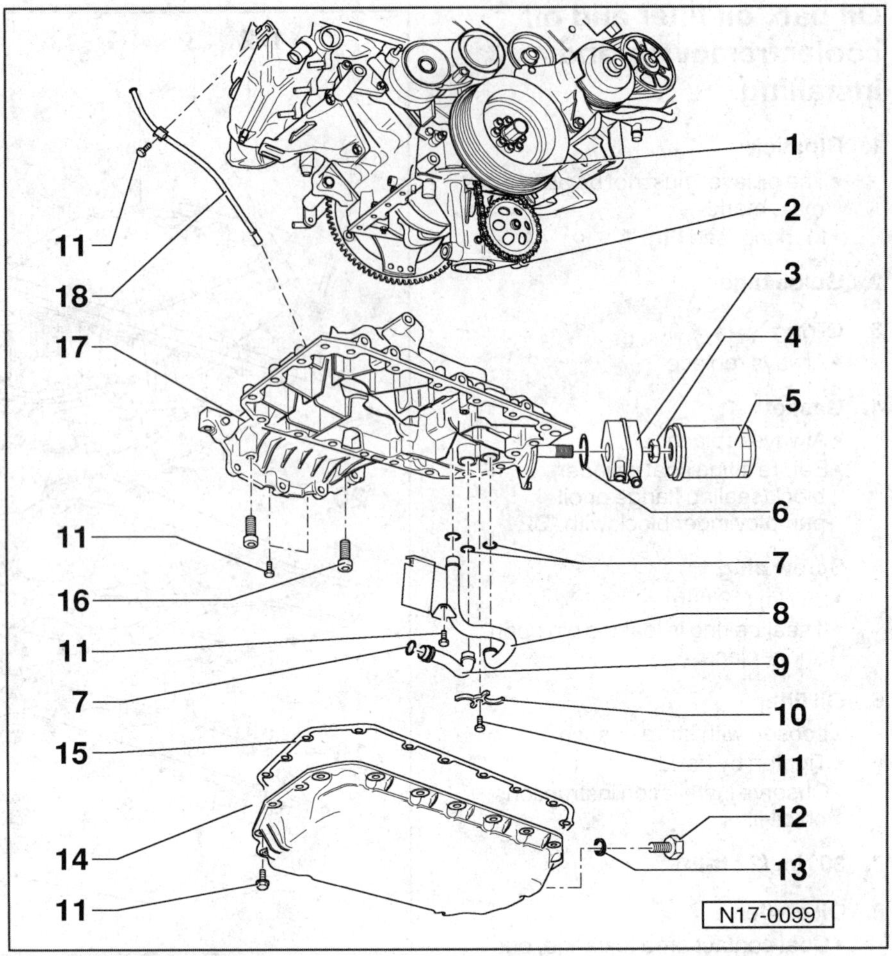

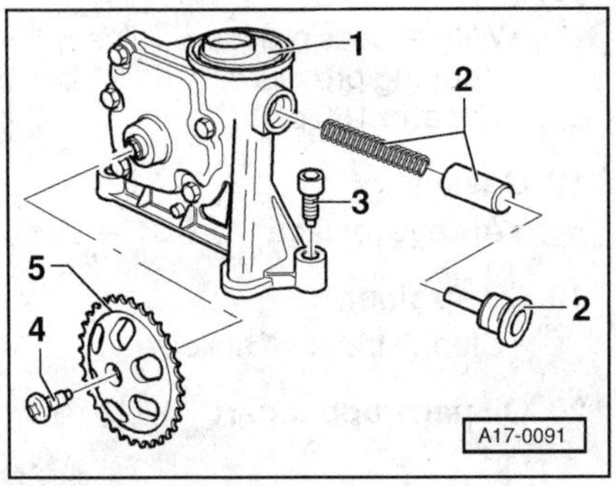

Fig. 1 Oil pump components

1. Oil pump, cannot be disassembled
2. Pressure regulation valve, 40 Nm (30 ft-lb)
3. 20 Nm (15 ft-lb)
4. 25 Nm (18 ft-lb)
5. Chain sprocket

Oil pump chain tensioner, checking and oil pump, removing

Special tools and equipment

- Hand drill with plastic brush
- Silicone sealant D 176 404 A2
- Flat scraper

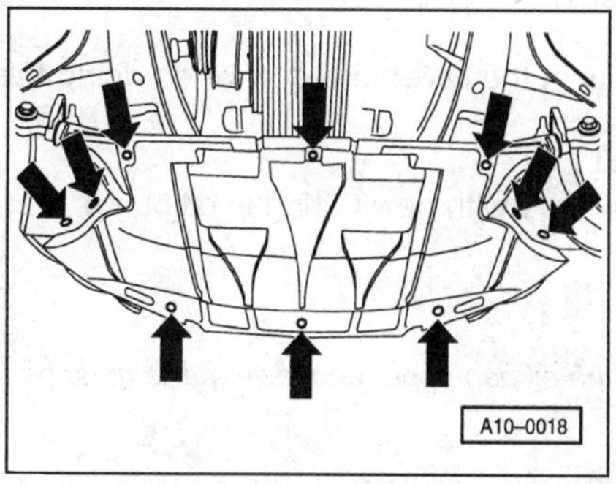

Removing

◄ Remove noise insulation (arrows).

– Drain engine oil.

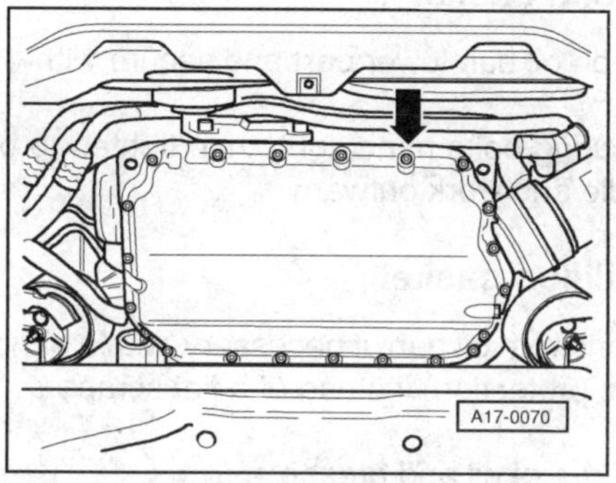

◄ Remove oil pan lower part (arrows).

Checking chain tensioner

◄ Insert a screwdriver between chain and chain tensioner and pry carefully against chain tensioner, in direction of arrow.

– If there is no spring tension or drive chain is slack, chain tensioner is malfunctioning and must be replaced.

NOTE —

To replace the chain tensioner, the toothed belt and then the front sealing flange must be removed, see **Repair Group 13**.

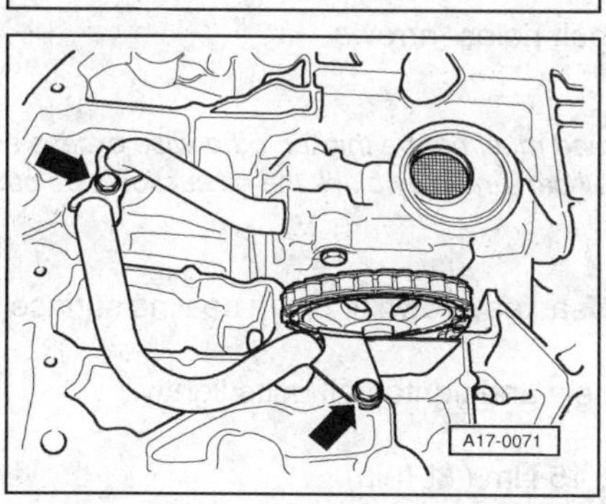

Oil pump, removing

◄ Remove both brackets (arrows) for oil supply pipe.

– Pull longer oil supply pipe out of oil pan upper part.

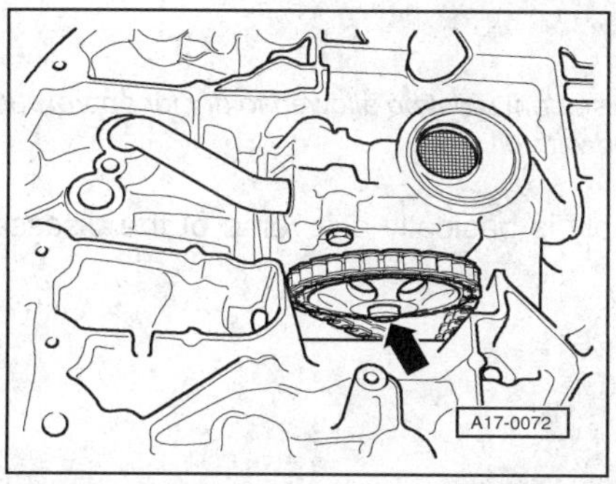

◄ Unbolt chain sprocket from oil pump (arrow) and pull down off oil pump drive shaft.

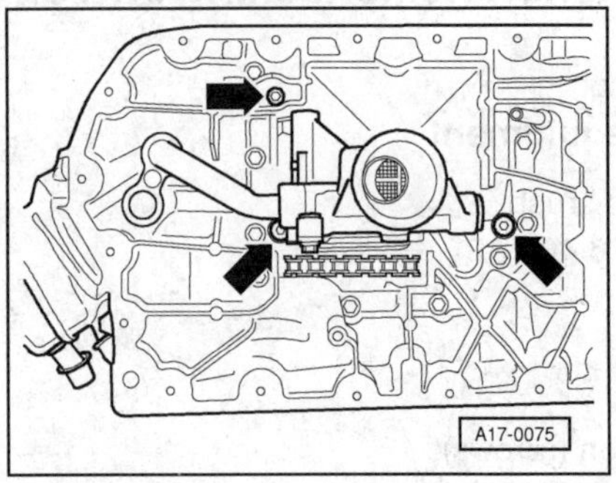

Remove oil pump together with shorter oil supply pipe.

Oil pump, installing

Installation is carried out in the reverse order, when doing this note the following:

– Insert shorter oil supply pipe with new O-ring in oil pump, then install oil pump.

NOTE —
The sealing surfaces on the oil pan upper and lower parts must be free of oil and grease.

For oil pans with solid gasket

– Position a new gasket on oil pan lower part and secure with 4 bolts.

– Tighten all oil pan securing bolts hand tight, then tighten to prescribed torque starting at middle and work outward.

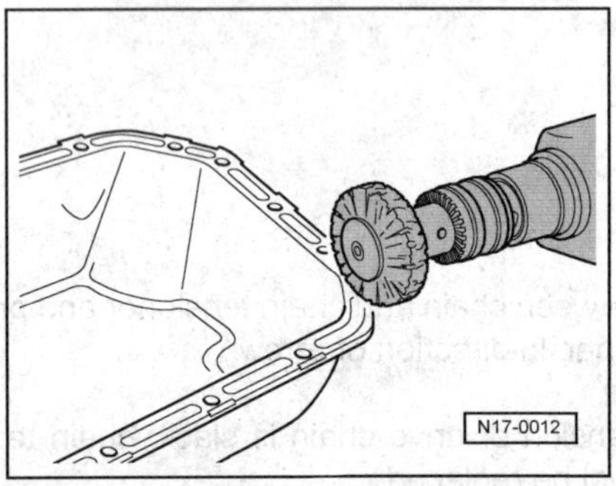

For oil pans with silicon gasket

Remove residual sealant from oil pan, if necessary, with a hand drill and plastic brush (wear eye protection) instead of a flat scraper.

– Clean sealing surface free of oil and grease.

NOTE —
- *Note the use by date of the sealant.*
- *The oil pan must be installed within 5 minutes of applying silicone sealing compound.*

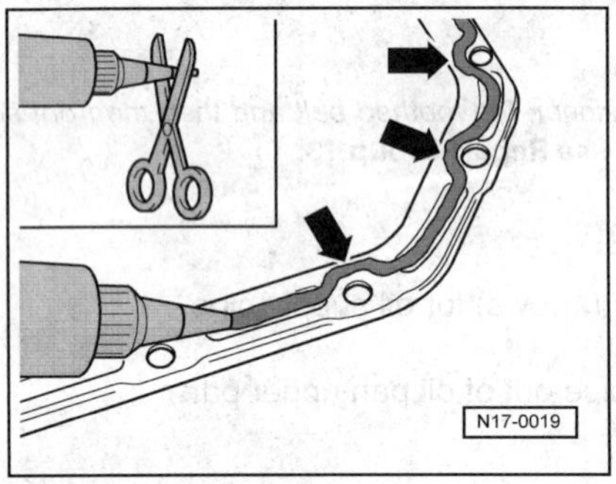

Cut-off tube nozzle at forward marking (nozzle approx. 3 mm dia.).

– Apply silicone sealing compound, as shown, to clean oil pan sealing surface. Sealing compound bead must be:

- 2 - 3 mm thick
  and run on inside of bolt holes -arrows-

NOTE —
The sealing compound bead must not be thicker, otherwise excess sealing compound will enter the oil pan and may block the oil suction pipe strainer.

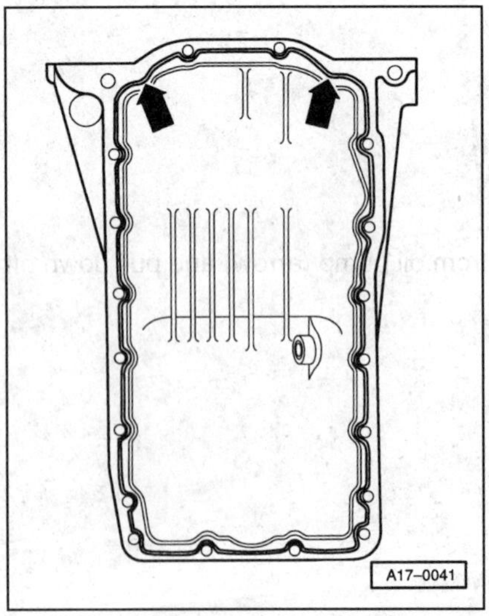

Apply silicone sealant bead as shown to clean sealing surface of oil pan.

– Install oil pan immediately and tighten all bolts lightly.

– Tighten oil pan bolts to 15 Nm (11 ft-lb).

NOTE —
After installing oil pan the sealant must be allowed to dry for approx. 30 minutes before engine is filled with oil.

The rest of the assembly is basically a reverse of the disassembly sequence.

19 Engine–Cooling System

Repair Group

D 19

Cooling System

NOTE —

- *When the engine is warm the cooling system is under pressure. If necessary release pressure before commencing repair work.*
- *Hoses are secured with spring-type clips. In cases of repair only use spring-type clips.*
- *Assembly tool VAS 5024 or hose clip pliers VAG 1921 are recommended for fitting spring-type clips.*
- *When installing coolant hoses route stress-free, so that they do not come into contact with other components (observe markings on coolant connection and hose).*

Perform cooling system leakage test with cooling system tester VAG 1274 and adapter 1274/9.

EDITOR'S NOTE —

Volkswagen identifies electrical components by a letter and/or a number in the electrical schematics. See **Wiring Diagrams, Fuses and Relays**. *These electrical identifiers are listed -between dashes- as an aid to electrical troubleshooting.*

Cooling system components (body side), overview

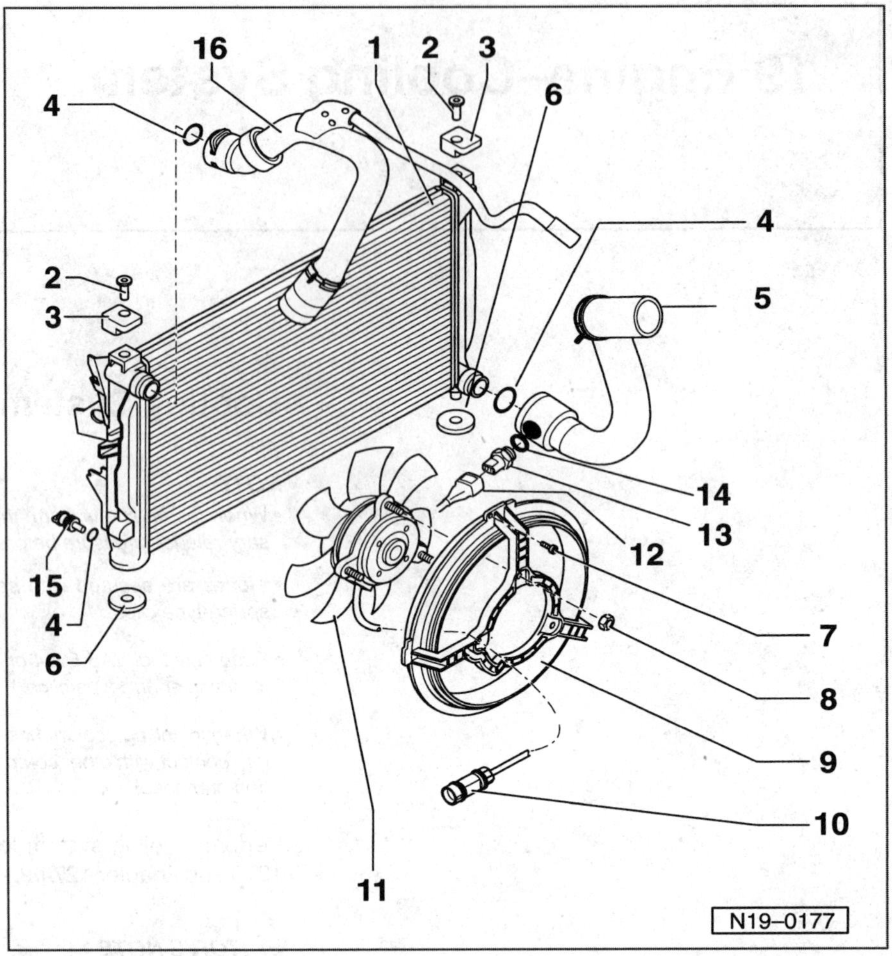

1. **Radiator**
   - See **Radiator, removing and installing**
   - After replacing, replace all coolant
   - With ATF cooler on vehicles with automatic transmission, see **5-speed Automatic Transmission 01V section, Repair Group 37**
2. **Securing clip**
3. **Securing rubber**
4. **O-ring**
   - Replace if damaged
5. **Lower coolant hose**
   - Secured to radiator with retaining clip
   - Check securely seated
   - See **Coolant hose connection diagram**
6. **Rubber washer**
7. **Locking bolt**
8. **10 Nm (7 ft-lb)**
9. **Fan ring**
   - Clipped into air ducting and secured with locking bolt -item 7-
10. **2-pin connector**
11. **Radiator fan (V7)**
    - See **Radiator fan, checking**
12. **Connector**
    - Black, 3 pin
    - For radiator fan thermo switch
13. **Radiator fan thermo switch (F18)**
    - 35 Nm (26 ft-lb)
    - For electric fan
    - Switching temperatures:
      Stage 1 on: 92-97°C (198-207°F)
      off: 84-91°C (183-196°F)
      Stage 2 on: 99-105°C (210-221°F)
      off: 91-98°C (196-208°F)
14. **Seal**
    - Always replace
15. **Drain screw, 10 Nm (7 ft-lb)**
16. **Upper coolant hose**
    - Secured to radiator with retaining clip
    - Check securely seated
    - See **Coolant hose connection diagram**

Cooling system components (engine side), overview

NOTE—

An additional electric coolant pump is installed in vehicles in hot countries, see **Continued coolant circulation pump (V51), checking.**

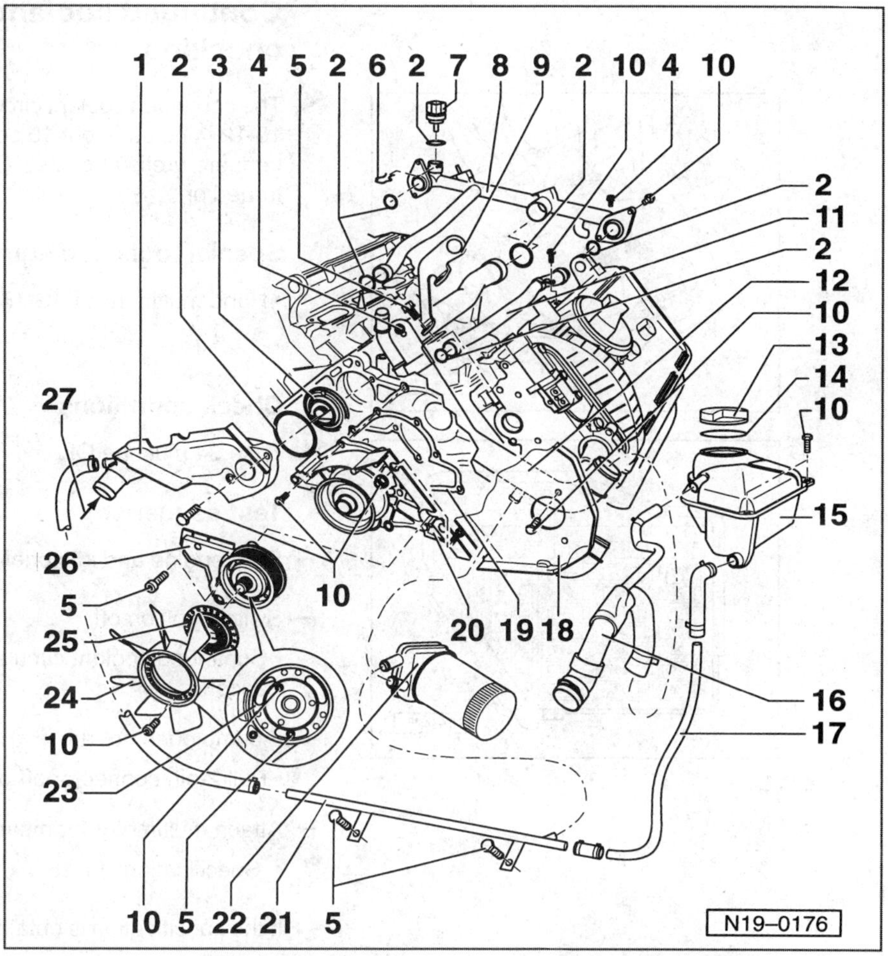

1. **Connection**
   - For thermostat
2. **O-ring**
   - Always replace
3. **Coolant thermostat**
   - Observe installation position, see **Thermostat, removing and installing**
   - Checking: heat-up thermostat in water
     Opening commences approx. 86°C (187°F)
     Opening lift min. 7 mm
4. **Breather bolt**
   - 20 Nm (15 ft-lb)
5. **25 Nm (18 ft-lb)**
6. **Retaining clip**
   - Check seated securely
7. **Coolant temperature sender (G62)**
   - For engine control module
   - Release pressure in cooling system before removing
   - Electrical testing, see **2.8 Liter Fuel Injection and Ignition section, Repair Group 28**
8. **Rear coolant hose**
   - See **Coolant hose connection diagram**
9. **Lifting eye**
10. **10 Nm (7 ft-lb)**
    - Also note -item 23-
11. **Coolant hose**
    - See **Coolant hose connection diagram**
12. **Front coolant hose**
    - See **Coolant hose connection diagram**
13. **Cap**
    - Check with cooling system tester VAG 1274 and adapter VAG 1274/9
    - Test pressure: 1.4-1.6 bar (20-23 psi)
14. **O-ring**
    - Replace if damaged
15. **Expansion tank**
    - Carry out leakage test of cooling system with cooling system tester VAG 1274 and adapter VAG1274/8
16. **Upper coolant hose**
    - Secured to radiator with retaining clip
    - Check securely seated
    - See **Coolant hose connection diagram**
17. **Coolant hose**
    - See **Coolant hose connection diagram**
18. **Rear toothed belt guard**
19. **Gasket**
    - Always replace
20. **Coolant pump**
    - Check for ease of movement
    - If damaged or leaking replace
    - See **Coolant pump, removing and installing**
21. **Lower coolant hose**
    - See **Coolant hose connection diagram**
22. **Oil cooler**
    - Removing and installing, see **Repair Group 17**
23. **Belt pulley**
    - For viscous fan coupling
    - Tightening torque for 3 securing bolts 30 Nm (22 ft-lb)
24. **Radiator fan**
25. **Viscous fan coupling**
    - Observe left-hand thread
    - Tightening torque only valid for open jaw spanner 3312
    - Tighten to 40 Nm (30 ft-lb)
    - See **Viscous fan coupling, removing and installing**
26. **Bracket**
    - For viscous fan coupling
27. **From bottom of radiator**
    - See **Coolant hose connection diagram**

Continued coolant circulation pump (V51), checking

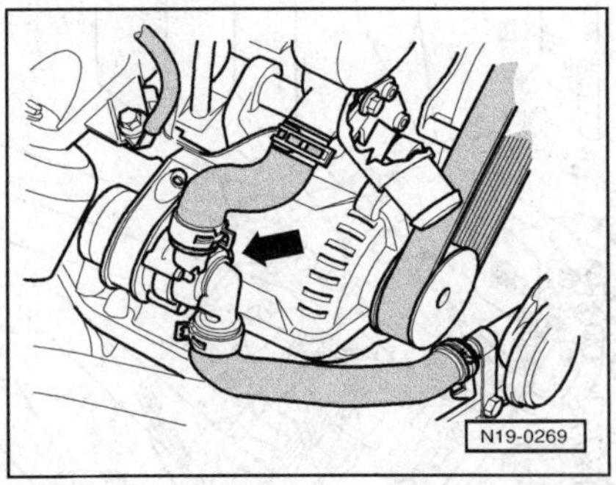

The continued coolant circulation pump -arrow- is generally activated for 10-12 minutes about 15 seconds after ignition is shut off. If the coolant temperature is above 92°C (198°F), the radiator fan (V7) will also be turned on.

Special tools and equipment

- Hand multimeter VAG 1526 or multimeter VAG 1715

Check conditions

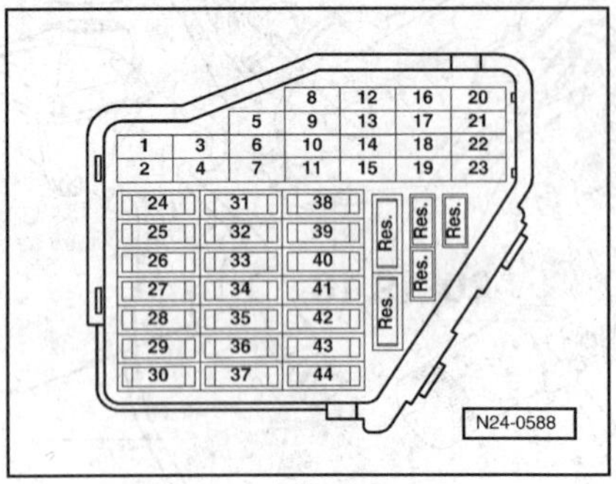

Fuse 34 must be OK.

Test sequence

– Start engine and idle briefly.

– Switch ignition off.

- Continued coolant circulation pump (V51) must start after about 15 seconds.

– If pump does not run:

- Pull 2-pin connector off pump.

– Attach multimeter for measuring voltage to contacts 1+2 of connector.

- Specification: 11-15 V

If the specification is obtained:

– Replace continued coolant circulation pump.

If no voltage is present:

– Check wiring from relay plate to continued coolant circulation pump using wiring diagrams.

Coolant hose connection diagram

1. **Lower coolant hose**
2. **Cylinder head/cylinder block**
3. **Intake manifold**
4. **Coolant hose**
   - From heat exchanger to engine block
5. **Heating system heat exchanger**
6. **Coolant hose**
7. **Rear coolant hose**
8. **Expansion tank**
9. **Oil cooler**
10. **Coolant hose**
    - From expansion tank to lower coolant hose
11. **Upper coolant hose**
12. **Cooler**
13. **Coolant pump**
14. **Lower coolant hose**
15. **Connection**
    - For thermostat

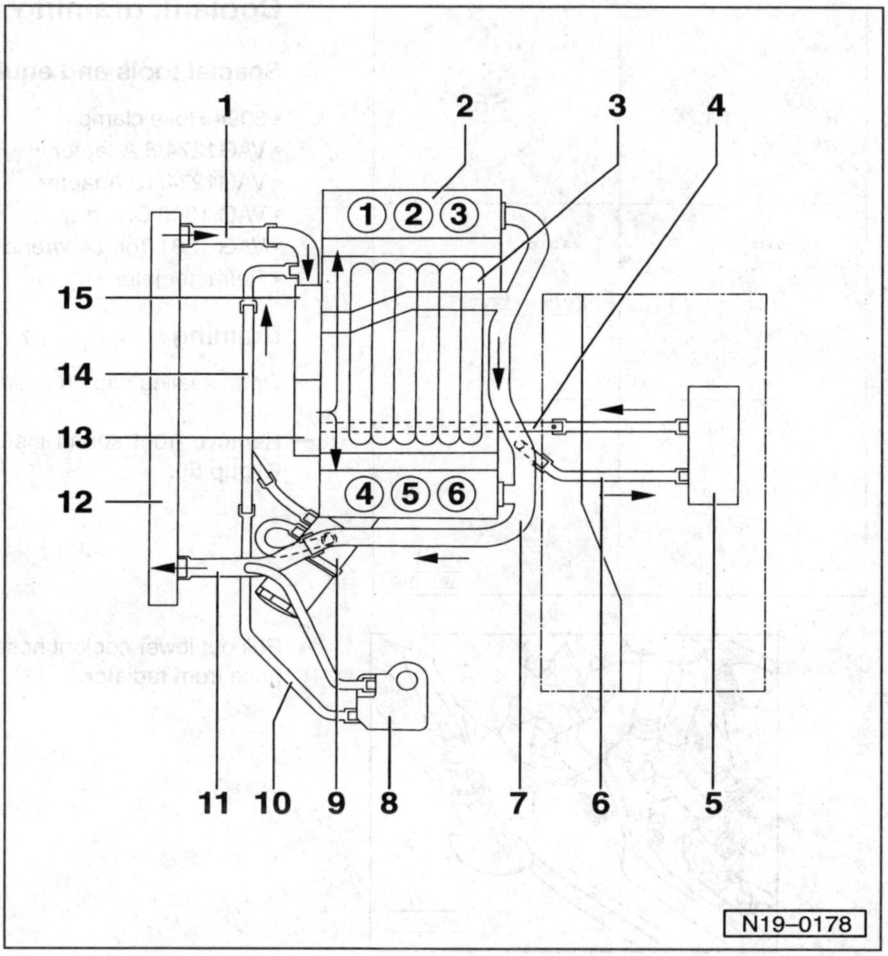

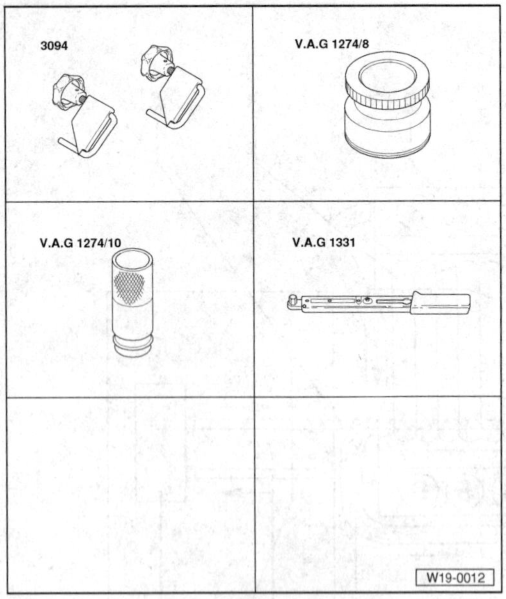

Coolant, draining and filling

Special tools and equipment

- 3094 Hose clamp
- VAG1274/8 Adapter
- VAG1274/10 Adapter
- VAG 1306 Drip tray
- VAG 1331 Torque wrench (5-50 Nm)
- Refractometer

Draining

– Open sealing cap on coolant expansion tank.

– Remove front sound insulation, see **Body–Exterior section, Repair Group 50.**

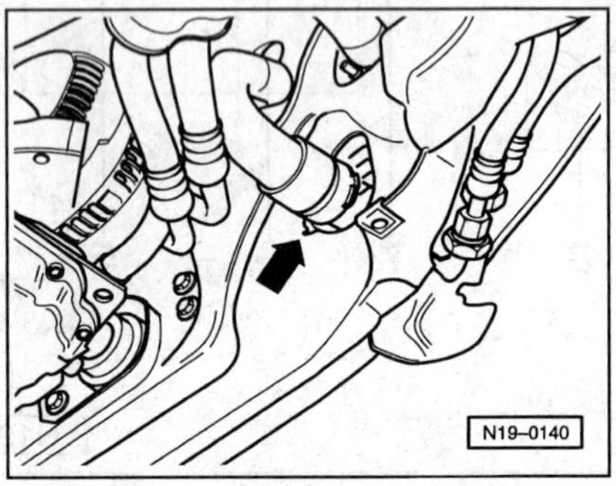

Pull out lower coolant hose retaining clip -arrow- and remove coolant hose from radiator.

Drain coolant from engine using drain plug -arrow-.

NOTE —
Observe disposal regulations for coolant!

Coolant, filling

NOTE —

- *Only use coolant additive G12 in accordance with TL VW 774 D. Identification: colored red*
- *On no account must G12 be mixed with other coolant additives!*
- *If the fluid in expansion tank is brown, G12 has been mixed with other coolant. In this case the coolant must be changed.*
- *G 12 and coolant additives marked "In accordance with TL VW 774 D" prevent frost and corrosion damage, scaling and also raise boiling point of coolant. For this reason the system must be filled all year round with frost and corrosion protection additives.*
- *Because of its high boiling point, the coolant improves engine reliability under heavy loads, particularly in countries with tropical climates.*
- *Protection against frost must be assured to about -25°C (-13°F) (in arctic climatic countries to about -35°C (-31°F).*
- *The coolant concentration must not be reduced by adding water even in warmer seasons and in warmer countries. The anti-freeze ratio must be at least 40%.*
- *If for climatic reasons a greater frost protection is required, the amount of G12 can be increased, but only up to 60% (frost protection to about -40°C (-40°F), as otherwise frost protection is reduced again and cooling effectiveness is also reduced.*
- *If radiator, heat exchanger, cylinder head or cylinder head gasket is replaced, do not reuse old coolant.*

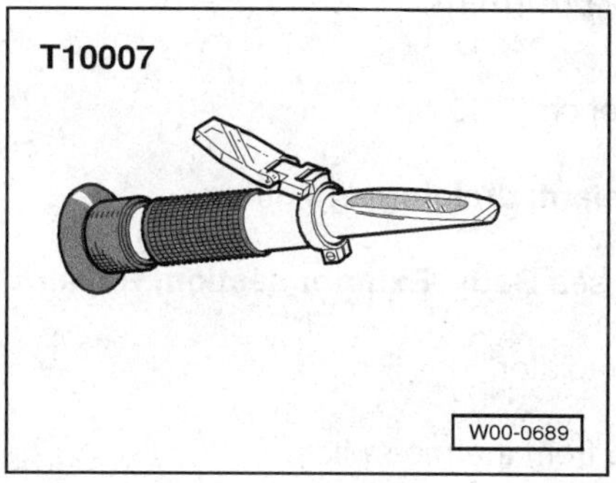

The refractometer T10007 is recommended for determining the current anti-freeze density.

Recommended mixture ratios:

Frost protection to	Anti-freeze amount	G12[1]	Water[1]
-25°C (-13°F)	40%	3.5 liters	4.5 liters
-35°C (-31°F)	50%	4.0 liters	4.0 liters

[1] The quantity of coolant can vary depending upon the vehicle equipment.

Work sequence

– Replace coolant drain plug O-ring and screw coolant drain plug into engine again.

Tightening torque: 20 Nm (15 ft-lb)

– Install lower coolant hose and secure.

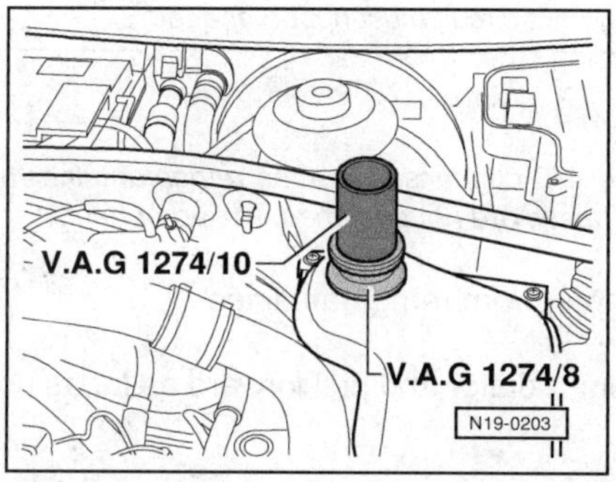

Screw adapter VAG 1274/8 with adapter VAG1274/10 onto expansion tank.

– Loosen both bleeder screws on coolant hoses.

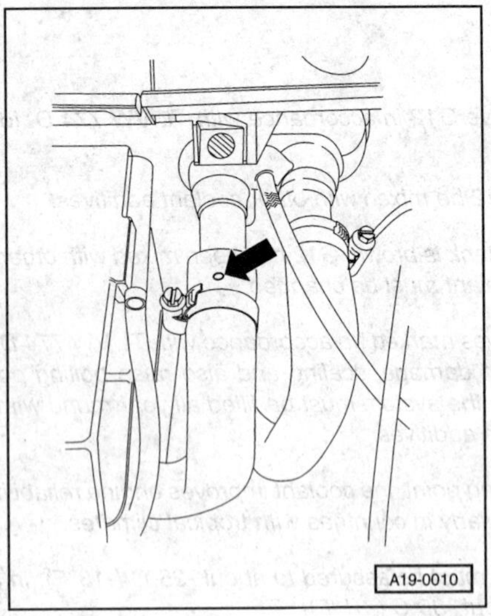

Loosen coolant hose on heat exchanger and pull back hose sufficiently so that bleeder hole -arrow- is no longer sealed by connection.

– Top up coolant until it flows out of bleeder hole in coolant hose and bleeder screws on coolant hoses.

– Push coolant hose on connection and tighten.

– Screw bleeder screws tight on coolant hoses.

• Tightening torque: 20 Nm (15 ft-lb)

– Fit expansion tank cap.

– Switch off heater and air conditioner.

– Start engine and maintain an engine speed of about 2000 rpm for approx. 3 minutes.

– Allow engine to run at idling speed until lower hose on radiator becomes hot.

– Check coolant level and top-up if necessary. When engine is at normal operating temperature, coolant level must be on max. mark and between min. and max. marks when engine is cold.

Radiator, removing and installing

Special tools and equipment

• VAG 1306 Drip tray
• VAG 1331 Torque wrench (5-50 Nm)

– Drain coolant, see **Coolant, draining and filling**.

– Remove front bumper, see **Body–Exterior section, Repair Group 63**

– Pull coolant hoses off radiator.

– Disconnect connectors from thermo-switch.

– On vehicles with automatic transmission remove ATF lines/hoses from radiator, see **5-Speed Automatic Transmission 01V section, Repair Group 37.**

Unbolt power steering hydraulic oil cooling line -arrows- and ensure it is free. Do not open hydraulic oil circuit.

> ***WARNING —***
> *The air conditioner refrigerant circuit must not be opened.*

NOTE —
To prevent damage to the A/C condenser and the refrigerant lines/hoses, ensure that the lines and hoses are not stretched, kinked or bent.

– Remove retaining clamp(s) from refrigerant lines.

– Remove condenser from radiator and pull forward as far as possible.

– Secure condenser to body so that refrigerant lines/hoses are not stressed.

– Remove upper radiator securing clips and pull radiator out between condenser and lock carrier.

Radiator, installing

Installation is carried out in the reverse order of removal, noting the following:

– On vehicles with automatic transmission, check ATF level, see **Maintenace.**

– Fill engine with coolant, see **Coolant, draining and filling**.

– Check all electrical connections and routing.

– Install front bumper, see **Body–Exterior section, Repair Group 63.**

– Check headlight adjustment and adjust if necessary, see **Maintenance.**

Coolant pump, removing and installing

Special tools and equipment

- VAS 5024 Assembly tool for spring-type clips
- VAG 1306 Drip tray
- VAG 1331 Torque wrench (5-50 Nm)

NOTE —
- *Always replace seals and gaskets.*
- *Cover toothed belt with a cloth to protect it from coolant before removing coolant pump.*

– Drain coolant, see **Coolant, draining and filling**.

– Move lock carrier into service position, see **Maintenance**.

– Remove viscous fan coupling, see **Viscous fan coupling, removing and installing**.

– Remove ribbed belt, see **Maintenance**.

– Remove toothed belt guard.

– Loosen tensioner and remove toothed belt from camshaft sprockets, see **Repair Group 13.**

– Remove bracket for viscous fan coupling.

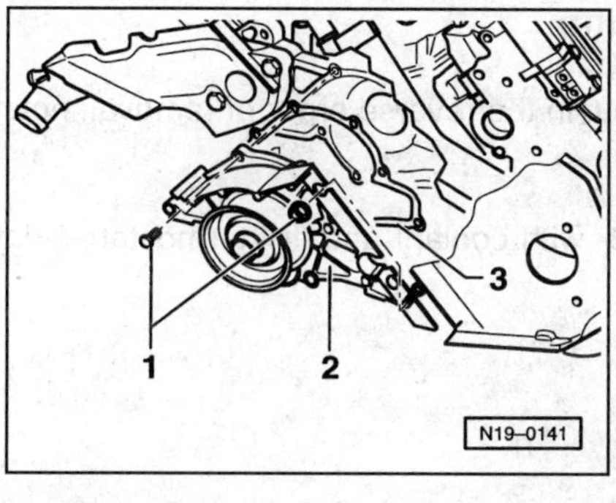

◄ Unscrew coolant pump securing bolts -1- and remove coolant pump -2-.

Coolant pump, installing

Installation is carried out in the reverse order of removal, noting the following:

– Fit new gasket -3- onto water pump flange.

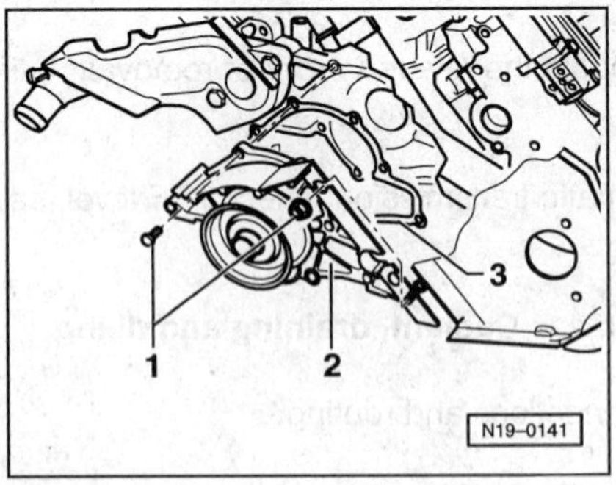

– Insert coolant pump -2- into cylinder block and tighten securing bolts -1-.
  - Tightening torque: 10 Nm (7 ft-lb)

– Install toothed belt and adjust timing, see **Repair Group 13.**

– Install toothed belt guard.

– Install ribbed belt, see **Maintenance**.

– Fill with coolant, see **Coolant, draining and filling**.

Thermostat, removing and installing

Special tools and equipment

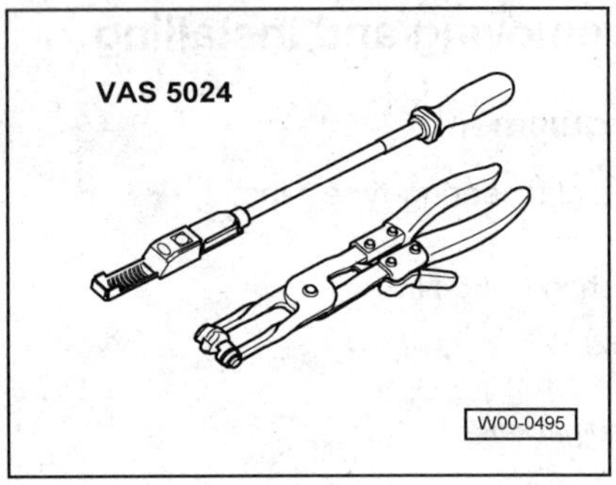

- VAS 5024 Assembly tool for spring-type clips
- VAG 1306 Drip tray
- VAG 1331 Torque wrench (5-50 Nm)

NOTE —
- *Always replace seals and gaskets.*
- *Cover toothed belt with a cloth to protect it from coolant before removing thermostat.*

– Drain coolant, see **Coolant, draining and filling.**

– Move lock carrier into service position, see **Maintenance**.

– Remove ribbed belt, see **Maintenance**.

– Remove toothed belt guard.

– Loosen tensioner and remove toothed belt from camshaft sprockets, see **Repair Group 13.**

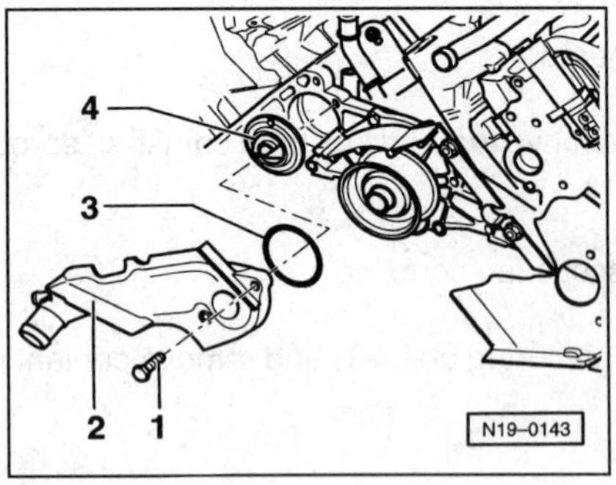

– Unscrew connection securing bolt -1- and remove connection -2-.

– Take out thermostat -4-.

Thermostat, installing

Installation is carried out in the reverse order of removal, noting the following:

– Moisten new O-ring -3- with coolant and fit thermostat -4- into engine block.

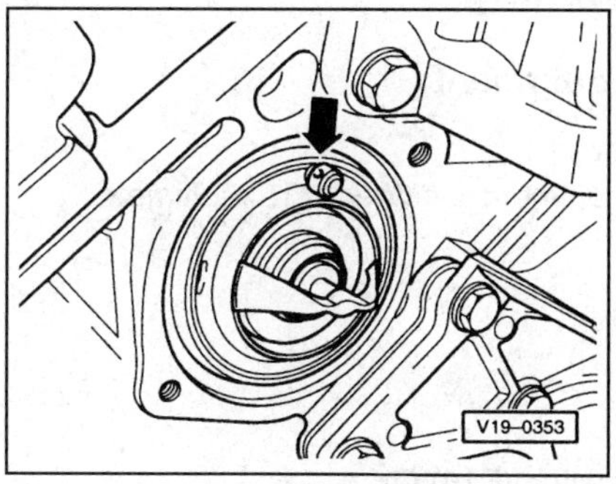

The thermostat bleeder valve -arrow- must be on top.

– Tighten thermostat securing bolt to 10 Nm (7 ft-lb)

– Install toothed belt and adjust timing, see **Repair Group 13.**

– Install toothed belt guard.

– Install ribbed belt, see **Maintenance**.

– Fill with coolant, see **Coolant, draining and filling**.

Viscous fan coupling, removing and installing

Special tools and equipment

- 3212 Two-hole pin wrench
- 3312 Open jaw spanner
- VAG 1331 Torque wrench (5-50 Nm)

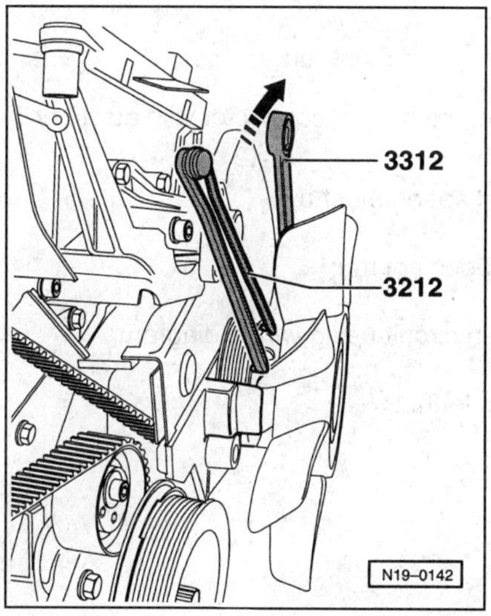

Counter hold radiator fan viscous coupling belt pulley with 2 hole pin wrench 3212 and unscrew radiator fan viscous coupling with open jaw spanner 3312 (left-hand thread).

– Remove fan viscous coupling carefully upward.

Installing

Installation is carried out in the reverse order, when doing this note the following:

– Tighten radiator fan viscous coupling to 40 Nm (30 ft-lb) (left-hand thread).

NOTE —
Tightening torque is only valid when used in conjunction with open jaw spanner 3312!

Radiator fan, checking

Special tools and equipment

- VAG 1594 Adapter set
- VAG 1526 Portable multimeter or VAG 1715 Multimeter
- Wiring diagram

Check conditions

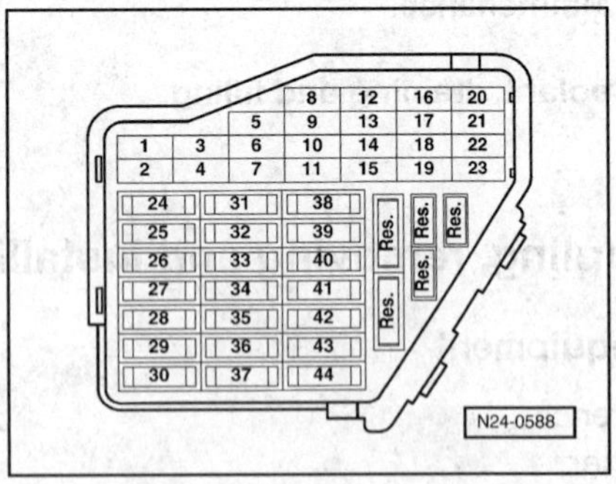

Fuse 34 must be OK.

- Radiator fan thermo switch (F18) OK.

Test sequence

– Pull 3 pin connector off radiator fan thermo switch (F18).

– Switch on ignition.

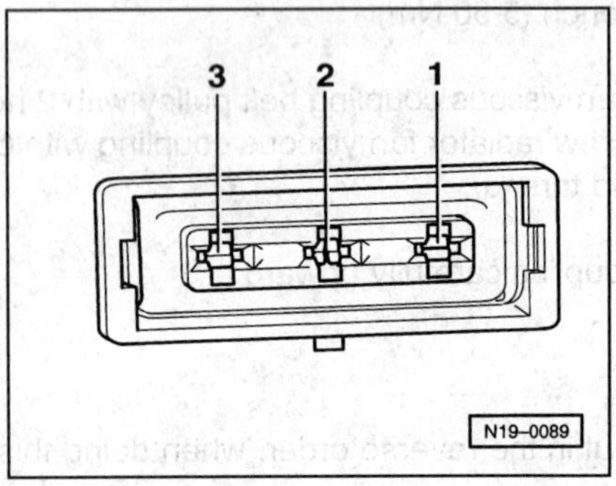

Bridge contact -1- and contact -2- of connector with auxiliary cables from VAG 1594.

- Radiator fan (V7) 1st speed must run.

– Bridge contact -1- and contact -3- of connector with auxiliary cables from VAG 1594.

- Radiator fan (V7) 2nd speed must run.

If 1st or 2nd fan speed does not run:

– Locate and correct open circuit using wiring diagram

If no open circuit can be found:

– Replace radiator fan (V7).

20 Fuel Supply

Repair Group

D 20

Fuel Supply System

NOTE—

- *All hose connections are secured with spring-type clamps or tensioning clamps.*
- *Always replace tensioning clamps with spring-type clamps.*
- *Fuel hoses at engine must only be secured with spring-type clamps. The use of clamping collars or screw clamps is not permitted.*
- *VAS 5024 Assembly tool or VAG 1921 Pliers are recommended for installing spring clips.*

Safety precautions when working on fuel supply

> ***WARNING—***
>
> *Fuel supply line is under pressure! Before loosening hose connections, place rags around the connection point. Then release pressure by carefully pulling on hose.*

Always observe the following when removing and installing the sensor for fuel gauge or the Fuel Pump (FP) from partially filled fuel tanks.

- Before starting work, switch on exhaust extraction system and place an extraction hose close to the sensor opening in fuel tank to extract escaping fuel fumes. If no exhaust extraction system is available, a radial fan (as long as motor is not in air flow) with a displacement greater than 15 m^3/h can be used.
- Prevent gasoline from contacting skin! Wear fuel-resistant gloves!

Rules of cleanliness

When working on the fuel supply or fuel injection systems, the following 5 rules should be carefully observed:

1. Thoroughly clean all connections and the surrounding area before disconnecting.
2. Place parts that have been removed on a clean surface and cover. Use lint-free cloths only!
3. Carefully cover over opened components or seal, if repairs are not carried out immediately.
4. Only install clean components: Only unpack replacement parts immediately prior to installation. Do not use parts that have been stored loose (in tool boxes etc.)
5. When the system is open: Do not work with compressed air if it can be avoided. Do not move vehicle unless absolutely necessary.

EDITOR'S NOTE—

Volkswagen identifies electrical components by a letter and/or a number in the electrical schematics. See **Wiring Diagrams, Fuses and Relays**. *These electrical identifiers are listed -between dashes- as an aid to electrical troubleshooting.*

Fuel tank with attachments and fuel filter (4WD)

1. **Sealing cap**
2. **Seal**
3. **Securing screw**
4. **Tank flap unit**
5. **O-ring**
   - Replace if damaged
6. **Vent valve**
   - Unclip sideways out of support
   - Before installing remove sealing cap -item 1-
   - Checking, see **Fig. 2**
7. **Connecting hose**
8. **Gravity valve**
   - Unclip upward out of support
   - Check flow-through valve vertical: open, valve tilted 45°: closed
9. **25 Nm (18 ft-lb)**
10. **Filler neck**
11. **Non-return valve**
    - White side of valve points in direction of sealing cap
12. **Connecting hose**
13. **Heat shield**
    - For fuel filler neck
14. **10 Nm (7 ft-lb)**
15. **Union nut**
    - 60 Nm (44 ft-lb)
    - Remove/install with spanner 3218
16. **Connector**
    - Black
    - For fuel gauge sender
17. **Flange**
    - Installation on fuel tank, see **Fig. 1**
18. **Sealing ring**
    - Insert seal "dry" into fuel tank opening
    - Replace if damaged
    - Moisten w/fuel when installing flange
19. **Vent line**
    - Black
    - Check for proper seating
    - Secure with spring-type clips
20. **Fuel gauge sender 3 -G237-**
    - Checking, see **Wiring Diagrams**

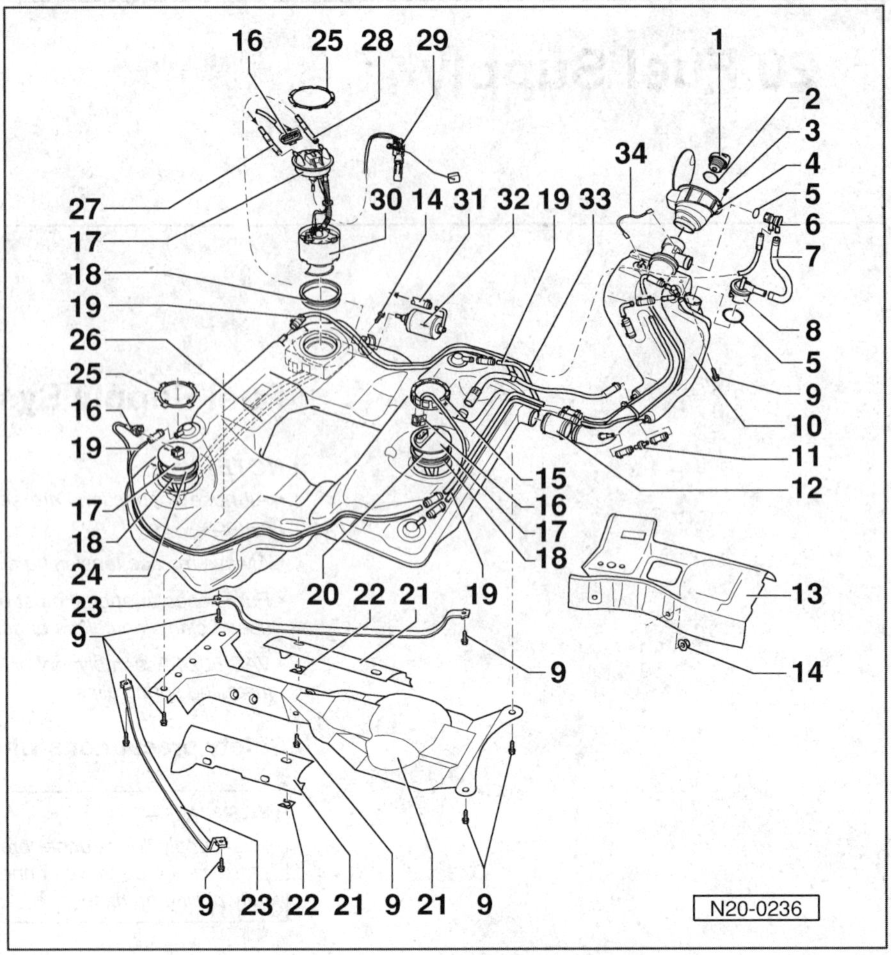

21. **Heat shield**
    - For fuel tank
22. **Securing clip**
    - Check for proper seating
23. **Securing strap**
    - Note installation position
24. **Fuel gauge sender 2 -G169-**
    - Checking, see **Wiring Diagrams**
25. **Locking ring**
    - Check for proper seating
    - Remove/install with spanner 3087
26. **Fuel tank**
    - Support with engine/transmission jack V.A.G 1383 A when removing
27. **Return line**
    - Blue or blue markings
    - Check for proper seating
    - Secure with spring-type clips
    - From fuel return line on fuel rail
28. **Supply line**
    - Black
    - Check for proper seating
    - Secure with spring-type clips
29. **Fuel gauge sender -G-**
    - Clipped onto reservoir housing in fuel tank
30. **Fuel delivery unit**
    - Removing and installing, see **Fuel delivery unit, removing and installing**
    - Checking fuel pump, see **Fuel pump, checking**
    - Clean strainer if soiled
31. **Supply line**
    - Black
    - Check for proper seating
    - Secure with spring-type clips
    - To fuel supply line on fuel rail
32. **Fuel filter**
    - Install position: Arrow points in through-flow direction
33. **Vent line**
    - To EVAP canister
    - Check for proper seating
    - Black
    - Clipped onto side of fuel tank
    - Secure with spring-type clips
34. **Ground connection**

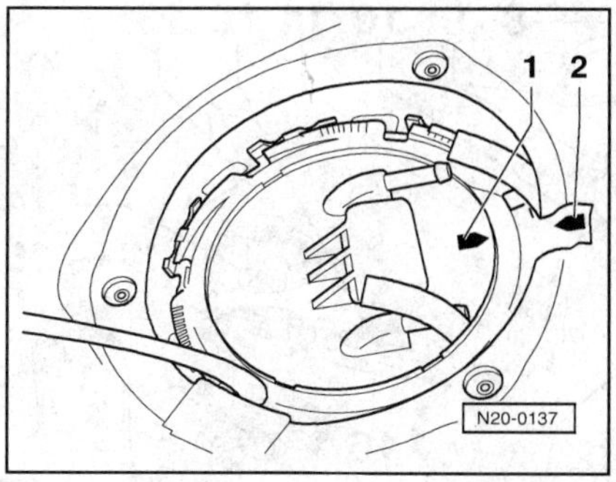

Fig. 1 Fitting position of fuel delivery unit flange

Mark on flange -1- must align with mark on fuel tank -2-.

NOTE—
After installing fuel delivery unit flange, check that the supply, return and breather lines are still clipped onto the fuel tank.

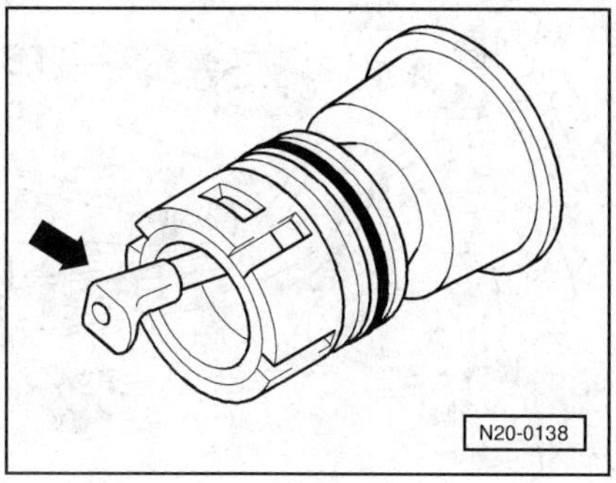

Fig. 2 Checking vent valve

Lever in rest position: Closed

Lever pushed in direction of arrow: Open

NOTE—
Before installing vent valve remove fuel tank cap.

Fuel tank with attachments and fuel filter (2WD)

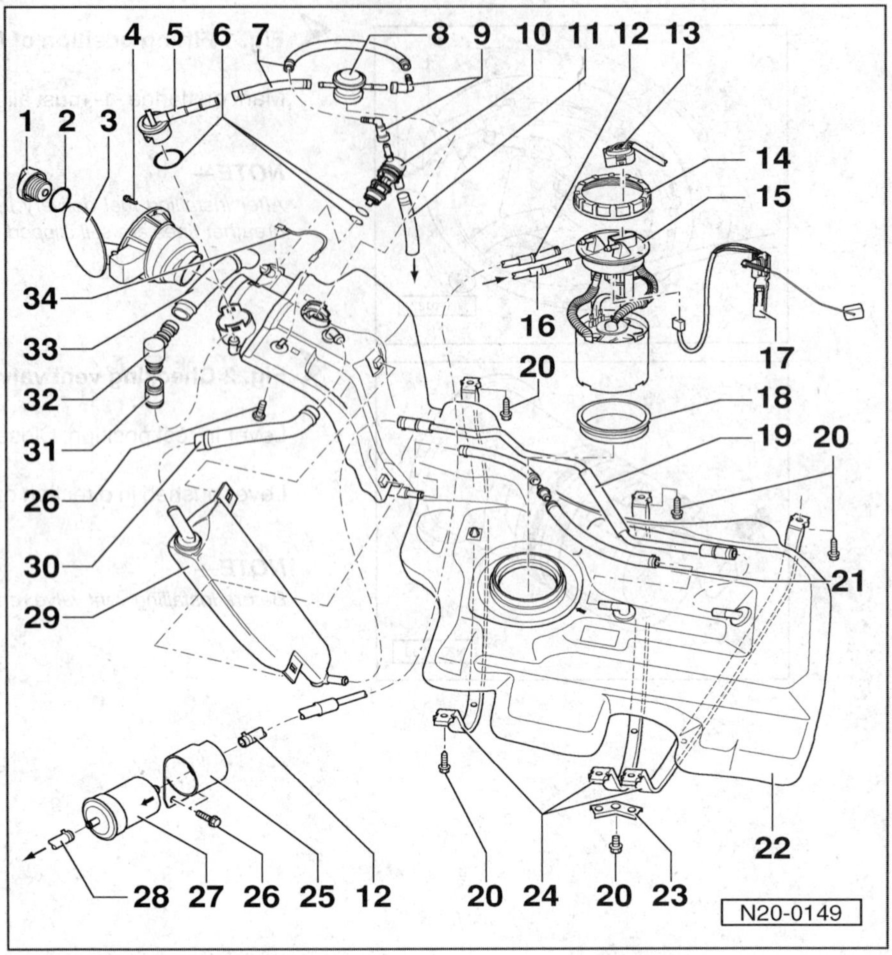

1. **Sealing cap**
2. **Seal**
3. **Tank flap unit**
   - With rubber cup
4. **Securing screw**
5. **Gravity valve**
   - Unclip upward out of support
   - Check flow-through-flow valve vertical: ope valve tilted 45°: closed
6. **O-ring**
   - Replace if damaged
7. **Vent line**
   - Operating vent
   - Black
   - Check for proper seating
   - Secure with spring-type clips
8. **Pressure retaining valve**
   - For fuel tank breathing
   - Checking, see **Fig. 3**
9. **Coupling**
   - Press together at front to release
10. **Change-over valve**
    - To remove valve unclip sideways out of support
    - Before installing remove sealing cap -item 1-
    - Checking, see **Fig. 2**
11. **Vent line**
    - Filling vent
    - Check for proper seating
    - Clipped onto top of fuel tank
    - To EVAP canister,
12. **Supply line**
    - Black
    - Clipped onto side of fuel tank
    - Check for proper seating
    - Secure with spring-type clips
13. **Connector**
    - Black, 4 pin
    - For fuel gauge sender -G- and fuel pump -G6-
14. **Union nut**
    - 60 Nm (44 ft-lb)
    - Remove and install with spanner 3217
15. **Fuel delivery unit**
    - Note installation position, see **Fig. 1**
    - Can only be replaced complete with flange
    - See **Fuel delivery unit, removing and installing**
    - See also **Fuel pump, checking**
    - Clean strainer if soiled
16. **Return line**
    - Blue or blue markings
    - Clipped onto side of fuel tank
    - Check for proper seating
    - Secure with spring-type clips
    - From fuel return line on fuel rail
17. **Fuel gauge sender (G)**
    - Clipped onto reservoir housing in fuel tank
18. **Sealing ring**
    - Replace if damaged
    - Coat with fuel when installing
19. **Vent line**
    - Operating vent
    - For droplet container see item**29**
20. **25 Nm (18 ft-lb)**
21. **Vent line**
    - Check for proper seating
    - Clipped onto top of fuel tank
22. **Fuel tank**
    - Support with engine/transmission jack V.A.G 1383 A when removing
23. **Bracket**
    - For securing straps
    - Note installation position
24. **Securing strap**
    - Note differing length
    - Installation position: Fixing point (holes) point in direction of travel
25. **Retaining ring**
    - Secured to body
26. **10 Nm (7 ft-lb)**
27. **Fuel filter**
    - Arrow points in flow direction
28. **Supply line**
    - Black
    - Check for proper seating
    - Secure with spring-type clips
    - To fuel supply line on fuel rail:

Fuel tank with attachments and fuel filter (2WD)

(continued from previous page)

29. Droplet container

30. Vent line
- For droplet container, see -item 29-

31. Connecting hose

32. Angle piece

33. Shaped hose

34. Ground connection
- Check for proper seating

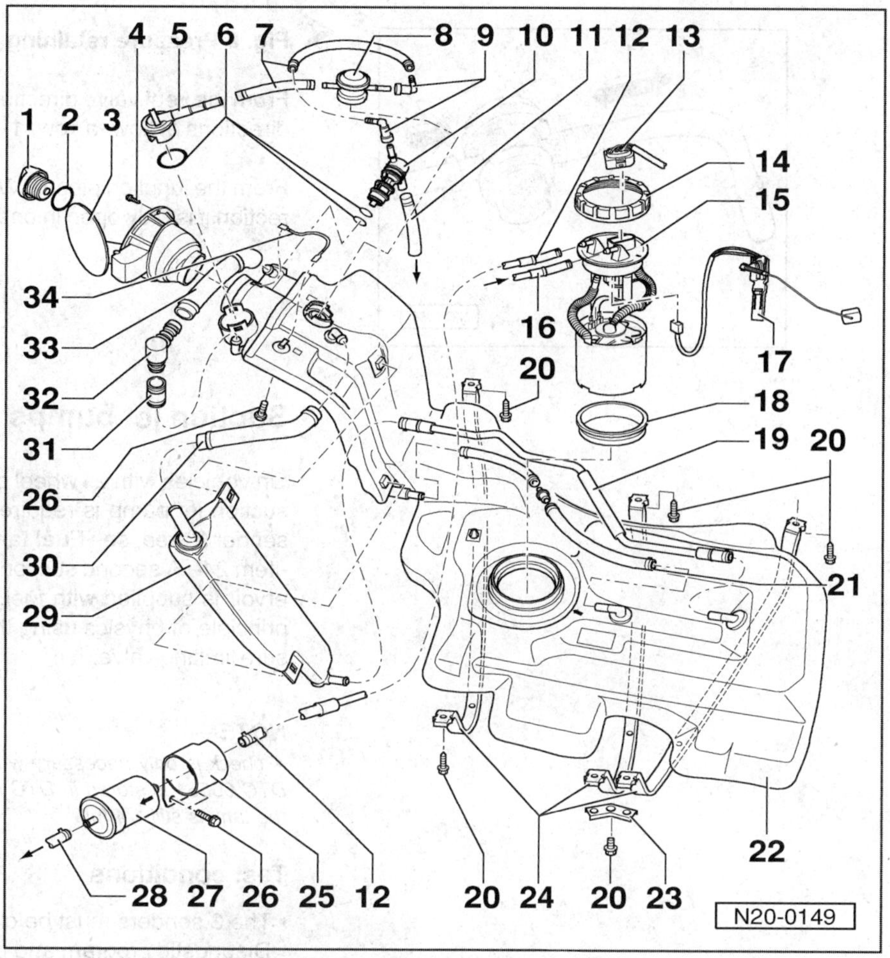

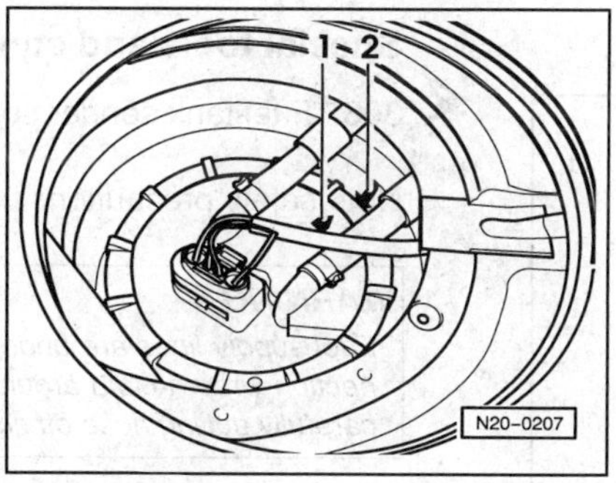

Fig. 1 Fuel delivery unit, installation position

- Mark on fuel delivery unit -1- must align with mark on fuel tank -2-.

NOTE —

After installing fuel delivery unit, check that the supply, return and vent lines are still clipped onto the fuel tank.

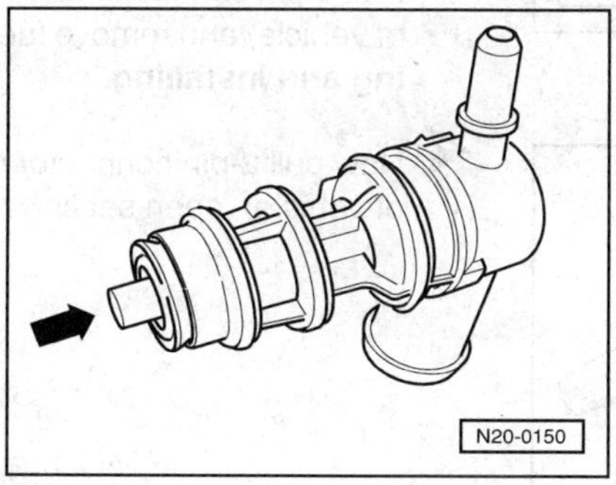

Fig. 2 Change-over valve, checking

- Valve in rest position: Closed
- Valve pushed in direction of arrow: Open

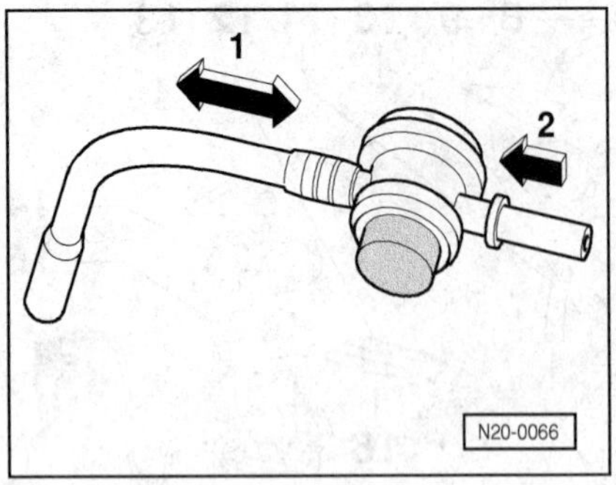

Fig. 3 Pressure retaining valve, checking

From the vent valve direction, the pressure retaining valve is open in both directions of flow -arrow -1-.

From the junction piece EVAP canister purge regulator solenoid valve direction it is only open in one direction of flow -arrow -2-.

Suction jet pumps (4WD), removing and installing

On vehicles with all wheel drive the shape of the fuel tank dictates that a suction jet pump is required to pump fuel to the fuel delivery unit from sender 2 area, see **Fuel tank with attachments and fuel filter (4WD),** -item 24-. A second suction jet pump ensures the delivery unit in the reservoir is supplied with fuel. The suction pumps function according to a principle of physics using the return flow of fuel via a junction with pressure limiting valve.

NOTE —
A check is only necessary when the engine stops due to lack of fuel or a DTC 00561 is stored in DTC memory, although the fuel gauge shows that the tank is still 1/4 full.

Test conditions

- The 3 senders must be checked using a scan tool and the On-Board Diagnostic program and be OK.
- Fuel tank must have been drained

Special tools and equipment

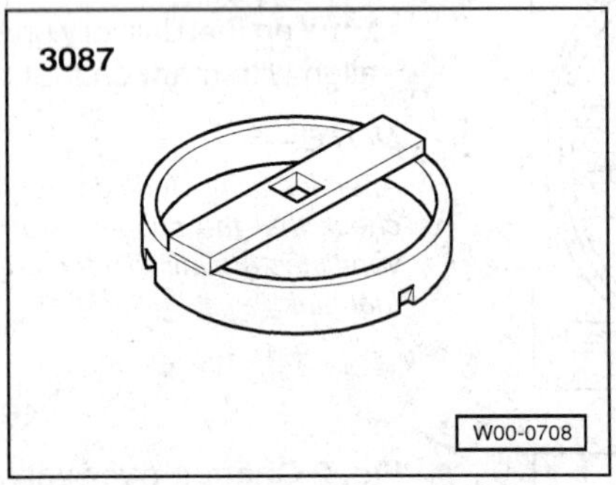

3087 Fuel tank sender spanner

– Note safety precautions before starting work

WARNING —
Fuel supply lines are under pressure! Before removing from hose connection wrap a cloth around the connection. Then release pressure by carefully pulling hose off connection.

– Pull 4-pin connector and fuel hoses off fuel delivery unit (right-hand side of vehicle) and remove fuel delivery unit, see **Fuel delivery unit, removing and installing**.

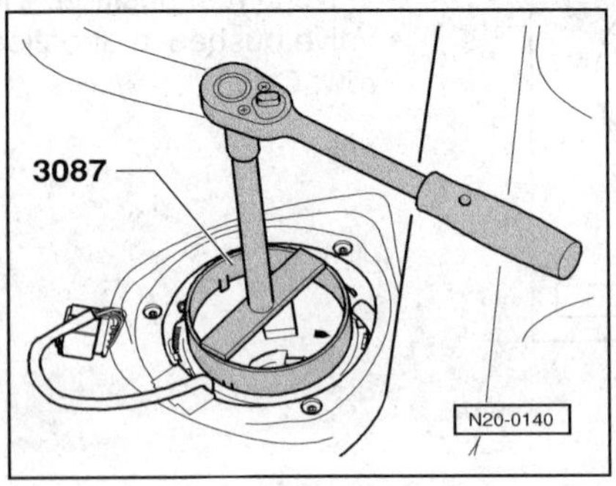

Now pull 2-pin connector off fuel gauge sender 2 (G169) (left-hand side of vehicle), open sealing ring with key 3087 and carefully lift the sender.

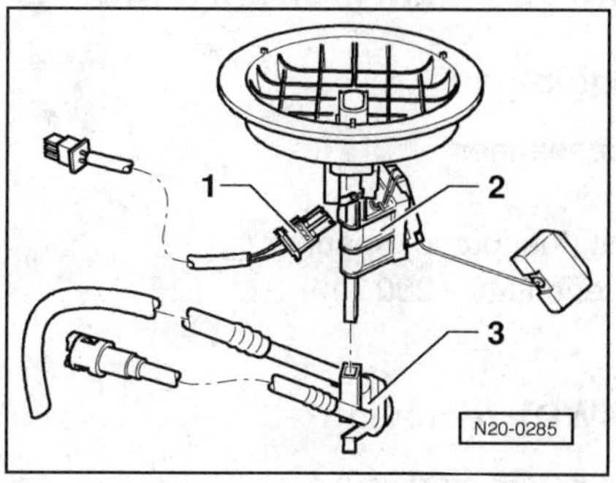

Now pull off connector -1- under sender 2 cover -2-.

– Lift sender 2 -2- and pull suction jet pump -3- out of locking device of sender.

– Take out sender 2.

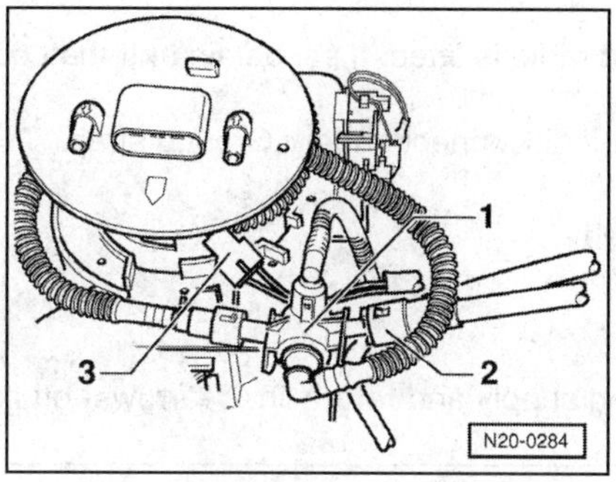

Now, on right-hand side of vehicle (fuel delivery unit), pull hose/line to suction jet pump -2- off junction -1- and pull connector -3- off fuel gauge sender -G- to sender 2 -G169-.

– Now suction jet pump can be pulled out from sender 2 side (left-hand side of vehicle).

– Check that fuel hoses/lines on suction jet pump are fitted securely and are not damaged.

– Additionally, check suction jet pump for contamination/dirt if necessary.

If no faults are found:

– Install suction jet pump and sender 2 again in reverse order.

NOTE —

In other work sequences the return supply line/hose must be pulled off junction. When doing this observe that all O-rings must be replaced.

Check junction with pressure relief valve as follows:

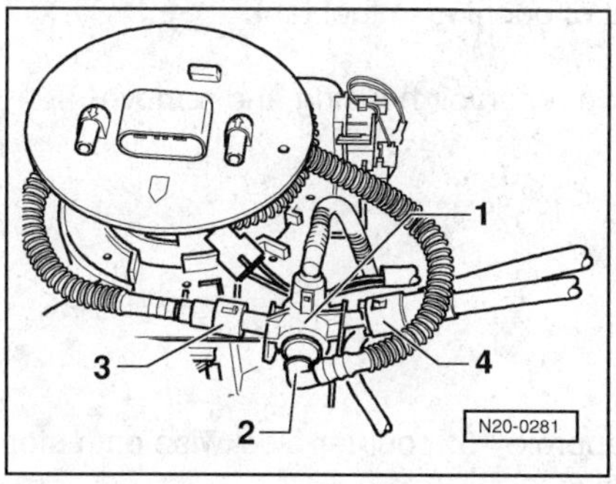

Pull off connections for return supply -2-, connections -3- and -4- for suction jet pump on junction with pressure relief valve -1-.

– Unclip junction which is secured on reservoir.

– Use a test hose and blow into connection, from direction of fuel return -2-, there must be through flow in directions -3- and -4-.

– Hold connections -3- and -4- closed, the pressure relief valve must open under slight opposing pressure.

– If no opposing pressure can be determined or pressure relief valve does not open, replace junction with pressure relief valve.

Fuel delivery unit, removing and installing

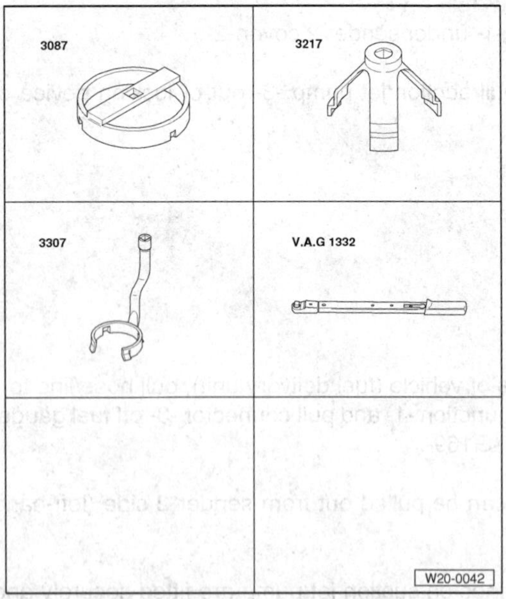

Special tools and equipment

- 3087 Fuel tank sender spanner
- 3217 Union nut spanner
- 3307 Fuel delivery unit (fuel pump) spanner
- V.A.G 1332 Torque wrench (40 - 200 Nm)

Fuel delivery unit (4WD), removing

– Note safety precautions before starting work

– Check whether a coded radio is fitted. If so, obtain anti-theft coding.

– With ignition switched off disconnect battery Ground strap.

– Remove rear seat bench.

– Remove cover in floor.

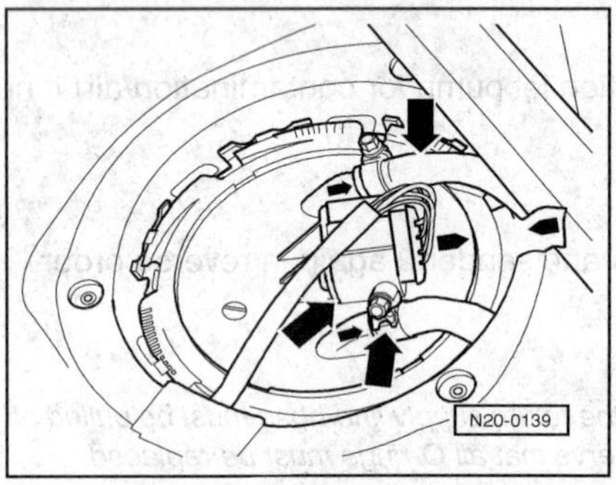

◀ Pull 4-pin connector and supply and return lines -arrows- off flange.

> ***WARNING —***
> *Fuel supply lines are under pressure! Before removing from hose connection wrap a cloth around the connection. Then release pressure by carefully pulling hose off connection.*

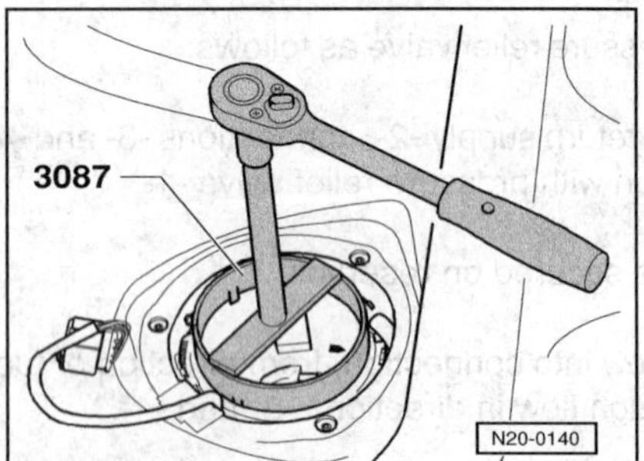

◀ Remove locking ring with key 3087.

– Pull flange and seal out of opening in fuel tank.

– Unclip fuel sender -G- on reservoir housing and remove.

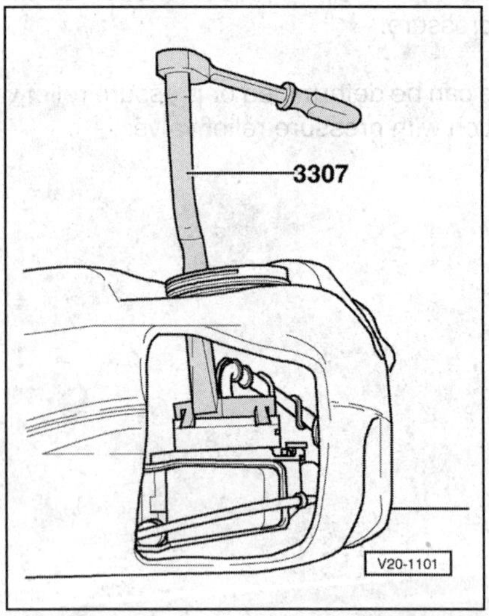

◀ Turn fuel delivery unit approx. 15° counter-clockwise onto stop with spanner 3307 and remove.

NOTE —
If the delivery unit is to be replaced then drain old delivery unit before disposal.

Fuel delivery unit (4WD), installing

Install fuel delivery unit in reverse order of removal, noting the following:

NOTE —

- *Do not bend fuel sender when installing.*
- *Coat flange seal with fuel when fitting.*
- *Do not interchange supply and return hoses (return hose -blue-).*
- *Secure fuel hoses with spring-type clamps.*

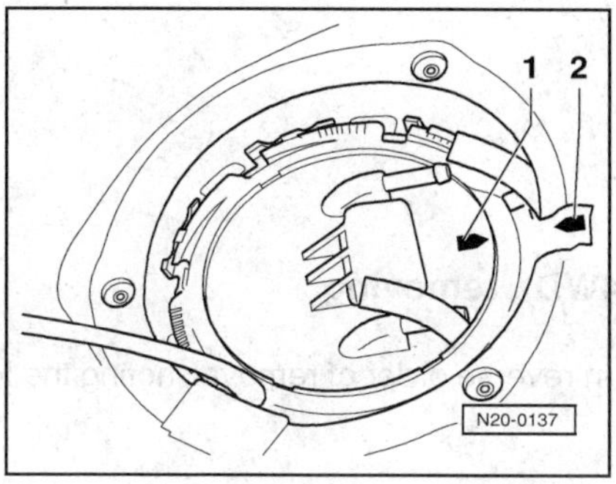

◄ Note installation position of fuel delivery unit flange:

- Mark on flange -1- must align with mark -2- on fuel tank.

– After installing fuel delivery unit flange, check that the supply, return and breather lines are still clipped onto fuel tank.

Fuel delivery unit (4WD), removing

– Note safety precautions before starting work.

– Check whether a coded radio is fitted. If not, obtain anti-theft coding.

– With ignition switched off disconnect battery Ground strap.

– Remove floor cover.

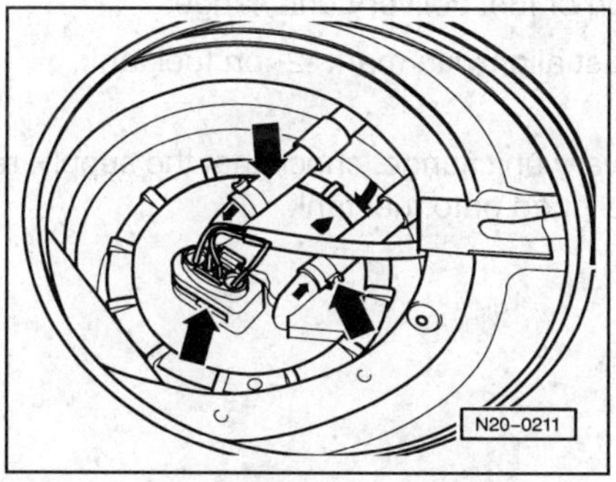

◄ Pull 4-pin connector and supply and return lines -arrows- off flange.

> ***WARNING —***
> *Fuel supply lines are under pressure! Before removing from hose connection wrap a cloth around the connection. Then release pressure by carefully pulling hose off connection.*

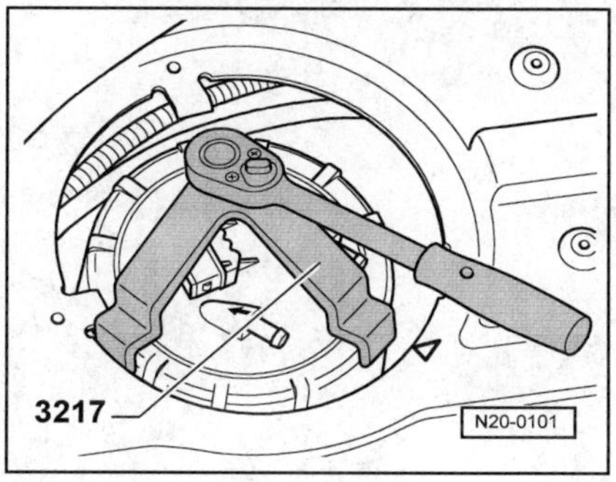

◄ Remove union nut with spanner 3217.

– Pull flange and seal out of the opening in fuel tank.

– Unclip fuel sender -G- on reservoir housing and remove.

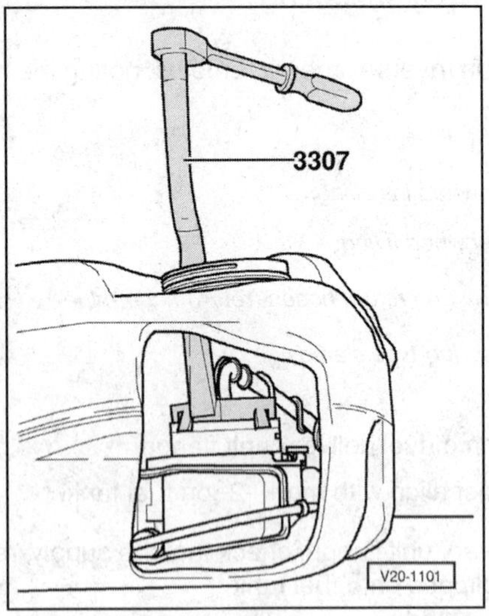

Turn fuel delivery unit approx. 15° counter-clockwise onto stop with spanner 3307 and remove.

NOTE—
If the delivery unit is to be replaced then drain old delivery unit before disposal.

Fuel delivery unit (4WD), removing

Install fuel delivery unit in reverse order of removal, noting the following:

NOTE—
• Do not bend fuel sender when installing.

• Coat flange seal with fuel when fitting.

• Ensure fuel hoses are seated securely.

• Do not interchange supply and return hoses (return hose -blue-).

• Secure fuel hoses with spring-type clamps.

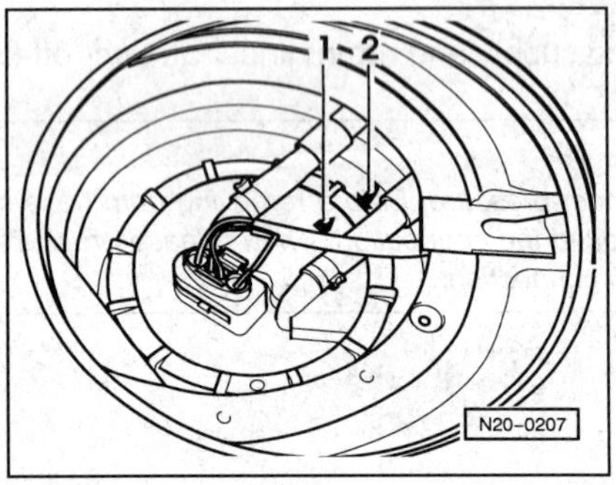

Note installation position of fuel delivery unit flange:
- Mark on flange -1- must align with mark -2- on fuel tank.

– After installing fuel delivery unit flange, check that the supply, return and breather lines are still clipped onto fuel tank.

Fuel pump, checking

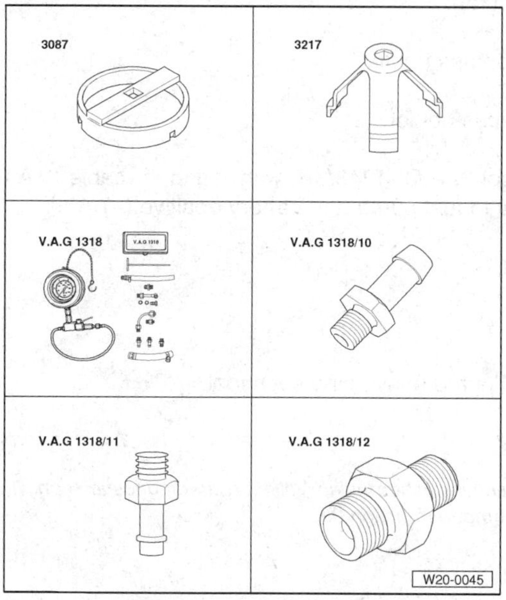

Special tools and equipment

- 3087 Fuel tank sender spanner
- 3217 Union nut spanner
- V.A.G 1318 Pressure tester
- V.A.G 1318/10 Adapter
- V.A.G 1318/11 Adapter
- V.A.G 1318/12 Adapter

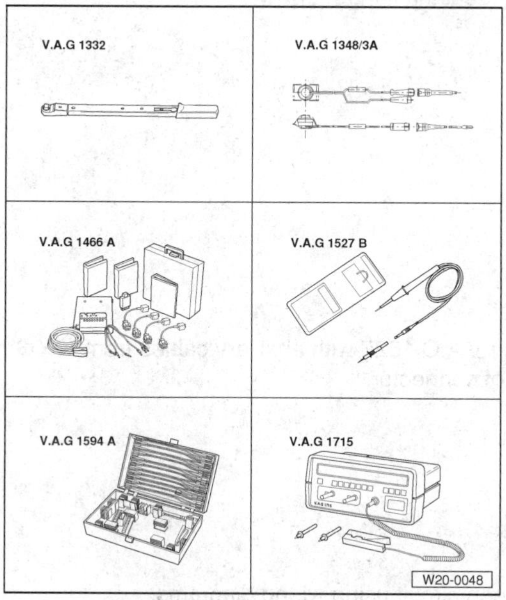

- V.A.G 1332 Torque wrench (40-200 Nm)
- V.A.G 1348/3A Remote control with V.A.G 1348/3-2 Adapter cable
- V.A.G 1466 Test system
- V.A.G 1527 B Diode test lamp
- V.A.G 1594 A Adapter set
- V.A.G 1715 Multimeter
- Measuring container
- Wiring diagram

Vehicles with all wheel drive (4WD)

- Fuel tank sender spanner 3087

Vehicles with front wheel drive (2WD)

- Union nut spanner 3217

Checking conditions

- Battery voltage at least 11.5 V

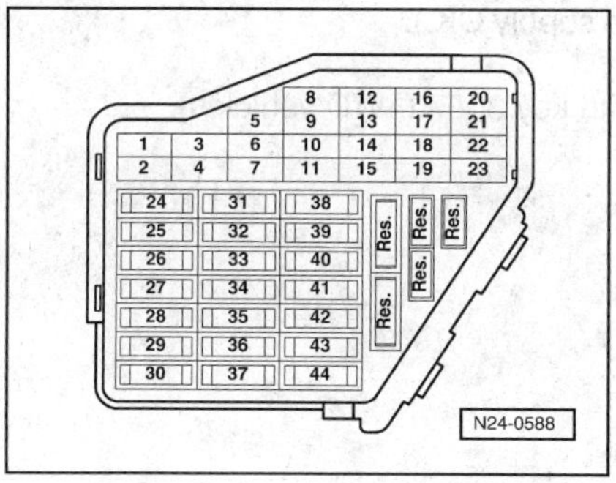

◄ Fuse No. 28 OK.

Checking function and voltage supply

– Remove rear seat bench (4WD vehicles).

– Remove floor cover.

– Briefly operate starter. Fuel pump must be audible.

– Switch ignition off.

If the fuel pump does not run:

– Remove cover in front of fuse holder.

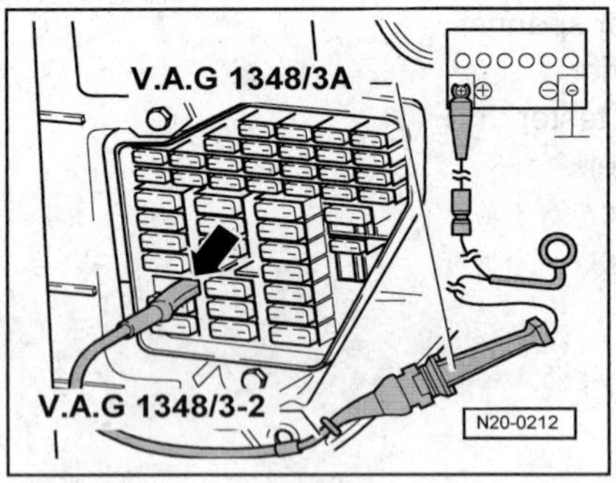

◄ Pull fuse No. 28 out of fuse holder.

– Connect remote control V.A.G 1348/3A with adapter cable V.A.G 1348/3-2 to contact 28a to fuel pump and battery positive (+).

– Operate remote control.

If fuel pump runs:

– Check activation of fuel pump relay using scan tool.

EDITOR'S NOTE —
Component activation is accomplished using Volkswagen diagnostic scan tool, output diagnostic test mode (function 03).

If fuel pump does not run:

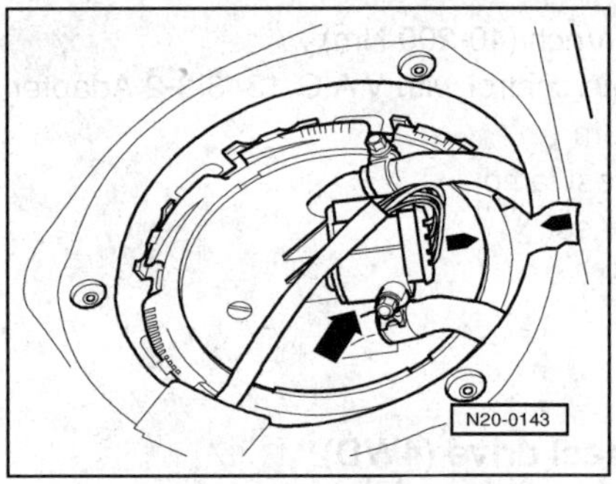

◄ Pull 4-pin connector off fuel tank flange -arrow-.

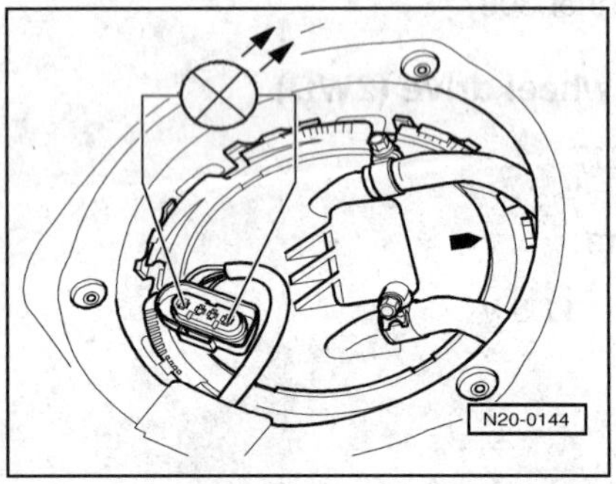

◄ Connect diode test lamp V.A.G 1527 with auxiliary cables from V.A.G 1594 to outer contacts of connector.

– Operate remote control.

- LED must light up

If LED does not light up:

– Locate and eliminate open circuit using wiring diagrams.

If LED lights up (voltage supply OK.):

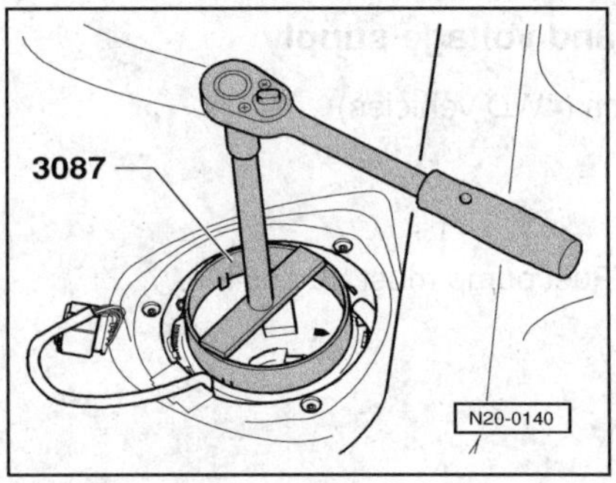

◄ Remove locking ring with key 3087 (4WD vehicles).

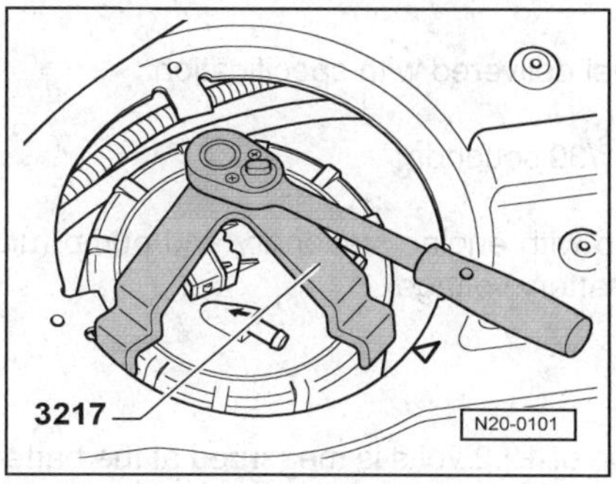

Remove union nut with spanner 3217 (2WD vehicles).

Fuel pump, checking, continued for all vehicles

– Check that electrical wires between flange and fuel pump are connected.

If no open circuit can be found:

– Fuel pump malfunctioning. Replace fuel delivery unit, see **Fuel delivery unit, removing and installing**.

Fuel pump, checking delivery rate

Check conditions

• Voltage supply OK.
• Remote control V.A.G 1348/3A connected.

Test sequence

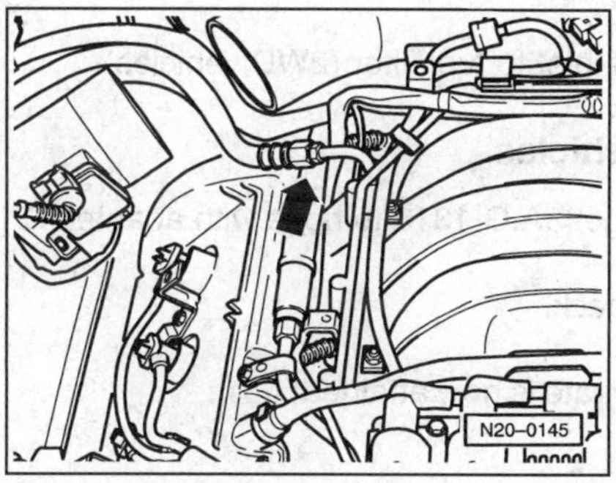

– Remove filler cap from fuel tank filler neck.

– Remove intake manifold cover.

Open threaded union -arrow- and catch escaping fuel with a cloth.

> ***WARNING —***
> *Fuel supply lines are under pressure! Before removing from hose connection wrap a cloth around the connection. Then release pressure by carefully pulling hose off connection.*

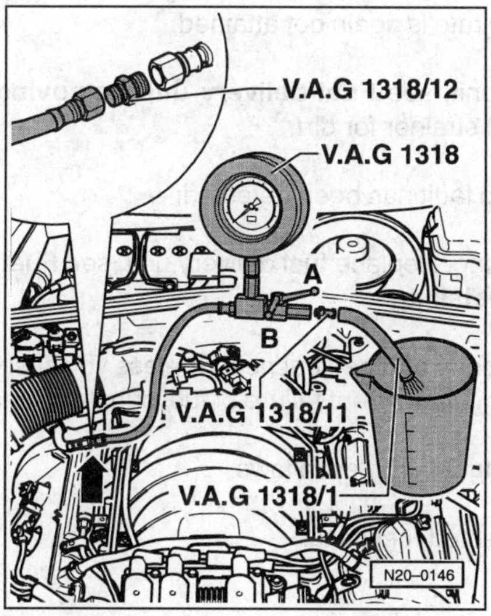

Connect pressure gauge V.A.G 1318 with adapter 1318/12 to fuel supply line.

– Push hose (commercially available) onto adapter V.A.G 1318/11 of pressure gauge and hold hose in a measuring container.

– Open pressure gauge shut-off tap. Lever then points in throughflow direction -A-.

– Operate remote control V.A.G 1348/3A and slowly close shut-off tap, until pressure gauge shows 4 bar. From this point on do not move position of shut-off tap.

– Empty measuring container.

– Fuel pump delivery range is dependent upon battery voltage. Therefore connect multimeter with adapter cables from V.A.G 1594 to vehicle battery.

– Operate remote control for 30 seconds and measure battery voltage

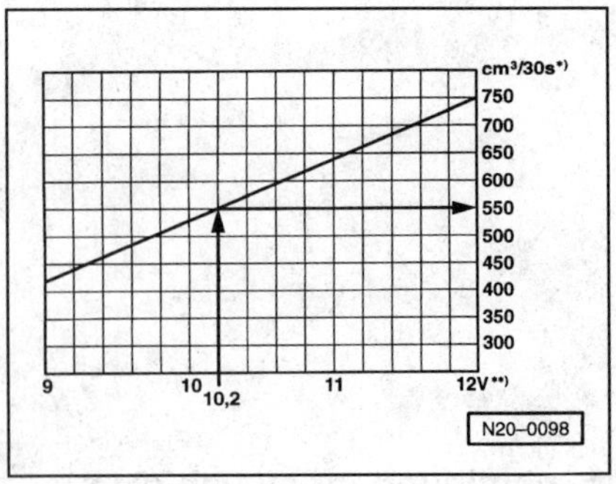

Compare quantity of fuel delivered with specification.

*) Minimum delivery cm^3/30 seconds

**) Voltage at fuel pump with engine stationary and pump running (approx. 2 volts less than battery voltage).

Example:

During the test a voltage of 12.2 volts is measured at the battery. As the voltage at the pump is approx. 2 volts less than the battery voltage then this will equate to a minimum delivery of 550 cm^3/30 s.

If the minimum delivery rate is not attained:

– Check fuel lines for possible restrictions (kinks) or blockages.

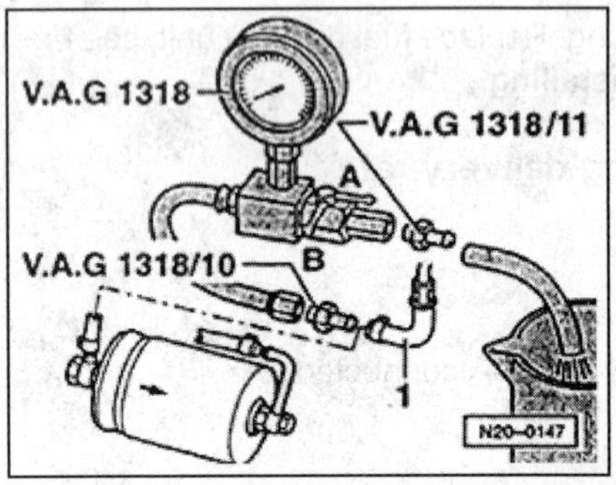

Pull supply hose -1- off inlet to fuel filter (4WD vehicles).

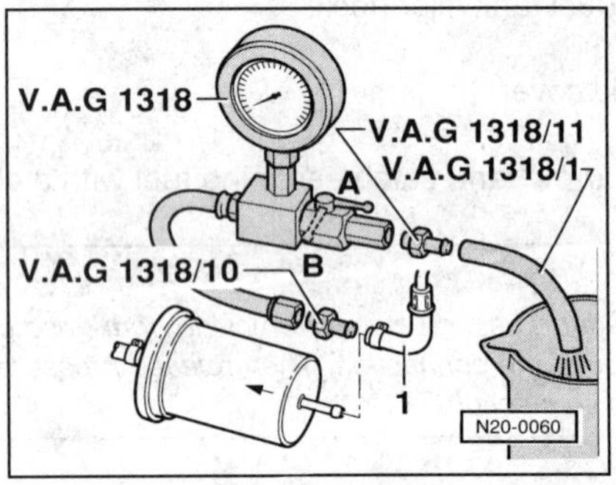

Pull supply hose -1- off inlet to fuel filter (2WD vehicles).

Continued for all vehicles

– Connect pressure gauge V.A.G 1318 to hose with adapter 1318/10.

– Repeat delivery rate check.

If the minimum delivery rate is now attained:

– Replace fuel filter.

If the minimum delivery rate is again not attained:

– Remove fuel delivery unit, see **Fuel delivery unit, removing and installing** and check fuel strainer for dirt.

Only when up to now no fault has been detected:

– Fuel pump malfunctioning. Replace fuel delivery unit, see **Fuel delivery unit, removing and installing**.

If the delivery rate has been attained but nevertheless you suspect a fuel supply system fault (e.g. intermittent failure of fuel supply system):

– Check current draw of fuel pump as follows:

– Reconnect all disconnected fuel lines.

– Remove rear seat bench (4WD vehicles).

Continued for all vehicles

– Remove floor cover.

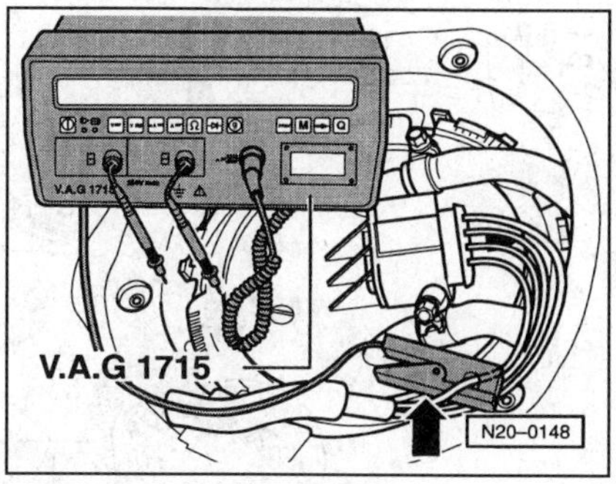

◄ Connect multimeter V.A.G 1715 with current clamp on green/yellow wire in wiring harness.

– Start engine and run at idling speed.

– Measure current draw of fuel pump.

– Specification: max. 8 amps.

NOTE —
If the fuel system malfunction is intermittent, you can perform the check during a test drive, to do this a 2nd person is necessary.

If the current draw is exceeded:

– Fuel pump malfunctioning, replace fuel delivery unit.

Checking fuel pump non-return valve

Check conditions

• Remote control V.A.G 1348/3A connected
• Pressure gauge V.A.G 1318 connected to hose on fuel supply line.

Test sequence

NOTE —
With this check the fuel supply line connections from the fuel delivery unit to the point at which the pressure gauge V.A.G 1318 is connected will be checked for leaks at the same time.

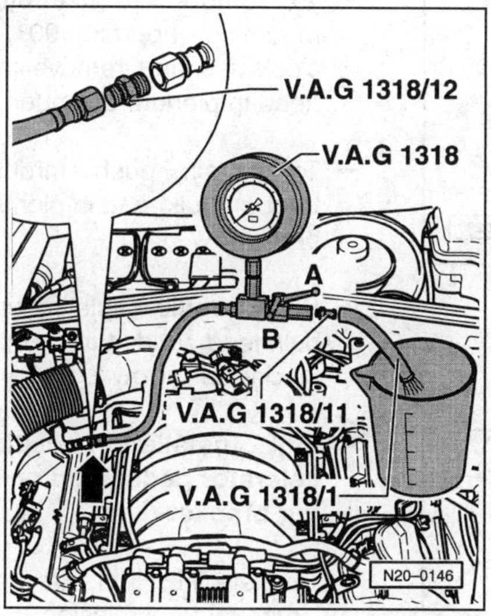

◄ Close pressure gauge shut-off tap (lever at right angles to throughflow direction - position -B-).

– Operate remote control at short intervals, until a pressure of approx. 4 bar (58 psi) has built-up.

– If pressure builds up too high, lower excess pressure by carefully opening shut-off tap.

> ***WARNING —***
> *Danger of spray when opening the shut-offtap, hold container over the pressure gauge free connection.*

– Watch pressure drop on gauge. After 10 minutes pressure must not drop below 2.2 bar (32 psi) (engine cold) or 3.0 bar (43 psi) (engine warm).

If the pressure drops further:

– Check line connections for leaks.

If no wiring fault is detected:

– Fuel pump malfunctioning, replace fuel delivery unit.

Throttle control

Accelerator mechanism, servicing (AHA engine)

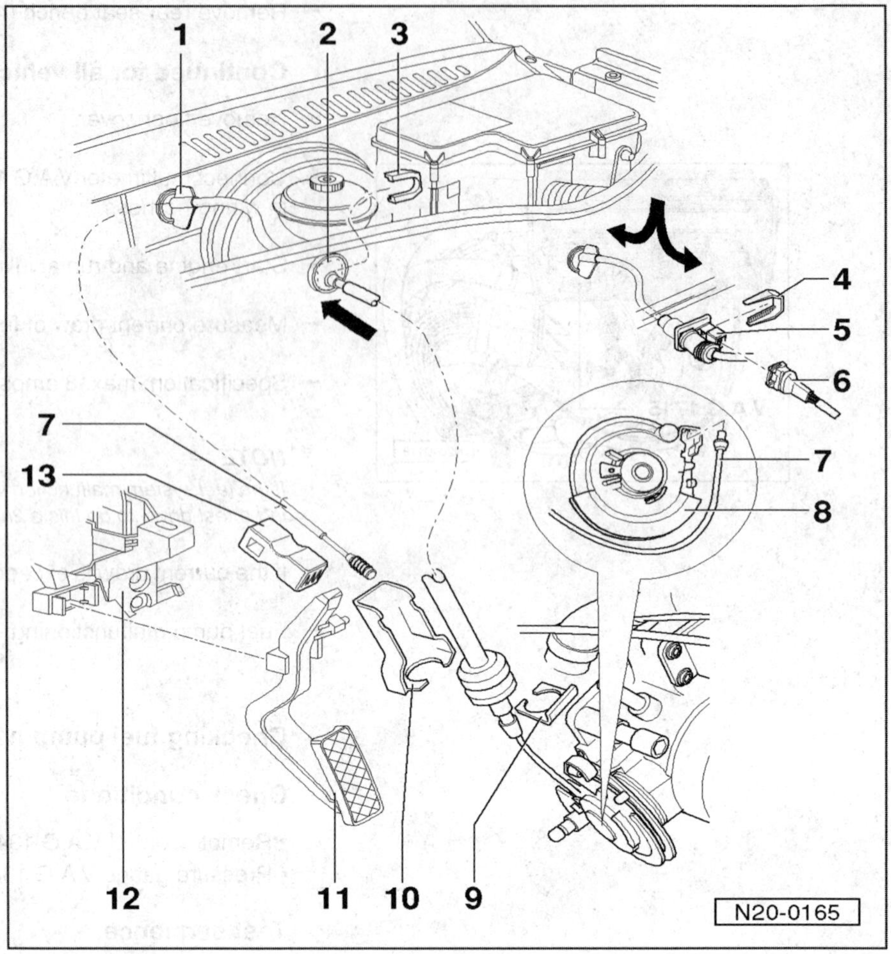

1. **Throttle cable mounting**
   - On bulkhead in plenum chamber
   - Removing and installing, see **Fig. 1**
2. **Throttle cable mounting**
   - On bulkhead
   - Only vehicles with manual transmission
3. **Retaining clip**
   - Only vehicles with manual transmission
   - Check seated securely
   - Secured behind bulkhead
4. **Retaining clip**
   - Only vehicles with automatic transmission
   - Check seated securely
   - Secured before bulkhead
5. **Kickdown switch (F8)**
   - On bulkhead
   - Checking, see **Throttle cable, adjusting**
6. **Connector**
   - Black, 2-pin
7. **Accelerator cable**
   - Adjusting, see **Throttle cable, adjusting**
   - Do not reuse damaged accelerator cable, replace
8. **Cable cam**
   - On throttle valve control part
9. **Retainer**
   - Adjusting throttle cable, see **Throttle cable, adjusting**
10. **Support bracket**
    - For accelerator cable
11. **Accelerator pedal**
12. **Mounting bracket**
    - Removing and installing, see **Brake System section, Repair Group 46**
13. **Accelerator pedal stop**
    - Installation position of accelerator pedal stop on mounting bracket, see **Fig. 2**

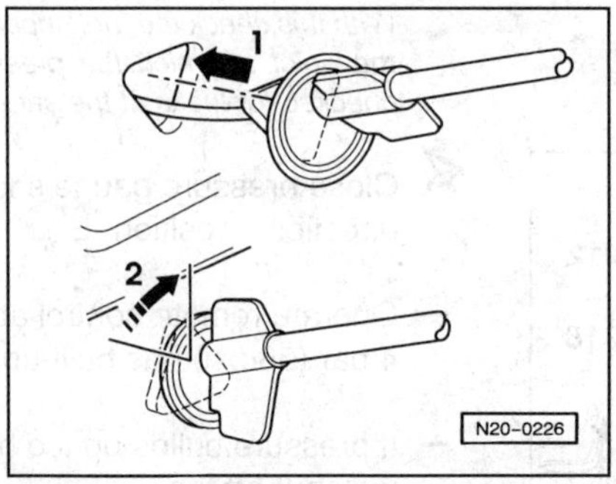

Fig. 1 Throttle cable mounting, removing and installing

– To remove, turn throttle cable mounting approx. 90° counter-clockwise and remove from bulkhead in plenum chamber.

– To install, push throttle cable through bulkhead in plenum chamber -arrow 1-.

– Push throttle cable mounting into bulkhead and turn approx. 90° clockwise -arrow 2-.

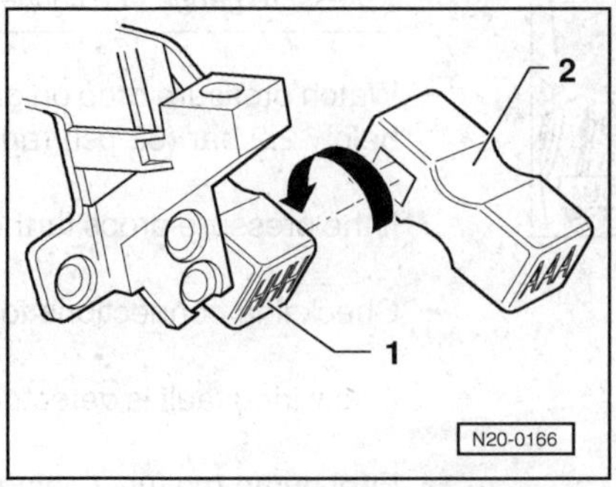

Fig. 2 Installation position of accelerator pedal stop on mounting bracket

- Letters "HHH" -1- are visible with manual transmission
- Letters "AAA" -2- are visible with automatic transmission.

Throttle cable, adjusting (AHA engine)

Vehicles with manual transmission

Work sequence

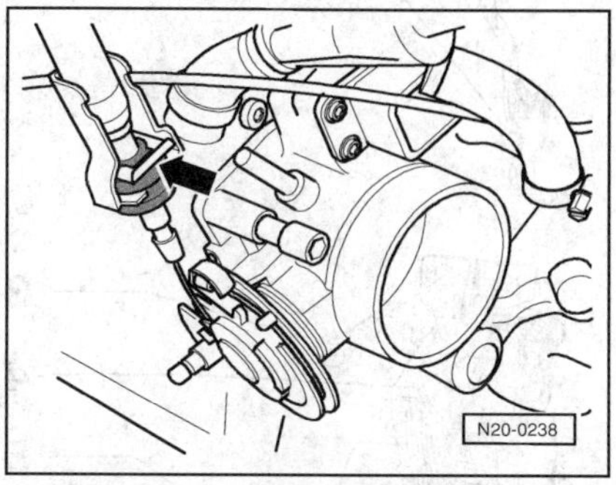

- Adjust accelerator cable by moving retainer -arrow- at support bracket, so that full throttle position is attained at throttle valve control module.

Vehicles with automatic transmission

Special tools and equipment

- Hand multimeter V.A.G 1526
- Adapter set V.A.G 1594

Work sequence

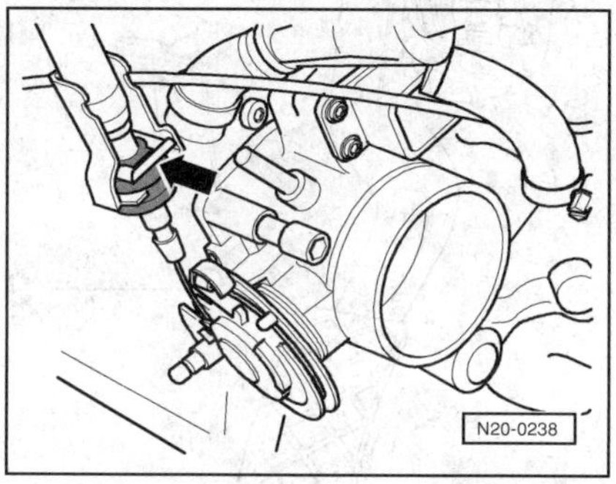

- Adjust accelerator cable by moving retainer -arrow- at support bracket, so that full throttle position is attained at throttle valve control module.

– Release accelerator pedal.

– Pull 2-pin connector off kick-down switch (F8) in engine compartment (on bulkhead).

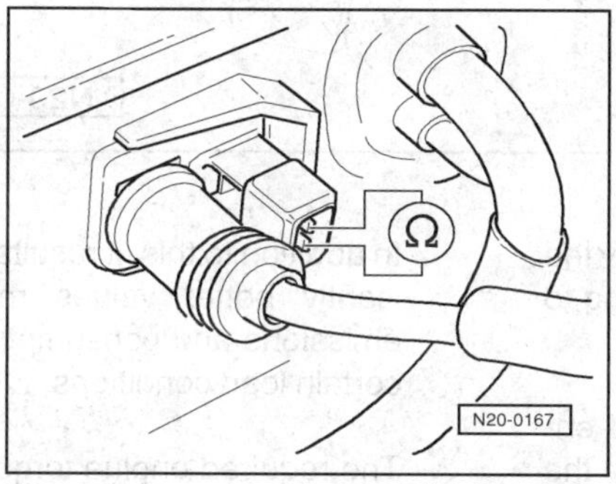

- Connect hand multimeter V.A.G 1526 with auxiliary cables from V.A.G 1594 to kick-down switch.

– Check resistance at switch.

Specification: ∞

– Slowly move accelerator pedal in direction of full throttle.

- Shortly after the kick-down pressure point the resistance must drop to 0. At this point the accelerator pedal must be just off the stop.

Electronic accelerator pedal, assembly (ATQ engine)

NOTE—
Vehicles with the engine code ATQ are equipped with a 121 pin control module.

1. **Mounting bracket**
   - Removing and installing, see **Brake System section, Repair Group 46, Brake pedal, assembly overview**
2. **Harness connector**
   - Black, 6-pin
3. **10 Nm (7 ft-lb)**
4. **Throttle Position (TP) sensor -G79-**
   - Checking, see **Throttle Position (TP) sensor, checking**
   - To remove, remove instrument panel cover and pedal assembly cover.
5. **Bracket**
   - For footwell cover
   - Clipped in at Throttle Position (TP) sensor

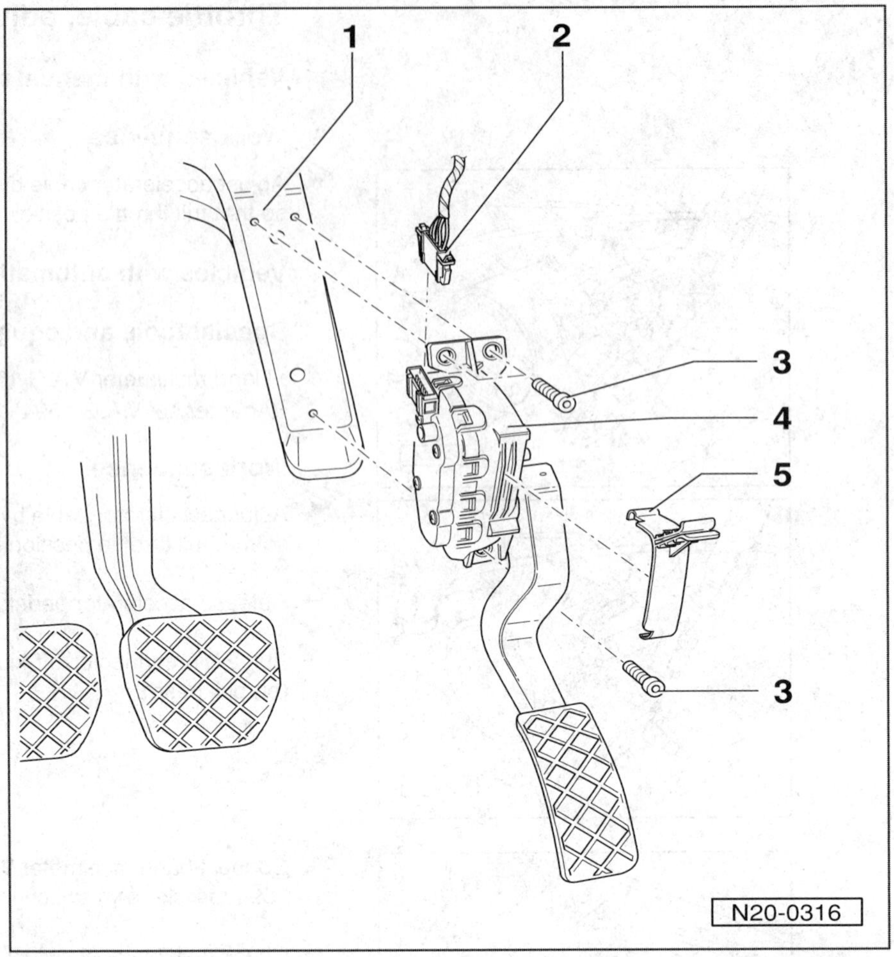

Electronic Power Control (EPC) system, function

For EPC or E-Gas, the throttle valve is not operated by a cable from the accelerator pedal. There is no mechanical connection between the accelerator pedal and the throttle valve.

The position of the accelerator pedal is communicated to the Engine Control Module (ECM) via two sensors for accelerator pedal position (variable resistances; stored in one housing) that are connected with the accelerator pedal.

The accelerator pedal position (driver's intention) is a main input unit for the ECM.

Operation of the throttle valve occurs via an electric motor (throttle valve actuator) in the throttle valve control module. This is true across the entire engine speed and engine load spectrum.

The throttle valve is operated by the throttle valve actuator according to the instructions of the ECM.

With ignition switched on and engine off, the ECM activates the throttle valve actuator precisely according to the sensor for accelerator pedal position. This means, if the accelerator pedal is depressed half way, the throttle valve actuator opens the throttle valve to the same degree; i.e. throttle valve would be opened half way.

With engine running (under load), the ECM can open or close the throttle valve independently of the sensor for accelerator pedal position.

This means, for example, that the throttle valve could be fully opened even though the accelerator pedal has only been depressed half way. This has the advantage of preventing torque losses at the throttle valve.

In addition to this, it results in significantly better values relative to emissions and consumption under certain load conditions.

The required engine torque can be achieved by the Engine Control Module (ECM) via the optimal combination of throttle valve opening and boost pressure.

It would be incorrect to think that "E-Gas" consists of only one or two components. E-Gas is much more of a system containing all components that contribute to recognizing, controlling and monitoring the position of the throttle valve, e.g. sensor for accelerator pedal position, throttle valve control module, EPC warning lamp, Engine Control Module (ECM).

Electronic power control (EPC) warning light -K132-, checking

Special tools and equipment

- VAG1598/31 test box
- VAG1526 or VAG1715 multimeter
- VAG1594 connector test kit
- Wiring diagram

Function test of EPC warning lamp

– Switch ignition on.

- EPC warning lamp must light up.

– Start engine and let run at idle.

- EPC warning lamp must go out if there are no E-Gas DTCs stored.

EPC warning lamp does not light up

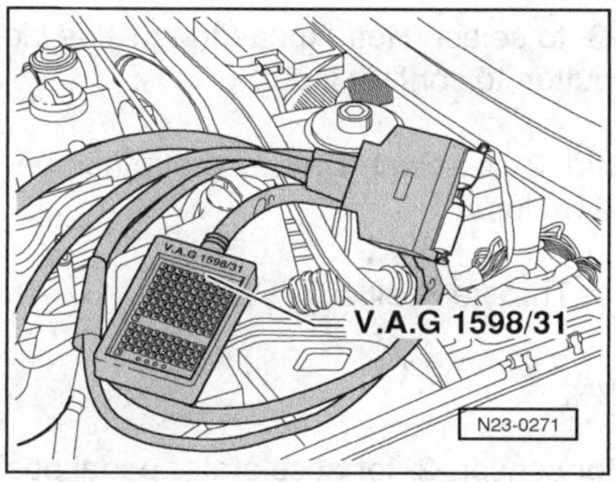

– Connect VAG1598/31 test box to control module wiring harness. Control module is not connected.

– Bridge test box sockets 1 + 48 using adapter cables from VAG1594 connector test kit.

– Switch ignition on.

- Warning lamp must light up.

If warning lamp does not light up:

– Switch ignition off.

– Remove instrument cluster and check bulb and its voltage supply, see **Electrical Equipment section, Repair Group 90.**

If no malfunctions are found:

– Check wiring between test box socket 48 and instrument cluster for open circuit according to wiring diagram.

If no malfunctions are found in wire:

– Replace Engine Control Module (ECM)

EDITOR'S NOTE —

ECM replacement requires coding of control module using Volkswagen factory scan tool. Consult an authorized Volkswagen service department for additional information.

EPC warning lamp is constantly lit

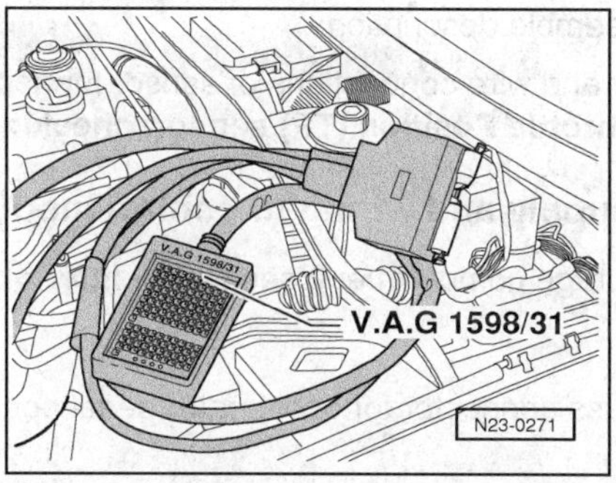

– Connect VAG1598/31 test box to control module wiring harness. Control module is not connected.

– Check wiring between test box socket 48 and instrument cluster for short circuit to Ground (GND) according to wiring diagram.

If no malfunctions are found in wire:

– Replace Engine Control Module (ECM)

Throttle Position (TP) sensor, checking

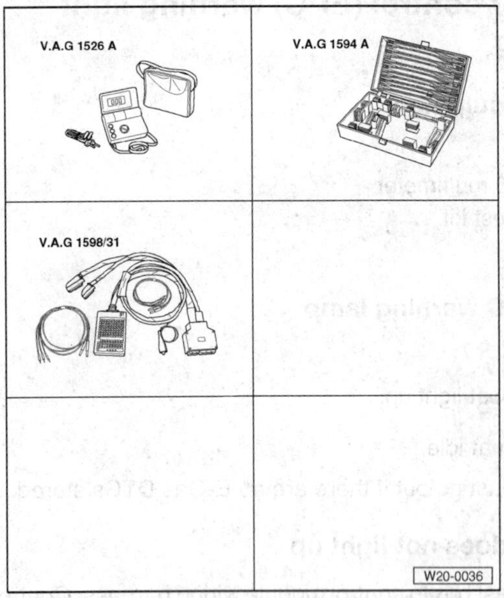

Special tools and equipment

- VAG 1526A Multimeter
- VAG 1594A Connector test kit
- VAG 1598/31 Test box
- VAG 1551 Scan tool (or VAG1552 Mobile scan tool) with VAG 1551/3 Cable
- Wiring diagram

Both the Throttle Position (TP) sensor -G79- and the Sensor -2- for accelerator pedal position -G185- are located at the accelerator pedal and communicate the driver's intentions to the ECM completely independently of each other. Both sensors are stored in one housing.

– Connect VAG 1551 Scan tool (or VAG 1552) and select the control module for engine electronics via address word 01. Ignition must be switched on for this, see **Maintenance.**

– Press buttons -0- and -8- to select "Read measuring value block" function 08. Press -Q- button to confirm input.

– Press buttons -0-, -6- and -2- to select "Display group number 62". Press -Q- button to confirm input.

– Check specified value for Throttle Position (TP) sensor -G79- at idle stop in display field 3.

- Specification: 12 to 97%

– Check specified value for sensor -2- for accelerator pedal position -G185- at idle stop in display field 4.

- Specification: 4 to 49%

NOTE—
Engine Control Module (ECM) calculates the voltage values of the angle sensors in percent relative to 5 volts and displays these percentage values. (5 volts supply voltage represents 100%).

– Depress accelerator pedal slowly until Wide Open Throttle (WOT), and observe percent indication in display fields 3 and 4:

- Percent indication in display field 3 must rise uniformly. The tolerance range 12 to 97% is not used completely during this.
- Percent indication in display field 4 must rise uniformly. In the process, the tolerance range of 4 to 49% is not completely utilized.

NOTE—
The indicated value in display field 3 must always be about twice as large as the value in display field 4.

– If indications do not resemble description:

- Check voltage supply and wire connections of sensor for accelerator pedal position, see **Throttle Position (TP) sensor, checking**.

Checking voltage supply and wiring to control module

– Remove driver-side storage compartment, see **Body–Interior section, Repair Group 68.**

– Disconnect 6-pin harness connector for accelerator pedal position sensor.

– Switch ignition on.

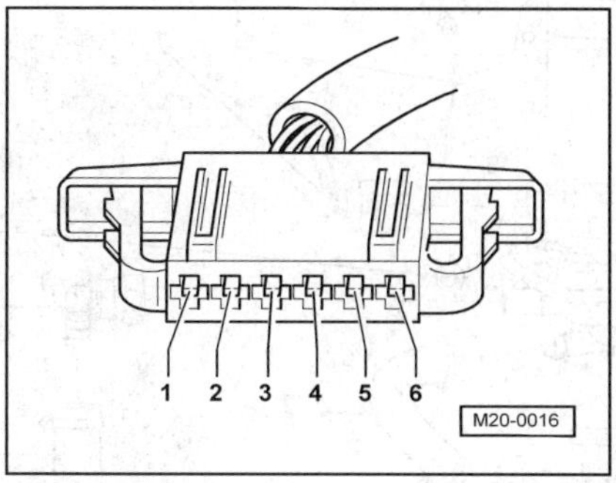

Connect multimeter to the following connector terminals for voltage measurement.

- Terminal 1 + Ground (GND)
- Terminal 1 + 5
- Terminal 2 + Ground (GND)
- Terminal 2 + 3
- Specification: at least 4.5 V

– Switch ignition off.

If specified values are not obtained:

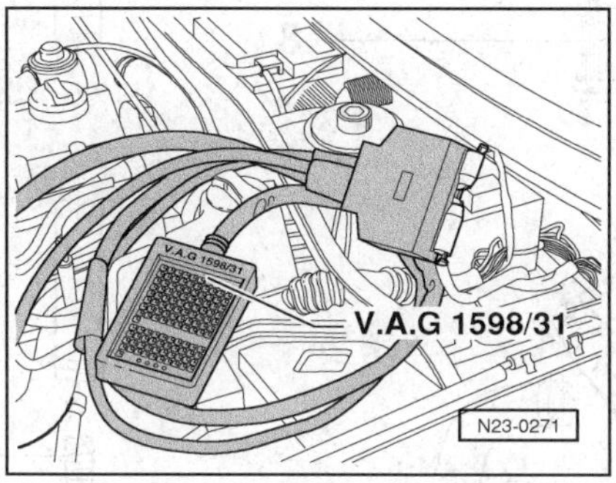

Connect VAG 1598/31 Test box to control module wiring harness. Engine Control Module (ECM) is not connected.

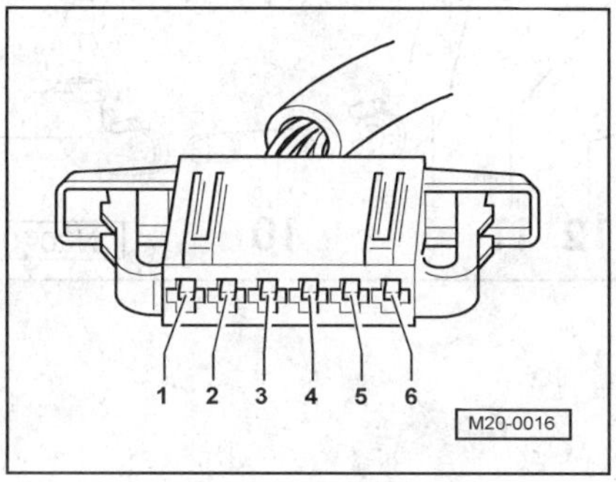

Check for open circuits in wires between test box and connector according to wiring diagram.

- Terminal 1 + socket 72
- Terminal 2 + socket 73
- Terminal 3 + socket 36
- Terminal 4 + socket 35
- Terminal 5 + socket 33
- Terminal 6 + socket 34
- Wire resistance: max: 1.5 Ω

– Also check wires for short circuits to B+ or Ground (GND).

– Also check wires for short circuit to each other.

– If no malfunctions are found in wires:

- Replace accelerator pedal position sensor, see **Electronic engine power control (EPC), checking**.

Evaporative Emissions (EVAP) system

Function description of leak diagnostic

The EVAP canister system (incl. fuel tank) is equipped with a leak diagnostic that detects leaks (improper seal) in the system.

The diagnostic is based on pressure characteristics and is intended to detect leaks with diameters greater than 1mm.

During the diagnostic, the Leak Detection Pump (LDP) -V144- generates a positive pressure of approx. 30 mbar in the EVAP canister system. The pump is switched off once the pressure is reached. If pressure falls below a specified value, the pump is switched on again. On Board Diagnostic monitors the switch condition and stores a DTC in DTC memory if the intervals are too short.

EVAP components, removing and installing

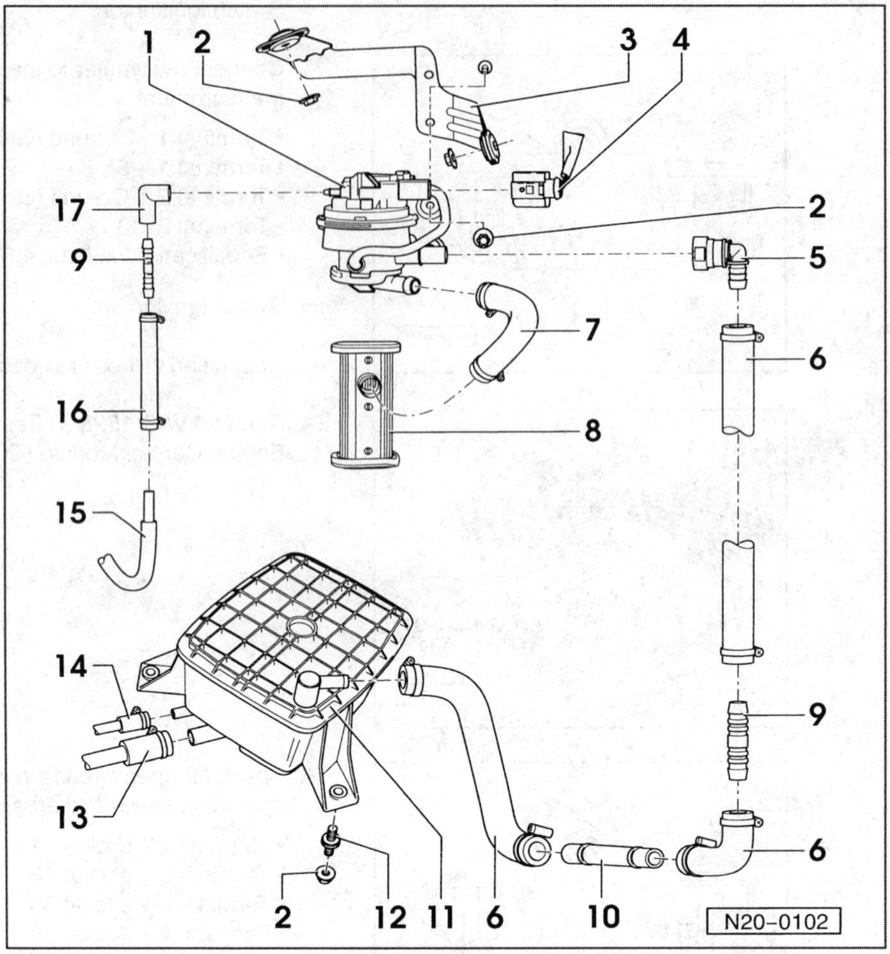

1. **Leak Detection Pump (LDP) -V144-**
   - In rear left wheel housing below wheel housing liner
   - Checking, see **Leak Detection Pump (LDP), checking**
2. **10 Nm (7 ft-lb)**
3. **Bracket for LDP**
4. **Connector**
   - 3 pin
5. **Coupling**
   - Press together at front to release
6. **Connecting hose**
   - Pressure side
7. **Connecting hose**
   - Suction side
8. **Air cleaner for LDP**
9. **Connector**
10. **Connecting line**
11. **EVAP canister**
    - In spare wheel well on vehicle floor
12. **10 Nm**
13. **To EVAP canister purge regulator valve 1 (N80)**
    - In engine compartment on air cleaner, see **EVAP system, overview**, item **-11-**
14. **Vent line**
    - From change-over valve on fuel tank
15. **Vacuum line**
    - To throttle valve control module (J388), see **EVAP system, overview**, item **-9-**
16. **Connecting hose**
    - Vacuum side
17. **Elbow**

EVAP system, overview

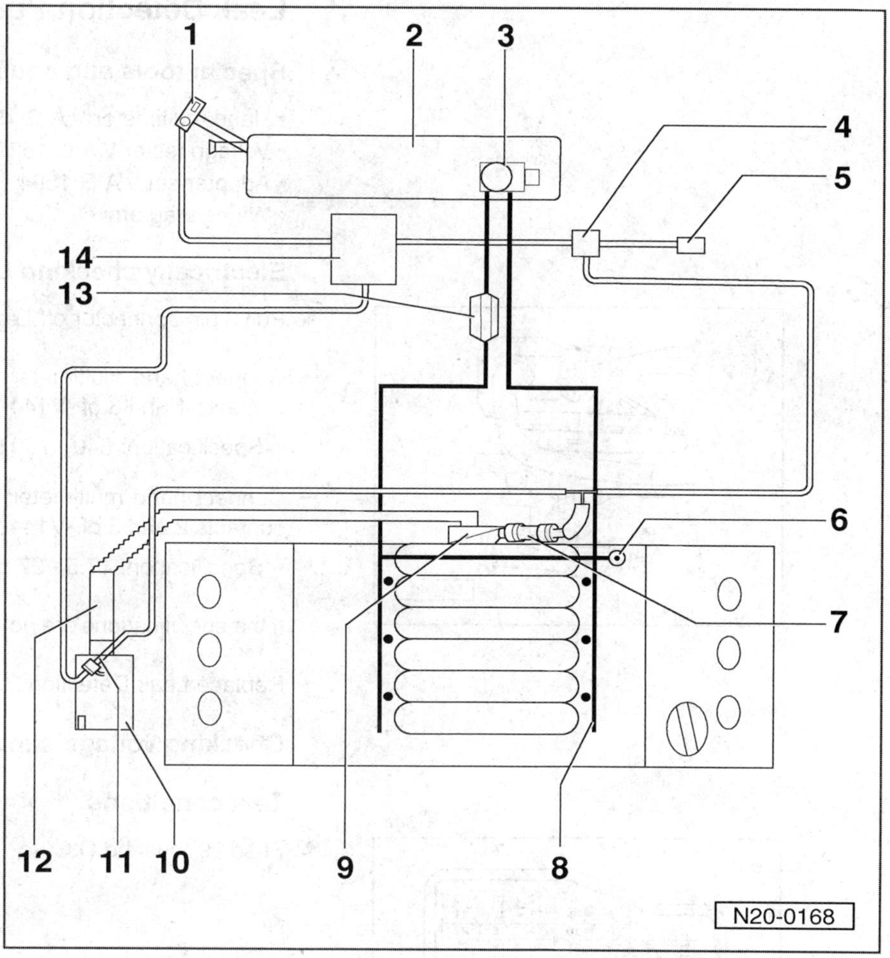

1. **Tank cap unit**
   - With rubber cup
2. **Fuel tank**
   - Support with engine and transmission jack V.A.G 1383 A when removing
3. **Fuel delivery unit**
   - Removing and installing, see **Fuel delivery unit, removing and installing**
4. **Leak Detection Pump (LDP) -V144-**
   - In rear left wheel housing below wheel housing liner
   - Checking, see **Leak Detection Pump (LDP), checking**
5. **Air cleaner for LDP**
6. **Fuel pressure regulator**
7. **Vacuum booster**
8. **Fuel rail with injectors**
9. **Throttle valve control module (J338)**
10. **Air cleaner housing**
11. **EVAP canister purge regulator valve 1 (N80)*/****
    - Checking activation using scan tool, see **Maintenance**
12. **Mass Air Flow meter (G70)***
13. **Fuel filter**
    - Fitting position: Arrow points in through-flow direction
14. **EVAP canister**

Leak Detection Pump (LDP), checking

Special tools and equipment

- Hand multimeter V.A.G 1526
- Voltage tester V.A.G 1527B
- Adapter set V.A.G 1594
- Wiring diagram

Electrically checking LDP

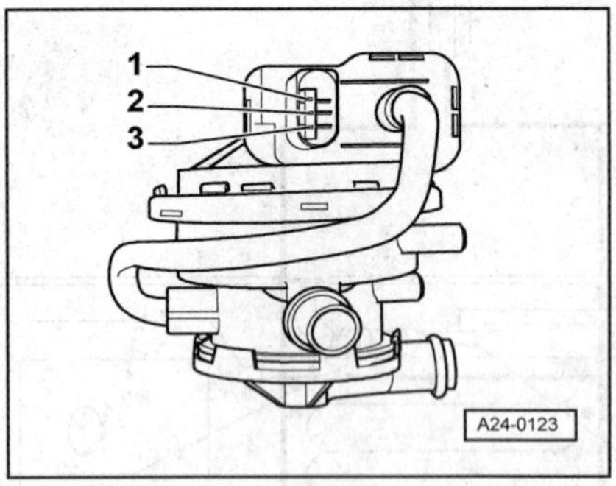

Pull 3 pin connector off Leak Detection Pump (V144).

– Connect hand multimeter V.A.G 1526 to measure resistance between contacts 1 and 3 of -V144-.

- Specification: 640 - 720 Ω.

– Connect hand multimeter V.A.G 1526 to measure resistance between contacts 2 and 3 of -V144-.

- Specification: 17.5 - 27.5 Ω.

If the specifications are not attained:

– Replace Leak Detection Pump -V144-.

Checking voltage supply

Test conditions

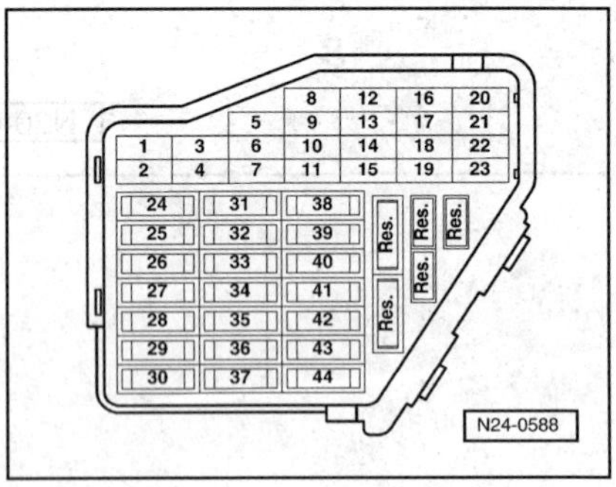

Fuse 29 must be OK.

Test sequence

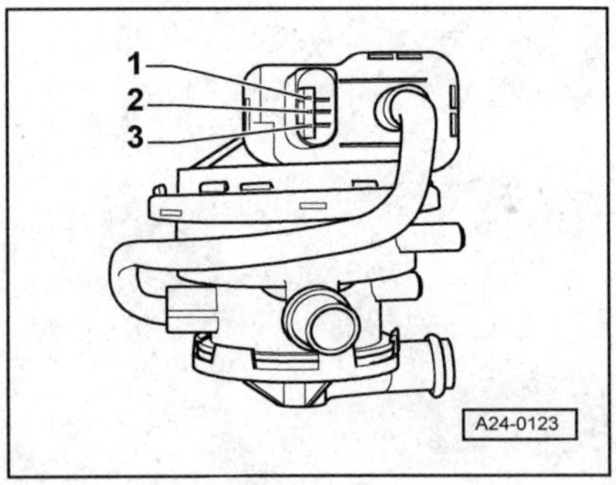

Pull 3 pin connector off Leak Detection Pump -V144-.

– Connect voltage tester V.A.G 1527B between connector contact 3 (positive) and engine Ground.

– Start engine and run at idling speed:

- LED must light up.

If the LED does not light up:

– Check wire between connector contact 3 and relay plate for open circuit using wiring diagrams.

26 Exhaust System and Emission Controls

Repair Group

D 26

Exhaust System

NOTE—

- *After working on the exhaust system ensure that the system is not under stress, and that it has sufficient clearance from the bodywork. If necessary, loosen double clamps and align muffler and exhaust pipe so that sufficient clearance is maintained to the bodywork and the support rings are evenly loaded.*
- *Replace self-locking nuts.*

EDITOR'S NOTE—

Volkswagen identifies electrical components by a letter and/or a number in the electrical schematics. See **Wiring Diagrams, Fuses and Relays**. *These electrical identifiers are listed -between dashes- as an aid to electrical troubleshooting.*

Exhaust manifold, front exhaust pipe and catalyst with attachments, overview

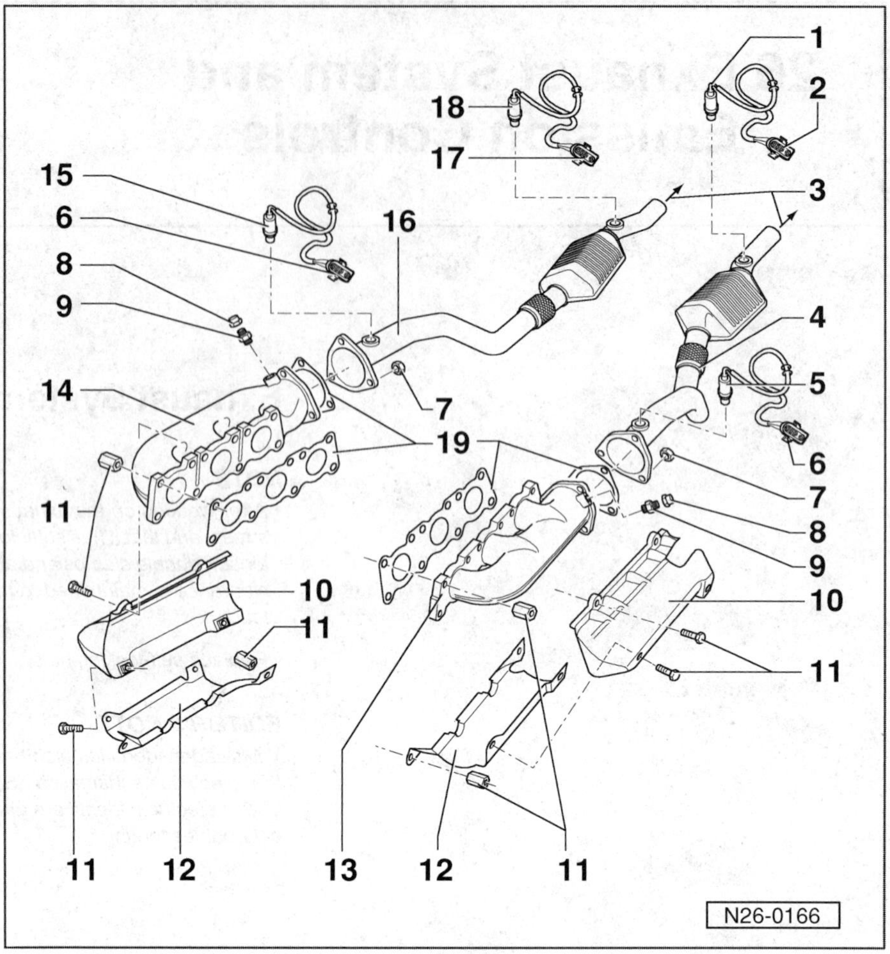

1. **Oxygen Sensor (O2S) 2 Behind Three Way Catalytic Converter -G131-,**
   - 50 Nm (37 ft-lb)
   - Cylinder bank 2
   - Grease only the threads with "G5"; "G5" must not get into the slots on the sensor body
   - Check using diagnostic scan tool
   - Electrical checks, see **2.8 Liter Fuel Injection & Ignition section, Repair Group 24**
2. **4-pin connector**
   - Brown
   - Secured to bulkhead
   - For Oxygen Sensor (O2S) 2 Behind Three Way Catalytic Converter -G131- and Heated Oxygen Sensor (O2S) -Z30-, cyl. bank 2
3. **To front muffler**
4. **Front left exhaust pipe w/catalyst**
5. **Heated Oxygen Sensor (HO2S) 2 -G108-,**
   - 50 Nm (37 ft-lb)
   - Grease only the threads with "G5"; "G5" must not get into the slots on the sensor body
   - Check using diagnostic scan tool
   - Electrical checks, see **2.8 Liter Fuel Injection & Ignition section, Repair Group 24**
6. **4-pin connector**
   - Black
   - Secured to bulkhead
   - For Heated Oxygen Sensor (HO2S) 2 -G108-* and Oxygen Sensor (O2S) Heater -Z28-, bank 2
7. **25 Nm (18 ft-lb)**
8. **Cover cap**
   - 10 Nm (7 ft-lb)
9. **Connection**
10. **Upper heat shield**
11. **25 Nm (18 ft-lb)**
12. **Lower heat shield**
13. **Left exhaust manifold**
14. **Right exhaust manifold**
15. **Heated Oxygen Sensor (HO2S) -G39-**
    - 50 Nm (37 ft-lb)
    - Grease only the threads with "G5"; "G5" must not get into the slots on the sensor body
    - Check using diagnostic scan tool
    - Electrical checks, see **2.8 Liter Fuel Injection & Ignition section, Repair Group 24**
16. **Front right exhaust pipe with catalyst**
17. **4-pin connector**
    - Green
    - Secured to bulkhead
    - For Oxygen Sensor (O2S) Behind Three Way Catalytic Converter (TWC) (G130) and Oxygen Sensor (O2S) Heater -Z29-, bank 1
18. **For Oxygen Sensor (O2S) Behind Three Way Catalytic Converter (TWC) -G130-**
    - 50 Nm (37 ft-lb)
    - Grease only the threads with "G5"; "G5" must not get into the slots on the sensor body
    - Check using diagnostic scan tool
    - Electrical checks, see **2.8 Liter Fuel Injection & Ignition section, Repair Group 24**
19. **Gasket**
    - Always replace

Muffler with mountings (4WD vehicles)

1. **40 Nm (30 ft-lb)**
2. **From catalyst**
3. **Carriage bolt**
4. **Double clamp**
5. **25 Nm (18 ft-lb)**
6. **Mounting**
   - With retaining ring
7. **Center muffler**
8. **Repair double clamp**
9. **Separating point**
   - As standard, center and rear mufflers are installed as a single component. In cases of repair the center and rear muffler are supplied individually and with a double clamp for connecting.
   - Cut through connecting pipe with body saw, e.g. V.A.G 1523, at right angles at the separating point see **Fig. 1**
10. **Right rear muffler**
11. **Mounting**
12. **Bracket**
13. **Left rear muffler**

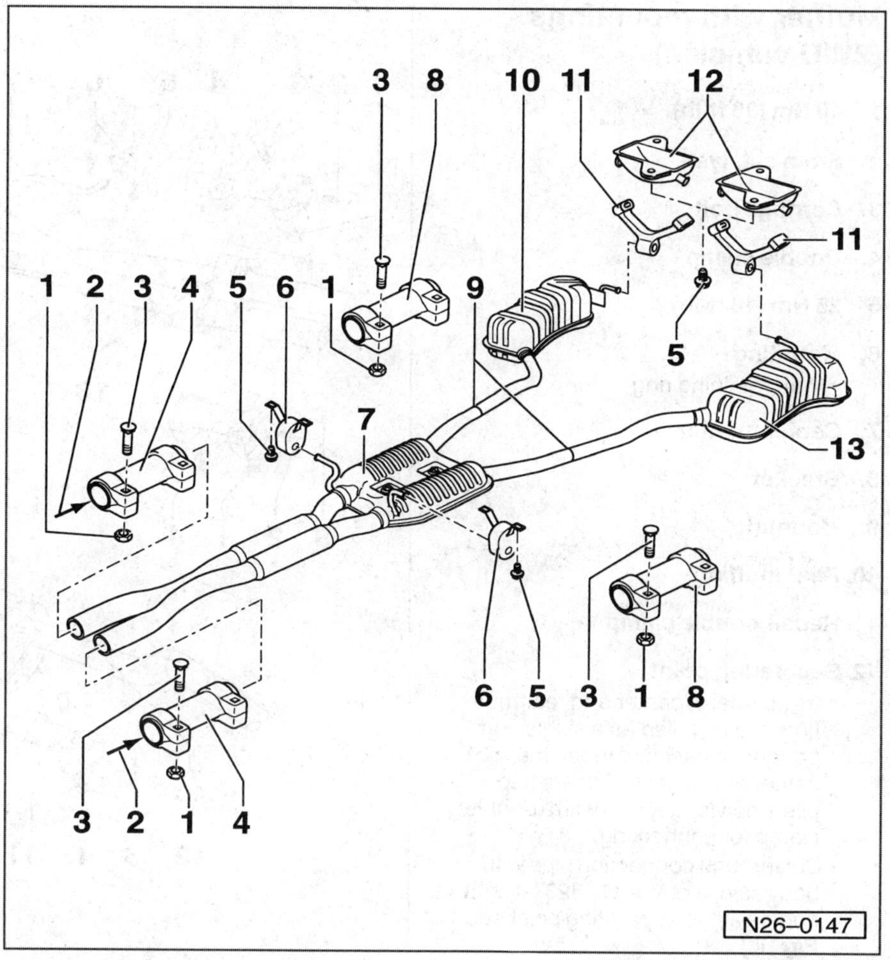

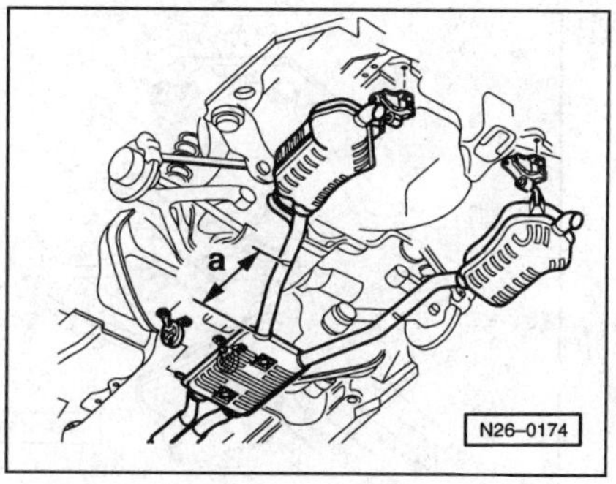

Fig. 1 Separating point between center and rear mufflers

Special tools and equipment

- Body saw e.g. V.A.G 1523

Work sequence

– Separate connecting pipe between center and rear mufflers at point marked with a groove:

Dimension -a- approx. 244 mm (9.5 in.)

Muffler with mountings (2WD vehicles)

1. 40 Nm (30 ft-lb)
2. From catalyst
3. Carriage bolt
4. Double clamp
5. 25 Nm (18 ft-lb)
6. Mounting
   - With retaining ring
7. Center muffler
8. Bracket
9. Mounting
10. Rear muffler
11. Repair double clamp
12. Separating point
    - As standard, center and rear mufflers are installed as a single component. In cases of repair the center and rear muffler are supplied individually and with a double clamp for connecting.
    - Cut through connecting pipe with body saw, e.g. V.A.G 1523, at right angles at the separating point see **Fig. 1**
13. Front muffler

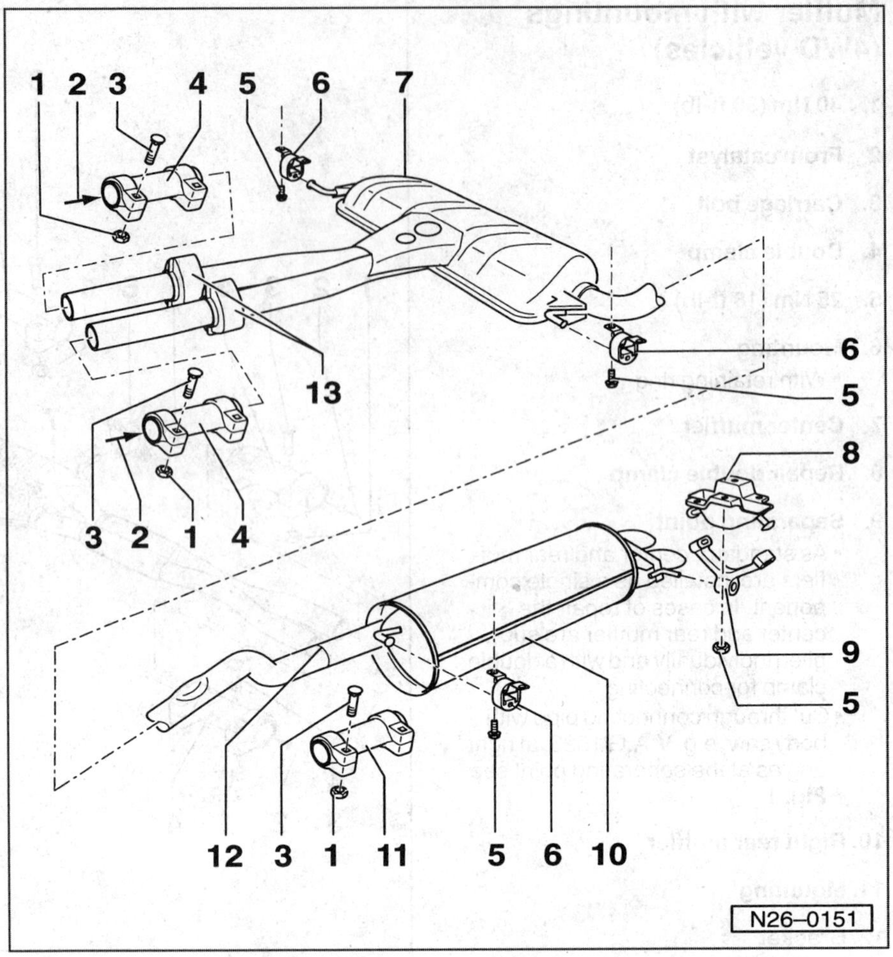

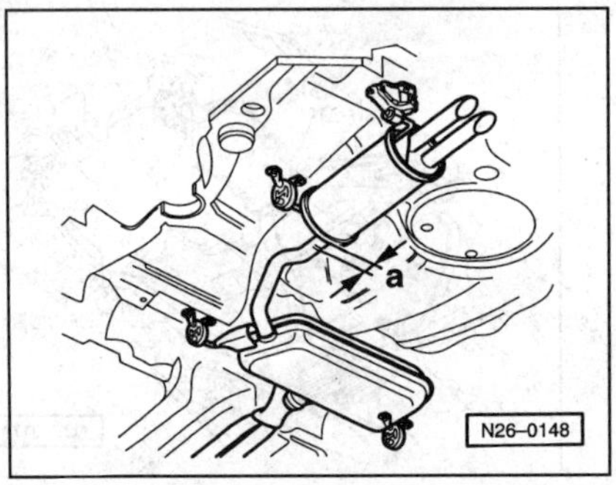

Fig. 1 Separating point between center and rear mufflers

Special tools and equipment

- Body saw e.g. V.A.G 1523

Work sequence

– Separate connecting pipe between center and rear mufflers at point marked with a groove:
  - Dimension -a- approx. 182 mm (7 in.)

Secondary Air System

Function

The secondary air system blows air in behind the exhaust valve for 65 seconds during the cold start (coolant temperature 15°C (59°F) to 35°C (95°F). This produces an oxygen rich exhaust gas, causes afterburning and reduces the heating-up phase of the catalyst. Activation occurs from the Motronic control module (J220) via the Secondary Air Injection (AIR) pump relay -J299- to Secondary Air Injection (AIR) valve -N112- and combi-valve. Additionally after each subsequent engine start (up to max. 85°C (185°F) engine temperature) the secondary air system will after a 20 second delay switch in for 5 seconds during idling and will be checked by the On Board Diagnostic (OBD).

Secondary air injection pump motor, assembly

NOTE—
- *Components marked with an asterisk* are checked by the On Board Diagnostic (OBD):*
- *Components marked with ** are checked by the output diagnostic test mode.*
- *Secondary Air Injection (AIR) pump relay -J299-**, see* **Fig. 1**

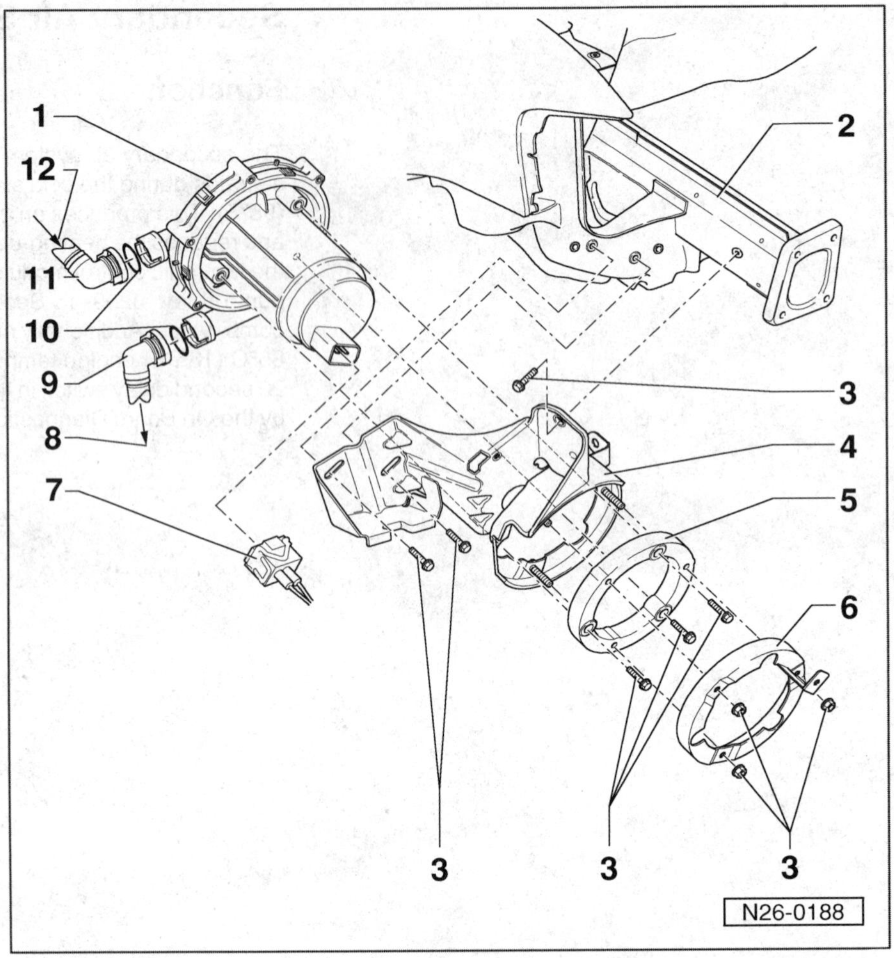

1. **Secondary Air Injection (AIR) pump motor V101*/****
2. **Front right longitudinal member**
3. **10 Nm (7 ft-lb)**
4. **Bracket**
   - For secondary air injection pump motor
   - Secured to intake manifold
5. **Insulation ring**
   - Securing collar must face to pump eye
6. **Protective cap**
7. **Connector**
   - For secondary air injection pump motor
   - Black, 2-pin
8. **To connecting pipe**
9. **Pressure hose**
   - Ensure seated tightly
   - Press together at front to release
10. **O-Ring**
    - Always replace
11. **Intake hose**
    - Ensure seated tightly
    - Press together at front to release
12. **From air cleaner upper part**

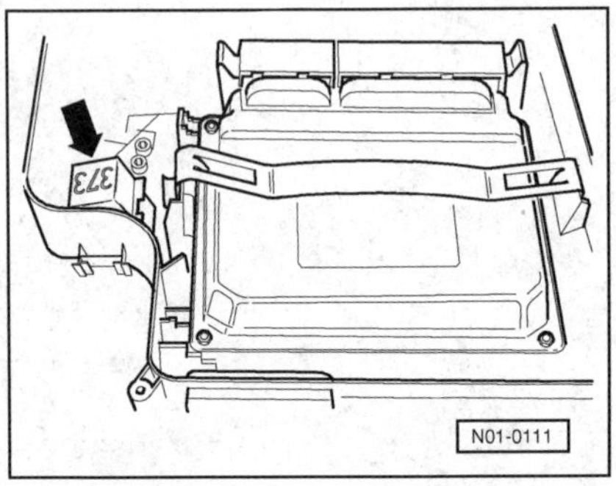

Fig. 1 Secondary Air Injection (AIR) pump relay -J299- -arrow-

Combination Valves

Connecting pipe and combi-valves, assembly overview

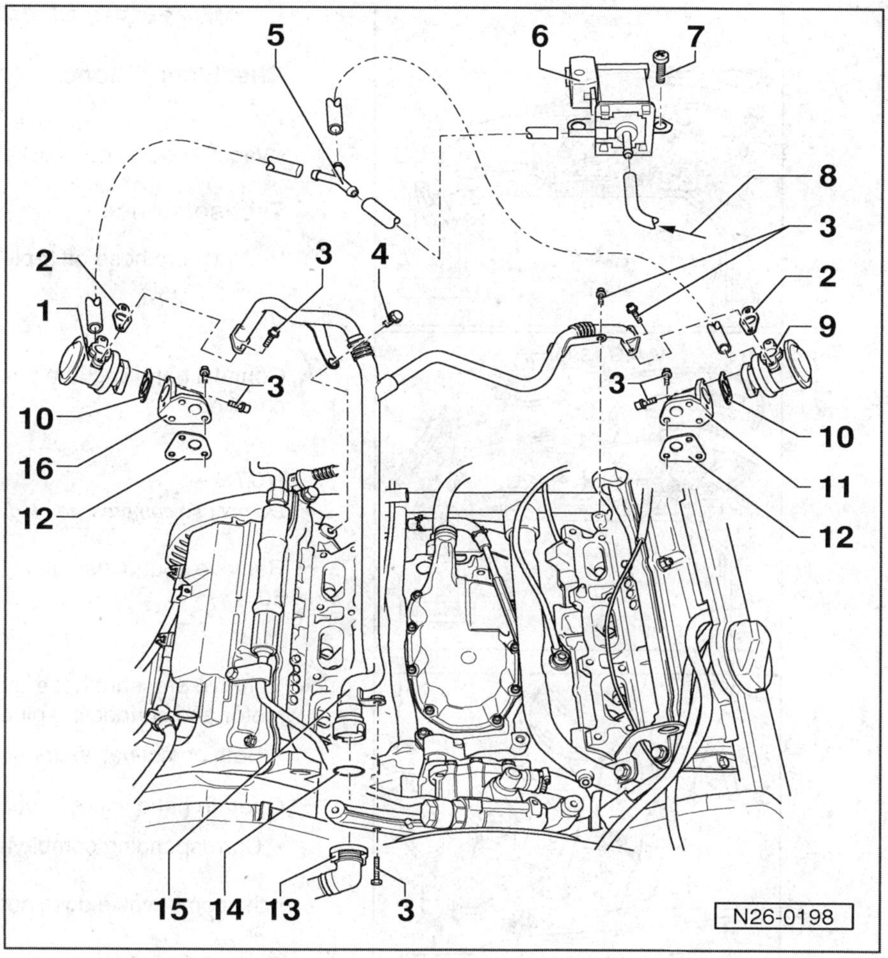

NOTE—
- *Components marked with an asterisk* are checked by the On*
- *Components marked with ** are checked by the output diagnostic test mode:*

1. **Combi-valve**
   - See **Combi-valve, checking**
   - See **Combi-valve, removing and installing**
2. **Gasket**
   - Always replace
3. **10 Nm (7 ft-lb)**
4. **25 Nm (18 ft-lb)**
5. **Y-piece**
6. **Secondary Air Injection (AIR) Solenoid valve -N112-*/****
7. **6 Nm (53 in-lb)**
8. **From vacuum reservoir**
   - Location: In left wheel housing
9. **Combi-valve**
   - See **Combi-valve, checking**
   - See **Combi-valve, removing and installing**
10. **Gasket**
    - Always replace
11. **Connection**
12. **Gasket**
    - Always replace
13. **Pressure hose**
    - Ensure seated tightly
    - Press together at front to release
14. **O-Ring**
    - Always replace
15. **Connecting pipe**
16. **Connection**

Combi-valve, checking

Check conditions

• Vacuum pipes and hose connections free of leaks.
• Vacuum pipes not blocked/kinked.

Test sequence

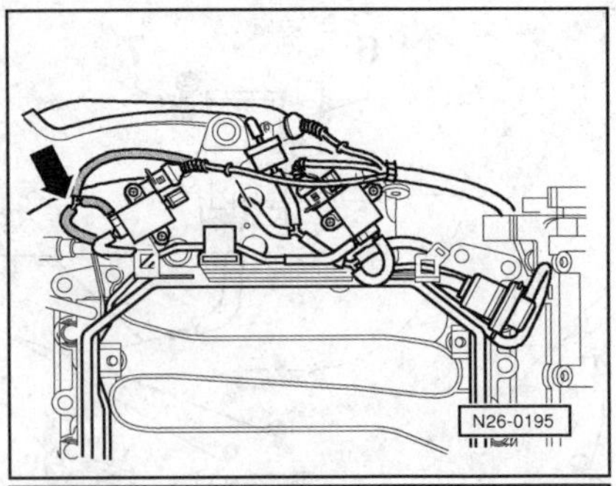

Pull vacuum hose off Y-piece -arrow- of combi-valve to be checked.

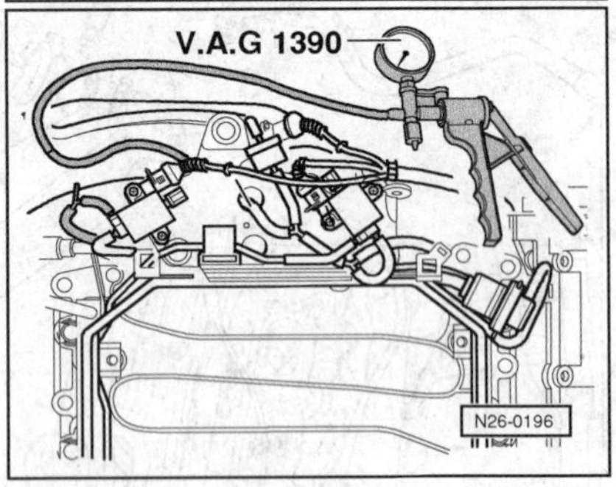

Connect hand vacuum pump V.A.G 1390 to vacuum hose of valve to be checked.

NOTE—
Do not use compressed air during the following check!

– Remove sound insulation, see **Body–Exterior section, Repair Group 50**

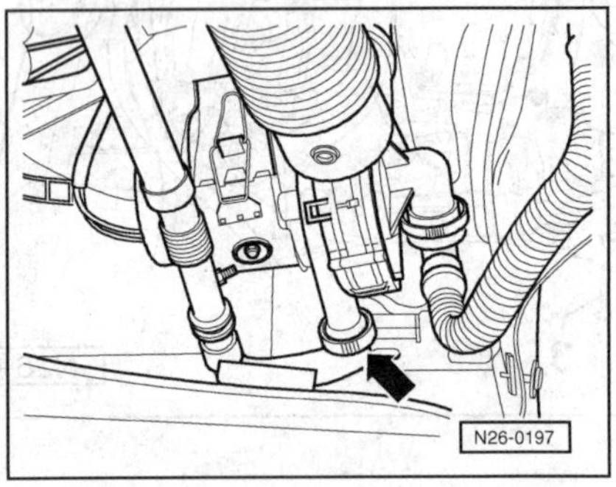

Remove pressure hose -arrow- between secondary air injection pump motor and connecting pipe and blow in with slight pressure.

• Both combi-valves must be closed.

– Operate hand vacuum pump.

• Corresponding combi-valve must open.

If the combi-valve does not open, replace corresponding combi-valve.

Combi-valve, removing and installing

NOTE—
Always replace all gaskets and seals.

Left combi-valve

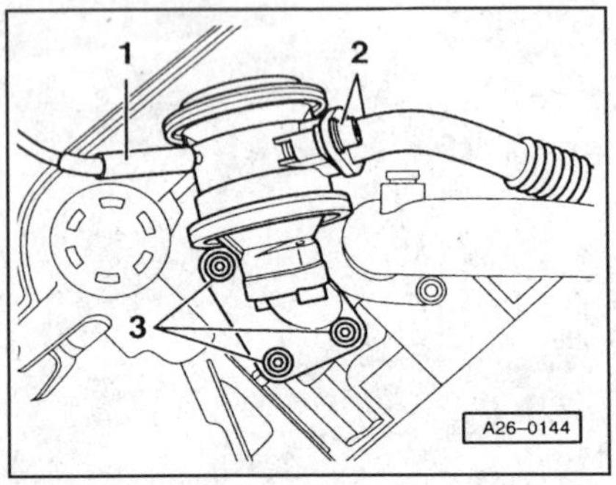

– Remove coolant system expansion reservoir and set to one side.

Pull vacuum hose -1- off combi-valve.

– Unscrew securing bolts -2- and remove connecting pipe.

– Unscrew securing bolts -3- and remove combi-valve from cylinder head.

Right combi-valve

– Remove connecting hose between air cleaner and intake air pipe:

– Remove intake air pipe from throttle valve control module:

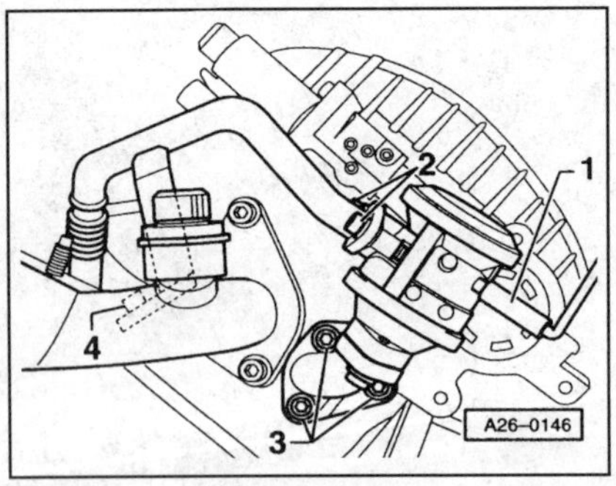

Pull vacuum hose -1- off combi-valve.

– Unscrew securing bolts -2- and remove connecting pipe.

– Unscrew securing bolts -3- and remove combi-valve from cylinder head.

– If necessary unscrew securing bolts -4- for connecting pipe retainer.

– Install in combi valves in reverse order of removal.

4.0 Liter W8 4V Engine Mechanical (code: BDP)

Section **E**

00 General, Technical Data

Repair Group

E 00

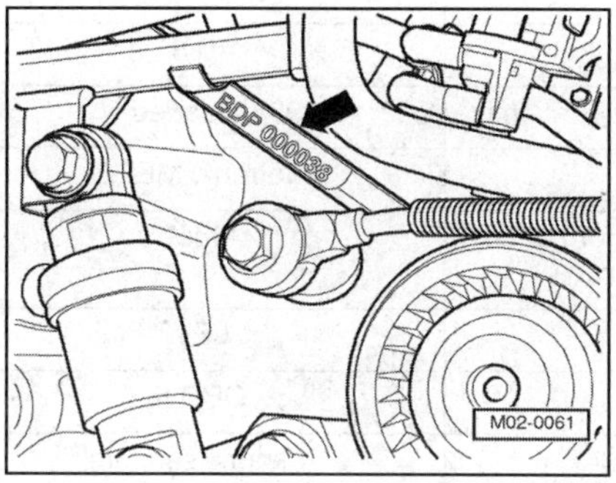

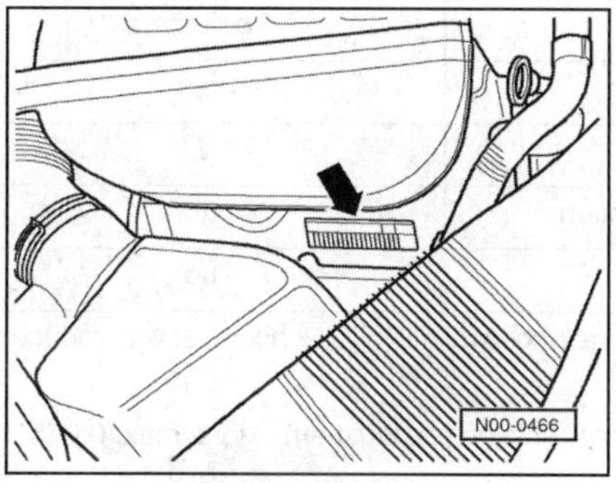

Technical data

Engine number

◀ The engine number -arrow- ("engine code" and "serial number") is located left on the cylinder block.

The engine number consists of up to nine characters (alphanumeric). The first part (maximum 3 characters) makes up the "engine code", and the second part (6 characters), the "serial number". If more than 999,999 engines with the same engine code are produced, the first of the six characters is replaced with a letter.

◀ Additionally there is a sticker -arrow- on the cylinder head cover with "engine code" and "serial number".

The engine code is also included on the vehicle data plate.

Engine data

Engine code		BDP
Manufactured		from 04.02
Capacity	ltr.	4.0
Output	hp (kw) at rpm	270 (202) /6000
Torque	ft-lb (Nm) at rpm	272 (370) /3250
Bore	diameter mm	84.0
Stroke	mm	90.2
Compression ratio		10.5
RON	min.	95 unleaded [2)]
System designation		Motronic ME7.1
Electronic engine performance control		yes
Emissions standard		LEV
Self-diagnosis		OBD II
Knock control		4 knock sensors
Oxygen sensor control		4 sensors
Variable valve timing		4 adjusters
Secondary air system		yes
Catalysts		2
Exhaust gas recirculation		no
Charging		no

[1)] In exceptional circumstances min. 93 RON, however with reduced performance

[2)] LEV = Low Emission Vehicles (exhaust emissions max. 0.075 g/mi HC)

10 Engine–Assembly

Repair Group

E 10

Engine, removing and installing

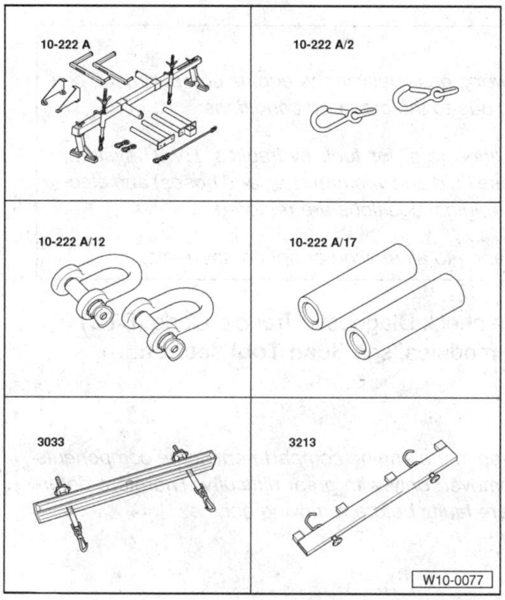

Special tools and equipment

- 10-222 A Support bracket with legs 10-222 A/1
- 10-222A/2 Additional hooks
- 10-222A/12 Shackle
- 10-222A/17 Spacer tube
- 3033 Lifting tackle
- 3213 Assembly device

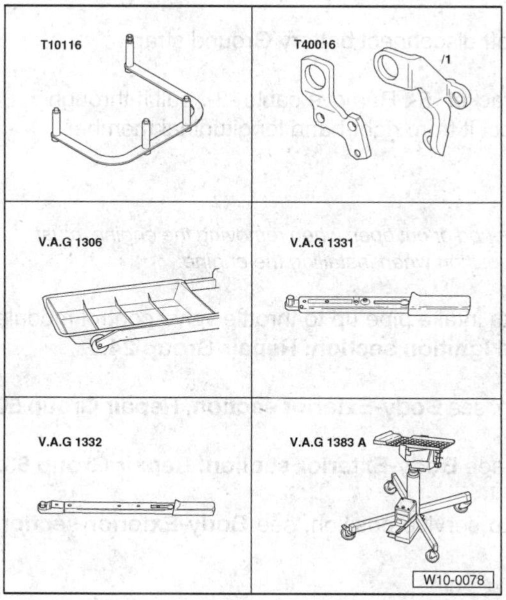

- T10116 Engine stand
- T40016 Transport bracket
- V.A.G 1306 Drip tray
- V.A.G 1331 Torque wrench (5...50 Nm)
- V.A.G 1332 Torque wrench (40...200 Nm)
- V.A.G 1383 A Engine/transmission jack

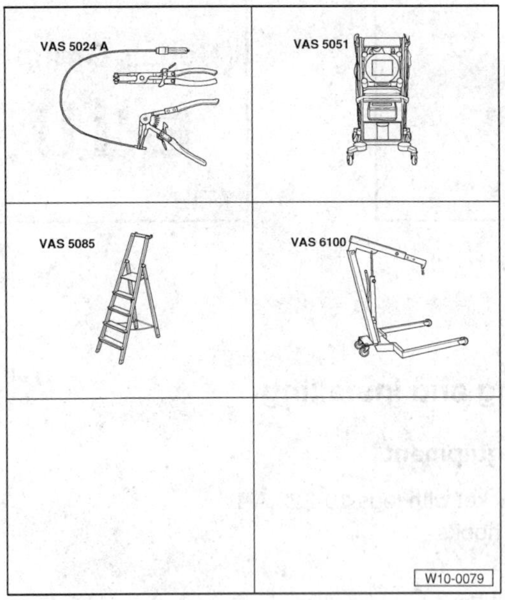

- VAS 5024 Fitting tool for spring-type clips
- VAS 5051 Vehicle Diagnostic, Testing and Information System
- VAS 5085 Step ladder
- VAS 6100 Workshop crane

Not illustrated:

- G 000 100 Grease (vehicles with manual transmission)
- Cable tie

Engine, removing

The engine with transmission is removed in a forward direction.

> ***WARNING —***
> *When doing any repair work, especially in the engine compartment, pay attention to the following due to the cramped conditions:*
>
> - *Route all the various lines (e.g. for fuel, hydraulics, EVAP system, coolant, refrigerant, brake fluid and vacuum lines and hoses) and electrical wiring so that the original positions are restored.*
> - *Ensure sufficient clearance to all moving or hot components.*

– Before removing engine check Diagnostic Trouble Code (DTC) memories of all control modules, see **Scan Tool section**.

NOTE —
When the engine is installed in the engine compartment some components cannot be removed, or removed only with great difficulty. Therefore determine which components are faulty before removing engine.

Work sequence

– Switch off ignition and all electrical consumers and remove ignition key.

– With ignition switched off disconnect battery Ground strap.

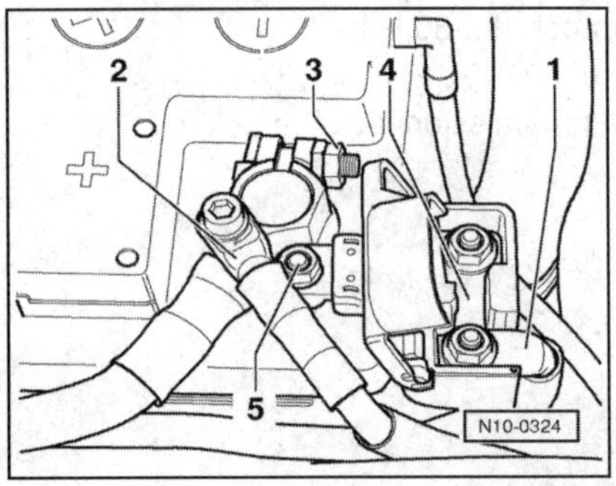

– Release line -2- and bracket -5-. Remove cable -3-, pull it through bulkhead and disconnect it from right-hand longitudinal member.

NOTE —
All cable ties which are opened or cut open when removing the engine, must be replaced in the same position when installing the engine.

– Remove air cleaner with intake pipe up to throttle valve control module, see **Fuel Injection and Ignition section, Repair Group 24**.

– Remove noise insulation, see **Body-Exterior section, Repair Group 50.**

– Remove front bumper, see **Body-Exterior section, Repair Group 63.**

– Position lock carrier into service position, see **Body-Exterior section, Repair Group 50.**

– Extract refrigerant from air conditioning system.

EDITOR'S NOTE —
Air conditioning refrigerant must be evacuated by a trained service professional using proper refrigerant recycling equipment.

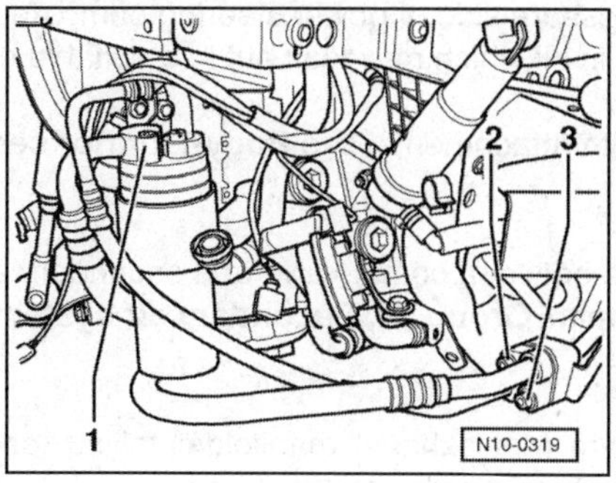

Separate air conditioning system lines -2- and -3- at condensor and line -1- at accumulator, see **Heating & Air Conditioning section, Repair Group 87.**

NOTE —
To prevent damage to the condenser and also to the refrigerant lines/hoses, ensure that the lines and hoses are not stretched, kinked or bent.

– Drain engine oil.

NOTE —
Observe disposal regulations!

– Remove oil filter housing complete with hoses, see **Repair Group 17, Oil filter housing, assembly**.

– Drain coolant, see **Repair Group 19, Cooling system, draining and filling**.

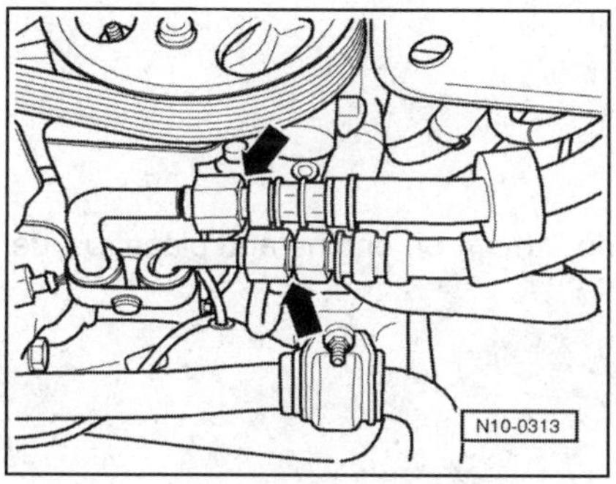

Separate hose connections -arrows- between transmission oil cooler and lines to manual/automatic transmission and collect oil which may leak out in a suitable container.

NOTE —
Observe disposal regulations!

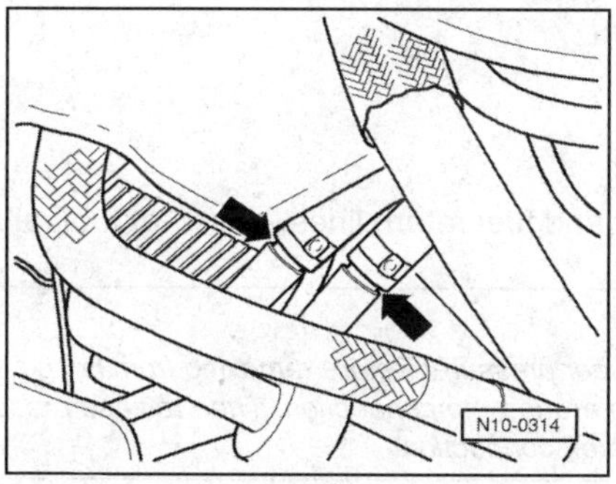

Separate lines to cooling coil at connections -arrows- and collect hydraulic oil which may leak out in a suitable container.

NOTE —
Observe disposal regulations!

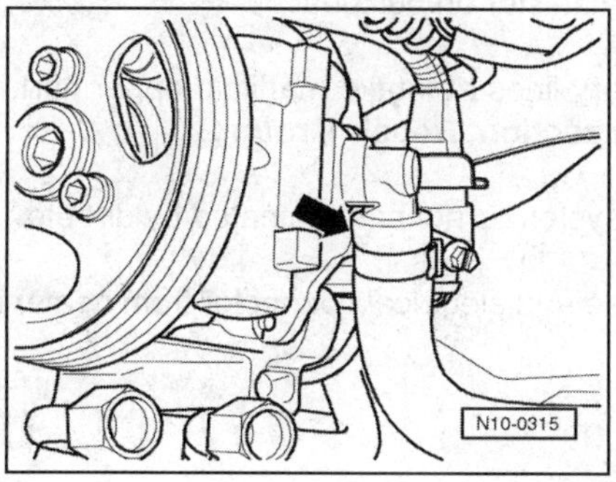

Remove hose -arrow- on suction side of power steering pump and collect hydraulic oil which may leak out in a suitable container.

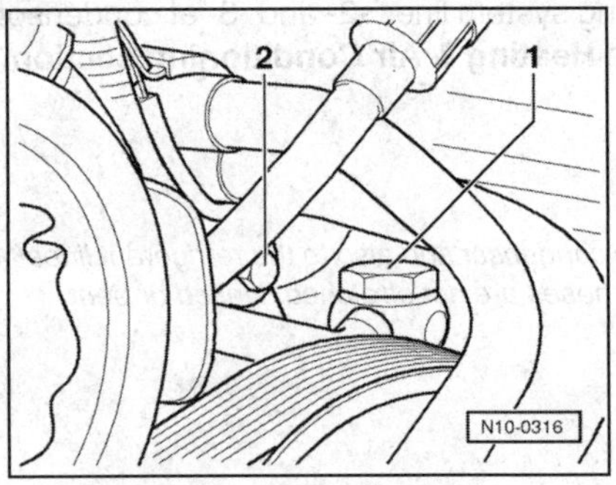

Separate line -1- on pressure side of power steering pump and remove line -2-. Collect hydraulic oil which may leak out in a suitable container.

– Remove lock carrier with attachments, see **Body-Exterior section, Repair Group 50.**

– Remove pressure hose between combi-valve and secondary air injection pump, see **Repair Group 26, Secondary air system components.**

– Remove exhaust system (not exhaust manifold/catalyst), see **Repair Group 26, Exhaust system, components.**

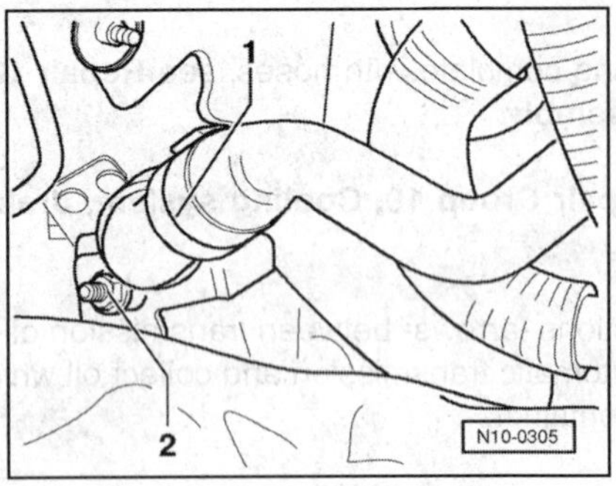

Disconnect coolant pipe -1- to Coolant Pump -V36- at longmember, left side.

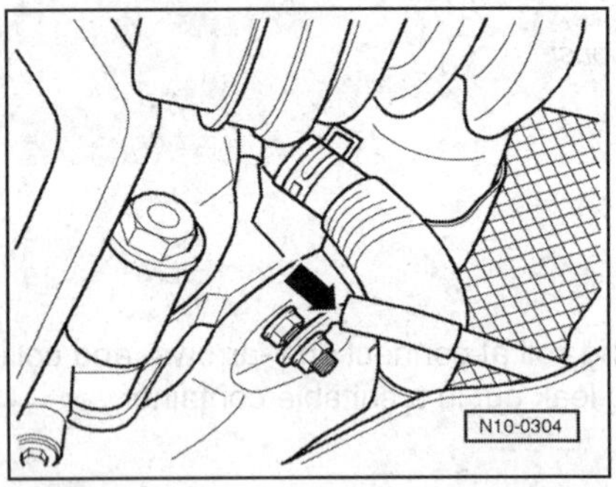

Disconnect Ground strap -arrow- on engine and place to side.

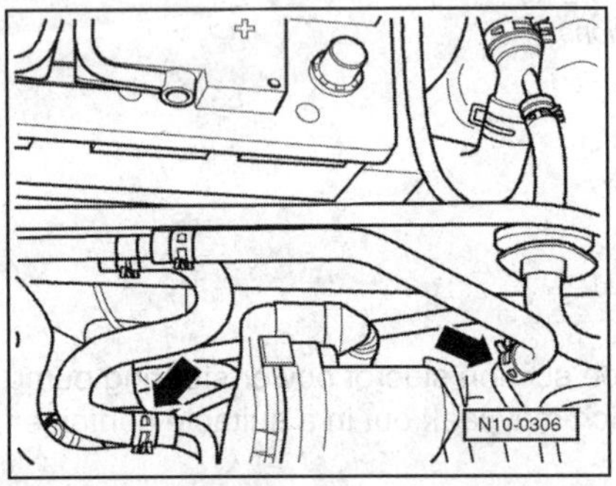

Disconnect fuel supply and fuel return lines -arrow- at fuel rail.

> ***WARNING —***
> *Fuel supply lines are under pressure! Before removing from hose connection wrap a cloth around the connection. Then release pressure by carefully pulling hose off connection.*

– Separate line to EVAP canister on bulkhead.

– Disconnect all necessary lines at intake manifold upper part, see **Fuel Injection and Ignition section, Repair Group 24.**

– Seal lines so that fuel system is not contaminated by dirt etc.

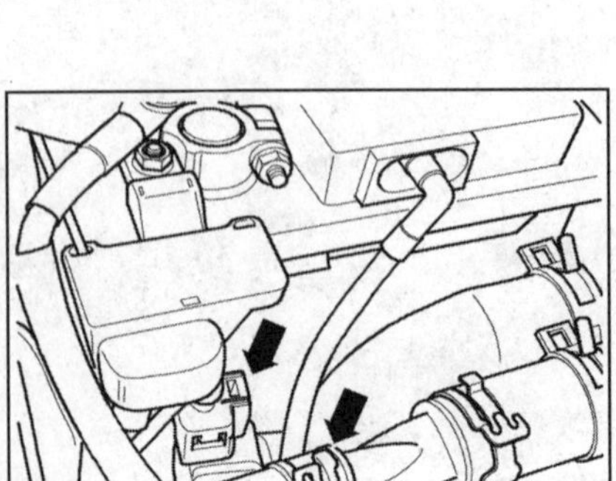

Disconnect coolant hose and electrical connector from heater regulator -arrows-.

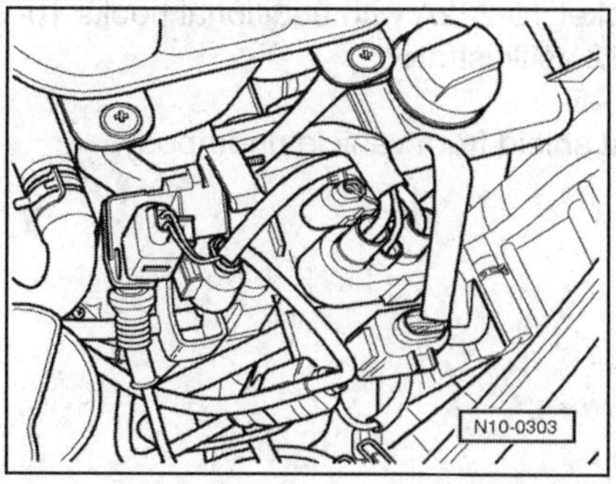

Disconnect connectors on left of engine compartment.

- Remove coolant system expansion tank, see **Repair Group 19, Cooling system components, engine side, top and rear,** item -10-.
- Remove wiper arms and cowl panel, see **Electrical Equipment section, Repair Group 92.**

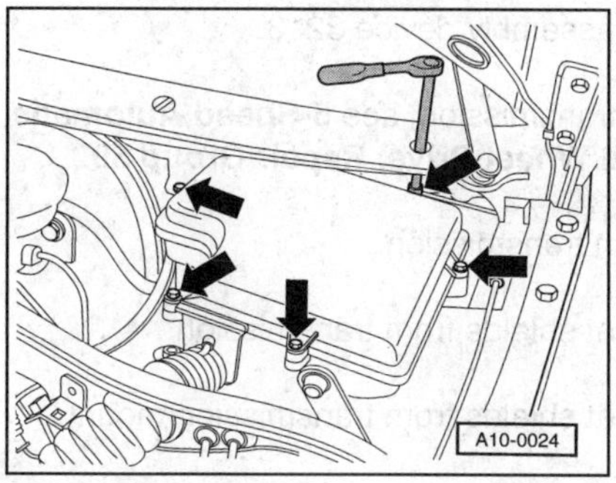

Remove cover of protective housing for control modules.

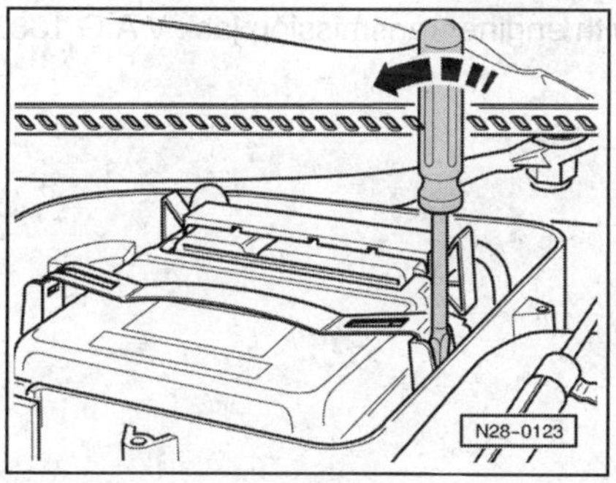

Carefully pry off retainer bar with a screwdriver -arrow-.

- Release connector from control module and pull off.
- Remove engine control module.

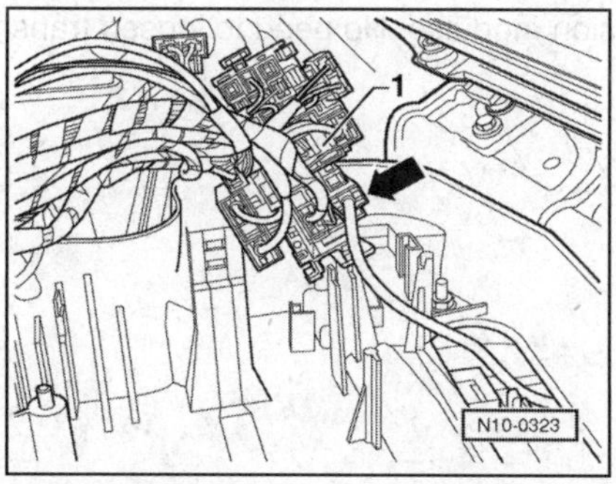

Remove relay carrier -1- from protective housing and pull out connector -arrow-.

- Disconnect engine wiring harness from central unit and place to one side.
- Disconnect all connectors on bulkhead and release fasteners.
- Disconnect all electrical connections as necessary from engine and place to one side.

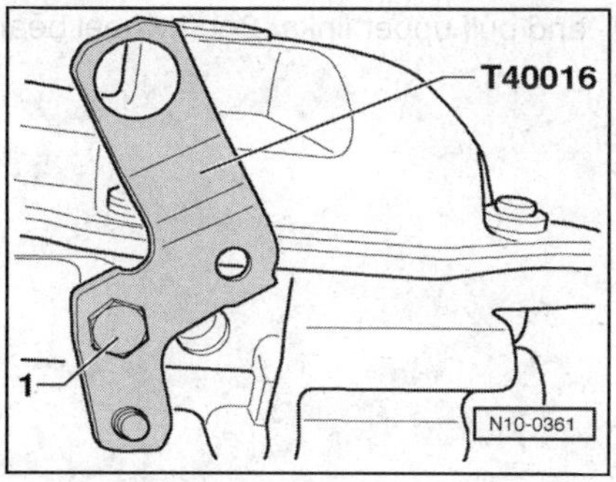

Secure T40016 transport bracket to rear left of cylinder head with securing screw -1-.

- Tightening torque: 25 Nm (18 ft-lb)

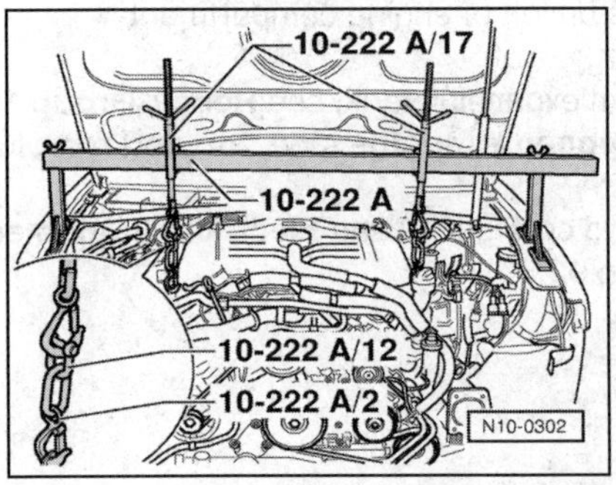

◀ Fit engine support bracket 10-222A with additional hooks 10-222A/2 and shackle 10-222A/12 as illustrated.

– Remove propshaft heat shield from vehicle underbody.

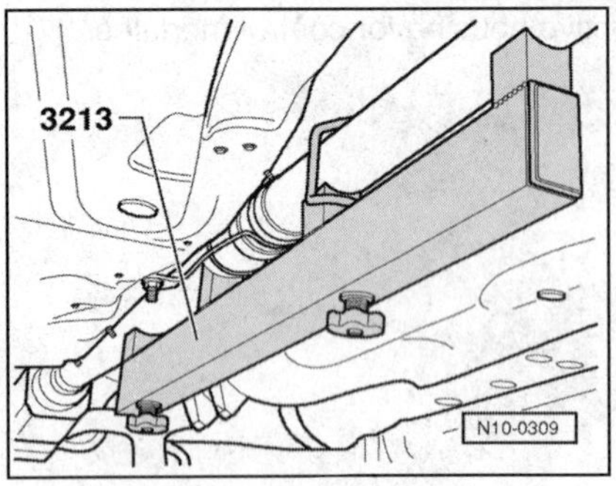

◀ Secure propshaft with assembly device 3213.

– Unbolt propshaft from transmission, see **5-Speed Automatic Transmission 01V, All Wheel Drive, Repair Group 39.**

– Unbolt drive shafts from transmission.

– Remove drive shaft heat shields from transmission.

– Remove drive shaft heat shields from transmission mount.

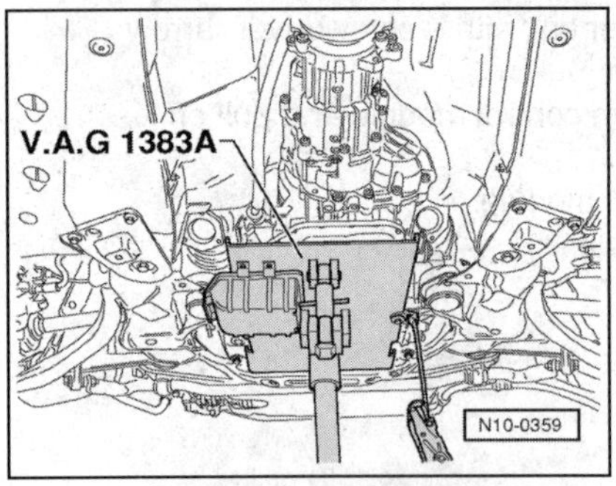

◀ Secure transmission with engine/transmission jack V.A.G 1383 A.

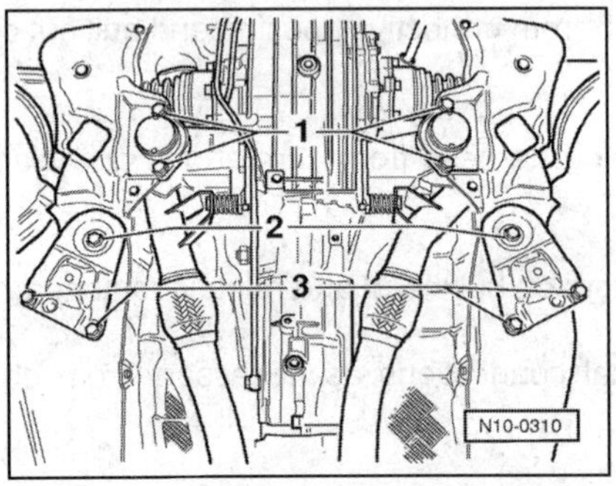

◀ Remove both transmission modules. No need to loosen transmission mounting -1-.

– Remove front wheels.

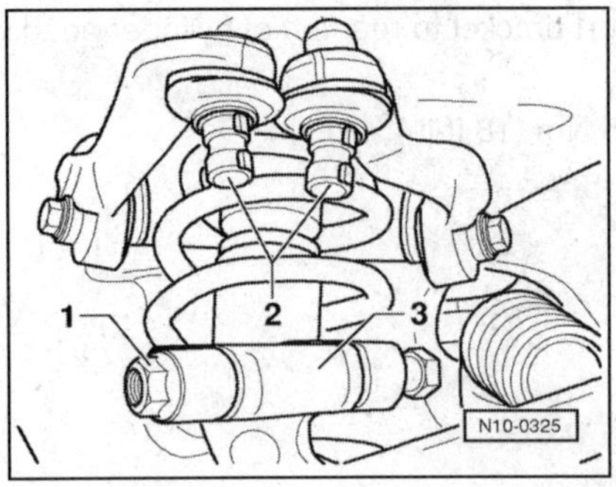

◀ Loosen bolted union -1- and pull upper links -2- off wheel bearing housing -3-.

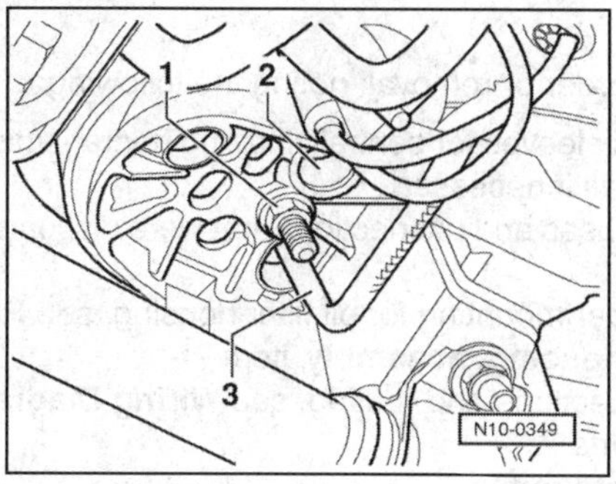

- Mark installation position of threaded connection -1- and lower locating sleeve -2- on left and right engine mountings.

– Disconnect vacuum lines -3- at Left Electro-Hydraulic Engine Mount Solenoid Valve -N144-/Right Electro-Hydraulic Engine Mount Solenoid Valve -N145-.

– Disconnect both lower engine mountings.

– Lower engine with transmission via support bracket 10-222A until following work sequences can be performed:

– Disengage selector lever cable at transmission, see **5-Speed Automatic Transmission 01V, All Wheel Drive, Repair Group 37.**

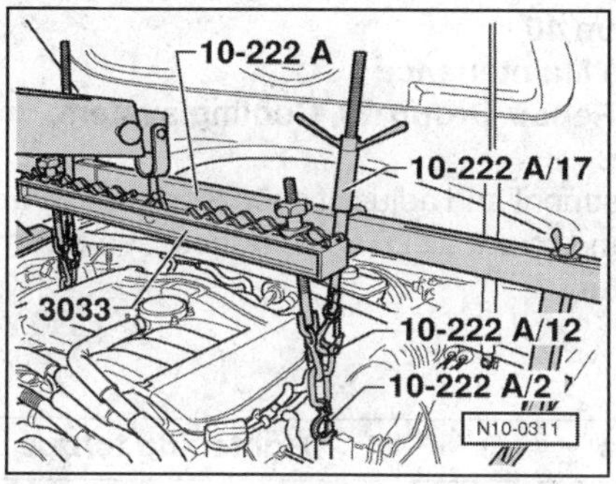

- Hook lifting tackle 3033 (item 2 and 11) into supplementary hooks 10-222A/2 and lift slightly using workshop crane VAS 6100.

– Lift engine with transmission until support bracket 10-222A/2 can be unhooked. To do this, turn stud out of shackle 10-222A/12.

– Pull engine with transmission forward slightly and remove coolant hose to heat exchanger.

– Check that all connections between bulkhead and engine have been separated.

– Lift out engine with transmission forward.

NOTE—

Guide the engine carefully when lifting out, to prevent damage to the bodywork.

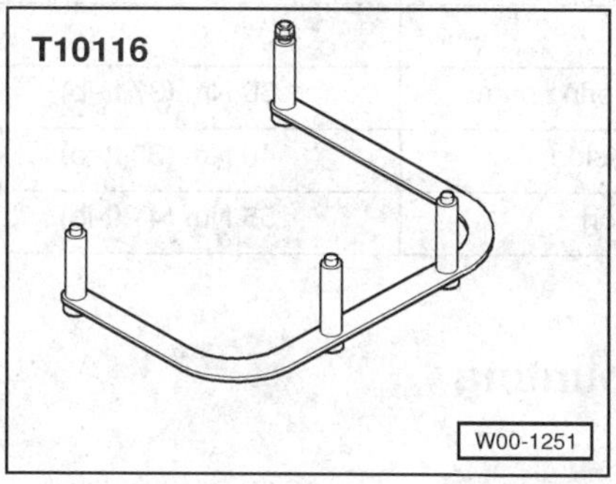

- Do not set the engine down on the sump or the liquid gasket between the upper and lower sump will be damaged. Use engine stand T10116 to hold engine.

NOTE—

Secure engine in engine and transmission stand VAS 6095 when working on engine.

Engine, installing

– Installation is reverse order or removal, noting the following:

- Check whether dowel sleeves for centralizing engine/transmission are in cylinder block, install if necessary.
- Reconnect all lines, hoses and connections that were disconnected for removal sequence.
- Replace bonded rubber mounting for oil filter housing, see **Repair Group 17, Oil filter housing, assembly**, item **-4-**.
- Check electrical connections and routing, see **Wiring Diagrams and Component Locations**
- Top up ATF, see **Maintenance**
- Connect A/C system lines, see **Heating & Air Conditioning section, Repair Group 87**
- Top up oil for power steering, see **Suspension, Wheels, Steering section, Repair Group 40**
- Top up engine oil, see **Maintenance**
- Fill with coolant, see **Repair Group 19, Cooling system, draining and filling.**
- Check headlight adjustment and adjust if necessary, see **Maintenance**
- Perform a test drive and check all DTC memories, see **Scan Tool section.**

Tightening torques

Bolted connections		Tightening torque
Nuts, bolts	M 6	10 Nm (7 ft-lb)
	M 7	15 Nm (11 ft-lb)
	M 8	25 Nm (18 ft-lb)
	M10	40 Nm (30 ft-lb)
	M 12	60 Nm (44 ft-lb)
Exceptions:		
Union bolt to power steering pump		50 Nm (37 ft-lb)
Drive shafts to transmission		40 Nm (30 ft-lb)
Propshaft to transmission		55 Nm (41 ft-lb)

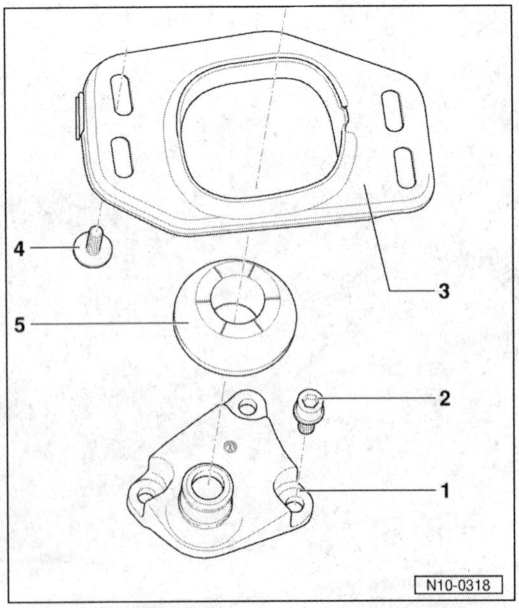

Front engine mounting

1. **Torque bracket**
2. **40 Nm (30 ft-lb)**
3. **Stop**
   - Allow to move onto stop buffer under its own weight and tighten securing bolts
4. **30 Nm (22 ft-lb)**
5. **Bump stop**

13 Engine–Crankshaft, Cylinder Block

Repair Group

E 13

Engine Block Assembly

NOTE —

If large quantities of metal particles or other deposits (caused, for example, by partial seizure of the crankshaft or connecting rod bearings) are found in the engine oil, the oil passages must be cleaned thoroughly and the oil cooler replaced in order to prevent further damage from re-occurring.

For accessory drive belt information, see **Maintenance**.

For cylinder block, engine bracket, and seal information, see **Engine block, components**.

Engine block, components

NOTE—
- *Do not set the engine down on the sump or the liquid gasket between the upper and lower sump will be damaged. Use engine stand T10116 to hold the engine.*
- *Secure engine in engine and transmission stand VAS 6095 when performing work on the engine.*

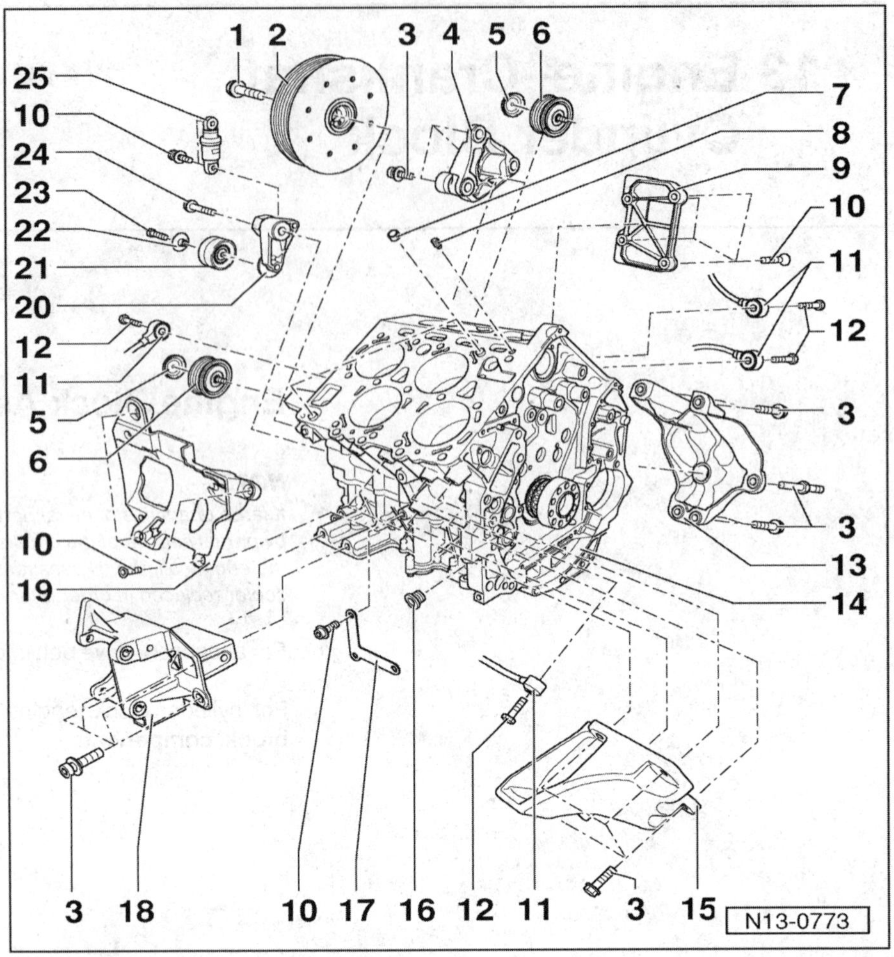

1. **100 Nm (74 ft-lb) + 1/4 turn (90 °) further**
   - Replace
   - Counter-hold with T10069 to loosen and tighten, see **Fig.1**
   - Tighten using torque wrench V.A.G 1601
2. **Vibration damper**
   - Removing and installing ribbed belt, see **Maintenance**
3. **40 Nm (30 ft-lb)**
4. **Torque bracket**
5. **Cover cap**
6. **Idler roller**
   - Tighten to 30 Nm (22 ft-lb)
7. **Dowel sleeve**
8. **Oil non-return valve**
   - Tighten to 6 Nm (53 in-lb)
   - Note fitting position
   - Clean if badly soiled
9. **Mount**
   - For alternator
10. **25 Nm (18 ft-lb)**
11. **Knock sensors**
    - See **Repair Group 28, Ignition system components**, item -6-
12. **20 Nm (15 ft-lb)**
    - Tightening torque influences function of knock sensor
13. **Right engine bracket**
14. **Cylinder block**
    - See **Sealing flange and dual mass flywheel, removing and installing**
15. **Left engine bracket**
16. **Sealing plug**
    - Tighten to 20 Nm (15 ft-lb)
17. **Bracket**
    - For wiring harness
18. **Retainer**
    - For power steering pump
19. **Retainer**
    - For air conditioning compressor
20. **Tensioning element**
    - For ribbed belt
    - See **Maintenance**
21. **Tensioner**
22. **Washer**
23. **20 Nm (15 ft-lb)**
24. **30 Nm (22 ft-lb)+ 1/4 turn (90 °) further**
25. **Tensioning element**

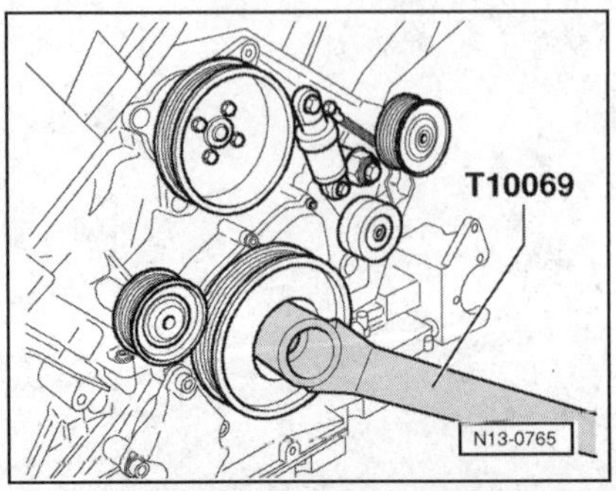

Fig. 1 To loosen and tighten securing bolt, hold vibration damper with counter-hold tool T10069

NOTE—
- *Vibration damper securing bolt must be replaced.*
- *Tighten securing bolt with torque wrench V.A.G 1601.*

Balancing shafts, disassembling and assembling

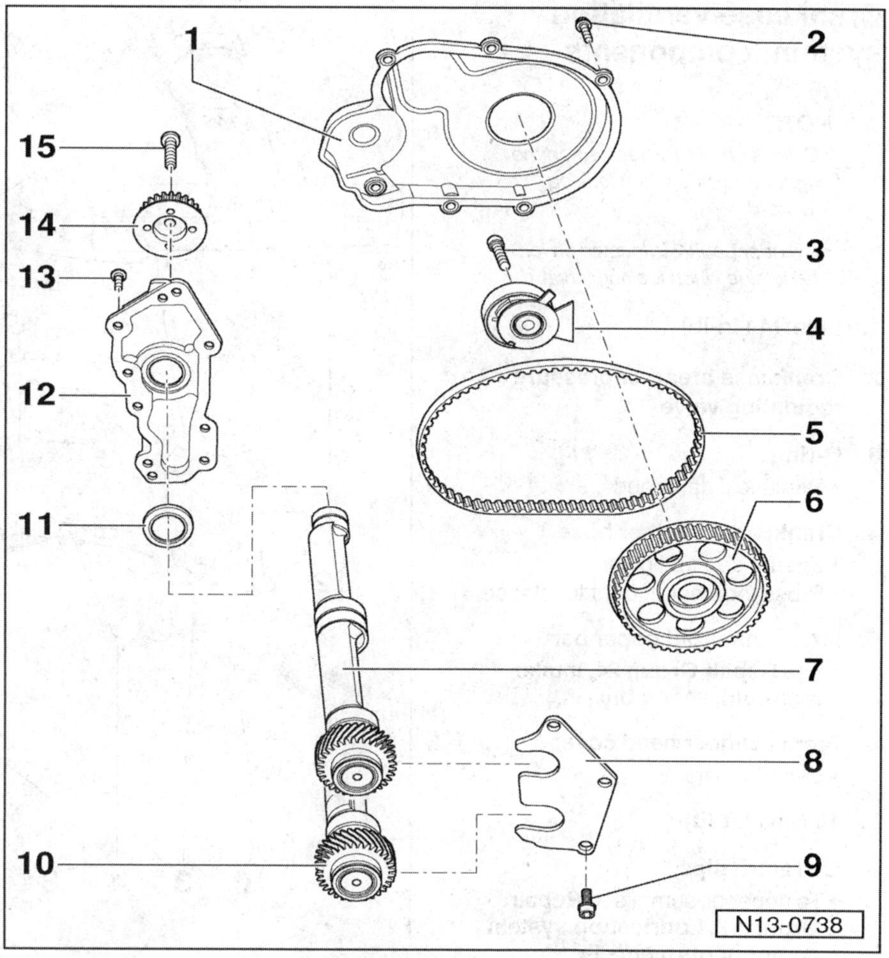

1. **Toothed belt guard**
2. **10 Nm (7 ft-lb)**
3. **30 Nm (20 ft-lb) + 1/4 turn (90 °) further**
4. **Tensioner**
5. **Toothed belt**
   - Mark direction of rotation before removing
   - Check for wear
   - Do not kink
   - See **Toothed belt drive for balancing shafts, removing, installing and adjusting**
6. **Crankshaft toothed belt sprocket**
7. **Upper balancing shaft**
   - Driven by toothed belt
   - Note position when fitting toothed belt, see **Toothed belt drive for balancing shafts, removing, installing and adjusting**
8. **Securing plate**
9. **10 Nm (7 ft-lb)**
   - Insert with locking fluid D 000 600 A2
10. **Lower balance shaft**
    - Note position when fitting toothed belt, see **Toothed belt drive for balancing shafts, removing, installing and adjusting**
11. **Seal**
    - Replace only in conjunction with front sealing flange , **-12-**
12. **Sealing flange, front**
    - Clean sealing surface before installing
    - Coat sealing surfaces to cylinder block with sealant D 188 800 A1
13. **8 Nm (71 in-lb)**
14. **Upper balancing shaft toothed belt sprocket**
    - Note position when fitting toothed belt, see **Toothed belt drive for balancing shafts, removing, installing and adjusting**
15. **25 Nm (18 ft-lb)**

Crankcase ventilation system, components

NOTE—
- *Cylinder bank 1 is located on the right-hand side when looking forward.*
- *Cylinder bank 2 is located on left-hand side when looking forward.*

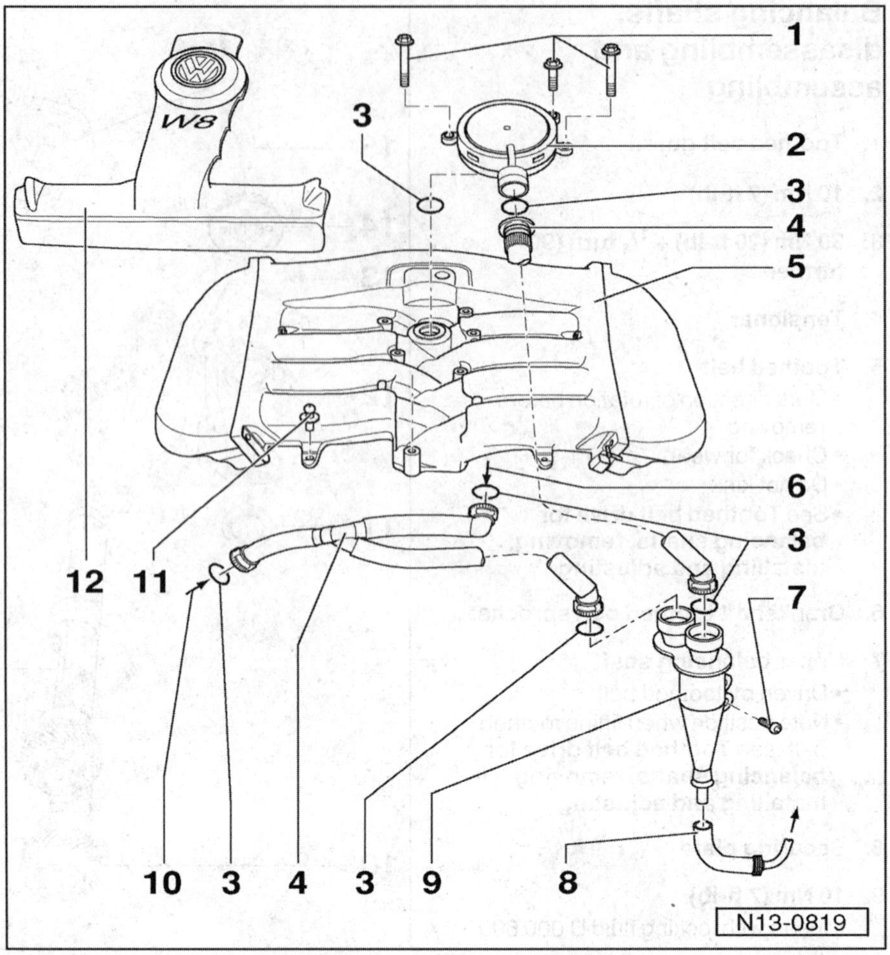

1. **5 Nm (44 in-lb)**
2. **Crankcase breather pressure regulating valve**
3. **O-ring**
   - Replace if damaged
4. **Crankcase breather hose**
   - Ensure seated tightly
   - Press together at front to release
5. **Intake manifold upper part**
   - See **Repair Group 24, Intake manifold, assembly**
6. **From cylinder head cover**
   - Cylinder bank 2
7. **10 Nm (7 ft-lb)**
8. **Oil return pipe**
   - To upper oil sump, see **Repair Group 17, Lubrication system components**, item **-5-**
9. **To oil separator**
   - Secured with retainer at connecting piece for combi-valve (cylinder bank 2), see **Repair Group 26, Secondary air system components**, item **-8-**
10. **From cylinder head cover**
    - Cylinder bank 1
11. **Ball head pin**
    - Tighten to 3 Nm (27 in-lb)
    - For engine cover
12. **Cover**

Toothed belt drive for balancing shafts, removing, installing and adjusting

Special tools and equipment

- T10069 Counter-holder
- V.A.G 1331 Torque wrench (5...50 Nm)
- V.A.G 1332 Torque wrench (40...200 Nm)
- V.A.G 1601 Torque wrench (150...800 Nm)
- 3242 Clamping bolt (with engine removed)
- Drill bit, diameter 2.0 mm

Removing

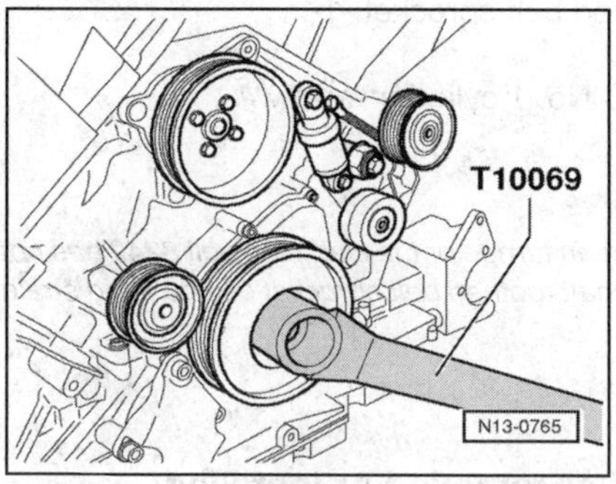

NOTE —
- *The toothed belt can be removed and installed when the engine is installed in the engine compartment.*
- *The upper and lower balancing shafts can only be removed and installed when the engine is removed from the engine compartment.*

– Remove ribbed belt, see **Maintenance**.

◄ Remove vibration damper securing bolt. To do this, lock vibration damper with counter-holder T10069.

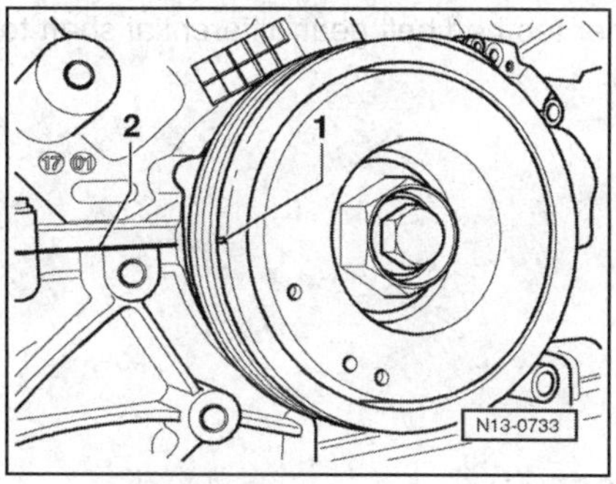

◄ Turn crankshaft to TDC No. 1 cylinder.

- Mark -1- must align with cylinder block joint -2-.

– Remove vibration damper.

With engine removed:

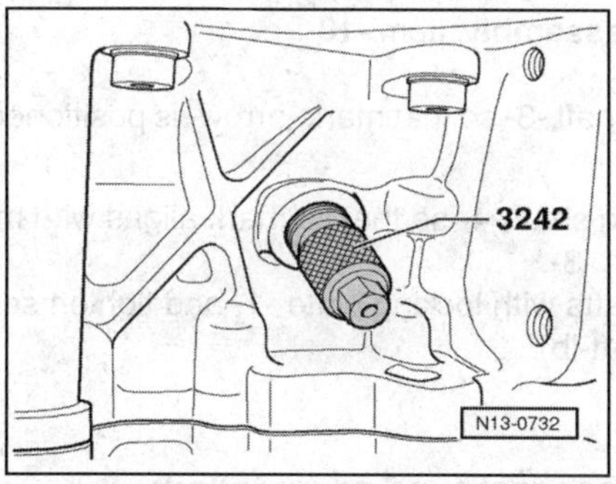

– Remove sealing plug from left side of cylinder block, see **Engine block, components,** item **-16-**.

◄ Screw clamping bolt 3242 into hole so that the crankshaft is secured against turning.

When engine is either removed or installed:

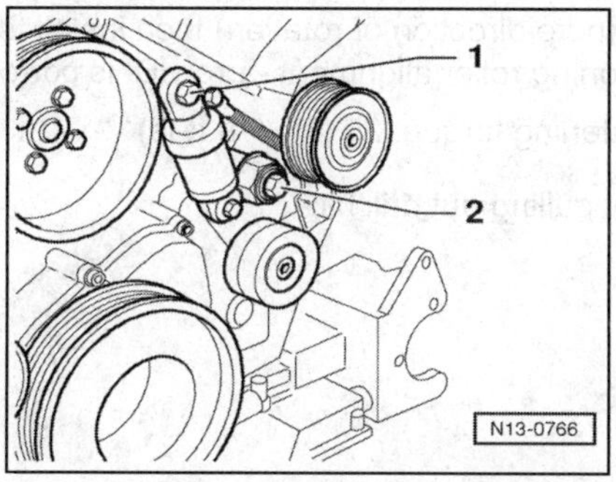

◄ Remove toothed belt tensioning device securing bolts -1- and -2- and take off tensioning device.

– Remove toothed belt guard.

– Mark direction of rotation of toothed belt.

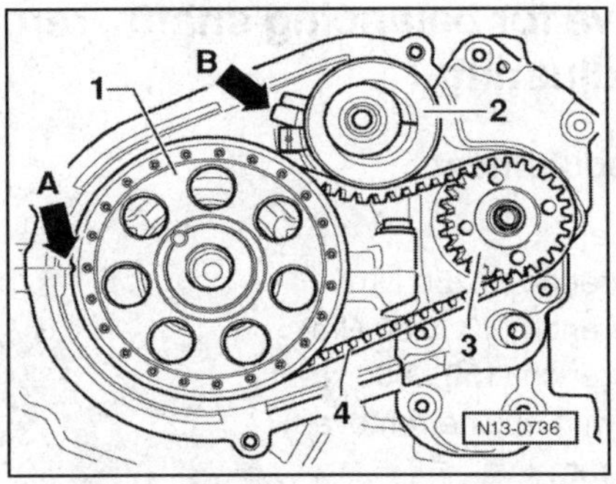

Release tension of tensioning roller -2- and lock tensioning roller with a drill bit (diameter: 2.0 mm).

- Remove tensioning roller -2- for toothed belt -4-.
- Remove toothed belt.
- Remove crankshaft toothed belt sprocket -1-.

Toothed belt drive for balancing shafts, installing

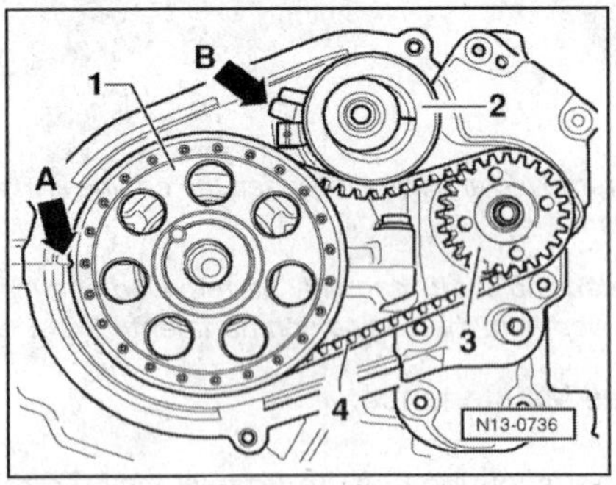

Install crankshaft toothed belt sprocket -1-.

– Turn crankshaft to TDC No. 1 cylinder -arrow A-.

NOTE—

If the crankshaft has not been turned or the clamping bolt 3242 has not been removed, then the crankshaft toothed belt sprocket can be fitted straight on.

When engine is either removed or installed:

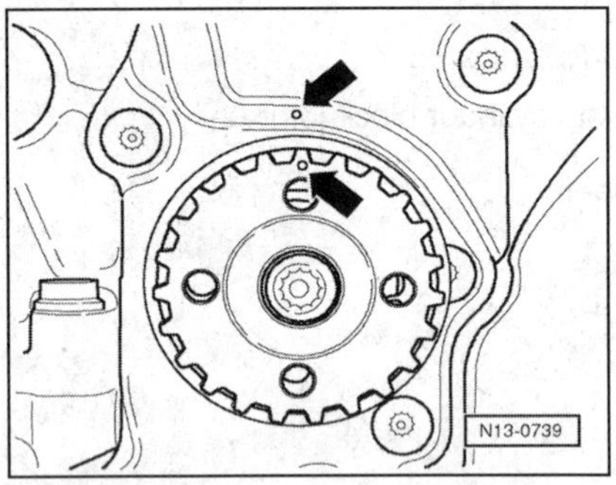

Line up marking of upper toothed belt gear differential shaft to cylinder block marking (arrows).

With engine removed:

– Remove flywheel/drive plate sealing flange, see **Sealing flange and dual mass flywheel, assembly**, item **-10-**.

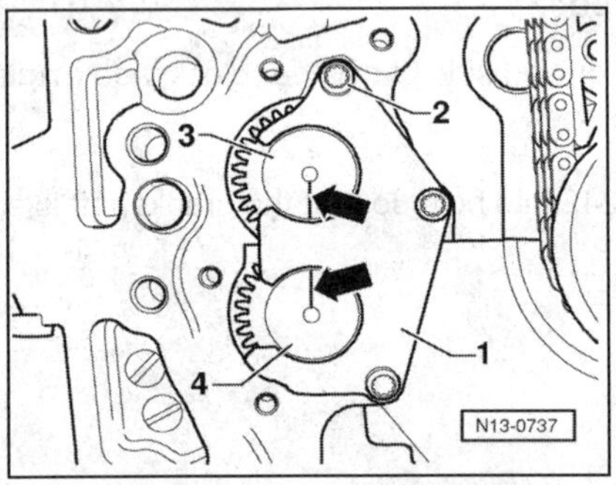

Turn upper balancing shaft -3- so that mark -arrow- is positioned straight down.

- Insert lower balancing shaft -4- so that its mark aligns with mark of upper balancing shaft -3-.
- Secure balancing shafts with locking plate -1- and tighten securing bolts -2- to 10 Nm (7 ft-lb).

When engine is either removed or installed:

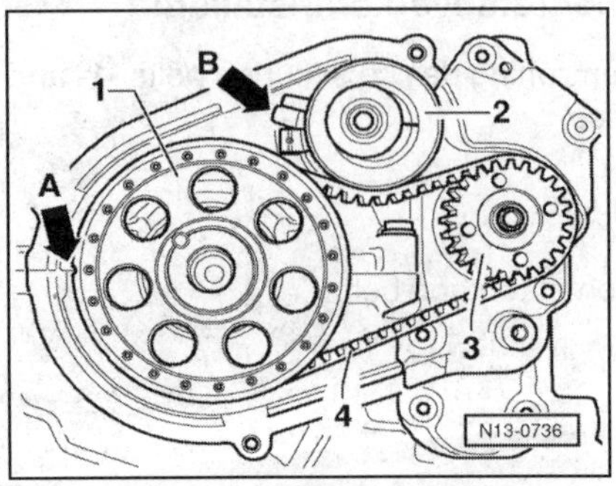

Install toothed belt -4- (note direction of rotation) then install tensioning roller -2-. Ensure tensioning roller alignment -arrow B- is correct.

- Tensioning roller tightening torque: 25 Nm (18 ft-lb)

– Tension toothed belt by pulling out drill bit.

When engine is installed in engine compartment:

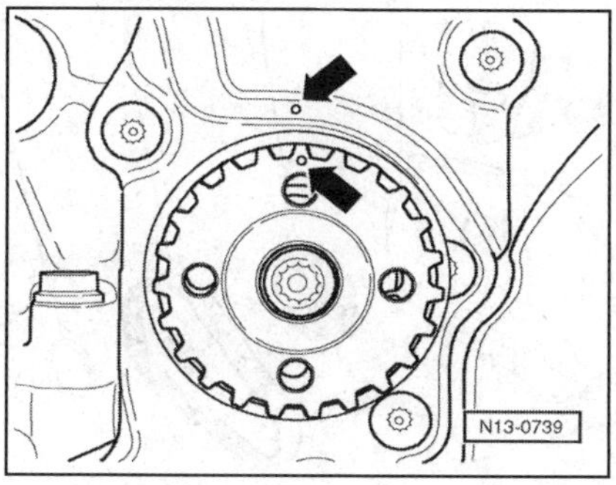

Check again position of marking of upper toothed belt gear differential shaft and cylinder block marking (arrows) to each other. (While doing so, crankshaft is at TDC cylinder 1.)

With engine removed:

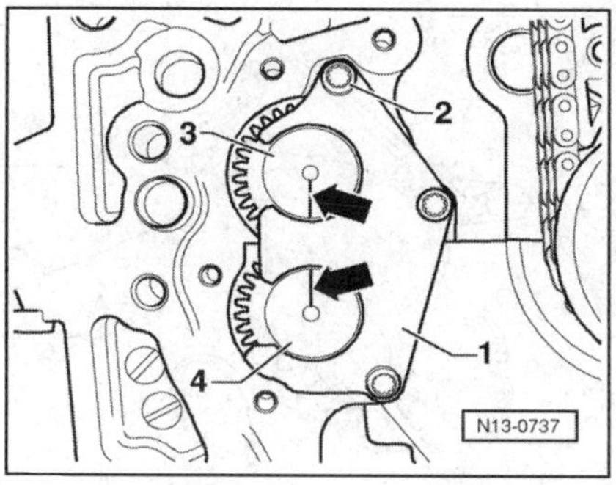

Check position of balancing shafts -3- and -4- to one another, again.

– Install flywheel/drive plate sealing flange, see **Sealing flange and dual mass flywheel, assembly**, item **-10-**.

– Replace seal on flywheel/drive plate sealing flange, see **Sealing flange and dual mass flywheel, assembly**, item **-13-**.

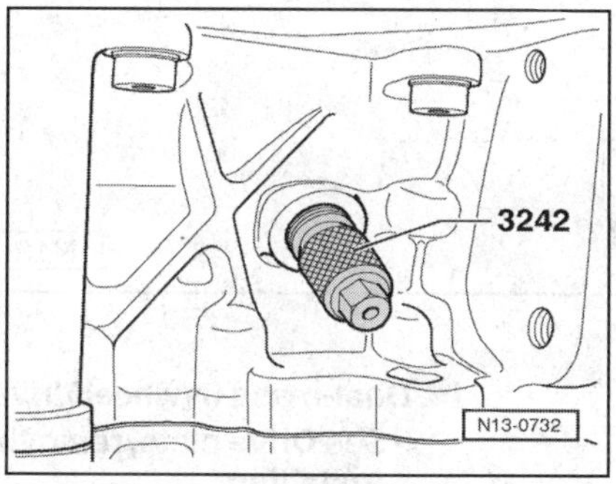

Remove clamping bolt 3242.

– Install sealing plug in cylinder block.
  - Tightening torque: 20 Nm (15 ft-lb)

When engine is either removed or installed:

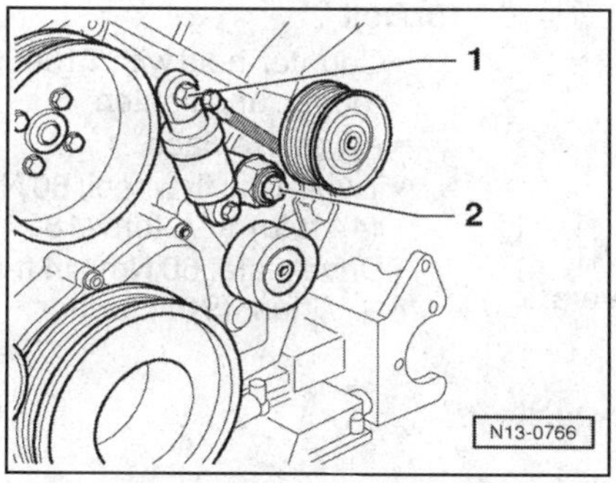

– Install toothed belt guard.
  - Torque setting: 10 Nm (7 ft-lb)

Install toothed belt tensioning device.
- Securing bolt M8 -1-: 25 Nm (18 ft-lb)
- Securing bolt M10 -2-: 30 Nm (22 ft-lb) + 90° (1/4 turn)

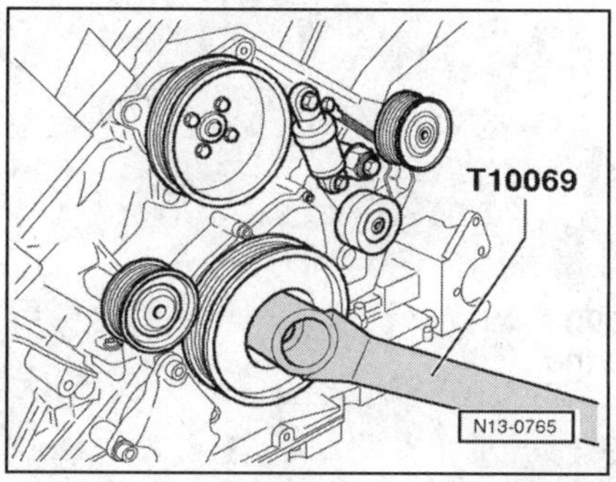

Install vibration damper and lock with counter-holder T10069.
- Tighten new securing bolt to 100 Nm (74 ft-lb)+ 90° (1/4 turn). (The additional turn may be performed in several steps.)

NOTE —
- *Vibration damper securing bolt must be replaced*
- *Tighten securing bolt with torque wrench V.A.G 1601.*

– Install ribbed belt, see **Maintenance**.

Sealing flange and dual mass flywheel, assembly

CAUTION—
Crankshaft, piston and conrods must not be removed.

NOTE—
- *The engine must not be set down on the oil sump or the liquid gasket between the upper and lower sump will be damaged. Use engine stand T10116 to hold engine.*
- *Secure engine in engine and transmission stand VAS 6095 when repairing.*
- *The sealing flange (item **-10-**) can be removed and installed with the cylinder head installed.*
- *Servicing clutch, see **Manual Transmission section, Repair Group 30.***

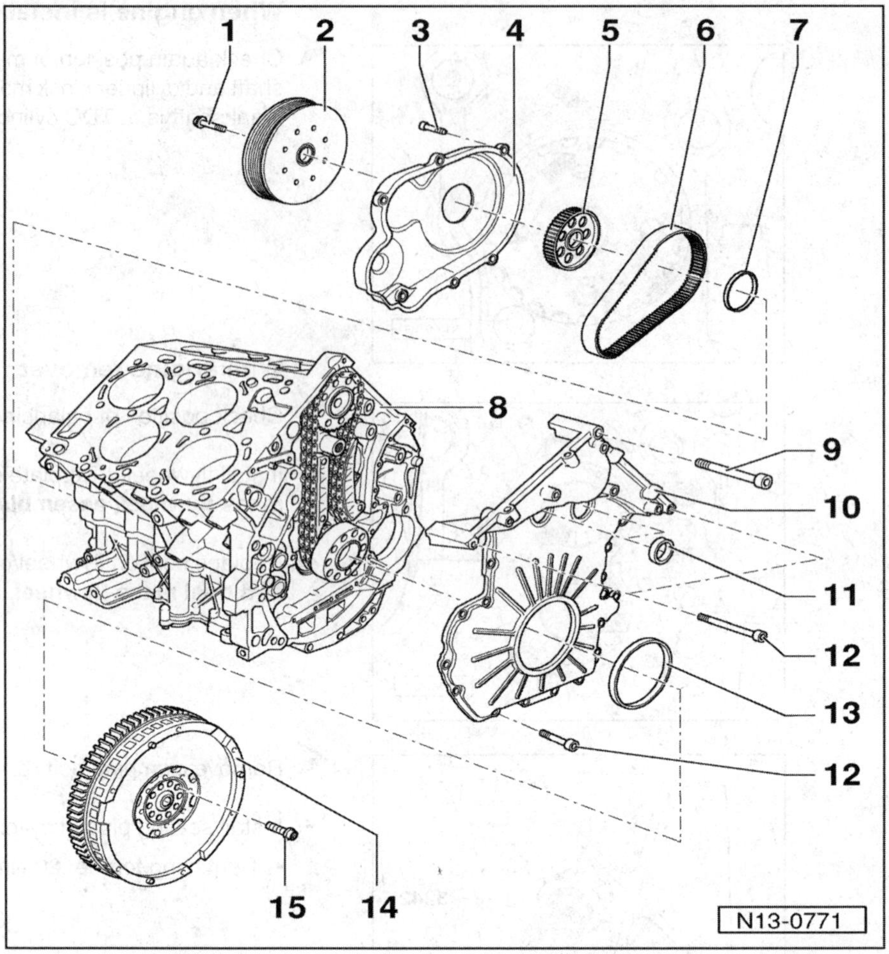

1. **100 Nm (74 ft-lb) + $^1/_4$ turn (90 °) further**
   - Alwaysreplace
   - Counter-hold with T10069 to loosen and tighten, see **Crankshaft oil seal (vibration damper end), replacing**
   - Tighten using torque wrench V.A.G 1601
2. **Vibration damper**
   - Removing and installing ribbed belt, see **Maintenance**
3. **10 Nm (7 ft-lb)**
4. **Toothed belt guard**
5. **Crankshaft toothed belt sprocket**
6. **Toothed belt**
   - Mark direction of rotation before removing
   - Check for wear
   - Do not kink
   - See **Toothed belt drive for balancing shafts, removing, installing and adjusting.**
7. **Oil seal**
   - PTFE version of oil seal
   - Identification: no inner coil spring
   - Do not additionally oil or grease the oil seal sealing lip
   - See **Crankshaft oil seal (vibration damper end), replacing**
8. **Aluminum cylinder block**
9. **20 Nm (15 ft-lb)**
   - Observe sequence when tightening, see **Fig. 1**
10. **Sealing flange**
    - Clean sealing surface before installing
    - Coat sealing surfaces to cylinder block with sealant D 188 800 A1
    - Sealing sealing surfaces to covers, see **Repair Group 15, Valve timing, adjusting**
11. **Seal**
    - Replace only in conjunction with rear sealing flange item**10**
12. **10 Nm (7 ft-lb)**
    - Observe sequence when tightening, see **Fig. 1**
13. **Oil seal**
    - PTFE version of oil seal
    - Identification: no inner coil spring
    - Do not additionally oil or grease the oil seal sealing lip
    - See **Crankshaft seal, flywheel end, replacing**
14. **Dual-mass flywheel/drive plate**
    - See **Drive plate, removing and installing**
15. **Bolt**
    - Counter-hold with T10069 to loosen and tighten
    - Always replace
    - Two-mass flywheel: 60 Nm (44 ft-lb) + $^1/_2$ turn (180 °) further
    - Drive plate: 60 Nm (44 ft-lb) + $^1/_4$ turn (90 °) further

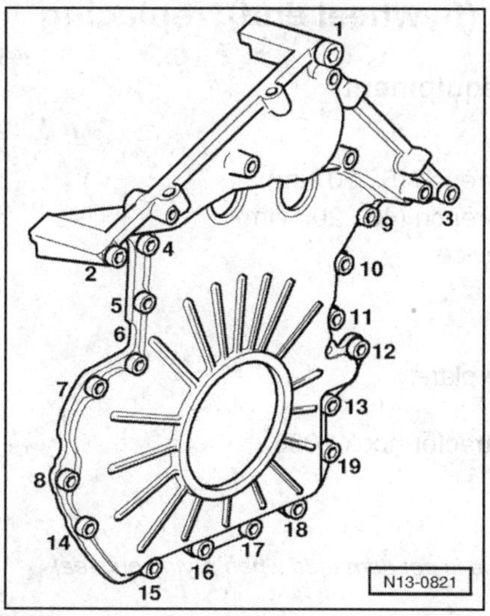

Fig. 1 Tightening sequence of securing screws for sealing flange

NOTE —
The cover must be installed within 5 minutes of applying sealing compound.

– Tighten screws in sequence indicated as follows and note different screw lengths

Position	Screw	Torque
1	M8	20 Nm (15 ft-lb)
2 – 19	M6	10 Nm (7 ft-lb)

Crankshaft oil seal (vibration damper end), replacing

Special tools and equipment

- VW 681 Extractor lever
- T10069 Counter-holder
- T10124 Fitting sleeve
- V.A.G 1332 Torque wrench (40...200 Nm)
- V.A.G 1601 Torque wrench (150...800 Nm)

Removing

– Remove ribbed belt, see **Maintenance**.

– Remove toothed belt, see **Toothed belt drive for balancing shafts, removing, installing and adjusting.**

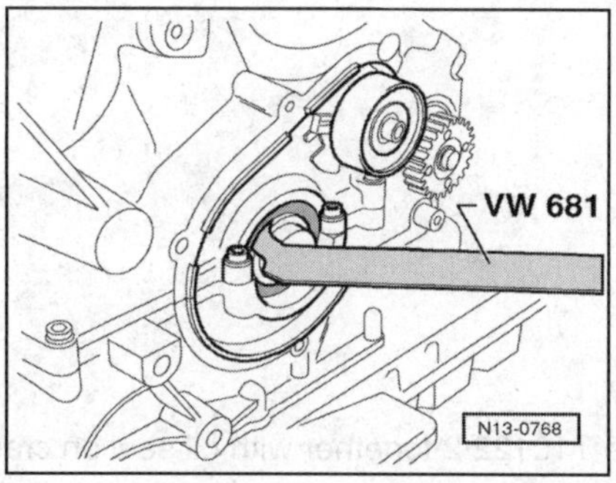

Pry out seal using extractor lever VW 681.

NOTE —
Ensure the sealing surface is not damaged when prying out the seal.

Installing

– Before installing, remove oil remains from crankshaft journal with a clean cloth.

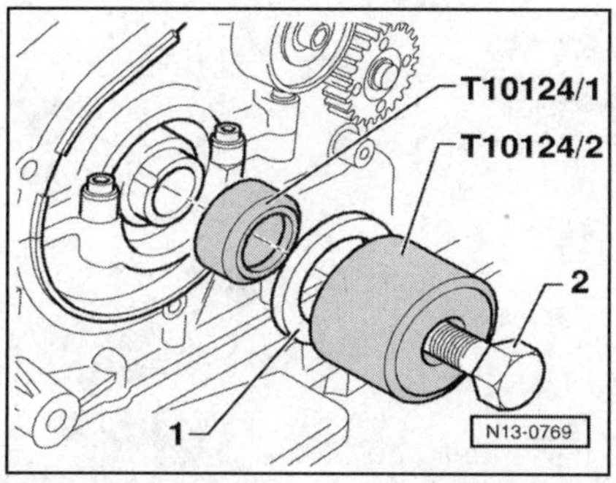

Place guide sleeve T10124/1 onto crankshaft journal.

- Slide dry seal -1- over guide sleeve.
- Press oil seal -1- in onto stop with press sleeve from T10124/2. To do this use an old vibration damper securing bolt -2-.

– Install toothed belt, see **Toothed belt drive for balancing shafts, removing, installing and adjusting**.

– Install ribbed belt, see **Maintenance**.

Crankshaft seal (flywheel end), replacing

Special tools and equipment

- 2086 Extractor hook
- V.A.G 1331 Torque wrench (5...50 Nm)
- V.A.G 1332 Torque wrench (40...200 Nm)
- T10122 Assembly device

Removing

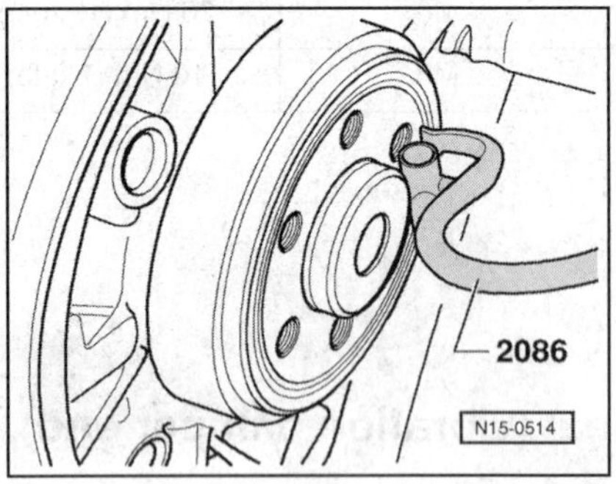

– Remove flywheel/drive plate.

◄ Remove seal using extractor hook 2086.

NOTE —
Ensure the sealing surface is not damaged when prying out seal.

Crankshaft seal (lywheel end), installing

– Before installing, remove oil remains from crankshaft journal with a clean cloth.

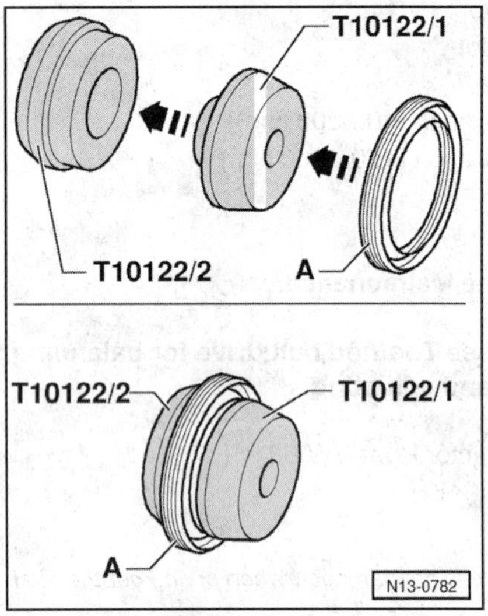

◄ Install assembly sleeves T10122/1 and T10122/2 together. Slide oil seal -A- until stop onto assembly sleeve T10122/2.

NOTE —
Do not remove support ring out of seal until immediately before installation.

– Separate both assembly sleeves.

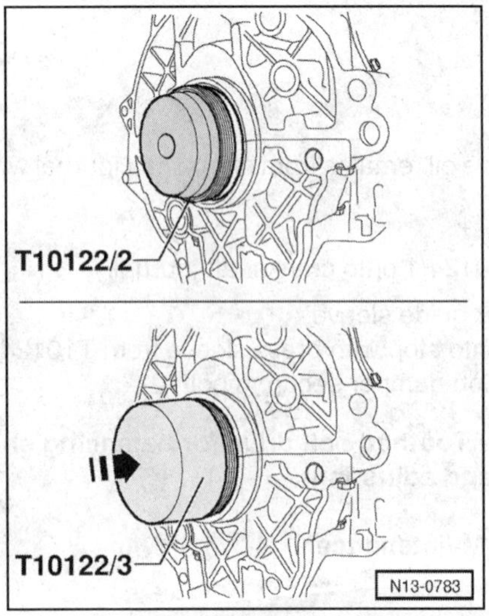

◄ Install assembly sleeve T10122/2 together with oil seal on crankshaft flange.

– Press in oil seal using assembly sleeve T10122/3 until stop.

– Installation is reverse of removal.

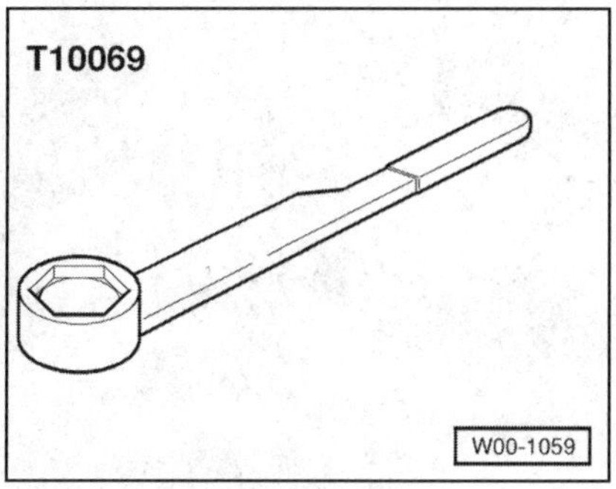

Drive plate, removing and installing

Special tools and equipment

◀ T10069 Counter-hold tool

- V.A.G 1332 Torque wrench (40...200 Nm)
- Depth gauge
- Straight edge

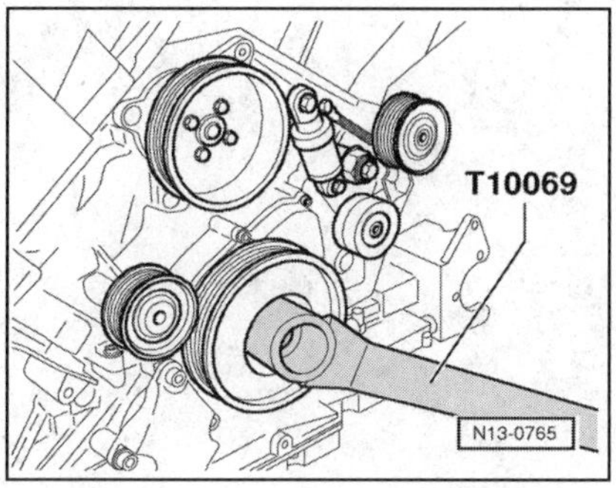

Removing

◀ Remove drive plate by locking vibration damper with counter-holder T10069.

– Loosen drive plate securing bolts diagonally and remove.

– Remove drive plate.

Drive plate, installing

– Install drive plate on crankshaft.

– Insert at least 3 old securing bolts and tighten to 30 Nm.

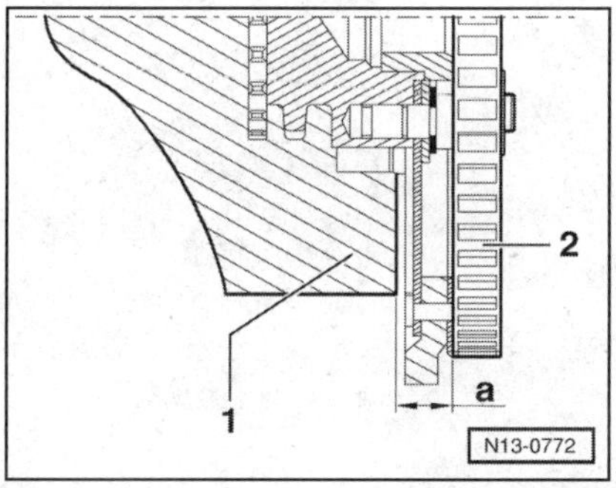

◀ Check dimension -a- through holes for securing torque converter using a straight edge and depth gauge and calculate average.

- 1 - Cylinder block
- 2 - Drive plate

– Compare average (measured distance + thickness of straightedge) with specification.

- Specification -a-: 18.1...19.7 mm

If the specification is not attained:

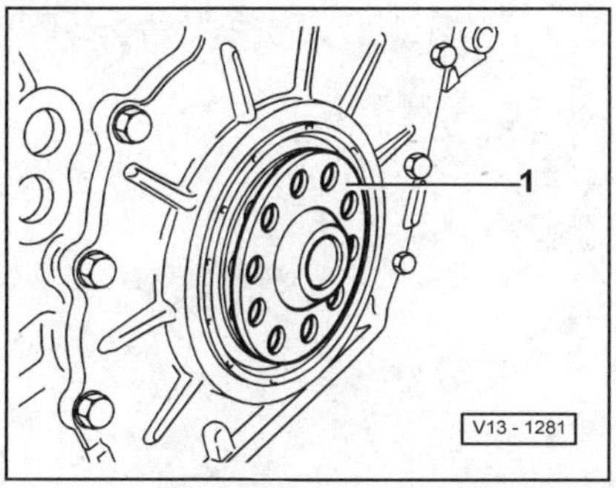

◀ Remove drive plate again and fit appropriate shim -1-.

NOTE —
Only one shim of the appropriate thickness may be used.

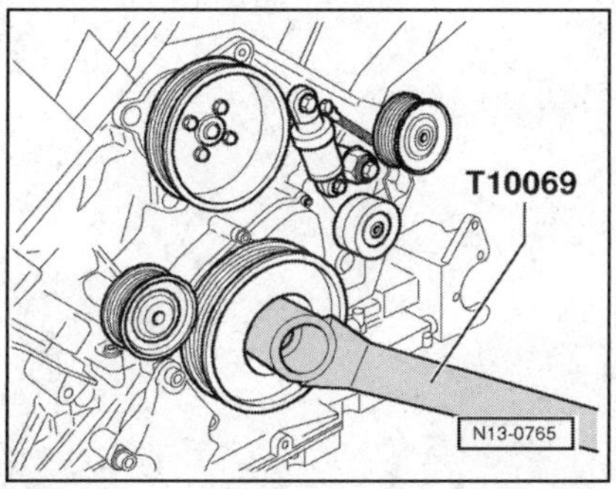

If the specification is attained:

– Insert new securing bolts and tighten by hand.

◀ Install drive plate. Lock vibration damper with counter-holder T10069.

– Tighten securing bolts to 60 Nm + 90° ($^1/_4$ turn).
(Turning further may be performed in several steps.)

15 Engine–Cylinder Head, Valve Gear

Repair Group

E 15

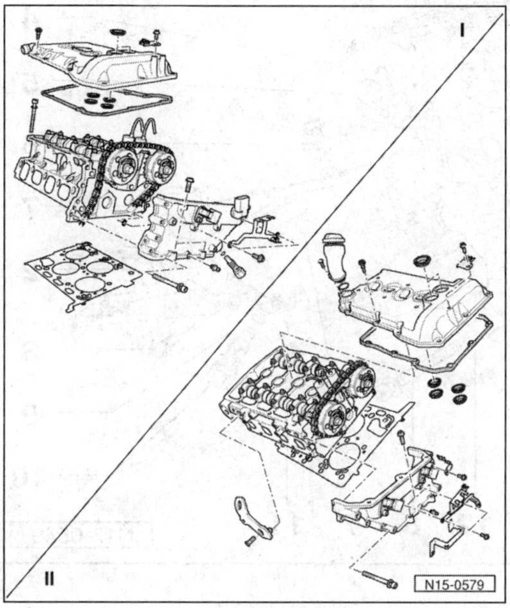

Cylinder heads, servicing

NOTE—

- *Secure the engine in the engine and transmission stand VAS 6095 when repairing.*
- *If you install an exchange cylinder head, all contact surfaces between bearing elements, roller rocker fingers and cam running surfaces of camshaft must be oiled before installing the cylinder head cover.*
- *The plastic protectors installed to protect the open valves must only be removed immediately before installing the cylinder head.*
- *When replacing the cylinder head the complete amount of engine coolant must be replaced.*

NOTE—

- *Cylinder bank 1 is located on the right-hand side when looking toward the front.*
- *Cylinder bank 2 is located on left-hand side when looking toward the front.*

I - Cylinder bank 1, see **Cylinder bank 1, assembly.**

II - Cylinder bank 2, see **Cylinder bank 2, assembly**.

Checking compression pressures, see **Compression pressures, checking.**

EDITOR'S NOTE—

Service procedures involving the timing chain drive system, valve timing, camshafts, valve train and cylinder head removal require that the engine assembly be removed from the vehicle.

Cylinder bank 1, assembly

NOTE —

- *Cylinder bank 1 is located on the right-hand side when looking toward the front.*
- *Disassembling and assembling intake manifold, see **Repair Group 24, Intake manifold, assembly**.*
- *Removing and installing parts of the chain drive system, see **Chain drive system, components.***

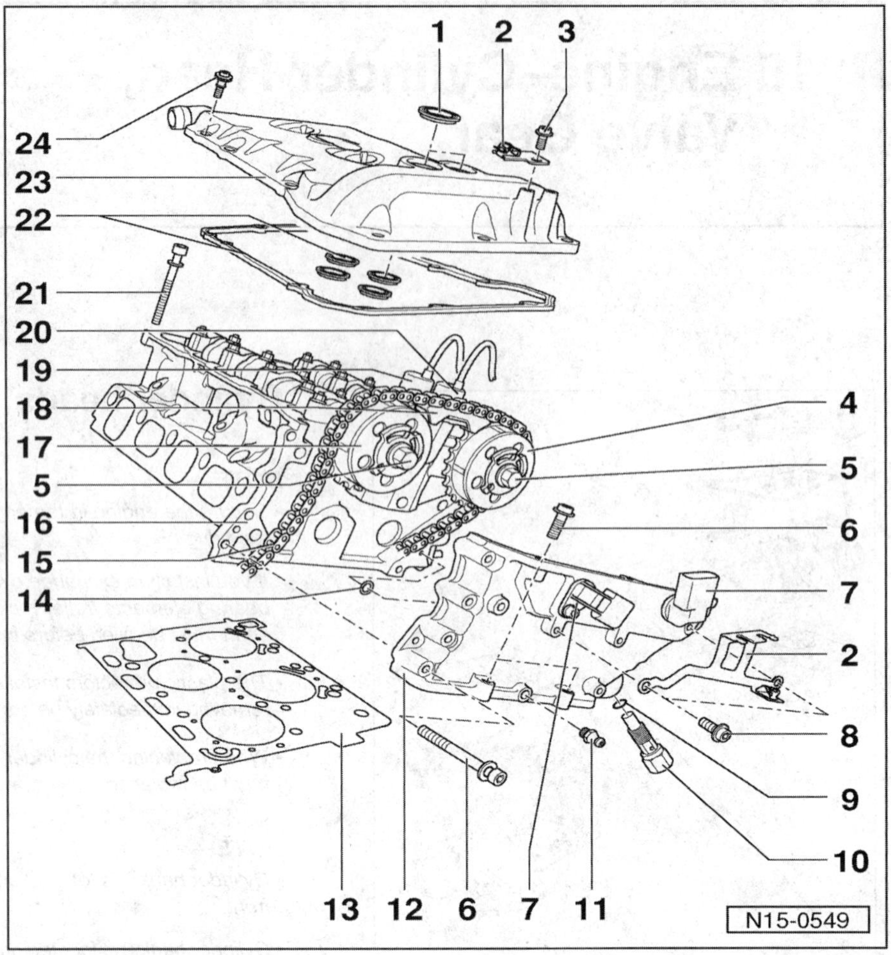

1. **Oil seal**
   - For valve 1 for camshaft adjustment -N205- **-19-** (intake) and camshaft adjustment valve -1- (exhaust) -N318-, **-20-**
   - Replace if damaged or leaking
   - Replace, see **Fig. 2**
2. **Bracket**
   - For wiring harness
3. **8 Nm (71 in-lb)**
4. **Exhaust camshaft adjuster**
   - Identification: B1 and A8
   - Rotate engine only when camshaft adjuster is installed
   - Installing, see **Valve timing, adjusting**
5. **60 Nm (44 ft-lb) + $^1/_4$ turn (90 °) further**
   - Replace
   - Contact surface of sender wheel must be dry around bolt head when assembled
   - To remove and install, counter-hold with 32 mm open jaw wrench on camshaft, see **Camshafts, removing and installing**
6. **10 Nm (7 ft-lb)**
   - Follow installation instructions and sequence when loosening and tightening, see **Valve timing, adjusting**
7. **Hall sender**
   - See **Repair Group 28, Ignition system components**, item **-3-**
   - Before removing, mark connector belonging to component
8. **10 Nm (7 ft-lb)**
9. **Seal**
   - Replace if damaged or leaking
10. **Chain tensioner**
    - Tighten to 40 Nm (30 ft-lb)
    - For camshaft roller chain **-15-**
    - Only rotate engine when chain tensioner is installed
11. **Sealing nipple**
    - Tighten to 4 Nm (35 in-lb)
    - Installed according to version of cover piece
12. **Cover**
    - Can only be removed and installed with engine removed
    - Can be removed and installed with cylinder head installed
    - Before removing, remove chain tensioner, item**-10-**
    - With O-ring for sealing oil gallery, item **-14-**
    - Installing, see **Valve timing, adjusting**
13. **Cylinder head gasket**
    - Replace, see **Cylinder head (cylinder banks 1 and 2), removing and installing**
    - If replaced, replace complete amount of engine coolant
14. **O-ring**
    - For oil channel seal
    - Replace
    - Oil before assembling
15. **Camshaft roller chain**
    - Mark direction of rotation before removing (installation position), see **Chain drive system components, Fig. 1**
    - Installing, see **Valve timing, adjusting**
16. **Cylinder head**
    - Check for distortion, see **Fig. 1**
    - Removing and installing, see **Cylinder head (cylinder banks 1 and 2), removing and installing**, Removing and installing cylinder head (cylinder banks 1 and 2)
    - If replaced, replace engine coolant

(continued on the following page)

Cylinder bank 1, assembly (continued)

17. **Intake camshaft adjuster**
    - Identification: B1 and E7
    - Rotate engine only when camshaft adjuster is installed
    - Installing, see **Valve timing, adjusting**
18. **Guide rail**
    - For camshaft roller chain **-15-**
    - Clipped to control housing
19. **Valve 1 for camshaft adjustment -N205-**
    - For intake camshaft
    - Before removing, mark connector belonging to component
20. **Camshaft adjustment valve -1- (exhaust) -N318-**
    - Before removing, mark connector belonging to component
21. **Cylinder head bolt**
    - Replace
    - Follow installation instructions and sequence when loosening and tightening, see **Cylinder head (cylinder banks 1 and 2), removing and installing**
22. **Seal for cylinder head cover**
    - Replace if damaged or leaking
    - Note installation position
23. **Cylinder head cover**
    - Removing and installing, see **Cylinder head cover (cylinder banks 1 and 2), removing and installing**
    - Replace if damaged
24. **10 Nm (7 ft-lb)**
    - With spacer sleeve and seal
    - Replace seal if damaged
    - Observe sequence when tightening, see **Cylinder head cover (cylinder banks 1 and 2), removing and installing**

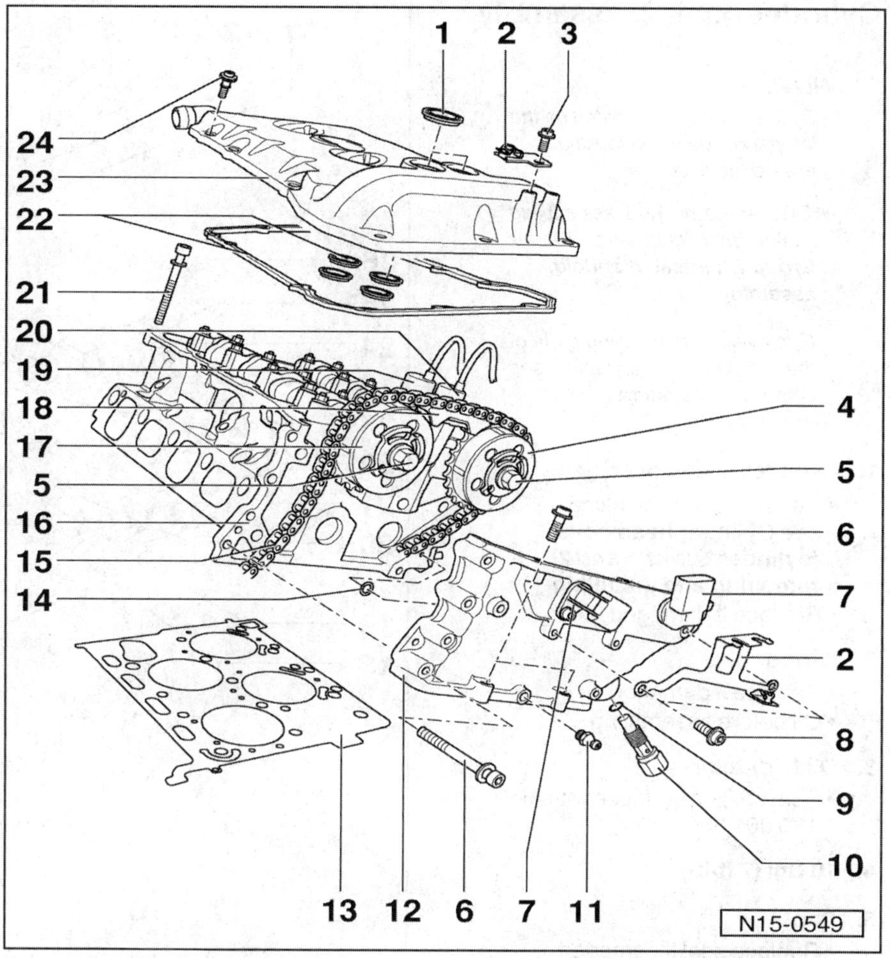

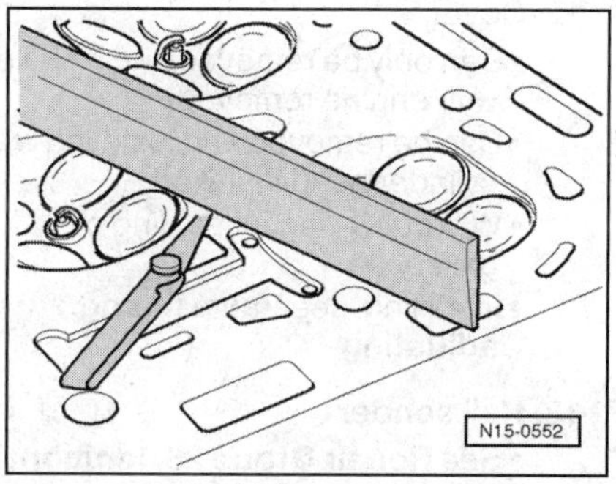

Fig. 1 Checking cylinder head for distortion

Special tools and equipment

- Straight edge
- Feeler gauges

Max. permissible distortion: 0.05 mm (0.00019 in.)

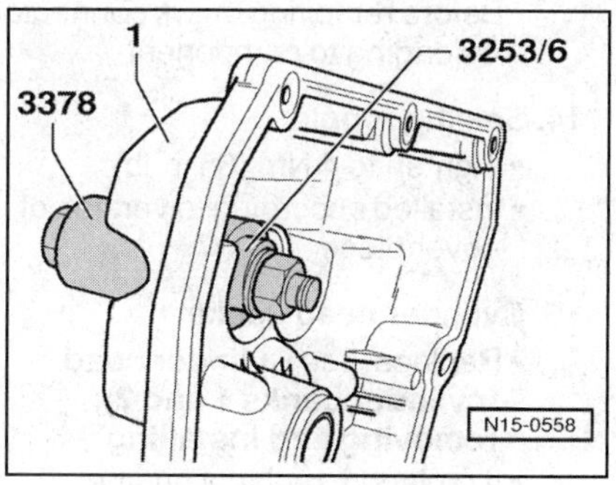

Fig. 2 Installing camshaft timing adjustment valve seal

Special tools and equipment

- 3378 Fitting sleeve
- 3253/6 From assembly device 3253

NOTE—
Do not lubricate seal.

– Install seal with sleeve 3378 in cylinder head cover -1- and pull in flush with sleeve 3253/6 from assembly device 3253.

Cylinder bank 2, assembly

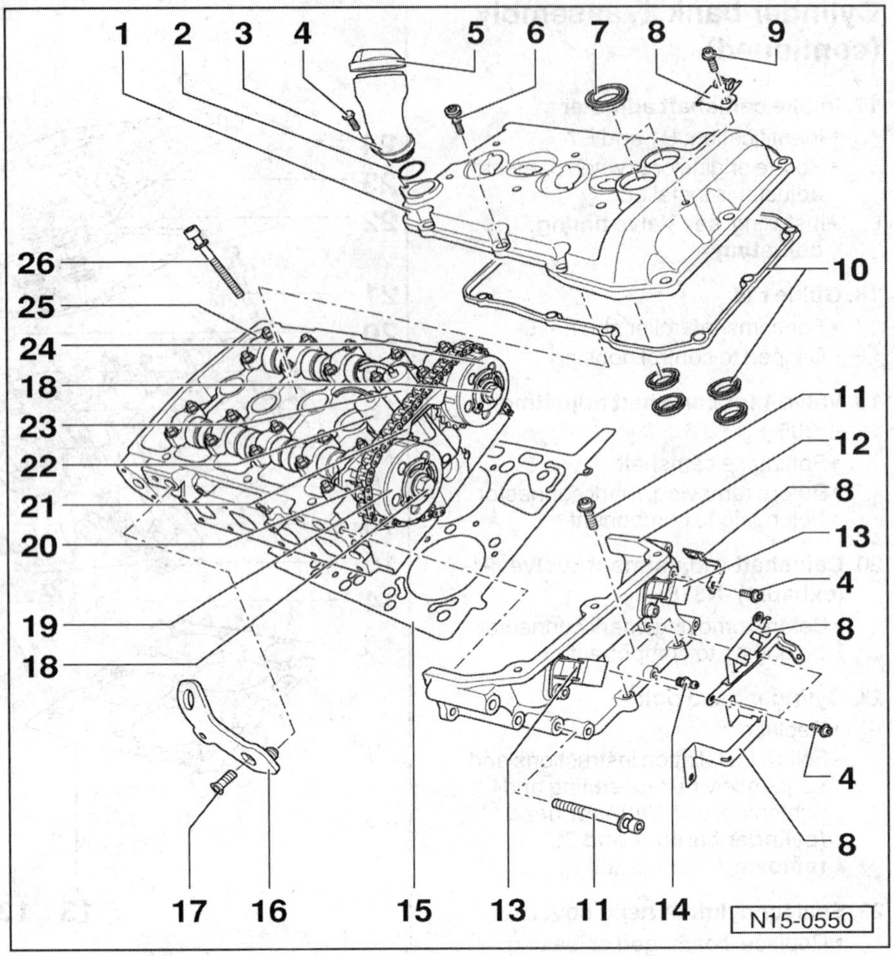

NOTE —

- *Cylinder bank 2 is located on the left-hand side when looking toward the front.*
- *Disassembling and assembling intake manifold, see **Repair Group 24, Intake manifold, assembly**.*
- *Removing and installing parts of the chain drive system, see **Chain drive system components.***

1. **Cylinder head cover**
   - Removing and installing, see **Cylinder head cover (cylinder banks 1 and 2), removing and installing**
   - Replace if damaged
2. **O-ring**
   - Replace if damaged
   - Oil before assembling
3. **Oil filler tube**
   - Coat sealing surfaces with AMV 188 001 02
4. **10 Nm (7 ft-lb)**
5. **Cap**
   - Replace seal if damaged
6. **10 Nm (7 ft-lb)**
   - With spacer sleeve and seal
   - Replace seal if damaged
   - Observe sequence when tightening, see **Cylinder head cover (cylinder banks 1 and 2), removing and installing**
7. **Seal**
   - For valve 2 for camshaft adjustment -N208- **-23-** (intake) and camshaft adjustment valve 2 (exhaust) -N319-, **-22-**
   - Replace if damaged or leaking
   - Replace, see **Cylinder bank 1, Fig. 2**
8. **Bracket**
   - For wiring harness
9. **8 Nm (71 in-lb)**
10. **Seal for cylinder head cover**
    - Replace if damaged or leaking
    - Note installation position
11. **10 Nm (7 ft-lb)**
    - Follow installation instructions and sequence when loosening and tightening, see **Valve timing, adjusting**
12. **Cover**
    - Can only be removed and installed with engine removed.
    - Can be removed and installed with cylinder head installed
    - Without O-ring for sealing oil gallery
    - Installing, see **Valve timing, adjusting**
13. **Hall sender**
    - See **Repair Group 28, Ignition system components**, item **-3-**
    - Before removing, mark connector belonging to component
14. **Sealing nipple**
    - Tighten to 4 Nm (35 in-lb)
    - Installed according to version of cover piece
15. **Cylinder head gasket**
    - Replace, see **Cylinder head (cylinder banks 1 and 2), removing and installing**
    - If replaced, replacef engine coolant
16. **Lifting eye**
17. **23 Nm (17 ft-lb)**
18. **60 Nm (44 ft-lb) + 1/4 turn (90 °) further**
    - Replace
    - Contact surface of sender wheel must be dry around bolt head when assembled
    - To remove and install, counter-hold with 32 mm open jaw wrench on camshaft, see **Camshafts, removing and installing**
19. **Exhaust camshaft adjuster**
    - Identification: B2 and A8
    - Rotate engine only when camshaft adjuster is installed
    - Installing, see **Valve timing, adjusting**
20. **Camshaft roller chain**
    - Mark direction of rotation before removing (installation position), see **Chain drive system components , Fig. 1**
    - Installing, see **Valve timing, adjusting**

(continued on following page)

Cylinder bank 2, assembly (continued)

21. Guide rail
- For camshaft roller chain **-20-**
- Clipped to control housing

22. Camshaft adjustment valve 2 (exhaust) -N319-
- For exhaust camshaft
- Before removing, mark connector belonging to component

23. Valve 2 for camshaft adjustment -N208-
- For intake camshaft
- Before removing, mark connector belonging to component

24. Intake camshaft adjuster
- Identification: B2 and E7
- Rotate engine only when camshaft adjuster is installed
- Installing, see **Valve timing, adjusting**

25. Cylinder head
- Check for distortion, see **Cylinder bank 1, Fig. 1**
- Removing and installing, see **Cylinder head (cylinder banks 1 and 2), removing and installing**
- If replaced, replace all engine coolant

26. Cylinder head bolt
- Replace
- Follow installation instructions and sequence when loosening and tightening, see **Cylinder head (cylinder banks 1 and 2), removing and installing**

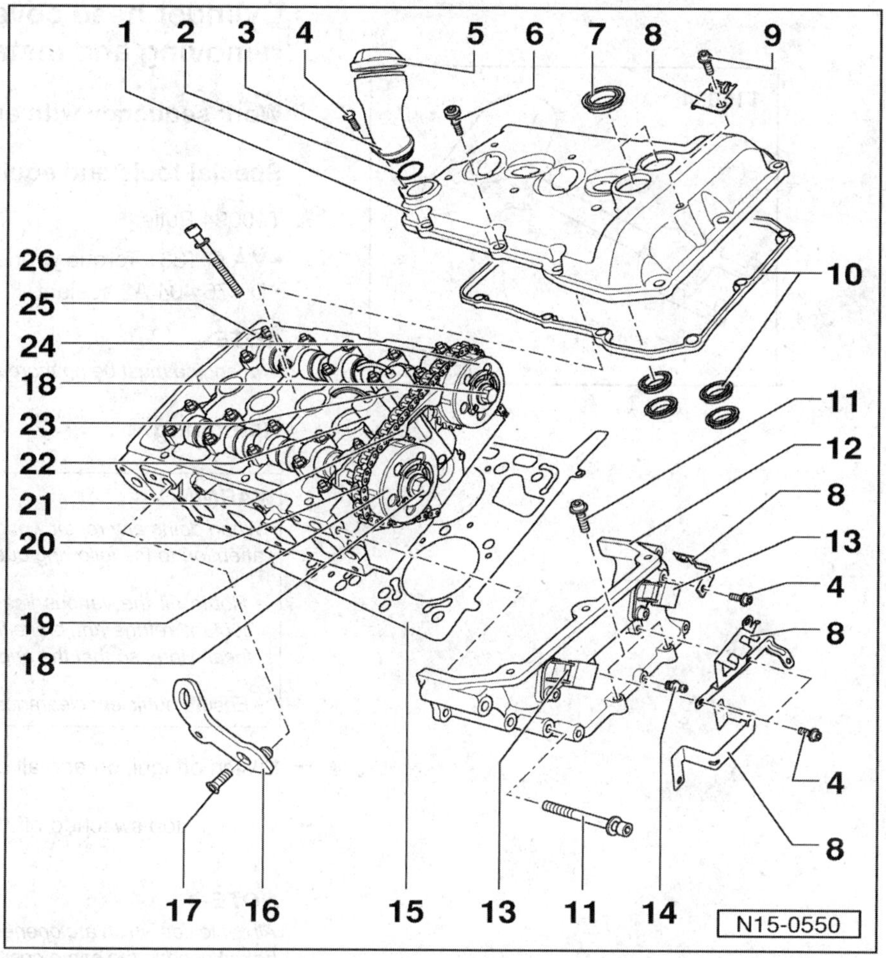

Cylinder head cover (cylinder banks 1 and 2), removing and installing

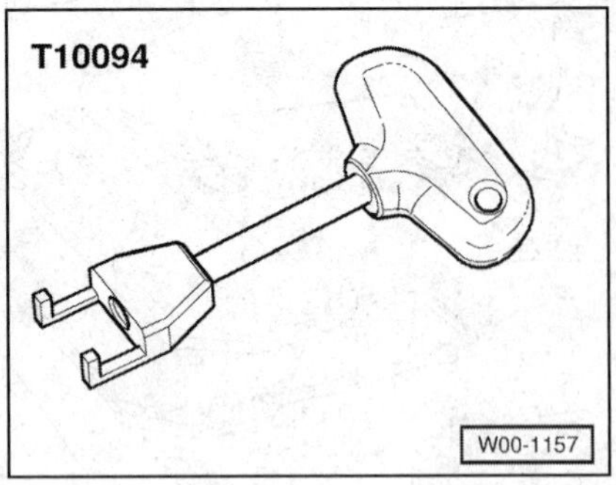

Work sequence with engine installed

Special tools and equipment

T10094 Puller

- V.A.G 1331 Torque wrench (5...50 Nm)
- D 176 404 A2 sealant

NOTE—
The engine must be no more than warm to touch.

Removing

WARNING—
When doing any repair work, especially in the engine compartment, pay attention to the following due to the cramped conditions:

- *Route all the various lines (e.g. for fuel, hydraulics, EVAP system, coolant, refrigerant, brake fluid and vacuum lines and hoses) and electrical wiring so that the original positions are restored.*
- *Ensure sufficient clearance to all moving or hot components.*

– Switch off ignition and all electrical consumers and remove ignition key.

– With ignition switched off disconnect battery Ground strap.

NOTE—
All cable ties which are opened or cut open when removing the engine, must be replaced in the same position when installing the engine.

– Remove engine cover.

– Disconnect hoses for crankcase ventilation at pressure regulator valve and at cylinder head covers, see **Repair Group 13, Crankcase ventilation system, components**, item **-4-**.

NOTE—
Press buttons on hose couplings to disconnect.

– Remove intake manifold upper part (along with both intake manifold side parts), see **Fuel Injection and Ignition section, Repair Group 24**.

NOTE—
Seal the intake ports in the lower part of the intake manifold with a clean cloth.

– Disconnect connectors from ignition coils.

– Remove ignition coils with output stages using puller T10094.

– Remove cylinder head cover.

Cylinder head cover (cylinder banks 1 and 2), installing

NOTE—

- *Replace cylinder head cover if damaged or leaking.*
- *Replace cylinder head cover seal if damaged or leaking.*

– Clean sealing surface. Sealing surface must be free of oil and grease.

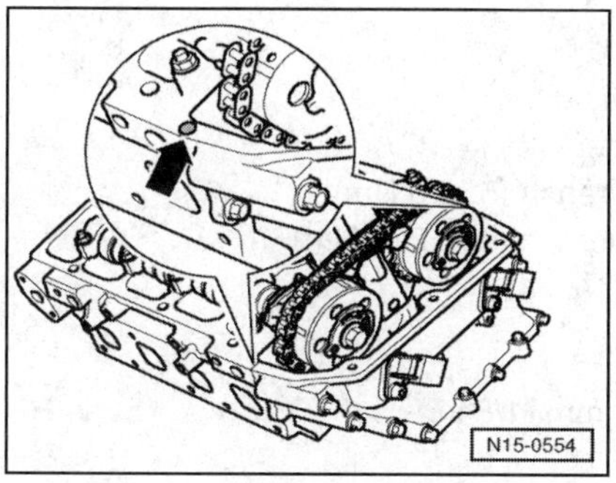

◄ Apply sealant D176 404 A2 to clean sealing surface between cylinder head and cover.

NOTE—

- *Note the use by date of the sealant.*
- *The cylinder head cover must be installed within 5 minutes of applying the sealant.*

– Install cylinder head cover immediately and tighten all securing bolts hand tight.

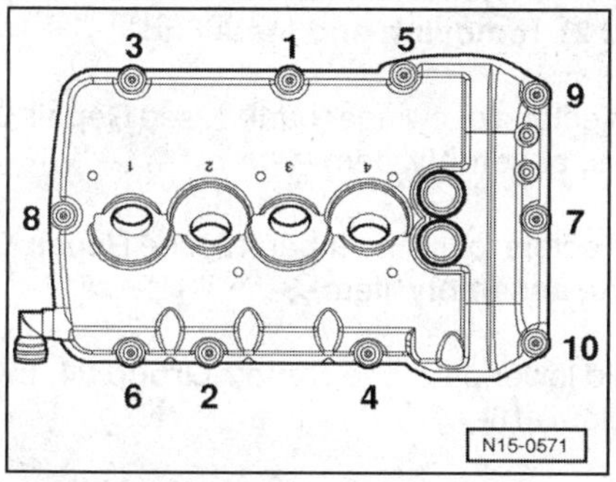

◄ **Cylinder bank 1**

- Tighten securing bolts in sequence as illustrated.
- Torque setting: 10 Nm (7 ft-lb)

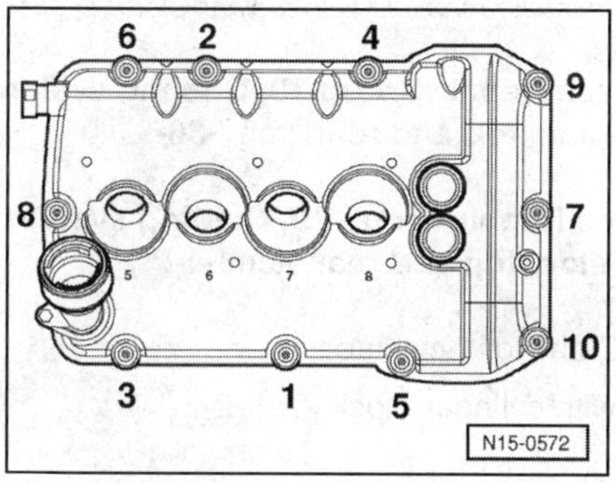

◄ **Cylinder bank 2**

- Tighten securing bolts in sequence as illustrated.
- Torque setting: 10 Nm (7 ft-lb)

Cylinder banks 1 and 2

– The rest of installation is reverse of the removal.

Cylinder head (cylinder banks 1 and 2), removing and installing

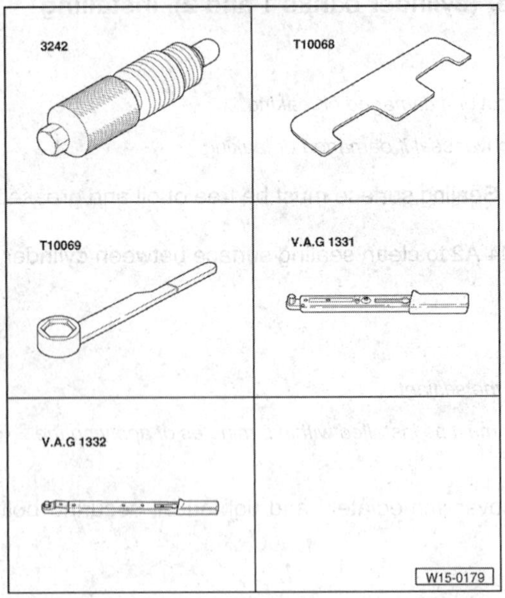

NOTE—
This work sequence is only possible when the engine is removed from the engine compartment.

Special tools and equipment

- 3242 Clamping bolt
- T10068 Camshaft bar
- T10069 Counter-holder
- V.A.G 1331 Torque wrench (5...50 Nm)
- V.A.G 1332 Torque wrench (40...200 Nm)

Not illustrated:

- Drill bit, diameter 3.5 mm (1/8 in.)

Removing

– Remove cylinder head cover, see **Cylinder head cover (cylinder banks 1 and 2), removing and installing**.

– Remove fuel rail with injectors for cylinder bank 1, see **Repair Group 24, Fuel rail with injectors, assembly**, item **-1-**.

– Remove fuel rail with injectors for cylinder bank 2, see **Repair Group 24, Fuel rail with injectors, assembly**, item **-3-**.

– Remove intake manifold lower part, see **Repair Group 24, Intake manifold, assembly**, item **-16-**.

NOTE—
Seal the intake ports in the intake manifold with a clean cloth.

– Remove coolant pipe, see **Repair Group 19, Cooling system components, engine side, top and rear**, item **-36-**.

– Remove intake pipe, see **Repair Group 19, Cooling system components, engine side, top and rear**, item **-4-**.

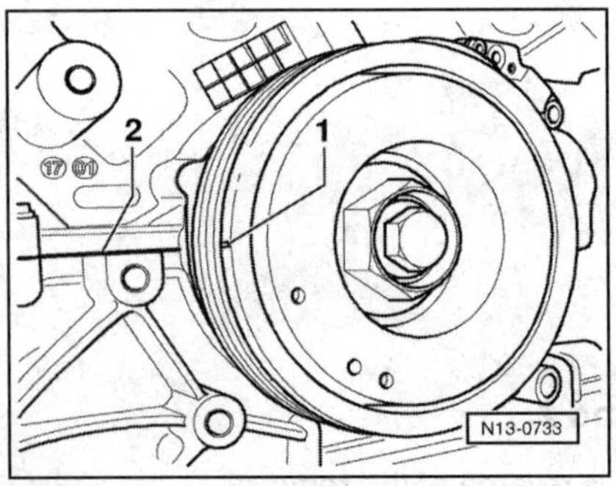

◀ Turn crankshaft to TDC on No. 1 cylinder.

- Mark -1- must align with cylinder block joint -2-.

Cylinder head, bank 1, removing

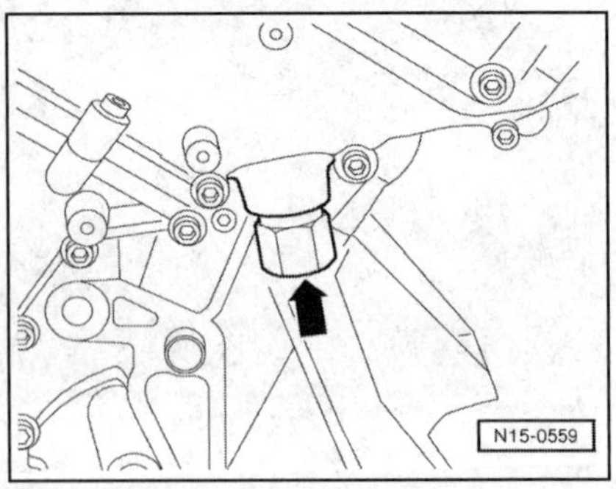

◀ Remove camshaft roller chain tensioner -arrow-.

– Remove camshaft cover part, see **Cylinder bank 1, assembly,** item **-12-**.

– Remove flywheel/drive plate sealing flange, see **Repair Group 13, Sealing flange and dual mass flywheel, assembly**, item **-10-**.

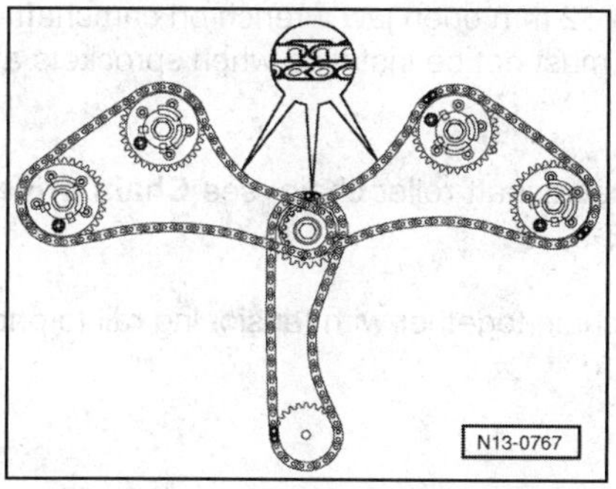

Mark roller chain before removing (e.g. with paint, arrow pointing in direction of rotation).

NOTE—
Do not mark chain with a center punch or any similar means!

– First remove exhaust timing camshaft adjuster and then remove intake camshaft timing adjuster.

– Remove both camshaft timing adjusters together with camshaft roller chain from camshafts.

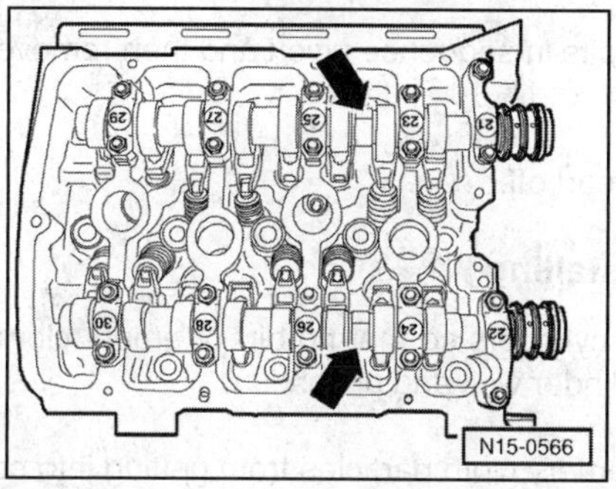

Counter-hold only with 32 mm open jaw wrench on camshaft -arrow-. Camshaft bar T10068 must not be installed when sprockets are loosened or tightened.

– Remove guide rail for camshaft roller chain, see **Chain drive system, components**, item **-7-**.

– Place camshaft roller chain to one side.

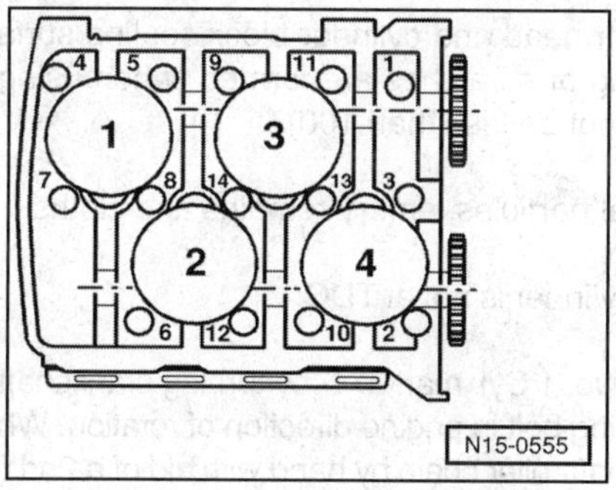

Loosen socket head bolts in sequence given and then remove completely.

– Carefully lift cylinder head off.

Cylinder head, bank 2, removing

– Remove camshaft cover, see **Cylinder bank 2, assembly** item -12-.

– Remove flywheel/drive plate sealing flange, see **Repair Group 13, Sealing flange and dual mass flywheel, assembly**, item **-10-**.

– Release tension on camshaft roller chain tensioner and lock chain tensioner using a drill bit (diameter: 3.5 mm, (1/8 in.), see **Chain drive system, components**, item **-31-**.

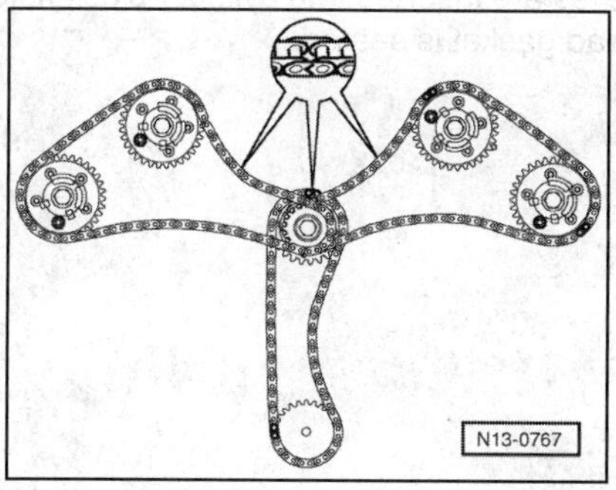

Mark roller chain before removing (e.g. with paint, arrow pointing in direction of rotation).

NOTE—
Do not mark chain with a center punch or any similar means!

– First remove exhaust timing camshaft adjuster and then remove intake camshaft timing adjuster.

– Remove both camshaft timing adjusters together with camshaft roller chain from camshafts.

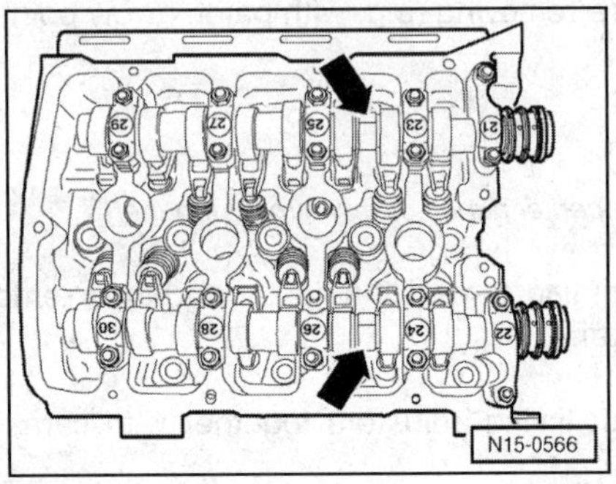

Counter-hold only with 32 mm open jaw wrench on camshaft -arrow-. Camshaft bar T10068 must not be installed when sprockets are loosened or tightened.

– Remove guide rail for camshaft roller chain, see **Chain drive system, components**, item **-28-**.

– Place camshaft roller chain together with tensioning rail to one side.

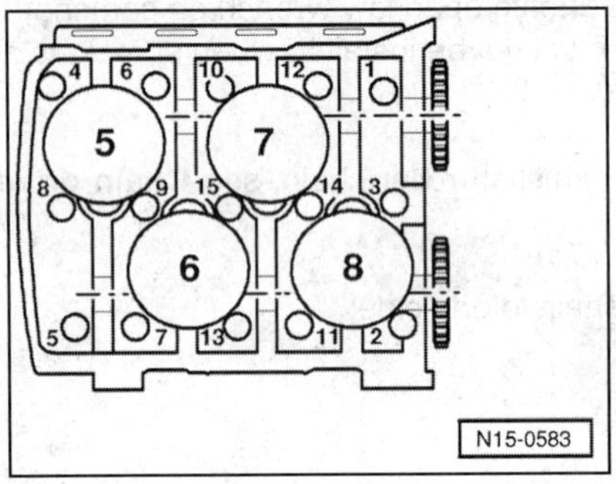

Loosen socket head bolts in sequence given and then remove completely.

– Carefully lift cylinder head off.

Cylinder heads, installing

– Place a clean cloth in cylinders so that no dirt or emery cloth particles can get in between cylinder wall and piston.

– Also prevent dirt and emery cloth particles from getting into coolant.

– Carefully clean cylinder head and cylinder block sealing surfaces. This ensures that no scoring or scratches are formed (when using abrasive paper the grade must not be less than 100).

– Carefully remove metal particles, emery remains and cloths.

If the piston for No. 1 cylinder is not at TDC:

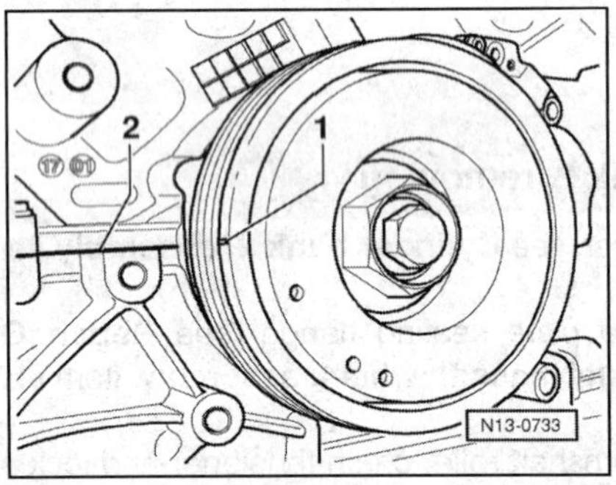

Set crankshaft to TDC No. 1 Cyl. marks -1- by turning crankshaft on vibration damper securing bolt in engine direction of rotation. When doing this, guide camshaft roller chain by hand with aid of a 2nd technician.

- Mark -1- must align with cylinder block joint -2-.

Cylinder head, bank 1, installing

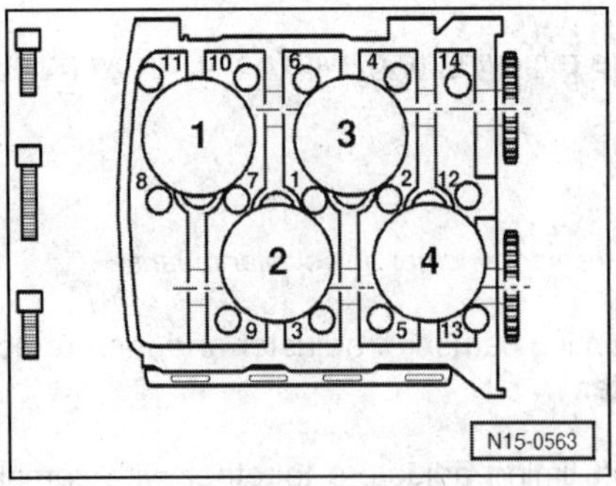

Ensure that dowel sleeves are inserted into cylinder block holes 9 and 14 and that cylinder head gasket is seated.

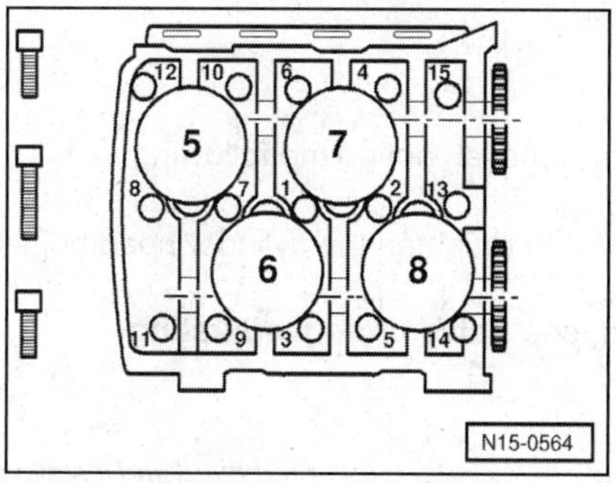

Cylinder head, bank 2, installing

Ensure that dowel sleeves are inserted into cylinder block holes 4 and 11 and that cylinder head gasket is seated.

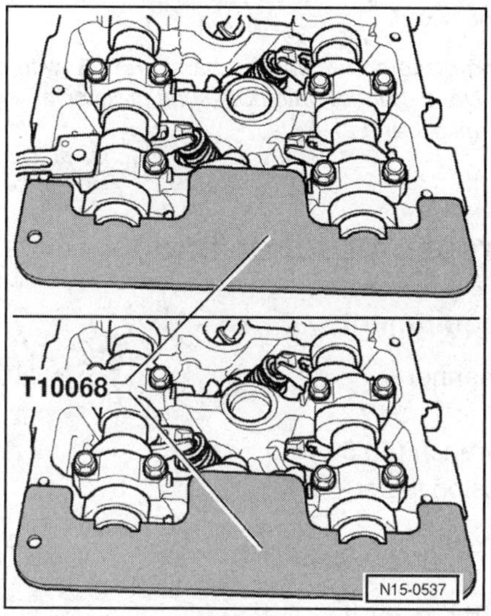

Cylinder banks 1 and 2

– Fit new cylinder head gasket.

NOTE —

- *Only remove the new cylinder head gasket from its packing immediately before installing.*
- *Handle the new gasket with extreme care. Damage to the gasket will lead to leaks.*

– Position camshafts in cylinder head to TDC No. 1 cylinder.

The camshaft bar T10068 must engage in both shaft grooves.

NOTE —

If the camshaft bar cannot be engaged in the camshafts, turn camshafts in direction of rotation past TDC No. 1 Cyl. and then back to TDC No. 1 Cyl.

– Install cylinder head, insert new cylinder head bolts and hand tighten.

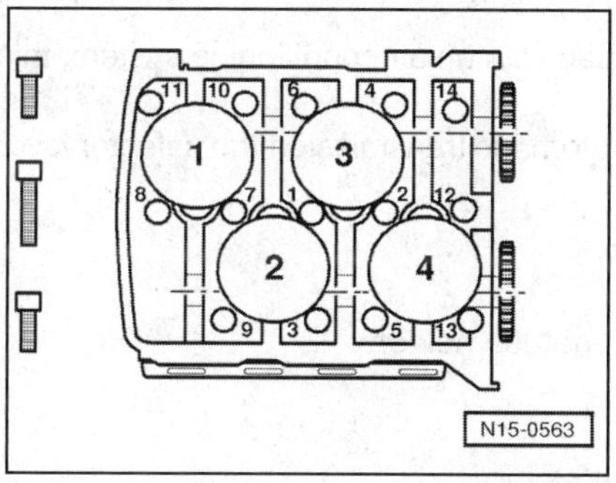

Cylinder head, bank 1

Tighten cylinder head in tightening sequence as follows:

NOTE —

The longer cylinder head bolts must be inserted in the middle holes of cylinder head.

- Pretighten all bolts to 30 Nm (22 ft-lb).
- Then tighten all bolts to 50 Nm (37 ft-lb).
- Then tighten all bolts $^1/_4$ turn (90 °) further using a rigid wrench.
- Then tighten all bolts again $^1/_4$ turn (90 °) further.

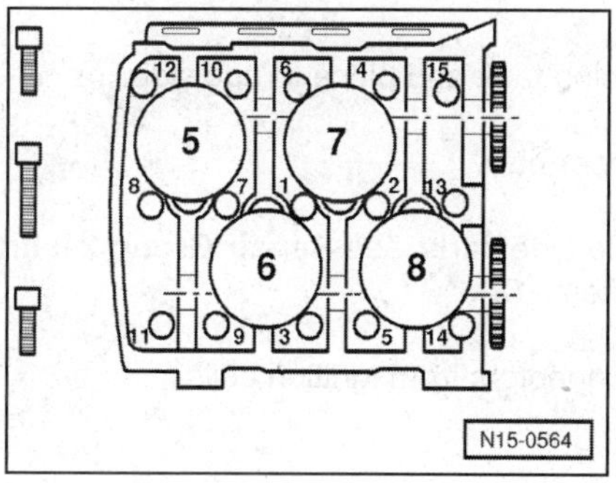

Cylinder head, bank 2

Tighten cylinder head in tightening sequence as follows:

NOTE —

The longer cylinder head bolts must be inserted in the middle holes of cylinder head.

- Pretighten all bolts to 30 Nm (22 ft-lb).
- Then tighten all bolts to 50 Nm (37 ft-lb).
- Then tighten all bolts $^1/_4$ turn (90 °) further using a rigid wrench.
- Then tighten all bolts again $^1/_4$ turn (90 °) further.

Cylinder banks 1 and 2

NOTE—
Installation is reverse of removal, noting the following:

- There is no requirement to retighten the cylinder head bolts after repairs.
- Adjusting valve timing, see **Valve timing, adjusting**.

NOTE—
- *If only the cylinder head of one cylinder bank is removed and installed, then the position of the copper links of the camshaft roller chain of the other cylinder bank need not be noted.*
- *After adjusting the timing place both camshaft bars T10068 (with crankshaft positioned at TDC No. 1 cylinder) into both shaft grooves of camshafts for cylinder banks 1 and 2.*

Compression pressures, checking

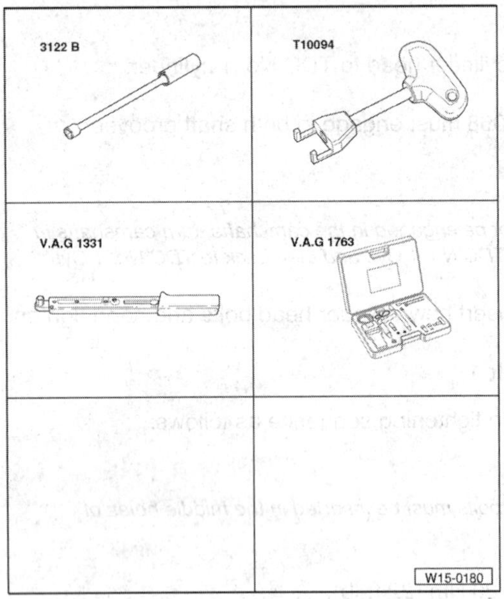

Special tools and equipment

- 3122 B Spark plug spanner
- T10094 Puller
- V.A.G 1331 Torque wrench (5...50 Nm)
- V.A.G 1763 Compression tester

Test conditions

- Engine oil temperature 20...30 °C.
- Battery voltage must be at least 11.5 V.
- All electrical consumers, e.g. lights and rear window defroster must be switched off.
- If the vehicle is equipped with an air conditioning system, this must be switched off.
- On vehicles with an automatic transmission the selector lever must be in "P" or "N" position

Test sequence

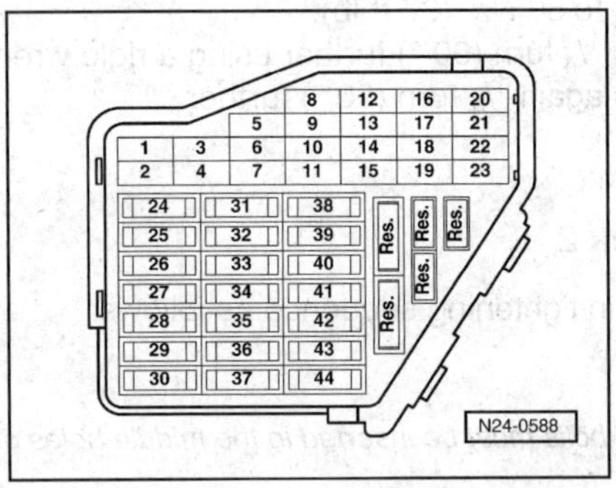

◄ Remove fuse No. 32 from fuse holder.

NOTE—
Removing fuse No. 32 interrupts the voltage supply to the injectors.

– Remove engine cover.

– Loosen coolant expansion tank and place to one side.

– Remove air cleaner upper part.

– Remove intake manifold side parts, see **Repair Group 24, Intake manifold, assembly**, item **-13-**.

– Disconnect all 4-pin connectors from ignition coils.

NOTE—

Before disconnecting the connector, mark the component to which it belongs.

- Remove ignition coils with output stages using puller T10094.
- Remove spark plugs with spark plug socket and extension 3122B.
- Check compression pressure with compression tester V.A.G 1763.
- Operate starter until tester shows no further pressure increase.

Compression pressure

Engine code	New	Wear limit
BDP	10 to 13 bar (145-190 psi)	9.0 bar (130 psi)

Permissible difference between all cylinders: 3 bar (45 psi)

- Check fault memory, if necessary, repair any malfunctions and then erase fault memory.

Chain drive system, components

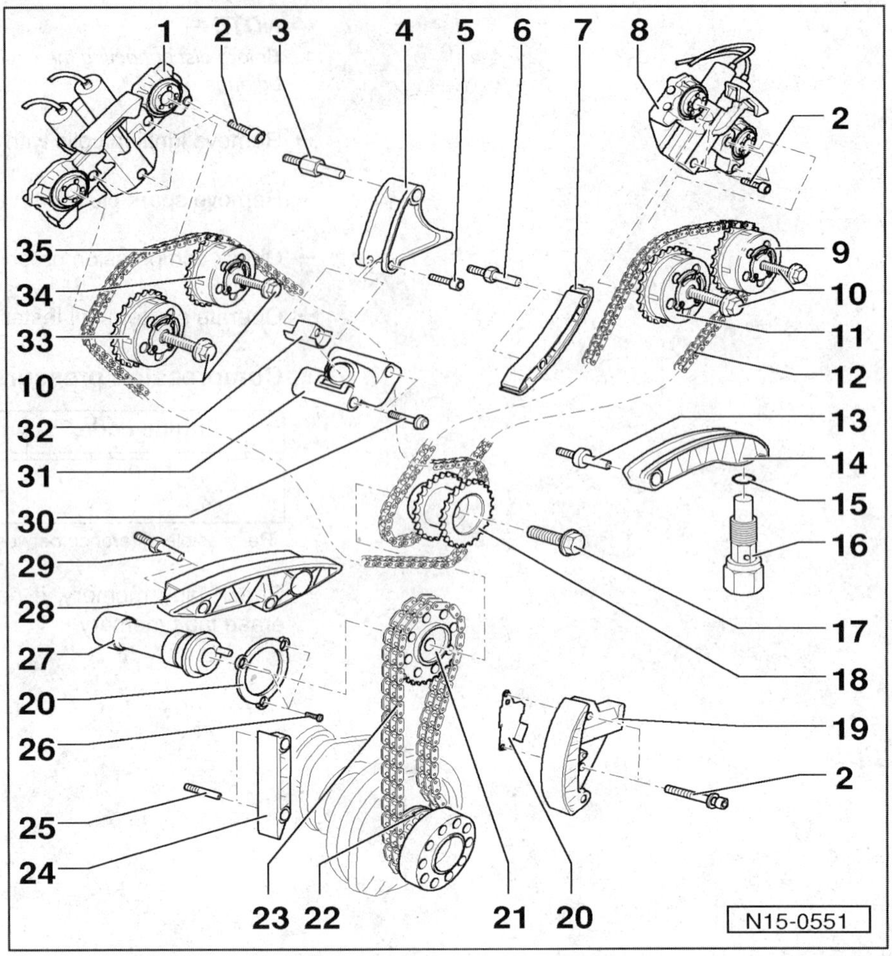

NOTE—
- *Cylinder bank 1 is located on the right-hand side when looking toward the front.*
- *Cylinder bank 2 is located on left-hand side when looking toward the front.*
- *Disassembling and assembling intake manifold, see* ***Repair Group 24, Intake manifold, assembly****.*

1. **Control housing**
   - Cylinder bank 2
   - Lightly lubricate contact surfaces of oil seals before installing
   - Removing and installing, see **Camshafts, removing and installing**
   - Disassembling and assembling, see **Valve train, assembly, Fig. 8**
2. **8 Nm (71 in-lb)**
3. **Pivot pin**
   - Tighten to 18 Nm (13 ft-lb)
   - For tensioning plate **-4-**
4. **Tensioning plate**
   - For camshaft roller chain **-35-**
5. **20 Nm (15 ft-lb)**
6. **Pivot pin**
   - Tighten to 18 Nm (13 ft-lb)
   - For guide rail item **-7-**
7. **Guide rail**
   - For camshaft roller chain **-12-**
8. **Control housing**
   - Cylinder bank 1
   - Lightly lubricate contact surfaces of oil seals before installing
   - Removing and installing, see **Camshafts, removing and installing**
   - Disassembling and assembling, see **Valve train, components, Fig. 8**
9. **Exhaust camshaft adjuster**
   - Cylinder bank 1
   - Identification: B1 and A8
   - Observe installation position to camshaft roller chain, see **Fig. 3**
   - Rotate engine only when camshaft adjuster is installed
   - Installing, see **Valve timing, adjusting**
10. **60 Nm (44 ft-lb) + $^1/_4$ turn (90 °) further**
    - Replace
    - Contact surface of sender wheel must be dry around bolt head when assembled
    - To remove and install, counter-hold with 32 mm open jaw wrench on camshaft, see **Camshafts, removing and installing**
11. **Intake camshaft adjuster**
    - Cylinder bank 1
    - Identification: B1 and E7
    - Observe installation position to camshaft roller chain, see **Fig. 2**
    - Rotate engine only when camshaft adjuster is installed
    - Installing, see **Valve timing, adjusting**
12. **Camshaft roller chain**
    - Cylinder bank 1
    - Mark direction of rotation before removing (installation position), see **Fig. 1**
    - Observe installation position to camshaft adjusters, see **Fig. 2** and, see **Fig. 3**
    - Installing, see **Valve timing, adjusting**
13. **Pivot pin**
    - Tighten to 12 Nm (9 ft-lb)
    - For tensioning rail **-14-**
14. **Tensioning rail**
    - For camshaft roller chain **-12-**
15. **Seal**
    - Replace if damaged or leaking
16. **Chain tensioner**
    - Tighten to 40 Nm (30 ft-lb)
    - For camshaft roller chain **-12-**
    - Only rotate engine when chain tensioner is installed
17. **60 Nm (44 ft-lb) + $^1/_4$ turn (90 °) further**
    - Replace
    - Counter-hold with T10069 to loosen and tighten, see **Valve timing, adjusting**
18. **Sprocket**
    - For camshaft roller chain items **-12-** and **-35-**
    - Installing, see **Valve timing, adjusting**

(continued on following page)

Chain drive system, components (continued)

19. **Chain tensioner with tensioning plate**
    - For roller chain **-23-**
    - Lock chain tensioner with drill bit (diameter: 3.5 mm (1/8 in.)
    - Only rotate engine with chain tensioner installed

20. **Securing plate**

21. **Sprocket**
    - For roller chain **-23-**
    - Installing, see **Valve timing, adjusting**

22. **Crankshaft sprocket**
    - Integral part of crankshaft

23. **Roller chain**
    - Mark direction of rotation before removing (installation position), see **Fig. 1**
    - Installing, see **Valve timing, adjusting**

24. **Guide rail**
    - For roller chain **-23-**

25. **Locating pin without collar**
    - Tighten to 10 Nm (7 ft-lb)
    - For guide rail **-24-**

26. **10 Nm (7 ft-lb)**
    - Insert with locking fluid D 000 600 A2

27. **Intermediate shaft**

28. **Guide rail**
    - For camshaft roller chain **-35-**

29. **Pivot pin**
    - Tighten to 10 Nm (7 ft-lb)
    - For guide rail **-28-**

30. **9 Nm (80 in-lb)**

31. **Chain tensioner**
    - For camshaft roller chain **-35-**
    - Lock chain tensioner with drill bit (diameter: 3.5 mm (1/8 in.)
    - Only rotate engine with chain tensioner installed

32. **Stop plate**
    - For chain tensioner **-31-**

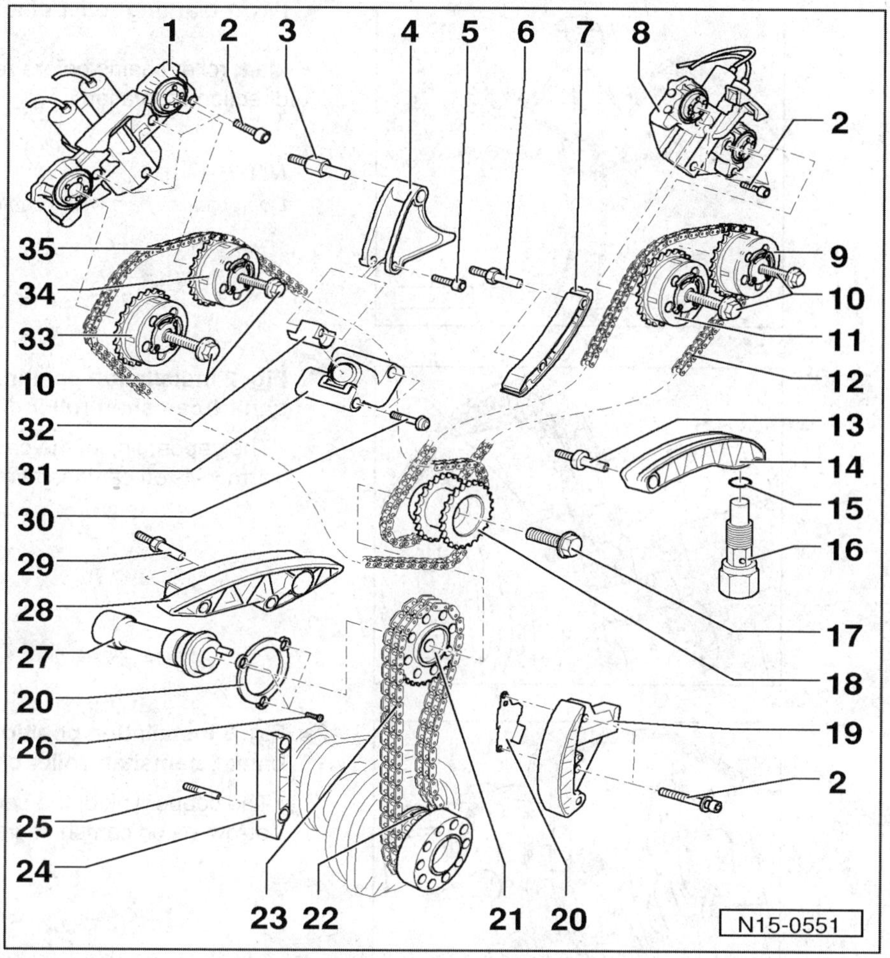

33. **Exhaust camshaft adjuster**
    - Cylinder bank 2
    - Identification: B2 and A8
    - Observe installation position to camshaft roller chain, see **Fig. 5**
    - Rotate engine only when camshaft adjuster is installed
    - Installing, see **Valve timing, adjusting**

34. **Intake camshaft adjuster**
    - Cylinder bank 2
    - Identification: B2 and E7
    - Observe installation position to camshaft roller chain, see **Fig. 4**
    - Rotate engine only when camshaft adjuster is installed
    - Installing, see **Valve timing, adjusting**

35. **Camshaft roller chain**
    - Cylinder bank 2
    - Mark direction of rotation before removing (installation position), see **Fig. 1**
    - Observe installation position to camshaft adjusters, see **Fig. 4** and, see **Fig. 5**
    - Installing, see **Valve timing, adjusting**

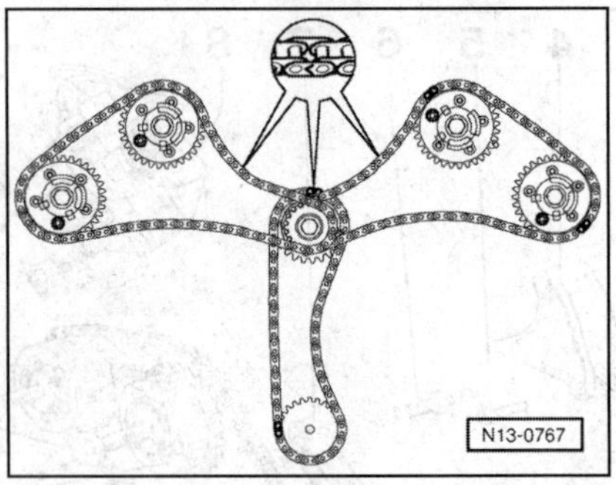

Fig. 1 Marking roller chains

– Mark roller chains before removing (e.g. with paint, arrow pointing in direction of rotation).

NOTE —
Do not mark chain with a center punch or any similar means!

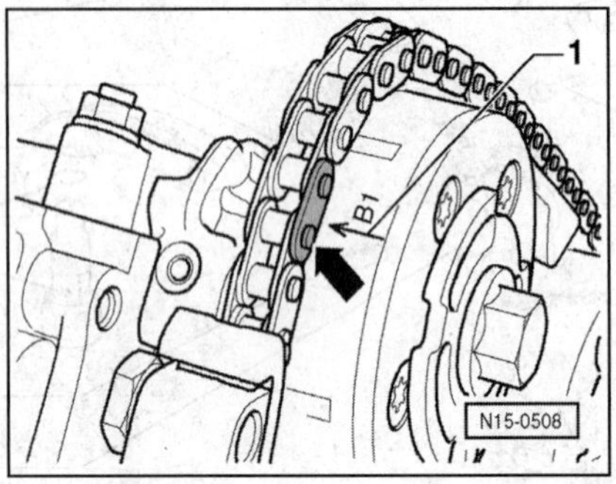

Fig. 2 Installation position: Intake camshaft adjuster to cylinder bank 1 camshaft roller chain

- The copper link of the camshaft roller chain -arrow- must align with arrow -1- on camshaft timing adjuster.

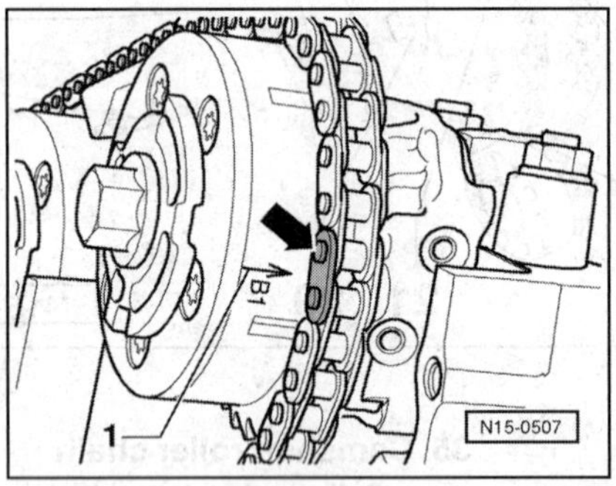

Fig. 3 Installation position: Exhaust camshaft adjuster to cylinder bank 1 camshaft roller chain

- The copper link of the camshaft roller chain -arrow- must align with arrow -1- on camshaft timing adjuster.

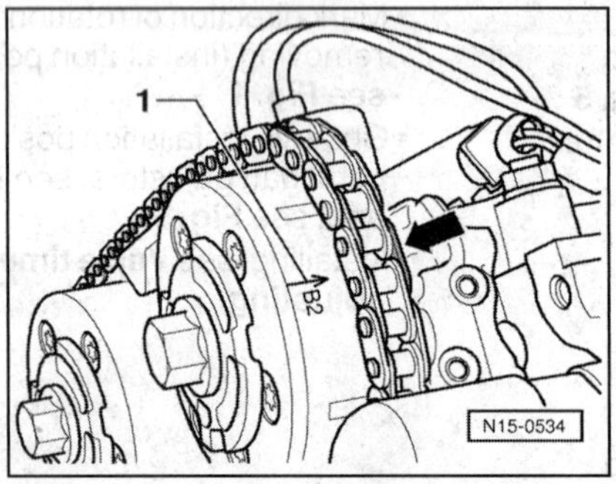

Fig. 4 Installation position: Intake camshaft adjuster to cylinder bank 2 camshaft adjuster roller chain

- The copper link of the camshaft roller chain -arrow- must align with arrow -1- on camshaft timing adjuster.

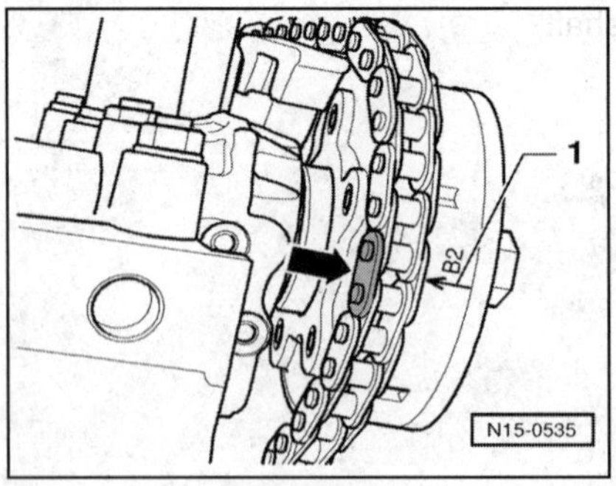

Fig. 5 Installation position: Exhaust camshaft adjuster to cylinder bank 2 camshaft roller chain

- The copper link of the camshaft roller chain -arrow- must align with arrow -1- on camshaft timing adjuster.

Valve timing, adjusting

NOTE—
Valve timing adjusting procedure requires that the engine is removed from the engine compartment.

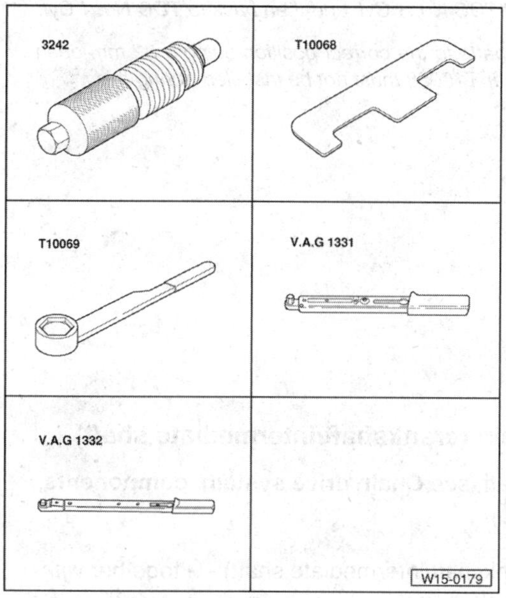

Special tools equipment

- 3242 Clamping bolt
- T10068 Camshaft bar
- T10069 Counter-holder
- V.A.G 1331 Torque wrench (5...50 Nm)
- V.A.G 1332 Torque wrench (40...200 Nm)
- LSE 020 100 A3 Silicone remover
- D 188 800 A1 Sealing compound
- 3.5 mm diameter (1/8 in.) drill bit

Work sequence

NOTE—
- *The following work sequence can only be performed when the engine is removed from the engine compartment. The adjustments required depend on how far the engine has been disassembled.*
- *Note direction of rotation of used chains, see* ***Chain drive system components****, see* **Fig. 1**.
- *Roller chains must run completely straight in guide and tensioner rails.*
- *Ensure that all chain sprocket securing bolts are replaced.*

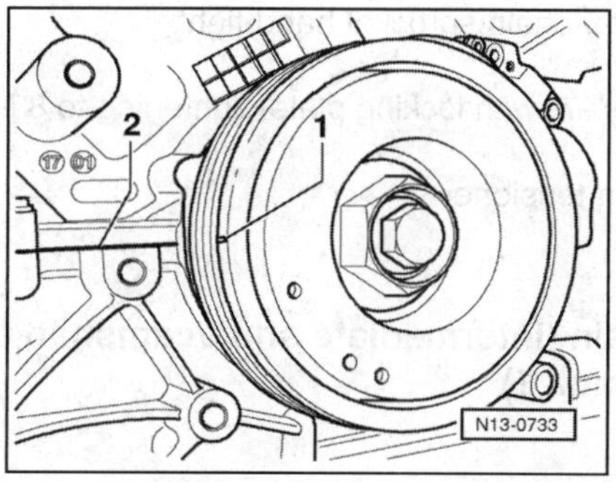

Turn crankshaft to TDC No. 1 cylinder.

- Mark -1- must align with cylinder block joint -2-.

– Remove sealing plug from left side of cylinder block.

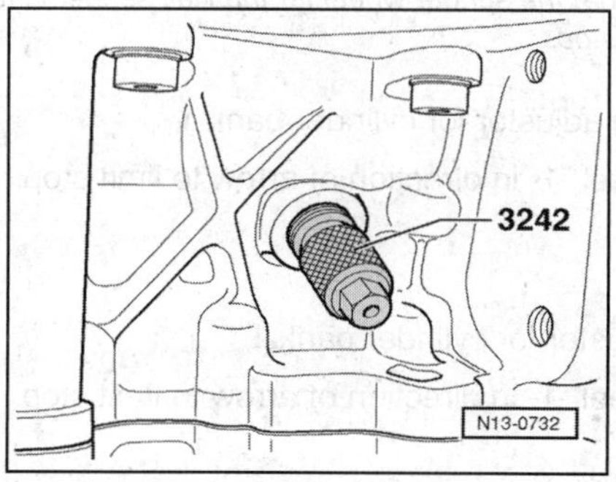

Screw clamping bolt 3242 into hole so that crankshaft is secured against turning.

– Position camshafts in cylinder head to TDC No. 1 cylinder.

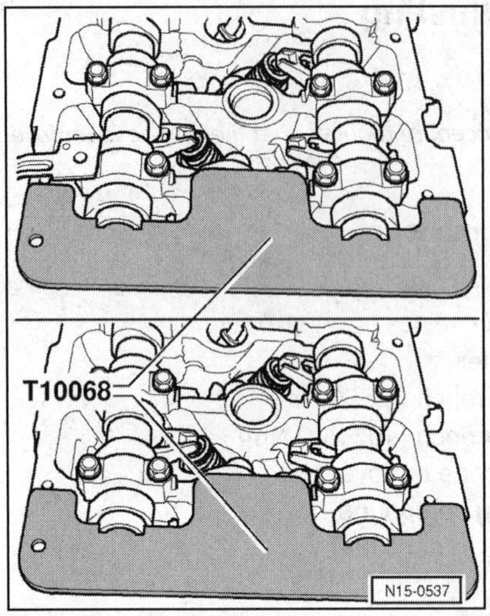

Camshaft bar T10068 must engage in both shaft grooves.

NOTE—
- *If the camshaft bar cannot be engaged in the camshafts, turn camshafts in direction of rotation past TDC No. 1 Cyl. and then back to TDC No. 1 Cyl.*
- *If necessary, turn camshafts to the correct position using a 32 mm open jaw wrench. Camshaft bar T10068 must not be installed during this process.*

Installing roller chain (crankshaft/intermediate shaft)

– Install intermediate shaft, see **Chain drive system, components**, item **-27-**.

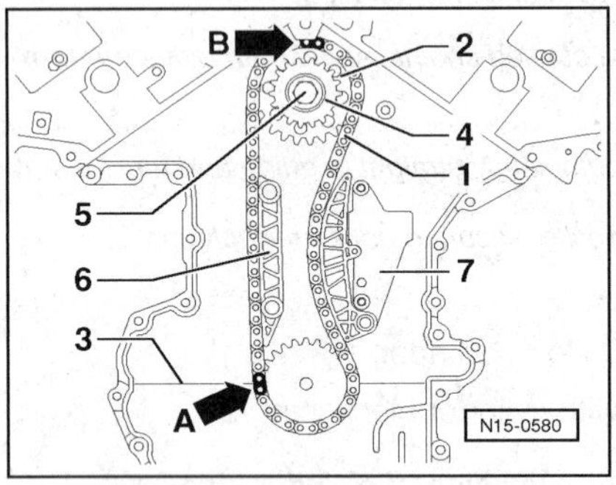

Install roller chain (crankshaft/intermediate shaft) -1- together with intermediate shaft sprocket -2-.

- Copper link -arrow A- must align with cylinder block joint -3-.
- Copper link -arrow B- must be positioned on top of intermediate shaft chain sprocket -2-.
- Install chain sprocket (intermediate shaft/camshaft adjuster) -4- onto intermediate shaft chain sprocket -2-.
- Tighten new bolt -5- for chain sprocket hand-tight.
- Install guide rail -6-.
- Install chain tensioner -7- with locking plate, tightening to 8 Nm (71 in-lb)
- Release lock for chain tensioner -7-.

Installing roller chain (intermediate shaft/camshaft timing adjuster cylinder bank 1)

NOTE—
The camshaft timing adjuster can be turned in two directions. When installing please ensure that the sender wheel for the Hall sender is turned on to the respective limit stop.

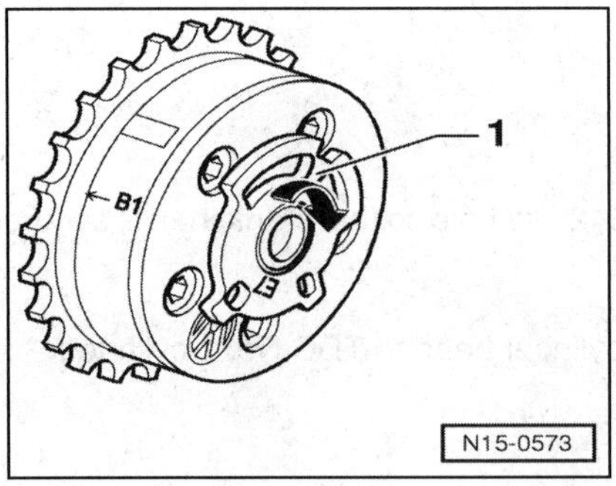

Intake camshaft timing adjuster for cylinder bank 1

- Turn Hall sender wheel -1- in direction of arrow to limit stop.

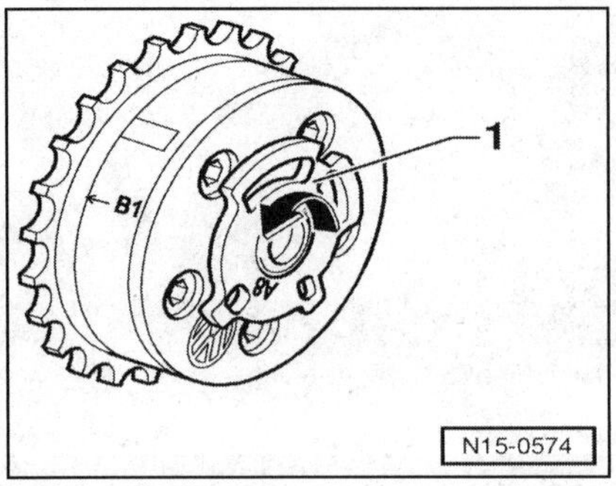

Exhaust camshaft adjuster for cylinder bank 1

- Turn Hall sender wheel -1- in direction of arrow to limit stop.

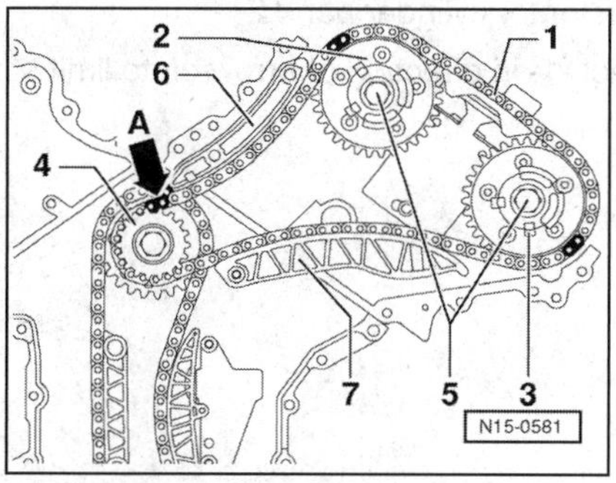

Install roller chain (intermediate shaft/camshaft adjuster cylinder bank 1) -1- together with camshaft adjusters.

- 2 - Intake camshaft timing adjuster, Identification: B1 and E7
- 3 - Exhaust camshaft timing adjuster, Identification: B1 and A8
- Copper link -arrow A- must be positioned on top of intermediate shaft chain sprocket -4-.

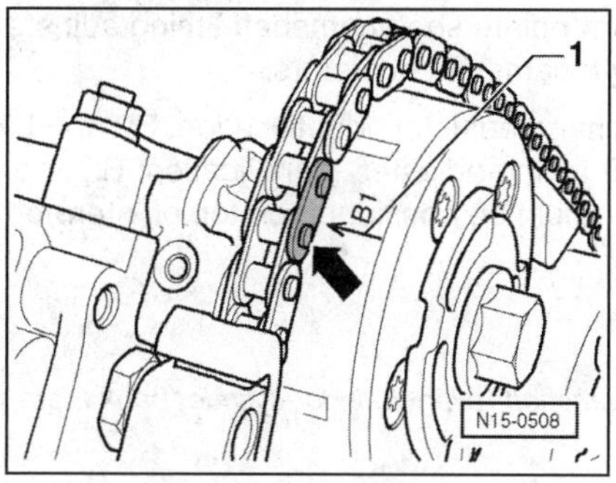

Copper link of camshaft roller chain -arrow- must align with arrow -1- on intake camshaft adjuster.

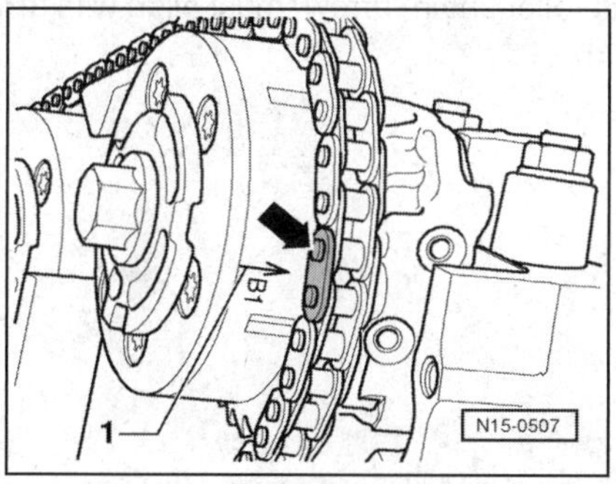

Copper link of camshaft roller chain -arrow- must align with arrow -1- on exhaust camshaft adjuster.

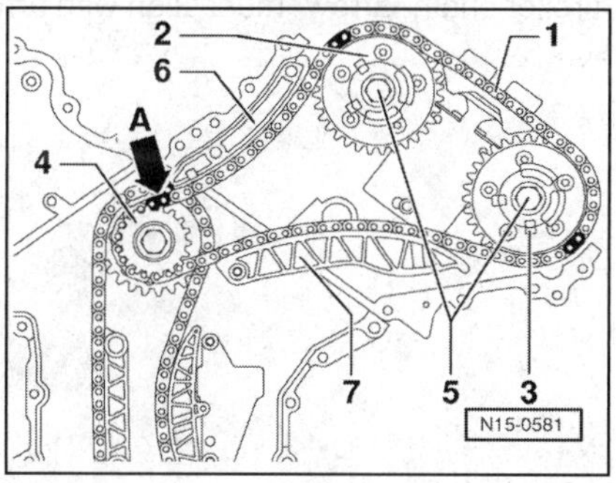

Tighten new bolts for camshaft adjuster -5- hand tight.

- Install guide rail -6-.
- Install tensioning rail -7-.

Installing roller chain (intermediate shaft/camshaft timing adjuster cylinder bank 2)

NOTE —

The camshaft timing adjuster can be turned in two directions. When installing please ensure that the sender wheel for the Hall sender is turned on to the respective limit stop.

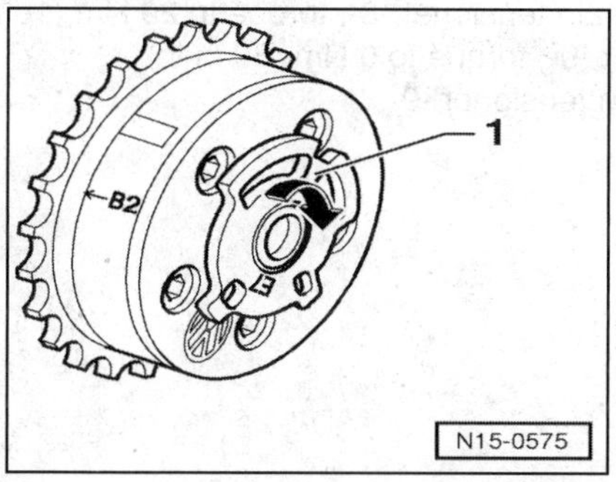

Intake camshaft adjuster for cylinder bank 2

- Turn Hall sender wheel -1- in direction of arrow on to limit stop.

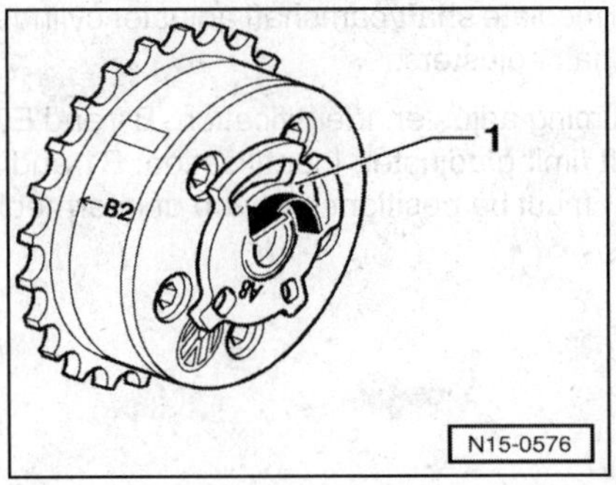

Exhaust camshaft adjuster for cylinder bank 2

- Turn Hall sender wheel -1- in direction of arrow on to limit stop.

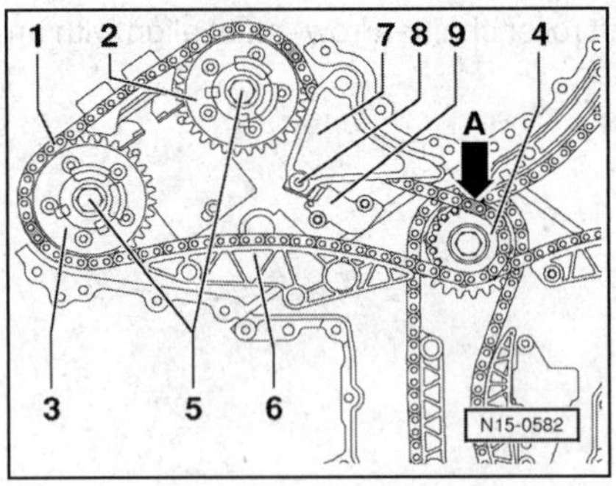

Install roller chain (intermediate shaft/camshaft timing adjuster cylinder bank 2) -1- together with camshaft adjusters.

- 2 - Intake camshaft timing adjuster, identification: B2 and E7
- 3 - Exhaust camshaft timing adjuster, identification: B2 and A8
- Copper link -arrow A- must be positioned on top of intermediate shaft chain sprocket -4-.

NOTE—

The copper links of the camshaft roller chain for cylinder bank 2 are visible from the back.

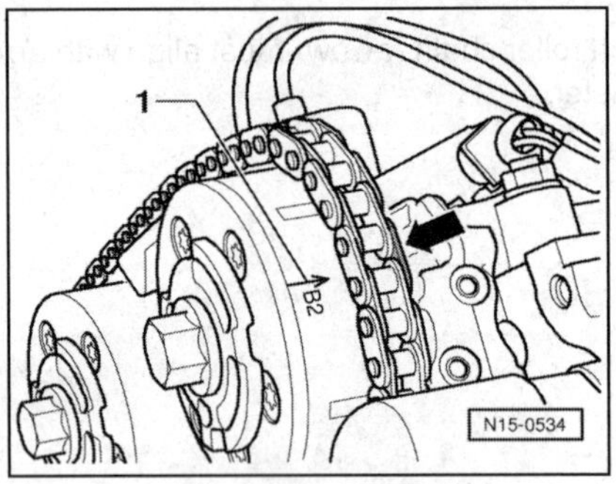

Copper link of camshaft roller chain -arrow- must align with arrow -1- on intake camshaft adjuster.

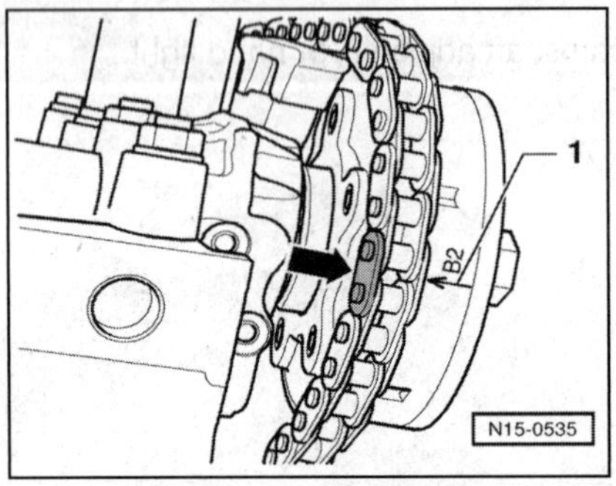

Copper link of camshaft roller chain -arrow- must align with arrow -1- on exhaust camshaft adjuster.

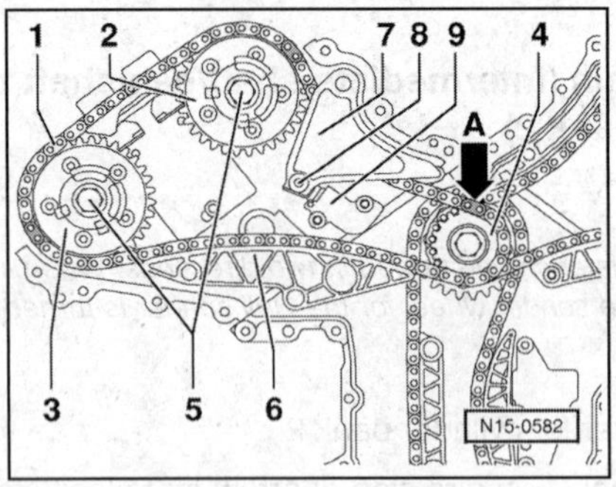

Tighten new bolts of camshaft adjuster -5-.

- Install guide rail -6-.
- Install tensioning rail -7-.
- Install limit stop for chain tensioner -8-, torque to 20 Nm (15 ft-lb)
- Install chain tensioner -9-, torque to 9 Nm (80 in-lb)
- Release lock on chain tensioner -9-.

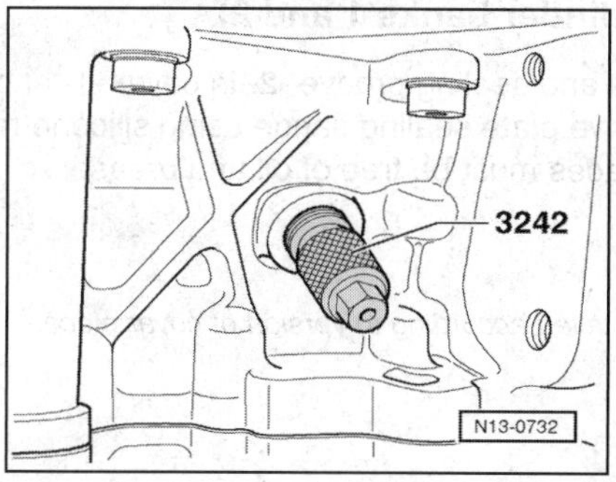

◀ Remove clamping bolt 3242.

– Install sealing plug in cylinder block, torque to 20 Nm (15 ft-lb)

– Remove camshaft bar T10068.

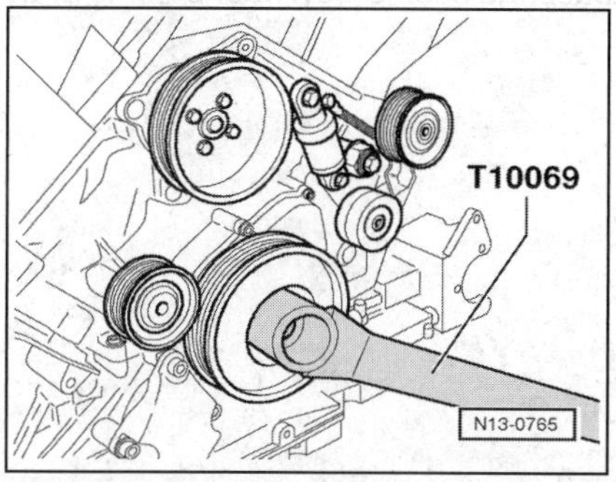

◀ Lock vibration damper with counter-holder T10069.

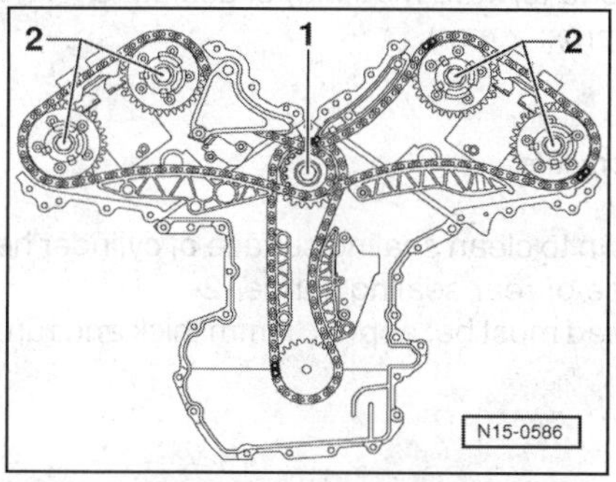

◀ Tighten new securing bolt for intermediate shaft sprocket -1- to 60 Nm (44 ft-lb) + 90° ($^1/_4$ turn). (Turning further can be performed in several steps.)

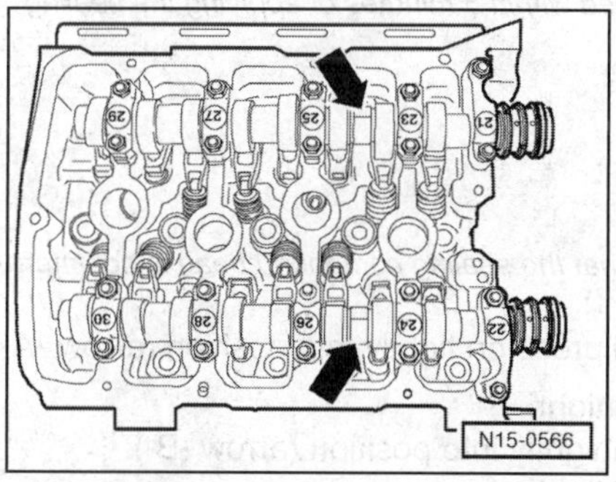

◀ Counter-hold camshafts using a 32 mm open jaw wrench.

NOTE —

Counter-hold only with 32 mm open jaw wrench on camshaft -arrow-. Camshaft bar T10068 must not be installed when sprockets are loosened or tightened.

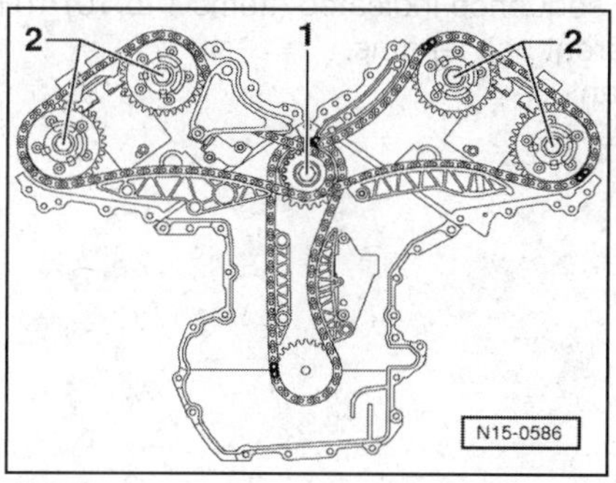

◀ Tighten new securing bolts for camshaft timing adjuster -2- to 60 Nm (44 ft-lb) + 90° ($^1/_4$ turn). (Additional turning can be performed in several steps.)

– Install flywheel/drive plate sealing flange, see **Repair Group 13, Sealing flange and dual mass flywheel, assembly**, item **-10-**.

– Replace seal on flywheel/drive plate sealing flange, see **Repair Group 13, Sealing flange and dual mass flywheel, assembly**, item **-13-**.

Installing cover (cylinder banks 1 and 2)

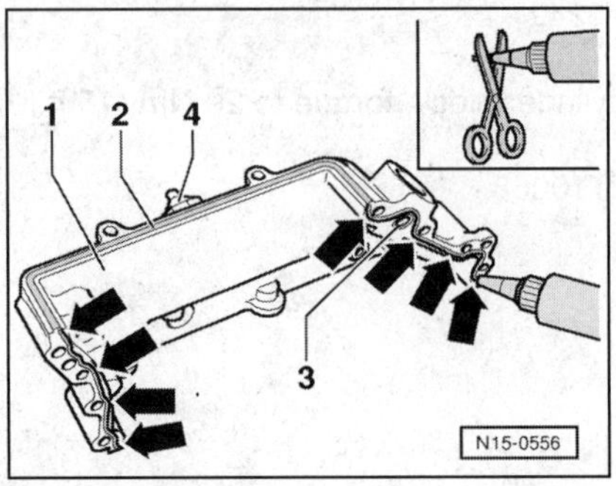

Clean sealing surfaces and sealing groove -2- in cover -1- and sealing surfaces of flywheel/drive plate sealing flange using silicone remover LSE 020 100 A3. Surfaces must be free of oil and grease.

NOTE —
A sealing nipple -4- is installed according to version of cover piece.

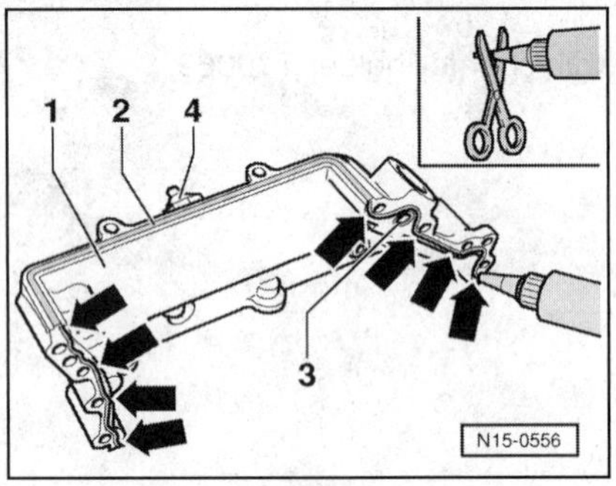

Lubricate O-ring for oil passage seal -3- (cylinder bank 1) and insert into cover -1-.

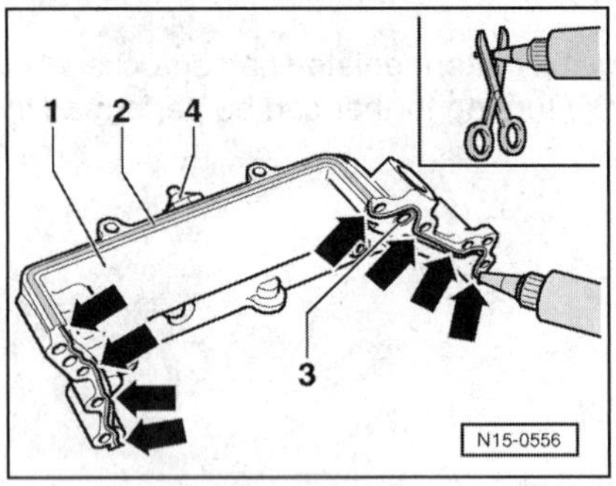

Cut through nozzle (along forward marking) of sealant D 176 501 A1 (diameter of nozzle approx. 2 mm).

NOTE —
Observe sealant expiration date.

- Apply sealant as shown to clean sealing surface of cylinder head cover and on sealing surface of rear sealing flange -2-.
- Sealing compound bead must be: approx. 2 mm thick and run on inside of bolt holes -arrows-.

NOTE —
The cover must be installed within 5 minutes of applying the sealing compound.

Cylinder bank 1

NOTE —
When installing, ensure that the sealing compound bead is not interrupted.

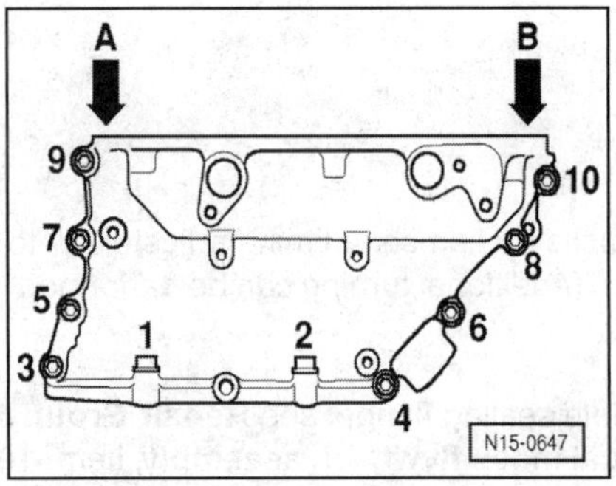

Install cover piece and press on tightly into position (arrow -A-).

- Tighten bolt -3- hand tight.
- Press cover piece on tightly into position (arrow -B-).
- Tighten bolt -4- hand tight.
- Then, tighten bolts in sequence indicated (items 1 to 10) 10 Nm (7 ft-lb) and note different bolt lengths.
- Remove excess sealant

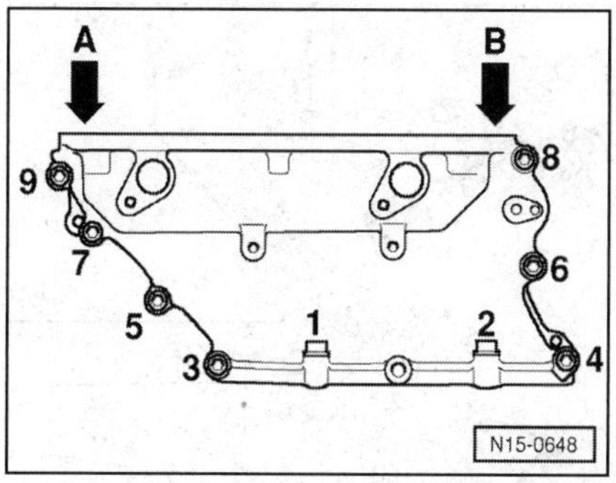

Cylinder bank 2

NOTE—
When installing, ensure that the sealing compound bead is not interrupted.

Install cover piece and press on tightly into position (arrow -A-).

- Tighten bolt -3- hand tight.
- Press cover piece on tightly into position (arrow -B-).
- Tighten bolt -4- hand tight.
- Then, tighten bolts in sequence indicated (items 1 to 10) 10 Nm (7 ft-lb) and note different bolt lengths.
- Remove excess sealant

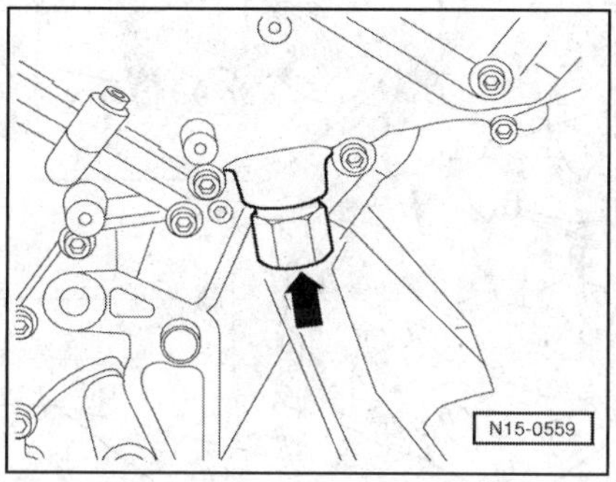

Cylinder bank 1

Install camshaft roller chain tensioner -arrow-.

- Tightening torque: 40 Nm (30 ft-lb)

Cylinder banks 1 and 2

– Install cylinder head cover, see **Cylinder head cover (cylinder banks 1 and 2), removing and installing**.

Valve train, components

NOTE —
- *Cylinder bank 1 is located on the right-hand side when looking toward the front.*
- *Cylinder bank 2 is located on left-hand side when looking toward the front.*
- *All bearing and running surfaces must be oiled before assembling.*

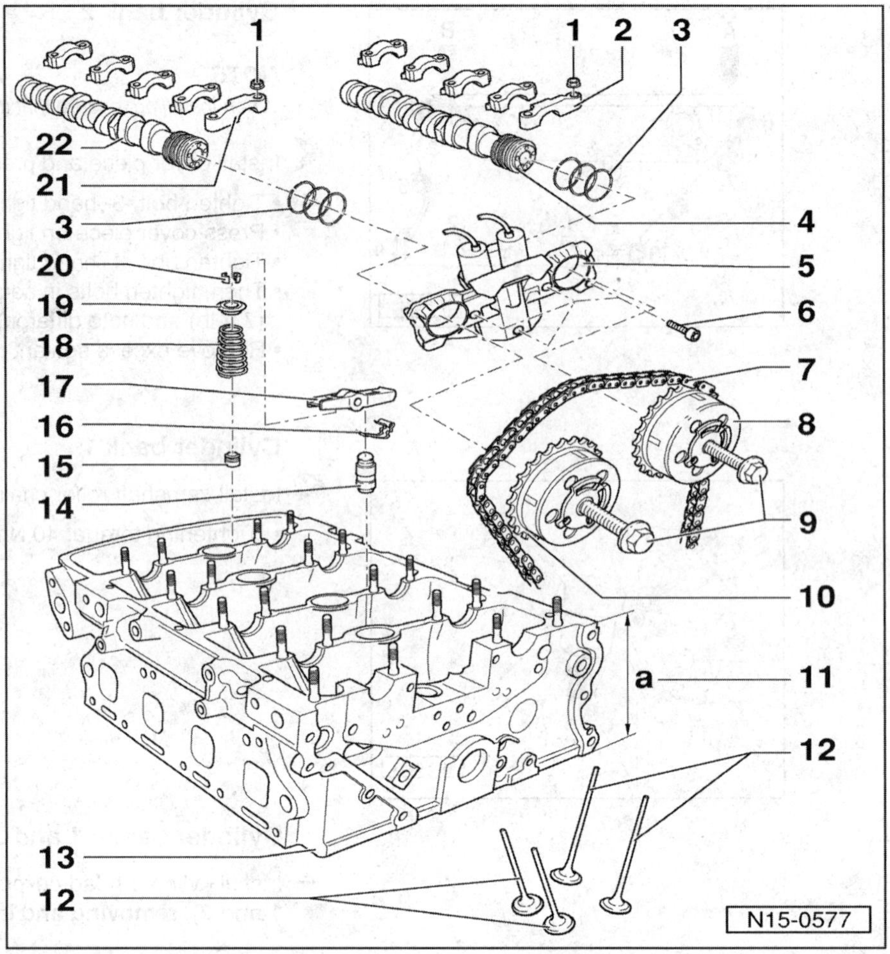

1. **5 Nm (44in-lb) + 1/8 turn (45 °) further**
2. **Intake camshaft bearing cap**
   - Lightly coat facing contact surfaces of bearing cap with G 052 723 A2 grease before installing, see **Fig. 2**:
     Cylinder bank 1: Bearing cap 7
     Cylinder bank 2: Bearing cap 27
   - Installation position:
     Cylinder bank 1, see **Fig. 3**
     Cylinder bank 2, see **Fig. 4**
   - Installation sequence, see **Camshafts, removing and installing**
3. **Seal**
   - If seal is leaking, replace complete
   - When installing control housing, lightly lubricate contact surfaces of seal
   - When replacing seals, do not spread too widely
   - Offset gaps by 120 °
4. **Intake camshaft**
   - Checking radial clearance with plastigage, Wear limit: 0.1 mm
   - Run-out: max. 0.01 mm
   - Checking axial clearance, see **Fig. 1**
   - Identification:
     Cylinder bank 1, see **Fig. 6**
     Cylinder bank 2, see **Fig. 7**
   - Removing and installing, see **Camshafts, removing and installing**
5. **Control housing**
   - Lightly lubricate contact surfaces of oil seals before installing
   - Disassembling and assembling, see **Fig. 8**
   - Removing and installing, see **Camshafts, removing and installing**
6. **8 Nm (71 in-lb)**
7. **Camshaft roller chain**
   - Mark direction of rotation before removing (installation position), see **Chain drive system, components , Fig. 1**
   - Installing, see **Valve timing, adjusting**
8. **Intake camshaft adjuster**
   - Identification:
     Cylinder bank 1: B1 and E7
     Cylinder bank 2: B2 and E7
   - Rotate engine only when camshaft adjuster is installed
   - Installing, see **Valve timing, adjusting**
9. **60 Nm (44 ft-lb) + 1/4 turn (90 °) further**
   - Replace
   - Contact surface of sender wheel must be dry around bolt head when assembled
   - To remove and install, counterhold with 32 mm open jaw wrench on camshaft, see **Camshafts, removing and installing**
10. **Exhaust camshaft adjuster**
    - Identification:
      Cylinder bank 1: B1 and A8
      Cylinder bank 2: B2 and A8
    - Rotate engine only when camshaft adjuster is installed
    - Installing, see **Valve timing, adjusting**
11. **Cylinder head height**
    - Minimum height: a = 139.9 mm
12. **Valves**
    - Do not rework; only lapping is permitted
    - Valve dimensions, see **Fig. 5**
13. **Cylinder head**
    - Cylinder bank 2
    - If replaced replace complete amount of engine coolant
    - Check for distortion, see **Cylinder bank 1, Fig. 1**
    - Removing and installing, see **Cylinder head (cylinder banks 1 and 2), removing and installing)**

(continued on following page)

Valve train, components (continued)

14. Support element
- Before installing check camshaft axial clearance, see **Fig. 1**
- Do not interchange
- With hydraulic valve clearance compensation

15. Valve stem seal

16. Retaining clip
- Check if seated securely

17. Roller rocker finger
- Before installing check camshaft axial clearance, see **Fig. 1**
- Do not interchange
- Check roller for easy movement
- Oil contact surface
- Use securing clip to clip onto support element when installing

18. Valve spring
- Note installation position

19. Valve spring plate

20. Valve cotters

21. Exhaust camshaft bearing cap
- Lightly coat facing contact surface of bearing cap with G 052 723 A2 grease, see **Fig. 2**:
  Cylinder bank 1: Bearing cap 8
  Cylinder bank 2: Bearing cap 28
- Installation position:
  Cylinder bank 1, see **Fig. 3**
  Cylinder bank 2, see **Fig. 4**
- Installation sequence, see **Camshafts, removing and installing**

22. Exhaust camshaft
- Checking radial clearance with plastigage, Wear limit: 0.1 mm
- Run-out: max. 0.01 mm
- Checking axial clearance, see **Fig. 1**
- Identification:
  Cylinder bank 1, see **Fig. 6**
  Cylinder bank 2, see **Fig. 7**
- See **Camshafts, removing and installing**

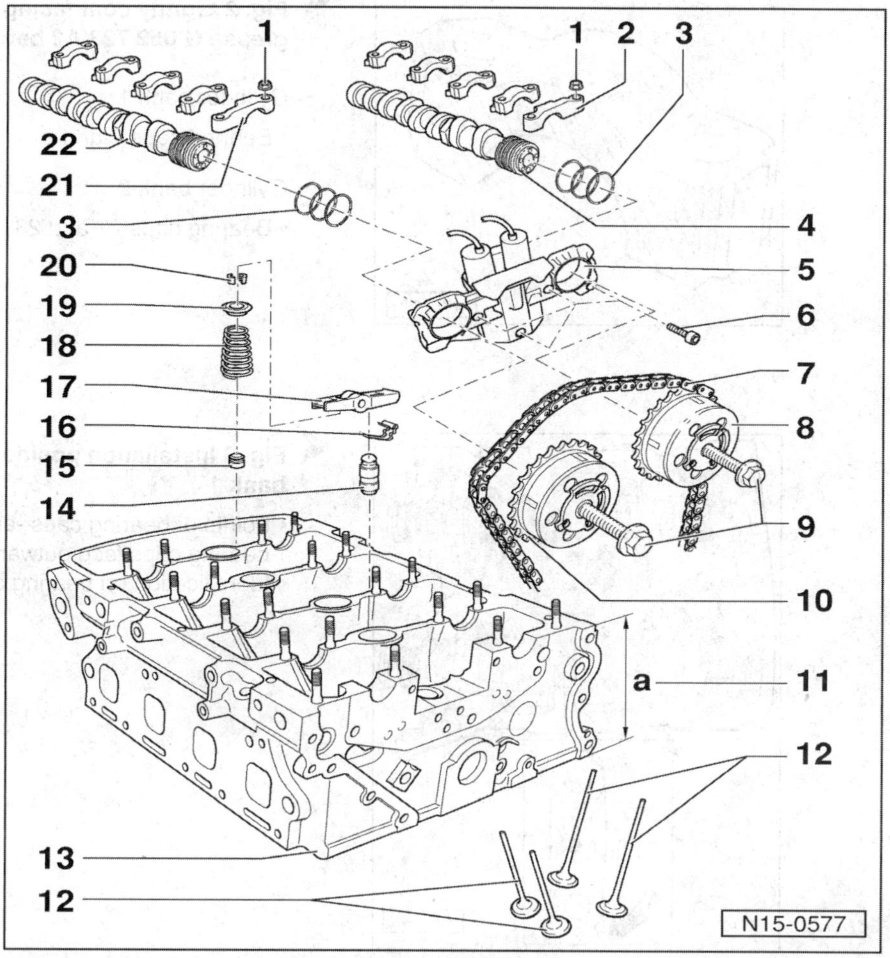

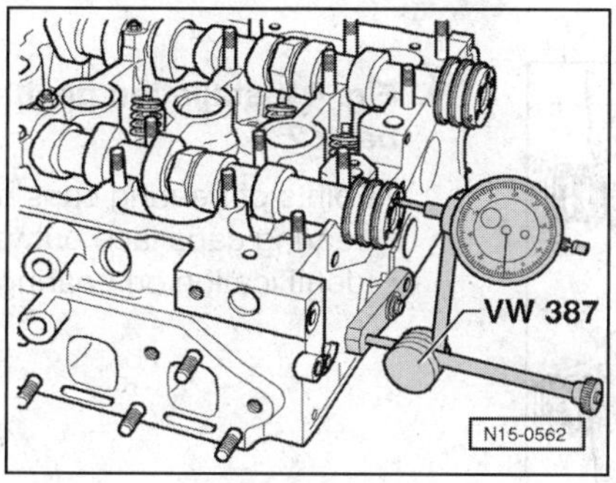

Fig. 1 Camshafts, checking axial clearance

Special tools and equipment

- VW 387 Universal dial gauge bracket
- Dial gauge

Perform measurements with support elements and roller rocker fingers removed.

Wear limit: max. 0.10 mm

– Cylinder bank 1
  - Bearing cap 7 installed on intake camshaft
  - Bearing cap 8 installed on exhaust camshaft

– Cylinder bank 2
  - Bearing cap 27 installed on intake camshaft
  - Bearing cap 28 installed on exhaust camshaft

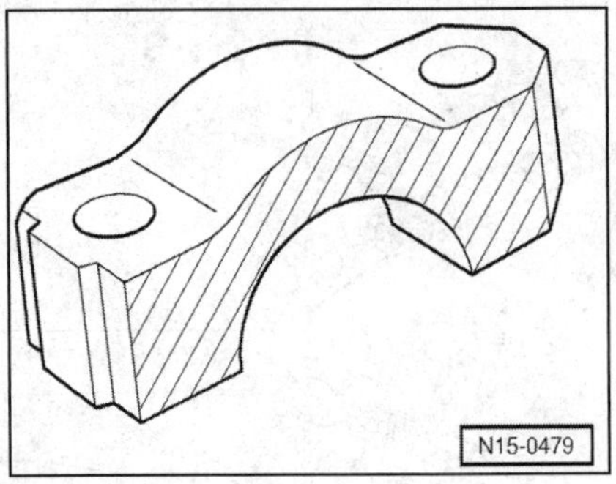

Fig. 2 Lightly coat facing contact surfaces of bearing caps with grease G 052 723 A2 before installing

– Cylinder bank 1
  - Bearing caps 7 and 8
– Cylinder bank 2
  - Bearing caps 27 and 28

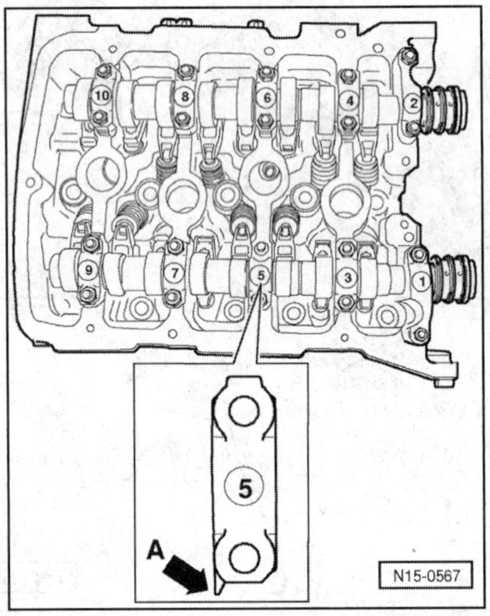

Fig. 3 Installation position of camshaft bearing caps for cylinder bank 1

- Points of bearing caps -arrow A- of intake and exhaust camshaft bearing caps face outward.
- Identification on bearing caps is legible when read from intake side.

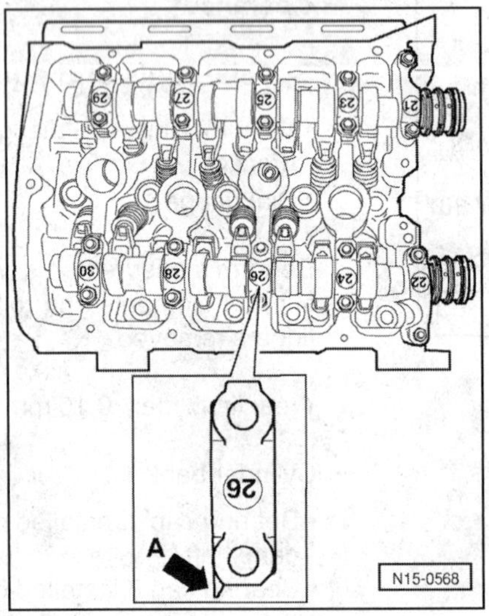

Fig. 4 Installation position of camshaft bearing caps for cylinder bank 2

- Points of bearing caps -arrow A- of intake and exhaust camshaft bearing caps face outward.
- Identification on bearing caps is legible when read from intake side.

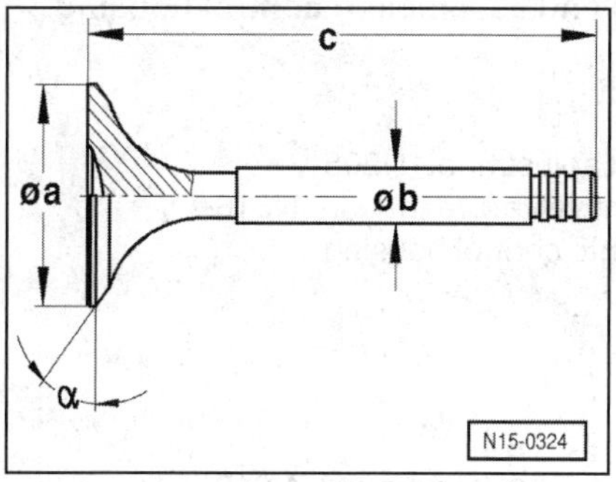

Fig. 5 Valve dimensions (cylinder banks 1 and 2)

NOTE—

Valves must not be reworked. Only lapping is permitted.

Valve dimensions for intake valves

Dimension	Short intake valve	Long intake valve
dia. a	32.50 mm	32.50 mm
dia. b	5.96 mm	5.96 mm
c	102.20 mm	136.10 mm
	45°	45°

Valve dimensions for exhaust valves

Dimension	Short exhaust valve	Long exhaust valve
dia. a	27.00 mm	27.00 mm
dia. b	5.94 mm	5.94 mm
c	102.50 mm	136.40 mm
〈	45°	45°

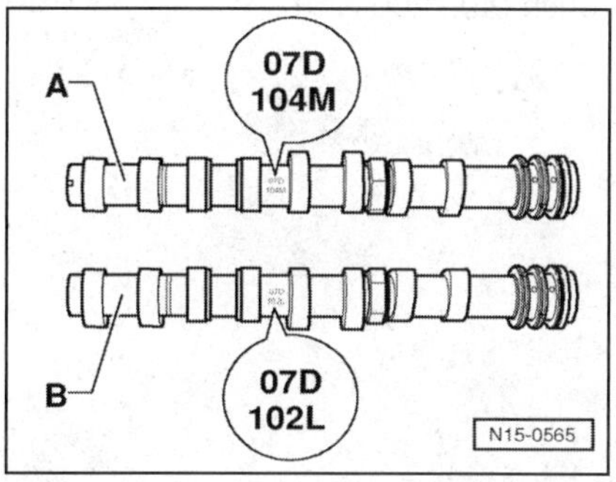

Fig. 6 Camshaft identification (cylinder bank 1)

- Identification between No. 2 cyl. and No. 3 cyl. cam pair of respective camshaft

Camshaft	Identification
A - Exhaust camshaft	07D 104M
B - Intake camshaft	07D 102L

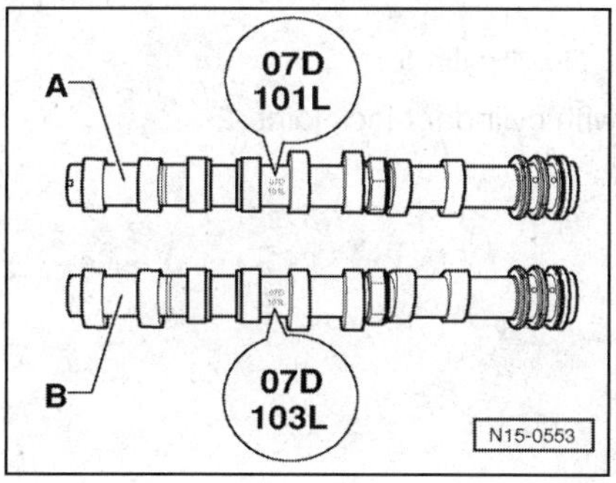

Fig. 7 Camshaft identification (cylinder bank 2)

- Identification between No. 6 cyl. and No. 7 cyl. cam pair of respective camshaft

Camshaft	Identification
B - Intake camshaft	07D 101L
A - Exhaust camshaft	07D 103L

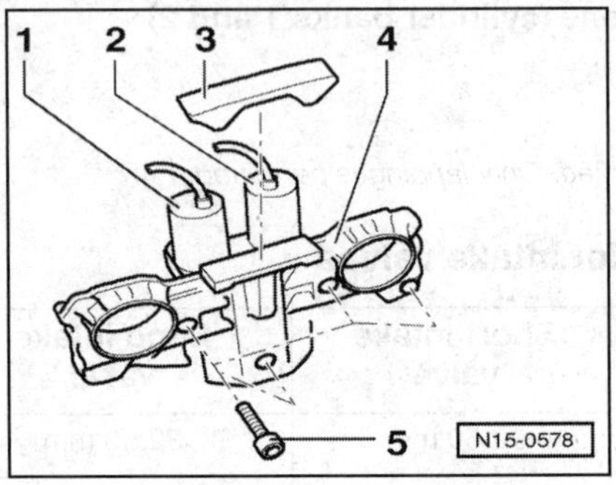

Fig. 8 Disassembling and assembling control housing

Cylinder bank 1

1. Valve 1 for camshaft adjustment -N205-
2. Camshaft adjustment valve 1 (exhaust) -N318-
3. Guide rail: clipped onto control housing
4. Control housing
5. 8 Nm (71 in-lb)

Cylinder bank 2

1. Camshaft adjustment valve 2 (exhaust) -N319-
2. Valve 2 for camshaft adjustment -N208-
3. Guide rail: clipped onto control housing
4. Control housing
5. 8 Nm (71 in-lb)

Camshafts, removing and installing

Work sequence is only possible when the engine is removed from the engine compartment

Special tools and equipment

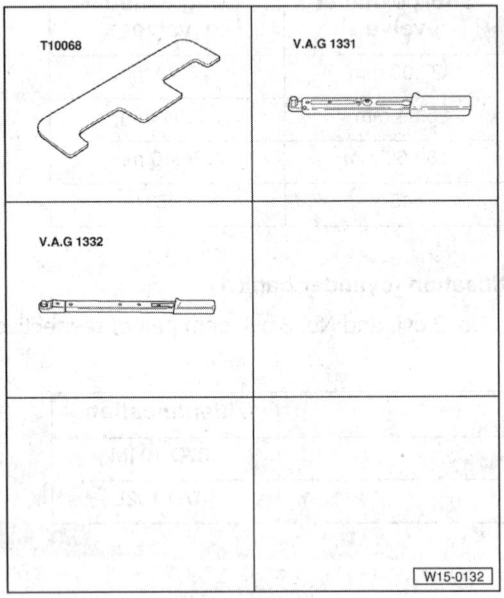

- T10068 Camshaft bar
- V.A.G 1331 Torque wrench (5...50 Nm)
- V.A.G 1332 Torque wrench (40...200 Nm)

Not illustrated:

- G 052 723 A2 Grease

Removing

– Remove cylinder head cover, see **Cylinder head cover (cylinder banks 1 and 2), removing and installing**.

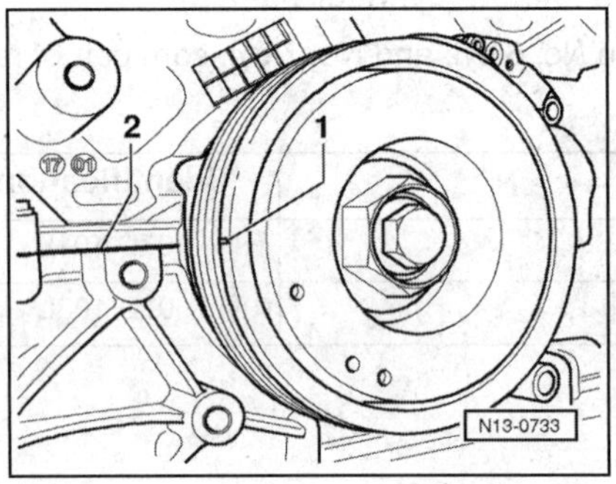

Turn crankshaft to TDC No. 1 cylinder.

- Mark -1- must align with cylinder block joint -2-.

Cylinder bank 1

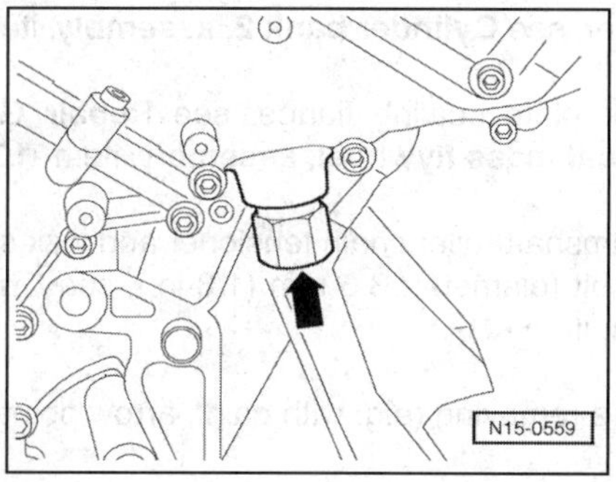

Remove camshaft roller chain tensioner -arrow-.

- Remove camshaft cover part, see **Cylinder bank 1,assembly,** item **-12-**.
- Remove flywheel/drive plate sealing flange, see **Repair Group 13, Sealing flange and dual mass flywheel, assembly**, item **-10-**.

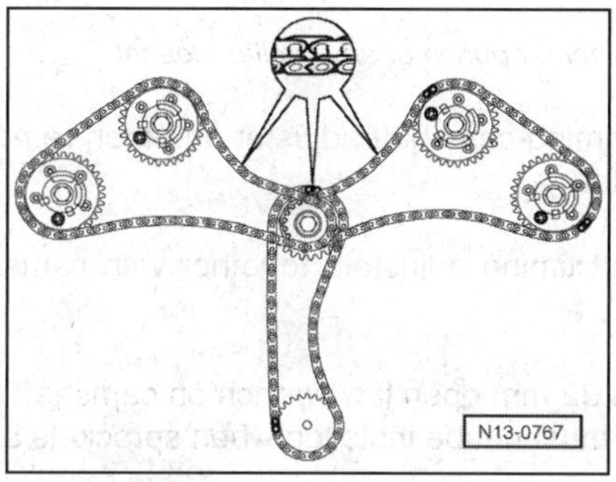

Mark roller chain before removing (e.g. with paint, arrow pointing in direction of rotation).

NOTE —
Do not mark chain with a center punch or any similar means!

- First remove exhaust timing camshaft adjuster and then remove intake camshaft timing adjuster.
- Remove both camshaft timing adjusters together with camshaft roller chain from camshafts.

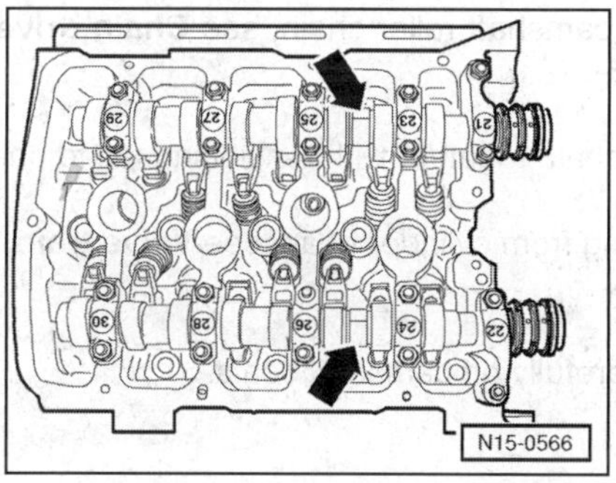

Counter-hold only with 32 mm open jaw wrench on camshaft -arrow-. Camshaft bar T10068 must not be installed when sprockets are loosened or tightened.

- Remove guide rail for camshaft roller chain, see **Chain drive system components**, item **-7-**.
- Place camshaft roller chain to one side.
- Remove control housing from cylinder head, see **Valve train, components**, item **-5-**.
- Pull control housing carefully off camshaft seals.

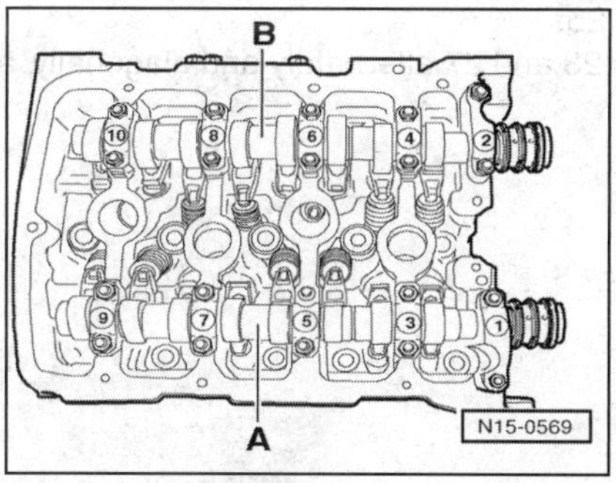

A - Intake camshaft

- First remove bearing caps 1 and 9.
- Remove bearing cap 5.
- Loosen bearing caps 3 and 7 alternately and diagonally, and remove.

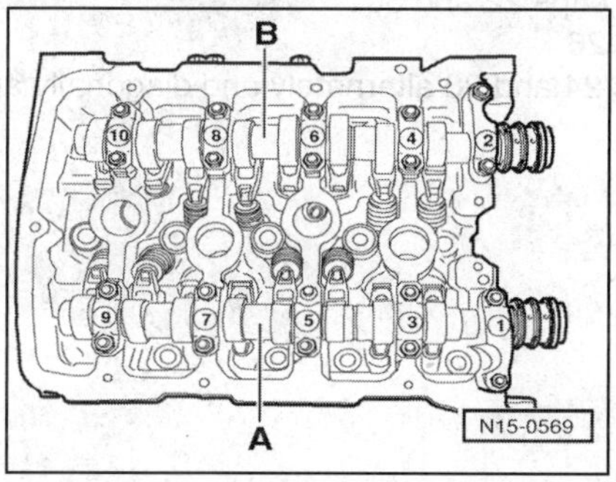

B - Exhaust camshaft

- First remove bearing caps 2 and 10.
- Remove bearing cap 6.
- Loosen bearing caps 4 and 8 alternately and diagonally and remove.

Cylinder bank 2

– Remove camshaft cover, see **Cylinder bank 2, assembly**, item **-12-**.

– Remove flywheel/drive plate sealing flange, see **Repair Group 13, Sealing flange and dual mass flywheel, assembly**, item **-10-**.

– Release tension on camshaft roller chain tensioner and lock chain tensioner using a drill bit (diameter: 3.5 mm (1/8 in.), see **Chain drive system, components**, item **-31-**.

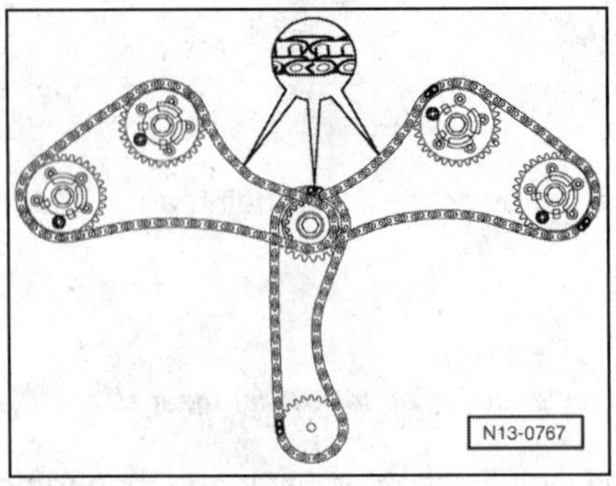

◀ Mark roller chain before removing (e.g. with paint, arrow pointing in directin of rotation).

NOTE—
Do not mark chain with a center punch or any similar means!

– First remove exhaust timing camshaft adjuster and then remove intake camshaft timing adjuster.

– Remove both camshaft timing adjusters together with camshaft roller chain from camshafts.

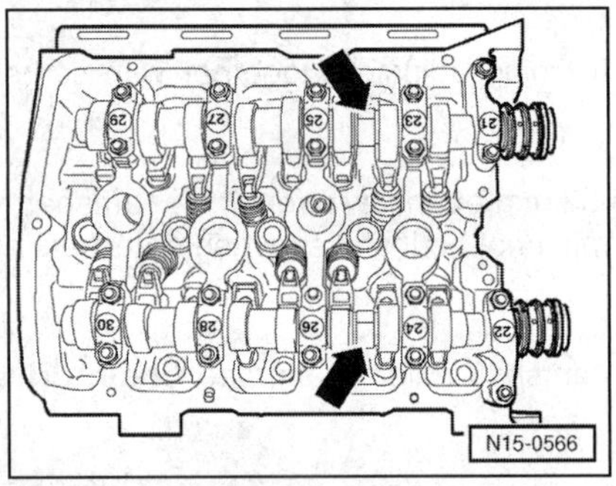

◀ Counter-hold only with 32 mm open jaw wrench on camshaft -arrow-. Camshaft bar T10068 must not be installed when sprockets are loosened or tightened.

– Remove guide rail for camshaft roller chain, see **Chain drive system, components**, item **-28-**.

– Place camshaft roller chain together with tensioning rail to one side.

– Remove control housing from cylinder head, see **Valve train, components**, item **-5-**.

– Pull control housing carefully off camshaft seals.

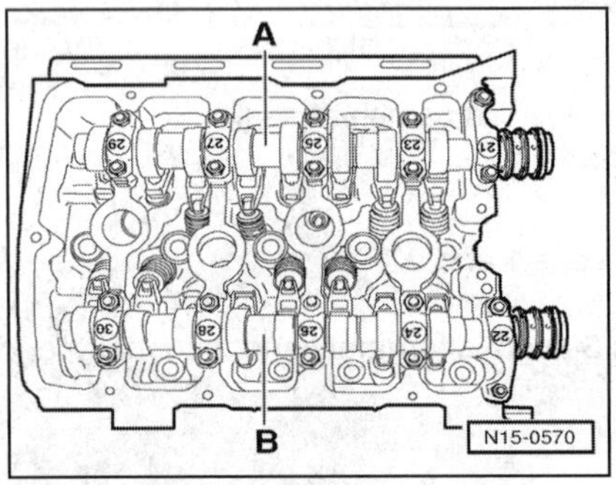

◀ A - Intake camshaft

- First remove bearing caps 21 and 29.
- Remove bearing cap 25.
- Loosen bearing caps 23 and 27 alternately and diagonally, and remove.

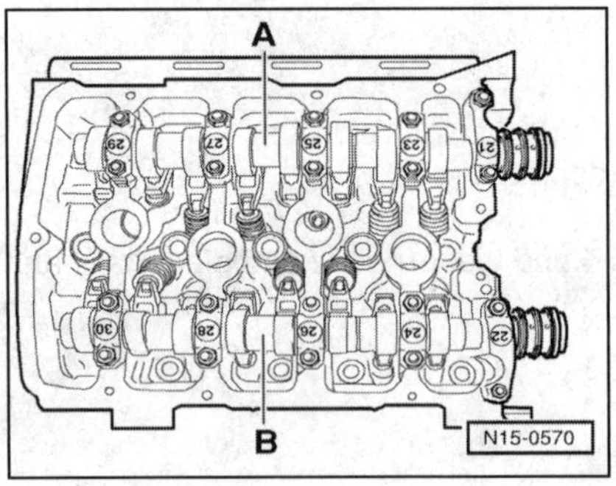

◀ B - Exhaust camshaft

- First remove bearing caps 22 and 30.
- Remove bearing cap 26.
- Loosen bearing caps 24 and 28 alternately and diagonally and remove.

Cylinder banks 1 and 2

– Carefully remove camshafts and place on a clean surface.

– Remove roller rocker finger together with support elements and place on a clean surface.

– Ensure that roller rocker fingers and support elements are not interchanged.

Installing

Requirements

- When installing, the camshafts must be positioned at TDC No. 1 cylinder.
- The pistons must not be positioned at TDC.

– Insert support element in cylinder head and install roller rocker finger onto respective valve stem end and support element.

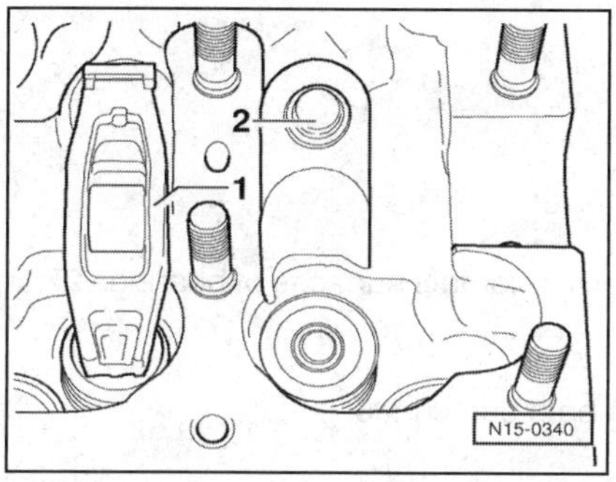

◀ Ensure that all roller rocker fingers seat properly on valve stem ends -1- and are clipped onto respective support elements -2-.

Cylinder bank 1

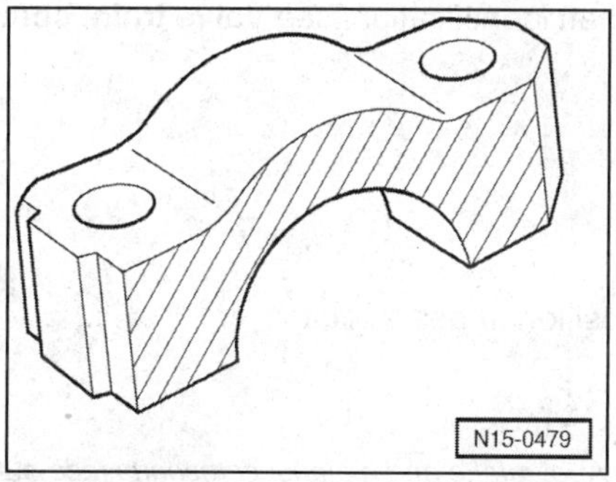

◀ Before installing coat facing contact surfaces of bearing caps 7 and 8 lightly with grease G 052 723 A2.

– Oil running surfaces of both camshafts.

– Place respective camshaft carefully in camshaft bearings of cylinder head, observing camshaft identification, see **Valve train, components, Fig. 6**.

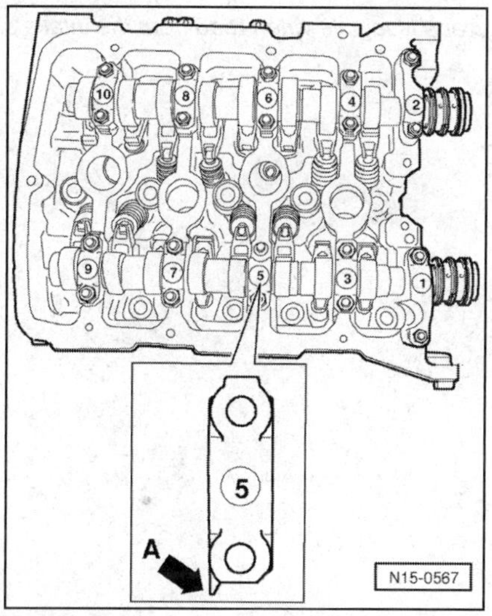

◀ Observe installation position of bearing caps:

NOTE —
- *Points of bearing caps -A- of intake and exhaust camshafts face outward.*
- *Identification on bearing caps is legible when read from the intake side.*

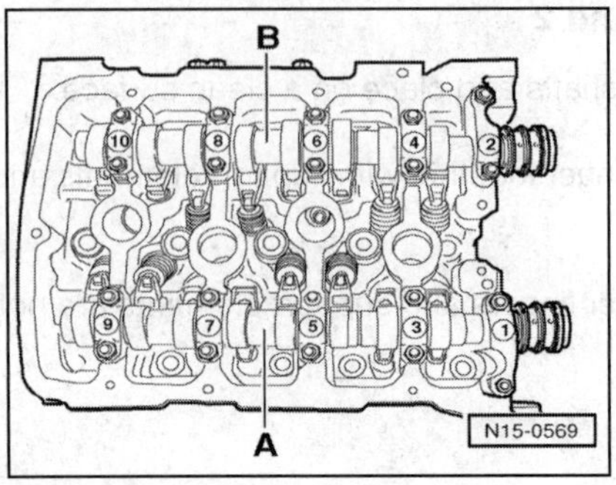

A - Intake camshaft

- Tighten bearing caps 3 and 7 alternately and diagonally to 5 Nm (44 in-lb) and $^{1}/_{8}$ turn (45°) further.
- Install bearing cap 5 and tighten to 5 Nm (44 in-lb) and $^{1}/_{8}$ turn (45°) further.
- Install bearing caps 1 and 9 and tighten to 5 Nm (44 in-lb) and $^{1}/_{8}$ turn (45°) further.

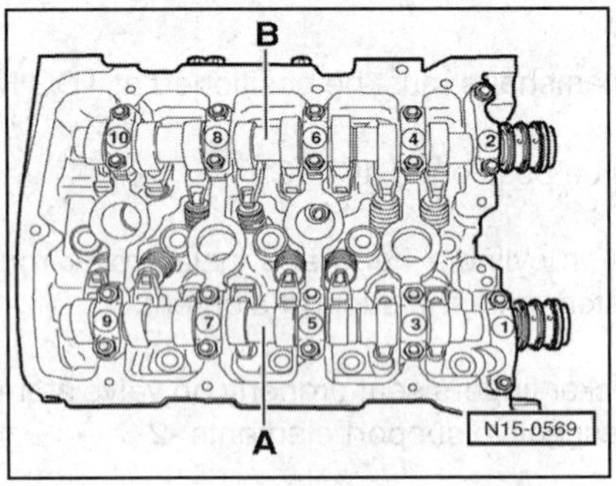

B - Exhaust camshaft

- Tighten bearing caps 4 and 8 alternately and diagonally to 5 Nm (44 in-lb) and $^{1}/_{8}$ turn (45°) further.
- Install bearing cap 6 and tighten to 5 Nm (44 in-lb) and $^{1}/_{8}$ turn (45°) further.
- Install bearing caps 2 and 10 and tighten to 5 Nm (44 in-lb) and $^{1}/_{8}$ turn (45°) further.

Cylinder bank 2

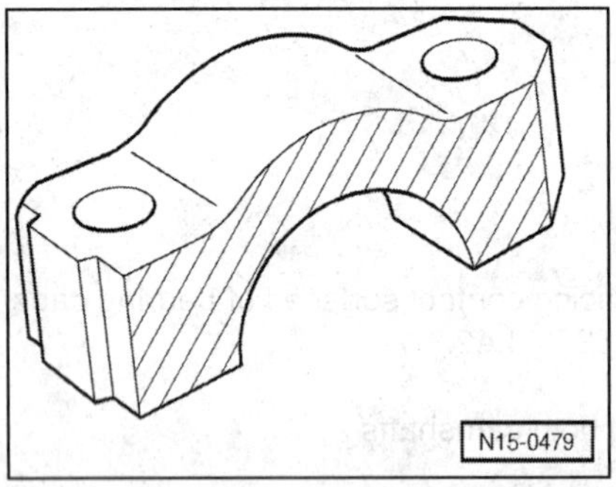

Before installing coat facing contact surfaces of bearings 27 and 28 with grease G 052 723 A2.

– Oil running surfaces of both camshafts.

– Place respective camshaft carefully in camshaft bearings of cylinder head, observing camshaft identification, see **Valve train, components, Fig. 7**.

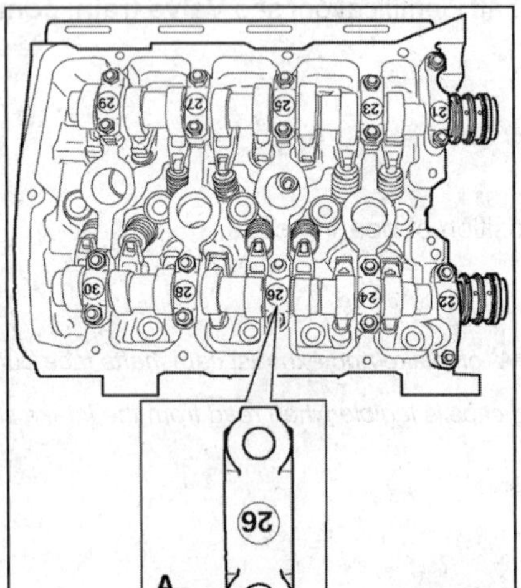

Observe installation position of bearing caps:

NOTE —

- *Points of bearing caps -A- of intake and exhaust camshafts face outward.*
- *Identification on bearing caps is legible when read from the intake side.*

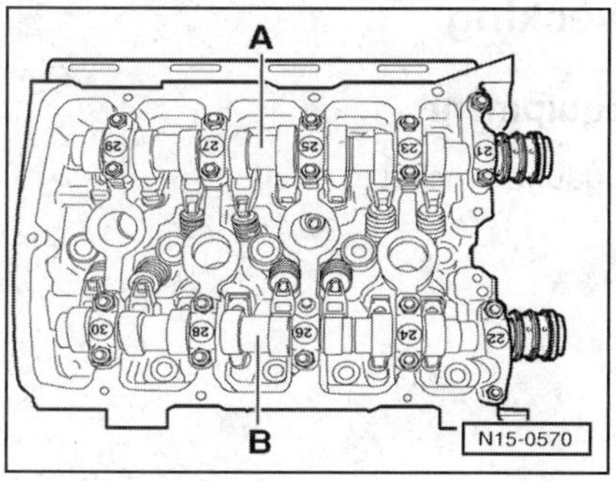

A - Intake camshaft

- Tighten bearing caps 23 and 27 alternately and diagonally to 5 Nm (44 in-lb) and $^1/_8$ turn (45°) further.
- Install bearing cap 25 and tighten to 5 Nm (44 in-lb) and $^1/_8$ turn (45°) further.
- Install bearing caps 21 and 29 and tighten to 5 Nm (44 in-lb) and $^1/_8$ turn (45°) further.

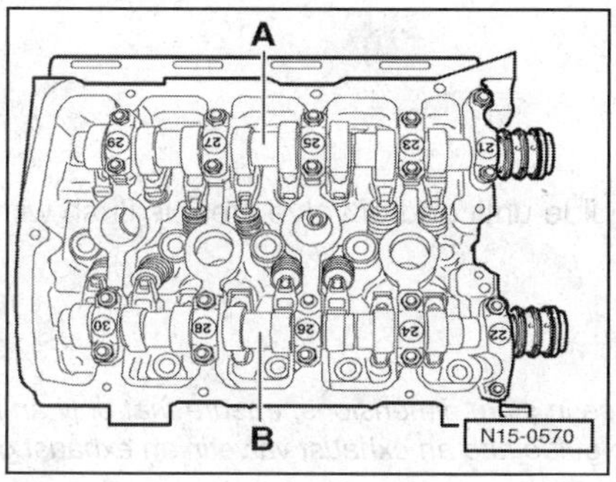

B - Exhaust camshaft

- Tighten bearing caps 24 and 28 alternately and diagonally to 5 Nm (44 in-lb) and $^1/_8$ turn (45°) further.
- Install bearing cap 26 and tighten to 5 Nm (44 in-lb) and $^1/_8$ turn (45°) further.
- Install bearing caps 22 and 30 and tighten to 5 Nm (44 in-lb) and $^1/_8$ turn (45°) further.

Cylinder banks 1 and 2

– Position camshafts in cylinder heads to TDC No. 1 cylinder.

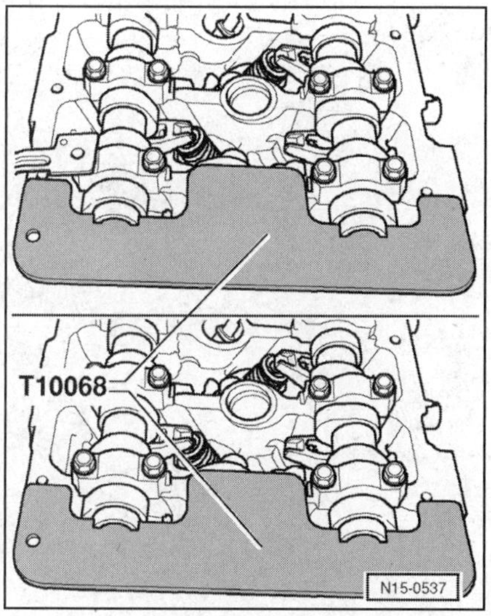

Camshaft bar T10068 must engage in both shaft grooves.

– Before installing control housing, lightly lubricate surfaces in control housing contacting camshaft seals.

– Lightly lubricate contact surfaces of camshaft seals and slowly push control housing over camshaft seals.

– Install control housing.

- Tightening torque: 8 Nm (71 in-lb)

– Adjust valve timing, see **Valve timing, adjusting**.

NOTE —

- *If only the cylinder head of one cylinder bank is removed and installed, then the position of the copper links of the camshaft roller chain of the other cylinder bank need not be noted.*
- *After adjusting timing place both camshaft bars T10068 (with crankshaft positioned at TDC No. 1 cylinder) into both shaft grooves of camshafts for cylinder banks 1 and 2.*

Valve guides, checking

Special tools and equipment

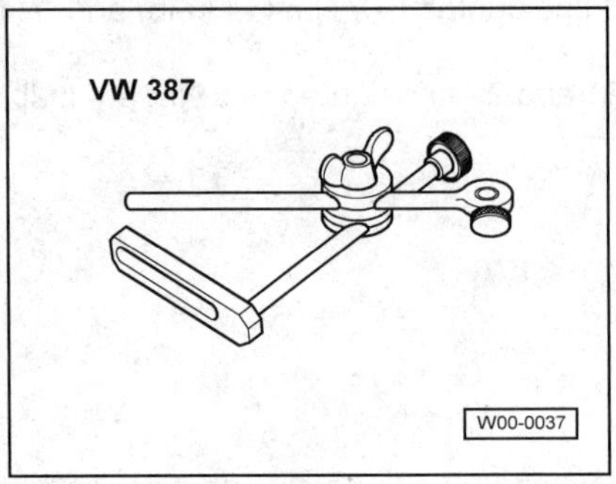

VW 387 Universal dial gauge bracket

- Dial gauge

Test sequence

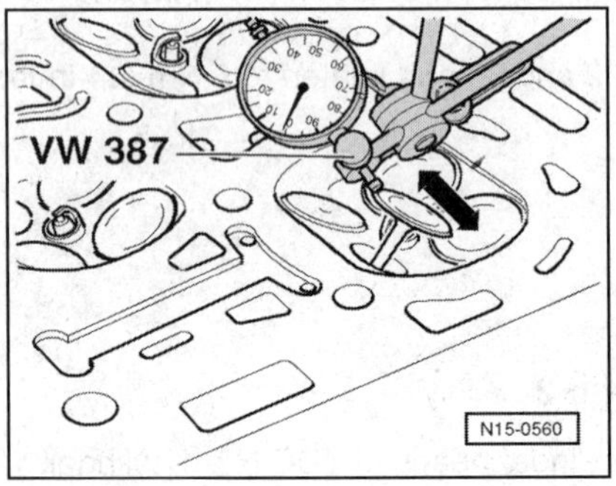

Insert new valve into guide until end of valve stem is flush with end of guide.

NOTE—
Due to the slight difference in stem dimensions, ensure that only an intake valve is used in an intake guide and an exhaust valve in an exhaust guide.

– Determine valve rock.
  - Wear limit: 0.8 mm

– If valve rock tolerance is exceeded:
  - Replace cylinder head.

17 Engine–Lubrication System

Repair Group
E 17

Lubrication system

> ***WARNING —***
> *When doing any repair work, especially in the engine compartment, pay attention to the following due to the cramped conditions:*
>
> - *Route all the various lines (e.g. for fuel, hydraulics, EVAP system, coolant, refrigerant, brake fluid and vacuum lines and hoses) and electrical wiring so that the original positions are restored.*
> - *Ensure sufficient clearance to all moving or hot components.*

NOTE —

- *For engine disassembly and assembly, mount the engine to VAS6095 holding fixture.*
- *The engine must not be put down on the sump, otherwise the fluid seal between the lower and upper sump will be damaged. Use T10016 engine stand.*
- *The oil level must not be above the max. mark - danger of damage to catalyst! Marks, see* ***Lubrication system, components, Fig. 1.***

NOTE —

If when repairing an engine, metal shavings or a large quantity of small metal particles are found, this could indicate that the crankshaft and conrod bearings are damaged. To prevent consequential damage perform the following repairs:

- *Thoroughly clean oil passages*
- *Replace oil cooler*
- *Replace oil filter*

NOTE —

Oil system capacity, see **Maintenance**.

Engine oil specifications:

The following terms must appear on the oil container singly or in combination with other designations:

- ACEA A3
- SAE 5W-40, SAE 10W-40, or SAE 15W-50

If the above described oils are not available, the following oil may be used for topping up only:

- ILSAC GF3
- SAE 5W-30

Lubrication system, components

1. **Cylinder block/bearing block**
2. **Gasket**
   - Replace
3. **Banjo bolt**
   - Tighten to 30 Nm (22 ft-lb)
4. **Seal**
   - Replace
5. **Oil return line**
   - From oil separator, see **Repair Group 13, Crankcase ventilation system, components**, Item **-8-**
6. **Retainer**
   - For power steering pump
7. **40 Nm (30 ft-lb)**
8. **20 Nm (15 ft-lb)**
9. **Banjo bolt**
   - 20 Nm (15 ft-lb)
10. **Lower sump**
    - Clean sealing surface before installing
    - Coat sealing surfaces with D 176 404 A2
    - See **Oil sump, removing and installing**
11. **Oil drain plug**
    - Tighten to 30 Nm (22 ft-lb)
    - If sealing ring is leaking cut open and replace.
12. **8 Nm (71 in-lb)**
13. **Oil pump**
    - Replace if running surfaces and gears are scored
    - See **Oil pump, removing and installing**
    - See **Oil pump, assembly**
14. **Chain tensioner with tensioning rail**
    - Individual components, see **Fig. 2**
15. **10 Nm (7 ft-lb)**
16. **Union for oil return line from oil filler bracket**
17. **Upper sump**
    - Clean sealing surface before installing
    - See **Oil sump, removing and installing**
18. **Guide tube**
    - For dipstick
    - Secured on cylinder head

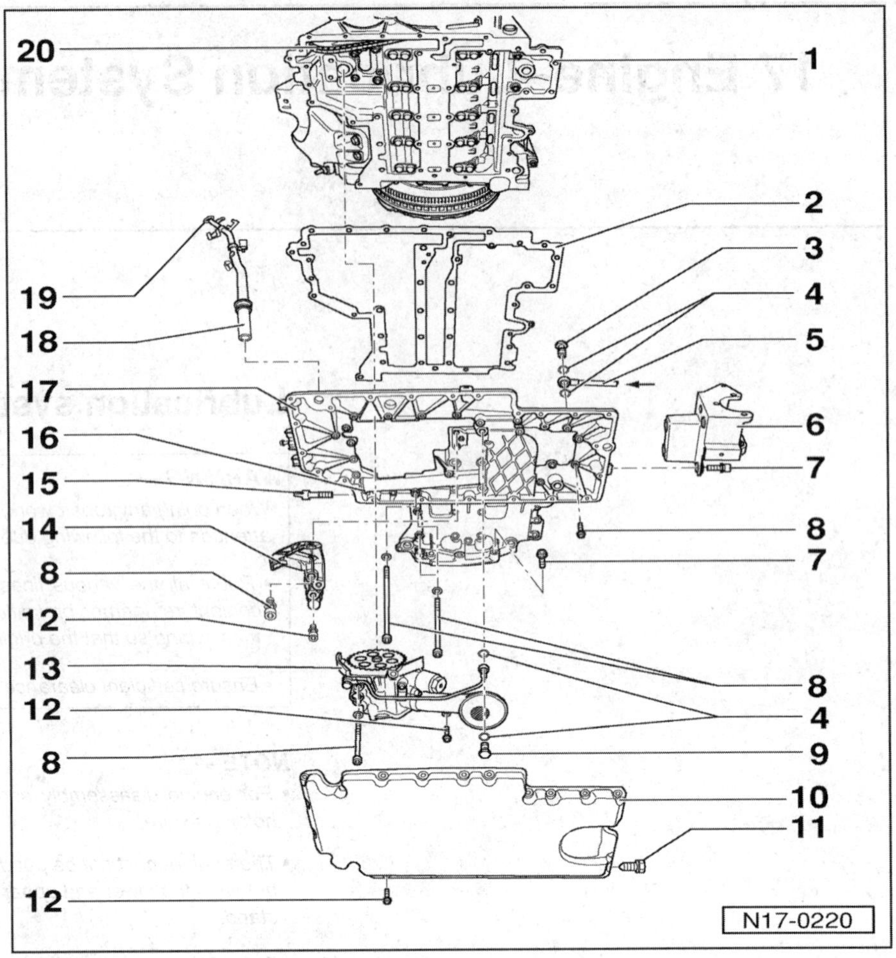

19. **Dipstick**
    - Oil level must not be above the max. mark!
    - Markings, see **Fig. 1**
20. **Chain**

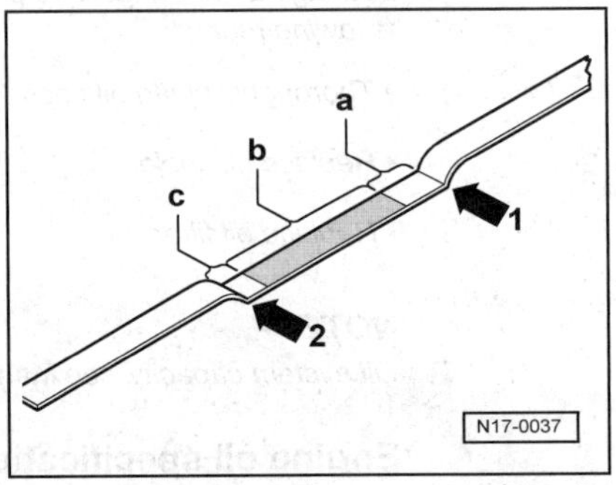

Fig. 1 Markings on oil dipstick

1. max. mark
2. min. mark

a. Area above hatched field up to max. mark: Do not replenish with engine oil!
b. Oil level within hatched field: can be replenished with engine oil
c. Area from min. mark up to hatched field: replenish with max. 0.5 ltr. of engine oil!

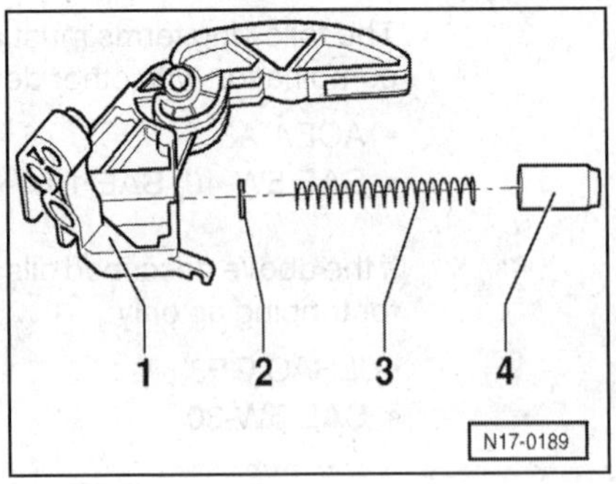

Fig. 2 Chain tensioner with tensioning rail

1. Chain tensioner
2. Disc
3. Spring
4. Piston

Oil filter housing, assembly

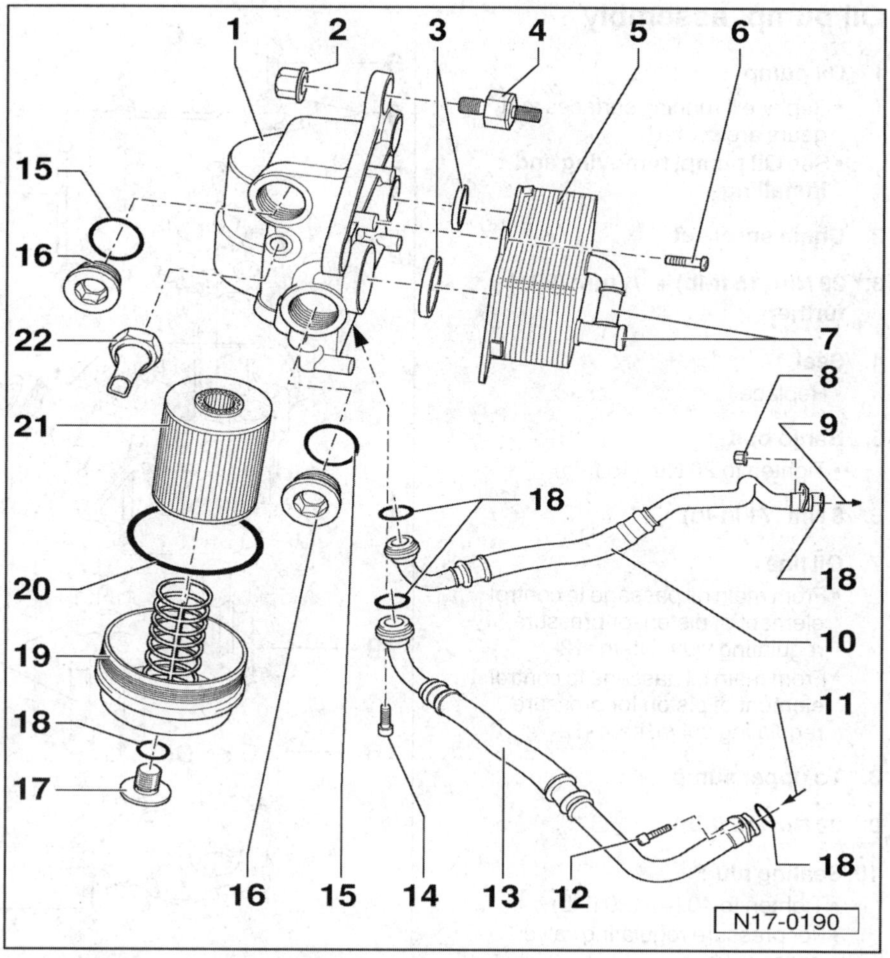

1. **Oil filter bracket**
   - With bypass valve for oil cooler
   - If oil filter bracket is removed, the bonded rubber mounting, item **-4-** must be replaced
2. **9 Nm (80 in-lb)**
   - Install with sealant AMV 200 002
3. **Gasket**
   - Replace
4. **Bonded rubber mounting**
   - Screwed in on longitudinal member, right
   - Replace
5. **Oil cooler**
   - Replace in cases of bearing or internal engine failure.
6. **10 Nm (7 ft-lb)**
7. **Connection for coolant hoses**
   - See **Repair Group 19, Coolant hose connection diagram**
8. **10 Nm (7 ft-lb)**
9. **To upper sump**
   - See **Lubrication system, components**, item **-16-**
10. **Oil return line**
11. **From union on front bearing block**
12. **10 Nm (7 ft-lb)**
13. **Oil supply line**
14. **20 Nm (15 ft-lb)**
15. **Seal**
    - Replace
16. **Sealing plug**
    - Tighten to 70 Nm (52 ft-lb)
17. **Oil drain plug**
    - Tighten to 10 Nm (7 ft-lb)
18. **O-ring**
    - Replace
    - Oil before assembling
19. **Screw cap**
    - Tighten to 25 Nm (18 ft-lb)
    - With bypass valve for oil filter
20. **O-ring**
    - Replace
    - Oil before assembling
21. **Oil filter element**
    - Observe change intervals
22. **1.4 bar oil pressure switch -F1-**
    - Tighten to 20 Nm (15 ft-lb)
    - If sealing ring is leaking cut open and replace.
    - See **Oil pressure and oil pressure switch, checking**

Oil pump, assembly

1. **Oil pump**
   - Replace if running surfaces and gears are scored
   - See **Oil pump, removing and installing**
2. **Chain sprocket**
3. **20 Nm (15 ft-lb) + 1/4 turn (90 °) further**
4. **Seal**
   - Replace
5. **Banjo bolt**
   - Tighten to 20 Nm (15 ft-lb)
6. **8 Nm (71 in-lb)**
7. **Oil line**
   - From main oil passage to control element of piston for pressure regulating valve, item **-12-**
   - From main oil passage to control element of piston for pressure regulating valve, item **-12-**
8. **To upper sump**
9. **20 Nm (15 ft-lb)**
10. **Sealing plug**
    - Tighten to 40 Nm (30 ft-lb)
    - For pressure regulating valve, approx. 4 bar
11. **Spring**
    - For pressure regulating valve, approx. 4 bar
12. **Piston**
    - For pressure regulating valve, approx. 4 bar
13. **8 Nm (71 in-lb)**
14. **Suction line**
    - Clean strainer if soiled
15. **O-ring**
    - Replace
    - Oil before assembling
16. **Gears**
    - Checking backlash, see **Fig. 1**
    - Checking axial clearance, see **Fig. 2**
17. **Cover**
18. **8 Nm (71 in-lb)**
19. **Drive gear**

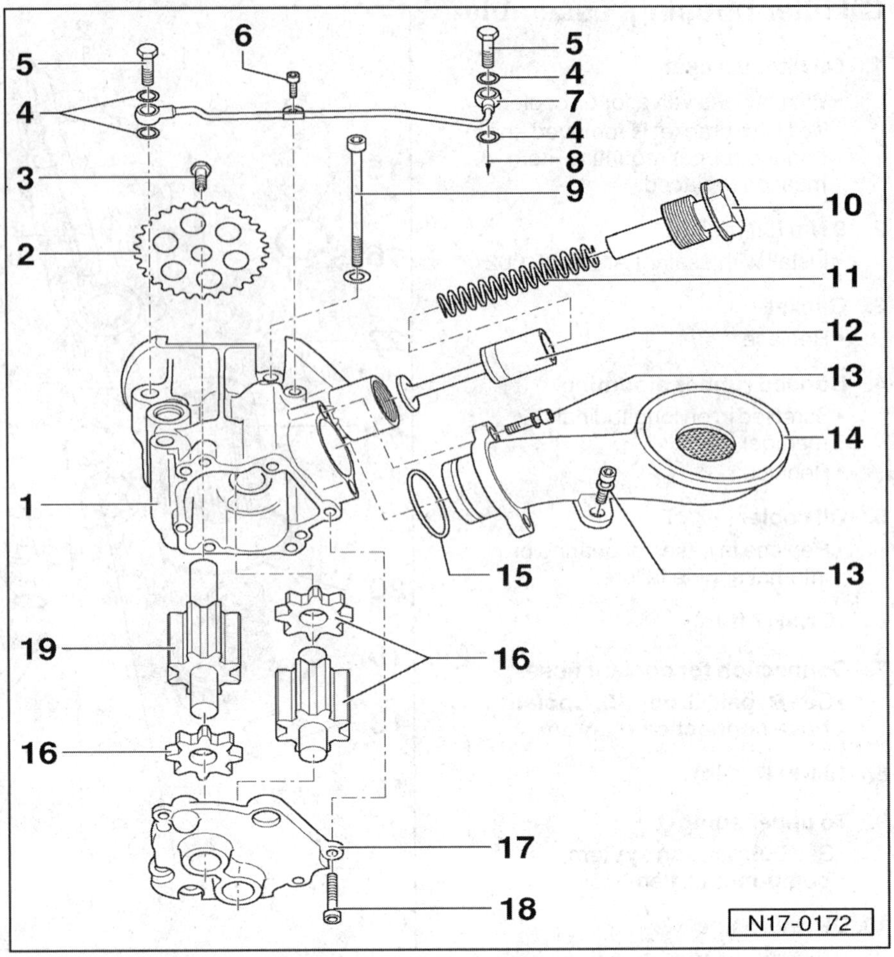

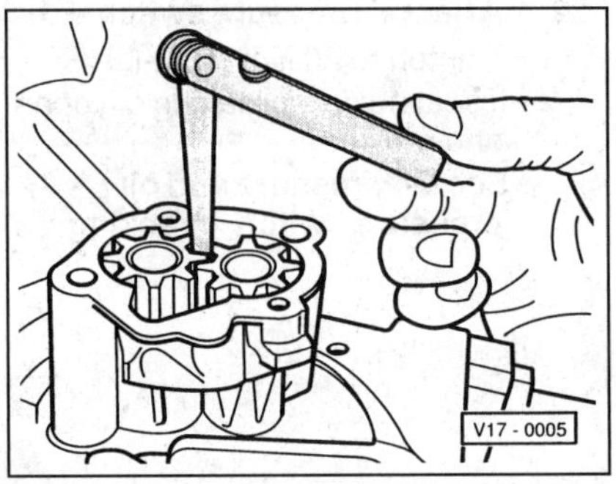

Fig. 1 Checking oil pump backlash

Wear limit: 0.20 mm

NOTE —
To prevent malfunctions, press toothed gears into housing during measurement.

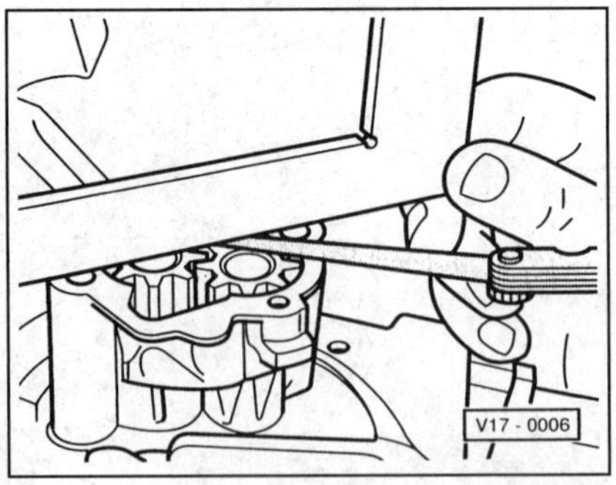

Fig. 2 Checking oil pump axial clearance

Wear limit: 0.10 mm

Oil sump, removing and installing

Removing lower sump

Special tools and equipment

• V.A.G 1331 Torque wrench (5...50 Nm)
• D 176 404 A2 Silicone sealant
• Hand drill with plastic brush
• Protective glasses
• Flat scraper

Removing

– Remove sound insulation, see **Body-Exterior section, Repair Group 50.**

– Remove air conditioning line bracket on upper sump.

– Drain engine oil.

NOTE —
Observe disposal regulations!

– Remove lower sump bolts.

– Loosen sump with light blows of a rubber headed hammer if necessary.

– Remove sealant residue on upper sump with a flat scraper.

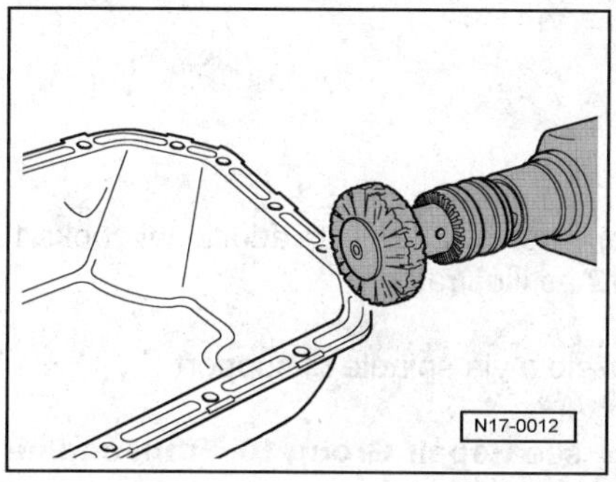

◄ Remove sealant residue on sump with a rotating brush, e.g. a hand drill with a plastic brush (wear eye protection).

– Clean sealing surface, must be free of oil and grease.

Removing upper sump

Special tools and equipment

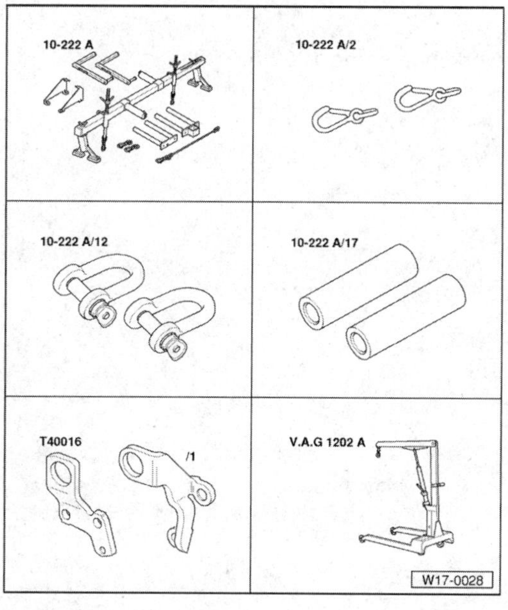

• 10-222 A Support bracket with legs 10-222 A/1
• 10-222A/2 Additional hooks
• 10-222A/12 Shackle
• 10-222A/17 Spacer tube
• T40016 Transport bracket
• V.A.G 1202 A Workshop crane
• V.A.G 1331 Torque wrench (5...50 Nm)

– Remove sound insulation, see **Body-Exterior section, Repair Group 50**

– Remove ribbed belt, see **Maintenance**.

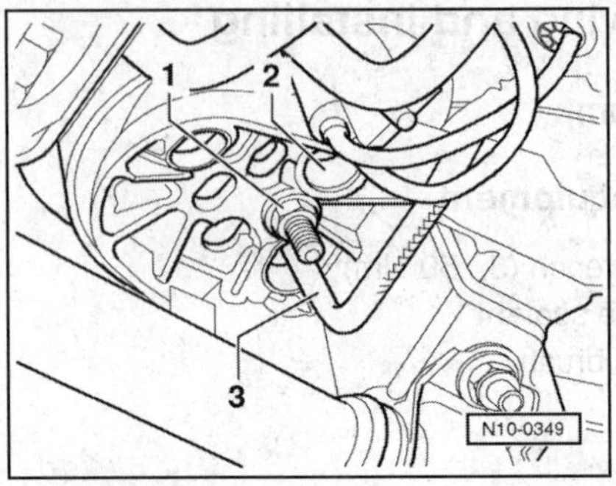

Mark installation position of threaded connection -1- and lower locating sleeve -2- on left and right engine mountings.

– Disconnect vacuum lines -3- at Left Electro-Hydraulic Engine Mount Solenoid Valve -N144-/Right Electro-Hydraulic Engine Mount Solenoid Valve -N145-.

– Disconnect both lower engine mountings.

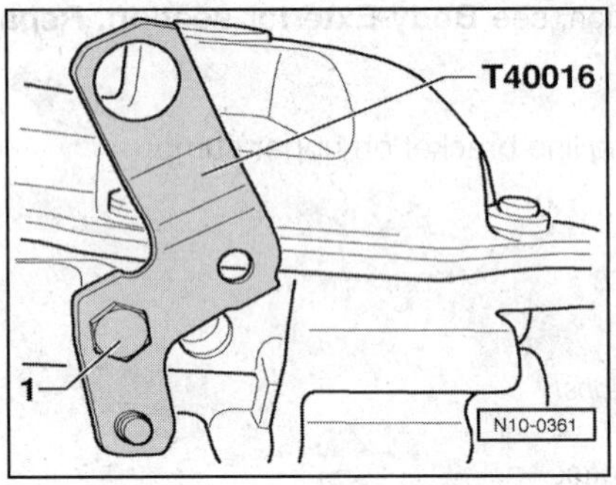

Secure T40016 transport bracket to rear left of cylinder head with securing screw -1-. Tighten to 25 Nm (18 ft-lb).

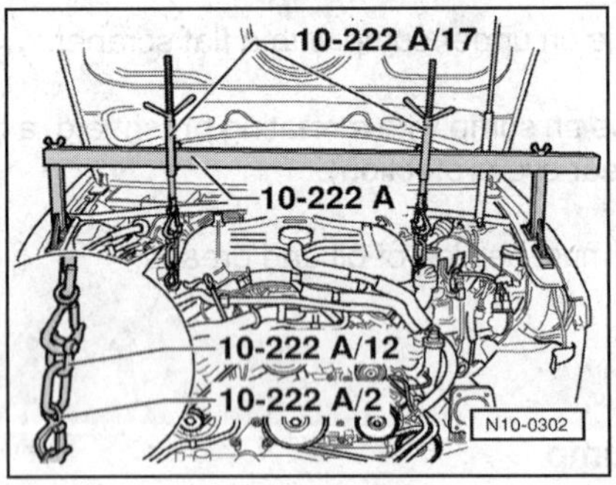

Install engine support bracket 10-222A with additional hooks 10-222A/2 and shackle 10-222A/12 as illustrated.

– Lift engine as far as possible via spindle of support.

– Remove torque bracket, see **Repair Group 10, Front engine mounting**, item **-1-**.

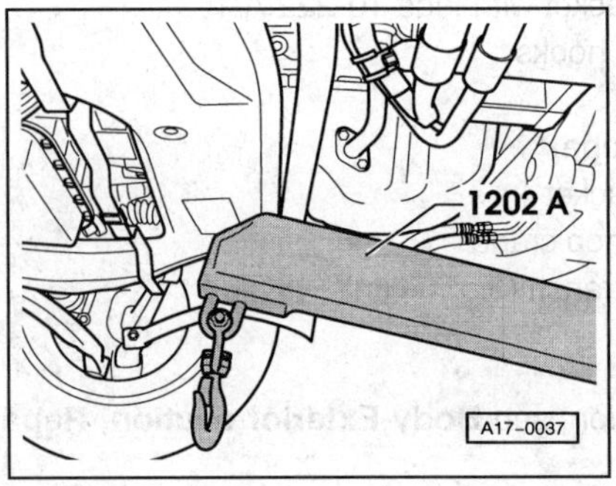

Support subframe with workshop crane 1202 A.

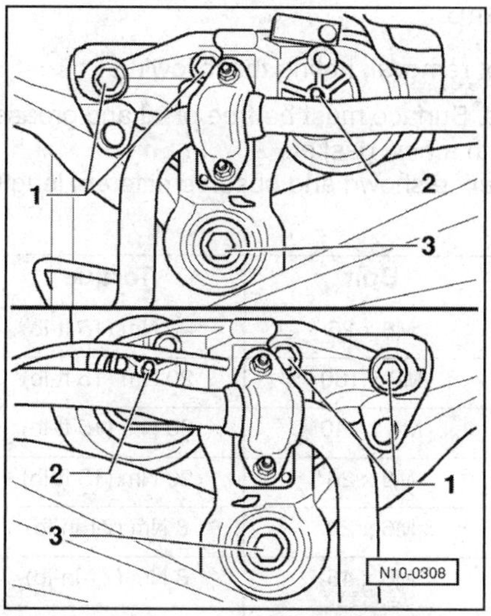

Remove subframe front bolts -1- on left and right-hand sides. Then remove bolts -3-.

NOTE —

To avoid having to measure the wheel alignment, the subframe must only be loosened or lowered at the front end.

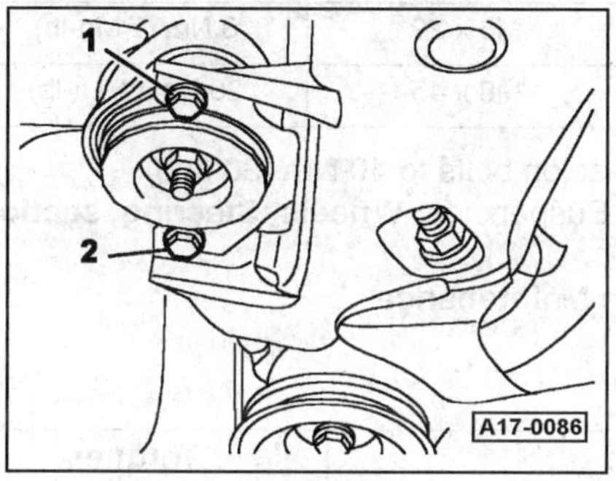

Loosen rear bolt -2- of left-hand transmission mounting a few turns. Remove front bolt -1- of left-hand transmission mounting.

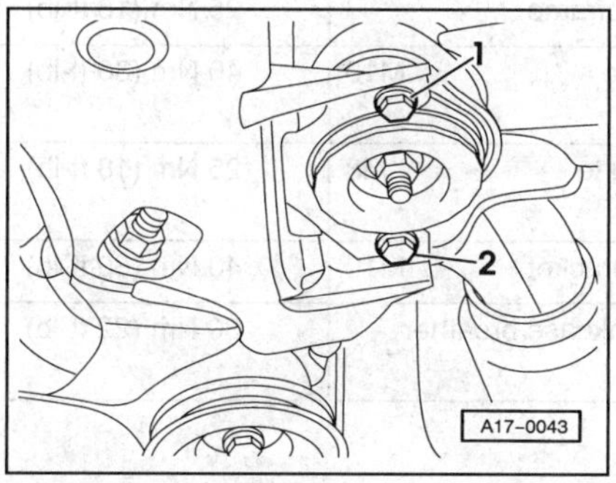

Loosen rear right transmission mounting rear bolt -2- several turns, remove right transmission mounting front bolt -1-.

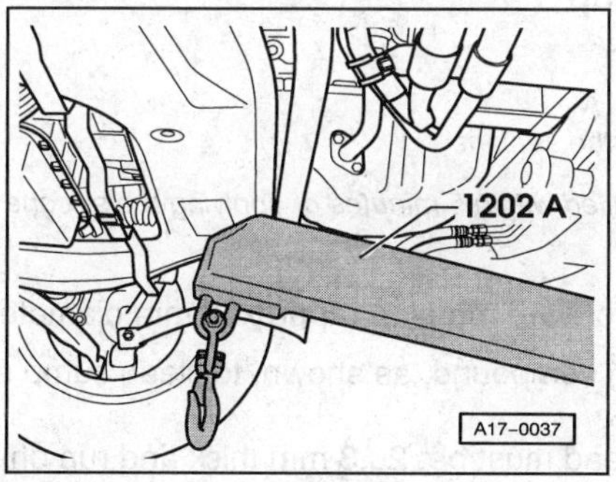

Lower subframe slowly with workshop crane.

- Remove workshop crane, swing stabilizer bar down.
- Remove retainers for coolant and refrigerant lines on upper sump.
- Remove oil pump oil lines, see **Oil pump, assembly**, item **-7-**.
- Remove crankcase breather oil return line, see **Lubrication system, components**, item **-5-**
- Remove power steering pump from bracket and remove bracket, see **Lubrication system, components**, item **-6-**.
- Remove oil return lines, see **Oil filter housing, assembly**, item **-9-**.
- Remove upper sump.

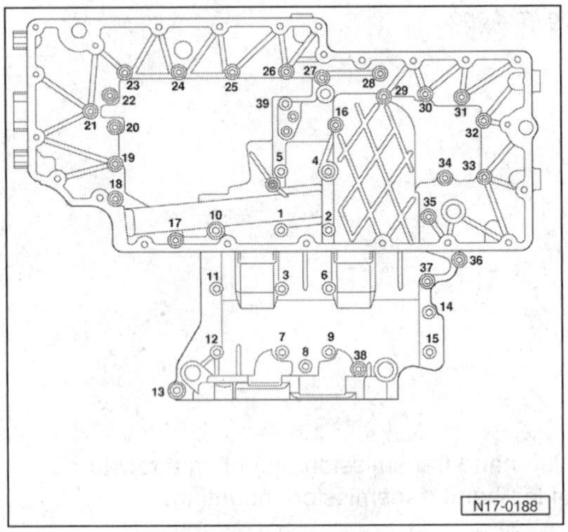

Installing upper sump

– Installation is reverse of removal, noting the following:

- Clean sealing surface. Surface must be free of oil and grease.
- Install upper sump with a new gasket.
- Tighten bolts in sequence shown and observe different lengths of bolts:

Position	Bolt	Torque
1...9	M8 x 26	20 Nm (15 ft-lb)
10	M8 x 160	20 Nm (15 ft-lb)
11+12	M8 x 140	20 Nm (15 ft-lb)
13...15	M8 x 26	20 Nm (15 ft-lb)
16...26	M6 x 28	8 Nm (71 in-lb)
27...29	M6 x 45	8 Nm (71 in-lb)
30...35	M6 x 28	8 Nm (71 in-lb)
36...38	M6 x 22	8 Nm (71 in-lb)
39	M8 x 45	20 Nm (15 ft-lb)

- Tighten sump/transmission bolts to 40 Nm (30 ft-lb).
- Install subframe, see Suspension, **Wheels, Steering, section, Repair Group 40**
- Install ribbed belt, see **Maintenance.**

Tightening torques

Bolted connections		Torques
Engine mounting to subframe		25 Nm (18 ft-lb)
Transmission support to transmission mounting	M10	40 Nm (30 ft-lb)
Transmission mounting to subframe	M8	25 Nm (18 ft-lb)
Power steering pump bracket	M10	40 Nm (30 ft-lb)
Oil return line from crankcase breather housing to sump		30 Nm (22 ft-lb)

Installing lower sump

NOTE —

- *Note the use by date of the sealant.*
- *The sump must be installed within 5 minutes of applying the silicone sealing compound.*

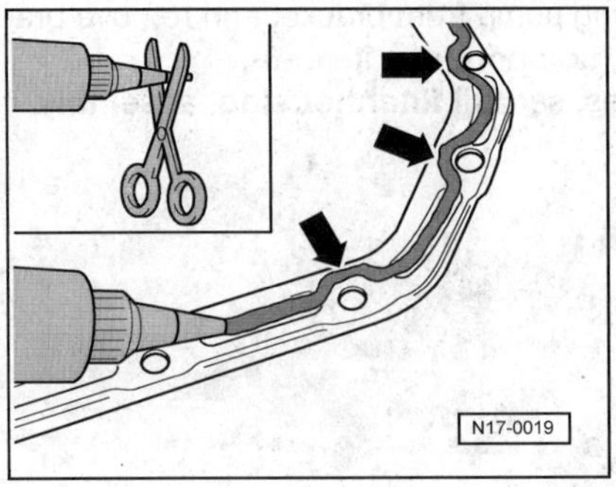

◄ Cut off tube nozzle at forward marking (approx. 3 mm diameter).

- Apply silicone sealing compound, as shown, to clean sump sealing surface.
- Sealing compound bead must be: 2...3 mm thick and run on inside of bolt holes -arrows-.

NOTE —

The sealing compound bead must not be thicker, otherwise excess sealing compound will enter the sump and may block the oil suction pipe strainer.

– Install sump immediately and tighten all sump bolts lightly. Tighten sump bolts to 8 Nm (71 in-lb).

– Tighten coolant / refrigerant line retainers on upper sump to 8 Nm (71 in-lb).

NOTE —
After installing the sump the sealant must be allowed to dry for approx. 30 minutes before engine oil is replenished.

– Remaining installation is reverse of removal.

Oil pump, removing and installing

Special tools and equipment

- V.A.G 1331 Torque wrench (5...50 Nm)
- 3.5 mm (1/8 in.) diameter drill bit

Removing

– Remove lower sump, see **Oil sump, removing and installing**

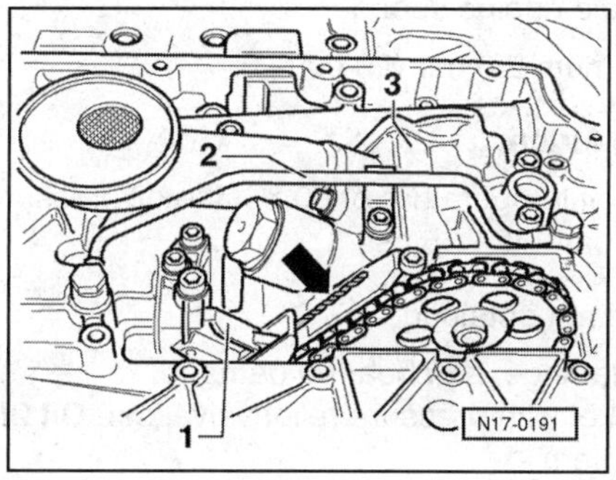

◄ Remove oil line -2-.

- Lock chain tensioner -1- using a 3.5 mm (1/8 in.) drill bit -arrow- and remove tensioner.
- Remove oil pump -3-.

Installing

– Installation is reverse of removal, noting the following:

- Tighten oil pump bolts to 20 Nm (15 ft-lb).
- Tighten chain tensioner bolts and remove drill bit: M6 to 8 Nm (71 in-lb), M8 to 20 Nm (15 ft-lb).
- Tighten oil line banjo bolts to 20 Nm (15 ft-lb) and securing bolts to 8 Nm (71 in-lb).
- Install lower sump, see **Oil sump, removing and installing**

Oil pressure and oil pressure switch, checking

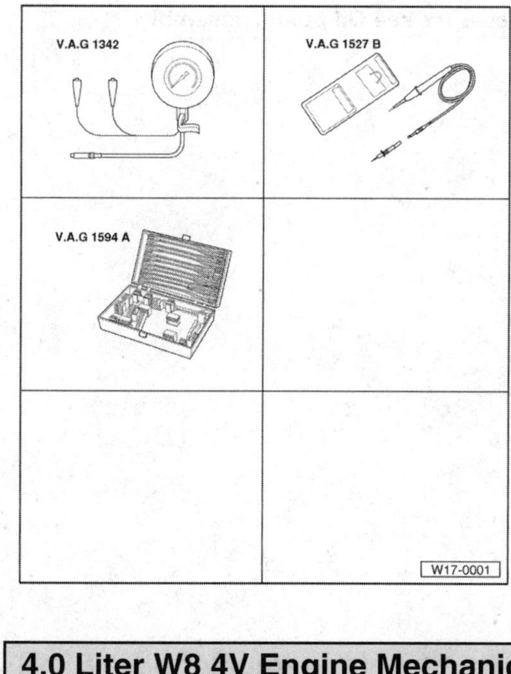

Special tools and equipment

- V.A.G 1342 Oil pressure tester
- V.A.G 1527 B Diode test lamp
- V.A.G 1594 A Adapter set

Test conditions

- Engine oil level OK, checking, see **Lubrication system, components, Fig. 1**.
- On vehicles with automatic transmission selector lever in "P" or "N" position

Checking oil pressure switch

– Remove front bumper, see **Body-Exterior section, Repair Group 63; Front bumper.**

– Position lock carrier into service position, see **Maintenance**.

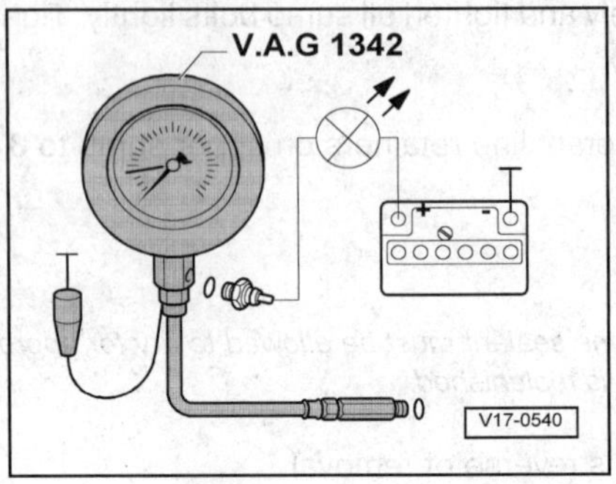

Remove oil pressure switch -F1- and thread into tester.

– Thread tester into oil filter bracket in place of oil pressure switch.

– Connect brown wire of tester to Ground (-).

– Connect diode test lamp V.A.G 1527 B using auxiliary cables from V.A.G 1594 A to battery positive (+) and oil pressure switch.
  - LED must not light up.

If the LED lights up:

– Replace oil pressure switch -F1-, see **Oil filter housing, assembly**, item **-22-**.

If the LED does not light up:

– Start engine and run at idling speed.
  - At 1.2...1.6 bar (17 to 23 psi) pressure LED must light up; otherwise replace oil pressure switch -F1-, see **Oil filter housing, assembly**, item **-22-**.

Check oil pressure

– Check oil pressure at idle engine speed.
  - Specified value: minimum 2.0 bar (29 psi)

– Increase engine speed further.
  - At 2000 rpm and an oil temperature of 80 °C, the oil pressure should be between 3.0 - 4.5 bar. (43 - 65 psi)

– If the specifications are not obtained:
  - Correct mechanical damage, e.g. bearing damage.
  - Replace oil filter bracket with pressure relief valve, see **Oil filter housing, assembly**, item **-1-**.
  - Replace oil pump if necessary, see **Oil pump, assembly** .

At higher engine speeds oil pressure must not exceed 7.0 bar (100 psi)

– If the specification is exceeded:
  - Check oil galleries.
  - Replace oil pump if necessary, see **Oil pump, assembly**.

19 Engine–Cooling System

Repair Group

E 19

Cooling system

> ***WARNING —***
> *When doing any repair work, especially in the engine compartment, pay attention to the following due to the cramped conditions:*
>
> - *Route all the various lines (e.g. for fuel, hydraulics, EVAP system, coolant, refrigerant, brake fluid and vacuum lines and hoses) and electrical wiring so that the original positions are restored.*
> - *Ensure sufficient clearance to all moving or hot components.*

NOTE —

- *When the engine is warm the cooling system is under pressure. If necessary release pressure before commencing repair work.*
- *Hoses are secured with spring-type clips. In cases of repair only use spring-type clips.*

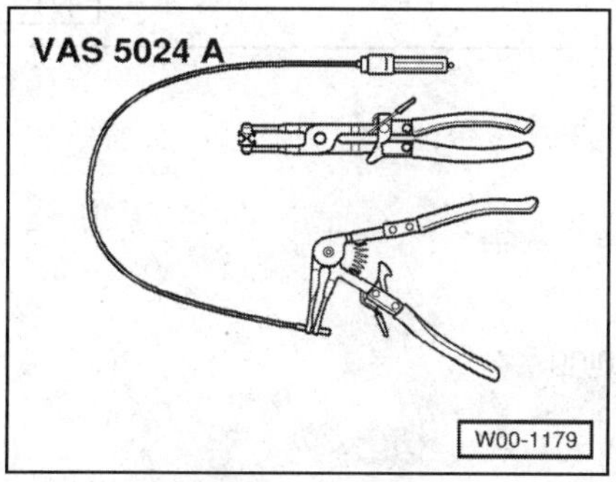

VAS 5024 pliers are recommended when installing spring-type clips.

- When installing coolant hoses route stress-free so that they do not come into contact with other components (observe markings on coolant connection and hose).

Perform cooling system leakage test with cooling system tester V.A.G 1274 and adapters V.A.G 1274/8 and V.A.G 1274/9.

Draining and filling with coolant, see **Cooling system, draining and filling.**

Coolant mixture ratios, see **Cooling system, draining and filling**

Cooling system components, body side

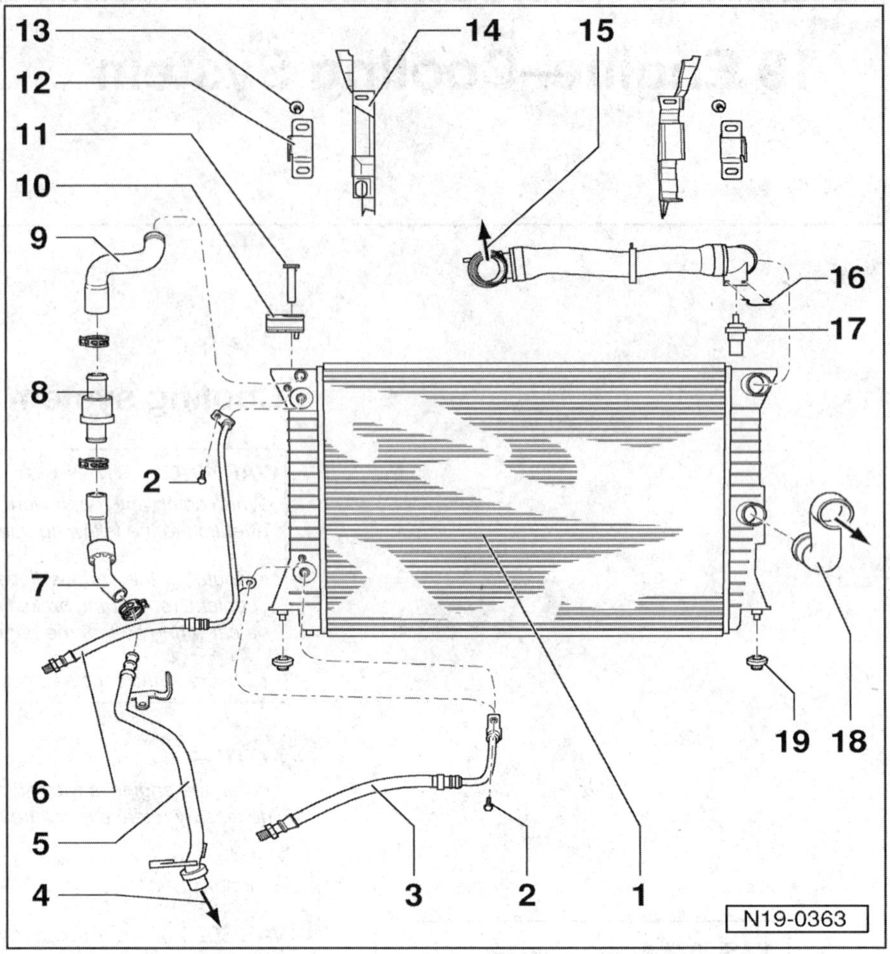

1. **Radiator**
    - With drain plug
    - See **Radiator, removing and installing**
    - After replacing replace entire amount of engine coolant
    - With ATF cooler on vehicles with automatic transmission
    - Removing and installing coolant fan, see **Coolant fan, assembly**
2. **10 Nm (7ft-lb)**
3. **Transmission oil line**
4. **To coolant pump -V36-**
5. **Left coolant pipe**
    - On left longitudinal member
6. **Transmission oil line**
7. **Coolant hose**
    - Check for secure seating
8. **Non-return valve**
    - Note installation position
9. **Coolant hose**
    - Check for secure seating
10. **Securing rubber**
11. **Securing pin**
12. **Retainer**
    - For air conditioning condenser
13. **10 Nm (7 ft-lb)**
14. **Air deflector plate**
15. **Upper coolant hose**
    - To coolant regulator housing, see **Cooling system components, engine side, top and rear**, item **-24-**
    - Secured to radiator and thermostat housing with a retaining clip
    - Check for secure seating
16. **Retaining clip**
    - Check for secure seating
17. **Engine Coolant Temperature (ECT) sensor (on radiator) -G83-**
    - If necessary release pressure in cooling system before removing
18. **Lower coolant hose**
    - To upper coolant hose, see **Cooling system components, engine side, top and rear**, item **-38-**
    - Secured to radiator with a retaining clip
    - Check for secure seating
19. **Rubber mounting**

Coolant fan, assembly

1. Lock carrier
2. Coolant fan 2 -V177-
3. Securing sleeve
4. Fan ring
5. 10 Nm (7 ft-lb)
6. 10 Nm (7 ft-lb)
7. 10 Nm (7 ft-lb)
8. Coolant fan control (FC) control module -J293-
9. Coolant fan -V7-

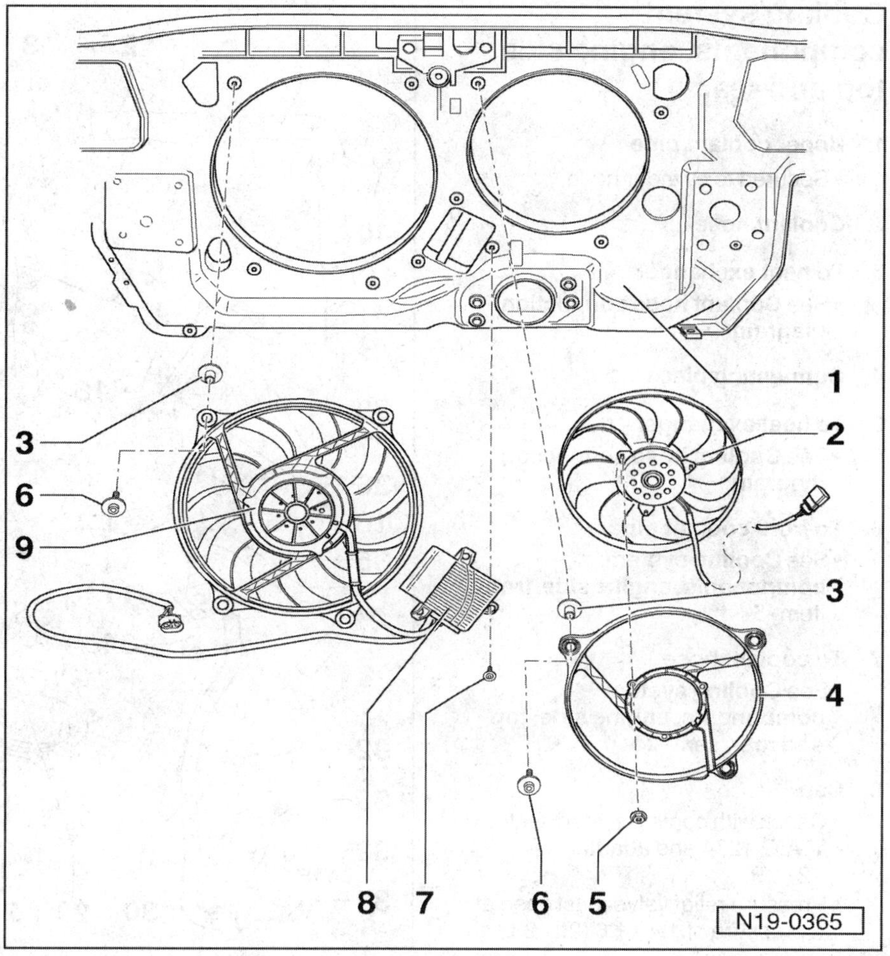

Cooling system components, engine side, top and rear

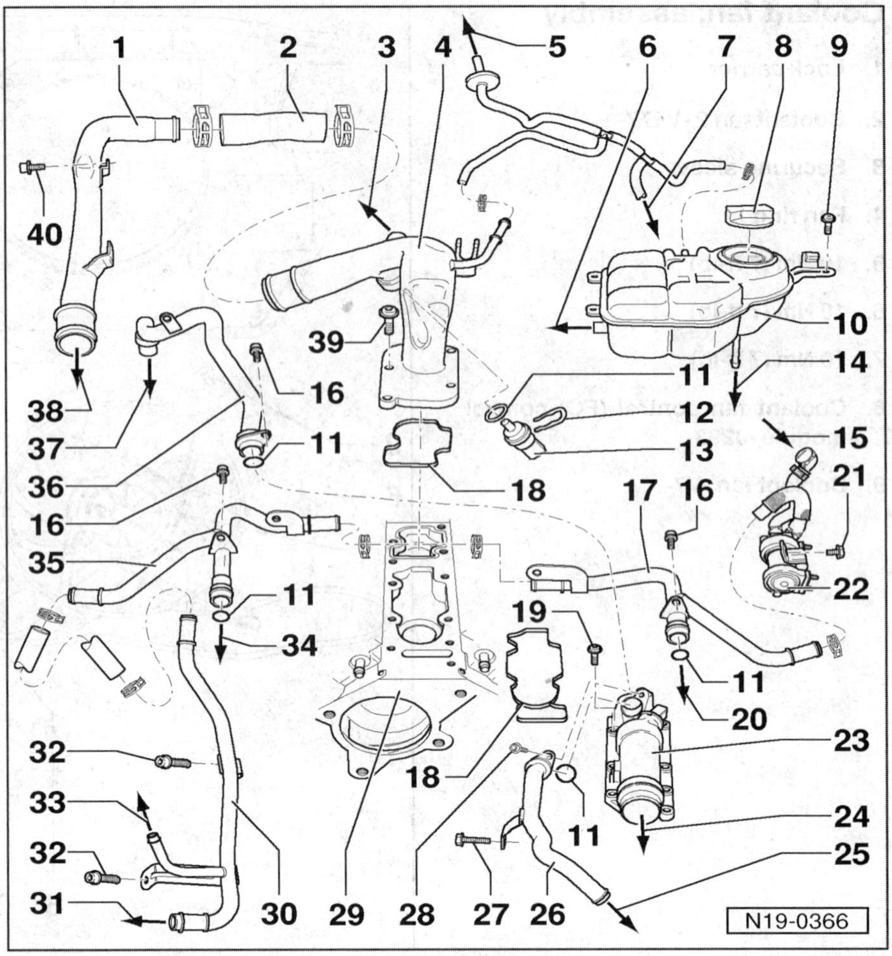

1. **Upper coolant pipe**
   - Secured to cylinder head
2. **Coolant hose**
3. **To heat exchanger**
   - See **Coolant hose connection diagram**
4. **Connecting piece**
5. **To heat exchanger**
   - See **Coolant hose connection diagram**
6. **To front coolant pipe**
   - See **Cooling system components, engine side, front**, item **-5-**
7. **To coolant hose**
   - See **Cooling system components, engine side, top and rear**, item **-24-**
8. **Cap**
   - Check with cooling system tester V.A.G 1274 and adapter V.A.G 1274/9
   - Pressure relief valve must open at a pressure of 1.4 - 1.6 (20 - 23 psi)
9. **10 Nm (7ft-lb)**
10. **Expansion tank**
    - Perform leak test of cooling system with cooling system tester V.A.G 1274 and adapter V.A.G1274/8
    - Test pressure 1.4...1.6 bar (20 - 23 psi)
11. **O-ring**
    - Replace
12. **Retaining clip**
    - Check for secure seating
13. **Engine Coolant Temperature (ECT) sensor -G62-**
    - For engine control module
    - If necessary release pressure in cooling system before removing
14. **Over-flow connection**
15. **From left coolant pipe**
    - See **Cooling system components, body side**, item **-4-**
16. **10 Nm (7 ft-lb)**
17. **Rear left coolant pipe**
18. **Gasket**
    - Replace if damaged
19. **10 Nm (7 ft-lb)**
20. **To left cylinder block**
21. **10 Nm (7 ft-lb)**
22. **Coolant pump -V36-**
23. **Coolant thermostat housing**
    - With Map controlled engine cooling thermostat -F265-
    - Checking: heat-up thermostat in water
    - Opening begins: approx. 86 °C (186 °F)
    - Opening lift: min. 7 mm
24. **To upper coolant hose**
    - See **Cooling system components, body side**, item **-15-**
25. **To front coolant pipe**
    - See **Cooling system components, engine side, front**, item **-4-**
26. **Coolant pipe**
27. **10 Nm (7 ft-lb)**
28. **10 Nm (7 ft-lb)**
29. **Cylinder head/cylinder block**
30. **Right coolant pipe**
31. **To oil cooler**
    - See **Repair Group 17, Oil filter housing, assembly**, item **-7-**
32. **10 Nm (7 ft-lb)**
33. **To alternator**
    - See **Cooling system components, engine side, front**, item **-15-**
34. **To right cylinder block**
35. **Rear right coolant pipe**
36. **Coolant pipe**
37. **From heat exchanger**
38. **Lower coolant hose**
    - See **Cooling system components, body side**, item **-18-**
39. **10 Nm (7 ft-lb)**
40. **10 Nm (7 ft-lb)**

Cooling system components, engine side, front

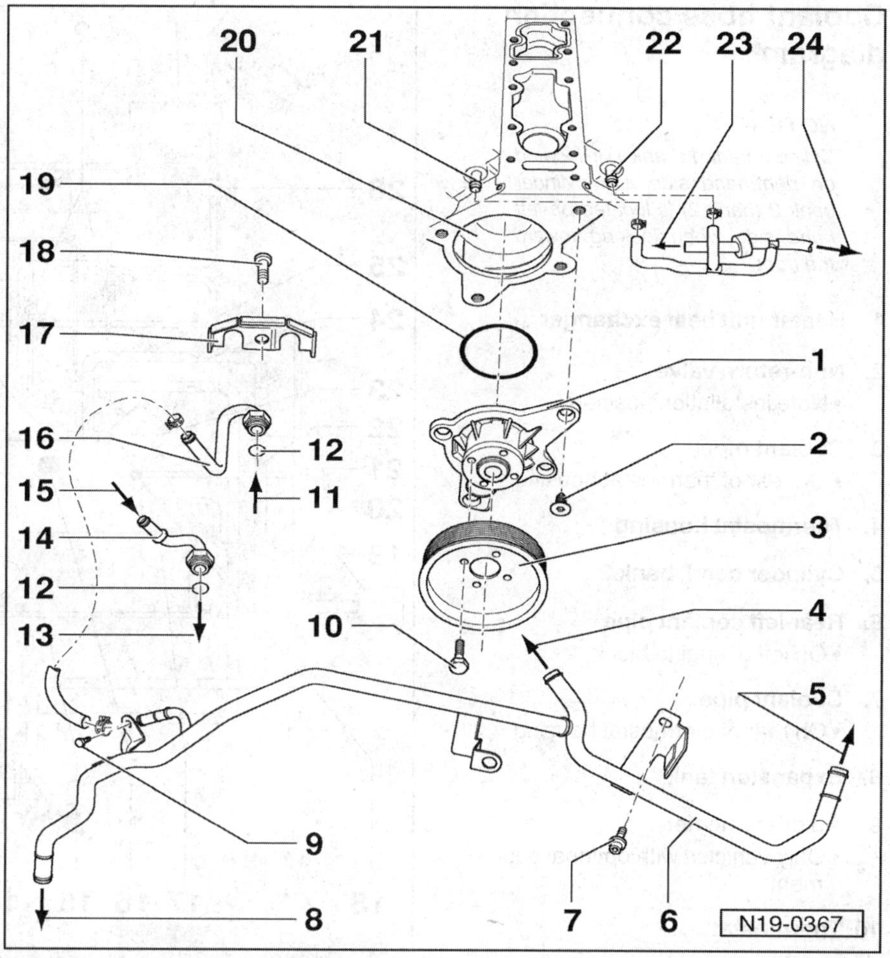

1. **Coolant pump**
   - Check for ease of movement
   - If damaged or leaking replace complete
   - See **Coolant pump, removing and installing**
2. **20 Nm (15 ft-lb)**
3. **Pulley**
   - For coolant pump
   - Removing and installing ribbed belt, see **Maintenance**
4. **To coolant pipe**
   - On front of thermostat housing, see **Cooling system components, engine side, top and rear**, item **-25-**
5. **To expansion tank**
   - See **Cooling system components, engine side, top and rear**, item **-6-**
6. **Front coolant pipe**
7. **10 Nm (7 ft-lb)**
8. **To oil cooler**
   - See **Repair Group 17, Oil filter housing, assembly**, item **-7-**
9. **10 Nm (7 ft-lb)**
10. **15 Nm (11 ft-lb)**
11. **From alternator**
12. **O-ring**
    - Replace
13. **To alternator**
14. **Coolant pipe**
    - On alternator
15. **From right coolant pipe**
    - See **Cooling system components, engine side, top and rear**, item **-33-**
16. **Coolant pipe**
    - On alternator
17. **Securing plate**
    - For coolant pipe on alternator
18. **10 Nm (7 ft-lb)**
19. **O-ring**
    - Replace
20. **Cylinder block with flange for coolant pump**
21. **Coolant connecting piece**
    - On right of cylinder head
22. **Coolant connecting piece**
    - On left of cylinder head
23. **To upper coolant hose**
    - See **Cooling system components, body side**, item **-15-**
24. **To coolant hose**
    - See **Cooling system components, engine side, top and rear**, item **-7-**

Coolant hose connection diagram

> ***NOTE —***
> *Cylinder bank 1 (bank 1) is located on right-hand side and cylinder bank 2 (bank 2) is located on left hand side when looking toward the front.*

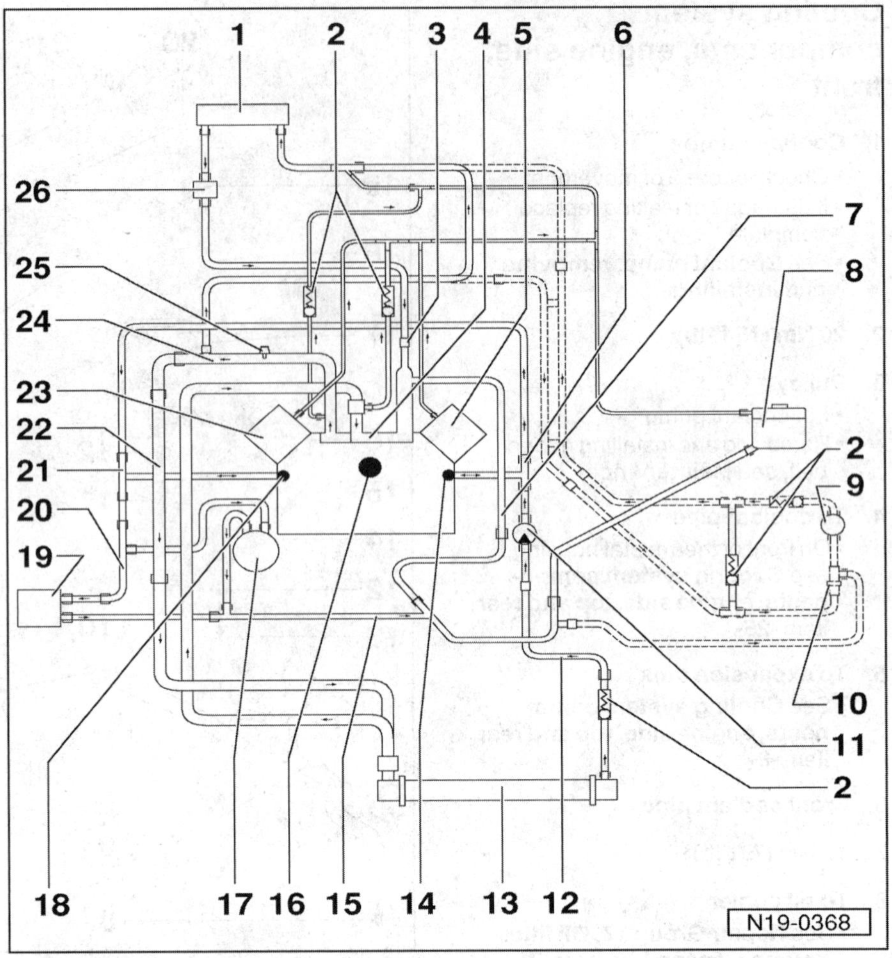

1. **Heater unit heat exchanger**
2. **Non-return valve**
   - Note installation position
3. **Coolant pipe**
   - On rear of thermostat housing
4. **Thermostat housing**
5. **Cylinder head, bank 2**
6. **Rear left coolant pipe**
   - On left of engine block
7. **Coolant pipe**
   - On rear of thermostat housing
8. **Expansion tank**
9. **Auxiliary heater**
   - Only vehicles with optional equipment
10. **Thermostat**
11. **Coolant pump -V36-**
12. **Left coolant pipe**
    - On left longitudinal member
13. **Radiator**
    - With ATF cooler on vehicles with automatic transmission
14. **Flange for rear left coolant pipe**
    - On left of engine block
15. **Front coolant pipe**
16. **Coolant pump**
17. **Alternator**
18. **Flange for rear right coolant pipe**
    - On left of engine block
19. **Oil cooler**
20. **Right coolant pipe**
    - On right longitudinal member
21. **Rear right coolant pipe**
    - On right of engine block
22. **Upper coolant pipe**
    - On cylinder bank 1
23. **Cylinder head, bank 1**
24. **Connection**
25. **Engine Coolant Temperature (ECT) sensor -G62-**
26. **Thermostat**

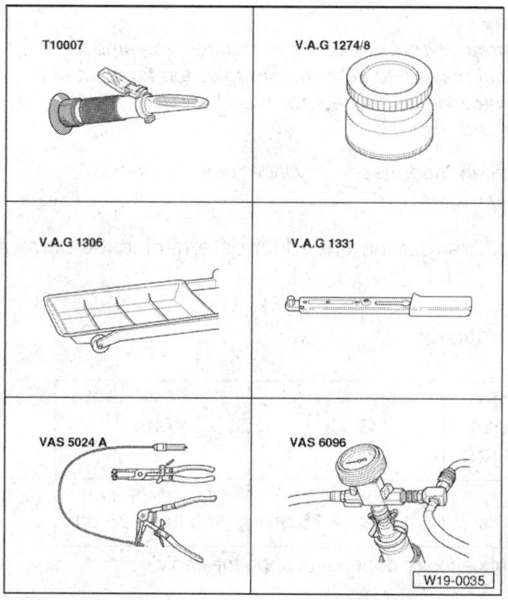

Cooling system, draining and filling

Special tools and equipment

- T10007 Refractometer
- V.A.G1274/8 Adapter
- V.A.G 1306 Drip tray
- V.A.G 1331 Torque wrench (5...50 Nm)
- VAS 5024 Installation tool for spring-type clips
- VAS 6096 Cooling system charging unit

Draining

– Open cap on coolant expansion tank.

> ***WARNING —***
> *Steam can be released when the cap is removed from the expansion tank. Cover cap with a cloth and open carefully.*

– Remove sound insulation, see **Body-Exterior section, Repair Group 50**

– Remove front bumper, see **Body-Exterior section, Repair Group 63**

– Position lock carrier into service position, see **Body-Exterior section, Repair Group 50**

– Drain coolant from radiator through drain plug.

NOTE —
Observe disposal regulations for coolant!

– Pull coolant hose retaining clip off downward and remove coolant hose from radiator.

– Disconnect coolant hose from oil cooler.

Filling

NOTE —

- *Only use coolant additive G 12 in accordance with TL VW 774 F. Distinguishing feature: violet color*
- *G 12 and coolant additives marked "In accordance with TL VW 774 F" prevent frost and corrosion damage, scaling and also raise the boiling point of the coolant. For this reason the system must be filled all year round with frost and corrosion protection additives.*
- *Because of its high boiling point, the coolant improves engine reliability under heavy loads, particularly in countries with tropical climates.*
- *Protection against frost must be assured to about -25 °C (in arctic climatic countries to about -35 °C).*
- *The coolant concentration must not be reduced by adding water even in warmer seasons and in warmer countries. The anti-freeze ratio must be at least 40 %.*

NOTE—

- *If for climatic reasons a greater frost protection is required, the amount of G 12 can be increased, but only up to 60 % (frost protection to about -40 °C), as otherwise frost protection is reduced again and cooling effectiveness is also reduced.*
- *If radiator, heat exchanger, cylinder head or cylinder head gasket is replaced, do not reuse old coolant.*

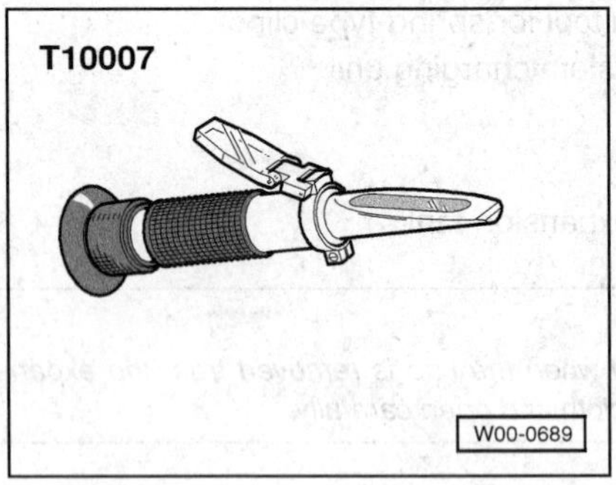

The refractometer T10007 is recommended for determining the current anti-freeze density.

Recommended mixture ratios:

Frost protection to	Anti-freeze portion	G 12[1]	Water[1]
-25° C -35° C	40 % 50 %	4.0 ltr. (4.2 qt) 4.5 ltr. (4.75 qt)	5.0 ltr. (5.3 qt) 4.5 ltr. (4.75 qt)

[1] The quantity of coolant can vary depending upon the vehicle equipment.

Work sequence

- Install lower coolant hose and secure.
- Connect coolant hoses on oil cooler.
- Replace O-ring of coolant drain plug and insert coolant drain plug in cooler again.
  - Tightening torque: 20 Nm (15 ft-lb)

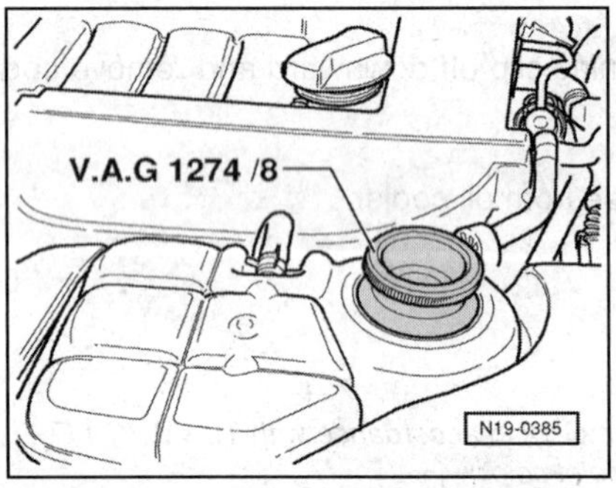

Screw adapter V.A.G 1274/8 onto expansion tank.

- Fill coolant circuit using cooling system charging unit VAS 6096.
- Start engine and maintain an engine speed of about 2000 rpm for approx. 3 minutes.
- Allow engine to run at idling speed until lower hose on radiator becomes hot.
- Switch ignition off.

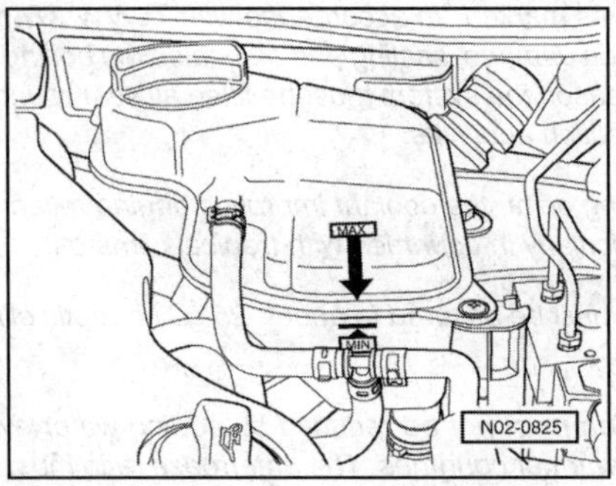

Check coolant level and top up as necessary:

- When engine is at normal operating temperature coolant level must be on max. mark.
- When engine is cold coolant level can be between min. and max. marks.

- Check headlight adjustment and adjust if necessary, see **Maintenance**

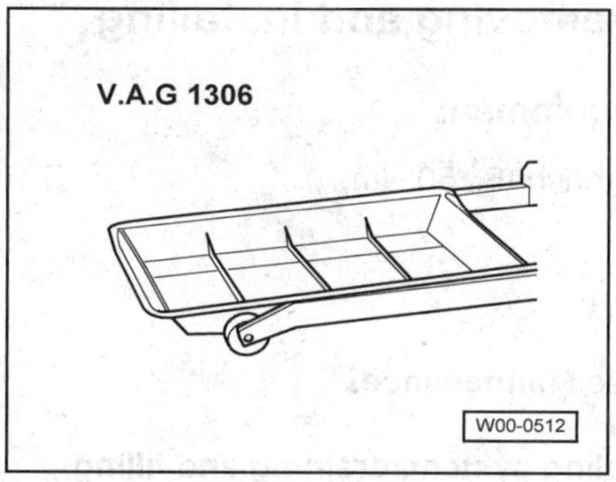

Radiator, removing and installing

Special tools and equipment

- V.A.G 1331 Torque wrench (5...50 Nm)

Drip tray V.A.G 1306

Removing

– Drain coolant, see **Cooling system, draining and filling**.

– Pull out retaining clip for upper coolant hose and remove coolant hose from quick-release coupling of radiator.

– Disconnect transmission oil line at transmission oil cooler and collect oil which may leak out in a suitable container.

NOTE —
Observe disposal regulations!

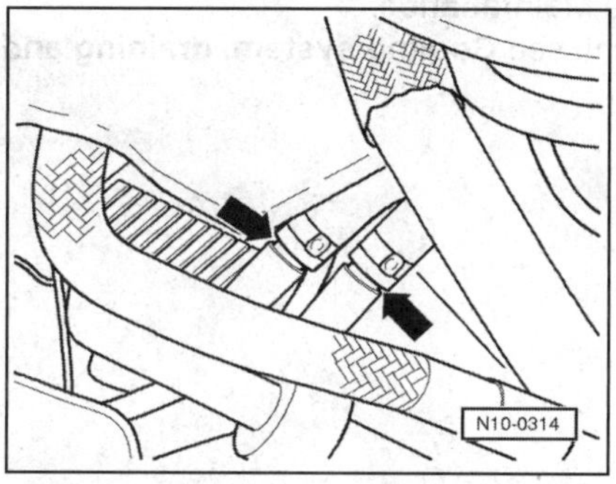

Separate lines to oil cooler at unions -arrows- and collect ATF which may leak out in a suitable container.

NOTE —
Observe disposal regulations!

– Remove radiator upper securing pin and take radiator out forward.

Installing

– Installation is reverse of removal, noting the following:

- Check electrical connections and routing, see **Wiring Diagrams and Component Locations**

Vehicles with manual transmission

- Top up transmission oil.

Vehicles with automatic transmission

- Check ATF level, see Maintenance.

Continuation for all vehicles

- Top up oil for power assisted steering system, see **Maintenance**.
- Fill system with coolant, see **Cooling system, draining and filling**.

Coolant pump, removing and installing

Special tools and equipment

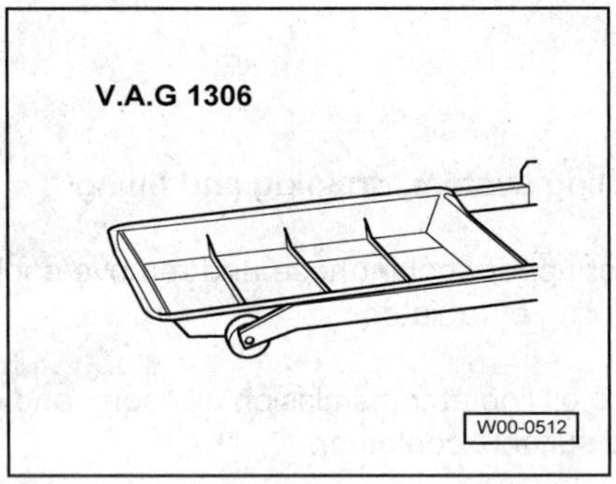

- V.A.G 1331 Torque wrench (5...50 Nm)

Drip tray V.A.G 1306

Removing

– Remove ribbed belt, see **Maintenance.**

– Drain coolant, see **Cooling system, draining and filling**.

– Unscrew coolant pump securing bolts and remove coolant pump.

Installing

– Installation is reverse of removal, noting the following:

- Moisten new O-ring with coolant.
- Install coolant pump. Tighten securing bolts to 20 Nm (15 ft-lb).
- Install ribbed belt, see **Maintenance**.
- Fill system with coolant, see **Cooling system, draining and filling**.

20 Fuel Supply

Repair Group

E 20

Fuel supply system

NOTE —

- *Hose connections are secured with either spring-type or clamp-type clips.*
- *Always replace clamp-type clips with spring-type clips.*
- *Fuel hoses in engine compartment must only be secured with spring type clips. The use of clamp or screw type clips is not permissible.*

Safety precautions when working on the fuel supply system

WARNING —

When doing any repair work, especially in the engine compartment, pay attention to the following due to the cramped conditions:

- *Route all the various lines (e.g. for fuel, hydraulics, EVAP system, coolant, refrigerant, brake fluid and vacuum lines and hoses) and electrical wiring so that the original positions are restored.*
- *Ensure sufficient clearance to all moving or hot components.*

WARNING —

Fuel supply lines are under pressure! Before removing from hose connection wrap a cloth around the connection. Then release pressure by carefully pulling hose off connection.

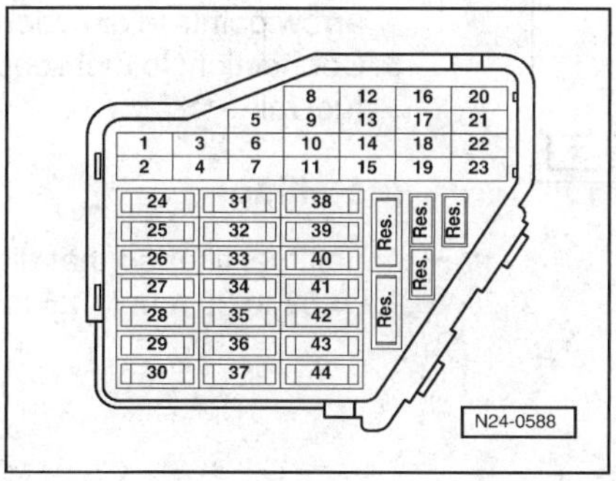

◄ For reasons of safety, remove fuse No. 28 from the fuse holder before opening the fuel system. This is because the fuel pump can be activated by the driver's door contact switch.

Rules for cleanliness

When working on the fuel supply/injection system, pay careful attention to the following 5 rules:

- Thoroughly clean all unions and the adjacent areas before disconnecting.
- Place parts that have been removed on a clean surface and cover. Do not use fluffy cloths!
- Carefully cover opened components or seal if the repair cannot be performed immediately.
- Only install clean components: Only unpack replacement parts immediately prior to installation. Do not use parts that have been stored loose (e.g. in tool boxes etc.).
- When the system is open: Do not work with compressed air if this can be avoided. Do not move vehicle unless absolutely necessary.

Fuel filter, components

1. **Banjo bolt**
   - Tighten to 30 Nm (22 ft-lb)
2. **Seal**
   - Replace
3. **Supply pipe**
   - Black
   - Check for secure seating
   - Secure with spring-type clips
   - From fuel delivery unit, see **Fuel tank with attachments, components**, item **-7-**
4. **Direction of travel**
5. **Retaining clip**
6. **Fuel tank**
7. **2 Nm (18 in-lb)**
8. **Banjo bolt**
9. **Tighten to 23 Nm (17 ft-lb)**
10. **Connection**
11. **Supply pipe**
    - Black
    - Check for secure seating
    - Secure with spring-type clips
    - To fuel supply pipe on fuel rail
12. **Fuel filter**
    - Note installation position, see **Fig. 1**

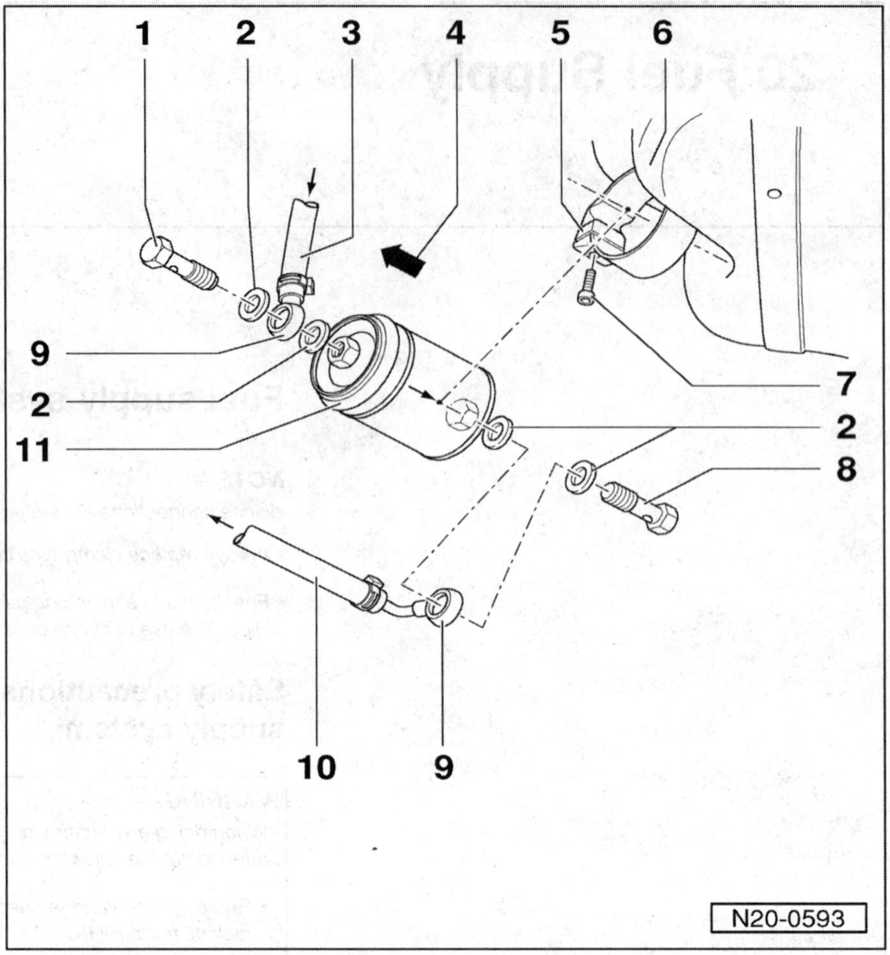

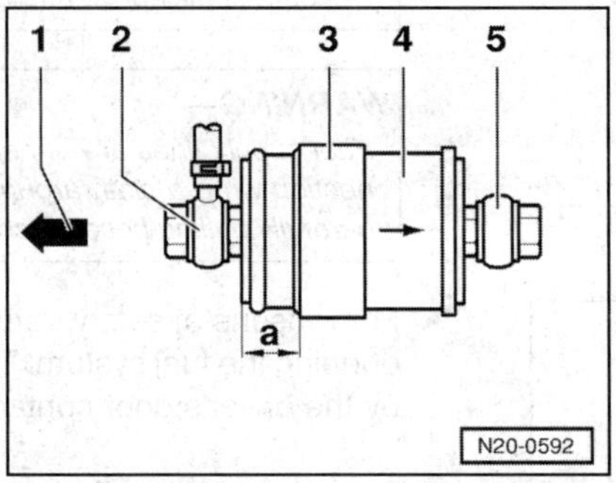

Fig. 1 Installation position of fuel filter

1. Direction of travel
2. Connection: for fuel delivery unit
3. Retaining clip
4. Fuel filter, installation position: arrow points in direction of flow
5. Connection: to fuel supply pipe at fuel rail

Installing

– Install fuel filter so that dimension -a- = 32 to 36 mm (1.25 to 1.42 in.).

Fuel tank with attachments, assembly

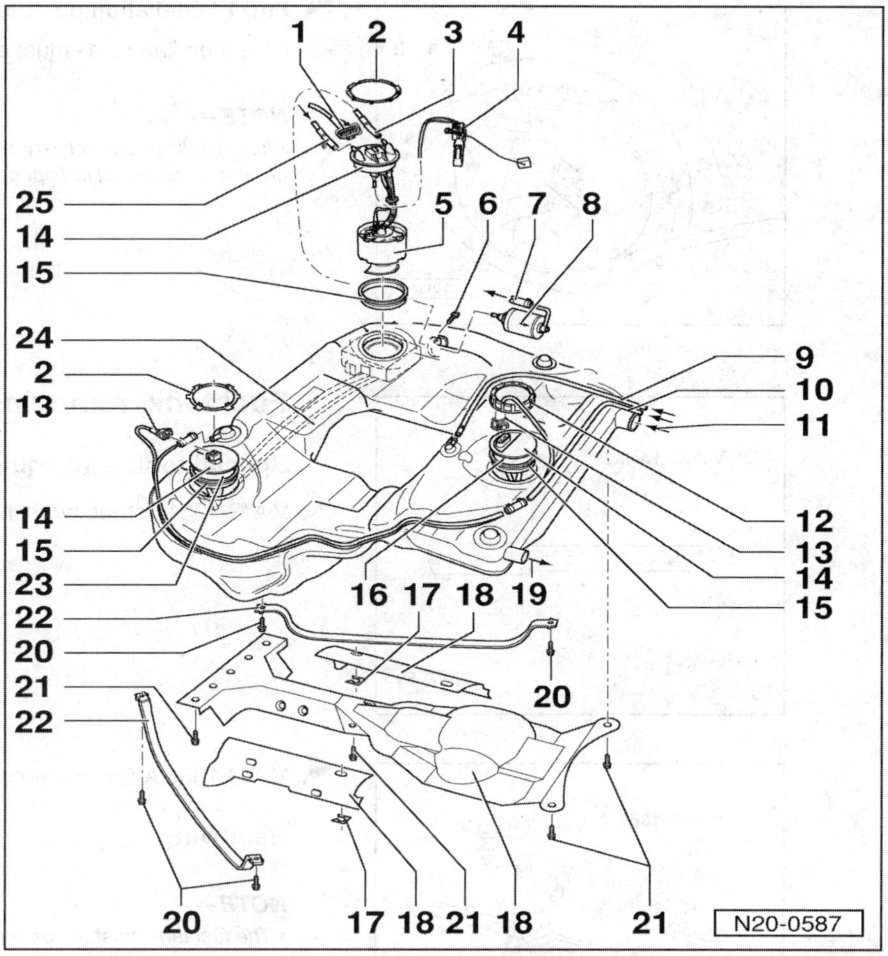

1. **Connector**
   - Black, 4-pin
   - For fuel gauge sender -G- and fuel pump -G6-
2. **Locking ring**
   - Tighten to 145 Nm (107 ft-lb)
   - Check for secure seating
   - Remove and install with spanner 3087
3. **Supply pipe**
   - Black
   - Check for secure seating
   - Secure with spring-type clips
4. **Sender for fuel gauge -G-**
   - See **Sender for fuel gauge -G-, removing and installing**
5. **Fuel delivery unit**
   - See **Fuel delivery unit, removing and installing**
   - Checking fuel pump, see **Fuel pump, checking**
   - Clean strainer if soiled
6. **2 Nm (18 in-lb)**
7. **Supply pipe**
   - Black
   - Check for secure seating
   - Secure with spring-type clips
   - To fuel supply pipe on fuel rail
8. **Fuel filter**
   - See **Fuel filter, assembly**
9. **Vent pipe**
   - See **Fuel filler neck, assembly,** Item **-16-**
   - Check for secure seating
   - Secure with spring-type clips
10. **Vent pipe**
    - See **Fuel filler neck, assembly,** Item **-17-**
    - Check for secure seating
    - Secure with spring-type clips
11. **Connecting hose**
    - See **Fuel filler neck, assembly,** Item **-18-**
12. **Union nut**
    - Tighten to 60 Nm (44 ft-lb)
    - Remove and install with spanner 3218
13. **Connector**
    - Black, 2-pin
    - For fuel gauge sender
14. **Flange**
    - Note fitting position on fuel tank, see **Fig. 1**
15. **Sealing ring**
    - Replace if damaged
    - When installing, insert seal "dry" into fuel tank opening
    - Moisten with fuel only when installing flange item **-14-**
16. **Fuel supply sensor 3 -G237-**
17. **Securing clip**
    - Check for secure seating
18. **Heat shield**
    - For fuel tank
19. **Connection**
    - For connecting hose to additional tank
20. **23 Nm (17 ft-lb)**
21. **10 Nm (7 ft-lb)**
22. **Securing strap**
    - Note installation position
23. **Fuel level sensor 2 -G169-**
24. **Fuel tank**
    - Support with engine and transmission jack V.A.G 1383 A when removing
    - See **Fuel tank, removing and installing**
25. **Return pipe**
    - Blue or with blue marking
    - Check for secure seating
    - Secure with spring-type clips
    - From fuel return pipe on fuel rail

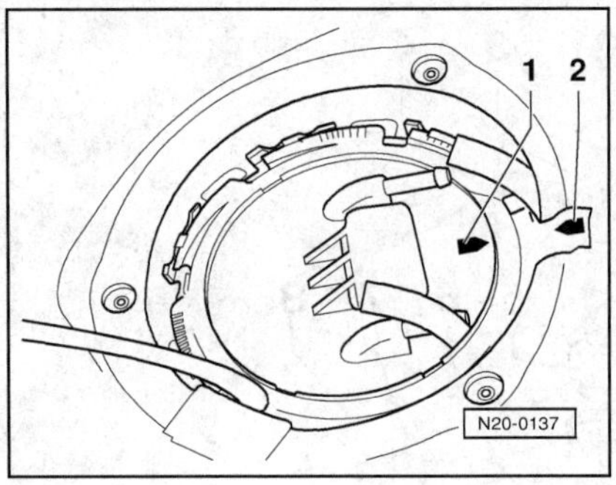

Fig. 1 Installation position of fuel delivery unit flange

- Mark on flange -1- must align with mark on fuel tank -2-.

NOTE —
After installing fuel delivery unit flange, check that the supply, return and breather lines are still clipped onto the fuel tank.

Fuel tank, removing and installing

Special tools and equipment

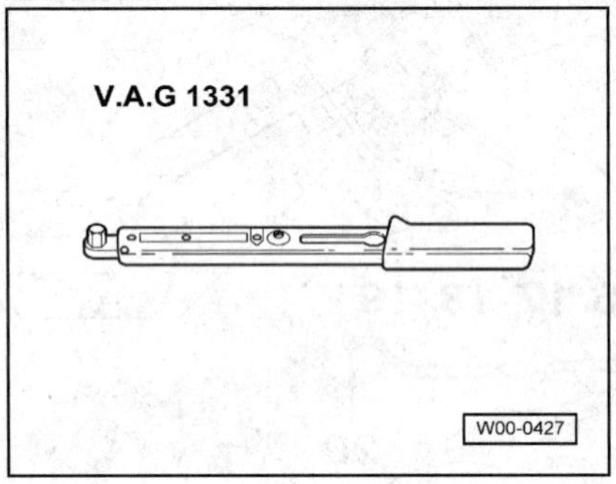

V.A.G 1331 Torque wrench (5...50 Nm)

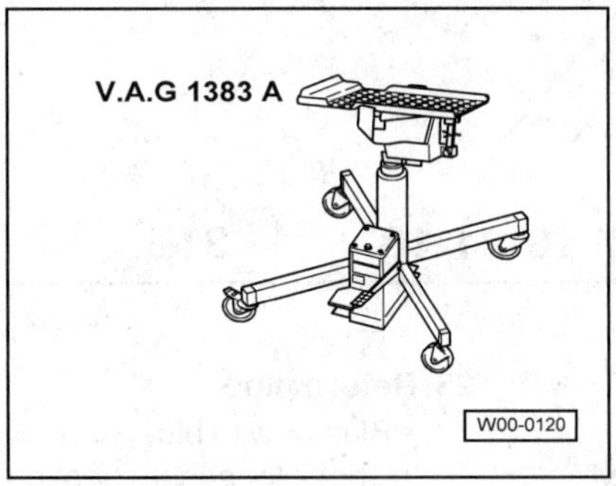

V.A.G 1383 A Engine/transmission jack

Removing

NOTE —
- *The fuel tank must not be more than 1/3 full.*
- *Empty the fuel tank using fuel pump unit VAS 5190 if necessary.*

– Note safety precautions before starting work.

– Switch off ignition and all electrical consumers and remove ignition key.

– With ignition switched off disconnect battery Ground strap.

NOTE —
All cable ties which are opened or cut open when removing the engine, must be replaced in the same position when installing the engine.

– Remove rear right wheel housing liner, see **Body-Exterior section, Repair Group 66.**

– Remove center and rear muffler (left and right-hand sides) of exhaust system, see **Repair Group 26, Exhaust system, components**.

– Remove driveshaft, see **5-Speed Automatic Transmission 01V, All Wheel Drive, Repair Group 39.**

– Remove rear axle, see **Suspension, Wheels, Steering section, Repair Group 42.**

– Drain fuel tank and clean fuel filler neck and surrounding area.

– Fold rear seat bench forward.

– Remove cover from fuel delivery unit.

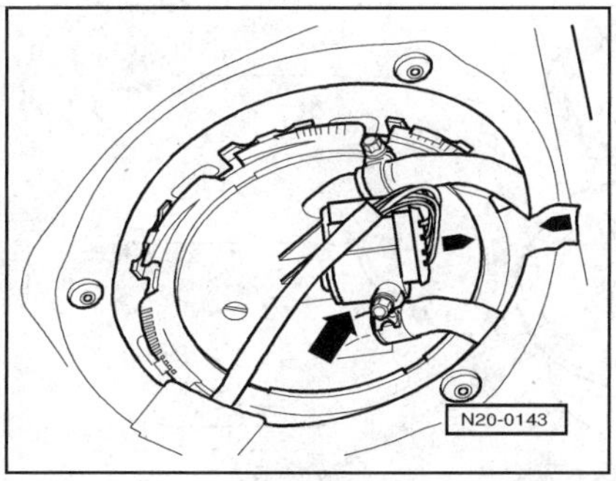

- Remove 4-pin connector -arrow- from flange on fuel tank.
- Pull connecting hose off union to additional fuel tank, see **Additional fuel tank, components**, item **-7-**.

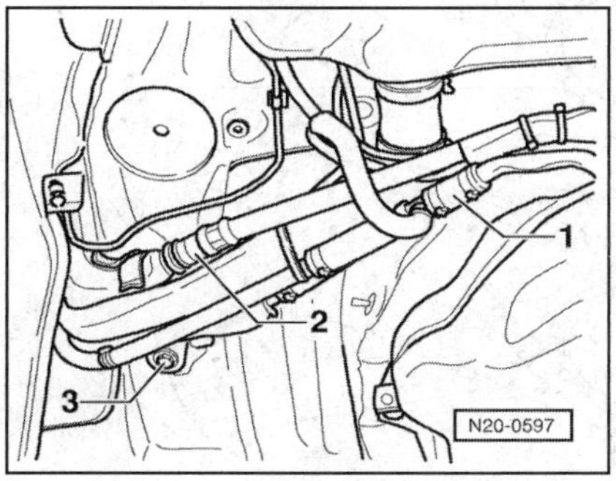

- Remove vent pipe from EVAP canister -1- at junction piece and vent pipe from additional fuel tank -2- at filler neck.
- Seal lines so that fuel system is not contaminated by dirt etc.
- Remove securing bolts on filler neck.
- Remove filler cap from fuel tank filler neck.
- Unscrew securing bolts and remove tank flap unit with rubber cup.
- Remove securing bolts on filler neck.

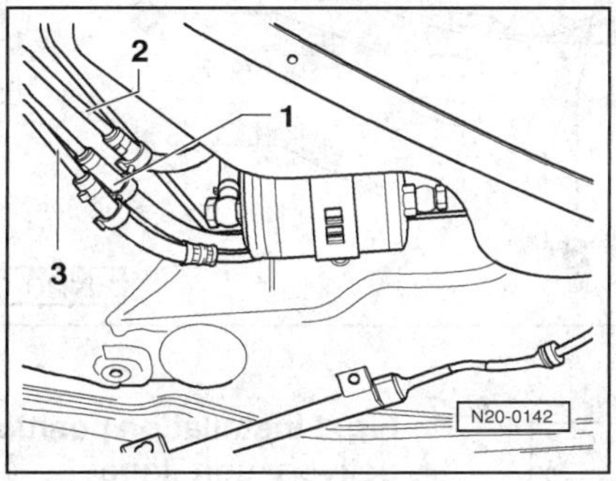

- Remove supply hose -1- from fuel filter.
  - Mark return hose -2- and breather hose -3- and separate at connecting point.

> ***WARNING —***
> *Fuel supply lines are under pressure! Before removing from hose connection wrap a cloth around the connection. Then release pressure by carefully pulling hose off connection.*

- Seal lines so that fuel system is not contaminated by dirt etc.
- Remove left and right protective plates on fuel tank and floor.
- Remove tensioning strap. When doing this support fuel tank with engine/transmission jack V.A.G 1383 A.
- Lower using engine/transmission jack V.A.G 1383 A but only sufficiently to allow connector to be pulled off flange.

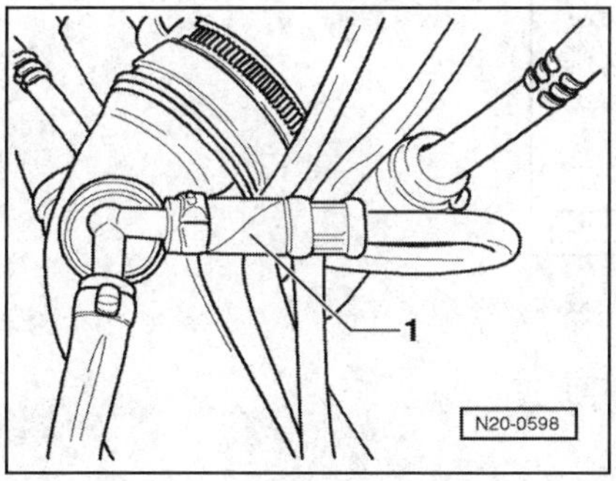

- Remove supply pipe -1- from fuel delivery unit in additional fuel tank at filler neck.
- Seal lines so that fuel system is not contaminated by dirt etc.
- Lower fuel tank.

Installing

- Installation is performed in the reverse order, noting the following:
  - Breather and fuel hoses are to be installed kink-free.
  - Ensure fuel hoses are seated securely.
  - Secure fuel hoses with spring-type clamps.
  - Do not interchange supply and return hose (return hose blue or blue markings, supply hose black).
  - After installing fuel gauge sender, check that the supply, return and breather lines are still clipped onto the fuel tank.

Additional fuel tank, assembly

1. **Additional fuel tank**
   - Support with engine and transmission jack V.A.G 1383 A when removing
2. **Locking ring**
   - Tighten to 145 Nm (107 ft-lb)
   - Check for secure seating
   - Remove and install with spanner 3087
3. **Supply pipe**
   - To fuel filler neck, see **Fuel filler neck, assembly**, item **-19-**
   - Ensure seated tightly
   - Press together at front to release
4. **Fuel delivery unit**
   - Note installation position on fuel tank, see **Fig. 1**
   - With Transfer Fuel Pump (FP) -G23-, additional fuel tank
   - With Auxiliary Tank Fuel Level Sensor -G292-
   - See **Auxiliary Tank Fuel Level Sensor -G292-, removing and installing**
   - 6-pin connector
5. **Sealing ring**
   - Replace if damaged
6. **Sleeve**
7. **Connecting hose**
   - To connection on fuel tank, see **Fuel tank with attachments, components**, item **-19-**
   - Check for secure seating
8. **Vent pipe**
   - To connection at fuel filler neck, see **Fuel filler neck, assembly**, item **-20-**
   - Check for secure seating
   - Secure with spring-type clips
9. **23 Nm (17 ft-lb)**

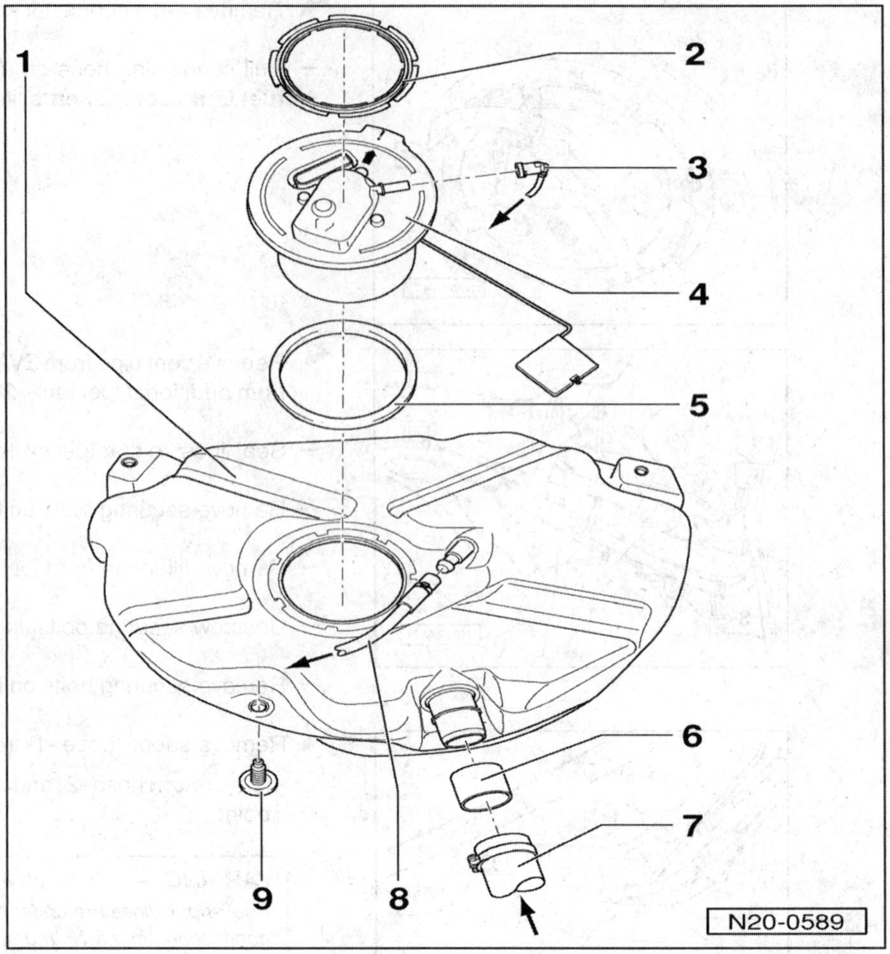

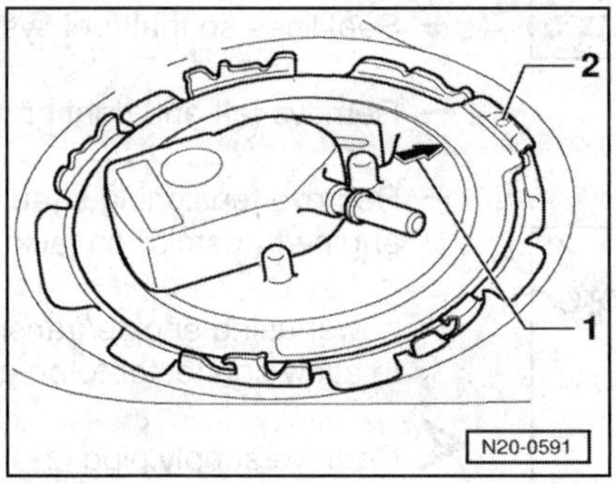

Fig. 1 Installation position of fuel delivery unit flange

- Mark on flange -1- must align with mark on fuel tank -2-.

Fuel filler neck, assembly

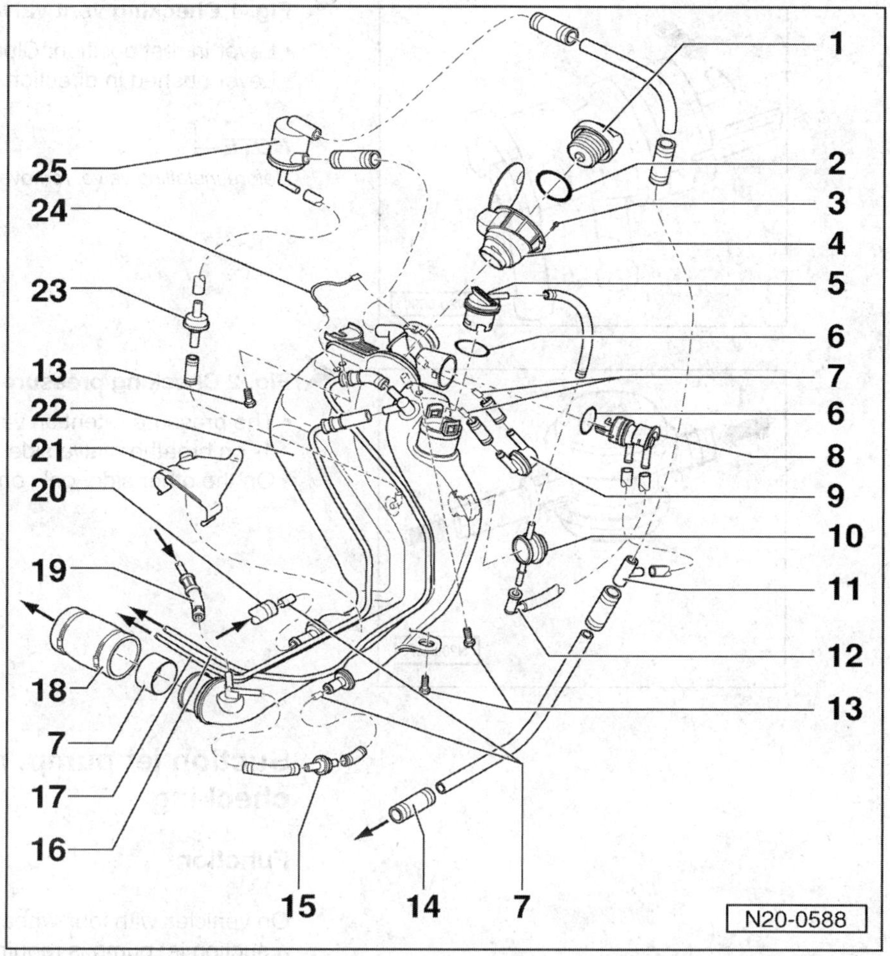

1. **Sealing cap**
2. **Seal**
   - Replace if damaged
3. **Securing screw**
4. **Tank flap unit**
   - With rubber cup
5. **Gravity valve**
   - To remove valve unclip upward out of support
   - Check valve for through-flow valve vertical: open, valve tilted 45 °: closed
6. **O-ring**
   - Replace if damaged
7. **Sleeve**
8. **Breather valve**
   - To remove valve unclip sideways out of support
   - Before installing remove sealing cap item **-1-**
   - Checking, see **Fig. 1**
9. **Combi-valve**
   - For tank filling ventilation
   - Note installation position: white side of valve points in direction of sealing cap item **-1-**
10. **Pressure retention valve**
    - Checking, see **Fig. 2**
11. **Junction piece**
12. **Connection piece**
    - Check securely seated
    - Press together at front to release
13. **15 Nm (11 ft-lb)**
14. **Vent pipe**
    - To junction piece at EVAP canister, see **EVAP system components**, item **-18-**
    - Check for secure seating
    - Secure with spring-type clips
15. **Non-return valve**
    - Note installation position: white side of valve points in direction of sealing cap item **-1-**
16. **Vent pipe**
    - See **Fuel tank with attachments, components**, item **-9-**
    - Check for secure seating
    - Secure with spring-type clips
17. **Vent pipe**
    - See **Fuel tank with attachments, components**, item **-10-**
    - Check for secure seating
    - Secure with spring-type clips
18. **Connecting hose**
    - To connection on fuel tank,see **Fuel tank with attachments, components**, item **-11-**
    - Check for secure seating
19. **Supply pipe**
    - From fuel delivery unit in additional fuel tank, see **Additional fuel tank, components**, item **-3-**
    - Check for secure seating
    - Secure with spring-type clips
20. **Vent pipe**
    - From connection at additional fuel tank, see **Additional fuel tank, components**, item **-8-**
    - Check for secure seating
    - Secure with spring-type clips
21. **Trim strip**
22. **Fuel filler neck**
23. **Non-return valve**
    - Note installation position: gray side of valve points in direction of vent valve item **-25-**
24. **Ground connection**
    - Check for secure seating
25. **Vent valve**
    - For tank filling ventilation

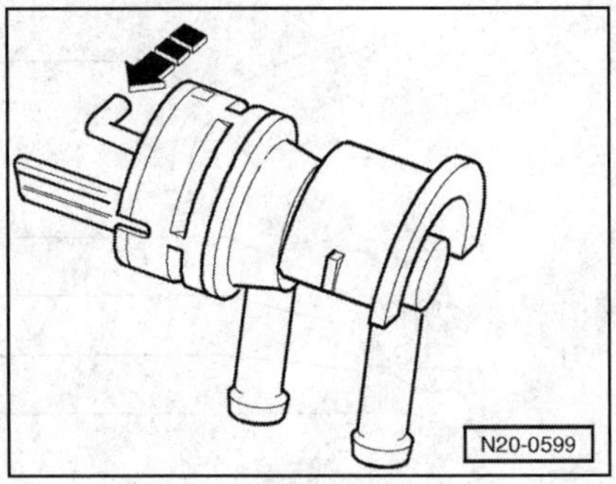

Fig. 1 Checking vent valve

- Lever in rest position: Closed
- Lever pushed in direction of arrow: Open

NOTE —

Before installing valve, remove fuel tank cap.

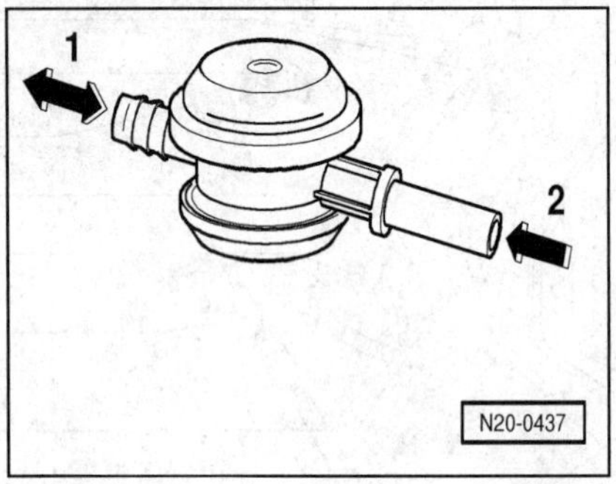

Fig. 2 Checking pressure retention valve

- The pressure retention valve is open for both directions of flow -arrow 1- on breather valve side.
- On the other side, only one direction of flow is open, -arrow 2-.

Suction jet pump, removing, installing and checking

Function

On vehicles with four-wheel drive the shape of the fuel tank dictates that a suction jet pump is required to pump fuel to the fuel delivery unit from the area of Fuel level sensor 2 -G169-, see **Fuel tank with attachments, components,** item **-23-**. A second suction jet pump ensures the delivery unit in the reservoir is supplied with fuel. These suction pumps function according to a principle of physics using the return flow of fuel via a junction with the pressure limiting valve.

NOTE —

A check is only necessary if the engine stops due to lack of fuel.

Special tools and equipment

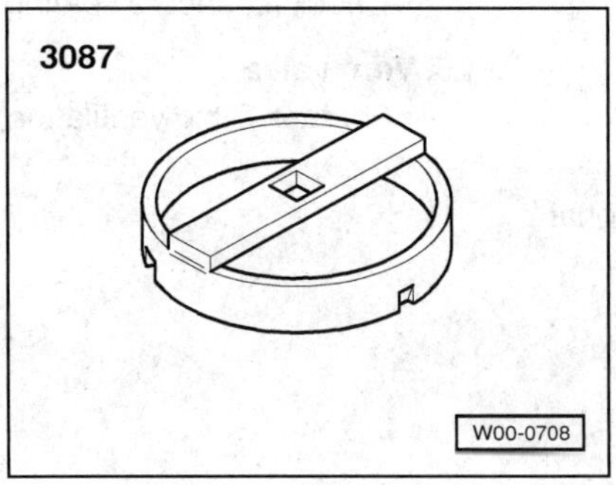

3087 Fuel tank sender spanner

- V.A.G 1332 Torque wrench (40...200 Nm)

Test conditions

- The 3 fuel gauge sensors (-G-, -G169- and -G237-) must be checked using the fault finding program and be OK
- Fuel tank must have been drained

Suction jet pump, removing and installing

NOTE—

• Empty the fuel tank using fuel pump unit VAS 5190 if necessary.

• Observe safety measures before performing repair work..

> ***WARNING—***
> *Fuel supply lines are under pressure! Before removing from hose connection wrap a cloth around the connection. Then release pressure by carefully pulling hose off connection.*

– Switch off ignition and all electrical consumers and remove ignition key.

– With ignition switched off disconnect battery Ground strap.

– Remove fuel delivery unit, see **Fuel delivery unit, removing and installing.**

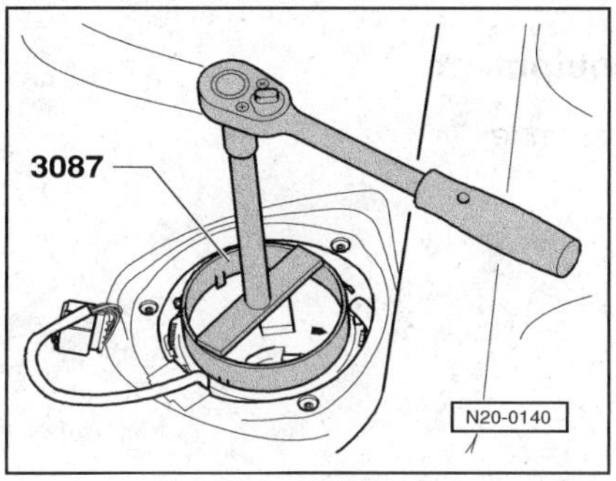

◄ Disconnect 2-pin connector from Fuel level sensor 2 -G169- (left-hand side of vehicle), open sealing ring with key 3087 and carefully lift sensor.

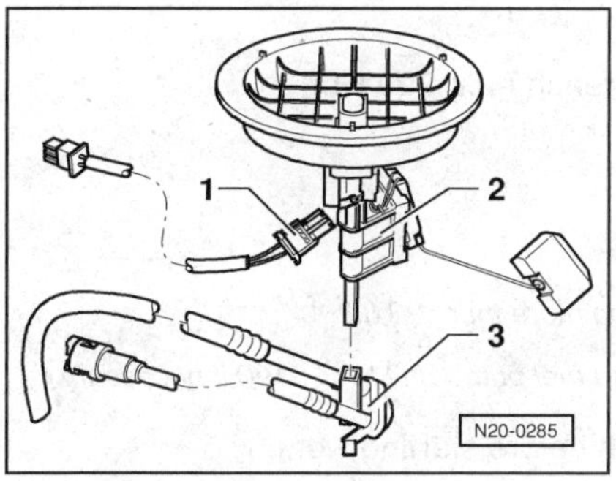

◄ Disconnect connector -1- under fuel level sensor 2 cover -2-.

- Lift fuel level sensor 2 -2- and pull suction jet pump -3- down out of sensor locking device.
- Take out fuel level sensor 2.

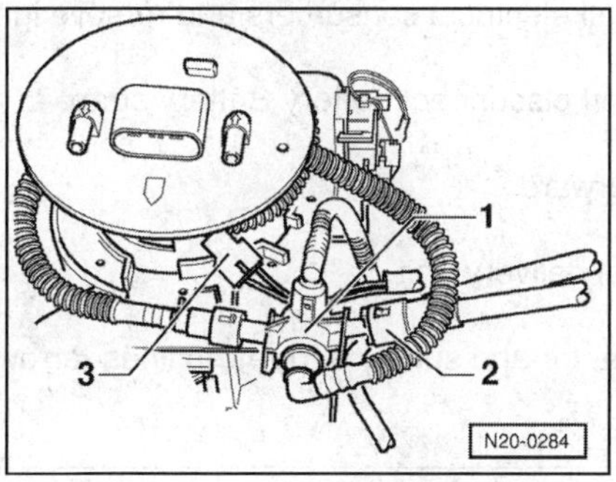

◄ Remove hose/pipe to suction jet pump -2- from junction -1- and disconnect connector -3- from fuel gauge sender -G- to Fuel level sensor 2 -G169-, right-hand side of vehicle (fuel delivery unit).

– Now suction jet pump can be pulled out from fuel level sensor 2 side (left-hand side of vehicle).

– Check that fuel hoses/lines on suction jet pump are installed securely and are not damaged.

– Additionally check suction jet pump for contamination/dirt if necessary.

If no faults are found:

– Install suction jet pump and fuel level sensor 2 again in reverse order.

NOTE—

In other work sequences the return supply line/hose must be pulled off junction. When doing this observe that all O rings must be replaced.

Suction jet pump, checking

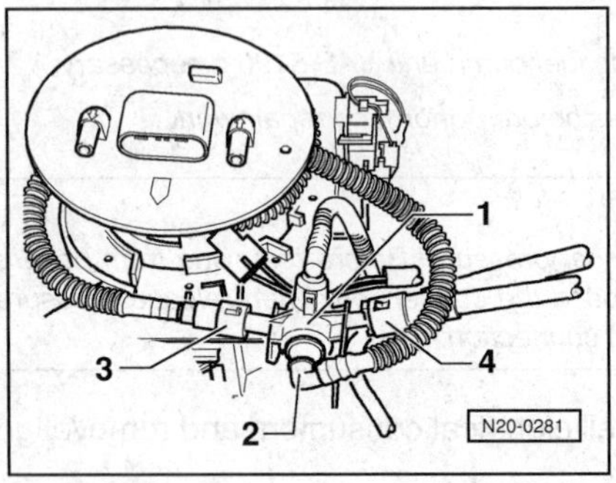

Check junction with pressure relief valve as follows:

◀ Pull off connections for return supply -2-, connections -3- and -4- for suction jet pump on junction with pressure relief valve -1-.

- Unclip junction which is secured on reservoir.
- Use a test hose and blow into connection, from direction of fuel return -2-; there must be throughflow in directions -3- and -4-.
- Hold connections -3- and -4- closed; pressure relief valve must open under slight opposing pressure.

– If no opposing pressure can be determined or pressure relief valve does not open, replace junction with pressure relief valve.

Fuel delivery unit, removing and installing

Special tools and equipment

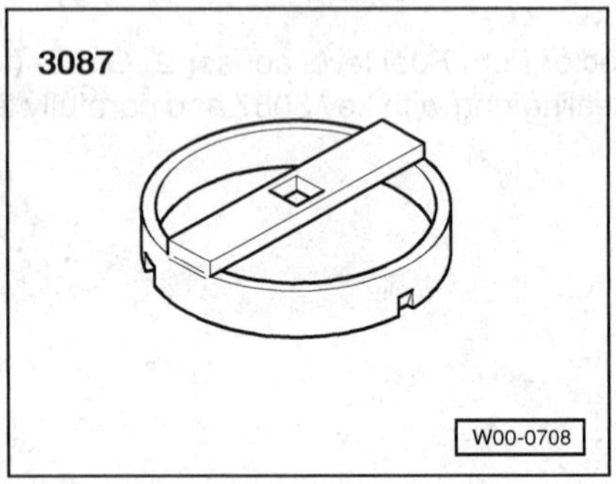

◀ 3087 Fuel tank sender spanner

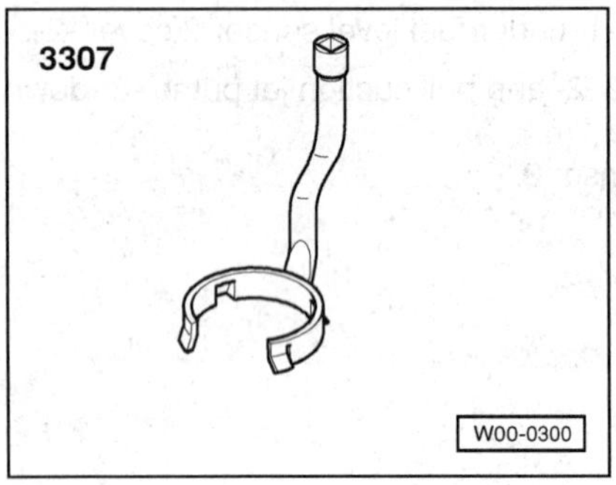

◀ 3307 Fuel pump tool

- V.A.G 1332 Torque wrench (40...200 Nm)

Removing

NOTE —

- *The fuel tank must not be more than 1/3 full.*
- *Empty the fuel tank using fuel pump unit VAS 5190 if necessary.*

– Note safety precautions before starting work.

– Switch off ignition and all electrical consumers and remove ignition key.

– With ignition switched off disconnect battery Battery strap.

– Fold rear seat bench forward.

– Remove cover from fuel delivery unit.

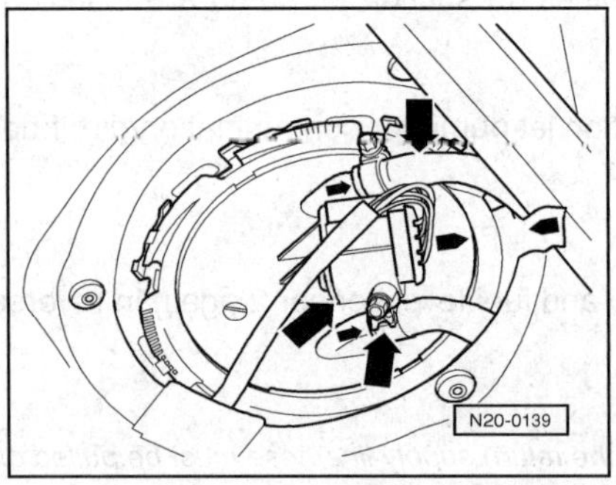

◀ Disconnect 4-pin connector and supply and return lines -arrows- from flange.

> ***WARNING —***
> *Fuel supply lines are under pressure! Before removing from hose connection wrap a cloth around the connection. Then release pressure by carefully pulling hose off connection.*

– Seal lines so that fuel system is not contaminated by dirt etc.

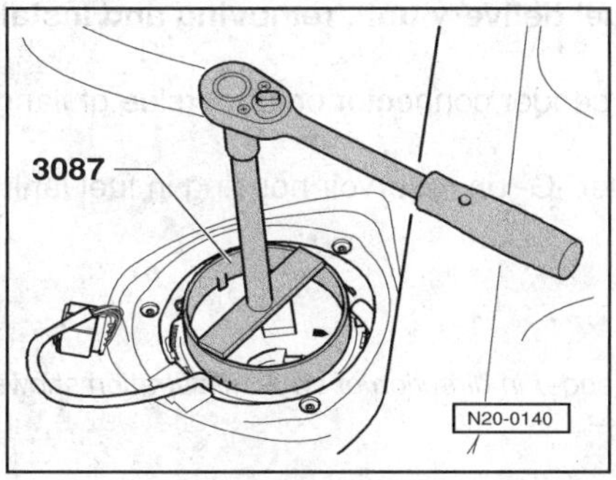

Remove locking ring with key 3087.

– Pull flange and seal out of opening in fuel tank.

– Unclip fuel gauge sender -G- on reservoir housing and remove.

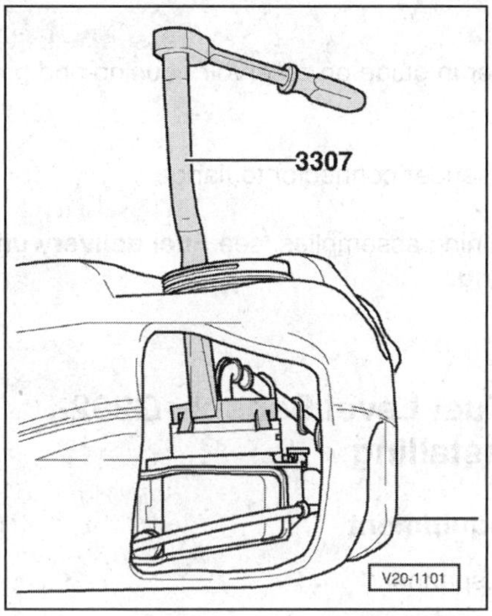

Turn fuel delivery unit approx. 15° counter-clockwise onto stop using spanner 3307 and remove with flange.

NOTE—
If the delivery unit is to be replaced then drain old delivery unit before disposal.

Installing

– Installation of the fuel delivery unit is the reverse of removal, noting the following:
 • Do not bend the fuel sender when installing.
 • Insert the seal for the flange or fuel delivery unit dry into opening of fuel tank.
 • Moisten seal with fuel only when installing the flange or fuel delivery unit.
 • Ensure that fuel hoses are seated securely.
 • Do not interchange supply and return hoses.
 • Secure fuel hoses with spring-type clamps.
 • Note the installation position of the fuel delivery unit flange: Mark on flange -1- must align with mark -2- on fuel tank.
 • After installing the fuel delivery unit flange, check that the supply, return and breather lines are still clipped onto fuel tank.

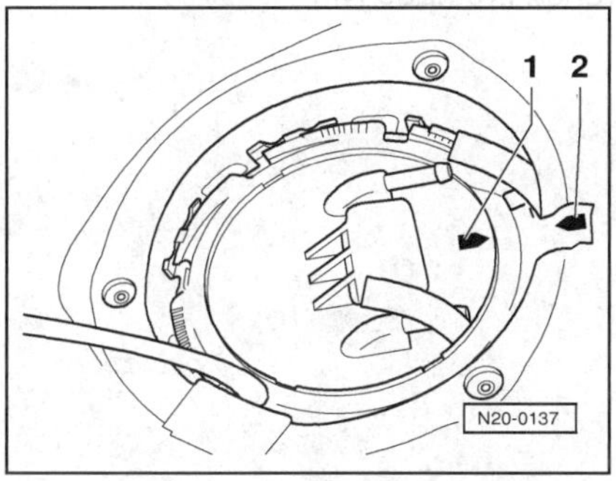

Sender for fuel gauge -G-, removing and installing

Special tools and equipment

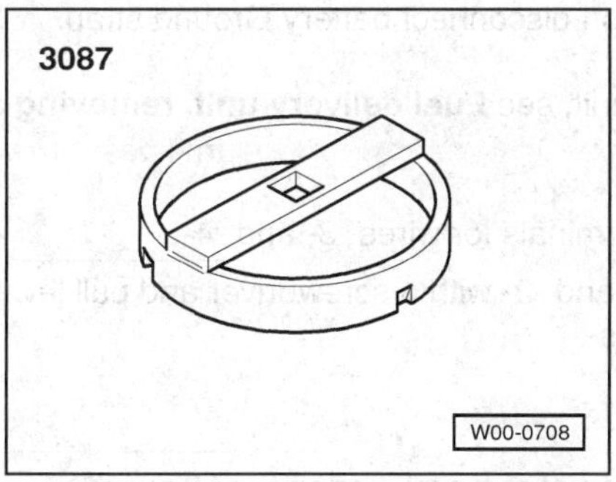

3087 Fuel tank sender spanner

• V.A.G 1332 Torque wrench (40...200 Nm)

Removing

NOTE—
• The fuel tank must not be more than 1/3 full.

• Empty the fuel tank using fuel pump unit VAS 5190 if necessary.

– Note safety precautions before starting work.

– Switch off ignition and all electrical consumers and remove ignition key.

– With ignition switched off disconnect battery Ground strap.

– Remove flange, see **Fuel delivery unit, removing and installing**.

– Disconnect fuel gauge sender connector on underside of flange.

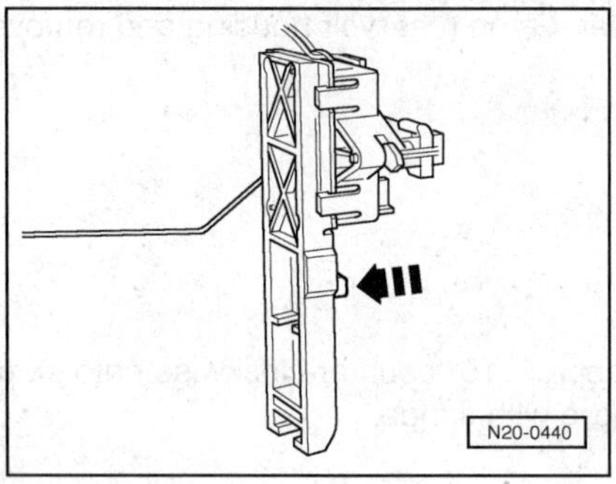

Unclip fuel gauge sender -G- on reservoir housing in fuel tank and remove.

NOTE—
Press locking device on sender in direction of arrow (illustration shows sender removed).

Installing

– Insert fuel gauge sender in guide on reservoir housing and press down until it engages.

– Reconnect fuel gauge sender connector to flange.

– Install flange and remaining assemblies, see **Fuel delivery unit, removing and installing**.

Auxiliary Tank Fuel Level Sensor -G292-, removing and installing

Special tools and equipment

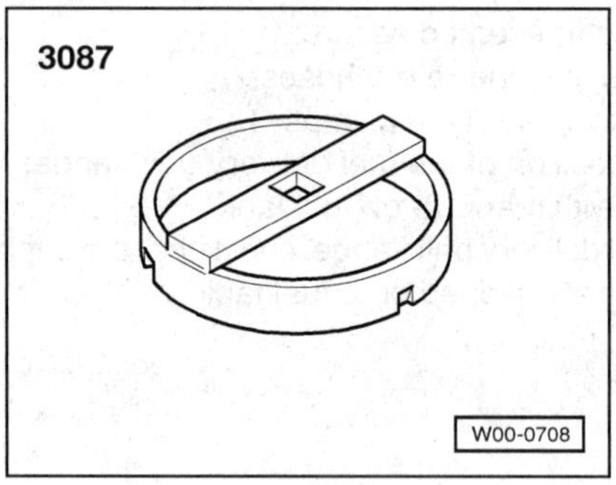

3087 Fuel tank sender spanner

- V.A.G 1332 Torque wrench (40...200 Nm)

Removing

– Note safety precautions before starting work.

– Switch off ignition and all electrical consumers and remove ignition key.

– With ignition switched off disconnect battery Ground strap.

– Remove fuel delivery unit, see **Fuel delivery unit, removing and installing**.

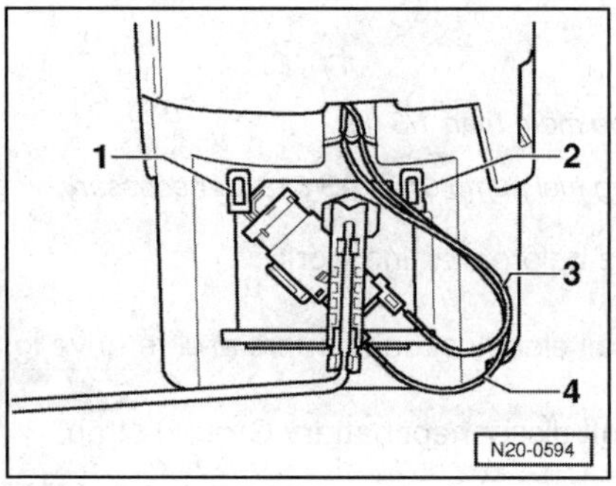

Release and pull off terminals for wires -3- and -4-.

- Lift retaining tabs -1- and -2- with a screwdriver and pull level sensor out downward -arrows-.

Installing

– Installation is the reverse of removal, noting the following:

- Insert level sensor in guides on fuel delivery unit and push up until it engages.

Crash fuel shut-off

Function

Vehicles with an airbag are equipped with a crash fuel shut-off system. It reduces the danger of a fire in a crash as the fuel pump is switched off via the fuel pump relay.

When opening the fuel system observe safety precautions.

Fuel pump, checking

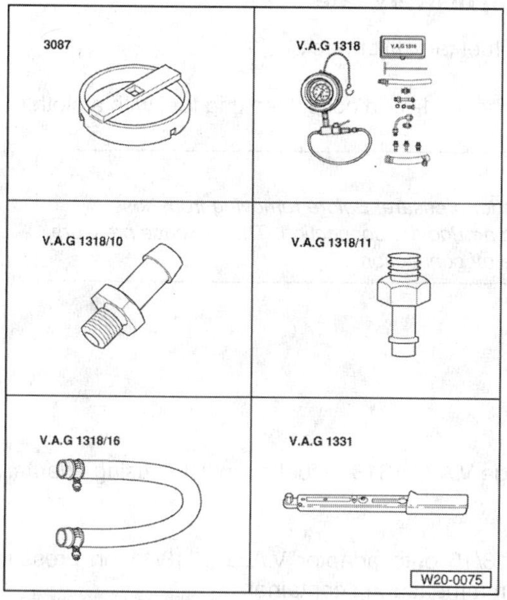

Special tools and equipment

- 3087 Fuel tank sender spanner
- V.A.G 1318 Pressure gauge
- V.A.G 1318/10 Adapter
- V.A.G 1318/11 Adapter
- V.A.G 1318/16 Adapter
- V.A.G 1331 Torque wrench (5...50 Nm)

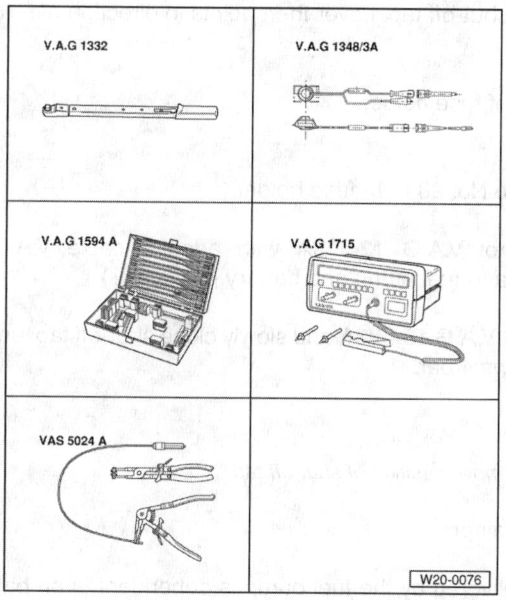

- V.A.G 1332 Torque wrench (40...200 Nm)
- V.A.G 1348/3A Remote control with V.A.G 1348/3-2 Adapter cable
- V.A.G 1594 A Adapter set
- V.A.G 1715 Multimeter
- VAS 5024 Installation tool for spring-type clips

Not illustrated:

- Measuring container
- Wiring diagram

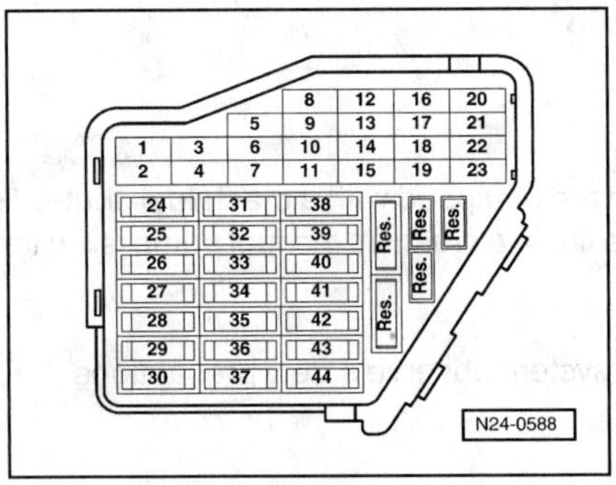

Fuel pump, checking conditions

◄ Fuse 28 must be OK

- Battery voltage must be at least 11.5 V
- All electrical consumers, e.g. lights and rear window defroster must be switched off.
- If the vehicle is equipped with an air conditioning system, this must be switched off.
- Voltage supply OK

Fuel pump, checking delivery rate

– Remove filler cap from fuel tank filler neck.

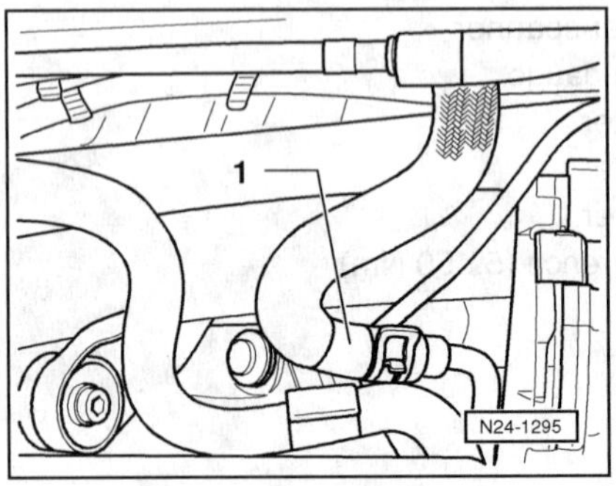

◄ Disconnect supply line/hose -1- and collect leaking fuel with a cloth.

> ***WARNING—***
> *Fuel supply lines are under pressure! Before removing from hose connection wrap a cloth around the connection. Then release pressure by carefully pulling hose off connection.*

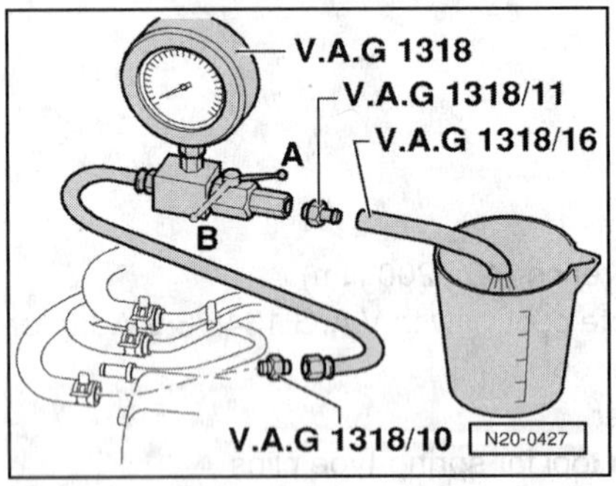

◄ Connect pressure gauge V.A.G 1318 to fuel supply line using adapter 1318/10.

– Install hose V.A.G 1318/16 onto adapter V.A.G 1318/11 on pressure gauge and hold hose in a measuring container.

– Open pressure gauge shut-off tap. Lever then points in direction of flow -A-.

– Remove cover in front of fuse holder.

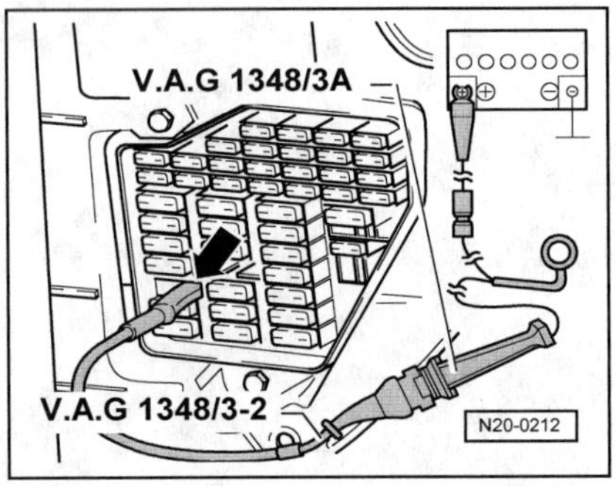

◄ Remove fuel pump fuse No. 28 from fuse holder.

– Connect remote control V.A.G 1348/3A with adapter cable V.A.G 1348/3-2 to contact 28a to fuel pump and battery positive (+).

– Operate remote control V.A.G 1348/3A and slowly close shut-off tap, until pressure gauge shows 4 bar.

NOTE—
From this point on do not move position of shut-off tap.

– Empty measuring container.

– The quantity of fuel delivered by the fuel pump is dependent upon battery voltage. Therefore,

- Connect multimeter with adapter cables from V.A.G 1594 A to vehicle battery.
- Operate remote control for 30 seconds and measure battery voltage

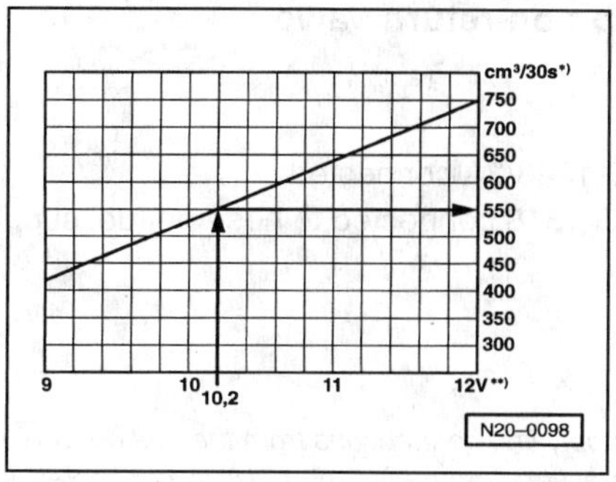

Compare quantity of fuel delivered with specification.

*) Minimum delivery cm^3/30 seconds

**) Voltage at fuel pump with engine stopped and pump running (approx. 2 volts less than battery voltage).

Example:

During the test a voltage of 12.2 volts is measured at the battery. As the voltage at the pump is approx. 2 volts less than the battery voltage then this will equate to a minimum delivery of 550 cm^3/30 s.

– If the minimum delivery rate is not attained:

• Check fuel lines for possible restrictions (kinks) or blockages.

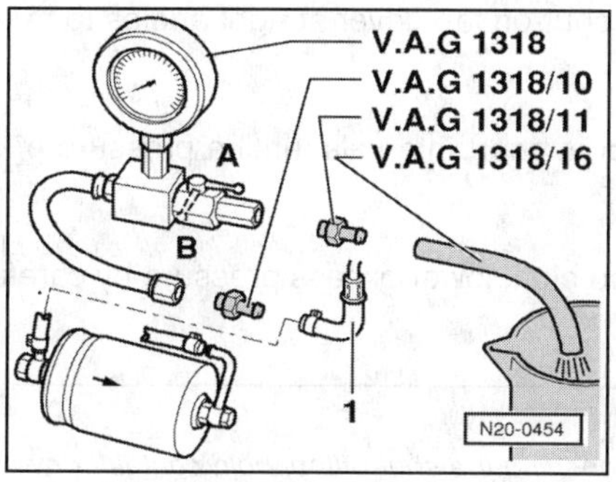

Remove supply hose -1- from inlet to fuel filter.

– Connect pressure gauge V.A.G 1318 to hose with adapter 1318/10.

– Repeat delivery rate check.

– If the minimum delivery rate is now attained:

• Replace fuel filter.

– If the minimum delivery rate is again not attained:

• Remove fuel delivery unit, see **Fuel delivery unit, removing and installing**.

– Only when up to now no fault has been detected:

• Fuel pump faulty. Replace fuel delivery unit, see **Fuel delivery unit, removing and installing**.

If the delivery rate has been attained but nevertheless you suspect a fuel supply system fault (e.g. intermittent failure of fuel supply system):

– Check current draw of fuel pump as follows:

– Reconnect all disconnected fuel lines.

– Fold rear seat bench forward.

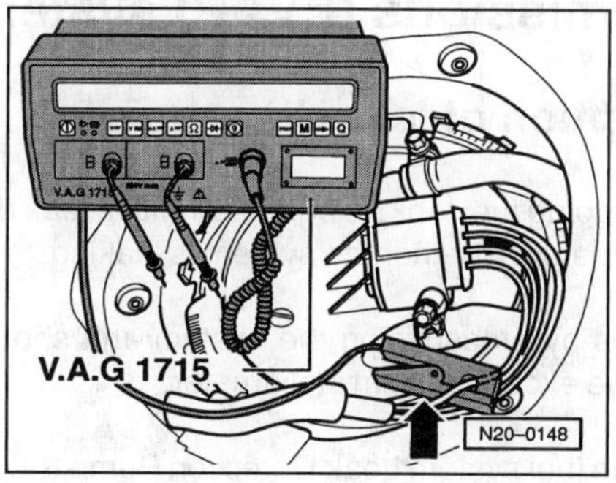

Connect multimeter V.A.G 1715 with current clamp on green/yellow wire in wiring harness.

– Start engine and run at idling speed.

– Measure current draw of fuel pump.

• Specification: max. 8 amps.

NOTE —

If the fuel system malfunction is intermittent, you can perform the check during a test drive with the help of a 2nd person.

– If the current draw is exceeded:

• Fuel pump faulty. Replace fuel delivery unit, see **Fuel delivery unit, removing and installing**.

Checking fuel pump non-return valve

Check conditions

• Remote control V.A.G 1348/3A connected

• Pressure gauge V.A.G 1318 connected to hose on fuel supply line.

Test sequence

NOTE—

With this check the fuel supply line connections from the fuel delivery unit to the point at which the pressure gauge V.A.G 1318 is connected will be checked for leaks at the same time.

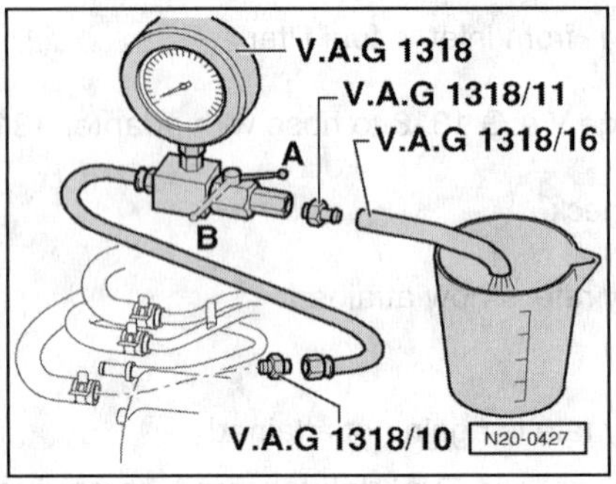

< Close pressure gauge shut-off tap (lever at right angles to throughflow direction - position -B-).

– Operate remote control at short intervals, until a pressure of approx. 4 bar (58 psi) has built-up.

– If pressure builds up too high, lower excess pressure by carefully opening shut-off tap.

WARNING—

Danger of spray when opening the shut-offtap, hold container over the free connection of the pressure gauge.

– Watch pressure drop on gauge. After 10 minutes pressure must not drop below 2.2 bar (31 psi) (engine cold) or 3.0 bar (43 psi) (engine warm).

– If the pressure drops further:

• Check line connections for leaks.

– If no fault is detected in the wiring:

• Fuel pump faulty. Replace fuel delivery unit, see **Fuel delivery unit, removing and installing**.

Evaporative emissions (EVAP) system

Function description of leak diagnosis

The EVAP system (including fuel tank) is equipped with a leak diagnosis. The leak diagnosis will detect whether the system is leaking.

The diagnosis operates by pressurizing the system and should detect leaks where the damage exceeds 1 mm in diameter.

During the diagnosis the fuel system Leak Detection Pump (LDP) -V144- builds up a pressure of approx. 30 mbar in the EVAP system. The pump will switch off when the pressure is attained. When the pressure falls to below a certain figure, the pump will switch on again. The self-diagnosis monitors the time between the switched on periods and if it is too short it will register a Diagnostic Trouble Code (DTC) in the DTC memory.

NOTE—

• Hose connections are secured with either spring-type or clamp-type clips.

• Always replace clamp-type clips with spring-type clips.

EVAP system components

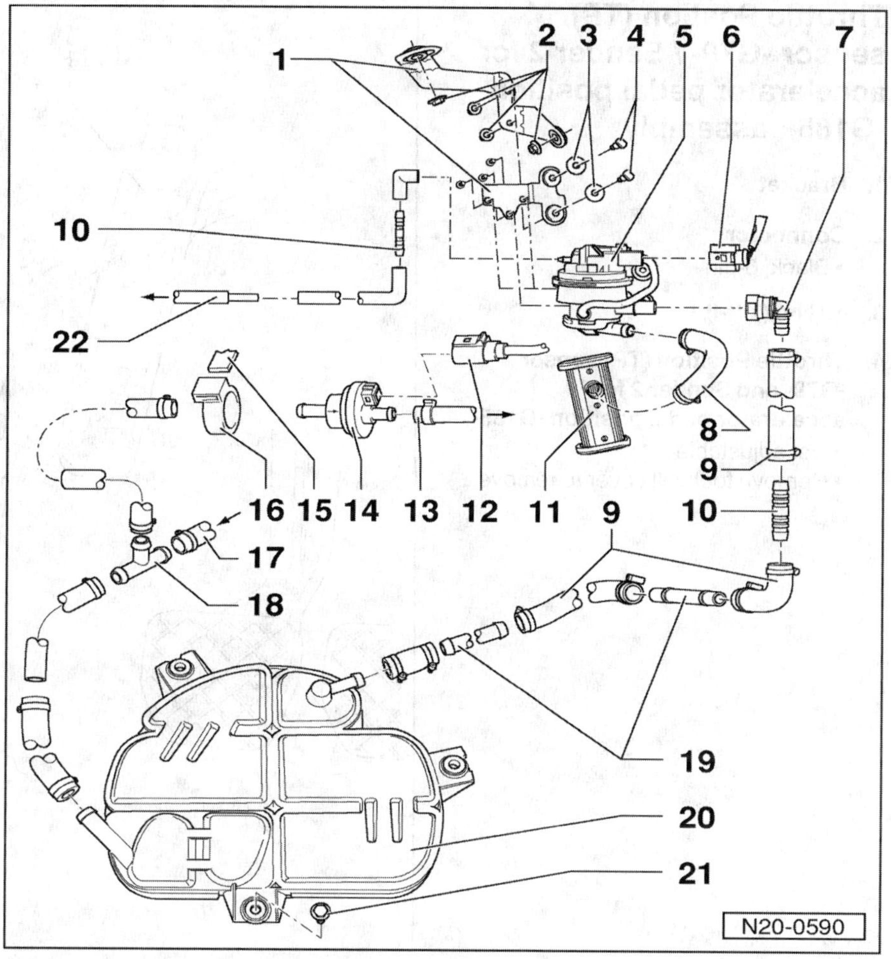

1. **Bracket**
2. **8 Nm (71 in-lb)**
3. **Washer**
4. **Securing screw**
5. **Leak Detection Pump (LDP) -V144-**
   - In left rear wheel housing behind inner fender liner
6. **Connector**
   - Black, 3-pin
7. **Connection piece**
   - Check for secure seating
   - Press together at front to release
8. **Connecting hose**
   - Suction side
   - Check for secure seating
   - Secure with spring-type clips
9. **Connecting hose**
   - Pressure side
   - Check for secure seating
   - Secure with spring-type clips
10. **Connector**
11. **Air cleaner for LDP**
12. **Connector**
   - Black, 2-pin
13. **Vacuum pipe**
   - Check for secure seating
   - Secure with spring-type clips
   - On intake manifold upper part, see **Repair Group 24, Intake manifold, assembly**
14. **EVAP canister purge regulator valve -N80-**
   - Installation position: Arrow points in direction of flow
15. **Retainer**
   - At upper part of air filter, see **Repair Group 24, Air cleaner, assembly**, item **-14-**
16. **Retaining ring**
17. **Vent pipe**
   - From junction piece at fuel filler neck, see **Fuel filler neck, assembly**, item **-11-**
   - Check for secure seating
   - Secure with spring-type clips
18. **Junction piece**
19. **Connecting pipe**
20. **EVAP canister**
   - In spare wheel well on vehicle floor
21. **15 Nm (11 ft-lb)**
22. **Vacuum pipe**
   - Check for secure seating
   - On intake manifold upper part, see **Repair Group 24, Intake manifold, assembly**

Throttle Postion (TP) sensor -G79- / Sender 2 for accelerator pedal position -G185-, assembly

1. **Bracket**
2. **Connector**
   - Black, 6-pin
3. **10 Nm (7 ft-lb)**
4. **Throttle Position (TP) sensor -G79- and Sender 2 for accelerator pedal position -G185-**
   - Not adjustable
   - Remove footwell cover to remove

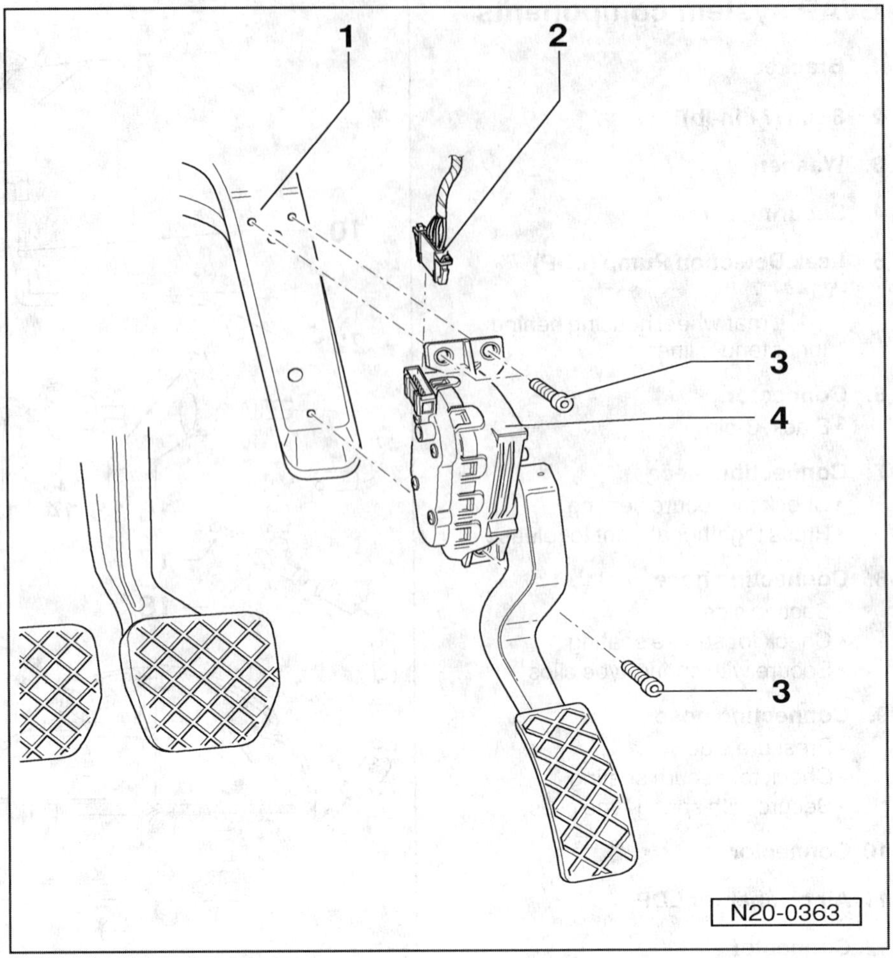

26 Exhaust System and Emission Controls

Repair Group

E 26

Exhaust system

WARNING —

When doing any repair work, especially in the engine compartment, pay attention to the following due to the cramped conditions:

- *Route all the various lines (e.g. for fuel, hydraulics, EVAP system, coolant, refrigerant, brake fluid and vacuum lines and hoses) and electrical wiring so that the original positions are restored.*
- *Ensure sufficient clearance to all moving or hot components.*

NOTE —

- *After working on the exhaust system ensure that the system is not under stress, and that it has sufficient clearance from the bodywork. If necessary, loosen double and single clamps and align muffler and exhaust pipe so that sufficient clearance is maintained to the bodywork and the support rings are evenly loaded.*
- *Replace self-locking nuts.*

Exhaust manifold/catalyst module with front exhaust pipe and ancillaries

NOTE —

- *Cylinder bank 1 is located on the right-hand side when looking toward the front.*
- *Cylinder bank 2 is located on left-hand side when looking toward the front.*

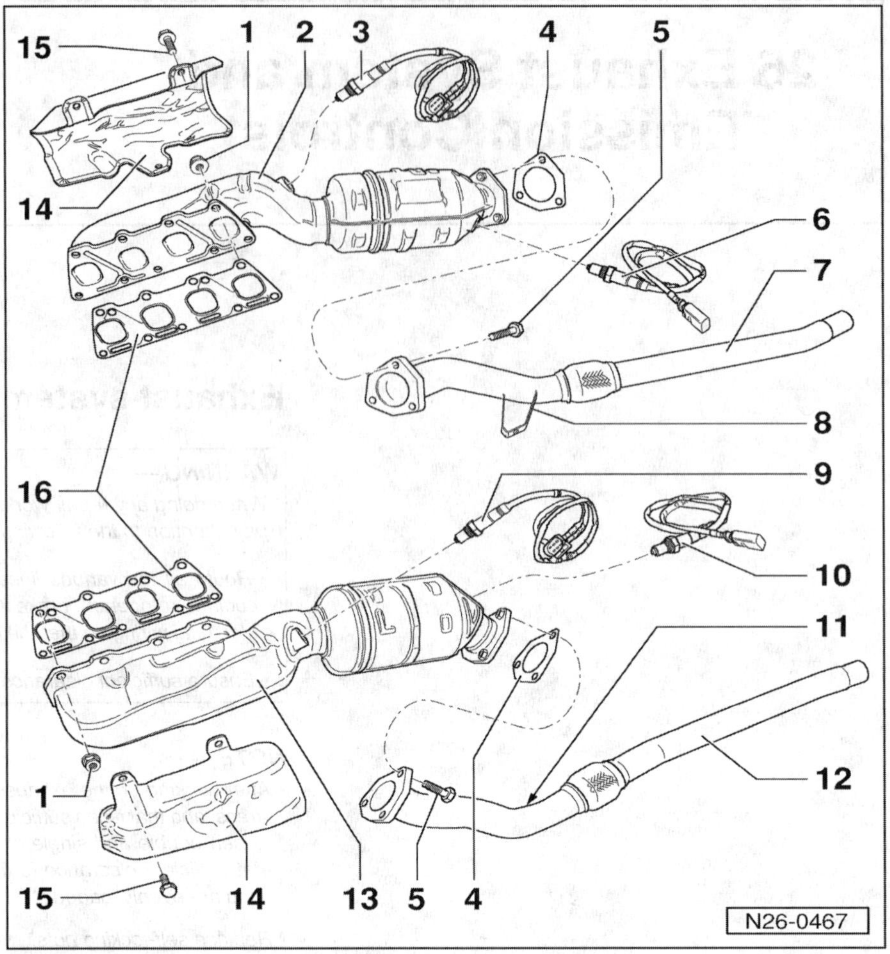

1. **25 Nm (18 ft-lb)**
   - Replace
   - Coat threads with G 052 112 A3
2. **Right exhaust manifold/catalyst module**
   - Can only be replaced when engine is removed from engine compartment, see **Repair Group 10, Engine removing and installing**
3. **Heated Oxygen sensor (H02S) -G39-**
   - Tighten to 50 Nm (37 ft-lb)
   - For cylinder bank 1
   - Grease only the threads with G 052 112 A3; G 052 112 A3 must not get into slots on sensor body
   - Can only be replaced when engine is removed from engine compartment, see **Repair Group 10, Engine removing and installing**
4. **Gasket**
   - Replace
5. **30 Nm (22 ft-lb)**
   - Replace
   - Coat threads with G 052 112 A3
6. **Oxygen sensor (02S) Behind Three Way Catalytic Converter (TWC) -G130-, 50 Nm**
   - For cylinder bank 1
   - Grease only the threads with G 052 112 A3; G 052 112 A3 must not get into slots on sensor body
   - Can only be replaced when engine is removed from engine compartment, see **Repair Group 10, Engine removing and installing**
7. **Front right exhaust pipe**
8. **Mounting**
   - Individual components see **Fig. 1** and **3**
9. **Heated Oxygen sensor (H02S) 2 -G108-**
   - Tighten to 50 Nm (37 ft-lb)
   - For cylinder bank 2
   - Grease only the threads with G 052 112 A3; G 052 112 A3 must not get into slots on sensor body
   - Can only be replaced when engine is removed from engine compartment, see **Repair Group 10, Engine removing and installing**
10. **Oxygen Sensor (O2S) 2 Behind Three Way Catalytic Converter -G131-**
    - Tighten to 50 Nm (37 ft-lb)
    - For cylinder bank 2
    - Grease only the threads with G 052 112 A3; G 052 112 A3 must not get into slots on sensor body
    - Can only be replaced when engine is removed from engine compartment, see **Repair Group 10, Engine removing and installing**
11. **Mounting**
    - Individual components see **Fig. 2** and **4**
12. **Front left exhaust pipe**
13. **Left exhaust manifold/catalyst module**
    - Can only be replaced when engine is removed from engine compartment, see **Repair Group 10, Engine removing and installing**
14. **Upper heat shield**
15. **10 Nm (7 ft-lb)**
16. **Gasket**
    - Replace

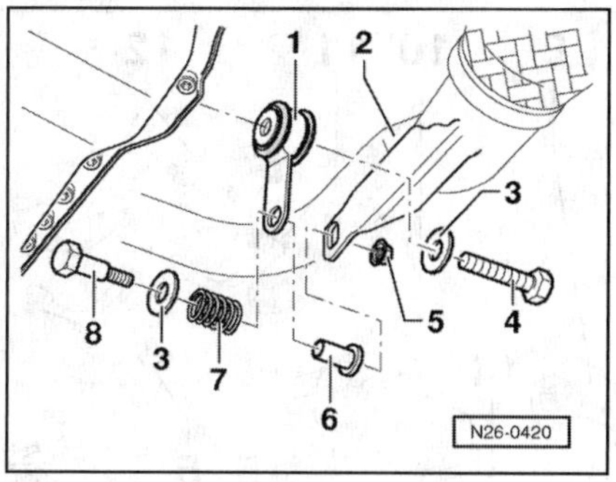

Fig. 1 Individual components for right mounting, vehicles with automatic transmission

1. Tongue with buffer
2. Exhaust pipe
3. Washer
4. Hex bolt, 25 Nm (18 ft-lb)
5. Self-locking nut, 25 Nm (18 ft-lb)
6. Spacer sleeve
7. Spring
8. Hex bolt

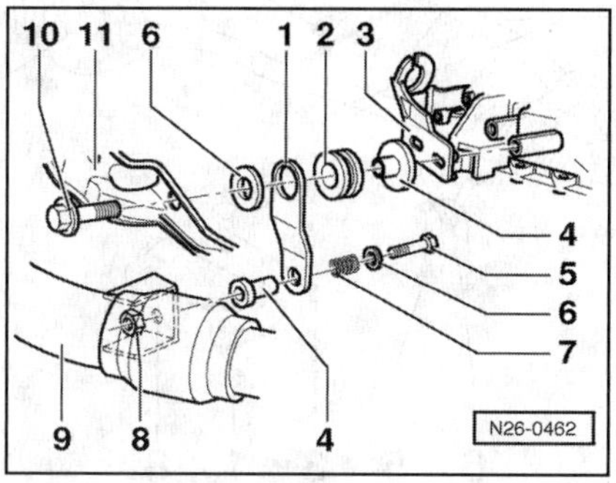

Fig. 2 Individual components for left mounting, vehicles with automatic transmission

1. Tongue
2. Buffer
3. Bracket on transmission
4. Spacer sleeve
5. Hex bolt, 25 Nm (18 ft-lb)
6. Washer
7. Spring
8. Self-locking nut, 25 Nm (18 ft-lb)
9. Exhaust pipe

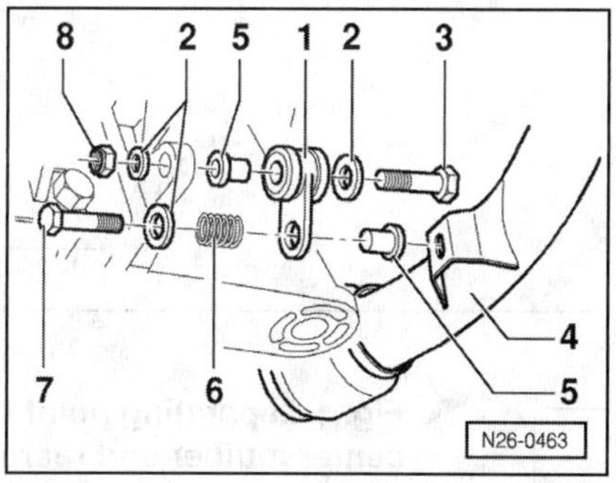

Fig. 3 Individual components for right mounting, vehicles with manual transmission

1. Tongue with buffer
2. Washer
3. Hex bolt, 25 Nm (18 ft-lb)
4. Exhaust pipe
5. Spacer sleeve
6. Spring
7. Hex bolt
8. Self-locking nut, 25 Nm (18 ft-lb)

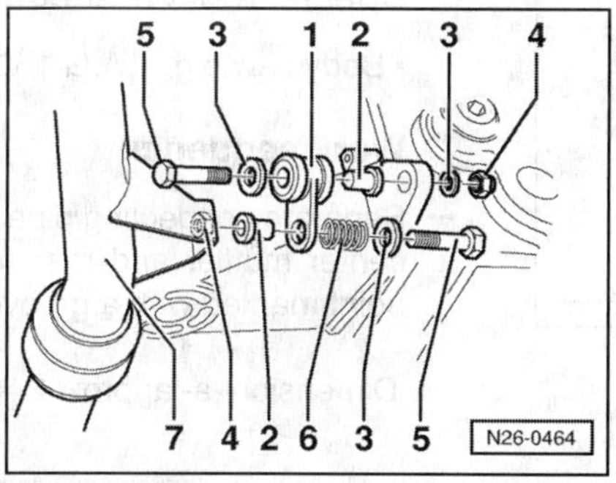

Fig. 4 Individual components for left mounting, vehicles with manual transmission

1. Tongue with buffer
2. Spacer sleeve
3. Washer
4. Hex bolt, 25 Nm (18 ft-lb)
5. Self-locking nut, 25 Nm (18 ft-lb)
6. Spring
7. Exhaust pipe

Exhaust system, components

1. **40 Nm (30 ft-lb)**
2. **From front exhaust pipe**
3. **Carriage bolt**
4. **Double clamp**
   - Installation position, see **Fig. 2**
5. **25 Nm (18 ft-lb)**
6. **Mounting**
   - With retaining ring
7. **Center muffler**
8. **Repair double clamp**
   - Installation position, see **Fig. 3**
9. **Separating point**
   - As standard, center and rear muffers are installed as a single component. In cases of repair center and rear muffler are supplied individually and with a double clamp for connecting
   - Cut through connecting pipe with body saw e.g. V.A.G 1523 at right angles at separating point, see **Fig. 1**
10. **Right rear muffler**
11. **Mounting**
12. **Bracket**
13. **Left rear muffler**
14. **25 Nm (18 ft-lb)**
15. **Connecting piece**

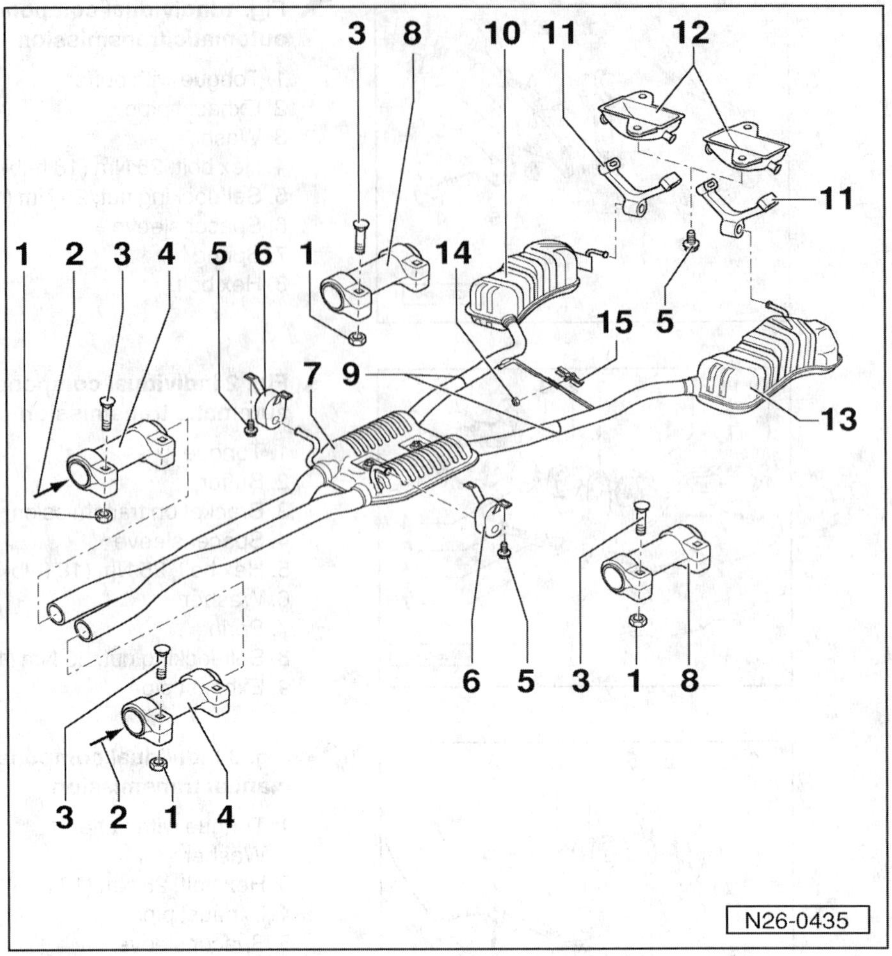

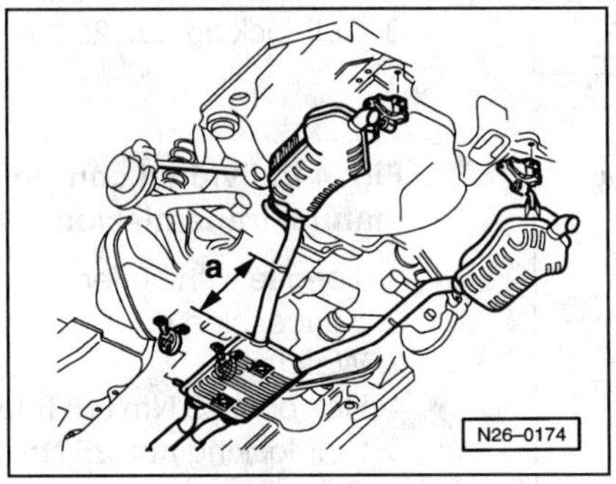

Fig. 1 Separating point between center muffler and rear muffler

Special tools and equipment

- Body saw e.g. V.A.G 1523

Work sequence

– Separate connecting pipe between center muffler and rear muffler at point marked with a groove:

Dimension -a- approx. 244 mm.

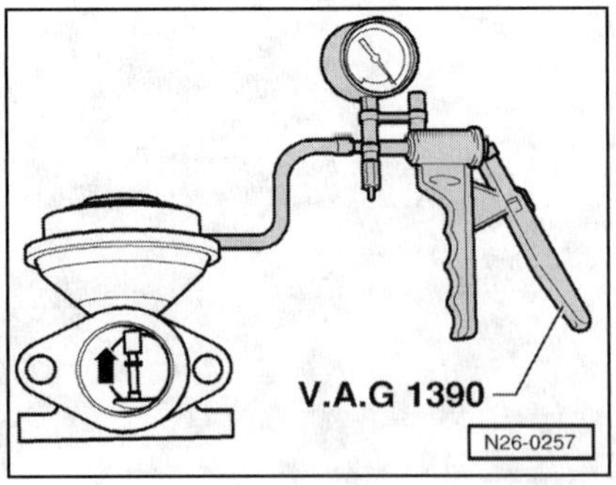

Fig. 2 Installation position of double clamps

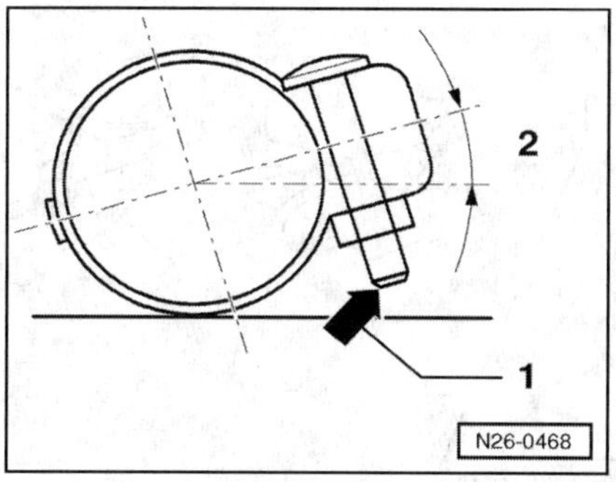

Fig. 3 Installation position of repair double clamps (viewed looking forward)

1. End of bolt must not project beyond lower edge of clamp
2. Angle 10°...15°

Secondary air system

> ***WARNING —***
> *When doing any repair work, especially in the engine compartment, pay attention to the following due to the cramped conditions:*
>
> - *Route all the various lines (e.g. for fuel, hydraulics, EVAP system, coolant, refrigerant, brake fluid and vacuum lines and hoses) and electrical wiring so that the original positions are restored.*
> - *Ensure sufficient clearance to all moving or hot components.*

Function

The secondary air system blows air in behind the exhaust valve for 100 seconds during the cold start (+5 °C...+33 °C coolant temperature). This produces an oxygen rich exhaust gas, causes afterburning and reduces the heating-up phase of the catalyst. Activation occurs from the Motronic engine control module -J220- via the secondary air injection pump relay -J299- to the secondary air injection solenoid valve -N112-, change-over valve and combi-valve. Additionally after each subsequent engine start (up to max. 96 °C engine temperature) the secondary air system will switch in for 10 seconds during idling and will be checked by the self-diagnosis. The oxygen sensor regulation must then be active.

Secondary air system components

NOTE —
- *Cylinder bank 1 is located on the right-hand side when looking toward the front.*
- *Cylinder bank 2 is located on left-hand side when looking toward the front.*

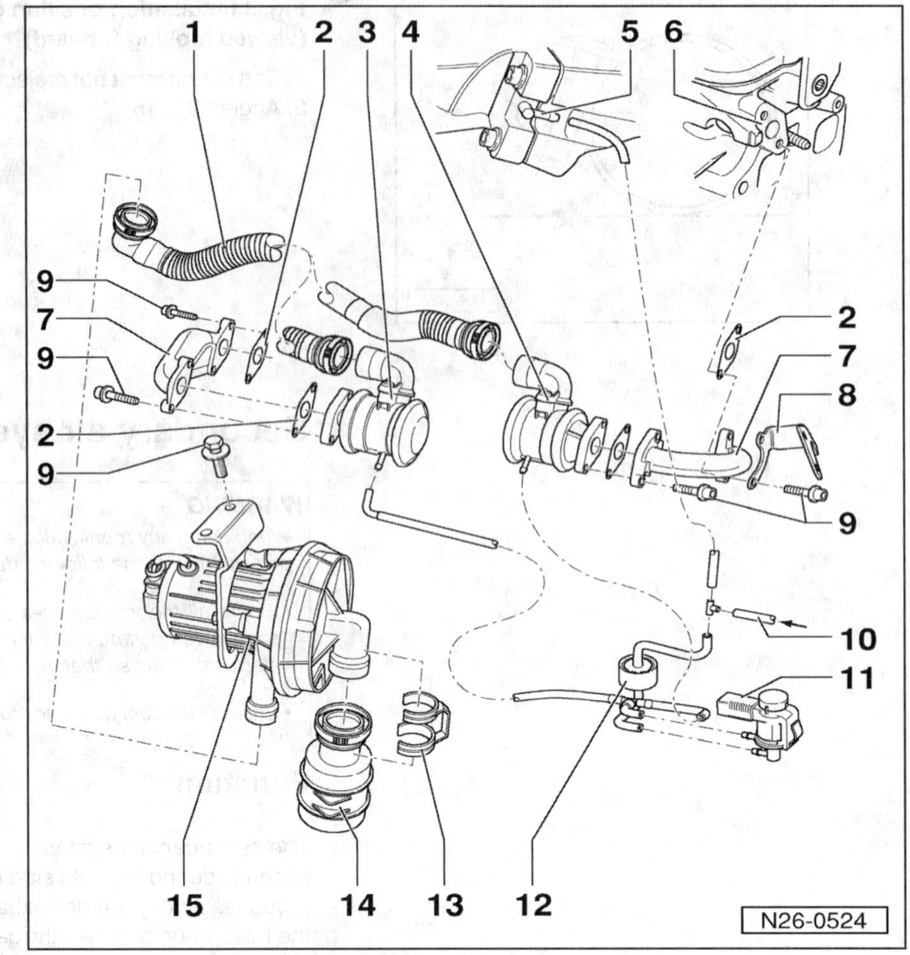

1. **Pressure hose**
   - Ensure seated tightly
   - Press together at front to release
2. **Gasket**
   - Replace
3. **Combi-valve**
   - Cylinder bank 1
4. **Combi-valve**
   - Cylinder bank 2
5. **Vacuum connection**
   - Left on intake pipe
6. **Flange**
   - On cylinder head
7. **Connection**
8. **Bracket**
   - For oil separator, see **Repair Group 13, Crankcase ventilation system, components**, item -9-
9. **10 Nm (7 ft-lb)**
10. **Vacuum hose**
    - From Left Electro-Hydraulic Engine Mount Solenoid Valve -N144- / Right Electro-Hydraulic Engine Mount Solenoid Valve -N145-
11. **Secondary Air Injection (AIR) solenoid valve -N112-**
12. **Non-return valve**
    - Note installation position
13. **Securing clip**
14. **Air cleaner**
15. **Secondary Air Injection (AIR) pump motor -V101-**

1.8 Liter Fuel Injection and Ignition (code: AEB)

Section

F

24 Multiport Fuel Injection (MFI)

Page

28 Ignition

Page

24 Multiport Fuel Injection (MFI)

Repair Group

F 24

Fuel Injection System

NOTE—

- *The Engine Control Module (ECM) is equipped with On Board Diagnostic. Before carrying out repairs and Troubleshooting; DTC memory must be checked. Also check the vacuum hoses and connections for leaks or unmeasured air.*
- *Fuel hoses in engine compartment must only be secured with spring clips. The use of clamp or screw clips is not permissible.*
- *VAG 1921 pliers are recommended for installing spring clips.*
- *Disconnecting and connecting the battery must only be done with the ignition switched off, otherwise the Engine Control Module could be damaged.*
- *Components marked with * are checked via the On Board Diagnostic system, see* **Maintenance**.
- *Components marked with ** can be checked via the Output Diagnostic Test Mode (DTM), see* **Maintenance**.
- *For trouble-free operation of the electrical components, 11.5 Volts minimum is required.*
- *Do not use sealants containing silicone. Particles of silicone drawn into the engine, will not be burned can impair Oxygen sensor function.*
- *During some checks it is possible that the Control Module will recognize and store a malfunction. Therefore after completing all checks and repairs DTC memory must be checked and if necessary erased, see* **Maintenance**.

CAUTION—

- *Do not use any gasket sealers or other automotive chemical compounds that contain silicone. Traces of silicone drawn into the engine pass unburned into the exhaust and can damage the oxygen sensor.*
- *Do not use any electrical contact sprays or equivalent near oxygen sensor connections. Proper oxygen sensor function requires reference to outside air via the cable. If any contact spray reaches the oxygen sensor, it can cause damage or improper function.*

EDITOR'S NOTE—

- *Checking components is accomplished using Volkswagen diagnostic scan tool VAG 1551 or equivalent, via Output Diagnostic Test Mode (function 03).*
- *Volkswagen identifies electrical components by a letter and/or a number in the electrical schematics. See* **Wiring Diagrams and Component Locations**. *These unique electrical component identifiers are listed -between dashes-, for example Mass Air Flow Sensor -G70-.*

EDITOR'S NOTE —
Functional testing of most fuel injection and ignition components is done using a VAG 1551, 1552, 5051 or equivalent aftermarket scan tool. In addition, the factory repair information calls for the use of various special tools including electronic test boxes and remote control switches. Many fuel injection components can be tested electrically using a digital multimeter and/or an LED test light. The following pages include many of these basic electrical tests as well as some of the testing that relies on the use of a diagnostic scan tool and specialized diagnostic equipment.

Safety precautions

> *WARNING —*
> *The fuel system is pressurized! Before loosening hose connections or opening the test connection, wrap a cloth around the connection. Then release pressure by carefully pulling off the hose or carefully unscrewing sealing plug.*

To prevent injuries to persons and/or damage to the fuel injection and ignition systems, observe the following precautions:

- Do not touch or disconnect ignition wiring when the engine is running or being turned over via the starter.
- Switch OFF the ignition before connecting or disconnecting fuel injection or ignition system wiring or tester cables.

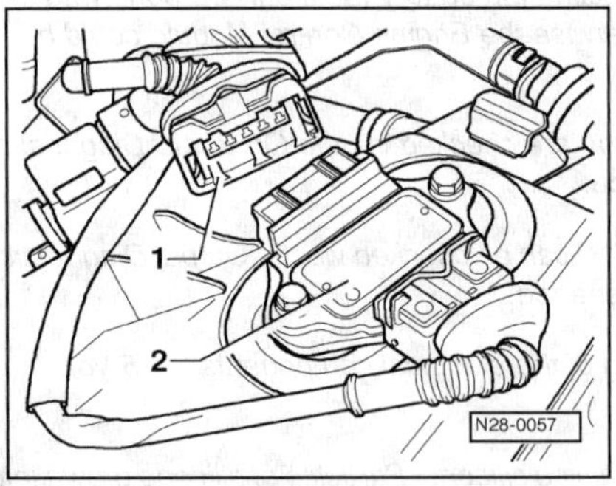

If the engine is to be turned at starter speed, without starting:

- Disconnect 5-pin harness connector -1- from ignition coil Power Output stage -2-.
- Test and measuring instruments must be secured to rear seat and operated by a 2nd person from this location. If test and measuring instruments are operated from front passenger's seat and the vehicle is involved in an accident, there is a possibility that the person sitting in this seat may receive serious injuries when the airbag is triggered.

Rules for cleanliness

When working on the fuel supply/injection system, pay careful attention to the following 5 rules:

1. Thoroughly clean all unions and the adjacent areas before disconnecting.
2. Place parts that have been removed on a clean surface and cover. Do not use fluffy cloths!
3. Carefully cover opened components or seal, if the repair cannot be carried out immediately.
4. Only install clean components: Only unpack replacement parts immediately prior to installation. Do not use parts that have been stored loose (e.g. in tool boxes etc.).
5. When the system is open: Do not work with compressed air if this can be avoided. Do not move vehicle unless absolutely necessary.

Technical data

Engine code		AEB
Idle speed check	rpm	820 to 920
Engine Control Module		Motronic M 5.9.2
Engine governed speed	rpm	from approx. 6500

Component locations, overview

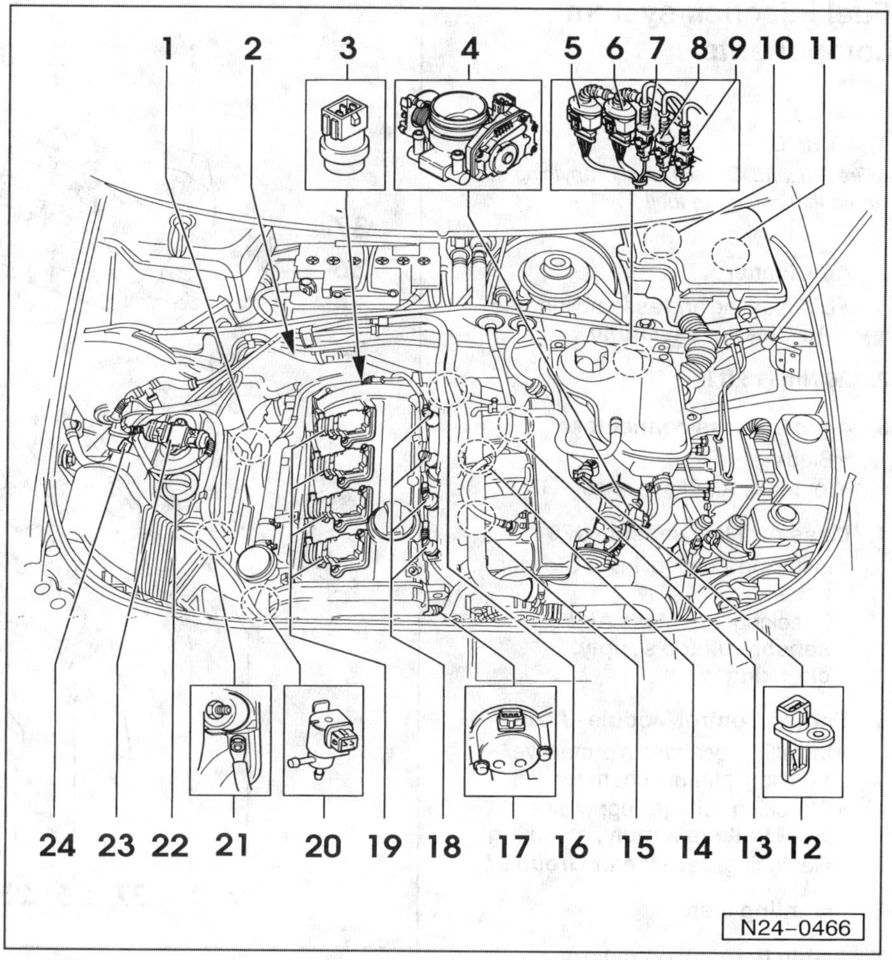

1. **Oxygen sensor 1 before Three Way Catalyst -G39-**
2. **Oxygen sensor 2 after Three Way Catalyst -G130-**
3. **Engine Coolant Temperature sensor -G62-**
4. **Throttle Valve Control Module -J338-**
5. **4-pin harness connector**
   - Brown for Oxygen sensor 1 -G39-
6. **4-pin harness connector**
   - Black for Oxygen sensor 2 -G130-
7. **3-pin harness connector**
   - Grey for engine speed sensor -G28-
8. **3-pin harness connector**
   - Green for knock sensor 1 -G61-
9. **3-pin harness connector**
   - Blue for knock sensor 2 -G66-
10. **Barometric sensor -F96-**
    - Installing location: In protective housing, plenum chamber, left
11. **Engine Control Module -J220-**
    - Installing location: In protective housing, plenum chamber, left
12. **Intake Air Temperature sensor -G42-**
13. **Engine speed sensor -G28-**
14. **Knock sensor 2 -G66-**
    - See **Repair Group 28, Ignition system components**
15. **Knock sensor 1 -G61-**
    - See **Repair Group 28, Ignition system components**
16. **Fuel pressure regulator**
17. **Camshaft Position sensor - G40-**
    - See **Repair Group 28, Ignition system components**
18. **Fuel Injector -N30-, -N31-, -N32-, -N33-**
19. **Ignition coils -N-, -N128-, -N158- and -N163-**
    - See **Repair Group 28, Ignition system components**
20. **Wastegate Bypass Regulator Valve -N75-**
    - Boost pressure system, see **1.8 Liter 4-Cyl. 5V Turbo Engine Mechanical, (code: AEB, ATW) section, Repair Group 21**
21. **Ground connection**
    - On engine support, right
22. **Mass Air Flow sensor -G70-**
23. **Power Output stage -N122-**
    - See **Repair Group 28, Ignition system components**
24. **EVAP canister purge regulator valve 1 -N80-**
    - On Air Cleaner
    - Evaporative emissions system (EVAP), see **1.8 Liter 4-Cyl. 5V Turbo Engine Mechanical (code: AEB, ATW) section, Repair Group 20**

Fuel injection system components

WARNING —
Fire Hazard! Do NOT have anything in area that can ignite fuel!

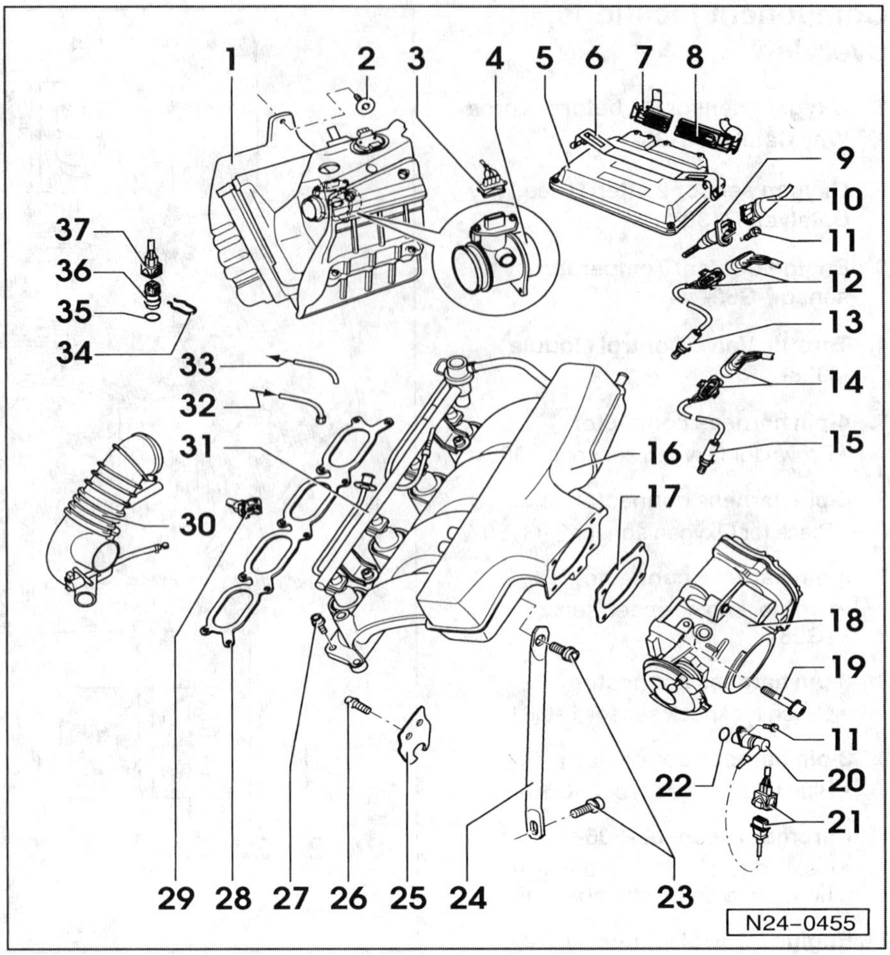

1. **Air Cleaner**
   - Dismantling and assembling, see **Air Cleaner assembly**
2. **20 Nm (15 ft lb)**
3. **4-pin harness connector**
   - Black
   - for Mass Air Flow sensor -G70-
4. **Mass Air Flow sensor -G70-***
   - Installing location: In Air Cleaner housing
   - Checking, see **Mass Air Flow sensor voltage supply, checking**
5. **Engine Control Module -J220-***
   - Installing location: In protective housing, plenum chamber, left
   - Checking voltage supply, see **ECM voltage supply, checking**
   - Replacing, see **Repair Group 97**
6. **Mounting plate**
7. **28-pin harness connector**
   - Only disconnect or connect with ignition switched OFF
   - Unlatch to disconnect
8. **Connector**
   - Only disconnect or connect with ignition switched OFF
   - Unlatch to disconnect
9. **Intake Air Temperature sensor -G42-***
   - Checking, see **Fig.1** and **Fig.3**
10. **2-pin harness connector**
    - Black
    - For Intake Air Temperature sensor G42
11. **10 Nm (7 ft lb)**
12. **4-pin harness connector**
    - Brown
    - For Oxygen sensor and Oxygen sensor heating before Three Way Catalyst
13. **Oxygen sensor 1 -G39-***
    - 50 Nm (37 ft lb)
    - Grease threads with "G5"only; it must not get into the slots on the sensor body
    - See **Oxygen sensor control, checking**
    - See **Oxygen sensor heating, checking**
14. **4-pin harness connector**
    - Black
    - For Oxygen sensor and Oxygen sensor heating after Three Way Catalyst
15. **Oxygen sensor -G130-* (after Three Way Catalyst)**
    - 50 Nm (37 ft lb)
    - Grease only the threads with "G5"; "G5" must not get into the slots on the probe body
    - See **Oxygen sensor control, checking**
    - See **Oxygen sensor heating, checking**
16. **Intake manifold**
17. **Seal**
    - Always replace
18. **Throttle Valve Control Module -J338-***
    - Checking, see **Throttle valve control module, checking.**
    - If replaced, match to ECM, see **Engine control module to throttle valve control module, matching.**
19. **10 Nm (7 ft lb)**
20. **Engine speed sensor -G28-***
    - Checking, see **Engine Speed sensor (RPM), checking**
21. **3-pin harness connector**
    - Grey
    - For engine speed sensor
22. **O-ring**
23. **20 Nm (15 ft lb)**
24. **Support**
    - Between intake manifold and assembly bracket

(continued on following page)

Fuel injection system components, (continued)

25. Support bracket
- for throttle cable

26. 10 Nm (7 ft lb)

27. 10 Nm (7 ft lb)

28. Seal
- Replace

29. 2-pin harness connector
- Black
- for Fuel injectors -N30- to -N33-

30. Intake hose

31. Fuel manifold with injectors
- Disassembling and assembling, see **Fuel injection manifold assembly**

32. Supply line
- Black

33. Return line
- Secure with spring clips
- Ensure it is securely seated
- Blue
- to fuel delivery unit in fuel tank

34. Retaining clip
- Check seated securely

35. O-ring
- Replace if damaged

36. Engine Coolant Temperature sensor -G62-*
- Blue
- For Engine Control Module
- With coolant temp gauge sensor
- Checking, see **Fig. 2** and **Fig.3**
- If necessary, release cooling system pressure before removing

37. 4-pin harness connector
- Blue
- For Engine Coolant Temperature sensor -G62-

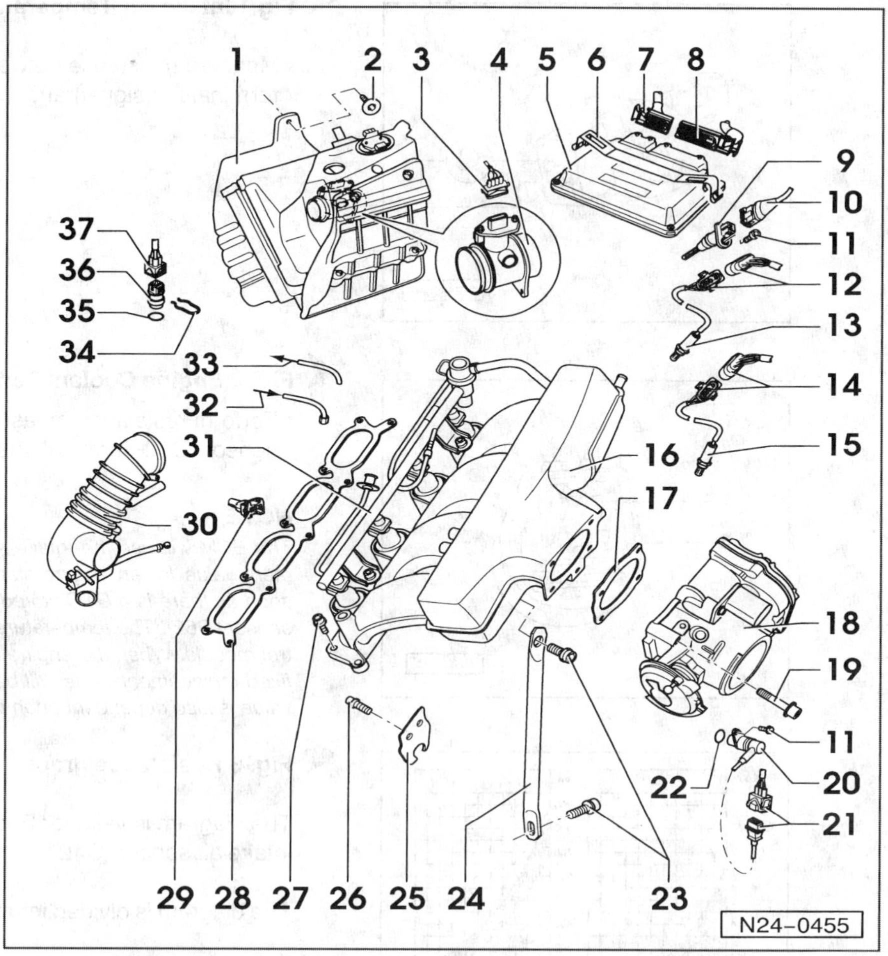

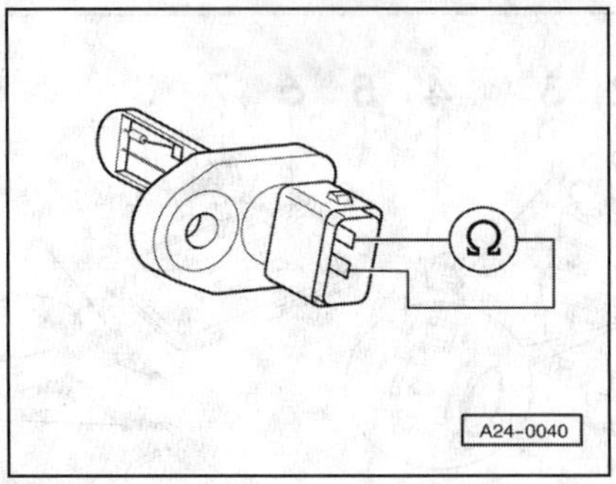

Fig.1 Intake Air Temperature sensor -G42-

- Measure resistance between Intake Air Temperature sensor G42 terminals 1 (signal) and 2 (Ground)

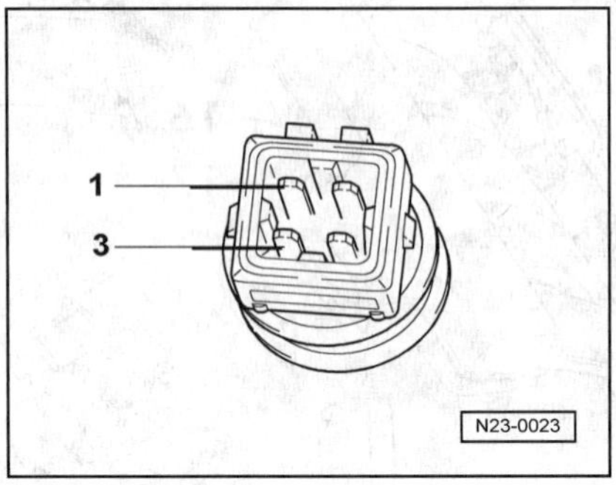

Fig. 2 Engine Coolant Temperature sensor -G62-, checking

- Perform resistance measurement at Engine Coolant Temperature sensor -G62- terminal 1 (signal) and 3 (Ground).

NOTE —
The ECM will use the Intake Air Temperature sensor signal as a replacement value for an engine start (start temperature replacement value) as soon as there is a DTC stored in the DTC memory, which effects the ECT sensor -G62-. The temperature then rises using a model stored in the control module. When the engine has reached normal working temperature, a fixed replacement value will be displayed after a certain period. This fixed value is also dependent upon the Intake Air Temperature.

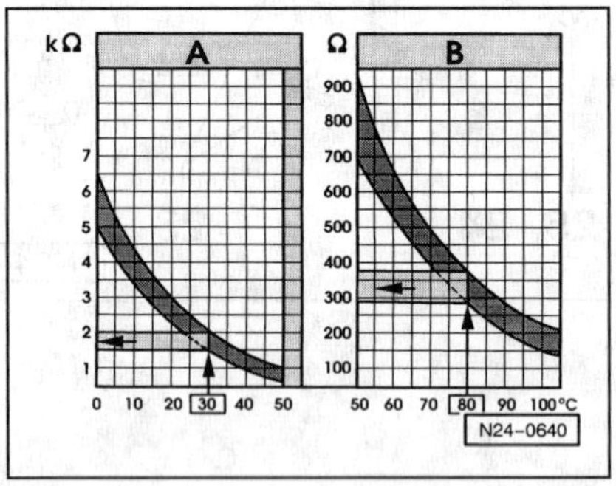

Fig. 3 Resistance graph

The diagram is valid for Engine Coolant Temperature sensor -G62- and intake air sensor -G42-.

The diagram is divided into two temperature ranges:

A - From 0 to 50 °C (32 to 122°F)

B - From 50 to 105 °C (122 to 302°F)

Examples:

- 30 °C is range A and corresponds to a resistance of 1.5 to 2.0 kΩ
- 80 °C is range B and corresponds to a resistance of 275 to 375Ω

Air cleaner assembly

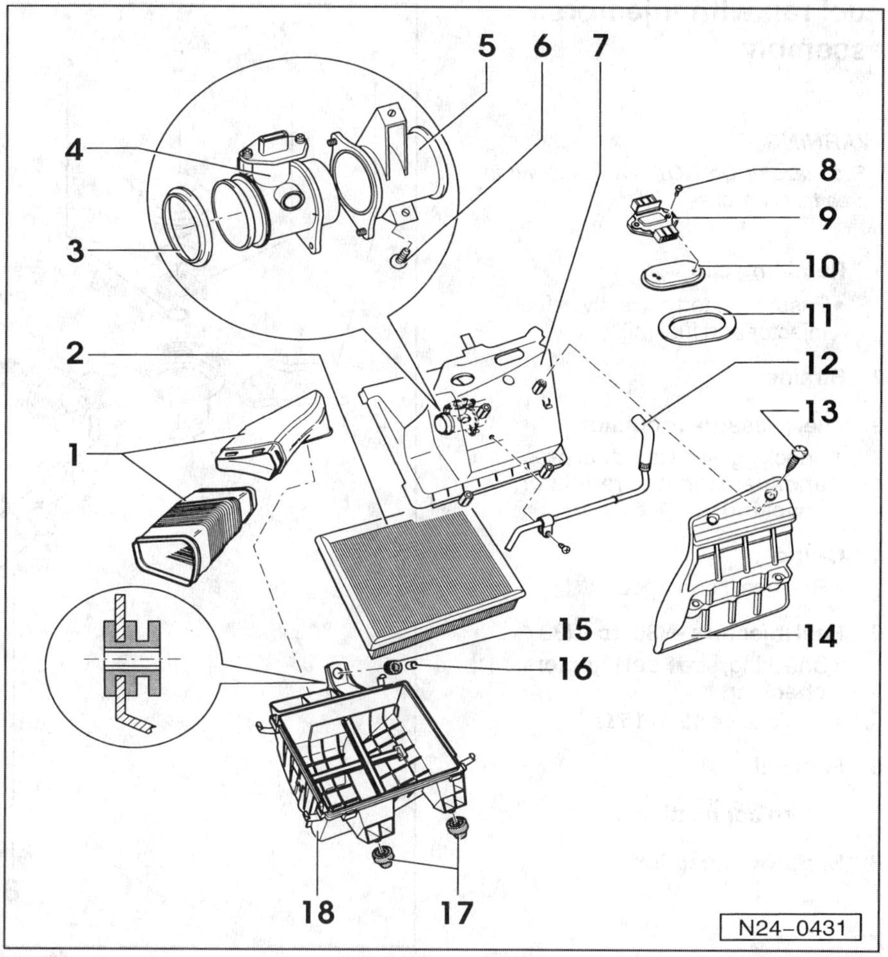

1. **Air ducting**
   - to lock carrier
2. **Filter element**
3. **Oil seal**
4. **Mass Air Flow sensor -G70-***
   - Checking, see **Mass air flow sensor voltage supply, checking**
5. **Air duct**
6. **6 Nm (53 in-lb)**
7. **Air Cleaner upper part**
8. **6 Nm (53 in-lb)**
9. **Power Output stage -N122-**
   - See **Repair Group 28, Ignition system components**
10. **Cooling element**
11. **Rubber grommet**
12. **Pipe**
    - for Evaporative emissions system
13. **10 Nm (7 ft-lb)**
14. **Heat shield**
15. **Rubber grommet**
16. **Spacer sleeve**
17. **Rubber grommet**
18. **Air Cleaner lower section**

Fuel rail with injectors, assembly

WARNING —

Fire Hazard! Do NOT have anything in area that can ignite fuel!

1. **Retaining clip**
   - Ensure seated correctly at fuel injector and fuel rail
2. **Strainer**
3. **Fuel pressure regulator**
   - Checking, see **Residual pressure and fuel pressure regulator, checking**
4. **O-ring**
   - Replace if damaged
5. **Fuel injectors -N30- to -N33-***
   - Checking, see **Fuel injectors, checking**
   - Resistance 12 to 15 Ω
6. **Fuel rail**
7. **Return connection**
8. **Supply connection**

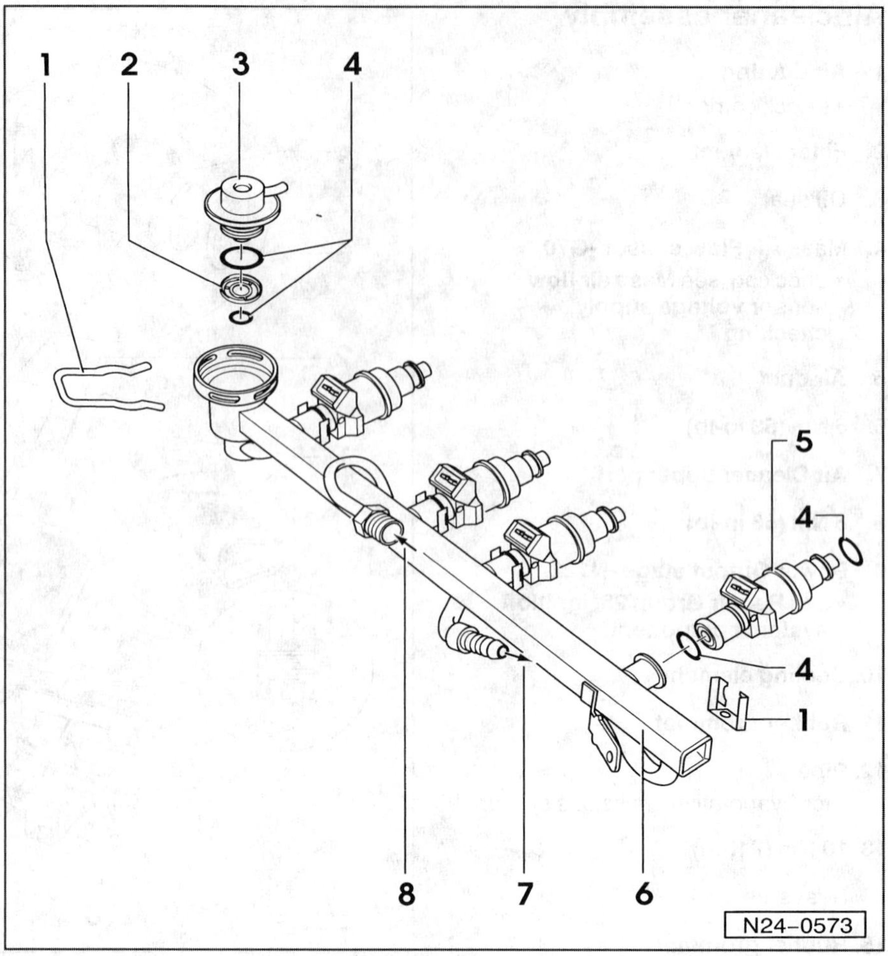

Fuel Injection Components, checking

Oxygen sensor heater, checking

Special tools

- Fluke 83 Multimeter
- Wiring diagram

Check conditions

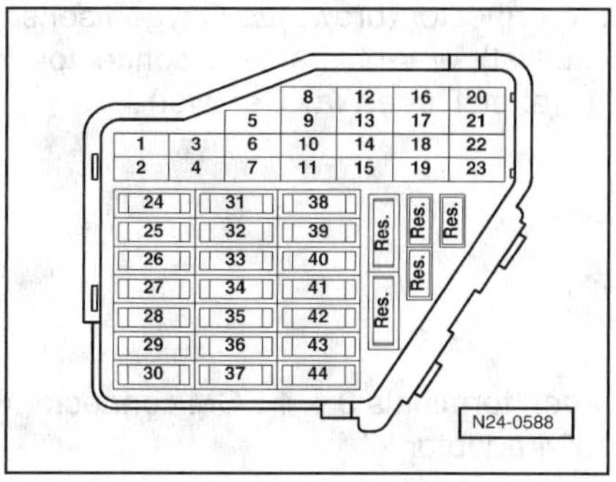

Fuse 29 OK.

- Battery Voltage 11.5 Volts minimum
- Fuel pump relay OK

Test sequence

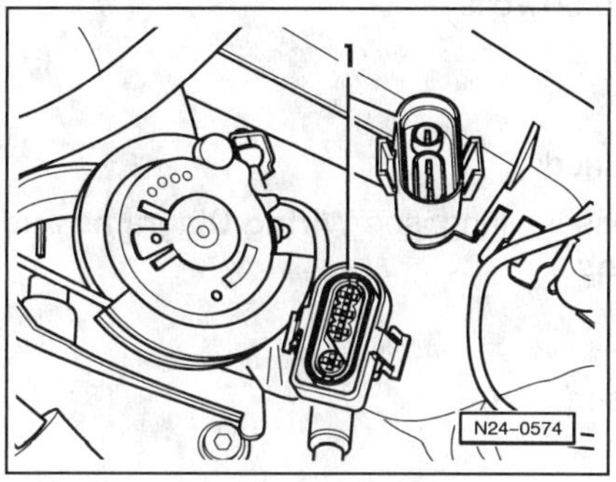

Disconnect 4-pin harness connector (brown) from Oxygen sensor -G39- (before Three Way Catalyst -1-), or 4-pin harness connector (black) to Oxygen sensor -1--G108- (after Three Way Catalyst).

– Switch Multimeter to "Resistance measurement" range.

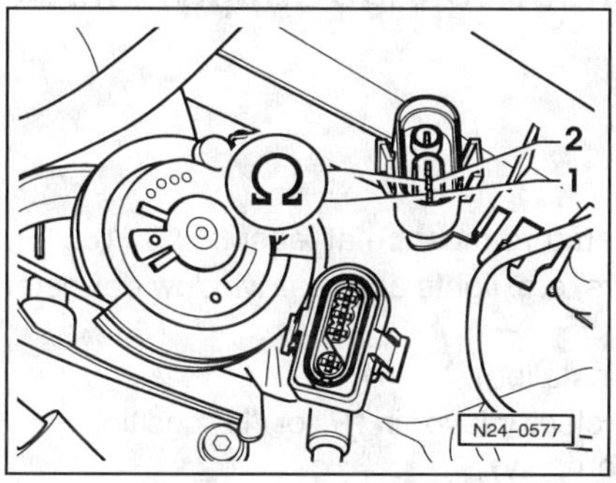

Connect Multimeter between terminals 1 + 2 of -G39- harness connector using jumper wires from VW 1594 adaptor kit

– Measure resistance:
  - Specification: 4.5 to 19.9 Ω

– If specification not obtained:
  - Replace Oxygen sensor -G39- (before Three Way Catalyst).

– If specification obtained:
  - Switch Multimeter to "Voltage measurement" range.

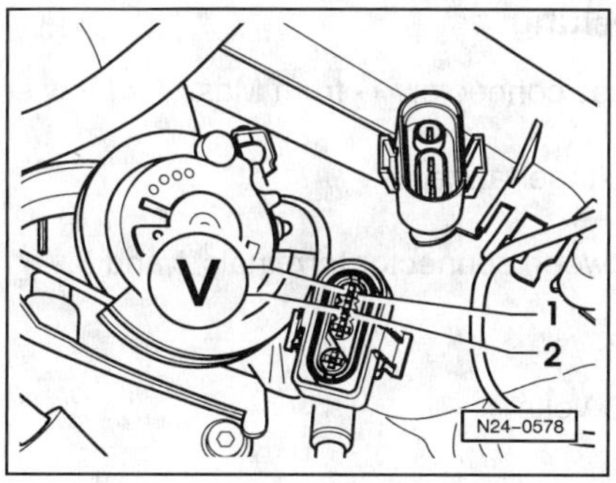

Connect Multimeter between terminals 1 + 2 of ECM harness connector using jumper wires from VW 1594 adaptor kit.

– Start engine and let idle.

– Measure the Voltage supply:
  - Specification with heater on: 11.0 to 14.5 Volts.
  - Specification with heater alternating: Between 0.0 and 12.0 Volts, fluctuating.

– Switch OFF ignition.

If NO Voltage present, check wiring diagrams, See **Wiring Diagrams and Component Locations.**

Oxygen sensor control, checking

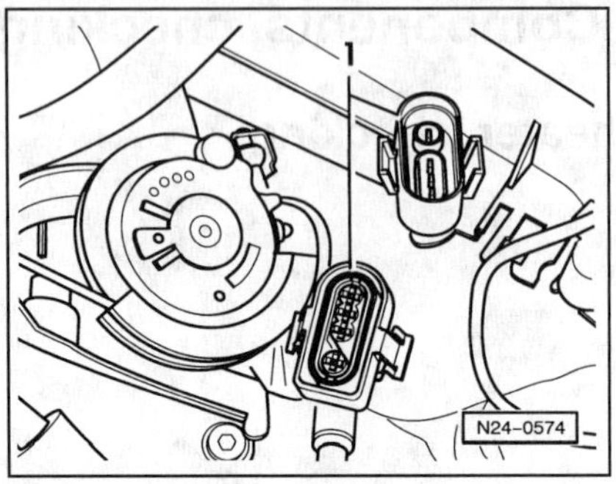

Basic Voltage, check

Disconnect 4-pin harness connector (brown) to Oxygen sensor -G39--1- (before Three Way Catalyst), or 4-pin harness connector (black) to Oxygen sensor -G108--1- (after Three Way Catalyst).

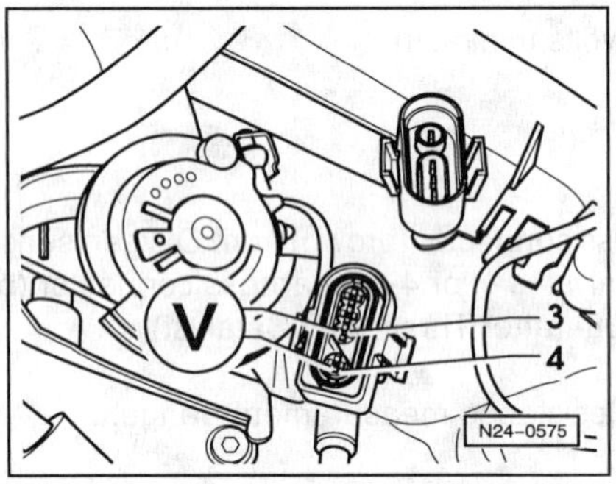

Connect Multimeter between terminals 3 + 4 ECM connector using jumper wires from VW 1594 adaptor kit.

– Start engine and measure basic Voltage.
  - Specification: 0.40 to 0.50 Volts
– Switch OFF ignition.
– If specification not obtained:
  - Check the Oxygen sensor wiring, see **Wiring Diagrams and Component Locations**
– If specification obtained:
  - Replace Oxygen sensor.

Mass air flow sensor voltage supply, checking

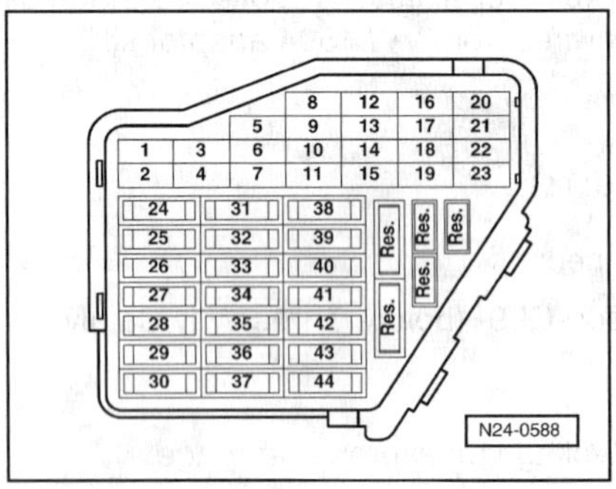

Check conditions

Fuse 29 OK.

- Engine Coolant Temperature must be at least 85 °C (185°F).
- All electrical consumers, e.g. lights and rear window defroster must be switched OFF.
- A/C switched OFF (if installed).
- Transmission Range selector lever in "P" or "N" position (automatic transmission only).

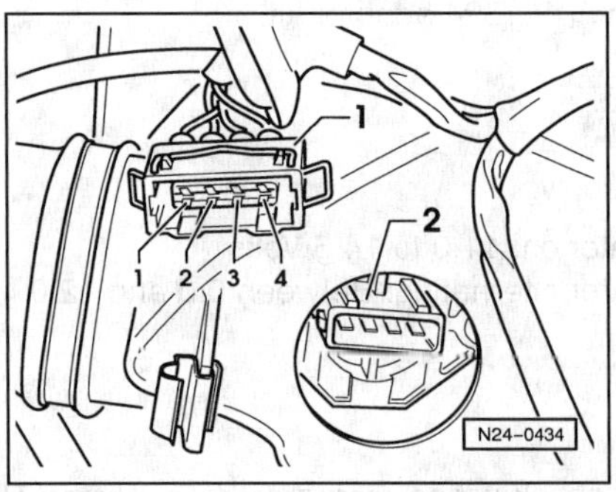

Voltage supply, checking

Disconnect 4-pin harness connector -1- from Mass Air Flow sensor -2-.

– Switch Multimeter to Voltage range.
– Connect Multimeter between connector terminals 1 and 3.
– Start engine and let idle.
  - Specification: 11 to 15 Volts
– Switch OFF ignition.
– If Voltage supply OK
  - Test signal and Ground wires.
– If NO Voltage present:
  - Check wiring between connector and fuel pump relay -J17-, see **Wiring Diagrams and Component Locations.**

Throttle valve control module, checking

NOTE —

- *If the throttle valve control module is replaced, the new assembly MUST BE matched to the engine control module, see* **Throttle valve control module, adaptation to Engine Control Module (ECM)**.
- *Components of throttle valve control module -J338-: Throttle valve drive -G186-, throttle valve drive angle sensor -G187-, throttle position sensor -G69- and closed throttle switch -F60-*
- *All settings are made using VAG 1551 Scan Tool in Function 04*
- *If the complete throttle valve control module has no current (e.g. connector disconnected) the throttle valve moves to a pre-determined mechanically determined position which will increase idle speed to 1200 - 1300 rpm when engine is at normal operating temperature.*
- *If no current reaches the throttle valve drive -G186,- the throttle will move to the mechanically predetermined position (emergency running gap), but because the closed throttle position switch -F60- is still recognized an almost normal idling speed of approx. 900 rpm is achieved by relatively retarding the ignition.*
- *If the engine control module recognizes a DTC for the throttle valve drive angle sensor -G187- the throttle valve drive -G186- is switched to receive no current by the ECM and the throttle moves again to the mechanically predetermined position (emergency running gap).*
- *The displayed value is dependent on the tolerances of the Throttle Position sensor and does not correspond to the actual opening angle.*

Voltage supply and wiring to throttle valve control module, checking

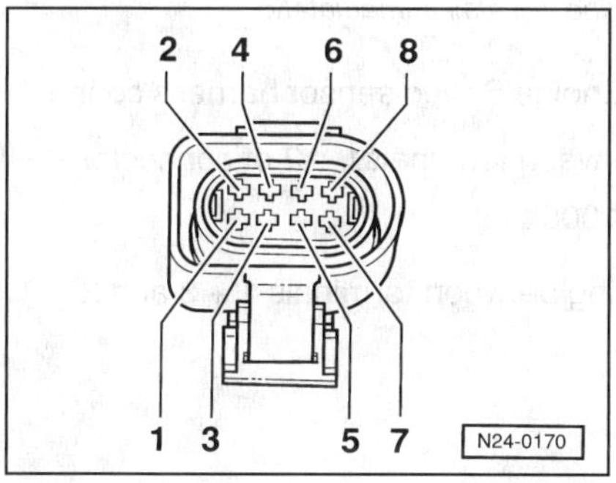

◄ Disconnect 8-pin harness connector to throttle valve control module.

– Switch Multimeter to Voltage range.

– Connect Multimeter between connector terminals 4 + 7.

– Switch ON ignition.

- Specification: 4.5 Volts minimum (5.0 Volts nominal)

– Switch OFF ignition.

– Connect Multimeter between connector terminals 3 + 7.

– Switch ON ignition.

- Specification: 9 Volts minimum (battery voltage nominal)

– Switch OFF ignition.

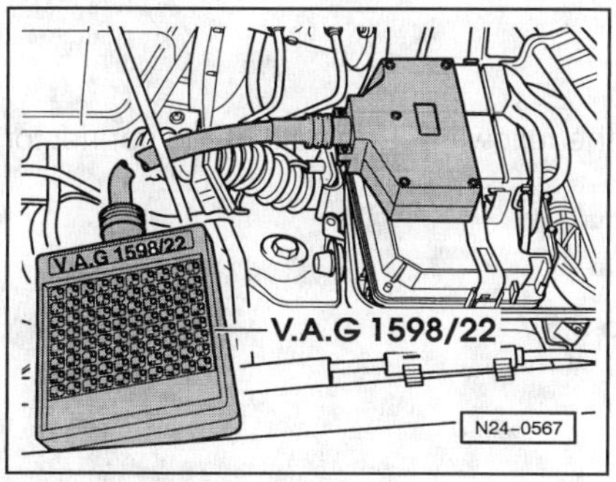

◄ Connect VAG 1598/22 Test Box to Engine Control Module harness connector.

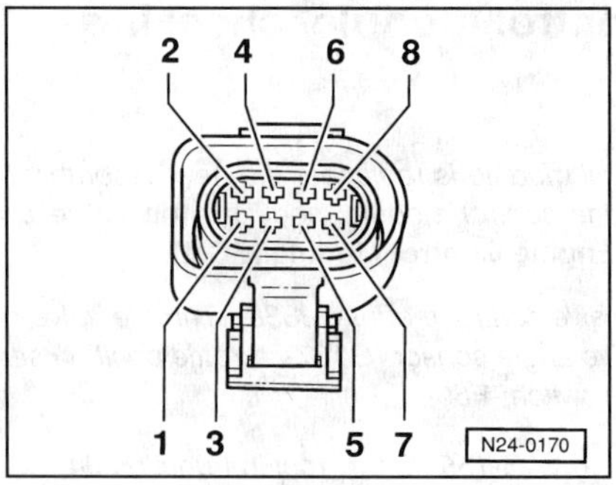

Check wiring for open circuit between Test Box and harness connector using Wiring diagram.

- Terminal 1 + socket 66
- Terminal 2 + socket 59
- Terminal 3 + socket 69
- Terminal 4 + socket 62
- Terminal 5 + socket 75
- Terminal 7 + socket 67
- Terminal 8 + socket 74
- Wire resistance: Max. 1.5 Ω

– Check wires for shorting to one another.

- Specification: ∞Ω

– If wiring OK

- Check ECM voltage supply, see **ECM voltage supply, checking.**

EDITOR'S NOTE —

If test box is not available, use wiring diagrams to check for open circuit between ECM harness connector and throttle valve control module harness connector.

Engine speed sensor (RPM), checking

NOTE —

Engine Speed sensor -G28- is a speed and reference mark sensor. The engine will not start without a speed signal. If the speed signal fails while the engine is running, the engine will stall immediately.

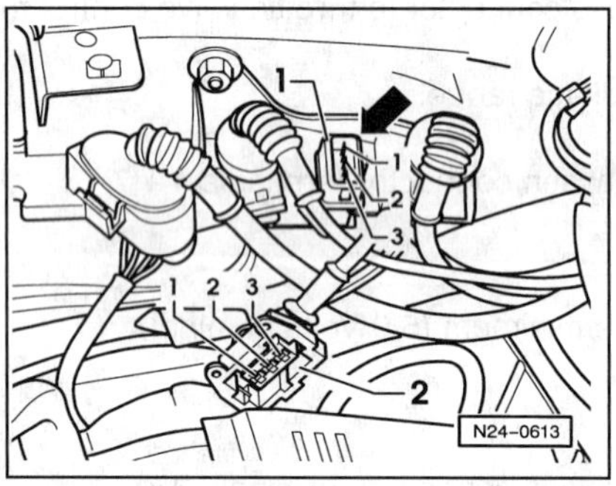

Disconnect gray 3-pin Engine Speed sensor harness connector -arrow-.

– Measure resistance between terminals1 + 2 of connector -1-

- Specification: 400 to 1000 Ω

– Check sensor for shorting between terminals 1 + 3 and 2 + 3

- Specification: ∞Ω

Barometric pressure (BARO) sensor -F96- reference voltage, checking

– Switch ignition on.

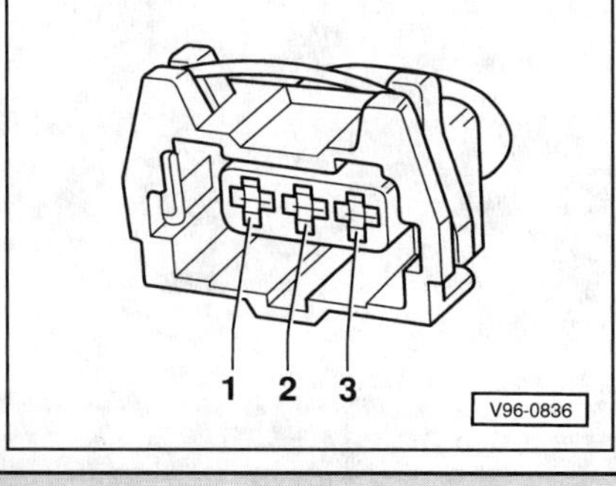

Connect multimeter to the following terminals of the connector for a voltage measurement:

- Terminal 1 + 3
- Terminal 2 + 3
- Specified value: 4.5 to 5.5V

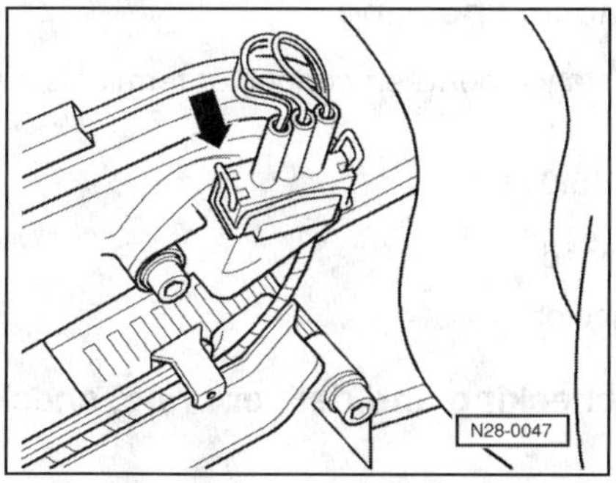

Fuel injectors, checking

– First check activation of Fuel injectors via using Volkswagen diagnostic scan tool, Output Diagnostic Test Mode (Function 03).

– If one or more injectors does not click, continue activation as follows.

Disconnect 5-pin harness connector -1- from ignition coil power output stage -2-.

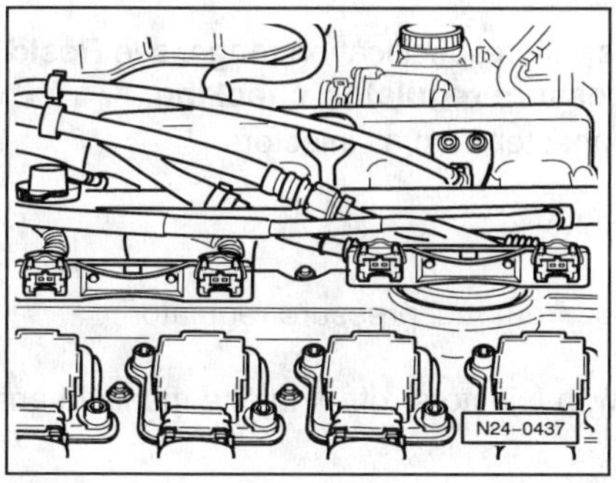

Disconnect all Fuel injector harness connectors

– Connect VAG 1527 B Voltage tester to cyl. 1 connector terminals using jumper wires from VW 1594 adaptor kit.

– Operate starter and check Voltage supply for cyl. 1 Fuel injector

- LED must light up

– Repeat check at remaining injector connectors

– Switch OFF ignition.

If LED does not flicker for any cylinder checked:

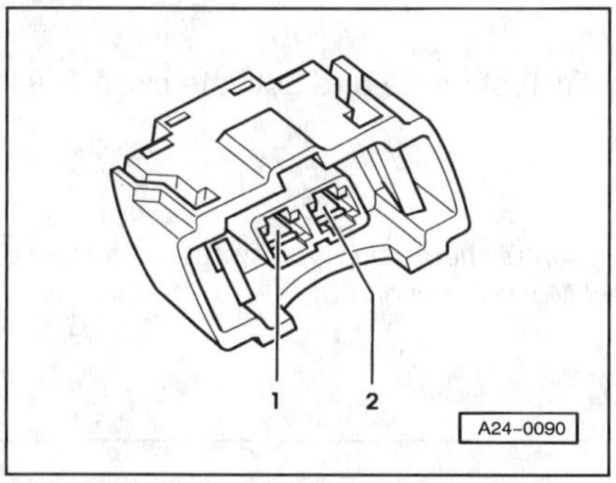

Connect VAG 1527B Voltage tester between cyl. 1 harness connector terminal 1 and Ground, 2.

– Operate starter.

- LED tester must light up

– If LED does not light up:

- Switch OFF ignition

– Check wiring for open circuit between 2-pin harness connector terminal 1 and fuel pump relay -J17- using Wiring diagram.

- Wire resistance: max. 1.5 Ω

If LED flickers on one or several cylinders:

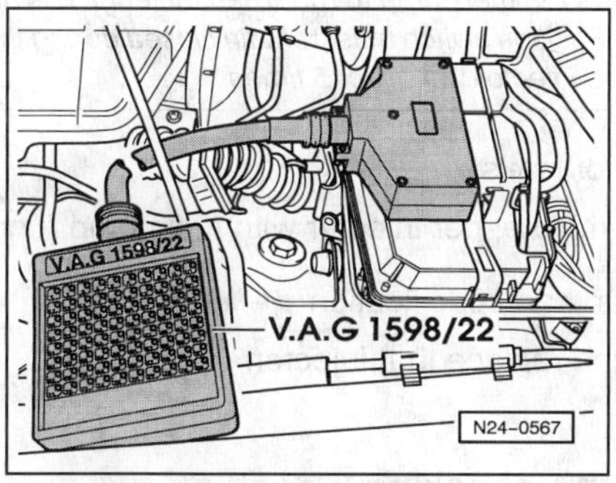

Connect VAG 1598/22 Test Box to Engine Control Module harness connector.

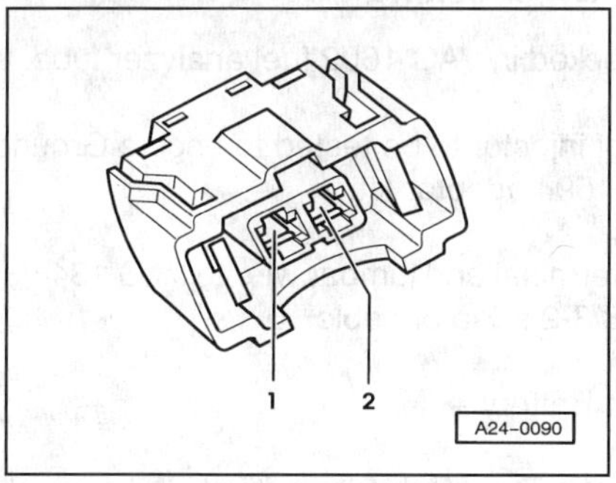

Check wiring for open circuit between Test Box and connector using Wiring diagram.

- Injector 1: Terminal 2 + socket 73
- Injector 2: Terminal 2 + socket 80
- Injector 3: Terminal 2 + socket 58
- Injector 4: Terminal 2 + socket 65
- Wire resistance: Max.1.5 Ω

– Check wires for shorting to one another.

- Specification: ∞Ω

– Check wiring for open circuit between connector terminals 1 of fuel injectors.

- Wire resistance: Max. 1.5 Ω

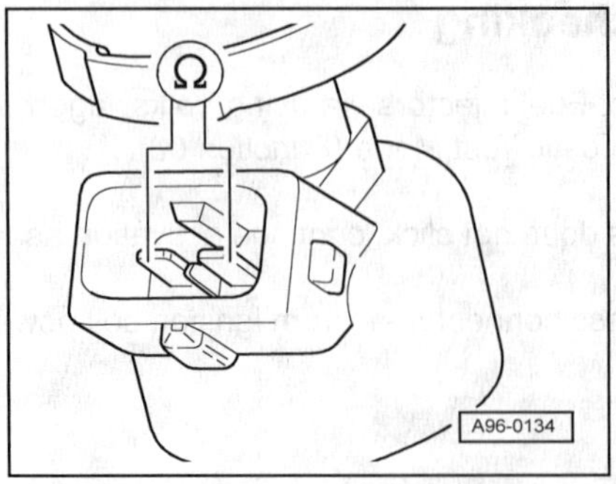

Fuel injector resistance, checking

➤ Check fuel injector resistance between connector terminals on the fuel injector.

- Specification: 12.0 to 15.0 Ω

– If specification not obtained

- Replace faulty fuel injector

Fuel injection leak checking and delivery rate, checking

Test conditions

- Fuel pressure must be within specification range, see **Residual pressure and fuel pressure regulator, checking**.

– Disconnect harness connector from all injectors.

– Remove fuel rail assembly.

– Disconnect vacuum hose from fuel pressure regulator.

– Lift fuel rail (complete with injectors) off of intake manifold and support as required.

Leak-checking

– Perform Output Diagnostic Test mode and activate cyl. 1 Fuel injector -N30-.

EDITOR'S NOTE —
Component activation is accomplished using Volkswagen diagnostic scan tool, Output Diagnostic Test Mode (function 03).

- Fuel pump must run.

> ***CAUTION —***
> *This work sequence allows the fuel pump to run when the engine is not running. Closed Throttle Position switch must remain closed during this check, otherwise selected injector will inject 5 times.*

– Visually check injector for leaks

- Only 1 to 2 drops per minute, per injector with fuel pump running

– If fuel leakage is greater than specification

- Switch OFF ignition and replace that injector.
- Always replace seals

Fuel injection quantity, checking

– Insert injector to be checked in VAG 1602 fuel analyzer tube.

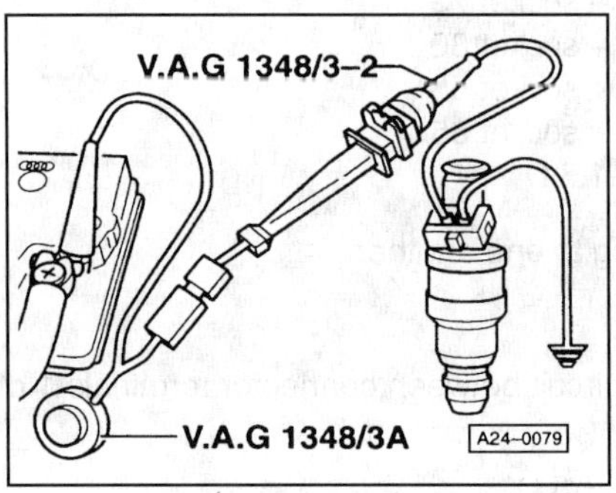

➤ Connect one terminal of injector to be tested to engine Ground using jumper wires from VW 1594 adaptor kit.

– Connect other injector terminal and jumper wire to VAG 1348/3A remote control using VAG 1348/3-2 adaptor cable.

– Connect alligator clip to Battery +.

– Perform Output Diagnostic Test Mode and activate cyl.1 injector -N30-.

EDITOR'S NOTE—
Component activation is accomplished using Volkswagen diagnostic scan tool, Output Diagnostic Test Mode (function 03).

- Fuel pump must run

– Operate VAG 1348/3A Remote control for 30 seconds.

– Repeat check on remaining injectors using new measuring tubes.

– After all injectors have been activated, place measuring tubes on horizontal surface and compare injected quantities.

- Specification: 110 to 130 ml per injector

If the measured values of one or more injectors are above or below the prescribed specifications:

– Replace faulty injector.

– Perform fuel injector installation in reverse order.

- Replace O-rings on all injectors and lightly moisten with clean oil.
- Insert injectors vertically and in correct position into fuel rail and secure with retaining clips.
- Install fuel rail assembly onto intake manifold and press in evenly.

Residual pressure and fuel pressure regulator, checking

The fuel pressure regulator controls the fuel pressure dependent upon intake manifold pressure.

Special tools, testers, measuring instruments and auxiliary items required

- VAG 1318 pressure gauge
- 1318/6 adaptor
- 1318/7 adaptor

WARNING—
Fire Hazard! Do NOT have anything in area that can ignite fuel!

Test conditions

- Fuel pump delivery rate OK, see **1.8 Liter 5V Turbo Engine Mechanical section, Repair Group 20**

Test sequence

WARNING—
Fuel system is under pressure! Before opening the system place a cloth around the connection. Then release pressure by carefully loosening the connection.

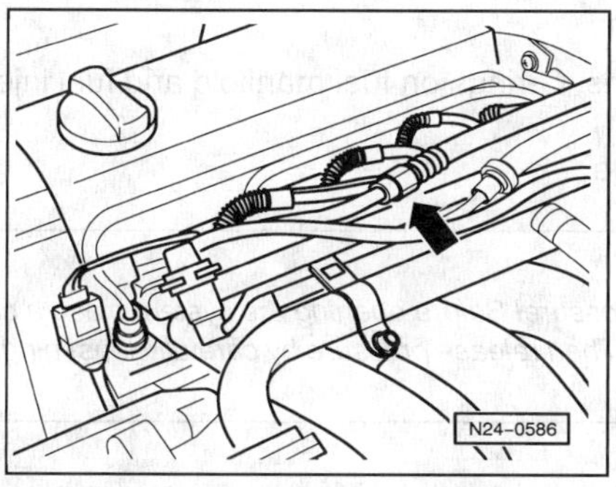

◀ Open the union -arrow- and catch escaping fuel with a cloth.

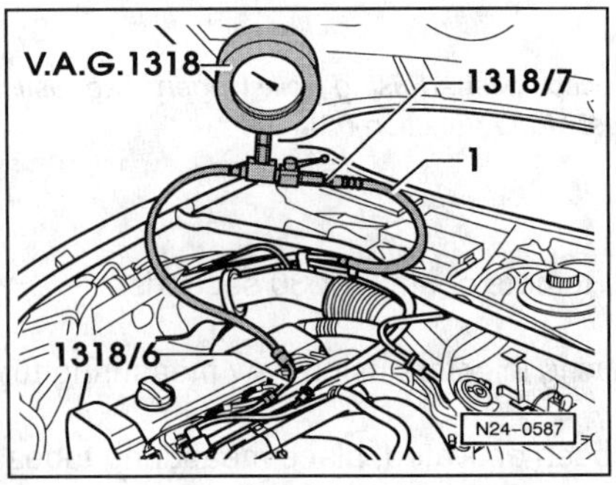

Connect pressure gauge VAG 1318 to fuel supply pipe -1- and to fuel rail using adaptors 1318/6 and 1318/7.

– Open shut-off tap on pressure gauge.
  - Handle points in direction of flow
– Start the engine and run at idling speed.
– Measure fuel pressure.
  - Specification: Approx. 3.5 bar (50 psi)

If specification obtained:

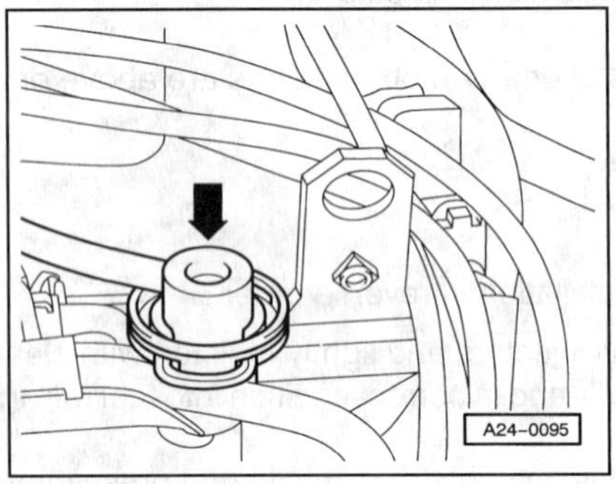

Disconnect the vacuum hose off fuel pressure regulator -arrow-.
  - Fuel pressure must increase to approx. 4.0 bar (58 psi).

– Switch OFF ignition.
– Check for leakage and residual pressure by observing pressure drop on gauge.
  - After 10 minutes residual pressure must be 2.0 bar (29 psi) minimum.

If the holding pressure drops below 2 bar (29 psi):

– Start engine and let idle.

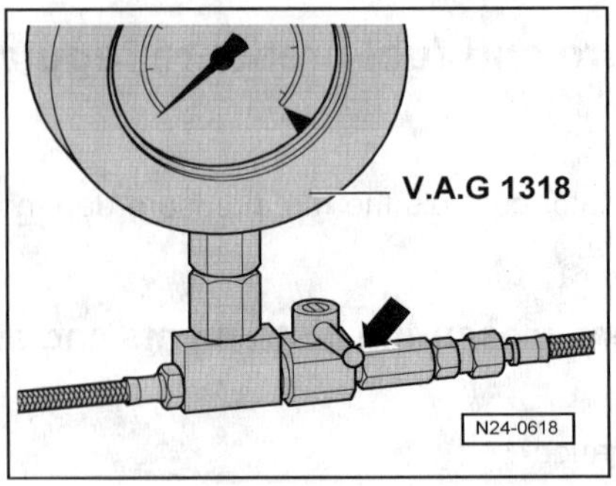

Switch OFF ignition after pressure has built-up, at same time close pressure tap.
  - VAG 1318 handle perpendicular to direction of flow -arrow-.

– Observe pressure drop on gauge.

If pressure does not drop:

– Check fuel pump check valve, see **1.8 Liter 5V Turbo Engine Mechanical (code: AEB, ATW) section, Repair Group 20**

If pressure drops again:

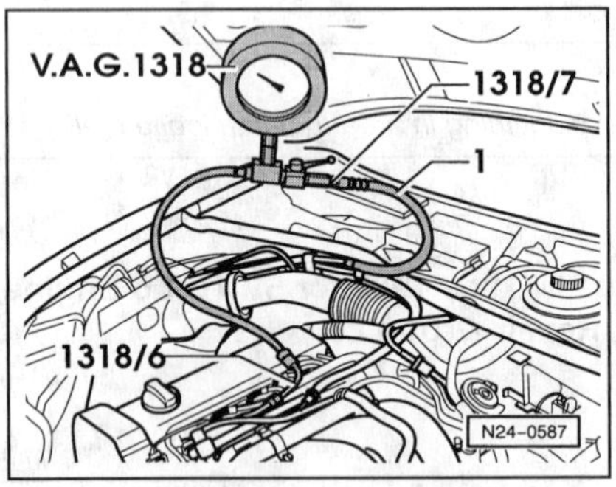

Open pressure tester VAG 1318 shut-off tap.
  - VAG 1318 handle in direction of flow.

– Start engine and let idle.
– Switch OFF ignition after pressure has built-up, at same time pinch/seal return hose.
– If pressure does not drop:
  - Replace fuel pressure regulator.
– If pressure drops again:
  - Check line connections, O-rings on fuel manifold and fuel injectors for leakage.
  - Check pressure gauge

> ***WARNING —***
> *Fuel system is under pressure! Before opening the system place a cloth around the connection. Then release pressure by carefully loosening the connection.*

Engine Control Module

ECM voltage supply, checking

Special tools, testers, measuring instruments and auxiliary items required

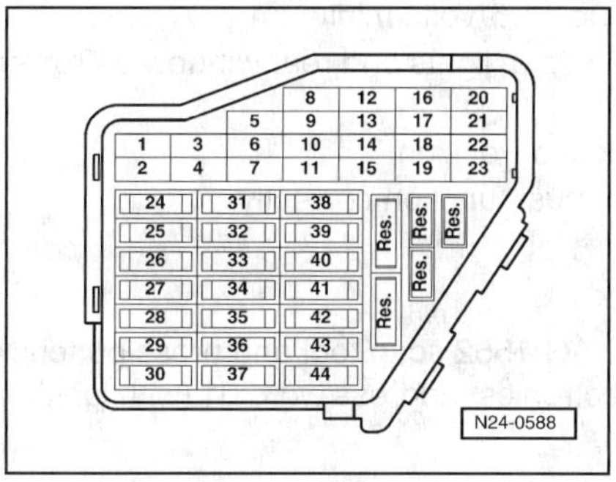

- VAG 1551/1552 Scan Tool with VAG 1551/3B adaptor cable
- VAG 1598/22 Test Box
- Fluke 83 Multimeter
- VW 1594 Adaptor kit
- Wiring diagram

Check conditions

Fuse 32 OK.

- Generator OK, see **Wiring Diagrams and Component Locations.**

Checking Voltage supply terminal 30

– Measure supply Voltage between Test Box sockets 2 + 3 using Multimeter and jumper wires from VW 1594 adaptor kit.

- Specification: 11.5 Volts minimum

– If specification not obtained:

- Check wiring connections fuse/relay panel using Wiring diagram, see **Wiring Diagrams and Component Locations.**

Checking Voltage supply terminal 15

– Measure supply Voltage between Test Box sockets 1 + 2 using Multimeter and jumper wires from VW 1594 adaptor kit.

– Switch ON ignition.

- Specification: 11.5 Volts minimum

– If specification not obtained:

- Check wiring connections to fuse/relay panel using Wiring diagram, see **Wiring Diagrams and Component Locations.**

Throttle valve control module, adaptation to Engine Control Module (ECM)

NOTE —

When adapting Throttle Valve Control Module -J338- to Engine Control Module with the ignition switched on and engine not running, the Engine Control Module learns the Throttle Valve Control Module -J338- stop positions and a comparison between Throttle Position sensor -G69- and throttle valve angle sensor (G187).

Control Module Matching must be performed if:

- Voltage supply is interrupted (Battery disconnected)
- Throttle Valve Control Module is removed and installed
- Throttle Valve Control Module is replaced
- If as a result of installing another engine, another Throttle Valve Control Module is installed

Special tools, testers, measuring instruments and auxiliary items required

- Scan Tool VAG 1551 with cable VAG 1551/3B

Test conditions

- No DTC's stored in DTC memory, see **Maintenance**
- Battery Voltage must be 11.5 Volts minimum
- All electrical consumers, e.g. lights and rear window defroster must be switched OFF
- Throttle must be in Closed Position.
- Cruise control system must function correctly

Work sequence

– Connect VAG 1551 or VAG 1552 scan tool and press buttons -0- and -1- to insert "Engine Electronics" address word 01 (with ignition switched on; engine not running).

```
Rapid data transfer          HELP
Select function  XX
```

◄ Indicated on display

– Press buttons -0- and -4- to select "Basic Setting" function 04, and press -Q- button to confirm input.

```
System in Basic Setting      HELP
Input display group number  XXX
```

◄ Indicated on display

– Press buttons -0-, -6- and -0- to input display group number 60, and press -Q- button to confirm input.

```
System in Basic Setting  60
1 2 3 4
```

◄ Indicated on display (1-4 = display fields)

NOTE —

- *After confirming Display group 60 the adaption is introduced and the Throttle Valve Drive is moved to min. max. and emergency running positions. Then using intermediate steps, a comparison is made between Throttle Position sensor -G69- and Throttle Valve Drive angle sensor -G187-.*
- *The Control Module stores these learned values in its permanent memory. Then the throttle stops in the start position.*

– Check specifications (Display zones 1 to 4, chart on next page).

	Display fields			
	1	2	3	4
Display group 60: Basic setting-throttle valve control module				
Display	xxx%	xxx %	Idle Part throttle Full throttle Decel Enrich	ADP runs ADP. OK ADP. ERROR
Indicates	Throttle valve angle	Throttle drive angle	Operating condition	Adaptation condition
Range	0-90 %	0-20 %	---	---
Specified value	1-10 % (automatic trans) 1-20 % (manual trans)		Idle	ADP. runs
	0-5 %	0-5 %	Idle	ADP. OK (ADP. ERROR indicates unsuccessful adaptation)

If displayed values are NOT OK:

- Check DTC memory
- Check throttle valve control module

NOTE —
If the Control Module Basic setting is interrupted, the cause could be one of the following:

- *Throttle cannot reach mechanical stop because of contamination e.g. oil deposits or an incorrectly adjusted throttle cable.*
- *Battery Voltage too low.*
- *Throttle Valve Control Module or wiring is faulty.*
- *After an abort, DTC "17967" is stored in DTC memory. When next switching ON the ignition Basic setting is automatically performed again.*

— End engine basic setting at the earliest after 30 seconds by pressing → button.

— Press 0 and 6 buttons to select Function 06: "End data transfer"

— Switch ignition off.

Additional Signals, checking

Vehicle speed signal, checking

Special tools, testers, measuring instruments and auxiliary items required

- Scan Tool VAG 1551 with cable VAG 1551/3B
- VAG 1598/22 Test Box
- Fluke 83 Multimeter
- VW 1594 Adaptor kit

Test sequence

NOTE —
To check the Vehicle speed signal, the vehicle must be driven, thus requiring a second technician.

WARNING —
- *When driving or riding in an airbag-equipped vehicle, NEVER hold scan tool or other test equipment in your hands or lap while in motion. Objects between you and the airbag increase the risk of injury in an accident.*
- *During a test drive in an airbag-equipped vehicle, test equipment must always be fastened to and operated from the rear seat by a second technician.*

— Connect VAG 1551/1552 Scan Tool, see **Maintenance.**

— Start engine and let idle.

— Select "Address word" 01 "Engine electronics".

— Select Function 08 "Read measuring value block".

— Select "Display group 5".

— Perform test drive with a 2nd technician to observe display.

– Observe figure shown in display zone 3.

• Specification: approx. driven speed

	Display zones			
	1	2	3	4
Display group 5: General engine data				
Display	xxx rpm	xx.xx ms	xxx km/h	Idle Part load Full load Overrun Enrichment
Indicated	Engine speed (in 40 rpm steps)	Engine load	Vehicle speed	Operating mode
Working range	0 to 6800 rpm	---	---	---
Specification	---	---	Approx. vehicle speed	---

If NO driving speed is displayed, see **Vehicle Speed Signal, checking**

– Press →button.

– Select Function 06: "End data transfer"

– Switch OFF ignition.

If NO speed indicated or displayed values reduce when driving slowly:

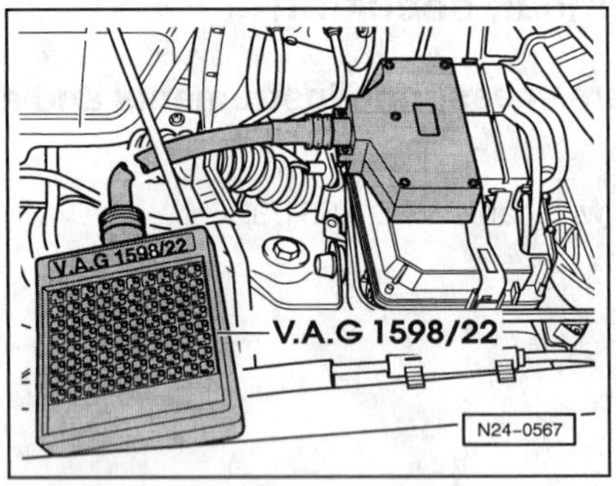

– Connect VAG 1598/22 Test Box to Engine Control Module harness connector.

– Switch Multimeter to Voltage measurement range.

– Connect Multimeter between Test Box sockets 2 + 20.

– Switch ON ignition.

– Lift vehicle at front left.

– Rotate front wheel and observe Voltage Display.

• Specification: between 0 and at least 4 Volts fluctuating

– If value does not fluctuate:

• Check wiring to speedometer / speedometer sensor, see **Wiring Diagrams and Component Locations**

A/C compressor signal, checking

A/C compressor signal: Voltage is supplied to the Engine Control Module (ECM) shortly before switching ON the A/C compressor. This signal ensures that the ECM maintains a constant idle speed even when differing A/C system power requirements are present.

A/C cut-off: The A/C compressor is switched OFF for approx. 12 seconds by the ECM when accelerating from a standstill and low speeds to ensure that the full engine power is available when accelerating. If the accelerator is released earlier, the time is shortened to a minimum of 3 seconds. At Wide Open Throttle the A/C is switched off by the Transmission Control Module (Kick-down switch).

Special tools, testers, measuring instruments and auxiliary items required

- Scan Tool VAG 1551 with cable VAG 1551/3B
- VAG 1598/22 Test Box
- Fluke 83 Multimeter
- VW 1594 Adaptor kit

Test conditions

- A/C functioning OK.
- A/C must be switched OFF
- No DTC's stored in DTC memory, see **Maintenance.**
- Vehicle at + 15 °C minimum

Test sequence

– Connect VAG 155/1552 scan tool, see **Maintenance.**

– Start engine and let idle.

– Select "Address word" 01 "Engine electronics".

– Select Function 08: "Read measuring value block".

– Press Q button to enter input.

– Select "Display group 50".

– Check A/C specifications (Display zones 3 and 4).

	Display zones			
	1	2	3	4
Display group 8: Signals to Engine Control Module				
Display	xxxx rpm	xxxx rpm	A/C LOW	Compr. ON Compr. OFF
Indicated	Actual engine speed (in 10 rpm steps)	Specified engine speed (in 10 rpm steps)	A/C status	A/C compressor status
Working range	0 to 2550 rpm	---	---	---
Specification	820 to 900 rpm	860 rpm	A/C ON: A/C-High A/C OFF: A/C-Low	Compressor ON: Compr. ON Compressor OFF: Compr. OFF

If specification is not obtained, see **A/C compressor signal, checking**

– Observe Display in Display zone 4.
  - It must show: "Compr. OFF"

– Switch ON A/C system.

– Select lowest temperature and highest blower speed.
  - Display in zone 4 must change to: Compr. ON.

– Floor accelerator pedal quickly and release again (short burst of throttle).
  - When quickly pressing the accelerator pedal down the Display in Display zone 4 must change from "ON" to "OFF" for a few seconds (compressor cut-off under vehicle acceleration).

– Select Function 06: "End data transfer".

– Switch OFF ignition.

NOTE —
A/C compressor is switched OFF by the Engine Control Module when accelerating from standstill and from low speeds. At Wide Open Throttle the A/C compressor is switched OFF by the Transmission Control Module (TCM) kick-down switch.

If display does not appear as described:

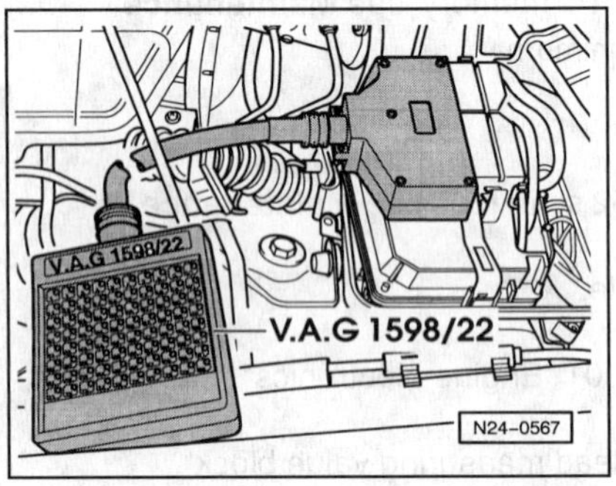

◄ Connect VAG 1598/22 Test Box to Engine Control Module harness connector.

– Check wiring for open circuit between Test Box sockets 8 and A/C using Wiring diagram.

- Wire resistance: Max. 1.5 Ω

– If wiring OK

- Check function of A/C Control Module.

NOTE —
If DTC memory has been erased or the Engine Control Module was disconnected from its Voltage supply, the Readiness Code must be created again.

Ignition retardation during transmission range selection, checking

NOTE —
- *The engine control module receives the gear change signal via the transmission control module. The engine control module then, depending on road and engine speeds, retards the ignition briefly. This reduces the torque and therefore the gear jolt.*
- *The VAG 1551 does not always recognize and display the signal for torque reduction because it is sent for an extremely short period.*

NOTE —
To check ignition retardation, the vehicle must be driven which requires a second technician to read the test equipment.

> ***WARNING —***
> - *When driving or riding in an airbag-equipped vehicle, NEVER hold scan tool or other test equipment in your hands or lap while in motion. Objects between you and the airbag increase the risk of injury in an accident.*
> - *During a test drive in an airbag-equipped vehicle, test equipment must always be fastened to and operated from the rear seat by a second technician.*

Special tools, testers, measuring instruments and auxiliary items required

- Scan Tool VAG 1551 with cable VAG 1551/3B
- VAG 1598/22 Test Box
- Fluke 83 Multimeter
- VW 1594 Adaptor kit

Test sequence

– Connect VAG 1551/1552 scan tool, see **Maintenance.**

– Start engine and let idle.

– Select "Address word" 01 "Engine electronics".

– Select Function 08: "Read measuring value block".

– Select "Display group 122".

– Perform test drive with 2nd technician to observe display.

– Check specifications in Display zone 4:

	Display zones			
	1	2	3	4
Display group 122: Torque reduction during gear change				
Display	xxxx rpm	xxx Nm	xxx Nm	No inter Eng. inter
Indicated	Engine speed (in 40 rpm steps)	Engine torque required (specified torque)	Actual engine torque (Actual torque)	Timing retardation active or not active
Working range	0 to 6800 rpm	100 to 410 Nm	100 to 310 Nm	---
Specification	---	---	---	"Eng. inter" must be displayed briefly when changing gears

NOTE—
It is possible that several gear changes must be performed before the display in Display zone 3 jumps to 0 because the signal from the automatic transmission is very brief.

If specification obtained:

– Select Function 06: "End data transfer".

– Switch OFF ignition.

If display does not change:

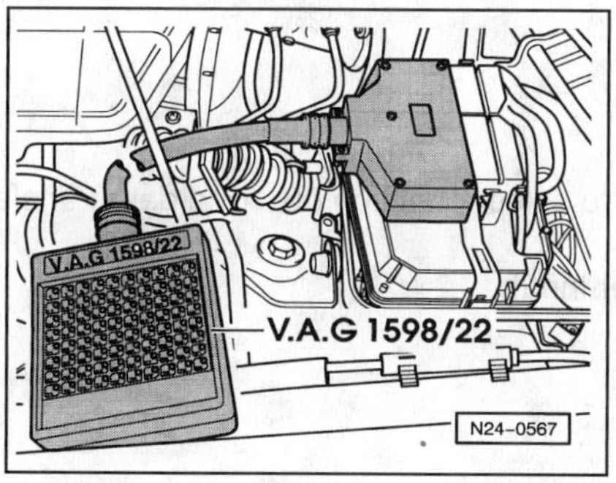

◄ Connect VAG 1598/22 Test Box to Engine Control Module harness connector.

– Check wiring for open circuit between Test Box socket 23 and Automatic transmission using Wiring diagram.

- Wire resistance: Max. 1.5 Ω

Transmission range signal, checking

NOTE—
Engine Control Module (ECM) receives following data from (Automatic) Transmission Control Module (TCM): Driving range selected (selector lever in 2/3/4/R/D) or No driving range selected (selector lever in P or N)

Special tools, testers, measuring instruments and auxiliary items required

- Scan Tool VAG 1551 with cable VAG 1551/3B
- VAG 1598/22 Test Box
- Fluke 83 Multimeter
- VW 1594 Adaptor kit

Test conditions

- Transmission Range selector must be in "P" or "N" position.

Test sequence

– Connect VAG 1551/1552 Scan Tool, see **Maintenance**

– Switch ON ignition

– Select "Address word" 01 "Engine electronics"

– Select Function 08: "Read measuring value block"

– Select "Display group 56"

– Check specifications in Display zone 4 on chart, next page

	Display zones			
	1	2	3	4
Display group 56: Idle speed check				
Display	xxxx rpm	xxxx rpm	x.xx g/s	xxxx
Indicated	Actual engine speed (in 10 rpm steps)	Specified engine speed (in 10 rpm steps)	Idle Air Control valve (Idle speed control)	Operating mode
Working range	0 to 2550 rpm	---	---	---
Specification	0 rpm	0 rpm	---	0000 = Selector lever in P or N 0010 = Selector lever in: 2 / 3 / 4 / R / D

If spec. not obtained, see **Transmission Range signal, checking**

If specification obtained

– Press →button.

– Press 0 and 6 buttons to select Function 06: "End data transfer"

– Press Q button to enter input

– Switch OFF ignition.

Significance of figures in 4 digit number block in Display zone 4, Display group 56

Significance when Display = 0				
x	x	x	x	Operating conditions Display group 56 (for Automatic Transmission equipped vehicles)
			0	A/C compressor OFF (0 = A/C compressor OFF) (1 = A/C compressor ON)
		0		Transmission Range selector lever in P or N (0 = selector lever in P or N) (1 = selector lever in 2/3/4/R/D)
	0			Always "0"
0				Always "0"

NOTE—

If DTC memory has been erased or the engine control module was disconnected from its voltage supply, the readiness code must be created again.

If display does not change:

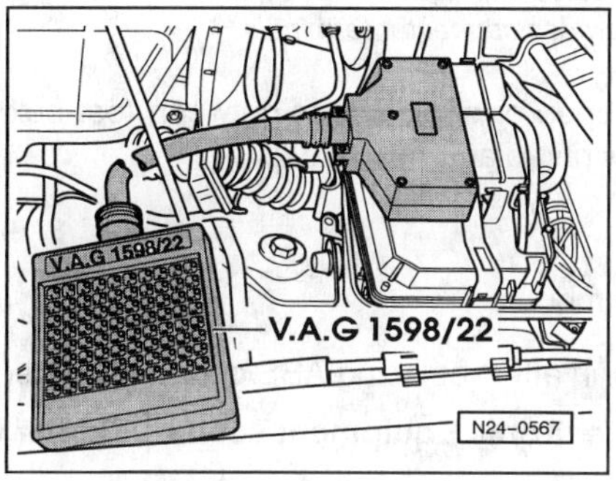

– Connect VAG 1598/22 Test Box to Engine Control Module harness connector.

– Check wiring for open circuit between Test Box socket 22 and Automatic transmission using Wiring diagram.

- Wire resistance: Max. 1.5 Ω

Rough Road conditions signal from ABS/EDL control module, checking

NOTE—

- *If a spinning wheel is detected by the ABS/EDL control module the ABS/EDL control module produces a rough road condition signal. If the engine control module recognizes a rough road signal, misfire detection is switched OFF in the ECM.*
- *The rough road condition signal should only be checked if DTC "18014 rough road information, engine specified torque, ABS-SG electrical malfunction" is stored in DTC memory.The possible DTC: "Misfire" which may become stored; is a resulting malfunction and can be disregarded.*
- *The wiring connections and/or the rough road condition signal are monitored by the engine control module.*
- *If an exhaust emissions test is performed on a front wheel drive vehicle with ABS/EDL where the front axle / wheels are rotated while the rear axle / wheels remain stationary, this can cause DTC code "18014" to be stored in DTC memory. Therefore, first remove the ABS/EDL control module fuse when performing emissions tests on these vehicles. Check and erase ABS/EDL control module and engine control module DTC memories after performing an emissions test.*

Special tools, testers, measuring instruments and auxiliary items required

- Scan Tool VAG 1551 with cable VAG 1551/3B
- VAG 1598/22 Test Box

Test conditions

- Control Module coding OK

Test sequence

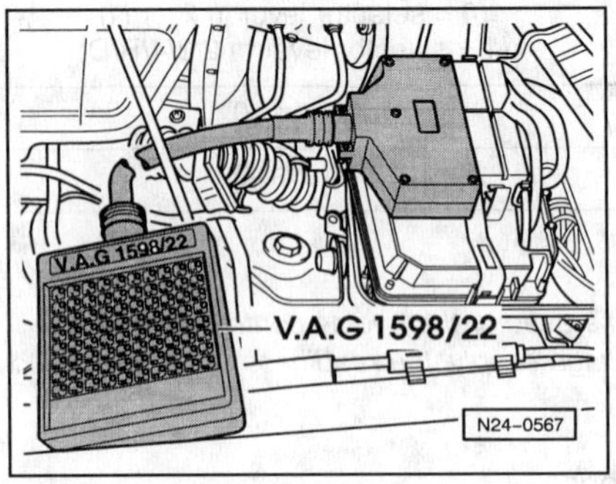

Connect VAG 1598/22 test box to engine control module harness connector.

EDITOR'S NOTE —

Volkswagen recommends using the VAG 1598/22 test box to prevent damage to ECM harness connectors during electrical testing. However the harness connectors can be tested without the test box if extra care is used to prevent damage to the harness connector terminals.

– Check signal wire for open or short circuit between test box and ABS control module using wiring diagram.

– If wiring OK:

- Replace ABS control module.

– If DTC is displayed again after replacing ABS control module:

- Replace ECM, see **Electronic Equipment section, Repair Group 97**

NOTE —

If DTC memory has been erased or the Engine Control Module was disconnected from its Voltage supply, the Readiness Code must be created again.

28 Ignition System

Repair Group

F 28

Ignition system, servicing

- Only components specifically related to the ignition system are contained in this Repair Group. For the other Fuel injection components, see **Repair Group 24**.
- Always switch OFF the ignition when disconnecting and connecting the battery, otherwise the Engine Control Module (ECM) could be damaged.
- The Engine Control Module is equipped with an On Board Diagnostic system.
- Components marked with an * are checked via On Board Diagnostic.
- 11.5 Volts minimum is required for trouble-free operation of the electrical components.
- While performing certain checks it is possible that the ECM will recognize and store a Diagnostic Trouble Code (DTC). Therefore after completing all checks and repairs check and erase the DTC memory as necessary, see **Maintenance.**

Safety precautions

To prevent injuries to persons and/or damage to the fuel injection and ignition system, the following must be noted:

- Do not touch or disconnect ignition wiring when the engine is running or being turned at starter speed.
- The ignition must be switched off before connecting or disconnecting injection or ignition system wiring or tester cables.

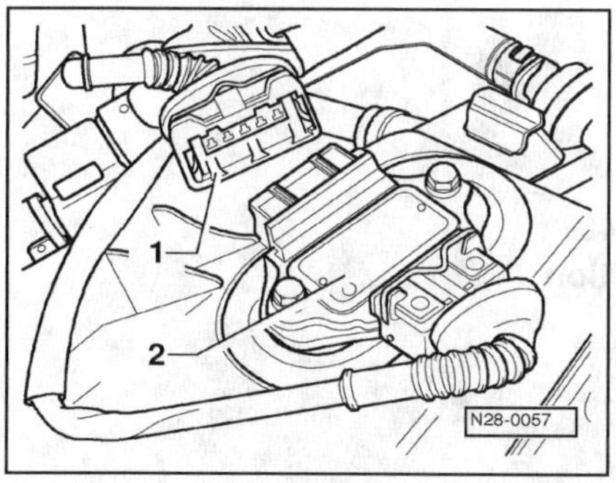

If the engine is to be turned at starter speed, without starting:

- Disconnect 5 pin connector -1- off output stage for ignition coils -2-.

Observe following if test and measuring instruments are required during a test drive:

- Test and measuring instruments must be secured to rear seat and operated by a 2nd person from this location.
- If test and measuring instruments are operated from front passenger's seat and the vehicle is involved in an accident, there is a possibility that the person sitting in this seat may receive serious injuries when the airbag is triggered.

EDITOR'S NOTE —

Volkswagen identifies electrical components by a letter and/or a number in the electrical schematics. See **Wiring Diagrams, Fuses and Relays**. *These electrical identifiers are listed -between dashes- as an aid to electrical troubleshooting.*

Ignition system components

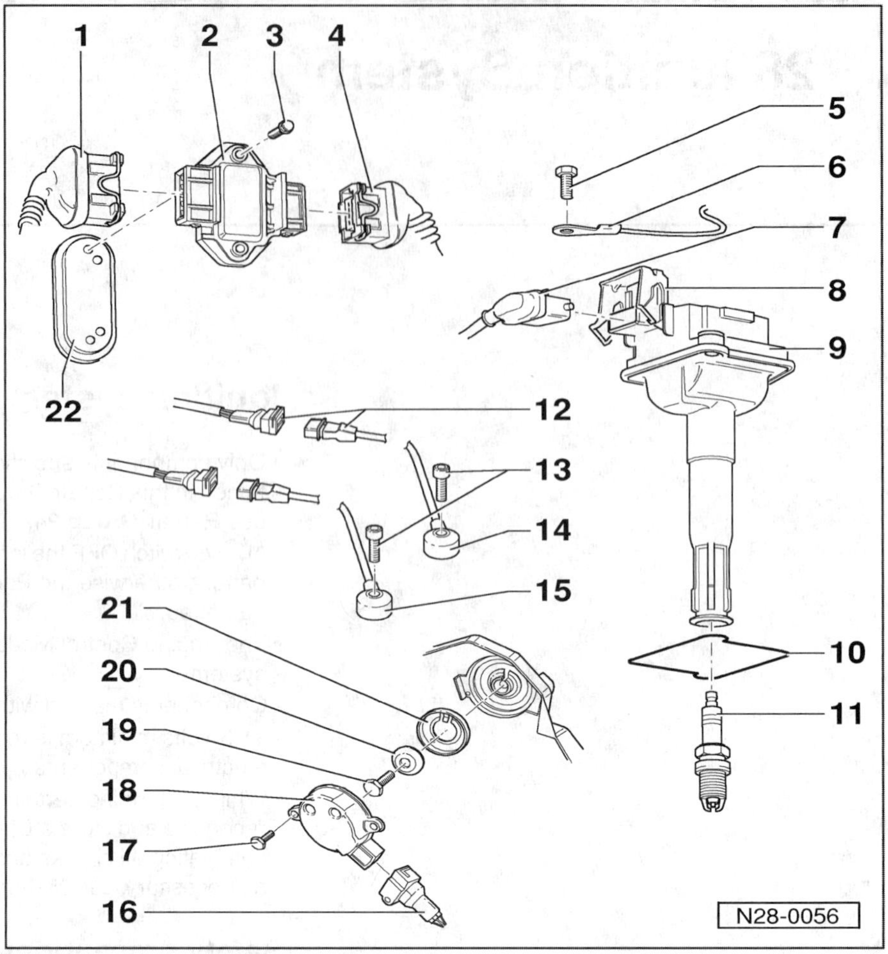

1. **5-pin harness connector**
   - Black
2. **Ignition coil power output stage -N122-**
   - Checking, see **Ignition coil power output stage, checking**
   - Coat lower part with heat conductive paste G 052 170 A2
3. **6 Nm (53 in. lb)**
4. **4-pin harness connector**
   - Black
5. **10 Nm (7 ft lb)**
   - Only loosen or tighten with ignition switched off
6. **Ground wire**
   - Only loosen or tighten with ignition switched off
7. **3-pin harness connector**
   - Black
8. **Connector lock**
9. **Ignition coils -N-, -N128-, -N158- and -N163-**
   - Spark plug connector can be pulled off
   - Checking, see **Ignition coil power output stage, checking**
10. **Oil seal**
    - Replace if damaged
11. **Spark plug**
    - 30 Nm (22 ft lb)
    - Remove and install using 3122B
    - Type and electrode gap, see **Test data, spark plugs**
12. **3-pin harness connector**
    - Gold plated terminals
    - Green for knock sensor 1 -G61-
    - Blue for knock sensor 2 -G66-
13. **20 Nm (15 ft lb)**
    - Tightening torque influences knock sensor function
14. **Knock sensor 1 -G61-**
    - Green
    - Gold-plated sensor and connector terminals
    - Checking, see **Knock sensor, checking**
15. **Knock sensor 2 -G66-**
    - Blue
    - Gold-plated sensor and connector terminals
    - Checking, see **Knock sensor, checking**
16. **3-pin harness connector**
    - Black
    - for Camshaft Position sensor -G40-
17. **10 Nm (7 ft lb)**
18. **Camshaft Position sensor -G40-**
    - Checking, see **Camshaft position sensor, checking**
19. **25 Nm (18 ft lb)**
20. **Conical washer**
21. **Hood**
    - for Camshaft Position sensor -G40-
    - When installing, note installation arrangement
22. **Heat sink**

Technical data, spark plugs

Engine code	AEB
Firing order	1-3-4-2
Spark plugs: VW/Audi Manufacturer's number	 101 000 051 AA F 7 LTCR
Electrode gap	0.9 to 1.1 mm (0.35 to 0.43 in.)
Tightening torque	30 Nm (22 ft lb)

Camshaft position sensor, checking

Special tools, testers, measuring instruments and auxiliary items required

- VAG 1598/22 Test Box
- Fluke 83 multimeter
- VW 1594 Adaptor kit
- Wiring diagram

Test conditions

- Battery voltage must be 11.5 Volts minimum

Test sequence

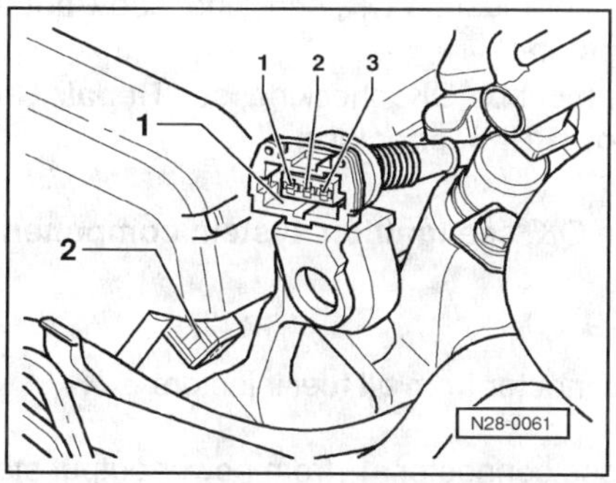

◄ Disconnect 3-pin harness connector -1- off Camshaft Position sensor -2-.

– Switch multimeter to Voltage measurement range

– Connect multimeter between outer terminals of connector using jumper wires from VW 1594

– Switch ON ignition.

- Specification: min. 4.5 Volts

– Switch OFF ignition.

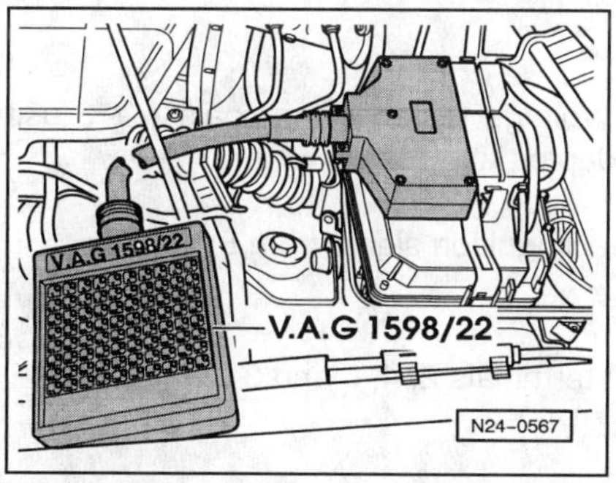

◄ Connect test box VAG 1598/22 to Engine Control Module harness connector.

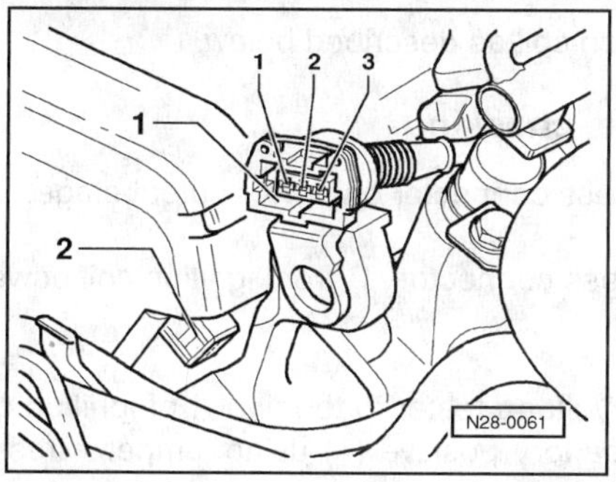

◄ Check wiring between Test Box and harness connector using wiring diagram:

- Terminal 1 + socket 11
- Terminal 2 + socket 76
- Terminal 3 + socket 67
- Wire resistance: Max. 1.5 Ω

– Check wires for shorting to one another.

- Specification: ∞Ω

– If wiring OK and voltage present between terminals 1 + 3
  - Replace Camshaft Position sensor.

– If wiring OK and no voltage present between terminals 1 + 3
  - Replace Engine Control Module, see **Electrical Equipment section, Repair group 97**.

EDITOR'S NOTE —

*ECM replacement requires coding of control module using a diagnostic scan tool, see **Scan Tool section**. Consult an authorized Volkswagen service department for additional information.*

Ignition coil power output stage, checking

Special tools, testers, measuring instruments and auxiliary items required

- VAG 1598/22 Test Box
- Fluke 83 multimeter
- VW 1594 Adaptor kit
- VAG 1527 B Voltage tester
- Wiring diagram

Checking requirements

- Battery voltage must be 11.5 Volts minimum
- Camshaft Position sensor must be OK, checking, see **Camshaft position sensor, checking**.
- Engine Speed sensor must be OK, checking, see **Repair Group 24, Engine Speed sensor (RPM), checking**.
- Fuse 32 OK.
- Ground wires must be OK, see **Ignition system components** (-6-).

Activation, checking

– Disconnect harness connector from all fuel injectors.

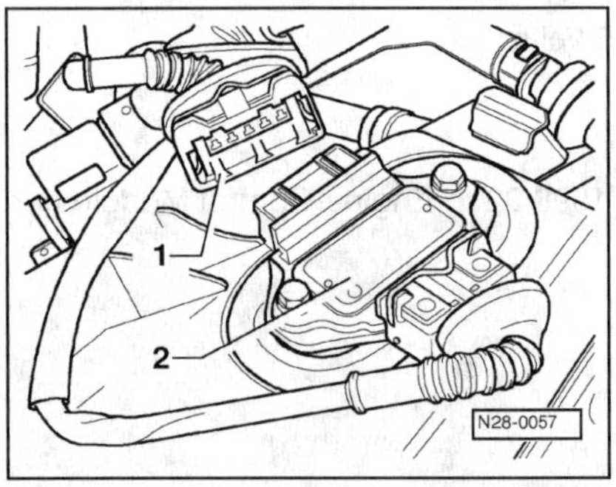

◄ Disconnect 5-pin harness connector -1- from power output stage for ignition coils -2-.

– Connect VAG 1527 B Voltage tester to terminals 1 + 3 using jumper wires from VW 1594 adaptor kit.

– Operate starter and check ignition signal from ECM.
  - LED must flicker

– Repeat check between terminals 2, 4, 5 and 3 (Ground).

– If LED does not flicker:
  - Check wiring to ignition coil as described below.

Power Output stage, checking

– Re-connect 5-pin harness connector on power output stage.

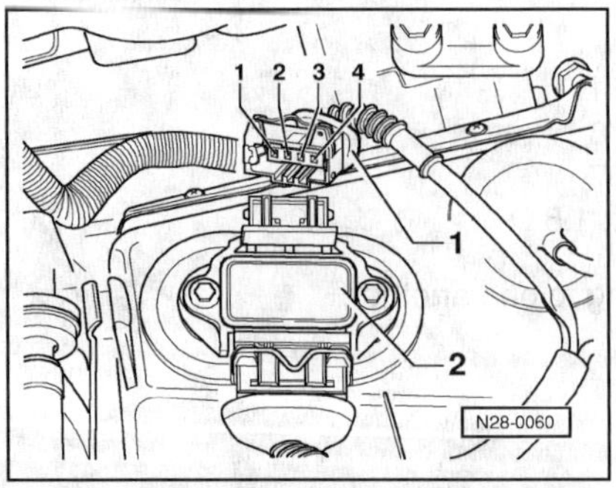

◄ Disconnect 4-pin harness connector -1- from ignition coil power output stage -2-

– Connect VAG 1527 B Voltage tester to terminal 1 of ignition coil power output stage -2- and battery positive (+) using jumper wires from VW 1594 adaptor kit.

– Operate starter.
  - LED must flicker

– Repeat check between terminals 2, 3 and 4.

– If LED does not flicker
  - Replace power output stage.

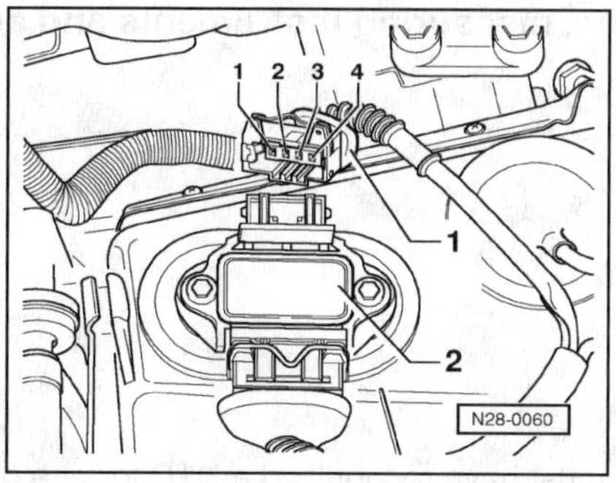

Checking ignition coils

– Switch ON ignition.

– Connect diode test lamp VAG 1527 with aux. cables from VAG 1594 to terminal 1 of 4 pin connector and to battery negative (-)
  - LED must light up

– Repeat check between terminals 2, 3 and 4.

If LED does not flicker on one terminal:

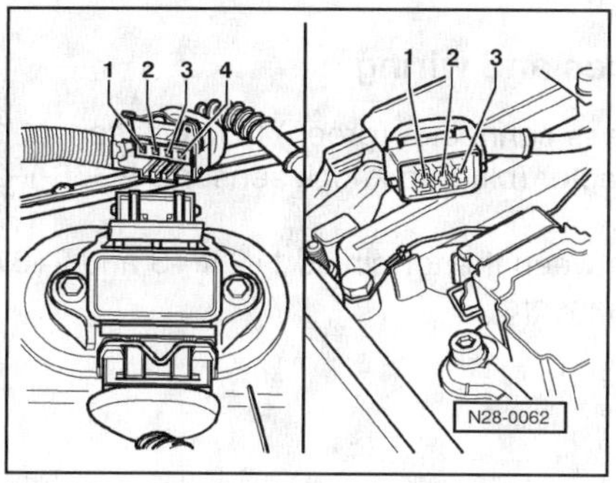

– Check for open circuit between 4-pin harness connector and (relevant ignition coil) 3-pin harness connector (terminal 1) using wiring diagram.
  - Wire resistance: Max. 1.5 Ω

– If wiring OK
  - Replace ignition coil.

Ignition wiring, checking

– Switch OFF ignition.

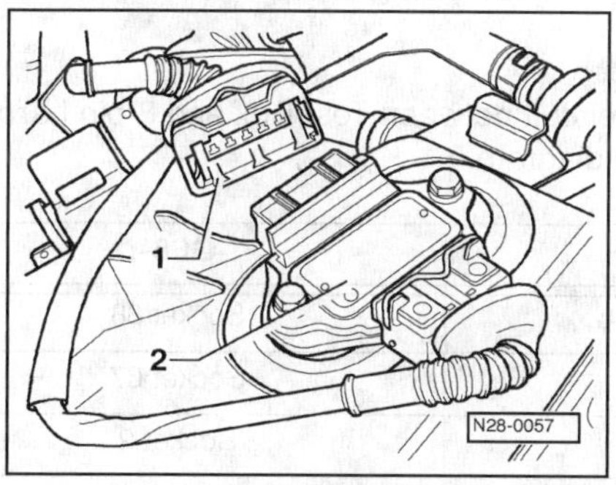

– Disconnect 5-pin harness connector -1- from ignition coil power output stage -2-.

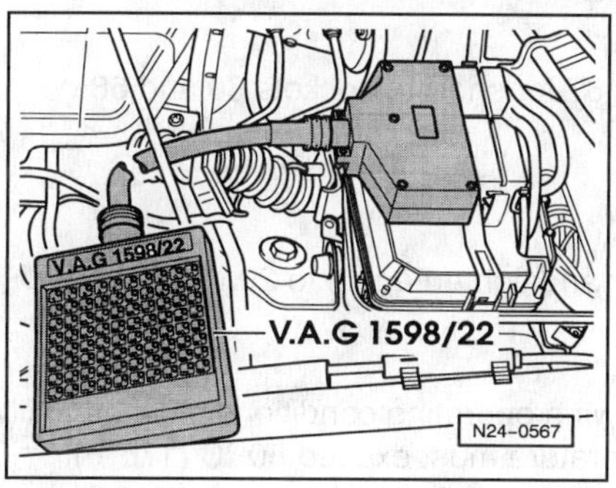

– Connect VAG 1598/22 Test Box to Engine Control Module harness connector.

– Check wiring between Test Box and 5-pin harness connector for open circuit using wiring diagram.
  - Terminal 1 + socket 70
  - Terminal 2 + socket 78
  - Terminal 3 + socket 2
  - Terminal 4 + socket 77
  - Terminal 5 + socket 71
  - Wire resistance: Max. 1.5 Ω

Knock sensor, electrically checking

NOTE —
- *Knock sensor torque must be maintained at 20 Nm (15 ft lb) to ensure the knock sensors function correctly.*
- *Only gold-plated terminals may be used when servicing the knock sensor connector terminals.*

Special tools, testers, measuring instruments and auxiliary items required

- VAG 1598/22 Test Box
- Fluke 83 multimeter
- VW 1594 Adaptor kit
- VAG 1527 B Voltage tester
- Wiring diagram

Test conditions

- On Board Diagnostic must have recognized a DTC for one or both knock sensors.

Checking resistances and wiring

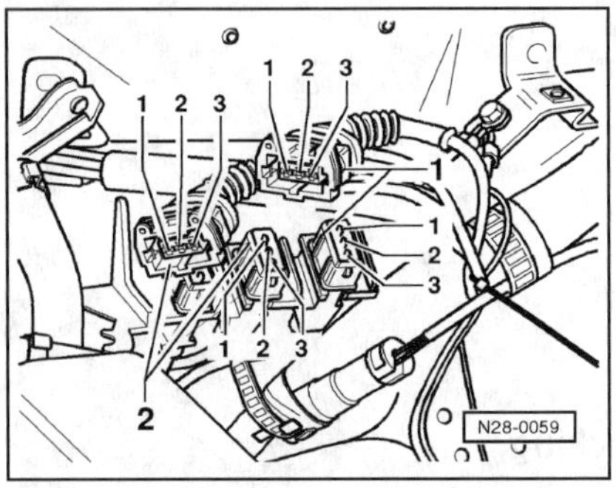

◄ Disconnect 3-pin harness connector (green) to knock sensor 1 (-G61-) -1- and / or 3-pin connector (blue) to knock sensor 2 (-G66-) -2-.

– Measure resistance between the terminals 1+2, 1+3 and 2+3 at connections to knock sensors.

- Specification ∞Ω

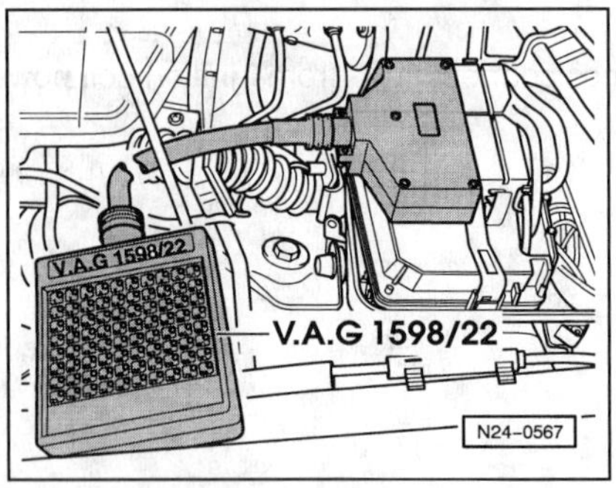

◄ Connect test box VAG 1598/22 to Engine Control Module wiring harness.

– Check wiring for open circuit between Test Box and 3-pin harness connector using wiring diagram.

-G61-:	-G66-:
Terminal 1	Socket 68
Terminal 2	Socket 67
Terminal 3	Socket 2

Wire resistance: Max. 1.5Ω

– Check wiring at socket 67 for short to sockets 60 and 68.

- Specification: ∞Ω

– If wiring OK

- Loosen knock sensor and tighten again to 20 Nm (15 ft lb).

– Perform road test.

- During road test following operating conditions must be fulfilled:
- Engine coolant temperature must exceed 80 °C (176°F).

After Engine Coolant Temperature is reached, the following operating conditions must be attained several times: Closed throttle (idle), Partial throttle, Wide Open Throttle Overrun

– At Wide Open Throttle, engine speed must exceed 3500 rpm.

– Check ECM Diagnostic Trouble Code memory using scan tool, see **Maintenance**.

– If knock sensor DTC still stored

- Replace knock sensor(s)

Misfiring detection, checking

Special tools, testers, measuring instruments and auxiliary items required

- VAG 1551/1552 Scan Tool with VAG 1551/3 cable

– Connect VAG 1551/1552 Scan Tool, see **Maintenance**.

– Start engine and let idle

– Press 0 and 1 buttons to select Address word 01: "Engine electronics"

– Press 0 and 8 buttons to select Function 08: "Read measuring value blocks"

– Press Q button to enter input

– Press 0, 1 and 4 buttons to select "Display group 014"

– Press Q button to enter input

– Check specification for misfire detection (display zones 3 to 4):

	Display zones			
	1	2	3	4
Display group 14: Misfiring recognition				
Display	xxx rpm	xx.xx ms	xxx	Active Blocked
Indicated	Engine speed (in 40rpm steps)	Engine load	Misfiring adversely affecting exhaust (total)	Misfiring recogni-tion status
Working range	0 to 6800 rpm	0.00 to 8.50 ms	0 to 500	---
Specific.	820 to 900 rpm	0.5 to 1.5 ms	0 to 20	Activated

– If specification is attained:

- Press →button.
- Press 0 and 6 buttons to select Function 06: "End data transfer"
- Press Q button to enter input

– If specifications are not obtained:

- Press C button.
- Press 0, 1 and 5 buttons to select "Display group 015"
- Press Q button to enter input

– Check specification for misfire detection (display zones 1 to 4):

	Display zones			
	1	2	3	4
Display group 15: Misfire recognition				
Display	xxx	xxx	xxx	Active Blocked
Indicated	Misfiring adversely affecting No. 1 Cyl.	Misfiring adversely affecting No. 2 Cyl.	Misfiring adversely affecting No. 3 Cyl.	Misfire recognition status
Working range	0 to 500	0 to 500	0 to 500	---
Specification	0 to 5	0 to 5	0 to 5	Activated
If specification not obtained, see **Evaluating display groups 15 and 16**				

– If specification attained:

- Press C button.
- Press 0, 1 and 6 buttons to select "Display group 016"
- Press Q button to enter input

– Check specification for misfire recognition, display zones 1 and 4

	Display zones			
	1	2	3	4
Display group 16: Misfiring recognition				
Display	xxx	---	---	Active Blocked
Indicated	Misfiring adversely affecting No. 4 Cyl.	---	---	Misfire detection status
Working range	0 to 500	---	---	---
Specification	0 to 5	---	---	Active
If specification not obtained, see **Evaluating display groups 15 and 16**				

Evaluating display groups 15 and 16

Display groups: 15 and 16 Display zone: 1 to 4	Possible cause of malfunction	Malfunction elimination
More than 5	Ignition coil faulty Spark plug connector faulty Spark plug faulty Ignition coil output stage faulty	- Check ignition coil Power Output stage, see **Ignition coil power output stage, checking**
	Injector faulty Fuel shortage	- Check fuel injectors, see **Repair Group 24** - Check quantity of fuel in tank

1.8 Liter Fuel Injection and Ignition (code: ATW, AUG, AWM)

Section

G

24 Multiport Fuel Injection (MFI)

Page

28 Ignition

Page

Section 24

1.8 Liter Fuel Injection and Ignition (code: ATW, AUG, AWM)

1 Multiport Fuel Injection (MPI) Page

2 Ignition

24 Multiport Fuel Injection (MFI)

Repair Group

G 24

Fuel Injection System

NOTE—

- *The Engine Control Module (ECM) is equipped with an On Board Diagnostic (OBD) Always check DTC memory before performing repairs or troubleshooting. Vacuum hoses and connections should also be checked (unmetered air).*
- *Fuel hoses in engine compartment may only be secured with spring-type clamps. The use of clamping collars or screw clamps is not permitted.*
- *Components marked with * are checked via On Board Diagnostic (OBD).*
- *Components marked with ** can be checked via output Diagnostic Test Mode (DTM).*
- *For the electric components to work properly, a voltage of at least 11.5 V is required.*
- *Do not use any silicone sealant. Traces of silicone components which are sucked into the engine are not burned there and damage the oxygen sensor.*
- *It is possible that the control module will recognize a malfunction and store a DTC during some tests. Therefore, after all tests and repairs have been performed, DTC memory should be checked and erased if necessary, see* **Maintenance**.

Safety precautions

WARNING—

Fuel system is under pressure! Before loosening hose connections or opening the test connection, place rags around the connection point. Then release pressure by carefully pulling on hose or unscrewing locking screw.

To reduce the risk of personal injury and/or damage to the fuel injection and ignition system, always observe the following:

- Do not touch or disconnect ignition wires when engine is running or turning at starting RPM.
- Only disconnect and reconnect wires for injection and ignition system, including test leads, when ignition is turned off.

If special testing equipment is required during road test, note the following:

- Test equipment must always be secured to the rear seat and operated from there by a second person.

If test and measuring equipment is operated from the passenger seat, the person seated there could be injured in the event of an accident involving deployment of the passenger-side airbag.

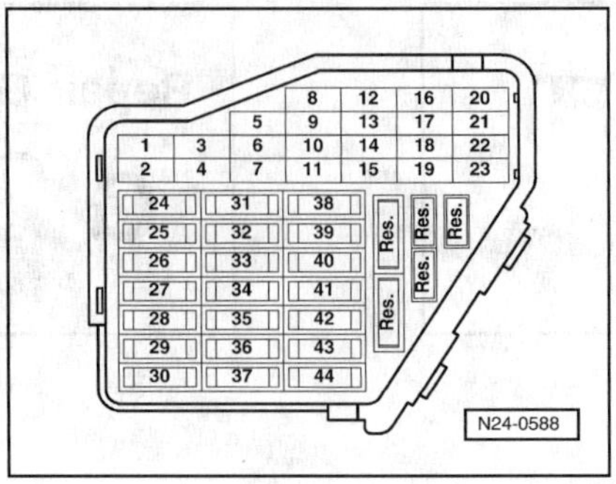

If engine is to be cranked at starting RPM without starting, remove fuses 29 and 32.

- By removing fuse 29, voltage supply to the ignition coils is interrupted.
- By removing fuse 32, voltage supply to the fuel injectors is interrupted.

> ***WARNING —***
>
> - *Fire hazard! Do not smoke, work near heaters or open flame, or have anything nearby that could ignite fuel. The fuel system is under pressure. Before opening, cover connections with a rag, and open carefully to release pressure.*
> - *When driving or riding in an airbag-equipped vehicle, NEVER hold the scan tool or other test equipment in your hands or lap while in motion. Objects between you and the airbag increase the risk of injury in an accident.*
> - *During a test drive in an airbag-equipped vehicle, test equipment must always be fastened to and operated from the rear seat by a second technician.*
> - *Be alert when working on or near the engine and ignition wires. High secondary ignition voltage can cause serious personal injury and damage vehicle components.*
> - *DO NOT touch or disconnect ignition system wires when the engine is running or being cranked at starting RPM.*
> - *DO NOT operate the starter if fuel injectors are removed.*

> ***WARNING —***
>
> *BEFORE cranking the engine at starting RPM (e.g. for compression testing) disable ignition and fuel injection systems:*
>
> - *Disconnect connector for ignition coil power output stage, and fuse for fuel pump.*
> - *Disconnect harness connectors from all fuel injectors.*

Rules of cleanliness

When working on the fuel supply or fuel injection systems, the following 5 rules of cleanliness should be carefully observed:

1. Thoroughly clean all connections and the surrounding area before disconnecting.
2. Place parts that have been removed on a clean surface and cover. Use lint-free cloths only!
3. Carefully cover over opened components or seal, if repairs are not carried out immediately.
4. Only install clean components: Only unpack replacement parts immediately prior to installation. Do not use parts that have been stored loose (in tool boxes etc.)
5. When the system is open: Do not work with compressed air if it can be avoided. Do not move vehicle unless absolutely necessary.

Technical data

Engine Code		**ATW**	**AUG**	**AWM**
Idle speed	RPM	740 to 920		
System identification		Motronic ME7.5		
ECM part number		4B0 906 018 P	4B0 906 018 BQ	4B0 906 018 CL
Engine speed limitation	RPM	from approx. 6800		

Component locations, overview

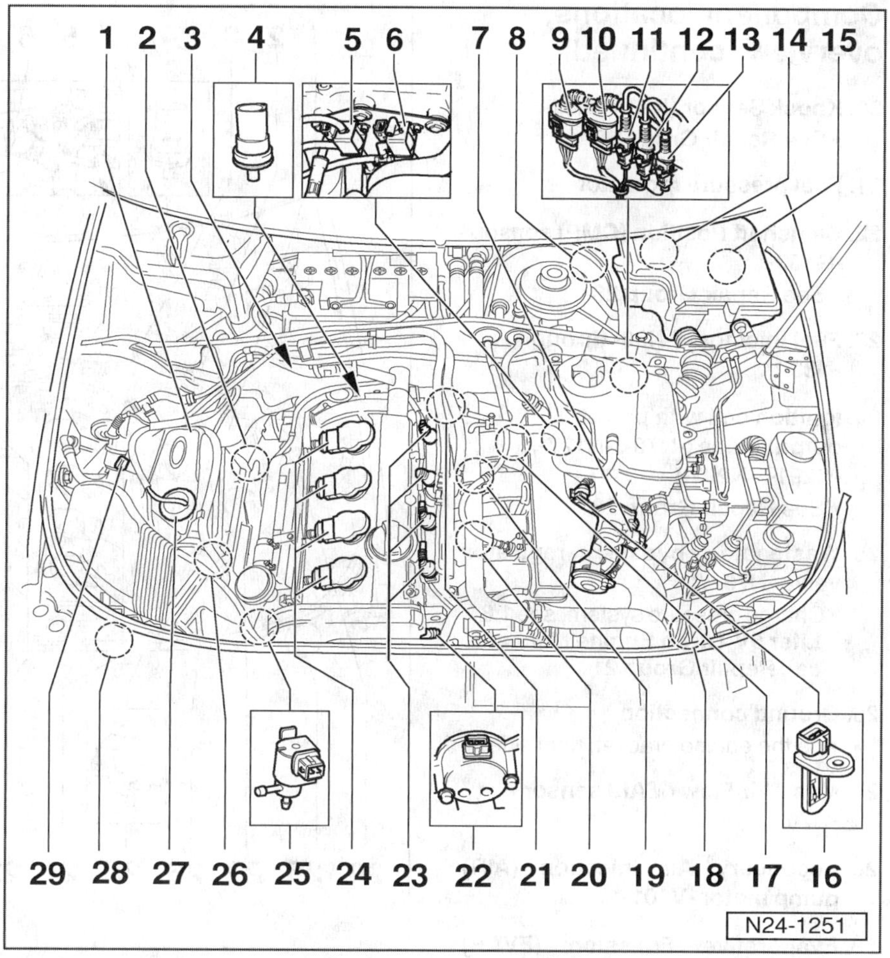

1. **Air filter housing**
   - Disassembling and assembling, see **Air filter, assembly**
2. **Heated Oxygen Sensor (HO2S) -G39-* before catalytic converter**
   - Tighten to 50 Nm (37 ft-lb)
3. **Oxygen Sensor (O2S) behind Three Way Catalytic Converter (TWC) -G130-***
   - Tighten to 50 Nm (37 ft-lb)
4. **Engine Coolant Temperature (ECT) sensor -G62-***
   - See **Fig. 2** and **Fig. 3**
5. **Recirculating valve for turbocharger -N249-*/****
6. **Secondary Air Injection (AIR) solenoid valve -N112-*/****
7. **Throttle valve control module -J338-***
   - Tighten to 10 Nm (7 ft-lb)
   - When replacing, adapt Engine Control Module (ECM), see **Throttle valve control module, adaptation to Engine Control Module (ECM)**
8. **Clutch vacuum vent valve switch -F36-*, brake light switch -F-*, brake pedal switch -F47-* and Throttle Position (TP) sensor -G79-* and sender -2- for accelerator pedal position -G185-***
   - In driver-side footwell
9. **4-pin harness connector**
   - Engine code ATW: Black for Heated Oxygen Sensor (HO2S) 1 before Three Way Catalytic Converter (TWC) -G39-
   - Engine codes AUG, AWM: Brown for Heated Oxygen Sensor (HO2S) 2 behind Three Way Catalytic Converter (TWC) -G130-

Engine code ATW:

10. **4-pin harness connector**
    - Brown for Heated Oxygen Sensor (HO2S) 2 behind Three Way Catalytic Converter (TWC) -G130- and oxygen sensor heater

Engine codes AUG, AWM:

10. **6-pin harness connector**
    - Black for Heated Oxygen Sensor (HO2S) 1 before Three Way Catalytic Converter (TWC) -G39- and oxygen sensor heater

Continuation for all:

11. **3-pin harness connector**
    - Gray for Engine Speed (RPM) sensor -G28-
12. **3-pin harness connector**
    - Green for Knock Sensor (KS) 1 -G61-
13. **3-pin harness connector**
    - Blue for Knock Sensor (KS) 2 -G66-
14. **Secondary Air Injection (AIR) pump relay -J299-*/****
15. **Engine Control Module (ECM)***
    - Checking voltage supply, see **ECM voltage supply, checking**
    - Procedure following interruption of voltage supply, see **Electrical Equipment section, Repair Group 27**
    - Replacing, see **Electrical Equipment section, Repair Group 97**
16. **Intake Air Temperature (IAT) sensor -G42-***
    - Tighten to 10 Nm (7 ft-lb)
    - See **Fig.1** and **Fig.3**
17. **Engine speed sensor -G28-***
    - Inductive sensor
    - Tighten to 10 Nm (7 ft-lb)
18. **Charge air pressure sensor -G31-***
19. **Knock Sensor (KS) 2 -G66-***
    - See **Repair Group 28**

(Continued on following page)

Component locations, overview (continued)

20. **Knock Sensor (KS) 1 -G61-***
    - See **Repair Group 28**
21. **Fuel pressure regulator**
22. **Camshaft Position (CMP) sensor -G163-***
    - See **Repair Group 28**
23. **Fuel injectors (-N30- through -N33-)***
24. **Ignition coil with power output stage (-N70-, -N127-, -N291-, -N292-)**
    - See **Repair Group 28**
25. **Wastegate bypass regulator valve -N75-*/****
    - Charge pressure system, see **1.8 Liter 5V Turbo Engine Mechanical, Repair Group 21**
26. **Ground connection**
    - on the engine bracket, right side
27. **Mass Air Flow (MAF) sensor -G70-***
28. **Secondary Air Injection (AIR) pump motor -V101-***
29. **Evaporative Emission (EVAP) canister purge regulator valve -N80-*/****
    - See **1.8 Liter 5V Turbo Engine Mechanical, Repair Group 20**

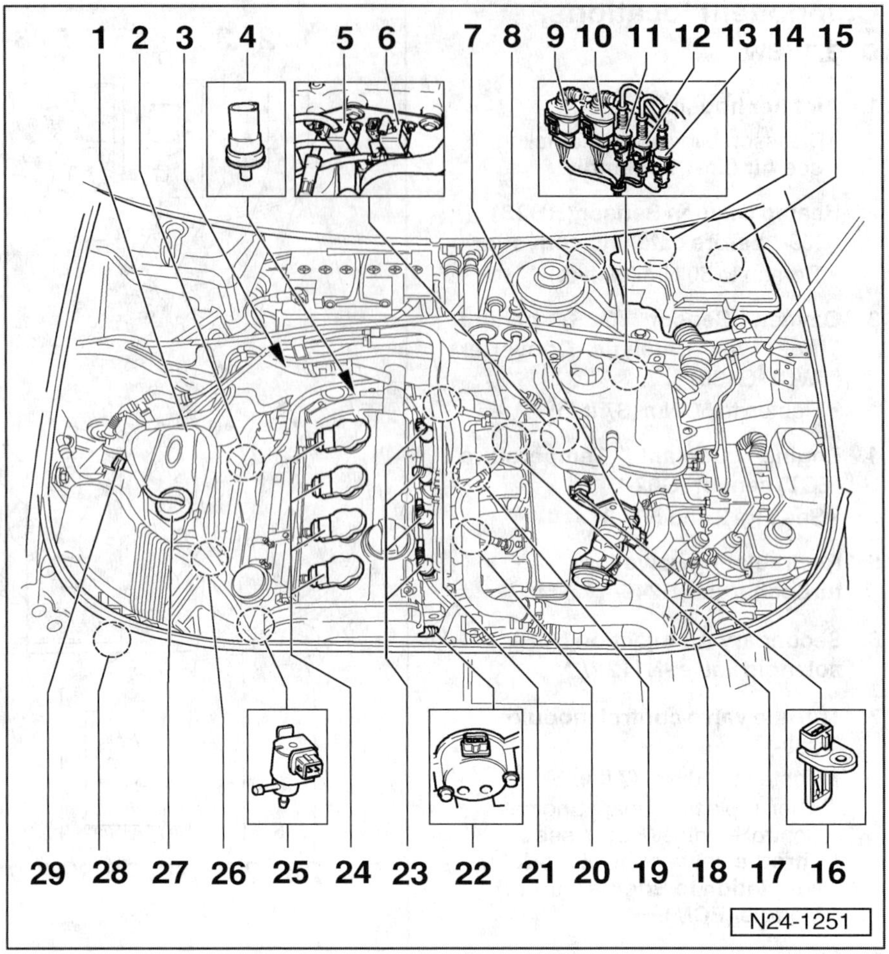

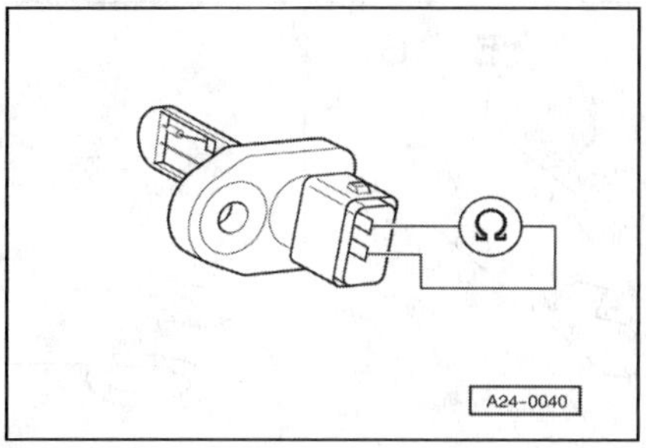

Fig.1 Intake Air Temperature sensor -G42-

- Measure resistance between IAT sensor G42 terminals 1 (signal) and 2 (Ground)

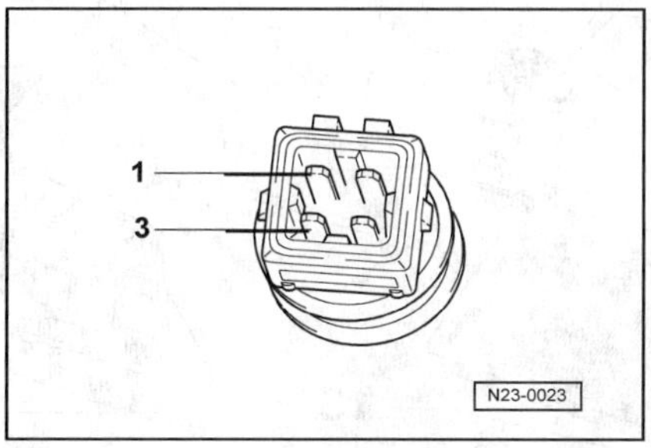

Fig. 2 Engine Coolant Temperature sensor -G62-, checking

- Perform resistance measurement at ECT sensor -G62- terminal 1 (signal) and 3 (Ground).

NOTE—

The ECM will use the Intake Air Temperature sensor signal as a replacement value for an engine start (start temperature replacement value) as soon as there is a DTC stored in the DTC memory, which effects the ECT sensor -G62-. The temperature then rises using a model stored in the control module. When the engine has reached normal working temperature, a fixed replacement value will be displayed after a certain period. This fixed value is also dependent upon the Intake Air Temperature.

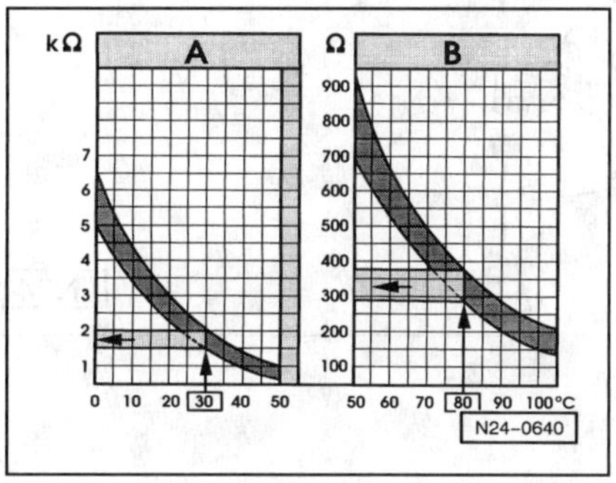

Fig. 3 Resistance graph

The diagram is valid for Engine Coolant Temperature sensor -G62- and intake air sensor -G42-.

The diagram is divided into two temperature ranges:

A - From 0 to 50 °C (32 to 122°F)

B - From 50 to 105 °C (122 to 302°F)

Examples:

- 30 °C is range A and corresponds to a resistance of 1.5 to 2.0 kΩ
- 80 °C is range B and corresponds to a resistance of 275 to 375Ω

Air filter, assembly

1. **Air ducts**
   - to metering unit
2. **Filter element**
3. **Seal**
4. **Mass Air Flow (MAF) sensor -G70-***
   - Checking, see **Mass Air Flow (MAF) sensor, checking**
5. **Air duct**
6. **6 Nm (53 in-lb)**
7. **Air filter, upper portion**
8. **Cooling body**
   - Not for engine code AWM
9. **Rubber ring**
   - Not for engine code AWM
10. **Wire**
    - For EVAP canister system
11. **10 Nm (7 ft-lb)**
12. **Heat shield**
13. **Rubber grommet**
14. **Spacer sleeve**
15. **Rubber grommet**
16. **Air filter, lower portion**

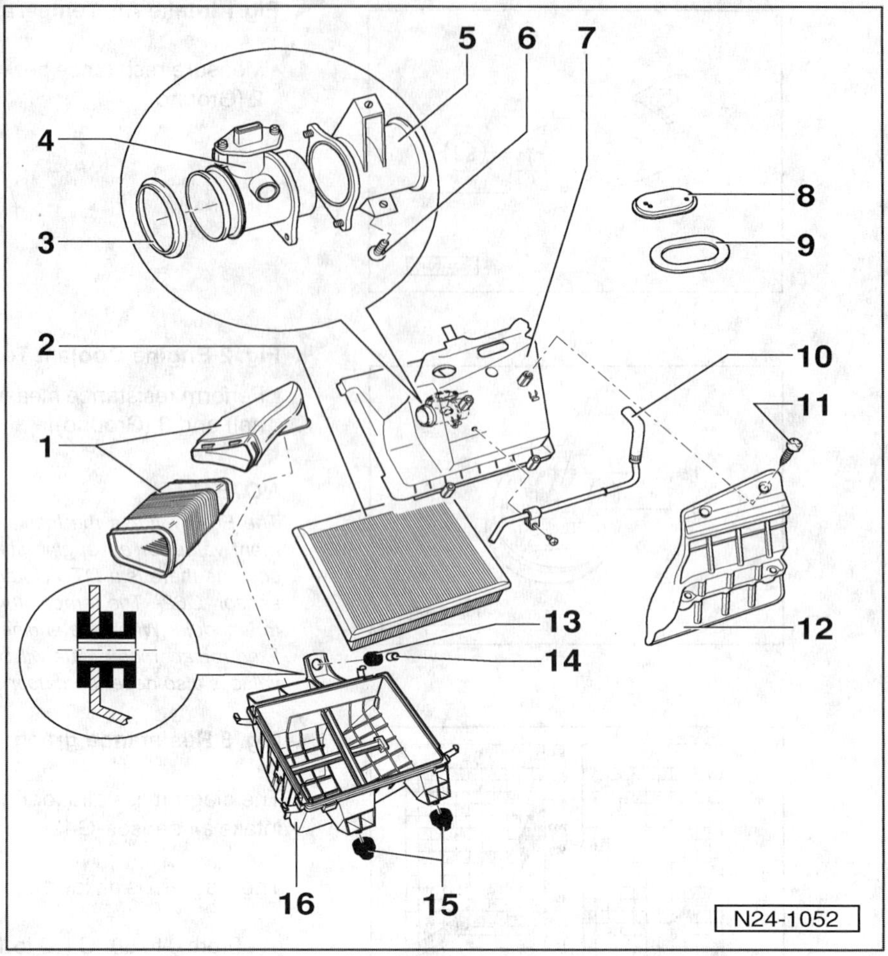

Fuel rail with fuel injectors, assembly

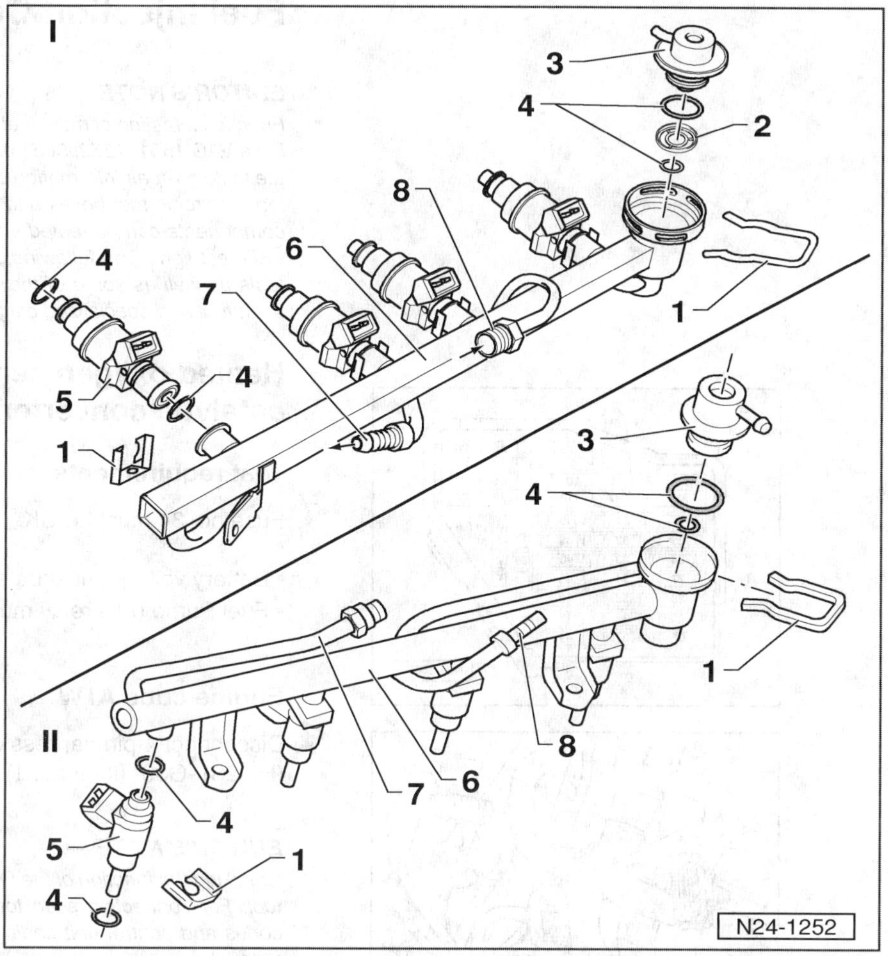

I -Engine code ATW

II -Engine codes AUG, AWM

1. **Retaining clip**
   - Make sure clip is correctly seated on fuel injector and fuel rail
2. **Sieve**
3. **Fuel pressure regulator**
   - Checking, see **Residual pressure and fuel pressure regulator, checking**
4. **O-ring**
   - Replace if damaged
5. **Fuel injectors (-N30- through -N33-)***
   - See **Fuel injectors, checking**
   - Resistance 12 to 17Ω
6. **Fuel distributor**
   - Tighten to 10 Nm (7 ft-lb)
7. **Connection, return line**
8. **Connection, supply line**

Fuel Injection Components, checking

EDITOR'S NOTE —

Functional testing of most fuel injection and ignition components is done using a VAG 1551, 1552, 5051 or equivalent aftermarket scan tool. In addition, the factory repair information calls for the use of various special tools including electronic test boxes and remote control switches. Many fuel injection components can be tested electrically using a digital multimeter and/or an LED test light. The following pages include many of these basic electrical tests as well as some of the testing that relies on the use of a diagnostic scan tool and specialized diagnostic equipment.

Heated oxygen sensor (HO2S) -G39- before catalytic converter, electrically checking

Test requirements

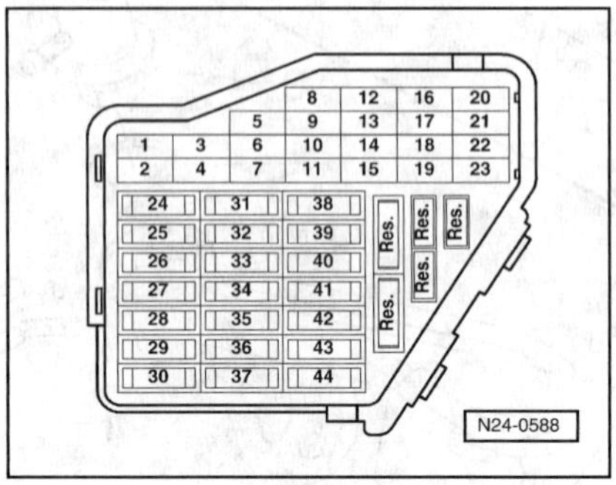

Fuse no. 34 must be OK.

- Battery voltage must be at least 11.5 volts.
- Fuel Pump (FP) relay must be OK.

Engine code ATW

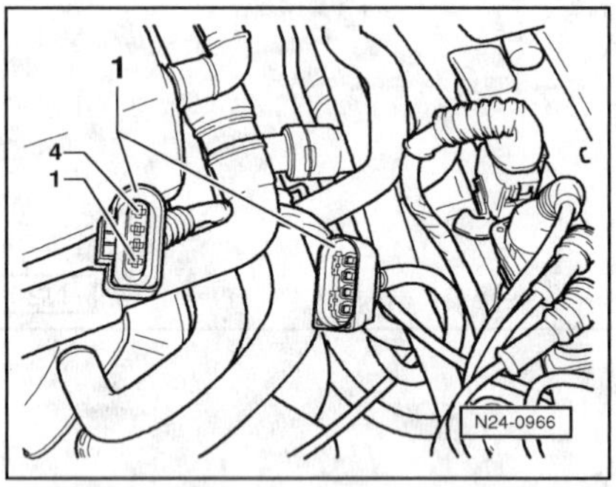

Disconnect 4-pin harness connector (black) to Heated Oxygen Sensor (HO2S) -G39- (item no. 1).

EDITOR'S NOTE —

Checking the function of the O2 sensor requires the use of a diagnostic scan tool. For connecting scan tool see, **Maintenance**. *For diagnostic trouble codes and control unit coding, see* **Scan Tool Codes** *at the back of this manual.*

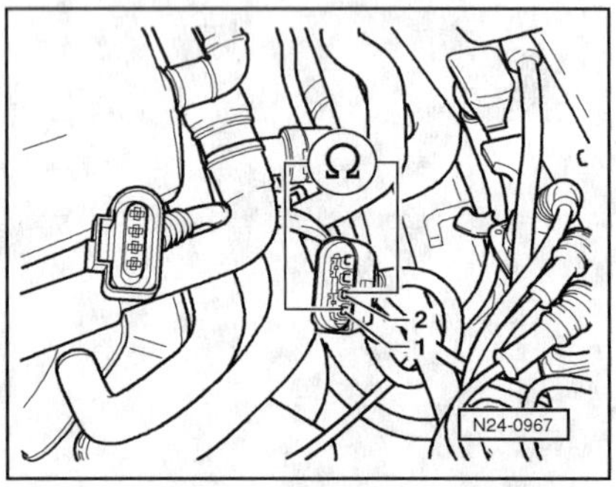

Check oxygen sensor heater for continuity at connector to Heated Oxygen Sensor (HO2S) terminals 1 + 2.

Engine codes AUG, AWM

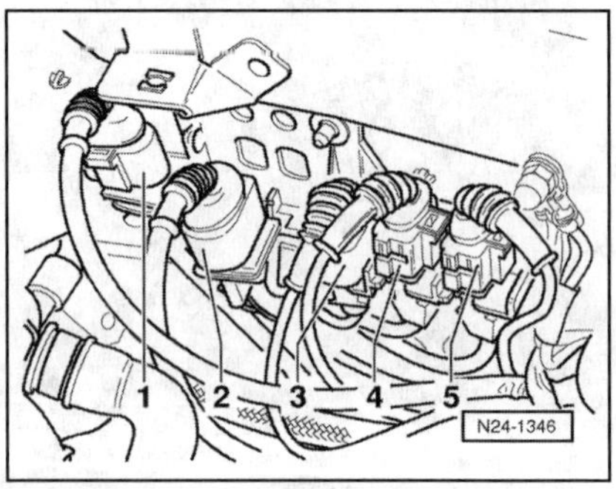

Disconnect 6-pin harness connector (black) to Heated Oxygen Sensor (HO2S) -G39- (item no. 2).

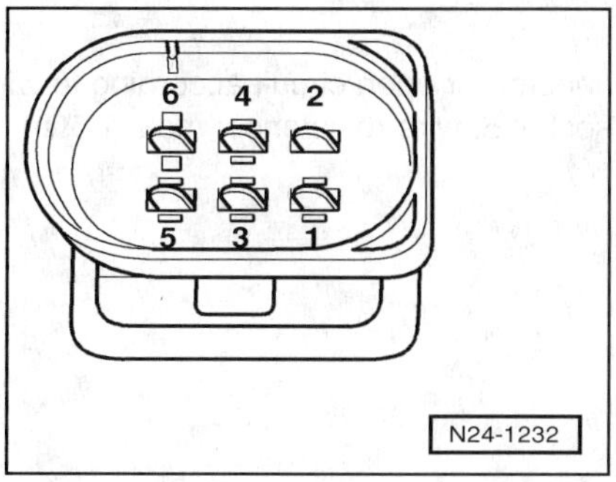

➤ Set multimeter for resistance measurement and connect it to terminals 3 + 4 at connector for oxygen sensor. Check oxygen sensor heater for continuity.

- Specification: 2.5 to 10Ω (at room temperature)

Continuation for all engine codes

– If a short circuit is found in the oxygen sensor heater:
  - Replace Heated Oxygen Sensor (HO2S) -G39-.

– If oxygen sensor heater has continuity:
  - Check voltage supply and wiring, see following procedure.

Checking voltage supply and wiring

Engine code ATW

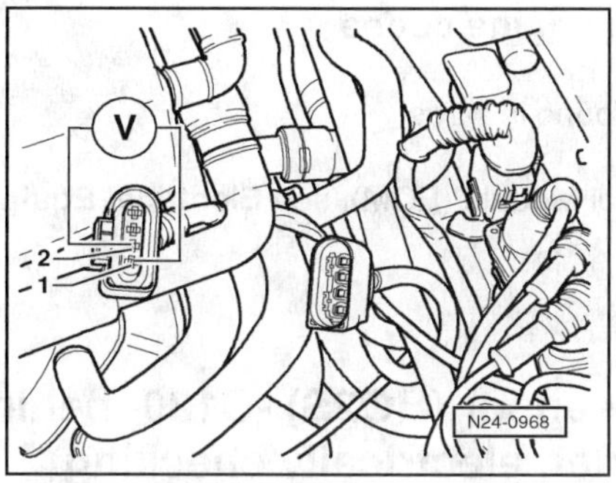

➤ Set multimeter for voltage measurement and connect it to terminals 1 + 2 at connector to control module.

Engine codes AUG, AWM

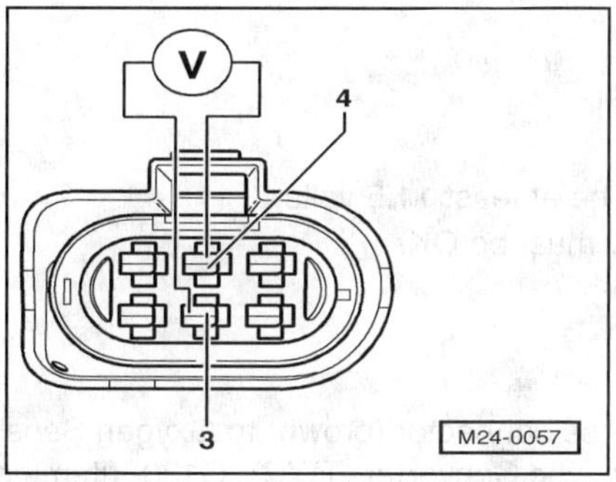

➤ Set multimeter for voltage measurement and connect it to terminals 3 + 4 at connector to control module.

Continuation for all engine codes

– Start engine, and let run at idle.

– Measure voltage supply
  - Specification: 11.0 to 14.5 V
  - Specification: oscillating between 0.0 and 12.0 V

EDITOR'S NOTE —
Oxygen sensor heater voltage will turn on and off based on engine operating parameters.

– If there is no voltage:
  - Connect multimeter to terminal 1 (engine ATW) terminal 3 (engines AUG, AWM) of connector and to engine Ground (GND) for a voltage measurement.
  - Specification: 11 to 15 V

– If there is no voltage again:
  - Check wire connection from terminal 1 (engine ATW) terminal 3 (engines AUG, AWM) to Fuel Pump (FP) relay -J17- according to wiring diagram, see **Wiring Diagrams and Component Locations.**

– If voltage was approximately same as battery voltage:
  - Switch ignition off.

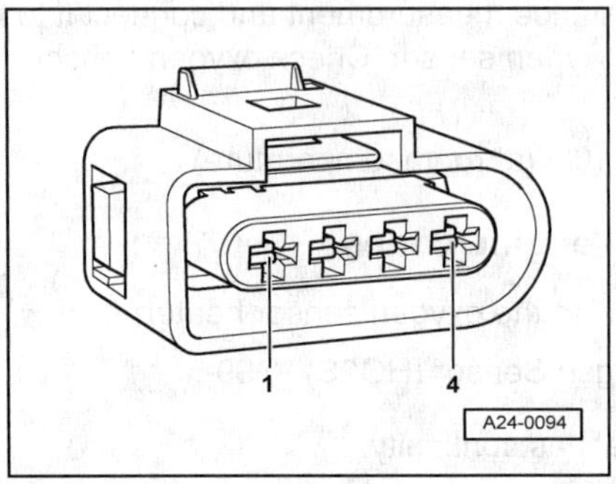

Engine code ATW

Check wire at 4-pin connector for open circuit according to wiring diagram. Terminal 2 + socket 5, wire resistance: max: 1.5Ω

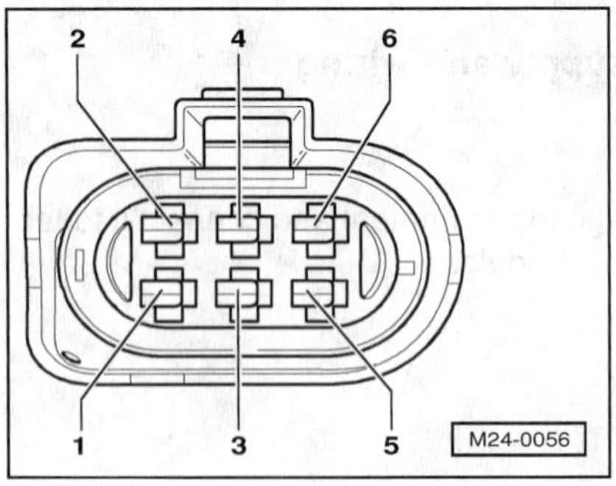

Engine codes AUG, AWM

Check wire at 6-pin connector for open circuit according to wiring diagram. Terminal 4 + socket 5, wire resistance: max: 1.5Ω

Continuation for all engine codes

If no malfunctions are found in wires:

– Replace Engine Control Module (ECM), see **Electrical Equipment section, Repair Group 97**.

Heated oxygen sensor (HO2S) -G130- behind catalytic converter, electrically checking

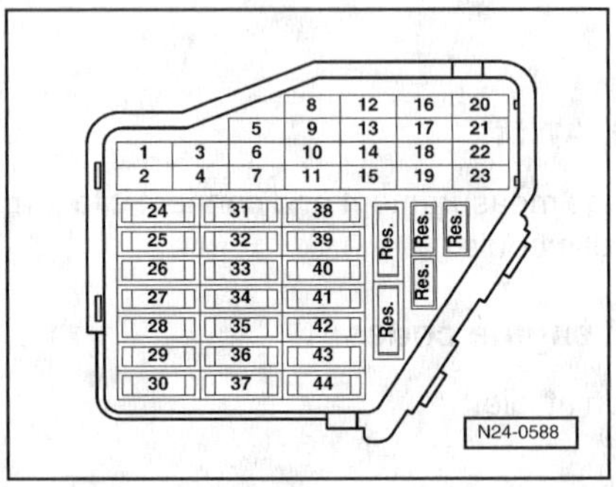

Test requirements

Fuse no. 34 must be OK.

- Battery voltage must be at least 11.5 volts.
- Fuel Pump (FP) relay must be OK.

Disconnect 4-pin harness connector (brown) to Oxygen Sensor (O2S) behind Three Way Catalytic Converter (TWC) -G130- (item no. 1).

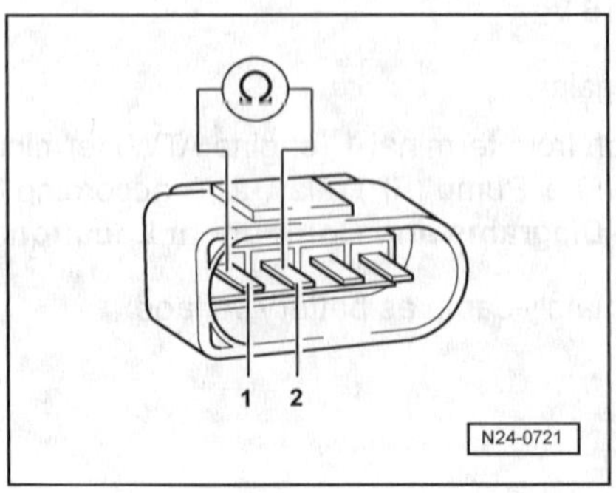

Set multimeter for resistance measurement and connect it to terminals 1 + 2 at connector for oxygen sensor. Check oxygen sensor heater for continuity.

- Specification: 8 to 20 Ω

– If a short circuit is found in the oxygen sensor heater:

- Replace Oxygen Sensor (O2S) Behind Three Way Catalytic Converter (TWC) -G130-.

– If oxygen sensor heater has continuity:

- Check voltage supply and wiring, see following procedure.

Checking voltage supply and wiring

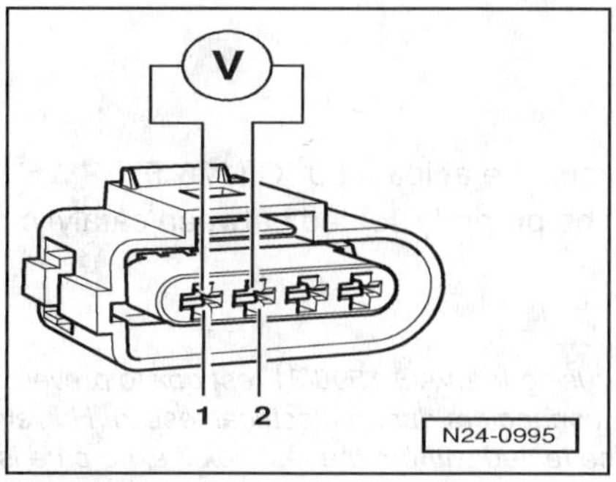

◄ Set multimeter for voltage measurement and connect it to terminals 1 + 2 at connector to control module.

– Start engine, and let run at idle.

– Measure voltage supply
  - Specification: 11.0 to 14.5 V
  - Specification: oscillating between 0.0 and 12.0 V

EDITOR'S NOTE —

Oxygen sensor heater voltage will turn on and off based on engine operating parameters.

– If there is no voltage:
  - Connect multimeter to terminal 1 of connector and engine Ground (GND) for voltage measurement.
  - Specification: 11 to 15 V

– If there is no voltage again:
  - Check wire connection from terminal 1 to Fuel Pump (FP) relay -J17- according to wiring diagram, see **Wiring Diagrams and Component Locations.**

If voltage was approximately same as battery voltage:

– Switch ignition off.

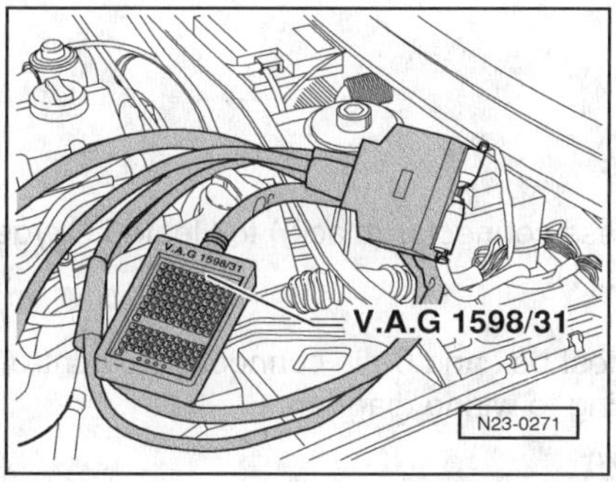

◄ Connect VAG1598/31 adapter cable to control module wiring harness. Engine Control Module (ECM) is not connected.

EDITOR'S NOTE —

Volkswagen recommends using the VAG 1598/31 test box to prevent damage to ECM harness connectors during electrical testing. However the harness connectors can be tested without the test box if extra care is used to prevent damage to the harness connector terminals.

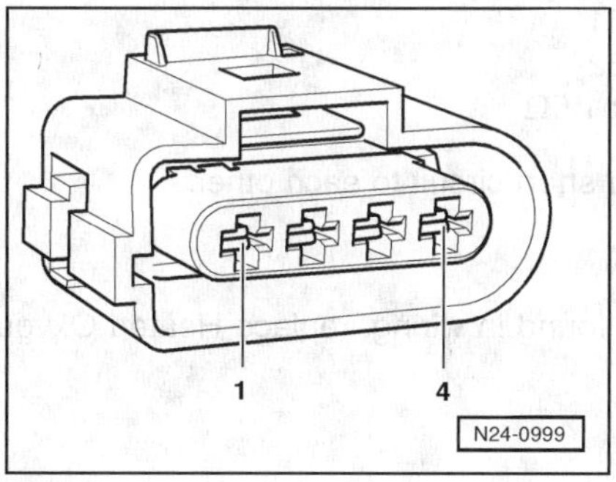

◄ Check wire between test box and 4-pin connector for open circuit according to wiring diagram.
  - Terminal 2 + socket 63
  - Wire resistance: max: 1.5Ω

If no malfunctions are found in wires:

– Replace Engine Control Module (ECM), see **Electrical Equipment section, Repair Group 97**.

Oxygen sensor control before catalytic converter, checking

Test requirements

- Coolant temperature must be at least 80 °C (176°F)
- Exhaust system must be properly sealed between catalytic converter and cylinder head.

EDITOR'S NOTE —

Volkswagen recommends using the VAG 1598/31 test box to prevent damage to ECM harness connectors during electrical testing. However the harness connectors can be tested without the test box if extra care is used to prevent damage to the harness connector terminals.

ATW engines

– Disconnect 4-pin harness connector (black) to Heated Oxygen Sensor (HO2S) -G39-.

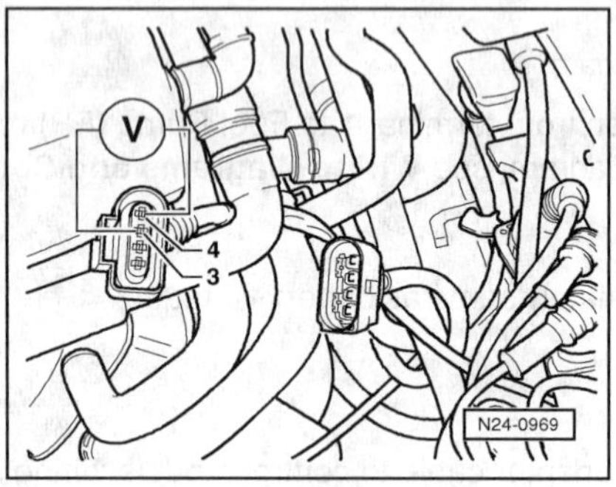

◄ Check wires between test box and 6-pin connector (to control module) for open circuit according to wiring diagram.

- Terminal 4 + socket 70
- Terminal 3 + socket 51
- Wire resistance: max 1.5Ω

– Check wires at connector terminal 3+4 for short circuit to terminal 1+2 according to wiring diagram.

If no malfunctions are found in wiring, replace Heated Oxygen Sensor (HO2S) -G39-.

AUG, AWM engines

– Disconnect 6-pin harness connector (black)) to Heated Oxygen Sensor (HO2S) -G39- (item no. 1).

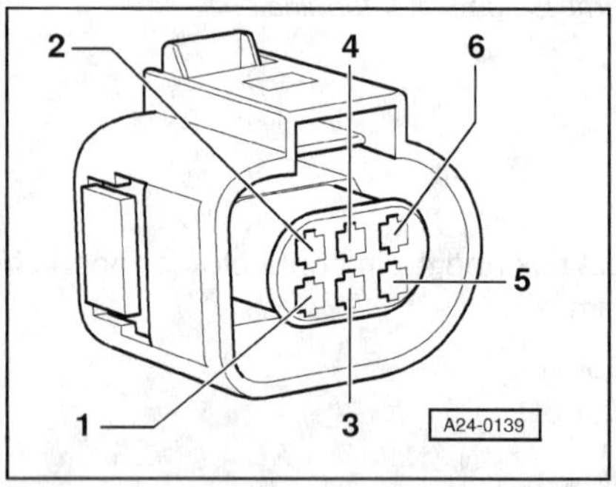

◄ Check wires between test box and 6-pin connector (to control module) for open circuit according to wiring diagram.

- Terminal 1 + socket 70
- Terminal 2 + socket 71
- Terminal 4 + socket 05
- Terminal 5 + socket 51
- Terminal 6 + socket 52
- Wire resistance: max 1.5Ω

– Also check all wires for short circuit to each other.

- Specification: ∞Ω

If no malfunctions are found in wiring, replace Heated Oxygen Sensor (HO2S) -G39-.

Oxygen sensor control behind catalytic converter, checking

Special tools and equipment

- VAG1788/10 RPM adjuster
- VAG1526 multimeter or VAG1715 multimeter
- VAG1594 connector test kit
- Wiring diagram

Test requirements

- Exhaust system must be properly sealed between catalytic converter and cylinder head.
- Coolant temperature must be at least 85 °C (185°F)

Checking primary voltage

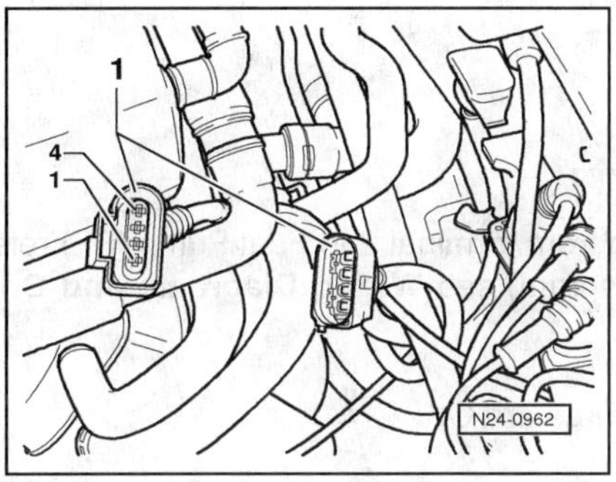

Disconnect 4-pin harness connector (brown) to Oxygen Sensor (O2S) behind Three Way Catalytic Converter (TWC) -G130- (-1-).

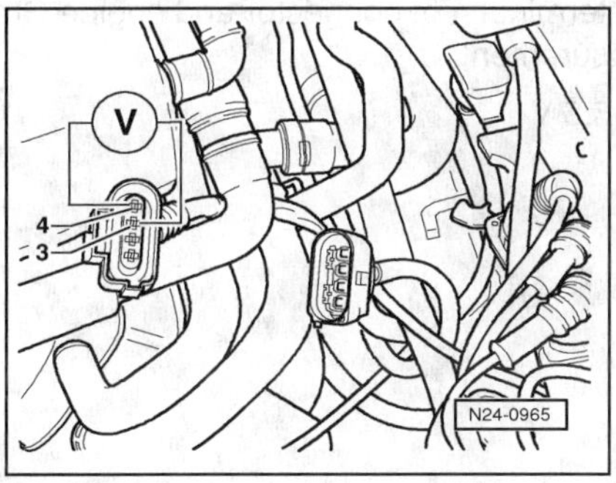

Use adapter cables from VAG1594 to connect multimeter to terminals 3 + 4 (Motronic ECM connector) for voltage measurement.

– Start engine and measure primary voltage.
  - Specification: 0.40 to 0.50 volts

– Switch ignition off.

– If specified value is not obtained:
  - Test oxygen sensor wires, see **Wiring Diagrams and Component Locations**.

– If no malfunctions are found in wires:
  - Replace Engine Control Module (ECM), see **Electrical Equipment section, Repair Group 97**.

EDITOR'S NOTE —

ECM replacement requires coding of control module using Volkswagen factory scan tool. Consult an authorized Volkswagen service department for additional information.

– If specified value is obtained:
  - Replace Oxygen Sensor (O2S) behind catalytic converter -G130-.

Mass Air Flow (MAF) sensor voltage supply, checking

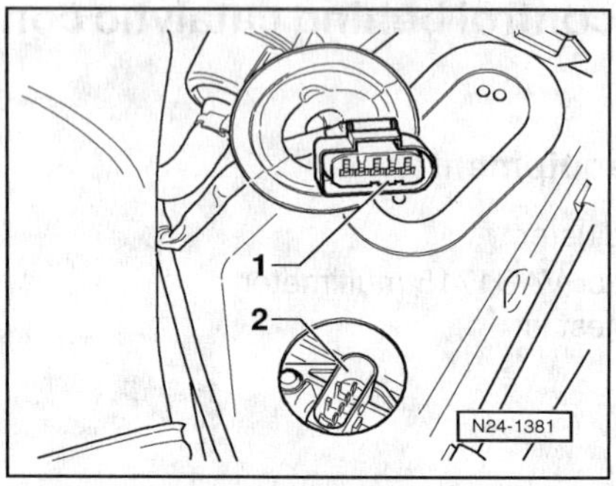

EDITOR'S NOTE —

Checking the function of the MAF sensor requires the use of a diagnostic scan tool. For connecting scan tool see, **Maintenance**. *For diagnostic trouble codes and control unit coding, see* **Scan Tool Codes** *at the back of this manual.*

◄ Disconnect 5-pin connector -1- from Mass Air Flow (MAF) sensor -2-.

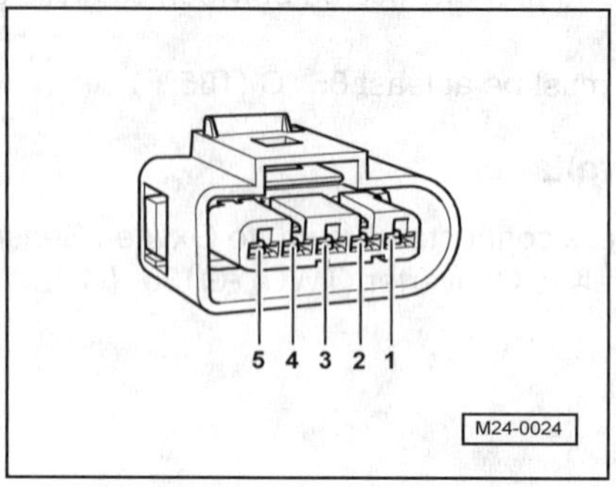

◄ Connect multimeter to terminal 2 of connector and engine Ground (GND) for voltage measurement.

– Start engine, and let run at idle.
  - Specification: 11 to 15 V

  If there is no voltage:

– Switch ignition off.

– Check wire connection from terminal 2 to Fuel Pump (FP) relay -J17- according to wiring diagram, see **Wiring Diagrams and Component Locations.**

– If voltage supply and wires are OK:
  - Switch ignition on.

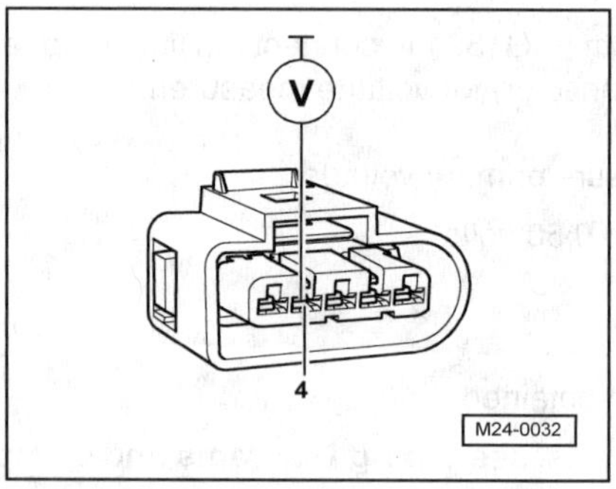

◄ Connect multimeter to terminal 4 of connector and engine Ground (GND) for voltage measurement:
  - Specification: 4.5 to 5.5 V

– If there is no voltage:
  - Switch ignition off.

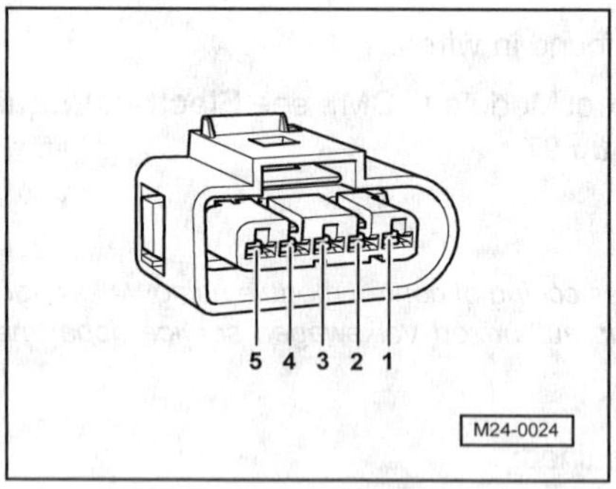

◄ Check wires between ECM and 5-pin connector for open circuits according to wiring diagram.

EDITOR'S NOTE —

Volkswagen recommends using the VAG 1598/31 test box to prevent damage to ECM harness connectors during electrical testing. However the harness connectors can be tested without the test box if extra care is used to prevent damage to the harness connector terminals.

- Terminal 5 + socket 29
- Terminal 4 + socket 53
- Terminal 3 + socket 27
- Wire resistance: max: 1.5Ω

– Also check wires for short circuit to each other.

– Also check wires for short circuits to B+ or Ground (GND).

– If no DTCs are found in wires:
  - Replace Mass Air Flow (MAF) sensor -G70-

Throttle valve control module, checking

NOTE—

When the throttle valve control module is replaced, it is absolutely necessary to adapt the new control module to the Engine Control Module (ECM), see **Throttle valve control module, adaptation to Engine Control Module (ECM)**.

Test requirements

- Throttle valve must not be damaged or dirty.
- Coolant temperature must be at least 80 °C (176°F)
- Ignition switched on, engine not running

Checking voltage supply and wiring to control module

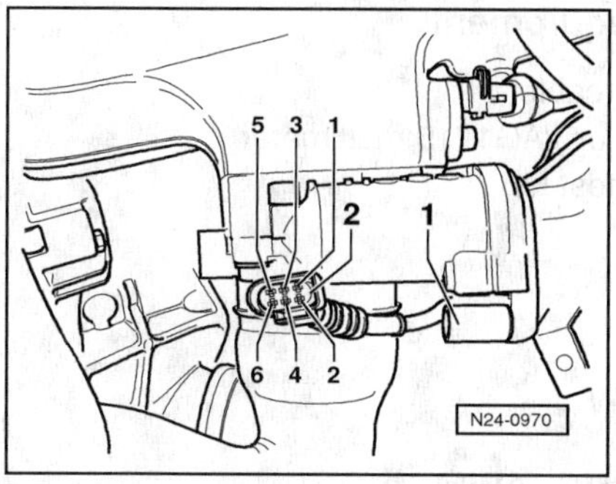

< Disconnect 6-pin connector -2- at the throttle valve control module.

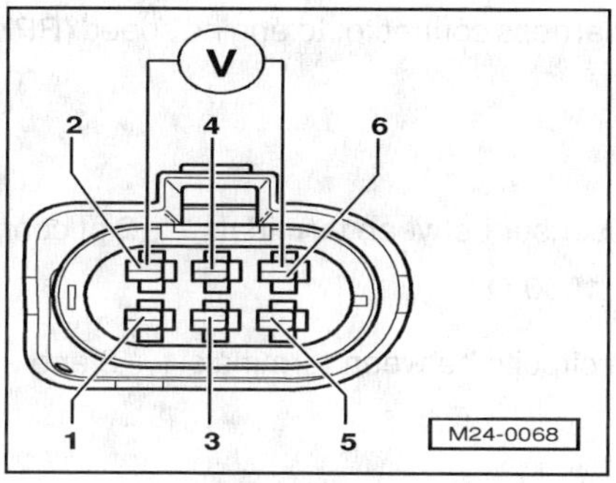

< Connect multimeter to terminals 2 + 6 of connector to make voltage measurement.

– Switch ignition on.

- Specification: at least 4.5 V

– Switch ignition off.

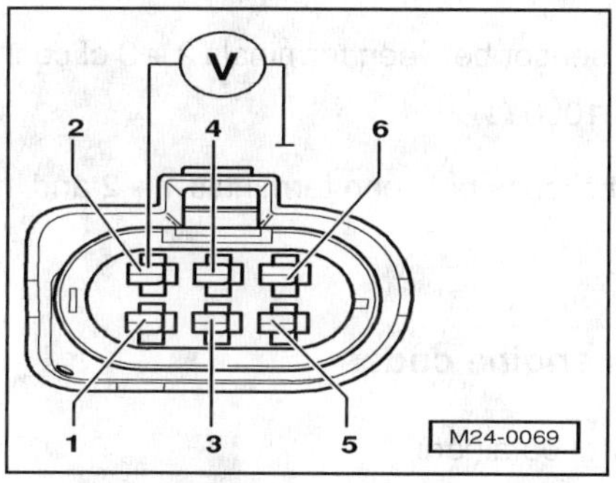

< Connect multimeter to terminal 2 of connector and Ground (GND) for voltage measurement.

– Switch ignition on.

- Specification: at least 4.5 V

– Switch ignition off.

If specified values are not obtained:

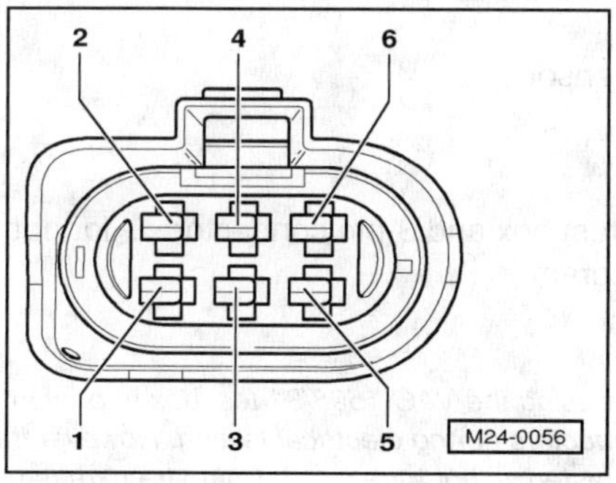

< Check for open circuits in wires between test box and connector according to wiring diagram:

- Terminal 1 + socket 92
- Terminal 2 + socket 83
- Terminal 3 + socket 117
- Terminal 4 + socket 84
- Terminal 5 + socket 118
- Terminal 6 + socket 91
- Wire resistance: max: 1.5Ω

– Also check wires for short circuit to each other.

– Also check wires for short circuits to B+ or Ground (GND).

– If no malfunctions are found in wires:

- Check voltage supply of Engine Control Module (ECM), see **ECM voltage supply, checking**.

Engine speed sensor, checking

The engine speed (RPM) sensor -G28- detects engine speed and reference marks. Without an engine speed signal, the engine will not start. If the engine speed signal fails while the engine is running, the engine will stop immediately.

Special tools and equipment

- VAG1598/31 adapter cable
- VAG1526 multimeter or VAG1715 multimeter
- VAG1594 connector test kit
- Wiring diagram

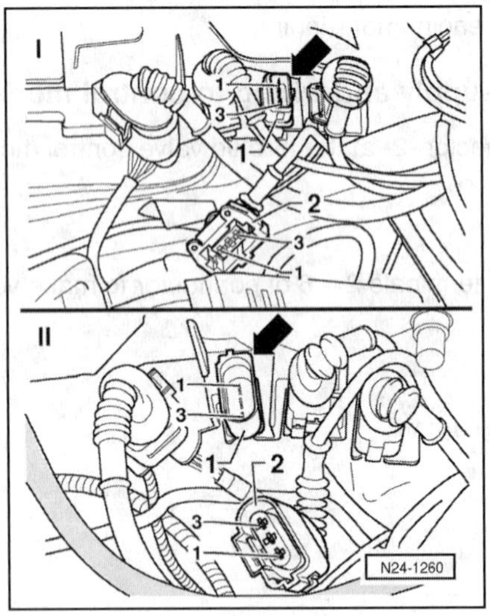

Test sequence

I - Engine code ATW

II - Engine codes AUG, AWM

Disconnect gray 3-pin harness connector to engine speed (RPM) sensor (arrow).

Engine code ATW

– Measure resistance of sensor between terminals 1 + 2 of connector -1-.

- Specification: 400 to 1000 Ω

– Check sensor for short circuits between terminals 1 + 3 and 2 + 3.

- Specification: ∞Ω

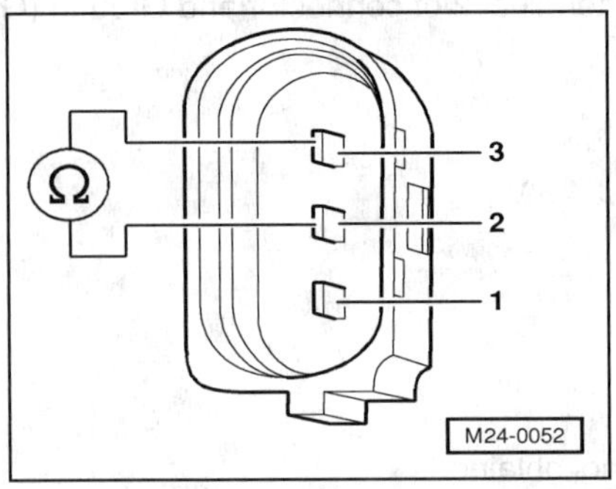

Engine codes AUG, AWM

Measure resistance of sensor between terminals 2 + 3 of connector -1-.

- Specification: 400 to 1000 Ω

– Check sensor for short circuits between terminals 1 + 2 and 1 + 3.

Specification: ∞ Ω

Continuation for all engine codes

If specified values are not obtained:

– Replace engine speed (RPM) sensor -G28-.

If no DTC is found on sensor:

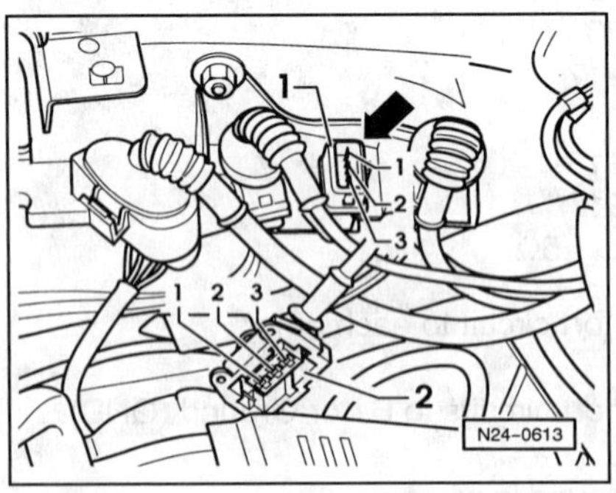

Engine code ATW

Check wires between test box and 3-pin connector -2- for open circuits according to wiring diagram.

EDITOR'S NOTE —

Volkswagen recommends using the VAG 1598/31 test box to prevent damage to ECM harness connectors during electrical testing. However the harness connectors can be tested without the test box if extra care is used to prevent damage to the harness connector terminals.

- Terminal 3 + socket 108
- Terminal 2 + socket 90
- Terminal 1 + socket 82
- Wire resistance: max: 1.5Ω

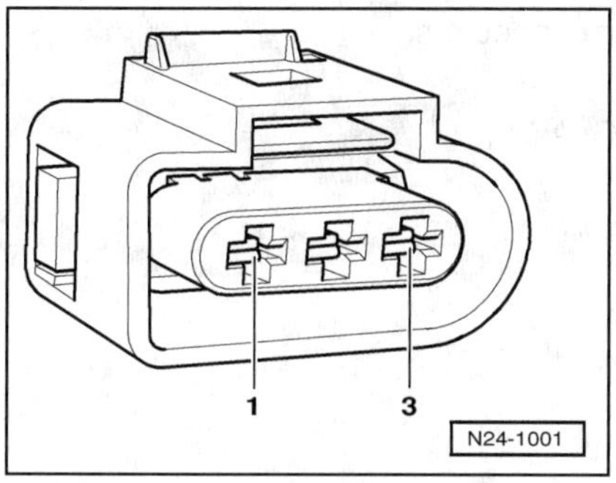

Engine codes AUG, AWM

◀ Check wires between ECM connector and 3-pin connector for open circuits according to wiring diagram.

- Terminal 3 + socket 82
- Terminal 2 + socket 90
- Terminal 1 + socket 108
- Wire resistance: max: 1.5Ω

Continuation for all engine codes

– Also check wires for short circuit to each other.

– If no malfunctions are found in wires:

- Remove sensor and check sensor wheel for secure fit, damage, and run-out.

***NOTE*—**

There is a larger-sized gap on the sensor wheel. This gap is a reference mark, not damage.

– If nothing seems to be wrong with the sensor wheel:

- Replace Engine Control Module (ECM), see **Electrical Equipment section, Repair Group 97**.

Fuel injectors, checking

Special tools and equipment

- VAG1598/31 adapter cable
- VAG1526 multimeter or VAG1715 multimeter
- VAG1527A voltage tester
- VAG1594 connector test kit
- Wiring diagram

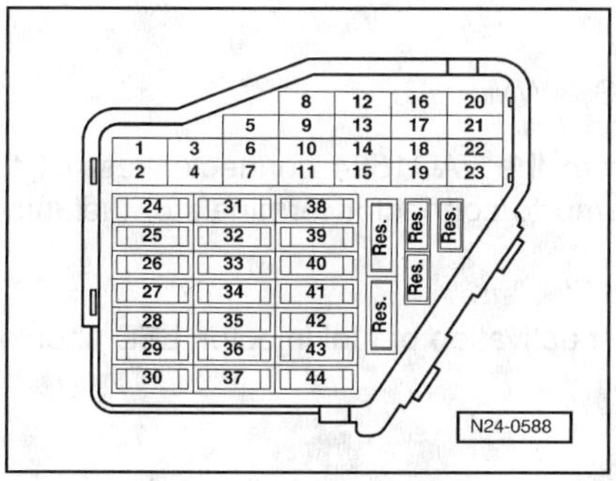

Test requirements

- Engine speed (RPM) sensor must be OK
- Fuel Pump (FP) relay must be OK.

◀ Fuses no. 28 and 32 must be OK.

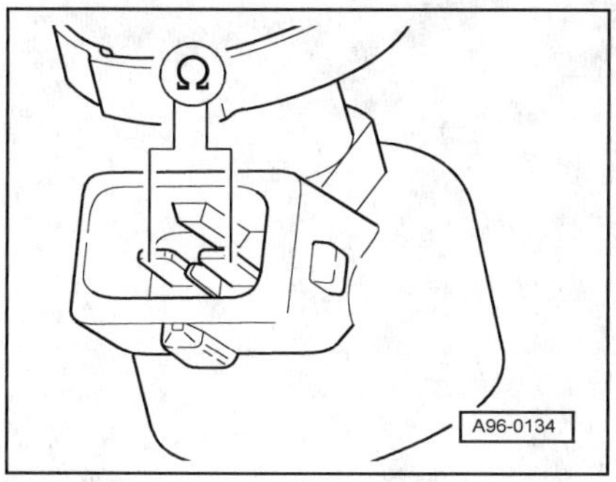

Resistance values of fuel injectors, checking

◀ Test resistance of fuel injectors between terminals.

- Specification: 12 to 17Ω

***NOTE*—**

Resistance value is valid at approx. 20 °C (68°F). At higher temperatures, resistance value increases.

– If specified value is not obtained:

- Replace faulty fuel injector

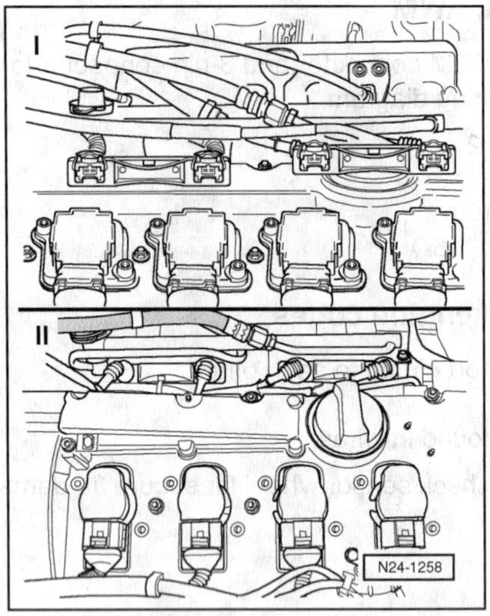

Checking voltage supply

Disconnect fuel injector connectors.

- I - Engine code ATW
- II - Engine codes AUG, AWM

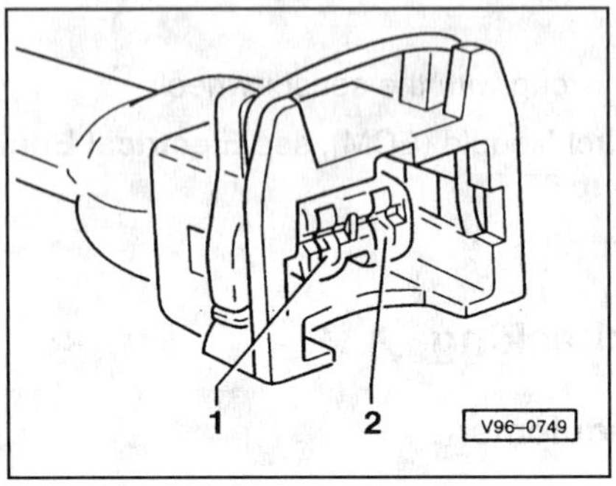

Connect LED test lamp to connector of tested fuel injector terminal 1 and Ground (GND).

– Operate starter and test voltage supply for fuel injector. LED must light up.

– If LED does not light up:

- Switch ignition off.
- Check wire between 2-pin connector terminal 1 and Fuel Pump (FP) relay -J17- for open circuit according to wiring diagram.
- Wire resistance: max: 1.5Ω

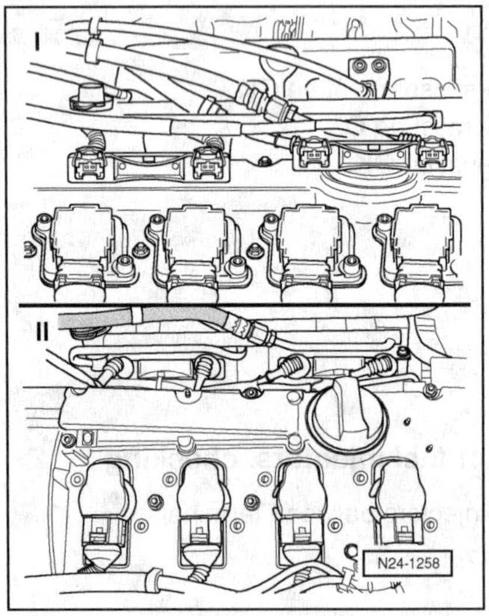

Activation and voltage supply, checking

Disconnect fuel injector connectors.

- I - Engine code ATW
- II - Engine codes AUG, AWM

– Use adapter cables from the VAG1594 connector test kit to connect VAG1527 diode test lamp to connector terminals of fuel injector to be tested.

– Operate starter and test activation of fuel injector. LED should flicker.

If LED does not flicker:

– Switch ignition off.

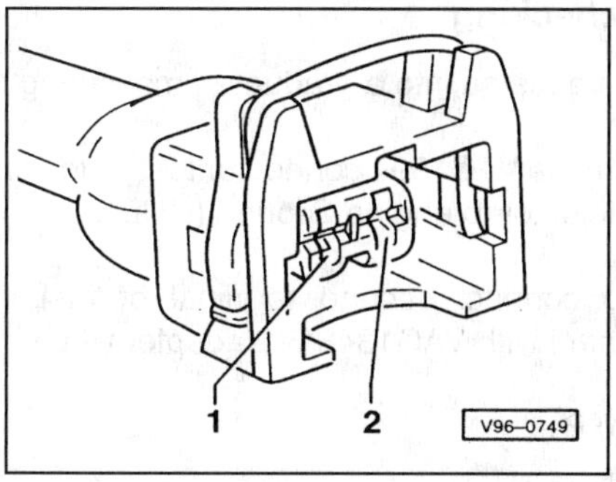

Use wiring diagram to check wires between ECM connector and harness connector for open circuits or short circuits to Ground (GND) or B+.

EDITOR'S NOTE—

Volkswagen recommends using the VAG 1598/22 test box to prevent damage to ECM harness connectors during electrical testing. However the harness connectors can be tested without the test box if extra care is used to prevent damage to the harness connector terminals.

- Fuel injector 1: Terminal 2 + socket 96
- Fuel injector 2: Terminal 2 + socket 89
- Fuel injector 3: Terminal 2 + socket 97
- Fuel injector 4: Terminal 2 + socket 88
- Wire resistance: max: 1.5Ω

– If there is no malfunction in wire connection:

- Replace Engine Control Module (ECM), see **Electrical Equipment section, Repair Group 97**.

Fuel injection quantity and injector leakage, checking

Test requirement

- Fuel pressure must be OK, checking, see **Fuel pressure regulator and residual pressure, checking**.

WARNING—

Fuel system is under pressure! Before opening system, place rags around the connection point. Then release pressure by carefully loosening connection.

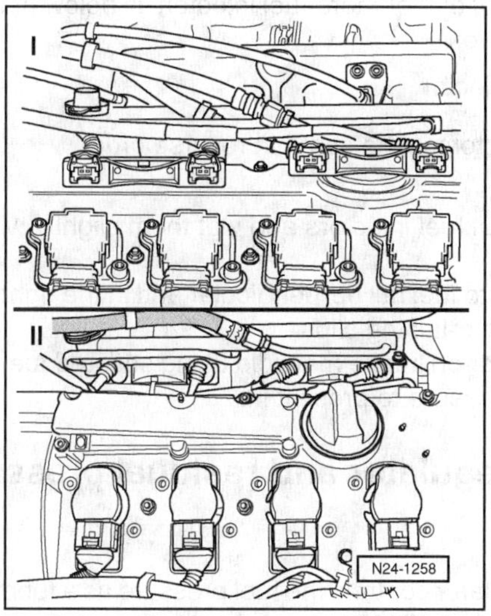

Disconnect fuel injector connectors.

- I - Engine code ATW
- II - Engine codes AUG, AWM

– Remove mounting bolts from fuel distributor.

– Remove fuel distributor completely.

– Disconnect vacuum hose from fuel pressure regulator.

– Lift fuel rail with fuel injectors off intake manifold and support it.

– Bridge test box pins 1 and 65 using auxiliary wires from VAG1594 connector test kit.

NOTE—

The only purpose of this step is to allow the Fuel Pump (FP) to run with the engine at standstill.

– Switch ignition on.

Injector seal, checking

– Check proper seal of fuel injectors (visual inspection). With Fuel Pump (FP) running, only 1 to 2 drops per minute should escape from each valve.

– If fuel loss is greater:

- Switch off ignition and replace leaking fuel injector.

NOTE—

Always use new sealing rings.

Injection quantity, checking

– Place a fuel injector to be tested into a graduated measuring glass.

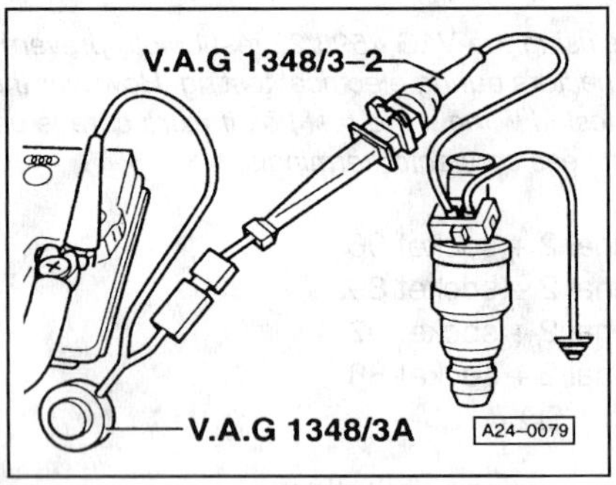

– ◄ Use adapter cables from the VAG1594 connector test kit to connect one terminal of tested fuel injector to engine Ground (GND).

– Use adapter cable to connect second terminal of fuel injector to VAG1348/3 remote control with VAG1348/3-2 adapter wire.

– Connect alligator clip to B+.

– Bridge ECM connector pins 1 and 65 using auxiliary wires from VAG1594 connector test kit.

NOTE —
The only purpose of this step is to allow the Fuel Pump (FP) to run with the engine at standstill.

– Switch ignition on.

– Operate fuel pump for 30 seconds.

– Repeat test on other fuel injectors. Use new graduated measuring glasses for this.

– After all fuel injectors have been activated, place graduated measuring glasses on a level surface and compare quantity of injected fuel.

• Specification: 133 to 157 ml. per injector

– If the measured value of one or more fuel injectors is below or above the indicated specified value:

• Replace faulty fuel injector

– Installation of fuel injectors is the same in reverse order. Note the following:

• Replace O-rings on all fuel injectors and wet them slightly with clean motor oil.
• Insert fuel injectors into fuel rail perpendicular and in the right position, and secure them with retaining clips.
• Position fuel distributor on intake manifold using secured fuel injectors and apply uniform pressure to press it in.

Fuel pressure regulator and residual pressure, checking

The fuel pressure regulator controls the fuel pressure as a function of the Manifold Absolute Pressure (MAP).

Special tools and equipment

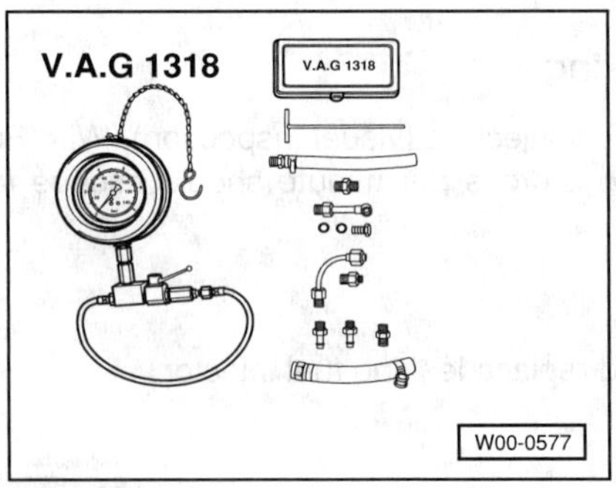

◄ VAG1318 pressure gauge

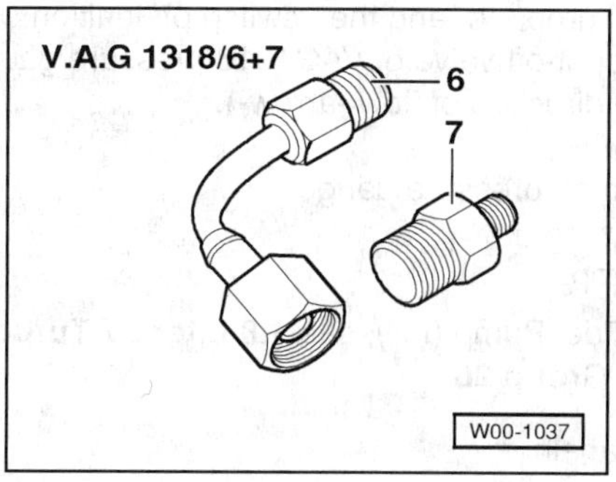

VAG1318/6+7 adapter

> ***WARNING —***
> *Fuel system is under pressure! Before opening system, place rags around the connection point. Then release pressure by carefully loosening connection.*

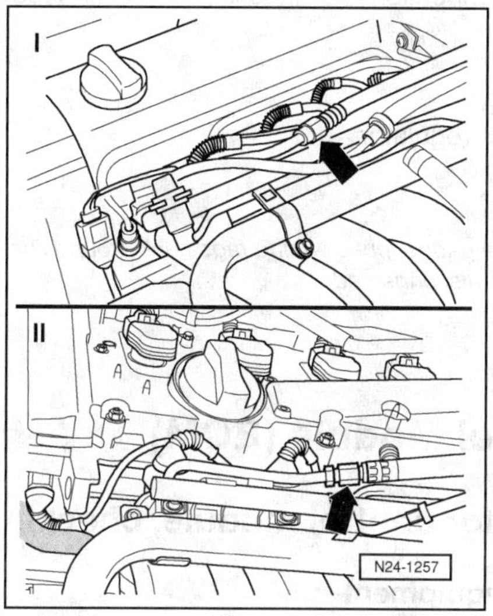

Test sequence

Open threaded connection (arrow) and catch exiting fuel using a rag.

- I - Engine code ATW
- II - Engine codes AUG, AWM

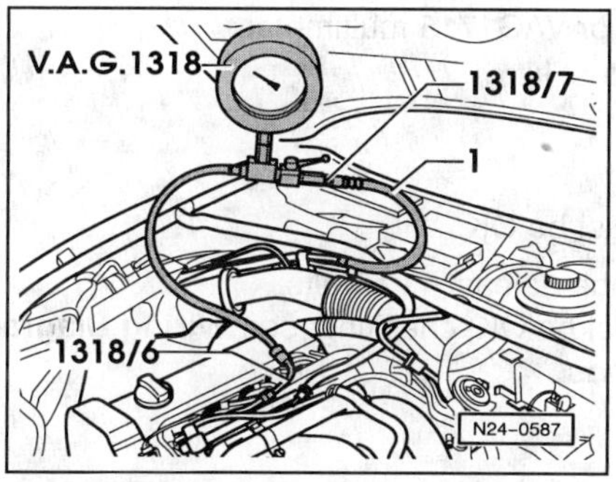

Connect VAG1318 pressure gauge to supply line -1- and fuel distributor using adapters VAG1318/6 and 1318/7.

– Open shut-off valve of pressure gauge. Lever points in direction of flow.

– Start engine and let run at idle.

– Measure fuel pressure.

- Specification: about 3.5 bar (50 psi) positive pressure

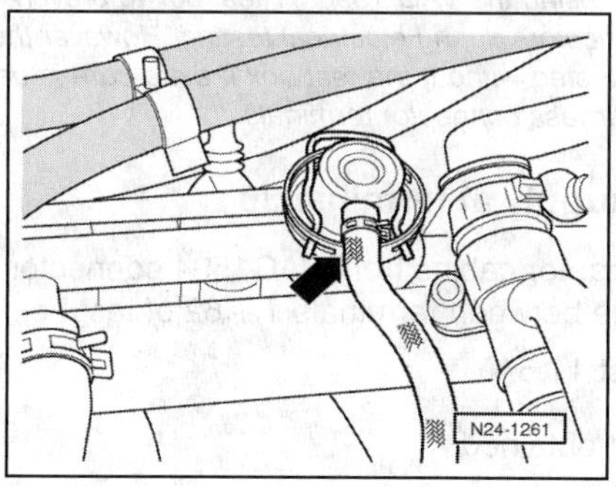

Disconnect vacuum hose from fuel pressure regulator (arrow). Fuel pressure must rise to about 4.0 bar (58 psi) positive pressure.

– If specified value is not obtained:

- Check flow quantity of Fuel Pump (FP), see **1.8 Liter 5V Turbo Engine Mechanical section, Repair Group 20**.

– If specified value is obtained:

- Switch ignition off.
- Now check for proper seal and residual pressure. Observe pressure drop on pressure gauge for this.
- After 10 minutes, there must be at least 2.0 bar (29 psi) positive pressure.

– If residual pressure falls below 2 bar (29 psi) positive pressure:

- Start engine and let run at idle.

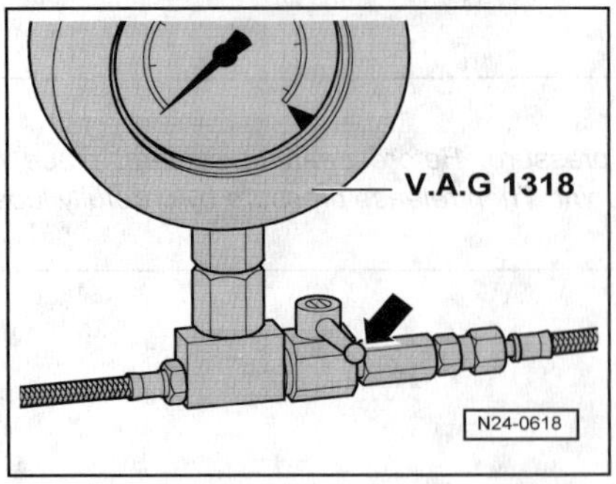

Wait until pressure has dropped, and then switch off ignition. Simultaneously close shut-off valve of VAG1318 pressure gauge (lever perpendicular to direction of flow -arrow-).

– Observe pressure drop on pressure gauge.

– If pressure does not drop:
  - Test check-valve on Fuel Pump (FP), see **1.8 Liter 5V Turbo Engine Mechanical, Repair Group 20,**

– If pressure does drop again:
  - Check whether pressure gauge is properly sealed.
  - Check whether line connections, O-rings in fuel rail, and fuel injectors are properly sealed.

– If no leaks are found:
  - Replace fuel pressure regulator

NOTE —
Before removing pressure gauge, put cleaning rags back around line connection which is to be disconnected.

Engine Control Module (ECM)

Voltage supply for control module, checking

Special tools and equipment

- VAG1598/31 adapter cable
- VAG1526 multimeter or VAG1715 multimeter
- VAG1594 connector test kit

Test requirements

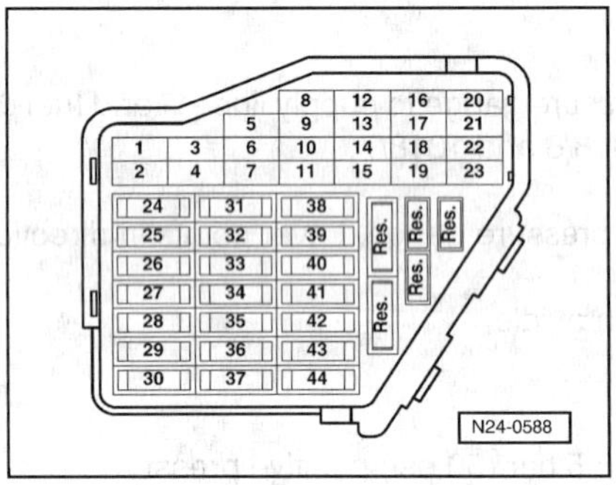

Fuses no. 7 and 29 must be OK.

- Battery voltage must be at least 11.5 volts.
- Generator (GEN) must be OK, checking, see **Wiring Diagrams and Component Locations.**

EDITOR'S NOTE —
Volkswagen recommends using the VAG 1598/31 test box to prevent damage to ECM harness connectors during electrical testing. However the harness connectors can be tested without the test box if extra care is used to prevent damage to the harness connector terminals.

Checking voltage supply at terminal 30

– Use multimeter and adapter cables from VAG1594 connector test kit to measure supply voltage between terminals 1 + 62 of test box:
  - Specification: at least 11.5 V

– If specified value is not obtained:
  - Check wire connections to relay carrier according to wiring diagram, see **Wiring Diagrams and Component Locations.**

Checking voltage supply at terminal 15

– Use multimeter and adapter cables from VAG1594 connector test kit to measure supply voltage between terminals 2 + 3 of test box:
  - Switch ignition on.
  - Specification: at least 11.5 V

– If specified value is not obtained:
  - Check wire connections to relay carrier according to wiring diagram, see **Wiring Diagrams and Component Locations.**

Check voltage supply via main relay

– Use multimeter and adapter cables from VAG1594 connector test kit to measure supply voltage between terminals 2 + 121 of test box:
  - Switch ignition on.
  - Specification: at least 11.5 V

– If specified value is not obtained:
  - Check wire connections to relay carrier according to wiring diagram

– Check activation wire from Engine Control Module (ECM) pin 21 to voltage supply relay for Engine Control Module (ECM) voltage supply relay -J271- according to wiring diagram, see **Wiring Diagrams and Component Locations.**

Throttle valve control module, adaptation to Engine Control Module (ECM)

The adaptation teaches the Engine Control Module (ECM) various positions of the throttle valve when the ignition is on and the engine is stopped. These positions are stored in the control module Throttle valve position feedback takes place through the two throttle valve drive angle sensors.

The adaptation must be performed if:

- the throttle valve control module has been removed and reinstalled or was cleaned
- the throttle valve control module has been replaced
- another engine has been installed, and another throttle valve control module along with it
- the Engine Control Module (ECM) was replaced

Special tools and equipment

- V.A.G1551 scan tool with VAG1551/3B cable

Test requirements

- Ignition switched on, engine not running
- Battery voltage must be at least 11.5 volts.
- If a cruise control system is present, it must be properly adjusted.
- The throttle valve must be in idle position (accelerator pedal must not be operated during test).
- There must not be any DTCs stored in DTC memory.
- Coolant temperature must be 6 to 105 °C (42 to 221°F)
- Intake air temperature must be at least 6 °C (42°F)

Procedure

– Connect V.A.G1551 scan tool (VAG1552). Switch ignition on and select Engine Control Module (ECM) using address word 01.

– Select "Basic Setting" function 04.

– Select "Display group number 60".

- Throttle valve actuator will be switched on without current.
- Adaptation step 1: The Engine Control Module (ECM) checks whether all test requirements have been met.
- Adaptation step 2: The throttle valve control module will be switched to no voltage (emergency running position). The angle sensor values for emergency running position are stored in the control module.
- Adaptation step 3: The throttle valve is set to a value above emergency position.
- Adaptation step 4: The output stage of the throttle valve actuator is switched off. Now, the mechanical spring must set the throttle valve to the previously learned emergency position within a specified time (spring return test).
- Adaptation step 5: The throttle valve is closed by the throttle valve actuator (under mechanical impact).
- Adaptation step 6. The angle sensor values for lower mechanical impact are stored in the control module. Finally, the lower electrical impact is derived from this (calculated) and stored in control module.
- Adaptation step 7: The output stage of the throttle valve actuator is switched off at lower mechanical impact. Now the mechanical spring places the throttle valve in emergency position (spring raising test).
- Adaptation step 8: Angle sensor values are checked.

Only engine code ATW

- Adaptation step 9. End of adaptation.

Continuation for all engine codes

The adaptation has been successfully completed.

– Select "End Output" function 06.

- To store the values:

– Switch ignition off.

NOTE—
The control module can abort basic adjustment for the following reasons:

- *Test requirements are not met*
- *Throttle valve control module or wire connection is faulty.*

NOTE—
After a cancellation, a DTC is entered in DTC memory. The next time ignition is switched on, basic adjustment is automatically repeated.

Terminal resistance for CAN-bus, checking

The Engine Control Module (ECM) communicates with all databus control modules via the CAN bus.

The control modules are connected via two data bus wires which are twisted together (CAN_High and CAN_Low), and exchange information (messages). Missing information on the data bus is recognized as an error by both the Engine Control Module (ECM) and by other CAN data bus control modules.

Trouble-free operation of the CAN data bus requires that it have a termination resistor. This central terminal resistor is located in the Engine Control Module (ECM).

- A DTC has been recognized by the CAN data bus On Board Diagnostic (OBD)

Special tools and equipment

- VAG1598/31 adapter cable
- VAG1526 multimeter or VAG1715 multimeter
- VAG1594 connector test kit

– Switch ignition off.

– Remove battery.

– Disengage connector from Engine Control Module (ECM) and then disconnect it.

– Connect VAG1598/31 adapter cable to Engine Control Module (ECM). Control module wiring harness is not connected.

– Check termination resistor in Engine Control Module (ECM), measure resistance between test box sockets 58 + 60:

- Specification: 60 to 72 Ω

– If resistance value is not in expected range, replace Engine Control Module (ECM), see **Electrical Equipment section, Repair Group 97**.

– If resistance value is in expected range, repair malfunction, see **Wiring Diagrams and Component Locations.**

Cruise Control System (CCS), activating and deactivating

Activating cruise control system

– Connect V.A.G 1551/1552 Scan Tool (ST), see **Maintenance**

– Press buttons -0- and -1- to insert "Engine electronics" address word "01".

– Press buttons -1- and -1- to select "Log-in procedure" function 11.

– Press -Q- button to confirm input.

– Enter code number 11463.

– Press -Q- button to confirm input.

Deactivating cruise control system

– Connect V.A.G 1551/1552 Scan Tool (ST), see **Maintenance.**

– Press buttons -0- and -1- to insert "Engine electronics" address word "01".

– Press buttons -1- and -1- to select "Log-in procedure" function 11.

– Press -Q- button to confirm input.

– Enter code number 16167.

– Press -Q- button to confirm input.

Additional signals, checking

Speed signal, checking

Special tools and equipment

- V.A.G1551 scan tool with VAG1551/3B cable
- VAG1598/31 adapter cable
- VAG1526 multimeter or VAG1715 multimeter
- VAG1594 connector test kit
- Wiring diagram

Test requirements

- Battery voltage must be at least 11.5 volts.
- Speedometer must be OK. Checking speedometer

Test sequence

NOTE —
To test speed signal, vehicle must be driven. Doing this requires a second person.

WARNING —
Secure Scan Tool (ST) on the rear seat and have operated from there by second technician.

– Connect V.A.G1551 scan tool. Start engine and select Engine Control Module (ECM) using address word 01.

– Press buttons -0- and -8- to select "Read measuring value block" function 08. Press -Q- button to confirm input.

– Press buttons -0-, -0- and -5- to select "Display group number 5". Press -Q- button to confirm input.

– Perform a road test, and have the 2nd person observe indication on display.

– Read value indicated in display field 3.

- Specification: approx. vehicle speed

– Press → button.

– Press buttons -0- and -6- to select "End Output" function 06. Press -Q- button to confirm input.

– Switch ignition off.

If no speed is indicated:

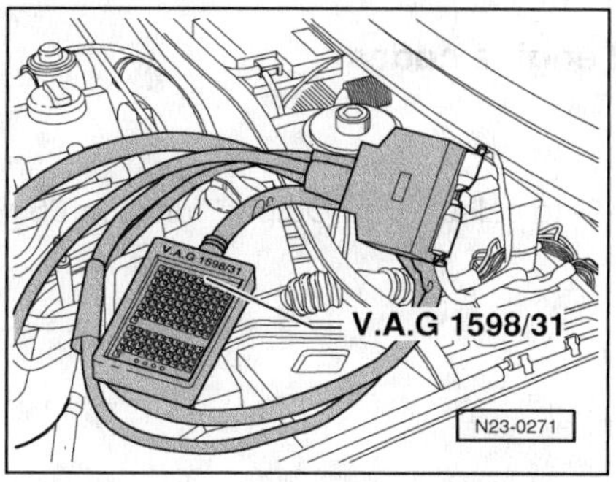

– Connect VAG1598/31 adapter cable to control module wiring harness. Engine Control Module (ECM) is not connected.

– To measure voltage, connect multimeter between sockets 3 (terminal 15) + 54 (speed signal) of test box.

– Switch ignition on.

– Jack up front left wheel.

– Rotate front wheel, and observe voltage indicated while doing so.

- Specification: oscillating between 0 and at least 4 volts

– If indicated voltage does not oscillate:

- Check wires going to speedometer and speed sensor, see **Wiring Diagrams and Component Locations.**

Signals to and from A/C System, checking

Special tools and equipment

- V.A.G1551 scan tool with VAG1551/3B cable
- VAG1598/31 adapter cable
- VAG1526 multimeter or VAG1715 multimeter
- VAG1594 connector test kit
- Wiring diagram

Test requirements

- The A/C system must function properly.
- A/C system must be switched off.
- There must not be any DTCs stored in DTC memory, see **Maintenance**.
- Vehicle must be at room temperature (warmer than + 15 °C).

Test sequence

– Connect V.A.G1551 scan tool (VAG1552). Start engine and select Engine Control Module (ECM) using address word 01.

– Press buttons -0- and -8- to select "Read measuring value block" function 08. Press -Q- button to confirm input.

– Press buttons -0-, -5- and -0- to select "Display group number 50". Press -Q- button to confirm input.

– Observe indication in display field 4.

- Compr.OFF must be indicated.

– Switch A/C system on.

– Set A/C to lowest temperature and highest fan setting.

- Indication in display field 4 must switch to compr.ON.

Not for engine code AWM

– Depress accelerator pedal completely (swiftly) and then release it (brief acceleration).

Indication in display field 4 must switch from compr.ON to compr.OFF for a few seconds (compressor shut-off for vehicle acceleration).

Continuation for all engine codes

– Press → button.

– Press buttons -0- and -6- to select "End Output" function 06. Press -Q- button to confirm input.

– Switch ignition off.

If reading does not change:

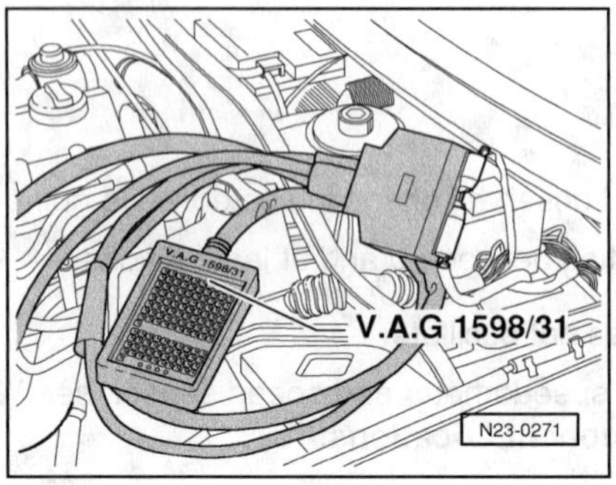

Connect VAG1598/31 adapter cable to control module wiring harness. Engine Control Module (ECM) is not connected.

– Check for open circuits in wires between test box socket 41 and A/C system according to wiring diagram.

- Wire resistance: max: 1.5Ω

28 Ignition

Repair Group

G 28

Ignition, servicing

NOTE *—*

- *This section deals especially with components having to do with ignition. For more information on components of the injection and ignition system, see* **Repair Group 24**
- *The Engine Control Module (ECM) is equipped with an On Board Diagnostic (OBD)*
- *Components marked with * are checked via On Board Diagnostic (OBD).*
- *For the electric components to work properly, a voltage of at least 11.5 V is required.*
- *It is possible that the control module will recognize a malfunction and store a DTC during some tests. Therefore, after all tests and repairs have been performed, DTC memory should be checked and erased if necessary, see* **Maintenance**.

Safety precautions

To reduce the risk of personal injury and/or damage to the fuel injection and ignition system, always observe the following:

- Do not touch or disconnect ignition wires when engine is running or turning at starting RPM.
- Only disconnect and reconnect wires for injection and ignition system, including test leads, when ignition is turned off.

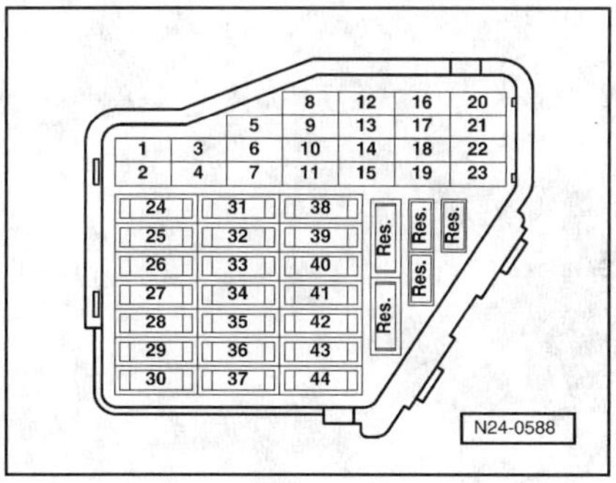

If engine is to be cranked at starting RPM without starting, remove fuses 29 and 32.

NOTE *—*

- *By removing fuse 29, voltage supply to the ignition coils is interrupted.*
- *By removing fuse 32, voltage supply to the fuel injectors is interrupted.*

Technical Data, Spark Plugs

Engine identification	**ATW, AUG, AWM**
Firing sequence	1-3-4-2
Spark plugs[1)] VW/Audi part # Manufacturer code Spark plug gap Tightening torque	 101 000 063 AA PFR 6 Q max. 0.8 mm 30 Nm

[1)] Use 3122B spark plug removal tool to remove and install spark plugs

Ignition system components, removing and installing

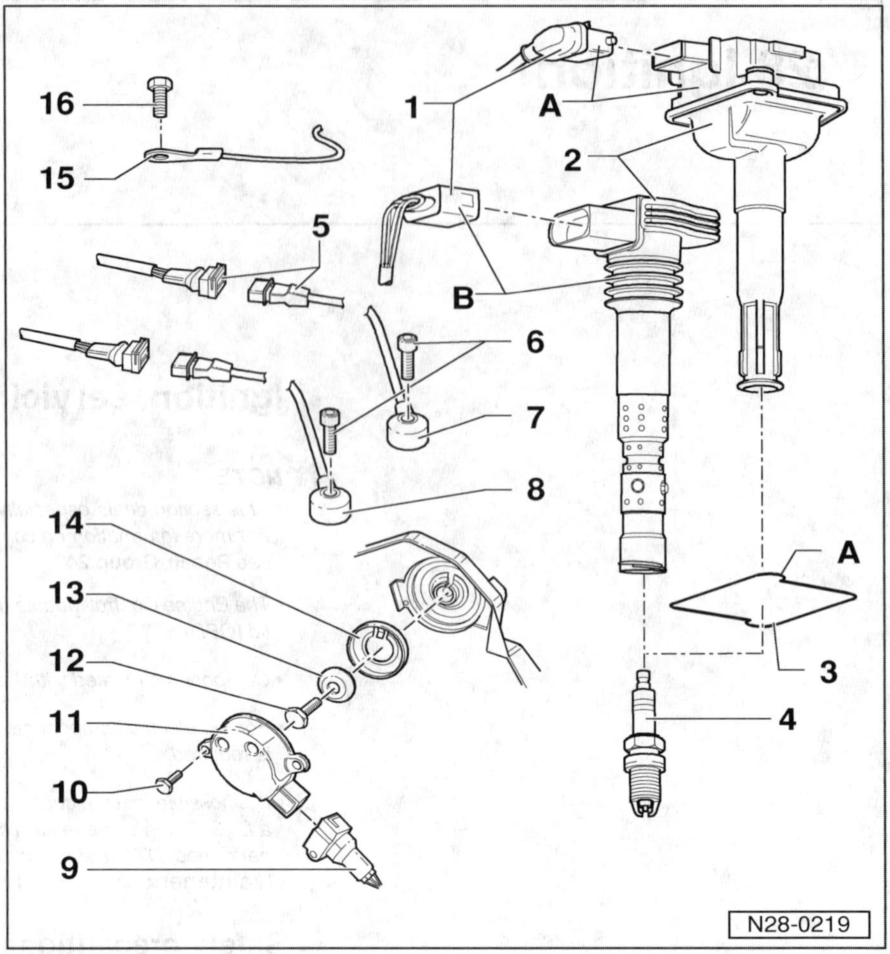

A parts: Only for engine code ATW

B parts: Only for engines AUG, AWM

1. **Connector**
   - Black, 4-pin
2. **Ignition coil with power output stage (-N70-, -N127-, -N291-, -N292-)**
   - See **Ignition coils with power output stages, checking**
   - Tighten mounting bolts to 10 Nm (7 ft-lb) (ATW engine)
   - Use T10094 or T40039 removal tool to remove (AUG/AWM engines)
3. **Seal**
   - ATW engine only
   - Replace if damaged
4. **Spark plug**
   - Tighten to 30 Nm ((22 ft-lb)
   - Remove and install with 3122 B
   - Type and spark plug gap, see **Technical Data, Spark Plugs**
5. **3-pin harness connector**
   - Terminals are gold-plated
   - Brown for Knock Sensor (KS) 1 -G61-
   - Black for Knock Sensor (KS) 2 -G66-
6. **20 Nm (15 ft-lb)**
   - Tightening torque affects function of Knock Sensor (KS)
7. **Knock Sensor (KS) 1 -G61-***
   - Terminals of sensor and connector are gold-plated
   - See **Knock sensors, checking**
8. **Knock Sensor (KS) 2 -G66-***
   - Terminals of sensor and connector are gold-plated
   - See **Knock sensors, checking**
9. **Connector**
   - Black, 3-pin
10. **10 Nm (7ft-lb)**
11. **Camshaft Position (CMP) sensor -G163-***
    - See **Camshaft Position (CMP) sensor, checking**
12. **25 Nm (18 ft-lb)**
13. **Washer**
    - Conical
14. **Face plate**
    - For Camshaft Position (CMP) sensor 2 -G163-
    - Note position when installing
15. **Ground (GND) wire**
    - Do not loosen or tighten unless ignition is switched off.
16. **10 Nm (7 ft-lb)**
    - Do not loosen or tighten unless ignition is switched off.

Camshaft Position (CMP) sensor, checking

Special tools and equipment

- VAG1598/31 adapter cable
- VAG1526 multimeter or VAG1715 multimeter
- VAG1594 connector test kit
- Wiring diagram

Test requirement

- Battery voltage must be at least 11.5 volts.

Test sequence

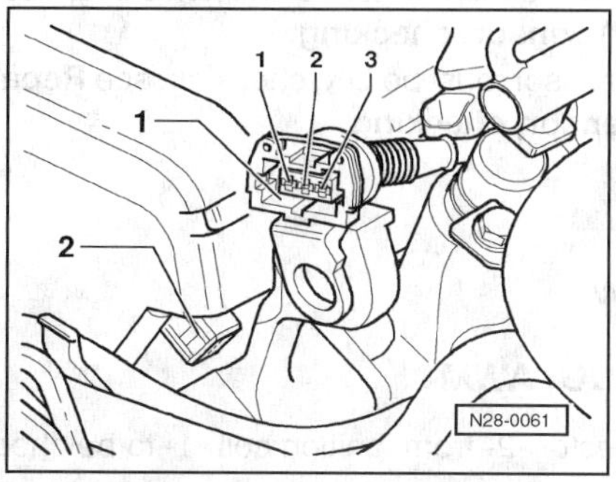

◄ Disconnect 3-pin connector -1- from Camshaft Position (CMP) sensor -2-.

– Use adapter cables from VAG1594 to connect multimeter to outer terminals of connector for voltage measurement.

– Switch ignition on.

- Specification: at least 4.5 V

– Switch ignition off.

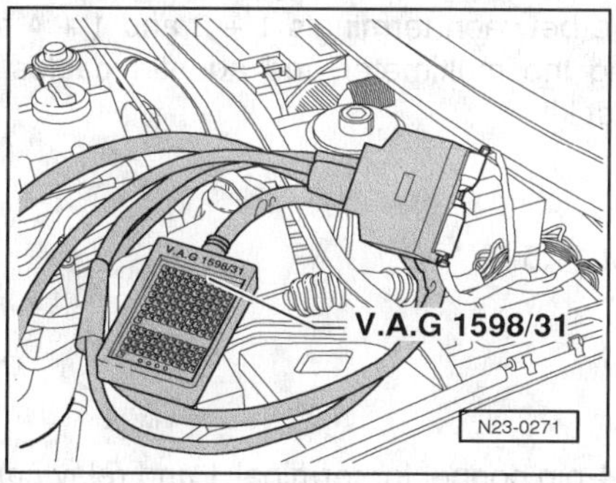

◄ Connect VAG1598/31 adapter cable to control module wiring harness. Engine Control Module (ECM) is not connected.

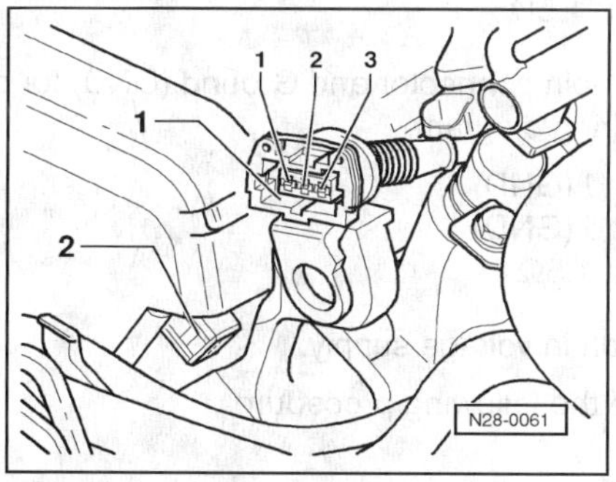

◄ Check for open circuits in wires between test box and connector according to wiring diagram.

- Terminal 1 + socket 98
- Terminal 2 + socket 86
- Terminal 3 + socket 108
- Wire resistance: max: 1.5Ω

– Check wires for shorts between one another.

– If no malfunctions are found in the wires and there was voltage between terminals 1 + 3:

- Replace Camshaft Position (CMP) Sensor 2 -G163-.

– If no errors are found in wires and there was no voltage between terminals 1 + 3, replace Engine Control Module (ECM).

- ECM removal and installation, see **Electrical Equipment section, Repair Group 97**

EDITOR'S NOTE—

ECM replacement requires coding of control module using Volkswagen factory scan tool. See **Scan Tool Codes** *at the back of this manual or consult an authorized Volkswagen service department for additional information.*

Ignition coils with power output stages, checking

Special tools and equipment

- VAG1598/31 adapter cable
- VAG1526 multimeter or VAG1715 multimeter
- VAG1594 connector test kit
- VAG1527 voltage tester
- Wiring diagram

Test requirements

- Battery voltage must be at least 11.5 volts.
- Camshaft Position (CMP) sensor must be OK, checking, see **Camshaft Position (CMP) sensor, checking**.
- Engine speed (RPM) sensor must be OK, checking, see **Repair Group 24, Engine Speed sensor, checking**.

Test sequence

I - Engine code ATW

II - Engine codes AUG, AWM

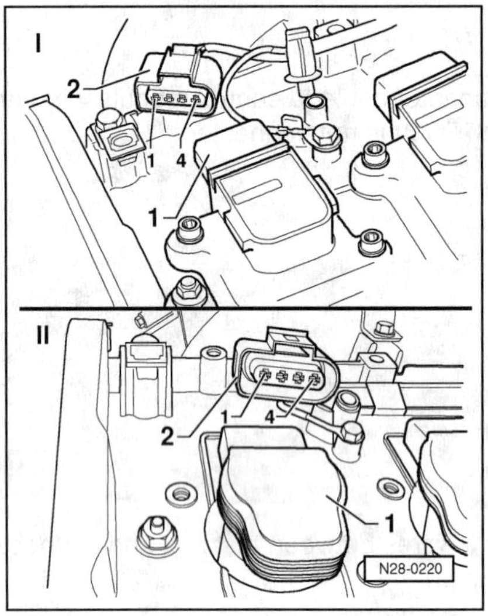

◄ Disconnect 4-pin connector -2- from ignition coil -1- to be checked.

– Measure supply voltage between terminals 1 + 2 and 1 + 4 of disconnected connector using the multimeter and adapter cables from the VAG1594 connector test kit.

– Switch ignition on.

- Specification: at least 11.5 V

– Switch ignition off.

– If there is no voltage:

- Check wire between 4-pin connector terminal 1 and relay carrier for open circuits according to wiring diagram.
- Wire resistance: max: 1.5Ω

– Check wires between 4-pin connector and Ground (GND) for open circuits according to wiring diagram.

- Terminal 2 and Ground (GND)
- Terminal 4 and Ground (GND)
- Wire resistance: max: 1.5Ω

– If there is no malfunction in voltage supply:

- Check activation, see the following procedure.

Checking activation

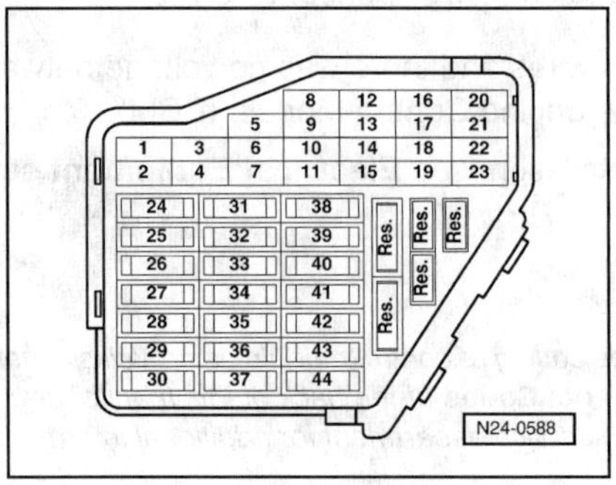

◄ Remove fuse 32.

NOTE—
By removing fuse 32, voltage supply to the fuel injectors is interrupted.

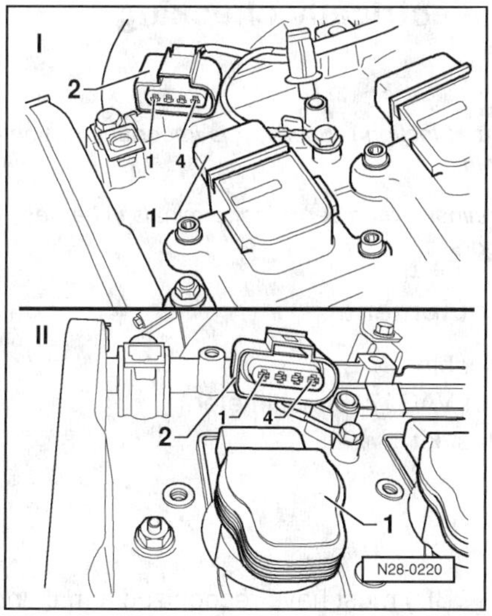

I - Engine code ATW

II - Engine codes AUG, AWM

◀ Use auxiliary wires from VAG1594 to connect diode test lamp VAG1527 to terminals 2 and 3 of disconnected connector.

– Operate starter and test ignition signal from Engine Control Module (ECM).
  - LED should flicker.

– Switch ignition off.

– If LED flickers and voltage supply is OK:
  - Replace ignition coils with power output stage

– If LED does not flicker:
  - Check wiring, see the following procedure.

Checking wiring

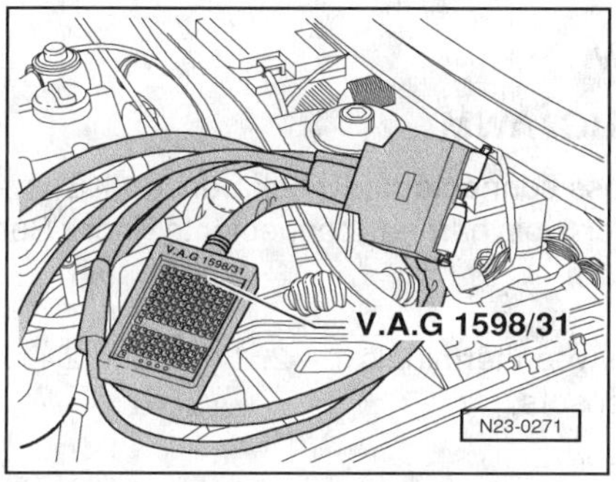

◀ Connect VAG1598/31 adapter cable to control module wiring harness. Engine Control Module (ECM) is not connected.

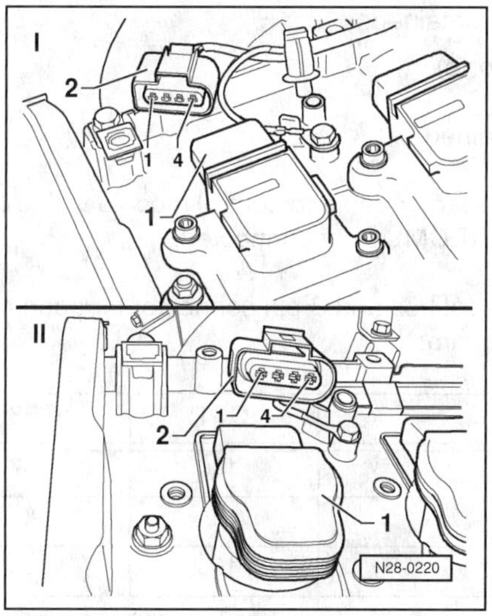

I - Engine code ATW

II - Engine codes AUG, AWM

◀ Check wires between test box and 4-pin connector for open circuits according to wiring diagram.
  - Cylinder 1 terminal 3 + socket 102
  - Cylinder 2 terminal 3 + socket 95
  - Cylinder 3 terminal 3 + socket 103
  - Cylinder 4 terminal 3 + socket 94
  - Wire resistance: max: 1.5Ω

– Check wires for shorts among one another.

– If no errors are found in wires, replace Engine Control Module (ECM).
  - ECM removal and installation, see **Electrical Equipment section, Repair Group 97**

EDITOR'S NOTE —

ECM replacement requires coding of control module using Volkswagen factory scan tool. See **Scan Tool Codes** *at the back of this manual or consult an authorized Volkswagen service department for additional information.*

Knock sensors, electrically checking

NOTE —

- *For the Knock Sensors to function properly, it is important for tightening torque to be exactly 20 Nm.*
- *Use only gold-plated terminals when servicing terminals in harness connector of knock sensors.*

Special tools and equipment

- VAG1598/31 adapter cable
- VAG1526 multimeter or VAG1715 multimeter
- VAG1594 connector test kit
- Wiring diagram

Test requirement

- On Board Diagnostic (OBD) must have recognized a malfunction at one or both of the Knock Sensors (KS).

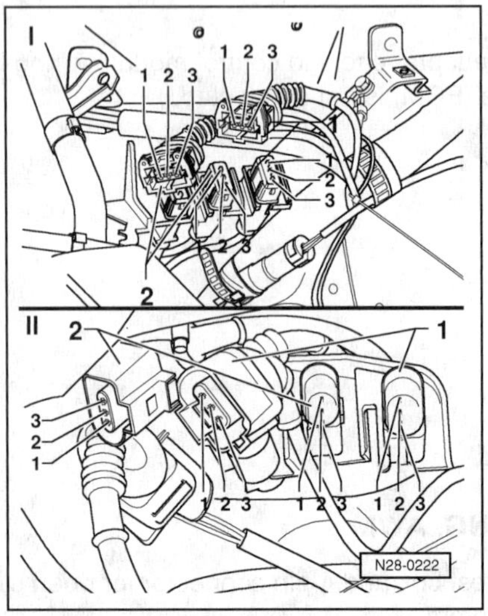

Checking resistances and wires

I - Engine code ATW

II - Engine codes AUG, AWM

Disconnect 3-pin harness connector to Knock Sensor (KS) 1 -G61- (green, item no. 1) and/or 3-pin harness connector to Knock Sensor (KS) 2 -G66- (blue, item no. 2).

– Measure resistance between terminals 1 + 2, 1 + 3, and 2 + 3 at connectors to knock sensors.

- Specification: ∞Ω

– If specified value is not obtained:

- Replace Knock Sensor(s)

If specified value is obtained:

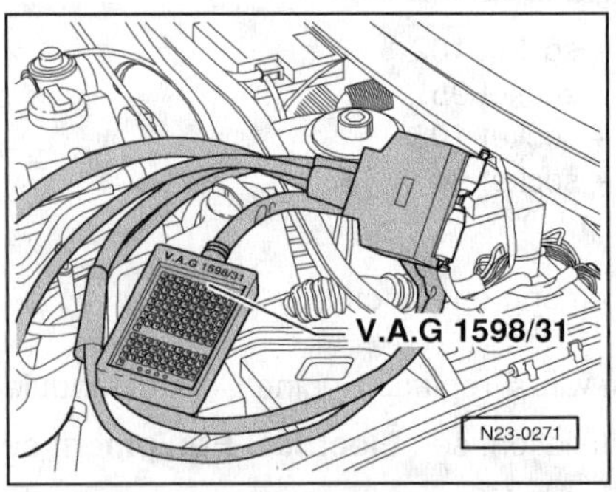

Connect VAG1598/31 adapter cable to control module wiring harness. Engine Control Module (ECM) is not connected.

– Check wiring between test box and 3-pin connector for open circuit according to wiring diagram.

Test Pin Connection	G61:	G66:
Terminal 2 + socket	106	107
Terminal 2 + socket	99	99
Terminal 3 + socket	108	108
Wire resistance:	max: 1.5	

– Check wires for shorts between one another.

– If no malfunctions are found in wires:

- Loosen Knock Sensor (KS) and then retighten to 20 Nm (15 ft-lb).

– Perform a road test.

- Check control module DTC memory again.

– If complaint persists:

- Replace Knock Sensor(s)

2.8 Liter Fuel Injection and Ignition (code: AHA)

Section H

24 Multiport Fuel Injection (MFI)

Page

28 Ignition System

Page

24 Multiport Fuel Injection (MFI)

Repair Group

H 24

Fuel Injection System, servicing

NOTE—

- *The Engine Control Module (ECM) is equipped with On board Diagnostic. Before carrying out repairs and Troubleshooting; DTC memory must be checked. Also check the vacuum hoses and connections for leaks or unmeasured air.*
- *Fuel hoses in engine compartment must only be secured with spring clips. The use of clamp or screw clips is not permissible.*
- *VAG 1921 pliers are recommended for installing spring clips.*
- *Disconnecting and connecting the battery must only be done with the ignition switched off, otherwise the Engine Control Module could be damaged.*
- *For trouble-free operation of the electrical components, 11.5 Volts minimum is required.*
- *During some checks it is possible that the Control Module will recognize and store a malfunction. Therefore after completing all checks and repairs DTC memory must be checked and if necessary erased. For connecting diagnostic scan tool and checking and erasing DTC memory, see* **Maintenance.**

CAUTION—

- *Do not use any gasket sealers or other automotive chemical compounds that contain silicone. Traces of silicone drawn into the engine pass unburned into the exhaust and can damage the oxygen sensor.*
- *Do not use any electrical contact sprays or equivalent near oxygen sensor connections. Proper oxygen sensor function requires reference to outside air via the cable. If any contact spray reaches the oxygen sensor, it can cause damage or improper function.*

EDITOR'S NOTE—

Volkswagen identifies electrical components by a letter and/or a number in the electrical schematics. See **Wiring Diagrams, Fuses and Relays***. These electrical identifiers are listed -between dashes- as an aid to electrical troubleshooting.*

EDITOR'S NOTE—

Functional testing of most fuel injection and ignition components is done using a VAG 1551, 1552, 5051 or equivalent aftermarket scan tool. In addition, the factory repair information calls for the use of various special tools including electronic test boxes and remote control switches. Many fuel injection components can be tested electrically using a digital multimeter and/or an LED test light. The following pages include many of these basic electrical tests as well as some of the testing that relies on the use of a diagnostic scan tool and specialized diagnostic equipment.

Safety precautions

> **WARNING —**
> - *Fire hazard! Do not smoke, work near heaters or open flame, or have anything nearby that could ignite fuel. The fuel system is under pressure. Before opening, cover connections with a rag, and open carefully to release pressure.*
> - *When driving or riding in an airbag-equipped vehicle, NEVER hold the scan tool or other test equipment in your hands or lap while in motion. Objects between you and the airbag increase the risk of injury in an accident.*
> - *During a test drive in an airbag-equipped vehicle, test equipment must always be fastened to and operated from the rear seat by a second technician.*
> - *Be alert when working on or near the engine and ignition wires. High secondary ignition voltage can cause serious personal injury and damage vehicle components.*
> - *DO NOT touch or disconnect ignition system wires when the engine is running or being cranked at starting RPM.*
> - *DO NOT operate the starter if fuel injectors are removed.*

> **WARNING —**
> *Be sure the ignition is switched OFF, when:*
> - *Disconnecting or connecting the battery, ignition wires, or test equipment leads.*
> - *Disconnecting fuel injection system wiring.*
> - *Washing the engine or engine compartment.*

> **WARNING —**
> *BEFORE cranking the engine at starting RPM (e.g. for compression testing) disable ignition and fuel injection systems:*
> - *Disconnect connector for ignition coil power output stage, and fuse for fuel pump.*
> - *Disconnect harness connectors from all fuel injectors.*
> - *Afterward, erase Diagnostic Trouble Code (DTC) memory.*

If the engine is to be turned at starter speed, without starting:

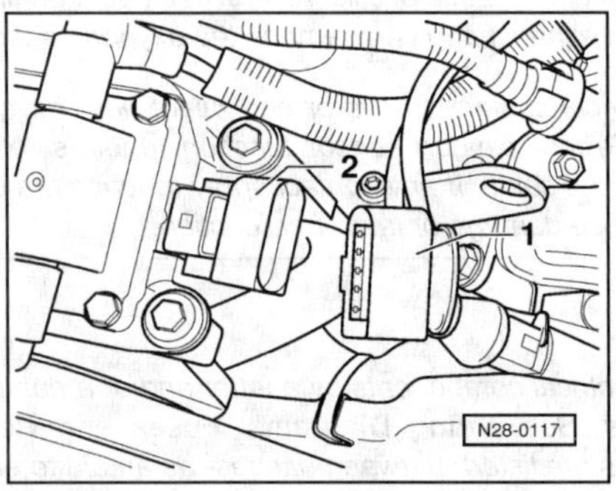

- Disconnect 5-pin connector -1- from power output stage for ignition coils -2-.

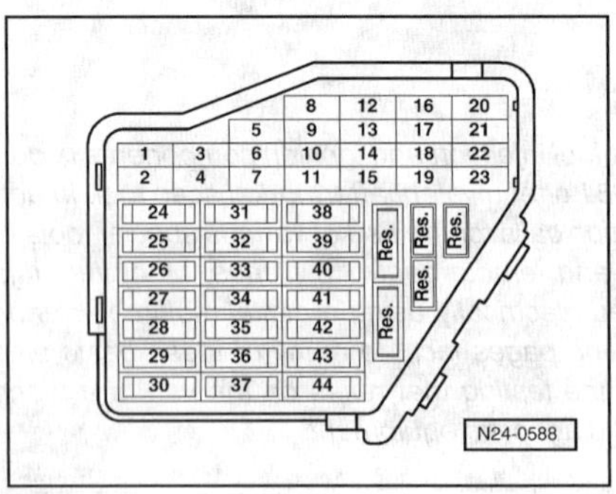

- Remove fuse 28 for fuel pump.

> **WARNING —**
> *BEFORE disconnecting the battery:*
> - *Stop the engine.*
> - *Be sure the ignition is switched OFF (also applies when connecting the battery). Failure to do so may damage the ECM.*
> - *Be sure of the proper radio code (for vehicles equipped with coded anti-theft radio).*

> ***WARNING —***
> *When connecting and disconnecting electrical test equipment (LED voltage tester, multimeter, etc.):*
>
> - *Be sure the ignition is switched OFF.*
> - *Use correct adapters from the VW1594 connector test kit.*

> ***WARNING —***
> *For any work affecting the Engine Control Module (ECM):*
>
> - *BEFORE disconnecting the ECM harness connector, switch the ignition OFF and WAIT at least 20 seconds. Failure to do so may damage the ECM.*
> - *DO NOT connect any outside voltage source to stimulate an output signal at the ECM.*

Rules of cleanliness

Whenever carrying out work on the fuel supply and fuel injection systems, carefully observe the following five rules of cleanliness.

1. Thoroughly clean fuel system line and hose connections and the surrounding area before disconnecting.
2. Place removed components on a clean surface and cover. Use plastic sheeting or paper. Do not use fluffy rags that could leave lint!
3. Carefully cover over or seal any components that have been opened if repairs are not carried out immediately.
4. Install only clean parts:
   - Do not remove replacement parts from the packaging until immediately before they are to be installed.
   - Do not use parts that have been stored without packaging (e.g. in toolboxes, etc.).
5. When the fuel system is opened:
   - Avoid working with compressed air whenever possible.
   - Avoid moving the vehicle if possible.

Technical data

Engine code	**AHA**
Idle check Idle speed[1] All wheel drive Front wheel drive	 620 - 740 RPM 740 - 860 RPM
Engine control module[2] System Part number	 Motronic M 5.9.2 F see factory parts information
Engine governed speed (RPM)	from approx. 6500

[1] Not adjustable

[2] Removing and installing engine control module, see **Electrical Equipment section, Repair Group 97**

Fuel injection components, overview

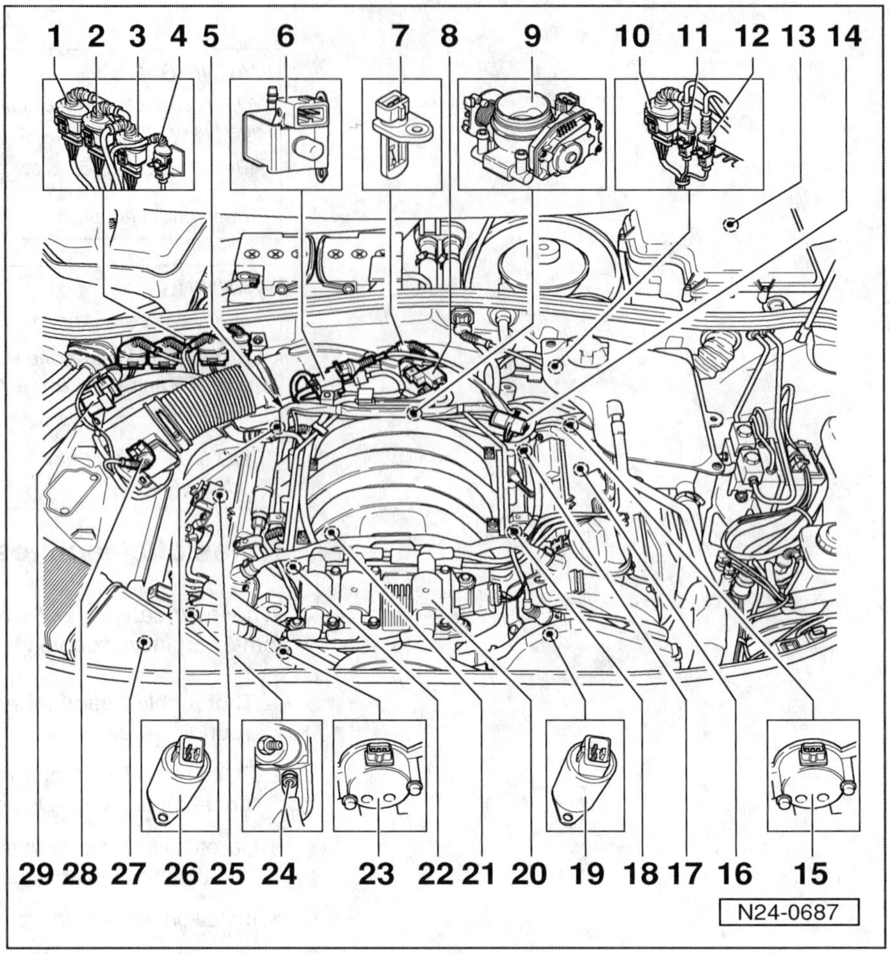

1. **Harness connector**
   - 4-pin connector, green
   - For O2S behind TWC -G130- and O2S heater -Z29-
   - For bank 1, sensor 2
2. **Harness connector**
   - 4-pin connector, brown
   - For O2S 2 behind TWC -G131- and O2S heater -Z30-
   - For bank 2, sensor 2
3. **Harness connector**
   - 4-pin connector, black
   - For HO2S -G39- and O2S heater -Z19-
   - For bank 1, sensor 1
4. **Harness connector**
   - 3-pin connector, blue
   - For Knock Sensor 1 -G61-, bank 1
5. **Engine Coolant Temperature (ECT) sensor -G62-**
6. **Secondary Air Injection (AIR) solenoid valve -N112-**
   - Removing and installing, see **2.8 Liter V6 5V Engine Mechanical section, Repair Group 26**
7. **Intake Air Temperature (IAT) sensor -G42-**
8. **Intake Manifold Tuning (IMT) valve -N156- (change-over valve)**
9. **Throttle valve control module -J338-**
10. **Harness connector**
    - 4-pin connector, black
    - For HO2S 2 -G108- and O2S heater -Z28-
    - Bank 2, sensor 1
11. **Harness connector**
    - 3-pin connector, blue
    - For Knock Sensor (KS) 2 -G66-, bank 2
12. **Harness connector**
    - 3-pin connector, gray
    - For Engine Speed (RPM) sensor -G28-
13. **Motronic Engine Control Module (ECM) -J220-**
    - Located in plenum, left side
14. **Fuel pressure regulator**
15. **Camshaft Position (CMP) sensor -G40-**
    - Cylinder bank 2
    - Location, see **Repair Group 28**
16. **Heated Oxygen Sensor (HO2S) 2 -G108-**
    - Tightening torque: 50 Nm (37 ft lb)
    - Cylinder bank 2
17. **Engine Speed (RPM) sensor -G28-**
18. **Knock Sensor (KS) 2 -G66-**
    - Cylinder bank 2
    - Removing and installing, see **Repair Group 28**
19. **Valve -2- for camshaft adjustment -N208-**
    - Cylinder bank 2
    - Removing and installing see **2.8 Liter V6 5V Engine Mechanical section, Repair Group 15**
20. **Ignition coils -N-, -N128-, -N158-**
    - With power output stage -N122-
    - Removing and installing, see **Repair Group 28**
21. **Knock Sensor (KS) 1 -G61-**
    - Cylinder bank 1
    - Removing and installing, see **Repair Group 28**
22. **Fuel injectors**
    - Cylinder 1: -N30-
    - Cylinder 2: -N31-
    - Cylinder 3: -N32-
    - Cylinder 4: -N33-
    - Cylinder 5: -N83-
    - Cylinder 6: -N84-
23. **Camshaft Position (CMP) sensor 2 -G163-**
    - Cylinder bank 1
    - Removing and installing, see **Repair Group 28**
24. **Ground connection**
    - On engine support, right
25. **Heated Oxygen Sensor (HO2S) -G39-**
    - Tightening torque: 50 Nm (37 ft lb)
    - Cylinder bank 1

(continued on following page)

Fuel injection components, overview (continued)

26. Valve -1- for camshaft adjustment -N205-
- Cylinder bank 1
- Removing and installing see **2.8 Liter V6 5V Engine Mechanical section, Repair Group 15**

27. Secondary Air Injection (AIR) pump motor -V101-
- Removing and installing see **2.8 Liter V6 5V Engine Mechanical section, Repair Group 26**

28. Mass Air Flow (MAF) sensor -G70-

29. Evaporative Emission (EVAP) canister purge regulator valve -N80-
- On air cleaner
- Removing and installing see **2.8 Liter V6 5V Engine Mechanical section, Repair Group 20**

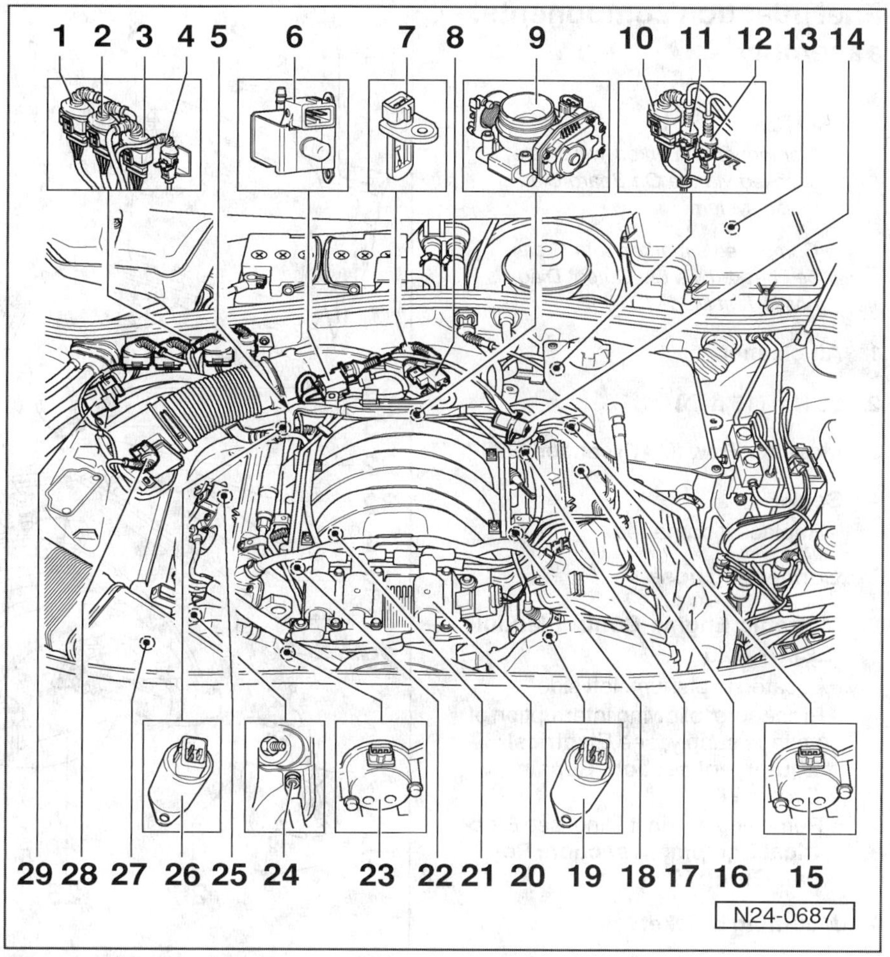

Fuel injection components, assembly

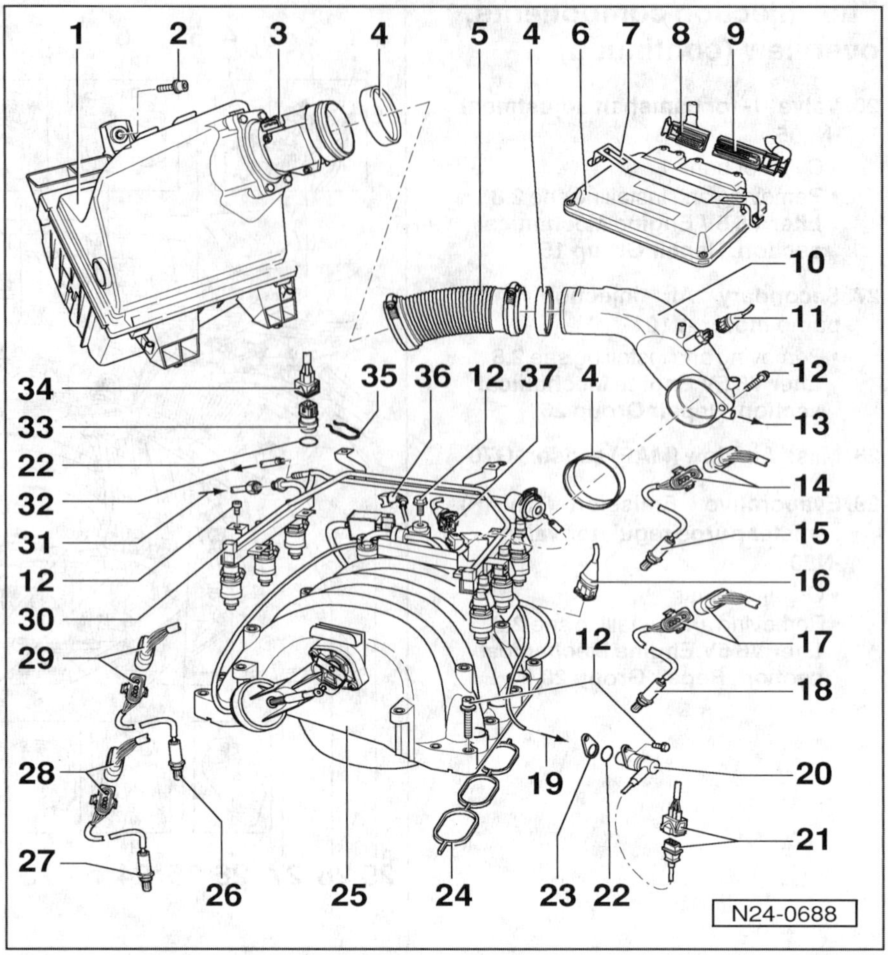

NOTE—

- *Components marked with * are checked via the On Board Diagnostic system.*
- *Components marked with ** can be checked via the Output Diagnostic Test Mode (DTM).*

1. **Air cleaner**
2. **20 Nm (15 ft lb)**
3. **Mass Air Flow (MAF) sensor -G70-***
4. **Seal**
   - Replace if damaged
5. **Connecting hose**
6. **Motronic Engine Control Module (ECM) -J220-***
   - Located in plenum, left side
   - Procedure following interruption of voltage supply, see **Electrical Equipment section, Repair Group 27**
   - Removing and installing, see **Electrical Equipment section, Repair Group 97**
7. **Mounting bracket**
8. **Harness connector (28-pin)**
   - Disconnect/connect w/ignition off
9. **Harness connector**
   - Only disconnect or connect with ignition switched off
10. **Intake air elbow**
11. **Harness connector for IAT sensor**
    - 2-pin, black
    - Checking IAT sensor resistance, see **Fig. 1**
12. **10 Nm (7 ft lb)**
13. **To positive crankcase vent valve**
14. **Harness connector**
    - 4-pin connector, brown
    - For Oxygen Sensor (O2S) 2 behind Three Way Catalytic converter (TWC) -G131- and O2S heater -Z30-
    - For bank 2, sensor 2
15. **Oxygen Sensor (O2S) 2 behind Three Way Catalytic Converter (TWC) -G131-***
    - Tightening torque: 50 Nm (37 ft lb)
    - Bank 2, sensor 2
    - Use special tool 3337 to remove
16. **Harness connector**
    - 2-pin, black
    - For fuel injectors -N30-, -N31-, -N32-, -N33-, -N83-, -N84-
17. **Harness connector**
    - 4-pin, black
    - For Heated Oxygen Sensor (HO2S) 2 -G108- and Oxygen Sensor (O2S) 2 heater -Z28-
    - For bank 2, sensor 2
18. **Heated Oxygen Sensor (HO2S) 2 -G108-***
    - Tightening torque 50 Nm (37 ft lb)
    - Bank 2, sensor 1
19. **To vacuum reservoir**
20. **Engine Speed (RPM) sensor -G28-***
    - In transmission housing above starter ring gear
21. **Harness connector**
    - 3-pin connector, gray
    - For Engine Speed sensor -G28-
22. **O-ring**
    - Replace if damaged
23. **Spacer**
24. **Gasket**
    - Replace if damaged
25. **Intake manifold**
    - See **Intake Manifold Tuning (IMT) (change-over) components**
    - See also **Throttle valve control module, assembly**
26. **Oxygen Sensor (O2S) behind Three Way Catalytic Converter (TWC) -G130-***
    - Tightening torque: 50 Nm (37 ft lb)
    - Bank 1, sensor 2
27. **Heated Oxygen Sensor (HO2S) -G39-***
    - Tightening torque: 50 Nm (37 ft lb)
    - Location: In front exhaust pipe
    - Grease only threads of sensor with "G5"; "G5" must not get into the slots of the sensor body
    - Oxygen Sensor heater voltage supply via fuel pump relay -J17-
    - Bank 1, sensor 1

Fuel injection components, assembly (continued)

28. Harness connector
- 4-pin connector, black
- For Heated Oxygen Sensor (HO2S) -G39- and Oxygen Sensor (O2S) heater -Z19-
- For bank 1, sensor 1

29. Harness connector
- 4-pin connector, green
- For Oxygen Sensor (O2S) behind Three Way Catalytic Converter (TWC) -G130- and Oxygen Sensor (O2S) heater -Z29-
- For bank 1, sensor 2

30. Fuel rail with fuel injectors
- See **Fuel rail with injectors**

31. Fuel supply line
- Black
- From fuel filter

32. Fuel return line
- Blue
- Secure with spring-type clamps
- Check for secure seating
- To the fuel delivery unit in fuel tank

33. Engine Coolant Temperature (ECT) sensor -G62-*
- Blue
- With coolant temperature gauge sensor -G2-
- Check resistance of sensor at terminals 1 (signal) and 3 (ground), compare to chart, see **Fig. 1**
- Before removing, release pressure from cooling system

34. Harness connector
- 4-pin connector, blue
- For Engine Coolant Temperature (ECT) sensor -G62-

35. Retaining clip
- Check for secure seating

36. Harness connector
- 2-pin connector, green
- For Secondary Air Injection (AIR) solenoid valve -N112-

37. Harness connector
- 2-pin connector, blue
- For Intake Manifold Tuning (IMT) valve -N156-

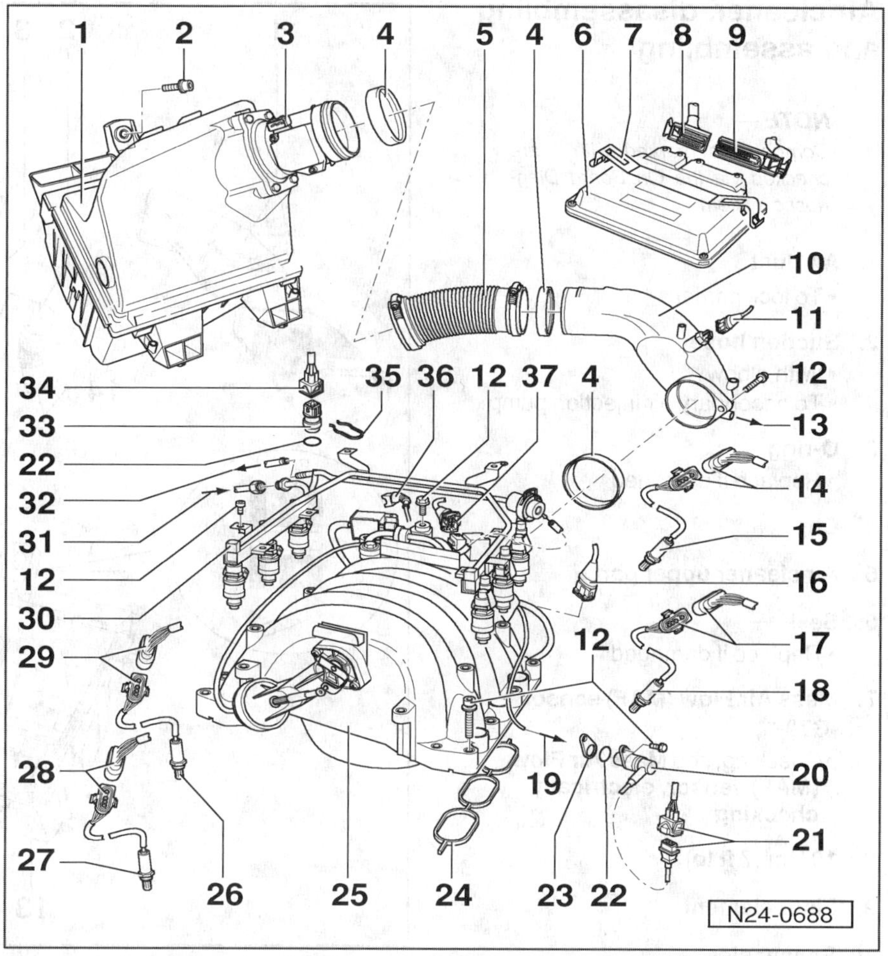

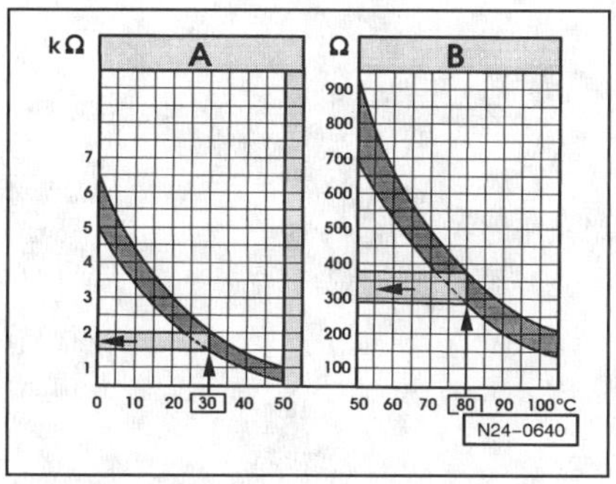

NOTE—
The ECM will use intake air temperature as a replacement value for an engine start (start temperature replacement value) as soon as there is a Diagnostic Trouble Code (DTC) stored in DTC memory for the Engine Coolant Temperature (ECT) sensor -G62-. The temperature then rises according to a program stored in the ECM. When the engine has reached normal operating temperature a fixed replacement value will be displayed. This fixed value is also dependent upon intake air temperature.

Fig. 1 Resistance graph

NOTE—
The graph is valid for the Engine Coolant Temperature (ECT) sensor -G62- and Intake Air Temperature (IAT) sensor -G42-.

The graph is divided into two temperature ranges:
- Range A: 0-50°C (32-122°F)
- Range B: 50-105°C (122-221°F)

Examples:
- 30°C (86°F) is in range A and corresponds to a resistance of 1.5-2.0 kΩ
- 80°C (176°F) is in range B and corresponds to a resistance of 275-375Ω

Air cleaner, disassembling and assembling

NOTE —
*Components marked with * are checked via the On Board Diagnostic system.*

1. **Air duct**
   - To lock carrier
2. **Suction hose**
   - With elbow
   - To secondary air injection pump
3. **O-ring**
   - Replace if damaged
4. **Clip**
5. **Air cleaner upper part**
6. **Seal**
   - Replace if damaged
7. **Mass Air Flow (MAF) sensor -G70-***
   - Checking, see **Mass Air Flow (MAF) sensor, electrically checking**
8. **10 Nm (7 ft lb)**
9. **Filter element**
10. **Spacer sleeve**
11. **Rubber grommet**
    - Note installation position
12. **Rubber grommet**
13. **Air cleaner lower part**
14. **To Secondary Air Injection (AIR) pump motor -V101-**
    - Removing and installing, see **2.8 Liter V6 5V Engine Mechanical section, Repair Group 26**

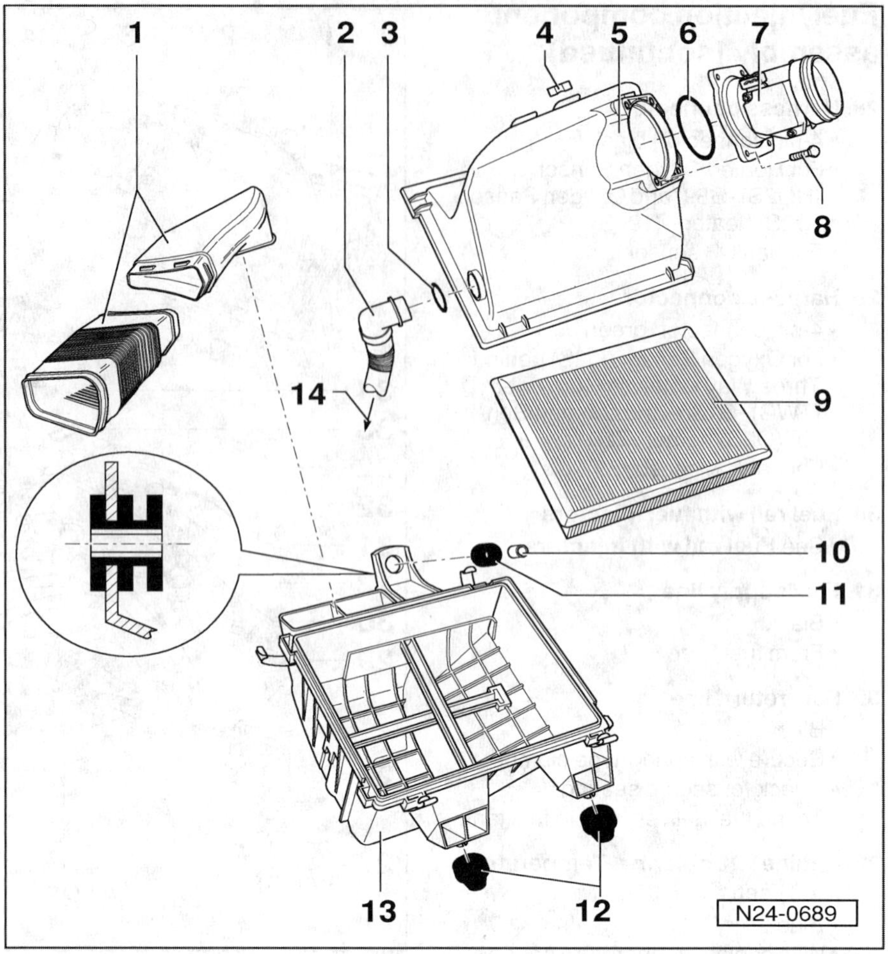

Fuel rail with injectors

NOTE —
- *Components marked with * are checked via the On Board Diagnostic system.*
- *Components marked with ** can be checked via the Output Diagnostic Test Mode (DTM).*

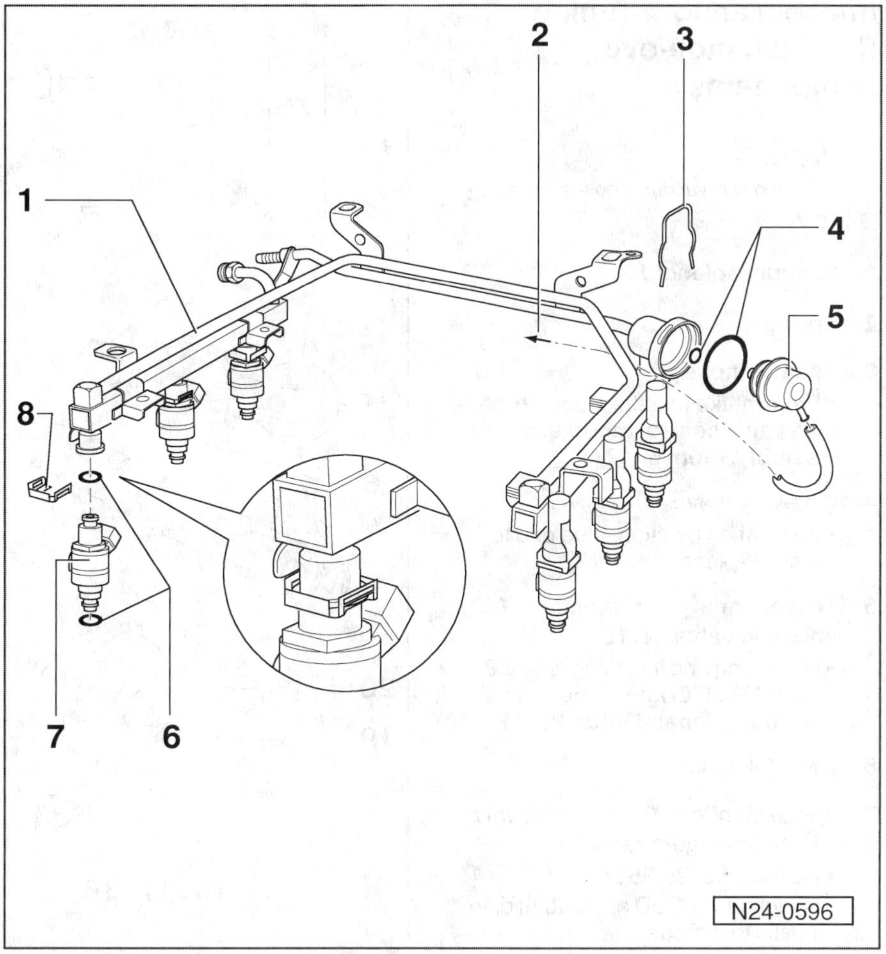

1. **Fuel rail**
2. **To T-piece**
   - Location, see **Throttle valve control module, assembly**
3. **Retaining clip**
   - Check for secure seating
4. **O-ring**
   - Replace
5. **Fuel pressure regulator**
   - See **Fuel pressure, fuel pressure regulator and residual fuel pressure, checking**
6. **O-ring**
   - Replace
7. **Fuel injector -N30-, -N31-, -N31-, -N33-, -N83-, -N84-*/****
   - Checking, see **Fuel injectors, checking**
   - Resistance: 13.5-15.5 Ω
8. **Retaining clip**
   - Check for secure seating at injector and fuel rail

Intake Manifold Tuning (IMT) (change-over) components

> ***NOTE —***
> *Routing of vacuum hoses, see* **Fig. 1** *and* **Fig. 2.**

1. **Vacuum solenoid**
2. **Spring**
3. **Vacuum hose**
   - From intake manifold, see **Throttle valve control module, assembly, -item 6-**
4. **One-way valve**
   - Installation position: blue end toward Y-piece
5. **To Secondary Air Injection (AIR) solenoid valve -N112-**
   - Removing and installing, see **2.8 Liter V6 5V Engine Mechanical section, Repair Group 26**
6. **6 Nm (53 in lb)**
7. **Intake Manifold Tuning (IMT) valve -N156- (change-over valve)**
   - Resistance: 25-35 Ω
   - Checked via OBD and output diagnostic test mode
8. **Y-piece**
9. **To fuel pressure regulator**
   - Location, see **Fuel rail with injectors**
10. **T-piece**
11. **10 Nm (7 ft lb)**
12. **Securing plate**
13. **Rubber grommet**
14. **Spacer sleeve**
15. **Vacuum reservoir**
    - Location: In left wheel housing
    - For intake manifold change-over with engine at full load
16. **Intake manifold**
17. **Washer**
    - Conical side to intake manifold
18. **Seal**
    - Replace if damaged
19. **Change-over valve**
20. **Seal**
    - For change-over valve

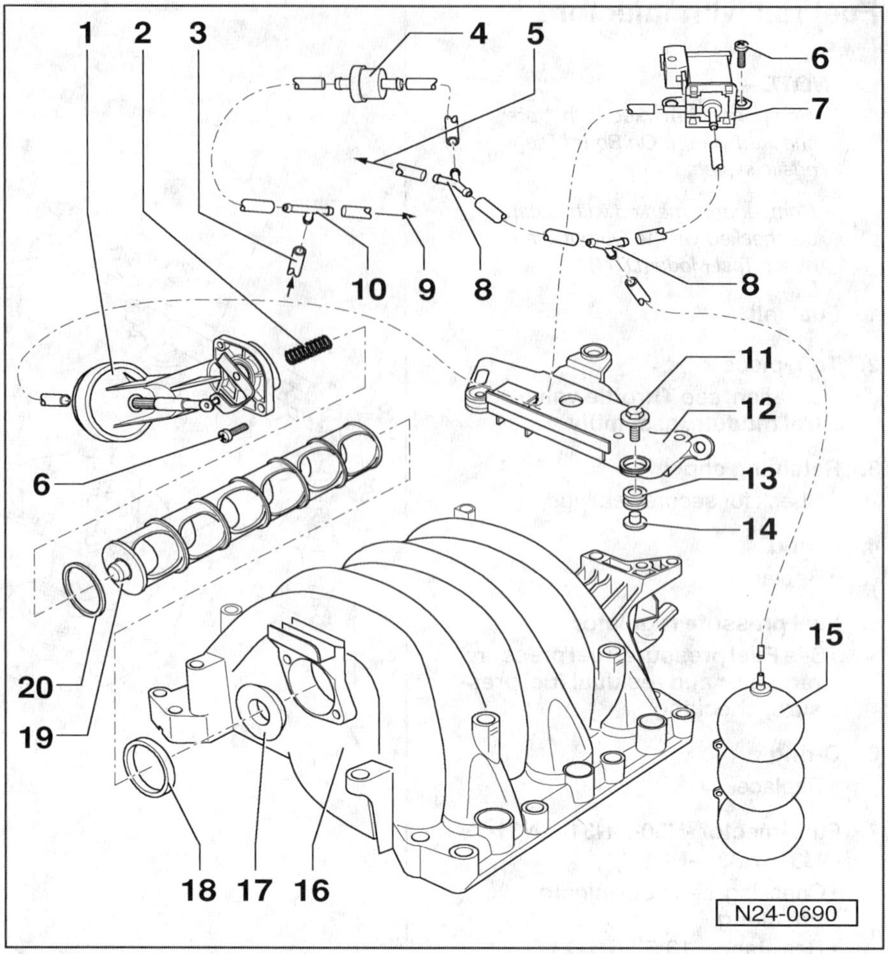

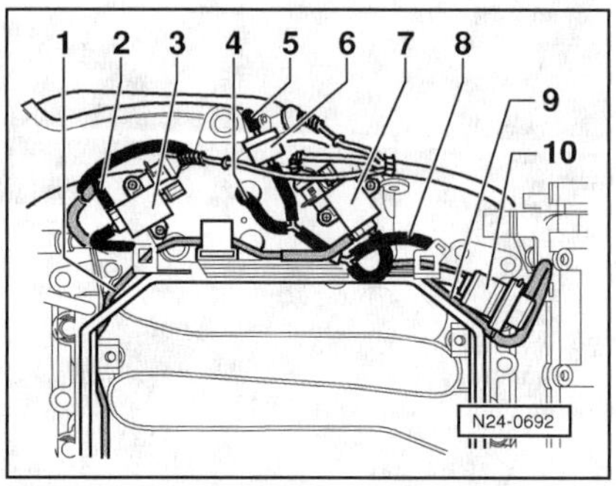

Fig. 1 Vacuum connections, upper

1. To intake manifold tuning (change-over) vacuum solenoid
2. To combination valves
3. Secondary Air Injection (AIR) solenoid valve -N112-
4. To secondary air injection valve
5. To T-piece on intake manifold
6. One-way valve
7. Intake manifold tuning change-over valve
8. To vacuum reservoir in left wheel housing
9. To T-piece on intake manifold
10. Fuel pressure regulator

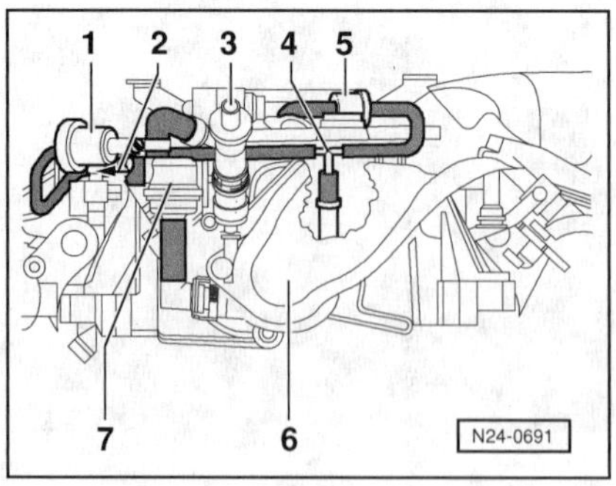

Fig. 2 Vacuum connections, rear

1. Fuel pressure regulator
2. To crankcase breather valve
3. Vacuum booster
4. T-piece
5. One-way valve
6. Intake air elbow
7. Crankcase breather valve

Throttle valve control module, assembly

> *NOTE—*
> *Components marked with * are checked via the On Board Diagnostic system.*

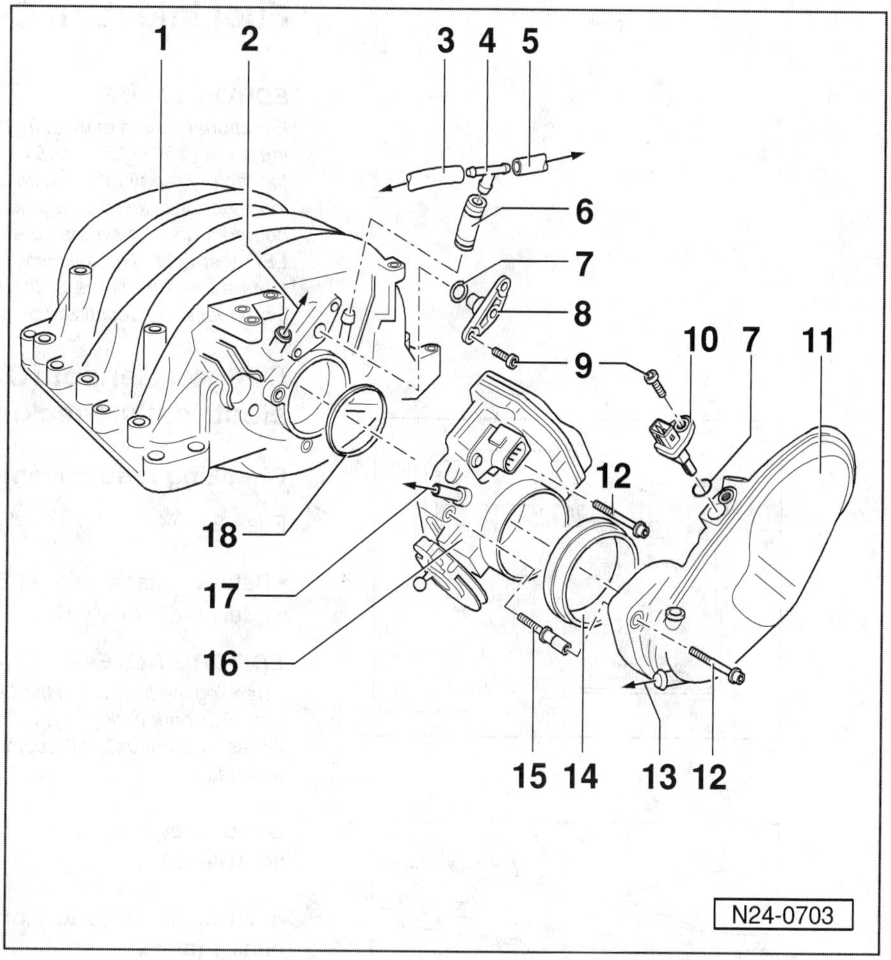

1. **Intake manifold**
2. **To vacuum booster**
3. **To fuel pressure regulator**
   - Location, see **Fuel rail with injectors**
4. **T-piece**
5. **To one-way valve**
   - Location, see **Intake Manifold Tuning (IMT) (change-over) components**
6. **Vacuum hose**
7. **O-ring**
   - Replace if damaged
8. **Support bracket**
   - For change-over valve
9. **6 Nm (53 in lb)**
10. **Intake Air Temperature (IAT) sensor -G42-***
    - Checking resistance, see **Fuel Injection components, assembly, Fig. 1**
11. **Intake air elbow**
12. **10 Nm (7 ft lb)**
13. **To positive crankcase ventilation valve**
14. **Seal**
    - Replace if damaged
15. **10 Nm (7 ft lb)**
    - Install with "D6" locking fluid
16. **Throttle valve control module -J338-***
    - If replaced, adapt to ECM, see **Throttle valve control module, adaptation to Engine Control Module (ECM)**
17. **To Evaporative Emission (EVAP) canister purge regulator valve -N80-**
    - Removing and installing see **2.8 Liter V6 5V Engine Mechanical section, Repair Group 20**
18. **Seal**
    - Replace if damaged

Fuel Injection Components, checking

EDITOR'S NOTE —

Functional testing of most fuel injection and ignition components is done using a VAG 1551, 1552, 5051 or equivalent aftermarket scan tool. In addition, the factory repair information calls for the use of various special tools including electronic test boxes and remote control switches. Many fuel injection components can be tested electrically using a digital multimeter and/or an LED test light. The following pages include many of these basic electrical tests as well as some of the testing that relies on the use of a diagnostic scan tool and specialized diagnostic equipment.

Oxygen Sensor (O2S) heater before TWC, electrically checking

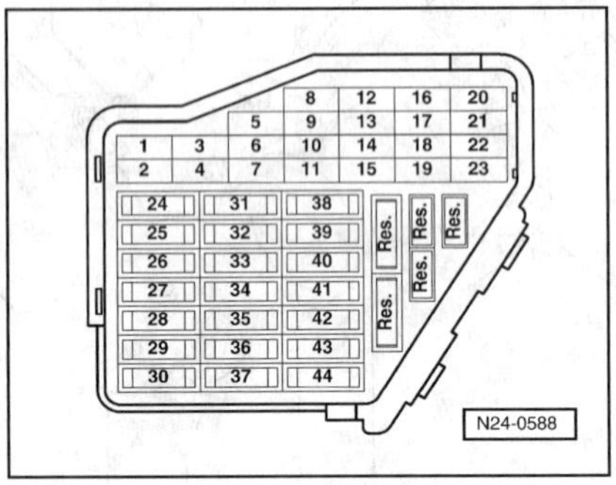

Checking requirements

◄ Fuse 29 OK

- Battery voltage: min. 11.5 volts
- Fuel pump relay OK

EDITOR'S NOTE —

Checking the function of the O2 sensor requires the use of a diagnostic scan tool. For connecting scan tool see, **Maintenance**. *For diagnostic trouble codes and control unit coding, see* **Scan Tool Codes** *at the back of this manual.*

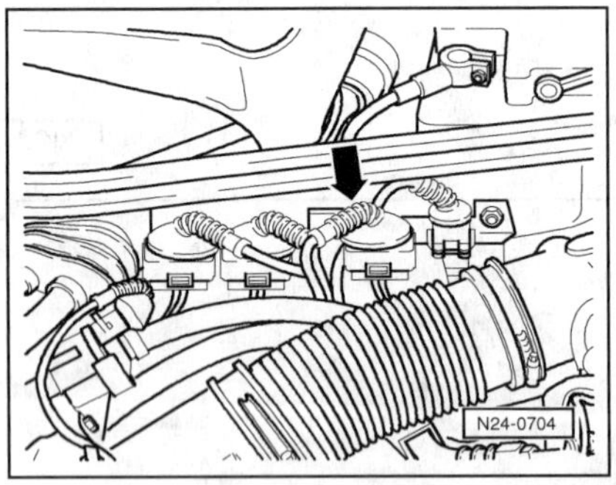

– Disconnect 4-pin connector (black) for respective Heated Oxygen Sensor (HO2S).

◄ 4-pin connector (black) for Heated Oxygen Sensor (HO2S) -G39- for bank 1 (arrow).

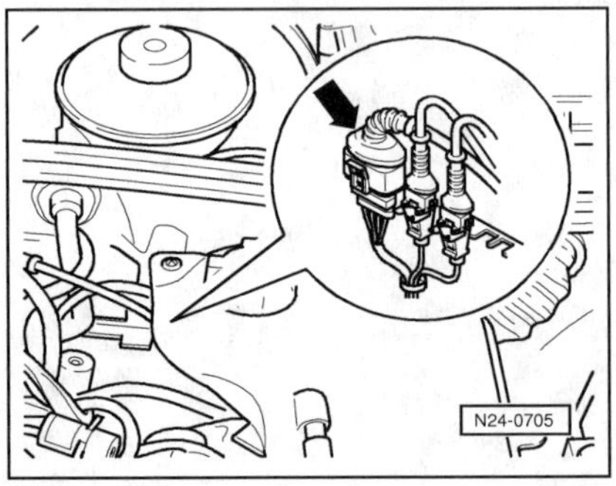

◄ 4-pin connector (black) for Heated Oxygen Sensor (HO2S) 2 -G108- for bank 2 (arrow).

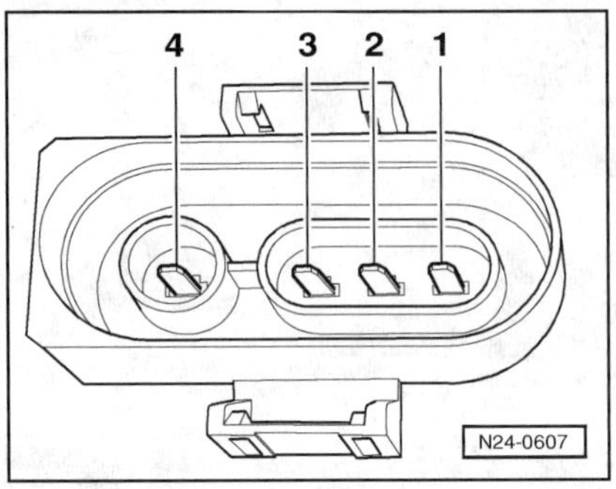

◄ Connect multimeter US1119 (Fluke 83 or equivalent) between HO2S connector terminals 1 and 2, using VW1594 connector test kit.

– Measure resistance.

- Specified value: 0.0-19.9 Ω

If resistance is NOT OK:

– Replace respective Heated Oxygen Sensor (HO2S).

– Check readiness code using diagnostic scan tool. If Diagnostic Trouble Code (DTC) memory has been erased, or ECM was disconnected, generate new readiness code.

If resistance is OK:

– Switch multimeter to 20 VDC measuring range.

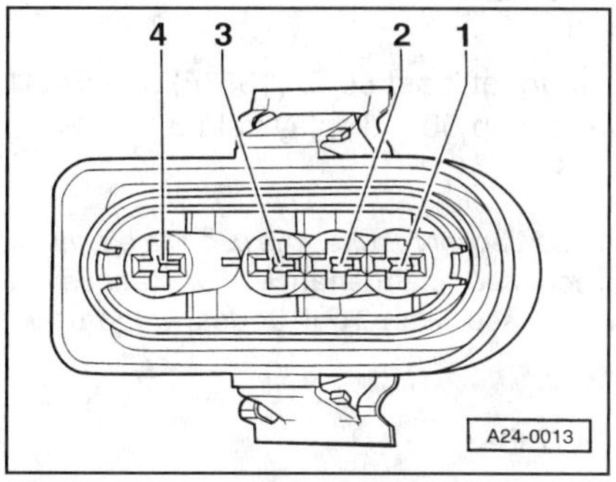

◄ Connect multimeter between harness connector terminals 1 and 2 (to ECM).

– Start engine and run at idle.

– Measure voltage supply and check scan tool display in display group 40, display field 2.

– If display = Htg.bC.ON:

• Specified value: 11.0-14.5 volts

– If display switches between "Htg.bC.ON" and "Htg.bC.OFF":

• Specified value: between 0.0-12.0 volts (oscillating)

– Switch ignition off.

If there is no voltage:

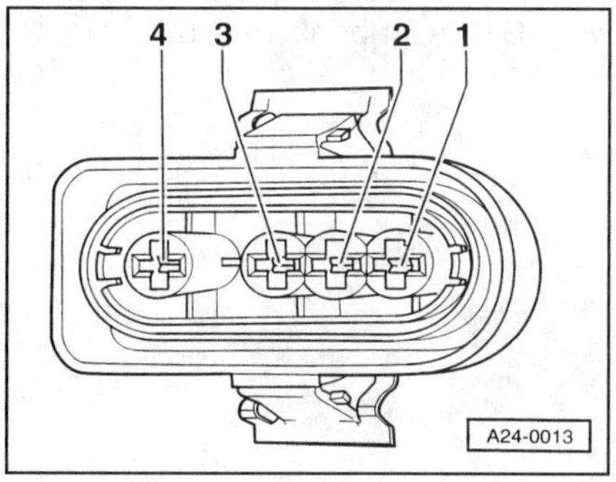

◄ Check wiring for open circuit between ECM and 4-pin HO2S harness connector (to ECM) using wiring diagrams.
Connector terminal 2 to ECM/test box terminal 27

• Specified value: max. 1.5 Ω

EDITOR'S NOTE —

Volkswagen recommends using the VAG 1598/22 test box to prevent damage to ECM harness connectors during electrical testing. However the harness connectors can be tested without the test box if extra care is used to prevent damage to the harness connector terminals.

– Check wiring at HO2S harness connector terminals 3 and 4 for short circuit to terminals 1 and 2 using wiring diagrams.

• Specified value: ∞

If the specification is attained:

– Check wiring between terminal 1 and fuel pump relay -J17- using appropriate wiring diagram.

– Check readiness code using diagnostic scan tool. If Diagnostic Trouble Code (DTC) memory has been erased, or ECM was disconnected, generate new readiness code.

If wiring is OK:

– Replace Motronic ECM -J220, see **Electrical Equipment section, Repair Group 97**.

EDITOR'S NOTE —

ECM replacement requires coding of control module using Volkswagen factory scan tool. See **Scan Tool Codes** *at the back of this manual or consult an authorized Volkswagen service department for additional information.*

Oxygen Sensor (O2S) heater behind TWC, electrically checking

Checking requirements

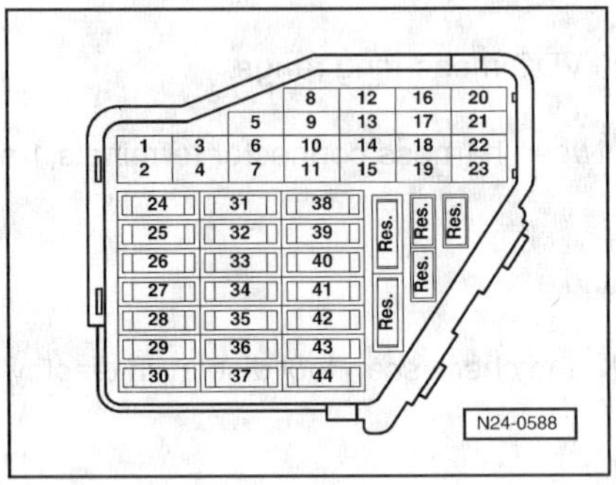

Fuse 29 OK

- Battery voltage: min. 11.5 volts
- Fuel pump relay OK
- Engine coolant temperature at least 85°C (185°F) see VAG1551 scan tool function 08, display group 004, display field 3

EDITOR'S NOTE —

Checking the function of the O2 sensor requires the use of a diagnostic scan tool. For connecting scan tool see, **Maintenance***. For diagnostic trouble codes and control unit coding, see* **Scan Tool Codes** *at the back of this manual.*

If displayed values are NOT OK:

– Switch ignition off.

– Disconnect 4-pin connector for respective Heated Oxygen Sensor (HO2S).

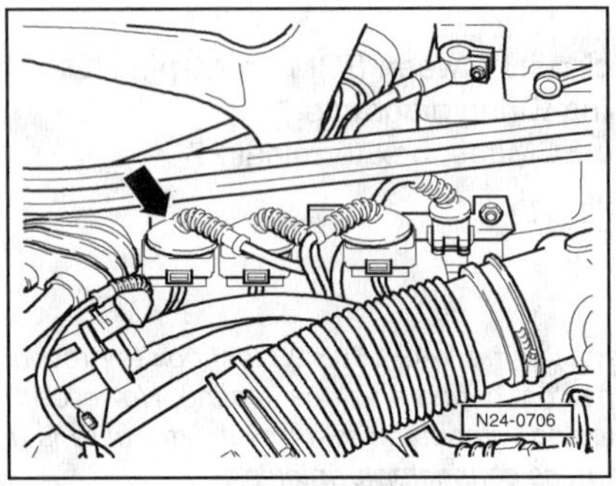

4-pin connector (green) for Oxygen Sensor (O2S) behind Three Way Catalytic Converter (TWC) -G130- for bank 1 (arrow).

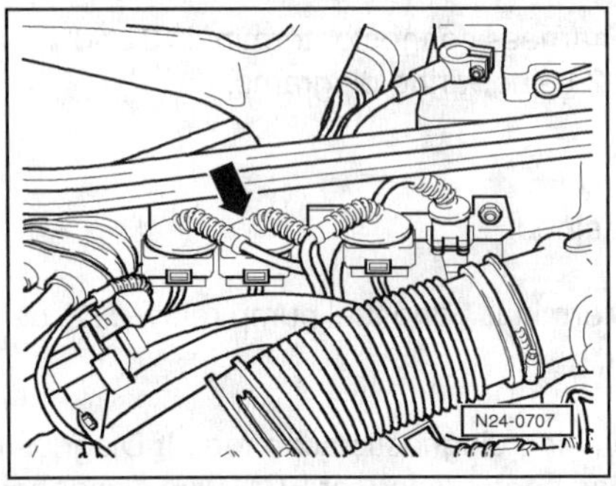

4-pin connector (brown) for Oxygen Sensor (O2S) 2 behind three Way Catalytic Converter (TWC) -G131- for bank 2 (arrow).

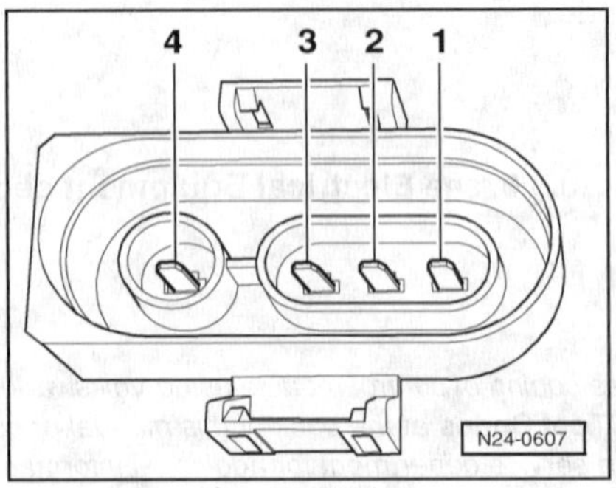

Connect multimeter US1119 (Fluke 83 or equivalent) between O2S connector terminals 1 and 2, using VW1594 connector test kit.

– Measure resistance.
  - Specified value: 0.0-19.9 Ω

If resistance is NOT OK:

– Replace respective Oxygen Sensor (O2S).

– Check readiness code using diagnostic scan tool. If Diagnostic Trouble Code (DTC) memory has been erased, or ECM was disconnected, generate new readiness code.

If the specification is attained:

- Switch multimeter to 20 VDC measuring range.

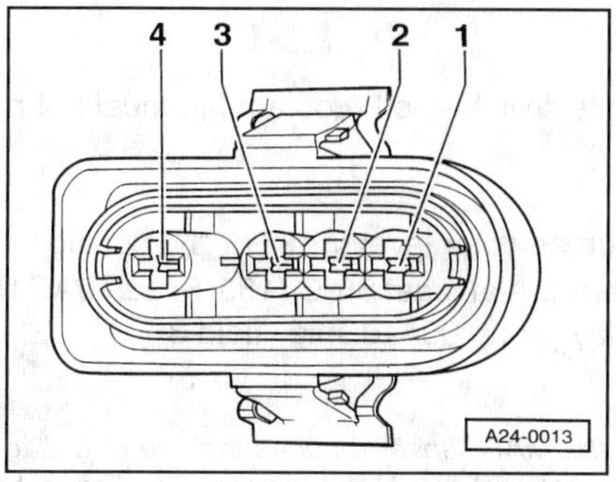

- Connect multimeter between harness connector terminals 1 and 2 (to ECM), using VW1594 connector test kit.

- Start engine and run at idle.

- Measure voltage supply and check display in display group 40, display field 4.

- If display = Htg.aC.ON:
  - Specified value: 11.0-14.5 volts

- If display switches between "Htg.aC.ON" and "Htg.aC.OFF":
  - Specified value: between 0.0-12.0 volts (oscillating)

- Switch ignition off.

If there is no voltage:

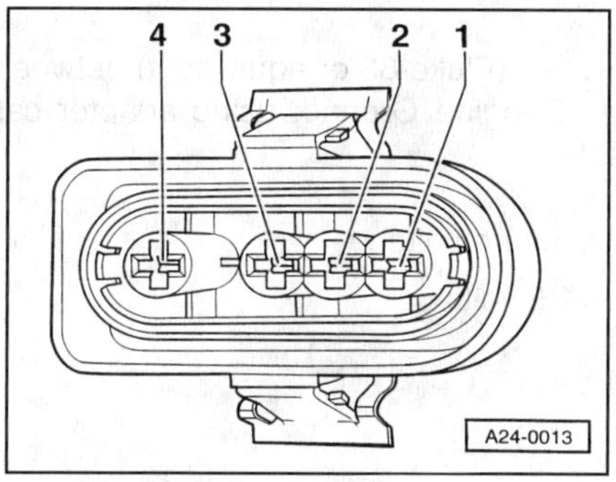

- Check wiring for open circuit between ECM/test box and 4-pin O2S harness connector (to ECM).
  Connector terminal 2 to ECM/test box socket 28
  - Specified value: max. 1.5 Ω

EDITOR'S NOTE —

Volkswagen recommends using the VAG 1598/22 test box to prevent damage to ECM harness connectors during electrical testing. However the harness connectors can be tested without the test box if extra care is used to prevent damage to the harness connector terminals.

- Check wiring at O2S harness connector terminals 3 and 4 for short circuit to terminals 1 and 2 using wiring diagram.
  - Specified value: ∞

If the specification is attained:

- Check wiring between terminal 1 and fuel pump relay -J17- using appropriate wiring diagram.

- Check readiness code using diagnostic scan tool. If Diagnostic Trouble Code (DTC) memory has been erased, or ECM was disconnected, generate new readiness code.

If wiring is OK:

- Replace Motronic ECM -J220, see **Electrical Equipment section, Repair Group 97**.

EDITOR'S NOTE —

ECM replacement requires coding of control module using Volkswagen factory scan tool. See **Scan Tool Codes** *at the back of this manual or consult an authorized Volkswagen service department for additional information.*

Mass Air Flow (MAF) sensor, electrically checking

Checking requirements

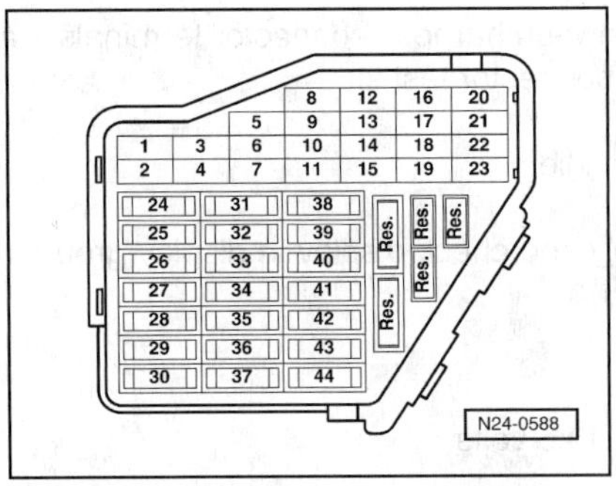

Fuses No. 31 and 32 OK

- All electrical consumers switched off (coolant fan must not run during test)
- A/C switched off
- Automatic transmission selector lever in "P" or "N" position
- Engine coolant temperature at least 85°C (185°F) see VAG1551 scan tool function 08, display group 004, display field 3

EDITOR'S NOTE —
Checking the function of the MAF sensor requires the use of a diagnostic scan tool. For connecting scan tool see, **Maintenance**. *For diagnostic trouble codes and control unit coding, see* **Scan Tool Codes** *at the back of this manual.*

Voltage supply, checking

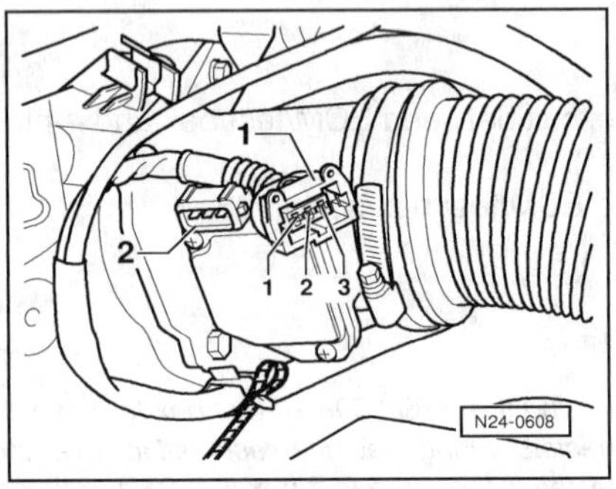

Disconnect 3-pin harness connector -1- from Mass Air Flow sensor (MAF) -2-.

– Connect multimeter US1119 (Fluke 83 or equivalent) between harness connector terminal 3 and engine Ground, using adapter cables from VW1594 connector test kit.

– Switch ignition on and measure voltage.

- Specified value: min. 11.5 volts

– Switch ignition off.

If the specification is not attained:

– Check wiring from terminal 3 to fuse panel using applicable wiring diagram.

Wiring, checking

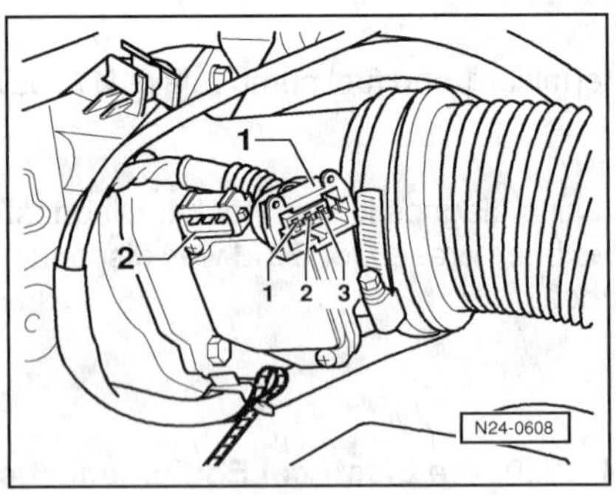

Check wiring for open circuits, between ECM/test box and 3-pin MAF sensor harness connector.
Connector terminal 1 to ECM/test box socket 13
Connector terminal 2 to ECM/test box socket 12

- Specified value: max. 1.5 Ω

EDITOR'S NOTE —
Volkswagen recommends using the VAG 1598/22 test box to prevent damage to ECM harness connectors during electrical testing. However the harness connectors can be tested without the test box if extra care is used to prevent damage to the harness connector terminals.

– Check wiring for short circuits between terminals of 3-pin connector.

- Specified value: ∞

If the voltage supply and wiring is OK:

– Replace Mass Air Flow (MAF) sensor -G70-.

– Check readiness code using diagnostic scan tool. If Diagnostic Trouble Code (DTC) memory has been erased, or ECM was disconnected, generate new readiness code.

Throttle valve control module, checking

NOTE—

If the throttle valve control module -J338- is removed and installed, or replaced, adaptation of the throttle valve control module to the Motronic ECM is required, see **Throttle valve control module, adaptation to Engine Control Module (ECM)**.

Checking requirements

• Throttle valve at idle position
• Accelerator Pedal (AP) cable adjustment OK, see **2.8 Liter V6 5V Engine Mechanical section, Repair Group 20**
• A/C switched off
• Automatic transmission selector lever in "P" or "N" position
• Cruise control system functioning correctly

Function

Idle speed stabilization is integrated in the throttle valve control module -J338-. In the housing opposite the throttle valve are Throttle Position (TP) sensor -G69-, Closed Throttle Position (CTP) switch -F60-, throttle drive -G186- and angle sensor -1- for throttle drive -G187-.

The individual components cannot be replaced or mechanically adjusted. All adjustments are accomplished by the VAG1551 Scan Tool (ST) using the "Basic Setting" function 04, or an equivalent aftermarket diagnostic scan tool.

If the complete throttle valve control module is without power (i.e. disconnected), the throttle valve assumes a mechanical closed throttle position that results in an increased idle speed of 1200-1300 RPM (with engine at operating temperature).

If only the Throttle Position (TP) actuator -V60- is without power, the throttle valve also assumes a mechanical closed throttle position (limp-home position), but because the signal from the CTP switch -F60- can still be recognized, a nearly "normal" idle speed of approx. 900 RPM is maintained with a corresponding reduction of ignition timing advance.

If the Engine Control Module (ECM) recognizes a malfunction in the angle sensor -1- for throttle drive -G187-, it switches the power off for throttle drive -G186-, and the throttle valve again assumes the mechanical closed throttle position (limp-home position).

Closed throttle position switch -F60-, electrically checking

EDITOR'S NOTE—

Checking the function of the closed throttle position switch requires the use of a diagnostic scan tool. For connecting scan tool see, **Maintenance**. *For diagnostic trouble codes and control unit coding, see* **Scan Tool Codes** *at the back of this manual.*

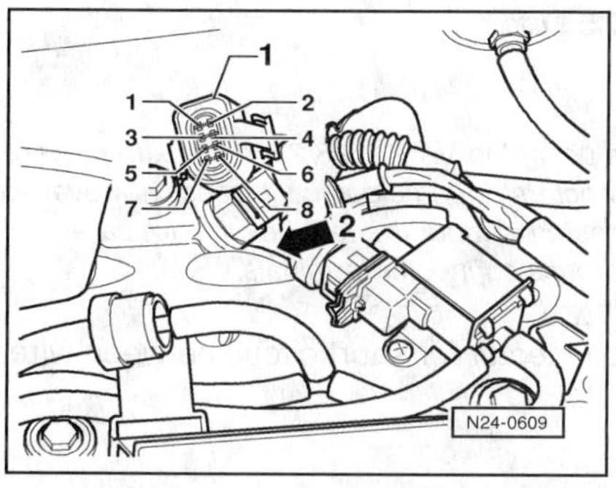

◄ Disconnect 8-pin harness connector -1- from throttle valve control module (arrow -2-).

– Measure Throttle Position (TP) actuator -V60- resistance between terminals 1 and 2 of throttle valve control module -J338-.

  • Specified value: 3-200 Ω

– Check voltage supply to throttle valve control module and wiring to ECM, using wiring diagrams.

If no malfunction is detected:

– Replace throttle valve control module.

– Carry out adaptation of throttle valve control module to ECM, see **Throttle valve control module, adaptation to Engine Control Module (ECM)**.

Throttle Position (TP) sensor -G69-, checking

EDITOR'S NOTE—

Checking the function of the closed throttle position switch requires the use of a diagnostic scan tool. For connecting scan tool see, ***Maintenance****. For diagnostic trouble codes and control unit coding, see* ***Scan Tool Codes*** *at the back of this manual.*

Checking conditions

• Engine coolant temperature at least 85°C (185°F) see VAG1551 scan tool function 08, display group 004, display field 3

Voltage supply and wiring to ECM, checking

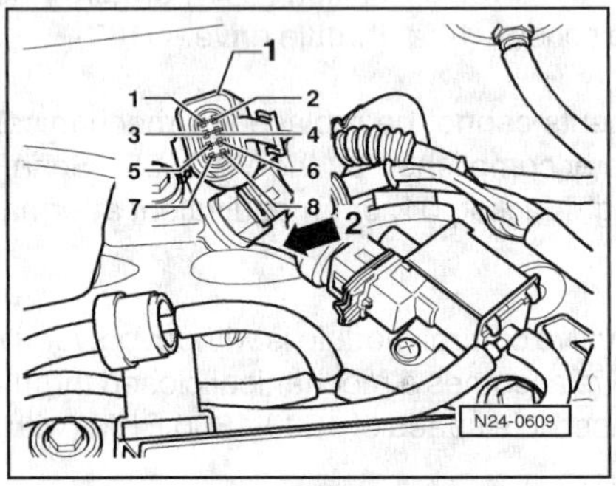

– Disconnect 8-pin harness connector -1- for throttle valve control module (arrow -2-).

– Connect multimeter US1119 (Fluke 83 or equivalent) between harness connector terminals 4 and 7.

– Switch ignition on and measure voltage.

- Specified value: min. 4.5 volts (5.0 volts nominal)

– Switch ignition off.

– Connect multimeter to measure voltage between harness connector terminals 3 and 7.

– Switch ignition on and measure voltage.

- Specified value: min. 9 volts (battery voltage nominal)

– Switch ignition off.

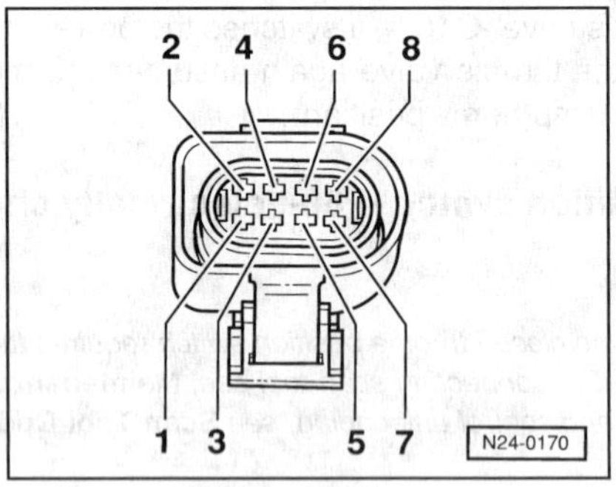

– Check wiring for open circuit between ECM/test box and 8-pin harness connector.

• Connector terminal 1 to ECM/test box socket 66
• Connector terminal 2 to ECM/test box socket 59
• Connector terminal 3 to ECM/test box socket 69
• Connector terminal 4 to ECM/test box socket 62
• Connector terminal 5 to ECM/test box socket 75
• Connector terminal 7 to ECM/test box socket 67
• Connector terminal 8 to ECM/test box socket 74

Specified value: max. 1.5 Ω

EDITOR'S NOTE—

Volkswagen recommends using the VAG 1598/22 test box to prevent damage to ECM harness connectors during electrical testing. However the harness connectors can be tested without the test box if extra care is used to prevent damage to the harness connector terminals.

– Check wiring of 8-pin connector for short circuit between wires, according to wiring diagram.

Specified value: ∞

If wiring is OK:

– Check ECM voltage supply, see **Engine Control Module (ECM) voltage supply, checking**.

– Carry out adaptation of throttle valve control module to ECM, see **Throttle valve control module, adaptation to Engine Control Module (ECM)**.

Engine Speed (RPM) sensor, electrically checking

The Engine Speed (RPM) sensor -G28- is a speed and reference mark sensor. The engine will not start if there is no speed signal. If the speed signal fails when the engine is running, it will cause the engine to stall immediately.

EDITOR'S NOTE—

Checking the function of the engine speed (RPM) sensor requires the use of a diagnostic scan tool. For connecting scan tool see, ***Maintenance****. For diagnostic trouble codes and control unit coding, see* ***Scan Tool Codes*** *at the back of this manual.*

Checking conditions

• Battery voltage: min. 11.5 volts

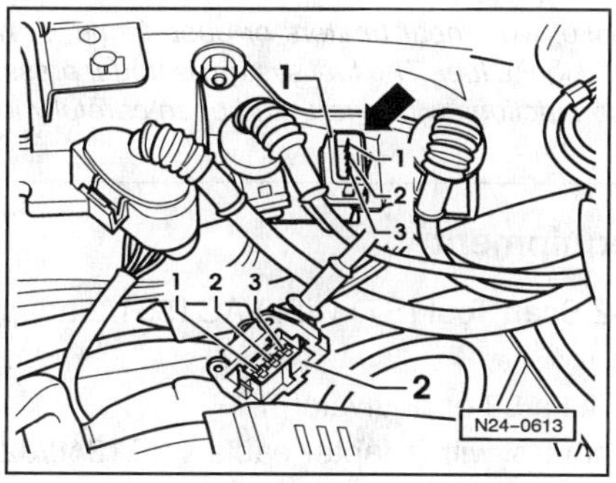

< Disconnect gray 3-pin harness connector from Engine Speed (RPM) sensor (arrow).

– Measure sensor resistance between terminals 2 and 3 of connector -1-.

• Specified value: 480-1000 Ω

– Check sensor for short circuit between terminals 1 and 2 and terminals 1 and 3.

• Specified value: ∞

If the specifications are not obtained:

– Replace Engine Speed (RPM) sensor -G28-.

If no sensor malfunction is detected:

– Check wiring for open circuit between ECM/test box and 3-pin harness connector.

• Connector terminal 1 to ECM/test box socket 67
• Connector terminal 2 to ECM/test box socket 63
• Connector terminal 3 to ECM/test box socket 56

Specified value: max. 1.5 Ω

EDITOR'S NOTE—

Volkswagen recommends using the VAG 1598/22 test box to prevent damage to ECM harness connectors during electrical testing. However the harness connectors can be tested without the test box if extra care is used to prevent damage to the harness connector terminals.

– Check wiring for short circuit between ECM/test box and 3-pin harness connector, according to wiring diagram.

If wiring is OK:

– Remove sensor and check sensor wheel for secure fit, damage and run-out.

– Check that spacer for sensor is seated securely.

NOTE —
There is one large gap on the sensor wheel. This gap is the reference mark and does not mean that the sensor wheel is damaged.

– Check readiness code, using diagnostic scan tool. If Diagnostic Trouble Code (DTC) memory has been erased, or ECM was disconnected, generate new readiness code.

If sensor wheel is OK:

– Replace Motronic ECM -J220-, see **Electrical Equipment section, Repair Group 97**.

EDITOR'S NOTE —
ECM replacement requires coding of control module using Volkswagen factory scan tool. See **Scan Tool Codes** *at the back of this manual or consult an authorized Volkswagen service department for additional information.*

Fuel injectors, checking

> ***WARNING —***
> *Fire hazard! Do not smoke, work near heaters or open flame, or have anything nearby that could ignite fuel. The fuel system is under pressure. Before opening, cover connections with a rag, and open carefully to release pressure.*

Special tools and equipment

- VAG1551 or VAG1552 Scan Tool (ST) with VAG1551/3 adapter cable
- VAG1598/22 test box
- Multimeter US1119 (Fluke 83 or equivalent)
- VAG1348/3A remote control with adapter cable VAG1348/32
- VAG1527B LED voltage tester
- VW1594 connector test kit
- VAG1602 measuring glass from calibration tester
- Wiring diagram

Checking conditions

- Engine Speed (RPM) Sensor OK, checking, see **Engine Speed (RPM) sensor, electrically checking**
- Fuel pump relay OK

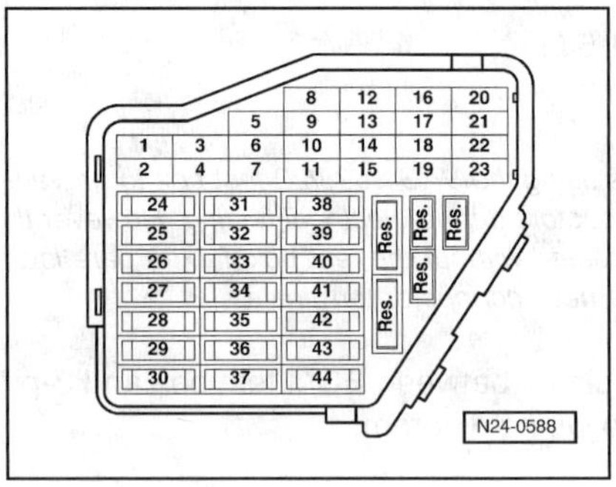

◄ Fuse 34 OK

Checking

– Use diagnostic scan tool to conduct output Diagnostic Test Mode (DTM) actuating fuel injectors.

If one or more fuel injectors do not click:

– Check actuation of fuel injectors.

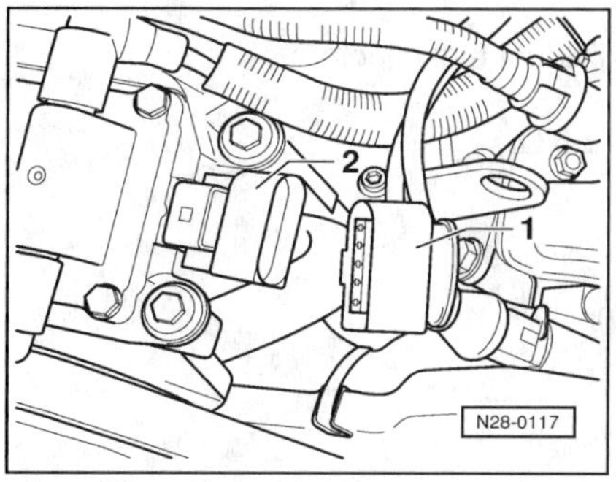

Disconnect 5-pin harness connector -1- from ignition coil -2-.

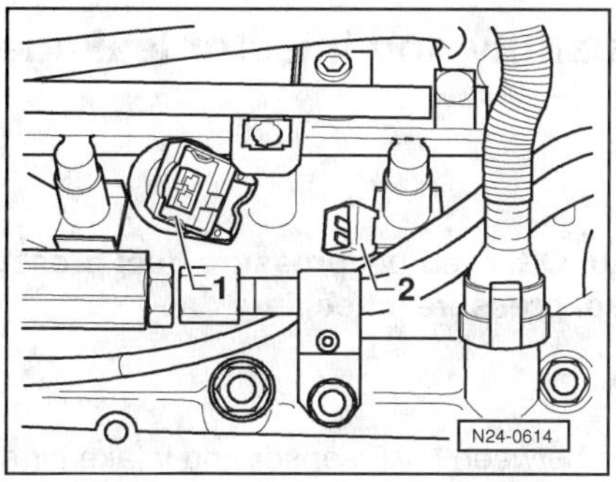

Disconnect harness connectors -1- from all fuel injectors -2-.

– Connect VAG1527B LED voltage tester to connector terminals for cylinder 1 fuel injector using adapter cables from VW1594 connector test kit.

– Operate starter and check voltage supply for cylinder 1 fuel injector.
  - LED must flicker.

– Repeat test for fuel injectors for cylinders 2-6.

– Switch ignition off.

If the LED does not flicker on any cylinder:

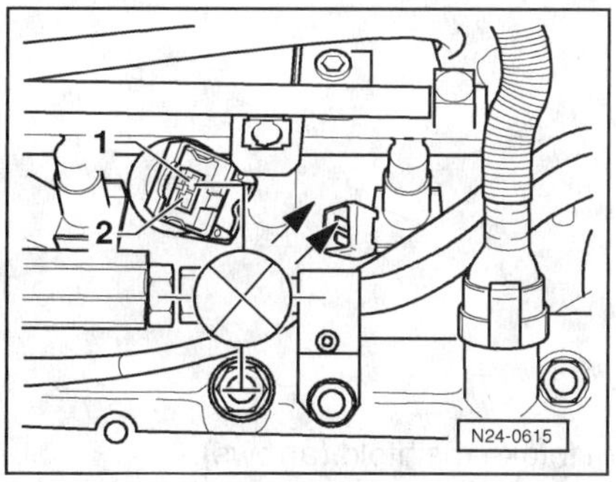

Connect VAG1527B LED voltage tester to cylinder 1 harness connector terminal 1 and Ground.

– Operate starter.
  - LED must light.

If the LED does not light up:

– Switch ignition off.

– Check wire between 2-pin connector terminal 1 and fuel pump relay -J17- for open circuit using wiring diagram.
  - Specified value: max. 1.5 Ω

If the LED flickers on one or several cylinders:

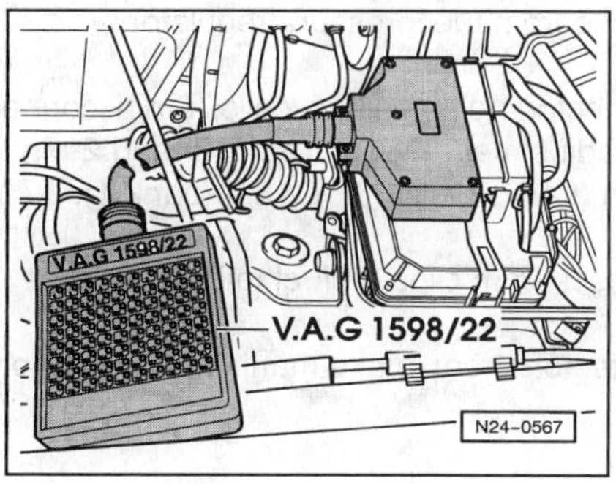

Connect VAG1598/22 test box to ECM harness connector.

– Check wiring for open circuits between ECM/test box and fuel injector harness connectors:
  - Cyl. 1 fuel injector terminal 2 to ECM/test box socket 73
  - Cyl. 2 fuel injector terminal 2 to ECM/test box socket 80
  - Cyl. 3 fuel injector terminal 2 to ECM/test box socket 58
  - Cyl. 4 fuel injector terminal 2 to ECM/test box socket 65
  - Cyl. 5 fuel injector terminal 2 to ECM/test box socket 72
  - Cyl. 6 fuel injector terminal 2 to ECM/test box socket 79

Specified value: max. 1.5 Ω

– Check wiring for short circuits between wires and connector terminals.
  - Specified value: ∞

– Check wiring between connector terminal 1 of all fuel injectors for open circuit.
  - Specified value: max. 1.5 Ω

Fuel injector resistance, checking

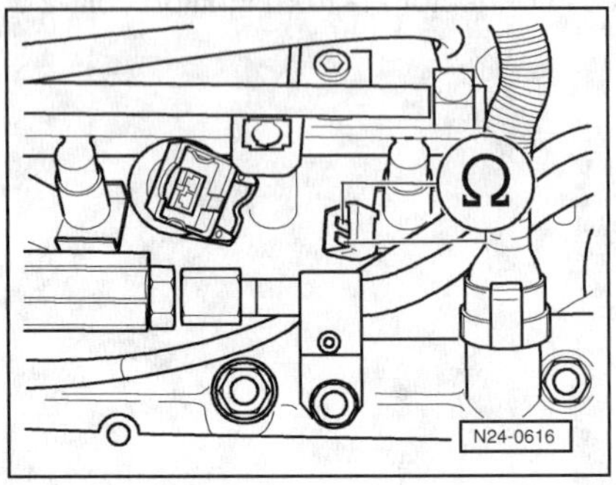

Check resistance of individual fuel injectors.

Specified value: 13.5-15.5 Ω

If resistance is NOT OK:

– Replace affected fuel injector.

Fuel injection quantity and injector leakage, checking

Checking conditions

- Fuel pressure regulator OK, see **Fuel pressure, fuel pressure regulator and residual fuel pressure, checking**

Checking

– Remove intake air hose between MAF sensor and intake air elbow.

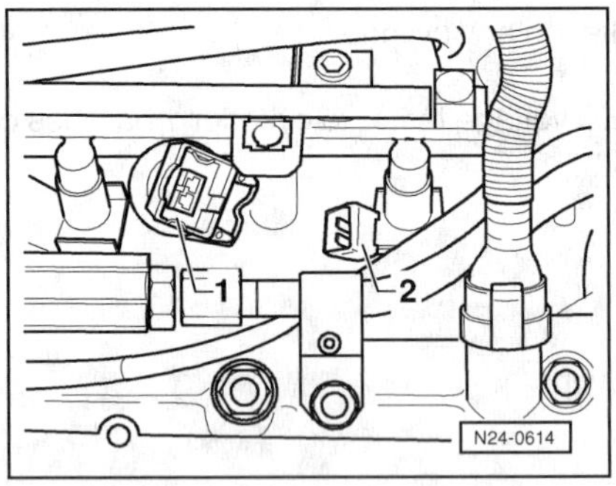

Remove harness connectors -1- from all fuel injectors -2-.

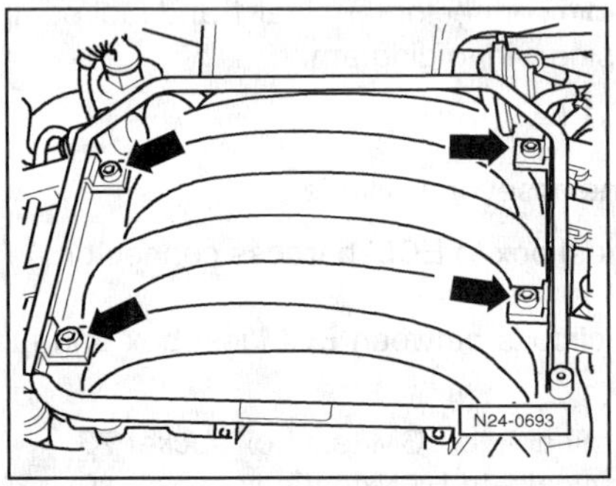

Disconnect fuel rail from intake manifold (arrows).

– Disconnect vacuum hose from Intake Manifold Tuning (IMT) valve -N156- (change-over valve).

– Disconnect vacuum hose from fuel pressure regulator.

– Disconnect 5-pin connector from ignition coils, 2-pin connector from valve -2- for camshaft adjustment -N208- (bank 2) and 2-pin connector from Camshaft Position (CMP) sensor 2 -G163- (bank 1).

– Disconnect wiring harness from clips on fuel rail.

– Lift fuel rail with fuel injectors from intake manifold and support it above manifold.

Checking for leakage

– Use diagnostic scan tool to conduct output DTM, actuating cylinder 1 fuel injector -N30- (fuel pump must be running).

NOTE—
This step lets the fuel pump run without the engine running. The closed throttle position switch must be closed during the test, or the selected fuel injector will be activated 5 times, discharging fuel.

– Visually check fuel injector for leaks.

- With fuel pump running, only 1 to 2 drops can drip per minute (per injector).

If fuel loss is greater:

– Switch ignition off and replace leaking injector.

NOTE—
Always replace seals.

Checking injection quantity

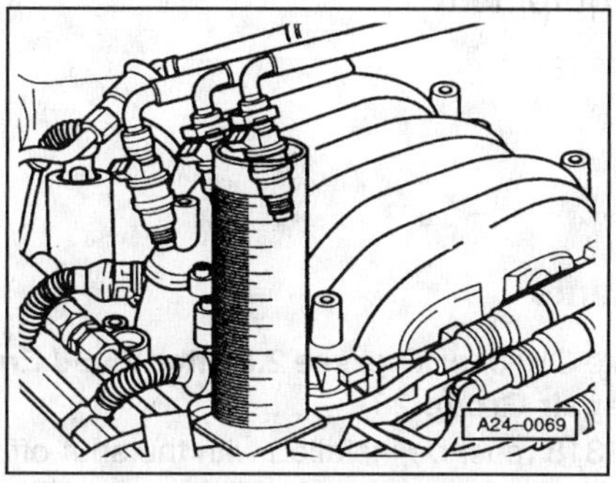

◄ Insert fuel injector to be tested into measuring glass of VAG1602 fuel analyzer.

– Connect one terminal of fuel injector being tested to engine Ground using test lead from VAG1594.

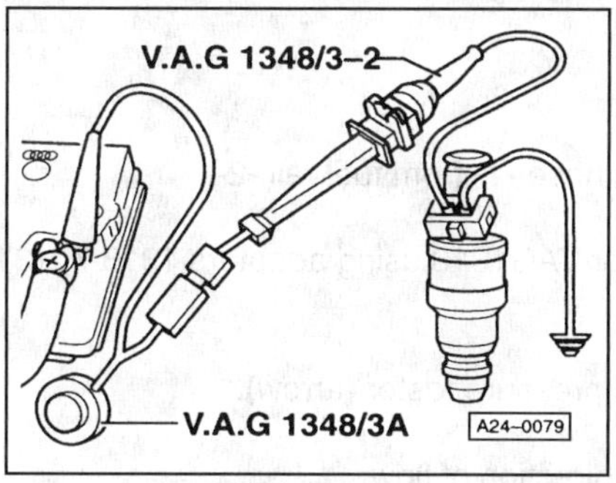

◄ Connect second terminal of fuel injector to B+ using VAG1348/3A remote control, VAG1348/3-2 adapter wire and an auxiliary wire.

– Connect alligator clip to battery positive (B+) terminal.

– Use diagnostic scan tool to conduct output DTM, actuating cylinder 1 fuel injector -N30- (fuel pump must be running).

– Operate remote VAG1348/3A for 30 seconds.

– Repeat check for other fuel injectors on cylinder bank. Use separate measuring containers.

– After all three fuel injectors on one cylinder bank have been activated, place three measuring containers on level surface and compare injected quantities.

- Specified volume per injector: 90-125 ml (3.04-4.2 fl oz.)

If the measured volume of one or more fuel injectors is below or above the specified volume:

– Replace malfunctioning fuel injector.

– Repeat test on fuel injectors in second bank of cylinders.

– Install fuel injectors in reverse order of removal.

NOTE—
When installing fuel injectors note the following:

- *Replace O-rings on all fuel injectors and coat O-rings lightly with clean engine oil.*
- *Slide fuel injectors (vertically) into position in fuel rail and secure injectors with retaining clips.*
- *Place fuel rail together with secured fuel injectors on manifold and press fuel rail and secured injectors in evenly.*

Fuel pressure, fuel pressure regulator and residual fuel pressure, checking

> ***WARNING —***
> *Fire hazard! Do not smoke, work near heaters or open flame, or have anything nearby that could ignite fuel. The fuel system is under pressure. Before opening, cover connections with a rag, and open carefully to release pressure.*

The fuel pressure regulator controls the fuel pressure depending on intake manifold pressure.

Special tools and equipment

- Pressure tester VAG1318
- Adapter 1318/7
- Adapter 1318/10
- Adapter 1318/13

Checking requirements

- Fuel pump delivery rate OK; checking, see **2.8 Liter V6 5V Engine Mechanical section, Repair Group 20**
- Pressure tester VAG1318 must be installed with the shut-off valve between the fuel supply line and the gauge for proper evaluation of residual pressure tests.

Checking

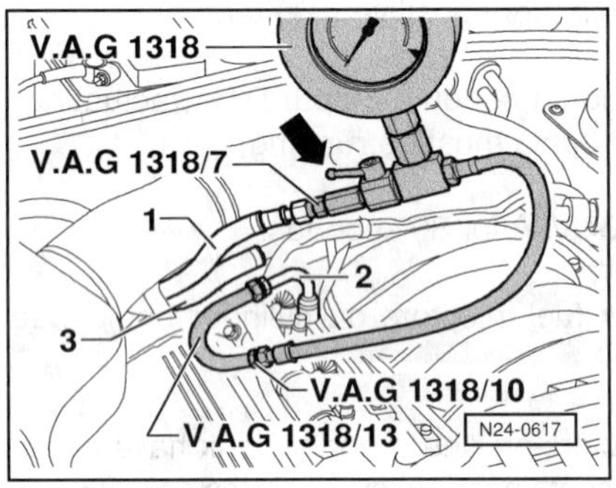

◄ Disconnect fuel supply hose -1- from fuel rail -2-.

– Connect pressure tester VAG1318 using adapters 1318/7, 1318/10 and 1318/13.

– Open shut-off valve on pressure tester (arrow).

The lever is parallel to direction of flow.

– Start engine and let idle.

– Measure fuel pressure.

- Specified value: approx. 3.5 bar (51 psi)

If the specification is attained:

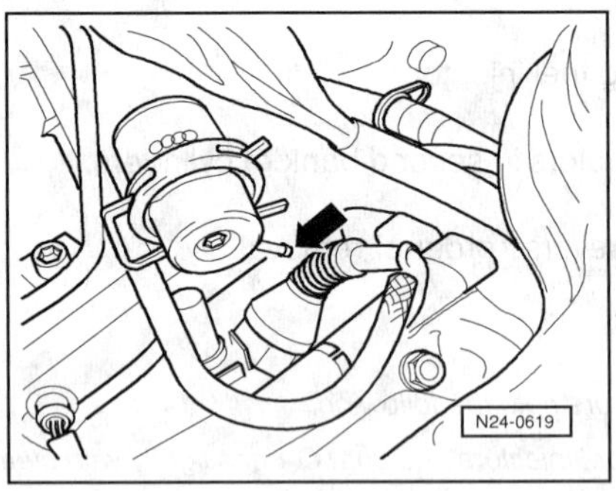

◄ Disconnect vacuum hose from fuel pressure regulator (arrow).

The fuel pressure must increase to approx. 4.0 bar (58.8 psi).

– Switch ignition off.

– Check for leaks and residual pressure by observing pressure drop on gauge.

- After 10 minutes there must be a residual pressure of at least 2.0 bar (29 psi).

If the residual pressure drops below 2 bar (29 psi):

– Start engine and let idle.

– Switch ignition off after pressure has built-up.

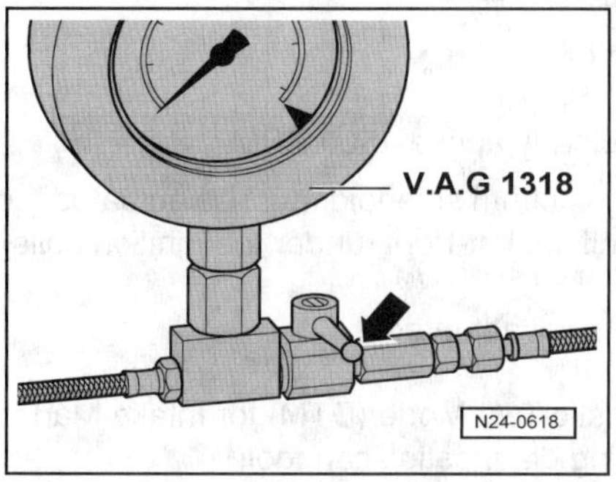

- Close shut off valve on VAG1318 pressure tester.
  - The lever is perpendicular to direction of flow.
- Observe pressure drop on gauge.

If the pressure does not drop:

- Check the check valve in electric fuel pump, see **2.8 Liter V6 5V Engine Mechanical section, Repair Group 20**

If the pressure drops again:

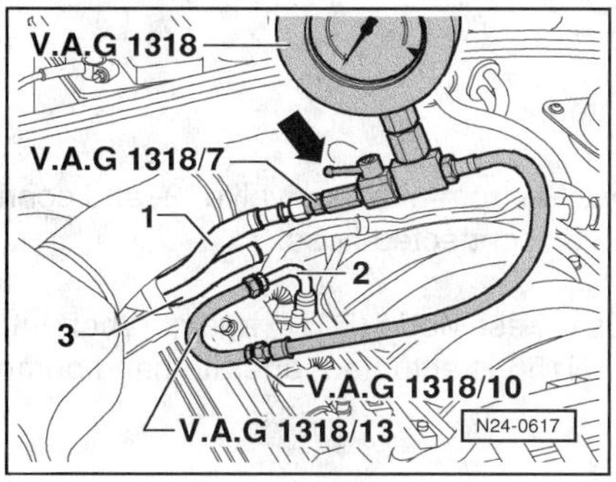

- Open shut-off valve on pressure tester (arrow).
  - The lever is parallel to direction of flow.
- Start engine and run at idle.
- Switch ignition off after pressure build-up.
- At same time pinch/seal return hose -3- together.

If the pressure does not drop:

- Replace fuel pressure regulator, see **Fuel rail with injectors.**

If the pressure drops again:

- Check pipe connections, O-rings on fuel rail and fuel injectors for leaks.
- Check pressure gauge for leaks.

NOTE—
Before removing pressure gauge place a cloth around the connections to be loosened.

Intake Manifold Tuning (IMT) valve, checking

NOTE—
- *Carry out this check for complaints of poor performance only.*
- *The tuning (change-over) of the manifold from long to short intake path occurs at approx. 4500 RPM.*

Special tools and equipment

- VAG1551 or VAG1552 Scan Tool (ST) with VAG1551/3 adapter cable
- VAG1598/22 test box
- Multimeter US1119 (Fluke 83 or equivalent)
- VW1594 connector test kit
- VAG1527B LED voltage tester
- VAG1390 hand pump
- Wiring diagrams

Check requirements

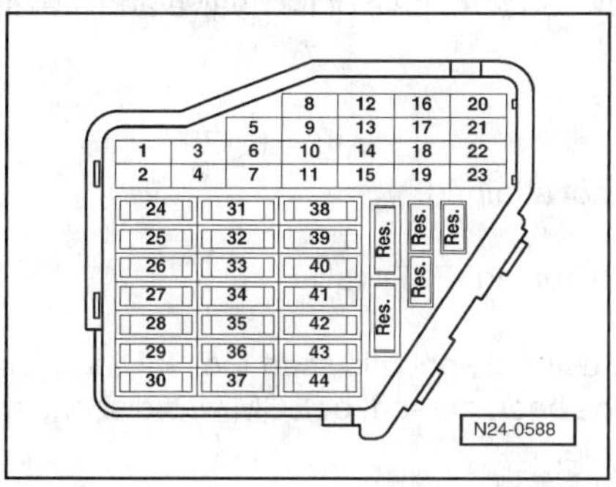

- Fuse 29 OK
- Fuel pump relay OK

IMT valve, checking function

– Start engine and run at idle.

– Increase engine RPM clearly above 4500 RPM.

  • Intake manifold tuning vacuum solenoid must be actuated (solenoid is located on intake manifold, left-front, under the ignition coils).

If the vacuum solenoid is not actuated:

– Conduct output Diagnostic Test Mode (DTM) for Intake Manifold Tuning (IMT) valve -N156-, using diagnostic scan tool.

If actuation is OK:

– Leave scan tool connected.

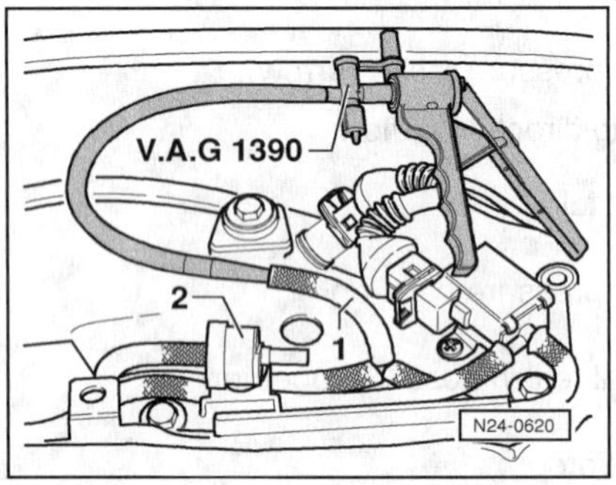

◄ Disconnect vacuum hose -1- from one-way valve -2- and connect VAG1390 hand pump to disconnected hose.

– Conduct output Diagnostic Test Mode (DTM) again. When Intake Manifold Tuning (IMT) valve -N156-is actuated, operate hand pump.

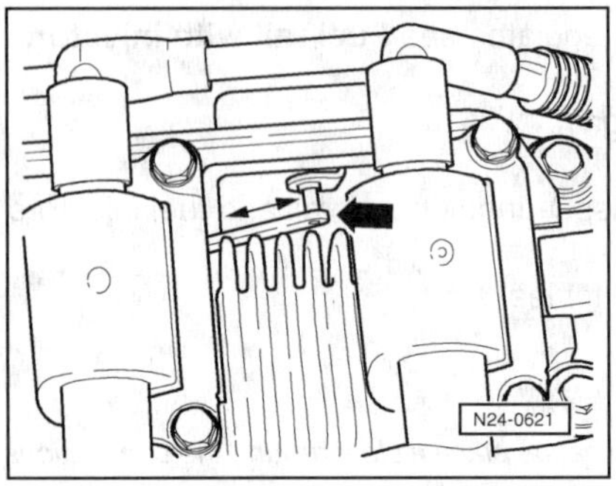

◄ The actuator must switch (change-over) between open and close (arrow).

If the change-over does not function as described:

– Check change-over mechanics for freedom of movement. To do this operate rods by hand.

– Check vacuum hoses for correct routing.

– Check vacuum system including vacuum reservoir in left wheel housing for leaks.

IMT valve, electrical check

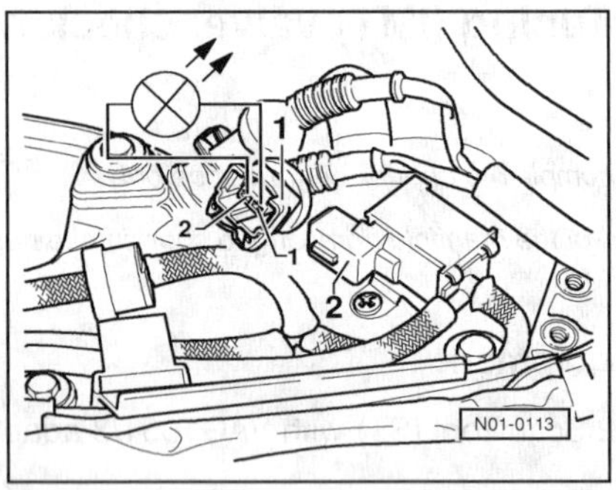

◄ Disconnect harness connector -1- from Intake Manifold Tuning (IMT) valve -2-.

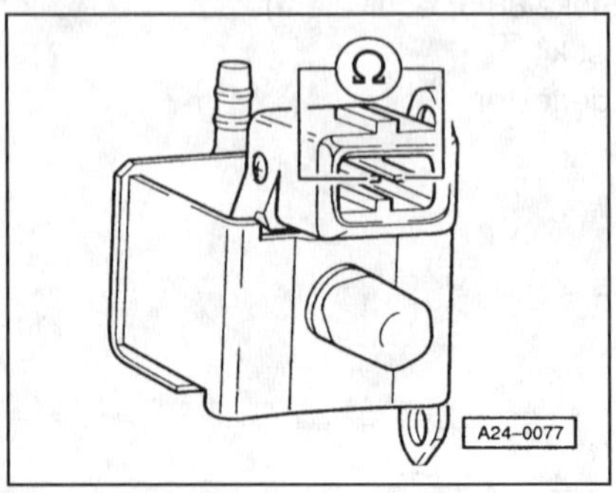

◄ Connect multimeter between terminals of IMT valve and measure resistance.

  • Specified value: 25-35 Ω

If the specified value is not attained:

– Replace Intake Manifold Tuning (IMT) valve.

– Check readiness code, using diagnostic scan tool. If Diagnostic Trouble Code (DTC) memory has been erased, or ECM was disconnected, generate new readiness code.

Throttle valve control module, adaptation to Engine Control Module (ECM)

During "adaptation" with ignition switched on and the engine not running, the Motronic ECM "learns" the closed throttle position (CTP), wide open throttle (WOT) and other positions of the throttle valve control module, and a characteristic curve of the relationship between Throttle Position (TP) sensor -G69- and angle sensor for throttle drive -G187-.

Adaptation must be performed if:

- The voltage supply is interrupted
- The throttle valve control module is removed and installed
- The throttle valve control module is replaced
- If when installing another engine another throttle valve control module is installed

Special tools and equipment

VAG1551 or VAG1552 Scan Tool (ST) with VAG1551/3 adapter cable

Test conditions

- No malfunctions stored in DTC memory, check DTC Memory (scan tool function 02) and Erase DTC Memory (scan tool function 05).
- Battery voltage: min. 11.5 volts
- All electrical consumers switched off (radiator coolant fan must not run during test)
- A/C switched off
- Throttle valve in closed throttle position; if necessary adjust accelerator pedal cable, see **2.8 Liter V6 5V Engine Mechanical section, Repair Group 20**
- Cruise control adjustment OK, see **Electrical Equipment section, Repair Group 27**

Adapting throttle valve control module to ECM, test procedure

– Connect VAG 1551 or VAG 1552 scan tool and press buttons -0- and -1- to insert “Engine Electronics” address word 01 (with ignition switched on; engine not running).

```
Rapid data transfer        HELP

Select function  XX
```

◀ Indicated on display

– Press buttons -0- and -4- to select “Basic Setting” function 04, and press -Q- button to confirm input.

```
System in Basic Setting    HELP

Input display group number  XXX
```

◀ Indicated on display

– Press buttons -0-, -6- and -0- to input display group number 60, and press -Q- button to confirm input.

```
System in Basic Setting  60

1 2 3 4
```

◀ Indicated on display (1-4 = display fields)

NOTE—

- *After confirming input of display group 60, adaptation is started. The Throttle Position (TP) sensor is positioned in the closed throttle, wide open throttle and limp-home positions. Following this procedure, other intermediate steps will create a characteristic curve of the relationship between Throttle Position (TP) sensor -G69- and angle sensor for throttle drive -G187-.*
- *The Motronic ECM stores these values in permanent memory. Afterward, the throttle valve assumes the starting position.*

– Compare display with specified values (display fields 1-4):

	Display fields			
	1	2	3	4
Display group 60: Basic setting-throttle valve control module				
Display	xxx %	xxx %	Idle Part throttle Full throttle Decel Enrich	ADP runs ADP. OK ADP. ERROR
Indicates	Throttle valve angle	Throttle drive angle	Operating condition	Adaptation condition
Range	0-90 %	0-20 %	---	---
Specified value	1-10 % (automatic trans) 1-20 % (manual trans)		Idle	ADP. runs
	0-5 %	0-5 %	Idle	ADP. OK (ADP. ERROR indicates unsuccessful adaptation)
	If displayed values are NOT OK: - Check DTC memory - Check throttle valve control module			

NOTE—

If the basic setting of the throttle valve control module is interrupted, the cause could be one of the following:

- *Throttle valve cannot reach the mechanical stop because of contamination e.g. oil deposits or an incorrectly adjusted accelerator pedal cable.*
- *Battery voltage is too low.*
- *Throttle valve control module or wiring is faulty. Checking, see* **Throttle valve control module, checking**.

NOTE—

After termination, DTCs P1559, P1564 or P1565 are stored in DTC memory. When next switching on ignition adaptation is automatically performed again.

– End adaptation as soon as possible after 30 seconds by pressing → button.

– Press buttons -0- and -6- to select "End Output" function 06, and press -Q- button to confirm input.

– Switch ignition off.

Additional signals, checking

Vehicle speed signal, checking

Special tools and equipment

- VAG1551 or VAG1552 Scan Tool (ST) with VAG1551/3 adapter cable
- VAG1598/22 test box
- Multimeter US1119 (Fluke 83 or equivalent)
- VW1594 connector test kit
- Wiring diagrams

Checking conditions

- Battery voltage: min. 11.5 volts
- Speedometer OK

Checking

– Connect VAG1551 or VAG1552 scan tool and press buttons -0- and -1- to insert "Engine Electronics" address word 01 (with engine running at idle).

– Press buttons -0- and -8- to select "Read Measuring Value Block" function 08, and press -Q- button to confirm input.

– Press buttons -0-, -0- and -5- to input display group number 5 (005), and press -Q- button to confirm input.

– Road test and compare indicated vehicle speed with value in display field 3 (2nd technician required).

> ***WARNING —***
> - *When driving or riding in an airbag-equipped vehicle, NEVER hold scan tool or other test equipment in your hands or lap while in motion. Objects between you and the airbag increase the risk of injury in an accident.*
> - *During a test drive in an airbag-equipped vehicle, test equipment must always be fastened to and operated from the rear seat by a second technician.*

– Observe value displayed in display field 3.

	Display fields			
	1	2	3	4
Display group 5: General engine data				
Display	xxx rpm	xx.xx ms	xxx km/h	Idle Part Throt Full Throt Decel Enrichment
Indicates	Engine speed (in 40 rpm steps)	Engine load	Vehicle speed	Operating condition
Range	0 - 6800 rpm	00.0 - 12.75 ms	0 - 326 km/h	---
Specification	---	---	Approx. vehicle speed	---
If NO driving speed is displayed, see **Continuation**, below				

If the specification is attained:

- Press →button.
- Press buttons -0- and -6- to select "End Output" function 06, and press -Q- button to confirm input.
- Switch ignition off.

Continuation

If no speed is indicated or the display values decrease when driving slowly:

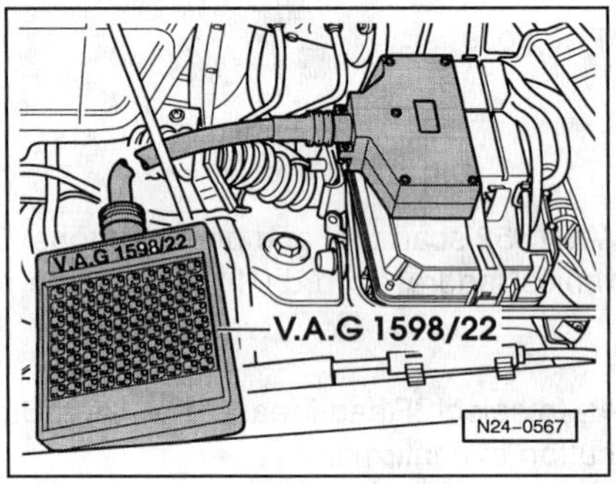

◄ Connect VAG1598/22 test box to ECM harness connector.

EDITOR'S NOTE —
Volkswagen recommends using the VAG 1598/22 test box to prevent damage to ECM harness connectors during electrical testing. However the harness connectors can be tested without the test box if extra care is used to prevent damage to the harness connector terminals.

- Connect multimeter to measure voltage between test box sockets 2 and 20.
- Switch ignition on.
- Lift vehicle at front left.
- Rotate front wheel and observe voltage display.
  - Specification: between 0 and at least 4 volts fluctuating

If the display does not fluctuate:

- Check wiring between speedometer and vehicle speed sensor using wiring diagrams.

Air conditioner (A/C) compressor signal, checking

Voltage is supplied to the engine control module (ECM) shortly before switching on the air conditioner compressor. The signal ensures that the control module keeps the idle speed constant for the differing power requirements of the A/C system.

The A/C compressor is switched off for approx. 12 seconds by the ECM when accelerating from standstill and from low speeds to ensure that the full engine power is available when accelerating. If the accelerator is released earlier the time is shortened to a minimum of 3 seconds. At full throttle the air conditioner is switched off by the Transmission Control Module (TCM) kick-down switch.

Special tools and equipment

- VAG1551 or VAG1552 Scan Tool (ST) with VAG1551/3 adapter cable
- VAG1598/22 test box
- Multimeter US1119 (Fluke 83 or equivalent)
- VW1594 connector test kit
- Wiring diagrams

Checking conditions

- A/C functioning OK
- A/C switched off
- No DTCs stored in DTC memory, check DTC Memory (scan tool function 02) and Erase DTC Memory (scan tool function 05).
- Vehicle at room temperature; warmer than +15°C (59°F).

Checking

– Connect VAG1551 or VAG1552 scan tool and press buttons -0- and -1- to insert "Engine Electronics" address word 01 (with engine running at idle).

– Press buttons -0- and -8- to select "Read Measuring Value Block" function 08, and press -Q- button to confirm input.

– Press buttons -0-, -5- and -0- to input display group number 50 (050), and press -Q- button to confirm input.

– Compare display with specified values for A/C (display fields 3 and 4).

– Check A/C specifications (Display zones 3 and 4)

	Display zones			
	1	2	3	4
Display group 50: Signals to Engine Control Module (ECM)				
Display	xxxx rpm	xxxx rpm	A/C LOW	Compr. on Compr. off
Indicated	Actual engine speed (in 10 rpm steps)	Specified engine speed (in 10 rpm steps)	A/C status	A/C compressor status
Range	0 to 2550 rpm	---	---	---
Specification	620 to 740 rpm (4WD) or 740 to 860 rpm (FWD)	680 rpm (4WD) or 800 rpm (FWD)	A/C ON: A/C-High A/C OFF: A/C-Low	Compressor ON: Compr. ON Compressor OFF: Compr. OFF
If specification is not obtained, see **Continuation**, below				

– Observe display in display field 4.

- Specified value: Compr. OFF

– Switch A/C system on.

– Select lowest temperature and highest blower speed.

– Observe display in display field 4.

- Specified value: Compr. ON

– Briefly press accelerator pedal to floor and release (short throttle burst).

When quickly pressing the accelerator pedal down the value in display field 4 must change from "Compr. ON" to "Compr. OFF" for a few seconds (compressor cut-off for vehicle acceleration).

– Press →button.

– Press buttons -0- and -6- to select "End Output" function 06, and press -Q- button to confirm input.

– Switch ignition off.

NOTE—
The air conditioner compressor will be switched off (Compr. OFF) when the engine control module recognizes full load and the air conditioning is switched on (A/C-HIGH).

Continuation

If the displayed values are NOT OK:

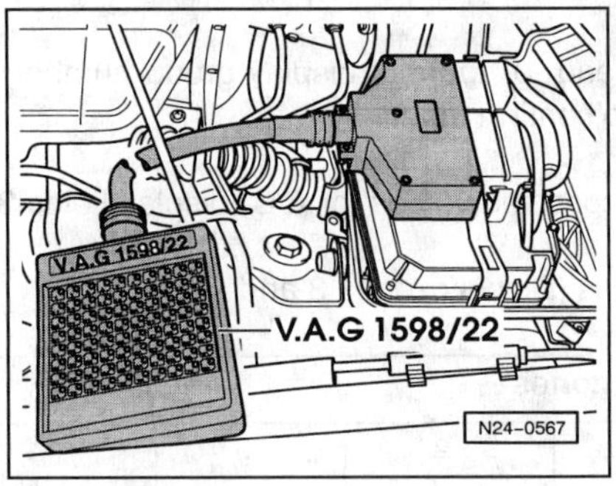

◄ Connect VAG1598/22 test box to ECM harness connector.

EDITOR'S NOTE—
Volkswagen recommends using the VAG 1598/22 test box to prevent damage to ECM harness connectors during electrical testing. However the harness connectors can be tested without the test box if extra care is used to prevent damage to the harness connector terminals.

– Check wiring for open circuit between ECM/test box terminal 8 and A/C control head using wiring diagrams.

If the wiring is OK:

– Check function of A/C control head -E87-.

– Check readiness code using diagnostic scan tool. If Diagnostic Trouble Code (DTC) memory has been erased, or ECM was disconnected, generate new readiness code.

Ignition retardation during gear shifts, checking

NOTE—
- *The Engine Control Module receives the gear change signal via the transmission Control Module. The Engine Control Module then, depending on road and engine speeds, retards the ignition briefly. This reduces the torque and therefore the gear jolt.*
- *The VAG 1551 does not always recognize and display the signal for torque reduction because it is sent for an extremely short period.*

NOTE —

To check ignition retardation, the vehicle must be driven which requires a second technician to read the test equipment.

> ***WARNING —***
>
> - *When driving or riding in an airbag-equipped vehicle, NEVER hold scan tool or other test equipment in your hands or lap while in motion. Objects between you and the airbag increase the risk of injury in an accident.*
> - *During a test drive in an airbag-equipped vehicle, test equipment must always be fastened to and operated from the rear seat by a second technician.*

Special tools, testers, measuring instruments and auxiliary items required

- Scan Tool VAG 1551 with cable VAG 1551/3B
- VAG 1598/22 Test Box
- Fluke 83 Multimeter
- VW 1594 Adaptor kit
- Wiring diagram

Test sequence

– Connect VAG 1551/1552 Scan Tool, see **Maintenance**

– Start engine and let idle

– Press 0 and 1 buttons to select "Address word" 01 "Engine electronics"

– Press 0 and 8 buttons to select Function 08: "Read measuring value block"

– Press Q button to enter input

– Press 1, 2 and 2 buttons to select "Display group 122"

– Press Q button to enter input

– Perform test drive with 2nd technician to observe display.

– Check specifications in Display zone 4:

	Display fields			
	1	2	3	4
Display group 122: Torque reduction during gear shift				
Display	xxxx rpm	xxx Nm	xxx Nm	TCS active Motor enga
Indicated	Engine speed (in 40 rpm steps)	Engine torque required (specified torque)	Actual engine torque (Actual torque)	Ignition angle retardation active or not active
Working range	0 to 6800 rpm	-100 to 410 Nm	-100 to 310 Nm	---
Specification	---	---	0 Nm	"Motor enga" must be displayed briefly when shifting gears

NOTE —
It is possible that several gear changes must be performed before the display in display zone 3 jumps to 0 because the signal from the automatic transmission is very brief.

If specification obtained

– Press →button.

– Press 0 and 6 buttons to select Function 06: "End data transfer"

– Press Q button to enter input

– Switch OFF ignition.

If display does not change:

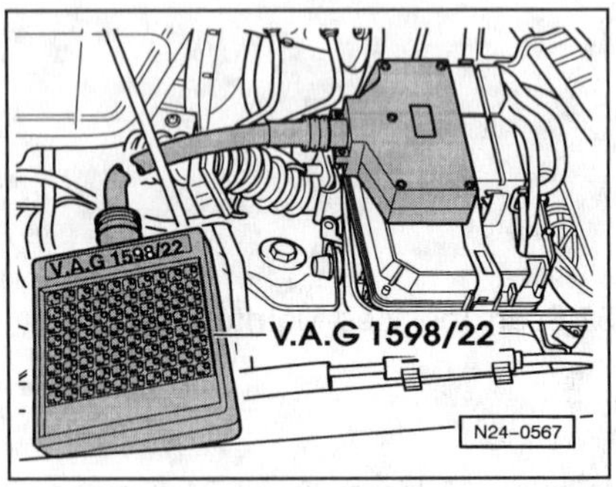

– Connect VAG 1598/22 Test Box to Engine Control Module harness connector.

EDITOR'S NOTE —
Volkswagen recommends using the VAG 1598/22 test box to prevent damage to ECM harness connectors during electrical testing. However the harness connectors can be tested without the test box if extra care is used to prevent damage to the harness connector terminals.

– Check wiring for open circuit between ECM/Test Box (socket 29 and 41) and TCM using appropriate wiring diagram.

• Wire resistance: Max. 1.5 Ω

Transmission range signal, checking

NOTE —
Engine Control Module (ECM) receives following data from (Automatic) Transmission Control Module (TCM): Driving range selected (selector lever in 2/3/4/R/D) or No driving range selected (selector lever in P or N)

Special tools, testers, measuring instruments and auxiliary items required

• Scan Tool VAG 1551 with cable VAG 1551/3B
• VAG 1598/22 Test Box
• Fluke 83 Multimeter
• VW 1594 Adaptor kit
• Wiring diagram

Test conditions

• Automatic transmission range selector must be in "P" or "N" position.

Test sequence

– Connect VAG 1551/1552 Scan Tool.

– Switch ON ignition

– Press 0 and 1 buttons to select "Address word" 01 "Engine electronics"

– Press 0 and 8 buttons to select Function 08: "Read measuring value block"

– Press Q button to enter input

– Press 0, 5 and 6 buttons to select "Display group 56"

– Press Q button to enter input

– Check specifications in Display zone 4 on chart, next page

	Display zones			
	1	2	3	4
Display group 56: Idle speed check				
Display	xxxx rpm	xxxx rpm	x.xx g/s	xxxx
Indicated	Actual engine speed (in 10 rpm steps)	Specified engine speed (in 10 rpm steps)	Idle mass air flow control regulation	Operating conditions
Range	---	---	---	---
Specification	0 rpm	0 rpm	---	0000 = Selector lever in P or N 0010 = Selector lever in: 2/3/4/R/D
If specification is not obtained, see **Continuation**, below				

If specification obtained

– Press →button.

– Press 0 and 6 buttons to select Function 06: "End data transfer"

– Press Q button to enter input

– Switch OFF ignition.

				Explanation when display field 4 = 0
x	x	x	x	**Display group 56**
			0	A/C compressor 0 = A/C compressor OFF 1 = A/C compressor ON)
		0		Selector lever position 0 = selector lever in P or N 1 = selector lever in 2/3/4/R/D)
	0			RPM increased 0 = no idle speed boost (vehicles with 4WD) 1 = idle speed boost (vehicles with front wheel drive)
0				Always "0"

If display does not change:

– Connect VAG 1598/22 Test Box to Engine Control Module harness connector.

– Check wiring for open circuit between ECM/Test Box (socket 29 and 41) and TCM using appropriate wiring diagram.

- Wire resistance: Max. 1.5 Ω

Rough road conditions signal from ABS/EDL control module, checking

NOTE —

- *If a spinning wheel is detected by the ABS/EDL control module the ABS/EDL control module produces a rough road condition signal. If the engine control module recognizes a rough road signal, misfire detection is switched OFF in the ECM.*
- *The rough road condition signal should only be checked if DTC "18014 rough road information, engine specified torque, ABS-SG electrical malfunction" is stored in DTC memory. The possible DTC: "Misfire" which may become stored; is a resulting malfunction and can be disregarded.*
- *The wiring connections and/or the rough road condition signal are monitored by the engine control module.*
- *If an exhaust emissions test is performed on a front wheel drive vehicle with ABS/EDL where the front axle/wheels are rotated while the rear axle/wheels remain stationary, this can cause DTC code 18014 (P1606) to be stored in DTC memory. Therefore, first remove the ABS/EDL control module fuse when performing emissions tests on these vehicles. Check and erase ABS/EDL control module and engine control module DTC memories after performing an emissions test.*
- *The fuse for the ABS/EDL control module must be removed when performing emissions tests on these vehicles. Check and erase DTC memory for the ABS/EDL control module and the engine control module after an emissions test.*

Special tools, testers, measuring instruments and auxiliary items required

- Scan Tool VAG 1551 with cable VAG 1551/3B
- VAG 1598/22 Test Box

Test conditions

- Control module coding OK, see **Scan Tool section**

Test sequence

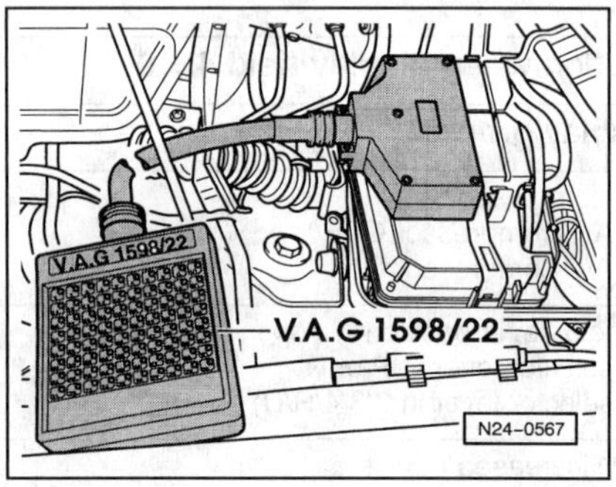

Connect VAG 1598/22 test box to engine control module harness connector.

EDITOR'S NOTE —

Volkswagen recommends using the VAG 1598/22 test box to prevent damage to ECM harness connectors during electrical testing. However the harness connectors can be tested without the test box if extra care is used to prevent damage to the harness connector terminals.

– Check signal wire for open or short circuit between test box and ABS control module using wiring diagram.

If wiring OK

– Replace ABS control module.

If DTC is displayed again after replacing ABS control module:

– Replace engine control module.

NOTE —

If DTC memory has been erased or the engine control module was disconnected from its voltage supply, the readiness code must be created again.

28 Ignition System

Repair Group

H 28

Ignition system, servicing

General notes on the ignition system

- Only the components which specifically relate to the ignition system are described here. For other components in the fuel injection and ignition system, see **Repair Group 24**.
- Disconnecting and connecting the battery must only be done with the ignition switched off, otherwise the Engine Control Module (ECM) could be damaged.
- The ECM is equipped with On Board Diagnostic (OBD).
- Components marked with * are checked by on board diagnostic.
- For trouble-free operation of the electrical components a voltage of at least 11.5 volts is necessary.
- During some checks it is possible that the ECM will recognize and store a DTC. After completing all checks and repairs the DTC memory must be checked and if necessary erased, see **Maintenance.**

Safety precautions

WARNING —

Be sure the ignition is switched OFF, when:

- *Disconnecting ignition wires*
- *Disconnecting fuel injection system wiring*
- *Connecting or disconnecting test equipment leads*
- *Disconnecting the battery*
- *Washing the engine or engine compartment.*

WARNING —

- *When driving or riding in an airbag-equipped vehicle, NEVER hold the scan tool or other test equipment in your hands or lap while in motion. Objects between you and the airbag increase the risk of injury in an accident.*
- *During a test drive in an airbag-equipped vehicle, test equipment must always be fastened to and operated from the rear seat by a second technician.*

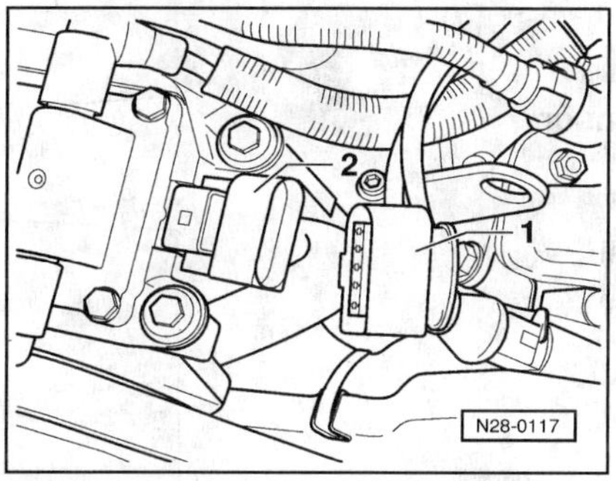

If the engine is to be turned at starter speed, without starting:

◄ Disconnect 5-pin connector -1- from power output stage -N122- for ignition coils -2-.

– Remove fuse 28 for fuel pump.

Ignition system components

> ***EDITOR'S NOTE—***
> *Volkswagen identifies electrical components by a letter and/or a number in the electrical schematics. See* **Wiring Diagrams, Fuses and Relays**. *These electrical identifiers are listed -between dashes- as an aid to electrical troubleshooting.*

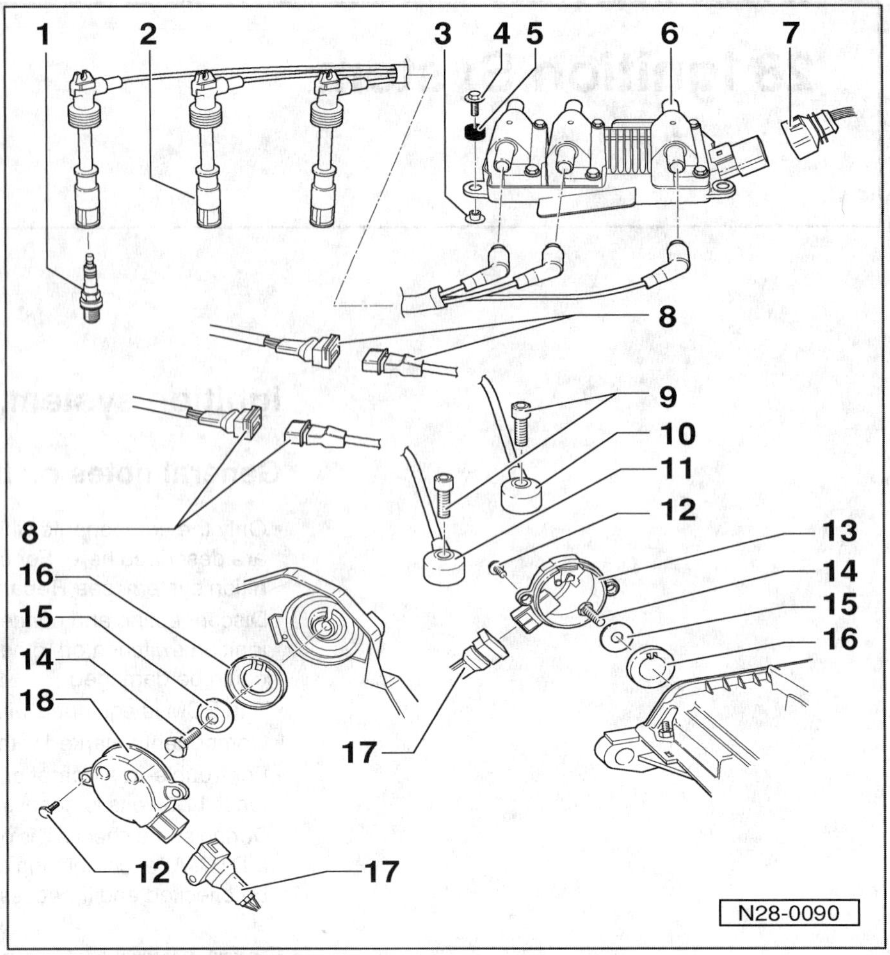

1. **Spark plug, 30 Nm (22 ft lb)**
   - Remove and install using 3122B
   - Type and electrode gap, see **Technical data, spark plugs**
2. **Spark plug connector**
   - 4 - 6 kΩ resistance
3. **Spacer sleeve**
4. **10 Nm (7 ft lb)**
5. **Rubber grommet**
6. **Ignition coils -N-, -N128-, -N158-**
   - With power output stage -N122-
   - With ignition cable identification, do not interchange
   - Checking, see **Ignition coils and power output stage, checking**
7. **Harness connector**
   - 5-pin, black
8. **Harness connector**
   - 3-pin connector, blue
   - Terminals are gold plated
   - For Knock Sensor (KS) 1 or 2 -G61- or -G66-
9. **20 Nm (15 ft lb)**
   - Tightening torque influences the function of the knock sensor
10. **Knock Sensor (KS) 2 -G66-***
    - Cylinder bank 2
    - Sensor and connector terminals are gold-plated
    - Checking, see **Knock sensor, checking**
11. **Knock Sensor (KS) 1 -G61-***
    - Cylinder bank 1
    - Sensor and connector terminals are gold-plated
    - Checking, see **Knock sensor, checking**
12. **10 Nm (7 ft lb)**
13. **Camshaft Position (CMP) sensor -G40-***
    - Cylinder bank 2
    - Checking, see **Camshaft Position (CMP) sensor, checking**
14. **25 Nm (18 ft lb)**
15. **Washer**
    - Conical
16. **Rotor**
    - For camshaft position sensor
    - Note position when installing
17. **Harness connector**
    - 3-pin, black
    - For camshaft position sensor -G40- or -G163-
18. **Camshaft Position (CMP) sensor 2 -G163-***
    - Cylinder bank 1
    - Checking, see **Camshaft Position (CMP) sensor, checking**

Technical data, spark plugs

Engine code	AHA
Firing order	1-4-3-6-2-5
Spark plugs[1)2)] VW/Audi Manufacturer's designation	 101 000 035 (NGK) BKR 6 EKUB (NGK)
Electrode gap	1.6 mm (0.063 in.)
Tightening torque	30 Nm (22 ft lb)

[1)] Spark plug change intervals see **Maintenance**

[2)] Remove and install spark plugs using special tool 3122B

> ***WARNING —***
> *Part numbers are listed here for reference only. Always check with your Parts department for the latest information.*

Camshaft Position (CMP) sensor, checking

Special tools and equipment

- VAG1598/22 test box
- Multimeter US1119 (Fluke 83 or equivalent)
- VW1594 connector test kit
- Wiring diagrams

Checking conditions

- Battery voltage: min. 11.5 volts

Checking

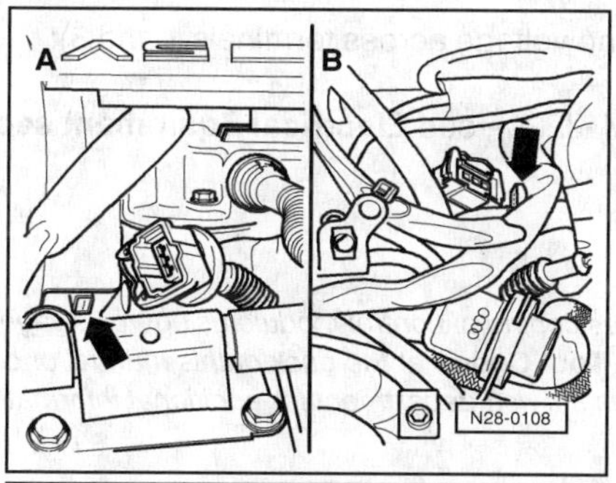

◀ Disconnect 3-pin harness connectors from Camshaft Position (CMP) sensors.

- A - Camshaft Position (CMP) sensor -G40-, cylinder bank 1, right front
- B - Camshaft Position (CMP) sensor -G163-, cylinder bank 2, left rear

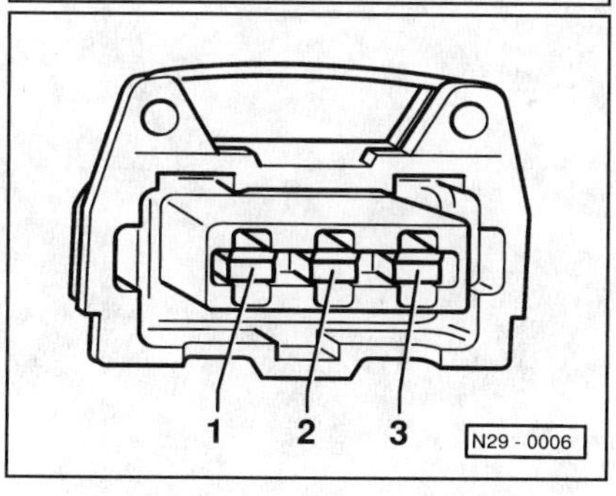

◀ Connect multimeter US1119 (Fluke 83 or equivalent) to measure voltage across CMP connector terminals 1 (B+) and 3 (Ground), using adapters from VW1594 connector test kit.

– Switch ignition on.

– Measure voltage between terminals 1 and 3.

- Specified value: min. 9.0 volts

– Switch ignition off.

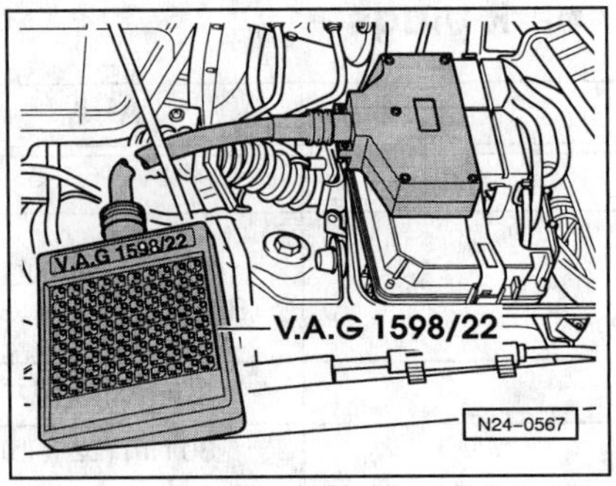

Connect VAG1598/22 test box to Motronic ECM harness connector.

EDITOR'S NOTE —
Volkswagen recommends using the VAG 1598/22 test box to prevent damage to ECM harness connectors during electrical testing. However the harness connectors can be tested without the test box if extra care is used to prevent damage to the harness connector terminals.

– Check wiring for open circuit between ECM/test box and 3-pin harness connector, according to wiring diagram.

CMP 2 -G163-, right cylinder head, front

- Connector terminal 1 to ECM/test box socket 11
- Connector terminal 2 to ECM/test box socket 76
- Connector terminal 3 to ECM/test box socket 67
- Specified value: max. 1.5 Ω

CMP 2 -G40-, left cylinder head, rear

- Connector terminal 1 to ECM/test box socket 11
- Connector terminal 2 to ECM/test box socket 44
- Connector terminal 3 to ECM/test box socket 14
- Specified value: max. 1.5 Ω

– Check wiring for short circuits to one another.

Specified value: ∞

If wiring is OK (and voltage across terminals 1 and 3 is OK):

– Replace Camshaft Position (CMP) sensor

If wiring is OK (but still no voltage across terminals 1 and 3):

– Replace Motronic ECM -J220-, see **Electrical Equipment section, Repair Group 97**.

EDITOR'S NOTE —
ECM replacement requires coding of control module using Volkswagen factory scan tool. See **Scan Tool Codes** *at the back of this manual or consult an authorized Volkswagen service department for additional information.*

Ignition coils and power output stage, checking

Special tools and equipment

- VAG1598/22 test box
- Multimeter US1119 (Fluke 83 or equivalent)
- VW1594 connector test kit
- VAG1527B voltage tester
- Wiring diagrams

Checking conditions

- Battery voltage: min. 11.5 volts
- Camshaft Position (CMP) sensors OK, checking, see **Camshaft Position (CMP) sensor, checking.**
- Engine Speed (RPM) sensor OK, checking, see **Repair Group 24.**

NOTE —

- *Ignition coils and power output stages are a combined unit.*
- *The ignition coil primary resistance cannot be measured.*
- *The secondary resistance can be measured with the ignition cables connected to the ignition coil. Measure between the spark plug connectors of the relevant ignition coil. For this measurement the suppression resistance of the ignition cables are also measured.*

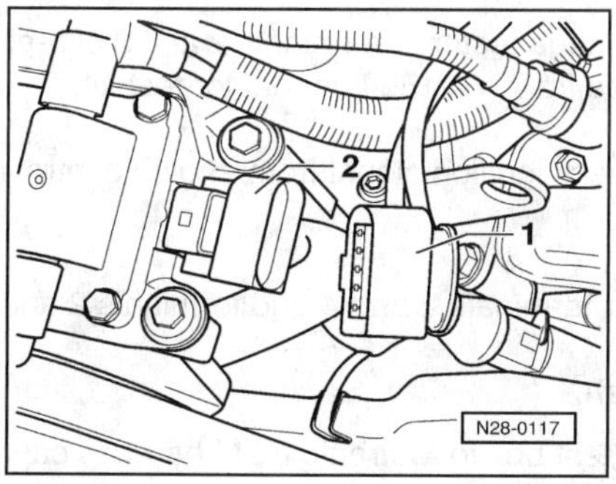

Ignition coils and ignition cables, checking

◀ Disconnect 5-pin connector -1- from ignition coil assembly -2-.

– Remove ignition cables from spark plugs.

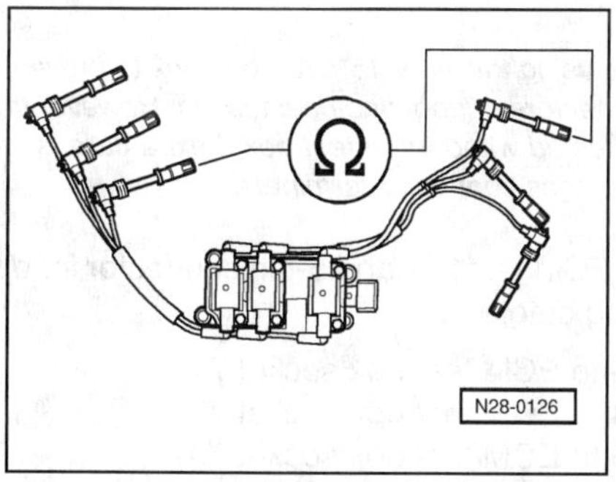

◀ Measure resistance between both spark plug connectors of an ignition coil.

- Specified value: approx. 18-25 kΩ.

– Repeat check at other two ignition coils.

If the specifications are not attained:

– Remove ignition cables from ignition coils.

– Check resistance of ignition cables.

- Specification: approx. 4.0-6.0 kΩ

– Check secondary resistance of ignition coils.

- Specification: 8.0-14.0 kΩ

If the specifications are again not attained:

– Replace ignition coils and power output stage.

If the specifications are attained:

– Check actuation signal for power output stage as described below.

Ignition coil voltage supply, checking

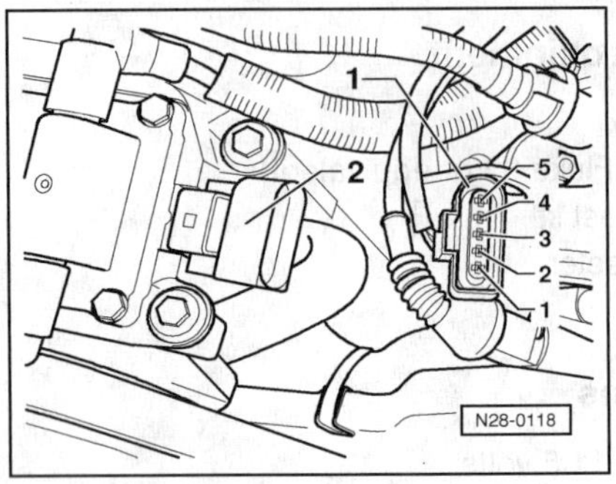

◄ Disconnect 5-pin connector -1- from ignition coil assembly -2-.

– Connect multimeter US1119 (Fluke 83 or equivalent) to measure voltage across terminals 1 and 2.

– Switch ignition on.

- Specified value: min. 11.5 volts

If no voltage is present:

– Switch ignition off.

– Check wire between 5-pin connector terminal 2 and Ground for open circuit using appropriate wiring diagram.

- Specified value: max. 1.5 Ω

– Check wire between 5-pin connector terminal 1 and relay panel for open circuit using appropriate wiring diagram.

- Specified value: max. 1.5 Ω

Checking actuation

– Remove fuse 28 for fuel pump or disconnect harness connectors from all fuel injectors.

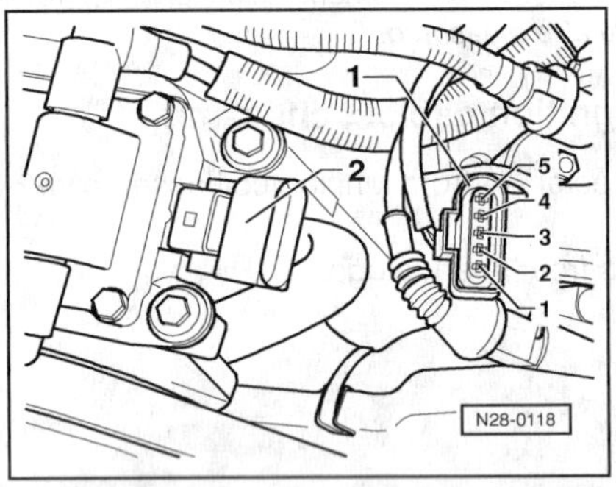

◄ Connect VAG1527B voltage tester to terminals 2 and 3 of 5-pin connector using adapter cables from VW1594 connector test kit.

– Operate starter and check ignition signal from engine control module.

- LED must flash.

– Repeat check between terminals 2 and 4 and terminals 2 and 5.

If the LED does not flash:

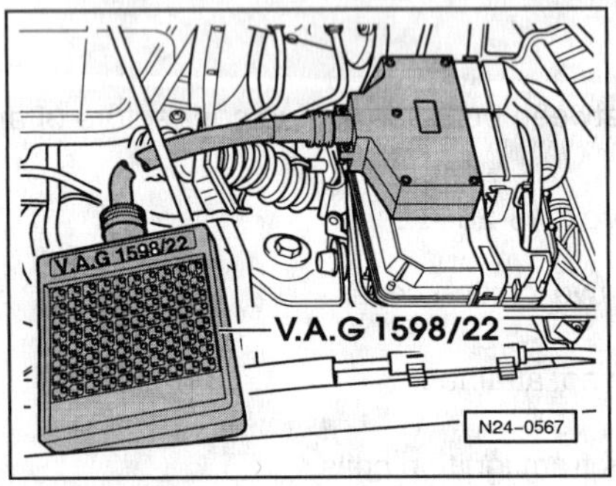

◄ Connect VAG1598/22 test box to Motronic ECM harness connector.

EDITOR'S NOTE —

Volkswagen recommends using the VAG 1598/22 test box to prevent damage to ECM harness connectors during electrical testing. However the harness connectors can be tested without the test box if extra care is used to prevent damage to the harness connector terminals.

– Check wiring between ECM/test box and 5-pin connector for open circuit using appropriate wiring diagram.

- Connector terminal 3 to ECM/test box socket 71
- Connector terminal 4 to ECM/test box socket 78
- Connector terminal 5 to ECM/test box socket 70

– Check wires for short circuits between using wiring diagrams.

- Specified value: max. ∞

– If wiring is OK and voltage was present between contacts 1 and 2, replace ECM, **Electrical Equipment section, Repair Group 97**

EDITOR'S NOTE —

ECM replacement requires coding of control module using Volkswagen factory scan tool. See **Scan Tool Codes** *at the back of this manual or consult an authorized Volkswagen service department for additional information.*

– If voltage supply and activation is OK, replace ignition coils and power output stage.

Knock sensor, electrically checking

NOTE—

- *The intake manifold must be removed to remove and install the knock sensors.*
- *It is extremely important to maintain a tightening torque of 20 Nm (15 ft lb) to ensure that the knock sensors perform adequately.*
- *Only gold-plated contacts may be used when servicing the knock sensor connector terminals.*

Special tools and equipment

- VAG1598/22 test box
- Multimeter US1119 (Fluke 83 or equivalent)
- VW1594 connector test kit
- VAG1527B voltage tester
- Wiring diagrams

Checking conditions

- On Board Diagnostic (OBD) must have recognized a malfunction for one or both knock sensors.

Resistances and wiring, checking

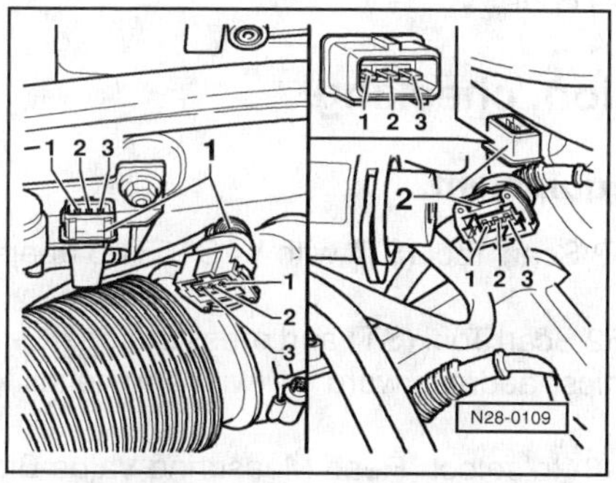

◄ Disconnect 3-pin connector from Knock Sensor (KS) 1 -G61- -1- and/or 3-pin connector from Knock Sensor (KS) 2 -G66- -2-.

– Connect multimeter US1119 (Fluke 83 or equivalent) to measure resistance between connector terminals, using adapters from VW1594 connector test kit.

- Knock sensor terminals 1 and 2
- Knock sensor terminals 1 and 3
- Knock sensor terminals 2 and 3
- Specified value: ∞

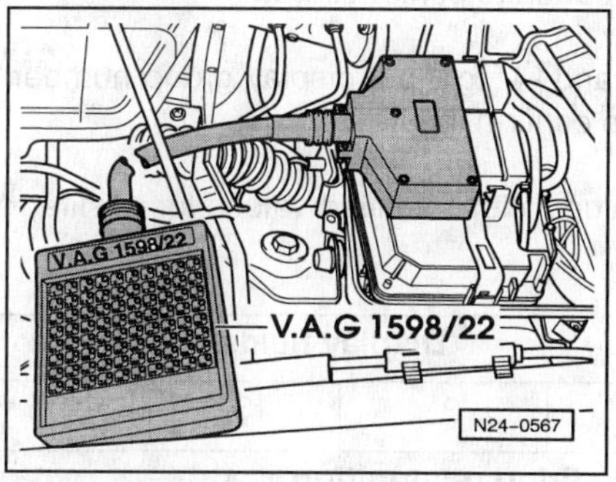

◄ Connect VAG1598/22 test box to Motronic ECM harness connector.

– Check wiring for open circuit between ECM/test box and 3-pin harness connector using appropriate wiring diagram.

Knock Sensor (KS) 1 -G61-:

- Connector terminal 1 to ECM/test box socket 68
- Connector terminal 2 to ECM/test box socket 67
- Connector terminal 3 to ECM/test box socket 67
- Specified value: max. 1.5 Ω

Knock Sensor (KS) 2 -G66-:

- Connector terminal 1 to ECM/test box socket 60
- Connector terminal 2 to ECM/test box socket 67
- Connector terminal 3 to ECM/test box socket 67
- Specified value: max. 1.5 Ω

– Check wires at test box socket 67 for short to test box sockets 60 and 68.

- Specified value: ∞

If wiring is OK:

– Loosen knock sensor and tighten again to 20 Nm (15 ft lb).

– Road test again and check knock sensor control specifications on display using diagnostic scan tool.

> ***WARNING —***
> - *When driving or riding in an airbag-equipped vehicle, NEVER hold the scan tool or other test equipment in your hands or lap while in motion. Objects between you and the airbag increase the risk of injury in an accident.*
> - *During a test drive in an airbag-equipped vehicle, test equipment must always be fastened to and operated from rear seat by a second technician.*

NOTE —
The test must be carried out during a road test because the knock sensor system is only active with engine speed greater than 2600 RPM and engine load greater than 3 ms.

– Check DTC memory again.

If DTC is still present:

– Replace knock sensor(s).

– Check readiness code using diagnostic scan tool. If Diagnostic Trouble Code (DTC) memory has been erased, or ECM has been disconnected, generate new readiness code.

Misfire recognition, checking

Special tools and equipment

- VAG1551 or VAG1552 Scan Tool (ST) with VAG1551/3 adapter cable

– Connect VAG1551/1552 Scan Tool (ST) and press buttons -0- and -1- to insert "Engine Electronics" address word 01 (with engine idling).

– Press buttons -0- and -8- to select "Read Measuring Value Block" function 08, and press -Q- button to confirm input.

– Press buttons -0-, -1- and -4- to input display group number 14 (014), and press -Q- button to confirm input.

– Compare displayed values with specified values for misfire recognition (display fields 3-4):

	Display fields			
	1	2	3	4
Display group 14: Misfiring recognition				
Display	xxx rpm	xx.xx ms	xxx	activated deact
Indicated	Engine speed (in 40rpm steps)	Engine load	Misfiring harmful to TWC(total)	Misfiring recognition status
Working range	0 to 6800 rpm	0.00 to 12.75 ms	---	---
Specified value	620 - 740 rpm (4WD) 740 - 860 rpm (2WD)	1.0 to 2.5 ms	0 to 30	Activated

– If displayed values are OK:

- Press → button.
- Press buttons -0- and -6- to select "End Output" function 06, and press -Q- button to confirm input.
- Switch ignition off.

If displayed values are NOT OK:

– Change to display group 15 as follows:

- VAG1551: Press -3- button
- VAG1552: Press ↑ button

– Compare displayed values with specified values for misfire recognition (display fields 1-4):

	Display fields			
	1	2	3	4
Display group 15: Misfire recognition				
Display	xxx	xxx	xxx	activated deact
Indicated	Misfiring adversely affecting No. 1 Cyl.	Misfiring adversely affecting No. 2 Cyl.	Misfiring adversely affecting No. 3 Cyl.	Misfire recognition status
Working range	0 to 500	0 to 500	0 to 500	---
Specification	0 to 5	0 to 5	0 to 5	Activated
If specification not obtained, see **Evaluating display groups 14, 15 and 16**				

If displayed values are OK:

– Change to display group 16 as follows:

- VAG1551: Press -3- button
- VAG1552: Press ↑button

– Compare displayed values with specified values for misfire recognition (display fields 1-4):

	Display fields			
	1	2	3	4
Display group 16: Misfiring recognition				
Display	xxx	---	---	activated deact
Indicated	Misfiring adversely affecting No. 4 Cyl.	Misfiring adversely affecting No. 5 Cyl.	Misfiring adversely affecting No. 6 Cyl.	Misfire detection status
Working range	0 to 500	0 to 500	0 to 500	---
Specification	0 to 5	0 to 5	0 to 5	activated
If specification not obtained, see **Evaluating display groups 14, 15 and 16**				

– Press → button.

- Press buttons -0- and -6- to select "End Output" function 06, and press -Q- button to confirm input.
- Switch ignition off.

Evaluation of display groups 14, 15 and 16

Display group: 14 Display zone: 3 and 4	Possible malfunction cause	Corrective action
greater than 30	Spark plug faulty Spark plug connector faulty Ignition coil faulty Ignition coil power output stage faulty	- Ignition coils and power output stage, checking, see **Ignition coils and power output stage, checking**
	Injector faulty Fuel shortage	- Fuel injectors, checking, see **Repair Group 24,** - Check amount of fuel in tank
Display groups: 15 and 16 **Display zones: 1 - 4**	**Possible malfunction cause**	**Corrective action**
greater than 5	Spark plug faulty Spark plug connector faulty Ignition coil faulty Ignition coil final output stage faulty	- Ignition coils and power output stage, checking, see **Ignition coils and power output stage, checking**
	Injector faulty Fuel shortage	- Fuel injectors, checking, see **Repair Group 24** - Check amount of fuel in tank

2.8 Liter Fuel Injection and Ignition (code: ATQ)

Section

J

24 Multiport Fuel Injection (MFI)

Page

28 Ignition

Page

24 Multiport Fuel Injection (MFI)

Repair Group

J 24

Fuel Injection System, servicing

NOTE —

- *The Engine Control Module (ECM) is equipped with On board Diagnostic. Before carrying out repairs and Troubleshooting; DTC memory must be checked. Also check the vacuum hoses and connections for leaks or unmeasured air.*
- *Fuel hoses in engine compartment must only be secured with spring clips. The use of clamp or screw clips is not permissible.*
- *VAG 1921 pliers are recommended for installing spring clips.*
- *Disconnecting and connecting the battery must only be done with the ignition switched off, otherwise the Engine Control Module could be damaged.*
- *For trouble-free operation of the electrical components, 11.5 Volts minimum is required.*
- *During some checks it is possible that the Control Module will recognize and store a malfunction. Therefore after completing all checks and repairs DTC memory must be checked and if necessary erased. For connecting diagnostic scan tool and checking and erasing DTC memory, see* **Maintenance.**

CAUTION —

- *Do not use any gasket sealers or other automotive chemical compounds that contain silicone. Traces of silicone drawn into the engine pass unburned into the exhaust and can damage the oxygen sensor.*
- *Do not use any electrical contact sprays or equivalent near oxygen sensor connections. Proper oxygen sensor function requires reference to outside air via the cable. If any contact spray reaches the oxygen sensor, it can cause damage or improper function.*

EDITOR'S NOTE —

Volkswagen identifies electrical components by a letter and/or a number in the electrical schematics. See **Wiring Diagrams and Component Locations section***. These electrical identifiers are listed -between dashes- as an aid to electrical troubleshooting.*

EDITOR'S NOTE —

Functional testing of most fuel injection and ignition components is done using a VAG 1551, 1552, 5051 or equivalent aftermarket scan tool. In addition, the factory repair information calls for the use of various special tools including electronic test boxes and remote control switches. Many fuel injection components can be tested electrically using a digital multimeter and/or an LED test light. The following pages include many of these basic electrical tests as well as some of the testing that relies on the use of a diagnostic scan tool and specialized diagnostic equipment.

Safety precautions

WARNING —

- *Fire hazard! Do not smoke, work near heaters or open flame, or have anything nearby that could ignite fuel. The fuel system is under pressure. Before opening, cover connections with a rag, and open carefully to release pressure.*
- *When driving or riding in an airbag-equipped vehicle, NEVER hold the scan tool or other test equipment in your hands or lap while in motion. Objects between you and the airbag increase the risk of injury in an accident.*
- *During a test drive in an airbag-equipped vehicle, test equipment must always be fastened to and operated from the rear seat by a second technician.*
- *Be alert when working on or near the engine and ignition wires. High secondary ignition voltage can cause serious personal injury and damage vehicle components.*
- *DO NOT touch or disconnect ignition system wires when the engine is running or being cranked at starting RPM.*
- *DO NOT operate the starter if fuel injectors are removed.*

WARNING —

Be sure the ignition is switched OFF, when:

- *Disconnecting or connecting the battery, ignition wires, or test equipment leads.*
- *Disconnecting fuel injection system wiring.*
- *Washing the engine or engine compartment.*

WARNING —

BEFORE cranking the engine at starting RPM (e.g. for compression testing) disable ignition and fuel injection systems:

- *Disconnect connector for ignition coil power output stage, and fuse for fuel pump.*
- *Disconnect harness connectors from all fuel injectors.*
- *Afterward, erase Diagnostic Trouble Code (DTC) memory.*

If the engine is to be turned at starter speed, without starting:

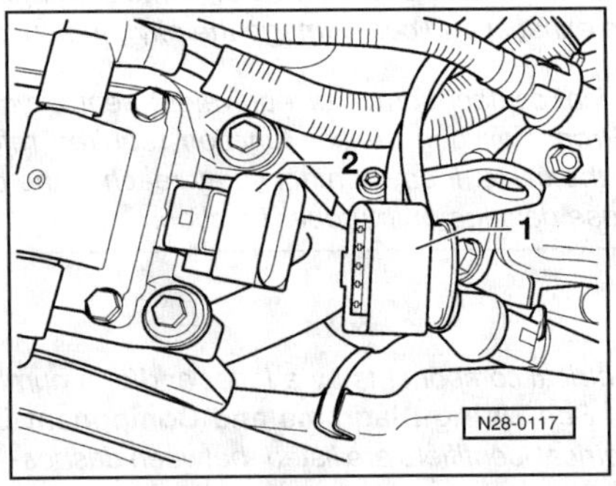

- Disconnect 5-pin connector -1- from power output stage for ignition coils -2-.

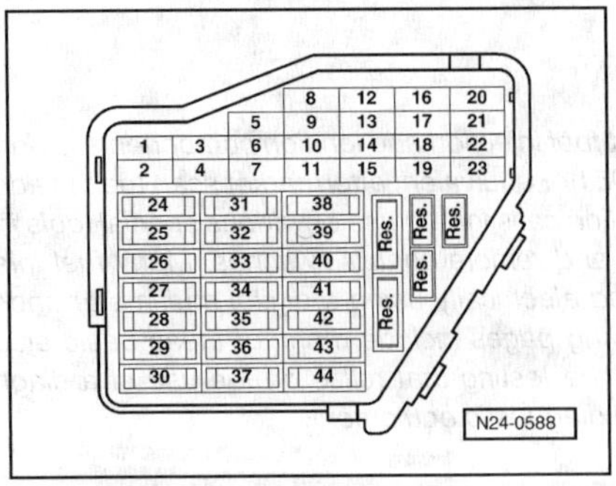

- Remove fuse 28 for fuel pump.

WARNING —

BEFORE disconnecting the battery:

- *Stop the engine.*
- *Be sure the ignition is switched OFF (also applies when connecting the battery). Failure to do so may damage the ECM.*
- *Be sure of the proper radio code (for vehicles equipped with coded anti-theft radio).*

> ***WARNING —***
> *When connecting and disconnecting electrical test equipment (LED voltage tester, multimeter, etc.):*
> - *Be sure the ignition is switched OFF.*
> - *Use correct adapters from the VW1594 connector test kit.*

> ***WARNING —***
> *For any work affecting the Engine Control Module (ECM):*
> - *BEFORE disconnecting the ECM harness connector, switch the ignition OFF and WAIT at least 20 seconds. Failure to do so may damage the ECM.*
> - *DO NOT connect any outside voltage source to stimulate an output signal at the ECM.*

Rules of cleanliness

Whenever carrying out work on the fuel supply and fuel injection systems, carefully observe the following five rules of cleanliness.

1. Thoroughly clean fuel system line and hose connections and the surrounding area before disconnecting.
2. Place removed components on a clean surface and cover. Use plastic sheeting or paper. Do not use fluffy rags that could leave lint!
3. Carefully cover over or seal any components that have been opened if repairs are not carried out immediately.
4. Install only clean parts:
   - Do not remove replacement parts from the packaging until immediately before they are to be installed.
   - Do not use parts that have been stored without packaging (e.g. in toolboxes, etc.).
5. When the fuel system is opened:
   - Avoid working with compressed air whenever possible.
   - Avoid moving the vehicle if possible.

Technical data

Engine code	**ATQ**
Idle check	
Idle speed[1]	
All wheel drive	620 - 740 RPM
Front wheel drive	710 - 860 RPM
Engine control module[2]	
System	Motronic ME 7.1
Part number	see factory parts information
Engine governed speed (RPM)	from approx. 6500

[1] Not adjustable

[2] Removing and installing engine control module, see **Electrical Equipment section, Repair Group 97**

Fuel injection components, overview

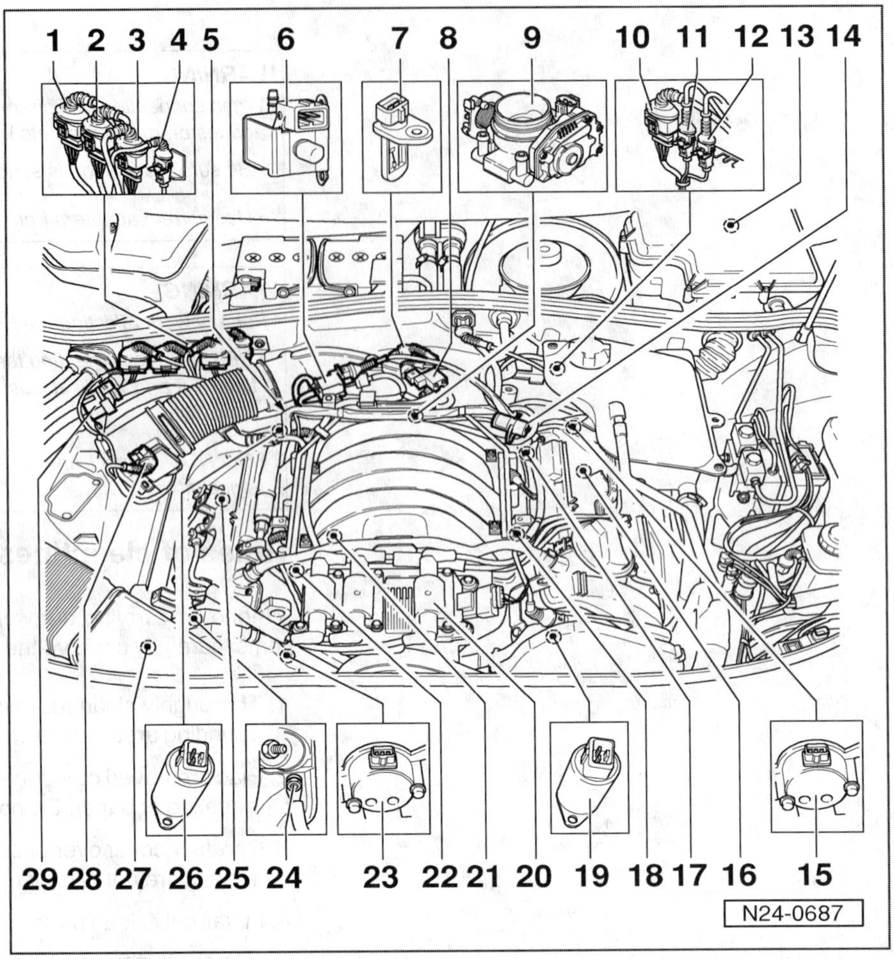

1. **Harness connector**
   - 4-pin connector, green
   - For O2S behind TWC -G130- and O2S heater -Z29-
   - For bank 1, sensor 2
2. **Harness connector**
   - 4-pin connector, brown
   - For O2S 2 behind TWC -G131- and O2S heater -Z30-
   - For bank 2, sensor 2
3. **Harness connector**
   - 4-pin connector, black
   - For HO2S -G39- and O2S heater -Z19-
   - For bank 1, sensor 1
4. **Harness connector**
   - 3-pin connector, blue
   - For Knock Sensor 1 -G61-, bank 1
5. **Engine Coolant Temperature (ECT) sensor -G62-**
6. **Secondary Air Injection (AIR) solenoid valve -N112-**
   - Removing and installing, see **2.8 Liter V6 5V Engine Mechanical section, Repair Group 26**
7. **Intake Air Temperature (IAT) sensor -G42-**
8. **Intake Manifold Tuning (IMT) valve -N156- (change-over valve)**
9. **Throttle valve control module -J338-**
10. **Harness connector**
    - 4-pin connector, black
    - For HO2S 2 -G108- and O2S heater -Z28-
    - Bank 2, sensor 1
11. **Harness connector**
    - 3-pin connector, blue
    - For Knock Sensor (KS) 2 -G66-, bank 2
12. **Harness connector**
    - 3-pin connector, gray
    - For Engine Speed (RPM) sensor -G28-
13. **Motronic Engine Control Module (ECM) -J220-**
    - Located in plenum, left side
14. **Fuel pressure regulator**
15. **Camshaft Position (CMP) sensor 2 -G160-**
    - Cylinder bank 2
    - Location, see **Repair Group 28**
16. **Heated Oxygen Sensor (HO2S) 2 -G108-**
    - Tightening torque: 50 Nm (37 ft lb)
    - Cylinder bank 2
17. **Engine Speed (RPM) sensor -G28-**
18. **Knock Sensor (KS) 2 -G66-**
    - Cylinder bank 2
    - Removing and installing, see **Repair Group 28**
19. **Valve -2- for camshaft adjustment -N208-**
    - Cylinder bank 2
    - Removing and installing see **2.8 Liter V6 5V Engine Mechanical section, Repair Group 15**
20. **Ignition coils -N-, -N128-, -N158-**
    - With power output stage -N122-
    - Removing and installing, see **Repair Group 28**
21. **Knock Sensor (KS) 1 -G61-**
    - Cylinder bank 1
    - Removing and installing, see **Repair Group 28**
22. **Fuel injectors**
    - Cylinder 1: -N30-
    - Cylinder 2: -N31-
    - Cylinder 3: -N32-
    - Cylinder 4: -N33-
    - Cylinder 5: -N83-
    - Cylinder 6: -N84-
23. **Camshaft Position (CMP) sensor -G40-**
    - Cylinder bank 1
    - Removing and installing, see **Repair Group 28**
24. **Ground connection**
    - On engine support, right
25. **Heated Oxygen Sensor (HO2S) -G39-**
    - Tightening torque: 50 Nm (37 ft lb)
    - Cylinder bank 1

(continued on following page)

Fuel injection components, overview (continued)

26. **Valve -1- for camshaft adjustment -N205-**
    - Cylinder bank 1
    - Removing and installing see **2.8 Liter V6 5V Engine Mechanical section, Repair Group 15**

27. **Secondary Air Injection (AIR) pump motor -V101-**
    - Removing and installing see **2.8 Liter V6 5V Engine Mechanical section, Repair Group 26**

28. **Mass Air Flow (MAF) sensor -G70-**

29. **Evaporative Emission (EVAP) canister purge regulator valve -N80-**
    - On air cleaner
    - Removing and installing see **2.8 Liter V6 5V Engine Mechanical section, Repair Group 20**

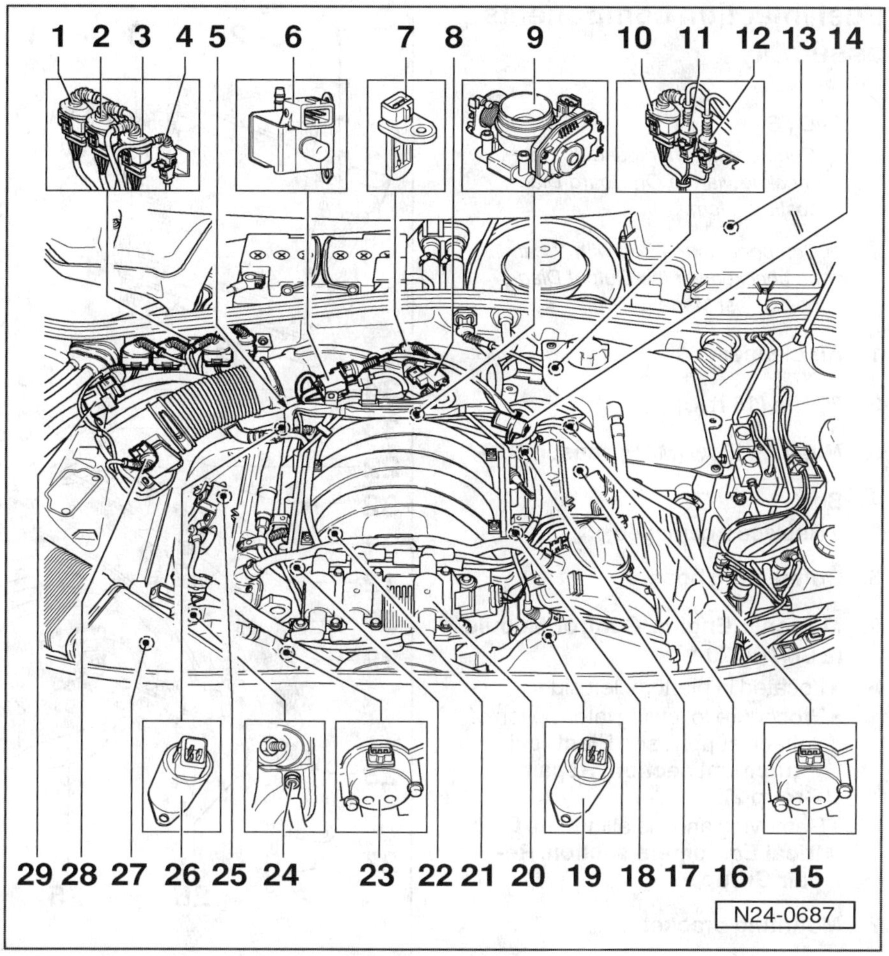

Fuel injection components, assembly

NOTE—

- *Components marked with * are checked via the On Board Diagnostic system.*
- *Components marked with ** can be checked via the Output Diagnostic Test Mode (DTM).*

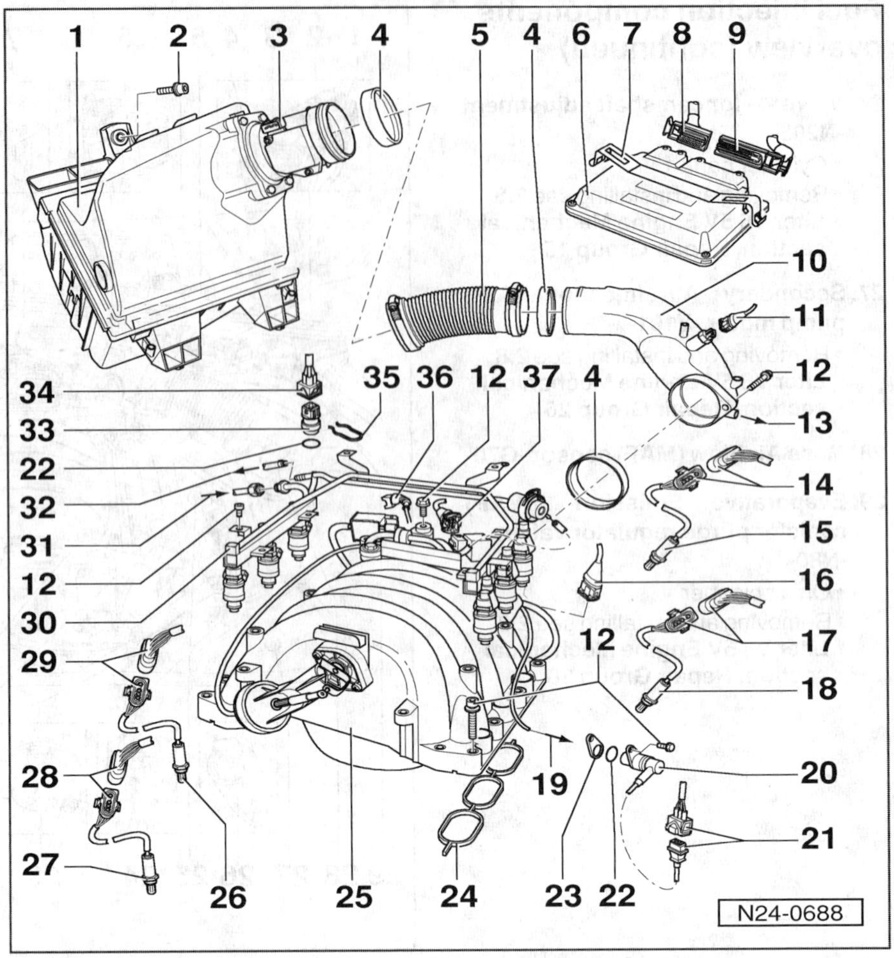

1. **Air cleaner**
2. **20 Nm (15 ft lb)**
3. **Mass Air Flow (MAF) sensor -G70-***
4. **Seal**
   - Replace if damaged
5. **Connecting hose**
6. **Motronic Engine Control Module (ECM) -J220-***
   - Located in plenum, left side
   - Procedure following interruption of voltage supply, see **Electrical Equipment section, Repair Group 27**
   - Removing and installing, see **Electrical Equipment section, Repair Group 97**
7. **Mounting bracket**
8. **Harness connector (28-pin)**
   - Disconnect/connect w/ignition off
9. **Harness connector**
   - Only disconnect or connect with ignition switched off
10. **Intake air elbow**
11. **Harness connector for IAT sensor**
    - 2-pin, black
    - Checking IAT sensor resistance, see **Fig. 1**
12. **10 Nm (7 ft lb)**
13. **To positive crankcase vent valve**
14. **Harness connector**
    - 4-pin connector, brown
    - For Oxygen Sensor (O2S) 2 behind Three Way Catalytic converter (TWC) -G131- and O2S heater -Z30-
    - For bank 2, sensor 2
15. **Oxygen Sensor (O2S) 2 behind Three Way Catalytic Converter (TWC) -G131-***
    - Tightening torque: 50 Nm (37 ft lb)
    - Bank 2, sensor 2
    - Use special tool 3337 to remove
16. **Harness connector**
    - 2-pin, black
    - For fuel injectors -N30-, -N31-, -N32-, -N33-, -N83-, -N84-
17. **Harness connector**
    - 4-pin, black
    - For Heated Oxygen Sensor (HO2S) 2 -G108- and Oxygen Sensor (O2S) 2 heater -Z28-
    - For bank 2, sensor 2
18. **Heated Oxygen Sensor (HO2S) 2 -G108-***
    - Tightening torque: 50 Nm (37 ft lb)
    - Bank 2, sensor 1
19. **To vacuum reservoir**
20. **Engine Speed (RPM) sensor -G28-***
    - In transmission housing above starter ring gear
21. **Harness connector**
    - 3-pin connector, gray
    - For Engine Speed sensor -G28-
22. **O-ring**
    - Replace if damaged
23. **Spacer**
24. **Gasket**
    - Replace if damaged
25. **Intake manifold**
    - See **Intake Manifold Tuning (IMT) (change-over) components**
    - See also **Throttle valve control module, assembly**
26. **Oxygen Sensor (O2S) behind Three Way Catalytic Converter (TWC) -G130-***
    - Tightening torque: 50 Nm (37 ft lb)
    - Bank 1, sensor 2
27. **Heated Oxygen Sensor (HO2S) -G39-***
    - Tightening torque: 50 Nm (37 ft lb)
    - Location: In front exhaust pipe
    - Grease only threads of sensor with "G5"; "G5" must not get into the slots of the sensor body
    - Oxygen Sensor heater voltage supply via fuel pump relay -J17-
    - Bank 1, sensor 1

Fuel injection components, assembly (continued)

28. Harness connector
- 4-pin connector, black
- For Heated Oxygen Sensor (HO2S) -G39- and Oxygen Sensor (O2S) heater -Z19-
- For bank 1, sensor 1

29. Harness connector
- 4-pin connector, green
- For Oxygen Sensor (O2S) behind Three Way Catalytic Converter (TWC) -G130- and Oxygen Sensor (O2S) heater -Z29-
- For bank 1, sensor 2

30. Fuel rail with fuel injectors
- See **Fuel rail with injectors**

31. Fuel supply line
- Black
- From fuel filter

32. Fuel return line
- Blue
- Secure with spring-type clamps
- Check for secure seating
- To the fuel delivery unit in fuel tank

33. Engine Coolant Temperature (ECT) sensor -G62-*
- Blue
- With coolant temperature gauge sensor -G2-
- Check resistance of sensor at terminals 1 (signal) and 3 (ground), compare to chart, see **Fig. 1**
- Before removing, release pressure from cooling system

34. Harness connector
- 4-pin connector, blue
- For Engine Coolant Temperature (ECT) sensor -G62-

35. Retaining clip
- Check for secure seating

36. Harness connector
- 2-pin connector, green
- For Secondary Air Injection (AIR) solenoid valve -N112-

37. Harness connector
- 2-pin connector, blue
- For Intake Manifold Tuning (IMT) valve -N156-

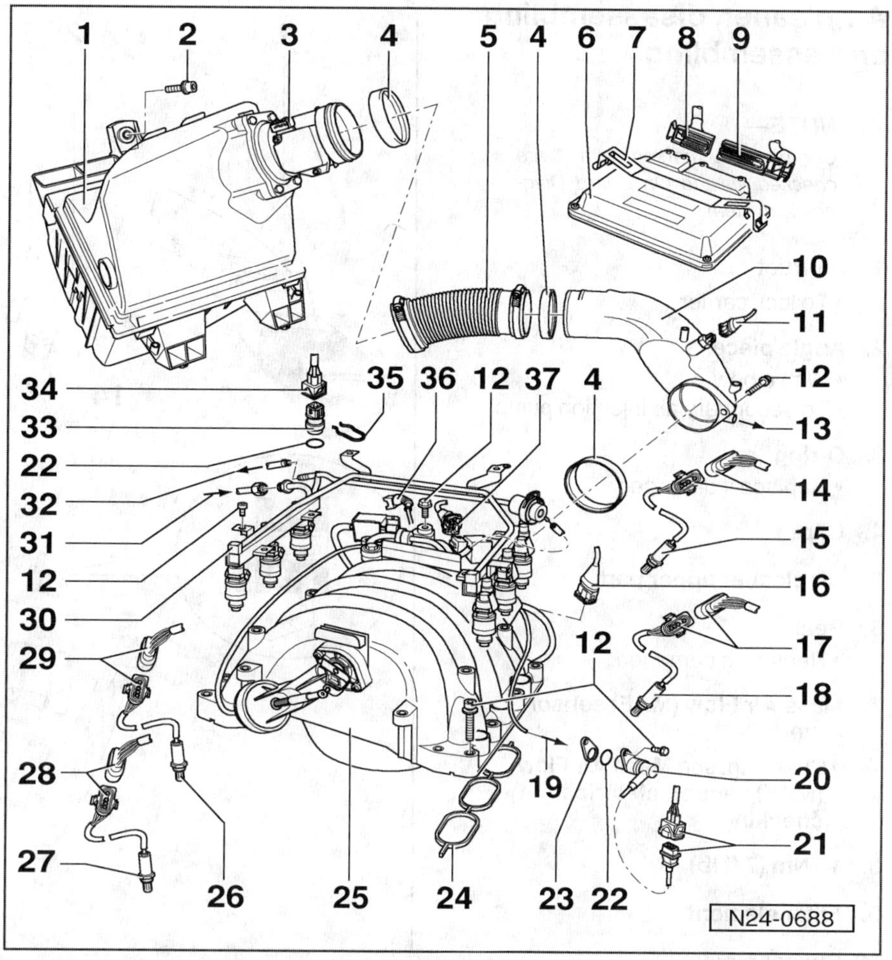

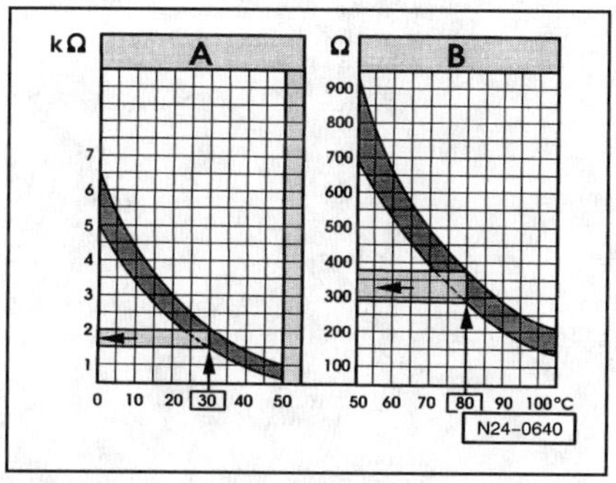

Fig. 1 Resistance graph

NOTE —
The graph is valid for the Engine Coolant Temperature (ECT) sensor -G62- and Intake Air Temperature (IAT) sensor -G42-.

The graph is divided into two temperature ranges:
- Range A: 0-50°C (32-122°F)
- Range B: 50-105°C (122-221°F)

Examples:
- 30°C (86°F) is in range A and corresponds to a resistance of 1.5-2.0 kΩ
- 80°C (176°F) is in range B and corresponds to a resistance of 275-375Ω

NOTE —
The ECM will use intake air temperature as a replacement value for an engine start (start temperature replacement value) as soon as there is a Diagnostic Trouble Code (DTC) stored in DTC memory for the Engine Coolant Temperature (ECT) sensor -G62-. The temperature then rises according to a program stored in the ECM. When the engine has reached normal operating temperature a fixed replacement value will be displayed. This fixed value is also dependent upon intake air temperature.

Air cleaner, disassembling and assembling

NOTE—
*Components marked with * are checked via the On Board Diagnostic system.*

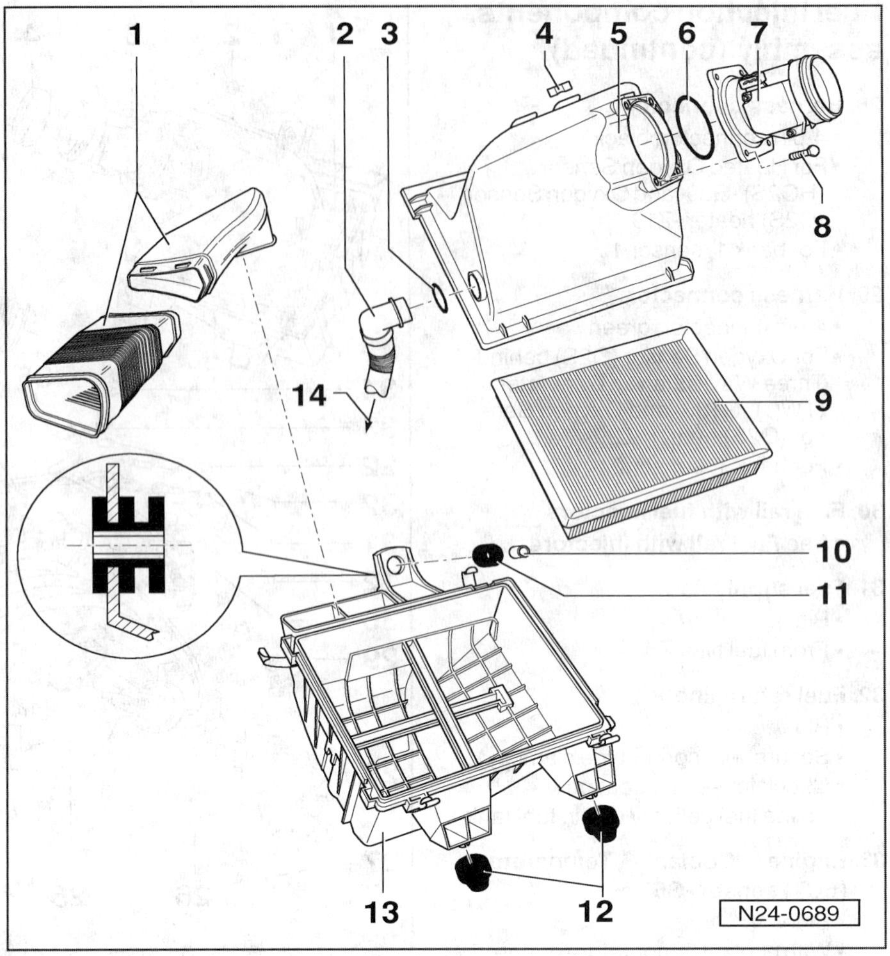

1. **Air duct**
   - To lock carrier
2. **Angle piece**
   - With elbow
   - To secondary air injection pump
3. **O-ring**
   - Replace if damaged
4. **Clip**
5. **Air cleaner upper part**
6. **Seal**
   - Replace if damaged
7. **Mass Air Flow (MAF) sensor -G70-***
   - Checking, see **Mass Air Flow (MAF) sensor, electrically checking**
8. **10 Nm (7 ft lb)**
9. **Filter element**
10. **Spacer sleeve**
11. **Rubber grommet**
    - Note installation position
12. **Rubber grommet**
13. **Air cleaner lower part**
14. **To Secondary Air Injection (AIR) pump motor -V101-**
    - Removing and installing, see **2.8 Liter V6 5V Engine Mechanical section, Repair Group 26**

Fuel rail with injectors

NOTE—

- *Components marked with * are checked via the On Board Diagnostic system.*
- *Components marked with ** can be checked via the Output Diagnostic Test Mode (DTM).*

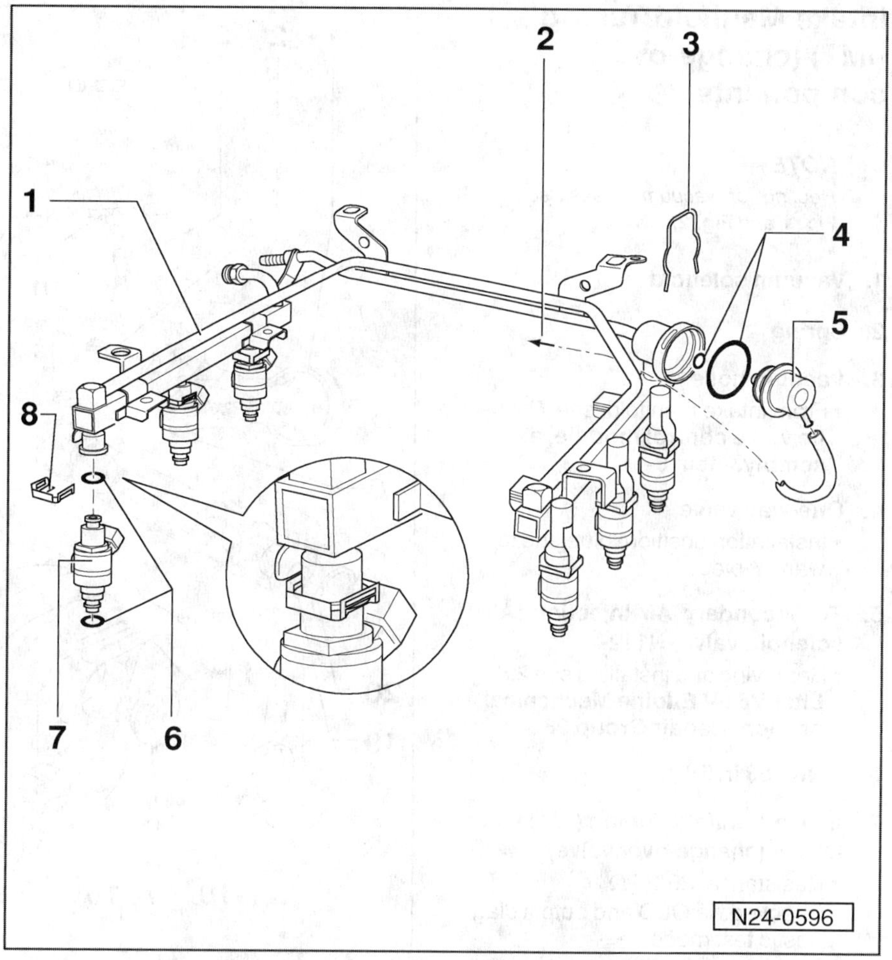

1. **Fuel rail**
2. **To T-piece**
   - Location, see **Throttle valve control module, assembly**
3. **Retaining clip**
   - Check for secure seating
4. **O-ring**
   - Replace
5. **Fuel pressure regulator**
   - See **Fuel pressure, fuel pressure regulator and residual fuel pressure, checking**
6. **O-ring**
   - Replace
7. **Fuel injector -N30-, -N31-, -N31-, -N33-, -N83-, -N84-*/****
   - Checking, see **Fuel injectors, checking**
   - Resistance: 13.5-15.5 Ω
8. **Retaining clip**
   - Check for secure seating at injector and fuel rail

Intake Manifold Tuning (IMT) (change-over) components

NOTE—

Routing of vacuum hoses, see **Fig. 1** *and* **Fig. 2.**

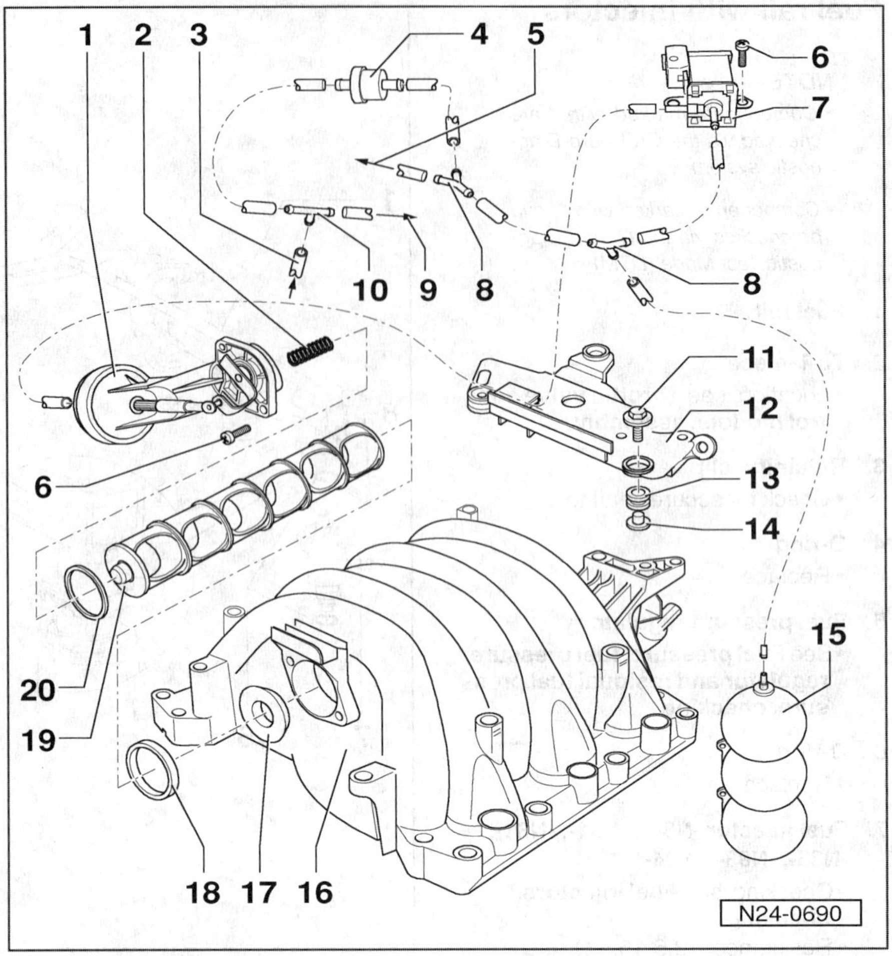

1. **Vacuum solenoid**
2. **Spring**
3. **Vacuum hose**
   - From intake manifold, see **Throttle valve control module, assembly, -item 6-**
4. **One-way valve**
   - Installation position: blue end toward Y-piece
5. **To Secondary Air Injection (AIR) solenoid valve -N112-**
   - Removing and installing see **2.8 Liter V6 5V Engine Mechanical section, Repair Group 26**
6. **6 Nm (53 in lb)**
7. **Intake Manifold Tuning (IMT) valve -N156- (change-over valve)**
   - Resistance: 25-35 Ω
   - Checked via OBD and output diagnostic test mode
8. **Y-piece**
9. **To fuel pressure regulator**
   - Location, see **Fuel rail with injectors**
10. **T-piece**
11. **10 Nm (7 ft lb)**
12. **Securing plate**
13. **Rubber grommet**
14. **Spacer sleeve**
15. **Vacuum reservoir**
    - Location: In left wheel housing
    - For intake manifold change-over with engine at full load
16. **Intake manifold**
17. **Washer**
    - Conical side to intake manifold
18. **Seal**
    - Replace if damaged
19. **Change-over valve**
20. **Seal**
    - For change-over valve

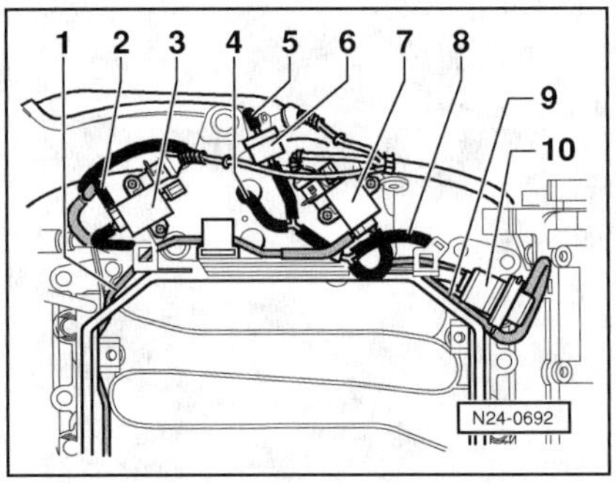

Fig. 1 Vacuum connections, upper

1. To intake manifold tuning (change-over) vacuum solenoid
2. To combination valves
3. Secondary Air Injection (AIR) solenoid valve -N112-
4. To secondary air injection valve
5. To T-piece on intake manifold
6. One-way valve
7. Intake manifold tuning change-over valve
8. To vacuum reservoir in left wheel housing
9. To T-piece on intake manifold
10. Fuel pressure regulator

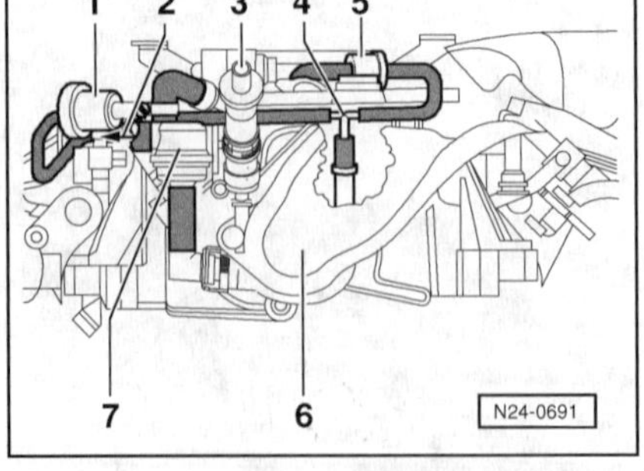

Fig. 2 Vacuum connections, rear

1. Fuel pressure regulator
2. To crankcase breather valve
3. Vacuum booster
4. T-piece
5. One-way valve
6. Intake air elbow
7. Crankcase breather valve

Throttle valve control module, assembly

NOTE—
*Components marked with * are checked via the On Board Diagnostic system.*

1. **Intake manifold**
2. **To vacuum booster**
3. **To fuel pressure regulator**
   - Location, see **Fuel rail with injectors**
4. **T-piece**
5. **To one-way valve**
   - Location, see **Intake Manifold Tuning (IMT) (change-over) components**
6. **Vacuum hose**
7. **O-ring**
   - Replace if damaged
8. **Support bracket**
   - For change-over valve
9. **6 Nm (53 in lb)**
10. **Intake Air Temperature (IAT) sensor -G42-***
    - Checking resistance, see **Fuel Injection components, assembly, Fig. 1**
11. **Intake air elbow**
12. **10 Nm (7 ft lb)**
13. **To positive crankcase ventilation valve**
14. **Seal**
    - Replace if damaged
15. **10 Nm (7 ft lb)**
    - Install with "D6" locking fluid
16. **Throttle valve control module -J338-***
    - If replaced, adapt to ECM, see **Throttle valve control module, adaptation to Engine Control Module (ECM)**
17. **To Evaporative Emission (EVAP) canister purge regulator valve -N80-**
    - Removing and installing see **2.8 Liter V6 5V Engine Mechanical section, Repair Group 20**
18. **Seal**
    - Replace if damaged

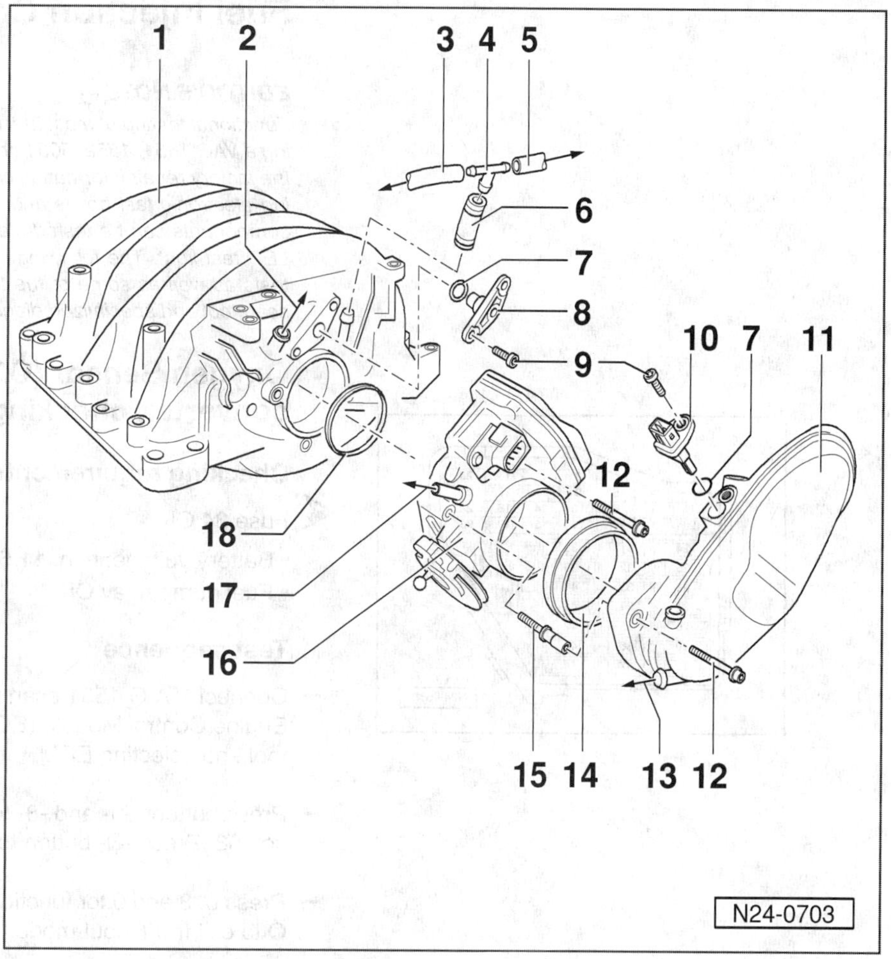

Fuel Injection Components, checking

EDITOR'S NOTE —
Functional testing of most fuel injection and ignition components is done using a VAG 1551, 1552, 5051 or equivalent aftermarket scan tool. In addition, the factory repair information calls for the use of various special tools including electronic test boxes and remote control switches. Many fuel injection components can be tested electrically using a digital multimeter and/or an LED test light. The following pages include many of these basic electrical tests as well as some of the testing that relies on the use of a diagnostic scan tool and specialized diagnostic equipment.

Oxygen Sensor (O2S) heater before catalytic converter, checking

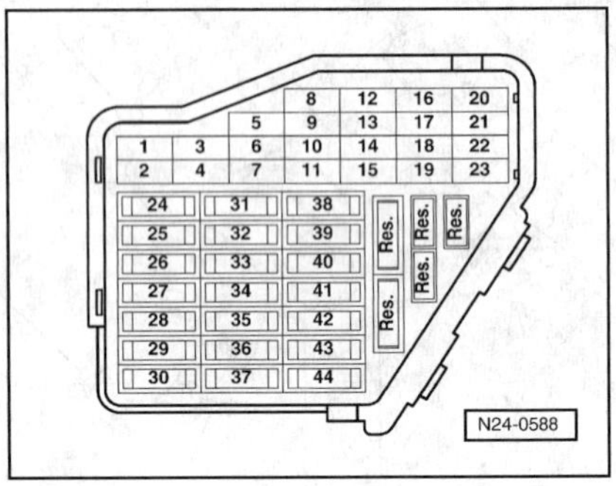

Checking requirements

Fuse 34 OK

- Battery voltage: min. 11.5 volts
- Fuel pump relay OK

Test sequence

– Connect V.A.G 1551 scan tool (VAG1552). Start engine and select the Engine Control Module (ECM) using address word 01 (connecting scan tool and selecting ECM.

– Press buttons -0- and -8- to select "Read measuring value block" function 08. Press -Q- button to confirm input.

– Press 0, 3 and 0 for function "display group number 30", and then press Q to exit from input mode.

– Check the status of oxygen sensor heater in display fields 1 and 3:

- Specification: 111
- 011 also briefly, since the oxygen sensor heater is clocked.

If specified value is not obtained:

– Press →button.

– Press buttons -0- and -6- to select "End Output" function 06. Press -Q- button to confirm input.

– Switch ignition off.

– Disconnect 4-pin connector (black) for respective Heated Oxygen Sensor (HO2S).

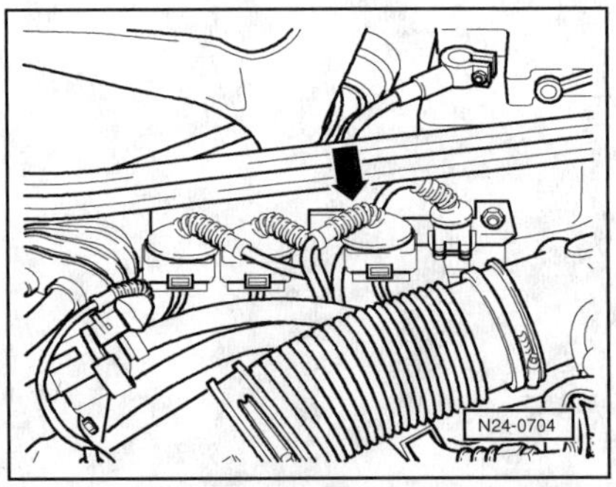

4-pin connector (black) for Heated Oxygen Sensor (HO2S) -G39- for bank 1 (arrow).

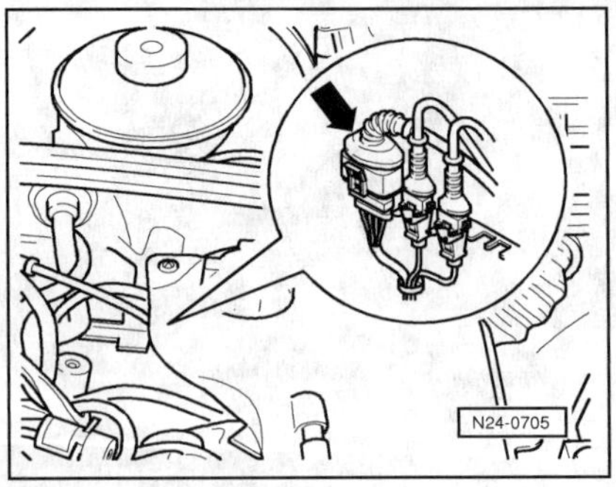

4-pin connector (black) for Heated Oxygen Sensor (HO2S) 2 -G108- for bank 2 (arrow).

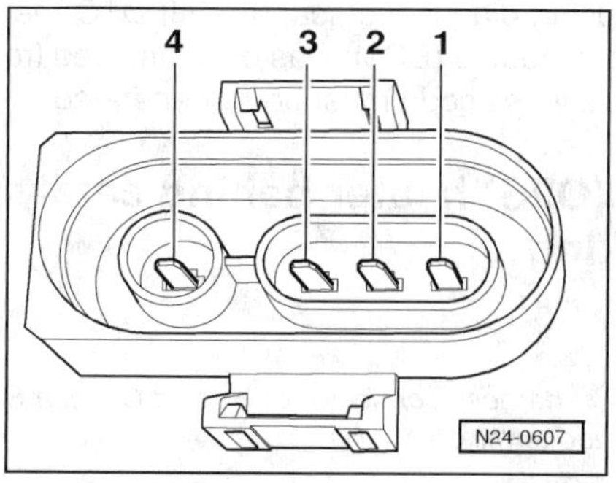

◄ Connect multimeter US1119 (Fluke 83 or equivalent) between HO2S connector terminals 1 and 2, using VW1594 connector test kit.

– Measure resistance.

- Specified value: 200 Ω

If resistance is NOT OK:

– Replace respective Heated Oxygen Sensor (HO2S).

If expected value is not obtained:

– Check Ground (GND) connection of appropriate oxygen sensor heater.

– Switch ignition off.

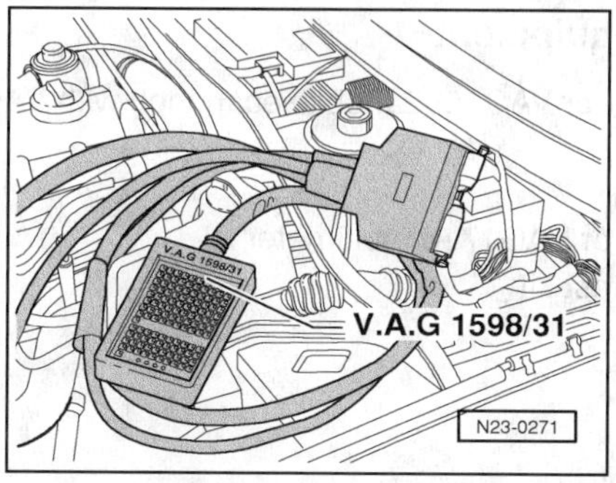

◄ Connect VAG1598/31 test box to control module wiring harness. Engine Control Module (ECM) is not connected.

EDITOR'S NOTE —

Volkswagen recommends using the VAG 1598/31 test box to prevent damage to ECM harness connectors during electrical testing. However the harness connectors can be tested without the test box if extra care is used to prevent damage to the harness connector terminals.

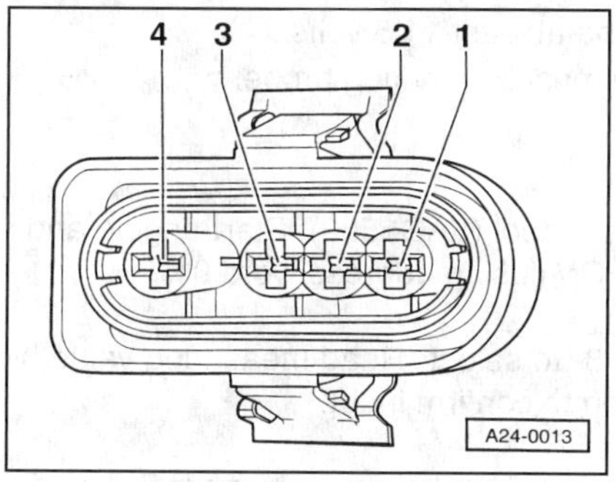

◄ Check wire connection of terminal 2 of appropriate connector of Heated Oxygen Sensor (HO2S) for continuity to test box.

- Specification: max. 1.5 Ω

Connector to Heated Oxygen Sensor (HO2S) -G39-, bank 1 before catalytic converter terminal	VAG1598/31 test box socket
2	5
Connector to Heated Oxygen Sensor (HO2S) 2 -G108-, bank 2 before catalytic converter terminal	**VAG1598/31 test box socket**
2	4

– If expected value is not obtained, switch ignition on.

– Bridge sockets 2 + 65 of test box using auxiliary wires from VAG 1594.

- Transfer Fuel Pump (FP) must be running.

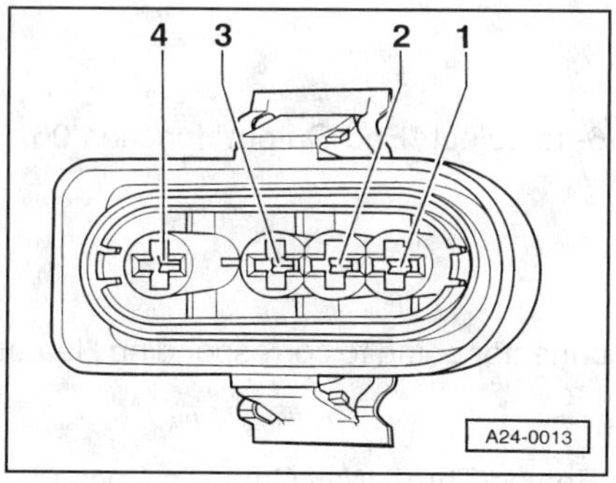

◄ Connect multimeter to terminals 1 + 2 of appropriate connector to make voltage measurement.

- Specification: at least 11.5 V

If there is no voltage:

– Switch ignition off.

– Check wire connection from terminal 1 to Fuel Pump (FP) relay -J17- (see wiring diagrams).

If voltage supply and wires are in order:

– Replace Engine Control Module (ECM), see **Electrical Equipment section, Repair Group 97**.

EDITOR'S NOTE —

ECM replacement requires coding of control module using Volkswagen factory scan tool. See **Scan Tool Codes** *at the back of this manual or consult an authorized Volkswagen service department for additional information.*

– Read readiness code using diagnostic scan tool. If DTC memory was erased or Engine Control Module (ECM) was disconnected from continuous voltage supply, readiness code must be re-generated.

Oxygen Sensor (O2S) heater behind catalytic converter, checking

NOTE —
When servicing terminals in harness connector of Heated Oxygen Sensor (HO2S), use only gold-plated terminals.

Special tools and equipment

- V.A.G1551 Scan Tool (or VAG1552 mobile scan tool) with VAG1551/3 cable
- VAG1598/31 test box
- VAG1526 multimeter or VAG1715 multimeter
- VAG1594 connector test kit
- Wiring diagram

Test requirements

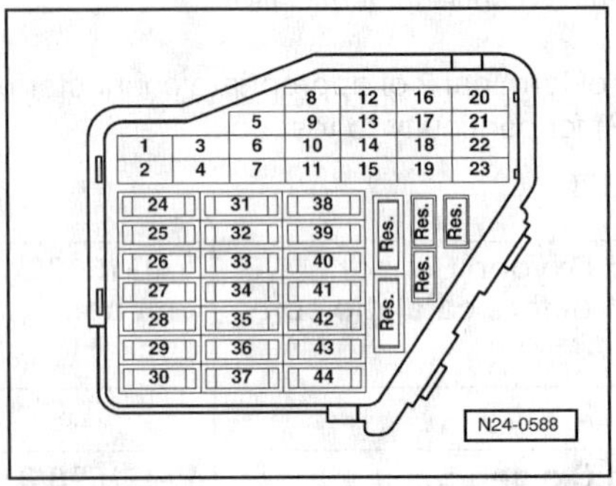

Fuse no. 34 must be OK.

- Battery voltage must be at least 11.5 volts.
- Fuel Pump (FP) relay must be working properly.

Test sequence

– Connect V.A.G 1551 scan tool (VAG1552). Start engine and select Engine Control Module (ECM) using address word 01.

– Press buttons -0- and -8- to select "Read measuring value block" function 08. Press -Q- button to confirm input.

– Press 0, 3 and 0 for function "display group number 30", and then press Q to exit from input mode.

– Check status of oxygen sensor heater in display fields 2 and 4:

- Specification: 111
- 011 also briefly, since the oxygen sensor heater is clocked.

If specified value is not obtained:

– Press →button.

– Press buttons -0- and -6- to select "End Output" function 06. Press -Q- button to confirm input.

– Switch ignition off.

– Unplug 4-pin harness connector going to corresponding Heated Oxygen Sensor (HO2S):

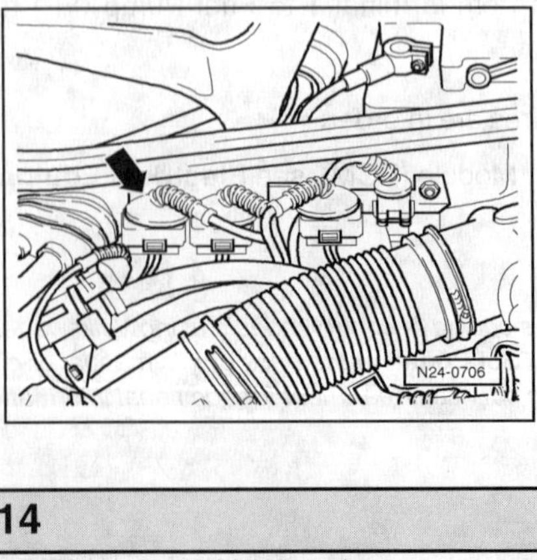

Oxygen Sensor (O2S) Behind Three Way Catalytic Converter (TWC) -G130-, bank 1, 4-pin harness connector (green) (arrow).

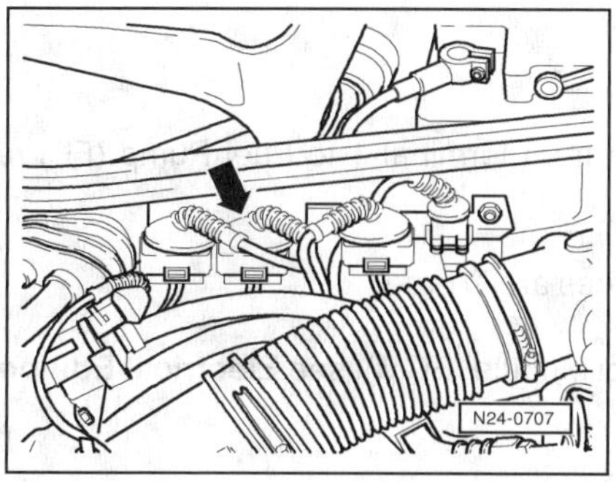

Oxygen Sensor (O2S) 2 Behind Three Way Catalytic Converter -G131- bank 2, 4-pin harness connector (brown) (arrow).

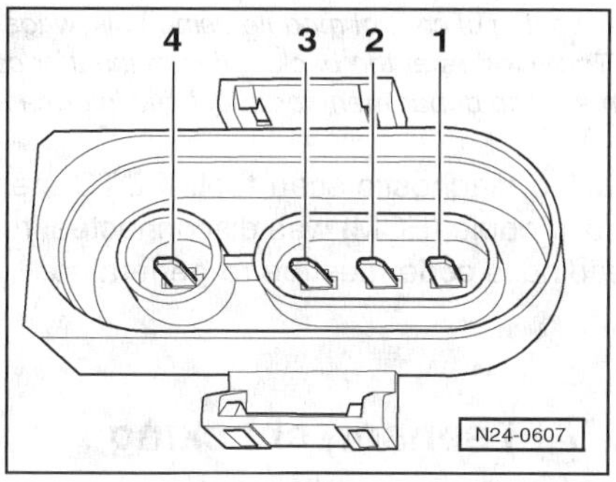

Check oxygen sensor heater on connector to Heated Oxygen Sensor (HO2S) terminals 1 + 2 for continuity

- Specification: max. 200 Ω

If specified value is not obtained:

– Replace appropriate Heated Oxygen Sensor (HO2S).

If expected value is not obtained:

– Check Ground (GND) connection of appropriate oxygen sensor heater.

– Switch ignition off.

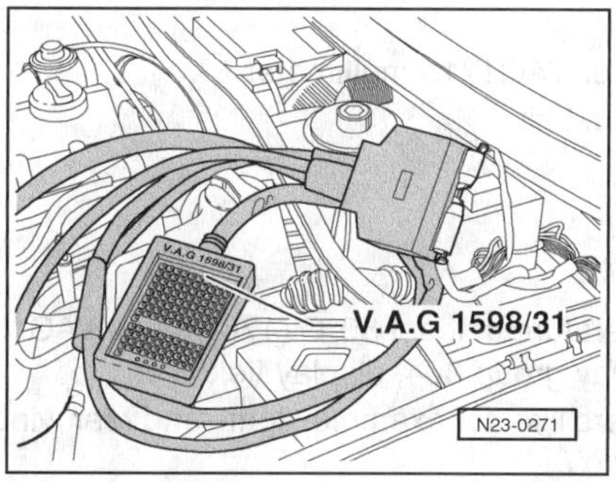

Connect VAG1598/31 test box to control module wiring harness. Engine Control Module (ECM) is not connected.

EDITOR'S NOTE—

Volkswagen recommends using the VAG 1598/31 test box to prevent damage to ECM harness connectors during electrical testing. However the harness connectors can be tested without the test box if extra care is used to prevent damage to the harness connector terminals.

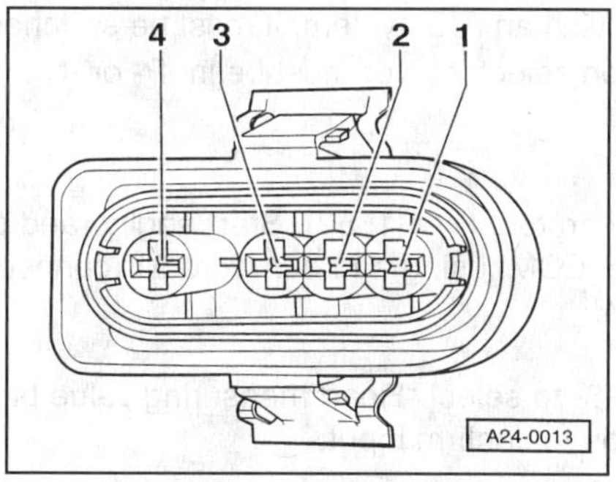

Check wire connection of terminal 2 of appropriate connector of Heated Oxygen Sensor (HO2S) for continuity to test box.

- Specification: max. 1.5 Ω

Connector to Oxygen Sensor (O2S), behind catalytic converter, -G130-, bank 1 terminal	VAG1598/31 test box socket
2	63
Connector to Oxygen Sensor (O2S) 2, behind catalytic converter, -G131-, bank 2 terminal	**VAG1598/31 test box socket**
2	6

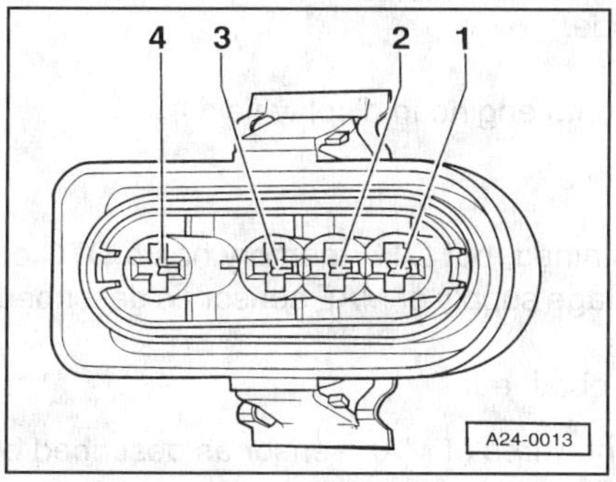

If expected value is not obtained:

– Connect together sockets 62 + 65 of test box using auxiliary wires from VAG 1594. Transfer Fuel Pump (FP) must be running.

Connect multimeter to terminals 1 + 2 of appropriate connector to make voltage measurement.

- Specification: at least 11.5 V

If there is no voltage:

– Switch ignition off.

– Check wire connection from terminal 1 to Fuel Pump (FP) relay (J17) (see wiring diagram).

If voltage supply and wires are in order:

– Replace Engine Control Module (ECM), see **Electrical Equipment section, Repair Group 97**.

EDITOR'S NOTE —

ECM replacement requires coding of control module using Volkswagen factory scan tool. See **Scan Tool Codes** *at the back of this manual or consult an authorized Volkswagen service department for additional information.*

– Read readiness code using diagnostic scan tool. If DTC memory was erased or Engine Control Module (ECM) was disconnected from continuous voltage supply, readiness code must be re-generated.

Mass Air Flow (MAF) sensor, checking

Special tools and equipment

- V.A.G1551 Scan Tool with VAG1551/3 cable
- VAG1598/31 test box
- VAG1526 multimeter or VAG1715 multimeter
- VAG1594 connector test kit

Test requirements

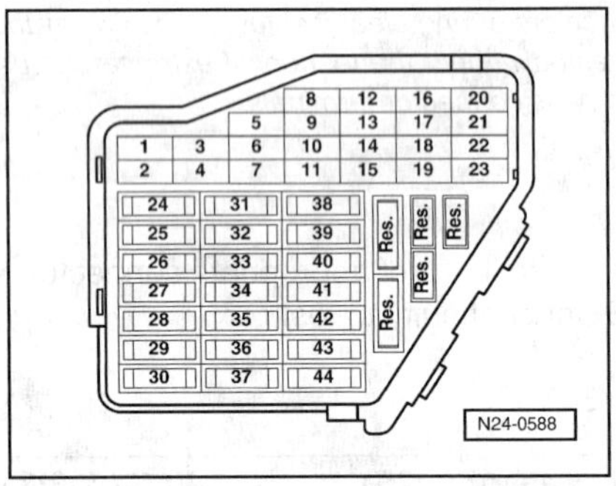

◄ Fuse no. 34 must be OK.

- Engine coolant temperature at least 80°C (176°F) see VAG1551 scan tool function 08, display group 004, display field 3
- All electrical consumers like, for example, lights and rear window heating must be switched off.
- If vehicle is equipped with an A/C system, it must be switched off.
- Automatic transmission selector lever must be in -P- or -N-.

Test sequence

– Connect V.A.G 1551 scan tool (VAG1552). Start engine and select the Engine Control Module (ECM) using address word 01 (connecting scan tool and selecting ECM.

– Press buttons -0- and -8- to select "Read measuring value block" function 08. Press -Q- button to confirm input.

– Press 0, 0 and 2 for function "display group number 2", and then press Q to exit from input mode.

– Check mass flow of air into engine in display field 4

- Specification: 3.0 - 5.0 g/s

If expected value is obtained, but DTC memory has a DTC concerning MAF sensor, check voltage supply of MAF sensor as described below.

If specified value is not obtained:

– Check signal and ground wires of MAF sensor as described below.

If the specified value is met:

- Press →button.
- Press buttons -0- and -6- to select "End Output" function 06. Press -Q- button to confirm input.
- Switch ignition off.

MAF sensor, checking voltage supply

- Start engine, and let it idle.

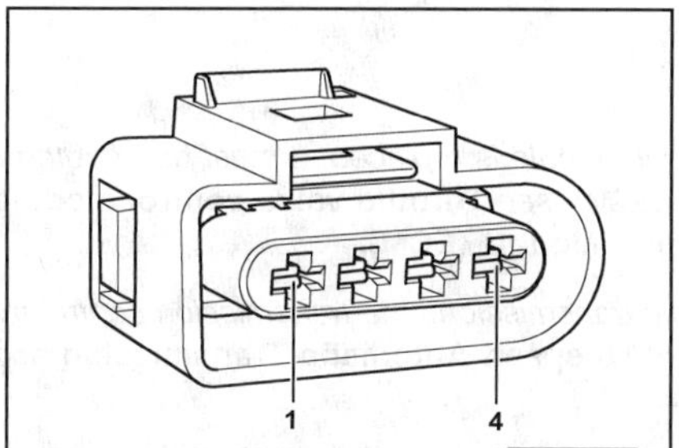

A24-0094

◄ Pull 4-way connector off Mass Air Flow (MAF) sensor.

- Connect multimeter to terminal 3 of connector and engine Ground (GND) to make voltage measurement.
    - Specification: 11-15 V

NOTE —

Voltage is supplied to MAF sensor from Fuel Pump (FP) relay.

If there is no voltage:

- Switch ignition off.
- Check wire connection from terminal 3 to Fuel Pump (FP) relay (J17) (see wiring diagram).

If voltage supply and wires are in order:

- Switch ignition on.
- Connect multimeter to terminals 2 and 3 of connector to make voltage measurement.
    - Specification: 4.5-5.5 V

If there is no voltage:

- Switch ignition off, check signal and ground wires as described below.

MAF sensor, checking signal and ground wires

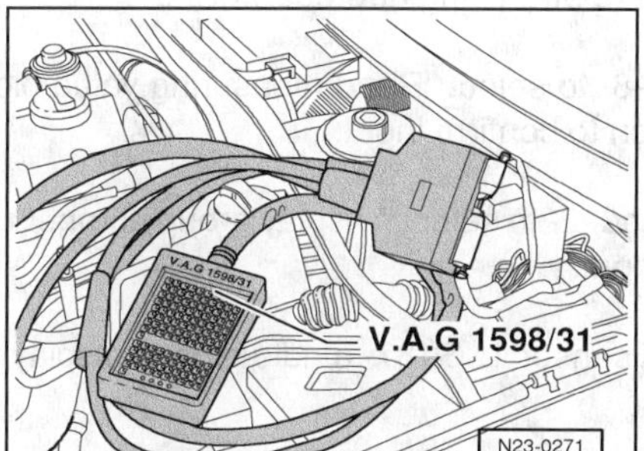

N23-0271

◄ Connect VAG1598/31 test box to control module wiring harness. Engine Control Module (ECM) is not connected.

EDITOR'S NOTE —

Volkswagen recommends using the VAG 1598/31 test box to prevent damage to ECM harness connectors during electrical testing. However the harness connectors can be tested without the test box if extra care is used to prevent damage to the harness connector terminals.

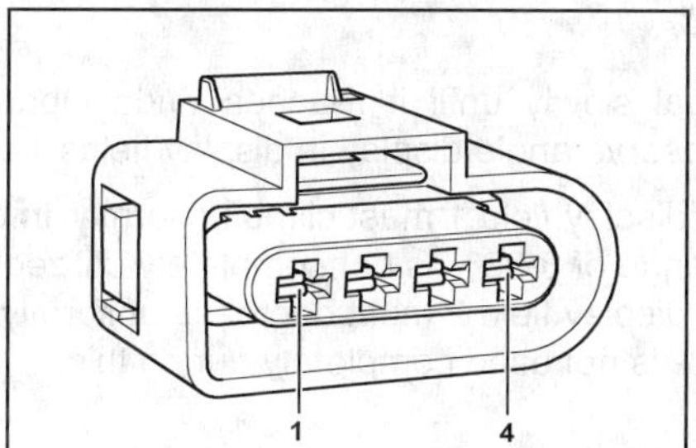

A24-0094

◄ Check for breaks in wires between test box and 4-way connector (see wiring diagram).

- Terminal 1 + socket 53
- Terminal 2 + socket 27
- Terminal 4 + socket 29
- Wire resistance: max. 1.5 Ω

- Check wires for shorts among one another, to battery B+ or Ground.
- If no DTCs are found in wires replace Mass Air Flow (MAF) sensor.

Throttle valve control module, checking

The engine identification ATQ indicates electronic gas activation. This means that the throttle valve is activated by an electric motor in the throttle valve control module. This is activated by the Engine Control Module (ECM) based on the information from the Throttle Position (TP) sensor -G79- and Sender 2 for accelerator pedal position -G185-.

Components of throttle valve control module -J338-:
Throttle drive -G186-
Angle sensor 1 for throttle drive -G187-
Angle sensor 2 for throttle drive -G188-

NOTE—

- *If the throttle valve control module is replaced, it must be adapted to the Engine Control Module (ECM), see* **Throttle valve control module, adaptation to Engine Control Module (ECM).**
- *In vehicles with automatic transmissions the transmission control module must also be adapted, see* **5-speed Automatic Transmission section, Repair Group 37**.

Special tools and equipment

- V.A.G1551 (or VAG1552 mobile scan tool) with VAG1551/3 cable
- VAG1598/31 test box
- VAG1526 multimeter or VAG1715 multimeter
- VAG1594 connector test kit
- Wiring diagram

Test condition

- Throttle valve must not be damaged or dirty.
- Coolant temperature must be at least 80°C (176 °F), see VAG 1551 scan tool function 08, display group 004, display field 3

Test sequence

– Connect V.A.G 1551 scan tool (VAG1552). Switch ignition on and select Engine Control Module (ECM) using address word 01.

– Press buttons -0- and -8- to select "Read measuring value block" function 08. Press -Q- button to confirm input.

– Press buttons -0-, -6- and -2- to select "Display group number 62". Press -Q- button to confirm input.

– Check throttle angle of sensor 1 -G187- at idle stop in display field 1

- Specification: 3 - 93 %

– Check throttle angle of sensor 2 -G188- at idle stop in display field 2

- Specification: 97 - 3 %

– Press accelerator pedal slowly until it reaches Wide Open Throttle (WOT) position, and observe angle display in display fields 1 and 2:

- Percent indication in display field 1 must climb uniformly. In the process, the tolerance range of 3-93% is not completely utilized.
- Percent indication in display field 2 must decrease uniformly. The tolerance range 97 - 3 % is not used completely during this.

NOTE—

- *The reason why the value indicated in display field 1 climbs, while the value indicated in display field 2 decreases has to do with the fact that the potentiometers (angle sensors) in the throttle valve control module run in opposite directions.*
- *This means that the voltage tap of angle sensor 1 runs from 0 to 5 volts. (The more the throttle valve is opened, the larger the voltage; percent indication climbs).*
- *The voltage tap of angle sensor 2 runs from 5 to 0 volts. (The more the throttle valve is opened, the lower the voltage; the percentage decreases).*

If indications do not resemble description:

– Press→button.

– Press buttons -0- and -6- to select "End data transfer" function 06. Press -Q- button to confirm input.

– Switch ignition off.

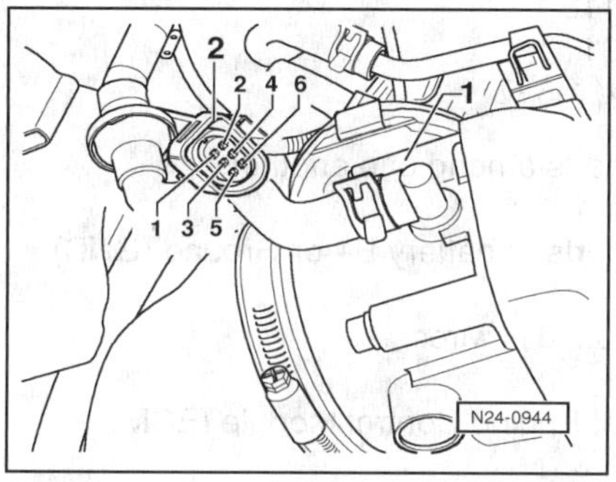

◄ Disconnect 6-way connector (2) from throttle valve control module.

– Measure resistance of throttle drive in throttle valve control module between terminals 3 + 5

- Specification: 3 - 200 Ω

If specified value is not obtained:

– Replace throttle valve control module -J338-.

If the specified value is met:

– Check voltage supply of throttle valve control module and wires to control module as described below.

– Test Throttle Position (TP) sensor and sensor 2 for accelerator pedal position, see **2.8 Liter V6 5V Engine Mechanical section, Repair Group 20**

If voltage supply and wires are ok:

– Replace Engine Control Module (ECM), see **Electrical Equipment section, Repair group 27**.

EDITOR'S NOTE—

ECM replacement requires coding of control module using Volkswagen factory scan tool. See **Scan Tool Codes** *at the back of this manual or consult an authorized Volkswagen service department for additional information.*

Throttle valve control module, checking voltage supply and wiring

– Pull 6-way connector (2) off throttle valve control module.

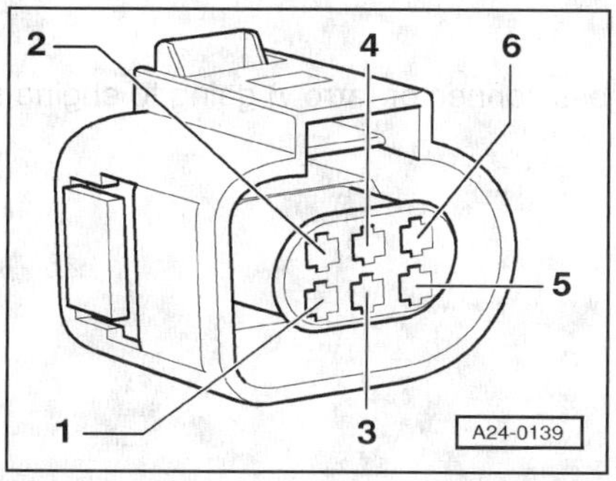

◄ Connect multimeter to terminals 2 + 6 of connector to make voltage measurement.

– Turn ignition on.

- Specification: at least 4.5 V

– Switch ignition off.

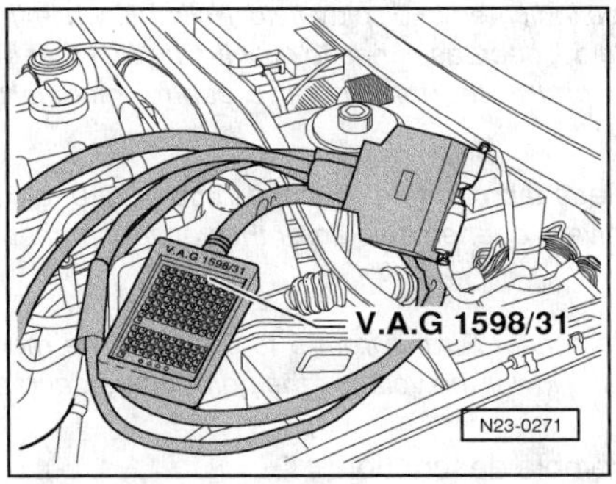

Connect VAG1598/31 test box to control module wiring harness. Engine Control Module (ECM) is not connected.

EDITOR'S NOTE —
Volkswagen recommends using the VAG 1598/31 test box to prevent damage to ECM harness connectors during electrical testing. However the harness connectors can be tested without the test box if extra care is used to prevent damage to the harness connector terminals.

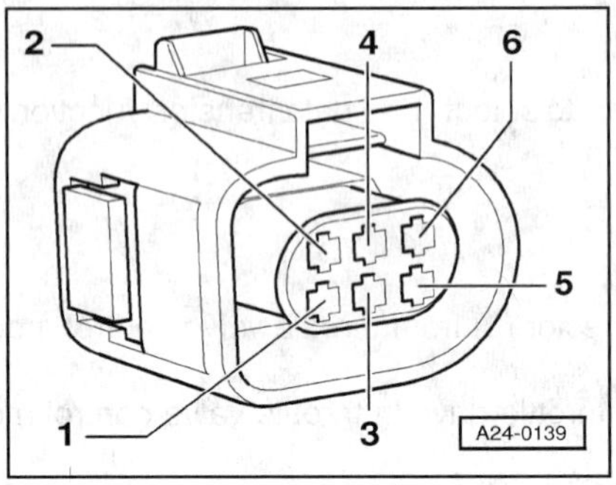

Check wires between test box and connector for breaks (see wiring diagram):

- Terminal 1 + socket 92
- Terminal 2 + socket 83
- Terminal 3 + socket 117
- Terminal 4 + socket 84
- Terminal 5 + socket 118
- Terminal 6 + socket 91
- Wire resistance: max. 1.5 Ω

– Also check wires for shorts among one another.

– Also check wires for shorts to battery B+ or Ground (GND).

If no malfunctions are found in wires:

– Check voltage supply of Engine Control Module (ECM).

Engine Speed (RPM) sensor, electrically checking

The engine speed (RPM) sensor -G28- detects engine speed and reference marks. Without an engine speed signal, the engine will not start. If the engine speed signal fails while the engine is running, this immediately causes the engine to stop.

EDITOR'S NOTE —
Checking the function of the engine speed (RPM) sensor requires the use of a diagnostic scan tool. For connecting scan tool see, **Maintenance**. *For diagnostic trouble codes and control unit coding, see* **Scan Tool Codes** *at the back of this manual.*

Checking conditions

- Battery voltage: min. 11.5 volts

Test sequence

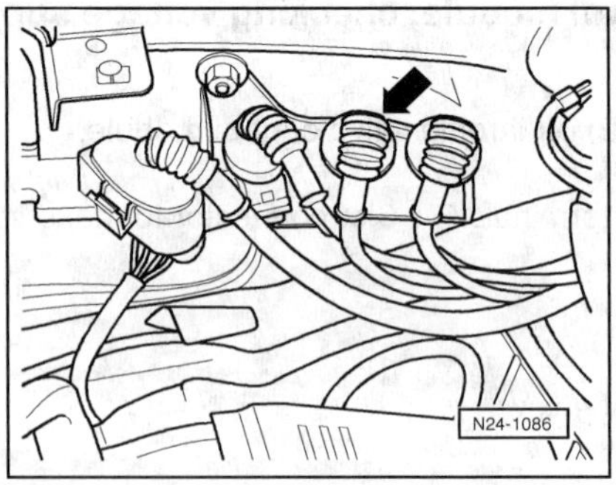

Unplug gray 3-pin harness connector (arrow) going to engine speed (RPM) sensor.

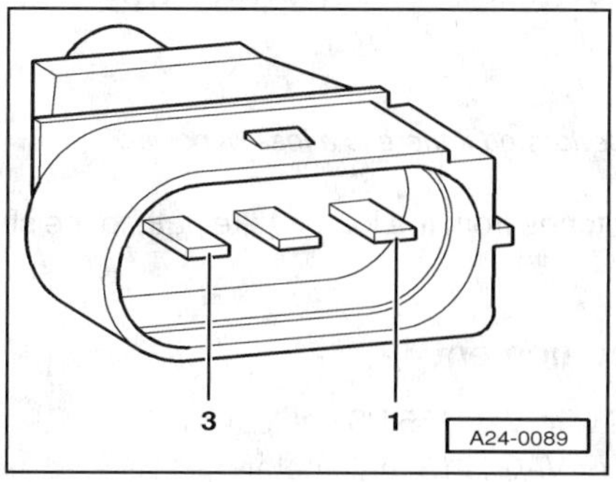

- Measure resistance of sensor between terminals 2 + 3 at connector to sensor.
  - Specification: 730 - 1000 Ω
- Check sensor for shorts between terminals 1 + 2 and 1 + 3.
  - Expected value: ∞

If specified values are not obtained:

- Replace engine speed (RPM) sensor -G28-.

If no DTC is found on sensor:

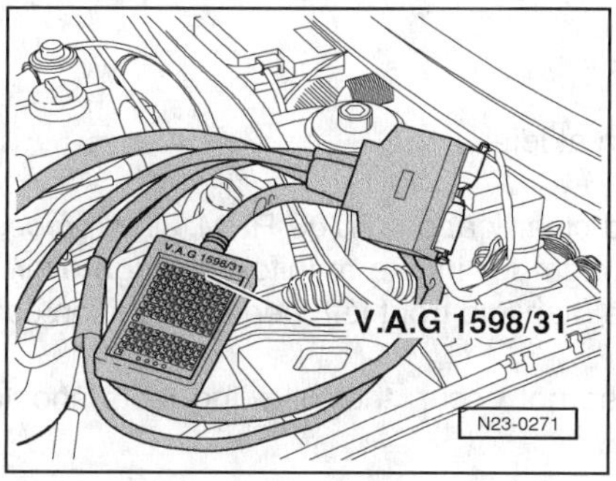

- Connect VAG1598/31 test box to control module wiring harness. Engine Control Module (ECM) is not connected.

EDITOR'S NOTE —
Volkswagen recommends using the VAG 1598/31 test box to prevent damage to ECM harness connectors during electrical testing. However the harness connectors can be tested without the test box if extra care is used to prevent damage to the harness connector terminals.

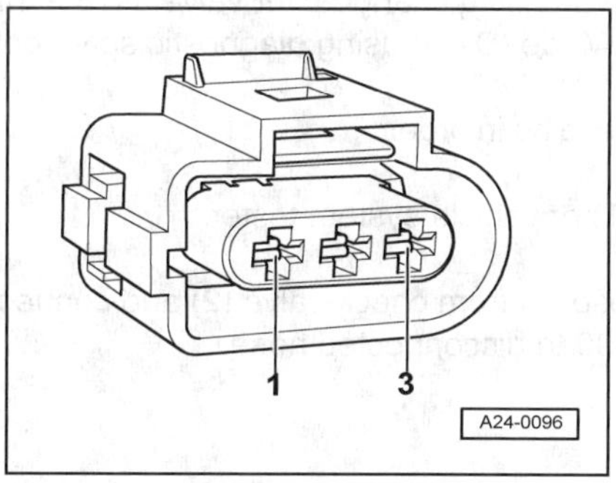

- Check for breaks in wires between test box and 3-way connector (see wiring diagram).
  - Terminal 1+socket 108
  - Terminal 2+socket 90
  - Terminal 3 + socket 82
  - Wire resistance: max. 1.5 Ω
- Also check wires for shorts among one another.

If no malfunctions are found in wires:

- Remove sensor and check sensor wheel for secure fit, damage, and run-out.

NOTE —
There is a larger-sized gap on the sensor wheel. This gap is a reference mark, not damage.

If nothing seems to be wrong with the sensor wheel, replace Engine Control Module (ECM), see **Electrical Equipment section, Repair group 27**.

EDITOR'S NOTE —
ECM replacement requires coding of control module using Volkswagen factory scan tool. See **Scan Tool Codes** *at the back of this manual or consult an authorized Volkswagen service department for additional information.*

Intake manifold change-over, checking

NOTE—
This test should only be performed if there is a loss of power.

The intake manifold switches from the long intake path to the short one at approx. 4700 RPM.

Special tools and equipment

- V.A.G1551 Scan Tool with VAG1551/3 cable
- VAG1526 multimeter or VAG1715 multimeter
- VAG1594 connector test kit

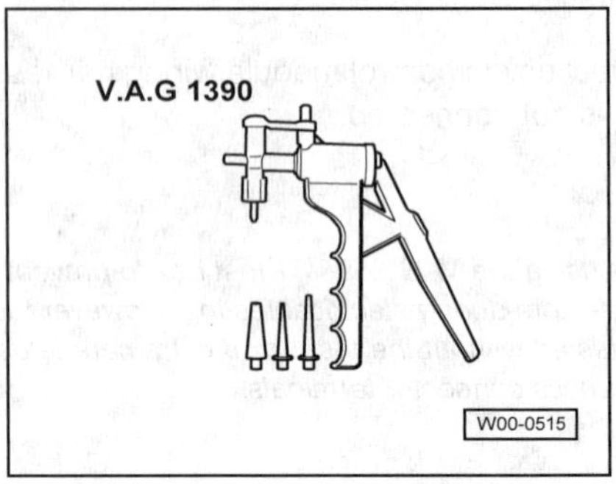

◄ VAG1390 Hand Vacuum Pump

Functional test

– Start engine and let run at idle.

– Increase engine speed clearly above 4700 RPM. This process should actuate vacuum diaphragm for intake manifold change-over. (Component location: on intake manifold front left, beneath ignition coils).

If the change-over does not occur, the following tests should be performed.

– Test activation of intake manifold change-over valve -N156- through the output Diagnostic Test Mode (DTM) using diagnostic scan tool.

If the activation appears to be in order:

– Connect Scan Tool (ST) or vehicle system tester.

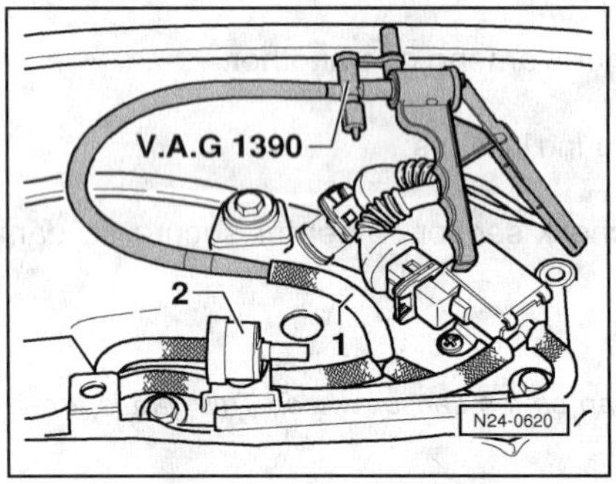

◄ Disconnect vacuum hose (1) from check-valve (2) and connect hand vacuum pump VAG 1390 to disconnected hose.

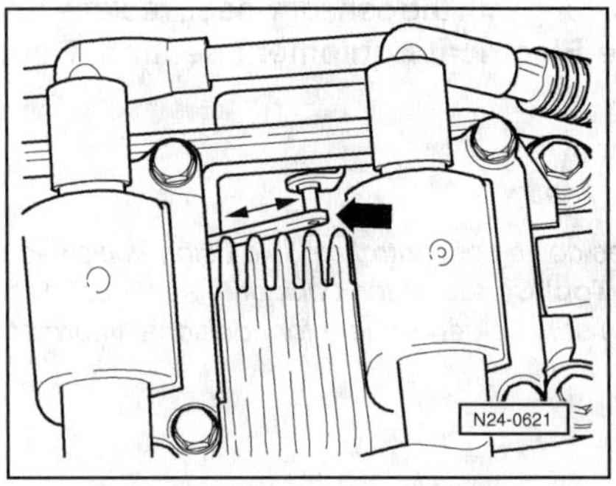

◄ Perform output Diagnostic Test Mode (DTM) again. If intake manifold change-over valve is activated, uniformly operate hand pump.

- The setting element must, between the two positions, switch back and forth

If the change-over does not function as described:

– Check whether change-over mechanism moves easily. To do this, operate linkage by hand.

– Check whether vacuum lines are correctly connected.

– Check whether vacuum system, including vacuum reservoir in left wheel well, is properly sealed.

Fuel injectors, checking

Special tools and equipment

- V.A.G1551 or VAG1552 mobile scan tool, with VAG1551/3 cable
- VAG1598/31 test box
- VAG1526 multimeter or VAG1715 multimeter
- VAG1527 voltage test lamp
- VAG1594 connector test kit
- Wiring diagram

Test requirements

- Engine speed (RPM) sensor must be working properly, see **Engine speed (RPM) sensor -G28-, checking**.
- Fuel Pump (FP) relay must be working properly.

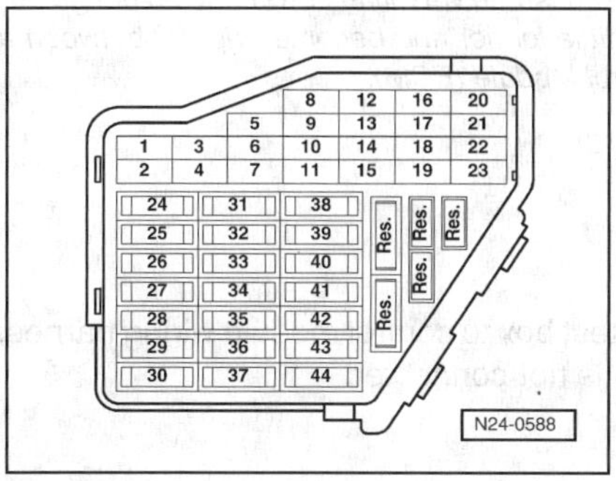

Fuses no. 28 and 32 must be OK.

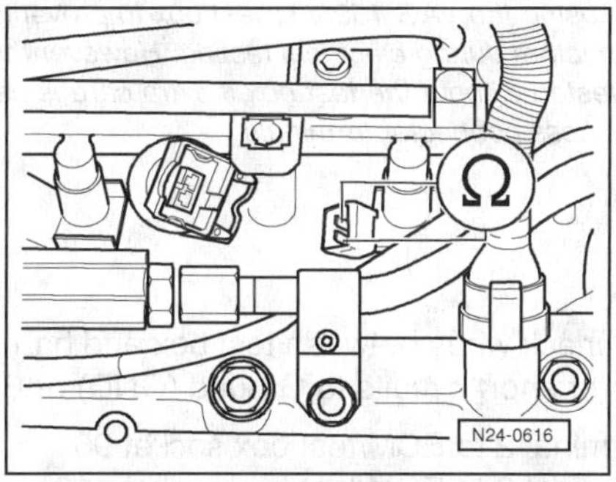

Test resistance of fuel injectors between terminals.

- Specification: 13.5 - 15.5 Ω

NOTE —

Resistance value is valid at approx. 20 °C (68 °F). At higher temperatures, the value will rise approx. 4 - 6 Ω

If specified value is not obtained:

– Replace faulty fuel injector.

Fuel injectors, checking voltage supply

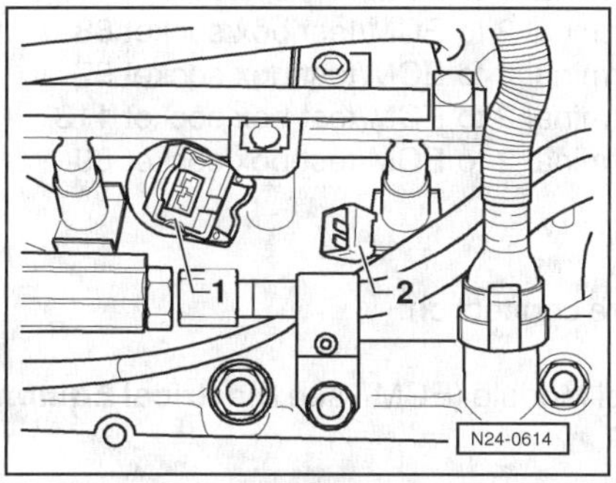

Pull connectors (1) off connection (2) of all fuel injectors.

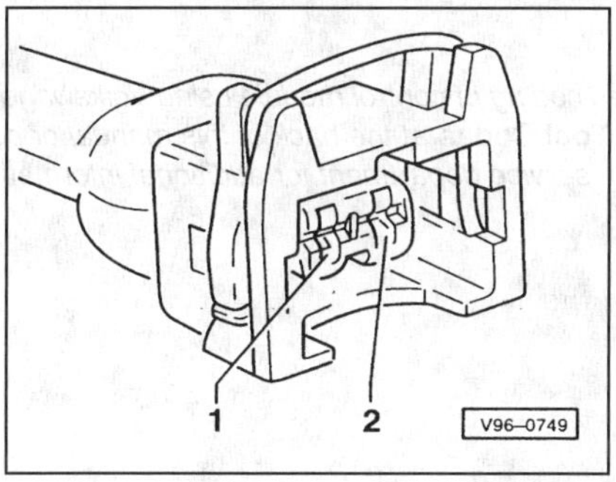

Connect diode test lamp to connector of tested fuel injector terminal 1 and Ground (GND).

– Switch on starter and test voltage supply for fuel injector.

- LED must light.

If LED does not light up:

– Switch ignition off.

– Check for breaks in wire between 2-way connector terminal 1 and Fuel Pump (FP) relay (J17) (see wiring diagram).

- Wire resistance: max. 1.5 Ω

Fuel injectors, checking activation and voltage supply

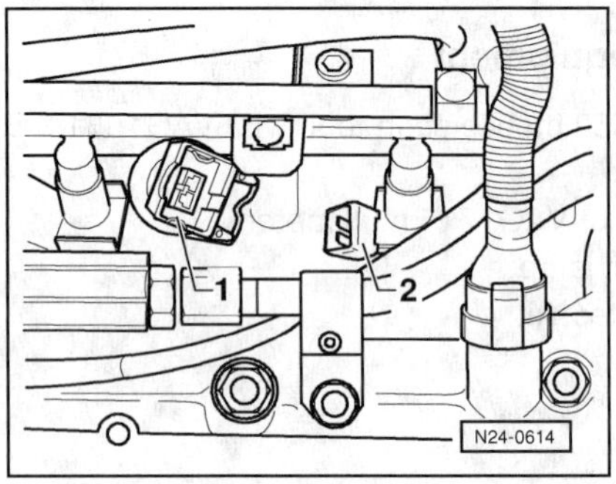

◄ Pull connectors (1) off connection (2) of all fuel injectors.

– Use auxiliary wires from VAG 1594 to connect diode test lamp VAG 1527 to connector terminals of fuel injector to be tested.

– Switch on starter and test activation of fuel injector.

- LED should flicker.

NOTE—

The LEDs in the test lamp consume very little power and do not go out completely but rather glow a little longer and become brighter between activations by the Engine Control Module (ECM).

If LED does not flicker:

– Switch ignition off.

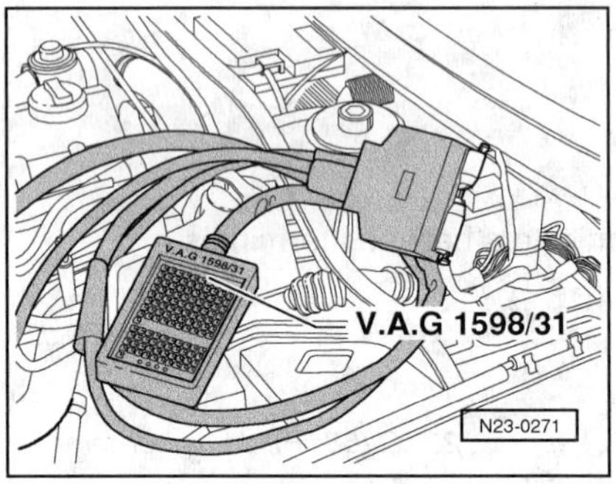

◄ Connect VAG1598/31 test box to control module wiring harness. Engine Control Module (ECM) is not connected.

EDITOR'S NOTE—

Volkswagen recommends using the VAG 1598/31 test box to prevent damage to ECM harness connectors during electrical testing. However the harness connectors can be tested without the test box if extra care is used to prevent damage to the harness connector terminals.

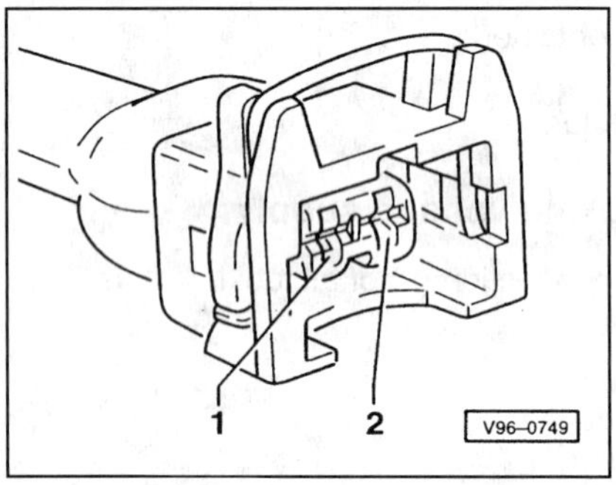

◄ Use wiring diagram to check wires between test box and harness connector for open circuits or short circuits to Ground (GND) or B+:

- Cyl. 1 fuel injector terminal 2 to ECM/test box socket 96
- Cyl. 2 fuel injector terminal 2 to ECM/test box socket 112
- Cyl. 3 fuel injector terminal 2 to ECM/test box socket 88
- Cyl. 4 fuel injector terminal 2 to ECM/test box socket 97
- Cyl. 5 fuel injector terminal 2 to ECM/test box socket 113
- Cyl. 6 fuel injector terminal 2 to ECM/test box socket 89

Specified value: max. 1.5 Ω

If there is no DTC in wire connection:

– Replace Engine Control Module (ECM), see **Electrical Equipment section, Repair group 27**.

EDITOR'S NOTE—

*ECM replacement requires coding of control module using Volkswagen factory scan tool. See **Scan Tool Codes** at the back of this manual or consult an authorized Volkswagen service department for additional information.*

Fuel injection quantity and injector leakage, checking

Special tools and equipment

- VAG 1348/3 remote control
- VAG 1348/3-2 adapter
- VAG 1602 tester for quantity of injected fuel

Test condition

- Fuel pressure must be working properly, see **Fuel pressure regulator and residual pressure, checking**.

> ***WARNING —***
> *Fuel system is under pressure! Before opening the fuel system, place rags under and around the connection then slowly release the fuel pressure.*

Test sequence

– Remove suction hose between Mass Air Flow (MAF) sensor and intake air scoop.

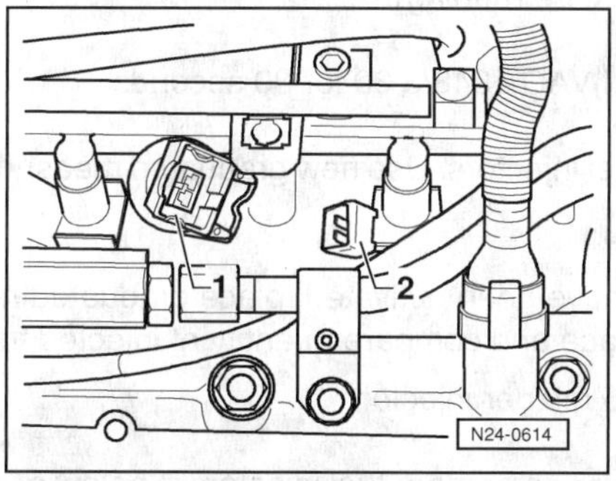

< Pull connectors (1) off connection (2) of all fuel injectors.

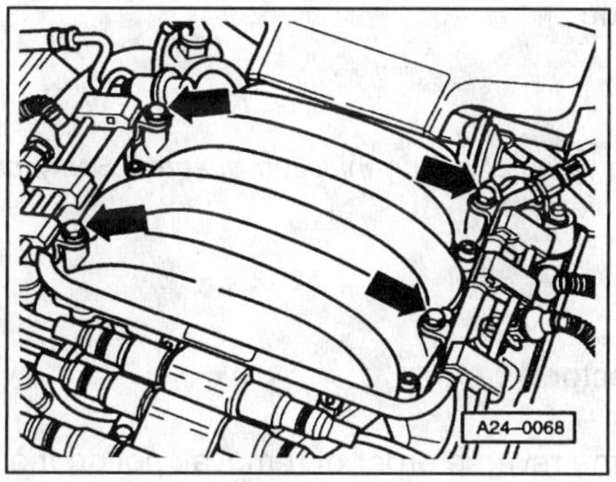

< Unscrew fuel rail from intake manifold (arrows).

– Pull vacuum hose off fuel pressure regulator.

– Lift fuel rail with fuel injectors off intake manifold and support it.

Checking proper seal

– Perform output Diagnostic Test Mode (DTM) using diagnostic scan tool (transfer fuel pump must be running).

NOTE —
The only purpose of this step is to allow the transfer Fuel Pump (FP) to run with the engine stationary.

– Check proper seal of fuel injectors (visual inspection). With transfer Fuel Pump (FP) running, only 1 to 2 drops per minute should escape from each valve.

If fuel loss is greater:

– Switch off ignition and replace leaking fuel injector.

NOTE —
Always use new sealing rings.

Injection quantity, checking

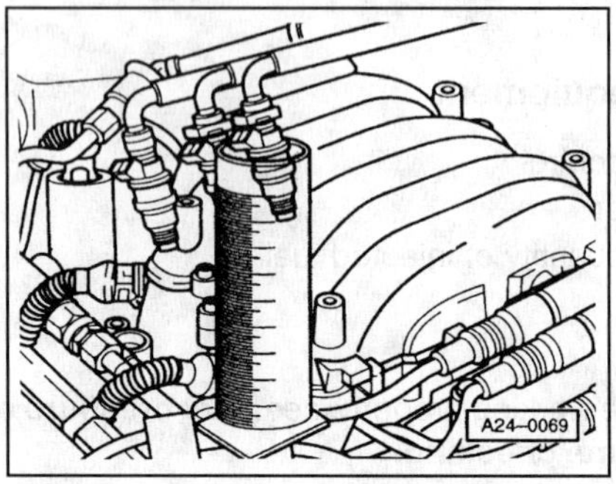

◀ Place a fuel injector to be tested into a graduated measuring glass of tester for quantity of injected fuel VAG 1602.

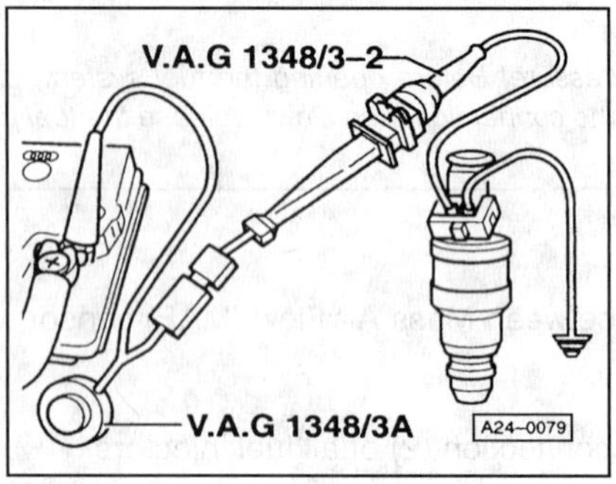

◀ Use auxiliary wires from VAG 1594 to connect one terminal of tested fuel injector to engine Ground (GND).

– Use auxiliary wire to connect second terminal of fuel injector to remote control VAG 1348/3 with adapter wire VAG 1348/3-2.

– Connect alligator clip to B+.

– Perform output Diagnostic Test Mode (DTM) using diagnostic scan tool (transfer fuel pump must be running).

– Operate remote control VAG 1348/3 30 for 30 seconds.

– Repeat test on other fuel injectors. Use new graduated measuring glasses for this.

– After all fuel injectors have been activated, place graduated measuring glasses on a level surface and compare quantity of injected fuel.

- Specification: 90-125 ml. per injector

If the measured value of one or more fuel injectors is below or above the indicated expected value:

NOTE —
Also check spray pattern when checking injection quantity. Spray pattern must be the same for all injectors

– Switch ignition off.

– Replace faulty fuel injector.

– Install fuel injectors is the reverse order of removal, noting the following:

- Replace O-rings on all fuel injectors and wet them slightly with clean motor oil.
- Insert fuel injectors into fuel rail perpendicular and in the right position, and secure them with retaining clips.
- Place fuel rail with secured fuel injectors onto intake manifold, and apply uniform pressure to press it in.

Fuel pressure regulator and residual pressure, checking

The fuel pressure regulator controls the fuel pressure as a function of the Manifold Absolute Pressure (MAP).

Special tools and equipment

- VAG 1318 pressure gauge
- VAG 1318/7 adapter
- VAG 1318/10 adapter
- VAG 1318/13 adapter

Test conditions

- Capacity of transfer Fuel Pump (FP) must be OK

> ***WARNING —***
> *Fuel system is under pressure! Before opening the fuel system, place rags under and around the connection then slowly release the fuel pressure.*

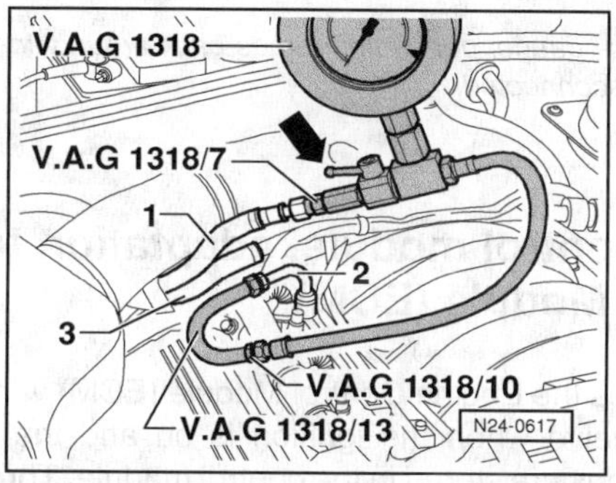

◀ Unscrew fuel supply hose (1) from fuel rail (2).

– Connect pressure gauge VAG 1318 with adapters 1318/7, /10 and /13.

– Open shut-off valve of pressure gauge. The lever points in the direction of flow (arrow).

– Start engine and let run at idle.

– Measure fuel pressure.
  - Specification: about 3.5 bar positive pressure

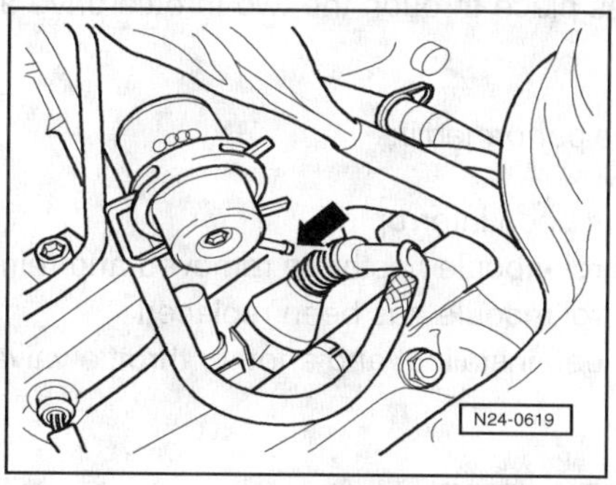

◀ Pull vacuum hose off fuel pressure regulator (arrow).
  - Fuel pressure should rise to about 4.0 bar positive pressure overpressure.

NOTE —
If fuel should escape at the vacuum connection of the pressure regulator, replace pressure regulator.

– Switch ignition off.

– Check for proper seal and residual pressure by observing pressure drop on pressure gauge.
  - After 10 minutes at least 2.2 bar overpressure should still be present.

If residual pressure falls below 2 bar overpressure:

– Start engine and let run at idle.

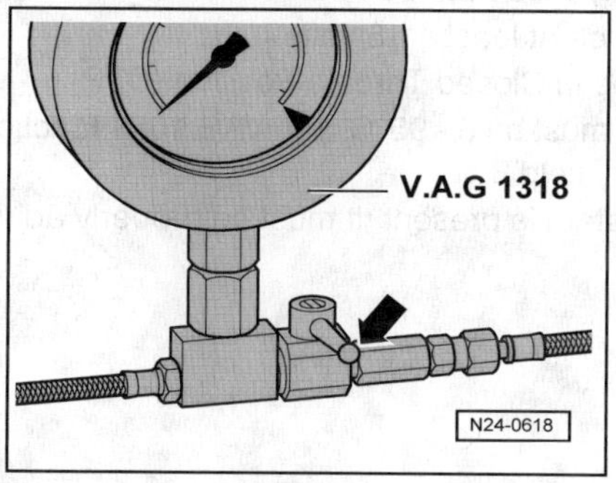

◀ Wait until pressure has dropped, and then switch off ignition. Simultaneously close stop valve of pressure gauge VAG 1318 (lever at right angles to direction of flow (arrow)).

– Observe pressure drop on manometer.

If pressure does not drop:

– Test check-valve on transfer Fuel Pump (FP).

If pressure does drop again:

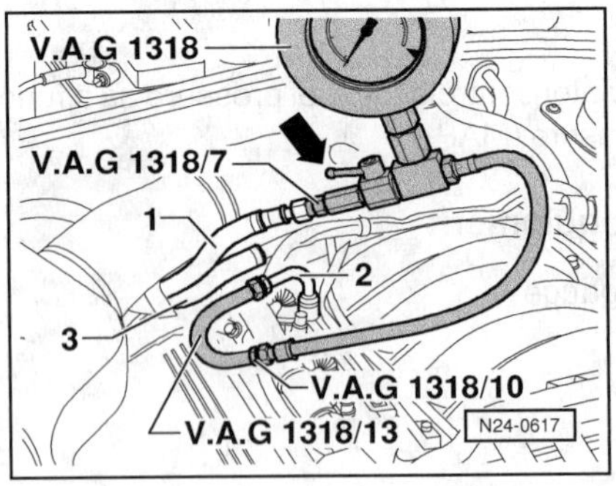

◄ Open stop valve of pressure gauge VAG 1318 (lever in direction of flow (arrow).

– Start engine and let run at idle.

– Wait until pressure has dropped, and then switch off ignition. Simultaneously pinch return hose (3) tightly together.

If pressure does not drop:

– Replace fuel pressure regulator.

If pressure does drop again:

– Check whether line connections, O-rings in fuel rail, and fuel injectors are properly sealed.

– Check whether pressure gauge is properly sealed.

NOTE —
Before removing pressure gauge, put cleaning rags back around line connection which is to be disconnected.

Throttle valve control module, adaptation to Engine Control Module (ECM)

The adaptation teaches the Engine Control Module (ECM) various positions of the throttle valve when the ignition is on and the engine is stopped. These positions are stored in the control module. Throttle valve position feedback takes place through the two throttle drive angle sensors.

The adaptation must be performed if:

- the supply voltage has been interrupted
- the throttle valve control module has been removed and reinstalled
- the throttle valve control module has been replaced
- another engine has been installed, and another throttle valve control module along with it

Special tools and equipment

- V.A.G1551 or VAG1552 mobile scan tool, with VAG1551/3 cable

Test condition

- There must not be any DTCs stored in DTC memory.
- Battery voltage must be at least 11.5 volts.
- Throttle valve must be in Closed Throttle Position (CTP)
- Coolant temperature must be 5 - 95°C, see VAG 1551 function 08, display group 04, display field 3
- If a cruise control system is present, it must be properly adjusted.

Adaptation procedure

– Connect V.A.G 1551 scan tool (VAG1552). Switch ignition on and select Engine Control Module (ECM) using address word 01 (connecting scan tool and selecting ECM.

– Press buttons -0- and -4- to select "Basic Setting" function 04. Press -Q- button to confirm input.

– Press buttons -0-, -6- and -0- to select "Display group number 60". Press -Q- button to confirm input.

After pressing the Q-button, the throttle valve actuator will be switched on without current.

In this condition, the throttle valve will be drawn into an emergency position by a mechanical spring, which is located in the throttle valve control module. The values supplied by the two angle sensors in this emergency position are stored by the Engine Control Module (ECM).

Next, the throttle valve is opened by a certain value. If this value is reached, the throttle valve actuator is once again de-energized. Now, the mechanical spring must close the throttle valve to the previously learned emergency position within a specified time (spring test).

Then, the throttle valve is closed by the throttle valve actuator; the values supplied by the angle sensors in the throttle valve control module are stored by the Engine Control Module (ECM).

If Engine Control Module (ECM) should de-energize throttle valve actuator during driving operation, this is expressed by an increased, surging idle speed (RPM). The engine accepts gas only after great delay.

```
System in Basic Setting  60
  xxx %   xxx % 9    ADP. OK
```

◄ Indicated on display (1 - 4 = display fields)

• The adaptation has been successfully completed.

To store the values:

– Switch ignition off.

NOTE —
If the basic adjustment is canceled by the control module, this can have the following causes:

- *The throttle valve does not reach the mechanical idle stop due to dirt, for example carbon buildup, or due to binding.*
- *Battery voltage is too low.*
- *Throttle valve control module or wire connection is defective.*

NOTE —
After a cancellation, a DTC is entered in the DTC memory. The next time the ignition is switched on, the basic adjustment is automatically performed again.

– End basic adjustment of the engine by pressing the→ key.

– Press buttons -0- and -6- to select "End Output" function 06. Press -Q- button to confirm input.

Additional signals, checking

Speedometer Vehicle Speed Sensor (VSS) -G22-, checking

Special tools and equipment

- VAG1551 or VAG1552 Scan Tool (ST) with VAG1551/3 adapter cable
- VAG1598/22 test box
- Multimeter US1119 (Fluke 83 or equivalent)
- VW1594 connector test kit
- Wiring diagrams

Checking conditions

- Battery voltage: min. 11.5 volts
- Speedometer OK

Checking

– Connect VAG1551 or VAG1552 scan tool and press buttons -0- and -1- to insert "Engine Electronics" address word 01 (with engine running at idle).

– Press buttons -0- and -8- to select "Read Measuring Value Block" function 08, and press -Q- button to confirm input.

– Press 0, 0 and 5 for function "display group number 5", and then press Q to exit from input mode.

– Perform a road test, and have 2nd person observe indication on display.

> ***WARNING —***
>
> - *When driving or riding in an airbag-equipped vehicle, NEVER hold scan tool or other test equipment in your hands or lap while in motion. Objects between you and the airbag increase the risk of injury in an accident.*
> - *During a test drive in an airbag-equipped vehicle, test equipment must always be fastened to and operated from the rear seat by a second technician.*

– Read value indicated in display field 3.

- Specification: approx. vehicle speed

– Press →button.

– Press buttons -0- and -6- to select "End Output" function 06. Press -Q- button to confirm input.

– Switch ignition off.

If no speed is indicated:

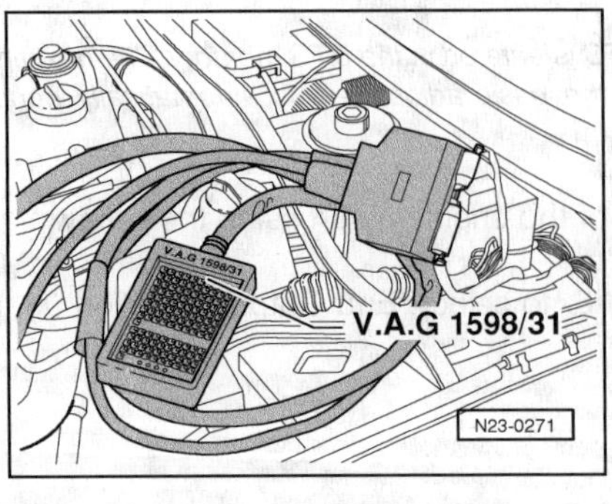

◄ Connect VAG1598/31 test box to control module wiring harness. Engine Control Module (ECM) is not connected.

EDITOR'S NOTE —

Volkswagen recommends using the VAG 1598/31 test box to prevent damage to ECM harness connectors during electrical testing. However the harness connectors can be tested without the test box if extra care is used to prevent damage to the harness connector terminals.

– To measure voltage, connect multimeter between sockets 3 (terminal 15) + 54 (speed signal) of test box.

– Switch ignition on.

– Jack up front left wheel.

– Rotate front wheel, and observe voltage indicated while doing so.

- Specification: between 0 and at least 4 volts, oscillating

If indicated voltage does not oscillate:

– Check wires going to speedometer and speed sensor using wiring diagrams.

A/C system signals, checking

Special tools and equipment

- V.A.G1551 or VAG1552 mobile scan tool, with VAG1551/3 cable
- VAG1598/31 test box
- VAG1526 multimeter or VAG1715 multimeter
- VAG1594 connector test kit
- Wiring diagram

Test requirements

- The A/C system must function properly.
- A/C system must be switched off.
- There must not be any DTCs stored in DTC memory.
- Vehicle must be at room temperature; warmer than 15°C (59°F).

Test sequence

– Connect V.A.G 1551 scan tool (V.A.G1552). Start engine and select Engine Control Module (ECM) using address word 01.

– Press buttons -0- and -8- to select "Read measuring value block" function 08. Press -Q- button to confirm input.

– Press 0, 5 and 0 for function "display group number 50", and then press Q to exit from input mode.

– Observe reading in display field 4.

– Switch A/C system on.

– Set it to lowest temperature and highest fan setting. Reading in display field 4 should change to:

- Compr.ON

– Suddenly press accelerator pedal to wide open throttle and then release it (brief acceleration).

– The reading in display field 4 should change from

- Compr.ON (for a few seconds to)
- Compr. OFF (compressor is switched off when vehicle accelerates).

– Press →button.

– Press buttons -0- and -6- to select "End Output" function 06. Press -Q- button to confirm input.

– Switch ignition off.

If reading does not change:

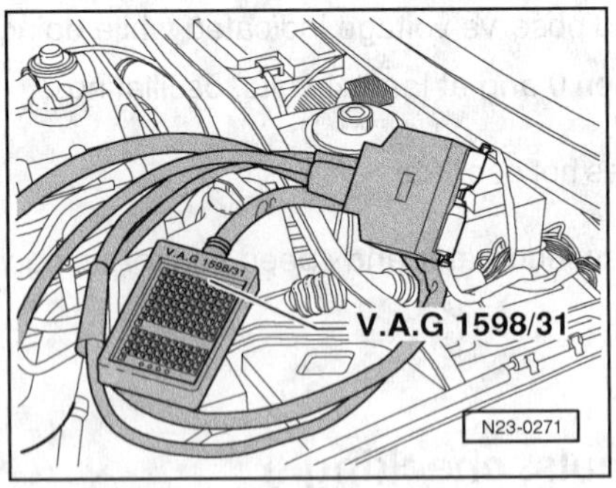

– Connect V.A.G1598/31 test box to control module wiring harness. Engine Control Module (ECM) is not connected.

– Check for breaks in wires between test box sockets 40 + 41 and A/C system (see wiring diagram).

- Wire resistance: max. 1.5 Ω

Clutch vacuum vent valve switch signal, checking

Special tools and equipment

- V.A.G1551 or VAG1552 mobile scan tool, with VAG1551/3 cable
- VAG1598/31 test box
- VAG1526 multimeter or VAG1715 multimeter
- VAG1594 connector test kit
- Wiring diagram

NOTE—
The cruise control system needs this signal to avoid tip-in shock.

Test sequence

– Connect V.A.G 1551 scan tool (VAG1552). Switch ignition on and select Engine Control Module (ECM) using address word 01.

– Press buttons -0- and -8- to select "Read measuring value block" function 08. Press -Q- button to confirm input.

– Press buttons 6-, -6- and -0- to select "Display group number 66". Press -Q- button to confirm input.

– Observe reading in display field 2. (2nd position from left)

- Specification: x 0 x x

– Press in clutch pedal and observe reading in display field 2.

- Specification: x 1 x x

– Press →button.

– Press buttons -0- and -6- to select "End Output" function 06. Press -Q- button to confirm input.

– Switch ignition off.

If specified value is not obtained:

– Remove driver-side storage compartment, see **Body–Interior section, Repair Group 68**

– Disconnect harness connector at Clutch Pedal Position (CPP) switch.

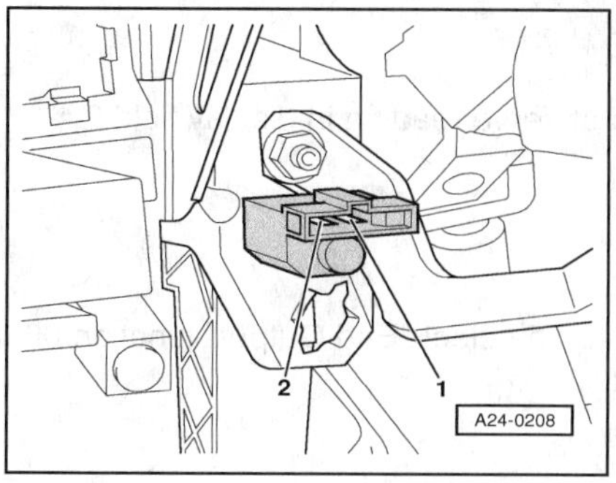

- Connect multimeter between terminal 1 and 2 for resistance measurement.
  - Specification: approx. 0 Ω
- Press in clutch pedal.
  - Expected value: ∞ (no flow of electricity)

If specified value is not obtained:

- Replace Clutch Pedal Position (CPP) switch.

If the specified value is met:

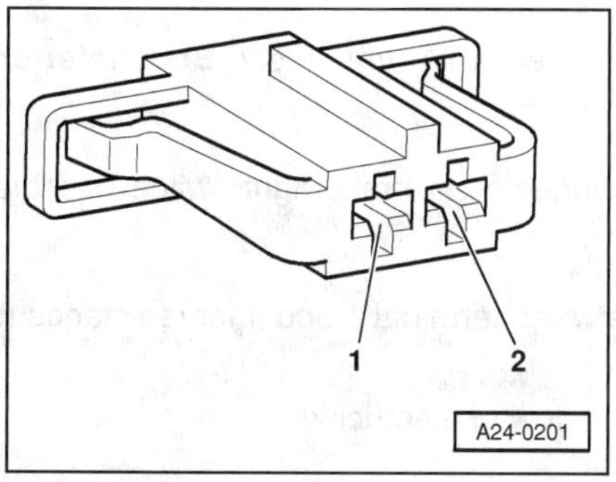

- Connect multimeter between terminal 1 and Ground (GND) to make voltage measurement.
  - Specification: at least 11.5 V

If specified value is not obtained:

- Check wire between 2-pin connector terminal 1 and central electrical system (see wiring diagram).

If the specified value is met:

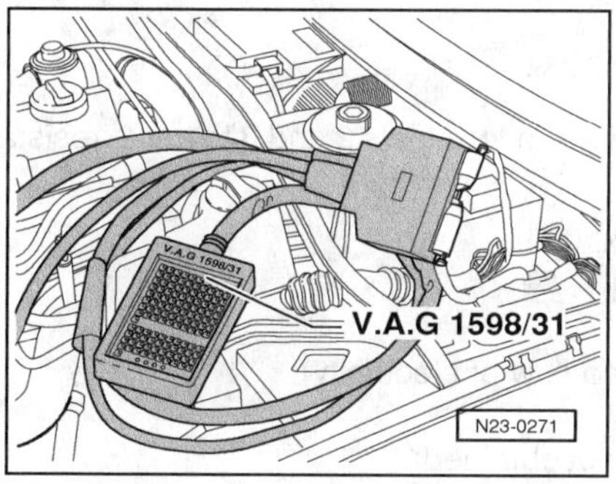

- Connect VAG1598/31 test box to control module wiring harness. Engine Control Module (ECM) is not connected.
- Check for breaks in wire between test box and 2-way connector (see wiring diagram).
  - Terminal 2 + socket 39
  - Wire resistance: max. 1.5 Ω
- Check wires for shorts among one another.
- Also check wires for shorts to battery B+ or Ground (GND).

Brake light switch and brake pedal switch signals, checking

Special tools and equipment

- V.A.G1551 or VAG1552 mobile scan tool, with VAG1551/3 cable
- VAG1598/31 test box
- VAG1526 multimeter or VAG1715 multimeter
- VAG1594 connector test kit
- Wiring diagram

Test sequence

- Connect V.A.G 1551 scan tool (VAG1552). Switch ignition on and select the Engine Control Module (ECM) using address word 01.
- Press buttons -0- and -8- to select "Read measuring value block" function 08. Press -Q- button to confirm input.
- Press buttons -0-, -6- and -6- to select "Display group number 66". Press -Q- button to confirm input.

– Observe reading in display field 2. (3rd and 4th position from left)

- Specification: x x 0 0

– Press brake pedal and observe reading in display field 2.

- Specification: x x 1 1

– Press →button.

– Press buttons -0- and -6- to select "End Output" function 06. Press -Q- button to confirm input.

– Switch ignition off.

If specified value is not obtained:

– Remove driver-side storage compartment, see **Body Interior section, Repair Group 68.**

– Disconnect harness connector at brake light switch and brake pedal switch.

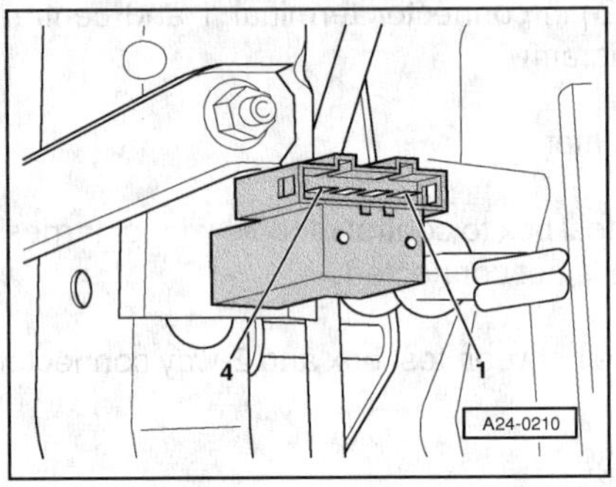

– Connect multimeter between terminal 1 and 4 for resistance measurement.

- Expected value: ∞ (no flow of electricity)

– Press brake pedal.

- Specification: approx. 0 Ω

– Connect multimeter between terminals 2 and 3 to make resistance measurement.

- Specification: approx. 0 Ω

– Press brake pedal.

- Expected value: ∞ (no flow of electricity)

If specified values are not obtained:

– Replace brake light switch / brake pedal switch.

If specified values are obtained:

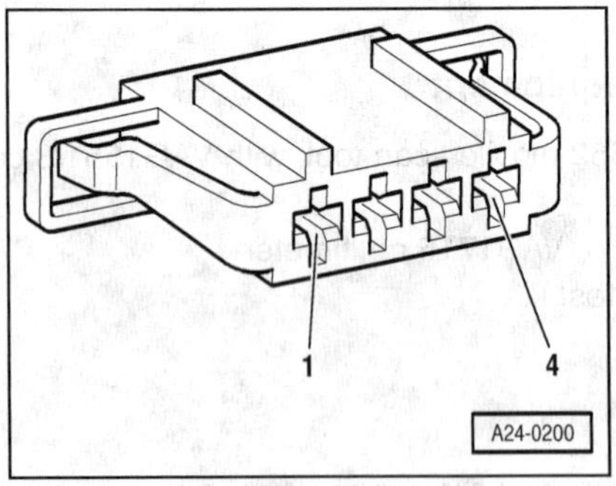

– Connect multimeter between terminal 1 and Ground (GND) to make voltage measurement.

- Specification: at least 11.5 V

– Switch ignition on.

– Connect multimeter between terminal 2 and Ground (GND) to make voltage measurement.

- Specification: at least 11.5 V

If specified values are not obtained:

– Check wire between 4-pin connector terminal 1 and main fuse (2) and central electronics according to wiring diagram.

If the specified value is met:

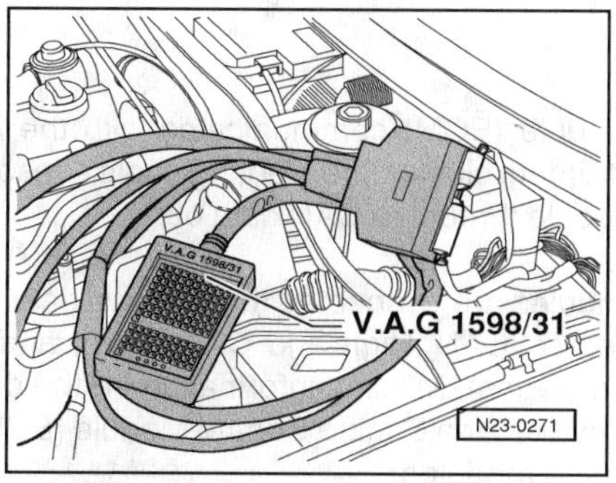

◄ Connect VAG1598/31 test box to control module wiring harness. Engine Control Module (ECM) is not connected.

EDITOR'S NOTE —

Volkswagen recommends using the VAG 1598/31 test box to prevent damage to ECM harness connectors during electrical testing. However the harness connectors can be tested without the test box if extra care is used to prevent damage to the harness connector terminals.

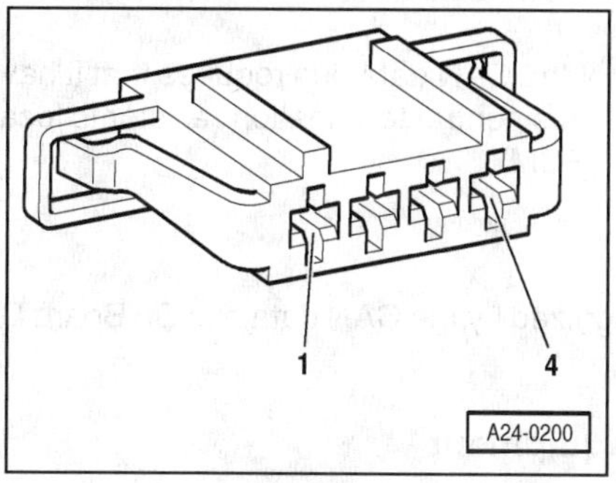

◄ Check for breaks in wire between test box and 4-way connector (see wiring diagram).

- Terminal 2 + socket 56
- Terminal 4 + socket 55
- Wire resistance: max. 1.5 Ω

– Check wires for shorts among one another.

– Also check wires for shorts to battery B+ or Ground (GND).

Cruise Control System (CCS), activating and deactivating

Activating cruise control system

– Connect V.A.G 1551/1552 Scan Tool (ST).

– Press buttons -0- and -1- to insert "Engine electronics" address word "01".

– Press buttons -1- and -1- to select "Log-in procedure" function 11.

– Press -Q- button to confirm input.

– Enter code number 11463.

– Press -Q- button to confirm input.

Deactivating cruise control system

– Connect V.A.G 1551/1552 Scan Tool (ST).

– Press buttons -0- and -1- to insert "Engine electronics" address word "01".

– Press buttons -1- and -1- to select "Log-in procedure" function 11.

– Enter code number 16167.

– Press -Q- button to confirm input.

Data Bus, checking

Function

The Engine Control Module (ECM) communicates with the ABS/ESP control module, the instrument cluster, the steering angle sensor, and the transmission control module through a CAN data bus.

The busable control modules are connected through two data bus wires which are twisted together (CAN_High and CAN_Low), and they exchange information (messages). Missing information on the data bus is recognized as an error by both the Engine Control Module (ECM) and by other CAN data bus control modules.

Trouble-free operation of the CAN data bus requires that it have a termination resistor of 66 Ω. This central termination resistor is located in the Engine Control Module (ECM).

Test condition

- A DTC has been recognized by the CAN data bus On Board Diagnostic (OBD)

Special tools and equipment

- VAG 1598 test box
- VAG1598/31 test box
- VAG 1598/33 test box
- VAG1526 multimeter or VAG1715 multimeter
- VAG1594 connector test kit
- Wiring diagram

Vehicles with automatic transmission:

- VAG 1598/18 test box

Test sequence

– Switch ignition off.

– Disengage connector from Engine Control Module (ECM) and then pull it off.

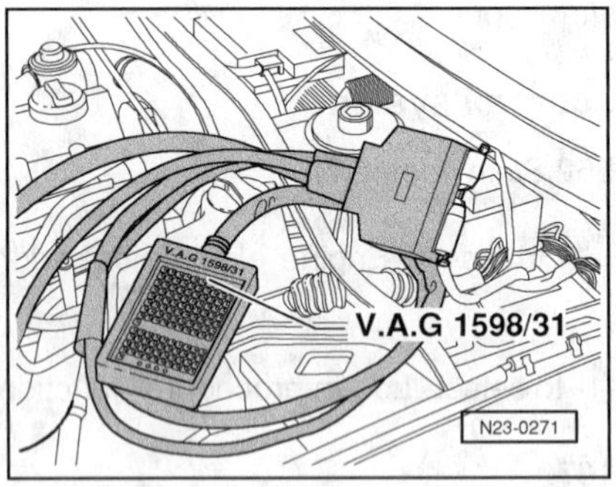

– Connect VAG 1598/31 test box to Engine Control Module (ECM). Control module wiring harness is not connected.

– Test central termination resistor in Engine Control Module (ECM) by measuring resistance between test box:

- Sockets 58 + 60 and
- Sockets 77 + 79
- Specification: 60 - 72 Ω

If resistance value is not in expected range:

– Replace Engine Control Module (ECM), see **Electrical Equipment section, Repair group 27**.

EDITOR'S NOTE —

ECM replacement requires coding of control module using Volkswagen factory scan tool. See **Scan Tool Codes** *at the back of this manual or consult an authorized Volkswagen service department for additional information.*

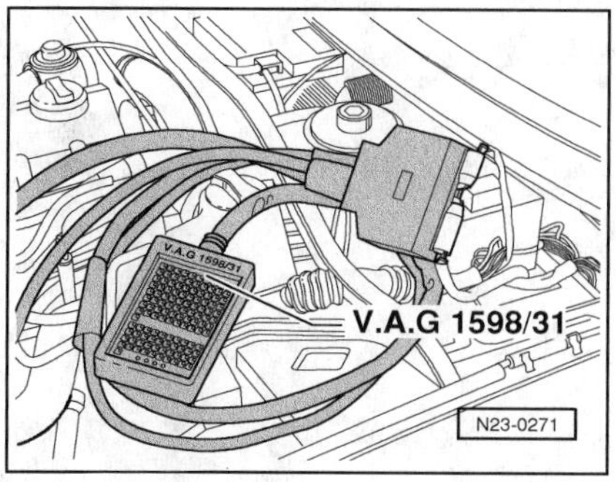

If resistance value is in expected range:

– Remove VAG 1598/31 test box from Engine Control Module (ECM).

◀ Connect VAG1598/31 test box to control module wiring harness.

– Check CAN data bus wires for shorts among one another by measuring resistance between test box sockets 58 + 60:

- Expected value: ∞

If expected value is obtained (no shorts in wiring):

– Eliminate error in wiring (see wiring diagram)

If expected value is obtained (shorts in wiring):

– Eliminate error in wiring (see wiring diagram).

28 Ignition System

Repair Group

J 28

Ignition system, servicing

General notes on the ignition system

- Only the components which specifically relate to the ignition system are described here. For other components in the fuel injection and ignition system, see **Repair Group 24**.
- Disconnecting and connecting the battery must only be done with the ignition switched off, otherwise the Engine Control Module (ECM) could be damaged.
- The ECM is equipped with On Board Diagnostic (OBD).
- Components marked with * are checked by on board diagnostic.
- For trouble-free operation of the electrical components a voltage of at least 11.5 volts is necessary.
- During some checks it is possible that the ECM will recognize and store a DTC. After completing all checks and repairs the DTC memory must be checked and if necessary erased, see **Maintenance.**

Safety precautions

WARNING —

Be sure the ignition is switched OFF, when:

- *Disconnecting ignition wires*
- *Disconnecting fuel injection system wiring*
- *Connecting or disconnecting test equipment leads*
- *Disconnecting the battery*
- *Washing the engine or engine compartment.*

WARNING —

- *When driving or riding in an airbag-equipped vehicle, NEVER hold the scan tool or other test equipment in your hands or lap while in motion. Objects between you and the airbag increase the risk of injury in an accident.*
- *During a test drive in an airbag-equipped vehicle, test equipment must always be fastened to and operated from the rear seat by a second technician.*

If the engine is to be turned at starter speed, without starting:

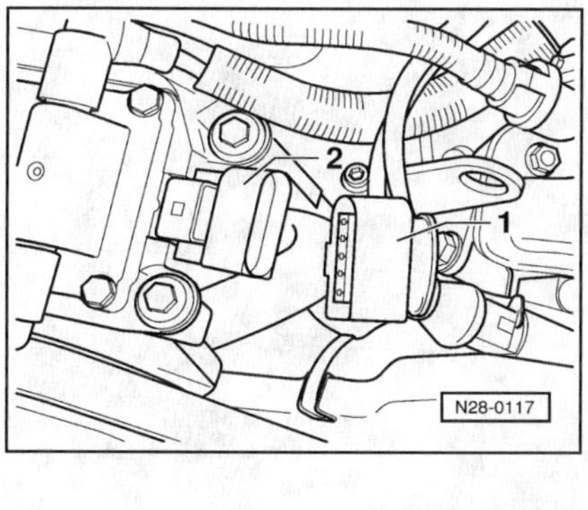

◄ Disconnect 5-pin connector -1- from power output stage -N122- for ignition coils -2-.

– Remove fuse 28 for fuel pump.

Ignition system components, removing and installing

1. **Spark plug, 30 Nm**
   - Remove and install with 3122 B
   - Type and spark plug gap, see **Technical data, spark plugs**
2. **Spark plug connectors with ignition wires**
   - 4 - 6 kΩ resistance
3. **Spacer sleeve**
4. **Rubber grommet**
5. **10 Nm (7 ft-lb)**
6. **Ignition coils (-N-, -N128-, -N158-)**
   - With power output stage -N122-
   - With marking for ignition wires, do not interchange
   - Checking, see **Ignition coils with power output stage, checking**
7. **Connector**
   - Black, 5-pin
8. **3-way connector**
   - Terminals are gold-plated
9. **20 Nm**
   - Tightening torque affects function of Knock Sensor (KS)
10. **Knock Sensor (KS) 2 -G66-***
    - Engine bank 2
    - Terminals of sensor and connector are gold-plated
11. **Knock Sensor (KS) 1 -G61-***
    - Engine bank 1
    - Terminals of sensor and connector are gold-plated
12. **10 Nm (7 ft-lb)**
13. **Camshaft Position (CMP) sensor 2 -G163-***
    - Engine bank 2
    - Checking, see **Camshaft Position (CMP) sensor, checking**
14. **25 Nm**
15. **Washer**
    - Conical
16. **Face plate**
    - For Camshaft Position (CMP) sensor
    - Note position when installing
17. **Connector**
    - Black, 3-pin
    - For Camshaft Position (CMP) sensor (-G40- or -G163-)

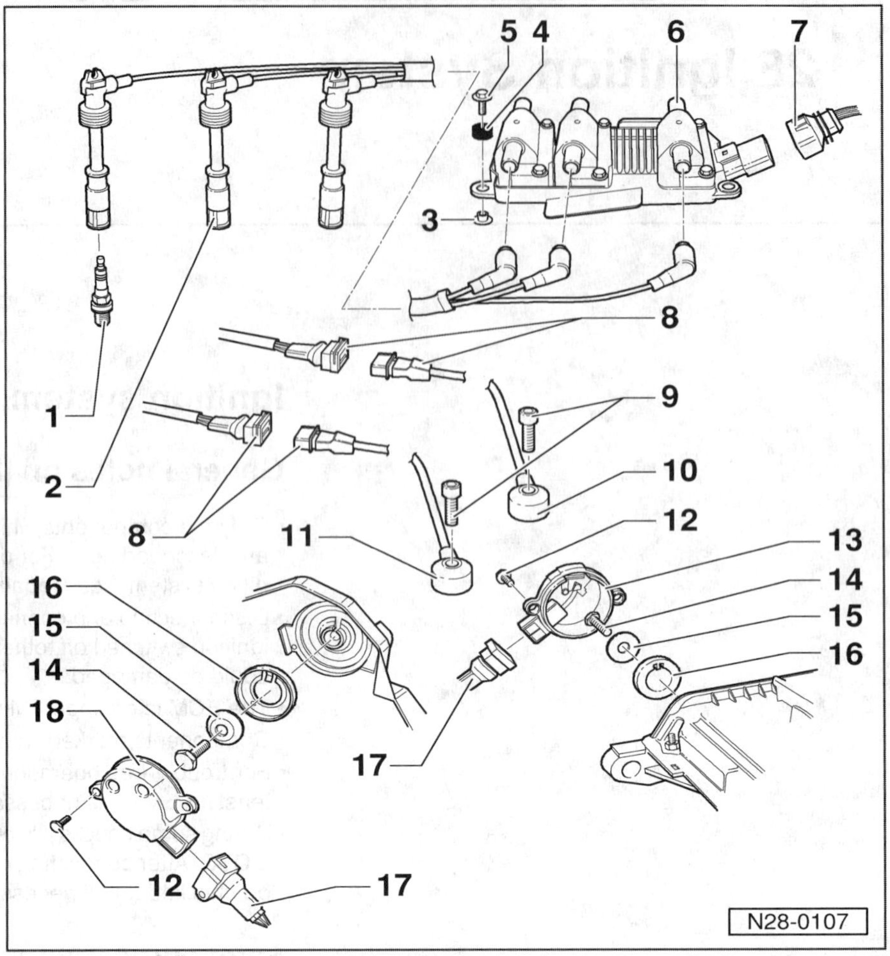

18. **Camshaft Position (CMP) sensor - G40-***
    - Engine bank 1
    - Checking, see **Camshaft Position (CMP) sensor, checking**

Technical data, spark plugs

Engine code	ATQ
Firing order	1-4-3-6-2-5
Spark plugs[1)2)] VW/Audi Manufacturer's designation	101 000 035 (NGK) BKR 6 EKUB (NGK)
Electrode gap	0.9 - 1.1 mm (0.035 - 0.043in.)
Tightening torque	30 Nm (22 ft lb)

[1)] Spark plug change intervals see **Maintenance**

[2)] Remove and install spark plugs using special tool 3122B

> ***WARNING —***
> *Part numbers are listed here for reference only. Always check with your Parts department for the latest information.*

Camshaft Position (CMP) sensor, checking

Special tools and equipment

- VAG1598/31 test box
- VAG1526 multimeter or VAG1715 multimeter
- VAG1594 connector test kit
- Wiring diagram

Test condition

- Battery voltage must be at least 11.5 volts.

Test sequence

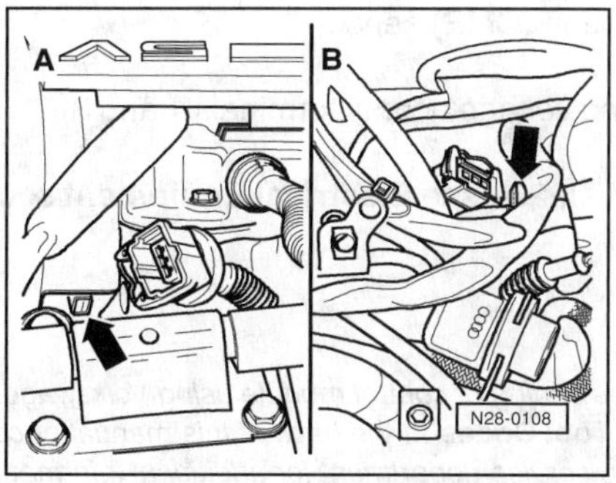

◄ Pull off connectors on Camshaft Position (CMP) sensors. The locations of the Camshaft Position (CMP) sensors are as follows:

- A - Cylinder bank 1 Camshaft Position (CMP) sensor -G40- Right front cylinder head
- B - Cylinder bank 2 Camshaft Position (CMP) sensor -G163- Left rear cylinder head

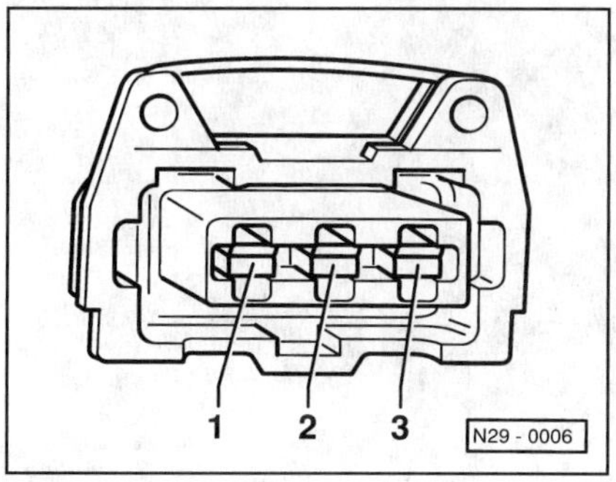

◄ Use auxiliary wires from VAG 1594 to connect multimeter with outer terminals of appropriate connector to make voltage measurement.

– Switch ignition on.

  - Specification: at least 5 V

– Switch ignition off.

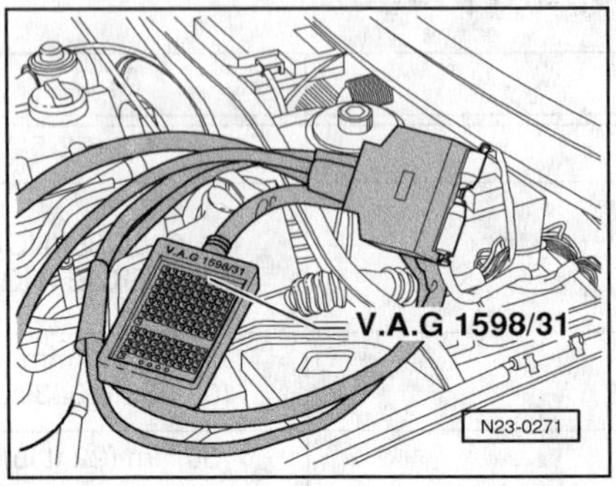

- Connect VAG1598/31 test box to control module wiring harness. Engine Control Module (ECM) is not connected.

- Connect VAG1598/22 test box to Motronic ECM harness connector.

EDITOR'S NOTE —
Volkswagen recommends using the VAG 1598/22 test box to prevent damage to ECM harness connectors during electrical testing. However the harness connectors can be tested without the test box if extra care is used to prevent damage to the harness connector terminals.

– Check wiring for open circuit between ECM/test box and 3-pin harness connector, according to wiring diagram.

CMP 2 -G163-, right cylinder head, front

- Connector terminal 1 to ECM/test box socket 98
- Connector terminal 2 to ECM/test box socket 86
- Connector terminal 3 to ECM/test box socket108
- Specified value: max. 1.5 Ω

CMP 2 -G40-, left cylinder head, rear

- Connector terminal 1 to ECM/test box socket 98
- Connector terminal 2 to ECM/test box socket 87
- Connector terminal 3 to ECM/test box socket 108
- Specified value: max. 1.5 Ω

– Check wiring for short circuits to one another.

Specified value: ∞

If wiring is OK (and voltage across terminals 1 and 3 is OK):

– Replace Camshaft Position (CMP) sensor

If wiring is OK (but still no voltage across terminals 1 and 3):

– Replace Motronic ECM -J220-, see **Electrical Equipment section, Repair Group 97**.

EDITOR'S NOTE —
ECM replacement requires coding of control module using Volkswagen factory scan tool. See **Scan Tool Codes** *at the back of this manual or consult an authorized Volkswagen service department for additional information.*

Ignition coils with power output stage, checking

NOTE—

- *Ignition coils and power output stage are a common part*
- *Primary resistance of ignition coils cannot be measured.*
- *Secondary resistance should be tested only with ignition wires connected to ignition coils. To do this, measurements must be made between spark plug connectors of appropriate ignition coil. This measurement also measures the interference-suppression resistors of all ignition wires.*

Special tools and equipment

- VAG1598/31 test box
- VAG1526 multimeter or VAG1715 multimeter
- VAG1594 connector test kit
- VAG1527 voltage test lamp
- Wiring diagram

Test requirements

- Battery voltage must be at least 11.5 volts.
- Camshaft Position (CMP) sensors must be in order, testing, see **Camshaft Position (CMP) sensor, checking.**
- Engine speed (RPM) sensor must be in order, testing, see **Repair Group 24.**

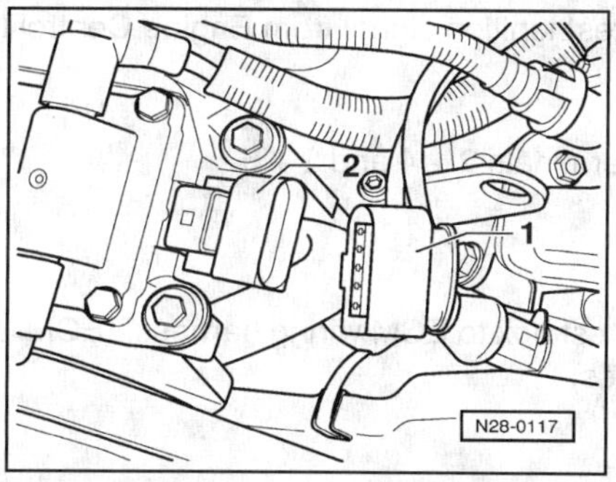

Ignition coils and ignition wires, checking

◄ Disconnect 5-way connector from ignition coils (N, N128, N158).

– Disconnect spark plug connectors.

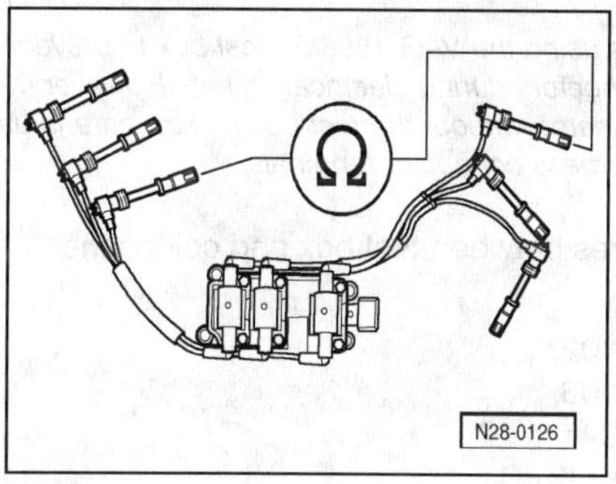

◄ Measure resistance between two spark plug connectors of an ignition coil.

- Specification: 16 - 27 kΩ

– Repeat test on other two ignition coils.

If specified values are not obtained:

– Disconnect ignition wires from ignition coils.

– Test resistance of ignition wires.

- Specification: 3 - 7 kΩ

– Test secondary resistance values of ignition coils.

- Specification: 8 - 14 kΩ

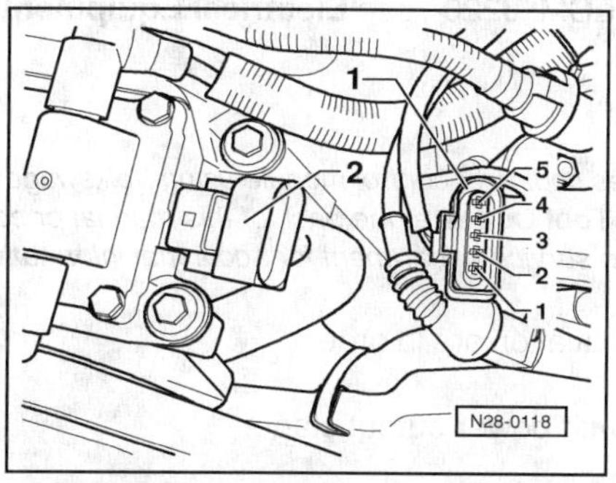

Ignition coil voltage supply, checking

◄ Pull 5-way connector (1) off output stage for ignition coils (2).

– Use auxiliary wires from VAG 1594 to connect multimeter between terminals 1 + 2 of the connector which has been pulled off to make voltage measurement.

– Operate starter.

- Specification: at least 11.5 V

If there is no voltage:

– Switch ignition off.

– Check for breaks in wire between 5-way connector terminal 2 and ground (see wiring diagram).
  - Wire resistance: max. 1.5 Ω

– Check for breaks in wire between 5-way connector terminal 1 and relay carrier (see wiring diagram).
  - Wire resistance: max. 1.5 Ω

NOTE —
Voltage is supplied to the ignition coils from the Fuel Pump (FP) relay through fuse 29.

Ignition coils, checking activation

– Disconnect connectors from fuel injectors.

– Use auxiliary wires from VAG 1594 to connect diode test lamp VAG 1527 to terminals 2 + 3 of connector.

– Switch on starter and test ignition signal from Engine Control Module.
  - LED should flicker.

– Repeat test between terminals 2 + 4 and 2 + 5.

If LED does not flicker:

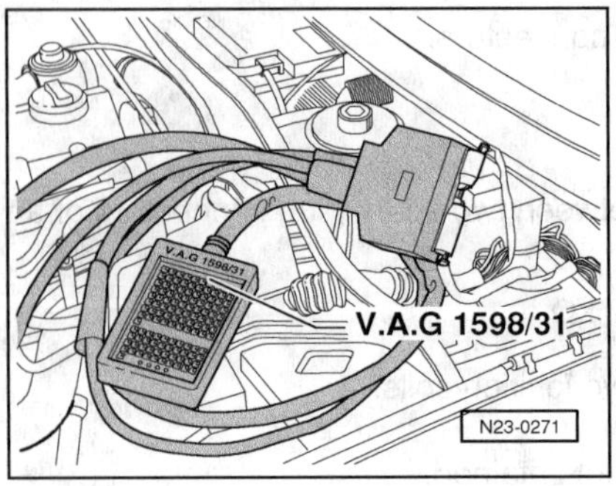

◄ Connect VAG1598/31 test box to ECM wiring harness. (ECM is not connected to wiring harness).

EDITOR'S NOTE —
Volkswagen recommends using the VAG 1598/31 test box to prevent damage to ECM harness connectors during electrical testing. However the harness connectors can be tested without the test box if extra care is used to prevent damage to the harness connector terminals.

– Check for breaks in wires between test box and coil connector (see wiring diagram).
  - Terminal 3 + socket 102
  - Terminal 4 + socket 103
  - Terminal 5 + socket 94
  - Wire resistance: max. 1.5 Ω

– Also check wires for shorts among one another.
  - Expected value: ∞

– If no errors are found in wires and there was voltage between terminals 1 + 2 replace Motronic ECM -J220-, see **Electrical Equipment section, Repair Group 97**.

EDITOR'S NOTE —
ECM replacement requires coding of control module using Volkswagen factory scan tool. See ***Scan Tool Codes*** *at the back of this manual or consult an authorized Volkswagen service department for additional information.*

If voltage supply and activation are in order:

– Replace ignition coils with power output stage.

Knock sensors, electrically checking

NOTE —

• To remove and install Knock Sensors, intake manifold must be removed.

• For the Knock Sensors to function properly, it is important for tightening torque to be exactly 20 Nm (15 ft-lb).

• When servicing terminals in harness connector of Knock Sensors, use only gold-plated terminals.

Special tools and equipment

- VAG1598/31 test box
- VAG1526 multimeter or VAG1715 multimeter
- VAG1594 connector test kit
- Wiring diagram

Test condition

- The On Board Diagnostic (OBD) must have identified a DTC on one or both knock sensors.

EDITOR'S NOTE —

Functional test of knock control and knock sensors is done using VAG 1551 diagnostic scan tool to read measuring value block, display groups 22 - 28.

Resistances and wiring, checking

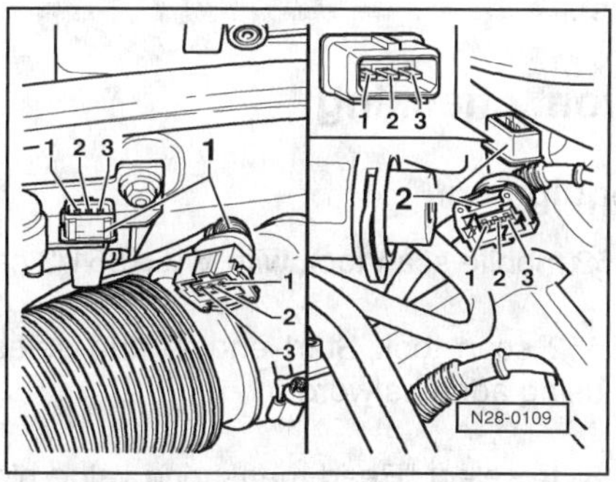

◄ Disconnect 3-pin harness connector going to Knock Sensor (KS) 1 (-G61-, bank 1) (1) and/or 3-pin harness connector going to Knock Sensor (KS) 2 (-G66-, bank 2) (2).

– Connect multimeter US1119 (Fluke 83 or equivalent) to measure resistance between connector terminals, using adapters from VW1594 connector test kit.

- Knock sensor terminals 1 and 2
- Knock sensor terminals 1 and 3
- Knock sensor terminals 2 and 3
- Specified value: ∞

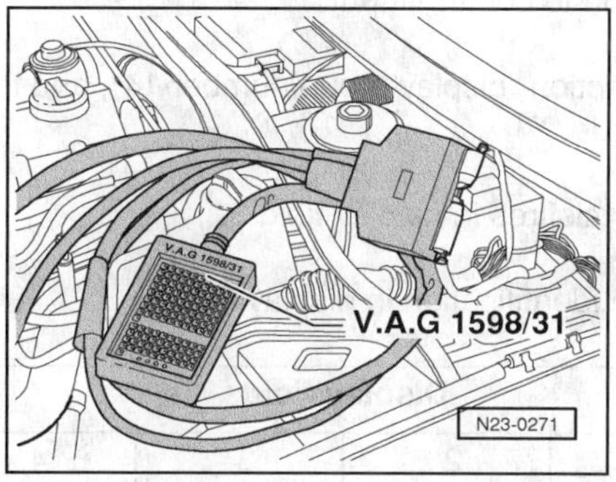

◄ Connect VAG1598/31 test box to control module wiring harness. Engine Control Module (ECM) is not connected.

EDITOR'S NOTE —

Volkswagen recommends using the VAG 1598/31 test box to prevent damage to ECM harness connectors during electrical testing. However the harness connectors can be tested without the test box if extra care is used to prevent damage to the harness connector terminals.

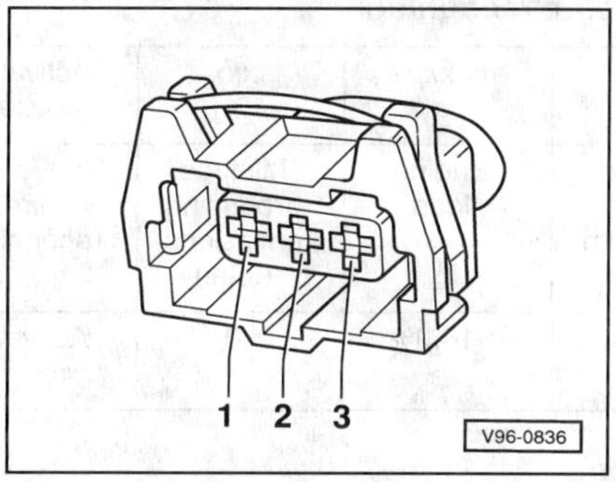

◄ Check for breaks in wires between test box and 3-way connector (see wiring diagram).

– Check wiring for open circuit between ECM/test box and 3-pin harness connector using appropriate wiring diagram.

Knock Sensor (KS) 1 -G61-:

- Connector terminal 1 to ECM/test box socket 106
- Connector terminal 2 to ECM/test box socket 99
- Connector terminal 3 to ECM/test box socket 108
- Specified value: max. 1.5 Ω

Knock Sensor (KS) 2 -G66-:

- Connector terminal 1 to ECM/test box socket 107
- Connector terminal 2 to ECM/test box socket 99
- Connector terminal 3 to ECM/test box socket 108
- Specified value: max. 1.5 Ω

– Also test lines at socket 108 for shorts to sockets 106 and 107.

- Expected value: ∞

If no malfunctions are found in wires:

– Loosen Knock Sensor (KS) and then retighten it to 20 Nm (15 ft-lb).

– Perform a road test. The following operating conditions must be obtained during road test:

- Coolant temperature must rise to above 80°C (176 °F).
- Once the temperature is reached, the following conditions must be reached several times each:
  idle, partial throttle, wide open throttle, deceleration mode
- Engine speed must be increased to above 3500 RPM during Wide Open Throttle (WOT).

– Check control module DTC memory again.

If DTC is still present:

– Replace knock sensor(s).

Misfire recognition, checking

Special tools and equipment

- V.A.G1551 or VAG1552 mobile scan tool, with VAG1551/3 cable

– Connect V.A.G 1551/1552 scan tool. Start engine and select Engine Control Module (ECM) using address word 01.

– Press buttons -0- and -8- to select "Read measuring value block" function 08. Press -Q- button to confirm input.

– Press 0, 1 and 4 for function "display group number 14", and then press Q to exit from input mode.

– Check total number of misfires in display field 3.

– Check status of misfire identification in display field 4.

	Display fields			
	1	2	3	4
Display group 14: Misfire recognition				
Display	xxx	xx.xx %	xxx	activated locked
Indicated	Engine speed (in 40rpm steps)	Engine load	Misfires producing emissions (total)	Status misfire recognition
Specified value	620 - 6800 rpm	>14.0%	0	Activated

If specified values are obtained:

- Press →button.
- Press buttons -0- and -6- to select "End Output" function 06. Press -Q- button to confirm input.
- Switch ignition off.

If specified values are not obtained:

- Change to display group 15 as follows:
- Press button -3- (V.A.G1551) or ↑button (VAG1552).
- Check misfires in display fields 1 - 3.
- Check status of misfire identification in display field 4.

	Display fields			
	1	2	3	4
Display group 15: Misfire recognition				
Display	xxx	xxx	xxx	activated locked
Indicated	Misfiring adversely affecting No. 1 Cyl.	Misfiring adversely affecting No. 2 Cyl.	Misfiring adversely affecting No. 3 Cyl.	Misfire recognition status
Specification	0	0	0	Activated
If specification not obtained, see **Evaluating display groups 15 and 16**				

If specified values are obtained:

- Change to display group 16 as follows:
- Press button -3- (V.A.G1551) or ↑button (VAG1552).
- Check misfires in display fields 1 - 3.
- Check status of misfire identification in display field 4.

	Display fields			
	1	2	3	4
Display group 16: Misfiring recognition				
Display	xxx	---	---	activated deact
Indicated	Misfiring adversely affecting No. 4 Cyl.	Misfiring adversely affecting No. 5 Cyl.	Misfiring adversely affecting No. 6 Cyl.	Misfire detection status
Specification	0	0	0	activated
If specification not obtained, see **Evaluating display groups 15 and 16**				

- Press →button.

– Press buttons -0- and -6- to select "End Output" function 06. Press -Q- button to confirm input.

– Switch ignition off.

Evaluation of display group 15 and 16

Indicated on display	Possible cause	Corrective action
greater than 0	Spark plug faulty Spark plug connector faulty Ignition coil faulty	-Test spark plugs, ignition wires and ignition coil
	Injector faulty	- Fuel injectors, checking, see **Repair Group 24**

4.0 Liter Fuel Injection and Ignition (code: BDP)

Section

K

24 Multiport Fuel Injection (MFI)

28 Ignition System

24 Multiport Fuel Injection (MFI)

Repair Group

K 24

Fuel Injection System, servicing

NOTE—

- *Fuel hoses in engine compartment must only be secured with spring type clips. The use of clamp or screw type clips is not permissible.*
- *Do not use sealants containing silicone. Particles of silicone drawn into the engine, will not be burned in the engine and will damage the oxygen sensors.*
- *The engine control module is equipped with On Board Diagnostic (OBD). Before performing repairs and fault finding the Diagnostic Trouble Code (DTC) memory must be checked. Also, the vacuum hoses and connections must be checked (unmetered air).*
- *For trouble-free operation of the electrical components, a voltage of at least 11.5 V is necessary.*
- *Disconnecting and connecting the battery must only be done with the ignition switched off, otherwise the engine control module could be damaged.*
- *During some checks it is possible that the control module will detect and store a fault. Check DTC memory and, if necessary, repair any faults and then erase DTC memory, see* **Scan Tool section**.

CAUTION—

- *Do not use any gasket sealers or other automotive chemical compounds that contain silicone. Traces of silicone drawn into the engine pass unburned into the exhaust and can damage the oxygen sensor.*
- *Do not use any electrical contact sprays or equivalent near oxygen sensor connections. Proper oxygen sensor function requires reference to outside air via the cable. If any contact spray reaches the oxygen sensor, it can cause damage or improper function.*

EDITOR'S NOTE—

Volkswagen identifies electrical components by a letter and/or a number in the electrical schematics. See **Wiring Diagrams, Fuses and Relays**. *These electrical identifiers are listed -between dashes- as an aid to electrical troubleshooting.*

EDITOR'S NOTE—

Functional testing of most fuel injection and ignition components is done using the VAS 5051 vehicle diagnosis, testing, and information system. Use of this system is beyond the scope of this manual.

Safety precautions

> ***WARNING —***
>
> *When doing any repair work, especially in the engine compartment, pay attention to the following due to the cramped conditions:*
>
> - *Route all the various lines (e.g. for fuel, hydraulics, EVAP system, coolant, refrigerant, brake fluid and vacuum pipes and hoses) and electrical wiring so that the original positions are restored.*
>
> - *Ensure sufficient clearance to all moving or hot components.*

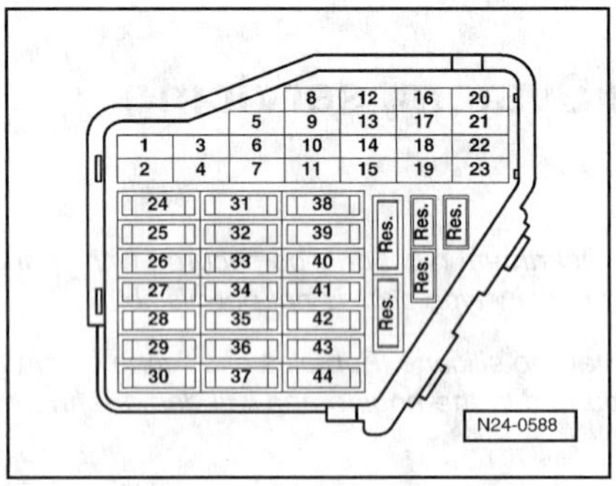

For reasons of safety, remove fuse No. 28 from the fuse holder before opening the fuel system. This is because the fuel pump can be activated by the driver's door contact switch.

> ***WARNING —***
>
> *Fuel system is under pressure! Before opening the system place a cloth around the connection. Then release pressure by carefully loosening the connection.*

To prevent injuries to persons and/or damage to the fuel injection and ignition system, the following must be noted:

- The ignition must be switched off before connecting or disconnecting injection or ignition system wiring or tester cables.
- If the engine is to be turned at starter speed, without starting, remove fuse No. 28.

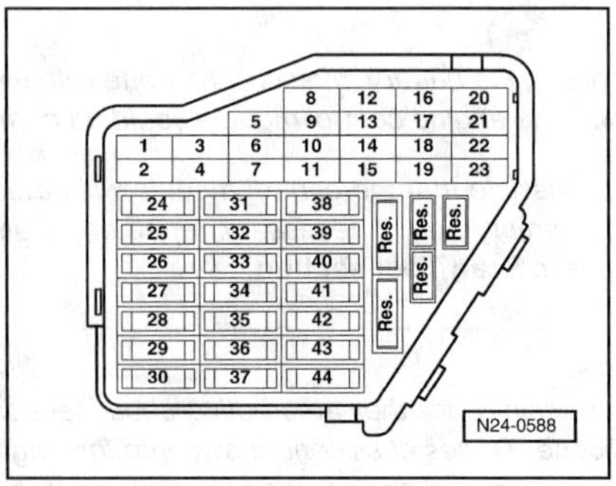

Removing fuse No. 32 from fuse holder

NOTE —

Removing fuse 32 interrupts the voltage supply to the injectors.

- Remove engine cover.
- Loosen coolant expansion tank and place to one side.
- Remove air cleaner upper part.
- Remove intake manifold side parts, see **Intake manifold, assembly**.
- Disconnect all 4-pin connectors from ignition coils.

NOTE —

Before disconnecting connector, mark which component that it belongs to.

Observe the following if test and measuring instruments are required during a test drive:

- Test and measuring instruments must be secured to the rear seat and operated by a 2nd person from this location.

If test and measuring instruments are operated from the front passenger's seat and the vehicle is involved in an accident, there is a possibility that the person sitting in this seat may receive serious injuries when the airbag is triggered.

Rules for cleanliness

When working on the fuel supply/injection system, pay careful attention to the following 5 rules:

- Thoroughly clean all unions and the adjacent areas before disconnecting.
- Place parts that have been removed on a clean surface and cover. Do not use fluffy cloths!
- Carefully cover opened components or seal, if the repair cannot be performed immediately.
- Only install clean components: Only unpack replacement parts immediately prior to installation. Do not use parts that have been stored loose (e.g. in tool boxes etc.).
- When the system is open: Do not work with compressed air if this can be avoided. Do not move the vehicle unless absolutely necessary.

Technical data

Engine code	**BDP**
Idling check Engine idle speed	 600 to 800 rpm (not adjustable)
Engine control module[1)] System Part number	 Motronic ME7.1.1 see factory parts information
Engine speed (RPM) limitation	from approx. 6800 rpm

[1)] Replacing engine control module, see **Electrical Equipment section, Repair Group 97**

Intake manifold, assembly

NOTE —

- *Removing and installing components of crankcase ventilation system, see **Repair Group 13, Crankcase ventilation system, components**.*
- *Cylinder bank 1 (bank 1) is located on right side in direction of travel, cylinder bank 2 (bank 2) on left side in direction of travel.*

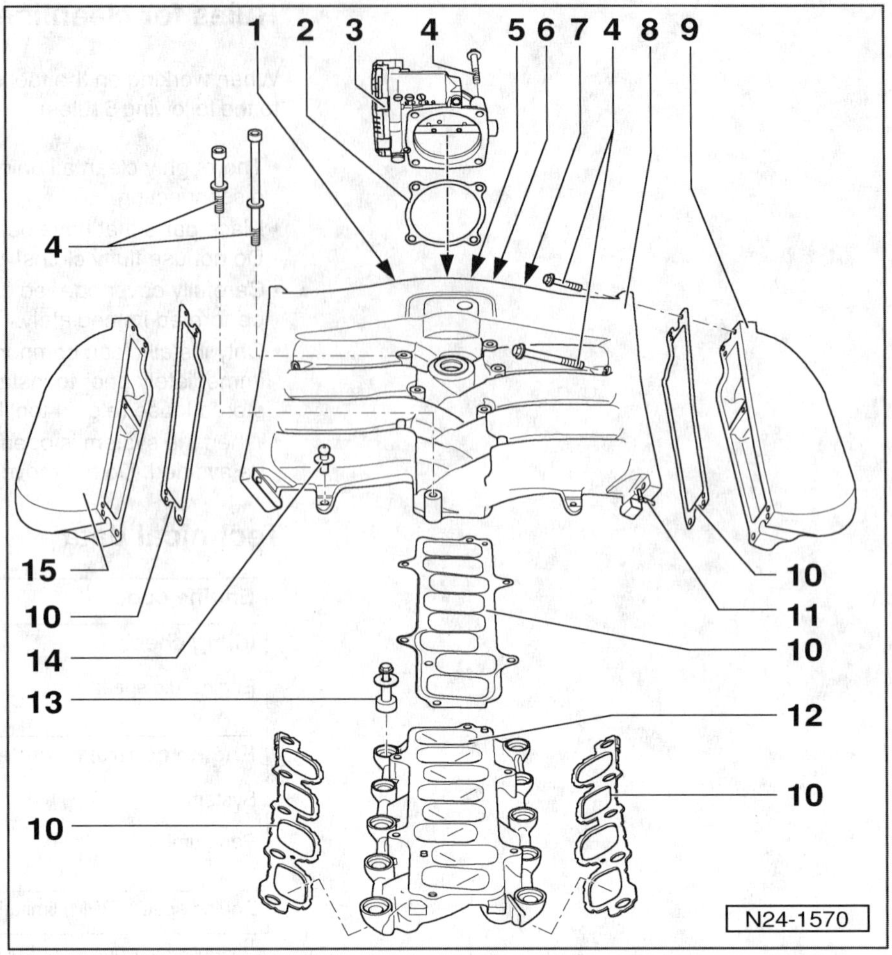

1. **Vacuum connection**
   - Below throttle valve control module
   - From Evaporative Emission (EVAP) Canister Purge Regulator Valve -N80-, see **Repair Group 20, EVAP system components**, item **-14-**
2. **Gasket**
   - Replace if damaged
3. **Throttle valve control module -J338-**
   - 6-pin connector
4. **8 Nm (71 in-lb)**
5. **Vacuum connection**
   - For Leak Detection Pump (LDP) -V144-, see **Repair Group 20, EVAP system components**, item **-5-**
6. **Vacuum connection**
   - See **Fuel rail with injectors, assembly**, item **-6-**
7. **Vacuum connection**
   - For brake servo
8. **Intake manifold upper section**
9. **Intake manifold - side section**
   - Cylinder bank 2
10. **Gasket**
    - Replace if damaged
11. **Vacuum connection**
    - For Secondary Air Injection (AIR) Solenoid Valve -N112- and Left Electro-Hydraulic Engine Mount Solenoid Valve -N144-/Right Electro-Hydraulic Engine Mount Solenoid Valve -N145-, see **Repair Group 26, Secondary air system components**, item **-5-**
12. **Intake manifold - lower section**
    - With two (2) locating pins (dowels) for aligning the intake manifold upper section to the intake manifold lower section.
    - Locating pins are not secured in lower manifold part. Use care to prevent pins from falling into the open intake manifold. Possible serious engine damage will result.
13. **9 Nm (80 in-lb)**
14. **Ball head pin**
    - Tighten to 3 Nm (27 in-lb)
    - For engine cover
15. **Intake manifold - side section**
    - Cylinder bank 1

Air cleaner, assembly

1. **Intake manifold**
2. **To throttle valve control module**
3. **6 Nm (53 in-lb)**
4. **Seal**
5. **Air cleaner upper part**
6. **Air duct**
   - Clipped in and screwed into air cleaner upper part
7. **Air ducts**
   - To lock carrier
8. **Filter element**
9. **Rubber grommet**
10. **Air cleaner lower part**
11. **10 Nm (7 ft-lb)**
12. **6 Nm (53 in-lb)**
13. **6 Nm (53 in-lb)**
14. **Retainer**
    - For retaining ring of Evaporative Emission (EVAP) Canister Purge Regulator Valve -N80-, see **Repair Group 20, EVAP system components**, item **-15-**
15. **Mass Air Flow (MAF) sensor -G70- with Intake Air Temperature (IAT) sensor -G42-**
    - 6-pin connector

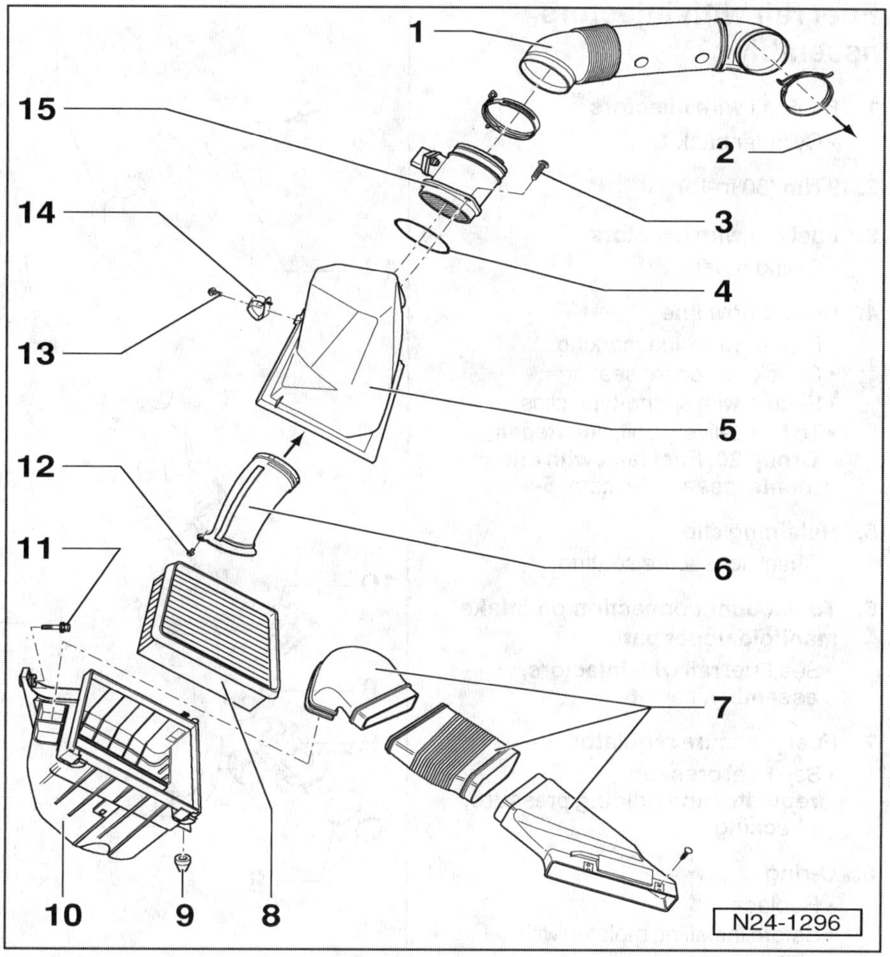

Fuel rail with injectors, assembly

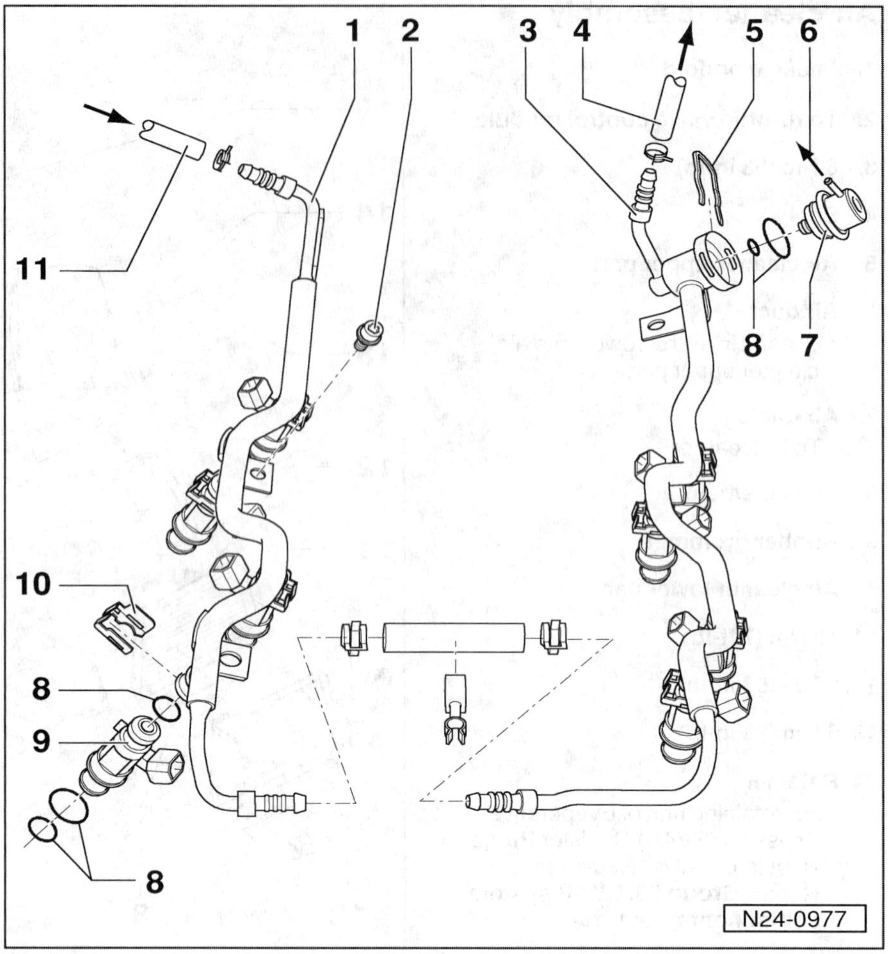

1. **Fuel rail with injectors**
   - Cylinder bank 1
2. **9 Nm (80 in-lb)**
3. **Fuel rail with injectors**
   - Cylinder bank 2
4. **Return flow line**
   - Blue or with blue marking
   - Check for secure seating
   - Secure with spring-type clips
   - To fuel delivery unit, see **Repair Group 20, Fuel tank with attachments, assembly**, item **-5-**
5. **Retaining clip**
   - Check for secure seating
6. **To vacuum connection on intake manifold upper part**
   - See **Fuel rail with injectors, assembly**, item **-6-**
7. **Fuel pressure regulator**
   - See **Fuel pressure regulator and holding pressure, checking**
8. **O-ring**
   - Replace
   - Before installing moisten with clean engine oil
9. **Injectors -N30- through -N33-, -N83- through -N86-***
   - Checking, see **Fuel pressure regulator and holding pressure, checking**
10. **Retaining clip**
    - Ensure seated correctly at injector and fuel rail
11. **Supply connection**
    - Black
    - Check for secure seating
    - Secure with spring-type clips
    - From fuel filter, see **Repair Group 20, Fuel tank with attachments, assembly**, item **-8-**

Fuel injection components, checking

Injectors, checking for leaks and quantity injected

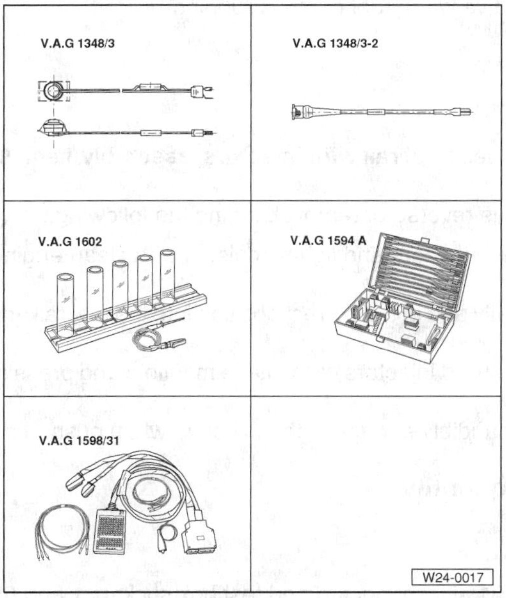

Special tools and equipment

- V.A.G 1348/3 Remote control
- V.A.G 1348/3-2 Adapter
- V.A.G 1602 Calibration tester
- V.A.G 1594 A Adapter set
- V.A.G 1598/31 Test box

Checking conditions

- Fuel pressure must be in correct; checking, see **Fuel pressure regulator and holding pressure, checking**.

> ***WARNING —***
> *Fuel system is under pressure! Before opening the system place a cloth around the connection. Then release pressure by carefully loosening the connection.*

Test sequence

– Remove intake manifold upper part (along with both intake manifold side parts), see **Intake manifold, assembly**.

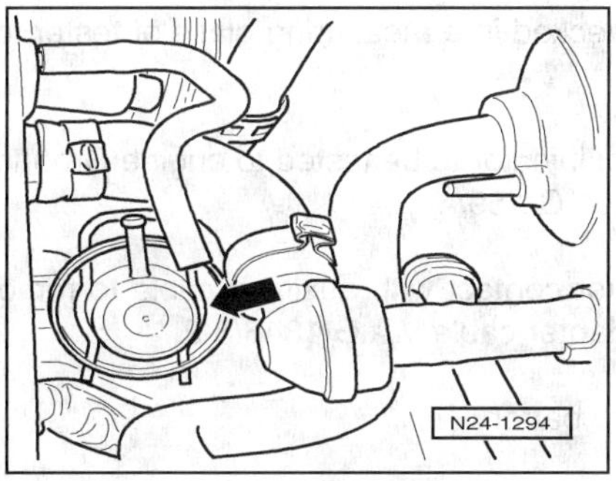

◀ Remove vacuum hose from fuel pressure regulator -arrow-.

– Disconnect connectors from injectors.

– Remove fuel rail securing bolts.

– Lift fuel rail with injectors off intake manifold and support.

– Remove wiper arms and cowl panel, see **Electrical Equipment section, Repair Group 92**

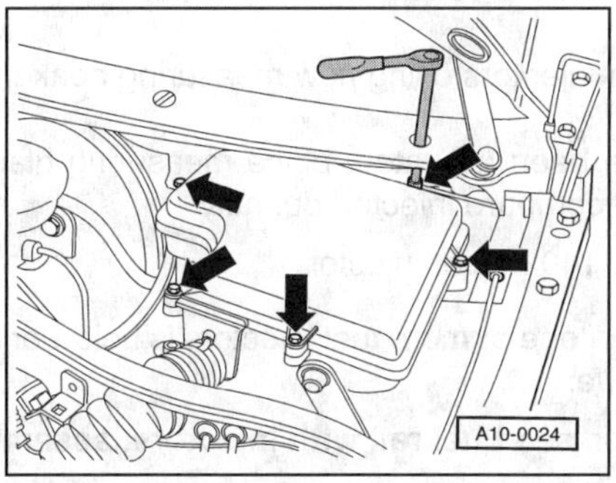

◀ Remove cover of protective housing for control modules.

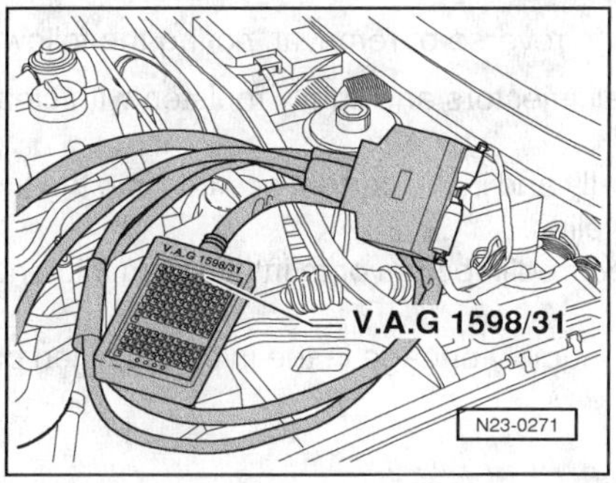

◀ Connect test box V.A.G 1598/31 to control module wiring harness. Engine control module remains disconnected.

– Bridge test box sockets 1 and 65 using adapter cables from adapter set V.A.G 1594.

> ***NOTE —***
> *This step is needed only to enable the fuel pump to run while the engine is shut off.*

– Switch ignition on.

Checking for leaks

– Check injector for leaks (visual check). Only 1 to 2 drops per minute must emit from each valve when fuel pump is running.

– If the fuel loss is greater:
  • Switch ignition off.

– Replace faulty injector, see **Fuel rail with injectors, assembly**, item **-9-**.

– Installation of injectors is reverse of removal, noting the following:
  • Replace O-rings on all injectors and lightly moisten with clean engine oil.
  • Insert injectors vertically and in the correct position into the fuel rail and secure with retaining clips.
  • Install fuel rail with secured injectors onto intake manifold and press in evenly.
  • The connectors must audibly engage in the injectors when pushed on.

Checking injected quantity

Test conditions

• Test box V.A.G 1598/31 still connected and test box sockets 1 and 65 bridged with adapter cable from adapter set V.A.G 1594.

Test sequence

– Insert injector to be checked in a measuring glass of tester for injected quantity V.A.G 1602.

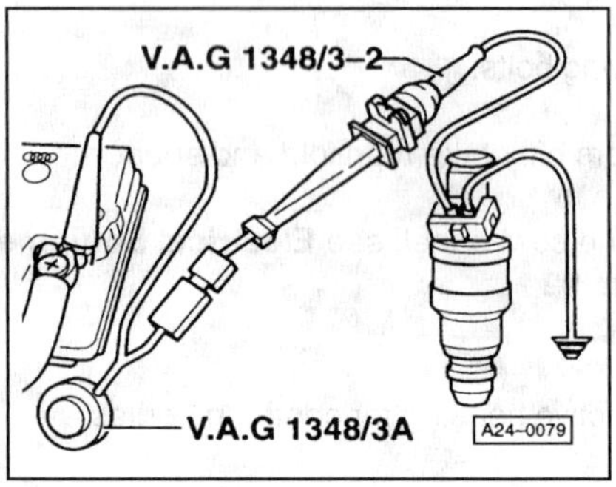

◄ Connect one contact of injector to be tested to engine Ground using auxiliary cables from V.A.G 1594.

– Connect second injector contact with auxiliary cable to remote control V.A.G 1348/3 using adapter cable V.A.G 1348/3-2.

– Connect alligator clip to battery (+).

– Operate remote control V.A.G 1348/3 for 30 seconds.

– Repeat check on other injectors using new measuring beakers.

– After all injectors have been activated, place measuring glasses on a horizontal surface and compare injected quantity.
  • Specification: 128 to 140 ml per injector

– If the measured value of one or more fuel injectors is below or above the indicated specified value:
  • Replace faulty injector, see **Fuel rail with injectors, assembly**, item **-9-**.

– Installation of injectors is reverse of removal, noting the following:
  • Replace O-rings on all injectors and lightly moisten with clean engine oil.
  • Insert injectors vertically and in the correct position into the fuel rail and secure with retaining clips.
  • Install fuel rail with secured injectors onto intake manifold and press in evenly.
  • The connectors must audibly engage in the injectors when pushed on.

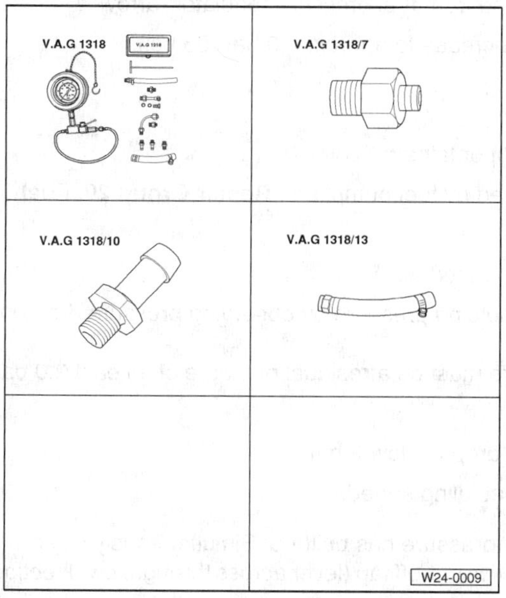

Fuel pressure regulator and holding pressure, checking

Function

The fuel pressure regulator controls the fuel pressure dependent from intake manifold pressure.

Special tools and equipment

- V.A.G 1318 Pressure gauge
- V.A.G 1318/7 Adapter
- V.A.G 1318/10 Adapter
- V.A.G 1318/13 Adapter

Not illustrated:

- VAS 5024 A Installation tool for spring-type clips

Test sequence

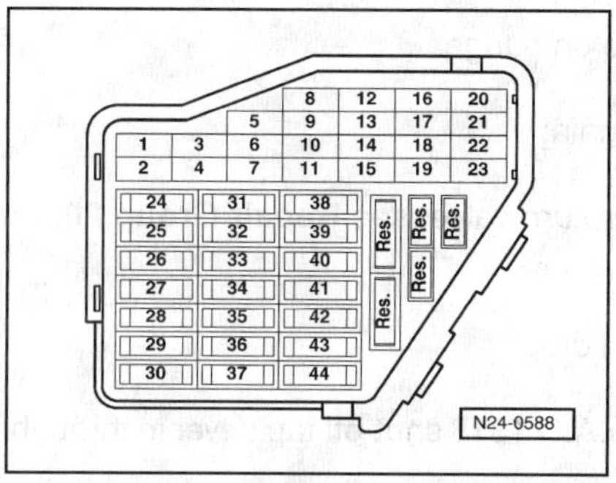

◀ Remove fuel pump fuse No. 28 from fuse holder.

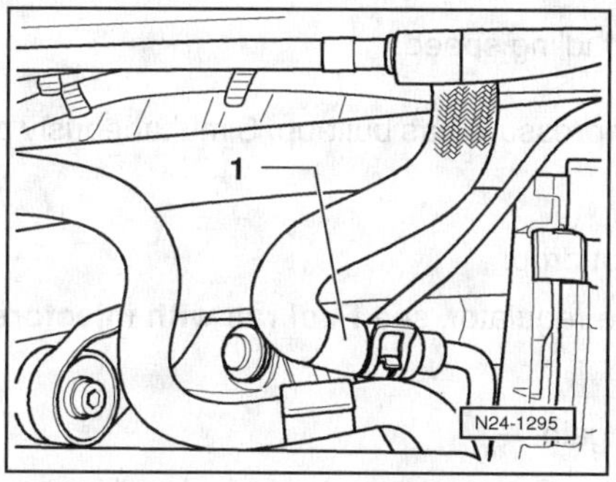

◀ Disconnect supply line/hose -1- and collect leaking fuel with a cloth.

> ***WARNING —***
> *Fuel supply lines are under pressure! Before removing from hose connection wrap a cloth around the connection. Then release pressure by carefully pulling hose off connection.*

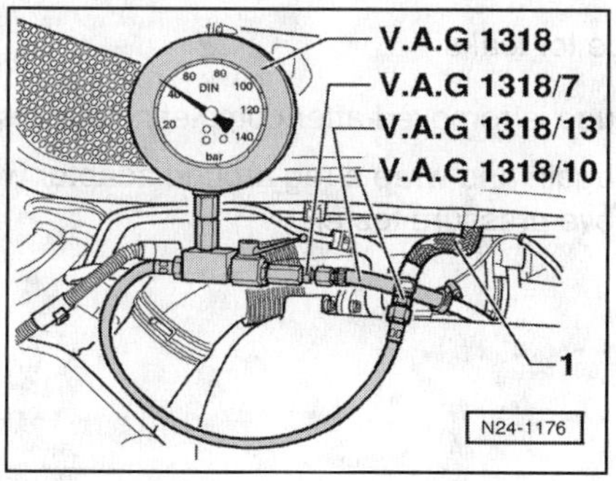

◀ Connect pressure gauge V.A.G 1318 to fuel supply line/hose and to fuel rail using adapters 1318/7, 1318/10 and 1318/13.

– Open shut-off tap on pressure gauge. Lever points in direction of flow.

– Reinstall fuse 28 (fuel pump) in fuse holder.

– Start engine and run at idling speed.

– Measure fuel pressure.

- Specification: Approx. 3.5 bar (50 psi)

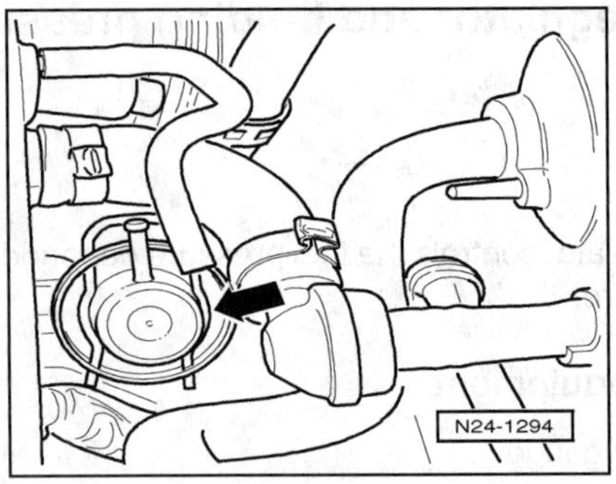

- Disconnect vacuum hose from fuel pressure regulator -arrow- .
  - Fuel pressure must increase to approx. 4.0 bar (58 psi).
- Switch ignition off.
- If the specification is not obtained:
  - Check quantity supplied by fuel pump, see **Repair Group 20, Fuel pump, checking**.
- If the specification is obtained:
  - Check for leaks and holding pressure by observing pressure drop on gauge.
  - After 10 minutes there must be a residual pressure of at least 2.0 bar (29 psi).
- If the holding pressure drops below 2 bar:
  - Start engine and run at idling speed.

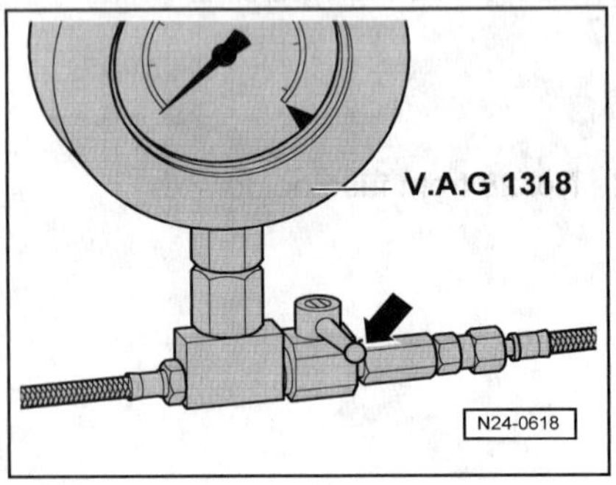

- Switch ignition off after pressure has built-up. Simultaneously close pressure measured device cut-off tap (lever across throughflow direction -arrow-).
- Observe pressure drop on gauge.

If the pressure drops again:

- Check fuel pump non-return valve, see **Repair Group 20, Fuel pump, checking**.

If the pressure does not drop:

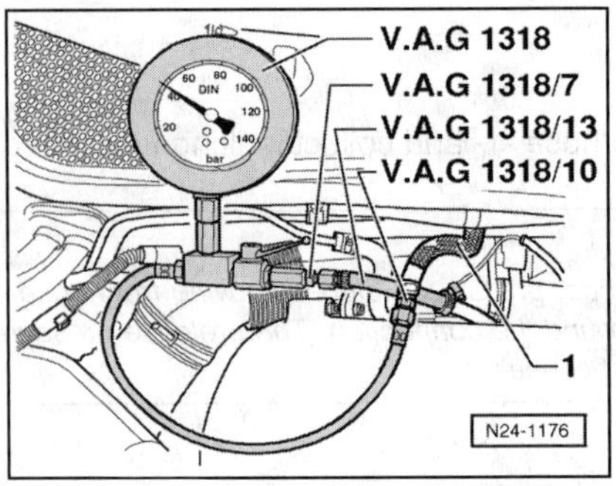

- Open pressure tester V.A.G 1318 shut-off tap (lever in throughflow direction).
- Start engine and run at idling speed.
- Switch ignition off after pressure has built-up. Simultaneously pinch/seal return hose together.
- If the pressure does not drop:
  - Replace fuel pressure regulator, see **Fuel rail with injectors, assembly,** item **-7-**.
- If the pressure drops again:
  - Check line connections, O-rings on fuel manifold and injectors for leaks.
  - Check pressure gauge for leaks.
- When the pressure gauge is removed after completing the test:
  - Remove fuse No. 28 again and wrap a rag around adapter which is to be removed and remove pressure tester.

28 Ignition

Repair Group

K 28

Ignition System, servicing

NOTE—

• The engine control module is equipped with self diagnosis.

• For trouble-free operation of the electrical components a voltage of at least 11.5 V is necessary.

• Disconnecting and connecting the battery must only be done with the ignition switched off, otherwise the engine control module could be damaged.

*• During some checks it is possible that the control module will detect and store a fault. Check fault memory and, if necessary, repair any faults and then erase fault memory, see **Scan Tool section**.*

Safety precautions

To prevent injuries to persons and/or damage to the fuel injection and ignition system, the following must be noted:

- The ignition must be switched off before connecting or disconnecting injection or ignition system wiring or tester cables.
- If the engine is to be turned at starter speed, without starting:

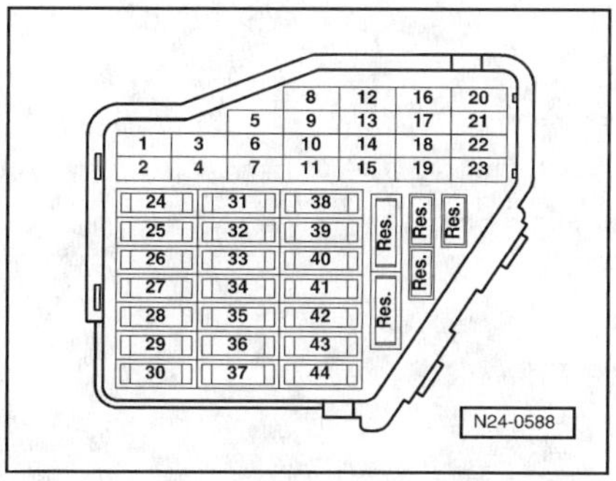

Removing fuse No. 32 from fuse holder

NOTE—

Removing fuse 32 interrupts the voltage supply to the injectors.

- Remove engine cover.
- Loosen coolant expansion tank and place to one side.
- Remove air cleaner upper part.
- Remove intake manifold side parts, see **Repair Group 24, Intake manifold, assembly**.
- Disconnect all 4-pin connectors from ignition coils.

NOTE—

Before disconnecting a connector, mark the component to which it belongs.

– Observe the following if test and measuring instruments are required during a test drive:

- Test and measuring instruments must be secured to the rear seat and operated by a 2nd person from this location.

If test and measuring instruments are operated from the front passenger's seat and the vehicle is involved in an accident, there is a possibility that the person sitting in this seat may receive serious injuries when the airbag is triggered.

Technical data, spark plugs

Engine code	BDP
Firing order	1-5-2-6-4-8-3-7
Spark plugs[1)] VW Manufacturer's designation	 101 000 062 AA BKR 6 EKUE
Electrode gap	max. 0.8 mm (0.031 in.)
Tightening torque	30 Nm (22 ft-lb)

[1)] Remove and install spark plugs with plug spanner 3122 B

Ignition system components

NOTE—

Cylinder bank 1 (bank 1) is located on right-hand side and cylinder bank 2 (bank 2) is located on left hand side when looking toward the front.

1. **Connector**
   - Black, 4 pin
   - Before removing, mark connector belonging to component
2. **Hood**
   - For Camshaft Position (CMP) sensor
3. **Camshaft Position (CMP) sensor**
   - -G40- = Bank 1 intake camshaft
   - -G163- = Bank 1 exhaust camshaft
   - -G300- = Bank 2 intake camshaft
   - -G301- = Bank 2 exhaust camshaft
   - Before removing, mark connector belonging to component
4. **10 Nm (7 ft-lb)**
5. **20 Nm (15 ft-lb)**
   - Tightening torque influences function of knock sensor
6. **Knock sensors**
   - Location, see **Fig. 1, 2** and **3**
   - Contact surfaces between knock sensor and cylinder block must be free of corrosion, dirt and grease
7. **3-pin connector**
   - Contacts gold plated
   - Black for Knock sensor (KS) 1 -G61-
   - Brown for Knock sensor (KS) 2 -G66-
   - Orange for Knock sensor 3 -G198-
   - Gray for Knock sensor 4 -G199-
8. **Spark plug**
   - Tighten to 30 Nm (22 ft-lb)
   - Remove and install with spark plug spanner 3122B
   - Type and electrode gap, see **Maintenance**
9. **Ignition coil with power output stage (-N70-, -N127-, -N291-, -N292-, -N323-, -N324-, -N325-, -N326-)**
   - Remove with puller T10095

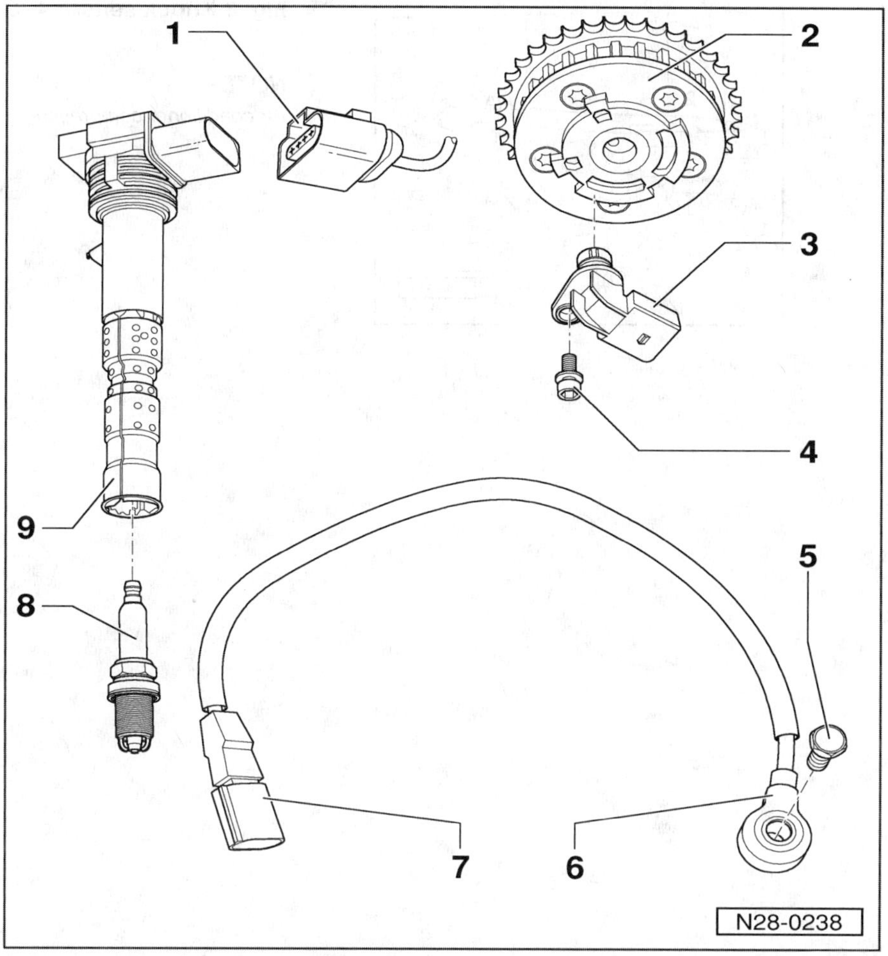

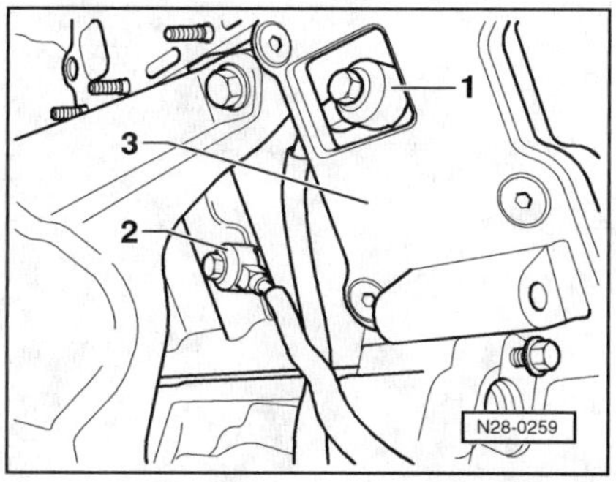

Fig. 1 Knock sensor (KS) 1 -G61- (black) -1- for bank 1, cylinders 1 + 2 and Knock sensor (KS) 2 -G66- (brown) -2- for bank 1, cylinders 3 + 4

NOTE—

Alternator mount -3- shown with alternator removed.

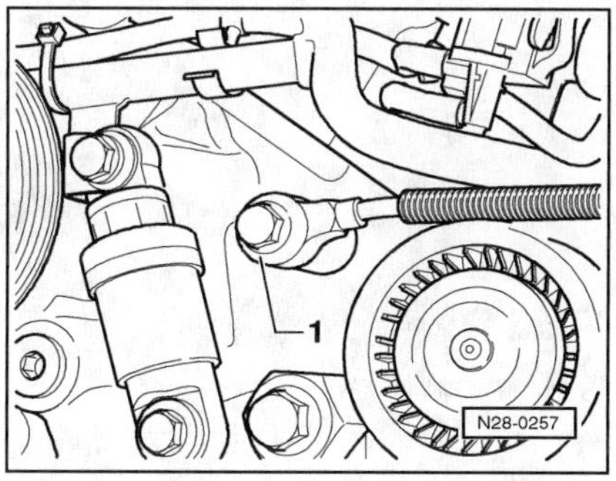

Fig. 2 Knock sensor 3 -G198- (orange) -1- for bank 2, cylinders 5 + 6

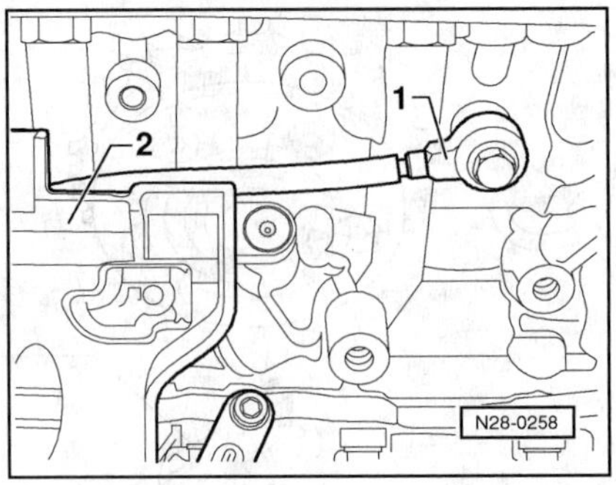

Fig. 3 Knock sensor 4 -G199- (gray) -1- for bank 2, cylinders 7 + 8

NOTE —
Air conditioner compressor mount -2- shown with compressor removed.

5-Speed Manual Transmission 012/01W

Section

M

00 General Information

30 Clutch

34 Controls, Housing

39 Final Drive, Differential

00 General Information

Repair Group
M 00

Transmission identification

The manual transmission 012 / 01W is installed in the Passat 1998 ➤as a 5 speed transmission in conjunction with the 4-Cyl. and 6-Cyl. engines.

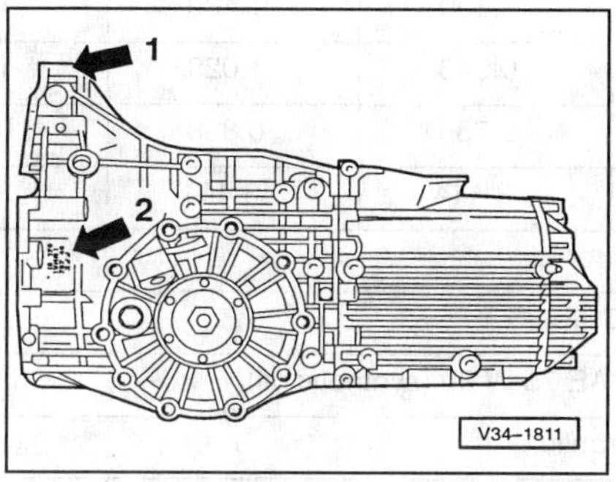

◄ Location on transmission

Code letters and date of manufacture (arrow 1)
Manual transmission 012 / 01W (-arrow 2-)

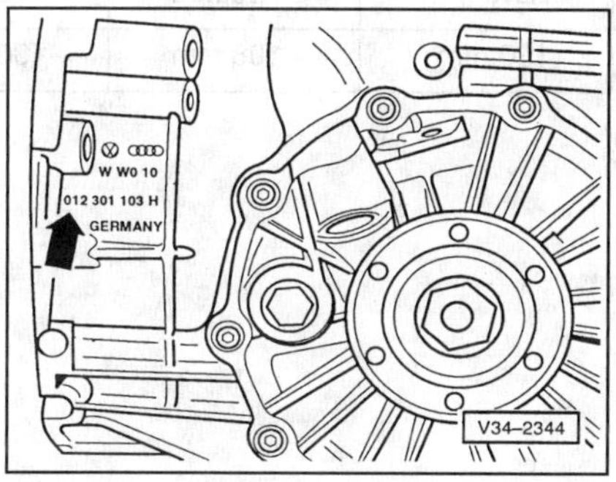

◄ Manual transmission 012 / 01W (-arrow 2-)

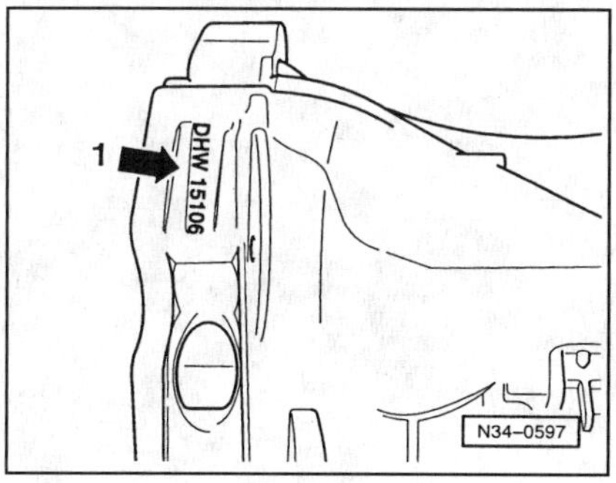

◄ Code letters and date of manufacture of transmission (-arrow 1-)

Transmission code letters:	Day	Month	Year
DHW	15	10	6
	Date (e.g. 15 OCT. 1996)		

Additional data depends on manufacture.

***NOTE*—**
The transmission code letters are also shown on the vehicle identification plates.

Code letters, transmission application, ratios, capacities

Transmission 012/01W front wheel drive Code letters		DHW	DHY	DVX	DVZ	EHV	EZG
Manufactured	from to	09/97 08/98	10/97 08/98	08/98 08/99	08/98 08/00	05/00 08/00	08/99 05/00
Application	Model	Passat 1998 ➤	Passat 1998 ➤	Passat 1998 ➤	Passat 1998 ➤	Passat 1998 ➤	Passat 1998 ➤
	Engine	1.8 ltr. - 127 kW	2.8 ltr. - 142 kW	1.8 ltr. - 127 kW	2.8 ltr. - 142 kW	1.8 ltr. - 127 kW	1.8 ltr. - 127 kW
Ratio:	Final drive	3.700	3.700	3.700	3.700	3.700	3.700
	1st Gear	3.500	3.500	3.500	3.500	3.778	3.778
	2nd Gear	2.118	1.944	2.118	1.944	2.176	2.176
	3rd Gear	1.300	1.300	1.300	1.300	1.429	1.429
	4th Gear	1.029	0.943	1.029	0.943	1.029	1.029
	5th Gear	0.838	0.789	0.838	0.789	0.838	0.838
	Reverse Gear	3.444	3.444	3.444	3.444	3.444	3.444
Speedometer		electronic					
Capacity		2.25 liter					
Specification		Gear oil G 052 911 A 1 SAE 75 W 90 (synthetic oil)					
Clutch control		hydraulic					
Clutch plate diameter		228 mm	240 mm	228 mm	240 mm	228 mm	228 mm
Drive axle flange diameter		108 mm	130 mm [1)]	108 mm	130 mm [1)]	108 mm	130 mm [1)]

[1)] Drive axle with triple rotor joint.

Clutch

Clutch system, bleeding

EDITOR'S NOTE —
Brake fluid is used for the clutch hydraulic system.

Special tools and misc. equipment required

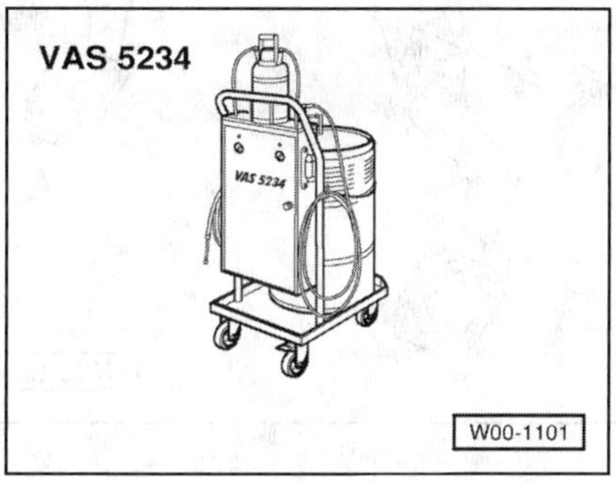

- VAS 5234 Brake filler and bleeder unit

– Brake fluid specifications, see **Brake System section, Repair Group 47.**

NOTE —
- *Open bleeder valve before operating bleeder unit.*
- *Clutch pedal is not operated.*

– Connect brake filler and bleeder unit VAS 5234.

– Remove insulation tray below engine/transmission, if installed.

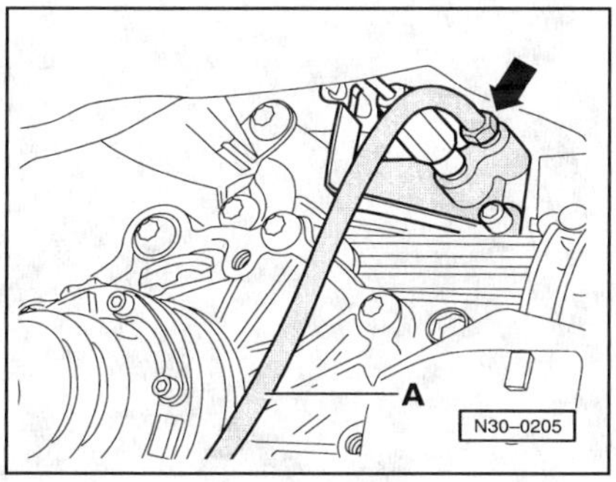

- Connect bleed hose -A- to slave cylinder (-arrow-) and open bleed valve.

– Now switch on bleeder unit and let about 100 cm^3 (3.4 oz.) brake fluid run out.

NOTE —
Brake fluid must not be allowed to get onto the transmission.

– Depress clutch pedal several times after completion of bleeding process.

Pedal cluster, assembly overview

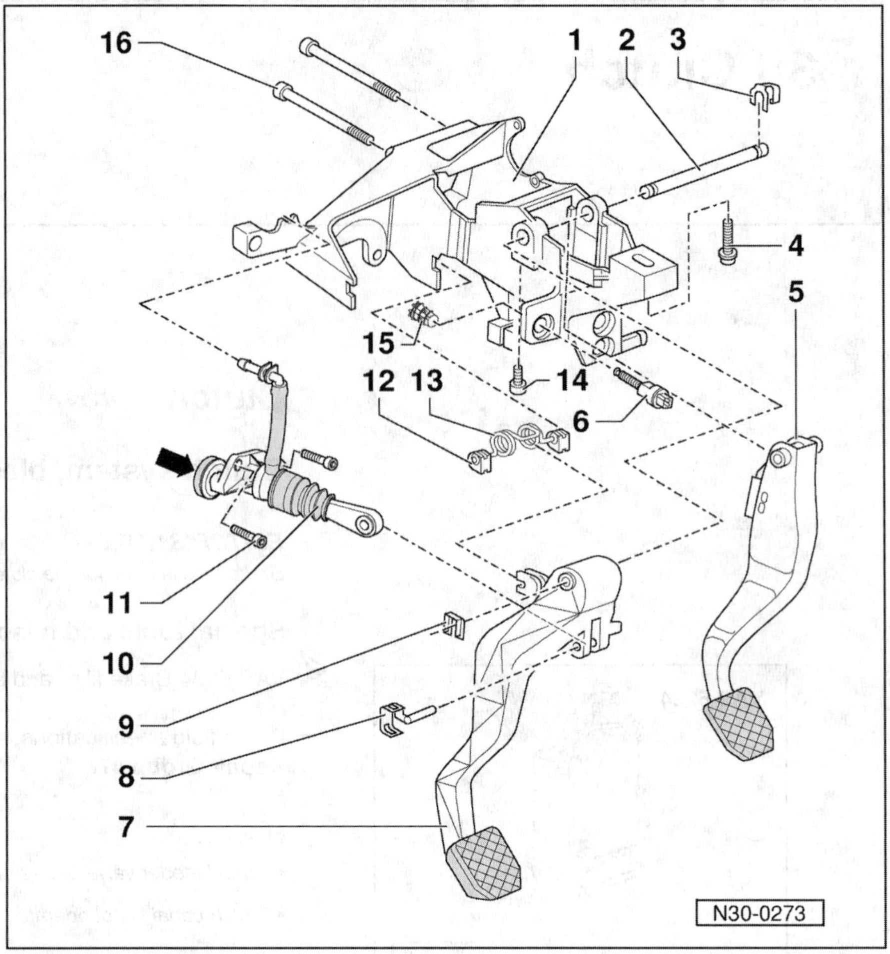

1. **Mounting bracket**
   - Before removing, detach steering column from steering gear, see **Suspension, Wheels, Steering section, Repair Group 48**
2. **Pivot pin for clutch and brake pedal**
3. **Locking clip**
4. **Hex socket head bolt**
   - Tighten to 25 Nm (18 ft-lb)
5. **Brake pedal**
   - Maintenance-free bearing; do not lubricate
6. **Clutch pedal switch**
   - Checking setting, see **Master cylinder, removing and installing**
7. **Clutch pedal**
   - Removing and installing, see **Clutch pedal and over-center spring, removing and installing**
8. **Pin**
   - Always replace
9. **Locking clip**
10. **Master cylinder**
    - Removing and installing, see **Master cylinder, removing and installing**
    - Seal -arrow- must not be removed
11. **Hex socket head bolt,**
    - Tighten to 20 Nm (15 ft-lb)
12. **Mounting**
    - Insert in mounting bracket with over-center spring
    - Installation position: deep groove faces downward
    - Allocate using parts catalog
13. **Over-center spring**
    - Allocate using parts catalog
    - Remove and install with clutch pedal, see **Clutch pedal and over-center spring, removing and installing**
14. **Hex socket head bolt**
    - Tighten to 5 Nm (44 in-lb)
    - To secure clutch and brake pedal pivot pin
    - Tighten only after locking clips -**3**- and -**9**- are in position
15. **Securing clip**
    - Holds clutch switch in adjusted position
16. **Torx screw**
    - Tighten to 25 Nm (18 ft-lb)
    - Also secures the brake master cylinder with brake servo

Clutch pedal and over-center spring, removing and installing

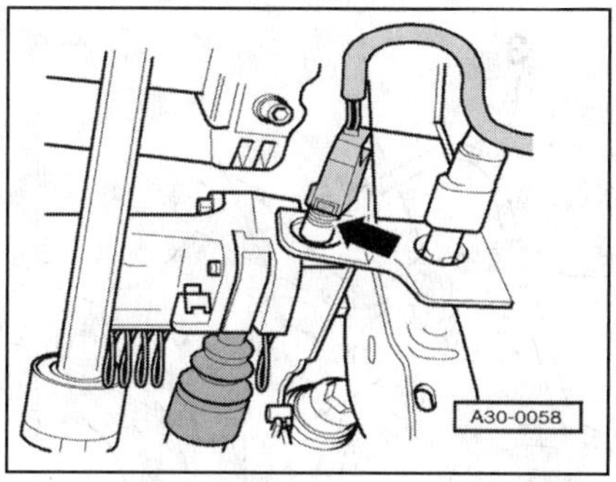

Removing

– Remove cover below instrument panel, see **Body-Interior section, Repair Group 70.**

◄ Make sure that the clutch pedal switch -arrow- is not pressed out of the securing clip. The thread of the switch will be damaged and the switch must be replaced.

NOTE —

In order to guarantee sufficiently secure seating, you can install the clutch pedal switch -arrow- one time only.

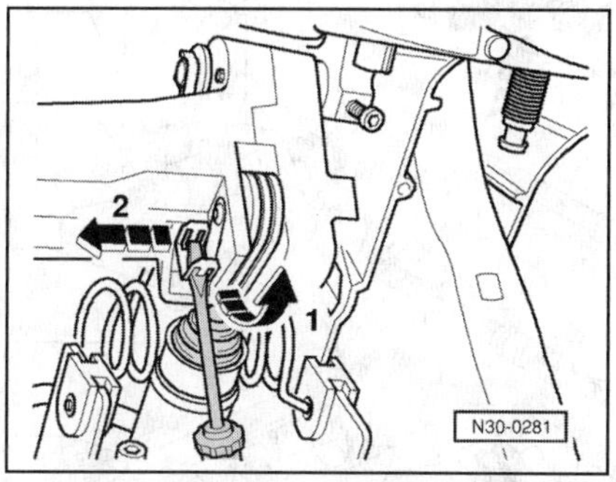

◄ Detach clutch pedal from master cylinder by prying out securing clip for pin with screw driver -arrow 1- and pulling out pin -arrow 2-.

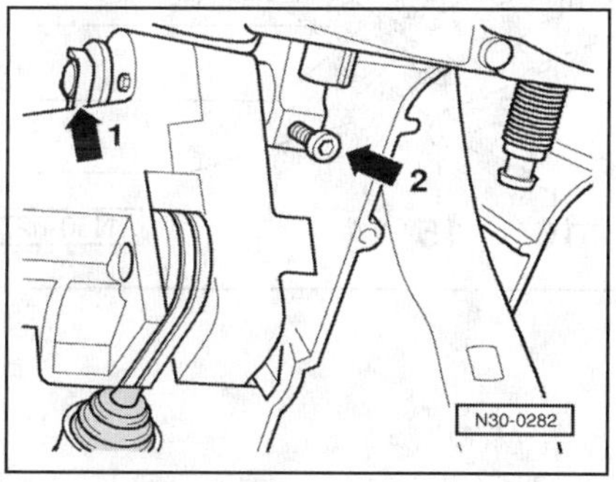

◄ Press clutch pedal securing clip -arrow 1- off pivot pin with a screwdriver.

– Remove plug -arrow 2-.

– Press clutch and brake pedal pivot pin out to the right, until the clutch pedal can be removed.

– If necessary press brake pedal securing clip off pivot pin with a screwdriver.

– Take out clutch pedal and over-center spring.

Installing

Installation is carried out in the reverse sequence, when doing this note the following:

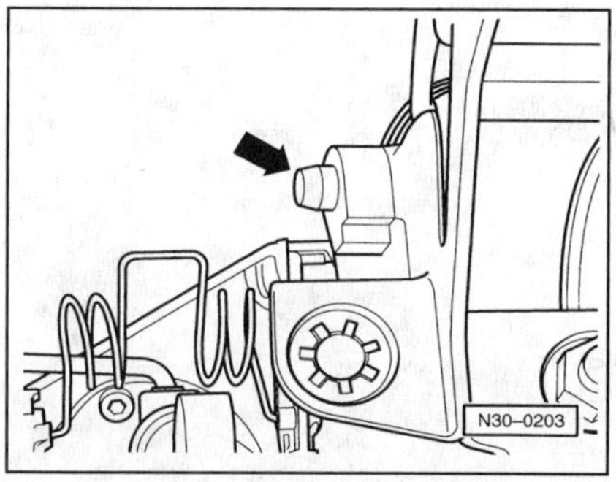

◄ Align the clutch pedal/brake pedal pivot pin so that it protrudes out of mounting bracket at the clutch pedal side (arrow).

– Press clutch pedal onto stop and insert over-center spring in its seat on clutch pedal.

– Now put clutch pedal on the pivot pin and push it to the left until securing clip can be attached.

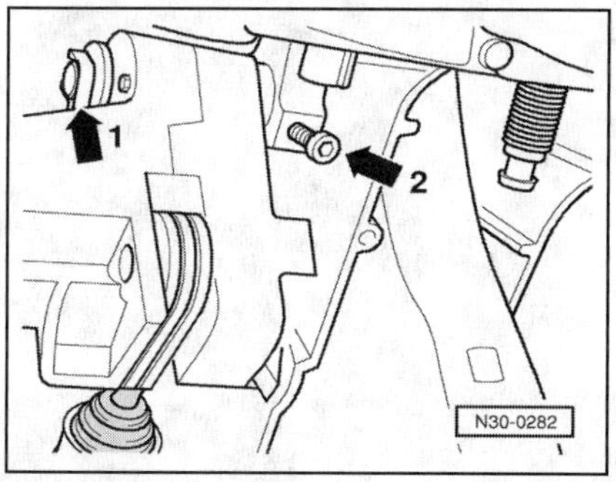

– Connect master cylinder to clutch pedal and engage securing clip for pin in clutch pedal.

◄ Do not tighten bolt -arrow 2- until both securing clips (-arrow 1- for clutch pedal; clip for brake pedal not visible) have been installed on the clutch pedal/brake pedal pivot pin.

– Adjust clutch pedal switch above clutch pedal, see **Master cylinder, removing and installing**.

Hydraulic system, assembly overview

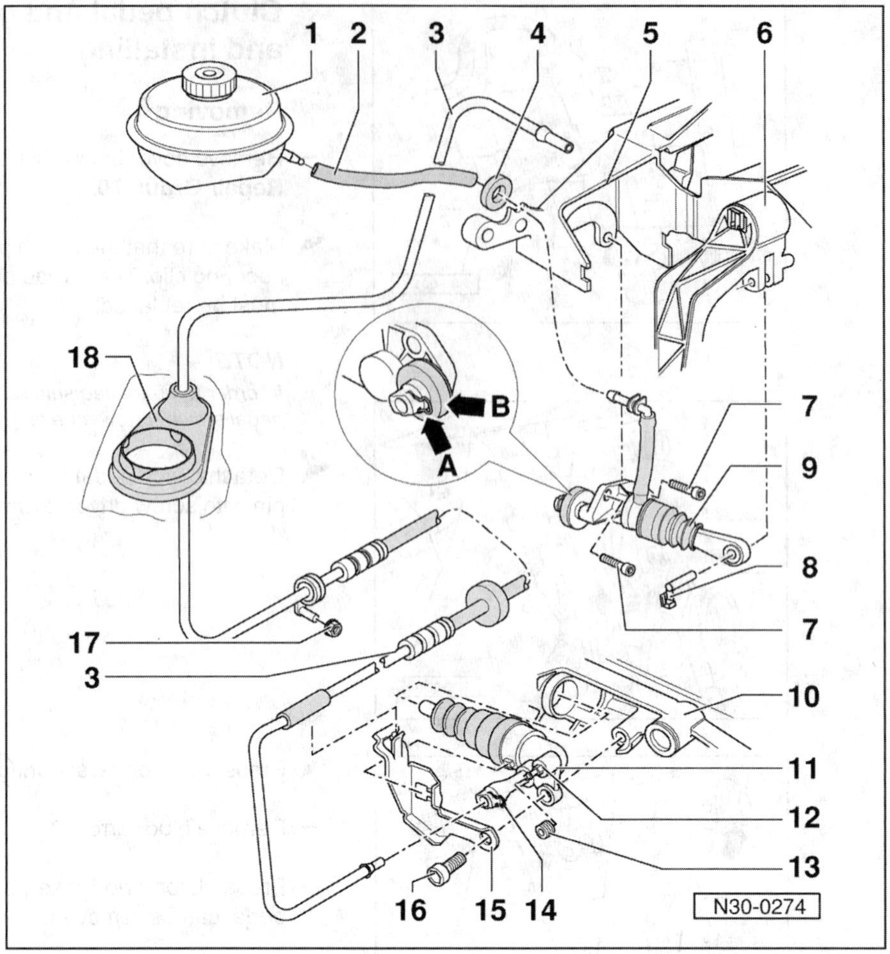

1. **Brake fluid reservoir**
2. **Supply hose**
3. **Line/hose assembly**
   - For connectors on master and slave cylinders
   - Separating and attaching to master cylinder, see **Master cylinder, removing and installing**
   - Separating from slave cylinder, see **Fig. 2**
   - Allocate in parts catalog
4. **Bush**
5. **Mounting bracket**
6. **Clutch pedal**
   - Removing and installing, see **Clutch pedal and over-center spring, removing and installing**
7. **Hex socket head bolt**
   - Tighten to 20 Nm (15 ft-lb)
8. **Pin**
   - Always replace
   - Lubricate with polyurea grease
9. **Master cylinder**
   - With connectors for line/hose assembly
   - Removing and installing, see **Master cylinder, removing and installing**
   - Clip (-arrow A-) and seal (-arrow B-) must not be dismantled
   - To remove line/hose assembly (-arrow A-) pull out to stop, see **Master cylinder, removing and installing**
10. **Transmission**
11. **Slave cylinder**
    - Do not operate clutch pedal after slave cylinder has been removed.
    - Brake fluid must not be allowed to get onto the transmission
    - To install, lubricate lightly, see **Fig. 1**
    - When installing, push in until the securing bolt **-16-** can be installed.
    - With connectors for line/hose assembly
    - Separate from line/hose assembly, see **Fig. 2**
    - Removing and installing, see **Clutch release mechanism, servicing**
    - Bleeding clutch system, see **Clutch system, bleeding**
12. **Bleeder valve**
    - Tighten to 4 Nm (35 in-lb)
    - Bleeding clutch system, see **Clutch system, bleeding**
13. **Dust cap**
14. **Clip**
    - To remove hose/line assembly, pull out clip (arrow) to stop, see **Fig. 2**
15. **Bracket**
    - For line/hose assembly
    - Secured to transmission with slave cylinder
16. **Hex socket head bolt**
    - Tighten to 20 Nm (15 ft-lb)
    - Slave cylinder on transmission
    - Self-locking
    - Always replace
17. **Hex nut**
    - Tighten to 2 Nm (18 in-lb)
18. **Seal**

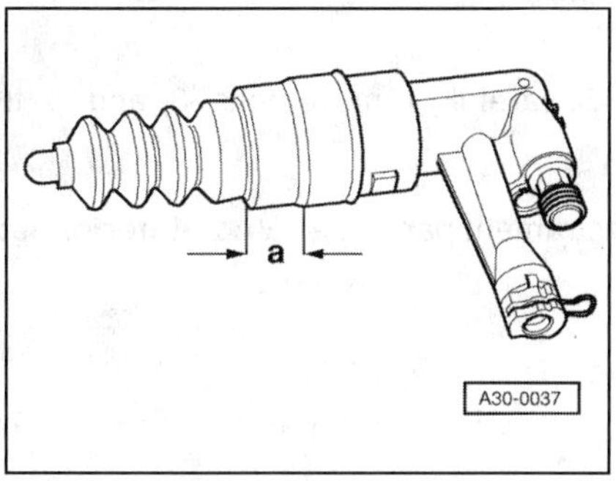

Fig. 1 Installing slave cylinder

– Before installing slave cylinder in transmission housing, you must coat the area -a- of the boot with Lithium grease G 052 150 A2.

– Apply a thin coat of copper grease to surface of clutch release lever where it contacts plunger.

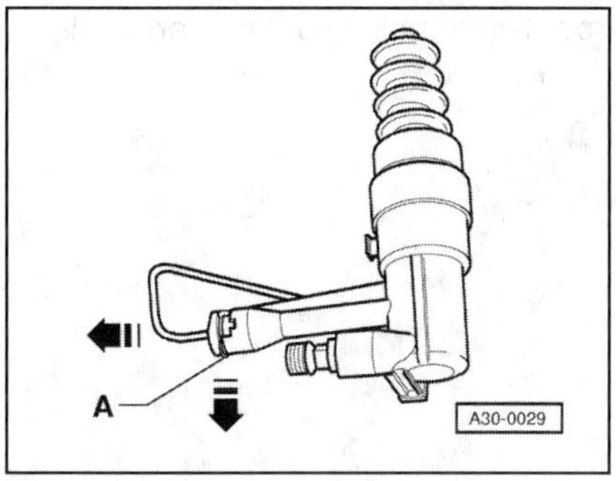

Fig. 2 Separating slave cylinder from line/hose assembly

Removing

– To remove line/hose assembly, you must pry clip -A- out to the stop, using a screwdriver. Only then can you pull out the line/hose assembly.

Installing

– Press in clip -A- to the stop.

– Push line/hose assembly into slave cylinder until assembly audibly engages.

– Test line/hose assembly by tugging on it.

Master cylinder, removing and installing

Special tools and misc. equipment required

- 3094 Hose clamps
- VAG 1331 Torque wrench

Removing

– First check whether a coded radio is installed. If this is the case obtain the radio anti-theft coding.

– With the ignition switched off disconnect the battery Ground (GND) strap.

– Lay a sufficiently lint-free cloth in plenum chamber below return hose to brake fluid reservoir.

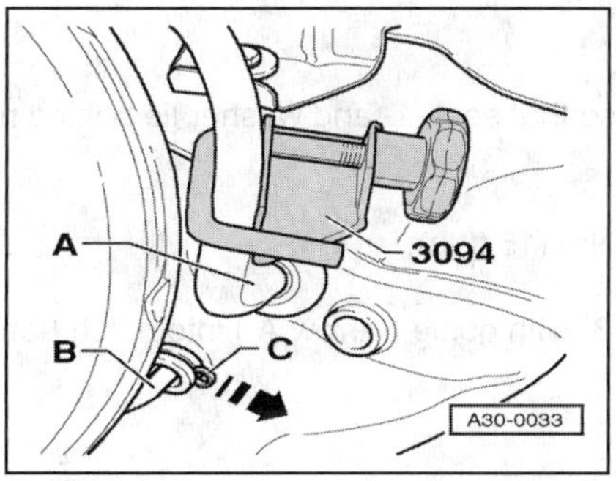

– Clamp shut the return hose -A- to brake fluid reservoir using hose clamp 3094, pull connecting line off master cylinder and seal off.

– Remove grommet for return hose -A- from bulkhead.

– Pull out clip -C- to the stop using screwdriver and then pull out line/hose assembly -B- slightly.

– Disconnect electrical connector for brake fluid level sensor from brake fluid reservoir.

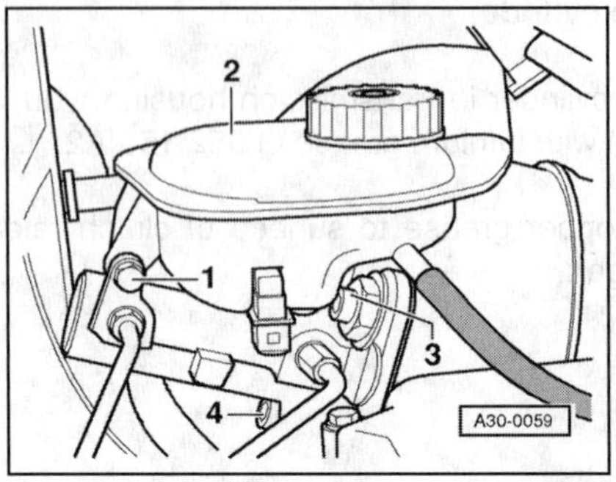

- Remove bolt -1-.

- Using inside Torx bit T45, carefully remove bolts -3- and -4- from brake servo.

- Remove cover below instrument panel, see **Body-Interior section, Repair Group 70.**

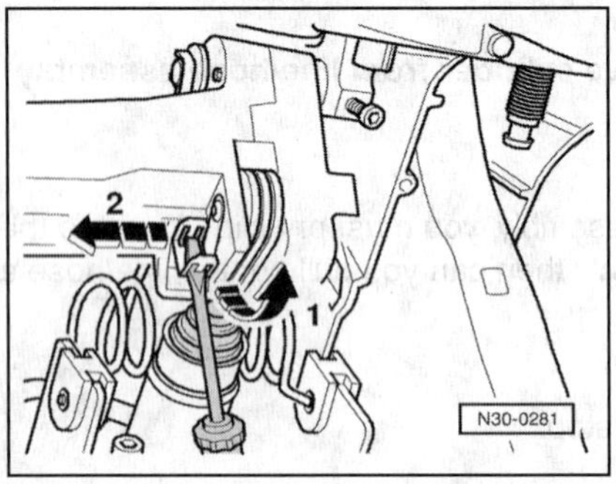

- Detach clutch pedal from master cylinder by prying out securing clip for pin with screwdriver -arrow 1- and pulling out pin -arrow 2-.

- Make sure that clutch pedal switch -1- is not pressed out of securing clip. This will damage the thread of the switch and it would have to be replaced.

NOTE —

To make sure sufficiently secure seating, a clutch pedal switch -1- may be installed only once.

- Unbolt (-arrows 2-) master cylinder -4- from within passenger compartment.

- Remove mounting bracket securing bolt (arrow 3).

- Pull complete foot pedal cluster slightly into passenger compartment.

- Remove master cylinder.

Master cylinder, installing

Installation is carried out in the reverse sequence, when doing this note the following:

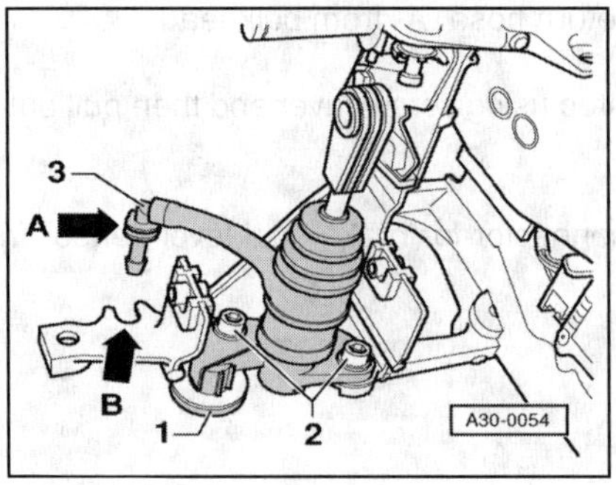

- Install master cylinder so that seal -1- and washer lie behind mounting bracket.

- Tighten bolts -2- to 20 Nm (15 ft-lb).

- Press connecting line -3- with guide (-arrow A-) into notch (-arrow B-) in mounting bracket.

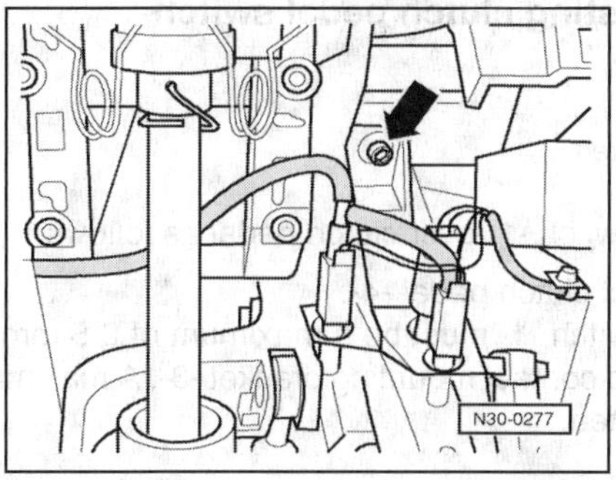

A second mechanic is required for the next work step.

- Push complete foot pedal cluster forward to bulkhead and insert bolt (-arrow-).

During this step, from plenum chamber, the second mechanic must guide line/hose assembly into master cylinder.

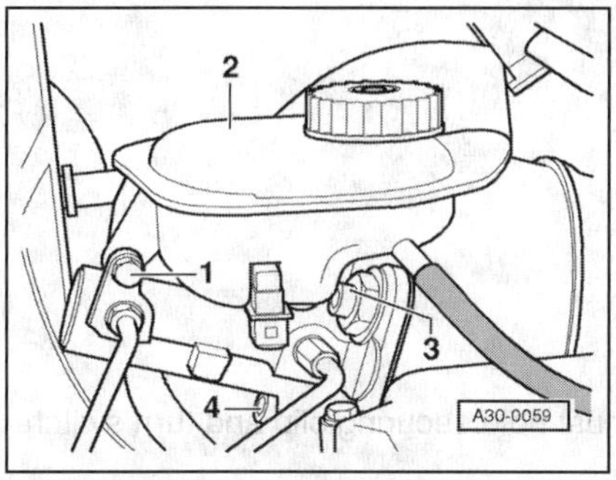

- Tighten brake servo securing bolts -3- and -4- to 25 Nm (18 ft-lb).
- Then tighten securing bolt -1- for brake fluid reservoir -2-.
- Attach connector to brake fluid level sensor on brake fluid reservoir.

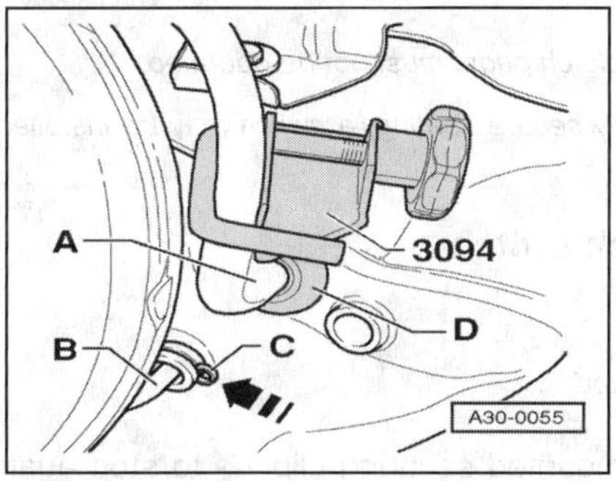

- Press in clip -C- to stop.
- Push line/hose assembly -B- into master cylinder until it audibly engages.
- Test line/hose assembly by tugging on it.
- Insert rubber grommet -D- for return hose -A- into bulkhead.
- Push return hose -A- to stop on connecting line of master cylinder.
- Remove hose clamp 3094.

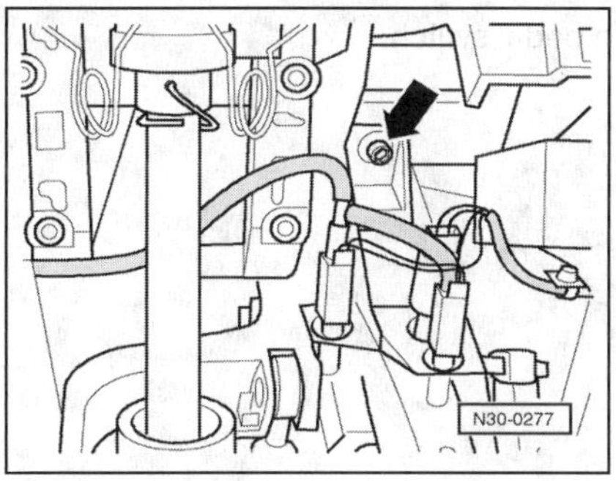

- Tighten securing bolt (-arrow-) for foot pedal cluster to instrument panel carrier to 25 Nm (18 ft-lb).
- Connect master cylinder to clutch pedal. Engage new pin in clutch pedal.
- Bleed clutch system after installing master cylinder, see **Clutch system, bleeding**.

Checking and adjusting clutch pedal switch

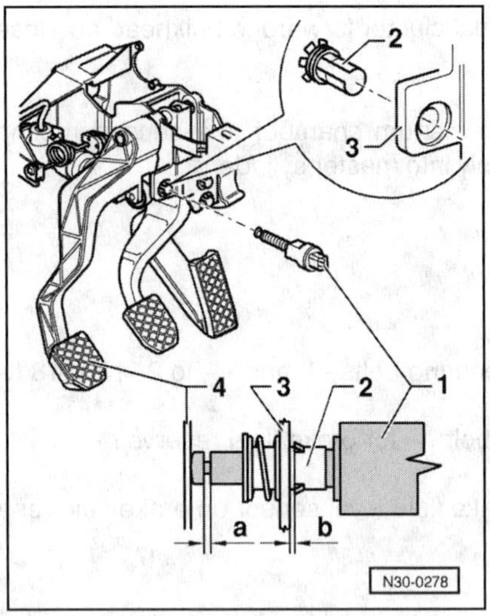

NOTE —
Clutch pedal is not operated.

Check adjustment of switch -1- over clutch pedal as follows:
Switch -1- must contact clutch pedal -4-.
Required play -a- at switch -1- must be a maximum of 0.5 mm.
Securing clip -2- should contact mounting bracket -3-. A maximum of 0.5 mm gap -b- is permitted.

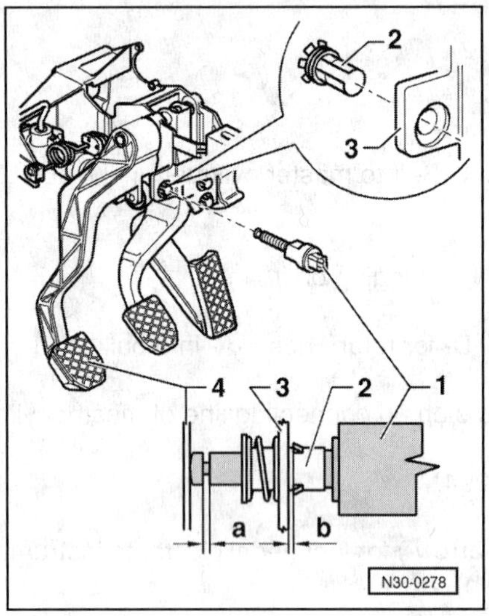

To adjust switch, you must hold securing clip and turn switch.

NOTE —
- *During adjustment, the clutch pedal must not be operated.*
- *To make sure sufficiently secure seating, a switch may be installed only one time.*

Installing new clutch switch

NOTE —
Clutch pedal is not operated.

- Press switch -1- into mounted securing clip -2- to stop against clutch pedal.
- Check and adjust clutch pedal switch.
- Attach connector.

Clutch release mechanism, servicing

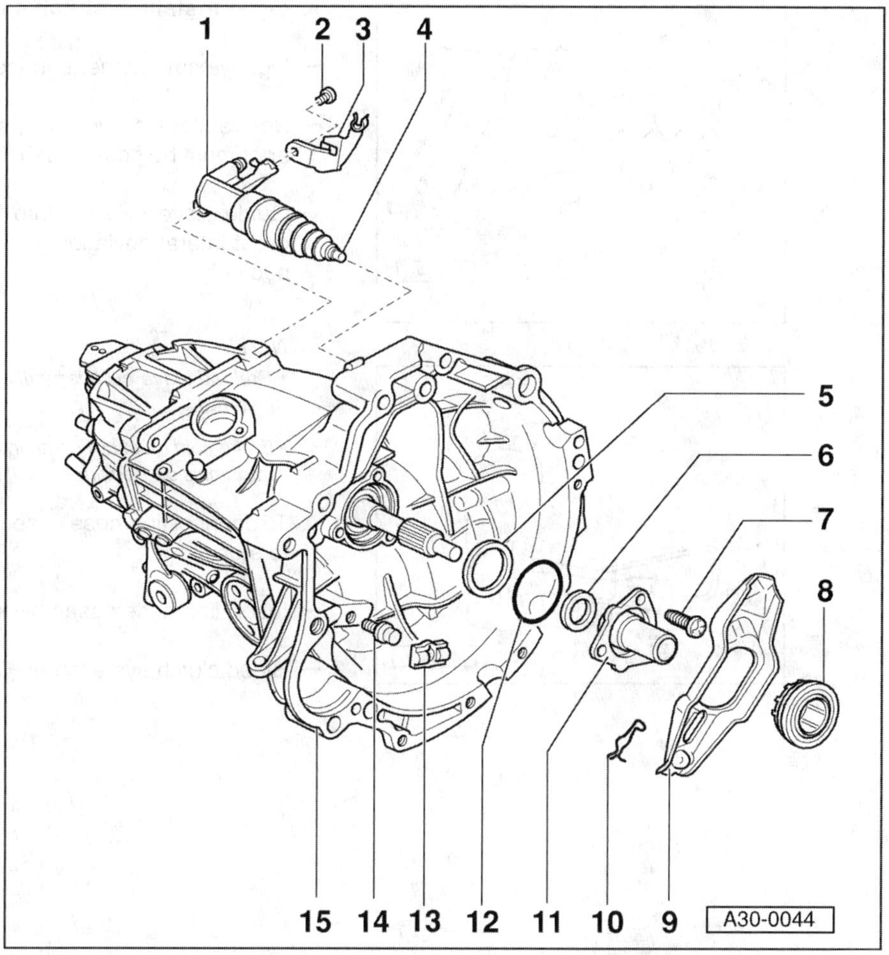

1. **Slave cylinder**
   - After removing, do not operate clutch pedal
   - Installing, see **Fig. 1**
   - When installing, push in until the securing bolt **-2-** can be installed
   - Tighten bleeder valve to 4 Nm (35 in-lb)
   - Bleeding clutch system, see **Clutch system, bleeding**
2. **Hex socket head bolt**
   - Tighten to 20 Nm (15 ft-lb)
   - Slave cylinder to transmission
   - Self-locking
   - Always replace
3. **Bracket**
   - For line/hose assembly
   - Secured to transmission with slave cylinder
4. **Plunger**
   - Coat end of plunger with copper grease
5. **Dished washer**
   - Smaller diameter (convex side) to guide sleeve
6. **Input shaft oil seal**
   - Pull out of guide sleeve with oil seal extractor lever VW 681
   - Knock in onto stop with fitting sleeve VW 192
7. **Torx socket head bolt**
   - Self-locking
   - Always replace
   - Black; only for aluminium transmissions
   - Bright; for aluminium and magnesium transmissions
   - For aluminium transmissions, 35 Nm (26 ft-lb)
   - For magnesium transmissions, 25 Nm (18 ft-lb)
8. **Release bearing**
   - Do not wash-out bearing, only wipe
   - Replace noisy bearings
9. **Clutch release lever**
   - Before installing, thinly coat contact surface of slave cylinder plunger with copper grease
10. **Retaining spring**
    - Secure to clutch release lever
11. **Guide sleeve**
    - Before pulling off or pressing on, pull shrink tube over input shaft splines to protect oil seal **-6-**
    - Installation position: oil drain hole facing downwards
    - Guide sleeve made of different materials for aluminium and magnesium transmissions
12. **O-ring**
    - Always replace
13. **Intermediate piece**
14. **Ball stud**
    - Lubricate with MoS_2 grease
    - Ball studs made from different materials for aluminium and magnesium transmissions
    - For aluminium transmissions, 25 Nm (18 ft-lb)
    - For magnesium transmissions, 20 Nm (15 ft-lb)
15. **Transmission**

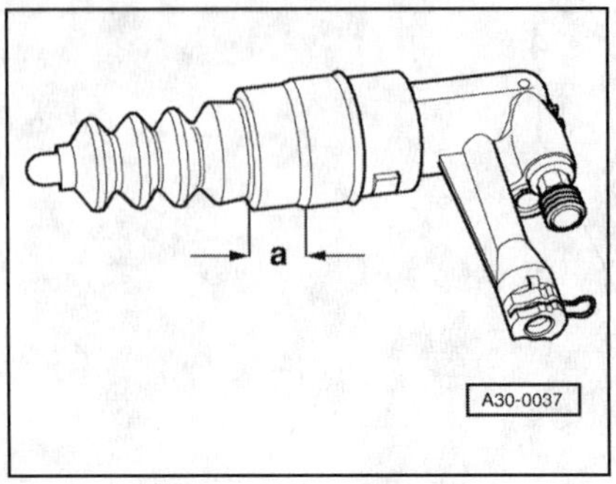

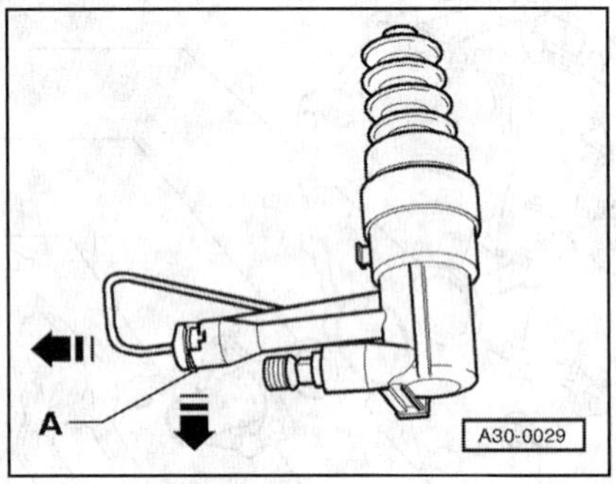

Fig. 1 Installing clutch slave cylinder

– Lightly coat plunger surface contacting clutch release lever.

– Before slave cylinder is installed in transmission housing, region -a- of boot must be coated with lithium grease G 052 150 A2.

– Guide slave cylinder into fitting hole of transmission housing without great lateral deviation from working direction of operating rod and bolt securely.

NOTE —

- *Preload slave cylinder sufficiently so that the securing bolt can easily be inserted.*
- *Brake fluid must not come in contact with transmission. If necessary, clean transmission housing.*

To connect line/hose assembly to slave cylinder, you must press clip -A- to stop.

– Insert line/hose assembly in slave cylinder until line audibly engages.

– Bleed clutch system, see **Clutch system, bleeding**.

Clutch, servicing

Special tools and misc. equipment required

- 3067 Counter hold tool
- 3176 Centering mandrel
- VAG 1331 Torque wrench
- (Transmission removed)

NOTE —

- *Replace clutch plates and pressure plates with damaged or loose rivets*
- *Select correct clutch plate and pressure plate according to parts list and engine code.*
- *Clutch illustrated in conjunction with dual-mass flywheel*

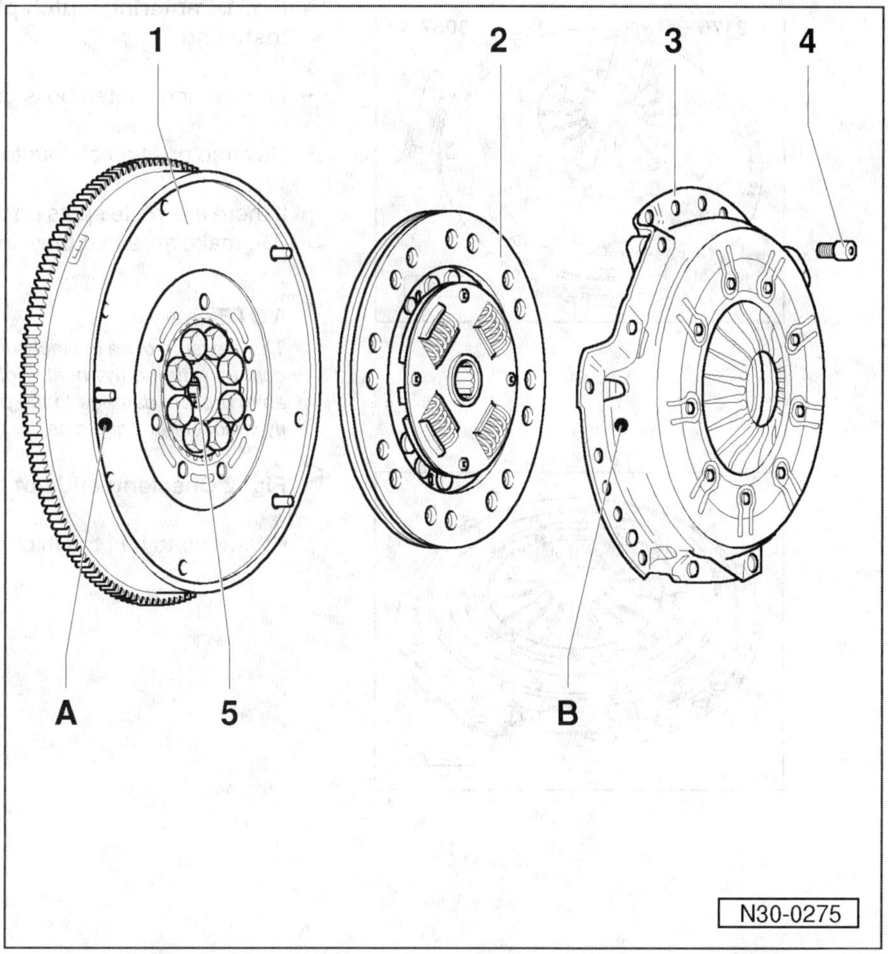

1. **Flywheel**
   - Make sure centering pins are a tight fit
   - Contact surface for clutch lining must be free of grooves, oil and grease
   - If white spots are present on flywheel -A- and pressure plate -B-, align them during assembly
   - Removing and installing, see **Engine Mechanical section, Repair Group 13**
2. **Clutch plate**
   - Diameter, see **Repair Group 00, Code letters, transmission allocation, ratios, capacities**
   - Installation position with two-part flywheel: the clutch plate lining must contact the flywheel
   - Installation position without two-part flywheel: spring cage faces the pressure plate
   - Centering, see **Fig. 1**
   - Lightly grease splines

   NOTE —

   Clean input shaft splines and on used clutch plates, the hub splines, remove corrosion and apply only a very thin coating of grease Part No: G 000 100 to the input shaft splines. Then move clutch plate to and fro on input shaft until hub moves freely on shaft. Excess grease must be removed.
3. **Pressure plate**
   - Removing and installing, see **Fig. 1**
   - Check ends of diaphragm spring, see **Fig. 2**
   - If white spots are present on flywheel -A- and pressure plate -B-, align them during assembly

   NOTE —

   Pressure plates are protected against corrosion and greased. Only the contact surface may be cleaned, otherwise the service life of the clutch will be considerably reduced.
4. **Bolt**
   - Tighten to 25 Nm (18 ft-lb)
   - Loosen and tighten gradually and diagonally
5. **Needle roller bearing**
   - Driving in and out, see **Engine Mechanical section, Repair Group 13**

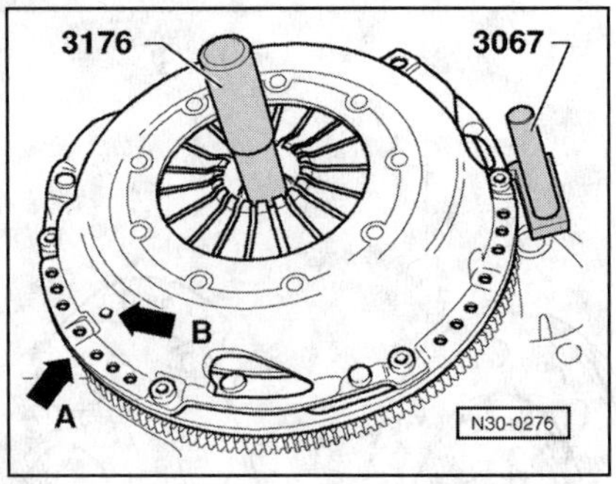

Fig. 1 Centering clutch plate and pressure plate, removing and installing

– Loosen and tighten bolts gradually and diagonally.

– Reverse position of counter-hold when removing.

– If there are white spots on the dual mass flywheel -A- and pressure plate -B-, make sure that they line up when you assemble them.

NOTE —
The pressure plate contact surface and the clutch plate lining must make full contact with the flywheel. Tighten securing bolts diagonally and evenly so as to prevent damage to the pressure plate centralizing hole and the flywheel centralizing pins.

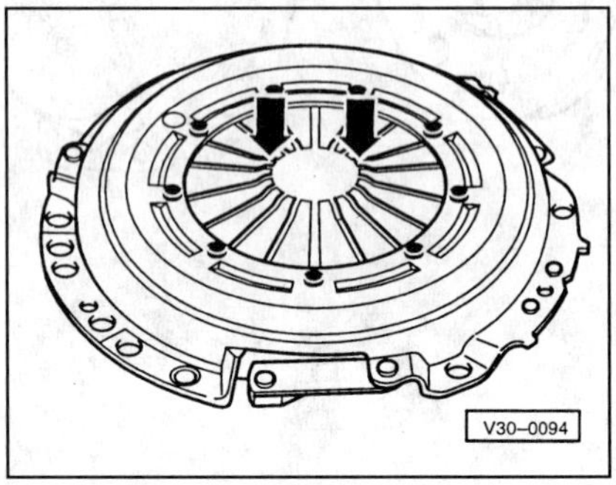

Fig. 2 Checking ends of the diaphragm spring

• Wear up to half the thickness of the diaphragm spring is permitted.

34 Controls, Housing

Repair Group

M 34

Shift Linkage

EDITOR'S NOTE —

There are two different types of selector mechanisms offered in Passats with manual transmissions; selector mechanisms that require the shift lever be pressed down to select reverse, and selector mechanisms that do not require that the shift lever be pressed to select reverse. These two mechanisms share some components, see exploded view assemblies for details.

NOTE —

To remove the complete shift linkage, remove the exhaust system, see **Engine Mechanical section, Repair Group 26**.

Selector mechanism (shift lever must be pressed down to select reverse), assembly

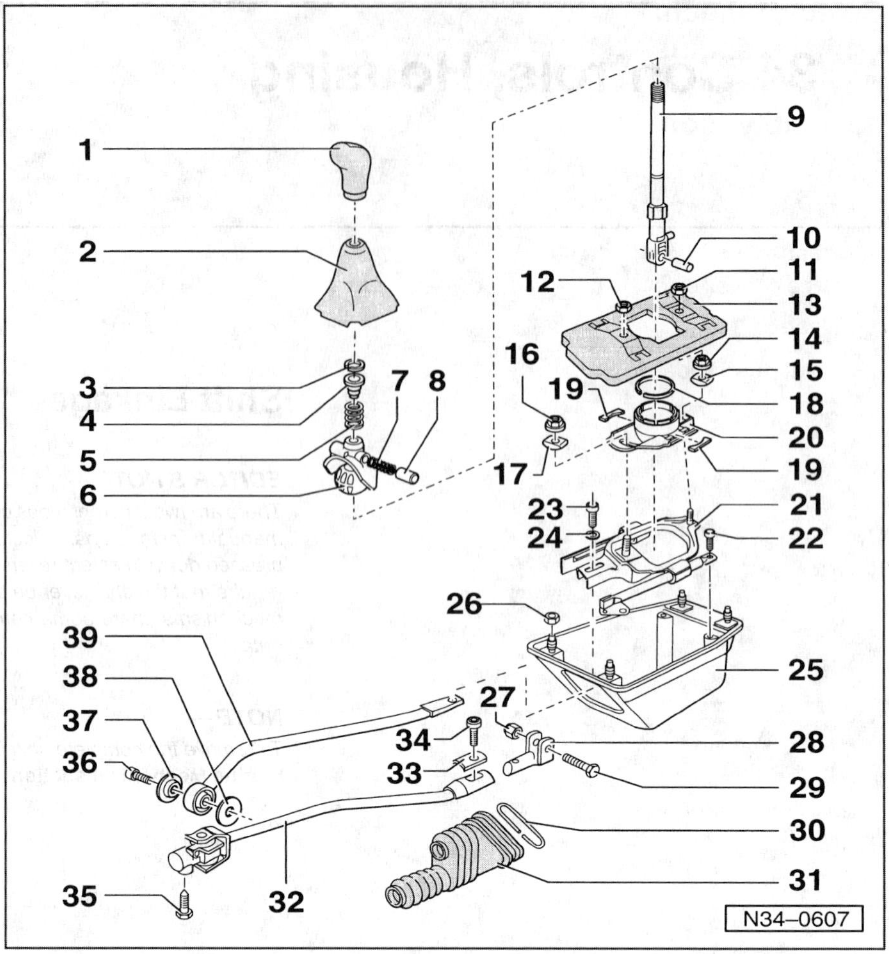

NOTE—

Lubricate all mountings and sliding surfaces with Polyurea grease Part No. G 052 142 A2.

1. **Shift lever knob**
   - Screw on and off together with cover
2. **Cover**
   - Separating from shift lever knob, see **Fig. 1**
   - Connect to shift lever knob before installing
3. **Securing clip**
   - Always replace
4. **Spacer sleeve**
5. **Spring**
6. **Ball stop**
   - Insert spring and bush in the ball stop and assemble on the shift lever so that the spring and bush are on the right when looking forwards.
   - Install ball stop before installing circlip **-18-**.
7. **Spring**
8. **Bush**
   - Installing position: The rounded end faces the shift lever
9. **Shift lever**
   - Only installs in one position in the ball housing
10. **Spacer tube**
11. **Hex nut**
    - Tighten to 10 Nm (7 ft-lb)
12. **Hex nut**
    - Tighten to 10 Nm (7 ft-lb)
13. **Cover**
14. **Hex nut**
    - Tighten to 25 Nm (18 ft-lb)
15. **Connecting piece**
16. **Hex nut**
    - Tighten to 25 Nm (18 ft-lb)
17. **Connecting piece**
18. **Circlip**
    - Always replace
    - Remove before removing the ball stop
    - Installation position: the rounded side faces to ball housing **-20-**
19. **Buffer**
20. **Ball housing**
21. **Rear push rod**
22. **Hex socket head bolt**
    - Tighten to 10 Nm (7 ft-lb)
23. **Hex socket head bolt**
    - Tighten to 25 Nm (18 ft-lb)
24. **Washer**
25. **Shift lever housing**
26. **Hex nut**
    - Tighten to 10 Nm (7 ft-lb)
27. **Hex nut**
    - Tighten to 10 Nm (7 ft-lb)
    - Self-locking
    - Always replace
28. **Selector fork**
29. **Bolt**
30. **Tensioning ring**
31. **Bellows**
    - Remove Three Way Catalytic Convertor (TWC) and front heat shield before removing and installing

NOTE—

- *Carefully pull bellows over selector rod*
- *Place bellows on the marks on selector rod and push rod*

32. **Selector rod**
    - With selector joint
33. **Clamping piece**
34. **Hex socket head bolt**
    - Tighten to 25 Nm (18 ft-lb)
35. **Hex bolt**
    - Tighten to 23 Nm (17 ft-lb)
    - Self-locking
    - Always replace

(continued on following page)

Selector mechanism (shift lever must be pressed down to select reverse), assembly (continued)

36. Hex socket head bolt
- Tighten to 40 Nm (30 ft-lb)

37. Washer
- The convex side faces to push rod

38. Washer

39. Front push rod

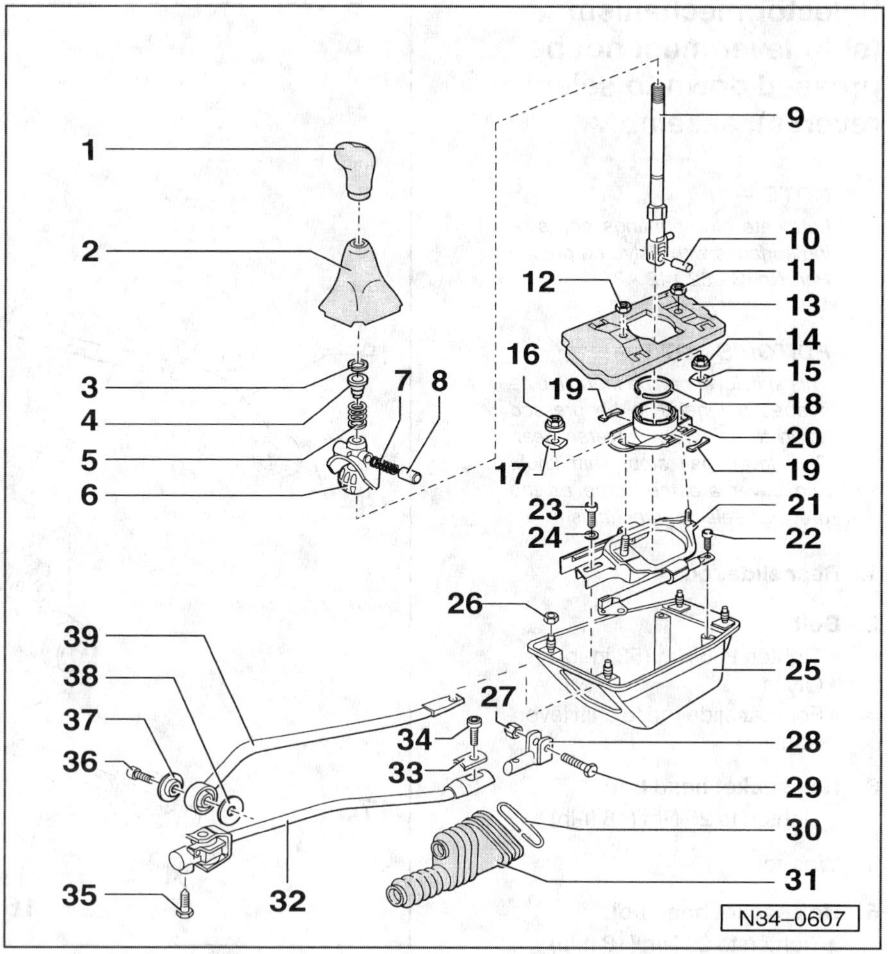

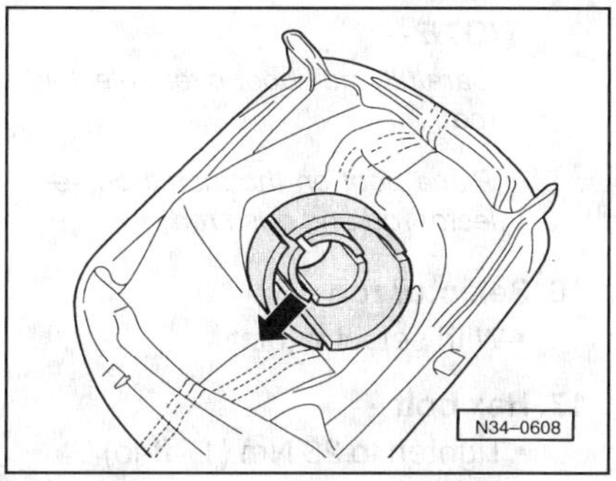

Fig. 1 Separating cover from shift lever knob

– Turn cover inside out.

– Carefully pry segment with a screwdriver in direction of arrow; while at the same time pulling the shift lever knob out.

Selector mechanism (shift lever must not be pressed down to select reverse), assembly

NOTE—
Lubricate all mountings and sliding surfaces with Polyurea grease Part No. G 052 142 A2.

EDITOR'S NOTE—
This shift lever can be identified as it does not need to be pressed down to select the reverse gear. Shift lever, ball stop, shift knob, and cover are the same as the previous selector mechanism

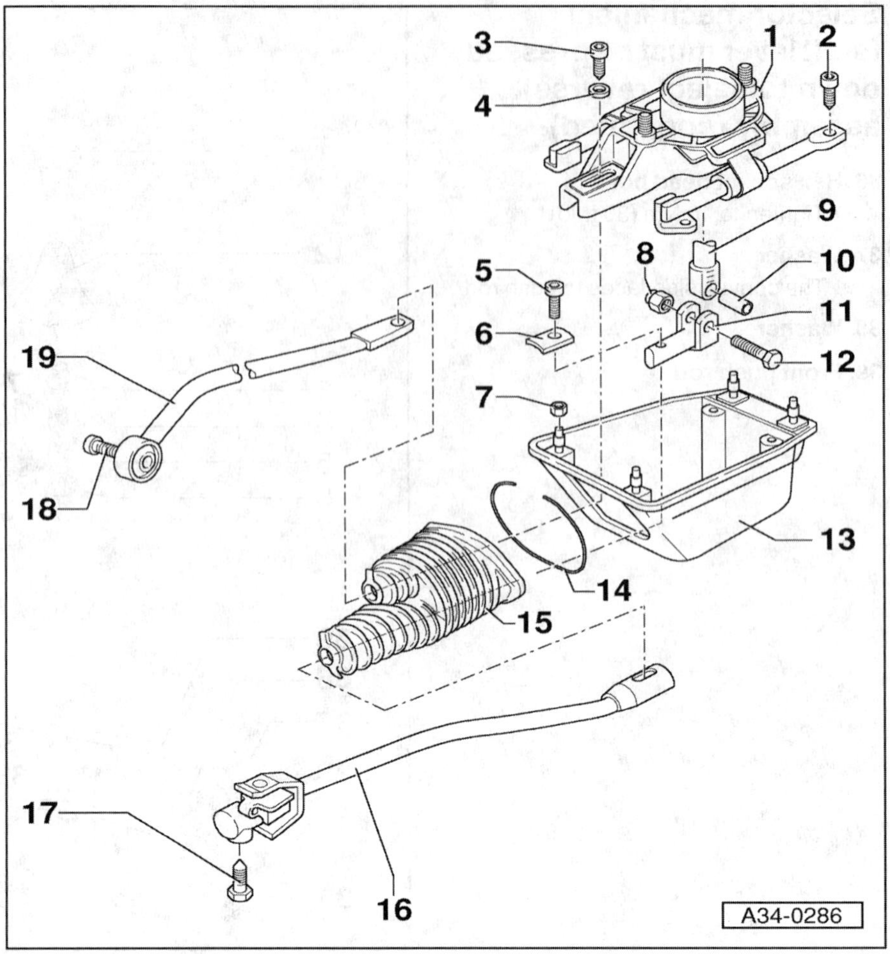

1. **Rear slide rod**
2. **Bolt**
   - Tighten to 6 Nm (53 in-lb)
   - Qty. 4
   - For rear slide rod to shift lever housing
3. **Hex socket head bolt**
   - Tighten to 25 Nm (18 ft-lb)
4. **Washer**
5. **Hex socket head bolt**
   - Tighten to 25 Nm (18 ft-lb)
6. **Clamping piece**
7. **Hex nut**
   - Tighten to 10 Nm (7 ft-lb)
   - Qty. 4
   - For shift lever housing to body
8. **Hex nut**
   - Tighten to 10 Nm (7 ft-lb)
   - Self-locking
   - Always replace
9. **Shift lever**
10. **Spacer tube**
11. **Selector fork**
12. **Bolt**
13. **Shift lever housing**
14. **Tensioning ring**
15. **Boot**
   - Remove Three Way Catalytic Convertor (TWC) and front heat shield before removing and installing

NOTE—
- *Carefully pull boot over selector rod*
- *Place boot on the marks on selector rod and push rod*

16. **Selector rod**
   - With selector joint
17. **Hex bolt**
   - Tighten to 23 Nm (17 ft-lb)
   - Self-locking
   - Always replace
18. **Hex socket head bolt**
   - Tighten to 40 Nm (30 ft-lb)
   - Cannot be removed from front slide rod bearing bush
19. **Front push rod**

Selector mechanism, removing and installing

– Carefully pry the cover off the center console.

– Then remove the shift lever knob with cover.

Shift lever must be pressed down to select reverse

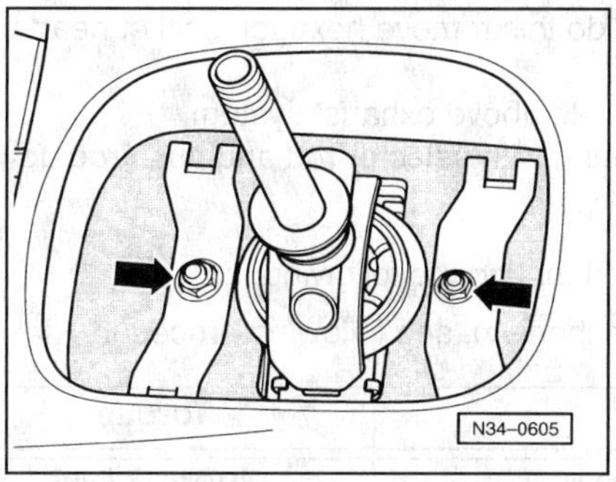

The rubber cover for the shift lever housing must be removed (arrows).

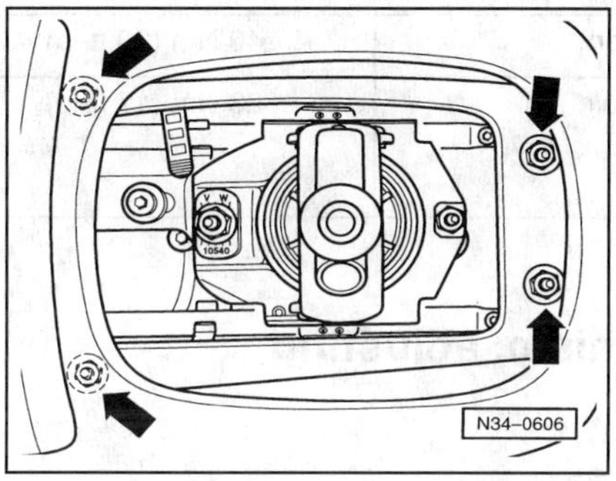

Now remove the shift lever housing securing nuts at outer corners (arrows).

Shift lever must not be pressed down to select reverse

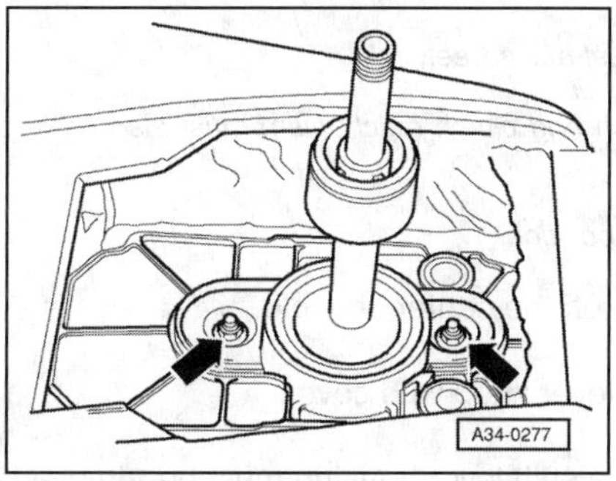

The rubber cover for the shift lever housing must be removed (arrows).

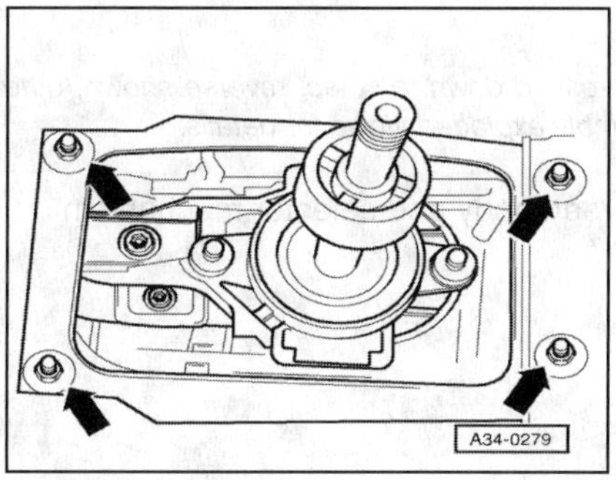

Now remove the shift lever housing securing nuts (arrows).

All vehicles

– Separate exhaust system behind the Three Way Catalytic Convertor (TWC), if necessary remove front exhaust system, see **Engine Mechanical section, Repair Group 26.**

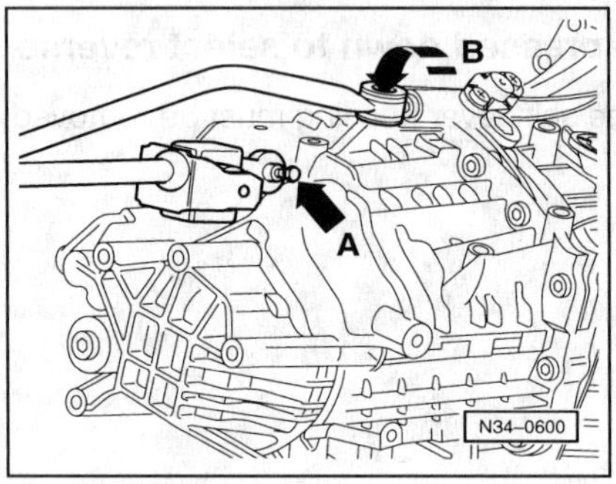

◄ Remove selector mechanism from transmission as follows:

- Remove selector rod (arrow -A-).
- Remove push rod, to do this remove hexagon socket head bolt (arrow -B-).
- Remove front heat shield above exhaust system.
- Swing shift lever housing with selector rod and push rod down and remove.

– Installation is or removal, noting the following:

- Adjusting selector mechanism, see following procedure.

Tightening torque	Torque
Shift lever housing to body	10 Nm (7 ft-lb)
Push rod to transmission	40 Nm (30 ft-lb)
Selector rod (selector joint) to transmission Use new bolts!	23 Nm (17 ft-lb)

Selector mechanism, adjusting

NOTE —

The following points are essential to ensure correct selector adjustment:

- *Moving parts of selector mechanism and transmission elements must be in proper condition.*
- *Selector mechanism must move freely.*
- *The transmission, clutch and clutch mechanism must also be in proper condition.*
- *Transmission in neutral position.*

– Carefully pry the cover off the center console.

– Then remove the shift lever knob with cover.

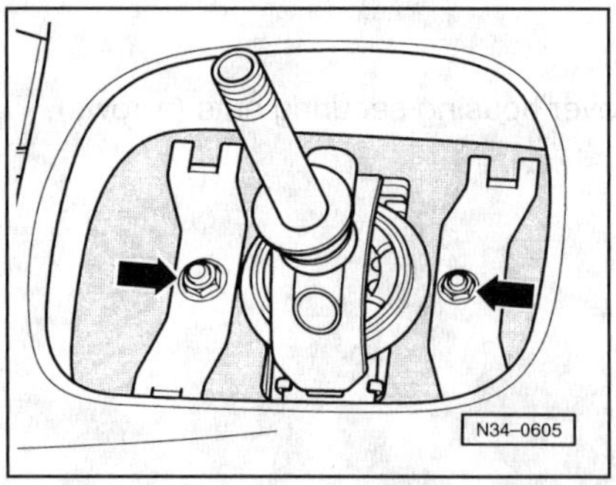

◄ The rubber cover for the shift lever housing must be removed (arrows).

EDITOR'S NOTE —

Shift lever that must be pressed down to select reverse shown, other shift lever is similar. See assembly exploded views for details.

– Measure distance between body and selector mechanism.

Shift lever must be pressed down to select reverse

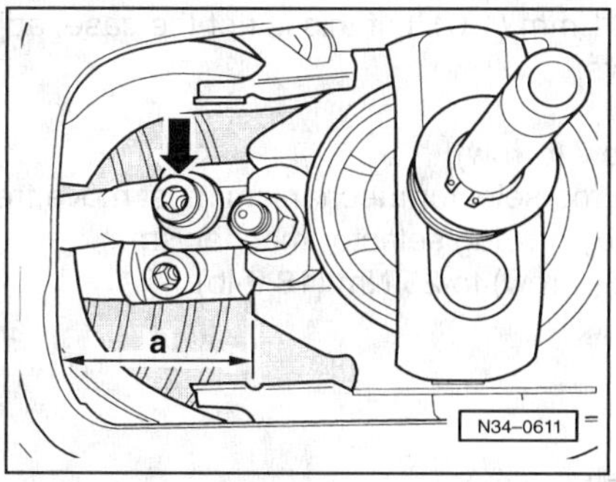

- Distance -a- must be 37 mm (1.45").
- If this is not the case, adjust dimension -a- as follows:
- Loosen push rod screw (arrow).
  - The connection push rod/selector mechanism must move freely.
- Set to dimension -a- by moving selector mechanism.
- Tighten slide rod bolt (arrow) to 25 Nm (18 ft-lb).

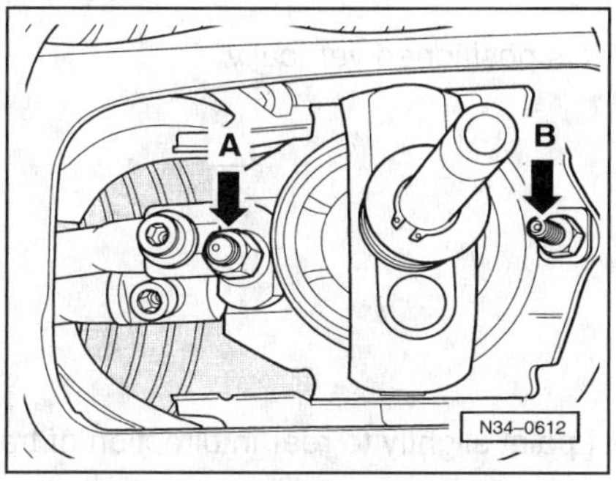

- Loosen ball housing nuts (arrow -A- and arrow -B-).
- The ball housing must be set so that it is horizontal.
- Then tighten ball housing nuts to 25 Nm (18 ft-lb).

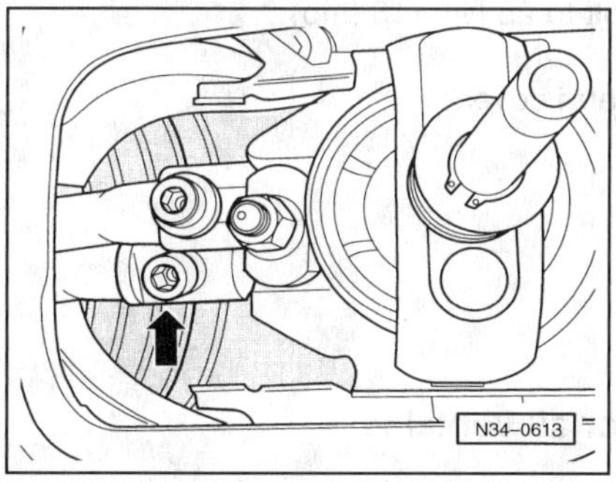

- Loosen selector rod bolt (arrow).
  - The connection selector rod/selector mechanism must move freely.
- Align shift lever so that it is positioned slightly to the rear.

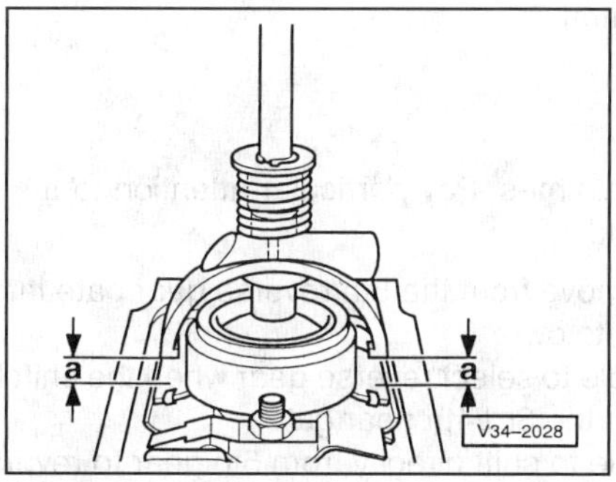

- The distance -a- for both ball stop lugs to ball housing must be equal.
- Tighten selector rod bolt to 25 Nm (18 ft-lb).

Should a gear fail to engage smoothly after repeated selection then continue as follows:

Fine adjustment

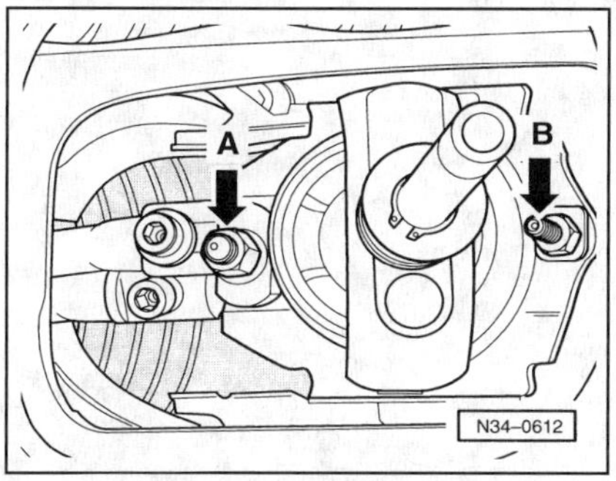

- Loosen ball housing nuts (arrow -A- and arrow -B-).
- Press the shift lever to right onto transmission stop.
- When doing this push the ball housing to left against the shift lever.
- Then hold shift lever and ball housing in this position and tighten to 25 Nm (18 ft-lb).
- Install shift lever knob and cover.

Shift lever must be pressed down to select reverse

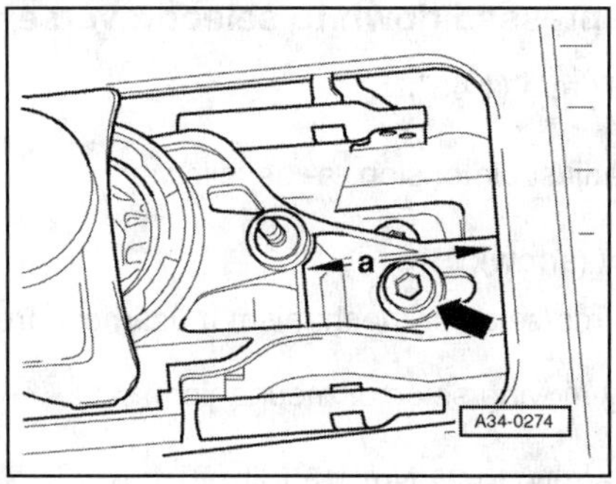

Distance -a- must be 41 mm (1.61"), if this is not the case, adjust dimension -a- as follows:

- Loosen push rod screw (arrow).
- The connection push rod/selector mechanism must move freely.
- Set to dimension -a- by moving selector mechanism.
- Tighten slide rod bolt (arrow) to 25 Nm (18 ft-lb).

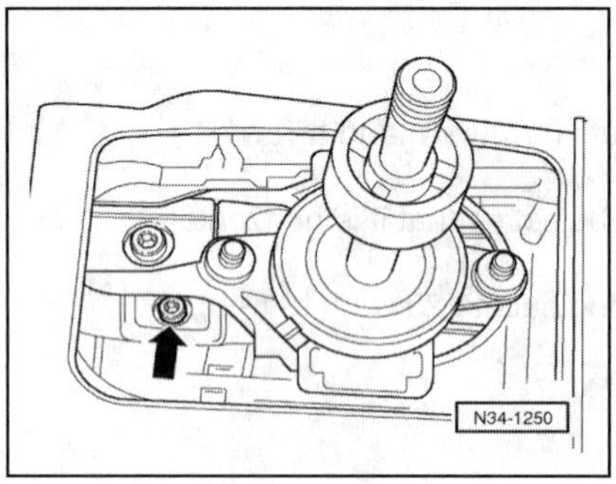

Loosen selector rod bolt (arrow).

- The connection selector rod/selector mechanism must move freely.

– Align shift lever so that it is positioned vertically.

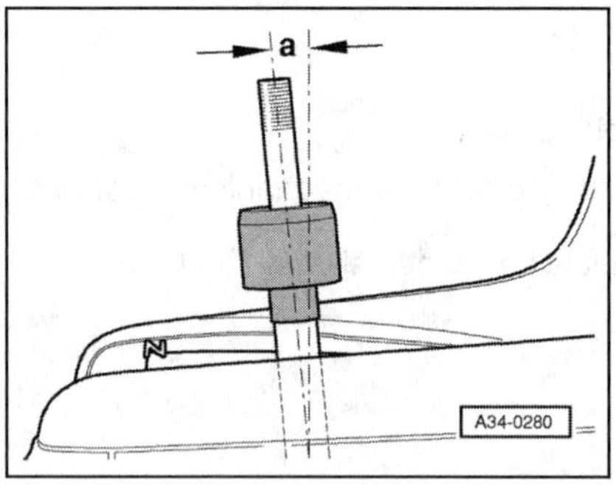

When doing this, it must point slightly to rear in direction of travel, -a-.

– Tighten selector rod bolt to 25 Nm (18 ft-lb).

– Install shift lever knob and cover.

Functional check for all vehicles

- The shift lever must rest in the 3rd/4th gear gate when transmission is in neutral.

– Operate clutch.

– Select all gears several times. Pay particular attention to the operation of the reverse gear lock.

- The shift lever must move from the 5th/reverse gear gate into the 3rd/4th gear gate on its own.
- It must only be possible to select reverse gear when the shift lever is in the neutral position in the 3rd/4th gear gate.
- It must not be possible to shift directly from 5th gear to reverse gear.

Transmission, removing and installing

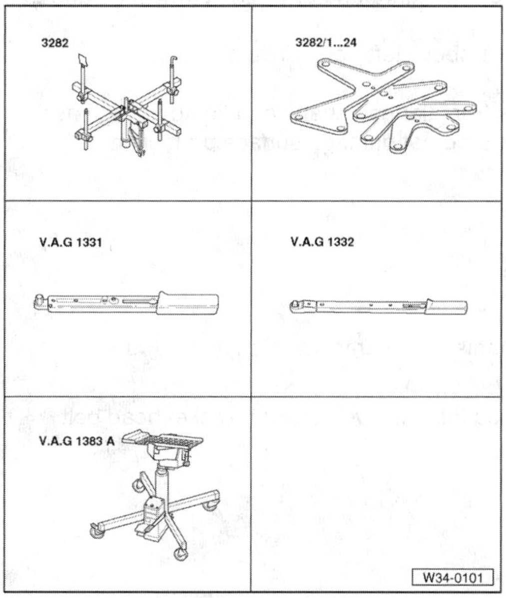

Special tools and misc. equipment required

- 3282 Transmission support
- 3282/10 Adjustment plate
- VAG 1331 Torque wrench
- VAG 1332 Torque wrench
- VAG 1383 A Engine/transmission jack
- 3336 Transmission lifting tackle

Removing

– First check whether a coded radio is installed. If this is the case obtain the anti-theft coding.

– With the ignition switched off disconnect the battery Ground (GND) strap.

– Remove coolant system expansion reservoir and lay to one side.

– Remove the intake manifold cover.

– Remove intake hose from air mass meter, see **Fuel Injection & Ignition section, Repair Group 24.**

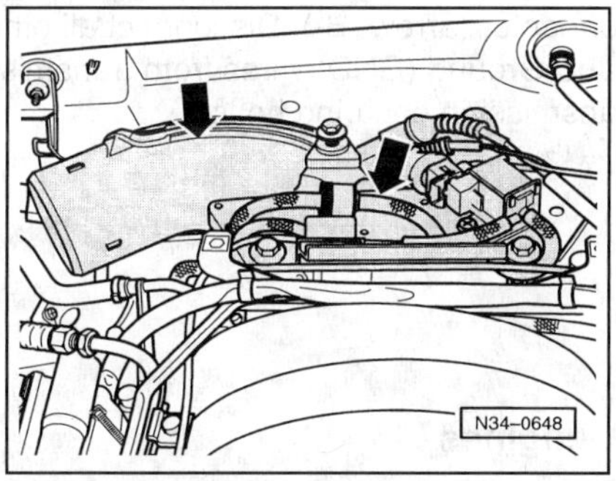

◄ Remove intake elbow (if installed) and lay to one side, see **Fuel Injection & Ignition section, Repair Group 24.**

– Remove upper engine/transmission securing bolts.

– Remove front wheels.

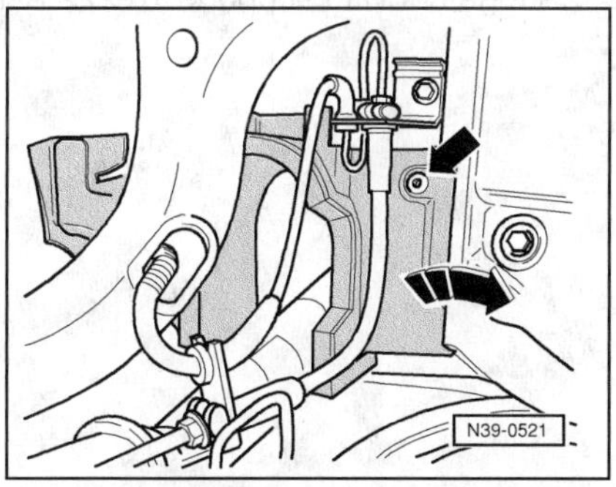

◄ Remove covers from wheel housing (arrows).

– Remove insulation tray below engine/transmission, if installed.

◄ Remove insulation tray bracket, if installed (arrow).

– Remove front exhaust system with Three Way Catalytic Convertor (TWC), see **Engine Mechanical section, Repair Group 26.**

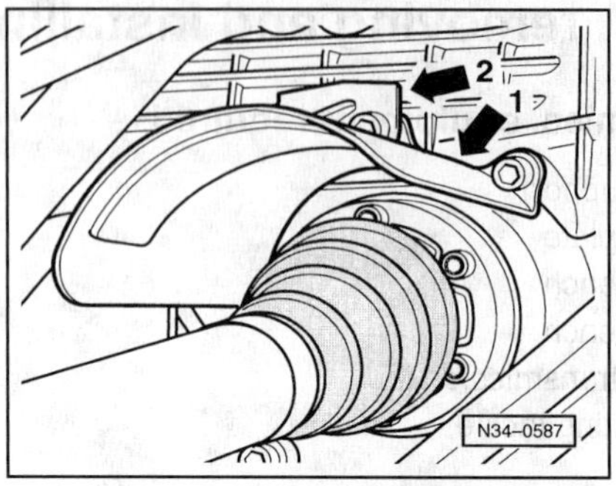

Remove cover plate over the right-hand drive shaft (arrows 1 and 2).

– If installed, remove cover above left-hand drive shaft.

– Disconnect drive shafts from flanged shafts and tie up as high as possible, when doing this do not damage surface protection.

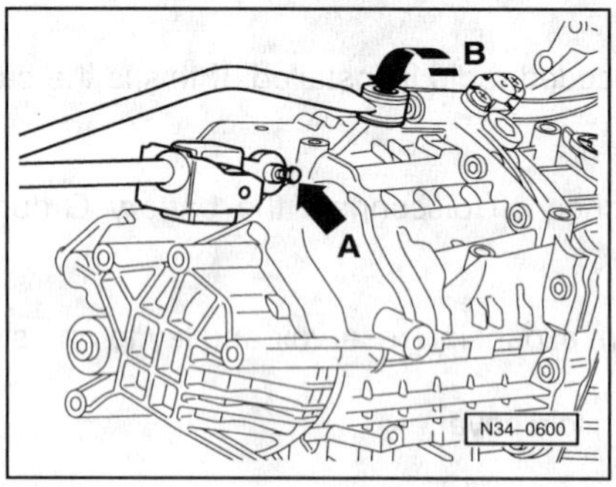

Remove selector mechanism from transmission as follows:

- Remove selector rod (arrow -A-).
- Remove push rod, to do this remove hexagon socket head bolt (arrow -B-).

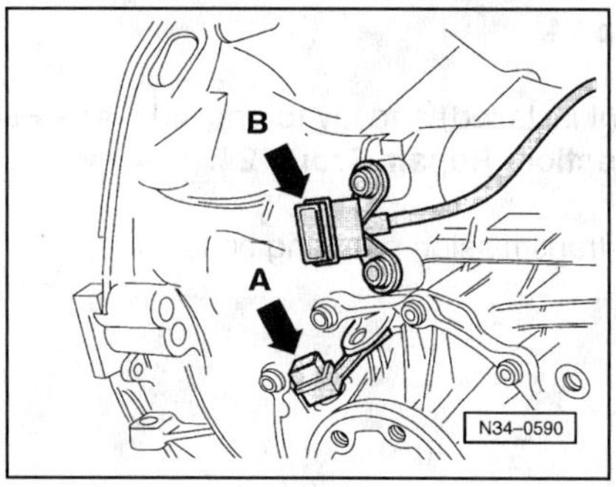

- Pull connector off road speed sender (arrow -A-).

– Pull off back-up light connector (arrow -B-). Disconnect all other electrical connections and Ground (GND) wires from transmission and from the engine/transmission securing bolts.

Vehicles with 6-Cyl. engines

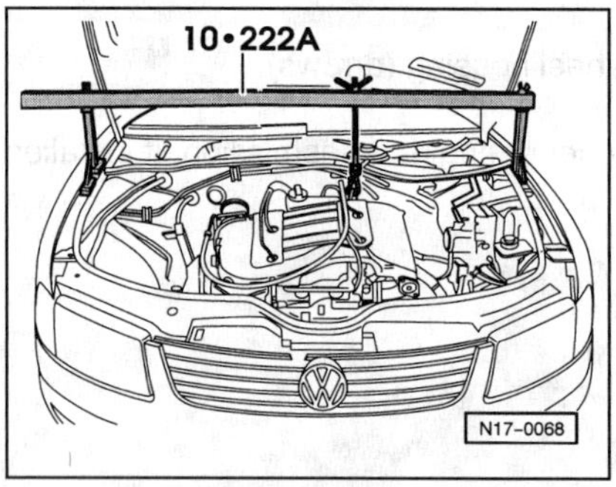

Install support bar 10-222A together with supports 10-222 A/1.

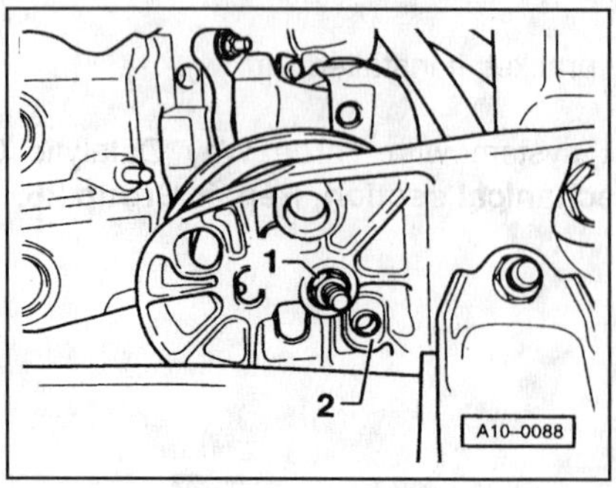

Mark installation position (items -1- and -2-) of both engine mountings from below.

– Remove nuts for both engine mountings (arrows).

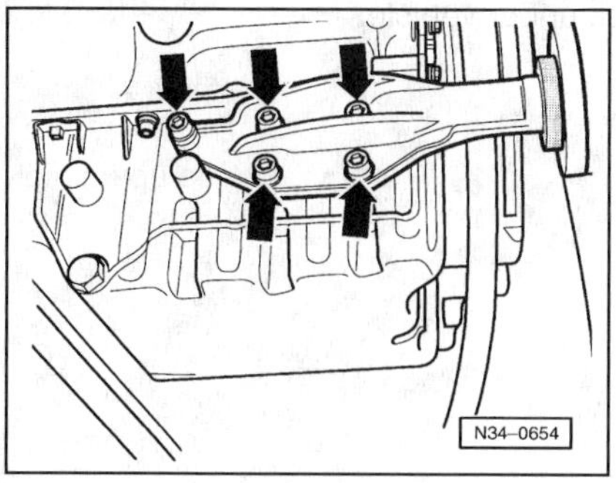

Remove torque rod support on engine (arrows) and take out, see **Engine Mechanical section, Repair Group 17.**

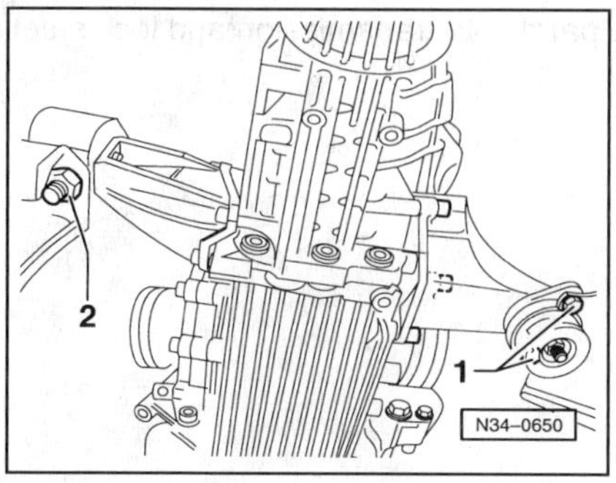

Then remove right transmission support bolts -1-.

– Remove left transmission support nut -2- from sub-frame.

– Lift engine/transmission assembly via the spindles of support bar 10-222A until the engine radiator fan lies against the underside of the bumper.

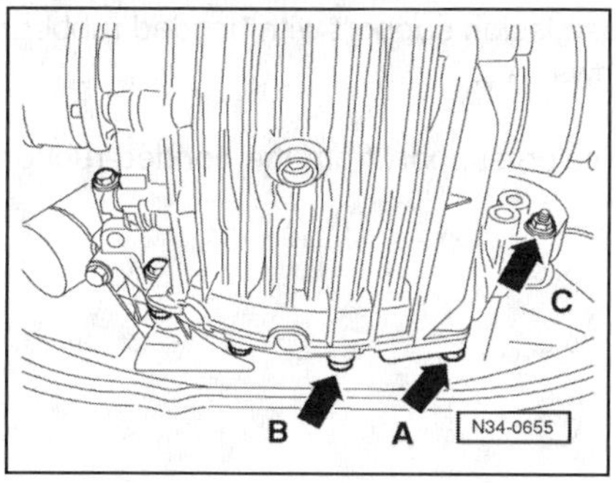

Remove engine/transmission connecting bolts from engine side (arrows -A- and -B-).

– Do not remove bolt (arrow -C-) at present.

– Lower engine/transmission assembly back to installation position again using spindles.

– Tighten torque rod support, engine mounting and left transmission support again hand tight.

All vehicles

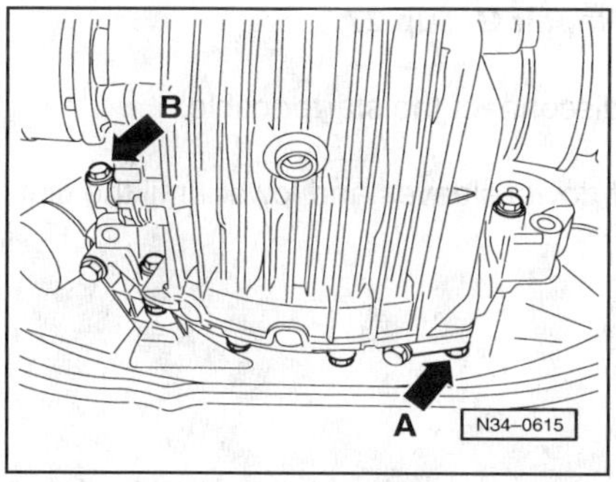

With the exception of bolts (arrows -A- and -B-) remove lower engine/transmission securing bolts.

– Install transmission support 3282, adjustment plate 3282/10 for "012" "01A" transmissions and support elements as follows to transmission jack:

– Place adjustment plate 3282/10 on transmission support 3282. (Adjustment plate only installs in one position).

– Align arms of transmission support according to holes in adjustment plate.

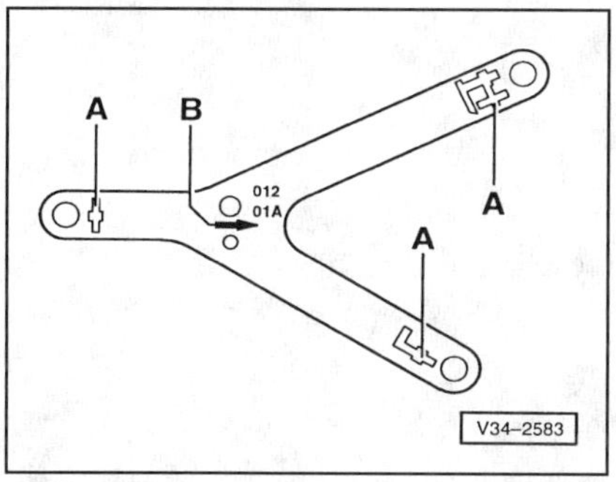

Bolt on support elements -A-, as illustrated on adjustment plate.

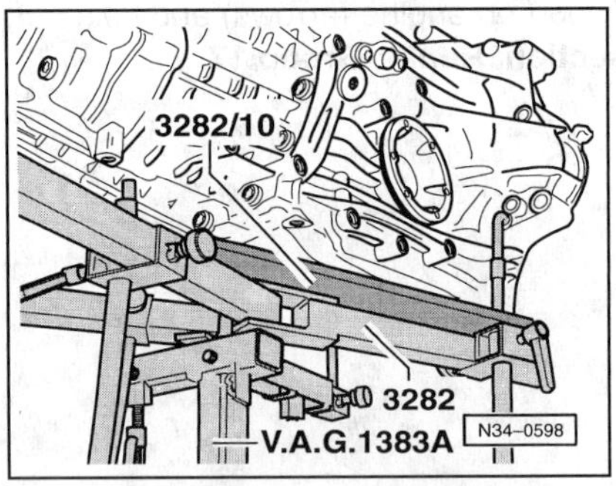

◀ Place transmission jack under vehicle.

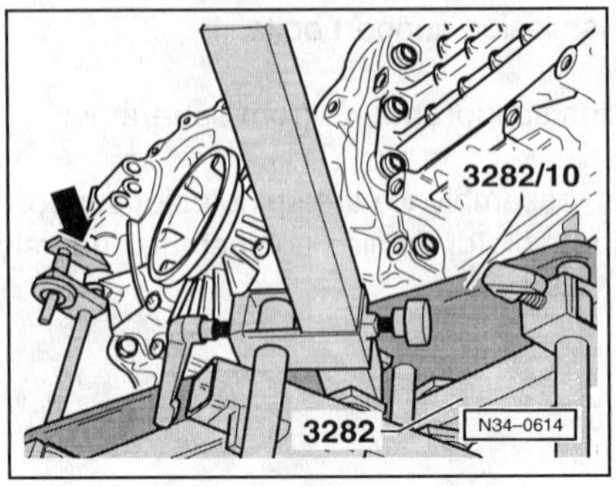

◀ Align adjustment plate parallel to transmission and lock safety support (arrow) on transmission.

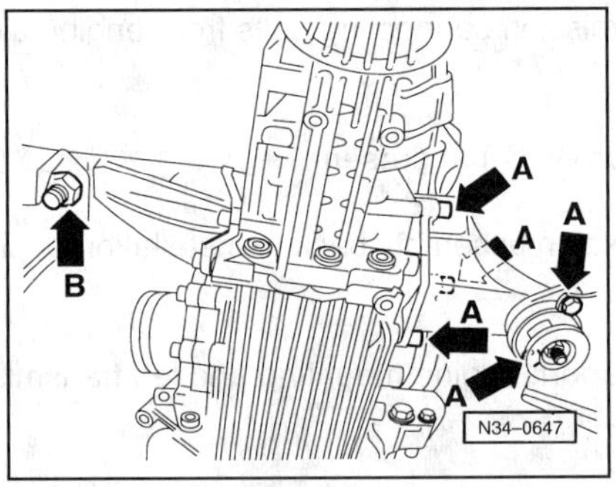

◀ Then remove right transmission support with bonded rubber mounting from transmission (arrows -A-).

– Detach the left transmission support from the bonded rubber mounting (arrow -B-).

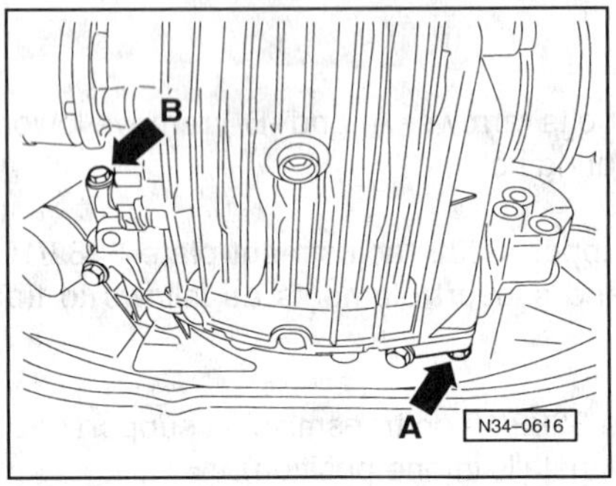

◀ Remove engine/transmission connecting bolts (arrows -A- and -B-) remove starter motor when doing this if necessary, see **Electrical Equipment section, Repair Group 27.**

– It is not necessary to disconnect the starter cable.

– Press transmission off dowel sleeves and lower until the clutch slave cylinder is accessible.

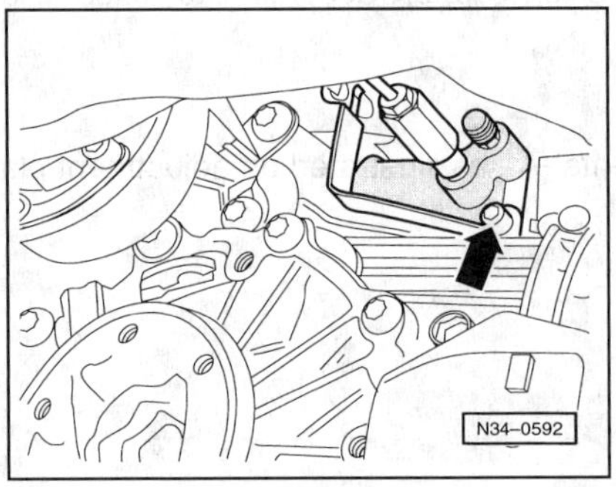

◀ Remove clutch slave cylinder (arrow) and secure with wire, do not disconnect pipe.

Then remove engine speed sender from clutch housing (if installed) (arrow).

– Lower transmission carefully.

– When lowering the transmission change position of transmission via the transmission jack spindle when in the area of the drive shafts.

When working on the transmission, it must be secured to the repair stand with transmission support VW 353 and the retaining plate VW 309.

Installing

– Clean input shaft splines and lightly grease with G000 100 grease.

The clutch plate must be able to slide lightly to and fro on the input shaft.

– Check clutch release bearing for wear, replace if necessary.

– Check whether dowel sleeves for aligning engine/transmission are installed in cylinder block, install if necessary.

– Ensure that the intermediate plate is correctly positioned on engine (if installed).

– Thinly coat clutch release lever surface contacting plunger.

– The threads of all threaded holes, in which self-locking bolts are screwed, should be cleaned of locking fluid remains using a thread cutter.

– Assemble transmission jack with transmission support 3282, adjustment plate 3282/10 for "012" "01A" transmission and support elements as follows:

– Place adjustment plate 3282/10 on transmission support 3282. (Adjustment plate only installs in one position).

– Align arms of transmission support according to holes in adjustment plate.

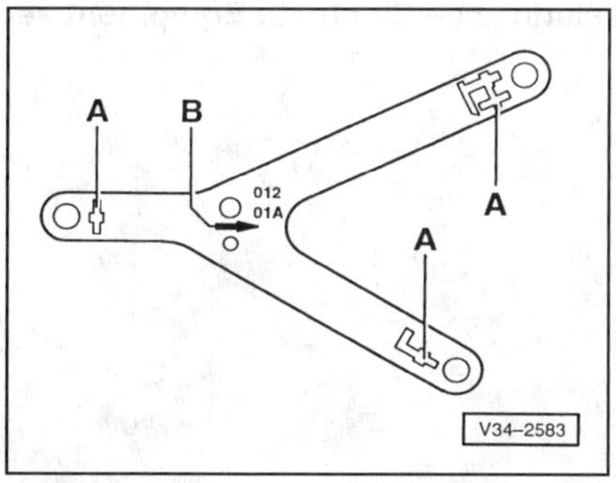

Bolt on support elements -A-, as illustrated on adjustment plate.

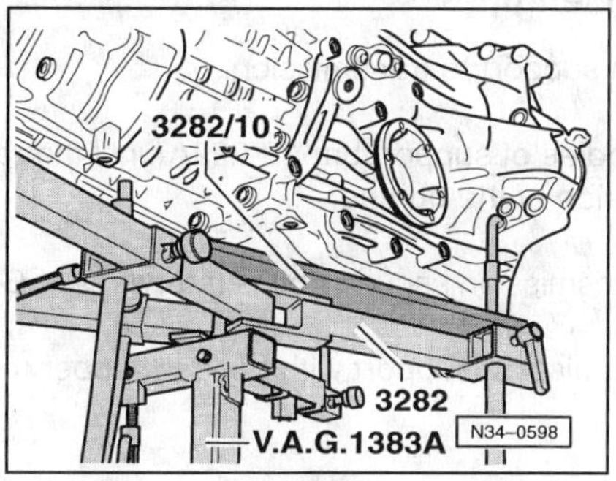

Place transmission on transmission jack.

– Align adjustment plate and transmission parallel with one another.

◀ Lock safety support (arrow) on transmission.

– Place transmission jack under vehicle.

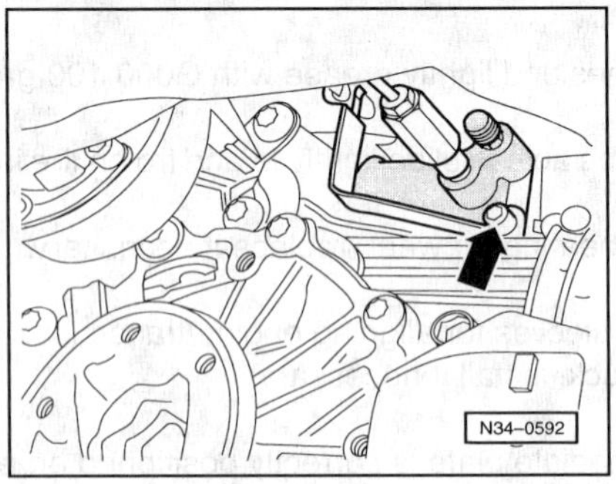

◀ Lift transmission until the clutch slave cylinder with the bracket for the pipe/hose can be installed (arrow).

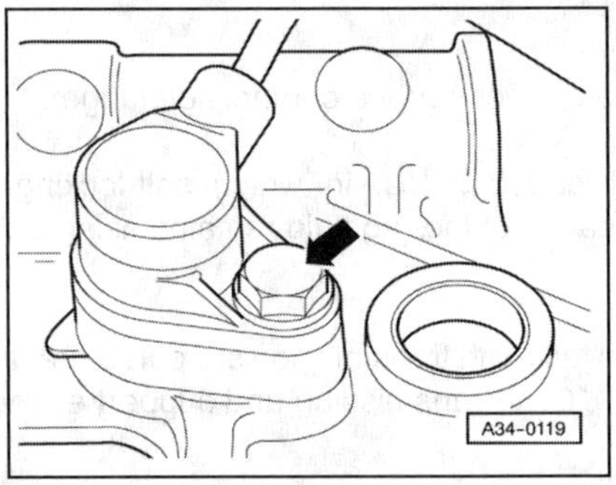

◀ Then install engine speed sender -G28-, if previously installed, on transmission housing (arrow).

– Install intermediate plate (if installed) onto engine dowel sleeves between transmission and engine.

– Ensure when inserting the transmission that the drive shaft flanges are located in the transmission flange shafts.

– Insert transmission.

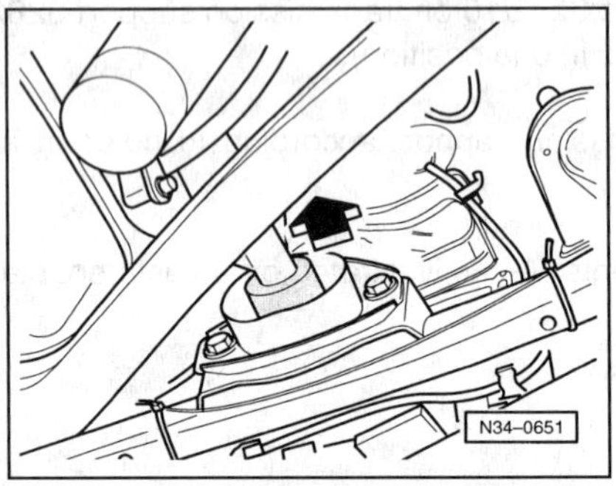

◀ When installing, press engine up at torque rod support (in direction of arrow) if necessary (2nd mechanic).

– Install lower engine/transmission securing bolts and bolts for starter when doing this install starter, see **Electrical Equipment section, Repair Group 27.**

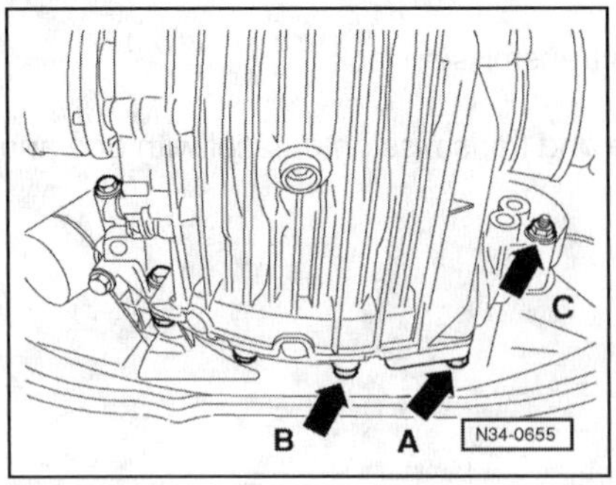

Vehicles with 6-Cyl. engines

– Bolt right transmission support to transmission.

– Lift engine via the spindles of support bar 10-222A until the radiator fan lies against the underside of the bumper.

◀ Install lower engine/transmission securing bolts (arrows -A-, -B- and -C-).

– Bolt left and right transmission support with bonded rubber mounting to sub-frame.

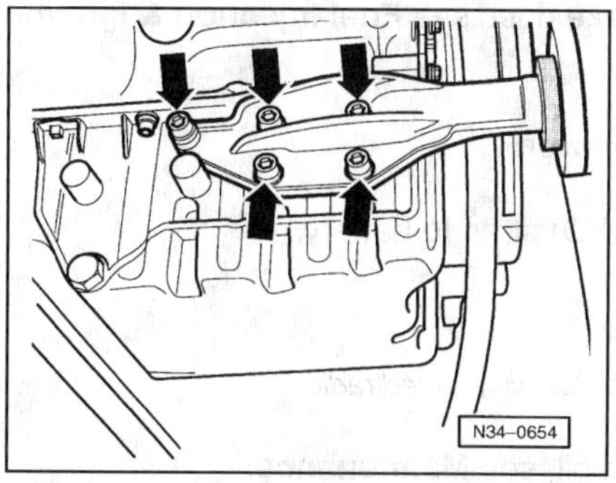

Bolt torque rod support to engine, see **Engine Mechanical section, Repair Group 17.**

– Tighten new nuts for both engine mountings to 20 Nm (15 ft-lb).

NOTE —
Install engine/transmission mountings stress-free.

All vehicles

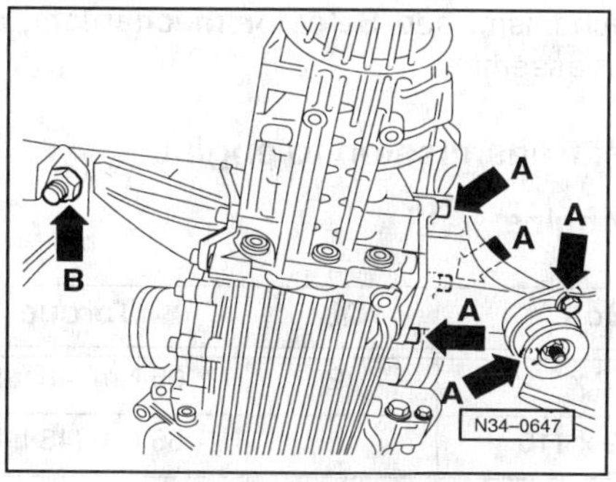

Bolt right transmission support with bonded rubber mounting to transmission and sub-frame (arrows -A-).

– Bolt left transmission support to bonded rubber mounting (arrow -B-) (illustrated in conjunction with 6-cyl. engine).

NOTE —
Install engine/transmission mountings stress-free.

– Install drive shafts to flange shafts.

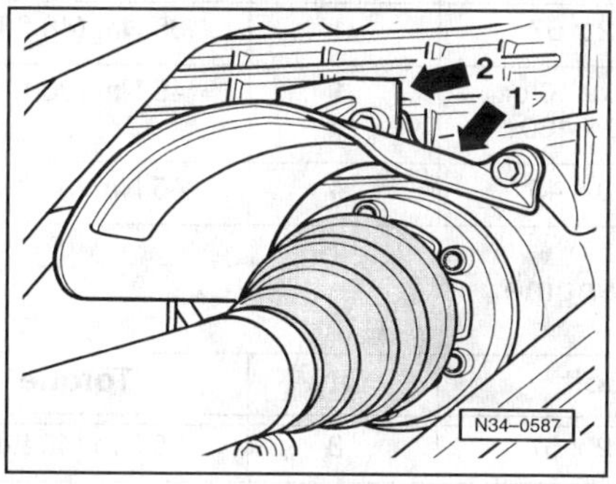

Install the cover plate above the right drive shaft (arrows 1 and 2).

– If installed install the cover plate above the left drive shaft.

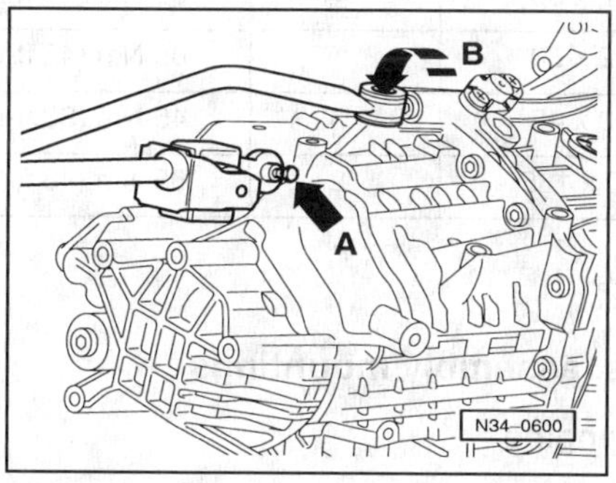

Push selector rod on so that the securing bolt installs in the recess in the selector shaft.

– Then tighten the selector rod first (arrow -A-).

– Then tighten push rod to transmission (arrow -B-).

– Install front exhaust system with Three Way Catalytic Convertor (TWC), see **Engine Mechanical section, Repair Group 26.**

– Install insulation tray below engine/transmission, if installed.

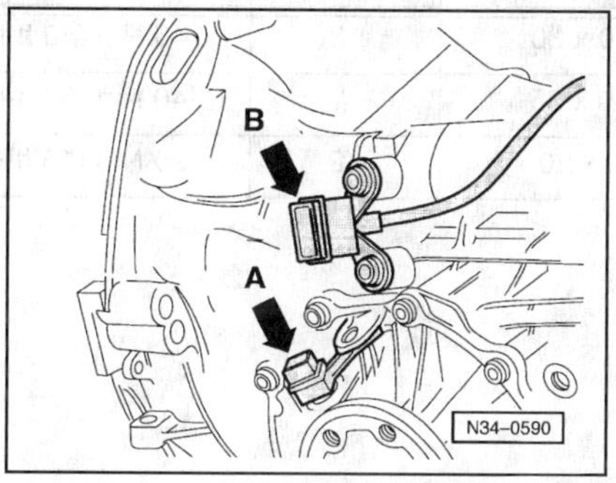

Connect reversing light connector (arrow -B-).

– Connect speedometer sender connector (arrow -A-).

– Install upper engine/transmission securing bolts.

– Secure coolant system expansion tank.

– Install intake hose to air mass meter, see **Fuel Injection & Ignition section, Repair Group 24.**

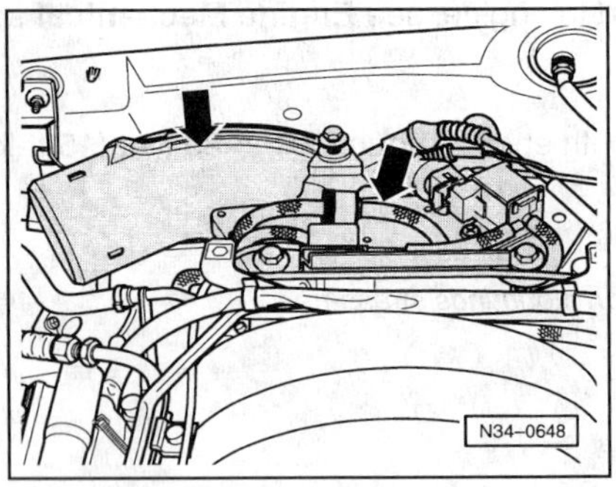

< Then assemble air intake duct, see **Fuel Injection & Ignition section, Repair Group 24.**

– Install intake manifold cover.

– Reconnect Ground (GND) strap to battery.

NOTE —

Note radio coding for vehicles with coded radio.

– Checking transmission oil, see **Maintenance**.

– Adjusting selector mechanism, see **Selector mechanism, adjusting** for appropriate selector assembly.

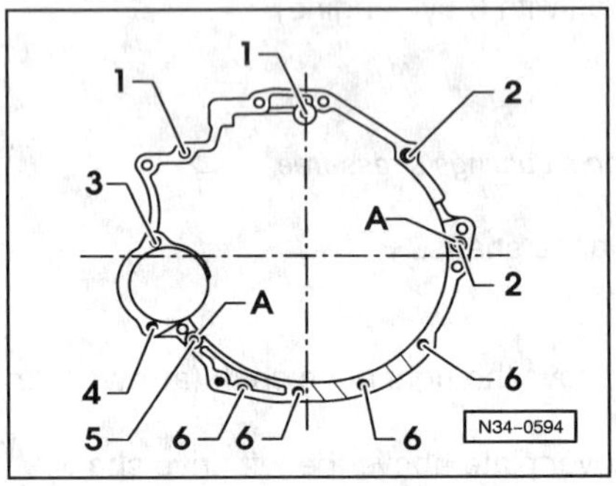

Tightening torques, transmission to engine

< **Vehicles with 4-Cyl. engine**

Item	Bolt	Qty.	Torque
1	M 12 x 75	2	65 Nm (48 ft-lb)
2	M 12 x 110	2	65 Nm (48 ft-lb)
3	M 12 x 90	1	65 Nm (48 ft-lb)
4	M 12 x 67	1	65 Nm (48 ft-lb)
5	M 10 x 110 or M 10 x 130	1	45 Nm (33 ft-lb)
6	M 10 x 45	4	45 Nm (33 ft-lb)

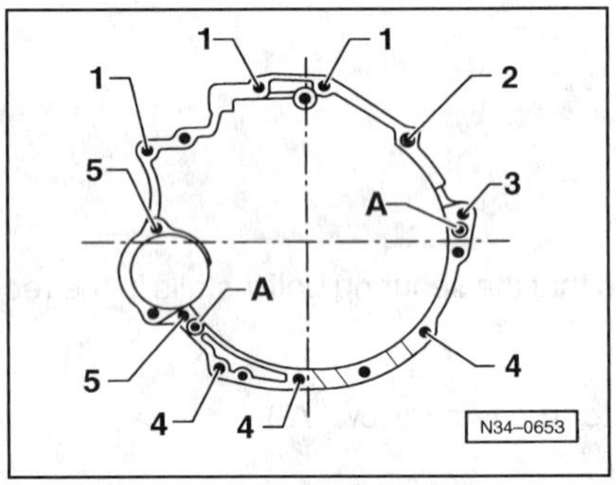

< **Vehicles with 6-Cyl. engine**

Item	Bolt	Qty.	Torque
1	M 12 x 67	3	65 Nm (48 ft-lb)
2	M 12 x 90	1	65 Nm (48 ft-lb)
3	M 12 x 80	1	65 Nm (48 ft-lb)
4	M 10 x 45	3	45 Nm (33 ft-lb)
5	M 10 x 135	2	65 Nm (48 ft-lb)

Item A = Dowel sleeves

Tightening torques, assembly mountings

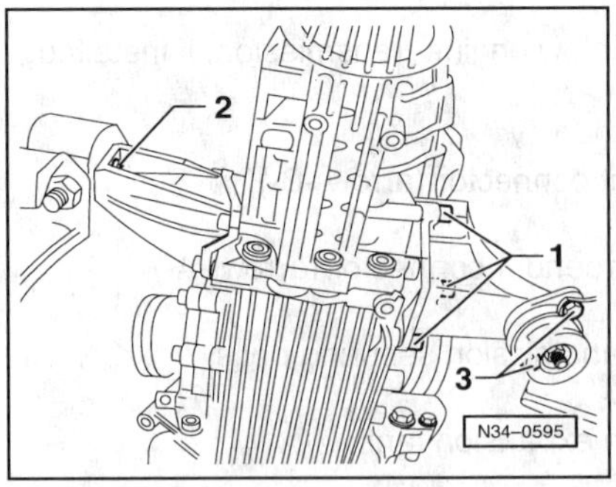

< **Vehicles with 4-Cyl. engine**

Item	Bolt	Qty.	Torque
1	M 10 x 30	3	40 Nm (30 ft-lb)
2	M 10 x 35	1	40 Nm (30 ft-lb)
3	M 8 x 20	2	20 Nm (15 ft-lb)

Vehicles with 6-Cyl. engine

Item	Bolt / nut	Qty.	Torque
1	M 10 x 30	3	40 Nm (30 ft-lb)
2	M 8 x 20	2	20 Nm (15 ft-lb)
3	M 10	1	45 Nm (33 ft-lb)

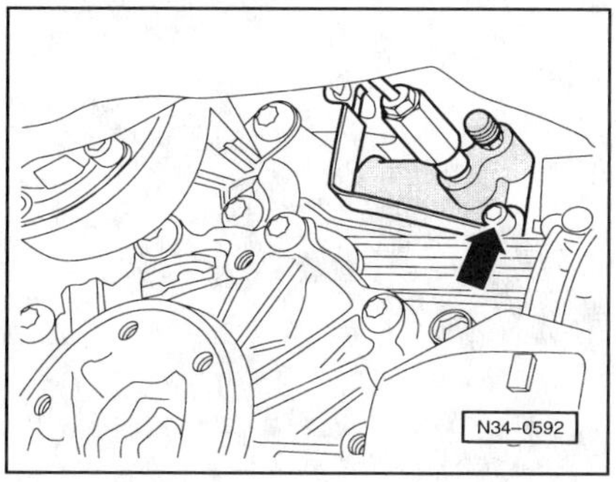

Additional torques	
Drive shaft to flanged shaft: M 8 M10	 45 Nm (33 ft-lb) 70 Nm (52 ft-lb)
Cover plate over the drive shafts to transmission	20 Nm (15 ft-lb)
Clutch slave cylinder to transmission (arrow)	20 Nm (15 ft-lb)
Selector rod (selector joint) to transmission (arrow -A-)	23 Nm (17 ft-lb)
Push rod to transmission (arrow -B)-	40 Nm (30 ft-lb)

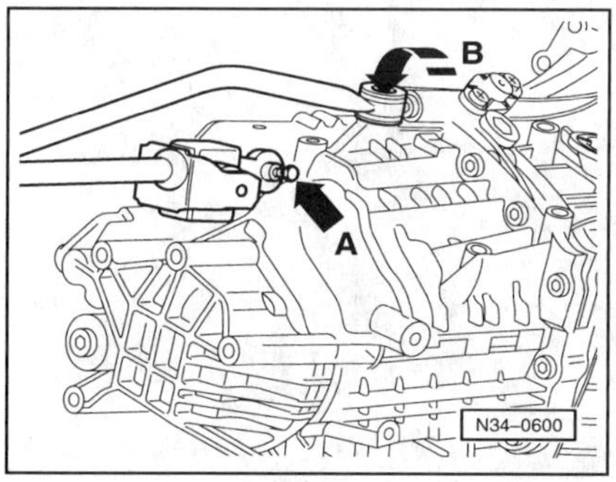

Bolts and other components which have direct contact with the transmission casing have a surface which is matched to the transmission casing depending upon whether it is manufactured from aluminium or magnesium.

39 Final Drive, Differential

Repair Group

M 39

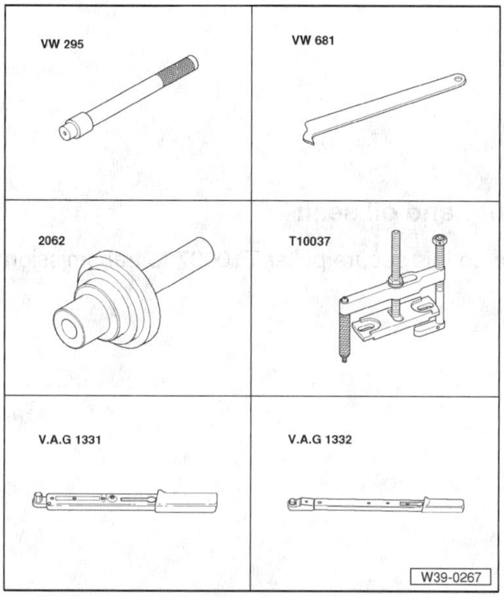

Drive axle flange seal, replacing

(Transmission in vehicle)

Special tools and workshop equipment required

- VW 295 Drift
- VW 681 Extractor
- 2062 Punch
- T10037 Puller
- VAG 1331 Torque wrench
- VAG 1332 Torque wrench

Removing

– Turn steering to left onto full lock stop.

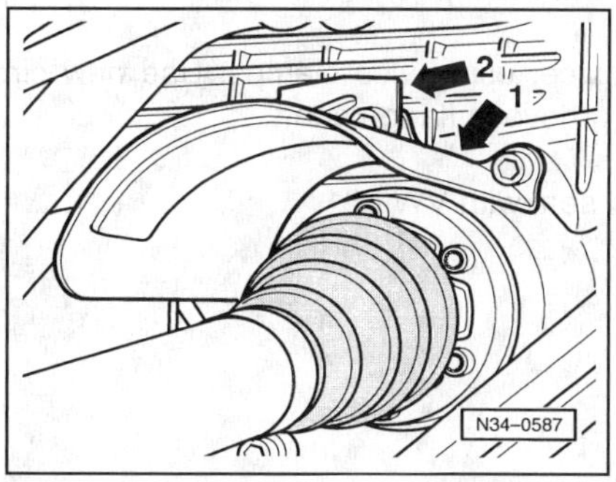

< Remove cover plate over the right-hand drive shaft (arrows 1 and 2).

– If fitted, remove cover above left-hand drive shaft.

– Place drip tray underneath.

– Disconnect drive shaft from flanged shaft.

– Tie-up drive shaft as high as possible. Do not damage paint on drive shaft.

When replacing left-hand oil seal:

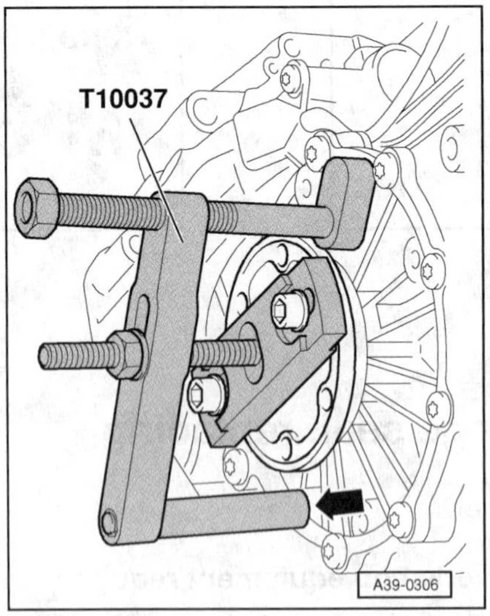

Remove flange shaft, to do this screw puller T10037 onto differential cover (arrow).

When replacing right-hand oil seal:

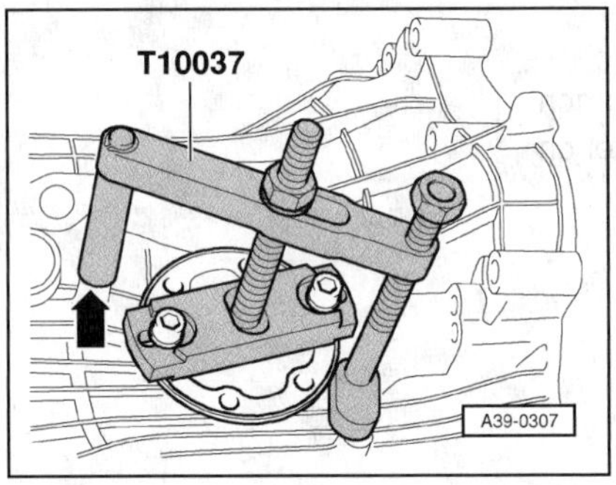

Remove flange shaft, to do this secure puller T10037 to transmission housing (arrow).

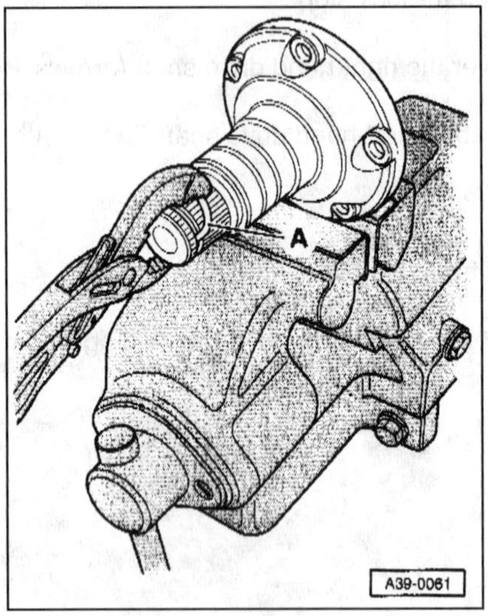

Always replace flange shaft circlip.

– Clamp flange shaft in vice, using vice clamps. Use new circlip -A- to press old circlip out of groove in flange shaft.

– Pull out flange shaft oil seal with VW 681.

Drive axle flange seal, installing

NOTE —
- *Repair pressing depth 6.5 mm if same flanged shaft is used again.*
- *Seal can also be driven in with punch 2062.*

– Lightly oil outer circumference of oil seal.

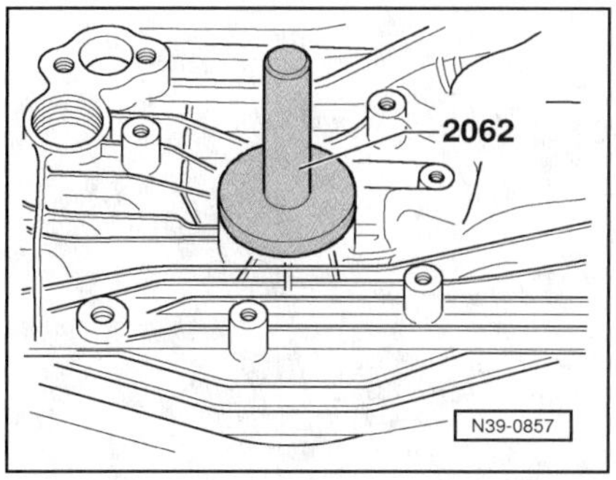

◄ Knock new right-hand flange shaft oil seal in 5.5 mm below housing outer surface, do not cant seal when doing this.

– Fill space between sealing lip and dust lip with grease G 052 128 A1.

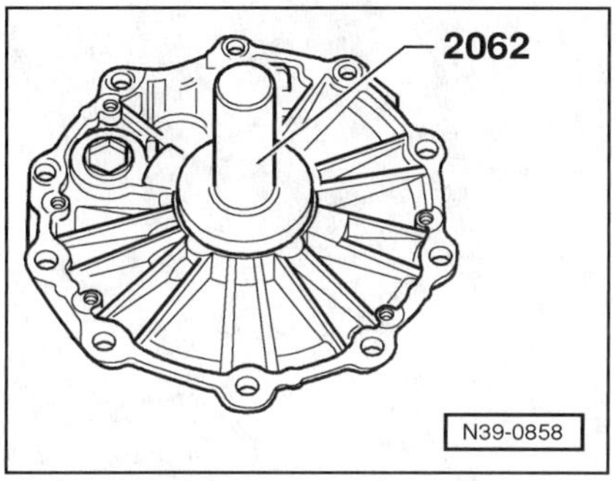

◄ Knock new left-hand flange shaft oil seal in 5.5 mm below housing outer surface, do not cant seal when doing this.

– Knock flange shaft in with drift VW 295.

– Bolt drive shaft to flange shaft.

– If part of vehicle equipment refit drive shaft cover plates.

– Check transmission oil, see **Maintenance**.

Tightening torques	
Drive shaft to flanged shaft M 8	45 Nm (33 ft-lb)
Drive shaft to flanged shaft M10	70 Nm (52 ft-lb)
Cover plates above the drive shafts to transmission	20 Nm (15 ft-lb)

Speedometer sender, replacing

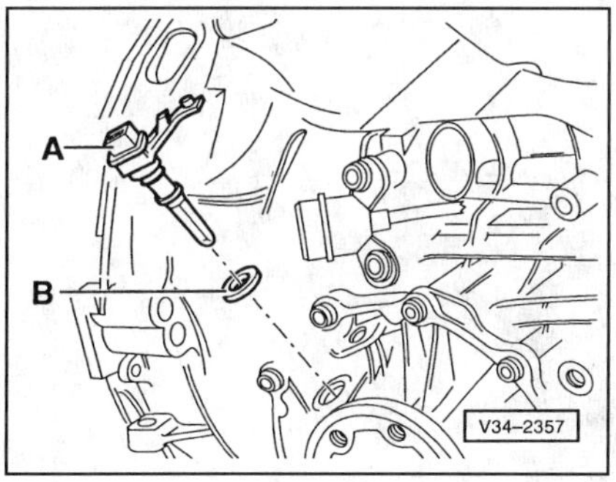

◄ Pull connector off sender -A-.

– Press retaining clip down, turn sender sideways and pull out upwards.

– When installing replace seal -B-.

– Handle sender carefully. If damaged it is possible that the speedometer will not indicate exactly.

5-Speed Automatic Transmission 01V, Front Wheel Drive

Section

N

5-Speed Automatic Transmission 01V Front Wheel Drive

Section N

00 General Information

Repair Group

N 00

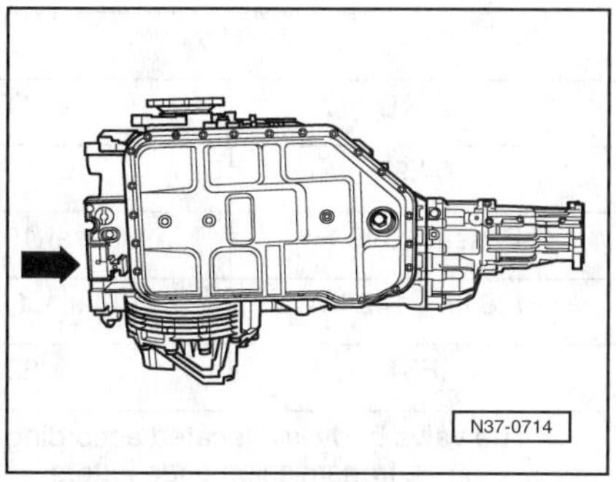

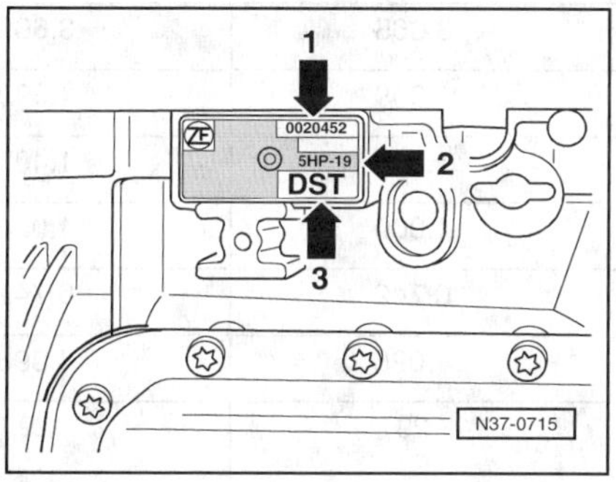

Transmission, identifying

The 5-speed automatic transmission 01V is installed in the Passat from 1998 in combination with 4 and 6-cylinder engines.

Location of code letters on transmission

The transmission code letters are given on the type plate at the bottom of the transmission -arrow-.

Transmission code letters and transmission serial number

Transmission Serial No. (arrow 1)
Transmission identification (arrow 2)
Code letters (arrow 3)

The code letters of the transmission are also given on the vehicle data plates.

5-speed automatic transmission 01V, notes

Torque converter

The torque converter is equipped with a lock-up clutch. The closing of the lock-up clutch is determined by the engine load and speed and produces very little vibration. The lock-up clutch drives all forward gears mechanically without slip.

Transmission

The 5-speed automatic transmission 01V has five hydraulically activated forward gears. When the lock-up clutch is closed these forward gears become mechanically driven gears by cutting out the torque converter slip.

Control module -J217- with driving situation gear changes (Fuzzy Logic)

Determining the driving situation and driving resistance dependent gear change point happens automatically.

Advantages:

- Gear changes will be consumption orientated
- Maximum engine output is always available
- Individual adaption of gear change points in all driving situations
- Gear change points are infinitely variable

Gear change point variations for gradients

An additional gear change map automatically selects gear changes for gradients dependent upon accelerator pedal position and driving speed.

- Gear change map for extreme uphill stretches is matched to engine output
- Gear change map for extreme downhill stretches is matched to the braking effect of the engine

Code letters, allocation, ratios, equipment

Automatic transmission			**01V**	
Transmission	Code letters		**DRD**	**DDT**
	Manufactured	from	04.97	04.97
Allocation	Model		Passat 1998 ➤	Passat 1998 ➤
	Engine		2.8 ltr. - 142 kW	1.8 ltr. - 110 kW
Torque converter	Code letters		F31	M28
Valve body	Code letters		The valve body is allocated according to the transmission code letters	
Ratios	1st gear		3.665	3.665
	2nd gear		1.999	1.999
	3rd gear		1.407	1.407
	4th gear		1.000	1.000
	5th gear		0.742	0.742
	Reverse gear		4.096	4.096
Intermediate drive	No. of teeth	Input gear	29	29
		Output gear	35	35
	Ratio		1.207	1.207
Final drive	No. of teeth	Drive pinion	11	11
		Crown wheel	30	34
	Ratio		2.727	3.091
Data bus			yes	no
Hydraulic control			E17	E17

Automatic transmission		01V		
Transmission	Codes	EZZ	EFP	EKC
	Manufactured from	06.00	08.99	08.99
Allocation	Model	Passat 1998 ➤	Passat 1998 ➤	Passat 1998 ➤
	Engine	2.8 ltr. - 142 kW	1.8 ltr. - 110 kW	2.8 ltr. - 142 kW
Torque converter	Codes	N28	M28	N28
Valve body	Code letters	The valve body is allocated according to the transmission code letters		
Ratios	1st gear	3.665	3.665	3.665
	2nd gear	1.999	1.999	1.999
	3rd gear	1.407	1.407	1.407
	4th gear	1.000	1.000	1.000
	5th gear	0.742	0.742	0.742
	Reverse gear	4.096	4.096	4.096
Intermediate drive	No. of teeth Input gear	29	29	29
	Output gear	35	35	35
	Ratio	1.207	1.207	1.207
Final drive	No. of teeth Drive pinion	11	11	11
	Crown wheel	34	34	34
	Ratio	3.091	3.091	3.091
Data bus		yes	no	no
Hydraulic control		E18/2	E18/2	E18/2

Capacities

Planetary transmission

Capacities	Planetary transmission	Automatic transmission
Initial filling	9.0 ltr.	01V
Changing	approx. 2.6 - 3.0 ltr.[1)]	
Lubricant	VW ATF	

1) Filled for life: Oil change required only following repair work or removal of oil pan up to 3 liters.

VW ATF is obtainable as a replacement part.

Container size 1.0 ltr. -Part No. G 052 162 A2

Only VW ATF (available as a replacement part) may be used in the planetary transmission of automatic transmission 01V.

Front final drive

Capacities	Front final drive	Automatic transmission
Initial filling	approx. 0.75 ltr.	01V
Oil change	Filled for life No change	
Lubricant	Gear oil SAE75 W 90 (synthetic oil)	

Gear oil SAE 75 W90 (synthetic oil) is available as a replacement part.

Size 1.0 ltr. - Replacement part No. G 052 145 S2

Checking oil level in front final drive, see **Maintenance.**

Transmission servicing, notes

The maximum possible care and cleanliness and proper tools are essential to ensure satisfactory and successful transmission repairs. The usual basic safety precautions also, naturally apply when carrying out vehicle repairs.

- If the automatic transmission has been exchanged, fluid levels are to be checked.
- Engine must not run if oil pan or ATF has been removed. Do not tow vehicle.
- After removing transmission, prevent torque converter from falling out.
- Thoroughly clean connecting points and surrounding areas and then loosen bolts.
- Before installing transmission, check fitting aperture for torque converter, see **Repair Group 32, Torque convertor, installing**.
- When installing ensure that the dowel sleeves between the engine and transmission are correctly located.
- Place removed parts on a clean surface. Cover parts to prevent soiling. Use plastic sheeting and paper. If necessary, use only lint-free cloth!
- Only install clean parts: Remove packing from replacement parts immediately prior to installation and not before.
- If repair work cannot be performed immediately, cover replacement parts from which packing has been removed.

Gaskets, sealing rings

- Always replace O-rings, oil seals and gaskets.
- Once oil seal/gasket has been removed, check contact surface on housing or shaft for burrs and damage. Rectify as necessary.
- Thoroughly clean housing connecting surfaces and apply sealant AMV 188 200 03.

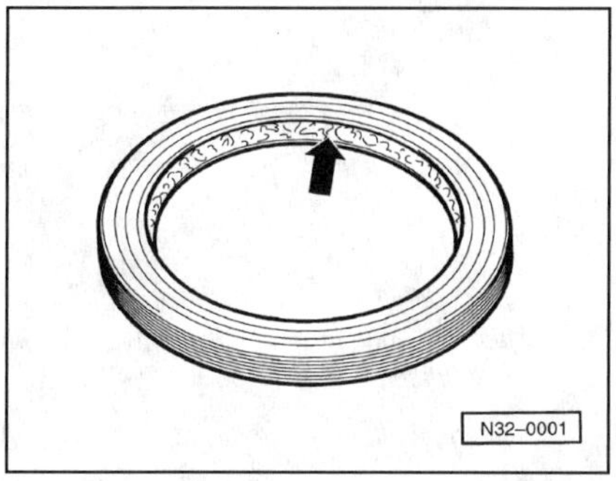

Before installing shaft-type oil seal, half fill space between sealing lips with grease G 052 128 A1 -arrows- and then install.

- Open side of oil seal faces oil.
- Before installing oil seals, moisten circumference of seal and sealing lip with ATF or gear oil depending on fitting location.
- Moisten O-rings with ATF to prevent damage occurring to the rings during assembly.
- Only use ATF for automatic transmission components. Other types of lubrication cause faults to occur in the transmission hydraulics.
- Check ATF level in planetary transmission or gear oil level in final drive following installation. Top up as necessary.

Guided fault finding, vehicle self-diagnosis and measuring technology

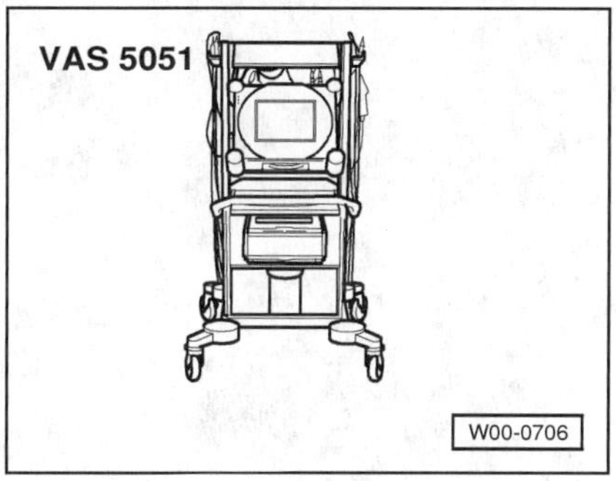

Before repair work is carried out on the automatic transmission, the fault cause should be determined as accurately as possible using vehicle diagnosis, testing and information system VAS 5051 in modes of operation "Guided fault finding", "Vehicle self-diagnosis" and "Test instruments".

32 Torque Converter

Repair Group

N 32

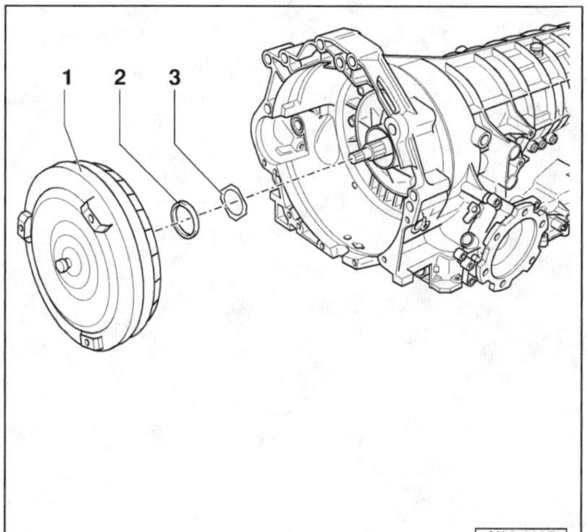

N32-0037

Torque converter

1. **Torque converter**
   - Secure in place when transmission is removed
   - Identification, see **Torque converter, identifying**
   - Installing, see **Torque convertor, installing**
2. **Oil seal**
   - Replacing, see **Torque converter oil seal, removing and installing**
3. **Bearing ring**
   - Behind oil seal
   - Replace if damaged

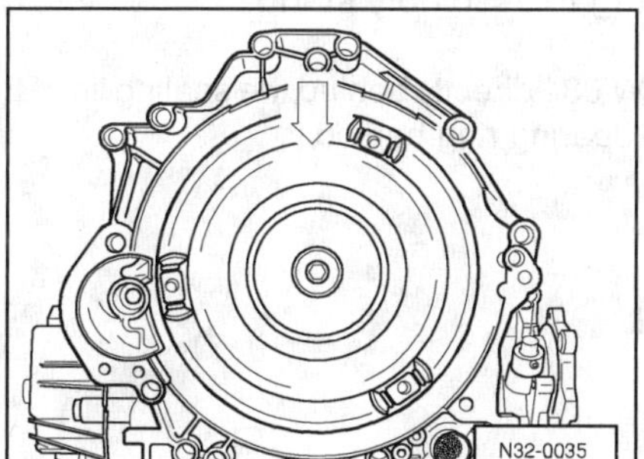

N32-0035

Torque converter, identifying

◄ There are various torque converters. They are identified by code letters (arrow).

Transmission/torque converter application, see **General Information.**

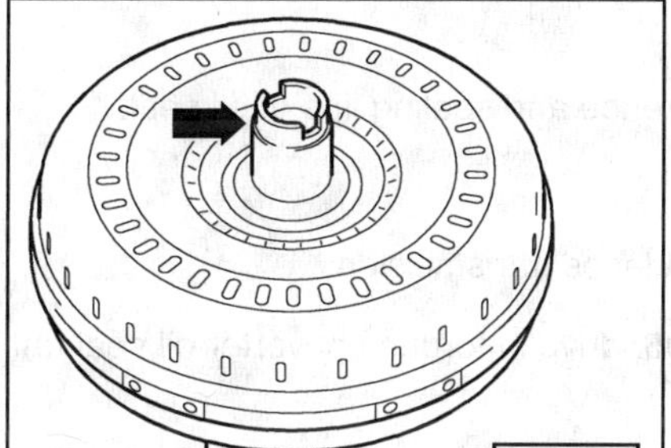

V32-0048

Torque converter, checking

◄ Check torque converter hub for irregular grooves -arrow-.

The torque converter is a sealed unit (welded) and must be replaced if damaged.

Torque converter oil seal, removing and installing

Special tools and equipment

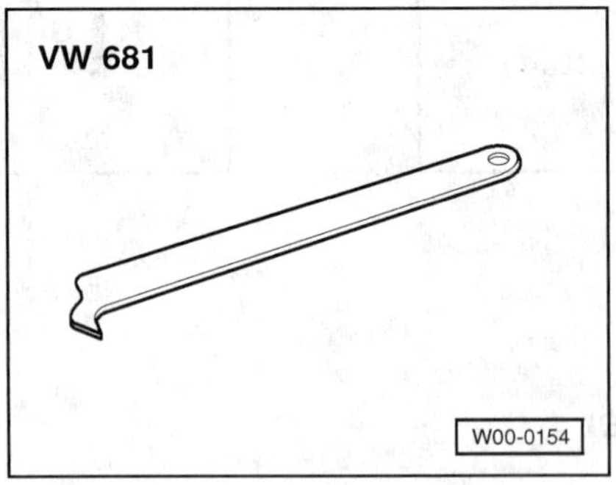

- VW 681 Extractor lever

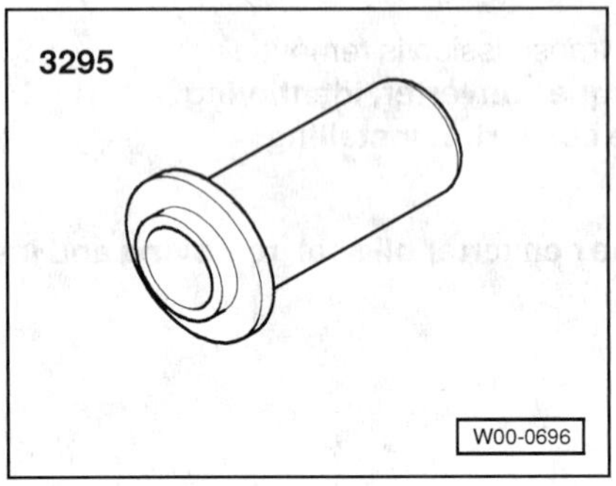

- 3295 Thrust piece

Removing

– Remove transmission, see **Repair Group 37, Transmission, removing and installing.**

– Secure transmission on repair stand, see **Repair Group 37, Transmission, securing to assembly stand.**

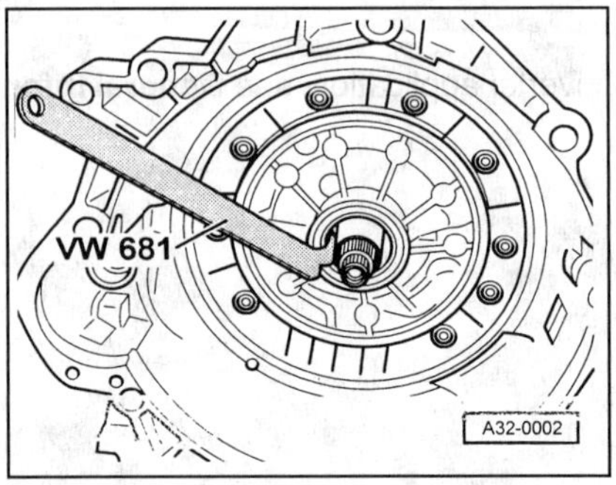

- Apply extractor lever VW 681 directly behind the sealing lip of the oil seal to avoid damaging the bearing ring behind it.

Installing

– Insert bearing ring.

– Apply ATF to circumference and sealing lips of oil seal.

– Installation position:

• Open side of oil seal faces transmission.

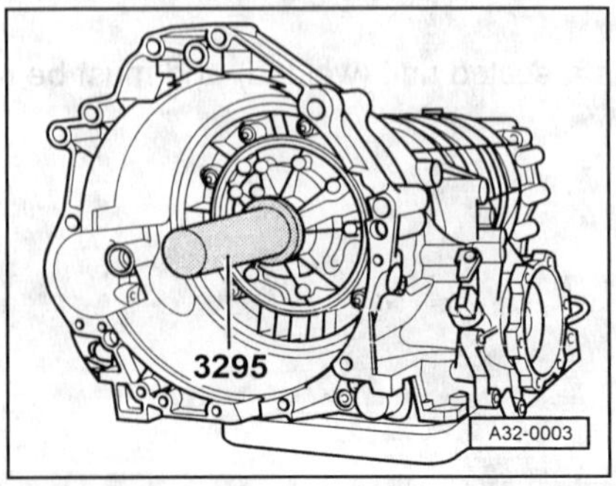

- Using thrust piece 3295, drive in torque converter oil seal until stop of thrust piece.

– Install transmission, see **Repair Group 37, Transmission, removing and installing**.

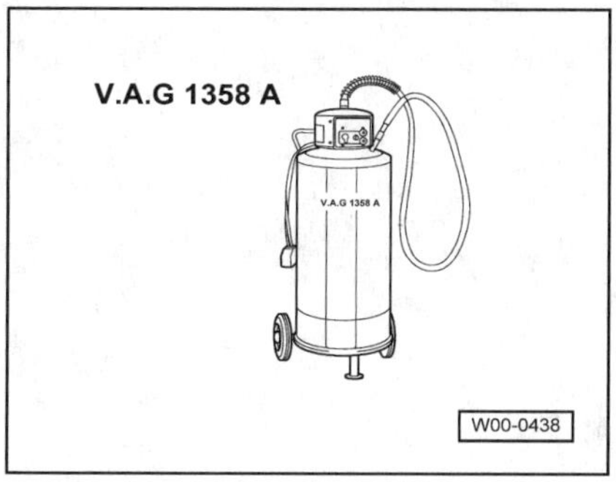

Torque converter, draining

Special tools and equipment

V.A.G 1358 A Oil extractor

• V.A.G 1358 A/1 Oil extractor probe

If the torque converter has become fouled by abrasion or when performing a major overhaul of the transmission, drain the torque converter as follows:

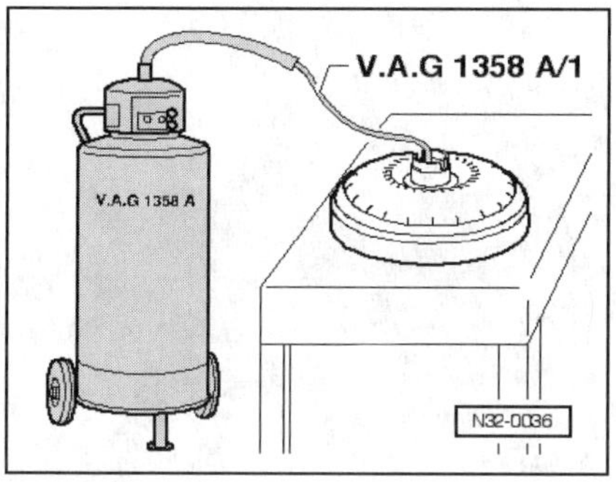

Extract ATF from torque converter using V.A.G 1358 A and probe V.A.G 1358 A/1.

Torque convertor, installing

Special tools and equipment

• Depth gauge

– Push on torque converter hub to first stop via oil seal.

– Turn torque converter inward using light pressure until opening in hub engages with follower of pump sprocket and until torque converter can be felt to slip in place.

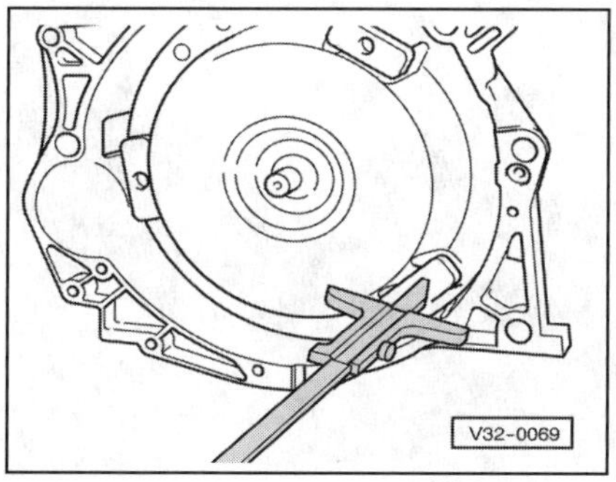

If the converter is correctly positioned, the distance between the contact surfaces below at the tapped holes in the converter and the contact surface of the converter bell of automatic transmission 01V is about 23 mm.

If the converter is not correctly positioned, the distance is about 11 mm.

> ***WARNING —***
> *If the torque converter is left in the wrong installation position, the follower of the converter or ATF pump will be destroyed when the transmission is connected to the engine.*

37 Controls, Housing

Repair Group

N 37

Shift Mechanism

Selector lever mechanism, checking

In selector lever positions "2", "3", "4", "D" and "R", do not operate starter motor.

The starter motor may only be operated in selector lever positions "P" and "N", providing the selector lever handle lock button is not pressed.

If selector lever position "N" is selected at speeds greater than 5 km/h, the selector lever lock solenoid must not engage and block the selector lever. Selector lever can be shifted in a driving gear.

If selector lever position "N" is selected at speeds less than 5 km/h (almost stationary), the selector lever lock solenoid should engage after approx. 1 second. Selector lever can only be shifted out of position "N" once foot brake has been depressed.

Selector lever in "P" position and ignition switched on:

– Brake pedal not depressed:
  - Selector lever is locked and cannot be shifted out of position "P" with button pressed. Selector lever lock solenoid blocks selector lever.

– Brake pedal depressed:
  - Selector lever lock solenoid releases selector lever. It is possible to shift into a driving gear. Shift selector lever slowly from "P" to "2" and, whilst doing this, check whether the display in the instrument cluster shows the correct selector lever position.

Selector lever in "N" position and ignition switched on:

– Brake pedal not depressed:
  - Selector lever is locked and cannot be shifted out of "N" position with button pressed. Selector lever lock solenoid blocks selector lever.

– Brake pedal depressed:
  - Solenoid for selector lever lock releases selector lever. It is possible to shift into a driving gear.

Selector lever cable, checking and adjusting

Special tools, workshop equipment, testers, measuring instruments and auxiliary items required

- 80-200 Removal lever
- V.A.G 1331 Torque wrench

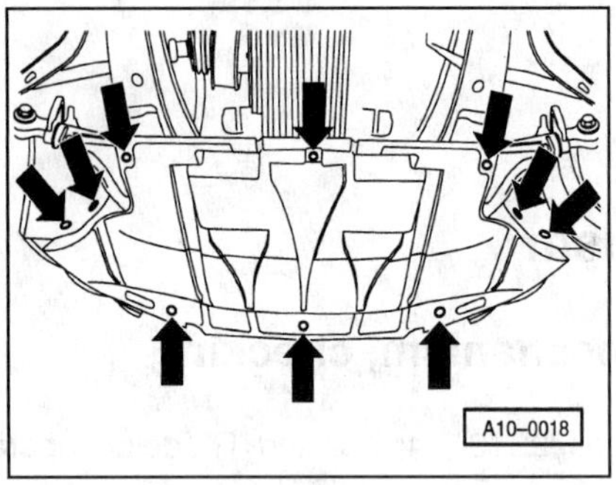

Checking

– Shift selector lever into "P" position.

– Raise vehicle.

◄ Remove noise insulation tray -arrows-.

Vehicles with V6 engine

– Remove front exhaust system with catalyst, see **Engine Mechanical section, Repair Group 26.**

– Remove selector lever cable protective plate.

Continued for all vehicles

– Remove left-hand transmission bracket, see **Transmission brackets, removing and installing**.

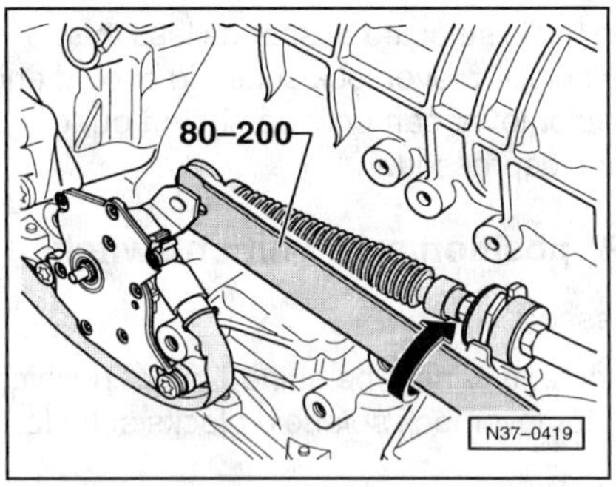

◄ Remove selector lever cable from selector shaft lever -arrow-.

– Shift selector lever from "P" to "2".

- Selector lever mechanism and selector lever cable must move freely. If necessary replace selector lever cable or service selector lever mechanism.

– Shift selector lever into "P" position.

– Bring selector shaft lever in position "P" on transmission. Locking lever must engage in park lock wheel, both front wheels are blocked (cannot be turned together in a direction).

- It should be possible to push the selector lever cable onto the selector shaft lever. Adjust selector lever cable if necessary.

– Using a pair of pliers, carefully push the selector lever cable onto the selector shaft lever. When doing this, take care not to bend the selector shaft lever, otherwise the selector mechanism will not be able to be adjusted correctly.

Adjusting

- Selector lever cable disconnected from selector shaft lever.

– Shift selector lever into "P" position.

– Bring selector shaft lever in position "P" on transmission. Locking lever must engage in park lock wheel, both front wheels are blocked (cannot be turned together in a direction).

◄ Loosen bolts -arrows- securing support bracket to transmission.

– Using a pair of pliers, carefully push the selector lever cable onto the selector shaft lever. When doing this, take care not to bend the selector shaft lever, otherwise the selector mechanism will not be able to be adjusted correctly.

– Align selector lever cable, free of stress, on the support bracket using the oblong holes.

– Tighten securing bolts -arrows- for support bracket.

– Check selector lever mechanism, see **Selector lever mechanism, checking**.

– Remove left-hand transmission bracket, see **Transmission brackets, removing and installing**.

Vehicles with V6 engine

– Fit selector lever cable protective plate.

– Install front exhaust system with catalyst, see **Engine Mechanical, section, Repair Group 26.**

Continued for all vehicles

– Install noise insulation.

Selector lever position signal, checking

The Engine Control Module (ECM) receives the following information from the transmission control module: Driving mode selected (selector lever in 2/3/4/R/D) or no driving mode selected (selector lever in P or N)

Test requirement

• Selector lever of automatic transmission must be in "P" or "N".

Test sequence

– Connect V.A.G1551 scan tool (VAG1552). Start engine and select Engine Control Module (ECM) using address word 01.

– Press buttons -0- and -8- to select "Read measuring value block" function 08. Press -Q- button to confirm input.

– Press buttons -0-, -5- and -6- to select "Display group number 56". Press -Q- button to confirm input.

– Observe display field 4.

– Press foot brake pedal and select driving position.

– Press → button.

– Press buttons -0- and -6- to select "End Output" function 06. Press -Q- button to confirm input.

– Switch ignition off.

– If reading does not change:

  • Check terminal resistance of CAN-bus, see **Terminal resistance for CAN-bus, checking.**

Shift mechanism, assembly

EDITOR'S NOTE —

Volkswagen identifies electrical components by a letter and/or a number in the electrical schematics. See **Wiring Diagrams, Fuses and Relays**. *These electrical identifiers are listed -between dashes- as an aid to electrical troubleshooting.*

Lubricate mounting and sliding surfaces with Polyurea grease, Part No. G 052 142 A2.

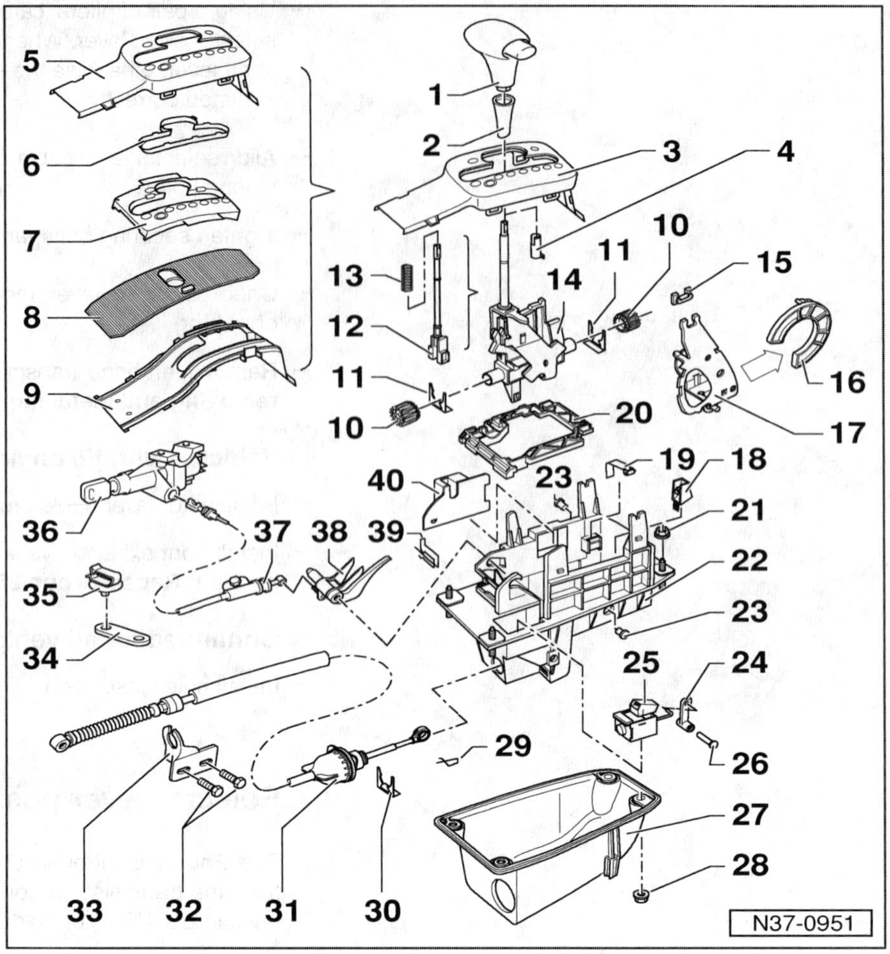

1. **Selector lever handle**
   - See **Removing and installing selector lever handle.**
2. **Sleeve**
   - Clips into selector lever handle
   - Installation position: Angled section faces inward
3. **Cover**
   - With trim piece **-6-**, symbol panel **-7-**, cover strip **-8-** and guide **-9-**
   - Lift up to remove
4. **Guide sleeve**
   - Slide onto selector lever **-14-**
   - Installation position: Engage foot to rear
5. **Cover molding**
   - With trim piece **-6-** and symbol panel **-7-**
   - Cannot be replaced as single item
6. **Trim piece**
   - Integrated in symbol panel **-7-** in cover molding **-5-**
   - Cannot be replaced as single item
7. **Symbol panel**
   - With printed circuit
   - With Tiptronic switch -F189- integrated in printed circuit
   - Integrated in trim piece **-6-** in cover molding **-5-**
   - Cannot be replaced as single item
8. **Cover strip**
   - With 2 solenoids for Tiptronic switch -F189- and gear indicator
   - Solenoids must be fitted securely on cover strip
9. **Guide**
   - Guides masking panel
   - Unclipping from cover molding **-5-**
10. **Mounting bush**
    - Secure in mounting bracket **-22-** with locking clip **-11-**
11. **Locking plate**
    - Secures mounting bush **-10-** in mounting bracket **-22-**
    - Install with angled ends pointing toward inside of mounting bracket
12. **Pull rod**
13. **Spring**
14. **Selector lever**
    - With pull rod, coil spring, selector gate and Tiptronic spring mechanism
15. **Stop buffer**
    - Clip into cable lever **-17-**
    - Install with large contact surface toward selector lever
16. **Detent**
    - Clip into cable lever **-17-**
17. **Cable lever**
    - Secured in mounting bracket **-22-** with fulcrum pin **-23-**
18. **Spring clip with roller**
    - For gearshift positions
19. **Spring clip with roller**
    - For Tiptronic position
20. **Frame**
    - Clip onto mounting bracket
    - Install with ribbed side upward
21. **Nut**
    - Tighten to 10 Nm (7 ft-lb)
22. **Mounting bracket**
23. **Fulcrum pin**
    - Secures cable lever **-17-** in mounting bracket **-22-**
24. **Locking pawl**
    - For selector lever lock
25. **Shift lock solenoid -N110-**
26. **Fulcrum pin**
    - For locking pawl
27. **Cover**
    - With seal for mounting bracket
28. **Nut**
    - Tighten to 10 Nm (7 ft-lb)

(continued on following page)

Shift mechanism, assembly (continued)

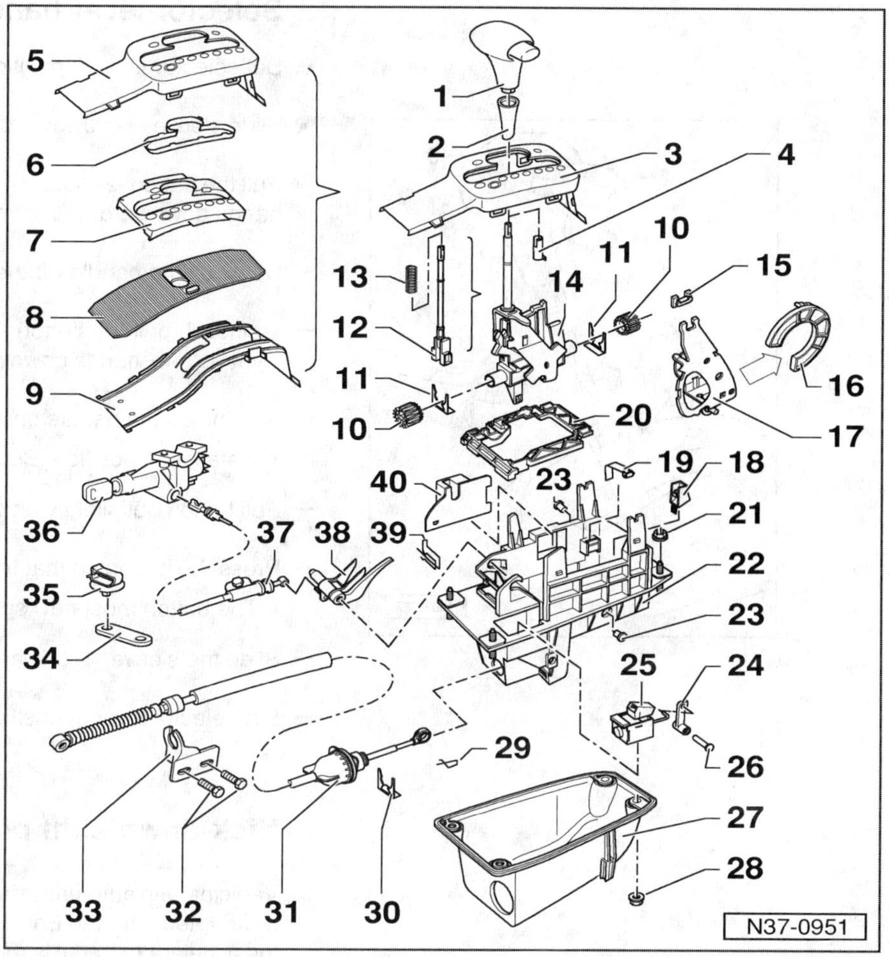

29. Securing clip
- Secures selector lever cable to cable lever **-17-**
- Install with angled end forward and upward

30. Locking plate
- Secures selector lever cable to mounting bracket
- Angled ends points to end of selector lever cable

31. Selector lever cable
- Do not bend or kink, lightly grease cable eye and ball socket before installing.
- Adjusting, see **Selector lever cable, checking and adjusting**
- Selector lever cable must be replaced if rubber sleeves are damaged
- When installing, make sure that rubber sleeve on transmission side is not twisted

32. Hex bolt
- Tighten to 23 Nm (17 ft-lb)

33. Support bracket
- For mounting selector lever cable on transmission

34. Bracket
- For locking cable

35. Cable clip
- For locking cable

36. Ignition/starter switch

37. Locking cable
- For ignition key removal lock
- Must not be kinked

38. Locking cable lever
- For ignition key removal lock

39. Securing spring
- For locking cable

40. Mounting
- For selector lever in Tiptronic position

Selector lever handle, removing and installing

– Set selector lever in position "3".

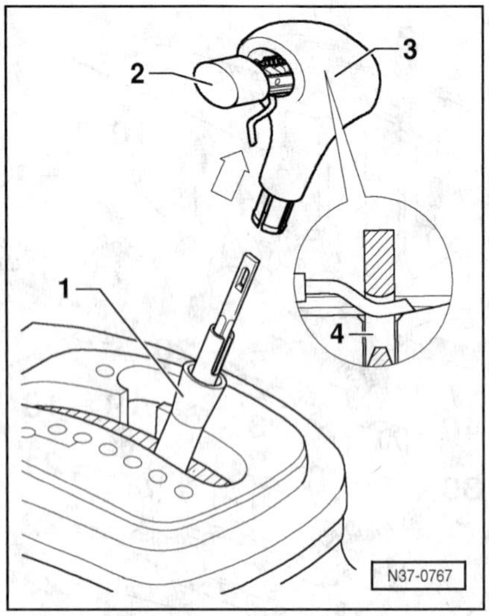

– Slide sleeve -1- downward.

– Pull the button -2- out of handle -3- just enough that the pull lever -arrow- hangs downward.

– Now pull the handle off the selector lever.

– To install, pull the button -2- out of handle -3- just enough that the pull lever -arrow- hangs downward.

– Now press the handle on selector lever onto stop.

- Installation position: Button points to driver.

– Pull button out slightly and guide the pull lever -arrow- into the handle.

– Press the button so that the pull lever engages in the pull rod -4-.

- The button must now spring back onto stop.

– Slide the sleeve -1- upward.

– Set selector lever in position "P".

Kick-down shift point, adapting

Vehicles with automatic transmission must have the kick-down shift point re-adapted after the Engine Control Module (ECM) is replaced and after the accelerator pedal is replaced.

Test requirement

- Ignition switched on, engine not running

Test sequence

– Connect V.A.G1551 scan tool (VAG1552). Switch ignition on and select Engine Control Module (ECM) using address word 01.

– Select "Basic Setting" function 04.

– Select "Display group number 63".

– Observe reading in display fields 3 and 4:

– Press accelerator pedal down past kick-down pressure point, as far as it goes.

– Hold accelerator pedal at impact for at least 2 seconds.

– Observe reading in display fields 3 and 4:

- Specification: Kick-down ADP.running, then Kick-down ADP.OK

– If display fields 3 and 4 indicate: Kick-down ERROR

- Check Throttle Position (TP) sensor, see appropriate **Engine Mechanical section, Repair Group 20**

Electrical/electronic components, locations

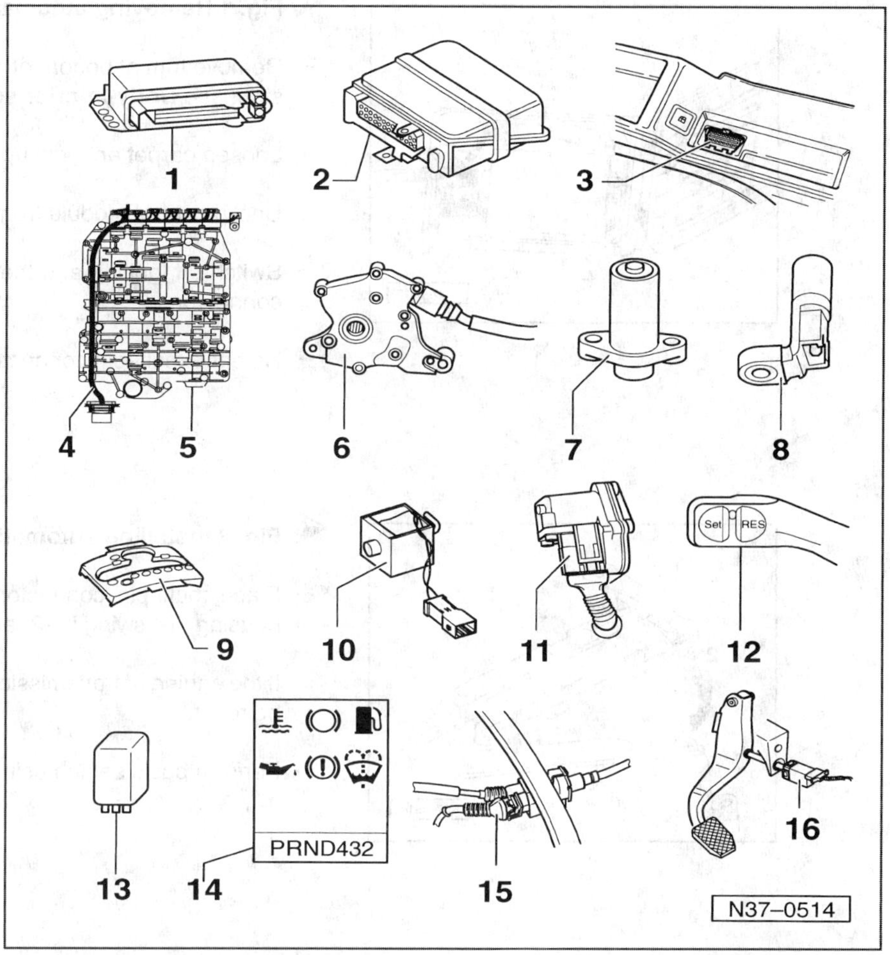

1. **Automatic transmission control module -J217-**
   - Removing, see **Fig. 1**
   - Installing, see **Fig. 2**
2. **Engine control module**
   - Located in left plenum chamber
   - If ecm is replaced, the system must be reset and the ecm coded using diagnostic scan tool
   - Control module coding, see **Scan Tool section**
3. **Data Link Connector (DLC)**
   - Installation location, see **Fig. 3**
4. **Conductor strip with -G93- Transmission fluid temperature sensor**
   - Location, see **Fig. 5**
5. **Valve body**
   - Location, see **Fig. 4**
   - With hydraulic control Type E17 or Type E18/2
   - Can be checked using "guided fault finding" on VAS 5051
6. **Multifunction switch -F125-**
   - Installation location, see **Fig. 7**
   - Removing and installing, see **Repair Group 38**
   - Can be checked using "guided fault finding" with the VAS 5051
7. **Sensor for Transmission RPM -G182-**
   - Records the transmission input speed
   - Installation location, see **Fig. 8**
   - Can be checked using "guided fault finding" with the VAS 5051
8. **Transmission vehicle speed sensor -G38- and Sender for transmission output RPM-G195-**
   - Records the transmission output speed
   - Installation location, see **Fig. 6**
   - Can be checked using "guided fault finding" with the VAS 5051
9. **Tiptronic switch -F189-**
   - Location, see **Fig. 12**
10. **Shift lock solenoid -N110-**
    - Location, see **Fig. 9**
11. **Throttle position (TP) sensors -G69- and -G79- and sender for accelerator pedal position -G185-**
    - Further information, see **Fig. 11**
12. **Cruise control switch -E45-**
    - Located on steering column switch
13. **Starting interlock relay -J207-**
    - Location, see **Wiring Diagrams and Component Loactions**
14. **Transmission Range (TR) selector lever display -Y6-**
    - If all segments of selector lever display light up together or
    - If selector lever display does not light up, transmission is in emergency running mode
15. **Kickdown switch -F8-**
    - Vehicles without EPC, switch is integrated into accelerator cable and is located on bulkhead in engine compartment (as shown)
    - Vehicles with EPC, see **Fig. 10**
16. **Brake light switch -F- or brake pedal switch -F47-**
    - Located in bracket at brake pedal

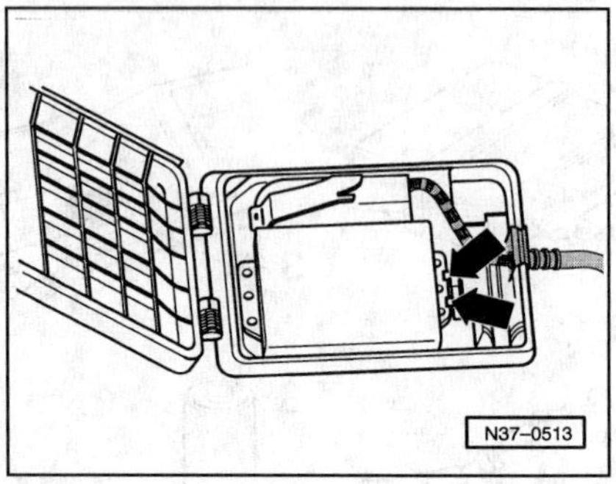

Fig. 1 Removing automatic transmission control module -J217-

– Remove trim at bottom of A pillar and remove right-hand door entry trim strip, see **Body-Interior section, Repair Group 70**

– Loosen carpet and fold up.

– Unclip control module from box -arrows-

– Switch off ignition and then wait 30 seconds before removing multi-pin connector.

– Release multi-pin connector, then pull connector off control module.

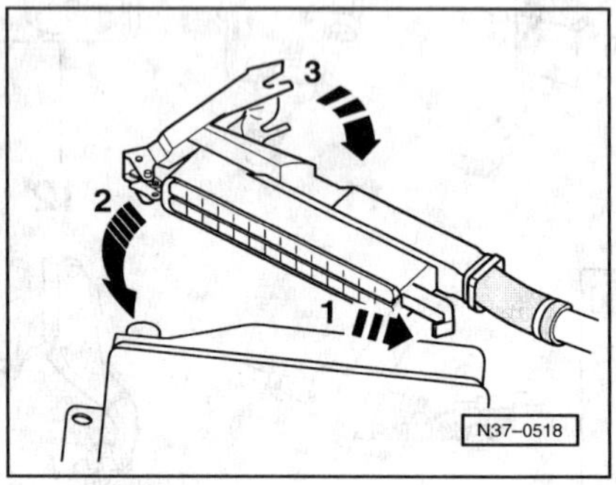

Fig. 2 Installing automatic transmission control module -J217-

– Place multi-pin connector on transmission control module connector housing -1-, swing in -2- and lock -3-.

If the engine or transmission control module is replaced, the system must be reset.

– Perform basic setting using "guided fault finding" on VAS 5051.

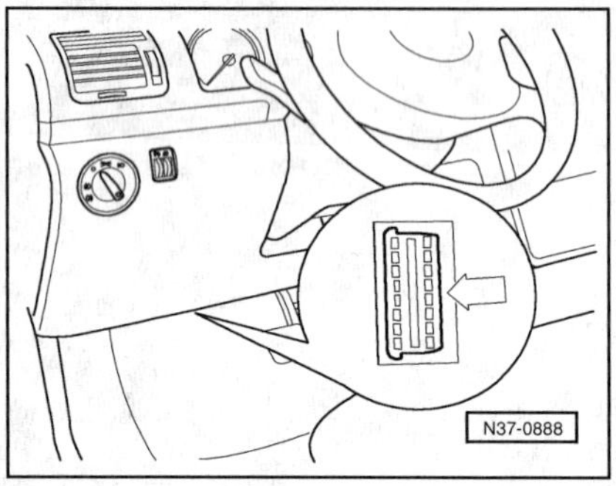

Fig. 3 DLC connector

Location: The diagnostic connection -arrow- is located under trim panel on driver's side.

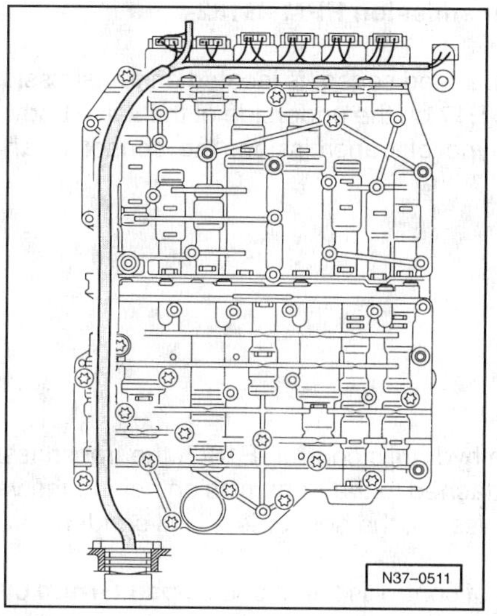

Fig. 4 Valve body

Location: The valve body is bolted to the underside of transmission housing inside the pan.

Solenoid valves -N88-, -N89-, -N90-, -N91-, -N92-, -N93-, -N94- are secured to the valve body of hydraulic control Type E17.

Solenoid valves -N91-, -N92-, -N93-, -N94- are known as pressure regulator valves 1 (-N215-), 2 (-N216-), 3 (-N217-) and 4 (-N218-) on the valve body of hydraulic control Type E18/2.

There are two different types of hydraulic control:

- Transmissions with hydraulic control Type E17, on which the sender for the transmission input seed (inductive sender) is attached to the underside of the valve body.
- Transmissions with hydraulic control Type E18/2 on which the sender for the transmission input speed (Hall sender) is attached to the transmission housing behind the valve body.

To find out which type of control is fitted, see **Repair Group 00, Code letters, allocation, ratios, equipment**

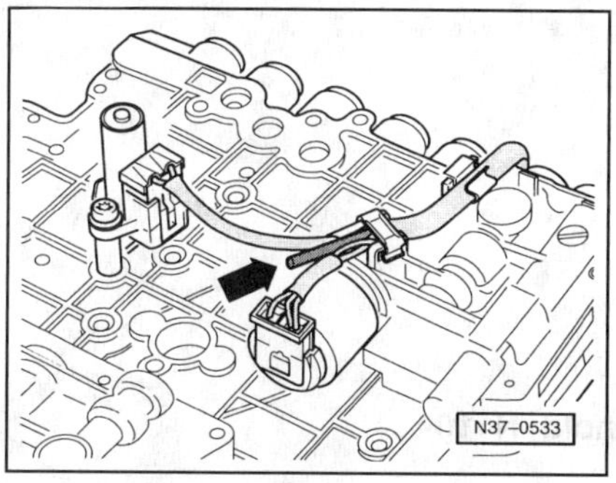

Fig. 5 Wiring harness with integral transmission fluid temperature sensor -G93-

Location: The transmission fluid temperature sender -arrow- is located in the wiring harness on valve body.

The valve body must be removed to replace the wiring harness. It is not necessary to remove the transmission.

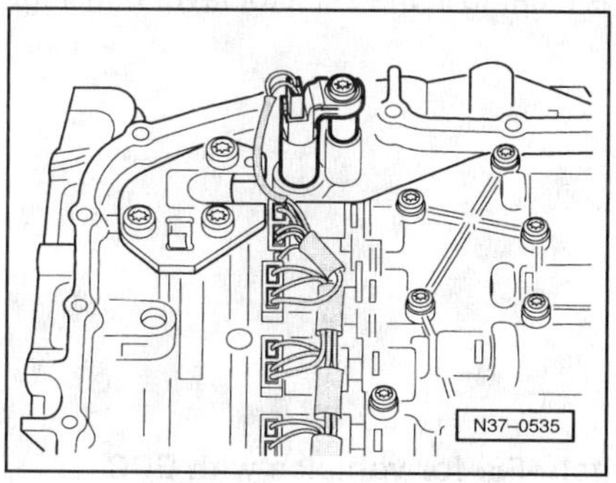

Fig. 6 Transmission vehicle speed sensor -G38- and Sender for transmission output RPM -G195-

Location: The transmission speed sensor is located on rear right of transmission, see **Repair Group 38.**

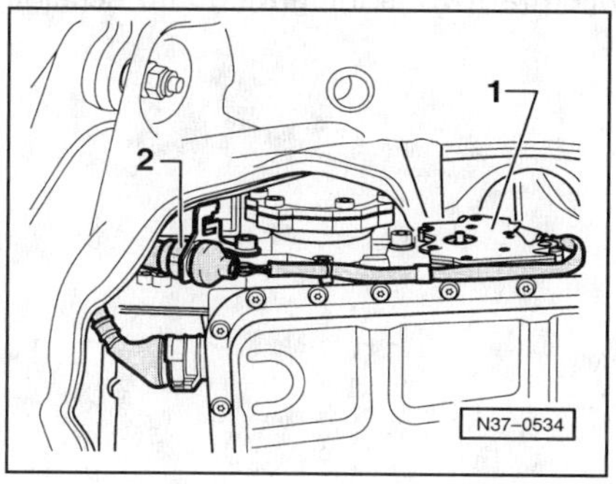

Fig. 7 Multi-function TR switch -F125-

Location: The multi-function switch -1- is located on left of transmission; Multi-pin connector for multi-function switch -2-.

Vehicles to 09.00

Multi-pin connector for 8-pin multi-function switch.

Vehicles from 10.00

Multi-pin connector for 10-pin multi-function switch.

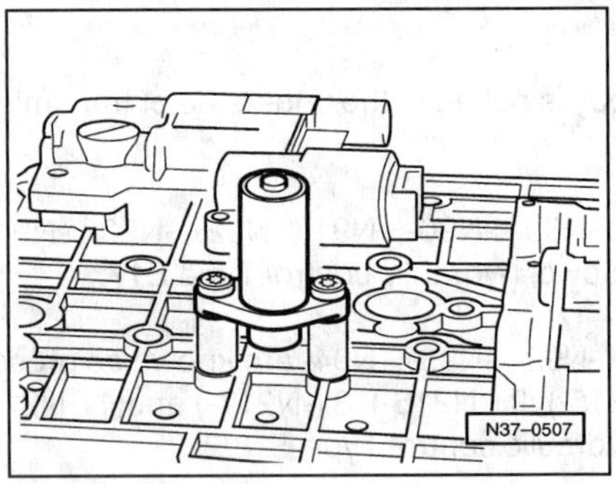

Fig. 8 Sensor for Transmission RPM -G182-

The transmission input speed sensor is located on transmissions with hydraulic control Type E17 on the underside of the valve body (see illustration). For this type of transmission, the sensor is an inductive sender.

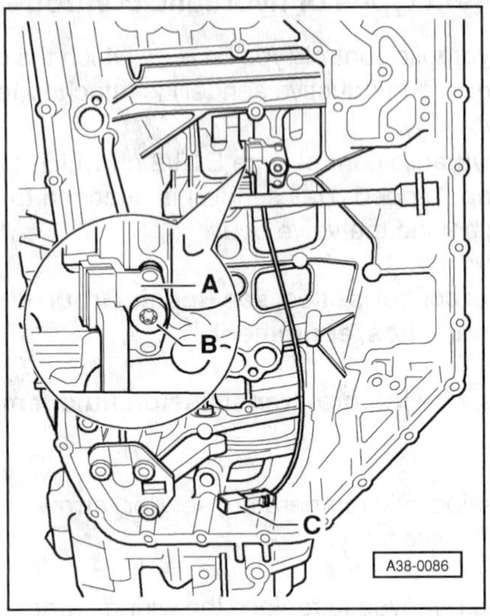

On transmissions with hydraulic control -E18/2, the transmission input speed sensor -A- is attached to the transmission behind the valve body. For this type of transmission, the sensor is a Hall sender.

To find out which type of control is fitted, see **Repair Group 00, Code letters, allocation, ratios, equipment**

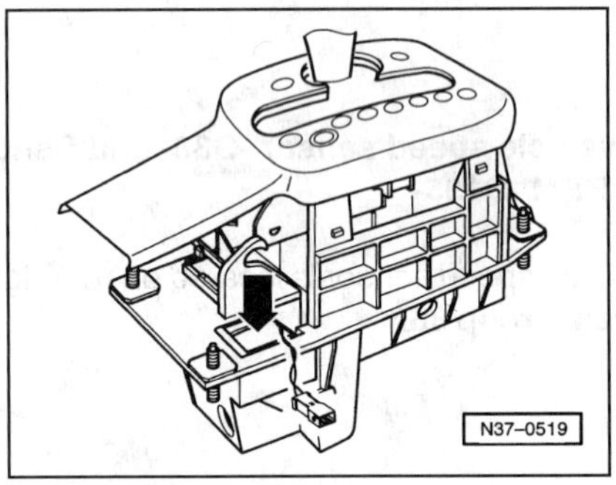

Fig. 9 Shift lock solenoid -N110-

The shift lock solenoid is located in the selector lever housing -arrow-.

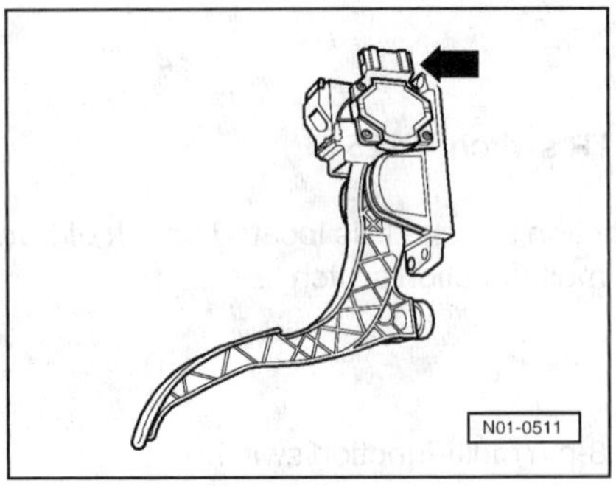

Fig. 10 Kickdown switch -F8- for vehicles with EPC

Location: Kickdown switch (arrow) is integrated into accelerator pedal position sender.

Fig. 11 Throttle position (TP) sensors -G69- and -G79- and accelerator pedal position sender -G185-

If the transmission control module signals are shown as faulty in the vehicle self-diagnosis, self-diagnosis of the engine control module must also be performed at all times.

Perform initial setting of engine control module following replacement or readjustment of throttle valve potentiometer or accelerator pedal position sender.

Locations:

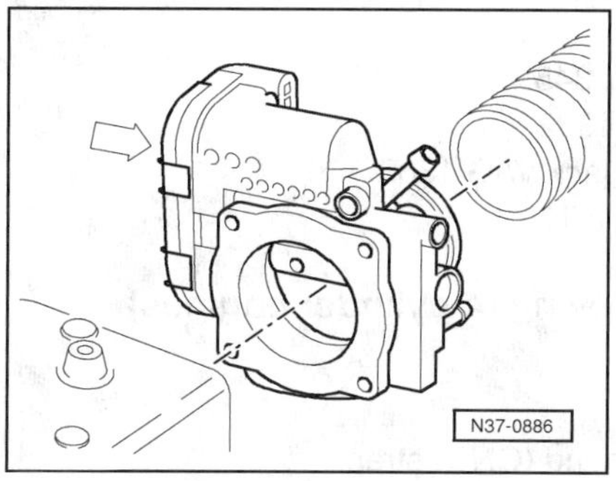

The potentiometer is part of the throttle valve control part -J338-. It is located on the throttle valve housing/intake manifold.

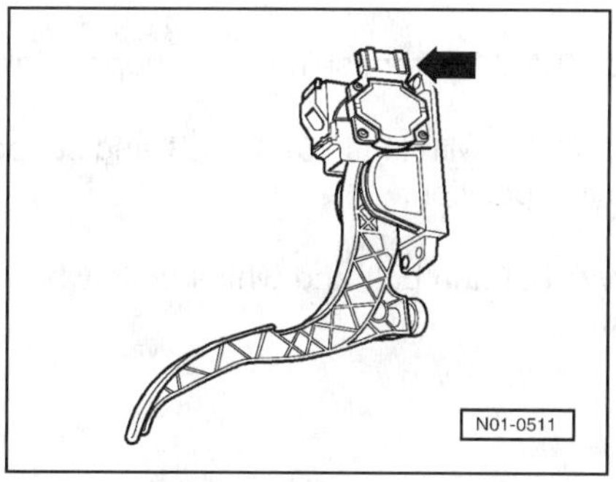

The throttle position sensor -G79- -arrow- is part of the electronic power control (EPC) and is located on the pedal cluster.

Fig. 12 Tiptronic switch -F189-

Location: The Tiptronic switch is integrated in the printed circuit of the symbol insert, which can be found in the selector mechanism cover -arrow-.

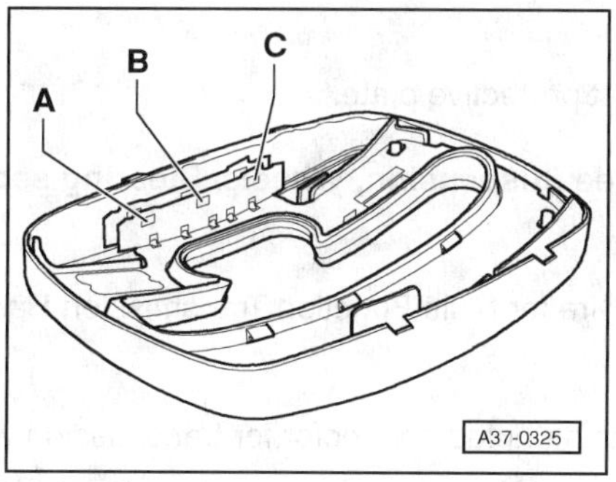

It consists of 3 Hall senders -A-, -B- and -C-, which are activated by a solenoid on the masking panel of the cover strip.

A - Tip down sensor

B - Tiptronic recognition sensor

C - Tip up sensor

Transmission, removing and installing

Transmission, removing and installing (4-cylinder engine)

Special tools and equipment

• Engine support bracket 10-222A with supports10-222A/1
• Adapter 10-222A/3
• Removal lever 80-200
• Transmission support 3282
• Alignment plate 3282/19
• Mandrel 3393
• Transmission support 3457
• Torque wrench VAG 1332
• Engine/transmission jack VAG 1383 A
• Special tool (socket) V 175

Transmission, removing (4-cylinder engines)

– Obtain radio code.

– Disconnect battery ground (GND) strap.

– Remove engine cover.

– Assemble adapter 10-222 A/3 for front spindle on support bar.

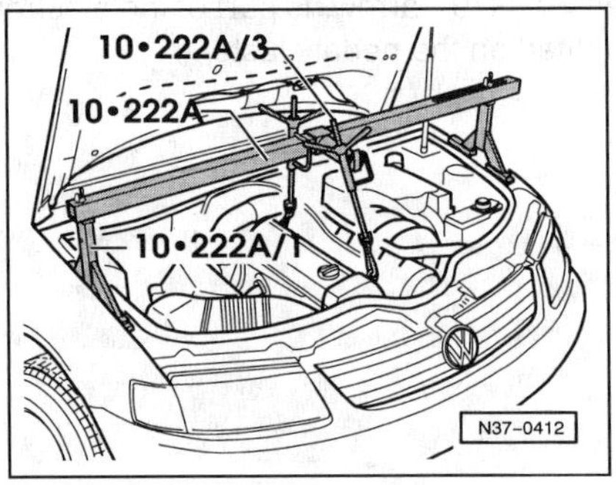

◄ Position support bar 10-222 A with legs 10-222 A/1 and support engine/transmission in this position.

– Only loosen drive axle/wheel hub bolt and wheel bolts when vehicle is standing on its wheels.

– Raise vehicle.

– Remove front wheels.

– Remove noise insulation tray.

– Remove noise insulation tray bracket.

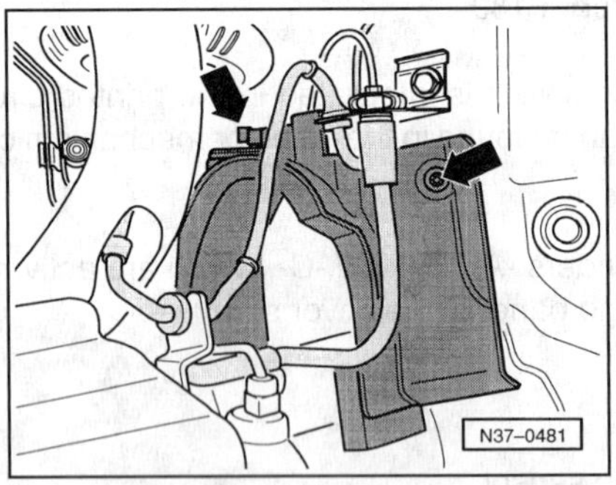

◄ Remove left and right axle shaft noise insulation from wheel housing (arrows).

– Remove right axle shaft protective plate.

– Remove drive axles, see **Suspension, Wheels, Steering section, Repair Group 40**

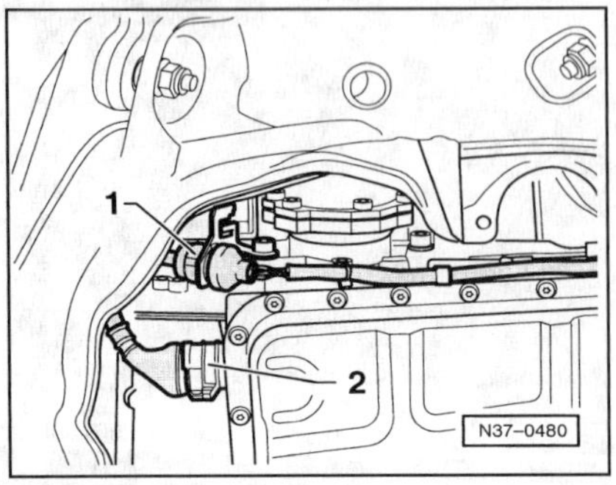

◄ Pull connector -1- off wire for Multi-Function Transmission Range (TR) -F125- Switch.

– Turn locking lever -2- and unplug connector for transmission wiring harness.

– Disconnect speedometer connector at transmission.

- Mark position of support bracket for securing selector lever cable to transmission housing (position must be correct when installing). Unbolt support bracket (arrows).

– Remove front exhaust pipe, see appropriate **1.8 Liter Engine Mechanical section, Repair Group 26**

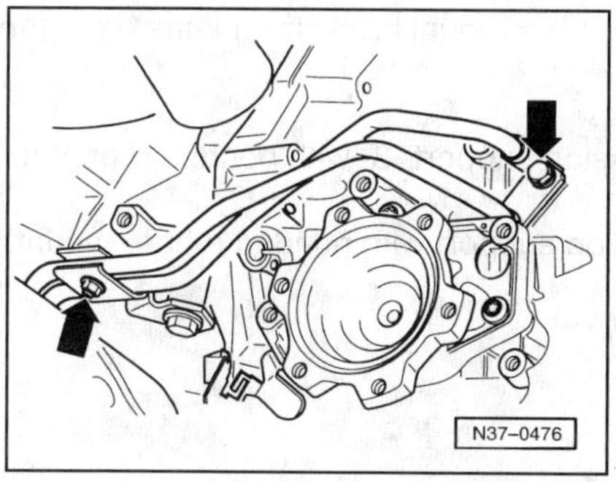

- Remove ATF lines from engine/transmission assembly (arrows) and tie-up.

– Plug ATF lines with clean plugs.

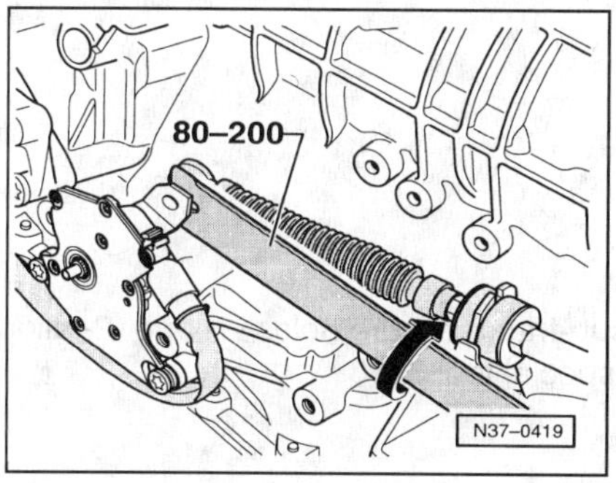

- Remove selector lever cable from selector shaft lever (arrow).

– Remove starter, see **Electrical Equipment section, Repair Group 27.**

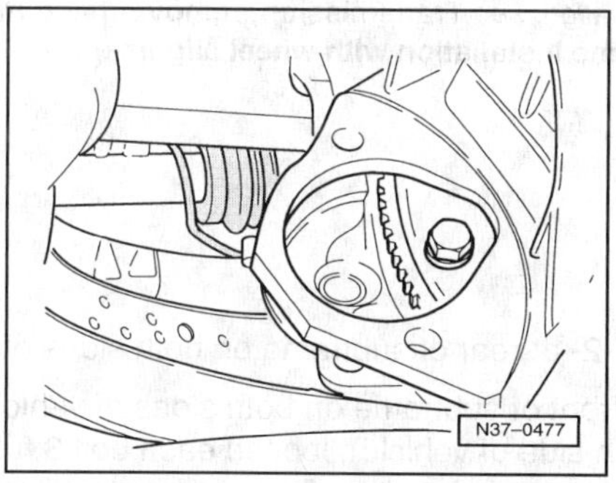

- Unbolt torque converter from drive plate using special tool V 175.

– Remove lower engine/transmission bolts.

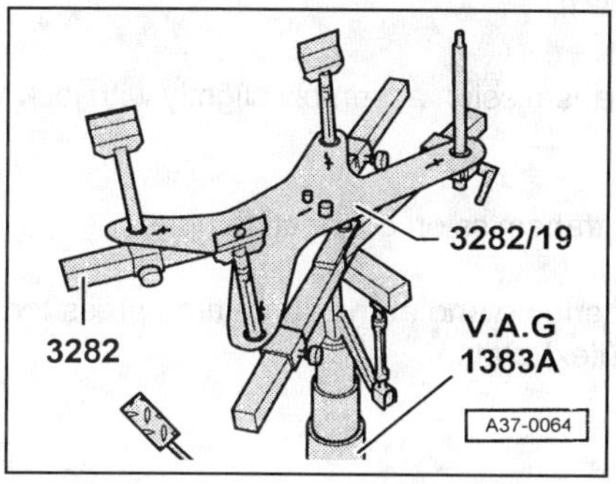

- Set up transmission mount 3282.
  - Transmission support 3282 is set up with adjusting plate 3282/19 for automatic transmission 01V.
  - Symbols on adjusting plate indicate mounts required, arrow points forward.

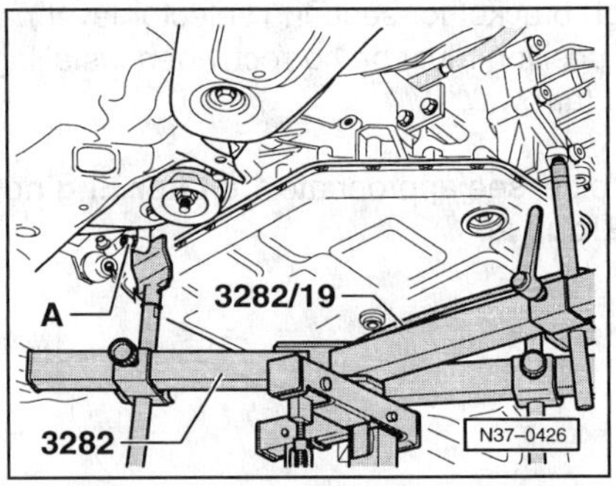

< Move jack VAG 1383 A with transmission mount 3282 under the transmission and support transmission.

– Align adjustment plate parallel with transmission.

– Secure transmission to transmission support 3282 with bolt -A-.

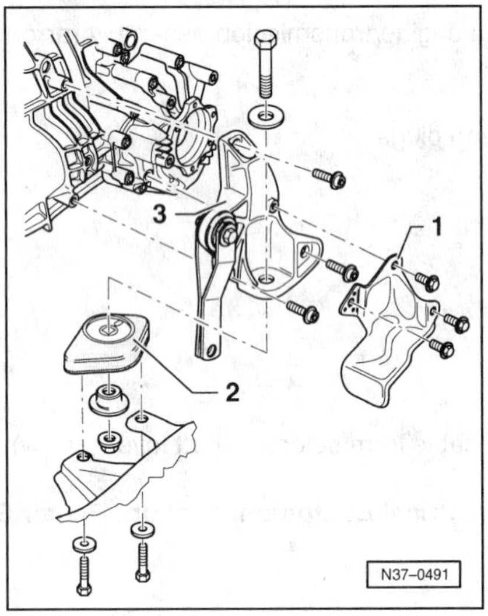

< Remove right bonded rubber mount protective plate -1- on transmission support.

– Remove right transmission support -3- with bonded rubber mount -2-.

– Remove left transmission support with bonded rubber mount.

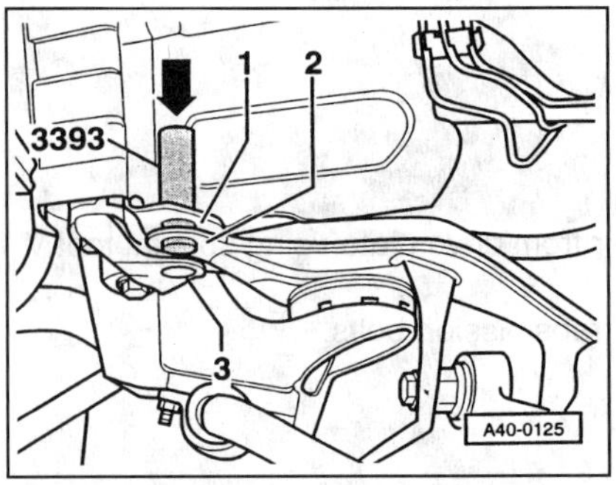

< Before loosening the sub-frame ensure holes -1- and -2- align on both sides of vehicle using special tool 3393.

EDITOR'S NOTE —

If subframe holes do not align, see **Transmission, removing and installing (V6 engine); Subframe installation with wheel alignment**.

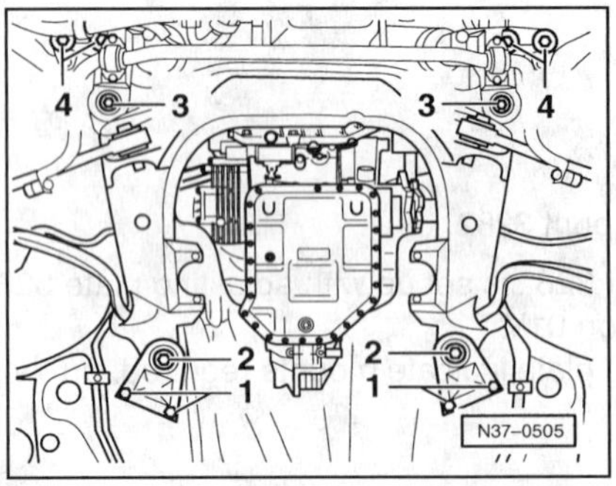

< Remove bolts -1- and -2- at rear of subframe on both sides of vehicle.

- Remove bolts -3- at front of subframe on both sides of vehicle.
- Bolts -4- (two on each side of vehicle), loosen each bolt 3 full turns.
- When the bolts are loosened, the rear of the subframe will drop about 150 mm (5.9 in.).

– Lower rear of engine/transmission assembly slightly with jack VAG 1383 A.

– Remove upper engine/transmission connecting bolts.

– Press transmission off engine and at the same time press torque converter out of drive (flex) plate.

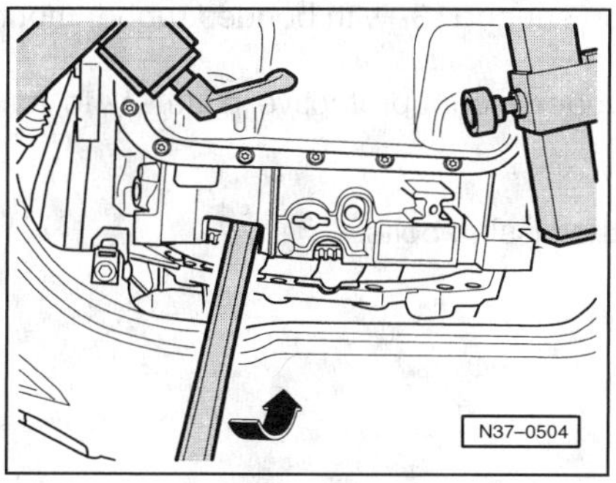

- Press torque converter against ATF pump (arrow).
- Secure torque converter to keep it from falling out.

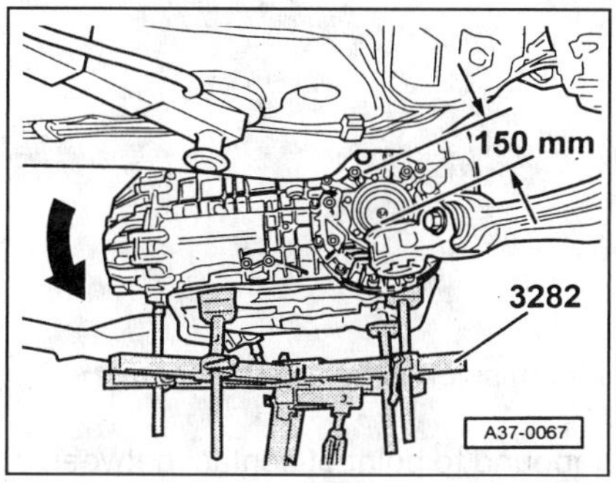

- Swivel transmission (arrow) and carefully guide out between subframe and underbody.
- Lower transmission.

Transmission, installing (4-cylinder engines)

Before installing:

- Clean ATF lines.
- Install torque converter in transmission, see **Repair Group 32, Torque convertor, installing.**
- Make sure dowel sleeves and intermediate plate are correctly positioned on engine.

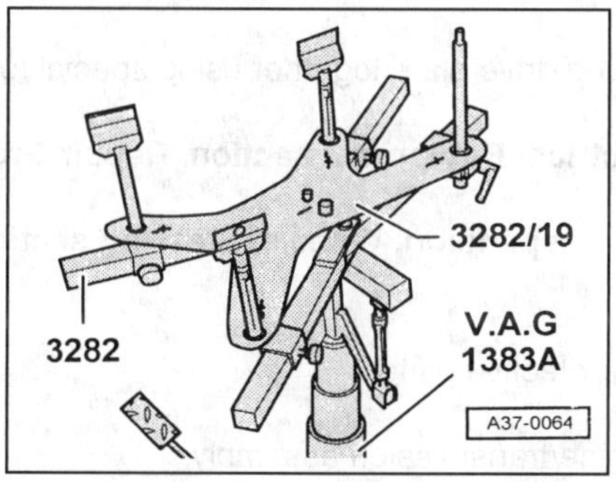

- Set up transmission mount 3282.
  - Transmission support 3282 is set up with adjusting plate 3282/19 for automatic transmission 01V.
  - Symbols on adjusting plate indicate mounts required, arrow points forward.
- Place transmission on transmission jack.

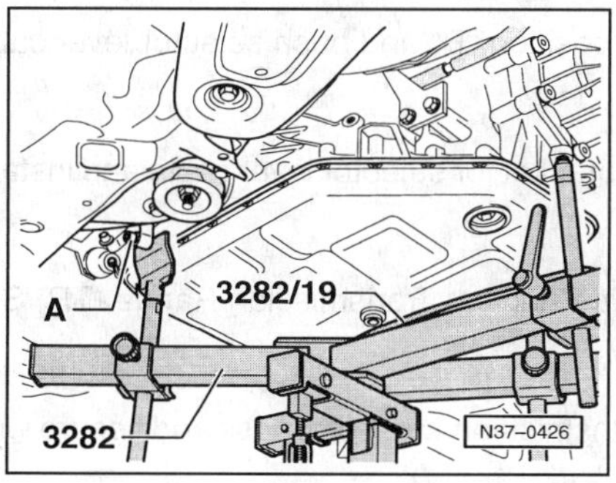

- Secure transmission to transmission support 3282 with bolt -A-
- Place transmission jack under vehicle.
- Guide transmission in between lowered subframe and underbody.
  - At this stage the subframe is lowered about 150 mm (5.9 in.) at the rear.
- Align transmission to engine.
- Install upper engine/transmission bolts.
- Raise rear engine/transmission assembly slightly using jack VAG 1383 A.
- Secure subframe, see **Suspension, Wheels, Steering section, Repair Group 40**
- Install left transmission support with bonded rubber mount.

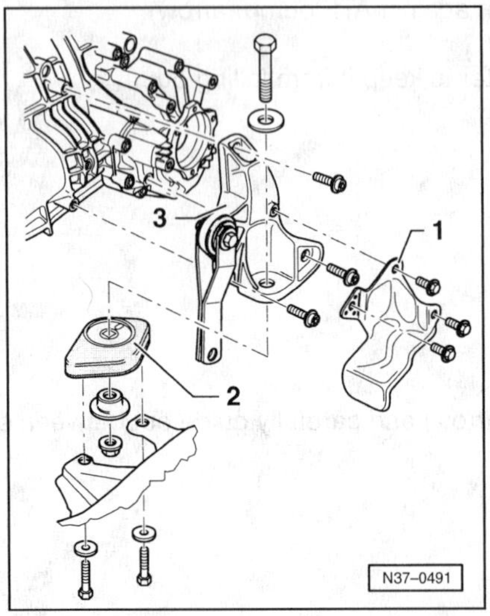

Install right transmission support -3- with bonded rubber mount -2-.

– Install right bonded rubber mount/protective plate -1- to transmission support.

– Install lower engine/transmission bolts.

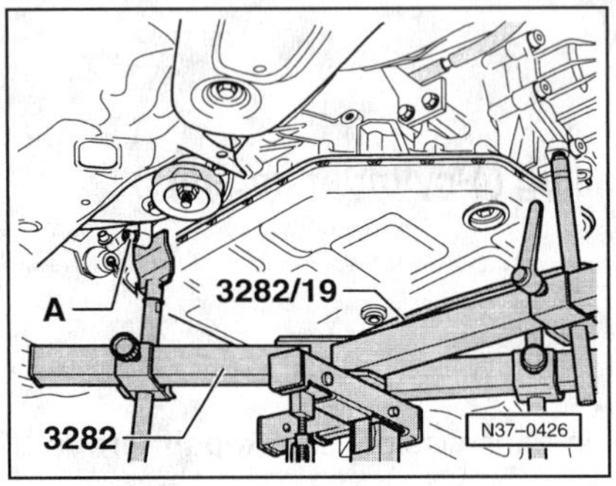

Remove bolt -A- after transmission is bolted to engine.

– Apply anti-corrosion compound to point of contact between bolt -A- and oil pan.

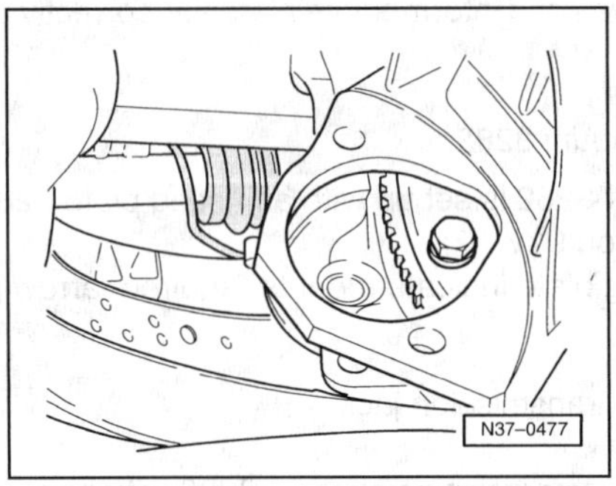

Bolt torque converter and drive plate together using special tool V 175.

– Install starter, see **Electrical Equipment section, Repair Group 27**

– Install drive axles, see **Suspension, Wheels, Steering section, Repair Group 40**

– Install right drive axle protective plate.

– Install ATF lines to engine/transmission assembly.

– Move selector lever to position "P" and push selector lever cable on selector lever shaft.

– Install bolts to support bracket for selector lever cable and install selector lever cable.

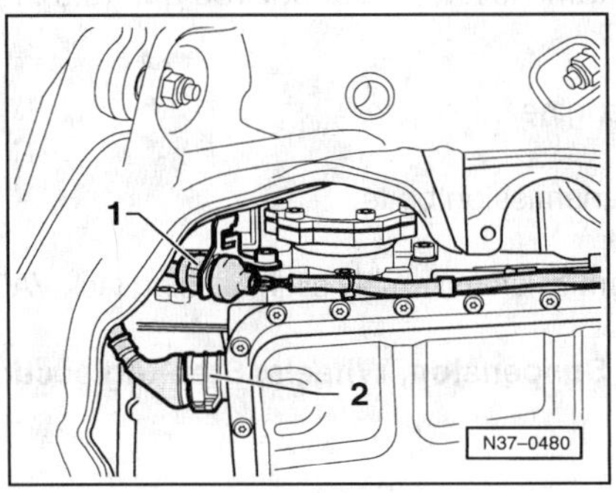

Plug connector on Multi-Function Transmission Range (TR) Switch -F125- -1-.

– Install connector for transmission wiring harness and secure with locking lever -2-.

– Reconnect speedometer sender connector.

– Install front exhaust pipe, see **1.8 Liter 4-Cyl. 5V Turbo Engine Mechanical section, Repair Group 26**

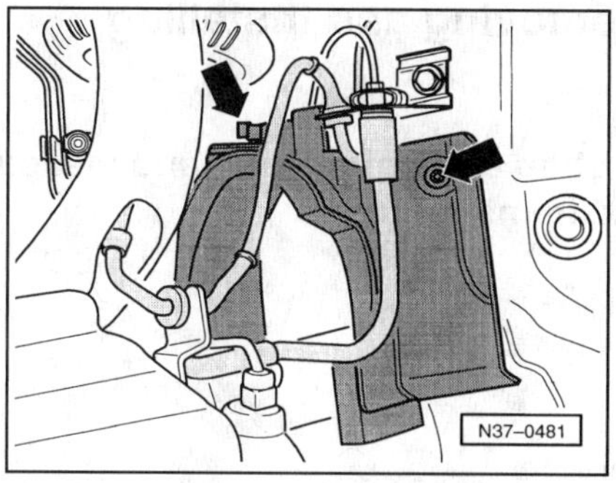

◄ Install left and right axle shaft noise insulation to wheel house.

– Install noise insulation tray bracket.

– Install noise insulation tray.

– Install front wheels.

– Remove support device 10-222 A.

– Install engine cover.

– Reconnect ground (GND) strap to battery.

– Check selector lever cable adjustment, see **Selector lever cable, checking and adjusting**.

– With transmission installed, check final drive oil level, see **Maintenance.**

– Check and top up ATF level, see **Maintenance**.

Tightening torques (4-cylinder)

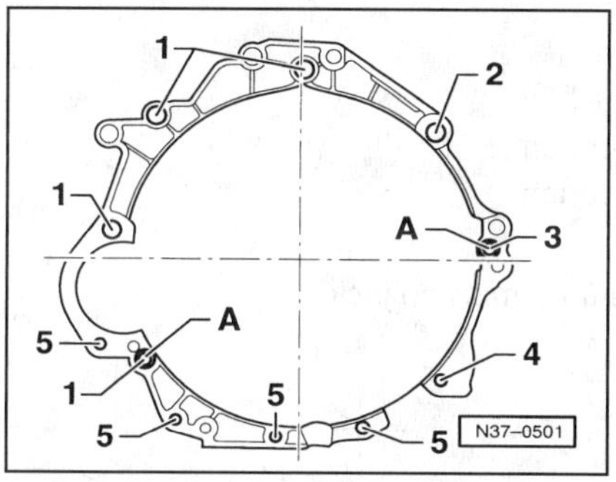

◄ Transmission to engine

Pos.	Bolt	Qty.	Torque
1	M 12 x 67	4	65 Nm (48 ft-lb)
2	M 12 x 110	1	65 Nm (48 ft-lb)
3	M 12 x 90	1	65 Nm (48 ft-lb)
4[1)]	M 10 x 60	1	45 Nm (33 ft-lb)
5	M 10 x 45	4	45 Nm (33 ft-lb)

[1)] with nut

Pos. A = Dowel sleeves

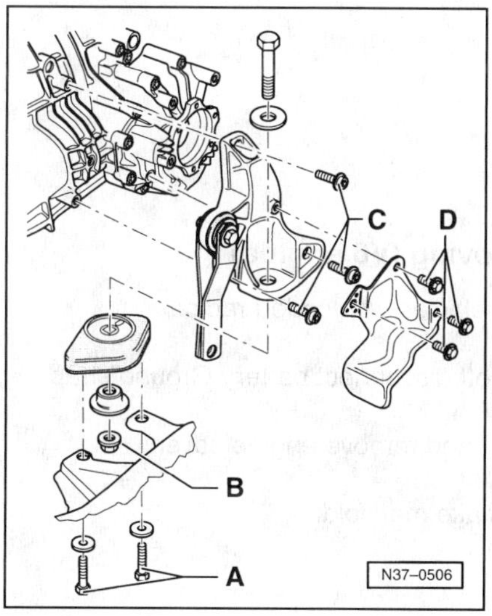

◄ Same torques for rear right (illustration) and left transmission mountings.

A. 23 Nm (17 ft lb)
B. 40 Nm (30 ft lb)
C. 40 Nm (30 ft lb)
D. 10 Nm (7 ft lb)

Tightening Torques	
Torque converter to drive plate	85 Nm (63 ft lb)
ATF line to transmission	20 Nm (15 ft lb)
Bracket for ATF line	10 Nm (7 ft lb)
Drive axle to transmission	80 Nm (59 ft lb)
Wheel bolts to wheel hub	120 Nm (89 ft lb)
Subframe to body: - Always replace bolts M8 M10 M12	 23 Nm (17 ft lb) 60 Nm (44 ft lb) 110 Nm (81 ft lb) + 90°
Axle shaft to wheel hub: - Always replace bolt M14 bolt M16 bolt	 115 Nm (85 ft lb) + 180° 190 Nm (140 ft lb) +180°

Transmission, removing and installing (V6 engine)

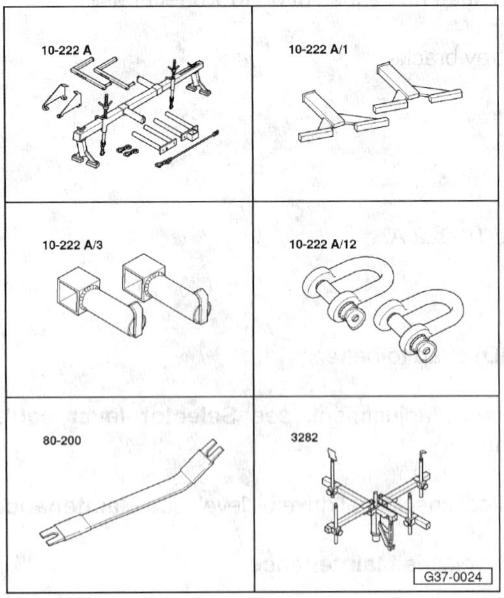

Special tools, workshop equipment, test and measuring appliances and aux. items required

- 10-222 A Support
- 10-222 A/1 Support
- 10-222 A/3 Adapter
- 10-222 A/12 Shackles
- 80-200 Removal lever
- 3282 Transmission support

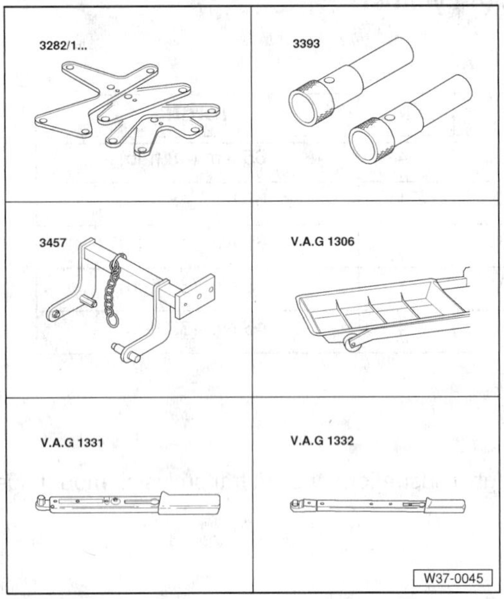

- 3282/19 Adjustment plate
- 3393 Test mandrel
- 3457 Transport device
- V.A.G 1306 Drip tray
- V.A.G 1331 Torque wrench
- V.A.G 1332 Torque wrench

- V.A.G 1383 A Engine/transmission jack

Transmission, removing (V6 engine)

– Obtain radio code on vehicles with coded radio.

– With ignition switched off disconnect battery Ground strap.

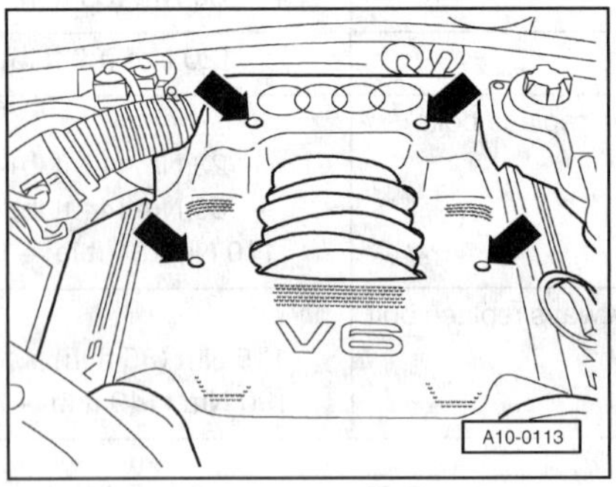

◄ Unscrew bolts -arrows- and remove engine cover.

– Take intake hose off intake manifold.

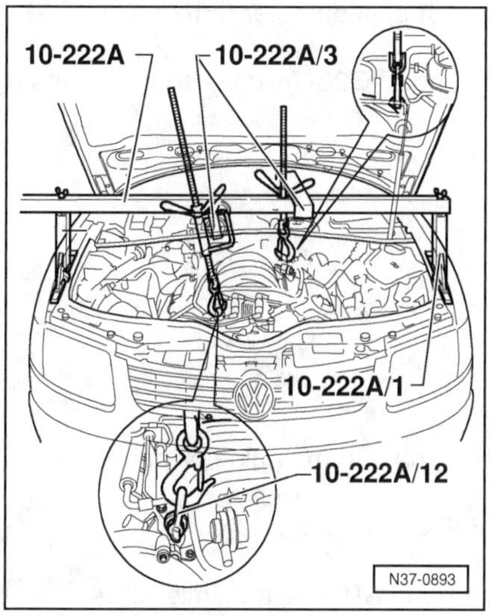

◀ Fit support bar 10-222A together with legs 10-222 A/1 and adapters 10-222 A/3.

– Secure spindles to engine. In addition, use a shackle 10-222 A/12 on the front of the engine to avoid damage.

– Take weight of engine/transmission assembly on the spindles.

– Loosen wheel bolts.

– Raise vehicle.

– Remove front wheels.

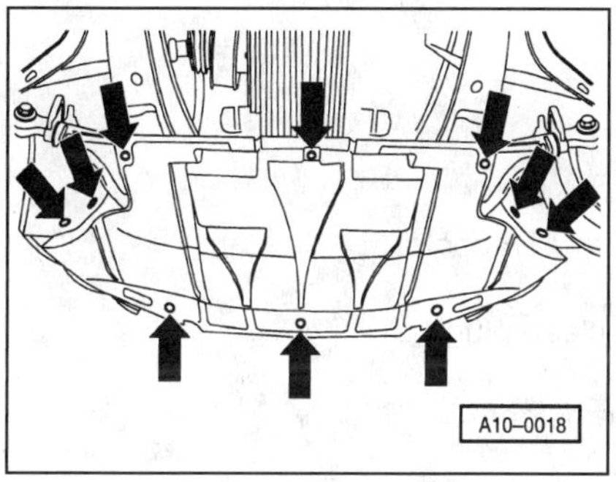

◀ Remove noise insulation tray -arrows-.

– Remove front exhaust system with catalyst, see **2.8 Liter V6 5V Engine Mechanical (code: AHA, ATQ) section, Repair Group 26**

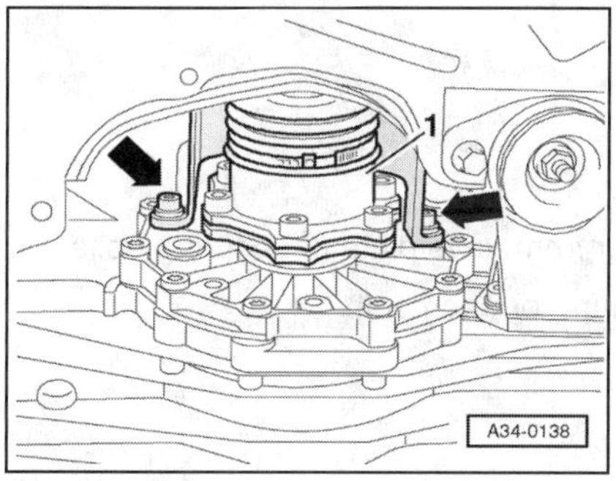

◀ Remove protective plates for left and right-hand drive axles -arrows-. The third bolt cannot be seen in the illustration.

– Disconnect left and right-hand drive axles -1- from flanges and tie-up.

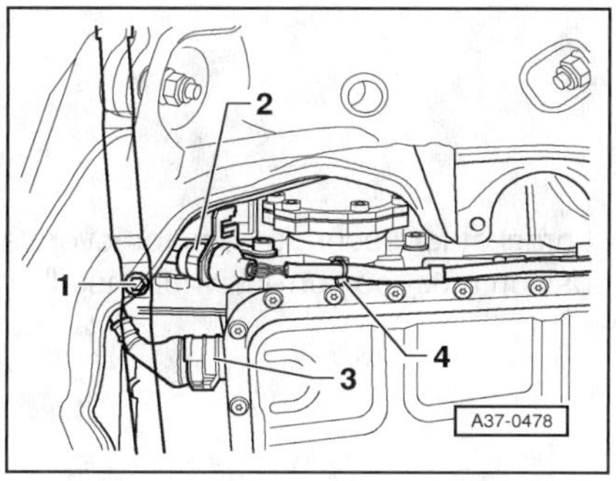

◀ Remove noise insulation tray bracket -1-.

– Pull connector -2- off multi-function switch lead.

– Turn locking lever -3- and remove connector for transmission wiring harness.

– Cut cable tie -4-.

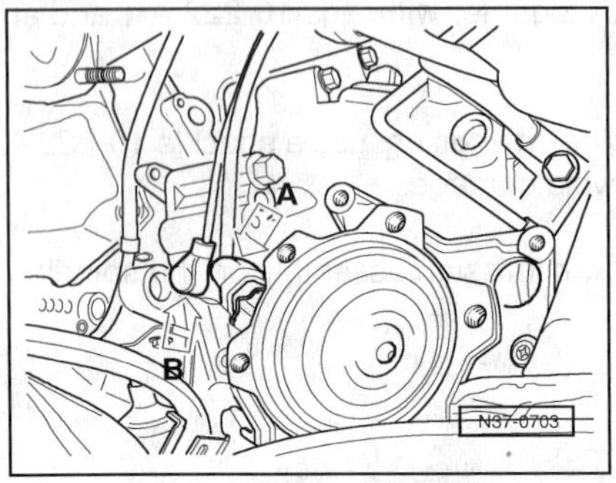

Pull connector off sender for speedometer (arrow -A-).

– Unscrew engine speed sender -G28- (arrow -B-) from transmission.

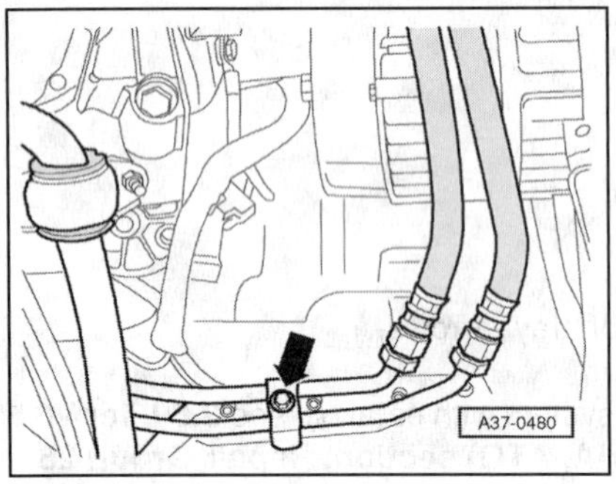

Remove ATF line holder -arrow- from engine oil pan.

– Place drip tray V.A.G 1306 underneath.

– Thoroughly clean area around ATF line/transmission.

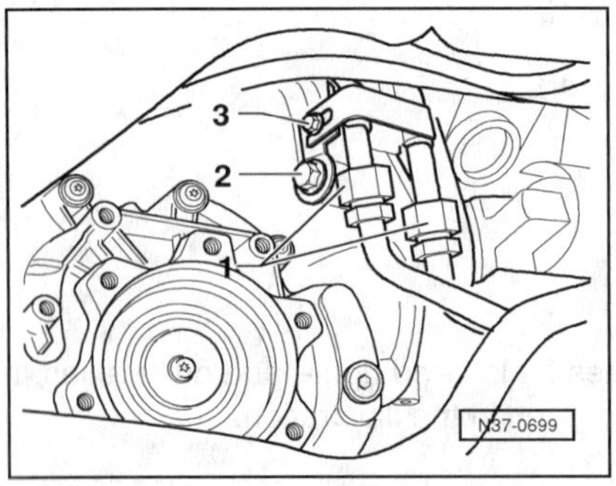

Separate ATF lines -1- to ATF cooler.

– Seal off ATF lines with clean plugs.

– Do not remove ATF line holder -3-.

– Remove transmission side starter motor bolt -2-.

– Remove starter.

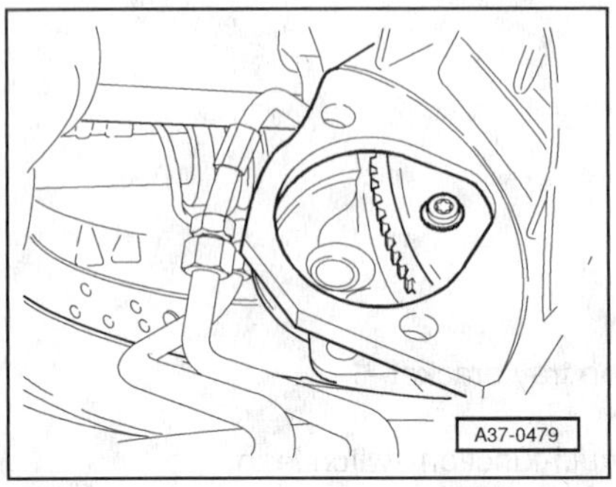

Unbolt torque converter from drive plate (3 securing bolts).

- Turn crankshaft 1/3 turn for each bolt.

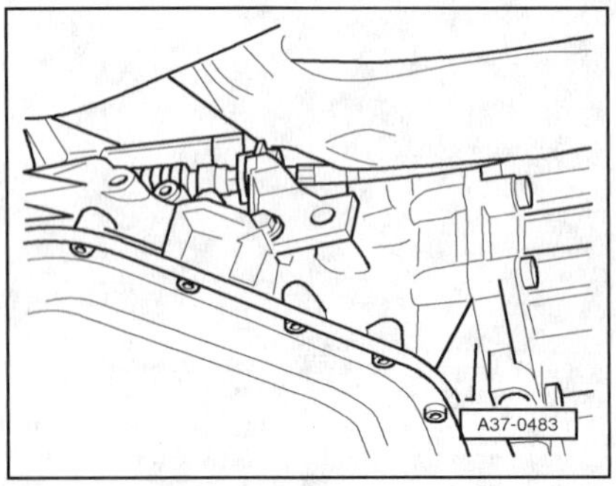

Mark position of support bracket for securing selector lever cable to transmission housing (position must be correct when installing). Unbolt support bracket -arrow-.

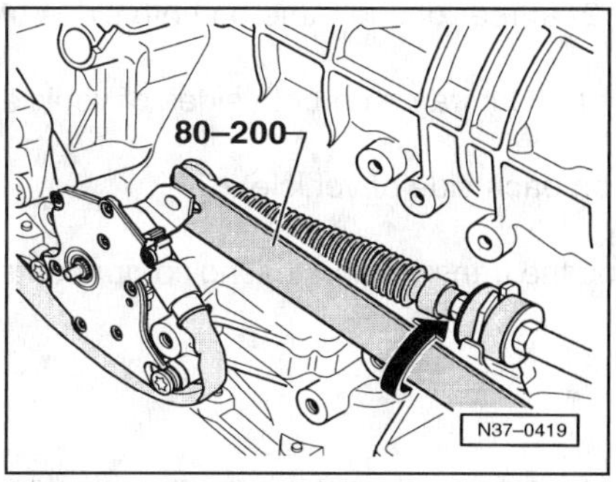

- Remove selector lever cable from selector shaft lever -arrow-.
- Remove side and lower engine/transmission connecting bolts.

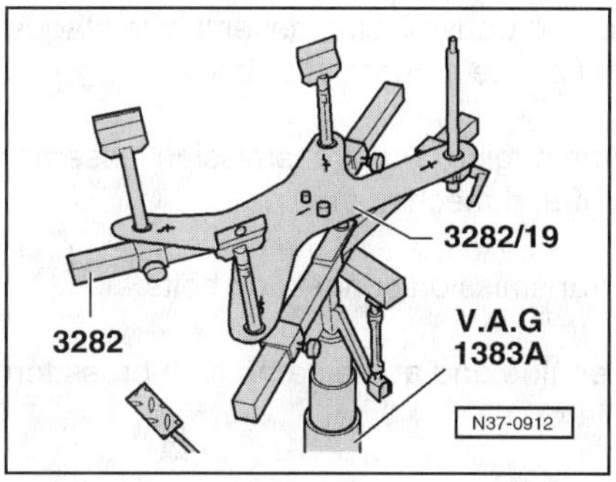

- Set up transmission support 3282.
  - Transmission support 3282 is set up with adjusting plate 3282/19 for automatic transmission 01V.
  - The symbols on the adjusting plate indicate the mounts required, the arrow points forward.

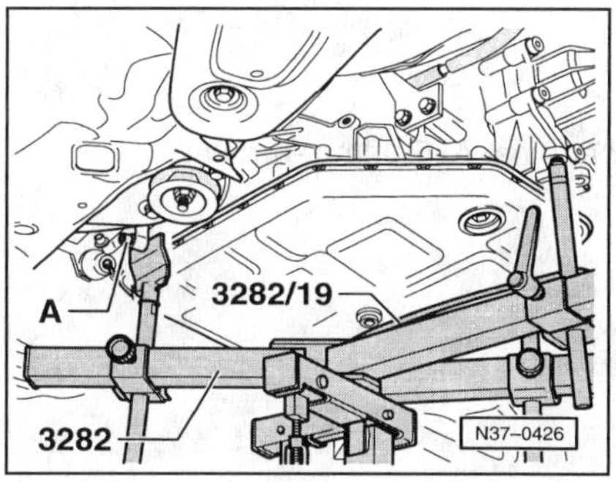

- Move jack V.A.G 1383 A with transmission support 3282 under the transmission, align adjusting plate parallel to transmission and support transmission.
- Secure transmission to transmission support 3282 with bolt -A-.
- Chock space between transmission support and oil pan using washers.

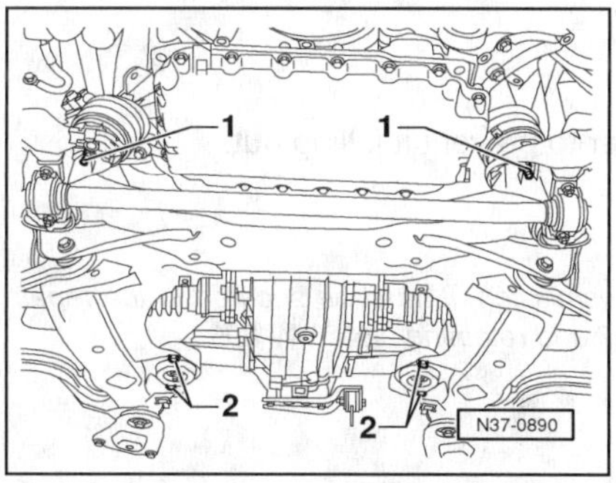

- Remove left and right-hand engine mounting nuts -1-.
- Remove bolts -2- from transmission mounting to subframe.

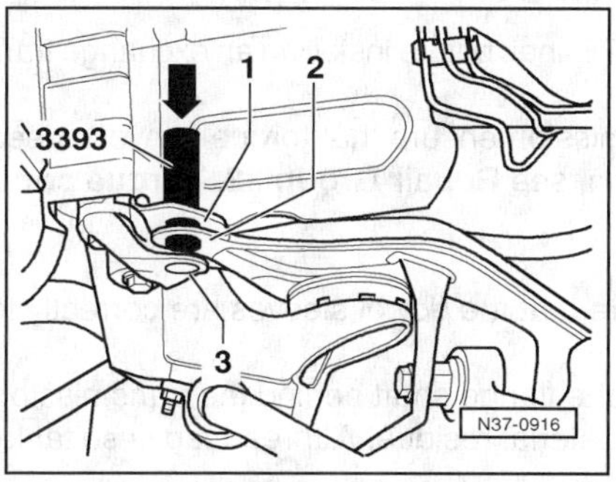

- Before loosening the subframe, it is essential to check whether at least holes -1- and -2- are aligned on both sides of the vehicle using test drifts 3393.
  - If this is not the case, perform steering alignment measurements after tightening the sub-frame!

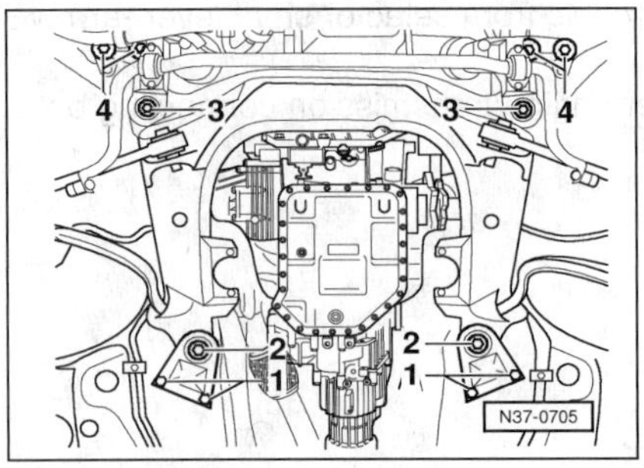

< Unscrew bolts -1- and -2- at rear of subframe on both sides of vehicle.

– Remove bolts -3- at front of subframe on both sides of vehicle.

– Remove bolts -4- (two on each side of vehicle):

Take note when lowering the transmission that no components are damaged.

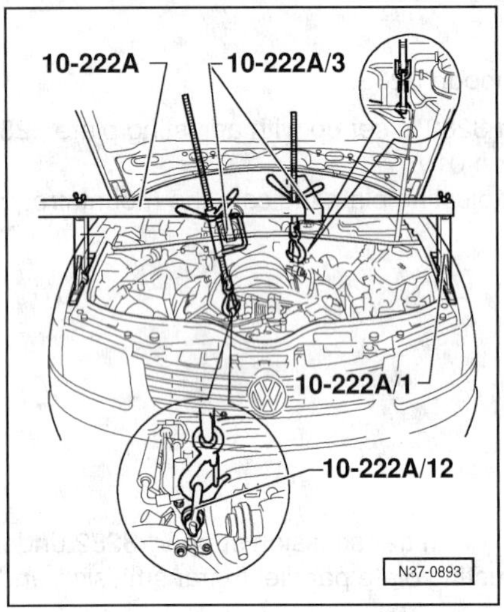

< Lower engine and transmission assembly 8 turns on spindles.

– To do this, lower engine and transmission assembly in stages alternatively on each spindle i.e. 2 turns at a time.

– At the same time, lower engine and transmission assembly on jack V.A.G 1383 A (with aid of 2nd mechanic).

– Remove upper engine/transmission connecting bolts.

– Press transmission off engine and at the same time press torque converter out of drive plate.

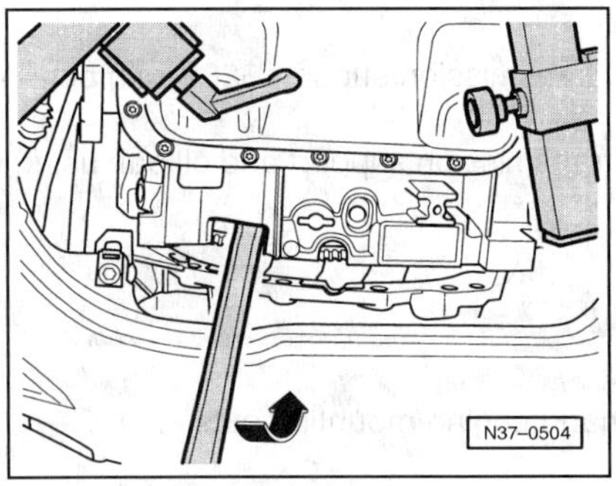

< Press torque converter against ATF pump -arrow-.

– Lower transmission.

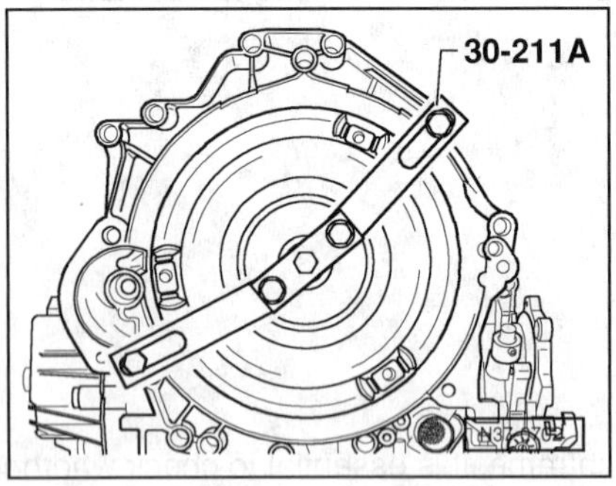

< Secure torque converter to prevent it falling out.

NOTE —
When working on the transmission, it must be secured to the repair stand with transport device 3457 and retaining plate VW 309.

Transmission, installing (V6 engine)

– Clean ATF cooler and ATF lines before installing an exchange transmission.

– Before installing transmission ensure that torque converter seats correctly in transmission, see **Repair Group 32, Torque convertor, installing**.

– Before installing, ensure that the dowel sleeves are correctly located.

– The threaded holes in the flange shaft behind the transmission must be cleaned of all locking material residue. A thread tap is suitable for this.

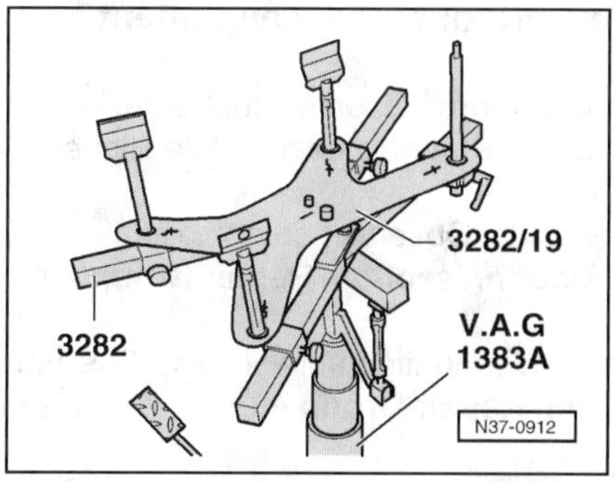

Set up transmission support 3282.

- Transmission support 3282 is set up with adjusting plate 3282/19 for automatic transmission 01V.
- The symbols on the adjusting plate indicate the mounts required, the arrow points forward.

– Place transmission on transmission jack.

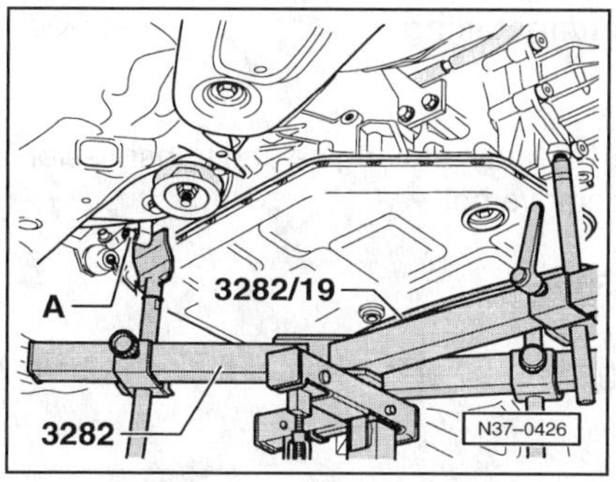

Secure transmission to transmission support 3282 with bolt -A-

– Place transmission jack under vehicle.

– Guide transmission in between lowered subframe and underbody.

– Align transmission to engine.

– Screw engine/transmission connecting bolt hand-tight in each side.

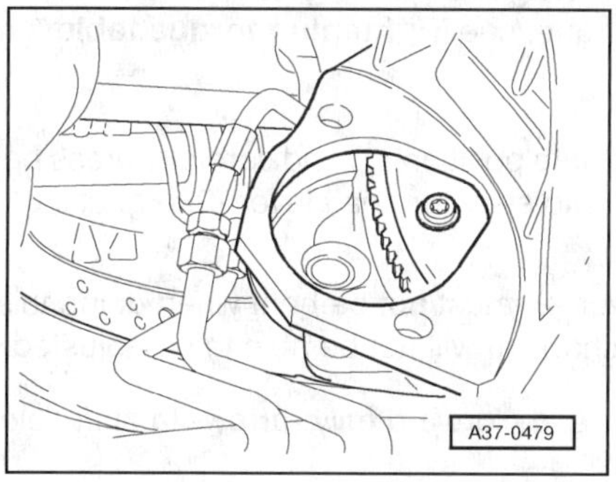

Bolt torque converter and drive plate together.

- To do this, screw in bolts by hand and then tighten to 85 Nm (63 ft-lb).

– Install upper engine/transmission securing bolts and tighten to prescribed torque.

NOTE —
Holders and Ground connections are fitted to some bolts see Tightening torque table.

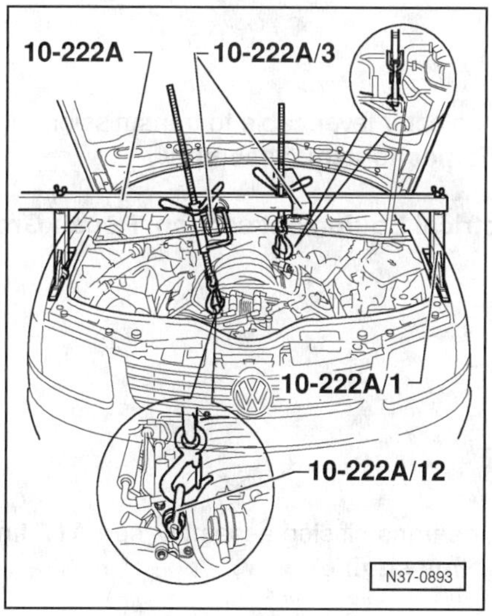

Lower engine and transmission assembly 8 turns on the spindle.

- To do this, lower engine and transmission assembly in stages alternatively on each spindle i.e. 2 turns at a time.
- At the same time, raise engine and transmission assembly on jack V.A.G 1383 A (with aid of 2nd mechanic).

Subframe installation with wheel alignment

If holes -1- and -2- did not align before loosening subframe, see illustration N37-0916, given Previously, the wheels must be aligned.

Install subframe using bolts originally removed. These must be replaced following wheel alignment.

When doing this, tighten original bolts only to prescribed torques, see **Suspension, Wheels, Steering section, Repair Group 40**

WARNING —
The vehicle must not be driven in this state.

Subframe installation without wheel alignment

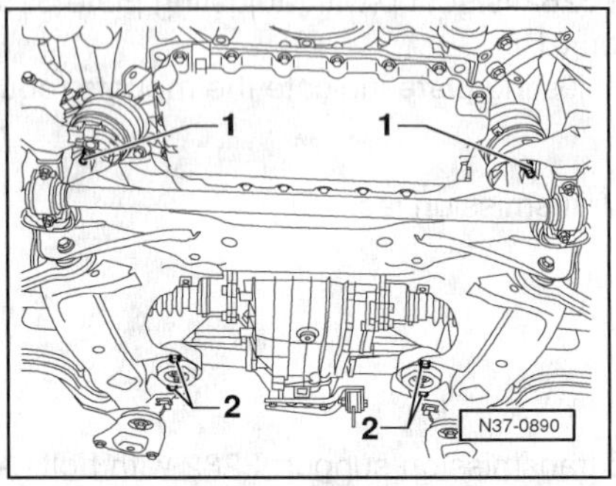

If holes -1- and -2- did align before loosening subframe, see illustration N37-0916, given previously, the wheels need not be aligned.

Install subframe with new bolts. Observe tightening sequence, see **Suspension, Wheels, Steering section, Repair Group 40.**

◀ Fit left and right-hand engine mounting nuts -1-, see **Suspension, Wheels, Steering section, Repair Group 40.**

– Fit bolts -2- for transmission mounting to subframe, see **Transmission brackets, removing and installing**.

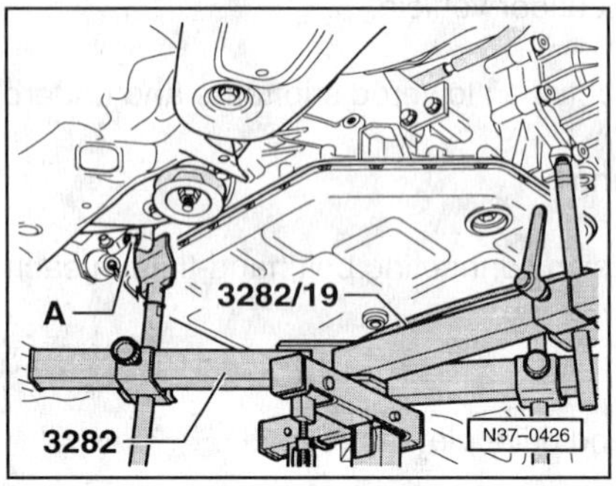

◀ Once subframe is bolted to body and engine and transmission assembly is bolted to subframe, remove bolt -A-.

– Remove jack V.A.G 1338 A.

– Apply anti-corrosion agent to point of contact between bolt -A- and oil pan.

– Fit side and lower connecting bolts for engine to transmission and tighten to prescribed torque, see **Tightening torque table,** given later.

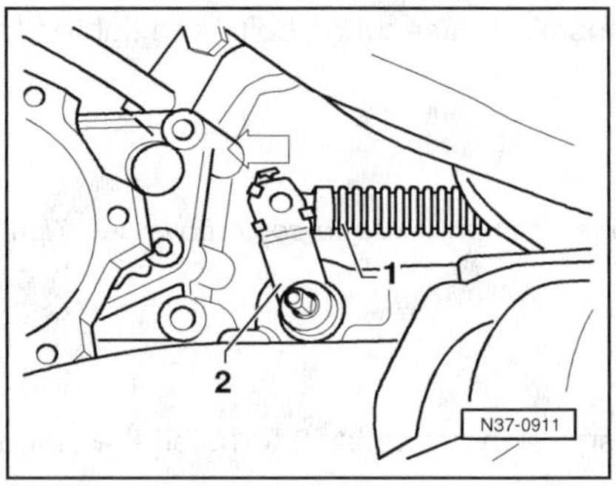

◀ Shift selector shaft lever into position "2" and carefully press ball head of selector lever cable -1- into selector shaft lever -2- e.g. using a pair of pliers.

- The selector shaft lever -2- must not be bent when doing this, otherwise the selector mechanism will not be able to be adjusted correctly.

– Push selector lever cable -1- forward fully -arrow- to shift selector lever shaft -2- into position "2".

◀ Bolt support bracket of selector lever cable to transmission -arrows- at place marked, see **Shift mechanism, assembly**.

– Install starter, see **Electrical Equipment section, Repair Group 27.**

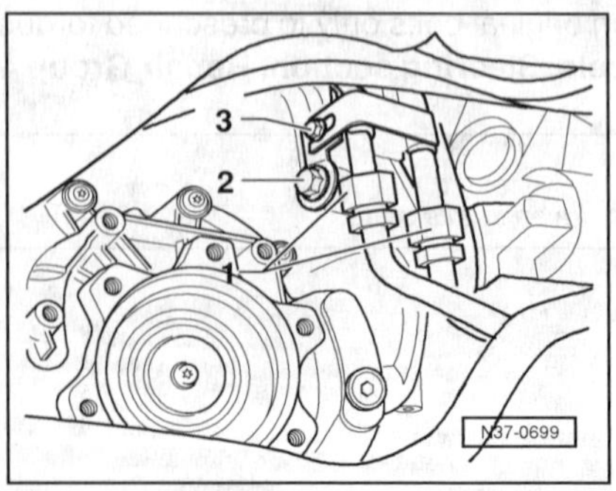

◀ Fit ATF lines -1- to engine/transmission assembly, see **ATF lines (vehicles with V6 gasoline engine)**.

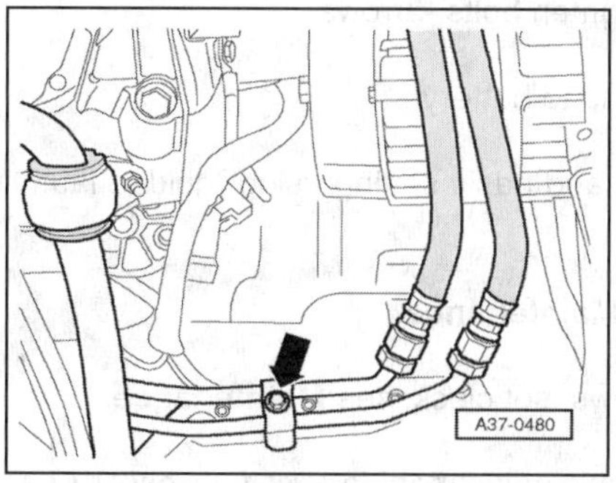

Remove ATF line holder -arrow- from engine oil pan, see **ATF lines**.

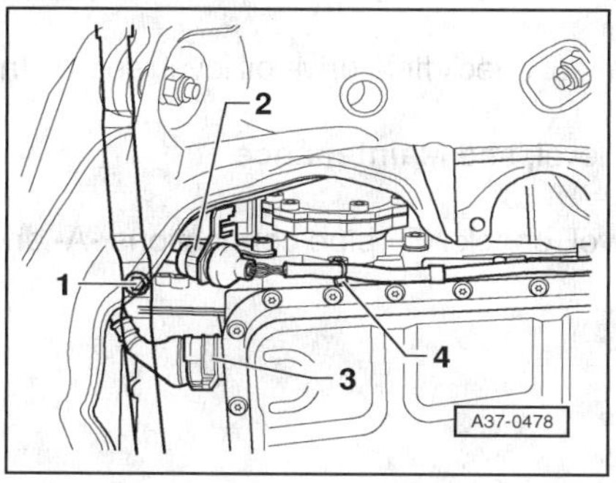

Fit connector to multi-function switch -2-.

– Fit connector for transmission wiring harness and secure with locking lever -3-.

– Install noise insulation tray holder -1-, see **Body Exterior section, Repair Group 50.**

– Replace cable tie -4-.

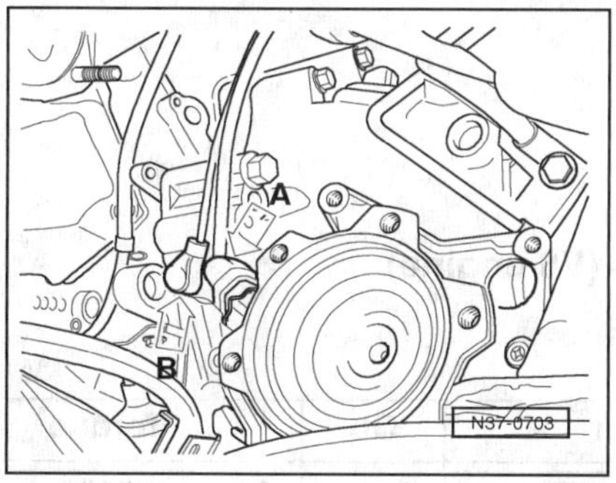

Reconnect speedometer sensor connector -arrow A-.

– Screw engine speed sensor -G28- (arrow -B-) into transmission.

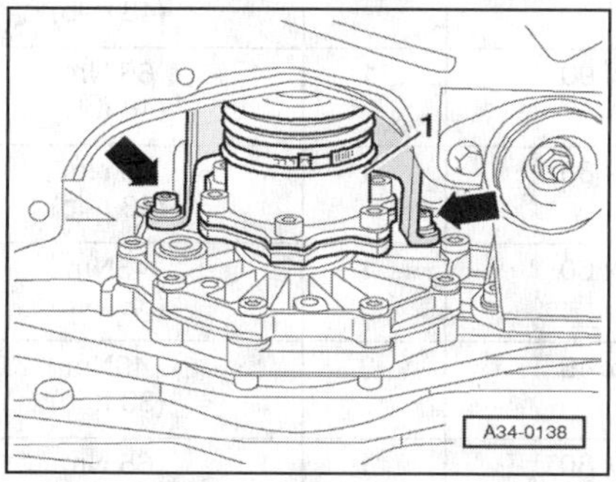

Fit drive axles -1- to flanged shafts, see **Suspension, Wheels, Steering section, Repair Group 40.**

– Fit protective plates for left and right-hand drive axles -arrows-. The third bolt cannot be seen in the illustration, see **Suspension, Wheels, Steering section, Repair Group 40.**

– Install front exhaust system with catalyst, see **2.8 Liter V6 5V Engine Mechanical, Repair Group 26**

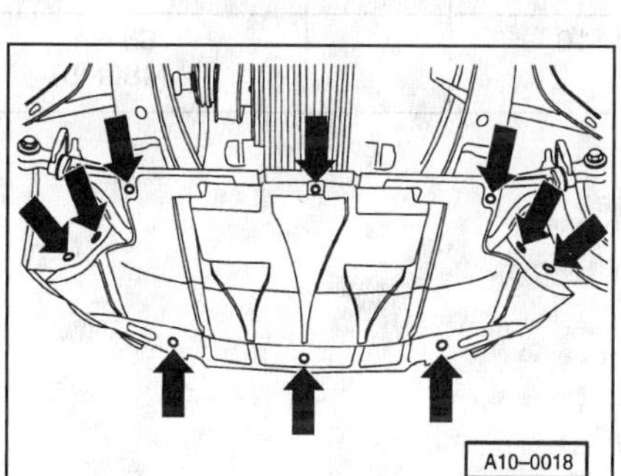

Install noise insulation tray -arrows-.

– Install front wheels, see **Suspension, Wheels, Steering section, Repair Group 40.**

– Remove support device 10-222 A.

– Fit intake hose to intake manifold.

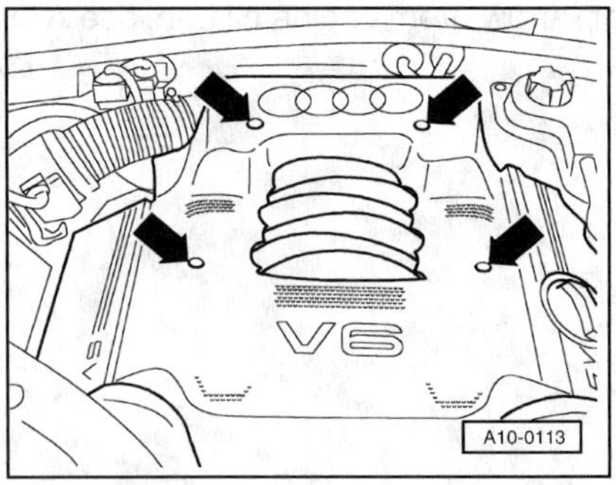

- Fit engine cover and tighten bolts -arrows-.
- Reconnect Ground strap to battery.
- Perform self-diagnosis and basic setting using "guided fault finding" on VAS 5051.
- Enter radio code, see **Maintenance.**
- Initialize electric windows, set clock, see **Maintenance.**
- Check selector lever cable adjustment, adjust if necessary, see **Selector lever cable, checking and adjusting**.
- With transmission installed, check final drive oil level, see **Maintenance.**
- Check and top up ATF level, see **Maintenance**.

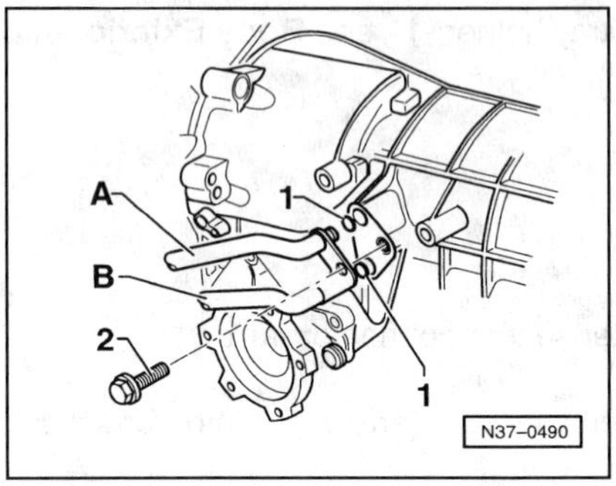

When checking ATF level, check ATF pipe connections -A- and -B- for leaks.

Tightening torques (V6 engine)

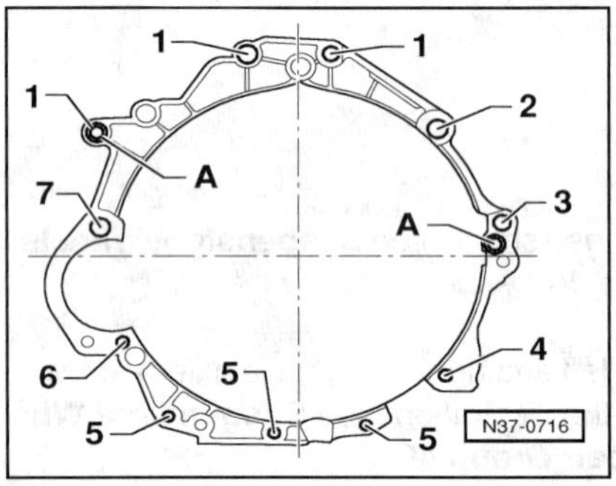

Transmission to engine

Item	Bolt	Qty.	Torque
1	M 12 x 67	3	65 Nm (48 ft-lb)
2	M 12 x 90	1	65 Nm (48 ft-lb)
3	M 12 x 80	1	65 Nm (48 ft-lb)
4[1]	M 10 x 60	1	45Nm (33 ft-lb)
5	M 10 x 45	3	45Nm (33 ft-lb)
6	M 10 x 80	1	65 Nm (48 ft-lb)
7[2]	M 12 x 110	1	65 Nm (48 ft-lb)

[1] with nut

[2] with bracket for ATF lines

Pos. A = Dowel sleeves

Transmission brackets, removing and installing

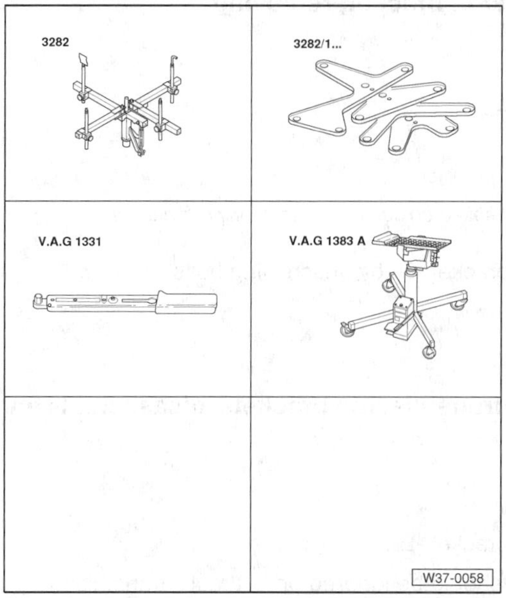

Special tools, workshop equipment, test and measuring appliances and aux. items required

- 3282 Transmission support
- 3282/19 Adjustment plate
- V.A.G 1331 Torque wrench
- V.A.G 1383 A Engine/transmission jack

Removing

If both transmission brackets are to be removed, the transmission must be supported as described. If only one bracket is to be removed, it is not necessary to support the transmission.

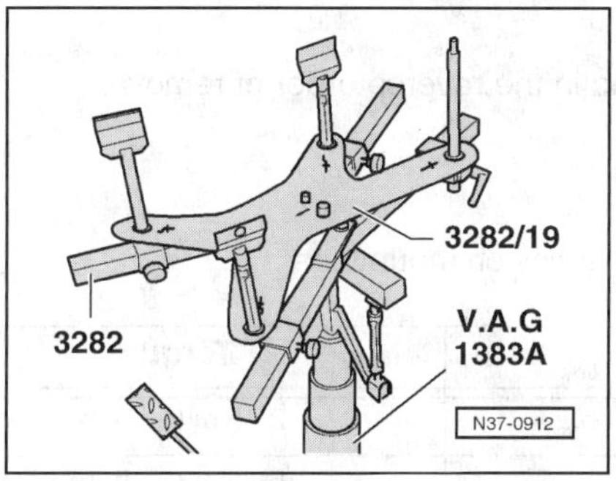

Set up transmission support 3282.

- Transmission support 3282 is set up with adjusting plate 3282/19 for automatic transmission 01V.
- The symbols on the adjusting plate indicate the mounts required, the arrow points forward.

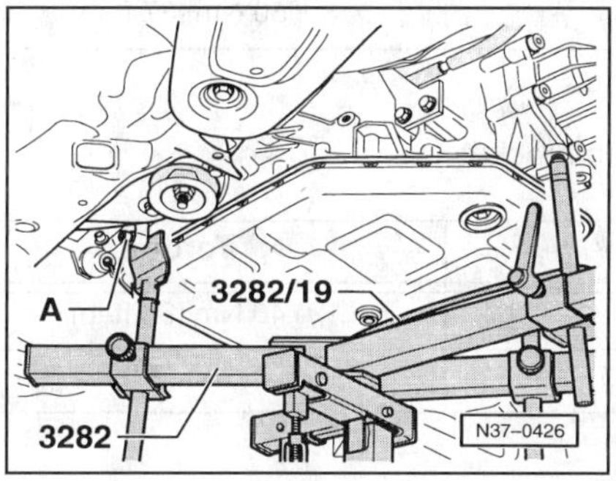

Move jack V.A.G 1383 A with transmission support 3282 under the transmission, align adjusting plate parallel to transmission and support transmission.

– Secure transmission to transmission support 3282 with bolt -A-.

– Chock space between transmission support and oil pan using washers.

Right-hand transmission bracket, removing

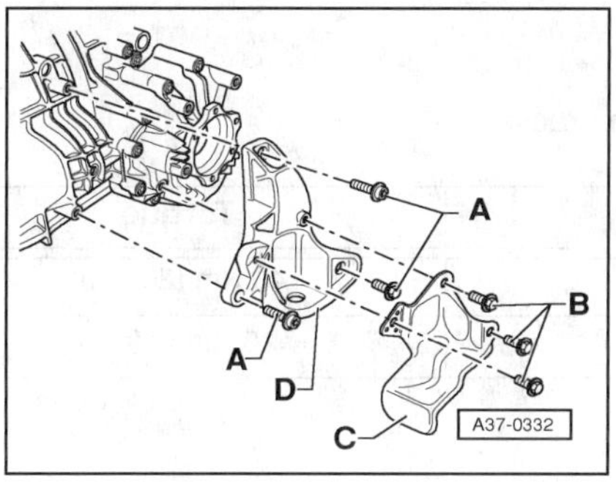

Remove bolts -B-.

– Remove heat shield -C-.

– Remove transmission bracket -D- by unscrewing bolts -A- from transmission.

Left-hand transmission bracket, removing

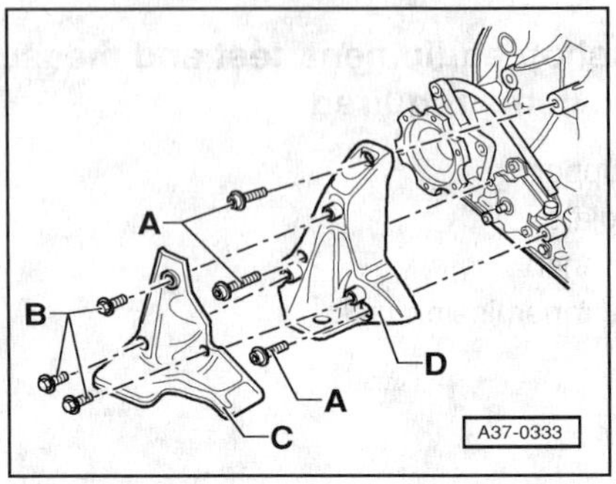

- Remove bolts -B-.
- Remove heat shield -C-.

NOTE —
There is no left-hand heat shield on vehicles with 4-cylinder engine.

- Remove transmission bracket -D- by unscrewing bolts -A- from transmission.

Left and right-hand transmission brackets, disassembly and installation

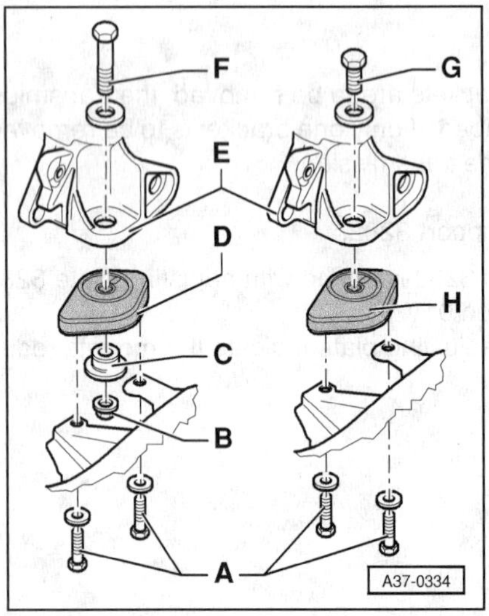

- Remove bolts -A-.
- Remove transmission bracket -E-.
  - Positions -C- and -B- should be ignored on hydraulic transmission mountings -H- as bolt -G- is screwed directly in the transmission mounting.
- Installation is performed in the reverse order of removal.

Tightening torques

Left and right-hand transmission mountings

Component	Torque
A	23 Nm (17 ft-lb)
B with F	50 Nm (37 ft-lb)
G	50 Nm (37 ft-lb)

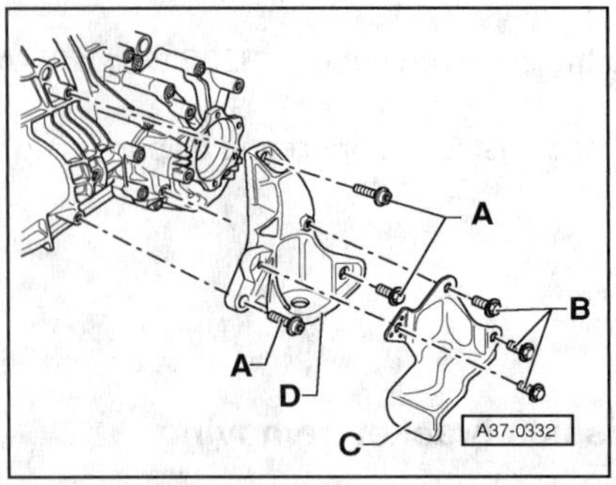

Right-hand transmission bracket

Component	Torque
A	40 Nm (30 ft-lb)
B	10 Nm (7 ft-lb)

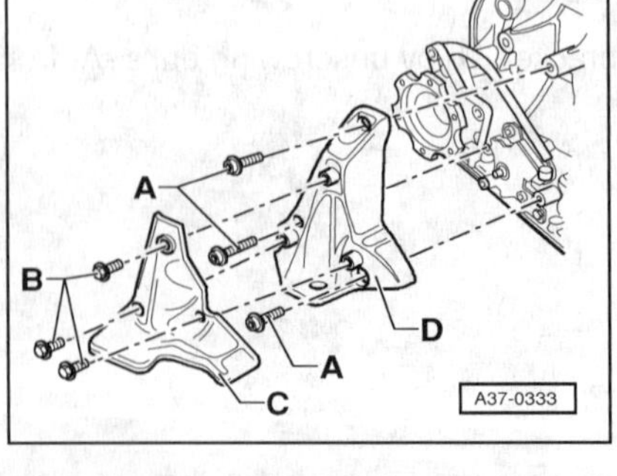

Left-hand transmission bracket

Component	Torque
A	40 Nm (30 ft-lb)
B	10 Nm (7 ft-lb)

ATF Lines

ATF lines (4-cylinder engine)

1. **Hex bolt**
   - Tighten to 20 Nm, (15 ft-lb)
   - Before tightening bolt, push in pipe onto stop by hand
2. **ATF pipe**
   - To cooler
   - Note inscription on transmission
3. **ATF pipe**
   - From cooler
   - Note inscription on transmission
4. **O-ring**
   - Lubricate with ATF
   - Always replace
5. **O-ring**
   - Lubricate with ATF
   - Always replace
6. **Hex bolt**
   - Tighten to 5 Nm (44 in-lb)
   - Before tightening bolt, push in pipe onto stop by hand
7. **O-ring**
   - Lubricate with ATF
   - Always replace
8. **Hex bolt**
   - Tighten to 5 Nm (44 in-lb)
   - Before tightening bolt, push in pipe onto stop by hand
9. **Union nut**
   - Tighten to 30 Nm (22 ft-lb)
10. **Bracket**
11. **Hex bolt**
    - To remove and install ATF lines it is not necessary to remove bolt.
12. **Hex nut**
    - Tighten to 5 Nm (44 in-lb)
13. **Clamp**
14. **Bracket**
    - On engine/transmission support

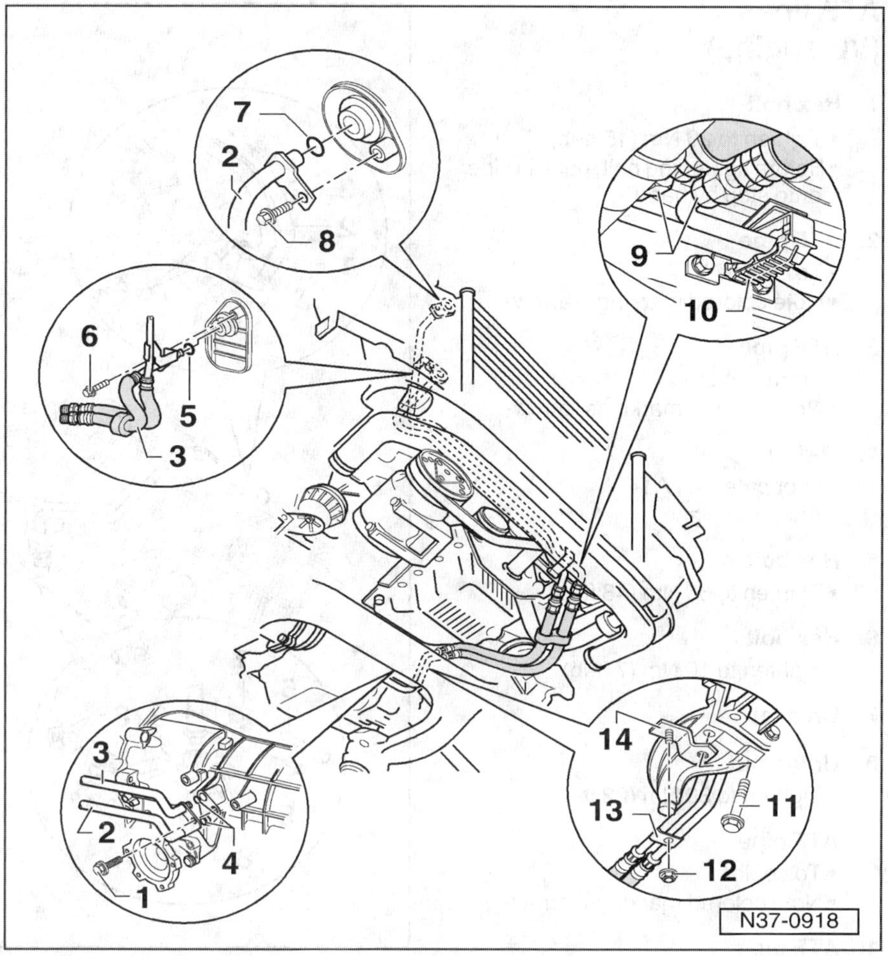

ATF lines (V6 engine)

1. **Hex bolt**
   - Tighten to 20 Nm (15 ft-lb)
   - Before tightening bolt, push in pipe onto stop by hand
2. **ATF pipe**
   - To cooler
   - Note colored markings -arrow-
3. **ATF pipe**
   - From cooler
   - Note colored markings -arrow-
4. **O-ring**
   - Lubricate with ATF
   - Always replace
5. **Hex bolt**
   - Tighten to 65 Nm (48 ft-lb)
6. **Hex bolt**
   - Tighten to 10 Nm (7 ft-lb)
7. **Bracket**
8. **Union nut**
   - Tighten to 30 Nm (22 ft-lb)
9. **ATF pipe**
   - To cooler
   - Note colored markings -arrow-
10. **ATF pipe**
    - From cooler
    - Note colored markings -arrow-
11. **Clamp**
12. **Hex bolt**
    - Tighten to 10 Nm (7 ft-lb)
13. **ATF pipe**
    - To cooler
    - Note colored markings -arrow-
14. **ATF pipe**
    - From cooler
    - Note colored markings -arrow-
15. **Hex bolt, 5 Nm**
    - Before tightening bolt, push in pipe onto stop by hand

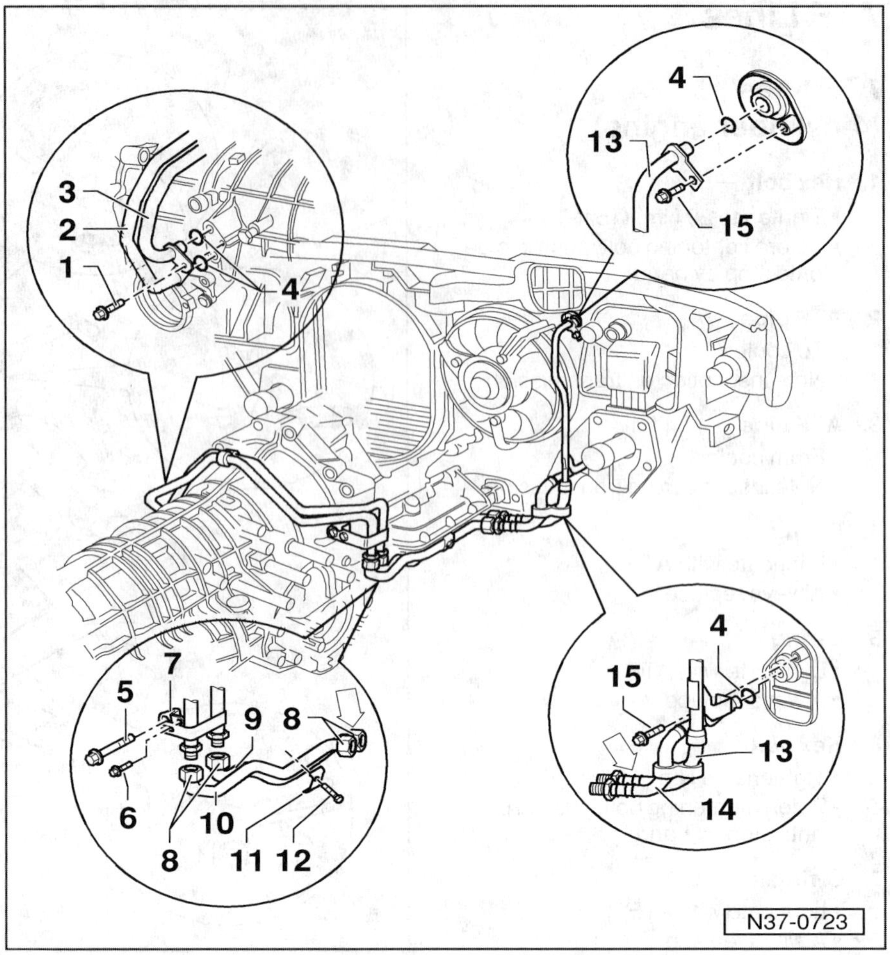

38 Gears, Hydraulics

Repair Group

N 38

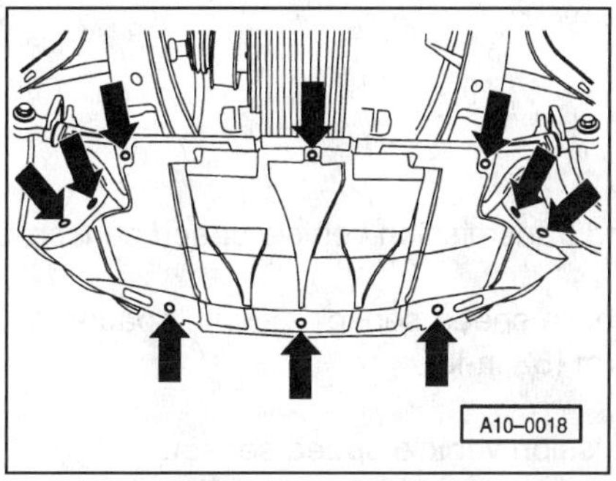

Oil pan, removing and installing

Special tools and equipment

- V.A.G 1306 Drip tray
- V.A.G 1331 Torque wrench

< Remove noise insulation tray -arrows-.

– Place drip tray V.A.G 1306 underneath.

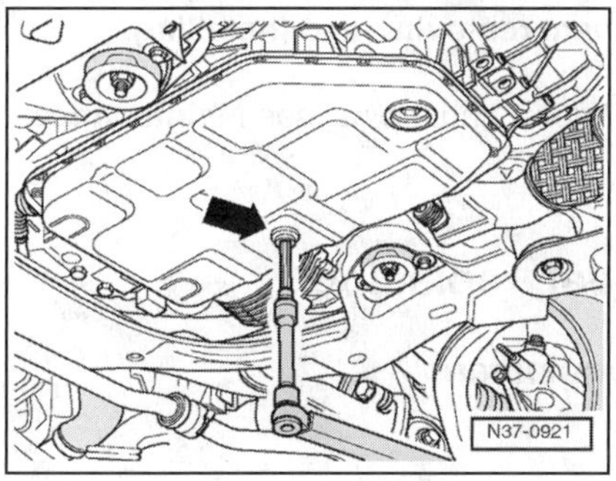

< Remove drain plug -arrow 1-.

– Drain ATF.

– Unscrew bolts on oil pan in diagonal sequence and remove oil pan.

– Install in reverse order but note following:

- The oil drain plug and seal must be replaced.

– Clean all four magnets in the recesses of the oil pan. Ensure magnets are seated correctly in the oil pan.

– Install oil pan with new gasket.

– Tighten oil pan bolts 10 Nm (7 ft-lb) in stages and in diagonal sequence.

– Tighten new drain plug to 40 Nm (30 ft-lb).

– Fill with ATF, then check ATF level and top up, see **Maintenance**.

Automatic Transmission Fluid (ATF) screen, removing and installing

– Remove oil pan, see **Oil pan, removing and installing**.

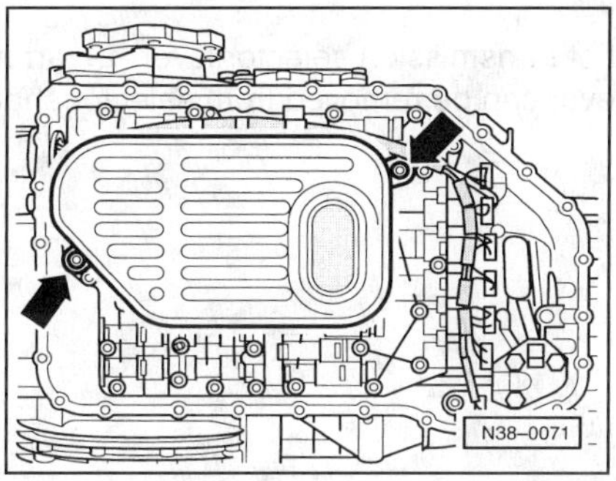

< Unscrew 2 bolts on ATF screen -arrows-.

– Pull ATF screen off valve body.

– Lubricate seal on intake neck of ATF screen with ATF.

– Carefully push intake neck of ATF screen into opening of valve body.

– Install ATF screen -arrows- and tighten bolts to 6 Nm (53 in-lb).

– Fit oil pan, see **Oil pan, removing and installing**.

– Fill with ATF, then check ATF level and top up, see **Maintenance**.

Transmission vehicle speed sensor -G38- and sender for transmission output RPM -G195-, replacing

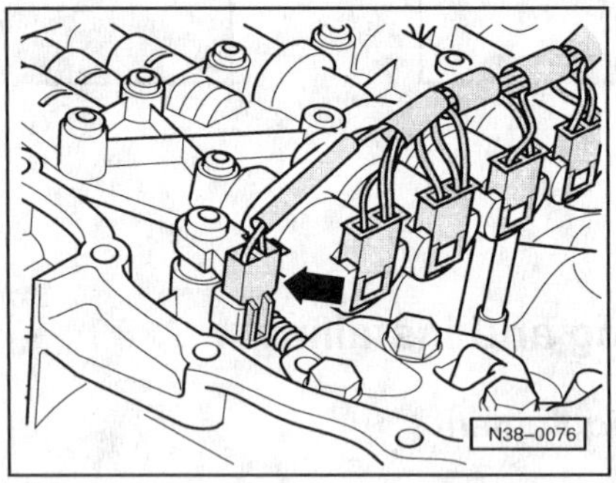

– Remove oil pan, see **Oil pan, removing and installing**.

◄ Pull connector -arrow- off transmission vehicle speed sensor.

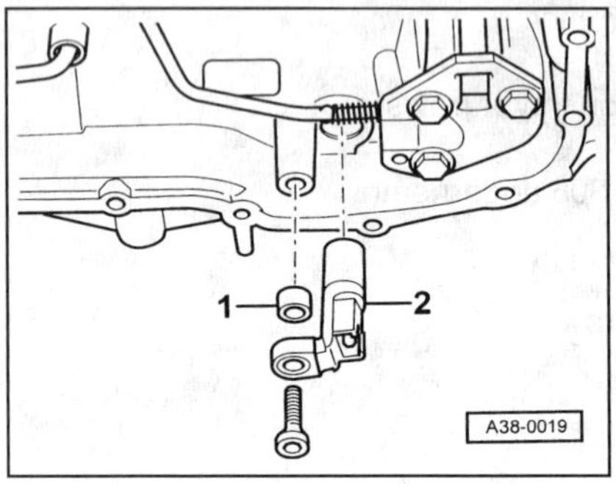

◄ Remove spacer -1-, and transmission vehicle speed sensor -2-.

– Install transmission vehicle speed sensor -2- with spacer -1- (height: 8 mm) and tighten to 6 Nm (53 in-lb).

– Fit connector to transmission vehicle speed sensor.

– Install oil pan, see **Oil pan, removing and installing**.

– Fill with ATF; check ATF level and top up, see **Maintenance**.

Selector shaft seal, replacing

Special tools and equipment

• 80-200 Removal lever
• 3385 Thrust piece

– Remove multi-function TR switch -F125-, see **Multi-function Transmission Range (TR) switch -F125-, replacing**.

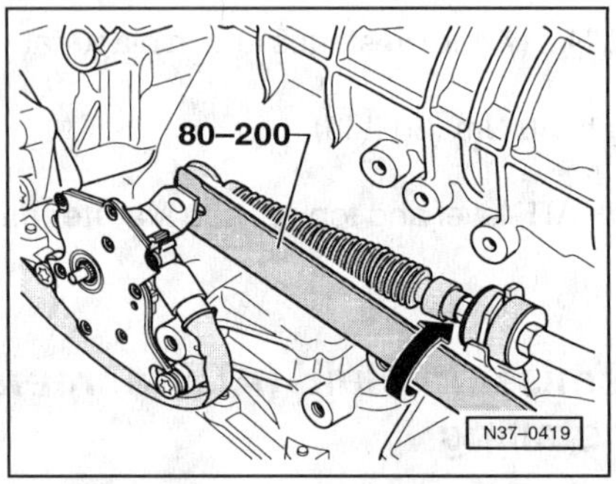

◄ Remove selector lever cable from selector shaft lever -arrow-.

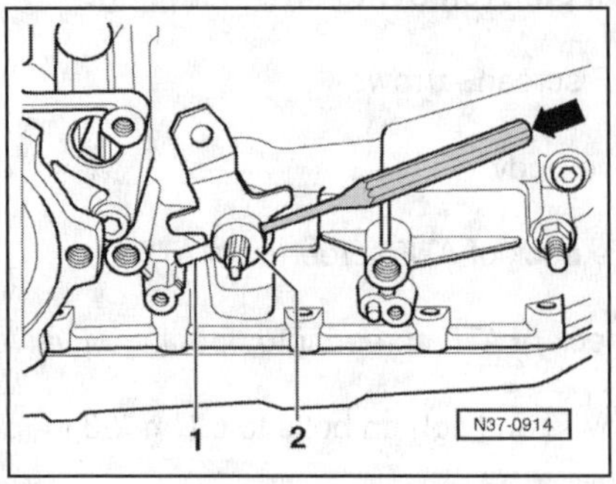

◄ Drive spring pin -1- out of transmission selector lever -2- -arrow- until transmission selector lever can be removed from selector shaft.

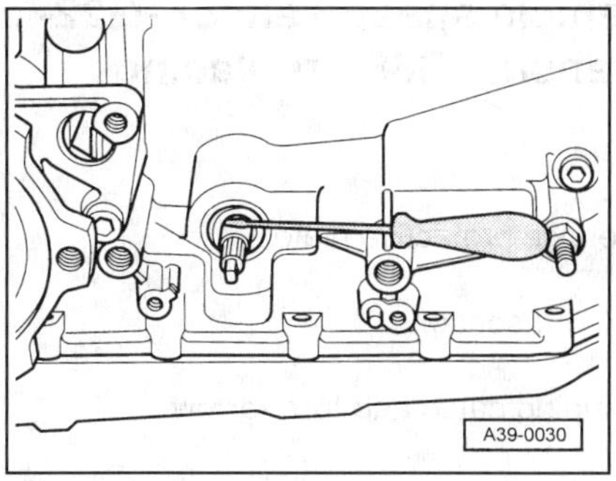

◄ Pry seal out with a screwdriver.

– Apply ATF to circumference and sealing lips of new seal.

NOTE —
Installation position: open side of seal faces transmission.

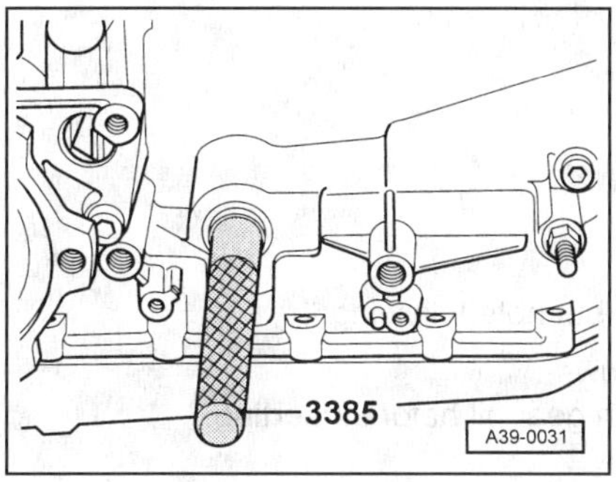

◄ Drive in new seal until thrust piece contacts stop. Ensure that seal remains straight when doing this.

– Push transmission selector lever onto selector shaft and install roll pin.

– Carefully press selector lever cable onto selector shaft lever using a pair of pliers. When doing this, take care not to bend the selector shaft lever, otherwise the selector mechanism will not be able to be adjusted correctly.

– Install multi-function switch, see following procedure.

Multi-function Transmission Range (TR) switch -F125-, replacing

– Remove left front wheel.

– Remove left-hand transmission bracket, see **Repair Group 37, Left and right-hand transmission brackets, removing and installing**.

– Unscrew bolts and detach multi-function TR switch -F125- from transmission; pull switch off selector shaft.

– Install multi-function TR switch -F125- centrally on selector shaft. Do not tilt or use force when installing. Danger of damaging switch contacts.

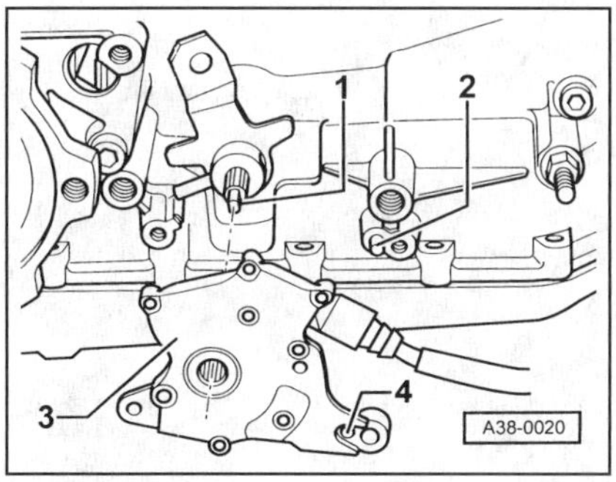

◄ Fit multi-function TR switch -F125- on selector shaft: the flat section in the splines in switch -3- must coincide with the flat section on selector shaft -1-.

– Turn switch so that drilling -4- on switch housing can be fitted on locating pin -2- on transmission housing.

– Bolt multi-function TR switch -F125- to transmission and tighten.

– Install left-hand transmission bracket, see **Repair Group 37, Left and right-hand transmission brackets, removing and installing**.

– Install left front wheel.

Speedometer vehicle speed sensor -G22- and Vehicle speed sensor -G68-, replacing

– Remove left front wheel.

– Remove left-hand drive axle protective plate.

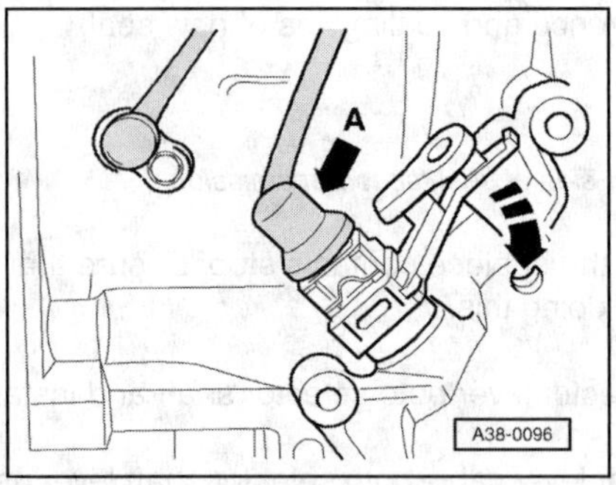

– Pull connector -arrow A- off sensor.

– Press down sensor retaining catch and turn -arrow-.

– Pull out sensor.

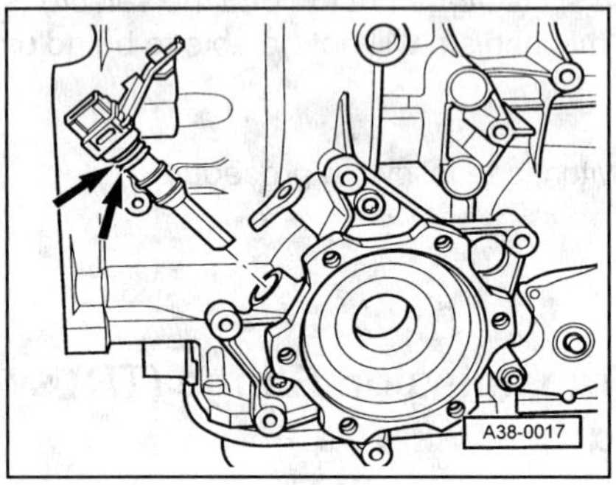

– Install in reverse order but note following:

- Replace O-rings -arrows-.
- Lubricate O-rings with gear oil before inserting.
- Insert sensor.
- Engage retaining catch on bearing bracket for flange shaft.
- Install left front wheel.

39 Final Drive, Differential

Repair Group

N 39

Gear oil in front final drive, checking

Special tools, workshop equipment, testers, measuring instruments and auxiliary items required

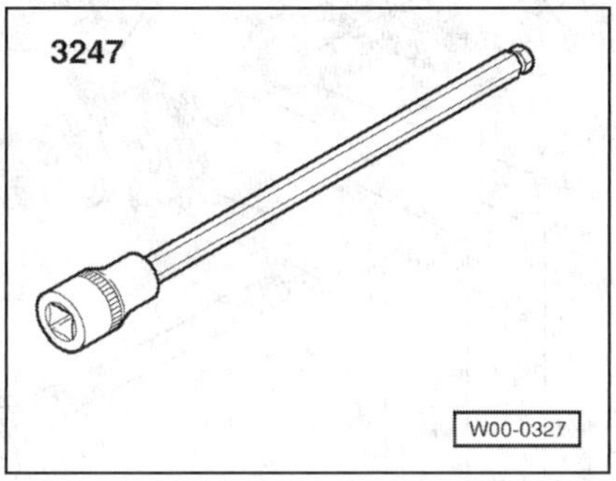

3247 Socket AF 8

Oil level is checked with transmission installed

– Drive the vehicle a short distance to warm up the gear oil.

– Position vehicle on hard, level standing e.g. on 4 pillar hoist or over pit.

– Switch off engine and let gear oil drain for 5 minutes.

– Place drip tray V.A.G 1306 underneath.

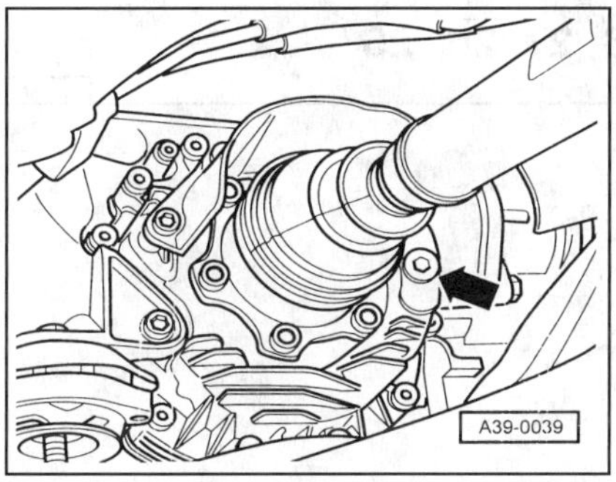

Remove sealing plug -arrow-.

The oil level is correct when the front final drive is full to the lower edge of the filler hole.

– Allow excess gear oil to pour out or top up as necessary.

• Specification, see **Maintenance**

– If gear oil needs to be replenished following repair work, fill front final drive slowly and evenly over a period of at least 5 minutes until the gear oil drips out over the lower edge of the oil filler hole.

A filling time of at least 5 minutes is necessary to allow the oil to settle evenly in the differential and cavity.

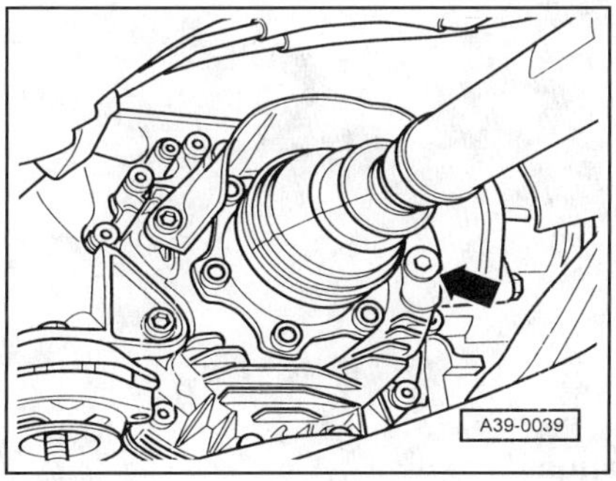

Screw in original sealing plug -arrow-.

– Drive the vehicle a short distance to warm up the gear oil.

– Check oil level in front final drive.

– If the oil level is OK, replace front final drive sealing plug -arrow-, see **Front final drive, assembly**.

Front final drive, assembly

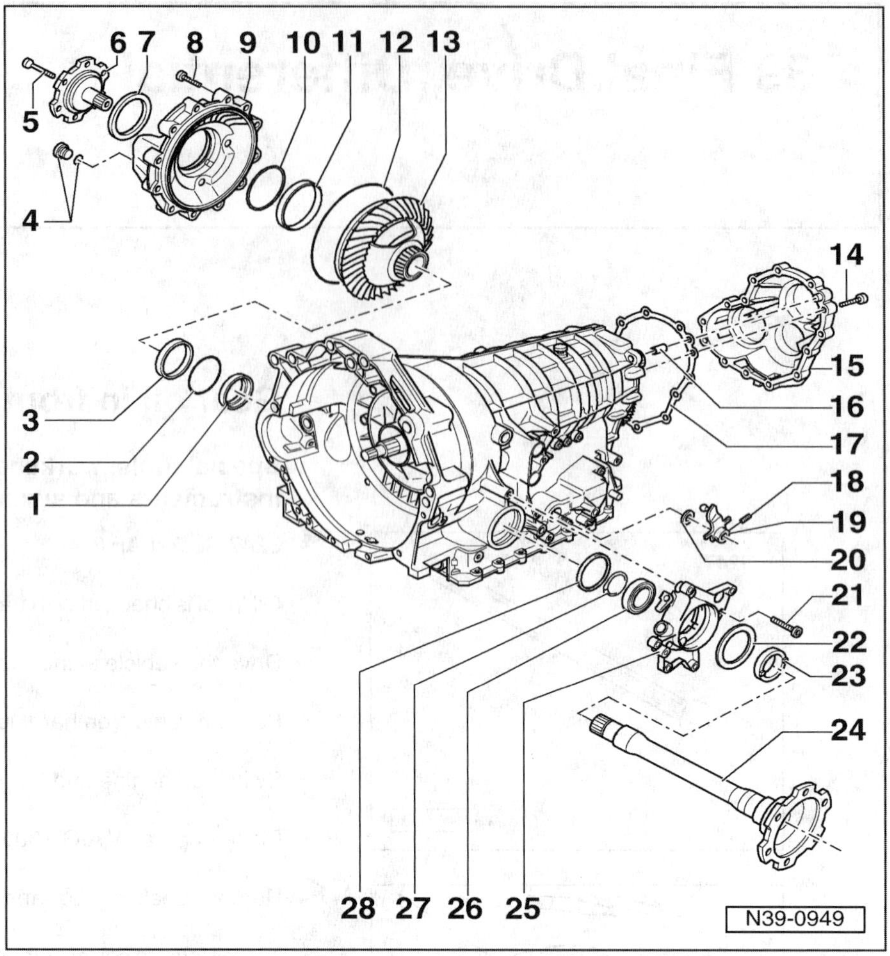

1. **Oil seal**
   - Between final drive and transmission housing
2. **Shim**
   - Behind tapered roller bearing outer race
3. **Tapered roller bearing outer race**
4. **Sealing plug**
   - Tighten to 30 Nm (22 ft-lb)
   - With seal, always replace
5. **Bolt**
   - Tighten to 23 Nm (17 ft-lb)
6. **Flange shaft, right**
7. **Oil seal for flange shaft, right**
   - Replacing, see **Seal for right-hand flange shaft, replacing**
8. **Bolt**
   - Tighten to 23 Nm (17 ft-lb)
9. **Cover for front final drive**
10. **Shim**
    - Behind tapered roller bearing outer race
11. **Tapered roller bearing outer race**
12. **O-ring**
    - Always replace
13. **Differential**
14. **Bolt**
    - Tighten to 23 Nm (17 ft-lb)
15. **Housing cover for front axle drive**
16. **Dowel sleeve**
17. **Seal**
18. **Spring pin**
19. **Lever**
    - For selector shaft
20. **Oil seal**
    - For selector shaft
21. **Bolt**
    - Tighten to 23 Nm (17 ft-lb)
22. **Oil seal for flange shaft, left**

    EDITOR'S NOTE —
    Replacement requires special tools and procedures that are beyond the scope of this manual.

    Requires special tools:
    - VW 401 Thrust plate
    - VW 402 Thrust plate
    - VW 407 Press tool
    - VW 412 Press tool
    - VW 681 Oil seal extractor
    - 32-109 Tube
    - 2040 Tube
    - 3384 Drift
23. **Drive gear**
    - For speedometer sender -G22-
24. **Flange shaft, left**
25. **Bearing bracket for flange shaft**
26. **Grooved ball bearing for flange shaft**
27. **Circlip**
28. **O-ring**
    - Square section

Seal for right-hand flange shaft, replacing

Special tools, workshop equipment, test and measuring appliances and aux. items required

- VW 681 Oil seal extractor
- 3382 Drift
- V.A.G 1331 Torque wrench
- V.A.G 1332 Torque wrench

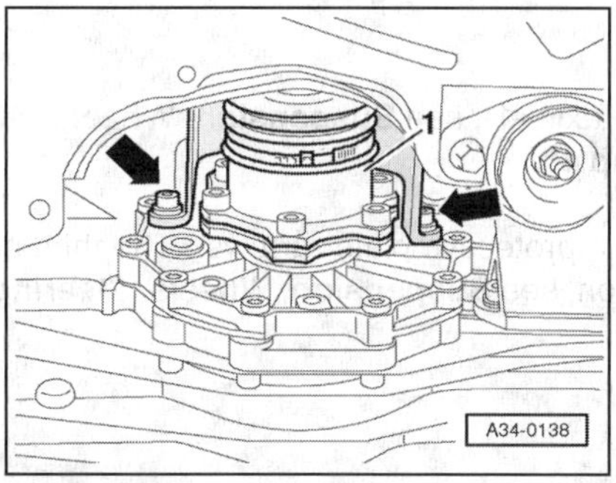

Removing

– Loosen drive shaft/wheel hub bolt and wheel bolts while vehicle is standing on its wheels.

– Remove wheel.

◄ Remove right-hand drive axle protective plate -arrows-. The third bolt cannot be seen in the illustration.

– Remove right-hand drive axle -1-, see **Suspension, Wheels, Steering section, Repair Group 40**

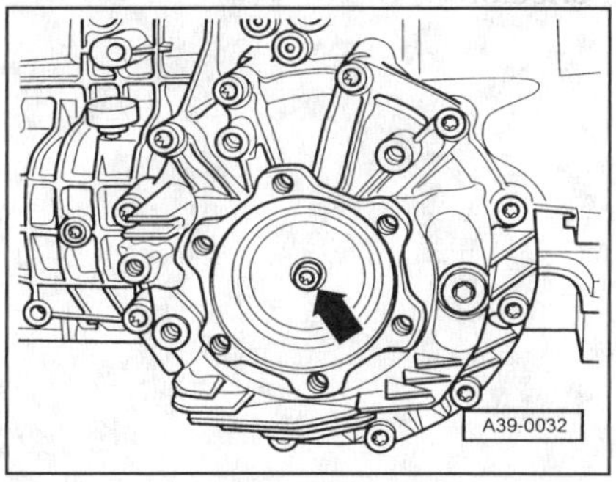

◄ Remove flange shaft bolt -arrow- (use pin to prevent flange turning).

– Place drip tray V.A.G 1306 or equivalent underneath.

– Pull out right flange shaft.

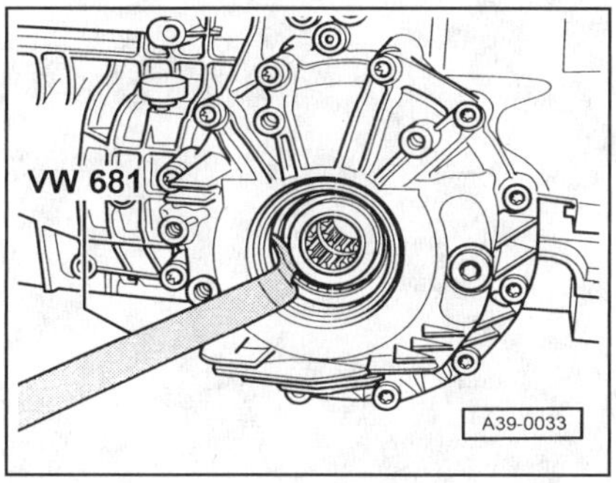

◄ Remove flange shaft oil seal.

Installing

– Apply gear oil to circumference and sealing lips of seal.

– Half fill space between sealing lip and dust lip with grease G 052 128 A1.

- Installation position: open side of seal faces transmission.

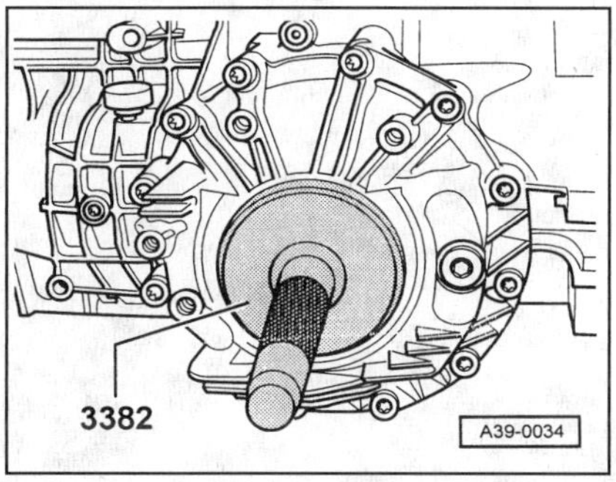

◄ Drive new seal in onto stop, when doing this do not cant seal.

– Install right-hand flanged shaft.

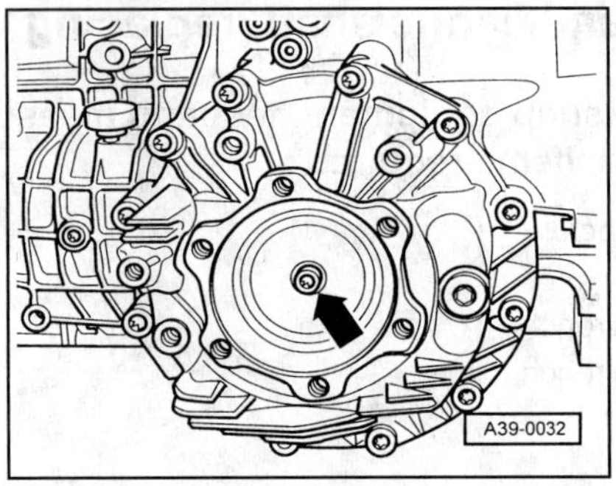

- Fit flanged shaft bolt -arrow- and secure against turning using a punch.

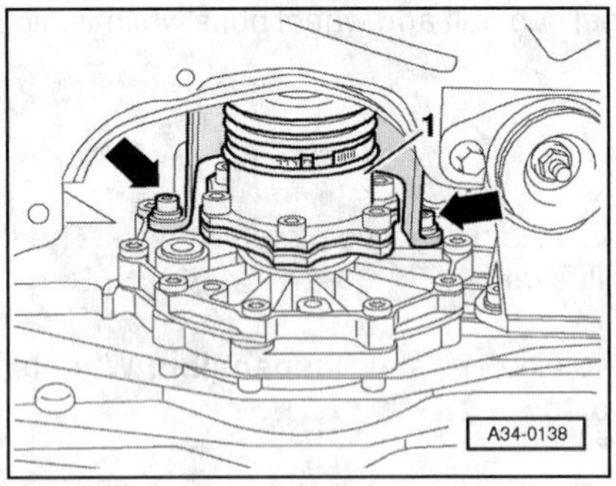

- Install right-hand drive axle -1-, see **Suspension, Wheels, Steering section, Repair Group 40**

– Fit right-hand drive axle protective plate -arrows-. The third bolt cannot be seen in the illustration, see **Suspension, Wheels, Steering section, Repair Group 40**

– Install front wheel.

– With transmission installed, check gear oil in front final drive, see **Gear oil in front final drive, checking**.

5-Speed Automatic Transmission 01V, All Wheel Drive

Section **P**

00 General Information

Repair Group

P 00

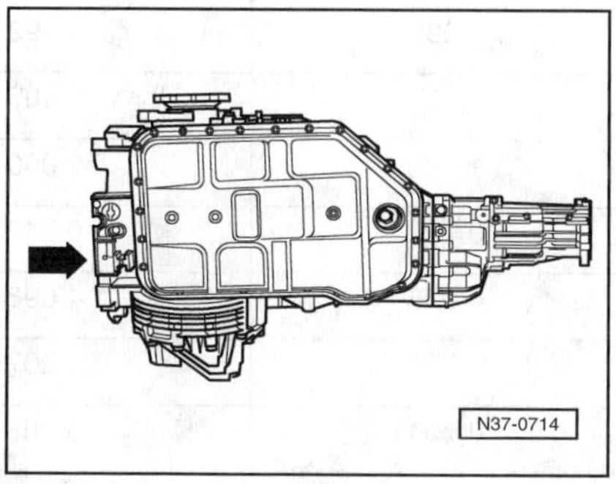

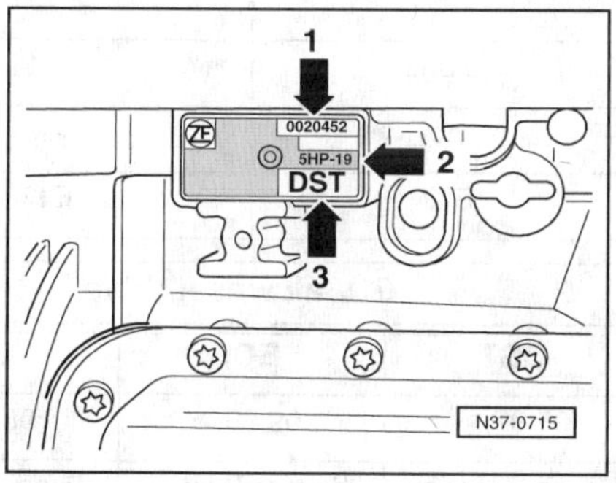

Transmission identification

The "5-speed automatic transmission 01V all wheel drive" is installed in the Passat 1998 ➤ in combination with a 6 or 8-cylinder engine.

Location of code letters on transmission

The transmission code letters are given on the type plate at the bottom of the transmission -arrow-.

Transmission code letters and transmission serial number

Transmission Serial No. -arrow 1-
Transmission identification -arrow 2-
Code letters -arrow 3-

The code letters of the transmission are also given on the vehicle data plates.

5-speed automatic transmission 01V, notes

Torque converter

The torque converter is equipped with a lock-up clutch. The closing of the lock-up clutch is determined by the engine load and speed and produces very little vibration. The lock-up clutch drives all forward gears mechanically without slip.

Transmission

The 5-speed automatic transmission 01V all wheel drive has five hydraulically activated forward gears. When the lock-up clutch is closed these forward gears become mechanically driven gears by cutting out the torque converter slip.

Control module -J217- with driving situation gear changes (Fuzzy Logic)

Determining the driving situation and driving resistance dependent gear change point happens automatically.

Advantages:

- Gear changes will be consumption orientated
- Maximum engine output is always available
- Individual adaption of gear change points in all driving situations
- Gear change points are infinitely variable

Automatic transmission		01V all wheel drive	
Transmission	Code letters	**CJR**	**DRN**
	Manufactured from/to	03.97 / 05.97	08.97 ➤
Application	Model	Passat 4Motion 1998 ➤	Passat 4Motion 1998 ➤
	Engine	2.8 ltr. - 142 kW	2.8 ltr. - 142 kW
Torque converter	Code letters	N28	F31
Valve body	Code letters	The valve body is allocated according to the transmission code letters	
Ratios	1st gear	3.665	3.665
	2nd gear	1.999	1.999
	3rd gear	1.407	1.407
	4th gear	1.000	1.000
	5th gear	0.742	0.742
	Reverse gear	4.096	4.096
Intermediate drive for front final drive	Ratio	1.000	1.207
Intermediate drive for rear final drive	Ratio	0.791	0.805
Front final drive	Ratio	3.091	2.727
Rear final drive	Ratio required	3.889	4.111
Data bus		no	yes
Hydraulic control		E17	E17

Automatic transmission		01V all wheel drive		
Transmission	Code letters	**DST**	**ECB**	**EKD**
	Manufactured from	05.97 ➤	09.99 ➤	08.99 ➤
Application	Model	Passat 4Motion 1998 ➤	Passat 4Motion 1998 ➤	Passat 4Motion 1998 ➤
	Engine	2.8 ltr. - 142 kW	2.8 ltr. - 142 kW	2.8 ltr. - 142 kW
Torque converter	Code letters	F31	F31	N28
Valve body	Code letters	The valve body is allocated according to the transmission code letters		
Ratios	1st gear	3.665	3.665	3.665
	2nd gear	1.999	1.999	1.999
	3rd gear	1.407	1.407	1.407
	4th gear	1.000	1.000	1.000
	5th gear	0.742	0.742	0.742
	Reverse gear	4.096	4.096	4.096
Intermediate drive for front final drive	Ratio	1.000	1.000	1.207
Intermediate drive for rear final drive	Ratio	0.791	0.791	0.902
Front final drive	Ratio	3.091	3.091	3.091
Rear final drive	Ratio required	3.889	3.889	4.111
Data bus		yes	yes	yes
Hydraulic control		E17	E18/2	E18/2

Automatic transmission			01V all wheel drive	
Transmission	Code letters		**FAR**	**FAS**
	Manufactured	from	06.00 ➤	08.00 ➤
Application	Model		Passat 4Motion 1998 ➤	Passat 4Motion 1998 ➤
	Engine		2.8 ltr. - 142 kW	2.8 ltr. - 142 kW
Torque converter	Code letters		F31	N28
Valve body	Code letters		The valve body is allocated according to the transmission code letters	
Ratios	1st gear		3.665	3.665
	2nd gear		1.999	1.999
	3rd gear		1.407	1.407
	4th gear		1.000	1.000
	5th gear		0.742	0.742
	Reverse gear		4.096	4.096
Intermediate drive for front final drive		Ratio	1.000	1.207
Intermediate drive for rear final drive		Ratio	0.791	0.902
Front final drive		Ratio	3.091	3.091
Rear final drive		Ratio required	3.889	4.111
Data bus			yes	yes
Hydraulic control			E18/2	E18/2

Automatic transmission			01V all wheel drive		
Transmission	Code letters		**EMQ**	**FVE**	**GAK**
	Manufactured	from / to	03.00 / 06.00	06.00 ➤	06.01 ➤
Application	Model		Passat 4Motion 1998 ➤	Passat 4Motion 1998 ➤	Passat 4Motion 1998 ➤
	Engine		4.0 ltr. - 202 kW	4.0 ltr. - 202 kW	4.0 ltr. - 202 kW
Torque converter	Code letters		Q56	Q56	Q56
Valve body	Code letters		The valve body is allocated according to the transmission code letters		
Ratios	1st gear		3.665	3.665	n/a
	2nd gear		1.999	1.999	n/a
	3rd gear		1.407	1.407	n/a
	4th gear		1.000	1.000	n/a
	5th gear		0.742	0.742	n/a
	Reverse gear		4.096	4.096	n/a
Intermediate drive for front final drive		Ratio	1.207	1.000	n/a
Intermediate drive for rear final drive		Ratio	0.902	0.711	n/a
Front final drive		Ratio	2.909	2.909	n/a
Rear final drive		Ratio required	4.111	4.111	n/a
Data bus			yes	yes	yes
Hydraulic control			E18/2	E18/2	E18/2

Rear final drive identification

Final drive unit 01R is used with the All Wheel Drive automatic transmission 01V.

> ***WARNING —***
> *If the rear final drive is to be replaced, it is absolutely necessary to note the code letters of not only the automatic transmission but also the engine of the vehicle in see Parts catalog. Only then is a correct application guaranteed.*

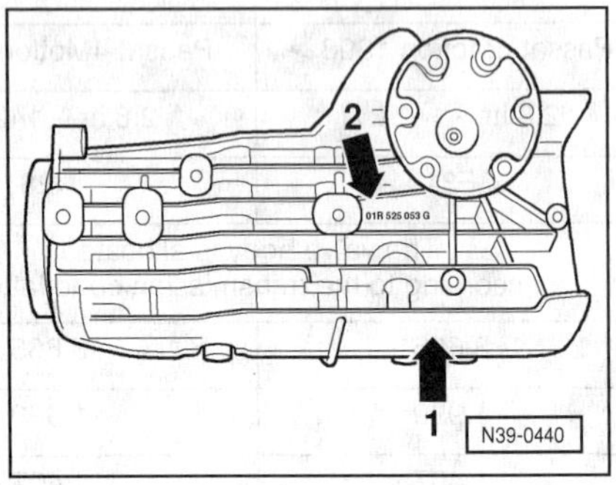

◄ location on final drive

Code letters and date of manufacture (arrow 1)
Final drive 01R (arrow 2)

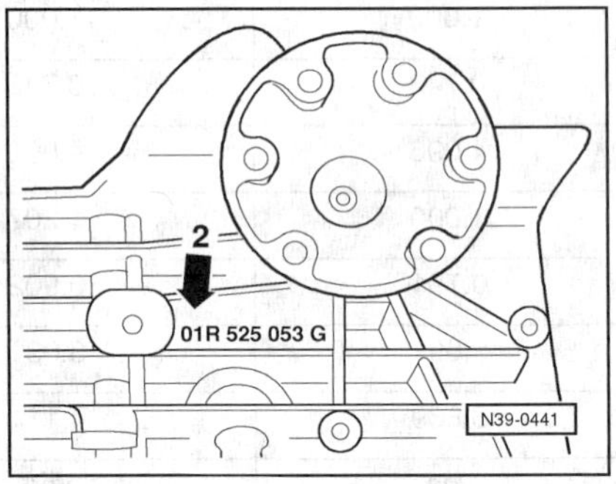

◄ Final drive 01R (arrow 2)

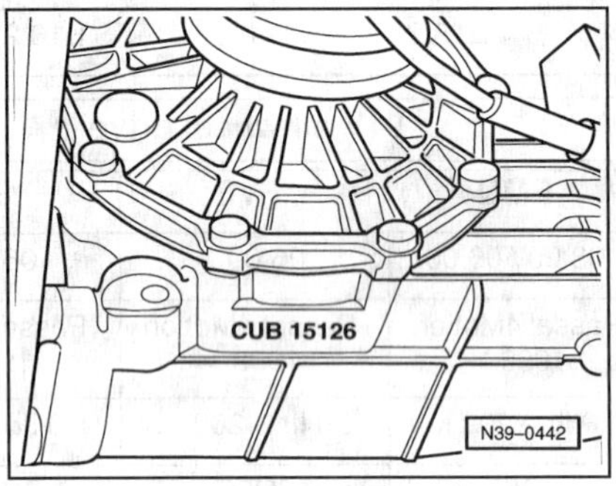

◄ Code letters and date of manufacture)

Transmission code letters:	Day	Month	Year
CUB	15	12	6
	Date (e.g. 15 DEC. 1996)		

Further data need not be regarded.

Code letters, assembly application, ratios, capacities

Rear final drive		01R		
Code letters		**CUC**	**CUB**	**EUU**
Manufactured	from / to	05.97	08.97	05.97
Application	Model	Passat 4Motion 1998 ➤	Passat 4Motion 1998 ➤	Passat 4Motion 1998 ➤
	Engine	2.8 ltr. - 142 kW	2.8 ltr. 142 kW	4.0 ltr. 202 kW
Final drive ratio		3.889	4.111	4.111
Drive shaft flange diameter		100 mm	100 mm	108 mm

Capacities

Planetary transmission

Capacities	Planetary transmission	Automatic transmission
Initial filling	9.0 ltr.	01V All Wheel Drive
Changing	ca. 2.6 - 3.0 ltr.[1]	
Lubricant	VW ATF	

[1] Filled for life: Oil change required only following repair work or removal of oil pan up to 3 liters.

VW ATF is obtainable as a replacement part.

Container size 1.0 ltr. - Replacement Part No. G 052 162 A2

Only VW ATF available as a replacement part may be used in the planetary transmission of the automatic transmission 01V all wheel drive.

Checking ATF level and topping up, see **Maintenance.**

Front final drive

Capacities	Front final drive	Automatic transmission
Initial filling	approx. 0.75 ltr.	01V All Wheel Drive
Oil change	Filled for life No change	
Lubricant	Gear oil SAE75 W 90 (synthetic oil)	

Gear oil SAE 75 W 90 (synthetic oil) is available as a replacement part.

Container size 1.0 ltr. - Replacement Part No. G 052 145 S2

Checking oil level in front final drive, see **Repair Group 39**.

Center differential

Capacities	Center differential	Automatic transmission
Initial filling	approx. 0.8 ltr.	01V All Wheel Drive
Oil change	Filled for life No change	
Lubricant	Gear oil SAE75 W 90 (synthetic oil)	

Gear oil SAE 75 W 90 (synthetic oil) is available as a replacement part.

Container size 1.0 ltr. - Replacement Part No. G 052 145 S2

Checking gear oil in center differential and topping up, see **Repair Group 39**.

Rear final drive

Capacities	Final drive 01R
Initial filling	1.5 ltr.
Oil change	Filled for life No change
Lubricant	Gear oil SAE75 W 90 (synthetic oil)

SAE 75 W 90 gear oil (synthetic oil) is available as a replacement part.

Container size 1.0 ltr. - Replacement Part No. G 052 145 S2

Checking gear oil in rear final drive, see **Repair Group 39**.

General repair instructions

The maximum possible care and cleanliness and proper tools are essential to ensure satisfactory and successful transmission repairs. The usual basic safety precautions also, naturally apply when carrying out vehicle repairs.

A list of special tools and workshop equipment required for repair procedures in a manual can be found at the start of each repair group section.

Transmission

- If the automatic transmission has been replaced, the following fluid levels are to be checked:
  ATF in planetary transmission, see **Maintenance**.
  Gear oil in front final drive, see **Repair Group 39**.
  Gear oil in center differential, see **Repair Group 39**.
  Fill levels and specifications, see **Capacities.**
- If rear final drive is replaced, fill it with gear oil to lower edge of filling hole, see **Repair Group 39**
- Engine must not run if oil pan or ATF has been removed.
  Do not tow vehicle.
- Before installing transmission, check torque converter position, see **Repair Group 32**.
- Place removed parts on a clean surface. Cover parts to prevent soiling. Use plastic sheeting and paper. If necessary, use only lint-free cloth!
- Only install clean parts: Remove packing from replacement parts immediately prior to installation and not before.
- If repair work cannot be performed immediately, cover replacement parts from which packing has been removed.

Gaskets, sealing rings

- Always replace O-rings, oil seals and gaskets.
- Once oil seal/gasket has been removed, check contact surface on housing or shaft for burrs and damage. Repair as necessary.
- Thoroughly clean housing connecting surfaces and apply sealant AMV 188 200 03.

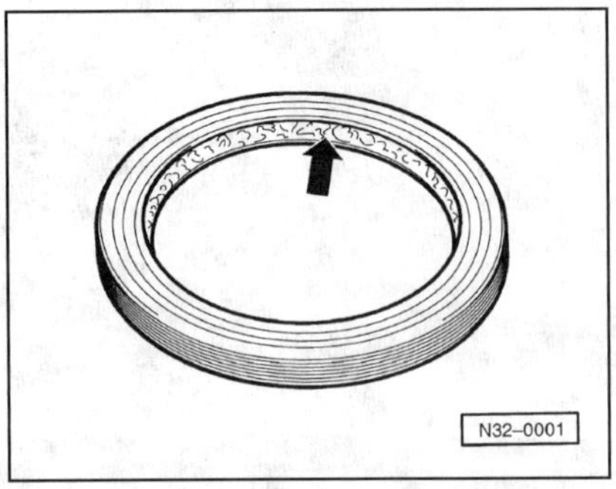

Before installing shaft-type oil seal, half fill space between sealing lips with grease G 052 128 A1 -arrow- and then install.

- Open side of oil seal faces oil.
- Before installing oil seals, moisten circumference of seal and sealing lip with ATF or gear oil depending on installation location.
- Moisten O-rings with ATF to prevent damage occurring to the rings during assembly.

37 Controls, Housing

Repair Group

P 37

Shift Mechanism

The following procedures are covered in **Automatic Transmission 01V, Front Wheel Drive section, Repair Group 37**:

- Kick-down shift point, adapting
- Selector lever cable, checking and adjusting
- Selector lever handle, removing and installing
- Selector lever mechanism, checking
- Selector lever position signal, checking
- Shift mechanism, assembly

EDITOR'S NOTE —
The shift mechanisms used in the 5-Speed Automatic Transmission 01V, Front Wheel Drive, and the 5-Speed Automatic Transmission 01V, All Wheel Drive are identical.

Electrical/electronic components

For information on Electrical/electronic components and locations see **Automatic Transmission 01V, Front Wheel Drive section, Repair Group 37**.

Transmission, removing and installing

Transmission, removing and installing (V6 engine)

Special tools and equipment

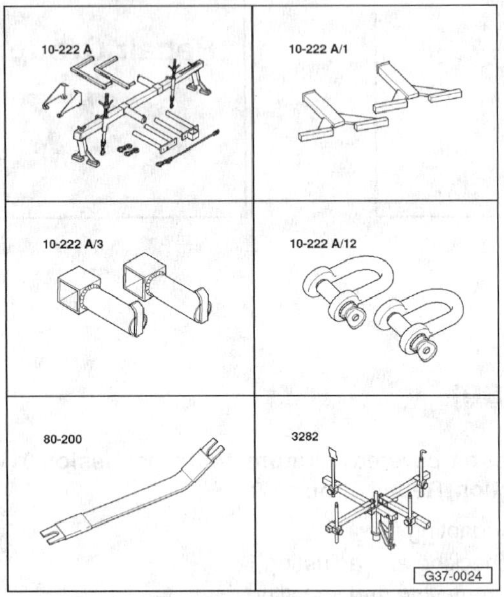

- 10-222A Engine support bridge
- 10-222A/1 Bracket for engine
- 10-222A3 Engine support adapter
- 10-222A12 Shackle
- 80-200 Pry lever
- 3282 Transmission support

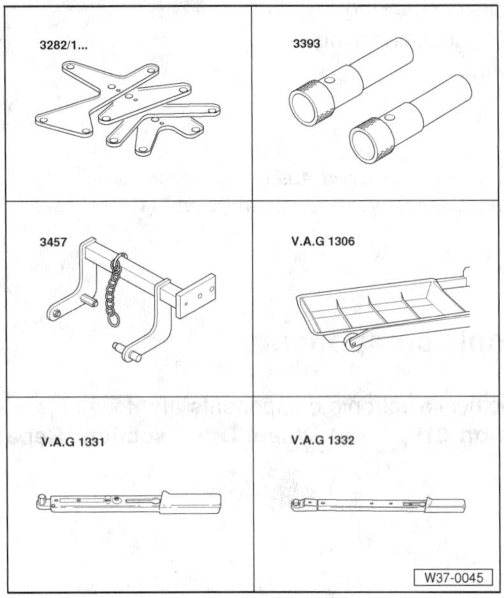

- 3282/19 Adjustment plate
- 3393 Testing mandrel
- 3457 Attachment and holding rig.
- VAG1306 Drip tray
- VAG1331 Torque wrench
- VAG1332 Torque wrench

- VAG1383A Engine/transmission jack
- VAS5085 Step ladder

Removing

– Obtain radio code on vehicles with coded radio.

– With ignition switched off disconnect battery Ground (GND) strap.

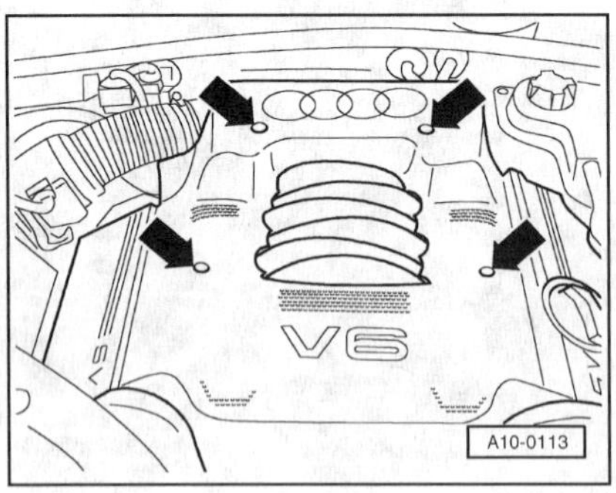

Unscrew bolts -arrows- and remove engine cover.

– Take intake hose off intake manifold.

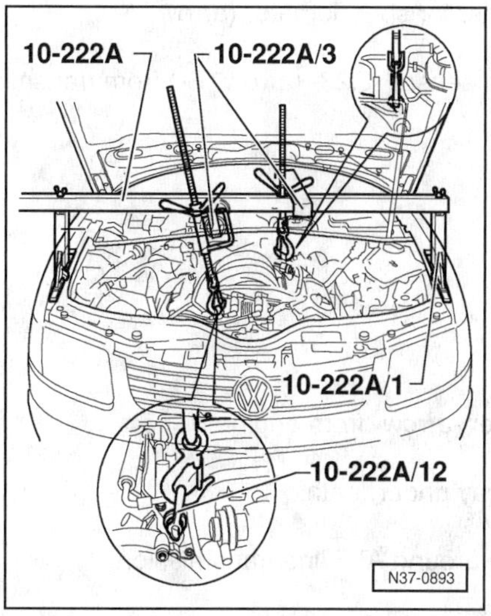

Install 10-222A Engine support bridge together with 10-222A/1 Bracket for engine and 10-222A3 Engine support adapter.

- Secure spindles to engine. In addition, use a 10-222A12 Shackle on the front of the engine to avoid damage.
- Take weight of engine/transmission assembly on the spindles.
- Loosen wheel bolts.
- Raise vehicle.
- Remove front wheels.

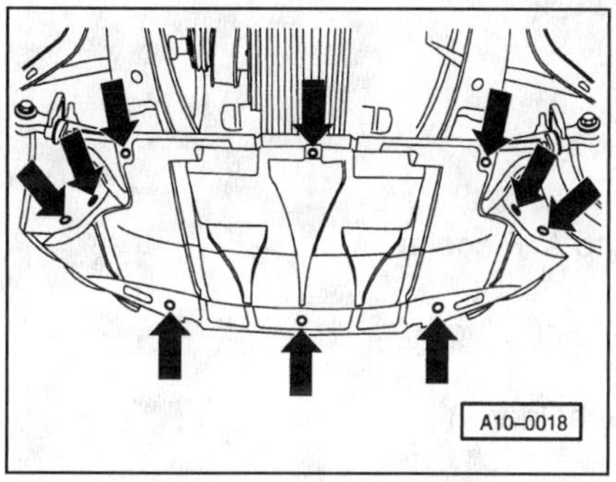

Remove noise insulation tray -arrows-.

- Remove front exhaust system with catalyst, see **2.8 Liter V6 5V Engine Mechanical (code: AHA, ATQ) section, Repair Group 26**

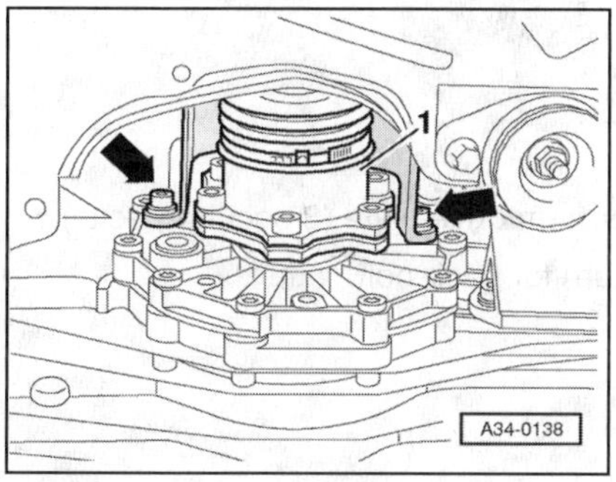

Remove protective plates for left and right-hand drive axles -arrows-. The third bolt cannot be seen in the illustration.

- Disconnect left and right-hand drive axles -1- from flanges and tie-up.

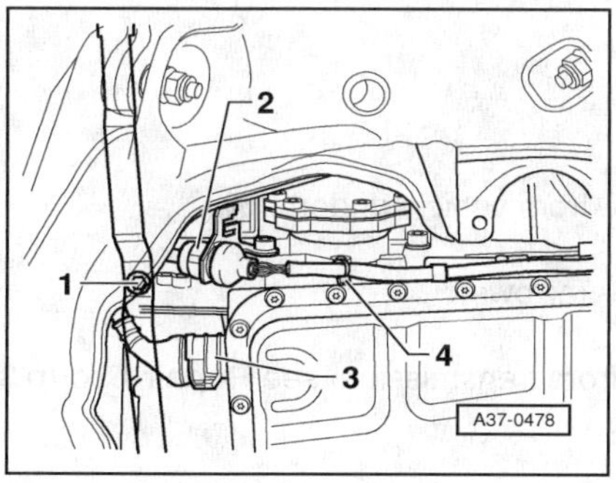

Remove noise insulation tray bracket -1-.

- Pull connector -2- off multi-function switch lead.
- Turn locking lever -3- and remove connector for transmission wiring harness.
- Cut cable tie -4-.

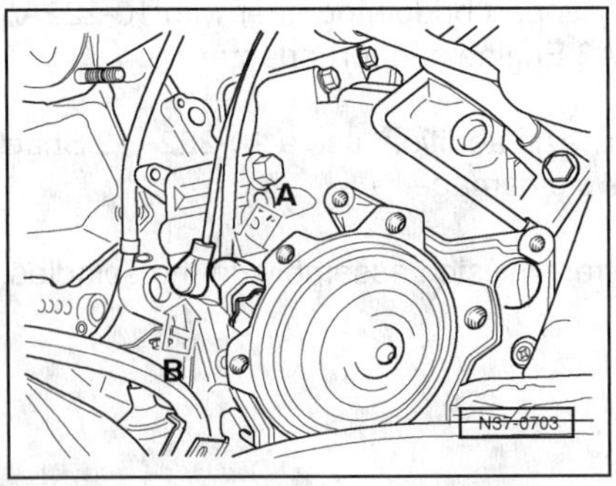

Pull connector off sender for speedometer (arrow -A-).

– Unscrew engine speed sensor -G28- (arrow -B-) from transmission.

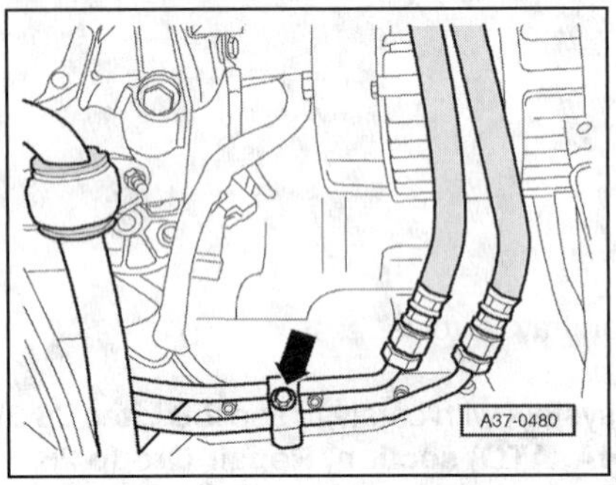

Remove ATF line holder -arrow- from engine oil pan.

– Place VAG1306 Drip tray underneath.

– Thoroughly clean area around ATF line/transmission.

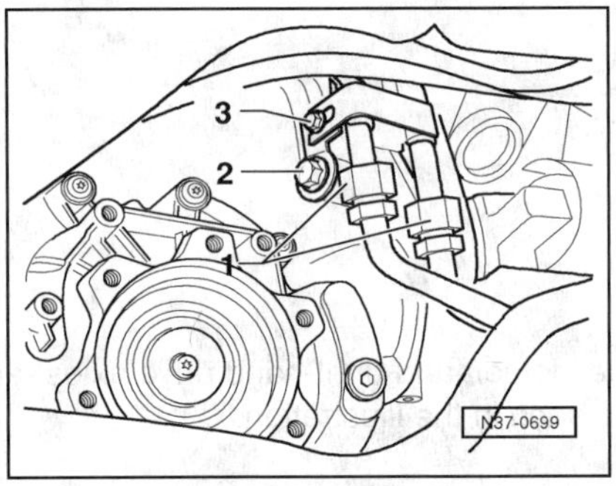

Separate ATF lines -1- to ATF cooler.

– Seal off ATF lines with clean plugs.

– Do not remove ATF line holder -3-.

– Remove transmission side starter motor bolt -2-.

– Remove starter.

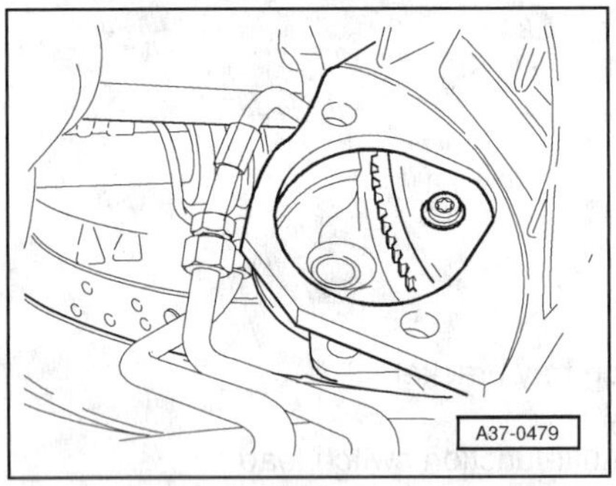

Unbolt torque converter from drive plate (3 securing bolts).

- Turn crankshaft 1/3 turn for each bolt.

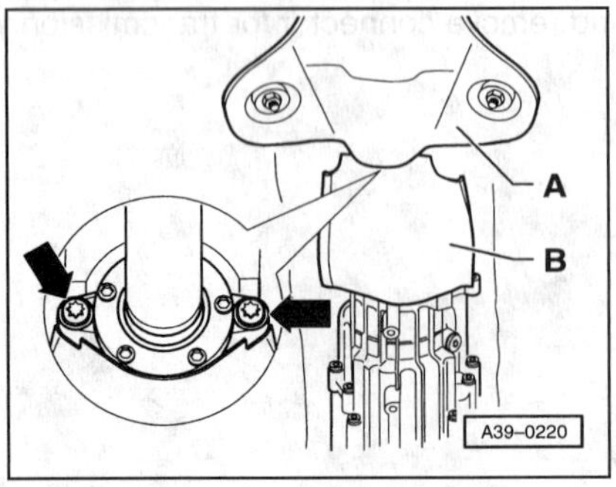

Remove heat shield -A- from vehicle underbody.

– Remove cover plate -B- (arrows).

– Disconnect driveshaft from transmission, see **Repair Group 39, Driveshaft, servicing**.

– Push driveshaft back onto stop and tie up.

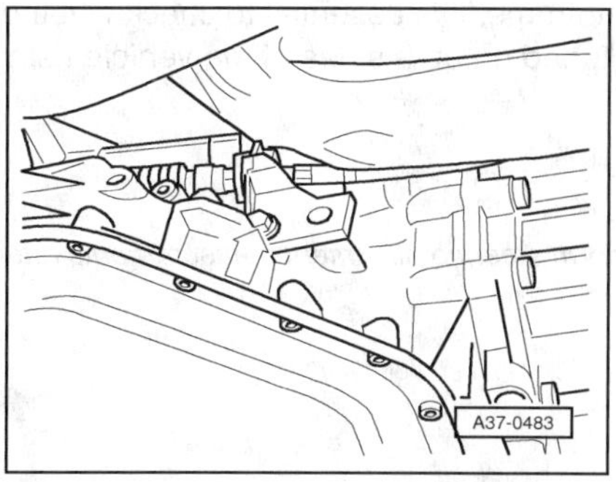

Mark position of support bracket for securing selector lever cable to transmission housing (position must be correct when installing). Unbolt support bracket -arrow-.

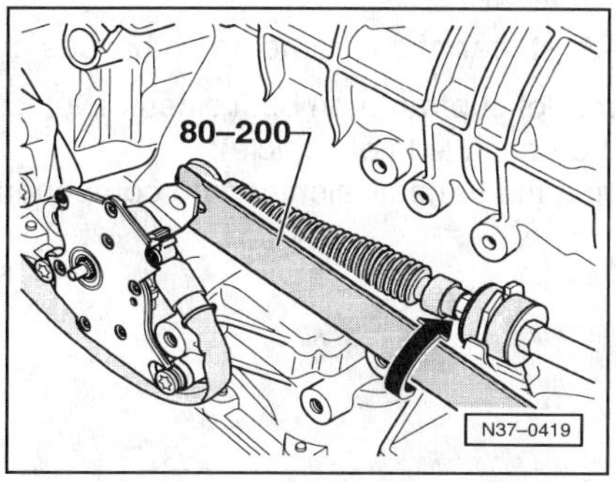

Remove selector lever cable from selector shaft lever -arrow-.

– Remove side and lower engine/transmission connecting bolts.

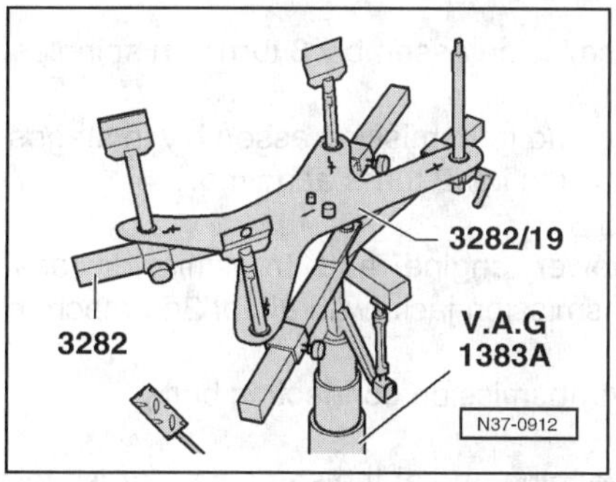

Set up 3282 Transmission support.

- 3282 Transmission support is set up with 3282/19 Adjustment plate for automatic transmission 01V.
- The symbols on the adjusting plate indicate the mounts required, the arrow points forward.

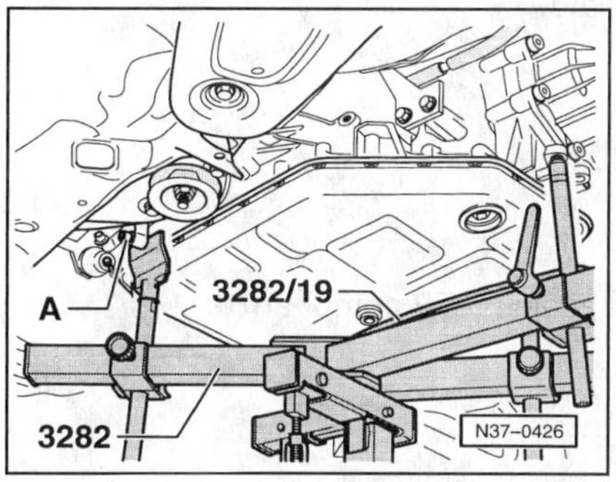

Move VAG1383A Engine/transmission jack with 3282 Transmission support under the transmission, align adjusting plate parallel to transmission and support transmission.

– Secure transmission to 3282 Transmission support with bolt -A-.

– Chock space between transmission support and oil pan using washers.

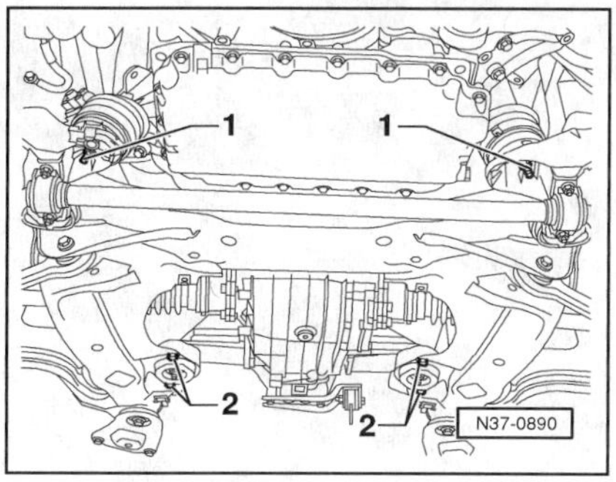

Remove left and right-hand engine mount nuts -1-.

– Remove bolts -2- from transmission mount to subframe.

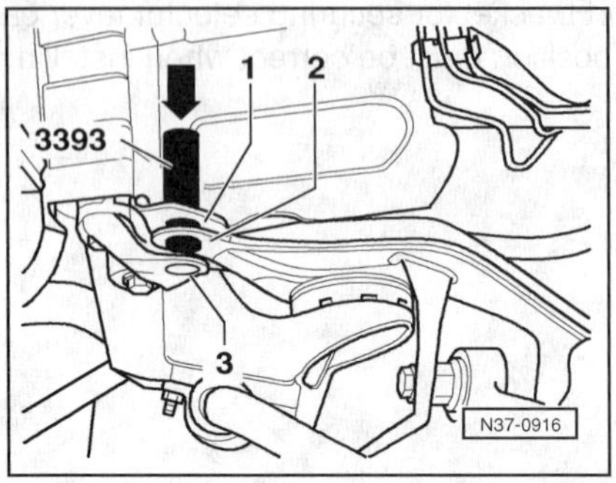

Before loosening the subframe, it is essential to check whether at least holes -1- and -2- are aligned on both sides of the vehicle using 3393 Testing mandrel.

NOTE—
If this is not the case, perform steering alignment measurements after tightening the sub-frame!

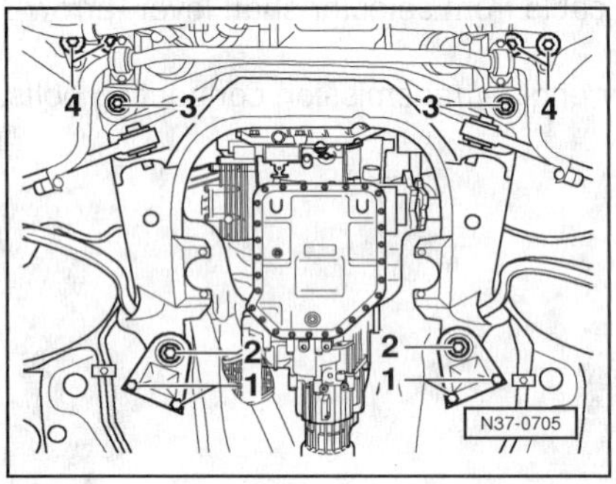

Unscrew bolts -1- and -2- at rear of subframe on both sides of vehicle.

- Remove bolts -3- at front of subframe on both sides of vehicle.
- Remove bolts -4- (two on each side of vehicle):
- Take note when lowering the transmission that no components are damaged.

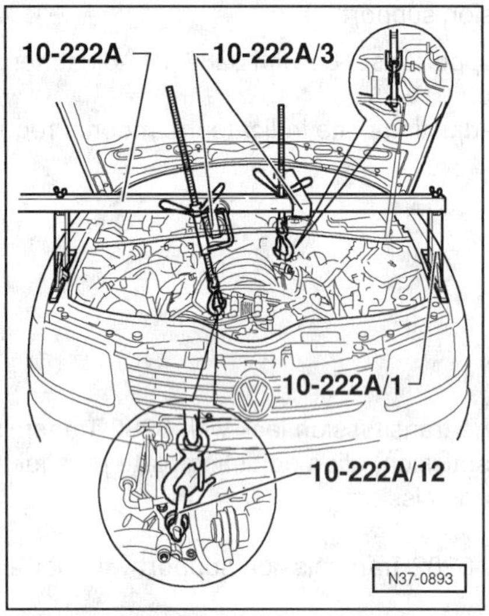

Lower engine and transmission assembly 8 turns on spindles.

– To do this, lower engine and transmission assembly in stages alternatively on each spindle i.e. 2 turns at a time.

– At the same time, lower engine and transmission assembly on VAG1383A Engine/transmission jack (with aid of 2nd mechanic).

– Remove upper engine/transmission connecting bolts.

– Press transmission off engine and at the same time press torque converter out of drive plate.

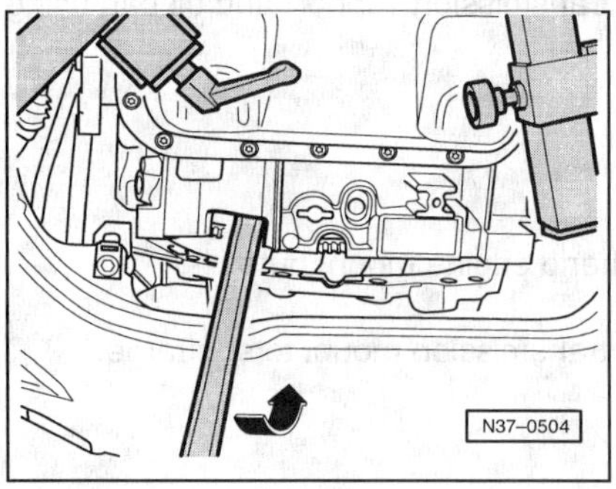

Press torque converter against ATF pump -arrow-.

– Lower transmission.

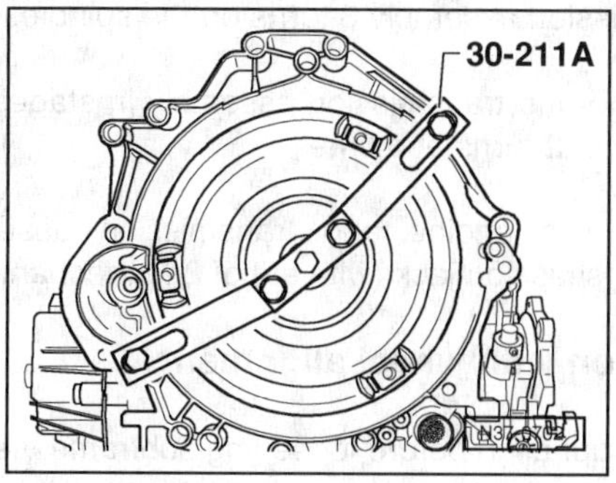

Secure torque converter to prevent it falling out.

When working on the transmission, it must be secured to the repair stand with 3457 Attachment and holding rig. and VW309 Holding plate.

Installing

– Clean ATF cooler and ATF lines before installing an exchange transmission.

– Before installing transmission ensure that torque converter seats correctly in transmission, see **Repair Group 32, Torque convertor, installing**.

– Before installing, ensure that the dowel sleeves are correctly located.

– The threaded holes in the flange shaft behind the transmission must be cleaned of all locking material residue. A thread tap is suitable for this.

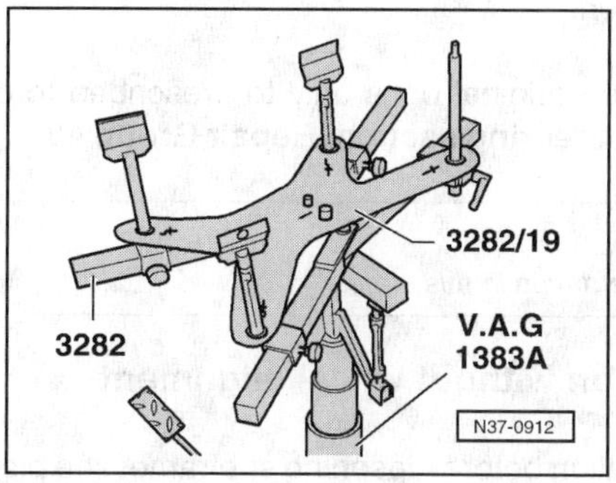

Set up 3282 Transmission support.

- 3282 Transmission support is set up with 3282/19 Adjustment plate for automatic transmission 01V.
- The symbols on the adjusting plate indicate the mounts required, the arrow points forward.

– Place transmission on transmission jack.

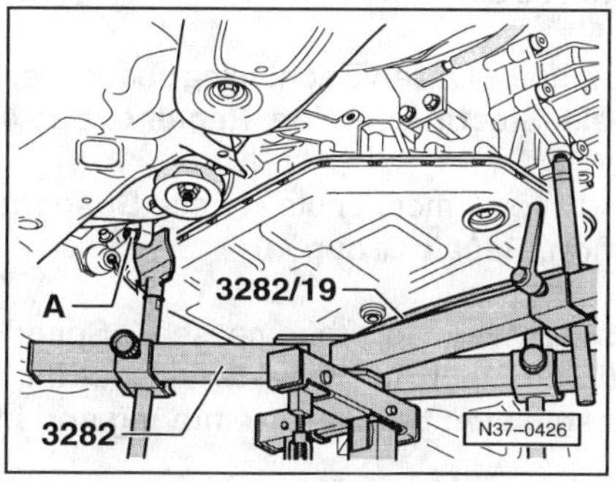

Secure transmission to 3282 Transmission support with bolt -A-

– Place transmission jack under vehicle.

– Guide transmission in between lowered subframe and underbody.

– Align transmission to engine.

– Screw engine/transmission connecting bolt hand-tight in each side.

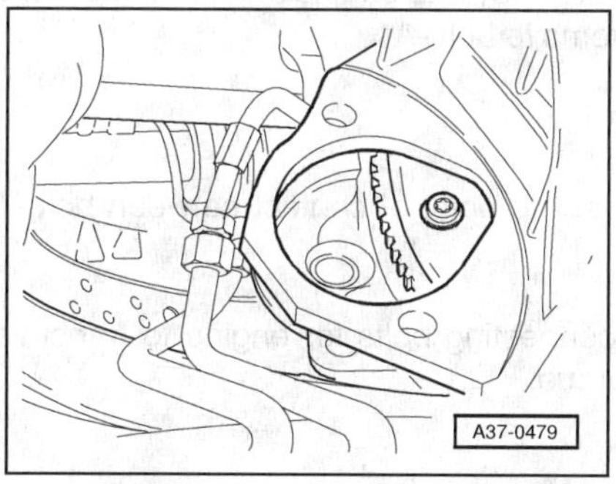

Bolt torque converter and drive plate together.

- To do this, screw in bolts by hand and then tighten to 85 Nm.

– Install upper engine/transmission securing bolts and tighten to prescribed torque.

– Holders and Ground (GND) connections are installed with some bolts see table on.

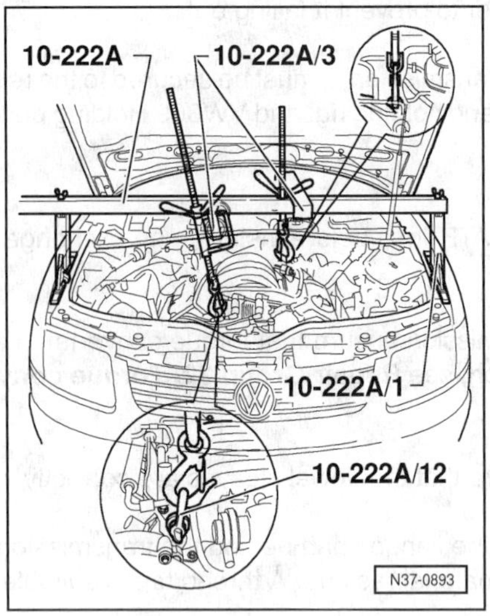

Lift engine and transmission assembly 8 turns on the spindle.

– To do this, lower engine and transmission assembly in stages alternatively on each spindle i.e. 2 turns at a time.

– At the same time, raise engine and transmission assembly on VAG1383A Engine/transmission jack (with aid of 2nd mechanic).

Subframe installation with wheel alignment

If holes -1- and -2- did not align before loosening subframe, see picture N37-0916, the wheels must be aligned.

Install subframe using bolts originally removed. These must be replaced following wheel alignment.

When doing this, tighten original bolts only to prescribed torques, see **Suspension, Wheels, Steering section, Repair Group 40**

> ***WARNING*** —
> *The vehicle must not be driven in this state.*

Subframe installation without wheel alignment

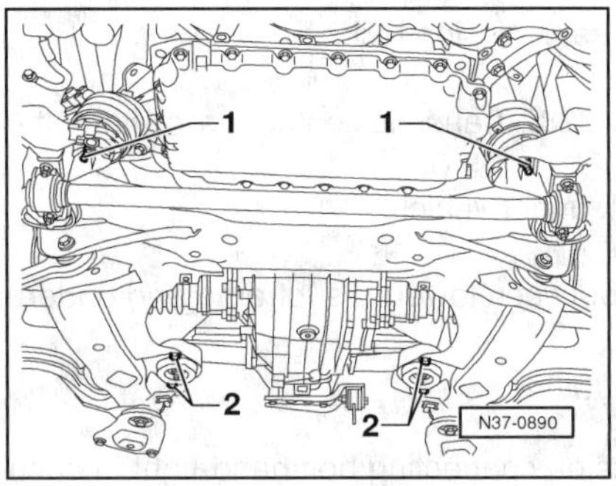

If holes -1- and -2- did align before loosening subframe, see picture N37-0916, the wheels need not be aligned.

Install subframe with new bolts. Observe tightening sequence, see **Suspension, Wheels, Steering section, Repair Group 40**

Install left and right-hand engine mount nuts -1-, see **Suspension, Wheels, Steering section, Repair Group 40**

– Fit bolts -2- for transmission mount to subframe, see **5-Speed Automatic Transmission 01V, Front Wheel Drive section, Repair Group 37, Transmission brackets, removing and installing**.

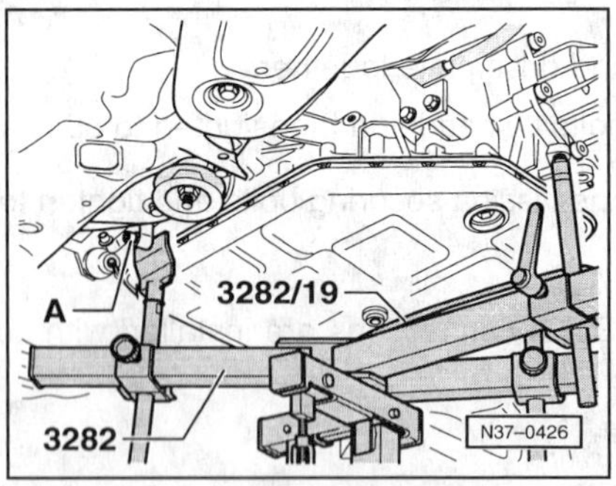

Once subframe is bolted to body and engine and transmission assembly is bolted to subframe, remove bolt -A-.

– Remove jack VAG1338 A.

– Apply anti-corrosion agent to point of contact between bolt -A- and oil pan.

– Install side and lower connecting bolts for engine to transmission and tighten to prescribed torque.

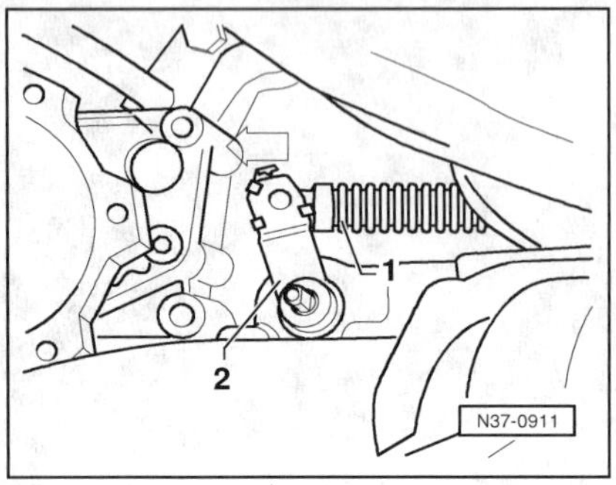

Shift selector shaft lever into position "2" and carefully press ball head of selector lever cable -1- into selector shaft lever -2- e.g. using a pair of pliers.

- The selector shaft lever -2- must not be bent when doing this, otherwise the selector mechanism will not be able to be adjusted correctly.

– Push selector lever cable -1- forward fully -arrow- to shift selector lever shaft -2- into position "2".

Bolt support bracket of selector lever cable to transmission -arrows- at place marked.

– Install starter, see **Electrical Equipment section, Repair Group 27**

– Connect driveshaft, see **Repair Group 39.**

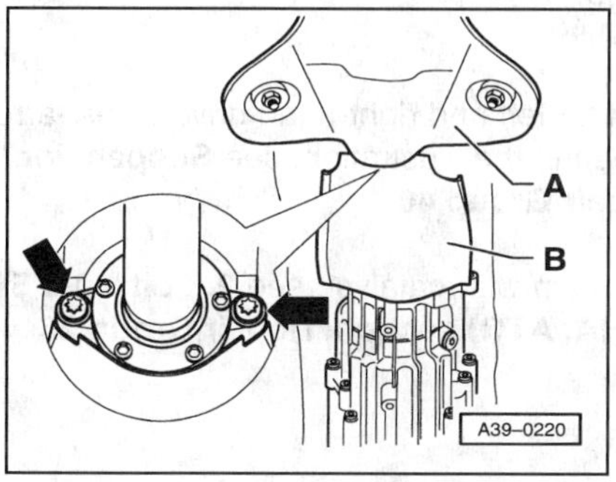

Install cover plate -B- (arrows), see **Repair Group 39**.

– Fit heat shield -A- to vehicle underfloor.

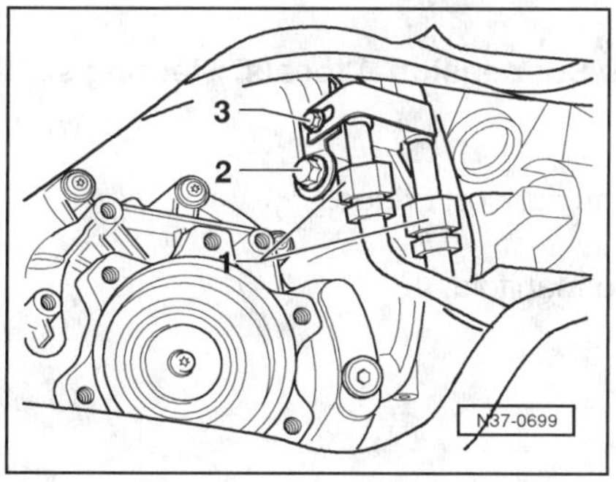

Fit ATF lines -1- to engine/transmission assembly, see **ATF lines, removing and installing (V-6 engine)**.

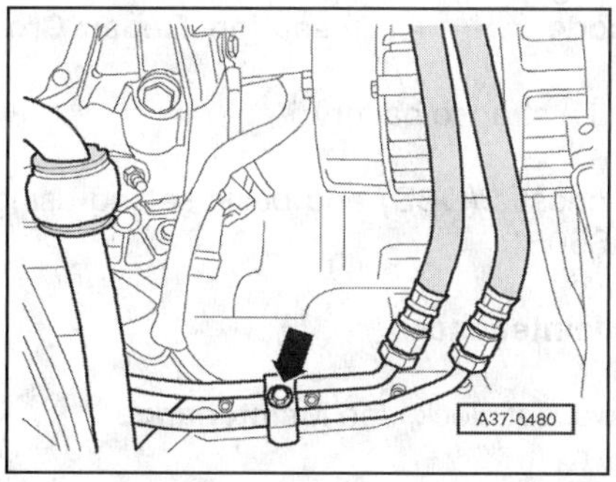

Remove ATF line holder -arrow- from engine oil sump, see **ATF lines, removing and installing (V-6 engine)**.

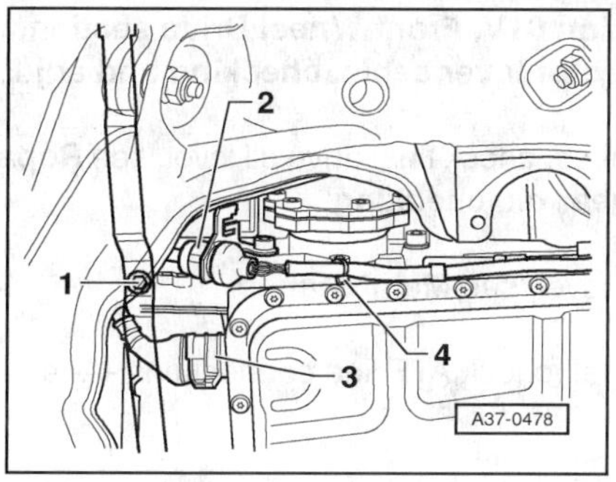

Fit connector to multi-function switch -2-.

– Install connector for transmission wiring harness and secure with locking lever -3-.

– Install noise insulation tray holder -1-, see **Body Exterior section, Repair Group 50**

– Replace cable tie -4-.

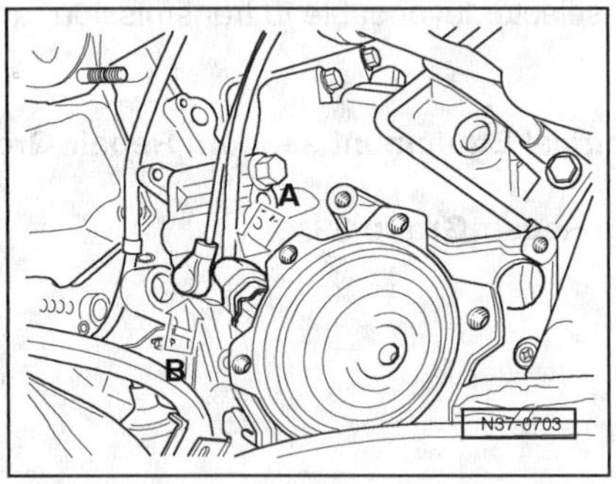

- Reconnect speedometer sensor connector -arrow A-.
- Screw engine speed sensor -G28- (arrow -B-) into transmission.

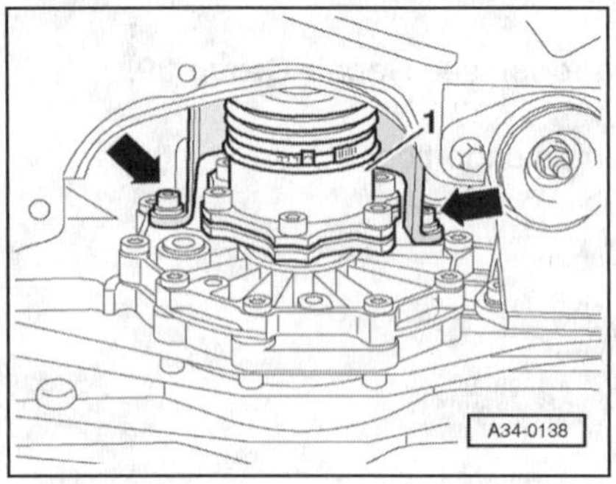

- Fit drive axles -1- to flanged shafts, see **Suspension, Wheels, Steering section, Repair Group 40**
- Install protective plates for left and right-hand drive axles -arrows-. The third bolt cannot be seen in the illustration, see **Suspension, Wheels, Steering section, Repair Group 40**
- Install front exhaust system with catalyst, see **2.8 Liter V6 5V Engine Mechanical (code: AHA, ATQ) section, Repair Group 26**

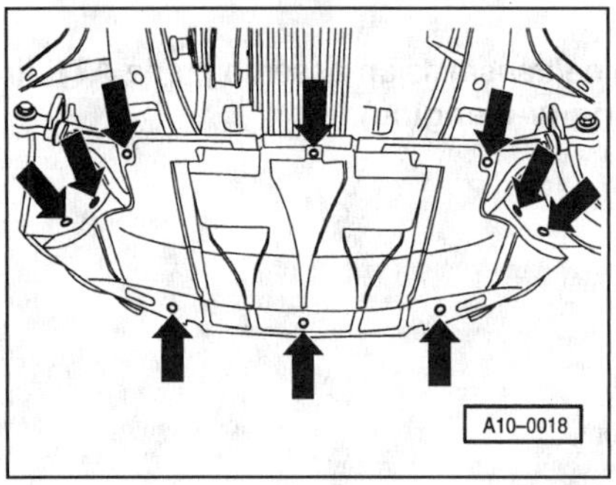

- Install noise insulation tray -arrows-.
- Install front wheels, see **Suspension, Wheels, Steering section, Repair Group 40**
- Remove 10-222A Engine support bridge.
- Fit intake hose to intake manifold.

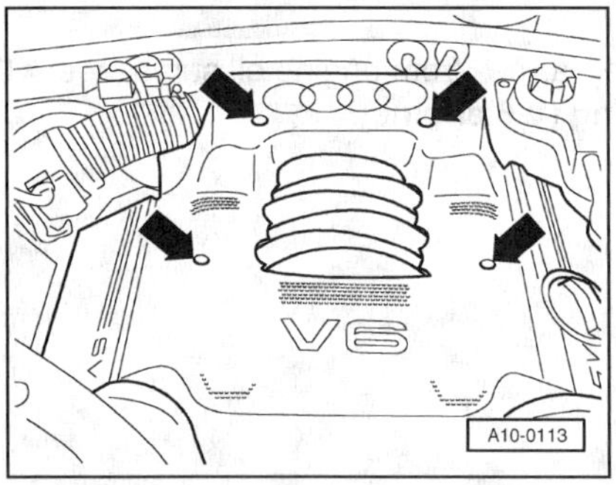

- Install engine cover and tighten bolts -arrows-, see **2.8 Liter V6 5V Engine Mechanical (code: AHA, ATQ) section, Repair Group 10**
- Reconnect Ground (GND) strap to battery.
- Perform On Board Diagnostic (OBD) and basic setting using "guided troubleshooting" on VAS5051.
- Enter radio code, see **Maintenance**
- Initialize electric windows, set clock, see **Maintenance**
- Check selector lever cable adjustment, adjust if necessary, see **5-Speed Automatic Transmission 01V, Front Wheel Drive section, Repair Group 37, Selector lever cable, checking and adjusting**.
- With transmission installed, check final drive oil level, see **Repair Group 39, Front final drive gear oil, checking**.
- Check and top up ATF level, see **Maintenance**.

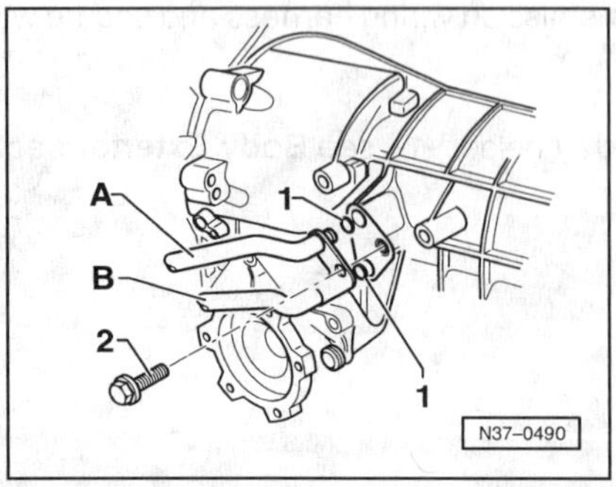

- When checking ATF level, check ATF line connections -A- and -B- for leaks.

Tightening torques (V6 engine)

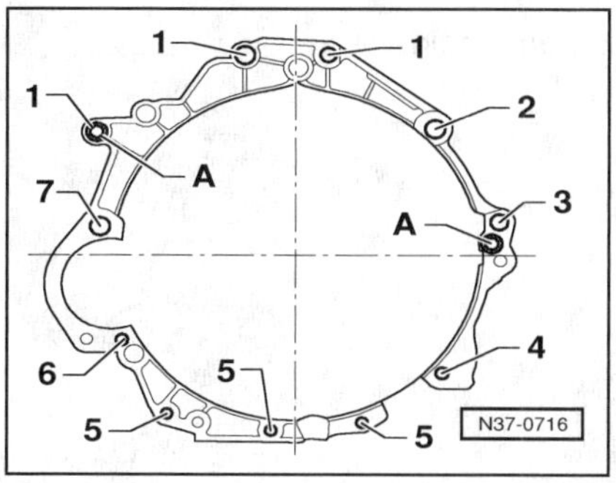

Transmission to engine

Item	Bolt	Qty.	Torque
1	M 12 x 67	3	65Nm (48ft-lb)
2	M 12 x 90	1	65Nm (48ft-lb)
3	M 12 x 80	1	65Nm (48ft-lb)
4[1)]	M 10 x 60	1	45Nm (33 ft-lb)
5	M 10 x 45	3	45Nm (33 ft-lb)
6	M 10 x 80	1	65Nm (48ft-lb)
7[2)]	M 12 x 110	1	65Nm (48ft-lb)

[1)] with nut

[2)] with bracket for ATF lines

Item A = Dowel sleeves

Transmission, removing and installing (8-cylinder engine)

Special tools and equipment

- 10-222 A Pull-out device
- 10-222 A/1 Frame
- 10-222 A/2 Extra hooks
- 10-222 A/10 Spindle
- 10-222 A/18 Adapter
- 80-200 Release lever
- 3282 Transmission mounting
- 3282/19 Adjustment plate
- 3393 Test mandrel
- 3457 Transport device
- V.A.G 1306 Drip tray
- V.A.G 1331 Torque wrench
- V.A.G 1332 Torque wrench
- V.A.G 1383 A Engine/transmission hoist
- VAS 5085 Ladder

Removing

– Obtain code for radio.

– Switch ignition off and disconnect battery ground strap.

– Remove intake air duct, upper part of air filter and intake hose, see **Repair Manual, 4.0 Liter W8 4V Fuel Injection & Ignition, Engine Mechanical (code: BDP), Repair Group 24.**

– Drain coolant, see **Repair Manual, 4.0 Liter W8 4V Engine Mechanical (code: BDP), Repair Group 19.**

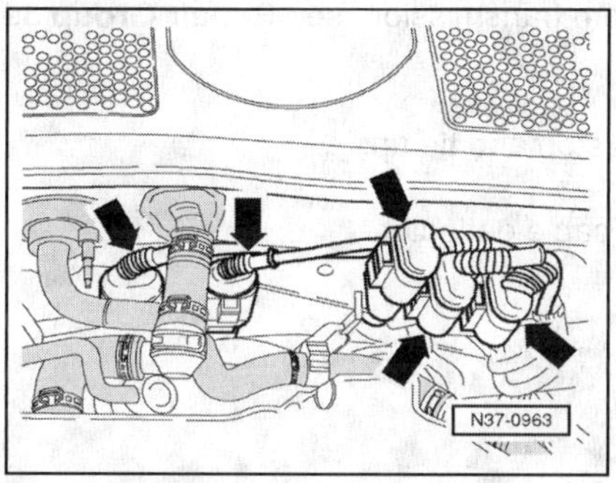

Unclip connectors -arrows- from holder.

– Move lock carrier in service position, see **Maintenance**

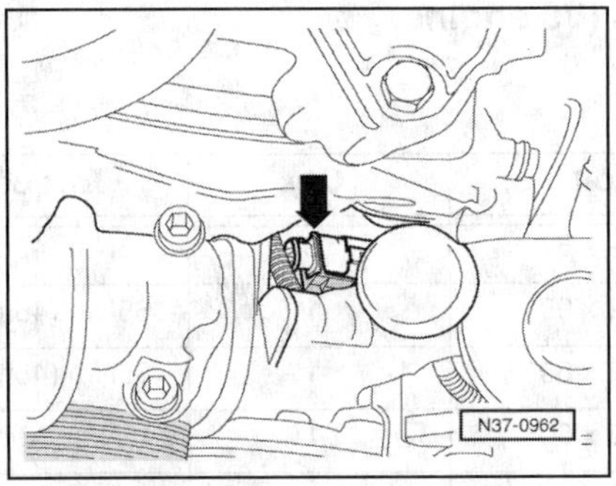

Unclip connector -arrow- for air conditioning compressor from holder.

– Disconnect oil filler neck from engine.

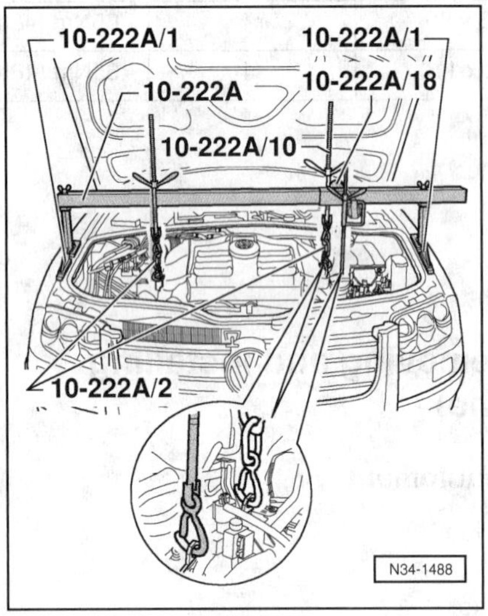

Install pull-out device 10-222A with stands 10-222 A/1 and adapter 10-222 A/18. Long edge of adapter 10-222 A/18 faces forward.

– Attach third spindle 10-222 A/10 with extra hook 10-222 A/2 to suspension eyelet on left rear of engine.

– Attach second hook 10-222 A/2 to eyelet on right rear of engine.

– Tension engine/transmission assembly slightly using spindles.

– Loosen wheel bolts.

– Raise vehicle.

– Remove front wheels.

– Remove front right wheel arch liner, see **Body–Exterior section, Repair Group 50**

– Remove front exhaust pipes, see **Repair Manual, 4.0 LiterW8 4V Engine Mechanical (code: BDP), Repair Group 26**

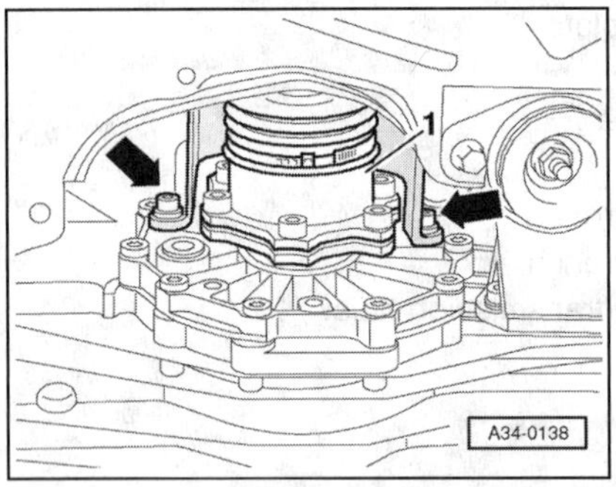

Remove right and left drive shaft heat shields -arrows-. Third screw is not visible in illustration.

– Unbolt drive axles -1- from flanges on left and right and tie up.

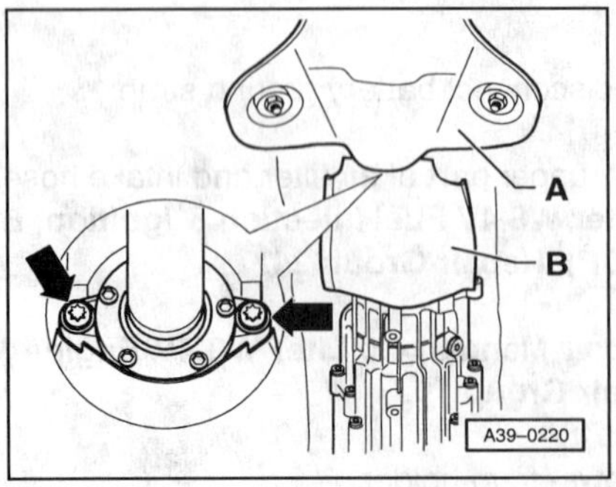

Remove heat shield -A- from floor of vehicle.

– Remove deflector -B- -arrows-.

– Separate drive shaft from transmission, see **Repair Group 39, Driveshaft, servicing**.

– Pull drive shaft back to stop and tie up.

– Remove selector lever cable deflector.

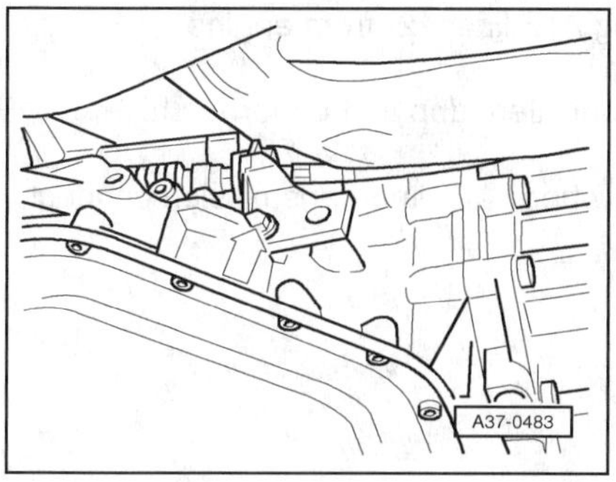

To facilitate re-installation, mark position of mounting for selector lever cable to transmission, remove bolt -arrow-.

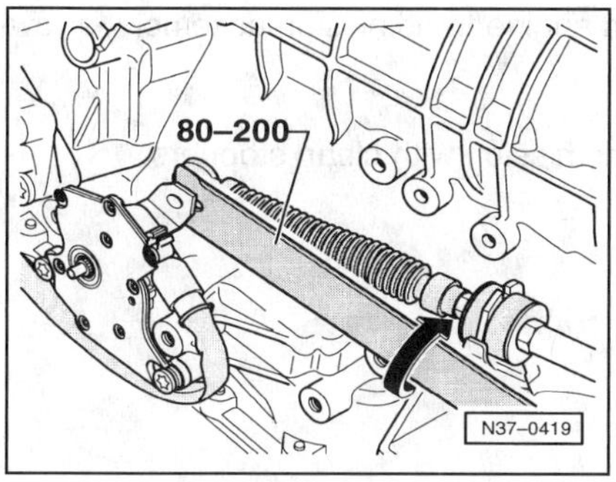

Disconnect selector lever cable from lever/third motion shaft -arrow-.

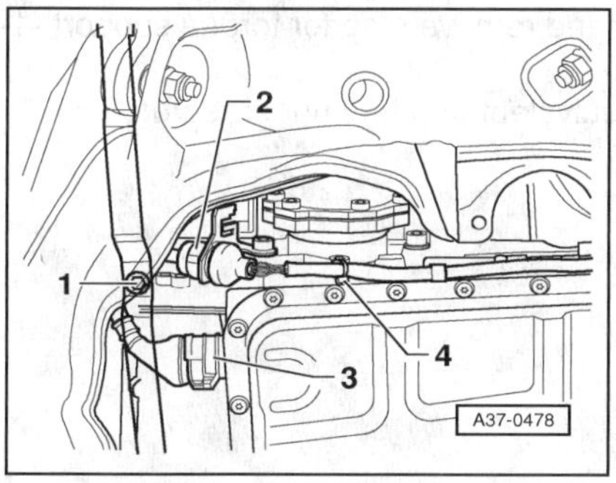

Disassemble holder for sound insulator -1-.

– Disconnect wiring connector -2- for multi-function switch.

– Move safety lever -3- aside; disconnect connector for transmission wiring harness.

– Cut cable tie -4-.

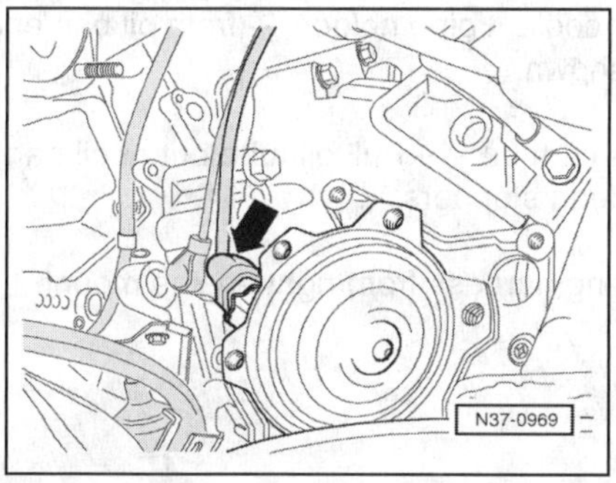

- Disconnect connector from speedometer sensor -G22- -arrow-.

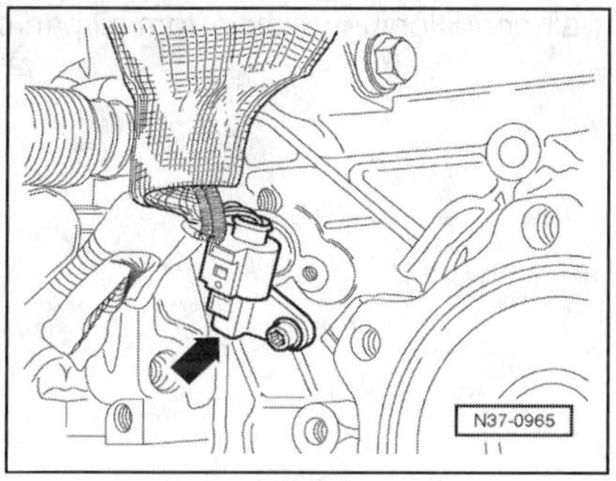

Remove engine speed sensor -G28- from lower mounting -arrow-.

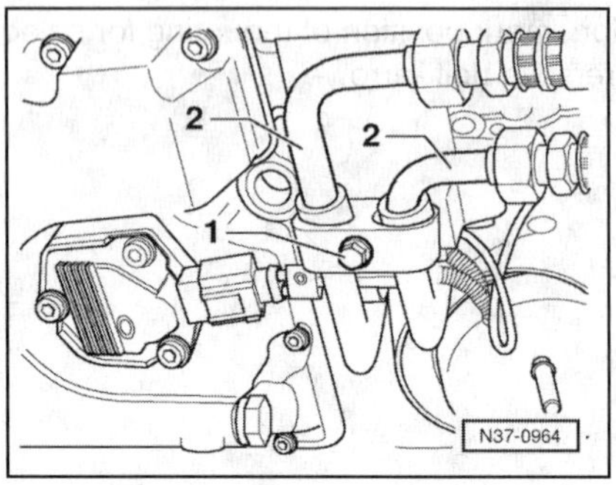

◄ Remove bolt -1- holding ATF lines -2- from engine.

– Place V.A.G 1306 or equivalent drip tray underneath.

– Thoroughly clean area where ATF line/transmission connect.

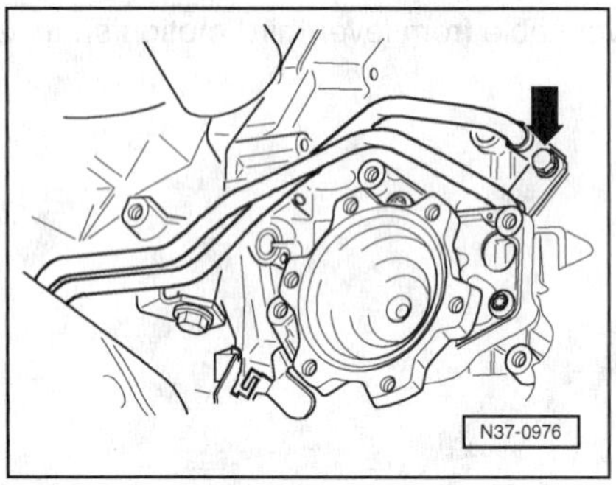

◄ Remove ATF lines from engine/transmission assembly -arrow- and tie up.

– Plug ATF lines and transmission with clean stoppers.

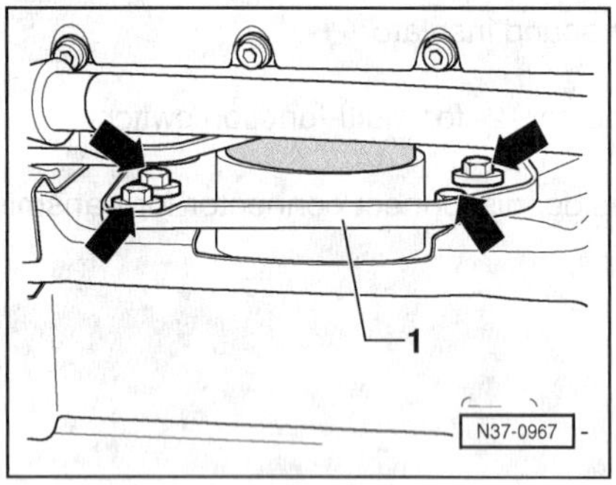

◄ Remove bolts -arrows- and remove stop for torque support -1-.

– Place V.A.G 1306 or equivalent drip tray underneath.

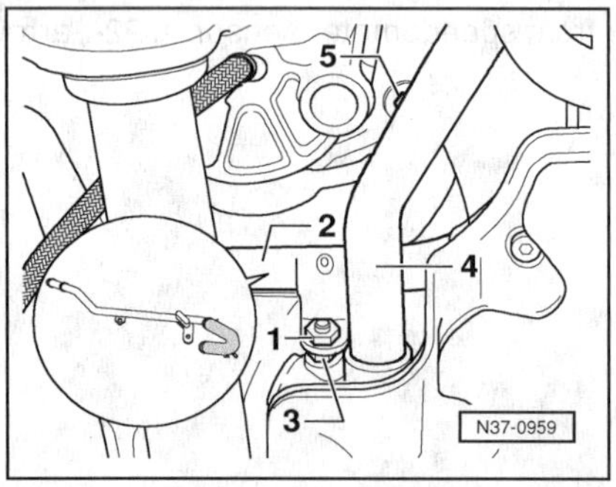

◄ Remove hex nut -1- for coolant pipe holder -2- from oil pan/engine. Remove coolant pipe as shown.

– Remove hex nut -3- for oil tube -4-. Pull oil tube out of oil pan and plug oil pan and tube with clean stoppers.

– Remove bolt -5- for wiring harness from right engine mount.

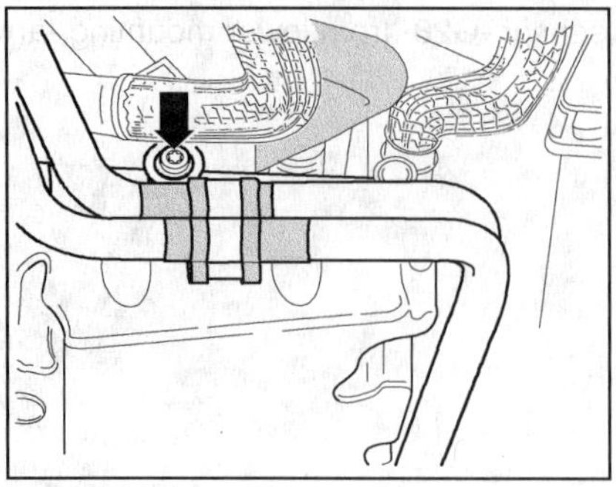

◄ Remove bolt -arrow- for air conditioning ducting from oil pan/engine.

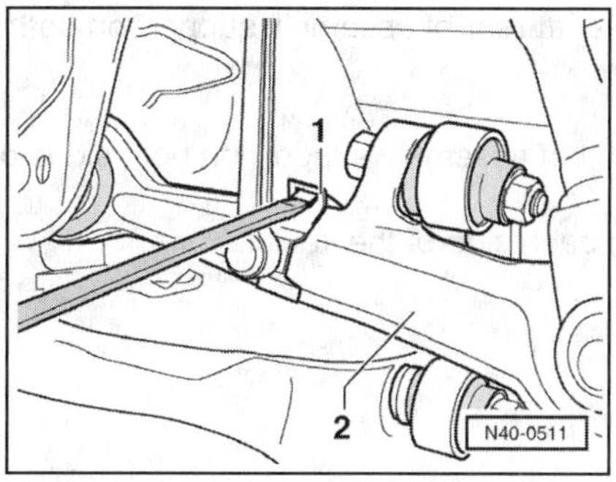

Unclip retaining bracket for coupling rod -1- of Left Front Level Control System Sensor -G78- from front left support guide -2-.

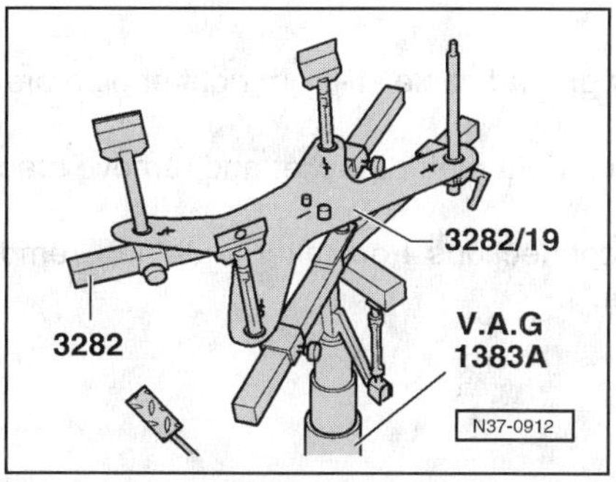

Install transmission mounting 3282.

- The transmission mounting 3282 for the 01V automatic transmission is installed with the 3282/19 adjustment plate.
- The symbols on the adjustment plate show the mountings required; the arrow points in the direction of travel.

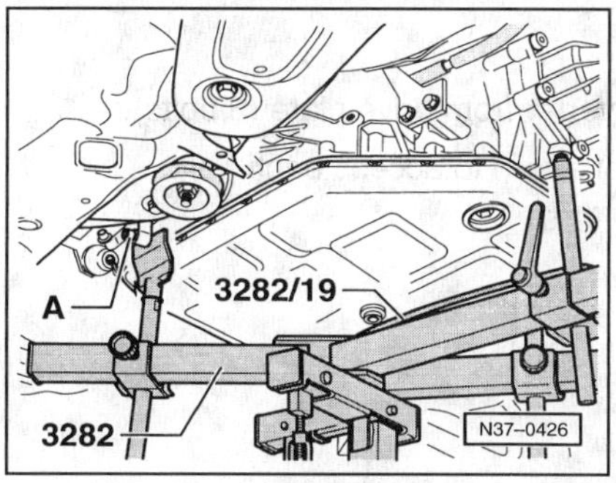

Position V.A.G 1383 A hoist and 3282 transmission mounting under transmission, align adjustment plate parallel to transmission and support transmission on it.

– Fasten transmission to 3282 transmission mounting using bolt -A-.

– Pack space between transmission mounting and oil pan with washers.

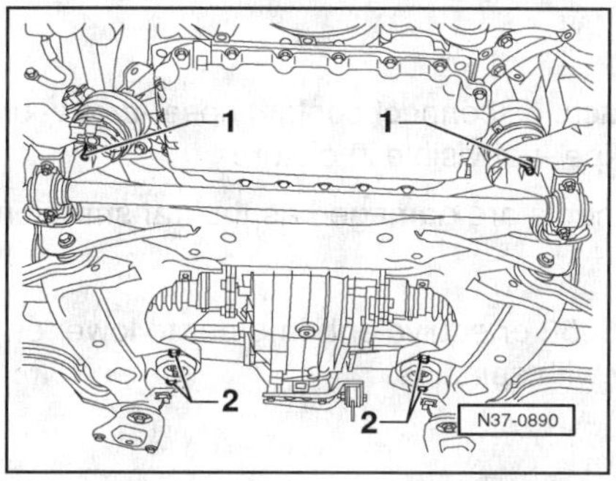

Remove nuts -1- on lower left and right of engine bracket.

– Remove bolts -2- on transmission mount/support.

– Disconnect lines from the engine mountings, unclip cables from assembly support and tie to one side.

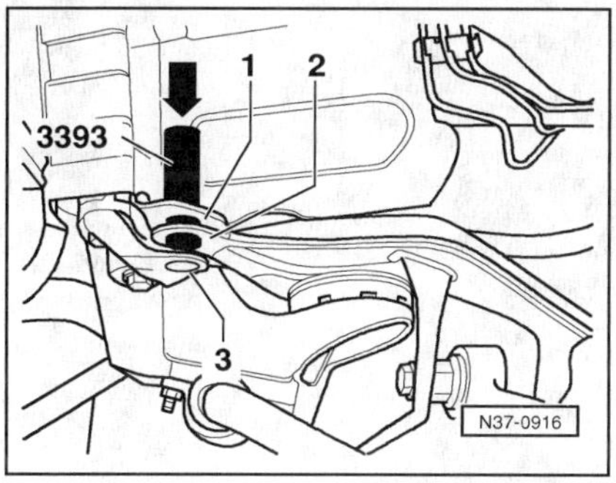

Before disconnecting assembly support, it is vital to inspect both sides of vehicle using 3393 test mandrels to check that holes -1- and -2- (at least) are aligned.

NOTE—

- *If this is not the case, a suspension alignment must be carried out after the assembly support is tightened!*
- *Hole -3- is of no relevance at this time.*

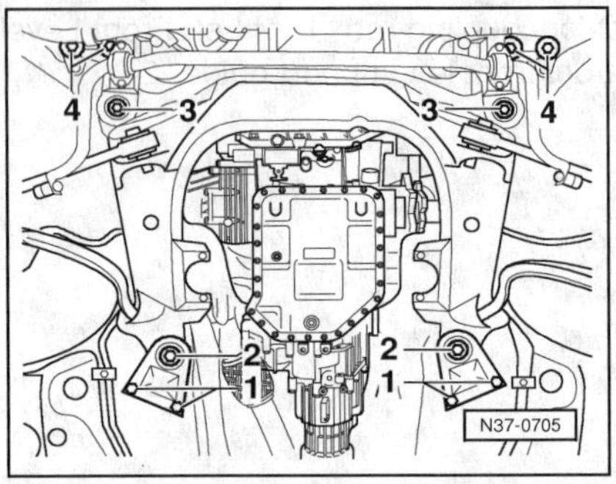

◄ Remove bolts -1- and -2- at rear of assembly support on both sides of vehicle.

– Remove bolts -3- at front of assembly support on both sides of vehicle.

– Remove bolts -4- (2 on each side of the vehicle).

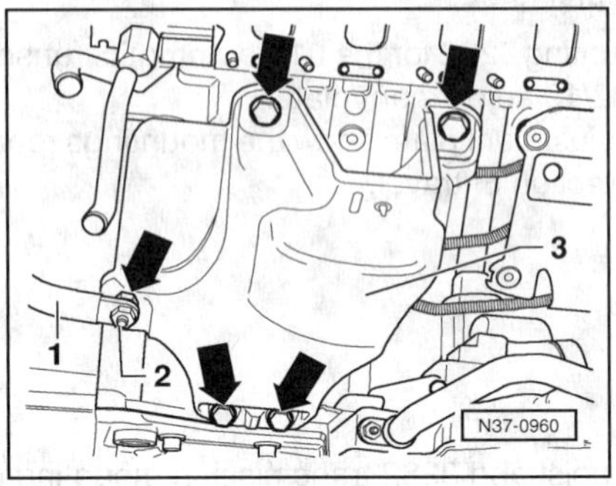

◄ Remove hex nut -2- for ground cable -1- from right engine bracket -3-.

– Loosen bolts -arrows- on right motor bracket and remove bracket.

– Disconnect electrical connections from alternator and remove wiring harness from alternator.

– Remove starter.

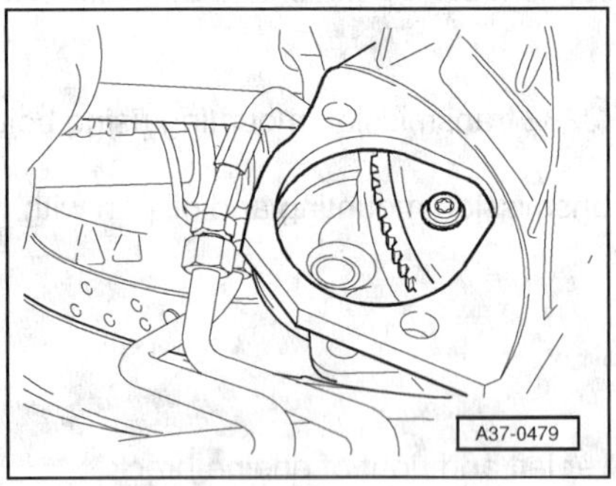

◄ Disconnect torque converter from drive plate (3 bolts).

- Turn crankshaft 1/3 of a turn to access each bolt.

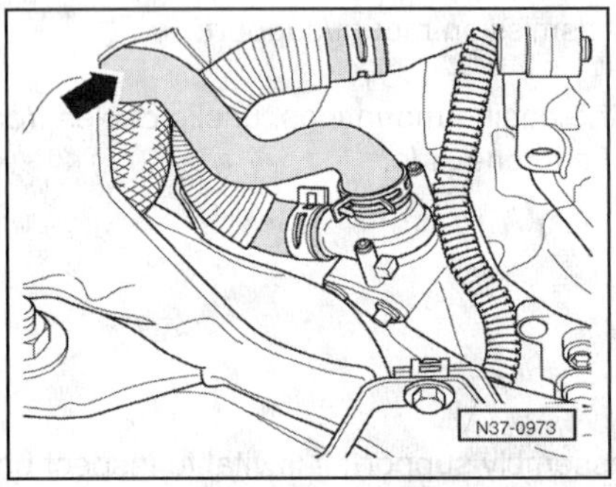

◄ Using quick-release catch, disconnect coolant hose -arrow- under left side rail from coolant pipe -not visible in picture-.

- Make sure no components are damaged as the transmission is lowered.

– Climb up ladder (VAS 5058 or equivalent) in order to lower engine/ transmission assembly with spindles.

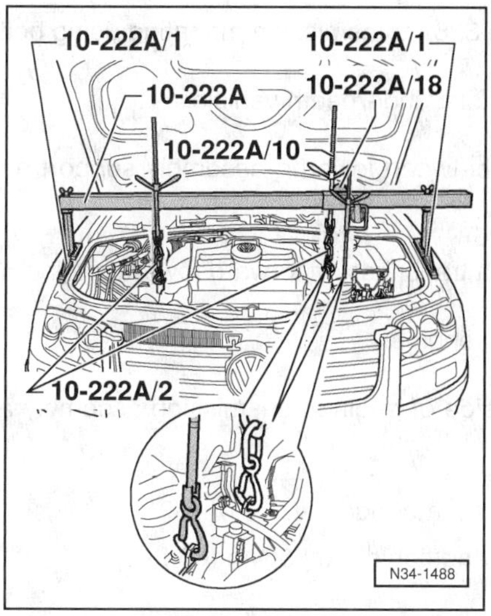

- Lower engine/transmission assembly approximately 4 turns using spindles.

– Lower assembly using spindles, lowering it alternately 2 turns.

– At same time, lower engine/transmission assembly using V.A.G 1383 hoist (2nd mechanic).

– Unclip wiring for oxygen sensors on transmission.

– Remove engine/transmission connecting bolts.

– While pushing transmission away from engine, press torque converter out of drive plate.

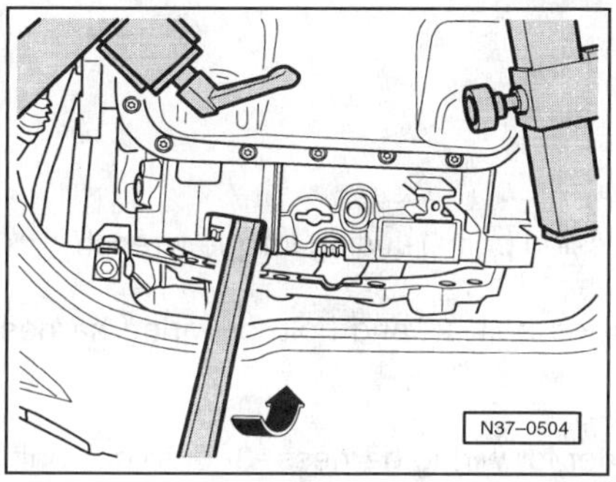

- Press torque converter against ATF pump -arrow-.

– Carefully retract transmission, and lower it.

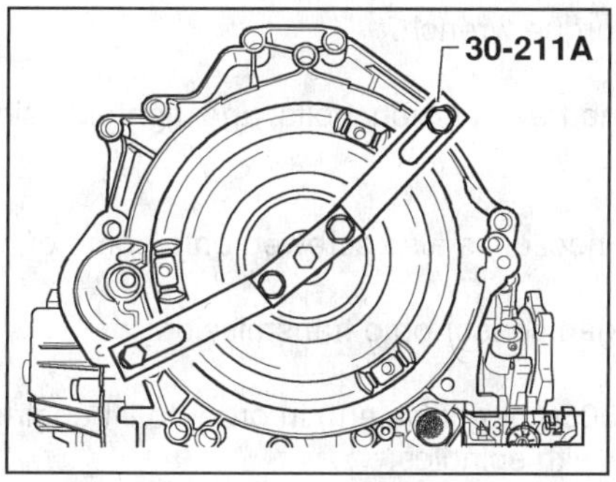

- Take steps to prevent converter falling out.
  - Use a 3457 transport device for transporting the automatic transmission unit.
  - To install the unit, the transmission together with the 3457 transport device and the VW 309 retaining plate is attached to the installation bracket.

Installing

– Before installing a replacement transmission unit, clean ATF cooler, secondary ATF cooler and ATF lines.

– Before installing transmission, make sure torque converter has been correctly inserted in transmission, see **Repair Group 32, Torque convertor, installing.**

– Ensure body-fit sleeves are seated correctly.

– Clean any adhesive residue from tapped holes in rear flanged shaft of transmission. This can be done using a tap thread.

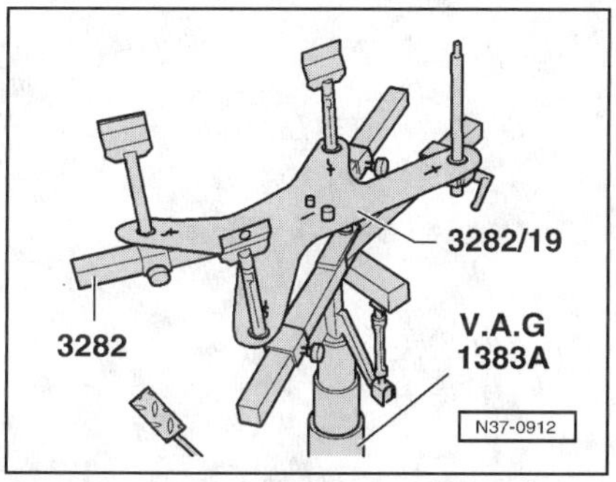

- Install transmission mounting 3282.
  - The transmission mounting 3282 for the 01V automatic transmission is installed with the 3282/19 adjustment plate.
  - The symbols on the adjustment plate show the mountings required; the arrow points in the direction of travel.

– Place transmission on transmission hoist.

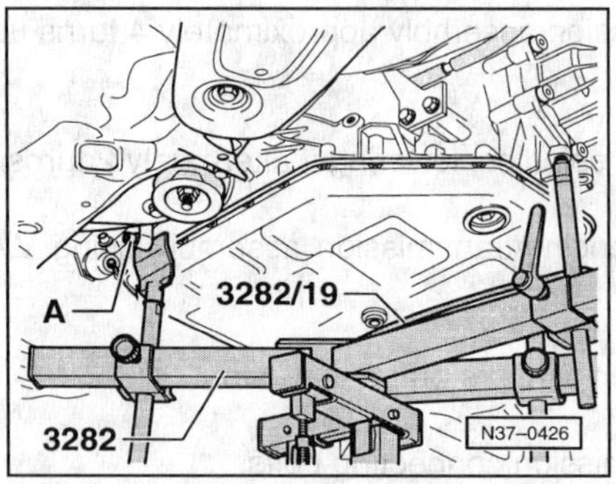

◀ Fasten transmission to 3282 transmission mounting using bolt -A-.

– Position transmission hoist underneath vehicle.

– Position transmission between lowered assembly support and vehicle floor.

– Alternately lift and push transmission forward toward engine.

– Align transmission with engine.

– Insert a bolt on each side of engine/transmission assembly and screw finger-tight.

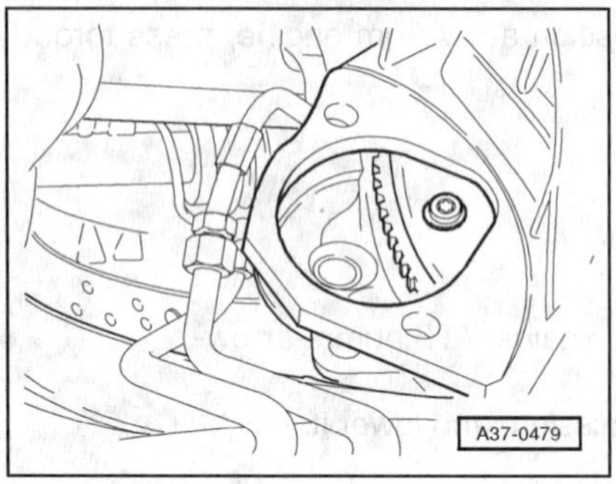

◀ Bolt converter and drive plate together.

- Insert bolts by hand and tighten to 85 Nm (63 ft-lb).

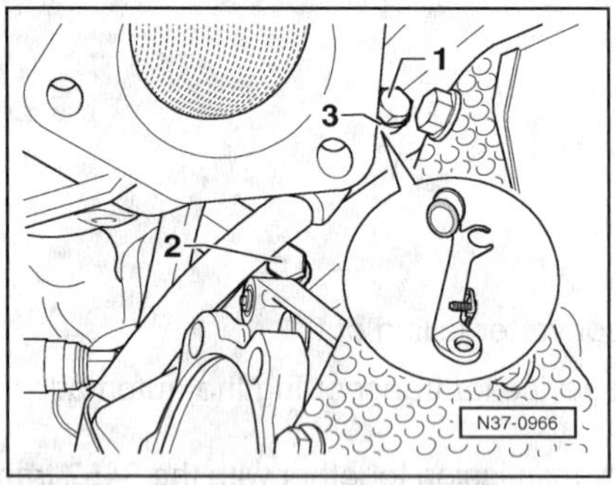

◀ Tighten engine/transmission bolt -1- using torque wrench.

– Insert engine/transmission bolt -2- and bolt for wiring harness -3- and thread in finger-tight.

– Place open side of holder for wiring harness -3- onto hex bolt head -1-.

– Tighten bolt -2- using torque wrench.

– Insert remaining engine/transmission bolts and tighten using torque wrench.

– A holder or ground connection is also screwed onto some of bolts.

– Clip harnesses for oxygen sensor onto transmission.

– Climb up ladder (VAS 5058 or equivalent) in order to lift engine/transmission assembly with spindles.

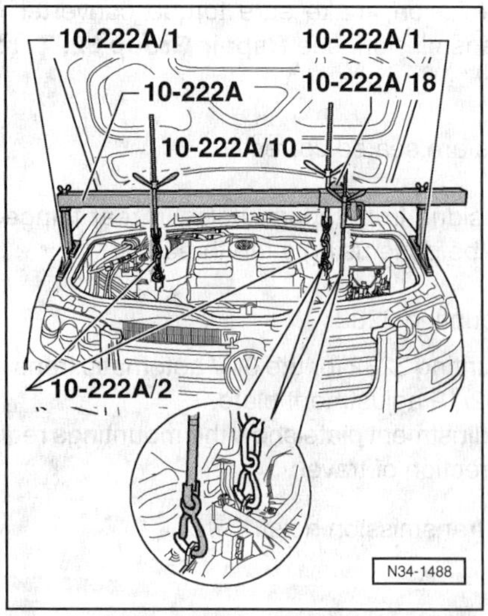

◀ Lift engine/transmission assembly approximately 4 turns using spindles.

– Lift assembly using spindles, raising it alternately 2 turns.

– At same time, raise engine/transmission assembly using V.A.G 1383 hoist (2nd mechanic).

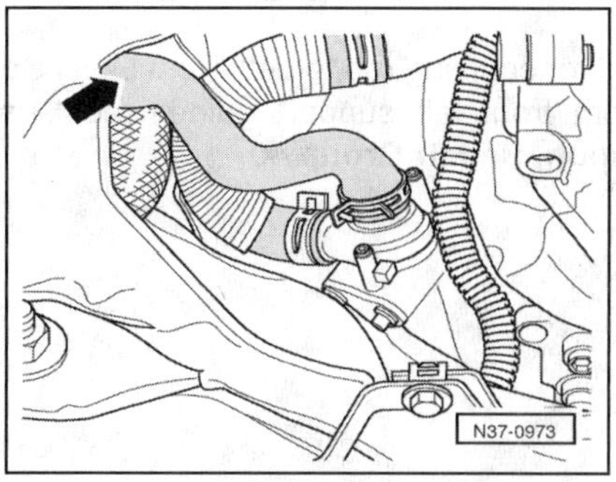

Using quick-release catch, connect coolant hose -arrow- under left side rail to coolant pipe -not visible in illustration-.

– Install starter, see **Electrical Equipment section, Repair Group 27**

– Connect electrical connections to alternator and fasten wiring harness to alternator.

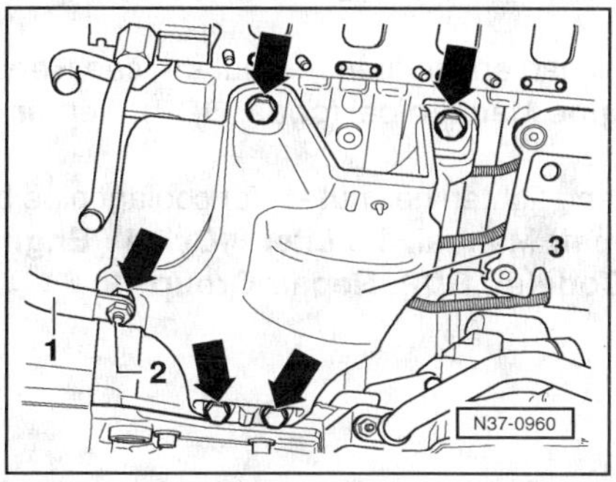

Install right engine bracket -3- and tighten bolts -arrows-, to 40 Nm (30 ft-lb)

– Attach ground cable -1- to right engine bracket using hex nut -2-.

Installing the assembly support with a suspension alignment measurement

If holes -1- and -2- were not aligned before the assembly support was removed, see Figure N37-0916, a suspension alignment must be performed.

Install assembly support using old bolts. These must be replaced by new bolts after suspension alignment has taken place.

Tighten old bolts only to specified tightening torque, see **Suspension, Wheels, Steering section, Repair Group 40**

> ***WARNING —***
> *The vehicle must not be driven when in this condition.*

Installing the assembly support without a suspension alignment measurement

If holes -1- and -2- were aligned before the assembly support was removed, see Figure N37-0916, a suspension alignment does not have to be performed.

Install assembly support using new bolts, observing tightening sequence, see **Suspension, Wheels, Steering section, Repair Group 40**

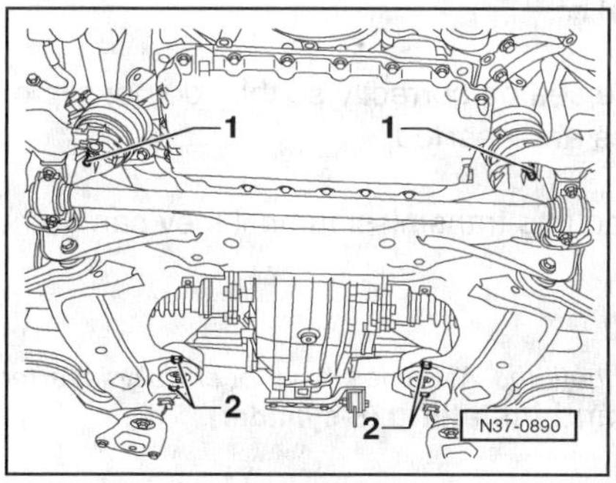

Tighten nuts -1- on lower left and right of engine bracket.

– Tighten transmission mount/assembly support bolts -2-, see **Transmission brackets, removing and installing**.

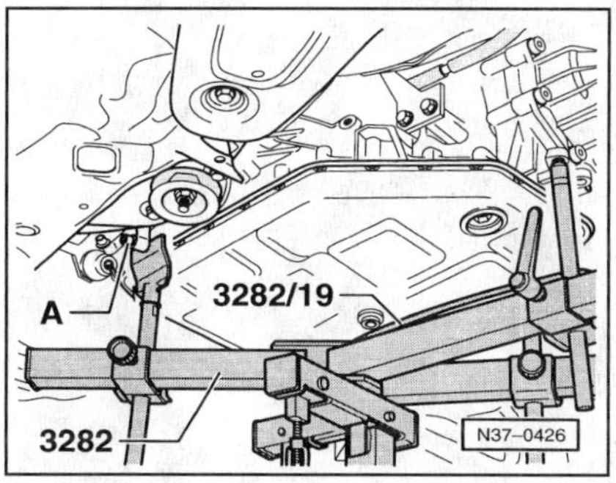

Remove bolt -A- when assembly support has been bolted to body and engine/transmission unit has been bolted to assembly support.

– Remove V.A.G 1338 A hoist.

– Apply an anti-corrosion agent at the point where bolt -A- touches the oil pan.

– Clip lines for engine mounted onto assembly support and attach lines to engine mount.

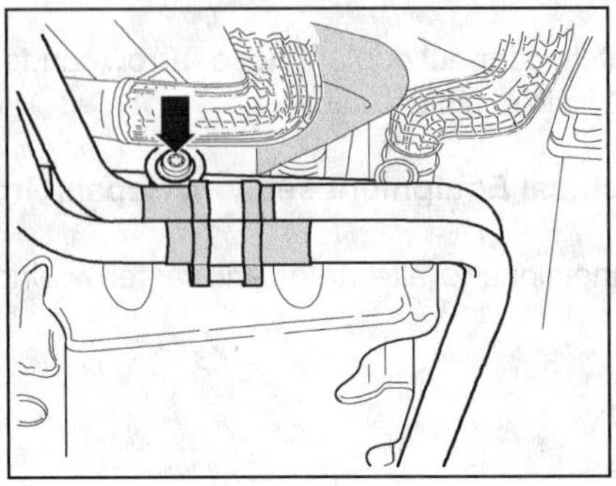

Install bolt -arrow- for air conditioning ducting to oil pan/engine.

- Attach retaining bracket for coupling rod to Left Front Level Control System Sensor -G78- from front left support guide, see **Suspension, Wheels, Steering section, Repair Group 40**

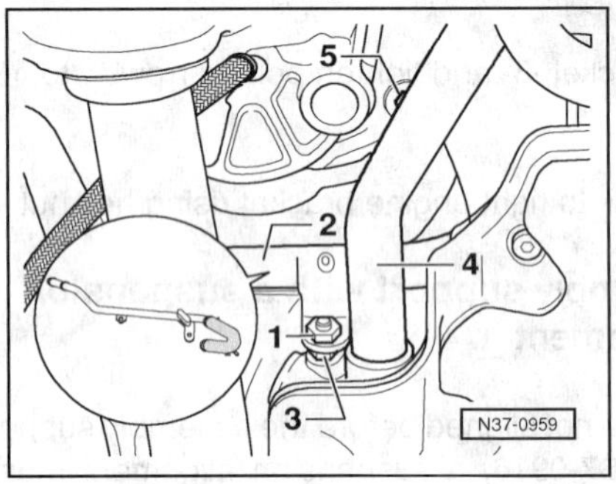

Install bolt -5- for wiring harness onto right engine mount.

- Insert oil tube -4- into oil pan and tighten hex nut -3-, see **Repair Manual, 4.0 Liter W8 4V Engine Mechanical (code: BDP), Repair Group 17**
- Install coolant pipe -2- and tighten hex nut -1- for coolant pipe bracket on oil pan/engine, see **Repair Manual, 4.0 Liter 8-Cyl. 4V Engine Mechanical, Engine Code(s): BDP, Repair Group 19.**

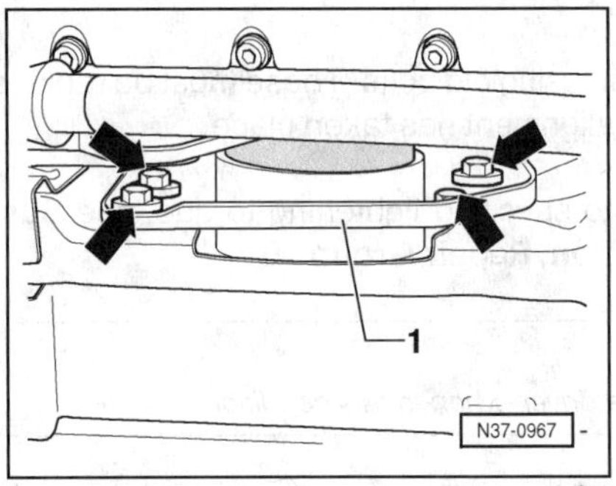

Attach stop for torque support -1- and tighten bolts -arrows- to specified tightening torque, see **Repair Manual, 4.0 Liter W8 4V Engine Mechanical (code: BDP), Repair Group 10.**

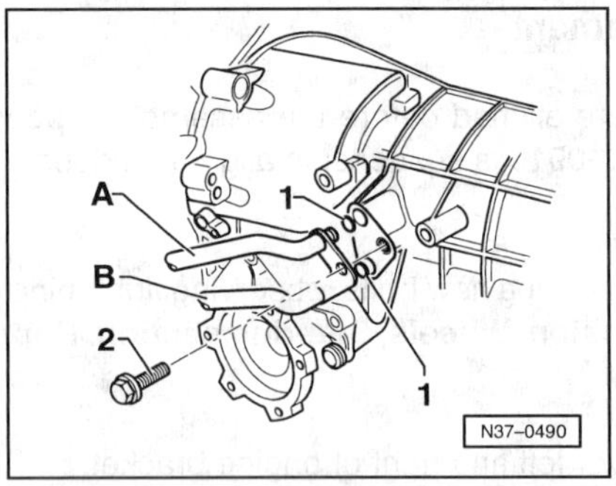

Always replace O-rings -1-.

- Moisten O-rings with ATF.
- Ensure that O-rings are seated correctly so they do not slip out of position when ATF lines are inserted.
- Insert ATF lines by hand into transmission until they can go no further.

NOTE—
Do not tighten screw -2- until the ATF lines have been fully inserted, see **ATF lines, removing and installing (8-cylinder)**.

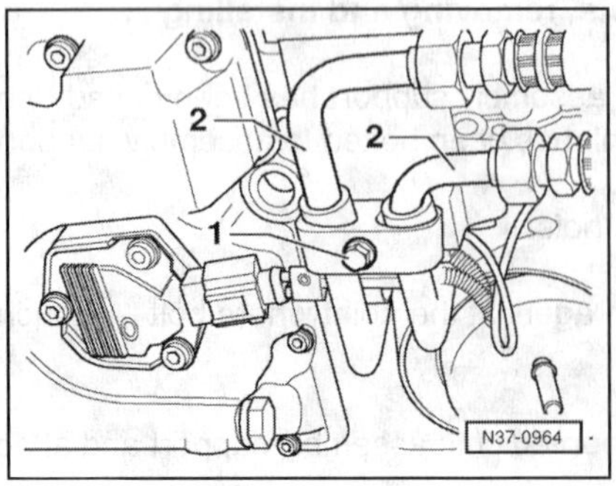

Screw ATF line holder -2- to engine, see **ATF lines, removing and installing (8-cylinder)**.

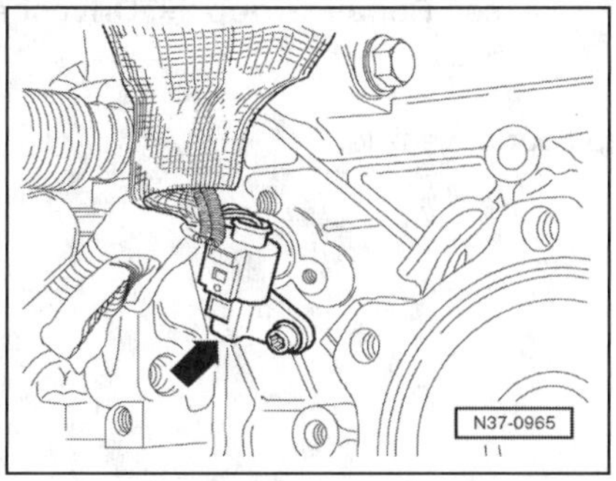

Install engine speed sensor -G28- to lower mounting -arrow-.

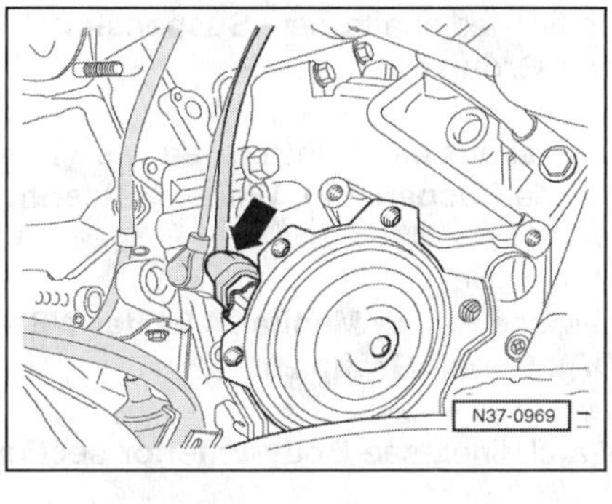

- Attach connector to speedometer sensor -G22- -arrow-.

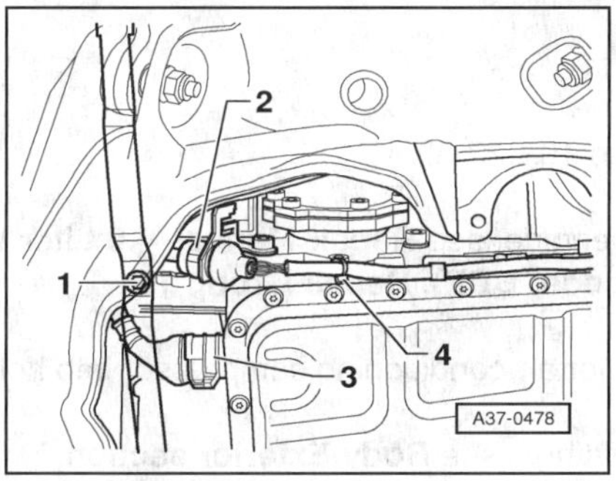

Attach multi-function switch connector -2-.

– Attach connector for transmission wiring harness and secure using safety lever -3-.

– Attach bracket for sound insulation -1-, see **Body Exterior section, Repair Group 50**

– Replace cable tie -4-.

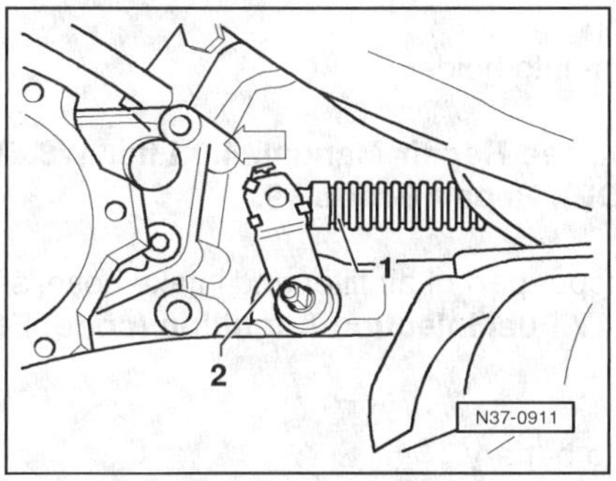

Move lever/third motion shaft to "2" and carefully press ball end of selector lever cable -1- onto lever of third motion shaft using forceps, for example.

NOTE —
Ensure the lever of the third motion shaft -2- is not bent during this operation as it will otherwise not be possible to adjust the gearshift exactly.

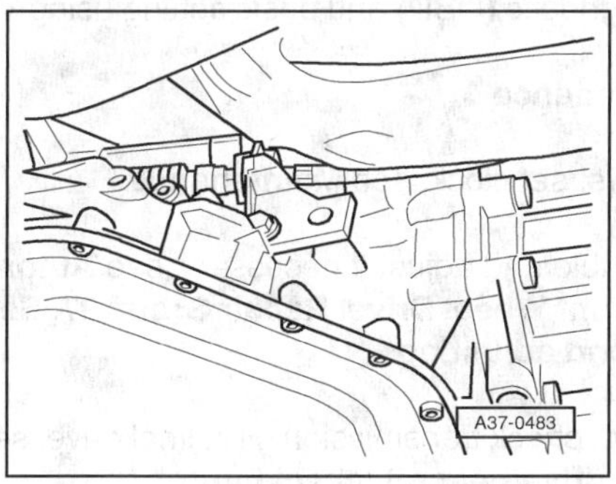

Install selector lever cable block tight according to marking on transmission -arrow-.

– Attach deflector/selector lever cable.

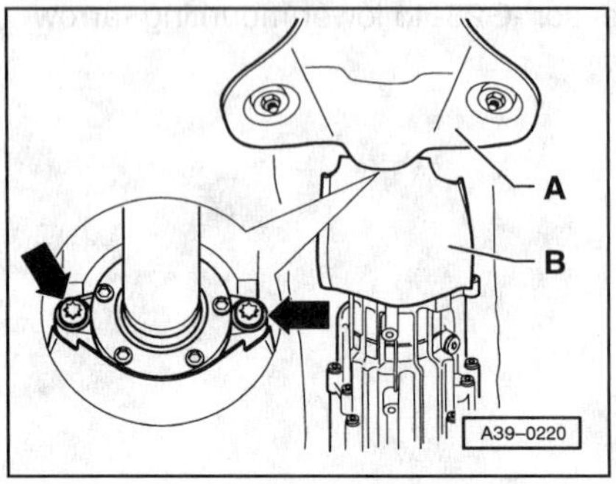

< Attach deflector -B- -arrows-, see **Repair Group 39, Driveshaft, servicing**.

– Attach heat shield -A- to floor of vehicle.

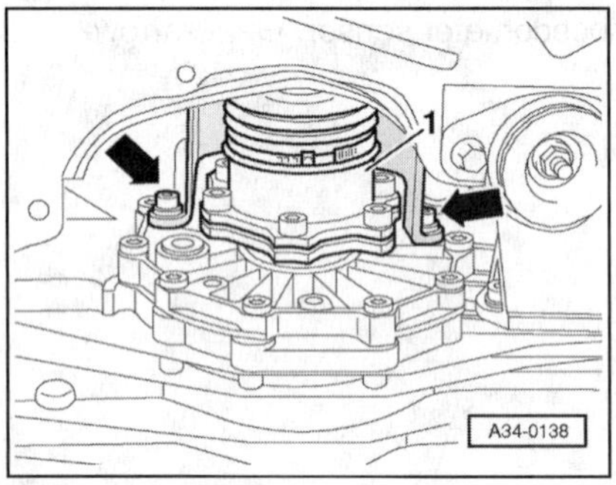

< Attach drive shafts -1- to flanged shafts, see **Suspension, Wheels, Steering section, Repair Group 40**

– Remove right and left drive shaft heat deflectors -arrows-. Third screw is not visible in illustration, see **Suspension, Wheels, Steering section, Repair Group 40**

– Install front exhaust pipes, see **Repair Manual, 4.0 Liter W8 4V Engine Mechanical (code: BDP), Repair Group 26.**

– Install right front wheel arch liner, see **Body-Exterior section, Repair Group 66**

– Install front wheels.

– Remove removal device 10-222 A.

– Attach oil filler neck to engine, see **Repair Manual, 4.0 Liter W8 4V Engine Mechanical (code: BDP), Repair Group 15.**

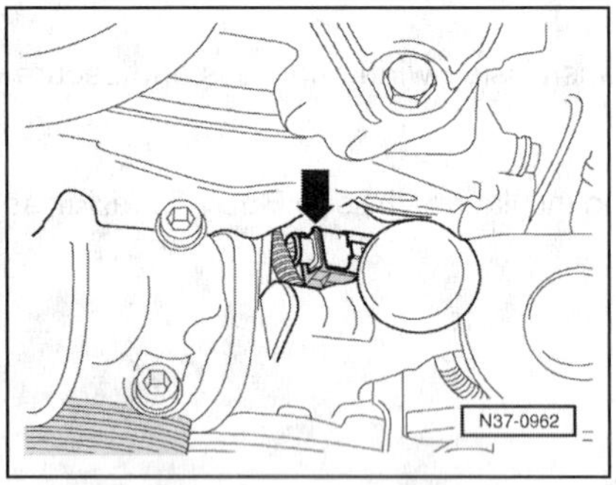

< Clip connector -arrow- for air conditioning compressor into holder.

– Install lock barrel and fittings, see **Body Exterior section, Repair Group 50**

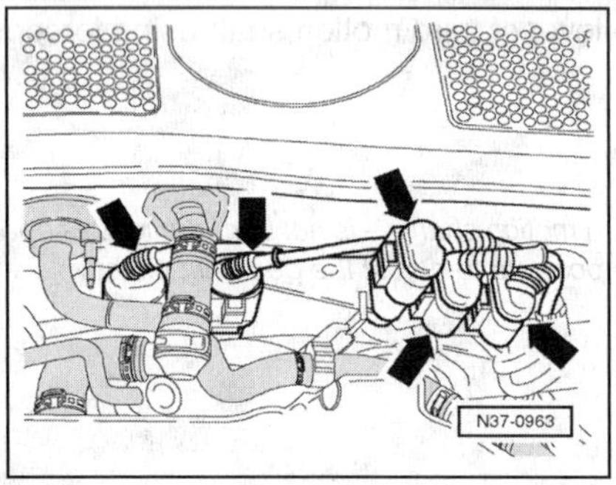

< Clip connectors -arrows- into holder.

– Fill engine with coolant, see **Repair Manual, 4.0 Liter W8 4V Engine Mechanical (code: BDP), Repair Group 19.**

– Install intake air duct, upper part of air filter and intake hose, see **Repair Manual, 4.0 Liter W8 4V Fuel Injection & Ignition (code: BDP), Repair Group 24.**

– Connect ground strap to battery.

– Perform On Board Diagnostic (OBD) and basic setting using VAS 5051.

– Code radio, see **Maintenance**

– Initialize power windows; set clock, see **Maintenance**

– Check selector lever cable and adjust if necessary, see **Automatic Transmission 01V, Front Wheel Drive, Repair Group 37, Selector lever cable, checking and adjusting**.

– If transmission installed, check transmission oil in final drive, see **Repair Group 39, Front final drive gear oil, checking**.

– Check ATF level and top up, see **Maintenance**.

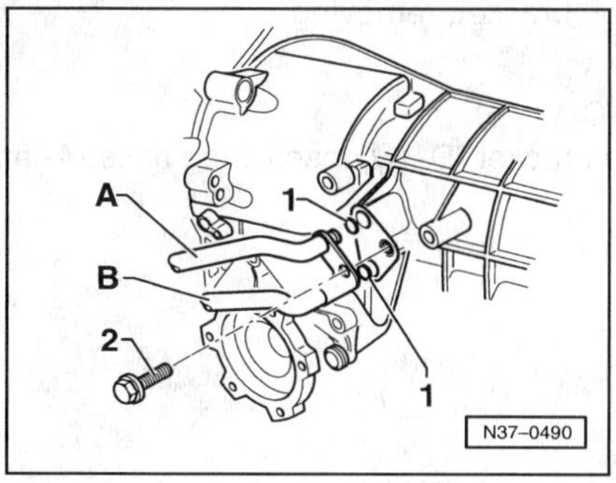

Check connections -A- and -B- of ATF lines while checking ATF level (look for leaks).

Tightening torques (8-cylinder)

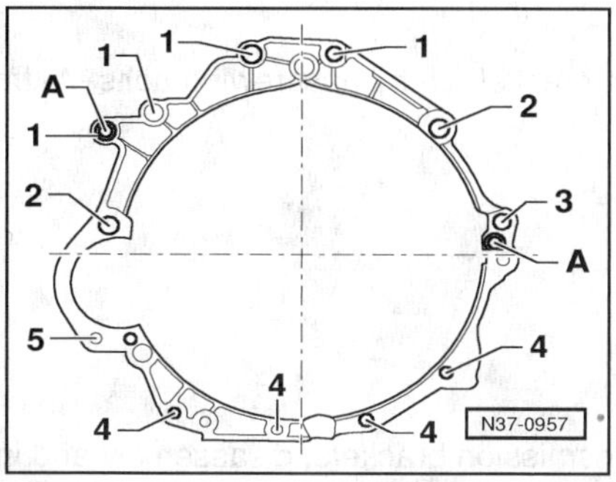

Transmission to engine

Item	Bolt	Qty.	Torque
1	M 12 x 80	4	65Nm (48 ft-lb)
2	M 12 x 110	2	65Nm (48 ft-lb)
3[1)]	M 12 x 110	1	65Nm (48 ft-lb)
4	M 10 x 60	4	45Nm (33 ft-lb)
5[2)]	M 10 x 80	1	65Nm (48 ft-lb)

[1)] with holder for wiring harness

[2)] with disc

Item A = dowel sleeves

Transmission brackets, removing and installing

Special tools, workshop equipment, test and measuring appliances and aux. items required

- 3282 Transmission support
- 3282/19 Adjustment plate
- V.A.G 1331 Torque wrench
- V.A.G 1383 A Engine/transmission jack

Removing

If both transmission brackets are to be removed, the transmission must be supported as described. If only one bracket is to be removed, it is not necessary to support the transmission.

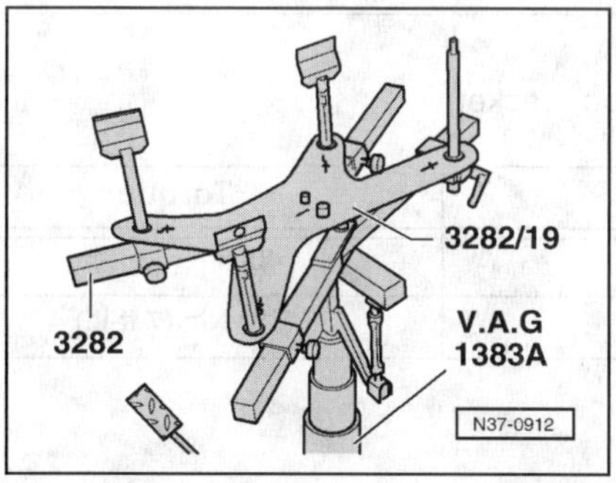

Set up transmission support 3282.

- Transmission support 3282 is set up with adjusting plate 3282/19 for automatic transmission 01V.
- The symbols on the adjusting plate indicate the mounts required, the arrow points forward.

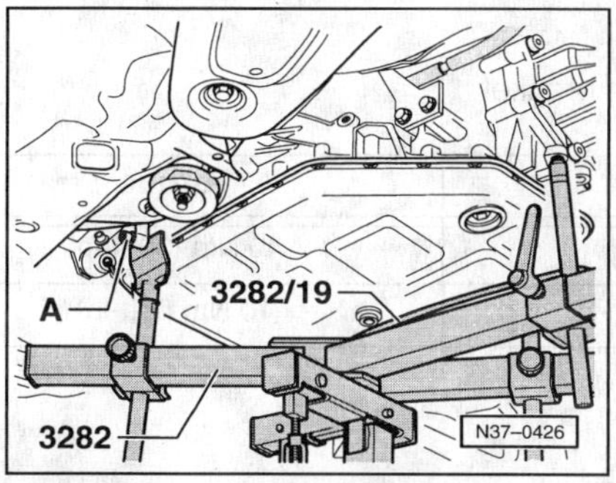

Move jack V.A.G 1383 A with transmission support 3282 under the transmission, align adjusting plate parallel to transmission and support transmission.

– Secure transmission to transmission support 3282 with bolt -A-.

– Chock space between transmission support and oil pan using washers.

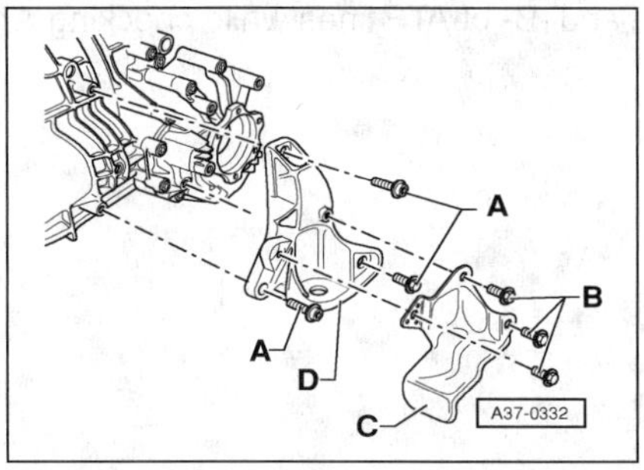

Right-hand transmission bracket, removing

- Remove bolts -B-.
- Remove heat shield -C-.
- Remove transmission bracket -D- by unscrewing bolts -A- from transmission.

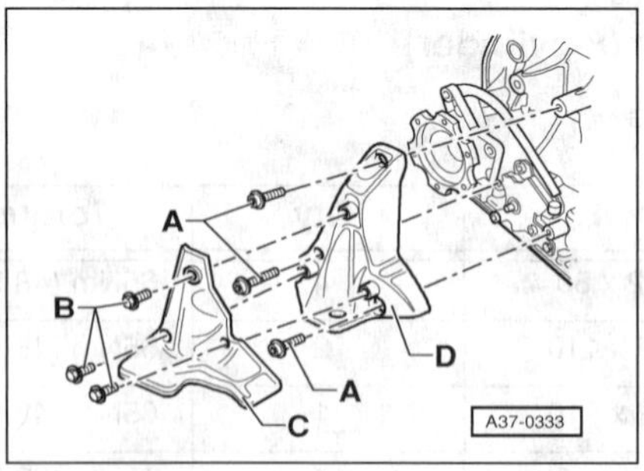

Left-hand transmission bracket, removing

- Remove bolts -B-.
- Remove heat shield -C-.
- Remove transmission bracket -D- by unscrewing bolts -A- from transmission.

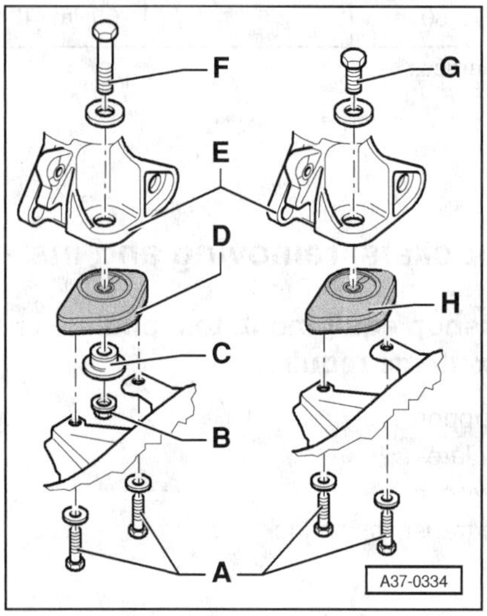

Left and right-hand transmission brackets, disassembly and installation

- Remove bolts -A-.
- Remove transmission bracket -E-.
- Positions -C- and -B- should be ignored on hydraulic transmission mountings -H- as bolt -G- is screwed directly in the transmission mounting.

- Installation is performed in the reverse order of removal.

Tightening torques

Component	Torque
A	23 Nm (17 ft-lb)
B with F	50 Nm (37 ft-lb)
G	50 Nm (37 ft-lb)

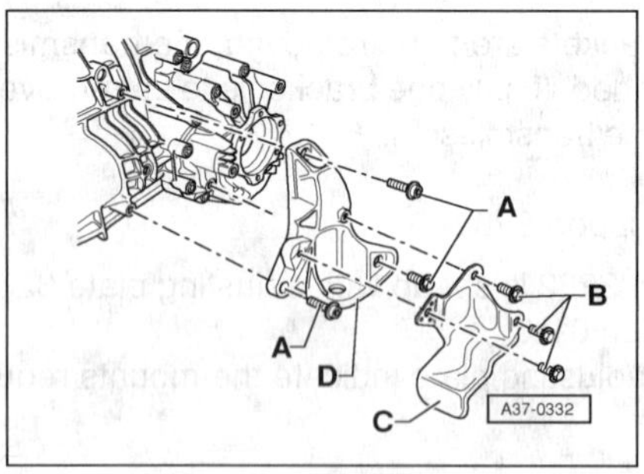

Right-hand transmission bracket

Component	Torque
A	40 Nm (30 ft-lb)
B	10 Nm (7 ft-lb)

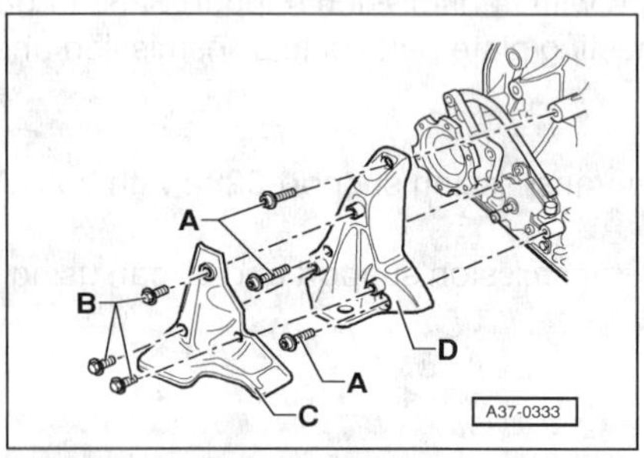

Left-hand transmission bracket

Component	Torque
A	40 Nm (30 ft-lb)
B	10 Nm (7 ft-lb)

ATF Lines

ATF lines, removing and installing (V6 engine)

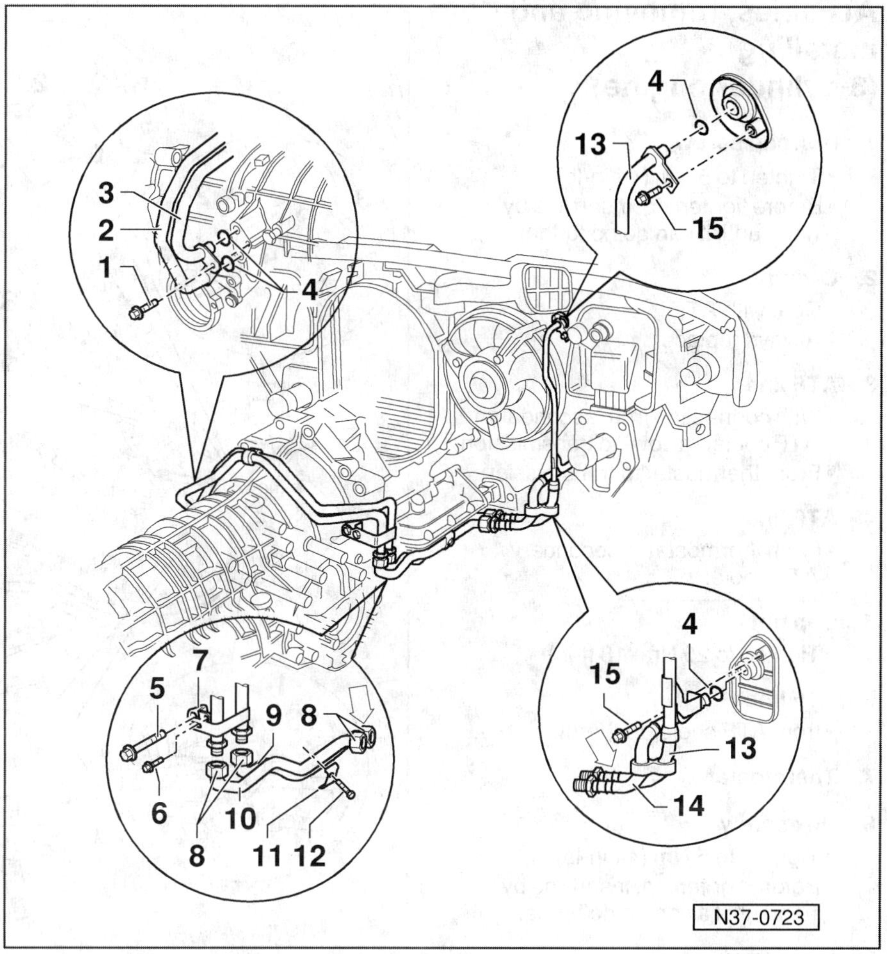

1. **Hex bolt**
   - Tighten to 20 Nm, (15 ft-lb)
   - Before tightening bolt, push in line onto stop by hand
2. **ATF line**
   - To cooler
   - Note colored markings -arrow-
3. **ATF line**
   - From cooler
   - Note colored markings -arrow-
4. **O-ring**
   - Insert with ATF
   - Always replace
5. **Hex bolt**
   - Tighten to 65 Nm, (48 ft-lb)
6. **Hex bolt**
   - Tighten to 10 Nm (7 ft-lb)
7. **Bracket**
8. **Union nut**
   - Tighten to 30 Nm, (22 ft-lb)
9. **ATF line**
   - To cooler
   - Note colored markings -arrow-
10. **ATF line**
    - From cooler
    - Note colored markings -arrow-
11. **Clamp**
12. **Hex bolt**
    - Tighten to 10 Nm (7 ft-lb)
13. **ATF line**
    - To cooler
    - Note colored markings -arrow-
14. **ATF line**
    - From cooler
    - Note colored markings -arrow-
15. **Hex bolt**
    - Tighten to 5 Nm (44 in-lb)
    - Before tightening bolt, push in line onto stop by hand

ATF lines, removing and installing (8-cylinder engine)

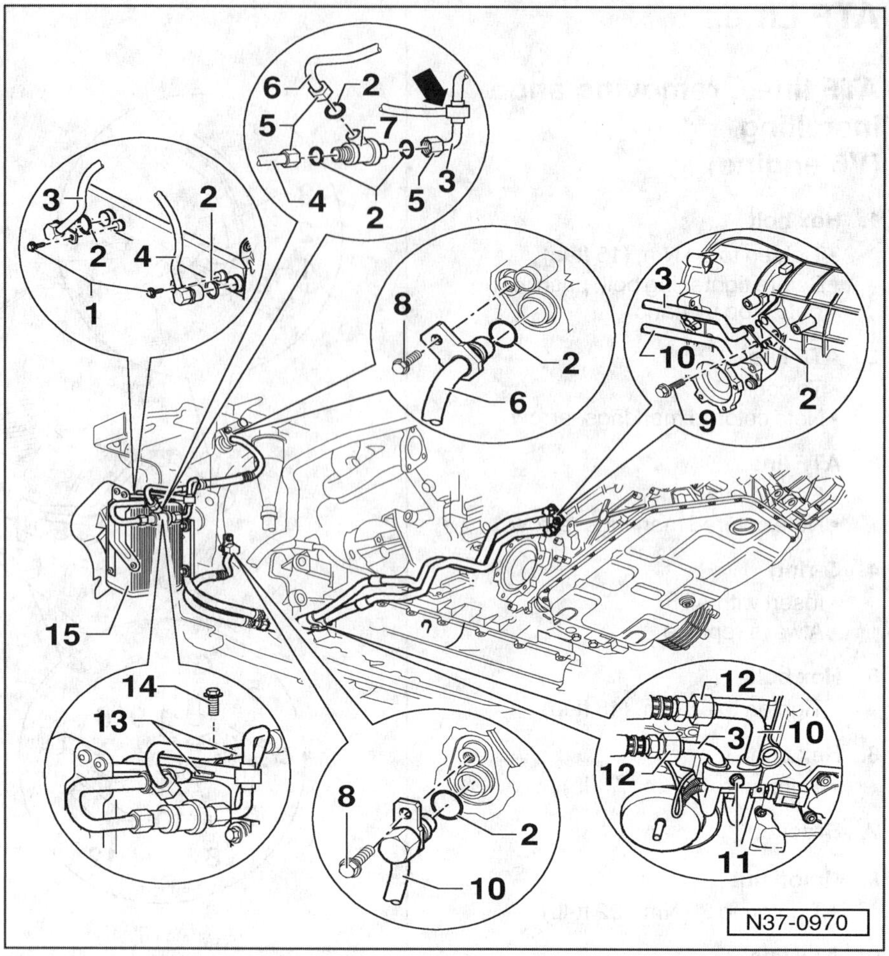

1. **Hex cap screw**
   - Tighten to 5 Nm (44 in-lb)
   - Before tightening, insert line by hand until it can go no further
2. **O-ring**
   - Coat with ATF
   - Always replace
3. **ATF line**
   - With connection from secondary ATF cooler -arrow- to transmission
   - From thermostat to transmission
4. **ATF line**
   - From thermostat to secondary ATF cooler
5. **Cap nut**
   - Tighten to 25 Nm (18 ft-lb)
6. **ATF line**
   - from ATF cooler to thermostat
7. **Thermostat**
8. **Hex screw**
   - Tighten to 5 Nm (44 in-lb)
   - Before tightening, insert line by hand until it can go no further
9. **Hex screw**
   - Tighten to 20 Nm (15 ft-lb)
   - Before tightening, insert line by hand until it can go no further
10. **ATF line**
    - From transmission to ATF cooler
11. **Hex cap screw**
    - Tighten to 9 Nm (18 in-lb)
12. **Cap nut**
    - Tighten to 30 Nm (22 ft-lb)
    - Thread connections cannot be crossed
13. **Clip**
14. **Hex screw**
    - Tighten to 5 Nm (44 in-lb)
15. **Secondary ATF cooler**

Secondary ATF cooler, removing and installing (8-cylinder engine)

Special tools and equipment

- V.A.G 1306 Drip tray
- V.A.G 1331 Torque wrench or equivalent

Removing

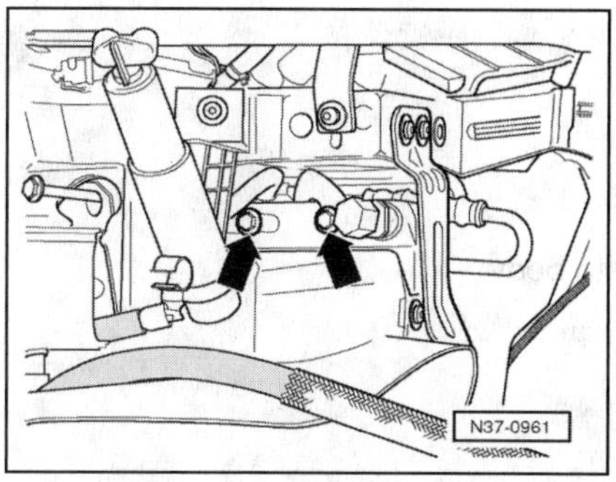

- Remove cover from front bumper, see **Repair Manual, Body Exterior, Repair Group 63**
- Place V.A.G 1306 or equivalent drip tray underneath.
- Thoroughly clean area where ATF line/secondary ATF cooler connect.
- Disconnect ATF lines from secondary ATF cooler -arrows-.
- Plug ATF lines and secondary ATF cooler with clean stoppers.

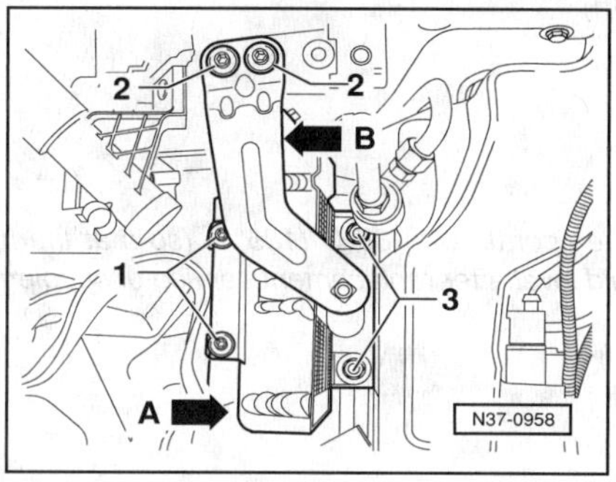

- Remove screw -1- from secondary ATF cooler -arrow A-.
- Remove screw -2- from holder -arrow B-.
- Remove bolt -3- from secondary ATF cooler -arrow A- and remove secondary ATF cooler.

Installing

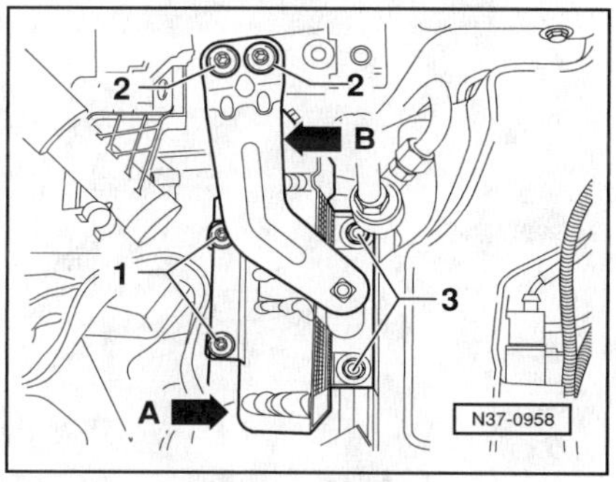

- Insert secondary ATF cooler -arrow A- and tighten screws -3-.
- Tighten screw -2- on holder -arrow B-.
- Tighten screw -1- on secondary ATF cooler -arrow A-.

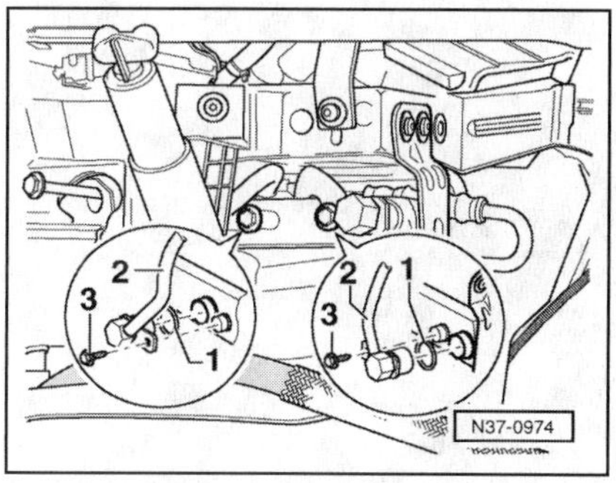

- Always replace O-rings -1-.
- Moisten O-rings -1- with ATF.
- Ensure that O-rings are seated correctly so they do not slip out of position when ATF lines are inserted.
- Insert ATF lines -2- by hand into secondary ATF cooler until they can go no further.
- Do not tighten screws -3- until ATF lines have been fully inserted, see **ATF lines, removing and installing (8-cylinder)**
- Check ATF level and top up if necessary, see **Maintenance**.

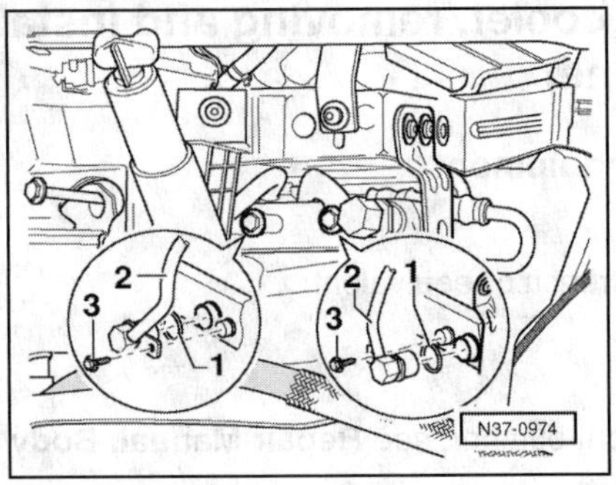

➤ Check connections of ATF lines -2- while checking ATF level (look for leaks).

– Replace front bumper cover, see **Body Exterior section, Repair Group 63**

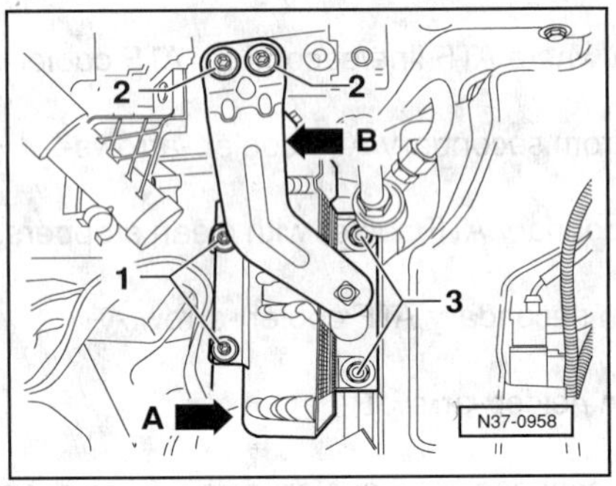

Tightening torques

➤ Secondary ATF cooler to body

1 - Torx screw, 5 Nm

2 - Torx screw, 5 Nm

3 - Hex cap screw, 5 Nm

NOTE—

Make sure to bring ATF temperature to 85°C (185°F) (so that thermostat opens) and check ATF fluid level after replacement/service of supplemental oil cooler (8-cylinder).

39 Final Drive, Differential

Repair Group

P 39

Front Final Drive

Gear oil in front final drive, checking and filling

Special tools, workshop equipment, testers, measuring instruments and auxiliary items required

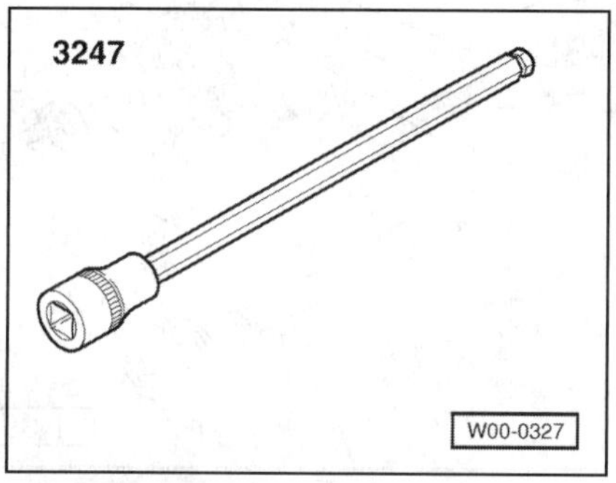

3247 Socket AF 8

Oil level is checked with transmission installed

– Drive the vehicle a short distance to warm up the gear oil.

– Position vehicle on hard, level standing e.g. on 4 pillar hoist or over pit.

– Switch off engine and let gear oil drain for 5 minutes.

– Place drip tray V.A.G 1306 underneath.

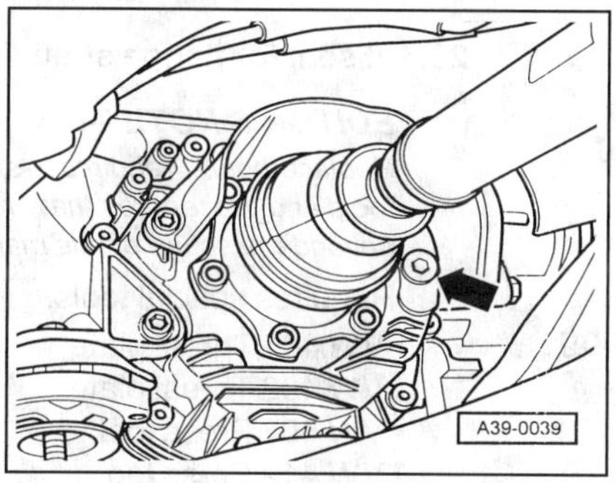

Remove sealing plug -arrow-.

The oil level is correct when the front final drive is full to the lower edge of the filler hole.

– Allow excess gear oil to pour out or top up as necessary.

• Specification, see **Maintenance**

– If gear oil needs to be replenished following repair work, fill front final drive slowly and evenly over a period of at least 5 minutes until the gear oil drips out over the lower edge of the oil filler hole.

A filling time of at least 5 minutes is necessary to allow the oil to settle evenly in the differential and cavity.

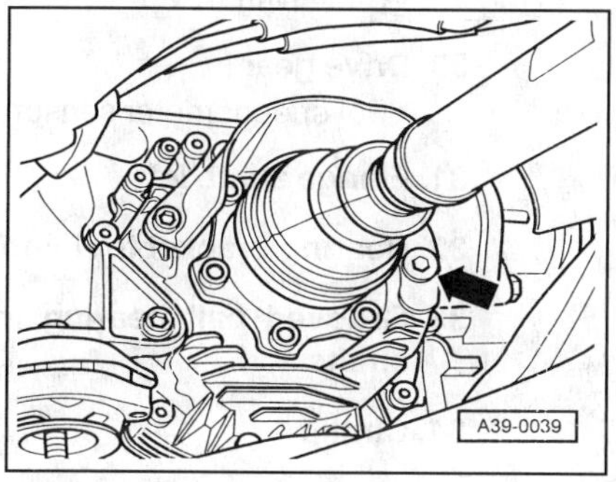

Screw in original sealing plug -arrow-.

– Drive the vehicle a short distance to warm up the gear oil.

– Check oil level in front final drive.

– If the oil level is OK, replace front final drive sealing plug -arrow-, see **Front final drive, assembly**.

Front final drive, servicing

Transmission in vehicle

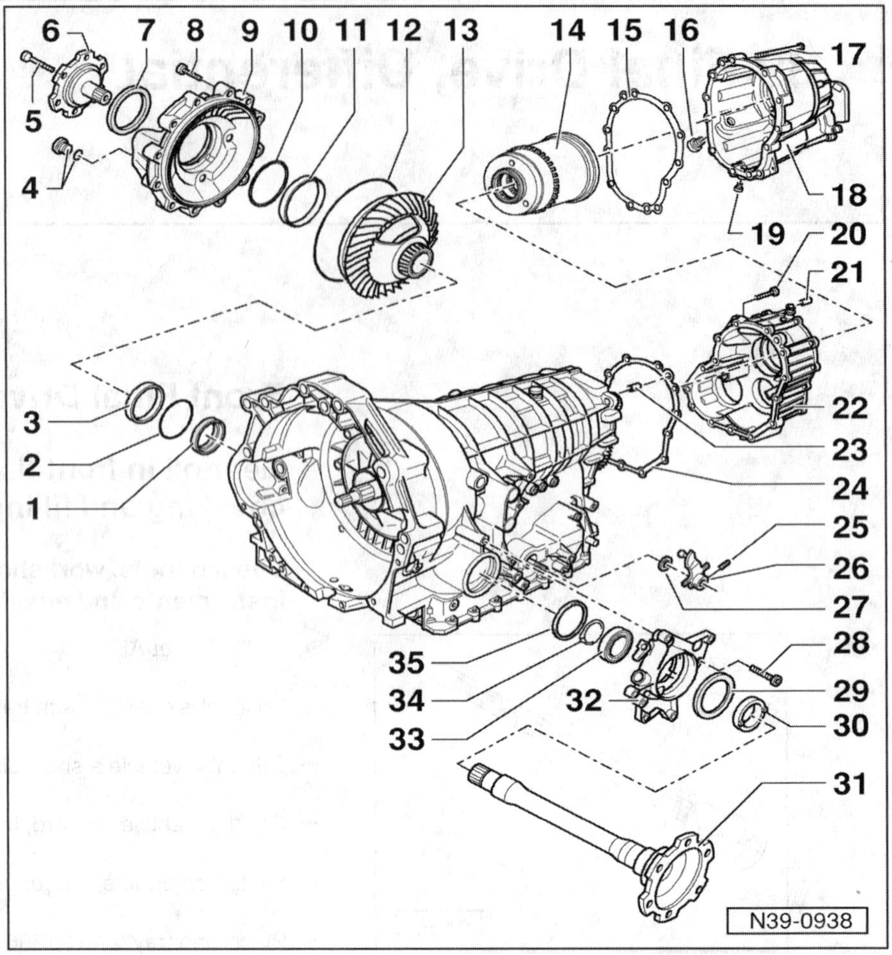

1. **Oil seal**
   - Between final drive and transmission housing
2. **Shim**
   - Behind tapered roller bearing outer race
3. **Tapered roller bearing outer race**
4. **Sealing plug,**
   - Tighten to 30 Nm (22 ft-lb)
   - With seal
   - Always replace
5. **Bolt**
   - Tighten to 23 Nm (17 ft-lb)
6. **Flange shaft, right**
7. **Oil seal for flange shaft, right**
   - Replacing, see **5-Speed Automatic Transmission 01V, Front Wheel Drive, repair group 37**
8. **Bolt**
   - Tighten to 23 Nm (17 ft-lb)
9. **Cover for final drive**
10. **Shim**
    - Behind tapered roller bearing outer race
11. **Tapered roller bearing outer race**
12. **O-ring**
    - Always replace
13. **Differential**
14. **Torsen differential**
    - Replace complete if damaged
    - Removing and installing, see **Center differential, removing and installing**
15. **Seal**
    - Replacing, see **Center differential, removing and installing**
16. **Plug,**
    - Tighten to 35 Nm (26 ft-lb)
    - See **Gear oil in center differential, checking, draining and filling**
    - Always replace
17. **Bolt**
    - Tighten to 23 Nm (17 ft-lb)
    - Observe tightening sequence, see **Center differential, removing and installing**
18. **Center differential**
    - Removing and installing, see **Center differential, servicing**
19. **Drain plug**
    - Tighten to 20 Nm (15 ft-lb)
    - See **Gear oil in center differential, checking, draining and filling**
20. **Bolt**
    - Tighten to 23 Nm (17 ft-lb)
21. **Dowel pin**
22. **Intermediate flange for front final drive**
23. **Dowel sleeve**
24. **Seal**
25. **Spring pin**
26. **Lever**
    - For selector shaft
27. **Oil seal**
    - For selector shaft
28. **Bolt**
    - Tighten to 23 Nm (17 ft-lb)
29. **Oil seal for flange shaft, left**

    EDITOR'S NOTE —
    Replacement requires special tools and procedures that are beyond the scope of this manual.

    Requires special tools:
    - VW 401 Thrust plate
    - VW 402 Thrust plate
    - VW 407 Press tool
    - VW 412 Press tool
    - VW 681 Oil seal extractor
    - 32-109 Tube
    - 2040 Tube
    - 3384 Drift
30. **Drive gear**
    - For speedometer sensor -G22-
31. **Flange shaft, left**
32. **Bearing bracket for flange shaft**
33. **Grooved ball bearing for flange shaft**
34. **Circlip**
35. **O-ring**
    - Square section

Center Differential

Gear oil in center differential, checking, draining and filling

Special tools and equipment

- VAG1331 Torque wrench
- VAG1306 Drip tray

– Drive vehicle onto a 4 pillar vehicle hoist or over a pit so that it stands level.

– Separate right exhaust pipe, if installed, from rear exhaust system at double clamp at front. Push front exhaust pipe in area of catalyst to right, see **Engine Mechanical section, Repair Group 26.**

– Place VAG1306 Drip tray underneath.

Checking

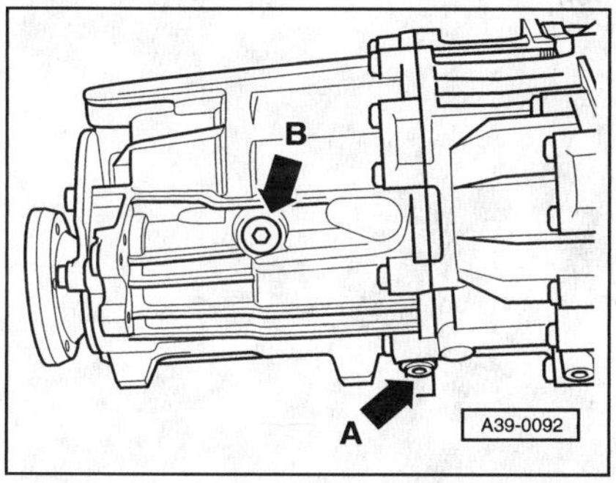

Remove plug -arrow B- to check gear oil.

- The oil level is correct when the transmission is full to the lower edge of the filler hole.

– Drain off excess gear oil or top up with gear oil, as necessary.

– Always replace bolts

– Tighten new plug -arrow B-, see **Front final drive, servicing**.

Draining gear oil

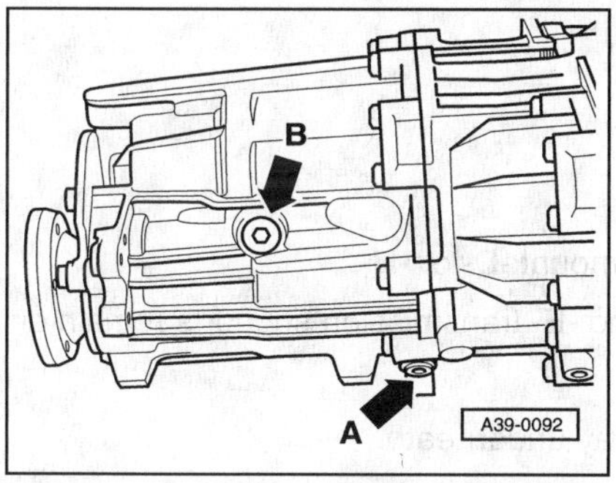

Remove drain plug -arrow A- and drain gear oil.

– Always replace drain plug.

– Screw in new drain plug -arrow A-, see **Front final drive, servicing**.

Filling gear oil

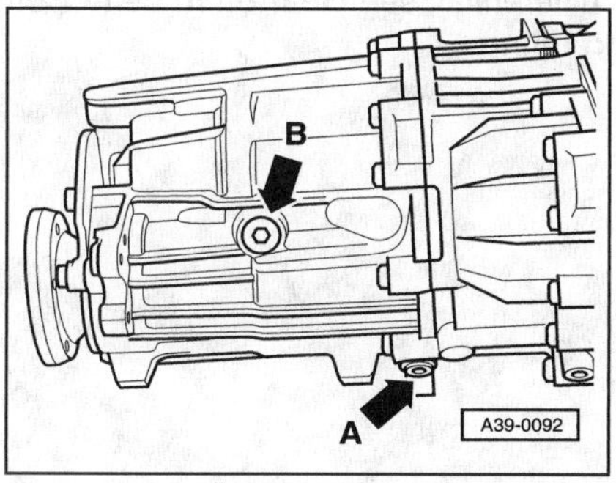

Remove plug -arrow B- to check gear oil and top up to lower edge of hole with gear oil.

- Screw in old plug.

NOTE —

If the exhaust system was loosened at the front or removed, it must be tightened or installed for the test drive, and then again removed and installed for checking the oil level.

- Carry out a test drive of at least 15 minutes pure driving time and then check gear oil in center differential and top up if necessary.
- Only after a test drive is it ensured that the newly added gear oil has reached the Torsen differential.
- The oil level is correct when the transmission is full to the lower edge of the filler hole.
- When oil level is correct, seal center differential with new plug -arrow B-.

Center differential, removing and installing

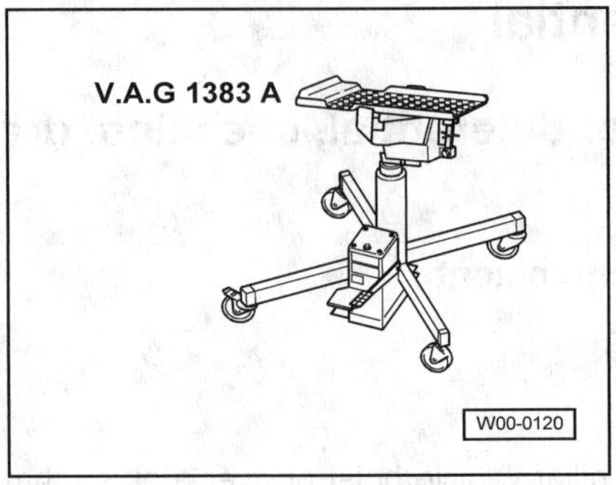

Special tools and equipment

- VAG1331 Torque wrench

VAG1383A Engine/transmission jack

Removing

– Remove front exhaust system with catalyst, see **Engine Mechanical section, Repair Group 26.**

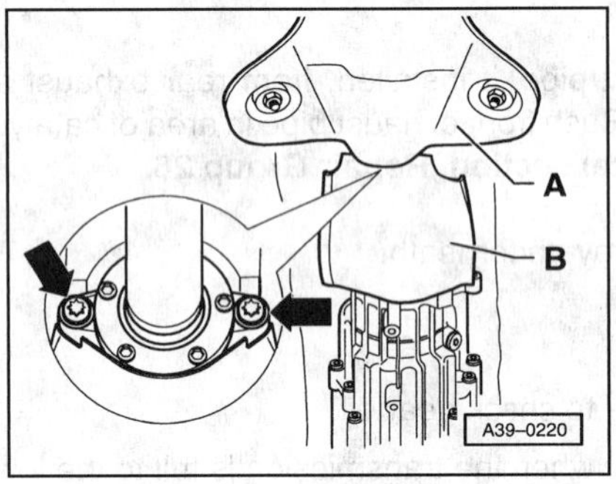

Remove heat shield -A- from vehicle underbody.

– Remove cover plate -B- (arrows).

– Disconnect driveshaft from transmission, see **Driveshaft, removing and installing**.

– Support transmission from below with transmission jack VAG1838 A and the respective plate 1359/2.

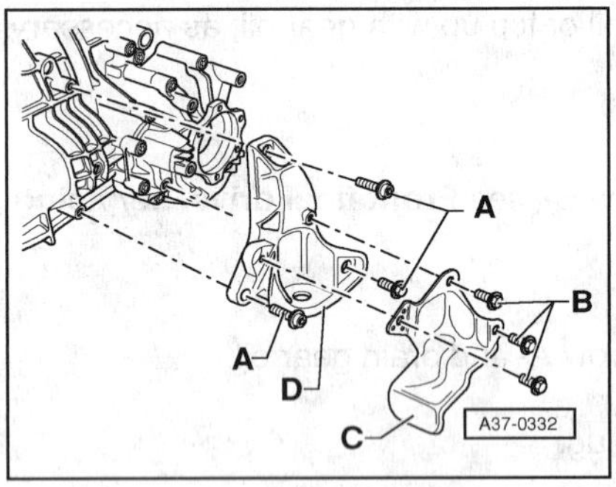

Remove bolts -B-.

– Remove heat shield -C-.

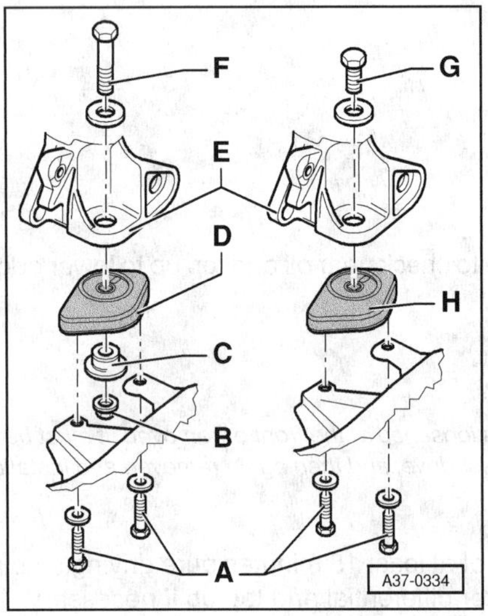

Remove transmission mount -D- or -H-.

- The left and right-hand -E- transmission supports remain on transmission.

– Place VAG1306 Drip tray underneath.

– Drain gear oil in center differential, see **Gear oil in center differential, checking, draining and filling.**

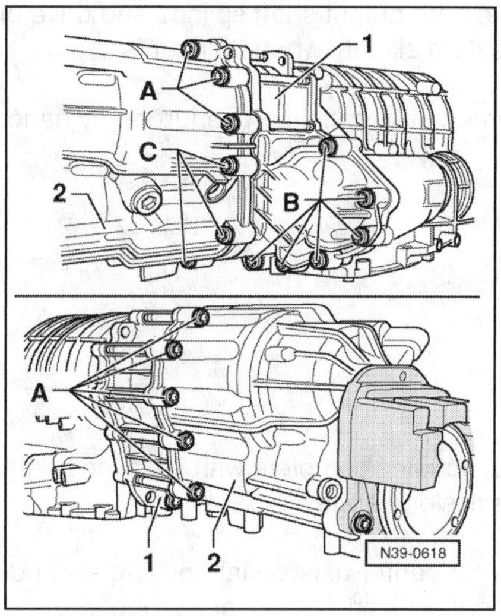

◀ Loosen bolts -A and C- for center differential housing -2- using diagonal sequence and remove.

– Do not loosen bolts -B-.

NOTE—

Pull center differential housing carefully to rear off transmission, otherwise the Torsen differential will fall out of the transmission.

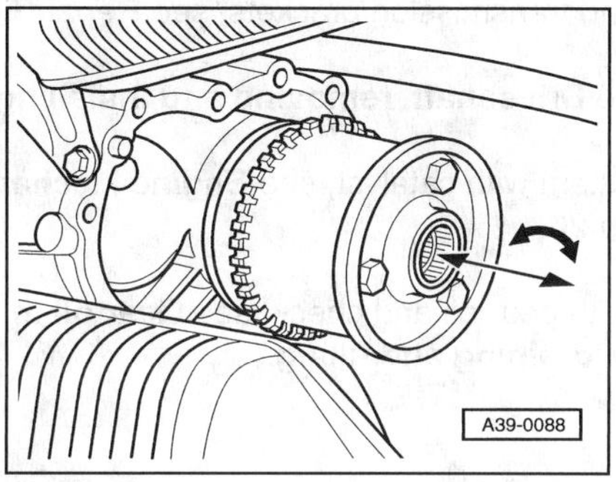

◀ Pull Torsen differential to rear off output shaft.

Center differential, installing

Install in reverse order but note following:

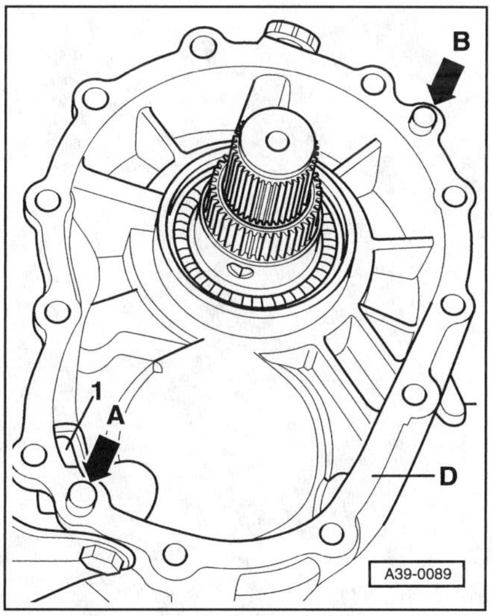

◀ Check dowel pins -arrows A and arrow B- are installed.

– Install gasket.

– Install cleaned magnets in the cast retainer -1- in intermediate flange.

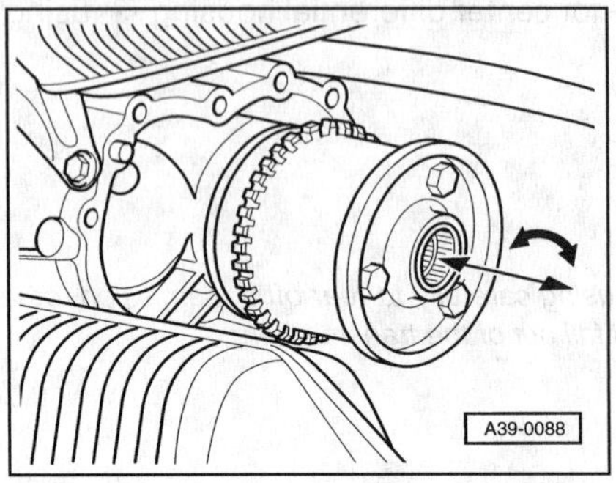

Install Torsen differential onto output shaft splines and drive pinion of front gear wheel, rotate item slightly when doing this.

– Check that installed Torsen differential can be turned by hand.

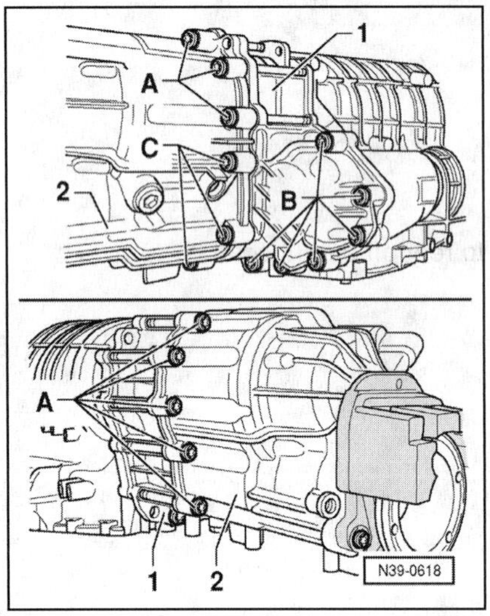

Install center differential housing complete with installed gears and rear flange shaft onto transmission flange.

– Insert bolts -A- and -C- for center differential housing -2- and tighten to 23 Nm (17 ft-lb) incrementally using diagonal sequence.

– Install left and right-hand transmission brackets, see **Repair Group 37**.

– Connect driveshaft, see **Driveshaft, removing and installing**.

– Install front exhaust system with catalyst, see **Engine Mechanical section, Repair Group 26.**

– Fill center differential with gear oil and check, see **Gear oil in center differential, checking, draining and filling**.

Driveshaft

Driveshaft, assembly

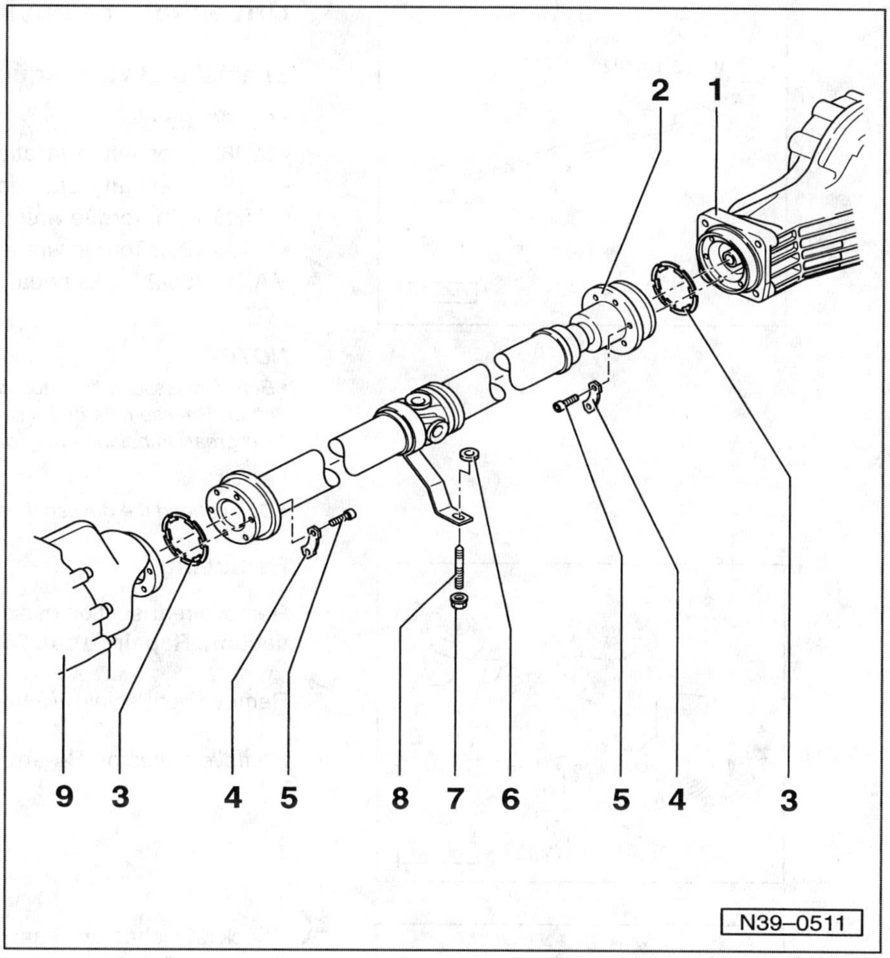

Sequence for removing and installing, see **Driveshaft, removing and installing**

- No repairs must be performed on the driveshaft
- If the driveshaft is disconnected from only the transmission or only the rear final drive, it must be tied up or supported.
- Remove and install the driveshaft only using 3298 or 3405 Driveshaft assembly appliance.
- Don't flex driveshaft; store and transport extended only.
- In the event of complaints (noise or vibration), check adjustment before replacing driveshaft, see **Driveshaft, adjusting**.

1. **Rear final drive**
2. **Driveshaft**
3. **Seal**
   - Replace
   - Pull off protective foil
   - Glue seal in flange shaft
4. **Lock plate**
5. **Allen head bolt**
   - Tighten to 55 Nm (41 ft-lb)
   - Self-locking
   - Always replace
   - Always clean threaded holes for bolts in flange shaft (e.g. with a thread tap)
6. **Differential**
   - Determining thickness, see **Driveshaft, adjusting**
7. **Nut**
   - Tighten to 25 Nm (18 ft-lb)
8. **Stud**
   - Tighten to 10 Nm (7 ft-lb)
   - Removing and installing, see **Fig. 1**
9. **Transmission**

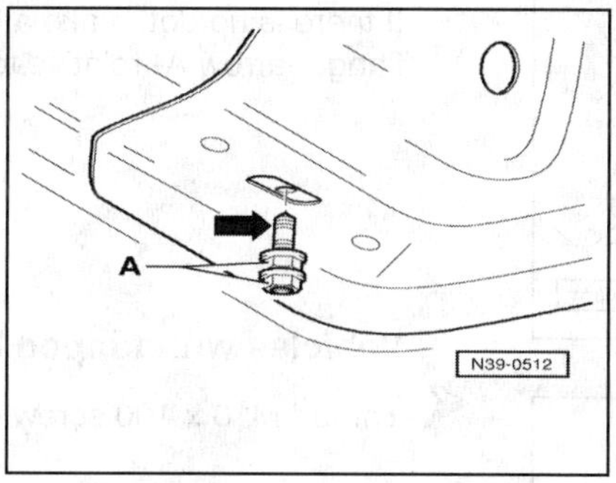

Fig. 1 Removing and installing stud

- Remove and install stud using two lock nuts -A-.
- Stud short thread is screwed into vehicle (arrow).
- Tighten to 10 Nm (7 ft-lb).

Driveshaft, removing and installing

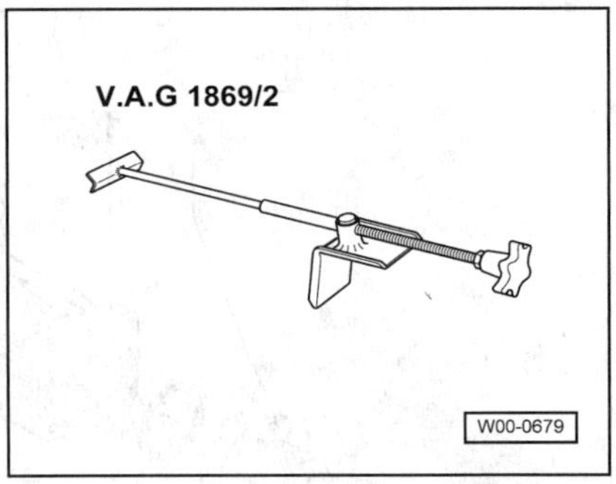

Special tools and equipment

- 3139/3 Spacers
- 3298 Driveshaft installation tool
- 3405 Driveshaft installation tool
- V.A.G 1331 Torque wrench
- V.A.G 1332 Torque wrench

◄ V.A.G 1869/2 Brake pedal applicator

NOTE —

- *Before disassembling, note the relative position of all the parts to one another. Reassemble in the same position, otherwise the imbalance will be too great. Imbalance may damage the bearings and generate humming noises.*
- *Do not bend the driveshaft; store and transport in extended state.*

Removing

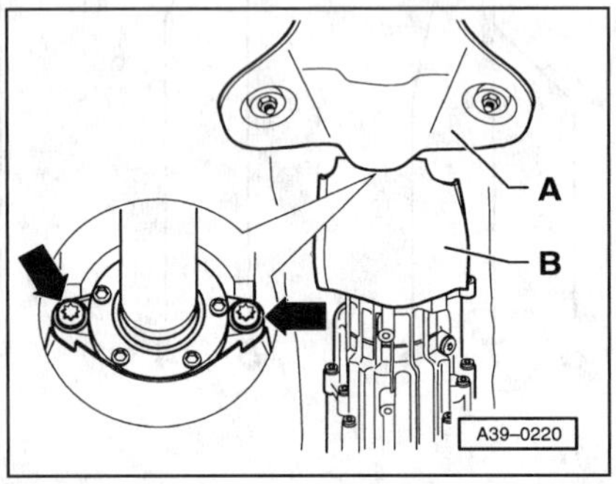

– Remove rear section of exhaust system, see **Engine Mechanical section, Repair Group 26**

◄ Remove heat shield -A- from floor of vehicle.

– Remove deflector -B- -arrows-.

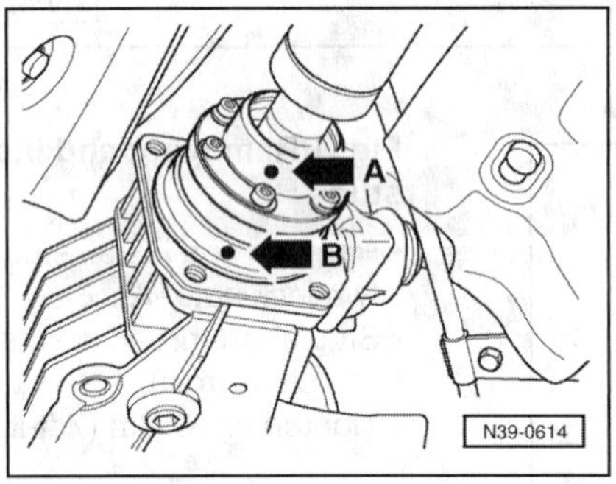

◄ Check driveshaft and flange/driveshaft on rear final drive to see whether a mark (red dot) is present.

– If there is no dot, make a colored mark to indicate position of driveshaft flange -arrow A- relative to rear final drive -arrow B-.

Vehicles with tapped holes in front of rear final drive

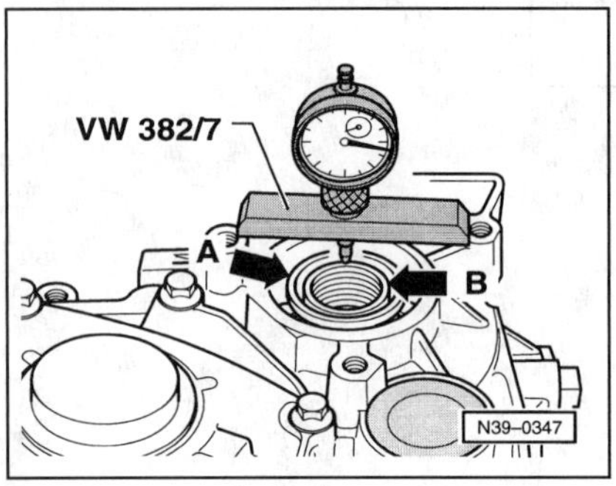

◄ Thread M10 x 100 screw -3- into rear final drive.

– To remove screws -1-, hold front and rear of driveshaft steady with a spanner -2-.

– At each constant-velocity joint on driveshaft, remove upper three bolts.

Vehicles without tapped holes in front of rear final drive

– Lower vehicle.

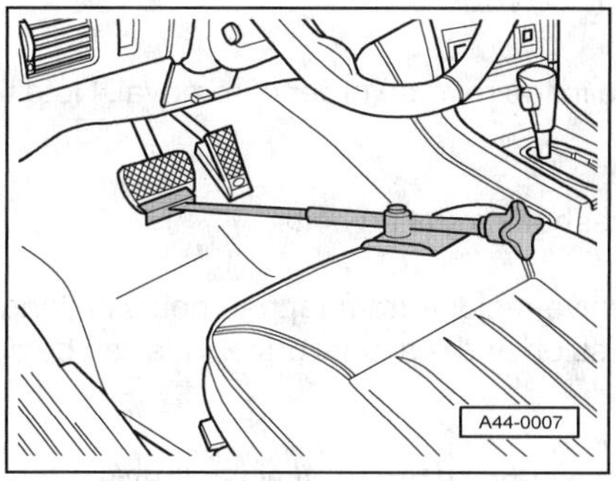

– Insert V.A.G 1869/2 brake pedal applicator.

– Operate brake pedal using brake pedal applicator.

– Raise vehicle.

– At each constant velocity joint on driveshaft, remove lower three bolts.

– Lower vehicle.

– Remove V.A.G 1869/2 brake pedal applicator.

– Raise vehicle.

– Turn two rear wheels in same direction to move driveshaft 1/2 turn (180°).

– Lower vehicle.

– Insert V.A.G 1869/2 brake pedal applicator.

– Operate brake pedal using brake pedal applicator.

– Raise vehicle.

All vehicles

Driveshaft with 60 mm diameter housing front and rear

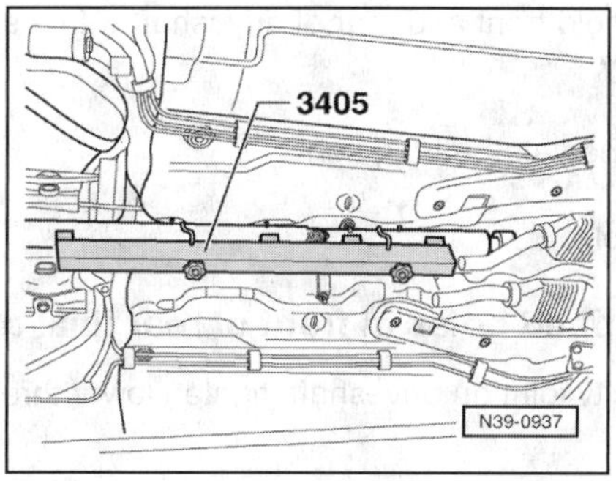

NOTE —
Do not place the installation tool on balance pieces.

– Insert 3405 installation tool and tighten plastic nuts.

Driveshaft with 60 mm diameter front housing and 70 mm rear housing

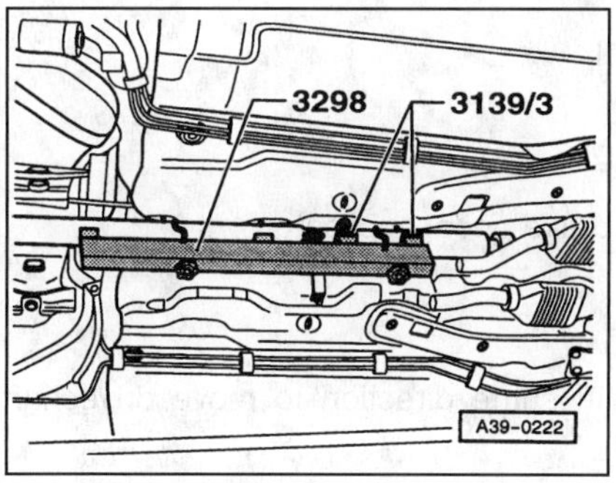

NOTE —
Do not place the installation tool on balance pieces.

– Insert 3298 installation tool and 3139/3 spacers and tighten plastic nuts.

All vehicles

– Undo remaining three bolts on front and rear constant velocity joints.

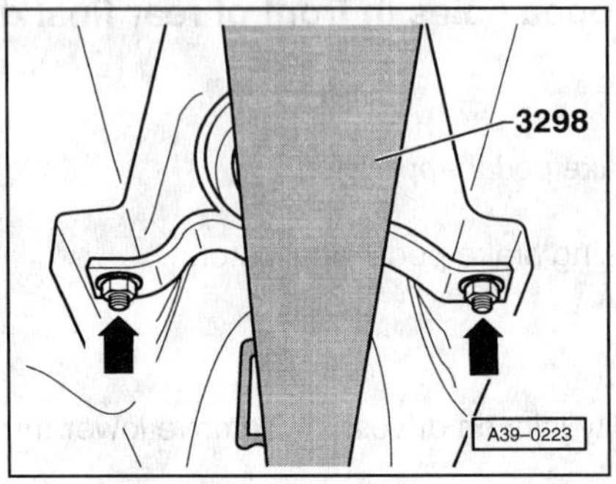

- Undo retaining nuts (arrows) for intermediate bearing and remove driveshaft, together with installation tool.

Driveshaft, installing

Installation is performed in the reverse order of removal. Note the following:

Always replace the driveshaft mounting bolts.

– Always clean any adhesive residue from tapped holes in flanged shafts for driveshaft of transmission and rear final drive. This can be done using a tap thread.

The mounting bolts may otherwise break off when they are threaded in.

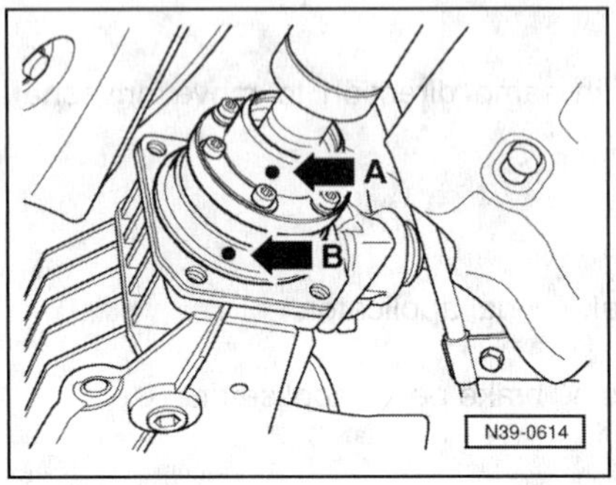

- Make certain that marks on driveshaft flange (arrow-A-) and transmission/rear final drive (arrow-B-) are in line.

If a new driveshaft is installed and the marks applied in the factory to the flange of the rear final drive are no longer visible, the radial run-out of the flange/driveshaft must be measured and the colored dot on the driveshaft aligned with the new mark on the flange/driveshaft.

– Replace driveshaft seals on transmission flange and/or rear final drive flange (remove protective film and stick seal in position).

Vehicles with tapped holes in front of rear final drive

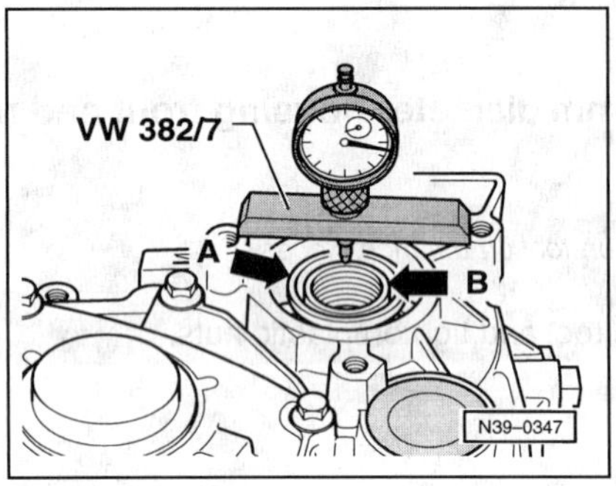

- Thread M10 x 100 screw -3- into rear final drive.

– To tighten screws -1-, hold front and rear of driveshaft with a spanner -2- to keep shaft steady.

– Screw driveshaft intermediate bearing tight.

– Remove installation tool.

Vehicles without tapped holes in front of rear final drive

– At each constant-velocity joint on driveshaft, tighten lower three bolts to 55 Nm (41 ft-lb).

– Screw driveshaft intermediate bearing to 25 Nm (18 ft-lb).

– Remove installation tool.

– Lower vehicle.

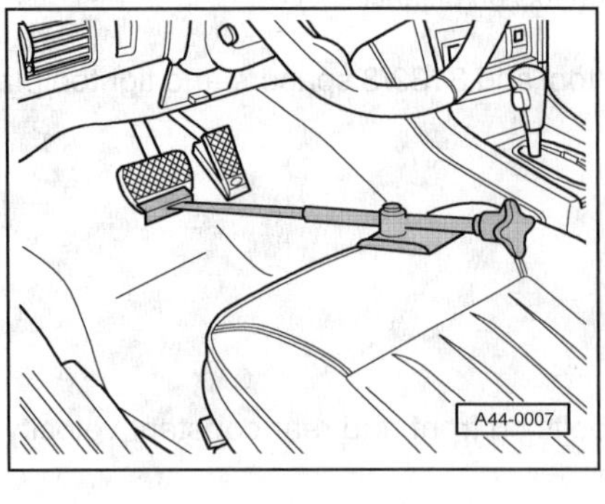

- Remove V.A.G 1869/2 brake pedal applicator.

– Raise vehicle.

– Turn two rear wheels in same direction to move driveshaft 1/2 turn (180°).

– Lower vehicle.

– Operate brake pedal using brake pedal applicator.

– Raise vehicle.

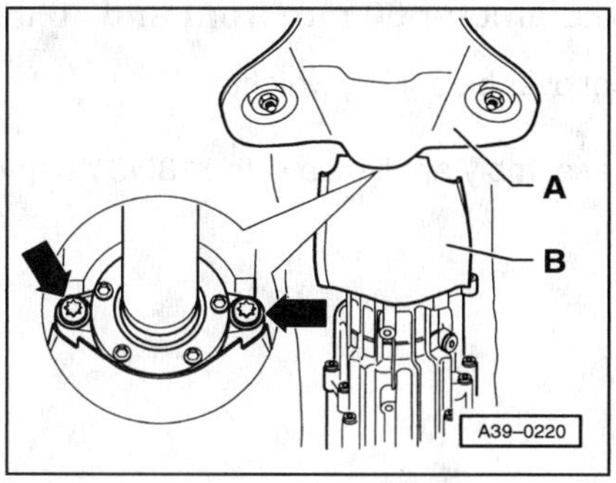

– At each of constant-velocity joints on driveshaft, tighten lower three mounting bolts to 55 Nm (41 ft-lb).

All vehicles

Tighten deflector -B- on transmission -arrows- to 25 Nm (18 ft-lb).

– Attach heat shield -A- to floor of vehicle.

– Install rear section of exhaust system, see **Engine Mechanical section, Repair Group 26.**

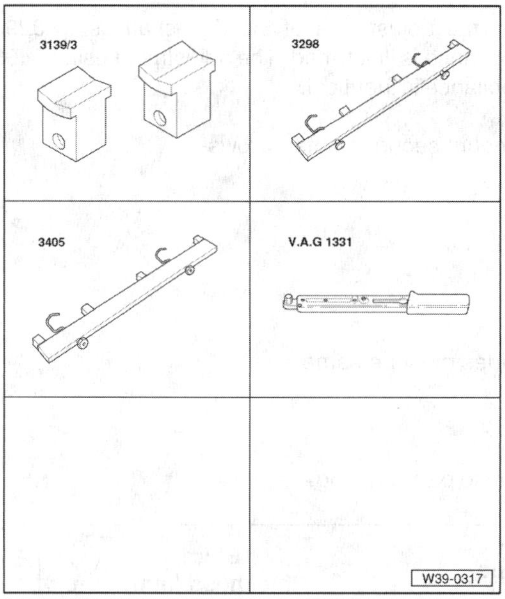

Driveshaft, adjusting

Special tools and equipment

- 3139/3 Adapters
- 3298 Driveshaft assembly appliance
- 3405 Driveshaft assembly appliance
- VAG1331 Torque wrench

– Remove rear section of exhaust system, see **Engine Mechanical section, Repair Group 26.**

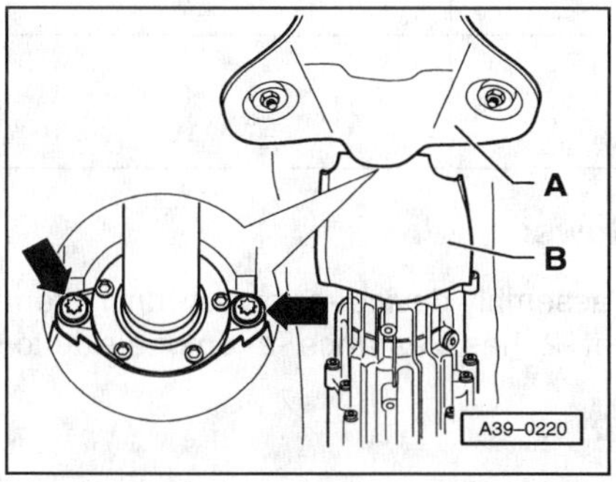

Remove heat shield -A- from vehicle underbody.

Driveshaft with tube diameter 60 mm front and rear

Do not set assembly tool over balancing weights.

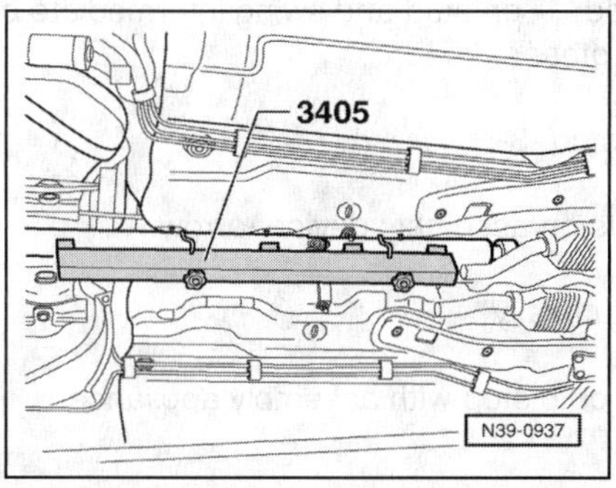

Install 3405 Driveshaft assembly appliance and tighten the plastic nuts.

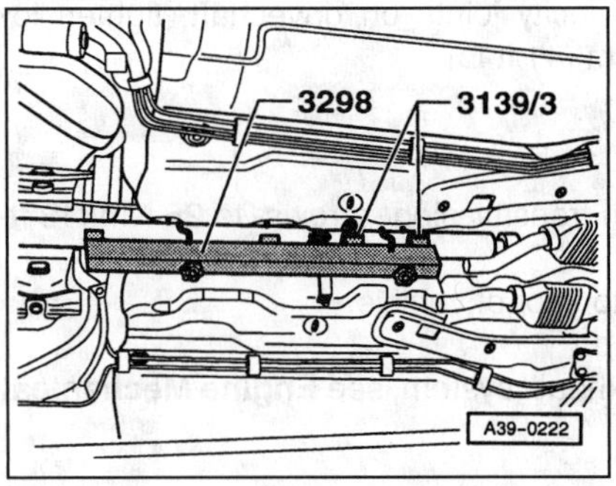

Driveshaft with tube diameter 60 mm front and 70 mm rear

– Do not set assembly tool over balancing weights.

Install 3298 Driveshaft assembly appliance with 3139/3 Adapters and tighten plastic nuts.

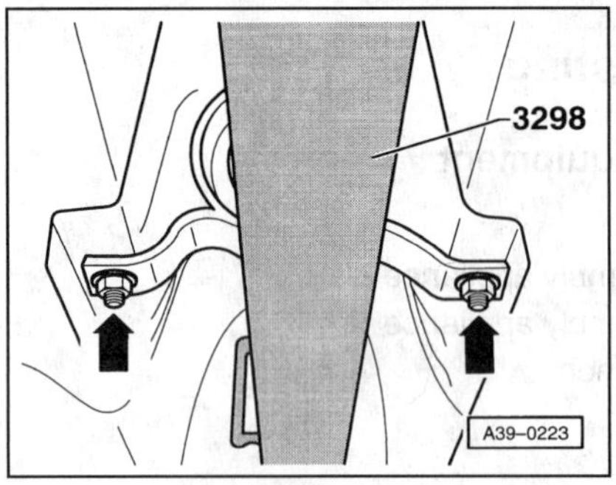

Continued for all vehicles

In the following figures, the adjustment of the driveshaft using 3298 Driveshaft assembly appliance is illustrated. The adjustment using 3405 Driveshaft assembly appliance is identical.

Remove intermediate mount securing nuts -arrows-.

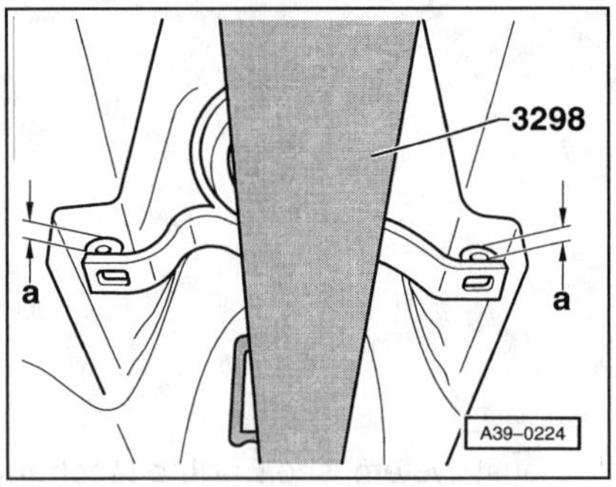

Measure dimension -a-.

- Left and right-hand sides must be same.

Table of shims

Allocate shims according to parts catalog.

Distance -a- measured value (mm)	Shim thickness (mm)
0 to 3.0 3.1 to 5.0 5.1 to 7.0	- 2 4
7.1 to 9.0 9.1 to 11.0 11.1 to 13.0	6 8 10

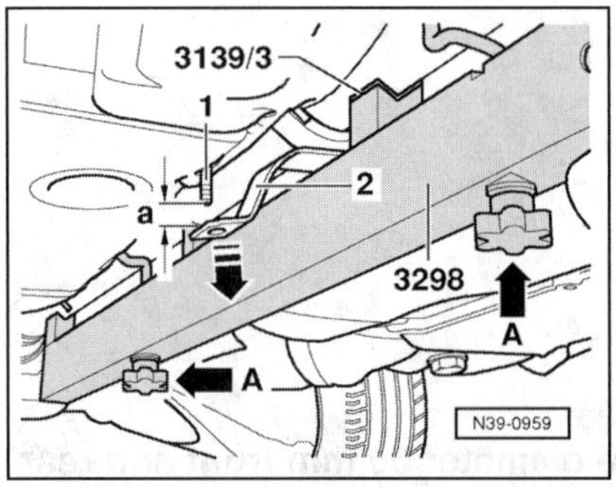

Install shims as follows:

Loosen plastic nuts on assembly device -arrows A- until intermediate bearing retaining bracket -2- has clearance -a- on one side to stud -1-.

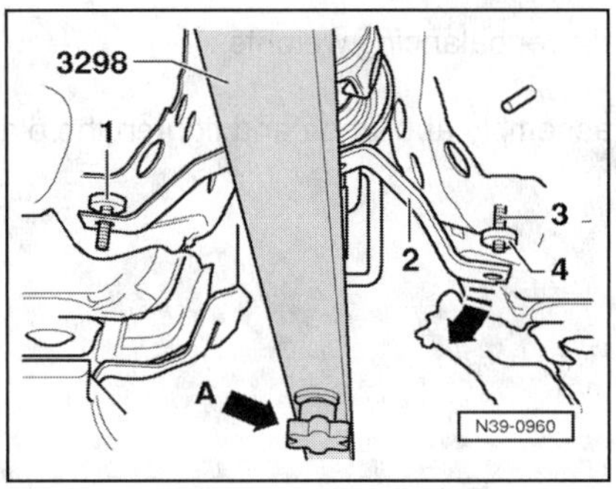

Place shim as measured -1- on stud and swing intermediate bearing retaining bracket -2- to other side.

– Place 2nd shim -4- on stud -3-.

– Tighten both plastic nuts on assembly device -arrow A-.

Center driveshaft longitudinally:

– Slide driveshaft to rear onto stop with assembly appliance.

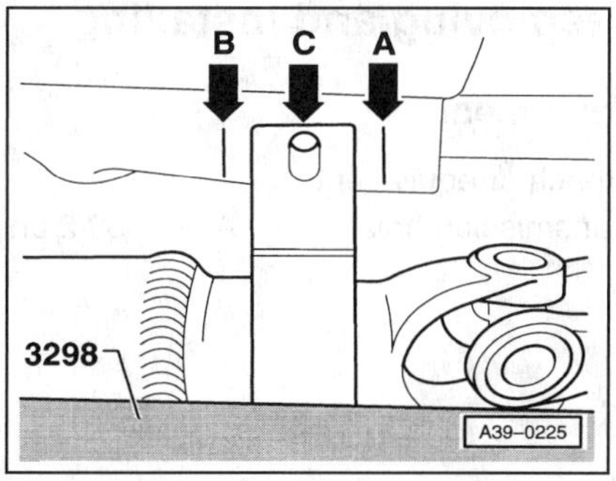

Mark position of intermediate mount on body (arrow -A-).

– Slide driveshaft forward onto stop with assembly appliance.

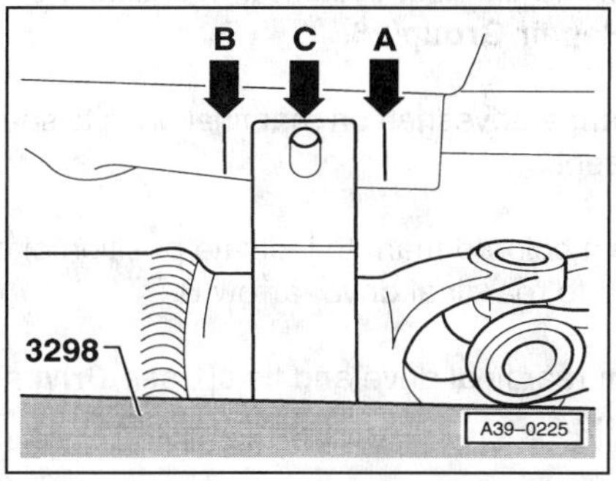

Mark position of intermediate mount on body (arrow -B-).

– Bolt intermediate mount in center position (arrow -C-), tighten to 25 Nm (18 ft-lb).

– Then remove assembly appliance.

– Install heat shield to vehicle floor.

– Install rear section of exhaust system, see **Engine Mechanical section, Repair Group 26.**

Rear Final Drive

Rear final drive oil, checking and filling

Special tools and equipment

- VAG1306 Drip tray
- VAG1331 Torque wrench

– Place VAG1306 Drip tray underneath.

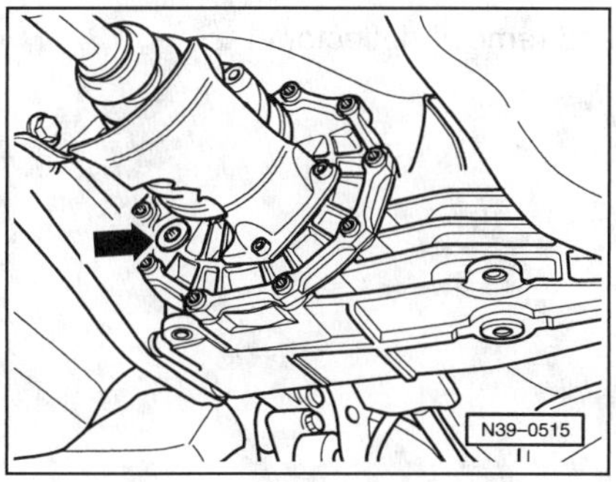

Remove plug -arrow- to check oil.

- The oil level is correct when the rear final drive is filled to the lower edge of the filler hole.

– Install plug -arrow-, tighten to 35 Nm (26 ft-lb).

– When filling with new oil note following:

- Remove plug -arrow-.
- Top up oil to lower edge of filler hole.
- Install plug -arrow-, tighten to 35 Nm (26 ft-lb)

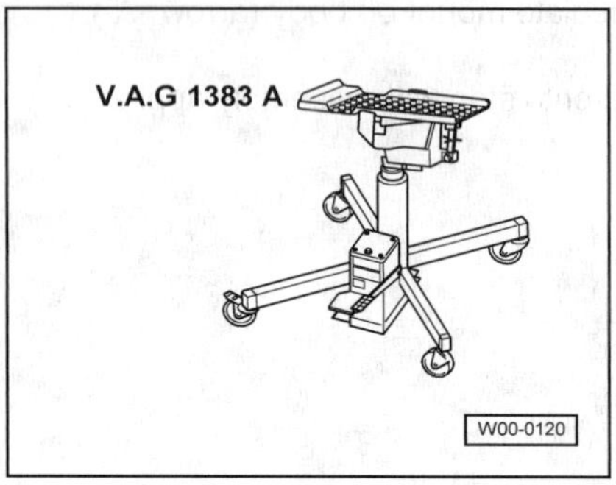

Rear final drive, removing and installing

Special tools and equipment

- V.A.G 1332 Torque wrench or equivalent

V.A.G 1383 A engine/transmission hoist with V.A.G 1359/2 universal transmission mounting

Removing

– Remove rear section of the exhaust system if necessary, see **Engine Mechanical section, Repair Group 26.**

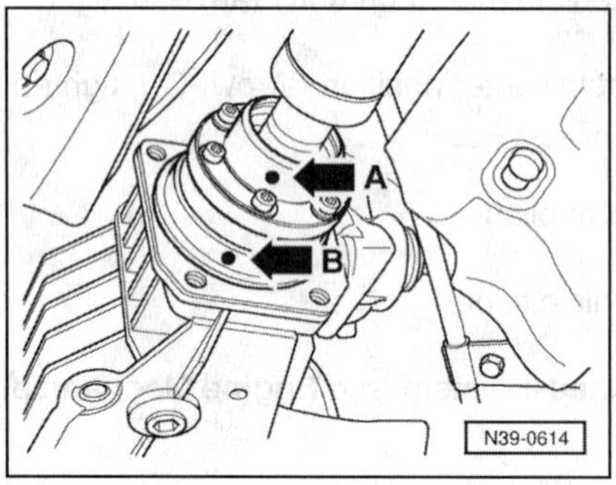

Check driveshaft and flange/driveshaft on rear final drive to see whether a mark (red dot) is present.

– If there is no dot, make a colored mark to indicate position of driveshaft flange -arrow A- relative to rear final drive -arrow B-.

– Remove driveshaft from rear final drive and tie up, see **Driveshaft, removing and installing**.

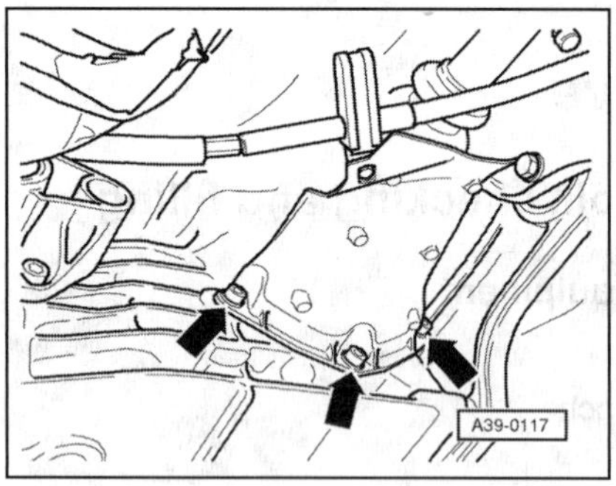

Remove deflectors on left and right of rear final drive -arrows-.

– Unscrew left and right drive shafts and tie up.

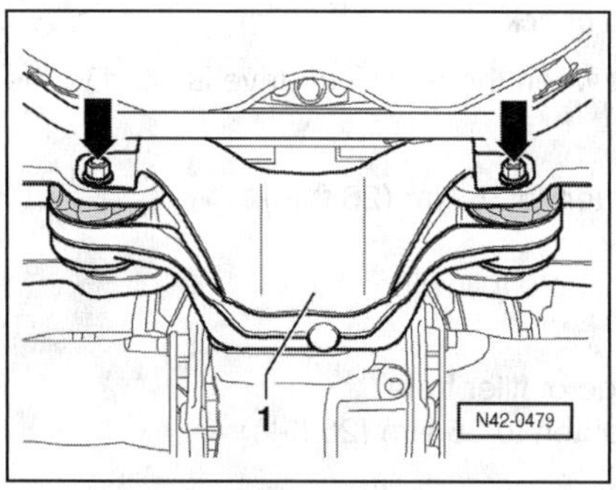

Remove nuts -arrows- and remove deflector -1-.

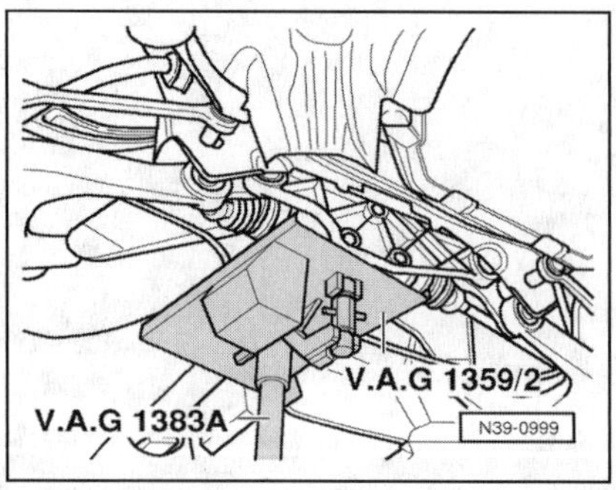

Position V.A.G 1383 A transmission hoist and 1359/2 universal transmission mounting under rear final drive.

– Strap rear final drive to universal transmission mounting.

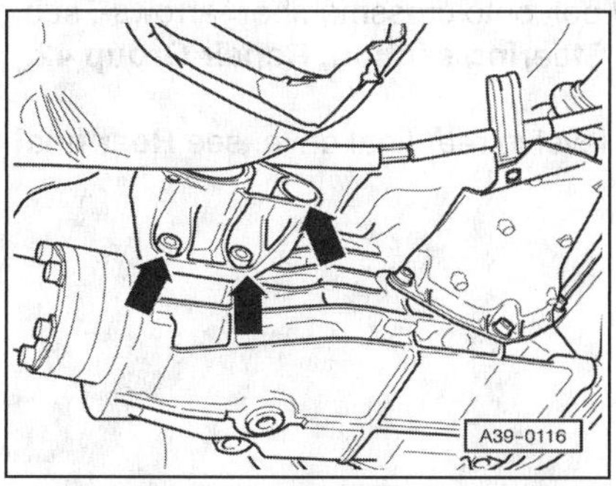

Remove transmission support mounting bolts -arrows-.

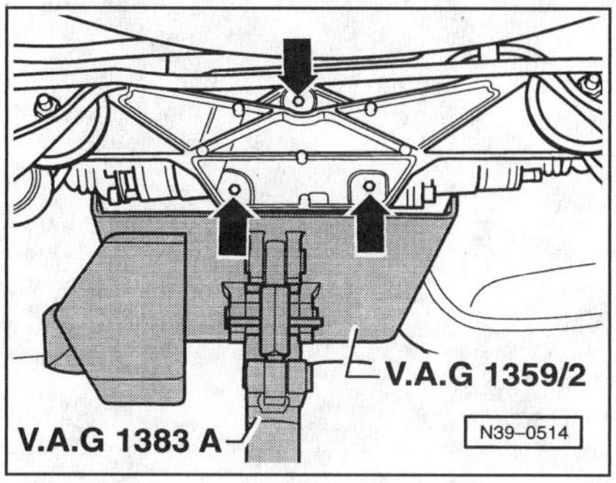

Remove crossmember mounting bolts -arrows-.

– Lower rear final drive.

Rear final drive, installing

Installation is performed in the reverse order of removal. Note the following:

Always replace the mounting bolts for the driveshaft.

– Clean any adhesive residue from tapped holes in flanged shaft on rear final drive. This can be done using a tap thread.

The mounting bolts may otherwise break off when they are threaded in.

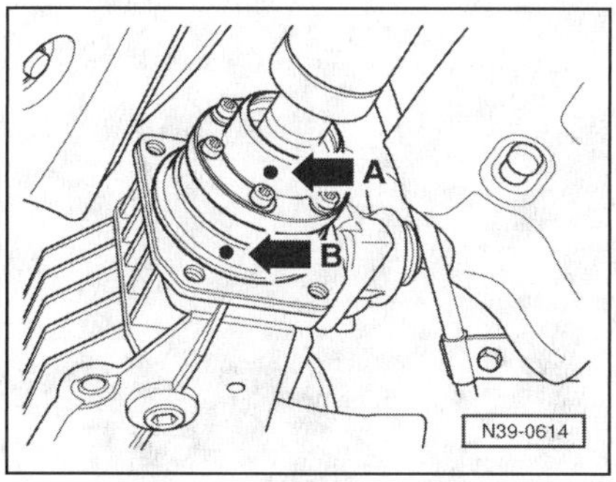

Ensure that marks on driveshaft flange (arrow-A-) and rear final drive flange (arrow-B-) are in line.

– Replace seals for drive shafts and driveshaft (remove protective film and stick seal in position).

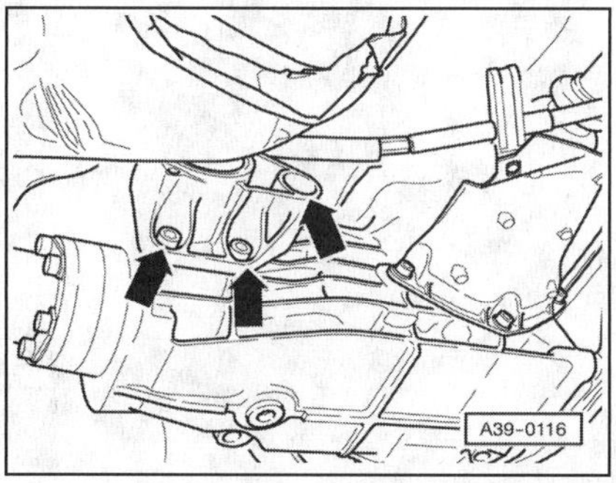

Tighten transmission support mounting bolts -arrows-, see **Suspension, Wheels, Steering section, Repair Group 42.**

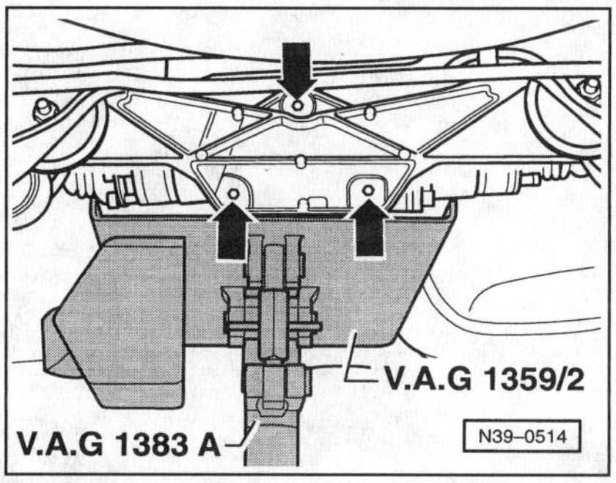

Bolt rear final drive to crossmember -arrows-, see **Suspension, Wheels, Steering section, Repair Group 42.**

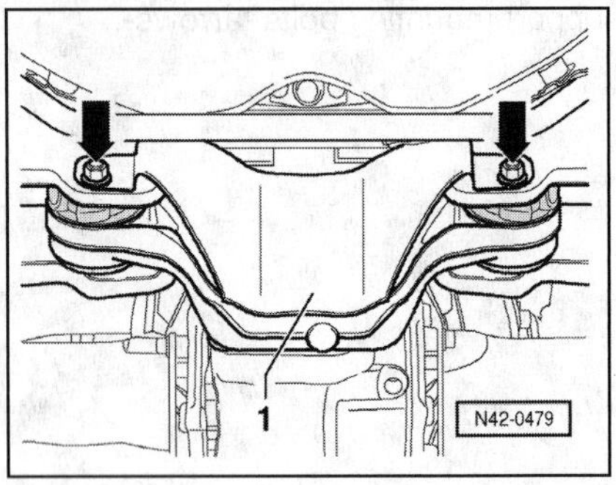

Install deflector -1- and bolt onto crossmember -arrows-, see **Suspension, Wheels, Steering section, Repair Group 42.**

– Check transmission oil level in rear final drive, see **Rear final drive oil, checking**.

6-Speed Manual Transmission 01E, All Wheel Drive

Section **Q**

39 Final Drive, Differential — Page

Transmission identification

The manual transmission 01E, all wheel drive (4 MOTION) is installed in the Passat 2000 ➤ as a 6-speed transmission in conjunction with the 4-cylinder, 6-cylinder Diesel and 8-cylinder engines.

EDITOR'S NOTE —
At the time of publication, the 01E 6-speed manual transmission was only available with the 8-cylinder (W8) engine for the North American market.

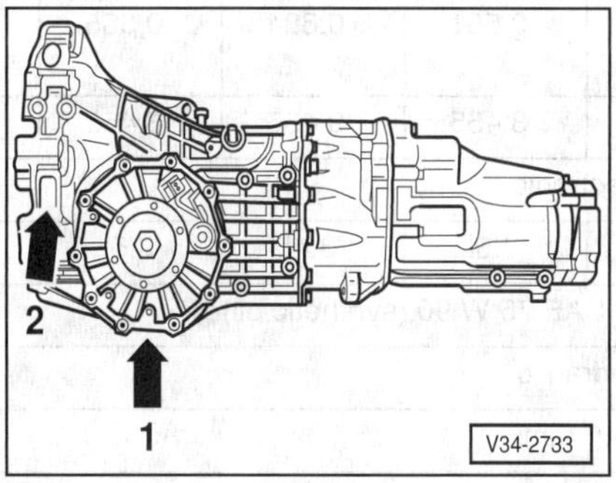

⊲ Location on transmission

Code letters and date of manufacture (**arrow 1**)
Manual transmission 01E (**arrow 2**)

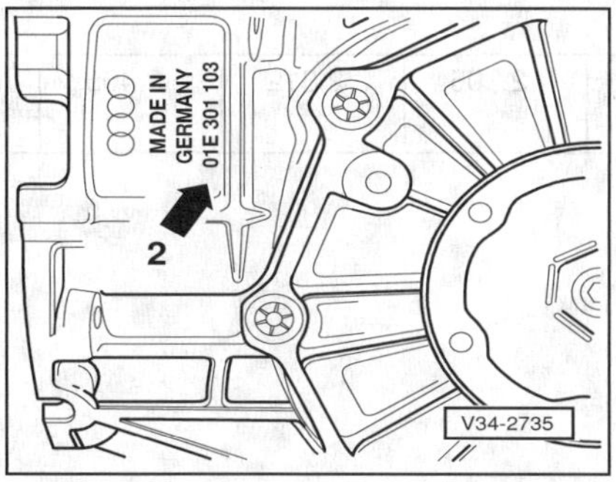

⊲ Manual transmission 01E (**arrow 2**)

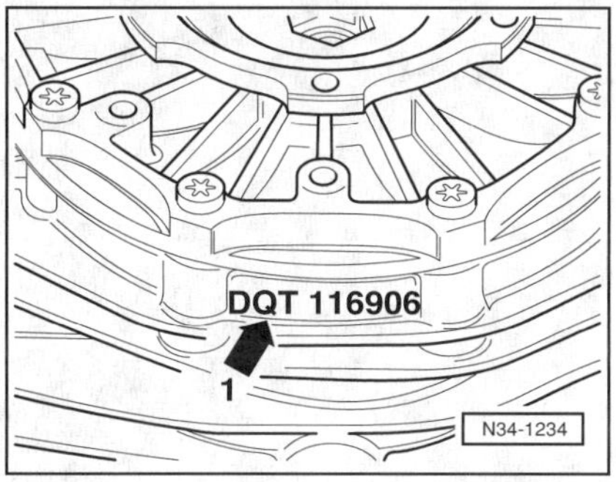

⊲ Code letters and transmission serial number (**arrow 1**)

Additional data depends on manufacture.

NOTE —
The transmission code letters are also shown on the vehicle identification plates.

Code letters, transmission application, ratios, capacities

Transmission 01E all wheel drive Code letters		DQT	EHW	FCP	ENK	FTF	FRR	FTS	GLX
Manufactured	from to	11/99 04/01	01/00 11/00	08/00 04/01	01/01 04/01	05/01 -	05/01 -	05/01 -	01/03 -
Application	Model	Passat 2000 ➤	Passat 2000 ➤	Passat 2000 ➤	Passat 2000 ➤	Passat 2000 ➤	Passat 2000 ➤	Passat 2000 ➤	Passat 2000 ➤
	Engine	2.5L - 110 kW V6 diesel	1.9L - 85 kW diesel	1.9L - 96 kW diesel	4.0L - 202 kW W8	2.5L - 110 kW diesel	4.0L - 202 kW W8	1.9L - 96 kW diesel	2.5L - 132 kW V6 diesel
Ratio:	Final drive	4.111	4.111	4.111	4.111	4.111	4.111	4.111	3.889
	1st Gear	3.500	3.500	3.500	3.500	3.500	3.500	3.500	3.750
	2nd Gear	1.889	1.889	1.889	1.889	1.889	1.889	1.889	2.059
	3rd Gear	1.231	1.231	1.231	1.231	1.231	1.231	1.231	1.320
	4th Gear	0.871	0.933	0.933	0.967	0.871	0.967	0.933	0.933
	5th Gear	0.667	0.730	0.730	0.806	0.667	0.806	0.730	0.711
	6th Gear	0.561	0.600	0.585	0.684	0.561	0.684	0.585	0.561
	Reverse Gear	3.455	3.455	3.455	3.455	3.455	3.455	3.455	3.455
Speedometer		electronic							
Capacity		2.5L	2.5L	2.5L	3.4L	2.5L	3.4L	2.5L	2.5L
Specification		Gear oil G 052 911 A 1 SAE 75 W 90 (synthetic oil)							
Clutch control		hydraulic							
Clutch plate diameter		240 mm							
Drive axle flange diameter		130 mm (triple roller axle shafts)							
Overall ratio in highest gear		2.306	2.467	2.405	2.812	2.306	2.812	2.405	2.182

30 Clutch

Repair Group

Q 30

Clutch

Clutch system, bleeding

EDITOR'S NOTE—
Brake fluid is used for the clutch hydraulic system.

Special tools and equipment

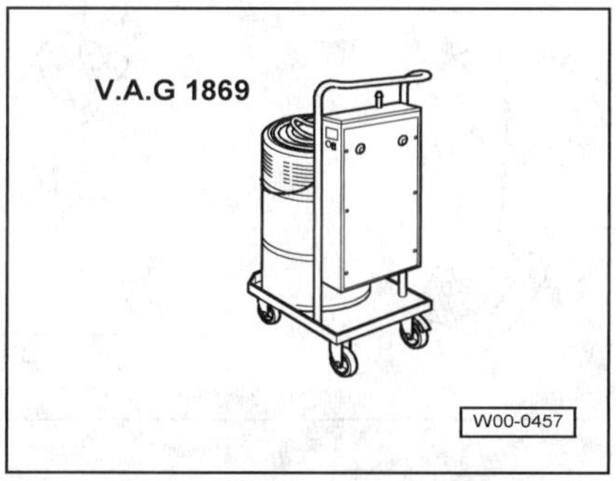

- VAS 5234 or V.A.G 1869 Brake charge and bleed unit.

– Brake fluid specifications, see **Brake System section, Repair Group 47**.

NOTE—
- *Open bleeder valve before operating bleeder unit.*
- *Clutch pedal is not operated.*

– Connect brake filling and bleeder unit VAG 1869.

– Remove sound insulation below transmission.

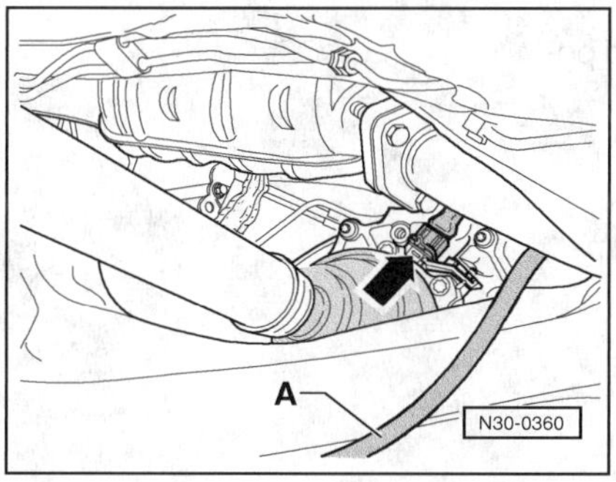

Vehicles with 6- or 8-cylinder engine

– Remove cover from left wheel housing.

– Remove left-hand axle shaft protective plate.

- Pull connector off speedometer sensor (**arrow**).

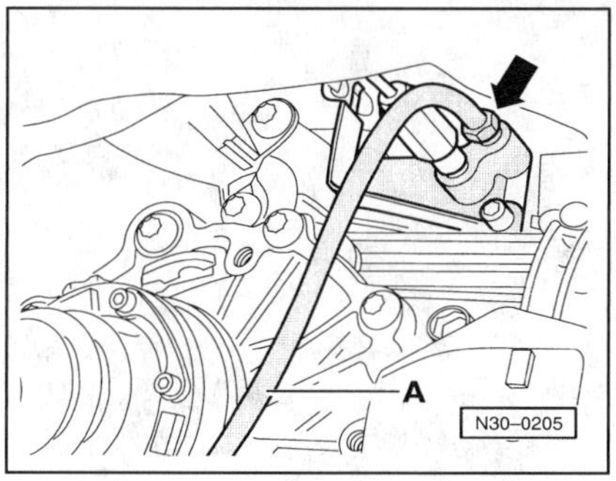

- Guide brake bleed hose -**A**- of fluid collection tank (e.g. V.A.G 1794) between front left exhaust pipe and transmission, and fit onto bleed valve.

Continued for all vehicles

– Connect brake bleed hose -A- of fluid collection tank (e.g. V.A.G 1794) to slave cylinder (arrow) and open bleed valve.

– Switch on bleeder unit and let about 100 cm^3 brake fluid run out.

NOTE—
Brake fluid must not be allowed to get onto the transmission.

– Depress clutch pedal several times after completion of bleeding process.

Pedal cluster, assembly overview

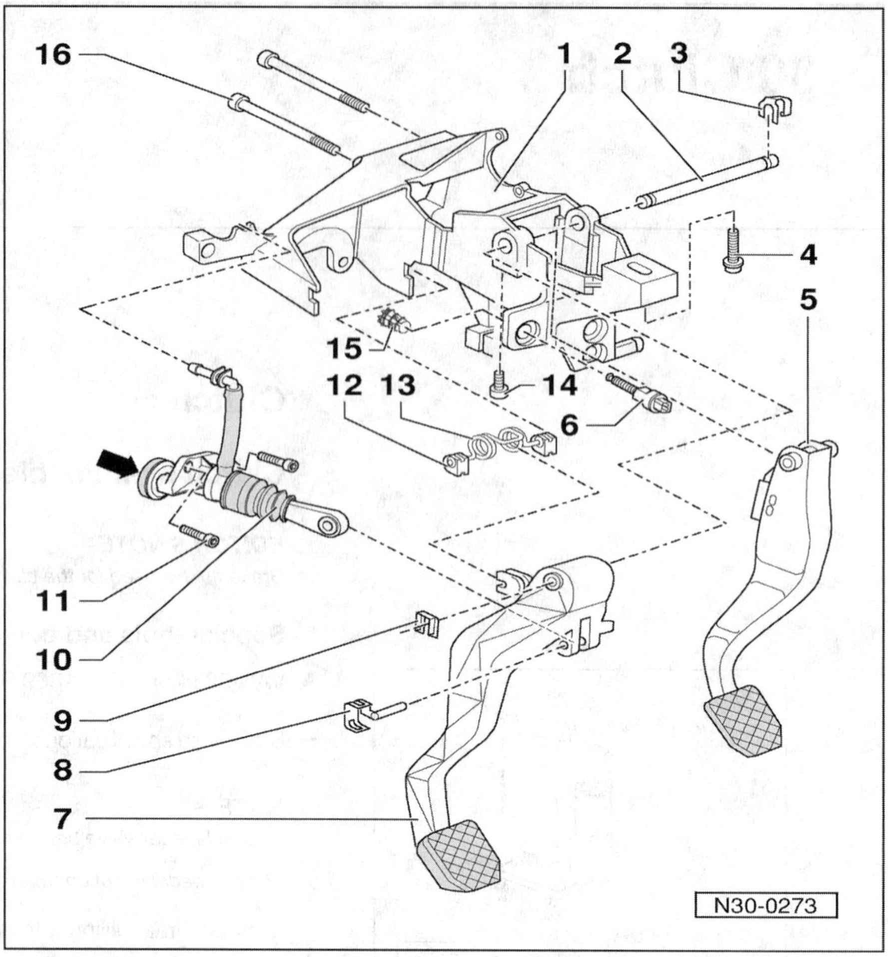

1. **Mounting bracket**
   - Before removing, disconnect steering column from steering box, see **Suspension, Wheels, Steering section, Repair Group 48**
2. **Pivot pin for clutch & brake pedal**
3. **Locking clip**
4. **Hex socket head bolt,**
   - Tighten to 25 Nm (18 ft-lb)
5. **Brake pedal**
   - Maintenance-free bearing; do not lubricate
6. **Clutch pedal switch**
   - Checking setting, see **Master cylinder, removing and installing**
7. **Clutch pedal**
   - Removing and installing, see **Clutch pedal and over-center spring, removing and installing**
8. **Pin**
   - Always replace
9. **Locking clip**
10. **Master cylinder**
    - Removing and installing, see **Master cylinder, removing and installing**
    - Do not operate clutch pedal after slave cylinder has been removed
    - Seal (**arrow**) must not be removed
11. **Hex socket head bolt**
    - Tighten to 20 Nm (15 ft-lb)
12. **Mounting**
    - Insert in mounting bracket with over-center spring
    - Installed position: deep groove faces downward
13. **Over-center spring**
    - Remove and install with clutch pedal, see **Clutch pedal and over-center spring, removing and installing**
14. **Hex socket head bolt**
    - Tighten to 5 Nm (44 in-lb)
    - Secures clutch and brake pedal pivot pin
    - Tighten only after locking clips, item -**3**- and item -**9**- are in position
15. **Securing clip**
    - Holds clutch switch in adjusted position
16. **Torx screw**
    - Tighten to 25 Nm (18 ft-lb)
    - Also secures brake master cylinder with brake servo

Clutch pedal and over-center spring, removing and installing

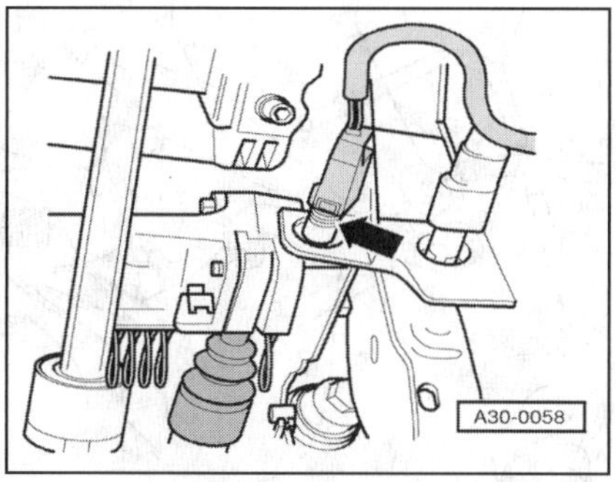

– Remove cover below instrument panel, see **Body Interior section, Repair Group 70**

Ensure that the clutch pedal switch (**arrow**) is not pressed out of the securing clip. If the switch is pressed out the threads will be damaged and the switch must be replaced.

NOTE—
In order to guarantee sufficiently secure seating, you can only install the clutch pedal switch one time only.

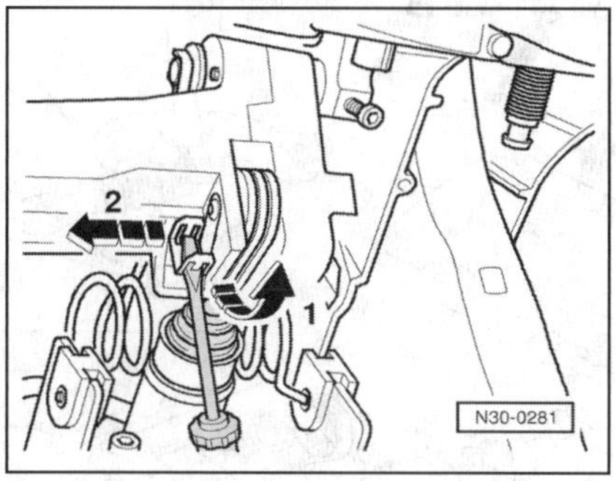

Disconnect clutch pedal from master cylinder by prying out securing clip for pin with screwdriver (**arrow 1**) and pulling out pin (**arrow 2**).

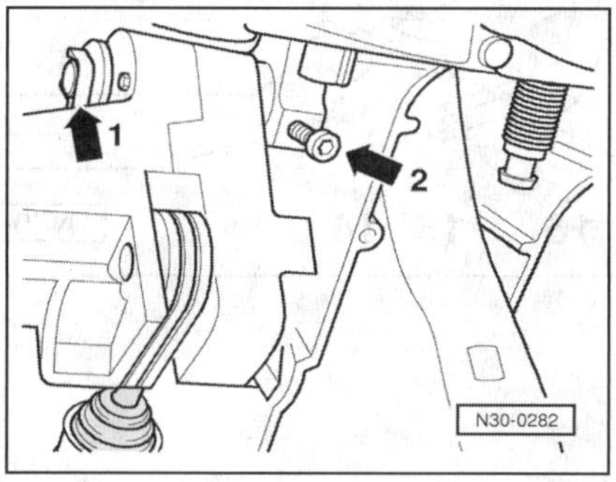

Press clutch pedal securing clip (**arrow 1**) off pivot pin with a screwdriver.

– Remove bolt (**arrow 2**).

– Press clutch and brake pedal pivot pin out to right, until clutch pedal can be removed.

– If necessary, press brake pedal securing clip off pivot pin with a screwdriver.

– Take out clutch pedal and over-center spring.

Installation is performed in the reverse sequence. Noting the following:

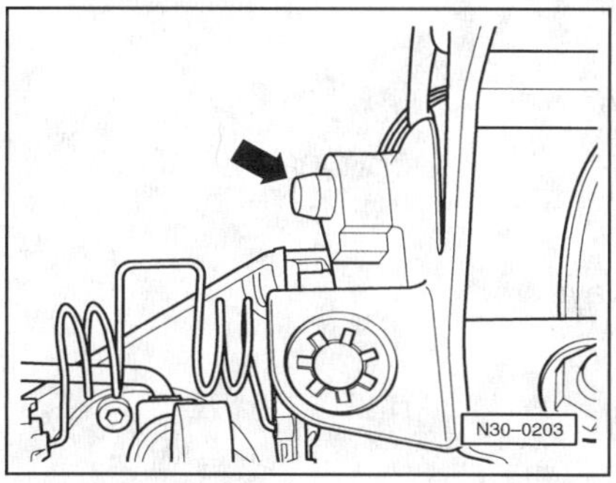

Align clutch pedal/brake pedal pivot pin so that it protrudes out of mounting bracket at clutch pedal side (**arrow**).

– Press clutch pedal onto stop and insert over-center spring in its seat on clutch pedal.

– Position clutch pedal on pivot pin and push it to left until securing clip can be attached.

– Connect master cylinder to clutch pedal and engage securing clip for pin in clutch pedal.

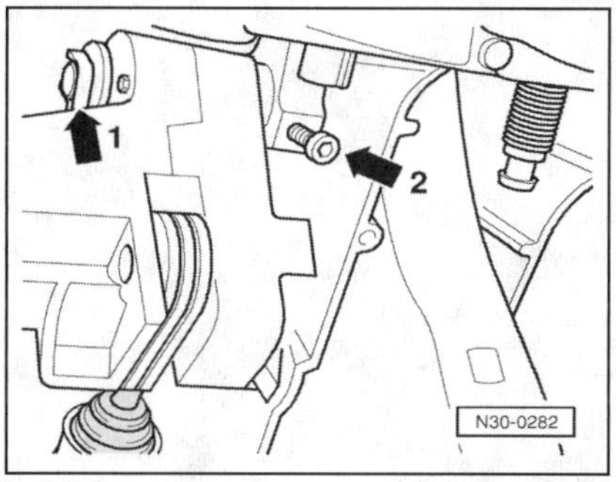

Tighten bolt (**arrow 2**) only after both securing clips (**arrow 1**) for clutch pedal; clip for brake pedal not visible) have been installed on clutch pedal/brake pedal pivot pin.

– Adjust clutch pedal switch above clutch pedal, see **Master cylinder, removing and installing**.

Hydraulic system, assembly overview

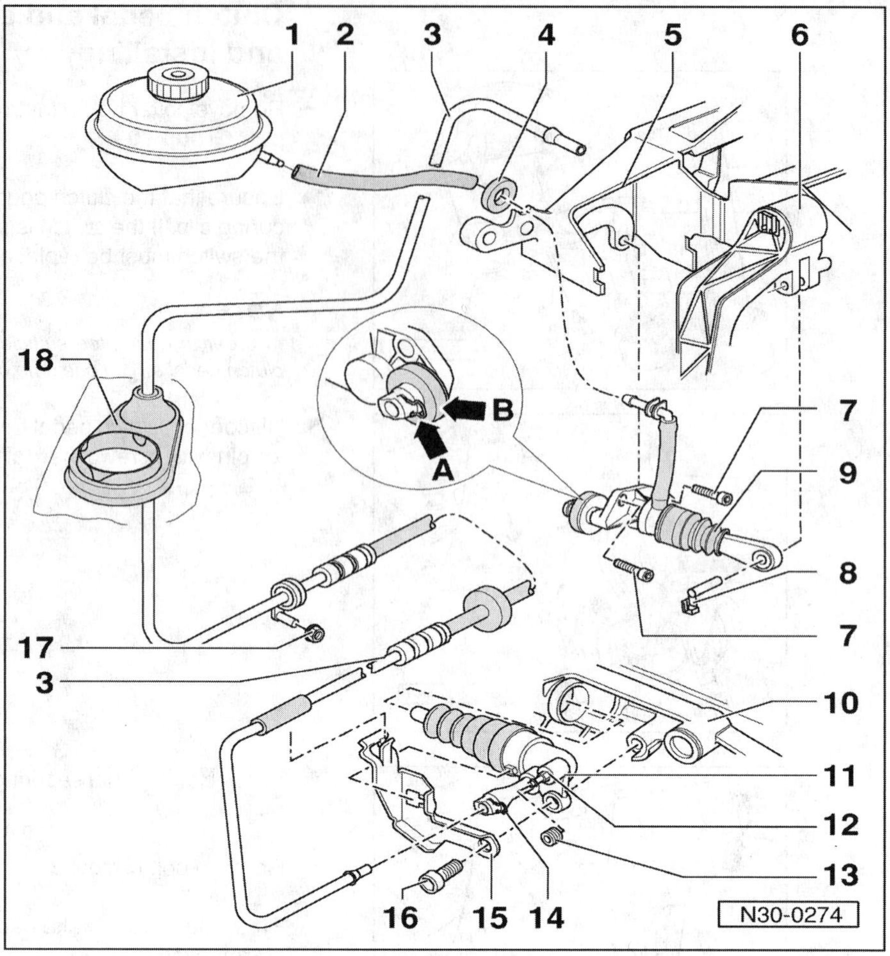

1. **Brake fluid reservoir**
2. **Supply hose**
3. **Line/hose assembly**
   - For connectors on master and slave cylinders
   - Separating and attaching to master cylinder, see **Master cylinder, removing and installing**
   - Separating from slave cylinder, see **Fig. 2**
4. **Bushing**
5. **Mounting bracket**
6. **Clutch pedal**
   - Removing and installing, see **Clutch pedal and over-center spring, removing and installing**
7. **Hex socket head bolt**
   - Tighten to 20 Nm (15 ft-lb)
8. **Pin**
   - Always replace
   - Lubricate with polyurea grease
9. **Master cylinder**
   - Removing and installing, see **Master cylinder, removing and installing**
   - Do not operate clutch pedal after slave cylinder has been removed.
   - Clip (**arrow A**) and seal (**arrow B**) must not be disassembled
   - To remove line/hose assembly (**arrow A**) pull out to stop, see **Master cylinder, removing and installing**
10. **Transmission**
11. **Slave cylinder**
    - Do not operate clutch pedal after slave cylinder has been removed.
    - Brake fluid must not be allowed to get onto transmission
    - To install, lubricate lightly, see **Fig. 1**
    - When installing, push in until securing bolt (item **16**) can be installed
    - Separate from line/hose assembly, see **Fig. 2**
    - See **Slave cylinder, removing and installing**
    - See **Clutch system, bleeding**
    - Vehicles with an 8-cyl. injection engine have an additional heat shield installed
12. **Bleeder valve**
    - Tighten to 4 Nm (35 in-lb)
    - See **Clutch system, bleeding**
13. **Dust cap**
14. **Clip**
    - To remove line/hose assembly, pull out clip to stop, see **Fig. 2**
15. **Bracket**
    - For line/hose assembly
    - Secured to transmission with slave cylinder
16. **Bolt**
    - Tighten to 25 Nm (18 ft-lb)
    - Slave cylinder on transmission
    - Self-locking
    - Always replace
17. **Hex nut**
18. **Seal**

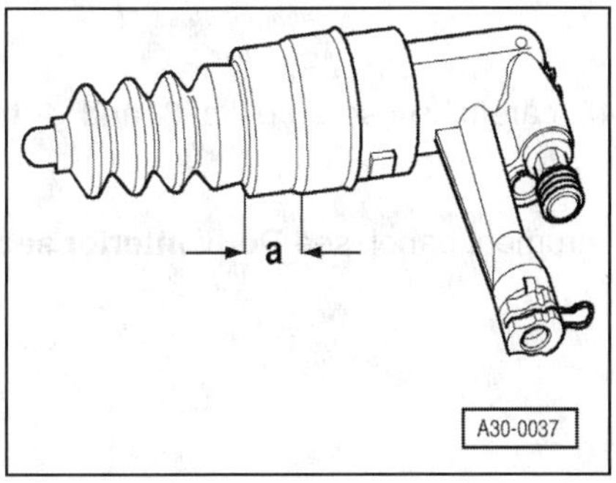

Fig. 1 Installing slave cylinder

– Before installing slave cylinder in transmission housing, coat area -a- of boot with Lithium grease G 052 150 A2.

– Apply a thin coat of copper grease to surface of clutch release lever where it contacts plunger.

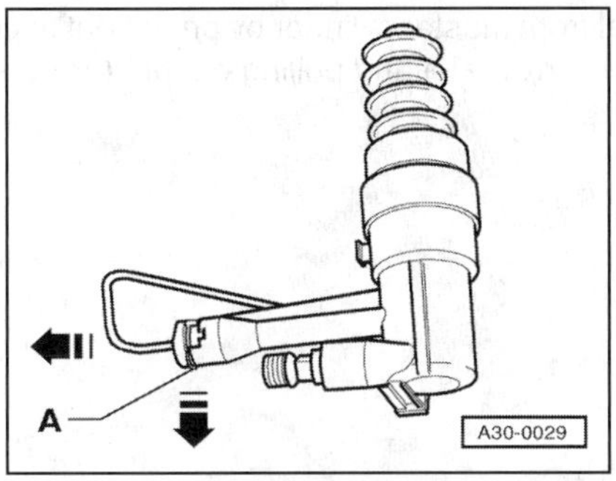

Fig. 2 Separating slave cylinder from line/hose assembly

Removing

– To remove line/hose assembly, pry out clip -A- to stop using a screwdriver. Pull out line-hose assembly.

Installing

– Press in clip -A- to stop.

– Push line/hose assembly into slave cylinder until assembly audibly engages.

– Test line/hose assembly for secure fit by tugging on it.

Master cylinder, removing and installing

Special tools and equipment

• 3094 Hose clamps
• V.A.G 1331 Torque wrench

Removing

– Check whether a coded radio is installed. If so, obtain anti-theft coding.

– With ignition switched off disconnect battery Ground strap.

– Lay a sufficiently lint-free cloth in plenum chamber below return hose to brake fluid reservoir.

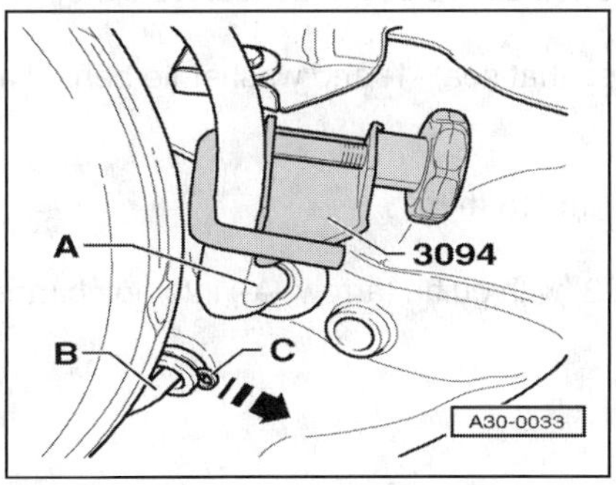

– Clamp shut return hose -A- to brake fluid reservoir using hose clamp 3094 and pull connecting line off master cylinder and seal off.

– Remove grommet for return hose -A- from bulkhead.

– Pull out clip -C- to stop using screwdriver and then pull out line/hose assembly -B- slightly.

– Remove harness connector for brake fluid level sensor from brake fluid reservoir.

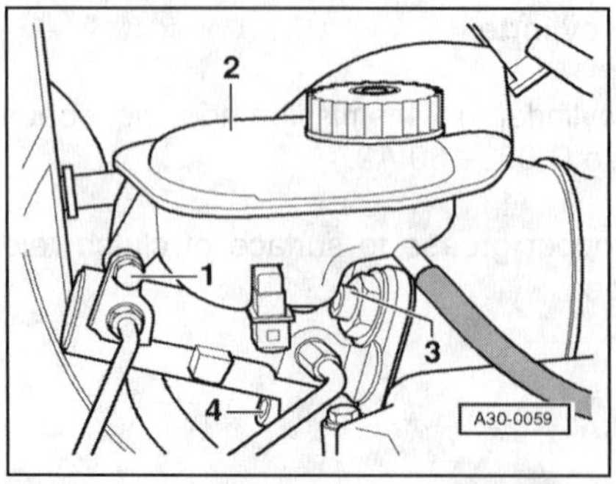

- Remove bolt -1-.

- Using inside Torx bit T 45, carefully unscrew bolts -3- and -4- from brake servo.

- Remove cover below instrument panel, see **Body Interior section, Repair Group 70.**

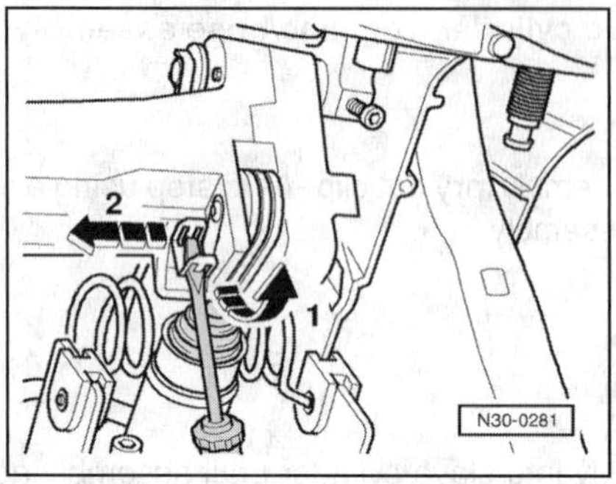

- Disconnect clutch pedal from master cylinder by prying out securing clip for pin with screw driver (arrow -1-) and pulling out pin (arrow -2-).

- Ensure that the clutch pedal switch (-1-) is not pressed out of the securing clip. If the switch is pressed out the threads will be damaged and the switch must be replaced.

NOTE —
To ensure sufficiently secure seating, a clutch pedal switch -1- may be installed only once.

- Unbolt (arrows -2-) master cylinder -4- from within passenger compartment.

- Unscrew mounting bracket securing bolt (arrow -3-).

- Pull complete pedal cluster slightly into passenger compartment.

- Remove master cylinder.

Master cylinder, installing

Installation is performed in the reverse sequence. Note the following:

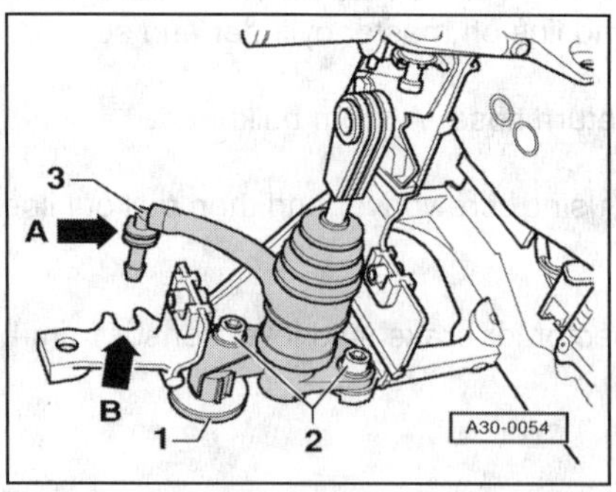

- Install master cylinder so that seal -1- and washer lie behind mounting bracket.

- Tighten bolts -2- to 20 Nm (15 ft-lb).

- Press connecting pipe -3- with guide (arrow -A-) into notch (arrow -B-) in mounting bracket.

A second technician is required for the next work step.

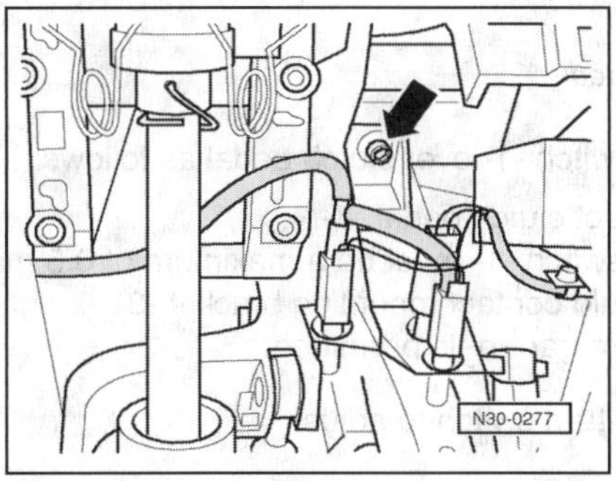

- Push complete pedal cluster forward to bulkhead and insert bolt (arrow).

NOTE —
During this step, from the plenum chamber, the second technician must guide line/hose assembly into master cylinder.

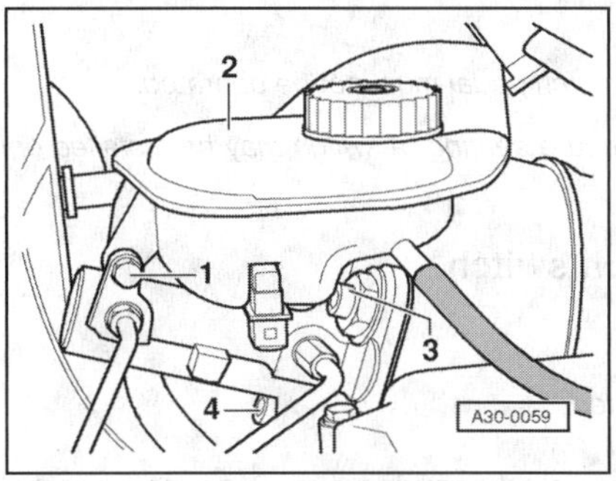

- Tighten brake servo securing bolts -3- and -4- to 25 Nm (18 ft-lb).

– Tighten securing bolt -1- for brake fluid reservoir -2-.

– Attach connector to brake fluid level sensor on brake fluid reservoir.

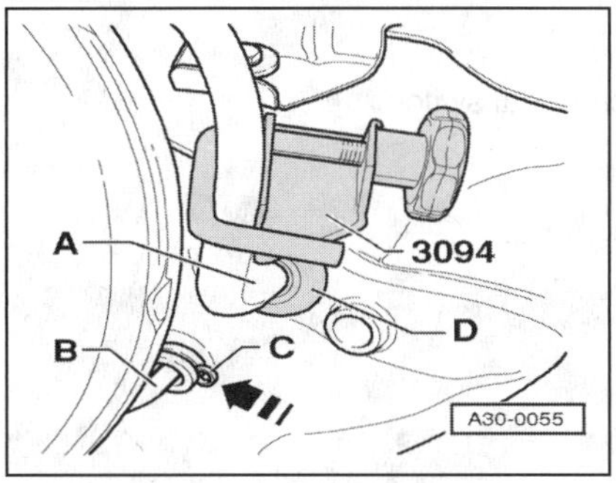

- Press clip -C- in direction of arrow onto limit stop.

– Push line/hose assembly -B- into master cylinder until it audibly engages.

– Test line/hose assembly by tugging on it.

– Insert rubber grommet -D- for return hose -A- into bulkhead.

– Push return hose -A- to stop on connecting pipe of master cylinder.

– Remove hose clamp 3094.

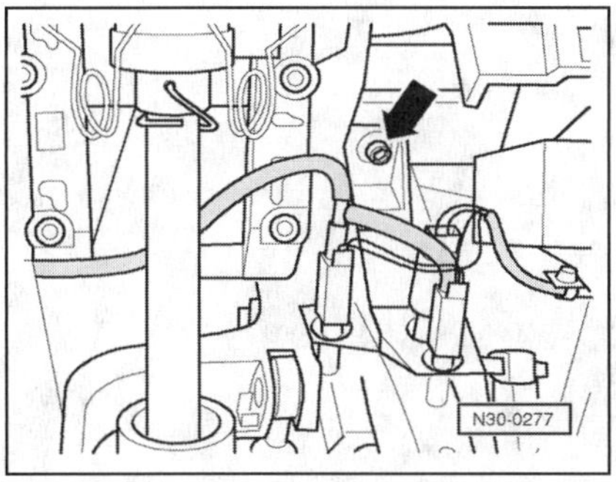

- Tighten bolt (arrow) for securing pedal cluster on instrument panel to 25 Nm (18 ft-lb).

– Connect master cylinder to clutch pedal. Engage new pin in clutch pedal.

– Bleed clutch system after installing master cylinder, see **Clutch system, bleeding**.

Checking and adjusting clutch pedal switch

NOTE—
Clutch pedal is not operated.

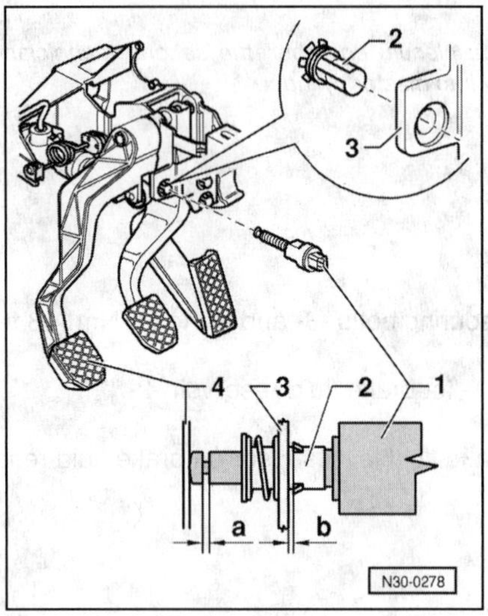

◄ Check adjustment of switch -1- over clutch pedal as follows:
- Switch -1- must contact clutch pedal -4-.
- Required play -a- at switch -1- must be a maximum of 0.5 mm.
- Securing clip -2- should contact mounting bracket -3-.
- A maximum of 0.5 mm gap -b- is permitted.

– Hold securing clip and turn switch to adjust.

NOTE—
- *During adjustment, the clutch pedal must not be operated.*
- *To ensure sufficiently secure seating, a switch may be installed only one time.*

Installing new clutch switch

NOTE—
Clutch pedal is not operated.

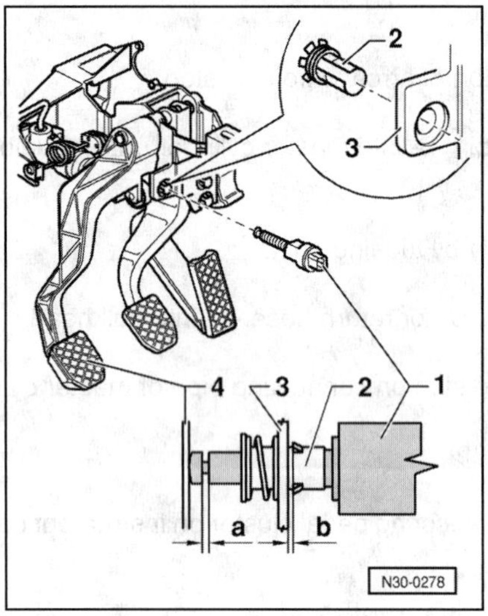

◄ Press switch -1- into mounted securing clip -2- to stop against clutch pedal.

– Check and adjust clutch pedal switch.

– Attach connector.

Clutch release mechanism, servicing

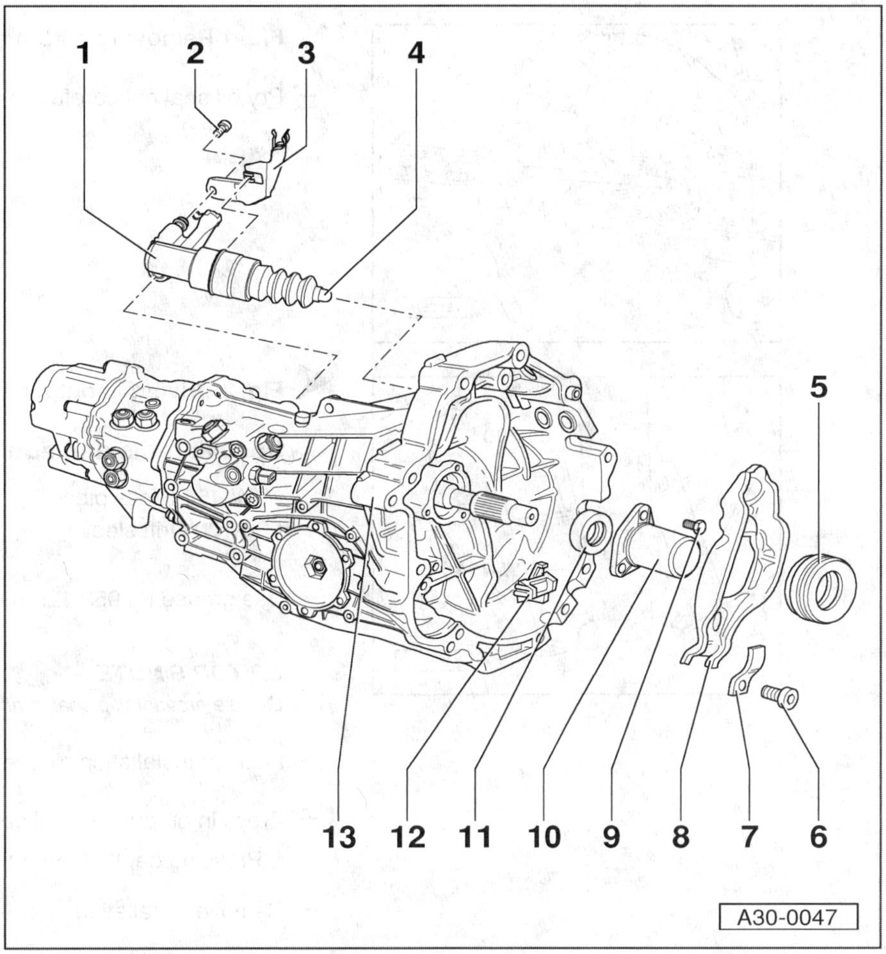

1. **Slave cylinder**
   - Removing and installing, see **Slave cylinder, removing and installing**
   - When installing, push in until securing bolt -item 2- can be installed.
   - After installing, see **Clutch system, bleeding**
   - Vehicles with an 8-cylinder engine have an additional heat shield installed
2. **Bolt**
   - Tighten to 25 Nm (18 ft-lb)
3. **Bracket**
   - For line/hose assembly
   - Secured to transmission with slave cylinder
4. **Plunger**
   - Coat end of plunger with copper grease
5. **Release bearing**
   - Do not wash out bearing, only wipe
   - Replace if noisy
   - Engage retaining tabs of release bearing in release lever
   - Re-glue plastic ring to bearing race with adhesive AMV 195 KD1 01 before installing transmission, see **Fig. 4**
   - If there are grooves deeper than 0.5 mm, release bearing must be replaced
   - Servicing release bearing, see **Fig. 4**
6. **Bolt**
   - Tighten to 25 Nm (18 ft-lb)
   - Self-locking
   - Always replace
7. **Leaf spring**
8. **Clutch release lever**
   - Removing and installing, see **Fig. 3**
   - Before installing, thinly coat contact surface of slave cylinder plunger with copper grease
9. **Bolt, 15 Nm**
   - Qty. 3
   - Tighten to 15 Nm (11 ft-lb)
10. **Guide sleeve**
11. **Input shaft oil seal**
    - Removing, see **Fig. 1**
    - Driving in, see **Fig. 2**
    - Pressing depth: 3.5 mm
12. **Intermediate piece**
    - Replace if damaged
13. **Transmission**

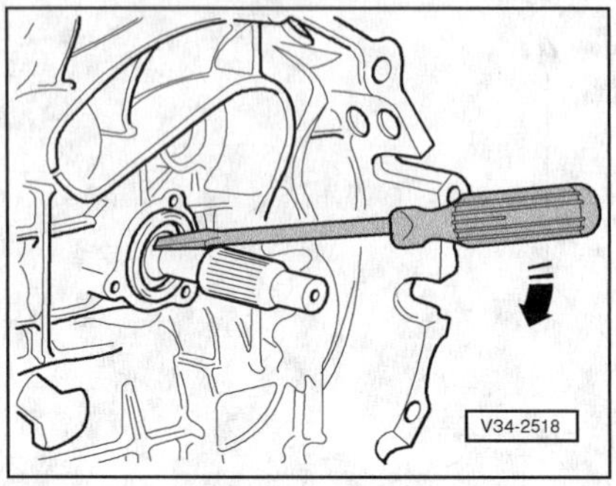

Fig. 1 Removing output shaft oil seal

– Pry oil seal out carefully with a screwdriver.

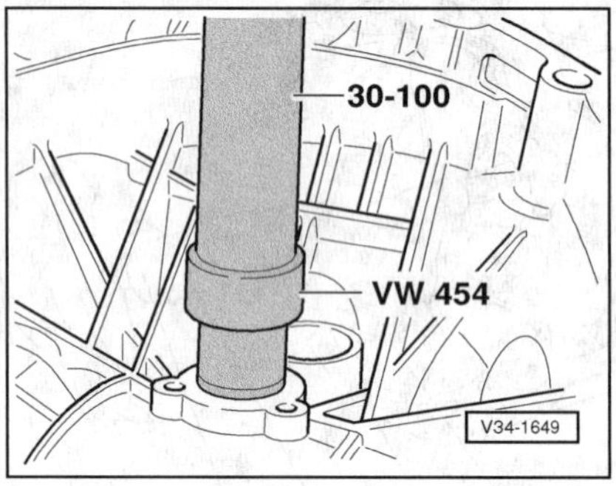

Fig. 2 Driving in output shaft oil seal

Special tools and equipment

- VW 454 Press piece
- 30-100 Drift sleeve

– Use grease (G 052 128 A1) to partially fill inside of new oil seal.

EDITOR'S NOTE —
Grease prevents lip seal spring from falling out during installation.

– Position installation sleeve over splines of output shaft.

– Press in output shaft oil seal using VW 454 and 30-100.
- Pressing depth: 3.5 mm

– Remove installation sleeve.

NOTE —
If the used output shaft is reused, repair pressing depth is 4.5 mm.

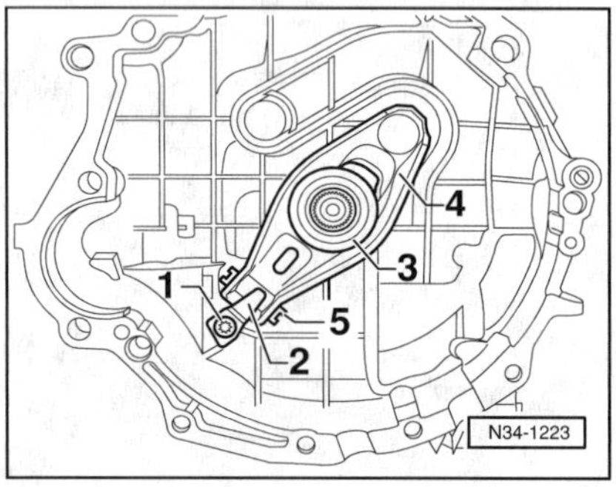

Fig. 3 Removing and installing clutch release lever

– Loosen bolt -1- and remove leaf spring -2-.

– Remove clutch release lever -4- together with release bearing -3- and intermediate piece -5-.

– Unhook retaining lugs of release bearing from release lever.

– Before installing clutch release lever, thinly coat contact surface of slave cylinder plunger with copper grease.

– Simultaneously set clutch release lever -4- and release bearing -3- on intermediate piece -5- as well as guide sleeve.

– Fit leaf spring -2- and tighten new bolt -1- to prescribed torque, see **Clutch release mechanism, servicing**, item -6-.

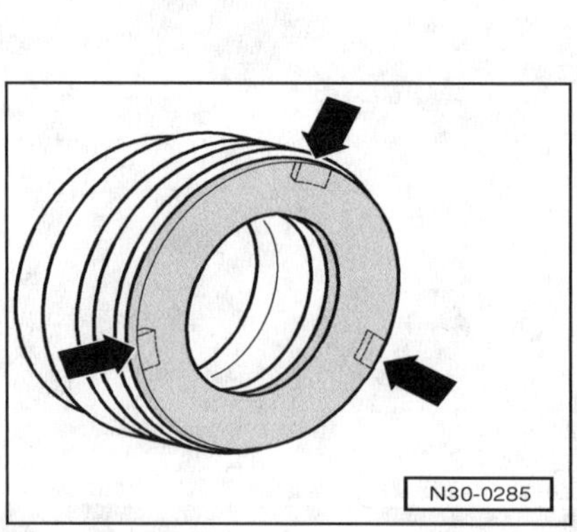

Fig. 4 Servicing release bearing

The three rectangular lugs (arrows) of the plastic ring engage in notches of the bearing race.

Slave cylinder, removing and installing

Removing

– Remove sound insulation below transmission.

– Remove front left exhaust pipe, see **Engine Mechanical section, Repair Group 26**.

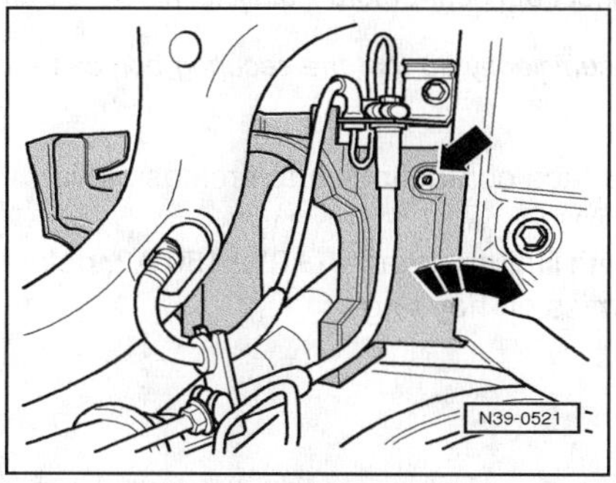

⊲ Remove cover from left wheel housing, if present (arrows).

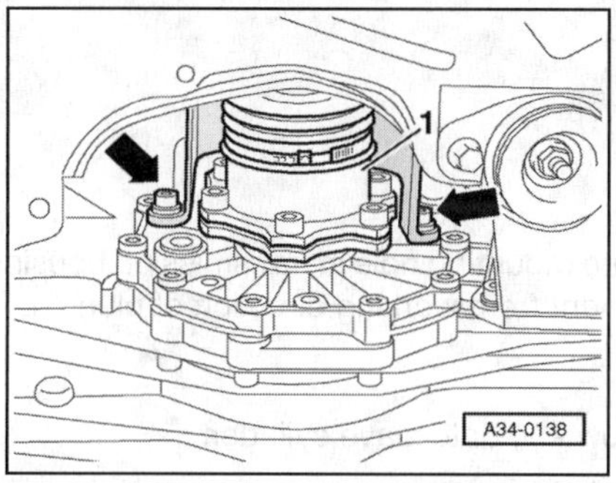

⊲ Remove left protective plate for axle shaft (arrows). Third bolt cannot be seen in picture.

– Disconnect left axle shaft -1- from transmission.

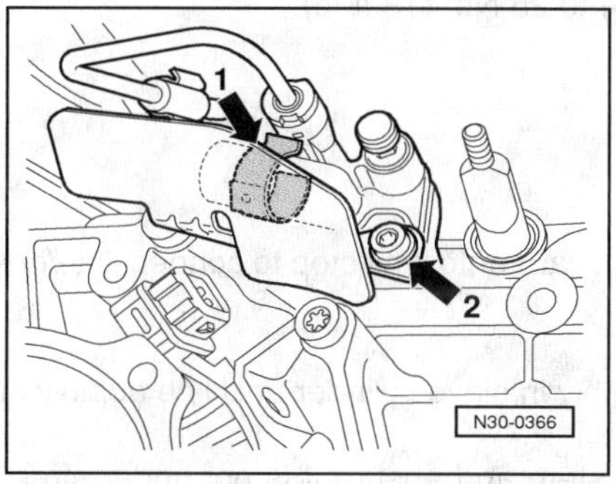

⊲ Remove bolt (arrow -2-) and pull out slave cylinder with heat shield (arrow -1- toward rear.

– Remove heat shield from slave cylinder.

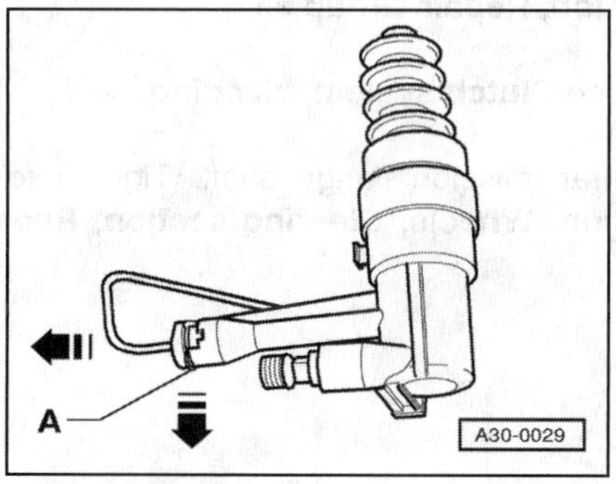

⊲ Pull out clip -A- to stop using screwdriver to remove line/hose assembly.

Slave cylinder, installing

Installation is performed in the reverse sequence. Note the following:

NOTE —

- *Brake fluid must not come in contact with the transmission. If necessary, clean the transmission housing.*
- *To simplify assembly, select 6th gear before installing the slave cylinder.*
- *Preload slave cylinder sufficiently so that the securing bolt can easily be inserted.*

– Lightly coat plunger surface contacting clutch release lever.

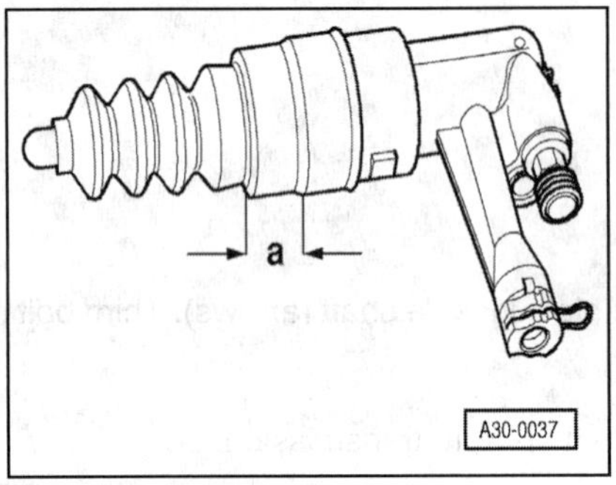

◄ Coat area -a- of boot with lithium grease, G 052 150 A2, before installing slave cylinder in transmission housing.

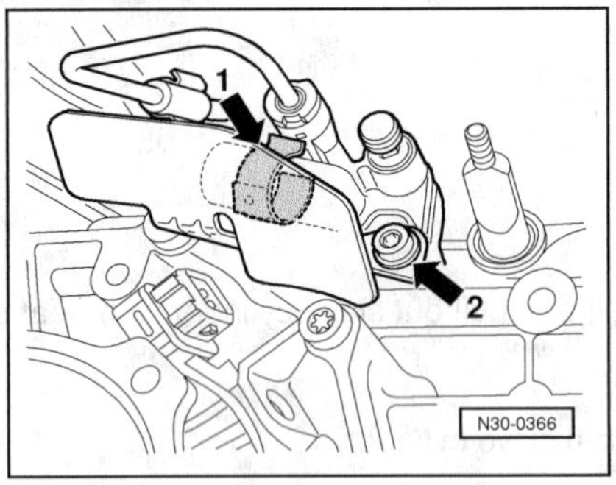

◄ Guide slave cylinder into mounting hole in transmission housing, without great sideways movement from working direction of plunger, and bolt tightly.

– Press heat shield (arrow -1-) onto slave cylinder.

– Tighten bolt (arrow -2-) to 25 Nm (18 ft-lb).

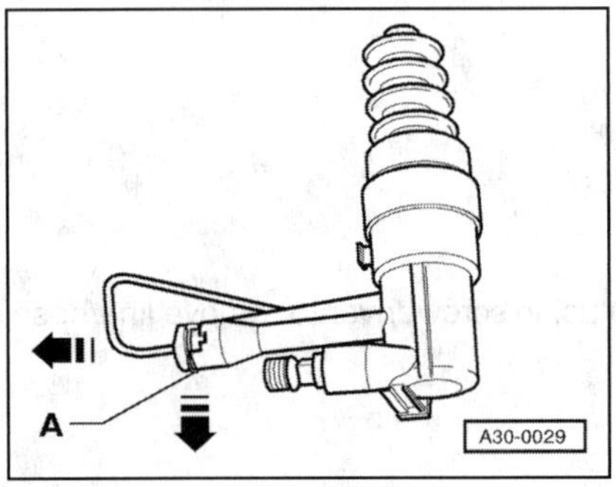

◄ Press clip (arrow -A-) inward onto limit stop to connect line/hose assembly to slave cylinder.

– Insert line/hose assembly in slave cylinder until line audibly engages.

– Reconnect exhaust system and ensure it is not under stress, see **Engine Mechanical section, Repair Group 26**.

– Bleed clutch system, see **Clutch system, bleeding**.

– Mount axle shaft to transmission flange shaft. Tighten to specified torque, see **Suspension, Wheels, Steering section, Repair Group 40.**

Clutch, servicing

Special tools and equipment

- 10-201 Counter-hold tool
- 3176 Centering mandrel

(Transmission removed)

NOTE —
- *Replace clutch plates and pressure plates with damaged or loose rivets*
- *Select correct clutch plate and pressure plate according to parts list and engine code.*
- *Clutch illustrated in conjunction with two-part flywheel*

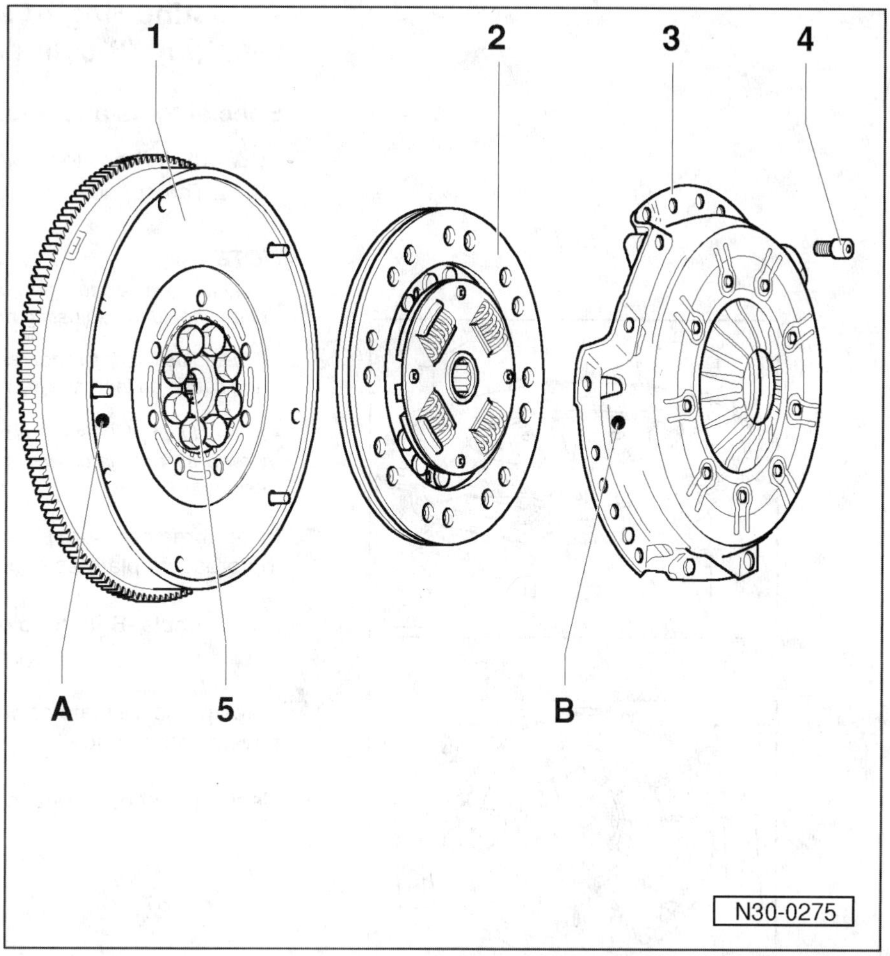

1. **Flywheel**
   - Make sure centering pins are a tight fit
   - Contact surface for clutch lining must be free of grooves, oil and grease
   - If white spots are present on flywheel -A- and pressure plate -B-, align them during assembly
   - Removing and installing, see **Engine Mechanical section, Repair Group 13**
2. **Clutch plate**
   - Diameter, see **Repair Group 00, Code letters, transmission allocation, ratios, capacities**
   - Installation position with two-part flywheel: the clutch plate lining must contact the flywheel
   - Installation position without two-part flywheel: spring cage faces the pressure plate
   - Centering, see **Fig. 1**
   - Lightly grease splines

   NOTE —
   Clean input shaft splines and on used clutch plates, the hub splines, remove corrosion and apply only a very thin coating of grease Part No: G 000 100 to the input shaft splines. Then move clutch plate to and fro on input shaft until hub moves freely on shaft. Excess grease must be removed.
3. **Pressure plate**
   - Removing and installing, see **Fig. 1**
   - Check ends of diaphragm spring, wear up to half the thickness of the diaphragm spring is permitted
   - If white spots are present on flywheel -A- and pressure plate -B-, align them during assembly

   NOTE —
   Pressure plates are protected against corrosion and greased. Only the contact surface may be cleaned, otherwise the service life of the clutch will be considerably reduced.
4. **Bolt**
   - Tighten to 25 Nm (18 ft-lb)
   - Loosen and tighten gradually and diagonally
5. **Needle roller bearing**
   - Driving in and out, see **Engine Mechanical section, Repair Group 13**

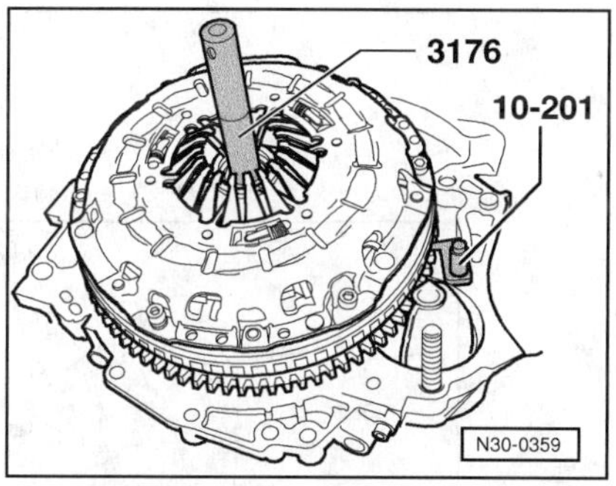

Fig. 1 Centering clutch and pressure plate (8-cylinder engine)

- Pressure plate contact surface and clutch plate lining must make full contact with the flywheel.
- Tighten securing bolts diagonally and evenly to prevent damage to pressure plate centralizing hole and flywheel centralizing pins.
- The adjusting ring on self-adjusting pressure plate must be reset when a new clutch plate is installed, as described below.

Adjusting ring of self-adjusting pressure plate, resetting (8-cylinder engine)

Special tools and equipment

- VW 408 A Press tool
- 3062 Thrust pad

NOTE—

- *Only when installing a new clutch plate with a used self-adjusting pressure plate should the adjusting ring be reset.*
- *If the adjusting ring is not reset, the clutch release pressure will be reduced and cause an increase in clutch plate wear.*
- *If the clutch plate is not replaced, the adjusting ring should not be reset. New pressure plates are pre-adjusted and resetting is therefore not necessary.*

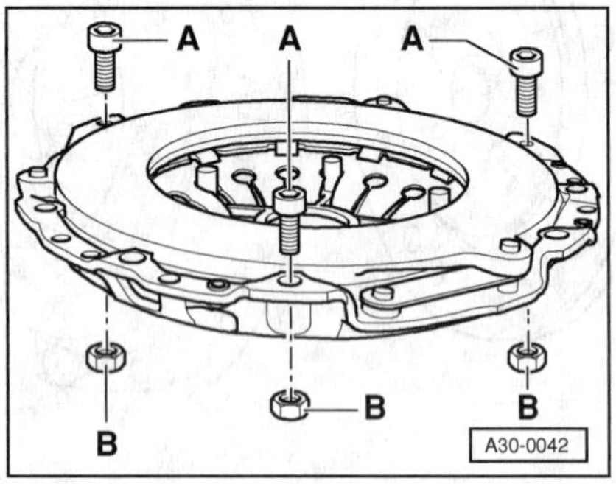

Insert three pressure plate securing bolts -A- at 120° in locating holes of the pressure plate as shown.

– Screw 3 nuts -B- (M8) to bolts -A- and tighten.

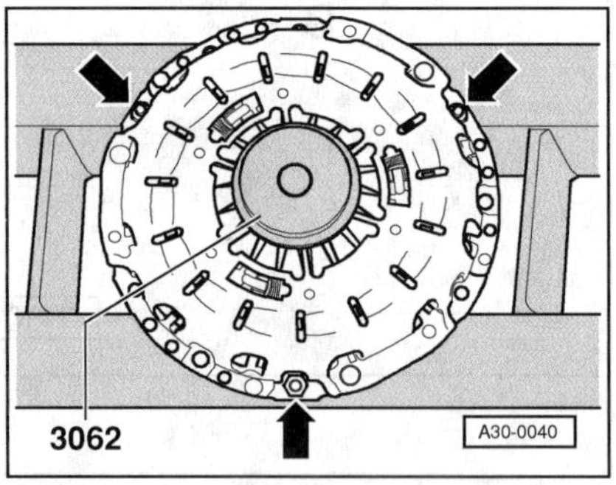

Place pressure plate on workshop press so that just 3 bolt heads (arrows) touch surface.

– Center thrust pad 3062 in middle of pressure plate.

CAUTION—
Do not use excessive force for the following work sequence otherwise the adjusting ring fork may break off

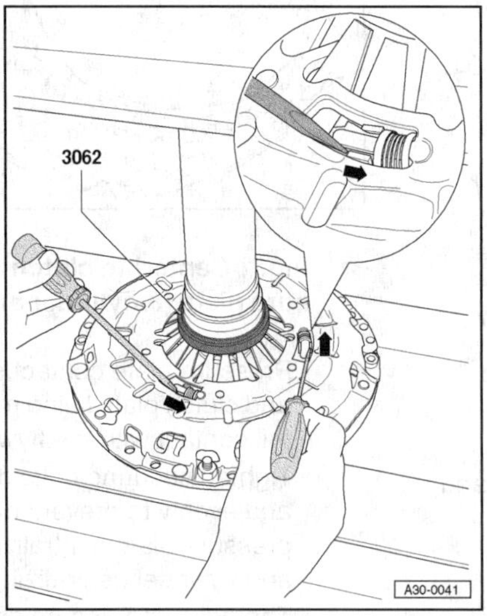

Position two screwdrivers on adjusting ring forks.

– Press down on pressure plate with press until adjusting ring can just be moved.

– Turn adjusting ring in direction of -arrow- back onto limit stop, using even pressure on both screwdrivers.

– Hold adjusting ring against limit stop and release pressure on press so that adjusting ring is held in this position.

34 Controls, Housing

Repair Group

Q 34

Shift Linkage

NOTE—
To remove the complete shift linkage, remove the exhaust system, see **Engine Mechanical section, Repair Group 26**.

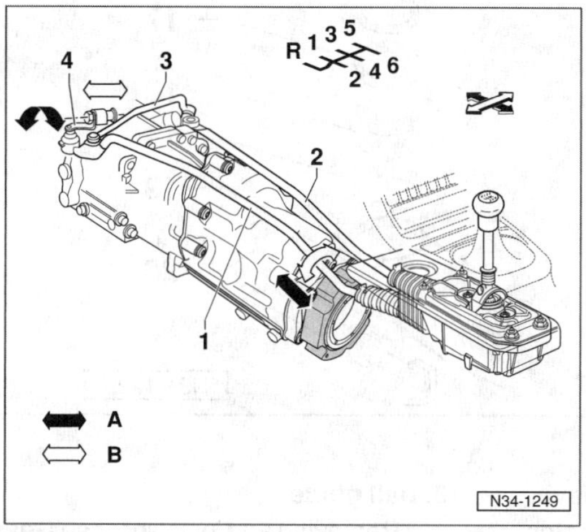

Gear selector mechanism, overview

1. **Selector rod**
2. **Front push rod**
3. **Connecting rod**
4. **Gear selector lever**

A. Gear selector movement arrow

B. Gate selector movement arrow

NOTE—
- *Before working on the selector mechanism, disconnect the Ground strap from the battery in the engine compartment.*
- *Note radio code on vehicles with coded radio.*

Gear selector lever and rear push rod, assembly

> ***NOTE —***
> *Lubricate all mountings and sliding surfaces with Polyurea grease Part No. G 052 142 A2.*

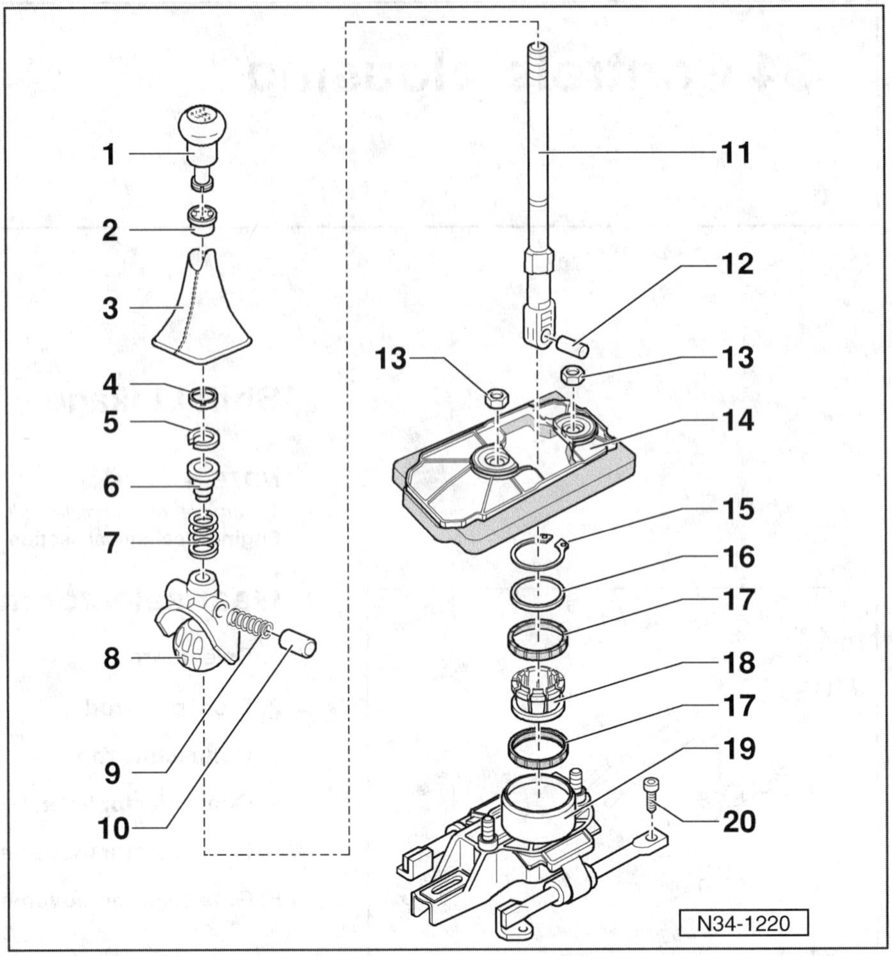

1. **Gear selector lever knob**
   - Screw on and off together with boot
   - As of 11/2000, secured on gear selector lever with a clamp, see **Fig. 2**
   - Determine correct application using parts catalog
2. **Bushing**
   - Install and remove along groove in gear selector lever knob
   - Discontinued as of 11/2000
3. **Cover**
   - Separating from gear selector lever knob, see **Fig. 1**
   - Connect to gear selector lever knob before installing, see **Fig. 1**
   - Remove and install together with gear selector lever knob
4. **Sleeve**
   - Always replace
   - Discontinued as of 11/2000
5. **Circlip**
   - Always replace
6. **Spacer sleeve**
7. **Spring**
8. **Ball stop**
   - Removing and installing, see **Gear selector lever with rear push rod, removing and installing**
   - Insert spring and bushing in ball stop and assemble on gear selector lever so that spring and bushing are on correctly when looking forward.
   - As of 11/2000, limit stop pins on both sides and mounting for spring have been discontinued, see **Fig. 3**
   - Determine correct application using parts catalog
9. **Spring**
   - Discontinued as of 11/2000
10. **Bushing**
    - Installed position: rounded side faces gear selector lever
    - Discontinued as of 11/2000
11. **Transmission gear selector lever**
    - Only fits in one position in ball stop (item -**8**-)
    - Removing and installing, see **Gear selector lever with rear push rod, removing and installing**
    - As of 11/2000, thread to secure gear selector lever has been discontinued
    - Determine correct application using parts catalog
12. **Spacer tube**
13. **Nut, hex**
    - Tighten to 10 Nm (7 ft-lb)
14. **Cover**
15. **Circlip**
    - Always replace
    - Remove before removing ball stop
16. **Washer**
    - Collar faces ball guide (item -**18**-)
17. **Rubber ring**
18. **Ball guide**
    - Removing and installing, see **Gear selector lever with rear push rod, removing and installing**
19. **Rear push rod**
    - With ball housing
    - Removing and installing, see **Gear selector lever with rear push rod, removing and installing**
20. **Bolt**
    - Tighten to 6 Nm (53 in-lb)
    - Rear push rod to gear selector lever housing, see **Selector rod, connecting rod and front push rod, assembly**

Fig. 1 Separating cover from gear selector lever knob and assembling to gear selector lever knob

– Turn cover inside out.

– Carefully press off clamping sleeve using screwdriver (**arrow**).

– Installation is the reverse.

– Insert gear selector lever knob in cover.

– Push clamping sleeve onto gear selector lever knob and engage.

– Install gear selector lever knob and boot together.

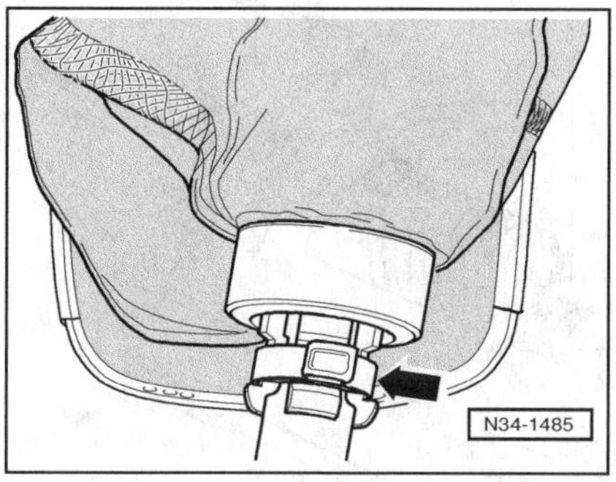

Fig. 2 Removing and installing cover with gear selector lever knob on gear selector lever

– Turn cover inside out.

– Destroy clip (**arrow**) and pull gear selector lever knob off with cover.

– Installation is the reverse.

– Install gear selector lever knob and cover together onto gear selector lever.

– Install new clip (**arrow**).

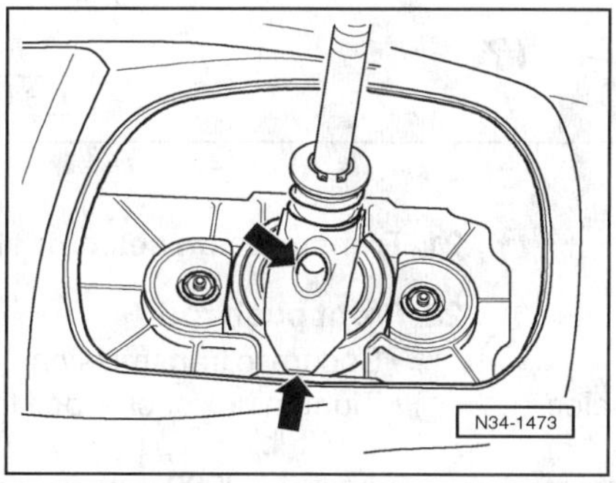

Fig. 3 Revised ball limit stop

– Limit stop pins on both sides and the mounting for the spring have been discontinued (**arrows**).

Selector rod, connecting rod and front push rod, assembly

NOTE—

Lubricate all mountings and sliding surfaces with Polyurea grease Part No. G 052 142 A2.

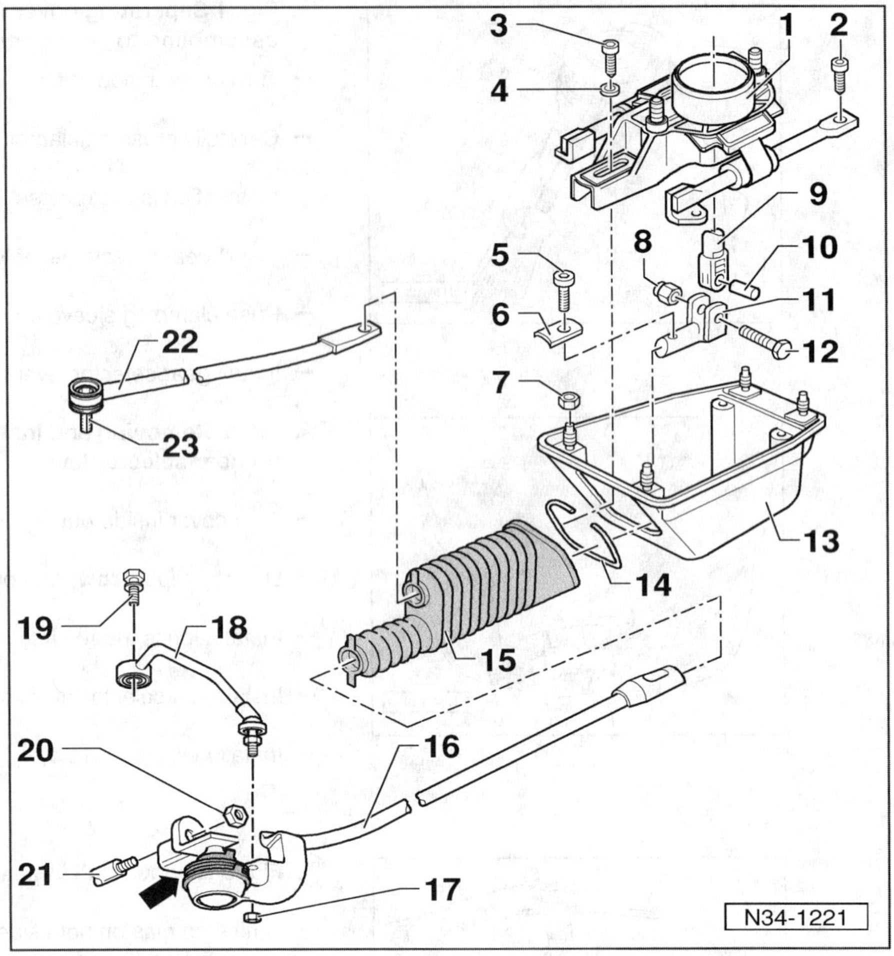

1. **Rear push rod**
   - With ball housing
   - Removing and installing, see **Gear selector lever with rear push rod, removing and installing**
2. **Bolt**
   - Tighten to 6 Nm (53 in-lb)
3. **Bolt, hex socket head**
   - Tighten to 25 Nm (18 ft-lb)
4. **Washer**
5. **Bolt, hex socket head**
   - Tighten to 25 Nm (18 ft-lb)
6. **Clamping piece**
7. **Nut, hex**
   - Tighten to 10 Nm (7 ft-lb)
8. **Nut, hex**
   - Tighten to 10 Nm (7 ft-lb)
   - Self-locking
   - Always replace
9. **Gear selector lever**
   - Only fits in one position in ball housing
10. **Spacer tube**
11. **Selector fork**
12. **Bolt**
13. **Selector housing**
14. **Tensioning ring**
15. **Boot**
    - Carefully pull boot over selector rod
    - Place boot on the marks on selector rod and push rod
16. **Selector rod**
    - As of 11/2000, revised dimensions
    - Determine correct application using parts catalog
17. **Nut, hex**
    - Tighten to 10 Nm (7 ft-lb)
18. **Connecting rod**
    - Remove with ball head from selector rod
    - As of 11/2000, secured to selector rod with a circlip, see **Fig. 1**
    - Remove from selector rod after removing circlip
    - Determine using parts catalog
    - Installed position on transmission, see **Gear selector mechanism, removing and installing**
19. **Bolt,**
    - Connecting rod to transmission
    - Tighten to 25 Nm (18 ft-lb)
20. **Nut, self-locking**
    - Self-locking
    - Always replace
    - Tighten to 25 Nm (18 ft-lb)
21. **Transmission selector shaft**
22. **Front push rod**
    - Secure to transmission
    - Tighten hex socket bolt to 40 Nm (30 ft-lb)
    - As of 11/2000, revised dimensions
    - Determine correct application using parts catalog
23. **Washer**
    - Peened over with front push rod
    - Reinstall front push rod with same number of washers

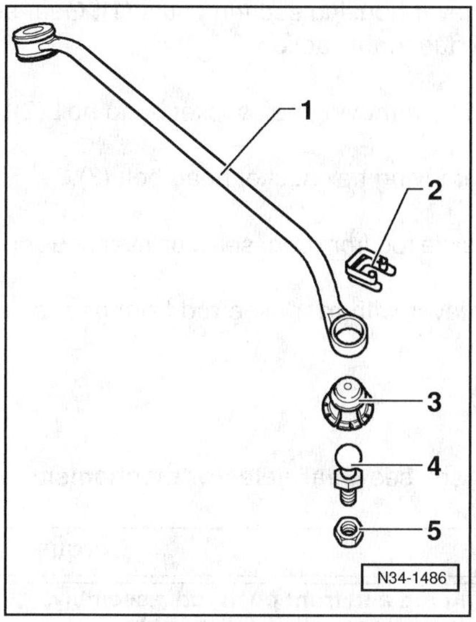

Fig. 1 Connecting rod with securing clip

1. Connecting rod
2. Securing clip
3. Bearing bushing
4. Ball head stud
5. Nut, hex head
   Tighten to 10 Nm (7 ft-lb)
   Self-locking
   Always replace

Gear selector lever with rear push rod, removing and installing

– Carefully pry cover off center console.

– For vehicles to 10/2000, unscrew gear selector lever knob with cover.

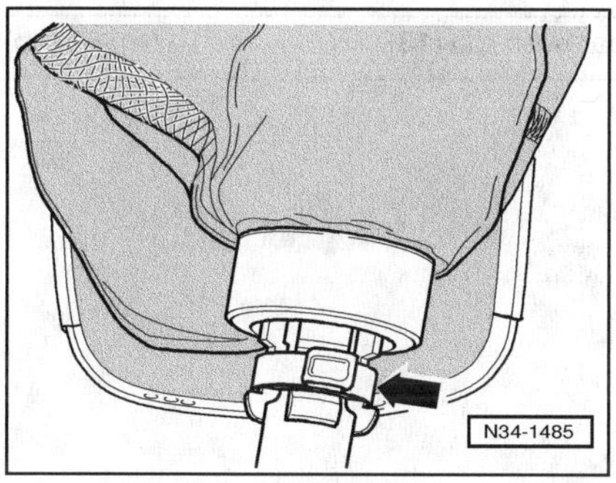

For vehicles from 11/2000, destroy clip (**arrow**) and pull gear selector lever knob off with cover.

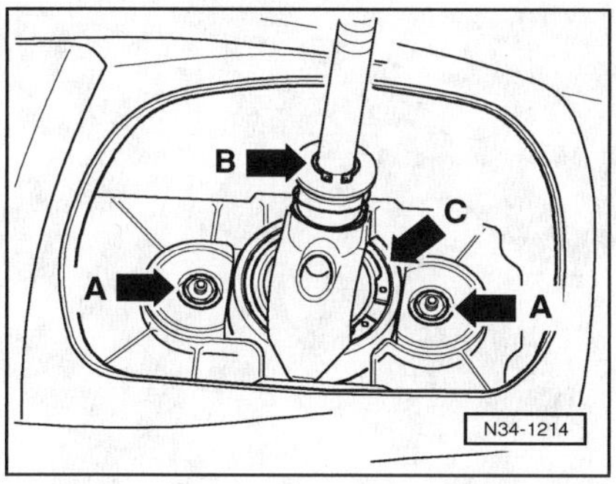

Unbolt gear selector lever housing cover (**arrows A**).

– Remove circlip (**arrow B**) while carefully relieving tension on spring.

– Take off spacer bushing and spring.

– Remove circlip (**arrow C**).

– Pull ball stop with ball guide and rubber rings out in upward direction.

– Remove ball stop from gear selector lever while (if part of original equipment) carefully relieving tension on spring and bushing.

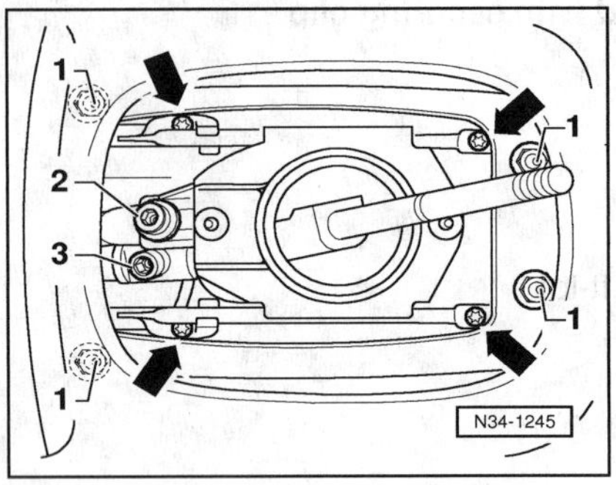

- Remove gear selector lever housing securing nuts (**1**). Gear selector lever housing is lowered due to this action.
- Remove front push rod by removing hex socket head bolt (**2**).
- Remove slide rod by removing hex socket head bolt (**3**).
- Release complete rear slide rod from gear selector lever housing (**arrows**).
- Remove gear selector lever with rear slide rod from gear selector lever housing.
- Install in reverse order.
- Adjust selector mechanism, see **Gear selector mechanism, adjusting**.

Tightening torques	Torque
Selector rod, connecting rod and front push rod, assembly, see Gear selector lever and rear push rod, assembly	
Bolt, rear push rod to selector housing, bolt (**2**)	6 Nm (53 in-lb)
Bolt, selector rod to gear selector lever selector fork (**5**)	25 Nm (18 ft-lb)
Bolt, front push rod to rear push rod (**3**)	25 Nm (18 ft-lb)
Nut, gear selector lever housing to body (**7**)	10 Nm (7 ft-lb)
Gear selector lever and rear push rod, see Gear selector lever and rear push rod, assembly	
Nut, push rod cover to rear push rod (**13**)	10 Nm (7 ft-lb)

Gear selector mechanism, removing and installing

– Carefully pry cover off center console.

– For vehicles to 10/2000, unscrew gear selector lever knob with cover.

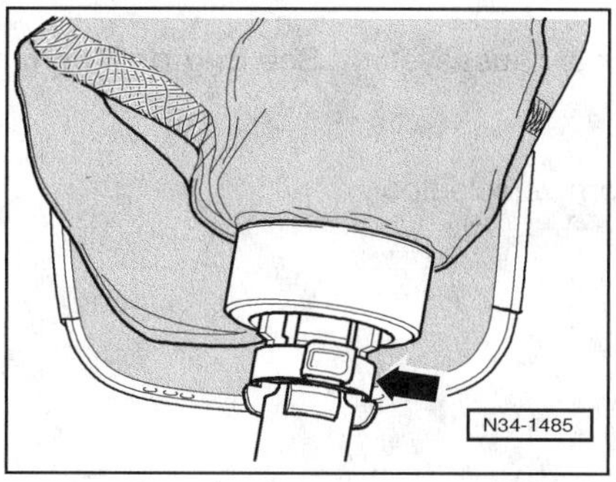

◄ For vehicles from 11/2000, destroy clip (**arrow**) and pull gear selector lever knob off with cover.

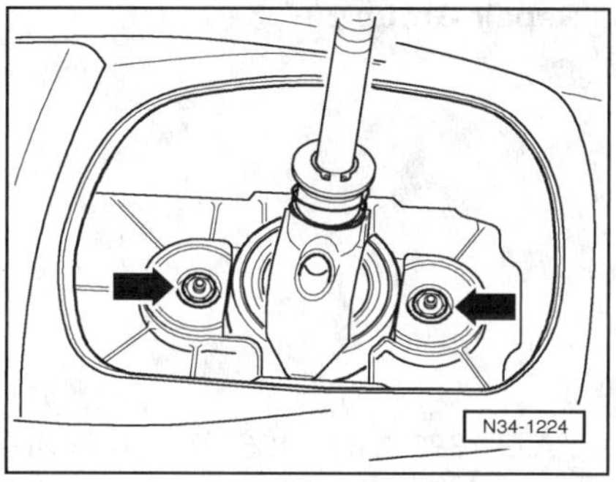

◄ Remove gear selector lever housing cover (**arrows**).

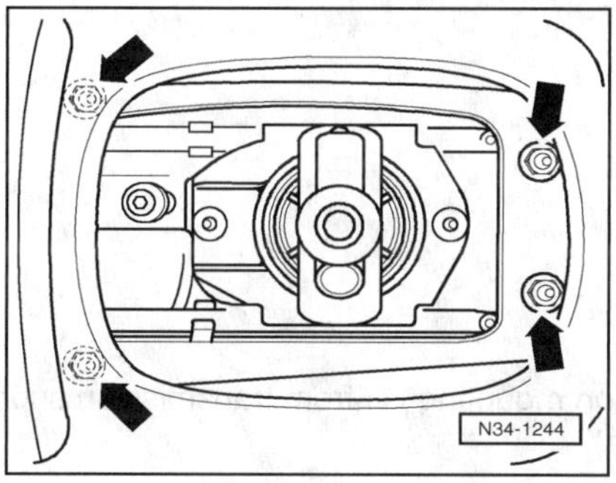

◄ Remove gear selector lever housing securing nuts (**arrows**).

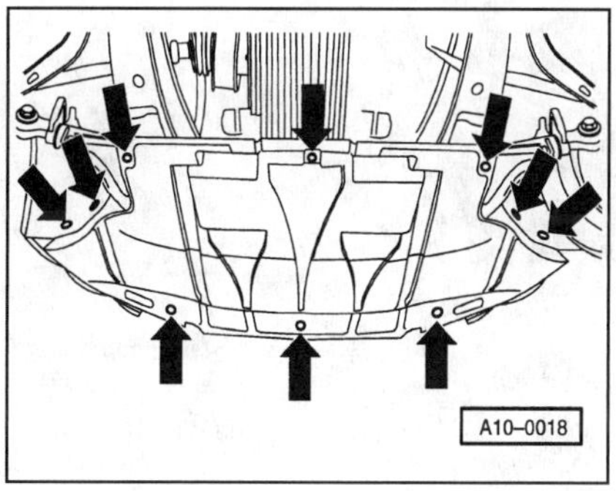

◄ Remove sound insulation tray (**arrows**)

– Remove sound insulation tray bracket.

– For 4-cylinder vehicles, remove intermediate pipe of exhaust system, see **Engine Mechanical section, Repair Group 26**

– For 8-cylinder vehicles, remove left front and right front exhaust pipes. See **Engine Mechanical section, Repair Group 26**

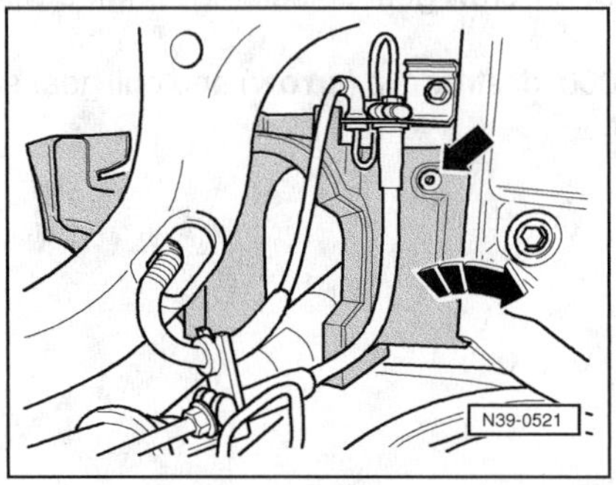

◄ Remove cover from left wheel housing, if present (**arrows**).

– Remove rear section of exhaust system. See **Engine Mechanical section, Repair Group 26.**

– Remove heat shield from vehicle floor.

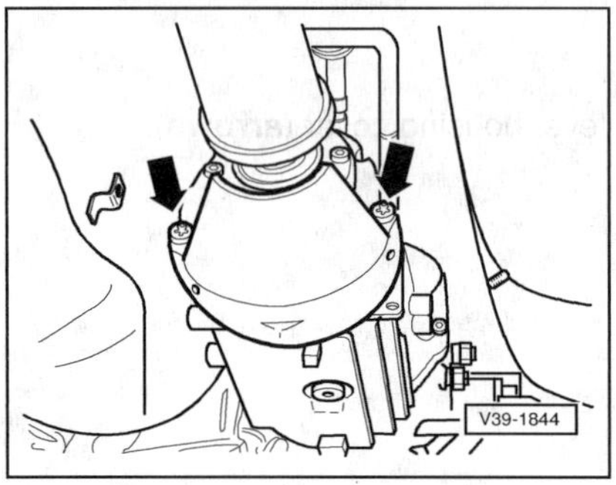

◄ Remove cover plate (**arrows**).

– Remove driveshaft, see **Repair Group 39**

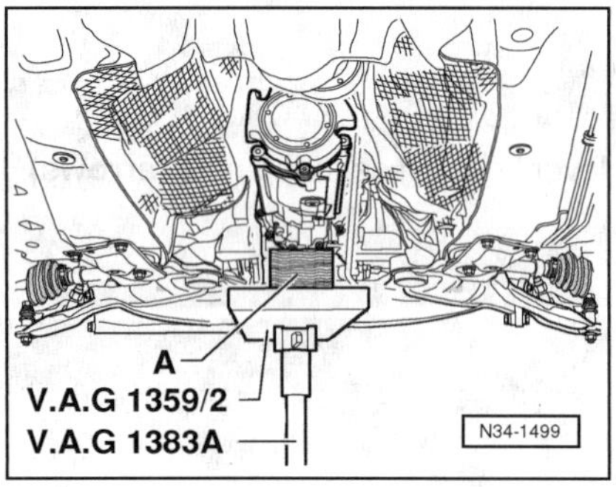

◄ Place transmission jack V.A.G 1383 A with 1359/2 (or equivalent) under subframe and exert slight counter-pressure.

A - Block of wood

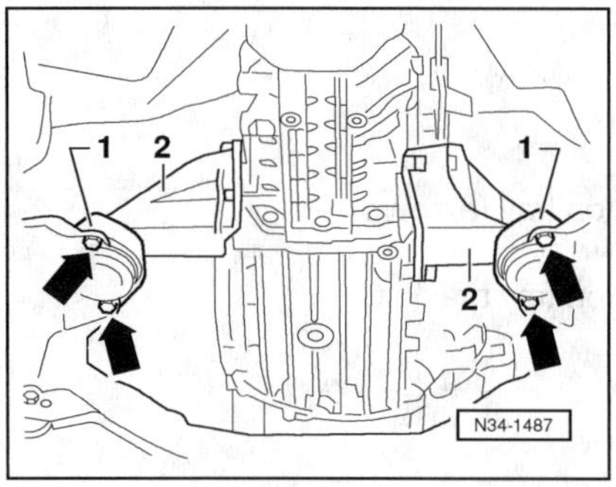

◄ Disconnect transmission mounting (**1**) from transmission support (**2**) and subframe (**arrows**).

– Using transmission jack V.A.G 1383A (or equivalent), lower transmission slightly.

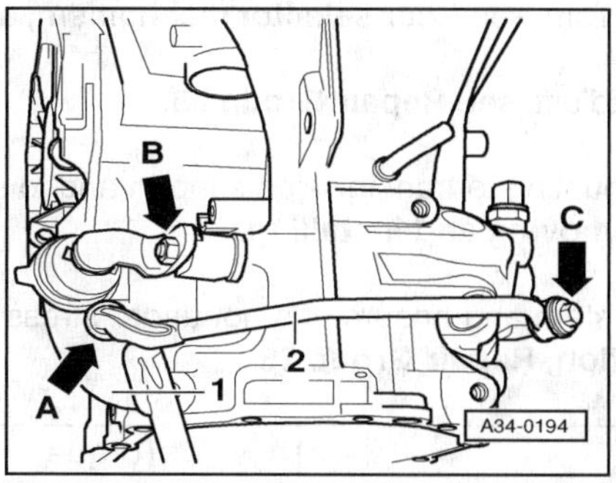

When removing gear selector mechanism, do not, under any circumstances, pull ball head (**arrow A**) of connecting rod (**2**) off selector rod (**1**) or the ball head will be destroyed.

– To remove selector rod, first remove nut (**arrow B**) and bolt (**arrow C**).

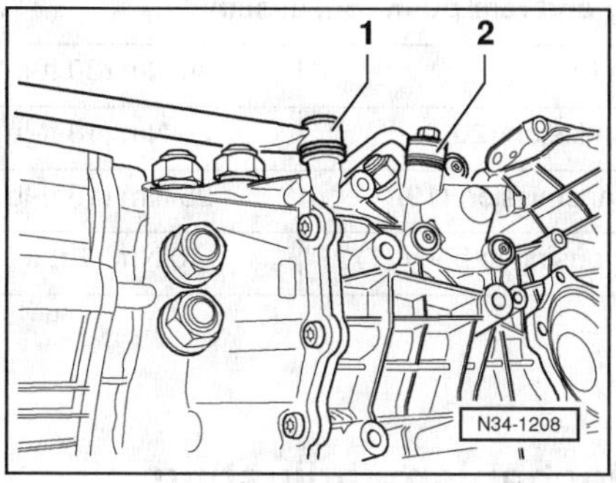

Remove selector mechanism from transmission by unbolting connecting rod (**2**) for selector rod from right side of transmission.

– Remove push rod (**1**) by unbolting hex socket head bolt.

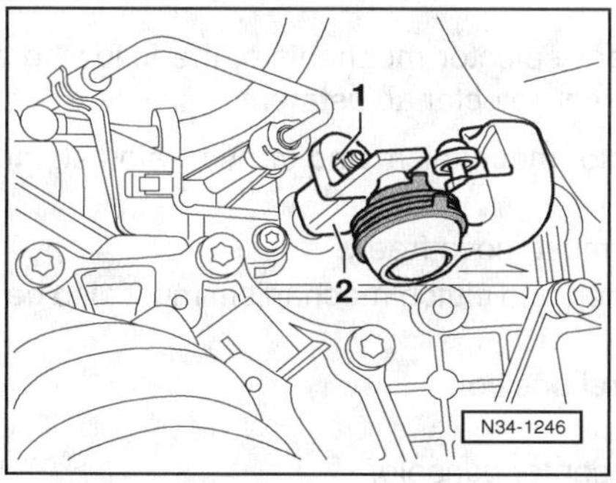

Remove nut (**1**) and pull gear selector lever (**2**) from selector shaft of transmission.

– Swing gear selector lever housing with selector rod and push rod down and remove.

– Installation is performed in the reverse sequence.

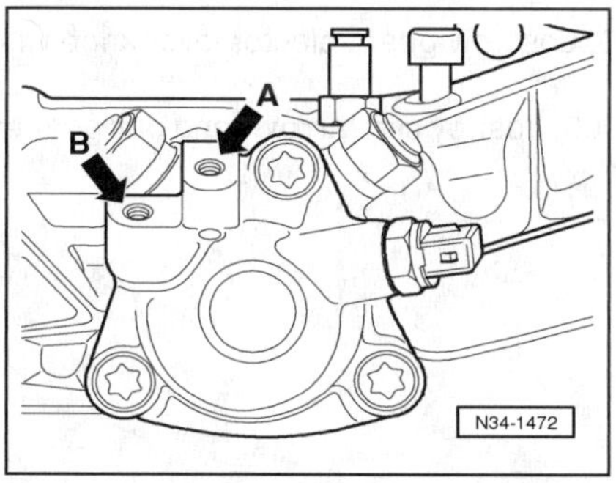

Note installed position of connecting rod between transmission and selector rod on transmission.

- Selector shaft cover has an additional threaded hole (**arrow A**) which is used to secure the new version connecting rod between selector rod and transmission.

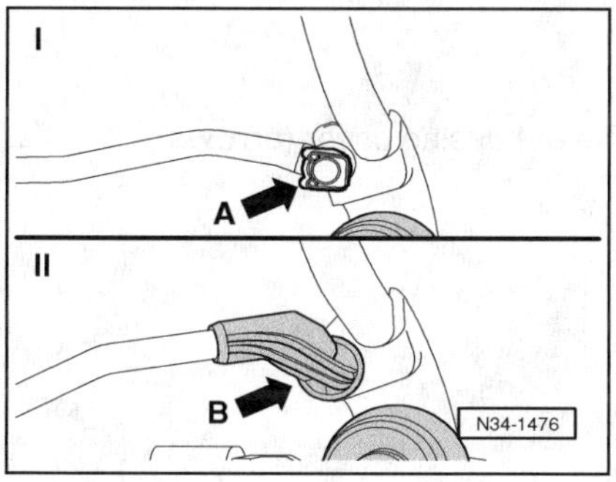

Note different securing methods when installing connecting rod

Connecting rod to selector rod version, illustration N34-1476 (left)	Connecting rod on selector shaft cover, illustration N34-1472 (above left)
I - Secured with a circlip arrow (**A**)	On threaded hole arrow (**A**)
II - Pressed onto ball head, arrow (**B**)	On threaded hole arrow (**B**)

– Adjust selector mechanism, see **Gear selector mechanism, adjusting**.

– Install driveshaft and adjust, see **Repair Group 39.**

– Install transmission mounting to transmission support and to subframe, see **Transmission, removing and installing.**

– Reconnect exhaust system and ensure it is not under stress. See **Engine Mechanical section, Repair Group 26**.

Tightening torques	Torque
Selector rod, connecting rod and front push rod, assembly, see Gear selector lever and rear push rod, assembly	
Push rod to transmission (**22**)	40 Nm (30 ft-lb)
Nut, selector rod to transmission (**20**)	25 Nm (18 ft-lb)
Bolt, connecting rod to transmission (**19**)	25 Nm (18 ft-lb)
Nut, gear selector lever housing to body (**7**)	10 Nm (7 ft-lb)

Gear selector mechanism, adjusting

When adjusting the gear selector mechanism, the following points are essential to ensure correct selector adjustment:

- Moving parts of selector mechanism and transmission elements must be in proper condition.
- Selector mechanism must move freely.
- The transmission, clutch and clutch mechanism must also be in proper condition.
- Transmission in neutral position.

– Carefully pry cover off center console.

– For vehicles to 10/2000, unscrew gear selector lever knob with cover.

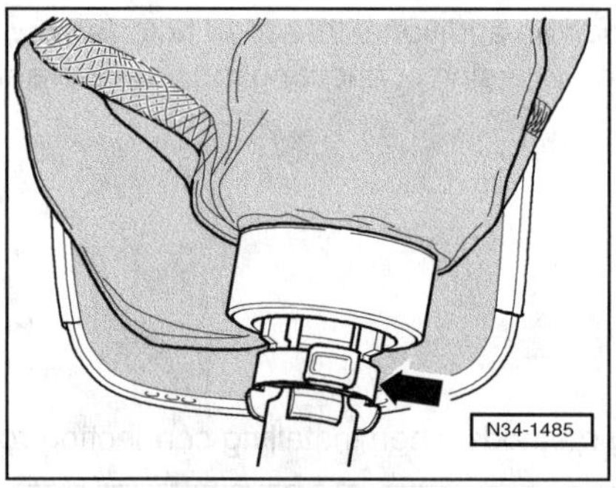

N34-1485

For vehicles from 11/2000, destroy clip (**arrow**) and pull gear selector lever knob off with cover.

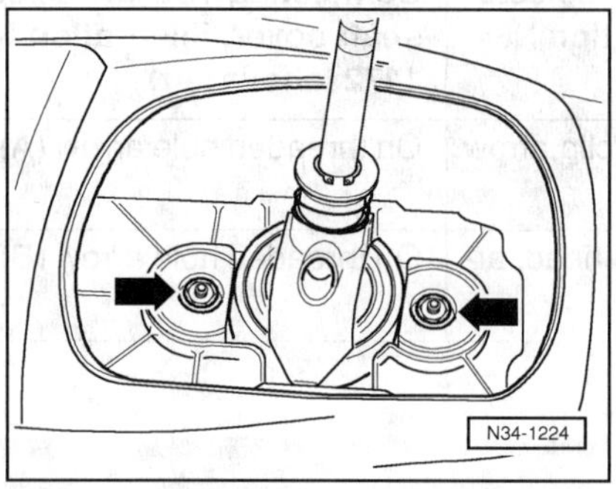

N34-1224

Remove gear selector lever housing cover (**arrows**).

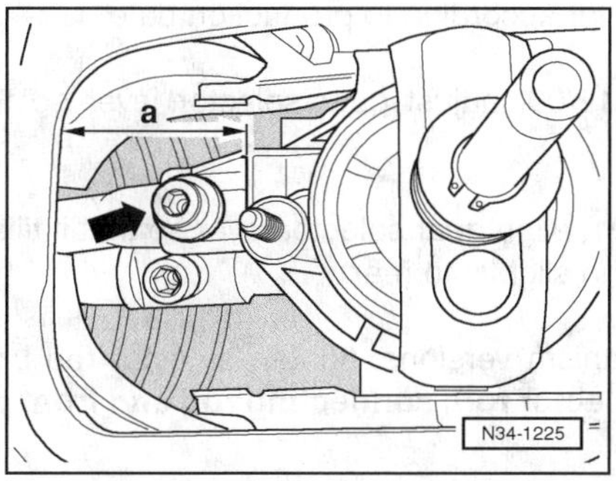

◀ Measure distance between body and gear selector mechanism.

• Distance (**a**) must be 43 mm (1.693 in).

– If not, adjust dimension (**a**). Loosen push rod screw (**arrow**).

• Connection of push rod/selector mechanism must move freely.

– Set to dimension (**a**) by moving selector mechanism.

– Tighten push rod bolt (**arrow**) to specified torque, see **Selector rod, connecting rod and front push rod, assembly** (item **3**).

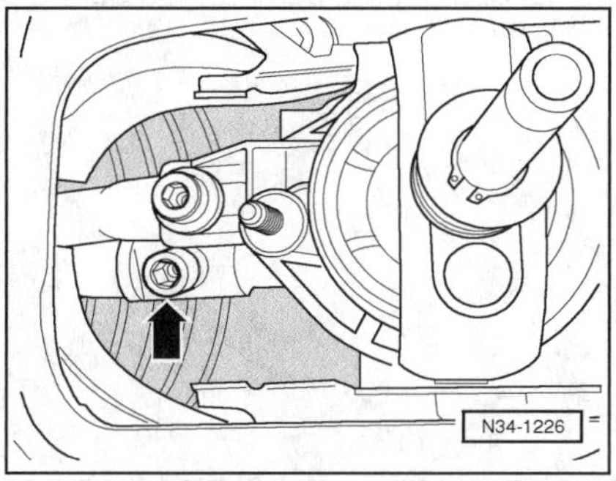

◀ Loosen selector rod bolt (**arrow**).

• Connection of selector rod/selector mechanism must move freely.

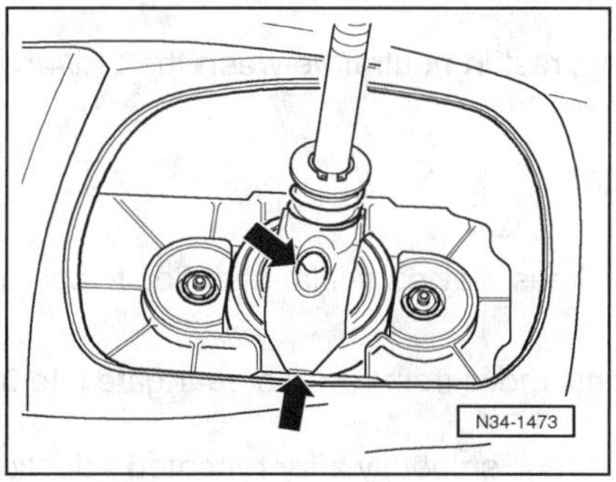

◀ Selector mechanism with limit stop pins of ball stop and mounting for spring (**arrows**).

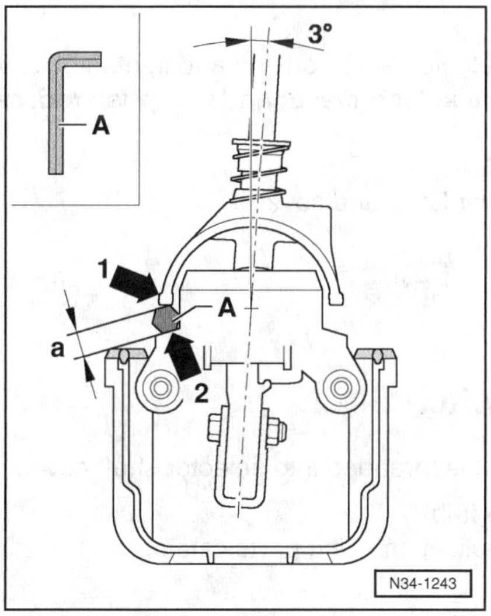

◀ Insert a 14 mm hex key (**A**) between the left tab (in direction of travel) (**arrow 1**) of ball stop and sloped shoulder alongside ball stop (**arrow 2**) and hold.

• Key must have contact along entire shoulder (**2**).

– Adjust gear selector lever according to production date

– For vehicles through 12/00, adjust gear selector lever so that it tilts slightly toward rear.

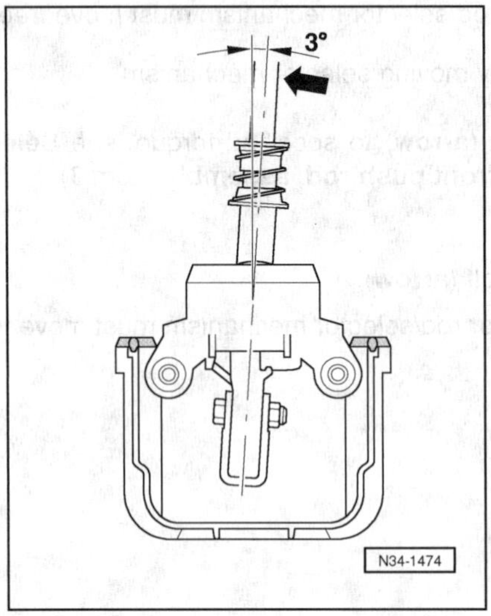

For vehicles as of 01/01, align gear selector lever so that it tilts slightly toward right (**arrow**) and slightly to rear.

– For all selector mechanism versions, tighten selector rod bolt to prescribed torque, see **Selector rod, connecting rod and front push rod, assembly** (item **5**).

• Position of gear selector lever must not be changed during procedure.

Gear selector mechanism, functional check

Gear selector lever must rest in neutral between the 3rd/4th gear gate when transmission is in neutral.

– Operate clutch.

– Select all gears several times. Pay particular attention to operation of reverse gear lockout.

• Gear selector lever must move from reverse gear gate into 3rd/4th gear gate on its own.
• Should a gear fail to engage smoothly after repeated selection, repeat selector adjustment

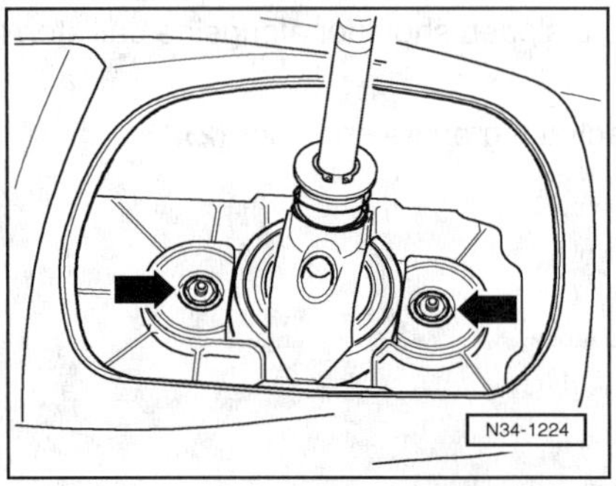

Install cover for gear selector lever housing and tighten bolts to prescribed torque, see **Gear selector lever and rear push rod, assembly, (**item **13)**.

– Install gear selector lever knob and cover.

Back-up light switch

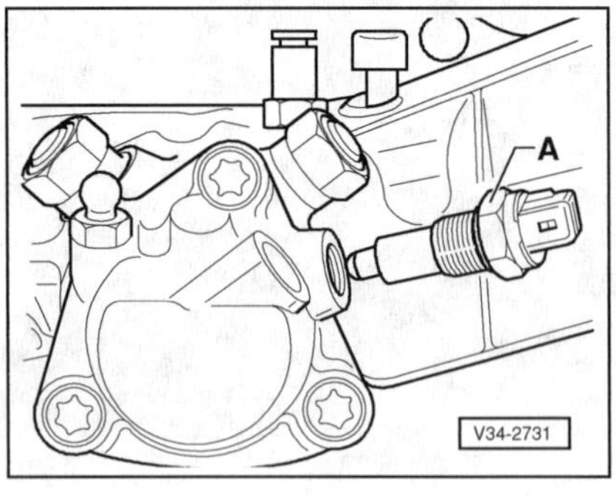

Back-up light switch (**A**) is threaded into selector shaft cover.

• Tighten to 20 Nm (15 ft-lb)
• Determine correct application using parts catalog

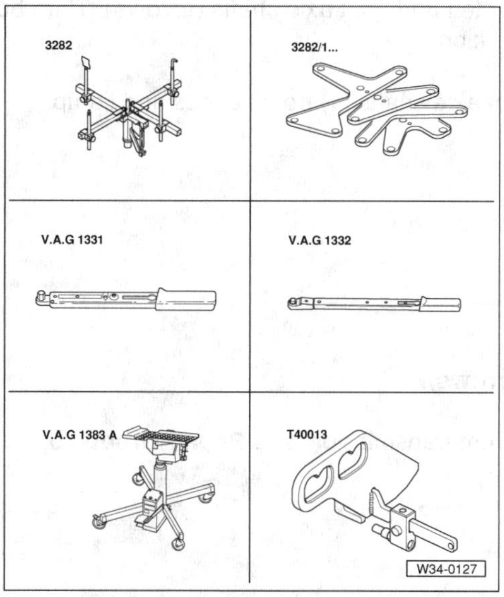

Transmission, removing and installing (vehicles with 4- or 6-cylinder engines)

NOTE —
For transmission removing and installing, vehicles with 8-Cyl. engines, see **Transmission, removing and installing (vehicles with 8-cylinder engines**.

Special tools and equipment

- 3282 Transmission support
- 3282/12 Adjustment plate
- V.A.G 1331 Torque wrench
- V.A.G 1332 Torque wrench
- V.A.G 1383 A Engine/transmission jack
- T40013 Transportation device

Additional information

- **Suspension, Wheels, Steering section**
- **Electrical Equipment section**
- **Engine Mechanical section**

Removing

– First, check whether a coded radio is installed. If so, obtain anti-theft coding.

– With ignition switched off, disconnect battery Ground (GND) strap.

– Remove cover from engine.

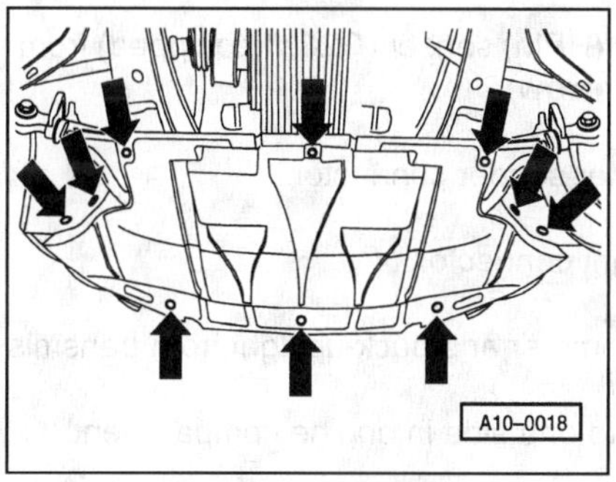

◄ Remove sound insulation tray (**arrows**).

– Remove sound insulation tray bracket.

– For 4-cylinder vehicles, remove intermediate pipe of exhaust system. See **Engine Mechanical section, Repair Group 26.**

Continuation for all vehicles

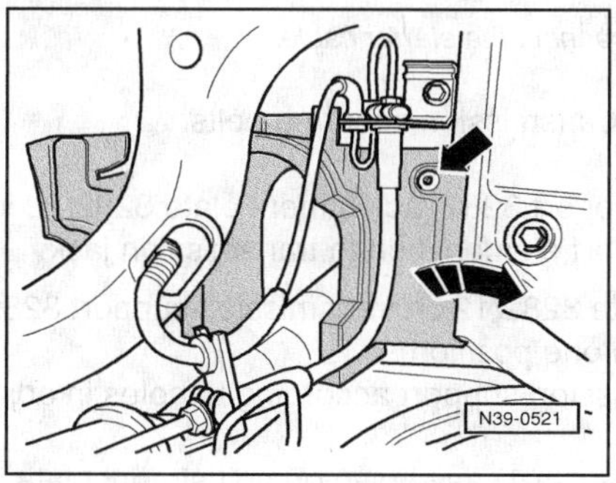

◄ Remove covers from wheelhousings, if present (**arrows**).

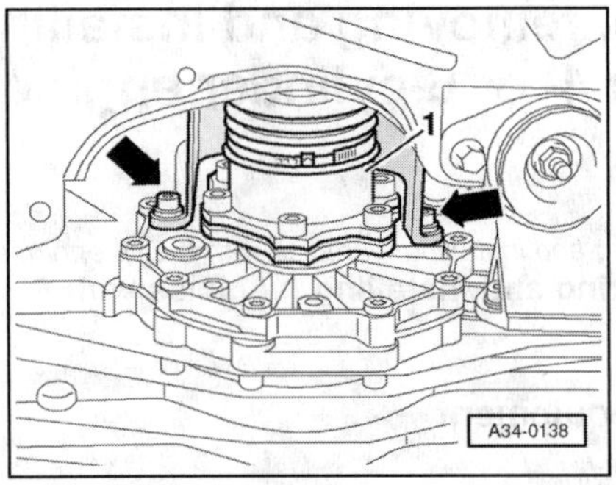

- Remove heat shields for left and right axle shafts (**arrows**). Third bolt cannot be seen in illustration.

– Disconnect left and right axle shafts (**1**) from flanges and tie-up.

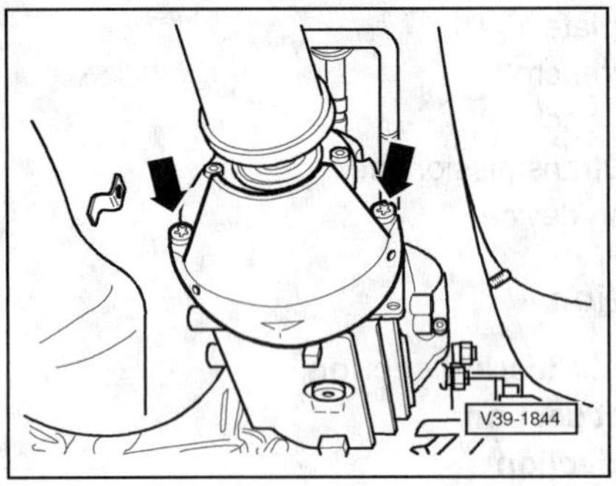

- Remove cover plate (**arrows**).

– Disconnect driveshaft from transmission, see **Repair Group 39**.

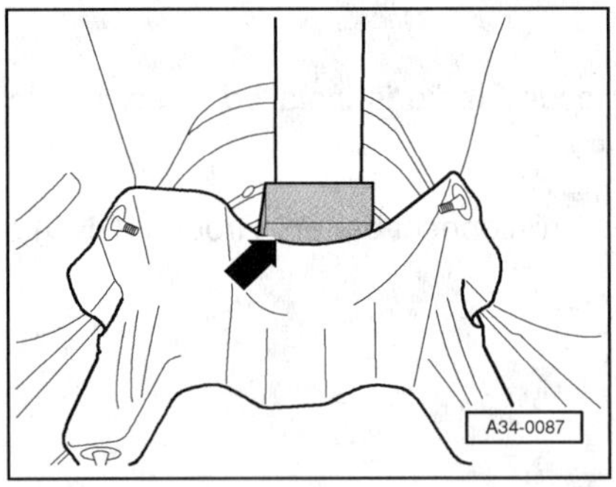

- Push driveshaft back onto limit stop and up onto heat shield with a wooden wedge (**arrow**).

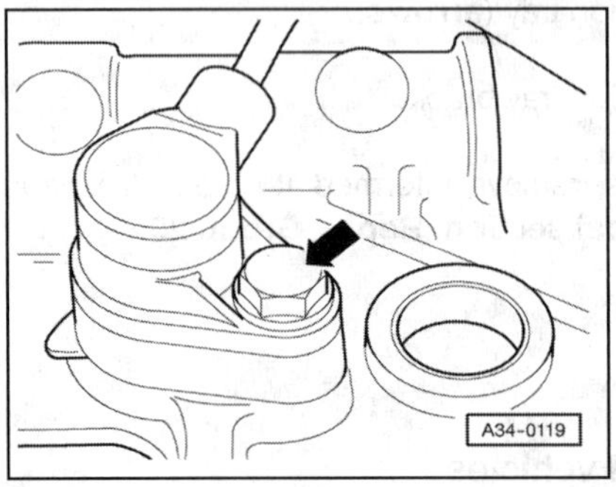

- Remove Engine speed (RPM) sensor (**G28**, if equipped) from top left of transmission housing (**arrow**).

– Disconnect speedometer sender connector.

– Disconnect back-up light connector.

– Unclip wires for speedometer and back-up light from transmission.

– Unbolt starter and set to one side in engine compartment.

NOTE —
It is not necessary to disconnect the starter cable.

– Remove lower engine/transmission securing bolts.

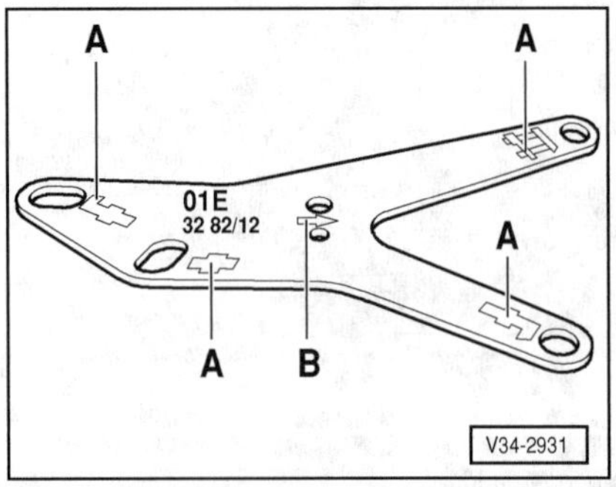

- Mount transmission support 3282, adjustment plate 3282/12 for "**01E**" transmission and support elements on to transmission jack.
  - Place adjustment plate 3282/12 on transmission support 3282 (adjustment plate fits only in one position).
  - Align arms of transmission support according to holes in adjustment plate.
  - Bolt on support elements (**A**) as shown on adjustment plate.
  - Arrow symbol (**B**) on adjustment plate points to front of vehicle.

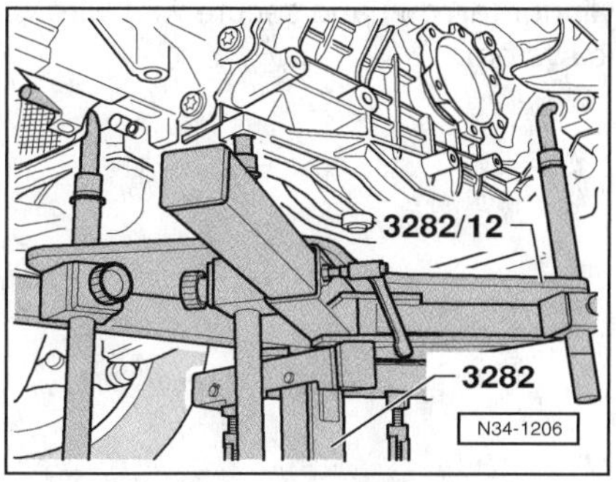

- Place transmission jack under vehicle with arrow symbol on adjustment plate pointing to front of vehicle.

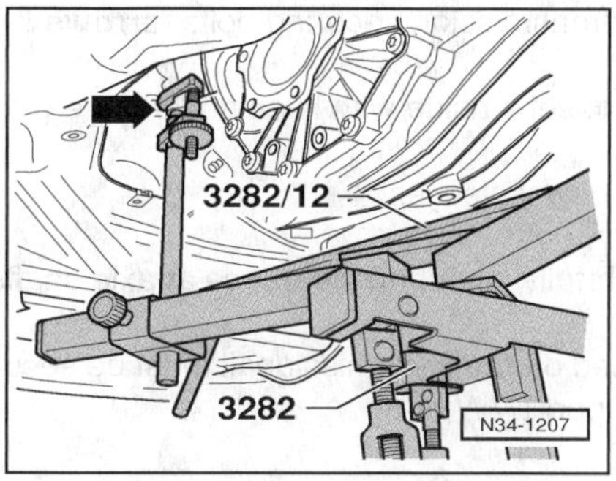

- Align adjustment plate parallel to transmission and lock safety support (**arrow**) on transmission.

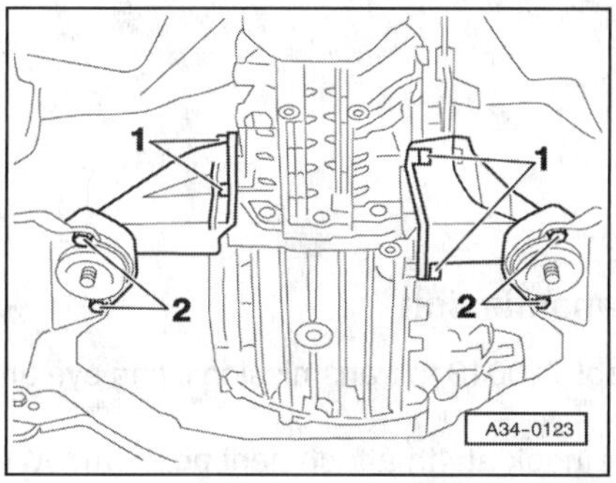

- Remove bolts (**1**) and (**2**) to disconnect left and right transmission supports (complete with transmission mounting) from transmission and subframe.

– Using transmission jack V.A.G 1383A, lower transmission slightly.

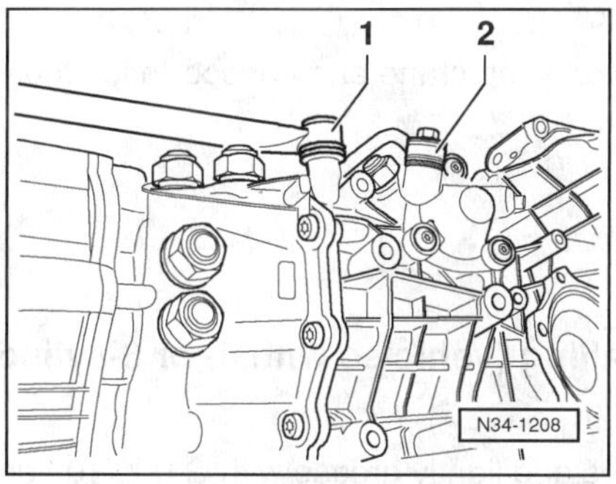

- Remove selector mechanism from transmission as follows:
  - Unbolt connecting rod (**2**) for selector rod from right side of transmission.

– Remove push rod (**1**) by unbolting hex socket head bolt.

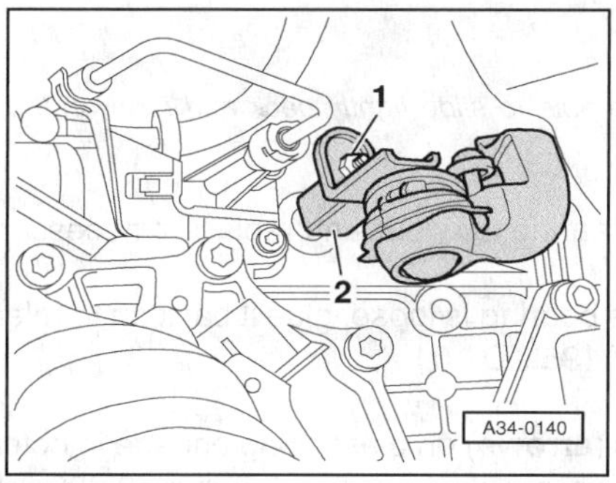

- Remove nut (**1**) and pull gear selector (**2**) from selector shaft of transmission.

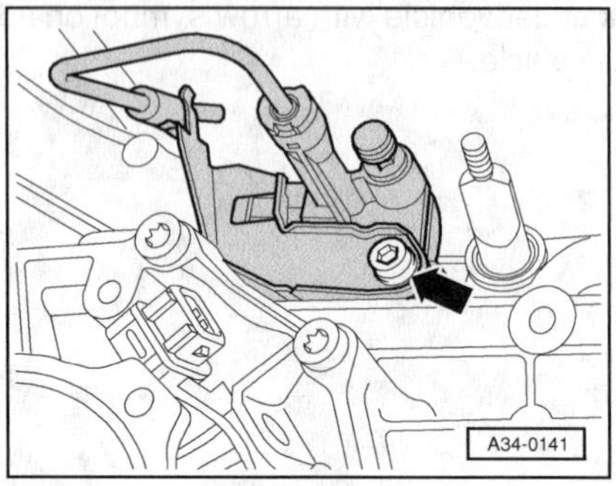

Remove clutch slave cylinder (**arrow**) and secure with wire without opening clutch hydraulic system.

NOTE—
Do not depress clutch pedal.

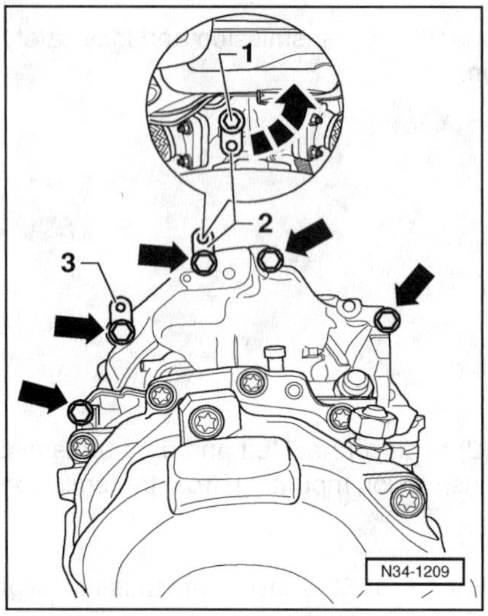

Remove upper engine/transmission securing bolts (**arrows**).

– If bracket (**3**) is installed, set it aside with lines and secure with wire.

– Press transmission off dowel sleeves.

– Lower transmission carefully, ensuring clearance at axle shafts.

– If work is to be performed on the transmission, it must be secured to the engine/transmission support VW 540.

Transporting the transmission

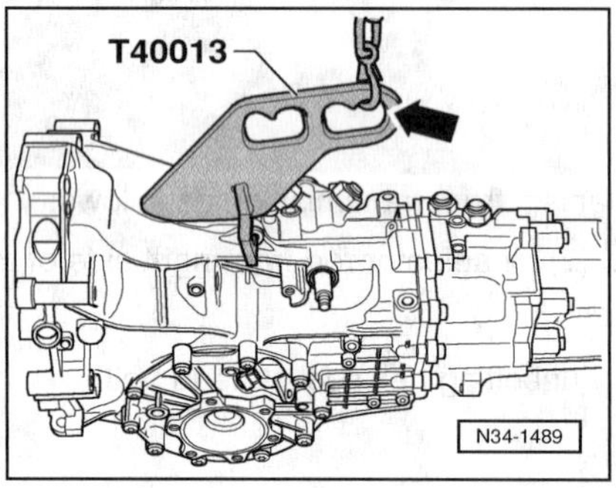

Mount transportation tool T40013 to transmission lifting eye and secure.

– Secure workshop crane hook at 4th attachment point (**arrow**) on transportation tool.

– Lift transmission with workshop crane and transportation tool T40013.

Transmission, installing (vehicles with 4- or 6-cylinder engines)

– Clean input shaft splines and lightly grease with G 000 100 grease.

NOTE—
The clutch plate must be able to slide lightly back and forth on the input shaft.

– Check clutch release bearing for wear and replace if necessary.

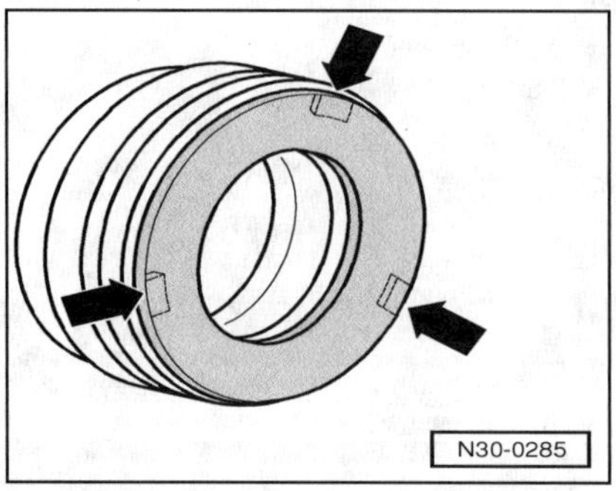

If plastic ring of release bearing is loose, glue it back on to release bearing with adhesive AMV 195 KD1 01.

– Three rectangular tabs (**arrows**) on plastic ring engage in notches of release bearing. If grooves are deeper than 0.5 mm (0.020 in), release bearing must be replaced.

– Check to insure that dowel sleeves for aligning engine/transmission are installed in cylinder block, install if necessary.

– Ensure that intermediate plate is correctly positioned on engine (if equipped).

– Thinly coat clutch release lever surface contacting plunger with copper grease.

– Before assembly, **without exception**, use a thread cutter/chaser to clean locking fluid residue from transmission flanged shaft threaded holes for driveshaft.

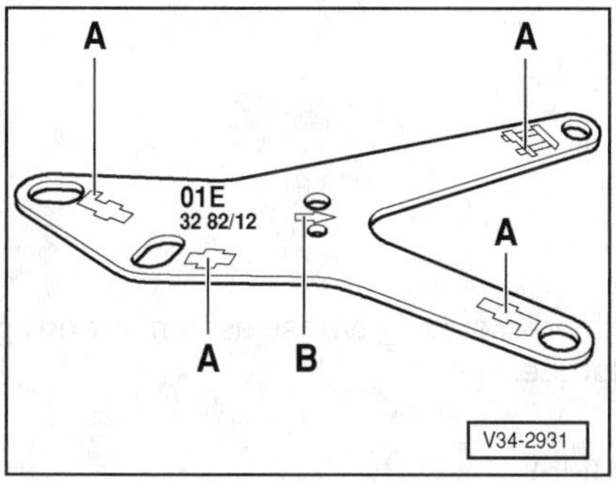

Mount transmission support 3282, adjustment plate 3282/12 for "**01E**" transmission and support elements on to transmission jack.

- Place adjustment plate 3282/12 on transmission support 3282 (adjustment plate fits only in one position).
- Align arms of transmission support according to holes in adjustment plate.
- Bolt on support elements (**A**) as shown on adjustment plate.
- Arrow symbol (**B**) on adjustment plate points to front of vehicle.

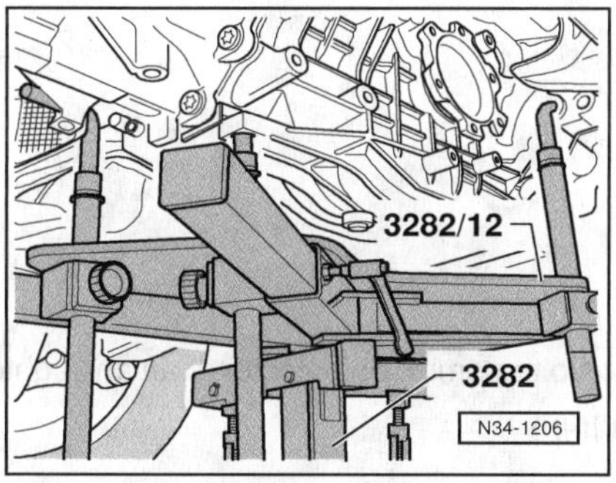

Place transmission on transmission jack. Align adjustment plate and transmission parallel with one another.

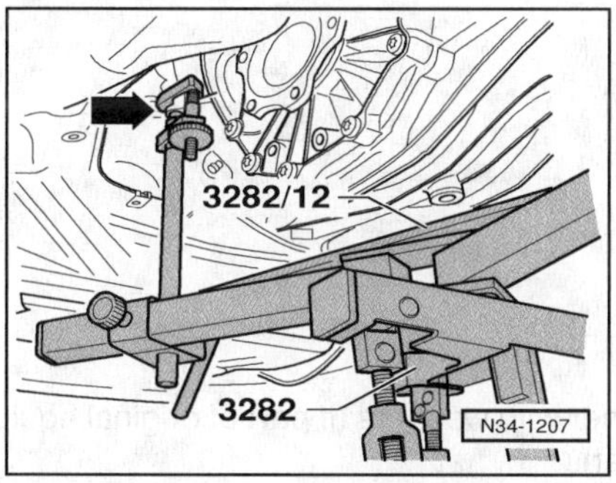

Lock safety support (**arrow**) on transmission. Place transmission jack under vehicle with arrow symbol on adjustment plate pointing to front of vehicle.

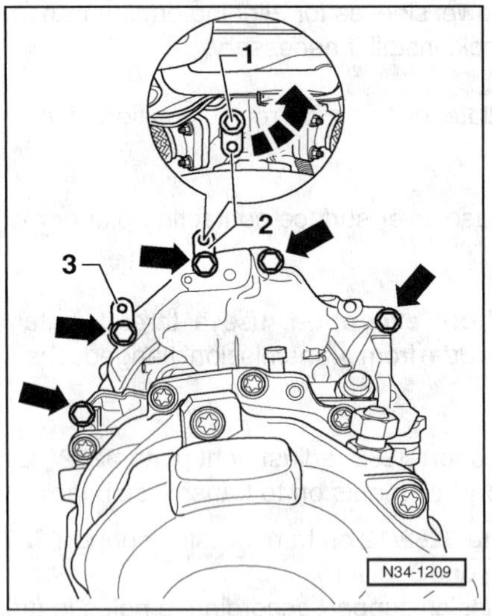

◀ Before installing transmission, ensure that brace (**2**) (if equipped) is swung to side (**direction of dashed arrow**).

– Install transmission, ensuring clearance at axle shafts.

– If bracket (**3**) is installed, push it along with any lines onto engine/transmission securing bolt.

– Tighten upper engine/transmission securing bolts to prescribed torque, see **Tightening Torques, transmission to engine.**

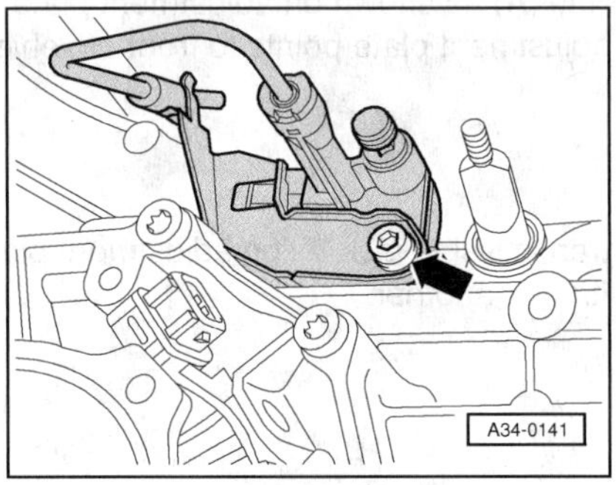

◀ Install slave cylinder with bracket for line/hose assembly and tighten bolt (**arrow**) to prescribed torque.

- Always replace
- Tighten to 25 Nm (18 ft-lb)

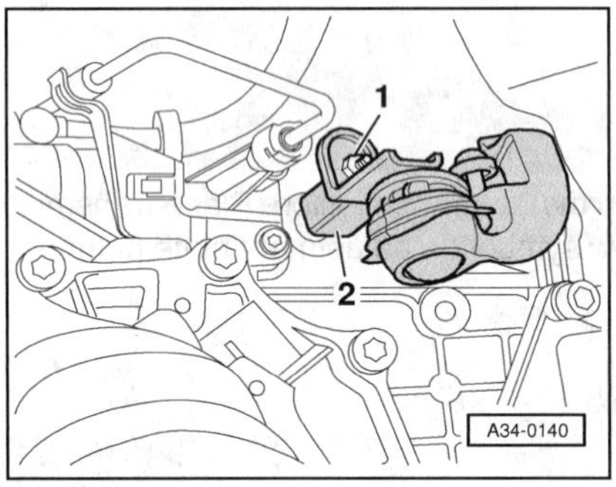

◀ Press gear selector (**2**) onto transmission selector shaft and tighten nut (**1**).

- Tighten to 25 Nm (18 ft-lb)

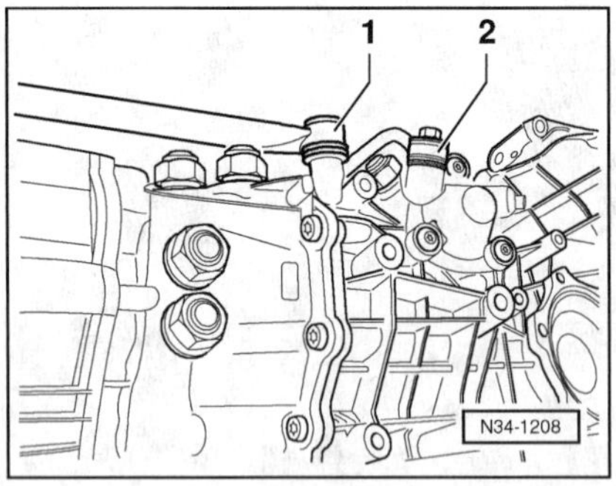

◀ Bolt push rod (**1**) together with washers (if part of original equipment) to transmission and tighten.

- Tighten to 40 Nm (30 ft-lb)

– Bolt connecting rod (**2**) for selector rod to right side of transmission.

- Tighten to 25 Nm (18 ft-lb)

– Raise transmission with transmission jack V.A.G 1383A until transmission supports can be inserted.

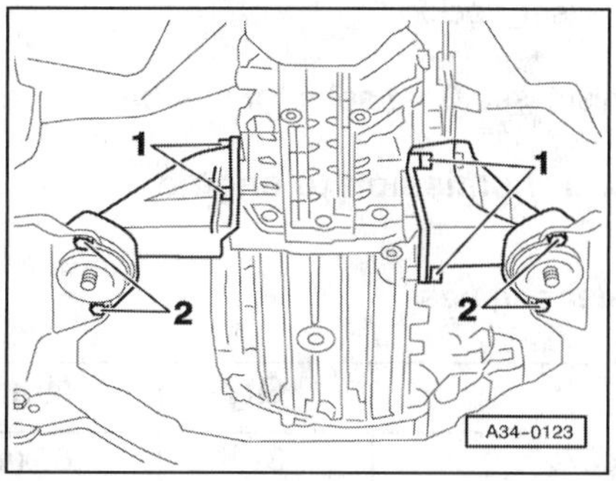

- Install right and left transmission supports complete with transmission mounting to transmission and subframe (**1** and **2**), see **Tightening torques, assembly mountings**.

– Install lower engine/transmission securing bolts and bolts for starter. When doing this, install starter. See **Tightening Torques, transmission to engine.**

– Connect back-up light connector.

– Connect speedometer sender connector.

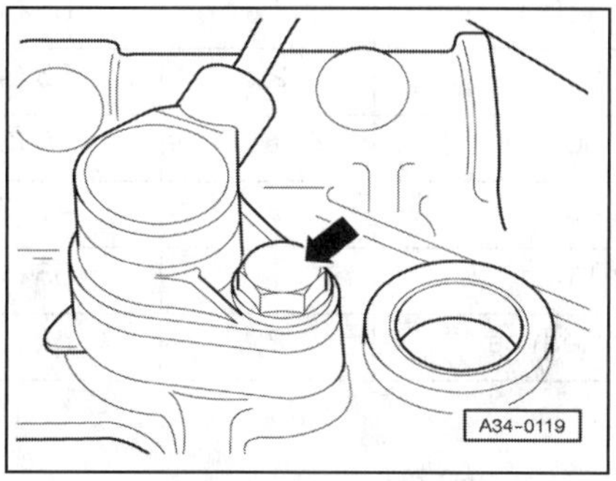

- Install Engine speed (RPM) sensor (**G28**), if previously removed, on transmission housing (**arrow**).

– Connect driveshaft, see **Repair Group 39.**

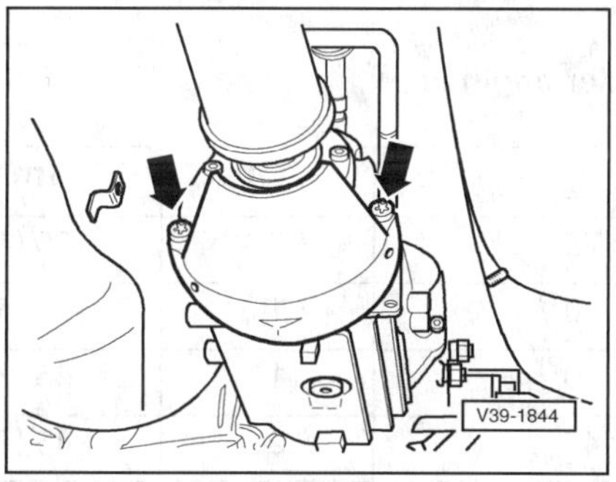

- Install protective plate (**arrows**), see **Repair Group 39.**

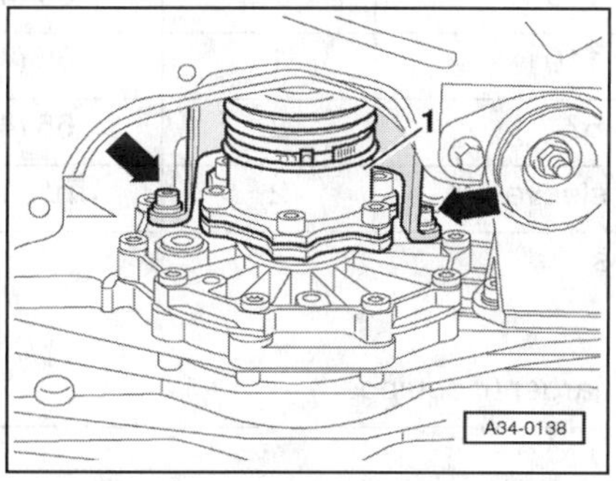

- Install axle shafts (**1**) to flange shafts.
  - Stage I:
    Tighten to 10 Nm (7 ft-lbs) diagonally
  - Stage II:
    M8 bolts, tighten to 40 Nm (30 ft-lb) diagonally
    M10 bolts, tighten to 77 Nm (57 ft-lb) diagonally

– Install heat shields for left and right axle shafts (**arrows**). Third bolt cannot be seen in illustration.
  - Tighten to 25 Nm (18 ft-lb)

– Install covers in wheel housings.

– Install front exhaust system with catalyst, see **Repair group 26.**

– Check transmission oil, see **Maintenance**.

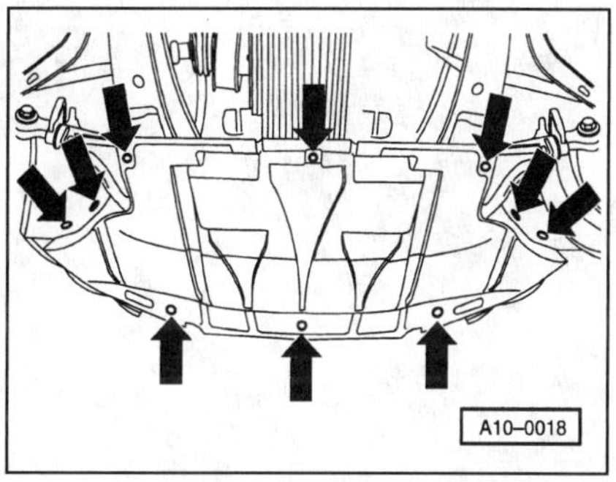

- Install sound insulation tray (**arrows**).

– Install engine cover.

– Reconnect Ground strap to battery.

– Recode radio.

– Initialize power windows, set clock.

– Adjust selector mechanism, see **Gear selector mechanism, adjusting**.

Tightening torques, transmission to engine

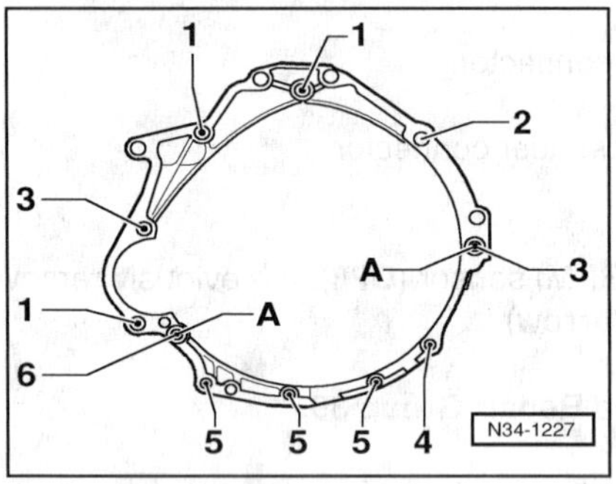

Vehicles with 4-cylinder engines

Item	Bolt	Qty.	Nm (ft-lb)
1	M 12 x 67	3	65 (48)
2	M 12 x 110	1	65 (48)
3	M 12 x 90	2	65 (48)
4	M 10 x 50[1)]	1	45 (33)
5	M 10 x 45	3	45 (33)
6	M 12 x 110[2)]	1	65 (48)
A	Dowel sleeves	2	n/a

1) Special bolt with nut
2) With nut

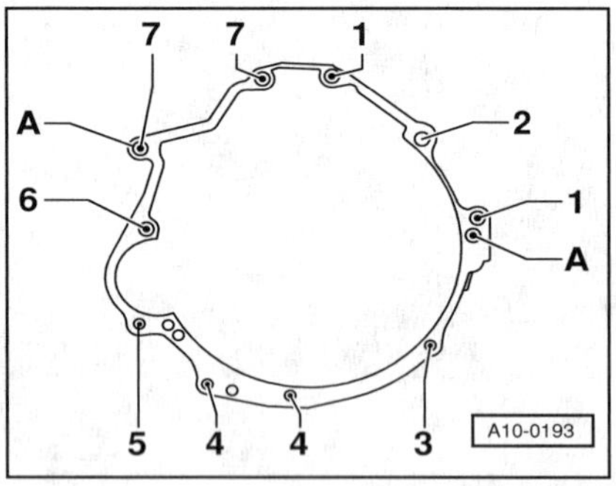

Vehicles with 6-cylinder engines

Item	Bolt	Qty.	Nm (ft-lb)
1	M 12 x 75	2	65 (48)
2	M 12 x 90[1)]	1	65 (48)
3	M 10 x 50[2)]	1	45 (33)
4	M 10 x 45	2	45 (33)
5	M 10 x 135[3)]	1	65 (48)
6	M 12 x 110	1	65 (48)
7	M 12 x 67[4)]	2	65 (48)
A	Dowel sleeves	2	n/a

1) With bracket for lines
2) Special bolt with nut
3) With nut
4) With strut for turbocharger (if equipped)

Tightening torques, assembly mountings

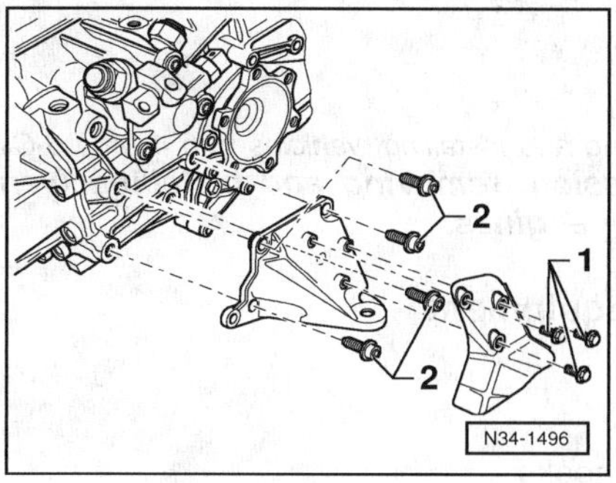

Left and right transmission supports to transmission (right side illustrated)

Bolts (**1**)	10 Nm (7 ft-lb)
Bolts (**2**)	40 Nm (30 ft-lb)

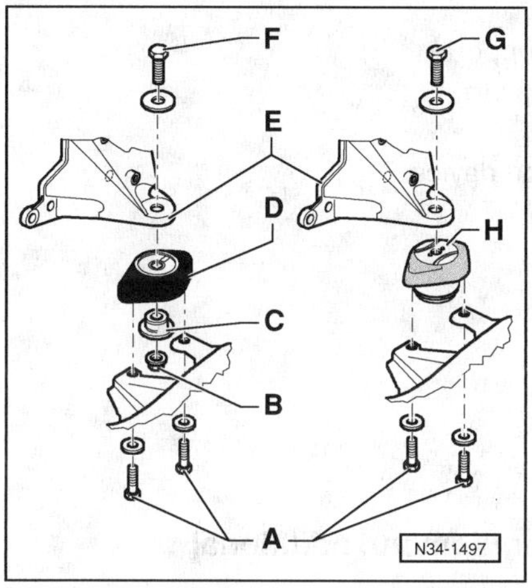

Left and right transmission supports to subframe (right side illustrated)

- Positions (**C**) and (**B**) should be ignored on hydraulic transmission mountings (**H**) as bolt (**G**) is screwed directly in the transmission mounting.

Bolts (**A**)	25 Nm (18 ft-lb)
Nut (**B**) to bolt (**F**)	50 Nm (37 ft-lb)
Bolt (**G**)	50 Nm (37 ft-lb)

Tightening torques, additional torque values

Tightening torques	Torque
Bolt, slave cylinder	25 Nm (18 ft-lb)
Nut, selector to transmission selector shaft	25 Nm (18 ft-lb)
Bolt, push rod to transmission	40 Nm (30 ft-lb)
Bolt, connecting rod for selector shaft to transmission	25 Nm (18 ft-lb)
Bolt, axle shaft to drive flange Stage 1 (diagonally) Stage 2 (diagonally) **M8** Stage 2 (diagonally) **M10**	 10 Nm (7 ft-lb) **M8** 40 Nm (30 ft-lb) **M10** 77 Nm (57 ft-lb)
Bolt, axle heat shields	25 Nm (18 ft-lb)

Transmission, removing and installing (vehicles with 8-cylinder engines)

NOTE—

For transmission removing and installing, vehicles with 4- and 6-Cylinder engines, see ***Transmission, removing and installing (vehicles with 4- and 6-cylinder engines.***

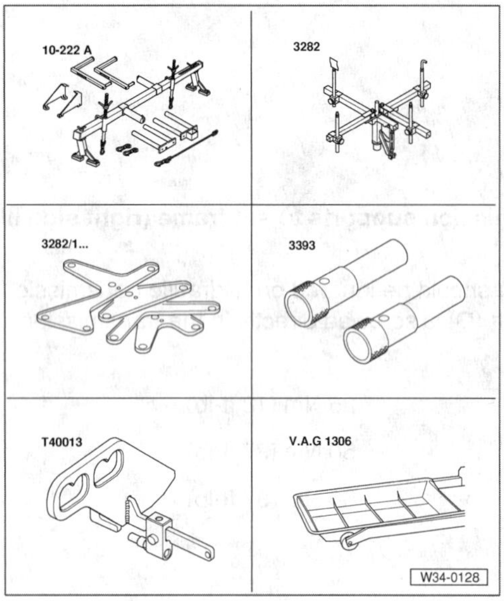

Special tools and equipment

- 10-222 A Support
- 10-222 A/1 Support
- 10-222 A/2 Auxiliary hooks
- 10-222 A/10 Spindle
- 10-222 A/18 Adapter
- 3282 Transmission support
- 3282/12 Adjustment plate
- 3393 Test mandrel
- T40013 Transportation device
- V.A.G 1306 Drip tray

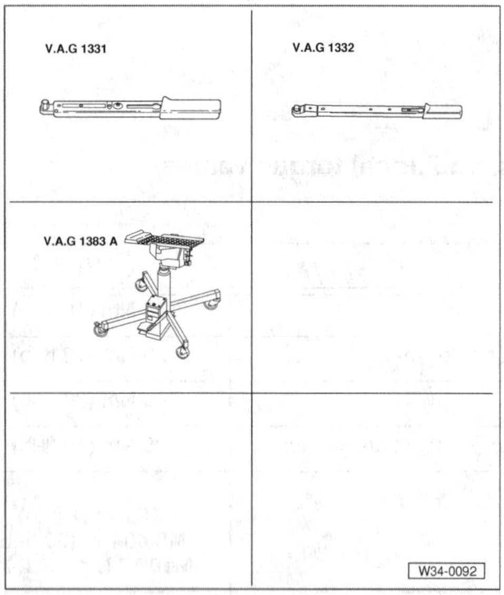

Special tools and equipment, additional

- V.A.G 1331 Torque wrench
- V.A.G 1332 Torque wrench
- V.A.G 1383 A Engine/transmission jack

Additional information

- **Suspension, Wheels, Steering section**
- **Electrical Equipment section**
- **Engine Mechanical section**

Removing

– First, check whether a coded radio is installed. If so, obtain anti-theft coding.

– With ignition switched off disconnect battery Ground (GND) strap.

– Remove intake air duct, air cleaner and intake hose. See **Fuel Injection & Ignition, section, Repair Group 24.**

– Bring lock carrier into service position. See **Body Exterior section, Repair Group 50**

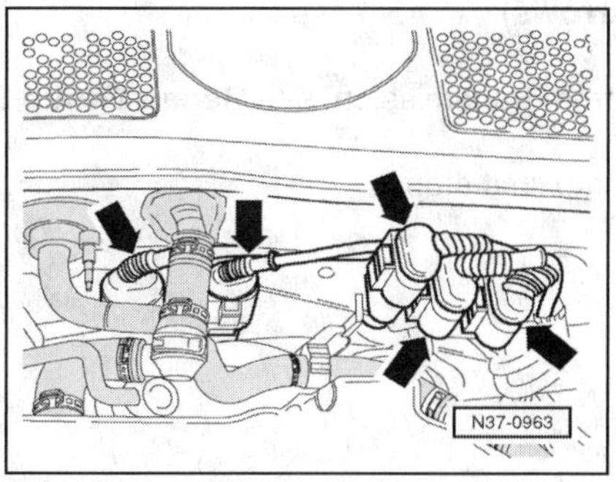

◄ Unclip connectors (**arrows**) from retainer.

– Remove connector from coolant expansion tank.

– Unbolt coolant expansion tank and place to one side. Do not open coolant system.

– Remove oil filler pipe from engine.

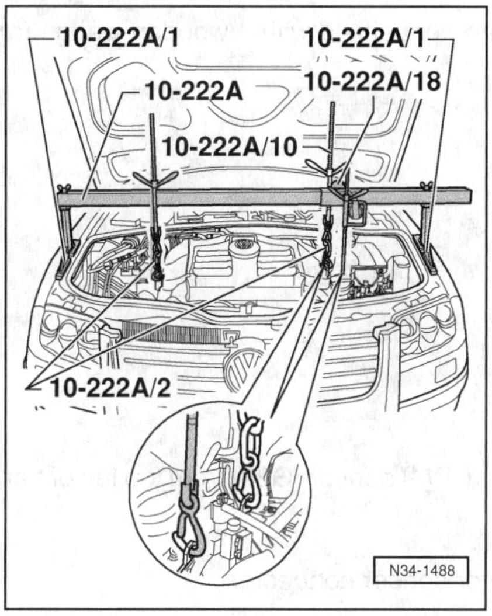

◄ Fit support bar 10-222A together with legs 10-222 A/1 and adapter 10-222 A/18. Longer side of adapter 10-222 A/18 faces forward.

– Secure third spindle 10-222 A/10 with additional hook 10-222 A/2 on lifting eye at rear left of engine.

– Hook second additional hook 10-222 A/2 on lifting eye at rear right of engine.

– Take weight of engine/transmission assembly onto spindles.

– Loosen wheel bolts.

– Raise vehicle.

– Remove front wheels.

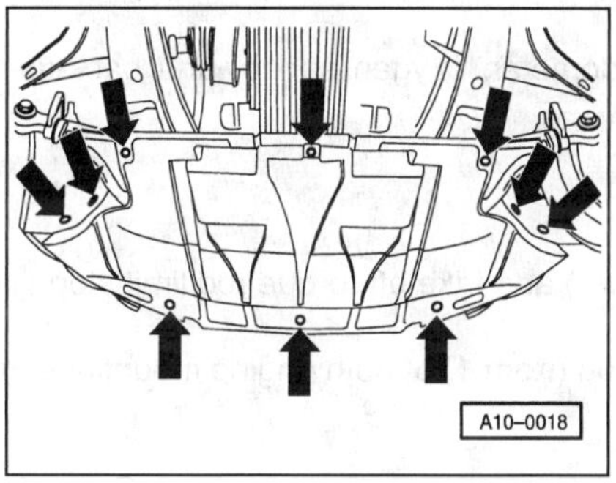

◄ Remove sound insulation tray (**arrows**).

– Remove sound insulation tray bracket.

– Remove front exhaust pipes. See **Engine Mechanical, section, Repair Group 26**.

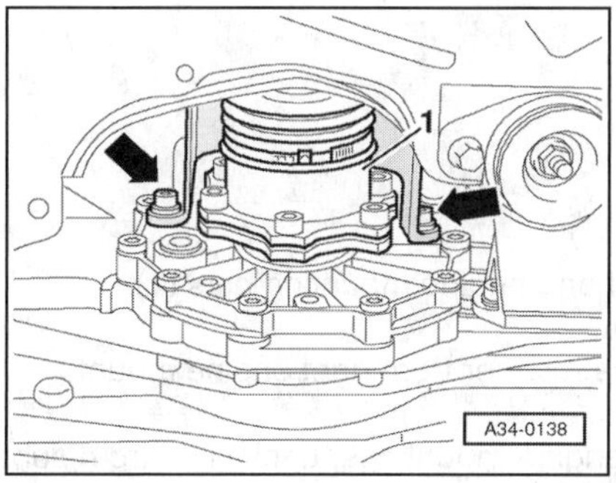

◄ Remove protective plates for left and right-hand axle shafts (arrows). Third bolt cannot be seen in illustration.

– Disconnect left and right-hand axle shafts (**1**) from flanges and tie-up.

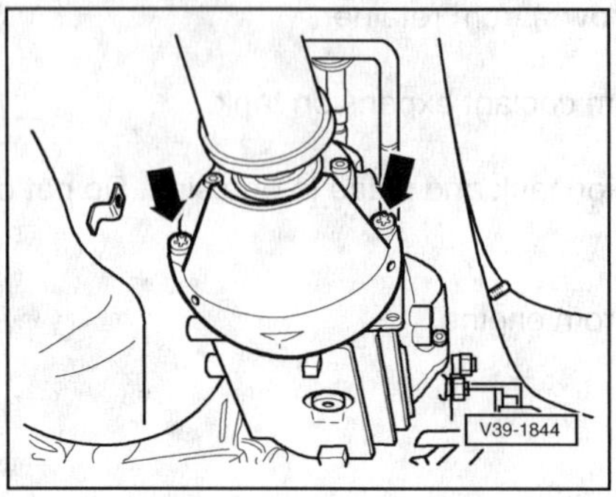

◀ Remove cover plate (**arrows**).

– Disconnect driveshaft from transmission, see **Repair Group 39**.

– Push driveshaft back onto limit stop.

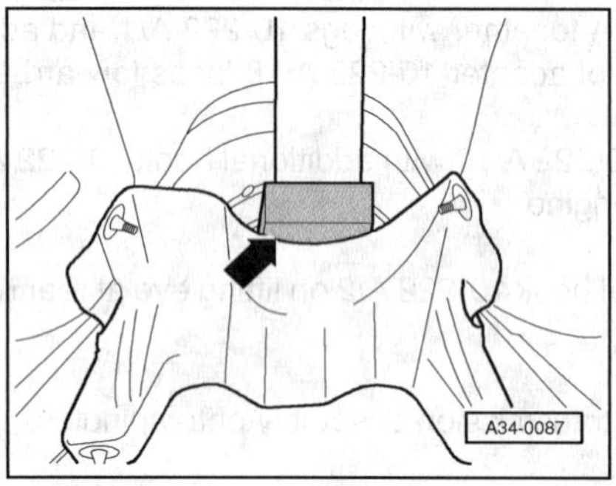

◀ Press driveshaft up onto heat shield with a wooden wedge (**arrow**).

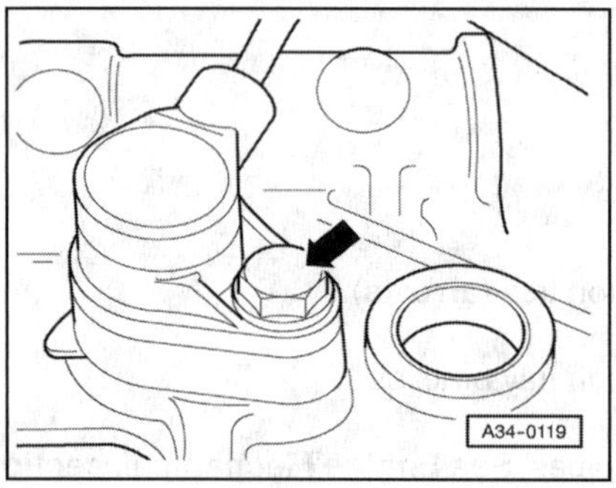

◀ Remove Engine speed (RPM) sensor -G28- from top left of transmission housing (**arrow**).

– Disconnect speedometer sensor connector.

– Disconnect back-up light connector.

– Unclip wires for speedometer, oxygen sensor and back-up light from transmission.

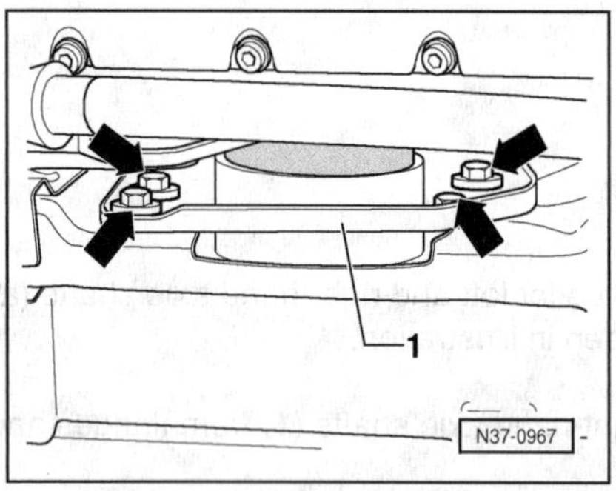

◀ Remove screws (**arrows**) and take off torque rod limit stop (**1**).

– Mark installation position (**item 1**) of both engine mountings from below.

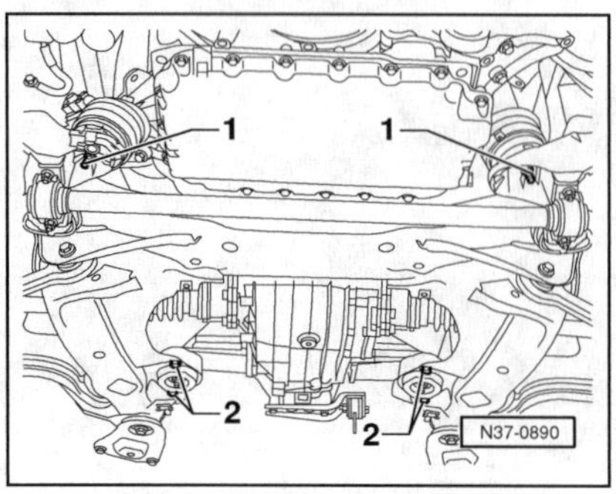

◀ Remove left and right-hand engine mounting nuts (**1**).

– Remove wiring harness retainer from right engine mount.

– Pull vacuum lines off engine mountings, unclip line from subframe and tie to side.

– Remove bolts (**2**) from transmission mounting to subframe.

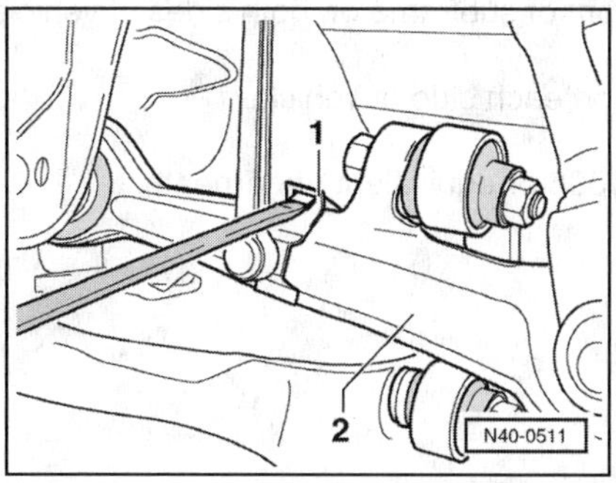

- Unclip coupling rod retaining clamp (**1**) of left front level control system sensor (G78) from left control arm (**2**).

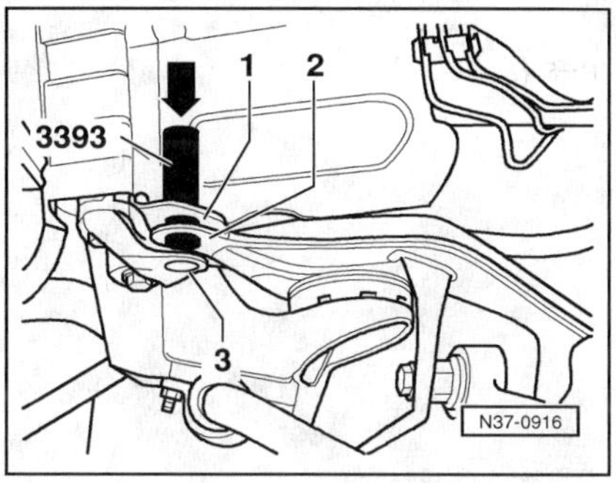

- Check whether holes (**1** and **2**) are aligned on both sides of vehicle using test drifts 3393. Disregard hole (**3**)

NOTE —

If the holes are not aligned, perform steering alignment measurements after tightening the subframe!

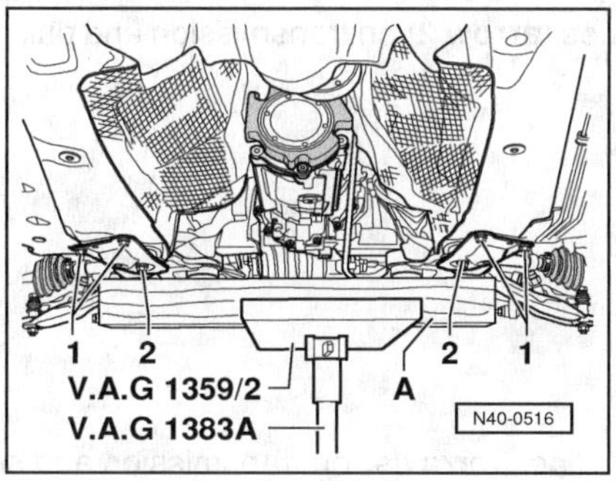

- Place transmission jack V.A.G 1383 A with 1359/2 under subframe with a block of wood (**A**) and exert slight counter-pressure.

– Unscrew bolts (**1** and **2**) at rear of subframe on both sides of vehicle.

– Using transmission jack V.A.G 1383A, lower subframe slightly.

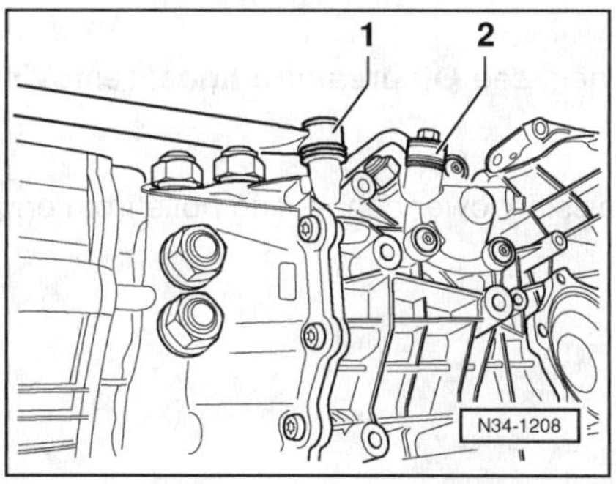

- Remove selector mechanism from transmission. Push rod (**1**) and connecting rod (**2**) securing bolts must be accessible.

– Unbolt connecting rod (**2**) for selector rod from right side of transmission.

– Remove push rod (**1**) by unbolting hex socket head bolt.

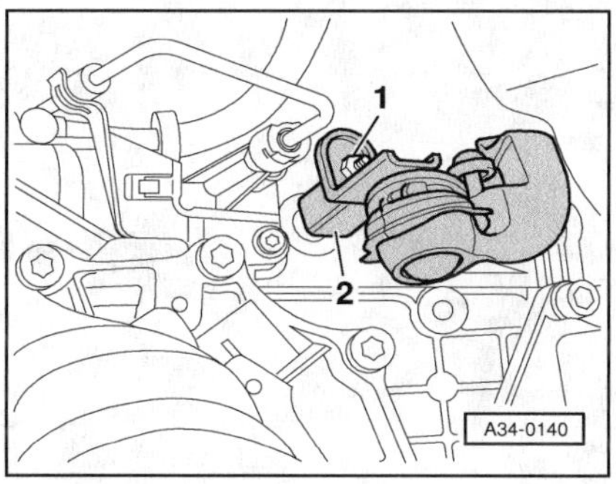

- Remove nut (**1**) and pull gear selector (**2**) from selector shaft of transmission.

– Remove transmission jack V.A.G 1383 A with 1359/2 from under subframe.

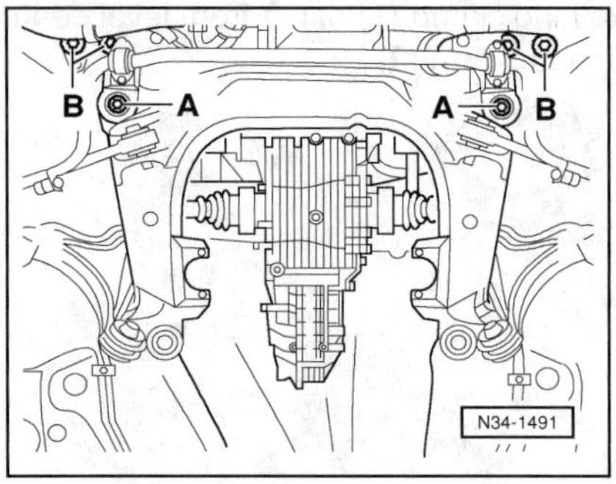

◀ Remove bolts (**A**) at front of subframe on both sides of vehicle.

– Remove bolts (**B**) two on each side of vehicle.

– Place drip tray V.A.G 1306 or equivalent underneath.

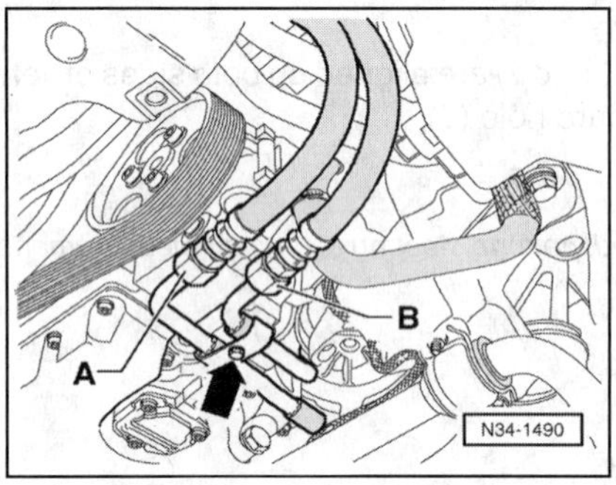

◀ Remove oil pressure line retainer (**arrow**) from engine.

– Separate oil pressure lines (**A** and **B**).

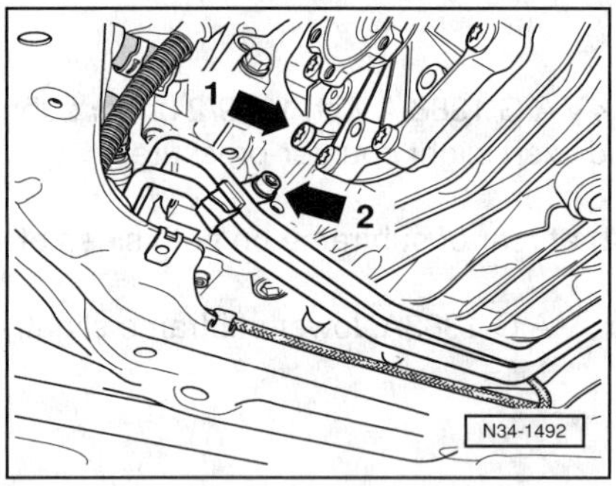

◀ Remove bolt (**arrow 1**) from final drive cover.

– Remove oil pressure lines (**arrow 2**) on transmission and pull out and off to the side.

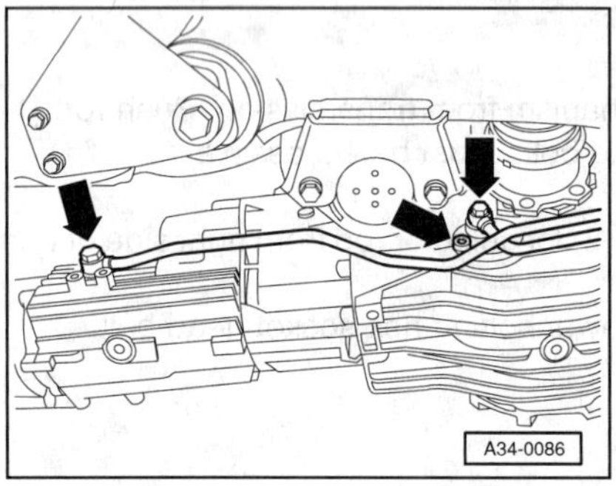

◀ Disconnect oil pressure lines (**arrows**) on transmission and seal holes with plugs.

– Remove oil pressure lines, see **Oil pressure lines, removing and installing.**

– Remove engine/transmission lower connecting bolts from engine side.

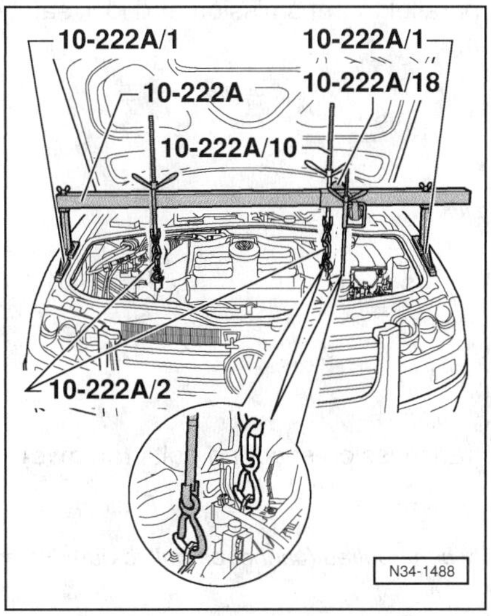

Lower engine and transmission assembly approximately 4 turns on spindles.

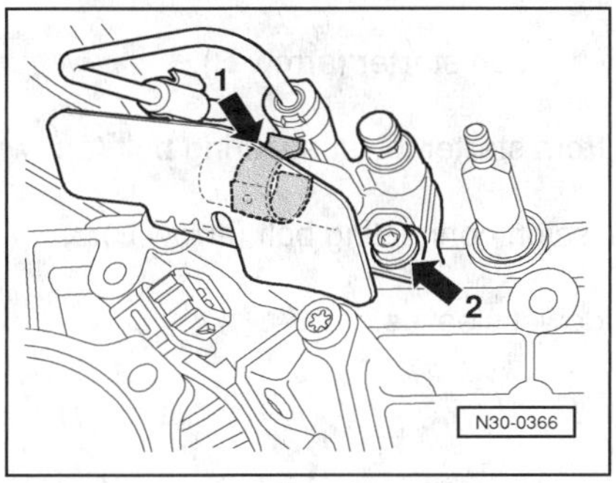

Remove clutch slave cylinder (**arrow 1**) with heat shield (**arrow 2**) and secure with wire without opening clutch hydraulic system.

NOTE —

Do not depress clutch pedal.

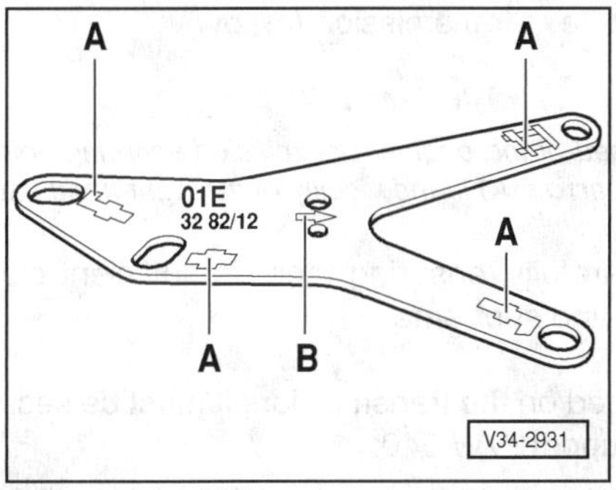

Mount transmission support 3282, adjustment plate 3282/12 for "**01E**" transmission and support elements on to transmission jack.

- Place adjustment plate 3282/12 on transmission support 3282 (adjustment plate fits only in one position).
- Align arms of transmission support according to holes in adjustment plate.
- Bolt on support elements (**A**) as illustrated on adjustment plate.
- Arrow symbol (**B**) on adjustment plate points to front of vehicle.

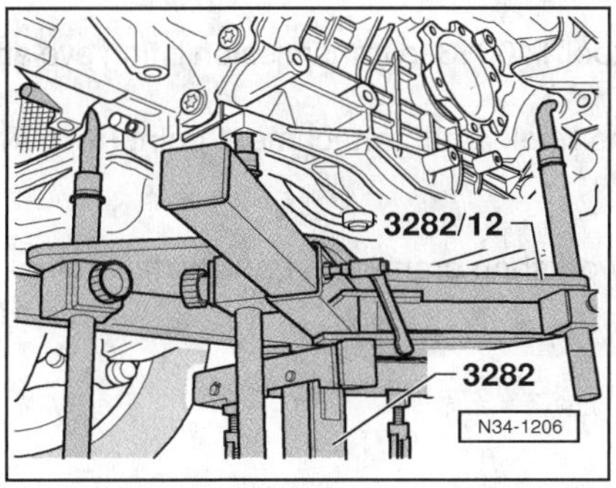

Place transmission jack under vehicle with arrow symbol on adjustment plate pointing to front of vehicle.

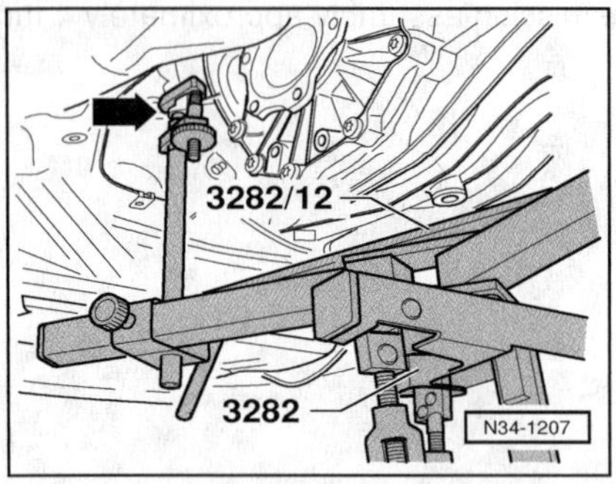

Align adjustment plate parallel to transmission and lock safety support (**arrow**) on transmission.

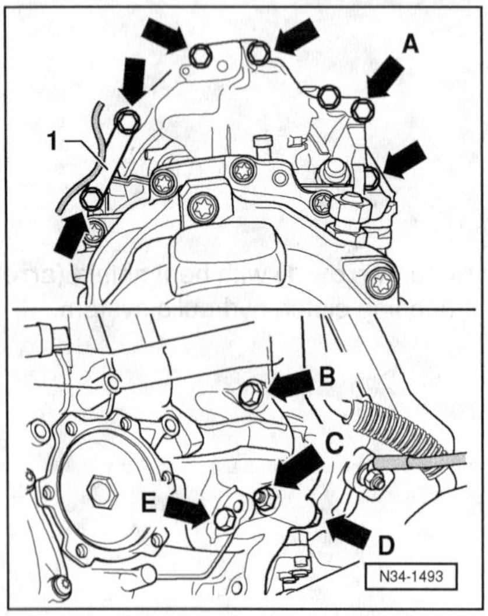

Remove upper engine/transmission securing bolts (**arrows**).

NOTE—

The bolt (arrow ***A****) can only be unbolted (cannot be pulled out). Remember this when reinstalling.*

- If bracket (**1**) is installed, set it aside with lines and secure with wire.
- Remove upper securing bolt on starter (arrow **B**).
- Remove nut (arrow **C**) from starter lower securing bolt (arrow **D**).
- Remove engine/transmission connecting bolt (arrow **E**).
- Press transmission off dowel sleeves.

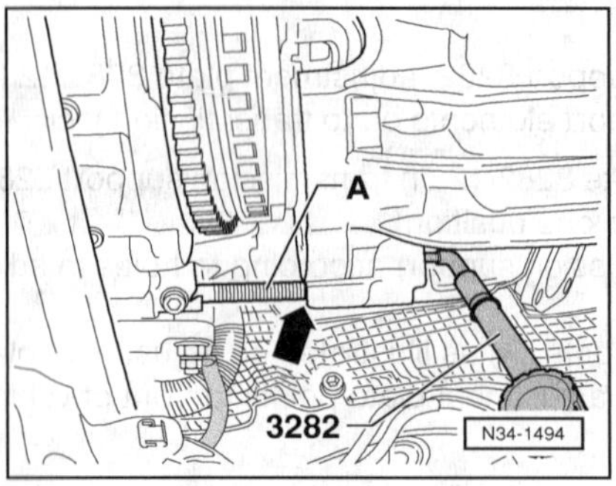

Guide transmission straight off engine until starter lower securing bolt (**A**) is no longer in its hole in transmission (**arrow**).

NOTE—

When doing this, take heed of the engine/transmission securing bolt (illustration N34-1493, above, ***arrow A****) in the vicinity of the right-hand catalyst.*

- Lower transmission carefully, ensuring there is sufficient clearance at axle shafts, driveshaft and subframe.
- If work is to be performed on the transmission, it must be secured to the engine/transmission support VW 540.

Transporting the transmission

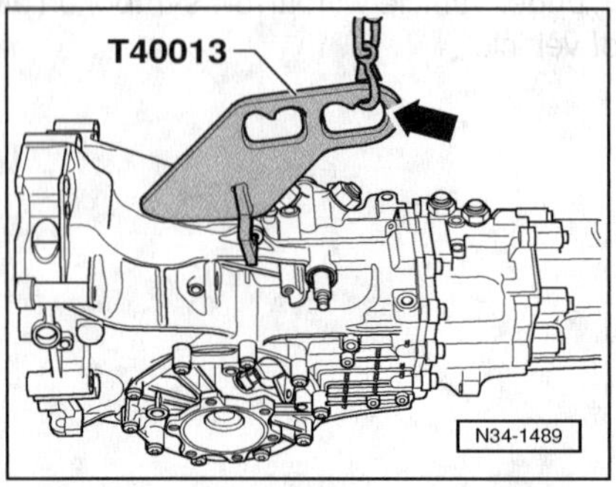

Mount transportation tool T40013 to transmission lifting eye and secure.

- Secure workshop crane hook at 4th attachment point (**arrow**) on transportation tool.
- Lift transmission with workshop crane and transportation tool T40013.

Transmission, installing (vehicles with 8-cylinder engines)

NOTE—
When replacing transmission, transfer transmission supports complete with transmission mounting.

– Clean input shaft splines and lightly grease with G 000 100 grease.

NOTE—
The clutch plate must be able to slide lightly back and forth on the input shaft.

– Check clutch release bearing for wear and replace if necessary.

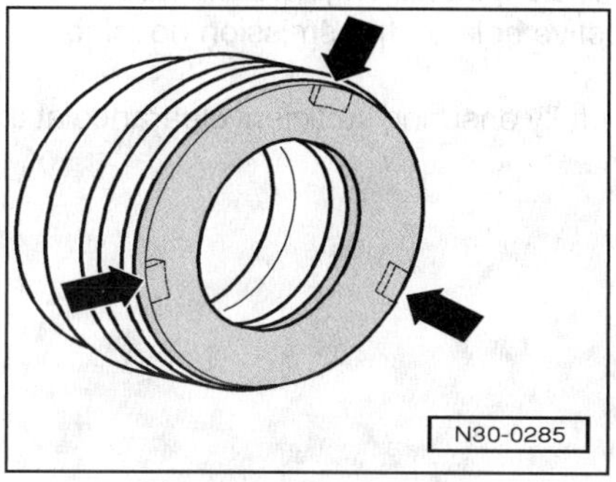

If plastic ring of release bearing is loose, glue it back on to release bearing with adhesive AMV 195 KD1 01.

– Three rectangular tabs (**arrows**) on plastic ring engage in notches of release bearing. If grooves are deeper than 0.5 mm, release bearing must be replaced.

– Check to insure that dowel sleeves for aligning engine/transmission are installed in cylinder block, install if necessary.

– Thinly coat clutch release lever surface contacting plunger with copper grease.

– Before assembly, **without exception**, use a thread cutter/chaser to clean locking fluid residue from transmission flanged shaft threaded holes for driveshaft.

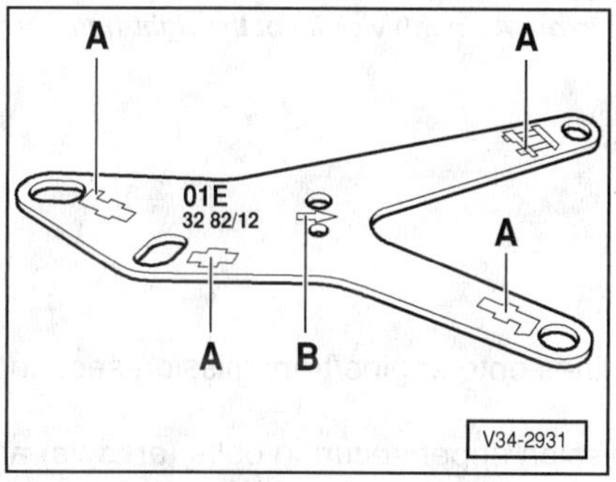

Mount transmission support 3282, adjustment plate 3282/12 for "**01E**" transmission and support elements on to transmission jack.

- Place adjustment plate 3282/12 on transmission support 3282 (adjustment plate fits only in one position).
- Align arms of transmission support according to holes in adjustment plate.
- Bolt on support elements (**A**) as shown on adjustment plate.
- Arrow symbol (**B**) on adjustment plate points to front of vehicle.

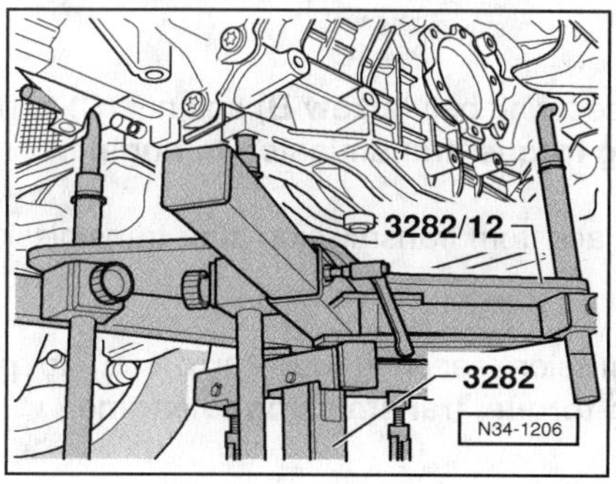

Place transmission on transmission jack. Align adjustment plate and transmission parallel with one another.

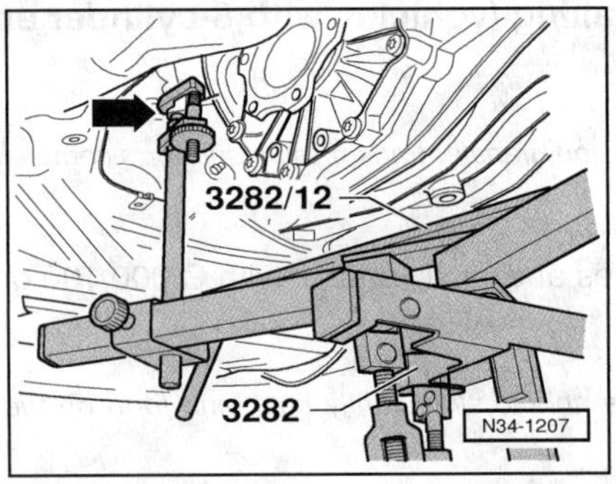

Lock safety support (**arrow**) on transmission. Place transmission jack under vehicle with arrow symbol on adjustment plate pointing to front of vehicle.

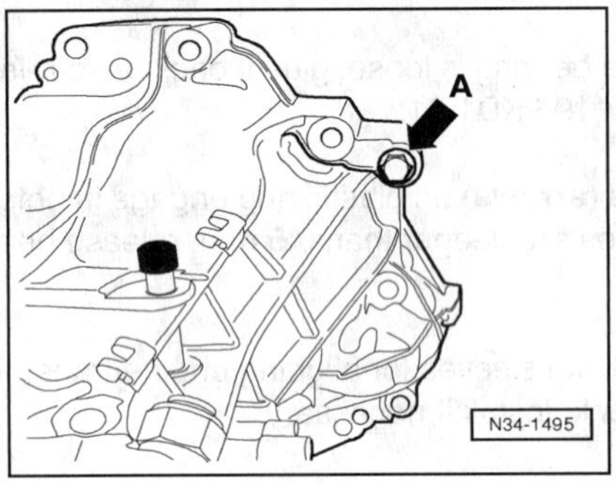

Before installing transmission, install engine/transmission connecting bolt (**arrow A**) in respective hole on transmission housing.

– Raise transmission carefully ensuring sufficient clearance at axle shafts, driveshaft and subframe.

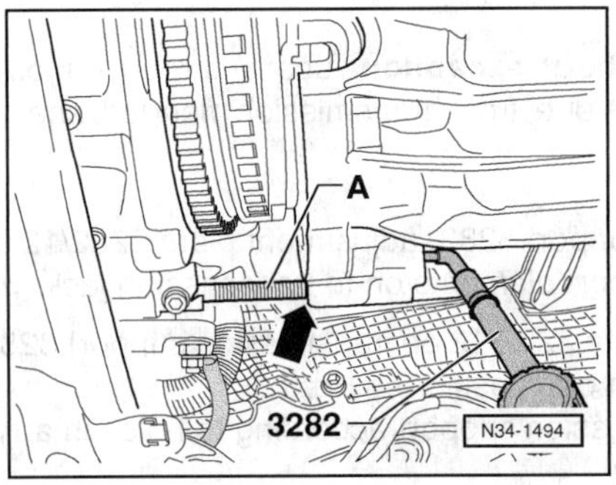

Align transmission to engine, and guide lower securing bolt (**A**) into respective hole of transmission housing (**arrow**).

NOTE —
*When doing this, take heed of the engine/transmission securing bolt (illustration N34-1493, above, **arrow A**) in the vicinity of the right-hand catalyst.*

– Install transmission.

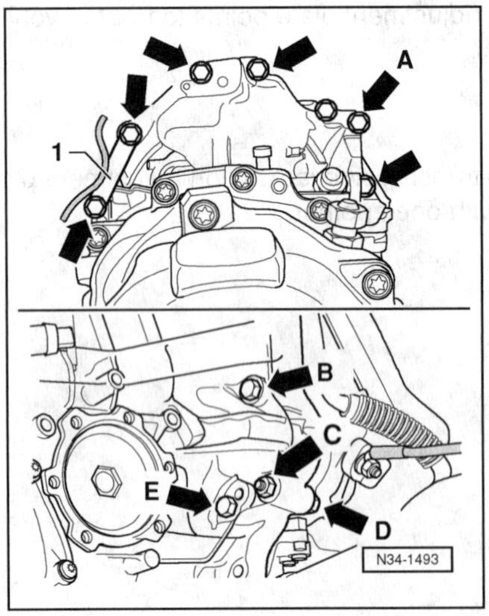

Install bracket (**1**) with lines onto engine/transmission securing bolts.

– Tighten engine/transmission upper securing bolts (**arrows**) and (**arrow A**) to prescribed torque, see **Tightening torque, transmission to engine**.

– Tighten starter upper securing bolt (**arrow B**) and nut (**C**) to prescribed torque, see **Tightening torque, transmission to engine**.

– Remove transmission jack from transmission after transmission is bolted to engine.

– Tighten engine/transmission securing bolt (**arrow E**) to prescribed torque see **Tightening torque, transmission to engine**.

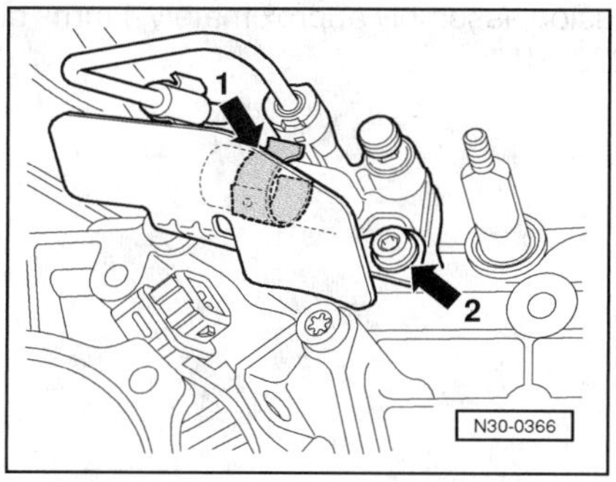

Install slave cylinder along with retainer for line/hose.

– Press heat shield (**arrow 1**) onto slave cylinder, install new bolt (arrow **2**), and tighten to prescribed torque.

  • Always replace
  • Tighten to 25 Nm (18 ft-lb)

– Install Engine speed (RPM) sensor (**G28**) on transmission housing.

– Install oil pressure lines between engine/transmission assembly and subframe.

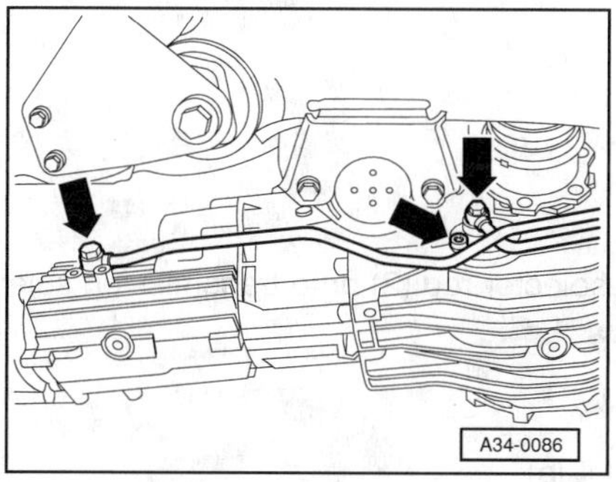

Install new seals on oil pressure lines and tighten oil pressure lines on transmission to prescribed torque (**arrows**).

  • Tighten to 25 Nm (18 ft-lb)

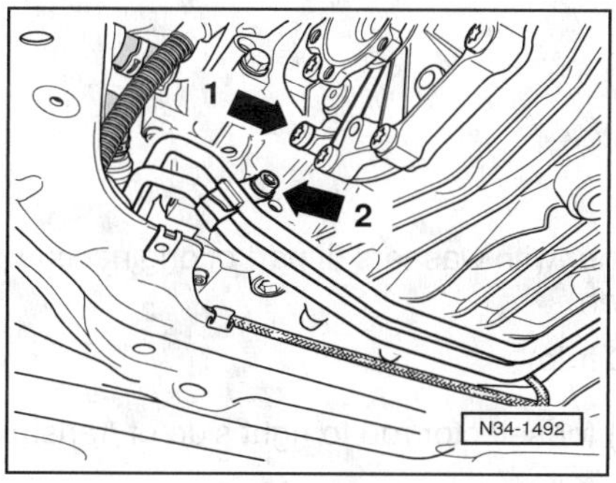

Install oil pressure line securing bolt to transmission (**arrow 2**) and tighten to prescribed torque, see **Oil pressure lines, removing and installing**.

– Install bolt (**arrow 1**) to final drive cover and tighten to prescribed torque.

  • Tighten to 25 Nm (18 ft-lb)

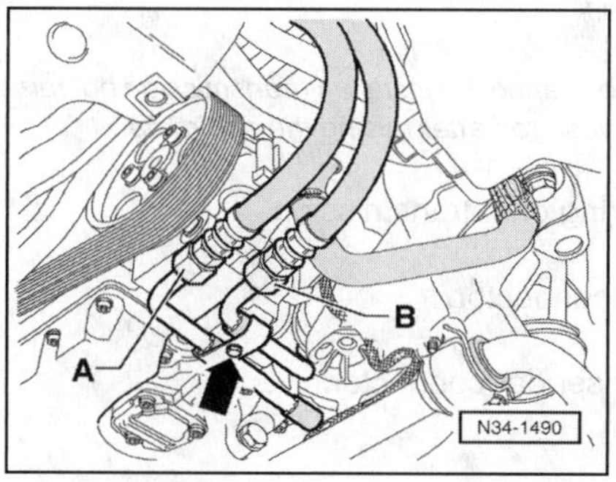

Install oil pressure line retainer (**arrow**) on engine.

  • Tighten to 10 Nm (7 ft-lb)

– Reconnect oil pressure lines (**A** and **B**) to cooler connecting point.

  • Tighten to 30 Nm (22 ft-lb)

– Install side and lower engine/transmission connecting bolts and tighten to prescribed torque, see **Tightening torques, additional values.**

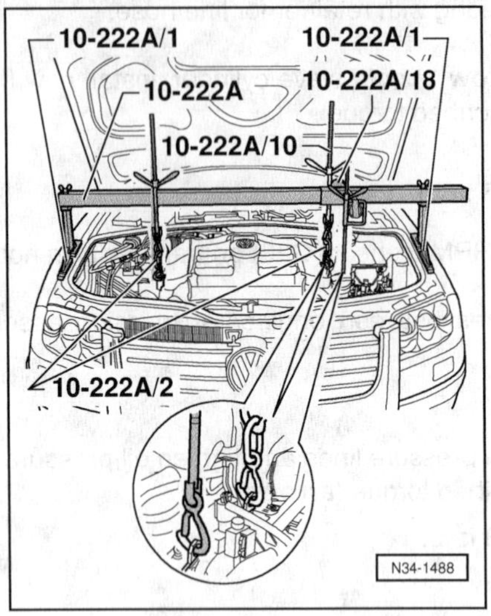

< Lift engine and transmission assembly approximately 4 turns on spindle. Start with spindle 10-222 A/10.

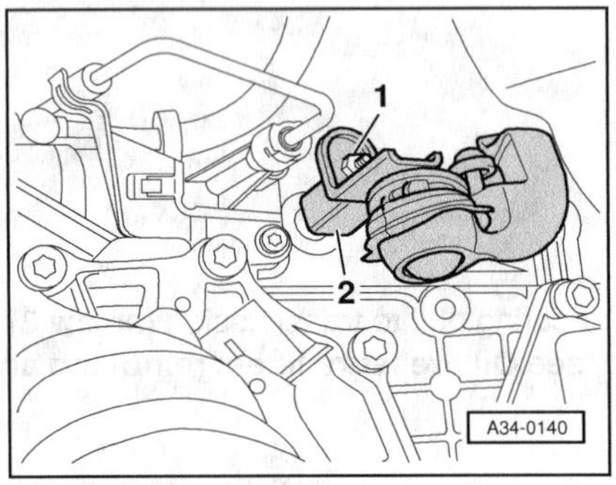

< Press gear selector of selector rod (**2**) onto transmission selector shaft.

– Tighten nut (**1**).

- Always replace
- Tighten to 25 Nm (18 ft-lb)

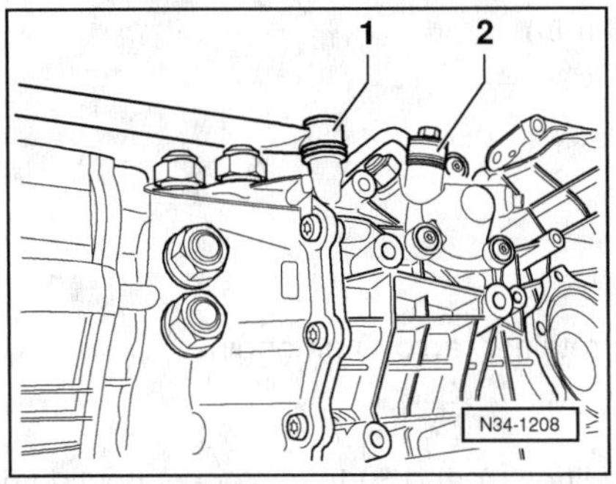

< Bolt push rod (**1**) together with washers (if part of original equipment) to transmission and tighten.

- Tighten to 40 Nm (30 ft-lb)

– Bolt connecting rod (**2**) for selector rod to right side of transmission.

- Tighten to 25 Nm (18 ft-lb)

NOTE —
If the push rod securing bolt cannot be installed (transmission not raised sufficiently), then install the push rod after raising the subframe.

– Clip oxygen sensor wiring onto transmission.

– Connect back-up light connector.

– Connect speedometer sender connector.

– Install subframe.

Subframe installation with alignment

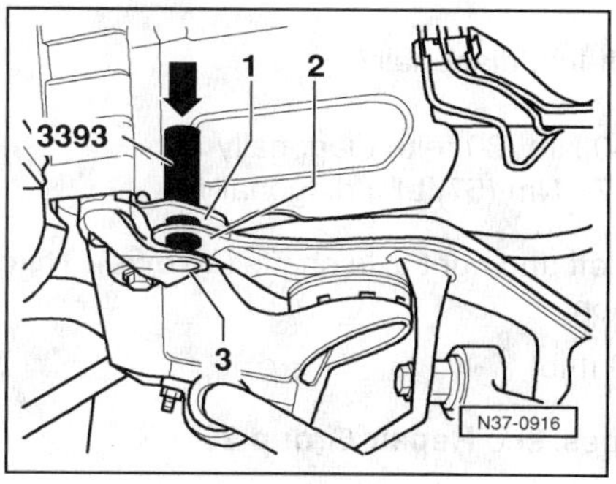

If holes (**1** and **2**) **did not** align before loosening subframe, the wheel alignment must be checked.

– Install subframe using bolts originally removed. These must be replaced following wheel alignment. See **Suspension, Wheels, Steering section, Repair Group 40; Removing and installing subframe**

NOTE—
When doing this, tighten original bolts only to prescribed torques.

WARNING—
*The vehicle **must not** be driven at this point.*

Subframe installation without alignment

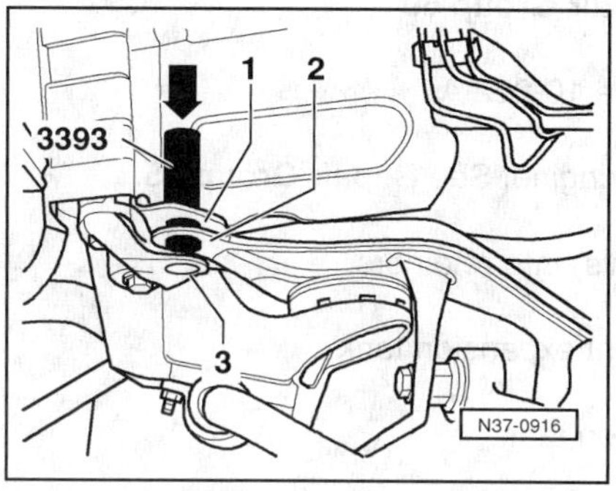

If holes (**1**) and (**2**) **did** align before loosening subframe, the wheel alignment need not be checked.

– Install subframe with new bolts. Observe tightening sequence. See **Suspension, Wheels, Steering section, Repair Group 40.**

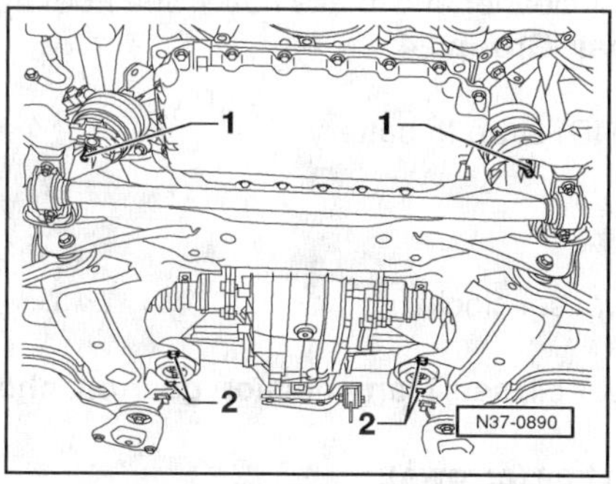

Tighten new M8 nuts (**1**) for both engine mountings.

- Always replace
- Tighten to 25 Nm (18 ft-lb)

– Install bolts (**2**) for transmission mounting/subframe.

- Tighten to 25 Nm (18 ft-lb)

– Clip in engine mounting vacuum line on subframe and connect lines to engine mounting.

– Install wiring harness retainer on right engine mount.

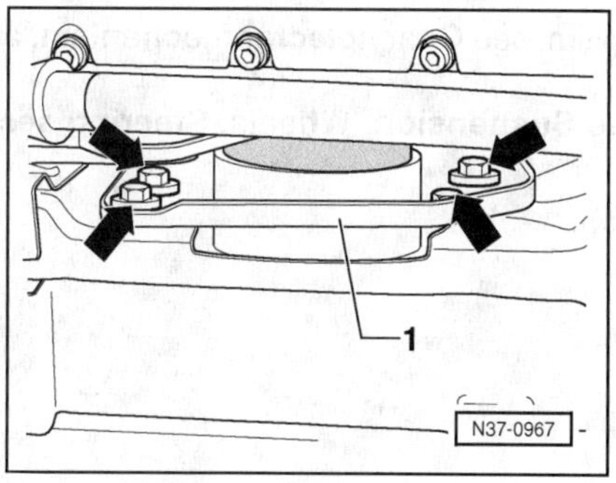

Fit torque rod limit stop (**1**) and tighten bolts (**arrows**) to prescribed torque.

- Tighten to 40 Nm (30 ft-lb)

– Install (clip) coupling rod retaining clamp for left front level control system sensor (**G78**) to left front control arm.

– Connect driveshaft, see **Repair Group 39**

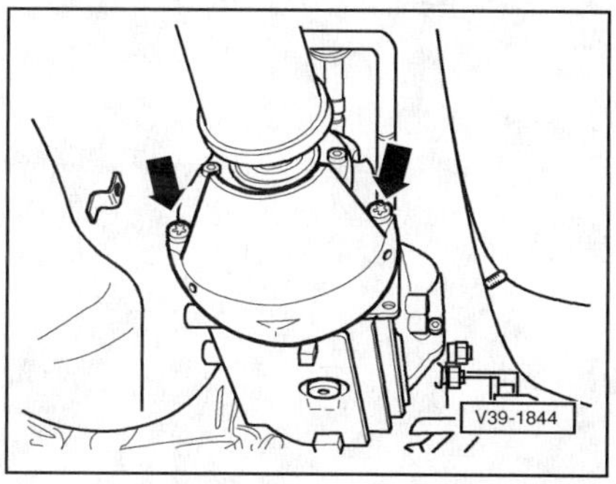

Install protective plate (**arrows**). see **Repair Group 39**

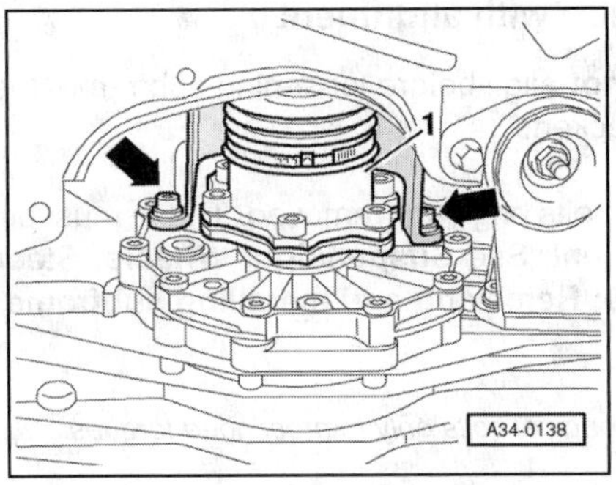

◀ Install axle shafts (**1**) to flange shafts.

- Stage I:
  Tighten to 10 Nm (7 ft-lbs) diagonally
- Stage II:
  M8 bolts, tighten to 40 Nm (30 ft-lb) diagonally
  M10 bolts, tighten to 77 Nm (57 ft-lb) diagonally

◀ Install heat shields for left and right axle shafts (**arrows**). Third bolt cannot be seen in illustration.

- Tighten to 25 Nm (18 ft-lb)

– Install front exhaust pipes, see **Repair Group 26**

– Return lock carrier to normal position from service position. See **Body Exterior section, Repair Group 50**

– Remove support device 10-222 A.

– Attach oil filler pipe to engine. See **Repair Group 15**

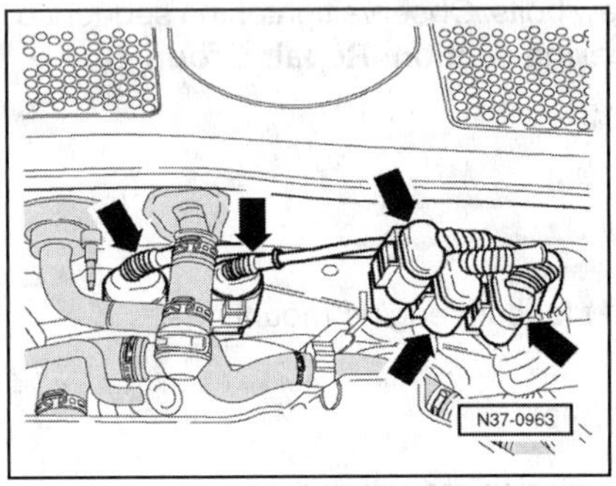

◀ Clip connectors (**arrows**) into retainers.

– Fit connector for coolant expansion tank.

– Install coolant expansion tank.

– Install intake air duct, air cleaner and intake hose. See **Fuel Injection & Ignition, section, Repair Group 24.**

– Reconnect Ground (GND) strap to battery.

– Re-code radio.

– Initialize power windows, set clock.

– Check transmission gear oil, see **Transmission gear oil, checking**.

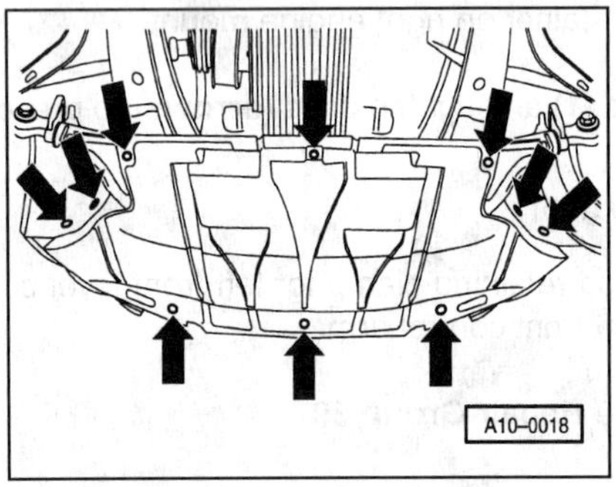

◀ Install sound insulation tray (**arrows**).

– Adjust selector mechanism, see **Gear selector mechanism, adjusting.**

– Install front wheels. See **Suspension, Wheels, Steering section, Repair Group 44**

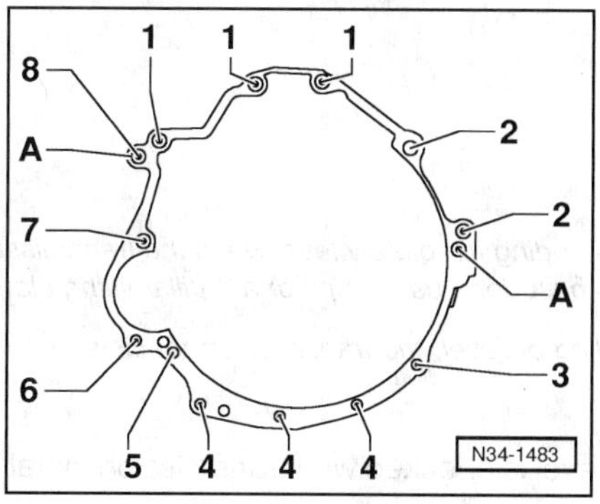

Tightening torques, transmission to engine

Item	Bolt	Qty.	Nm (ft-lb)
1	M 12 x 90	3	65 (48)
2	M 12 x 110[1)]	2	65 (48)
3	M 10 x 60[2)]	1	45 (33)
4	M 10 x 60	3	45 (33)
5	M 12 x 100	1	65 (48)
6	M 12 x 110[3)]	1	65 (48)
7	M 12 x 130	1	65 (48)
8	M 12 x 90[4)]	1	65 (48)
A	Dowel sleeves	2	n/a

1) With bracket for lines
2) With retainer for oil pressure lines
3) With nut, bolt can only be removed after first removing engine support
4) Bolt must remain in assembly when removing and installing transmission housing

Tightening torques, additional torque values.

Tightening torques	Torque
Bolt, slave cylinder	25 Nm (18 ft-lb)
Fitting, oil pressure line on transmission	25 Nm (18 ft-lb)
Bolt, final drive cover	25 Nm (18 ft-lb)
Bolt, oil pressure line retainer on engine	10 Nm (7 ft-lb)
Oil pressure lines to cooler connection point	30 Nm (22 ft-lb)
Nut, selector to transmission selector shaft (always replace)	25 Nm (18 ft-lb)
Bolt, push rod to transmission	40 Nm (30 ft-lb)
Bolt, connecting rod for selector shaft to transmission	25 Nm (18 ft-lb)
Nut, front engine mounting (always replace)	25 Nm (18 ft-lb)
Bolt transmission mounting, subframe	25 Nm (18 ft-lb)
Bolt, torque rod limit stop	40 Nm (30 ft-lb)
Bolt, wheel (lug bolt)	120 Nm (89 ft-lb)
Bolt, axle shaft to drive flange Stage 1 (diagonally) Stage 2 (diagonally) **M8** Stage 2 (diagonally) **M10**	 10 Nm (7 ft-lb) **M8** 40 Nm (30 ft-lb) **M10** 77Nm (57 ft-lb)
Bolt, axle heat shields	25 Nm (18 ft-lb)

Transmission gear oil

Checking

NOTE —
- *The vehicle should be standing absolutely level when the transmission oil level is checked. This is best done using a pit or a 4-pillar lifting platform.*
- *Adhere strictly to specified oil level; the transmission reacts sensitively to overfilling.*

Transmission gear oil level is checked with transmission installed.

Gear oil specification, see **Maintenance**

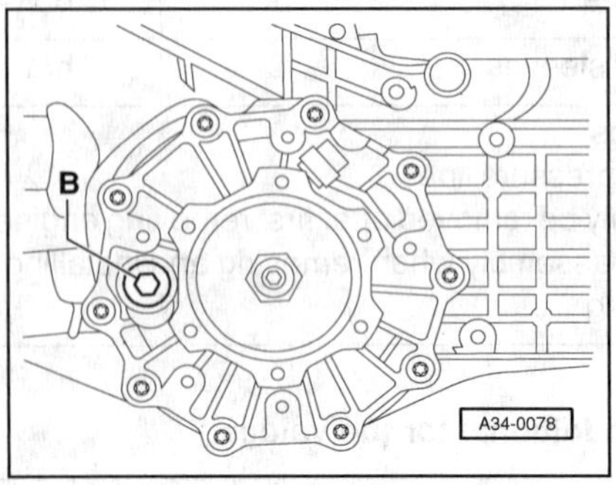

Remove plug (**B**) to check oil.

– Check oil level with home-made tool, e.g. a bent wire.
  - Transmission without oil pump:
    Oil level is correct when transmission is filled to 4 mm below oil filler hole.
  - Transmission with oil:
    Oil level is correct when transmission is filled to 9 mm below oil filler hole.

– Tighten bolt (**B**) to prescribed torque.
  - Tighten to 40 Nm (30 ft-lb)

Filling with new oil

– Remove plug (**B**).

– Fill transmission oil to 4 mm below lower edge of filler hole.

– Install plug (**B**).

– Start engine, engage gear and allow transmission to turn for approx. 2 minutes.

– Replenish transmission oil.
  - Approximate capacity 2.5 liters (2.6 qts).
  - Use synthetic oil, SAE 75W90, Part No. G 052 911 A1
  - Transmission without oil pump:
    Oil level is correct when transmission is filled to 4 mm below oil filler hole.
  - Transmission with oil:
    Oil level is correct when transmission is filled to 9 mm below oil filler hole.

– Tighten bolt (**B**) to prescribed torque.
  - Tighten to 40 Nm (30 ft-lb)

Oil pressure lines, removing and installing

> *NOTE —*
> *The subframe must be lowered to remove and install the oil pressure lines.*

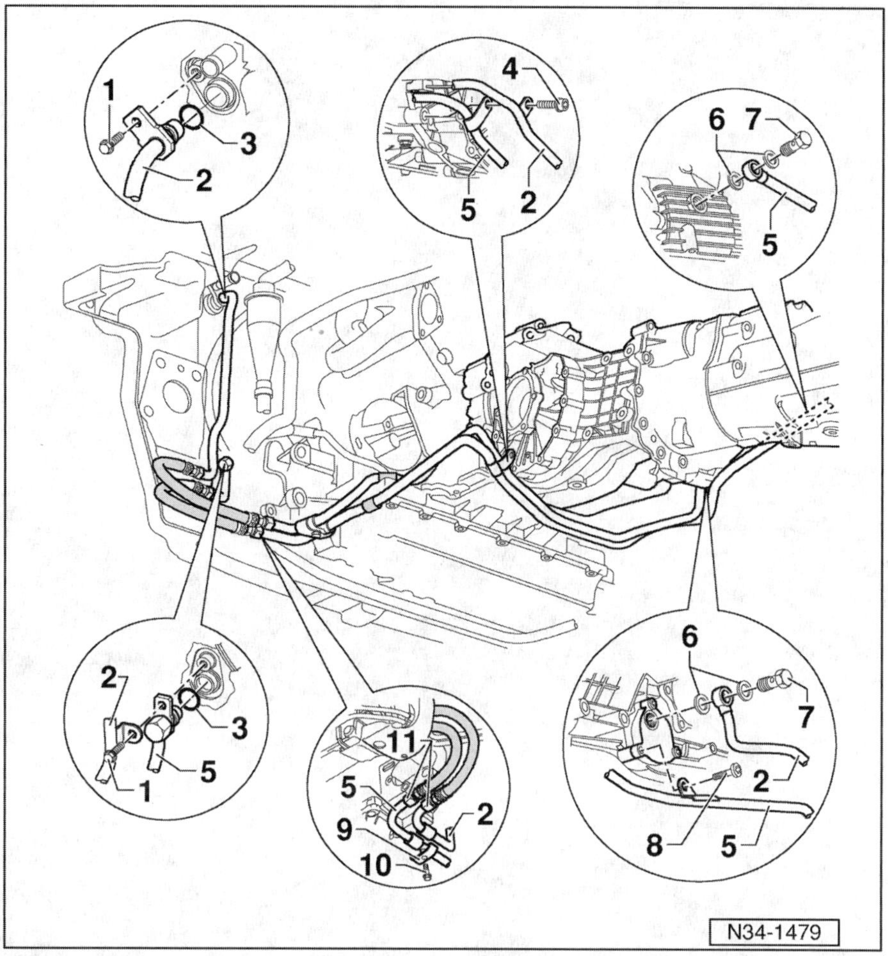

1. **Bolt, hex**
   - Before tightening bolt, push line in and onto stop by hand
   - Tighten to 5 Nm (44 in-lb)
2. **Oil pressure line**
   - From top of cooler to oil pump
   - Insert into cooler onto limit stop by hand
3. **O-ring**
   - Always replace
   - Coat with transmission oil when installing
4. **Bolt, hex socket head**
   - Tighten to 45 Nm (33 ft-lb)
5. **Oil pressure line**
   - From bottom of cooler to end of cover
   - Insert into cooler onto limit stop by hand
6. **Oil seal**
   - Always replace
7. **Bolt, banjo**
   - Tighten to 25 Nm (18 ft-lb)
8. **Screw, Torx**
   - Tighten to 25 Nm (18 ft-lb)
9. **Clamp**
10. **Bolt, hex,**
    - Tighten to 10 Nm (7 ft-lb)
11. **Union nut**
    - Tighten to 30 Nm (22 ft-lb)
    - For oil pressure line from bottom of cooler to end cover
    - For oil pressure line from top of cooler to oil pump
    - Oil pressure line distinguishing features, see **Fig. 1**

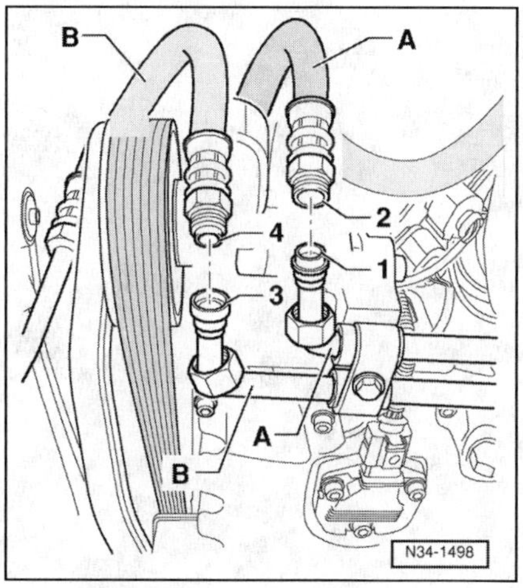

Fig. 1 Oil pressure line distinguishing features

A. Oil pressure line from top of cooler to oil pump
Secure pressure line (**1**) (male chamfer) to pressure hose union (**2**) (female chamfer).

B. Oil pressure line from bottom of cooler to end cover
Secure pressure line (**3**) (female chamber) to pressure hose union (**4**) (male chamfer).

39 Final Drive, Differential

Repair Group

Q 39

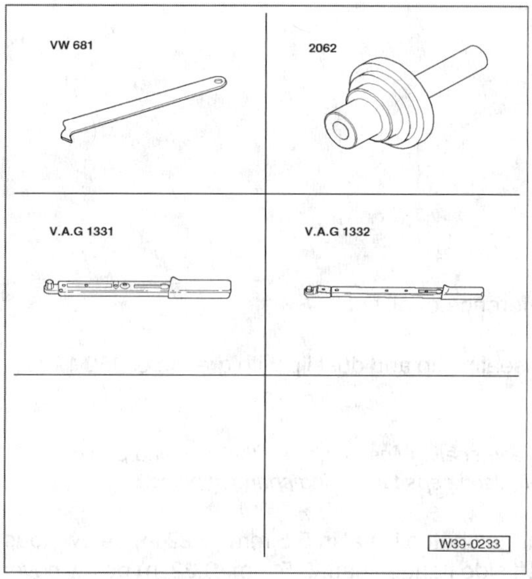

Front Final Drive

Axle shaft (drive flange) oil seals, replacing (transmission in vehicle)

Special tools and equipment

- VW 681 Oil seal extractor
- 2062 Drift
- V.A.G 1331 Torque wrench
- V.A.G 1332 Torque wrench

Removing

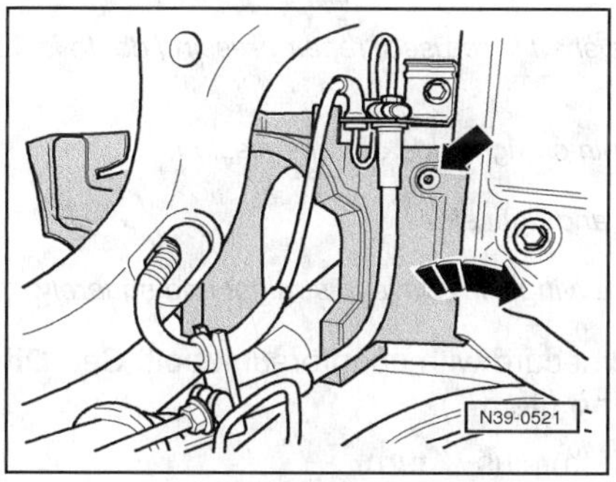

- Remove covers, if part of original equipment, from wheelhousing (**arrows**).
- Remove sound insulation below transmission if part of original equipment.

Left side oil seal

- Remove front exhaust system with catalyst on vehicles with 6-cylinder engine. See **Engine Mechanical section, Repair Group 26.**

Continued for both oil seals

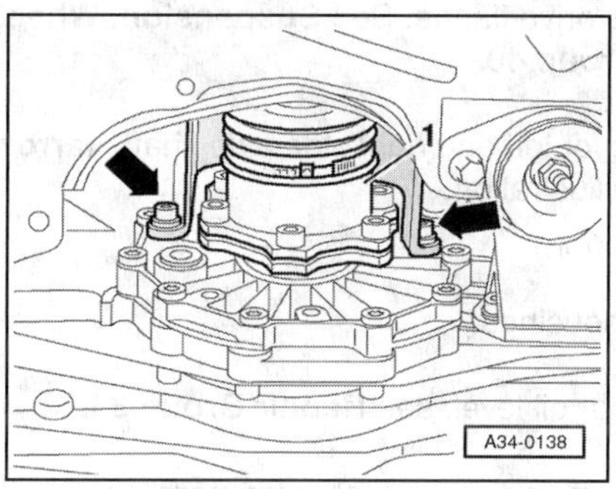

- Remove protective plates for left and right side axle shafts (**arrows**). Third bolt cannot be seen in illustration.
- Disconnect left and right side axle shafts (**1**) from drive flanges and tie-up.

Vehicles with 8-cylinder engine

- Remove axle shaft. See **Suspension, Wheels, Steering section, Repair Group 40.**

Continued for all vehicles

– Place drip tray underneath.

– Remove drive flange securing bolts by threading two bolts into flange and counter-holding flange shaft using a lever.

– Remove drive flange.

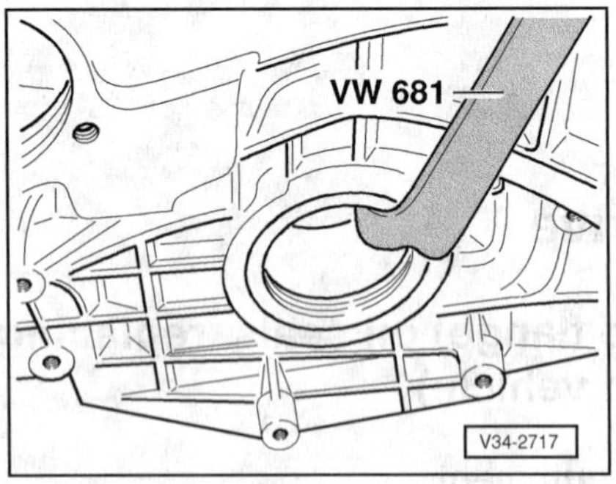

– Pull out drive flange oil seal using VW 681.

Installing

– Lightly oil outer circumference of oil seal.

– Half fill space between sealing lip and dust lip with grease G 052 128 A1.

EDITOR'S NOTE —

Greasing the space between sealing lips lubricates the seal and prevents damage due to dryness. It also keeps tensioning spring in place.

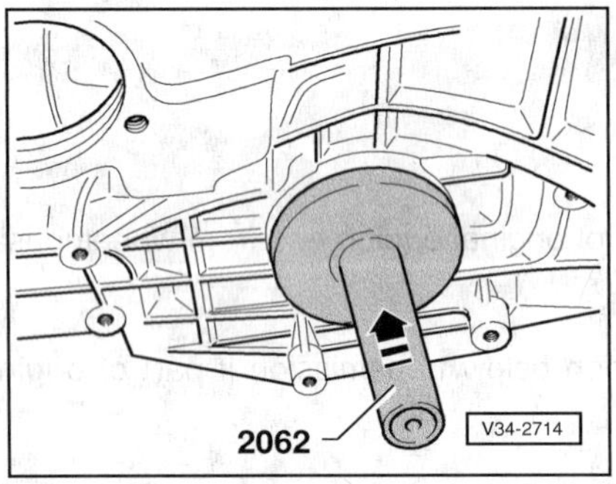

– Drive new right side flange shaft oil seal in 5.5 mm (0.22 in) below housing outer surface; for left side flange shaft, 5.5 mm (0.22 in) below cover outer surface except as noted below.

NOTE —

- *If previously used flange shaft is re-used, repair pressing depth is 6.5 mm (0.26 in).*
- *Figure illustrates driving-in on right side of transmission.*
- *Same procedure for left and right sides.*
- *Do not tilt oil seal when driving it in. Drive in straight and squarely.*

– Insert drive flange and secure with countersunk bolt. See **Differential, removing and installing** (item **1**).

• Tighten to 10 Nm (7 ft-lb) plus ¼ turn.

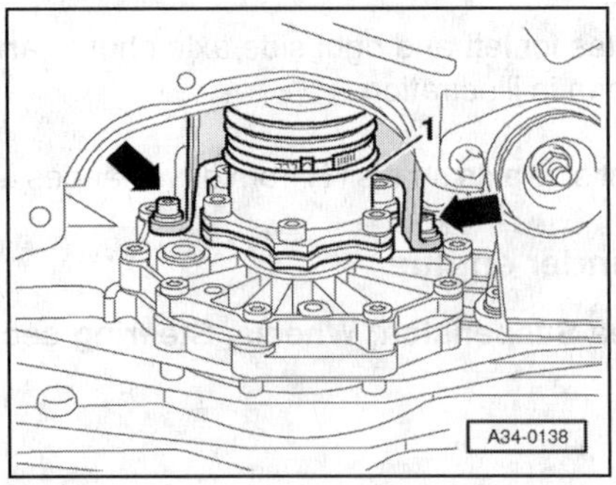

– Install axle shafts (**1**) to drive flange. See **Suspension, Wheels, Steering section, Repair Group 40.**

– Install protective plates for left and right side axle shafts (**arrows**). Third bolt cannot be seen in illustration.

• Tighten to 25 Nm (18 ft-lb)

– Install covers in wheelhousings.

– Check transmission gear oil level, see **Repair Group 34.**

– Install sound insulation, if part of original equipment.

Driveshaft flange oil seal and grooved ball bearing at transmission, assembly

1. **Transmission, rear section**
2. **Grooved ball bearing**
3. **Bearing plate with balance weight**
4. **Oil seal, rear**
   - Installing, see **Driveshaft flange oil seal and grooved ball bearing at transmission, replacing**
5. **Driveshaft flange**
6. **Bolt, socket head**
   - Qty. 6
   - Tighten to 25 Nm (18 ft-lb)
7. **O-ring**
   - Always replace
8. **Circlip**
9. **Dished spring**
   - Mark installed position before removing: larger diameter (concave side) faces shims (item **10**)
10. **Shims**
11. **Spring**
12. **Oil drain plug**
    - Tighten to 40 Nm (30 ft-lb)

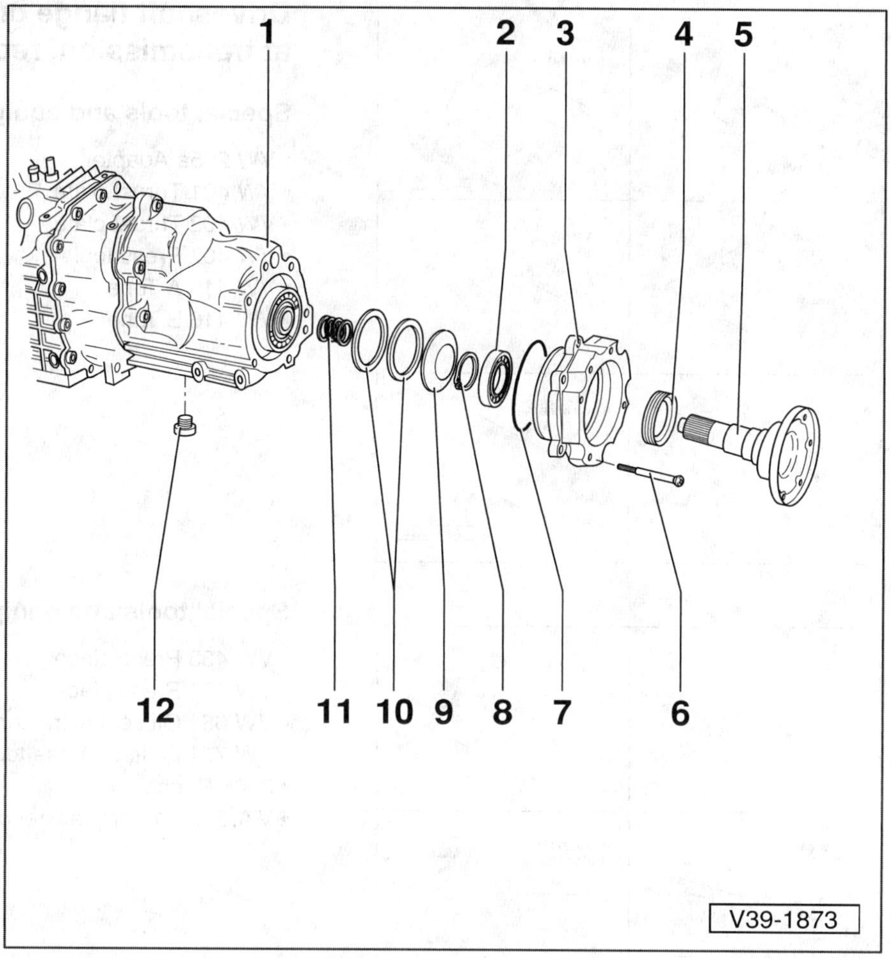

Driveshaft flange oil seal and grooved ball bearing at transmission, replacing (transmission installed)

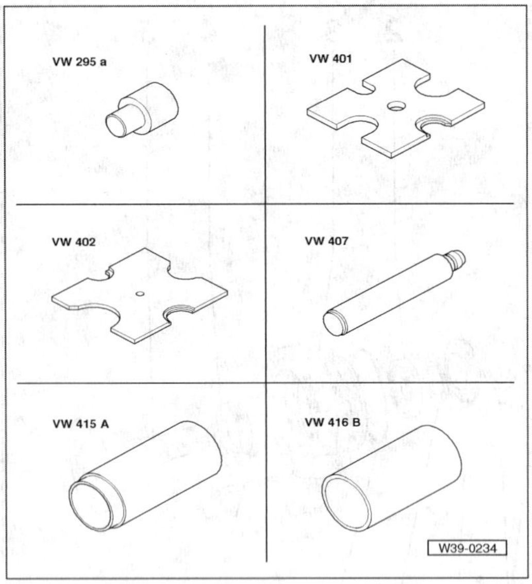

Special tools and equipment

- VW 295a Adapter
- VW 401 Thrust plate
- VW 402 Thrust plate
- VW 407 Press tool
- VW 415 A Tube
- VW 416 B Tube

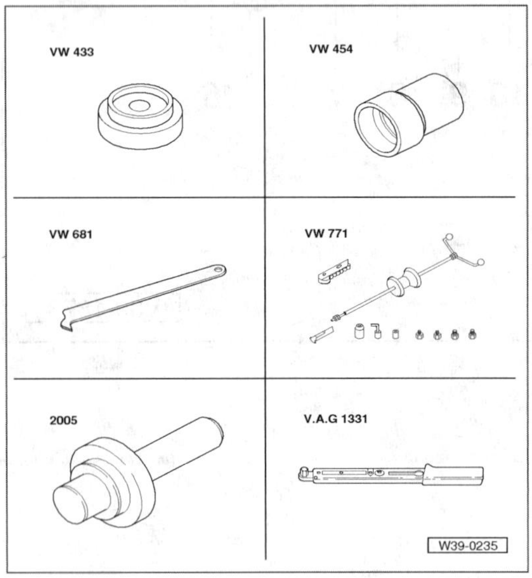

Special tools and equipment

- VW 433 Press piece
- VW 454 Press piece
- VW 681 Oil seal extractor
- VW 771 Multi-purpose tool
- 2005 Punch
- V.A.G 1331 Torque wrench

Driveshaft flange oil seal and grooved ball bearing at transmission, removing

– Disconnect driveshaft from transmission, see **Driveshaft, removing and installing, assembly overview**.

– Secure driveshaft to selector linkage.

– Place drip tray V.A.G 1306 or equivalent underneath.

– Remove rear oil drain plug (on end cover) and drain transmission oil.

– Remove bearing plate securing bolts.

NOTE—

The bearing plate will be pressed slightly away from the end cover by the coil spring as the securing bolts are loosened.

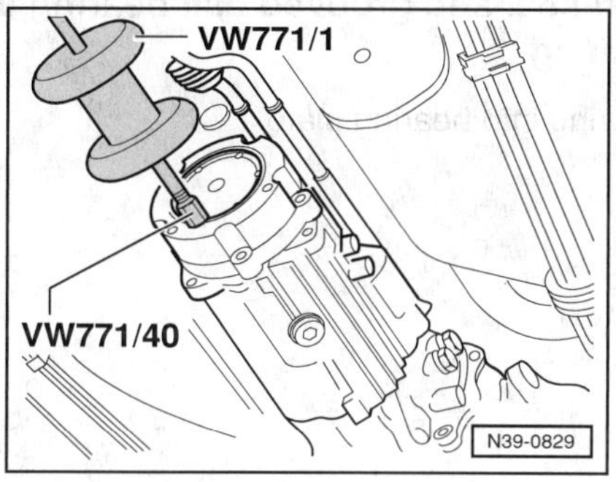

◄ Pull driveshaft flange together with bearing plate and vibration damper off end cover.

– Remove bearing plate.

– Remove circlip from driveshaft flange.

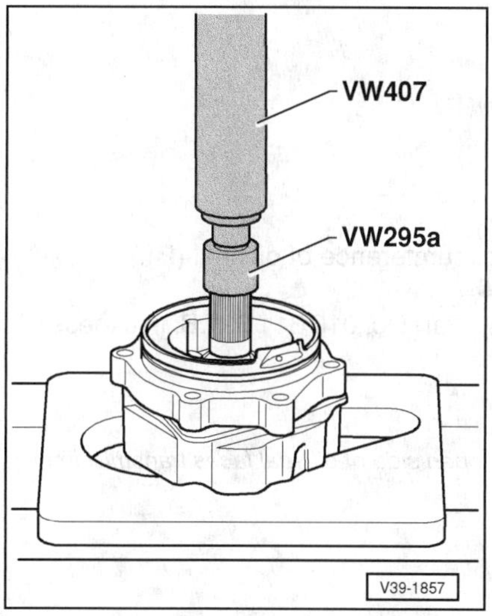

◄ Press out driveshaft flange.

– Pull out driveshaft flange oil seal.

– Thoroughly clean oil seal seat.

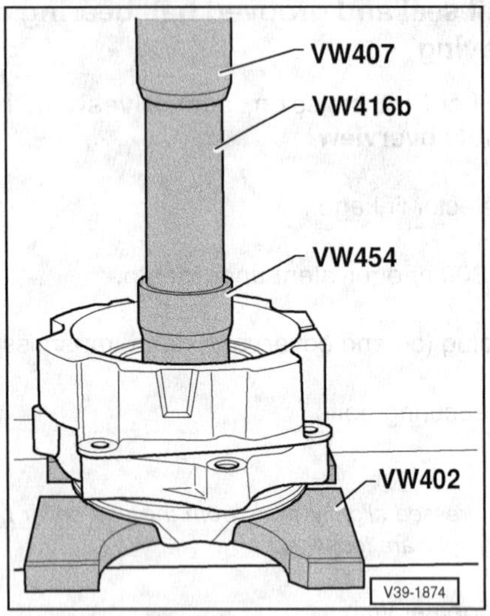

Press out grooved ball bearing.

Driveshaft flange oil seal and grooved ball bearing at transmission, installing

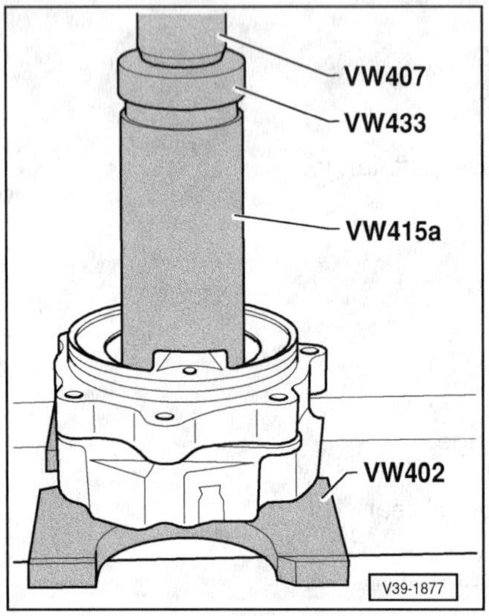

Press grooved ball bearing into bearing plate.

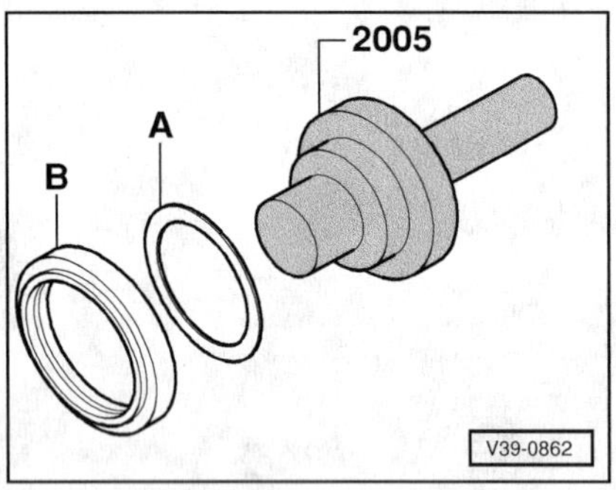

Lightly lubricate outer circumference of oil seal (**B**).

– Put oil seal with disk (**A**) (Part No. 016 311 391 B, thickness1.7 mm) onto drift 2005.

NOTE —

Installed position of seal: open side of oil seal faces transmission.

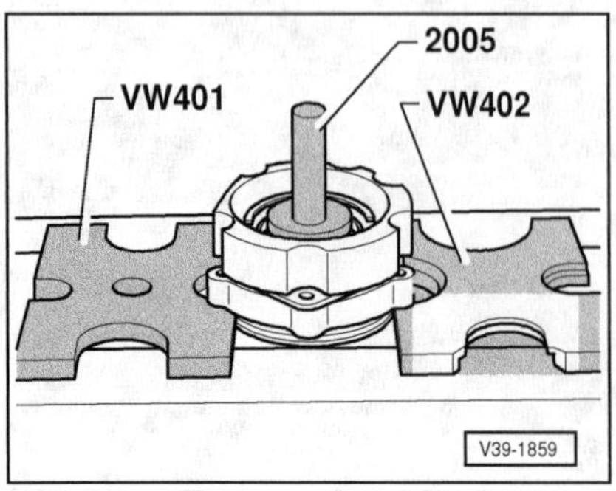

Drive in driveshaft flange oil seal.

– Remove disk after driving in.

– Half fill space between sealing lip and dust lip with grease G 052 128 A1.

EDITOR'S NOTE —

Greasing the space between sealing lips lubricates the seal and prevents damage due to dryness. It also keeps tensioning spring in place.

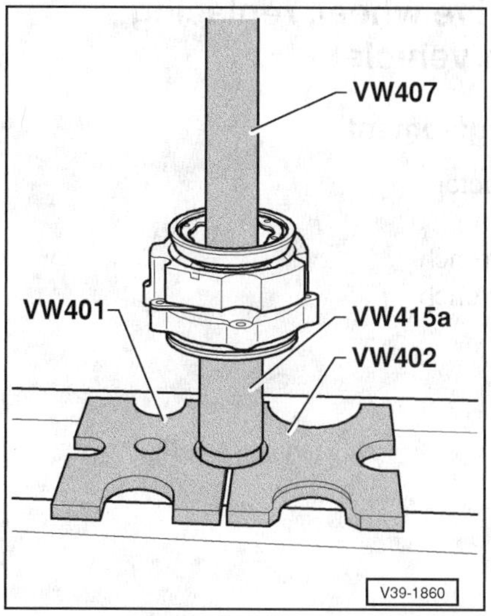

- Press in driveshaft flange.
- Put circlip on flange shaft.
- Lightly lubricate O-ring and fit into groove of bearing plate.
- Note previously marked dished spring and install together with shims in bearing plate.

NOTE —
Installed position of dished spring: larger diameter (concave side) faces shims

- Push coiled spring onto driveshaft flange.
- Tighten bearing plate securing bolts to prescribed torque in stages using diagonal sequence.
  - Tighten to 25 Nm (18 ft-lb)
- Connect driveshaft, see **Driveshaft, removing and installing, assembly overview.**
- Fill transmission with gear oil and check level, see **Repair Group 34.**

Speedometer sensor and drive wheel, replacing

Speedometer vehicle speed sensor (VSS), replacing (transmission in vehicle)

- Remove left front wheel.
- Remove left axle shaft protective plate (if equipped).

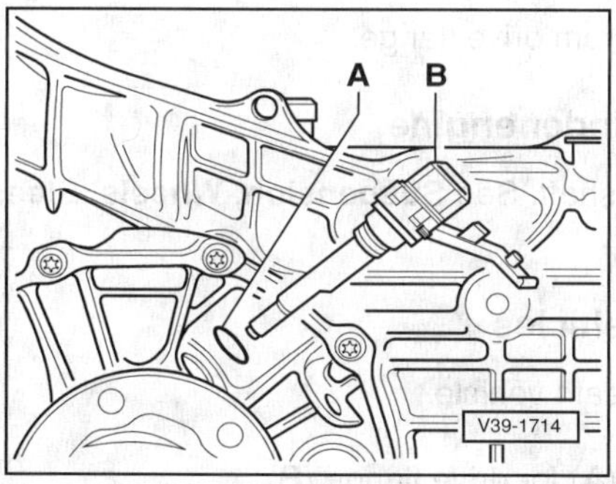

- Disconnect connector from VSS (G22) (**B**).
- Press retaining clip down, turn sensor sideways and pull out upward.
- When installing, replace O-ring (**A**).

NOTE —
Handle the sensor carefully. If damaged it is possible that the speedometer will not indicate exactly.

- Install protective plates for left side axle shaft.
  - Tighten to 25 Nm (18 ft-lb)
- Install front wheel.
  - Tighten to 120 Nm (89 ft-lb)

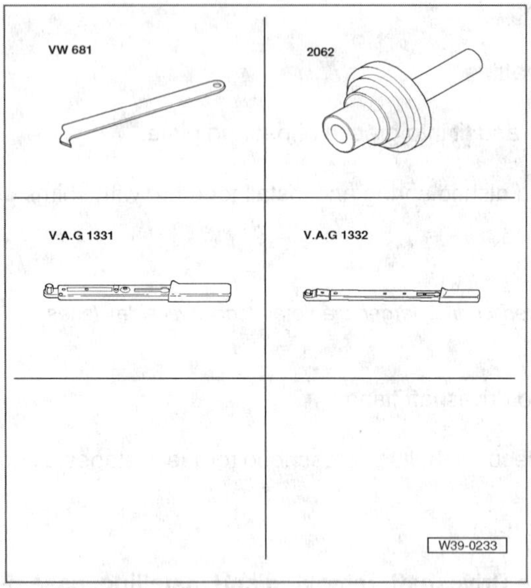

Speedometer drive wheel, replacing (transmission in vehicle)

Special tools and equipment

- VW 681 Oil seal extractor
- 2062 Drift
- V.A.G 1331 Torque wrench
- V.A.G 1332 Torque wrench

Removing

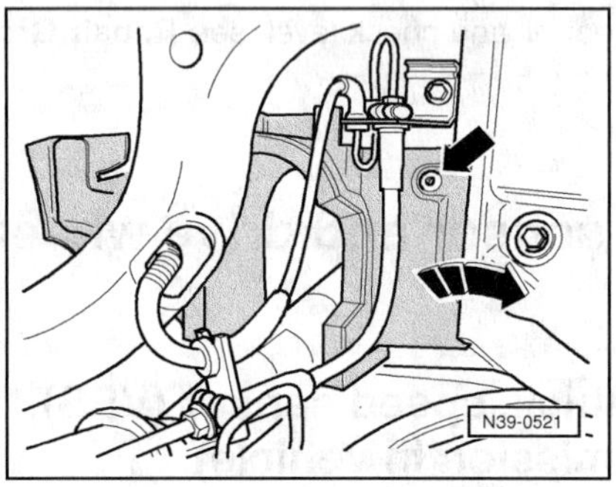

< Remove cover from left wheelhousing (**arrows**).

– Remove sound insulation below transmission if part of original equipment.

– Remove front exhaust system with catalyst on vehicles with a V6 Diesel engine. See **Engine Mechanical section, Repair Group 26.**

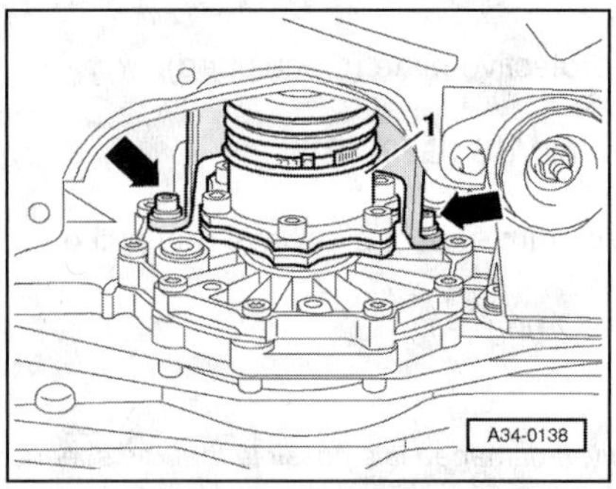

< Remove left side axle shaft protective plate (**arrows**) if part of original equipment. Third bolt cannot be seen in illustration.

– Disconnect axle shaft from drive flange.

Vehicles with 8-cylinder engine

– Remove left front axle shaft. See **Suspension, Wheels, Steering section, Repair Group 40.**

Continued for all vehicles

– Place drip tray underneath vehicle.

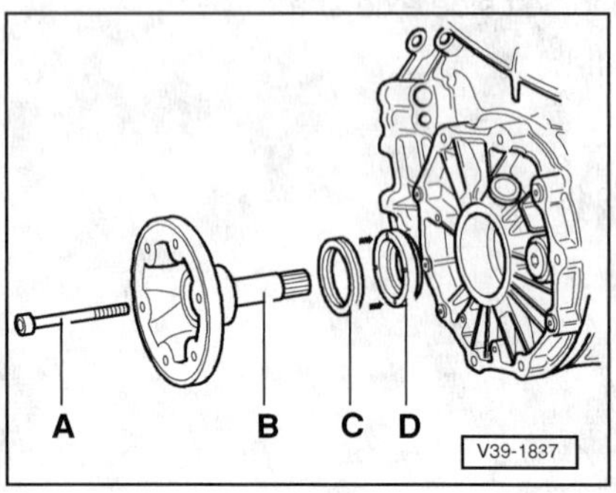

< Remove securing bolt (**A**) for drive flange (**B**).

– Remove drive flange.

– Pull out drive flange oil seal (**C**) using VW 681 or equivalent.

– Remove speedometer sensor, see **Speedometer vehicle speed sensor (VSS), replacing (transmission in vehicle).**

– Remove speedometer drive wheel (**D**) from grooves of differential by prying at connecting tabs (**small arrows**), alternating from side to side.

Speedometer drive wheel, installing

– Installation is the reverse.

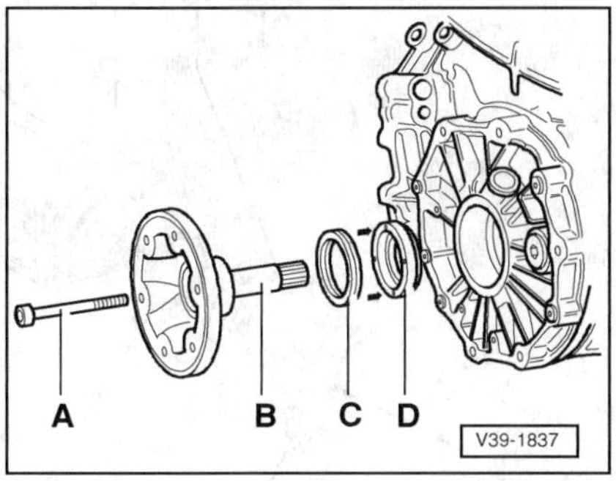

Speedometer drive wheel connecting tabs (**small arrows**) engage in grooves of differential.

A Bolt
B Drive flange
C Oil seal
D Speedometer drive wheel

– Drive in drive flange oil seal (**C**) and install drive flange, see **Axle shaft (drive flange) oil seals, replacing (transmission in vehicle)**

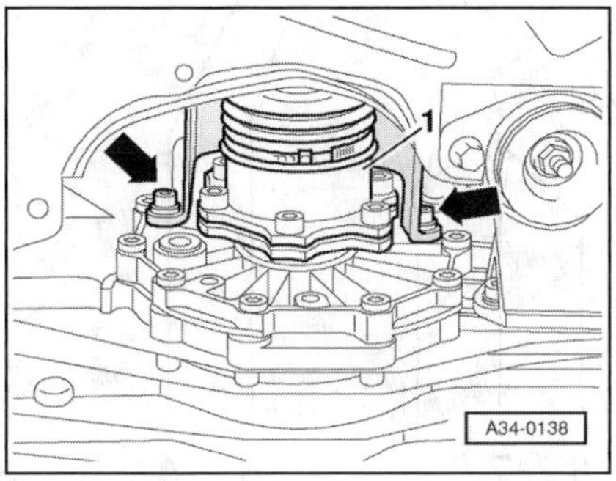

Install axle shafts (**1**) to drive flange. See **Suspension, Wheels, Steering section, Repair Group 40.**

– Install protective plates for left and right side axle shafts (**arrows**). Third bolt cannot be seen in illustration.

• Tighten to 25 Nm (18 ft-lb)

– Fill transmission with gear oil and check level, see **Repair Group 34.**

– Install sound insulation, if part of original equipment.

Driveshaft

Driveshaft, removing and installing, assembly overview

NOTE —

- *Must be removed and installed in correct sequence. See **Removing and installing driveshaft.***
- *Driveshaft cannot be repaired.*
- *The driveshaft must be tied up or supported if it is only disconnected at the transmission or final drive.*
- *Only remove and install the driveshaft with assembly device 3139.*
- *Keep the driveshaft straight and only store and transport it fully extended.*
- *Check adjustment before replacing the driveshaft for complaints concerning noise and/or vibrations., see **Driveshaft, adjusting.***

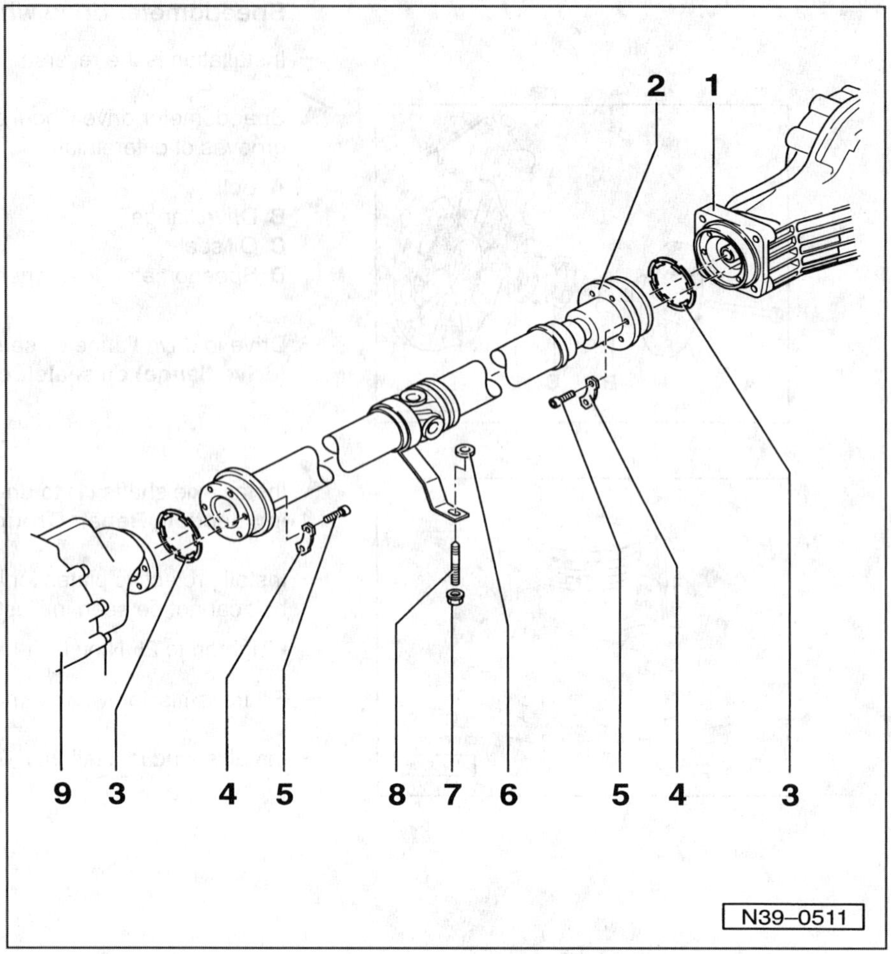

1. **Rear final drive (rear differential)**
2. **Driveshaft**
   - Two sections jointed by a universal joint
3. **Seal**
   - Always replace
   - Remove protective foil prior to installation
   - Glue seal in driveshaft flange
4. **Lock plate**
5. **Bolt, Allen head**
   - Tighten to 55 Nm (41 ft-lb)
   - Self-locking
   - Always replace
   - Always clean threaded holes for bolts in flange shaft (e.g. with a thread tap/chaser)
6. **Shim**
   - Determining thickness, see **Driveshaft, adjusting**
7. **Nut, hex**
   - Tighten to 25 Nm (18 ft-lb)
   - Check torque of stud (item **8**) each time nut is loosened.
8. **Stud,**
   - Tighten to 10 Nm (7 ft-lb)
   - Check torque of stud each time nut is loosened.
   - Removing and installing, see **Fig. 1**
9. **Transmission, rear housing**

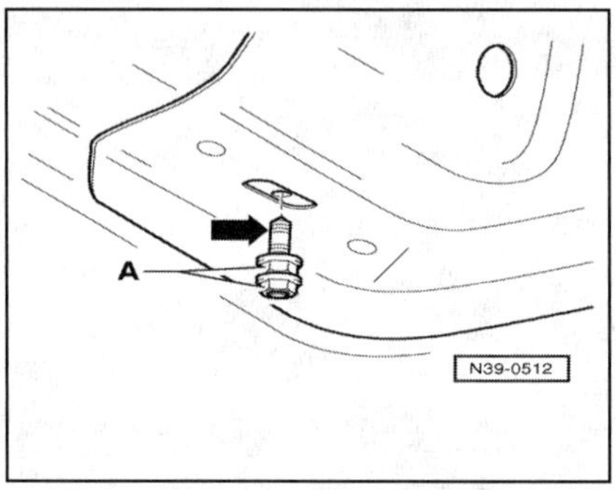

Fig. 1 Stud, removing and installing

- Remove and install stud using two lock nuts (**A**), (double-nut technique).
- Install short threaded end into vehicle body (**arrow**).
- Tighten to 10 Nm (7 ft-lb) using double-nut technique.

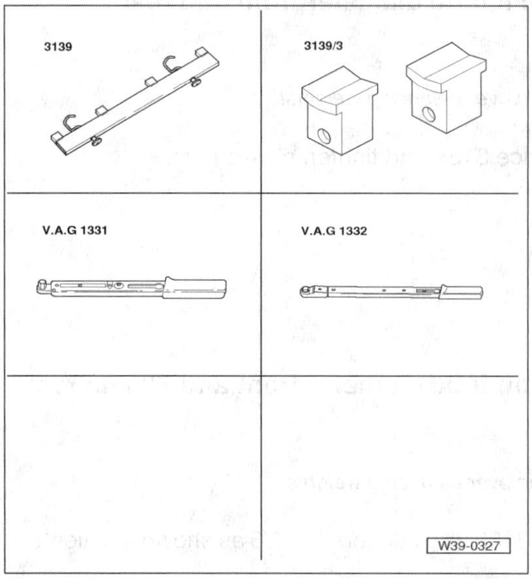

Driveshaft, removing and installing

Special tools and equipment

- 3139 Assembly device for driveshaft
- 3139/3 Spacers
- V.A.G 1331 Torque wrench
- V.A.G 1332 Torque wrench

NOTE—

- *Before removing, mark the positions of all parts (including transmission and rear differential driveshaft flanges) in relation to each other. Reinstall in the same position or imbalance will be excessive, the mountings could be damaged causing rumbling noises.*
- *Keep the driveshaft straight, only store and transport it fully extended.*
- *A twin post lifting platform should be used when working on the driveshaft.*

Removing

– Remove front cross support, below exhaust system, if equipped.

– Remove rear section of exhaust system. See **Engine Mechanical section, Repair Group 26.**

– Remove heat shield from vehicle floor.

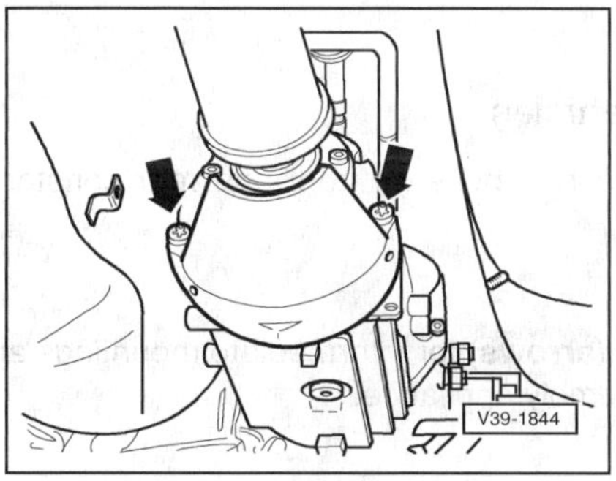

◀ Remove protective plate (**arrows**).

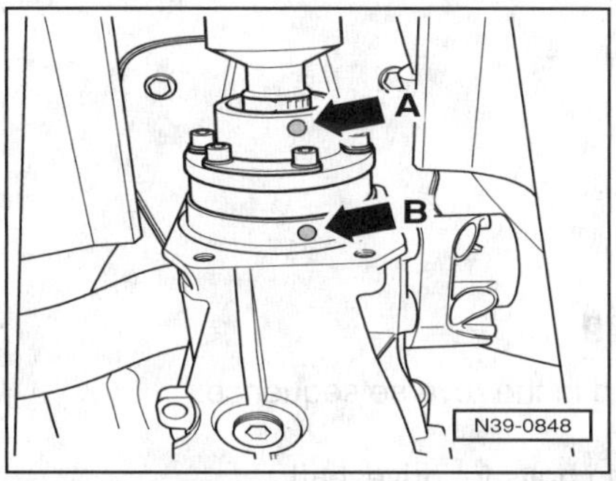

◀ Check whether there is a colored dot on driveshaft and driveshaft flange on rear final drive.

– If no markings can be found mark (**in color**) position of driveshaft flange (**arrow A**) to rear final drive (**arrow B**).

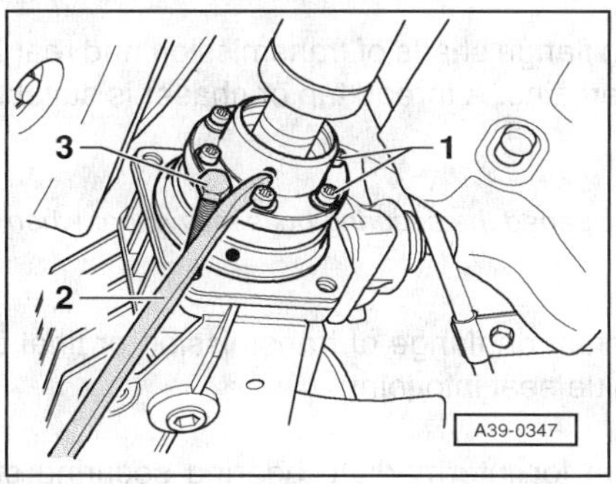

◀ Screw bolt (**3**) (M10 x 100) in rear final drive.

– To loosen front and rear driveshaft bolts (**1**), counter-hold with an assembly lever (**2**).

NOTE—

*If the threaded hole in the rear final drive for bolt (**3**) does not exist, then counter-hold the driveshaft with brake pedal actuator V.A.G 1869/2.*

– Remove upper 3 securing bolts in each driveshaft constant velocity joint.

Driveshaft with 60 mm tube diameter front and rear

NOTE—
Do not set assembly tool on/over balancing weights!

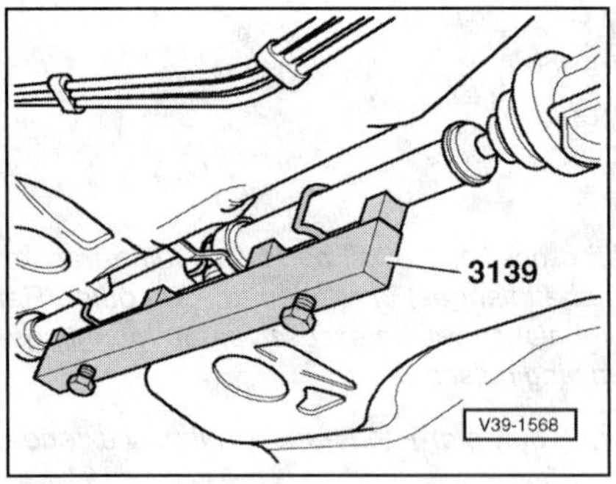

Mount assembly appliance 3139 and tighten plastic nuts.

Driveshaft with 60 mm tube diameter front and 70 mm rear

NOTE—
Do not set assembly tool on/over balancing weights!

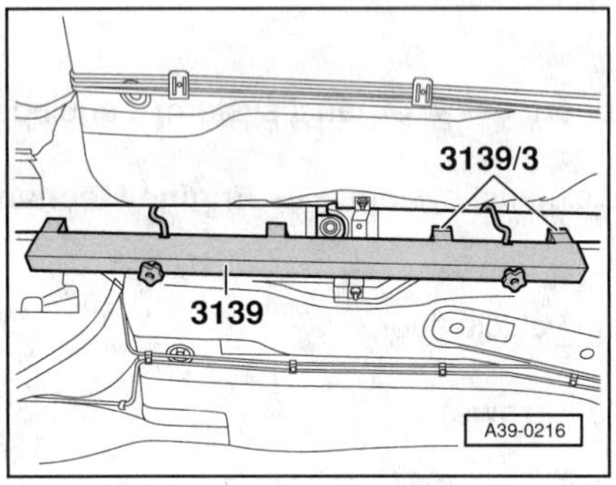

Mount assembly device 3139 with spacers 3139/3 as shown and tighten plastic nuts.

Continued for all vehicles

– Remove remaining securing bolts on front and rear constant velocity joints.

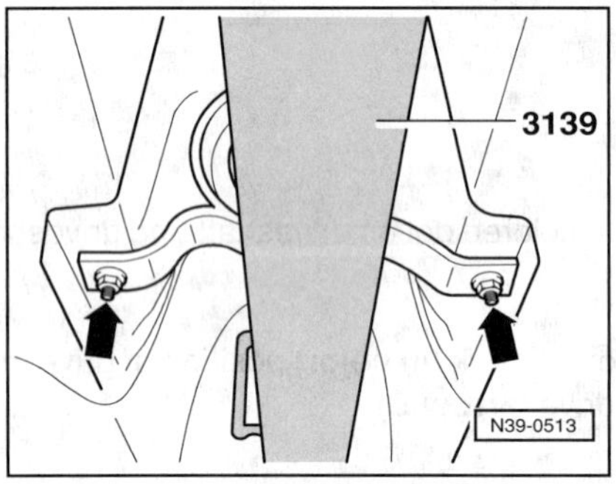

Remove securing nuts (**arrows**) for intermediate mountings and take out driveshaft with assembly appliance.

Driveshaft, installing

– Installation is performed in the reverse sequence.

– Always replace securing bolts for driveshaft.

– Clean threaded holes in flange shafts of transmission and rear final drive of all locking material remains. A thread tap or chaser is suitable for this.

NOTE—
If threaded holes are not cleaned the securing bolts may shear when properly tightened.

– Replace seal for driveshaft on flange of transmission or final drive (pull off protective foil and glue seal into joint).

– Check tightening torque for intermediate bearing securing studs, see **Driveshaft, removing and installing, assembly overview, (**item **8).**

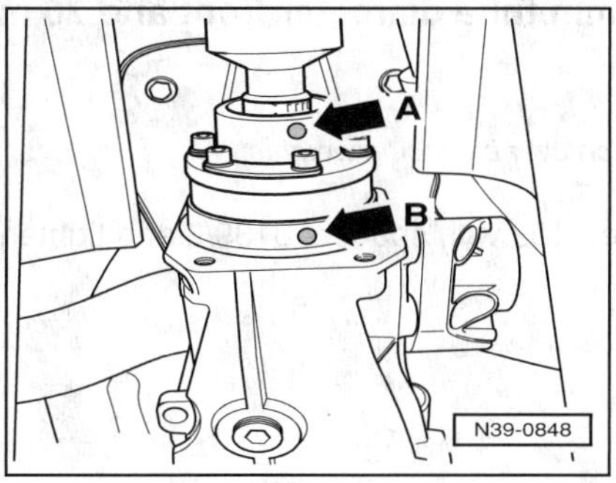

- Install driveshaft so that colored marks from factory or those subsequently made are in alignment in order to prevent imbalance.

– Driveshaft flange (**arrow A**) and rear final drive flange (arrow **B**) must be aligned.

NOTE —
If a new driveshaft is installed and the factory made colored mark on the rear final drive flange is no longer visible, the rear final drive flange run-out must be measured, see ***Driveshaft, adjusting****. The colored mark on the driveshaft must then be aligned with the new mark on the rear final drive flange.*

– Install rear section of exhaust system. See **Engine Mechanical, Repair Group 26.**

Tightening torques	Torque
Nut (on stud), intermediate bearing to body	25 Nm (18 ft-lb)
Bolt, protective plate to transmission	25 Nm (18 ft-lb)
Bolt, driveshaft to driveshaft flange on transmission and on rear differential (always replace)	55 Nm (41 ft-lb)

Driveshaft, adjusting

Special tools and equipment

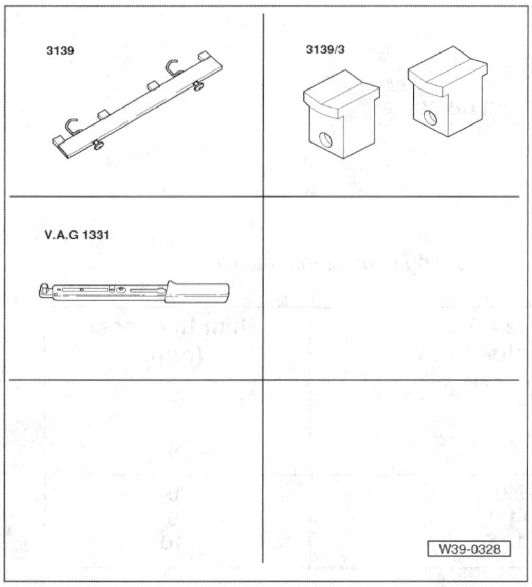

- 3139 Assembly device for driveshaft
- 3139/3 Spacers
- V.A.G 1331 Torque wrench

– Remove front cross support below exhaust system, if equipped.

– Remove rear section of exhaust system. See **Engine Mechanical Repair Group 26.**

– Remove heat shield from vehicle floor.

Driveshaft with 60 mm tube diameter front and rear

NOTE —
Do not set assembly tool on/over balancing weights!

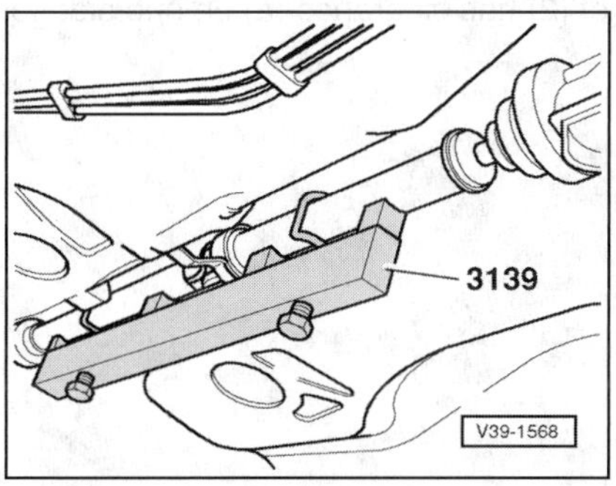

- Mount assembly appliance 3139 and tighten plastic nuts.

Driveshaft with 60 mm tube diameter front and 70 mm rear

NOTE—
Do not set assembly tool on/over balancing weights!

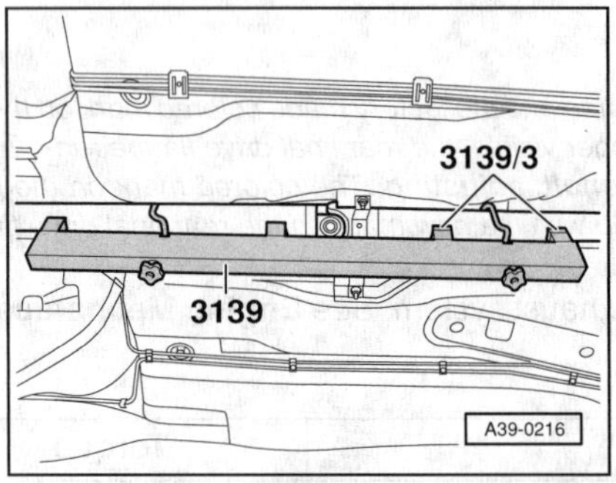

- Mount assembly device 3139 with spacers 3139/3 and tighten plastic nuts.

Continued for all vehicles

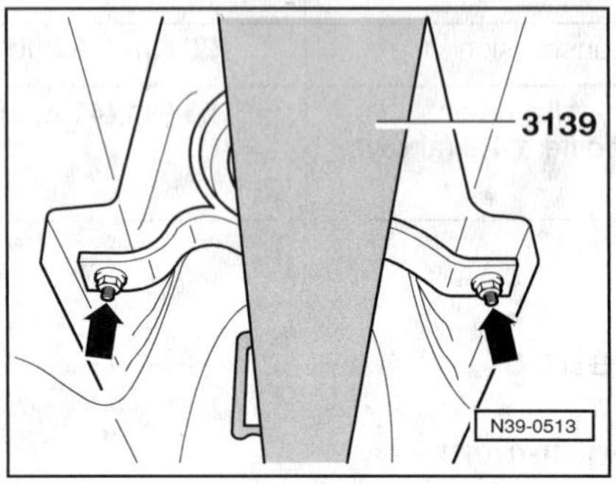

- Remove intermediate mounting securing nuts (**arrows**).
- Check tightening torque for intermediate bearing securing studs.
  - Tighten to 25 Nm (18 ft-lb)

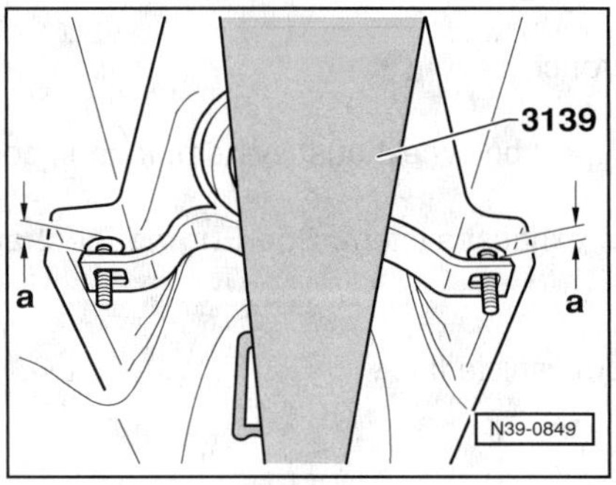

- Measure dimension (**a**).
  - Left and right sides must be same.

Table of shims

NOTE—
Determine shim application according to parts information.

Distance (a) measured value (mm)	Shim thickness (mm)
0 – 3.0 3.1 – 5.0 5.1 – 7.0	- 2 4
7.1 – 9.0 9.1 – 11.0 11.1 – 13.0	6 8 10

Install driveshaft shims

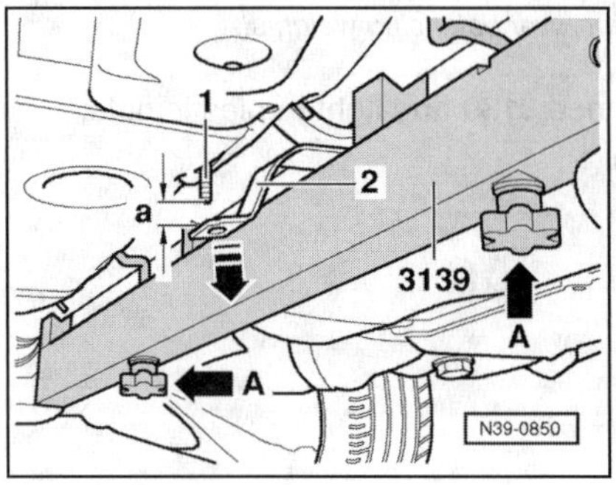

- Loosen plastic nuts (**arrows A**) on assembly device until intermediate bearing retaining bracket (**2**) has clearance (**a**) on one side to stud (**1**).

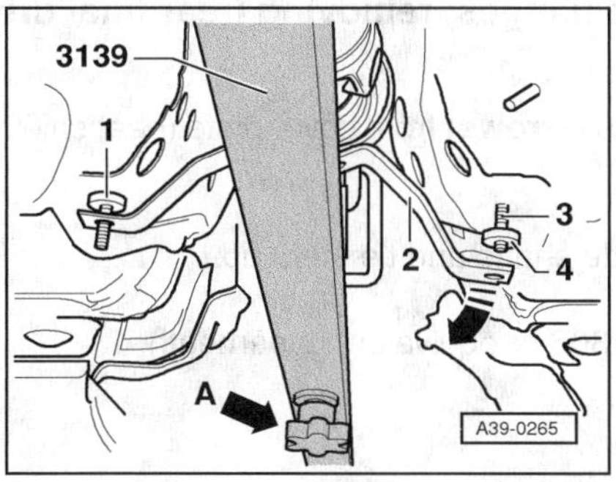

Place shim as previously measured (**1**) on stud and swing intermediate bearing retaining bracket (**2**) to other side.

– Place 2nd shim (**4**) on stud (**3**).

– Tighten both plastic nuts on assembly device (**arrow A**).

– Centralize driveshaft longitudinally (fore/aft). Slide driveshaft to the **REAR** onto stop with assembly appliance.

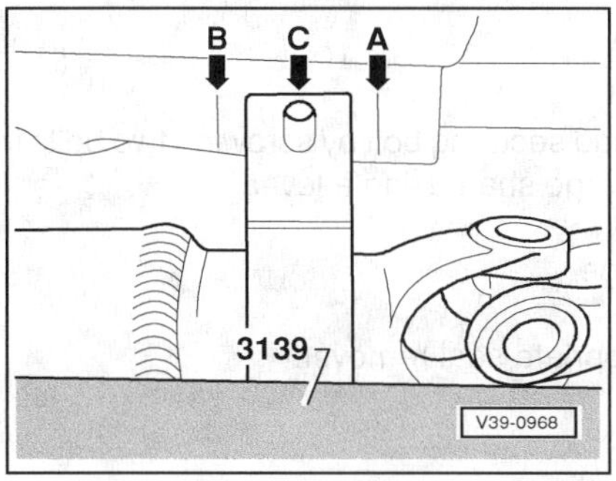

Mark position of intermediate mounting on body (**arrow A**).

– Slide driveshaft **FORWARD** onto stop with assembly appliance.

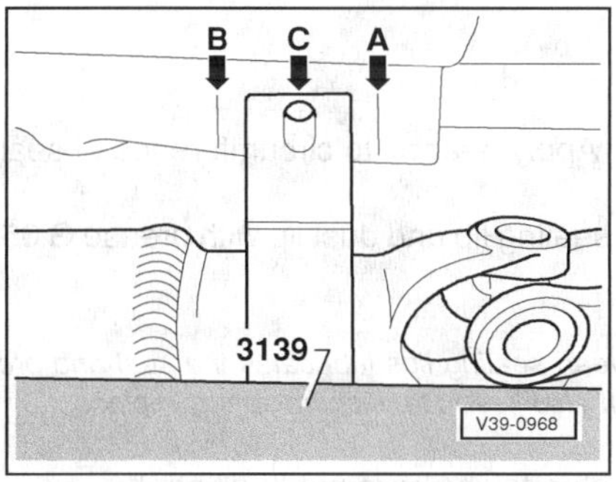

Mark position of intermediate mounting on body (**arrow B**).

– Install intermediate mounting in central position (**arrow C**).
 • Tighten to 25 Nm (18 ft-lb).

– Remove assembly appliance.

– Install heat shield to vehicle floor.

– Install rear section of exhaust system. See **Engine Mechanical, Repair Group 26.**

– Install cross support below exhaust system, if equipped.

Rear Final Drive

Seals for axle drive flanges, replacing (rear final drive installed)

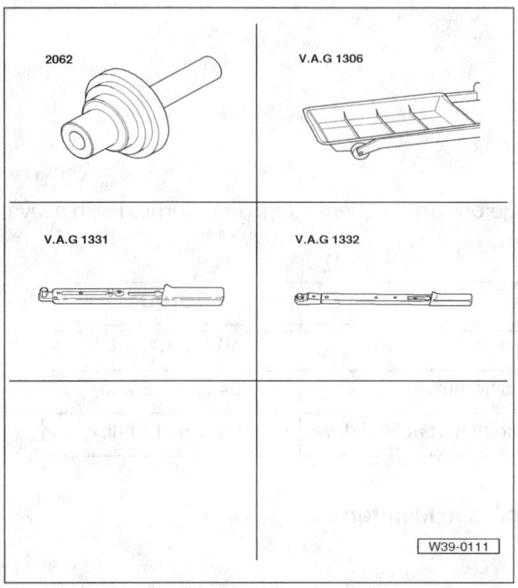

Special tools and equipment

- 2062 Drift
- V.A.G 1306 Drip tray
- V.A.G 1331 Torque wrench
- V.A.G 1332 Torque wrench

Additional information

- **Suspension, Wheels, Steering section**

NOTE —

All the illustrations show the left side of the final drive. Removal and installation of the flange shaft and seal on the right side is identical.

Seals for axle drive flanges, removing (rear final drive installed)

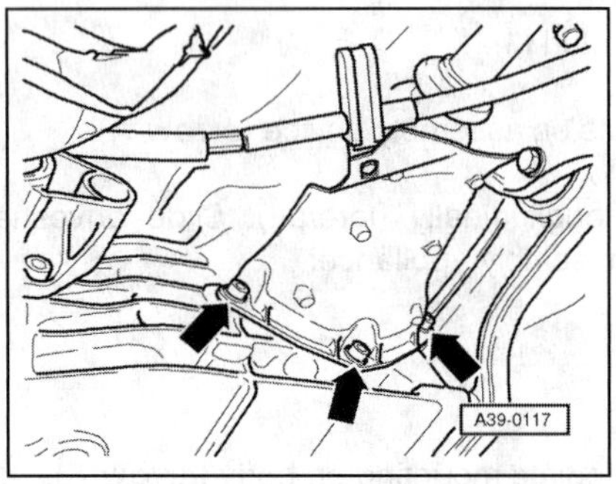

- Remove mounting bolts (**arrows**) from cover plate (heat shield) on final drive.

- Unbolt left and right axle shafts and tie them up.

- Place drip tray V.A.G 1306 or equivalent underneath.

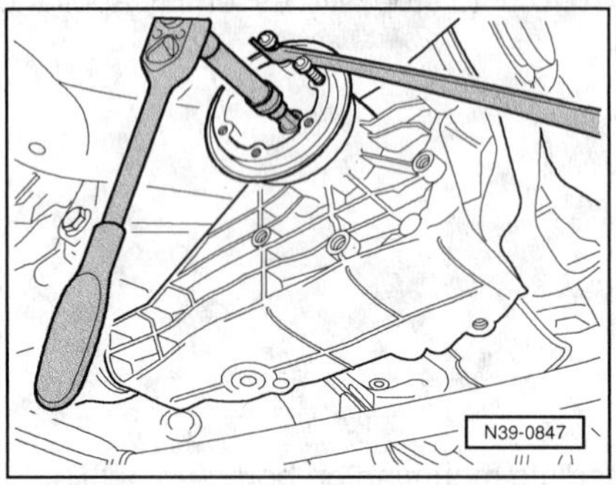

- Remove axle drive flange securing bolt by screwing two bolts into flange and counter-holding flange shaft using a lever.

- Pull out axle drive flange.

- Pry seal out with appropriate seal remover.

Installing

- Install in reverse order. Apply gear oil to circumference of seal.

- Half fill space between sealing lip and dust lip with grease G 052 128 A1.

EDITOR'S NOTE —
Greasing the space between sealing lips lubricates the seal and prevents damage due to dryness. It also keeps tensioning spring in place.

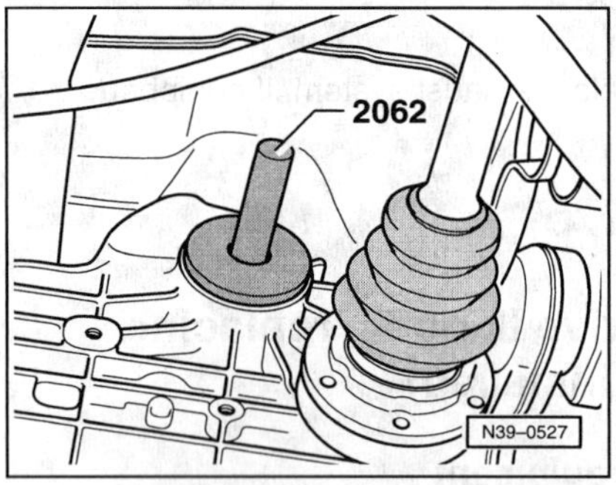

- Drive new seal in onto stop taking care not to tilt seal.

- Insert axle drive flange.

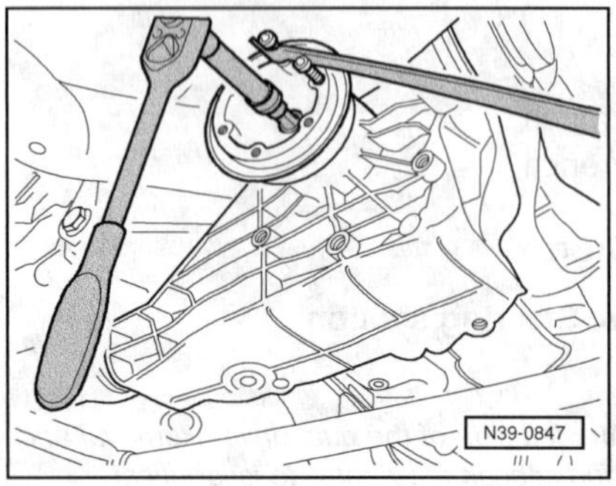

- Screw in axle shaft flange bolt and prevent flange from turning with a lever.
  • Tighten to 24 Nm (18 ft-lb)

Tightening torques	Torque
Bolt, rear axle to drive flange	40 Nm (30 ft-lb)
Bolt, axle drive flange to differential	24 Nm (18 ft-lb)
Bolt, cover plate (heat shield) to rear final drive	23 Nm (17 ft-lb)

- Check rear final drive oil, see **Maintenance**.

Seal for driveshaft flange, replacing (rear final drive installed)

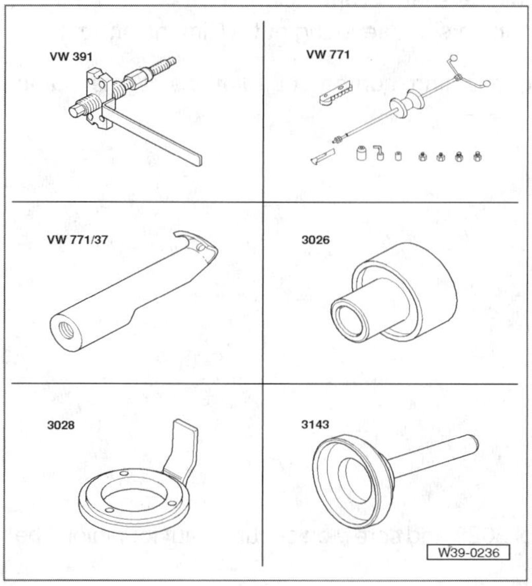

Special tools and equipment

- VW 391 Driveshaft installation appliance
- VW 771 Multi-purpose tool
- VW771/37 Puller hooks
- 3026 Drift
- 3028 Counter-hold
- 3143 Press tool

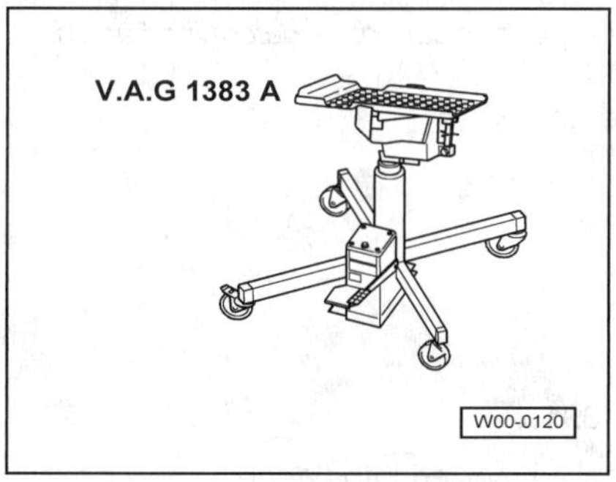

Special tools and equipment

- V.A.G 1383 A Engine/transmission jack with universal transmission mounting V.A.G 1359/2

Additional information

- **Suspension, Wheels, Steering section**

Removing

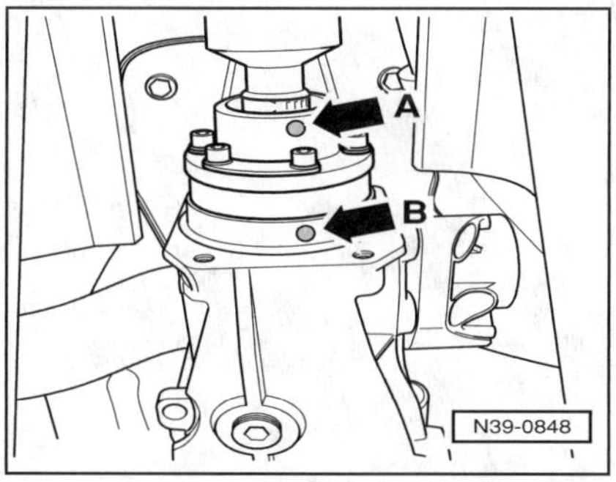

◄ Only if there are no marks, mark position of rear driveshaft sections on constant velocity joints with paint (**arrows A** and **B**).

– Disconnect driveshaft from rear final drive and tie up, see **Driveshaft, removing and installing, assembly overview.**

– Place transmission jack V.A.G 1383 A with universal transmission mounting 1359/2 under final drive and support final drive.

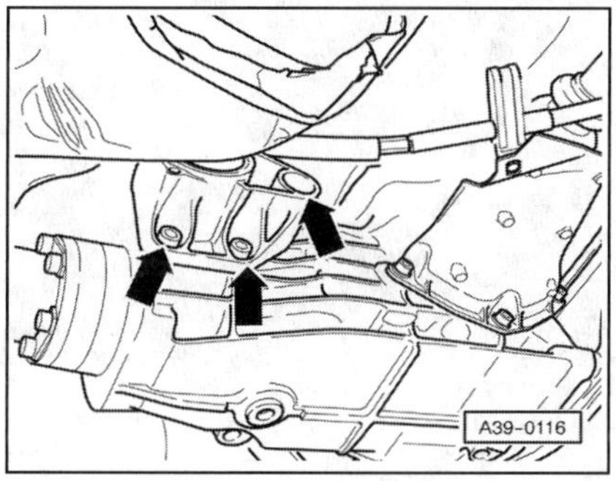

◄ Remove support securing bolts (**arrows**).

– Carefully lower final drive until flange is accessible.

– Mark position of pinion shaft securing nut with paint.

Determining dimension -a- and dimension -b-

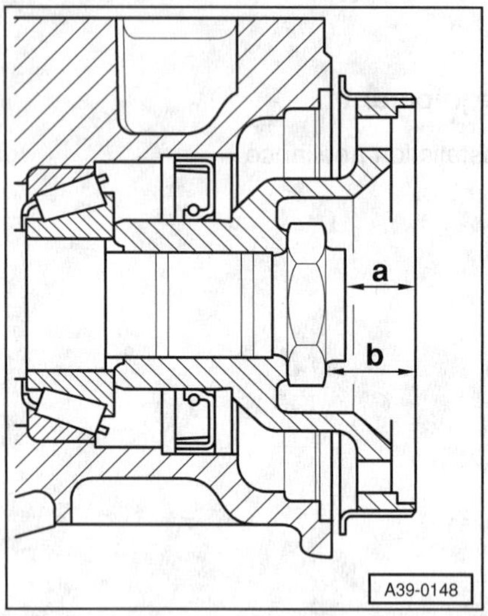

◀ Measure as indicated and note measurements.

- Distance of flange to pinion shaft = Dimension -a-
- Distance of flange to pinion shaft securing nut = Dimension -b-

– Remove securing nut, counting number of rotations. Write number down.

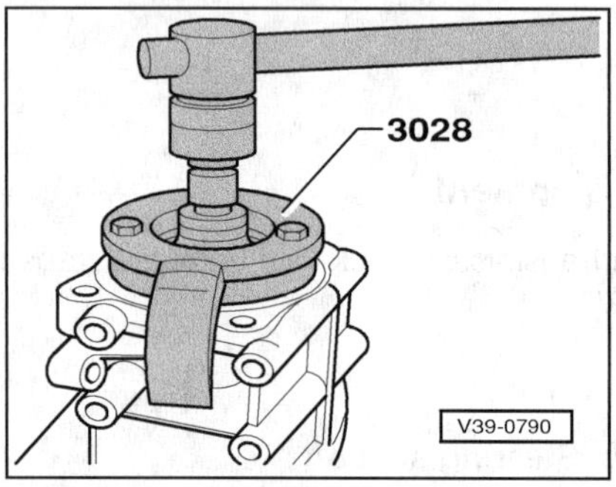

◀ Mount counter-hold tool 3028 and screw on securing nut for pinion shaft.

NOTE —

Re-use previous hex nut to secure flange to pinion shaft. If another nut is used the original installation position cannot be determined again.

– Place drip tray underneath.

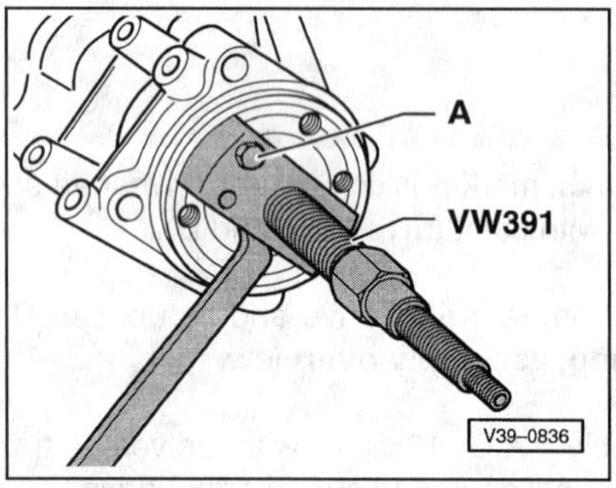

◀ Pull off flange with VW 391.

A - Two M8 x 30 hex bolts inserted into flange

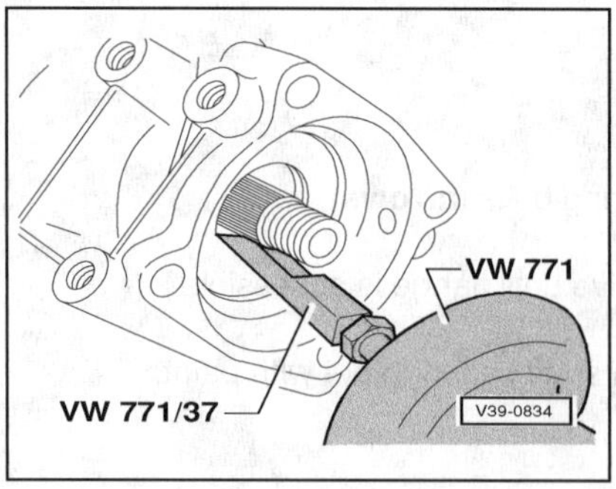

◀ Pull out seal.

Seal for driveshaft flange, installing (rear final drive installed)

– Installation is the reverse of removal noting the following:

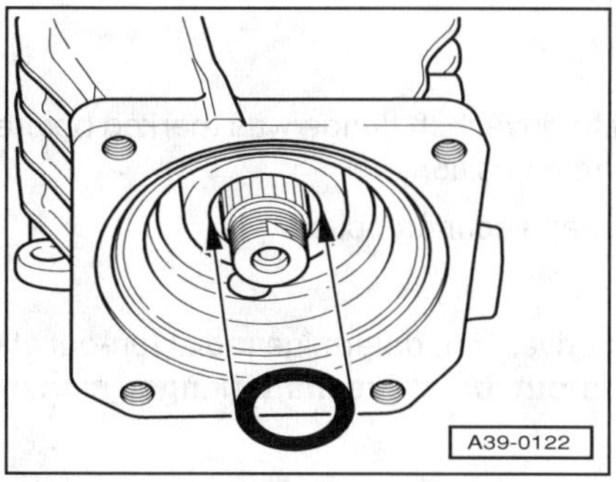

◄ Replace O-ring behind flange.

– Apply gear oil to circumference of seal.

– Half fill space between sealing lip and dust lip with grease G 052 128 A1.

EDITOR'S NOTE —

Greasing the space between sealing lips lubricates the seal and prevents damage due to dryness. It also keeps tensioning spring in place.

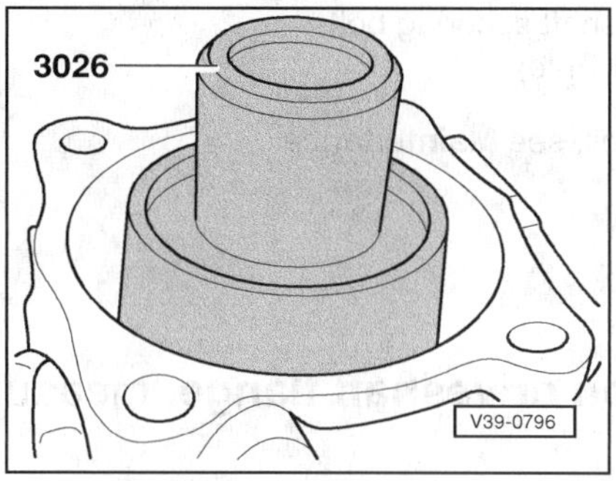

◄ Drive new seal in onto stop in rear axle housing taking care not to tilt seal.

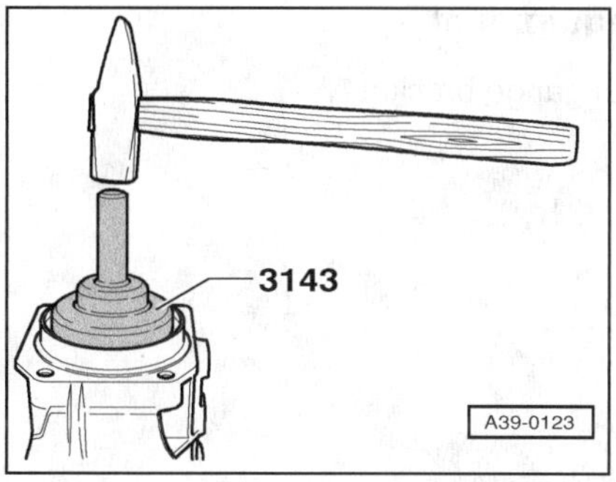

◄ Drive driveshaft flange onto pinion shaft until securing nut can be installed.

– Clean threads on pinion shaft and flange nut so that they are free of grease and coat thinly with locking compound D 000 600.

– Insert previous (original) hex nut and tighten to position marked.

NOTE —

*The same number of turns **MUST** be used when installing as was noted during removal.*

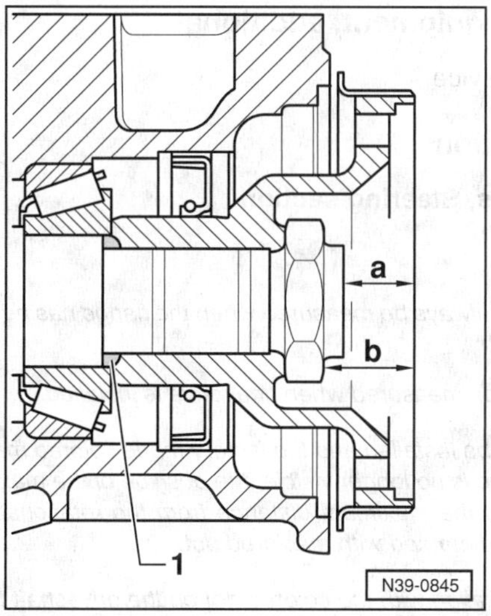

◄ Perform check measurement. Measure dimension -a- and dimension -b- in the same manner as previously noted.

NOTE —

- *Seal (**1**) must be installed when measuring.*
- *The hex nut must be exactly on the marked position again.*
- *Maximum deviation from the original measurement is ± 0.3 mm (0.012 in).*

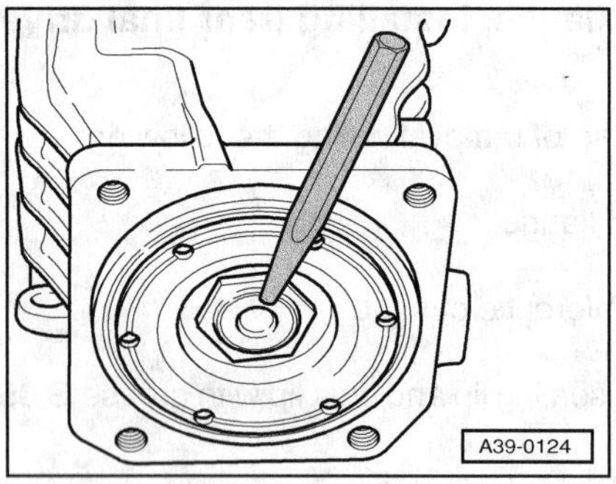

Secure hex nut by peening as shown.

– Position final drive in installed position again.

– Bolt final drive to support.
  • Tighten to 40 Nm (30 ft-lb)

– If position of driveshaft to driveshaft flange was marked before removal, reinstall driveshaft in same position.
  • Always replace driveshaft securing bolts
  • Tighten to 55 Nm (41 Ft-lb).

– If marks are present on driveshaft, determine radial run-out at driveshaft flange, see **Radial run-out on driveshaft flange, measuring and marking**.

– Reinstall driveshaft to rear final drive.
  • Always replace driveshaft securing bolts
  • Tighten to 55 Nm (41 Ft-lb).

– Check rear final drive oil, see **Maintenance**.

Radial run-out on driveshaft flange, measuring and marking

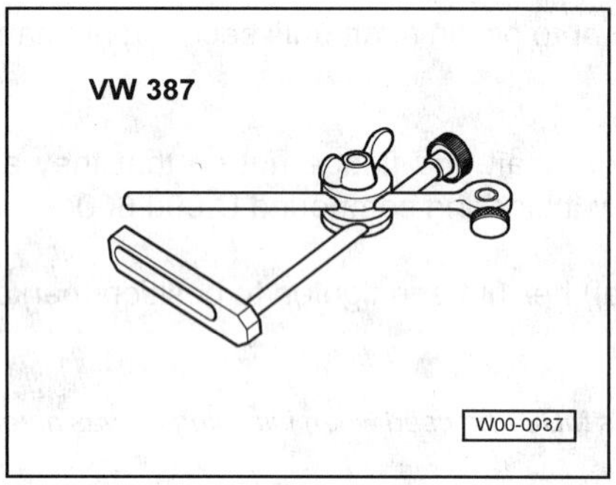

Special tools and equipment

• VW 387 Universal dial gauge bracket
• Dial gauge

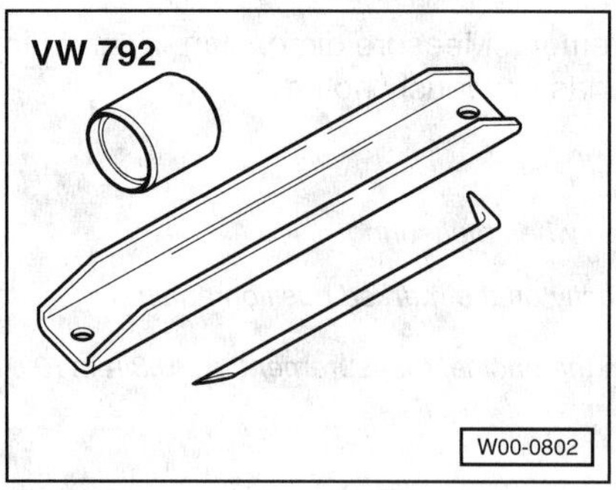

Special tools and equipment, additional

• VW 792 Assembly device

Additional information

• Suspension, Wheels, Steering section

NOTE —

• The radial run-out must always be measured when the flange has been removed.

• The radial run-out can be measured when final drive is installed.

• If a new driveshaft is to be installed and the factory made colored mark on the rear final drive flange is no longer visible, the position of the maximum run-out (corresponds to the maximum distance from the rotational axis) must be determined and marked with a colored dot.

*• This colored dot **MUST** align with the colored dot on the driveshaft flange as shown in illustration N39-0848 on following page.*

Vehicles with a threaded hole on front face of rear final drive

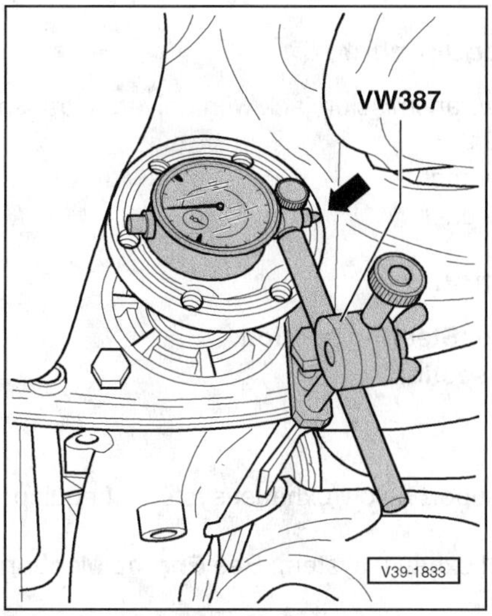

- Mount measuring tools as shown.

Vehicles without threaded hole in front side of rear final drive

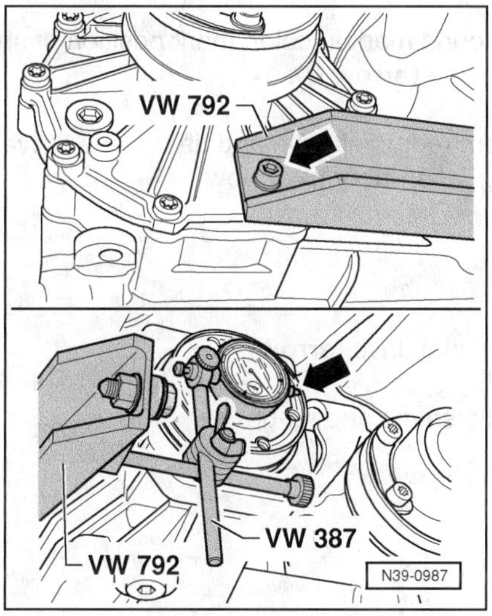

- Mount assembly appliance VW 792 in center hole (**arrow**) for right side axle shaft protective plate.

– Mount universal dial gauge bracket VW 387 with dial gauge to assembly appliance VW 792.

Continued for both versions

– Set dial gauge to zero with 1 mm preload and position on ground surface (**arrow**) in flange.

– Turn flange one complete rotation.

– Mark maximum run-out on edge of flange with a colored paint dot (corresponds to maximum distance from rotational axis).

– Remove previous marks on flange.

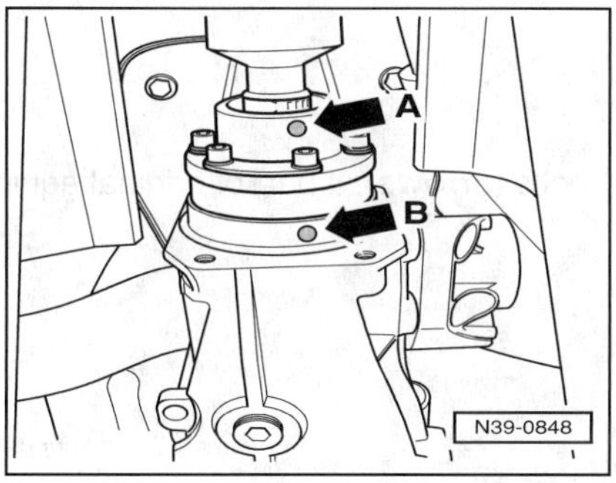

- When installing driveshaft, marks on driveshaft flange (**arrow A**) and on rear final drive flange (**arrow B**) must be aligned.

Rear final drive, removing and installing

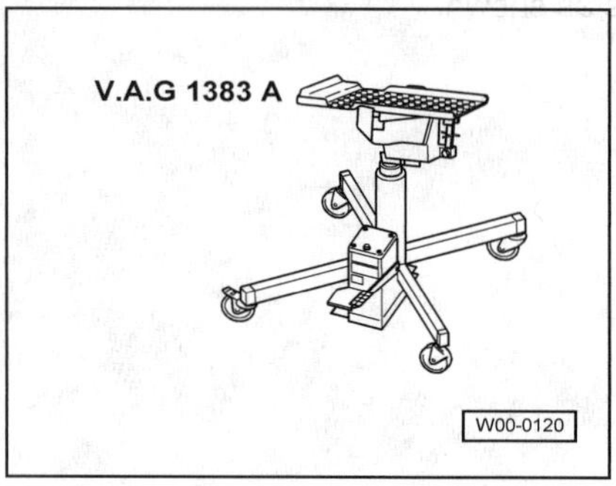

Special tools and equipment

- V.A.G 1383 A Engine/transmission jack with universal transmission mounting V.A.G 1359/2
- V.A.G 1331 Torque wrench
- V.A.G 1332 Torque wrench

Additional information

- **Suspension, Wheels, Steering section**
- **Engine Mechanical section**

Removing

– Remove front cross support below exhaust system, if equipped.

– Remove rear section of exhaust system: See **Engine Mechanical, section, Repair Group 26.**

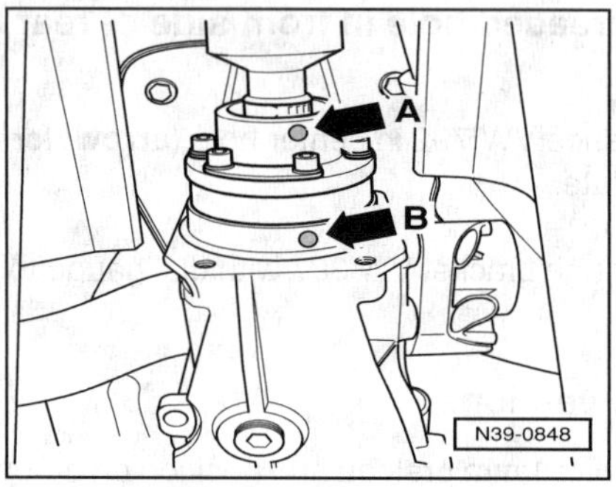

Check whether there is a colored dot on driveshaft and driveshaft flange on rear final drive.

– If no markings can be found mark in color, mark position of shaft flange (**arrow A**) to rear final drive (**arrow B**).

– Separate driveshaft from rear final drive and tie up, see **Driveshaft, removing and installing, assembly overview.**

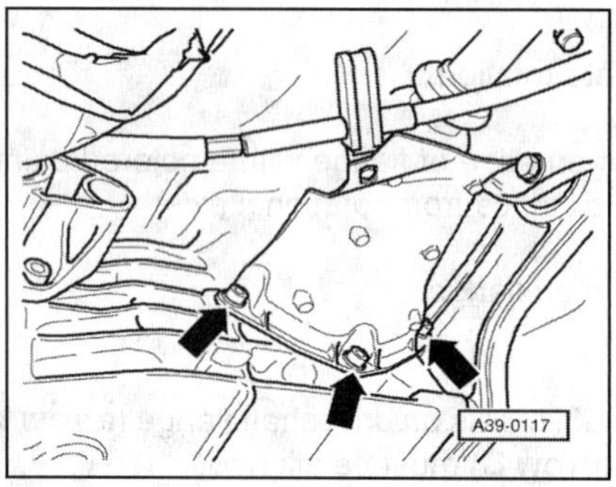

Remove cover plate on final drive (**arrow**).

– Unbolt left and right axle shafts and tie-up.

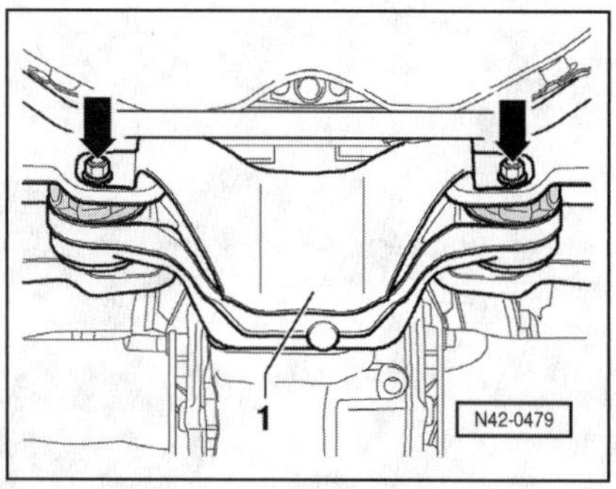

Remove cover plate (**1**) bolts (**arrows**), if part of original equipment.

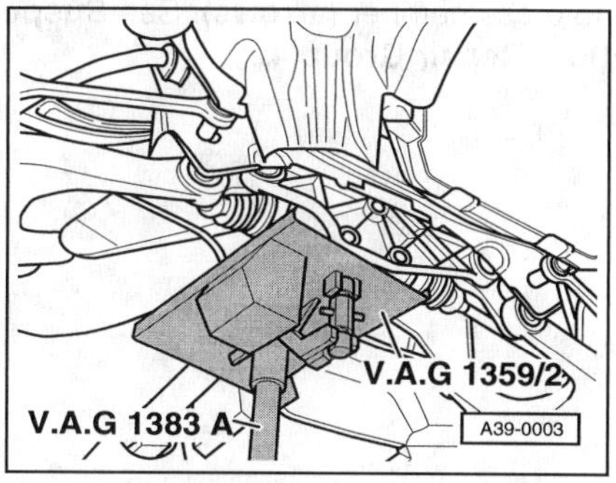

Place transmission jack V.A.G 1383 A with universal mounting 1359/2 under final drive and support final drive.

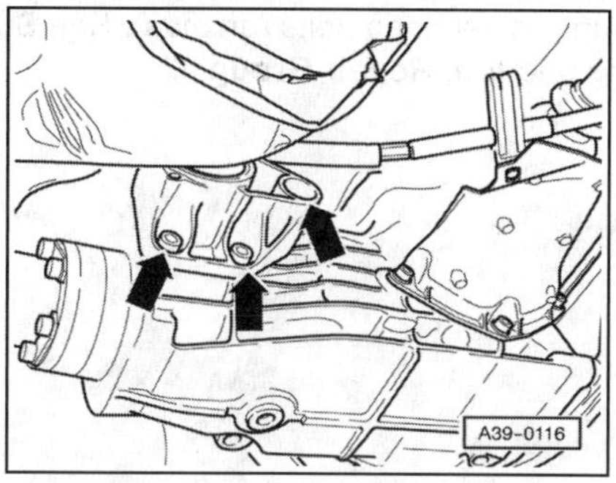

Remove transmission support securing bolts (**arrows**).

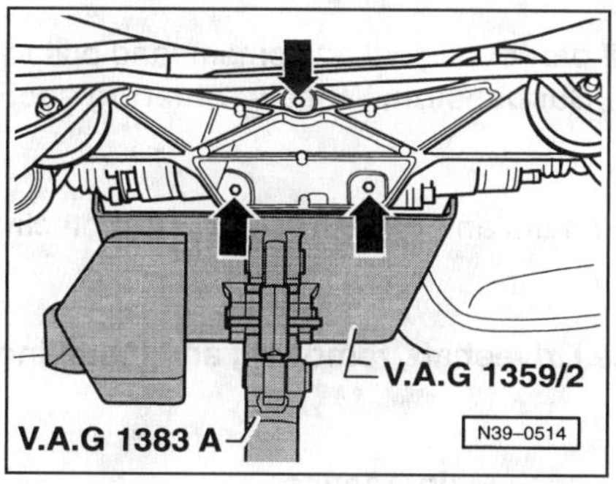

Remove securing bolts from crossmember (**arrows**).

– Lower final drive.

Rear final drive, installing

– Installation is performed in the reverse sequence. Note the following:

– Always replace securing bolts for driveshaft.

– Clean threads in flange shaft of rear final drive of locking fluid remains.

NOTE —

- *The threads can be cleaned with a thread tap.*
- *If threads in the flange shaft are not cleaned the securing bolts can shear when tightened.*

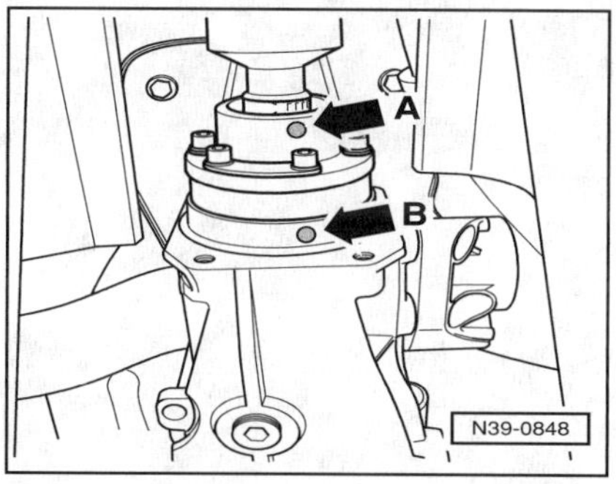

To prevent imbalance, driveshaft flange (**arrow A**) and rear final drive flange (**arrow B**) must be installed so that colored marks from factory, or those subsequently made, are in alignment.

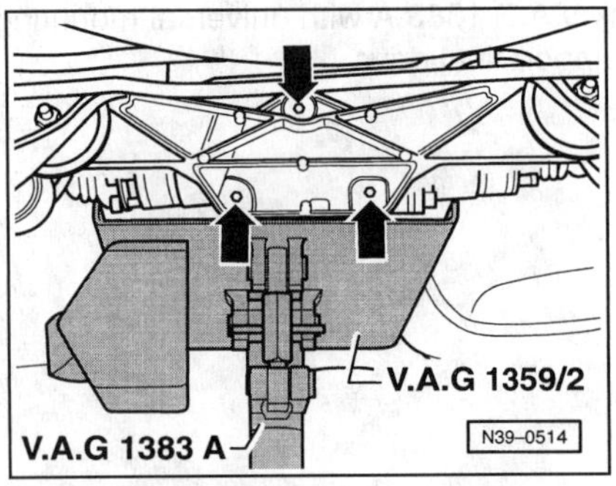

Tighten rear final drive to crossmember (**arrows**). See **Suspension, Wheels, Steering section, Repair Group 42**

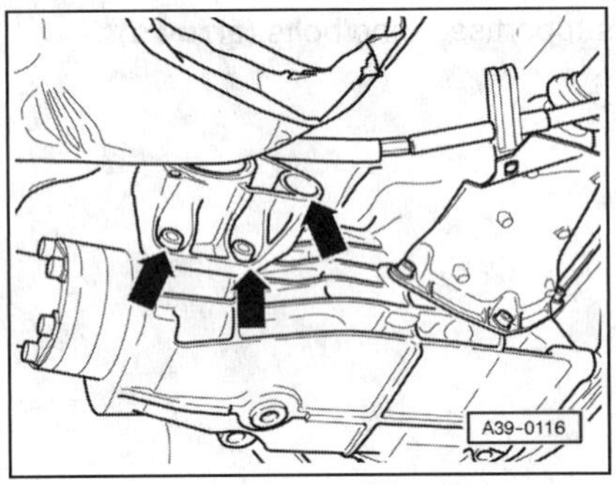

Tighten transmission support securing bolts (**arrows**). See **Suspension, Wheels, Steering section, Repair Group 42**

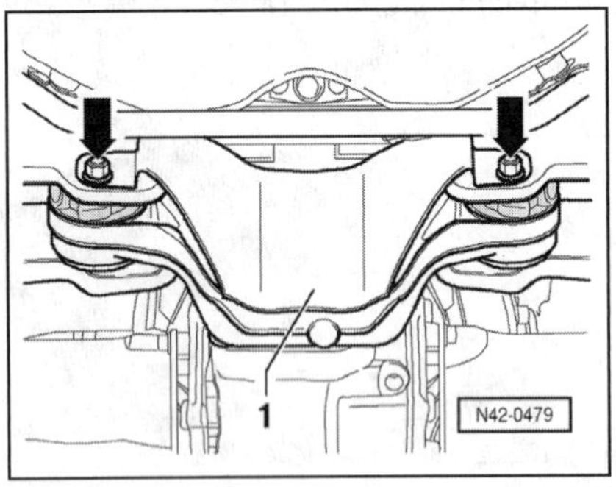

Install cover plate (**1**), if part of original equipment, and bolt to crossmember (**arrows**). See **Suspension, Wheels, Steering section, Repair Group 42**

– Replace seals for axle shafts and driveshaft (pull off protective foil and glue seal into joint).

– Connect driveshaft, see **Driveshaft, removing and installing, assembly overview.**

– Check rear final drive oil, see **Maintenance.**

Tightening torques	Torque
Bolt, drive shaft to driveshaft flange (always replace)	55 Nm (41 ft-lb)
Bolt, axle shaft to drive flange	40 Nm (30 ft-lb)
Bolt, cover plate (heat shield) to rear final drive	23 Nm (17 ft-lb)

Brake System

Section

R

00 General Information

Page

45 Anti-Lock Brake Systems

Page

46 Brakes, Mechanical — Page

47 Brakes, Hydraulic — Page

00 General Information

Repair Group

R 00

Technical data

1998 and 1999 model years

- 23.81 mm diameter brake master cylinder
- 10 inch diameter brake servo

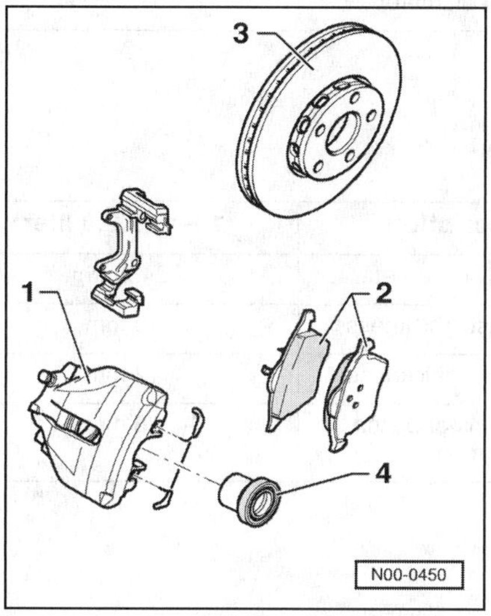

⮜ Front brakes, FN3 front brake calipers

Item	Description	1.8 and 2.8 liter
1	Brake caliper	FN3
2	Brake pad, thickness	14 mm
3	Brake disc diameter ➤VIN# 3b x 124 107 VIN# 3b x 124 108 ➤	 282.5 mm 288 mm
	Brake disc, thickness	25 mm
4	Brake caliper, piston diameter	57mm

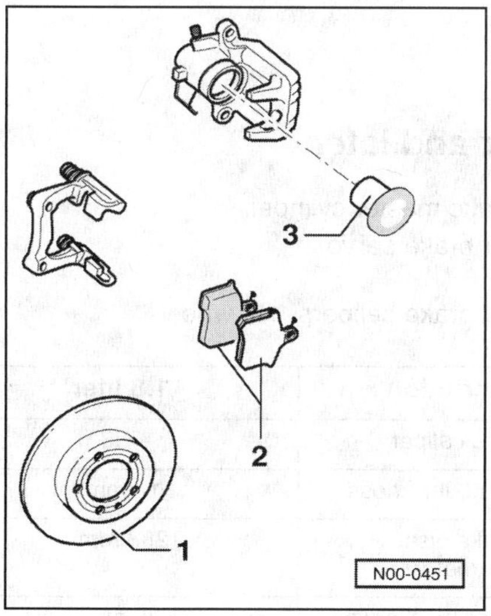

⮜ Rear brakes, 15" wheel

Item	Description	1.8 and 2.8 liter
1	Brake disc diameter	245 mm
	Brake disc, thickness	10 mm
2	Brake pad, thickness	12 mm
3	Brake caliper piston diameter	38 mm

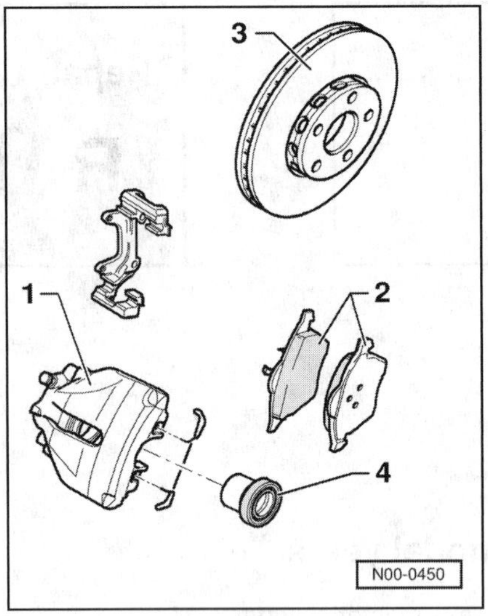

2000 model year

- 25.4 mm diameter brake master cylinder
- 7 and 8 inch diameter brake servo

< Front brakes, FN3 front brake calipers

Item	Description	1.8 and 2.8 liter
1	Brake caliper	FN3
2	Brake pad, thickness	14 mm
3	Brake disc diameter	288 mm
	Brake disc, thickness	25 mm
4	Brake caliper, piston diameter	57 mm

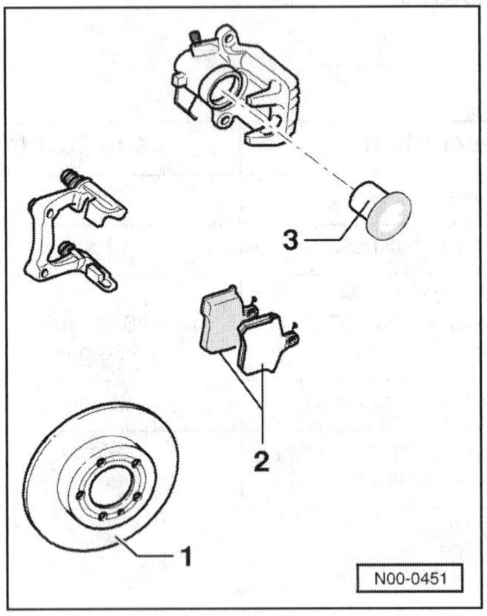

< Rear brakes, 15” wheel

Item	Description	1.8 and 2.8 liter
1	Brake disc diameter	245 mm
	Brake disc, thickness	10 mm
2	Brake pad, thickness	12 mm
3	Brake caliper piston diameter	38 mm

2001 model year and later

- 25.4 mm diameter brake master cylinder
- 7 and 8 inch diameter brake servo

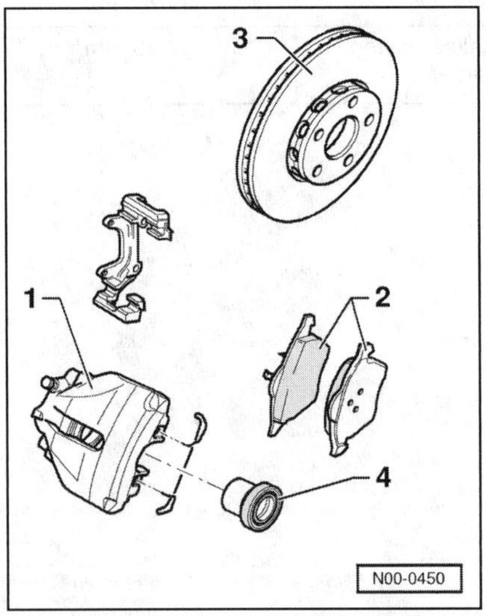

< Front brakes, FN3 front brake calipers, 15" wheel

Item	Description	1.8 liter
1	Brake caliper	FN3
2	Brake pad, thickness	14 mm
3	Brake disc diameter	288 mm
	Brake disc, thickness	25 mm
4	Brake caliper, piston diameter	57 mm

2001 model year and later (continued)

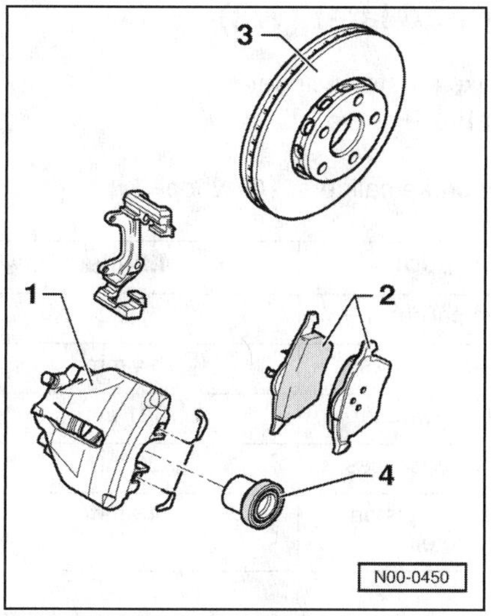

Front brakes, FN3 front brake calipers, 16" wheel

Item	Description	2.8 liter
1	Brake caliper	FN3
2	Brake pad, thickness	14 mm
3	Brake disc diameter	312 mm
	Brake disc, thickness	25 mm
4	Brake caliper, piston diameter	57 mm

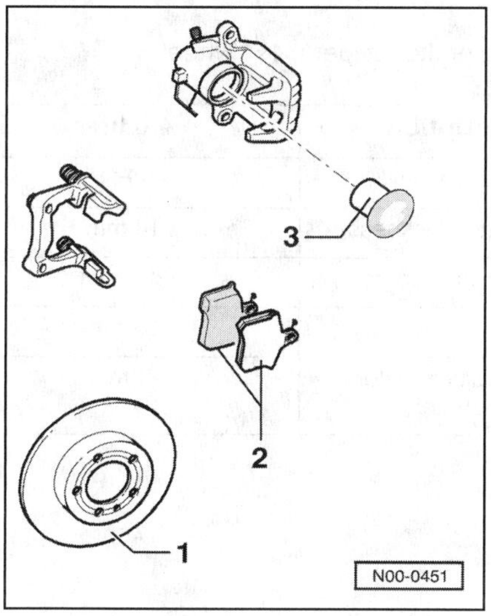

Rear brakes, front wheel drive, 15" and 16" wheels

Item	Description	1.8 and 2.8 liter
1	Brake disc diameter	245 mm
	Brake disc, thickness	10 mm
2	Brake pad, thickness	12 mm
3	Brake caliper piston diameter	38 mm

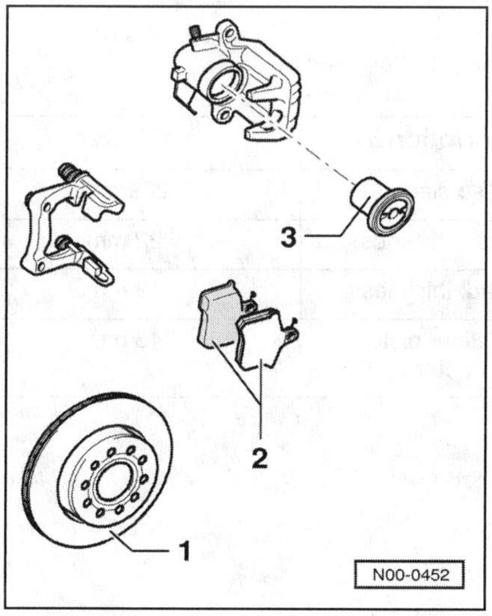

Rear brakes, 4Motion, 16" wheel

Item	Description	2.8 liter
1	Brake disc diameter	256 mm
	Brake disc, thickness	22 mm
2	Brake pad, thickness	12 mm
3	Brake caliper piston diameter	43 mm

2002 model year and later (W8)

- 25.4 mm diameter brake master cylinder
- 7 and 8 inch diameter brake servo

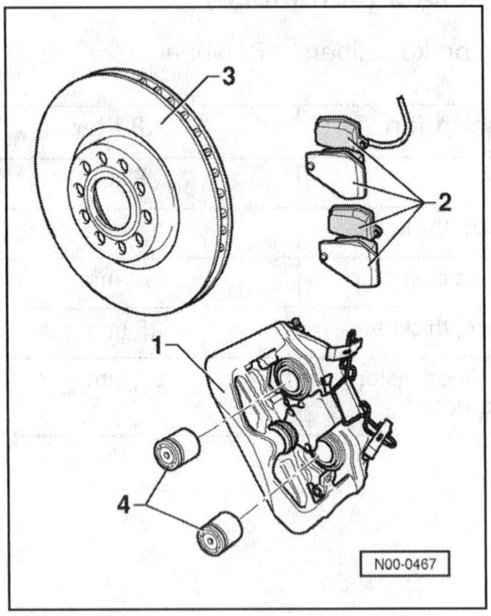

Front brakes, HP2 front brake calipers, 16" wheel

Item	Description	4.0 liter
1	Brake caliper	HP2
2	Brake pad, thickness	14 mm
3	Brake disc diameter	321 mm
	Brake disc, thickness	30 mm
4	Brake caliper, piston diameter	43 mm

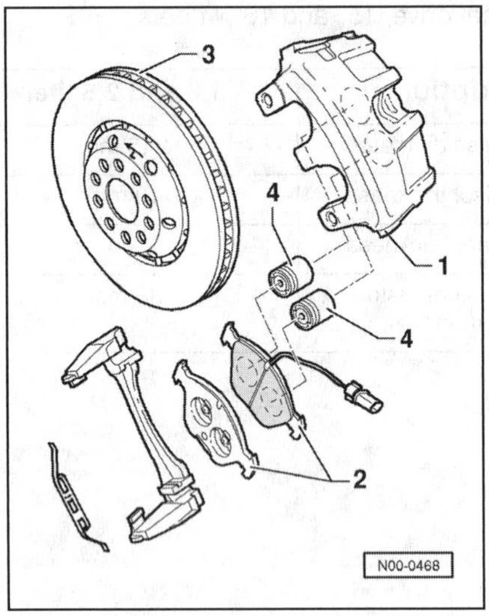

Front brakes, 2FN front brake calipers, 17" wheel

Item	Description	4.0 liter
1	Brake caliper	2FN
2	Brake pad, thickness	14 mm
3	Brake disc diameter	334 mm
	Brake disc, thickness	32mm
4	Brake caliper, piston diameter	42mm

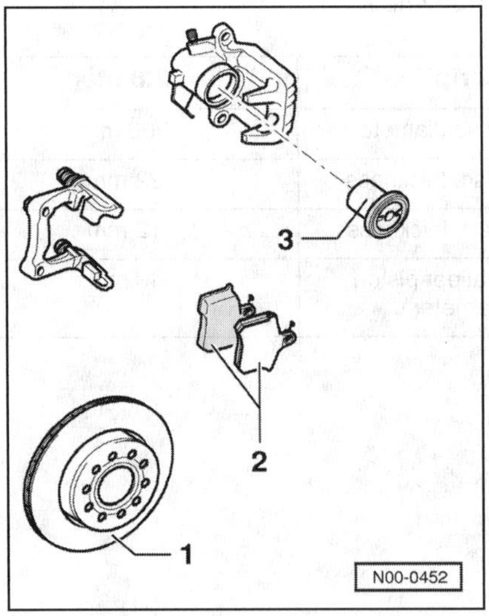

Rear disc brakes, 16" and 17" wheels

Item	Description	4.0 liter
1	Brake disc diameter	269 mm
	Brake disc, thickness	22 mm
2	Brake pad, thickness	12 mm
3	Brake caliper piston diameter	43 mm

45 Anti-Lock Brake Systems

Repair Group

R 45

Anti-Lock Brake System (ABS)

The following variants of ABS are possible:

- ABS
- ABS/EDL
- ABS/EDL with ASR

The ABS brake system is divided diagonally. The servo-assistance is effected pneumatically, by the vacuum brake servo unit.

Malfuncitons on the ABS influence the brake system and servo assistance. A braking behavior change should be checked. When the ABS warning lamp comes on, rear wheels can lock-up when braking!

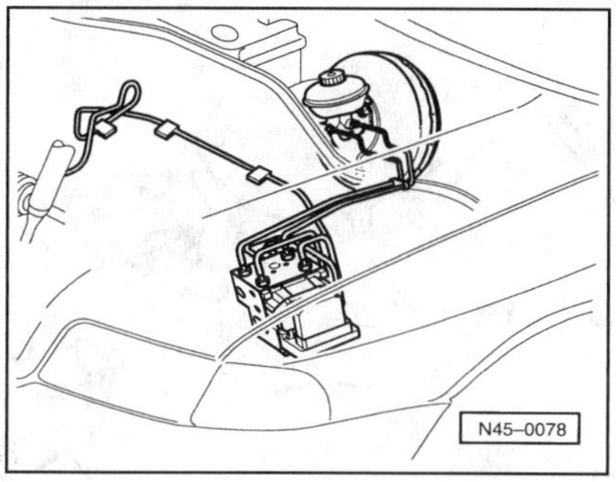

ABS layout.

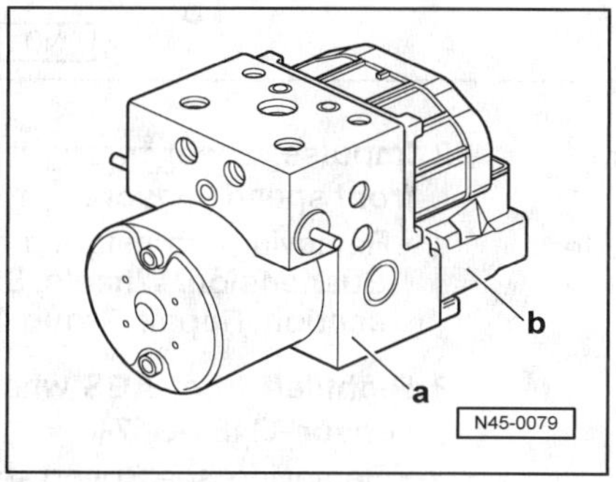

Hydraulic unit -a- and control module -b- are one unit and if malfunctions occur they should be replaced (apart from ABS/EDL/ASR Bosch 5.3).

Information for repair work on ABS

NOTE —

- *Disconnect battery ground strap with ignition switched off.*
- *When working with brake fluid, observe safety precautions, see* **Repair Group 47, Braking system, bleeding.**
- *After working on an open brake system, bleed the brake system with brake fluid and bleeding unit VAS 5234, see* **Repair Group 47, Braking system, bleeding.**
- *During the final road test, make sure that a controlled brake test is performed at least once (pulsations must be felt at the brake pedal).*
- *Absolute cleanliness is required when working on the anti-locking brake system. Do not use any products which contain mineral oil, such as oils, greases etc.*
- *Thoroughly clean all connections and the adjacent areas before loosening, do not use cleaning agents such as brake cleaner, gasoline, or thinners.*
- *Cover or seal open components if repairs cannot be completed immediately. (Use sealing plugs repair kit 1 H0 693 311 A).*
- *Do not use fluffy cloths.*
- *When the system is open do not work in areas with compressed air and do not move the vehicle.*
- *When painting, electronic control module can be exposed to a maximum temperature of 95°C (203°F) only briefly, and a maximum temperature of 85 °C (185°F) for longer periods of time (approx. 2 hours).*

Anti-Lock Brake System (ABS), Bosch 5.3

Electrical/electronic components, ABS/EDL/ASR

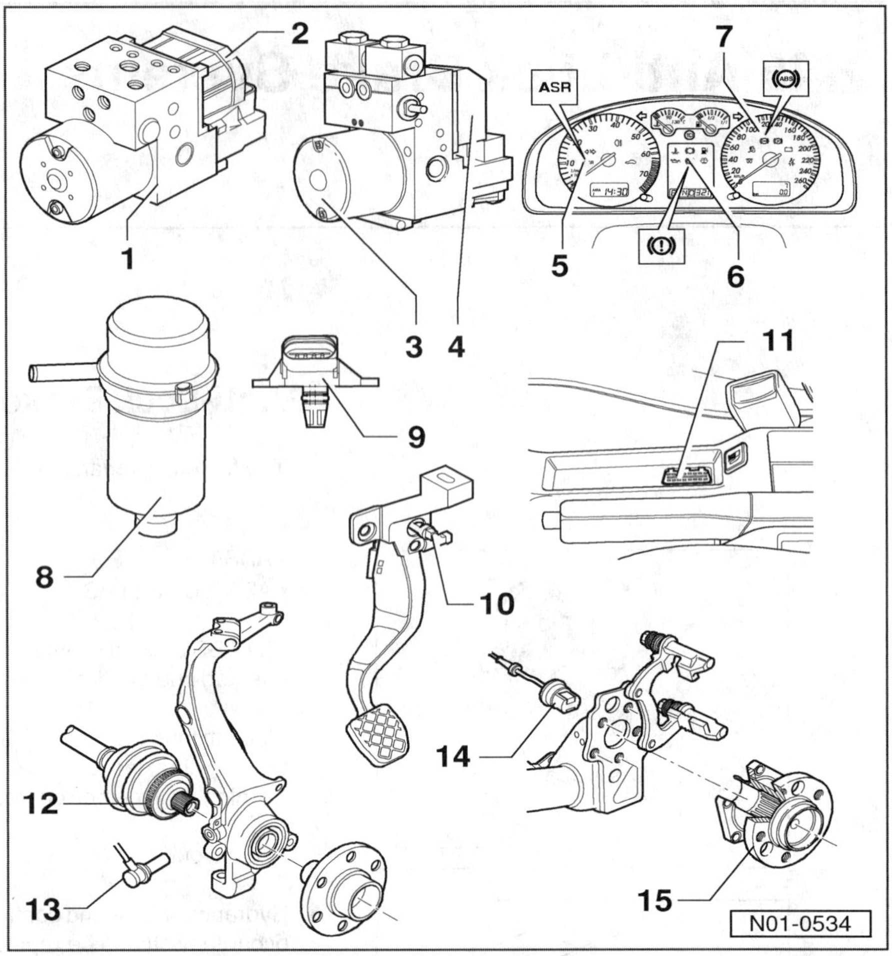

1. **ABS hydraulic unit -N55-**
   - Located on left side of engine compartment.
   - The hydraulic unit consists of: ABS return flow pump -V39- and Valve block (contains in/outlet valves).
   - The ABS return flow pump -V39- and valve block must not be separated from one another
2. **ABS control module (w/EDL) -J104-**
   - Located on the hydraulic unit on left side of engine compartment.
   - Do not disconnect connector(26 pin) before checking of Diagnostic Trouble Code (DTC) memory. First switch ignition off then separate connector.
3. **ABS hydraulic unit for ABS/EDL or ABS/EDL with ASR -N55-**
   - Located on left side of engine compartment.
   - The hydraulic unit consists of: ABS return flow pump -V39- and Valve block (contains the in/outlet valves).
   - The ABS return flow pump -V39- and valve block must not be separated from one another
   - Distinguishing features: Hydraulic unit can only be distinguished externally by different part numbers.
4. **Control module for ABS/EDL or ABS/EDL with ASR -J104-**
   - Located on hydraulic unit on left side of engine compartment.
   - Do not disconnect connector (26 or 31 pin) before checking of (DTC) memory. First switch ignition off then separate connector.
   - Hydraulic unit can be distinguished externally by different part numbers.
   - With ASR wiring harness connection for ABS control module (w/EDL) -J104- has 5 additional contacts (31 pin).
5. **Traction control indicator light -K86-**
   - Location: In the instrument cluster
6. **Warning light for brake system -K118-**
   - Location: In the instrument cluster
7. **ABS warning light -K47-**
   - Location: In the instrument cluster
8. **Brake system vacuum pump -V192-**
   - Not in all vehicles
   - Located on left side of engine compartment.
9. **Brake booster pressure sensor -G294-**
   - Not in all vehicles
   - Location: in vacuum hose on brake servo
10. **Brake light switch -F-**
    - Adjusting, see **Repair Group 46, Brake light switch -F- and Brake pedal switch (cruise control/Diesel Direct Fuel Injection) -F47-, adjusting**
11. **Data Link Connector (DLC)**
    - 2000m.y.: below cover next to parking brake lever
    - 2001m.y.: below stowage compartment,driver's side
12. **Impulse wheel for right and left front speed sensors**
    - Removing and installing, see **Suspension, Wheels, Steering section, Repair Group 40**
13. **Right/left front ABS wheel speed sensor -G45-/-G47-**
    - Removing, see **Speed sensor on front axle, removing and installing**
14. **Right/left rear ABS speed sensor -G44-/-G46-**
    - Figure for front wheel drive shown
    - Removing, see appropriate **Speed sensor on rear axle, removing and installing**
15. **Impulse wheel for right and left rear speed sensors**
    - Figure for front wheel drive shown
    - Removing and installing, see **Suspension, Wheels, Steering section, Repair Group 42**

Hydraulic unit, brake servo/brake master cylinder, overview (through 1999 m.y.)

NOTE—
The control module (item 6) and the hydraulic unit (item 8) must always be removed together. The control module and hydraulic unit are therefore not available as individual items.

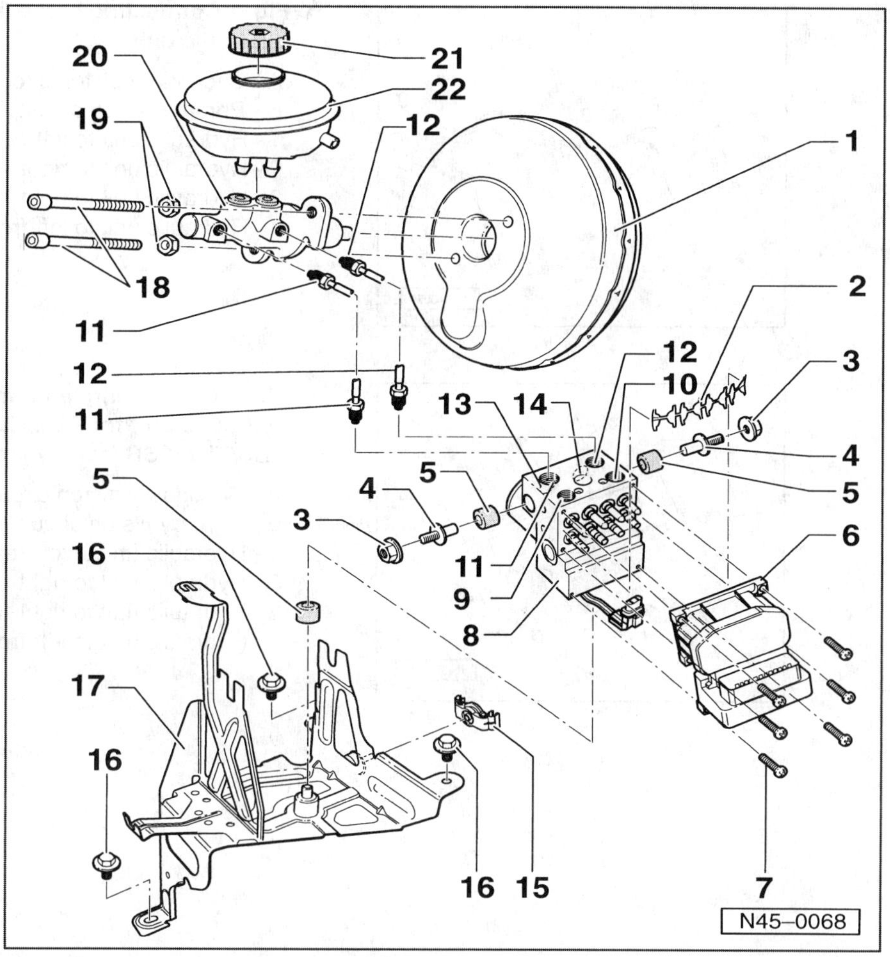

1. **Brake servo**
   - On gasoline engines the vacuum required is taken from the intake manifold.
   - Functional check
     With engine switched off, depress brake pedal firmly several times (to exhaust the vacuum in the module).
     Depress brake pedal with average foot pressure, hold and start engine. If the servo module is working properly, the pedal will drop slightly (servo assistance becomes effective).
     If faulty replace.
2. **Spring**
3. **Self-locking nut**
   - Tighten to 9 Nm (80 in-lb)
4. **Threaded stud**
5. **Rubber damper**
6. **Control module**
7. **T 20 torx socket head bolt**
   - Tighten to 8 Nm (71 in-lb)
8. **ABS hydraulic unit**
   - Note different versions:
   - ABS front wheel drive, see **Fig. 1**
   - ABS/EDL front and all wheel drive and ABS/EDL/ASR front wheel drive, see **Fig.2**
9. **Brake line connection**
   - Hydraulic unit to right-front brake caliper
10. **Brake line connection**
    - Hydraulic unit to left-front brake caliper
11. **Brake line**
    - Master brake cylinder/secondary piston circuit to hydraulic unit
12. **Brake line**
    - Master brake cylinder/primary piston circuit to hydraulic unit
13. **Brake line connection**
    - Hydraulic unit to left-rear brake caliper
14. **Brake line connection**
    - Hydraulic unit to right-rear brake caliper
15. **Retainer**
16. **Hex bolt**
    - Tighten to 9 Nm (80 in-lb)
17. **Retainer**
18. **T 45 torx socket head bolt**
    - Tighten to 25 Nm (18 ft-lb)
19. **Self-locking nut**
    - Tighten to 49 Nm (36 ft-lb)
20. **Tandem brake master cylinder**
    - With seal
21. **Cap**
22. **Brake fluid reservoir**

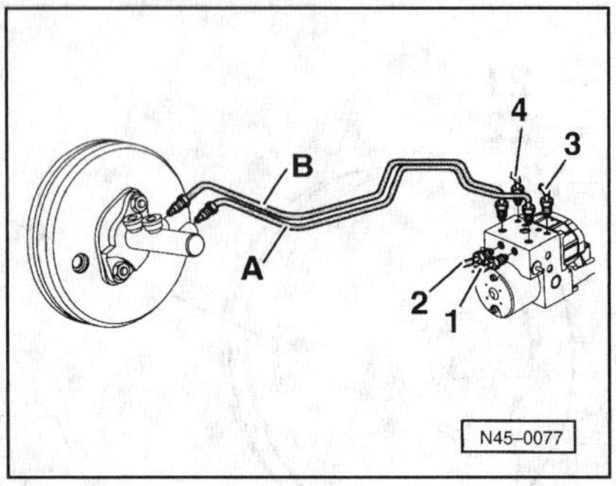

Fig. 1 Connecting brake lines from tandem master cylinder to ABS hydraulic unit

A - Secondary piston circuit
B - Primary piston circuit
1 - Hydraulic unit to left-rear parking brake caliper
2 - Hydraulic unit to right-rear parking brake caliper
3 - Hydraulic unit to right-front brake caliper
4 - Hydraulic unit to left-front brake caliper

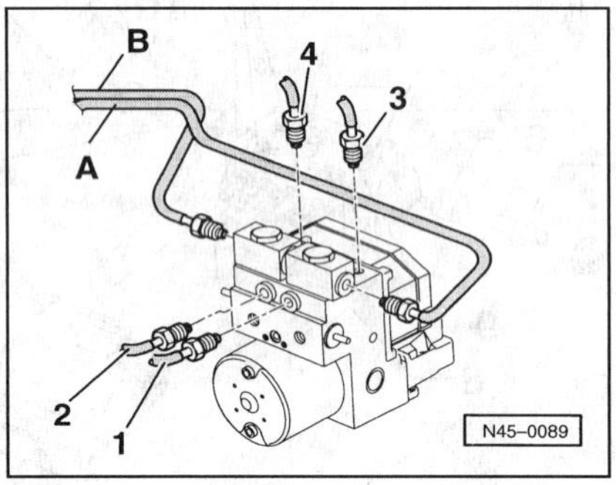

Fig. 2 Connecting tandem brake master cylinder brake lines to the hydraulic unit for ABS/EDL front and all-wheel-drive and ABS/EDL/ASR front wheel drive

A - Secondary piston circuit
B - Primary piston circuit
1 - Hydraulic unit to left-rear parking brake caliper
2 - Hydraulic unit to right-rear parking brake caliper
3 - Hydraulic unit to right-front brake caliper
4 - Hydraulic unit to left-front brake caliper

Hydraulic unit, brake servo/brake master cylinder, overview (from 2000 m.y.)

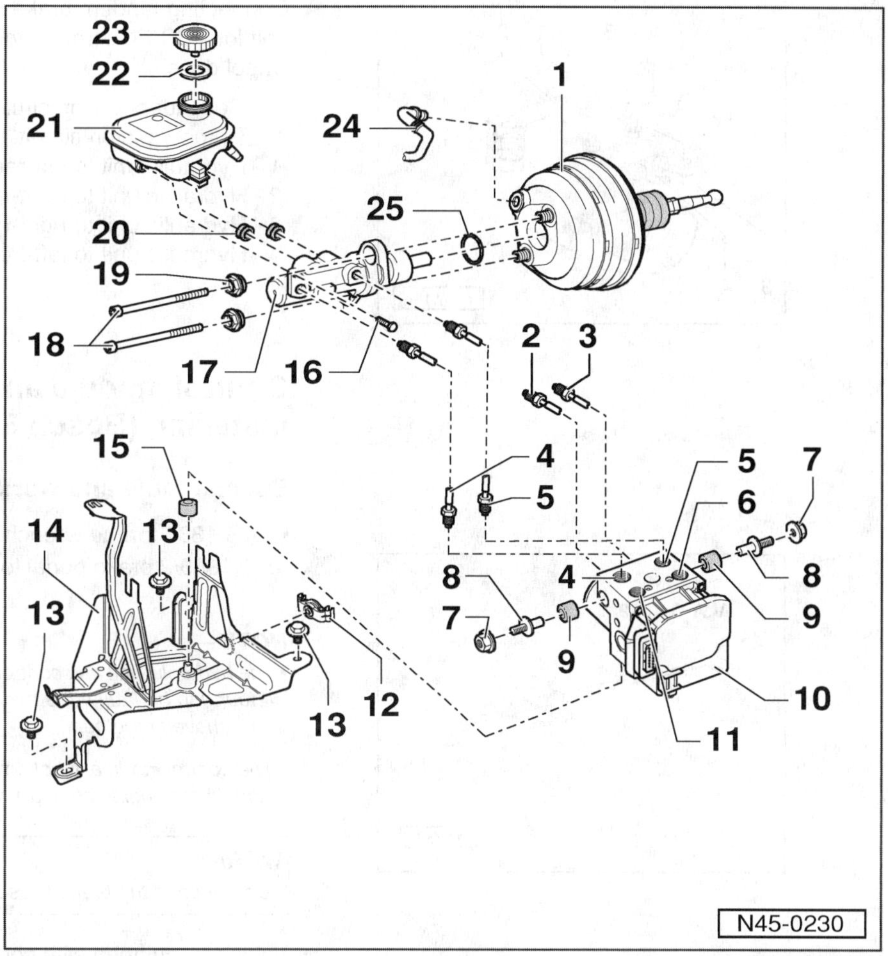

1. **Tandem brake servo**
   - On gasoline engines the vacuum required is taken from the intake manifold.
   - From model year 2001, vehicles with a gasoline engine and automatic transmission are equipped with a brake vacuum pump, see **Repair Group 47, Brake vacuum pump -V192-, assembly**
   - Function check
     With engine switched off, depress brake pedal firmly several times (to exhaust the vacuum in the module).
     Depress brake pedal with average foot pressure, hold and start engine. If the servo module is working properly, the pedal will drop slightly (servo assistance becomes effective).
     If faulty replace.
2. **Brake line connection**
   - To left-rear caliper
   - Marked on hydraulic unit with -HL-
3. **Brake line connection**
   - To right-rear caliper
   - Marked on hydraulic unit with -HR-
4. **Brake line**
   - Master brake cylinder/secondary piston circuit to hydraulic unit
5. **Brake line**
   - Brake master cylinder/primary piston circuit to hydraulic unit
6. **Brake line**
   - To left-front brake caliper
   - Marked on hydraulic unit with -VL-
7. **Hex nut**
   - Tighten to 12 Nm (9 ft-lb)
8. **Threaded stud**
9. **Rubber damper**
10. **ABS hydraulic unit**
11. **Brake line**
    - To right-front brake caliper
    - Marked on hydraulic unit with -VR-
12. **Retainer**
13. **Hex bolt**
    - Tighten to 9 Nm (80 in-lb)
14. **Retainer**
15. **Rubber damper**
16. **Hex bolt**
17. **Tandem brake master cylinder**
    - Cannot be repaired. If faulty, replace.
    - Removing and installing, see **Repair Group 47, Brake servo/ brake master cylinder, assembly (from 2000 m.y.)**
18. **T 45 torx socket head bolt**
    - Tighten to 25 Nm (18 ft-lb)
19. **Self-locking nut**
    - Tighten to 49 Nm (36 ft-lb)
20. **Sealing plug**
    - Coat with brake fluid and press into brake fluid reservoir
21. **Brake fluid reservoir**
22. **Oil seal**
23. **Cap**
24. **Vacuum hose**
    - Insert into brake servo module
25. **Oil seal**

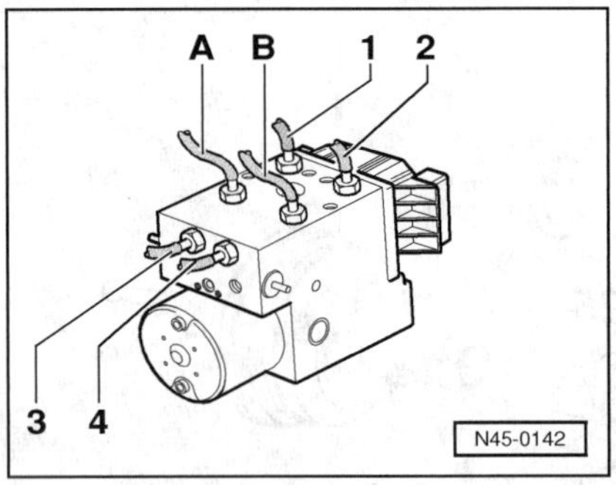

Connecting tandem brake master cylinder brake lines to the hydraulic unit for ABS/EDL front and all wheel drive and ABS/EDL/ASR front wheel drive.

A - Secondary piston circuit
B - Primary piston circuit
1 - Hydraulic unit to left-rear parking brake caliper
2 - Hydraulic unit to right-rear parking brake caliper
3 - Hydraulic unit to right-front brake caliper
4 - Hydraulic unit to left-front brake caliper

Control module and hydraulic unit, removing and installing (Bosch 5.3)

Special tools and workshop equipment required

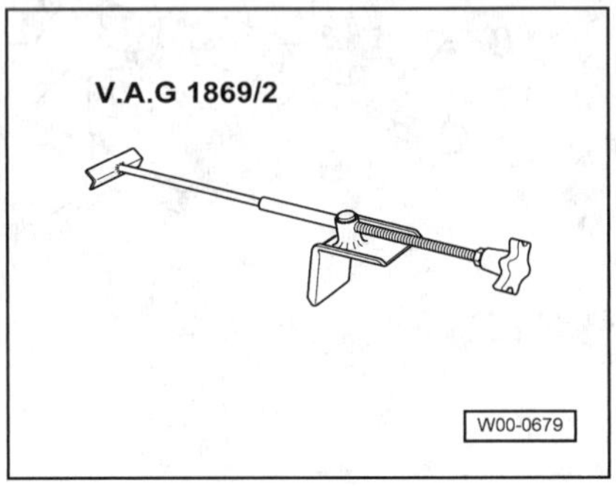

• VAG 1331 torque wrench (or equivalent 5 to 50 Nm)
VAG 1869/2 brake pedal loading device

NOTE —
• The hydraulic unit and control module are not available as individual items. If found to be faulty, the hydraulic unit and control module should always be removed together.

• The control module is bolted to the hydraulic unit and is located on the left side, in the engine compartment.

WARNING —
Do not bend the brake lines in the area of the hydraulic unit!

– Vehicles equipped with coded radio, note or request radio code, if necessary.

– Disconnect battery.

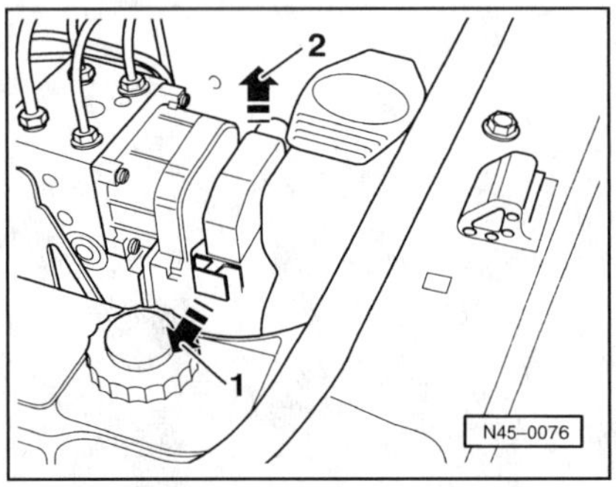

Release control module connector (arrow -1-) and pull off (arrow -2-).

– Connect bleeder bottle hose to left-front brake caliper bleed screw and open bleed screw.

– Install brake pedal loading device from VAS 5234

– Depress brake pedal with brake pedal depressor.

– Close left-front bleeder screw.

– Place lint-free cloths under the control module and hydraulic unit.

– Disconnect brake lines on the hydraulic unit.

– Seal brake lines and threaded holes with plugs from repair kit.

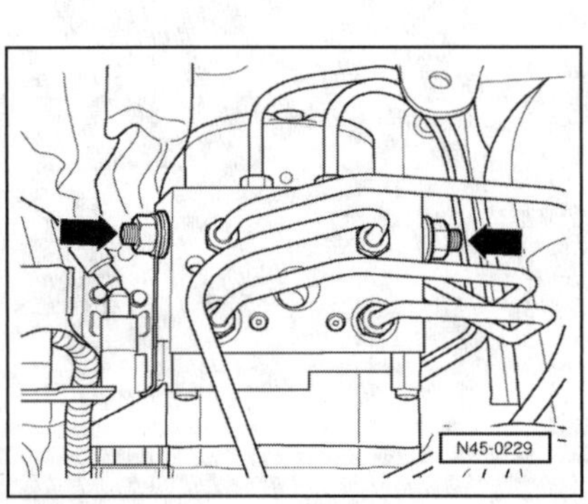

Remove hex nuts on hydraulic unit bracket -arrows-.

– Remove hydraulic unit.

Control module and hydraulic unit, installing

NOTE—

- *Remove sealing plugs on new hydraulic unit when the corresponding brake line is going to be connected.*
- *If the sealing plugs are removed too early, brake fluid can escape, and can no longer be guaranteed that the unit is filled or bled.*

– Installation is the reverse of removal, noting the following:

- Bleed the brake system, see **Repair Group 47, Braking system, bleeding**
- Enter radio code.
- Code ABS control module, see **Scan Tool section**

Tightening torques:	
Brake lines to ABS module: M 10 x 1 and M 12 x 1	4 Nm (35 in-lb)
Hydraulic unit hex nut to retainer	12 Nm (9 ft-lb)

Anti-Lock Brake System (ABS), Bosch 5.7

Hydraulic unit, brake servo/brake master cylinder, overview (from 2002 m.y.)

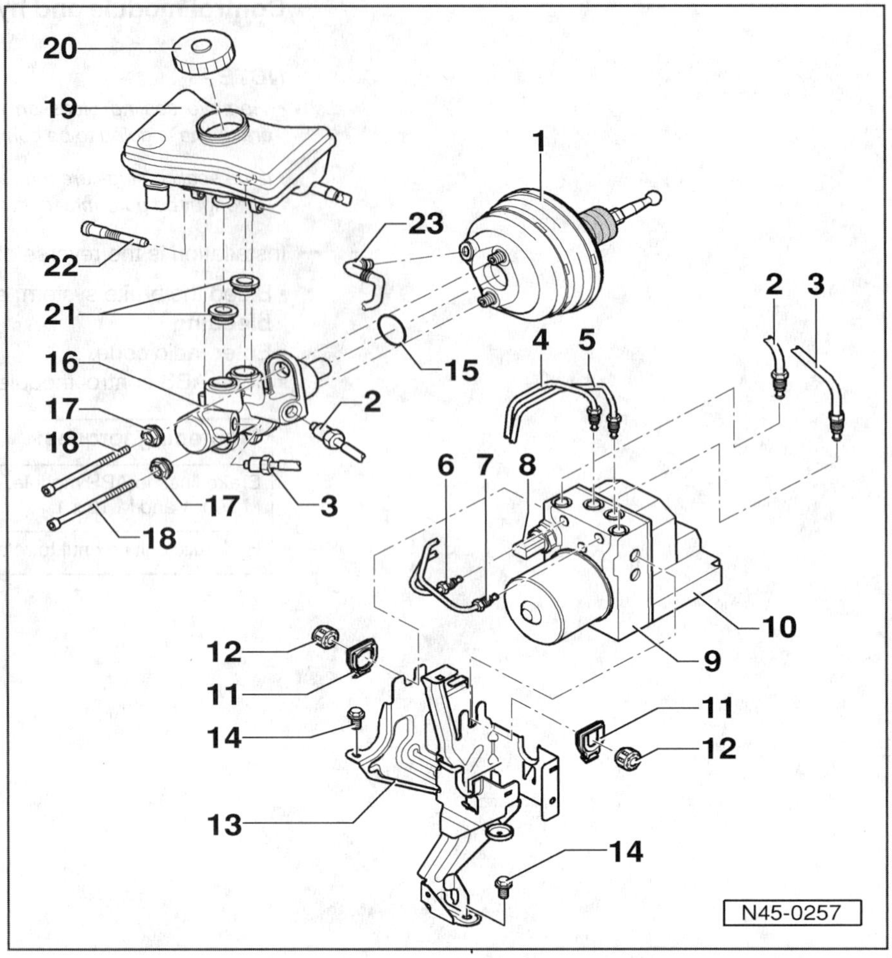

1. **Tandem brake servo**
   - On gasoline engines the vacuum required is taken from the intake manifold.
   - Vehicles with a gasoline engine and automatic transmission are equipped with a brake vacuum pump, see **Repair Group 47, Brake vacuum pump -V192-, assembly**
   - Functional check
     With engine switched off, depress brake pedal firmly several times (to exhaust the vacuum in the module).
     Depress brake pedal with average foot pressure, hold and start engine. If the servo module is working properly, the pedal will drop slightly (servo assistance becomes effective).
     If faulty replace.
   - With bonded seal
   - Removing and installing, see **Repair Group 47, Brake servo/ brake master cylinder, assembly (from 2002 m.y.)**
2. **Brake line**
   - Brake master cylinder/primary piston circuit to hydraulic unit
   - Marked on hydraulic unit with -HZ1-
3. **Brake line**
   - Master brake cylinder/secondary piston circuit to hydraulic unit
   - Marked on hydraulic unit with -HZ2-
4. **Brake line**
   - To left-front brake caliper
   - Marked on hydraulic unit with -VL-
5. **Brake line**
   - To right-front brake caliper
   - Marked on hydraulic unit with -VR-
6. **Brake line**
   - To right-rear caliper
   - Marked on hydraulic unit with -HR-
7. **Brake line**
   - To left-rear caliper
   - Marked on hydraulic unit with -HL-
8. **Brake pressure sender**
   - Removing and installing, see **Sender 1 for brake booster -G201-, removing and installing**
9. **ABS hydraulic unit**
10. **Control module**
11. **Locking clip**
12. **Buffer**
13. **Retainer**
14. **Hex bolt**
    - Tighten to 9 Nm (80 in-lb)
15. **Oil seal**
16. **Tandem brake master cylinder**
    - Cannot be repaired. If faulty, replace.
17. **Self-locking nut**
    - Tighten to 49 Nm (36 ft-lb)
18. **T 45 torx socket head bolt**
    - Tighten to 25 Nm (18 ft-lb)
19. **Brake fluid reservoir**
20. **Cap**
21. **Sealing plug**
    - Coat with brake fluid and press into brake fluid reservoir
22. **Trunnion bolt**
    - Tighten to 10 Nm (7 ft-lb)
23. **Vacuum hose**
    - Insert into brake servo module

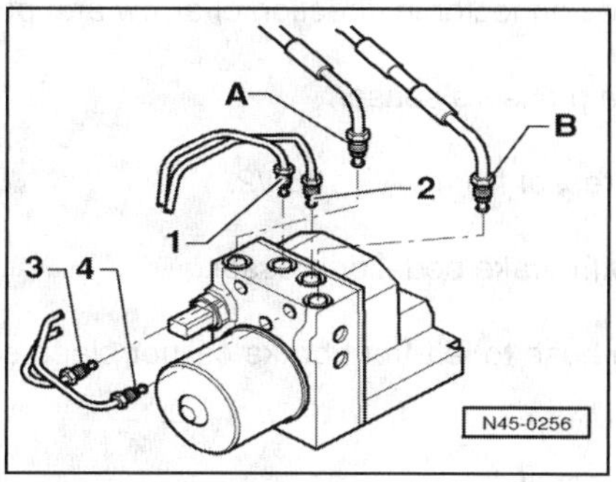

Connecting brake lines from tandem master cylinder to hydraulic unit

A - Hydraulic unit to master cylinder primary piston circuit
Marked -HZ1- on hydraulic unit
B - Hydraulic unit to master cylinder secondary piston circuit
Marked -HZ2- on hydraulic unit
1 - Hydraulic unit to left-front brake caliper
Marked -VL- on hydraulic unit
2 - Hydraulic unit to right-front brake caliper
Marked -VR- on hydraulic unit
3 - Hydraulic unit to right-rear brake caliper
Marked -HR- on hydraulic unit
4 - Hydraulic unit to left-rear brake caliper
Marked -HL- on hydraulic unit

Control module and hydraulic unit, removing and installing (Bosch 5.7)

Special tools and workshop equipment required

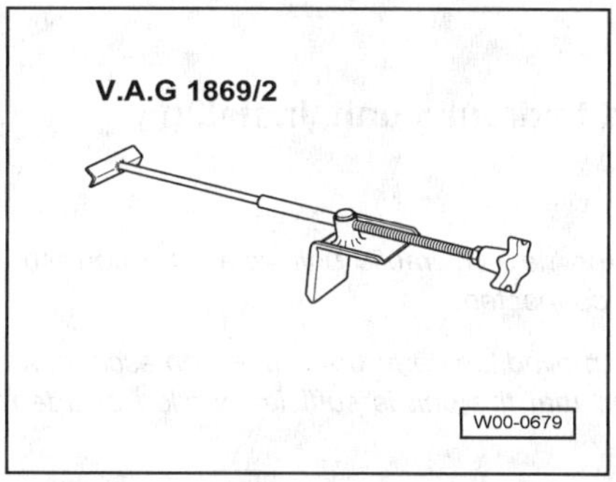

• VAG 1331 Torque wrench
VAG 1869/2 Brake pedal loading device

NOTE—
The hydraulic unit and control module are not available as individual items. If found to be faulty, the hydraulic unit and control module should always be removed together.

• The control module is bolted to the hydraulic unit and is located on the left side, in the engine compartment.

WARNING—
Do not bend the brake lines in the area of the hydraulic unit!

Removing

– Vehicles equipped with coded radio, note or request radio code if necessary.

– Disconnect battery.

NOTE—
The hydraulic unit -1- is turned 180° on the W8 engine.

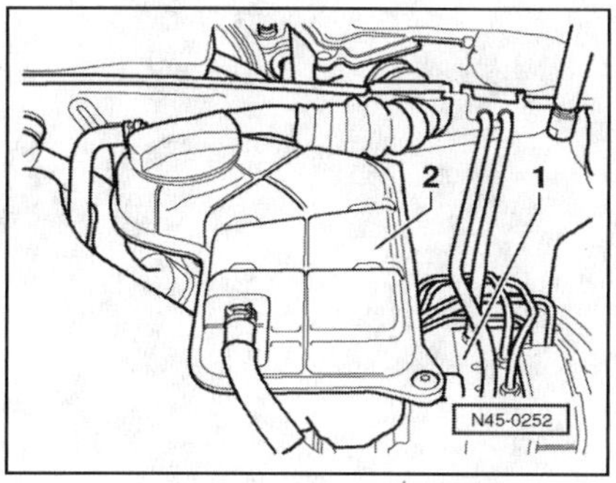

Unscrew coolant expansion tank -2- and place to side with coolant hoses still connected.

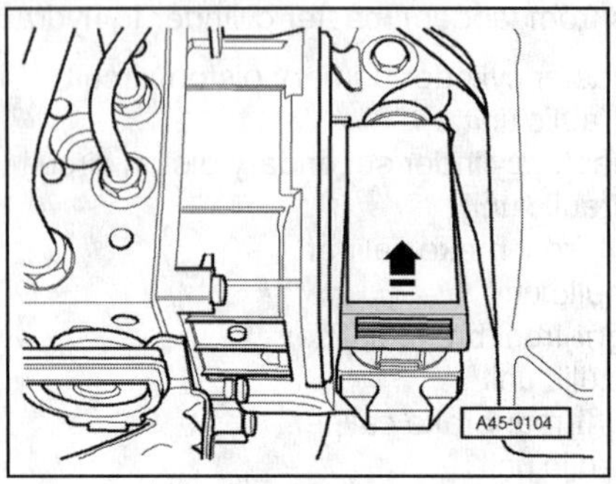

◀ Release control module connector in direction of arrow and pull off.

– Pull connector off brake pressure sensor.

– Install brake pedal depressor from VAG 1869/2.

– Depress brake pedal with brake pedal depressor.

– Connect bleeder bottle hose to left-front brake caliper bleed screw and open bleed screw.

– Close left-front bleeder screw.

– Place lint-free cloths under the control module and hydraulic unit.

• Make sure no brake fluid gets onto contacts.

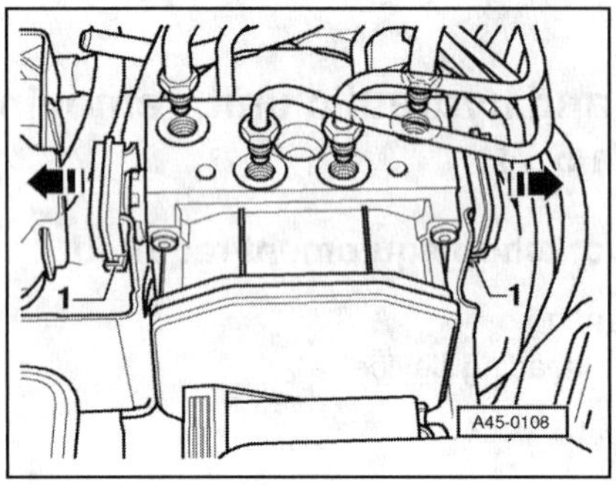

– Disconnect brake lines on the hydraulic unit.

– Seal brake lines and threaded holes with plugs from repair kit.

◀ Unclip locking clips -1- from below and remove in direction of arrow.

– Remove hydraulic unit.

Control module and hydraulic unit, installing

NOTE —

• Remove sealing plugs on new hydraulic unit when the corresponding brake line is going to be connected.

• If the sealing plugs are removed too early, brake fluid can escape, and can no longer be guaranteed that the unit is sufficiently filled or adequately bled.

– Installation is the reverse of removal noting the following:

• Bleed the brake system, see **Repair Group 47, Braking system, bleeding**
• Enter radio code.
• Code ABS control module, see **Scan Tool section**

Tightening torques:	
Brake lines to ABS module: M 12 x 1, AF 11 M 12 x 1, AF 12	 14 Nm (10 ft-lb) 24 Nm 18 ft-lb)

Sender 1 for brake booster -G201-, removing and installing

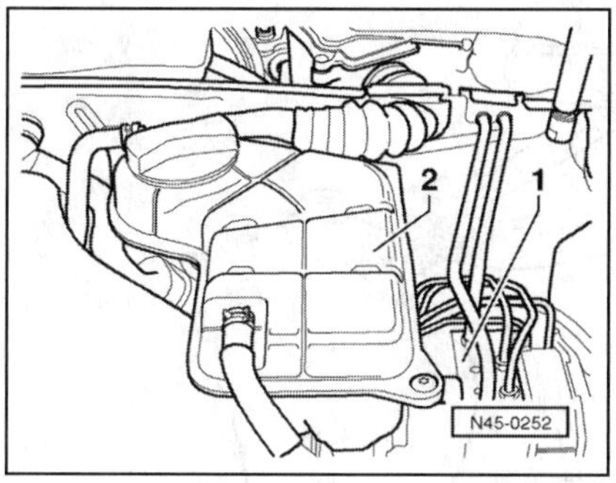

NOTE —
The brake pressure sender can be found on the hydraulic control unit -1-.

The hydraulic unit -1- is turned 180° on the W8.

This work sequence is for all vehicles, except with W8 engine:

– Unscrew coolant expansion tank -2- and place to one side with coolant hoses still connected.

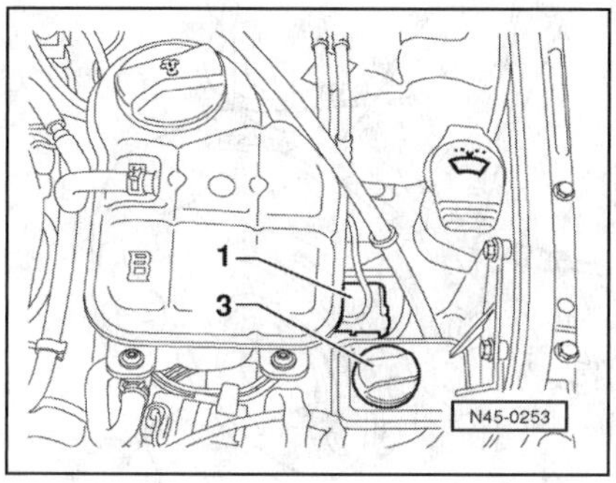

This work sequence is for vehicles w/W8 engine:

Unscrew power steering reservoir -3- and place to one side with hoses still connected.

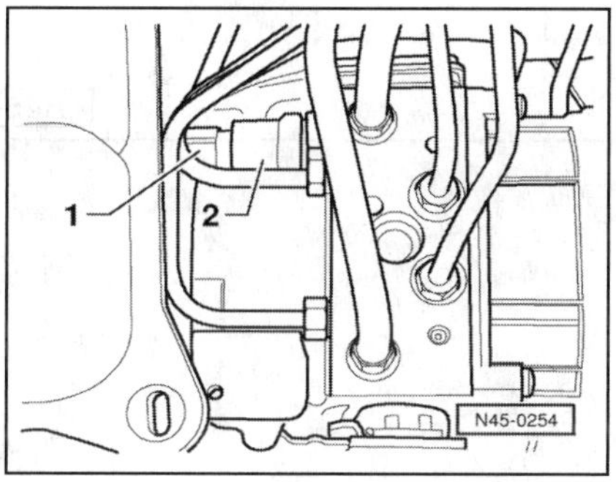

Continued for all vehicles

Pull connector -1- off brake pressure sensor -2-.

– Place lint-free cloths under the control module and hydraulic unit.

– Remove brake pressure sensor.

Sender 1 for brake booster -G201-, installing

– Installation is reverse of the removal, noting the following:

- Bleed the brake system, see **Repair Group 47, Braking system, bleeding**

Tightening torques:	
Brake pressure sensor	20 Nm (15 ft-lb)

Suspension mounted ABS components

ABS system on front axle

> ***NOTE—***
> *Removing and installing ABS system parts is identical for all brake calipers.*

1. **ABS speed sensor**
   - Before inserting sensor, clean inner surface bore hole and coat with lubricating paste G 000 650
   - Always replace
2. **Drive shaft with rotor**
   - Check rotor for damaged or dirt
3. **Hex bolt**
   - See **Suspension, Wheels, Steering, section Repair Group 40; Servicing drive axle**
4. **Clamping sleeve**
   - Always replace
   - Left and right side identical
   - Before inserting into wheel housing, coat circumference with grease G 000 650
   - Push into wheel bearing housing up to stops
5. **Sealing cuff**

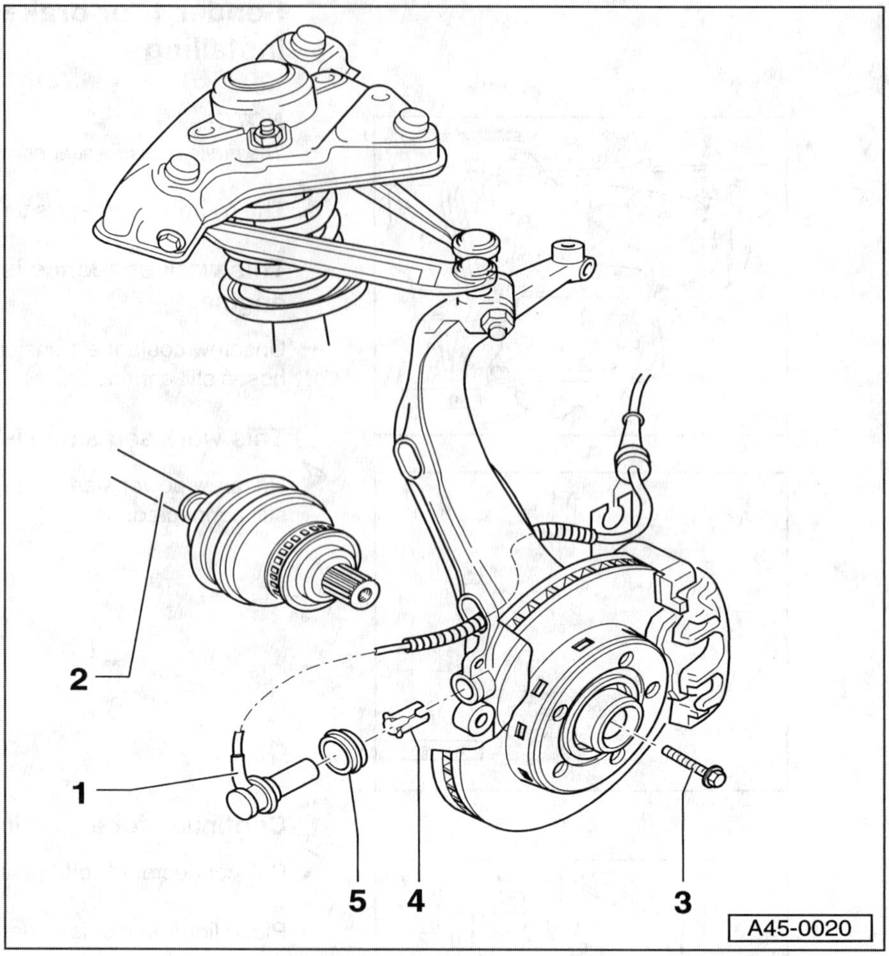

Speed sensor on front axle, removing and installing

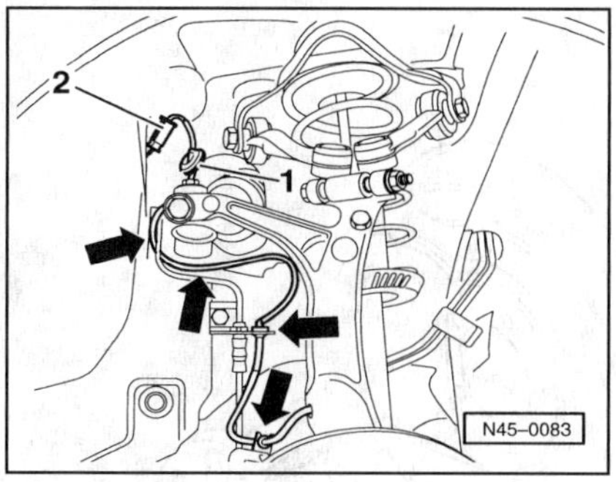

– Loosen wheel bolts.

– Raise vehicle.

– Remove wheel.

< Release grommet -1- in wheel housing.

- Disconnect ABS wheel speed sensor harness connectors -2-.
- Disconnect ABS wheel speed sensor wiring from retainers (arrows).

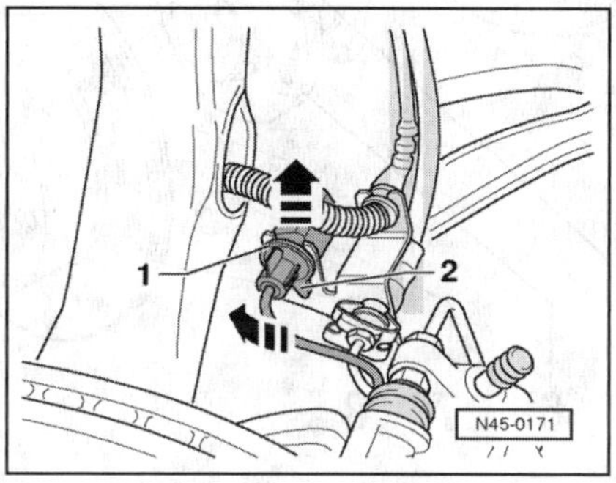

< For vehicles with brake wear indicator separate the connector -1-.

– Lift locking tab slightly on connector lower part -2- and then turn 90°.

– Pull connector lower part out of bracket.

– Pull ABS wheel speed sensor from wheel bearing housing.

– Before inserting ABS wheel speed sensor, clean inner surface of bore hole and coat speed sensor with solid lubricant paste G 000 650.

– Install ABS wheel speed sensor.

- Install ABS wheel speed sensor wire in retainers and install grommet in wheel housing.
- Insert brake wear indicator connector into bracket on brake caliper.
- The brake wear indicator wiring must be routed between the brake c arrier and brake line.

– Turn wheels fully until left and right-hand lock, and check clearance to ABS wheel speed sensor wiring.

Rotor on front axle, checking, removing and installing

– Raise vehicle.

– Remove wheel.

– Disconnect ABS wheel speed sensor out of wheel bearing housing.

– Check rotor for damaged or dirty parts by rotating drive axle.

– If rotor is damaged, remove drive axle with rotor and outer constant velocity joint, see **Suspension, Wheels, Steering section, Repair Group 40.**

ABS system on rear axle (front wheel drive)

1. **Clip**
2. **ABS speed sensor**
   - Before inserting sensor, clean hole inner surface and coat with lubricating paste G 000 650
   - Removing, see **ABS wheel speed sensor on rear axle (front wheel drive), removing and installing**
3. Axle beam
4. **Backplate**
5. **ABS speed sensor rotor**
   - If damaged or dirty, remove together with wheel bearing/hub module, see **Suspension, Wheels, Steering section, Repair Group 42**
6. **Wheel bearing/hub module**
   - Removing and installing, see **Suspension, Wheels, Steering section, Repair Group 42**
7. **Hex socket head bolt**
   - See **Suspension, Wheels, Steering section, Repair Group 42**
8. **Brake disc**

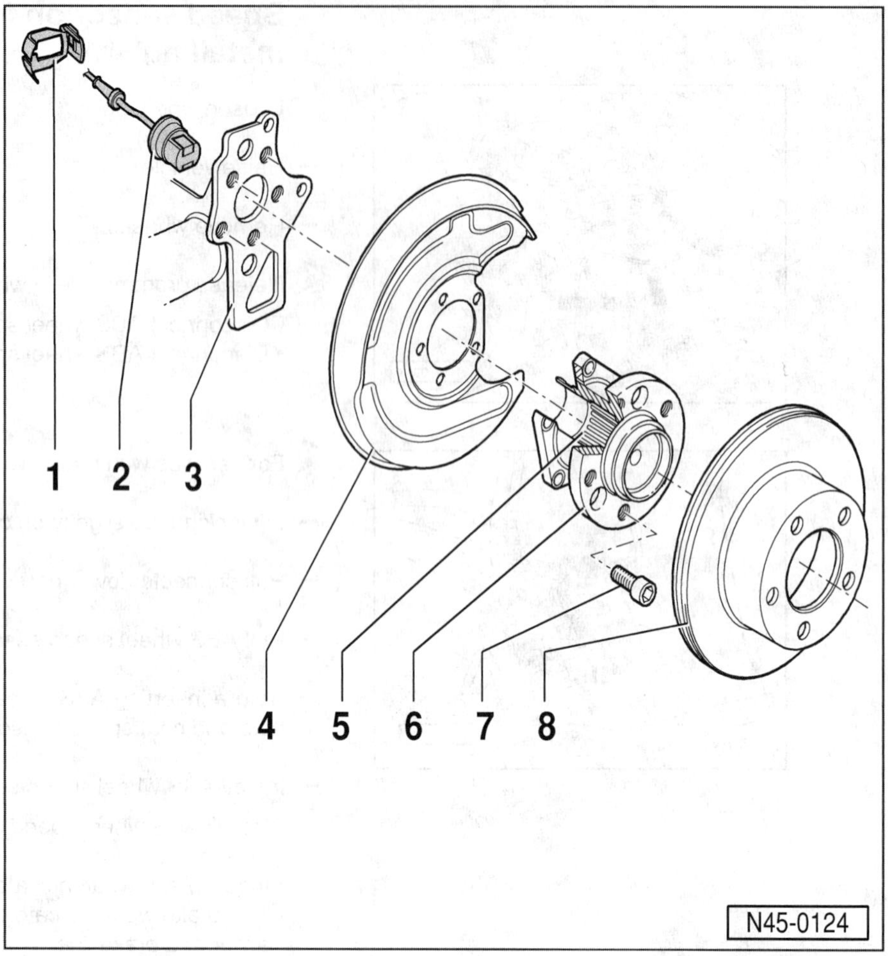

ABS wheel speed sensor on rear axle (front wheel wrive), removing and installing

Special tools and workshop equipment required

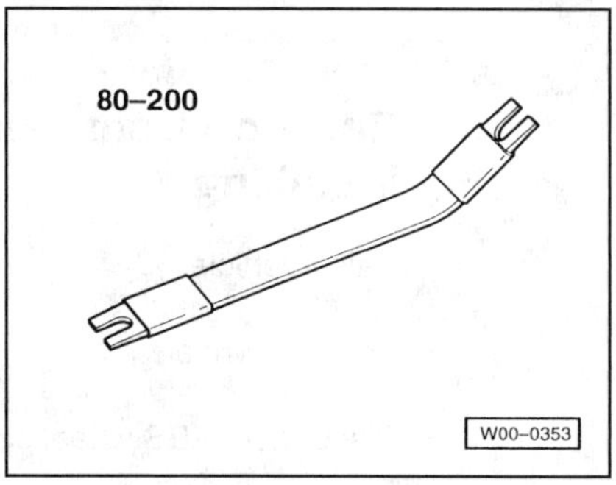

- 80-200 pry lever - removing outside mirror

– Remove rear seat, see **Body-Interior section, Repair Group 72**

– Remove lower C pillar trim, see **Body-Interior section, Repair Group 70.**

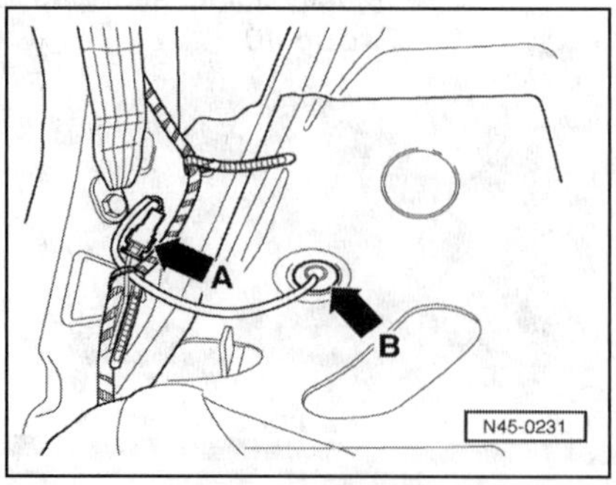

– Disconnect ABS wheel speed sensor harness connector -arrow A- and push out grommet -arrow B-.

– Loosen wheel bolts.

– Raise vehicle.

– Remove wheel.

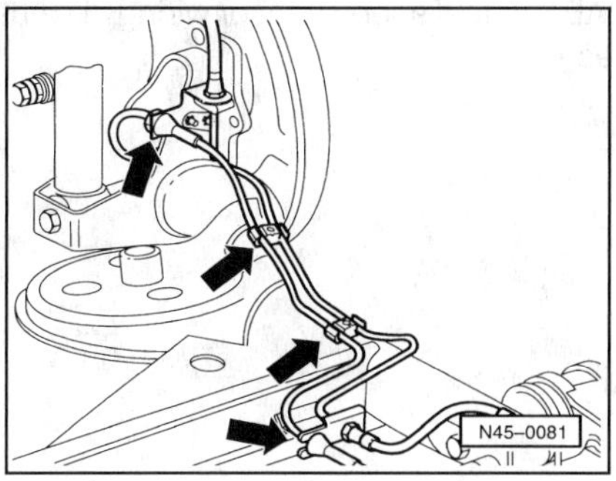

- Disconnect ABS wheel speed sensor wiring from retainers (arrows).
- Remove clip from ABS wheel speed sensor.

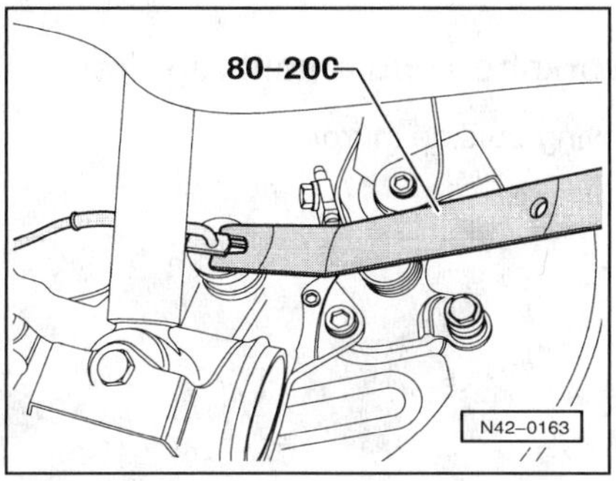

- Pry ABS wheel speed sensor out of wheel bearing module using special tool 80-200.

Speed sensor on rear axle (front wheel drive), installing

- Before inserting ABS wheel speed sensor, clean inner surface of bore hole and coat speed sensor with solid lubricant paste G 000 650.

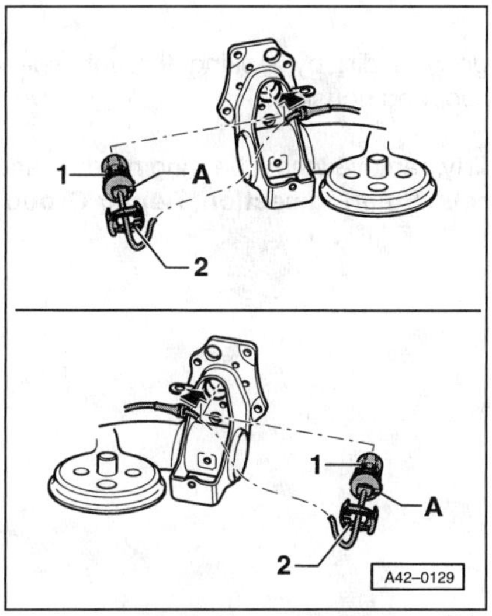

- Push ABS wheel speed sensor -1- into wheel bearing housing unit by hand until it stops .
  - Lug -A- faces toward front on the left-hand side and the rear on the right-hand side.
  - Insert clip -2- for speed sensor.

NOTE —
When routing ABS wheel speed sensor wiring note:

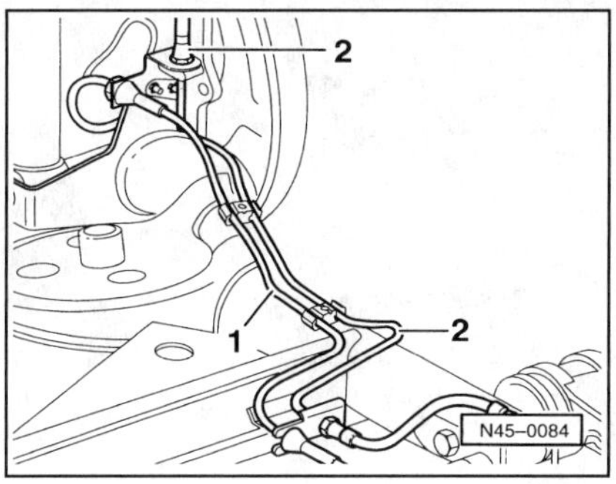

- The ABS wheel speed sensor wiring -1- is to be routed next to the brake line -2- on both sides as illustrated.
  - On the left-hand side there must be a clearance of approx. 2 cm between the exhaust system heat shield and the ABS wheel speed sensor wiring.

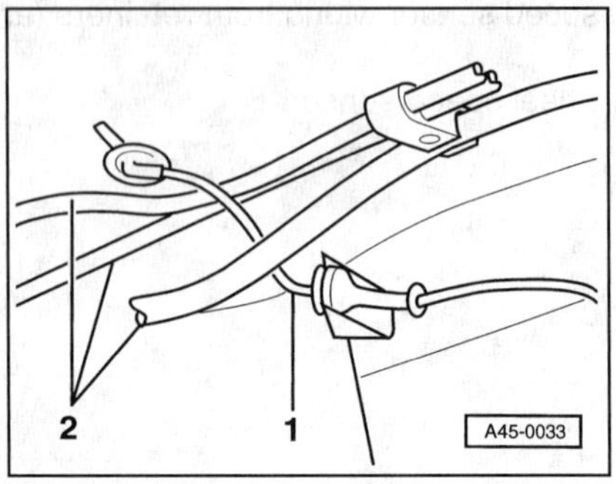

On the right-hand side, ABS wheel speed sensor wiring -1- must be routed between the fuel lines -2-.

Checking rotor on rear axle

Special tools and workshop equipment required

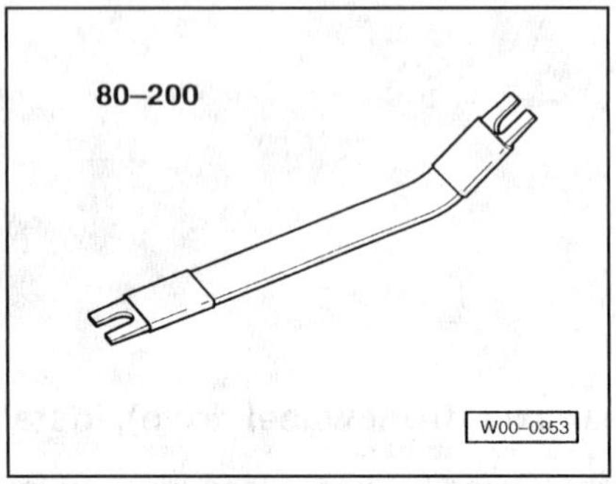

80-200 pry lever - removing outside mirror

– Raise vehicle.

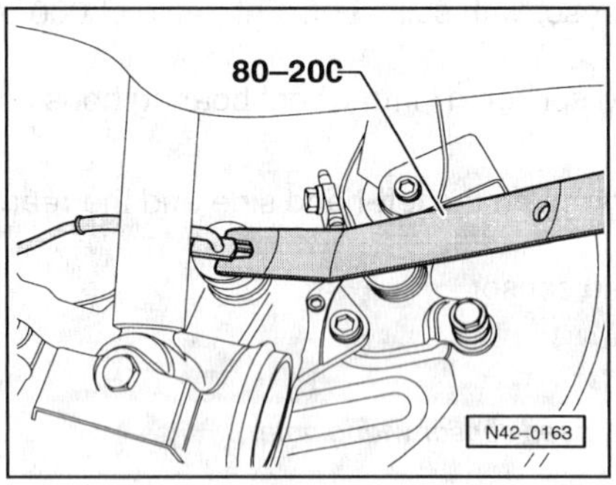

Pry ABS wheel speed sensor out of wheel bearing module using special tool 80-200.

– Check rotor for damage and dirt by looking through speed sensor mounting hole in wheel bearing housing.

– If rotor is damaged or dirty, remove wheel bearing module and replace, see **Suspension, Wheels, Steering section, Repair Group 42.**

ABS system on rear axle (all wheel drive)

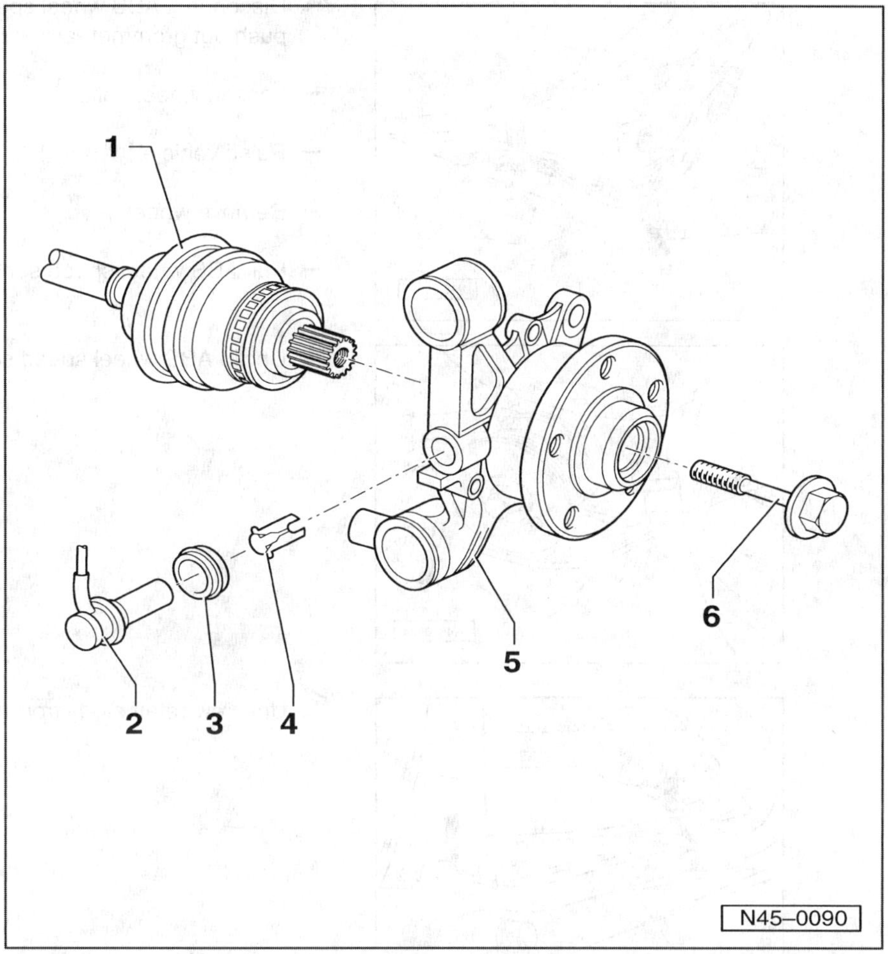

1. **Drive shaft with rotor**
   - Check for damaged or dirty rotor
2. **ABS speed sensor**
   - Before inserting sensor, clean inner surface and coat with lubricating paste G 000 650
   - Always replace
3. **Sealing cuff**
4. **Clamping sleeve**
   - Always replace
   - Left and right hand identical
   - Coat circumference with grease, before inserting into wheel housing G 000 650
   - Push into wheel bearing housing up to stops
5. **Wheel bearing/hub module**
6. **Hex bolt**
   - Always replace each time after removing, see **Suspension, Wheels, Steering section, Repair Group 42**

Speed sensor on rear axle (all wheel drive), removing and installing

– Remove rear seat, see **Body-Interior section, Repair Group 72; Rear seats**

– Remove lower C pillar trim, see **Body-Interior section, Repair Group 70; Lower C pillar trim.**

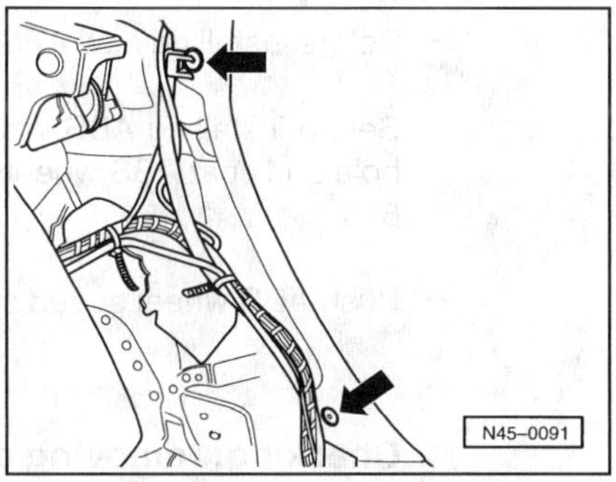

◀ Unscrew and remove screws from side trim (arrows).

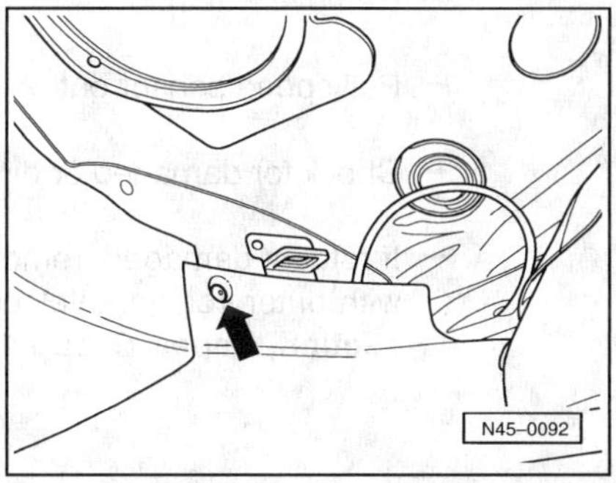

◀ Unscrew screw from side trim (arrow).

– Lift side trim and open cable tie behind it.

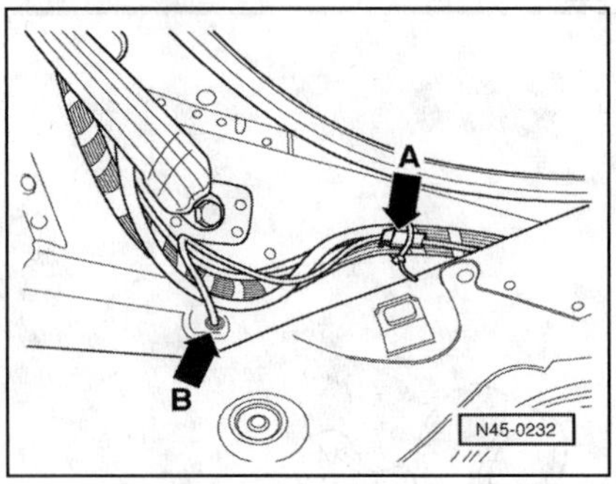

⊲ Disconnect ABS wheel speed sensor harness connector -arrow A- and push out grommet -arrow B-.

– Loosen wheel bolts.

– Raise vehicle.

– Remove wheel.

– Pull ABS wheel speed sensor from wheel bearing housing.

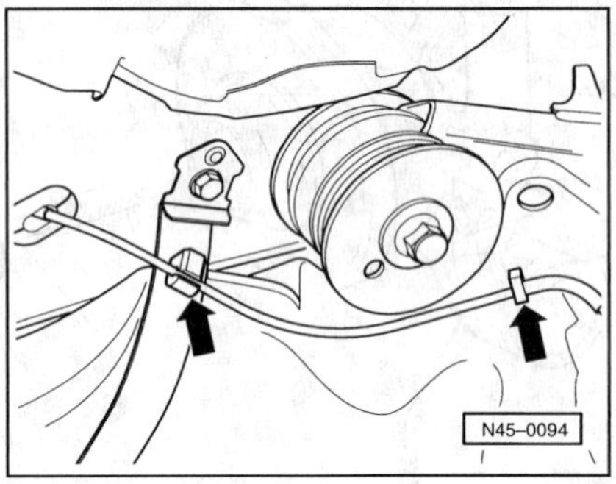

⊲ Unclip ABS wheel speed sensor wiring from retainers (arrows).

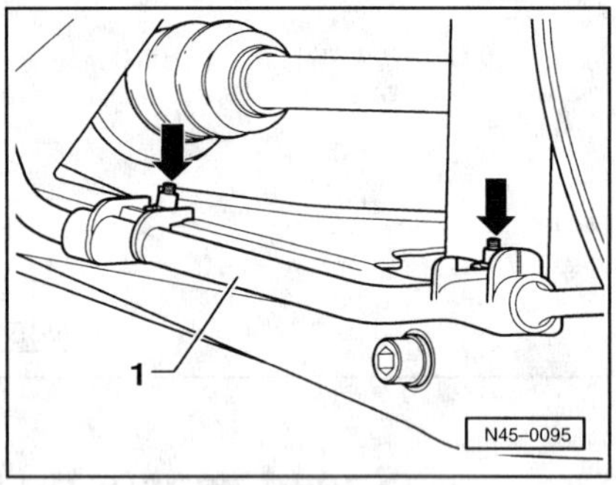

⊲ Unscrew retainer -1- from ABS wheel speed sensor wire (arrows).

Speed sensor on rear axle (all wheel drive), installing

– Before installing new ABS wheel speed sensor replace clamping sleeve.

– Before installing ABS wheel speed sensor, clean inner surface of bore hole and coat ABS wheel speed sensor with solid lubricant paste G 000 650.

– Push ABS wheel speed sensor into wheel bearing housing by hand.

Checking, removing and installing rotor on rear axle

– Raise vehicle.

– Remove wheel.

– Pull speed sensor out of wheel bearing housing.

– Check for damaged or dirty rotor.

– If rotor is damaged, remove drive axle with rotor and replace drive axle with outer constant velocity joint, see **Suspension, Wheels, Steering section, Repair Group 42**

46 Brakes, Mechanical

Repair Group

R 46

Front Brakes, Servicing

EDITOR'S NOTE —

See **Repair Group 00, General Information** *for details of brake caliper identification.*

- *When replacing brake pads, be sure to replace pad sets for calipers on both sides of the axle.*
- *Brake hoses do not need to be dosconnected for replacement of brake pads and rotors.*

Front brake calipers, removing and installing

Special tools

- VAG 1331 Torque wrench
- VAG 1410 Torque wrench
- VAG 1869/2 Brake peadal depressor

– Remove brake pads, see **Brake pads, removing and installing** for appropriate brake caliper assembly.

– Install brake pedal depressor VAG 1869/2 or equivalent.

– Unscrew brake hose from caliper.

– Install brake pads.

– Install brake hose to brake caliper, torque to proper setting, see appropriate caliper assembly.

– Remove brake pedal depressor.

– Bleed the brakes, see **Repair Group 47, Braking system, bleeding**.

C54 brake caliper, assembly

> ***NOTE—***
> *To draw off brake fluid from the reservoir, use a bleeder bottle which is used only for brake fluid. Brake fluid is poisonous, NEVER siphon brake fluid with your mouth!*

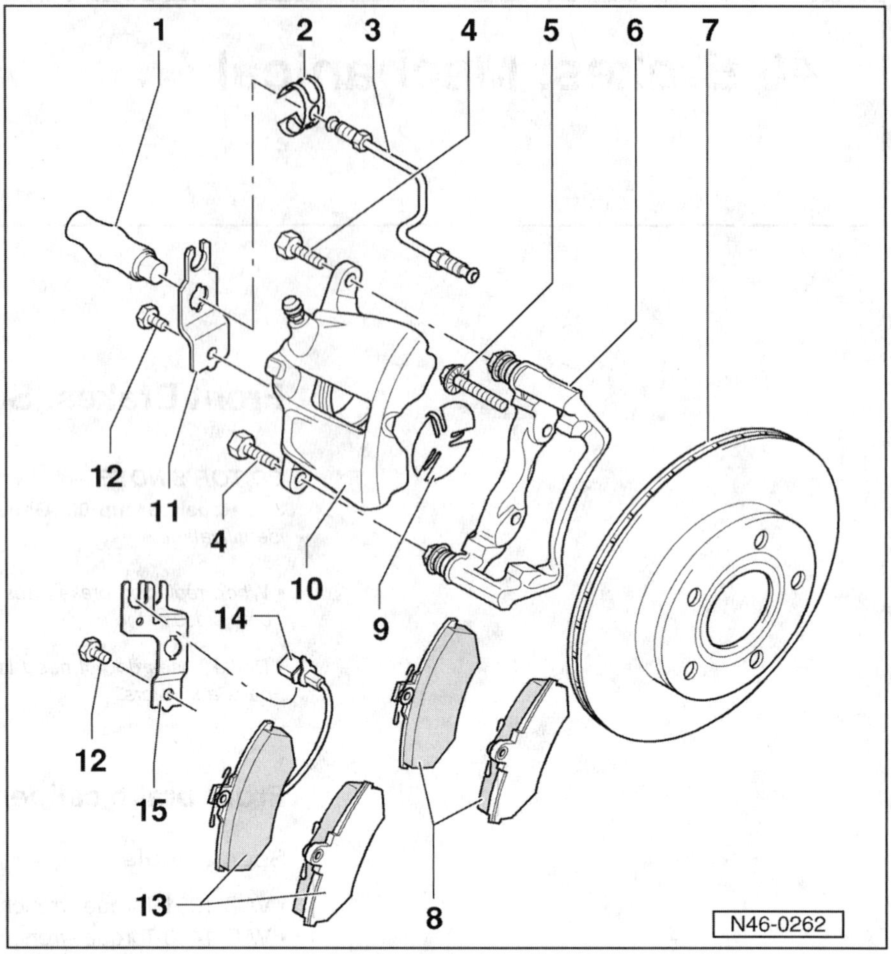

1. **Brake hose**
   - Do not disconnect brake hose when changing brake pads
2. **Spring clip**
3. **Brake pipe**
   - Tighten to 15 Nm (11 ft-lb)
   - Screw into brake caliper housing
   - Screw into brake hose, counter hold hex on brake hose.
   - Make sure that lugs are properly seated in retainer grooves.
4. **Self-locking hex bolt**
   - Tighten to 35 Nm (26 ft-lb)
   - Replace
   - When loosening and tightening counter hold on guide pin
5. **Ribbed combi bolt**
   - Tighten to 120 Nm (89 ft-lb)
   - Clean ribs if re-using
6. **Brake carrier with guide pins and protective caps**
   - Supplied as assembled replacement part with sufficient grease on guide pins
   - If protective caps or guide pins are damaged use repair kit. Use grease packet supplied to lubricate guide pins.
7. **Brake disc**
   - Vented brake disc diameter: 280 mm
   - Brake disc thickness: 22 mm
   - Wear limit: 20 mm
   - Always replace both sides
   - Never remove brake discs from hub by force. If necessary use penetrating fluid, otherwise brake discs can be damaged.
8. **Brake pads**
   - Without wear indicator
   - Thickness 14 mm
   - Wear limit: 7 mm including backing plate
   - Removing and installing, see **Brake pads, removing and installing (C54 brake caliper)**
9. **Heat shield**
   - Insert into piston
10. **Brake caliper**
    - Do not disconnect brake hose when changing brake pads
    - Removing and installing, see **Front brake calipers, removing and installing**
11. **Retainer**
    - Bolt to brake caliper housing
12. **Hex bolt**
    - Tighten to 10 Nm (7 ft-lb)
13. **Brake pads**
    - With wear indicator
    - Thickness 14 mm
    - Wear limit: 7 mm including backing plate
    - With wear indicator
    - When wear limit is reached (limit: 2 to 3 mm) warning lamp lights up on instrument cluster
    - Removing and installing, see **Brake pads, removing and installing (C54 brake caliper)**
14. **Connector**
    - When changing pads remove from bracket **-15-**
15. **Bracket**

Brake pads, removing and installing (C54 brake caliper)

Special tools and workshop equipment required

- VAG 1331 Torque wrench
- Piston resetting tool

Removing

NOTE —

- *When reusing brake pads mark location, install in same position otherwise uneven braking!*
- *After replacing brake pads with vehicle stationary, depress brake pedal firmly several times so that the brake pads are properly seated in their normal operating position.*
- *Wheel bolt tightening torque 120 Nm (89 ft-lb)*

– Remove wheels.

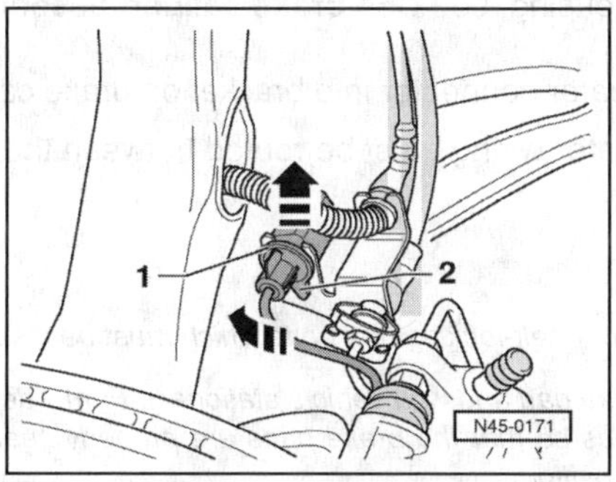

◄ For vehicles with brake wear indicator disconnect the connector -1-.

– Lift locking tab slightly on connector lower part -2- and then turn 90°.

– Pull connector lower part out of bracket.

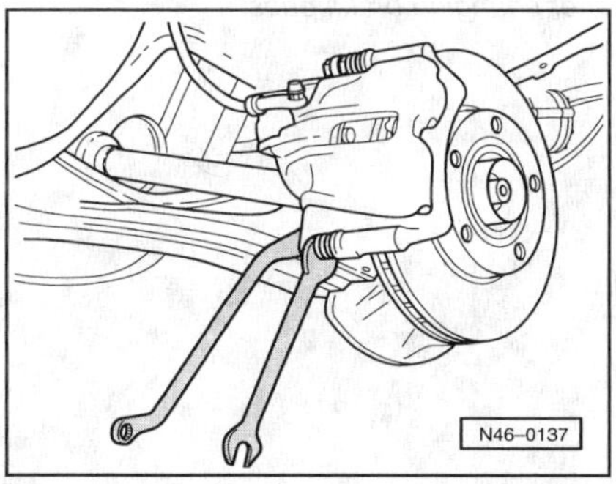

◄ Remove lower brake caliper housing securing bolt. Counter hold on guide pin.

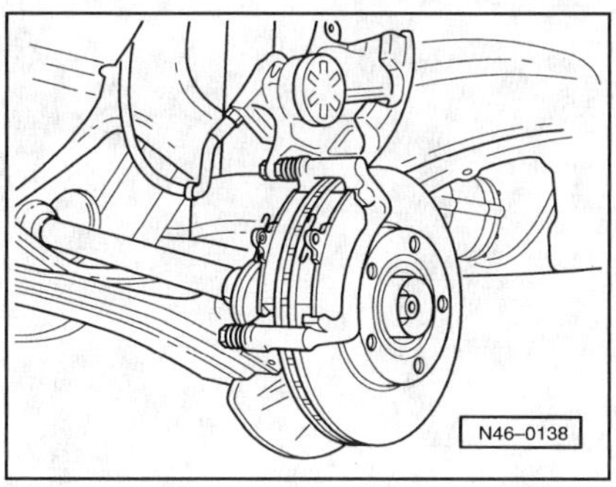

◄ Swing brake caliper housing upward.

– Remove brake pads.

Brake pads, installing

NOTE—

- *Press piston back into the cylinder with resetting tool. Before inserting new brake pads, draw off brake fluid from the reservoir with a bleeder bottle. Before pressing the piston back, make sure the reservoir has not been topped off, otherwise brake fluid will overflow and cause damage.*
- *Use only appropriate solvent for cleaning the brake caliper housing.*

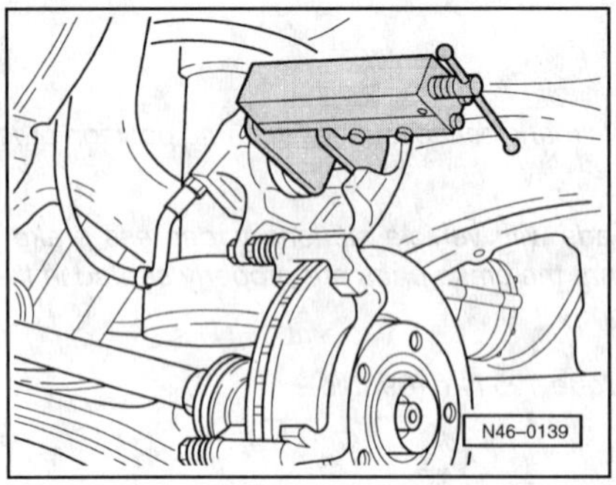
N46-0139

◀ Pressing piston back.

– Insert brake pads.

– Swing brake caliper housing down and secure with new self-locking bolt.

– Remove upper brake caliper housing securing bolt. Counter hold on guide pin.

– If removing the upper brake caliper housing securing bolt, it must always be replaced.

– Bolt on brake caliper housing, see **C54 brake caliper, assembly**; -4-.

– Insert brake wear indicator connector into bracket on brake caliper.

- The brake wear indicator wiring must be routed between the brake carrier and brake line.

NOTE—

- *The repair kit includes four self-locking hex bolts which must be installed.*
- *After changing each brake pad and with vehicle stationary. Firmly depress brake pedal several times so that the brake pads are properly seated in their normal operating position.*
- *Check brake fluid level, after changing brake pads.*

FN3 brake caliper, assembly

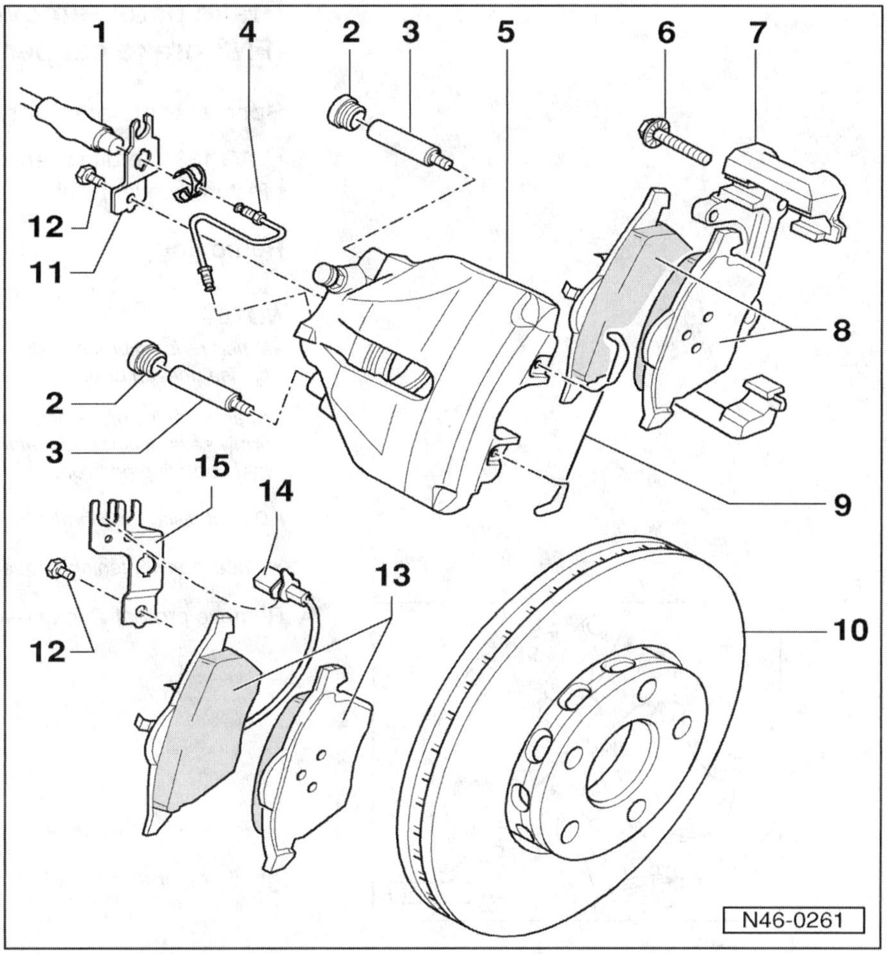

NOTE —

- *Install complete repair kit.*
- *To draw off brake fluid from the reservoir, use a bleeder bottle which is used only for brake fluid. Brake fluid is poisonous, NEVER siphon brake fluid with your mouth!*

1. **Brake hose**
   - Do not disconnect brake hose when changing brake pads
2. **Protective cap**
   - Remove
3. **Guide pins**
   - Tighten to 25 Nm (18 ft-lb)
4. **Brake line**
   - Tighten to 15 Nm (11 ft-lb)
   - Screw into brake caliper housing
   - Screw into brake hose, counter hold on brake hose hex.
   - Make sure that lugs are properly seated in retainer grooves.
5. **Brake caliper**
   - Removing and installing, see **Front brake calipers, removing and installing**
6. **Ribbed combi bolt**
   - Tighten to 120 Nm (89 ft-lb)
   - Clean ribs if re-using
7. **Brake carrier**
   - Bolt to wheel bearing housing
8. **Brake pads**
   - Without wear indicator
   - Thickness 14 mm
   - Wear limit: 7 mm including backing plate
   - Removing and installing, see **Brake pads removing and installing**
   - The arrow on the inner pad (with spring) must point in the direction of brake disc rotation when moving forward. May create noises if incorrectly installed.
   - There is an adhesive foil on the backplates of the outer brake pads. The protective foil must be pulled off before inserting.
9. **Retaining spring**
   - Insert in both holes of brake caliper housing

   NOTE —
   The retaining spring must be pressed under the brake carrier after inserting into both holes. Faulty installation will not allow compensation of outer pad wear. This will cause pedal travel to increase.
10. **Brake disc**
    - Vented brake disc to vehicle identity No. 3B X 124 107: diameter 282.5 mm
    - Vented brake disc from vehicle identity No. 3B X 124 108: diameter 282.5 mm
    - Brake disc thickness: 25 mm
    - Wear limit: 23 mm
11. **Retainer**
    - Bolt to brake caliper housing
12. **Hex bolt**
    - Tighten to 10 Nm (7 ft-lb)
13. **Brake pads**
    - With wear indicator
    - Warning lamp lights up on instrument panel when wear limit is reached (limit: 2 to 3 mm)
    - Thickness 14 mm
    - Wear limit: 7 mm including backing plate
    - Removing and installing, see **Brake pads, removing and installing (FN3 brake caliper)**
14. **Connector**
    - For pad replacement remove from bracket **-15-**
15. **Bracket**

Brake pads removing and installing (FN3 brake caliper)

Special tools and workshop equipment required

- VAG 1331 Torque wrench
- Piston resetting tool

Removing

NOTE—

- *When reusing brake pads mark location and install in same position, otherwise uneven braking will occur!*
- *After replacing brake pads and with vehicle stationary depress brake pedal firmly several times so that the brake pads are properly seated in their normal operating position.*
- *Do not disconnect brake hose when changing brake pads.*
- *Wheel bolt tightening torque 120 Nm (89 ft-lb)*

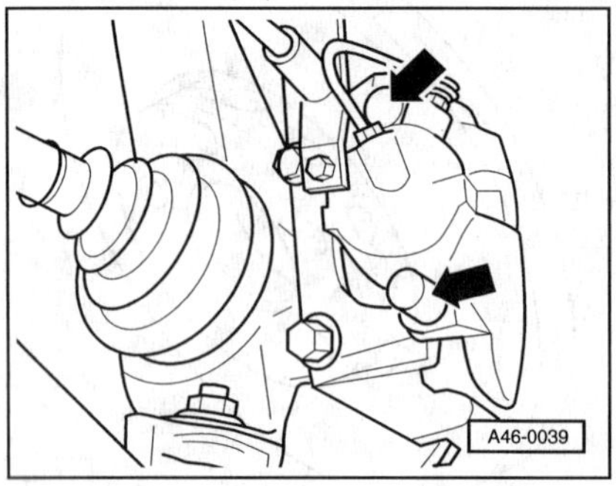

◀ Remove protective caps -arrows-.

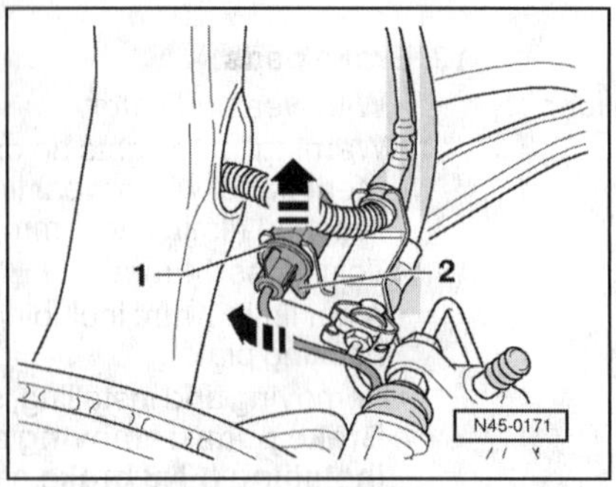

◀ For vehicles with brake wear indicator disconnect the connector -1-.

– Lift locking tab slightly on connector lower part -2- and then turn 90°.

– Pull connector lower part out of bracket.

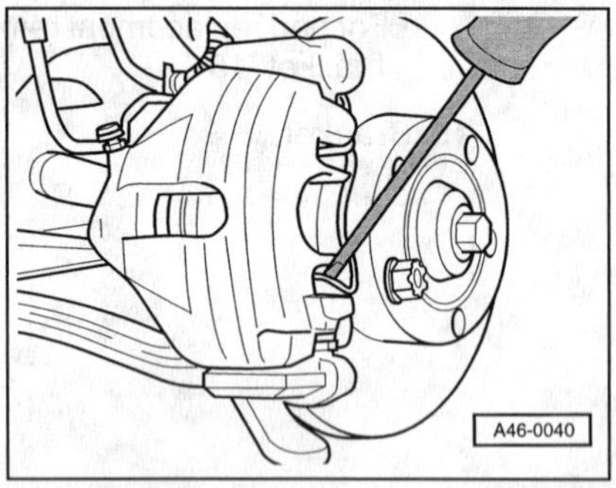

◀ Using a screwdriver pry off brake pad retaining spring from brake caliper housing and remove.

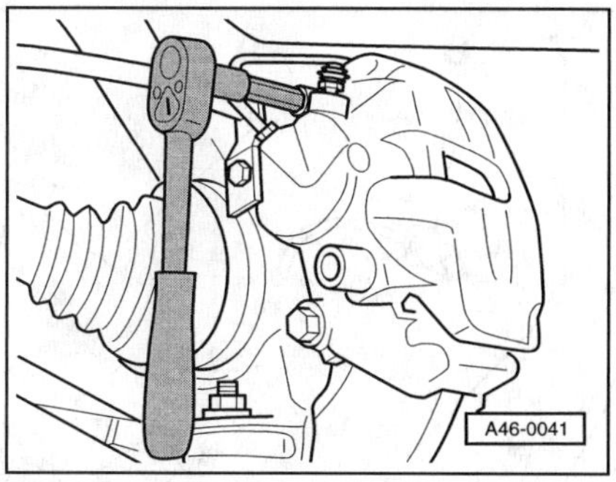

- Remove both guide pins from brake caliper.

– Remove brake caliper housing and lay to one side so that the weight of the brake caliper does not stress or damage the brake hose.

– Remove brake pads out of brake caliper housing or from brake carrier.

NOTE—
Use only appropriate solvent for cleaning the brake caliper housing.

Brake pads, installing

NOTE—
Press piston back into the cylinder with resetting tool. Before inserting new brake pads, draw off brake fluid from the reservoir with a bleeder bottle. Before pressing the piston back, make sure reservoir has not been topped off, otherwise brake fluid will overflow and cause damage.

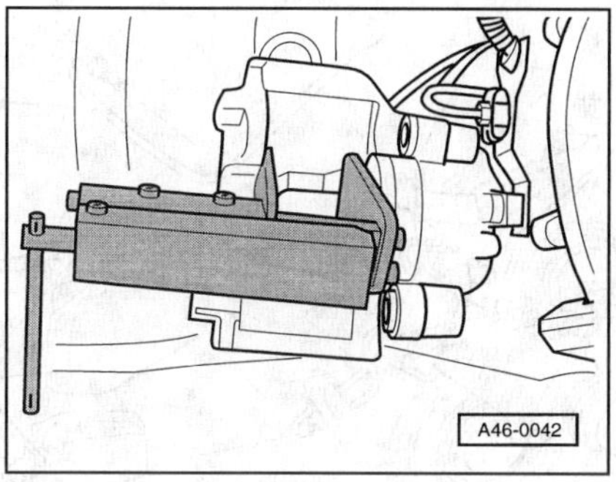

- Press piston back into brake caliper.

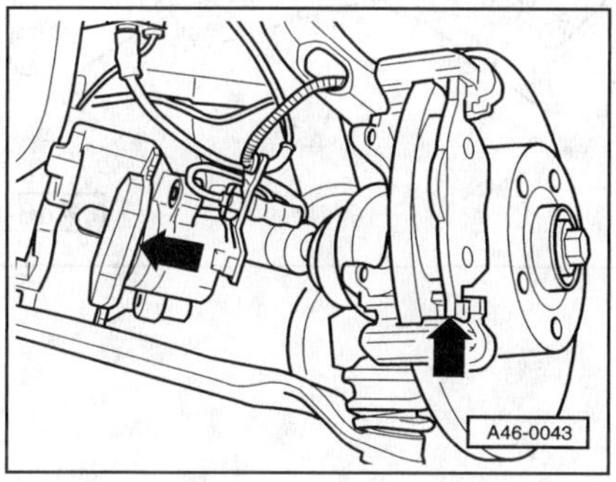

- Insert brake pad with retaining spring in brake caliper housing (piston).

NOTE—
The inner pad (with spring) has an arrow. The arrow must point in the direction of the forward rotation of the disc brake. May create noises if incorrectly installed.

– Remove protective foil on backing plate of outer brake pad.

– Install outer brake pad to brake carrier.

– Tighten brake caliper housing bolt to brake carrier with both guide pins, see **FN3 brake caliper, assembly**; -3-.

– Install both protective caps.

– Install retaining spring into brake caliper housing.

– Insert brake wear indicator connector into bracket on brake caliper.

- The wiring for the brake wear indicator must be routed between the brake carrier and brake line.

NOTE—
- *When inserting the outer brake pad, make sure that it does not stick to the brake caliper housing before achieving correct installation position.*
- *Do not damage adhesive surface.*
- *After replacing brake pads and with vehicle stationary, firmly depress brake pedal several times so that the brake pads are properly seated in their normal operating position.*
- *Check brake fluid level after changing brake pads.*

HP-2 brake caliper, assembly

Vehicles with W8 engines only

NOTE—
- *Install complete repair kit.*
- *To draw off brake fluid from the reservoir, use a bleeder bottle which is used only for brake fluid. Brake fluid is poisonous, NEVER siphon brake fluid by mouth!*

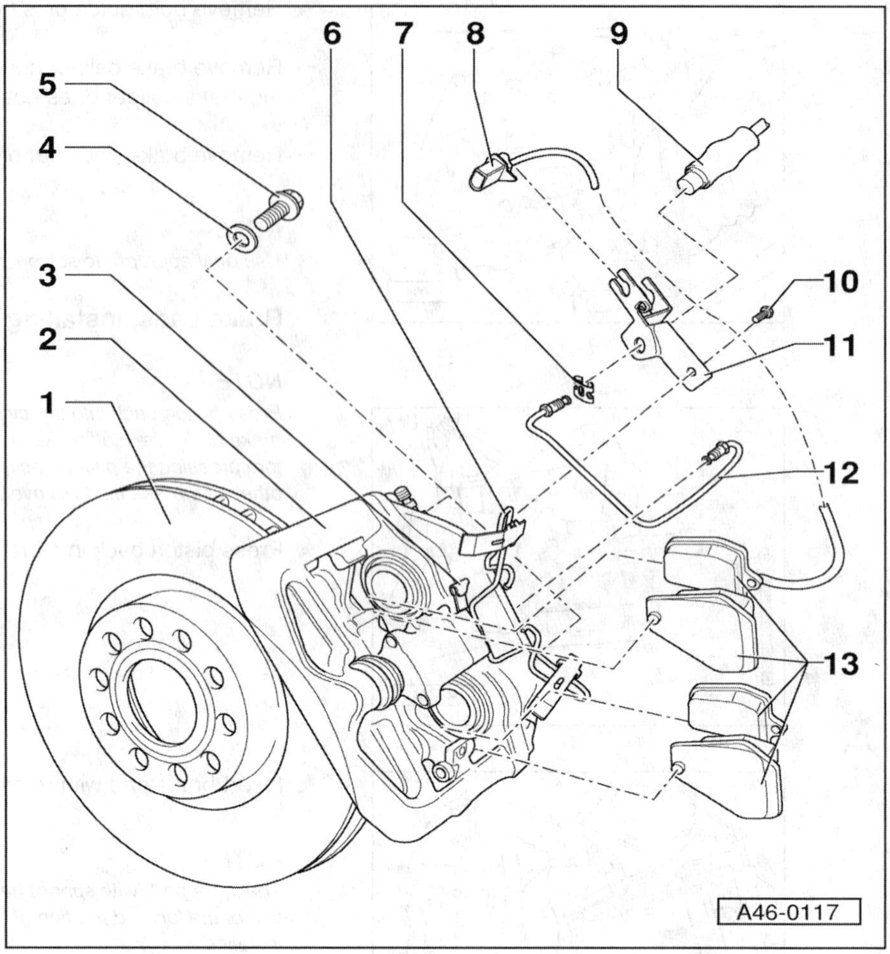

1. **Brake disc**
   - Ventilated brake disc: diameter 321 mm
   - Brake disc thickness: 30 mm
   - Wear limit: 28 mm (Some parts may be incorrectly stamped 27 mm for wear limit.)
   - Always replace both sides
2. **Brake caliper**
   - Removing, see **Front brake calipers,removing and installing**
3. **Cable holder for wear indicator**
4. **Shim**
5. **Ribbed combi bolt**
   - Tighten to 190 Nm (140 ft-lb)
   - Clean ribs if using again
6. **Retaining spring**
   - Unclip sideways
   - Complete removal not possible
7. **Spring clip**
8. **Connector**
   - Remove from bracket -**15**- when changing pads
9. **Brake hose**
   - Do not disconnect brake hose when changing brake pads
10. **Hex bolt**
    - Tighten to 25 Nm (18 ft-lb)
11. **Bracket**
12. **Brake pipe**
    - tighten to 15 Nm (11 ft-lb)
13. **Brake pads**
    - Warning light lights up on instrument cluster when wear limit is reached (limit: 2 to 3 mm), Wear limit: 7 mm including backing plate
    - Thickness 14 mm
    - Removing and installing, see **Brake pads, removing and installing (HP2 brake caliper)**

Brake pads, removing and installing (HP-2 brake caliper)

Removing

- *When reusing brake pads mark location and install in same position otherwise uneven braking will occur!*

- *Do not disconnect brake hose to remove brake pads.*

- *After replacing brake pads, and with vehicle stationary depress brake pedal firmly several times so that the brake pads are properly seated in their normal operating position.*

- *Wheel bolt tightening torque 120 Nm (89 ft-lb)*

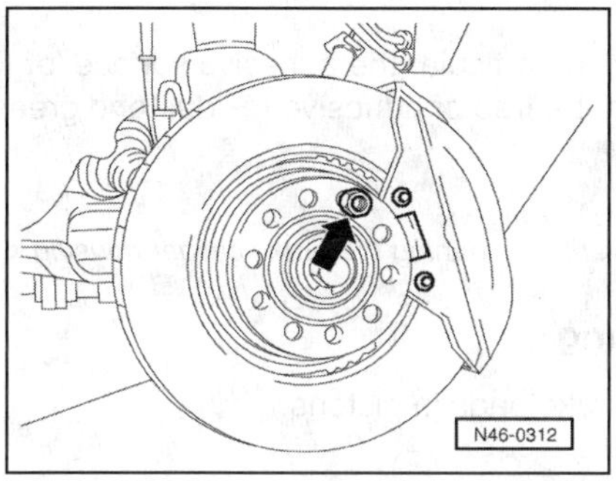

– Remove wheels.

◄ Secure brake disc by inserting a wheel bolt.

– Disconnect connector for brake pad wear indicator.

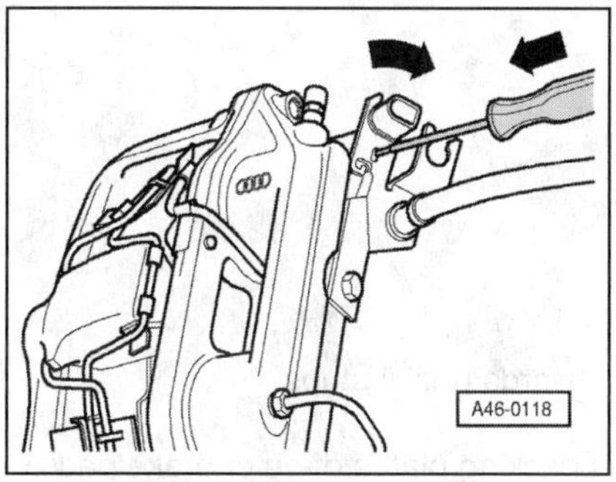

◄ Lift locking tab slightly on connector lower part and then turn 90°.

– Pull connector lower part out of bracket.

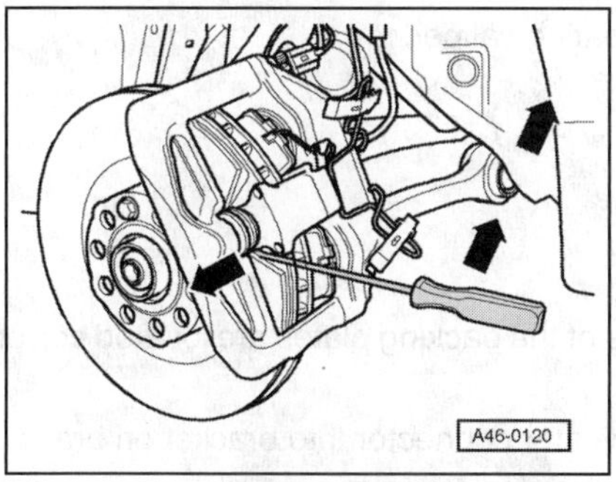

◄ Unclip upper and lower parts of retaining spring using a pair of pliers -arrows-.

– Push spring sideways.

NOTE—
Before pressing piston back, draw off brake fluid from reservoir using a bleeder bottle. Otherwise fluid will overflow and cause damage if the reservoir was topped off during repair work.

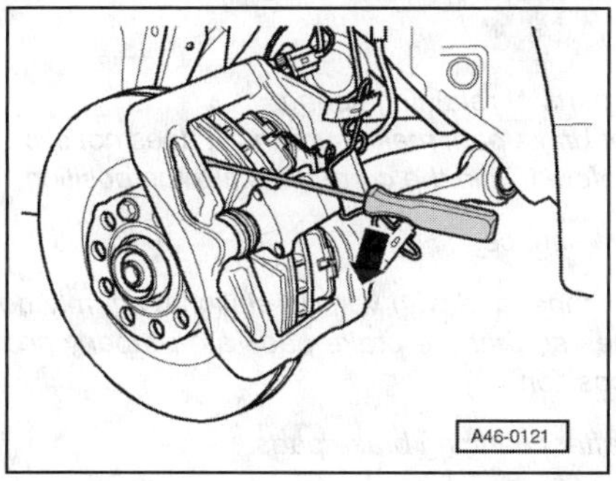

◄ Press piston back.

– To do this, press floating caliper outward using a screwdriver as shown in the diagram.

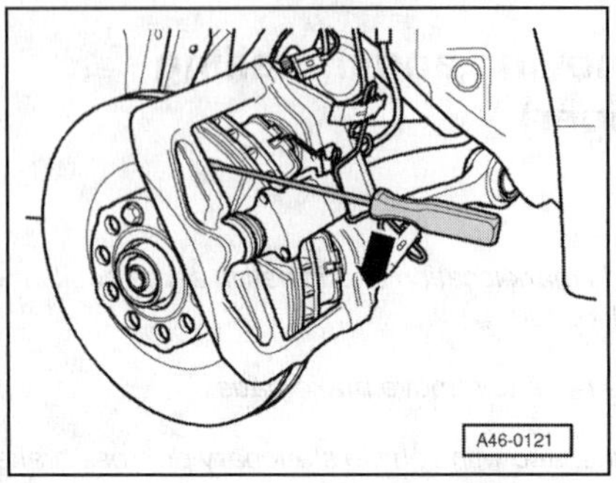

Using a screwdriver, first press outer brake pads off caliper and then remove.

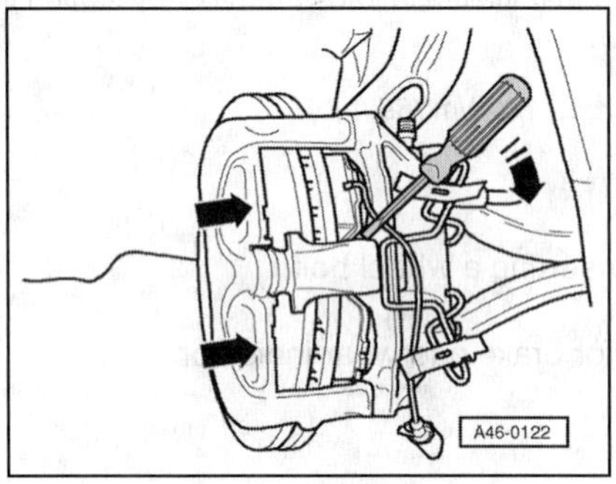

Push floating caliper in direction of arrow until it stops.

– Carefully press inner brake pads out of pistons and remove from floating caliper.

• Clean floating caliper, in particular the adhesive surface for the outer brake pads. This must be free of adhesive residue and grease.

NOTE —
Use only appropriate solvent for cleaning the brake caliper housing.

Brake pads, installing

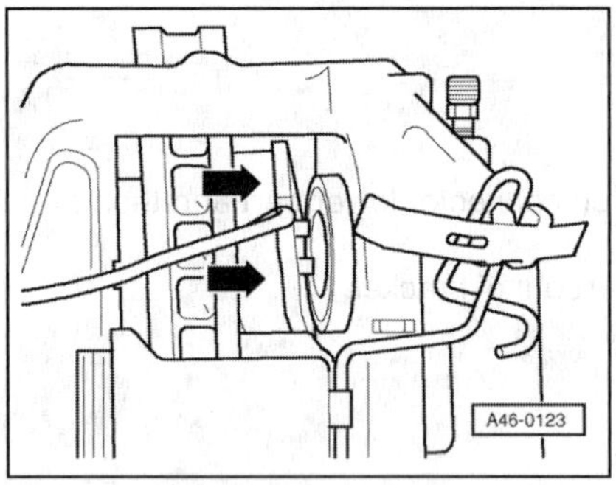

Carefully install inner brake pads in pistons.

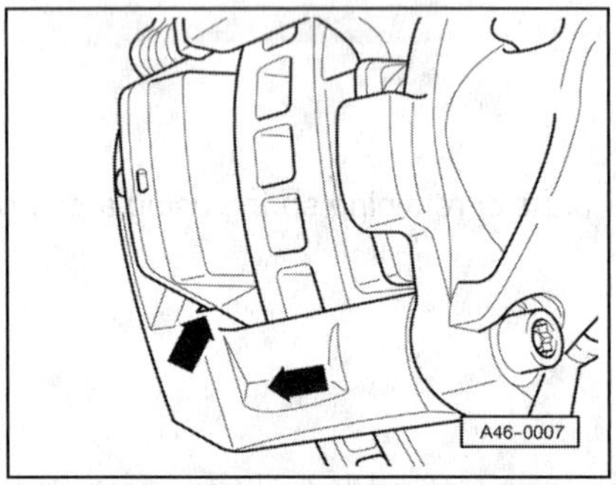

Press floating caliper outwards until it stops.

– Pull off protective foil on backing plates of outer brake pads.

– Install brake pads in floating caliper.

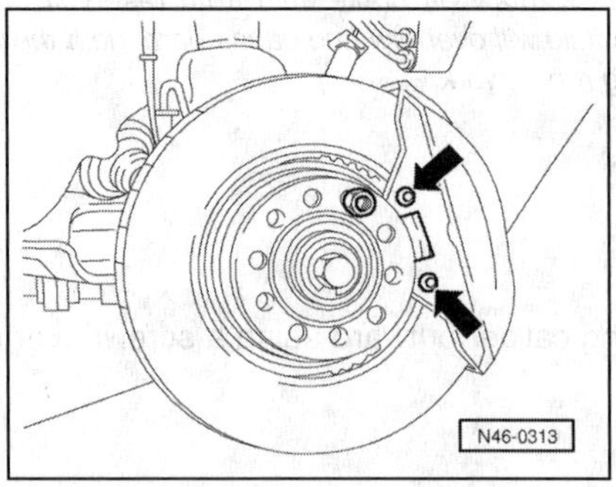

Make sure that the lugs of the backing plates are located correctly in the floating caliper.

• Install brake wear indicator connector into bracket on brake caliper.

NOTE —
• Clip in upper and lower parts of retaining spring.
When installing the outer brake pad, make sure that it does not stick to the brake caliper housing before it is in the correct installation position.

• Do not damage adhesive surface.

• After each brake pad change, and with vehicle stationary, firmly depress brake pedal several times so that the brake pads are properly seated in their normal operating position.

• Check brake fluid level after changing brake pads.

2FN brake caliper, assembly

NOTE—
- *Install repair kit complete.*
- *To draw off brake fluid from the reservoir, use a bleeder bottle which is used only for brake fluid. Brake fluid is poisonous, NEVER siphon brake fluid by mouth!*

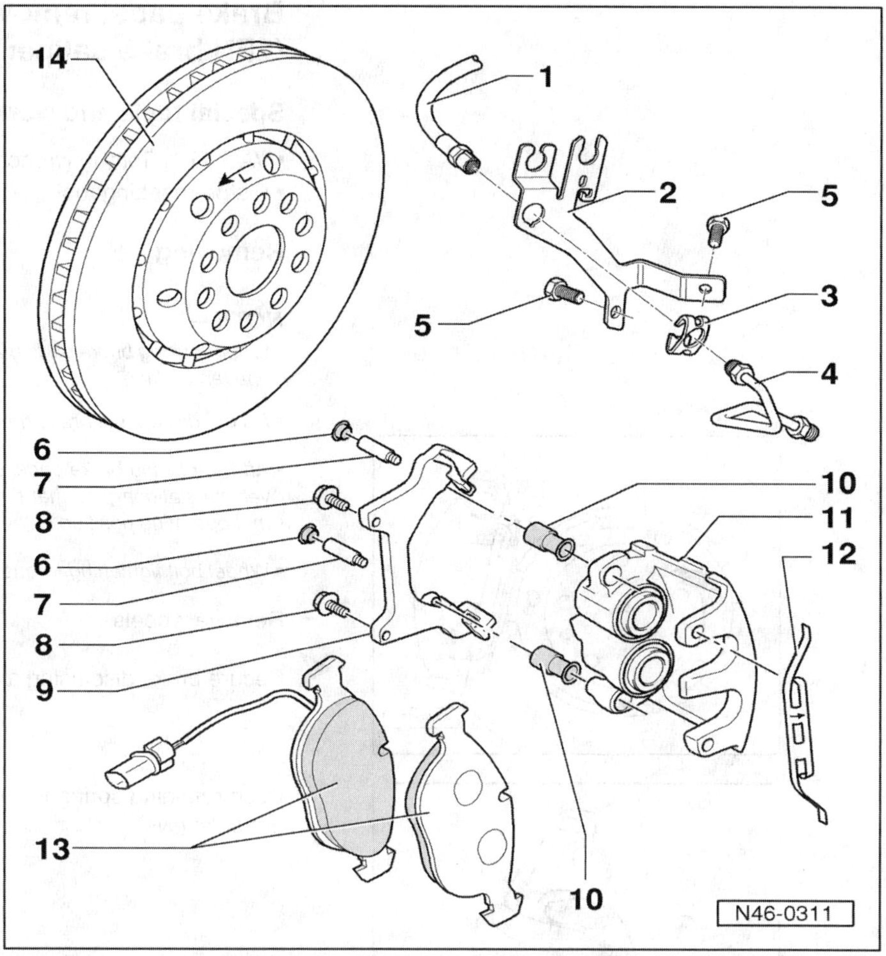

1. **Brake hose**
2. **Retainer**
   - Bolt to brake caliper housing
3. **Spring clip**
4. **Brake pipe**
   - Tighten 15 Nm (11 ft-lb)
   - Screw into brake caliper housing
   - Screw into brake hose, counter hold on brake hose hex.
   - Make sure that lugs are properly seated in retainer grooves.
5. **Hex bolt**
   - Tighten to 8 Nm (71 in-lb)
6. **Protective cap**
   - Take off
   - Upper protective cap with cable holder for brake pad wear indicator
7. **Guide pins**
   - Tighten to 25 Nm (18 ft-lb)
8. **Ribbed combi bolt**
   - Tighten to 190 Nm (140 ft-lb)
   - Clean ribs if using again
9. **Brake carrier**
   - Bolt to wheel bearing housing
10. **Bearing bushing**
11. **Brake caliper**
    - Removing and installing, see **Front brake pads, removing and installing**
12. **Retaining spring**
    - Insert in both brake caliper housing drillings

    NOTE—

    Inserting into both holes the retaining spring must be pushed under the brake carrier. Faulty installation will not allow compensation of outer pad wear and therefore the pedal travel i ncreases.
13. **Brake pads**
    - With wear indicator
    - Thickness 14 mm
    - Wear limit: 7 mm including backing plate
    - Warning lamp lights up on instrument panel when wear limit is reached (limit: 2 to 3 mm)
    - Removing and installing, see **Brake pads, removing and installing (2FN brake caliper)**
14. **Brake disc**
    - Ventilated brake disc: diameter 334 mm
    - Brake disc thickness: 32 mm
    - Wear limit: 30 mm
    - Always replace both sides
    - Left differs from right. Arrow points in direction of normal travel

Brake pads, removing and installing (2FN brake caliper)

Special tools and workshop equipment required

- VAG 1331 Torque wrench
- Piston resetting tool

Removing

NOTE —
- *When reusing brake pads mark location, install in same position otherwise uneven braking!*
- *Do not disconnect brake hose when changing brake pads*
- *After replacing brake pads, depress brake pedal firmly several times with vehicle stationary so that the brake pads are properly seated in their normal operating position.*
- *Wheel bolt tightening torque 120 Nm (89 ft-lb)*

– Remove wheels.

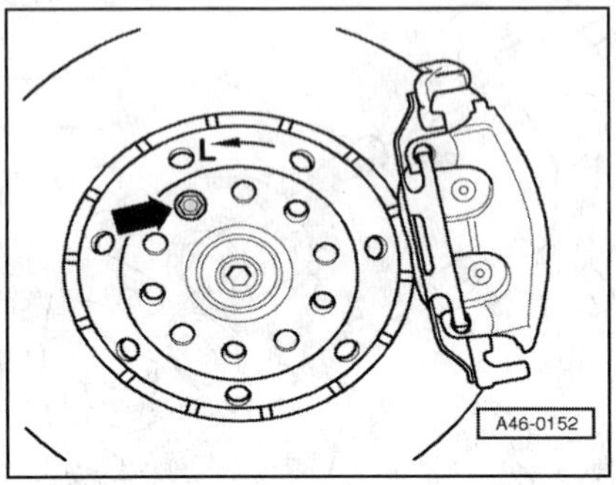

◄ Secure brake disc using a wheel bolt.

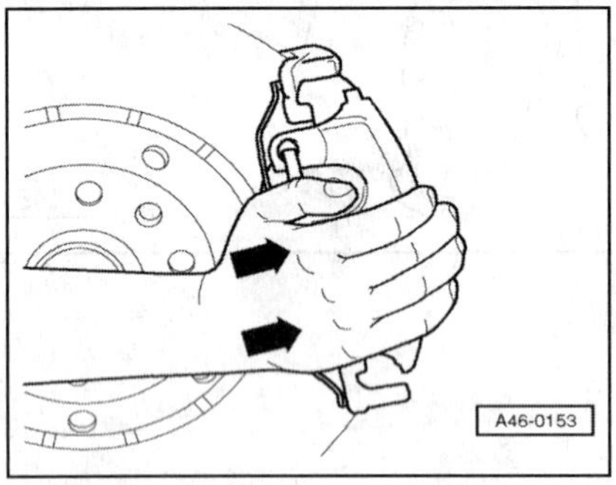

◄ Push retaining spring in direction of arrow and remove from brake caliper sideways.

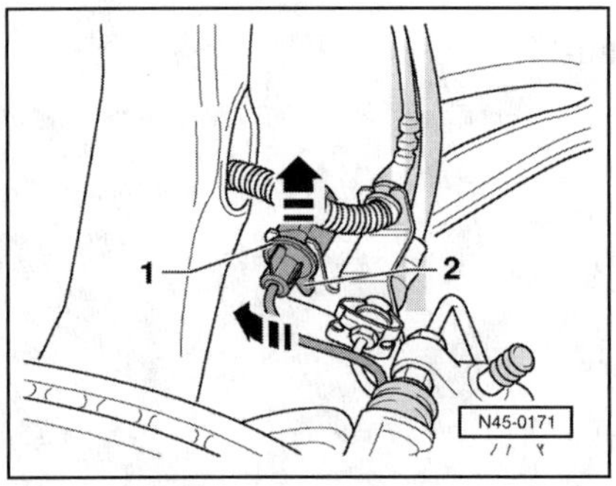

◄ Separate the connector -1- for brake pad wear indicator.

– Lift locking tab slightly on connector lower part -2- and then turn 90°.

– Pull connector lower part out of bracket.

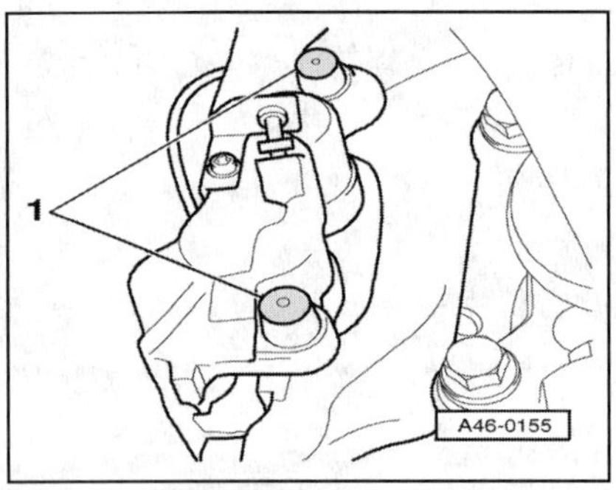

◄ Remove protective caps -1-.
- Upper protective cover with cable holder for brake pad wear indicator

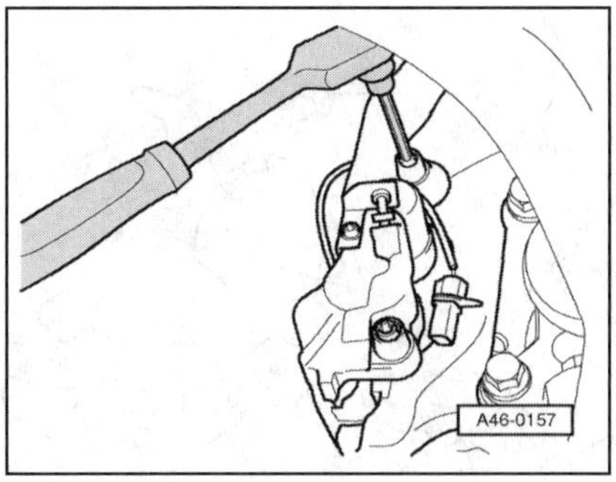

Remove both guide pins from brake caliper.

– Remove brake caliper housing and lay to one side so that the weight of the brake caliper does not stress or damage the brake hose.

– Remove brake pads out of brake caliper housing or brake carrier.

NOTE—
Use only appropriate solvent for cleaning the brake caliper housing.

Brake pads, installing

NOTE—
Before installing new brake pads, press piston back into the cylinder with resetting tool. Draw off brake fluid from the reservoir with a bleeder bottle. Before pressing the piston back, make sure reservoir has not been topped off, otherwise fluid will overflow and cause damage.

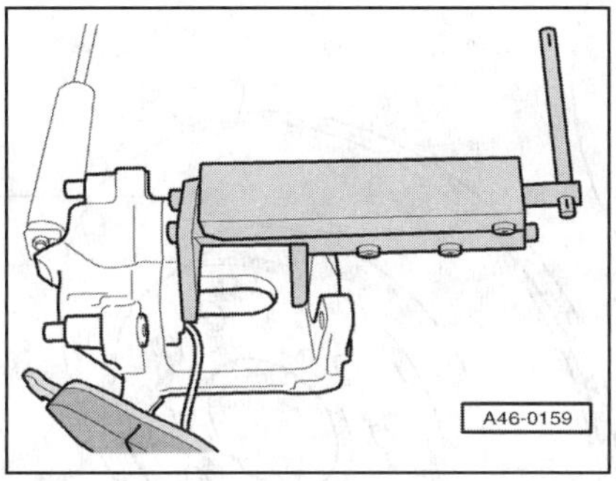

Press piston back into brake caliper.

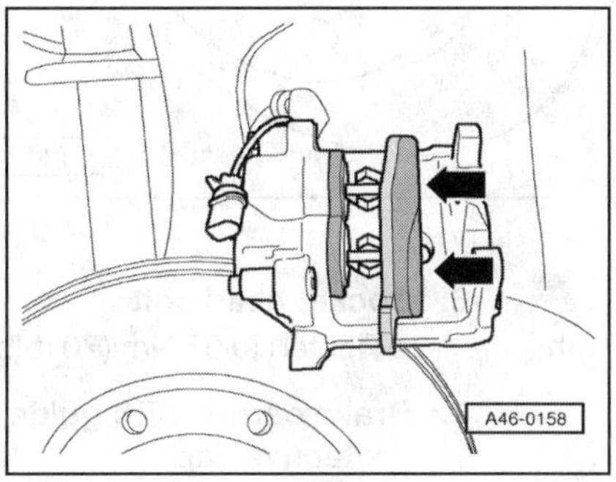

Insert brake pad with retaining spring in brake caliper housing (piston).

– Install outer brake pad to brake carrier.

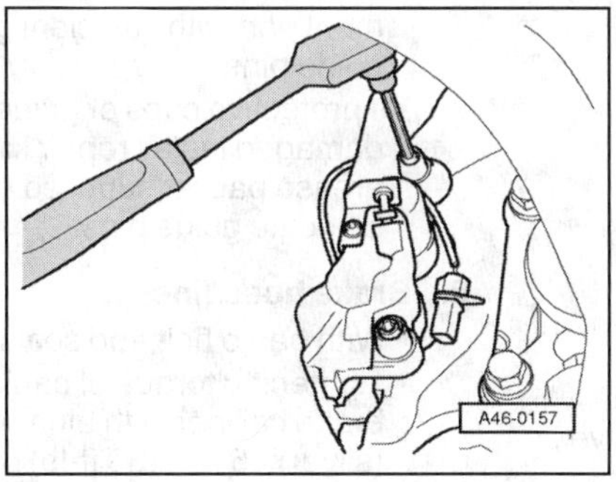

Bolt brake caliper housing to brake carrier with both guide pins.

– Install both protective caps.
  - The brake wear indicator wiring must be routed beneath the bracket on the brake caliper.

– Install brake pad wear indicator connector under dust cap of bleed valve, in upper protective cap and in bracket on brake caliper.

– Install brake wear indicator connector into bracket on brake caliper.

– Install retaining spring into brake caliper housing and then press under brake carrier.

NOTE—
- *After each brake pad change with vehicle stationary, firmly depress brake pedal several times so that the brake pads are properly seated in their normal operating position.*
- *Check brake fluid level, after changing brake pads.*

Rear Brakes, Servicing

Rear wheel brake, assembly

Special tools and workshop equipment required

- VAG 1331 Torque wrench (or equivalent)
- VAG 1332 Torque wrench (or equivalent)
- VAG 1410 Torque wrench (or equivalent)
- VAG 1869/2 Brake pedal depressor

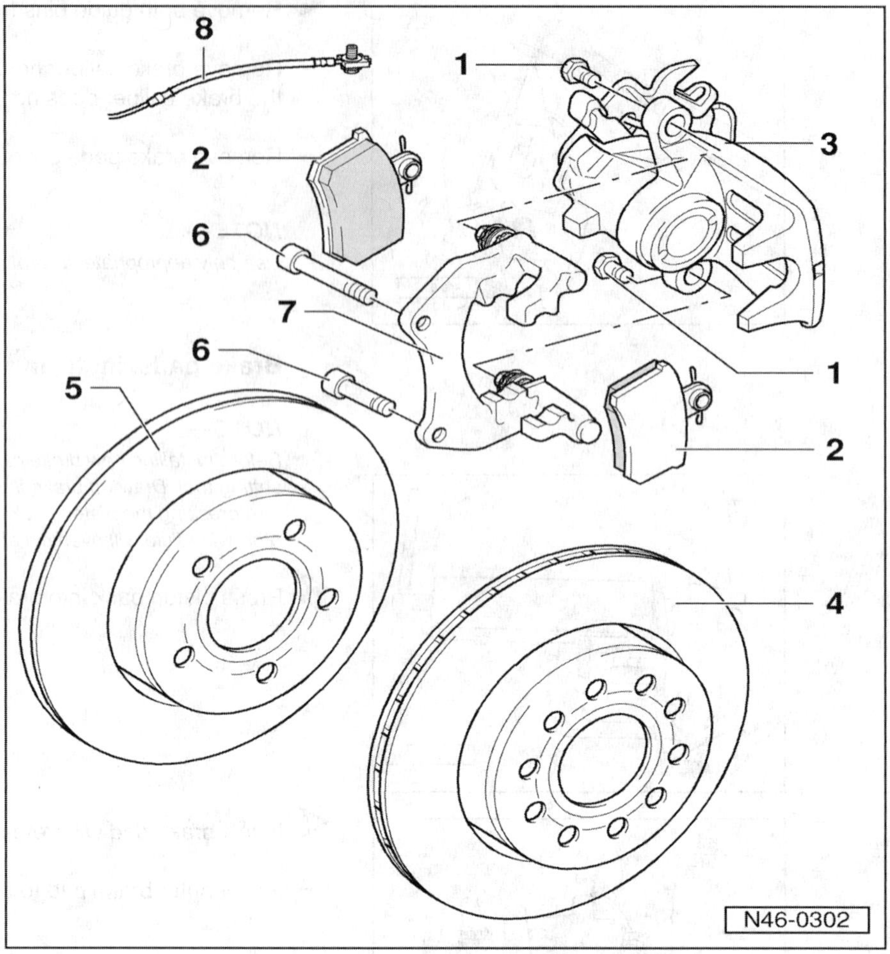

1. **Self-locking hex bolt**
   - Tighten to 35 Nm (26 ft-lb)
   - Always replace
   - When loosening and tightening counter hold on guide pin
2. **Brake pads**
   - Thickness 12 mm
   - Wear limit: 7 mm including backing plate
   - Always replace pads on both wheels
   - Removing and installing, see **Rear brake pads, removing and installing**
3. **Brake caliper**
   - Do not disconnect brake hose when changing brake pads
   - Removing:
     Remove brake pads -**2**-, see **Rear brake pads, removing and installing**
     Install brake pedal depressor VAG 1869/2
     Unscrew brake hose -**8**- from brake caliper
     Install brake pads -**2**-, see **Rear brake pads, removing and installing**
     Install brake hose -**8**- on brake caliper
     Remove brake pedal depressor VAG 1869/2
     Bleed brake system, see **Repair Group 47, Braking system, bleeding**
   - Adjust parking brake cable first after maintenance or replacement, see **Parking brake adjusting**
4. **Brake disc**
   - Vented disc brake for vehicles with 16" wheel all wheel drive: diameter 256 mm
   - Ventilated brake disc for vehicles with 16" and 17" wheels all wheel drive: diameter 269 mm
   - Brake disc thickness: 22 mm
   - Wear limit: 20 mm
   - When pads are worn always
   - replace both sides
   - Remove brake caliper prior to removing
5. **Brake disc**
   - Disc brake for vehicles with 15" and 16" wheels front wheel drive : diameter 245 mm
   - Disc brake for vehicles with 15" wheels all wheel drive: diameter 255 mm
   - Disc brake thickness: 10 mm
   - Wear limit: 8 mm
   - When pads are worn always replace both sides
   - Remove brake caliper prior to removing
6. **Socket head bolt**
   - Tighten to 95 Nm (70 ft-lb)
7. **Brake carrier with guide pins and protective cap**
   - Supplied as assembled replacement part with sufficient grease on guide pins
   - If protective caps or guide pins are damaged install repair kit. Use grease packet supplied to lubricate guide pins.
8. **Brake hose/line**
   - With banjo bolt and seals
   - Tightening torque of banjo bolt to brake calliper with aluminium seal (silver) 45 Nm (33 ft-lb)
   - Tightening torque of banjo bolt to brake calliper with titanium seal (black) 38 Nm (28 ft-lb)
   - Do not disconnect brake hose when changing brake pads

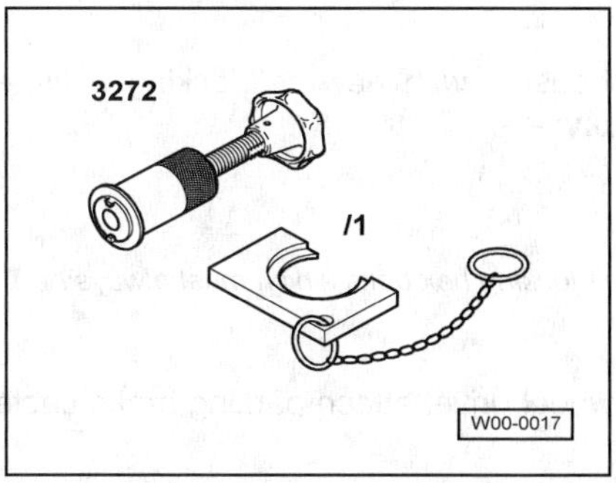

Rear brake pads, removing and installing

Special tools and workshop equipment required

3272 Reset and screwout tool

- 3272/1 Reset and screwout tool plate
- VAG 1331 Torque wrench (or equivalent)

Removing

NOTE—
When removing, mark brake pads if they are to be reused. Install in their original position to prevent uneven braking!

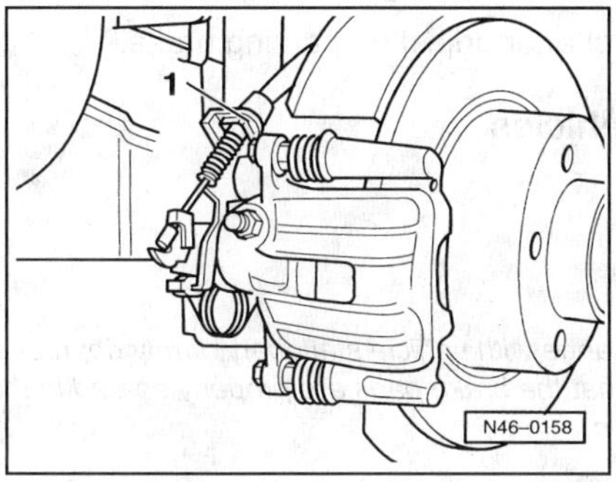

– Remove wheels.

On vehicles with front wheel drive, pull out securing clip -1- and unhook parking brake cable from brake caliper.

Continued for all vehicles

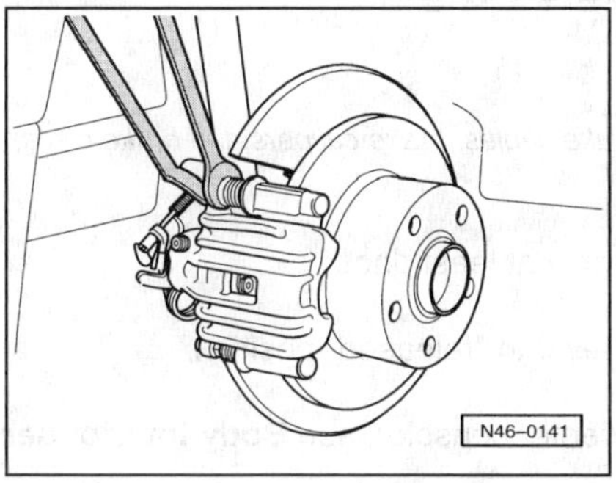

Remove securing bolts from brake caliper housing, when doing this counter-hold on guide pins.

– Remove brake caliper housing and secure with wire so that the weight of the brake caliper does not stress or damage the brake hose.

– Remove brake pads.

– Clean brake caliper housing, in particular the bonding surface for the brake pad. It must be free of adhesive and grease residues.

NOTE—
Use only appropriate solvent for cleaning the brake caliper housing.

Rear brake pads, installing

NOTE—
Draw off brake fluid from the reservoir before pressing the piston back. Use a bleeder bottle or a plastic bottle which is only used for brake fluid. Brake fluid is poisonous. NEVER siphon brake fluid with your mouth!

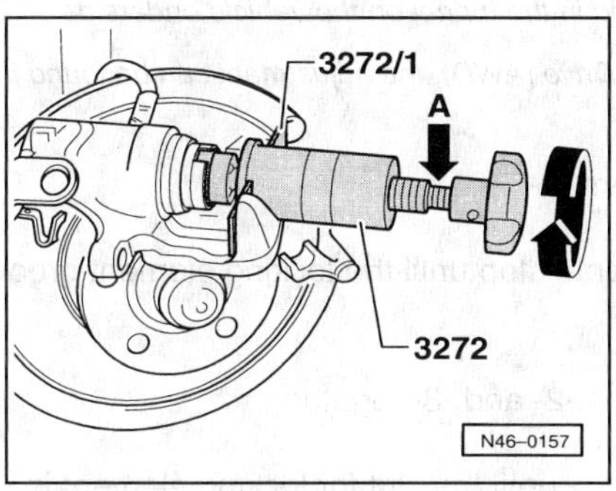

Install piston by turning knurled wheel of special tool 3272 clockwise. Do not damage protective cap when doing this.

– Use special tool 3272/1 for assistance when installing.

NOTE—
- *Install special tool 3272 so that the collar of the tool is seated on special tool 3272/1.*
- *For pistons that are difficult to move, use open-end wrench (use 13 mm) on the flats (arrow A).*

– Remove protective foil off backplate of new brake pads.

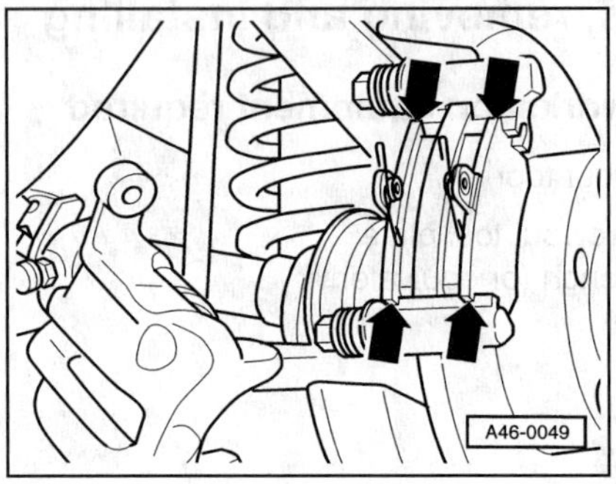

◀ Insert brake pads.

– Secure brake caliper housing with new self-locking bolts, see **Rear wheel brakes, assembly**; -1-.

NOTE —
Repair kit includes four self-locking hex bolts which must always be installed.

– On vehicles with front wheel drive, attach parking brake cable to brake caliper housing.

– Adjusting parking brake, see **Parking brake adjusting**.
  - Operate foot brake first after adjusting parking brake.

Continued for all vehicles

– Install wheels.

NOTE —
- *After each brake pad change with vehicle stationary, firmly depress brake pedal several times so that the brake pads are properly seated in their normal operating position.*
- *Check brake fluid level after changing brake pads.*

Parking brake, adjusting

NOTE —
After replacing parking brake cables, brake calipers and brake discs, adjustment is necessary.

– Depress brake pedal firmly at least once.

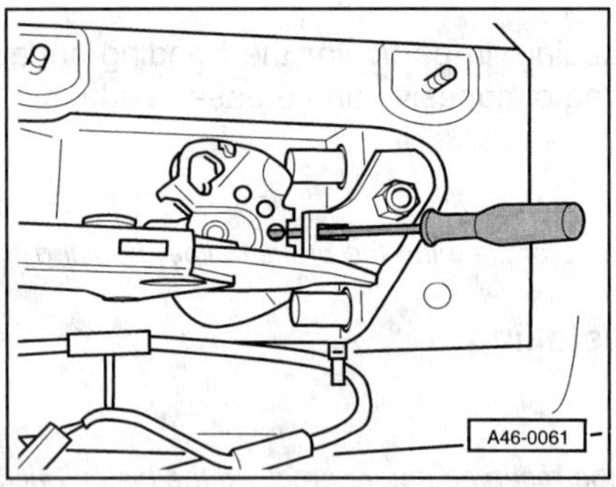

– Position parking brake lever in "released" position.

– Remove rear ashtray (center console), see **Body-Interior section, Repair Group 68**

◀ Remove any plastic parts on compensator; they do not need to be re-installed.

– To prevent compensator from turning, counterhold with a screwdriver (shown without center console for clarity).

NOTE —
- *For vehicles with Front Wheel Drive (FWD) the adjustment for the parking brake cable can be found in the tunnel on the vehicle underside.*
- *Vehicles with All Wheel Drive (AWD), the adjustment can be found in front of the lower control arm.*

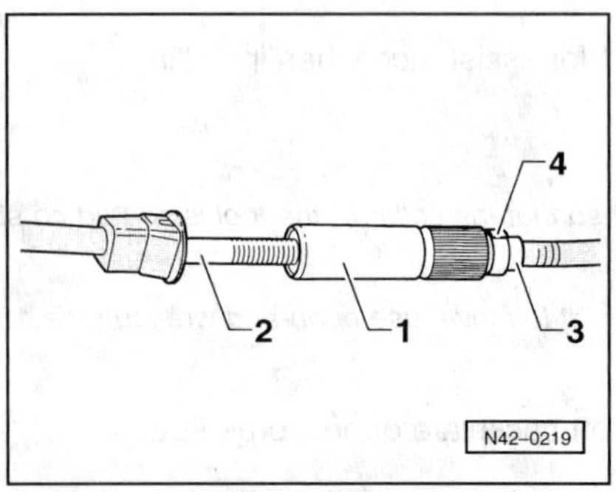

◀ Remove locking element -4-.

– Turn adjusting nut -1- onto stop until the locking element groove is covered.

– Push both cable sleeves -2- and -3- together.

– Unscrew adjusting nut -1- until the slot for locking element is visible.

– Insert locking element -4-.

- Pull both cable sleeves apart at the same time until the cables are tightened, doing this the lever on the brake caliper does not lift off.

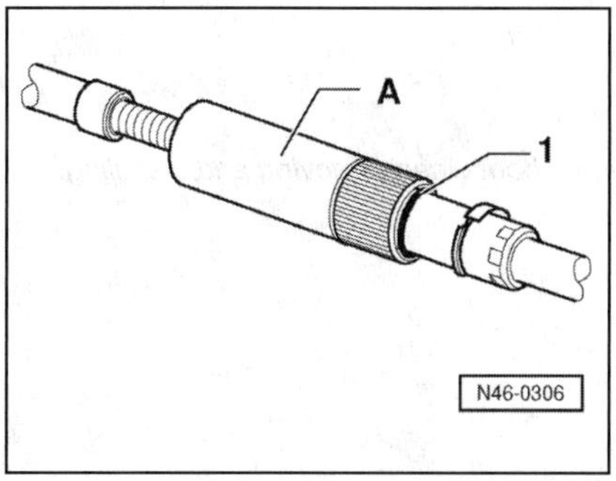

- Remove screwdriver from compensator and apply parking brake 3 times.

◄ The colored O-ring -1- must not be visible. Make sure that the mechanism of the parking brake cable adjustment element is sufficiently protected against dirt and water spray.

- Check tension of parking brake cable and, if necessary, turn adjusting nut -A- in/out...

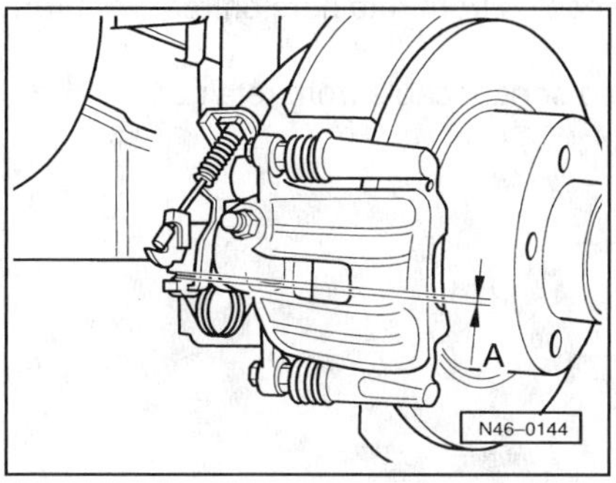

◄ ...until a distance -A- of max. 1.5 mm is set at the brake caliper.

NOTE—

Due to the automatic rear wheel brake adjustment there is no requirement to adjust the parking brake (after initial adjustment).

Parking brake cables (front wheel drive), removing and installing

Removing

- Remove both rear wheels.

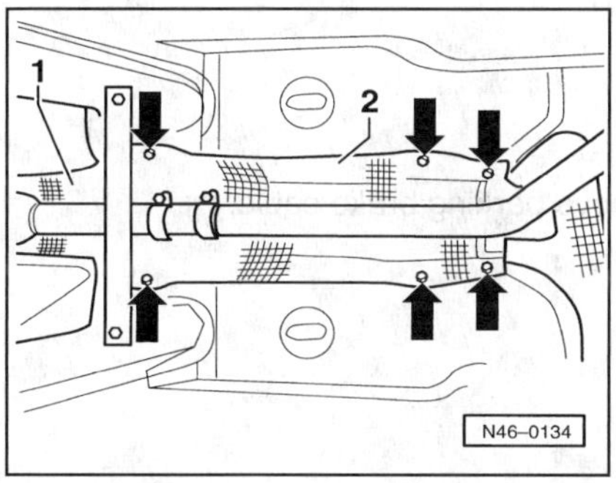

◄ Remove heat shield -1-.

- Remove retaining nuts -arrows- from heat shield -2- and slide heat shield forward.

NOTE—

Parking brake cable adjusting components are located in the tunnel on the underside of vehicle.

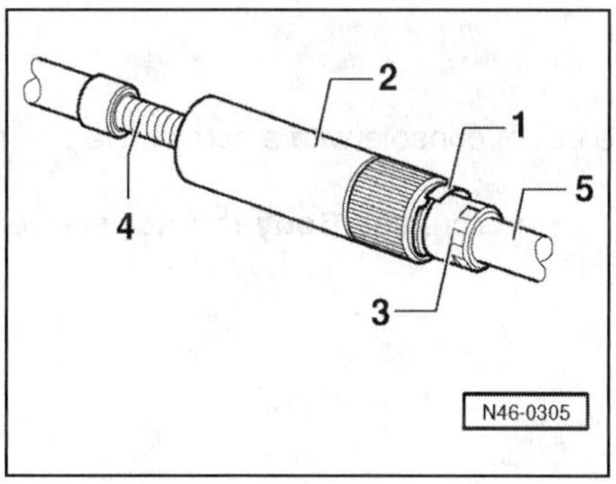

◄ Remove locking element -1-.

- Screw adjusting nut -2- in until it stops. To do this, counter-hold using an open-ended wrench (use 13 mm) in area -3-.

- Push both cable sleeves -4- and -5- together.

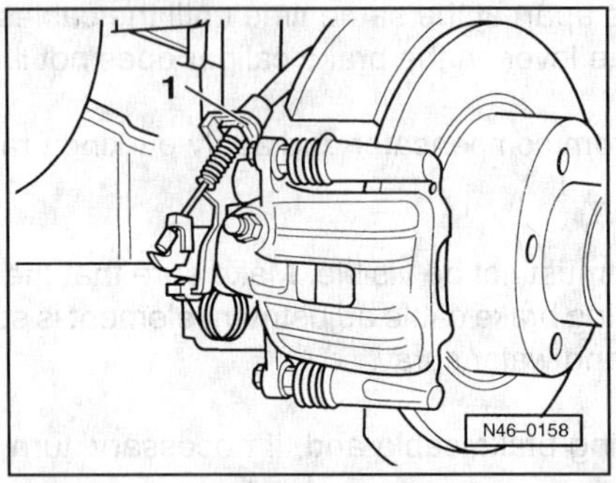

◄ Pull out securing clip -1- and unhook parking brake cable from brake caliper.

NOTE —
Do not damage parking brake boot when removing and installing.

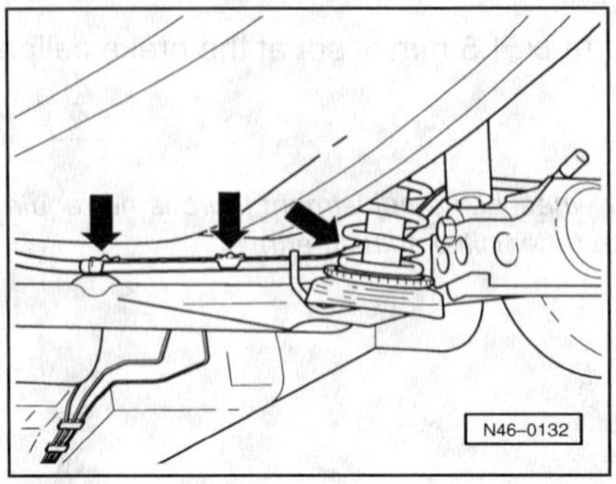

◄ Remove parking brake cables from retainers on rear axle (arrows).

– Unclip ABS wheel speed sensor cable from retainer.

◄ Loosen heat insulation above rear silencer (arrows) and place to one side.

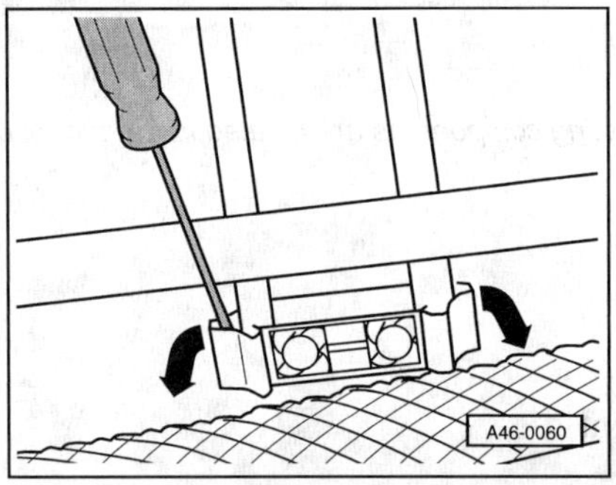

◄ Open retainer and remove parking brake cable.

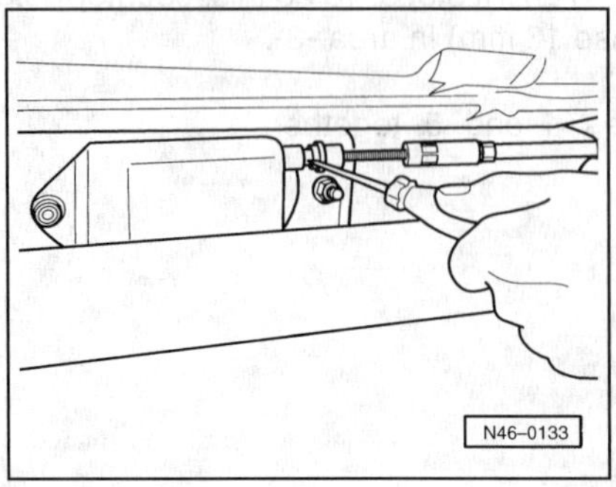

◄ Pry parking brake cable out of console with a screwdriver.

– Remove center console extension, see **Body-Interior section, Repair Group 68**

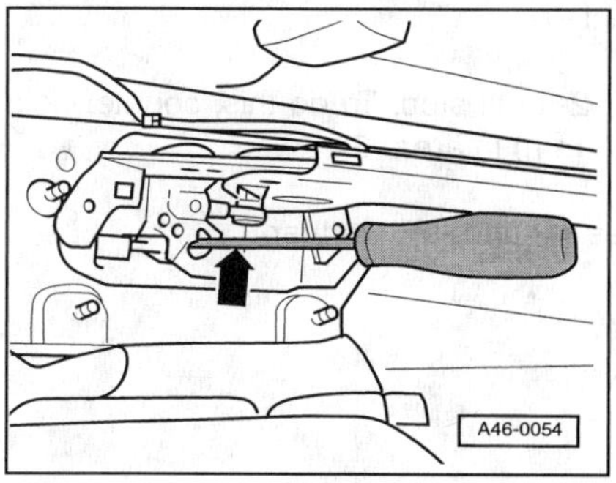

Using a screwdriver press parking brake cable in direction of parking brake lever and push ball out of retainer.

– Remove parking brake cable.

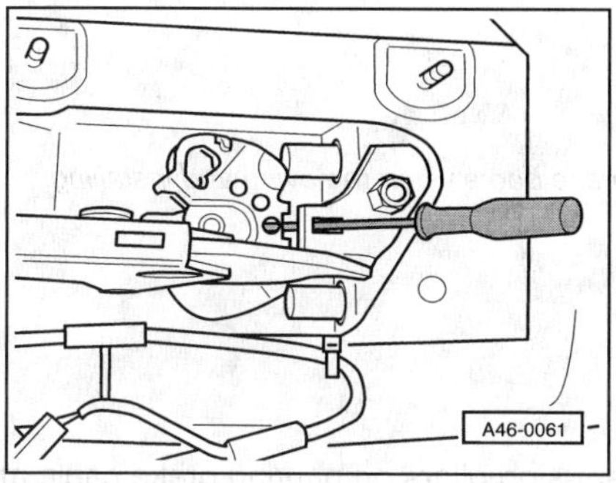

Parking brake cables (front wheel drive), installing

Remove any plastic parts on compensator; they do not need to be re-installed.

– Prevent compensator from turning by counterholding using a screwdriver.

– Insert ball head in compensator via console.

– Parking brake cable must engage in console.

– Secure parking brake cables to rear axle beam.

– Adjusting parking brake, see **Parking brake adjusting**.

– Install center console extension, see **Body-Interior section, Repair Group 68**

Parking brake cables (all wheel drive), removing and installing

Removing

– Remove rear wheels.

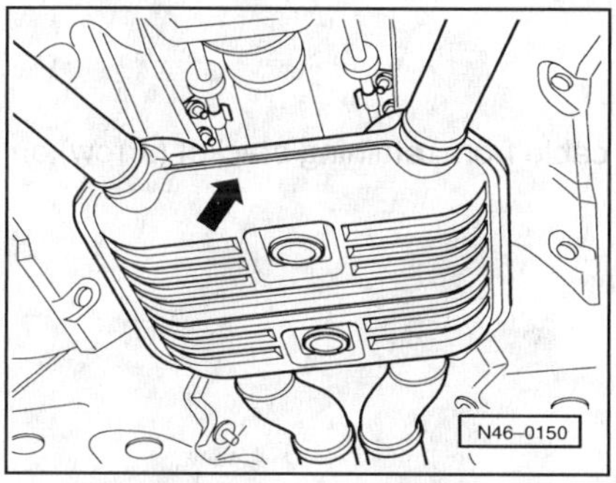

Remove exhaust system (arrow), see **Engine Mechanical section, Repair Group 26.**

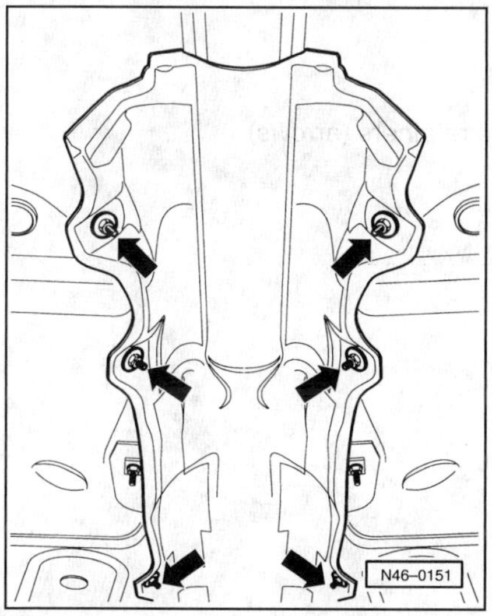

Unscrew retaining nuts (arrows) from heat shield and remove.

NOTE—
Adjusting the parking brake cables are located in front of the lower control arm.

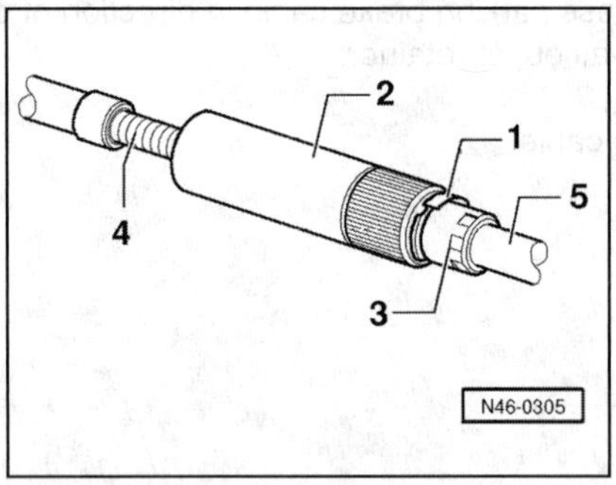

< Remove locking element -1-.

- Screw in adjusting nut -2- until stop. To do this, counter-hold using an open-ended wrench (13 mm) in area -3-.

- Push both cable sleeves -4- and -5- together.

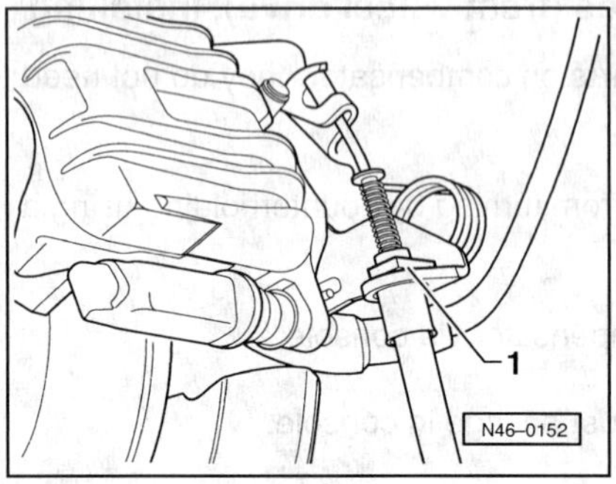

< Pull out securing clip -1- and unhook parking brake cable from brake caliper.

NOTE —
Do not damage parking brake boots when removing and installing.

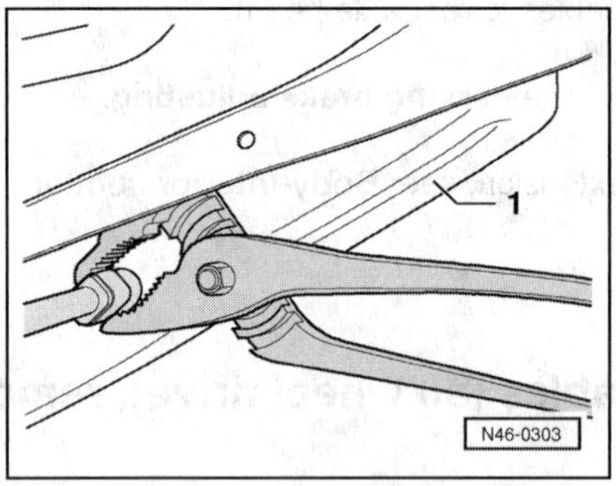

< Squeeze plugs together using pliers on parking brake cable and take parking brake cable out through hole in lower control arm -1-.

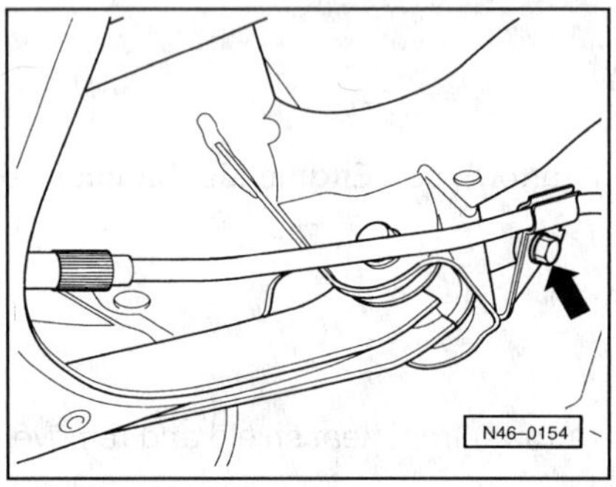

< Remove parking brake cable from retaining bracket (arrow) on rear axle.

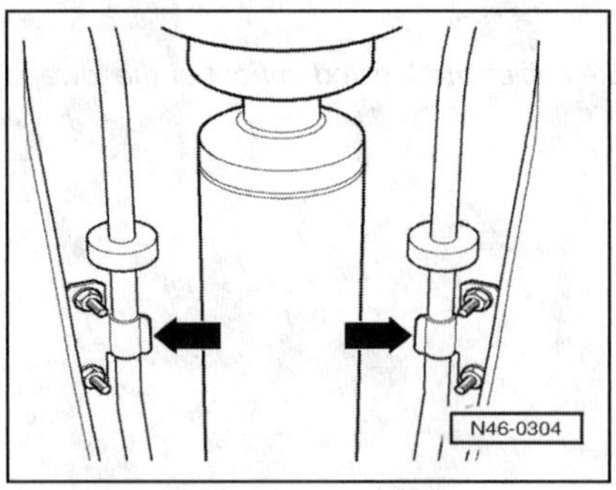

< Unclip brake pipes from retainers (arrows).

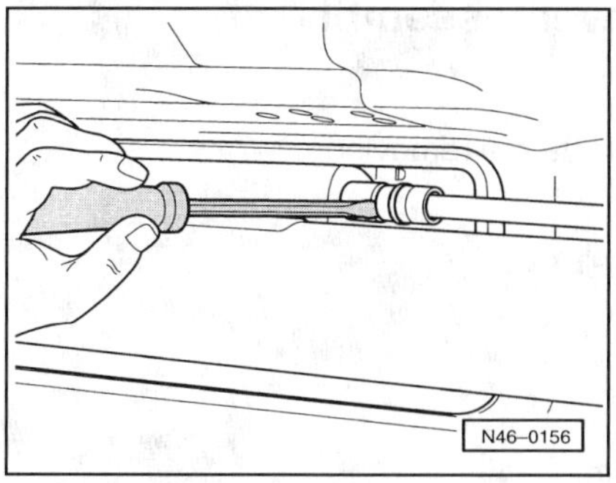

- Pry parking brake cable out of console with a screwdriver.
- Remove center console extension, see **Body-Interior section, Repair Group 68;**

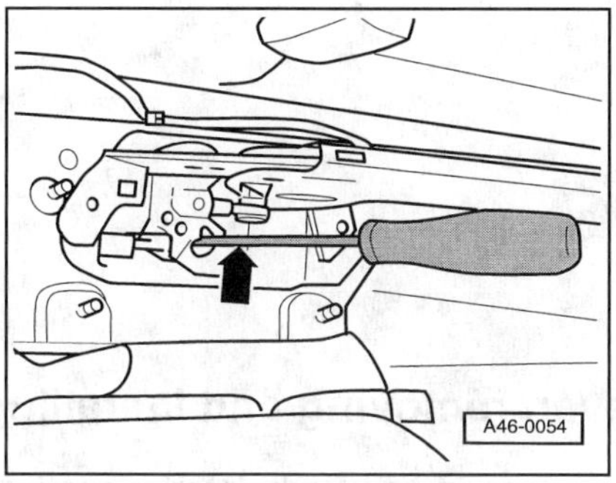

- Using a screwdriver press parking brake cable in direction of parking brake lever and push ball out of retainer.
- Remove parking brake cable.

Parking brake cables (all wheel drive), installing

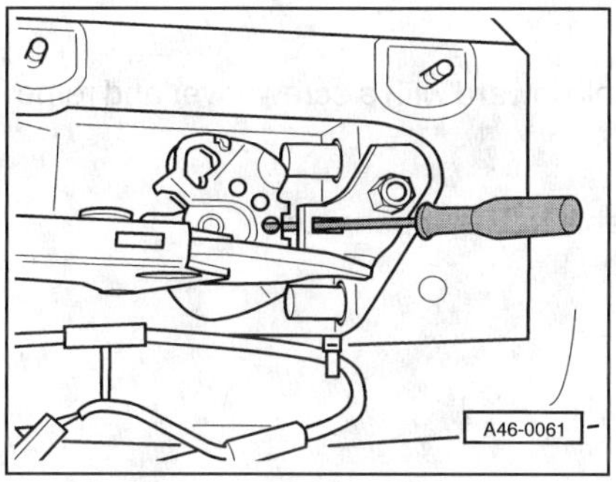

- Remove any plastic parts on compensator; they do not need to be re-installed.
- Prevent compensator from turning by counterholding using a screwdriver.
- Insert ball head in compensator via console.
  - Parking brake cable must engage in console.
- Secure parking brake cables to rear axle beam.
- Adjust parking brake, see **Parking brake adjusting**.
- Install center console extension, see **Body-Interior section, Repair Group 68**

Parking brake lever, assembly

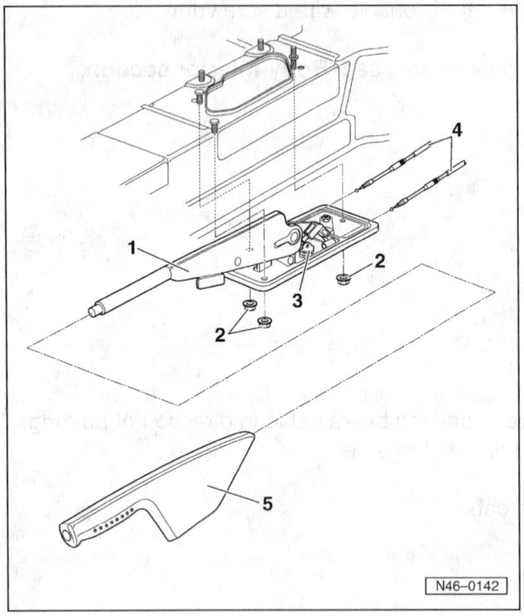

1. **Parking brake lever**
   - Remove center console before removing
2. **Hex nut**
   - Tighten to 23 Nm (17 ft-lb)
3. **Compensator**
4. **Parking brake cables**
5. **Parking brake lever trim**
   - Pull torward front

Parking brake lever, removing and installing

– Remove center console extension, see **Body-Interior section, Repair Group 68**

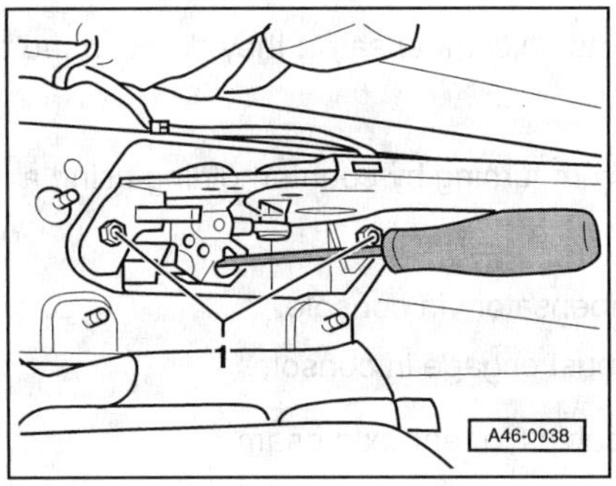

◄ Remove hex nuts -1-.

– Press parking brake cable inward with a screwdriver and remove parking brake lever.

– Install parking brake cable into compensator.

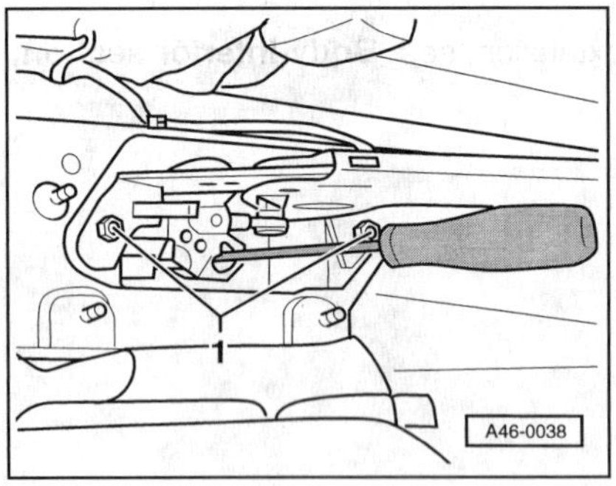

◄ Tighten hex nuts -1- to 25 Nm (18 ft-lb).

– Adjust parking brake, see **Parking brake adjusting**.

– Install center console extension, see **Repair Group, Body-Interior section, Repair Group 68**

Brake Pedal

Brake pedal, assembly

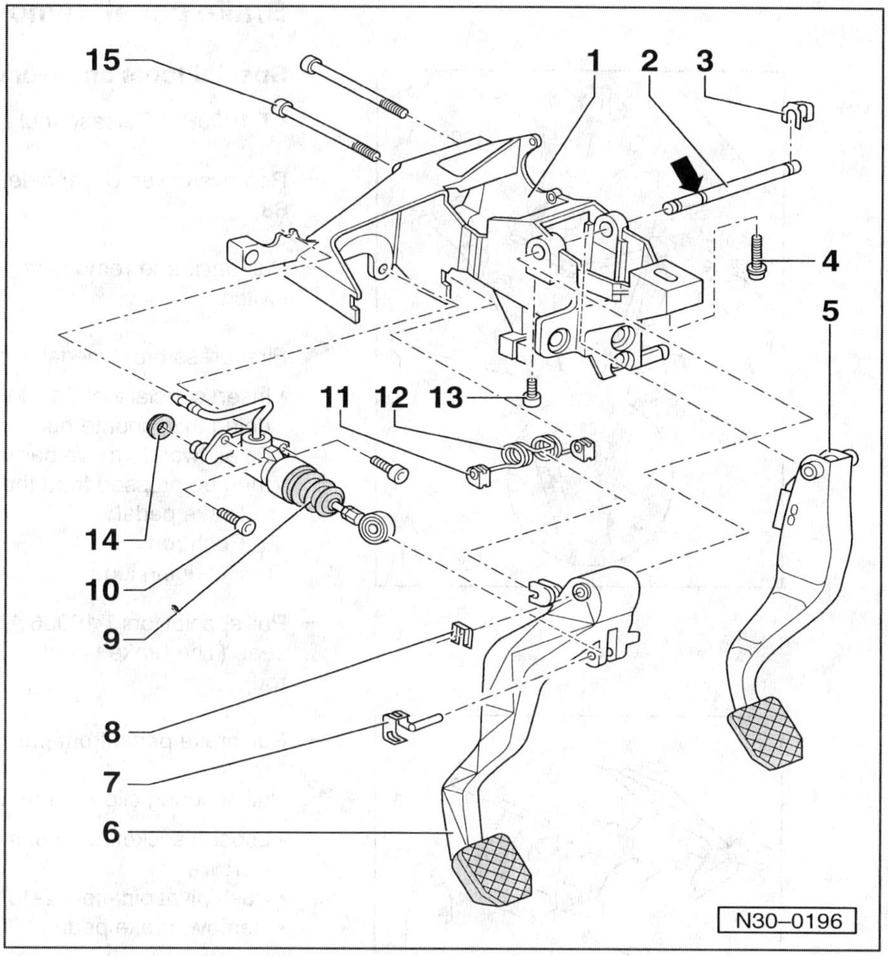

WARNING —
Brake pedal travel must not be restricted by additional floor covering.

1. **Mounting bracket**
2. **Pivot pin for clutch and brake pedal**
   - Installation position: The groove (arrow) faces towards clutch pedal.
3. **Locking clip**
4. **Hex socket head bolt**
   - Tighten to 25 Nm (18 ft-lb)
5. **Brake pedal**
   - Install onto pin
   - Removing and installing, see **Brake pedal, removing and installing**
6. **Clutch pedal**
   - Removing and installing, see **Manual Transmission section, Repair Group 30**
7. **Pin**
   - Always replace
8. **Locking clip**
9. **Master cylinder**
10. **Hex socket head bolt**
    - Tighten to 20 Nm (15 ft-lb)
11. **Mounting**
    - Install in mounting bracket with over-center spring
    - Installion position: Deep groove faces downward
12. **Over-center spring**
    - Remove and install with clutch pedal, see **Manual Transmission section , Repair Group 30**
13. **Hex socket head bolt**
    - Tighten to 5 Nm (44 ft-lb)
    - Securing clutch pedal and brake pedal pivot pin
14. **Seal**
15. **Torx screw**
    - Tighten to 25 Nm (18 ft-lb)
    - Secures the brake master cylinder and brake servo

Brake pedal, removing and installing

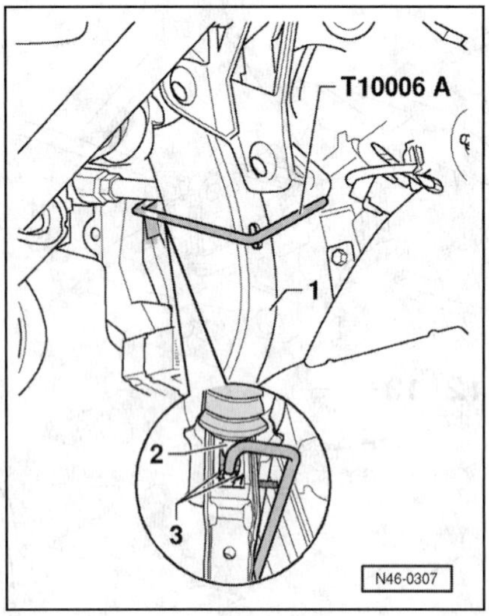

Special tools and workshop equipment required

- T 10006 A Release tool

– Remove cover, driver side, see **Body-Interior section, Repair Group 68**

– Pull back and remove brake light switch, if necessary, the brake pedal switch.

First press brake pedal in direction of brake servo and hold.

- Insert special tool T 10006 A and pull in direction of driver's seat. When doing this, counterhold brake pedal. (At this point the pedal must not be allowed to move backward). The mounting retaining lugs -3- will then be pressed from the ball head of the push rod -2-.

1 - Brake pedal
2 - Push rod
3 - Retaining lugs

– Pull special tool T 10006 A together with brake pedal towards the driver's seat. (The brake pedal will then be pulled off the ball head of the push rod).

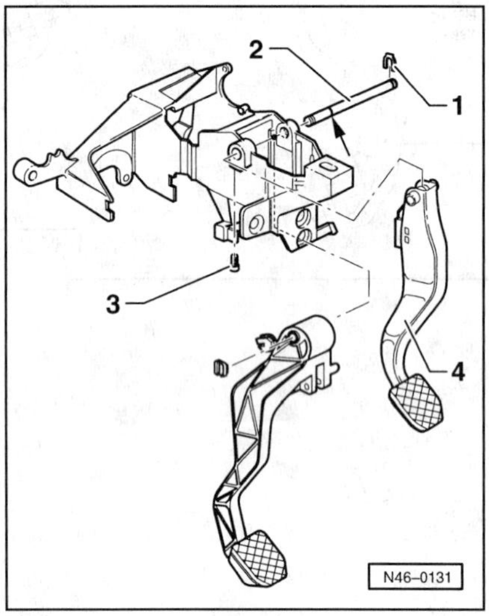

– Pull brake pedal from push rod.

Pull securing clip -1- from clutch and brake pedal pivot pin -2-.

- Loosen socket head bolt -3- until it is no longer seated in the groove -arrow-.
- Push pivot pin -item 2- to the left until the brake pedal can be removed.
- Remove brake pedal.

– Installation is the reverse order of removal, noting the following:

- Do not tighten socket head bolt -3- until retaining clip has been installed to clutch and brake pedal pivot pin.
- Hold push rod ball head in front of mounting and push brake pedal in direction of brake servo until ball head engage audibly.
- Adjust brake light switch and, if necessary, brake pedal switch, see the following procedure.

Brake light switch -F- and Brake pedal switch (cruise control/Diesel Direct Fuel Injection) -F47-, adjusting

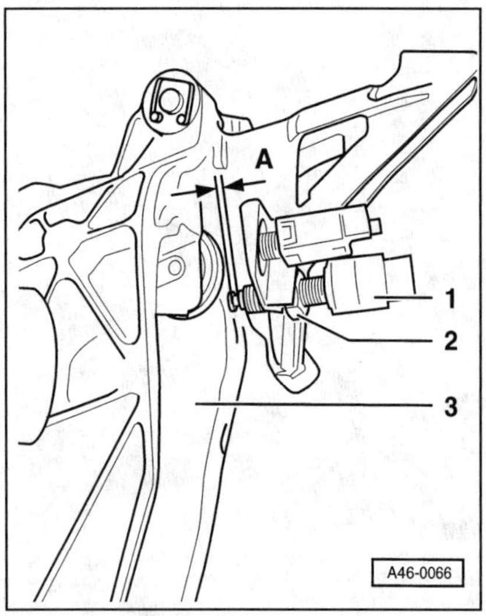

– Remove storage compartment on driver's side, see **Body-Interior section, Repair Group 68**

Mounting bracket with threaded hole for mounting brake light switch.

1 - Brake light switch
2 - Hex nut, M12 × 1.5
3 - Brake pedal

NOTE —

- *A spring washer (not visible in illustration) is installed between nut -2- and mounting bracket.*
- *Before adjusting brake light switch, check whether brake pedal is in end position (released position).*

– Rotate brake light switch in or out (as necessary) to obtain dimension -A- = 0.1 to 0.5 mm.

– Tighten nut -2- to 4.5 Nm.

– Check operation of brake light switch.

Mounting bracket with clip for brake light switch.

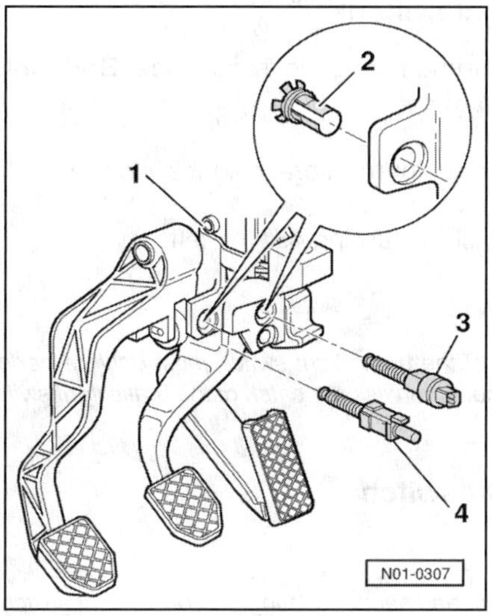

1 - Mounting bracket for pedal assembly

2 - Clip

3 - Brake Light Switch -F- with Brake Pedal Switch (cruise control/Diesel Direct Fuel Injection) -F47-

4 - Clutch Vacuum Vent Valve Switch -F36-

NOTE—

In order to assure secure installation, install brake light switch only once.

- Mounting procedure resembles procedure for Brake Light Switch -F-

Installing brake light switch

- Clip installed in mounting bracket
- Brake pedal not depressed

– Guide the switch -3- into the mounted clip -2-.

– Press switch in far enough that the plunger is resting at brake pedal.

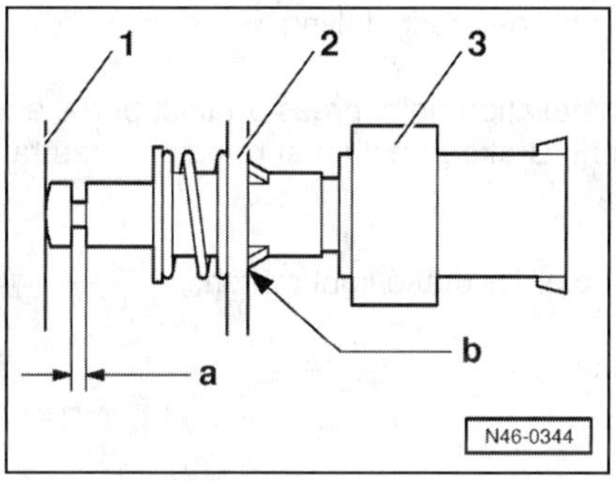

Checking adjustment after installation

1 - Brake pedal

2 - Mounting bracket

3 - Brake Light Switch -F- with Brake Pedal Switch (cruise control/Diesel Direct Fuel Injection) -F47-

- Dimension -a- = 0.5 to 1.2 mm
- Catches for clip -b- may not be lifted off mounting bracket -2-.
- Check adjustment dimension -a- and correct the adjustment if necessary by holding clip firmly and rotating the brake light switch inward or outward.

NOTE—

After adjusting brake light switch, check whether brake pedal is in end position (released position).

– Checking function of brake light switch

- Depress brake pedal.
  Brake lights light up.
- Remove foot from brake pedal.
  Brake lights must not light up.

– Install storage compartment on driver's side, see **Body-Interior section, Repair Group 68**

Brake light switch -F- and Brake pedal switch (cruise control/Diesel Direct Fuel Injection) -F47-, adjusting, from model year 2003

NOTE—
The brake light switch must be removed for adjustment.

Removing brake light switch

– Remove storage compartment on driver's side, see **Body-Interior section, Repair Group 68**

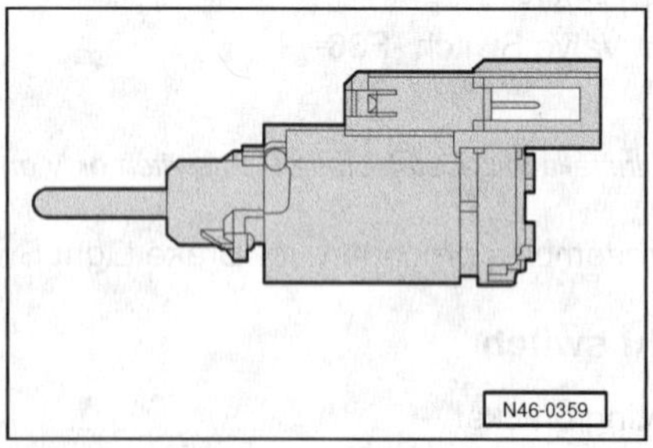

◄ Disconnect harness connector from brake light switch.

– Remove brake light switch by turning 45° the left.

NOTE—
Removal and installation of the brake light switch must only be performed with the plunger depressed, otherwise the catch of the brake light switch will be damaged.

Installing brake light switch

NOTE—
In order to ensure sufficient and secure seating, the brake light switch must only be installed once.

– Completely pull out plunger before installing

◄ Guide switch through installation hole, press against pedal and secure by turning 45° to the right. Brake pedal must remain in neutral position (not operated).

– Connect harness connector for brake light switch.

– Check operation of brake light.

NOTE—
After adjusting brake light switch, check whether the brake pedal is at final stop (neutral position, not operated).

– Install storage compartment on driver's side, see **Body-Interior section, Repair Group 68**

47 Brakes, Hydraulic

Repair Group

R 47

Brake system pressure leak test

Special tools and equipment

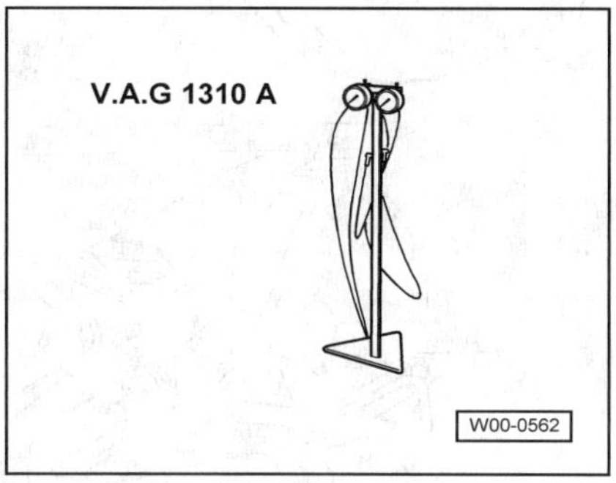

- VAG 1310 A Brake system test unit

Function and proper seal of brake system (hydraulic unit, brake hoses, brake lines and brake calipers).

– Remove bleed screw at one of the front brake calipers. Connect pressure gauge from VAG 1310 and bleed.

– Apply pressure to brake pedal until the gauge indicates a pressure of 50 bar. The pressure must not drop by more than 4 bar during the test period of 45 seconds. If pressure drops greatly replace master cylinder.

Braking system, bleeding

NOTE —

- *Bleeding the braking system on vehicles with ABS is carried out as for vehicles with conventional braking systems.*
- *Only use new brake fluid conforming to US standard FMVSS 116 DOT 4.*
- *Genuine VW/Audi brake fluid conforms to this specification.*
- *Brake fluid is poisonous. Due to its corrosive nature brake fluid must not come into contact with paint.*
- *Brake fluid is hygroscopic, which means it absorbs moisture from the ambient air and must therefore always be stored in air tight containers.*
- *Rinse off spillages, by using plenty of water.*
- *Volkswagen recommends replacing the brake fluid every two years.*
- *To bleed the hydraulic unit a pressure of at least 2 bar is necessary.*

– Build-up brake pressure in brake system by pumping the brake pedal.

– Following bleeding sequence, attach bleeder bottle hose and open bleed screw.

1 - Right-rear brake caliper
2 - Left rear brake caliper
3 - Right front brake caliper
4 - Left front brake caliper

– With brake pedal held down, close bleed screw.

– Repeat operation until no air is present.

Brake servo/brake master cylinder, assembly (through 1999 m.y.)

NOTE—
- *Brake master cylinders and brake servos are made by different manufacturers. Complete master cylinders and brake servos can be replaced independently.*
- *Brake master cylinders cannot be disassembled, i.e. they are not recommended to be serviced.*
- *Use only new brake fluid. Read information on brake fluid reservoir!*

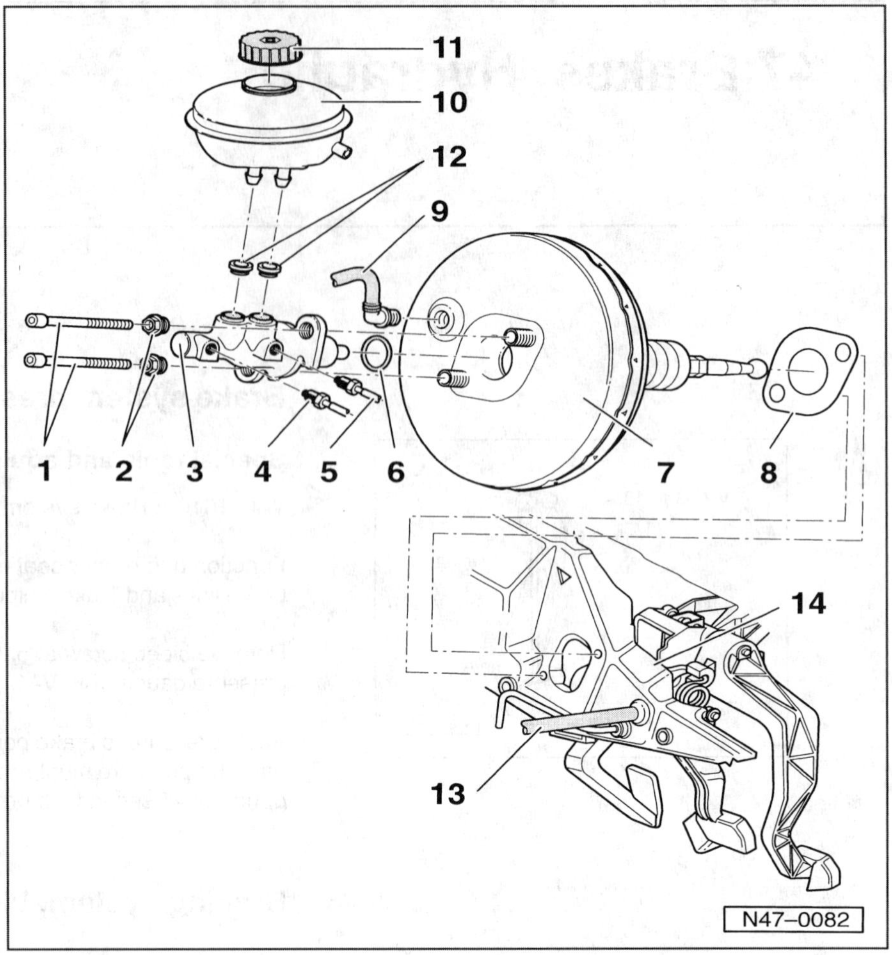

1. **T 45 torx socket head bolt**
   - Tighten to 25 Nm (18 ft-lb)
2. **Self-locking hex nut**
   - Tighten to 49 Nm (36 ft-lb)
3. **Brake master cylinder, diameter 23.81 mm**
   - Cannot be repaired. If faulty, replace complete.
4. **Brake line**
   - Master brake cylinder/secondary piston circuit to hydraulic unit
5. **Brake line**
   - Master brake cylinder/primary piston circuit to hydraulic unit
6. **Sealing ring**
   - Always replace
7. **Brake servo**
   - Adjusting ball plunger, see **Fig. 1**
   - On gasoline engines the vacuum required is taken from the intake manifold.
   - Functional check
     With engine switched off, depress brake pedal firmly several times (to exhaust the vacuum in the unit).
     Depress brake pedal with average foot pressure, hold and start engine. If the servo unit is working properly, the pedal will drop slightly (servo assistance becomes effective).
     If faulty replace complete.
8. **Seal**
9. **Vacuum hose**
   - Insert into brake servo unit
10. **Brake fluid reservoir**
11. **Cap**
12. **Sealing plug**
    - Coat with brake fluid and press into brake fluid reservoir
13. **Supply hose**
    - From brake fluid reservoir to clutch master cylinder
14. **Bulkhead**

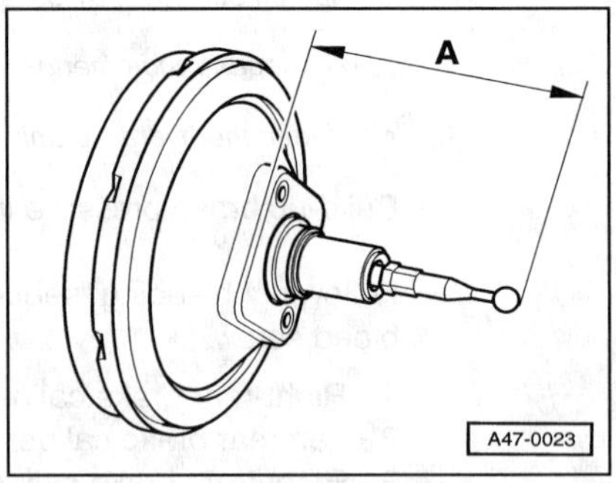

Fig. 1 Adjust ball plunger

A = 158.5 mm ± 0.5mm

NOTE—
- *When measuring the ball plunger it must be set at right angles to the surface of the brake servo.*
- *Measure to end of ball plunger without gasket installed.*

Brake servo/brake master cylinder, assembly (from 2000 m.y.)

NOTE—
- *Use only new brand brake fluid. Read information on brake fluid reservoir!*

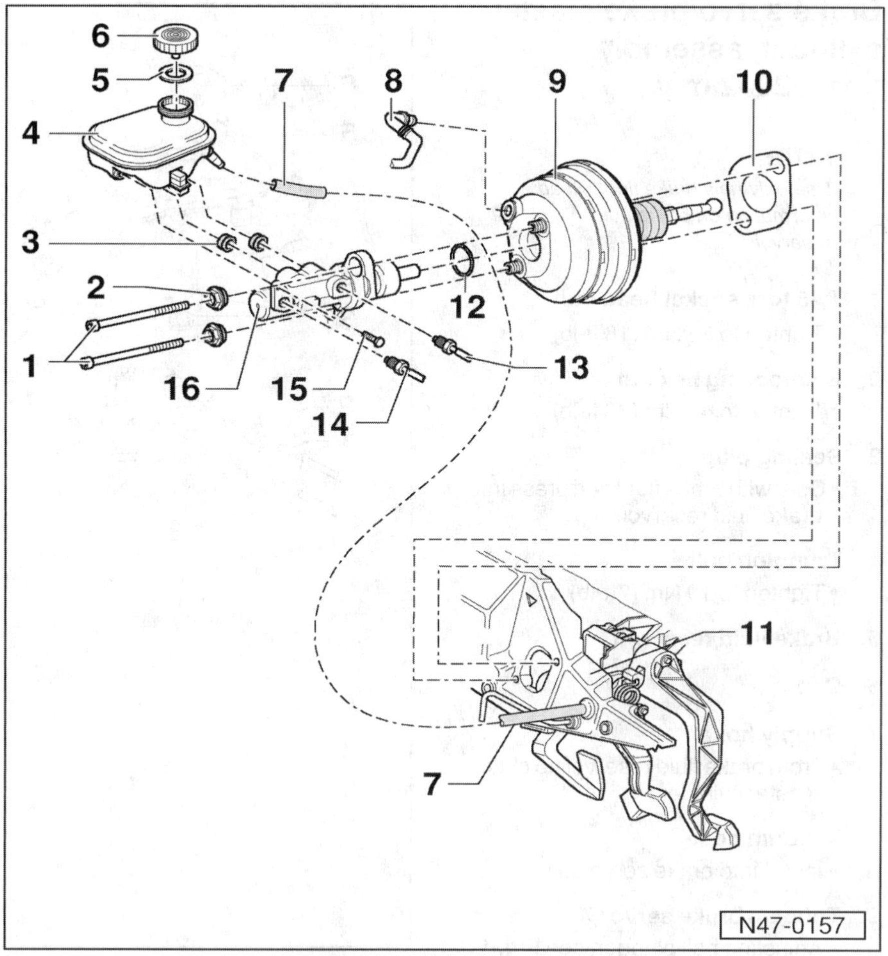

1. **T 45 torx socket head bolt**
   - Tighten to 25 Nm (18 ft-lb)
2. **Self-locking hex nut**
   - Tighten to 49 Nm (36 ft-lb)
3. **Sealing plug**
   - Coat with brake fluid and press into brake fluid reservoir
4. **Brake fluid reservoir**
5. **Oil seal**
6. **Cap**
7. **Supply hose**
   - From brake fluid reservoir to clutch master cylinder
8. **Vacuum hose**
   - Insert into brake servo unit
9. **Tandem brake servo**
   - Adjusting ball plunger, see **Fig. 1**
   - On gasoline engines the vacuum required is taken from the intake manifold.
   - From model year 2001, vehicles with a gasoline engine and automatic transmission are equipped with a brake vacuum pump, see **Brake vacuum pump -V192-, assembly**
   - Functional check

     With engine switched off, depress brake pedal firmly several times (to exhaust the vacuum in the unit).

     Depress brake pedal with average foot pressure, hold and start engine. If the servo unit is working properly, the pedal will drop slightly (servo assistance becomes effective).
   - If faulty replace complete.
10. **Seal**
11. **Bulkhead**
12. **Sealing ring**
    - Always replace
13. **Brake line**
    - Master brake cylinder/primary piston circuit to hydraulic unit
14. **Brake line**
    - Master brake cylinder/secondary piston circuit to hydraulic unit
15. **Hex bolt**
16. **Tandem brake master cylinder**
    - Cannot be repaired. If faulty, replace.

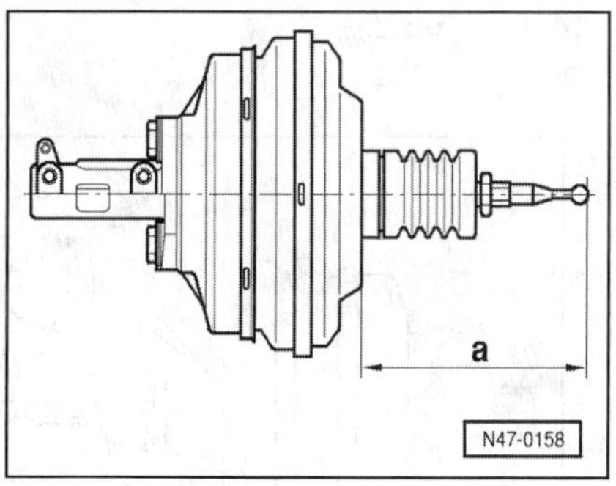

Fig. 1 Adjusting ball plunger

a = 159 mm ± 0.5 mm

NOTE—
- *When measuring the ball plunger it must be set at right angles to the surface of the brake servo.*
- *Measure to end of ball plunger without gasket installed.*

Brake servo/brake master cylinder, assembly (from 2002 m.y.)

NOTE—
Use only new brake fluid. Read information on brake fluid reservoir!

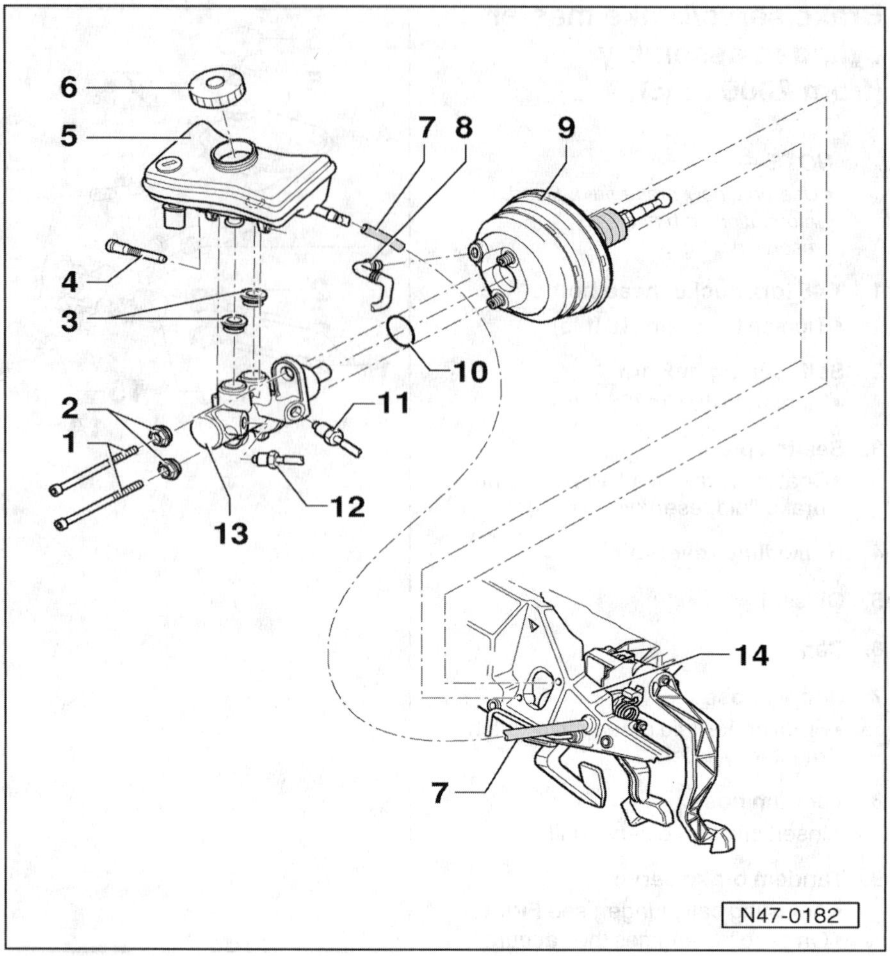

1. **T 45 torx socket head bolt**
   - Tighten to 25 Nm (18 ft-lb)
2. **Self-locking hex nut**
   - Tighten to 49 Nm (36 ft-lb)
3. **Sealing plug**
   - Coat with brake fluid and press into brake fluid reservoir
4. **Trunnion bolt**
   - Tighten to 10 Nm (7ft-lb)
5. **Brake fluid reservoir**
6. **Cap**
7. **Supply hose**
   - From brake fluid reservoir to clutch master cylinder
8. **Vacuum hose**
   - Insert into brake servo unit
9. **Tandem brake servo**
   - Adjusting ball plunger, see **Fig. 1**
   - Ball plunger to brake servo, tighten to 30 Nm (22 ft-lb)
   - On gasoline engines the vacuum required is taken from the intake manifold.
   - Vehicles with a gasoline engine and automatic transmission are equipped with a brake vacuum pump, see **Brake vacuum pump -V192-, assembly**
   - Functional check
     With engine switched off, depress brake pedal firmly several times (to exhaust the vacuum in the unit).
     Depress brake pedal with average foot pressure, hold and start engine. If the servo unit is working properly, the pedal will drop slightly (servo assistance becomes effective).
   - If faulty replace complete.
   - With bonded seal
10. **Sealing ring**
    - Always replace
11. **Brake line**
    - Tighten to 24 Nm (18 ft-lb)
    - Master brake cylinder/primary piston circuit to hydraulic unit
12. **Brake line**
    - Tighten to 24 Nm (18 ft-lb)
    - Master brake cylinder/secondary piston circuit to hydraulic unit
13. **Tandem brake master cylinder**
    - Cannot be repaired. If faulty, replace.
    - Remove and install together with brake servo.
14. **Bulkhead**

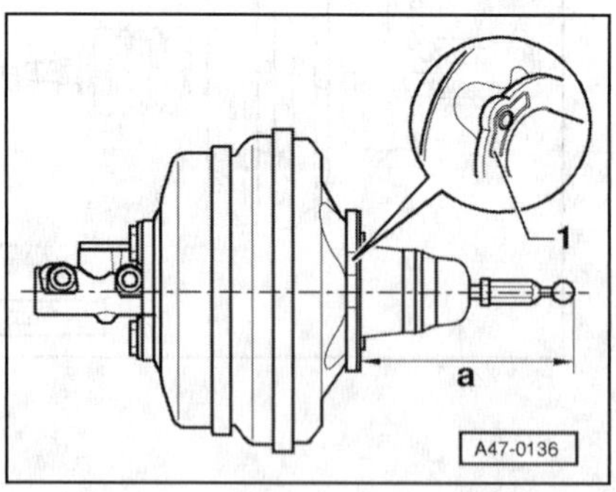

Fig. 1 Adjusting ball plunger

a = 159 mm ± 0.5 mm

NOTE—
- *Surface -1- for measuring*
- *When measuring the ball plunger it must be set at right angles to the surface of the brake servo.*
- *Measure to end of ball plunger without gasket installed.*

Brake vacuum pump -V192-, assembly

Gasoline engines with automatic transmission only

1. **Brake servo pressure sensor -G294-**
2. **Tandem brake servo**
3. **Brake servo relay -J569-**
4. **Brake vacuum pump -V192-**
5. **Vacuum hose**
6. **Check valve**
7. **Vacuum hose**
   - To engine

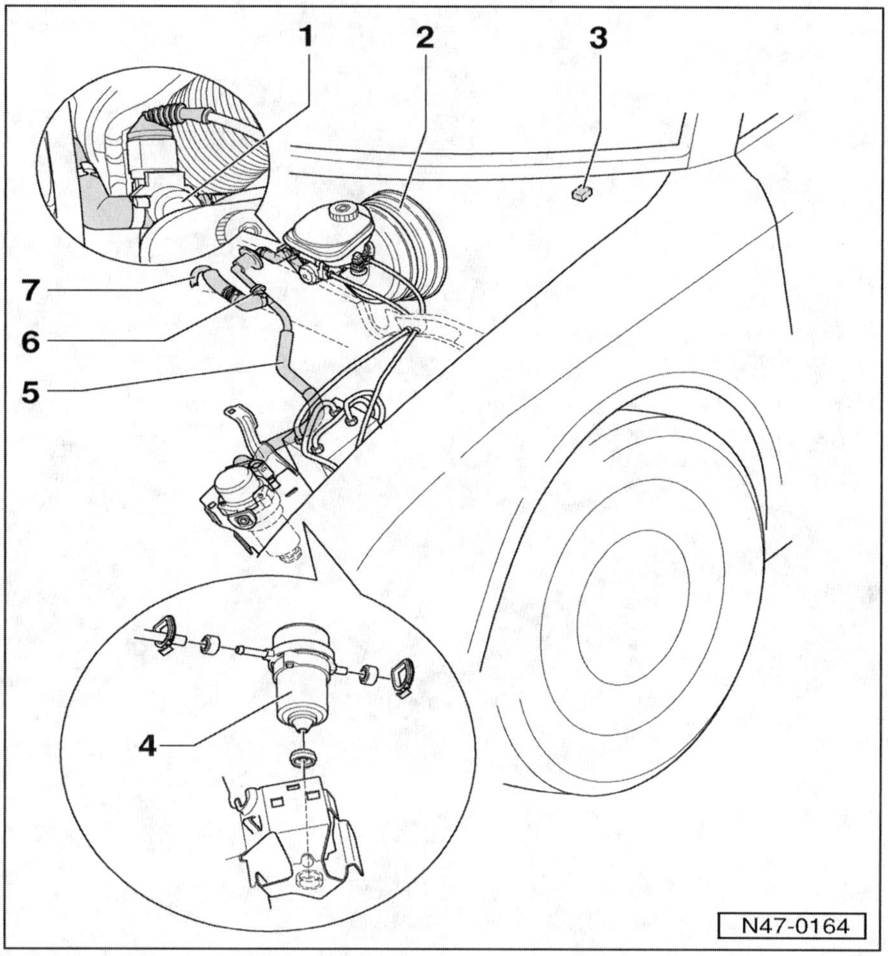

Suspension, Wheels, Steering

Section

S

00 General Information

40 Front Suspension

42 Rear Suspension

44 Wheels, Tires, Alignment — Page

48 Steering — Page

Technical data

Steering gear

Models: all except 8-cylinder engine			Front wheel drive	All wheel drive
Front axle			Independent, four-link suspension, tubular anti-roll bar, twin-tube gas shock absorbers and coil springs, tension stop springs, subframe, guide links with hydraulic damping mounts	
Rear axle			Torsion beam axle, anti-roll bar	Double wishbones, anti-roll bar
Wheelbase for Pr.-No.:[1]	1BB 1BA/BP 1BE	mm	2710 2708 2705	2700 2701 2702
Front track for Pr.-No.:[1]	1BB	mm	1489, rim ET 45	1489, rim ET 45 1499, rim ET 40 1505, rim ET 37
	1BA/1BP	mm	1495, rim ET 45	1495, rim ET 45 1505, rim ET 40 1511, rim ET 37
	1BE	mm	1498, rim ET 45	1498, rim ET 45 1508, rim ET 40 1514, rim ET 37
Rear track for Pr.-No.:[1]	1BB	mm	1500, rim ET 45	1491, rim ET 45 1501, rim ET 40 1517, rim ET 37
	1BA/1BP	mm	1500, rim ET 45	1493, rim ET 45 1503, rim ET 40 1509, rim ET 37
	1BE	mm	1500, rim ET 45	1494, rim ET 45 1504, rim ET 40 1510, rim ET 37

[1] For explanations of PR numbers, see page, see **Repair Group 44**

Models: 8-cylinder engine only			All wheel drive
Front axle			Independent, four-link suspension, tubular anti-roll bar, twin-tube gas shock absorbers and coil springs, tension stop springs, subframe, guide links with hydraulic damping mounts
Rear axle			Double wishbones anti-roll bar
Wheelbase for Pr.-No.:[1]	1BA 1BE	mm	2701 2702
Front track for Pr.-No.:[1]	1BA/1BE	mm	1507, rim ET 45 1517, rim ET 40 1523, rim ET 37
Rear track for Pr.-No.:[1]	1BA	mm	1505, rim ET 45 1515, rim ET 40 1521, rim ET 37
	1BE	mm	1507, rim ET 45 1517, rim ET 40 1523, rim ET 37

[1] For explanations of PR numbers, see page, see **Repair Group 44**

Steering

Model	All
Steering gear	Power assisted steering box
Maximum wheel lock angle	40°

40 Front Suspension

Repair Group

S 40

Front Suspension

WARNING—

- *Do not re-use any fasteners that are worn or deformed in normal use.*
- *Some fasteners are designed to be used only once, and are unreliable and may fail if used a second time. This includes, but is not limited to, nuts, bolts, washers, circlips and cotter pins. Always follow the recommendations in this manual-replace these fasteners with new parts where indicated, and any other time it is deemed necessary by inspection.*

CAUTION—

- *Do not attempt to weld or straighten the load-bearing components or parts of the suspension.*
- *Do not attempt to move a vehicle from which the drive axle has been removed. Otherwise wheel bearing damage will result. If the vehicle does have to be moved, install an outer CV joint in place of the drive axle and tighten to 50 Nm (37 ft lb).*
- *The bonded rubber bushings can only be turned to a limited extent. The bolted connections on the suspension links should therefore only be tightened when the vehicle is standing on the ground.*

EDITOR'S NOTE—

Special tools, equipment and procedures are required for most front suspension repair and component replacement. In addition, front wheel alignment is almost always disturbed when suspension components are removed or replaced.

Front suspension, assembly

NOTE —

- *If a vehicle has to be moved after removing the drive shaft, first install outer joint and tighten to 50 Nm (37 ft-lb) otherwise the wheel bearing will be damaged.*
- *It is not permitted to carry out welding and straightening operations on load bearing suspension components or those components which locate the wheels.*
- *Always replace self-locking nuts.Always replace corroded nuts/bolts.*

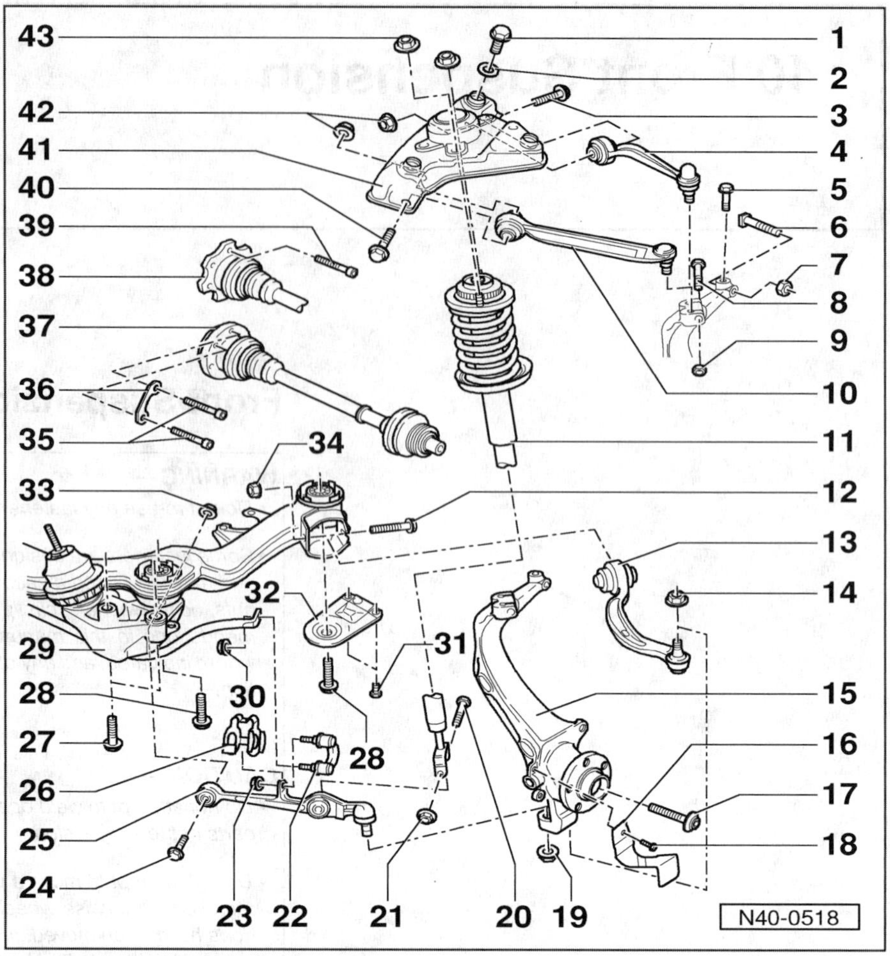

1. **Hex bolt**
   - Tighten to 75 Nm (55 ft-lb)
2. **Washer**
3. **Hex bolt**
   - Vehicles with steel mounting bracket; Bolt length M10 x 60
   - Vehicles with aluminium mounting bracket; Bolt length M10 x 62
   - Replace each time after removing
4. **Upper rear link**
   - Pressed in using VAG 3301 and 3348 assembly tools.
5. **Hex bolt**
   - Tighten to 7 Nm (62 in-lb)
   - Replace each time after removing
6. **Bolt**
7. **Self-locking nut**
   - Tighten to 50 Nm (37 ft-lb)
   - Replace each time after removing
8. **Hex bolt M 10 x 100**
9. **Self-locking nut**
   - Tighten to 40 Nm (30 ft-lb)
   - Replace each time after removing
10. **Upper front link**
    - Can only be removed together with mounting bracket
    - Removing and installing, see **Bearing bracket, removing and installing**
    - Replacing bearing, see **Mountings for upper front and rear links, replacing**
11. **Suspension strut**
12. **Hex bolt M 12 x 1.5 x 120**
    - Replace each time after removing
13. **Guide link**
    - Hydro-bushing must be replaced if leaking. Pressed in using VAG 3301, 3346, and 3347.
14. **Self-locking nut**
    - Tighten to 100 Nm (74 ft-lb)
    - Replace each time after removing
15. **Wheel bearing housing**
    - Removing and installing, see **Wheel bearing housing, removing and installing**
    - Modification, see **Wheel bearing housing/track rod ends, modified**
16. **Splash plate**
17. **Hex head bolt**
    - Changed to hex key head bolt
    - Replace each time after removing
18. **Socket head bolt**
    - Tighten to 10 Nm (7 ft-lb)
19. **Self-locking nut**
    - Tighten to 100 Nm (74 ft-lb)
    - Replace each time after removing
20. **Hex bolt M 12 x 1.5 x 85**
21. **Self-locking nut**
    - Tighten to 90 Nm (66 ft-lb)
    - Replace each time after removing
22. **Connecting link**
    - Arrow on connecting link points forward
    - Modification, see **Coupling rod for anti-roll bar, modified**
23. **Self-locking nut**
    - 40 Nm (30 ft-lb) and turn 90° further
    - Replace each time after removing
    - Use self-locking nut or nut with grooves on underside depending on type
24. **Hex bolt M 12 x 1.5 x 100**
    - Replace each time after removing
25. **Track control link**
    - Removing and installing, see **Track-control link, removing and installing**
    - Bearing pressed in using VAG 3301, 3346, and 3350.

(Continued on following page)

Front suspension, assembly (continued)

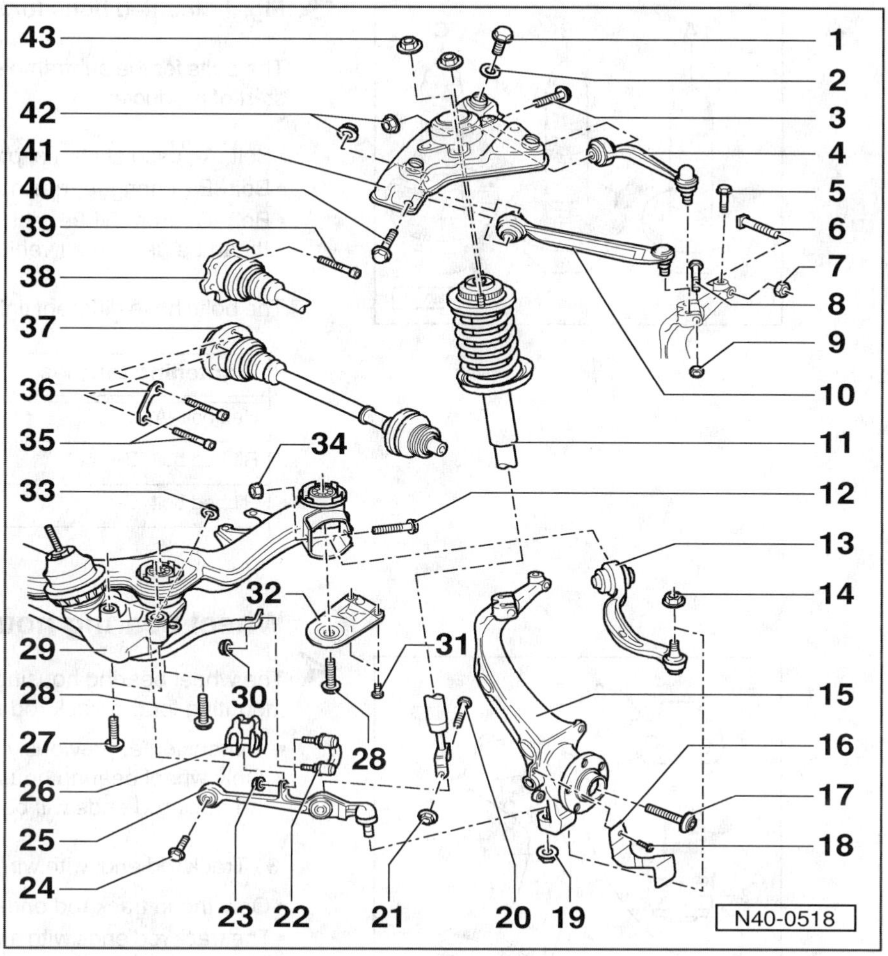

26. Clamp
- Replace each time after removing

27. Hex bolt
- Tighten to 75 Nm (55 ft-lb)
- M10 x 70
- Replace each time after removing
- If the welded nut threads in the cross member are damaged, the thread can be serviced with Heli-coil thread insert.

28. Hex bolt M 12 x 1.5 x 110
- 110 Nm (81 ft-lb) and turn 90° further
- Replace each time after removing
- If the welded nut threads in the cross member are damaged, the thread can be serviced with Heli-coil thread insert.

29. Subframe
- Removing and installing, see **Subframe, removing and installing**

30. Self-locking nut
- 40 Nm (30 ft-lb) and turn 90° further
- Replace each time after removing
- Modification, see **Coupling rod for anti-roll bar, modified**

31. Hex bolt
- M8 x 25
- Has been changed, see **Fig. 1**
- If the welded nut threads in the cross member are damaged, the thread can be serviced with Heli-coil thread insert.

32. Sub-frame support

33. Self-locking nut
- 80 Nm (59 ft-lb) and turn 90° further
- Replace each time after removing

34. Self-locking nut
- 80 Nm (59 ft-lb) and turn 90° further
- Replace each time after removing

35. Multi-point socket head bolt
- See **Drive shaft with constant velocity joint, servicing, -15-**

36. Backing plate

37. Drive shaft

38. Drive shaft with tripod roller joint

39. Multi-point socket head bolt
- See **Drive shaft with triple roller joint, repairing, -22-**

40. Hex bolt
- Vehicles with steel mounting bracket; Bolt length M10 x 60
- Vehicles with aluminium mounting bracket; Bolt length M10 x 62
- Replace each time after removing

41. Mounting bracket
- Material changed from steel to aluminium.
- Installation of different types of mounting brackets is not permitted.
- Removing and installing, see **Bearing bracket, removing and installing**

42. Self-locking nut
- 50 Nm (37 ft-lb) and turn 90° further
- Replace each time after removing

43. Self-locking nut
- Tighten to 22 Nm (16 ft-lb)
- Replace each time after removing

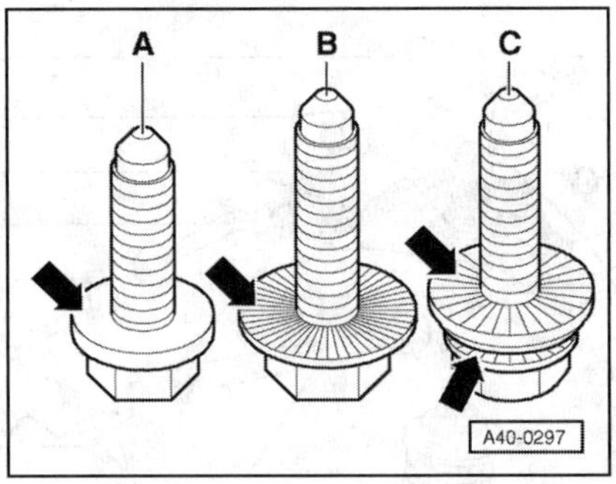

Fig. 1 Modified bolts for subframe support

The bolts for the subframe support have been changed three times since start of production.

- Bolt -A-, used at start of production, has no ribs on underside of washer.
- Bolt -B-, changed, ribs on underside -arrow-.
- Bolt -C-, from 04.99, ribs on underside of bolt head and washer. It can be used in all vehicles.

The bolts have different tightening torques.

Tightening torques	
Hex bolt -A-	25 Nm (18 ft-lb)
Ribbed bolt -B-	75 Nm (55 ft-lb)
Ribbed bolt -C-	30 Nm (22 ft-lb)

Wheel bearing housing/track rod ends, modified

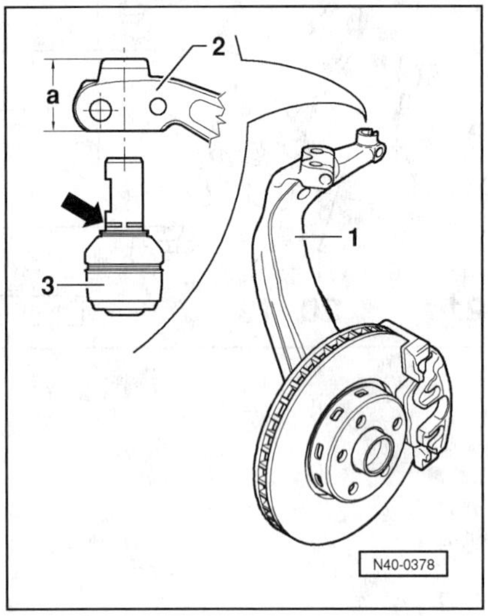

The wheel bearing housing -1- has been modified in the area of the mounting for the track rod ends -2-.

- Dimension a, previously: 32 mm, new: 30 mm
- Only wheel bearing housings with the modified mounting (30 mm) for the track rod ends will be supplied as a replacement part.

3 - Track rod end, with wire ring -arrow-

- Only these track rod ends will be supplied as a replacement part.
- The track rod ends with a wire ring can also be used with wheel bearing housing (dimension a: 32 mm).
- The toe adjustment range is reduced by approx. 2 mm with this combination.
- If the toe adjustment range for the track curve is not sufficient, the wire -arrow- ring can be removed.

Coupling rod for anti-roll bar, modified

The ball joint in the coupling has been changed to a rubber mounting, and the track-control link was also slightly modified.

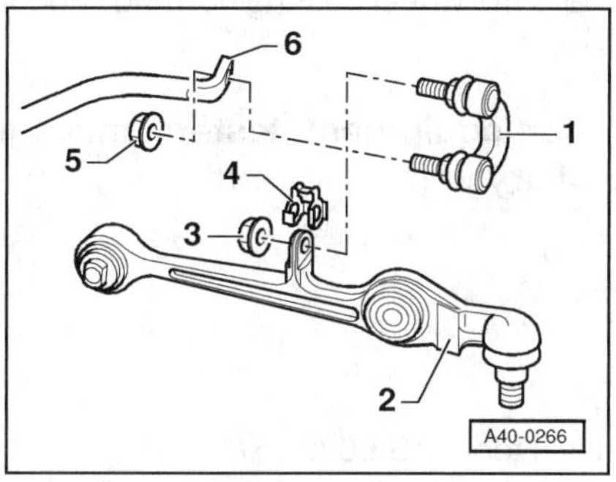

Coupling rod with ball joint

1 - Coupling with ball joint: no longer installed on vehicles beginning with vehicle identity no. 3-B-X P-213 727.
2 - Track-control link: was modified. Only the modified track-control link is still supplied as a spare part. This track-control link can also be installed in vehicles produced prior to VIN# 3-B-X P-213 727. In this case, the clamp -4- must also be used.
3 - Self-locking hex nut; 40 Nm (30 ft-lb).
4 - Clamp: Is not installed in vehicles beginning with VIN# 3-B-X P-213 727.
5 - Self-locking hex nut; 100 Nm (75 ft-lb)
6 - Anti-roll bar

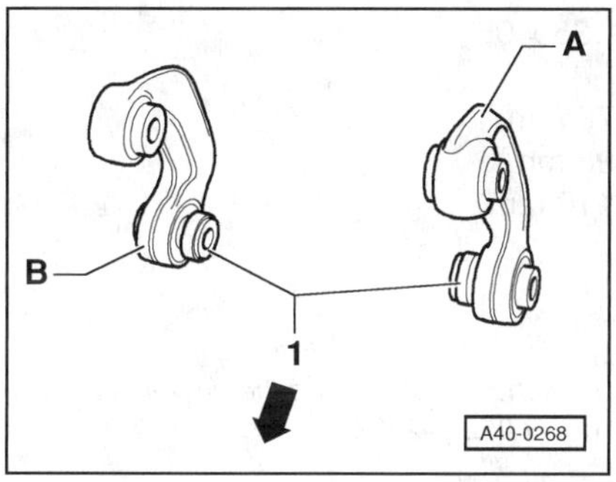

Coupling rod with bonded rubber mounting

- Can be installed in vehicles up to but not including VIN# 3-B-X P-213 727. In this case, the clip -4- from diagram A 40- 0266 is no longer installed.

Coupling -A- for left side of vehicle:

- The arrow on the coupling points in direction of travel.
- The permanently mounted shim -1- on the coupling must be bolted to the anti-roll bar.

Coupling -B- for the right side of vehicle:

- The arrow on the coupling points in direction of travel.
- The arrow in the figure symbolizes the direction of travel.

CAUTION—

Bonded rubber mountings can only be twisted to a limited degree. Therefore, you should only tighten the bolted connections of the suspension links when the vehicle is standing on its wheels.

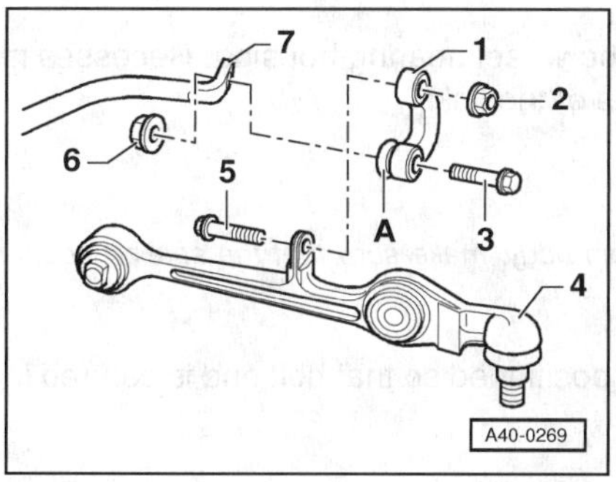

1 - Coupling with rubber mounting, rubber mounting with attached shim -A- must be bolted to anti-roll bar.
2 - Self-locking hex nut, 40 Nm (30 ft-lb) and turn 90° further
3 - Hex bolt
4 - Track-control link
5 - Hex bolt
6 - Self-locking hex nut, 40 Nm (30 ft-lb) and turn 90° further
7 - Anti-roll bar

Separating connections between upper links and wheel bearing housing

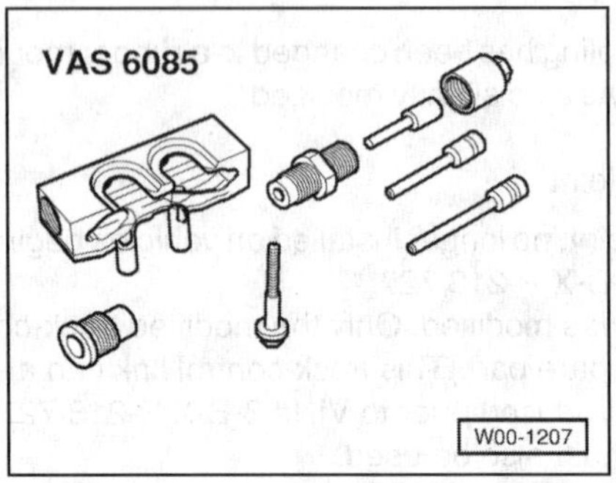

This procedure is only necessary if the hex bolt for the connections between the upper links and the wheel bearing housing cannot be removed.

Special tools, workshop equipment, testers, measuring instruments and auxiliary items required

- VAG 1332 Torque wrench
- VAS 6085 Removal tool

Removing

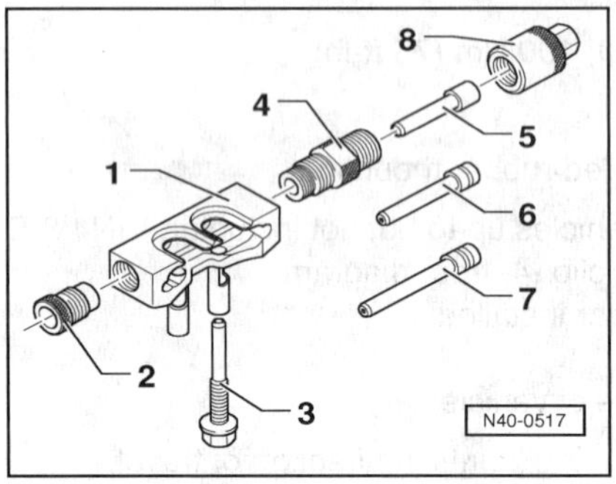

Overview of part of removal tool VAS 6085

1 - Main body
2 - Adapter
3 - Thrust bolt M10 x 1.25 x 95
4 - Threaded sleeve
5 - Thrust rod (length 78 mm)
6 - Thrust rod (length 98 mm)
7 - Thrust rod (length 118 mm)
8 - Nut

NOTE—
Before working, the thrust bolt, the thrust rod and the nut must be lubricated with polyurea grease part no. G 052 142 A2.

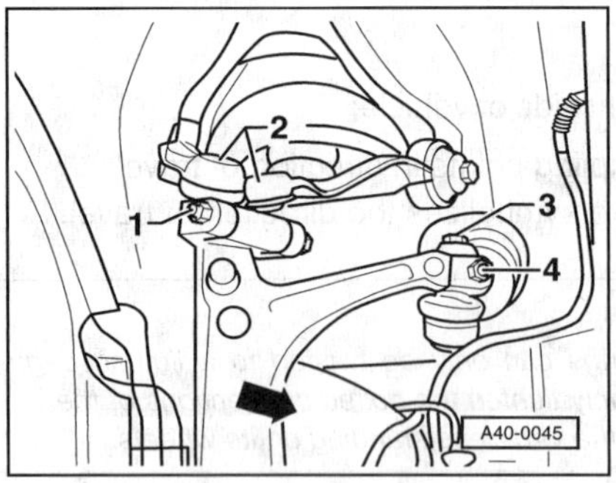

– Remove wheel.

– Disconnect ABS speed sensor wiring out from retainer on brake caliper.

Remove nut -1-.

- Do not loosen bolts -3- and -4-.

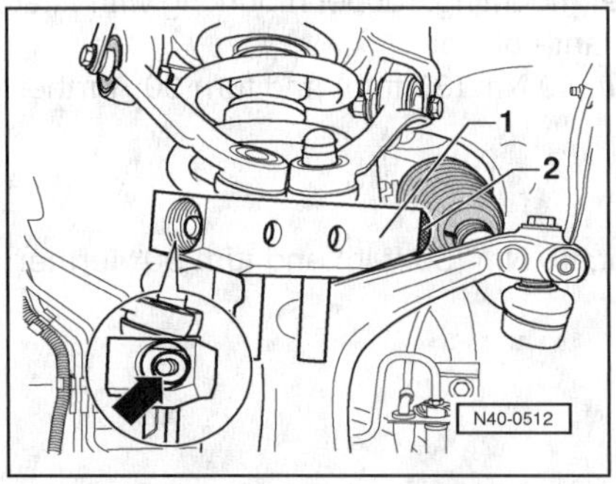

Set main body -1- on the wheel bearing housing. Recesses must engage under the joints of the links.

NOTE—
When positioning the main body, make sure that the sealing boots of the links are not damaged.

– Main body -1- must be positioned so that bolt end is centred in drilling of main body.

– Screw adapter -2- into main body -1- to stop so that the main body is locked in position.

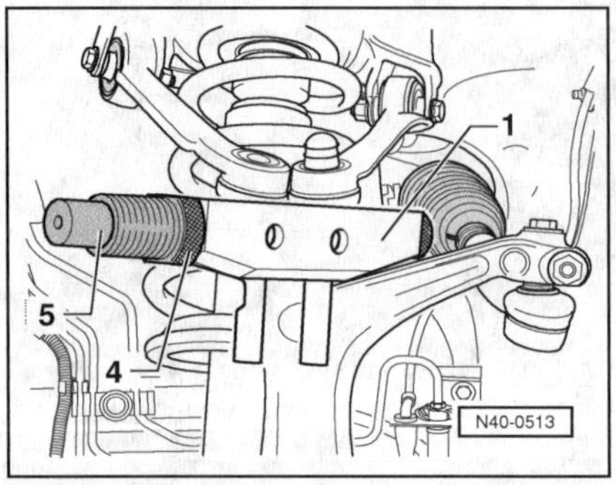

Screw threaded sleeve -4- into main body -1- to stop.

– Insert thrust rod -5- in threaded sleeve -4-.

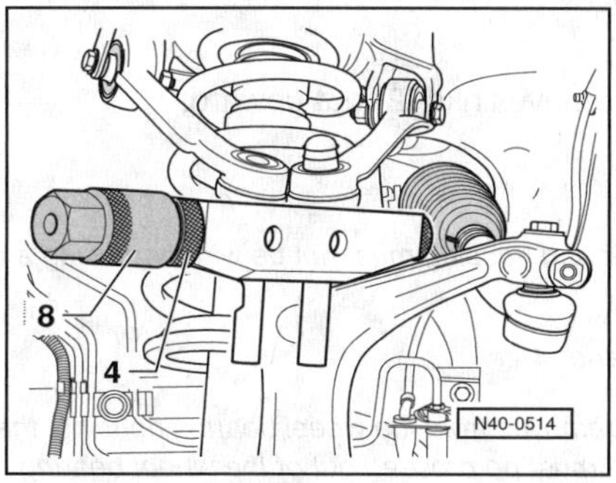

Screw nut -8- onto threaded sleeve -4- and turn to stop so that bolt is pressed out.

NOTE —
The nut -8- may be turned with a wrench or impact driver.

– Unscrew nut -8- from threaded sleeve -4-, remove 78 mm thrust rod and insert 98 mm thrust rod in threaded sleeve -4-.

– Repeat pressing procedure.

– If bolt still cannot be removed by hand, repeat pressing procedure with 118 mm thrust rod .

– Unscrew nut -8- from threaded sleeve -4-, remove thrust rod and bolt.

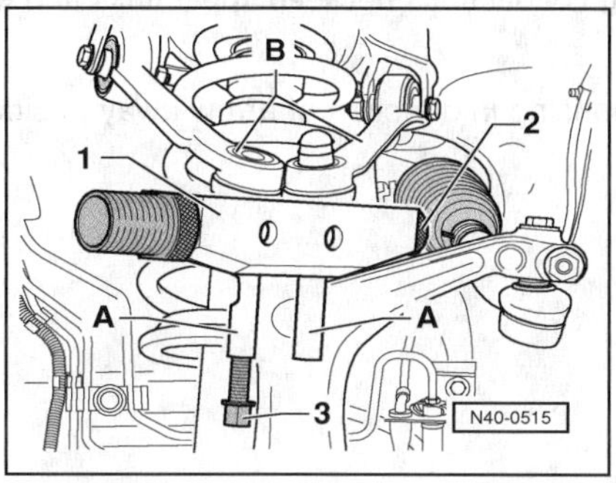

Screw thrust bolt -3- into one guide after the other -A- and press links -B- out of wheel bearing housing.

– Unscrew thrust bolt -3- out of guide -A- and loosen adapter -2-.

– Remove main body -1- from wheel bearing housing.

Suspension strut, removing and installing

Special tools, workshop equipment, testers, measuring instruments and auxiliary items required

• VAG 1331 Torque wrench
• VAG 1332 Torque wrench

Removing

– Remove wheel.

– Remove plenum chamber cover, see **Electrical Equipment section, Repair Group 27.**

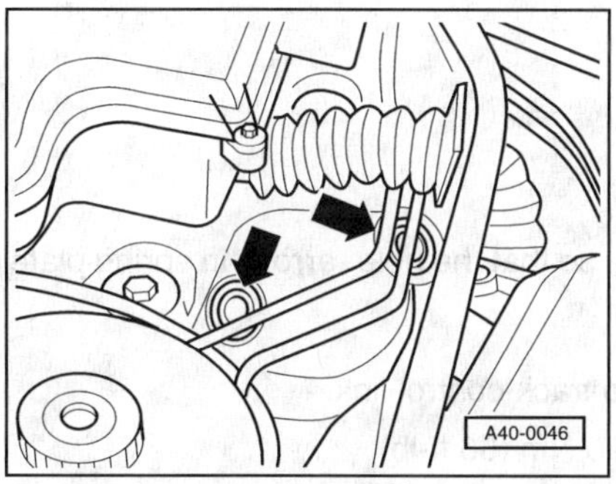

Take out rubber grommets in plenum chamber -arrows-.

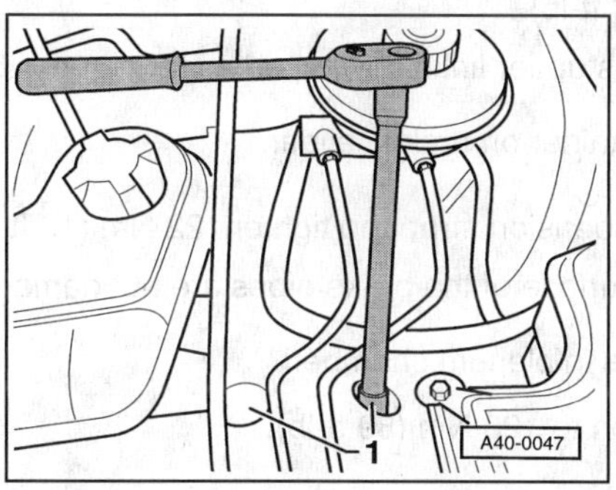

Unbolt suspension strut at body.
1 - Nuts

– Disconnect ABS speed sensor wiring out from retainer on brake caliper.

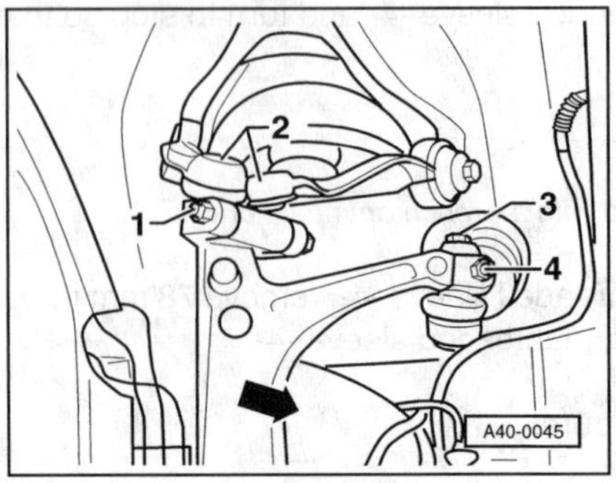

Remove nut -1-.

– Take out hex bolt and remove links -2- out upward.

NOTE —
- *The slits in the wheel bearing housing must not be widened using a chisel or similar tool!*
- *Do not loosen bolts -3- and -4-.*
- *If the hex bolt cannot be removed from the wheel bearing housing, then the hex bolt and the links -2- must be pressed out of the wheel bearing housing, see* **Separating connections between upper links and wheel bearing housing.**

– Swing wheel bearing housing in direction of arrow away to side.

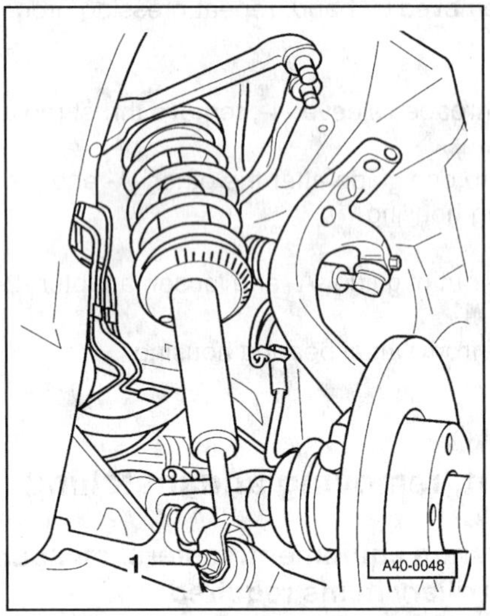

Remove nut -1-.

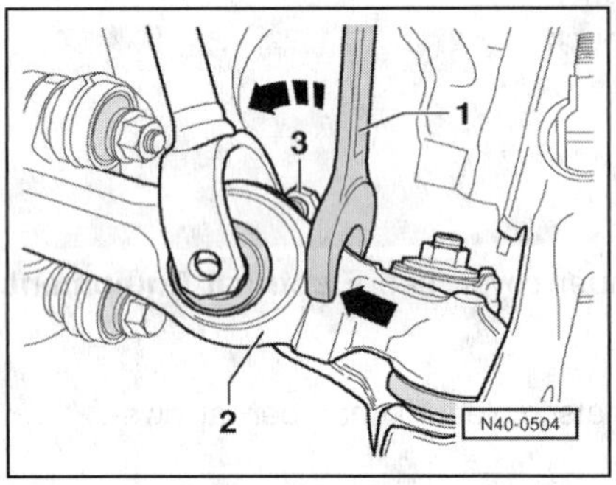

To facilitate removal of suspension strut, apply an open-jaw wrench -1- to wrench surfaces -arrow- of track-control link -2-.

– Turn wrench in direction of arrow (direction of travel) and remove bolt -3-.

– Take out suspension strut.

Installing

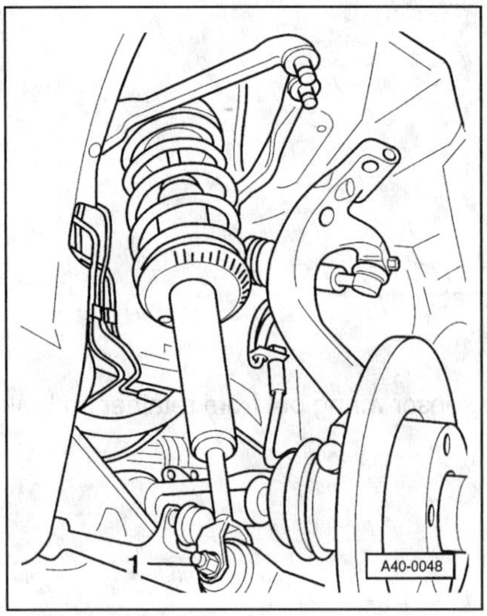

Insert suspension strut, so that the hole -arrow- in spring plate points to center of vehicle.

– Bolt suspension strut to track-control link.
  - Tighten new nut -1-, 90 Nm (66 ft-lb).

– Insert upper links into wheel bearing housing, insert new bolt and tighten new nut -2-, 50 Nm (37 ft-lb).
  - When tightening press upper links downward as far as possible!

– Insert ABS wiring in retainer on brake caliper.

– Install new nuts for suspension strut and tighten, 22 Nm (16 ft-lb).
  - Make sure that the surface of the brake pipes are not damaged!

– Insert rubber grommets in plenum chamber.

– Install wheel and tighten to 120 Nm (89 ft-lb).

Suspension strut, assembly

1. **Hex nut**
   - Tighten to 50 Nm (37 ft-lb)
2. **Suspension strut mounting**
3. **Washer**
4. **Upper spring plate**
5. **Coil spring**
   - Outer surface of spring must not be damaged
   - Observe color coding
   - The coil spring allocation for vehicles from vehicle identification No. 3B - WE 113 562 is performed via PR numbers. These numbers are indicated on the vehicle data plate/sticker.
6. **Buffer stop**
7. **Protective sleeve**
8. **Protective cap**
9. **Lower support**
   - Located by cut-out in ribbing on spring plate
10. **Lower spring plate**
11. **Shock absorber**

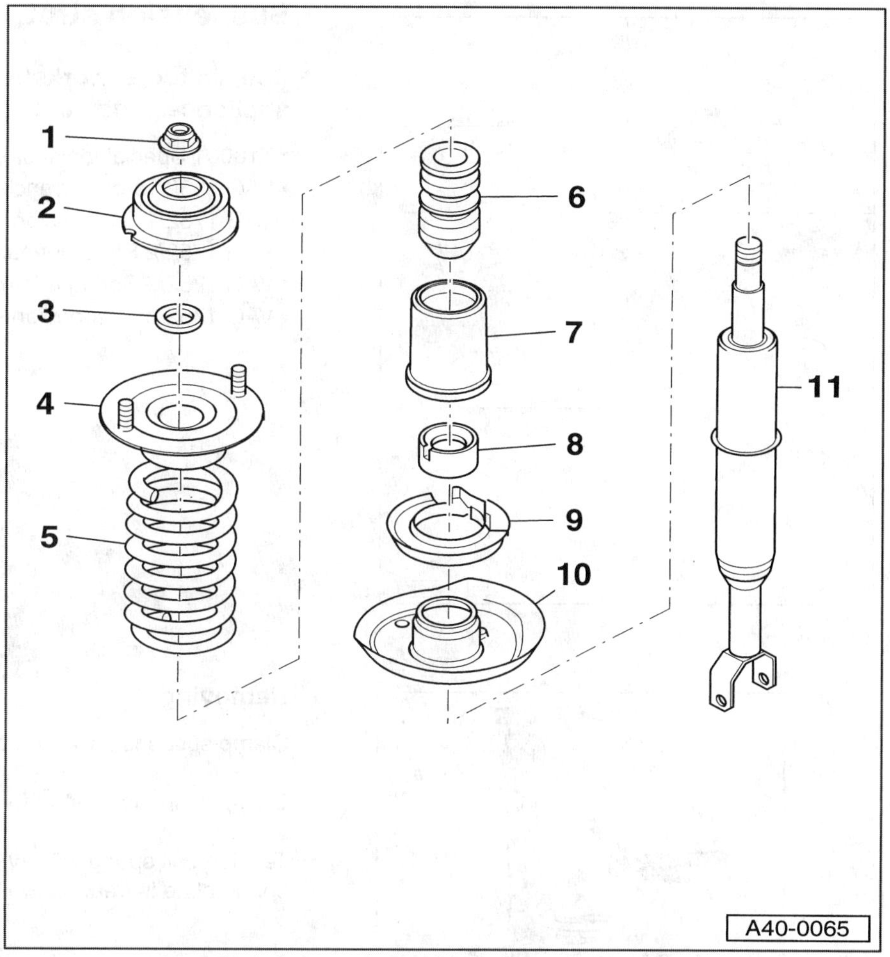

Fig. 1 Vehicle data sticker

The vehicle data sticker is located in the spare wheel recess and in service booklet.

The figure (arrows) gives information on coil springs and shock absorbers installed in vehicle.

The part number of the relevant component is also coded in the Production code.

The following table explains the Production code. This is relevant for the coil springs/shock absorber allocation for the vehicle.

	Production code	Component
1	0JH	Coil spring, front
2	0YF	Coil spring, rear
3	1BA	Standard shock absorber front and rear

EDITOR'S NOTE —

Always consult with your Volkswagen Parts Department or aftermarket parts specialist for the latest replacement parts information

Suspension strut, disassembly

Special tools, workshop equipment, test and measuring appliances and aux. items required

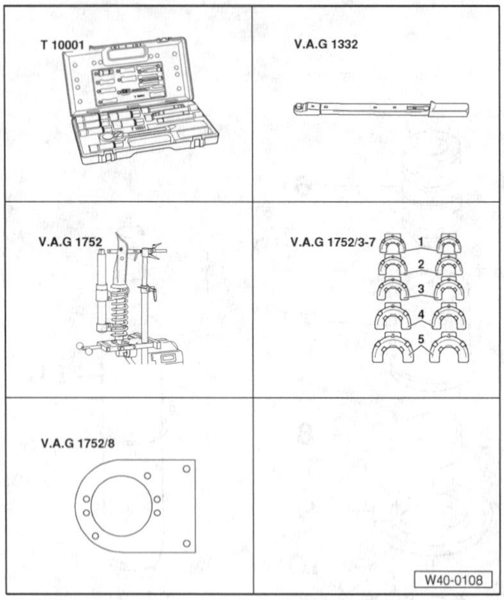

- T10001 Special tools for struts/shock absorbers
- VAG 1332 Torque wrench
- VAG 1752/1 Suspension strut clamp
- VAG 1752/2 Strut mounting
- VAG 1752/7 Spring retainer
- VAG 1752/8 Suspension strut alignment gauge

Removing

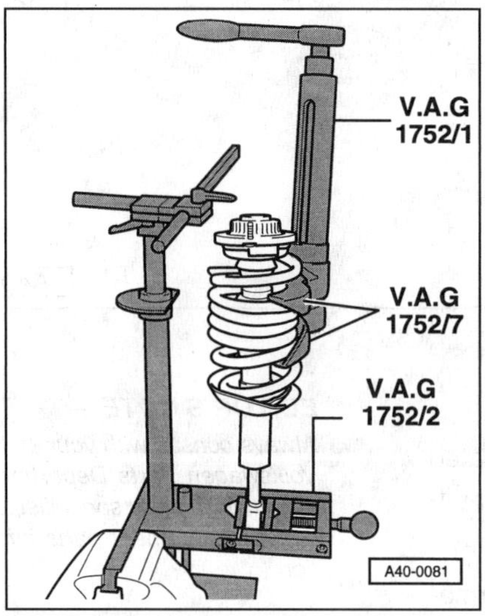

- Clamp suspension strut mounting VAG 1752/2 in a vice.
- Clamp suspension strut forked head in suspension strut mounting.
- Tension coil spring with -VAG 1752/1- tensioning device until the upper spring plate is free.

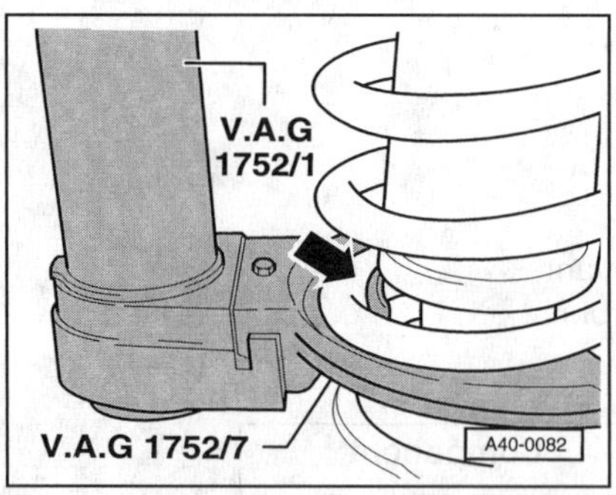

- Make sure that the coil spring is seated correctly in adapter VAG 1752/7 -arrow-.

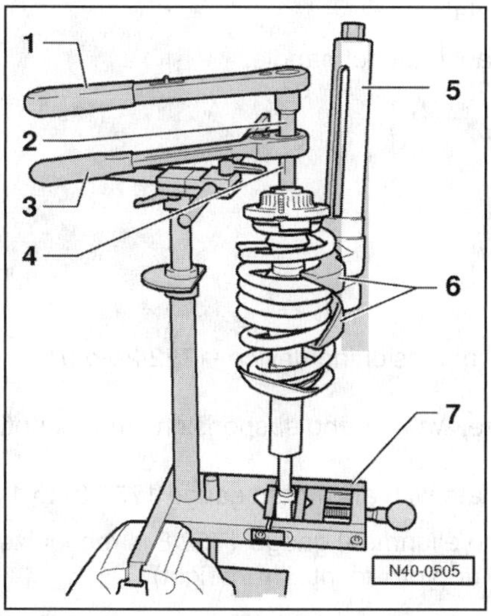

Unscrew hex nut from piston rod.

– Take off individual parts of suspension strut and coil spring with tensioning device -VAG 1752/1-.

1 - Commercially available ratchet handle
2 - T 10001/8
3 - T 10001/11
4 - T 10001/5
5 - VAG 1752/1
6 - VAG 1752/7, 1752/2

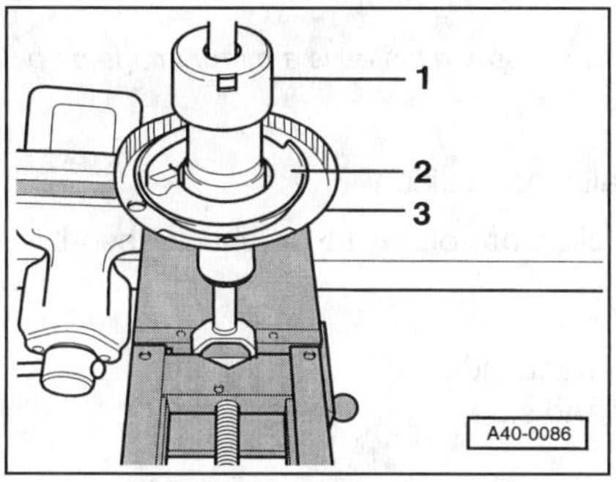

Replacing shock absorber

If the shock absorber requires replacing then additional work is necessary.

Take off protective cap -1- and lower spring support -2-.

– Loosen spring plate -3- with a synthetic head hammer and remove.

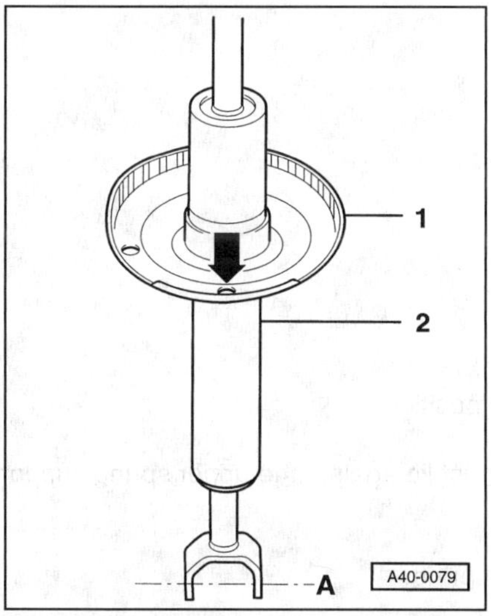

Installing

Before inserting coil spring make sure that the lower spring plate is in the correct installation position.

• The hole -arrow- in spring plate -1- should be offset 90° to shock absorber -2- bolt axis -A-.
• Permissible deviation of spring plate to shock absorber bolt axis; ± 2°.

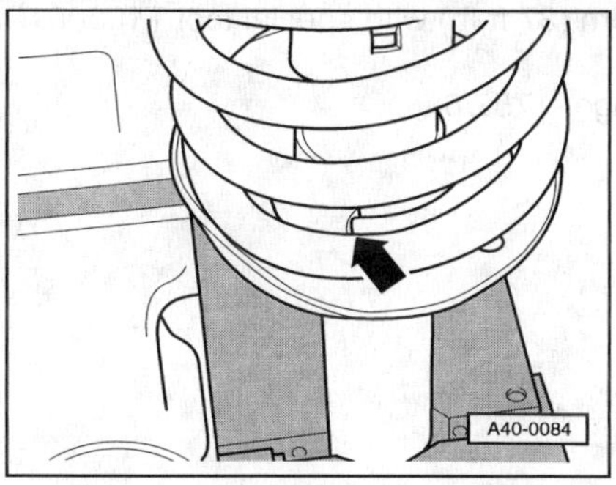

Install lower spring support, protective cap and stop buffer.

– Install coil spring with tensioning device -VAG 1752/1- onto lower spring support.

• The end of the coil spring must lie against stop -arrow-.

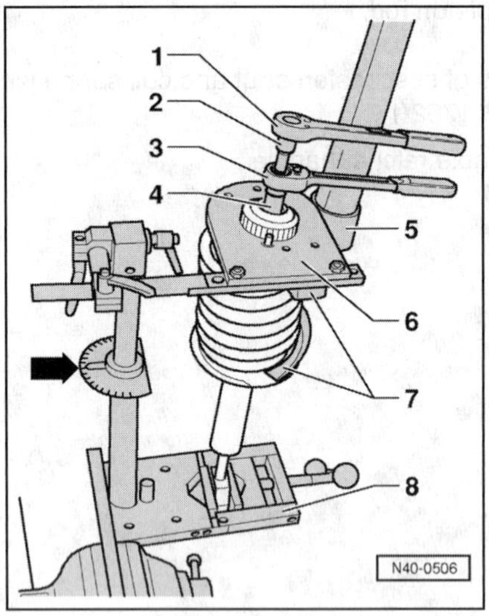

Installing upper spring plate

1 - Commercially available ratchet handle
2 - T 10001/8
3 - T 10001/11
4 - T 10001/5
5 - VAG 1752/1
6 - VAG 1752/8
7 - VAG 1752/7
8 - VAG 1752/2

– Set angle scale -arrow- of tensioning device -1752/2- to 0°.

– Install upper spring plate, washer and suspension strut mounting.

– Position upper spring plate with alignment gauge 1752/8- to 11°.
  - The 11° position on the alignment gauge 1752/8 is recognized with "vorn links (front left)" or "vorn rechts (front right)".

NOTE —
Note that the 11° position of the spring plates is a mirror image on both the left and right-hand side

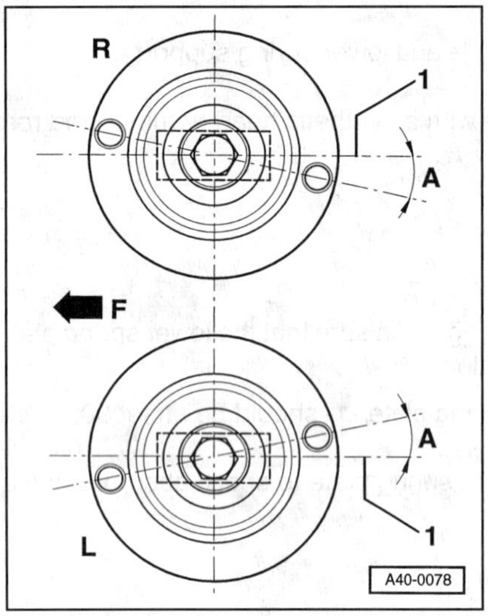

Upper spring plate installation position

- Studs in upper spring plate are offset 11° to shock absorber forked head bolt axis -1-.
- F = Forward
- R = Spring plate, right-hand side
- L = Spring plate, left-hand side
- A = 11° + 2°

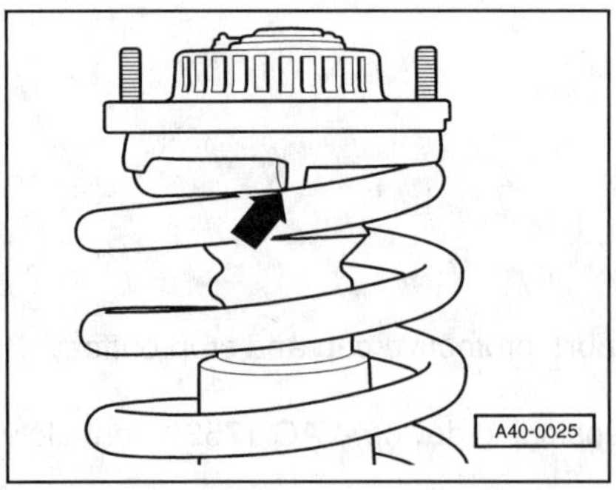

Rubber ring installation position

– The end of the spring must lie against the upper spring support stop -arrow-.

– Relieve tension on coil spring.

– Tighten hex nut to 50 Nm (37 ft-lb) with special tool T100001.

– Remove alignment gauge -1752/8-.

Wheel bearing housing, removing and installing

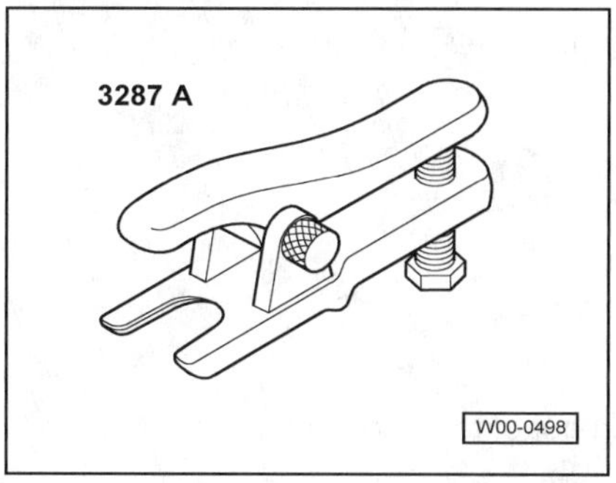

Special tools, workshop equipment, testers, measuring instruments and auxiliary items required

- VAG 1332 Torque wrench
- 3287A Ball joint puller

Removing

– Unbolt hex bolt for drive shaft. (Loosen only when vehicle is standing on wheels -danger of accident-).

– Remove wheel.

– Disconnect ABS speed sensor wiring out of retainer on brake caliper.

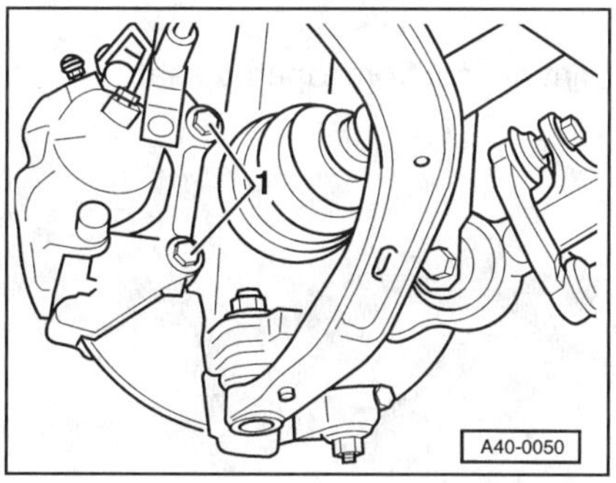

– Remove bolts -1- for brake caliper and take off brake caliper.

– Remove disc brake.

– Secure brake caliper to body with wire.

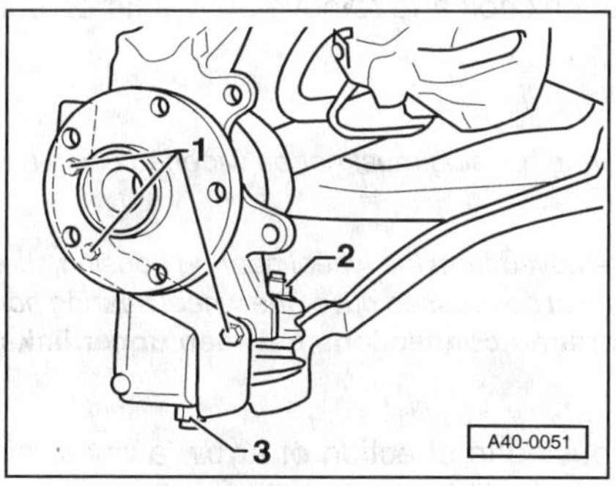

– Remove bolts -1- for cover plate.

– Disconnect ABS speed sensor from wheel bearing housing.

– Remove nuts -2- and -3-.

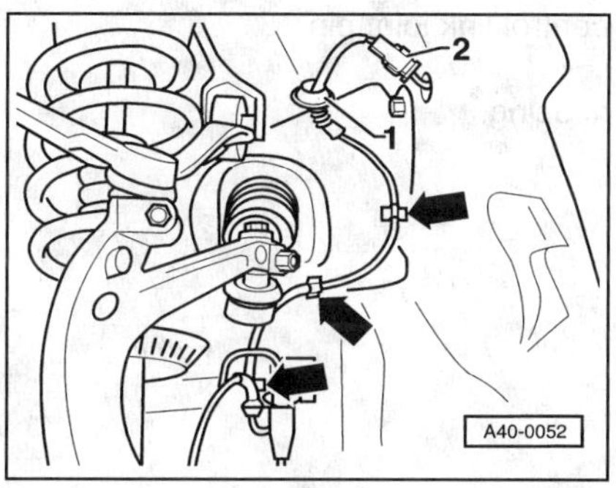

– Remove rubber grommet -1- and separate connector -2-.

– Disconnect ABS speed sensor wiring from retainers -arrows-.

– Guide wiring through opening in wheel bearing housing and take out.

NOTE —
Do not damage the rubber grommet!

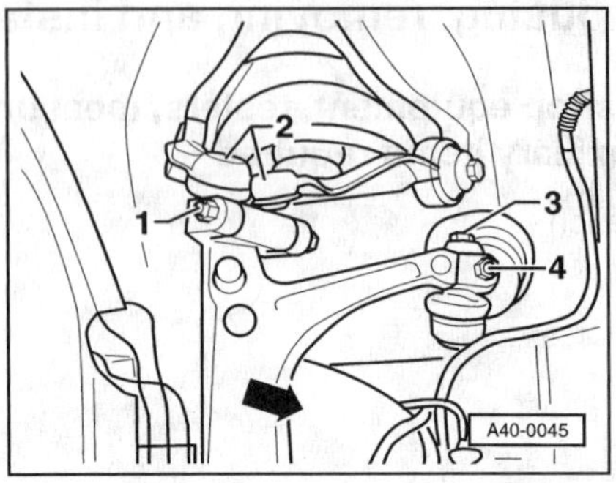

◄ Remove hex bolt -3- and nut -4-.

– Remove track rod ball joint.

– Remove nut from joint pin.

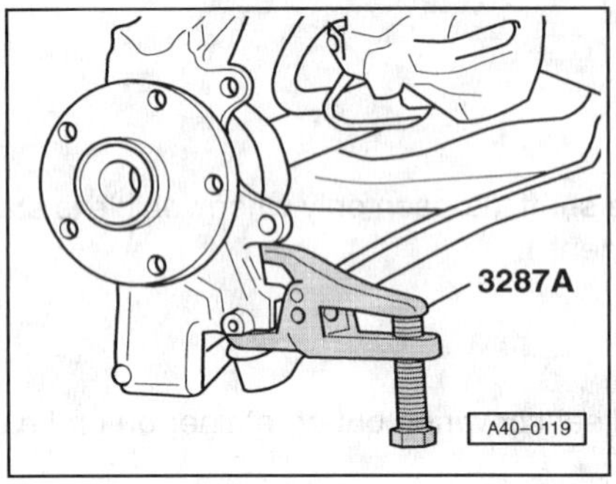

◄ Press guide link joint pin out of tapered seat.

NOTE —
Do not damage joint protective boot when doing this!

– Loosen joint pin nut, do not remove completely.

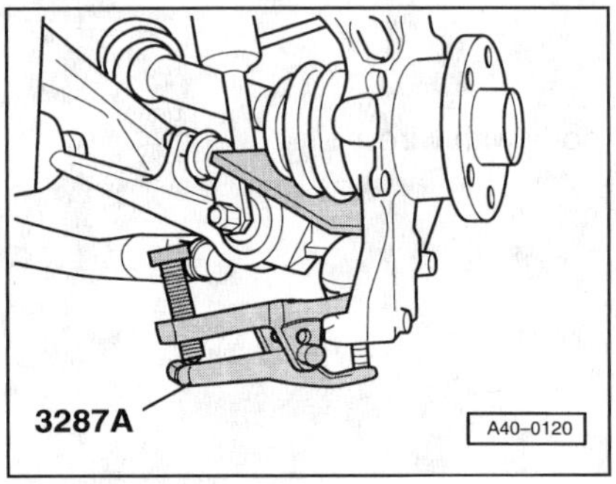

◄ Press track-control link joint pin out from tapered seat.

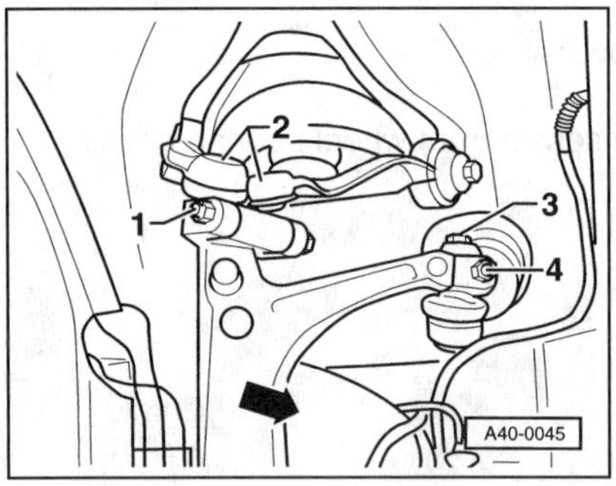

◄ Remove nut -1-, take out hex bolt and remove both links -2- out upward.

NOTE —

- *The slits in the wheel bearing housing must not be widened using a chisel or similar tool!*
- *If the hex bolt cannot be removed from the wheel bearing housing, then the hex bolt and the links -2- must be pressed out of the wheel bearing housing see from page, see* **Separating connections between upper links and wheel bearing housing**.

– Swing wheel bearing housing in direction of arrow away to side, when doing this remove drive shaft pin from wheel hub.

– Remove nut from track-control link joint pin.

– Take off wheel bearing housing.

Wheel bearing housing, installing

– Install wheel bearing housing.

– Install joint pins of track-control and guide links into wheel bearing housing.

– Screw on new self-locking nut and tighten to 100 Nm (74 ft-lb).
  - Counter hold joint pin with 4 mm AF hex key if necessary.

– Slide drive shaft outer joint pin into wheel hub and tighten new hex bolt hand tight.

– Insert upper links into wheel bearing housing, insert new bolt and tighten new nut to 40 Nm (30 ft-lb).
  - When tightening press upper links downward as far as possible!

– Install track rod.
  - Tighten hex bolt to 7 Nm (62 in-lb).
  - Screw on new self-locking nut and tighten to 45 Nm (33 ft-lb).

– Connect ABS vehicle speed sensor.

– Screw on cover plate and tighten to 10 Nm (7 ft-lb).

– Install disc brake, screw on brake caliper, see **Brake System section, Repair Group 46.**

– Install wheel and tighten to 120 Nm (89 ft-lb).

– Tighten drive shaft hex bolt, see **Removing and installing drive axles**.

– Check and if necessary adjust the front axle alignment.

Wheel bearing housing, servicing

Special tools required

- VW 401 Thrust plate
- VW 402 Thrust plate
- VW 407 Press tool
- VW 412 Press tool
- VW 416 B Tube
- VW 432 Press piece
- VW 447 i Thrust plate
- VW 519 Tube
- 30-100 Drift sleeve
- 3005 Press disk
- 3124 Press piece
- 3291 Assembly tool
- 3345 Wheel bearing tube
- Kukko 17/2 Separating device

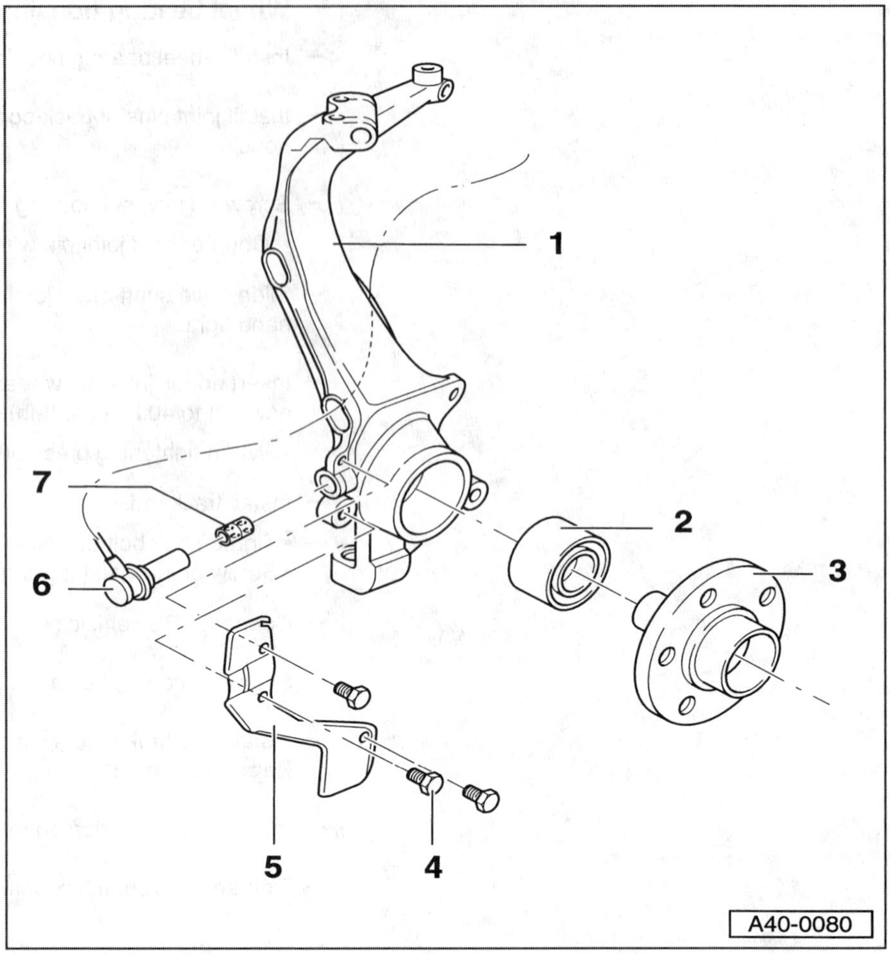

1. **Wheel bearing housing**
2. **Wheel bearing**
   - Installation position: Larger internal diameter of wheel bearing faces wheel hub
   - Pressing out diameter 75 mm, see **Fig. 3**
   - Pressing out diameter 82 mm, see **Fig. 4**
   - Pressing bearing inner race off, see Fig. **5** and **6**
   - Pressing in diameter 75 mm, see **Fig. 7**
   - Pressing in diameter 82 mm, see **Fig. 8**
3. **Hub**
   - Pressing out diameter 75 mm, see **Fig. 1**
   - Pressing out diameter 82 mm, see **Fig. 2**
   - Pressing in diameter 75 mm, see **Fig. 9**
   - Pressing in diameter 82 mm, see **Fig. 10**
4. **Bolt**
   - See **Front suspension, servicing**, -18-
5. **Splash plate**
6. **Speed sensor**
   - Disconnect out to remove
   - Before inserting sensor, clean inner surface of installation hole and coat with grease G 000 650
7. **Clamping sleeve**
   - Before inserting into wheel bearing housing, coat circumference with grease G 000 650
   - Press into wheel bearing housing onto stop

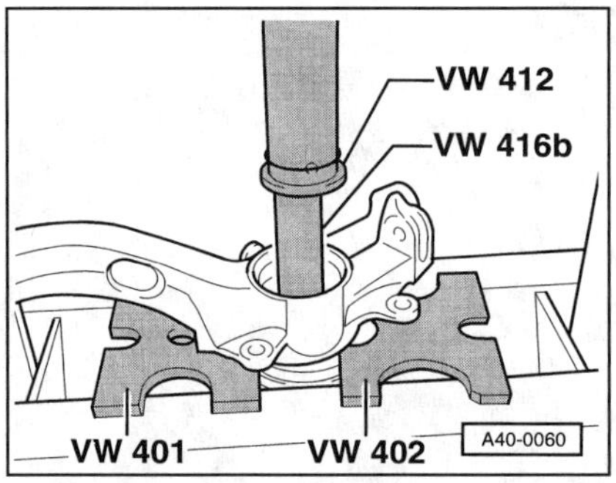

Fig. 1 Pressing wheel hub out of 75 mm diameter wheel bearing housing

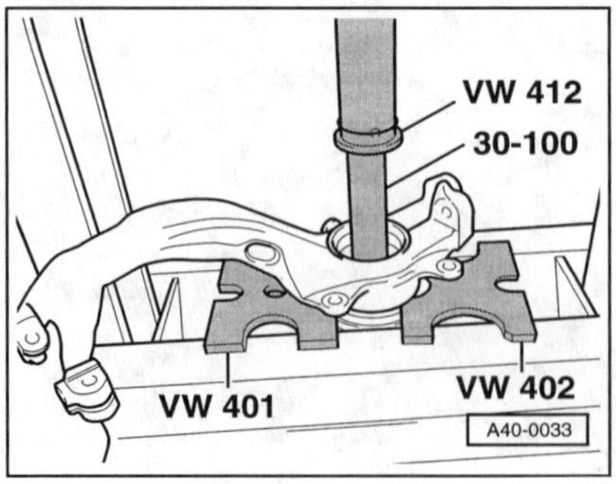

Fig. 2 Pressing wheel hub out of 82 mm diameter wheel bearing housing

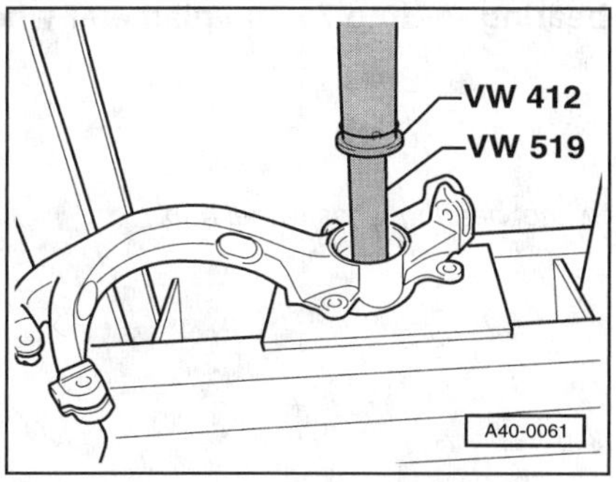

Fig. 3 Pressing wheel bearing out of 75 mm diameter wheel bearing housing

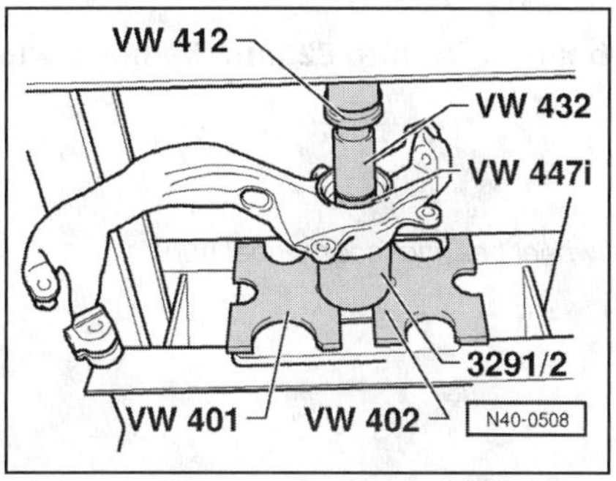

Fig. 4 Pressing wheel bearing out of 82 mm diameter wheel bearing housing

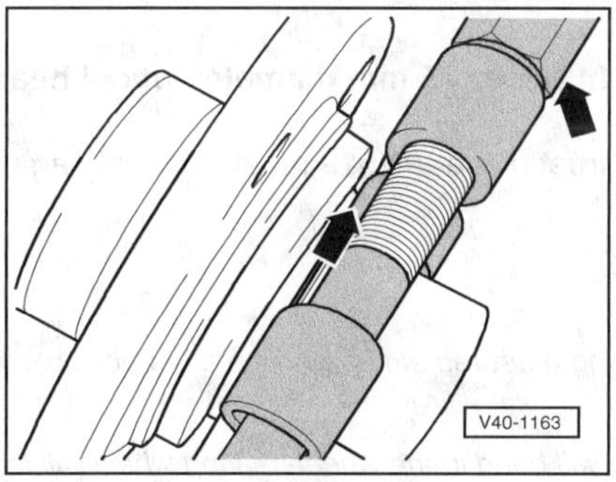

Fig. 5 Installing separating device

Install separating device into bearing inner race annular groove -arrow- and tension spindle.

NOTE —

Use commercially available separating device e.g. Kukko 17/2.

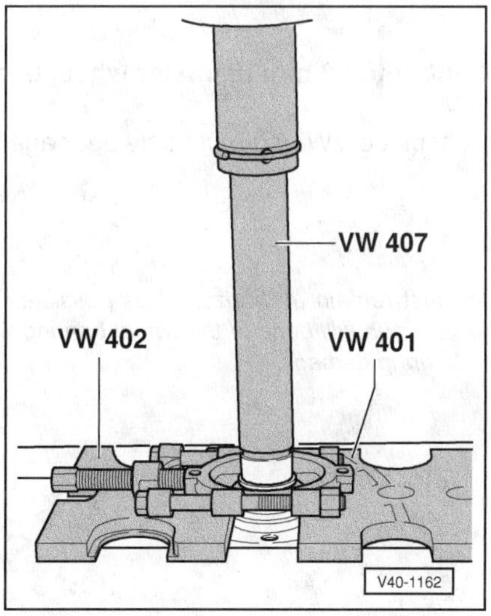

Fig. 6 Pressing bearing inner race off wheel hub

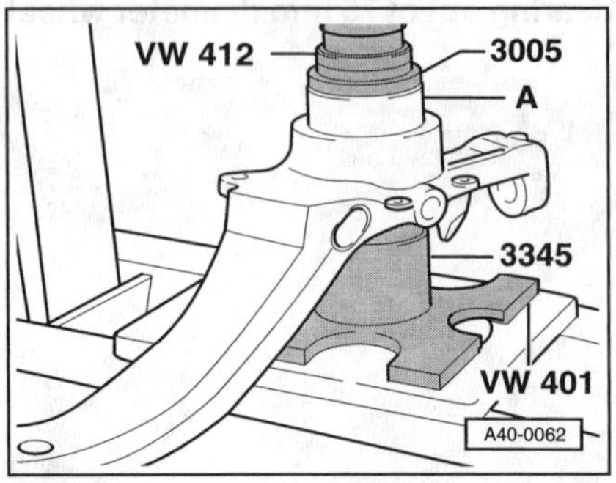

Fig. 7 Pressing wheel bearing -A- into 75 mm diameter wheel bearing housing

EDITOR'S NOTE —

Larger internal diameter of wheel bearing faces wheel hub.

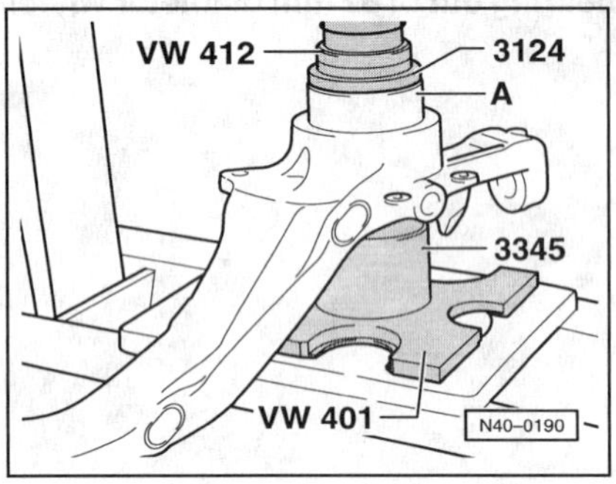

Fig. 8 Pressing wheel bearing -A- into 82 mm diameter wheel bearing housing

NOTE —

Larger internal diameter of wheel bearing faces wheel hub.

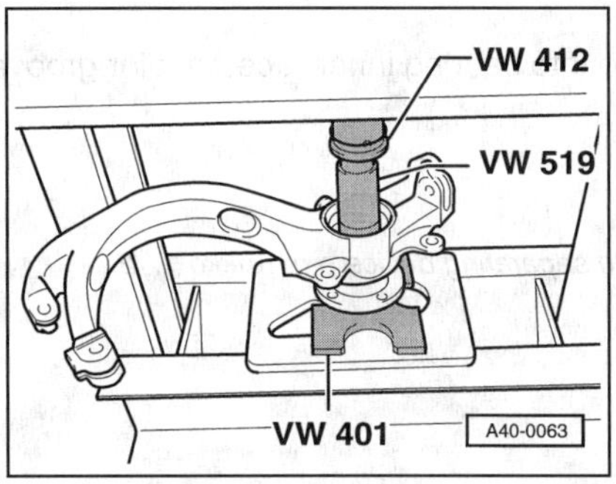

Fig. 9 Pressing wheel hub into 75 mm diameter wheel bearing

– When pressing in, the thrust piece VW519 must bear only against inner race.

NOTE —

- *The wheel bearing housing must remain as horizontal as possible when pressing in.*
- *Otherwise the wheel hub will bend in the wheel bearing which will lead to premature wheel bearing damage.*

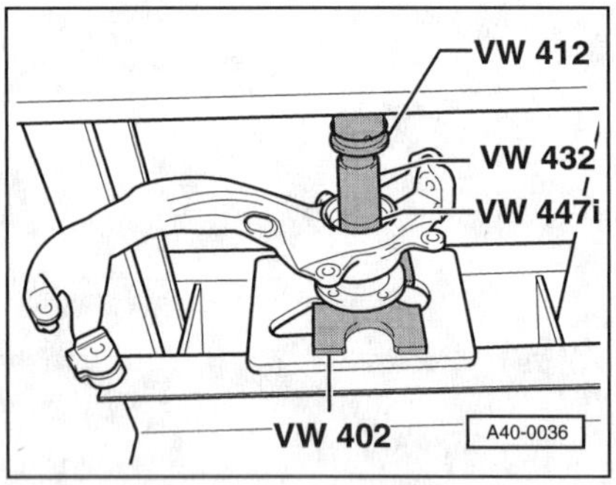

Fig. 10 Pressing wheel hub into 82 mm diameter wheel bearing

– When pressing in, the thrust piece VW447i must only bear against inner race.

NOTE —

The wheel bearing housing must remain as horizontal as possible when pressing in. Otherwise the wheel hub will bend in the wheel bearing which will lead to premature wheel bearing damage.

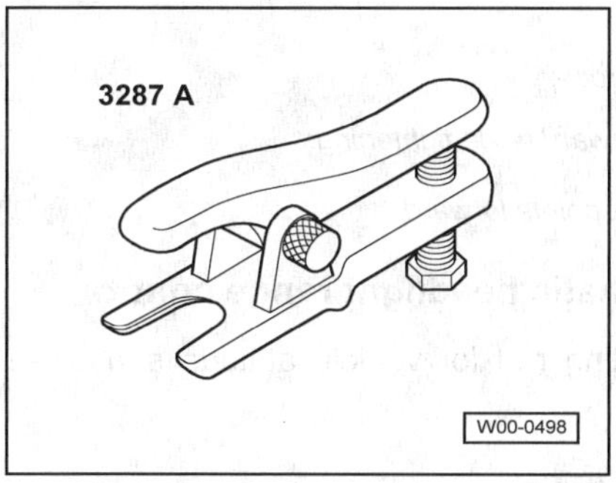

Track-control link, removing and installing

Special tools, workshop equipment, testers, measuring instruments and auxiliary items required

- VAG 1332 Torque wrench
- 3287A Ball joint puller

Removing

– Remove wheel.

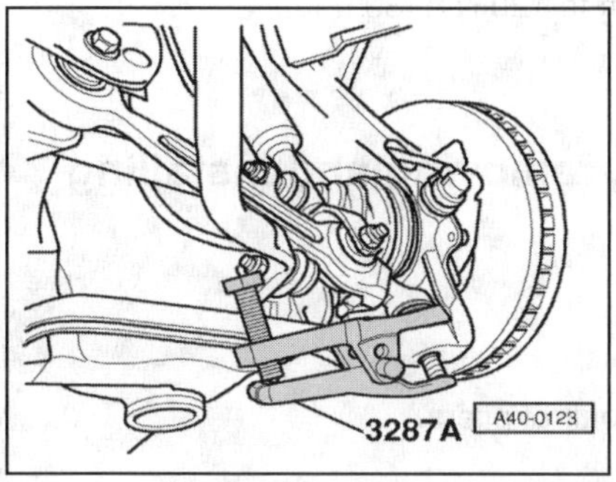

Remove nut from track-control link joint pin and press joint pin out of tapered seat.

NOTE—
Do not damage joint protective boot when doing this!

Vehicles with automatic headlight range control

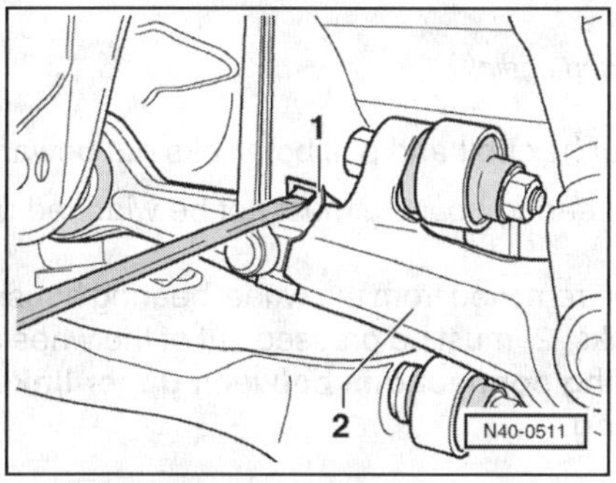

Unclip clamp for coupling rod -1- for vehicle attitude sensor -G78- from track-control link -2-.

Continued for all vehicles

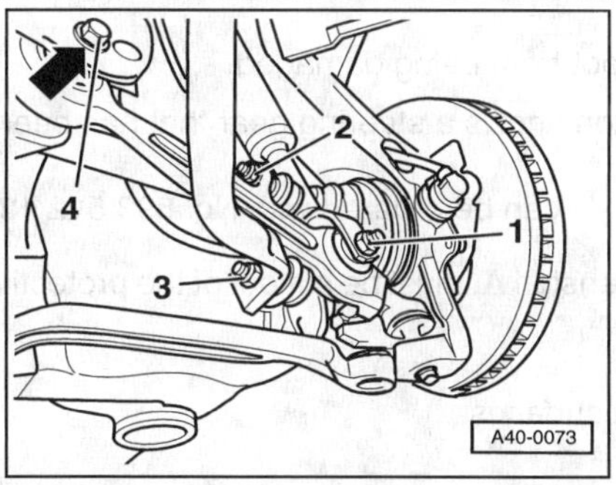

Remove hex bolt -1-.

– Remove ribbed nuts -2- and -3- and take out connecting rod.

NOTE—
To prevent the upper link joints from being damaged, support with e.g. VAG 1383-A to prevent spring extending too much.

– Unscrew hex bolt -4-.

– Take out track-control link.

Installing

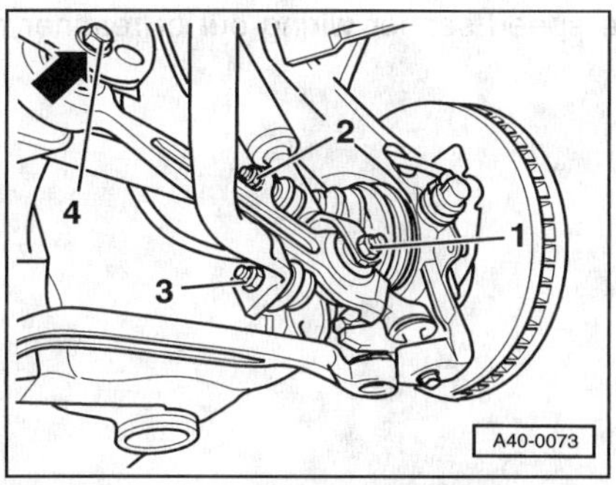

Insert track-control link.

- Tighten nut on joint pin,100 Nm (74 ft-lb).
- Tighten hex nut -1-, 90 Nm (66 ft-lb).
- Tighten nut -2-, 40 Nm (30 ft-lb) + turn 90° further.
- Tighten nut -3, 40 Nm (30 ft-lb) + turn 90° further.
- Install new hex bolt -4- and new hex nut and tighten, 80 Nm (59 ft-lb) + turn 90° further.

NOTE—
- *Only use inner holes -arrow-.*
- *Push track-control link inward while tightening.*
- *Arrow on connecting link points forward.*

Vehicles with automatic headlight range control

– Install clamp for coupling rod for vehicle attitude sensor -G78- from track-control link.

Continued for all vehicles

– Install wheel and tighten to 120 Nm (89 ft-lb).

Bearing bracket, removing and installing

Removing

– Remove wheels.

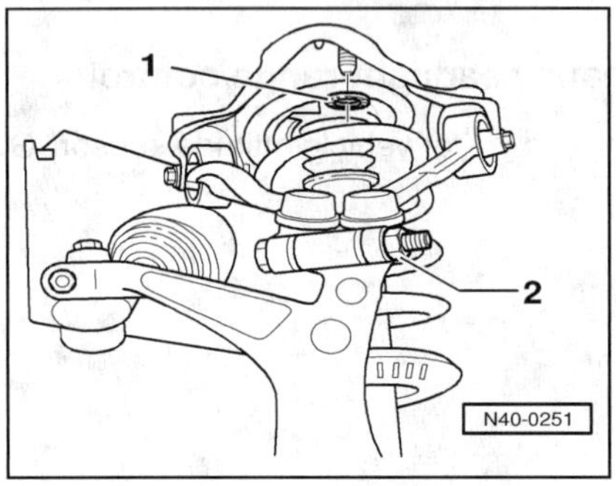

◄ Remove clip -1- with a pair of pliers.

NOTE—
This clip need not be inserted again!

– Remove nut -2-, take out hex bolt and pull both links out upward.

- The slits in the wheel bearing housing must not be widened using a chisel or similar tool!
- If the hex bolt cannot be removed from the wheel bearing housing, then the hex bolt and the links -2- must be pressed out of the wheel bearing housing, see **Separating connections between upper links and wheel bearing housing**.

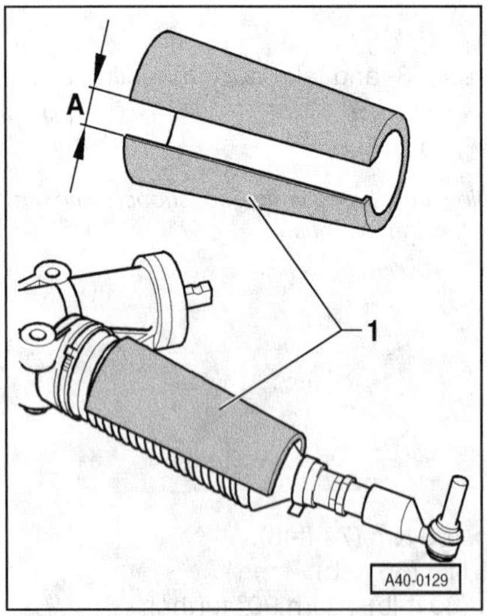

◄ Protect steering gear boot from being damaged.

- For clarity the illustration shows a steering gear that has been removed.
- The protective sleeve -1- can be ordered with No. 893 512 137.

– A 20... 25 mm strip, dimension A, must be cut out of the protective sleeve -1-.

– Carefully deburr the cut surfaces.

– Slide protective sleeve over the steering gear boot.

- The open side of the sleeve must point downward.

– Disconnect ABS vehicle speed sensor wiring out of retainer on brake caliper.

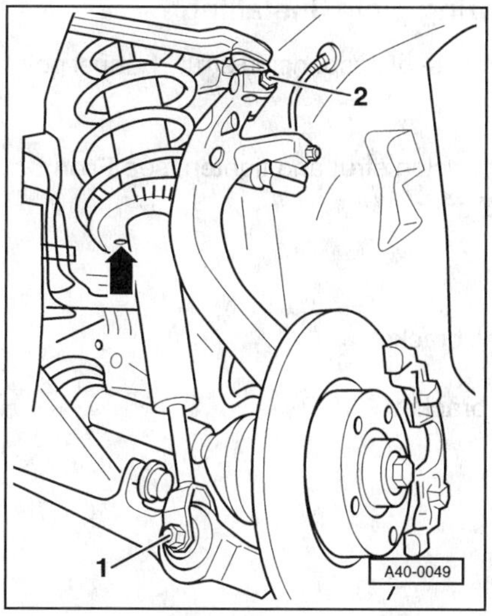

➤ Remove nut -1-.

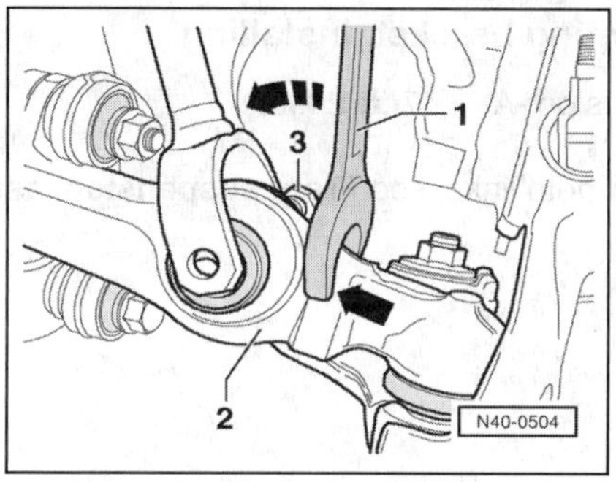

➤ To facilitate removal of suspension strut, apply an open-end wrench -1- to wrench surfaces -arrow- of track-control link -2-.

– Turn wrench in direction of arrow (direction of travel) and pull out bolt -3-.

– Swing wheel bearing housing to side.

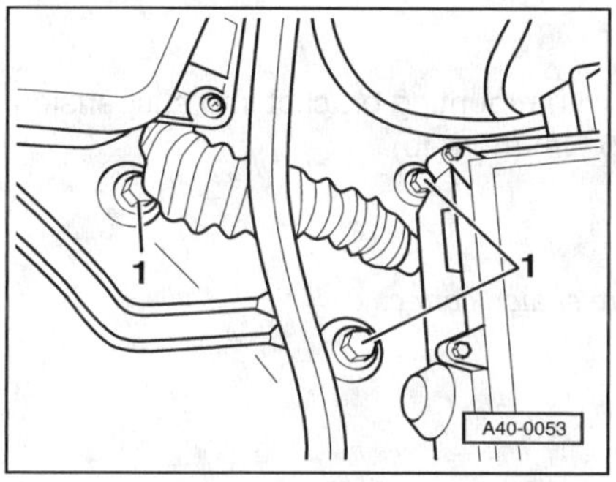

➤ Unscrew hex bolts -1- in plenum chamber.

– Take out suspension strut with mounting bracket.

NOTE —
Do not damage joint protective boot when doing this!

Upper link from mounting bracket, removing

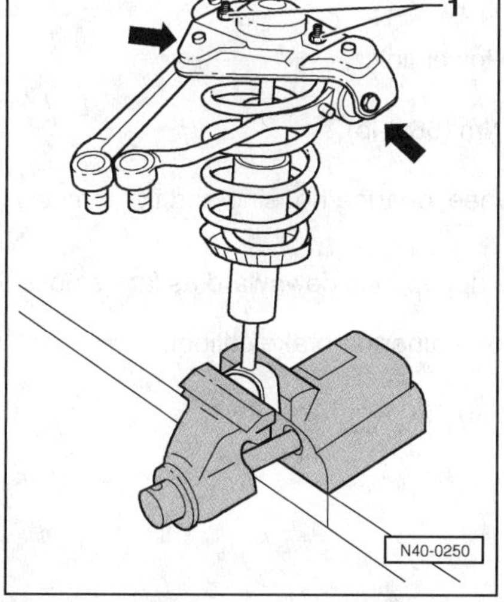

➤ Tension suspension strut in a vice (use protective clamps).

– Unscrew both links -arrows-.

– Unscrew nut -1- and take off mounting bracket.

Suspension strut spring plate, installing

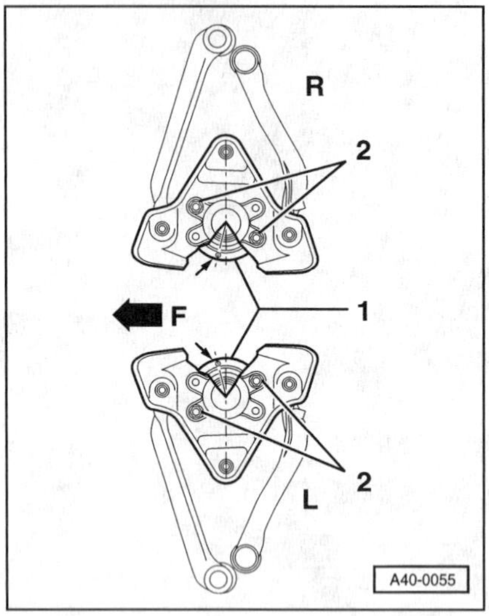

◄ Holes in spring plate -arrows- of suspension strut -1- must point towards center of vehicle.

– Fit new nuts -2- for suspension strut and tighten, see **Front suspension, assembly**, **-43-**.

F = Forward

R = Right-hand mounting bracket

L = Left-hand mounting bracket

Upper links to mounting bracket, installing

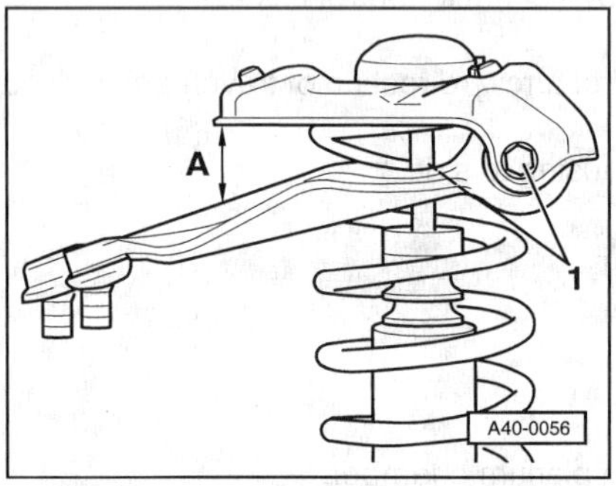

◄ Align upper links, dimension -A- = 47 ± 2 mm.

– Tighten hex bolts -1- for both links, see **Front suspension, assembly**, **-42-**.

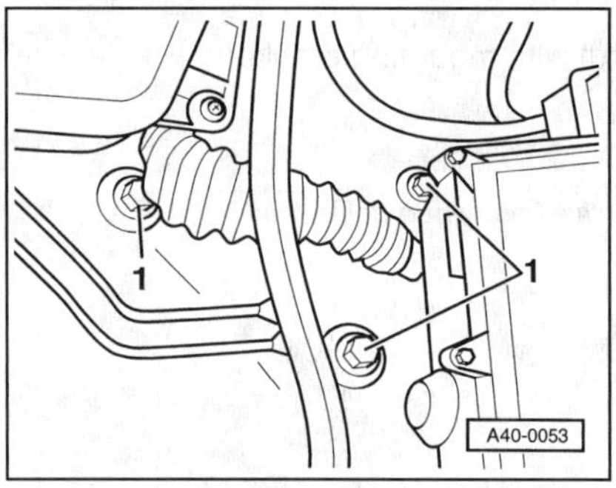

◄ Insert suspension strut with mounting bracket into suspension turret. Tighten hex bolts -1-, 75 Nm (55 ft-lb).

NOTE —
Make sure that washers are seated correctly.

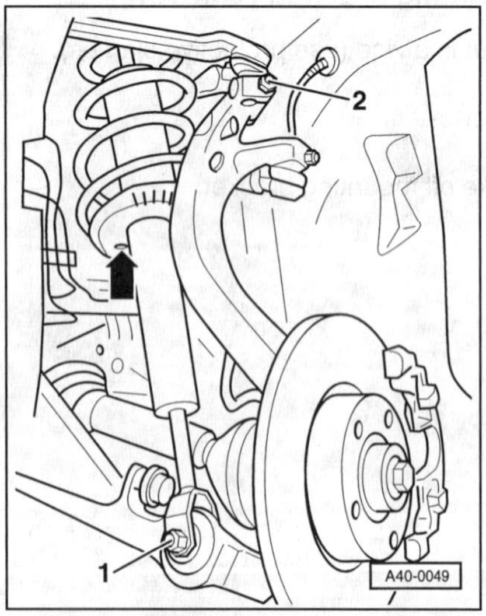

◄ Insert suspension strut forked head into link.

– Bolt suspension strut to lower link.

Tighten new nut -1-, 90 Nm (66 ft-lb).

– Insert upper links into wheel bearing housing and tighten new nut -2-, 40 Nm (30 ft-lb).

• When tightening press upper links downward as far as possible!

– Connect ABS wiring into retainer on brake caliper.

– Install wheel and tighten to 120 Nm (89 ft-lb).

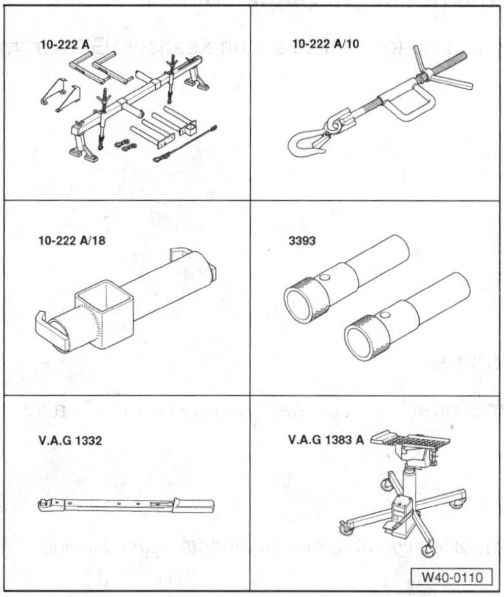

Subframe, removing and installing

Special tools, workshop equipment, test and measuring appliances and aux. items required

- 10-222 A Support
- 10-222 A/10 Additional spindle
- 10-222 A/18 Adapter
- 3393 Test mandrel
- VAG 1332 Torque wrench
- VAG 1383 A Engine/transmission jack

Removing

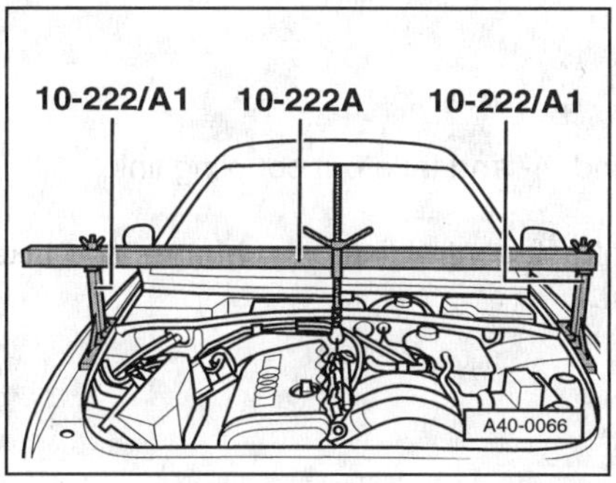

Install support bar 10-222A together with supports 10-222 A/1.

Vehicles with 8-cylinder engine

– Remove oil filler pipe from engine.

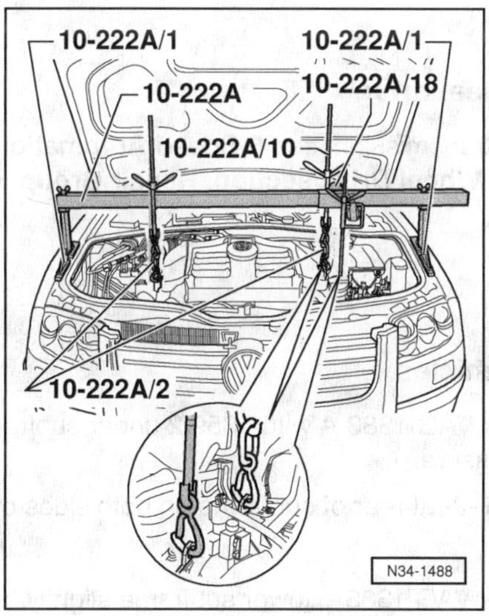

Install support bar 10-222A together with legs 10-222 A/1 and adapter 10-222 A/18.

- The longer side of adapter 10-222 A/18 faces forward.

– Secure third spindle 10-222 A/10 with additional hook 10-222 A/2 on lifting eye at rear left of engine.

– Hook second additional hook 10-222 A/2 on lifting eye at rear right of engine.

Continued for all vehicles

– Take weight of engine/transmission assembly on the spindles.

– Remove wheels.

– Remove noise insulation tray.

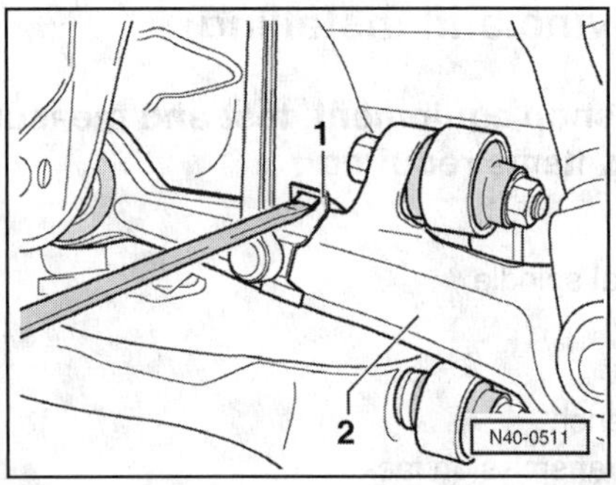

Vehicles with automatic headlight range control

Unclip clamp for coupling rod -1- for vehicle attitude sensor -G78- from track-control link -2-.

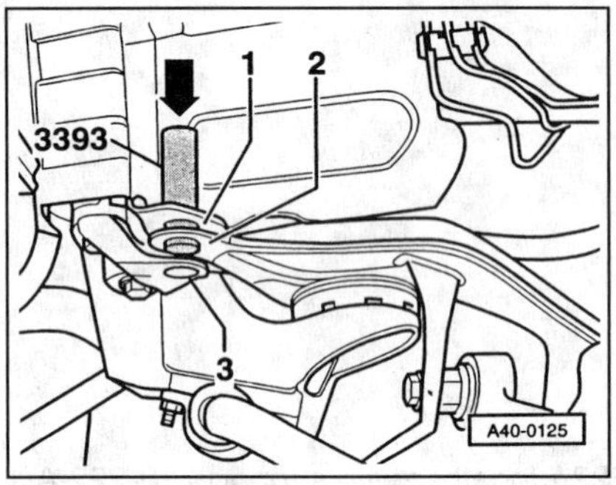

Continued for all vehicles

Before loosening subframe, check that at least holes -1- and -2- align using special tool 3393.

NOTE —

If this is not the case, perform axle alignment measurements after installing the sub-frame!

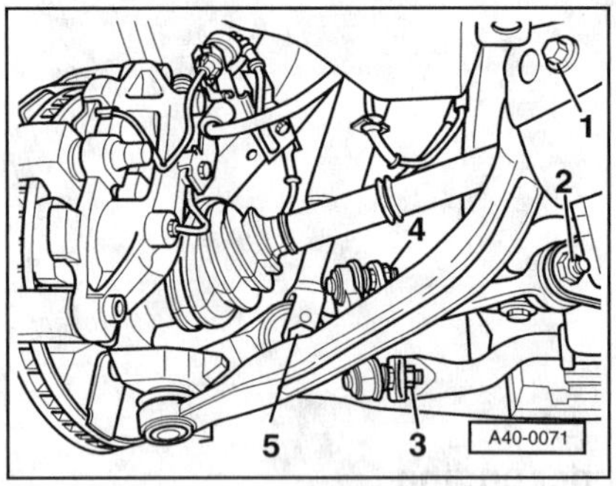

Unscrew hex bolt -5-.

- Unscrew hex nuts -3- and -4- and take out coupling link.
- Remove hex nut -2- for track-control link on subframe and take out hex bolt.
- Remove hex bolt -1-.

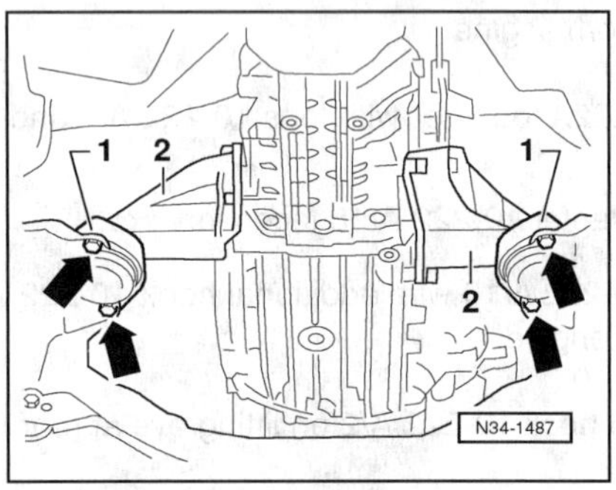

Unscrew hex bolts -arrows- from transmission mounting -1-.

Vehicles with all wheel drive

- Unbolt driveshaft from transmission, see **5-Speed Automatic Transmission 01V All Wheel Drive section, Repair Group 39.**

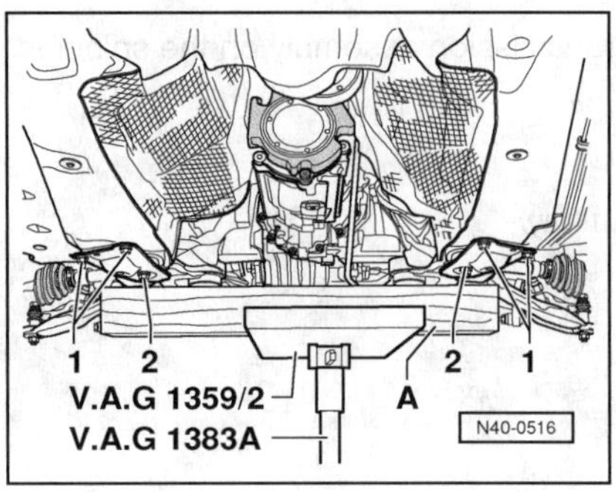

Continued for all vehicles

Place transmission jack VAG 1383 A with 1359/2 under subframe and exert slight counter pressure.

- Unscrew bolts -1- and -2- at rear of subframe on both sides of vehicle.
- A - Block of wood

- Using transmission jack VAG 1383A, lower subframe slightly.
  - Remove hex bolts for track-control and guide links.
- Disconnect ABS vehicle speed sensor wiring out of retainer on brake caliper.

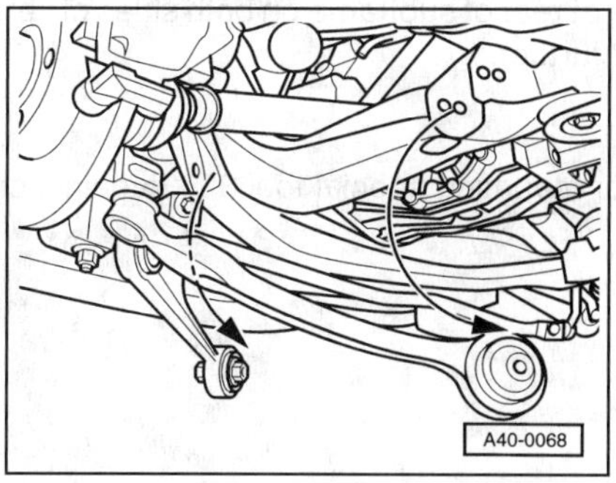

Pull both track-control and guide links out of subframe.

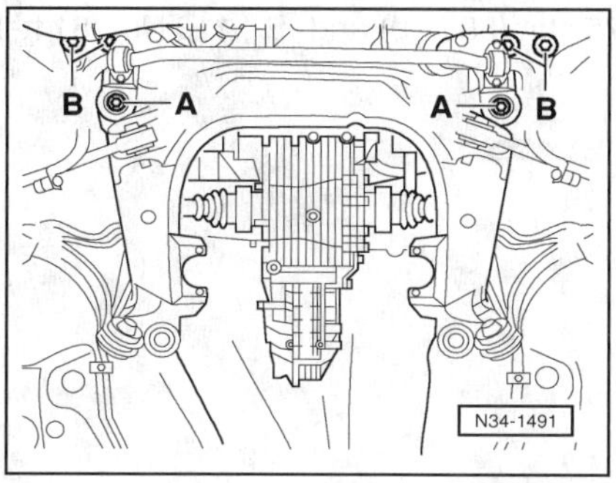

Unscrew hex bolts -A-.

– Loosen hex bolts -B- until the subframe can be taken out.

Subframe, installing

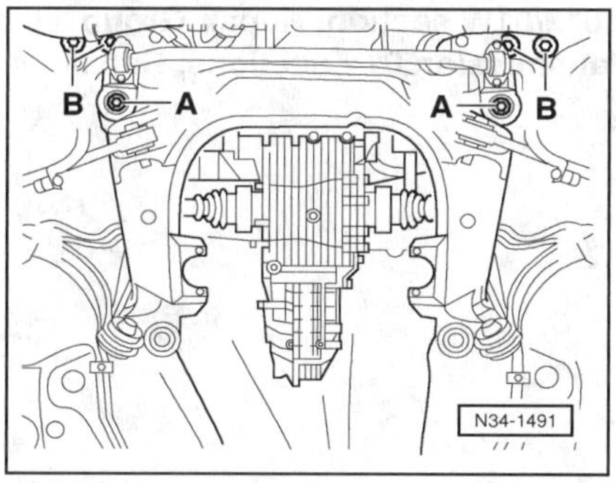

Install subframe and start hex bolts -A- and -B-.

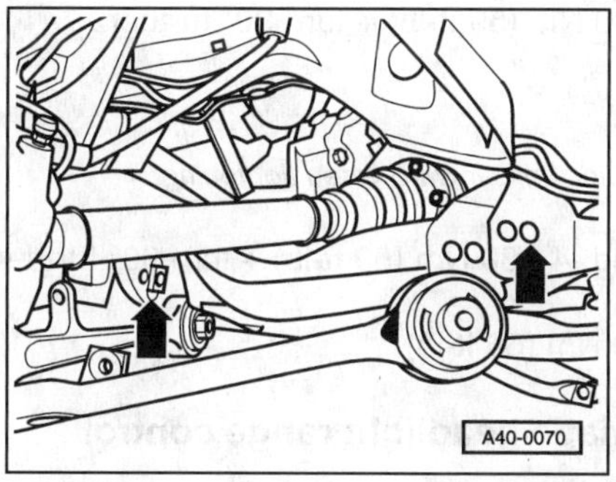

Guide both track-control and guide links into subframe.

- Install bolts into the inner holes -arrows-.

– Using transmission jack VAG 1383A, raise back of subframe slightly.

NOTE—

The subframe must be attached to the body with the help of special tool 3393 before tightening.

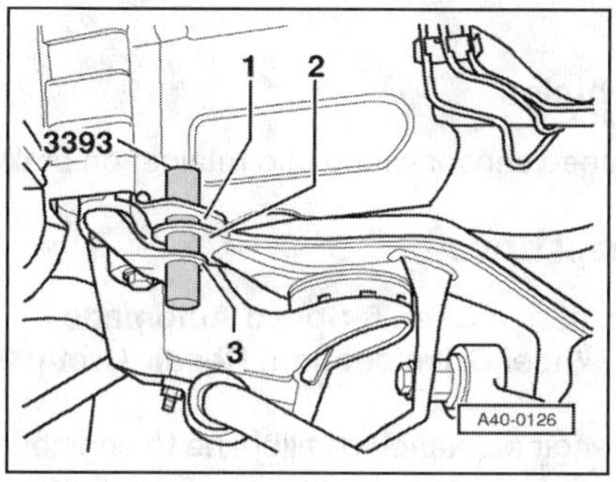

Hole -1-, sub-frame hole -2- and aluminium console hole -3- must align on both sides.

NOTE—

- *This attachment only replaces subsequent axle alignment measurements when:*
- *The sub-frame is removed completely and installed.*
- *Holes -1- and -2- align before repairs are made.*

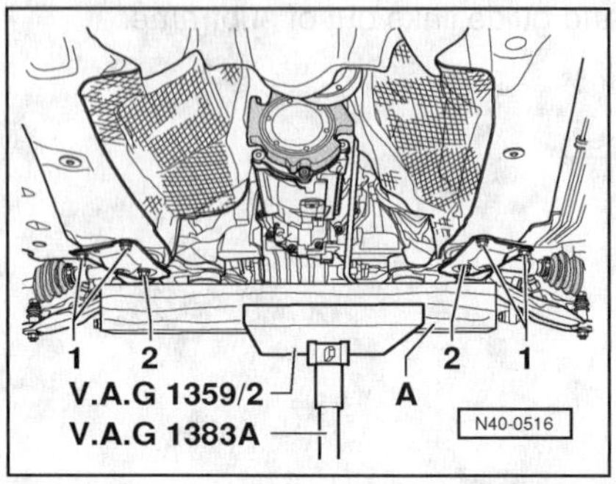

Tighten bolts -1- and -2- at rear of subframe on both sides of vehicle, 110 Nm (81 ft-lb) + turn 90° further.

- A - Block of wood

– Protect holes in body for mounting special tool 3393 against corrosion.

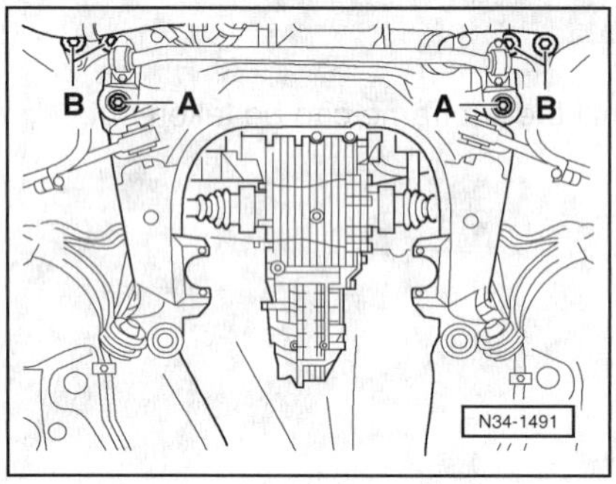

Tighten hex bolts -A- , 75 Nm (55 ft-lb) and -B-, 110 Nm (81 ft-lb) + turn 90° further.

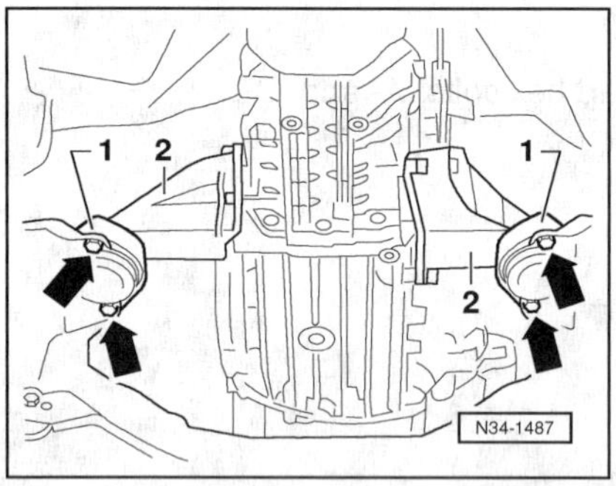

Tighten hex bolts -arrows- at transmission mounting -1-, see **5-Speed Manual Transmission 012/01W section, Repair Group 34, or 5-Speed Automatic Transmission 01V sections, Repair Group 37.**

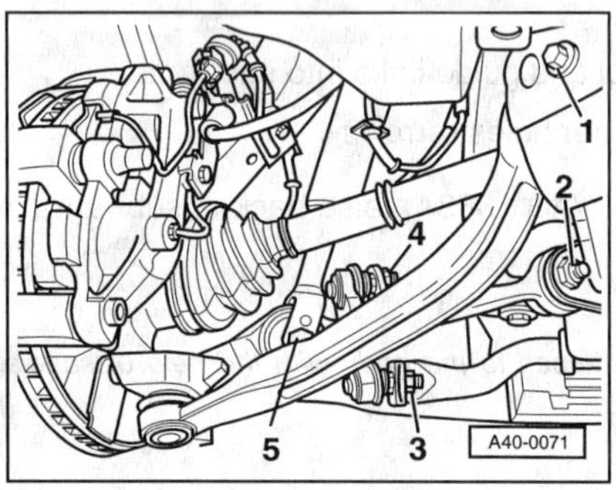

Insert new hex bolts for guide link -1- and track-control link -2- and tighten new hex nuts, 80 Nm (59 ft-lb) + turn 90° further..

NOTE —
Press track-control and guide links inward when tightening.

– Tighten hex nuts -3- and -4-, 80 Nm (59 ft-lb) + turn 90° further.

– Tighten hex bolt -5-, 90 Nm (66 ft-lb).

Vehicles with automatic headlight range control

– Install clamp for coupling rod for vehicle left front level control system sensor -G78-.

Continued for all vehicles

– Connect ABS vehicle speed sensor wiring into retainer on brake caliper.

Vehicles with all wheel drive

– Bolt propshaft onto transmission, see **5-Speed Automatic Transmission 01V All Wheel Drive section, Repair Group 39.**

– For vehicles with 8 cyl. engines, attach oil filler line to engine

– Install wheel and tighten, 120 Nm (89 ft-lb).

Drive axle, servicing

Removing and installing drive axles

Removing

– Unbolt hex bolt for drive shaftaxle. (Loosen only when vehicle is standing on wheels -danger of accident-).

– Remove wheel.

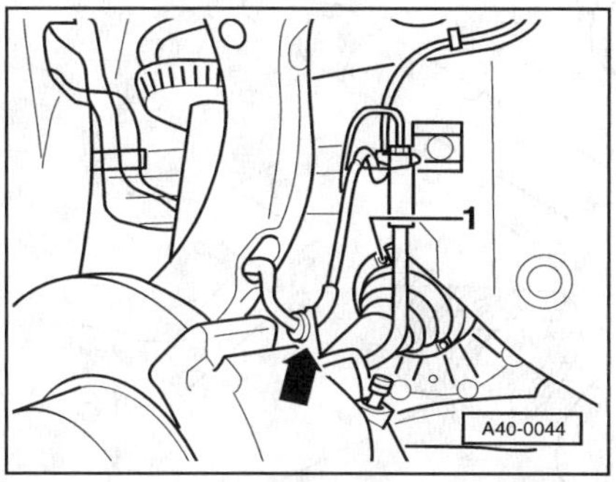

◄ Unbolt drive axle at flange shaft.

• 1 - hex key head bolt or multi-point key head bolt

– Disconnect ABS vehicle speed sensor wiring out of retainer on brake caliper -arrow-.

– Disconnect ABS vehicle speed sensor out of wheel bearing housing slightly.

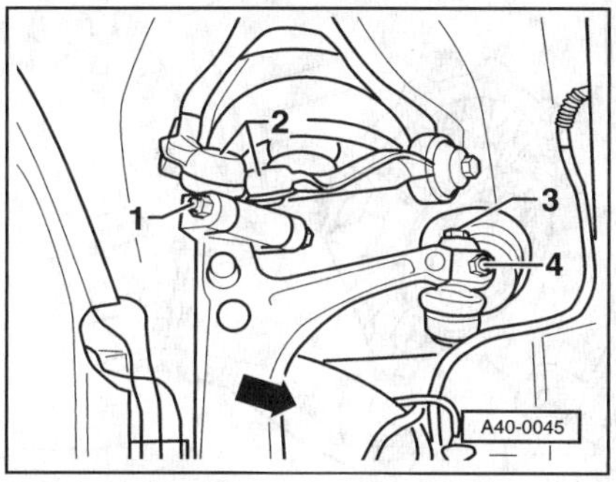

◄ Remove nut -1-, take out hex bolt and pull both links -2- out upward.

• The slits in the wheel bearing housing must not be widened using a chisel or similar tool!

If the hex bolt cannot be removed from the wheel bearing housing, then the hex bolt and the links -2- must be pressed out of the wheel bearing housing, see **Separating connections between upper links and wheel bearing housing**.

NOTE —
Do not loosen bolts -3- and -4- otherwise the front axle alignment is changed!

– Swing wheel bearing housing in direction of arrow away to side.

– Remove drive axle.

Installing

– Insert drive axle onto transmission shaft and into wheel bearing housing.

– Insert upper links into wheel bearing housing, insert new bolt and tighten new nut -2-, 40 Nm (30 ft-lb).

NOTE —
When tightening press upper links downward as far as possible!

– Bolt drive axle to transmission.

– Tighten multi-point socket head bolts to prescribed tightening torque using diagonally to 10 Nm (7 ft-lb).

• Multi-point socket head bolt M 8 x 48, 40 Nm (30 ft-lb).
• Multi-point socket head bolt M 10 x 48, 77 Nm (57 ft-lb).

– Push ABS vehicle speed sensor into wheel bearing housing onto stop and insert cable in retainer on brake caliper.

– Install wheel and tighten, 120 Nm (89 ft-lb).

– Tighten drive axle hex bolt only with vehicle standing on its wheels.

Drive axle with constant velocity joint, assembly

Special tools, required

- VW 401 Thrust plate
- VW 402 Thrust plate
- VW 408 A Press tool
- VW 411 Press tool
- VW 447 H Press plate
- VW 454 Press piece
- VW 161 A Circlip pliers
- VAG 1331 Torque wrench
- VAG 1332 Torque wrench
- VAG 1682 Tension clamp
- T10065 Assembly device

Grease quantity and type

Constant velocity joints are packed with grease G 000 603:

Outer CV joints

Outer joint dia.	Total grease required	Proportions:	
		Joint	Boot
88mm (3.5 in.)	90g (3.5oz)	40g (1.8oz)	50g (1.8oz)
98mm (3.9 in.)	120g (4.2oz)	80g (2.8oz)	40g (1.4oz)

Inner CV joints

Inner joint dia.	Total grease required	Proportions:	
		Joint	Boot
100mm (3.9 in.)	90g (3.8 oz)	40g (1.9 oz)	50g (1.9 oz)
108mm (4.3 in.)	120g (4.2 oz)	35g (1.2oz)	85g (3.0oz)

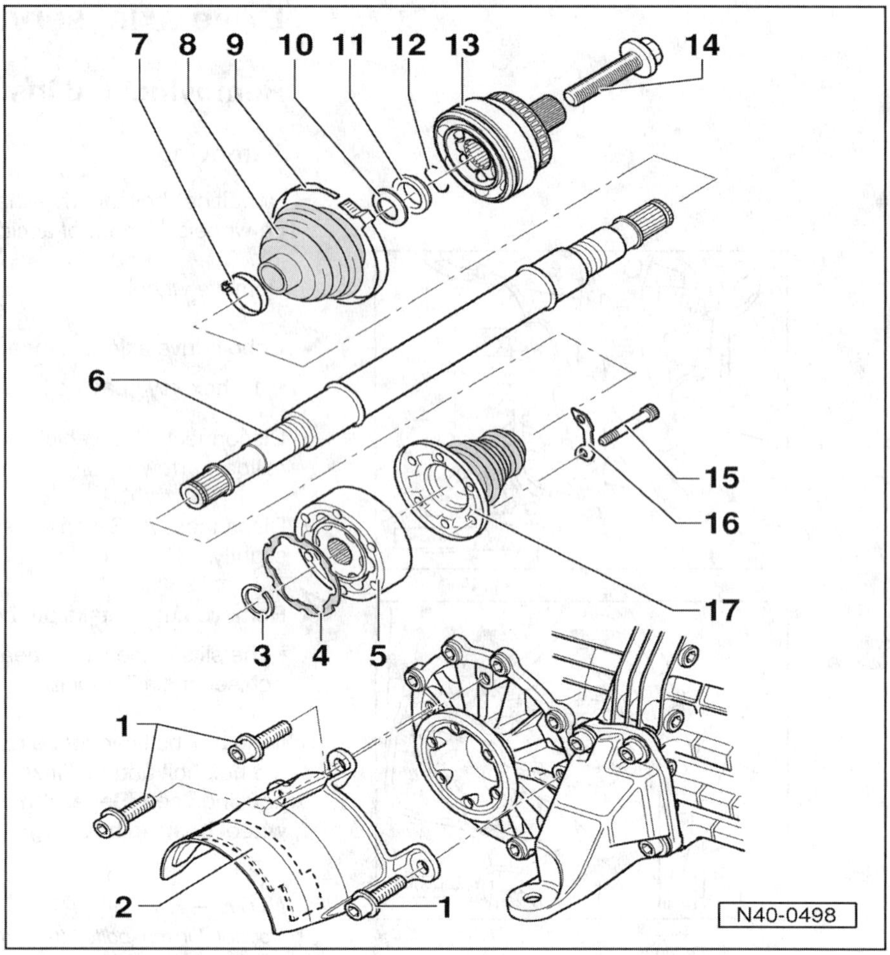

1. **Hex key head bolt**
   - Tighten to 25 Nm (18 ft-lb)
2. **Heat shield**
3. **Circlip**
   - Always replace
   - Remove and install with VW 161 a
4. **Gasket**
   - Replacing. Pull off protective foil and stick onto joint.
5. **Inner constant velocity joint**
   - Outer diameter 100 or 108 mm depending on engine/ transmission combination
   - Pressing off, see **Fig. 3**
   - Pressing on, see **Fig. 4**
6. **Drive axle (tube shaft)**
7. **Hose clip**
   - Always replace
8. **Protective boot**
9. **Hose clip**
   - Tensioning, see **Fig. 1** and **Fig. 5**
10. **Dished washer**
    - Outer diameter (concave side) contacts thrust washer
11. **Thrust washer**
12. **Circlip**
    - Insert in shaft groove
13. **Outer constant velocity joint**
    - Removing, see **Fig. 2**
    - Installing: drive onto shaft on to stop with plastic hammer
14. **Hex bolt**
    - Changed to hex key head bolt
    - Replace each time after removing
    - Tightening torque:
Bolt M14: 115 Nm (85 ft-lb) and turn 180° further
Bolt M16: 190 Nm (140 ft-lb) and turn 180° further
15. **Multi-point socket head bolt**
    - Tighten using diagonal sequence
M 8 x 48, 40 Nm (30 ft-lb)
M 10 x 48, 77 Nm (57 ft-lb)
16. **Plate**
17. **Protective boot for inner constant velocity joint**
    - Drive off with drift
    - Before installing on constant velocity joint, coat sealing surface with D-3
    - Joint protective boots made from rubber are tightened with pliers VAG 1275.

NOTE —

The instructions in Fig. 1 are only for hose clamps made from stainless steel!

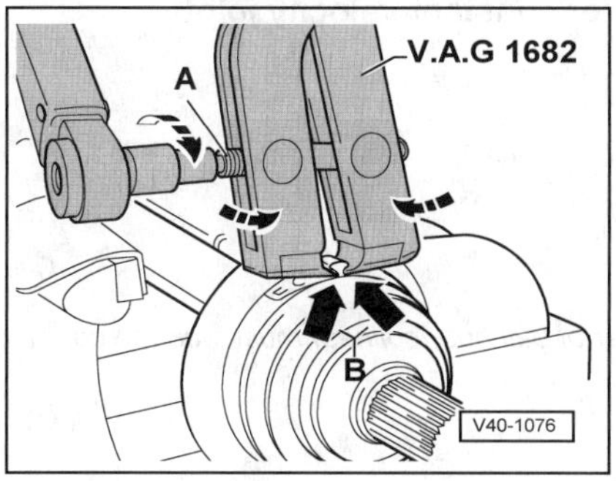

Fig. 1 Tightening hose clamp on outer joint

– Install tensioning clamp VAG 1682 as shown in illustration. Thereby ensure that the jaws of the tension clamp seat in the corners (arrow B) of the hose clip.

– Tighten hose clamp by turning the spindle with a torque wrench (do not bend tensioning clamp).

NOTE—

- *The hard material of the joint boot (compared to rubber) makes it necessary to use a stainless steel hose clamp, it is only possible to tighten the hose clamp with tensioning clamp VAG 1682.*
- *Tightening torque: 20 Nm (15 ft-lb).*
- *Use torque wrench with 5...50 Nm range (e.g. VAG 1331).*
- *Make sure the spindle thread is not tight. If necessary lubricate with MOS 2 grease*
- *If the thread is tight e.g. dirty, the required tensioning force for the hose clamp will not be achieved in spite of correct tightening torque settings.*

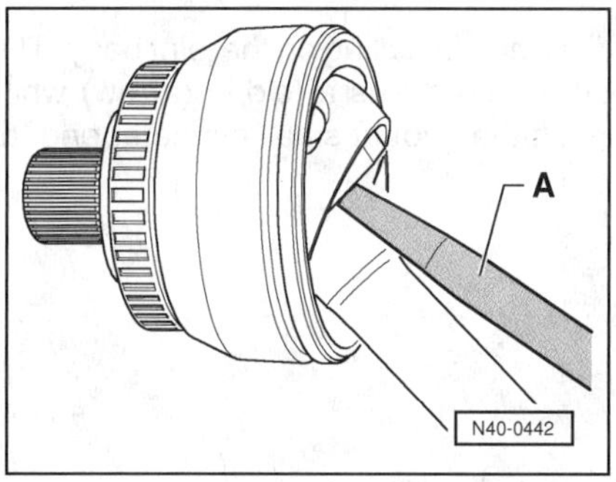

Fig. 2 Pressing off outer constant velocity joint

– Clamp drive shaft in vice using vice clamps.

– Remove hose clip and slide back boot.

– Drive constant velocity joint off drive shaft using drift -A-.

– Drive onto shaft with plastic hammer until securing ring engages.

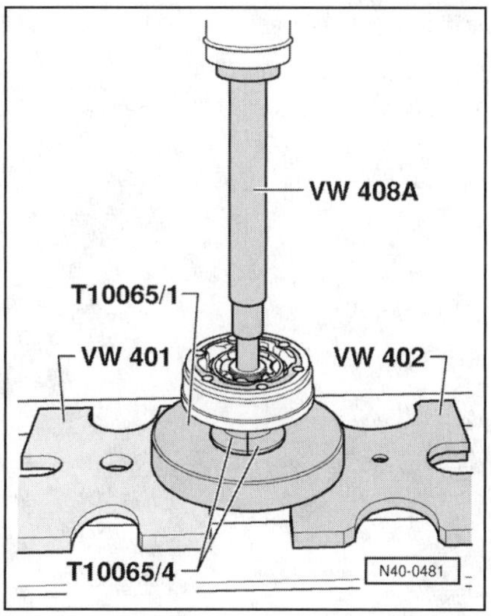

Fig. 3 Pressing off inner constant velocity joint

NOTE—
First drive boot off with drift

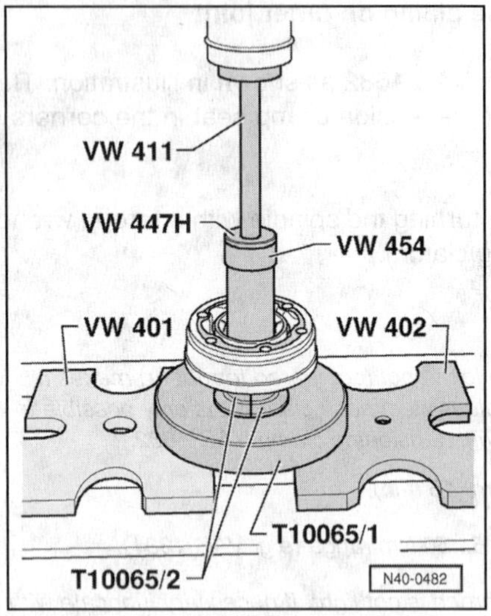

Fig. 4 Pressing on inner constant velocity joint

- Press on joint up to stop.
- Insert circlip.

NOTE —
Chamfer on inner diameter of ball hub (splines) must face the contact shoulder on the drive shaft.

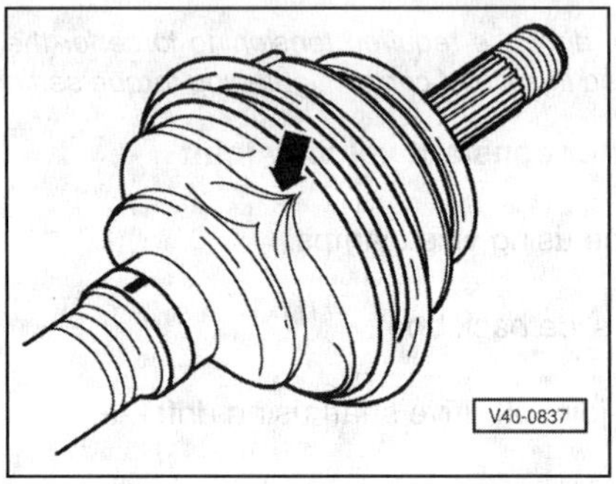

Fig. 5 Ventilating joint boot

• Only for joint protective boot of rubber

Often the boot is pushed in when installing on the joint body. This creates a vacuum inside the boot which draws a fold in (arrow) while driving. Therefore after installing, lift the boot at small diameter end to equalize pressure.

Drive shaft with triple roller joint, assembly

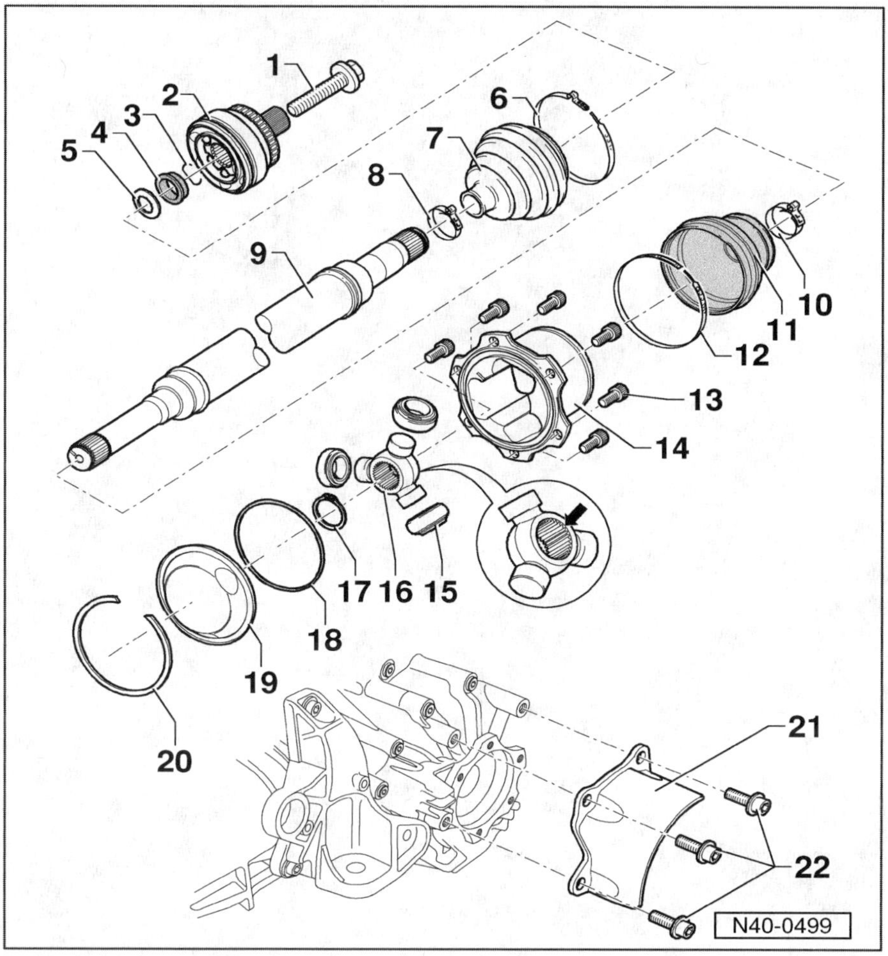

Special tools required

- VW 401 Thrust plate
- VW 402 Thrust plate
- VW 408 A Press tool
- VW 411 Press tool
- VW 416 B Tube
- VW 447 H Press plate
- VAG 1275 Pliers
- VAG 1331 Torque wrench
- VAG 1332 Torque wrench
- VAG 1682 Tension clamp
- T10065 Assembly device

1. **Hex bolt**
   - See **Drive shaft with constant velocity joint, assembly**, -14-
2. **Outer constant velocity joint**
   - Only replace complete joint
   - Installing: drive onto shaft with plastic hammer until compressed circlip seats
   - Greasing, see **Drive shaft with constant velocity joint, assembly**
3. **Circlip**
   - Insert in shaft groove
4. **Thrust washer**
5. **Dished washer**
   - Outer diameter (concave side) contacts thrust washer
6. **Hose clip**
   - Tightening, see **Drive shaft with constant velocity joint, servicing**
7. **Constant velocity joint boots**
   - Balance pressure, by briefly ventilating boot before tensioning the small clamp, see **Drive shaft with constant velocity joint, servicing**
8. **Hose clip**
9. **Drive shaft**
10. **Hose clip**
11. **Triple roller joint boot**
12. **Hose clip**
13. **Multi-point socket head bolt**
    - Tighten using diagonal sequenc M 10 x 20, 77 Nm (57 ft-lb)
14. **Joint piece**
15. **Rollers**
16. **Tripodic star**
    - The chamfer -arrow- points to drive shaft splines.
17. **Circlip**
18. **Seal**
19. **Cover**
    - Destroyed when dismantling.
    - No longer required when assembling and is therefore no longer supplied as a spare part.
    - A new cover is included in the repair kit.
20. **Circlip**
    - This circlip is not installed during production. It is only available as a replacement part.
    - The circlip is included in the repair kit.
21. **Heat shield**
22. **Allen head bolt**
    - M 8 x 22, 25 Nm (18 ft-lb)
    - M 8 x 28, 25 Nm (18 ft-lb)
    - M 10 x 70, 40 Nm (30 ft-lb)

42 Rear Suspension

Repair Group

S 42

WARNING —

- *Do not re-use any fasteners that are worn or deformed in normal use.*
- *Some fasteners are designed to be used only once, and are unreliable and may fail if used a second time. This includes, but is not limited to, nuts, bolts, washers, circlips and cotter pins. Always follow the recommendations in this manual-replace these fasteners with new parts where indicated, and any other time it is deemed necessary by inspection.*

CAUTION —

- *Do not attempt to weld or straighten the load-bearing components or parts of the suspension.*
- *Do not attempt to move a vehicle from which the drive axle has been removed. Otherwise wheel bearing damage will result. If the vehicle does have to be moved, install an outer CV joint in place of the drive axle and tighten to 50 Nm (37 ft lb).*
- *The bonded rubber bushings can only be turned to a limited extent. The bolted connections on the suspension links should therefore only be tightened when the vehicle is standing on the ground.*

EDITOR'S NOTE —

Special tools, equipment and procedures are required for most rear suspension repair and component replacement. In addition, wheel alignment can be disturbed when suspension components are removed or replaced.

Rear Axle, Vehicles with Front Wheel Drive, servicing

Rear axle with front wheel drive, assembly

NOTE—
- *Welding and straightening on axle beam and stub axle is not permissible.*
- *Always replace self-locking nuts.*

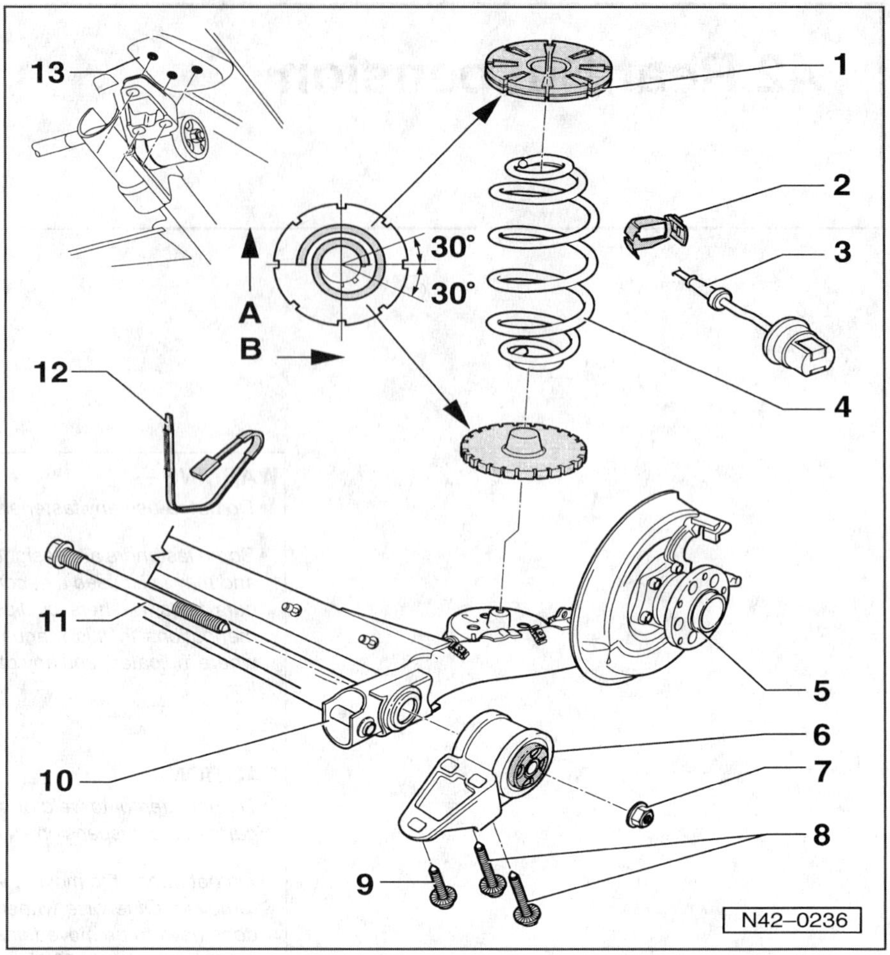

1. **Spring support**
2. **Vehicle speed sensor retainer**
3. **Vehicle speed sensor**
   - Only on vehicles with ABS
4. **Coil spring**
   - Removing: first remove upper bolt from shock absorber, see **Shock absorber/spring, removing and installing (front wheel drive)**
   - After installing, the ends of the spring must lie in area marked, as illustrated.
     Arrow A points forward
     Arrow B points right
     A tolerance of ± 30° is permissible
   - Examine for paint damage and rectify paint damage if necessary
   - For coil spring application see latest factory parts information and vehicle data sticker.
5. **Wheel bearing**
6. **Mounting bracket**
   - Check and if necessary adjust rear axle total track after installation
   - The bearing bracket contact surface is to be clean and free of wax, paint and soil
7. **Hex nut**
   - Replace each time after removing
8. **Hex bolt M 12 x 1.5 x 90**
   - 110 Nm (81 ft-lb) and turn 90° further
   - Replace each time after removing
9. **Hex bolt M 12 x 1.5 x 60**
   - 110 Nm (81 ft-lb) and turn 90° further
   - Replace each time after removing
10. **Axle beam**
    - Wheel bearing/wheel hub contact surfaces and threaded holes are to be free of paint and dirt
11. **Hex bolt M 14 x 1.5 x 190**
    - 120 Nm (89 ft-if) and turn 90° further
    - Replace each time after removing
12. **Retaining strap**
    - Installing, see **Fig.1**
13. **Threads in longitudinal member**
    - If threads in welded nut in longitudinal member are damaged, they can be serviced using a Heli-coil.

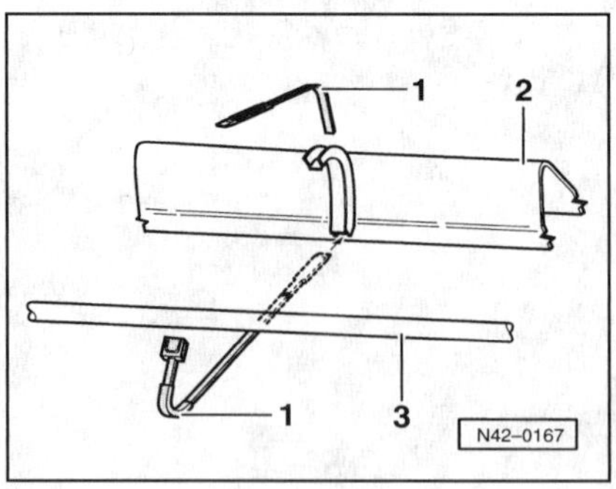

Fig. 1 Installing anti-roll bar retaining strap

- Thread retaining strap -1- behind the retainer on axle beam -2-.
- Place retaining strap around anti-roll bar -3- and close strap
- Push anti-roll bar approx. 8 mm toward axle beam and push retaining strap into lock as far as possible.

Shock absorbers on vehicles with front wheel drive, assembly

1. **Gas-filled shock absorber**
   - Can be replaced individually
   - Allocation
2. **Hex nut**
   - Replace each time after removing
3. **Hex bolt M 10 x 90**
   - 50 Nm (37 ft-lb) and turn 90° further
   - Replace each time after removing
   - Only tighten when vehicle is standing on wheels.
4. **Protective cap**
5. **Protective tube**
6. **Stop buffer**
7. **Hex bolt**
   - Tighten to 45 Nm (33 ft-lb)
8. **Shock absorber mounting**
9. **Hex self-locking nut**
   - Tighten to 25 Nm (18 ft-lb)
   - Loosening and tightening, see **Fig. 1**
   - Replace each time after removing
10. **Threads in wheel housing**
    - If thread in welded nut in wheel housing is damaged, then this can be serviced with a Heli-coil.

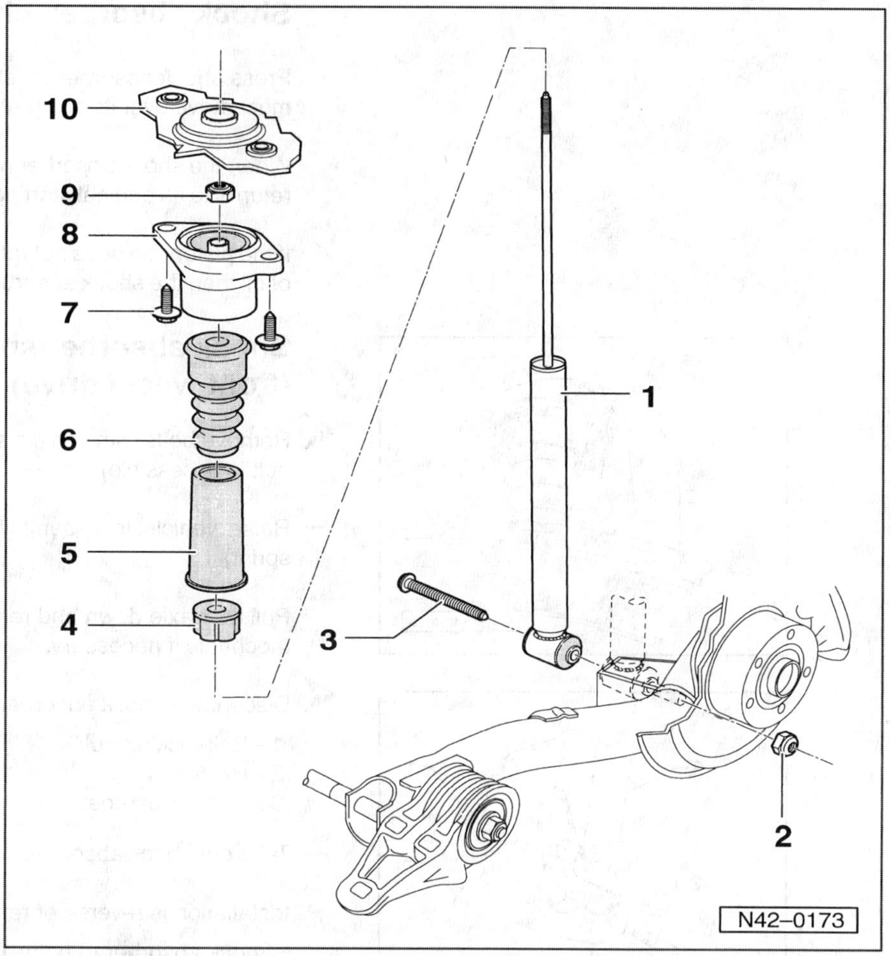

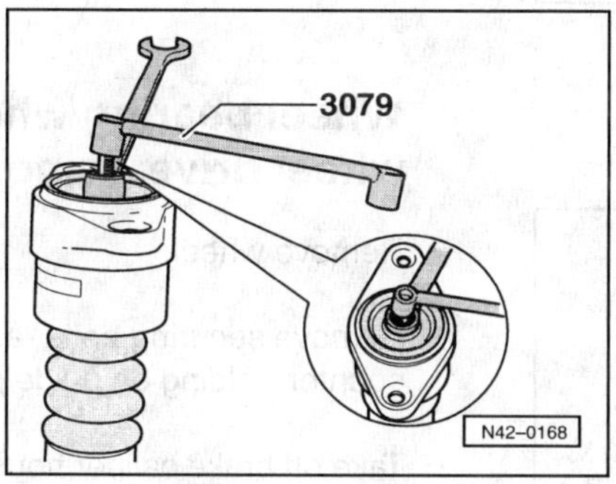

Fig. 1 Loosening and tightening shock absorber mounting threaded connection

Shock absorber, checking

Press shock absorber together by hand. When doing this the piston rod must move over its complete length smoothly and with even force.

When the shock absorber has sufficient gas pressure the piston rod returns to its original starting position.

If the piston rod does not return to its starting position and there is no loss of oil then the shock absorber is still OK.

Shock absorber/spring, removing and installing (front wheel drive)

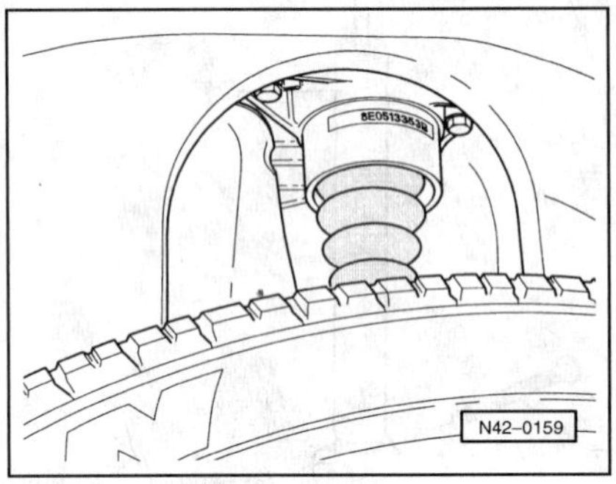

◄ Remove bolts with vehicle standing on its wheels (raise vehicle until the bolt is accessible).

– Raise vehicle to assembly height to relieve the pressure on the coil spring.

– Pull rear axle down and remove coil spring with the help of a second mechanic if necessary.

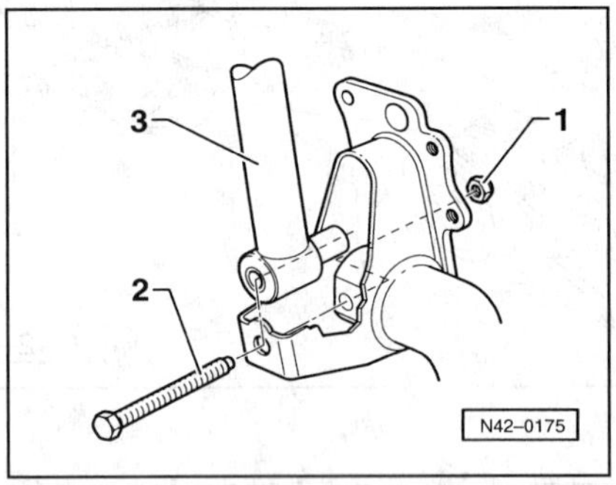

◄ Disconnect shock absorber from rear axle.

1 - Self-locking nut
2 - Hex bolt
3 - Shock absorber

– Take out shock absorber.

– Installation is reverse of removal, noting the following:

- Shock absorber to body, tighten bolt to 45 Nm (33 ft-lb).
- Shock absorber to rear axle, tighten nut to 50 Nm (37 ft-lb).

Wheel bearing/wheel hub, vehicles with front wheel drive, removing and installing

– Remove wheel.

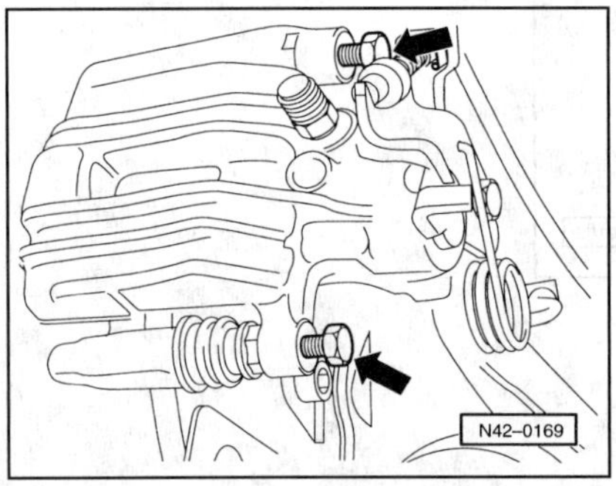

◄ Remove securing bolts -arrows- from brake caliper housing while counter-holding on guide pins.

– Take off brake caliper housing and secure with wire so that the weight of the brake caliper does not stress or damage the brake hose.

– Remove brake pads.

– Take off disc brake.

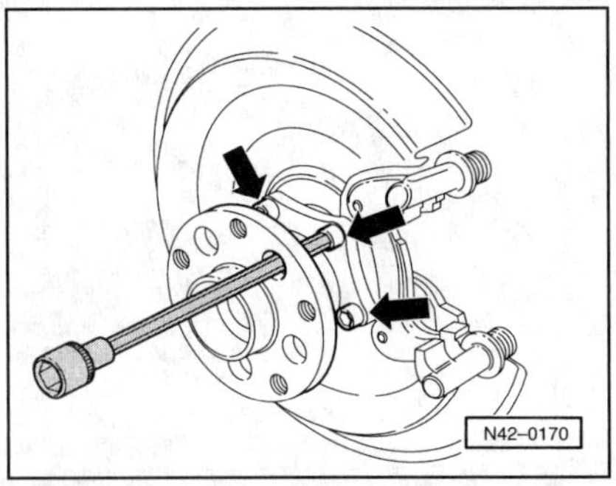

◄ Remove bolts -arrows- for wheel bearing/wheel hub unit.

– Remove wheel bearing/wheel hub unit.

– Installation is reverse order or removal:

- Install wheel and tighten to 120 Nm (89 ft-lb).
- Wheel bearing to rear axle, tighten bolts to 60 Nm (44 ft-lb).
- Brake caliper to brake carrier, tighten bolt to 35 Nm (26 ft-lb)

Wheel mounting for vehicles with front wheel drive, assembly

The wheel bearing and wheel hub are installed together in a housing.

This wheel bearing/hub unit is maintenance and adjustment free. Adjusting and servicing is not possible!

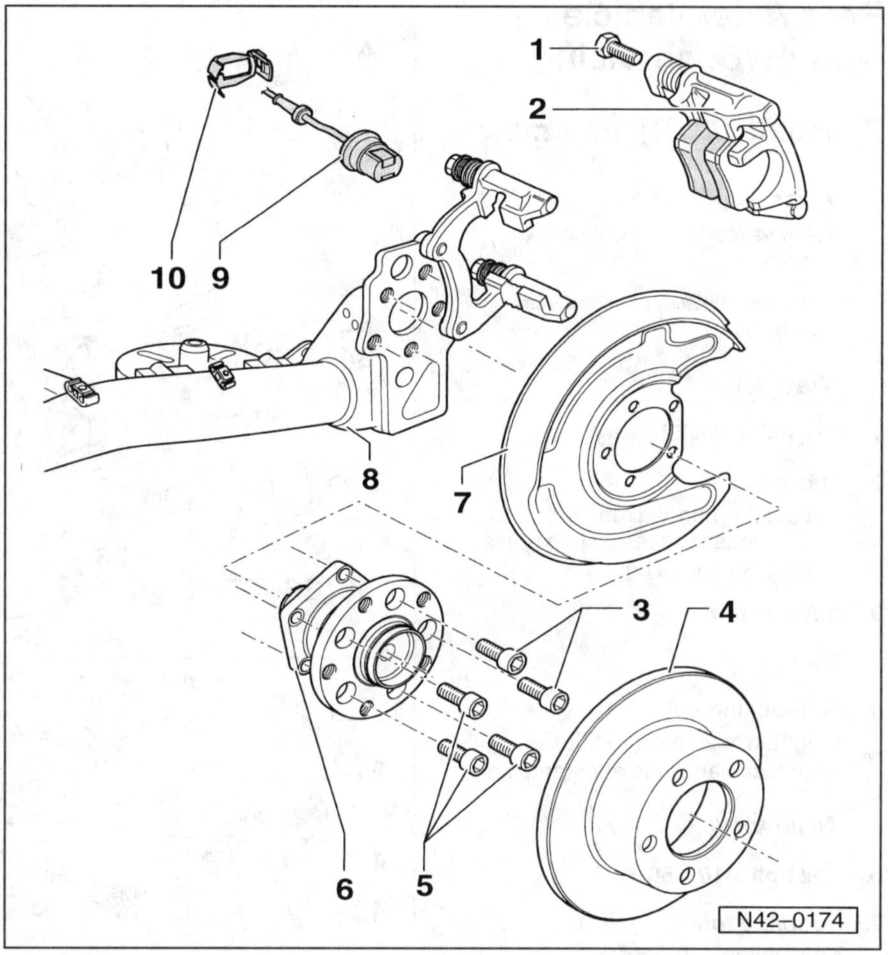

1. **Hex bolt**
   - Always replace
   - Tighten to 35 Nm (26 ft-lb)
2. **Brake caliper**
   - Do not disconnect hydraulic hose to remove caliper from assembly
   - See **Brake System section, Repair Group 46**
3. **Allen head bolt**
   - 60 Nm (44 ft-lb)
4. **Brake disc**
5. **Hex socket head bolt**
   - 60 Nm (44 ft-lb)
6. **Wheel bearing/hub unit**
7. **Splash shield**
8. **Axle beam**
9. **Vehicle speed sensor**
   - Only on vehicles with ABS
10. **Vehicle speed sensor retainer**
    - Only on vehicles with ABS

Rear Axle, Vehicles with 4WD, servicing

Rear axle (4WD), assembly

NOTE—
- *Always replace self-locking nuts and bolts.*
- *Always replace corroded nuts and bolts.*

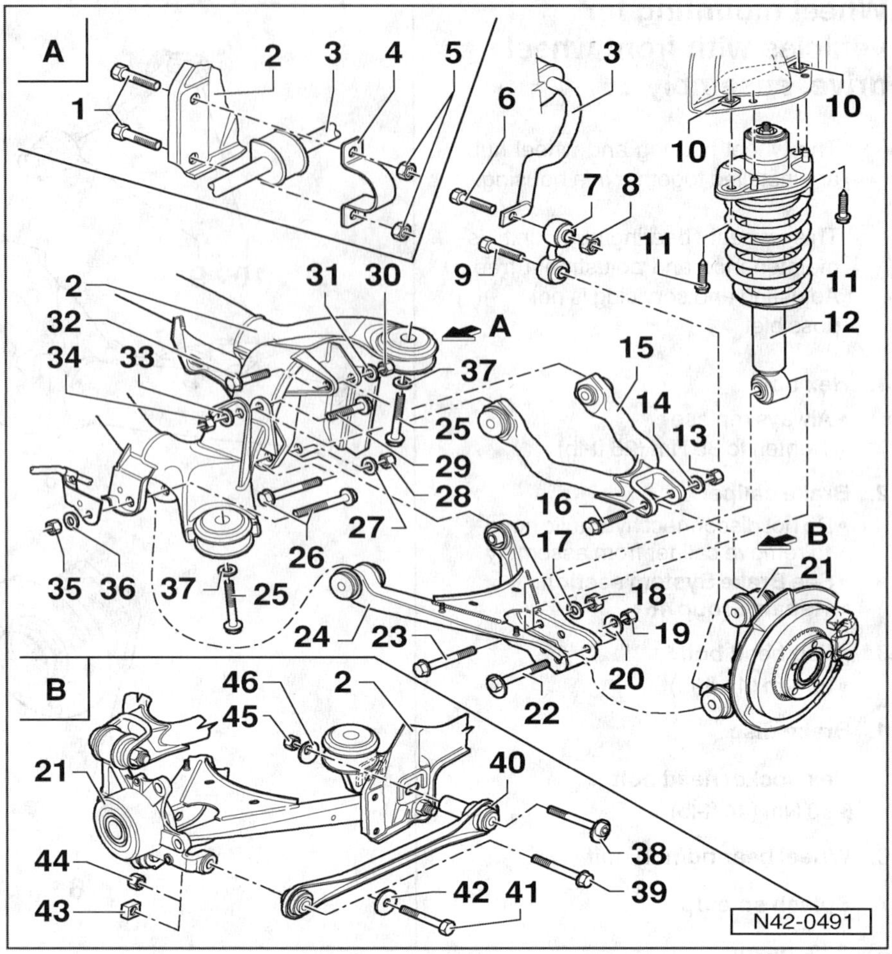

View A

1. **Hex bolt M8 x 22**
2. **Subframe**
   - Replacing bonded rubber mountings requires a shop press and assorted fittings
3. **Anti-roll bar**
4. **Clamp**
5. **Self-locking nut**
   - Tighten to 20 Nm (15 ft-lb)
   - Replace each time after removing

Main view

6. **Hex bolt M10 x 50**
7. **Coupling rod**
   - Allocation
   - Vehicles with steel wheel bearing housings, see **Wheel bearing housing (steel) and control arm, (4WD), assembly, -13-**
   - Vehicles with aluminium wheel bearing housings, see **Wheel bearing housing (aluminium) and control arm, (4WD), servicing, -27-**
8. **Self-locking nut**
   - Tighten to 50 Nm (37 ft-lb)
   - Replace each time after removing
9. **Hex bolt**
   - Vehicles with steel wheel bearing housings, see **Wheel bearing housing (steel) and control arm, (4WD), assembly, -12-**
   - Vehicles with aluminium wheel bearing housings, see **Wheel bearing housing (aluminium) and control arm, (4WD), assembly, -17-**
10. **Threads in wheel housing**
    - If thread in welded nut in wheel housing is damaged, it can be serviced with a Heli-coil.
11. **Hex head bolt**
    - Tighten to 45 Nm (33 ft-lb)
12. **Suspension strut**
    - See **Suspension strut, vehicles (4WD), assembly**
13. **Self-locking nut**
    - Replace each time after removing
14. **Washer**
15. **Upper control arm**
16. **Hex head bolt M 12 x 1.5 x 95**
    - Vehicles with steel wheel bearing housings, see **Wheel bearing housing (steel) and control arm, (4WD), assembly, -14-**
    - Vehicles with aluminium wheel bearing housings, see **Wheel bearing housing (aluminium) and control arm, (4WD), assembly, -24-**
17. **Washer**
18. **Self-locking nut**
    - Tighten to 70 Nm (52 ft-lb)
    - Replace each time after removing
19. **Self-locking nut**
    - Tighten to 95 Nm (70 ft-lb)
    - Vehicles with steel wheel bearing housings, see **Wheel bearing housing (steel) and control arm (4WD), assembly, -11-**
    - Vehicles with aluminium wheel bearing housings, see **Wheel bearing housing (aluminium) and control arm (4WD), assembly, -6-**
20. **Eccentric washer**
21. **Wheel bearing housing**
22. **Eccentric bolt**
23. **Hex head bolt M 12 x 1.5 x 95**
    - Replace each time after removing

(continued on following page)

Rear axle (4WD), assembly overview (continued)

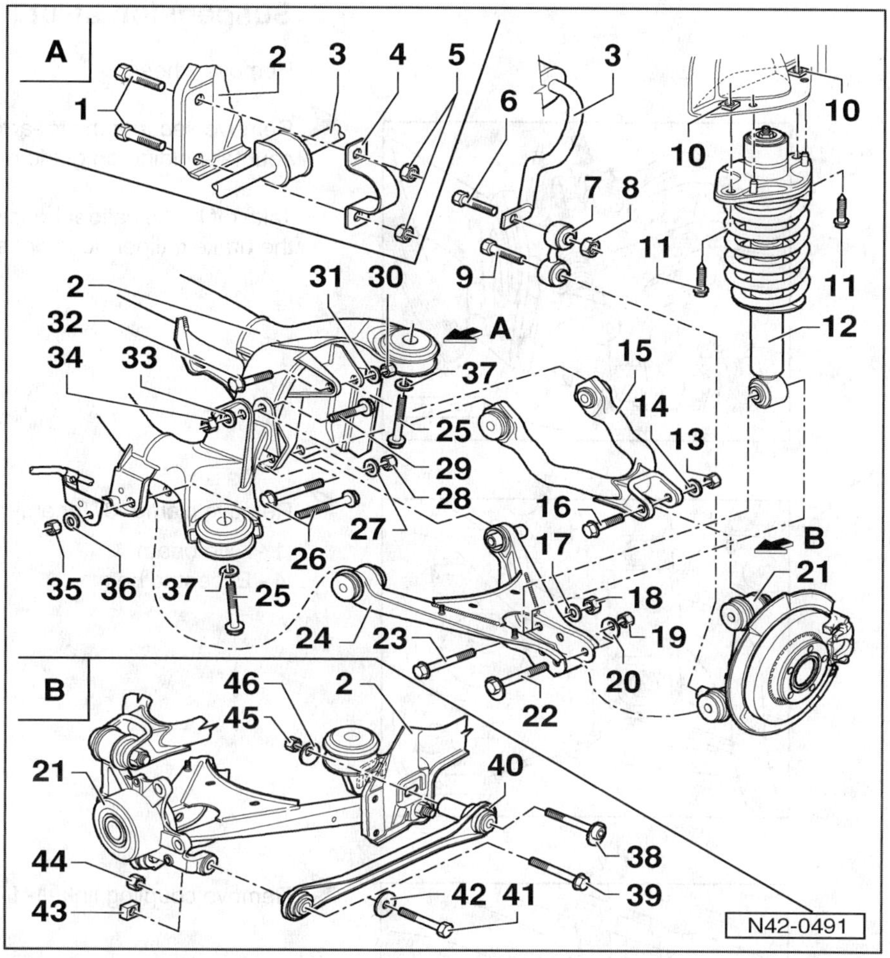

24. Lower control arm

25. Hex bolt M 12 x 1.5 x 110
- 110 Nm (81 ft-lb) and turn 90° further
- Replace each time after removing

26. Hex bolt M 12 x 1.5 x 100
- Turn 70 Nm (52 ft-lb)+ 90° further
- Replace each time after removing

27. Washer

28. Self-locking nut
- Replace each time after removing

29. Hex bolt M 12 x 1.5 x 85
- Tighten to 95 Nm (70 ft-lb)

30. Self-locking nut
- Tighten to 95 Nm (70 ft-lb)
- Replace each time after removing

31. Washer

32. Hex bolt M 12 x 1.5 x 80
- Tighten to 95 Nm (70 ft-lb)

33. Washer

34. Self-locking nut, 95 Nm
- Replace each time after removing

35. Self-locking nut
- Replace each time after removing

36. Washer

37. Washer

View B

38. Eccentric bolt

39. Hex bolt M 12 x 1.5 x 85
- Tighten to 95 Nm (70 ft-lb)
- Only for vehicles with 8-cylinder engines

40. Track rod
- The bonded rubber mountings are not available as replacement part

41. Hex bolt M 12 x 1.5 x 85
- Tighten to 95 Nm (70 ft-lb)

42. Washer

43. Square nut
- Only for vehicles with 8-cylinder engines

44. Self-locking nut
- Replace each time after removing

45. Self-locking nut, 95 Nm
- Replace each time after removing

46. Eccentric washer

Suspension strut (4WD), removing and installing

– Remove wheel.

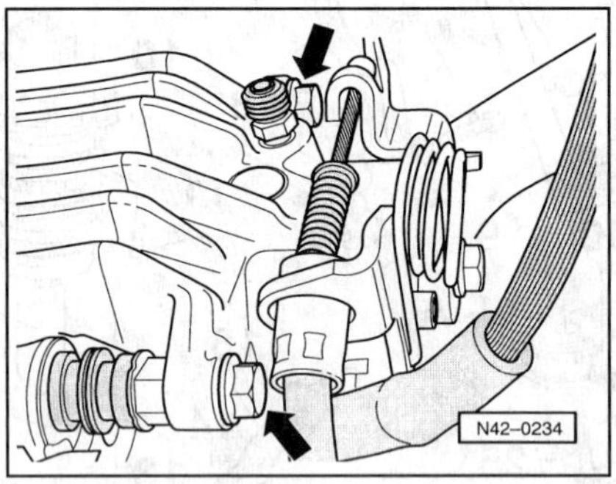

◄ Remove securing bolts -arrows- from brake caliper housing while counter-holding on guide pins.

– Take off brake caliper housing and secure with wire so that the weight of the brake caliper does not stress or damage the brake hose.

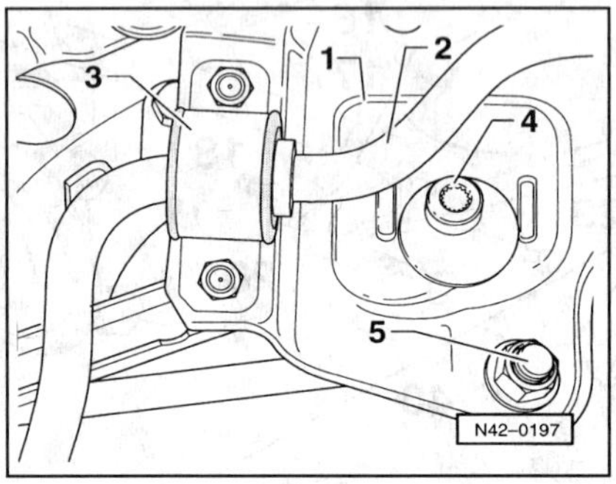

◄ Remove clamp -3- for anti-roll bar -2-.

1 - Axle beam
4 - Eccentric bolt

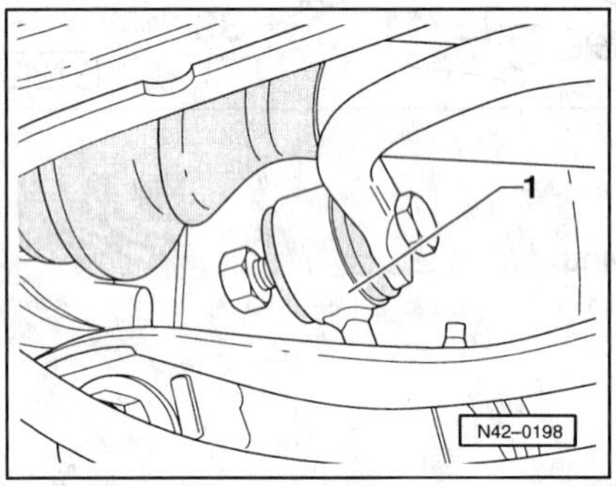

◄ Remove coupling link -1- from anti-roll bar.

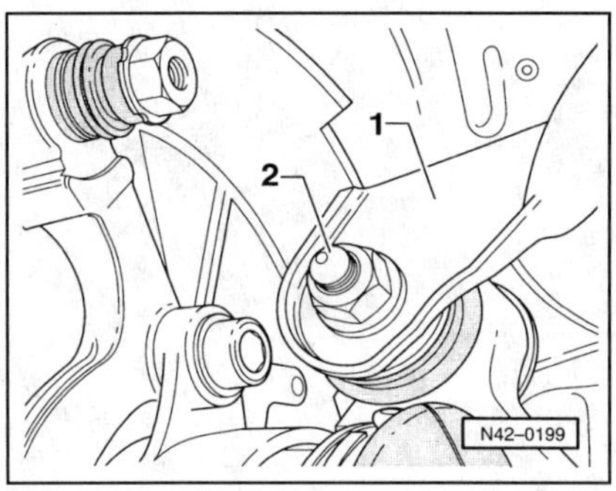

◄ Remove bolt -2- for control arm -1-.

Vehicles with automatic headlight range control

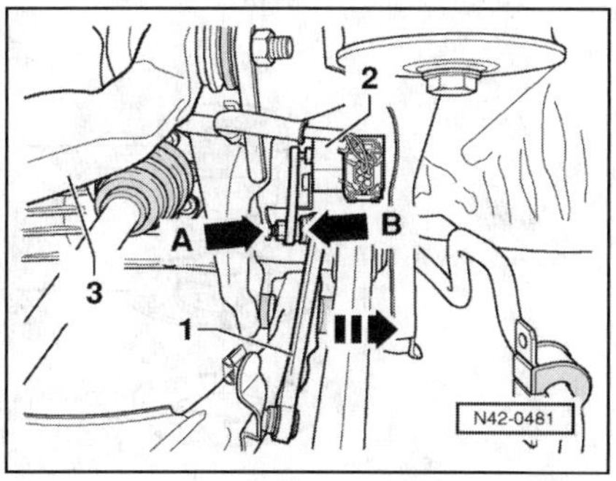

Remove nut -arrow A- from coupling rod -1- for vehicle attitude sensor -2- while counterholding on the wrench surfaces -arrow B-.

- Pull out coupling rod -1- for vehicle attitude sensor -2- in direction of arrow.

3 - Upper control arm

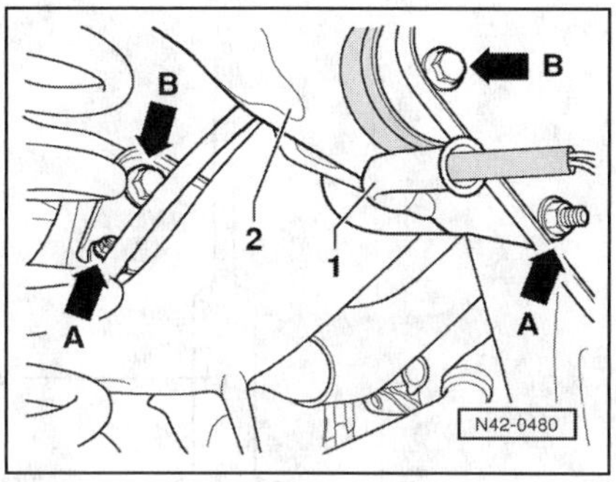

Unbolt -arrows A- protective tube -1- from upper control arm -2-.

Continued for all vehicles

– Remove securing bolts -arrows B- for upper control arm -2- and take off control arm upward.

Unbolt suspension strut mounting hex bolts (arrows) to body.

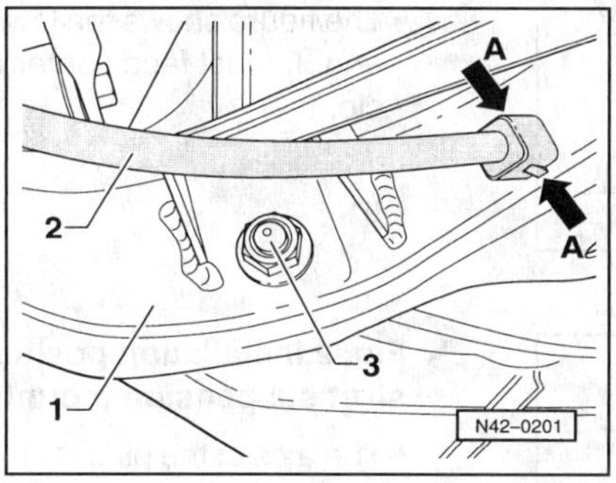

Unbolt suspension strut bolt -3- to control arm -1- and remove shock absorber.

– To remove shock absorber pull wheel bearing housing lightly downward.

– Installation is reverse order of removal, noting the following:

- Install wheel and tighten to 120 Nm (89 ft-lb).
- Anti-roll bar clamp to subframe, tighten nut to 20 Nm (15 ft-lb).
- Coupling link to anti-roll bar, tighten nut to 50 Nm (37 ft-lb).
- Upper control arm to wheel bearing housing, tighten bolt to 70 Nm (52 ft-lb) + turn 90° further.
- Shock absorber to body, tighten bolt to 45 Nm (33 ft-lb)
- Shock absorber to control arm, tighten nut to 70 Nm (52 ft-lb)
- Brake caliper to brake carrier, tighten bolt to 35 Nm (26 ft-lb).

Suspension strut (4WD), assembly

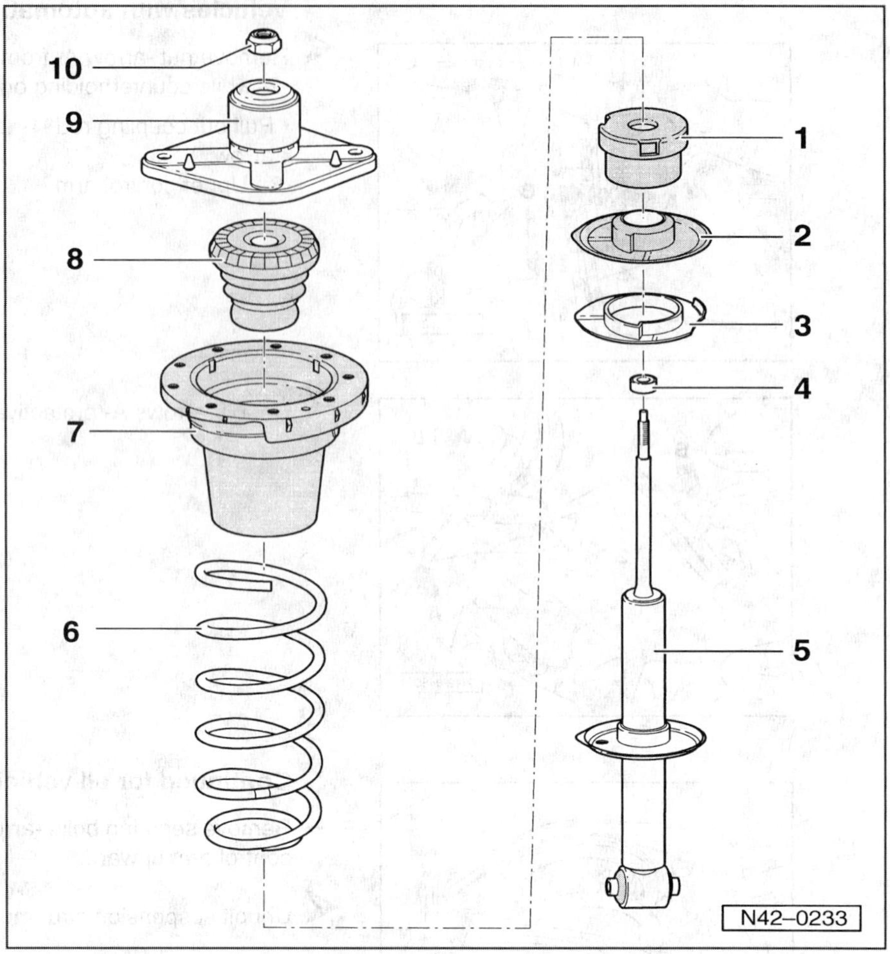

1. **Protective cap**
   - Pre assembled on shock absorber
2. **Rubber insert**
   - Pre assembled on shock absorber
3. **Spring seat**
4. **Spacer**
   - The spacer is only installed on shock absorbers from the manufacturer "Bilstein".
   - It must not be installed to shock absorbers of other manufacturers!
5. **Gas filled shock absorber**
   - Can be replaced individually
6. **Coil spring**
   - Examine for paint damage and rectify paint damage if necessary
   - Spring allocation
     For coil spring application see latest factory parts information and vehicle data sticker.
   - Removal of the coil spring requires the use of a spring compressor.
   - Assembling, see **Fig.1**
7. **Spring seat with protective tube**
   - Two-part
8. **Buffer stop**
9. **Suspension strut mounting**
   - Assembling, see **Fig. 2**
10. **Self-locking hex nut**
    - Tighten to 27 Nm (20 ft-lb)
    - Replace each time after removing

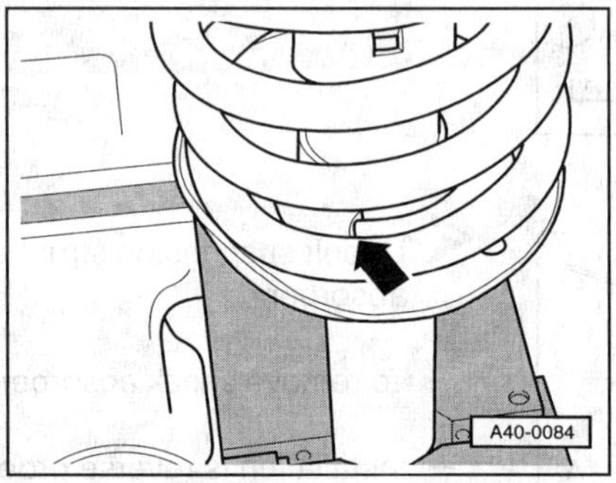

Fig. 1 Suspension strut, assembling

- The ends of the upper and lower of the coil spring must be in contact with the spring seat -arrow-.
- The figure shows the lower spring end. It must face center of vehicle.

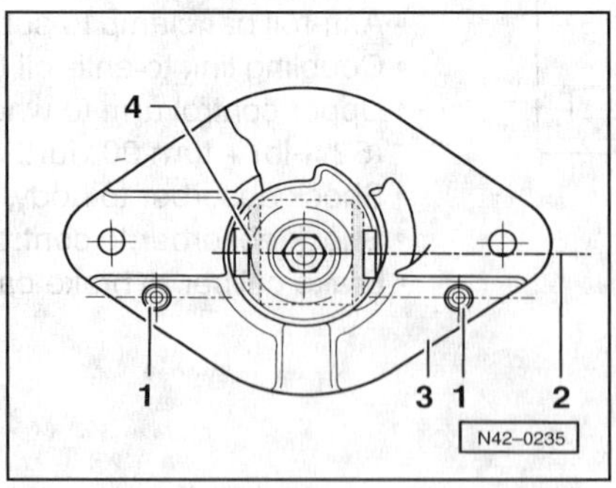

Fig. 2 Installation position of strut suspension mounting

- The axis of the pins -1- must align with the lower shock absorber bushing axis -2-.The pins -1- faces outer side of vehicle.

3 - Shock absorber mounting

Wheel bearing housing (steel) and control arm (4WD), servicing

NOTE—

If a vehicle has to be moved after removing the drive axle, install an outer joint instead of a drive axles and tighten to 50 Nm (37 ft-lb). Otherwise the wheel bearing will become damaged.

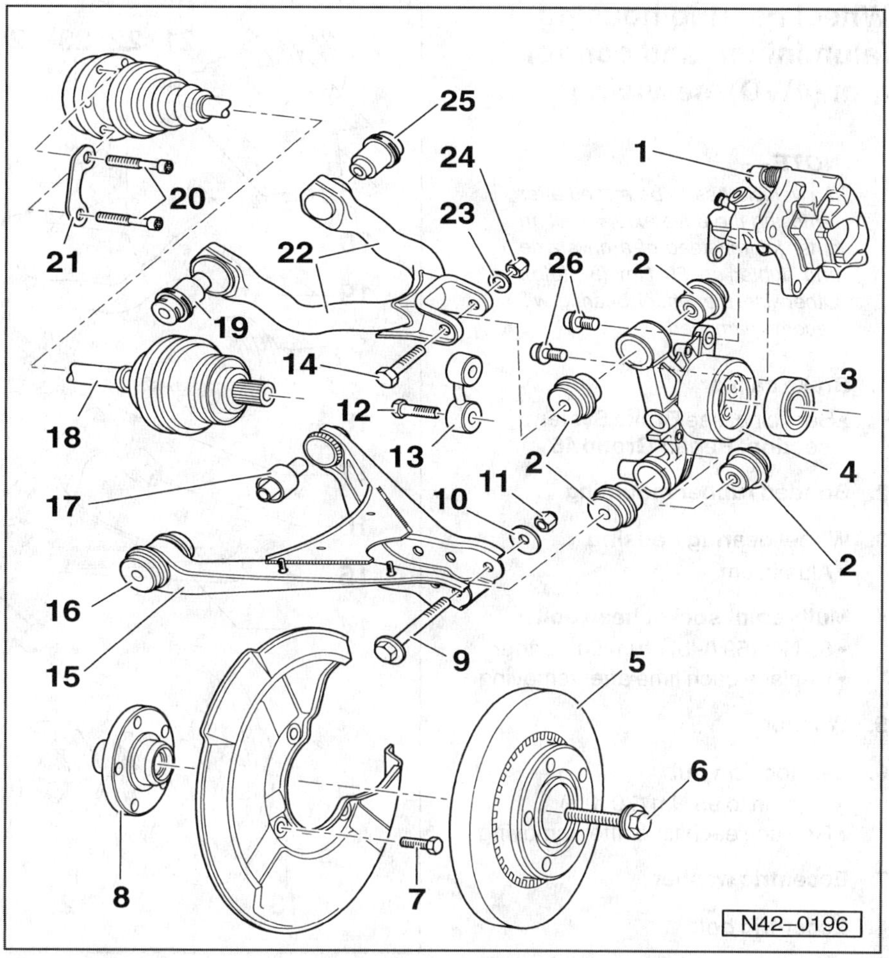

1. **Brake caliper**
   - Servicing, see **Brake System section, Repair Group 46**
2. **Bonded rubber mounting**
   - Removing and installing requires the use of a shop press and assorted press tubes and plates.
3. **Wheel bearing**
   - Removing and installing requires the use of a shop press and assorted press tubes and plates.
   - 75 mm diameter
   - Pressing out, see **Fig. 2**
   - Pressing in, see **Fig. 4**
4. **Wheel bearing housing**
5. **Brake disc**
   - See **Brake System section, Repair Group 46**
6. **Hex bolt**
   - See **Rear drive axle with constant velocity joint (4WD), servicing, -2-**
7. **Hex head bolt**
   - Tighten to 10 Nm (7 ft-lb)
8. **Hub**
   - Pressing out, see **Fig. 1**
   - Pulling off inner race, see **Fig. 3**
   - Pressing in, see **Fig. 5**
9. **Eccentric bolt**
10. **Eccentric washer**
11. **Self-locking nut**
    - Tighten to 95 Nm (70 ft-lb)
    - Replace each time after removing
12. **Hex bolt**
    - Tighten to 50 Nm (37 ft-lb)
13. **Coupling rod**
    - Installation position: Bend faces outward
14. **Hex bolt M 12 x 1.5 x 95**
    - 70 Nm (52 ft-lb)+ turn 90° further
    - Replace each time after removing
15. **Lower control arm**
16. **Front bonded rubber mounting for lower control arm**
    - Pressing out is extremely difficult or in some cases impossible!
    - If bonded rubber mounting is faulty replace complete control arm.
17. **Rear bonded rubber mounting for lower control arm**
18. **Drive axle**
    - Removing and installing, see **Rear drive axles (4WD), removing and installing**
    - Servicing, see **Rear drive axle with constant velocity joint (4WD), servicing**
19. **Front bonded rubber mounting for upper control arm**
    - Removing and installing requires the use of a shop press and assorted press tubes and plates.
20. **Multi-point socket head bolt**
    - See **Drive axle with constant velocity joint, servicing, -5-**
21. **Plate**
22. **Upper control arm**
23. **Washer**
24. **Self-locking nut**
    - Replace each time after removing
25. **Rear bonded rubber mounting for upper control arm**
    - Removing and installing requires the use of a shop press and assorted press tubes and plates.
26. **Hex socket head bolt**
    - See **Brake System section, Repair Group 46**

Wheel bearing housing (aluminium) and control arm (4WD), servicing

NOTE—
If a vehicle has to be moved after removing the drive axle, install an outer joint instead of a drive axle and tighten to 50 Nm (37 ft-lb). Otherwise the wheel bearing will become damaged.

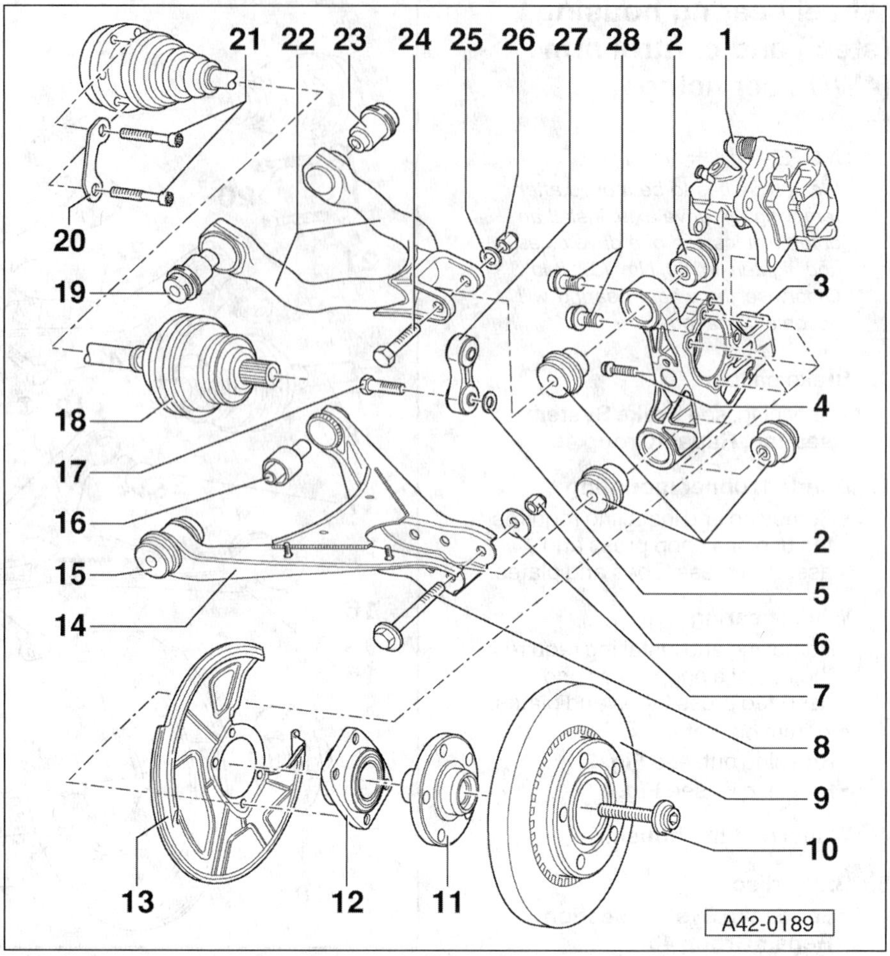

1. **Brake caliper**
   - Servicing, see **Brake System section, Repair Group 46.**
2. **Bonded rubber mounting**
3. **Wheel bearing housing**
   - Aluminium
4. **Multi-point socket head bolt**
   - 80 Nm (59 ft-lb)+ turn 90° further
   - Replace each time after removing
5. **Washer**
6. **Self-locking nut**
   - Tighten to 95 Nm (70 ft-lb)
   - Replace each time after removing
7. **Eccentric washer**
8. **Eccentric bolt**
9. **Brake disc**
   - See **Brake System section, Repair Group 46.**
10. **Hex socket head bolt**
    - See **Rear drive axle with constant velocity joint (4WD), servicing, -2-**.
11. **Hub**
12. **Wheel bearing**
    - Removing and installing requires the use of a shop press and assorted press tubes and plates.
13. **Backplate**
14. **Lower control arm**
15. **Front bonded rubber mounting for lower control arm**
    - Pressing out is extremely difficult or in some cases impossible!
    - If bonded rubber mounting is faulty replace complete control arm.
16. **Rear bonded rubber mounting for lower control arm**
    - Removing and installing requires the use of a shop press and assorted press tubes and plates.
17. **Hex bolt**
    - Tighten to 55 Nm (41 ft-lb).
18. **Drive axle**
    - See **Rear drive axles, removing and installing (4WD)**
    - Servicing, see **Rear drive axle with constant velocity joint (4WD), servicing**
19. **Front bonded rubber mounting for upper control arm**
    - Removing and installing requires the use of a shop press and assorted press tubes and plates.
20. **Plate**
21. **Multi-point socket head bolt**
    - See **Rear drive axle with constant velocity joint (4WD), servicing, -5-**
22. **Upper control arm**
23. **Rear bonded rubber mounting for upper control arm**
    - Removing and installing requires the use of a shop press and assorted press tubes and plates.
24. **Hex bolt M 12 x 1.5 x 95**
    - 70 Nm (52 ft-lb) + turn 90° further
    - Replace each time after removing
25. **Washer**
26. **Self-locking nut**
    - Replace each time after removing
27. **Coupling rod**
    - Installation position: Bend faces outward
28. **Hex socket head bolt**
    - See **Brake System section, Repair Group 46.**

Steel wheel bearing housing, removing and installing

Removing

– Remove hex bolt for drive axle.

– Remove wheels.

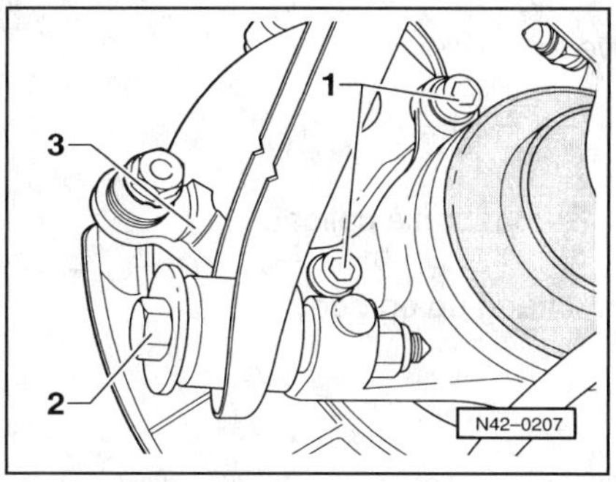

– Remove securing bolts -1- from brake caliper.

– Remove brake caliper and hang up on body.

– Take off disc brake.

– Remove bolts for splash plate.

– Take off splash plate.

– Disconnect ABS vehicle speed sensor out from wheel bearing housing.

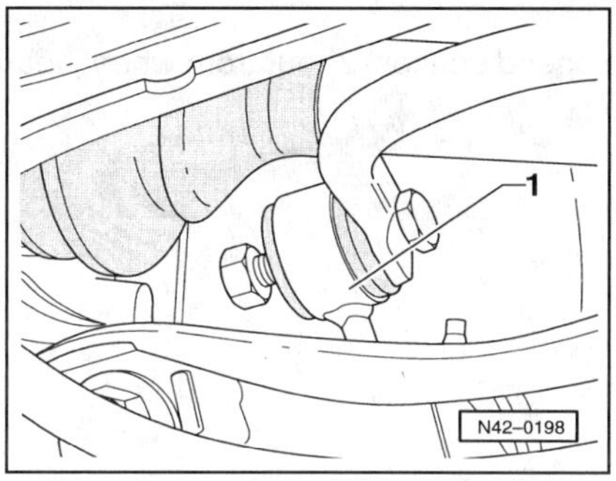

– Remove coupling rod from anti-roll bar.

• 1 - Coupling link

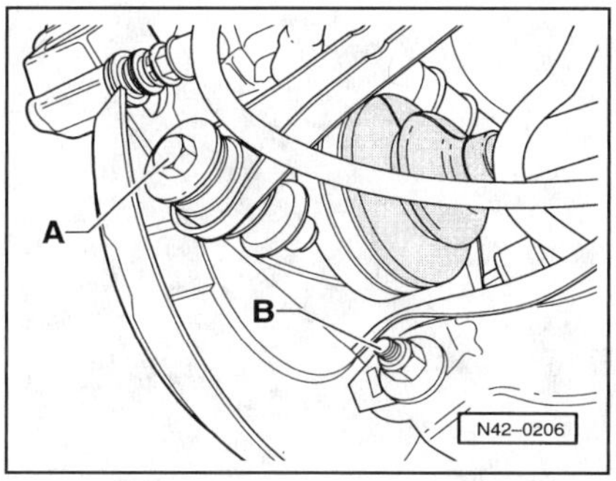

– Remove bolt -A- for track rod.

– Mark installation position of eccentric washer for bolt -B-.

– Remove bolt -B-.

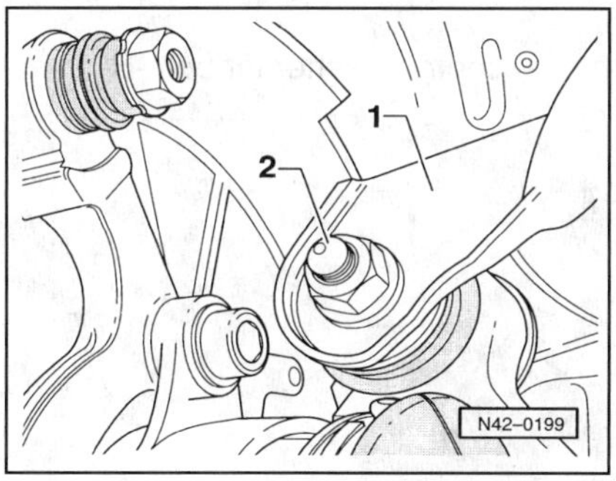

– Remove bolt -2- for control arm -1-.

– Pull wheel bearing housing off drive axle.

Steel wheel bearing housing, installing

– Installation is reverse order of removal, noting the following:

- Install wheels and tighten to 120 Nm (89 ft-lb).
- After installing check position of steering wheel during a test drive. If steering wheel is not in straight ahead position the front axle tracking must be checked!
- Upper control arm to wheel bearing housing, tighten bolt to 70 Nm (52 ft-lb)
- Lower control arm to wheel bearing housing, tighten nut to 95 Nm (70 ft-lb)
- Drive axle to wheel hub, see **Rear drive axle with constant velocity joint (4WD), servicing, -2-**
- Track rod to wheel bearing housing, tighten bolt to 95 Nm (70 ft-lb)
- Splash plate to wheel bearing housing, tighten bolt to 10 Nm (7 ft-lb)
- Brake caliper to wheel bearing housing, tighten bolt to 35 Nm (26 ft-lb)

Aluminium wheel bearing housing, removing and installing

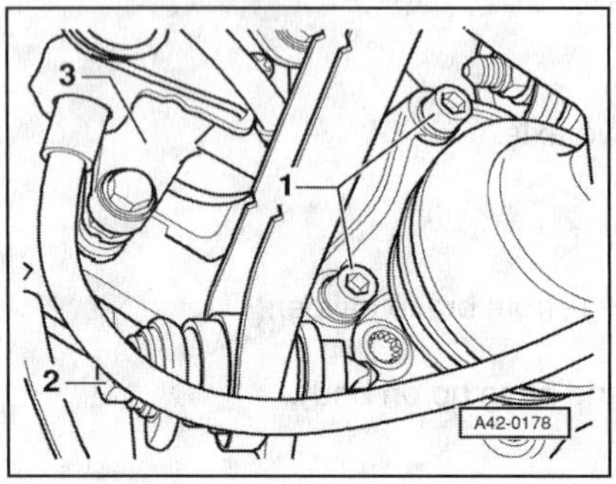

Removing

– Remove hex bolt for drive axle.

– Remove wheels.

◄ Remove securing bolts -1- from brake caliper.

– Remove brake caliper and hang up on body.

– Take off disc brake.

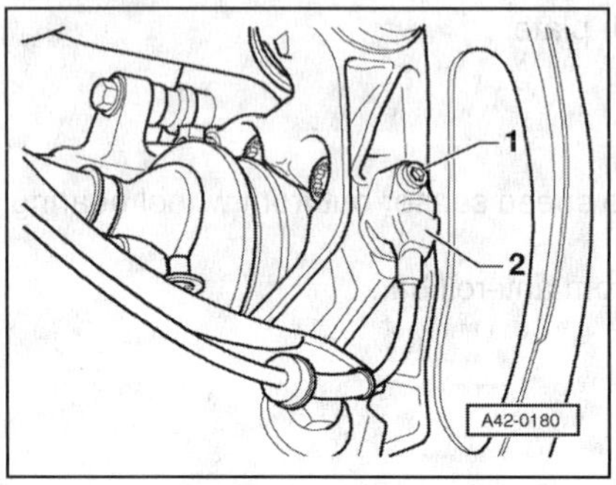

◄ Remove hex key head bolt -1-.

– Disconnect ABS vehicle speed sensor -2- out from wheel bearing housing.

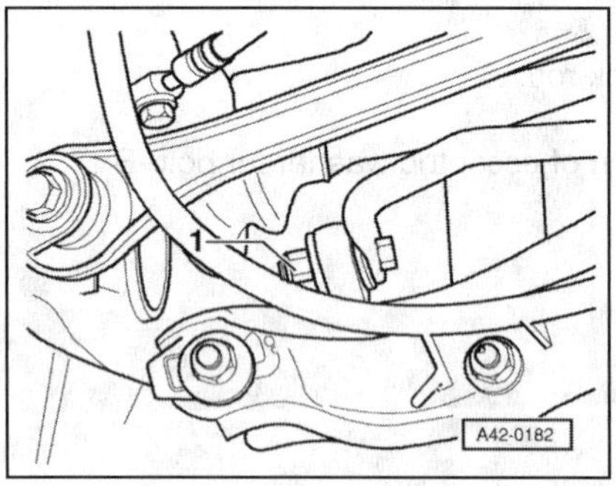

◄ Remove bolt -1- for coupling link.

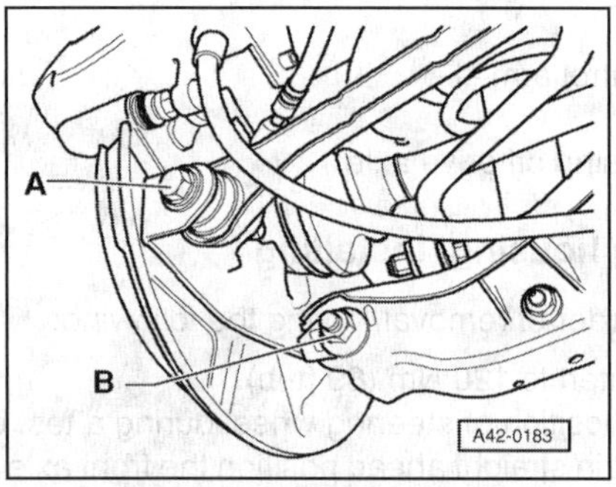

◄ Remove bolt -A- for track rod.

– Mark installation position of eccentric washer for bolt -B-.

– Remove bolt -B-.

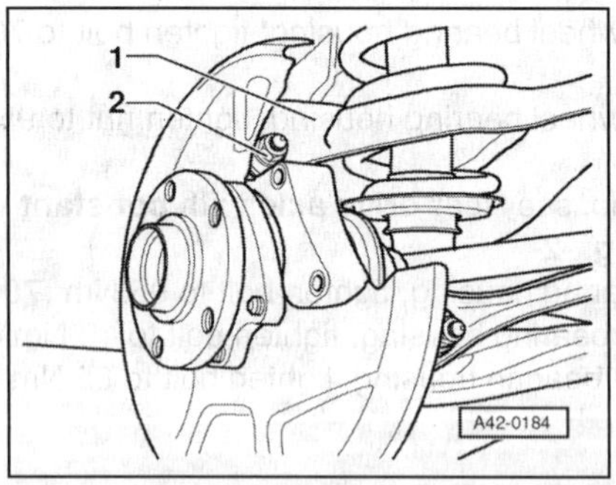

◄ Remove bolt -2- for control arm -1-.

– Pull wheel bearing housing off drive axle.

Installing

– Installation is reverse order of removal, noting the following:

- Install wheel and tighten to 120 Nm (89 ft-lb).
- After installing check position of steering wheel during a test drive. If steering wheel is not in straight ahead position the front axle tracking must be checked!
- Upper control arm to wheel bearing housing, Tighten bolt to 70 Nm (52 ft-lb) + turn 90° further.
- Lower control arm to aluminium wheel bearing housing, tighten nut to 95 Nm (70 ft-lb) Drive axle to wheel hub, see **Rear drive axle with constant velocity joint (4WD), servicing, item #2.**
- Track rod to wheel bearing housing, tighten bolt to 95 Nm (70 ft-lb)
- Coupling rod to anti-roll bar, tighten nut to 50 Nm (37 ft-lb)
- Brake caliper to wheel bearing housing, tighten bolt to 35 Nm (26 ft-lb)

Subframe and final drive (4WD), overview

1. **Subframe**
2. **Washer**
3. **Nut**
   - Tighten to 40 Nm (30 ft-lb)
4. **Protective plate**
5. **Washer**
6. **Hex bolt M 12 x 1.5 x 130**
   - Tighten bolt to 110 Nm (81 ft-lb) + 90° further
7. **Bonded rubber mounting**
8. **Hex bolt M 10 x 75**
9. **Hex socket head bolt M10 x 45**
   - Tighten to 55 Nm (41 ft-lb)
10. **Hex socket head bolt M10 x 25**
    - Tighten to 40 Nm (30 ft-lb)
11. **Final drive support**
12. **Final drive**
    - If the final drive is removed, the protective plate -4- must be removed on some vehicles, see **Fig. 1**
13. **Cross member**

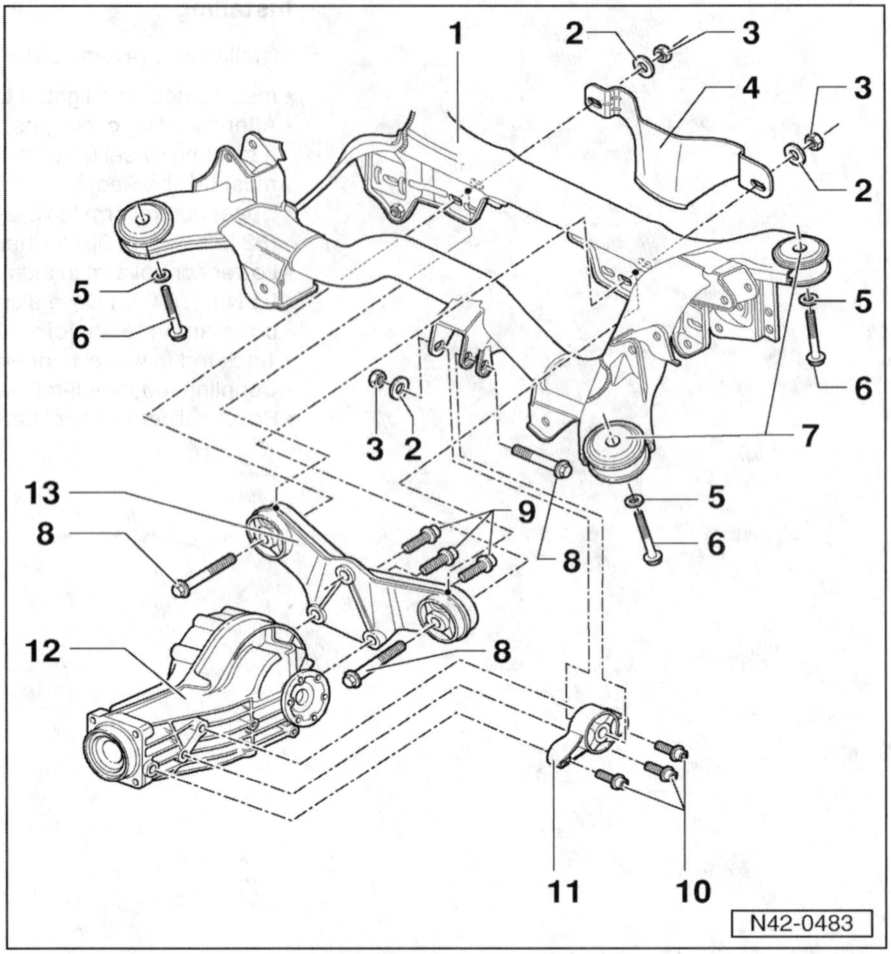

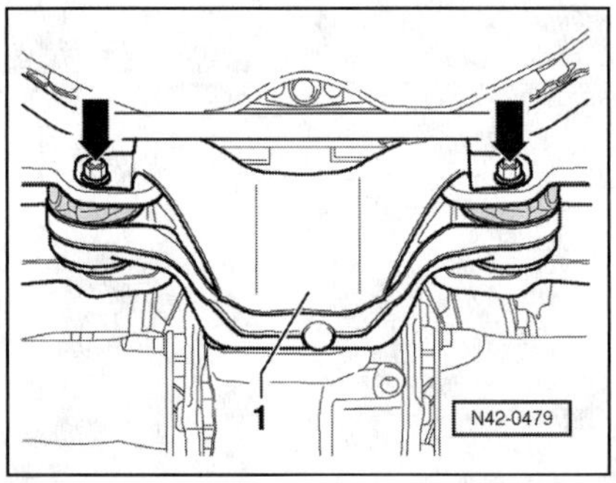

Fig. 1 Removing protective plate

- Remove protective plate -1- from cross member -arrows-, if installed.
- The cross member securing bolts remain installed on subframe.

Rear drive axles (4WD), removing and installing

Special tools, workshop equipment, testers, measuring instruments and auxiliary items required

• VAG 1383 A Engine/transmission jack

Removing

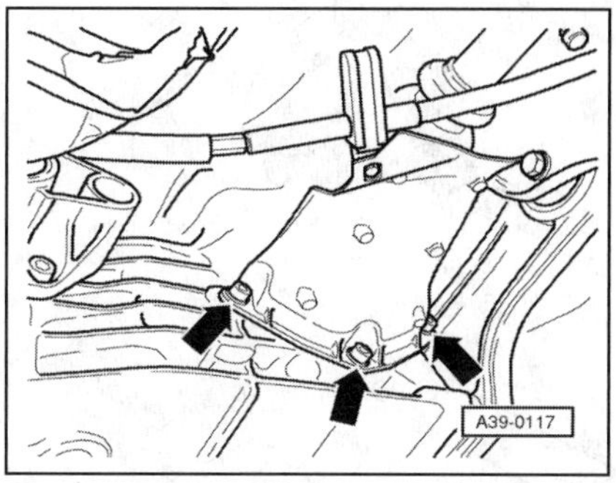

– Remove hex bolt for drive axle.

◄ Remove heat shield for drive axle -arrows-.

– Unbolt drive axle from drive flange.

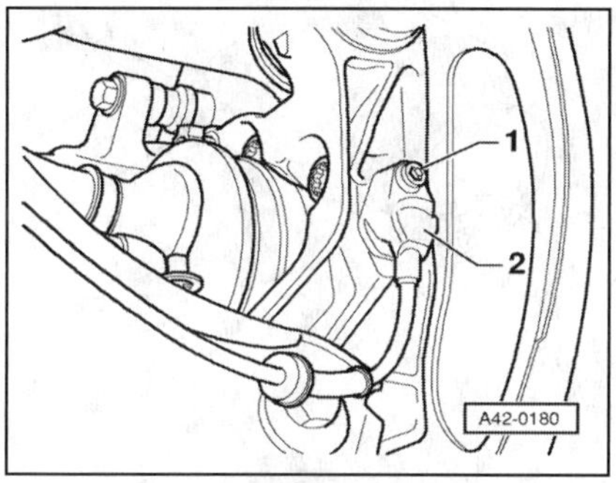

◄ Remove hex key head bolt -1-.

– Disconnect ABS vehicle speed sensor -2- out from wheel bearing housing.

– Disconnect connector behind Three Way Catalytic Converter (TWC) and remove rear part of exhaust system.

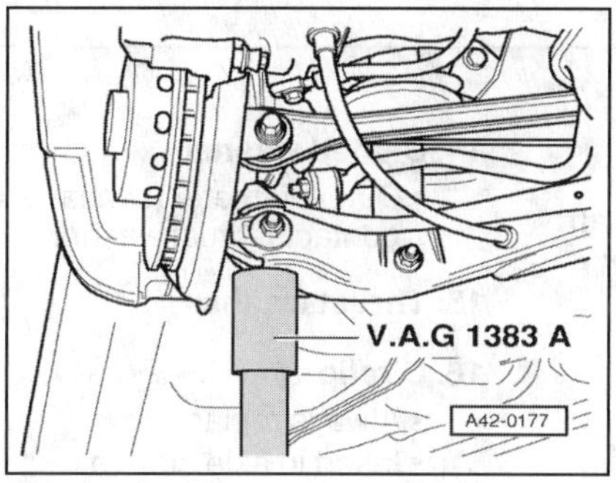

◄ Place transmission jack e.g. VAG 1383 A under control arm.

– Raise control arm slightly.

– Drive drive axle out of hub with a plastic head hammer if necessary.

– Remove drive axle.

Installing

– Installation is reverse order of removal, noting the following:

• Drive axle to wheel hub, see **Rear drive axle with constant velocity joint (4WD), servicing**, item #**2**
• Drive axle to flanged axle, tighten to 40 Nm (30 ft-lb).

Rear drive axle with constant velocity joint (4WD), servicing

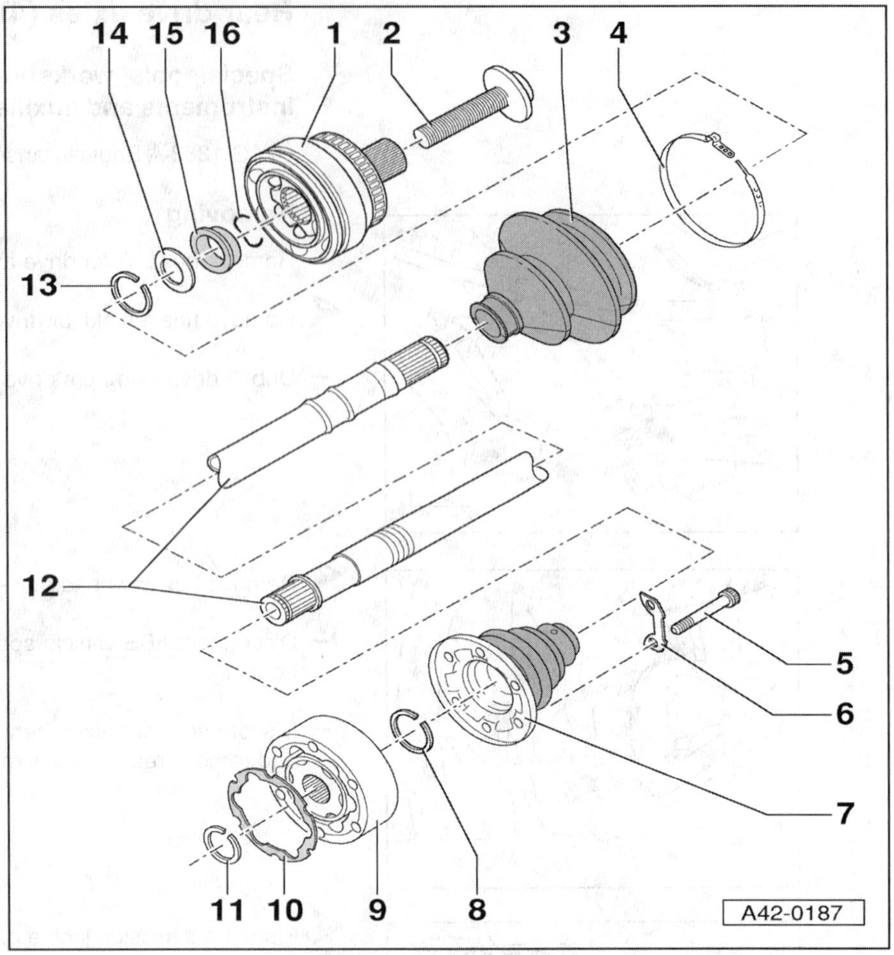

Special tools, workshop equipment, test and measuring appliances and aux. items required

- VW 401 Thrust plate
- VW 402 Thrust plate
- VW 408 A Press tool
- VW 411 Press tool
- VW 447 H Press plate
- VW 454 Press piece
- VW 161 A Circlip pliers
- VAG 1331 Torque wrench
- VAG 1332 Torque wrench
- VAG 1682 Tension clamp
- T10065 Assembly device

Grease quantity and type

Joint dia.	Total grease required	Proportions:	
		Joint	Boot
Outer 89mm (3.5 in.)	90g (3.8 oz.)	40g (1.9 oz.)	50g (1.9 oz.)
Inner 88mm (3.4 in.)	120g (4.2 oz.)	35g (1.2oz)	85g (3.0oz)

NOTE—
Constant velocity joints are packed with grease G 000 603:

1. **Outer constant velocity joint**
   - Only replace complete
   - Removing, see **Fig. 1**
   - Installing: drive onto axle on to stop with plastic hammer
2. **Hex bolt**
   - Tightening torque:
     Changed to hex key head bolt
     Bolt M14: 115 Nm (85 ft-lb) and turn 180° further
     Bolt M16: 190 Nm (140 ft-lb) and turn 180° further
   - Vehicle must be standing on ground when tightening
   - Replace each time after removing
3. **Protective boot**
   - Check for tears and chafing
4. **Hose clip**
   - Replace
5. **Multi-point socket head bolt**
   - M 8 x 48; 40 Nm (30 ft-lb)
6. **Plate**
7. **Protective boot for inner constant velocity joint**
   - Drive off with drift
   - Before installing on constant velocity joint, coat sealing surface with D-3
8. **Circlip**
   - Always replace
   - Remove and install with VW 161 a
9. **Inner constant velocity joint**
   - Pressing off, see **Fig. 2**
   - Pressing on, see **Fig. 3**
10. **Gasket**
    - Replacing. Pull off protective foil and stick onto joint.
11. **Circlip**
12. **Drive axle**
13. **Circlip**
    - Not installed to all drive axles
    - Always replace
    - Remove and install with VW 161 a
14. **Dished washer**
    - Outer diameter (concave side) contacts thrust washer
15. **Thrust washer**
16. **Circlip**
    - Always replace
    - Insert in axle groove

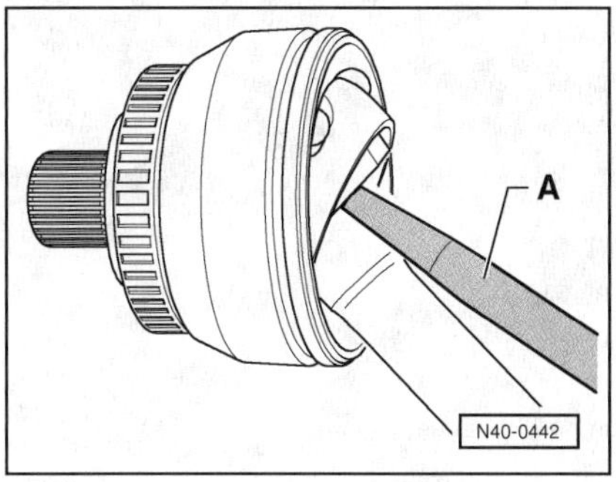

Fig. 1 Pressing off outer constant velocity joint

- Clamp drive axle in vice using vice clamps.
- Remove clamp and slide back boot.
- Drive constant velocity joint off drive axle using drift -A-.
- Drive must be applied exactly on star of constant velocity joint.

Driving joint on

– Drive onto axle with plastic hammer until securing ring engages.

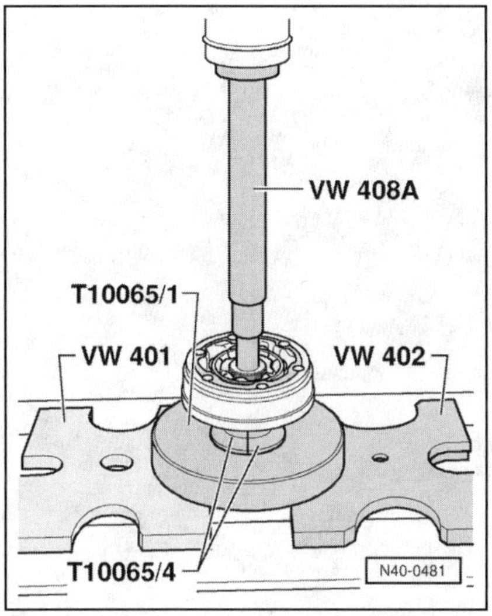

Fig. 2 Pressing off inner constant velocity joint

NOTE—

• First drive boot off with drift

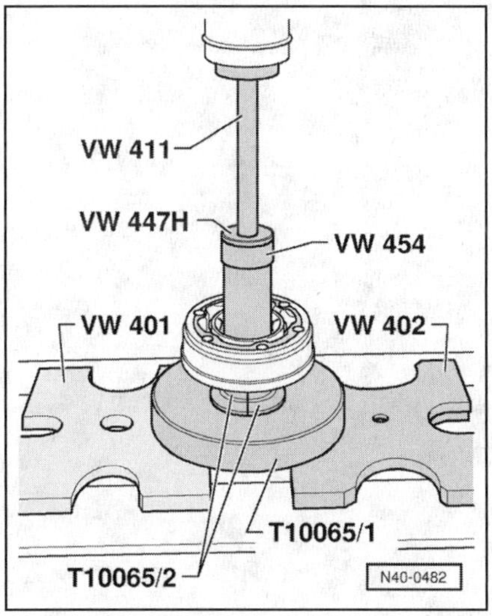

Fig. 3 Pressing on inner constant velocity joint

- Press on joint up to stop.
- Insert circlip.

NOTE—

Chamfer on inner diameter of ball hub (splines) must face the contact shoulder on the drive axle.

44 Wheels, Tires, Alignment

Repair Group

S 44

Tightening torques for wheel bolts

Wheel bolts to wheel hub for all vehicles: 120 Nm (89 ft-lb)

Vehicle alignment

General

NOTE —

- *Vehicle alignment should not be checked before the vehicle has completed 1000 to 2000 km (600 to 1200 miles) as the coil springs have had time to settle.*
- *The cause of vehicle imbalance may also be excessive residual wheel imbalance and/or vertical wheel runout.*
- *When making adjustments the relevant specifications are to be adhered to as close as possible.*
- *The vehicle alignment should be measured using VW/AUDI approved wheel alignment equipment!*
- *When measuring, the front and rear axles must be measured every time., otherwise the proper running characteristics of the vehicle cannot be ensured!*

EDITOR'S NOTE —

- *Front axle camber is only slightly adjustable by location of front subframe to vehicle body.*
- *Rear axle camber can be checked but not adjusted on vehicles with front wheel drive. If the readings are outside the permitted tolerances the axle beam must be checked for damage and replaced if necessary.*
- *The total toe for the rear axle on all wheel drive vehicles is not adjustable.*
- *It is possible to adjust the individual rear axle toe (vehicles with all-wheel-drive) by adjusting the mounting bracket*

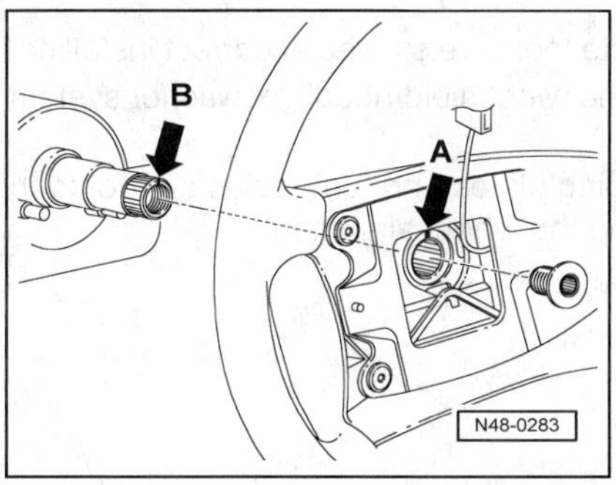

Steering wheel and steering column are marked.

These positions must not be changed!

A - Line on steering wheel

B - Punch mark on steering column

Only then the center position on the rack can be guaranteed!

Wheel alignment must be checked if:

- Not handling properly.
- Involved in an accident.
- Axle components have been removed.
- Tires wear on one side.

Front axle components	Wheel alignment check required	Toe constant "S" checking/ setting required
Upper rear link	No	No
Upper front link	Yes	No
Guide link with hydro-mounting	No	No
Toe control link	No	No
Mounting bracket	Yes	Yes
Wheel bearing housing	Yes	Yes
Tie rod	Yes	Yes
Steering gear	Yes	Yes
Subframe	Yes	Yes

Rear axle components	Wheel alignment check required
Shock absorber	No
Coil spring	No
Upper link[1)]	No
Lower link[1)]	Yes
Wheel bearing housing[1)]	Yes
Tie rod[1)]	Yes
Subframe[1)]	Yes
Torsion beam axle (front wheel drive)	Yes

[1)] Only affects vehicles with all-wheel-drive

Test requirements:

- Check suspension, wheel bearing, steering and steering linkage for excessive play and damage.
- Tread depth difference of no more than 2 mm on an axle.
- Tires inflated to correct pressure
- Vehicle unloaded
- Fuel tank must be full
- Spare wheel and vehicle tools are stowed at correct installing locations
- The fluid reservoir for the windshield/headlight washer system must be full.
- Make sure that the sliding plates and turn tables are not touching the end stop when checking the wheel alignment.

Explanation of the Production code numbers

Various suspensions are available as optional equipment. These are identified by the Production code numbers.

The suspensions installed in a vehicle is documented on the vehicle data plate with the relevant Production code.

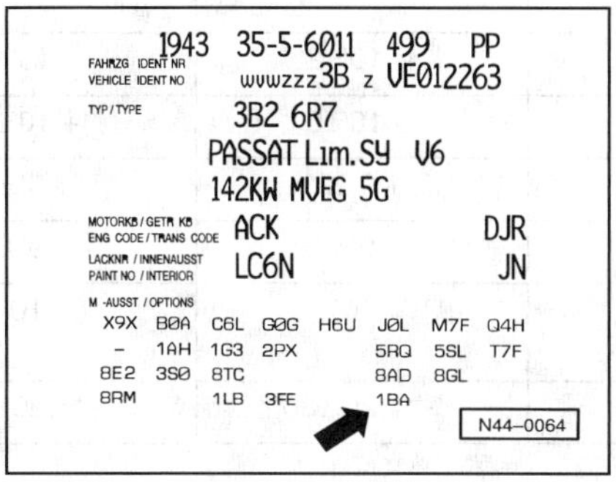

Example of a vehicle data sticker

- In this example the vehicle is equipped with the standard suspension 1BA -arrow-.
- The vehicle data sticker is located in the spare wheel recess and in service booklet.

Standard suspension	Sport suspension	Heavy duty suspension
1BA	1BE	1BB/1BP

Vehicle alignment specifications

Front wheel drive vehicles with 4-cylinder and 6-cylinder engines.

Vehicles with front wheel drive Front axle	Standard suspension (1BA)	Sport suspension (1BE)	Heavy duty suspension (1BP)	Heavy duty suspension (1BB)
Camber	- 35′±25′	- 50′±25′	- 35′±25′	- 20′±25′
Maximum permissible difference between both sides	30′	30′	30′	30′
Toe setting for each wheel (unladen)	+ 10′±2′	+ 10′±2′	+ 10′±2′	+ 10′±2′
Toe check value for each wheel (unladen)	+ 10′±5′	+ 10′±5′	+ 10′±5′	+ 10′±5′
Toe constant for each wheel (setting)	+ 9′±2′	+ 9′±2′	+ 9′±2′	+ 9′±2′
Toe constant for each wheel (check value)	+ 9′±7′	+ 9′±7′	+ 9′±7′	+ 9′±7′
Maximum permissible difference between both sides	10′	10′	10′	10′
Toe angle difference at 20 degrees	- 1° 20′±30′	- 1° 20′±30′	- 1° 20′±30′	- 1° 20′±30′
Rear axle				
Camber	- 90′±20′	- 90′±20′	- 90′±20′	- 90′±20′
Maximum permissible difference between both sides	30′	30′	30′	30′
Total toe	$+ 20'^{+15'}/_{-10'}$	$+ 28'^{+15'}/_{-10'}$	$+ 14'^{+15'}/_{-10'}$	$+ 20'^{+15'}/_{-10'}$
Maximum permissible deviation in direction of travel relative to vehicle longitudinal axis	±15′	±15′	±15′	±15′

4MOTION vehicles with 6-cylinder engines.

V6 with all-wheel drive (4MOTION) **Front axle**	**Standard suspension (1BA)**	**Sport suspension (1BE)**	**Heavy duty suspension (1BP)**	**Heavy duty suspension (1BB)**
Camber	- 42′±25′	- 50′±25′	- 35′±25′	- 20′±25′
Maximum permissible difference between both sides	30′	30′	30′	30′
Toe setting for each wheel (unladen)	+ 10′±2′	+ 10′±2′	+ 10′±2′	+ 10′±2′
Toe check value for each wheel (unladen)	+ 10′±5′	+ 10′±5′	+ 10′±5′	+ 10′±5′
Toe constant for each wheel (setting)	+ 9′±2′	+ 9′±2′	+ 9′±2′	+ 9′±2′
Toe constant for each wheel (check value)	+ 9′±7′	+ 9′±7′	+ 9′±7′	+ 9′±7′
Maximum permissible difference between both sides	10′	10′	10′	10′
Toe angle difference at 20 degrees	- 1° 20′±30′	- 1° 20′±30′	- 1° 20′±30′	- 1° 20′±30′
Rear axle				
Camber	- 40′±30′	- 40′±30′	- 40′±30′	- 40′±30′
Maximum permissible difference between both sides	30′	30′	30′	30′
Toe for each wheel	+ 8′±5′	+ 8′±5′	+ 8′±5′	+ 8′±5′
Total toe	+ 16′±10′	+ 16′±10′	+ 16′±10′	+ 16′±10′
Maximum permissible deviation in direction of travel relative to vehicle longitudinal axis	±10′	±10′	±10′	±10′

4MOTION vehicles with W8 engines.

W8 with all-wheel drive (4MOTION) **Front axle**	**Standard suspension (1BA)**	**Sport suspension (1BE)**
Camber	- 42′±25′	- 50′±25′
Maximum permissible difference between both sides	30′	30′
Toe setting for each wheel (unladen)	+ 10′±2′	+ 10′±2′
Toe check value for each wheel (unladen)	+ 10′±5′	+ 10′±5′
Toe constant for each wheel (setting)	+ 9′±2′	+ 9′±2′
Toe constant for each wheel (check value)	+ 9′±7′	+ 9′±7′
Maximum permissible difference between both sides	10′	10′
Toe angle difference at 20 degrees	- 1° 20′±30′	- 1° 20′±30′
Rear axle		
Camber	- 40′±30′	- 40′±30′
Maximum permissible difference between both sides	30′	30′
Toe for each wheel	+ 8′±5′	+ 8′±5′
Total toe	+ 16′±10′	+ 16′±10′
Maximum permissible deviation in direction of travel relative to vehicle longitudinal axis	±10′	±10′

Tire/rim/chain combinations

The use of chains on tires is permitted in some cases. Particular attention must be paid to the table below to prevent damage due to the proximity of bodywork and suspension components.

VW AG Approved Tire/Rim Combination

Passat, Passat 4Motion and Wagon

Wheel/Cover Part Number	Wheel Sizes	OEM Tire Sizes	Tire Chain Usage	Winter Tire Sizes	Tire Chain Usage
Wheel – 3B0601027D **Cover** – 3B0601147H *(Steel Wheel with Wheel Cover)*	6Jx15H2 ET37 (5x112) *(Factory Wheel)*	195/65R15 91H	**Yes**	195/65R15 91H *(with wheel 6Jx15H2 ET 37)*	**Yes**
Wheel – 3B7601025B **Cover** – 3B7601171 *(Steel Wheel with Wheel Cover)*	6Jx16H2 ET45 (5x112) *(Winter Wheel)*	205/55R16 91H	**Yes**	205/55R16 91H *(with wheel 6Jx16H2 ET 45)*	**Yes**
Wheel – 3B0601025S **Cover** – 3B0601149L *(Aluminum Wheel with Center Cap)*	7Jx15H2 ET37 (5x112) *(Factory Wheel)*	195/65R15 H	**No**	195/65R15 91H *(with wheel 6Jx15H2 ET 37)*	**Yes**
Wheel – 3B0601025T **Cover** – 3B7601171 *(Aluminum Wheel with Center Cap)*	7Jx16H2 ET37 (5x112) *(Factory Wheel)*	205/55R16 91H	**No**	205/55R16 91H *(with wheel 6Jx16H2 ET 45))*	**Yes**
Wheel – 3B0601025L **Cover** – 7D0601165 *(Aluminum Wheel with Center Cap)*	7Jx16H2 ET37 (5x112) *(Factory Wheel)*	205/55R16 91H	**No**	205/55R16 91H *(with wheel 6Jx16H2 ET 45)*	**Yes**
Wheel – 3B7601025F **Cover** – 3B7601171 *(Aluminum Wheel with Center Cap)*	7Jx16H2 ET37 (5x112) *(Factory Wheel)*	225/55R16 91H	**No**	205/55R16 91H *(with wheel 6Jx16H2 ET 45)*	**Yes**
Wheel – 3B7601025A **Cover** – 3B7601171 *(Aluminum Wheel with Center Cap)*	7Jx17H2 ET37 (5x112) *(Factory Wheel)*	225/45R17 91H	**No**	205/55R16 91H *(with wheel 6Jx16H2 ET 45)*	**Yes**

Passat W8

Wheel/Cover Part Number	Wheel Sizes	OEM Tire Sizes	Tire Chain Usage	Winter Tire Sizes	Tire Chain Usage
Wheel – 3B7601025A **Cover** – 3B7601171 *(Aluminum Wheel with Center Cap)*	7.5Jx17H2 ET45 *(Factory Wheel)*	225/45R17 H	**No**	205/50R17 93H *(with wheel 6Jx17H2 ET 42)*	**Yes**
Wheel – 3B7601025C **Cover** – *(Aluminum Wheel with Center Cap)*	6.0Jx17H2 ET42 *(Factory Wheel)*	205/50R17 93H	**Yes**	205/50R17 93H *(with wheel 6Jx17H2 ET 42)*	**Yes**
Wheel – 3B7601025 **Cover** – 3B7601149 *(Aluminum Wheel with Center Cap)*	7Jx16H2 ET45 () *(Factory Wheel)*	215/55 R16 H	**No**	205/55R16 94H *(with wheel 6Jx16H2 ET 45)*	**Yes**
Wheel – 3B7601025B **Cover** – 3B7601171 *(Steel Wheel with Wheel Cover)*	6Jx16H2 ET45 *(Winter Wheel)*	205/55R16 94H	**Yes**	205/55R16 94H	**Yes**

Tire chains must not be used on the compact spare wheel.
Ultra High performance tires are not suited for use on ice or snow.
4Motion vehicles snow chains are permitted on the front wheels only.
Part numbers are for reference only. Always check with your Parts Dept. for the latest parts information.

Part numbers that do not appear in this document are not suitable for tire chain usage.

0024499

48 Steering

Repair Group

S 48

EDITOR'S NOTE —
For information on steering wheel and airbag removal, see **Body-Interior section, Repair Group 69**.

Steering Column

Column tube, steering wheel with steering column, removing and installing

Removing

The steering column is supplied as a spare part. Servicing is not possible.

The steering lock housing can be transferred.

> ***WARNING —***
> *Before working on the electrical system and before removing the steering wheel the following conditions must be met:*
>
> *• Ground strap disconnected from battery.*
>
> *• Wheels must be in straight ahead position.*
>
> *• If these instructions are not observed it is possible that the airbag system may fail later!*

Steering wheel, removing

– Turn wheels to straight ahead position.

– Pull lever below steering column down.

– Pull steering column down and out as far as possible.

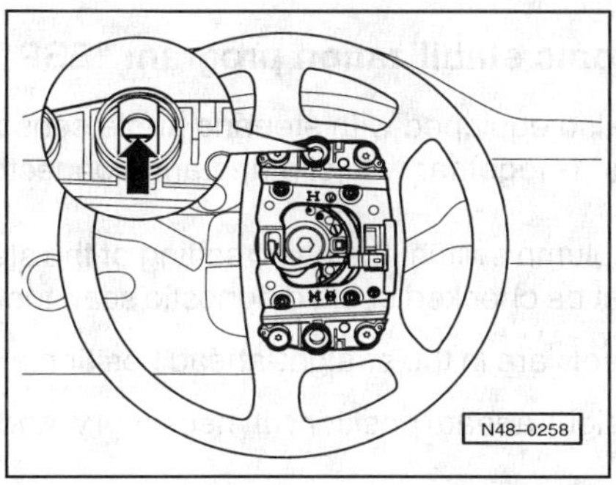

◄ Position steering wheel as shown in illustration.

– Remove airbag module, see **Body-Interior section, Repair Group 69**.

– Place airbag down so that it will not be damaged or cannot fall!

EDITOR'S NOTE —
Do not leave airbag where anyone else can access it. Store in a safe, out of the way location or specially labeled container.

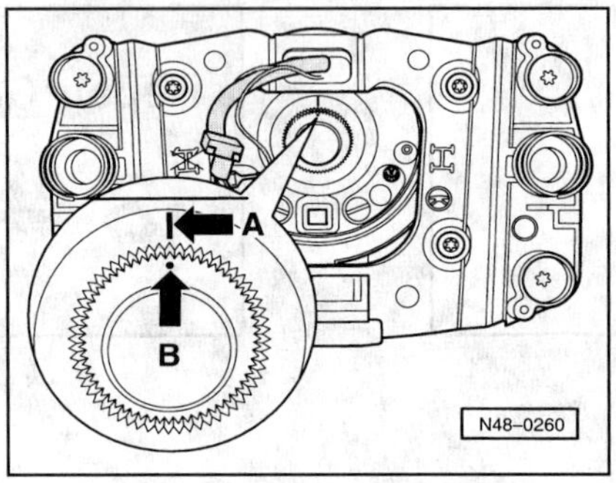

Before pulling off steering wheel, check the mark -arrow A- aligns with the punch mark -arrow B-.

- If this is not the case, mark installation position of steering wheel to steering column using a felt tip pen or scriber for example.

– Remove bolt for steering wheel and pull off.

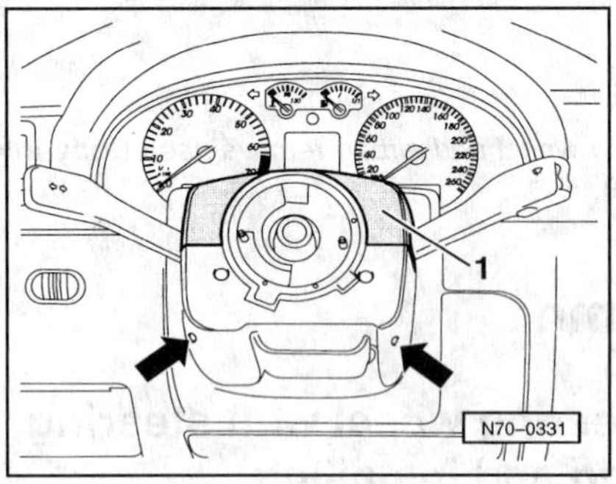

Trim for steering column switch, removing

Remove two cross-head screws -arrows-.

– Take off upper steering column switch trim -1-.

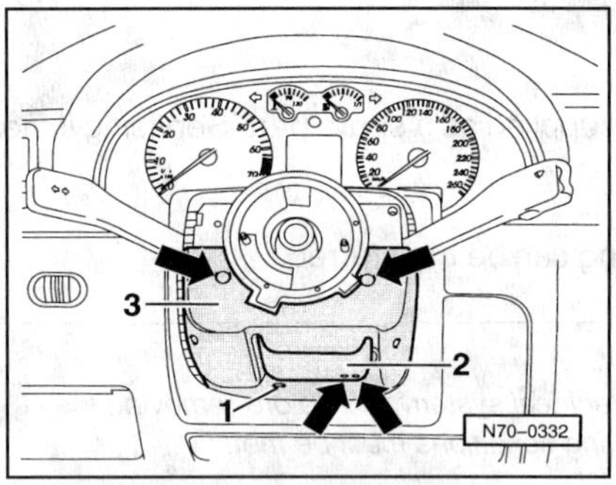

Remove four cross-head screws -arrows-.

– Remove hex socket head bolt -1-.

– Release steering wheel height adjustment -2-.

– Take off lower steering column switch trim -3-.

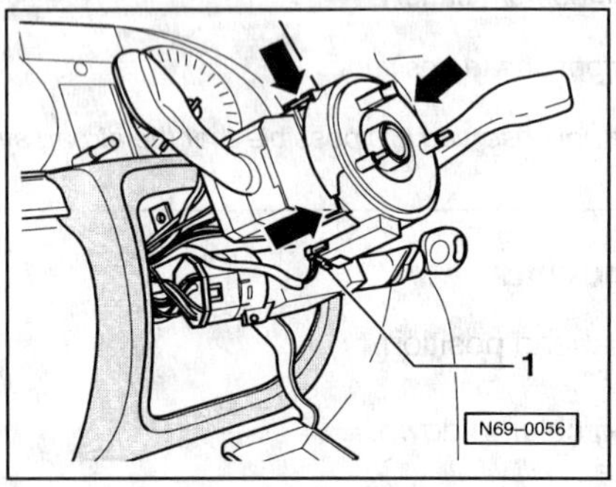

Separate connection -1-.

– Release locking lugs -arrows- and pull coil connector with slip ring off steering column switch.

NOTE —
The steering must be in the center position (wheels in straight-ahead position) when the coil connector and slip ring are removed and installed.

NOTE —
Coil connector with slip ring is secured as a new part in the center position with a cable tie.

Vehicles with electronic stabilization program "ESP"

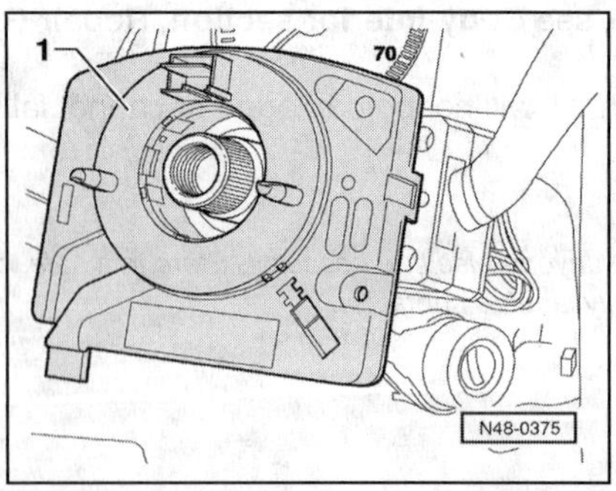

Vehicles with ESP are also equipped with steering angle sensor -G85-. It is installed in housing -1- together with slip ring and connector coil.

After work on steering column switch, the basic setting of the steering angle sensor -G85- must be checked using diagnostic scan tool.

Check that the front wheels are in the straight-ahead position.

– Position wheels to straight-ahead position, if necessary and remove steering wheel.

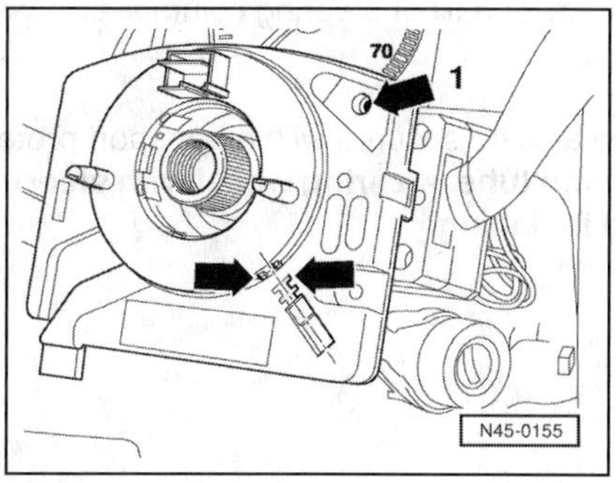

Position steering angle sensor -G85- in center position;

- A yellow spot must be visible in hole -1-.
- The marks -arrow- must align.

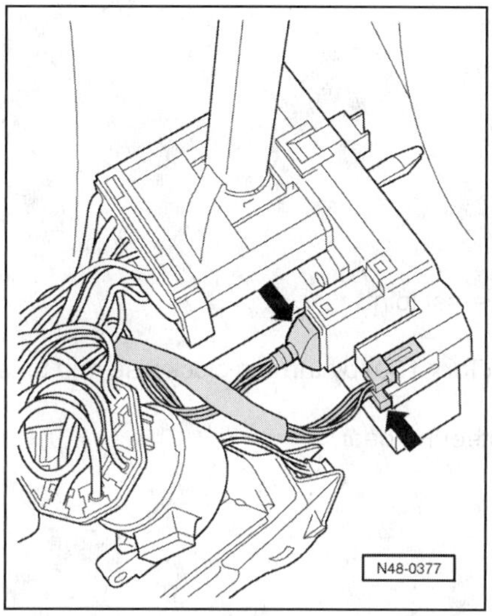

Pull off connectors -arrows-.

Lift hooks -arrows- carefully and pull off steering angle sensor.

The following steps apply to all vehicles.

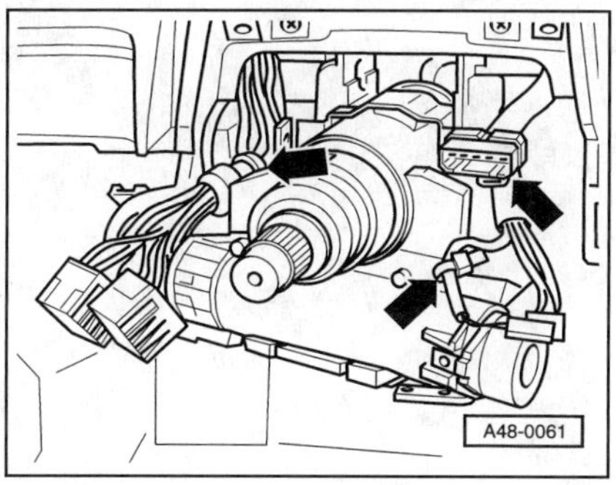

Pull connectors -arrows- off steering lock housing.

The ignition key removal lock cable must be unhooked on vehicles with an automatic transmission, see **Locking cable for ignition with key withdrawal lock, removing and installing.**

Securing steering column before removal

An assembly aid is required so that the upper and lower parts of the steering column do not pull apart when pulling off steering gear.

If the upper and lower parts of the steering column are pulled too far apart or pushed too close together the splines separate.

It is possible that rattling noises are created later if the splines are not in their original installation position.

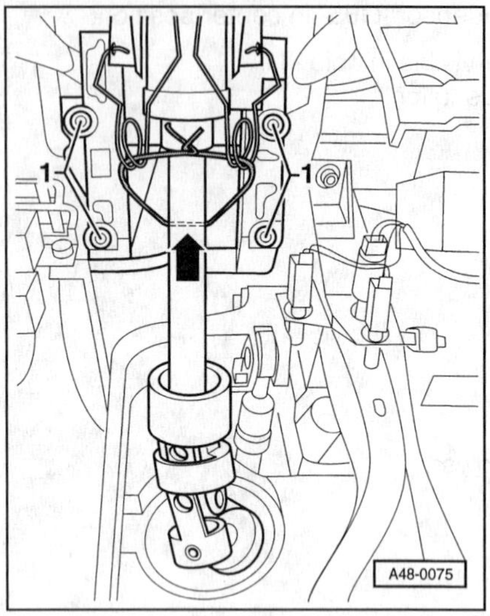

Pull wire through hole in lower part of steering column -arrow- and spring.

The steering column can also be secured with a transport protection device. Installing, see **Column tube, steering wheel with steering column, removing and installing**

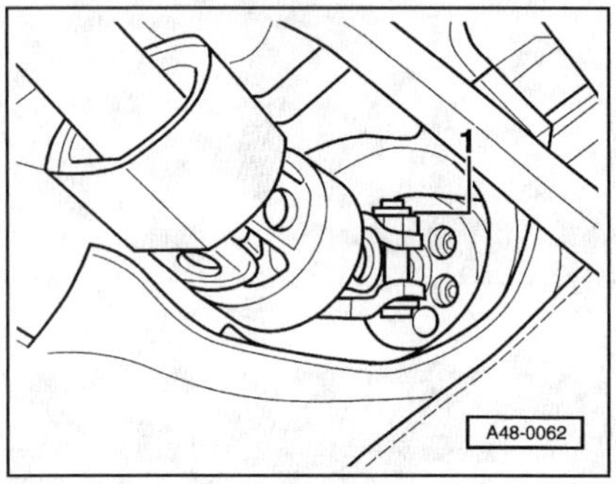

Remove nut -1- on universal joint.

– Relieve tension on bolt (Torx T50) by turning clockwise and take out bolt.

– Pull universal joint off steering gear.

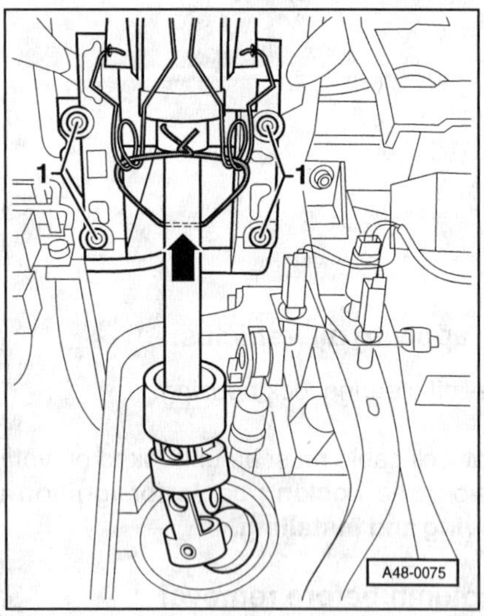

Remove steering column hex socket head bolts -1- and take out steering column.

– Removing and installing steering lock housing, see **Steering lock housing, removing and installing**.

Steering column, installing

NOTE —

- *New steering columns supplied as spare parts are secured with a transport protection device.*
- *This transport protection device must be removed after installing the steering column in the vehicle.*
- *We recommend that this part is retained as it can be reused when removing steering columns.*

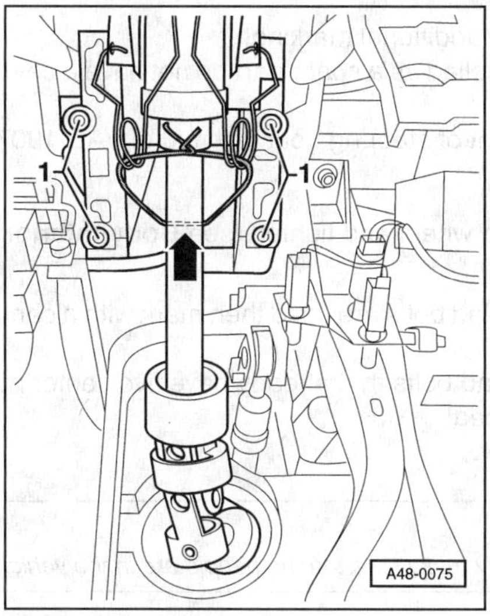

◄ Attach steering column to cross member with 4 hex socket head bolts -1- (do not tighten).

– Install universal joint onto steering gear pinion.

– Install clamping bolt (Torx T50) through lower part and tension counterclockwise.

– Install hex nut and tighten to 40 Nm (30 ft-lb).

– Remove wire/transport protection between upper and lower parts of steering column.

Following work sequence affects only vehicle with ESP

– Press on steering angle sensor -G85- until lugs engage.

- Remove transportation protection when a new steering angle sensor -G85- is installed.

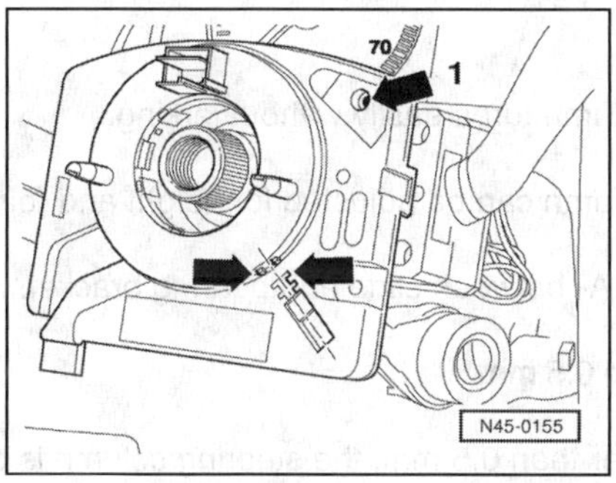

◄ Position the steering angle sensor -G85- in the center position;

- Yellow spot must be visible in hole -1-.
- The marks -arrow- must align.
- Make sure it does not move from center position.

– The basic setting for steering angle sensor must be checked after the following assembly work:

- When the steering angle sensor -G85- has been removed or replaced.
- After removing or replacing the steering column.
- After removing or replacing the steering column switch.
- After removing or replacing the steering lock housing.
- When steering wheel is not in straight-ahead position.

The continuation of the work sequence is valid for all vehicles

– Install steering lock housing.

– Install new shear bolts.

– Tighten shear head bolts until head shears off.

– Install steering lock switch with trim.

– Align steering column so that the gaps in the trim are even.

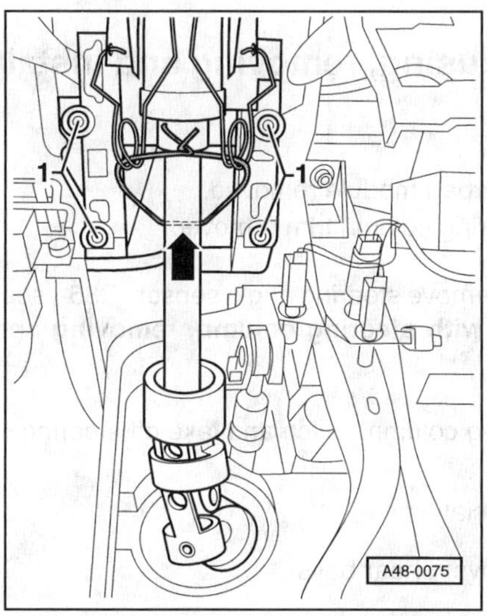

◄ Tighten steering column Allen head bolts -1- to 25 Nm (18 ft-lb).

– Install driver's side stowage compartment.

Installing steering wheel

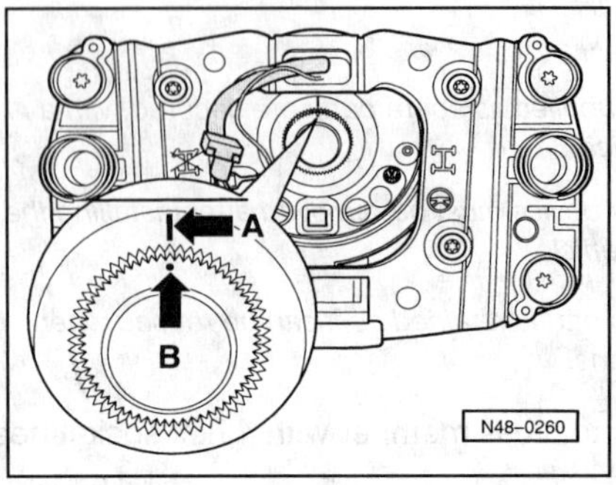

Install steering wheel so that the marked line -arrow A- on the steering wheel and the center punch mark -arrow B- on the steering column align.

- If necessary, observe additional marking!
- Steering columns supplied as a spare part do not have a center punch mark.
- Determine center point of steering gear with aid of VAG 1907 if necessary.

– Install bolts for steering wheel and tighten, see **Body-Interior section, Repair Group 69.**

- If multi point socket head bolts are used then mark with a center punch mark.
- Multi point socket head bolts that already have five center punch marks must be replaced!
- Install in reverse order.

> ***WARNING —***
> *When connecting battery make sure that no people are in the vehicle!*

Steering column, checking for damage

Visual check

– Check all steering column parts for damage.

Functional check

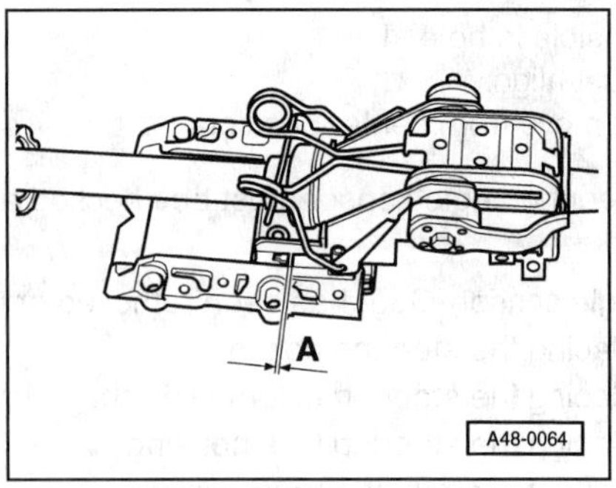

– Check that steering column turns easily without jerking.

– Check that steering column can be adjusted for height and reach.

Check gap dimension -A- between slide and bearing bracket.

Dimension A: Maximum 0.5 mm

– If dimension -A- is larger than 0.5 mm, the steering column is damaged and must be replaced.

Steering lock housing, removing and installing

Requirements

- Steering wheel and airbag module removed.
- Upper and lower steering column trim removed.

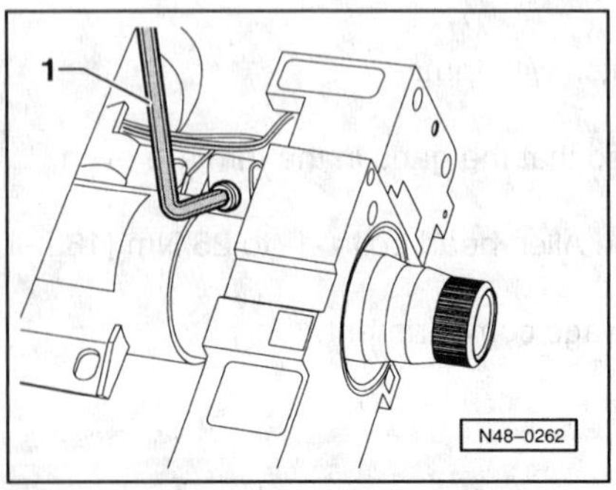

– On vehicles with ESP, remove steering angle sensor -G85-, see **Column tube, steering wheel with steering column, removing and installing.**

Remove bolt for steering column switch and take off steering column switch.

- 1 - Hex key (commercial type)

– Remove plastic cover over shear bolts.

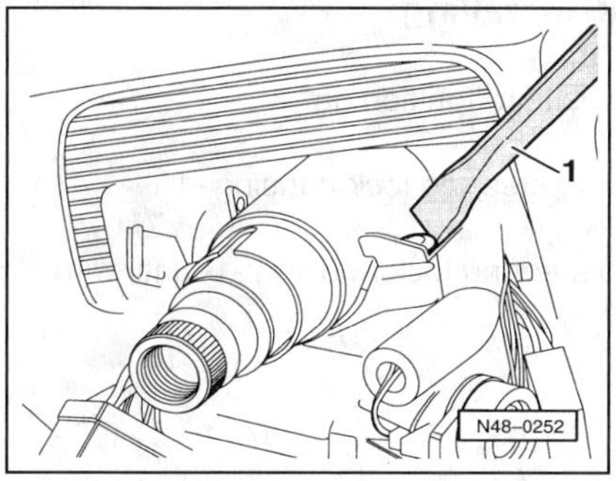

Drill out shear bolts or shear off using a suitable chisel -1-.

The ignition key removal lock cable must be additionally unhooked on vehicles with an automatic transmission, see **Locking cable for ignition with key withdrawal lock, removing and installing**

Removing and installing ignition switch / lock cylinder, see **Electrical Equipment section, Repair Group 94**

Steering lock housing, installing

– Reconnect connector for ignition starter switch.

– Attach ignition key removal lock cable only with automatic transmissions, if necessary, see **Locking cable for ignition with key withdrawal lock, removing and installing**.

– Install new shear bolts.

– Tighten shear head bolts until head shears off.

– On vehicles with ESP, install steering angle sensor -G85-, see **Column tube, steering wheel with steering column, removing and installing**.

– Installation is in reverse order of removal.

Automatic shift lock, checking

– Turn ignition key to position "Ignition on".

– Depress break pedal and hold

- Must be possible to move selector lever out of "Park" position.
- Must not be possible to remove ignition key with selector lever in any position other than "Park"
- If this is not the case, the locking cable must be adjusted.

NOTE —
- *When ignition key is in position "Ignition off" it must be impossible to move selector lever out of "Park" position!*

Automatic shift lock, removing and installing

Only vehicles with automatic transmission

Removing

Requirements

- Steering wheel and airbag module removed.
- Lower steering column trim removed.

– Move selector lever to position "P", if necessary.

– Turn ignition key to position "Ignition on".

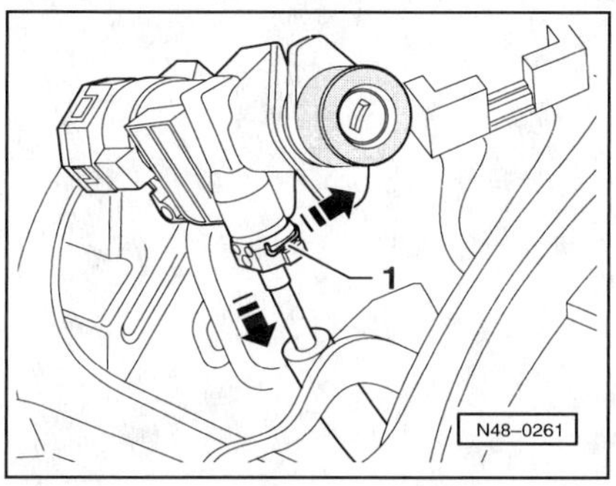

Press wire clip -1- upward or downward according to installation position and pull locking cable out at the same time.

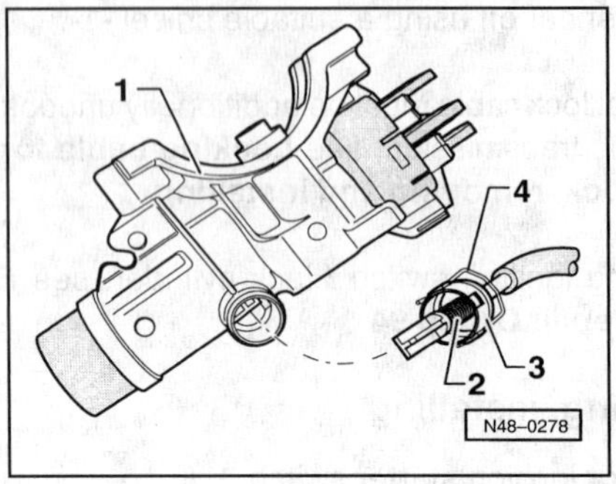

Automatic shift lock, installing

Ignition key must be in position "Ignition on".

< Push locking cable -2- into steering lock housing -1-.

– Push locking cable onto steering lock housing -1- until wire clip -4- engages.

Power Steering Gear

Tie rods, removing and installing

Special tools, workshop equipment, testers, measuring instruments and auxiliary items required

- VAG 1332 Torque wrench
- VAG 1923 Open end wrench attachment

NOTE—
Left and right-hand tie rods are identical. They can be removed and installed with steering gear remaining installed.

Removing

NOTE—
Air cleaner and air line must be removed.

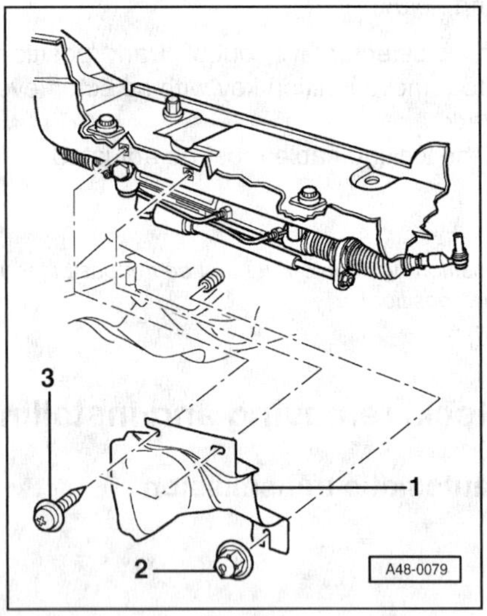

< To remove the left tie rod on vehicles with 4-cyl. engine, additionally remove heat shield.

– Remove nut -2- and cross head screws -3-.

– Remove heat shield -1-.

– Remove front wheel.

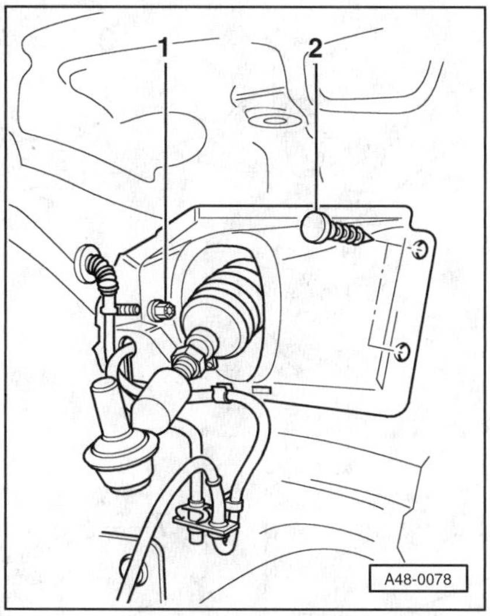

Remove plastic nut -1-.

– Pry out clips -2-.

– Take out cover for tie rod.

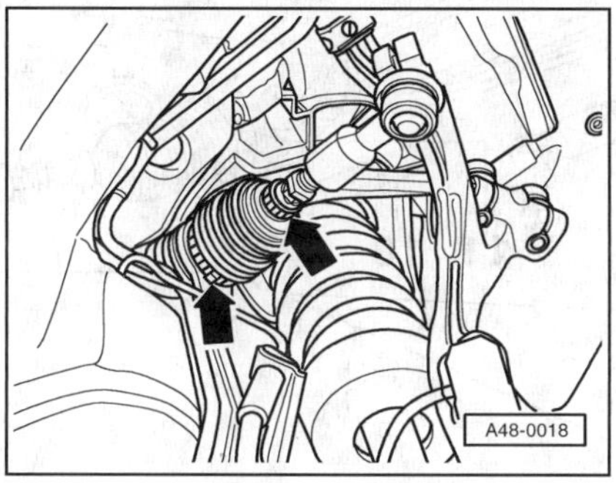

Open clips on boot.

• Illustration shows tie rod for vehicles with V6 engine.

– On vehicles with 4-cyl. engine, open clip on steering gear.

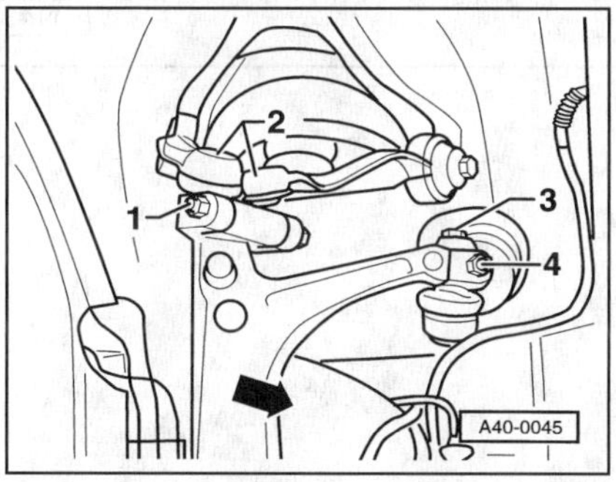

Loosen bolts -3- and -4- and pull tie rod ball joint off downward.

– Pull boot as far as possible outward.

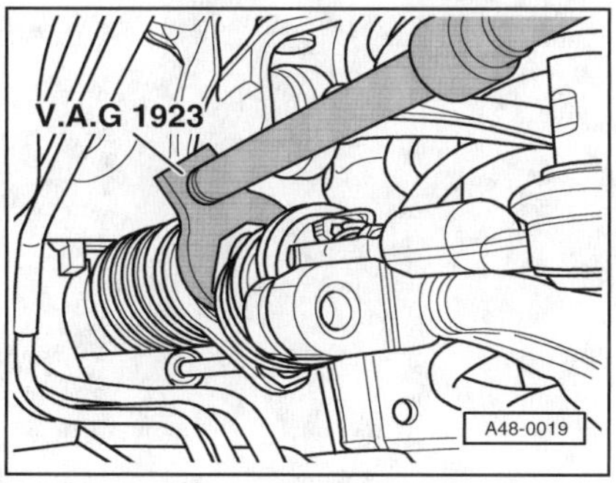

Remove tie rod with VAG 1923.

Tie rod, installing

– Installation is reverse order or removal, noting the following:

• Install tie rod to steering gear, tighten to 100 Nm (74 ft-lb).
• Checking suspension geometry, see **Repair Group 44, Vehicle alignment**.

Power steering gear, components

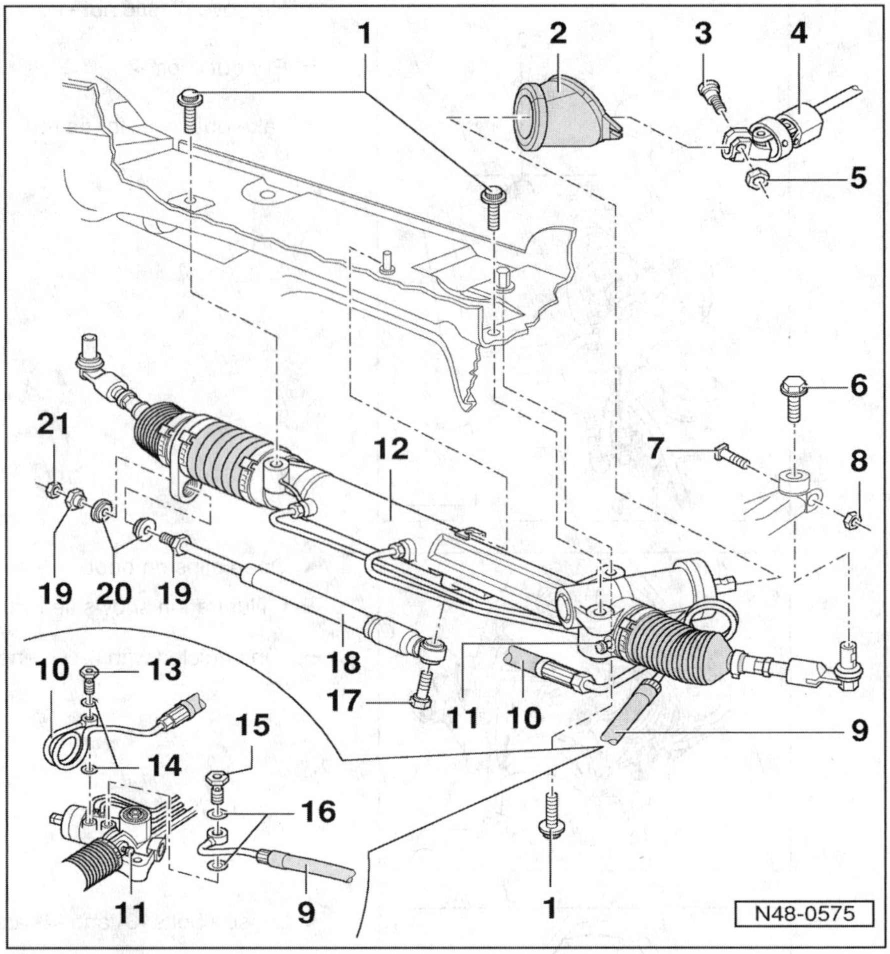

1. **Twelve point bolt**
   - Tighten to 65 Nm (48 ft-lb)
2. **Sealing boot**
3. **Eccentric bolt**
   - Torx T50
4. **Steering column**
5. **Self-locking hex nut,**
   - Tighten to 40 Nm (30 ft-lb)
   - Always replace
6. **Hex bolt,**
   - Tighten to 7 Nm (62 in-lb)
7. **Bolt**
8. **Self-locking hex nut**
   - Tighten to 45 Nm (33 ft-lb)
   - Always replace
9. **Return hose**
10. **Pressure hose**
11. **Hex key head bolt**
    - Tighten to 13 Nm (10 ft-lb)
12. **Power assisted steering gear with tie rods**
13. **Banjo bolt**
    - Tighten to 40 Nm (30 ft-lb)
    - With integrated non-return valve
14. **Oil seal, 14 x 20 mm**
    - Always replace
15. **Banjo bolt**
    - Tighten to 50 Nm (37 ft-lb)
16. **Oil seal, 16 x 22 mm**
    - Always replace
17. **Hex bolt**
    - Tighten to 35 Nm 26 ft-lb)
18. **Steering damper**
    - Introduced in vehicles with 4-cylinder engines at a later stage in production
19. **Bushing**
20. **Rubber mounting**
    - Two part
21. **Hex nut**
    - Tighten to 10 Nm (7 ft-lb)
    - To loosen and tighten, counter hold on steering damper piston rod hex

Steering gear, vehicles with 4 or 6-cylinder engines, removing and installing

Special tools, workshop equipment, testers, measuring instruments and auxiliary items required

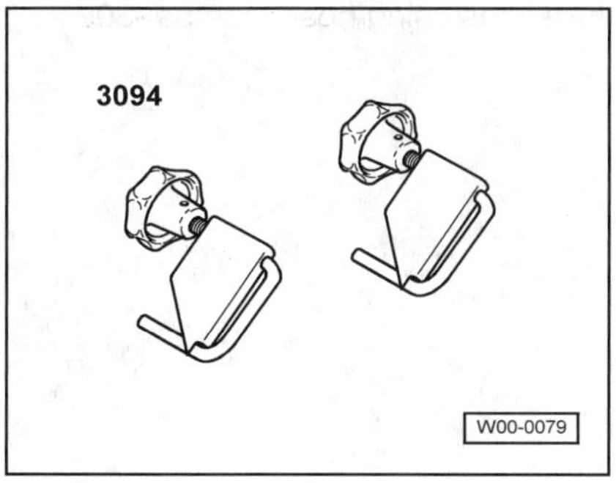

◀ 3094 Hose clamps

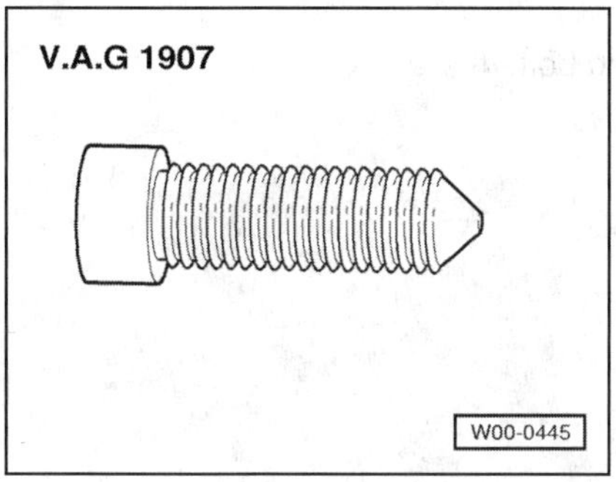

◀ VAG 1907 Steering centering bolt

Removing

– Remove plenum chamber cover.

– Remove battery.

– Remove driver's side stowage compartment.

– Bring steering wheel to center position.

– Remove ignition key with steering wheel in center position.

– Move steering wheel slightly so that the steering lock engages.

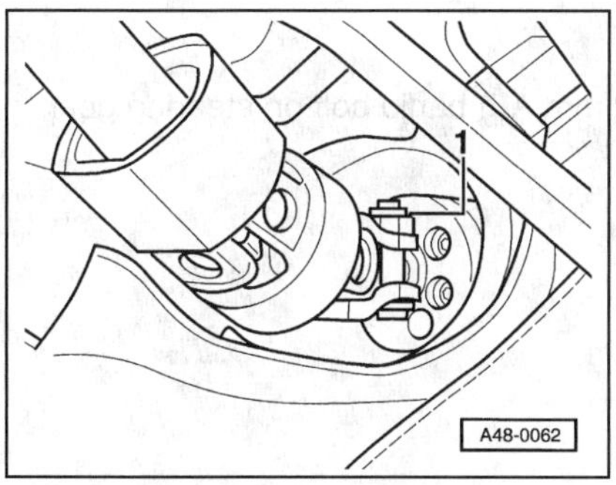

NOTE —
If the steering wheel is turned excessively with the steering gear removed, the spiral spring for the airbag system may become over-wound and damaged.

– Secure steering column against moving apart, see **Column tube, steering wheel with steering column, removing and installing**

◀ Remove nut -1- on universal joint.

– Relieve tension on eccentric by turning tensioning bolt (Torx T50) clockwise and remove bolt.

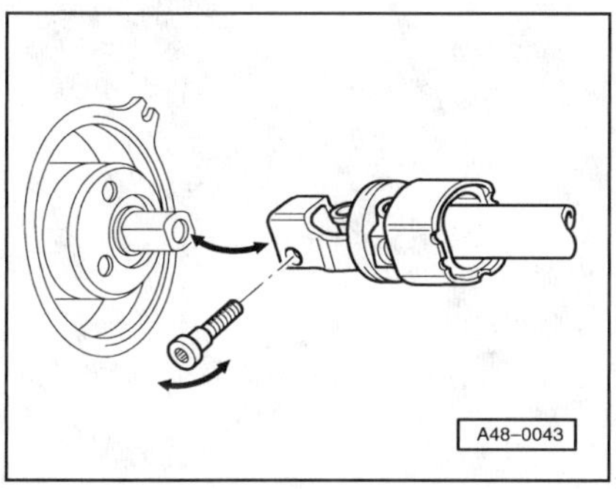

◀ Swing universal joint away downward.

– Pull boot out from bulkhead.

– Remove front wheels.

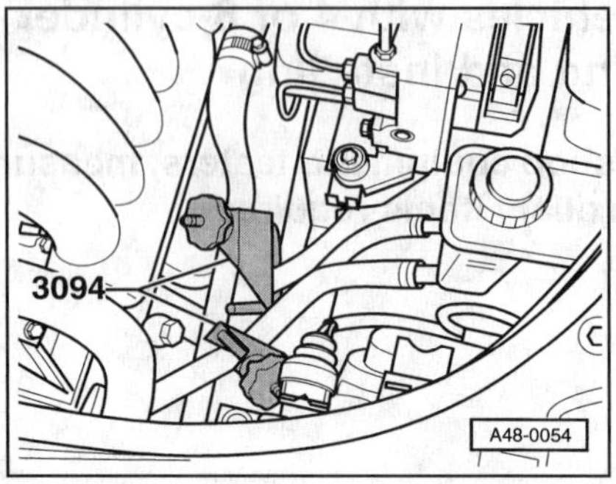

Clamp off intake and return lines using hose clamps -3094-.

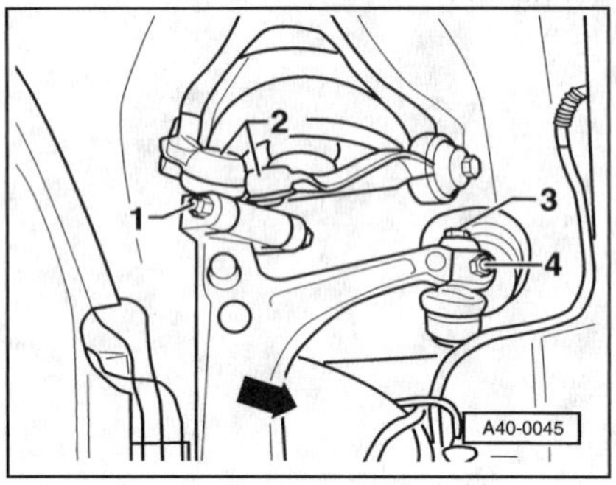

Remove hex bolt -3- and bolt -4-.

– Pull out tie rod.

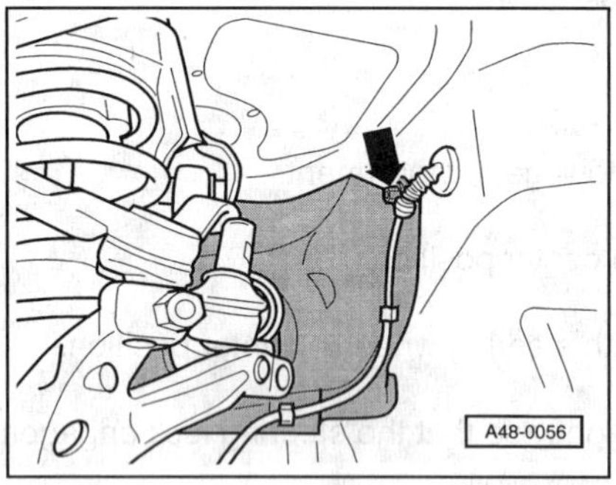

Remove plastic nut -arrow-.

– Pry out clips and take out cover.

– Place oil tray under vehicle.

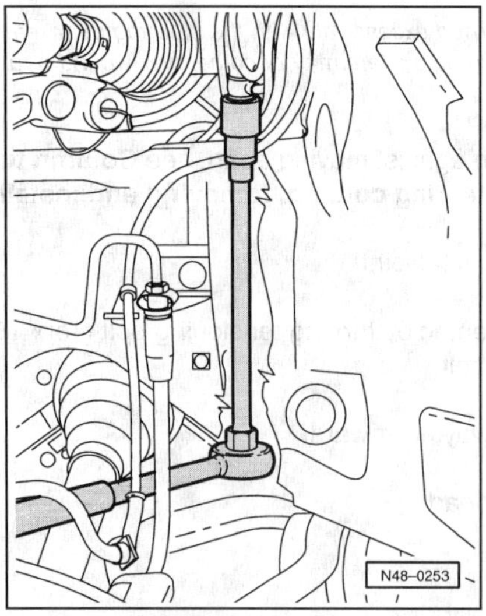

Remove return line (22 mm AF) banjo bolt on steering gear.

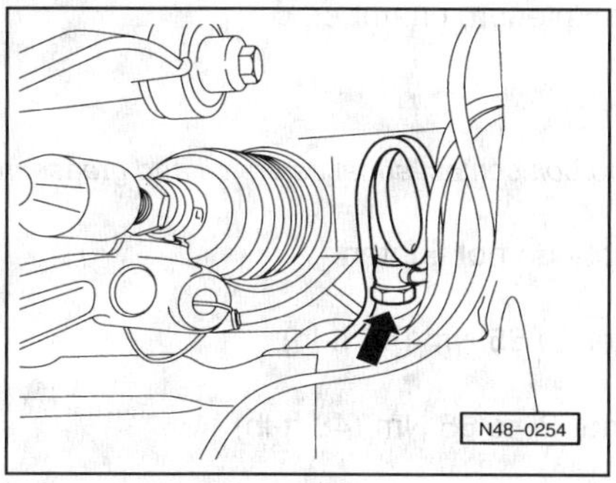

Remove pressure hose banjo bolt (19 mm AF) from steering gear.

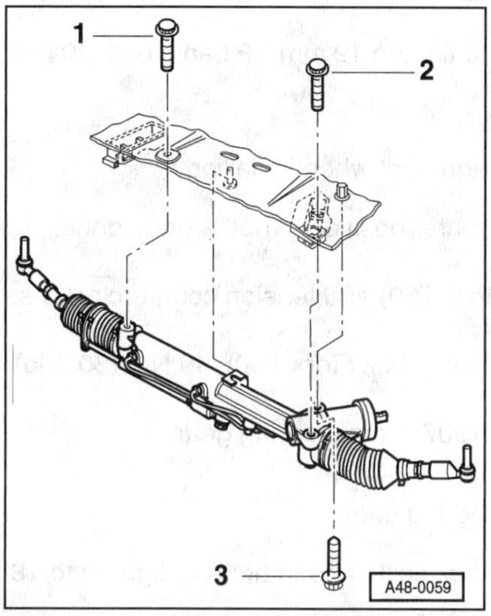

Remove steering gear bolts -1-, -2- and -3-.

– Take out steering gear through left-hand wheel housing (second mechanic required).

Steering gear, installing

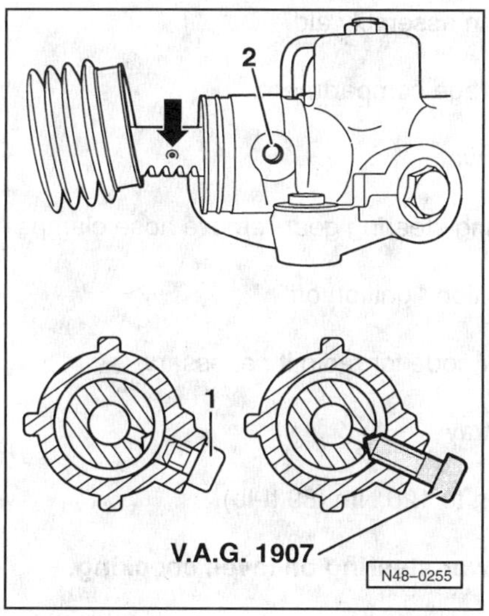

Centralize removed steering gear with -VAG 1907-.

- Centering is also possible when steering gear is installed. But access is poor.

– To do this remove hex socket head bolt -1- on steering gear.

– Move rack until center marking on rack -arrow- aligns with threaded hole -2-.

– Screw in VAG 1907.

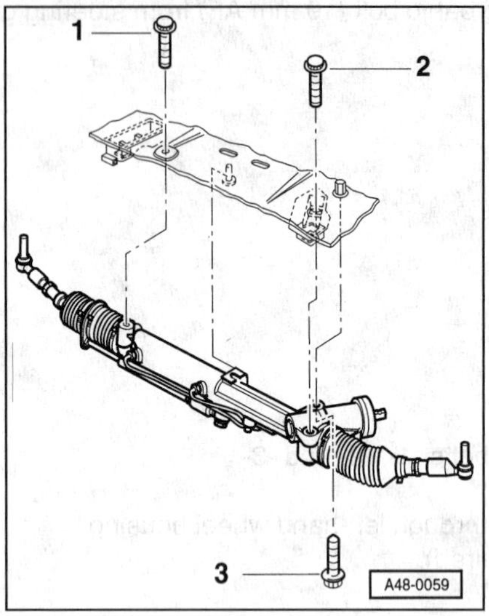

Position steering gear on plenum chamber.

NOTE—
Make sure that threads and bolt contact surfaces are oil and grease free!

- Insert bolts -1- and -2- but do not tighten.
- Install bolt -3- and tighten to 65 Nm (48 ft-lb)
- Now tighten bolts -1- and -2- to 65 Nm (48 ft-lb)
- Install return hose and tighten 22 mm AF banjo bolt, 50 Nm (37 ft-lb).
- Install pressure hose and tighten 19 mm AF banjo bolt, 40 Nm (30 ft-lb).
- Install boot in bulkhead.
  - Take care not to damage boot when installing.
- Turn steering wheel to center position, if not already done.
- Insert tensioning bolt (Torx T50) and tension counterclockwise.
- Tighten hex nut of tensioning bolt (Torx T50), 40 Nm (30 ft-lb).

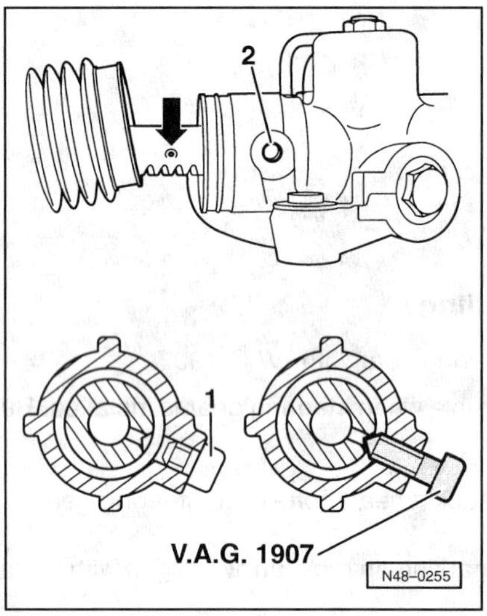

Now remove tool -VAG 1907- from steering gear.

1 - hex socket head bolt
2 - Threaded hole in steering gear

- Seal steering gear with hex socket head bolt -1-; tighten to 18 Nm (13 ft-lb).
- Install universal joint to steering pinion.
- Remove steering column assembly aid.
- Install driver's side stowage compartment.
- Install wheel housing cover.
- After completely installing steering gear remove hose clamps -3094-.
- Turn ignition key to position "ignition off".
- Install battery and enter code for radio if necessary.
- Install noise insulation tray.
- Install wheel and tighten to 120 Nm (89 ft-lb).
- Check oil level, see **Power steering oil level, checking**.
- Check steering system for leaks, see **Steering system checking for leaks**.
- Bleed steering system, see **Steering system, bleeding**.
- After installing check position of steering wheel during a test drive.
  - If steering wheel is not in straight ahead position the front axle tracking must be checked and if necessary adjusted!
  - Check toe adjustment, see **Repair Group 44, Vehicle alignment**.

Steering gear, vehicles with 8-cylinder engines, removing and installing

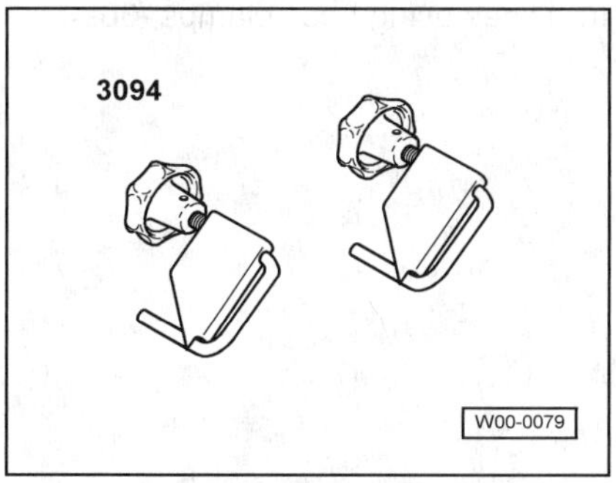

Special tools, workshop equipment, testers, measuring instruments and auxiliary items required

- 3094 Hose clamps

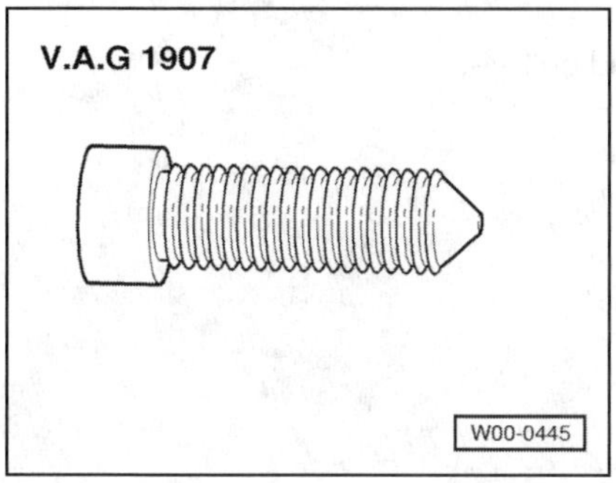

- VAG 1907 Steering centering bolt

Removing

- Remove plenum chamber cover.
- Remove battery.
- Remove driver's side stowage compartment.
- Bring steering wheel to center position.
- Remove ignition key with steering wheel in center position.
- Move steering wheel slightly so that the steering lock engages.

NOTE —
If the steering wheel is turned excessively with the steering gear removed, the spiral spring for the airbag system may become over-wound and damaged.

- Secure steering column against moving apart, see **Column tube, steering wheel with steering column, removing and installing**

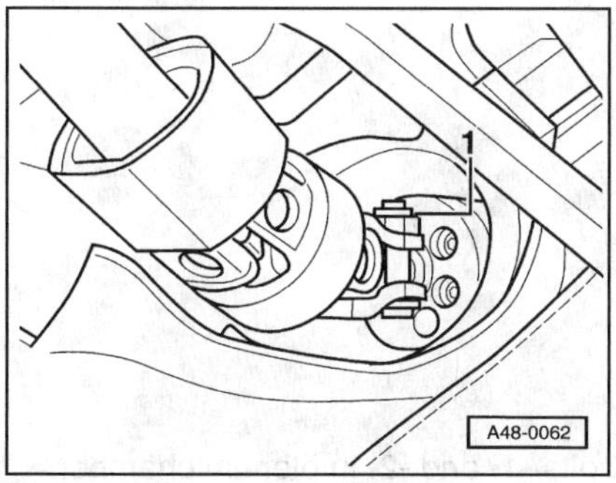

- Remove nut -1- on universal joint.
- Relieve tension on eccentric by turning tensioning bolt (Torx T50) clockwise and remove bolt.

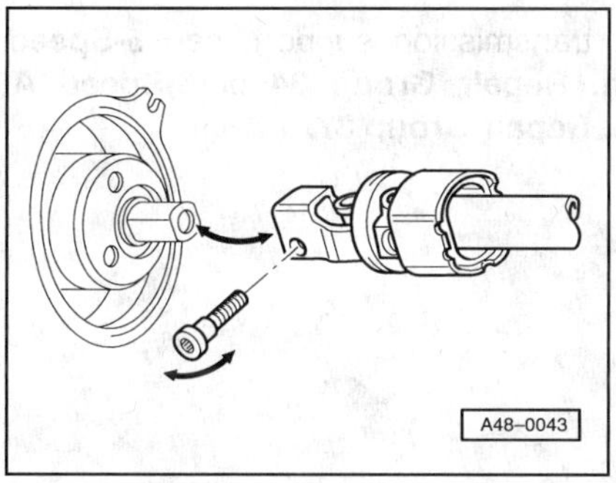

- Swing universal joint away downward.
- Pull boot out from bulkhead.
- Remove noise insulation tray.
- Remove front wheels.

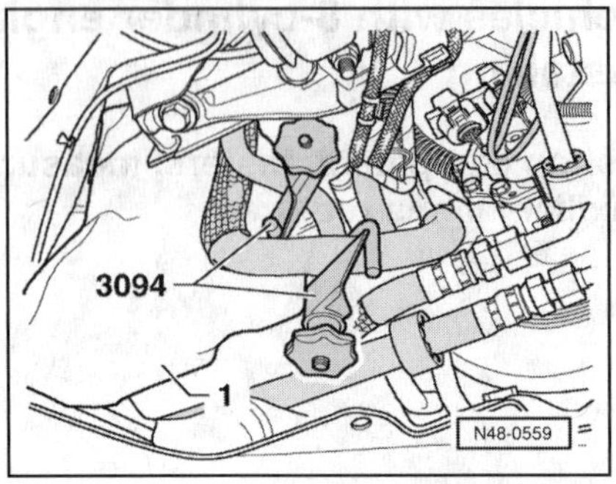

Clamp off intake and return lines using hose clamps -3094-.

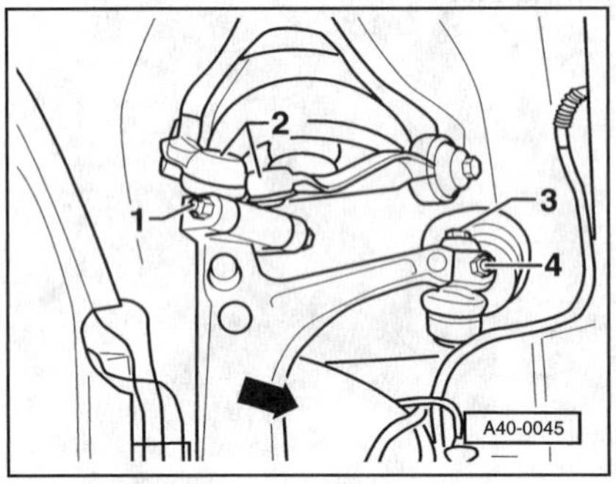

Remove hex bolt -3- and bolt -4-.

– Pull out tie rod ball joint.

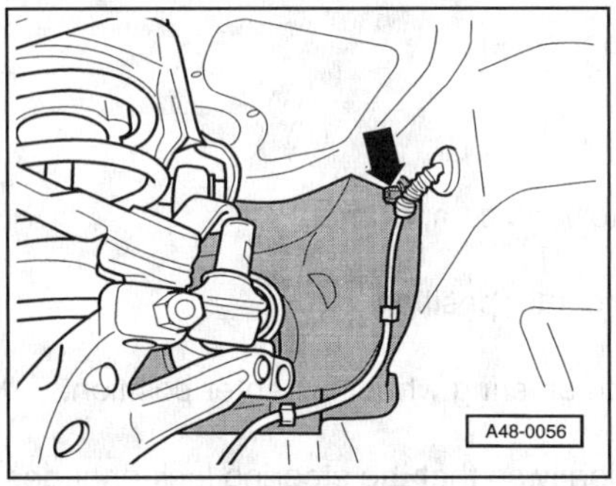

Remove panel nut -arrow-.

– Pry out clips and take out cover.

– Place oil tray under vehicle.

– Remove return line banjo bolt (22 mm AF) from steering gear.

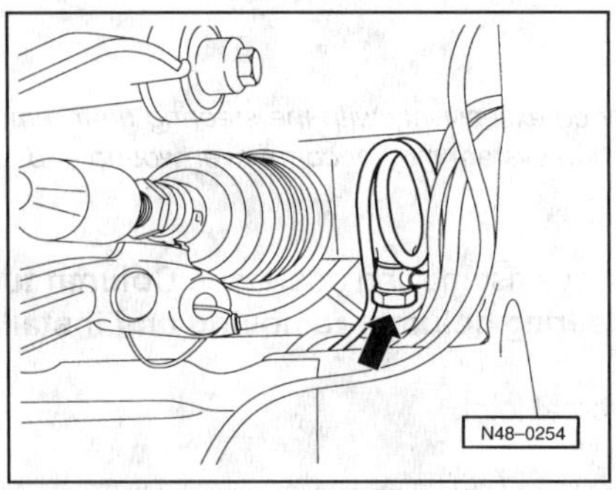

Remove pressure hose banjo bolt (19 mm AF) from steering gear.

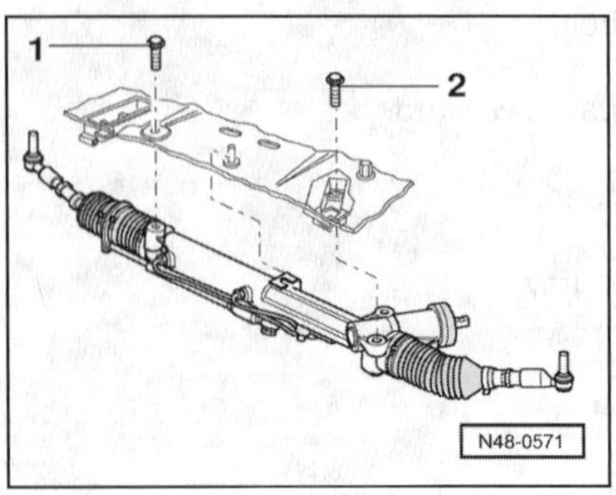

Remove steering gear bolts -1- and -2- in plenum chamber.

– Remove complete left transmission support, see **5-Speed Manual Transmission section, Repair Group 34** or **5-Speed Automatic Transmission section, Repair Group 37.**

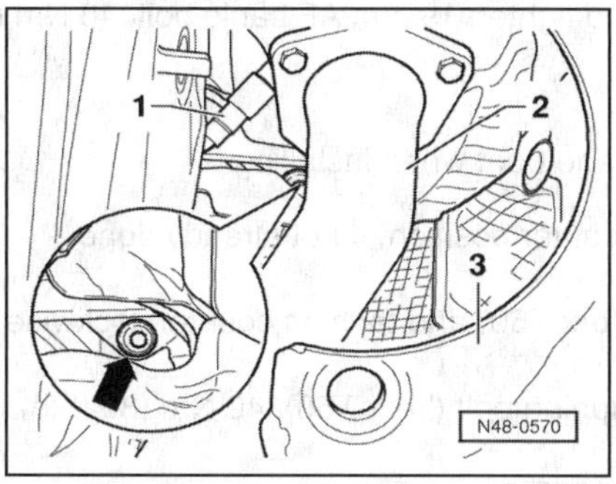

Remove steering gear securing bolt -arrow-.

NOTE—
The bolt -arrow- is accessible only through a recess in heat shield between oxygen sensor -1- and front exhaust pipe -2-.

3 - Subframe

Two mechanics are required for the next work steps

– Guide steering gear as far as possible to left over wheel bearing housing and turn 90° counterclockwise.

– Guide steering gear as far as possible to left and turn again 90° counterclockwise.

– Take out steering gear.

Steering gear, installing

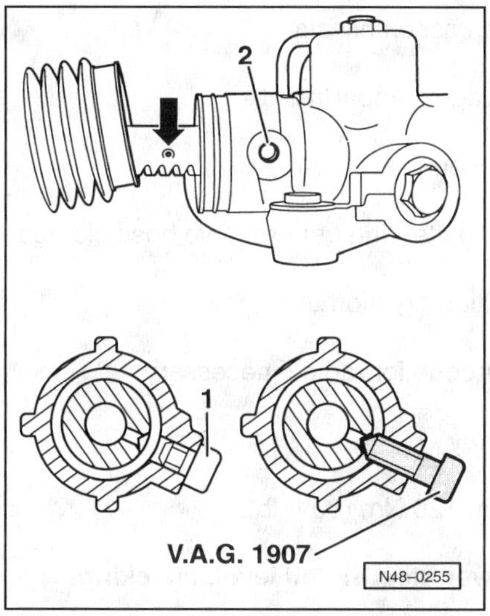

Centralize removed steering gear with -VAG 1907-.

• Centering is also possible when steering gear is installed. But access is poor.

– To do this remove hex socket head bolt -1- on steering gear.

– Move rack until center marking on rack -arrow- aligns with threaded hole -2-.

– Screw in VAG 1907.

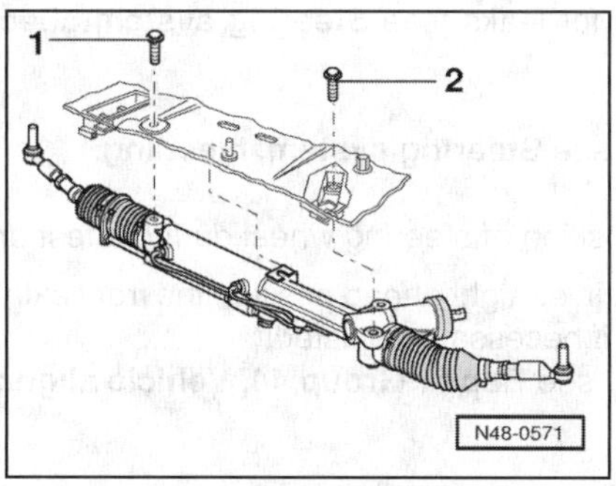

Position steering gear on plenum chamber.

NOTE—
Make sure that threads and bolt contact surfaces are oil and grease free!

– Insert bolts -1- and -2- but do not tighten.

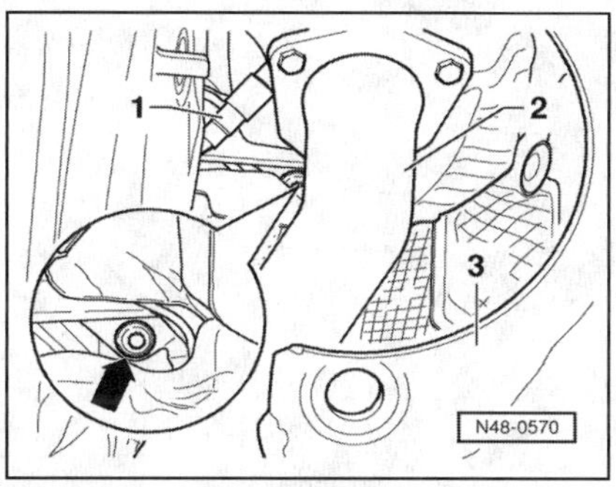

Install bolt -arrow- and tighten, 65 Nm (48 ft-lb).

NOTE—
The bolt -arrow- is accessible only through a recess in heat shield between Heated Oxygen Sensor (HO2S) -1- and front exhaust pipe -2-.

3 - Subframe

– Now tighten bolts -1- and -2-, 65 Nm (48 ft-lb).

– Install return hose and tighten 22 mm AF banjo bolt, 50 Nm (37 ft-lb).

– Install pressure hose and tighten 19 mm AF banjo bolt, 40 Nm (30 ft-lb).

– Install boot in bulkhead.
  • Take care not to damage boot when installing.

– Turn steering wheel to center position, if not already done.

– Insert tensioning bolt (Torx T50) and tension counterclockwise.

– Tighten hex nut of tensioning bolt (Torx T50), 40 Nm (30 ft-lb).

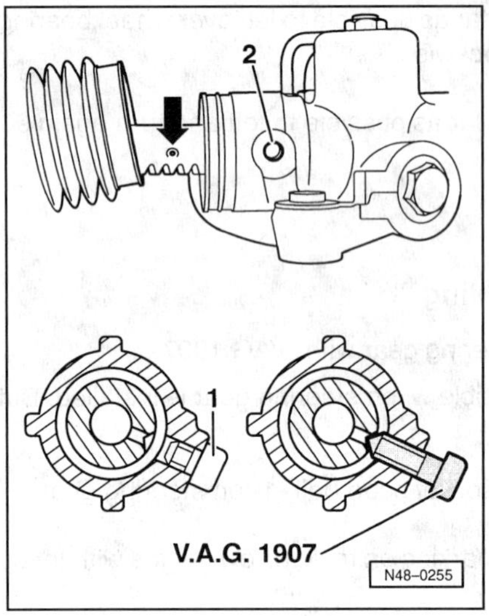

◄ Now remove tool -VAG 1907- from steering gear.
  1 - Hex socket head bolt
  2 - Threaded hole in steering gear

– Seal steering gear with hex socket head bolt -1-; tighten to 18 Nm (13 ft-lb).

– Install universal joint to steering pinion.

– Remove steering column assembly aid.

– Install driver's side stowage compartment.

– Install wheel housing cover.

– After completely installing steering gear remove hose clamps -3094-.

– Turn ignition key to position "ignition off".

– Install battery and enter code for radio if necessary.

– Install noise insulation tray.

– Install wheel and tighten, 120 Nm (89 ft-lb).

– Check oil level, see **Power steering oil level, checking**.

– Check steering system for leaks, see **Steering system checking for leaks**.

– Bleed steering system, see **Steering system, bleeding**.

– After installing check position of steering wheel during a test drive.
  • If steering wheel is not in straight ahead position the front axle tracking must be checked and if necessary adjusted!
  • Check toe adjustment, see **Repair Group 44, Vehicle alignment**.

Power Steering Hydraulic System

Power steering oil level, checking

– Do not start engine but bring front wheels to straight-ahead position.

Oil in cold condition:

– Remove cap.

– Clean dipstick with a clean cloth.

– Screw cap on hand tight and unscrew again.

NOTE —

The cap must be fully screwed in order to obtain an accurate fluid level reading.

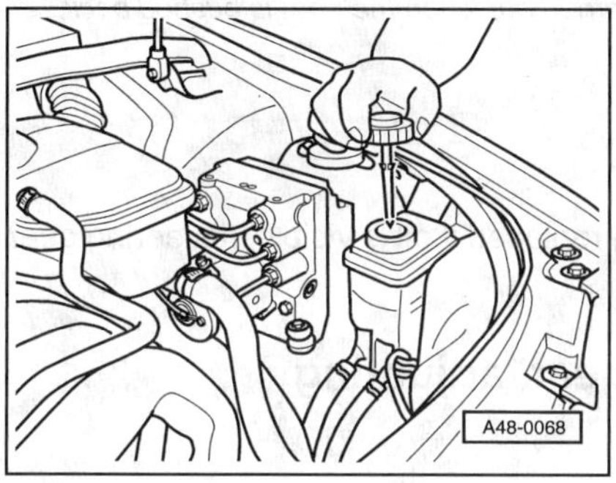

< Check oil level: The oil level must be in area of MIN. marking (up to 2 mm above or below marking).

Oil at operating temperature (above approx. 50°C):

– Check oil level: The oil level must be between MIN. and MAX. marks.

NOTE —

- *If the fluid level is above the area specified, the excess fluid must be extracted.*
- *If oil level is below afore mentioned area, then the hydraulic system must be checked for leaks. It is not sufficient to top-up with oil.*

Steering system, bleeding

– Check hydraulic oil level and top-off if necessary.

– Raise vehicle until front wheels are free.

– Turn steering wheel from lock to lock 10 times, with engine switched off.

– Check hydraulic oil level and top-off if necessary.

– Lower vehicle.

– Start engine.

– Turn steering wheel from lock to lock 10 times.

– Check hydraulic oil level and top-off if necessary.

- Any remaining air in steering system will dissipate when driving over the next 10 - 20 km (6 - 12 miles).

Steering system checking for leaks

NOTE —
After completing repairs if hydraulic oil level is low the steering system must be checked for leaks.

– Start engine.

– Turn steering from lock to lock and hold briefly.

– The maximum possible pressure is then produced. In this position the following components must be checked for leaks:

• Pinion seal on steering gear valve housing.
• All line connections
• Rack seals

NOTE —
This check can only be carried out when the boot is pushed back.

– Open boot hose clip.

– Push boot back.

• Steering gear must be replaced if oil is visible in steering gear housing and/or in the boot.

Power steering gear, adjusting

NOTE —
Two mechanics are required when adjusting. Adjust with engine switched off.

– Raise vehicle on hoist.

– Bring wheels to straight-ahead position.

– Turn steering wheel left and right (approx. 30° from center axis).

• Knocking noise will be heard if there is excessive play.

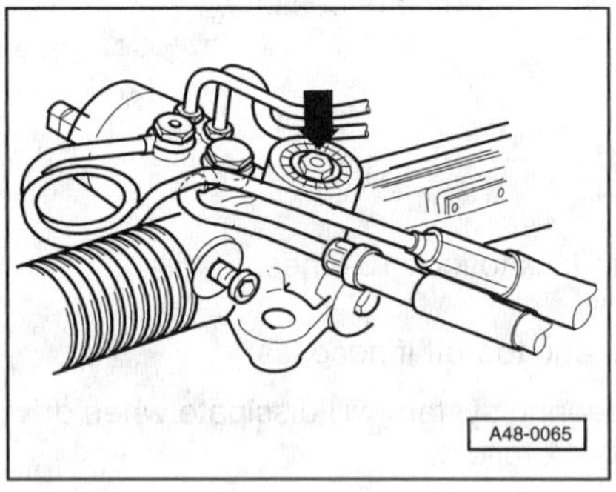

< The second mechanic must carefully screw the adjustment screw (arrow) until the knocking noise can no longer be heard inside the vehicle.

– Carry out a road test.

– Make sure that when maneuvering or negotiating corners the steering self-centers without sticking, re-adjust if necessary.

Power steering pump, reservoir and hydraulic lines, assembly (4-cylinder engine)

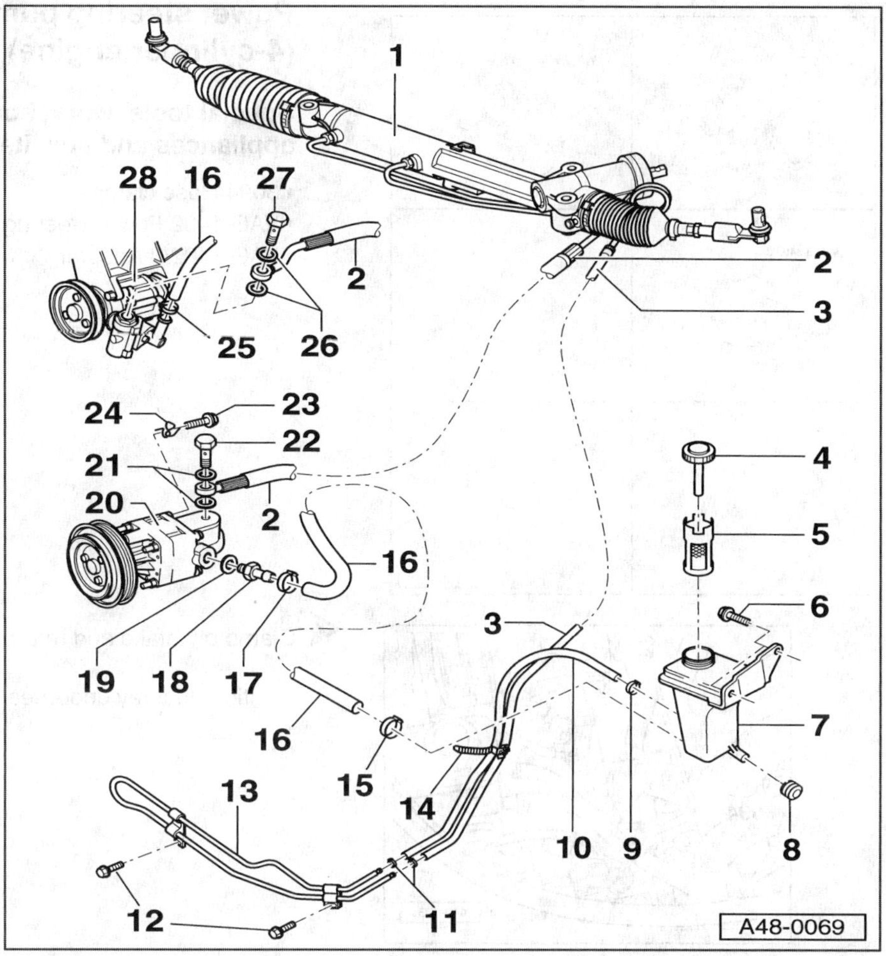

NOTE—

- *If reservoir fluid level is low then check steering system for leaks.*
- *In the case of leaks at the line connections first tighten connections and wipe dry then recheck for leaks.*
- *Pumps supplied as spare parts are not filled with oil. Therefore it is essential to fill the pump with hydraulic oil G 002 000 and to rotate pump by hand, otherwise the pump may be damaged or be noisy during operation.*
- *Type of oil: hydraulic oil G 002 000*
- *Oil system capacity: 0.7 to 0.9 ltr.*

1. **Power steering gear**
2. **Pressure hose**
3. **Return hose**
4. **Cap with dip stick**
   - Checking oil level, see **Maintenance**
5. **Strainer**
6. **Hex bolt**
   - Tighten to 9 Nm (80 in-lb)
7. **Reservoir**
8. **Rubber mounting**
9. **Hose clip**
   - Tension with VAG 1275
   - Always replace
10. **Return hose**
    - Between reservoir and cooling line
11. **Hose clip**
    - Tension with VAG 1275
    - Always replace
12. **Hex bolt**
    - Tighten to 18 Nm (13 ft-lb)
13. **Cooling line**
14. **Cable tie**
15. **Hose clip**
    - Tension with VAG 1275
    - Always replace
16. **Suction hose**
17. **Hose clip**
    - Tension with VAG 1275
    - Always replace
18. **Threaded union**
    - Tighten to 50 Nm (37 ft-lb)
19. **Oil seal, 18 x 24 mm**
    - Always replace
20. **Power steering pump**
    - Checking delivery pressure, see **Power steering pump delivery pressure, checking (4-cylinder engine)**
    - Fill with oil before installing.
21. **Oil seal, 16 x 22 mm**
    - Always replace
22. **Banjo bolt**
    - Tighten to 50 Nm (37 ft-lb)
23. **Hex bolt**
    - Tighten to 18 Nm (13 ft-lb)
24. **Securing clip**
25. **Hose clip**
    - Tension with VAG 1275
    - Always replace
26. **Oil seal, 16 x 22 mm**
    - Always replace
27. **Banjo bolt**
    - Tighten to 50 Nm (37 ft-lb)
28. **Power steering pump for TDI**

Power steering pump delivery pressure, checking (4-cylinder engine)

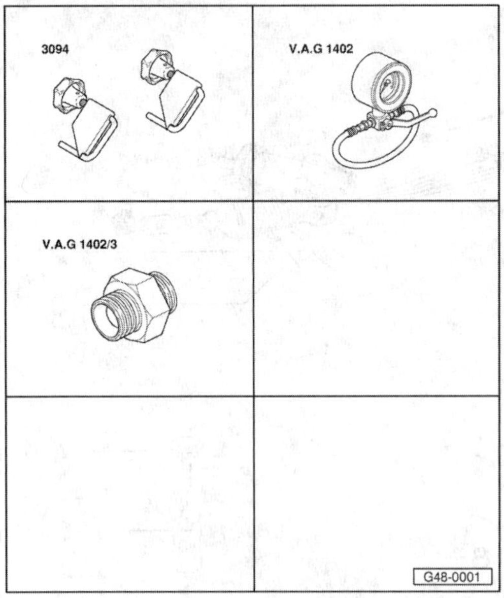

Special tools, workshop equipment, test and measuring appliances and aux. items required

- 3094 Hose clamps
- VAG 1402 Power steering test module
- VAG 1402/3 Adapter

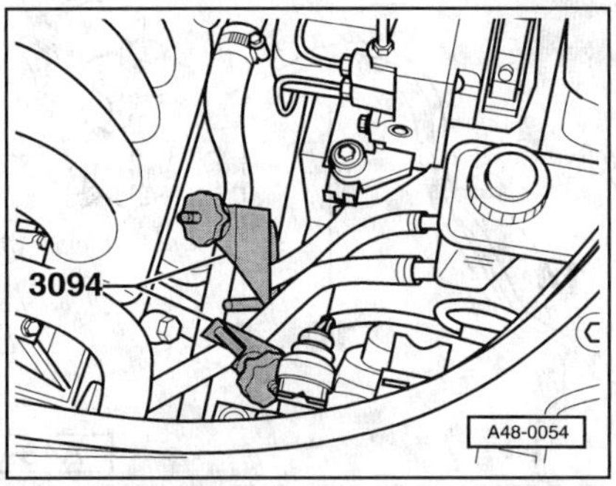

Clamp off intake and return lines using hose clamps -3094-.

– Position drip tray underneath.

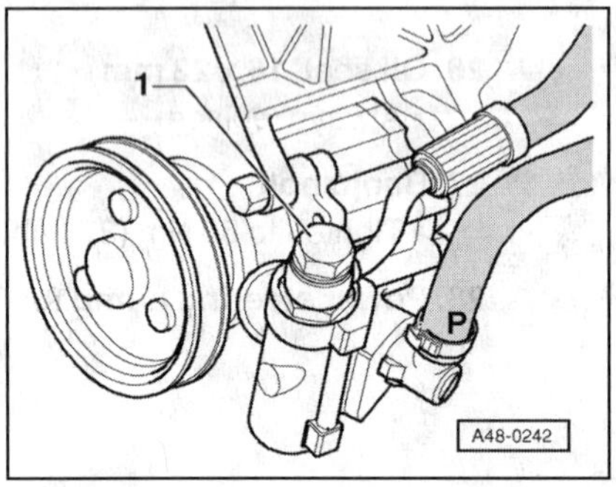

Remove banjo bolt -1- and place pressure line to side.

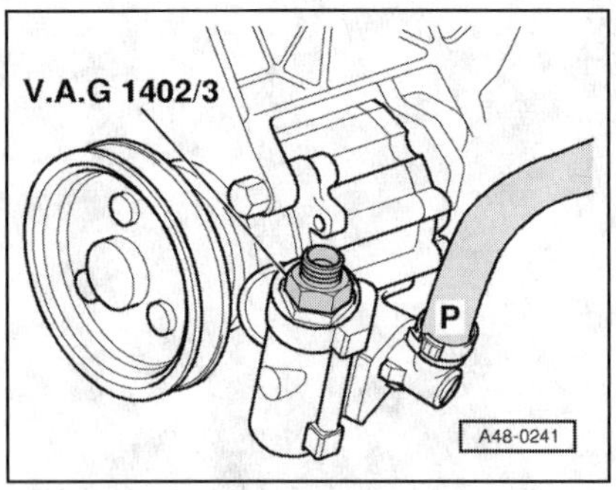

Screw adapter -VAG 1402/3- into power steering pump in place of banjo bolt.

– Screw gauge -VAG 1402- onto adapter -VAG 1402/3-.

– Close gauge shut-off valve (lever to left).

– Remove hose clamps -3094-.

– Start engine and if necessary add fluid to reservoir.

– Read off gauge pressure with engine at idling speed (do not measure longer than 10 seconds).

- Specification: 110 -120 bar. (1585 to 1740 psi).

NOTE—
If the specifications are not reached, replace power steering pump, see **Power steering pump, removing and installing (4-cylinder engine)**.

– Check oil level, see **Power steering oil level, checking**.

– Check steering system for leaks, see **Steering system checking for leaks**.

– Bleed steering system, see **Steering system, bleeding**.

Power steering pump, removing and installing (4-cylinder engine)

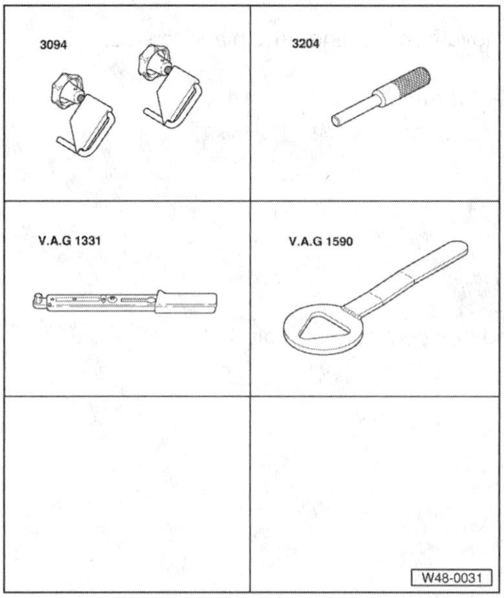

Special tools, workshop equipment, test and measuring appliances and aux. items required

- 3094 Hose clamps
- 3204 Locking pin
- VAG 1331 Torque wrench
- VAG 1590 Water pump wrench

Removing

– Remove engine noise insulation.

– Set lock carrier to service position, see **Maintenance**.

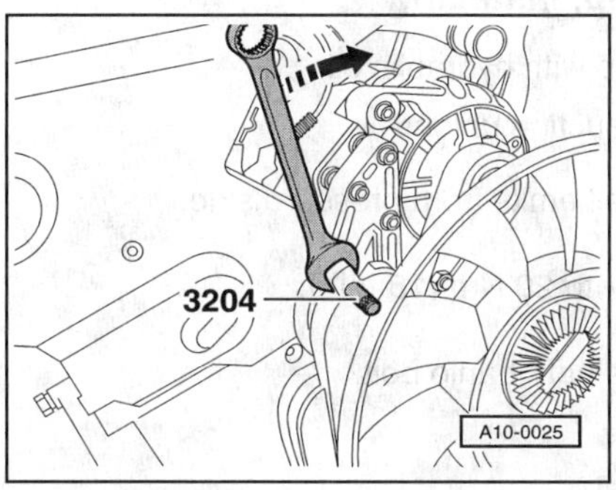

◄ Turn tensioning device in direction of arrow to relieve tension on ribbed belt.

– Secure with special tool -3204-.

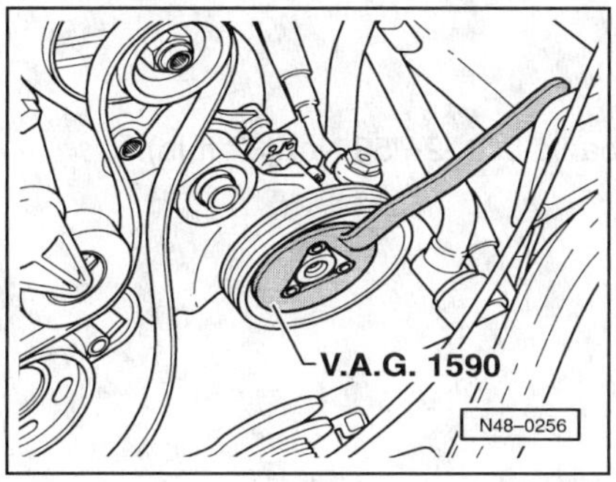

◄ Remove coolant pump ribbed belt pulley.

– Remove ribbed belt.

– Remove power steering pump ribbed belt pulley.

NOTE—
It may be necessary to file out corners of tool VAG 1590

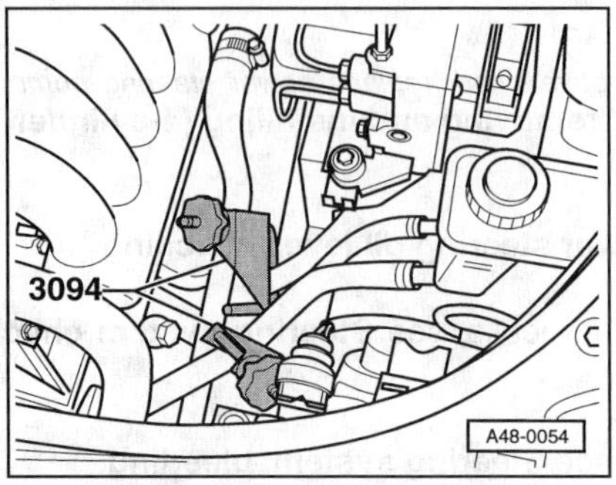

Clamp off intake and return lines using hose clamps -3094-.

– Position drip tray underneath.

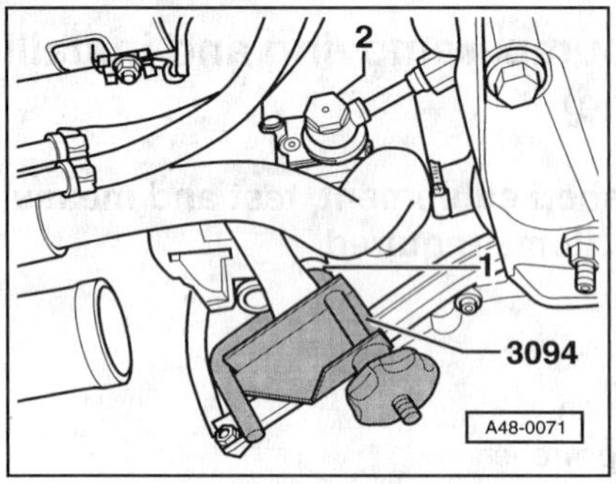

Remove hose clamp -1- from suction hose.

– Pull off suction hose, remove from retainer and place to side.

– Unbolt pressure hose -2-, 22 mm AF.

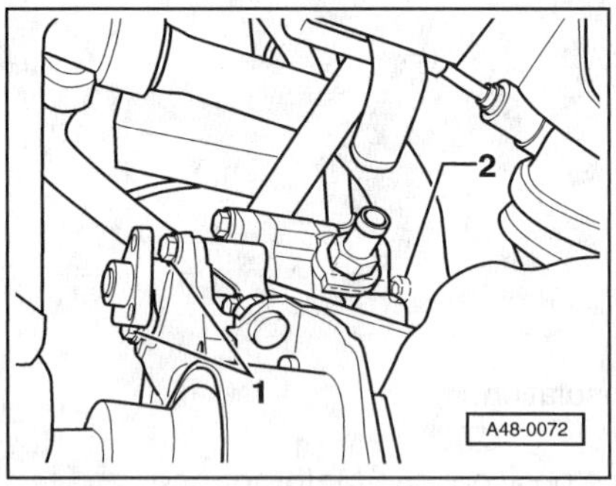

Remove hex bolts -1- and hex socket head bolt -2-.

– Take out pump.

Power steering pump, installing

– Fill power steering pump with hydraulic oil.

- Fill pump with oil via suction union.

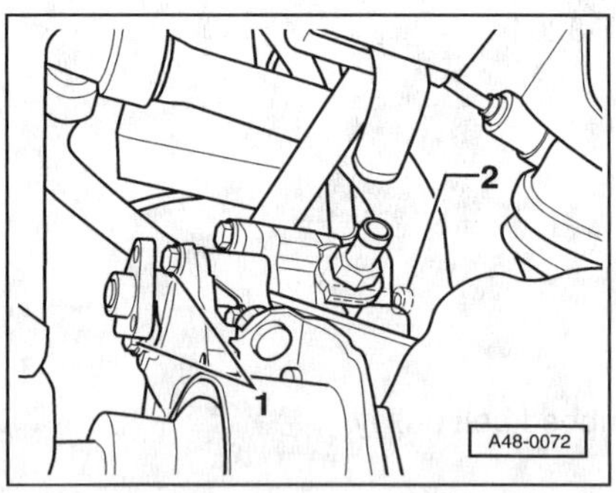

– Turn hub by hand until oil emits from pressure side.

Tighten bolts -1- and -2- to 20 Nm (15 ft-lb).

– Install new sealing rings onto banjo bolt.

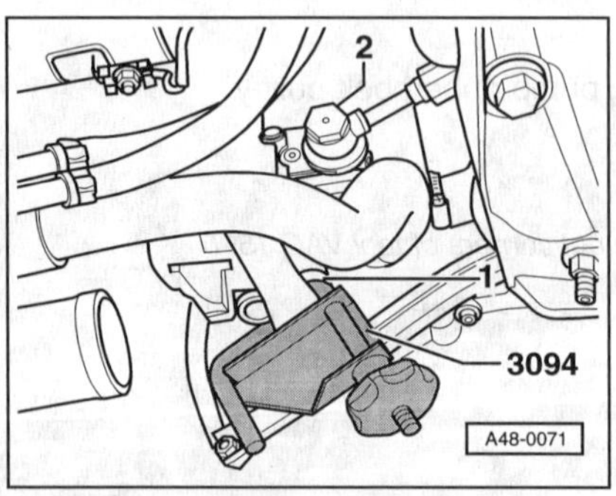

Tighten pressure hose banjo bolt -2-, 50 Nm (37 ft-lb).

– Install suction hose.

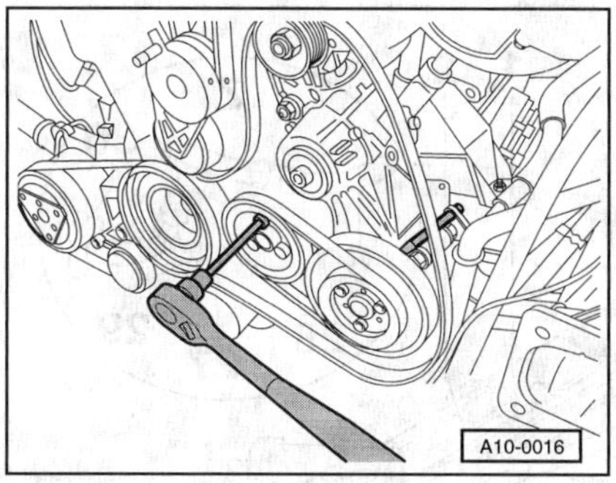

Tighten coolant/power steering pump pulley hex socket head bolts to 25 Nm (18 ft-lb).

– Install radiator fan to viscous coupling.

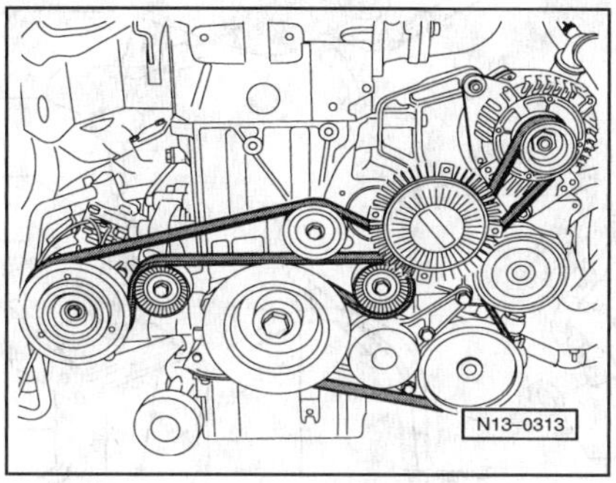

Install ribbed belt.

NOTE —
When installing the belt make sure it is correctly seated in the pulley.

For ease of illustration the routing is shown without the lock carrier being in position.

– Top up hydraulic oil.

– Install bumper.

– Start engine and check belt running.

– Check oil level, see **Power steering oil level, checking**.

– Check steering system for leaks, see **Steering system checking for leaks**.

– Bleed steering system, see **Steering system, bleeding**.

Power steering pump, reservoir and hydraulic lines, assembly (6-cylinder engine)

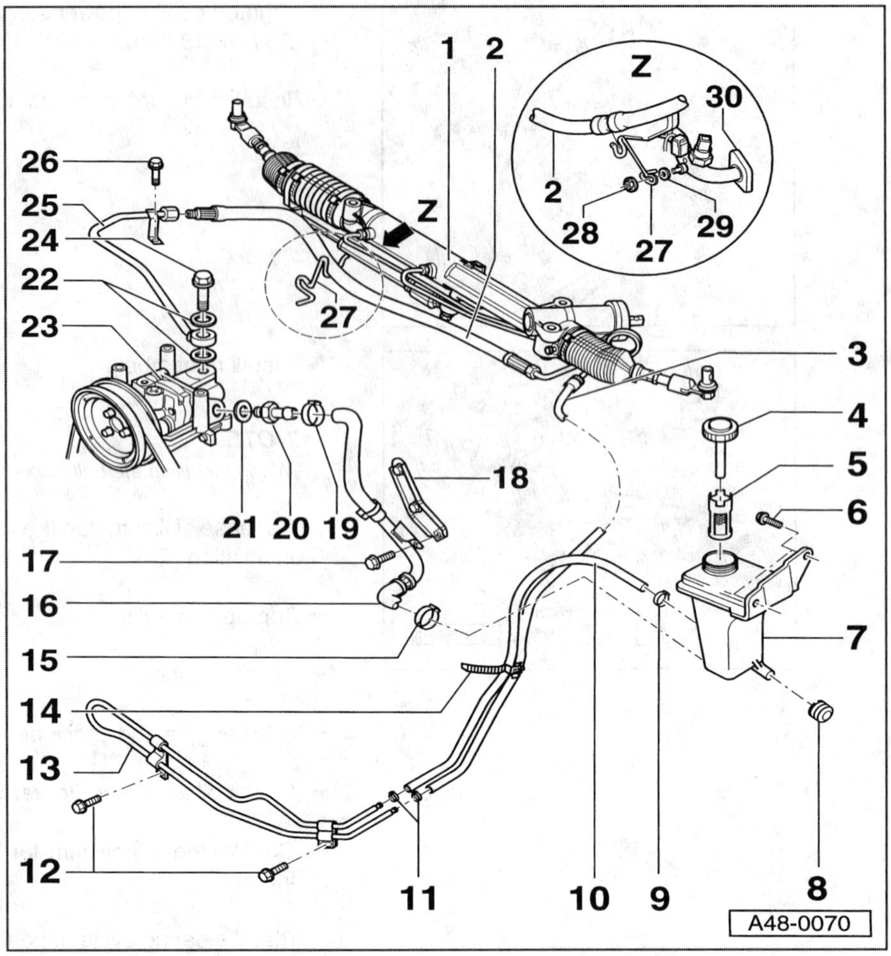

NOTE—

- *If reservoir fluid level is low then check steering system for leaks.*
- *In the case of leaks at the line connections first tighten connections and wipe dry then recheck for leaks.*
- *Pumps supplied as spare parts are not filled with oil. Therefore it is essential to fill the pump with hydraulic oil G 002 000 and to rotate pump by hand, otherwise the pump may be damaged or be noisy during operation.*
- *Type of oil: hydraulic oil G 002 000*
- *Oil system capacity: 0.7 - 0.9 liter.*

1. **Power assisted steering gear**
2. **Pressure hose**
3. **Return hose**
4. **Cap with dip stick**
   - Checking oil level, see **Power steering oil level, checking.**
5. **Strainer**
6. **Hex bolt**
   - Tighten to 9 Nm (80 in-lb)
7. **Reservoir**
8. **Rubber mounting**
9. **Hose clip**
   - Tension with VAG 1275
   - Always replace
10. **Return hose**
    - Between reservoir and cooling line
11. **Hose clip**
    - Tension with VAG 1275
    - Always replace
12. **Hex bolt**
    - Tighten to 10 Nm (7 ft-lb)
13. **Cooling line**
14. **Cable tie**
15. **Hose clip**
    - Tension with VAG 1275
    - Always replace
16. **Suction hose**
17. **Hex bolt**
    - Tighten to 8 Nm (71 in-lb)
18. **Retainer**
    - Bolted to cylinder head
19. **Hose clip**
    - Tension with VAG 1275
    - Always replace
20. **Threaded union**
    - Tighten to 50 Nm (37 ft-lb)
21. **Oil seal, 18 x 24 mm**
    - Always replace
22. **Oil seal, 16 x 22 mm**
    - Always replace
23. **Power steering pump**
    - Checking delivery pressure, see **Power steering pump delivery pressure, checking (6-cylinder engine)**
    - Fill with oil before installing.
24. **Banjo bolt**
    - Tighten to 50 Nm (37 ft-lb)
25. **Pressure hose**
    - Tighten to 40 Nm (30 ft-lb)
    - Screw onto pressure hose, when doing this counter-hold on pressure hose hex.
    - Bolted to intake line with hex bolt - 26-.
26. **Hex bolt**
    - Tighten to 8 Nm (71 in-lb)
27. **Bracket**
28. **Hex nut**
    - Tighten to 20 Nm (15 ft-lb)
29. **Washer**
30. **Coolant line**

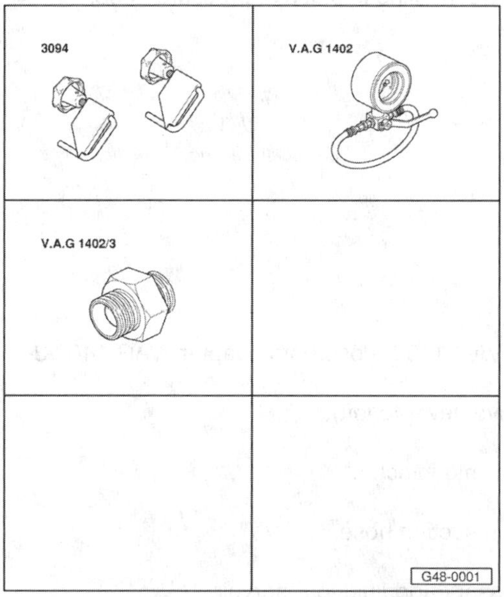

Power steering pump delivery pressure, checking (6-cylinder engine)

Special tools, workshop equipment, test and measuring appliances and aux. items required

- 3094 Hose clamps
- VAG 1402 Power steering test module
- VAG 1402/3 Adapter

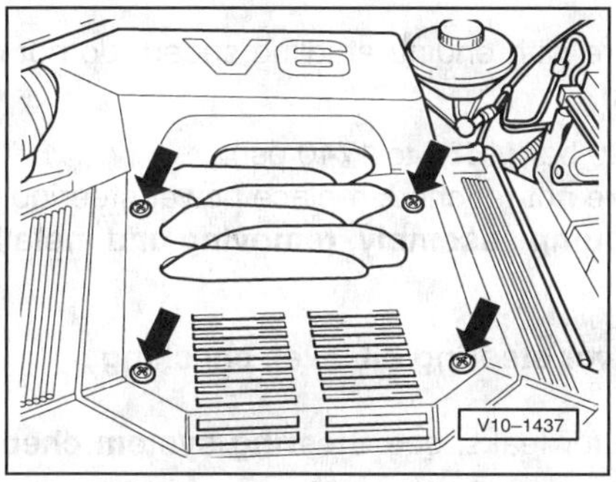

◀ Unclip engine cover and take off.

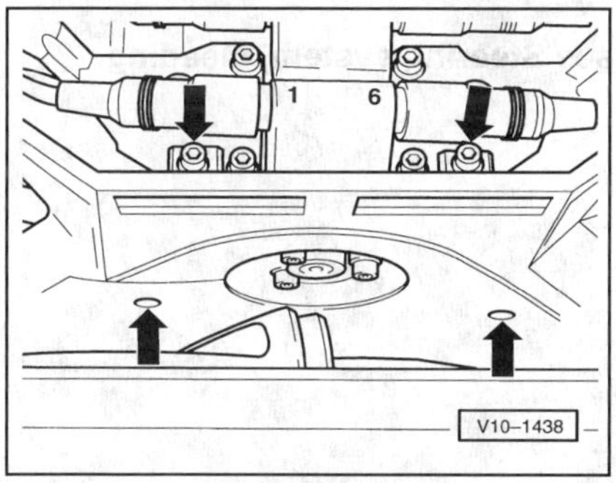

◀ Remove ribbed belt cover -arrows-.

– Clamp off suction hose with hose clamp -3094-.

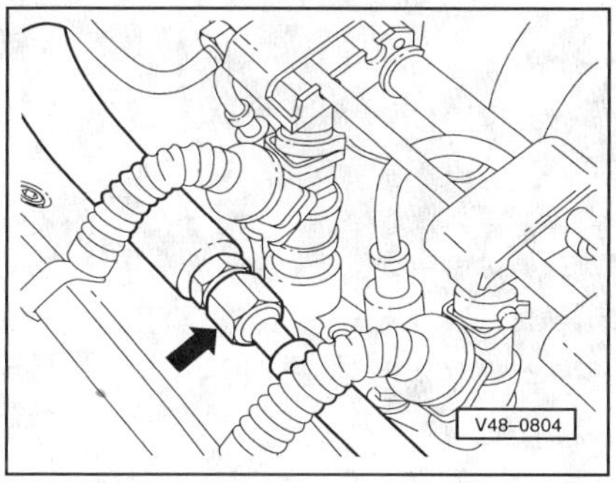

◀ Press connector spring clip and pull connector off injector.

– Remove line from pressure hose (arrow), when doing this counter-hold on pressure hose hex.

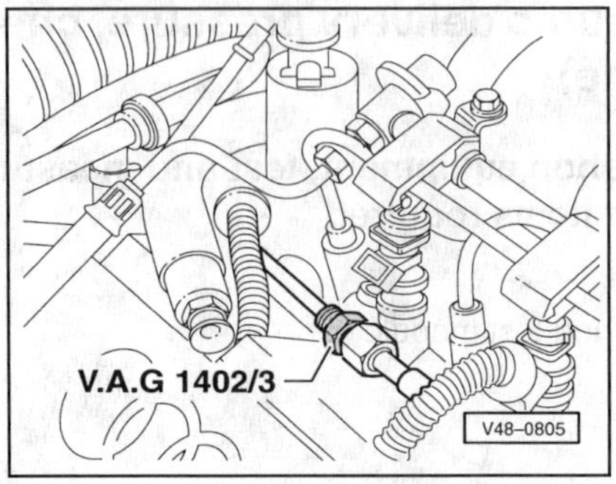

Screw adapter VAG 1402/3 into line in place of pressure hose.

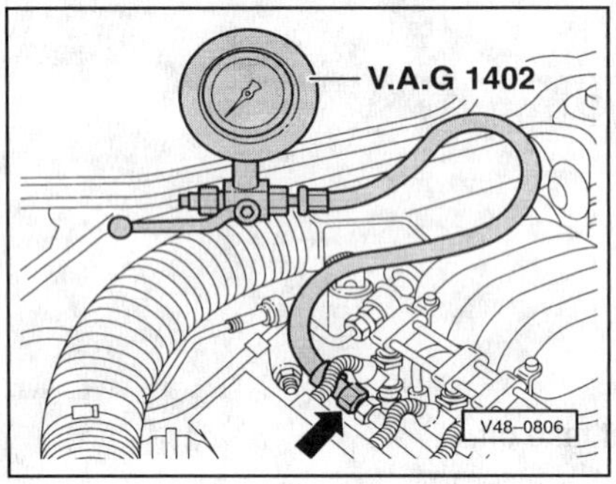

Screw pressure gauge -VAG 1402- hose onto adapter -VAG 1402/3-.

– Close gauge shut-off valve (lever to left).

– Install injector connector onto injector.

– Remove hose clamp from suction hose.

– Start engine and if necessary add fluid to reservoir.

– Read off gauge pressure with engine at idling speed (do not measure longer than 10 seconds).

  • Specification: 110 - 120 bar (1585 to 1740 psi).
  • If the specifications are not reached, replace power steering pump, see **Power steering pump assembly, removing and installing (6-cylinder engine)**.

– Check oil level, see **Power steering oil level, checking**.

– Check steering system for leaks, see **Steering system checking for leaks**.

– Bleed steering system, see **Steering system, bleeding**.

Power steering pump assembly, removing and installing (6-cylinder engine)

NOTE —
- *Replace sealing rings.*
- *Do not reuse drawn off hydraulic oil.*
- *Hydraulic oil: Part No. G 002 000*
- *Quantity of oil in system approx. 1 liter.*

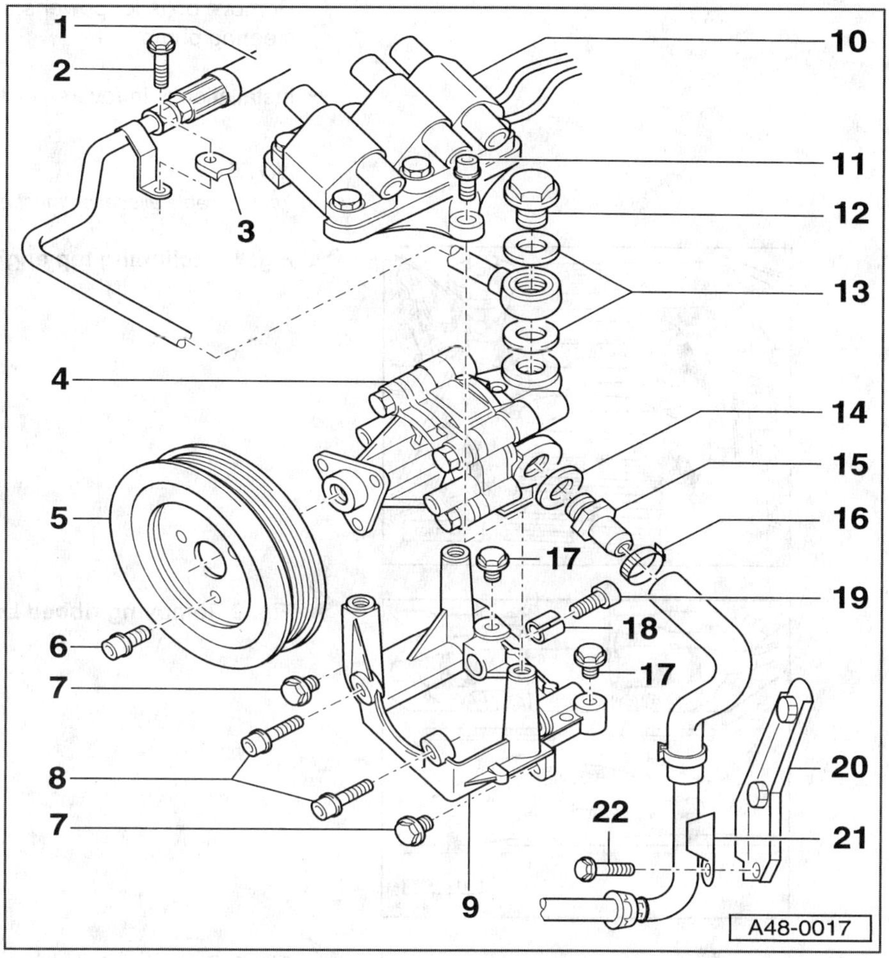

1. **Pressure line with hose**
2. **Hex bolt**
   - Tighten to 8 Nm (71 in-lb)
3. **Hole on intake line**
4. **Power steering pump**
   - Checking delivery pressure, see **Power steering pump delivery pressure, checking (6-cylinder engine)**
   - Fill with oil before installing.
5. **Belt pulley**
6. **Hex socket head bolt**
   - Tighten to 25 Nm (18 ft-lb)
   - When loosening and tightening counter-hold with VAG 1590 if necessary
7. **Hex bolt**
   - Tighten to 25 Nm (18 ft-lb)
8. **Hex socket head bolt**
   - Tighten to 25 Nm (18 ft-lb)
9. **Bracket**
10. **Ignition coil with bracket**
11. **Hex socket head bolt**
    - Tighten to 25 Nm (18 ft-lb)
12. **Banjo bolt**
    - Tighten to 50 Nm (37 ft-lb)
13. **Oil seal, 16 x 22 mm**
    - Always replace
14. **Oil seal, 18 x 24 mm**
    - Always replace
15. **Threaded union**
    - Tighten to 50 Nm (37 ft-lb)
16. **Hose clip**
    - Tension with VAG 1275
    - Always replace
17. **Hex bolt**
    - Tighten to 25 Nm (18 ft-lb)
18. **Sleeve, slotted**
19. **Hex socket head bolt**
    - Tighten to 25 Nm (18 ft-lb)
20. **Mounting bracket**
21. **Suction line with hose**
22. **Hex bolt**
    - Tighten to 8 Nm (71 in-lb)

Power steering pump assembly, removing and installing

- Remove upper engine cover, see **Fig. 1**.
- Remove ribbed belt guard, see **Fig. 2**.
- Relieve tension on ribbed belt, see **Fig. 3**.
- Remove belt pulley, counter-hold with VAG 1590 if necessary.
- Pull ignition cable out of ignition coils **-10-**.
- Clamp off suction hose with 3094.
- Remove suction and pressure lines.
- Remove bracket **-9-** and remove complete with power steering pump.

– Remove bolts for power steering pump from bracket and take off power steering pump.

– Installation is in reverse order or removal.

NOTE —
Ignition cable allocation must equate to numbers marked see Fig. **4**

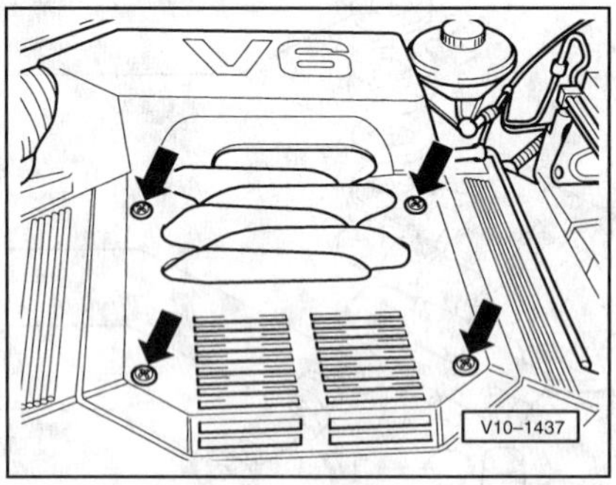

Fig. 1 Unclipping top engine cover and removing

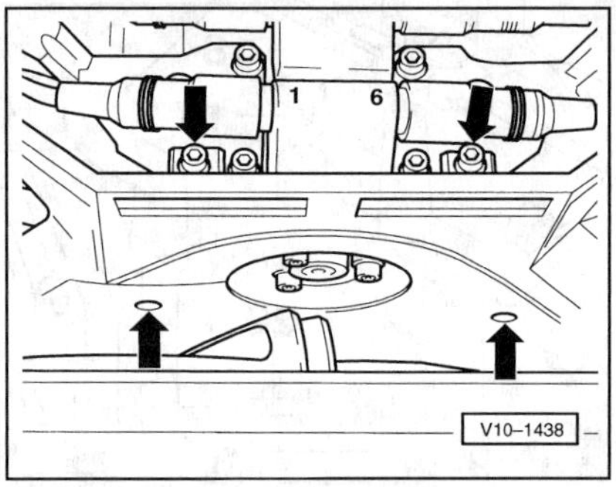

Fig. 2 Removing ribbed belt guard

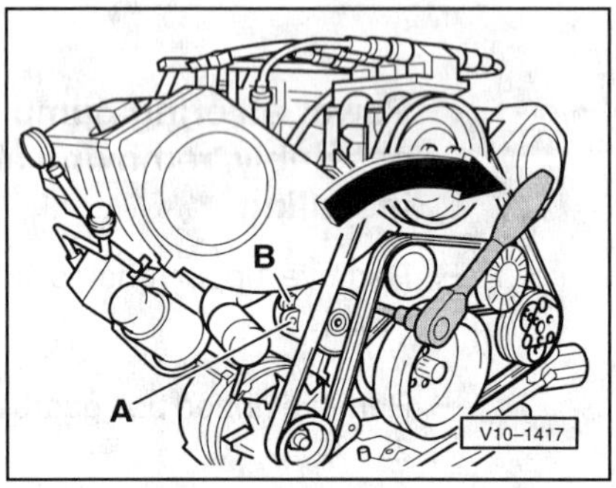

Fig. 3 Relieving tension and removing ribbed belt

– To relieve tension on ribbed belt, push hex socket head wrench (10 mm AF) downward.

– Align hole -A- with hole -B-.

– Insert pin -3204- to secure.

– Take ribbed belt off belt pulley.

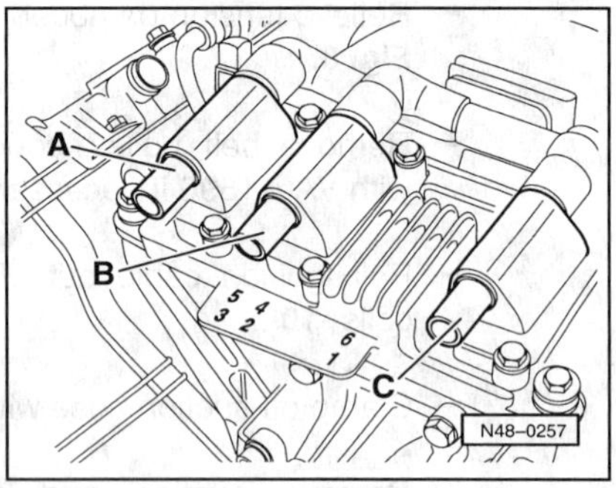

Fig. 4 Allocating ignition cables to ignition coils

A - Ignition coil for cylinders 5 and 3

B - Ignition coil for cylinders 4 and 2

C - Ignition coil for cylinders 6 and 1

• Ignition cables are marked with number 1. .. 6.

Power steering pump, reservoir and hydraulic lines, assembly (8-cylinder engine)

NOTE —

- *If reservoir fluid level is low then check steering system for leaks.*
- *In the case of leaks at the line connections first tighten connections and wipe dry then recheck for leaks.*
- *Pumps supplied as spare parts are not filled with oil. Therefore it is essential to fill the pump with hydraulic oil G 002 000 and to rotate pump by hand, otherwise the pump may be damaged or be noisy during operation.*
- *Type of oil: hydraulic oil G 002 000*
- *Oil system capacity: 0.7 to 0.9 ltr.*

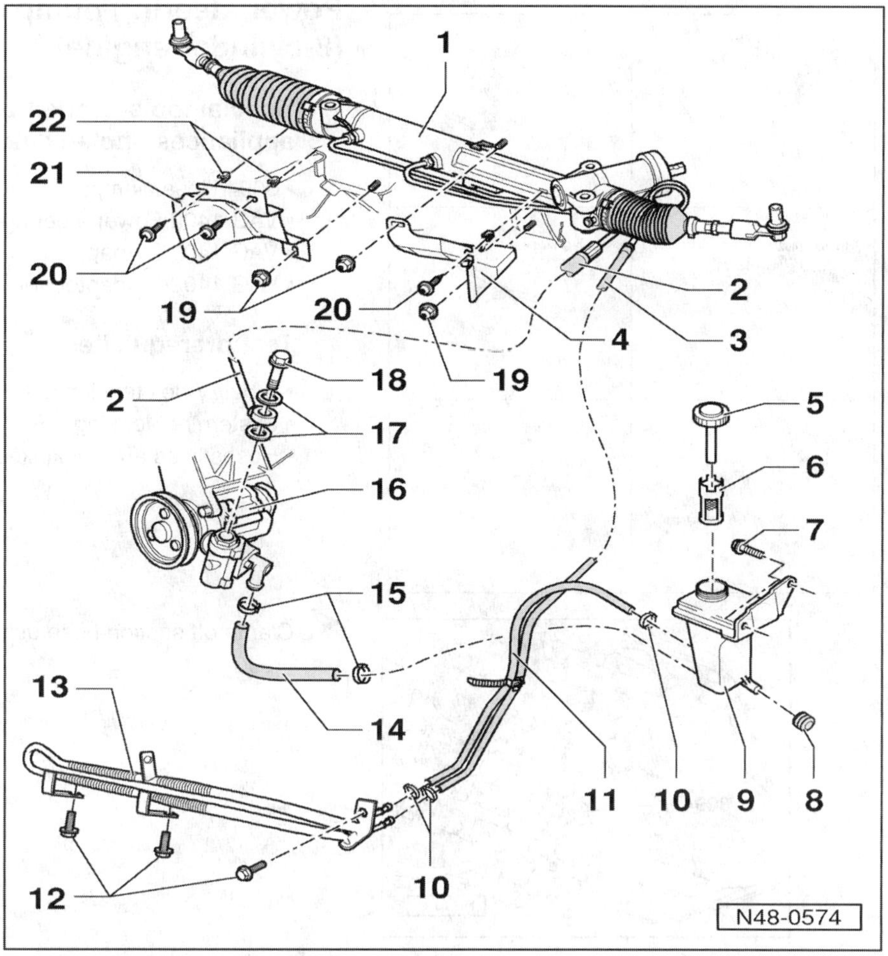

1. **Power assisted steering gear**
2. **Pressure hose**
3. **Return hose**
4. **Backplate**
5. **Cap with dip stick**
   - Checking oil level, see **Power steering oil level, checking**
6. **Strainer**
7. **Hex bolt**
   - Tighten to 9 Nm (80 in-lb)
8. **Rubber mounting**
9. **Reservoir**
10. **Hose clip**
    - Tension with VAG 1275
    - Always replace
11. **Return hose**
    - Between reservoir and ribbed cooling line
12. **Hex bolt**
    - Tighten to 10 Nm (7 ft-lb)
13. **Ribbed cooling line**
14. **Suction hose**
15. **Hose clip**
    - Tension with VAG 1275
    - Always replace
16. **Power steering pump**
    - Checking delivery pressure, see **Power steering pump delivery pressure, checking (8-cylinder engine).**
    - Fill with oil before installing.
17. **Oil seal, 16 x 22 mm**
    - Always replace
18. **Banjo bolt**
    - Tighten to 50 Nm (37 ft-lb)
19. **Hex nut**
    - Tighten to 1.3 Nm (11 in-lb)
20. **Hex bolt**
    - Tighten to 2 Nm (18 in-lb)
21. **Backplate**
22. **Spreader nut**
    - Always replace

Power steering pump delivery pressure, checking (8-cylinder engine)

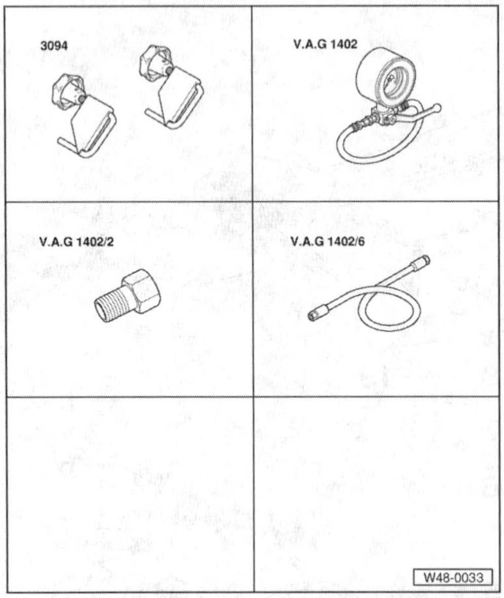

Special tools, workshop equipment, test and measuring appliances and aux. items required

- 3094 Hose clamps
- VAG 1402 Power steering test unit
- VAG 1402/2 Adapter
- VAG 1402/6 Adapter set

Test prerequisites:

- V-belt/V-belt tension OK.
- System not leaking.
- Hoses/lines are not kinked or restricted.

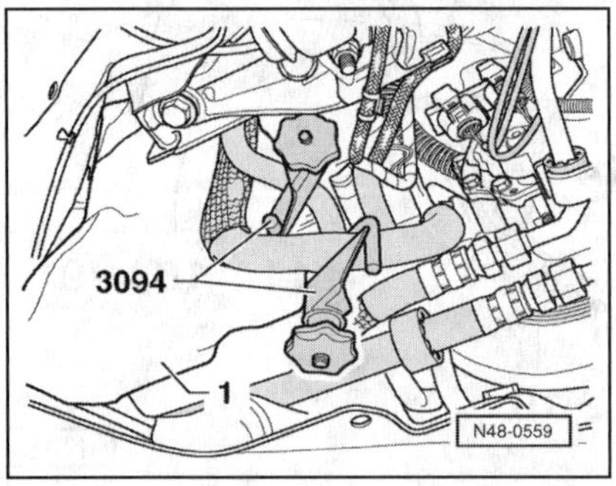

◄ Clamp off suction hose using hose clamp -3094-.

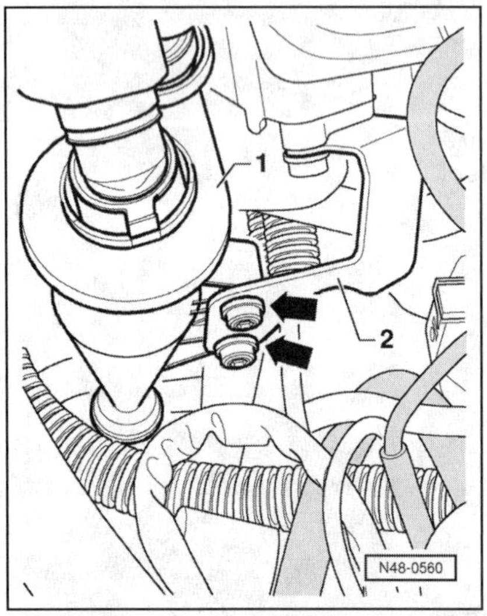

◄ Unbolt -arrows- crankcase breather -1- from bracket -2-.

– Place oil tray under vehicle.

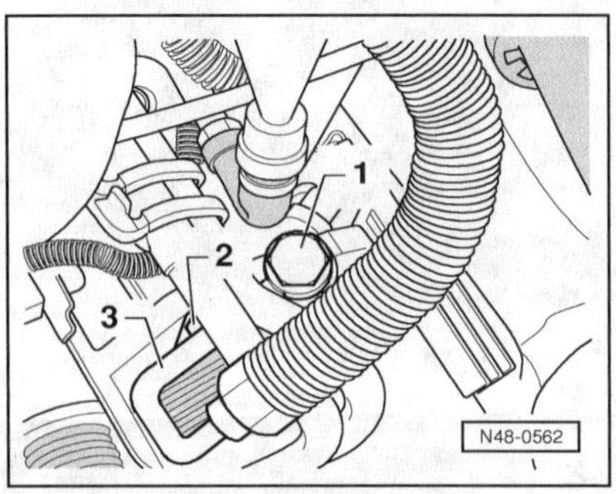

◄ Remove banjo bolt -1- and hex bolt -2-.

– Set pressure line -3- to side.

- Seal pressure line/hose with a plastic bag or similar.

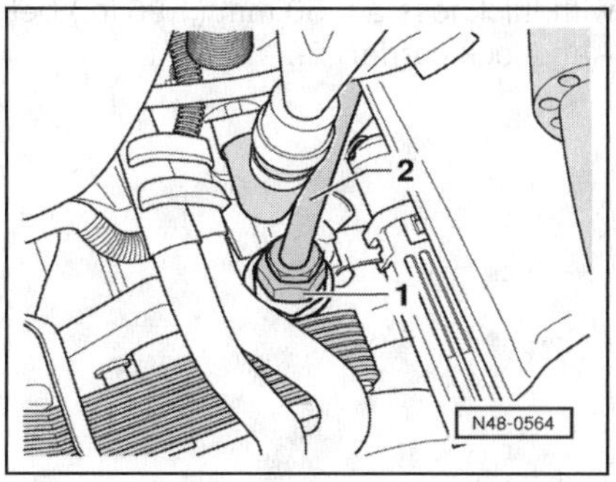

◀ Screw adapter -1- from VAG 1402/6 into power steering pump in place of banjo bolt.

– Screw adapter -2- of VAG 1402/6 into adapter -1- from VAG 1402/6.

– Screw hose of pressure gauge VAG 1402 onto adapter -2- of VAG 1402/6.

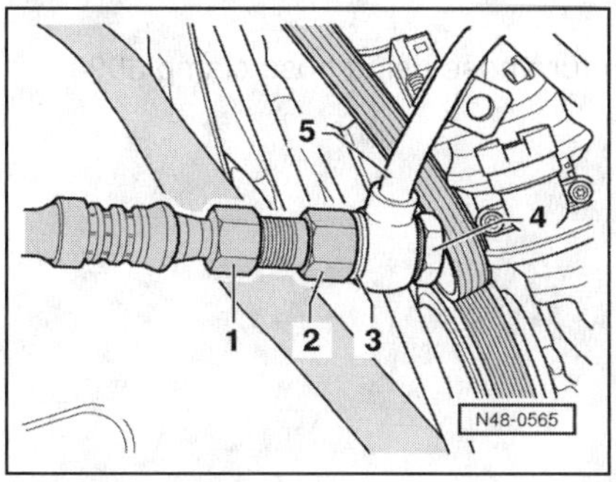

◀ Assemble VAG adapter, banjo bolt and pressure line as shown in diagram.

1 - VAG 1402/2
2 - VAG 1402/6
3 - Four seals
4 - Banjo bolt
5 - Pressure line

– Remove hose clamp 3094 from suction hose.

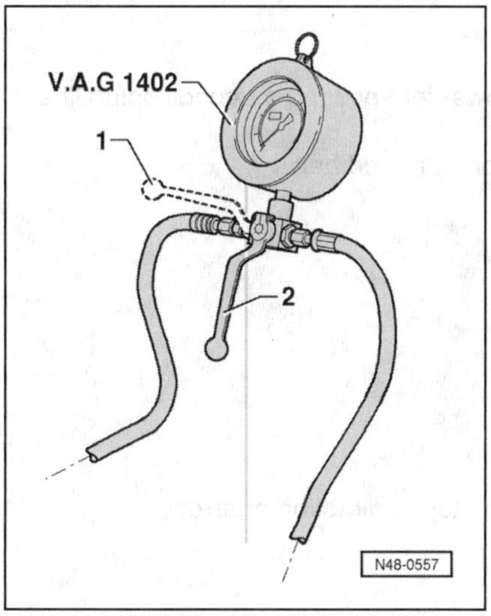

◀ Make sure that the lever on the pressure gauge remains in position -2-.

– Start engine and if necessary add fluid to reservoir.

– Turn steering wheel approx.10 times from lock to lock.

– With engine running at idling speed. Close cut-off valve (set lever to position -1-, not longer than 5 seconds) and read off pressure.

• Specifications:110 - 120 bar (1595 - 1740 psi)

Power steering pump, removing and installing (8-cylinder engine)

Special tools, workshop equipment, testers, measuring instruments and auxiliary items required

• 3094 Hose clamps
• VAG 1331 Torque wrench

Removing

– Remove engine noise insulation.

– Move lock carrier to service position, see **Maintenance**.

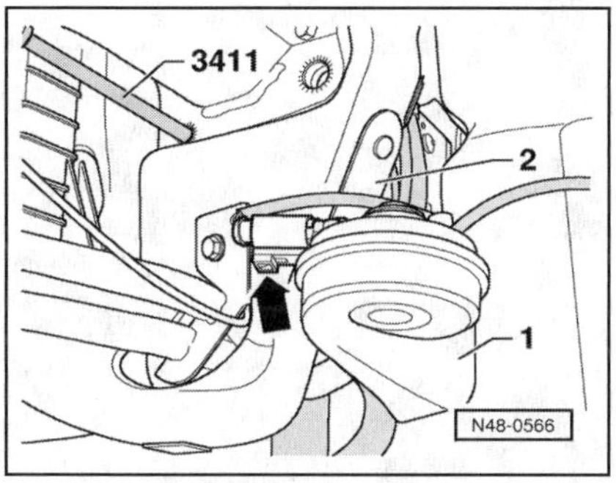

◀ Unclip connector -arrow- from horn -1-.

2 - Lock carrier

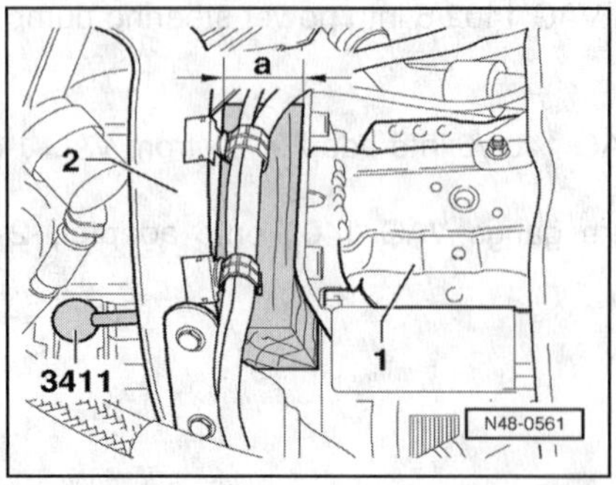

◄ Insert a piece of wood with thickness a = 50 mm (1.96 in.) between longitudinal member -1- and lock carrier -2-.

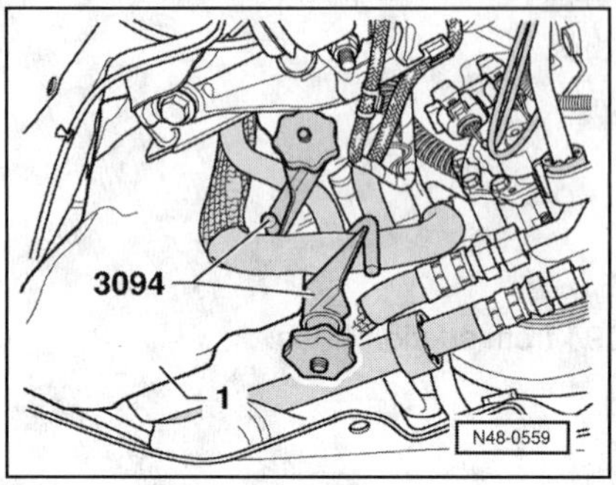

◄ Clamp off suction and return hose using hose clamp 3094.

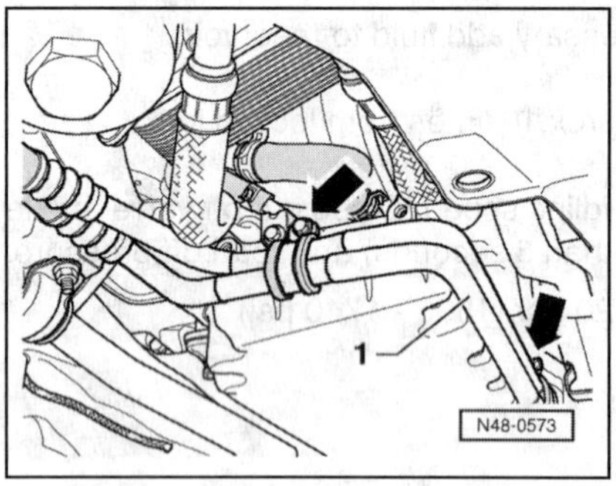

◄ Remove hex bolts -arrows- for securing air conditioning line -1-.

– Mark direction of rotation of ribbed belt.

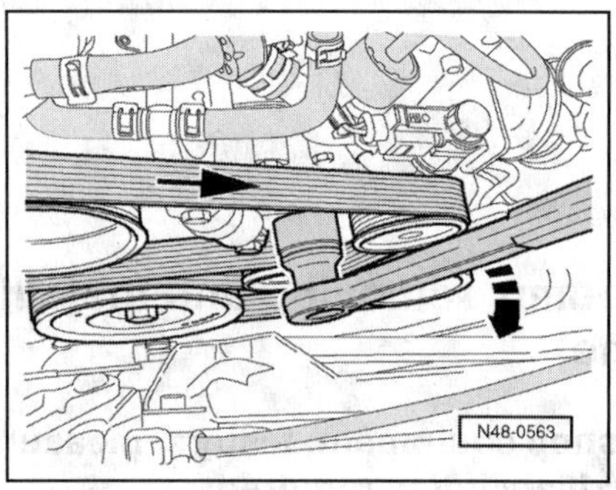

◄ Turn tensioning roller to stop in direction of arrow.

– Remove ribbed V-belt.

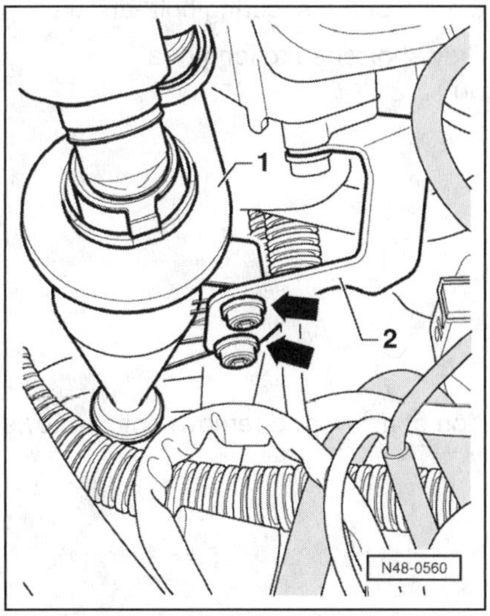

◀ Unbolt -arrows- crankcase breather -1- from bracket -2-.

– Place oil tray under vehicle.

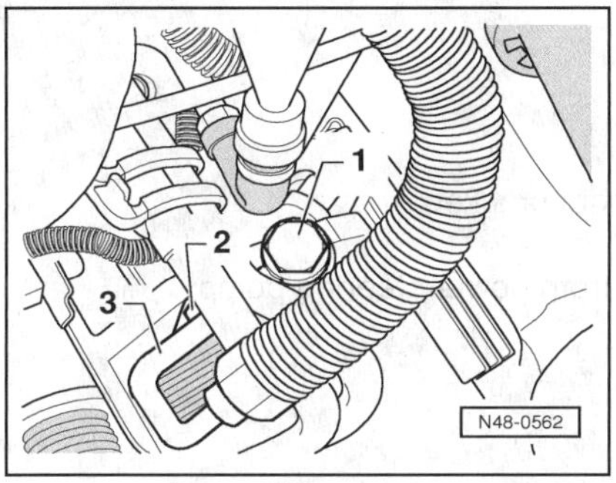

◀ Remove banjo bolt -1- and hex bolt -2-.

- Set pressure line -3- to side.
- Seal pressure line/hose with a plastic bag or similar.

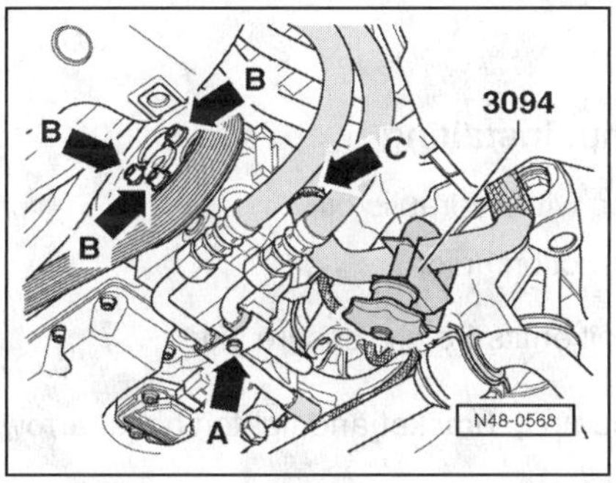

◀ Remove hex bolt -A-.

– Unbolt power steering pump pulley -arrows B-.

NOTE —

- *When removing power steering pump pulley, observe the shim located behind it. It is important for the measurement and adjustment of the power steering pump.*
- *Do not yet remove shim from power steering pump gear.*

– Open hose clip -arrow C- and pull suction hose off power steering pump.

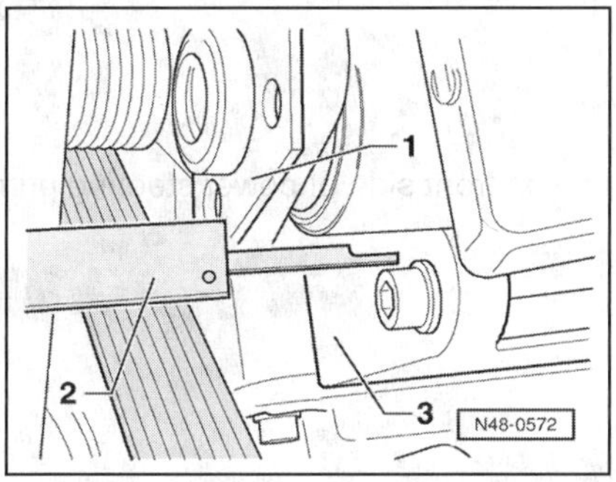

◀ Measure distance from front side of hub with shim -1- to power steering pump bracket -3- with depth gauge -2-.

– Note the measured value.

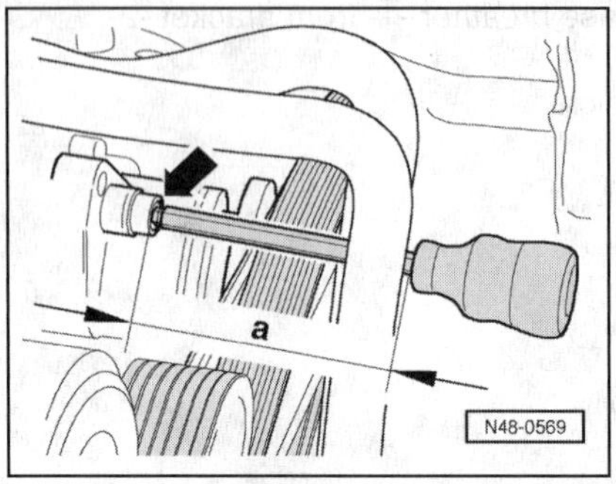

Remove upper power steering pump securing bolt -arrow-.

• To do so, use an Allen key shortened to length -a-.
a = about 75 mm (2.95 in.).

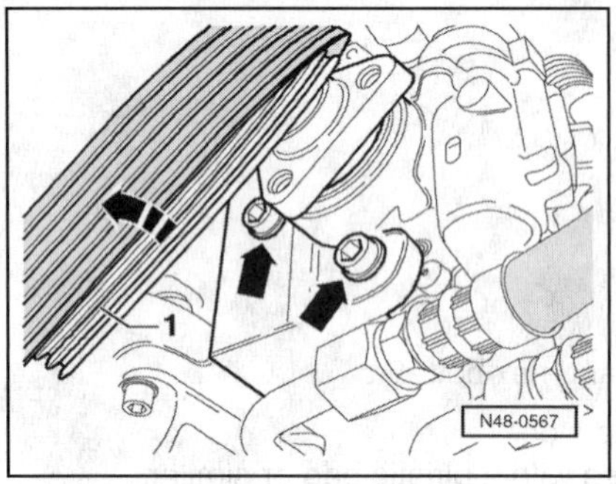

Press pulley -1- in direction of arrow and remove bolts -arrows-.

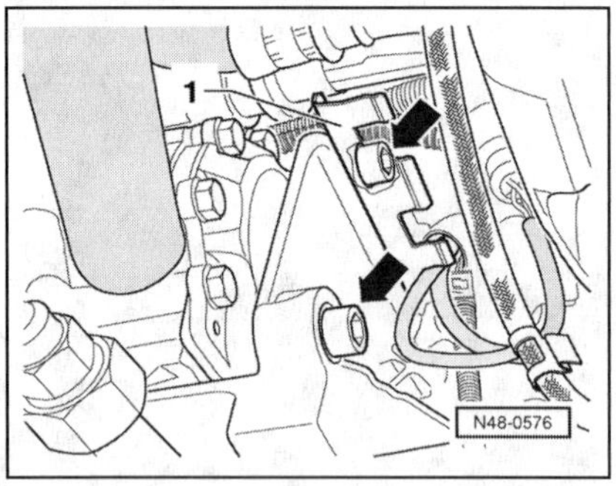

Unclip connectors -1- from bracket.

– Remove bolts -arrows- from power steering pump.

– Take out pump.

Power steering pump, installing

– Fill power steering pump with hydraulic oil.

• Fill pump with oil via suction union.

– Turn hub by hand until oil emits from pressure side.

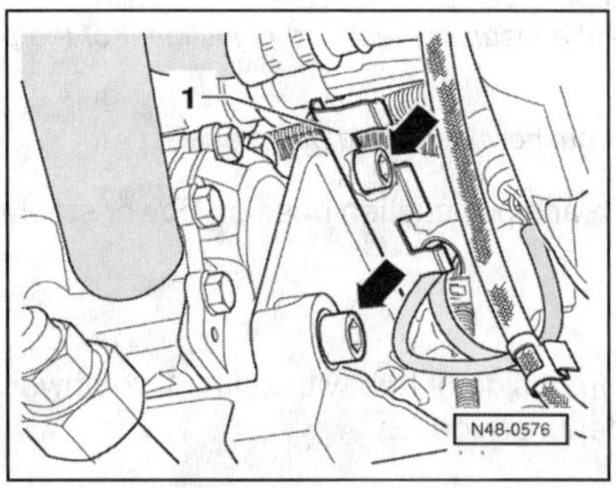

Install power steering pump in bracket and tighten bolts -arrows- to 23 Nm (17 ft-lb).

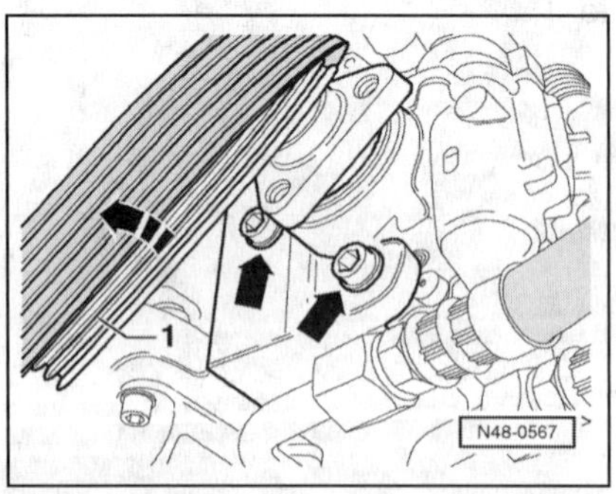

Then tighten bolts -arrows- on front side of power steering pump to 23 Nm (17 ft-lb).

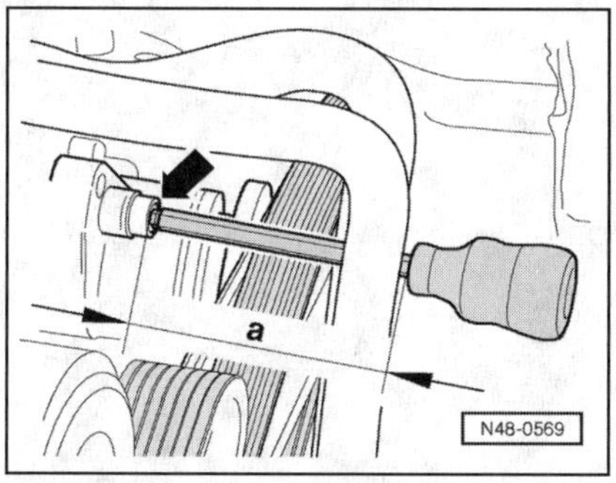

Tighten upper power steering pump securing bolt -arrow- to 23 Nm (17 ft-lb)

- To do so, use an Allen key shortened to length -a-.
  a = about 75 mm (2.95 in.)

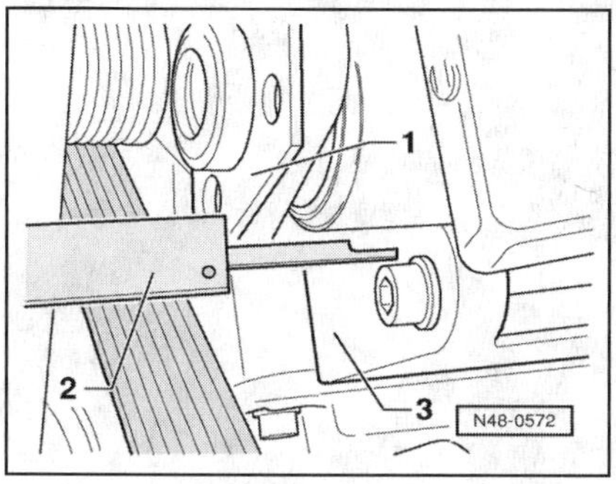

Determining shim thickness

Measure distance from front of hub without shim -1- to power steering pump bracket -3- with depth gauge -2-.

– Note the measured value.

– Now determine shim thickness using values measured before removing and after installing.

Example:

Value before removing power steering pump, with shim	33.65 mm
Value after installing power steering pump, without shim	32.85 mm
Shim thickness =	0.80 mm

The following shims are available: 0.6, 0.8, 1.0, 1.2, 1.4 mm.

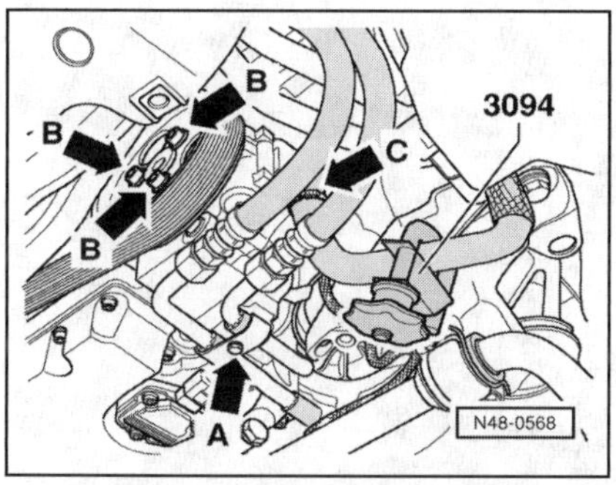

Install suction hose and install new hose clip -arrow C-.

– Bolt on vane pump pulley with torque setting 23 Nm (17 ft-lb) -arrows B-.

– Screw in hex bolt -arrow A-.

– Install pressure line to vane pump and tighten banjo bolt, 50 Nm (37 ft-lb).

– Remove hose clamp 3094 from suction and pressure lines.

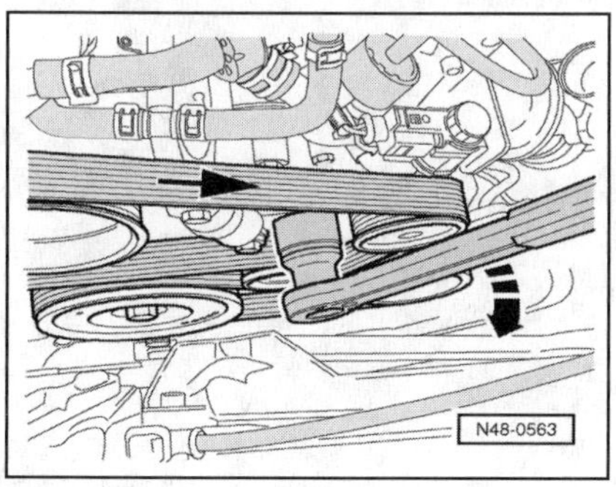

Turn tensioning roller to stop in direction of arrow.

– Install ribbed belt and turn tensioning roller opposite direction of arrow until ribbed belt is tensioned.

NOTE —
When installing the ribbed belt make sure that the direction of rotation and seating in the pulley are correct.

– Installation is continued in reverse order of removal, noting the following:

- Check oil level, see **Power steering oil level, checking.**
- Check steering system for leaks, see **Steering system checking for leaks**.
- Bleed steering system, see **Steering system, bleeding**.

Body–Exterior

Section

T

50 Body, Front

Page

55 Hoods, Lids

Page

57 Front Doors, Central Locking — Page

58 Rear Doors — Page

60 Sunroof — Page

63 Bumpers — Page

64 Window Regulators — Page

66 Exterior Equipment — Page

50 Body, Front

Body, Front

Tools

Special tools, workshop equipment, testers, measuring instruments and auxiliary items required

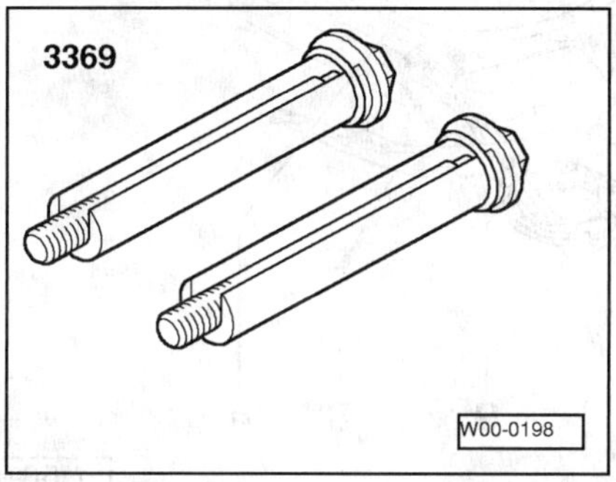

3369 Support tool

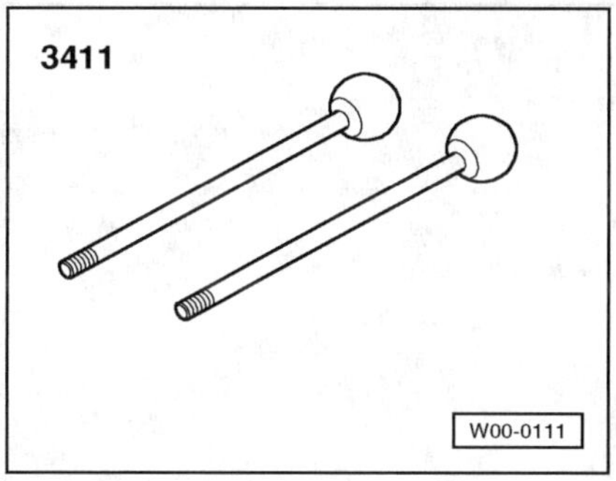

3411 Guide rods

Lock carrier and front bumper reinforcement, assembly (through 09/2000)

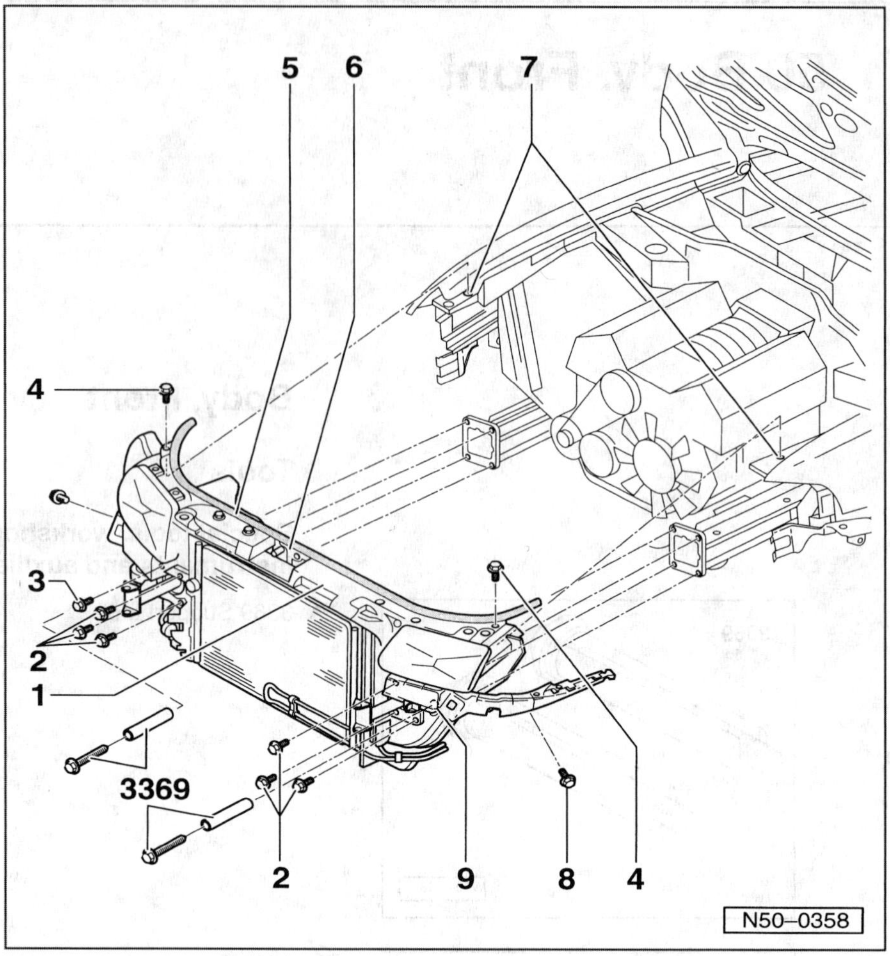

1. **Lock carrier and front bumper reinforcement**
   - Removing:
     Removing noise insulation, see **Noise insulation, assembly**
     Remove turn signals.
     Disconnect Bowden cable **-6-** from coupling and disconnect electrical connections.
     Remove front bumper, see **Repair Group 63, Front bumper, assembly (through 09/2000)** and remove side mounted guide from fender.
     Disconnect hoses for charge air cooler (only certain engine variants).
     Drain coolant and disconnect coolant hoses.
     Remove condenser from lock carrier, see **Heating & Air Conditioning section, Repair Group 87**
     Unscrew hydraulic oil cooler from lock carrier, see **Hydraulic oil cooler, unbolting**.

   NOTE —
   - *Do not hang up condenser and hydraulic oil cooler on pipes/ hoses.*
   - *Do not kink condenser and hydraulic pipe/hoses.*

   - Installing:
     Align lock carrier to longitudinal members and between fenders.
     Gaps/shut lines
     Adjust headlights.

2. **Bolt**
   - Qty. 6
   - Tighten to 50 Nm (37 ft-lb)
3. **Bolt**
   - Tighten to 50 Nm (37 ft-lb)
4. **Bolt**
   - Qty. 2
   - Tighten to 8 Nm (71 in-lb)
5. **Seal**
6. **Bowden cable**
7. **Hole in side panel**
8. **Bolt**
   - Qty. 2
9. **Hole for special tool**
   - Special tool 3369

Lock carrier and front bumper reinforcement, moving to service position (through 09/2000)

– Remove front bumper cover, see **Repair Group 63, Front bumper assembly (through 09/2000)**.

– Unbolt air guide between lock carrier and air cleaner on lock carrier.

– Unscrew bolt -**3**- and screw special tool 3369 onto right and left-hand longitudinal member.

– Unscrew bolts -**2**- and -**4**- and pull lock carrier in area of special tool 3369 forwards.

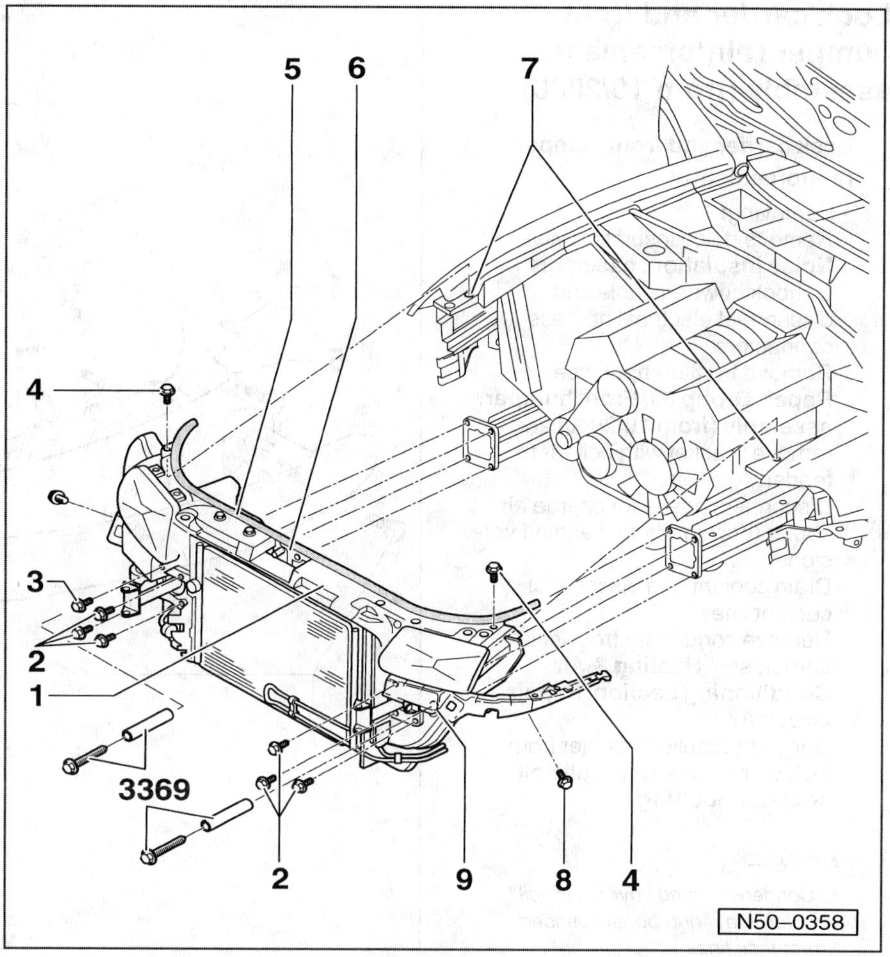

Lock carrier and front bumper reinforcement, assembly (from 10/2000)

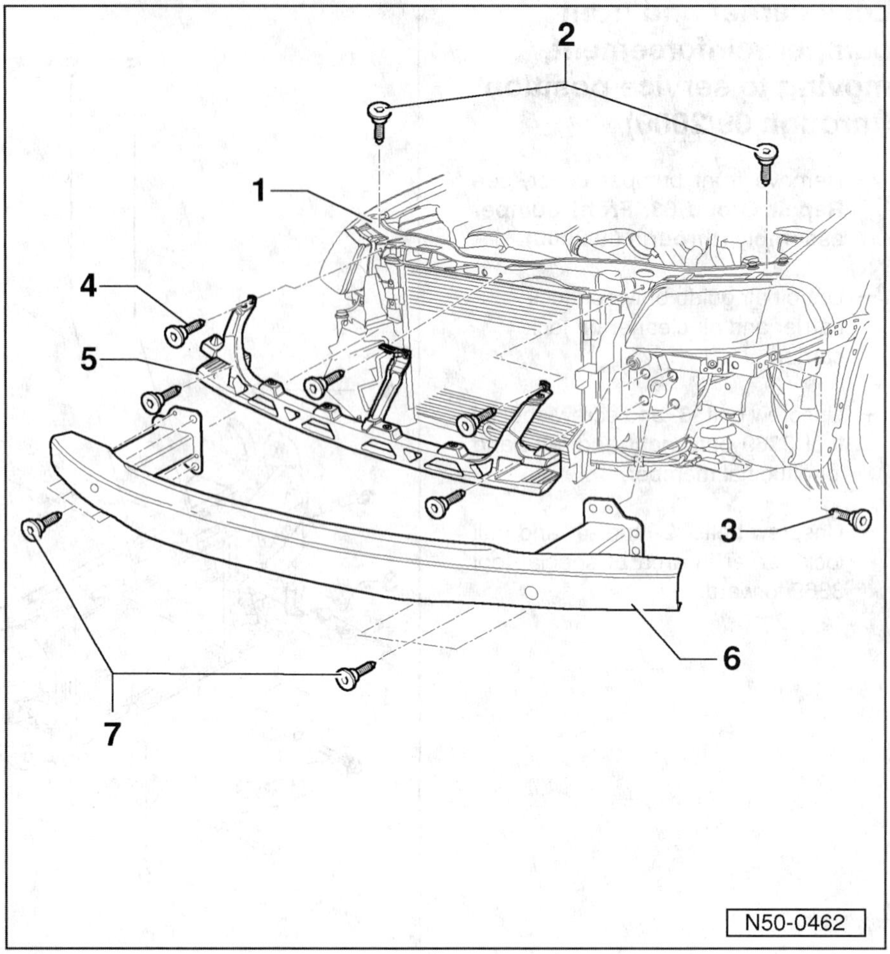

1. **Lock carrier and front bumper reinforcement**
   - Removing:
     Remove noise insulation, see **Noise insulation, assembly**
     Unhook bowden cable and disconnect electrical harness connectors
     Remove front bumper, see **Repair Group 63, Front bumper, assembly (from 10/2000)** and remove bracket with bolt -3- to fender.
     Disconnect hoses for charge air cooler (only for certain engine versions).
     Drain coolant and disconnect coolant lines
     Remove condenser from lock carrier, see **Heating & Air Conditioning section, Repair Group 87**
     Unbolt hydraulic oil cooler from lock carrier, see **Hydraulic oil cooler, unbolting**

   NOTE —
   - *Condenser and hydraulic oil cooler may not be suspended from their lines.*
   - *Condenser and hydraulic lines may not be kinked.*

   - Installing:
     Tape off fender points, to avoid damaging paint.Center lock carrier at longmembers and between fenders. Adjust headlights.

2. **Bolt**
   - Qty: 2
   - Tighten to 8 Nm (71 in-lb)
3. **Bolt**
   - Qty: 3
4. **Bolt**
   - Qty: 5
   - Tighten to 6 Nm (53 in-lb)
5. **Mounting frame**
6. **Bumper carrier**
7. **Bolt**
   - Qty: 7
   - Tighten to 50 Nm (37 ft-lb)

Lock carrier and front bumper reinforcement, moving to service position (from 10/2000)

– Remove front bumper cover, see **Repair Group 63, Front bumper , assembly (from 10/2000)**

– Remove noise insulation, see **Noise insulation, assembly**.

– Unbolt air guide between lock carrier and air filter at lock carrier.

– Remove bolts -5-, and bolt 3411 guide rods to right and left longmember, see **Fig. 1**.

– Remove bolts -2- and -3-.

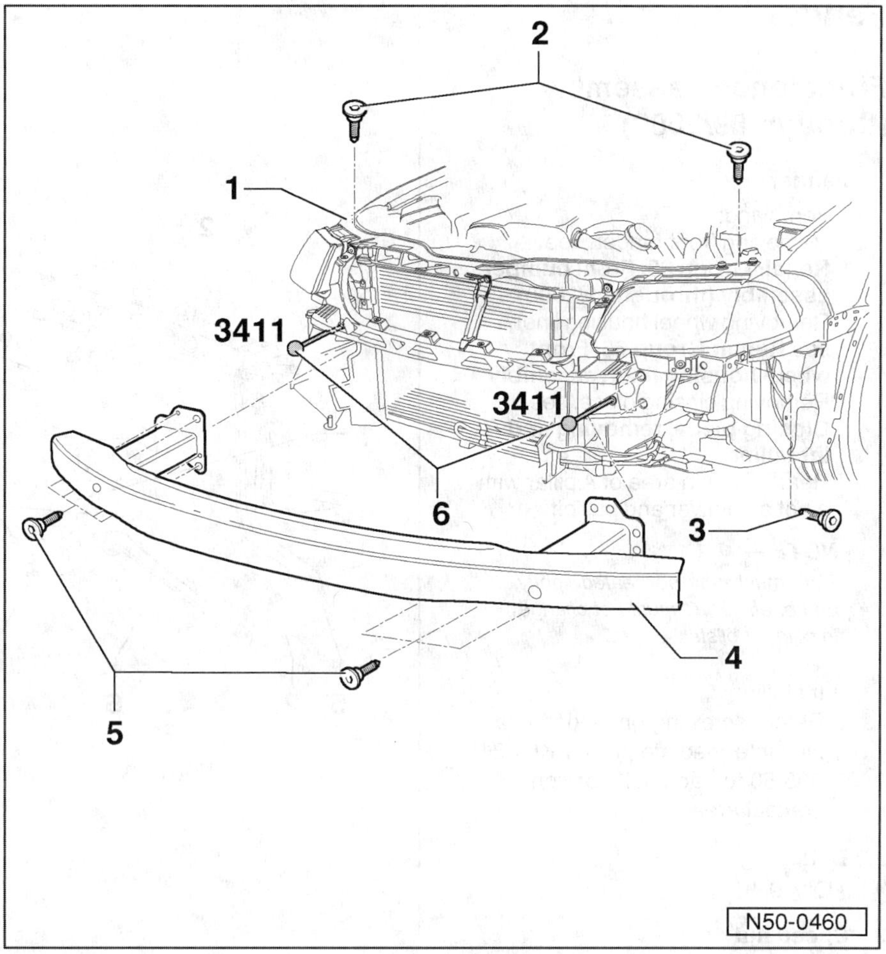

1. **Lock carrier with components**
2. **Bolt**
   - Qty: 2
   - Tighten to 8 Nm (71 in-lb)
3. **Bolt**
   - Qty: 3
4. **Bumper carrier**
5. **Bolt**
   - Qty: 7
   - Tighten to 50 Nm (37 ft-lb)
6. **3411 guide rods**

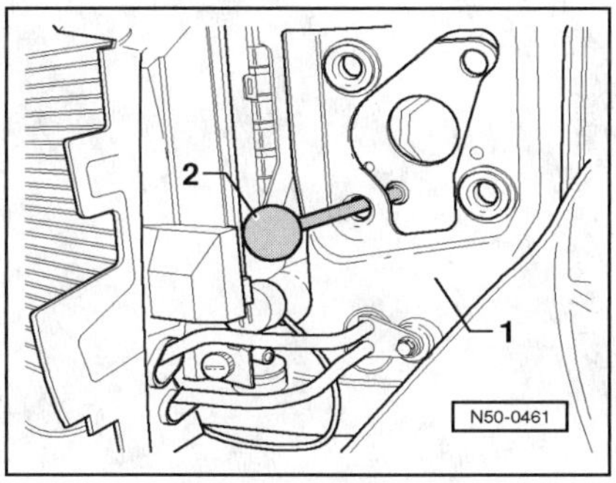

Fig. 1 Lock carrier -1- can be pulled forward approx. 10 cm (3.9 in.) on 3411 guide rods -2-.

Hydraulic oil cooler, unbolting

Unbolt hydraulic oil cooler from lock carrier -arrows-.

NOTE —

Do not hang oil cooler up on pipes/hoses and do not kink hydraulic pipes/hoses.

Fender

Front fender, assembly (through 09/2000)

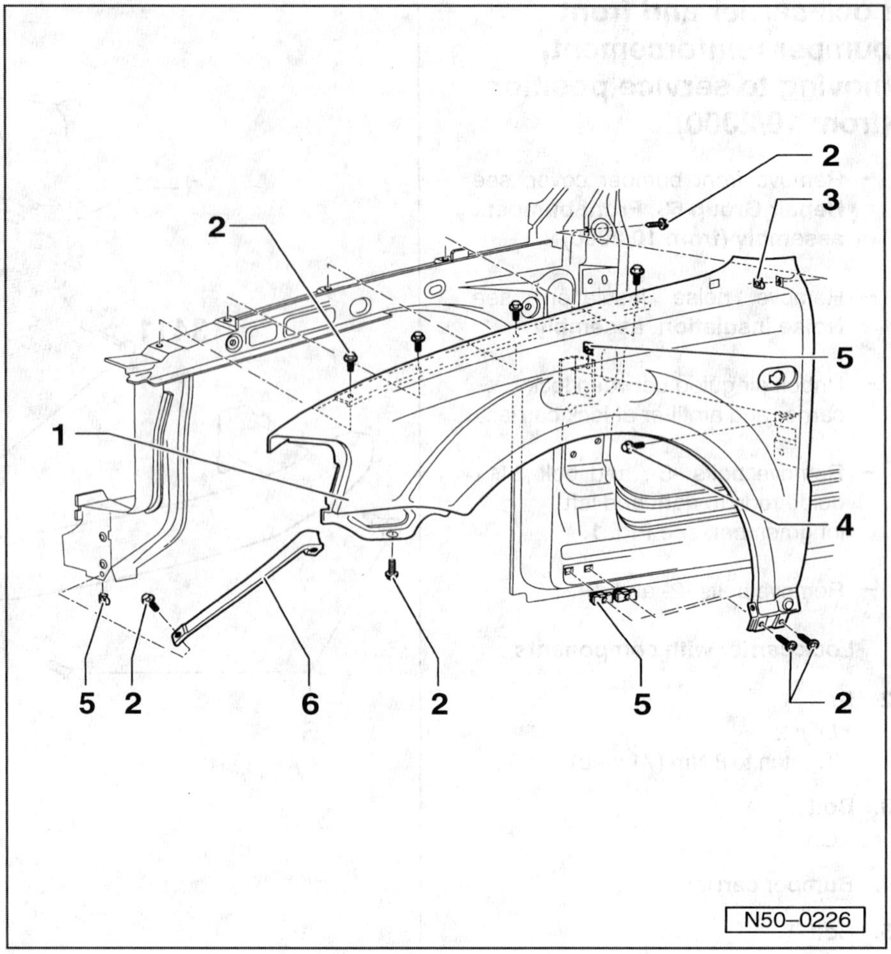

1. **Fender**
   - Removing:
     Removing front fender, see **Repair Group 63, Front bumper, assembly (through 09/2000)**.
     Removing wheel housing liner, see **Repair Group 66, Front wheel housing liner, assembly**.
     Removing closing piece, see **Closing piece, removing and installing**.
     Heat fender in area of A pillar with a hot air blower and pull off.

   NOTE —
   PVC must only be heated lightly and briefly. PVC must not change in color or blister.

   - Installing:
     Before screwing on fender, fit a zinc intermediate piece AKL 381 035 50 to each bolt position contact area.

2. **Bolt**
   - Qty. 9
3. **Speed nut**
4. **Torx® screw**
   - Torx® key T 25
   - Tighten to 6 Nm (53 in-lb)
5. **Spreader nut**
   - Qty. 4
6. **Bracket**

Front fender, assembly (from 10/2000)

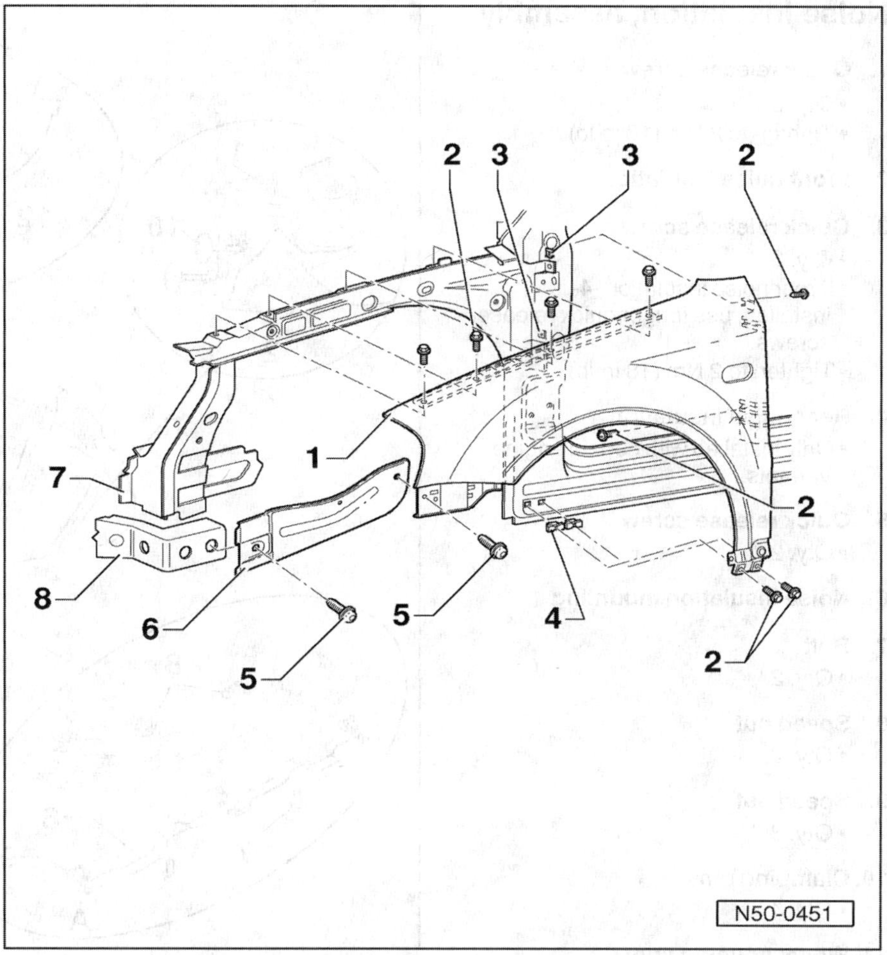

1. **Fender**
   - Removing:
     Remove front bumper, see **Repair Group 63, Front bumper, assembly (from10/2000)**.
     Remove wheel housing liner, see **Repair Group 66, Front wheel housing liner, assembly**.
     Remove closing piece, see **Closing piece, removing and installing**.
     Heat up fender in area of A-pillar using a hot air gun and pull off.

   NOTE —
   - *Only heat up PVC material slightly and briefly. PVC must not change color and must not form bubbles.*
   - *Installing: Before bolting on fender, each bolting point of contact areas must be coated with a zinc-layer AKL 381 035 50*

2. **Bolt**
   - Qty: 8
   - Tighten to 7.4 Nm (65 in-lb)
3. **Speed nut**
   - Qty: 2
4. **Expanding nut**
   - Qty: 2
5. **Torx bolt**
   - Qty: 2
   - Tighten to 6 Nm (53 in-lb)
6. **Bracket**
   - Connects fender to lock carrier
7. **Wheelhousing**
8. **Bracket**

Closing piece, removing and installing

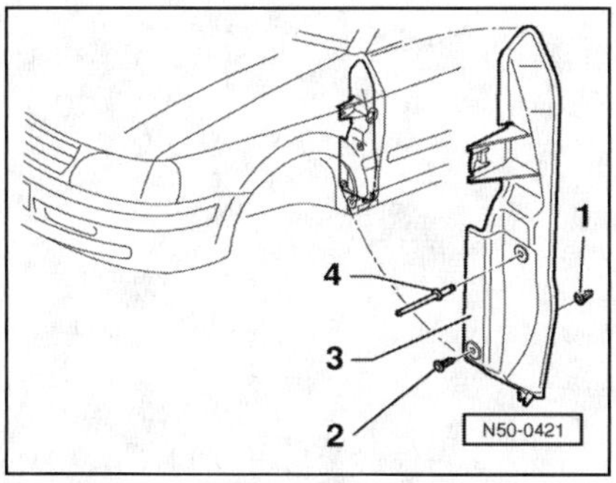

Removing

Remove wheel housing liner, see **Repair Group 66, Front wheel housing liner, assembly**.

– Remove bolts -2- from spreader clips -1-. Tightening torque 2.5 Nm (22 in-lb).

– Drill out plastic rivets -4-.

– Pull closing piece -3- out from fender.

Installing

– The plastic rivet can be pulled in with the rivet tongs V.A.G 1753 A if the mouthpiece is first removed from the rivet tongs.

Noise insulation, assembly

1. **Quick release screw**
   - Qty. 3
   - Tighten to 2 Nm (18 in-lb)
2. **Front noise insulation**
3. **Quick release screw**
   - Qty. 3
   - If rear noise insulation -**4**- is installed, use longer quick release screws.
   - Tighten to 2 Nm (18 in-lb)
4. **Rear noise insulation**
   - Only installed with certain engine variants.
5. **Quick release screw**
   - Qty. 2
6. **Noise insulation mounting**
7. **Bolt**
   - Qty. 2
8. **Speed nut**
   - Qty. 2
9. **Speed nut**
   - Qty. 3
10. **Clamping pin**
    - Qty. 4
11. **Wheel housing liner**
12. **Bumper**
13. **Lock carrier with attachments**

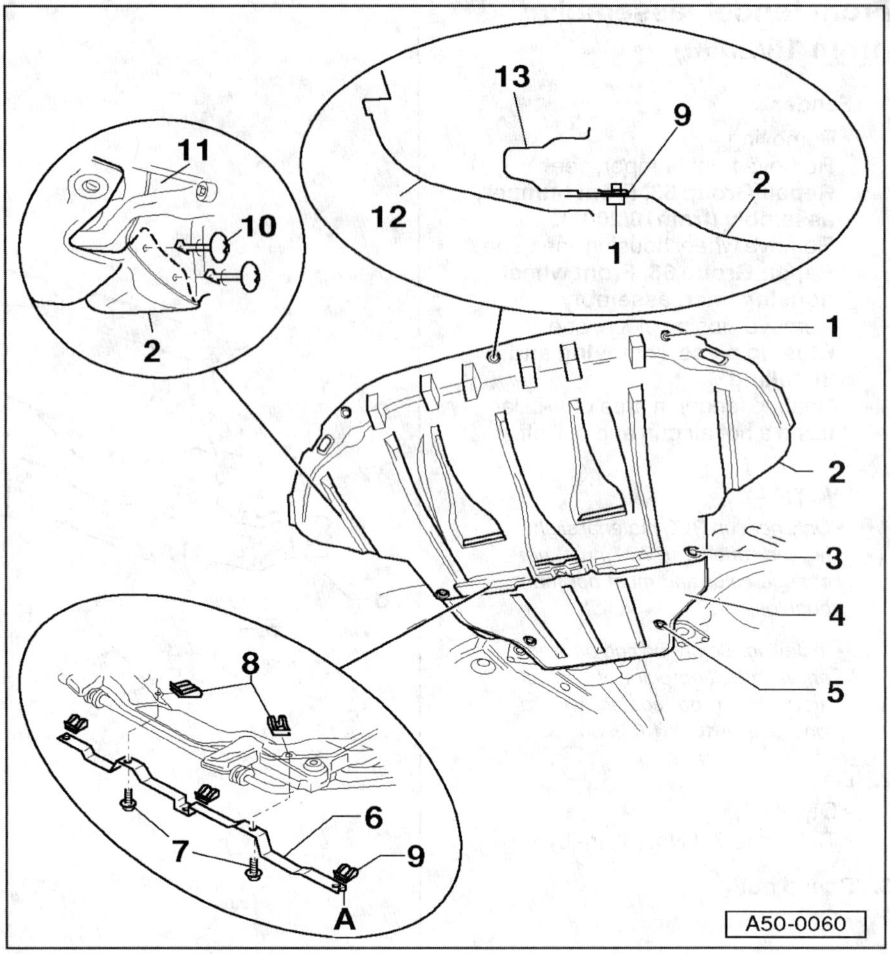

55 Hoods, Lids

Repair Group

T 55

Hood

Gas strut, hood, removing

– Support hood.

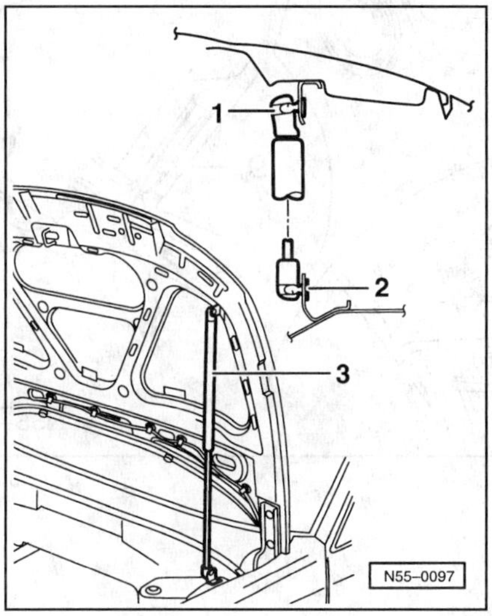

◀ Lift locking clip -1- with a screwdriver and pull gas-filled strut off ball-head stud -2-.

NOTE —

If it is intended to re-use gas-filled strut, then do not pry out locking clip -1- completely from the ball socket otherwise clip will be damaged.

1 - Locking clip
2 - Mounting bracket with ball head
3 - Gas strut

Hood, assembly

1. **Trunk lid/tailgate**
   - Removing:
     Remove cables/hoses for windscreen washer system and washer jets.
     Disconnect gas strut -2- at hood.
     Unscrew bolt -7- for hinge cover -6- and take off hinge cover.
     Unscrew bolts -9- and take off hood.
   - Adjusting:
     Align hood -1- between fenders by moving hood hinges.
     The hood can be adjusted for height relative to fenders using rubber stop -10-.
   - After reinstalling or adjusting carry out corrosion protection measures to hinges and bolts
2. **Gas-filled strut**
   - Removing and installing, see **Gas strut, hood, removing**
   - Releasing gas, see **Gas strut, releasing gas**
3. **Hose**
4. **Clip**
5. **Washer jet**
   - Removing, see **Electrical Equipment section, Repair Group 92**
6. **Hinge cover**
7. **Bolt**
   - Tighten to 6 Nm (53 in-lb)
8. **Hinge**
9. **Bolt**
   - Tighten to 23 Nm (17 ft-lb)
10. **Rubber stop**
    - The hood can be adjusted for height relative to fenders using rubber stop
11. **Rubber buffer**
12. **Guide piece**

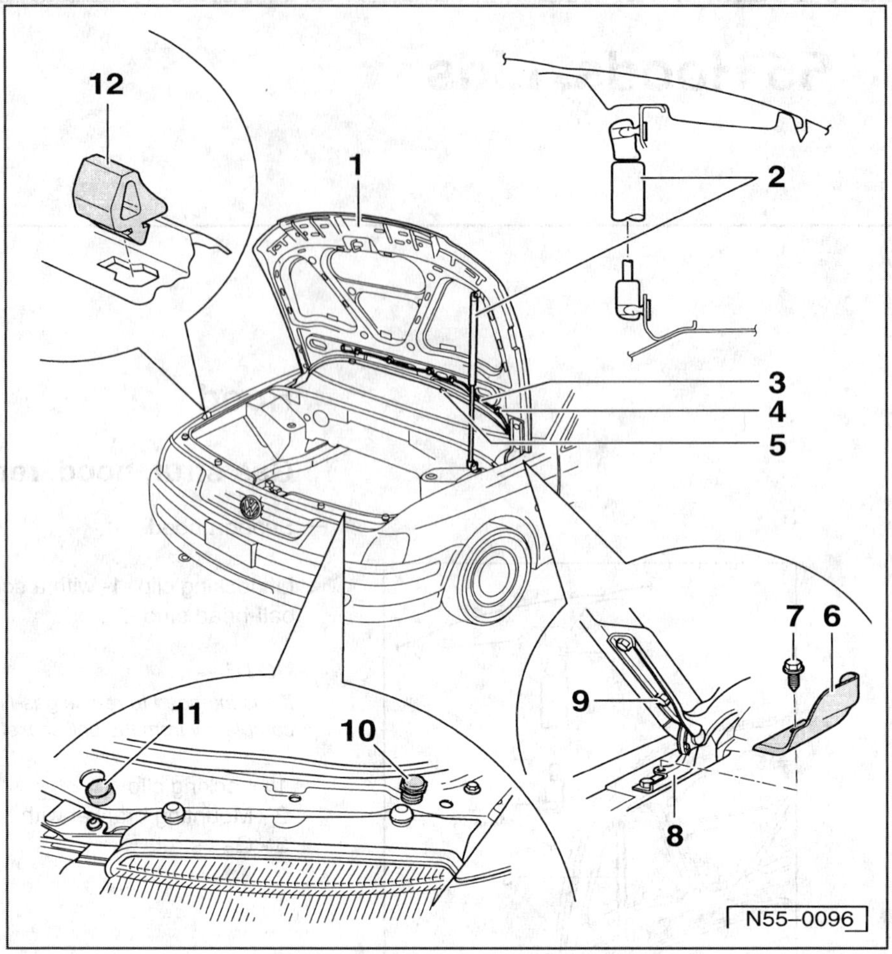

Hood lock cable and hood lock, removing and installing

1. **Cable**
   - Replacing, unscrew release lever and lock, detach cable.
   - Cable is split by Bowden cable coupling -6-.
   - Bowden cable can be detached at Bowden cable coupling to remove front end.
   - Separating, see **Bowden cable, disconnecting**
2. **Sealing grommet**
3. **Plenum chamber seal**
   - Side contour of plenum chamber seal must contact fender seal.
4. **fender seal**
5. **hood lock**
   - Removing and installing, see **Hood lock, removing and installing**
6. **Bowden cable coupling**
7. **Release lever**
   - Cable attached
8. **Self-tapping screw**

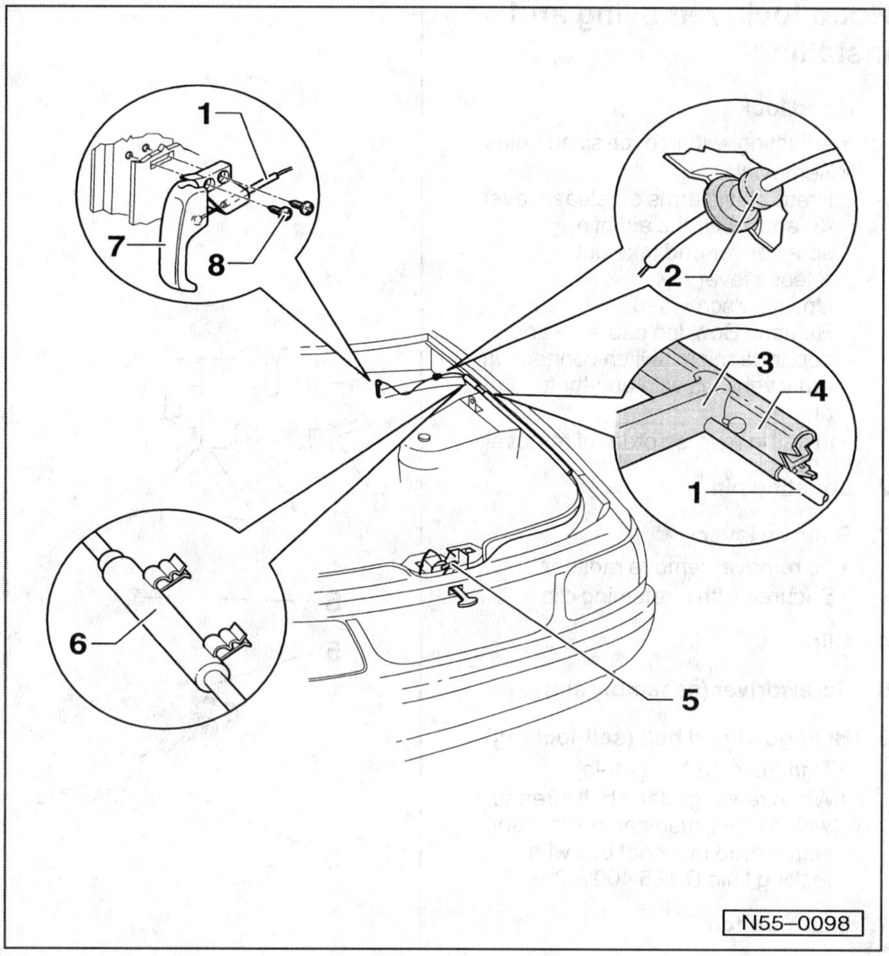

Bowden cable, disconnecting

– Pull Bowden cable coupling off fender.

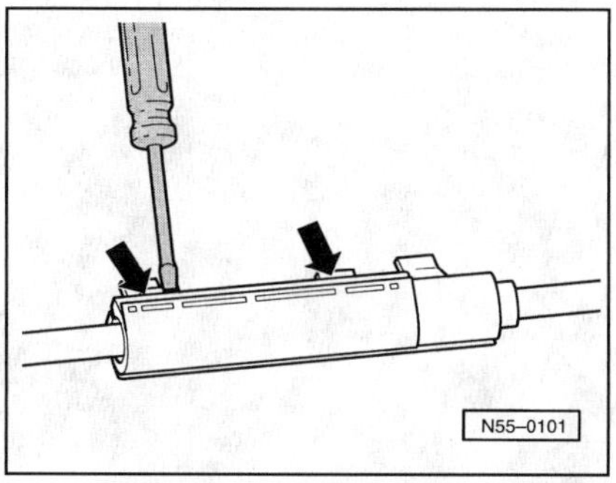

◀ Open Bowden cable coupling at locking devices with a small screwdriver -arrows-.

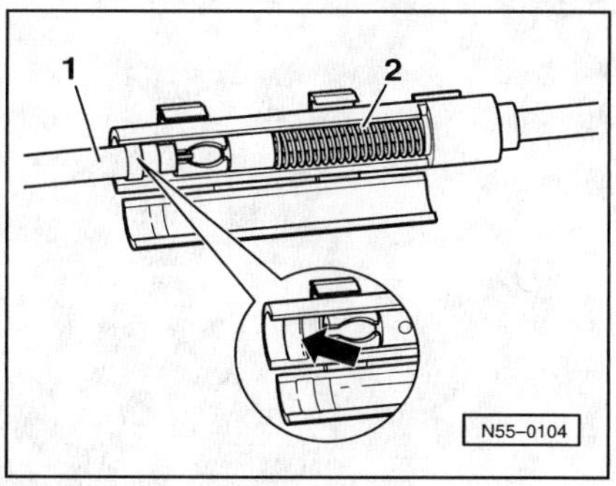

◀ Take front Bowden cable -1- out of coupling -2-.

NOTE —
When installing, ensure that the Bowden cable sleeve is inserted correctly -arrow- and the coupling engages.

Hood lock, removing and installing

1. **Hood lock**
   - Adjusting: within over-sized holes
   - Removing
     Press clevis arms of release lever -3- apart with the aid of a screwdriver and take out release lever -5-.
     Unscrew screws -6-.
     Release Bowden cable -8- and separate micro switch connection (only vehicles with anti-theft alarm) -7-.
   - Install in reverse order of removal.
2. **Locating pin**
3. **Release lever**
   - To remove, remove radiator
   - Secured with a retaining clip
4. **Clip**
5. **Screwdriver (assembly aid)**
6. **Hexagon head bolt (self-locking)**
   - Tighten to 12 Nm (9 ft-lb)
   - When reusing: clean bolt threads with a wire brush and nut threads with thread tap; coat bolt with locking fluid D 185 400 A 2.
7. **Micro switch**
8. **Bowden cable**

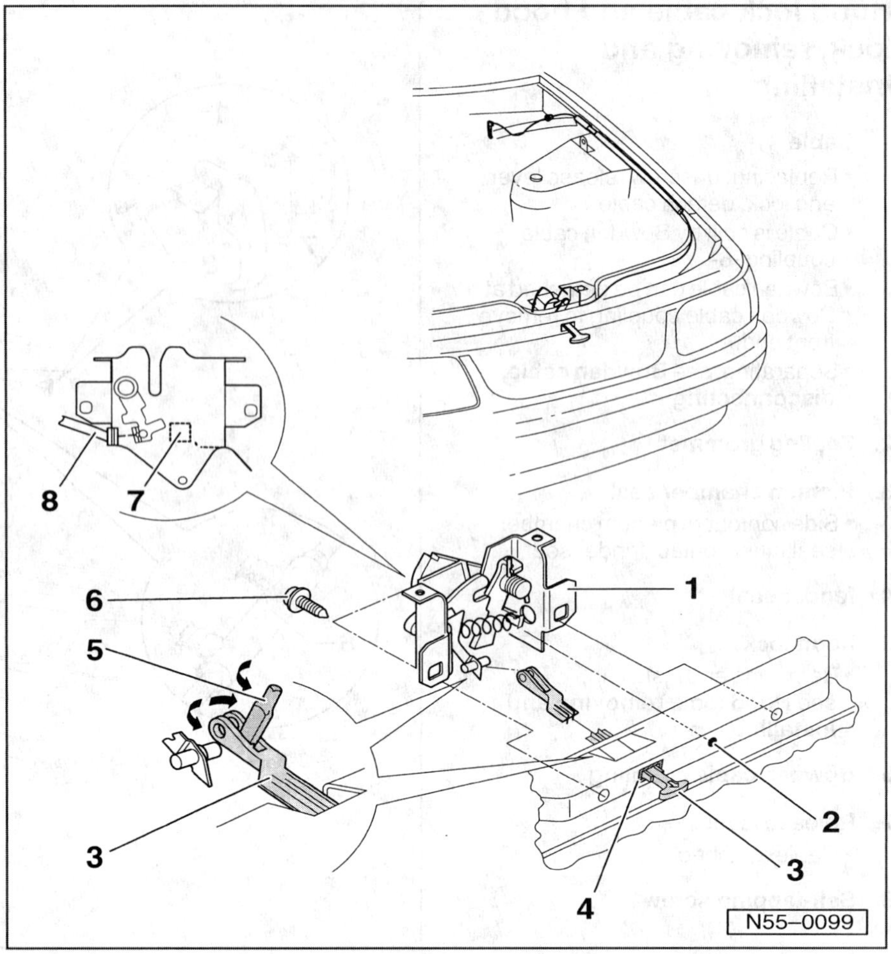

Hood seal, assembly

1. **Plenum chamber seal**
   - Side contour of plenum chamber seal must contact fender seal.
2. **Fender seal**
3. **Front end seal**
   - Connect into fender seal
4. **Plugs**
   - Seal open Bowden cable channel with plugs.
5. **Bowden cable**

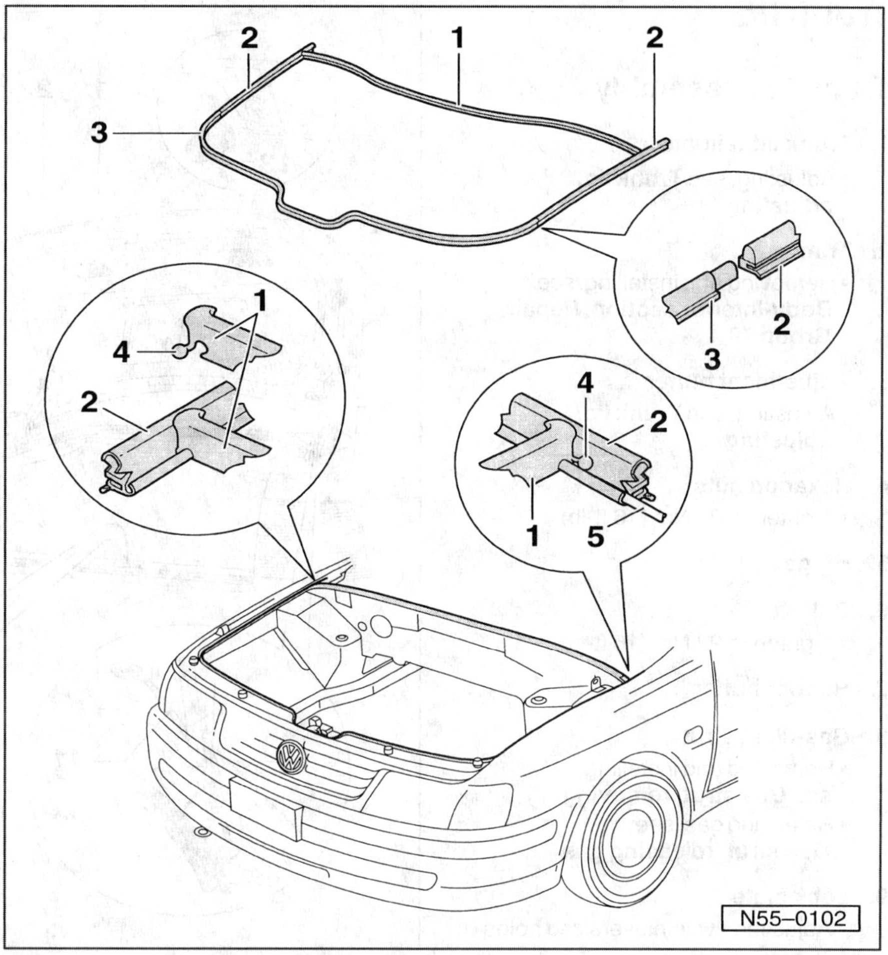

Trunk lid

Trunk lid, assembly

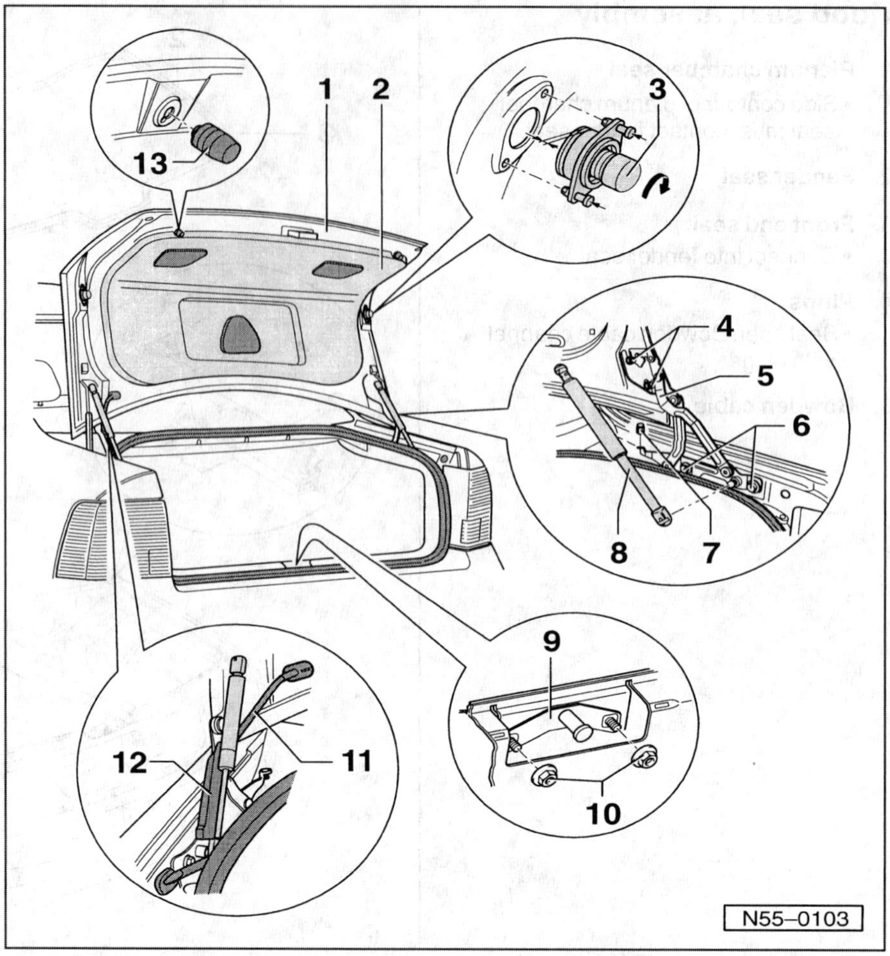

1. **Trunk lid/tailgate**
   - Adjusting, see **Trunk lid, adjusting**
2. **Trim**
   - Removing and installing, see **Body-Interior section, Repair Group 70**
3. **Adjustment buffer**
   - Adjusting, see **Trunk lid, adjusting**
4. **Hexagon nuts**
   - Tighten to 22 Nm (16 ft-lb)
5. **Hinge**
6. **Bolt**
   - Tighten to 22 Nm (16 ft-lb)
7. **Rubber buffer**
8. **Gas-filled strut**
   - Removing and installing, see **Gas strut, removing**
   - Releasing gas, see **Gas strut, releasing gas**
9. **Lock plate**
   - Adjusting - within oversized holes
10. **Hexagon nut**
    - Tighten to 7.2 Nm (64 in-lb)
11. **Cables/hoses**
12. **Hinge cover**
13. **Rubber buffer**

Gas strut, trunk lid /tailgate, removing

– Support trunk lid/tailgate.

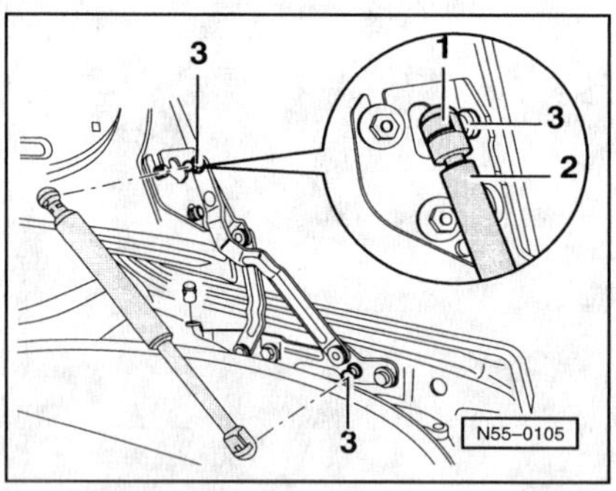

◄ Lift locking clip -1- with a screwdriver and pull gas-filled strut off ball-head stud -3-.

NOTE —

If it is intended to re-use gas-filled strut, then do not pry out locking clip -1- completely from the ball socket otherwise clip will be damaged.

1 - Locking clip
2 - Gas strut
3 - Ball head pin (part of trunk lid/tailgate hinge)

Gas strut, releasing gas

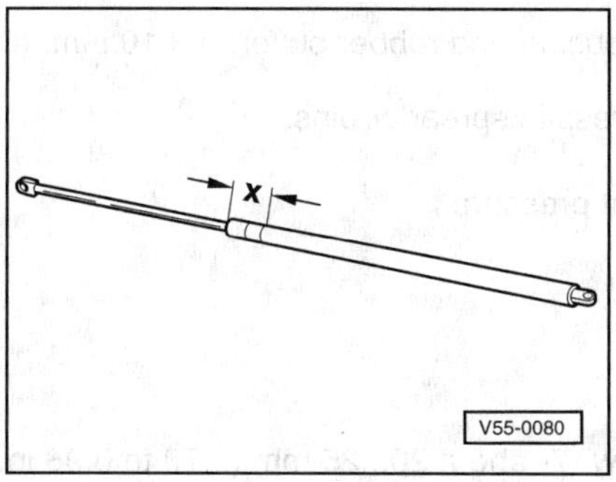

➤ Clamp gas-filled strut in vice in area x = 50 mm (1.9 in.).

> ***WARNING —***
> *Clamp only within this area otherwise danger of accident!*

– Saw through cylinder of gas-filled strut within first third of cylinder's overall length using piston rod end of cylinder as reference point.

NOTE —

• Protective goggles must be worn during sawing process.

• Cover area of saw cut with a cloth.

• Dispose of oil and cloth via existing disposal channels.

Trunk lid, adjusting (through 09/2000)

NOTE —

• The trunk lid is adjusted correctly, when the shut lines are even all-round, and it does not project too far in or out and all contours align when closed.

• Vehicle must be standing on its wheels when performing adjustments.

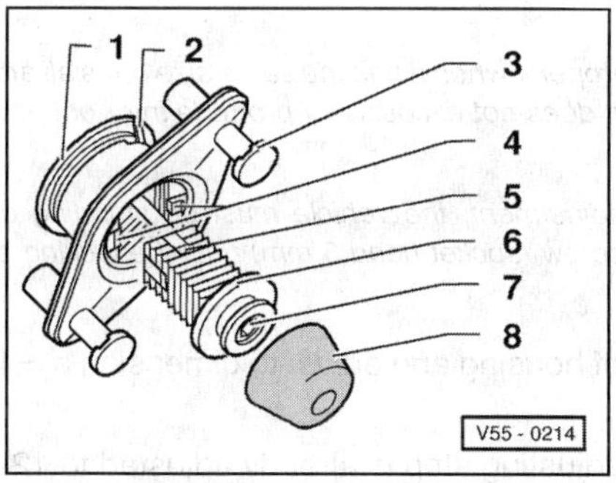

➤ 1-Threaded sleeve (with left-hand thread)
2 - Spacer web
3 - Spreader pins with dismantling slot
4 - Detent lugs for basic setting
5 - Recess for detent lugs
6 - Detent slide
7 - Clamping screw
8 - Rubber buffer

– Lever out spreader pin in dismantling slot with a narrow screwdriver and take adjusting stop out of tailgate.

– Take off rubber buffer and unscrew clamping screw (hexagon socket 3 mm) up to the edge of the detent slot.

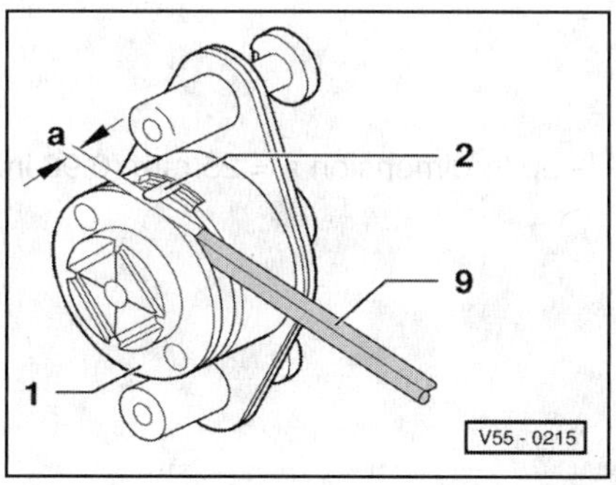

➤ Adjust threaded sleeve -1- (left-hand thread) to a clearance of a = 3 mm (0.11 in.). check clearance, e.g. with hexagon socket head wrench -9-. Reinstall rubber buffer.

NOTE —

When new, the adjusting stop should be set to a clearance of 3 mm (0.11 in.). The spacer web -2- rests against the threaded sleeve -1-.

- Pull out detent slide and press in again far enough until locking lugs -4- engage in the recess -5-.

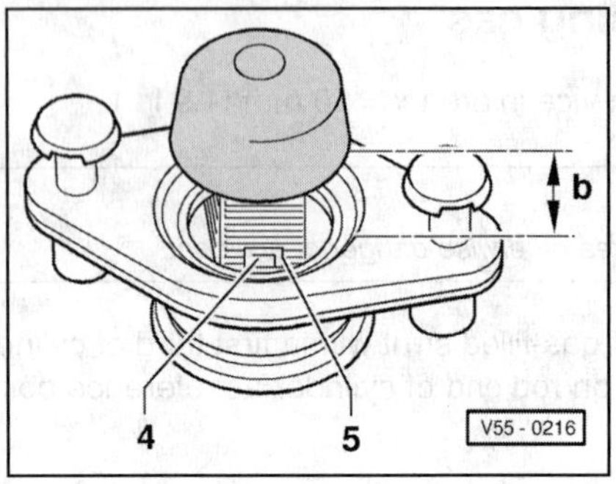

  - Clearance between housing and rubber buffer: b = 10 mm. (0.39 in.)

– Insert adjusting stop, press in spreader pins.

– Close tailgate with slight pressure.

– Open tailgate again.

- Screw in clamping screw -7- about 20...25 mm (0.78 to 0.98 in.).

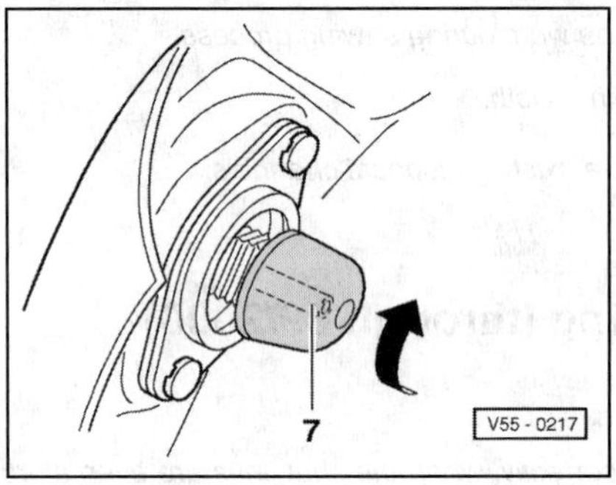

– Check adjustment.

Trunk lid, adjusting (from 10/2000)

NOTE —

- *The trunk lid is adjusted properly when it has the same size gaps all around in closed position and if it does not excessively protrude in or out and the contours are flush.*
- *To be able to perform adjustment, the vehicle must be standing on its wheels. Loosen locking screw (socket head 3 mm) until the locking slider can be pulled out.*

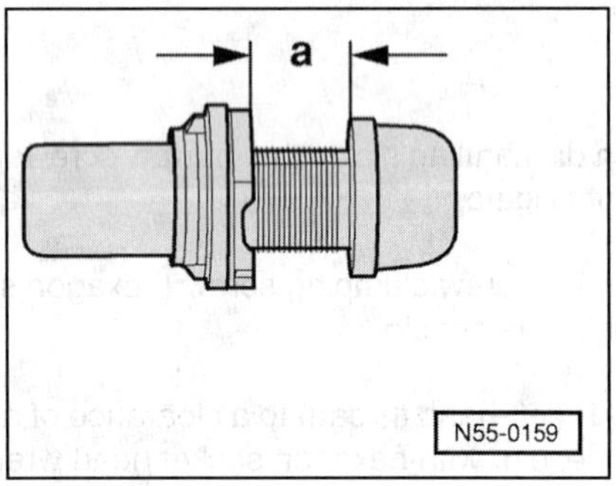

- Pull locking slider out of housing and adjust to dimension a = 12.5 mm (0.49 in.).

  - In new condition, the adjusting stop is already adjusted to 12.5 mm (0.49 in.) distance.

– With slight pressure, close trunk lid from center, while pulling on the handle.

– Open trunk lid again.

- Screw in locking screw -1- up to dimension a = 25 mm (0.98 in.).

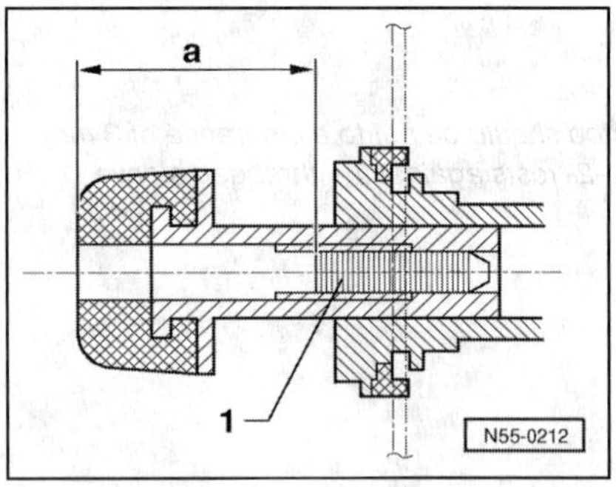

– Check adjustment.

Trunk lid lock, assembly (through 09/2000)

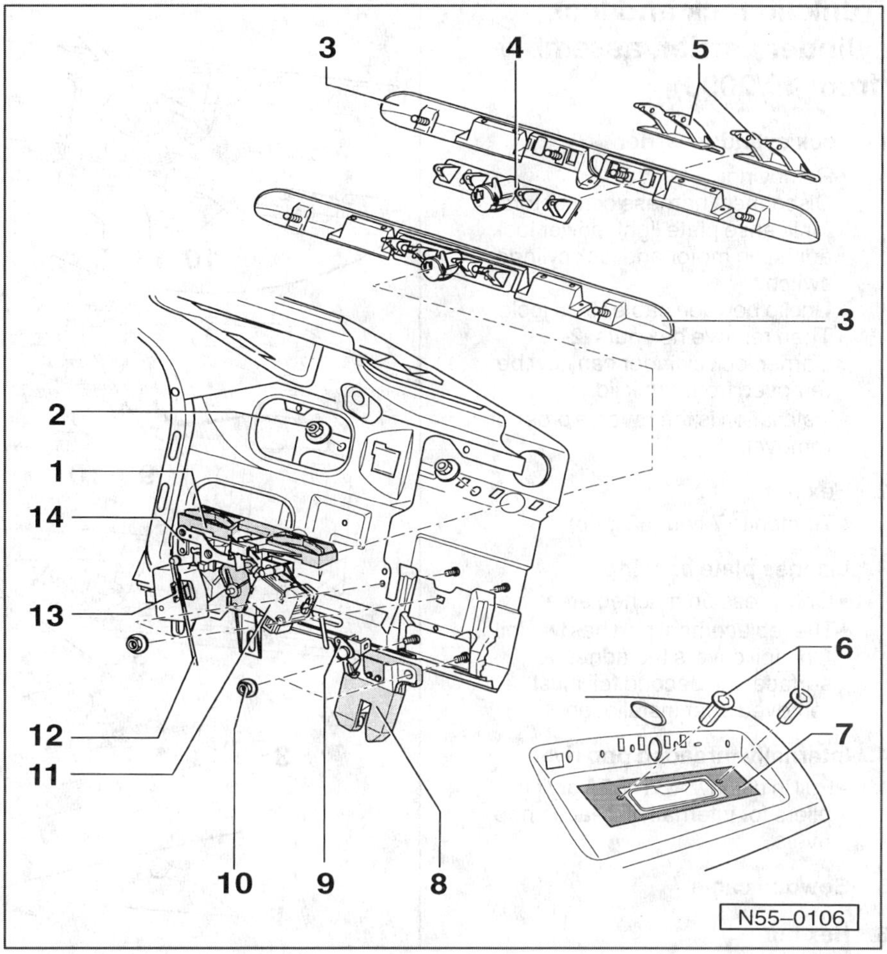

1. **Lock cylinder carrier**
   - The lock cylinder can only be removed in conjunction with the carrier.
   - See **Lock cylinder carrier, removing and installing (through 09/2000)**.
2. **Hexagon nut**
   - Tighten to 7 Nm (62 in-lb)
3. **Handle trim**
   - Removing:
     Remove lock cylinder carrier, see **Lock cylinder carrier, removing and installing (through 09/2000)**.
     Unscrew hexagon nuts -2- and take outer hand grip off trunk lid.
4. **Handle**
5. **Push-button**
6. **Pop rivet nut**
   - Pull in with pop rivet nut tool V.A.G 1765 A
7. **Base for registration plate**
   - Only press on the hatched area
   - The replacement part has two foils. The first foil covers the adhesive surface, the second must be pulled off after assembling.
8. **Lock**
   - Removing:
     Remove tailgate trim, see **Body-Interior section, Repair Group 70**
     Disconnect connector on lock.
     Unclip operating rod -12-, unscrew hexagon nuts -2- and take lock out of trunk lid.
9. **Relay lever**
10. **Hexagon nut**
    - Tighten to 7 Nm (62 in-lb)
11. **Central locking actuator**
    - Removing:
      Remove lock cylinder carrier, see **Lock cylinder carrier, removing and installing (through 09/2000).**
      Unbolt actuator and allow to hang.
12. **Operating rod**
    - Press relay lever -14- against stop and insert operating rod stress-free
13. **Connector sleeve**
14. **Relay lever**

Installing

– Engage securing clip again.

– Insert lock cylinder into carrier.

– Fit the complete carrier onto the outer grip and the studs, and engage into the outer grip (a clicking must be heard).

Trunk lid lock and lock cylinder carrier, assembly (from 10/2000)

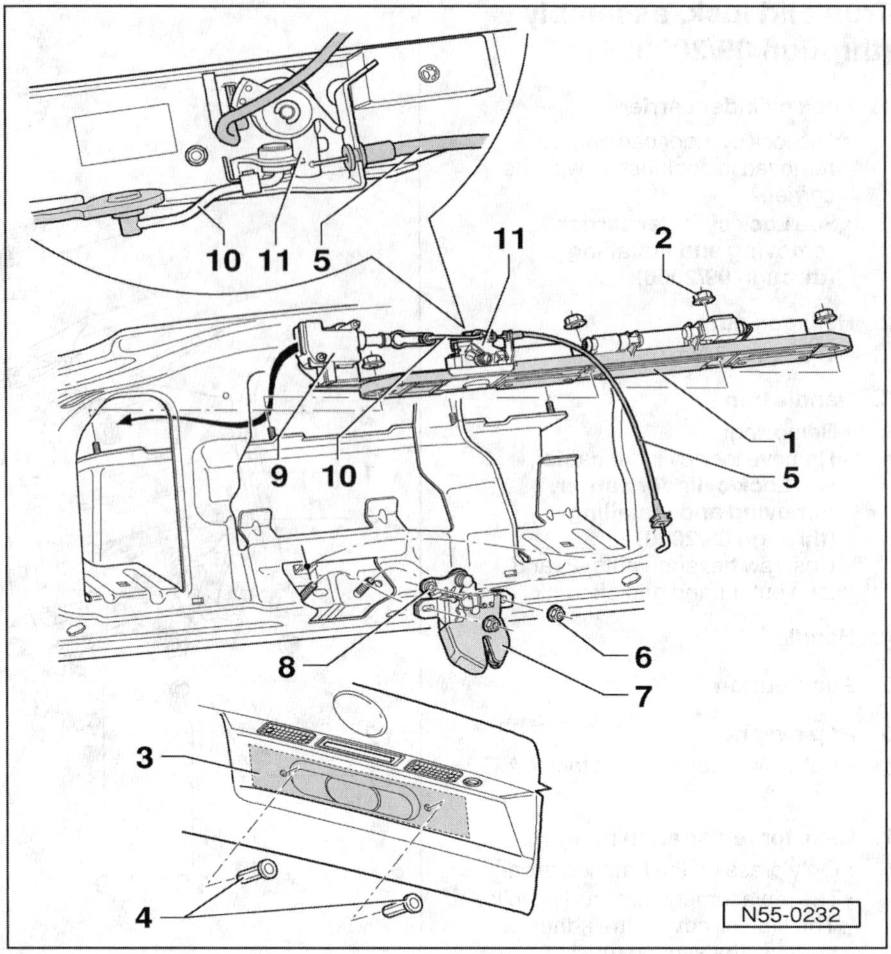

1. **Lock cylinder carrier**
   - Removing:
     Disconnect harness connectors for license plate light, power lock adjusting motor and lock cylinder switch.
     Unclip bowden cable -5- at lock.
     Then remove hex nuts -2-.
     Carrier lock cylinder can now be removed from trunk lid.
   - Installation is the reverse order of removal.
2. **Hex nut**
   - Tighten to 7 Nm (62 in-lb)
3. **License plate backing**
   - Only press on hatched area
   - The replacement part has two foils. One foil covers the adhesive surface, the second foil must be removed after installation.
4. **Internally threaded pop rivet**
   - Pull in using VAG1765A prying pliers for internally threaded pop rivets
5. **Bowden cable**
6. **Hex nut**
   - Tighten to 7 Nm (62 in-lb)
7. **Lock**
   - Removing:
     Remove trim for trunk lid, see **Body-Interior section, Repair Group 70**
     Disconnect harness connector at lock.
     Unclip bowden cable -5-, remove hex nuts -6- and remove lock from trunk lid.
8. **Reversing lever**
9. **Motor to unlock trunk lid -V139-**
   - Remove lock cylinder carrier.
   - Unbolt motor and unhook.
10. **Operating rod**
11. **Reversing lever**

Lock cylinder carrier, removing and installing (through 09/2000)

Removing

– Remove trunk lid trim, see **Body-Interior section, Repair Group 70.**

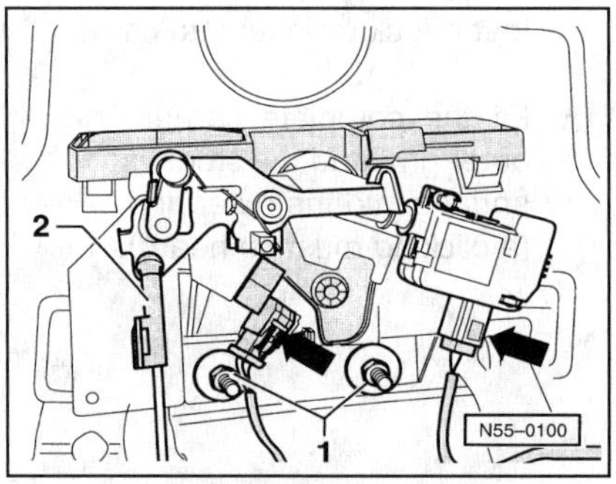

◄ Separate connectors -arrows-.

– Unclip operating rod -2- for trunk lid lock.

– Unscrew hexagon nuts -1-.

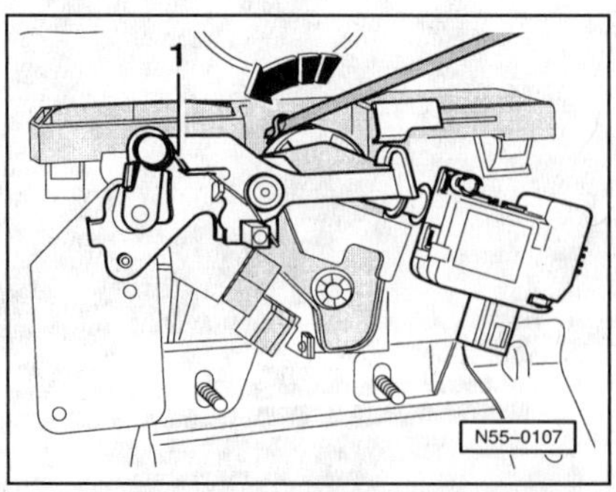

◄ Release securing clip -arrow- by turning anti-clockwise with a screwdriver and hold.

– Pull lock cylinder carrier -1- off trunk lid and therewith off outer hand grip.

– The lock cylinder can now be taken out of the carrier.

Lock cylinder housing, assembly (from 10/2000)

1. **Carrier lock cylinder**
2. **Guide piece**
3. **Guide pin**
4. **Handle**
5. **Seal**
6. **Lock cylinder housing**
   - Clipped into the lock cylinder carrier
7. **Motor to unlock rear lid -V139-**
   - Removing:
     Remove lock cylinder carrier -1-, see **Trunk lid lock, assembly (through 09/2000)**.
     Unbolt motor and unhook.
8. **Screw**
9. **Carrier plate**
   - Bolted onto the lock cylinder carrier
10. **Conductor housing**
    - Secured to carrier using circlip
11. **Trunk lid release switch -E165-**
    - Soldered onto lock cylinder carrier

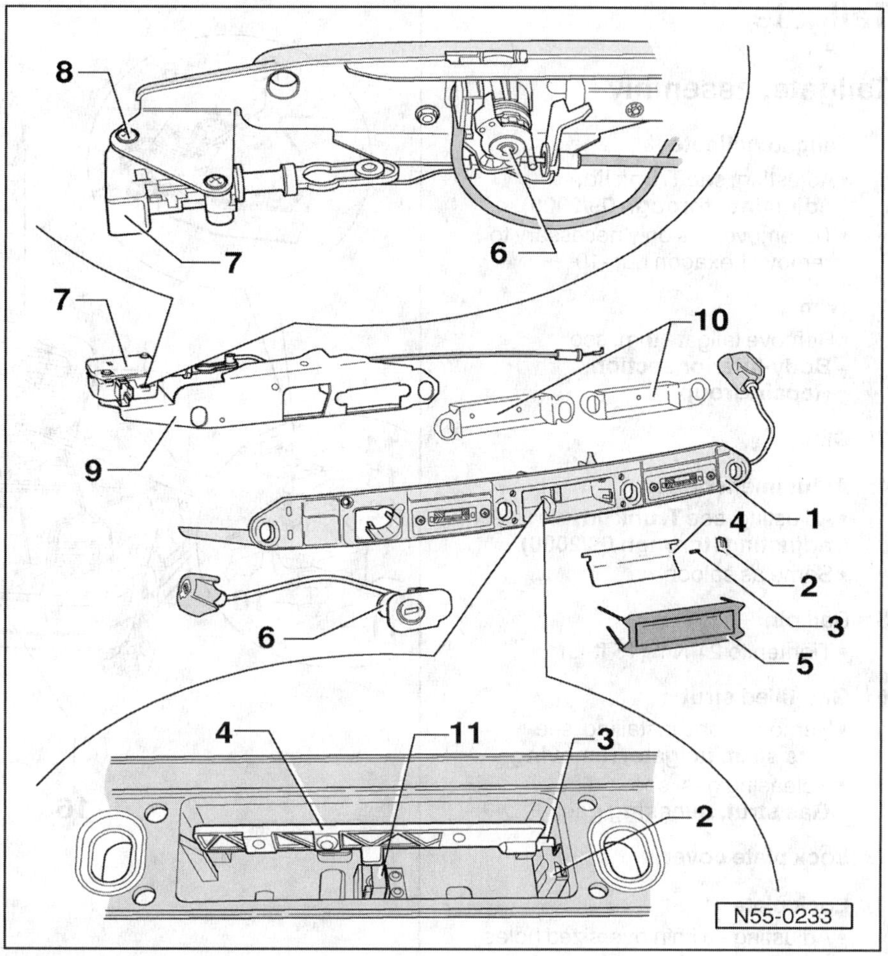

Tailgate

Tailgate, assembly

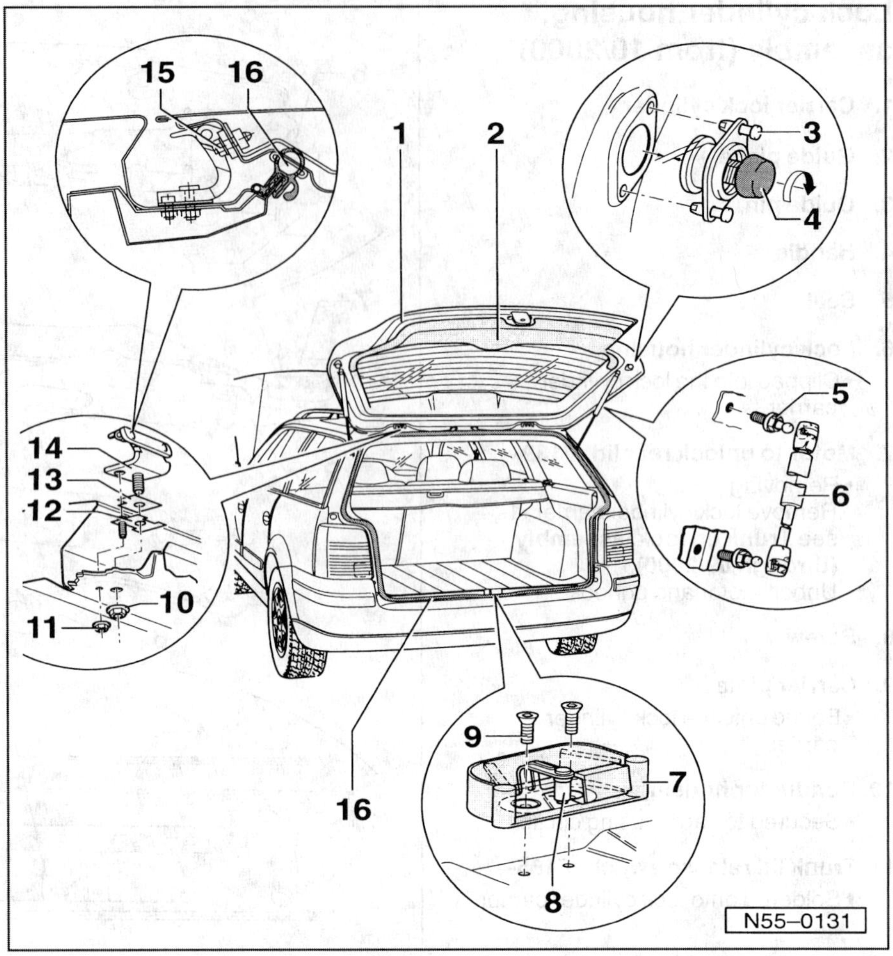

1. **Tailgate/tailgate**
   - Adjusting, see **Trunk lid, adjusting (through 09/2000)**
   - To remove, it is only necessary to remove hexagon nut -10-
2. **Trim**
   - Remove tailgate trim, see **Body-Interior section, Repair Group 70**
3. **Clip**
4. **Adjustment buffer**
   - Adjusting, see **Trunk lid, adjusting (through 09/2000)**
   - Same as saloon
5. **Ball pin**
   - Tighten to 21 Nm (15 ft-lb)
6. **Gas-filled strut**
   - Removing and installing, see **Gas strut, tailgate, removing**
   - Releasing gas, see **Gas strut, releasing gas**
7. **Lock plate cover**
8. **Lock plate**
   - Adjusting - within oversized holes
9. **Bolt**
   - Tighten to 23 Nm (17 ft-lb)
10. **Hexagon nut**
    - Tighten to 22 Nm (16 ft-lb)
    - Tailgate can be removed by unscrewing this hexagon nut
11. **Hexagon nut**
    - 10 Nm
12. **Adapter plate**
    - With the adapter plate the tailgate adjustment is discontinued
13. **Seal**
14. **Hinge**
15. **12 point head bolt**
    - Tighten to 10 Nm (7 ft-lb)
16. **Seal**

Tailgate lock and lock cylinder carrier, assembly (through 09/2000)

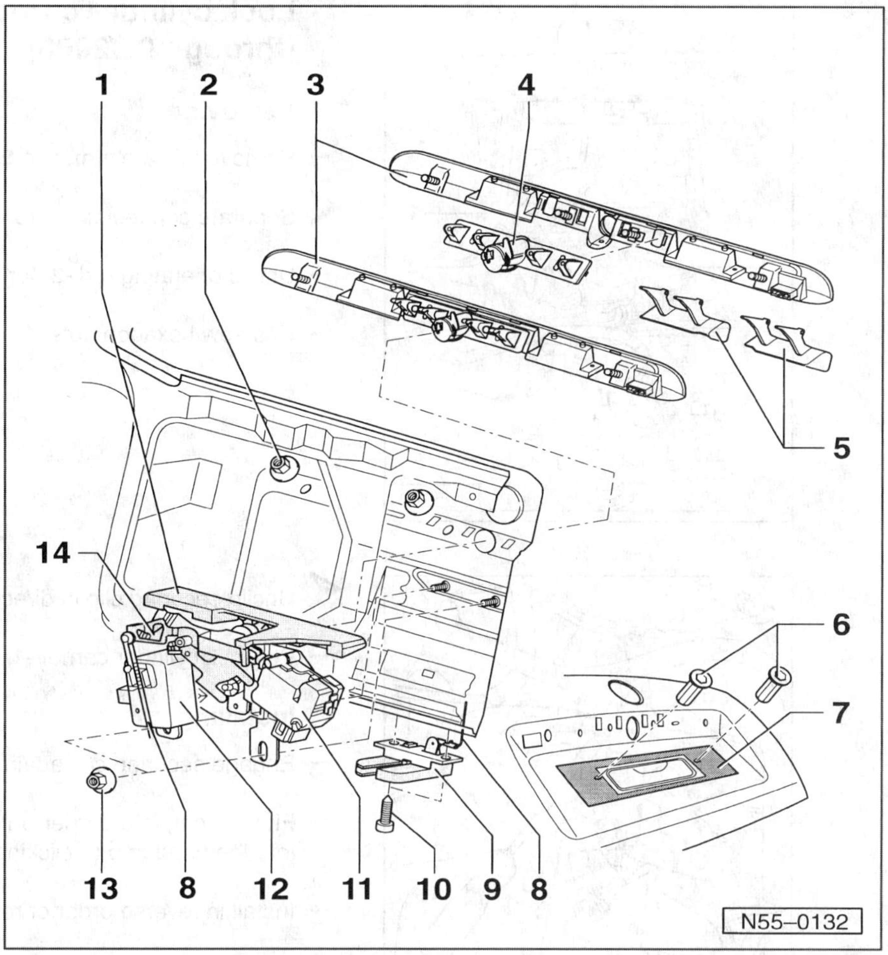

1. **Lock cylinder carrier**
   - The lock cylinder can only be removed in conjunction with the carrier.
   - Removing and installing, see **Lock cylinder carrier, removing and installing (through 09/2000)**.
2. **Hexagon nut**
   - Tighten to 7 Nm (62 in-lb)
3. **Handle trim**
   - Removing:
     Remove lock cylinder carrier, see **Lock cylinder carrier, removing and installing (through 09/2000)**.
     Separate connector for number plate light.
     Unscrew hexagon nuts -2- and take outer hand grip off tailgate.
4. **Handle**
5. **Push-button**
6. **Pop rivet nut**
   - Pull in with pop rivet nut tool V.A.G 1765 B
7. **Base for registration plate**
   - Only press on the hatched area
   - The replacement part has two foils. The first foil covers the adhesive surface, the second must be pulled off after assembling.
8. **Operating rod**
9. **Lock**
   - Removing:
     Remove tailgate trim, see **Body-Interior section, Repair Group 70**
     Disconnect connector on lock.
     Unclip operating rod -8-, remove bolt -10- and take lock off tailgate.
10. **Bolt**
    - Tighten to 23 Nm (17 ft-lb)
11. **Central locking actuator**
    - Removing:
      Remove lock cylinder carrier, see **Lock cylinder carrier, removing and installing (through 09/2000)**.
      Unbolt actuator and allow to hang.
12. **Tailgate release motor -V139-**
    - Only USA models
13. **Hexagon nut**
    - Tighten to 7 Nm (62 in-lb)
14. **Relay lever**

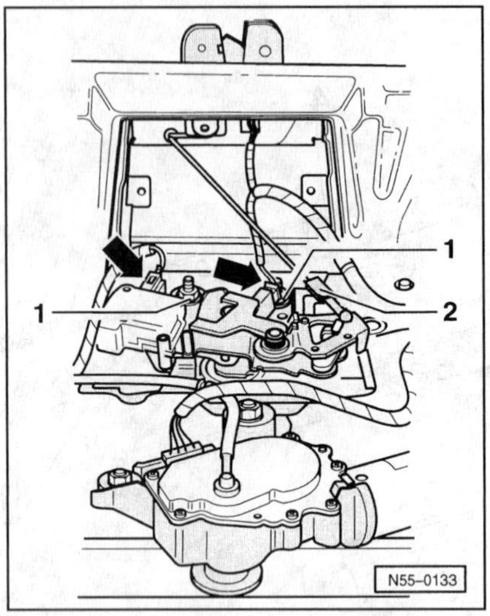

Lock cylinder carrier, removing and installing (through 09/2000)

Removing

– Remove tailgate trim, see **Body-Interior section, Repair Group 70.**

< Separate connectors -arrows-.

– Unclip operating rod -2- for tailgate lock.

– Unscrew hexagon nuts -1-.

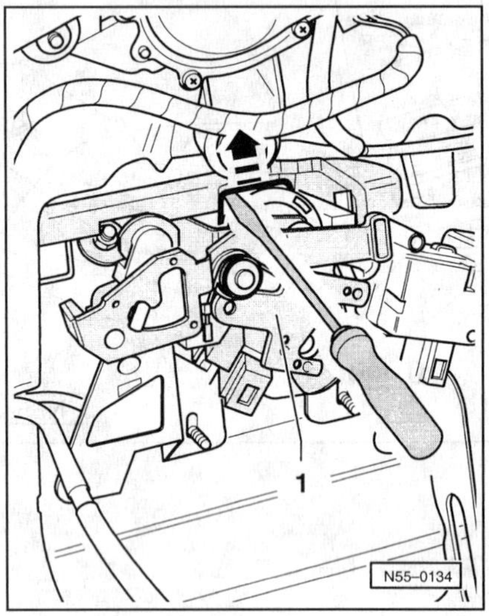

< Unclip securing clip in direction of arrow with a screwdriver and hold.

– Pull lock cylinder carrier -1- off tailgate and thereby off hand grip.

Installing

– Engage securing clip again.

– Fit the complete carrier onto the hand grip and the studs, and engage into the hand grip (a clicking must be heard).

– Install in reverse order of removal.

Tailgate lock and lock cylinder carrier, assembly (from 10/2000)

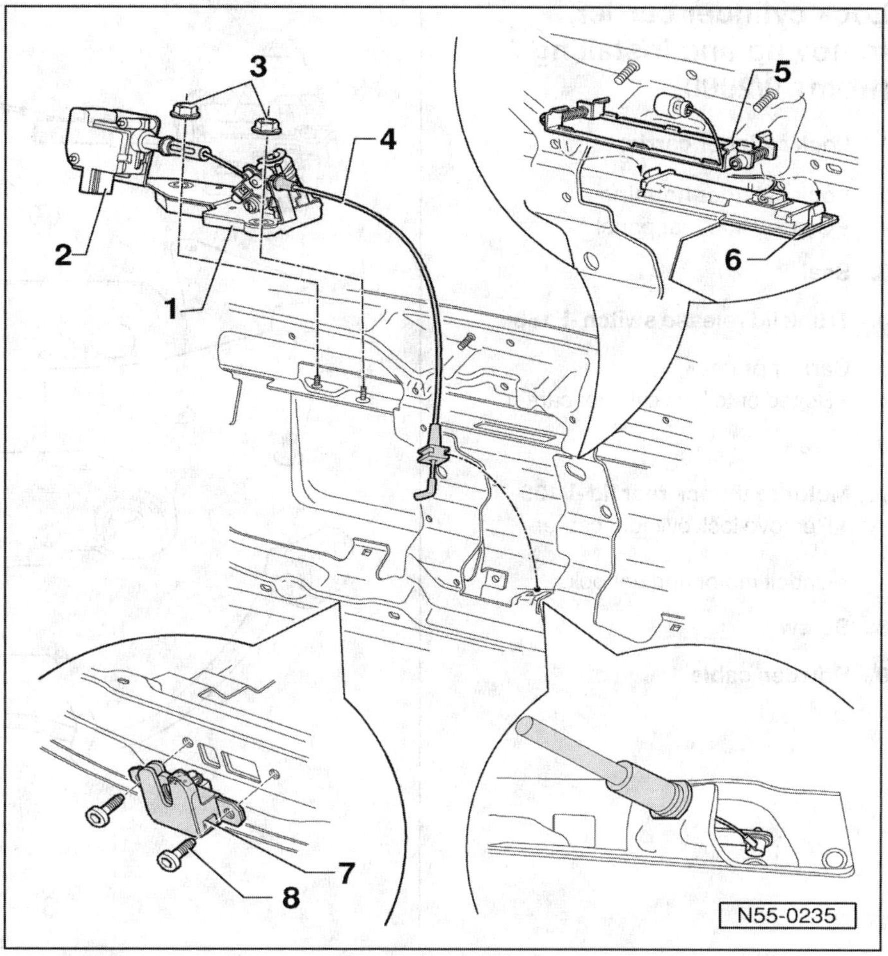

1. **Lock cylinder carrier**
   - Removing
     Disconnect harness connectors for license plate light, power lock adjusting motor and lock cylinder switch.
     Unclip bowden cable -4- at lock.
     Then remove hex nut -8-.
     Carrier lock cylinder can now be removed from tailgate.
   - Installation is the reverse order of removal.
2. **Motor to unlock -V139-**
3. **Hex nut**
   - Tighten to 7 Nm (62 in-lb)
4. **Bowden cable**
5. **Mounting frame**
6. **Pressure switch for electrical tailgate lock operation**
7. **Lock**
   - Removing:
     Remove trim for tailgate,
     see **Body-Interior section, Repair Group 70.**
     Disconnect harness connector at lock.
     Unclip bowden cable -4- at lock.
     Then remove hex nut -8-.
     Lock can now be removed.
8. **Bolt**
   - Tighten to 23 Nm (17 ft-lb)

Lock cylinder carrier, removing and installing (from 10/2000)

1. **Lock cylinder carrier**
2. **Lock cylinder housing**
   - Clipped into rear panel
3. **Seal**
4. **Trunk lid release switch -E165-**
5. **Carrier plate**
   - Bolted onto lock cylinder carrier
6. **Screw**
7. **Motor to unlock rear lid -V139-**
   - Remove lock cylinder carrier -1-.
   - Unbolt motor and unhook.
8. **Screw**
9. **Bowden cable**

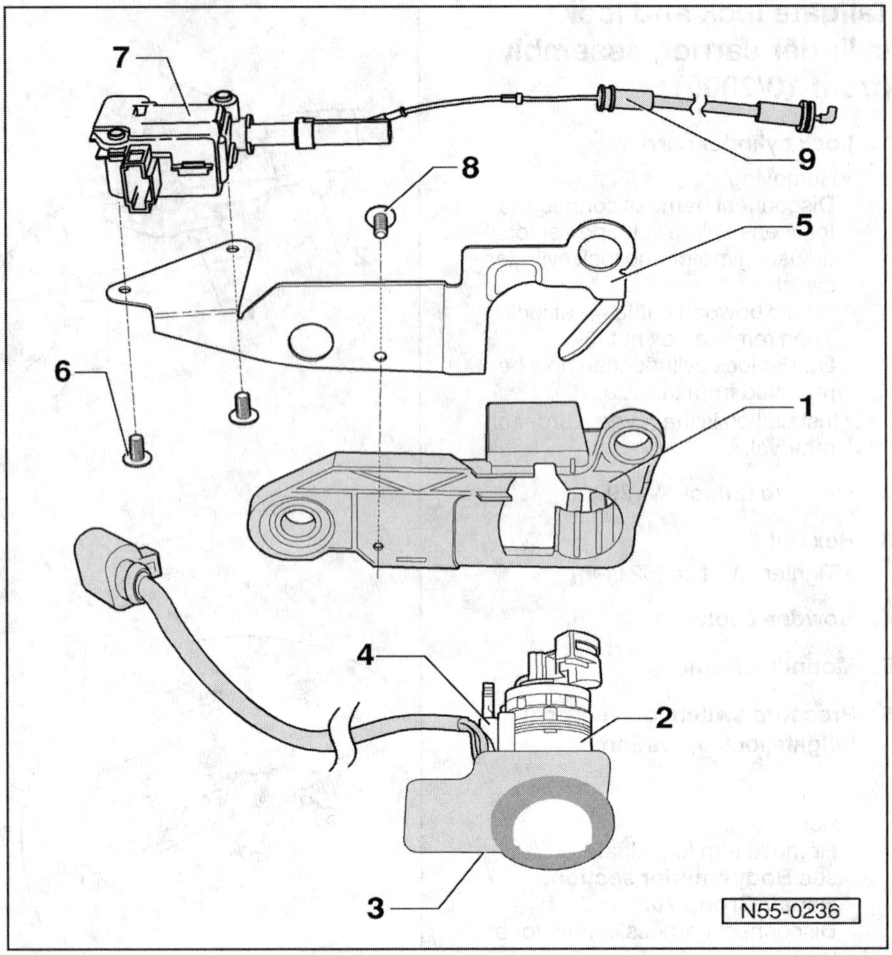

Lock cylinder, removing and installing

Removing

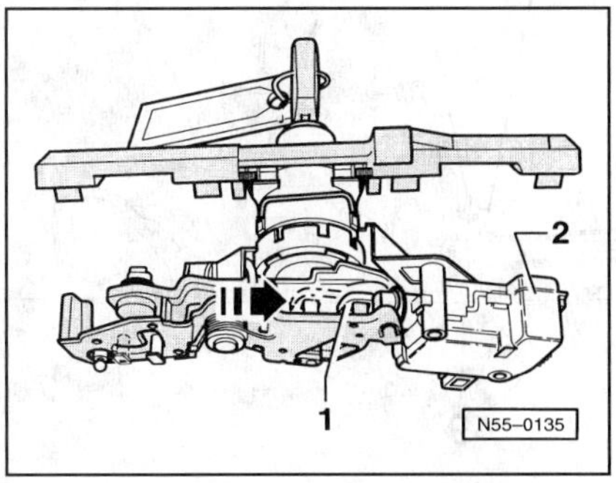

Slide lift rod -1- into positioning motor -2- -arrow-.

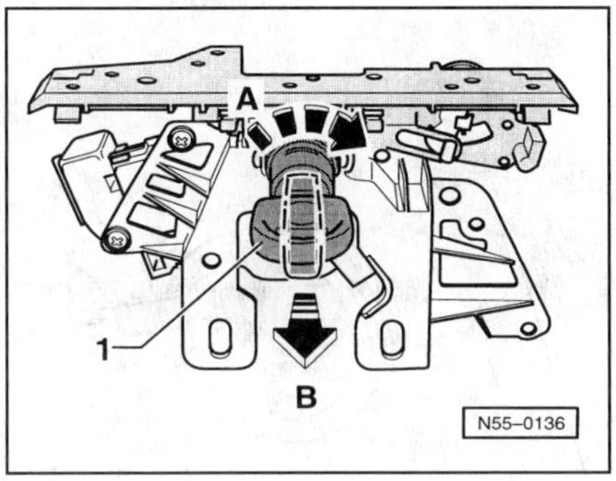

Turn lock cylinder -1- clockwise through 90° -arrow A- and then take out forwards off carrier.

Installing

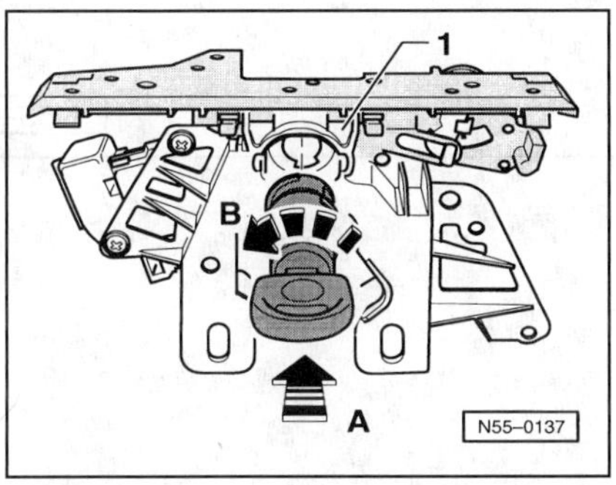

Insert lock cylinder in carrier -1- -arrow A-, turn anti-clockwise through 90° -arrow B-.

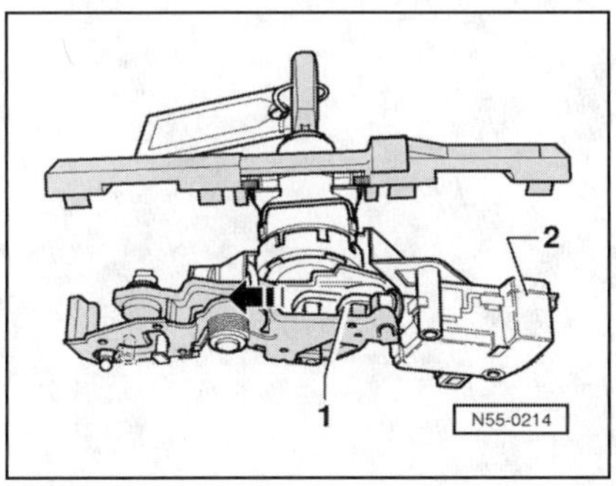

Lift rod -1- must be pulled out of positioning motor -2- in this position. Simultaneously the lock cylinder is engaged in the carrier with an audible check.

Tank Flap Unit

Tank flap unit, assembly

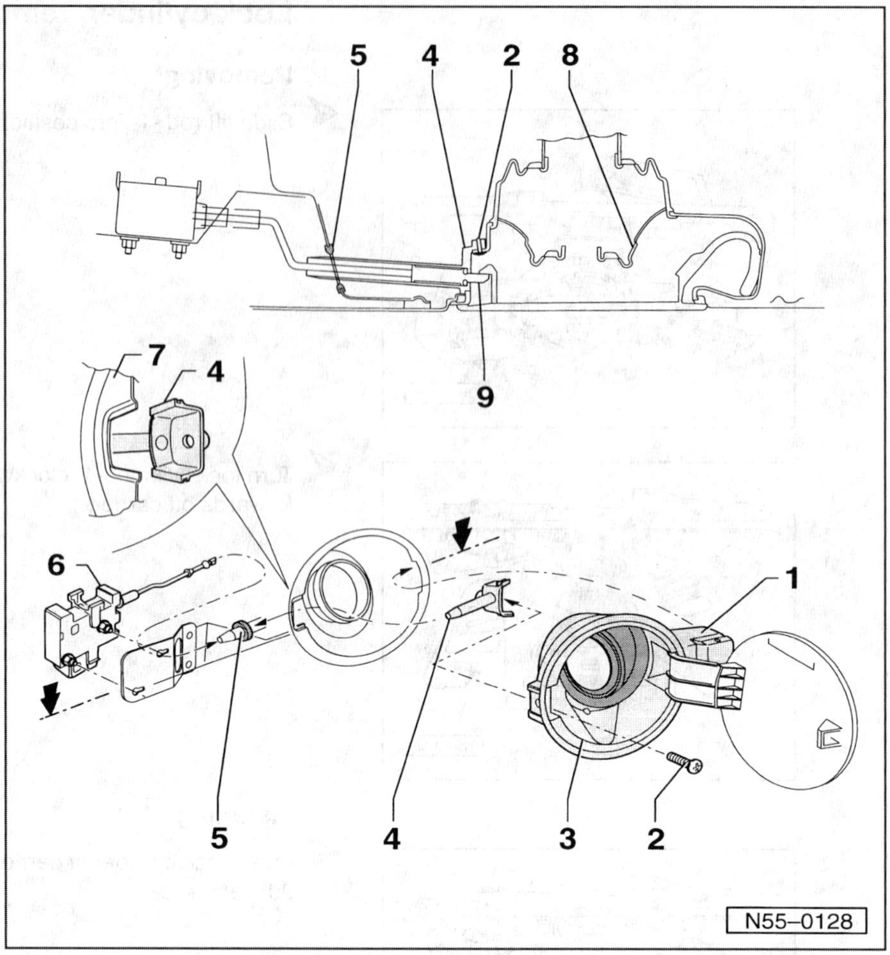

1. **Tank flap unit**
   - Removing:
     Unscrew tank filler cap.
     Press back lock release rod by operating remote release.
     Remove cross-head screw -2-.
     Pull rubber part off fuel tank filler and swing tank flap unit out of side panel.
   - Installing:
     Assembly part and positioning element are installed
     Install tank flap unit (rubber part rolled up) with hinge side leading.
     Secure tank flap unit and assembly part.
2. **Cross-head screw**
3. **Tank flap cup**
   - Removing and installing (can only be removed as part of tank flap unit)
   - Press out securing pin in tank flap unit to remove the tank flap
4. **Assembly piece**
   - Pushed into side panel
5. **Seal**
6. **Fuel filler flap positioner**
7. **Side panel**
8. **Rubber part**
   - Roll up rubber part to install
9. **Release rod**

57 Front Doors, Central Locking

Repair Group

T 57

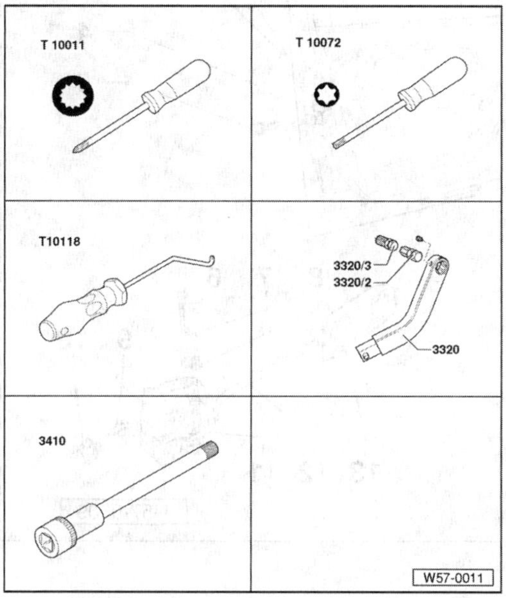

Front Door

Special tools

Special tools, workshop equipment, testers, measuring instruments and auxiliary items required

- T 10011 Socket insert
- T 10072 Socket wrench
- T10118 Assembly tool
- 3320/2 Box spanner
- 3320/3 Box spanner
- 3410 Socket insert

Front door, assembly

NOTE—

- *The hinge bolts must always be replaced if loosened.*
- *The dash panel must be removed to remove and adjust the upper door hinges.*

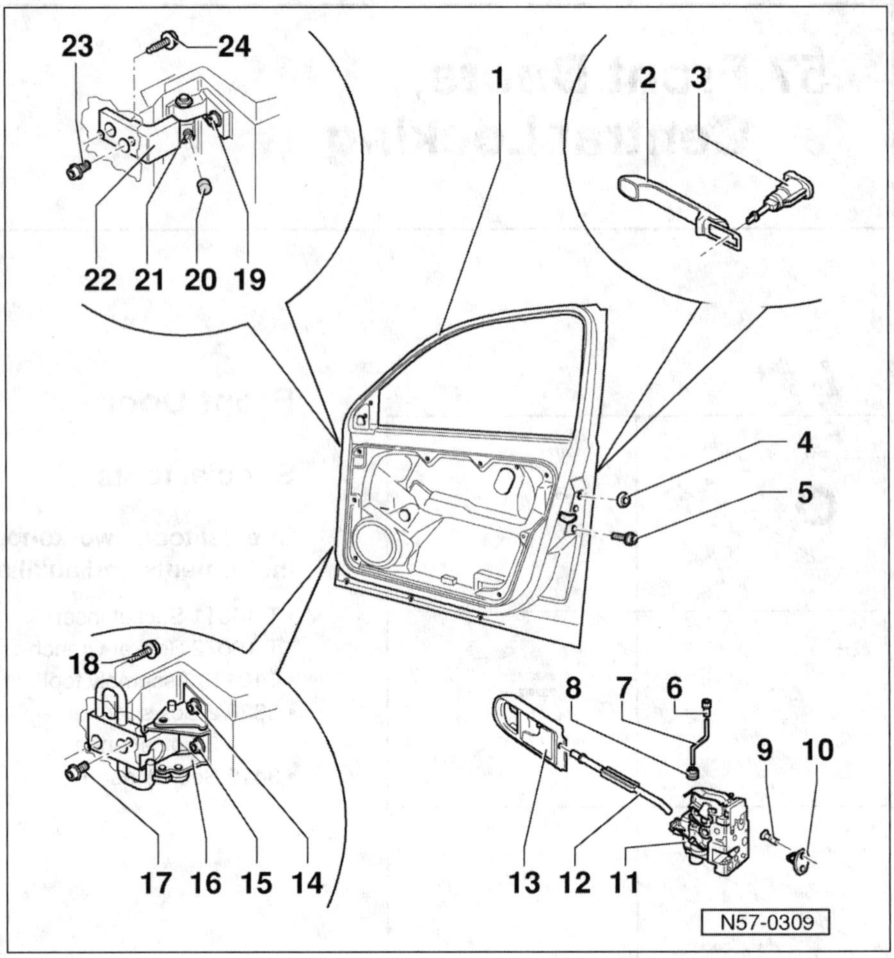

1. **Door**
   - See **Door, removing and installing**
2. **Door handle with base**
   - See **Door handle, removing and installing**
3. **Lock cylinder housing**
   - See **Lock cylinder housing, removing and installing**
4. **Cover cap**
5. **Screw**
   - Tighten to 20 Nm (15 ft-lb)
6. **Locking button**
   - See **Locking knob for locking rod, removing and installing**
7. **Locking rod**
8. **Bellows**
9. **Screw**
   - Tighten to 20 Nm (15 ft-lb)
10. **Lock plate**
11. **Door lock**
    - See **Door lock, removing and installing**
12. **Bowden cable**
13. **Interior release handle**
    - See **Inner release handle, removing**
14. **Multi-point socket head bolt**
    - M8 x 28
    - Tighten to 20 Nm (15 ft-lb) + $^1/_4$ turn further (90°)
    - Always replace bolt after loosening
15. **Multi-point socket head bolt**
    - Only remove this bolt from hinge to remove door
    - M8 x 28
    - Tighten to 20 Nm (15 ft-lb) + $^1/_4$ turn further (90°)
    - Always replace bolt after loosening
16. **Door hinge with arrester**
    - Hinge is split
17. **Multi-point socket head bolt**
    - M8 x 28
    - Install from inside vehicle
    - Remove A pillar lower trim, see **Body-Interior section, Repair Group 70**
    - 20 Nm + $^1/_4$ turn further (90°)
    - Always replace bolt after loosening
18. **Multi-point socket head bolt**
    - M8 x 28
    - Tighten to 20 Nm (15 ft-lb) + $^1/_4$ turn further (90°)
    - Always replace bolt after loosening
19. **Multi-point socket head bolt**
    - M8 x 28
    - Tighten to 20 Nm (15 ft-lb) + $^1/_4$ turn further (90°)
    - Always replace bolt after loosening
20. **Cover cap**
21. **Grub screw**
    - Tighten to 23 Nm (17 ft-lb)
22. **Door hinge**
    - Hinge is split
23. **Multi-point socket head bolt**
    - M8 x 22
    - Install from inside vehicle
    - To loosen or tighten the bolt, remove and install dash panel, see **Body-Interior section, Repair Group 70**
    - Tighten to 20 Nm (15 ft-lb) + $^1/_4$ turn further (90°)
    - Always replace bolt after loosening
24. **Multi-point socket head bolt**
    - M8 x 28
    - Tighten to 20 Nm (15 ft-lb) + $^1/_4$ turn further (90°)
    - Always replace bolt after loosening

Door, removing and installing

NOTE—
Various screws are used to secure the door hinge. There are two box spanners for this, bit 3320/2 (Emden production only) as well as 3320/3.

– Remove A pillar lower trim -5-, see **Body-Interior section, Repair Group 70**

– Pull bellows -3- off A pillar. Guide electrical cables out of pillar through hole -arrow-.

– Pry out cap -1- and remove grub screw -2- from upper hinge.
  - Tightening torque: 23 Nm (17 ft-lb)

– Remove lower bolt -4- from hinge.
  - Tightening torque: 20 Nm (15 ft-lb) + $^{1}/_{4}$ turn further (90°).

NOTE—
Always replace this bolt.

– Lift door out of door brackets upwards.

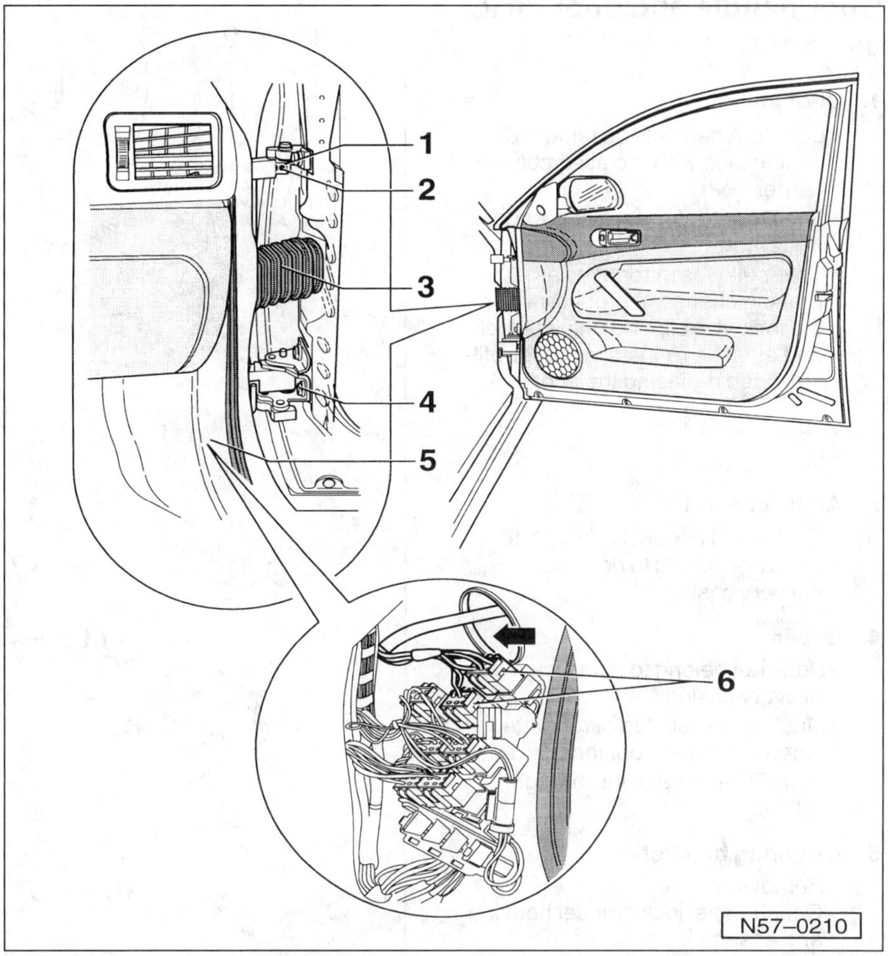

Door, adjusting

To ensure a correct door adjustment the door hinges on the pillar must be loosened. Other measures such as raising the door upwards have no effect. Over pressuring thereafter will again cause the door to sag.

Use door adjustment wrench 3320 with box spanner 3320/2 (Emden production only) or box spanner 3320/3.

If there is a requirement to loosen the door hinge on the upper A-pillar from the interior, socket 3410 can be used. The dash panel must be removed for this, see **Body-Interior section, Repair Group 70.**

Door handle and door lock, assembly

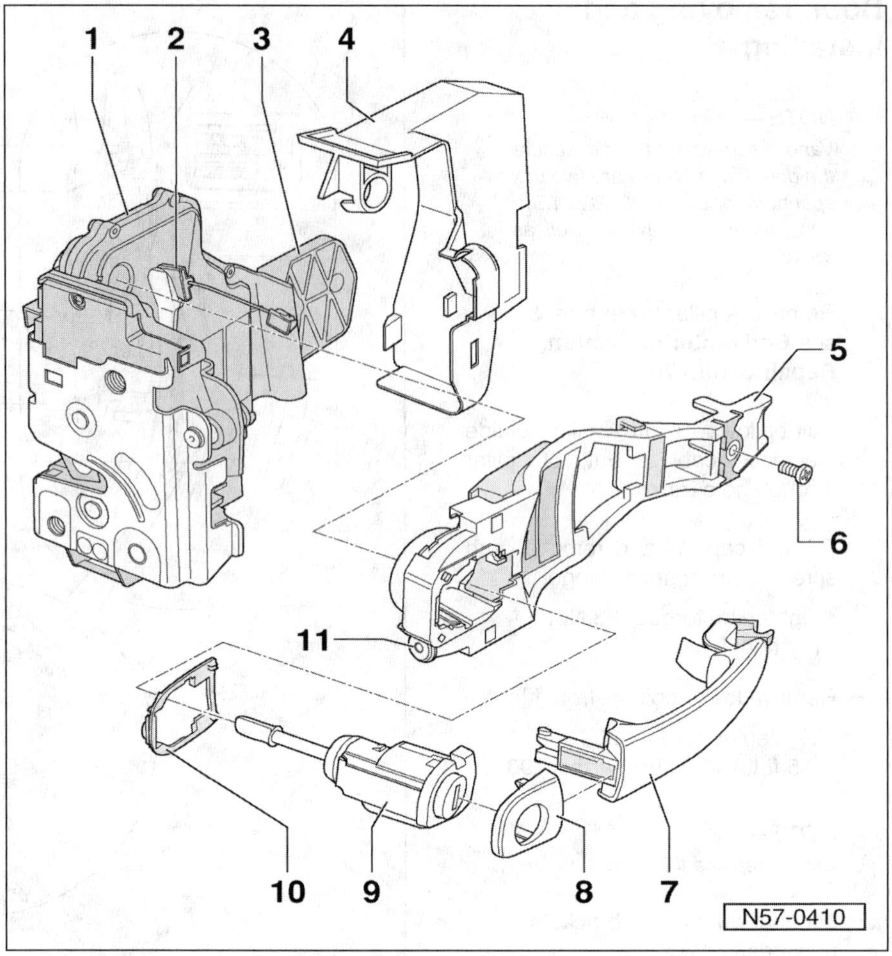

1. **Door lock**
   - Door lock can only be removed in conjunction with the assembly carrier
   - See **Door lock, removing and installing**
   - Cover -4- was introduced as a running change in production for the lock. It is clipped onto the lock and must be re-installed or retrofitted when replacing the door lock.
2. **Cable**
   - Lock release
3. **Angle bracket**
   - Bolted and riveted to door lock
   - Does not belong to door lock delivery casing
4. **Cover**
   - Does not belong to door lock delivery casing
   - Must be re-installed or retrofitted when replacing door lock
   - Retrofit for all vehicles through 09/1999
5. **Mounting bracket**
   - Removing:
     Door handle, lock cylinder housing and assembly carrier have been removed
     Remove Phillips-head screw -6-, slide mounting bracket slightly to rear and remove from door.
6. **Phillips-head screw**
7. **Door handle with backing plate**
   - See **Door handle, removing and installing**
8. **Concealment cap**
9. **Lock cylinder housing**
   - See **Lock cylinder housing, removing and installing**
   - Lock cylinder is not available as an individual replacement part
10. **Base plate**
11. **Socket-head bolt**
    - T 10011 Socket insert
    - T 10072 Socket wrench
    - By loosening this bolt, locking mechanism for lock cylinder housing -9- is disengaged and may be removed from mounting bracket -5-
    - Bolt -11- must not be threaded in without lock cylinder housing installed. Lock mechanism ring may fall into door

Lock cylinder housing, removing and installing

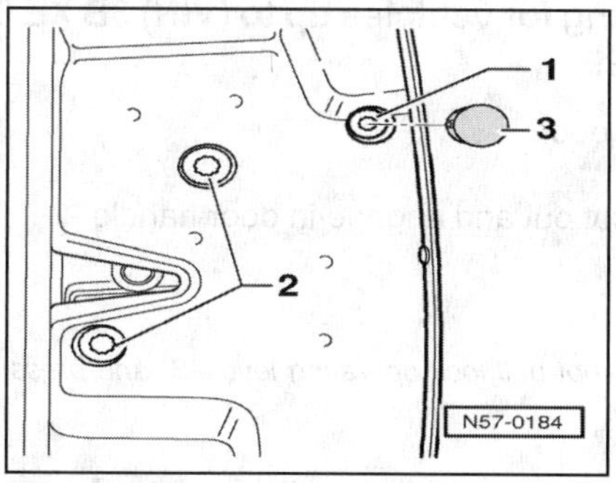

Removing

Pry out cover cap -3-.

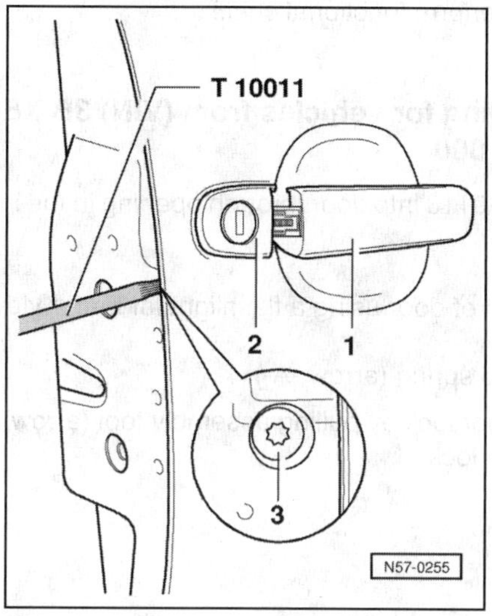

Pull door handle -1- and hold. Unscrew screw -3- onto stop with socket insert T 10011 (socket wrench T10072 has been introduced gradually). This releases the lock cylinder housing.

NOTE—
Unscrewing the bolt too far may result in the locking ring releasing from the mounting bracket and falling into the door.

– Pull lock cylinder housing -2- out of door handle mounting bracket at right angles.

Installing

– Fit lock cylinder housing in door handle mounting bracket at right angles.

– Fit screw in mounting bracket and tighten with socket wrench.
  - The door handle locates in the lock cylinder housing with an audible click.

NOTE—
Press door handle and lock cylinder housing against door panel when assembling.

– Perform remaining installation in reverse order.

– Always perform a functional check as an incorrectly fitted and adjusted Bowden cable will prevent the door from opening

Door handle, removing and installing

EDITOR'S NOTE—
There are two different prodcedures based upon location and date of production. See **Forward** *in front of book for VIN# information.*

Remove lock cylinder housing, see **Lock cylinder housing, removing and installing**.

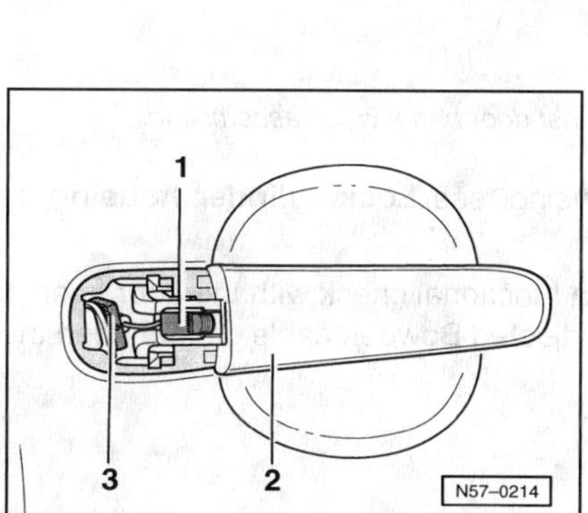

Removing

Unclip clip -1- from door handle.

– Swing door handle out of door.

Door handle, installing for vehicles up to (VIN) 3B XE 059 999 and to 3B XP 077 999

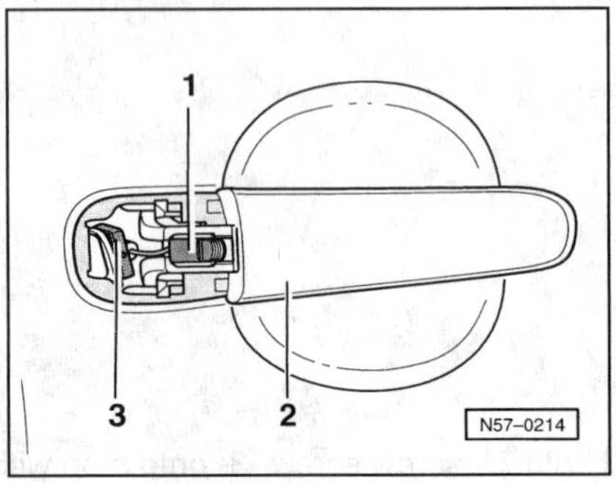

◄ Swing door handle into door.

– Pull clip -1- into panel cut out and engage in door handle -2-.

NOTE —
When assembling clip, do not pull lock operating lever -3- and press door handle -2- onto door panel.

– Install lock cylinder housing, see **Lock cylinder housing, removing and installing.** Then perform functional check.

Door handle, installing for vehicles from (VIN) 3B XE 060 000, and 3B XP 078 000

– Guide assembly tool T10118 into door through opening in the inner door plate.

– Illuminate the inner part of door using a flashlight for better view.

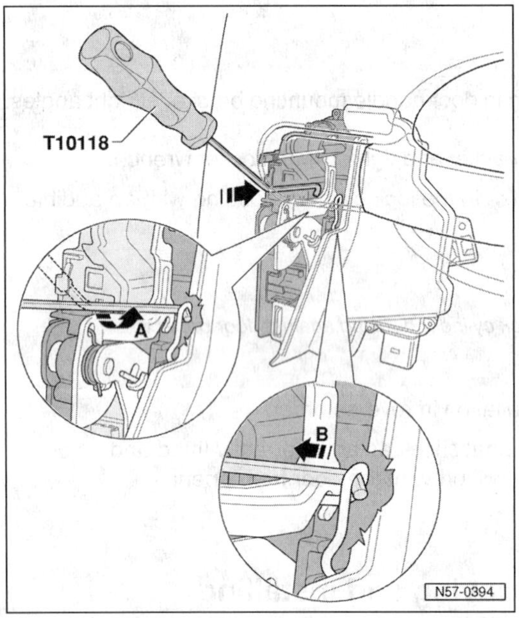

◄ Hook assembly tool into spring (arrow -A-).

- Engage spring into door lock by pulling assembly tool (arrow -B-).
- Activation lever is now locked.

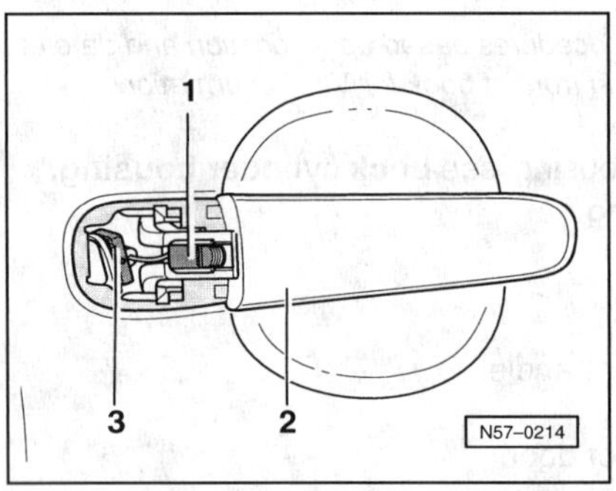

◄ Swing door handle into door.

– Pull clip -1- into panel cut out and engage in door handle -2-.

NOTE —
Press door handle -2- against door panel when assembling.

– Install lock cylinder housing, see **Lock cylinder housing, removing and installing**

- Then always perform a functional check with the door open as an incorrectly fitted and adjusted Bowden cable cannot open the door.

Door lock, removing and installing

The window lifters, the door lock and the loudspeaker are secured to the assembly carrier.

The door lock can only be removed in conjunction with the assembly carrier.

The assembly carrier can only be removed when the door window clamping brackets of the window lifter are unscrewed. To do this lower door window to the height of the assembly holes in assembly carrier and loosen clamping brackets.

Door lock, removing

– Remove front door trim, see **Body-Interior section, Repair Group 70.**

– Remove lock cylinder housing, see **Lock cylinder housing, removing and installing**.

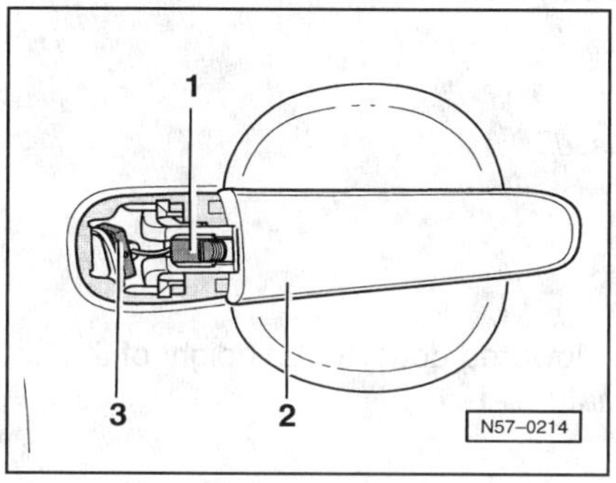

◄ Unclip clip -1- from door handle.

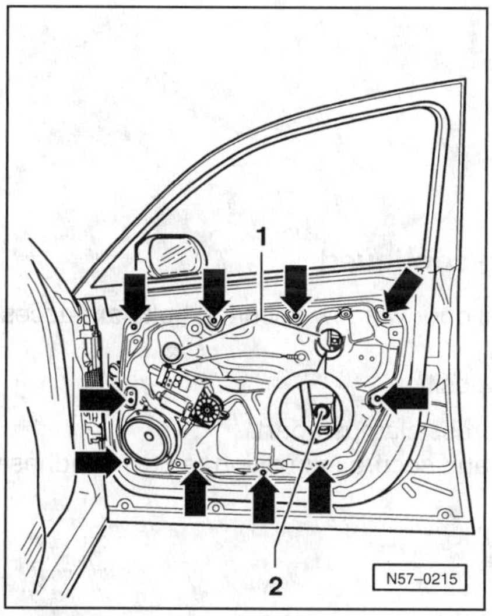

◄ Pry out cover cap -1-.

If electric window lifter motor has failed, proceed as follows

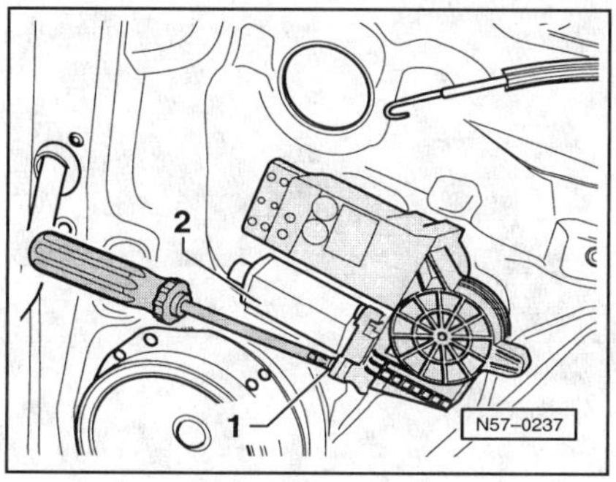

◄ Unscrew bolts -1- (Qty. 4).

– Pull housing -2- off window motor.

Manually lowering electric window (Bosch lifter motor)

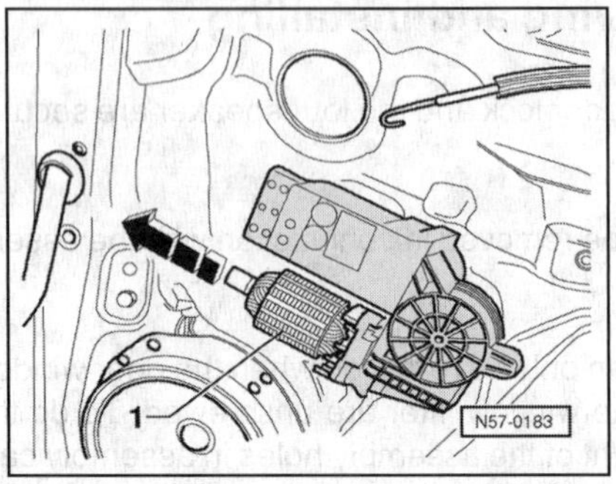

Pull armature -1- off window motor -arrow-.

– The door window can now be pushed down to the height of the assembly holes in the assembly carrier.

NOTE —
The motor must not be reused.

Manually lowering electric window (Siemens lifter motor)

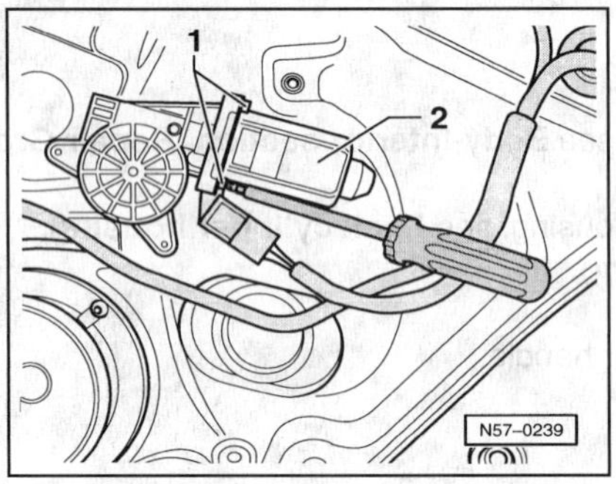

Unscrew bolts -1- (Qty. 4).

– Pull housing -2- off window motor.

NOTE —
The motor must not be reused.

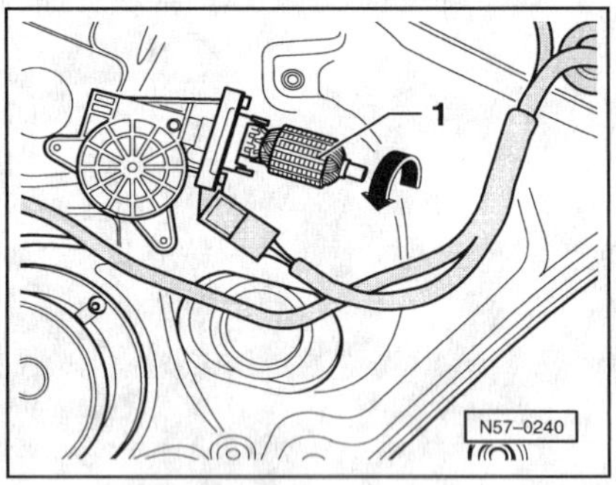

The door window can be lowered down to the height of the assembly holes by turning the armature -1-.

Door lock, removing continued

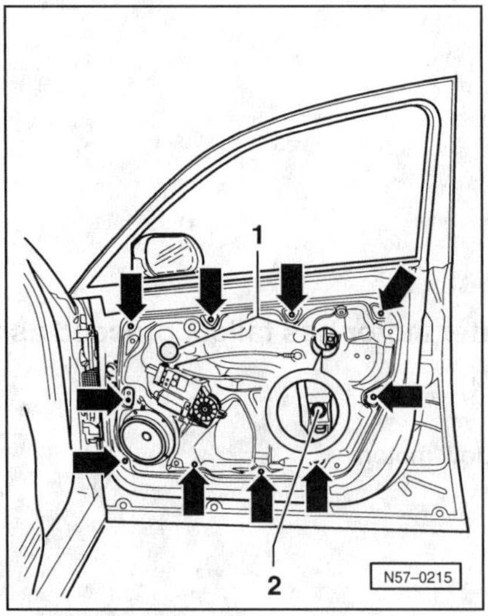

Lower door window until door window securing bolts are accessible.

– With window in service position

- Loosen bolts -2- and press clamps apart.
- Push window upward and secure in position (e.g. with adhesive tape).

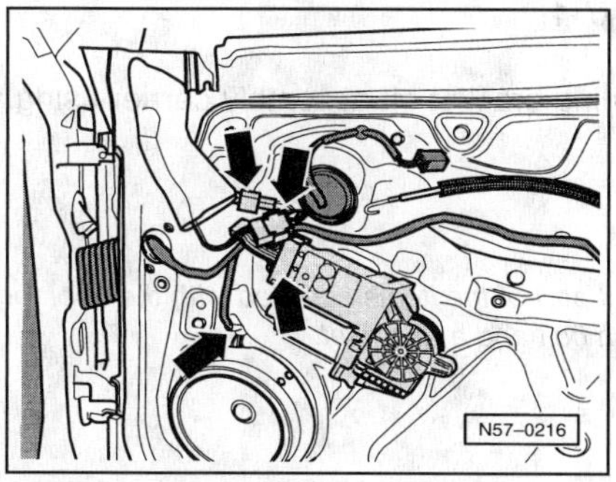

◄ Disconnect all connections -arrows-.

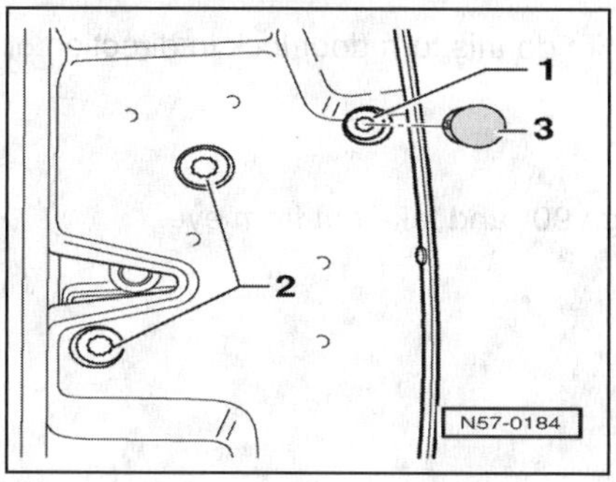

◄ Lever out cover cap -3-.

– Unscrew bolts -2- Tightening torque: 20 Nm (15 ft-lb).

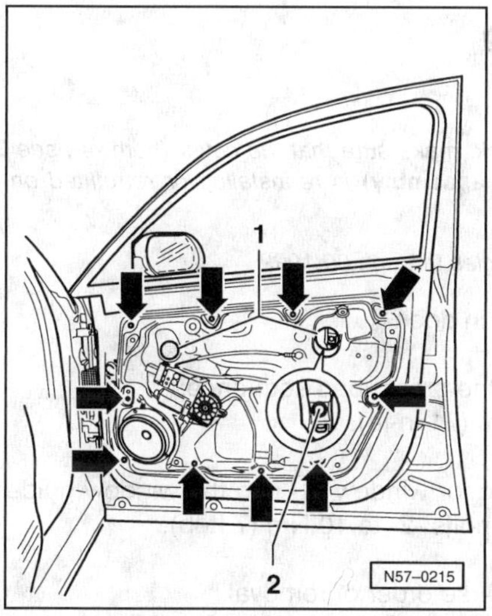

◄ Unscrew bolts -arrows-. Pull top of assembly carrier off door, lift and take out of door towards hinges.

– Turn assembly carrier and pull connector off door lock.

– Then unclip clips for electrical cables on back of assembly carrier.

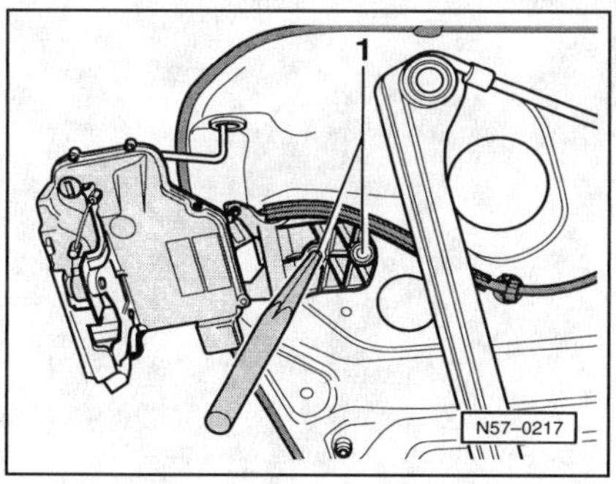

◄ Drive out clips -1- with a drift.

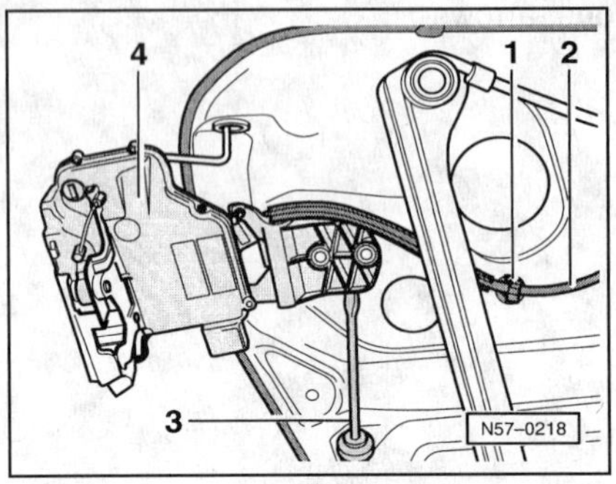

Unclip cable -2- from clip -1-.

– Pry door lock with retaining bracket off assembly carrier using a screwdriver -3-.

NOTE—
The retaining bracket is not part of the items supplied with the door lock. Secured to door lock with a bolt and a pop rivet.

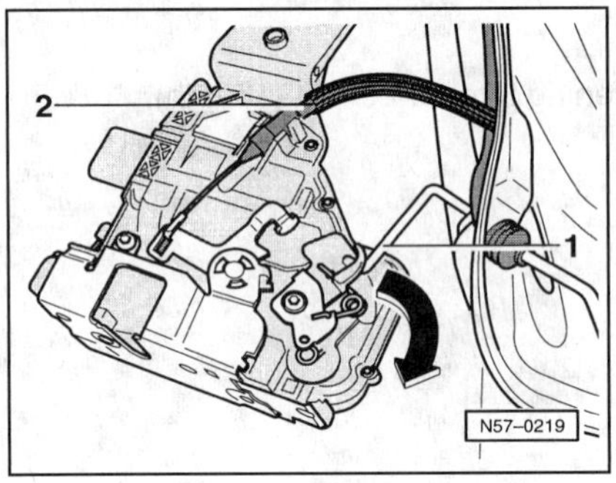

Unclip securing rod -1-. To do this turn door lock in direction of arrow.

– Unclip cable -2-.

– Turn cable nipple through 90° and take out from eye.

Door lock, installing

NOTE—

- *When replacing door lock, make sure that the cover (Item -4-, see* **Door handle and door lock, assembly)** *is re-installed or retrofitted on door lock.*
- *Retrofit cover for all vehicles through 09/1999*

– Insert assembly carrier in door.

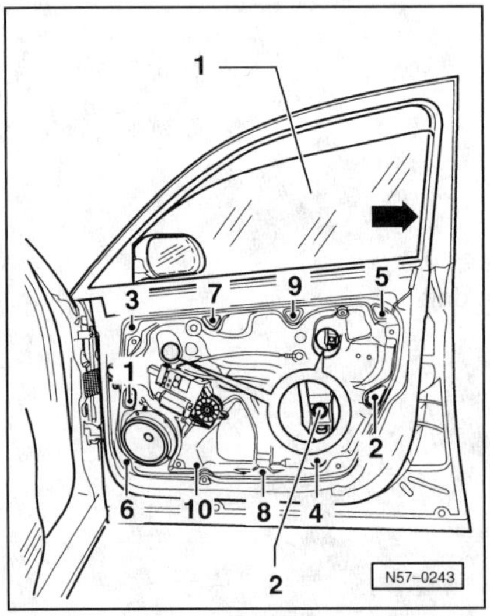

Fit bolts and tighten in the prescribed sequence.
Tightening torque: 8 Nm (71 in-lb).

– Loosen clamps, press door window -1- into the window guides -arrow- and tighten clamp with bolts -2- to 10 Nm (7 ft-lb).

– Install remainder in reverse order of removal.

- Then perform functional check.

Inner release handle, removing

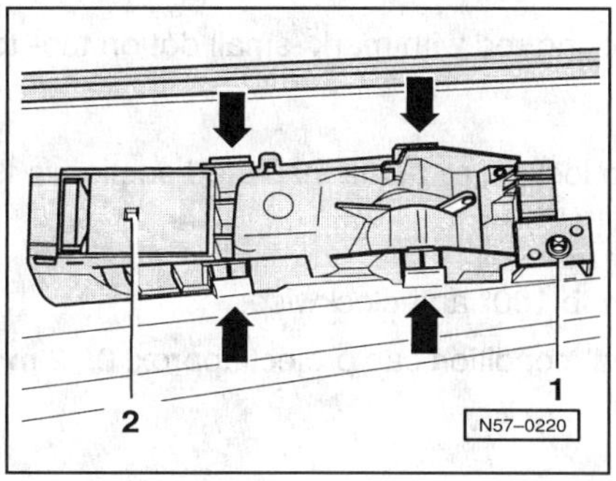

– Removing front door trim, see **Body-Interior section, Repair Group 70.**

◄ Remove cross head screw -1-.

– Unclip switch -2- for electric windows.

– Unclip door inner release -arrows- and take out from door trim.

Door inner seal, removing and installing

Door seals are fitted in series production with a sealant, they are placed on the door flange and then rolled on.

NOTE —
- *If they are removed the sealant will spread on the inside of the seal and the flanks will be bent out slightly. If the seal is then refitted there is no guarantee that it will fit tightly and seal.*
- *Therefore each seal which is removed completely should be replaced by a so called "hammer blow" seal.*
- *When seals have been partially removed squeeze the flanks together before refitting.*

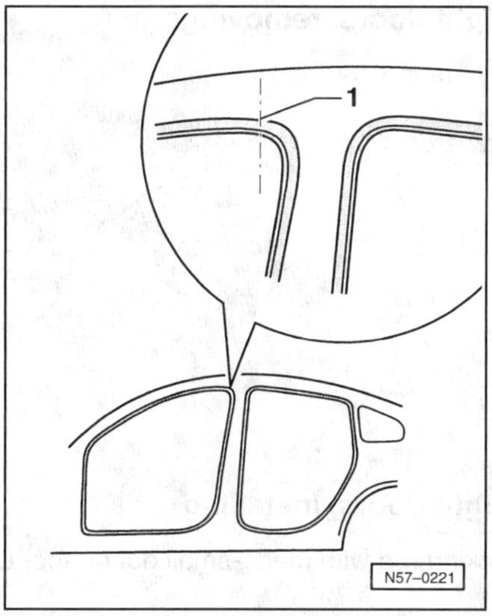

◄ Pull door inner seal off body flange.

– When installing the door inner seal, start at upper radius of the door cut-out -1-.

Locking knob for locking rod, removing and installing

The description for removing and installing the locking knob must be adhered to.

Not observing the sequence, may unscrew the locking rods in the front door locks and the relays in the rear doors or damage the mountings.

Front right and rear left doors, removing

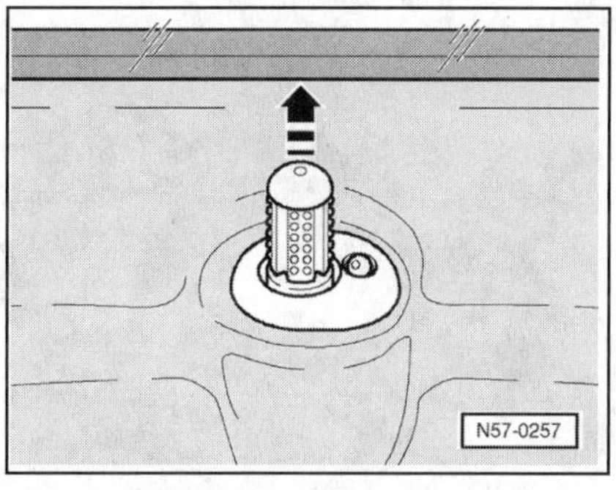

◄ Release locking knob by turning 180° anti-clockwise.

– Take locking knob off upwards.

Front right and rear left doors, installing

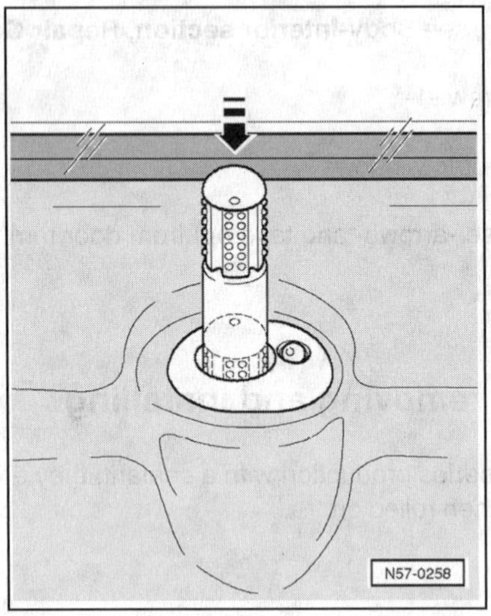

- Fit locking knob onto locking rod with mark -small dot on top- towards vehicle interior.
- Press locking knob onto locking rod until the fluted section is flush with the height of the door trim.
- Then turn the locking knob 180° anti-clockwise.
  - Locking knob in "locked" condition can project approx. 0...2 mm above door trim.

Front left and rear right doors, removing

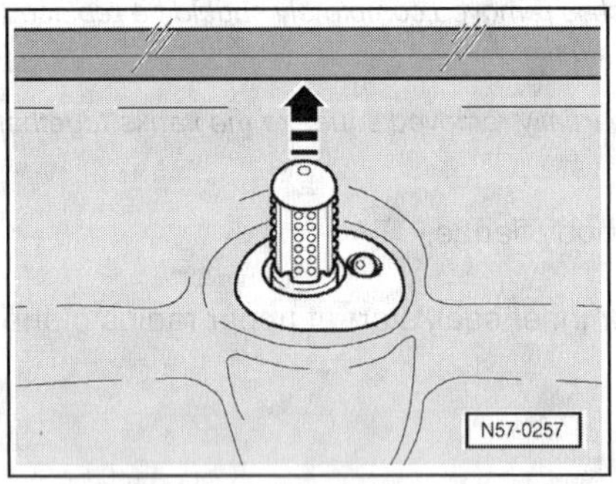

- Release locking knob by turning 180° clockwise.
- Take locking knob off upwards.

Front left and rear right doors, installing

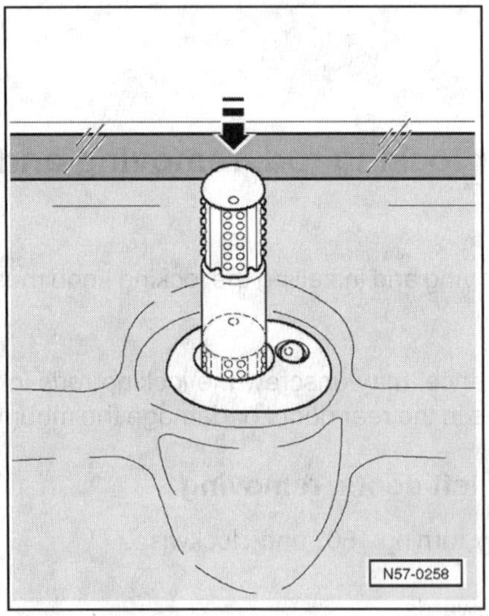

- Fit locking knob onto locking rod with mark -small dot on top- towards vehicle interior.
- Press locking knob onto locking rod until the fluted section is flush with the height of the door trim.
- Then turn the locking knob 180° clockwise.
  - Locking knob in "locked" condition can project approx. 0...2 mm above door trim.

Central Locking

Central locking and comfort system, assembly

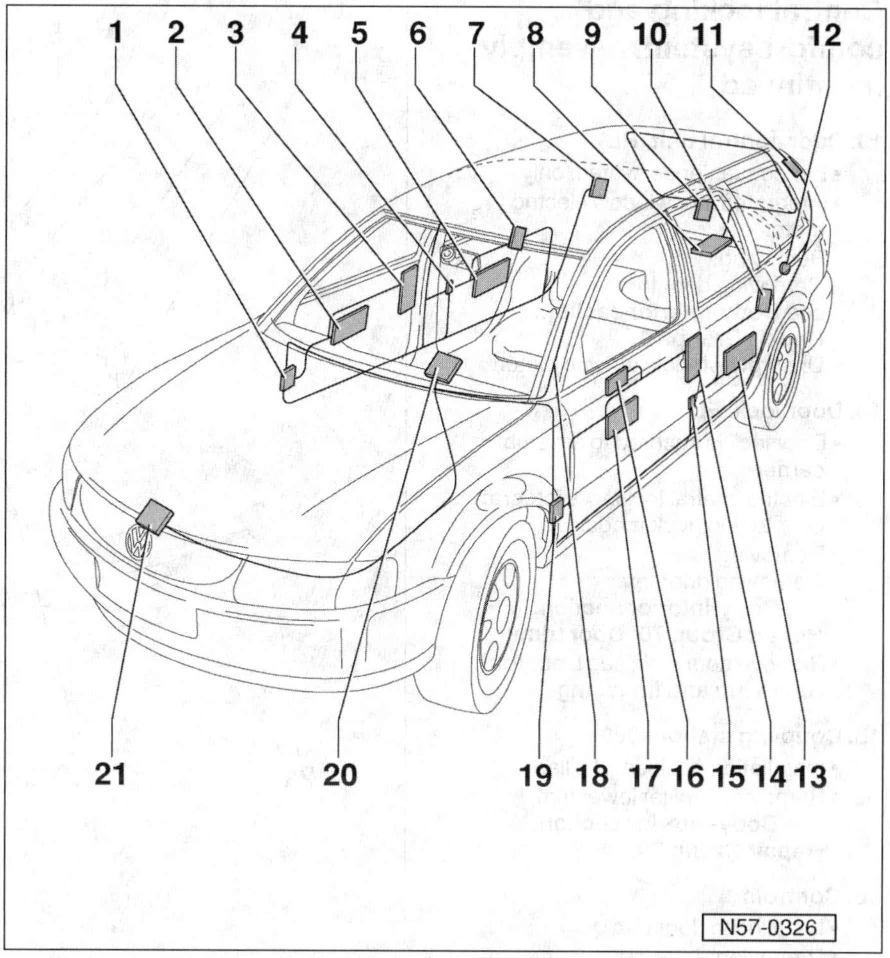

1. **Coupling station -080-**
   - Installation position: Lower A pillar covered by trim in footwell
   - Removing A pillar lower trim, see **Body-Interior section, Repair Group 70.**
2. **Door control unit, FR**
   - For convenience system only
   - Integrated into window electric motor
   - Removing:
     Removing door trim,
     see **Body -Interior section, Repair Group 70.**
     Unscrew window electric motor.
3. **Door lock, FR**
   - Door lock is secured to assembly carrier
   - Electric central locking is integrated into door lock module
   - Removing:
     Removing door trim,
     see **Body Interior section, Repair Group 70.**
     Remove door lock, see **Door lock, removing and installing.**
4. **Coupling station -080-**
   - Installation position: B pillar
   - Removing B pillar lower trim, see **Body-Interior section, Repair Group 70.**
5. **Door control unit, RR**
   - For convenience system only
   - Integrated into window electric motor
   - Removing:
     Removing door trim,
     see **Body-Interior section, Repair Group 70.**
   - Unscrew window electric motor.
6. **Door lock, RR**
   - Door lock is secured to assembly carrier
   - Electric central locking is integrated into door lock module
   - Removing:
     Removing door trim,
     see **Body-Interior section, Repair Group 70.**
   - Removing rear door lock, see **Repair Group 58, Door lock, removing and installing.**
7. **Motor for fuel tank lid unlock -V155-**
   - Installation position: Under C pillar trim
   - To remove, the trunk trim must be detached first of all up to wheel house, see **Body-Interior section, Repair Group 70.**
   - Can be removed with special socket insert T 10010
8. **Tailgate lock**
   - Bolted to tailgate
   - Removing:
     Removing tailgate trim,
     see **Body-Interior section, Repair Group 70**
     Remove tailgate lock, see **Repair Group 55, Tailgate lock and lock carrier assembly**
   - For trunk lid lock, see **Repair Group 55, Trunk lid lock assembly**
9. **Tailgate lock actuator**
   - Bolted to tailgate
   - Can be removed with special socket insert T 10010
10. **Door lock, RL**
    - Door lock is secured to assembly carrier
    - Electric central locking is integrated into door lock module
    - Removing:
      Removing door trim,
      see **Body-Interior section, Repair Group 70**
    - Removing rear door lock, see **Repair Group 58, Door lock, removing and installing.**
11. **Coupling station -080-**
    - Installation position: In area of rear roof carrier, covered by roof trim
    - Removing roof trim, see **Body-Interior section, Repair Group 70**
12. **Anti-theft alarm horn**
    - Installation position: In area of C pillar covered by C pillar trim

(continued on following page)

Central locking and comfort system, assembly (continued)

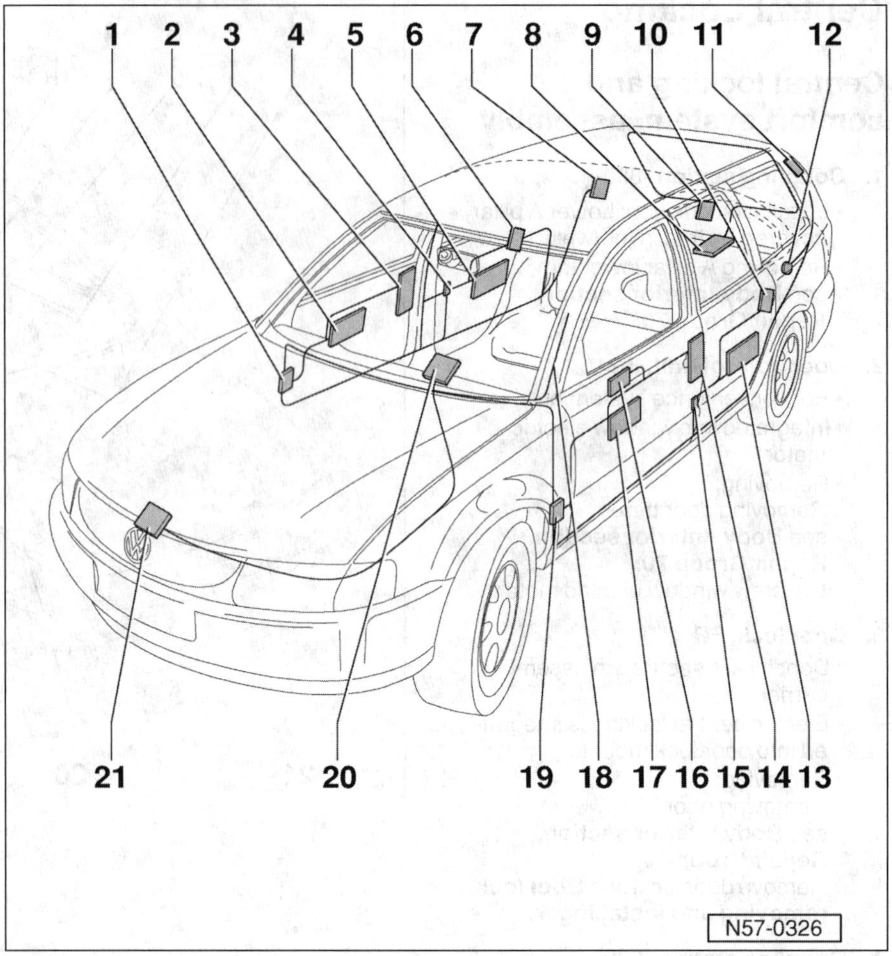

13. Door control unit, RL
- For convenience system only
- Integrated into window electric motor
- Removing:
  Removing door trim, see **Body-Interior section, Repair Group 70**
  Unscrew window electric motor.

14. Door lock, FL
- Door lock is secured to assembly carrier
- Electric central locking is integrated into door lock module
- Removing:
  Removing door trim, see **Body-Interior section, Repair Group 70; Door trims**
- Remove door lock, see **Door lock, removing and installing.**

15. Coupling station -080-
- Installation position: B pillar
- Removing B pillar lower trim, see **Body-Interior section, Repair Group 70.**

16. Controls
- Installed in door trim
- Removing:
  Removing door trim, see **Body-Interior section, Repair Group 70**

17. Door control unit, FL
- For convenience system only
- Integrated into window electric motor
- Removing:
  Removing door trim, see **Body-Interior section, Repair Group 70; Door trims**
  Unscrew window electric motor.

18. Aerial for radio remote control
- Under left A pillar trim
- Removing:
  Removing A pillar trim, see **Body-Interior section, Repair Group 70**

19. Coupling station -080-
- Installation position: Lower A pillar, covered by trim in footwell
- Removing left A pillar lower trim, see **Body-Interior section, Repair Group 70**

20. Central module
- Bolted to bracket on steering column under dash panel
- To remove, remove lower steering column trim first
- Comfort system control module under carpet left front floor, see **Body-Interior section, Repair Group 68**

21. Hood lock
- Installation position: In lock carrier
- See **Repair Group 55, Hood lock, removing and installing**

58 Rear Doors

Repair Group

T 58

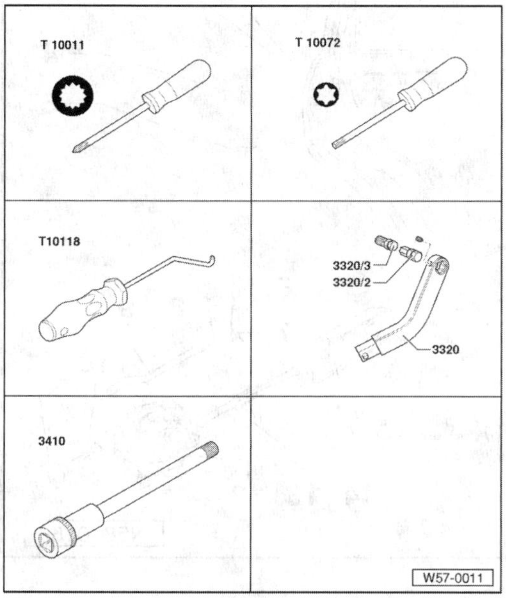

Rear Door

Special tools

Special tools, workshop equipment, testers, measuring instruments and auxiliary items required

- T 10011 Socket insert
- T 10072 Socket wrench
- T10118 Assembly tool
- 3320/2 Box spanner
- 3320/3 Box spanner
- 3410 Socket insert

Rear door, assembly

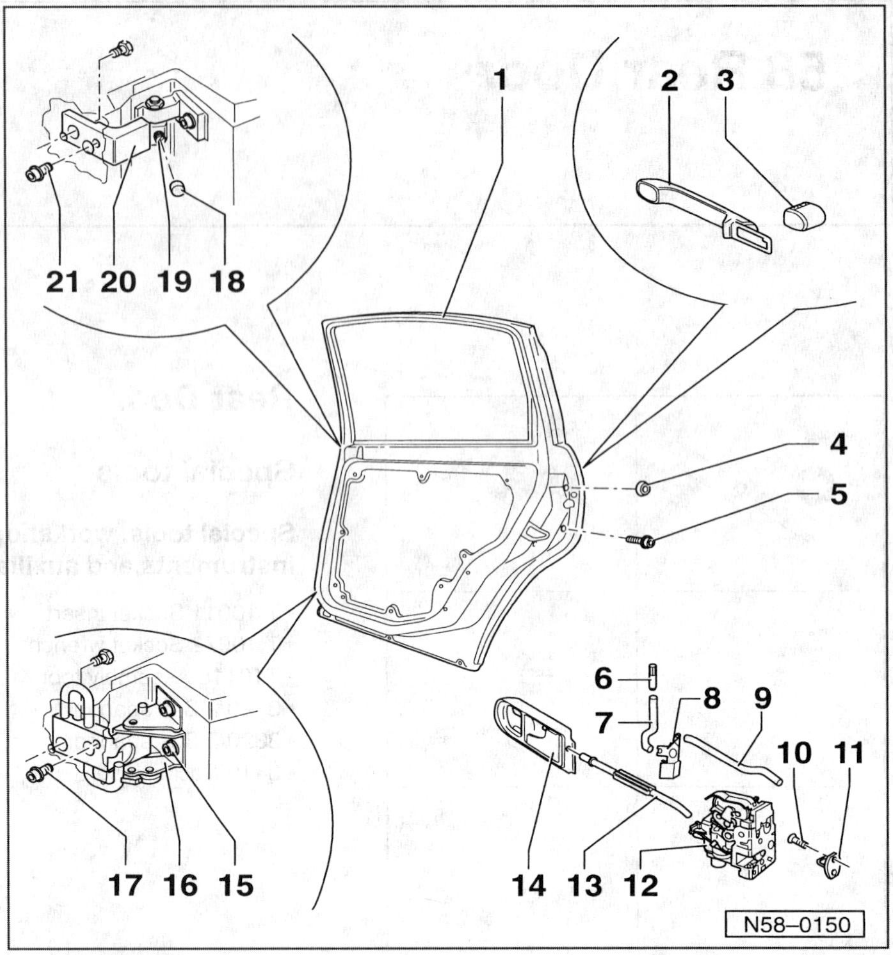

NOTE —

The hinge bolts must always be replaced if loosened.

1. **Door**
   - See **Door, removing and installing**
2. **Door handle with base**
   - See **Repair Group 57, Door handle, removing and installing**
3. **Housing**
   - See **Repair Group 57, Lock cylinder housing, removing and installing**
4. **Cover cap**
5. **Screw**
   - Tighten to 20 Nm (15 ft-lb)
6. **Locking button**
7. **Locking rod**
8. **Bell crank**
9. **Locking rod**
10. **Screw**
    - Tighten to 20 Nm (15 ft-lb)
11. **Lock plate**
12. **Door lock**
    - See **Door lock, removing and installing**
13. **Bowden cable**
14. **Interior release handle**
    - See **Inner release handle, removing**
15. **Bolt**
    - M8 x 28
    - Only remove this bolt from hinge to remove door
    - Tighten to 20 Nm (15 ft-lb) + $^1/_4$ turn further (90°)
    - Always replace after loosening
16. **Door hinge with arrester**
    - Hinge is split
17. **Bolt**
    - M8 x 28
    - Only remove this bolt from hinge to remove door
    - Tighten to 20 Nm (15 ft-lb) + $^1/_4$ turn further (90°)
    - Always replace after loosening
18. **Cover cap**
19. **Grub screw**
    - Tighten to 23 Nm (17 ft-lb)
20. **Door hinge**
    - Hinge is split
21. **Bolt**
    - Tighten to 20 Nm (15 ft-lb) + $^1/_4$ turn further (90°)
    - Always replace bolt after loosening

Door, removing and installing

NOTE—
Various screws are used to secure the door hinge. There are two box spanners for this, bit 3320/2 (Emden production only) as well as 3320/3.

– Remove B-pillar upper and lower trim, see **Body-Interior section, Repair Group 70**

◄ Disconnect multi-pin connector under rubber boot -3- on B pillar.

– Lever off cover cap -2- with a screwdriver.

– Remove grub screw -1- from upper hinge.
  - Grub screw tightening torque = 23 Nm (17 ft-lb).

– Remove lower bolt -4- from hinge.
  - Tightening torque = 20 Nm (15 ft-lb) + $^1/_4$ turn (90°) further

NOTE—
Always replace this bolt.

– Lift door upwards out of hinge brackets.

Door, adjusting

To ensure a correct door adjustment the door hinges on the pillar must be loosened. Other measures such as raising the door upwards have no effect. Over pressuring thereafter will again cause the door to sack.

Use door adjustment wrench 3320 with box spanner 3320/2.

If there is a requirement to loosen the door hinge on B-pillar from the interior, socket 3410 can be used. The B-pillar lower trim must be removed for this, see **Body-Interior section, Repair Group 70.**

Door handle and door lock, assembly

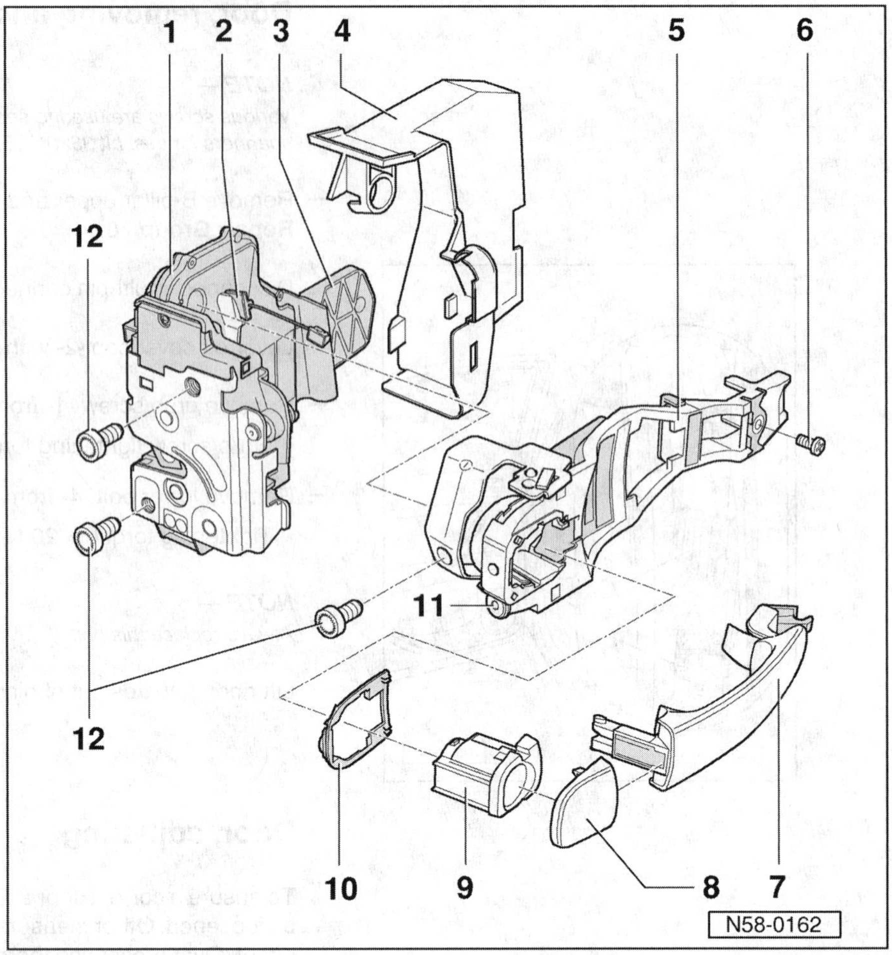

1. **Door lock**
   - Door lock can only be removed in conjunction with the subframe
   - See **Door lock, removing and installing**
   - Cover -4- was introduced as a running change in production for the lock. It is clipped onto the lock and must be re-installed or retrofitted when replacing the door lock.
2. **Cable**
   - Lock release
3. **Angle bracket**
   - Bolted and riveted to door lock
   - Does not belong to door lock delivery casing
4. **Cover**
   - Does not belong to door lock delivery casing
   - Must be re-installed or retrofitted when replacing door lock
   - Retrofit for all vehicles through 09/1999
5. **Mounting bracket**
   - Removing
     Door handle, lock cylinder housing and subframe have been removed
     Remove Phillips-head screw -6-, slide mounting bracket slightly to rear and remove from door.
6. **Phillips-head screw**
7. **Door handle with backing plate**
   - See **Door handle, removing and installing**
8. **Concealment cap**
9. **Housing**
   - See **Repair Group 57, Lock cylinder housing, removing and installing**
10. **Base plate**
11. **Socket-head bolt**
    - This bolt was replaced by a Torx bolt, and a new tool must be used for it
    - T 10011 Socket insert
    - T 10072 Socket wrench
    - By loosening this bolt, locking mechanism for housing -9- is disengaged and housing may be removed from mounting bracket -5-
    - Bolt -11- must not be threaded in without lock cylinder housing installed. Lock mechanism ring may fall into door.
12. **Bolts**
    - Tighten to 20 Nm (15 ft-lb)

Door handle, removing and installing

EDITOR'S NOTE —

There are two seperate procedures based on year and location of production, see Forward at the front of book for VIN# information.

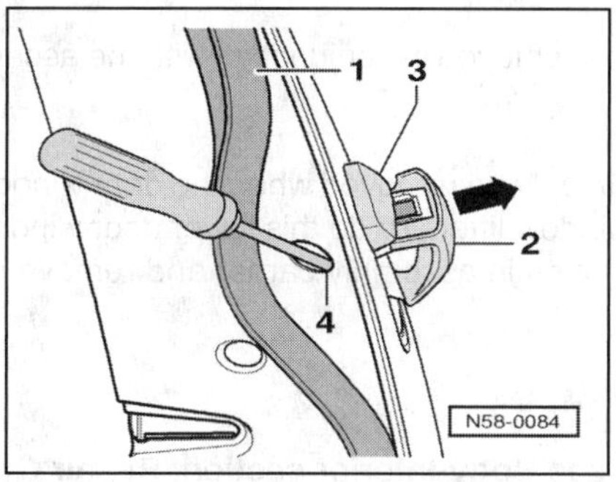

Removing

◄ Pull off door seal -1- in area of door handle.

– Pull door handle -2- in direction of arrow and hold. Unscrew bolt -4- onto stop using socket wrench T 10011 (T10072). This releases the lock cylinder housing.

NOTE —

Unscrewing the bolt too far may cause the locking ring to release from the support bracket and fall into the door.

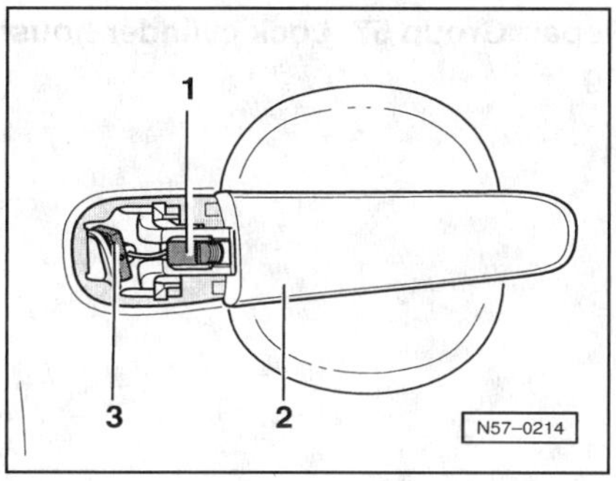

– Pull housing -3- at right angles to door, out from door handle bearing plate.

◄ Unclip clip -1- from door handle.

– Swing door handle out of door.

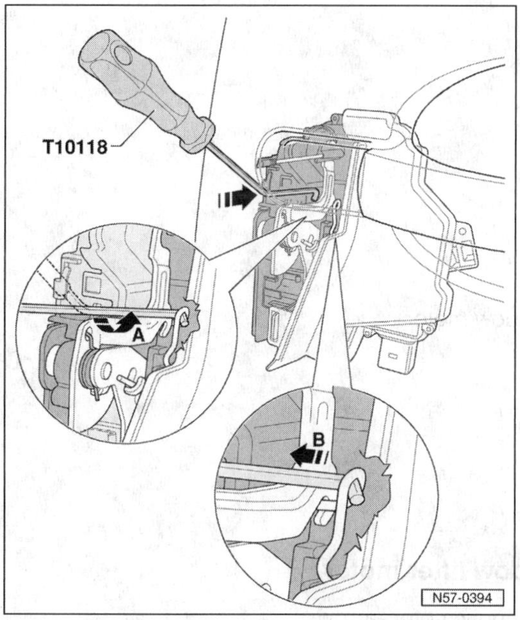

Door handle, installing on vehicles from (VIN) 3B XE 060 000, and 3B XP 075 000

– Guide assembly tool T 10118 into door through opening in the inner door plate.

– Illuminate the inner part of door using a flashlight for better view.

◄ Hook assembly tool into spring (arrow -A-).

• Engage spring into door lock by pulling assembly tool (arrow -B-).
• Activation lever is now locked.

– Swing door handle into door.

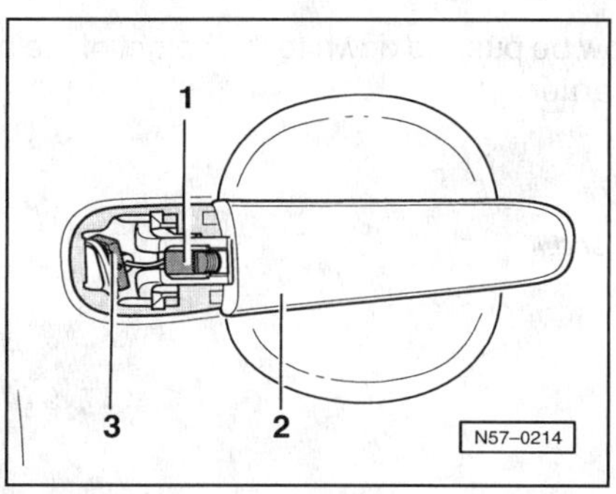

◄ Pull clip -1- into panel cut out and engage in door handle -2-.

NOTE —

When assembling clip, do not pull lock operating lever -3- and press door handle -2- onto door panel.

– Removing housing, see removing lock cylinder housing, see **Repair Group 57, Lock cylinder housing, removing and installing**.

• Perform a functional check as an incorrectly fitted and adjusted Bowden cable will prevent the door from opening.

Door lock, removing and installing

The window lifters, the door lock and the loudspeaker are secured to the assembly carrier.

The door lock can only be removed in conjunction with the assembly carrier.

The assembly carrier can only be removed when the door window is released from electric window lifter. To do this lower door window to the height of the assembly holes in assembly carrier and remove spreader pins.

Removing

– Remove rear door trim, see **Body-Interior section, Repair Group 70**

– Remove housing, see **Repair Group 57, Lock cylinder housing, removing and installing**

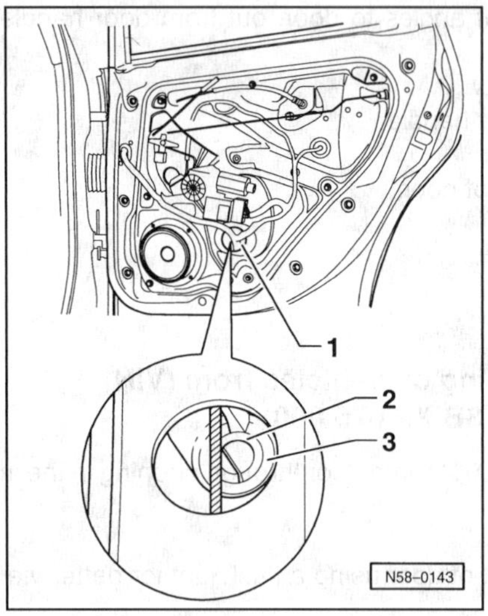

◄ Lever out cover cap -1-.

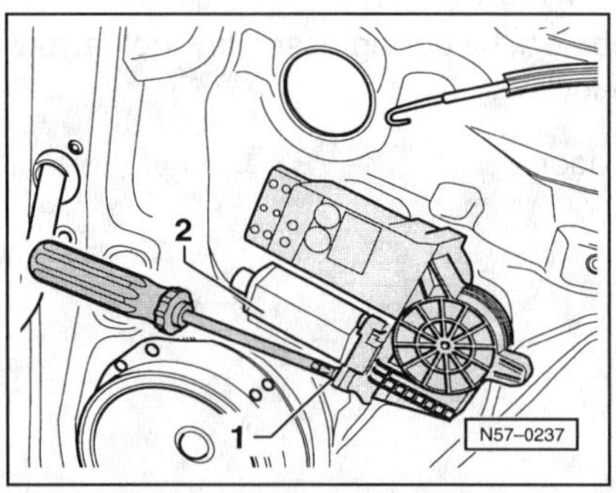

◄ Unscrew bolts -1- (Qty. 4).

– Pull housing -2- off window motor.

Bosch electric window lifter motor

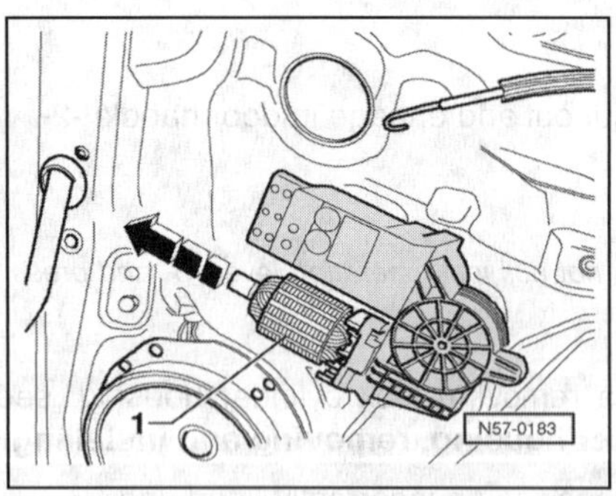

◄ Pull armature -1- off window motor -arrow-.

– The door window can now be pushed down to the height of the assembly holes in the assembly carrier.

NOTE *—*
The motor must not be reused.

Siemens electric window lifter motor

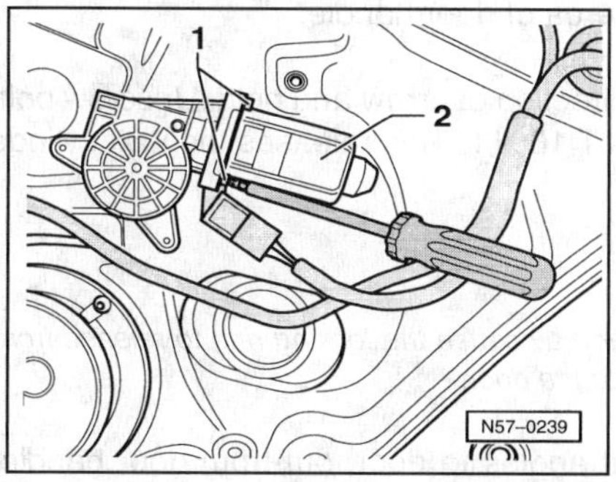

- Unscrew bolts -1- (Qty. 4).
- Pull housing -2- off window motor.

NOTE—
The motor must not be reused.

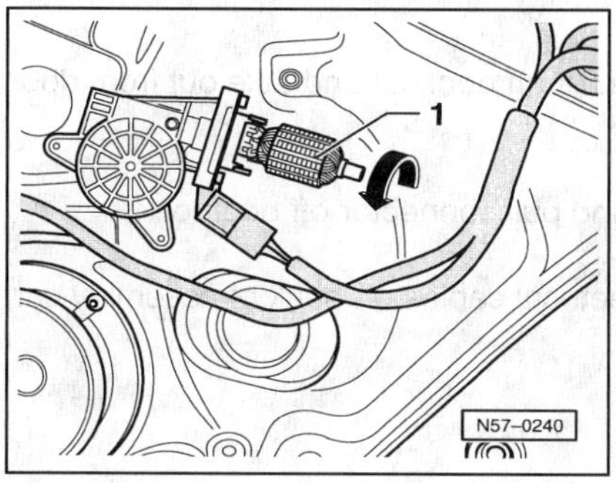

- The door window can be lowered down to the height of the assembly holes by turning the armature -1-.

All vehicles

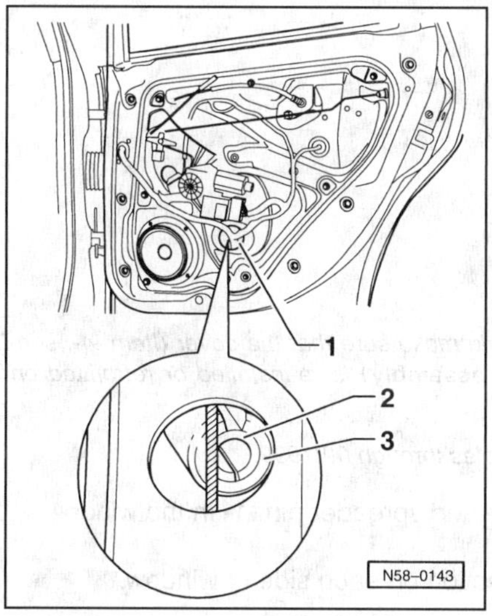

- Lower door window until spreader pin -2- and spreader plug -3- are accessible in window lifter cut-out.

NOTE—
When pressing out the spreader pin and the spreader plug ensure that the door outer panel is not damaged.

- Screw a 5 mm bolt (approx. 70 mm (2.76 in.) long) into the spreader pin -2- and pull out from spreader plug -3-.
- Now screw an 8 mm bolt (approx. 80 mm (3.15 in.) long) into spreader plug -3-.

NOTE—
Do not exert to great a pressure on the plug otherwise the plug will fall into the door.

- Pull spreader plug out from window guide and subsequently out from door window.
- When doing this press window lifter cable -2- to side with a screwdriver.
- Push window upward and secure in position (e.g. with adhesive tape).

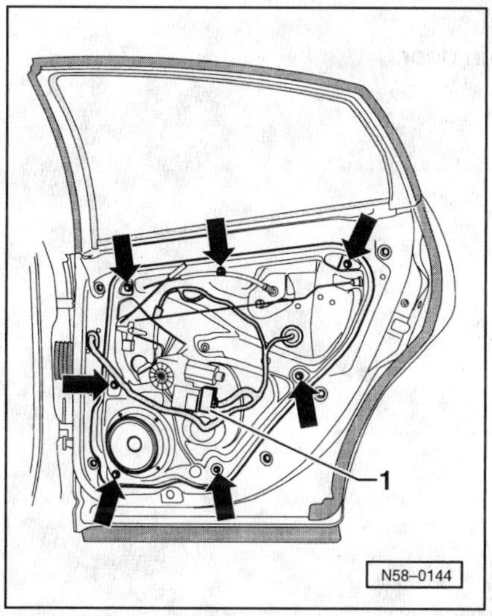

- Separate connector -1-.
- Unscrew bolts -arrows-.

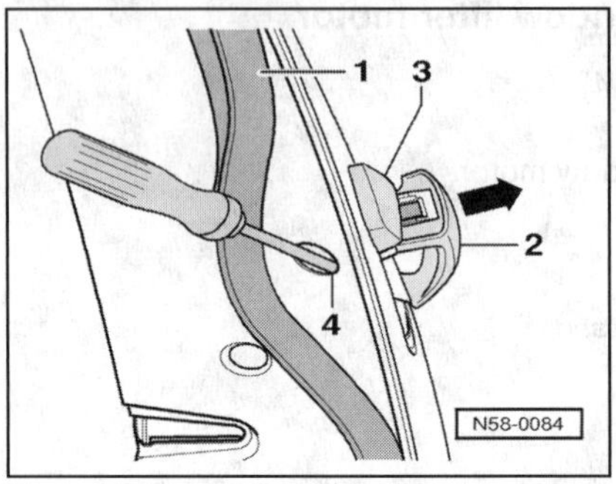

Pull off door seal -1- in area of door handle.

– Pull door handle -2- in direction of arrow and hold. Unscrew bolt -4- onto stop using socket insert T 10011. This releases the lock cylinder housing.

NOTE —
Unscrewing the bolt too far may cause the locking ring to release from the support bracket and fall into the door.

– Pull housing -3- at right angles to door, out from door handle bearing plate.

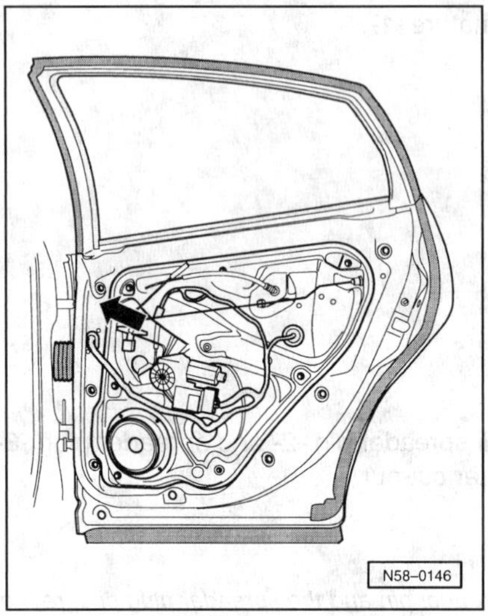

Pull top of assembly carrier off door, lift and take out from door towards hinges -arrow-.

– Turn assembly carrier and pull connector off door lock.

– Then unclip clips for electrical cables on back of assembly carrier.

Door lock, installing

NOTE —
- *When replacing door lock, make sure that the cover (Item -4-, see* **Door handle and door lock, assembly***) is re-installed or retrofitted on door lock.*
- *Retrofit cover for all vehicles through 09/1999.*

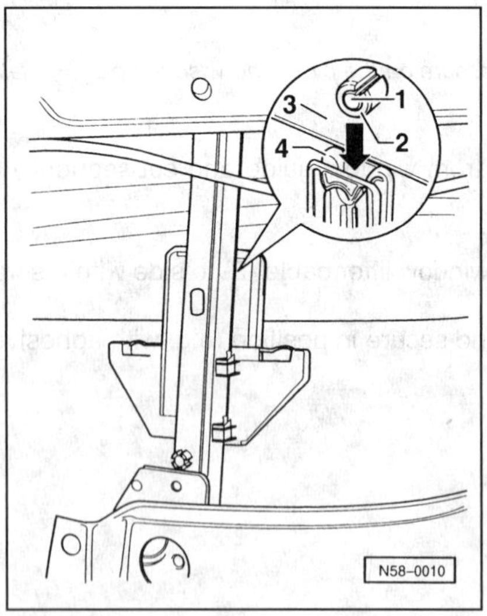

Insert spreader plug -2- and spreader pin -1- in the window.

– Set spreader plug projection on each side of window.

– Guide window into door.

– Insert assembly carrier in door.

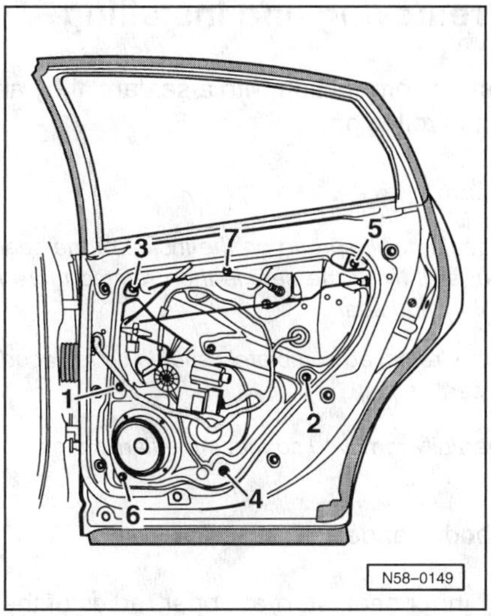

◀ Fit bolts and tighten in the prescribed sequence.

- Tightening torque: 8 Nm (71 in-lb).

– Install door window, see **Door window, fitting to window lifter**.

– Install remainder in reverse order or removal, then perform functional check.

Door window, fitting to window lifter

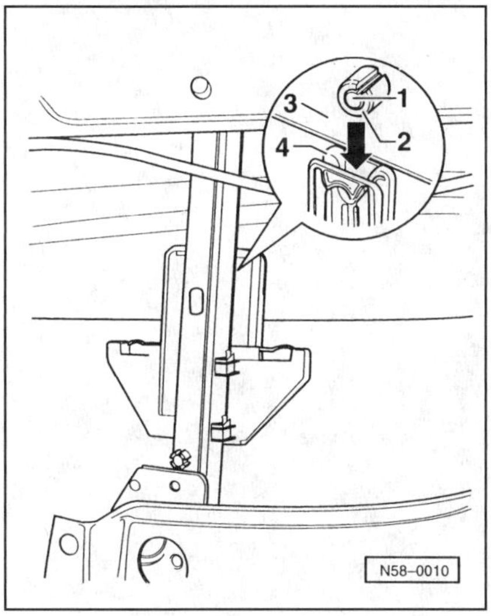

◀ Insert spreader plug -2- and spreader pin -1- in the window.

– Set spreader plug projection on each side of window.

– Guide window into door.

– Insert door -3- in the window lifting rail guide slot -4-.

– Lightly tap top of window, so that the window locates in lifting rail -arrow-.

Inner release handle, removing

– Remove rear door trim, see **Body-Interior section, Repair Group 70;**

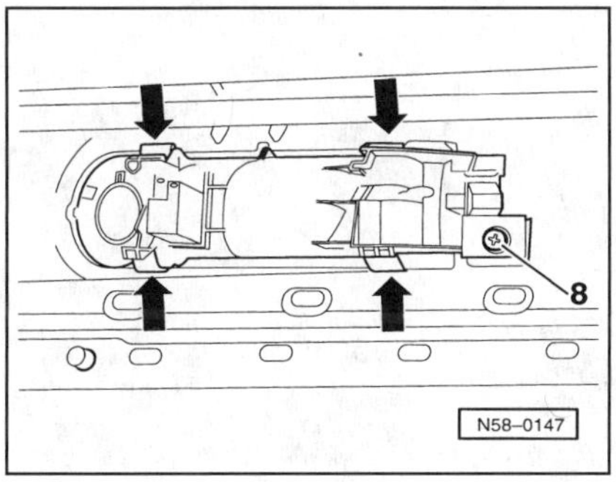

◀ Remove cross head screw -1-.

– Unclip door inner release -arrows- and take out from door trim.

Door inner seal, removing and installing

Door seals are fitted in series production with a sealant, they are placed on the door flange and then rolled on.

NOTE —

- *If they are removed the sealant will spread on the inside of the seal and the flanks will be bent out slightly. If the seal is then refitted there is no guarantee that it will fit tightly and seal.*
- *Therefore each seal which is removed completely should be replaced by a so called "hammer blow" seal.*
- *When seals have been partially removed squeeze the flanks together before refitting.*

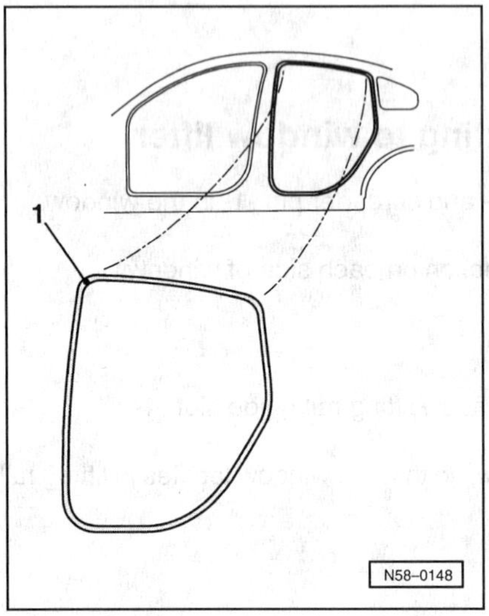

◄ Pull door inner seal off body flange.

– When installing the door inner seal, start at upper radius of the door cut-out -1-.

60 Sunroof

Repair Group

T 60

Function

With ignition switched on the sliding/tilting sunroof is opened and closed by turning the rotary switch.

After switching off the ignition the siding/tilting sunroof can be opened or closed until either the driver's or front passenger's door is opened.

On the rotary switch there are marks for preset opening positions of the glass panel. Any other intermediate opening sliding position can be selected by turning the switch clockwise.

The sliding/tilting sunroof is equipped with a excess force limitation system (roll-back) for the sliding function. The roof will open automatically when contacting an obstruction during the closure sequence.

In addition there is also an emergency closure function. For problems in closing the sunroof, the sunroof can be compulsory closed by pressing the preselector in position "close sunroof".

The "roll back" function is disengaged when using the compulsory closure function.

The sliding/tilting sunroof drive is protected against overheating by a time limiter. The protective mechanism operates when the sunroof is operated continuously for approx. 2 minutes. It is functional again after a cooling-off phase.

Sliding/tilting roof with glass panel (through 07/1998)

Sliding/tilting roof with glass panel assembly, (through 07/1998)

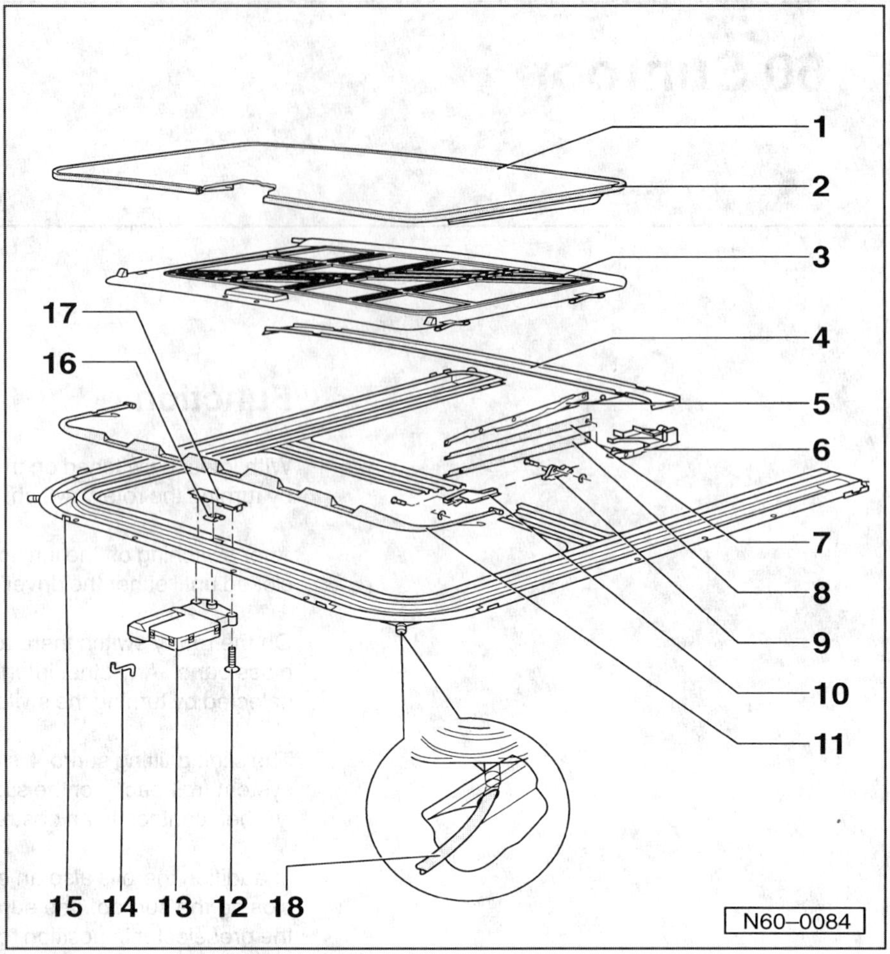

1. **Glass panel for sliding/tilting roof (single pane safety glass)**
   - Removing sliding/tilting roof glass panel, see **Sliding/tilting roof glass panel, removing**
   - Installing sliding/tilting roof glass panel, see **Sliding/tilting roof glass panel, installing**
   - Adjusting sliding/tilting roof glass panel, see **Sliding/tilting roof glass panel, adjusting height**
2. **Panel seal**
   - See **Panel seal, adjusting**
   - Seal joint central in rear area
   - Treat panel seal clamping channel with lubricant (soap solution) before fitting
3. **Sliding headlining**
   - See **Sliding headlining, removing**
4. **Water channel**
   - Remove glass panel for sliding/tilting roof
   - Press clip off forwards and pull water channel off upwards
5. **Guide rail**
   - See **Guide, removing**
6. **Trim frame**
   - Removed together with guide rail
7. **Upper trim**
   - See **Sliding/tilting roof glass panel, removing**
8. **Lower trim**
   - See **Sliding/tilting roof glass panel, removing**
9. **Center guide**
   - Removed together with guide rail
10. **Front guid**
    - Removed together with guide rail.
11. **Wind deflector**
    - See **Wind deflector, removing**
12. **Countersunk screws (micro encapsulated)**
    - Always use new countersunk screws
    - Tighten to 3.5 Nm (31 in-lb)
13. **Electric drive**
    - See **Drive for sliding/tilting roof, removing**
    - Checking cable drive (0 position), see **Drive for sliding/tilting roof, adjusting ("0" position)**
14. **Hexagon key for emergency operation**
    - Clipped into electric drive
15. **Carrier unit**
    - See **Sliding/tilting roof carrier unit, removing**
    - U frame with guide channels, if required, guide channels are to be greased with special grease G 000 450 02 only, otherwise correct operation cannot be guaranteed.
16. **Spring plate**
17. **Pinion cover**
18. **Water drain hose**
    - See **Water drain hoses, cleaning**

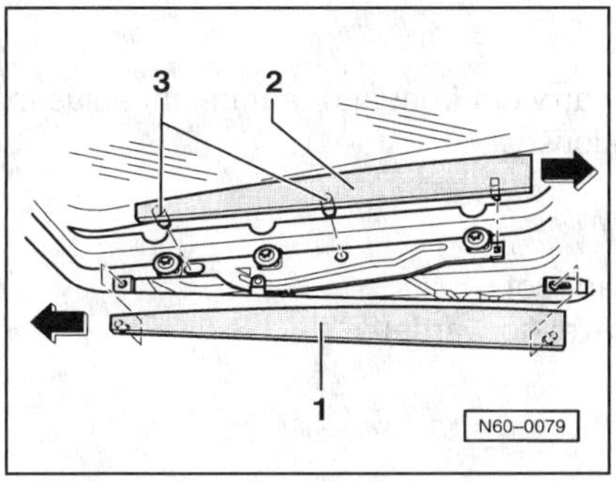

Sliding/tilting roof glass panel, removing

– Tilt sliding/tilting roof open.

– Slide sun blind to rear.

Unclip lower trim -1- in forward area, inwards to center of vehicle. Slide forwards out of elongated holes and take off.

– Unclip front and center of upper trim -2- (pins -3-) and take off in direction of arrow.

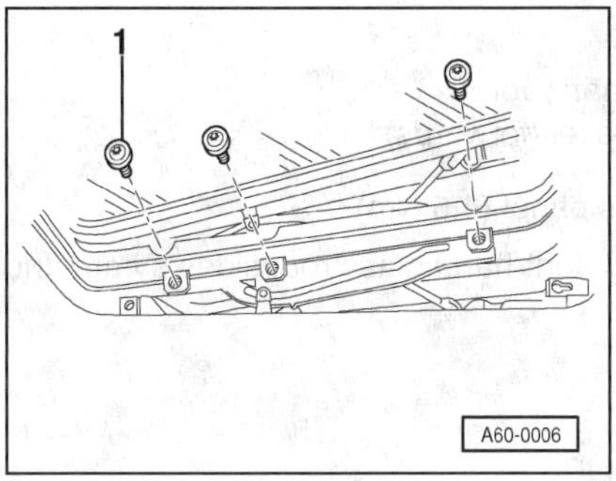

Remove securing screws -1- (Torx key T 25).

– Take sliding/tilting roof panel out upwards.

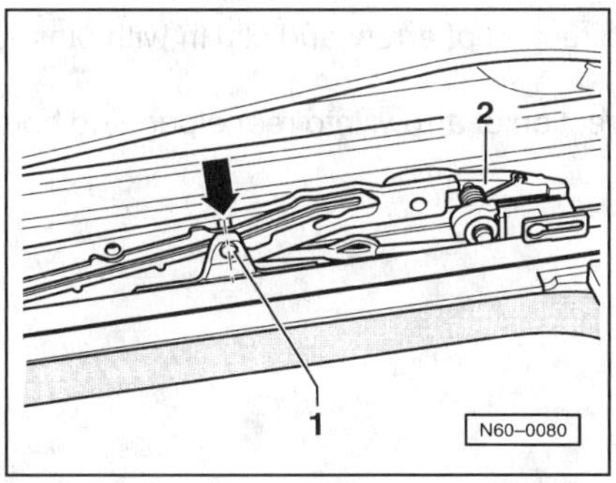

Sliding/tilting roof glass panel, installing

Panel must be installed in "0 position" (panel closed).

"0" position:

The guide pins -1- must lie between the markings on the guide rail -arrow-.

- Locating hooks -2- must be located in the guide rails.
- If this is not the case, adjust parallel running, see **Parallel running, adjusting**.

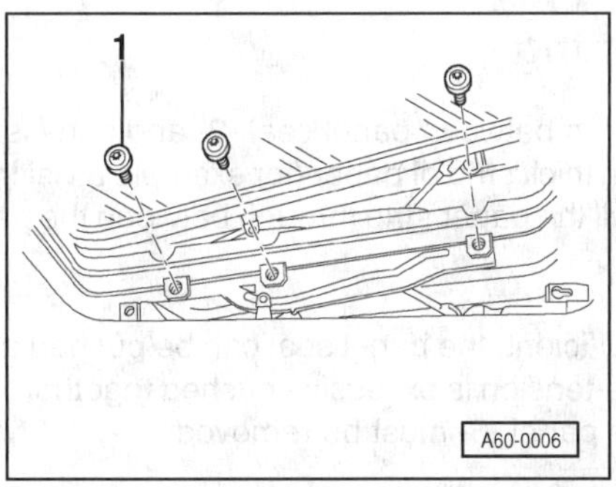

Place sliding/tilting roof on from above and screw in securing screws (guide rail/panel).

– Lightly tighten securing screws (Torx T 25).

NOTE —
After adjusting panel height tighten securing screws to 4.5 Nm (39 in-lb).

Sliding/tilting roof glass panel, adjusting height

Sliding/tilting roof "0" position OK.

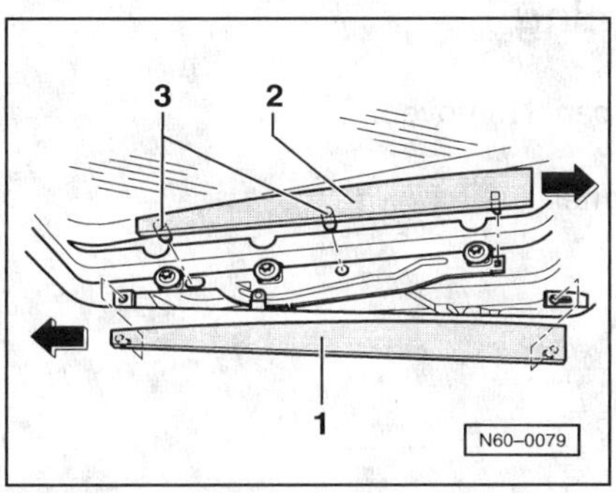

– Tilt glass panel open.

– Fully push back sliding headlining.

Unclip lower trim -1- in forward area, inwards to center of vehicle. Slide forwards out of elongated holes and take off.

– Unclip front and center of upper trim -2- (pins -3-) and take off in direction of arrow.

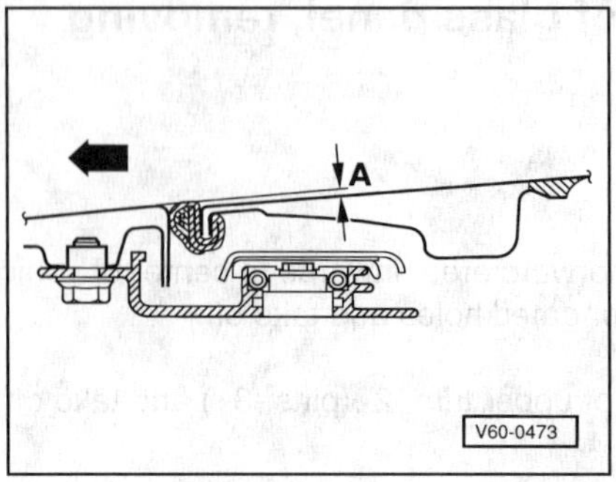

– Close glass panel.

– To prevent wind noises carry out front and rear height adjustment in accordance with illustrations.

Panel adjustment front:

- a = 0....1 mm lower than roof
- Arrow = Direction of travel (forwards)

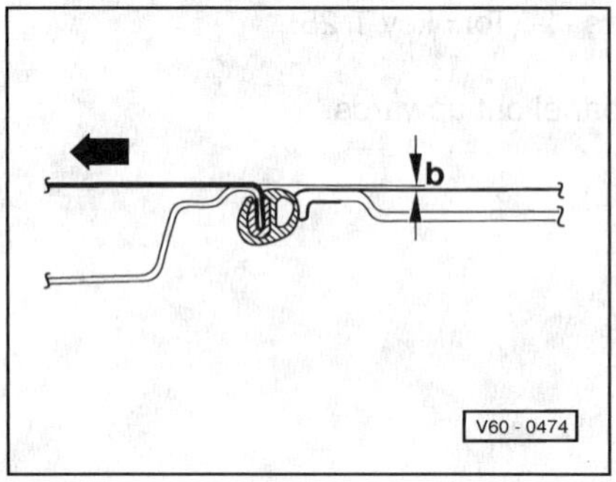

Panel adjustment rear:

- b = 0....1 mm higher than roof
- Arrow = Direction of travel (forwards)

– Tighten screws for glass panel (4.5 Nm).

- Always tighten left and right-hand glass roof sides symmetrically.

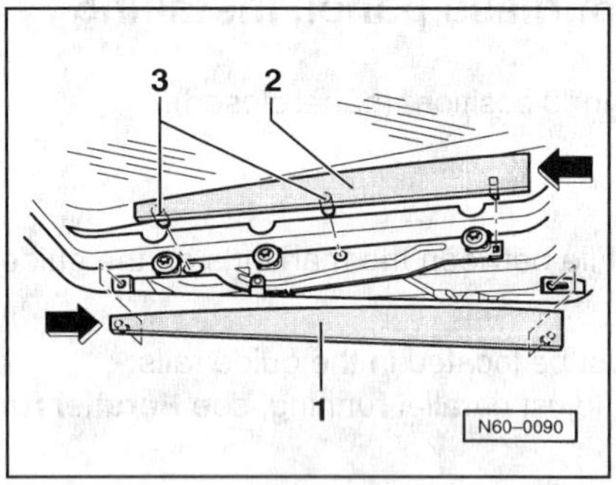

Fit on upper trim -2- in direction of arrow and clip in with pins -3-.

– Slide lower trim -1- in direction of arrow, into rear elongated hole and clip front section in.

Panel seal, adjusting

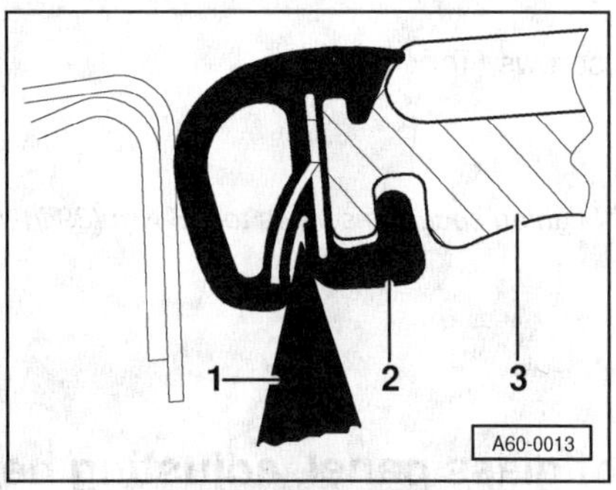

Check that the pre-tension between panel seal -2- and body is uniform all-round using a 0.3 mm thick strip of paper (for example a visiting card). It must be possible to pull the paper strip through between the panel seal and body under tension.

– If the pre-tension is insufficient, the panel seal can be pushed apart with a wedge -1-, or if the pre-tension is excessive pushed together. To adjust the panel seal the glass panel -3- must be removed.

Panel seal, replacing

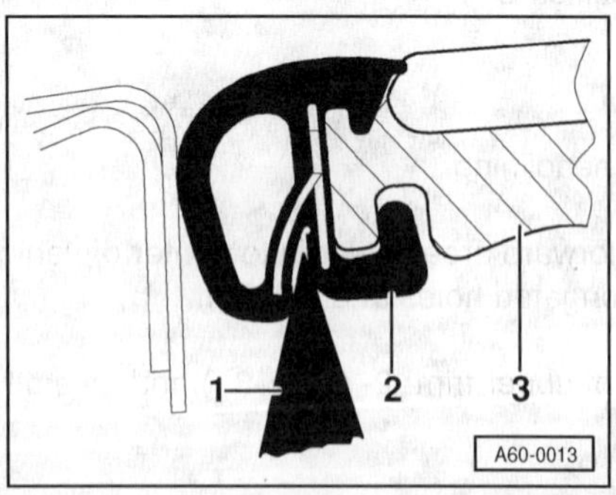

Sliding/tilting roof glass panel removed.

Pull seal -2- off glass panel.

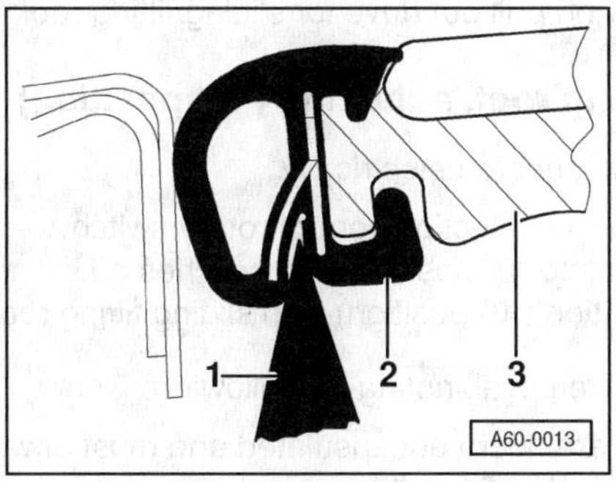

Pull new seal onto glass panel -3- from bottom to top starting at center of rear.

NOTE —
To ease fitting of seal, coat panel edge with a soapy solution.

– Cut panel seal to length after installing.

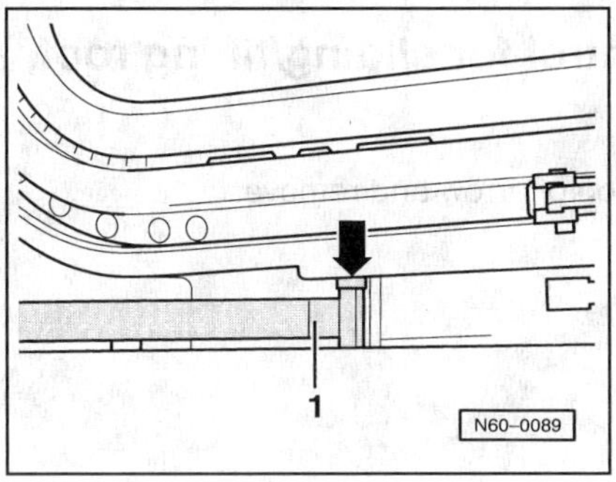

Wind deflector, removing

– Sliding/tilting roof completely open.

Press wind deflector -1- down and at same time pull out of both side guides -2- in direction of arrow.

– Take out wind deflector.

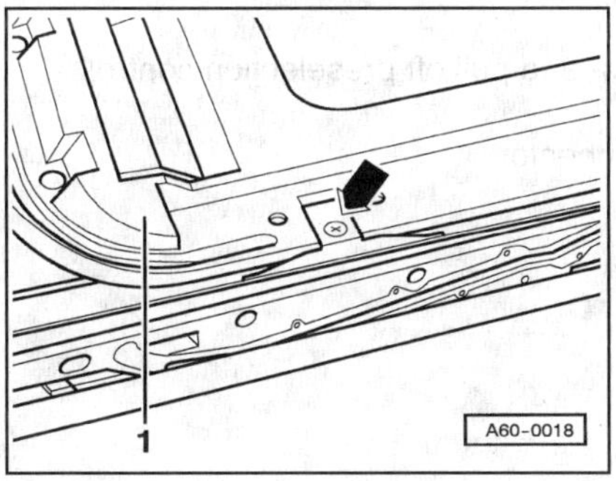

Sliding headlining, removing

Sliding/tilting roof glass panel removed.

– Slide sun blind to rear.

Remove forward bolts -arrow- on left and right-hand sides. Pull out slide forwards over the stop -1-.

– Slide sun blind forwards.

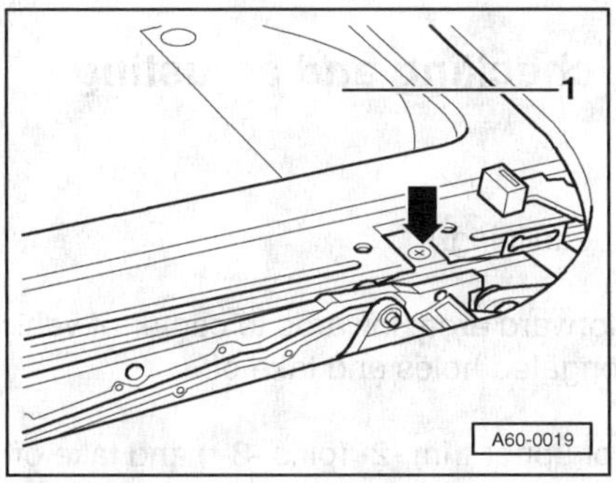

Remove rear bolts -arrow- on left and right-hand sides. Pull out slide to rear and take out sun blind -1-.

Drive for sliding/tilting roof, removing, adjusting, and installing

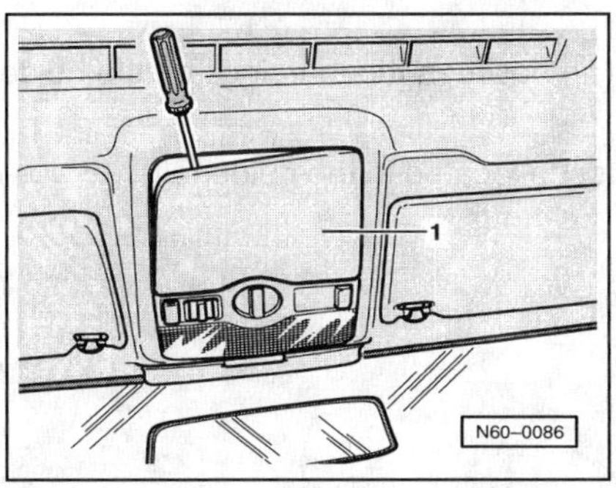

– Disconnect battery Ground (GND) strap from battery.

Unclip cover -1- for drive.

NOTE —
Remove and install drive for sliding/tilting roof only when roof is closed ("0" position).

– Unclip and separate connector.

Remove bolts -arrows- and pull out drive for sliding/tilting roof.

Drive for sliding/tilting roof, adjusting ("0" position)

– Drive removed but still connected electrically.

- Select "close roof" with preselection control rotary switch.
- Drive automatically runs to "0" position and switches off.
- Install drive in this position ("0" position) with sliding/tilting roof closed.

– Installation is reverse of removal, noting the following:

- The bolts for the drive are micro encapsulated and must always be replaced. Tighten to 3.5 Nm (31 in-lb).

Preselection control for sliding/tilting roof, removing

Unclip trim -1- in direction of arrow and remove.

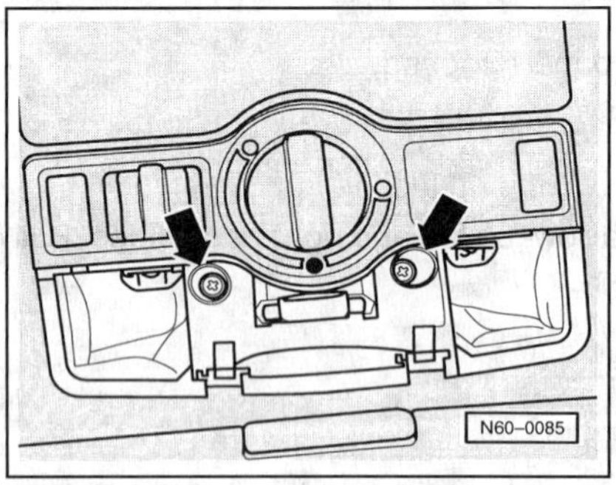

Remove screws -arrows- and pull off preselection control.

– Unclip and separate connector.

Parallel running, checking and adjusting

– Tilt glass panel open.

– Fully push back sliding headlining.

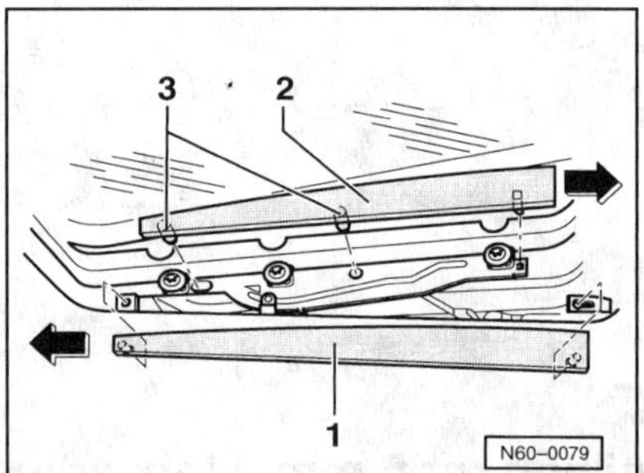

Unclip lower trim -1- in forward area, inwards to center of vehicle. Slide forwards out of elongated holes and take off.

– Unclip front and center of upper trim -2- (pins -3-) and take off in direction of arrow.

– Close glass panel.

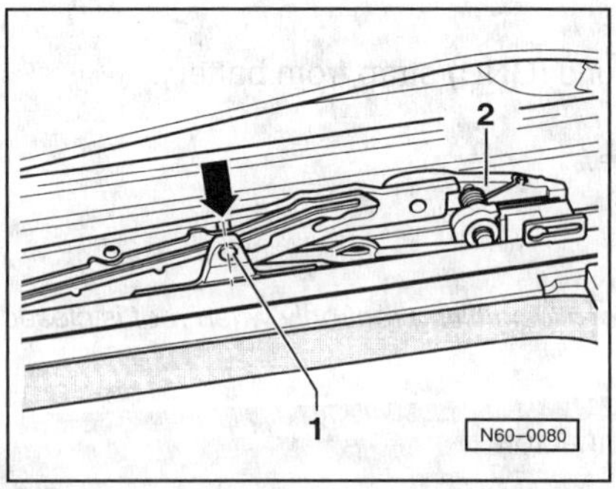

The guide pins -1- must lie between the markings on the guide rail -arrow-.

- Locating hooks -2- must be located in the guide rails.

Parallel running, adjusting

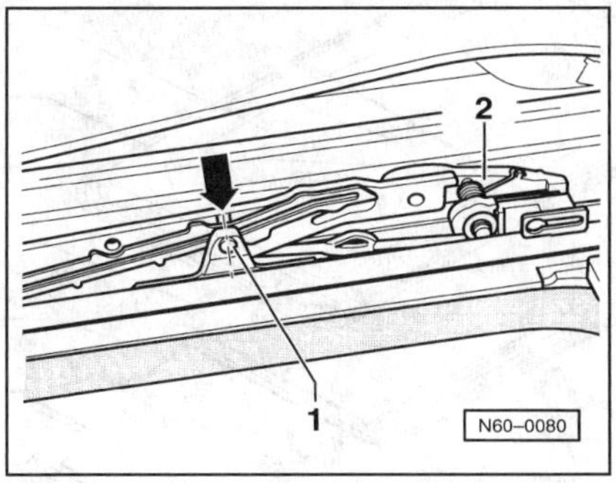

NOTE—
The parallel running adjustment can only be undertaken when the drive and glass panel are removed ("0" position).

– Slide guides from front to rear only.

The guide pins -1- must lie between the markings on the guide rail -arrow-.

• Locating hooks -2- must be located in the guide rails.

– Install drive in this position ("0" position).

• Then check "0" position.

Sliding/tilting roof carrier unit, removing

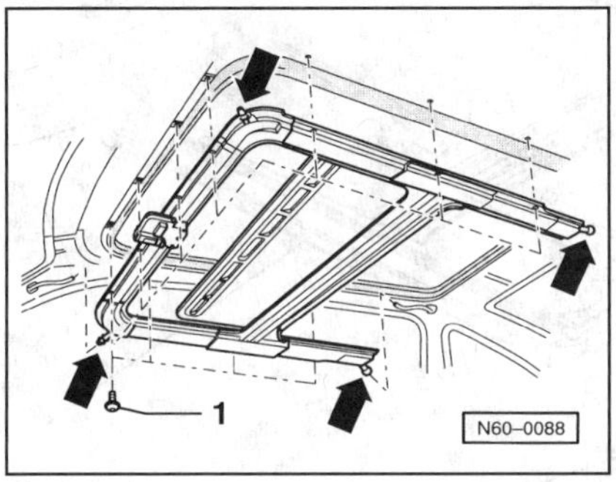

Headliner removed.

– Disconnect battery Ground (GND) from battery.

– Pull connections off drive.

Pull water drain hoses -arrows- from assembly unit.

– Unscrew bolts -1- and take assembly unit, with help of a second person, out from the vehicle.

Installing assembly unit

– Align assembly unit within roof cut-out (use centralizing holes).

– Tighten bolts -1- to 8 Nm (71 in-lb) from front to rear on left and right-hand sides starting at drive, in sequence.

Guide, removing

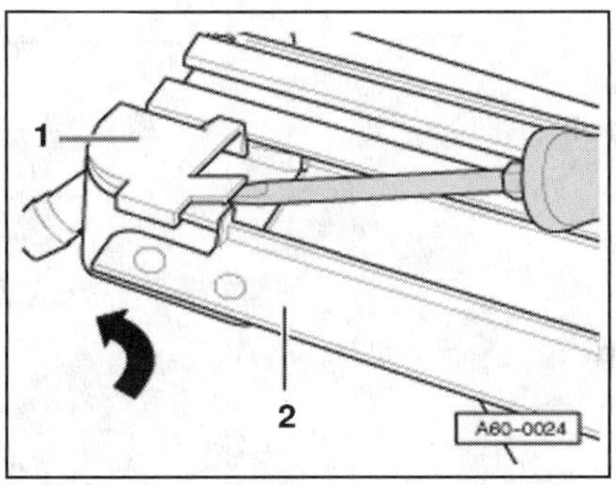

Assembly unit for sliding/tilting roof removed.

Pry cover cap -1- off guide rail -2- and pull off to rear in direction of arrow.

– Remove rear stop on guide rail.

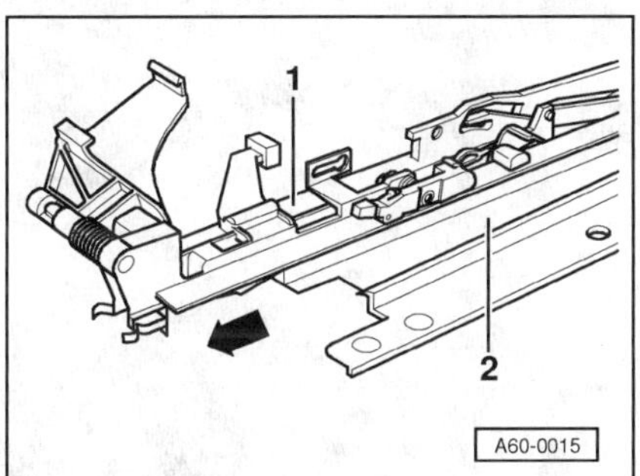

Pull rear guide -1- with center guide and front guide with cable out of guide rail -2- in direction of arrow.

NOTE—
Always replace guide with cable as a pair.

Sliding/tilting sunroof with glass panel (from 08/1998)

Sliding/tilting sunroof with glass panel, assembly (from 08/1998)

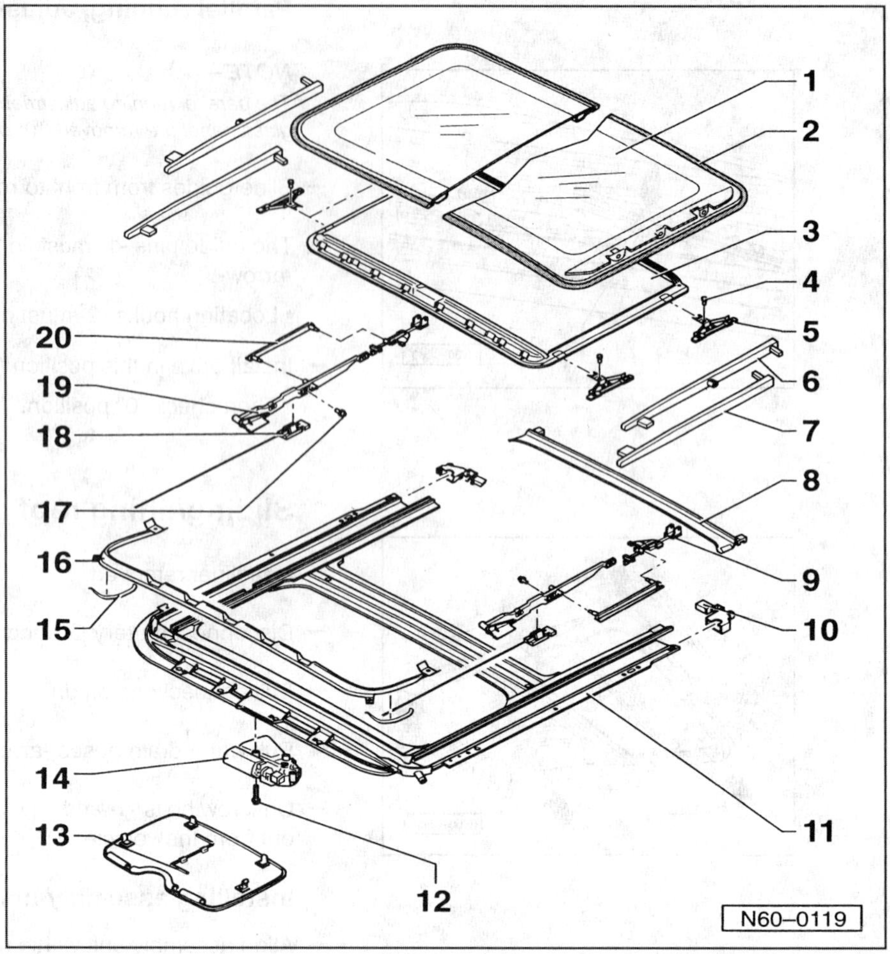

1. **Glass panel for sliding/tilting roof (single pane safety glass)**
   - Removing sliding/tilting sunroof glass panel, see **Sliding/tilting sunroof glass panel, removing**
   - Installing sliding/tilting sunroof glass panel, see **Sliding/tilting sunroof glass panel, installing**
   - Adjusting sliding/tilting roof glass panel, see **Sliding/tilting roof glass panel (height adjustment)**
2. **Panel seal**
   - See **Panel seal, adjusting**
   - Seal joint located centrally at rear
   - Replacing panel seal, see **Panel seal, adjusting**
3. **Sliding headlining**
   - Removing and installing sliding headlining, see **Sliding headlining, removing**
4. **Countersunk screws (micro encapsulated)**
   - Always use new countersunk screws
   - Tighten to 3.5 Nm (31 in-lb)
5. **Slide**
6. **Upper trim**
   - See **Sliding/tilting sunroof glass panel, removing**
7. **Lower trim**
   - See **Sliding/tilting sunroof glass panel, removing**
8. **Water channel**
   - Removed together with guide rail.
9. **Guide locking hook**
10. **End piece**
    - Use Butyl adhesive sealing cord AKL 450 005 05 to seal
11. **Carrier unit**
    - See **Carrier unit, removing and installing**
    - If required, guide channels are to be greased with special grease G 000 450 02 only, otherwise correct operation cannot be guaranteed.
12. **Bolt**
13. **Hexagon key for emergency operation**
    - Clipped into cover for electric drive
14. **Electric drive**
    - See **Drive for sliding/tilting sunroof, removing**
    - Adjusting drive (0 position), see **Drive for sliding/tilting sunroof, adjusting ("0" position)**
15. **Spring**
16. **Wind deflector**
    - See **Wind deflector, removing**
17. **Torx screw**
    - Torx key T 25
    - Tighten to 4.5 Nm (39 in-lb)
18. **Rear guide**
    - Removed together with guide rail.
19. **Guide rail**
    - See **Guide, removing**
20. **Spacer**
    - Removed together with guide rail.

Sliding/tilting sunroof glass panel, removing

– Slide sun blind to rear.

– Tilt sliding/tilting roof open.

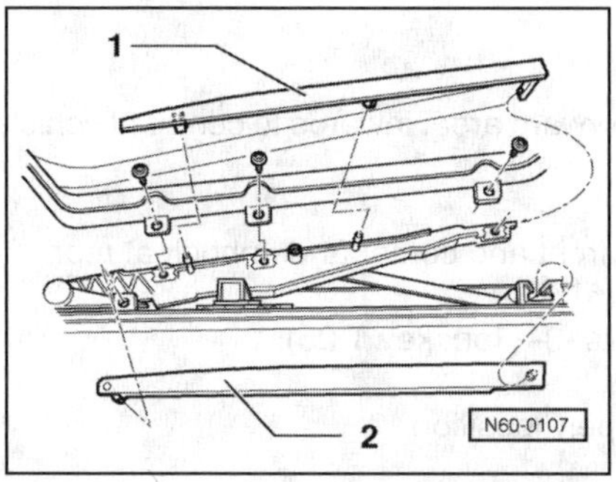

Unclip lower trim -2- in forward area, inwards to center of vehicle and unhook at rear.

– Unclip upper trim -1- at front and center and unhook at rear.

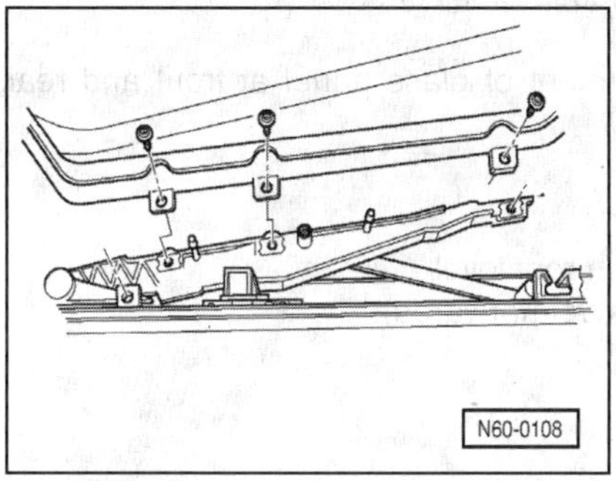

Remove securing screws (Torx insert T 25).

– Take sliding/tilting roof panel out upwards.

NOTE—

The sliding/tilting sunroof must not moved to position "Open" when removed, because the water channel will not be compressed by the glass panel and can cant inside the roof.

Sliding/tilting sunroof glass panel, installing

Panel must be installed in "0 position" (panel closed).

"0" position:

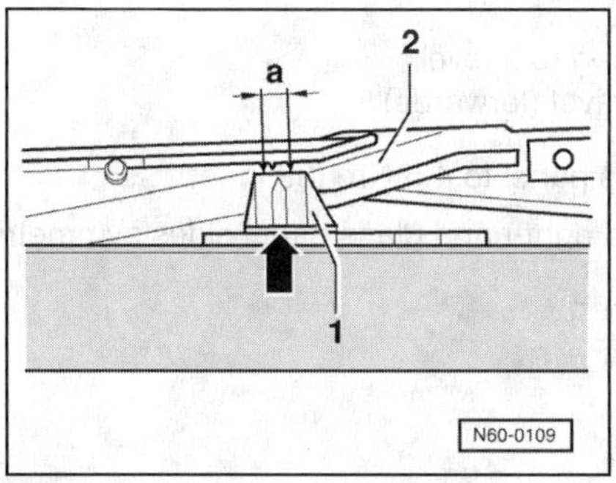

The mark -arrow- on the rear of the guide upper section -1- must, on both sides, lie within the marks -dimension a- on the guide rail gate -2-.

- The guide rail -2- must be located in the guide rails (cannot be moved by hand).
- If this is not the case, adjust parallel running, see **Parallel running, adjusting**.

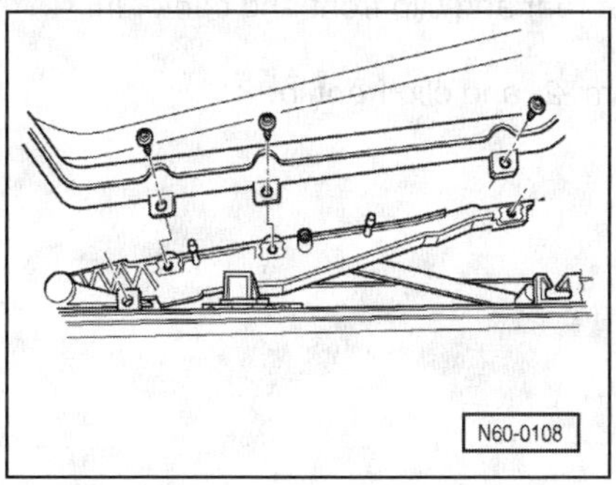

Place sliding/tilting roof on from above and screw in securing screws (guide rail/panel).

– Lightly tighten securing screws (Torx T 25).

NOTE—

After adjusting panel height tighten securing screws to 4.5 Nm (39 in-lb).

Sliding/tilting roof glass panel (height adjustment)

Sliding/tilting roof "0" position OK.

– Tilt sliding/tilting roof open.

– Slide sun blind to rear.

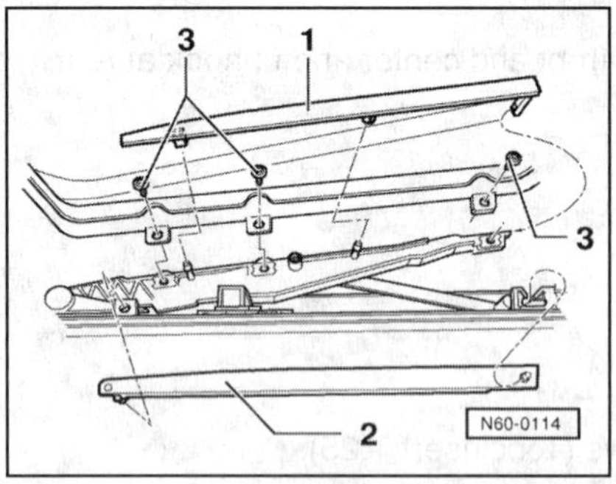

– Unclip lower trim -2- in forward area, inwards to center of vehicle and unhook at rear.

– Unclip upper trim -1- at front and center and unhook at rear.

– Remove securing screws -3- (Torx key T 25).

– Run glass panel into "open" position.

– Run glass panel into "closed" position.

– Carry out height adjustment of glass panel at front and rear on both sides as follows:

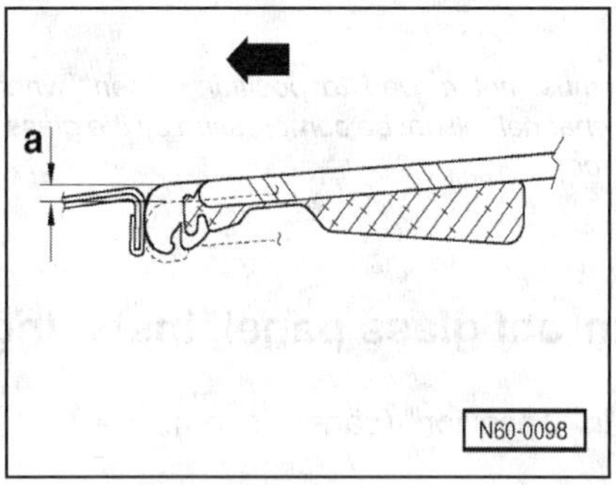

Panel adjustment front:

• a = 0...1 mm lower than roof level
• Arrow = Direction of travel (forwards)

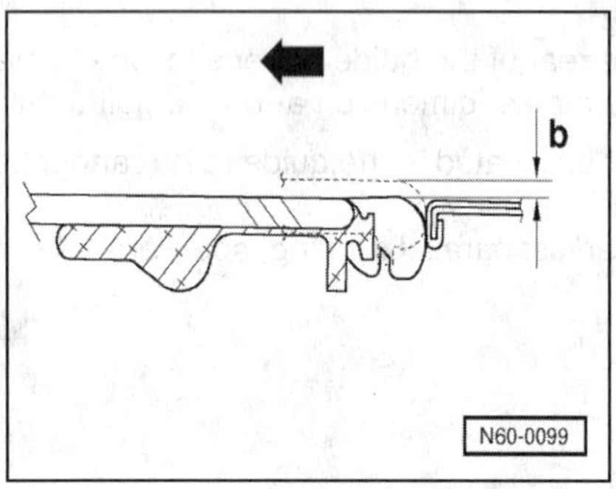

Panel adjustment rear:

• b = 0...1 mm higher than roof level
• Arrow = Direction of travel (forwards)

– Tighten screws for glass panel to 4.5 Nm (39 in-lb).

• Always tighten left and right-hand glass panel sides symmetrically.

– Tilt sliding/tilting roof open.

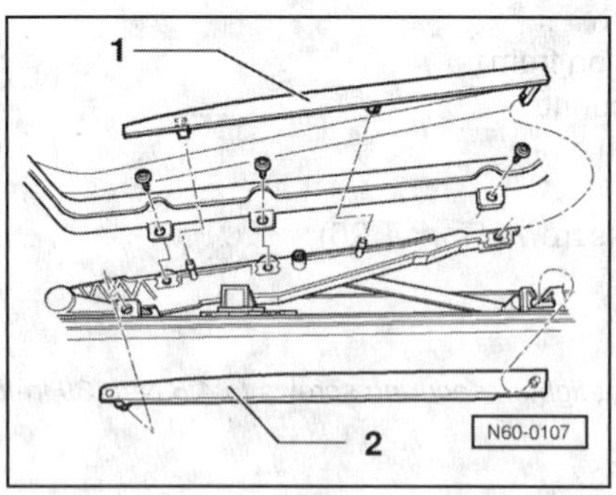

Hook in upper trim -1- at rear and clip front and center in.

– Hook in rear of lower trim -2- and clip front in.

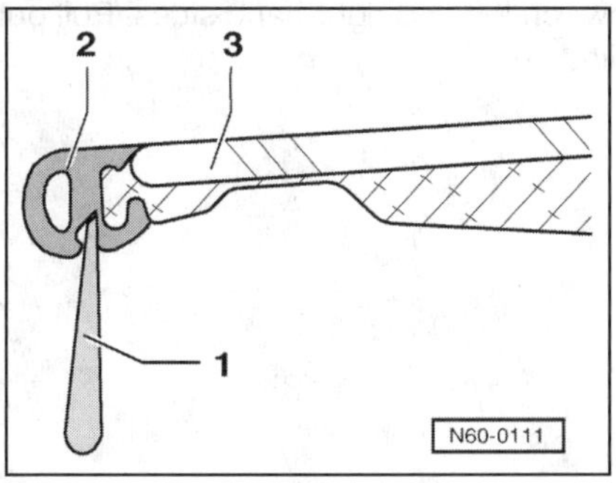

Panel seal, adjusting

Check that the pre-tension between panel seal -2- and body is uniform all-round using a 0.3 mm thick strip of paper (a business card for example). It must be possible to pull the paper strip through between the panel seal and body under tension.

– If the pre-tension is insufficient, the panel seal can be pushed apart with a wedge -1-, or if the pre-tension is excessive pushed together. To adjust the panel seal the glass panel -3- must be removed.

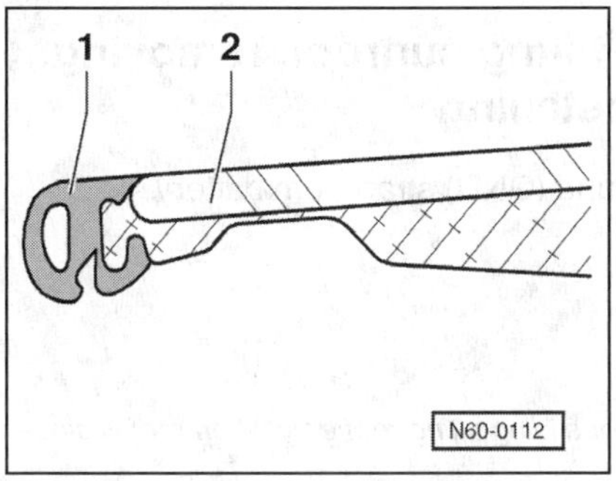

Panel seal, replacing

– Removing glass panel for sliding/tilting sunroof, see **Sliding/tilting sunroof glass panel, removing**.

Pull seal -1- off glass panel.

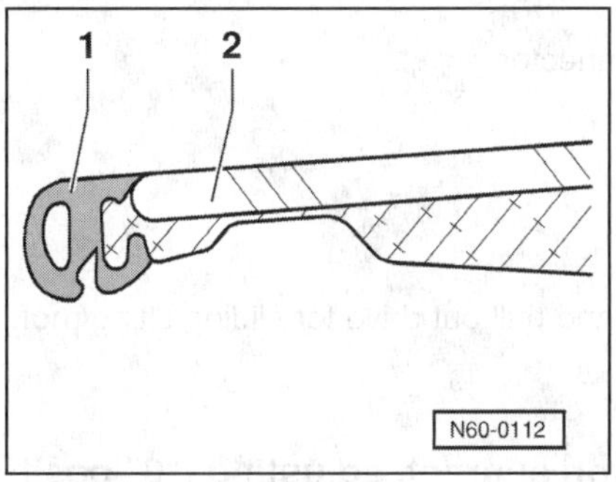

Pull new seal onto glass panel -2- from bottom to top starting at center of rear.

NOTE—
- *To ease fitting of seal, coat panel edge with a soapy solution.*
- *If the seal is not correctly installed the upper surface will be wavy.*
- *The panel seal must be shortened to suit the panel circumference.*

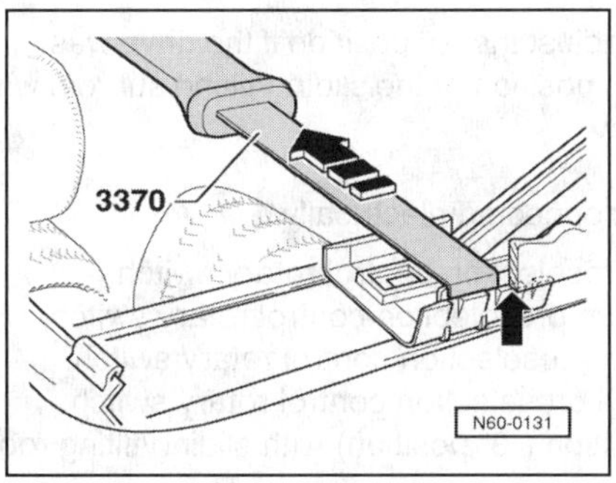

Wind deflector, removing

– Sliding/tilting roof completely open.

Insert special tool 3370 between edge of roof and wind deflector and release wind deflector side mountings from locking devices on left and right sides by pulling on special tool 3370.

– Lever wind deflector mounting out of guide rail (on left and right) with a screwdriver.

– Take out wind deflector.

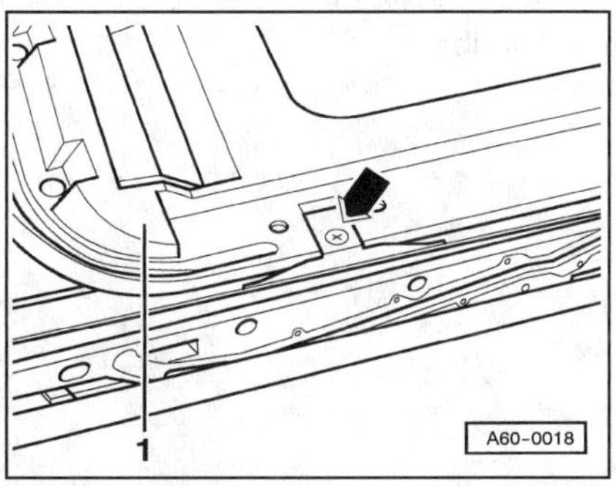

Sliding headlining, removing

– Removing glass panel for sliding/tilting sunroof, see **Sliding/tilting sunroof glass panel, removing**.

– Slide sun blind to rear.

Remove forward bolts -arrow- on left and right-hand sides. Pull out slide forwards over the stop -1-.

– Slide sun blind forwards.

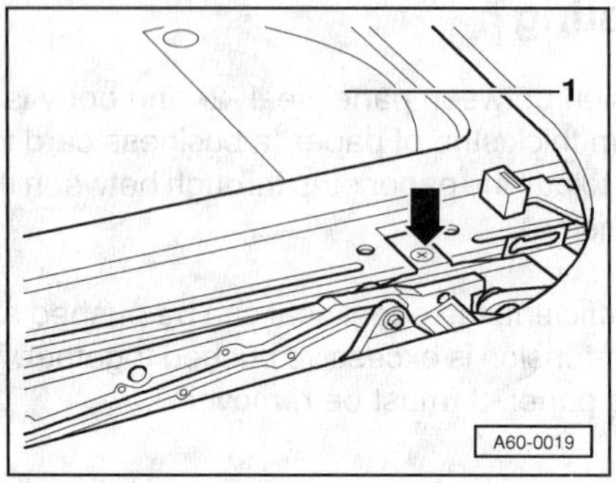

Remove rear bolts -arrow- on left and right-hand sides. Pull out slide to rear and take out sun blind -1-.

Drive for sliding/tilting sunroof, removing, adjusting, and installing

– Disconnect battery Ground (GND) strap from battery.

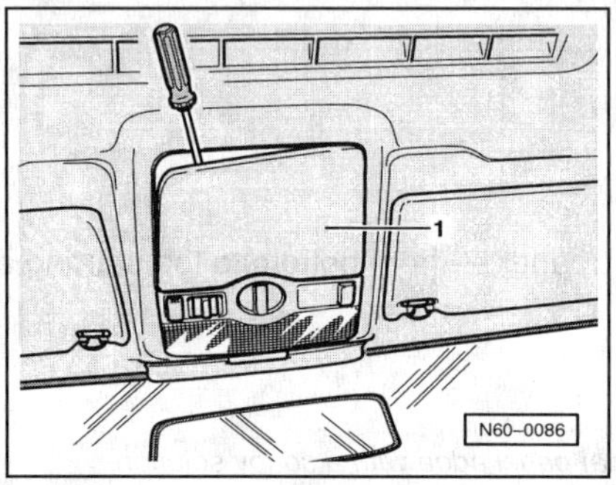

Unclip cover -1- for drive.

NOTE —

Remove and install drive for sliding/tilting roof only when roof is closed ("0" position).

– Unclip and separate connector.

Remove bolts -arrows- and pull out drive for sliding/tilting roof.

Drive for sliding/tilting sunroof, adjusting ("0" position)

It may be necessary to adjust the "0" position if the drive was removed when not in "0" position or the sliding/tilting sunroof was closed using the emergency key.

– Drive removed but still connected electrically.
  - Select "tilt roof" with preselection control rotary switch.
  - Select "closed roof" with preselection control rotary switch.
  - Select "open roof" with preselection control rotary switch.
  - Select "close roof" with preselection control rotary switch.
  - Install drive in this position ("0" position) with sliding/tilting roof closed.

– Installation is reverse of removal, noting the following:
  - The bolts for the drive are micro encapsulated and must always be replaced .
  - Tighten bolts to 3.5 Nm (31 in-lb).

Preselection control for sliding/tilting sunroof, removing

- ◄ Unclip trim -1- in direction of arrow and remove.

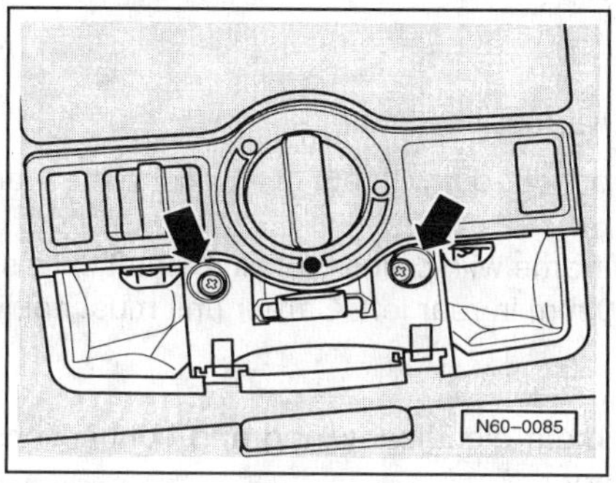

- ◄ Remove screws -arrows- and pull off preselection control.
- Unclip and separate connector.

Parallel running, checking

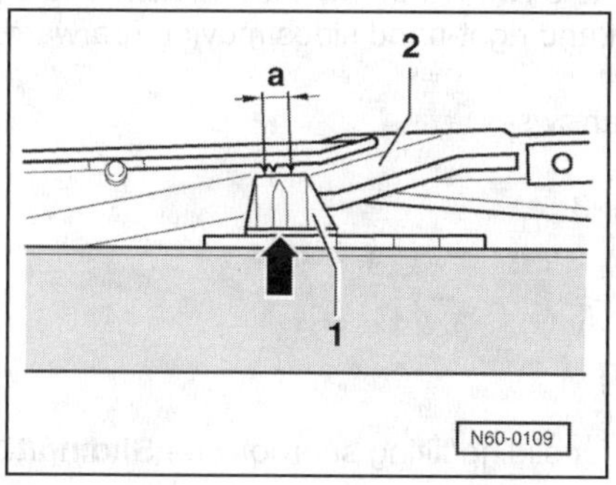

- Removing glass panel for sliding/tilting sunroof, see **Sliding/tilting sunroof glass panel, removing.**
- ◄ The mark -arrow- on the rear of the guide upper section -1- must, on both sides, lie within the marks -dimension a- on the guide rail gate -2-.
  - The guide rail -2- must be located in the guide rails (cannot be moved by hand).

Parallel running, adjusting

NOTE—

The parallel running adjustment can only be undertaken when the drive and glass panel are removed ("0" position).

- Removing sliding/tilting sunroof, see **Drive for sliding/tilting sunroof, removing**.

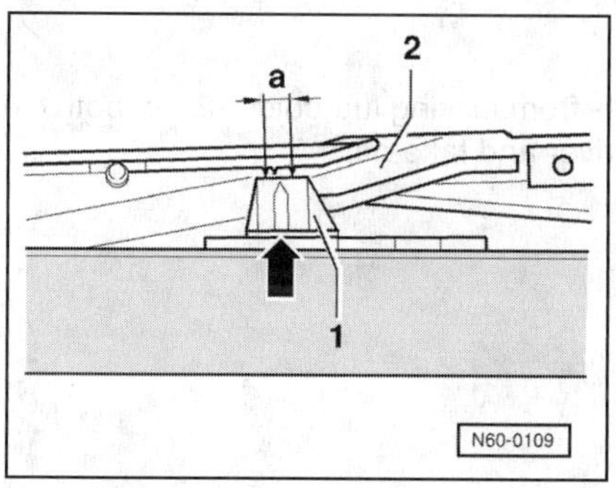

- ◄ The mark -arrow- on the rear of the guide upper section -1- must, on both sides, lie within the marks -dimension a- on the guide rail gate -2-.
  - The guide rail -2- must be located in the guide rails (cannot be moved by hand).
- Slide guide upper part -1- from front to rear only centrally between the marks.
- Install drive in this position ("0" position), then check "0" position.

Carrier unit, removing and installing

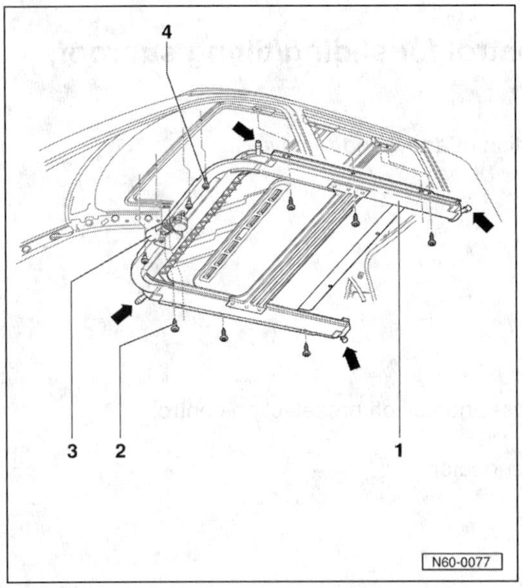

Removing

- Remove preselection control for sliding/tilting sunroof, see **Preselection control for sliding/tilting sunroof, removing**.
- Remove moulded headlining, see **Body-Interior section, Repair Group 70**
- Pull water drain hoses -arrows- off carrier unit -1-.
- Remove bolts -2- and -4- (with sleeves) and lift carrier unit out of vehicle with help of second mechanic.

Installing

- Fit carrier unit -1- into roof cut-out with aid of a second mechanic.
- Align carrier unit in roof frame with cylindrical pins (shaft of twist drill), 12 mm in front right and 10 mm in rear left. Carrier unit must not lie on roof frame.
- Check electrical consumers electrical wiring and connections on roof and reinstate if necessary.
- Tighten carrier unit -1-, to 8 Nm (71 in-lb), starting at sliding/tilting sunroof drive -3- on left and right-hand sides moving rearward.
- Fit water drain hoses -arrows-.
- Fit connection for drive -3-.

Guide, removing

- Removing glass panel for sliding/tilting sunroof, see **Sliding/tilting sunroof glass panel, removing**.
- Removing carrier unit, see **Carrier unit, removing and installing**.
- Removing sliding/tilting sunroof, see **Drive for sliding/tilting sunroof, removing**.

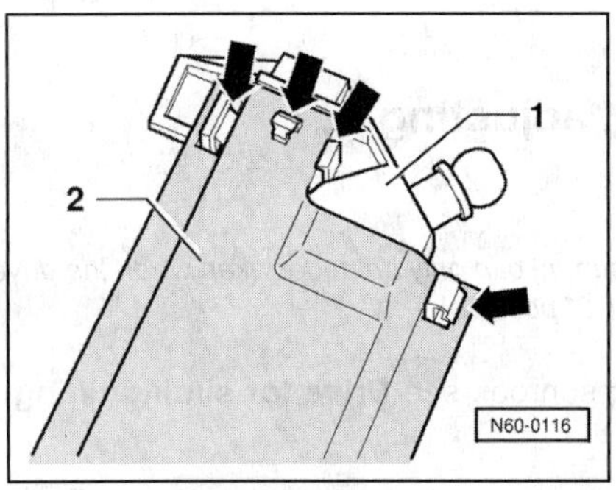

◀ Unclip end cap -1- -arrows- from guide rail -2- and then pull off.

NOTE —

When installing seal end piece -1- to guide rail with Butyl adhesive sealing cord AKL 450 005 05.

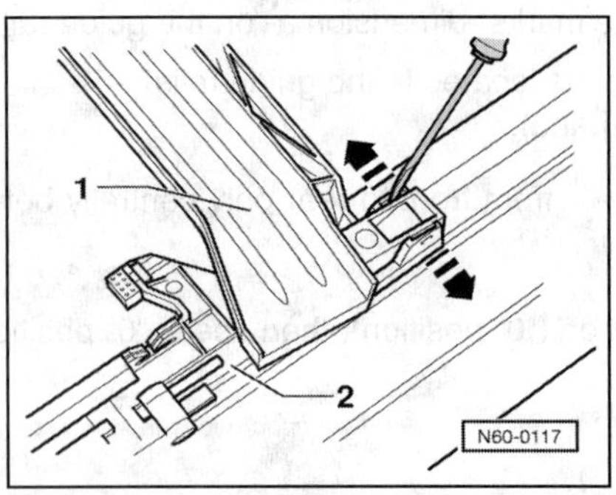

◀ Unclip water channel -1- from locking lug guide -2- on both sides carefully with a screwdriver and take off.

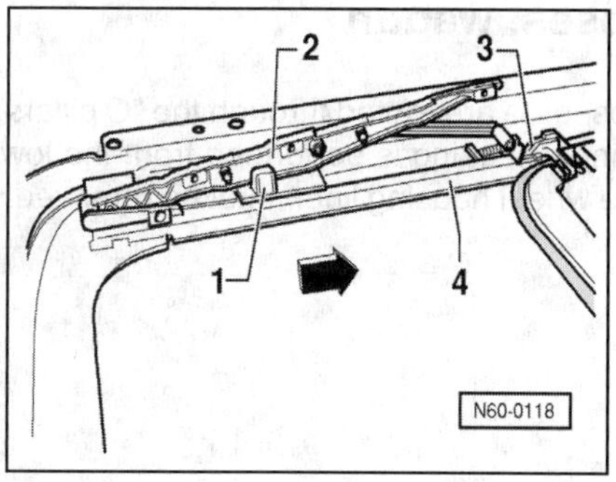

◄ Slide rear guide -1- in guide rail -4- in direction of arrow until the locking lug guide -3- is released.

– Pull rear guide -1-, locking lug guide -3- and guide -2- out of guide rail -4-.

NOTE—
Always replace guide with cable as a pair.

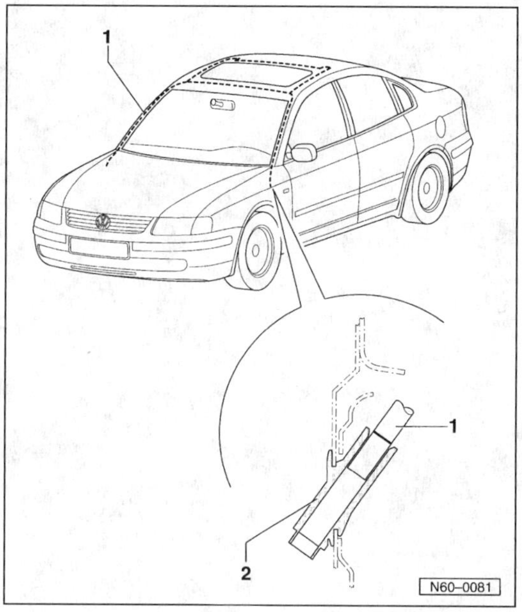

Water Drain Hoses

Front water drain hoses

The front water drain hoses -1- are routed in the "A pillars" and end between door and A pillar. Cleaning is done from the sliding/tilting roof panel aperture.

1 - Front water drain hoses

2 - Hose grommet

NOTE—
For cleaning, a locally manufactured tool made from a speedometer inner cable, approx. 2300 mm (90 in.) ong is recommended.

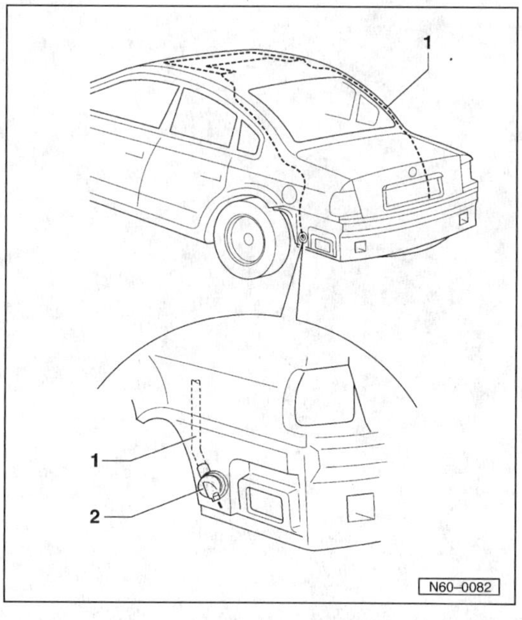

Rear water drain hoses

The rear water drain hoses -2- are routed in the "C pillars" and end at the side behind the bumper cover. Cleaning is performed from the lower end of the hose. To do this, the bumper must be removed.

1 - Water drain hose

2 - Hose grommet

NOTE—
For cleaning, a locally manufactured tool made from a speedometer inner cable, approx. 2300 mm (90 in.) long is recommended.

Rear water drain hoses, wagon

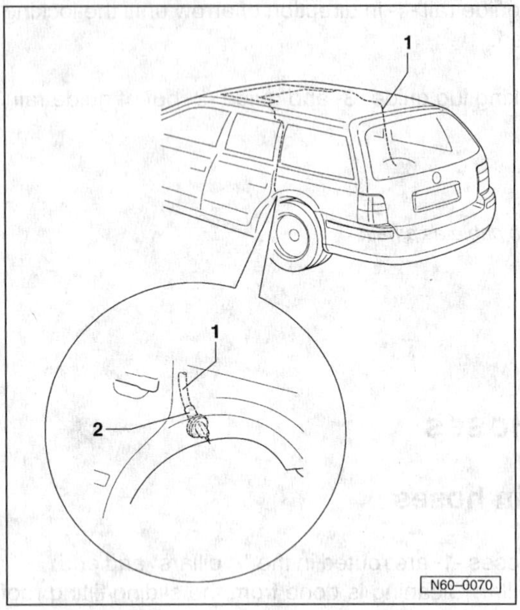

The rear water drain hoses -1- are routed through the "C pillars" and end in the rear wheel housing. Cleaning is performed from the lower end of the hose. To do this, the wheel housing liner must be removed.

1 - Water drain hose

2 - Hose grommet

NOTE —
For cleaning, a locally manufactured tool made from a speedometer inner cable, approx. 2300 mm (90 in.) long is recommended.

63 Bumpers

Repair Group

T 63

Front Bumper

Front bumper strip, removing and installing (from 10/2000)

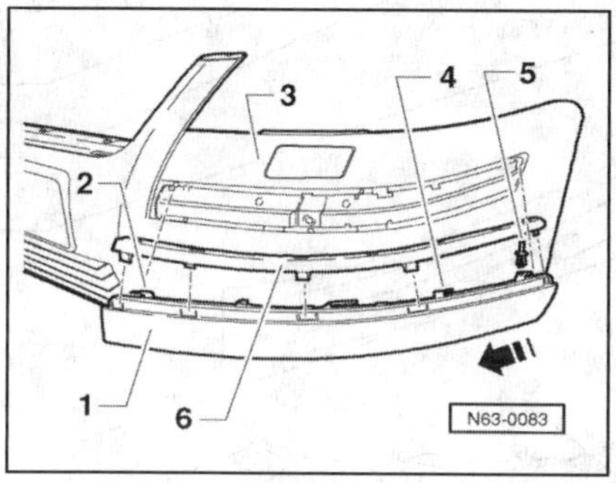

◄ Loosen clip -5- from inside of cover -3-.

– From center of vehicle, loosen bumper trim -1- from front retaining clip -2- and pull out of guides -4- toward front.

– Chrome strip -6- is only secured to bumper trim via retaining clips in certain versions. These retaining clips must be removed from the inside of the bumper trim.

Release lever, removing

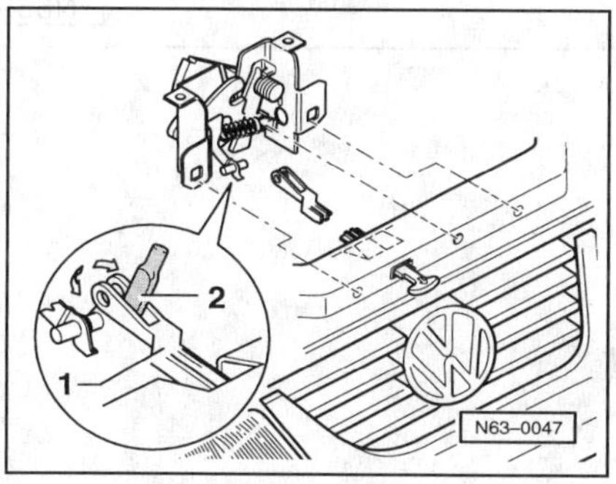

◄ Open release lever -1- with a screwdriver -2- and pull off hood lock.

– Pull release lever -1- with guide out from lock carrier.

Guide, adjusting

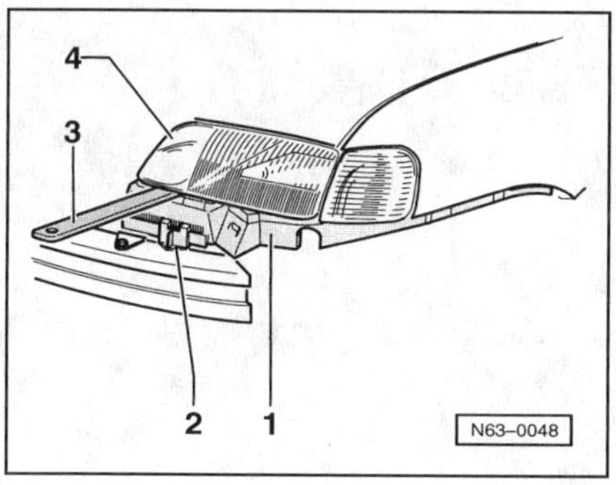

◄ Adjust slide -2- on guide -1- so that between headlight -4- and upper edge of guide -1- a gap of 2 mm exists.

– Check and adjust gaps/shut lines with special tool 3371 (-3-)

Front bumper, assembly (through 09/2000)

NOTE—
The following procedure is for removing the front bumper cover. For instructions on removal of the front bumper reinforcement please see, see **Repair Group 50, Lock carrier and front bumper reinforcement, assembly (through 09/2000)**

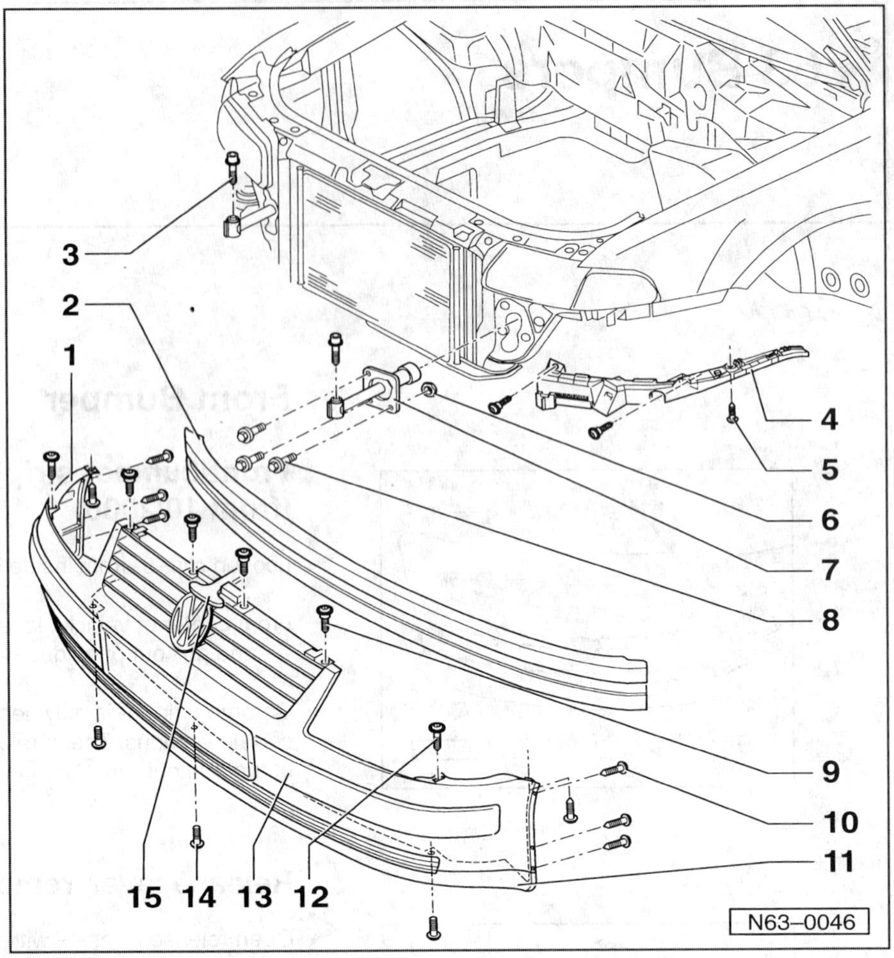

1. **Cover**
   - Material: PP/EPDM
   - Removing:
     Remove turn signals.
     Separate release lever -15- from trunk lock, see **Release lever, removing**.
     Unscrew screws -9-, -10- and -14-.
     Pull cover out from guides -4- (left and right-hand side).
2. **Bumper carrier**
3. **Bolt**
   - Qty. 2
   - Tighten to 20 Nm (15 ft-lb)
4. **Guide**
   - To remove/install bumper cover, pull out or push in parallel to guides at left and right.
   - See **Guide, adjusting**
5. **Bolt**
   - Qty. 6
   - Tighten to 2 Nm (18 in-lb)
6. **Hexagon nut**
   - Qty. 2
7. **Impact damper**
8. **Bolt**
   - Qty. 7
   - Tighten to 50 Nm (37 ft-lb)
9. **Bolt**
   - Qty. 4
   - Tighten to 3 Nm (27 in-lb)
10. **Bolt**
    - Qty. 8
    - 2 Nm
11. **Spoiler**
    - Clipped into cover
    - Is not necessary to remove when removing the cover
12. **Bolt**
    - Qty. 2
13. **Bumper strip**
    - Clipped into cover
14. **Quick release bolt**
    - Qty. 3
    - Tighten to 2 Nm (18 in-lb)
15. **Release lever**
    - See **Release lever, removing**

Front bumper, assembly (from 10/2000)

NOTE —

The following procedure is for removing the front bumper cover. For instructions on removal of the front bumper reinforcement please see, see **Repair Group 50, Lock carrier and front bumper reinforcement, assembly (from 10/2000)**.

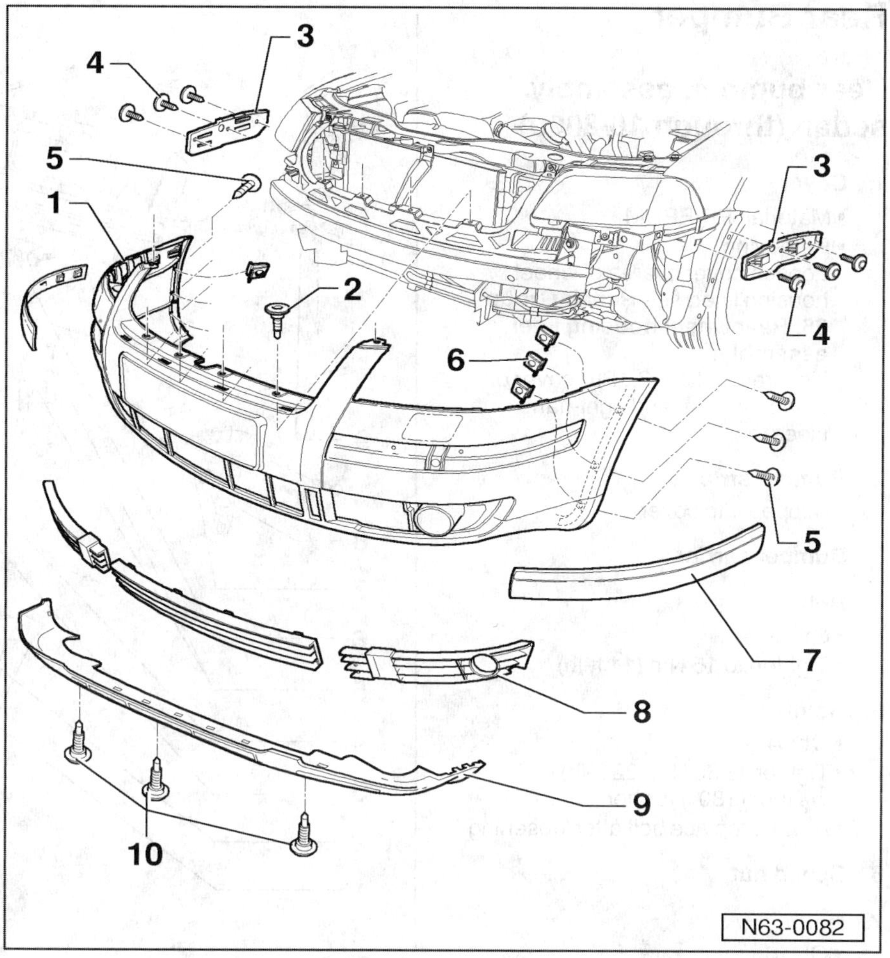

1. **Cover**
   - Material -PP/EPDM
   - Removing:
     Remove radiator grille, see **Repair Group 66, Radiator grill, assembly (from 10/2000)**
     Remove screws -2-, -5- and -10-.
     Pull cover off guide pieces -3- (left and right).
2. **Bolt**
   - Qty: 6
   - Tighten to 6 Nm (53 in-lb)
3. **Guide piece**
   - To remove and install bumper cover, pull out or slide into guide pieces in parallel manner (left and right).
4. **Screw**
   - Qty: 3
   - Tighten to 2 Nm (18 in-lb)
5. **Screw**
   - Qty: 8
   - Tighten to 2 Nm (18 in-lb)
6. **Speed nut**
   - Qty: 6
7. **Bumper strip**
   - Clipped into cover
8. **Ventilation grill**
   - Consists of three parts and is clipped into cover
9. **Spoiler**
   - Clipped into cover
   - Does not have to be removed when removing cover
10. **Quick-release screws**
    - Qty: 3
    - Tighten to 2 Nm (18 in-lb)

Rear Bumper

Rear bumper, assembly, sedan (through 10/2000)

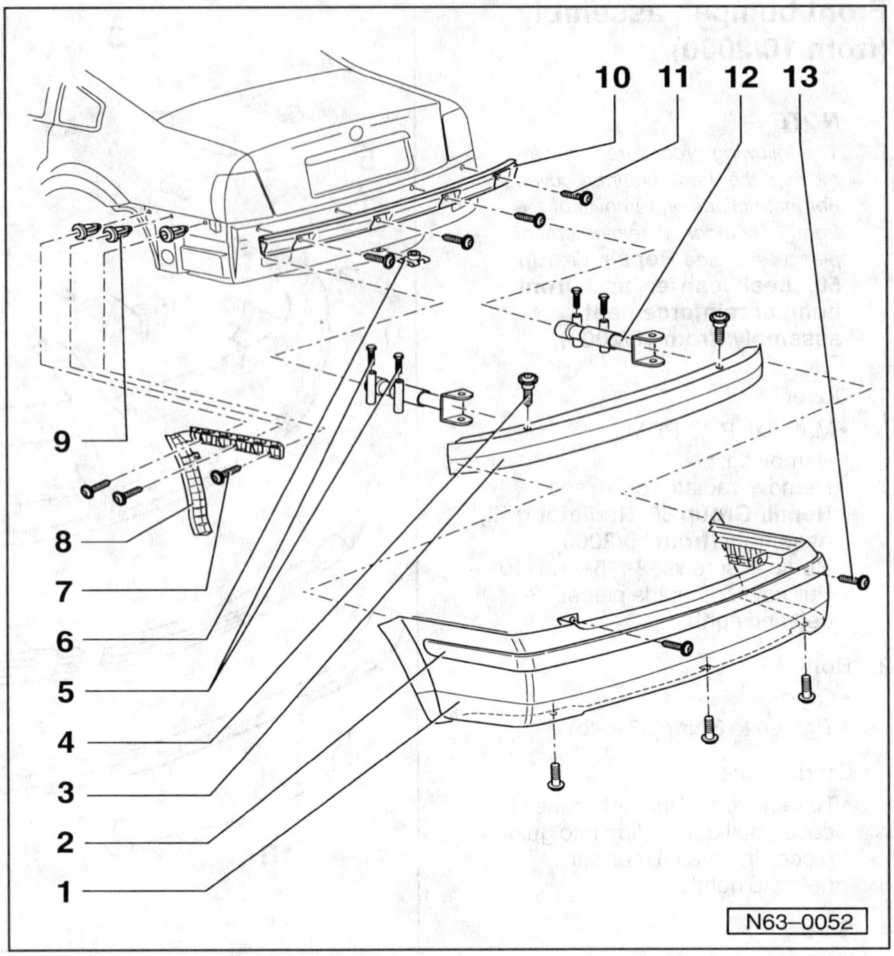

1. **Cover**
   - Material: PP/EPDM
   - Removing:
     Loosen cover in area of wheel housing liner, see **Repair Group 66, Rear wheel housing liner, assembly**.
     Unscrew bolts -13-. Pull cover off guides -8- (left and right-hand sides).
2. **Bumper strip**
   - Clipped into cover
3. **Bumper carrier**
4. **Bolt**
   - Qty. 2
   - Tighten to 15 Nm (11 ft-lb)
5. **Bolt**
   - Qty. 4
   - Tighten to 30 Nm (22 ft-lb) + $^1/_2$ turn (180°) further
   - Always replace bolt after loosening
6. **Speed nut**
7. **Bolt**
   - Qty. 6
   - Tighten to 2 Nm (18 ft-lb)
8. **Guide**
   - To remove/install bumper, pull out or push in parallel to guides on left and right.
9. **Spreader nut**
   - Qty. 6
10. **Securing strip**
11. **Bolt**
    - Qty. 4
    - Tighten to 1.2 Nm (11 in-lb)
12. **Impact damper**
13. **Bolt**
    - Qty. 5
    - Tighten to 6 Nm (53 in-lb)

Rear bumper, assembly, sedan (from 10/2000)

1. **Cover**
   - Material -PP/EPDM
   - Removing:
     Loosen cover in area of wheel-housing liner, see **Repair Group 66, Rear wheel housing liner, assembly**.
     Remove taillights, see **Electrical Equipment section, Repair Group 94**
     Remove bolts -4-, -19- and -20-.
     Pull off cover from guide pieces -8- (left and right).
2. **Bolts**
   - Qty: 6
   - Tighten to 2 Nm (18 ft-lb)
3. **Speed nut**
   - Qty: 6
4. **Screw**
   - Tighten to 1.2 Nm (11 in-lb)
   - Qty: 4
   - Below taillights
5. **Bolt**
   - Tighten to 1.2 Nm (11 in-lb)
   - Qty: 2
6. **Locking piece**
7. **Guide piece**
   - To remove and install bumper, pull out or slide into guide pieces in parallel manner (left and right).
8. **Bolt**
   - Tighten to 1.2 Nm (11 in-lb)
   - Qty: 6
9. **Expanding nut**
   - Qty: 6
10. **Securing strip**
11. **Bolt**
    - Tighten to 1.2 Nm (11 in-lb)
    - Qty: 4
12. **Hex nut**
    - Qty: 2
13. **Bolt**
    - Qty: 4
    - Tighten to 30 Nm (22 ft-lb) + $^1/_2$ turn (180°) further
    - Always replace bolts after removing
14. **Impact absorber**
15. **Bolt**
    - Qty: 2
    - Tighten to 15 Nm (11 ft-lb)

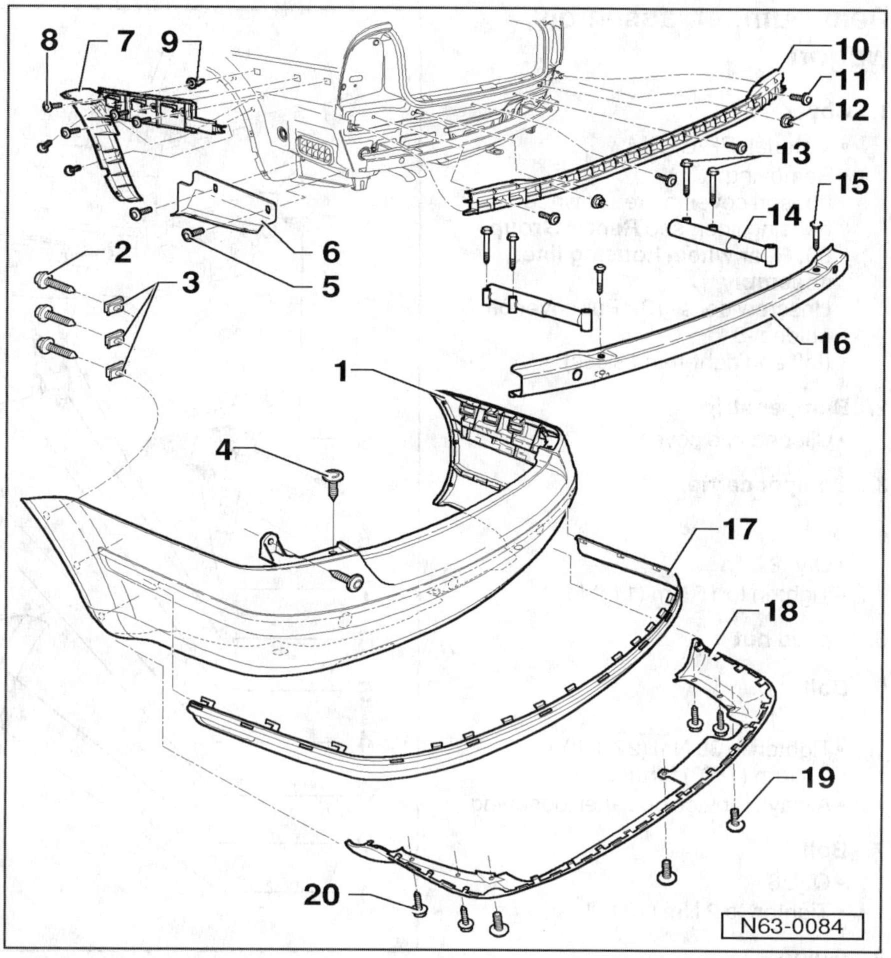

16. **Bumper carrier**
17. **Bumper strip**
    - Clipped into cover
    - See **Rear bumper strip, removing (from 10/2000)**
18. **Spoiler**
    - Clipped into cover
19. **Bolt**
    - Tighten to 6 Nm (53 in-lb)
    - Qty: 3
20. **Bolt**
    - Tighten to 1.2 Nm (11 in-lb)
    - Qty: 4

Rear bumper, assembly, wagon

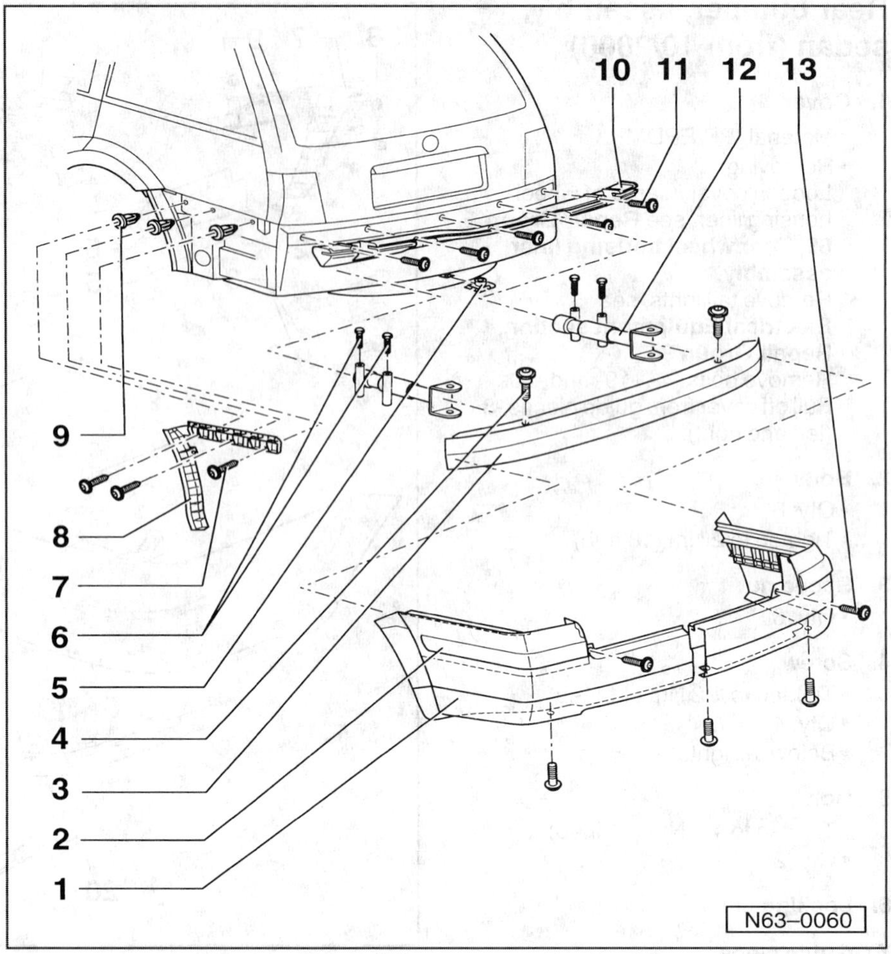

1. **Cover**
   - Material: PP/EPDM
   - Removing:
     Loosen cover in area of wheel housing liner, see **Repair Group 66, Rear wheel housing liner, assembly**.
     Unscrew bolts -13-. Pull cover off guides -8-
     (left and right-hand sides).
2. **Bumper strip**
   - Clipped into cover
3. **Bumper carrier**
4. **Bolt**
   - Qty. 2
   - Tighten to 15 Nm (11 ft-lb)
5. **Speed nut**
6. **Bolt**
   - Qty. 4
   - Tighten to 30 Nm (22 ft-lb) + $^1/_2$ turn (180°) further
   - Always replace bolt after loosening
7. **Bolt**
   - Qty. 6
   - Tighten to 2 Nm (18 ft-lb)
8. **Guide**
   - To remove/install bumper, pull out or push in parallel to guides on left and right.
9. **Spreader nut**
   - Qty. 6
10. **Securing strip**
11. **Bolt**
    - Qty. 4
    - Tighten to 1.2 Nm (11 in-lb)
12. **Impact damper**
13. **Bolt**
    - Qty. 5
    - Tighten to 6 Nm (53 in-lb)

Rear bumper strip, removing and installing (from 10/2000)

– Loosen inner clips -5- on left and right side of cover -1-.

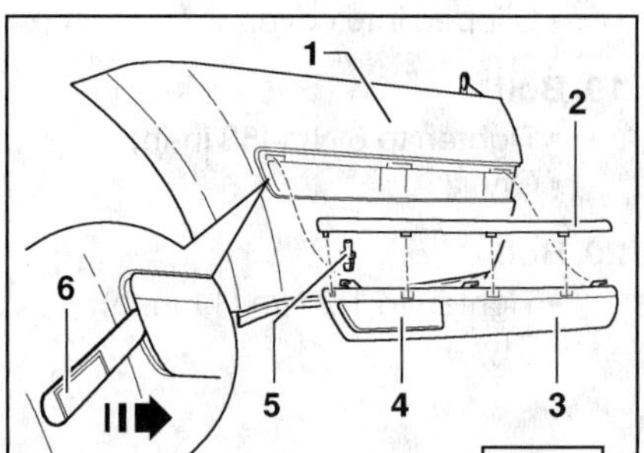

Pry chrome strip -2- out of guides at left and right in direction of arrow using a plastic wedge -6-. Pull bumper strip off of clips further in direction of arrow.

– Chrome strip -3- is secured to bumper strip with release catches. Loosen release catches from inside of bumper strip.

– Rear side marker light -4- is removed from the inside of bumper strip -3-.

64 Window Regulators

Repair Group

T 64

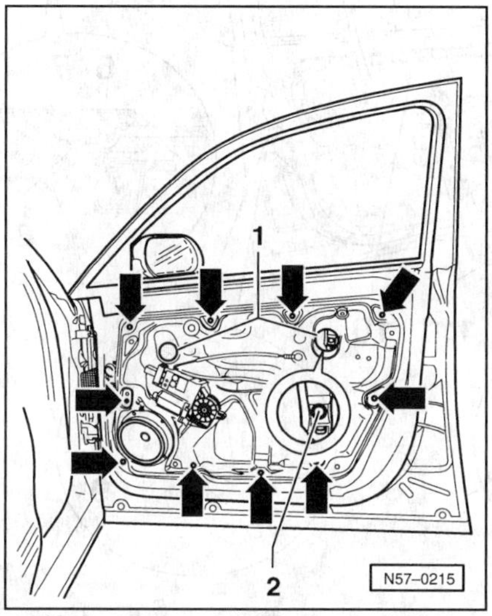

Front door window, removing

– Removing front door trim, see **Body-Interior section, Repair Group 70; Door trim**

◄ Pry out cover cap -1-.

– Lower door window until clamp securing bolts are accessible.

If the electrically driven door window will not lower sufficiently then proceed as described in procedure **Repair Group 57, Door lock, removing and iinstalling**.

– Loosen bolts -2- and press clamps apart.

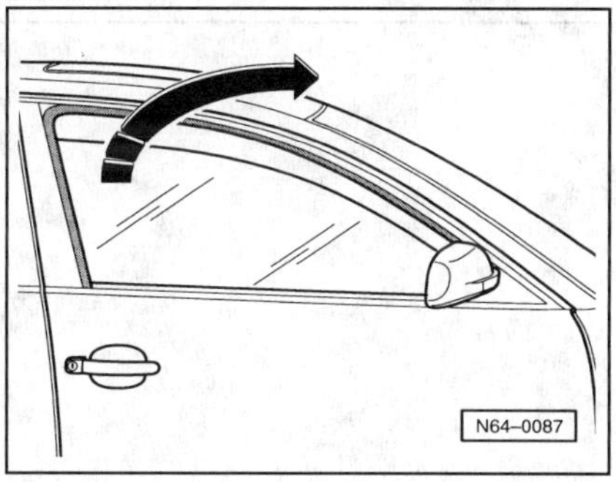

◄ Lift rear of door window and take out of door forwards, in direction of arrow.

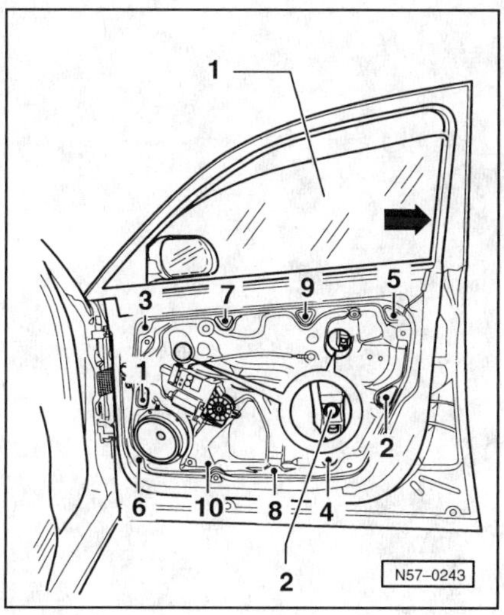

Front door window, adjusting

◄ The door window clamping jaws must be positioned in the assembly opening.

– Loosen clamping jaws, push door window -1- into window channel -arrow- and tighten clamping jaws bolts -2- to 10 Nm (7 ft-lb).

Front door window, assembly

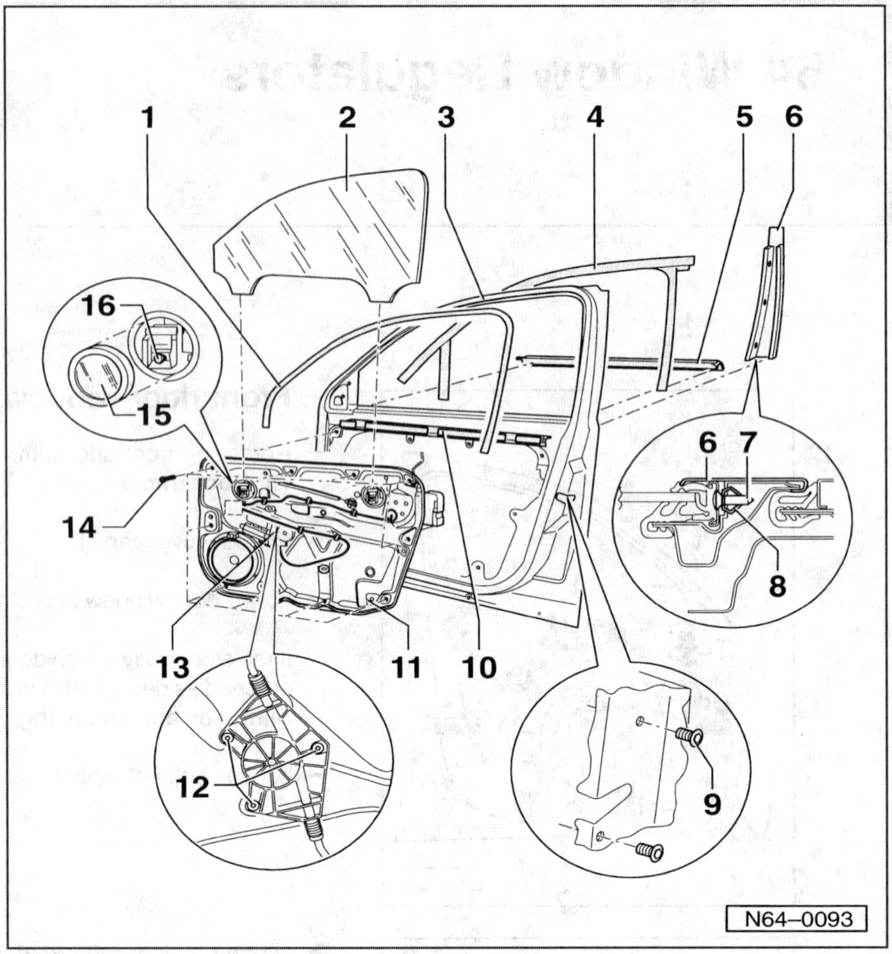

1. **Window channel**
   - Pushed onto flange
2. **Door window**
   - See **Front door window, removing**
   - See **Door window, adjusting**
3. **Door**
4. **Outer window channel**
   - Pushed onto flange
5. **Window slot seal**
   - Pushed onto flange
6. **Trim**
7. **Cross-head screw**
8. **Spring clip**
9. **Screw**
   - Tighten to 20 Nm (15 ft-lb)
10. **Window slot seal**
    - Pushed onto flange
11. **Assembly carrier**
    - See **Assembly carrier, removing and installing**
    - Window lifter is part of assembly carrier
12. **Bolt**
    - Tighten to 8 Nm (71 in-lb)
13. **Motor for window lifter**
    - Is screwed onto assembly carrier from the back with bolt -12-
    - See **Window lifter motor or assembly carrier with window lifter, removing and installing**
14. **Bolt**
    - Tighten to 8 Nm (71 in-lb)
15. **Cover cap**
16. **Bolt**
    - Tighten to 8 Nm (71 in-lb)

Rear door window, removing

– Removing rear door trim, see **Body-Interior section, Repair Group 70;**

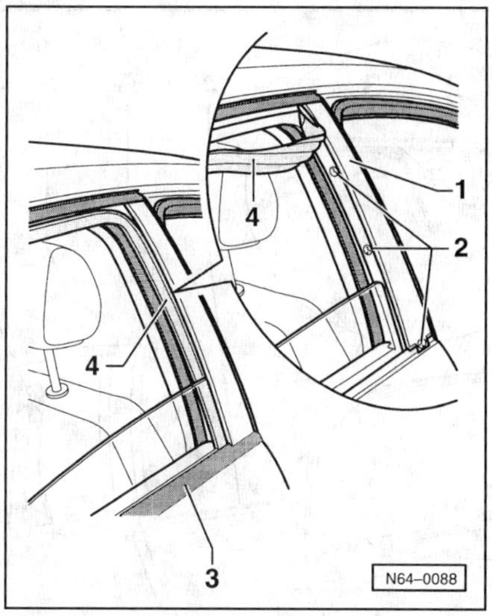

Pull outer window slot seal -3- off door flange.

- Pull window guide -4- out of trim -1-.
- Unscrew bolts -2- and take trim -1- off door frame.

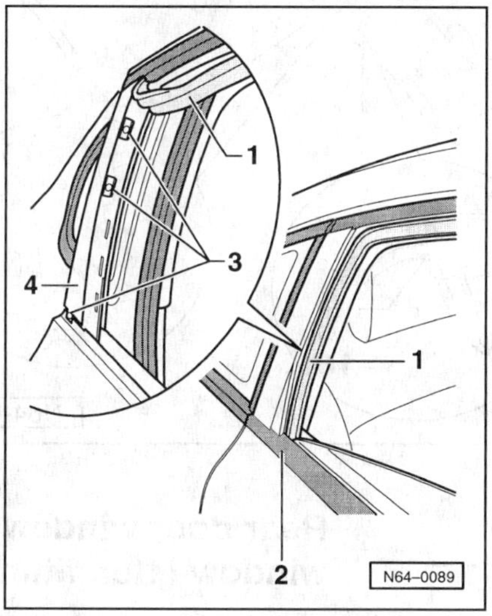

Pull outer window slot seal -2- off door flange.

- Pull window guide -1- out of trim -4-.
- Unscrew bolts -3- and take trim -4- off door frame.

– Remove housing from door handle, see **Repair Group 57, Lock cylinder housing, removing and installing**

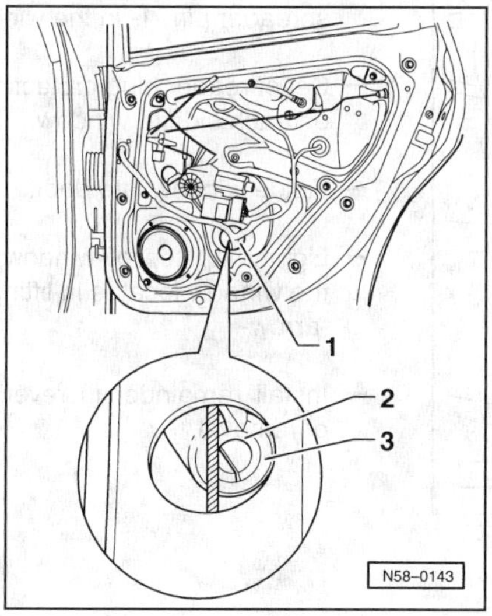

Pry out cover cap -1-.

- If a door window with electrical drive will not lower fully, proceed as described in procedure **Repair Group 57, Door lock, removing and installing**
- Lower door window until the spreader pin -2- and the spreader plug -3- are accessible in the window lifter cut-out.

NOTE —

- *When pressing out the spreader pin and the spreader plug ensure that the door outer panel is not damaged.*
- *Pressed out spreader pins must be removed from door.*

- Press spreader pin -2- out of spreader plug -3- with a 3 mm drift.
- Press spreader plug -3- out of door window with a 3 mm drift.

– Take door window out of door upwards.

Rear door window, assembly

1. **Door**
   - See **Repair Group 58, Door, removing and installing**
2. **Spring clip**
3. **Cross-head screw**
4. **Window slot seal**
   - Pushed onto flange
5. **Trim**
6. **Outer window channel**
   - Pushed onto flange
7. **Trim**
8. **Bolt**
9. **Seal**
10. **Clip**
11. **Assembly carrier**
    - See **Assembly carrier, removing and installing**
    - Window lifter is part of assembly carrier
12. **Bolt**
    - Tighten to 8 Nm (71 in-lb)
13. **Motor for window lifter**
    - Is screwed onto assembly carrier from the back with bolt -12-
    - See **Window lifter motor or assembly carrier with window lifter, removing and installing**
14. **Bolt**
    - Tighten to 8 Nm (71 in-lb)
15. **Window slot seal**
    - Pushed onto flange
16. **Inner window channel**
    - Pushed onto flange
17. **Door window**
    - See **Rear door window, removing**

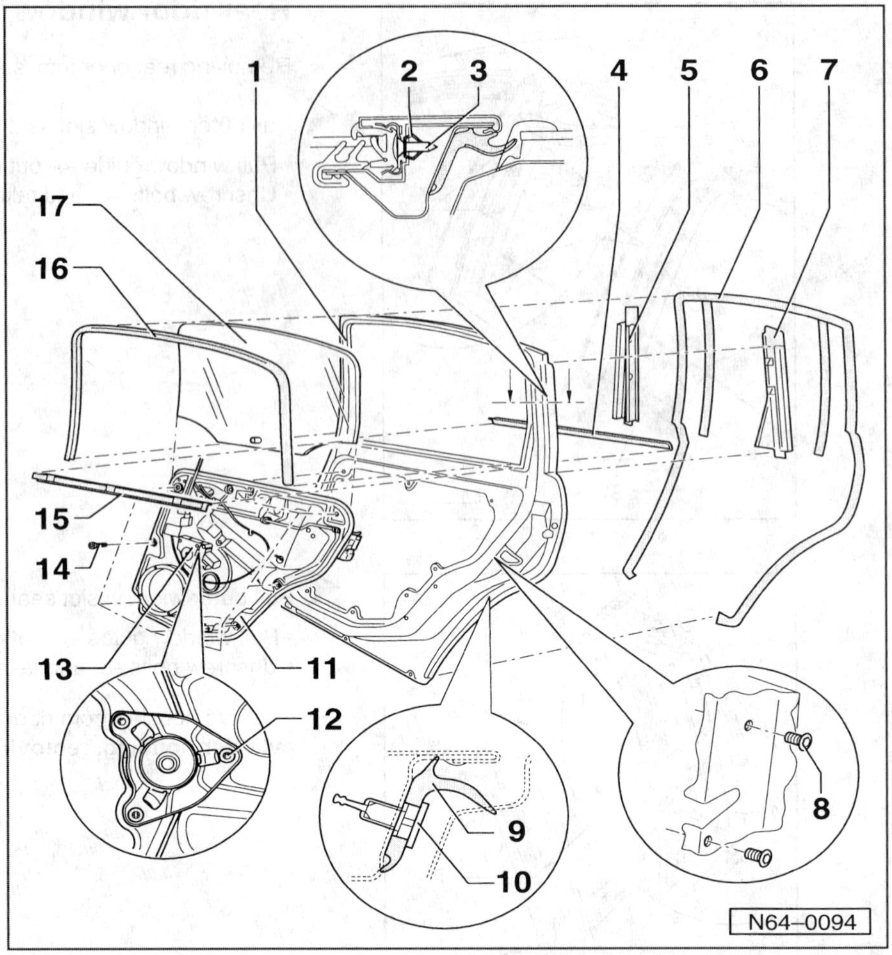

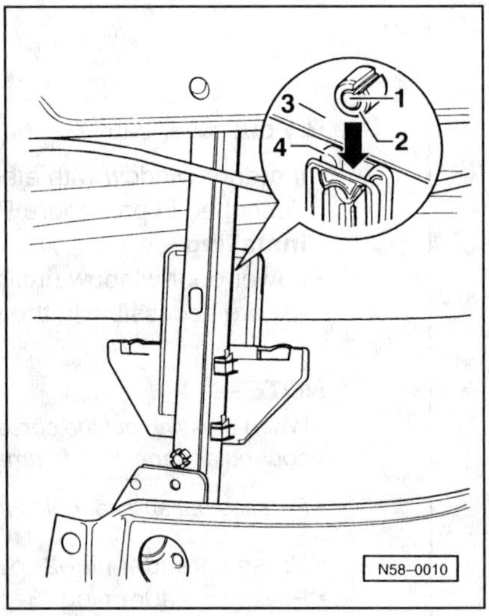

Rear door window to window lifter, fitting

◄ Insert spreader plug -2- and spreader pin -1- in the window.

– Set spreader plug projection on each side of window.

– Guide window into door.

– Lightly tap top of window, so that the window locates in lifting rail -arrow-.

– Install remainder in reverse order of removal.

Assembly carrier, removing and installing

The window lifters, the door lock and the loudspeaker are secured to the assembly carrier.

The window lifter is supplied with the assembly carrier.

The assembly carrier can only be removed when the door window clamping brackets of the window lifter are unscrewed. To do this lower door window to the height of the assembly holes in assembly carrier and loosen clamping brackets.

Front assembly carrier, removing

– Removing front door trim, see **Body-Interior section, Repair Group 70**

– Removing lock cylinder housing, see **Repair Group 57, Lock cylinder housing, removing and installing**.

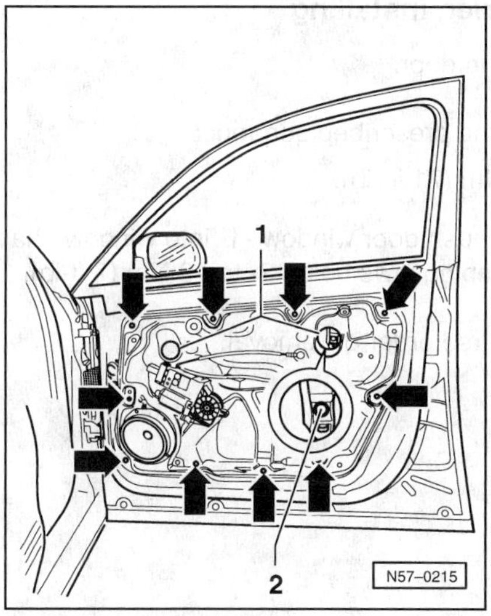

◄ Pry out cover cap -1-.

- Lower door window until securing bolts for door window are accessible.
- If the window will not lower sufficiently then proceed as described in procedure **Repair Group 57, Door lock, removing and installing**

– Loosen bolts -2- and press clamps apart.

– Push window upward and secure in position (e.g. with adhesive tape).

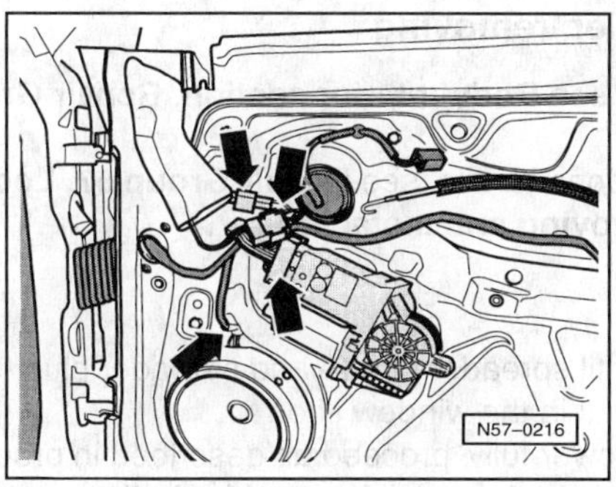

◄ Disconnect all connections -arrows-.

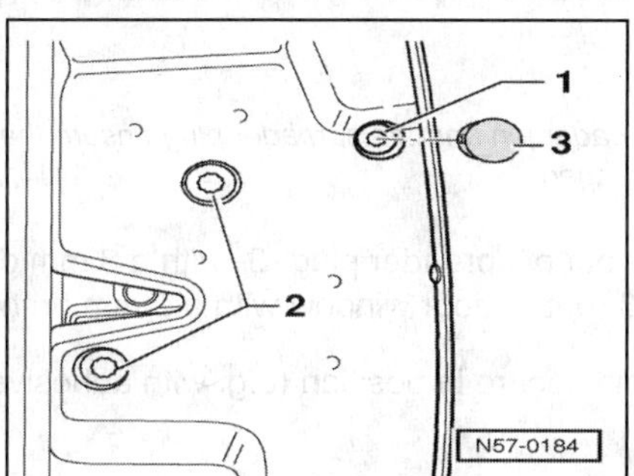

◄ Pryout cover cap -3-.

- Unscrew bolts -2-
- Tightening torque: 20 Nm (15 ft-lb).

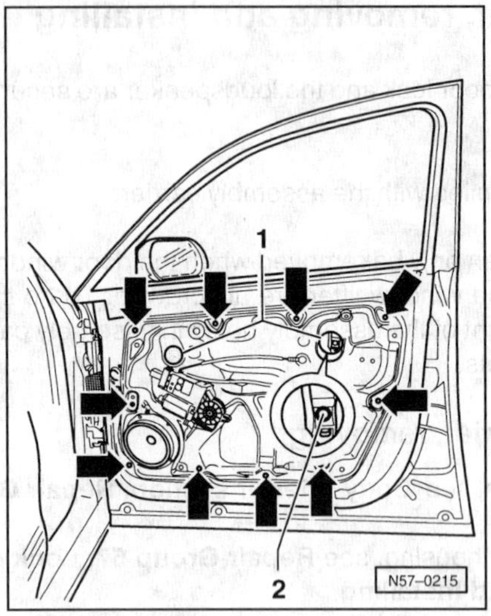

Unscrew bolts -arrows-. Pull top of assembly carrier off door, lift and take out of door towards hinges.

– Turn assembly carrier and pull connector off door lock.

– Then unclip clips for electrical cables on back of assembly carrier.

Front assembly carrier, installing

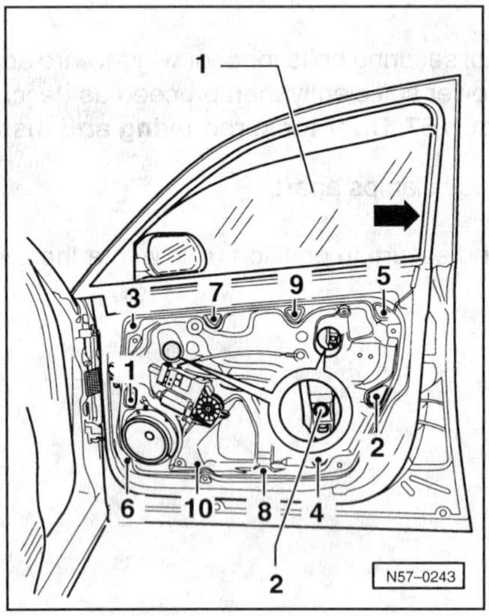

– Insert assembly carrier in door.

Fit bolts and tighten in the prescribed sequence

- Tightening torque: 8 Nm (71 in-lb).

– Loosen clamping jaws, push door window -1- into window channel -arrow- and tighten clamping jaws bolts -2- to 10 Nm (7 ft-lb).

– Install remainder in reverse order or removal.

Rear assembly carrier, removing

– Remove rear door trim, see **Body-Interior section, Repair Group 70**

– Remove housing from door handle, see **Repair Group 57, Lock cylinder housing, removing and installing,**

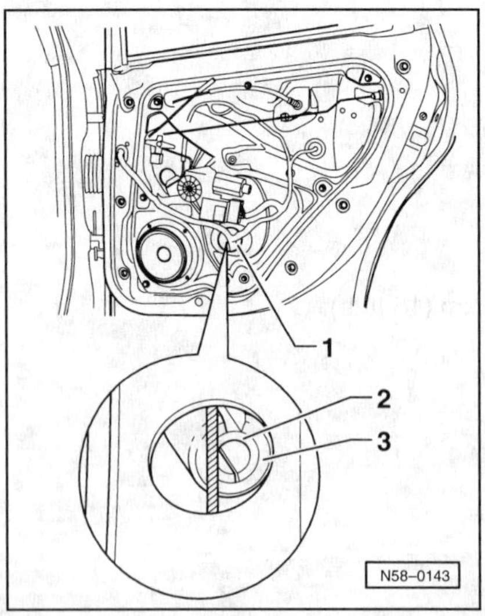

Pry out cover cap -1-.

- Lower door window until spreader pin -2- and spreader plug -3- are accessible in the cut-out in the window lifter.
- If the window will not lower fully, proceed as described in procedure **Repair Group 57, Door lock, removing and installing**

NOTE —
When pressing out the spreader pin and the spreader plug ensure that the door outer panel is not damaged.

- Press spreader pin -2- out of spreader plug -3- with a 3 mm drift.
- Press spreader plug -3- out of door window with a 3 mm drift.

– Push window upward and secure in position (e.g. with adhesive tape).

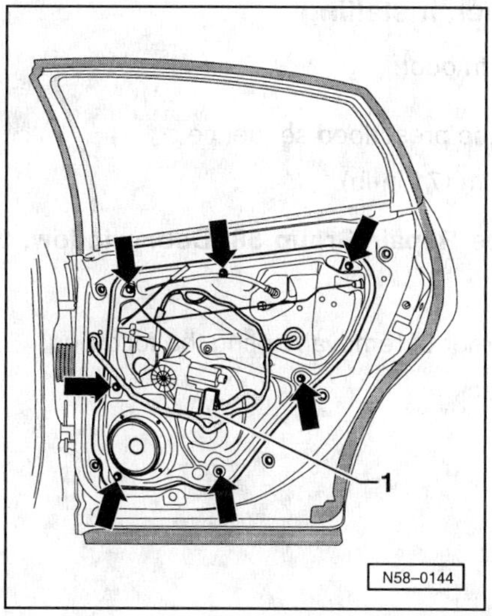

- Separate connector -1-.
  - Unscrew bolts -arrows-.

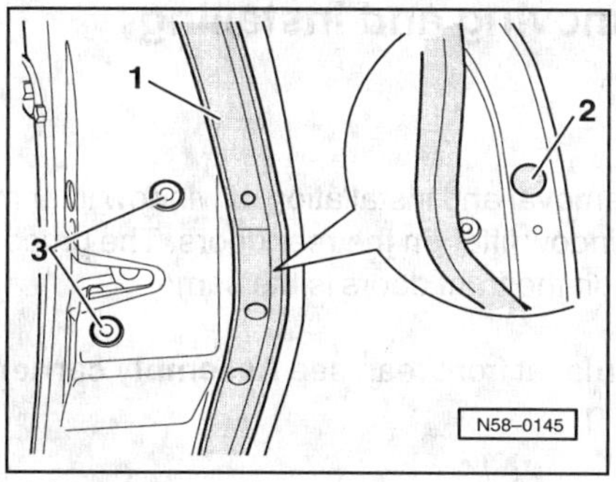

- Loosen door seal -1- in area of door lock.
  - Pry out cover cap -2-.
  - Unscrew bolts -3-.

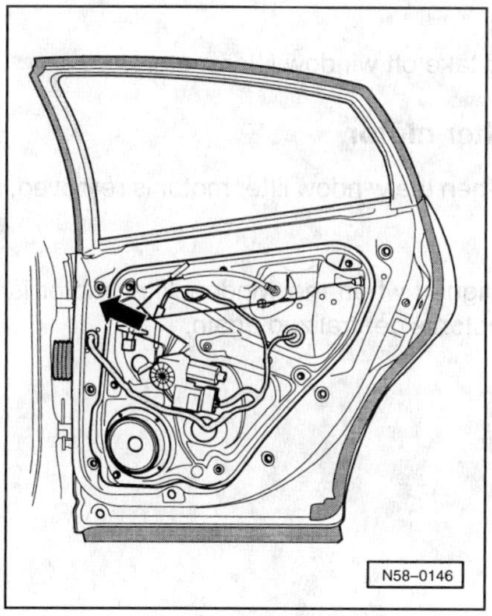

- Pull assembly carrier at top off door, lift and take off door towards hinges.

– Turn assembly carrier and pull connector off door lock.

– Then unclip clips for electrical cables on back of assembly carrier.

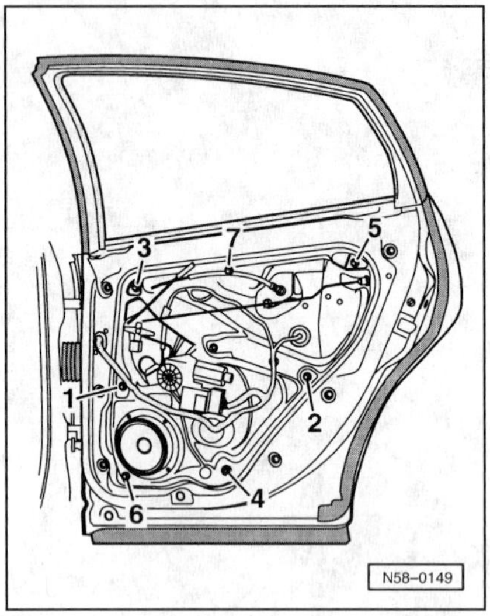

Rear assembly carrier, installing

– Insert assembly carrier in door.

< Fit bolts and tighten in the prescribed sequence

- Tightening torque: 8 Nm (71 in-lb).

– Install door window, see **Repair Group 58, Door window, fitting to window lifter**.

– Installation is reverse order of removal, noting the following:

- Perform functional check.

Window lifter motor or assembly carrier with window lifter, removing and installing

Removing

The illustration shows removal and installation of window lifter motor/ assembly carrier with window lifter on the rear doors. The procedure for removing and installing on the front doors is the same.

Removing assembly carrier at front/rear, see **Assembly carrier, removing and installing**

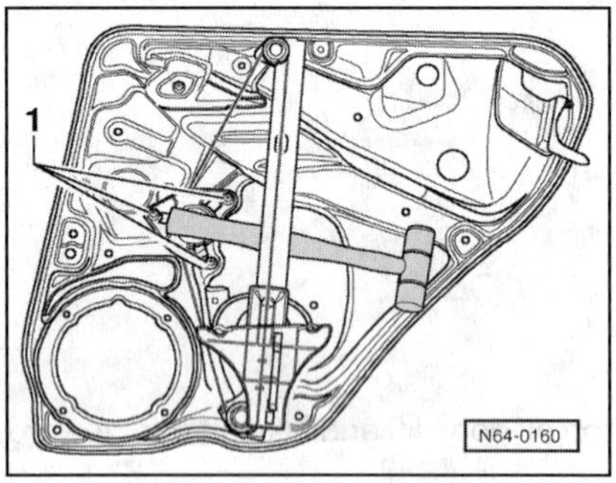

< Slide a hammer shaft between the window guide and the assembly carrier until the end of the hammer shaft presses against the assembly carrier.

– Remove screws -1- and take off window lifter motor.

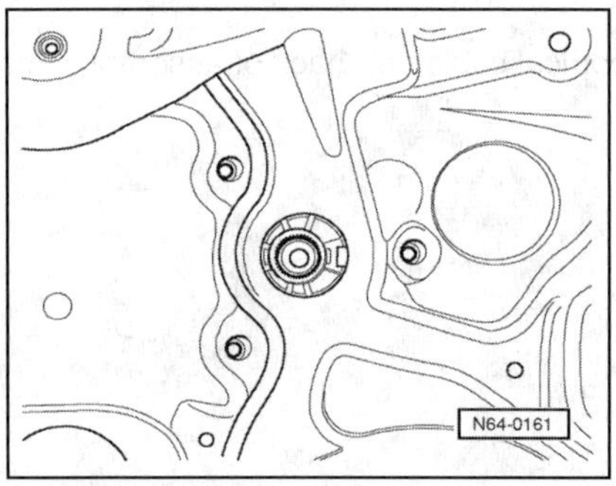

Installing window lifter motor

< The gearing will move when the window lifter motor is removed, it is then no longer centralized.

– The gearing must be aligned when the window lifter motor is installed until the window lifter motor is centralized again.

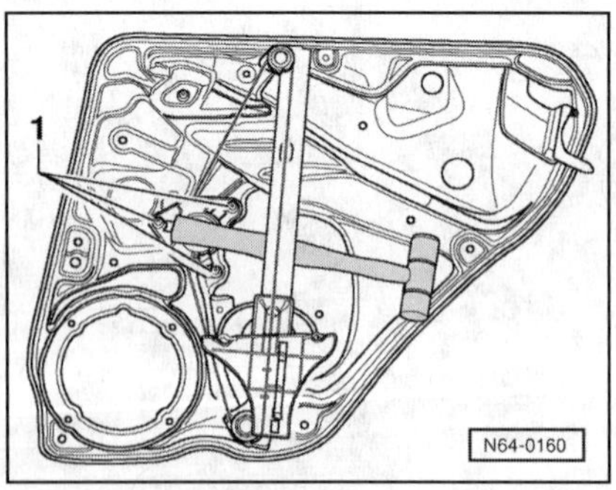

< Fit bolts -1- and tighten

- Tightening torque: 3.5 Nm (31 in-lb)
- The tightening torque must be strictly adhered to, otherwise the plastic sleeve on the window lifter could be damaged.

Installing assembly carrier with window lifter

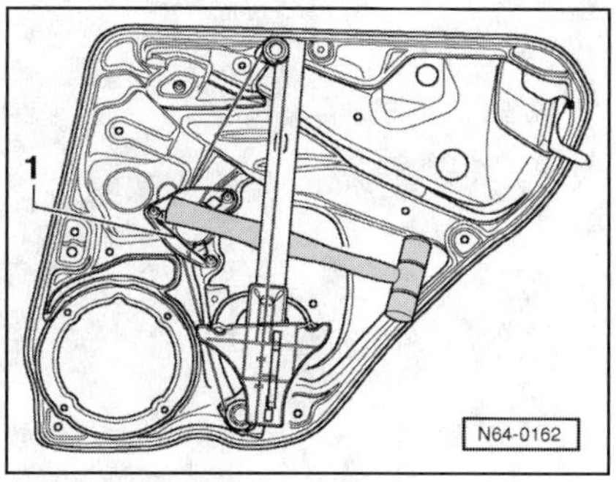

On a replacement assembly carrier the gearing is secured with a cable tie -1- to prevent it moving or falling out.

- Slide a hammer shaft between the window guide and the assembly carrier until the end of the hammer shaft presses against the assembly carrier.
- Remove cable tie -1-.

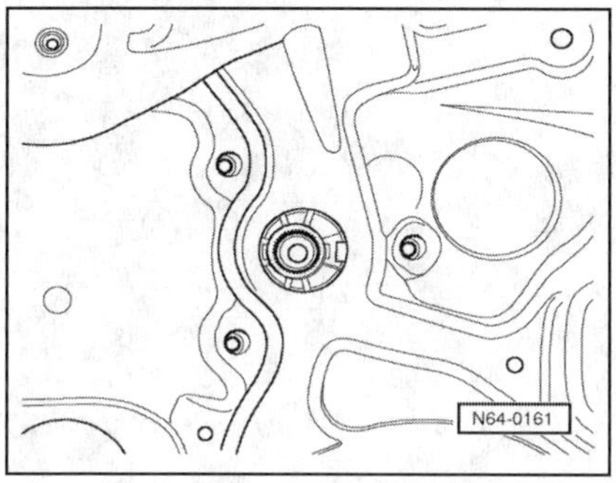

The gearing will move when the window lifter motor is removed, it is then no longer centralized.

– The gearing must be aligned when the window lifter motor is installed until the window lifter motor is centralized again.

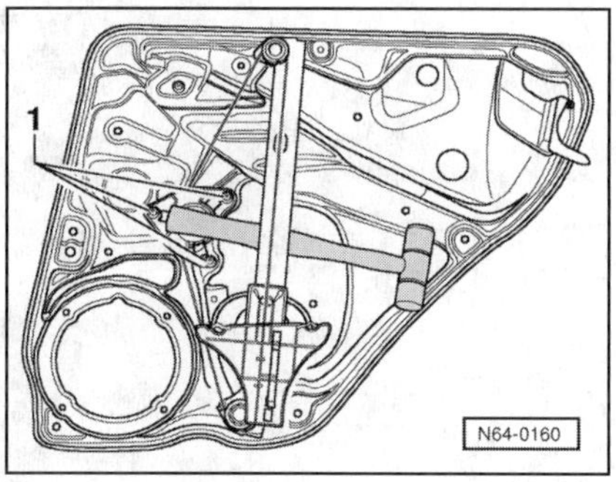

Fit bolts -1- and tighten

- Tightening torque: 3.5 Nm (31 in-lb).
- The tightening torque must be strictly adhered to, otherwise the plastic sleeve on the window lifter could be damaged.

66 Exterior Equipment

Repair Group

T 66

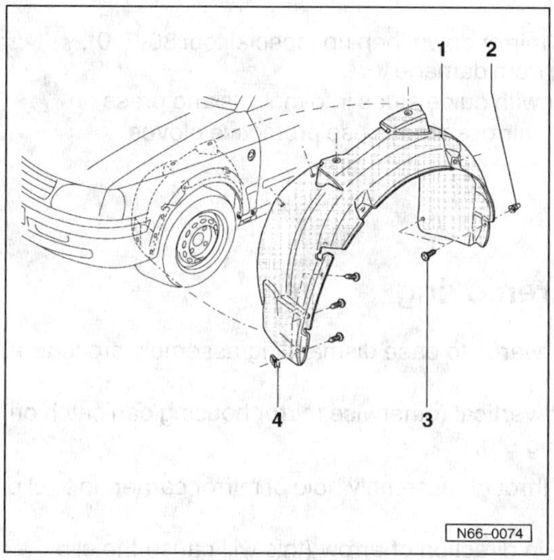

Front wheel housing liner, assembly

1. **Wheel housing liner**
   - Material: PP/EPDM
   - Removing:
     Remove wheel. Wheel bolt tightening torque = 120 Nm (89 ft-lb).
     Unscrew cross-head screws -3- (Qty. 9) and pull off wheel housing liner.
2. **Spreader nut**
3. **Cross-head screw**
4. **Speed nut**

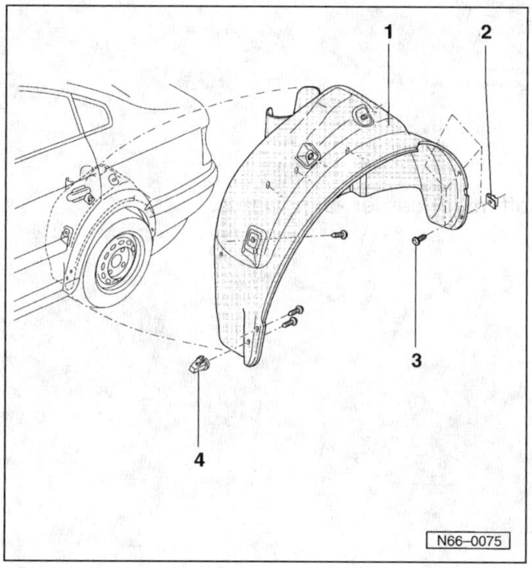

Rear wheel housing liner, assembly

1. **Wheel housing liner**
   - Material: PP/EPDM
   - Removing:
     Remove wheel. Wheel bolt tightening torque = 120 Nm (89 ft-lb).
     Unscrew cross-head screws -3- (Qty. 7) and pull off wheel housing liner.
2. **Speed nut**
3. **Cross-head screw**
4. **Spreader nut**

Exterior mirror housing, assembly

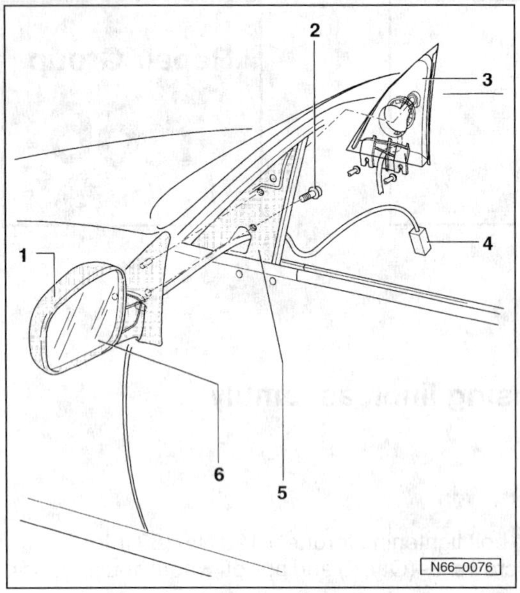

1. **Mirror housing**
   - See **Mirror housing, removing**.
   - Material: ABS
2. **Bolt**
   - Tighten to 10 Nm (7 ft-lb)
3. **Cover for exterior rear view mirror**
4. **Connector**
5. **Insulation**
6. **Mirror glass**
   - Removing - first, press mirror down then up (special tool 80-200). Protect mirror housing from damage
   - Installing - insert mirror with guide studs into mounts and press on
   - Press only on center of mirror - always use protective gloves

Mirror housing, removing

– Fold rear view mirror forwards to ease dismantling/assembly procedure.

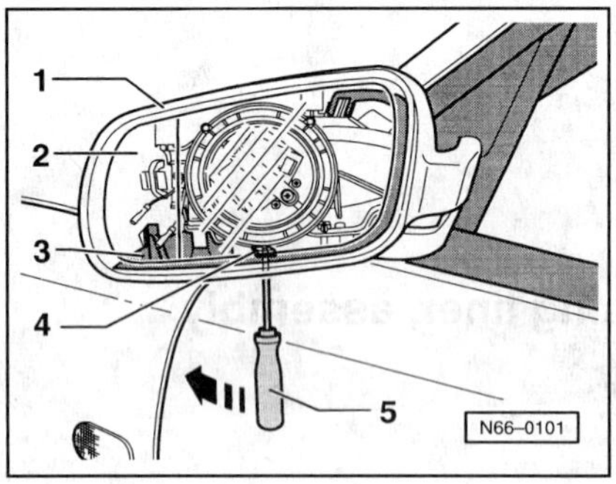

Position mirror glass -2- vertical (otherwise mirror housing can catch on mirror glass when pulling off).

- Insert screwdriver -5- through assembly hole of mirror carrier up to clip -4-.
- Press screwdriver -5- in direction of arrow (this will cause the clip -4- to release the mirror housing).

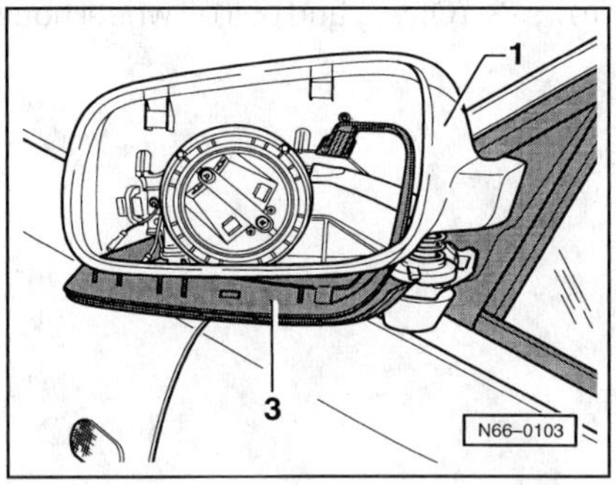

Pull mirror housing -1- off mirror carrier -3- upwards.

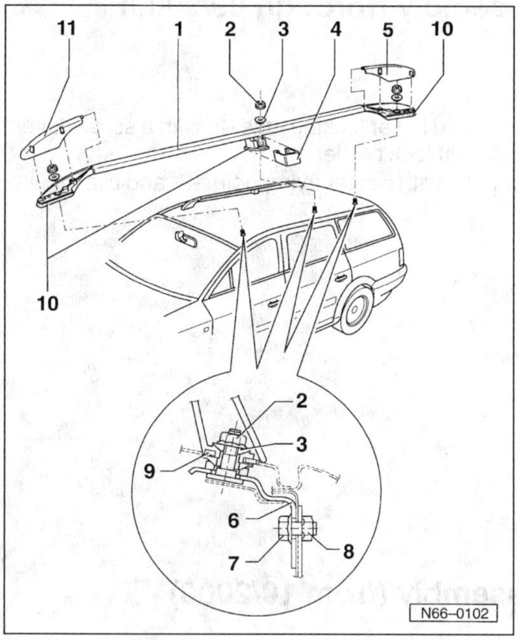

Roof rail, assembly (wagon)

1. **Roof railing**
   - Pry off cover caps -4-, -5- and -11-.
   - Remove hexagon nuts -2-, take off cone washers -3- and pull roof railing off retaining brackets.
2. **Hexagon nut**
   - Tighten to 20 Nm (16 ft-lb)
3. **Cone washer**
4. **Cover cap**
   - Center
5. **Cover cap**
   - Rear
   - When clipping in, insert cover cap guide lugs into roof railing first and then press cover cap onto roof railing foot (check clamping sleeve on roof railing foot is seated correctly)
6. **Retaining bracket**
   - With securing stud, cone washer and seal.
   - Assembling sequence: Press retaining bracket onto roof skin from below and align in hole in roof. Bolt retaining bracket to roof frame (tighten lower bolt first). Fit roof railing and bolt on. Fit cover caps -4-, -5- and -11-.
7. **Hexagon bolt**
   - Tighten to 20 Nm (15 ft-lb)
8. **Welded nut**
9. **Seal**
10. **Base**
11. **Cover cap**
    - Front
    - When clipping in, insert cover cap guide lugs into roof railing first and then press cover cap onto roof railing foot (check clamping sleeve on roof railing foot is seated correctly)

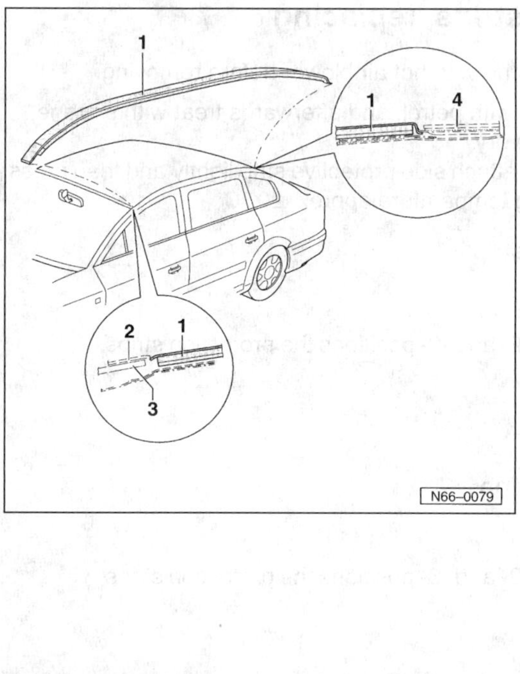

Roof moulding (sedan)

1. **Roof moulding**
   - Removing; starting at windscreen, lever off with soft plastic wedge.
   - Do not bend roof moulding
   - Installing; place roof moulding on end of rear roof channel and press in from rear to front up to stop.
2. **Water deflector**
3. **Windscreen**
4. **Rear window**

Radiator grill, assembly (through 09/2000)

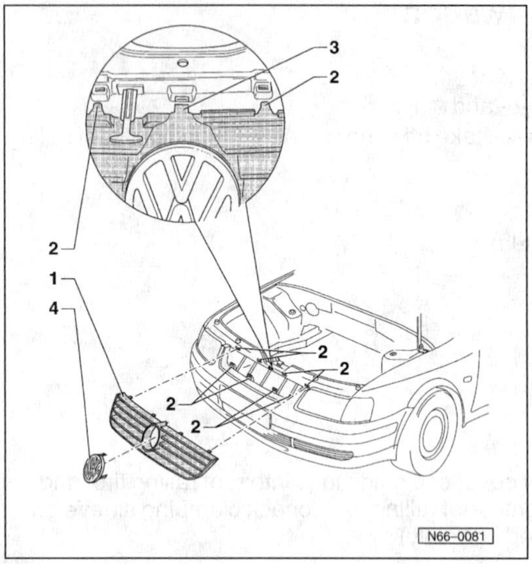

1. **Radiator grille**
   - Material: ASA
   - Removing: Release upper and lower locking lugs -2- with a screwdriver. Pull radiator grille upwards off lock carrier.
   - Installing: Align radiator grille with center fixing point -3- and clip in.
2. **Locking tab**
3. **Center fixing point**
4. **VW emblem**
   - Clipped into radiator grille

Radiator grill, assembly (from 10/2000)

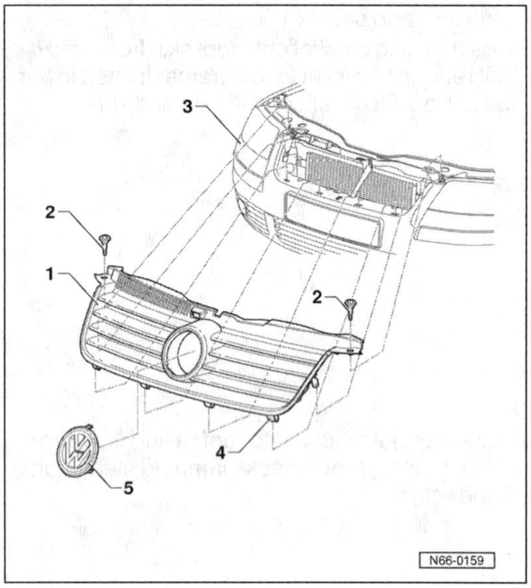

1. **Radiator grille**
   - Material ASA
   - See **Repair Group 63, Release lever, removing**
   - Remove screws -2-. Remove radiator grille upward out of retaining clips -4- of cover -3-.
   - Installing: Align radiator grille, clip in and bolt in.
2. **Bolt**
   - Qty: 2
3. **Cover**
4. **Retaining clips**
   - Qty: 6
5. **Company ornament**
   - Clipped into radiator grille.

Side protective strips, replacing

– Warm side protective strip with hot air blower before removing.
   - Clean exterior panels with petrol, and afterwards treat with silicone remover and then rub dry.
   - Pull off protective foil, attach side protective strip lightly and then press forcefully on. Working temperature approx. 20 °C.

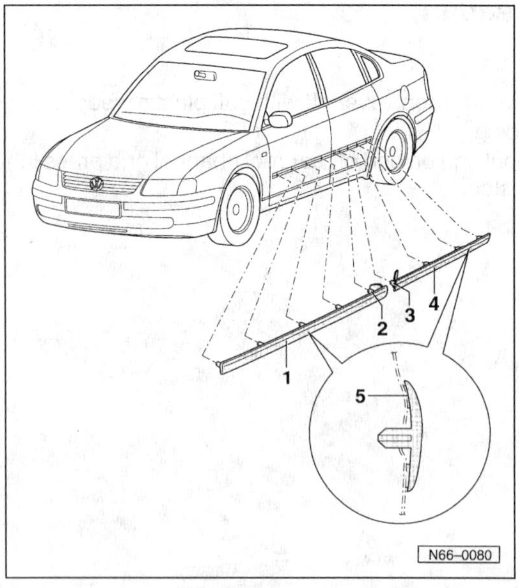

1. **Front door protective strip**
   - Self-adhesive
   - Material PP/EPDM
   - The centralizing lugs -2- and -3- positions the protection strips
2. **Centralizing lug**
3. **Centralizing lug**
4. **Rear door protective strip**
   - Self-adhesive
   - Material PP/EPDM
   - The centralizing lugs -2- and -3- positions the protection strips
5. **Protective foil**

Body-Interior

Section

U

68 Interior Equipment

69 Passenger Protection

Section U – Body-Interior

69 Passenger Protection (continued) Page

70 Interior Trim Page

72 Seats

Page

68 Interior Equipment

Repair Group

U 68

Automatic dimming interior rear-view mirror

The automatic dimming interior rear-view mirror dims in steps, if the driver is blinded by high beams from behind.

It consists of a mirror element and electronics with two photo-sensors.

The electronics recognizes via the photo sensors the light influence from the front and back.

If the light influence from behind is greater than from the front, the electronics emit a voltage to the coating.

The applied voltage changes the color of the electrolytes. The higher the voltage, the darker the electrolyte becomes and causing the light not to be reflected back as strongly.

When reverse is engaged, the dimming function is shut off. Thereby with the aid of the interior rear-view mirror e.g. one can drive out of a dark garage.

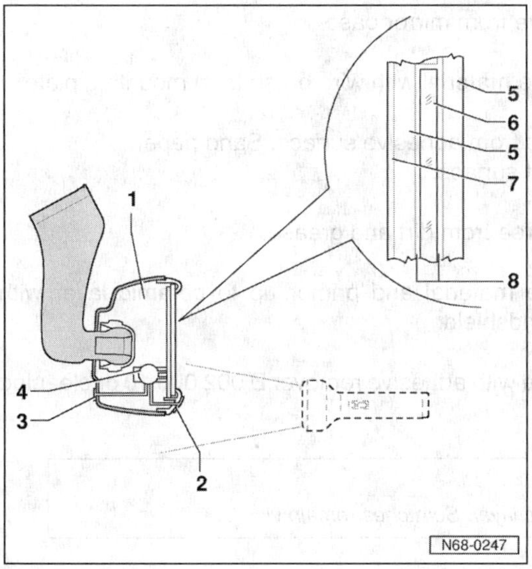

1. **Interior rear-view mirror**
   - Checking function: switch on ignition. Cover photo sensor -20- and light up second photo sensor -19- with flashlight. Mirror surface must dim (reverse gear must not be engaged).
2. **Photo sensor**
3. **Photo sensor**
4. **Electronics**
5. **Glass**
6. **Electrolytic coating**
7. **Silver-reflective coating**
8. **Conductive coating**

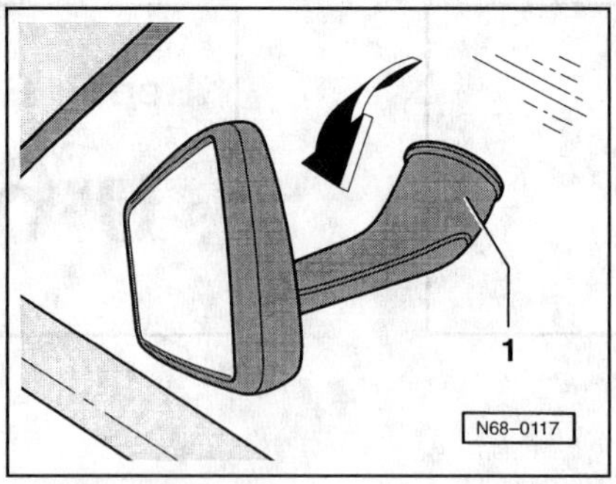

Interior rear-view mirror, removing and installing

– Press interior rear-view mirror -1- off the retaining plate (arrow) by pushing down at angle (locking springs in base of mirror).

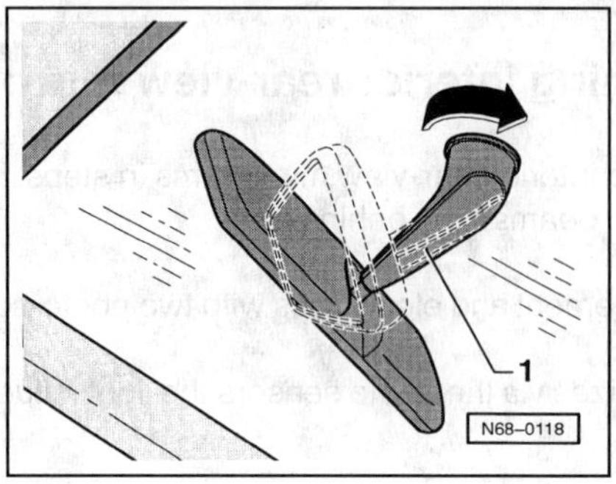

– When installing mirror -1- turn it 60 - 90° from installed position then turn to installed position (arrow), until locking springs engage.

Mounting plate for interior rear-view mirror, bonding

Materials

- Glass-metal-adhesive-set D 000 703 A1

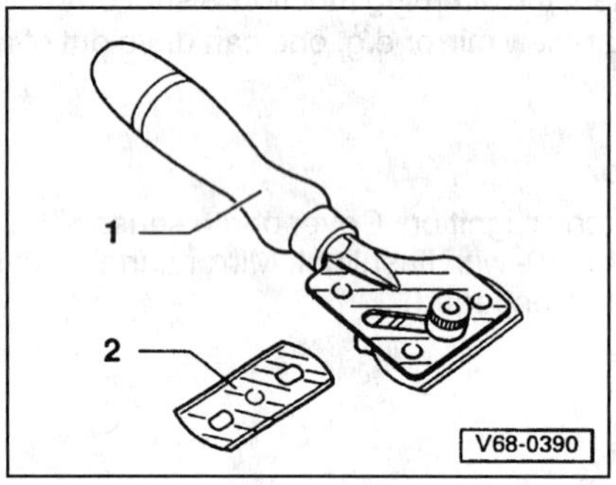

Tools

1 - Glass scraper (commercial)
2 - Blade (commercial)

– Remove mounting plate from mirror base.

– Remove PUR-adhesive material with wire brush from mounting plate.

– Grind off spacer knobs from adhesive surface. Sand paper (360/400 grade) on flat surface.

– Keep ground surface free from dirt and grease.

– Scrape PUR-adhesive material and primer up to ceramic layer with glass scraper off of windshield.

– Clean adhesive surface with adhesive remover D 002 000 10 or cleaning solution D 009 401 04.

> ***CAUTION—***
> *Do not damage ceramic layer. Scratches remain visible.*

– Cut nylon mesh weave to exact size of mirror base.

NOTE—
Hands should be protected (rubber gloves).

– Apply adhesive evenly on mirror base.

– Place nylon mesh weave on mirror base.

– With the tube, dab adhesive on nylon mesh adhesive.

- Only 30 seconds are available from applying the adhesive to the nylon mesh weave until it is pressed onto the windshield.

– Press mirror base securely for 15 seconds (not with force) to the windshield.

– Remove excess adhesive with a cloth.

NOTE —
After 15 minutes the interior rear-view mirror can be mounted.

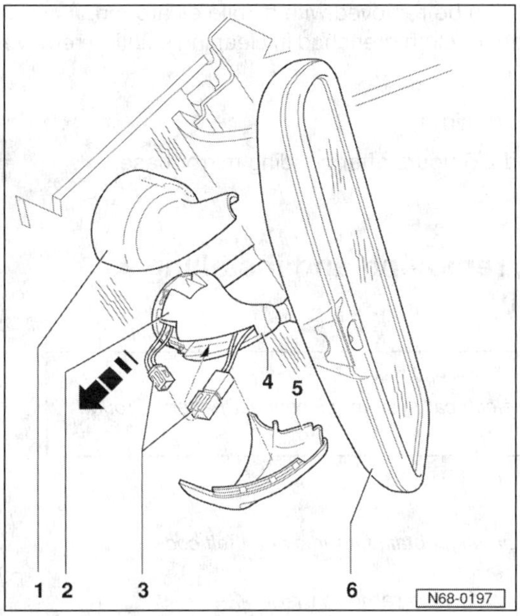

Interior rear-view mirror (with rain sensor), removing and installing

– Press apart left cover cap halves -1- and right -5- and remove.

– Pull connector -3- out of mirror base -4- and disconnect.

– Remove mirror base -4- with mirror -6-downward from retaining plate -2- (arrow).

– Installation is reverse order of removal.

Mirror base for interior rear-view mirror with rain sensor, bonding

The previously described repair procedure can be avoided by using the repair procedure below.

This makes it (simultaneously) possible to re-use the old mirror base.

– Preparation for windshield

- All adhesive and primer residue must be removed up to ceramic layer.
- Clean adhesive surface with cleaning solution D 009 401 04.
- Let dry for at least 10 minutes.
- Apply glass/paint primer D 009 200 02.
- Drying time until bonding: at least 10 minutes to max. 1 hour.

– Preparation of re-used and new mirror bases

- Mechanically remove adhesive residue (applies only for an removed mirror base).
- Sand wet adhesive surface with very fine sand paper (800-1200 grit) until adhesive surface is uniformly moist.
- Then clean with cleaning solution D 009 401 04.

– Bonding mirror base

• Apply 2K Polyurethane adhesive D 180 KD2 A1 directly on ground surface and clean adhesive surface of mirror base.
• Adhesive bead diameter: approx. 2.5 - 3 mm

• After adhesive application press mirror base on to the prepared windshield glass all around.

• Position mirror base in position onto the ceramic coating and hold with adhesive tape.

– Cleaning and removing any excess adhesive

• Carefully remove adhesive tape 0.5 - 1 hour after mirror base has set.
• Soft excess adhesive can be removed with a knife all around.
• With a primer applicator or cloth drenched in cleaning solution remove excess adhesive.

– Installing rain sensor and mirror

• Mirror can be installed 2.5 hours after bonding mirror base.

Center console, removing and installing (through 09/2000)

> ***CAUTION —***
> *Before working on the electrical system; disconnect battery Ground (GND) strap.*

NOTE —
Before disconnecting the battery, obtain for radio anti-theft code.

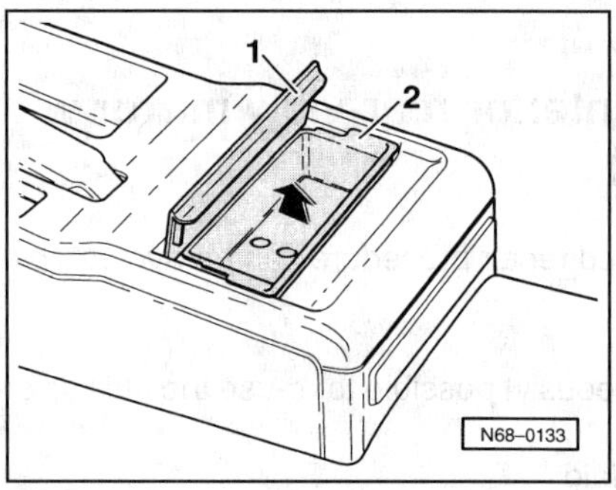

◄ Open cover from ashtray -1- and remove insert -2-.

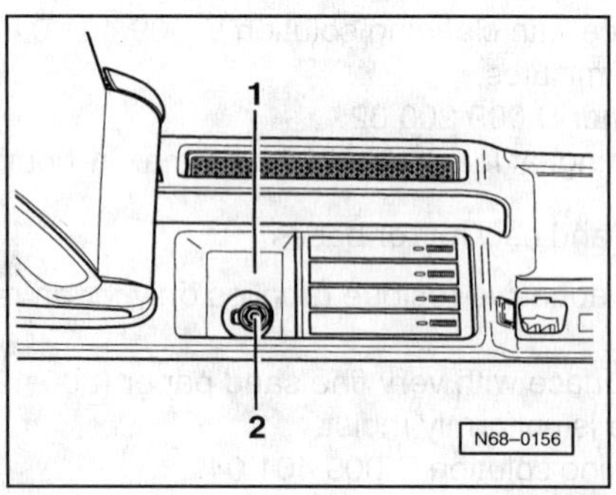

◄ Flip up center armrest.

– Remove rubber base from shelf -1-.

– Remove lock nut -2-.

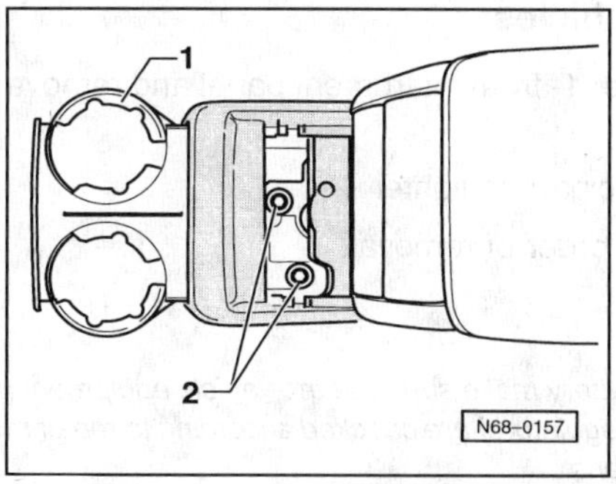

- Fold out cup holder -1-.
- Remove lock nuts -2-.

NOTE —

For vehicles without center armrest only: lock nut is installed under the ashtray.

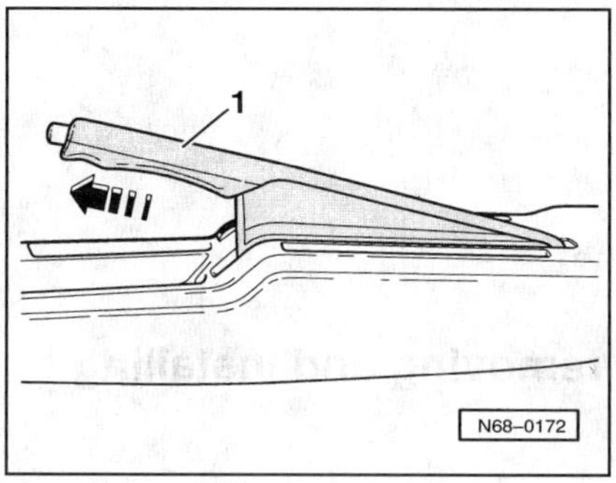

- Pull up parking brake lever.
- Remove trim -1- from parking brake lever (arrow).

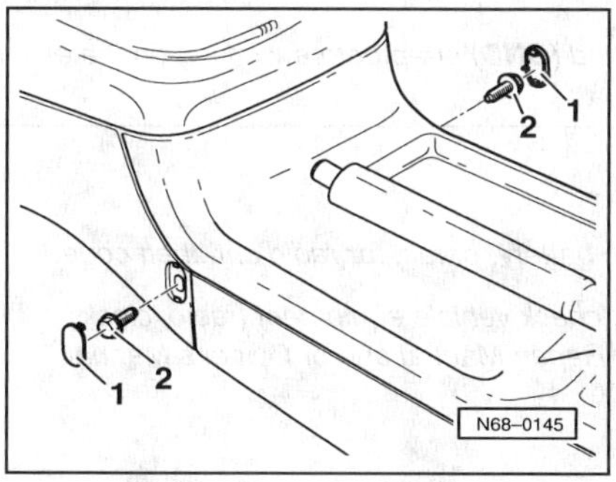

- Pry off cover caps -1-
- Unscrew hex head bolts -2-.

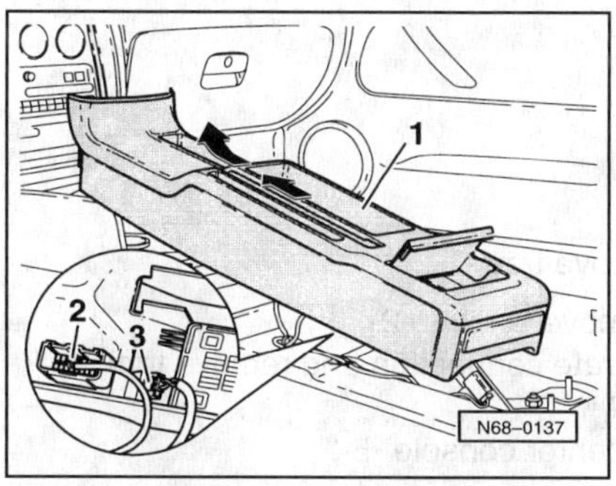

- Remove center console extension -1- from parking brake lever.
- Unclip diagnostic connector -2- (if applicable).
- Remove connector -3- from switch remote/fuel tank door -E204.

For vehicles with manual transmission only:

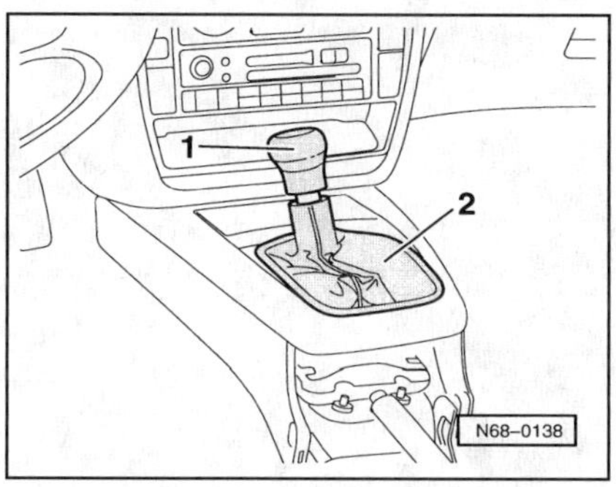

- Unclip cover -2-.
- Unscrew shift knob -1- together with cover -2- and remove from the top.

Continued for all vehicles:

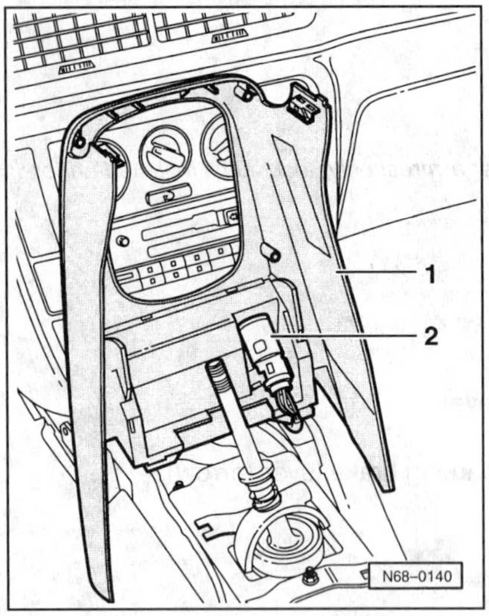

◄ Remove center console -1- from instrument panel and remove from the top out of shift lever.

– Remove connector for cigarette lighter -2-.

- Installation is reverse order of removal.

NOTE —

After disconnecting the battery make sure that the vehicle equipment (radio, clock and power window regulators) are checked according to repair manual and/or operating instructions.

Center console, removing and installing (from 10/2000)

CAUTION —

Disconnect battery Ground (GND) strap before working on the electrical system.

NOTE —

- *Before disconnecting the battery, obtain for radio anti-theft code.*
- *After connecting battery, check vehicle equipment (radio, clock, electric windows) as per Repair Manual and/or Owner's Manual.*

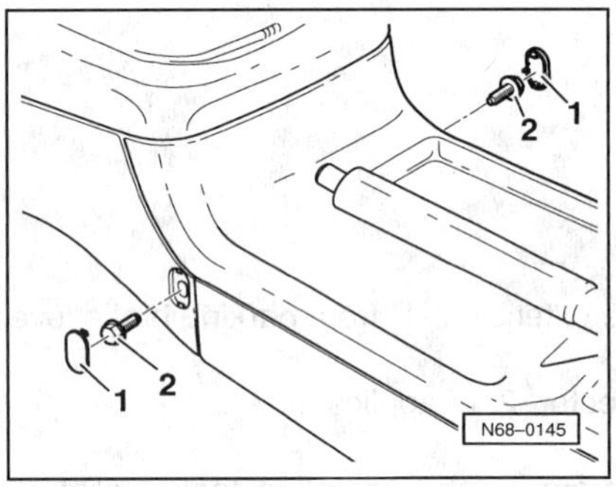

◄ Remove cover caps -1-.

- Remove hex bolts -2-.

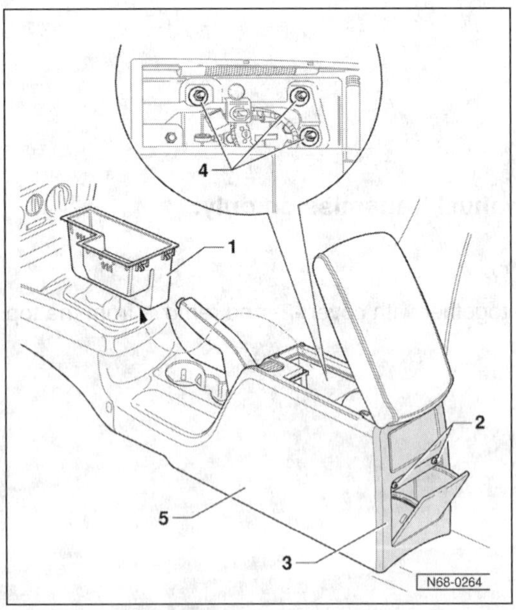

◄ Raise armrest and remove tray -1-.

- Open ashtray and remove screws -2-.
- Loosen trim -3-, separate connection and remove trim.
- Unscrew nuts -4-.
- Remove rear part of center console -5-.

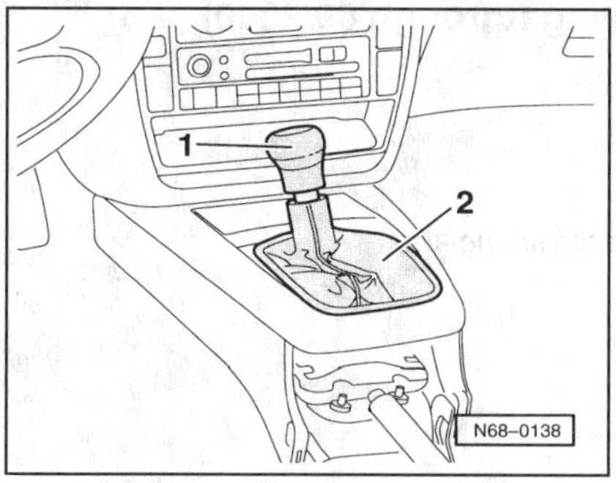

Vehicles with manual transmission

◄ Unclip cover -2-.

• Unscrew shift knob -1- together with cover -2- and remove upwards.

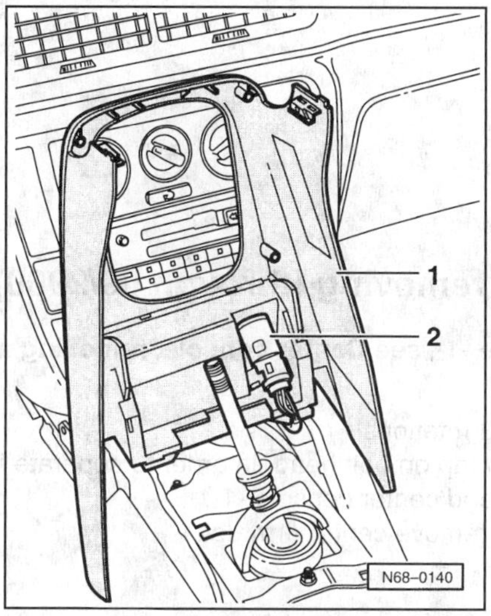

Continued for all vehicles

◄ Pull center console -1- off dash panel and remove shifter upwards.

• Pull connector off cigarette lighter -2-.
• Installation is reverse order of removal.

Front ashtray, removing

– Remove center console, see **Center console, removing and installing**.

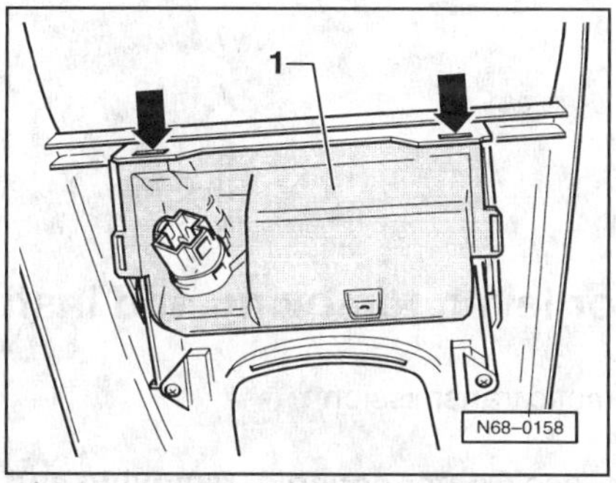

◄ Unhook tabs (arrows) with small screw driver and remove ashtray from center console.

Rear ashtray, removing and installing (through 09/200)

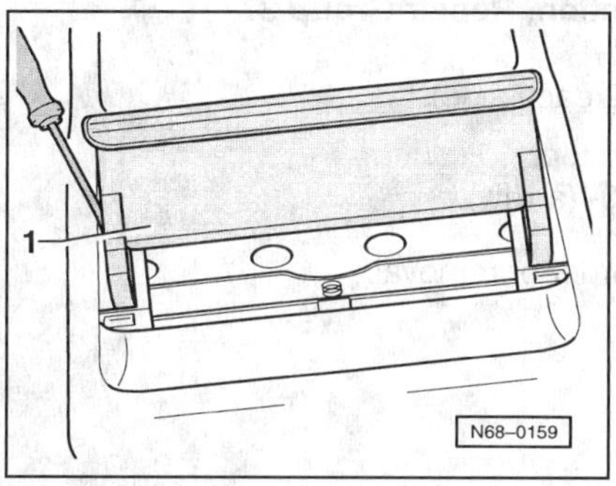

◄ Unhook cover for ashtray -1- from side holes and remove.

Rear ashtray, installing (through 09/2000)

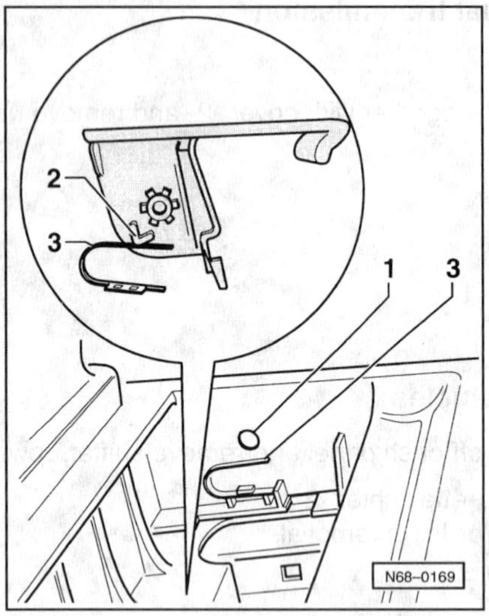

◄ Lock in cover in side holes -1-.

NOTE—

- *Lock stud -2- must lie on the spring -3-.*
- *Left lid hinge shown.*

Center armrest, removing (through 09/2000)

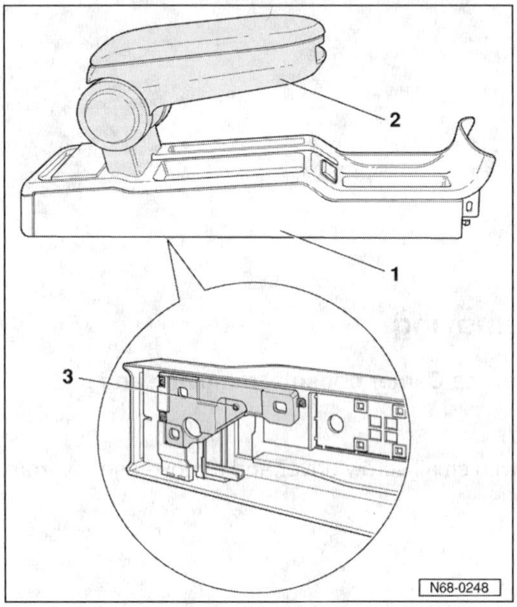

◄ Remove center console -1-, see **Center console, removing and installing**.

- Unscrew bolt -3-, 2 - 3 rotations.
- With a hammer lightly tap on bolt head, in order to separate fitting between armrest -2- and center console -1-.
- Unscrew bolt -3- and remove center armrest.

Cover for selector lever, removing and installing

Only vehicles with automatic transmission

– Remove center console, see **Center console, removing and installing.**

– Remove handle for selector lever, see **5-Speed Automatic Transmission 01V section, Repair Group 37**

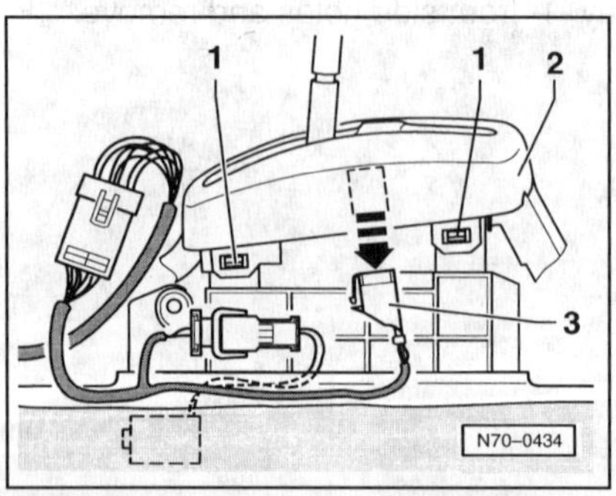

◄ Unhook clips -1- (Qty. 2 each side).

- Remove cover -2-from top.
- Separate connector -3- (arrow).

– Installation is reverse order of removal.

Front cup holder, removing

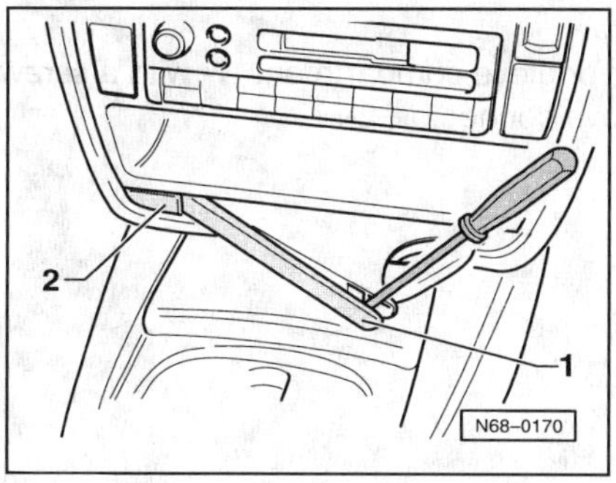

Pry out right trim -1- with screwdriver from cup holder.

- Pry out left trim -2-.

Drivers side shelf, removing

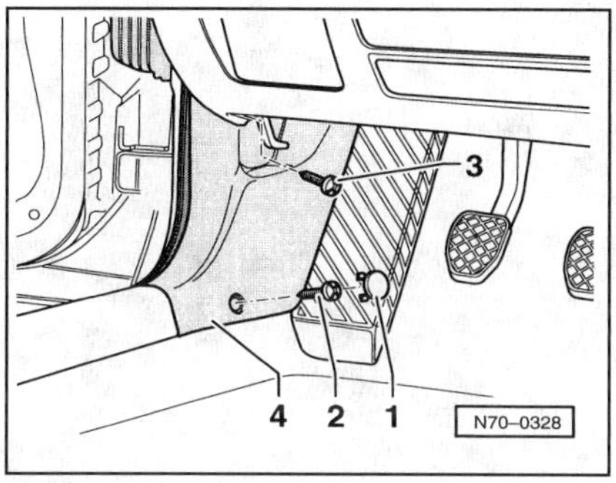

Pry off cover cap -1-.

- Unscrew bolts -2- and -3-.
- Unclip lower A-pillar trim -4- from sill plate and remove.

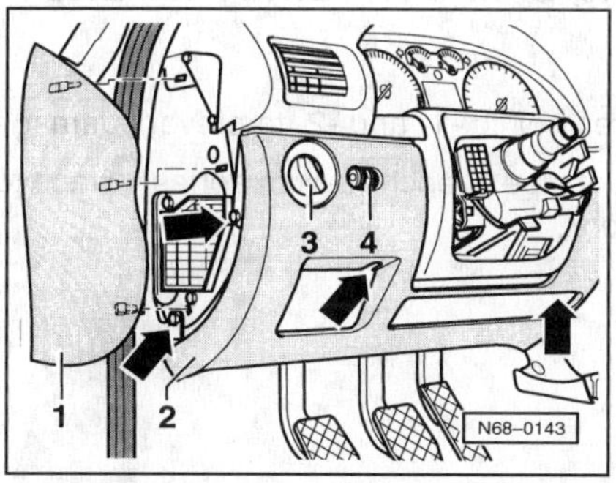

Unclip cover -1-.

- Unscrew bolts (arrows).
- Remove shelf driver side -2-.
- Remove connector from light switch -3- and from adjuster for automatic vertical aim control system -4-.

Passenger side glovebox, removing

> ***WARNING —***
> *Disconnect battery Ground (GND) strap before working on the electrical system.*

NOTE —
- *Before disconnecting the battery, obtain for radio anti-theft code.*
- *After connecting battery, check vehicle equipment (radio, clock, electric windows) as per Repair Manual and/or Owner's Manual.*

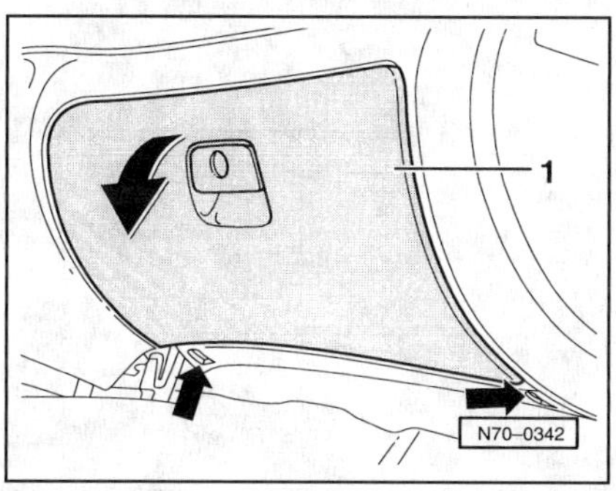

Unscrew bolts (arrows).

- Open glove compartment -1-.

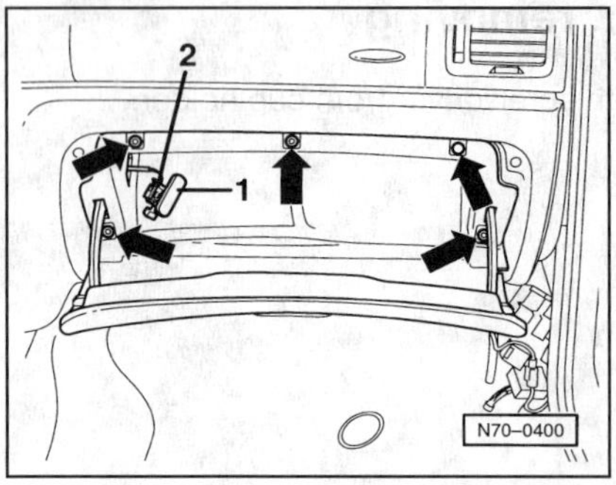

< Unscrew bolts (arrows)

- Carefully pry out light for glove compartment -1- with a screwdriver, from the top and remove connector -2-.

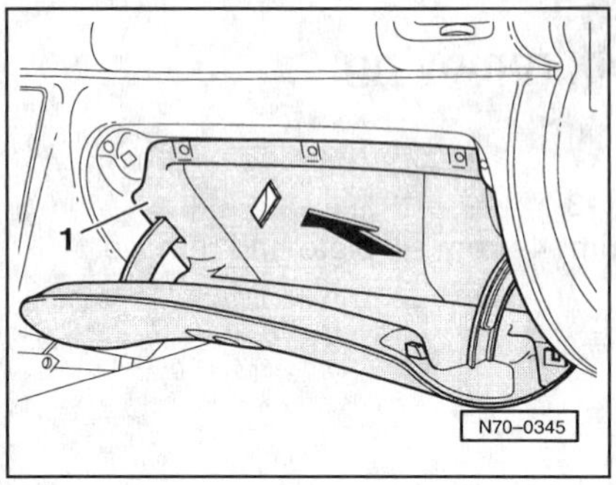

< Remove glove compartment liner -1- from dash board.

Glove box locking cylinder, removing and installing

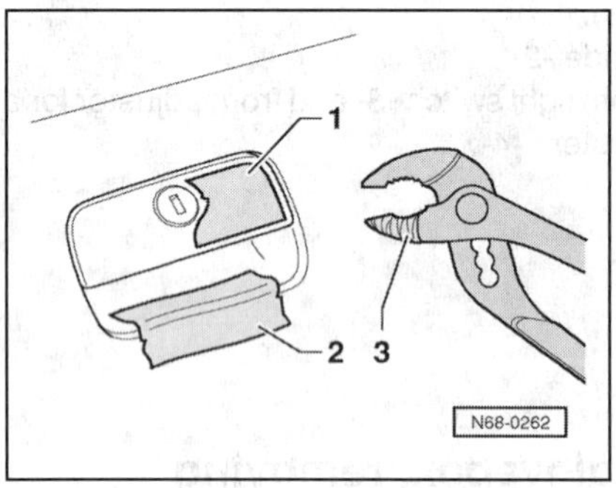

< Mask off the shaded areas of lid -1- and -2- to prevent damage.

- Mask off marked area of pipe wrench -3- with adhesive tape to prevent damage to the glove box.

Removing

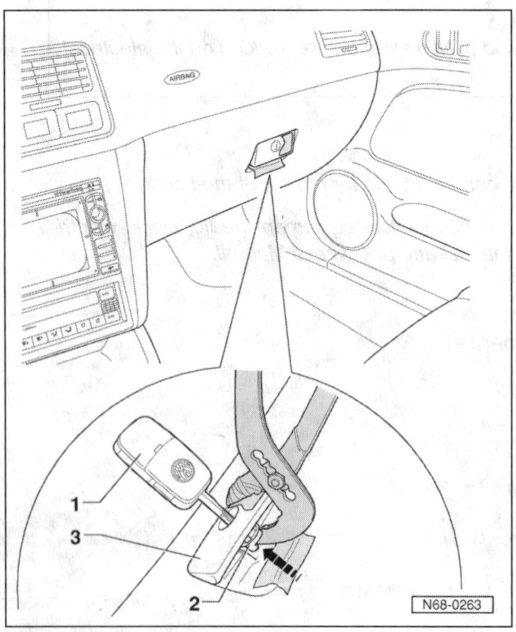

< Insert ignition key -1- into lock cylinder -2-.

- Using pipe wrench, press lock cylinder out of button -3- in direction of arrow.

Installing

– Press locking cylinder into opening in button until it locates audibly.

Cover for outside rear-view mirror, removing

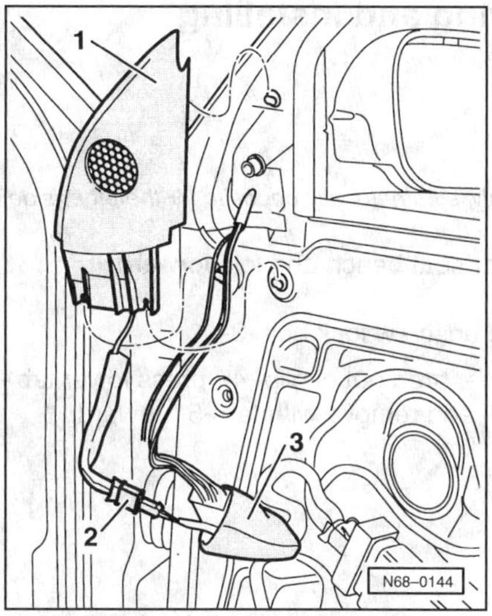

Cover without speaker

Remove cover for outside rear-view mirror -1- from the top.

Cover with speaker

- Remove front door trim, see **Repair Group 70, Front door trim driver side, removing and installing**.
- Remove foam sleeve -3- from connector and disconnect connector for high tones -2-.
- Remove cover for outside rear-view mirror -1- from the top.

Sun visor, removing

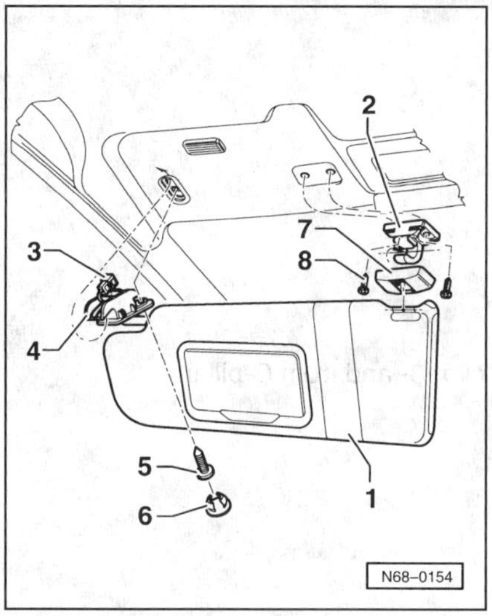

Unhook sun visor -1- from mount -2-.

- Pry off cap -6-.
- Remove screw -5-.
- Unhook sun visor bearing -4- from mount and disconnect connector -3-.
- Pry off cap -7-.
- Remove bolts -8- and remove mount -2-.

Handle (roof), removing

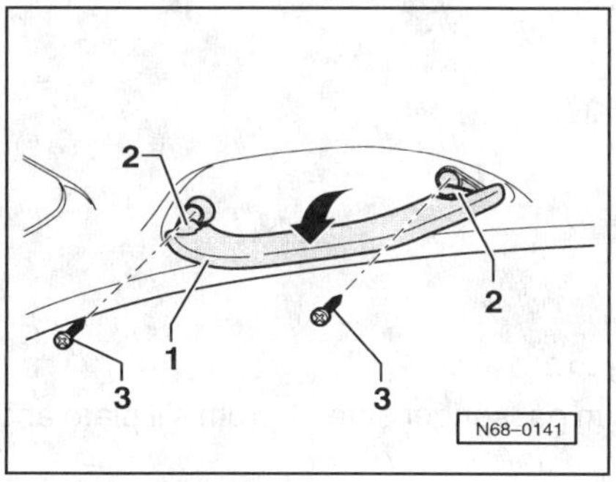

Fold handle -1- down.

- Pry off covers -2- with screwdriver and fold up.
- Unscrew Phillips-head screws -3- and remove handle -1-.

Sill plate, removing and installing

Removing

NOTE—
Illustration for one side only, symmetrically opposite for the other side.

– Unlock seatback for rear seat bench and fold forward.

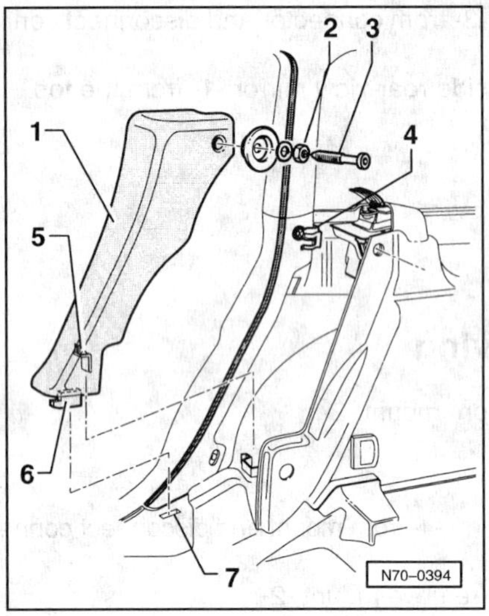

◄ Loosen hex nut -2- and unscrew lock bolt -3-.

- Pull cushion -1- in upper area out of clip -4-, press lower tab -6- to the back from sill plate -7- and remove with tab -5- upward.

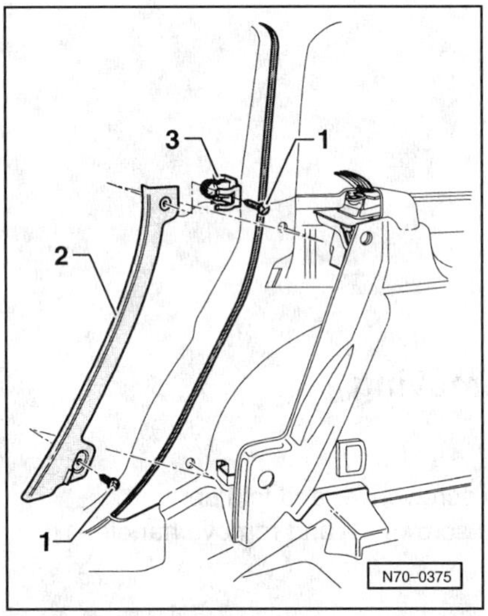

◄ Unscrew bolts -1-

- Remove trim -2- from clip -3- and from C-pillar.

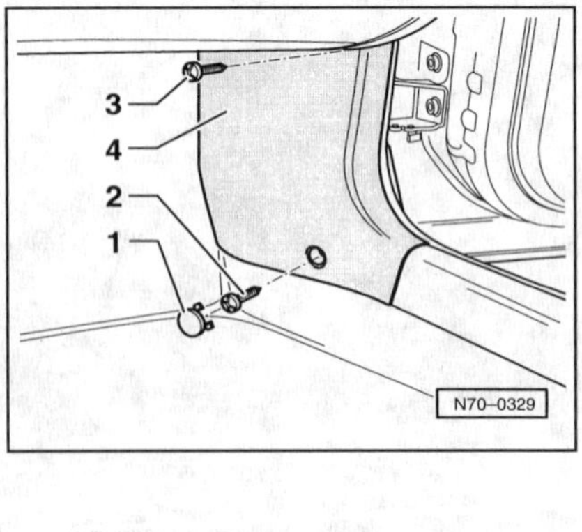

◄ Pry off cover cap -1-.

- Unscrew bolts -2- and -3-.
- Unclip lower A-pillar trim passenger side -4- from sill plate and remove.

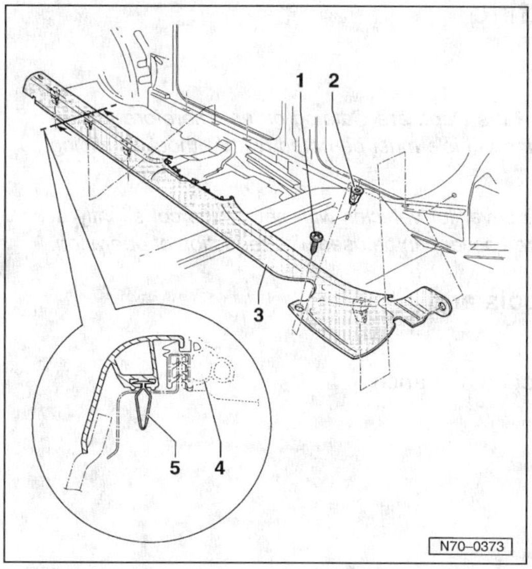

- Remove screw -1-.
  - Unclip sill plate -3-.
- Installation is reverse of removal, noting the following:
  - Before installing sill plate check expanding nut -2- and clips -5-, replace if necessary.
  - After installing sill plate check for correct fit of seal -4-.

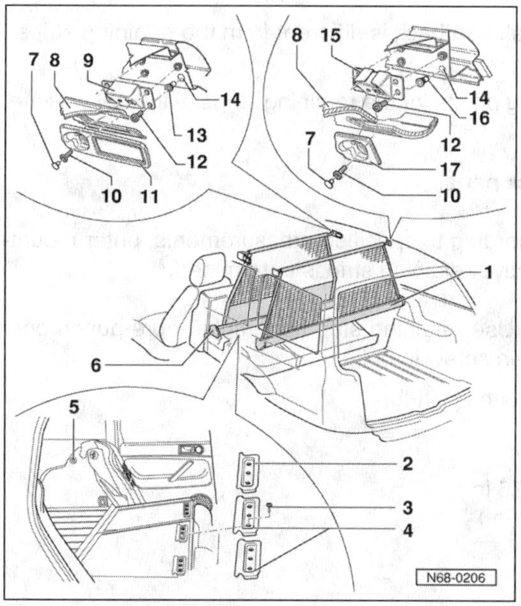

Net wall, assembly

1. **Net wall**
   - Rear position, seatback set up
2. **Retaining strip left**
3. **Bolt**
   - Qty.: 9
   - Tighten to 5 Nm (44 in-lb)
4. **Retaining strips center and right**
5. **Cover**
   - For locking bolt
6. **Net wall**
   - Front position, seatback folded over
7. **Cover cap**
8. **Molded headliner**
9. **Bracket front**
10. **Bolt**
    - Tighten to 23 Nm (17 ft-lb)
11. **Front eye hook**
12. **Bolt**
    - Qty.: 4
    - Tighten to 23 Nm (17 ft-lb)
13. **Blind rivet nut**
    - Qty.: 4
    - M 8 x11
14. **Hex hole**
    - For blind rivet nut
15. **Bracket rear**
16. **Blind rivet nut**
    - Qty.: 4
    - M 8 x9
17. **Rear eye hook**

Net wall, retrofitting

NOTE —

- *The required parts per Parts Dept. are Catalog order. Therefore the respective equipment for the vehicle must be noted (with/without tilt sliding sunroof).*
- *The edge of the loading cover for vehicles with net wall is cut slightly at overlap area. Loading area cover can be used with reduction of operation.*

Required special tools and equipment

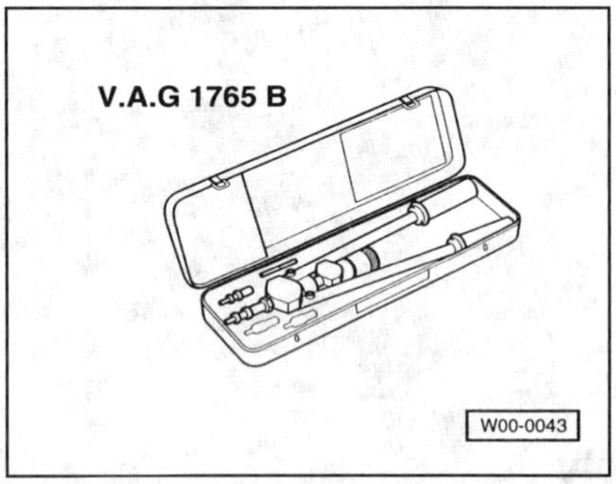

V.A.G 1765 B Blind rivet nuts-wrench

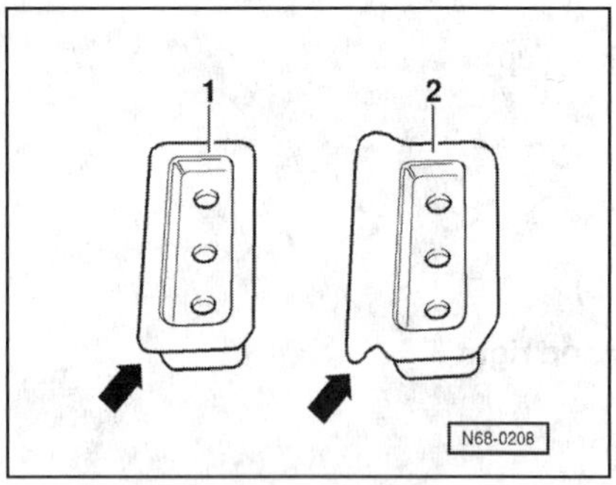

Retaining strip -1- for left seatback is different from the retaining strips -2- for the right (arrows).

- The required mounting points for the retaining strips of the net wall are on the seatback.

– Fold over both seatback parts.

– By pressing a tack according to specified measurements, outer mounting point for the respective retaining strip is determined.

– Bolt on retaining strip: Use retaining strip as template, pre-punch both mounting holes and then screw in bolts.

- Tightening torque: 5 Nm (44 ft-lb)

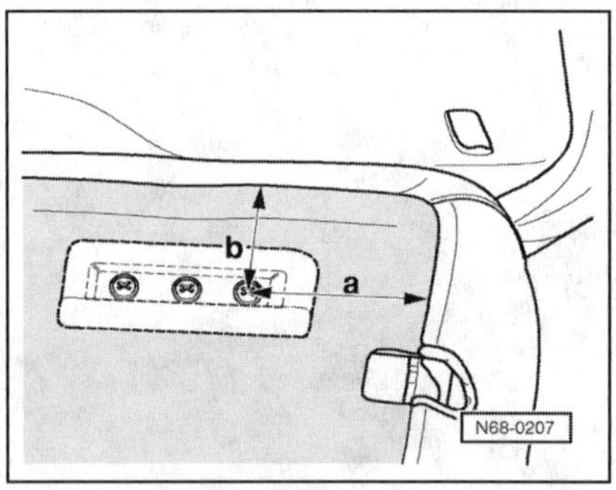

Right retaining strip

- Dim. -a- = 9.5 cm (3.75 in.)
- Dim. -b- = 5 cm (1.96 in.)

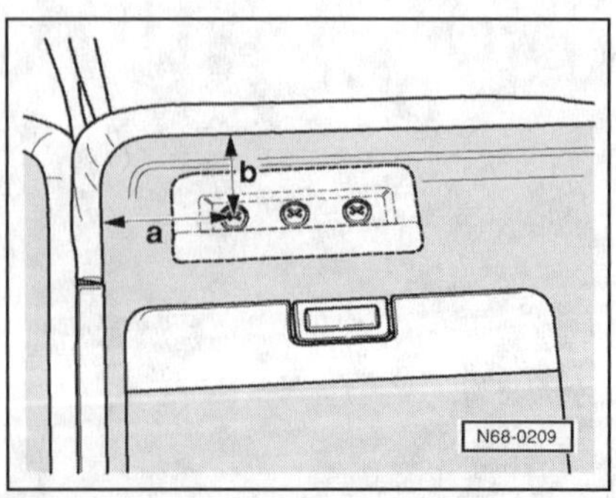

Retaining strip in the center

- Dim. -a- = 7 cm (2.75 in.)
- Dim. -b- = 5 cm (1.96 in.)

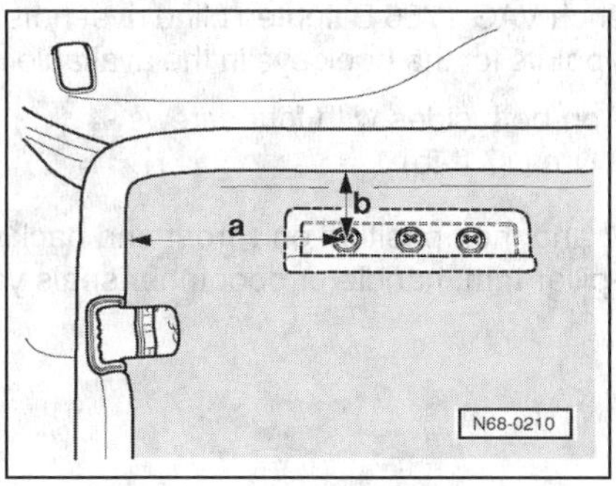

Retaining strip left

- Dim. -a- = 11.5 cm (4.5 in.)
- Dim. -b- = 5 cm (1.96 in.)

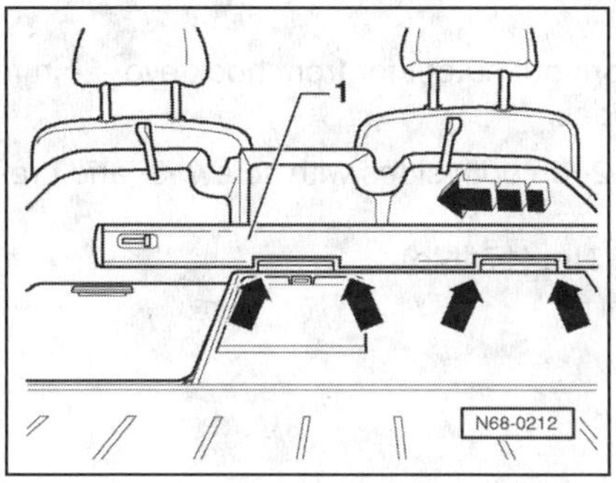

Place net wall -1- with openings (arrows) on the right and center bracket -2- and slide to the left until it locks in place.

– Both seatback parts folded up to lock.

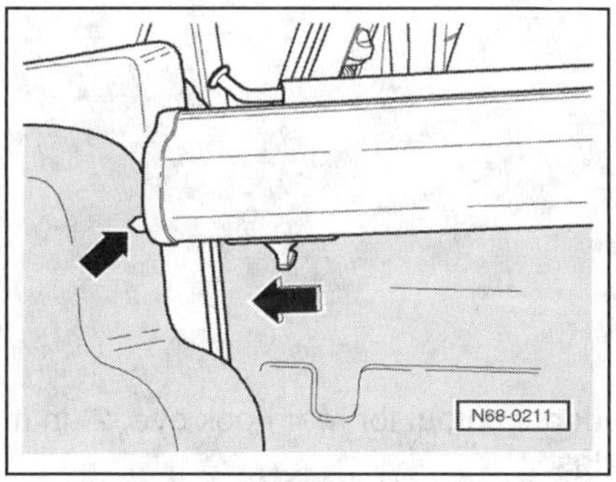

Slide locking bolt of net wall out to the left and mark contact point on the left and load trim (arrow).

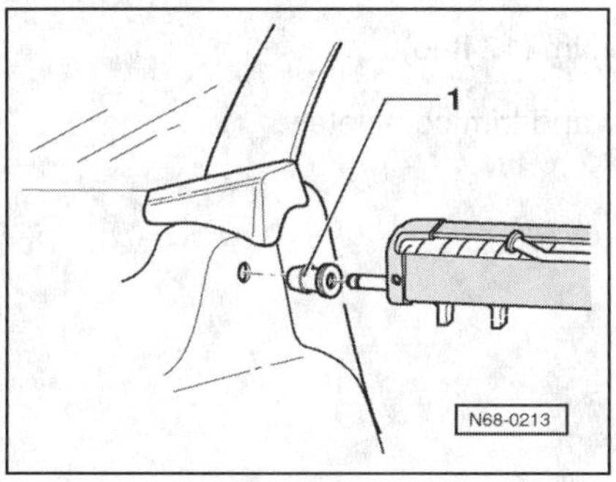

Place hole (diameter= 20 mm (0.78 in.)) in the trim (e.g. hole circle saw) and insert -1- in body hole.

– Remove headliner, see **Repair Group 70, Molded headliner with sunroof, removing and installing**.

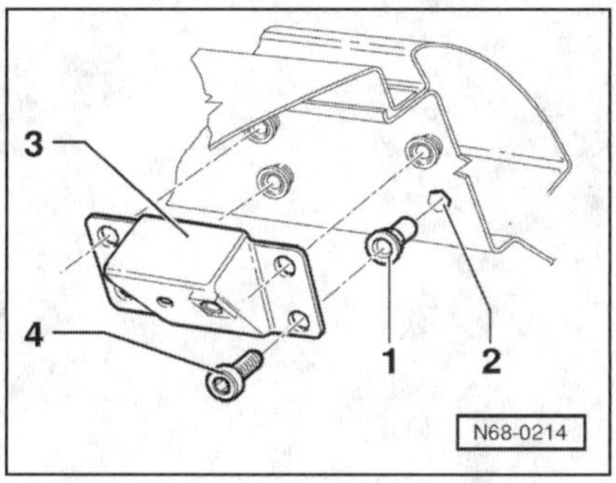

With blind rivet nuts wrench VAG 1765 B, tighten blind rivet nuts (M 8 x 11) -1- at the front mounting points (in front of interior light) for the hook eye in the available holes -2-.

– Install front bracket -3- on both sides with four screws -4-.

- Tightening torque: 23 Nm (17 ft-lb)

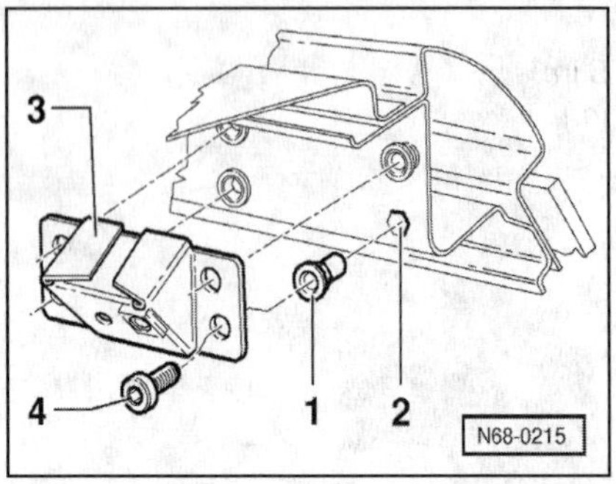

◄ With blind rivet nuts wrench VAG 1765 B tighten blind rivet nuts (M8 x 9) -1- at the rear mounting points for the hook eye in the available holes -2-.

- Install rear bracket -3- on both sides with four screws -4-.
- Tightening torque: 23 Nm (17 ft-lb)

– Install molded headliner and then position only front and back. (Do not install B and C-pillar trim, handle or door inner seals yet).

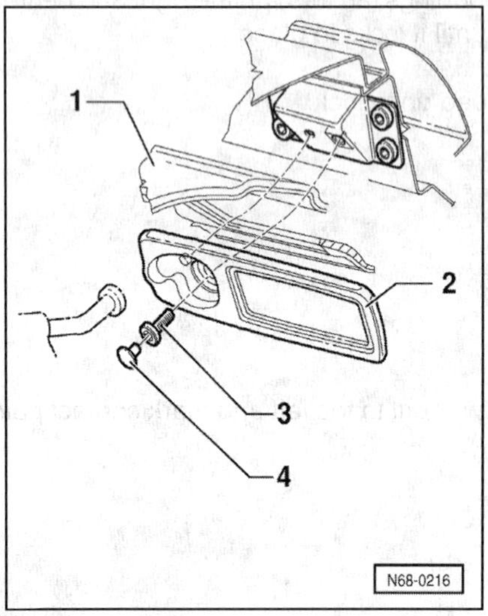

◄ With workshop tools work out cutout for front hook eye -2- in molded headliner -1- on both sides.

- Install front hook eye -2- on both sides with screw -3- and press in cap -4- in the bolt head.
- Tightening torque: 23 Nm (17 ft-lb)

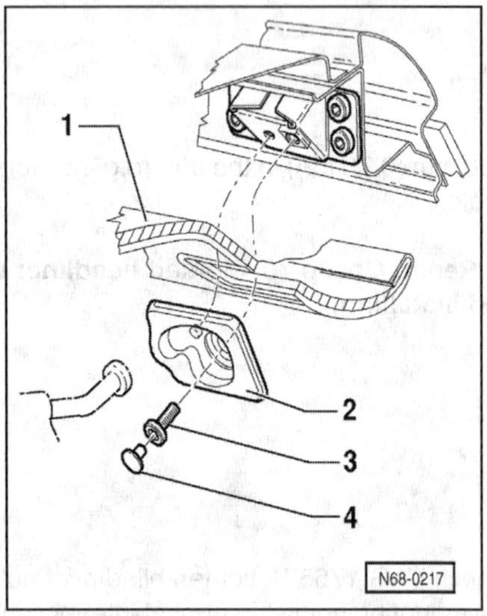

◄ With workshop tools work out cutout for rear hook eye -2- in molded headliner -1- on both sides.

- Install rear hook eye -2- on both sides with screw -3- and press in cap -4- in the bolt head.
- Tightening torque: 23 Nm (17 ft-lb)

– Install molded headliner and trim completely.

69 Passenger Protection

Repair Group

U 69

Seat belts

Safety requirements for belt tensioner

- Testing, assembly and repair work should only be conducted by trained personnel.
- Belt tensioner components should not be opened nor repaired; always use new parts (risk of injury).
- Belt tensioner units which take a hard fall or drop to the floor, must not be installed into the vehicle.
- Belt tensioner units, which are mechanically damaged (bubbles, tears) must always be replaced.
- Storage and transport fall under the laws governing explosive materials.
- Install belt tensioner units immediately after removing from transport container.
- For disassembly of belt tensioner do not use impact wrench.
- If work is interrupted place belt tensioner unit into transport container again.
- Do not leave un-installed belt tensioner unattended.
- Belt tensioner unit should not be handled with grease, cleaners or other solutions and do not expose to temperatures over 100°C (212°F) even briefly.
- The pyrotechnic propellant has no expiration date, this means it can be stored indefinitely and is maintenance free.

Front seat belts with belt tensioners, assembly (through 09/2000)

CAUTION—

Before starting any body disassembly, straightening or any other body repair work, the belt tensioner units must be removed.

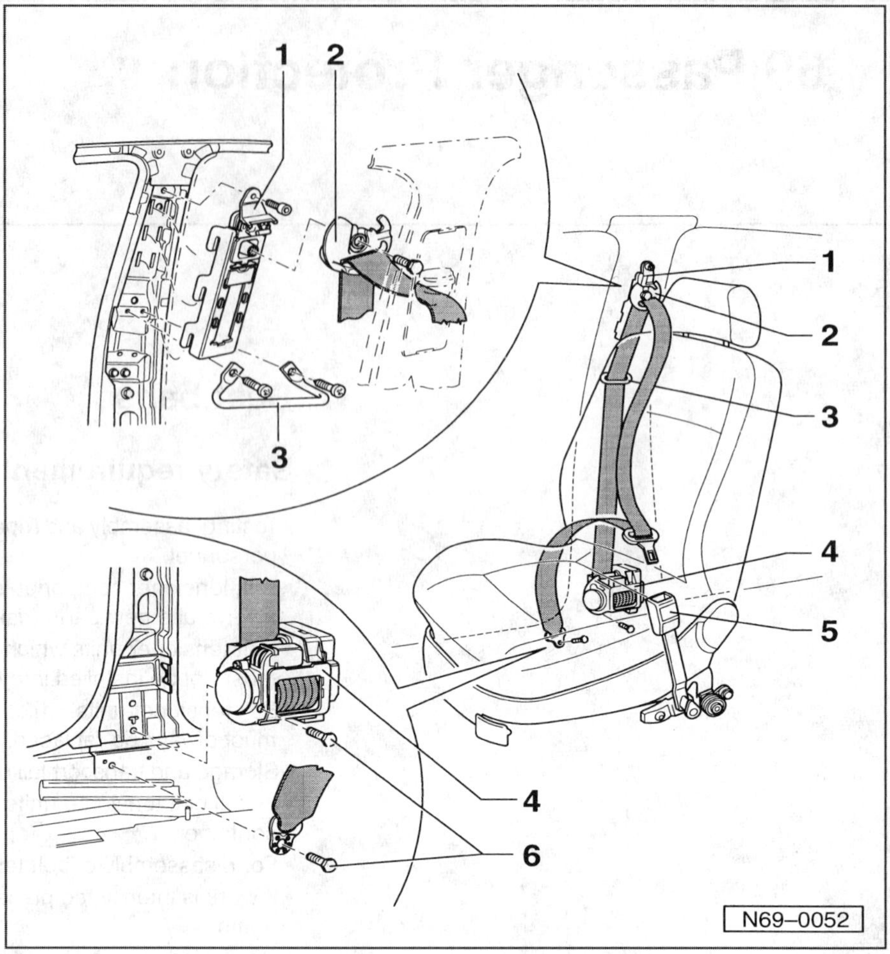

1. **Belt height adjustment**
   - To remove the height adjustment, remove B-pillar trim beforehand
2. **D-ring**
3. **Belt guide**
   - Fastened with 2 Phillips-head screws
   - Tighten to 1.5 Nm (13 in-lb)
4. **Seat belt with belt tensioner**
   - To remove, take out lower B-pillar trim beforehand and remove sill panel, see **Repair Group 70, Lower B-pillar trim, removing and installing** and **Repair Group 68, Sill plate, removing and installing**
   - Retaining tab determines the position of belt retractor
   - By loosening the belt retractor bolt -6- belt tensioner is secured, and in reverse it is functional again
   - Return old (removed) components which have not been ignited to Volkswagen of America for disposal.
5. **Buckle front**
6. **Hex head bolt**
   - Tighten to 40 Nm (30 ft-lb)

Front seat belt with belt tensioner, assembly (from 10/2000)

CAUTION —

Before starting any body disassembly, straightening or any other body repair work, the belt tensioner units must be removed.

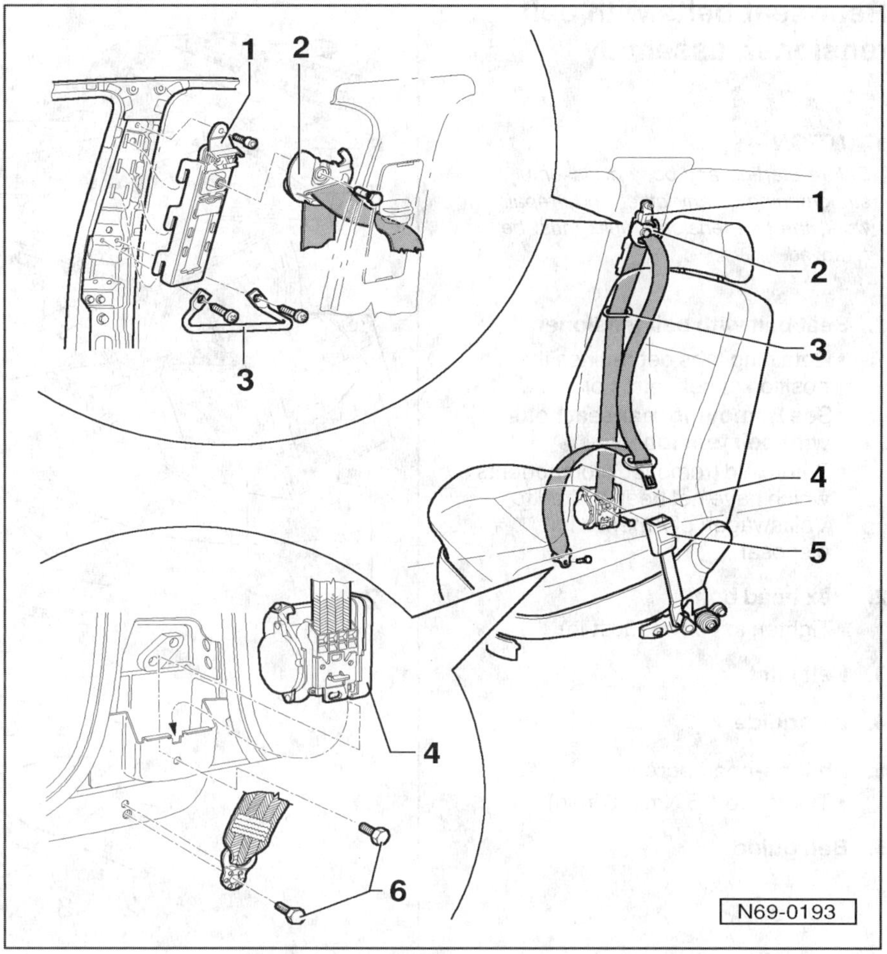

1. **Belt height adjustment**
   - To remove height adjustment remove B-pillar trim beforehand
2. **D-ring**
3. **Belt guide**
   - Secured with 2 Phillips-head screws
   - Tighten to 1.5 Nm (13 in-lb)
4. **Seat belt with belt tensioner**
   - To remove, take out lower B-pillar trim beforehand and remove sill panel trim, see **Repair Group 70, Lower B-pillar trim, removing and installing** and **Repair Group 68, Sill plate, removing and installing.**
   - Retaining tab determines position of belt retractor
   - Loosening the belt reel bolt -6- secures belt tensioner safe and the opposite returns belt tensioner to a functional condition
   - Return old (removed) components which have not been ignited to Volkswagen of America for disposal.
5. **Buckle front**
6. **Hex head bolt**
   - Tighten to 40 Nm (30 ft-lb)

Rear seat belts with belt tensioner, assembly

CAUTION —

Before starting any body disassembly, straightening or any other body repair work, the belt tensioner units must be removed.

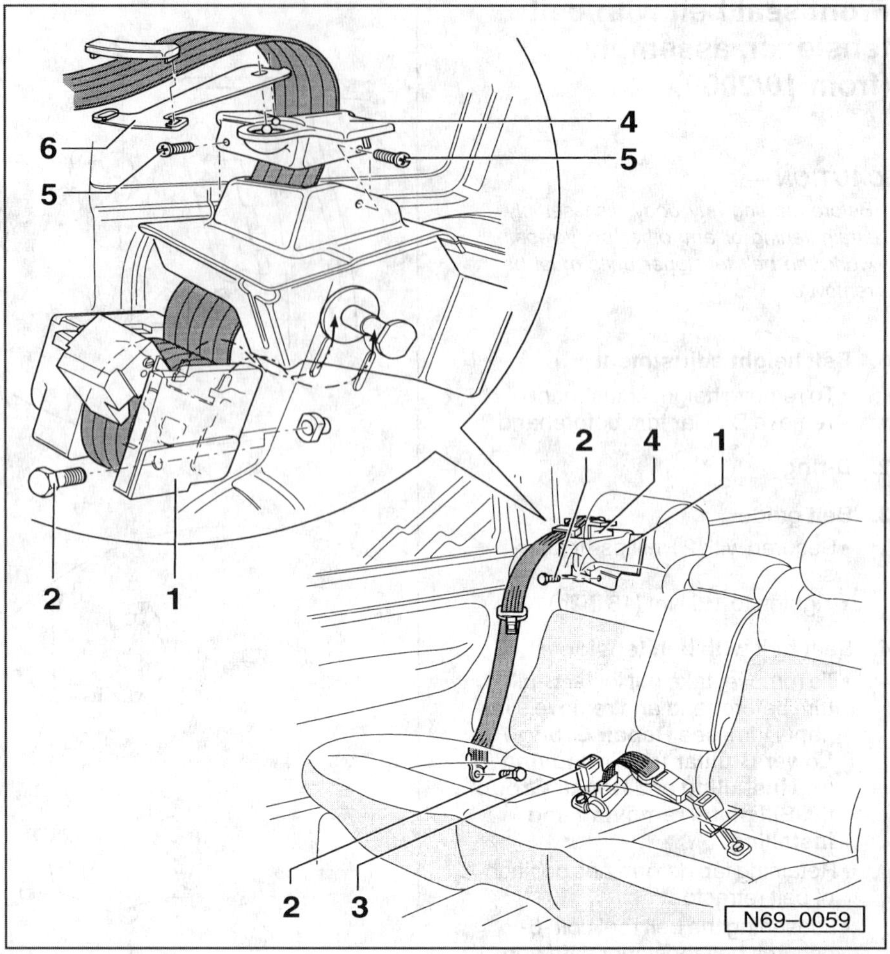

1. **Seat belt with belt tensioner**
   - Retaining tabs determines the position of belt retractor
   - See **Removing rear seat belts with belt tensioner**
   - Return old (removed) components which have not been ignited to Volkswagen of America for disposal
2. **Hex head bolt**
   - Tighten to 40 Nm (30 ft-lb)
3. **Belt latch**
4. **Belt guide**
5. **Phillips-head screw**
   - Tighten to 1.5 Nm (13 in-lb)
6. **Belt guide**

Removing rear seatbelt with belt tensioner

– Unlock seatback for rear seat bench and fold forward.

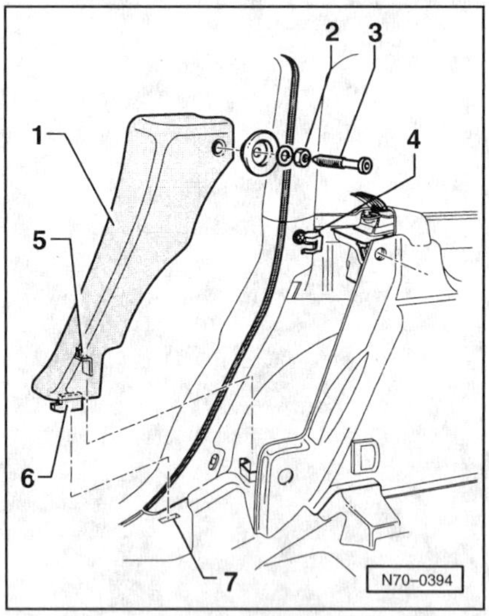

◄ Loosen hex nut -2- and unscrew lock bolt -3-.

- Pull cushion -1- in upper area out of clip -4-, press lower tab -6- to the back from sill plate -7- and remove with tab -5- upward.

◄ Unscrew bolt -1- (hereby belt tensioner secured).

- Unhook seatbelt with belt tensioner (arrows).
- Opening belt guide, see **Rear seat belts with belt tensioner, assembly**, item -4-.
- Unscrew belt strap -2- and remove from console.

Rear center seatbelt, assembly

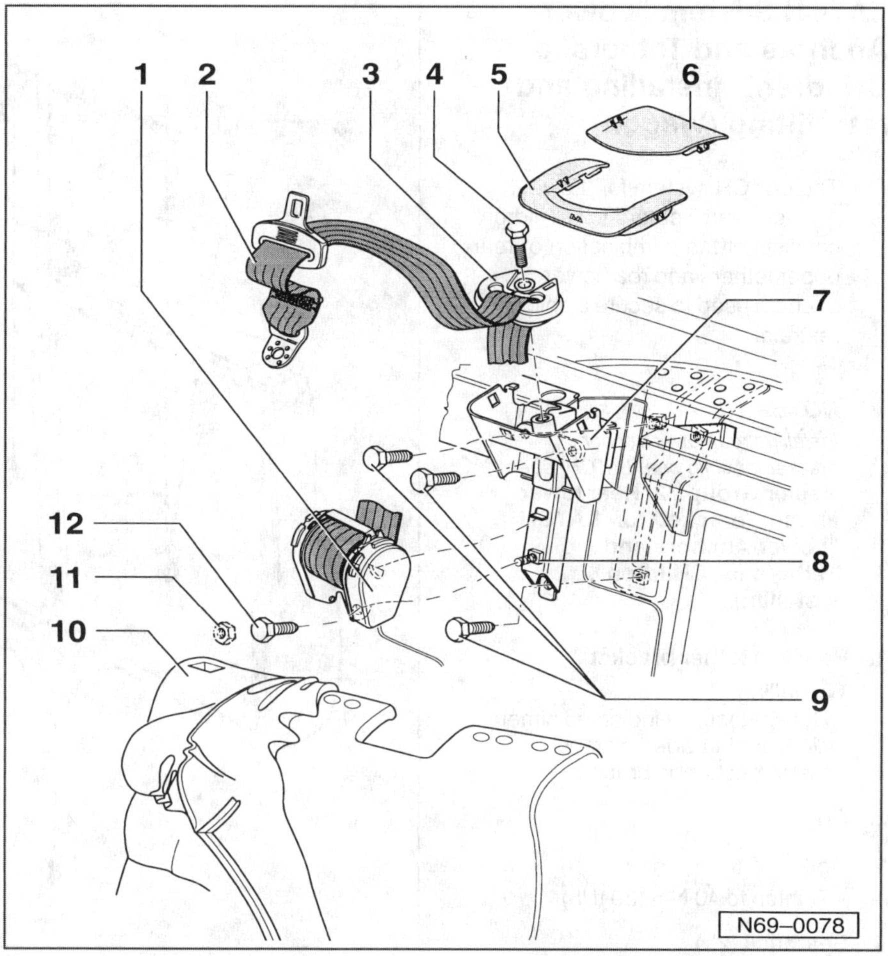

1. **Belt retractor**
   - See **Rear center seatbelt, assembly**
2. **Belt**
3. **Belt guide**
4. **Hex head bolt**
   - Tighten to 40 Nm (30 ft-lb)
5. **Belt guide**
6. **Cover cap**
7. **Support for seatbelt**
   - Bolted only with rigid seatback version, for all other versions the support is welded to seat frame
8. **Threaded plate**
   - Only for rigid seatback, for all other versions there are threaded bolts
9. **Hex head bolts**
   - Only for rigid seatback
   - Tighten to 40 Nm (30 ft-lb)
10. **Cushion**
11. **Hex nut**
    - not for rigid seatback
12. **Hex head bolt**
    - Only for rigid seatback
    - Tighten to 40 Nm (30 ft-lb)

Seatbelt retractor, removing

– Remove seat backrest.

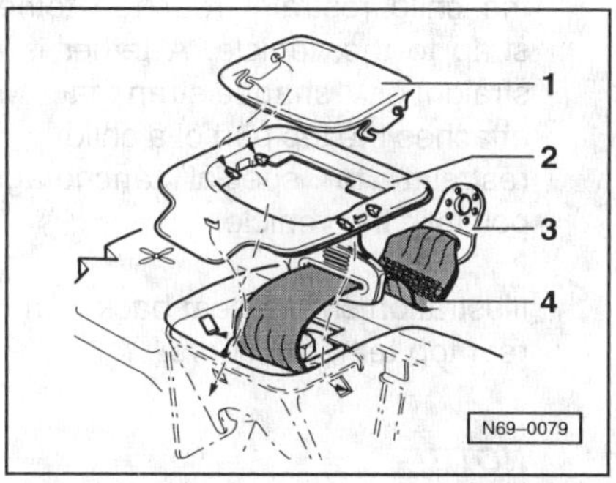

◀ Pry cover cap -1- from belt guide -2-.
- Unscrew belt strap -3- from attachment point.
- Unclip belt guide -2- and remove from belt -4-.

– Remove headrest guides, see **Repair Group 72, Headrest guides, removing.**

– Loosen cover and cushion for seat bench back, see **Repair Group 74, Cover for seat bench, removing and installing.**

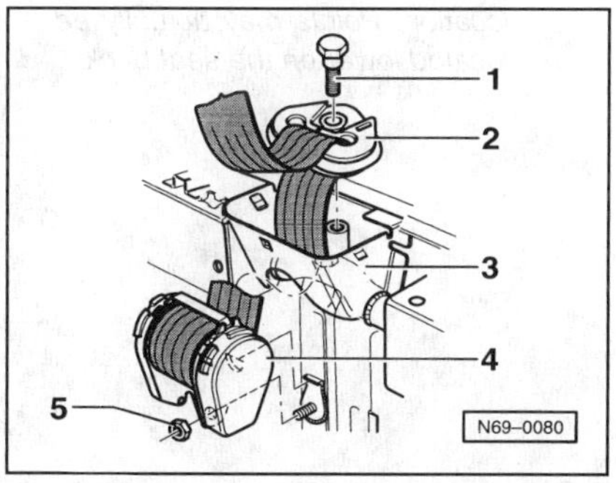

◀ Unscrew hex head bolt -1-, and remove belt guide -2- from support-3-.
- Unscrew hex nut -5- (or hex head bolt on rigid seatback) (and unhook belt retractor -4- from support -3-.
- Remove belt with belt guide from support -3- downward.

NOTE —
To uncoil belt, belt retractor must be in installed position.

LATCH System "Lower Anchors and Tethers for CHildren", installing and retrofitting (Wagon)

The LATCH system for "Lower Anchors and Tethers for CHildren" consists of the combination of rear upper tethers and rear lower anchors used to secure a child restraint.

NOTE —
Installing rear lower anchor bracket for child seat LATCH, see **Repair Group 72, Rear lower anchor bracket for LATCH "Lower Anchors and Tethers for CHildren", installing**.

1. **Rear top tether bracket**
   - Installing:
     Cut seatback at indicated dimensions for threaded plate -5-.
     Install anchoring bracket
2. **Cover**
3. **Bolt**
   - Tighten to 40 Nm (30 ft lb)
4. **Seatback (2x)**

   NOTE —
   - *Verify threaded plate is in place.*
   - *Cut material to allow bolt clearance.*
5. **Threaded plate (3x)**
   - Part of backrest

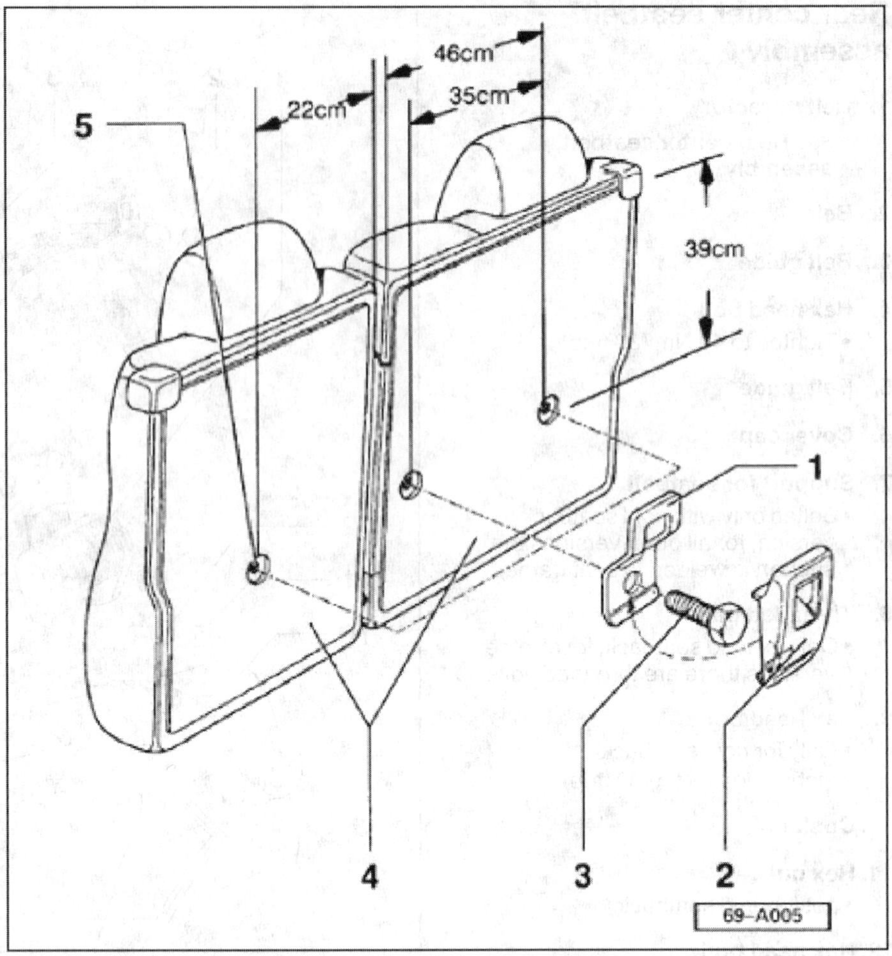

A tether anchor is used to connect the child restraint rear top tether strap to the vehicle. A tether is a straight or V-shaped strap that attaches the top part of a child restraint to special anchorage points in the vehicle.

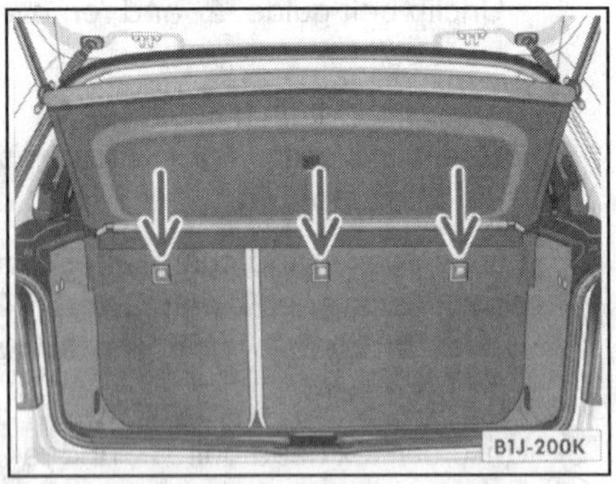

Illustration shows seat back with rear top tether anchors.

NOTE —
- *Some Passat Wagon models may require a seat back change for tether anchor installation.*
- *Anchor points shown in illustration at left are in the approximate location. Points may actually be located lower on the seat back.*

LATCH System "Lower Anchors and Tethers for CHildren", installing and retrofitting (Sedan)

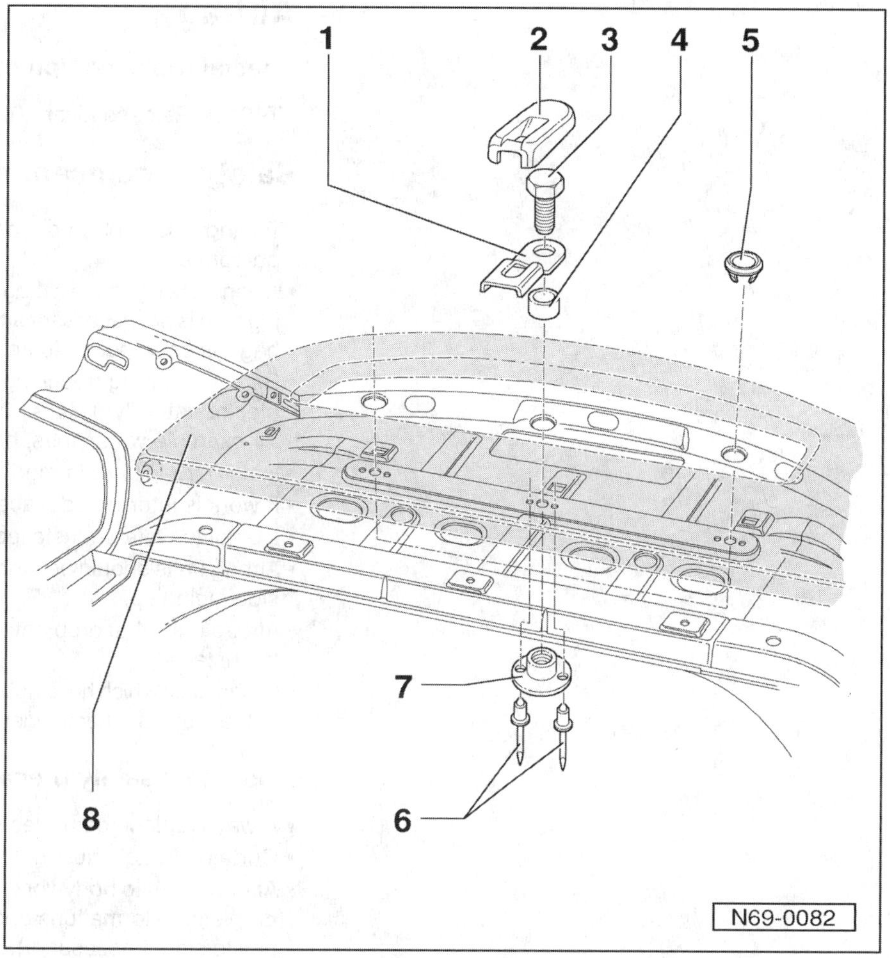

The LATCH system for "Lower Anchors and Tethers for CHildren" consists of the combination of rear upper tethers and rear lower anchors used to secure a child restraint.

NOTE —
Installing rear lower anchor bracket for child seat LATCH, see **Repair Group 72, Rear lower anchor bracket for LATCH "Lower Anchors and Tethers for CHildren", installing**.

1. **Rear top tether bracket**
   - Installing:
     Remove rear shelf -8- and knock out holes at marked locations
     Install threaded plate to inside of rear compartment with rivets -6-
     Install rear shelf into vehicle
     Install bracket -1- or plug -5-
2. **Cover**
3. **Bolt**
   - Tighten to 40 Nm (30 ft lb)
4. **Sleeve**
5. **Plug**
6. **Rivet**
7. **Threaded plate**
8. **Rear shelf**

A tether anchor is used to connect the child restraint rear top tether strap to the vehicle. A tether is a straight or V-shaped strap that attaches the top part of a child restraint to special anchorage points in the vehicle.

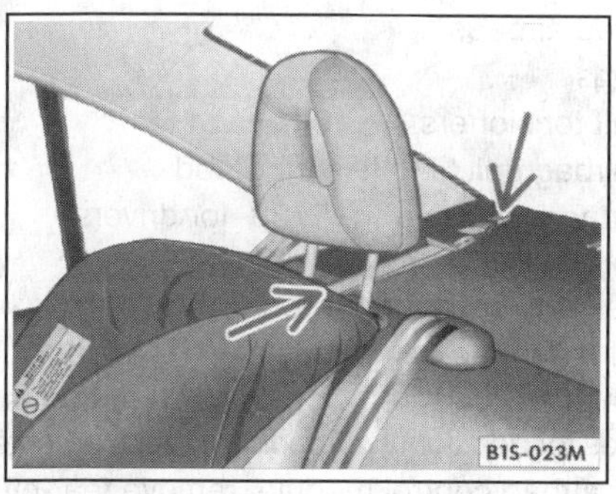

Illustration shows rear top tether strap attached to anchor.

Airbag

Special tools, equipment required

- T10119 Release lever

Safety specifications when working on airbag

- Testing, assembly and repair work should only be conducted by trained personnel.
- When working on the airbag system, disconnect battery ground strap. Waiting time is not necessary after disconnecting battery. When connecting airbag system to power source, no one should be in the interior of the vehicle.
- Before installing (touching) airbag unit technician must discharge electrostatically. This is attained, by touching Grounded metal parts, for example water lines, heater tubes or metal supports.
- Install airbag units immediately after removing from transport container.
- If work is interrupted place airbag unit into transport container again.
- Do not leave a un-installed airbag unattended.
- Airbag units should be stored in removed condition, with the cushioned side facing up.
- Airbag units that drop onto a hard surface or show damage, should not be installed.
- Airbag units which have not been activated should be marked and returned to the manufacturer for disposal (use airbag unit transport container).

Additional safety precautions for curtain airbags

- Always replace damaged pillar; never repair.
- Curtain airbags must not be kinked or twisted.
- After repairs to bodywork, check area of curtain airbags for weld spray droplets, deformations and areas which will abrade (compare with opposite side if necessary).

EDITOR'S NOTE —

Replacement of airbag components will require coding of airbag control module using a diagnostic scan tool, see **Scan Tool section**.

Airbag, replacing after an accident

Accident involving activation of airbag

Always replace the following:

- All triggered airbag units
- All control modules for triggered airbags
- All sensors for triggered airbags
- All seatbelts with triggered belt tensioners
- Support for front passenger airbag unit, if unit is triggered
- Airbag spiral spring/return spring with slip ring -F138- for drivers airbag, if unit is triggered
- Seat cover and cover clips if side airbag is triggered
- Visually check and replace all damaged components
- If side curtain airbags have been deployed, see **Component locations, overview** on the following pages for additional component replacements.
- After replacing airbag units or airbag control module, remove tear-off sticker and attach to registration card. Return card per instructions on card. Order registration card number 000 010 227 C separately.

Accident not involving activation of airbag

- Airbag Malfunction Indicator Light (MIL) -K75- displays no malfunction, replacement of airbag components is not required. Check seat belts separately.

Component locations, overview (through 09/2000)

EDITOR'S NOTE —
Replacement of airbag components will require coding of airbag control module using a diagnostic scan tool, see **Scan Tool section**.

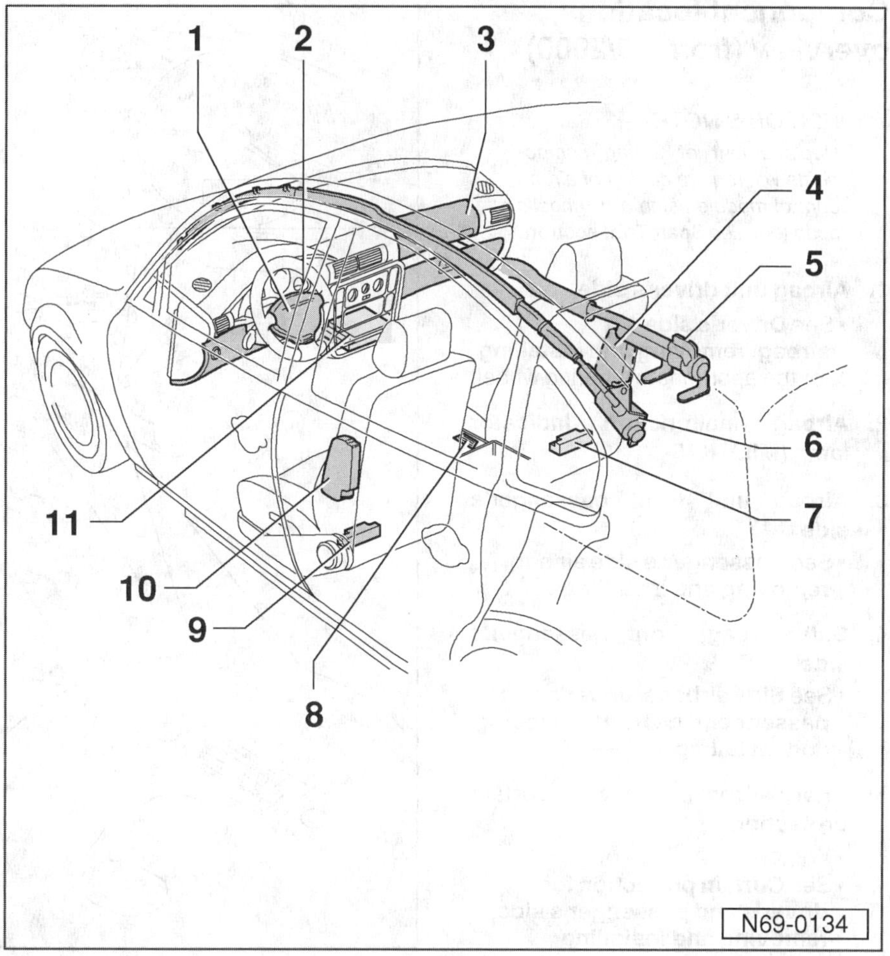

1. **Airbag unit driver's side**
   - See **Driver's side airbagirbag, removing and installing** for the appropriate steering wheel
2. **Airbag warning lamp (K 75)**
3. **Airbag unit front passenger's side**
   - See **Passenger side air bag, removing and installing**
4. **Side airbag, front passenger's side**
   - See **Side airbags, driver's and passenger side front, removing and installing**
5. **Driver's/front passenger's curtain protection**
   - Wagon
   - See **Curtain protection for driver's and passenger's side, removing and installing**
6. **Driver's/front passenger's curtain protection**
   - Sedan
   - See **Curtain protection for driver's and passenger's side, removing and installing**
7. **Sensor (side airbag, front passenger's side)**
   - See **Sensor for side airbag driver/passenger, removing and installing**
8. **Diagnostic socket**
9. **Sensor (side airbag, driver's side)**
   - See **Sensor for side airbag driver/passenger, removing and installing**
10. **Side airbag, driver's side**
    - See **Side airbags, driver's and passenger side front, removing and installing**
11. **Airbag control module -J234-**
    - See **Airbag control module -J234-, removing and installing**

In case of side curtain airbag deployment, replace the following additional components:

- Trim for A, B, C, and D pillars
- Door seal
- Guide piece for head airbag
- Headliner
- Headliner connecting strip
- Rear reading light*
- Grab handle*
- Bracket for grab handle*

NOTE —
Components identified with an (*) can be reused up to six (6) times.

Component locations, overview (from 10/2000)

EDITOR'S NOTE —

Replacement of airbag components will require coding of airbag control module using a diagnostic scan tool, see **Scan Tool section**.

1. **Airbag unit driver's side**
   - See **Driver's side airbag, removing and installing** for the appropriate steering wheel
2. **Airbag malfunction indicator lamp (MIL) -K75-**
3. **Airbag unit front passenger's side**
   - See **Passenger's side air bag, removing and installing**
4. **Side airbag, front passenger's side**
   - See **Side airbags, driver's and passenger side front, removing and installing**
5. **Driver's/front passenger's curtain protection**
   - Wagon
   - See **Curtain protection for driver's and passenger's side, removing and installing**
6. **Driver's/front passenger's curtain protection**
   - Sedan
   - See **Curtain protection for driver's and passenger's side, removing and installing**
7. **Sensor (side airbag, front passenger's side)**
   - See **Sensor for side airbag driver/passenger, removing and installing**
8. **Diagnostic socket**
9. **Sensor (side airbag, driver's side)**
   - See **Sensor for side airbag driver/passenger, removing and installing**
10. **Side airbag, driver's side**
    - See **Side airbags, driver's and passenger side front, removing and installing**
11. **Airbag control module -J234-**
    - See **Airbag control module -J234-, removing and installing**

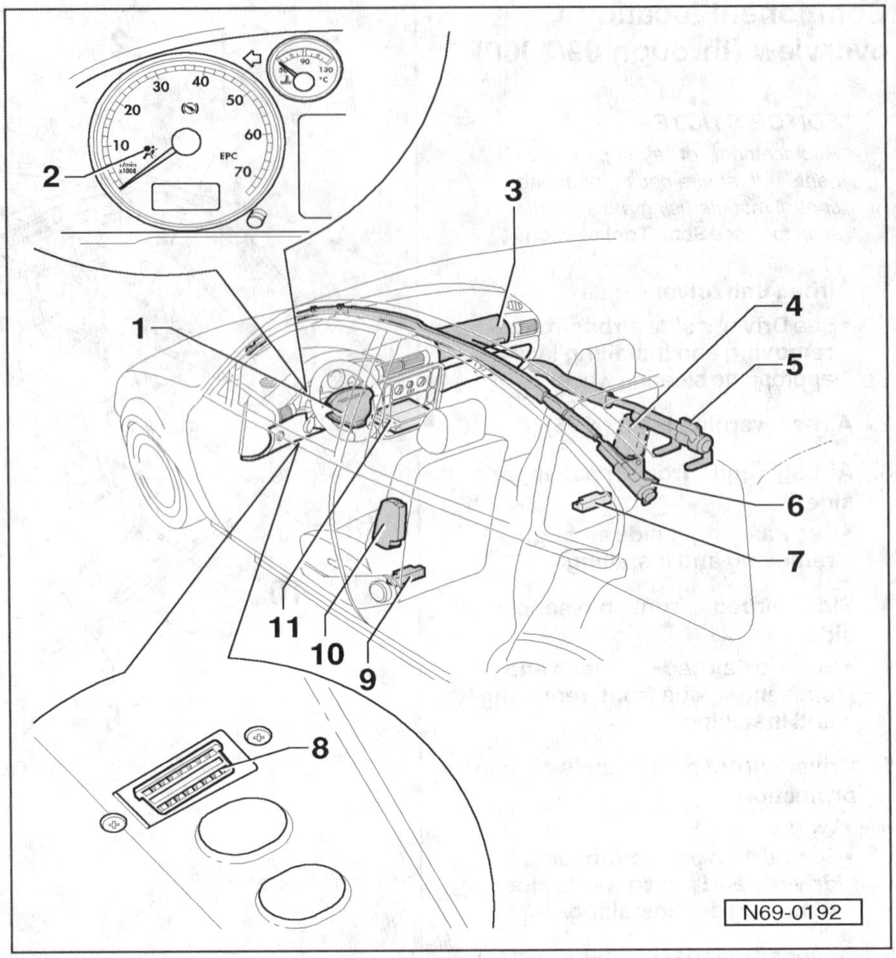

In case of side curtain airbag deployment, replace the following additional components:

- Trim for A, B, C, and D pillars
- Door seal
- Guide piece for head airbag
- Headliner
- Headliner connecting strip
- Rear reading light*
- Grab handle*
- Bracket for grab handle*

NOTE —

Components identified with an (*) can be reused up to six (6) times.

Driver's side airbag assembly, (four spoke steering wheel)

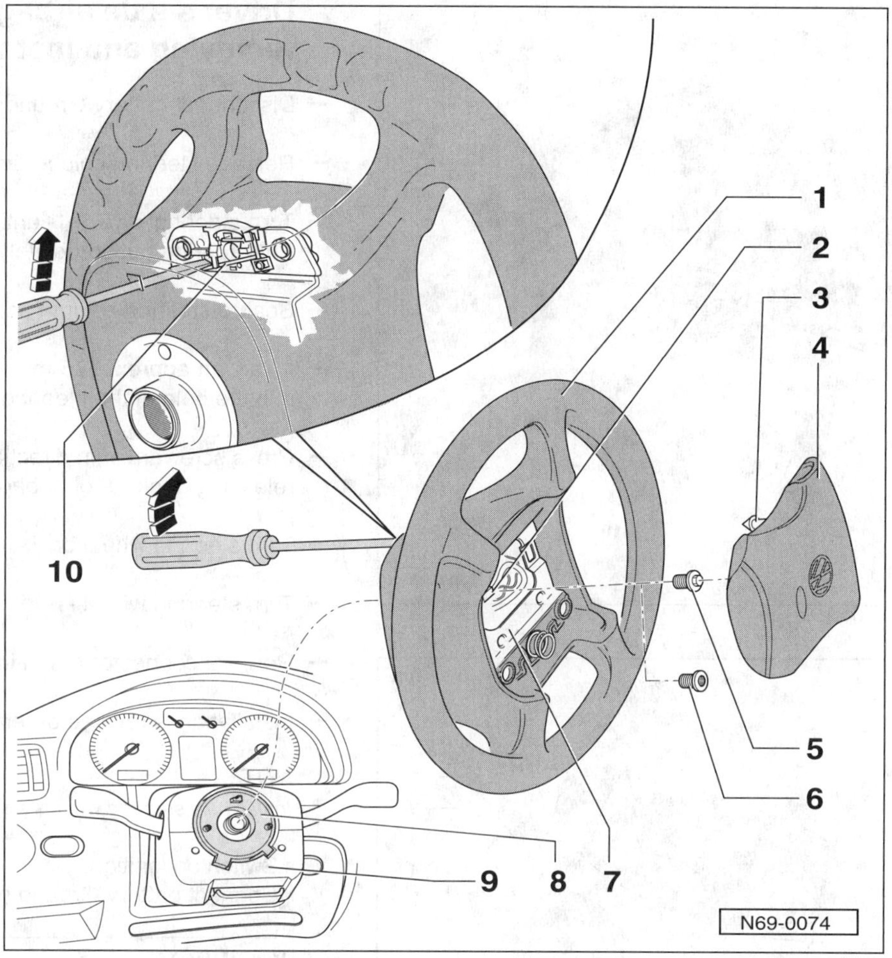

WARNING—

Before working on the electrical system, disconnect battery Ground (GND) strap.

1. **Steering wheel**
   - See **Steering wheel, removing and installing**
2. **Connector**
3. **Hook**
   - Release from steering wheel
4. **Airbag unit**
   - Due to different securing systems, only steering wheels and airbag units from the same manufacturer can be installed
   - For wood steering wheel, the airbag unit is bolted with steering wheel
   - See **Driver's side air bag (four spoke steering wheel), removing and installing**
5. **Hex head bolt**
   - Tighten to 75 Nm (55 ft-lb)
   - Deleted shortly after start of production
   - After stocks have been used up, only multi-point bolt -6- will be supplied as replacement part
6. **Multi-point bolt**
   - Tighten to 60 Nm (44 ft-lb)
   - Supplied with locking compound
   - Can be used up to five times. Mark with center punch after each assembly
7. **Mounting plate**
8. **Return ring with collector ring**
   - See **Return ring and collector ring, removing and installing**
9. **Trim**
10. **Clip**

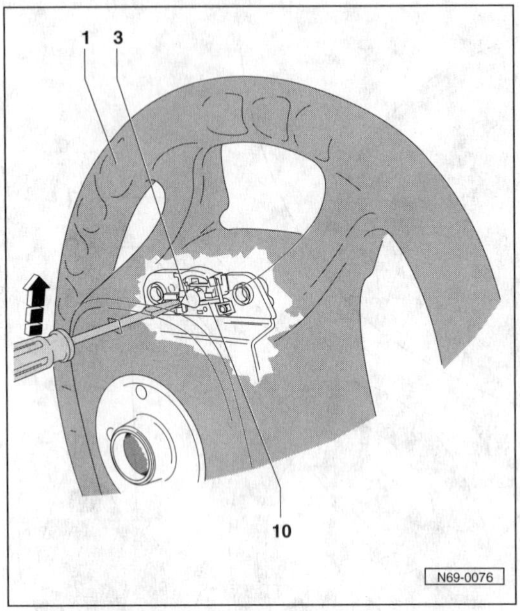

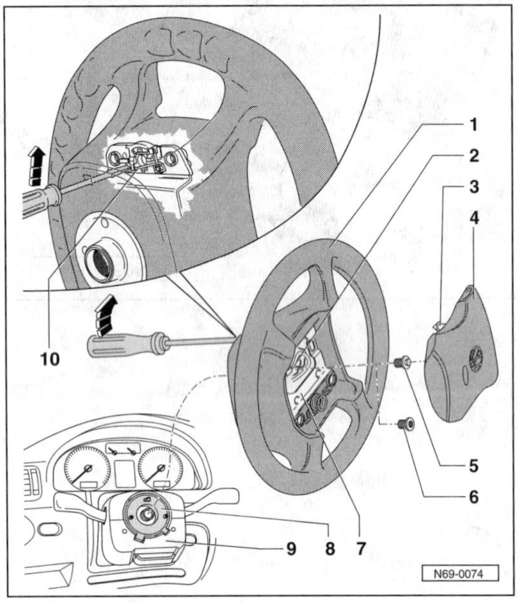

Driver's side airbag (four spoke steering wheel), removing and installing

– Disconnect battery Ground strap.

– Release steering column adjustment.

– Turn steering wheel -1- until steering wheel spokes are vertical. Pull steering out completely and press down to lowest position.

– Secure steering column adjustment.

– Insert an approx. 175 mm (6.8 in.) long screwdriver from the backside into the hole of the steering wheel approx. 45 mm (1.77 in.).

– Press screwdriver in direction of arrow. This presses back clip -10- releasing hook -3- of airbag.

– Turn steering wheel back 180° and release hook on the opposite side.

– Turn steering wheel -1- in center position (wheel straight ahead).

– Remove connector from airbag unit.

– Installation is reverse of removal, noting the following:

NOTE —
Install only steering wheels and airbags from the same manufacturer.

• Switch on ignition.
• Connect battery Ground (GND) strap.

WARNING —
Make sure that no one is in the vehicle.

Driver's side airbag assembly, (three spoke steering wheel)

1. **Steering wheel**
   - See **Steering wheel, removing and installing**
2. **Connector**
3. **Hook**
   - Release from steering wheel pressure
4. **Airbag unit**
   - Due to different securing systems, only steering wheels and airbag units from the same manufacturer can be installed.
   - See **Driver's side airbag (three spoke steering wheel), removing and installing**
5. **Hex head bolt**
   - Tighten to 75 Nm (55 ft-lb)
   - Deleted shortly after start of production
   - After stocks have been used up, only multi-point bolt -6- will be supplied as replacement part
6. **Multi-point bolt**
   - Tighten to 60 Nm (44 ft-lb)
   - Supplied with locking compound
   - Can be used up to five times.
   - Mark with center punch after each assembly
7. **Mounting plate**
8. **Return ring with collector ring**
   - See **Return ring and collector ring, removing and installing**
9. **Trim**
10. **Clip**

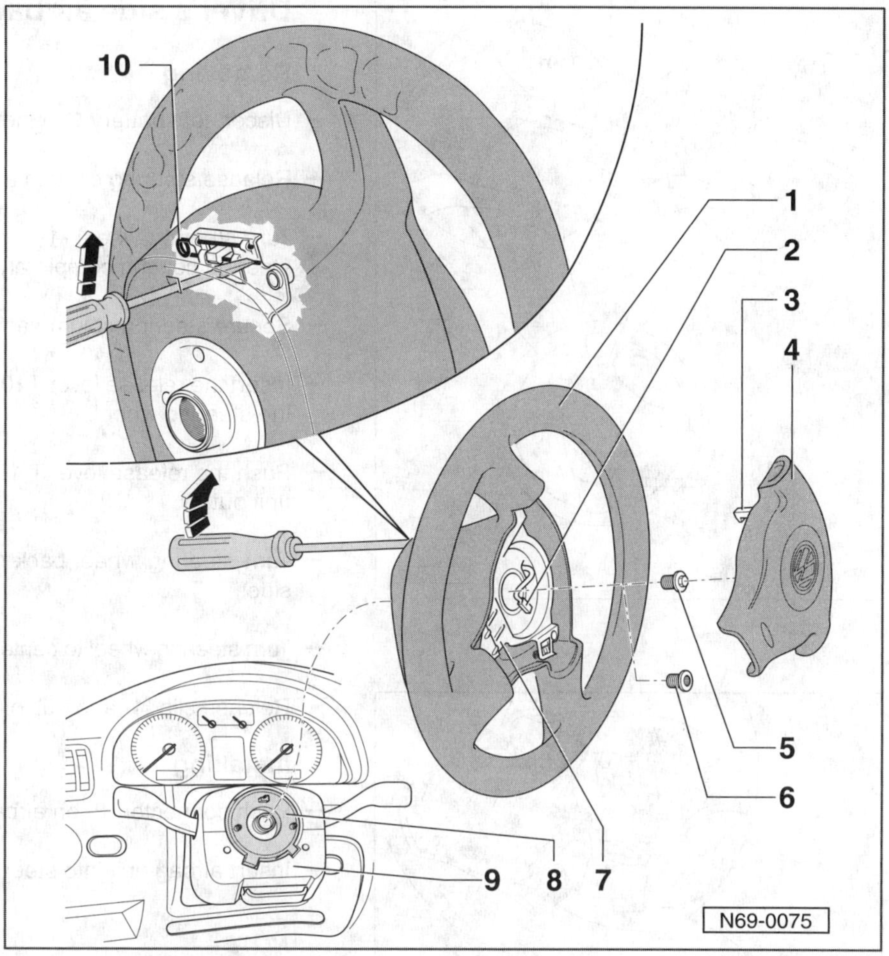

Driver's side airbag (three spoke steering wheel), removing and installing

– Disconnect battery Ground (GND) strap.

– Release steering column adjustment.

– Turn steering wheel -1- until steering wheel spokes are vertical. Pull steering out completely and press down to lowest position.

– Secure steering column adjustment.

– Insert an approx. 175 mm (6.8 in.) long screwdriver from the backside into the hole of the steering wheel approx. 45 mm (1.77 in.).

– Press screwdriver in direction of arrow. This presses back clip -10- and releases hook -3- of airbag.

– Turn steering wheel back 180° and release hook on the opposite side.

– Turn steering wheel -1- to center position (wheel straight ahead).

– Remove connector from airbag unit.

– Installation is reverse of removal, noting the following:

NOTE —
Install only steering wheels and airbags from the same manufacturer.

- Switch on ignition.
- Connect battery Ground (GND) strap.

WARNING —
Make sure that no one is in the vehicle.

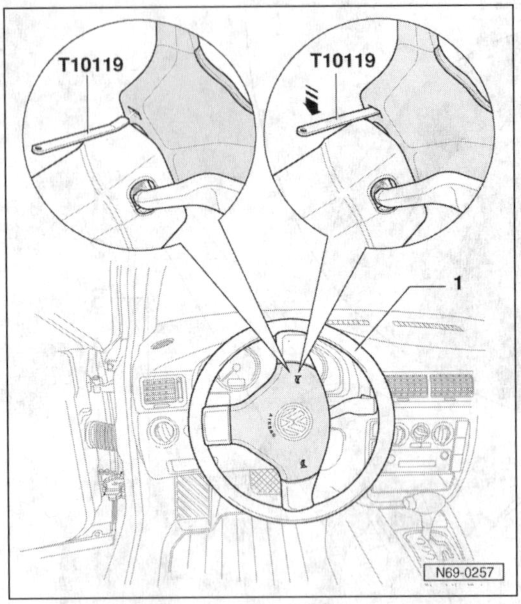

Driver's side air bag (W8), removing and installing

Removing

- Disconnect battery Ground (GND) strap.
- Release steering column adjustment.
- Turn steering wheel -1-, until steering wheel spoke is vertical. Pull out steering column completely and press up to upper position.
- Secure steering column adjustment.
- Insert the release lever T10119 in the rectangular opening at the back of the steering wheel.
- Push the release lever T 10119 in direction of arrow and pull the airbag unit out.
- Turn steering wheel back 180° and release second hook on opposite side.
- Turn steering wheel to center position (wheels in straight-ahead position).

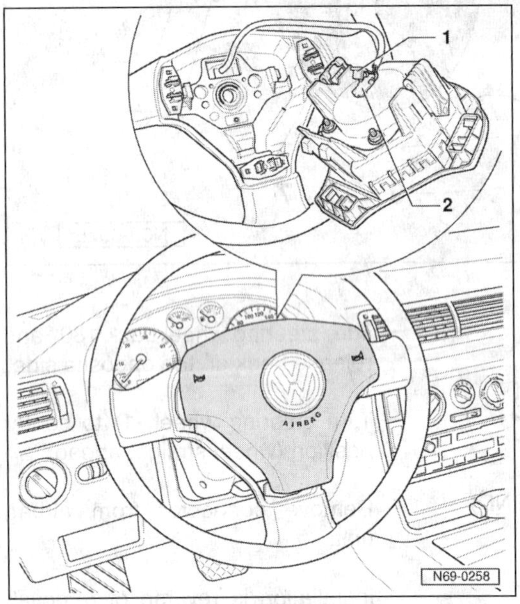

◄ Release clip -1- and pull off connector -2-.

Installing

- Push connector -2- on airbag unit and lock clip -1- in place.
- Insert airbag unit into steering wheel.

NOTE —
• The locking tabs of the airbag unit must be audibly engaged!

- Switch on ignition.
- Connect battery Ground (GND) strap.

WARNING —
When connecting the battery Ground (GND) strap make sure no person is in the vehicle.

Steering wheel, removing and installing

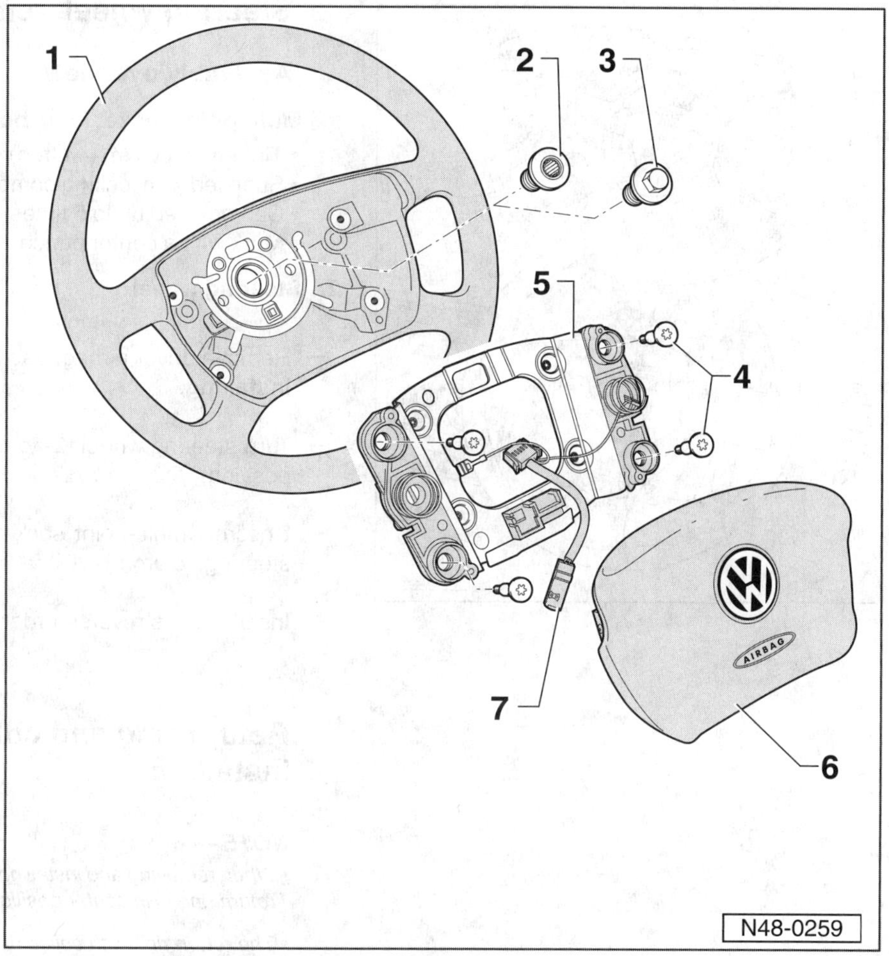

1. **Steering wheel**
2. **Multi-point bolt**
   - Tighten to 60 Nm (44 ft-lb)
   - Supplied with locking compound
   - Can be used up to five times.
   - Mark with a punch after each assembly
3. **Hex head bolt**
   - Tighten to 75 Nm (55 ft-lb)
   - Discontinued shortly after start of production
   - After stocks have been used up, only multi-point bolt -2- will be supplied as replacement part
4. **Torx bolt**
   - Tighten to 5 Nm (44 in-lb)
5. **Mounting plate**
6. **Airbag unit**
   - Due to different securing systems, only steering wheels and airbag units from the same manufacturer can be installed
7. **Connector**

– Remove driver airbag, see **Driver's side air bag, removing and installing** for the appropriate steering wheel.

– Turn steering wheel -1- to center position (wheels straight ahead).

– Remove hex head bolt -3- or multi-point bolt -2- and remove steering wheel from steering column.

– Installation is reverse order of removal.

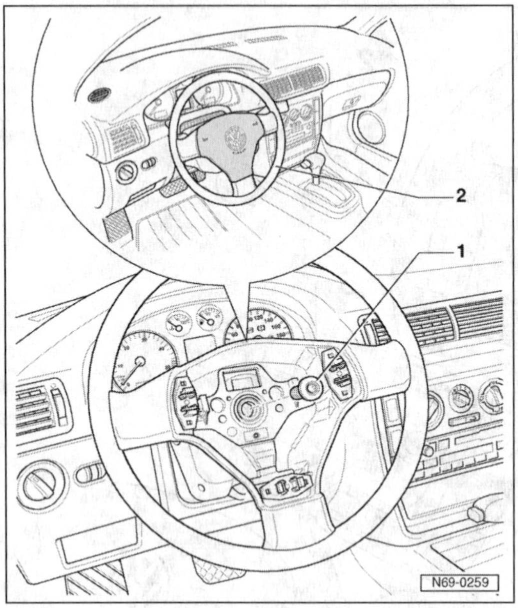

Steering wheel, removing and installing (W8)

Assembly overview

1. **Multi-point socket head bolt**
   - Tighetn to 60 Nm (44 ft-lb)
   - Supplied with locking compound
   - Can be used up to 5 times
   - Mark with a center punch after each assembly
2. **Steering wheel**

– Remove driver airbag, see **Driver's side airbag, removing and installing.**

– Turn steering wheel -2- to center position (wheels in straight-ahead position).

– Unscrew multi-point socket head bolt -1- and pull steering wheel off steering column.

– Installation is reverse order of removal.

Return ring and collector ring, removing and installing

NOTE —
- *When removing and installing return ring with collector ring, make sure that return ring is in center position.*
- *The return ring with collector ring is secured with a tie wrap in center position as a new part.*

Removing

– Remove airbag unit driver side and steering wheel.

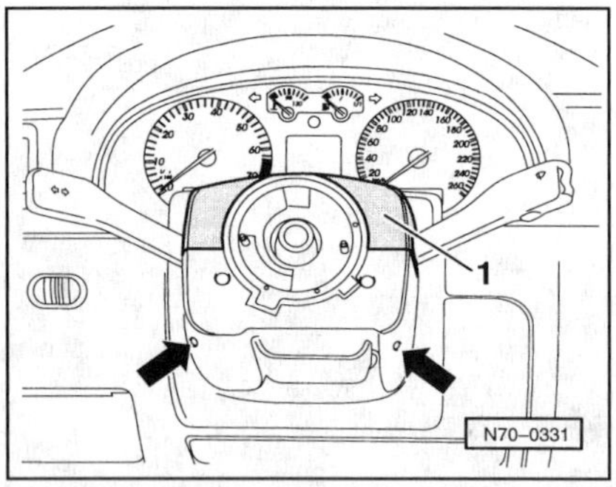

◀ Unscrew two Phillips-head screws (arrows).
- Remove upper trim for steering column switch -1-.

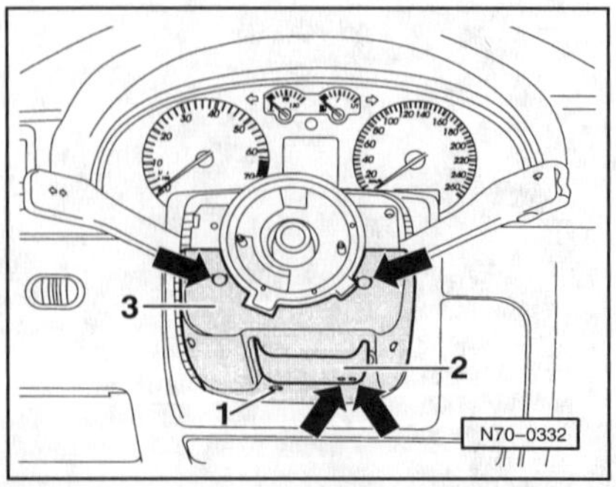

◀ Unscrew four Phillips-head screws (arrows)
- Unscrew inner hex head bolts -1-.
- Release height adjustment for steering wheel -2-.
- Remove lower trim for steering column switch -3-.

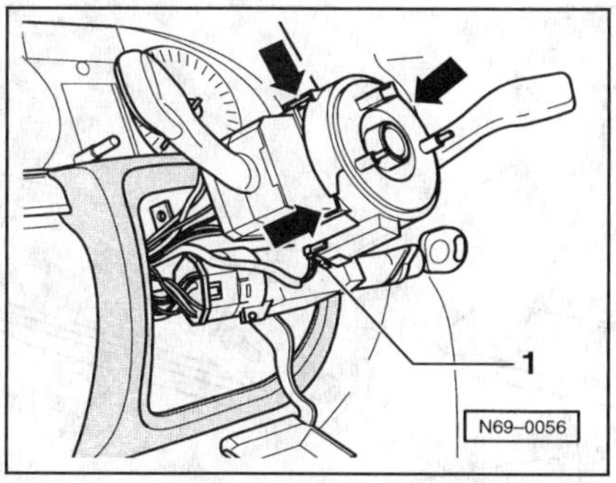

Disconnect connector -1-.

– Release hook (arrows) and remove return ring with collector ring from steering column switch.

Vehicles with Electronic Stabilization Program (ESP)

Vehicles with ESP are equipped with steering angle sensor -G85-. It is installed together with the collector and return ring in a housing.

NOTE—
Front wheels and steering wheel must be in straight ahead position

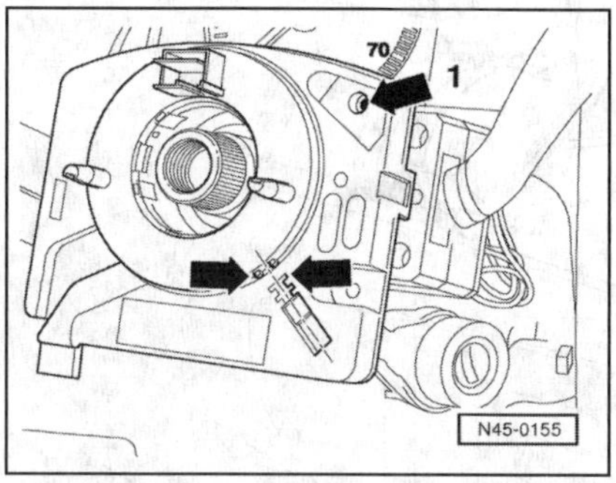

Place front wheels in straight ahead position and remove steering wheel.

- Position steering wheel sensor in center position.
- In the hole -1- a yellow dot must be visible.
- The markings (arrows) must align.

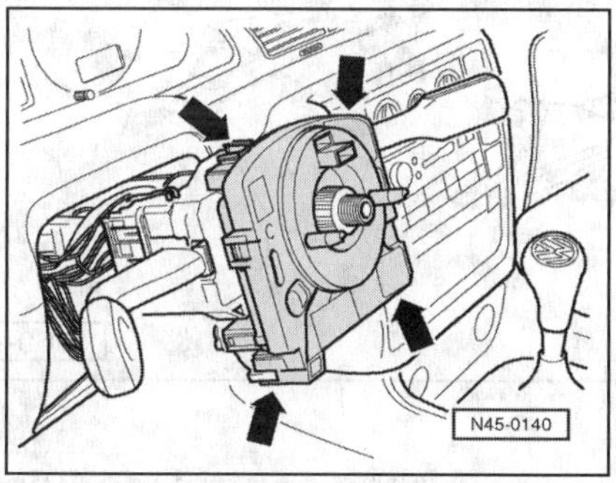

Carefully loosen four fastening clips (arrows) and remove sensor for steering angle in steering column direction.

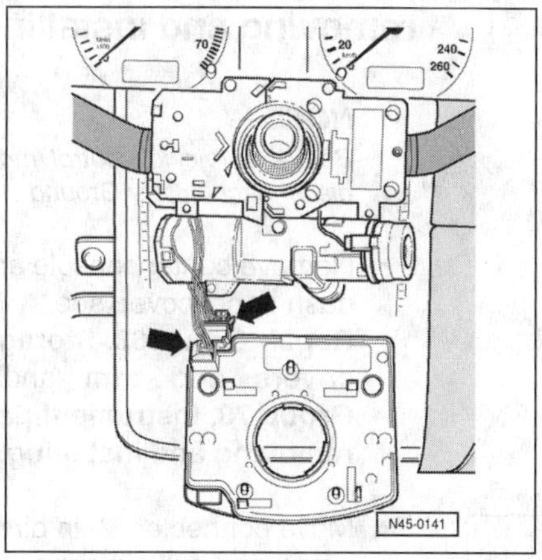

Loosen connectors (arrows).

– Installation is reverse order of removal, noting the following:
  - When installing make sure steering angle sensor is correctly seated and is free of stress.
  - Front wheel and steering wheel must be in straight-ahead position.
  - Remove transportation protection on markings before installing a new steering angle sensor.
  - Position steering angle sensor, until mounting clips lock in properly.

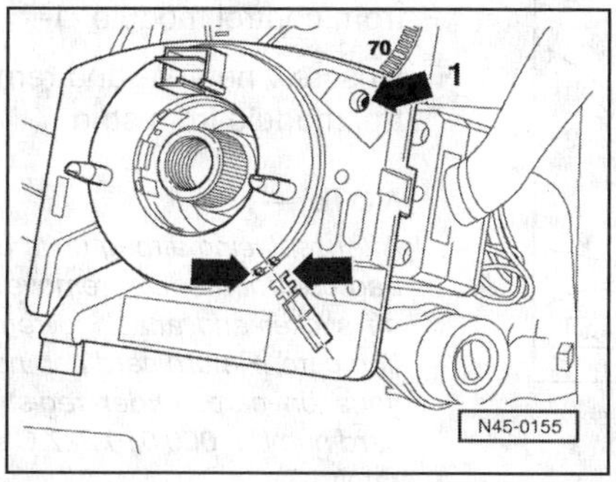

Position steering wheel sensor in center position.

- In the hole -1- a yellow dot must be visible
- The markings (arrows) must align

– Then adapt control module, see **Scan Tool section**.

Airbag unit passenger side, removing and installing

1. **Airbag unit passenger side**
2. **Support**
   - Must be replaced after airbag has been deployed
3. **Hex head bolt**
   - Tighten to 4 Nm (35 in-lb)
4. **Connector**
5. **Clip**
6. **Hex nut**
   - Tighten to 4 Nm (35 in-lb)
7. **Cover**

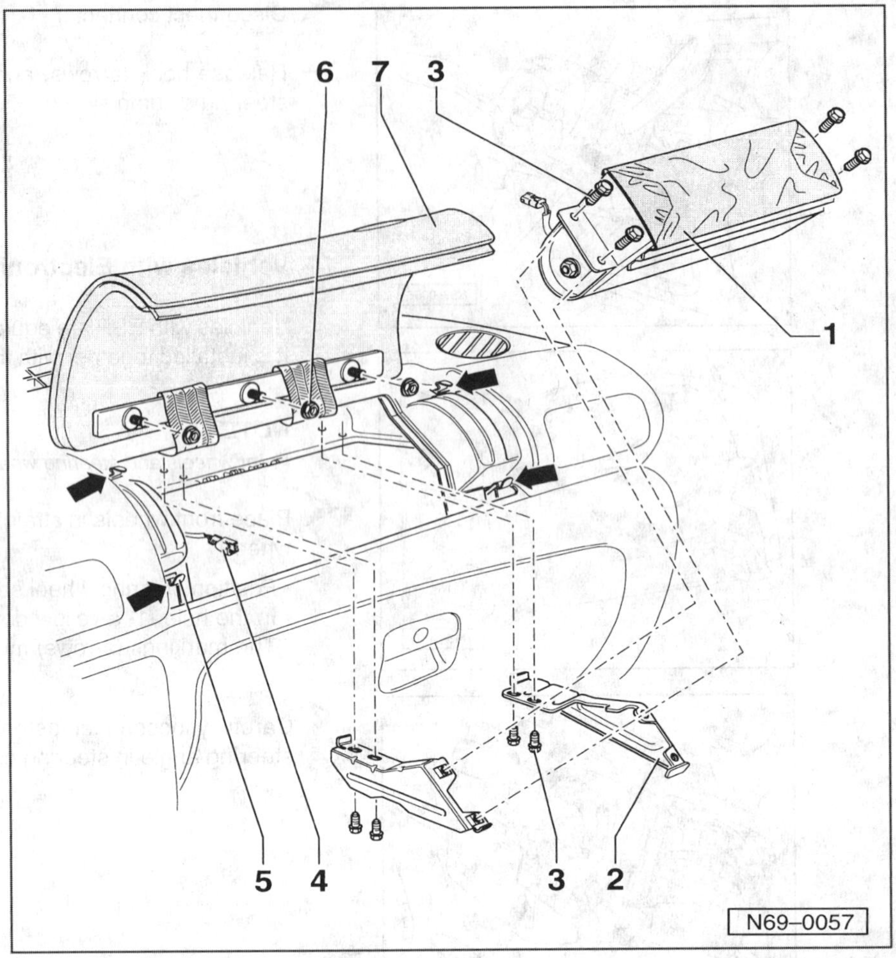

– Disconnect battery Ground (GND) strap.

– Release cover -7- for passenger airbag from dash board (cover held by 4 clips (arrows)).

– Swing cover upward and unscrew hex nuts -6-. Remove cover.

– Remove hex head mounting bolts from left sides of airbag. Remove airbag unit from supports -2- and disconnect connector -4-.

– Installation is reverse of removal, noting the following:
  - Switch on ignition.
  - Connect battery Ground (GND) strap.

__WARNING__ —

Make sure that no one is in the vehicle.

Airbag control module -J234- removing and installing

__NOTE__ —

Before servicing control module, disconnect battery Ground strap.

– Remove center console and center dash panel cover, see **Repair Group 68, Storage, covers and trim** and **Repair Group 70, Instrument panel, removing and installing**.

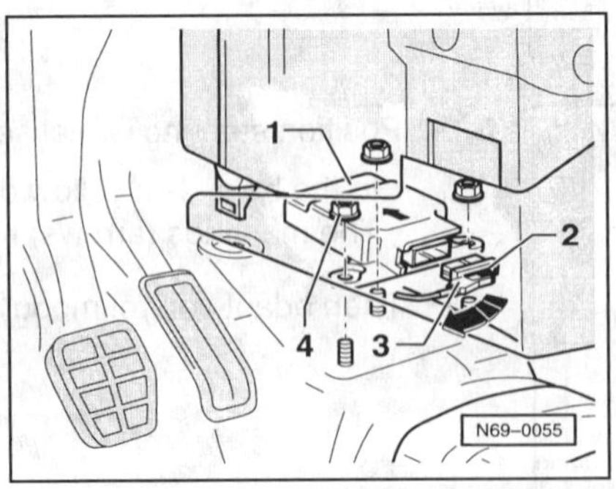

< Move connector -2- in direction of arrow and pull out connector -3- from control module -1-.

– Unscrew nuts -4- and remove control module from stud.

__NOTE__ —

After replacing airbag units or airbag control module, remove tear-off sticker and attach to registration card. Return card per instructions on card. Order registration card number 000 010 227 C separately.

Side airbags, driver's and passenger side front, removing and installing

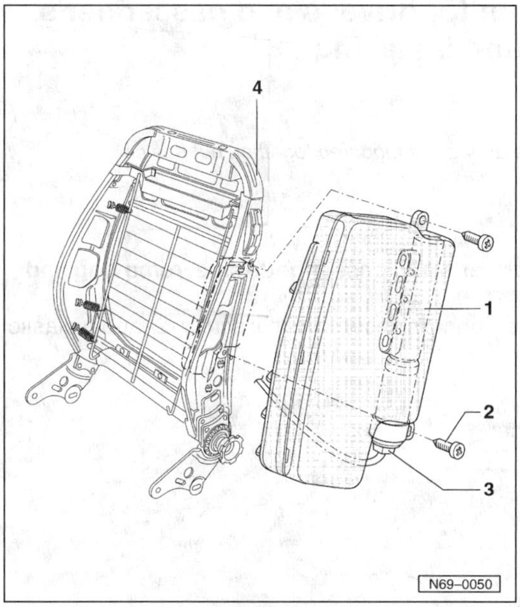

1. **Side airbag driver's/passenger side**
2. **Bolt**
   - Tighten to 7 Nm (62 in-lb)
3. **Connector**
4. **Seatback frame**
   - Disconnect battery Ground strap.
   - Remove seatback frame, see **Repair Group 72, Front backrest, removing and installing** and loosen seatback cover in area of airbag -1-.
   - Remove bolts -2-.
   - Disconnect connector -3- from airbag unit -1-.
   - Remove airbag unit -1- from seatback frame -4-.

Installation notes

– Switch on ignition.

– Close doors.

– Battery Ground strap connected.

> ***WARNING —***
> *Make sure that no one is in the vehicle.*

Sensor for side airbag driver/passenger, removing and installing

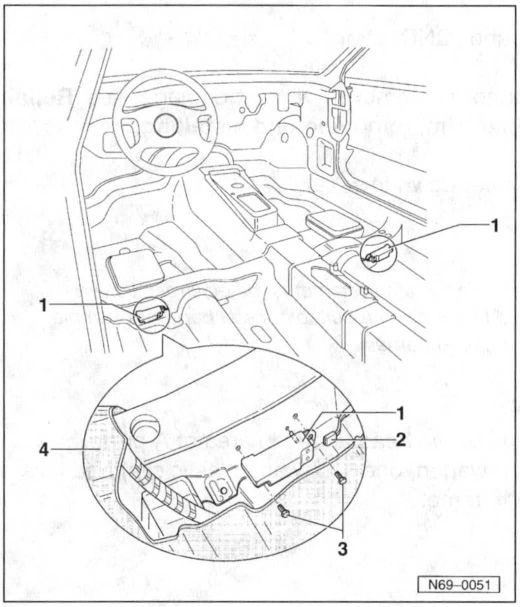

1. **Sensor driver-/passenger side**
2. **Connector**
3. **Bolt**
   - Tighten to 6 Nm (53 in-lb)
4. **Carpet**
   - Battery Ground strap disconnected.
   - Remove driver or passenger seat (Use Airbag - Adapter VAS 5061), see **Repair Group 72, Front seat, removing and installing**.
   - remove carpet -4- and loosen dampening mats in area of sensor.
   - Disconnect connector -2- and unscrew bolts -3-.

Installation notes:

– Switch on ignition.

– Close doors.

– Connect battery Ground strap.

> ***WARNING —***
> *Make sure that no one is in the vehicle.*

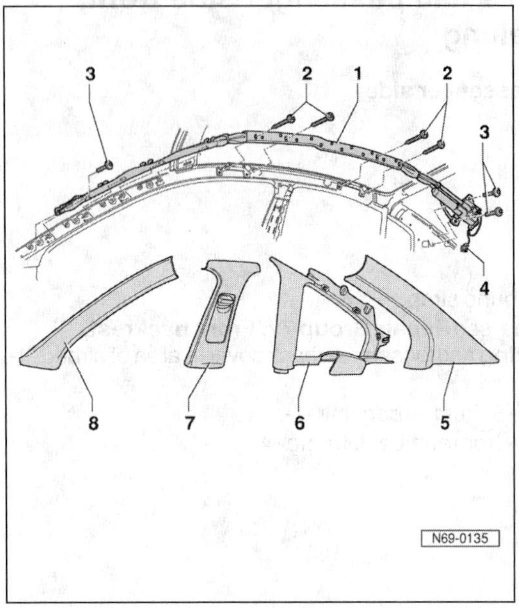

Curtain protection for driver's and passenger's side, removing and installing

NOTE —
The head airbag and side airbag are triggered together, but has no eparate crash sensor.

1. **Head airbag**
   - See **Head airbag on driver's and passenger's side, removing and installing**
   - The head airbag for the wagon is secured with a screw -2- and a plastic clip in the C pillar.
2. **Screw**
   - Panel screw with washer
3. **Bolt**
4. **Connector**
5. **C pillar trim (upper part)**
   - See **Repair Group 70**
6. **C pillar trim (lower part)**
   - See **Repair Group 70**
7. **B pillar trim**
   - See **Repair Group 70,**
8. **A-pillar trim**
   - See **Repair Group 70**

Removing

– Disconnect battery Ground (GND) strap.

– Remove pillar trim (cannot be removed without damage, see **Repair Group 70, Upper A-pillar trim, removing and installing**.

– Release roof trim and lower down to seats.

NOTE —
Before removing the airbag unit the technician must be electrostatically discharge. This is achieved by touching a suitable metal part, for example, a water pipe, heater pipe or metal frame.

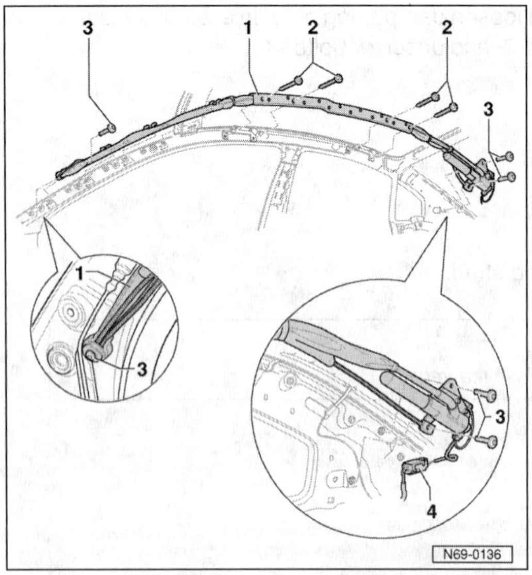

◀ Disconnect connector -4-.

- Remove screws -3- and unclip head airbag in area of A-pillar.
- Remove screws -2- (on wagon, one screw and plastic clip) and take head airbag -1- off roof frame.

Installing

– Switch on ignition.

– Connect battery Ground (GND) strap.

WARNING —
Make sure that nobody is in the vehicle.

70 Interior Trim

Repair Group

U 70

Instrument panel

Instrument panel, removing and installing

> ***WARNING —***
> *Before working on the electrical system; disconnect battery Ground (GND) strap.*

NOTE —
Before disconnecting the battery, obtain radio anti-theft code.

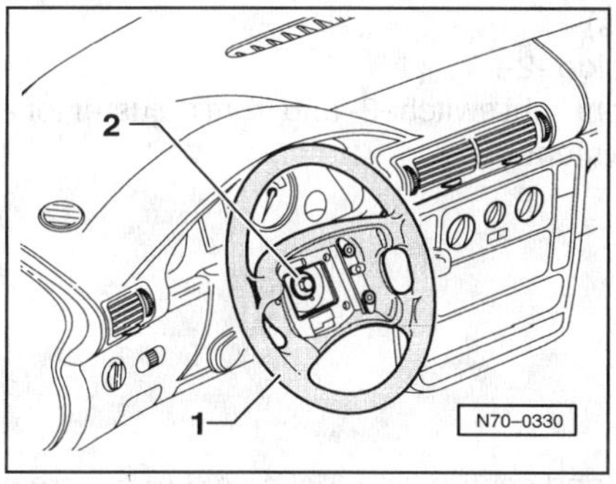

– Remove center console, see **Repair Group 68, Storage, covers and trim**.

– Remove airbag unit driver side, see **Repair Group 69, Airbag unit driver side, removing and installing**.

< Remove steering wheel bolt -2-.

- Place steering wheel-1- in center position (wheels straight ahead).
- Remove steering wheel -1- from steering column.

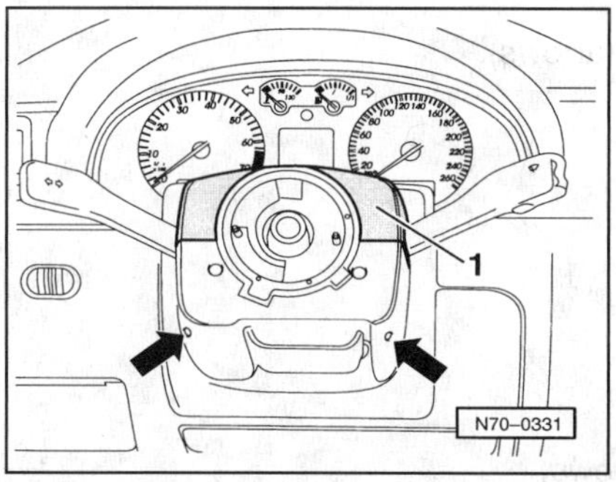

< Unscrew two Phillips-head screws (arrows)

- Remove upper trim for steering column switch -1-

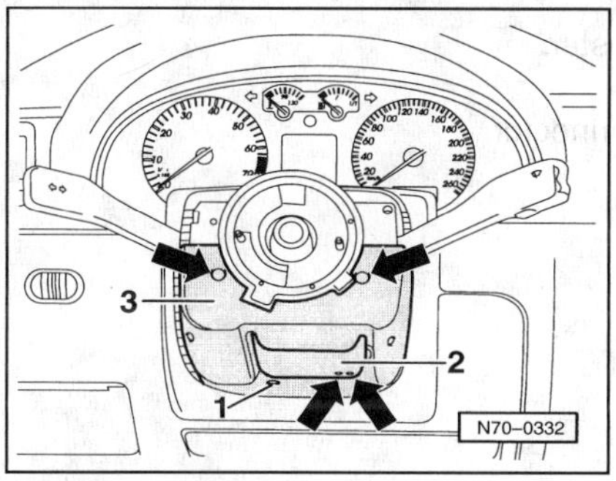

< Unscrew four Phillips-head screws (arrows)

- Unscrew inner hex head bolts -1-.
- Release height adjustment for steering wheel -2-.
- Remove lower trim for steering column switch -3-

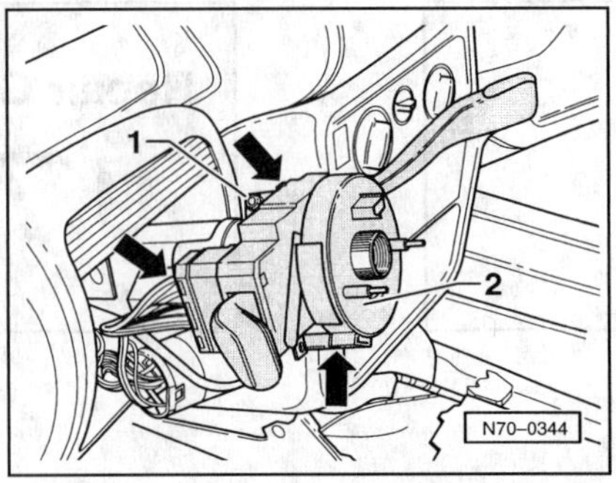

◄ Loosen inner hex head bolt -1-.

• Remove connectors (arrows) from steering column switch.

– Remove steering column switch

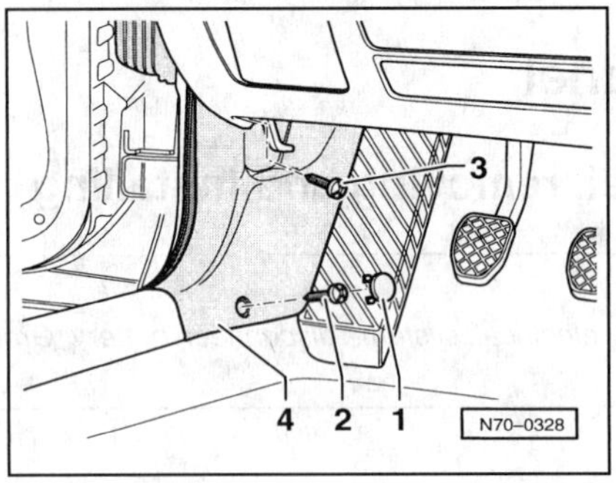

◄ Pry off cover cap -1-

• Unscrew bolts -2- and -3-.
• Unclip lower A-pillar trim -4- from sill plate and remove.

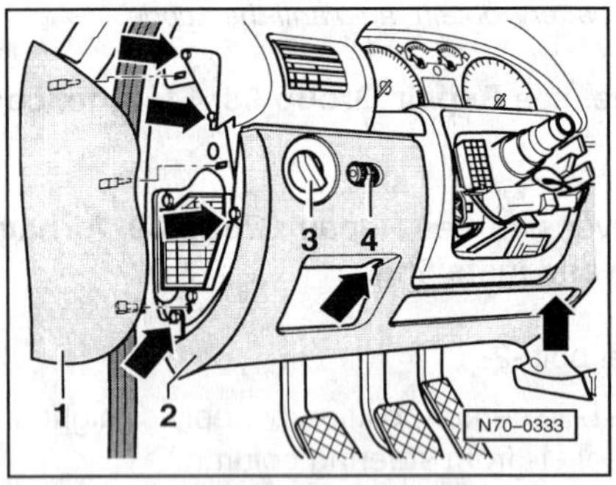

◄ Unclip -1- cover.

• Unscrew bolts (arrows).
• Remove shelf driver side -2-.
• Remove connector from light switch -3- and from adjuster for automatic vertical aim control system -4-.

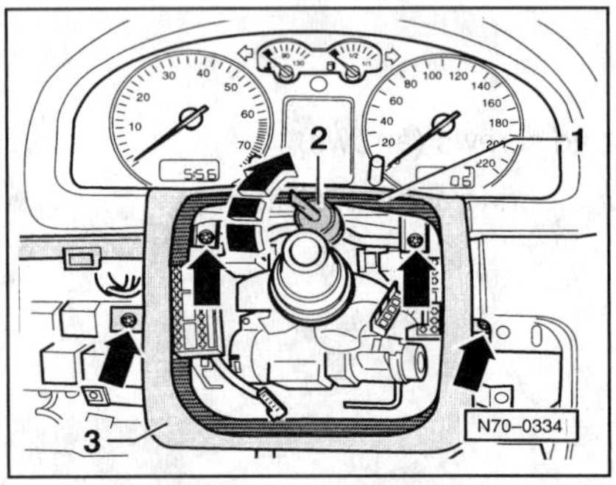

◄ Slide cover -1- upward and secure with tooling aid (e.g. screwdriver handle) -2-.

• Remove four screws (arrows).
• Remove cover -3-.

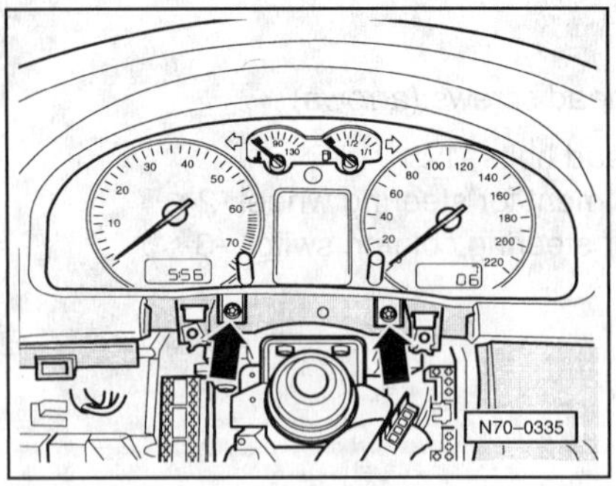

◄ Remove two bolts (arrows)

– Remove instrument cluster.

– Disconnect harness connector.

Remove two bolts (arrows)

Center dash panel cover, removing

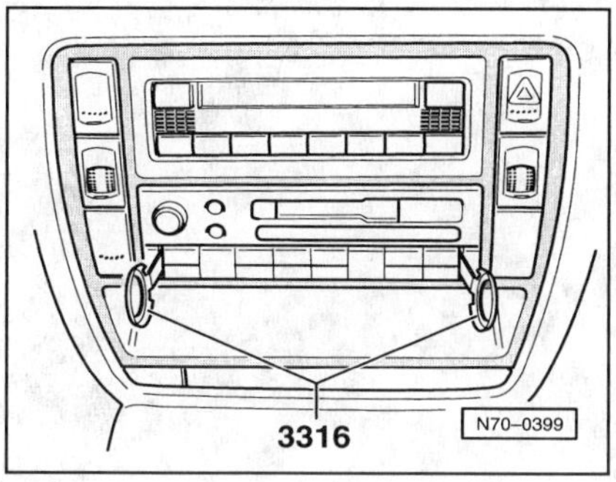

Remove radio at handles of tool from instrument panel.

NOTE —

- *Release tool for radio 3316 should not be pushed to the side or tilted.*
- *To remove the tool the side tabs on the radio must be pressed inward.*

– Disconnet harness connector.

EDITOR'S NOTE —

radio shown is for early models, see Radio section for more specific radio removal instructions.

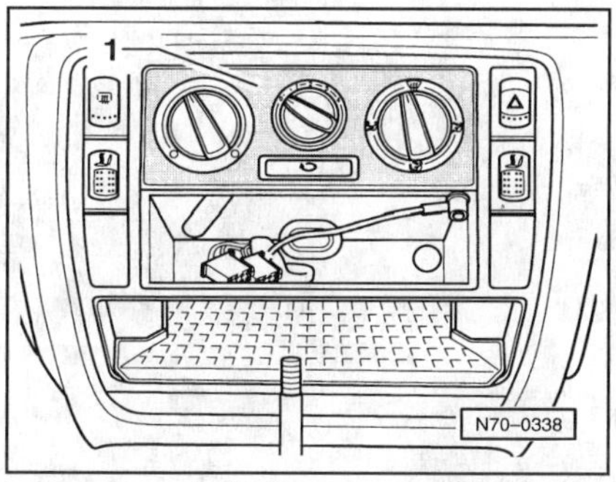

Unclip -1- trim.

EDITOR'S NOTE —

Manually operated climate control shown. For digital system (Climatronic) see Heating and Air Conditioning section for more specific removal instructions.

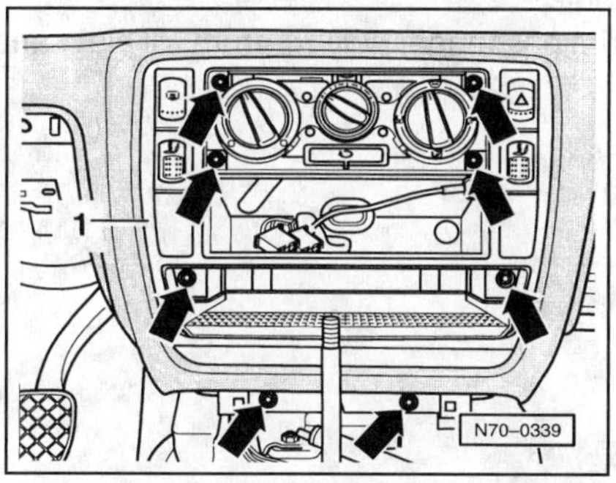

Unscrew eight bolts (arrows)

– Remove center dash panel cover -1- and take out connector.

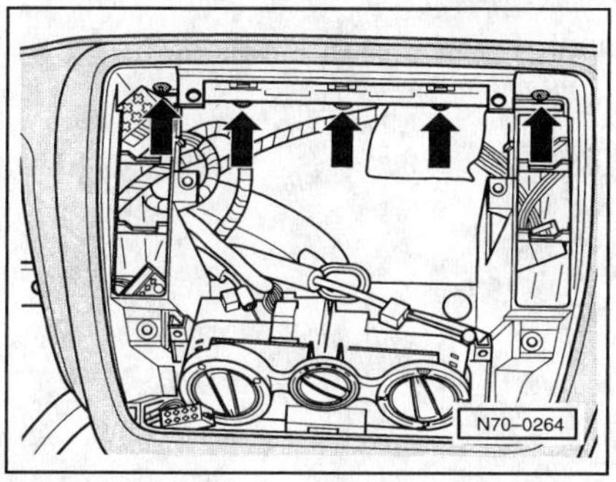

Remove five screws (arrows).

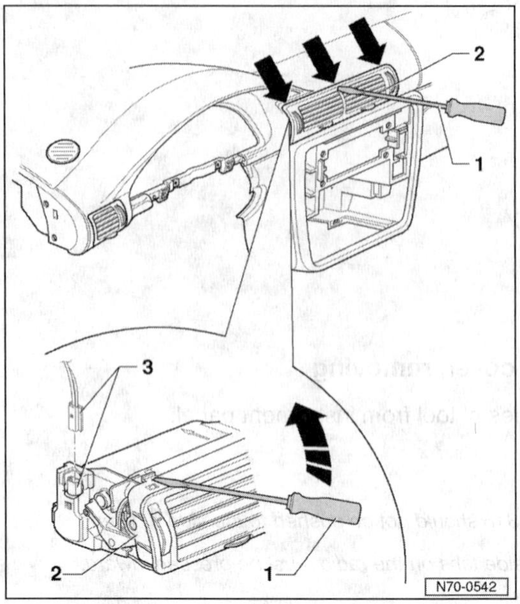

With a flat screwdriver -1- release three clips (arrows).

- Maintain tension of outlets -2-, in order to avoid clips from locking again.
- Remove outlets -2- from instrument panel and disconnect harness connector -3-.

NOTE—

If reinstalling the same instrument panel (dash board), disconnect connector -3- and leave outlets -2- and wiring installed.

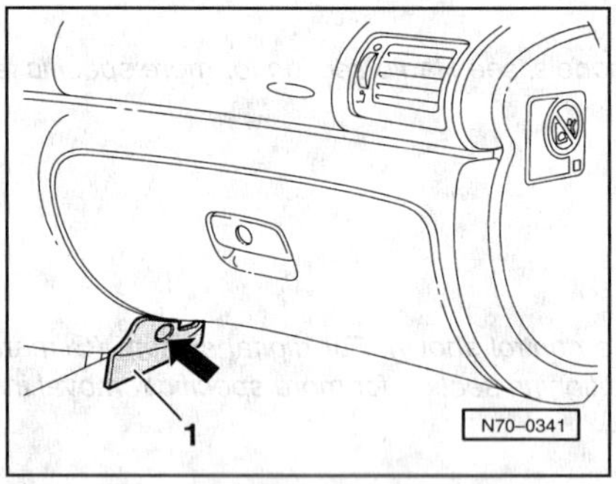

Unscrew bolt (arrow)

- Remove trim -1-.

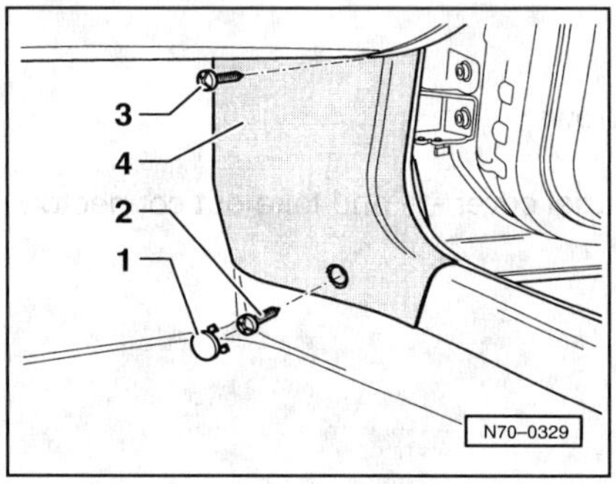

Pry off cover cap -1-

- Remove screws -2- and -3-.
- Unclip lower A-pillar trim passenger side -4- from sill plate and remove.

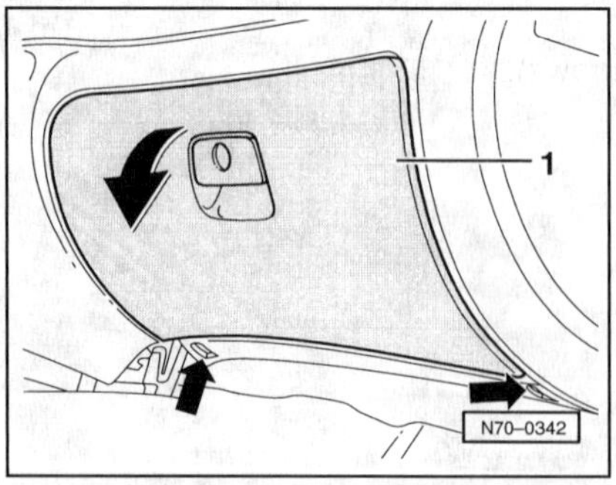

Unscrew bolts (arrows)

- Open glove compartment -1-.

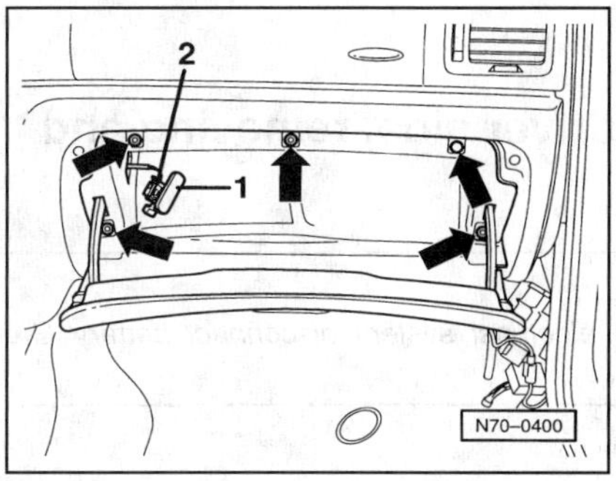

◄ Unscrew bolts (arrows)

- Carefully pry out light for glove compartment -1- with a screwdriver, from the top and remove connector -2-.

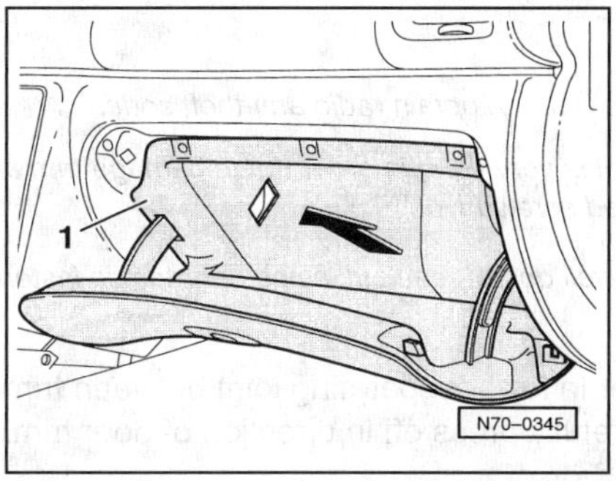

◄ Remove glove compartment liner -1- from instrument panel.

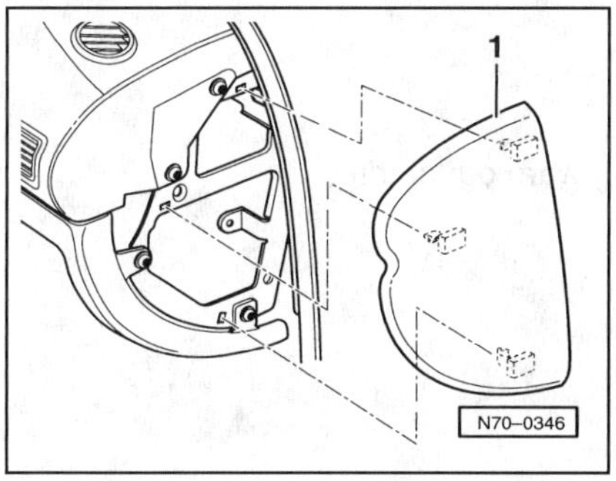

◄ Pry off cover -1-.

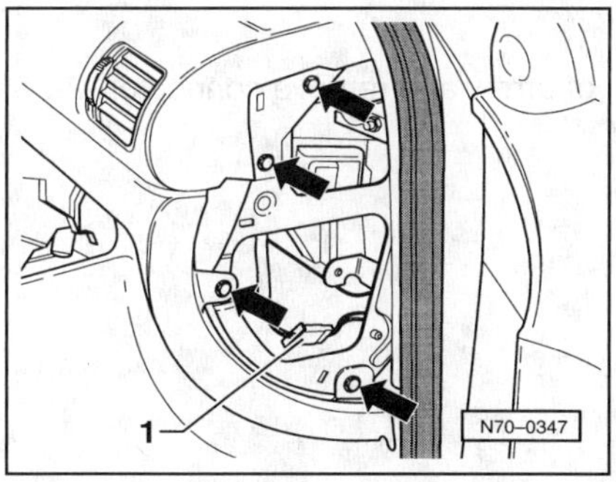

◄ Unscrew hex bolts (arrows).

- Disconnect connector -1-.

– Lift out instrument panel

– Installation is reverse order of removal.

NOTE—

- *After disconnecting the battery make sure that the vehicle equipment (radio, clock and power window regulators) are checked according to repair manual and/or operating instructions.*
- *If airbag malfunction indicator lamp (MIL) -K75- for airbag signals a malfunction after assembly, erase DTC memory with Scan Tool V.A.G 1551 and check again.*

Door trim

Front door trim driver side, removing and installing

> ***WARNING —***
> *Before working on the electrical system disconnect battery Ground (GND) strap*

Removing

NOTE —

- *Before disconnecting the battery, obtain radio anti-theft code.*
- *Since the separating joint is very tight, in order not to damage trim and handle well, use a pointed screwdriver.*
- *When removing handle well do not reach in, since door trim is fastened to strut underneath.*

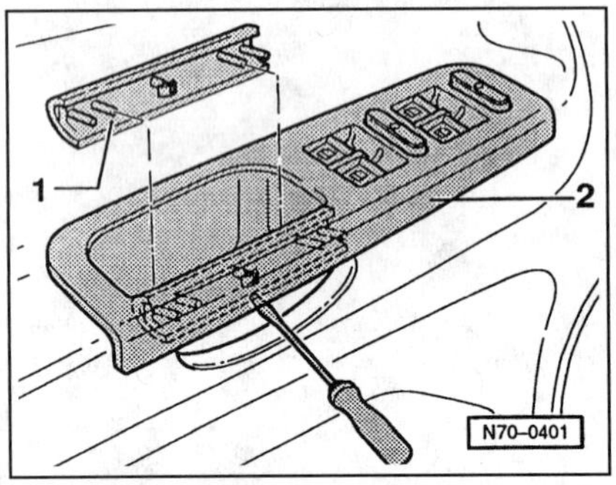

Insert small screwdriver in lower separating joint between trim -1- and handle well -2- and carefully press off in direction of door trim.

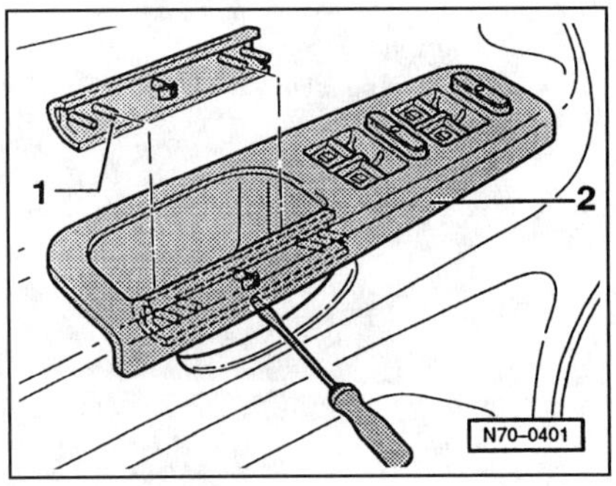

Unclip handle well -2- upward out of trim.

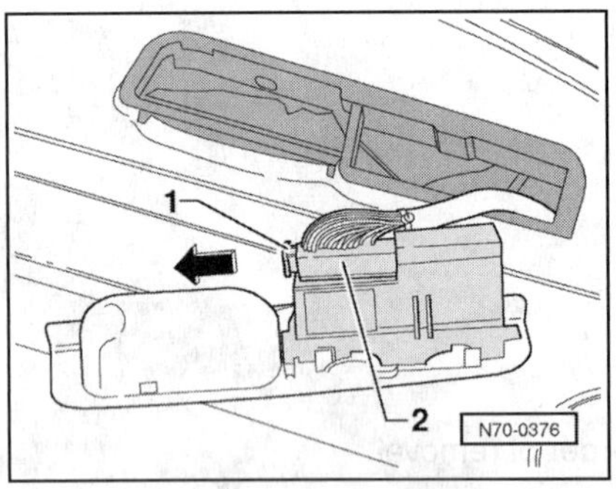

Pull lock -1- in direction of arrow and remove connector -2-.

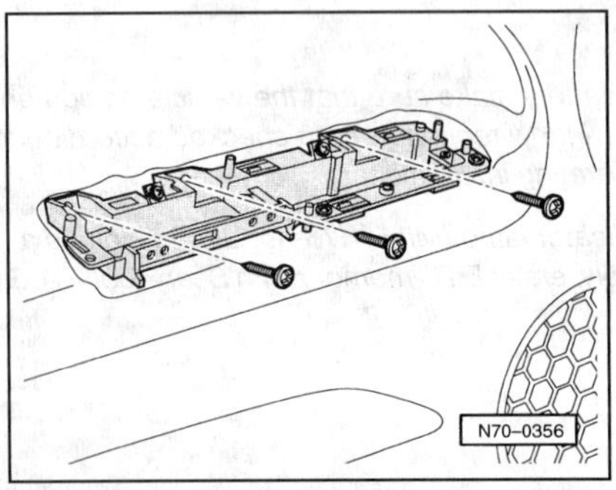

Unscrew 3 Phillips-head screws.

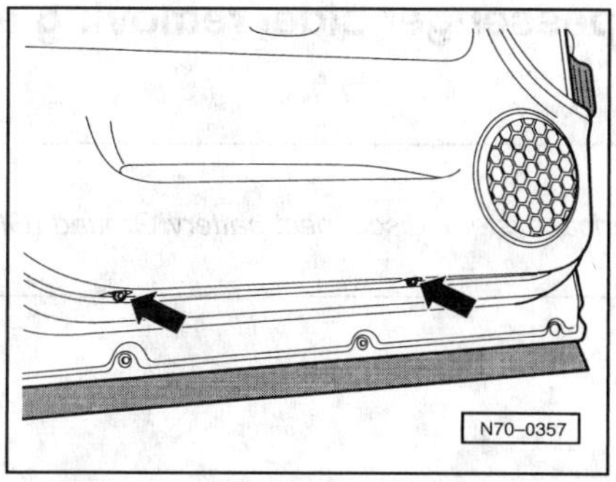

Unscrew bolts (arrows)

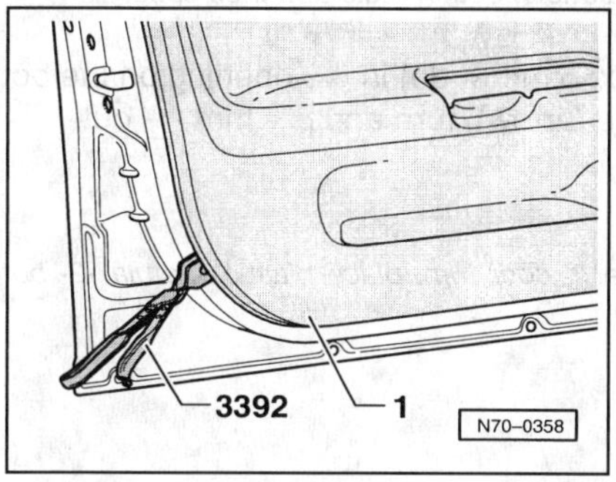

Unclip door trim -1- with aid of tool 3392 on the sides.

– Raise door trim upward out of window channel.

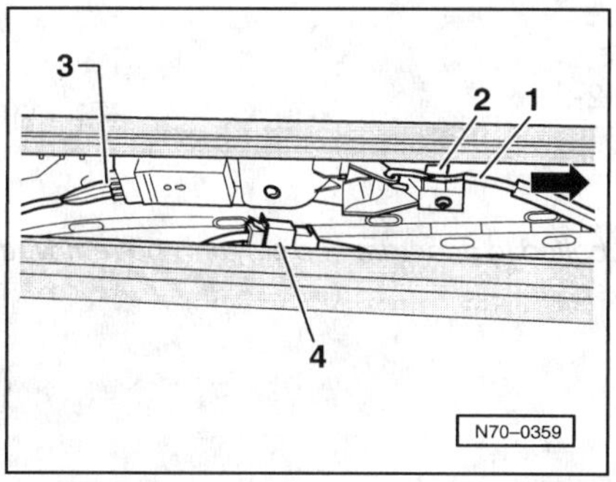

Pull cable -1- out in direction of guide -2- and unhook.

- Disconnect connectors -3- and -4-.

Front door trim driver side, installing

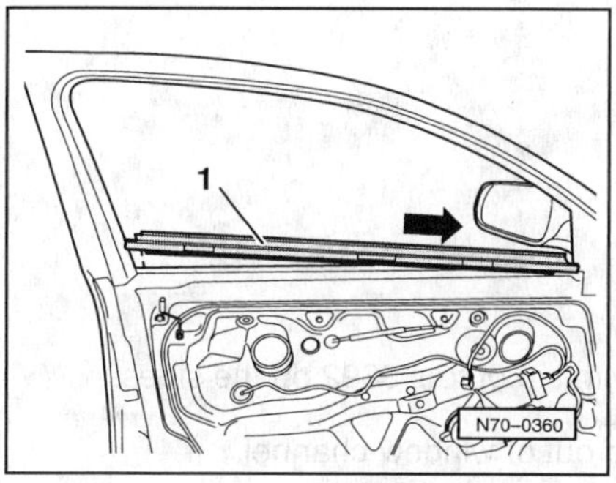

Remove window channel seal -1- from door trim and install in window channel.

NOTE—

- *Before installing the door trim check clips and if necessary replace.*
- *After connecting the battery make sure that the vehicle equipment (radio, clock and power window regulators) are checked according to repair manual and/or operating instructions.*

Front door trim passenger side, removing and installing

> ***WARNING —***
> *Before working on electrical system: disconnect battery Ground (GND) strap.*

Removing

NOTE —
Before disconnecting the battery, obtain radio anti-theft code.

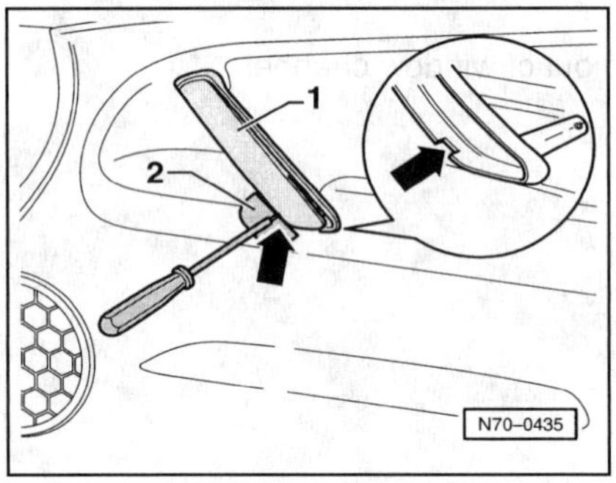

◄ Guide screwdriver (max. 5 mm wide) in the opening on the bottom side of the door inner handle (arrow), turn and pry trim -1- out.

NOTE —
In order to avoid damage to door trim, place plastic spatula -2- between screwdriver and door trim.

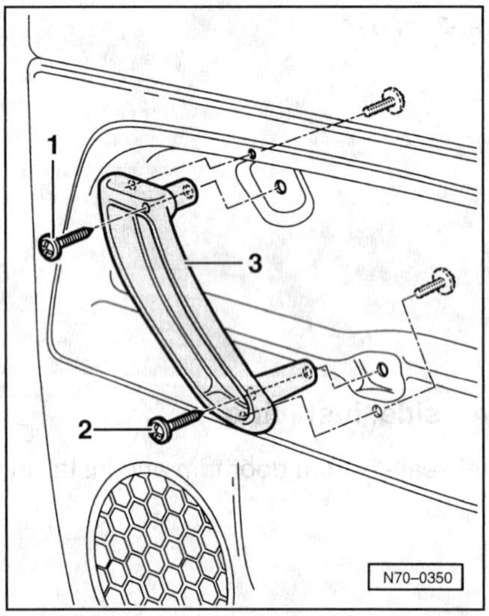

◄ Unscrew bolts -1- and -2-.

NOTE —
Door interior handle -3- is bolted to door trim and can only be removed after door trim has been removed.

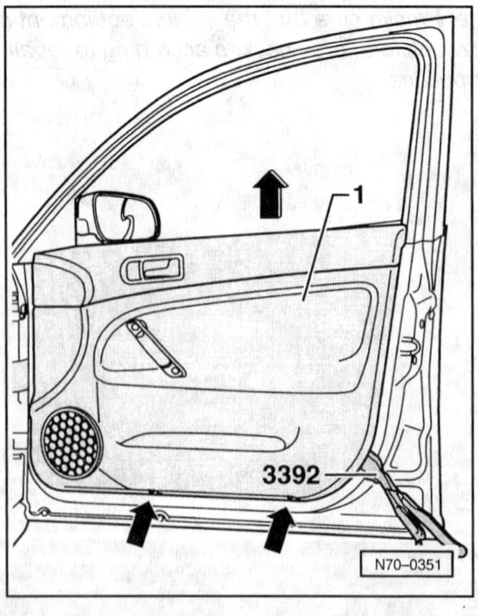

◄ Unscrew bolts (arrows)

• Unclip door trim -1- with aid of tool 3392 on the sides.

– Raise door trim upward out of window channel.

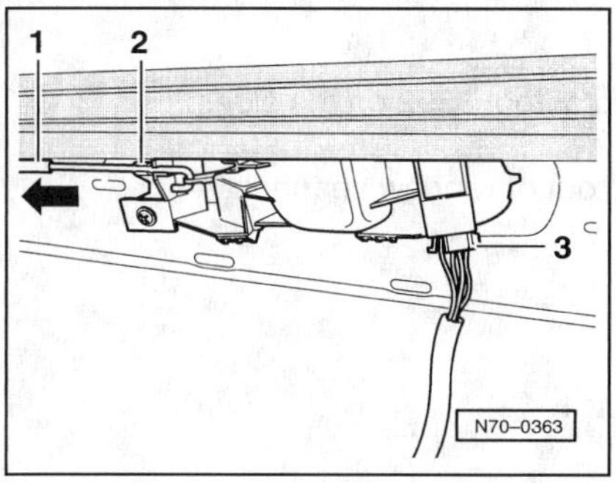

Pull cable -1- out in direction of guide -2- and unhook.

- Disconnect connector -3-.

Front door trim passenger side, installing

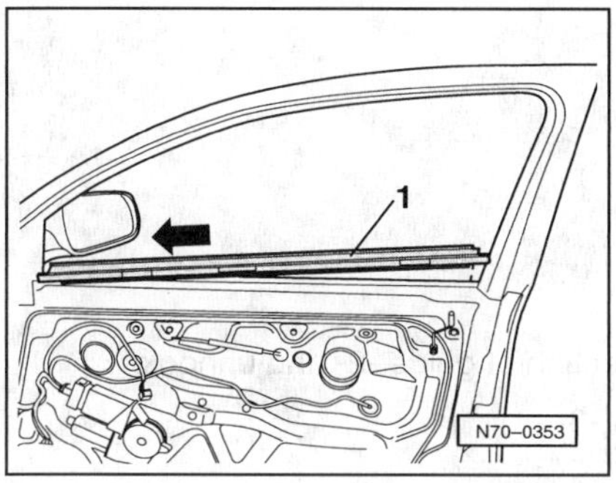

Remove window channel seal -1- from door trim and install in window channel.

NOTE—

- *Before installing the door trim check clips and if necessary replace.*
- *After connecting the battery make sure that the vehicle equipment (radio, clock and power window regulators) are checked according to repair manual and/or operating instructions.*

Rear door trim, removing and installing

> ***WARNING—***
> *Before working on electrical system: disconnect battery Ground (GND) strap.*

Removing

NOTE—

Before disconnecting the battery, obtain radio anti-theft code.

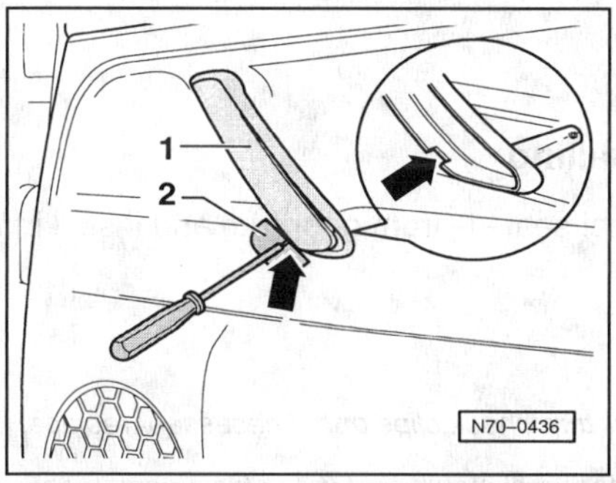

Guide screwdriver (max. 5 mm wide) in the opening on the bottom side of the door inner handle (arrow), turn and pry trim -1- out.

NOTE—

- *In order to avoid damage to door trim, place plastic spatula -2- between screwdriver and door trim.*

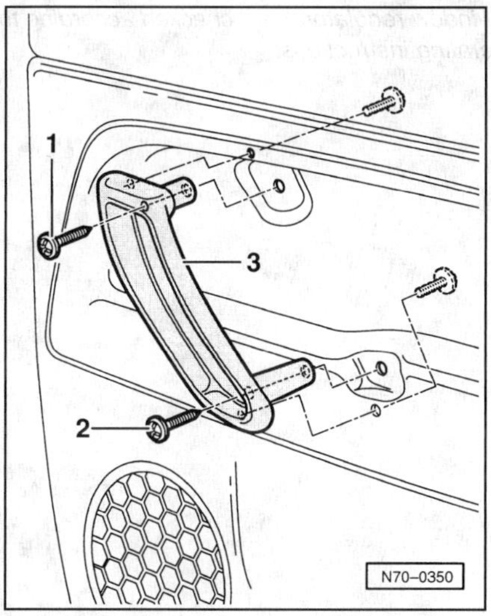

Unscrew bolts -1- and -2-.

NOTE—

Door interior handle -3- is bolted to door trim and can only be removed after door trim has been removed.

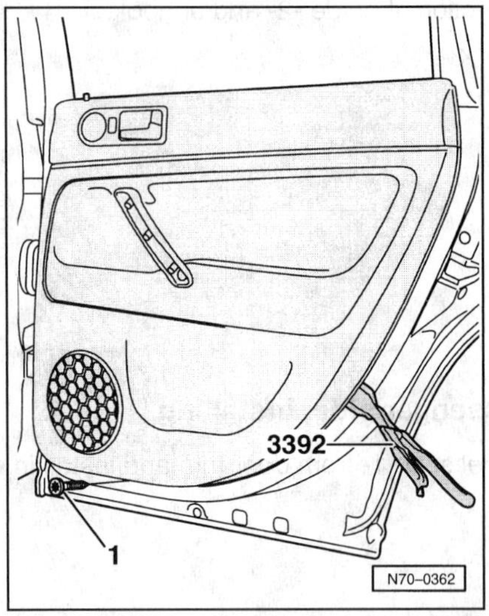

Unscrew bolt -1-

– Unclip door trim with aid of tool 3392 on the sides.

– Raise door trim upward out of window channel.

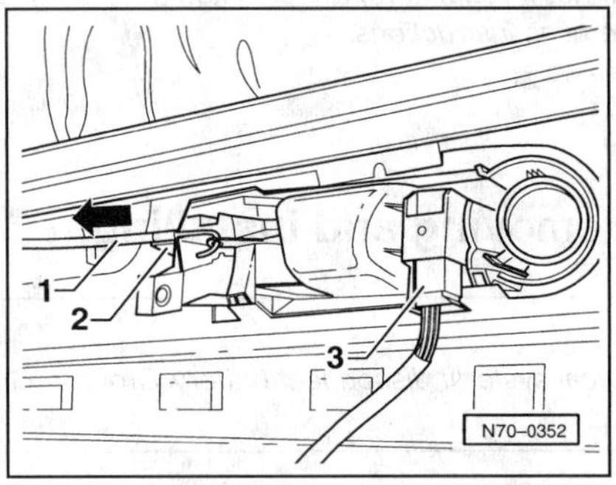

Pull cable -1- out in direction of guide -2- and unhook.

• Disconnect connector -3-.

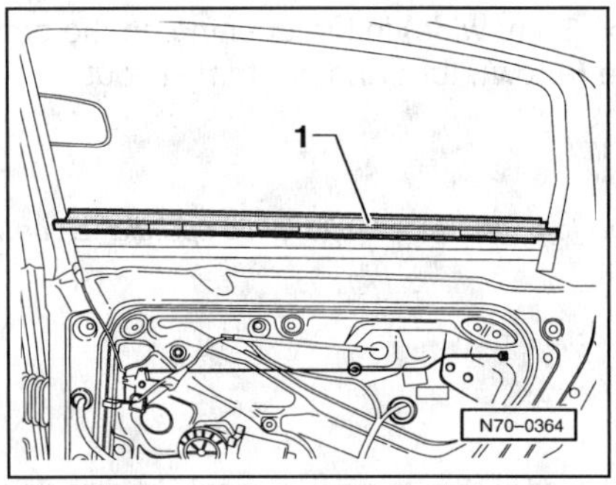

Rear door trim, installing

Remove window channel seal -1- from door trim and install in window channel.

NOTE —

• Before installing the door trim check clips and if necessary replace.

• After connecting the battery make sure that the vehicle equipment (radio, clock and power window regulators) are checked according to repair manual and/or operating instructions.

Pillars and side trim

Upper A-pillar trim, removing and installing

Vehicles without side curtain protection

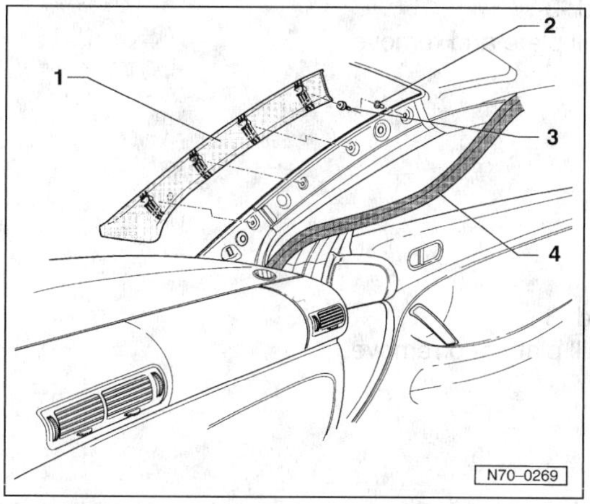

1. **Trim**
   - Removing:
     - Remove seal -4- in area of A-pillar trim.
     - Begin to unclip trim from the top.
     - Pull trim from the top.
   - Installing:
     - Check clips -3-, if necessary replace and install in trim mounting.
     - Clip trim in securely and check seal -4- is seated correctly.
2. **Sleeve**
   - Mounting for clip -**3**-
   - Qty.: 4
3. **Clip**
   - Qty.: 4
4. **Seal**
   - After installing the trim check for correct fit of seal.

Vehicles with side curtain protection

NOTE —

Removal of upper A-pillar trim cannot be performed without damage to mounting clips. Inspect A-pillar trim for integrity before reinstalling.

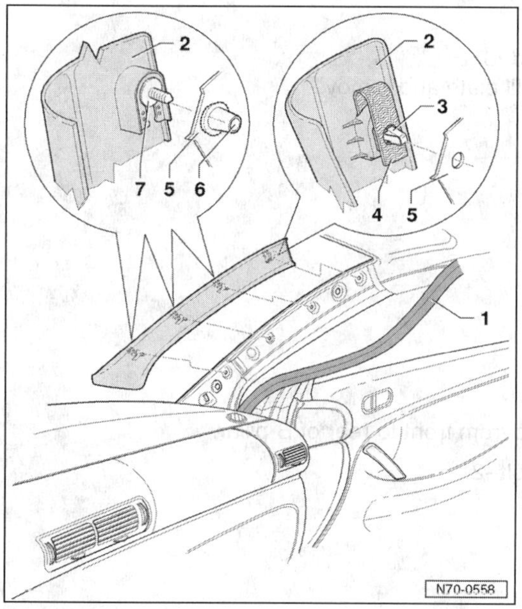

1. **Seal**
2. **Trim panel**
   - Removing:
     - Pull seal -1- off in area of A pillar trim.
     - Unclip trim beginning at top.
     - Pull trim out upwards.
     - If clip -3- remains in body, it must be squeezed together with pliers to remove without damage.
   - Installing:
     - Insert trim at same height as instrument panel, clip in securely and check that seal -**1**- is positioned correctly.
3. **Clip**
   - Damaged during removal.
4. **Support strap**
5. **A-pillar**
6. **Sleeve**
   - Clip mounting -**7**-
   - Qty. 3
7. **Clip**
   - Qty. 3

Lower A-pillar trim, removing

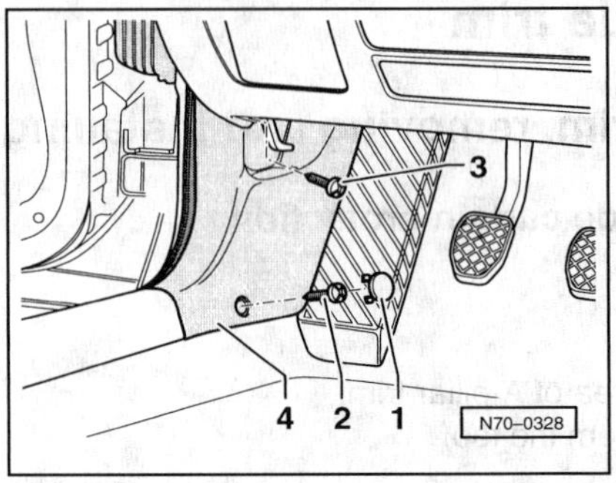

Driver side

- Pry off cover cap -1-
  - Unscrew bolts -2- and -3-.
  - Unclip trim -4- from sill plate and remove.

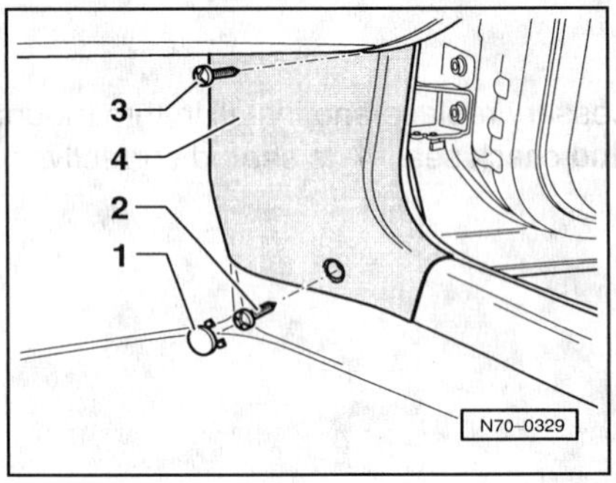

Passenger side

- Pry off cover cap -1-
  - Unscrew bolts -2- and -3-.
  - Unclip trim -4- from sill plate and remove.

Upper B-pillar trim, removing and installing

Removing

NOTE —
Illustration for one side only, symmetrically opposite for the other side.

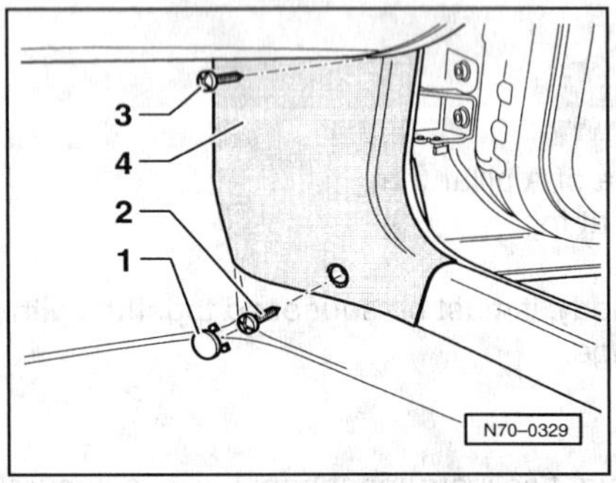

- Pry off cover cap -1-.
  - Unscrew bolts -2- and -3-.
  - Unclip trim -4- from sill plate and remove.

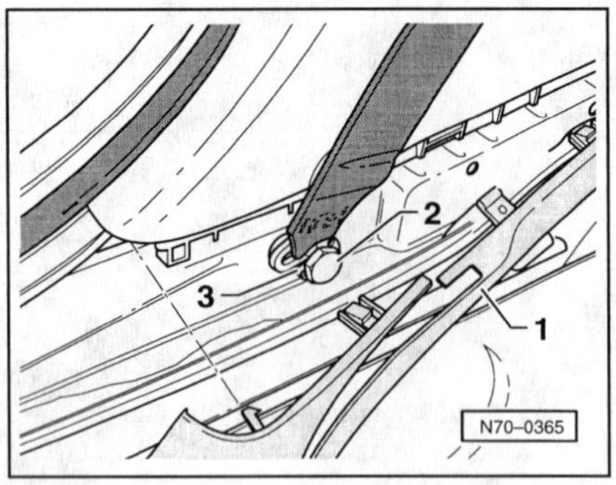

- Sill plate -1- un-clipped from front to rear of B-pillar.
  - Unscrew hex head bolt -2-.
  - Remove belt end -3-.

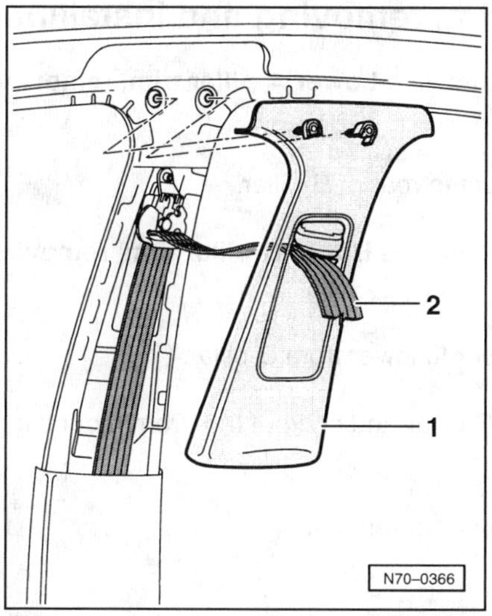

Vehicles without side curtain protection

◄ Unclip trim -1- first on the sides, then in upper area of B-pillar.

- Pull out seatbelt -2- from trim.

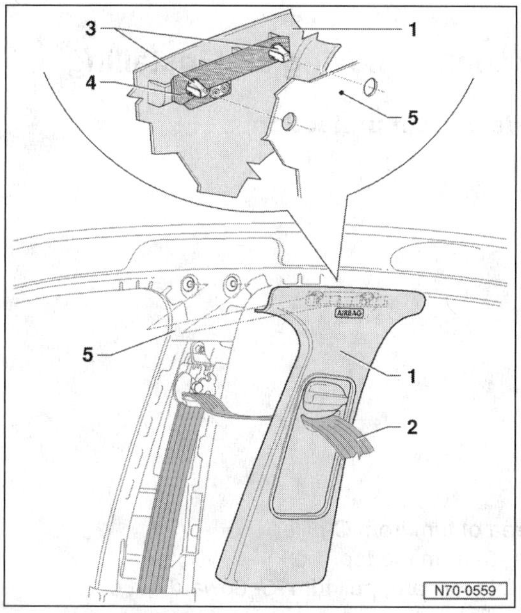

Vehicles with side curtain protection

NOTE —
Removal of upper B-pillar trim cannot be performed without damage to mounting clips. Inspect B-pillar trim for integrity before reinstalling.

◄ Unclip trim -1- from B pillar, first at sides and then at top.

- Pull seat belt -2- out of trim.
- If clip -3- remains in body, it must be squeezed together with pliers to remove without damage.

EDITOR'S NOTE —
Clips -3- are damaged during removal, be sure to have replacements on hand for reassembly

Installing (all vehicles)

– Pull seat belt with tongue and fitting through upper B-pilar trim.

– Position trim in installation position and clip into B-pillar. Make sure seat belt height adjustment fitting is correctly fitted.

- Replace upper B-pilar trim retaining clips if necessary.

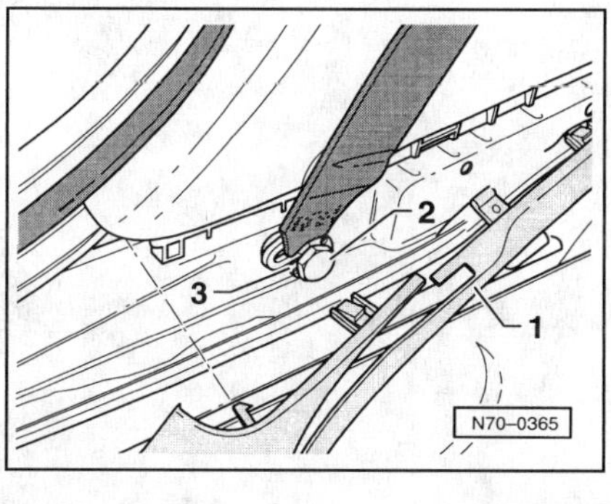

◄ Using bolt -2- screw seat belt fitting -3- onto longitudinal member (sill), tighten to 40 Nm (30 ft-lb).

- Clip sill panel strip -1- in.

– Install lower A pillar trim.

– Make sure seal is correctly positioned.

Lower B-pillar trim, removing and installing

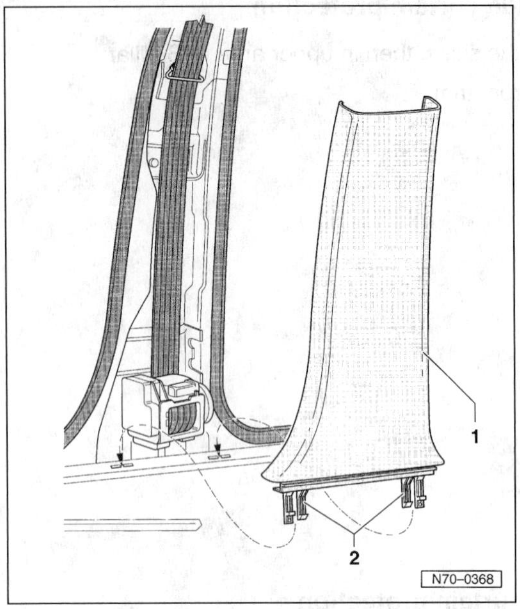

- Remove lower A-pillar trim, see **Lower A-pillar trim, removing and installing**.
- Unclip sill plate from front to rear of B-pillar.
- Remove upper B-pillar trim, see **Upper B-pillar trim, removing and installing**.
- Unclip trim -1- from upper to lower from B-pillar.
- Press hook -2- toward B-pillar and pull out trim from mounting in sill panel.
- Installation is reverse order of removal.

NOTE *—*
After installing the trim, check for correct fit of door seals.

Upper C/D-pillar trim, removing and installing

Vehicles without side curtain protection

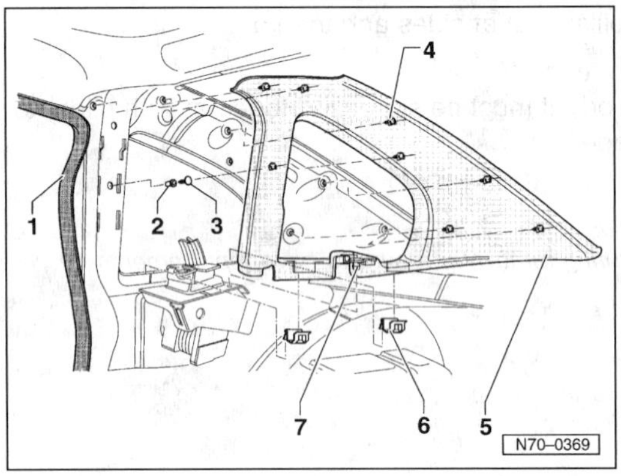

1. **Seal**
2. **Sleeve**
   - Qty.: 8
3. **Clip**
   - Qty.: 8
4. **Mounting**
   - Component of trim
5. **Trim**

   Removing:
   - Remove seal -1- in area of trim from C-pillar.
   - Begin un-clipping trim -5- from the top.
   - Unlock seatback and fold forward, pull trim -5- upward out of clips -6-.

   Installing:
   - Check and if necessary replace clips -3- and clip -7-, and install in trim -4- mounting.
   - Check and if necessary replace clips -6- and insert mounting to window flange.
   - Clip trim into clips -6- securely and make sure seal -1- is seated correctly.
6. **Clip**
   - Qty.: 2
7. **Clip**

Vehicles with side curtain protection

NOTE —

Removal of upper C/D pillar trim cannot be performed without damage.

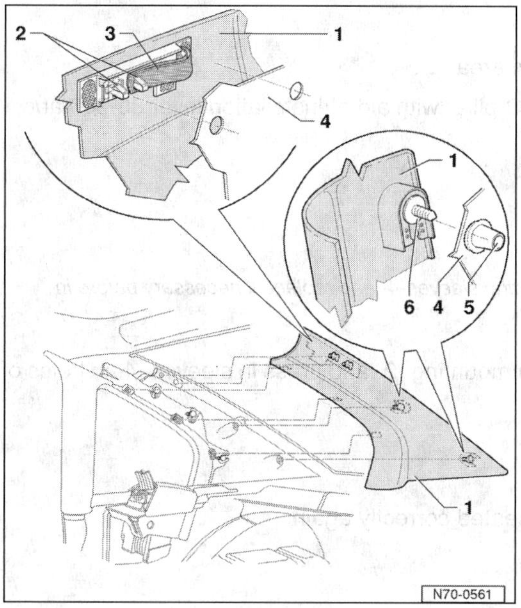

1. **Trim (upper)**

   Removing:
   - Unclip trim, starting at top.
   - If a clip -**2**- remains in the body, it must be pinched together using pliers to remove it from the body without causing damage.

   Installing:
   - Position trim on side panel and clip in securely.

2. **Clip**
   - Qty. 2

3. **Support strap**

4. **Side panel**

5. **Sleeve**
   - Mounting for clip -**6**-
   - Qty. 2

6. **Clip**
   - Qty. 2

Trim (lower), removing and installing

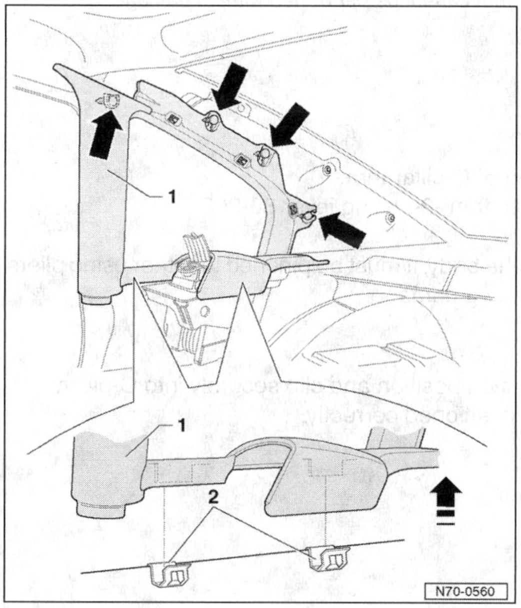

◄ Removing:
- Pull door seal off C pillar in area of trim.
- Unclip trim, starting at top.
- Release seat backrest and fold forward, pull trim -1- upwards out of clips -2-.

Installing:
- Position trim on side panel and clip in securely.
- Fit door seal.

Upper C-pillar trim, removing and installing (wagon)

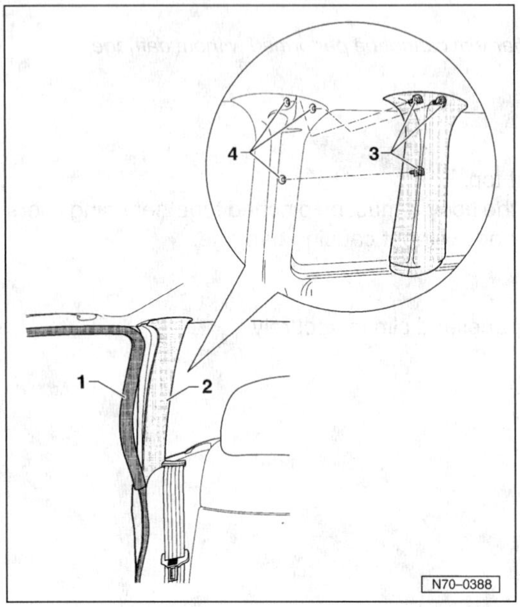

Vehicles without side curtain protection

Removing

Remove seal -1- in trim area -2-.

• Pry out trim -2- from C-pillar with aid of installation lever 80-200 and pull out from the top.

Installing

NOTE —

Check clips -3- and mounting sleeves-4- and replace if necessary before installing the trim -2-.

– Insert clips -3- into trim mounting -2- and press in sleeves -4- in holes of C-pillar.

– Clip trim -2- to C-pillar.

– Make sure seal -1- is seated correctly again.

Vehicles with side curtain protection

NOTE —

Removal of upper C-pillar trim cannot be performed without damage.

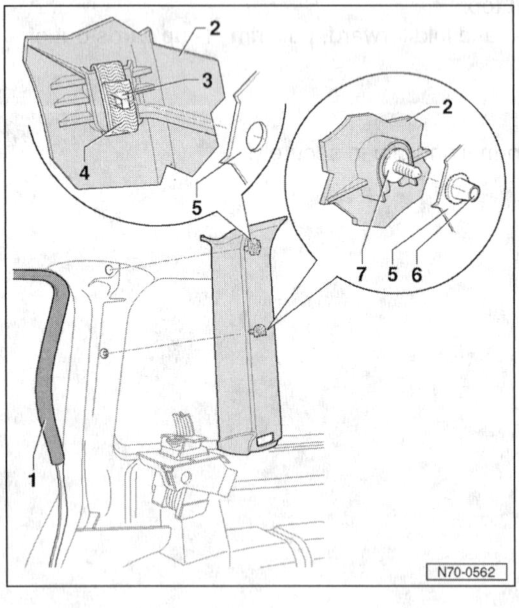

1. **Seal**
2. **Trim panel**

   Removing:

   • Pull seal -1- off in area of C-pillar trim.
   • Starting from top unclip trim -2-, using lever 80-200.
   • Pull trim out upwards.
   • If a clip -3- remains in the body, it must be pinched together using pliers to remove without causing damage.

   Installing:

   • Position trim in installation position and clip securely into C-pillar.
   • Make sure seal -1- is positioned correctly.

3. **Clip**
   • Damaged upon removal.
4. **Support strap**
5. **C-pillar**
6. **Sleeve**
   • Mounting for clip -7-
7. **Clip**

Lower C-pillar trim, removing and installing

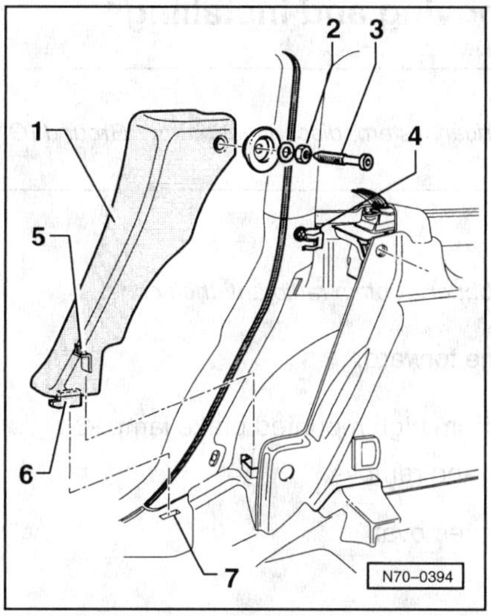

NOTE—
Illustration for one side only, symmetrically opposite for the other side.

– Unlock seatback for rear seat bench and fold forward.

Loosen hex nut -2- and unscrew lock bolt -3-.

- Pull cushion -1- in upper area out of clip -4-, press lower tab -6- to the back from sill plate -7- and remove with tab -5- upward.

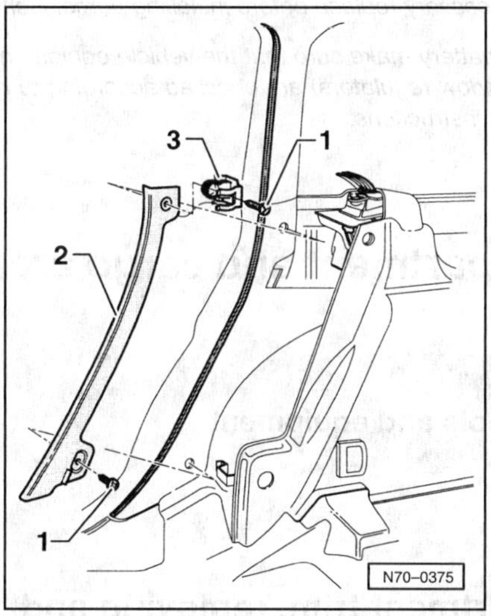

Unscrew bolts -1-

- Remove trim -2- from clip -3- and from C-pillar.

– Installation is reverse of removal.

NOTE—
After installing the trim, check for correct fit of door seals.

Upper D-pillar trim, removing and installing (wagon)

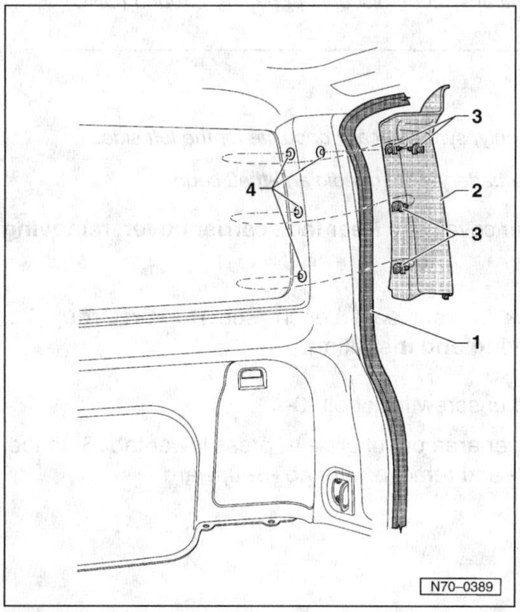

Remove seal -1- in trim area -2-.

- Pry out trim -2- from D-pillar with aid of installation lever 80-200 and pull out from the top.

NOTE—
Check and replace if necessary clips -3- and mounting sleeves-4- before installing the trim -2-.

Installing:

- Insert clips -3- into trim mounting -2- and press in sleeves -4- in holes of C-pillar.
- Insert trim -2- in lower area behind lateral trim and clip to D-pillar.
- Make sure seal -1- is seated correctly.

Parcel shelf, removing and installing

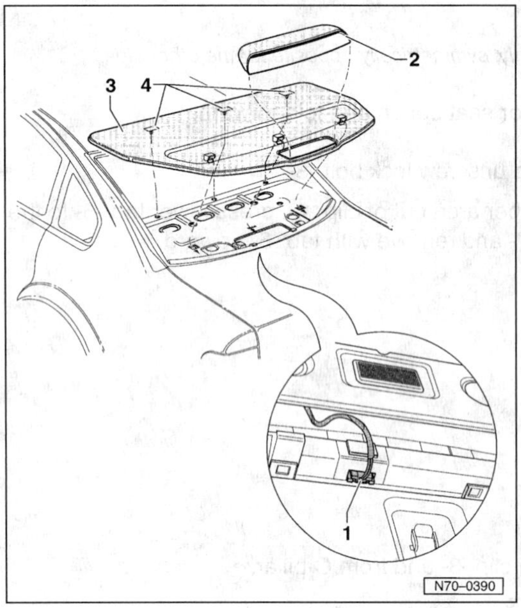

> ***WARNING —***
> *Before working on electrical system, disconnect battery Ground (GND) strap.*

NOTE —
Before disconnecting the battery, obtain radio anti-theft code.

– Fold rear seatback frame forward.

◀ Remove connector -1- from high mounted brake lamp -2-.

- Unclip parcel shelf -3- and remove.

– Installation is reverse of removal.

NOTE —

- *Check clips -4- and if necessary replace before installing parcelshelf -3-.*
- *After disconnecting the battery make sure that the vehicle equipment (radio, clock and power window regulators) are checked according to repair manual and/or operating instructions.*

Luggage compartment and cargo area trim

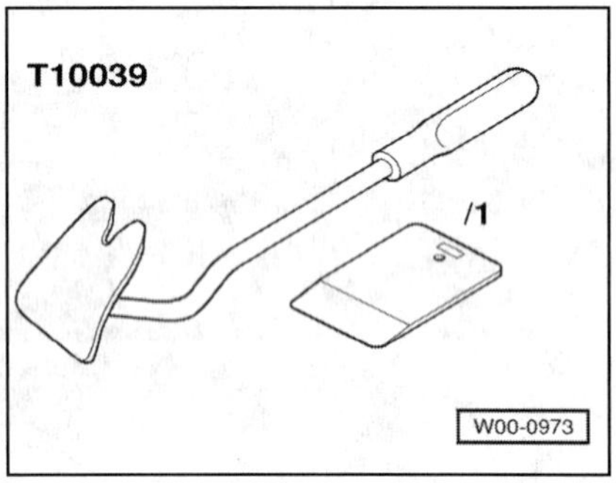

Tools

Required special tools and equipment

◀ T10039 Pry bar

Luggage compartment trim, removing and installing

> ***WARNING —***
> *Before working on electrical system, disconnect battery Ground (GND) strap.*

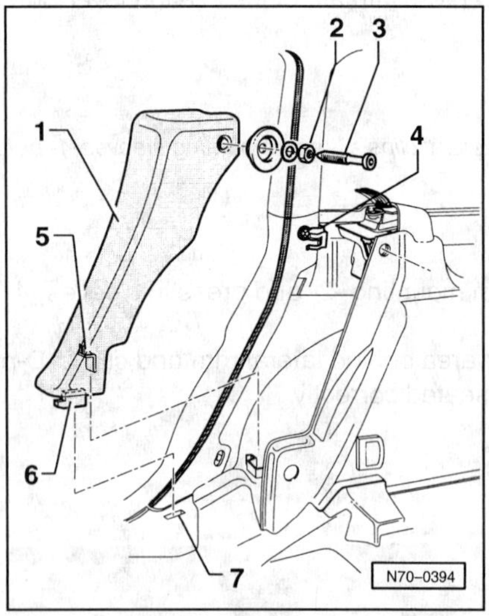

NOTE —

- *Illustration for right side only, symmetrically opposite for the left side.*
- *Before disconnecting the battery, obtain radio anti-theft code.*

– Remove rear lock carrier cover, see **Rear lock carrier cover, removing and installing**.

– Remove right seatback for rear seat bench, see **Repair Group 72, Right backrest, removing and installing.**

◀ Loosen hex nut -2- and unscrew lock bolt -3-.

- Pull cushion -1- in upper area out of clip -4-, press lower tab -6- to the back from sill plate -7- and remove with tab -5- upward.

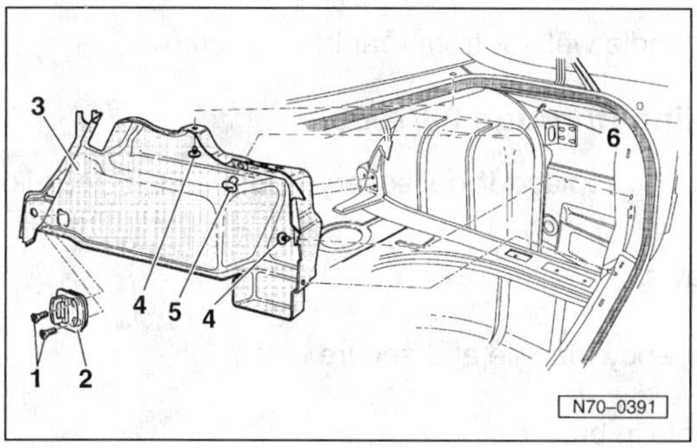

Remove screws -1-.

- Remove lashing eyes -2-.
- Remove clips -4-.
- Unscrew plugs -5- from threaded pin -6-.
- Remove trim -3-.

– Installation is reverse order of removal.

NOTE —
After disconnecting the battery make sure that the vehicle equipment (radio, clock and power window regulators) are checked according to repair manual and/or operating instructions.

Rear lock carrier cover, removing and installing

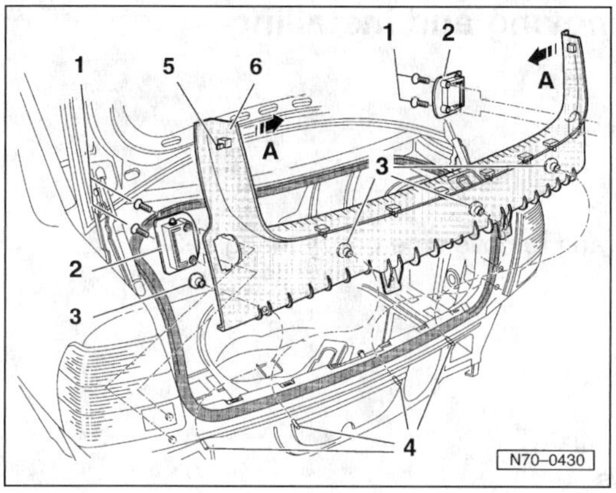

Remove screws -1-.

- Remove lashing eyes -2-.
- Unscrew four plugs -3- with screwdriver from threaded pins -4-.
- Unclip upper edge of cover -6- (arrows -A-).

– Unclip cover from lock carrier upward.

– Installation is reverse order of removal.

NOTE —
Check and if necessary replace plugs -3- and clips -5- before installing.

Rear lid trim, removing and installing (through 09/2000)

Removing

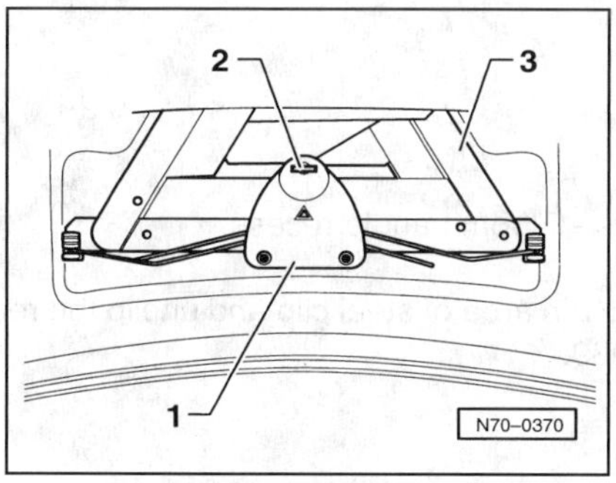

Unscrew lock -2-.

- Fold open bracket -1- and remove emergency triangle -3-.

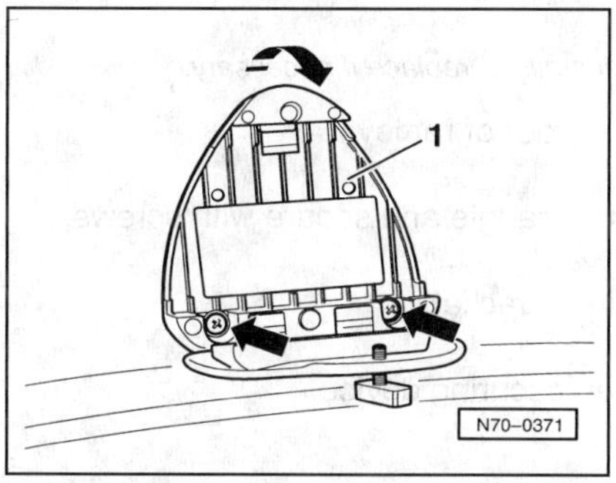

Unscrew Phillips-head screws (arrows).

- Slide bracket -1- upward and remove.

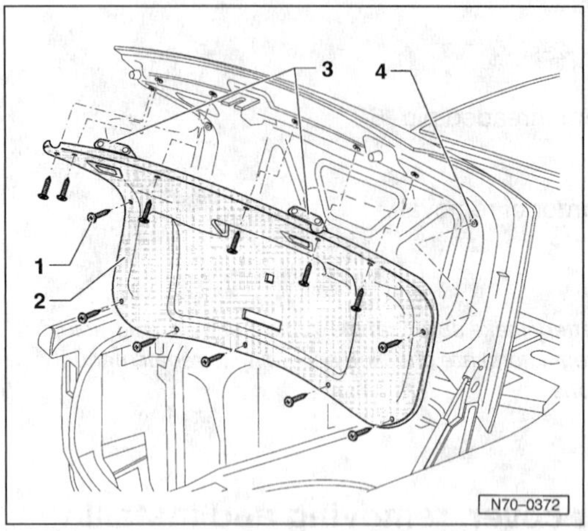

Unscrew mounting screws -1-.

- Pull out trim -2- with handle well -3- from rear lid and remove

Rear lid trim, installing (through 09/2000)

– Check expanding nuts -4-, replace if necessary and insert in rear lid holes.

– Install trim -2- with screws.

– Install bracket for emergency triangle and secure.

– Insert emergency triangle in bracket.

– Fold bracket and screw on lock.

Rear lid trim, removing and installing (from 10/2000)

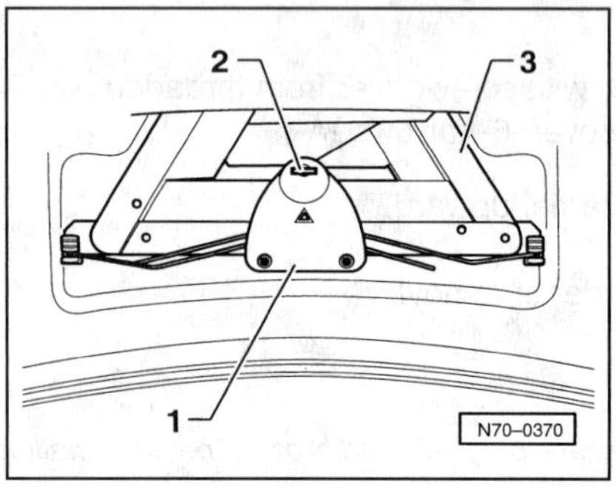

Removing

Unscrew lock -2-.

- Fold open bracket -1- and remove emergency triangle -3-.

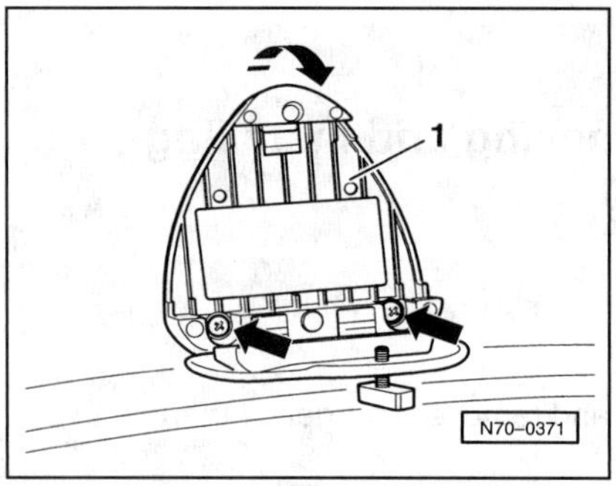

Remove screws -arrows-.

- Slide bracket -1- up and remove.

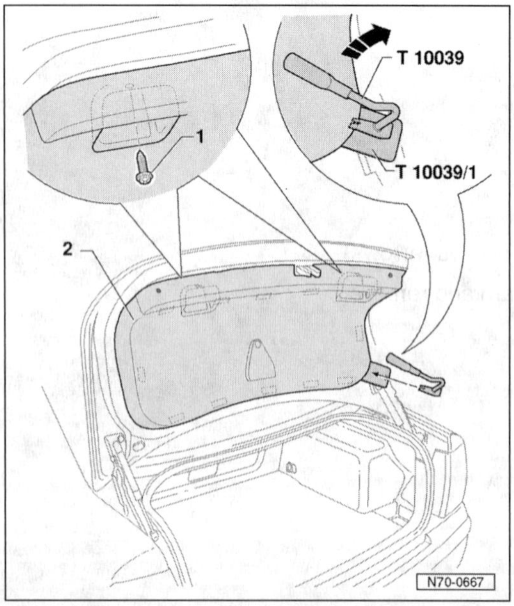

Unscrew the two screws -1- from handle recess.

– Position wedge T10039/1 in area of steel clip and unclip the rear lid trim with release lever T10039.

Installing

NOTE —
• Check steel clips before installing, replace if necessary.

– Install trim -2- in reverse order of removal.

– Insert bracket for warning triangle and secure with screws.

– Insert warning triangle into bracket.

– Close bracket and tighten securing device.

Cargo area trim, removing and installing (wagon)

> ***WARNING—***
> *Before working on electrical system, disconnect battery Ground (GND) strap.*

NOTE—
- *Illustration for right side only, symmetrically opposite for the left side.*
- *Before disconnecting the battery, obtain radio anti-theft code.*

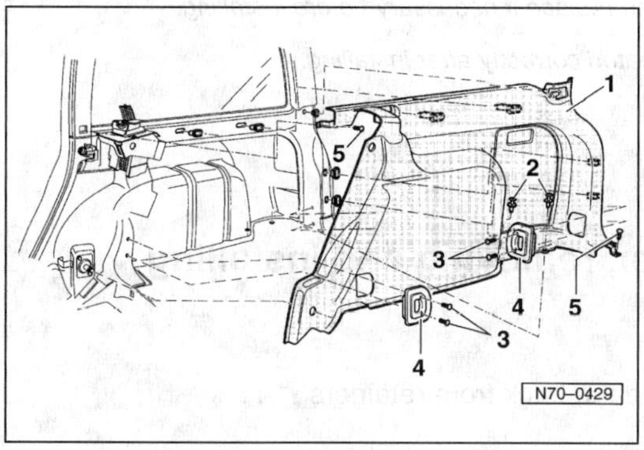

1. **Trim panel**
2. **Expanding clip**
3. **Bolts for lashing eyes**
4. **Lashing eyes**
5. **Bolts**

– Remove cargo area cover, see **Cargo area cover, removing and installing**.

– Remove cover for lock carrier, see **Rear lock carrier cover, removing and installing (wagon)**.

– Remove right seatback of rear seat bench, see **Repair Group 72, Right backrest, removing and installing.**

– Replace cargo area floor covering, see **Cargo area floor covering, removing (wagon)**.

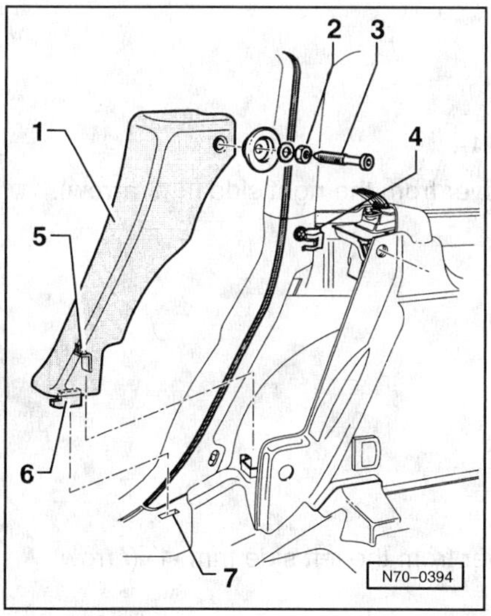

◄ Loosen hex nut -2- and unscrew lock bolt -3-.

- Pull cushion -1- in upper area out of clip -4-, press lower tab -6- to the back from sill plate -7- and remove with tab -5- upward.
- Expanding clip -2- pressed out with assembly lever 80-200.
- Remove bolts -3- and remove lashing eyes -4-.
- Remove screws -5-.
- Remove trim -1-.

– Installation is reverse order of removal.

NOTE—
After connecting the battery make sure that the vehicle equipment (radio, clock and power window regulators) are checked according to repair manual and/or operating instructions.

Rear lock carrier cover, removing and installing (wagon)

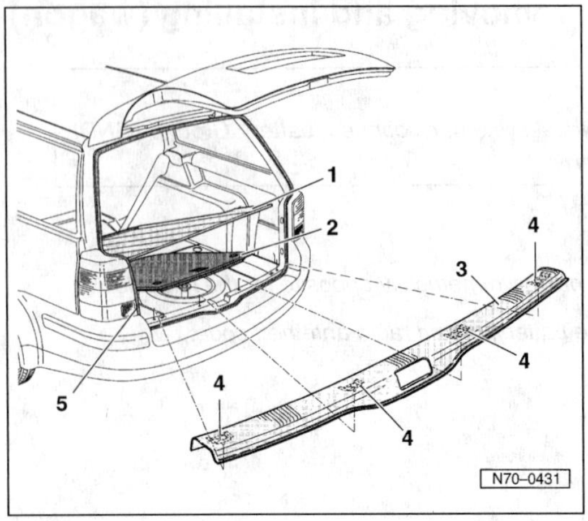

Raise cargo area floor -1-.

- Open spare tire well cover -2- pull up and support.
- Unclip lock carrier cover -3-.

– Installation is reverse order of removal.

NOTE —

- *Check cover clips -4- and replace if necessary before installing.*
- *Make sure seal -5- is seated correctly after installing.*

Cargo area cover, removing and installing

Removing

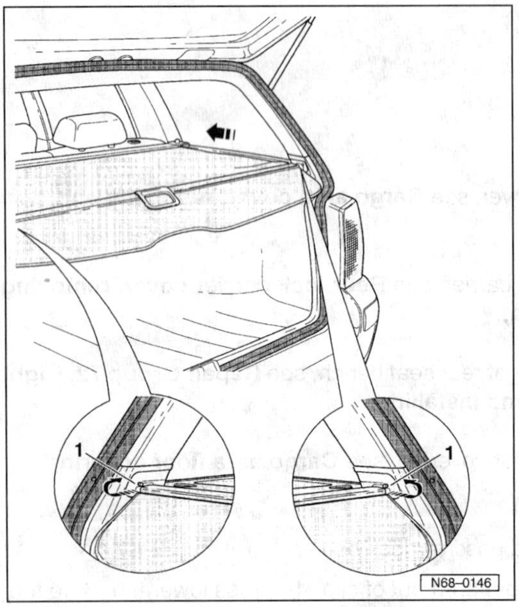

Pull cargo area cover to the back from retainers -1-.

– Carefully roll up in direction of arrow (automatic roll up).

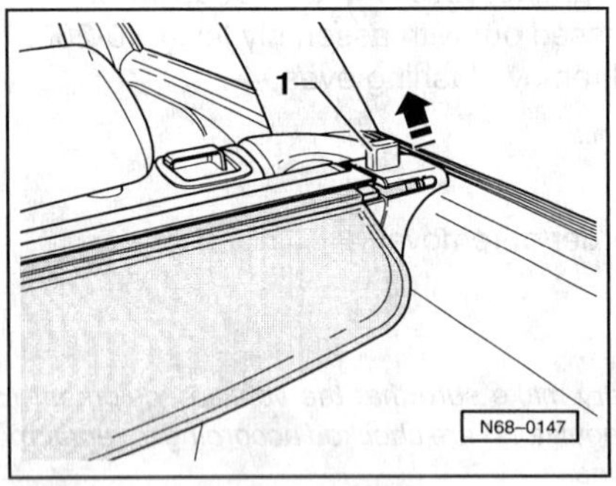

Push down lock button-1-.

- Pull out cargo area cover from the right side trim (arrow).

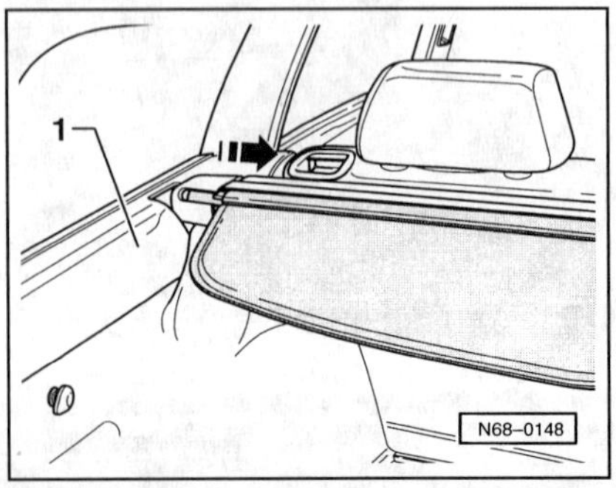

Pull out cargo area cover from the left side trim -1- (arrow).

Cargo area cover, Installing

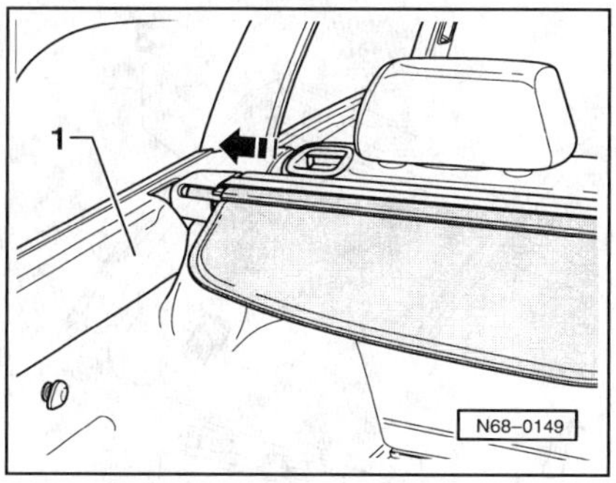

◄ Guide in cargo area cover in the left side trim -1- (arrow).

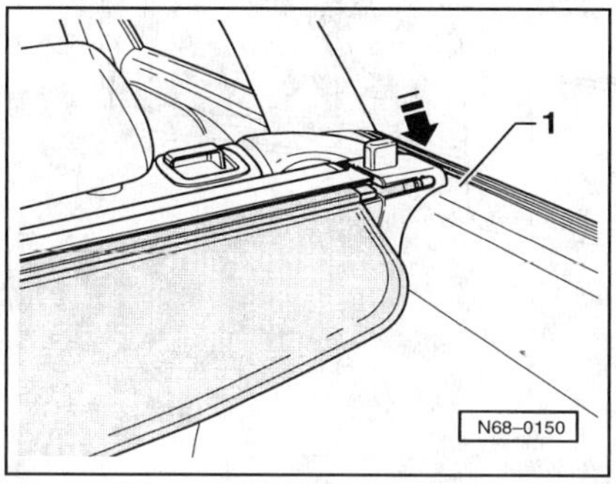

◄ Engage cargo area cover into the right side trim -1- (arrow).

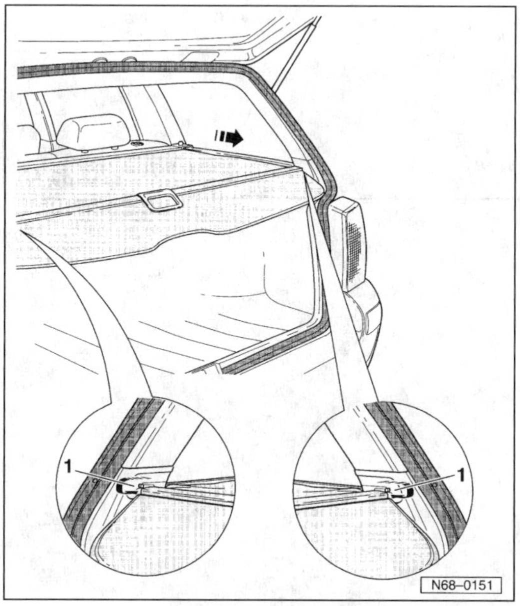

◄ Hook cargo area cover in rear retainers -1- (arrows).

Tailgate trim, removing (wagon) (through 09/2000)

1. **Clip**
   - Qty.: 4
2. **Upper trim**

   Removing:
   - Remove lower trim -7-.
   - Unclip upper trim -2- first laterally out of clips -3- and then from the upper area.
   - Check and if necessary replace clips -1- before installing trim and insert in mounting of trim.
3. **Clip**
   - Qty.: 4
   - Check for damage and replace if necessary before installing the upper trim.
4. **Expanding nut**
   - Qty.: 2
5. **Cover cap**
   - Qty.: 2
6. **Clip**
   - 12 pieces
7. **Lower trim**

   Removing:
   - Pry out caps -5-.
   - Remove screws -8-.
   - Begin un-clipping lower trim -7- from below.
   - Check and replace if necessary trim clips -6- before installing and insert in mounting of trim.
8. **Screw**
   - Qty.: 2

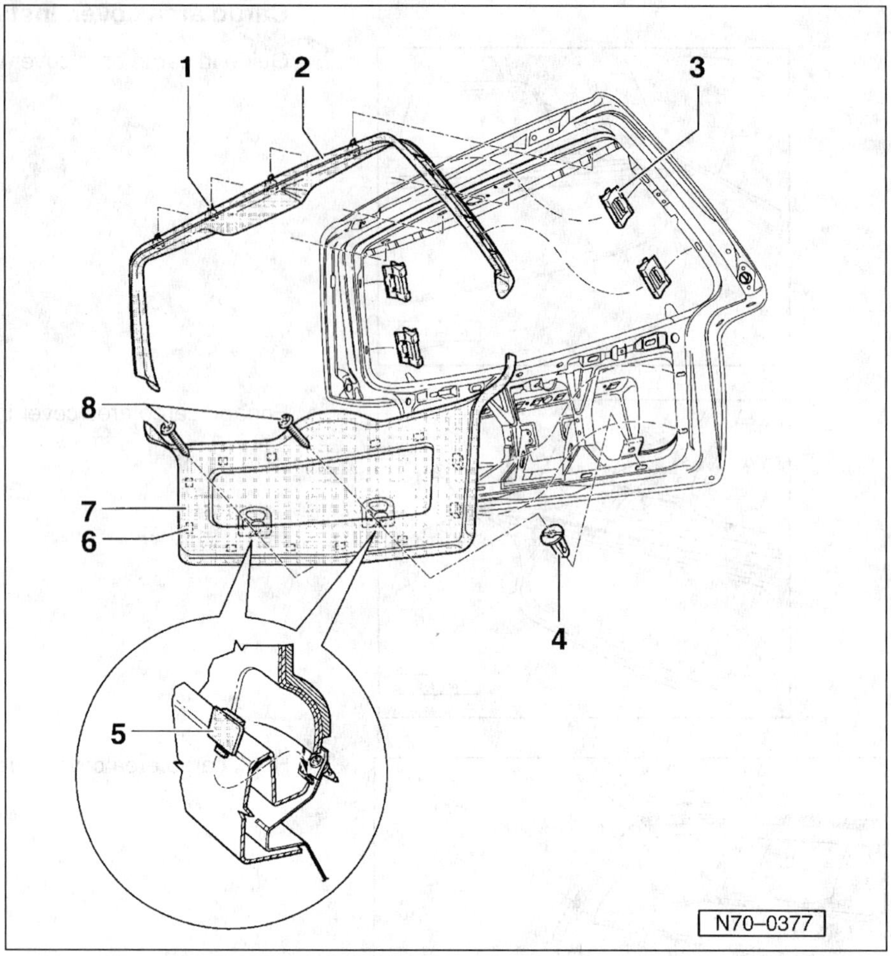

Tailgate trim, removing (wagon) (from 10/2000)

1. **Clip**
   - Qty. 4
2. **Upper trim**

   Removing:
   - Remove lower trim panel -7-.
   - Unclip trim panel -2- sides from clips -3- first, then unclip upper area.
   - Check clips -1- before installing trim, replace if necessary and insert into trim mountings.
3. **Clip**
   - Qty. 4
   - Check for damage and if necessary replace before installing upper trim panel.
4. **Expanding nut**
   - Qty. 2
5. **Cover Cap**
   - Qty. 2
6. **Clip**
   - Qty. 12
7. **Lower trim**

   Removing:
   - Pry out cover caps -5-.
   - Remove bolts -8-.
   - Unclip lower trim panel -7- starting at lower edge.
   - Check clips -6- before installing trim, replace if necessary and insert into trim mountings.
8. **Screw**
   - Qty. 4

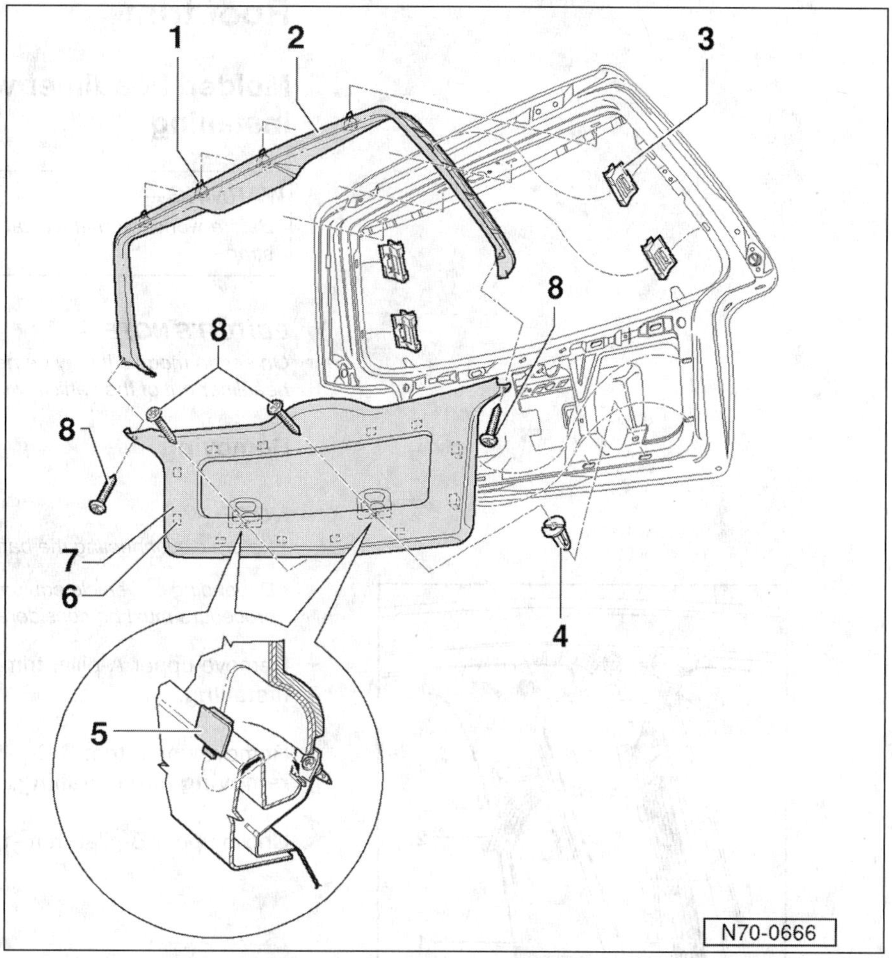

Cargo area floor covering, removing (wagon)

– Unlock rear seat back and fold forward.

◄ Remove welting strip -1- of pad -3- from fastening rail -2-.

- Remove pad -3- out of the vehicle toward rear.

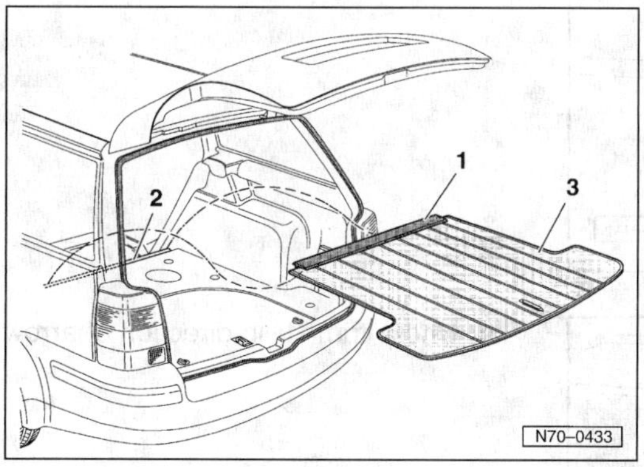

Roof trim

Molded headliner with sunroof, removing and installing

> ***WARNING —***
> *Before working on electrical system, disconnect battery Ground (GND) strap.*

EDITOR'S NOTE —
On sedan models it may be necessary to remove the rear windo to get the headliner out of the vehicle without damage.

Removing

NOTE —
- *Before disconnecting the battery, obtain radio anti-theft code.*
- *Depending on vehicle equipment, variations of the removing and installing procedure must be considered.*

– Remove upper A-pillar trim, see **Upper A-pillar trim, removing and installing**.

– Remove upper trim C/D-pillar trim, see **Upper C/D-pillar trim, removing and installing**.

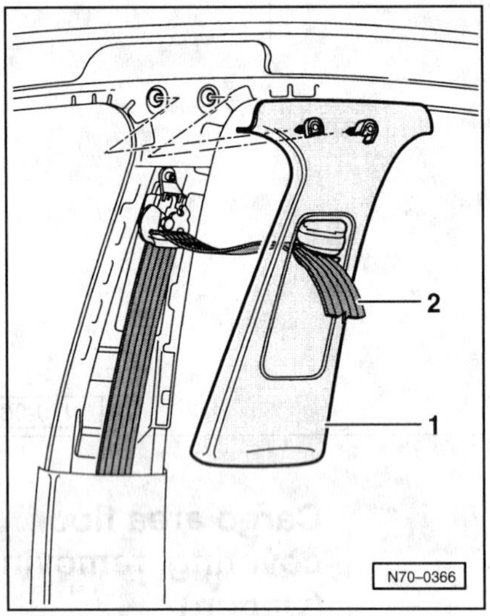

◄ Unclip upper B-pilar trim -1- at the sides, then at the upper area.

Roof handles, removing

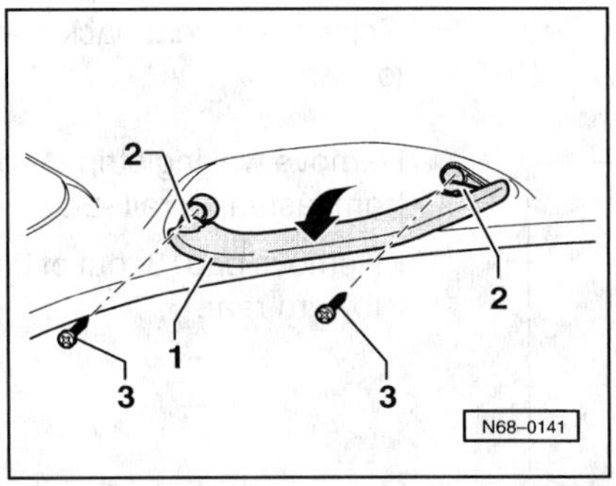

◄ Fold handle -1- down.

- Pry off covers -2- with screwdriver and fold up.
- Unscrew Phillips-head screws -3- and remove handle -1-.

◄ Unclip trim -1- in direction of arrow and remove.

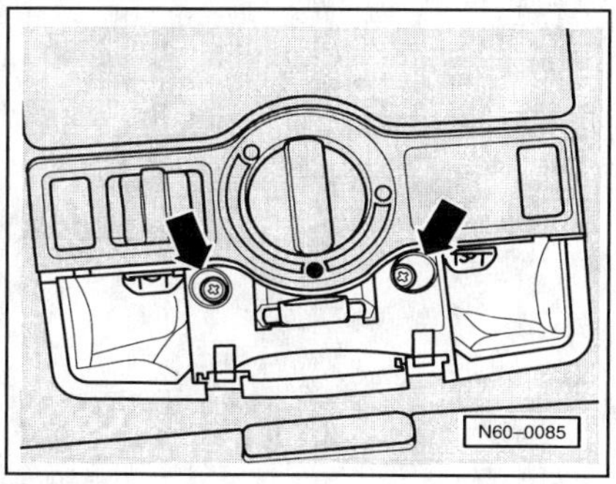

Unscrew screws (arrows) and remove controller.

– Disconnect harness connector.

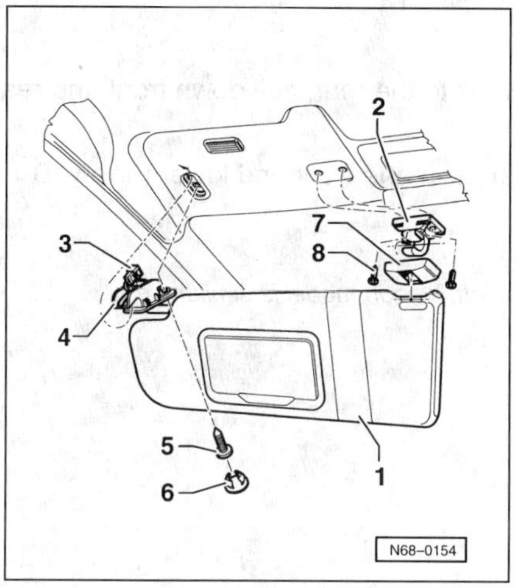

Sun visor, removing

Unhook sun visor -1- from mount -2-.

- Pry off cap -6-.
- Remove screw -5-.
- Unhook sun visor bearing -4- from mount and disconnect connector -3-.
- Pry off cap -7-.
- Remove bolts -8- and remove mount -2-.

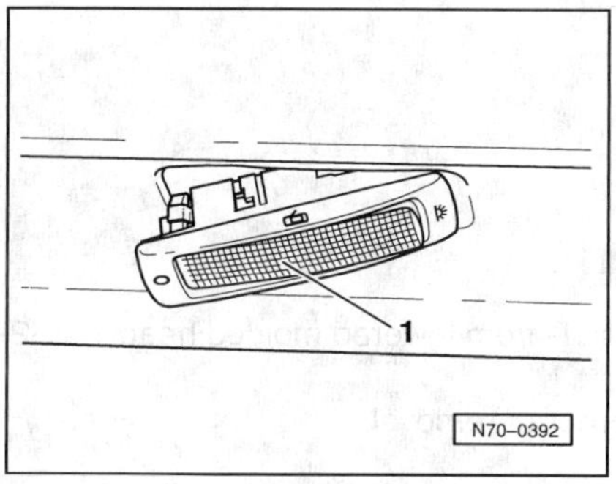

Interior light, removing

Pry out interior light -1- and separate connector.

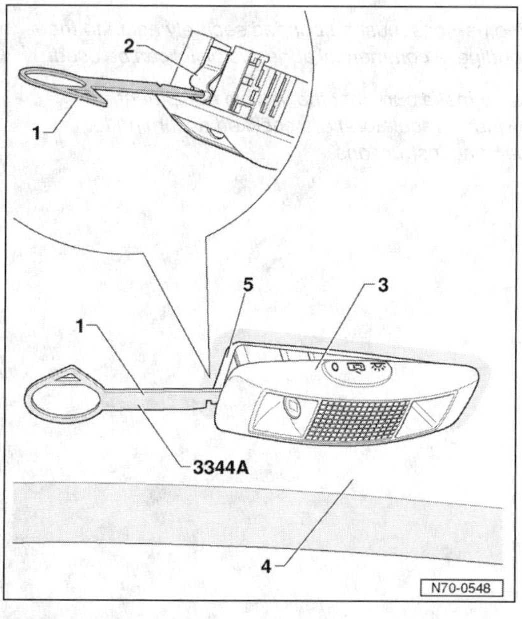

Reading light with interior monitoring, removing and installing

With the unlocking tool -1- press spring of reading light -2- to the back.

- Remove reading light -3- and disconnect connector.

NOTE—

Make sure that when pressing the spring -2- the release tool -1- does not enter between the headliner -4- and the metal frame in the headliner -5-.

Wagon only

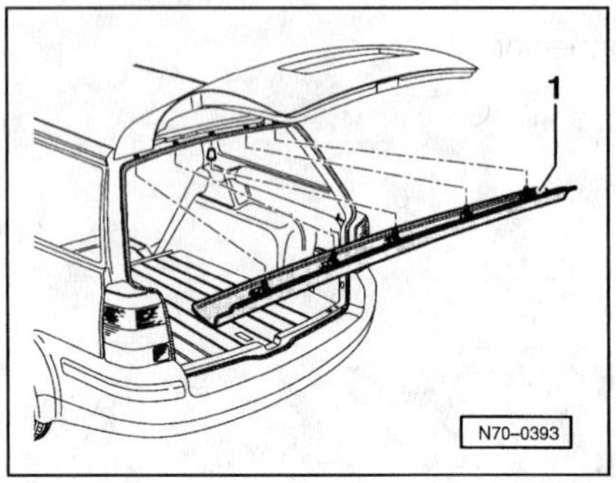

< Unclip cover strip -1-.

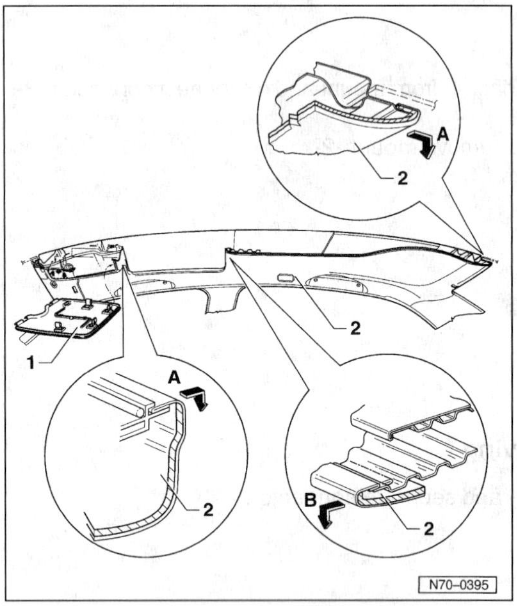

< Slide molded headliner -2- to the rear, pull down front and rear (arrows -A-).

- Push molded headliner -2- to the front and lower (arrow -B-).

EDITOR'S NOTE —

Sedan headliner shown iat left, wagon model is similar.

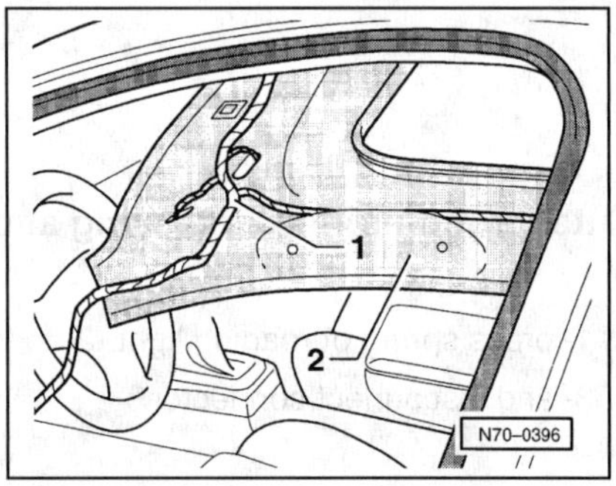

< Remove wiring harness -1- from lowered molded headliner -2-.

– Installation is reverse order of removal.

NOTE —

- *When installing, the wiring harness must be bonded securely again to the molded headliner. For bonding, a commercial adhesive pistol can be used.*
- *After connecting the battery make sure that the vehicle equipment (radio, clock and power window regulators) are checked according to repair manual and/or operating instructions.*

72 Seats, frames

Repair Group

U 72

Front seats

Tools

Required special tools and equipment

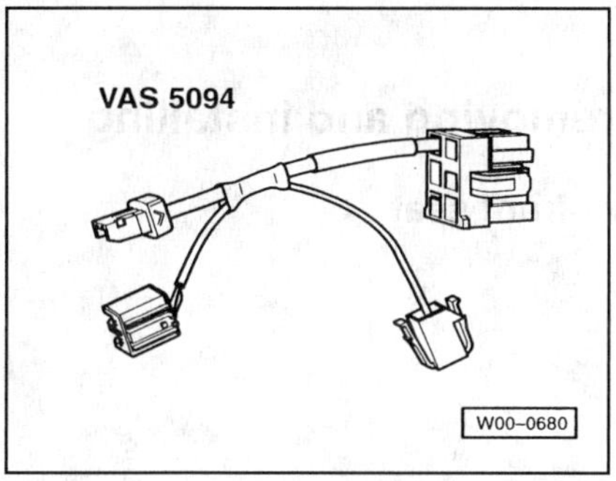

VAS 5094 Airbag adapter

Front seat, removing and installing

> ***WARNING —***
> *Before working on the electrical system, disconnect battery Ground (GND) strap.*

NOTE —

- *Depending on variations in vehicle equipment, slight deviations from the removing and installing procedure must be considered.*
- *Before disconnecting the battery, obtain radio anti-theft code.*
- *After disconnecting the battery make sure that the vehicle equipment (radio, clock and power window regulators) are checked according to repair manual and/or operating instructions.*

– Slide seat forward.

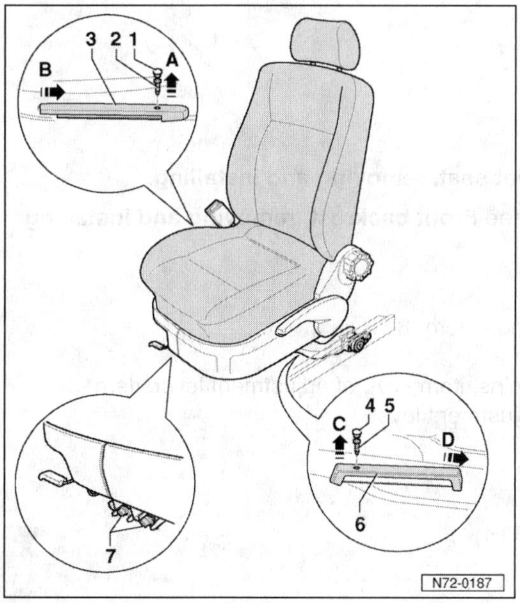

Unclip cap -1- from tunnel side of cover slip -3-, unscrew screw -2-, unclip cover strip from seat rail (arrow -A-) and remove (arrow -B-).

- Unclip cap -4- from rocker panel side cover strip -6-, unscrew screw -5-, unclip cover strip from seat rail (arrow -C-) and remove (arrow -D-).
- Slide seat backward.
- Unscrew bolts -7-.
- Slide seat to the back out of guide rails.

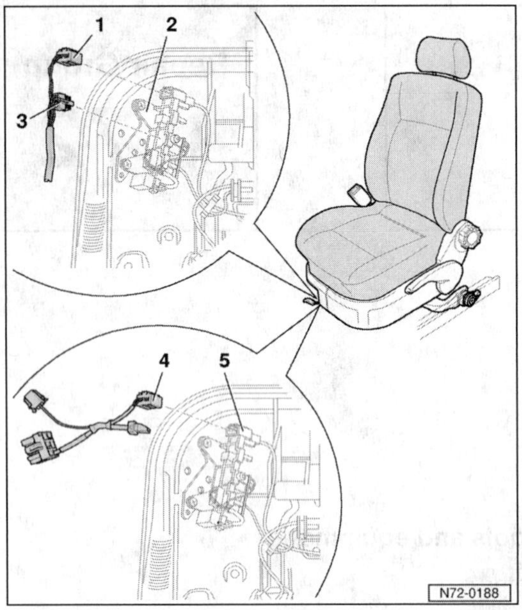

NOTE—

Before disconnecting ignition and Ground wires, technician must discharge from himself any electrostatic charge (to do this briefly touch striker for door or body).

< Disconnect connector for side airbag -1- from coupling station -2-.

- Disconnect connector for seat heater -3-, if available, from coupling station.
- Insert connector -4- (airbag adapter VAS 5094) in connector housing -5-.

– Installation is reverse of removal.

NOTE—

If airbag malfunction indicator lamp (MIL) -K75- for airbag signals a malfunction after assembly, erase DTC memory with Scan Tool VAG 1551 and check again.

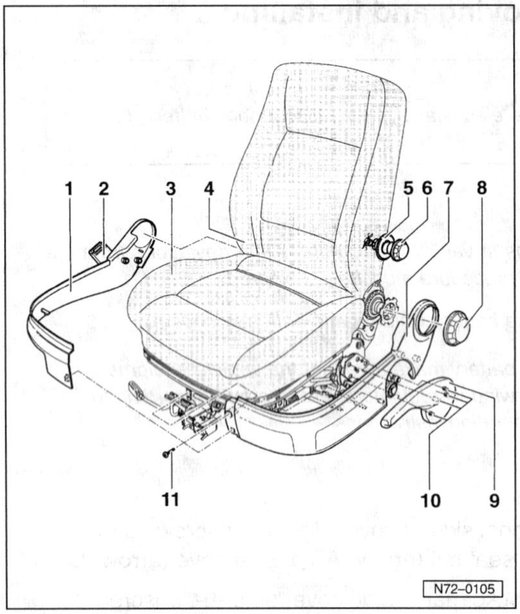

Front backrest, removing and installing

Assembly overview - front seat

1. **Seat trim, right**
2. **Belt lock**
3. **Seat**
4. **Backrest**
5. **Trim collar**
6. **Adjusting knob (lumbar support)**
7. **Seat trim, left**
8. **Adjustment knob (backrest)**
9. **Adjustment lever (Seat height)**
10. **Expanding pins**
11. **Screw for left trim**

– Remove seat, see **Front seat, removing and installing**.

- Assembly overview, see **Front backrest, removing and installing**.

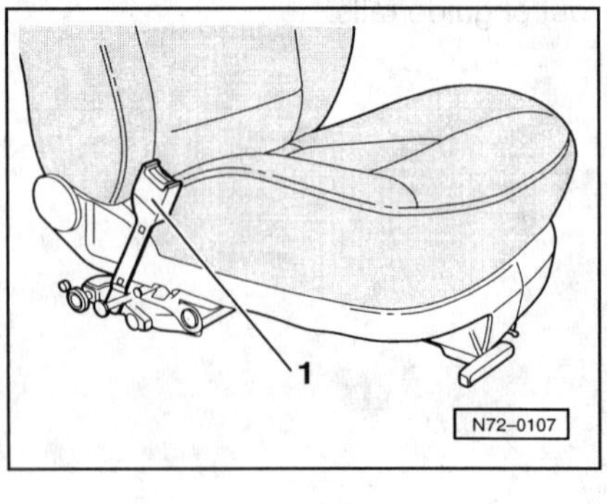

< Remove belt lock -1-.

– Remove adjustment knob, item -8-.

– Press four expanding pins, item -10-, of adjustment lever, item -9-, inward and remove adjustment lever.

– Remove trim, item -1-.

– Remove screw, item -11-.

– Remove trim, item-7-.

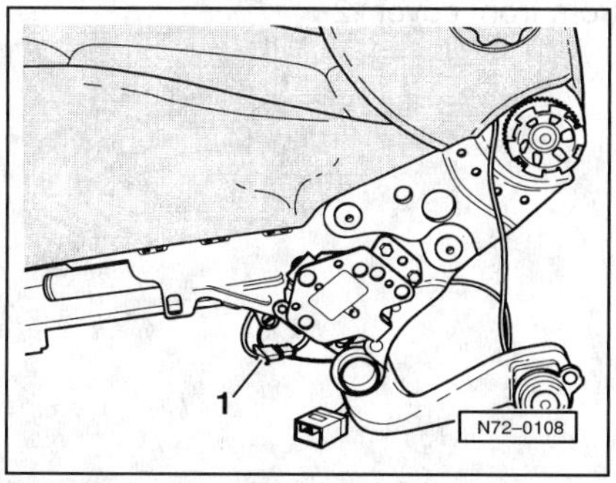

◀ Disconnect connector -1- for backrest heater.

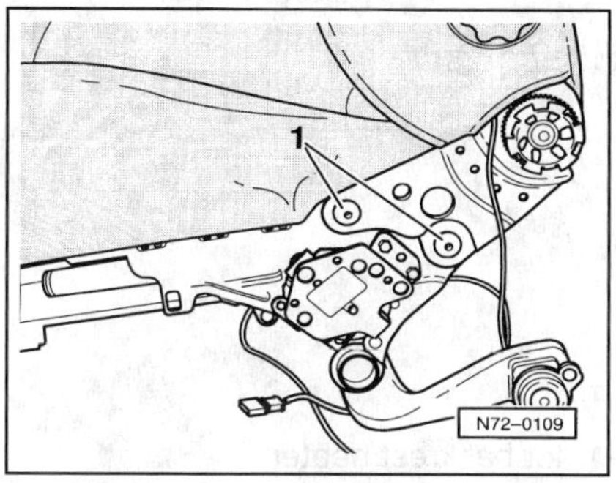

◀ Remove screws -1- on both sides of the seat and remove backrest.

– Installation is reverse of removal.

NOTE —
Screws -1- must be secured with locking fluid D 000 600 A2. Tighten to 20 Nm (15 ft-lb).

Lumbar support, removing and installing

NOTE —
Lumbar support permits backrest adjustment at lumbar area.

– Remove seat, see **Front seat, removing and installing**.

– Remove backrest, see **Front backrest, removing and installing**.

◀ Separate beading strip -1- of cover -2- with screwdriver (arrows).

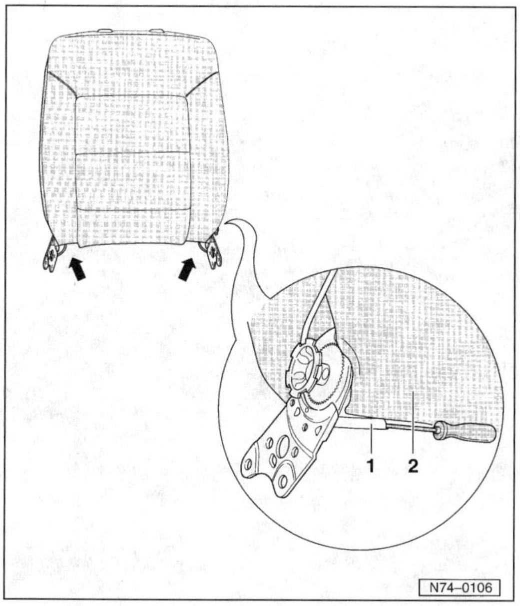

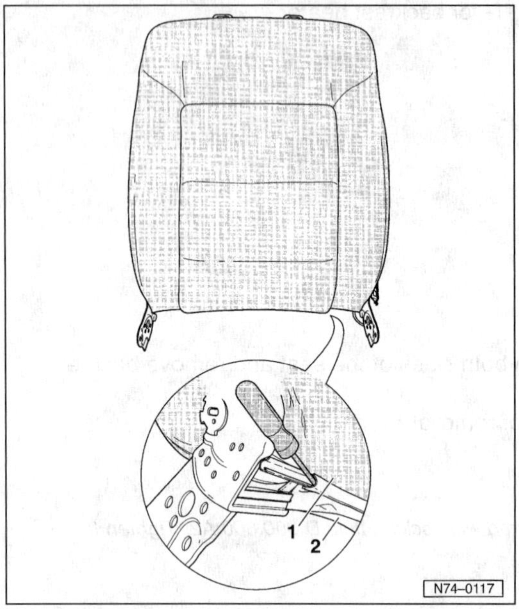

Loosen rear cover -1- from front cover -2-.

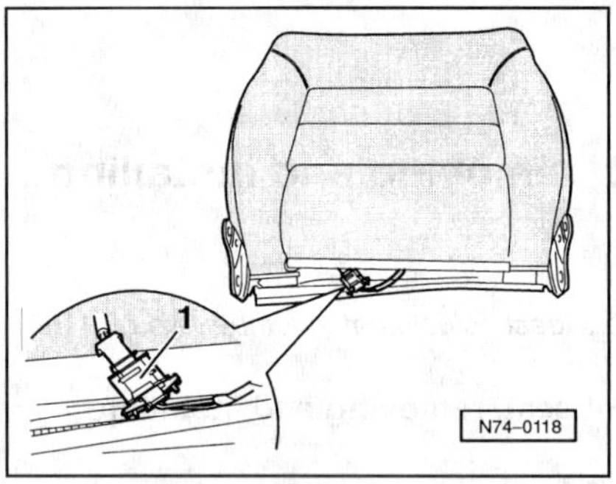

Disconnect connector -1- for backrest heater.

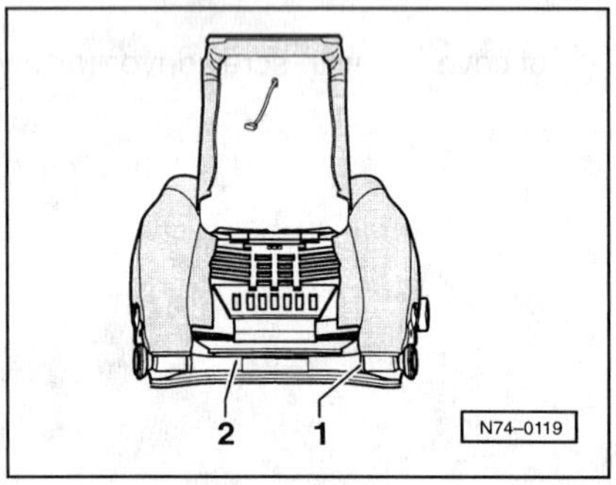

Remove tension wire -1- from seat frame -2-.

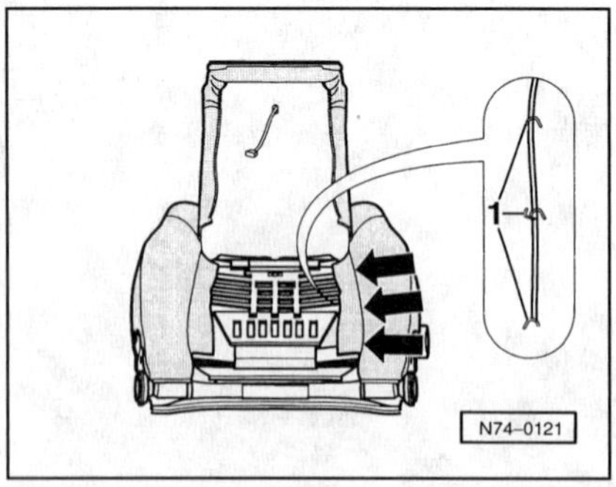

Cut through three upholstery clips-1-.

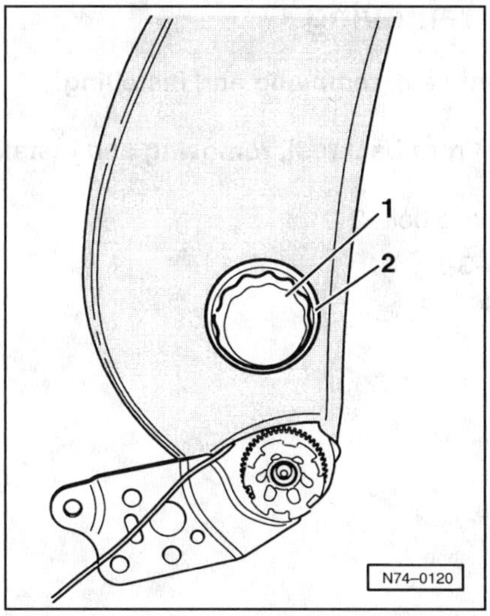

◄ Pry off adjustment knob -1- and trim collar -2-.

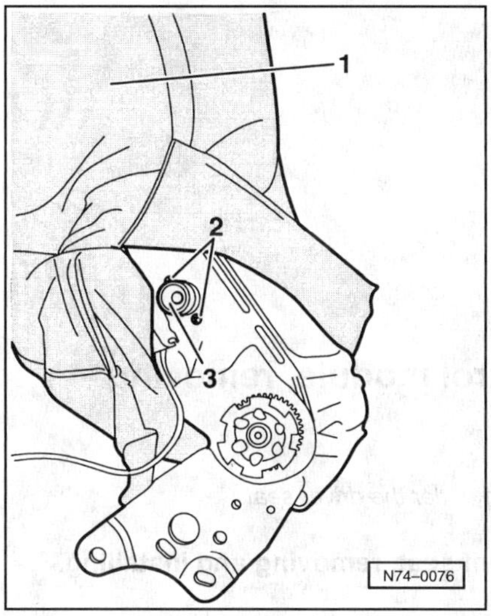

◄ Slide up cover -1- and remove screws -2- for lumbar support adjustment -3-.

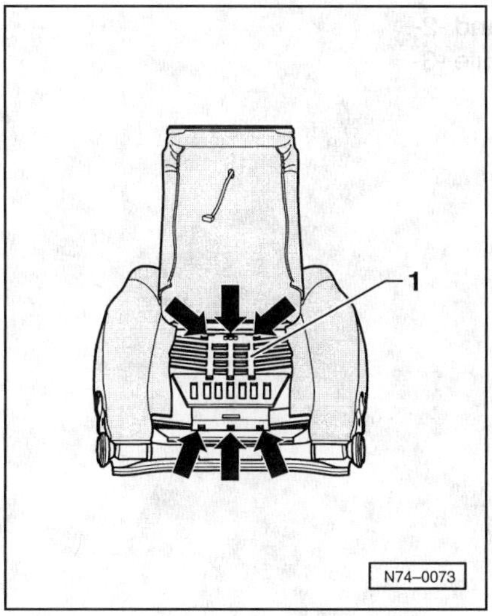

◄ Unclip lumbar support -1- (arrows) and remove backrest frame.

– Installation is reverse of removal.

8-way seat trim, removing

– Remove seat, see **Front seat, removing and installing**.

– Remove belt lock, see **Front backrest, removing and installing**.

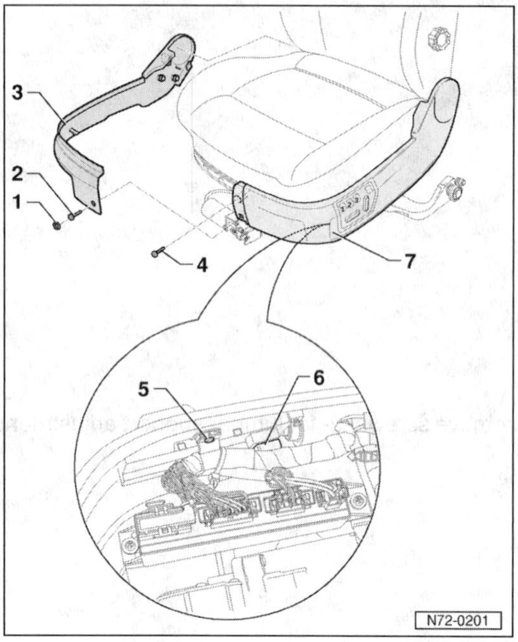

∢ Unclip cap -1- and remove bolt -2-.

- Unclip right seat trim -3-.
- Remove screws -4- and -5-.
- Disconnect connector -6-.
- Unclip left seat trim -7-.

8-way seat control module, removing

NOTE—
Control module is located under the driver seat.

– Remove seat, see **Front seat, removing and installing**.

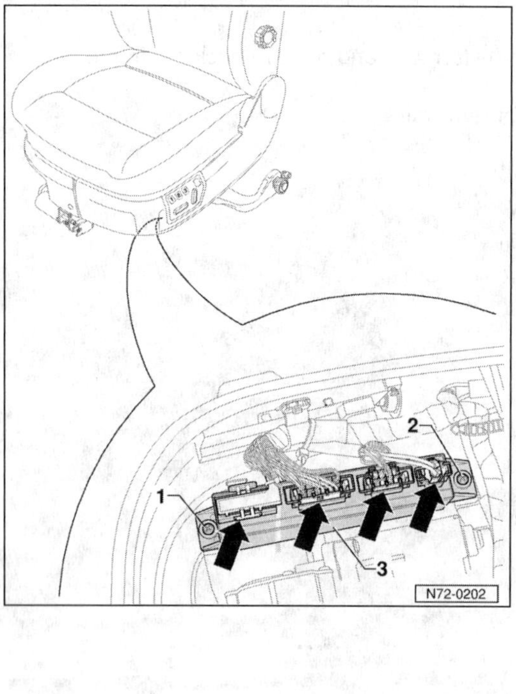

∢ Disconnect connector -6-.

- Remove screws -1- and -2-.
- Remove control module -3-.

8-way seat operating unit, removing

– Remove seat, see **Front seat, removing and installing**.

– Remove 8-way seat trim, see **8-way seat trim, removing**.

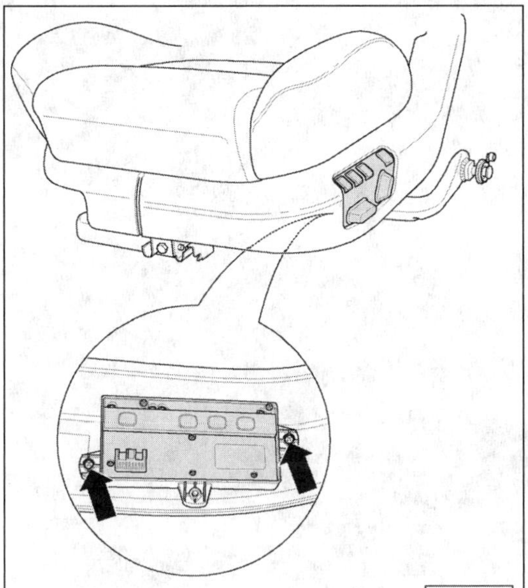

N72-0203

◄ Remove two screws (arrows) on the inner side of the left seat trim and remove operating unit.

8-way seat seat length adjuster, removing

– Remove seat, see **Front seat, removing and installing**.

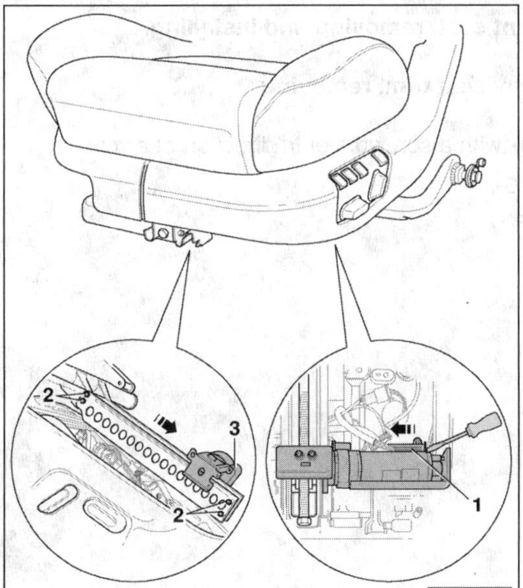

N72-0204

◄ Separate connector -1- with a screwdriver in direction of arrow.

- Remove eight screws -2- (four on each side).
- Remove seat length adjuster -3- in direction of arrow.

8-way seat front drive for seat height adjustment, removing

– Remove seat, see **Front seat, removing and installing**.

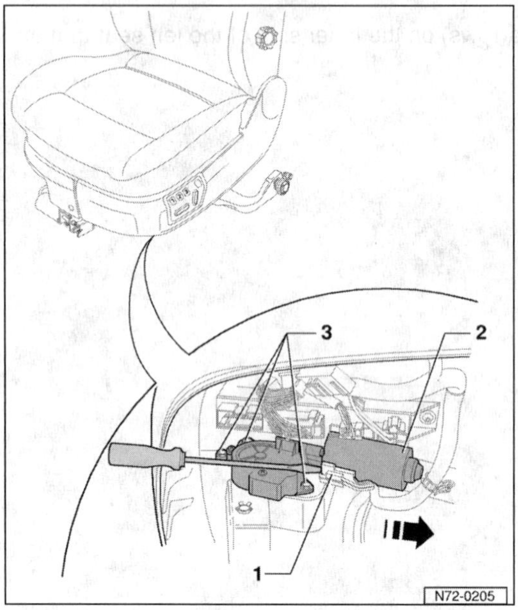

Separate connector -1- with a screwdriver in direction of arrow.
- Remove three bolts -3-.
- Remove front drive -2-.

8-way seat rear drive for seat height adjustment, removing

– Adjust seat to top most position.

– Remove seat, see **Front seat, removing and installing**.

– Remove trim, see **8-way seat trim, removing**.

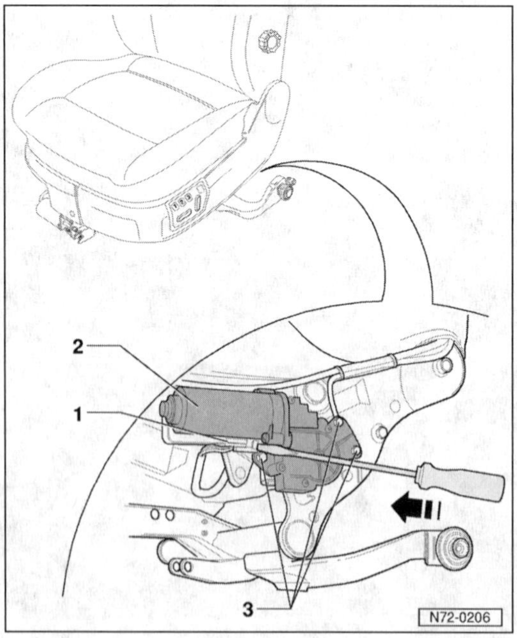

Separate connector -1- with a screwdriver in direction of arrow.
- Remove three bolts -3-.
- Remove rear drive -2-.

8-way seat drive for backrest adjustment, removing

– Remove seat, see **Front seat, removing and installing**.

– Remove trim, see **8-way seat trim, removing**.

– Loosen cover in lower area of backrest and roll up.

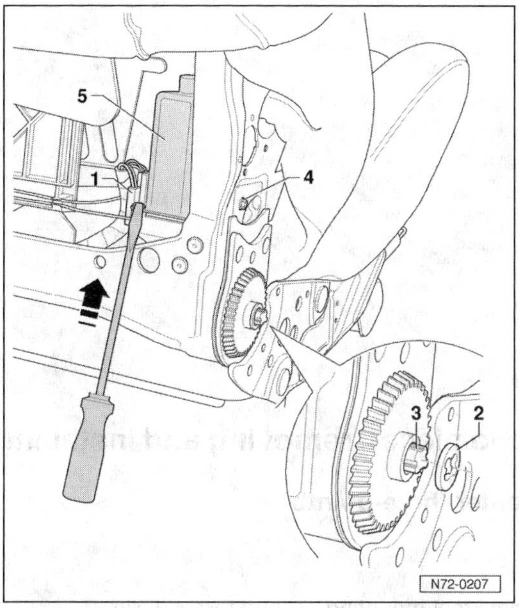

◄ Separate connector -1- with a screwdriver in direction of arrow.

- Pry off locking washer -2- from shaft -3- with a screwdriver.
- Drive out shaft -3- with hammer and mandrel approx. 15 cm (5.9") in the backrest frame.
- Unscrew nut -4-.
- Remove drive -5-.

Rear seats

Rear bench seat (sedan), removing and installing

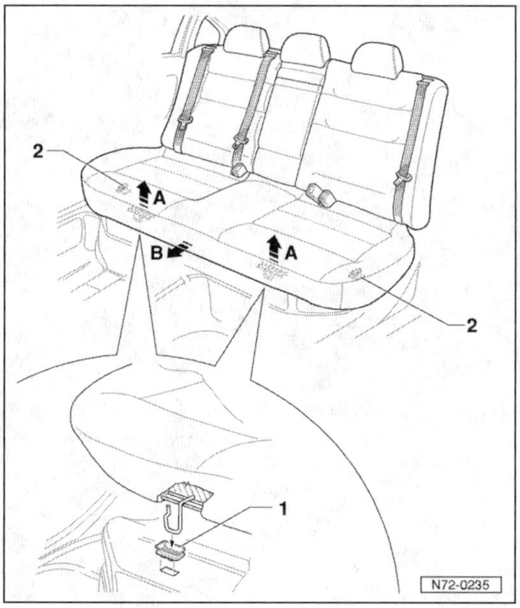

◄ Lift rear seat arrows -A- and pull forward arrow -B-.

– Installation is reverse of removal procedure.

NOTE—

- *Replace plastic sleeves -1- before installing bench seat.*
- *The bench seat is secured at the sides via two guides -2-.*

Rear bench seat (wagon), removing and installing

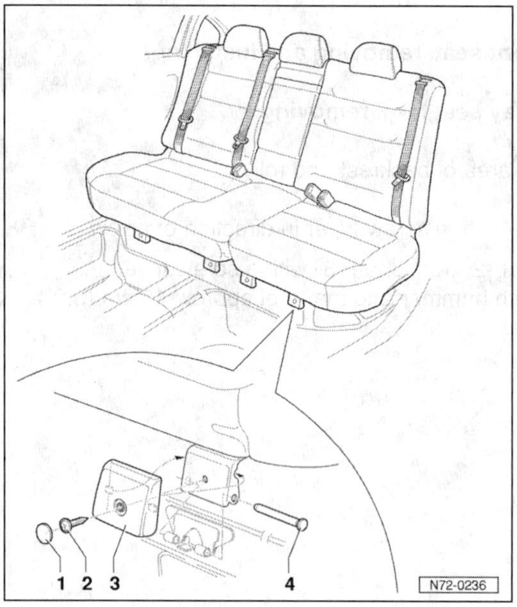

Unclip cover cap -1-, remove screw -2-.

• Remove cover -3- and pull out hinge pins -4-.

– Pull out remaining hinge pins and remove left and right-hand bench seat parts.

– Installation is reverse of removal procedure.

Rear backrest (two piece), removing and installing

Vehicles without center three-pointbelt

– Fold down backrests.

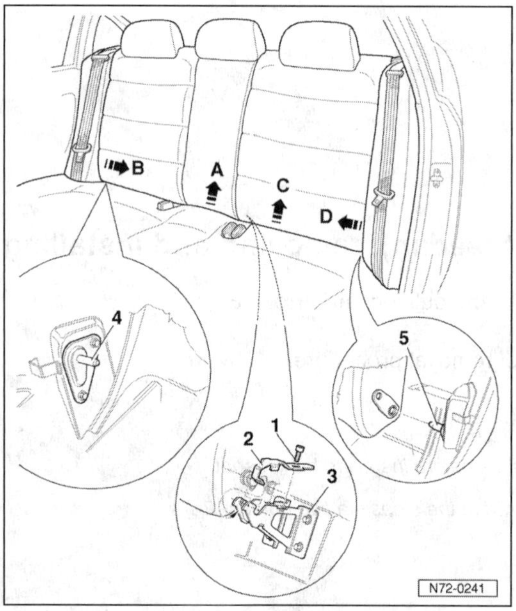

Unscrew bolt -1-from backrest mounting -3- and remove clip -2-.

• Remove right backrest upward out of backrest mounting (arrow -A-) and pull out of pin -4- (arrow -B-).

• Remove left backrest upward out of backrest mounting (arrow -C-) and pull out of pin -5- (arrow -D-).

– Installation is reverse of removal procedure.

Rear backrest (two piece), removing and installing, vehicles with center three-point belt

– Remove seat, see **Rear bench seat (sedan), removing and installing**.

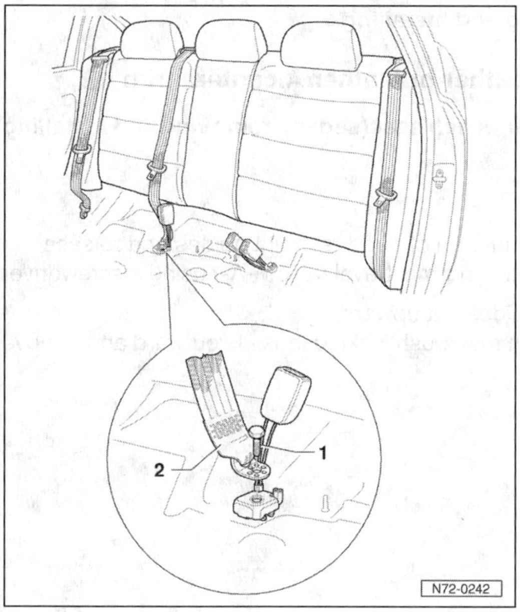

◄ Unscrew bolt -1- and remove belt anchor -2-.

– Fold down rear backrests.

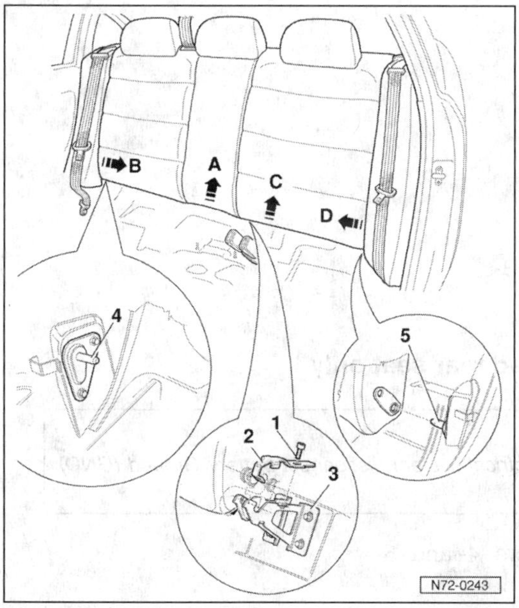

◄ Unscrew bolt -1- out of backrest mounting -3- and remove clip -2-.

- Remove right backrest upward out of backrest mounting (arrow -A-) and pull out of pin -4- (arrow -B-).
- Remove left backrest upward out of backrest mounting (arrow -C-) and pull out of pin -5- (arrow -D-).

– Installation is reverse of removal procedure.

Removing and installing rear backrest (W8 sedan)

Vehicles with part leather trim

– Removing and installing rear backrest (one piece), see **Rear backrest (one piece), removing and installing**.

Vehicles with full leather or leather/Alcantara trim

– Remove seat, see **Rear bench seat (sedan), removing and installing**.

– Remove head restraint.

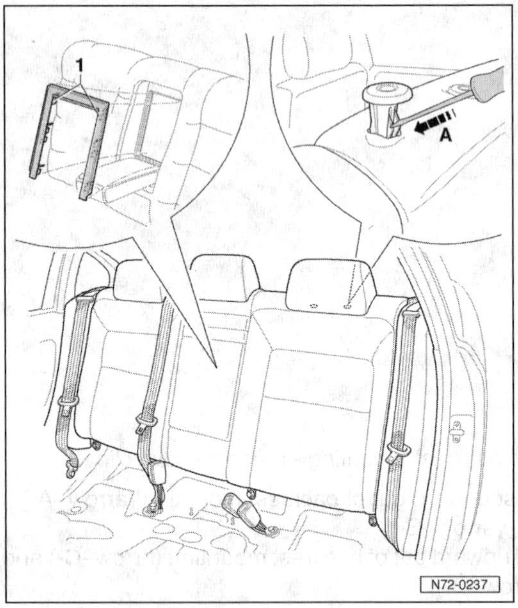

◄ Push upholstery down in area of head restraint guides and release locking tabs in direction of normal travel (arrow -A-) using a screwdriver.

- Pull head restraint guides out upward.
- Fold down center arm rest, push locking tabs -1- outward and remove frame.

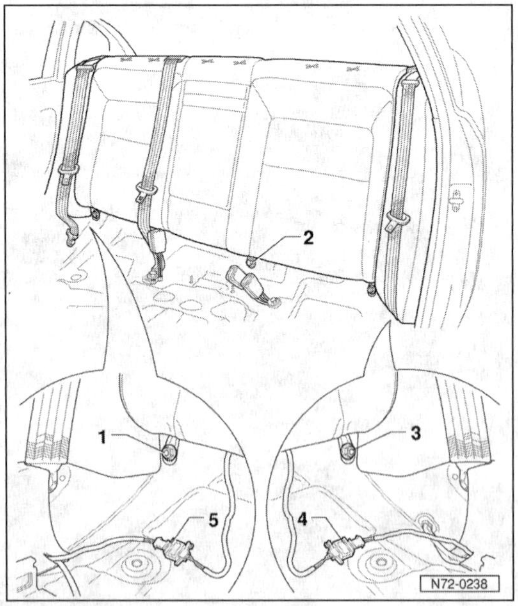

◄ Remove nuts -1- and -3-.

- Remove bolt -2-.

Vehicles with heated rear seat only

> ***WARNING —***
> *Before working on electrical system disconnect battery Ground (GND) strap.*

- Disconnect connectors -4- and -5-.

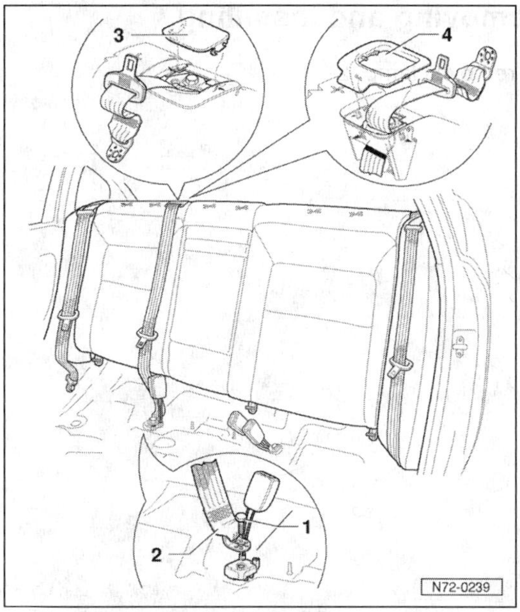

Vehicles with center three-point belt only

◀ Unscrew bolt -1- and remove belt anchor -2-.

- Unclip cover -3- and belt guide -4-.

NOTE—

- *When installing, make sure that the head restraint guides are in place before the belt guide -4- and cover -3- are installed.*

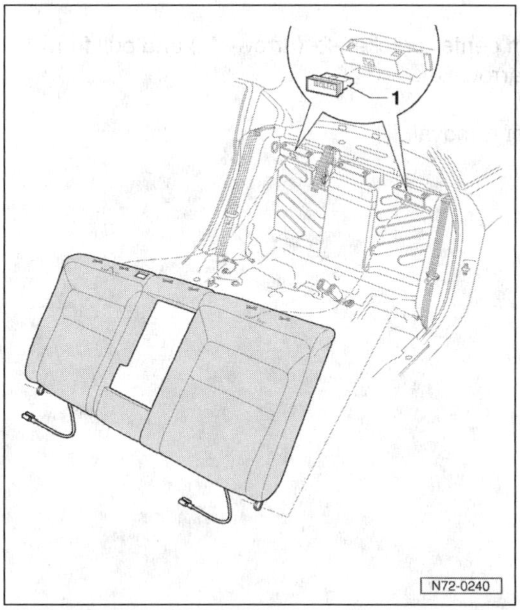

All vehicles

◀ Release rear backrest from clips in area of left and right head restraints.

Installation is reverse of removal procedure.

NOTE—

- *Replace mounting clips -1- before installing rear backrest.*

Left backrest, removing and installing

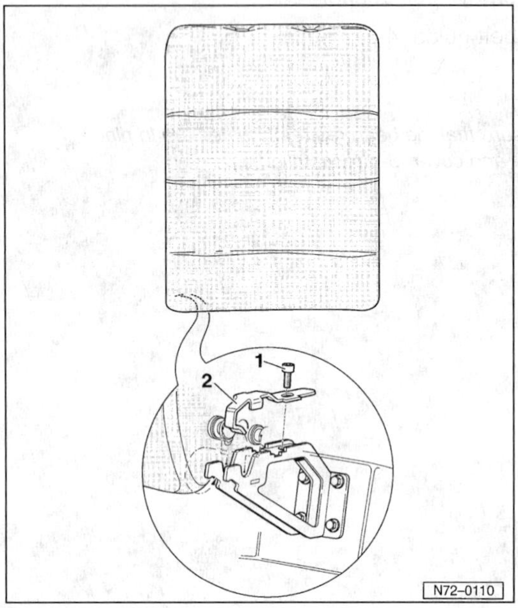

- Unscrew bolts -1- and remove clamp -2-.

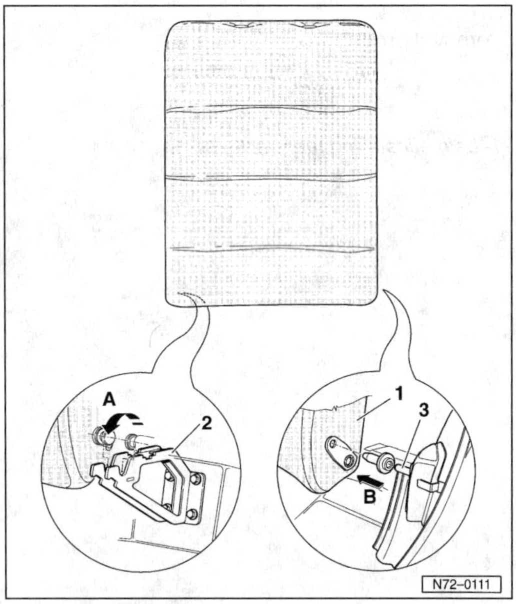

- Raise backrest -1- from center bracket -2- (arrow -A-) and pull from mounting bracket -3- (arrow-B-).

– Installation is reverse of removal.

Right backrest, removing and installing

Removing

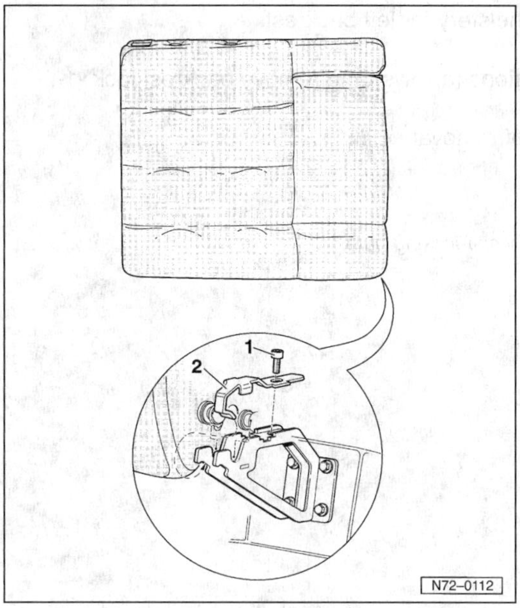

Unscrew bolts -1- and remove clamp -2-.

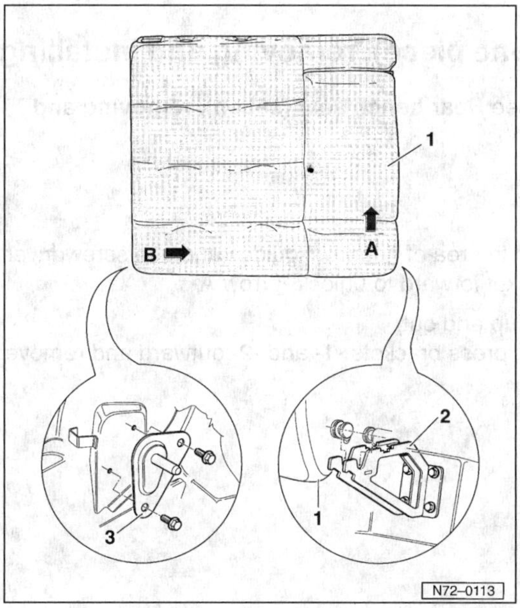

Raise backrest -1- from center bracket -2- (arrow -A-) and pull from mounting bracket -3- (arrow -B-).

– Installation is reverse of removal.

Left backrest lock, removing and installing

– Remove left backrest, see **Left backrest, removing and installing**.

– Remove cover and upholstery for left backrest.

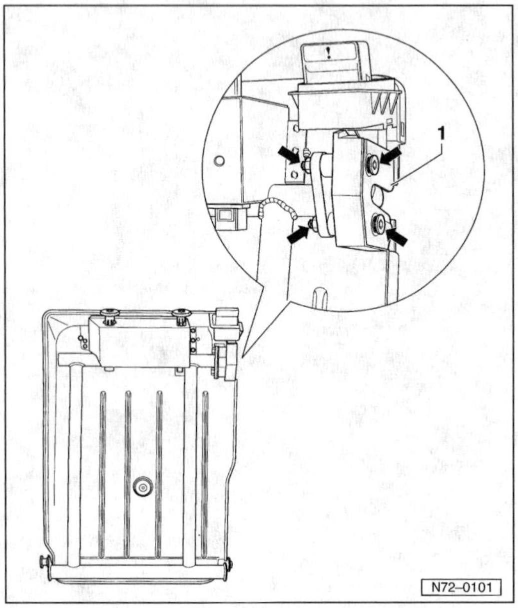

– Loosen bolted connections (arrows) and remove backrest lock -1-.

– Installation is reverse of removal.

NOTE —
When installation use new self-locking nuts.

Rear backrest (one piece), removing and installing

– Remove seat bench, see **Rear bench seat (sedan), removing and installing**.

– Remove headrests.

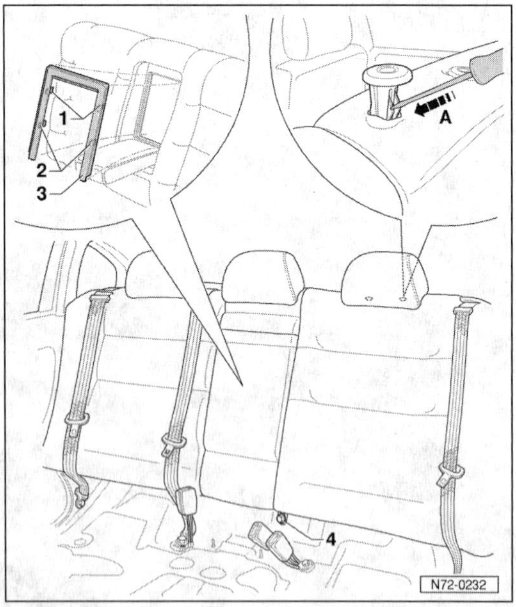

– Press upholstery down in area of headrest guides and use screwdriver to press headrest bracket forward to unlock -arrow A-.

- Pull headrest guides up and out.
- Open center armrest, press brackets -1- and -2- outward and remove frame
- Remove screw -4-.

Vehicles with three point tether anchor

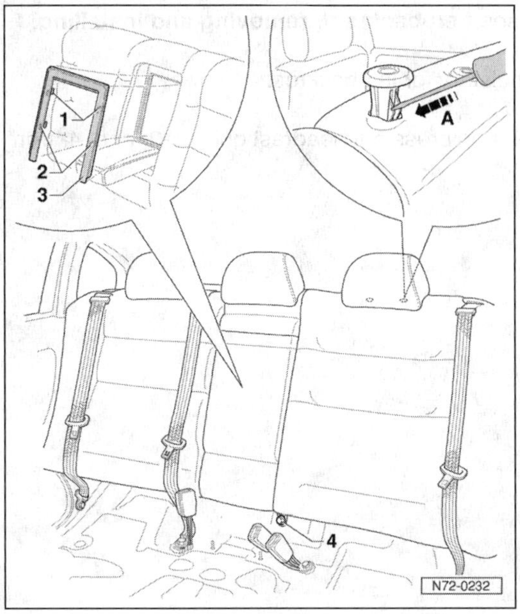

< Remove screw -1- and remove metal fitting -2- from seatbelt.

- Remove cover -3- and unclip belt guide -4-.

NOTE —
With installation, be sure to install headrest guides prior to mounting belt guide -4- and cover -3-.

All vehicles

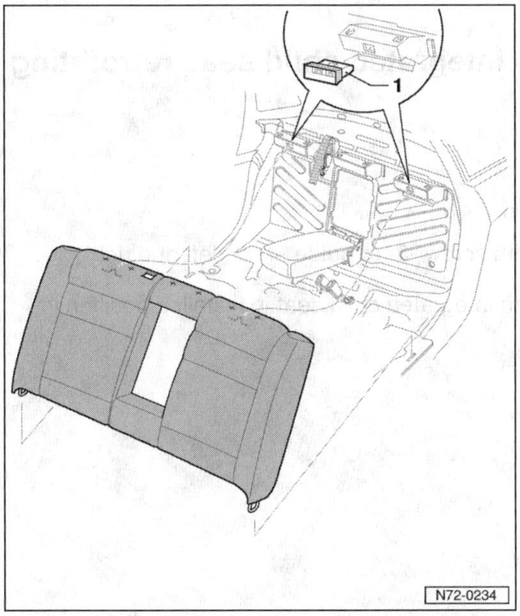

< Unfasten rear backrest in area of outer headrests.

– Remove seat backrest from lock and lift upward.

– Installation is reverse of removal

NOTE —
Replace guide grommets -1- prior to re-installing seat backrests.

Headrest guides, removing

– Remove left backrest, see **Left backrest, removing and installing**.

– Remove cover and upholstery for left backrest.

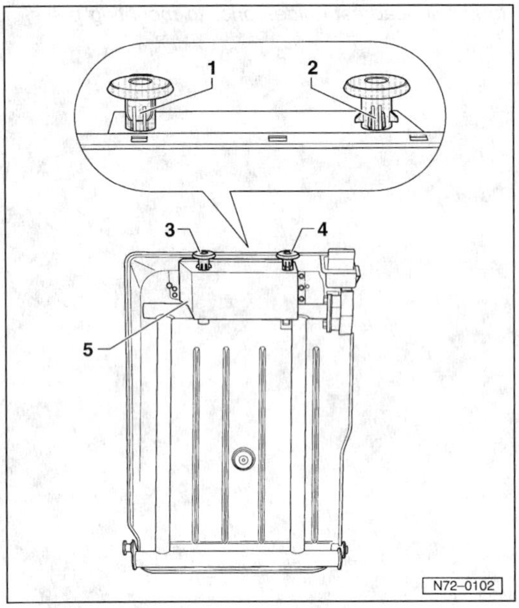

≮ Press in tabs -1- and -2- and press out headrest guides -3- and -4- from bracket -5-.

Seat bench with integrated child seat, retrofitting (1998 m.y.)

– Remove seat bench.

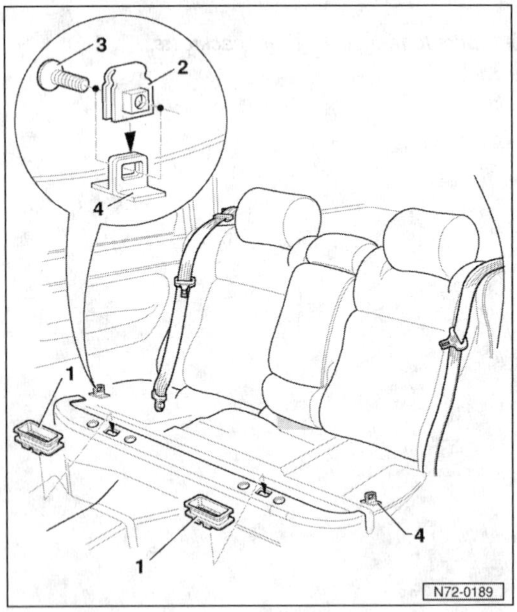

≮ Clip in two grommets -1-.

• Secure adapter -2- with screw -3- to guide -4- on left and right.

– Position seat bench with integrated child seat in installed position in vehicle.

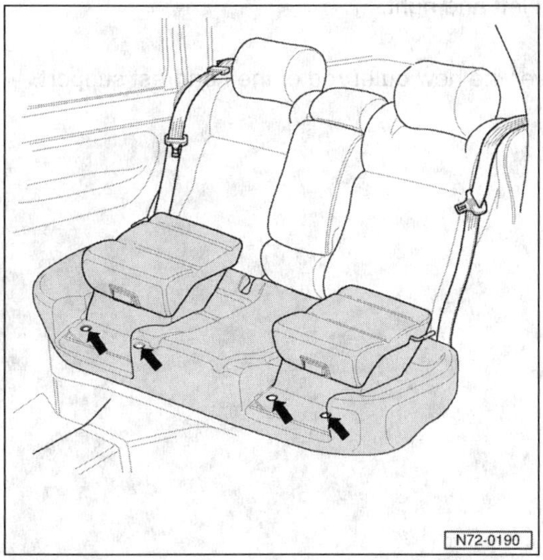

Mark the position of the holes to be drilled (arrows) with a punch.

– Remove seat bench.

– Drill four holes (arrows), use a 9 mm (0.35 in) drill bit.

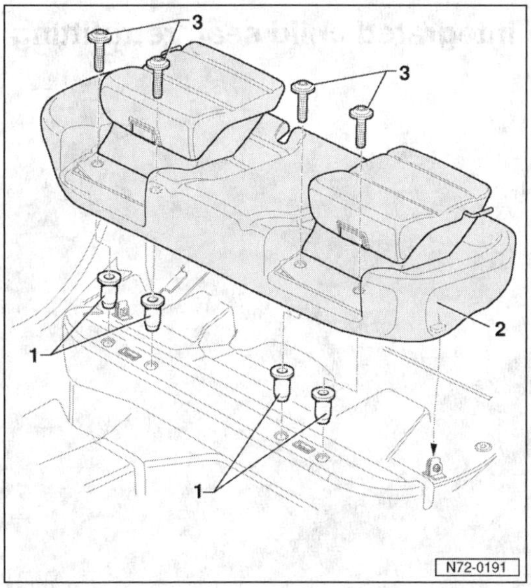

Tighten blind rivet nuts -1- with wrench VAS 5073.

• Install seat bench -2- and fasten bolts -3-.

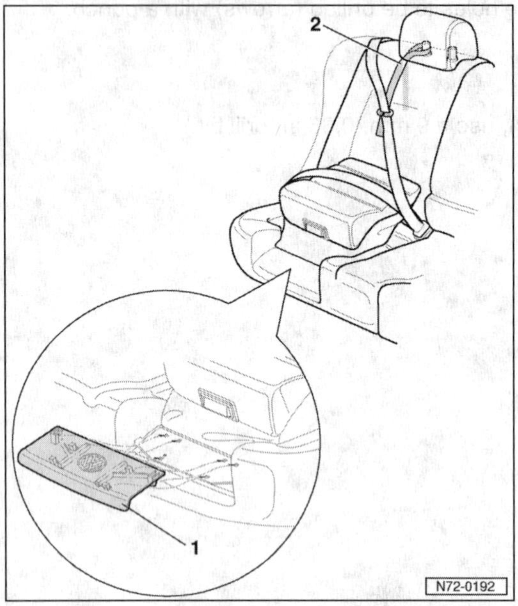

Clip in cover cap -1- on left and right.

- Removing headrest.
- Slide belt guide -2- over the new outer rod of the headrest support.
- Install headrest.

Seat bench with integrated child seat, retrofitting (from 1999 m.y.)

– Remove seat bench.

Clip in two grommets -1-.

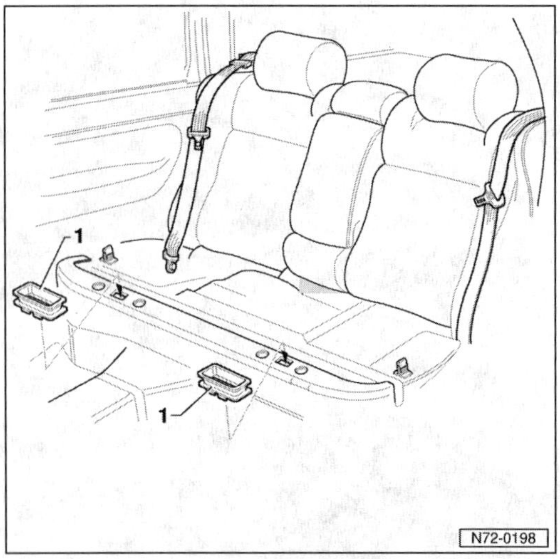

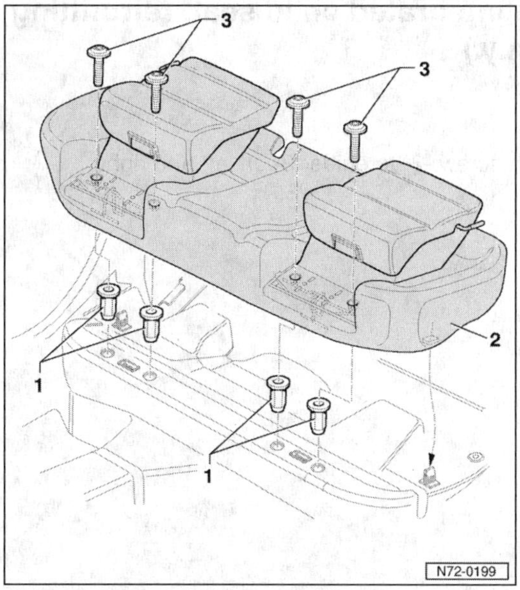

Tighten blind rivet nuts -1- with wrench VAS 5073.

- Install seat bench -2- and fasten bolts -3-.

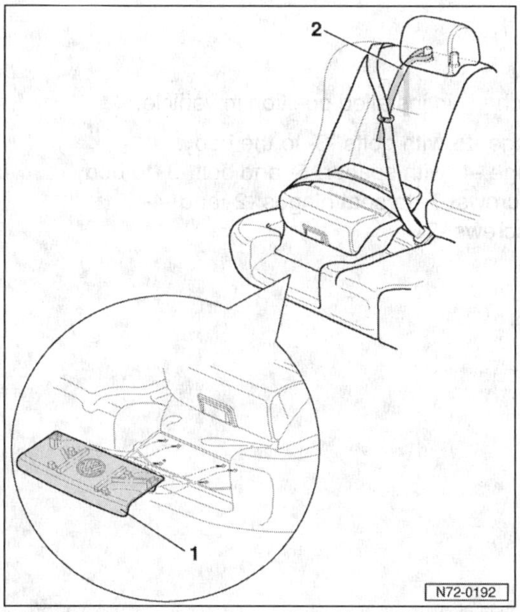

Clip in cover cap -1- on left and right.

- Remove headrest.
- Slide belt guide -2- over the new outer rod of the headrest support.
- Install headrest.

Seat bench with integrated child seat, retrofitting (wagon) (1998 m.y.)

– Remove seat bench.

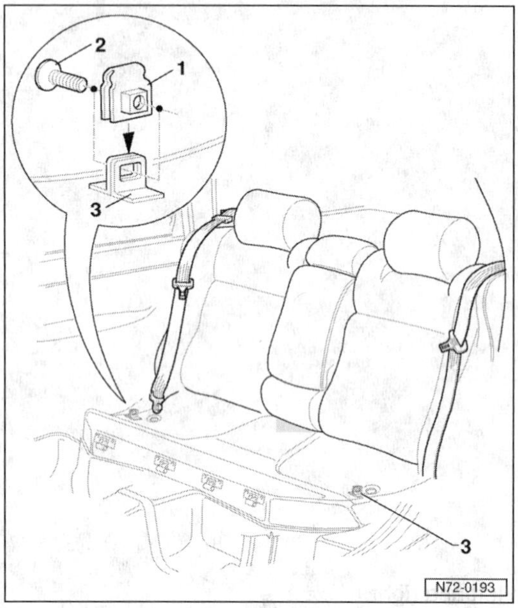

Secure adapter -1- with screw -2- to guide -3- on left and right.

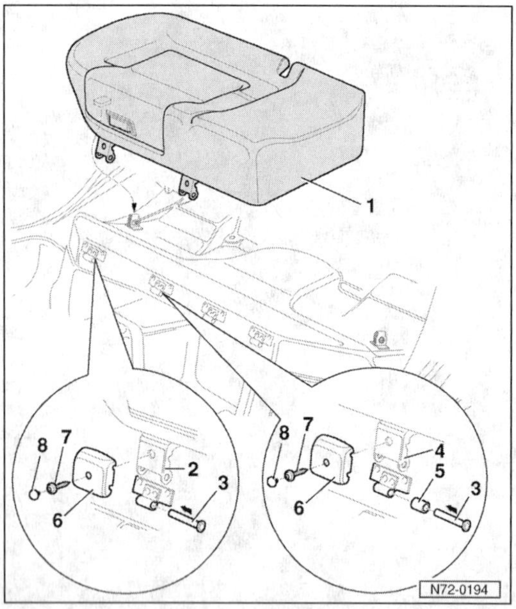

Position right seat bench -1- in installed position in vehicle.

- Secure outer seat hinge -2- with bolts -3- to the body.
- Secure inner seat hinge -4- with spacer -5- and bolt -3- to body.
- Install caps -6- with screws -7- to seat hinges -2- and -4-.
- Install caps -8- over screws -7-.

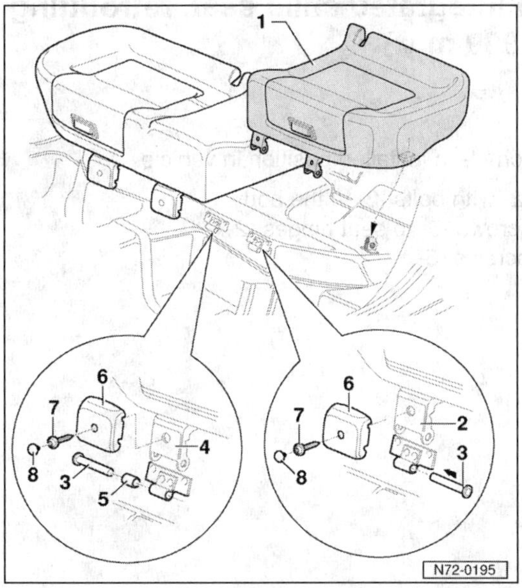

◀ Position left seat bench -1- in installed position in vehicle.

- Secure outer seat hinge -2- with bolts -3- to the body.
- Secure inner seat hinge -4- with spacer -5- and bolt -3- to body.
- Install caps -6- with screws-7- to seat hinges -2- and -4-.
- install caps -8- over screws -7-.

– Remove headrest.

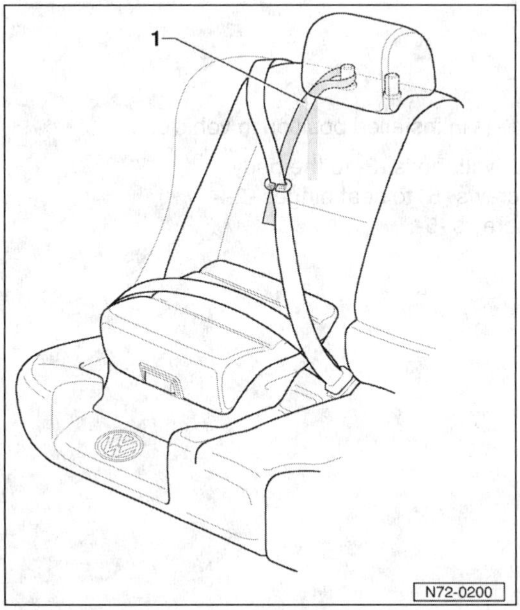

◀ Slide belt guide -1- over the new outer rod of the headrest support.

– Install headrest.

Seat bench with integrated child seat, retrofitting (wagon) (from 1999 m.y.)

– Remove seat bench.

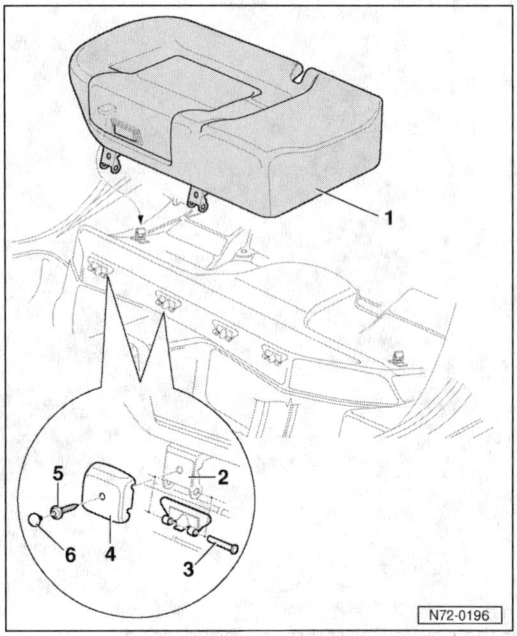

◄ Position right seat bench -1- in installed position in vehicle.

- Secure seat hinges -2- with bolts -3- to the body.
- Install caps -4- with screws -5- to seat hinges -2-.
- Install caps -6- over screws -5-.

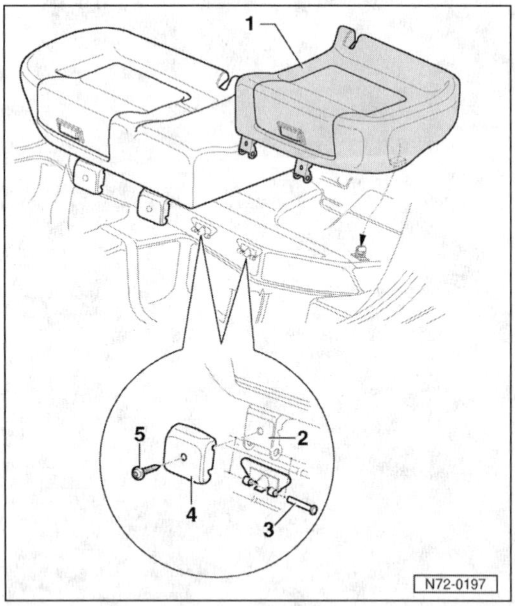

◄ Position left seat bench -1- in installed position in vehicle.

- Secure seat hinges -2- with bolts -3- to the body.
- Install caps -4- with screws -5- to seat hinges -2-.
- Install caps -6- over screws -5-.

– Remove headrest.

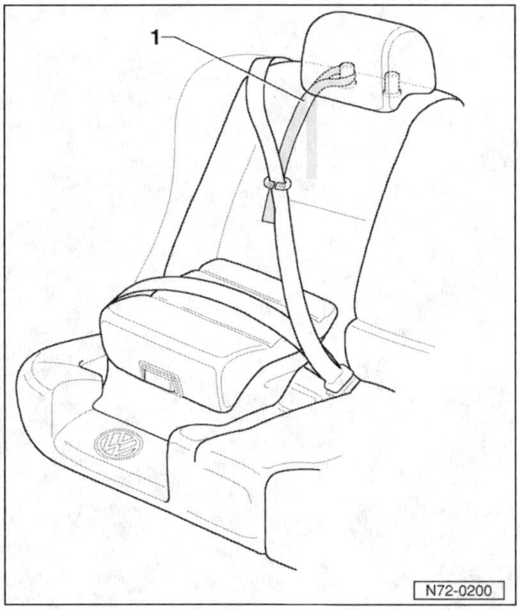

◀ Slide belt guide -1- over the new outer rod of the headrest support.

– Install headrest.

Rear lower anchor bracket for LATCH "Lower Anchors and Tethers for CHildren", installing

– Remove seat bench.

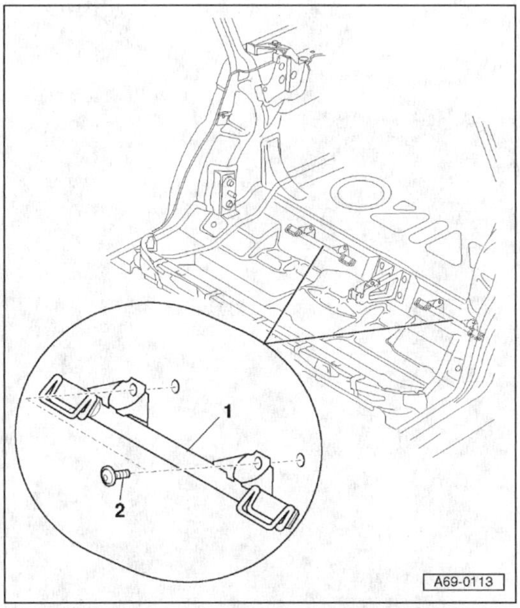

◀ Install bracket -1- with bolts -2-.

- Tighten to 30 Nm (22 ft-lb).

NOTE —
- *The anchorage bars shown in the illustration are for vehicles with front wheel drive. The anchorage bars for vehicles with 4Motion are similar.*

Rear lower anchors

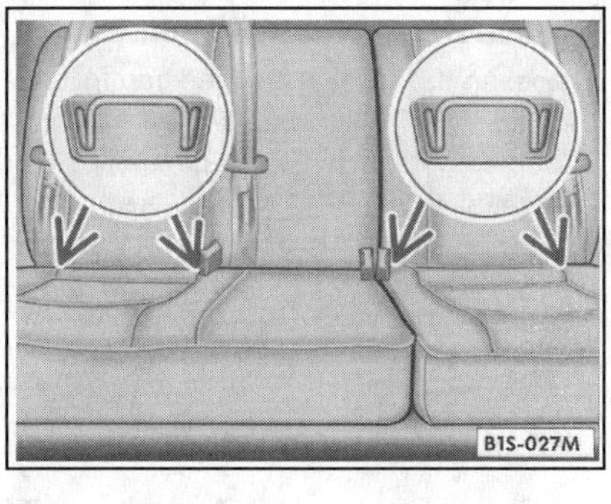

◀ Illustration shows location of rear lower anchors for rear seat.

NOTE —
- *Rear lower anchors secure the child restraint in the seat without using the vehicle's safety belts. Anchors provide a secure and easy-to-use attachment and minimize the possibility of improper child restraint installation.*
- *Rear lower anchor attachment points are located between the rear seat back and rear seat cushion for the outboard seating positions.*
- *Inside anchors are not recommended for attaching rear lower anchors to a LATCH seat in the middle position. Three-point lap/shoulder belt should be used.*

Heating and Air Conditioning

Section

V

87 Air Conditioning

Page

87 Air Conditioning

Repair Group

V 87

A/C System

Safety measures

The subassemblies and lines of the air conditioner are filled with Tetrafluoroethane also known as refrigerant R-134a.

A label on the lock carrier in the engine compartment indicates the refrigerant type and capacity.

WARNING —

- *Always use an Underwriter's Laboratory (UL) approved refrigerant recovery/recycling/recharging unit such as Kent-Moore ACR[4], or equivalent, whenever discharging an R-134a A/C system*
- *As of January, 1992 any person who services a motor vehicle air conditioner MUST, by law, be properly trained and certified and use approved refrigerant recycling equipment. Technicians must complete an EPA approved recycling course to be certified*
- *State and Provincial governments may have additional requirements regarding air conditioning servicing. Always comply with local laws*
- *A/C system is filled with refrigerant gas which is under pressure*
- *Always be careful that refrigerant does not come in contact with your skin*

WARNING —

If liquid refrigerant has come in contact with your skin or eyes:

- *Do not rub skin or eyes*
- *Immediately flush with cool water for 15 minutes*
- *Rush to a doctor or hospital*
- *Do not attempt to treat yourself*

CAUTION—

- *Work in a well ventilated area because refrigerant gases are heavier than air, displace oxygen and may cause suffocation in areas of poor air circulation, like under the car*
- *Avoid breathing refrigerant vapors. Exposure may irritate eyes, nose and throat*
- *Always wear hand and eye protection (gloves and goggles) when working around the A/C system*
- *Keep refrigerant containers stored below 50°C (122°F) and never drop from high places. DO NOT warm refrigerant containers with an open flame. If refrigerant needs to be warmed, place bottom of tank in warm water.*
- *Keep refrigerant away from open flames because poisonous gas will be produced if it burns. Do not smoke when refrigerant gases are present for the same reason*
- *Electric welding near refrigerant hoses causes R-134a to decompose from ultraviolet light. Discharge system before electric welding*
- *Pressurized R-134a refrigerant in the presence of oxygen may form a combustible mixture. Never introduce compressed air into any R-134a container (full or empty), A/C component or piece of service equipment*
- *Do not expose any component of the A/C system to high temperatures (for example, open flames). Excessive heat will cause a pressure increase which could burst the system*
- *Do not steam clean condensers or evaporators. Use only cold water or compressed air*

CAUTION—

- *Before working on electrical system, obtain radio security code and disconnect battery Ground (GND) strap*
- *When reconnecting battery, always check operation of vehicle equipment such as radio, clock, and power windows according to Repair Manual and/or Owner's Manual*
- *R-12 and R-134a refrigerant are NOT compatible. Never add R-12 refrigerant to an R-134a system or R-134a refrigerant to an R-12 system. If the refrigerants are mixed, total system contamination will occur and compressor failure may result*

WARNING—

- *Refrigerant oils used for the R-134a system and R-12 system are NOT compatible. Use only the specified synthetic oil (Polyalkylene Glycol/PAG) for the R-134a refrigerant system. DO NOT use R-12 system oil in an R-134a system or R-134a system oil in an R-12 system. If the refrigerant oils are mixed, system contamination will occur and compressor failure may result*
- *R-134a refrigerant system oil (PAG oil) absorbs moisture very rapidly. Moisture combines with the refrigerant to form acids which will damage the system. Use only the specified oil from a sealed container and ALWAYS reseal oil container immediately after use. DO NOT use oil if it has become contaminated with moisture*

CAUTION—

- *R-134a and R-12 A/C components including compressor, hoses, O-rings, evaporator, condenser, receiver-drier, etc. are NOT interchangeable. Components of the R-134a system are identified by lettering (R-134a) or by a green label (or stripe). In addition a label on the upper radiator support identifies which type refrigerant is used. Use only the correct system component for each refrigerant type*
- *Immediately plug open connections on A/C components to prevent dirt and moisture contamination. Likewise, DO NOT remove new components from packaging until ready to install. Immediately tighten component connections after installation*
- *Always use separate refrigerant recovery/recycling/recharging servicing equipment for R-12 and R-134a systems. DO NOT use one piece of equipment for both R-12 and R-134a systems. The residual traces of refrigerant will contaminate and damage the equipment. Servicing equipment includes recovery/recycling/recharging unit, charging station, vacuum pump, manifold gauges, etc.*
- *DO NOT use R-12 servicing equipment on R-134a systems or R-134a equipment on R-12 systems or damage to both the vehicle A/C system and servicing equipment may result. Use only equipment designed to meet Society of Automotive Engineers (SAE) standards*
- *R-134a and R-12 systems use different size service fittings. NEVER use adaptors to convert an R-12 fitting to R-134a size or R-134a fitting to R-12 size*
- *Always replace damaged and/or leaking A/C system components. Do not attempt repair by soldering or welding*
- *Work area must be extremely clean when working on A/C system components*
- *Discharge A/C system before removing any A/C system component*
- *Always reinstall caps over A/C service valves*
- *Always replace O-rings, DO NOT reuse. Use only the correct size and type of O-rings specified for use with R-134a refrigerant. Lubricate O-ring with refrigerant oil before installing*

A/C refrigerant R-134a, properties

Commercial designation

The following designations are used and appear on component labels and factor tags:

- R-134a
- Tetrafluoroethane
- CH_2F CF_3
- H-KW 134a
- SUVA® 134a
- ARCTON® 134a

Color

Refrigerant R-134a is colorless and is invisible as a gas. R-134a when viewed through the sight glass may appear milky due to the mixture of refrigerant and lubricating oil (PAG oil).

Temperature/pressure relationship

R-134a in an enclosed container will have a specific temperature/ pressure relationship.

Temperature	Pressure
-30°C (-22°F)	0.0 bar (0.0 psi)
-20°C (-4°F)	0.3 bar (4.4psi)
-10°C (14°F)	1.0 bar (14.5 psi)
0°C (32°F)	1.9 bar (27.5 psi)
10°C (50°F)	3.1 bar (45.0 psi)
20°C (68°F)	4.7 bar (68.2 psi)
30°C (86°F)	6.7 bar (97.2 psi)
40°C (104°F)	9.1 bar (132.0 psi)
50°C (122°F)	12.2 bar (177.0 psi)
60°C (140°F)	15.8 bar (229.2 psi)
70°C (158°F)	20.2 bar (293.0 psi)

Refrigerant oil

A special Polyalkylene Glycol (PAG) synthetic oil is used in R-134a systems. This oil is NOT compatible with mineral based oils used in R-12 systems. See safety measures.

Airborne properties

Escaped refrigerant gases are heavier than air and will gather first in low places, like under the car. R-134a refrigerant gas displaces oxygen and may cause suffocation in low areas of poor air circulation.

Refrigerant R-134a is not poisonous in any state (liquid or gas) and is safe when used properly.

Affects on plastic

Refrigerant R-134a will deteriorate some plastics. Therefore, when making system repairs, use only genuine VW replacement parts which are specified for use with R-134a refrigerant.

Affects on metal

In its pure state, refrigerant R-134a is chemically stable and will not attack iron, copper, brass or aluminum. However, the mixture of R-134a and PAG oil may deteriorate certain metals (copper). Therefore, when making system repairs, use only genuine VW replacement parts which are specified for use with R-134a refrigerant.

Leak detection

Use halogen leak detector Hitec HI400A-TELor equivalent to check for R-134a system leaks. This tool can also be used to detect leaks in R-12 systems. Many currently available R-12 leak detectors cannot detect R-134a refrigerant leaks.

Manual A/C System

Heating and A/C components in passenger compartment, assembly

CAUTION —
Before beginning repairs on the electrical system:

- *Obtain the anti-theft radio security code.*
- *Switch the ignition off.*
- *Disconnect the battery Ground (GND) strap.*
- *After reconnecting battery, re-code and check operation of anti-theft radio. Also check operation of clock and power windows according to Owner's Manual.*

NOTE —
*System components identified with an * must only be serviced or replaced by a qualified technician after discharging refrigerant system. Use Kent Moore ACR4 or equivalent refrigerant recycling equipment.*

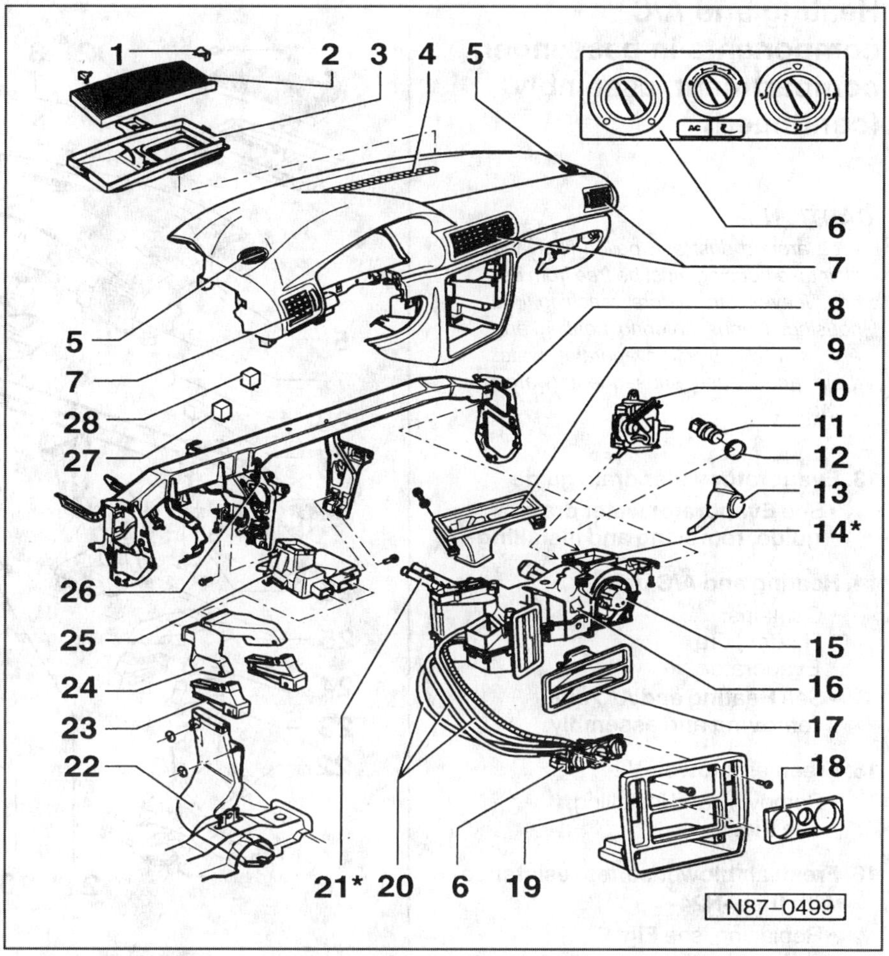

1. **Clip**
   - Lift out with screwdriver.
2. **Dust and pollen filter**
   - See **Maintenance**.
3. **Air intake housing**
4. **Defroster air outlet**
   - Removing and installing, see **Body-Interior section, Repair Group 70.**
5. **Side window air outlet**
   - Removing see **Fig. 1**
6. **Heating and A/C controls**
   - With:
     A/C switch -E35-
     Fresh air blower switch -E9-
     Switch for fresh air blower and recirculated air -E184-
   - See **Heating and A/C controls, removing and installing**
   - See **Heating and A/C controls, assembly**
7. **Air outlets**
   - Center and side air outlets, removing (early prod.), see **Fig. 2**
   - Center air outlet, removing (later prod.) removing, see **Fig. 3**
   - Side air outlets, removing (later prod.), see **Fig. 4**
8. **Instrument panel cross member**
   - Always follow installation instructions for bolt, see **Fig. 7**
9. **Defroster duct**
   - Replacing:
     Remove instrument panel, see **Body-Interior section, Repair Group 70**
10. **Servo motor for fresh/ recirculating air door -V154-**
    - See **Servo motor for fresh/ recircuating air door -V154-, removing and installing**
11. **Ambient temperature switch -F38-**
    - Switches off A/C clutch -N25- at low ambient temperatures
    - Switching point: 5°C (41.5°F)
    - Replacing:
      Remove glove box, see **Body-Interior section, Repair Group 68**
      Reach behind heating and A/C unit from below and pull out switch.
12. **Gasket**
    - Replace if damaged
    - Replacing:
      Remove ambient temperature switch -F38-
      Remove dust and pollen filter
      Set fresh air/recirculation flap in "fresh air" position
      Reach down into air intake housing and pull out gasket
      Remove instrument panel, see **Body-Interior section, Repair Group 70**
      Loosen cross member -10- . Always follow installation instructions for bolt, see **Fig. 7**

(Continued on following page)

Heating and A/C components in passenger compartment, assembly (continued)

CAUTION—

Area around dust and pollen filter and air intake housing must be free from debris, leaves etc. before reaching into housing. Debris entering heating and A/C unit will block evaporator water drain and cause subsequent (water) damage.

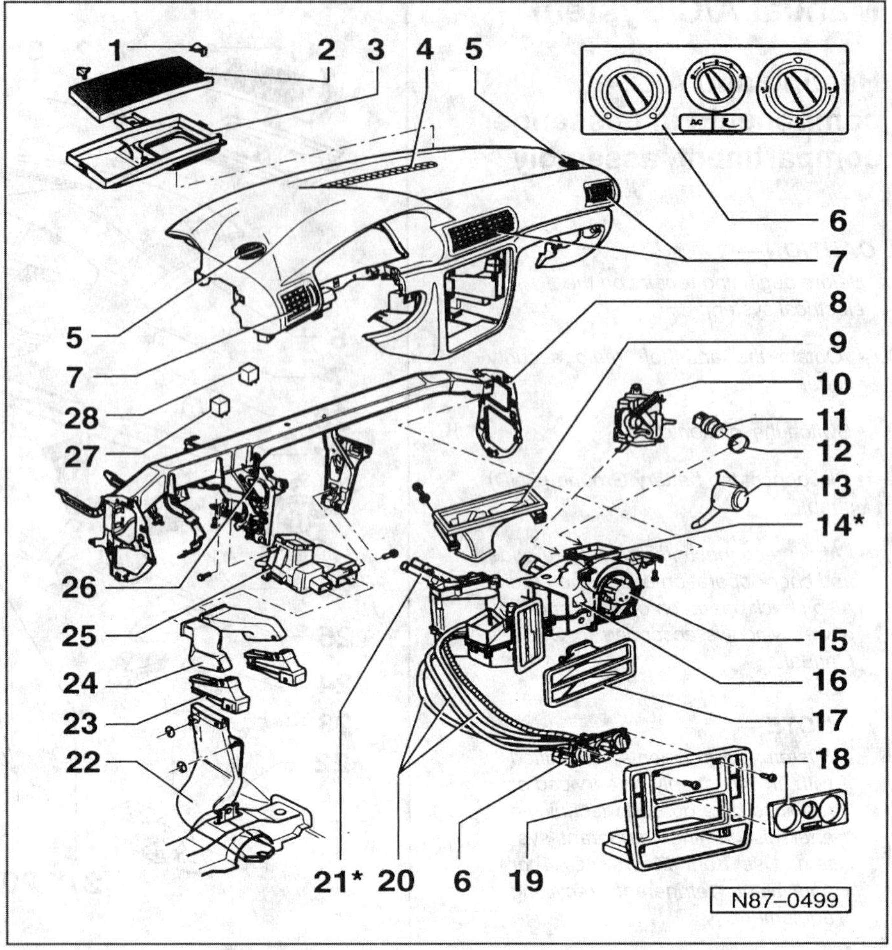

13. **Evaporator water drain guide**
    - See **Evaporator water drain guide, removing and installing**
14. **Heating and A/C unit***
    - Contains:
      Heater core
      Evaporator
    - See **Heating and A/C unit, removing and assembly**.
15. **Fresh air blower -V2-**
    - Removing and installing, see **Fig. 6**
16. **Fresh air blower series resistance with fuse -N24-**
    - Replacing, see **Fig. 5**
17. **Center air outlet duct**
18. **Control panel trim**
    - Unclip using screwdriver
19. **Center trim**
20. **Heating and A/C cables**
    - See **Heating and A/C cables, installing**
21. **Heater core***
    - Always use new coolant after replacement
22. **Left rear duct**
23. **Lower duct**
24. **Upper duct**
25. **Footwell air outlet**
26. **Instrument panel cross member and left side member securing bolt**
    - Always follow installation instruction for bolt, see **Fig. 7**
27. **Relay**
    - Coolant Fan Control (FC) Relay -J26-, Control number 214
    - First Speed Coolant Fan Control (FC) relay -J279-
      Vehicles through 09/2000: Control number 214
      Vehicles from 10/2000: Control number 373
    - Second Speed Coolant Fan Control (FC) Relay -J280-
      Vehicles through 09/2000: Control number 213
      Vehicles from 10/2000: Control number 370
    - Component locations, see **Wiring Diagrams and Component Locations.**
28. **A/C cut-off control module -J314-**
    - Control number 384
    - Component location, see **Wiring Diagrams and Component Locations.**

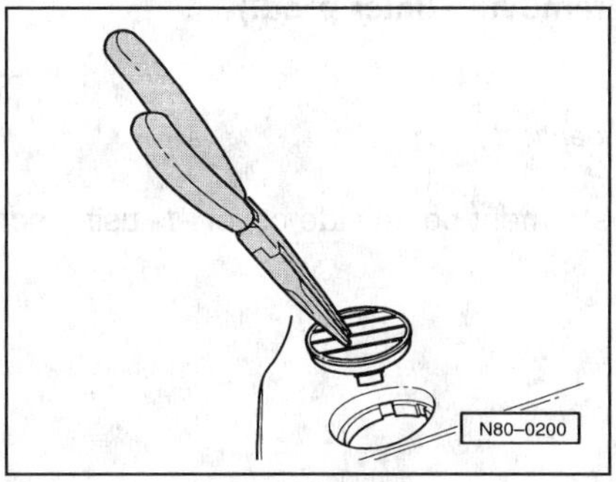

Fig. 1 Side window air outlet, removing

- Carefully pull out using needle-nose pliers.

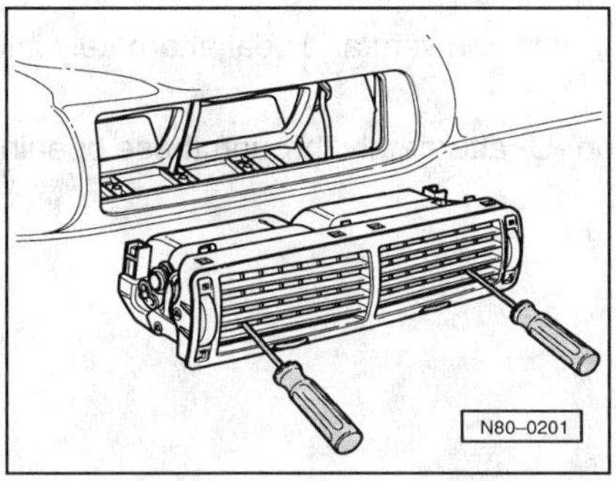

Fig. 2 Center and side air outlets, removing (early prod.)

- With flashlight, locate the two retaining tabs inside housing, lift tabs with two small screwdrivers and carefully pull outlet from instrument panel.

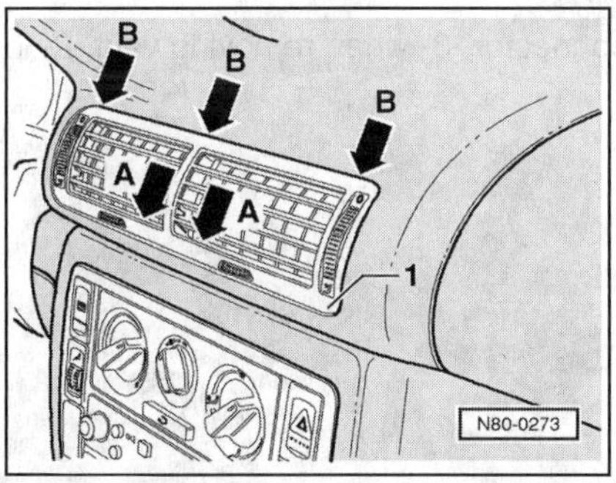

Fig. 3 Center air outlet, removing (later production)

- Insert fingers of both hands into lower row of grille -A- and push center vent -1- down suing force until locking lugs -B- are released from their mounts.

NOTE —

- *Sit between front seats to apply the force required.*
- *If outer locking lugs -B- will not release, grip at this point and push center vent down while simultaneously pulling out slightly.*

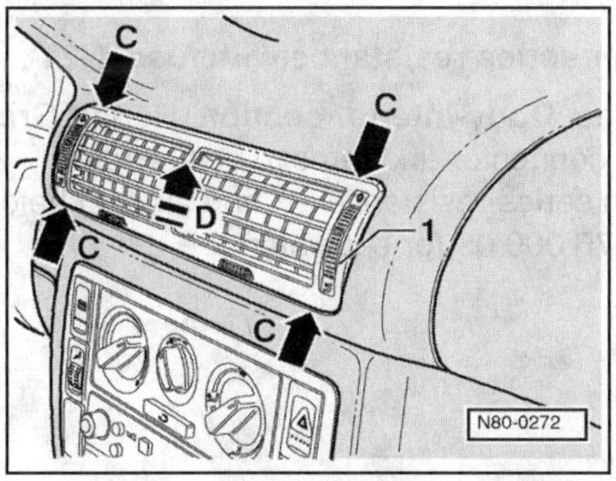

Grip center vent -1- at point -C- which has been raised up slightly from instrument panel opening by action of pushing down.

- Push center vent in direction -D- to release locking lugs and pull it outward slightly.

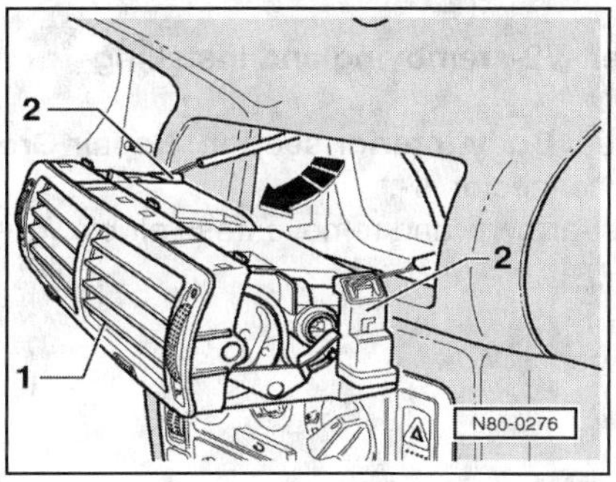

Disconnect electrical connectors -2- from center vent before removing in direction of -arrow-.

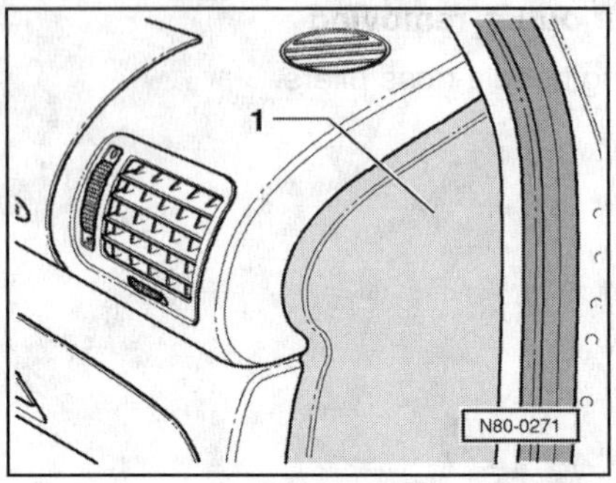

Fig. 4 Side air outlet, removing (later prod.)

NOTE —
Removal of both sides is identical.

- Unclip and remove instrument panel side cover -1- using screwdriver.

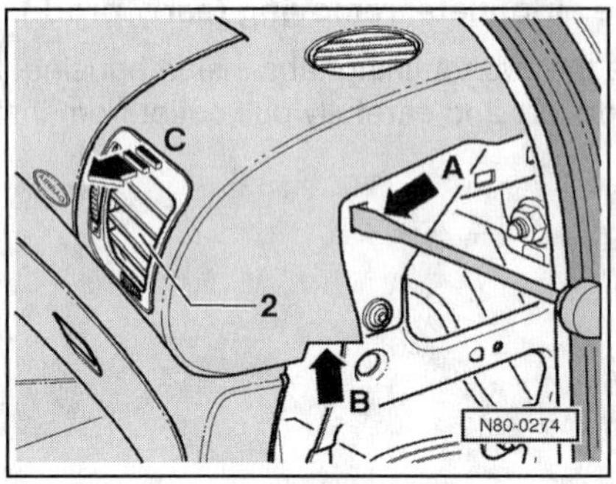

Guide a flat screwdriver between vent and seal alternately into side openings -A- and -B-.

- Pry vent -2- in direction -C- alternately through these openings.

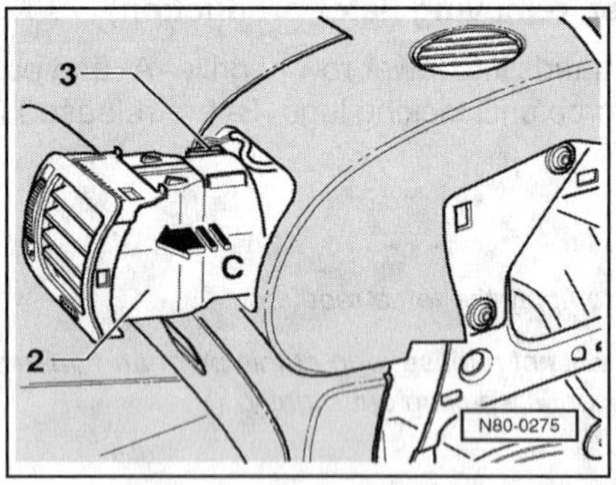

Disconnect electrical connector -3- when removing vent -2- in direction -C-.

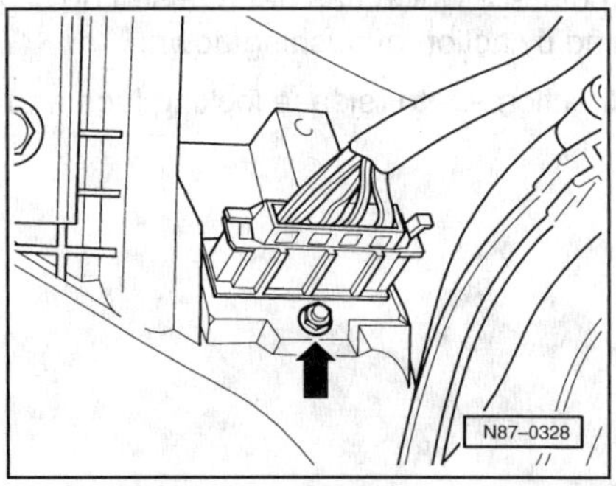

Fig. 5 Fresh air blower series resistance with fuse -N24-, replacing

- Remove glove box, see **Body-Interior section, Repair Group 68.**
- Disconnect electrical connector and remove bolt -arrow-.
- Seal surface between series resistance -1- and air duct before installing with AMV 176 000 05 (or equivalent).

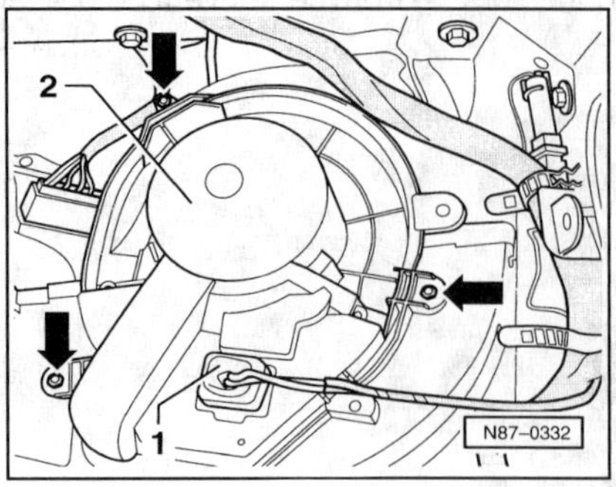

Fig. 6 Fresh air blower -V2-, removing and installing

- Remove glove box, see **Body-Interior section, Repair Group 68.**
- Disconnect electrical connector -1-.
- Remove three screws -arrows- and remove fresh air blower -2- downwards.

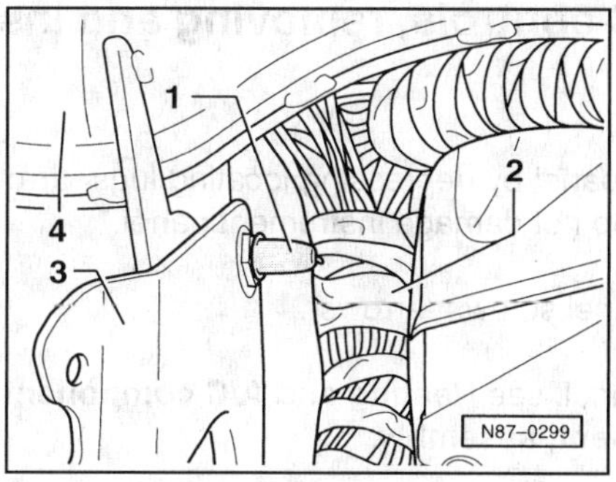

Fig. 7 Instrument panel cross member and left side member securing bolt, installation note:

- To avoid damage to instrument panel wiring harness -2-, ensure that the protective cap -1- is installed on the securing bolt for upper instrument panel cross member -4- at left side member -3-.

Evaporator water drain guide, removing and installing

– Remove instrument panel, see **Body-Interior section, Repair Group 70.**

– Release instrument panel cross member from bulkhead, see **Fig.7** (above).

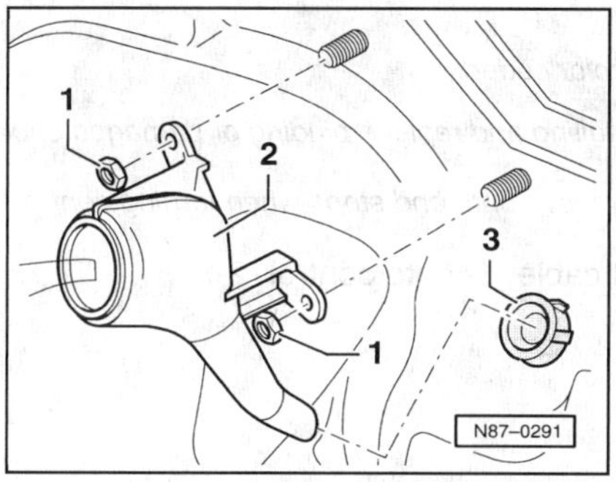

NOTE —
For the next step, an assistant must lift and hold the heating and A/C unit.

Remove nuts -1- (2.5 Nm / 23 in lb) and remove guide -2- from bulkhead.

When installing, ensure water drain valve -3- sealing lip completely encloses the guide -2-. If necessary, lubricate to aid installation, see **Heating and A/C components in engine compartment, assembly, Fig. 1**

Servo motor for fresh/recirculated air door -V154-, removing and installing

Removing

– Remove glove box, see **Body-Interior section, Repair Group 68**

– From the front passenger side, use a mirror to look behind heating and A/C unit.

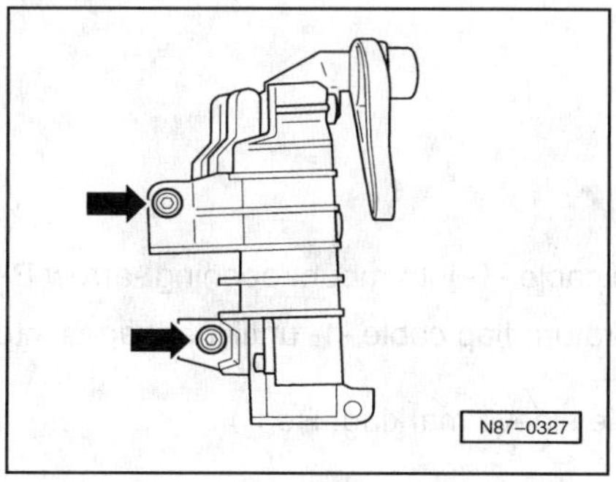

Remove screws (TORX T15) -arrows-.

– Un-clip lever for fresh/recirculated air flap.

– Pull off connector.

– Pull out motor downwards.

Installing

Position fresh/recirculated air flap lever by hand dependent on position of servo motor.

Heating and A/C controls, removing and installing

Removing

– Carefully remove trim panel by depressing locating lugs -arrows-with a covered screwdriver. Do not damage instrument panel.

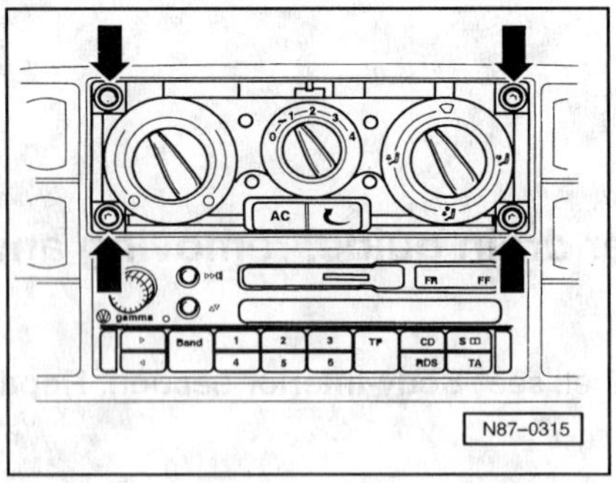

◀ Remove instrument panel screws -arrows-.

– Remove center trim panel, see **Heating and A/C components in passenger compartment, assembly**.

– Pull controls with cables attached from instrument panel.

– Release cables on controls.

– Disconnect electrical connector.

Installing

NOTE—

• Cable sleeve ends are color coded.

• Check cables before installing and replace binding or damaged cables.

• All air flaps must audibly move onto end stops when turning controls.

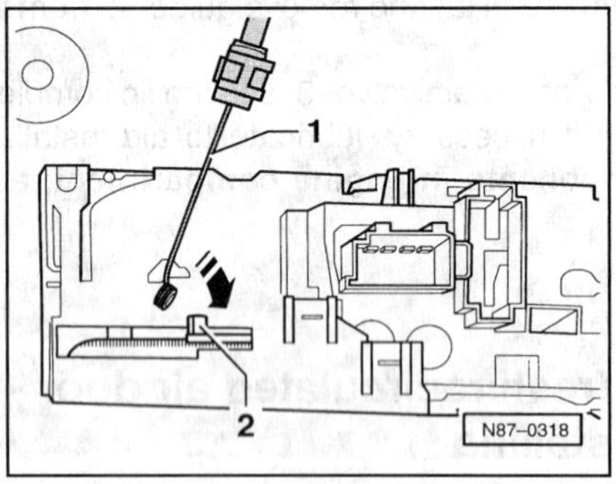

◀ Mount eyelet on end of cable -1- onto control -2-.

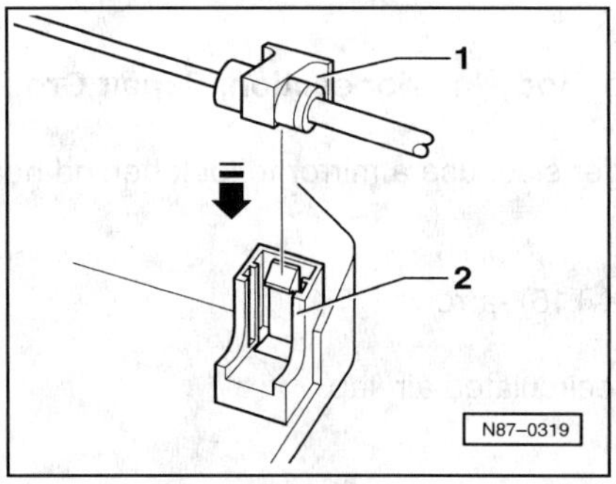

◀ Press footwell/defrost or central flap cable -1- into mount -2- until it engages.

- Footwell/defrost flap cable sleeve marking: White
- Central flap cable sleeve marking: Black

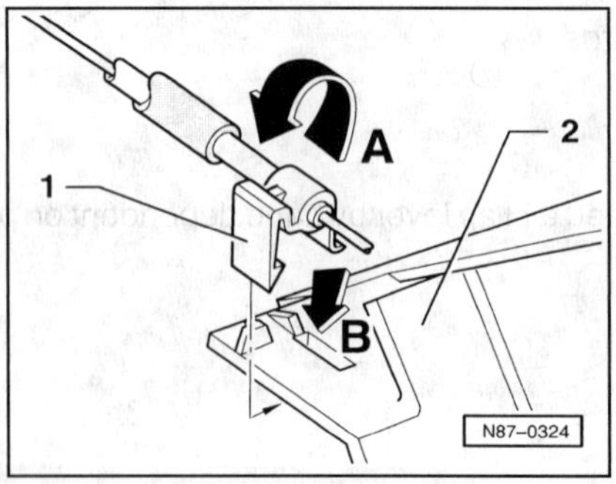

◀ Insert temperature flap cable -1- into mount opening -arrow B-.

- Turn -arrow A- temperature flap cable -1- until it engages into air distributor housing -2-.
- Temperature flap cable sleeve marking: Red

Heating and A/C controls, assembly

1. **Heating and A/C controls**
   - With:
     A/C switch -E35-
     Fresh air blower switch -E9-
     Switch for fresh air blower and recirculated air -E184-
   - See **Heating and A/C controls, removing and installing**
2. **Illumination filter**
3. **Trim panel**
4. **Fresh air control lever light -L16-**
   - 12V, 1.2W
   - Removing:
     Un-clip trim panel -7- and rotary control -5-.
     Remove bulb (if necessary, slide a short piece fuel hose over bulb and pull out).
5. **Rotary control**
   - For blower speed
   - To remove: Use pliers with suitable plastic or rubber protection on jaws
6. **Rotary control**
   - For air distribution
   - To remove: Use pliers with suitable plastic or rubber protection on jaws
7. **Trim panel**
8. **Rotary control**
   - For interior temperature
   - To remove: Use pliers with suitable plastic or rubber protection on jaws

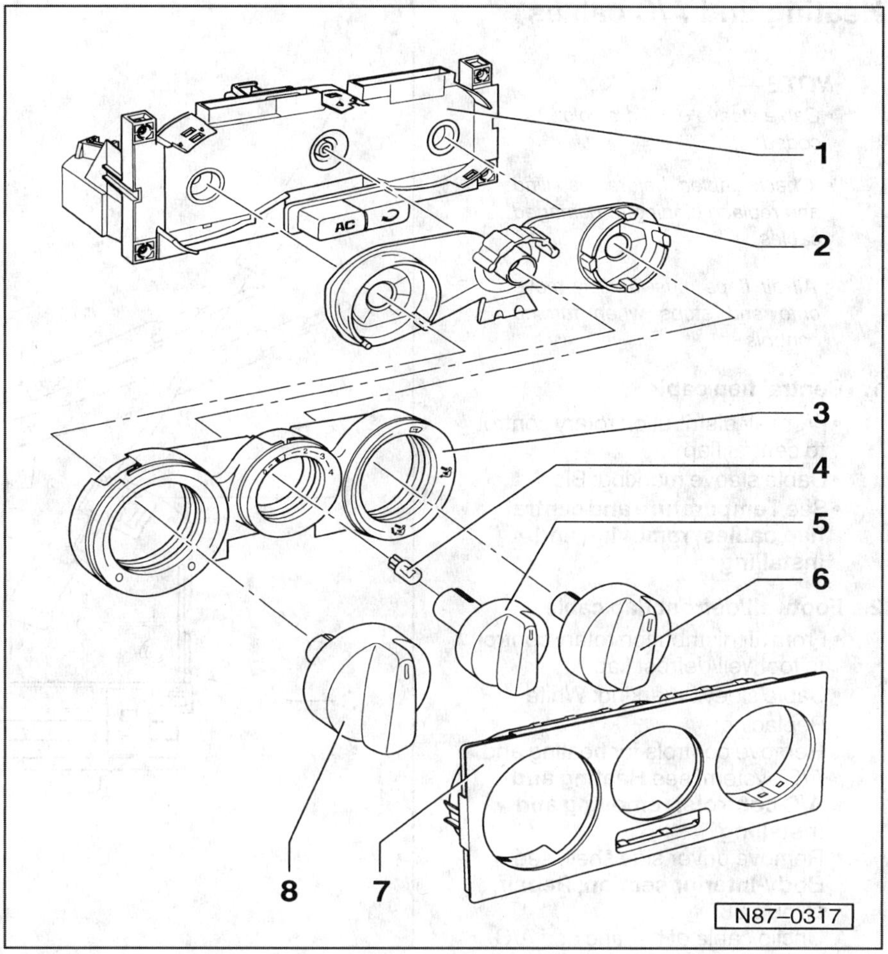

Heating and A/C cables

NOTE —
- *Cable sleeve ends are color coded.*
- *Check cables before installing and replace binding or damaged cables.*
- *All air flaps must audibly move onto end stops when turning controls.*

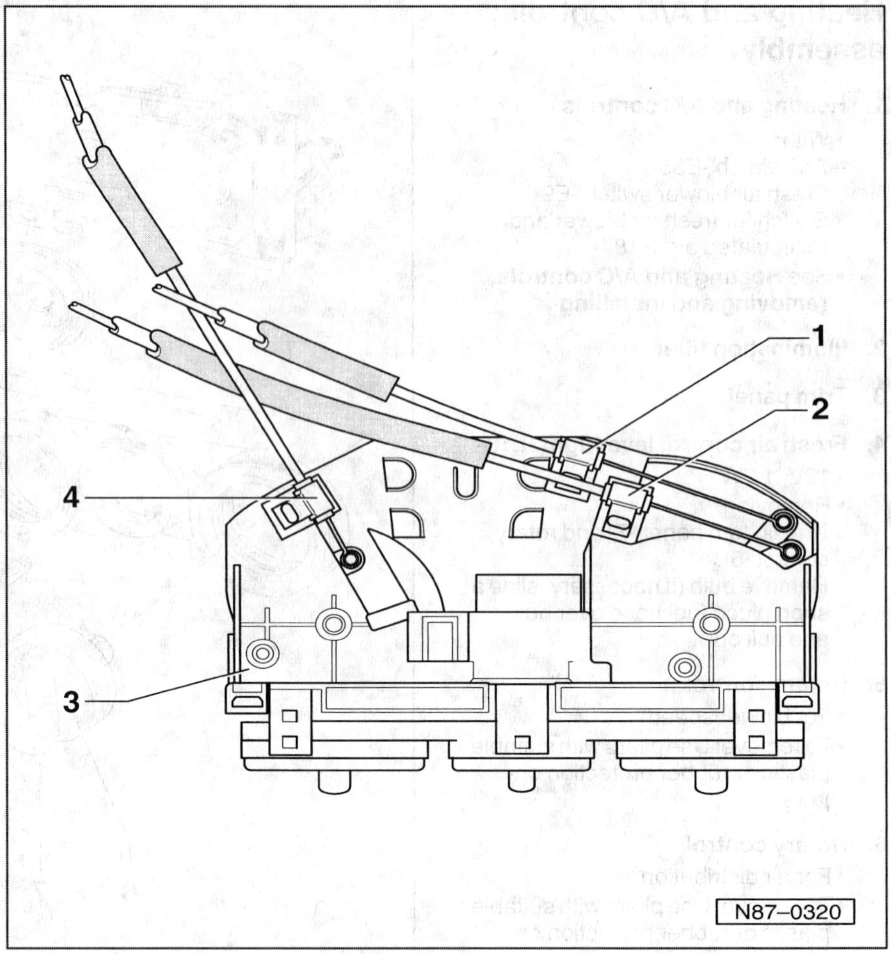

1. **Central flap cable**
   - From air distribution rotary control to central flap
   - Cable sleeve marking: Black
   - See **Temperature and central flap cables, removing and installing**
2. **Footwell/defrost flap cable**
   - From air distribution rotary control to footwell/defrost flap
   - Cable sleeve marking: White
   - Replacing:
     Remove controls for heating and A/C system, see **Heating and A/C controls, removing and installing**
     Remove driver side shelf, see **Body-Interior section, Repair Group 68**
     Unclip cable at heating and A/C unit.
     When installing, attach cable to heating and A/C unit first.
3. **Heating and A/C controls**
   - See **Heating and A/C controls, removing and installing**
4. **Temperature flap cable**
   - From temperature rotary control to temperature flap
   - Cable sleeve marking: Red
   - See **Temperature and central flap cables, removing and installing**

Temperature and central flap cables, removing and installing

NOTE —
• Cable sleeve ends are color coded.

• When installing secure cable to heating and A/C unit first.

• Check cables before installing and replace binding or damaged cables.

Removing

– Remove heating and A/C controls, see **Heating and A/C controls, removing and installing**

– Remove instrument panel, see **Body-Interior section, Repair Group 70.**

– Remove footwell air outlet, see **Heating and A/C components in passenger compartment, assembly**.

– Disconnect cable for central flap or temperature flap on heating and A/C unit.

Installing

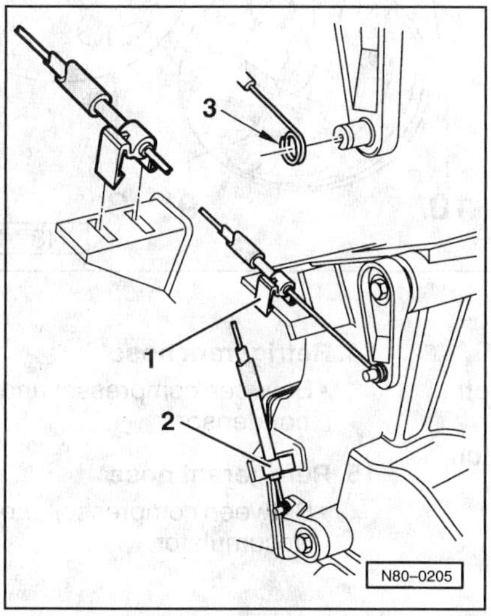

Illustration shows right-hand drive version.

1- Central flap cable, sleeve marking: Black
2- Temperature flap cable, sleeve marking: Red
3- When attaching inner cable note: Do not press end of wire (arrow) against lever.

Heating and A/C components in engine compartment, assembly

CAUTION—

Before beginning repairs on the electrical system:

- *Obtain the anti-theft radio security code.*
- *Switch the ignition off.*
- *Disconnect the battery Ground (GND) strap.*
- *After reconnecting battery, re-code and check operation of anti-theft radio. Also check operation of clock and power windows according to Owner's Manual.*

NOTE—

- *System components identified with an * must only be serviced or replaced by a qualified technician after discharging refrigerant system. Use Kent Moore ACR4 or equivalent refrigerant recycling equipment.*
- *Illustration shows vehicle with engine code AHA.*

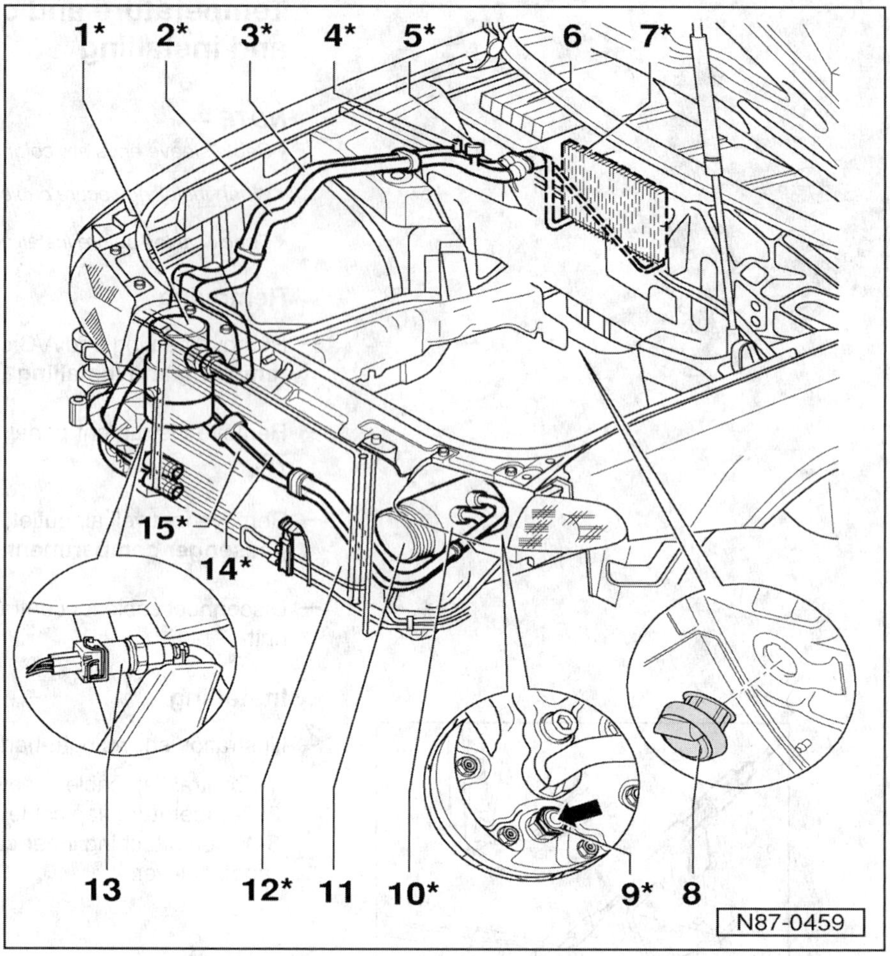

1. **Accumulator***
2. **Refrigerant hose***
   - Between evaporator and accumulator
3. **Refrigerant hose***
   - Between evaporator and condenser
4. **Low pressure service valve***
5. **High pressure service valve***
6. **Dust and pollen filter**
   - See **Maintenance**
   - Ambient temperature switch -F38- located below dust and pollen filter in lower air intake housing.
7. **Evaporator***
   - In heating and A/C unit.
8. **Evaporator water drain valve**
   - Checking, see **Fig. 1**.
9. **Pressure relief valve***
   - Zexel compressor illustrated
   - Checking, see **A/C refrigerant system assembly, Fig. 6 or Fig.7**
10. **Compressor***
11. **A/C clutch -N25-**
    - Servicing: Zexel, see **A/C clutch -N25-, servicing (Zexel)**
    - Servicing: Denso, see **A/C clutch -N25-, servicing (Denso)**
12. **Condenser***
13. **A/C pressure switch -F129-**
    - Tightening torque 8 Nm (71 in lb)
    - Replace O-ring (note Part No.).

    Function:
    - Switches coolant fan -V7- to next highest speed when pressure increases in refrigerant circuit.
    - Switches off A/C clutch -N25- when refrigerant pressure is excessive (e.g.: insufficient engine cooling) or if pressure is too low (e.g.: refrigerant loss).
    - Checking, see appropriate **A/C refrigerant system assembly**
    - Replacing, through 09/2000, see **A/C refrigerant system assembly (through 09/2000), Fig. 4**.
    - Replacing, from 10/2000, see **A/C refrigerant system assembly (from 10/2000), Fig. 5.**
14. **Refrigerant hose***
    - Between compressor and condensor
15. **Refrigerant hose***
    - Between compressor and accumulator.

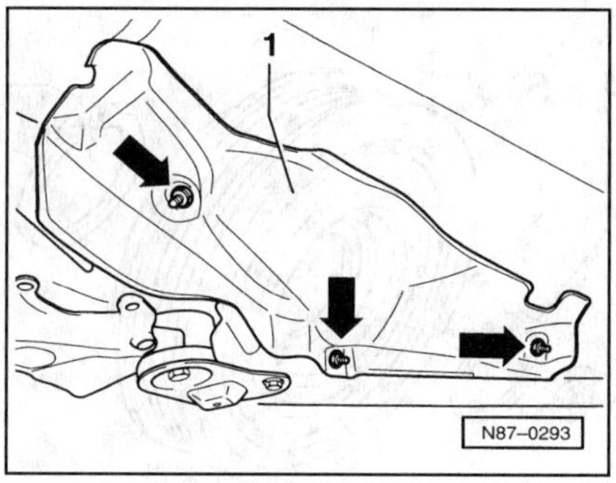

Fig. 1 Evaporator water drain valve, checking

– Check drain valve as follows:

- Remove nuts -arrows- on right catalytic converter heat shield -1-.
- Remove catalytic converter heat shield.

NOTE —

Catalytic converter heat shield from 10/2000 differs from illustration.

– Check valve as follows to ensure proper function:

- The valve must not be stuck together.
- The opening in the sealing cap of the valve must point downwards.
- The right catalytic converter heat shield plate must not be deformed or damaged in the area of the valve.

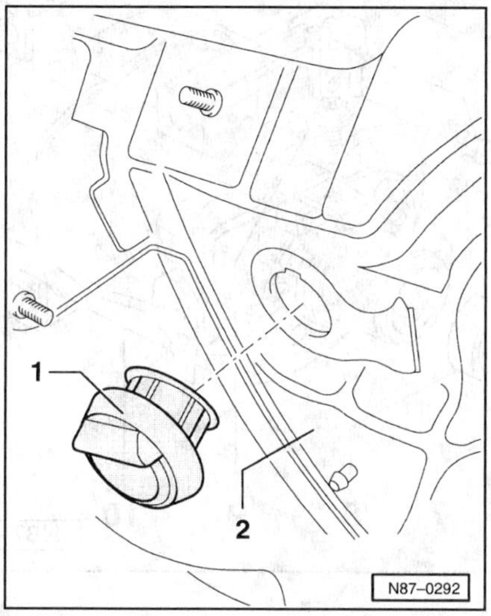

Remove valve -1- from bulkhead -2-.

- The valve sealing lip -1- must cover the water drain guide outlet completely, see **Heating and A/C components in passenger compartment, assembly**.

Ventilation system, checking

NOTE —

- *Passenger compartment air is vented to the outside trough openings in the luggage compartment trim.*
- *In order for the ventilation system to work properly, these openings must not be blocked.*
- *Vent frames are located in both rear quarter panels, behind the bumper cover.*

– Remove rear bumper cover, see **Body-Exterior section, Repair Group 63**

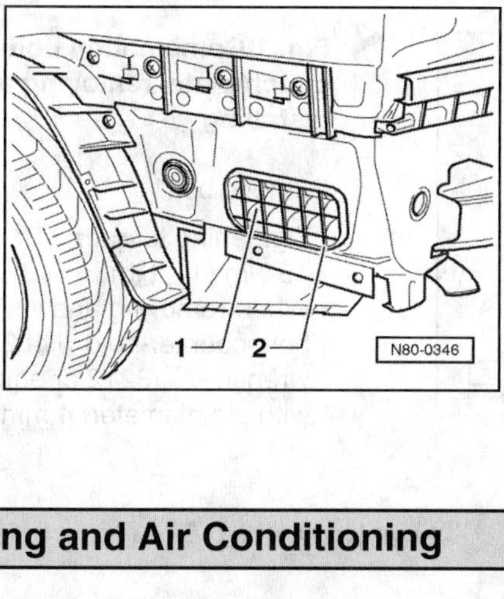

Sealing lips -1- in vent frame -2- must move and close freely.

A/C clutch -N25-, servicing (Zexel)

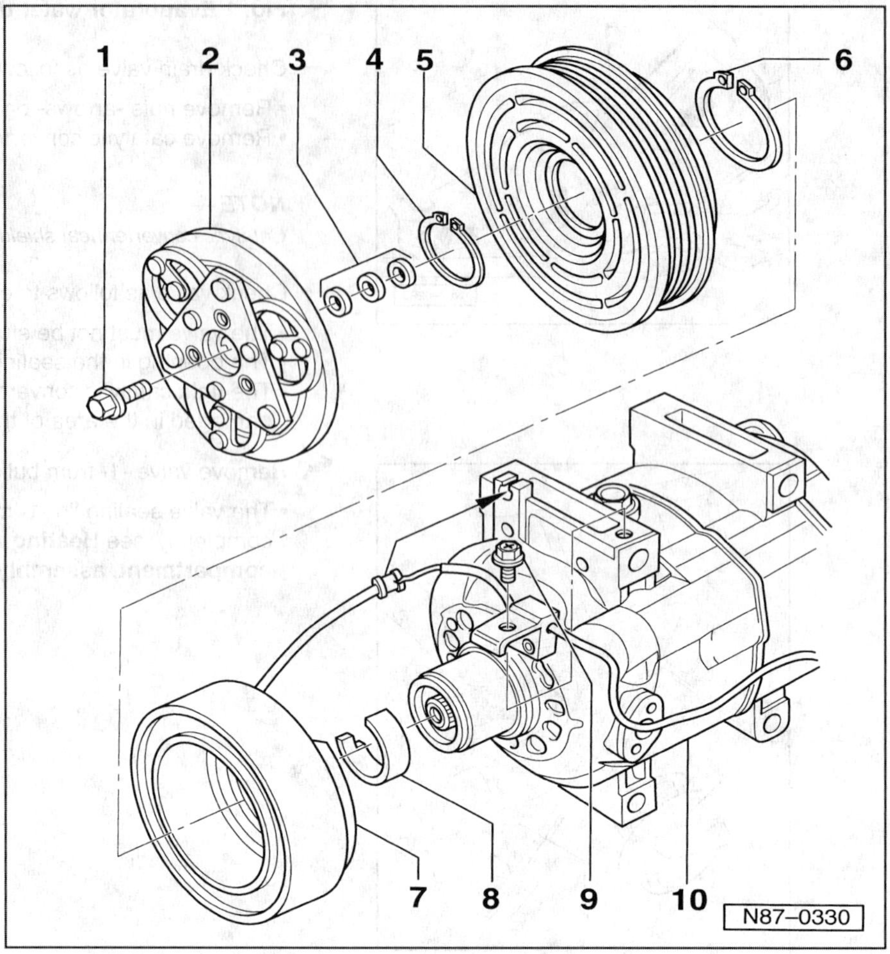

Special tools, workshop equipment, test and measuring appliances and auxiliary items required

- 3212 Two hole pin wrench
- VAG 1719 A/C Clutch set
- VAG 1719/1 Adapter
- VAG 1719/2 Adapter
- Torque wrench (5 - 50 Nm)
- Locally available two-arm puller (e.g. Kukko 20-10) with fixtures for a depth of 100 mm.
- Feeler gauge

NOTE —

- *Refrigerant circuit does not need to be opened to service A/C clutch.*
- *If unable to service A/C clutch on compressor installed in vehicle, discharge refrigerant and remove compressor from compressor bracket.*
- *System components identified with an * must only be serviced or replaced by a qualified technician after discharging refrigerant system. Use Kent Moore ACR4 or equivalent refrigerant recycling equipment.*

1. **Hex head combination bolt**
   - 15 Nm (11 ft lb)
   - Removing and installing, see **Fig. 1**
2. **Clutch plate**
   - Removing, see **Fig. 2**
   - Checking clutch plate gap, see **Fig. 6**
3. **Shims**
   - Used to adjust gap between clutch plate and pulley
   - Gap: 0.3 - 0.6 mm (0.011 - 0.023 in)
4. **Circlip**
   - Always replace
     Install correctly: thick side faces compressor
   - Ensure correct seating in groove
5. **Clutch pulley**
   - Removing, see **Fig. 3**
   - Installing, see **Fig. 4**
6. **Circlip**
   - Always replace
   - Install correctly: Flat side faces compressor
   - Ensure correct seating in groove
7. **Clutch coil**
   - Installing, see **Fig. 5**
8. **Felt ring**
   - Always replace
9. **Bolt**
10. **Compressor**
    - Manufacturer: Zexel

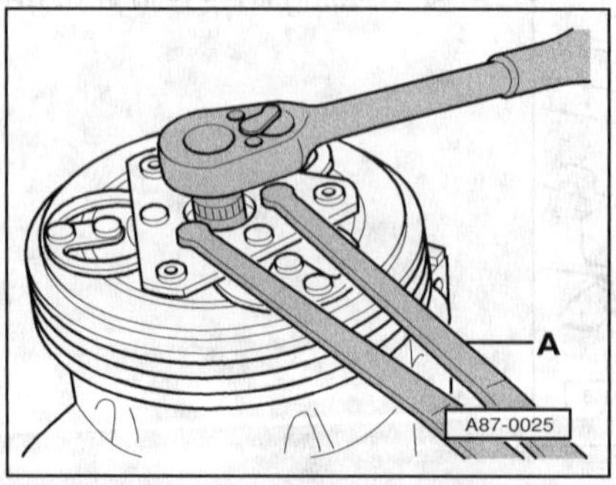

Fig. 1 Combination bolt for clutch plate, removing and installing

– Remove bolt.
- Tightening torque: 15 Nm (11 ft lb)
- When removing and installing bolt, counter-hold with 2 hole pin wrench -A- (3212 or equivalent) with pin diameter: 4 mm.

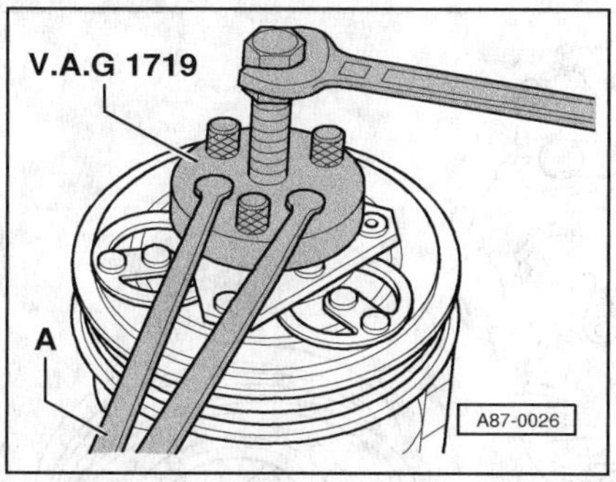

Fig. 2 Clutch plate, removing

– Mount VAG 1719 and turn in center bolt to extract clutch plate.
  - When removing, counter-hold with 2 hole pin wrench -A- (3212 or equivalent).

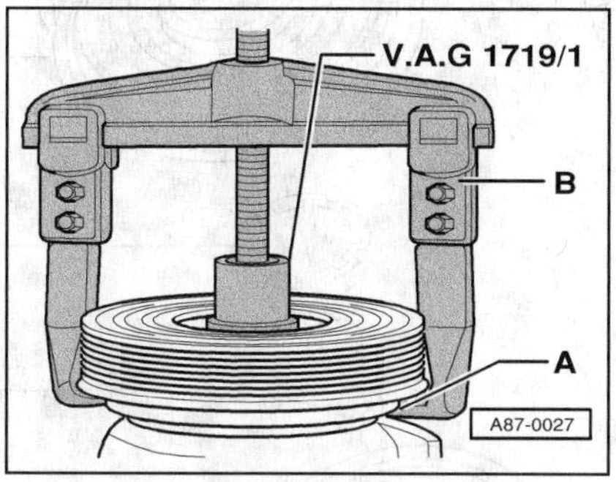

Fig. 3 Clutch pulley, removing

– Remove felt ring.

– Fit adapter VAG 1719/1 (or equivalent).

– Fit locally available 2 arm puller -B- on shoulder -A- and extract pulley.

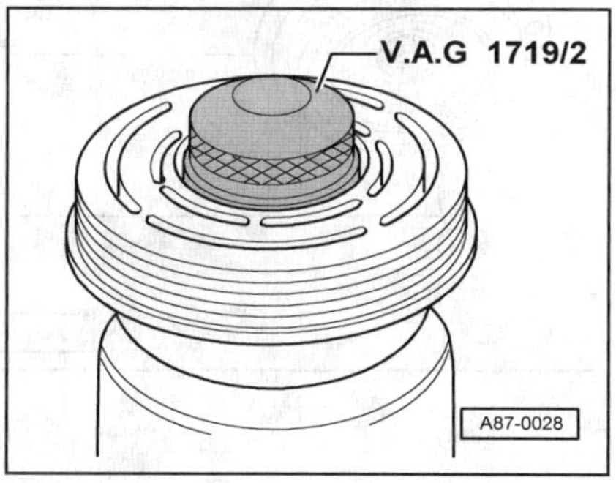

Fig. 4 Clutch pulley, installing

– Drive on belt pulley with plastic head hammer and VAG 1719/2 (or equivalent).

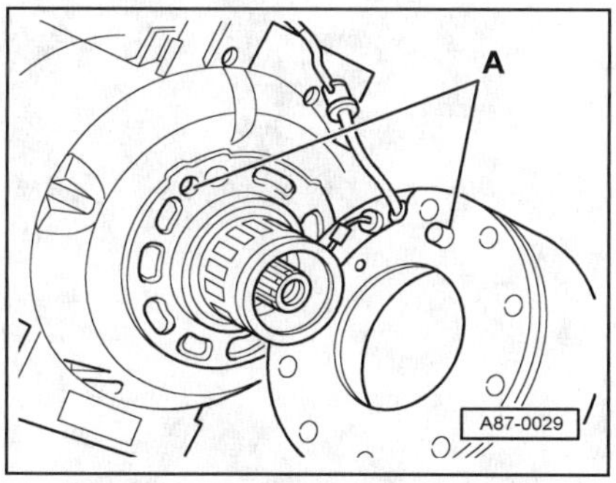

Fig. 5 Clutch coil, installing

– Install coil on compressor.

– Position pin -A- in recess.

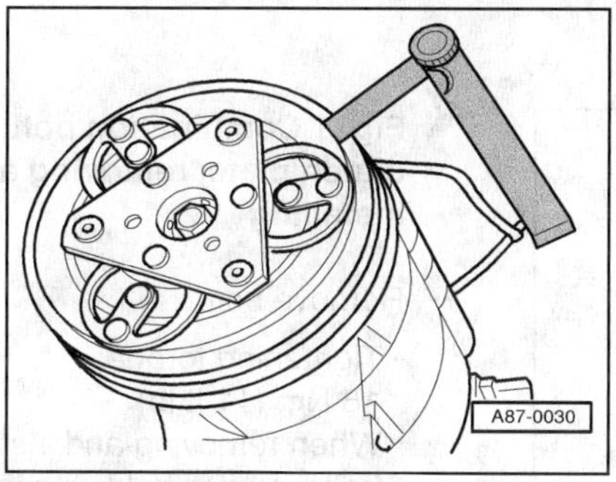

Fig. 6 Clutch plate gap, checking

– Check gap with feeler gauge.
  - Gap: 0.3 - 0.6 mm (0.011 - 0.023 in)

NOTE—
- *Gap must be within tolerance around entire circumference.*
- *If gap is outside tolerance range, remove clutch plate and adjust gap by adding or removing shims.*

A/C clutch -N25-, servicing (Denso)

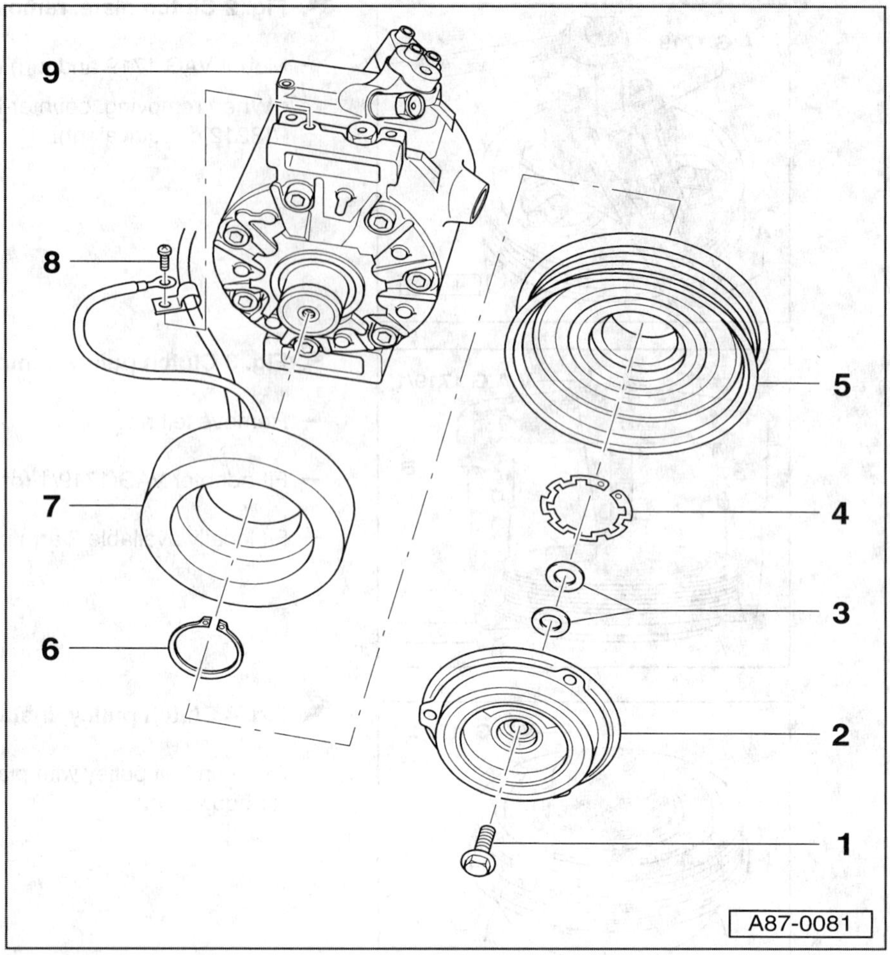

Special tools, workshop equipment, test and measuring appliances and auxiliary items required

- Torque wrench (5 - 50 Nm)
- Locally available two-arm puller (e.g. Kukko 20-10) with fixtures for a depth of 100 mm.
- Locally available strap wrench
- Depth gauge

NOTE —
- *Refrigerant circuit does not need to be opened to service A/C clutch.*
- *If unable to service A/C clutch on compressor installed in vehicle, discharge refrigerant and remove compressor from compressor bracket.*
- *System components identified with an * must only be serviced or replaced by a qualified technician after discharging refrigerant system. Use Kent Moore ACR4 or equivalent refirgerant recycling equipment.*

1. **Hex head combination bolt**
   - 15 Nm (11 ft lb)
   - Removing and installing, see **Fig. 1**
2. **Clutch plate**
   - Removing, see **Fig. 2**
   - Checking clutch plate gap, see **Fig. 4**
3. **Shims**
   - Used to adjust gap between clutch plate and pulley
   - Gap: 0.4 - 0.6 mm (0.015 - 0.023 in)
4. **Circlip**
   - Always replace
   - Install correctly: Thick side faces compressor.
   - Ensure correct seating in groove.
5. **Clutch pulley**
   - Remove with locally available 2 or 3 arm puller.
   - Ensure puller is seated correctly to prevent damage to pulley.
6. **Circlip**
   - Always replace
   - Install correctly: Flat side faces compressor.
   - Ensure correct seating in groove.
7. **Clutch coil**
   - Installing, see **Fig. 3**

   NOTE —
   A thermo-fuse is incorporated into the clutch coil. Current to the clutch coil is interrupted in the event of compressor overheating (e.g. a binding compressor).
8. **Screw**
9. **Compressor**
   - Manufacturer: Denso

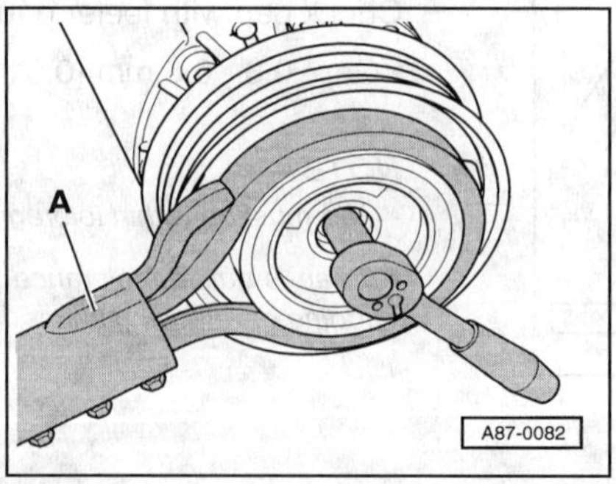

Fig. 1 Combination bolt for clutch plate, removing and installing

– Remove bolt.
- Tightening torque: 15 Nm (11 ft lb)
- When removing and installing bolt, counter-hold with locally available strap wrench -A-.

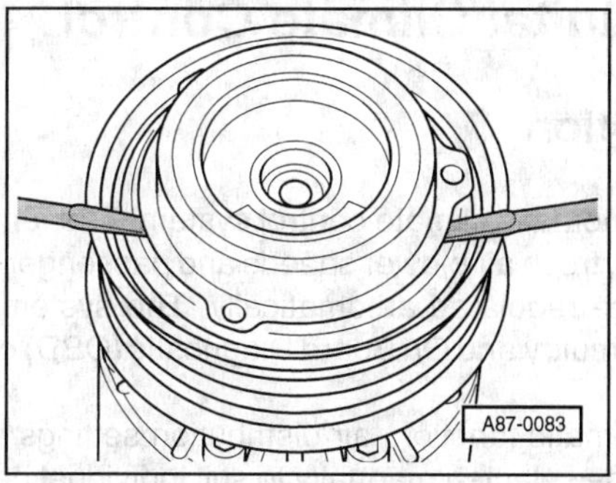

Fig. 2 Clutch plate, removing

– Pry off clutch plate carefully with two large screwdrivers.

- Use locally available 2 or 3 arm puller if plate is securely seated (difficult to remove).

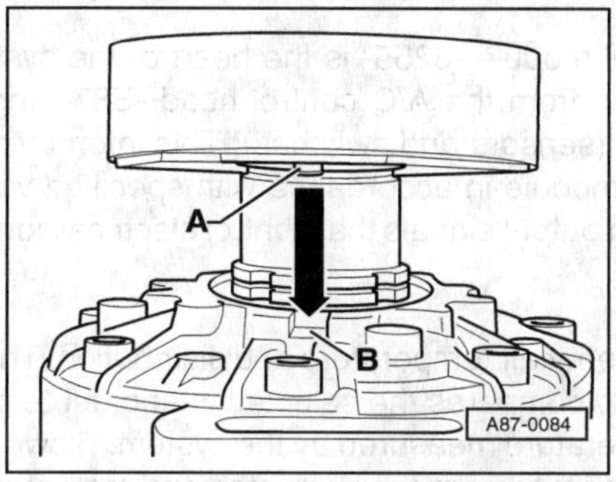

Fig. 3 Clutch coil, installing

– Install coil on compressor.

– Locate locking pin -A- in recess -B-.

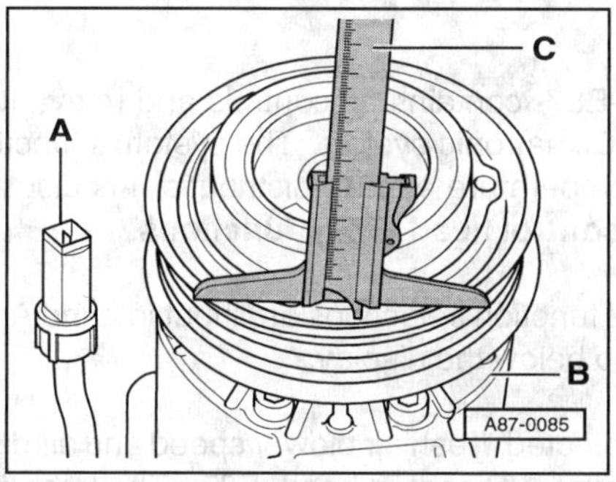

Fig. 4 Clutch plate gap, checking

- Gap: 0.4 - 0.6 mm (0.015 - 0.023 in)

NOTE —

- *In order to check gap with compressor installed in vehicle:*

- Move front end assembly (lock carrier) into service position, see **Body-Exterior section, Repair Group 50.**

Determine "Dimension 1"

– Using depth gauge -C-, measure gap between top of clutch plate and belt pulley at three equally spaced locations (measurement must not deviate).

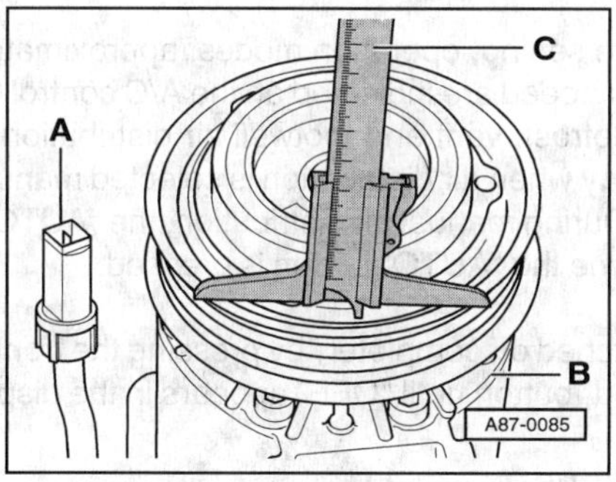

Determine "Dimension 2"

– If compressor is installed in vehicle, disconnect compressor electrical connector -A- from vehicle wiring harness.

– Using a jumper wire, energize clutch coil by applying 12V to connector -A- (battery terminal B+). If compressor is removed from vehicle, compressor -B- must be grounded.

– With clutch energized, measure gap between belt pulley and clutch plate with depth gauge -C-:

- The allowable gap (0.4 - 0.6 mm / 0.015 - 0.023 in) is the difference between "Dimension 1" and "Dimension 2".

NOTE —

- *Gap must be within tolerance around entire circumference.*
- *If gap is outside the permitted tolerance range, remove clutch plate and adjust gap by adding or removing shims.*

Climatronic Digital Climate Control

General information

Climatronic is a fully automatic climate control system. The temperature of the interior air supply, fresh air blower speeds and passenger compartment air distribution are regulated automatically. The system features fully digital electronic circuitry and On Board Diagnostic (OBD) capability.

In addition to fully automatic function, air distribution settings and fresh air blower speeds can be selected manually to suit individual needs.

The Climatronic control module -J255- is the heart of the system.It receives input information from the A/C control head -E87- and various electronic components (sensors and switches). This information is processed by the control module in accordance with specified values and provides corresponding output signals that control electrical components (actuators).

The system controls interior air temperature and distribution. The system continually samples and compares the selected interior air temperature with the actual air temperature measured by the system. Blower speeds, air temperature and distribution are then adjusted and regulated accordingly.

The A/C control head -E87- contains all controls and relevant displays needed to operate the Climatronic system. The system's functional status (mode selection, temperature and air distribution) is displayed in a LCD readout, see **A/C control head -E87-, functions**.

All automatic or manual function selections are input into the A/C control head via buttons located below the display.

When AUTO mode is selected, fresh air blower speed and air distribution signals are provided by the A/C control head -E87-. The fresh air blower speed is high when there is a large difference between the selected temperature and the actual interior temperature, and low when the difference is small or the same.

The interior temperature setting, operation modes, approximate outside temperature and blower speed are indicated on the A/C control head display at all times. The defrost, vent and footwell air distribution symbols only appear in the display when air distribution is selected manually (AUTO function override). During manual air distribution, the "AUTO" display remains off until such time the "AUTO" button is pressed.

The system can be switched off completely by pressing the front passenger "decrease fan speed" button until "OFF" appears in the display.

Climate control options

- Automatic temperature and fan speed mode
- Manual temperature and fan speed mode
- Manual selection of air recircultion mode
- Manual selection of air flow to istrument panel vents
- Manual selection of air flow to footwell vents
- Manual selection of air flow towindshield defroster

The Climatronic control module -J255- is controlled by an internal microprocessor with On Board Diagnostic (OBD) capability. If malfunctions, short or open circuits occur on monitored components, Diagnostic Trouble Codes (DTC) are stored in memory.

NOTE—

- *Pressing the button "AUTO" overrides any manually input settings.*
- *In ECON operation, only the compressor is switched off. Heating and A/C operation continues to be controlled electronically.*
- *Manually selected settings remain stored in memory after switching the ignition off and will be recalled on subsequent re-start. Only the function "Air recirculation" is erased after 20 minutes.*
- *Climatronic, manual operation see Owner's Manual.*
- *A malfunction is present if all display symbols on A/C control head -E87- flash. In this case check the Diagnostic Trouble Code (DTC) memory first, see* ***Scan Tool section.***

A/C control head -E87-, functions

CAUTION—

Before beginning repairs on the electrical system:

- *Obtain the anti-theft radio security code.*
- *Switch the ignition off.*
- *Disconnect the battery Ground (GND) strap.*
- *After reconnecting battery, re-code and check operation of anti-theft radio. Also check operation of clock and power windows according to Repair Manual and/or Owner's Manual.*

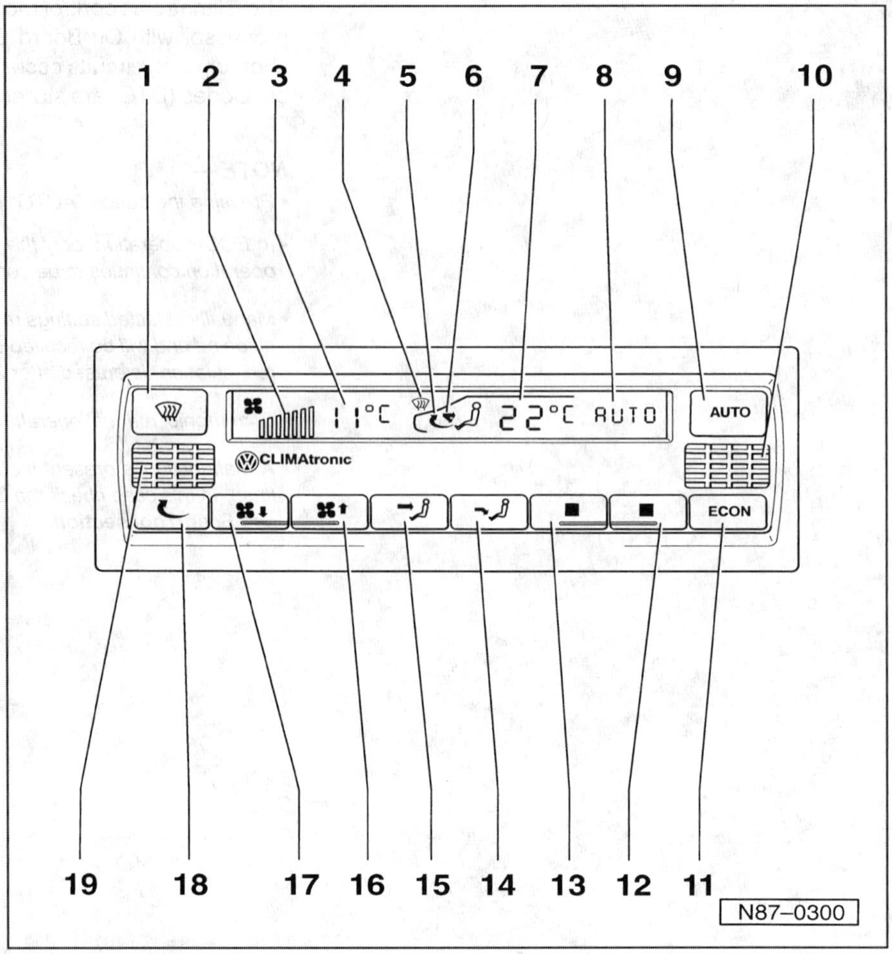

1. **Windshield defrost button**
2. **Blower speed display**
   - In automatic operation -8- a mid-range is displayed independent of actual blower speed.
3. **Ambient (outside) temperature display**
   - Display will not change when road speed is below 15 km/h (9.5 mph) and coolant temperature is above 70°C (158°F). Under these conditions, the actual measured value is not displayed because engine heat will adversely affect reading.
4. **Windshield defrost display**
5. **Air recirculation display**
6. **Air distribution display**
   - Function: "air flow to footwell" is selected and indicated when pressing button -14-.
   - Function: "air flow to upper body" is selected and indicated when pressing button -15-.
7. **Selected temperature display**
   - To change display from Celsius to Fahrenheit and vice versa: Press and hold button -11- and press button -9-.
   - Selected temperature scale appears in display.
8. **Operating mode display**

   "AUTO": Automatic mode
   - In automatic mode, the Climatronic maintains the selected interior temperature automatically.
   - With this setting, the air outlet temperature, blower speed and air distribution are controlled automatically.

   "ECON": A/C system off
   - In ECON mode only the compressor is switched off. Heating and A/C operation continues to be controlled automatically.

   "OFF": Complete system is switched off
   - Select by pressing and holding "decrease blower speed" button -17- until "OFF" appears in display.
   - In this mode, all air flaps remain in the last set position before switching off. (Setting should only be used in the case of a malfunction).
9. **Automatic operation button**
10. **Instrument panel interior temperature sensor -G56- and interior temperature sensor fan -V42-**
    - Installation location on right-hand drive vehicles.
11. **"ECON" button**
    - In ECON mode only the compressor is switched off. Heating and A/C operation continues to be controlled automatically.
12. **"Warmer" button**
13. **"Cooler" button**
14. **"Air flow to footwell" button**
15. **"Air flow to upper body" button**
16. **"Increase blower speed" button**
17. **"Decrease blower speed" button**
    - To switch off A/C system, press and hold button until "OFF" appears in display.
    - In this mode, all air flaps remain in the last set position before switching off.
      (Setting should only be used in the case of a malfunction).
18. **Recirculating air button**
19. **Instrument panel interior temperature sensor -G56- and interior temperature sensor fan -V42-**
    - Installation location on left-hand drive vehicles.

Climatronic Control Module -J255-, multi-pin connector assignments

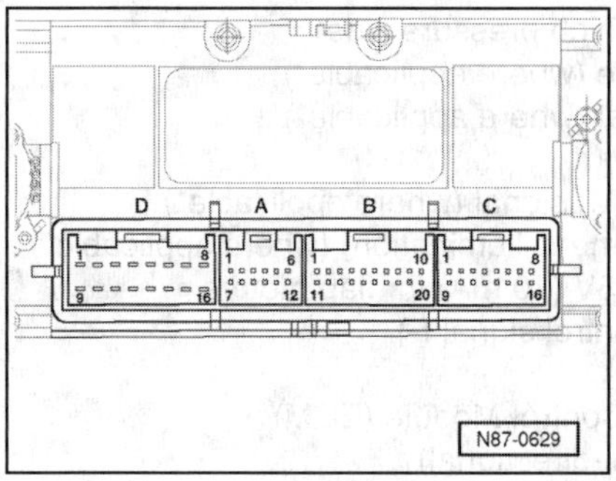

12-pin multi-connector -A- (black -T12/-)

1 - Data Link Connector (DLC) K-wire
2 - Motronic Engine Control Module -J220- (idle speed control)
3 - A/C readiness
4 - Open
5 - Terminal 58d (instrument illumination - dimmer)
7 - Fresh air intake duct temperature sensor -G89-
8 - Outside air temperature sensor -G17-
9 - Ground signal (sensors - motors)
10 - Open
11 - Engine Coolant Temperature (ECT) warning lamp (where applicable*)
12 - Ground signal (sensors - motors)

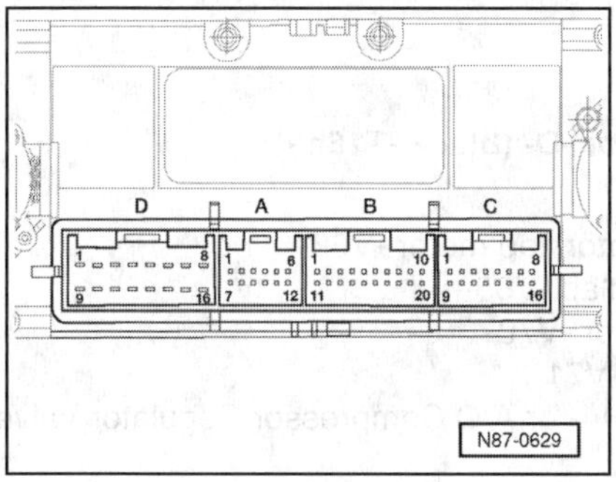

20-pin multi-connector housing -B- (red -T20/-)

1 - Outside temperature signal from instrument cluster (where applicable*)
2 - CAN Bus - high (where applicable*)
3 - CAN Bus - low (where applicable*)
4 - CAN Bus - shielding (where applicable*)
5 - Sender for outlet temperature, floor outlet -G192-
6 - Sender for outlet temperature, center -G191-
7 - Evaporator vent temperature sensor -G263-
8 - Temperature regulator flap motor position sensor -G92-
9 - Central flap motor position sensor -G112-
10 - Footwell/Defrost flap motor position sensor -G114-
13- Standing time - instrument cluster (where applicable*)
14- Temperature increase off/PTC off (P/N 3B1/2 907 044 F only - where applicable*)
15- Engine Coolant Shutoff Two-Way Vacuum Valve Relay -J585- (where applicable*)
17- Back pressure flap motor position sensor -G113-
18- Sunlight photo sensor -G107-

*) Confirm using applicable wiring diagram

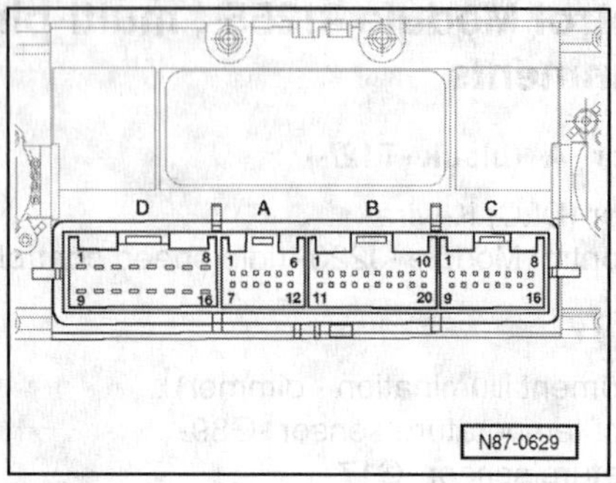

16-pin multi-connector housing -C- (brown -T16b/-)

1 - Kickdown (where applicable*)
2 - A/C pressure switch -F129- pin 2 (diagnosis 2/32 bar - high pressure side)
3 - Compressor voltage (where applicable*)
4 - Speedometer signal (where applicable*)
5 - Open
6 - Engine Speed (RPM) signal (where applicable*)
7 - Terminal 58 (instrument illumination) (where applicable*)
8 - Reference voltage 5V (10 mA) for flap motors
9 - Fresh air blower control signal (-)
10 - 11 Open
12 - Motronic Engine Control Module (ECM) (engine elecronics - bi-directional)
13 - Open
14 - Fresh air blower control signal (+)
15 - Wiring control (internal)
16 - Control module for fresh air blower -J126-

*) Confirm using applicable wiring diagram

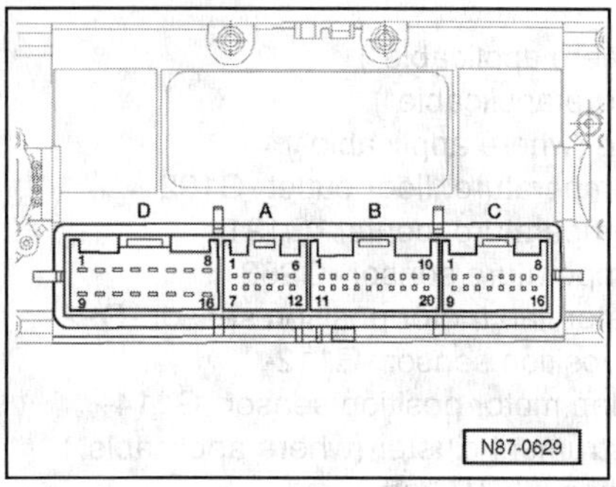

16-pin multi- connector -D- (black -T16a/-)

1 - Open
2 - Temperature regulator flap motor -V68-
3 - Footwell/Defroster flap -V85-
4 - Central air flap motor -V70-
5 - Air flow flap motor -V71-
6 - A/C Clutch relay -J44- or A/C Compressor Regulator Valve -N280- (where applicable*)
7 - Terminal 30 (where applicable*)
8 - A/C Clutch relay -J44- (where applicable*)
9 - Terminal 15
10 - Temperature regulator flap motor -V68-
11 - Footwell/Defroster flap motor -V85-
12 - Central air flap motor -V70-
13 - Air flow flap motor -V71-
14 - Terminal 31 (Ground - where applicable*)
15 - Terminal 31 (Ground)
16 - Coolant (FC) Fan Control Control Module -J293- or First Speed Engine Coolant Fan Control (FC) relay -J279- (where applicable**)

*) Confirm using applicable wiring diagram

**) Engine dependant - confirm using applicable wiring diagram

Climatronic components in passenger compartment, assembly

CAUTION—

Before beginning repairs on the electrical system:

- *Obtain the anti-theft radio security code.*
- *Switch the ignition off.*
- *Disconnect the battery Ground (GND) strap.*
- *After reconnecting battery, re-code and check operation of anti-theft radio. Also check operation of clock and power windows according to Owner's Manual.*

NOTE—

*System components identified with an * must only be serviced or replaced by a qualified technician after discharging refrigerant system. Use Kent Moore ACR4 or equivalent refrigerant recycling equipment.*

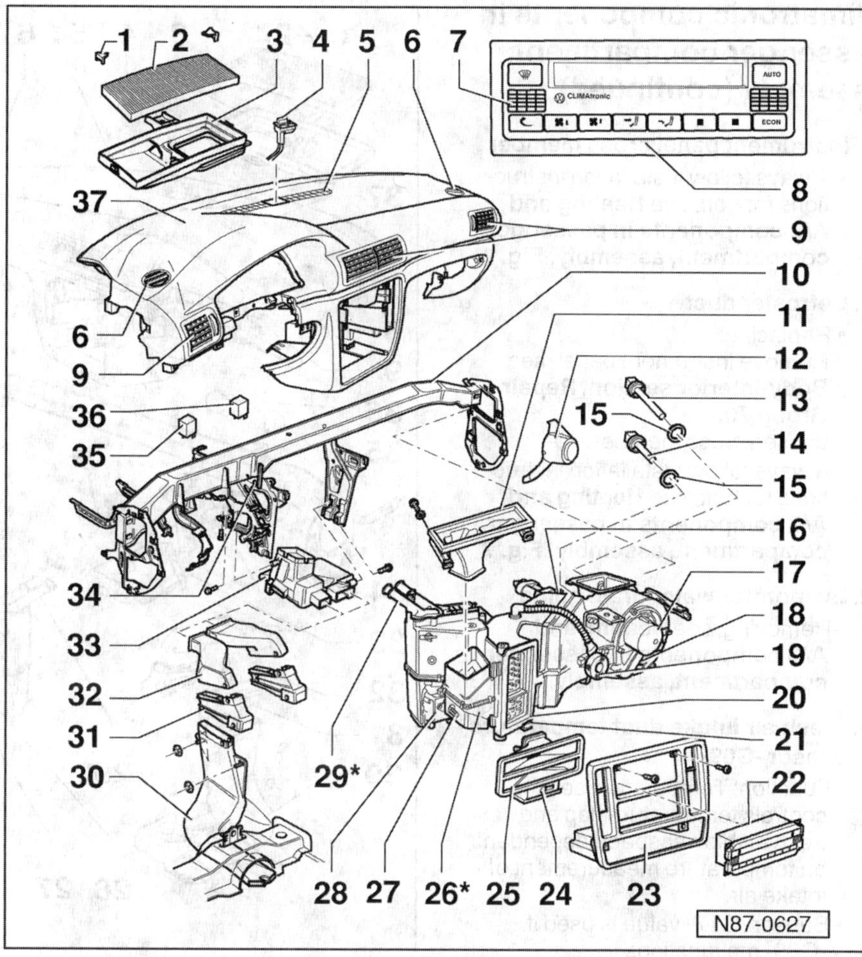

1. **Clip**
   - Lift out with screwdriver
2. **Dust and pollen filter**
   - See **Maintenance**
3. **Air intake housing**
4. **Sunlight photo sensor -G107-**
   - Function: Controls temperature flap and fresh air blower speed depending on light intensity
   - In event of failure: Climatronic control module -J255- assumes fixed value
   - Removing, see **Fig. 4**
5. **Defroster air outlet**
   - Removing and installing, see **Body-Interior section, Repair Group 70.**
6. **Side window air outlet**
   - Removing, see **Heating and A/C components in passenger compartment, assembly**
7. **Instrument panel interior temperature sensor -G56- with interior temperature sensor fan -V42-**
   - Climatronic control module -J255-, A/C control head -E87- and instrument panel interior temperature sensor -G56- with interior temperature sensor blower -V42- are integrated into a single unit and cannot be serviced separately.
   - Function: Temperature sensor controls temperature flap and fresh air blower speed dependant on temperature measurement of interior air.
   - In event of -G56- malfunction, operation continues using assumed value of 24°C (75°F)
   - Replacing and adjusting: Code Climatronic control module, function 07 and then set Basic Setting, function 04 using VAG 1551 Scan Tool (ST), see **Scan Tool section**.
8. **A/C control head -E87-**
   - Climatronic control module -J255-, A/C control head -E87- and instrument panel interior temperature sensor -G56- with interior temperature sensor blower -V42- are integrated into a single unit and cannot be serviced separately.
   - Functions, see **A/C control head -E87-, functions**
   - Removing and installing, see **A/C control head -E87- with Climatronic control module -J255-, removing and installing**
9. **Air outlets**
   - Center and side air outlets, removing (early prod.), see **Heating and A/C components in passenger compartment, assembly, Fig. 2**
   - Center air outlet, removing (later prod.) removing, see **Heating and A/C components in passenger compartment, assembly, Fig. 3**
   - Side air outlets, removing (later prod.), see **Heating and A/C components in passenger compartment, asssembly, Fig. 4**

(continued on following page)

Climatronic components in passenger compartment, assembly (continued)

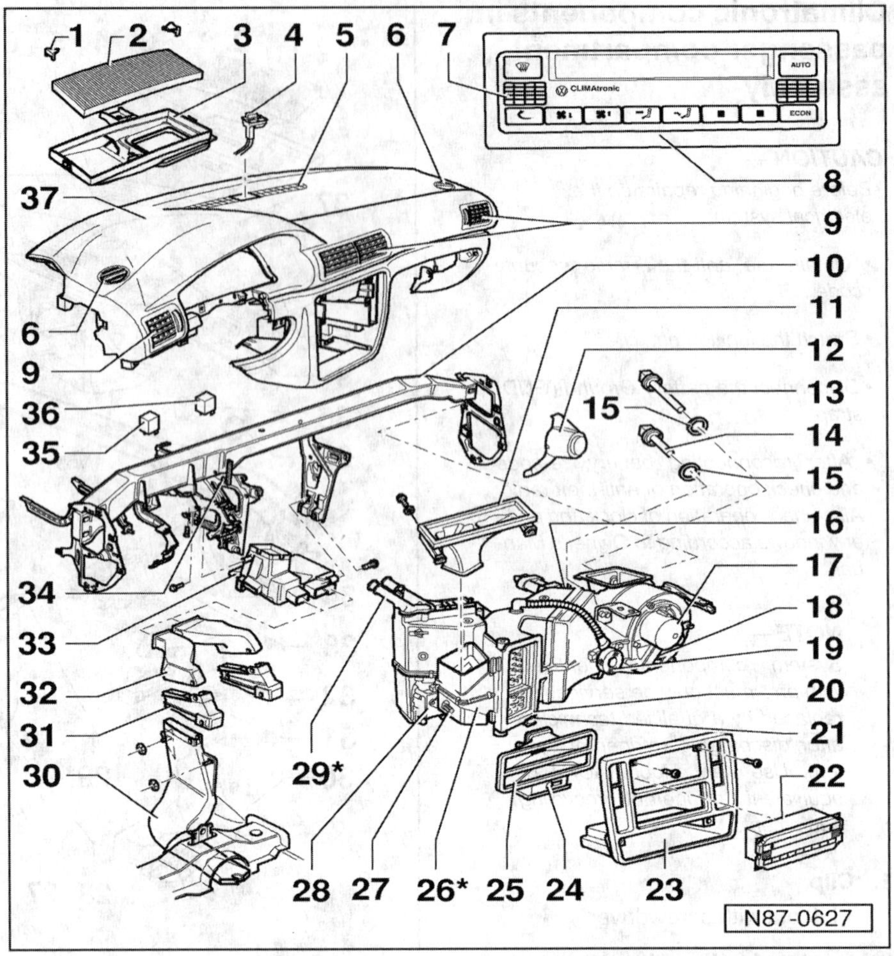

10. Instrument panel cross member
- Always follow installation instructions for bolt, see **Heating and A/C components in passenger compartment, assembly**, **Fig. 7**

11. Defroster duct
- Replacing:
  Remove instrument panel, see **Body-Interior section, Repair Group 70.**
  Loosen cross member -10- .
- Always follow installation instructions for bolt, see **Heating and A/C components in passenger compartment, assembly, Fig. 7**

12. Evaporator water drain guide
- Removing, see **Heating and A/C components in passenger compartment, assembly**.

13. Fresh air intake duct temperature sensor -G89-
- Function: Temperature sensor controls temperature flap and fresh air blower speed dependant on temperature measurement of intake air.
- Sensor -G17- value is used if -G89- malfunctions.
- Replacing:
  Remove glove box, see **Body-Interior section, Repair Group 68**
  Reach behind heat/A/C unit from below.
  Turn sensor 90° and pull out.
  Lubricate seal -15- with oil when installing.

NOTE —

If unable to access and remove sensor as described above, proceed as follows:

Remove instrument panel, see **Body-Interior section, Repair Group 70**
Loosen cross member -10- .
- Always follow installation instructions for bolt, see **Heating and A/C components in passenger compartment, assembly, Fig. 7**

14. Evaporator vent temperature sensor -G263-
- Vehicles from 10/2000 only
- Replacing
  Remove glove box, see **Body-Interior section, Repair Group 68**
  Reach behind heating and A/C unit from below.
  Turn sensor 90° and pull out.
- Lubricate seal -15- with oil when
- installing.

15. Gasket
- Lubricate with oil when installing.

CAUTION —

Area around dust and pollen filter and air intake housing must be free from debris, leaves etc. before reaching into housing. Debris entering heating and A/C unit will block evaporator water drain and cause subsequent (water) damage.

16. Air flow flap motor -V71-
- Also operates fresh and recirculating air flap
- Removing, see **Air flap motors (for climate control regulation), replacing**
- Replacing and adjusting: Initiate Basic Setting, function 04 with VAG 1551 Scan Tool (ST), see **Scan Tool section**.

17. Fresh air blower -V2-
- Removing and installing, see **Fresh air blower -V2- with control module for fresh air blower -J126-, removing and installing**

18. Control module for fresh air blower -J126-
- Function: Controls fresh air blower speed with variable voltage signal.
- Removing and installing, see **Fresh air blower -V2- with control module for fresh air blower -J126-, removing and installing**

(continued on following page)

Climatronic components in passenger compartment, assembly (continued)

19. **Air outlet for glove box**
    - Removing and installing, see **Fig. 2**

20. **Temperature regulator flap motor -V68-**
    - Removing, see **Air flap motors (for climate control regulation), replacing**
    - Replacing and adjusting: Initiate Basic Setting, function 04 with VAG 1551 Scan Tool (ST), see Scan Tool section.

21. **Central air flap motor -V70-**
    - Removing, installing and adjusting, see **Air flap motors (for climate control regulation), replacing**
    - Replacing and adjusting: Initiate Basic Setting, function 04 with VAG 1551 Scan Tool (ST), see Scan Tool section.

    NOTE —

    A modified flap (with cut-out at top) was introduced 11/1998. This design allows air flow to footwell to be blocked.

22. **Climatronic control module-J255-**
    - Climatronic control module -J255-, A/C control head -E87- and instrument panel interior temperature sensor -G56- with interior temperature sensor blower -V42- are integrated into a single unit and cannot be serviced separately.
    - Removing and installing, see **A/C control head -E87- with Climatronic control module -J255-, removing and installing**

23. **Center trim**

24. **Sender for outlet temperature, center -G191-**
    - Function: Controls center air distribution and fresh air blower speed, dependent on air outlet temperature measurement.
    - Removing and installing, see **Climatronic components in passenger compartment, assembly**

25. **Center air outlet duct**

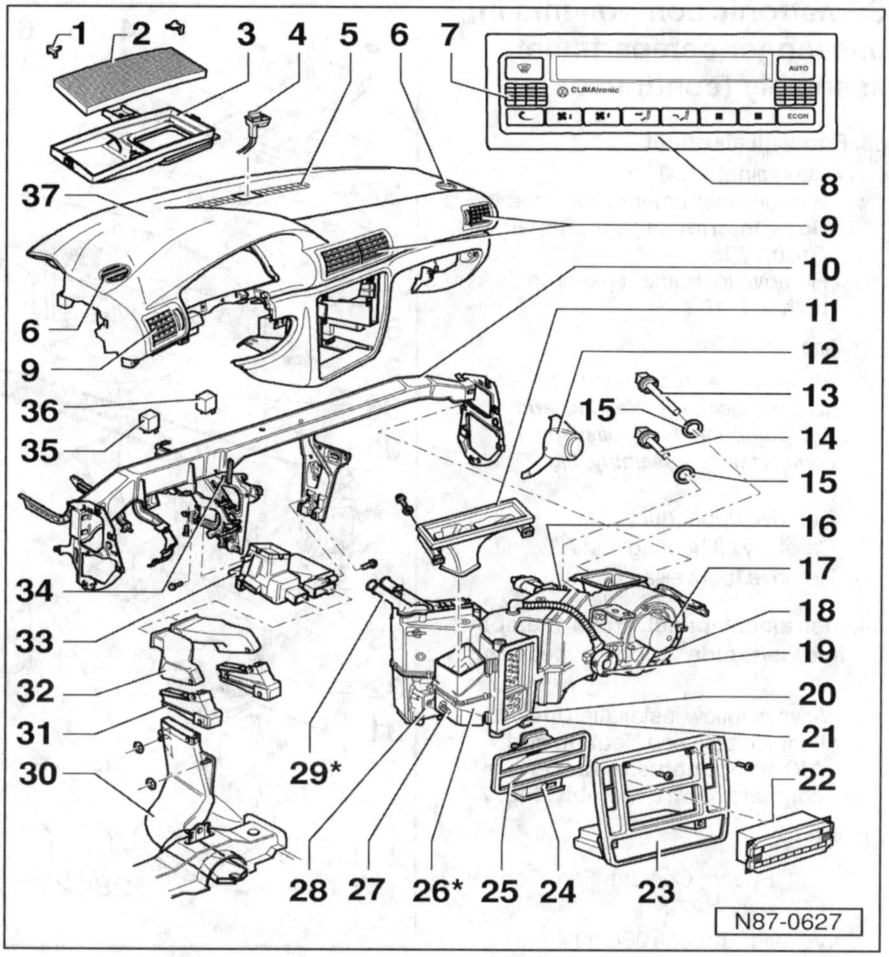

26. **Heating and A/C unit***

    Contains:
    - Heater core
    - Evaporator
    - Removing/assembly, see **Heating and A/C unit, removing and assembly.**

27. **Sender for outlet temperature, floor outlet -G192-**
    - Function: Controls air distribution to defrost/footwell and fresh air blower speed dependant on outlet temperature measurement.
    - In event of malfunction, operation continues using assumed value of 80°C (176°F)
    - Removing and installing, see **Fig. 1**

28. **Footwell/defroster flap motor -V85-**
    - Removing, see Air flap motors (for climate control regulation), replacing
    - Replacing and adjusting: Initiate Basic Setting, function 04 with VAG 1551 Scan Tool (ST), see **Scan Tool section**

29. **Heater core***
    - Always use new coolant after replacement
    - Removing: Remove heating and A/C unit, removing/assembly, see **Heating and A/C unit, removing and assembly.**

30. **Left rear duct**

31. **Lower duct**

32. **Upper duct**

(continued on following page)

Climatronic components in passenger compartment, assembly (continued)

33. **Footwell air outlet**
   - Replacing:
     Remove instrument panel, see **Body-Interior section, Repair Group 70.**
     Remove instrument panel cross member -10-.

     NOTE —
     *Always follow installation instructions for bolt, see **Heating and A/C components in passenger compartment, assembly, Fig. 7***

     Remove upper duct.
     Carefully lift heating and A/C unit.
     Remove footwell air outlet.

34. **Instrument panel cross member and left side support securing bolt**
   - Always follow installation instructions for bolt, see **Heating and A/C components in passenger compartment, assembly, Fig. 7**

35. **Relay**
   First Speed Coolant Fan Control (FC) relay -J279-
   - Vehicles through 09/200: Control number 214
   - Vehicles from 10/200: Control number 373

   Second Speed Coolant Fan Control (FC) Relay -J280-
   - Vehicles through 09/2000: Control number 213
   - Vehicles from 10/2000: Control number 370
   - Component locations see **Wiring Diagrams and Component Locations**.

36. **A/C clutch relay -J44-**
   - Control number 267
   - Component locations see Wiring Diagrams and Component Locations.

37. **Instrument panel**
   - Removing and installing, see **Body-Interior section, Repair Group 70.**

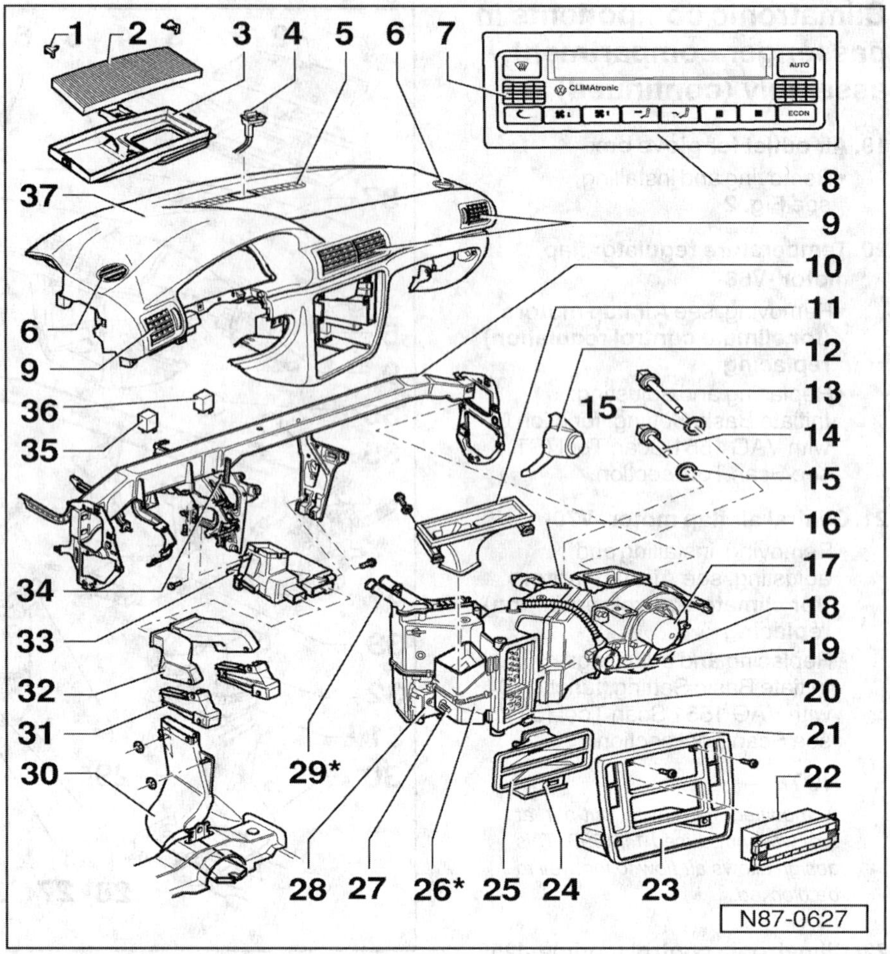

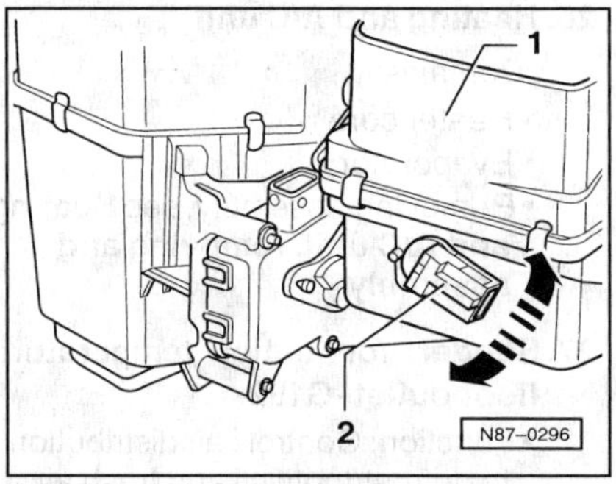

Fig. 1 Sender for outlet temperature, floor outlet -G192-, removing and installing

– Remove driver side shelf, see **Body-Interior section, Repair Group 68.**

– Disconnect electrical connector from sender -G192-.

– Rotate sender -2-, 90° -arrow- and remove from housing -1-.

– Lubricate rubber seal on footwell sender -G192- with oil when installing.

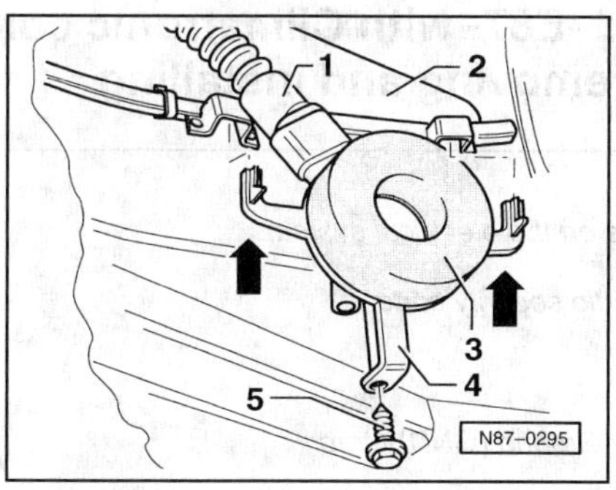

Fig. 2 Air outlet for glovebox, removing and installing

Removing

– Remove glove box, see **Body-Interior section, Repair Group 68.**

– Remove securing screw -5-.

– Pull air outlet -4- out from guides on heating and A/C unit -2-.

– Pull off hose -1-.

Installing

– Replace seal -3- if damaged.

- When installing, ensure that locking lugs engage in guides on heating and A/C unit -arrows-.

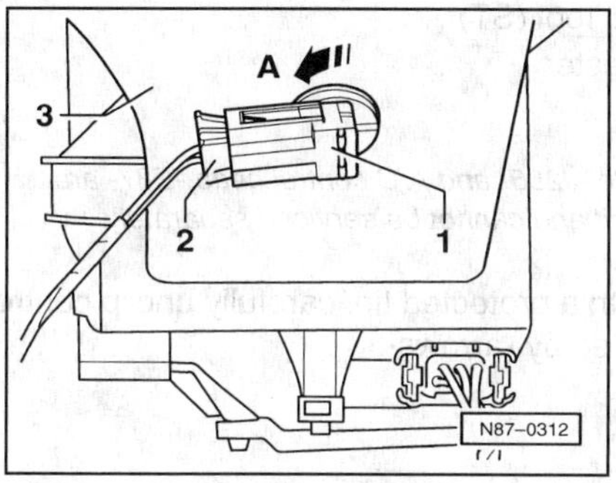

Fig. 3 Sender for outlet temperature, center -G191-, removing and installing

> ***CAUTION—***
>
> *Before beginning repairs on the electrical system:*
>
> - *Obtain the anti-theft radio security code.*
> - *Switch the ignition off.*
> - *Disconnect the battery Ground (GND) strap.*
> - *After reconnecting battery, re-code and check operation of anti-theft radio. Also check operation of clock and power windows according to Repair Manual and/or Owner's Manual.*

– Remove radio, see **Radio section, Repair Group 91.**

– Remove Climatronic control module -J255-, see **A/C control head -E87- with Climatronic control module -J255-, removing and installing**

– Remove center trim, see **Climatronic components in passenger compartment, assembly**

– Disconnect electrical connector -2- from sender -G191-.

– Rotate sender -1-, 90° -arrow A- and remove from housing -3-.

– Install in reverse order of removal, being sure to coat seal lightly with oil.

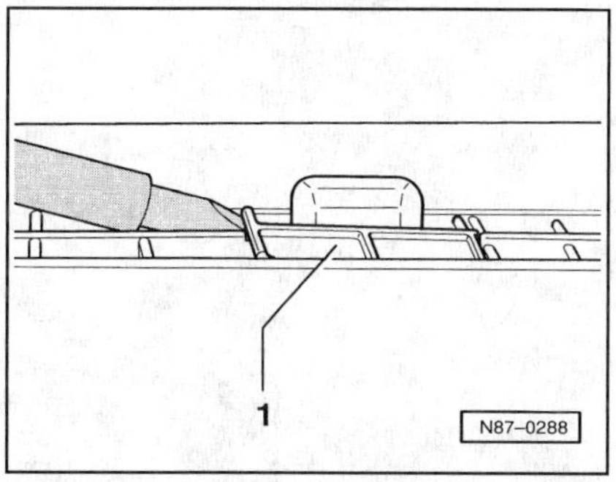

Fig. 4 Sunlight photo sensor -G107-, removing

– Carefully lift sender -1- at side with small screwdriver.

A/C control head -E87- with Climatronic control module -J255-, removing and installing

CAUTION—

Before beginning repairs on the electrical system:

- *Obtain the anti-theft radio security code.*
- *Switch the ignition off.*
- *Disconnect the battery Ground (GND) strap.*
- *After reconnecting battery, re-code and check operation of anti-theft radio. Also check operation of clock and power windows according to Repair Manual and/or Owner's Manual.*

Special tools, testers, measuring instruments and auxiliary items required

- VAG 1551/1552 Scan Tool (ST)
- Optional: VAS 5051 tester

NOTE—

Climatronic control module -J255- and A/C control head -E87- are integrated into a single unit and cannot be serviced separately.

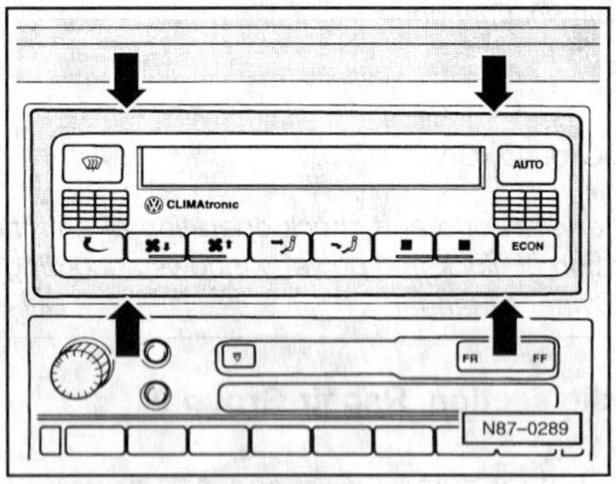

◄ Using a screwdriver with a protected tip, carefully unclip control head trim at locations indicated by -arrows-.

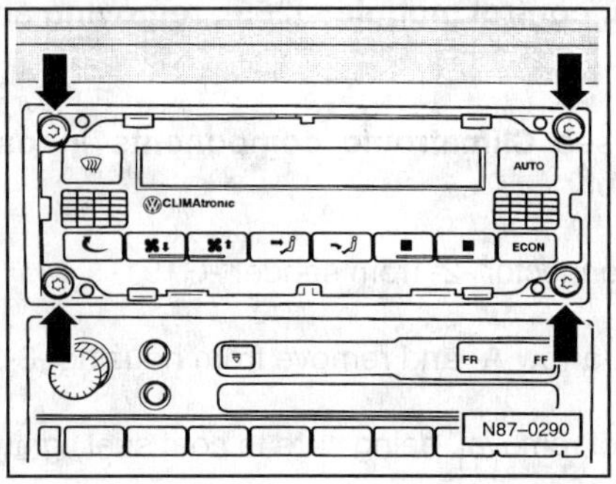

◄ Remove instrument panel screws -arrows-.

– Pull A/C control head/control unit from instrument panel opening.

– Disconnect electrical connections on rear of unit.

– Remove A/C control head with control unit.

– Installation is reverse order of removal, noting the following:

- Code Control Module, function 07, then initiate Basic Setting, function 04, with VAG 1551/1552 Scan Tool (ST), see **Scan Tool section**.

Fresh air blower -V2- with control module for fresh air blower -J126-, removing and installing

Removing

– Remove glove box, see **Body-Interior section, Repair Group 68.**

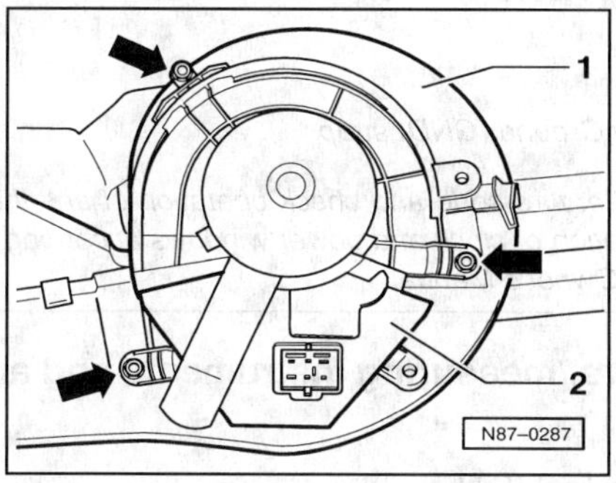

◄ Remove screws -arrows-.

– Remove blower with blower control unit -2- from housing -1-.

Control module for fresh air blower -J126-, removing (from blower)

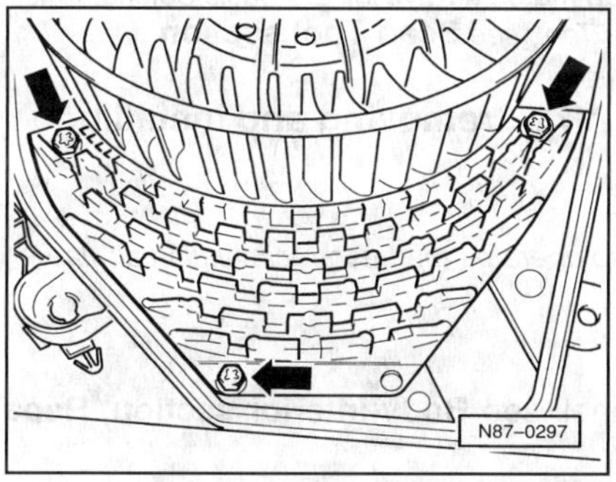

◄ Remove screws -arrows-.

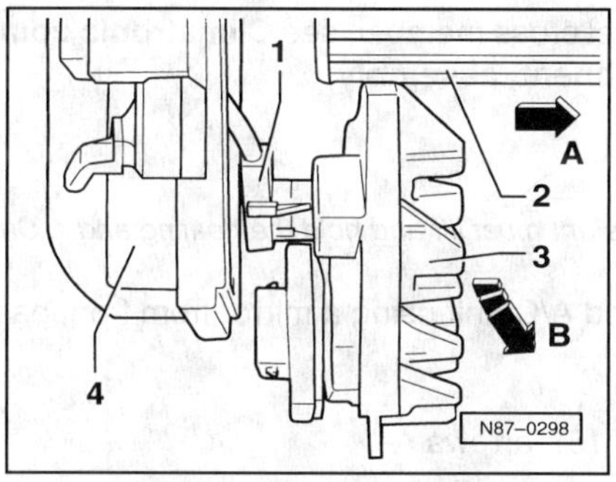

◄ Lift control module -3- and pull out forwards -arrow B-.

– Disconnect electrical connector -1-.

• Observe following when inserting module -J126-:

– Pull both components carefully apart in direction of -arrow A- to provide a slight clearance between fresh air blower wheel -2- and housing -4-.

– Insert control module -3- into housing -4- and secure.

Installing

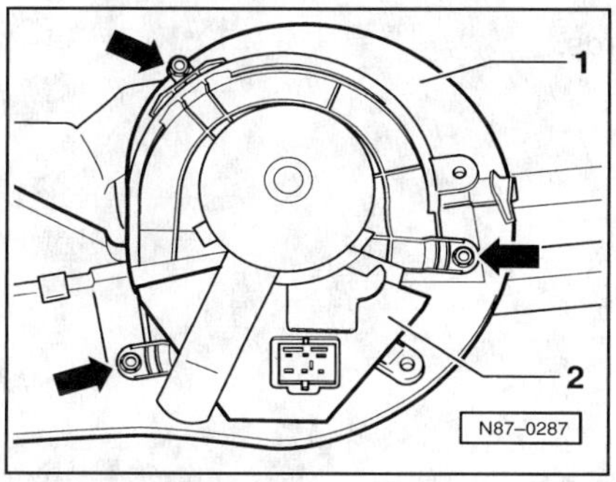

◄ Insert fresh air blower with blower control unit -2- into housing -1- and secure with screws -arrows-.

Air flap motors (for climate control regulation), replacing

> ***CAUTION—***
> *Before beginning repairs on the electrical system:*
>
> - *Obtain the anti-theft radio security code.*
> - *Switch the ignition off.*
> - *Disconnect the battery Ground (GND) strap.*
> - *After reconnecting battery, re-code and check operation of anti-theft radio. Also check operation of clock and power windows according to Repair Manual and/or Owner's Manual.*

Special tools, testers, measuring instruments and auxiliary items required

- VAG 1551/1552 Scan Tool (ST)
- Optional: VAS 5051 tester

NOTE—
After installing a new air flap motor, always initiate Basic Setting, function 04 with VAG 1551/1552 Scan Tool, see **Scan Tool section**.

Air flow flap motor -V71-, removing and installing

NOTE—
The air flow flap motor also operates the fresh/recirculating air flap.

Removing

– Remove instrument panel, see **Body-Interior section, Repair Group 70.**

– Loosen instrument panel cross member, see **Climatronic components in passenger compartment, assembly.**

NOTE—
For the next step, an assistant must lift and hold the heating and A/Cunit.

– Look behind heating and A/C unit using a mirror from front passenger's side.

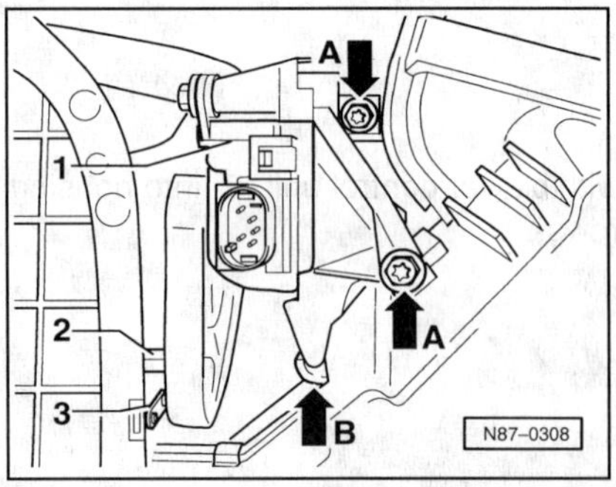

◄ Remove bolts (TORX T15) -arrows A-.

– Disconnect electrical connector.

– Pull out motor downwards.

Air flap motors (for climate control regulation), installing and adjusting

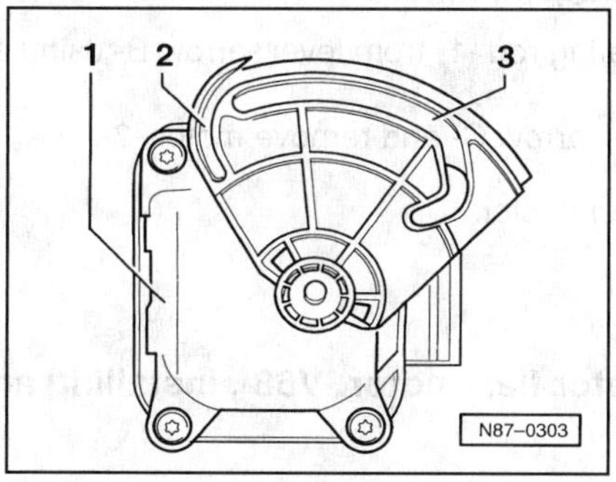

NOTE —

Observe correct orientation of flap levers to drive cam when installing flap motor.

1 - Air flow flap motor -V71-

- 2 - Guide for plastic lever
- 3 - Guide for metal lever

– Connect connector.

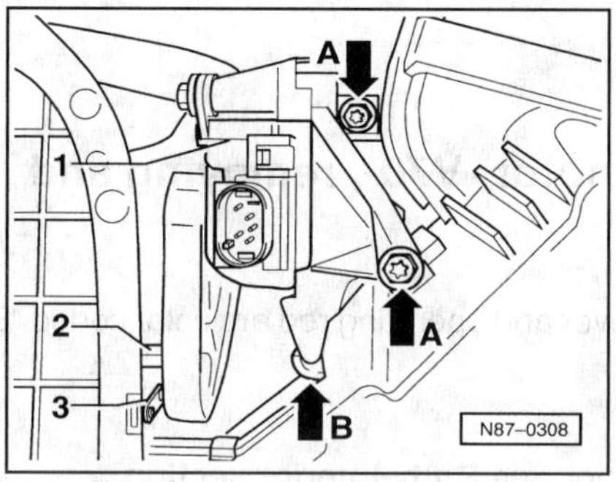

Insert flap motor -1- into guide -arrow B-.

- Position air flow flap and recirculated air flap levers -2 and 3- by hand according to position of flap motor.

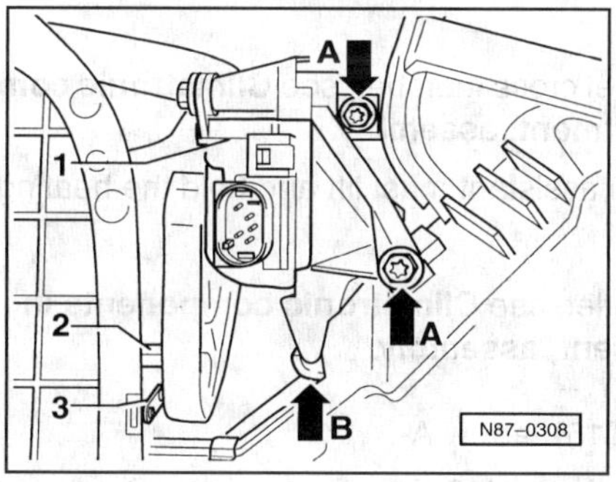

Insert flap levers -2- and -3- into drive cam guides.

- Plastic lever, position -2-
- Metal lever, position -3-

– Initiate Basic Setting, function 04, with VAG 1551 Scan Tool (ST), see **Scan Tool section.**

Temperature regulator flap motor -V68-, removing and installing

The temperature flap lever and operating rod are color coded "Red".

Removing

– Remove instrument panel, see **Body-Interior section, Repair Group 70**

– Remove dust and pollen filter.

– Loosen instrument panel cross member, see **Climatronic components in passenger compartment, assembly.**

- For the next steps, an assistant must lift and hold the heating and A/Cunit.

– Remove footwell air outlet, see **Climatronic components in passenger compartment, assembly**

– Remove air flow flap motor, see **Air flap motors (for climate control regulation), replacing**.

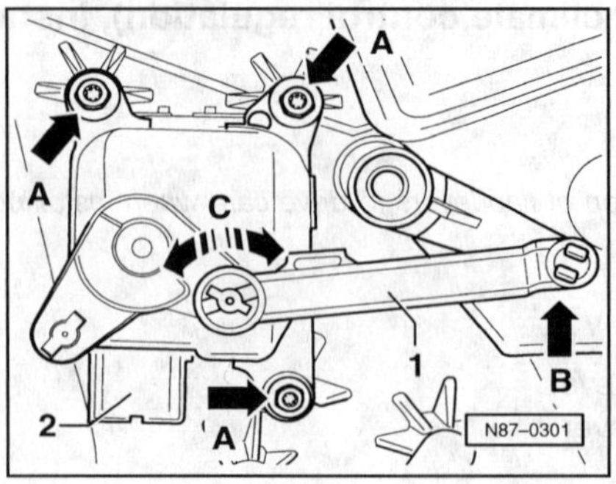

< Remove bolts (TORX T15) -arrows A-.

- Carefully unclip operating rod -1- from lever -arrow B- using a screwdriver.
- Turn operating rod -1- -arrow C- and remove motor -2-.

– Disconnect electrical connector.

– Remove flap motor.

Temperature regulator flap motor -V68-, installing and adjusting

– Initiate Basic Setting, function 04, with VAG 1551 Scan Tool (ST), see **Scan Tool section.**

Central air flap motor -V70-, removing and installing

The temperature flap lever and operating rod are color coded "Blue".

Removing

– Remove instrument panel, see **Body-Interior section, Repair Group 70**

– Loosen instrument panel cross member, see **Climatronic components in passenger compartment, assembly**.

- For the next steps, an assistant must lift and hold the heating and A/Cunit.

– Remove footwell air outlet, see **Climatronic components in passenger compartment, assembly**.

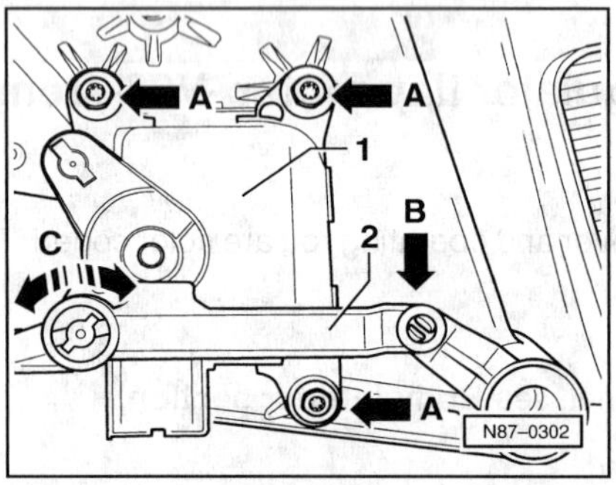

< Remove bolts (TORX T15) -arrow A-.

- Carefully unclip operating rod -2- from lever -arrow B- using a screwdriver.
- Turn operating rod -2- -arrow C- and remove motor -1-.

– Disconnect electrical connector.

– Remove motor.

Installing and adjusting

– Initiate Basic Setting, function 04, with VAG 1551 Scan Tool (ST), see **Scan Tool section**.

Footwell/defroster flap motor -V85-, removing and installing

Removing

– Remove Instrument panel, see **Body-Interior section, Repair Group 70.**

– Loosen Instrument panel cross member, see **Climatronic components in passenger compartment, assembly**.

- For the next steps, an assistant must lift and hold the heating and A/Cunit.

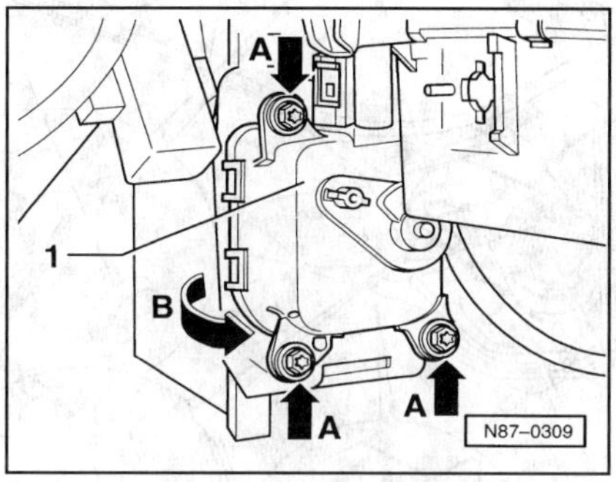

– Remove bolts (TORX T15) -arrows A-.

- Remove motor -3- from housing.

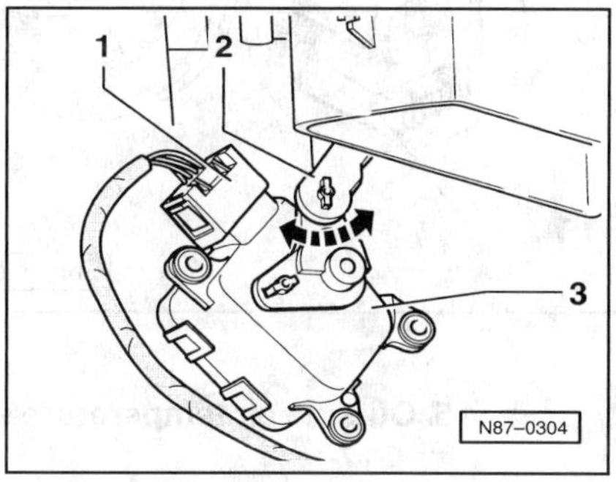

– Disconnect electrical connector -1-.

- Turn motor -3- and remove operating rod -2-.

Footwell/defroster flap motor -V85-,installing and adjusting

– Initiate Basic Setting, function 04, with VAG 1551 Scan Tool (ST), see **Scan Tool section**.

Climatronic components in engine compartment, assembly

CAUTION—

Before beginning repairs on the electrical system:

- *Obtain the anti-theft radio security code.*
- *Switch the ignition off.*
- *Disconnect the battery Ground (GND) strap.*
- *After reconnecting battery, re-code and check operation of anti-theft radio. Also check operation of clock and power windows according to Owner's Manual.*

NOTE—

- *System components identified with an * must only be serviced or replaced by a qualified technician after discharging refrigerant system. Use Kent Moore ACR4 or equivalent refrigerant recycling equipment.*
- *Illustration shows vehicle with engine code AHA.*

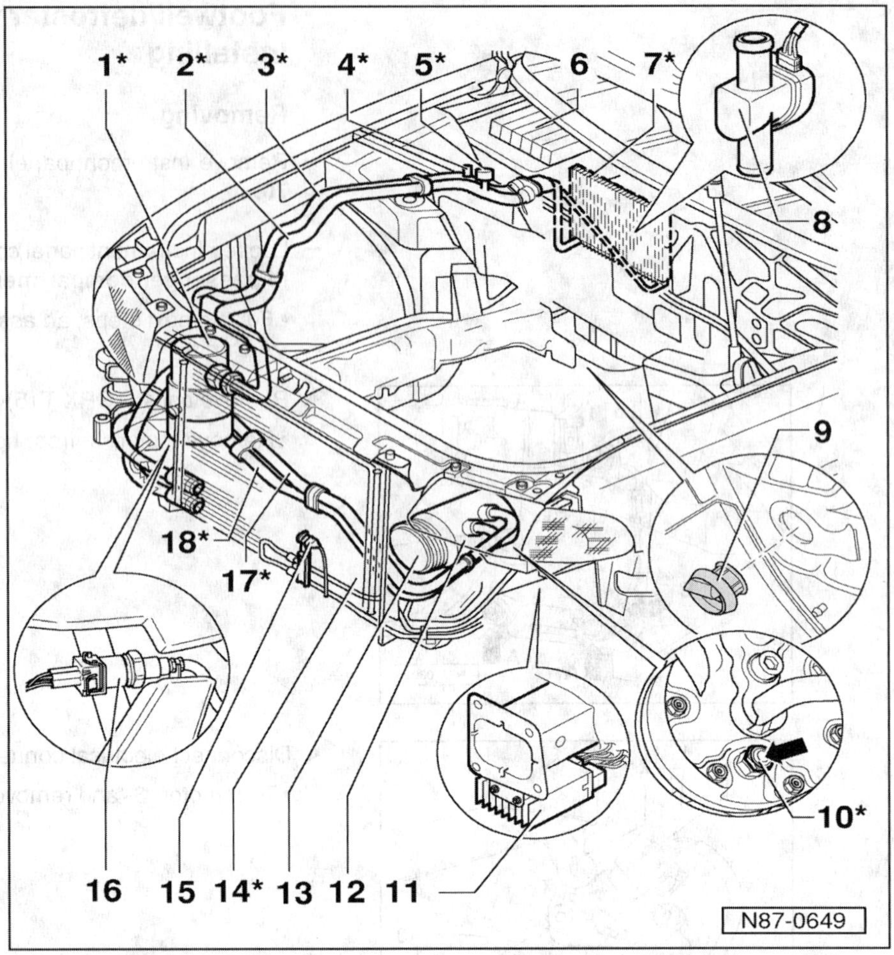

1. **Accumulator***
2. **Refrigerant hose***
   - Between evaporator and accumulator
3. **Refrigerant hose***
   - Between evaporator and condenser
4. **Low pressure service valve***
5. **High pressure service valve***
6. **Dust and pollen filter**
   - See **Heating and A/C components in engine compartment, assembly**
7. **Evaporator***
   - In heating and A/C unit
   - Removing/assembly, see **Heating and A/C unit, removing and assembly**.
8. **Engine Coolant (EC) Switch-off Valve (heater) -N279**
   - 8-cylinder engine only
9. **Evaporator water drain valve**
   - Checking, see **Heating and A/C components in engine compartment, assembly**
10. **Pressure relief valve***
    - Checking, Zexel compressor, see **A/C refrigerant system, assembly (through 09/2000)**
    - Checking, Denso compressor, see **A/C refrigerant system, assembly (from 10/2000)**
11. **Coolant FC (Fan Control) Control Module -J293-**
    - 6-cylinder engine, code ATQ
    - 8-cylinder engine, code BDP
12. **Compressor***
    - Externally regulated compressor without A/C clutch -N25- on 8-cylinder engine
13. **A/C clutch -N25-**
    - Servicing: Zexel, see **A/C clutch -N25-, servicing (Zexel)**
    - Servicing: Denso, see **A/C clutch -N25-, servicing (Denso)**
    - Not for 8-cylinder engine
14. **Condenser***
15. **Outside air temperature sensor -G17-**
    - Function: Temperature sensor controls temperature flap and fresh air blower speed dependant of temperature measurement.
    - In event of sensor malfunction, fresh air intake duct temperature sensor -G89- measurement value is used instead. If both sensors malfunction, operation continues with default value of 10°C (50°F). Air recirculation not possible. A/C control head -E87- display indicates: "--"

(continued on following page)

Climatronic components in engine compartment, assembly (continued)

16. A/C pressure switch -F129- or high pressure switch -G65-

NOTE —

- *Application is engine dependent.*
- *Always confirm system use of -F129- or -G65- and applicable wiring circuit using applicable "Engine" and "Climatronic" wiring diagrams.*

- -F129- function, see **A/C refrigerant system, component overview**
- -F129- checking, see **A/C refrigerant system, assembly (through 09/2000), Fig. 3**
- -F129- through 09/2000, replacing, see **A/C refrigerant system, assembly (through 09/2000)**
- -F129- from 10/2000, replacing, see **A/C refrigerant system, assembly (from 10/2000)**
- -G65- function, see **A/C refrigerant system, component overview**
- -G65-, replacing, see **A/C refrigerant system, assembly (from 10/2000)**
- Switch or sensor can be removed without discharging refrigerant system.
- Tightening torque 8 Nm (71 in lb)
- Always replace O-ring (note Part No.).

17. Refrigerant hose*

- Between compressor and evaporator.

18. Refrigerant hose*

- Between compressor and accumulator.

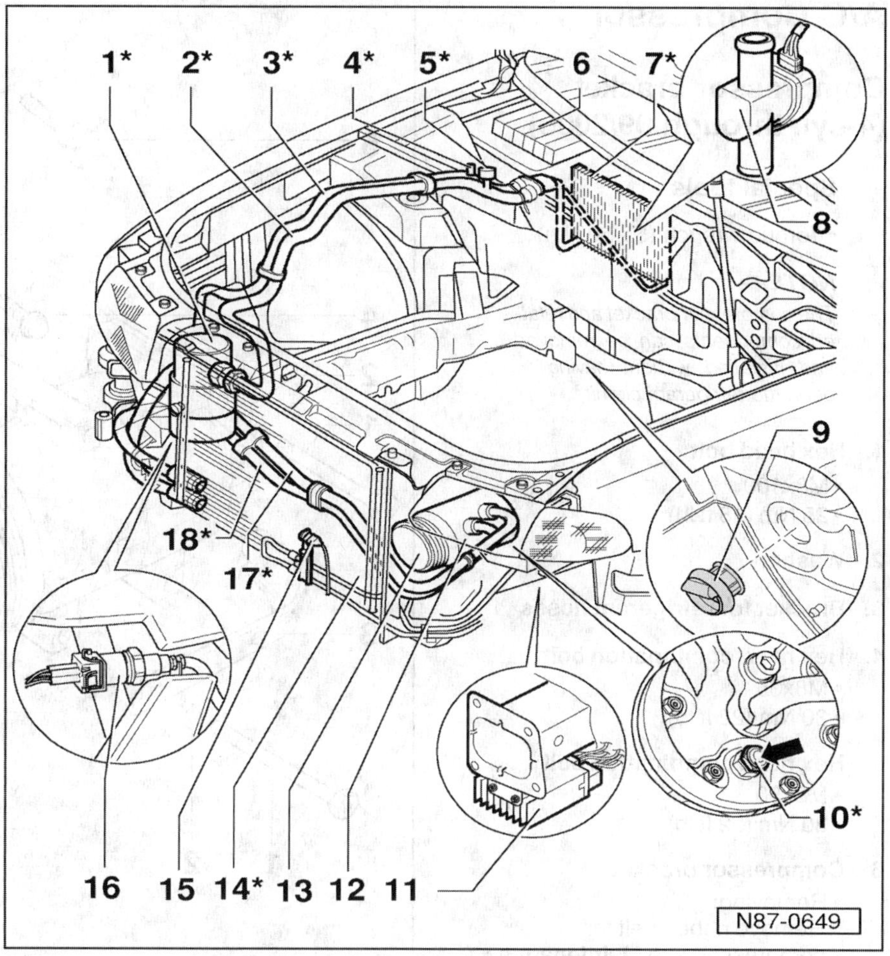

A/C Compressor

Compressor bracket (4-cyl. through 09/2000)

Special tools

- Torque wrench (5 - 50 Nm)

NOTE—
The compressor bracket and related components can be removed and installed without having to open the refrigerant circuit

1. **Hex head bolt**
   - M8x100
   - 25 Nm (18 ft lb)
2. **Washer**
3. **Bracket for refrigerant hoses**
4. **Hex head combination bolt**
   - M8x35
   - 30 Nm (22 ft lb)
5. **Hex head combination bolt**
   - M8x50
   - 30 Nm (22 ft lb)
6. **Compressor bracket**
   - Removing:
     Remove ribbed belt for compressor, see **Maintenance**
     Remove bolts -1-.
     Remove compressor from bracket, swing away (leave refrigerant hoses connected) and secure in engine compartment with wire, see **Fig. 1**.
     Remove bolts -4- and -5-.
     Remove compressor bracket from engine.

CAUTION—
- *Do not allow compressor to hang from refrigerant hoses unsupported.*
- *Note different length bolts and their installation locations.*

7. **Tensioning roller**
   - Removing and installing ribbed belt, see **Maintenance**
8. **Hex socket bolt**
   - M8 X 28
   - 23 Nm (17 ft lb)

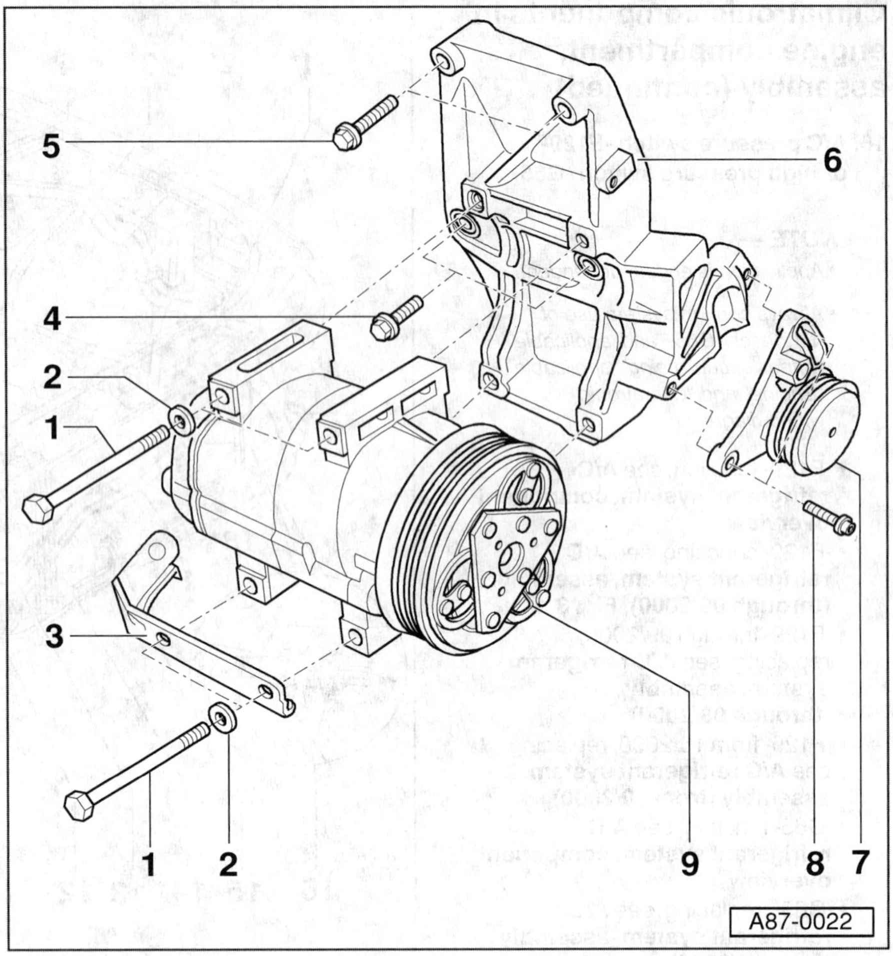

9. **Compressor**
   - Manufacturer through 09/2000: Zexel
   - With A/C clutch -N25-
   - Servicing A/C clutch -N25- (Zexel), see **A/C clutch -N25-, servicing (Zexel)**.

Compressor bracket, (4-cyl. from 10/2000)

Special tools required

- Torque wrench (5 - 50 Nm)

NOTE—
The compressor bracket and related components can be removed and installed without having to open the refrigerant circuit

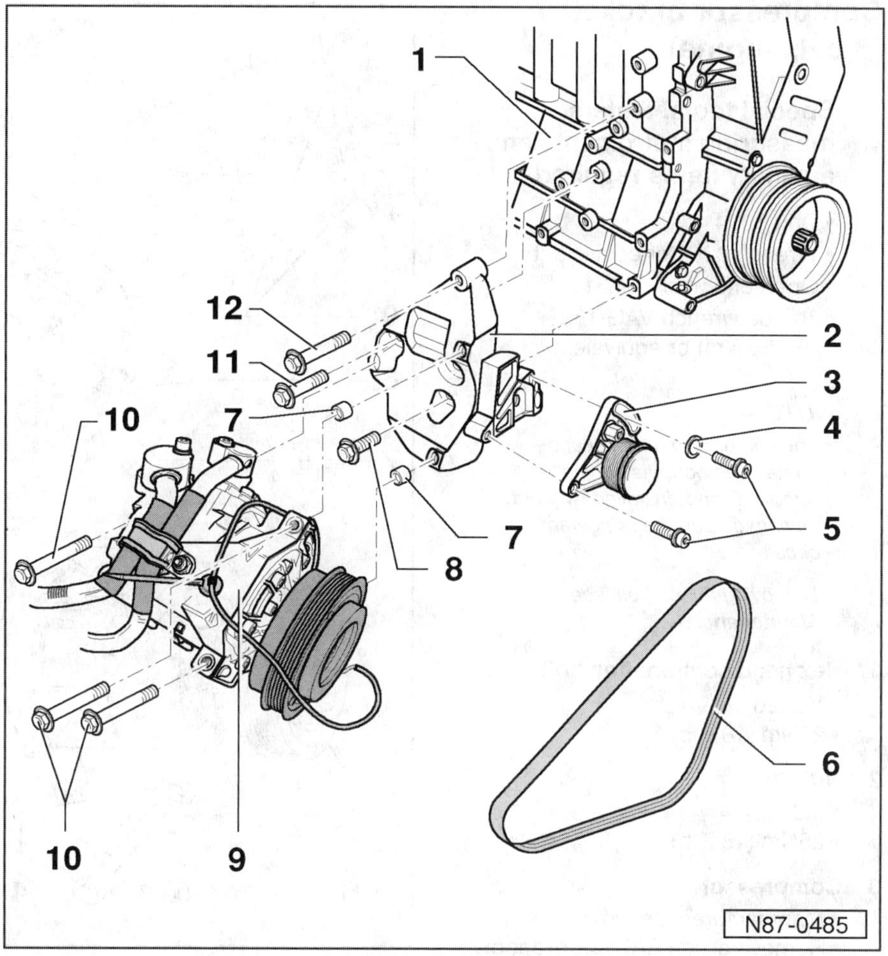

1. **Cylinder block**
2. **Compressor bracket**
   - Cast aluminium
   - Removing:
     Remove ribbed belt for compressor, see **Maintenance**
     Remove bolts -10-
     Remove compressor from bracket, swing away (leave refrigerant hoses connected) and secure in engine compartment with wire, see **Fig. 1**.
     Remove compressor bracket from engine.

CAUTION—
- *Do not allow compressor to hang from refrigerant hoses unsupported.*
- *Note different length bolts and their installation locations.*

3. **Tensioning roller**
   - Removing and installing ribbed belt, see **Maintenance**
4. **Washer**
5. **Hex socket bolt**
   - 23 Nm (17 ft. lb.)
6. **Ribbed belt**
   - Removing and installing, see **Maintenance**
7. **Spacer sleeve**
   - Qty. 2
   - Ensure proper fit in bracket
8. **Hex head combination bolt**
   - M8x25
   - 30 Nm (22 ft. lb.)
9. **Compressor**
   - Manufacturer from 10/2000: Denso
   - With A/C clutch -N25-
   - See **A/C clutch -N25-, servicing (Denso)**.
10. **Hex head combination bolt**
    - M8x100
    - 25 Nm (17 ft. lb.)
    - Qty. 3
11. **Hex head combination bolt**
    - M8x35
    - 35 Nm (26 ft. lb.)
12. **Hex head combination bolt**
    - M8x52
    - 35 Nm (26 ft. lb.)

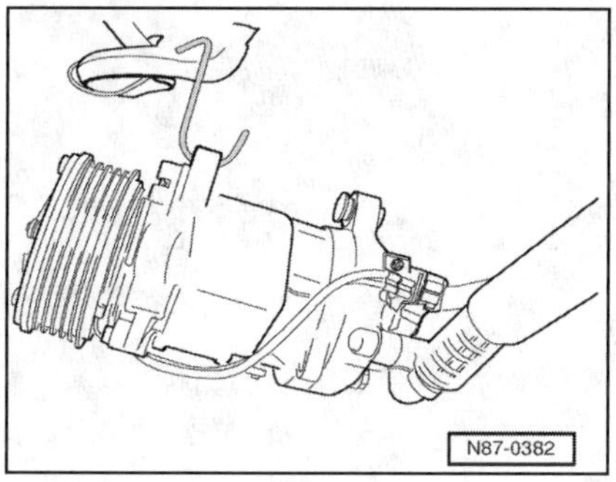

Fig. 1 Compressor, securing with wire
- When removing the compressor from it's bracket without opening the refrigerant system, secure the compressor to the lower frame member with a suitable wire (e.g.: welding rod or similar) as illustrated.

CAUTION—
Do not allow compressor to hang from refrigerant hoses unsupported.

Compressor bracket (6-cyl. engine)

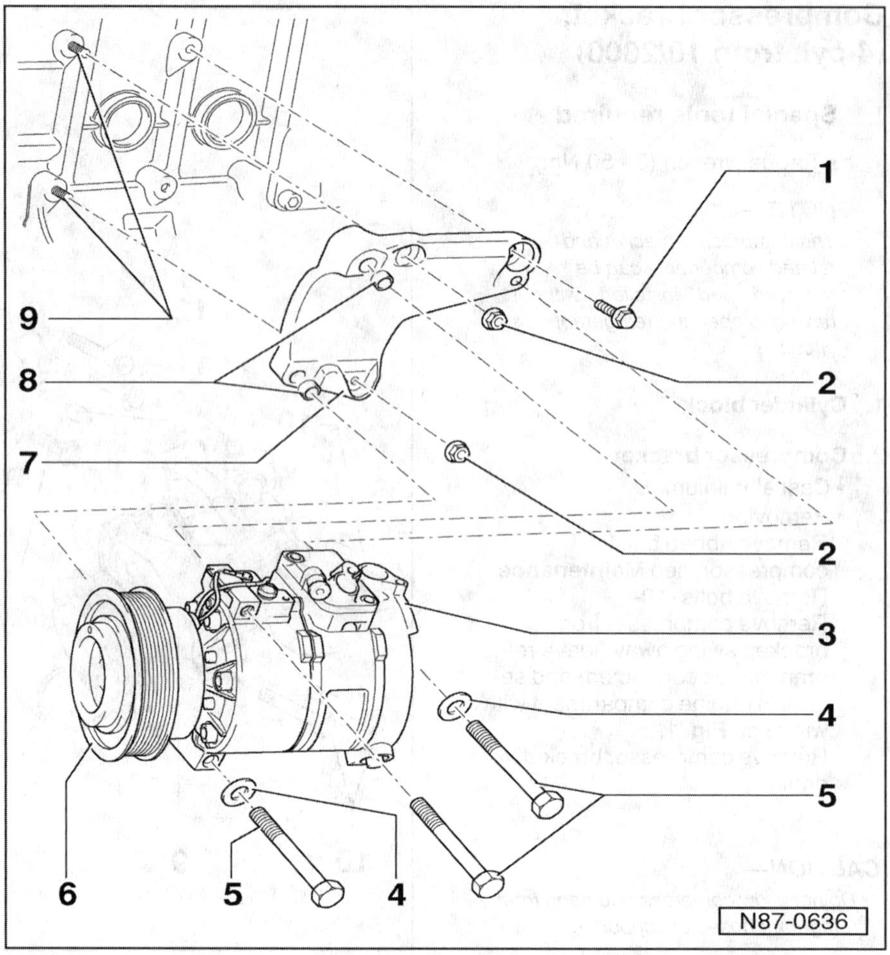

Special tools, testers, measuring instruments and auxiliary items required

- Drift 3204
- Wrench 3212 or equivalent
- Open end wrench 3312
- Torque wrench VAG 1331 (5 - 50 Nm) or equivalent

NOTE —

- *The compressor bracket and related components can be removed and installed without having to open the refrigerant circuit*
- *Removing ribbed belt, see **Maintenance**.*

1. **Hex head combination bolt**
   - M8x26
   - 25Nm (18 ft lb)
2. **Nuts**
   - M8
   - 25Nm (18 ft lb)
3. **Compressor**
   - Manufacturer: Denso
   - Removing, see **A/C compressor, removing and installing (6-cyl. engine)**
   - With A/C clutch -N25-
   - Servicing -N25-, see **A/C clutch -N25-, servicing (Denso)**
4. **Washer**
5. **Hex head bolt**
   - M8x88
   - 25Nm (18 ft lb)
6. **A/C clutch -N25-**
   - Servicing, see **A/C clutch -N25-, servicing (Denso)**
7. **Compressor bracket**
   - Remove compressor first, see **A/C compressor, removing and installing (6-cyl. engine)**
8. **Spacer sleeve**
   - Qty.: 2
   - Ensure proper fit in bracket
9. **Stud**
   - M8X20

A/C compressor, removing and installing (6-cyl. engine)

- Torque wrench (5 - 50 Nm)

– Move front end assembly (lock carrier) into service position, see **Body-Exterior section, Repair Group 50.**

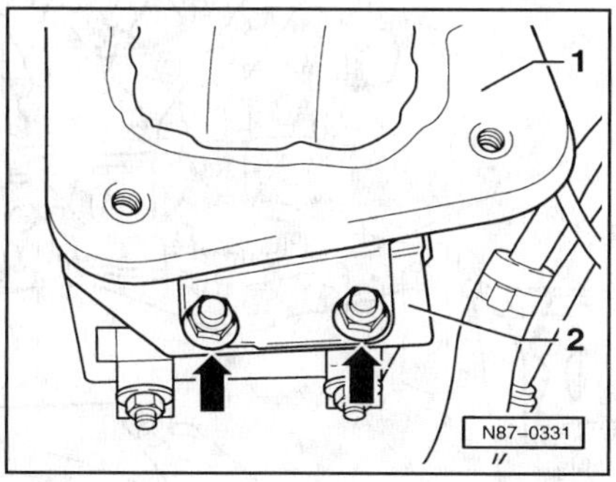

– Remove series resistance for radiator fan -2- from left longitudinal member -1-.

- Nut tightening torque: 8 Nm (71 in lb)

– Remove oil filter.

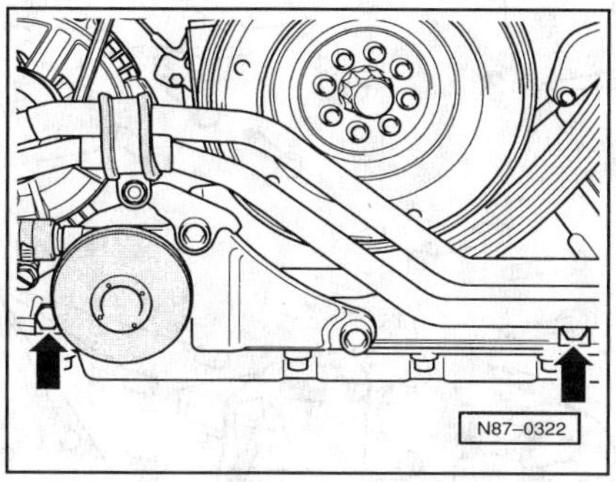

– Loosen coolant pipe securing bolts -arrows- on oil pan and remove coolant hoses from oil cooler.

- Bolt tightening torque: 10 Nm (69 in lb).

> ***WARNING —***
> *The cooling system is pressurized when the engine is warm. Wear gloves and other appropriate protection, slowly and carefully release system pressure (if necessary) before performing repairs.*

– Remove oil cooler, see **2.8 Liter V6 5V Engine Mechanical (code: AHA, ATQ) section, Repair Group 17.**

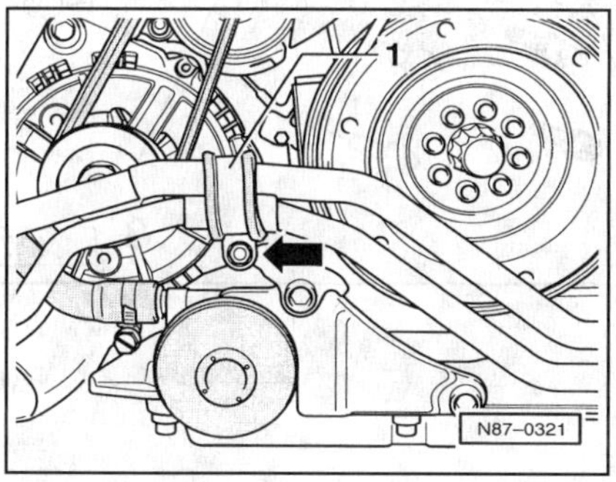

– Loosen nut -arrow- on bracket -1- for refrigerant hoses.

- Nut tightening torque: 8 Nm (71 in lb).

– Remove bolts for compressor, see **Compressor, (6-cyl. engine)**, item **-5-**.

– Remove compressor from bracket, swing away downward (leave refrigerant hoses connected) and secure with wire see below.

– When reinstalling compressor, note routing of ribbed belt, see **Fig. 2**.

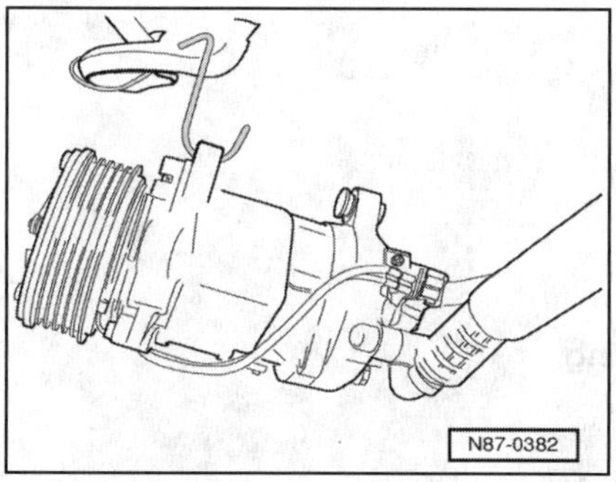

Fig. 1 Compressor, securing with wire

- When removing the compressor from it's bracket without opening the refrigerant system, secure the compressor to the lower frame member with a suitable wire (e.g.: welding rod or similar) as illustrated.

> ***WARNING —***
> *Do not allow compressor to hang from refrigerant hoses unsupported.*

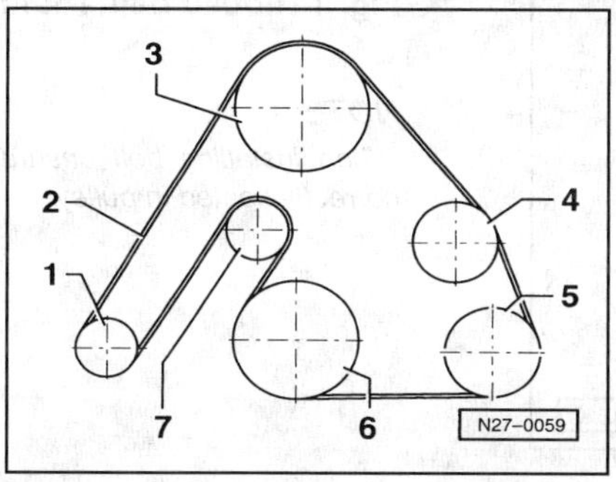

Fig. 2 Ribbed belt, routing

1 - Generator (GEN)
2 - Ribbed belt
3 - Power steering pump
4 - Viscous fan
5 - A/C compressor
6 - Crankshaft pulley
7 - Tensioner pulley

NOTE —
When installing belt ensure it is correctly seated in pulley.

Compressor bracket (8-cyl. engine)

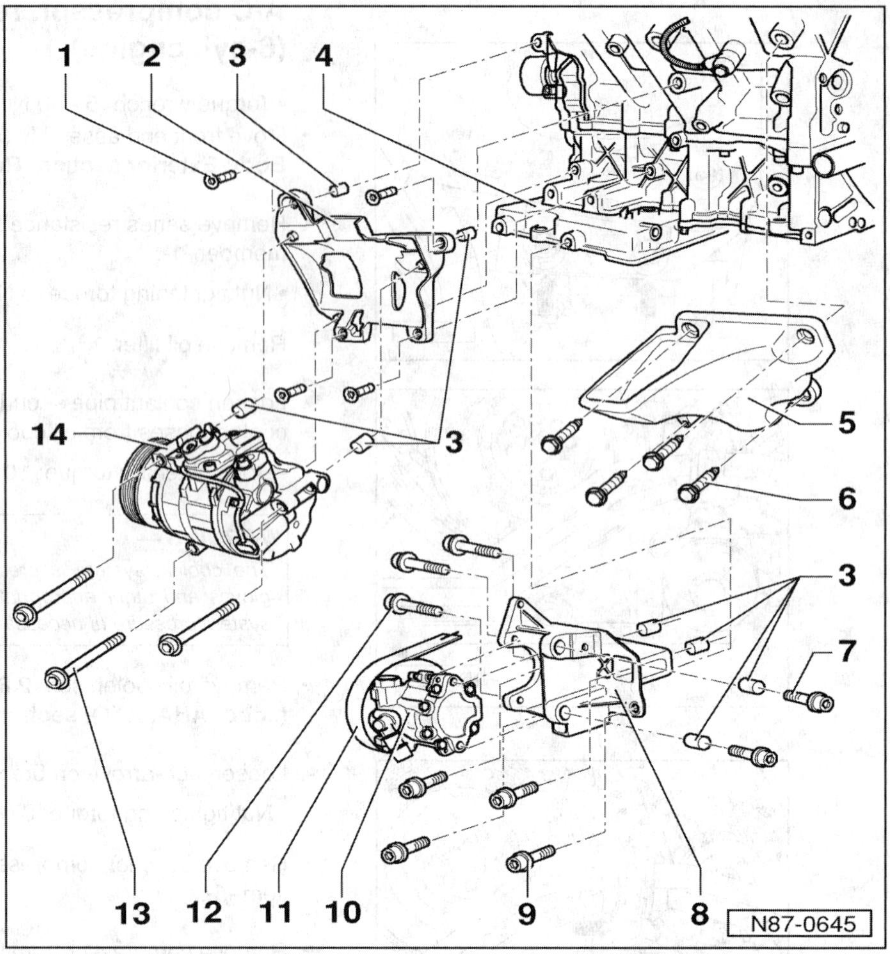

Special tools, testers, measuring instruments and auxiliary items required

- Torque wrench (5 - 50 Nm)
- Kent-Moore ACR4 (or equivalent)
- VAS 5532 Socket set

NOTE—

*System components identified with an * must only be serviced or replaced by a qualified technician after discharging refrigerant system. Use Kent Moore ACR4 or equivalent refrigerant recycling equipment.*

- *The cylinder block on 8-cylinder engines is made from aluminium.*
- *Removing ribbed belt, see* ***Maintenance***

1. **Screw**
   - M8x25
   - 23 Nm (17 ft. lb.)
   - Qty. 4
2. **Compressor bracket**
   - To remove, first remove A/C compressor, see **A/C compressor, removing and installing (8-cyl. engine)**
3. **Spacer sleeves**
   - Ensure proper fit in bracket
4. **Oil pan**
5. **Engine mount support**
6. **Hex head bolt**
   - M10x40
   - 50 Nm (37 ft. lb.)
   - Qty. 4
7. **Hex socket bolt**
   - M8x48
   - 23 Nm (17 ft. lb.)
   - Qty. 2
8. **Power steering pump bracket**
9. **Hex socket bolt**
   - M10x30
   - 32 Nm (24 ft. lb.)
   - Qty. 4
10. **Power steering pump**
11. **Ribbed belt**
    - Routing, see **Fig. 1**.
    - Removing, see **Maintenance**
12. **Hex socket bolt**
    - M8x34
    - 23 Nm (17 ft. lb.)
    - Qty. 3
13. **Hex socket bolt**
    - M8x95
    - 23 Nm (17 ft. lb.)
    - Qty. 3
14. **Compressor**
    - Manufacturer: Denso
    - Without A/C clutch
    - See **A/C compressor, removing and installing (8-cyl. engine)**

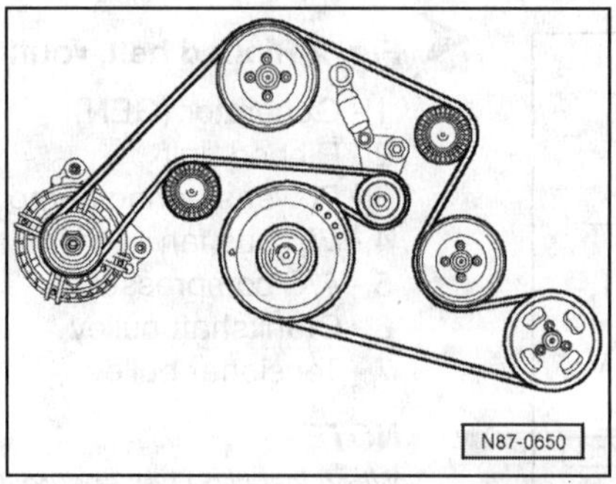

Fig. 1 Ribbed belt, routing

NOTE—

When installing belt ensure it is correctly seated in pulley.

A/C compressor, removing and installing (8-cyl. engine)

General information

8-cylinder models are equipped with Denso compressor 7SEU16. This compressor is driven continuously when the engine is running (no A/C clutch present).

Compressor output is regulated by the Climatronic control module -J255- based on input signals from high pressure sensor -G65- and control of signal to A/C compressor regulator valve -N280-.

Compressor regulator valve -N280- is not activated with empty refrigerant circuit (compressor idles with engine).

An internal oil circuit in the compressor ensures the compressor is not damaged when the refrigerant circuit is empty. This feature depends on a residual amout of refigerant oil being present in the circuit.

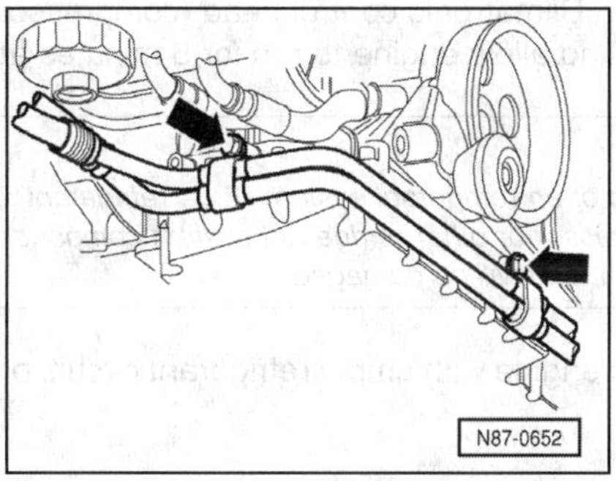

– Discharge refrigerant system.

– Remove ribbed belt, see **Maintenance.**

– Remove refrigerant line retainers from oil pan.

– Remove windshield washer reservoir.

– Remove power steering pump from bracket and lay to side; hoses remain connected, see **Suspension, Wheels, Steering section, Repair Group 48**

NOTE —
*Note installation position of two upper spacer sleeves when removing power steering pump bracket, see **Compressor bracket, removing and installing (8-cyl. engine)**; item -3-.*

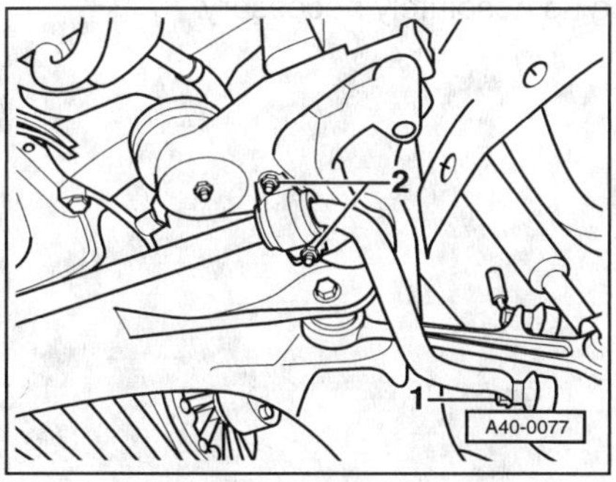

– Remove power steering pump bracket.

– Remove both front stabilizer bar brackets -2- and pull stabilizer bar downwards, see **Suspension, Wheels, Steering section, Repair Group 40**

NOTE —
Before removing engine mount, support subframe from below.

– Remove left-engine mount carrier from subframe, see **Suspension, Wheels, Steering section, Repair Group 40**

– Remove left hydraulic engine mount from carrier and engine mount support.

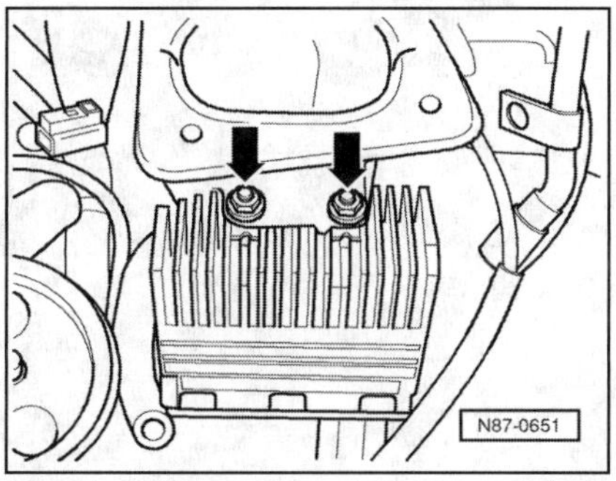

– Remove both fasteners -arrows- from cooling fan control module and remove.

• Tightening torque: 8 Nm (71 in. lb.)

– Disconnect electrical connectors and wiring as necessary.

– Remove securing bolts from compressor (leave refrigerant lines on bracket).

– Remove refrigerant lines from compressor and pull compressor out downwards.

– Installation is reverse order of removal, noting the following:
  - Always use new seals at compressor refrigerant lines.
  - Compressor refrigerant line hex socket bolts M8 - tightening torque: 25 Nm (18 ft. lb.)
  - Ensure proper fit of compressor bracket spacer sleeves, see **Compressor bracket (8-cyl. engine)**; item **-3-**.
  - Turn compressor 10 revolutions (clockwise) by hand.
  - Install ribbed belt, see **Maintenance.**
  - Install hydraulic engine mount to mount carrier and engine mount support. Tightening torque: 23 Nm (17 ft. lb.)
  - Install mount carrier to subframe, see **Suspension, Wheels, Steering section, Repair Group 40.**
  - Switch on ignition (do not start engine!)
  - Select "Econ" mode on Climatronic control head
  - Start engine and wait until idle speed stabilizes.
  - Open all instrument panel vents.
  - Select "Low" temperature setting on Climatronic control head.
  - Select "Auto" mode on Climatronic control head (compressor regulation activated) and allow engine to run for 5 minutes at idle.

WARNING —
Do not start engine with open refrigerant system (E.g.: refrigerant lines not connected to compressor or other refrigerant system components). Compressor will overheat and will be damaged.

If it is necessary to start engine with empty refrigerant circuit, proceed as follows:

– Refrigerant circuit must be fully assembled/closed.

– At least 1/4 of the prescribed refrigerant oil must be in the compresssor.

– Engine speed must not exceed 2500 rpm.

– Run engine only as long as absolutely necessary.

A/C refrigerant system

A/C refrigerant system, identification

NOTE —

- *Before carrying out any work on the A/C refrigerant system, refer to A/C system safety measures, see* **Safety measures**.

A/C refrigerant systems on Passat models are charged with refrigerant R-134a.

Labels specifying refrigerant type are located on the compressor or radiator support. Before proceeding with refrigerant system servicing or repairs, always confirm refrigerant type used.

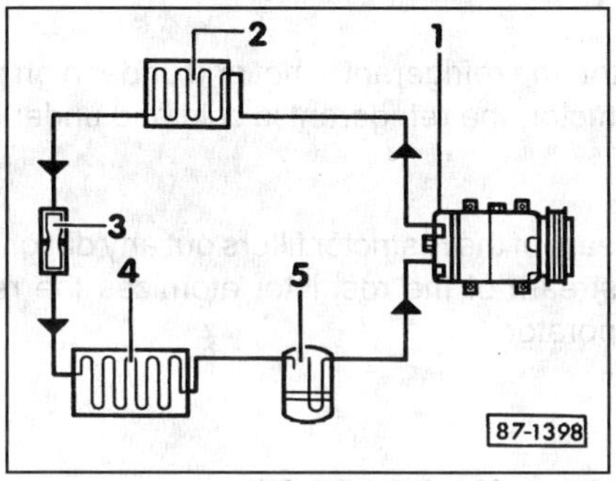

A/C refrigerant circuit with restrictor and accumulator

1 - Compressor
2 - Condenser
3 - Restrictor
4 - Evaporator
5 - Accumulator

NOTE —

Arrows indicate direction of refrigerant flow.

A/C refrigerant system, component overview

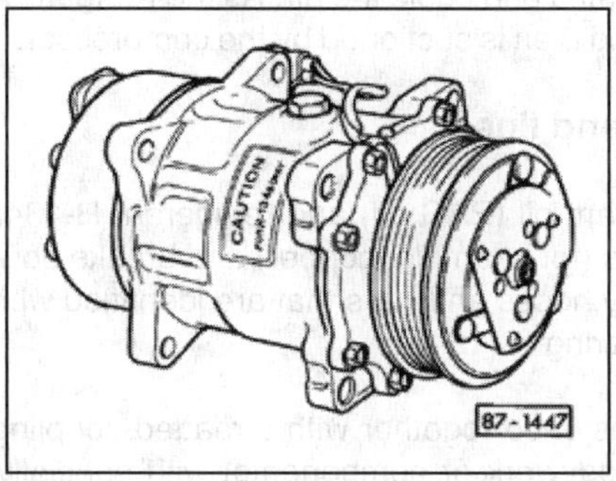

Compressor

On 4-cylinder and 6-cylinder models the compressor is driven via a belt on the engine when power is applied to the the A/C clutch -N25- (A/C "ON").

On 8-cylinder models with externally regulated Denso compressor 7SEU16, the compressor is driven continuously (no A/C clutch). Compressor output is regulated by the Climatronic control module -J255- based on input signals from the high pressure sensor -G65- and evaporator vent temperature sensor -G263-. -J255- provides an output signal to A/C compressor regulator valve -N280- in accordance with the interior temperature setting and refrigerant system pressures/engine load.

Low-pressure refrigerant gas from the evaporator is compressed by the compressor. After compression, the refrigerant gas (now high-pressure) flows to the condenser.

NOTE —

- *From start of production, several compressor and A/C clutch types (manufacturers) with specific servicing requirements are installed see from page, see* ***A/C refrigerant system, servicing.***
- *The compressor contains refrigerant oil that is mixable under all temperatures with the refrigerant.*
- *A label on the compressor indicates that compressor is for R-134a systems only.*

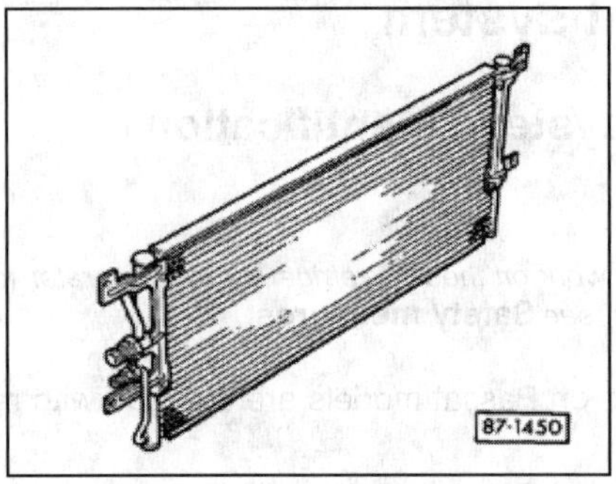

Condenser

The condenser transfers heat from the compressed refrigerant gas to the outside air which causes the refrigerant to change state from a gas to a liquid.

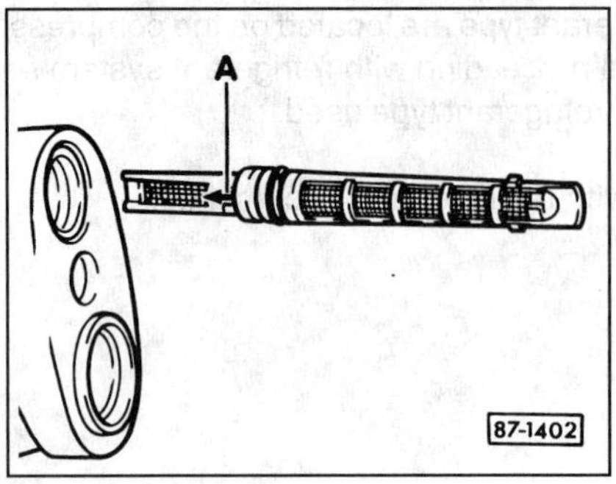

Restrictor

The restrictor restricts and controls refrigerant flow to the evaporator thus lowering refrigerant temperature and pressure.

Upstream of the restrictor, the refrigerant is hot and under high pressure. Downstream of the restrictor, the refrigerant is cold and under low pressure.

A strainer located upstream of the restrictor filters out any dirt or particles. Another strainer downstream of the restrictor atomizes the refrigerant before it enters the evaporator.

NOTE —
Insert restrictor so that arrow -A- points to evaporator.

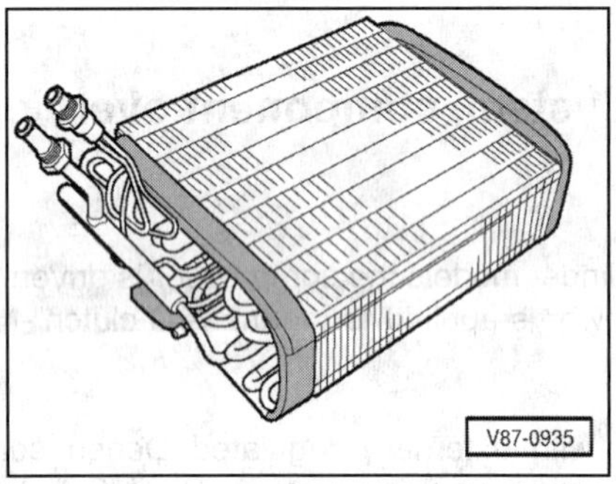

Evaporator

Liquid refrigerant entering the evaporator absorbs heat from air passing through the evaporator fins and cools the air. As the refrigerant absorbs heat it turns to vapor and then is suctioned by the compressor.

A/C system hoses and lines

The mixture of refrigerant oil (PAG oil) and refrigerant R-134a attacks some metals and alloys (for example, copper) and breaks down certain hose material. Use only hoses and lines that are identified with a green mark (stripe) or the lettering "R-134a".

Hoses and lines are fastened together with threaded couplings/fittings and are retained (to bodywork or components) with specially isolated hose clamps.

NOTE —
- *During servicing, all couplings, fittings and related fasteners must be torqued to specifications, see* **A/C refrigerant system, servicing**.
- *Ensure that only special tools (as specified) are used while servicing.*

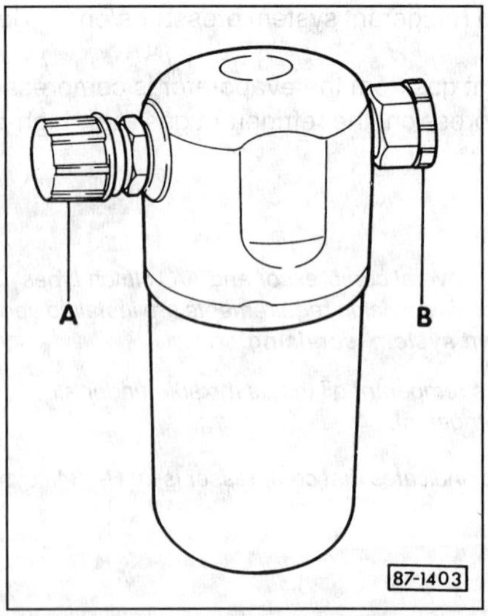

Accumulator

The accumulator traps the mixture of refrigerant vapor (liquid) and gas as coming from the evaporator. The accumulator acts as a refrigerant reservoir for the system and ensures that the compressor only suctions gaseous refrigerant. Any moisture in the refrigerant system is absorbed by drier desiccant in the accumulator.

NOTE—
- *Replace the accumulator each time the refrigerant circuit is opened.*
- *Do not remove sealing caps -A- and -B- until just prior to installation. The drier dessicant in non sealed accumulators becomes saturated with moisture and therefor unusable.*

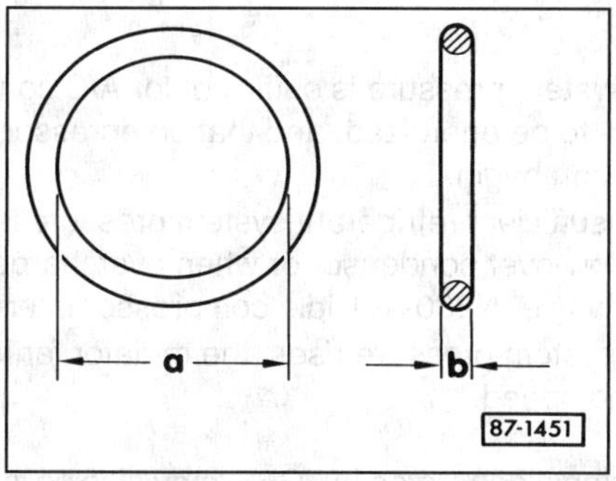

◀ O-rings

O-rings seal connections between A/C system components.

NOTE—
- *Always use correct size O-rings (dimensions -a- and -b-).*
- *Do not reuse O-rings, always replace. Use only new O-rings that are compatible with R-134a refrigerant and refrigerant (PAG) oil on R-134a systems.*
- *Lubricate O-rings with the appropriate refrigerant oil before installing (only use PAG oil).*

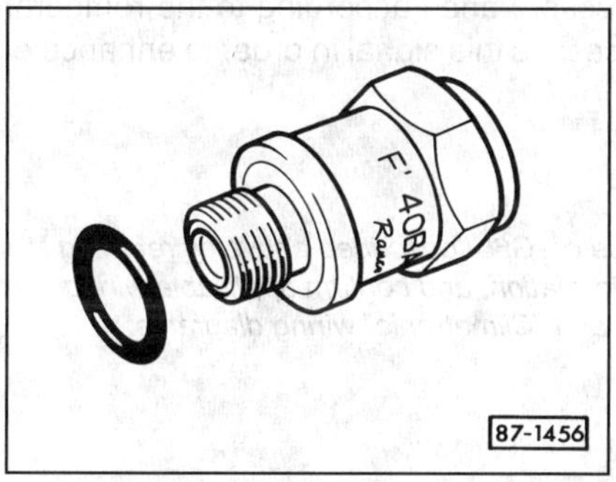

◀ Pressure relief valve

The pressure relief valve is mounted on the compressor or compressor inlet/outlet manifold. At pressures above 40 bar (580 psi), the pressure relief valve opens to vent excessive pressure. When the system pressure is reduced, the valve closes to prevent total refrigerant loss.

On some models, a cap on the pressure relief valve will pop out if the valve has opened.

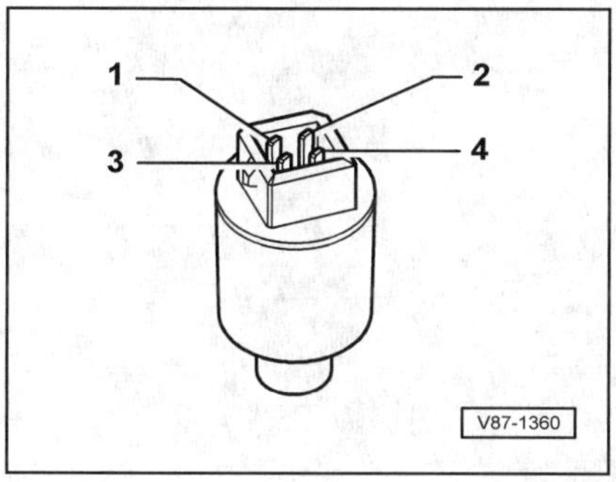

◀ A/C pressure switch -F129-

Where applicable, the A/C pressure switch is a combination switch (two switching components) that protects refrigerant circuit components from malfunctioning due to excessive or insufficient refrigerant system pressures.

- Switch part between terminals 1 and 2 switches off the A/C clutch -N25- when the system pressure is too high or when there is not enough refrigerant in the system .
- Switch part between terminals 3 and 4 switches coolant fan -V7- to the next higher fan speed when system pressures increase.

NOTE—
Always confirm system use of -F219- described above, and applicable wiring circuit using applicable "Engine" and "Climatronic" wiring diagrams.

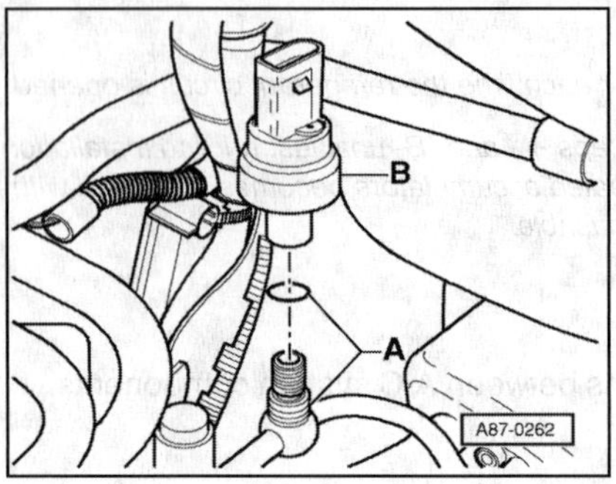

High pressure sensor -G65-

Where applicable, high pressure sensor -G65- -B- is used instead of A/C pressure switch -F129-. -G65- transmits a square wave signal to the Climatronic control module -J255- at a rate which varies according to the refrigerant system pressure. -J255- processes the signal/pulse rate from -G65- as follows:

- Confirms refrigerant system pressure is sufficient for A/C compressor regulator valve -N280- to be energized, and that compressor output regulation via -N280- can begin.
- Where excessive or insufficient refrigerant system pressure is present (E.g.: insufficient air flow over condensor or when overcharged), A/C compressor regulator valve -N280- will idle compressor internally.
- When the refrigerant system pressure rises, the radiator fan(s) are switched to the second speed.

Where applicable, the signal generated by -G65- may also be input to the Motronic Engine Control Module (ECM). As the amount of torque needed to drive the A/C compressor varies according to the refrigerant system pressure, the ECM processes this signal in order to enhance engine performance.

NOTE —

Always confirm system use of -G65- described above by referring to model specific Training (SSP) information, and confirm applicable wiring circuit using applicable "Engine" and "Climatronic" wiring diagrams.

A/C refrigerant system, assembly (through 09/2000)

NOTE —

- *Before carrying out any work on the A/C refrigerant system, refer to all CAUTIONS and WARNINGS.*
- *Service refrigerant system using Kent More ACR 4 or equivalent*
- *Only system components identified with an * can be serviced or replaced without discharging refrigerant system.*
- *During servicing, all couplings, fittings and related fasteners must be torqued to specifications.*

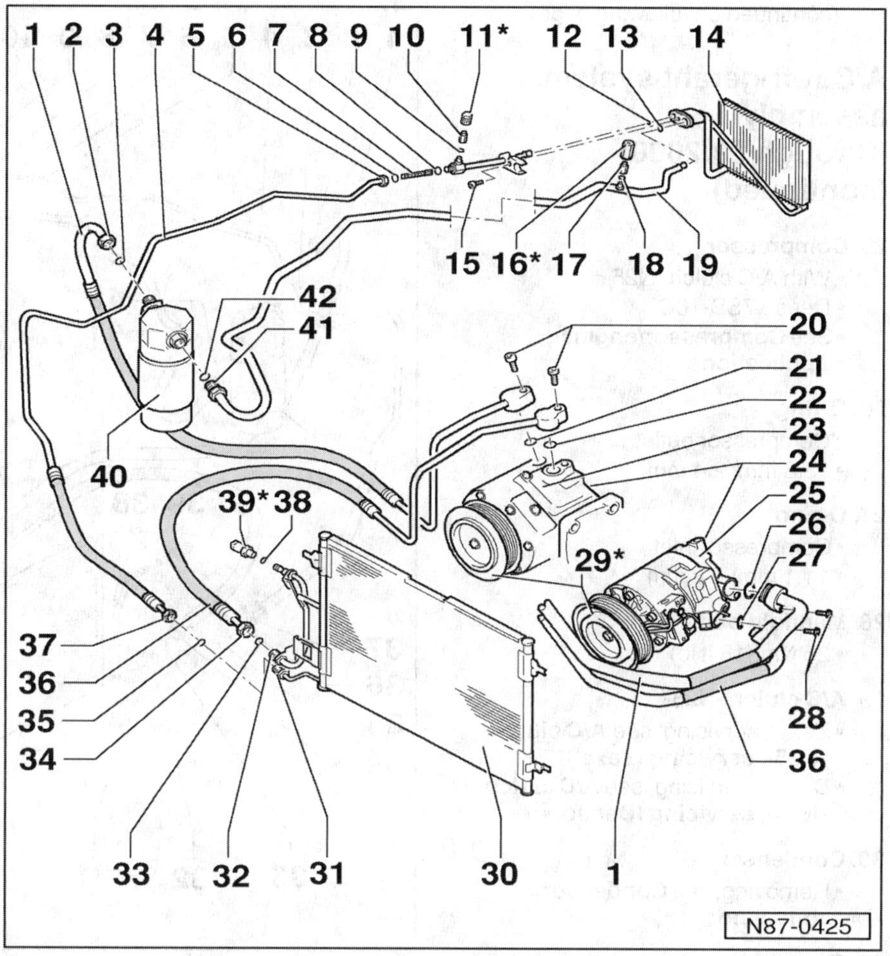

1. **Refrigerant pipe/hose**
   - Between accumulator and compressor
2. **Coupling**
   - Thread diameter ∅ 28 mm; pitch1.5 mm
   - 40 Nm (30 ft lb)
3. **O-ring**
   - 17.2 mm; 1.8 mm
4. **Refrigerant pipe/hose**
   - Between evaporator and condensor
5. **Coupling**
   - Thread diameter ∅ 20 mm; pitch1.5 mm
   - 15 Nm (11 ft lb)
6. **O-ring**
   - 10.8 mm; 1.8 mm
7. **Restrictor**
   - Removing and installing, see **Fig. 6**
8. **O-ring**
   - 7.5 mm; 1.5 mm
9. **O-ring**
   - 10.8 mm; 1.8 mm
10. **High pressure service valve**
    - Only use Kent Moore ACR4 or equivalent.
    - Removing and installing, see **Fig. 1**
11. **Cap***
12. **O-ring**
    - 10.8 mm; 1.8 mm
13. **O-ring**
    - 17.2 mm; 1.8 mm
14. **Evaporator**
    - In heating and A/C unit
15. **Hex socket bolt(s)**
    - At evaporator coupling
    - 10 Nm (7 ft lb)
16. **Cap***
17. **Low pressure service valve**
    - Only use Kent Moore ACR4 or equivalent.
    - Removing, see **Fig. 2**
18. **O-ring**
    - 7.6 mm; 1.8 mm
19. **Refrigerant hose**
    - Between evaporator and accumulator
20. **Manifold bolts**
    - 25 Nm (18 ft lb)
21. **O-ring**
    - Compressor inlet
    - 11.1 mm; 1.8 mm
22. **O-ring**
    - Compressor outlet
    - 23.8 mm; 2.4 mm
23. **Compressor**
    - With A/C clutch -N25-
    - Zexel DCW 17D
    - See **Compressor/engine applications**
24. **Pressure relief valve**

    Checking:
    - Zexel compressor, see **Fig. 6**
    - Denso compressor, see **Fig. 7**

(continued on following page)

A/C refrigerant system, assembly (through 09/2000) (continued)

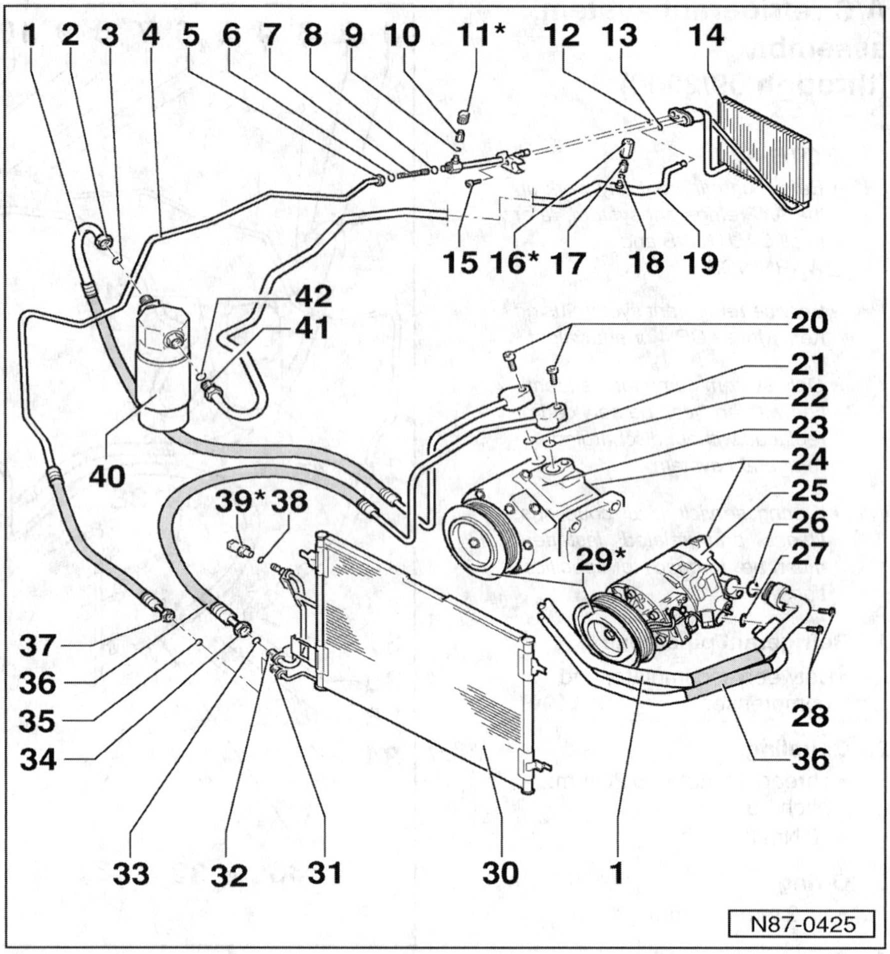

25. Compressor
- With A/C clutch -N25-
- Denso 7SB-16C
- See **Compressor/engine applications**

26. O-ring
- Compressor outlet
- 23.8 mm; 2.4 mm

27. O-ring
- Compressor inlet
- 11.1 mm; 1.8 mm

28. Manifold bolts
- 25 Nm (18 ft lb)

29. A/C clutch -N25-
- Zexel, servicing, see **A/C clutch -N25-, servicing (Zexel)**
- Denso, servicing, see **A/C clutch -N25-, servicing (Denso)**

30. Condenser
- Removing, see **Condensor, removing**

31. Coupling
- Thread diameter ∅ 20 mm; pitch1.5 mm

32. Coupling
- Thread diameter ∅ 24 mm; pitch1.5 mm

33. O-ring
- 14 mm; 1.8 mm

34. Coupling
- Thread diameter ∅ 24 mm; pitch1.5 mm

35. O-ring
- 10.8 mm; 1.8 mm

36. Refrigerant pipe/hose
- Between compressor and condensor

37. Coupling
- Thread diameter ∅ 20 mm; pitch1.5 mm

38. O-ring
- 7.6 mm; 1.8 mm

39. A/C pressure switch -F129-*
- Checking, see **Fig. 3**
- Inspect O-ring for damage and proper seating in recess of switch before installation.
- 8 Nm (71 in lb)

40. Accumulator
- Always replace whenever refrigerant system has been opened

41. Coupling
- Thread diameter ∅ 28 mm; pitch1.5 mm

42. O-ring
- 17.2 mm; 1.8 mm

A/C refrigerant system, assembly (from 10/2000)

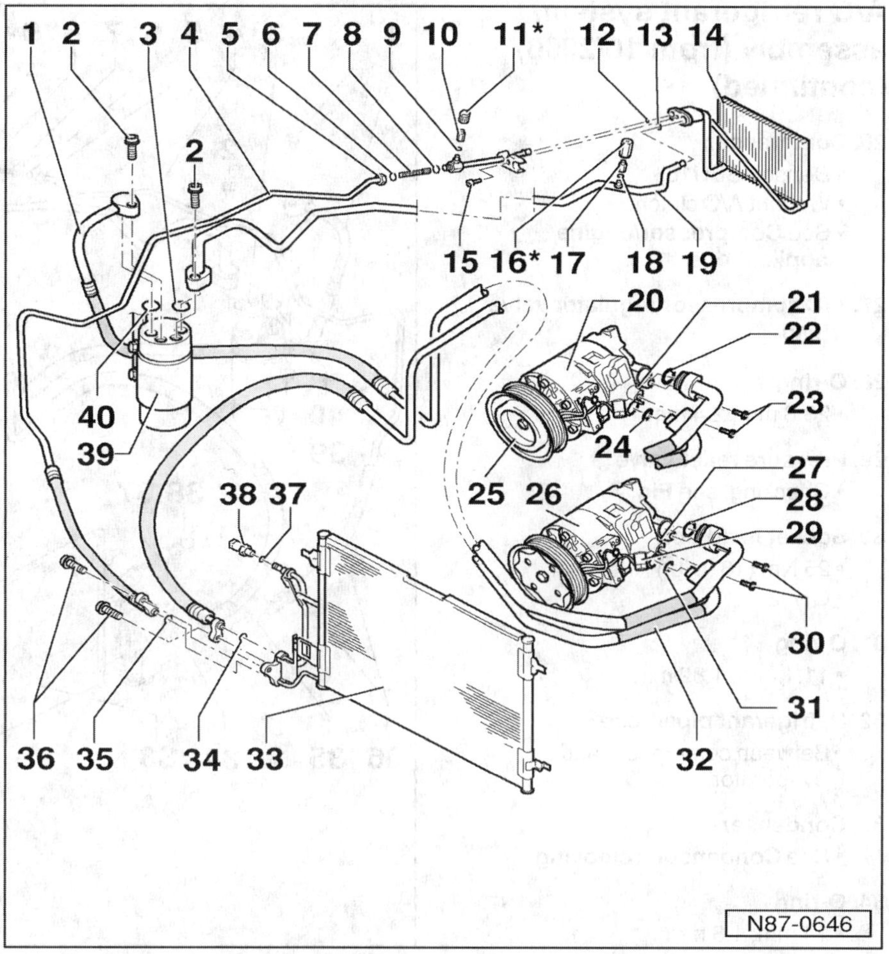

NOTE —

- *Before carrying out any work on the A/C refrigerant system, refer to all CAUTIONS and WARNINGS.*
- *Service refrigerant system using Kent More ACR 4 or equivalent*
- *Only system components identified with an * can be serviced or replaced without discharging refrigerant system.*
- *During servicing, all couplings, fittings and related fasteners must be torqued to specifications.*

1. **Refrigerant pipe/hose**
   - Between accumulator and compressor
2. **Hex socket bolt**
   - 15 Nm (11 ft. lb.)
3. **O-ring**
   - 17.2 mm; 1.8 mm
4. **Refrigerant pipe/hose**
   - Between evaporator and condenser
5. **Coupling**
   - Thread diameter ∅ 20 mm; pitch1.5 mm
   - 15 Nm (11 ft. lb.)
6. **O-ring**
   - 10.8 mm; 188 mm
7. **Restrictor**
8. **O-ring**
   - 7.5 mm; 1.5 mm
9. **O-ring**
   - 10.8 mm; 1.8 mm
10. **High pressure service valve**
    - Only use Kent Moore ACR4 or equivalent.
    - Removing and installing, see **Fig. 1**
11. **Cap***
12. **O-ring**
    - 10.8 mm; 1.8 mm
13. **O-ring**
    - 17.2 mm; 1.8 mm
14. **Evaporator**
    - In heating and A/C unit
    - Removing/assembly, see **Heating and A/C unit, removing and assembly**.
15. **Socket head bolt**
    - 10 Nm (89 in. lb.)
16. **Cap***
17. **Low pressure service valve**
    - Only use Kent Moore ACR4 or equivalent.
    - Removing, see **Fig. 2**
18. **O-ring**
    - 7.6 mm; 1.8 mm
19. **Refrigerant pipe/hose**
    - Between accumulator and evaporator
20. **Compressor**
    - Denso 7SB-16C
    - With A/C clutch -N25-
    - See **Compressor/engine applications**
21. **Pressure relief valve**
    - Checking see **Fig. 7**
22. **O-ring**
    - 23.8 mm; 2.4 mm
23. **Socket head bolt**
    - 25 Nm (18 ft. lb.)
    - Qty. 2
24. **O-ring**
    - 11.1 mm; 1.8 mm
25. **A/C clutch -N25-***
    - Manufacturer: Denso
    - Repairing A/C clutch with compressor installed is not recommended.
    - Servicing, see **A/C clutch -N25-, servicing (Denso)**

(continued on following page)

A/C refrigerant system, assembly (from 10/2000) (continued)

26. Compressor
- Denso 7SEU16
- Without A/C clutch
- See **Compressor/engine applications**

27. A/C compressor regulator valve -N280-

28. O-ring
- 23.8 mm; 2.4 mm

29. Pressure relief valve
- Checking, see **Fig. 7**

30. Socket head bolt
- 25 Nm (18 ft. lb.)
- Qty. 2

31. O-ring
- 11.1 mm; 1.82 mm

32. Refrigerant pipe/hose
- Between compressor and evaporator.

33. Condenser
- See **Condensor, removing**

34. O-ring
- 14 mm; 1.8 mm

35. O-ring
- 17.2 mm; 1.8 mm

36. Hex socket bolt
- 15 Nm (11 ft. lb.)

37. O-ring
- 4- & 6-cyl. engine: 7.6 mm; 1.8 mm
- 8-cyl. engine: 10.8 mm; 1.82 mm

38. A/C pressure switch -F129- or high pressure sensor -G65-

NOTE —
- *Application is engine dependent.*
- *Always confirm system use of -F129- or -G65- and applicable wiring circuit using applicable "Engine" and "Climatronic" wiring diagrams.*

- -F129- function, see **A/C refrigerant system, component overview**
- -G65- function, see **A/C refrigerant system, component overview**
- Switch or sensor can be removed without discharging refrigerant system.
- Tightening torque 8 Nm (71 in lb)
- Always replace O-ring (note Part No.)

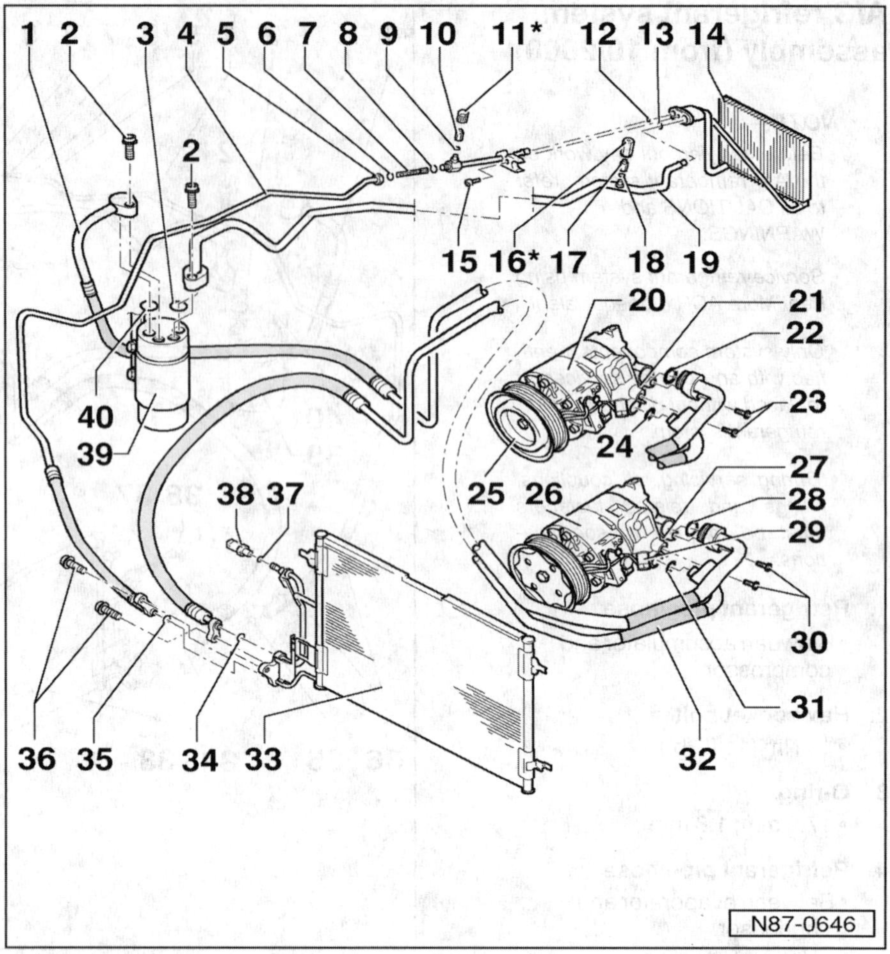

39. O-ring
- 17.2 mm; 1.82 mm

40. Accumulator
- Always replace whenever refrigerant system has been opened

Fig. 1 High pressure service valve, removing

- Discharge refrigerant system.
- Unscrew and remove high pressure service valve -3-.

1 - Connection with internal thread
2 - O-ring, 10.8 mm; 1.8 mm
3 - High pressure service valve, with groove for O-ring and internal thread for cap
4 - O-ring, 10.8 mm; 1.8 mm
5 - Cap

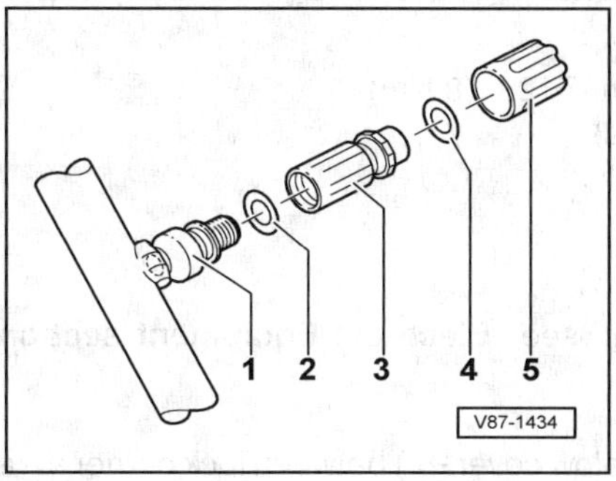

Fig. 2 Low pressure service valve, removing

– Discharge refrigerant system.

– Unscrew and remove low pressure service valve -3-.

1 - Connection with external thread and groove for O-ring
2 - O-ring, 7.6 mm; 1.8 mm
3 - Low pressure service valve
4 - O-ring, 7.6 mm; 1.8 mm
5 - Cap

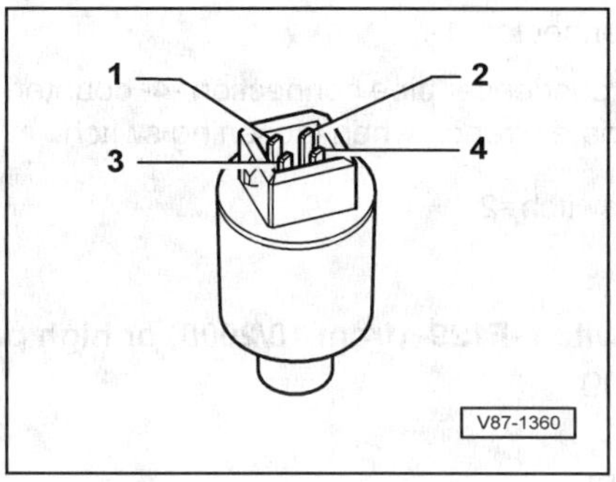

Fig. 3 A/C pressure switch -F129-, checking

- The pressure switch can be removed without discharging refrigerant system.
- Visually check that O-ring 10,8 mm x 1,8 mm is positioned correctly in the groove.

Function

Switch part between terminals 1 and 2 switches off the A/C clutch -N25- when the system pressure is too high or when there is not enough refrigerant in the system .

- Switch opens when the pressure is below 1.2 bar (17 psi) (and closes again above 2.4 bar (35 psi) (switch point).
- Switch opens above 32 bar (464 psi) and closes again below 24 bar (348 psi) (switch point).

Switch part between terminals 3 and 4 switches coolant fan -V7- (via the coolant fan control module -J293-) to the next higher fan speed when system pressures increase.

- Switch closes above 16 bar (232 psi) and opens again below 12.5 bar (181 psi)(switch point).

Checking:

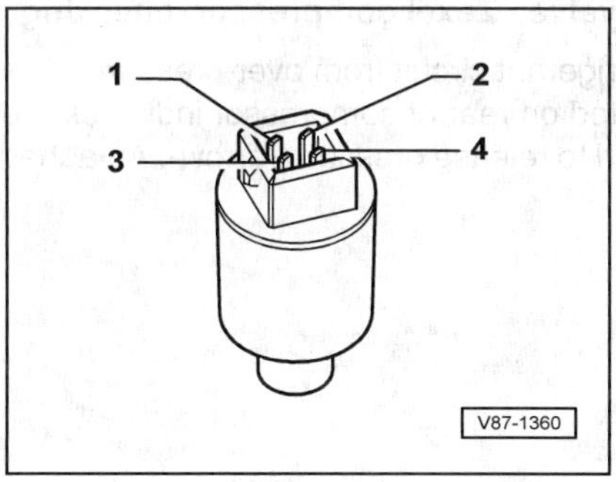

With the engine running, briefly bridge terminal 1 and 2 of connector. If the A/C clutch -N25- engages, the refrigerant circuit is empty.

-F129- functional checks

To determine proper A/C system performance, perform the following functional checks with the switch installed and connected.

– Check refrigerant capacity (fill). If system pressure is too low (refrigerant leakage), -F129- switches A/C clutch off.

– Check cooling fan function. When system pressure increases, -F129- causes coolant fan -V7- to switch to next higher fan speed.

– Check if A/C system switches off. At excessive system pressures (e.g.: caused by high ambient temperatures or if condensor is blocked), -F129- switches A/C clutch off.

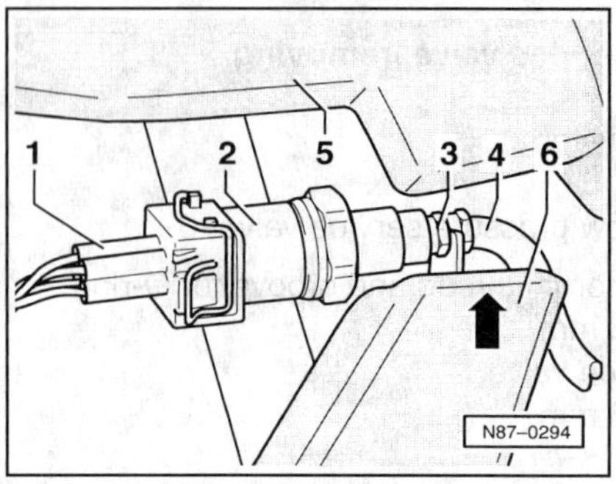

Fig. 4 A/C pressure switch -F129- (through 09/2000), replacing

1 - Connection
2 - A/C pressure switch -F129- (8 Nm)
3- Hex (to counter-hold)
4 - Condensor connection
5 - Lock carrier
6 - Cover

– Remove right headlight, see **Electrical Equipment section, Repair Group 94.**

– Remove bolt -arrow- (below cover -6-) between lock carrier -5- and cover -6-.

– Disconnect electrical connector -1-.

• To prevent damage to condenser pipe connection -4- counter-hold hex fitting -3- with appropriate wrench when loosening switch.

– Remove A/C pressure switch -2-.

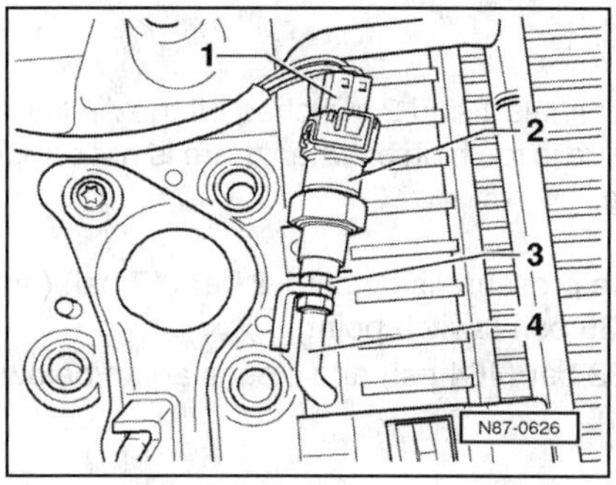

Fig. 5 A/C pressure switch -F129- (from 10/2000) or high pressure sensor -G65-, replacing

1 - Connector
2 - Switch/sensor (8 Nm)
3 - Hex (to counter-hold)
4 - Condensor connection

– Remove front bumper, see **Body-Exterior section, Repair Group 63**

– Move front lock carrier into service position, see **Body-Exterior section, Repair Group 50**

– Separate electrical connection -1-

• To prevent damage to condenser pipe connection -4- counter-hold hex fitting -3- with appropriate wrench when loosening switch.

– Remove switch/sensor -2-.

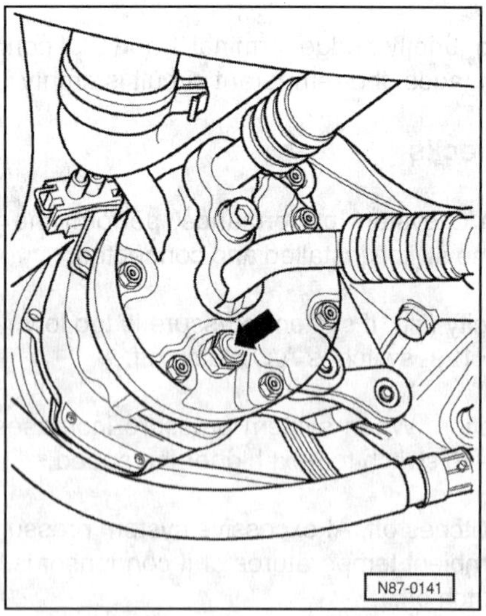

Fig. 7 Pressure relief valve - Zexel compressor, checking

• Function: Protects refrigerant circuit from over-pressure.
• Refrigerant oil deposited on rear of compressor indicates pressure relief valve has opened to release pressure -arrow-. An adhesive plate is pushed out.

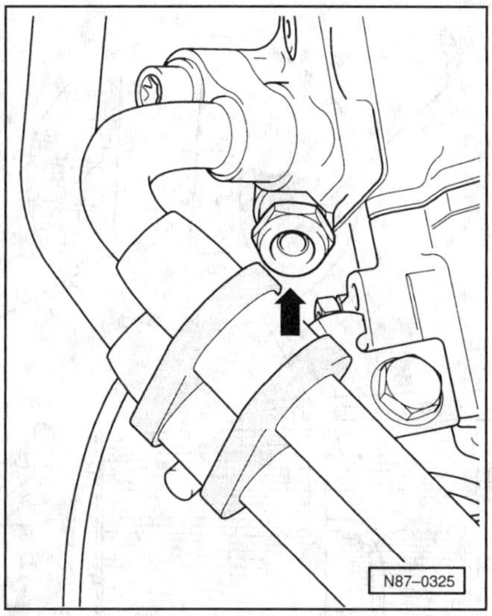

Fig. 8 Pressure relief valve - Denso compressor, checking

- Function: Protects refrigerant circuit from over-pressure.
- Refrigerant oil deposited in area of compressor manifold indicates pressure relief valve has opened to release pressure -arrow-. An adhesive plate is pushed out.

A/C refrigerant system, replacing components

Replacement A/C compressors, evaporators and condensers supplied by the Parts Dept. are filled with Nitrogen. If gas (Nitrogen) does not escape when component is first opened, the component may be faulty (leaking), do not install.

Replacement A/C compressors are filled with the total refrigerant oil quantity needed for the entire refrigerant system. When an additional, rear evaporator is present, additional compressor oil must be added after replacing the compressor.

Always replace the accumulator whenever the refrigerant system has been opened. Install immediately after opening to prevent moisture contamination of drier desiccant. Moreover, keep refrigerant system and other replacement components sealed for as long as possible to minimize the chance of dirt and moisture contamination.

Always plug open refrigerant line connections to prevent dirt and moisture contamination.

If the system has been discharged due to a damaged or leaking component (refrigerant hose/line, compressor, evaporator, condenser, etc.), flush the refrigerant system first with compressed air, then with nitrogen (available locally) and collect the oil that runs out. This will remove the refrigerant oil which may be saturated with moisture.

> ***CAUTION —***
> *DO NOT flush R-134a refrigerant system with R-12. R-12 is not compatible with refrigerant R-134a and PAG oil and will cause total system contamination.*

If the compressor is not replaced after flushing the system, fill compressor with the correct type and quantity of refrigerant oil as specified under A/C system capacities, see **A/C refrigerant system, capacities**. Fill compressor with the total system oil capacity.

If the compressor is replaced, do not add any additional oil as the total amount of oil required is already in the compressor.

Follow all applicable Warnings and Cautions when working on the A/C system.

Condensor, removing

CAUTION—
Before beginning repairs on the electrical system:

- *Obtain the anti-theft radio security code.*
- *Switch the ignition off.*
- *Disconnect the battery Ground (GND) strap.*
- *After reconnecting battery, re-code and check operation of anti-theft radio. Also check operation of clock and power windows according to Owner's Manual.*

– Discharge refrigerant system.

– Move lock carrier into service position, see **Maintenance**.

– Vehicles from 10/2000: Remove front bumper carrier, see **Body-Exterior section, Repair Group 63**

– Loosen and separate refrigerant pipes/couplings from evaporator.

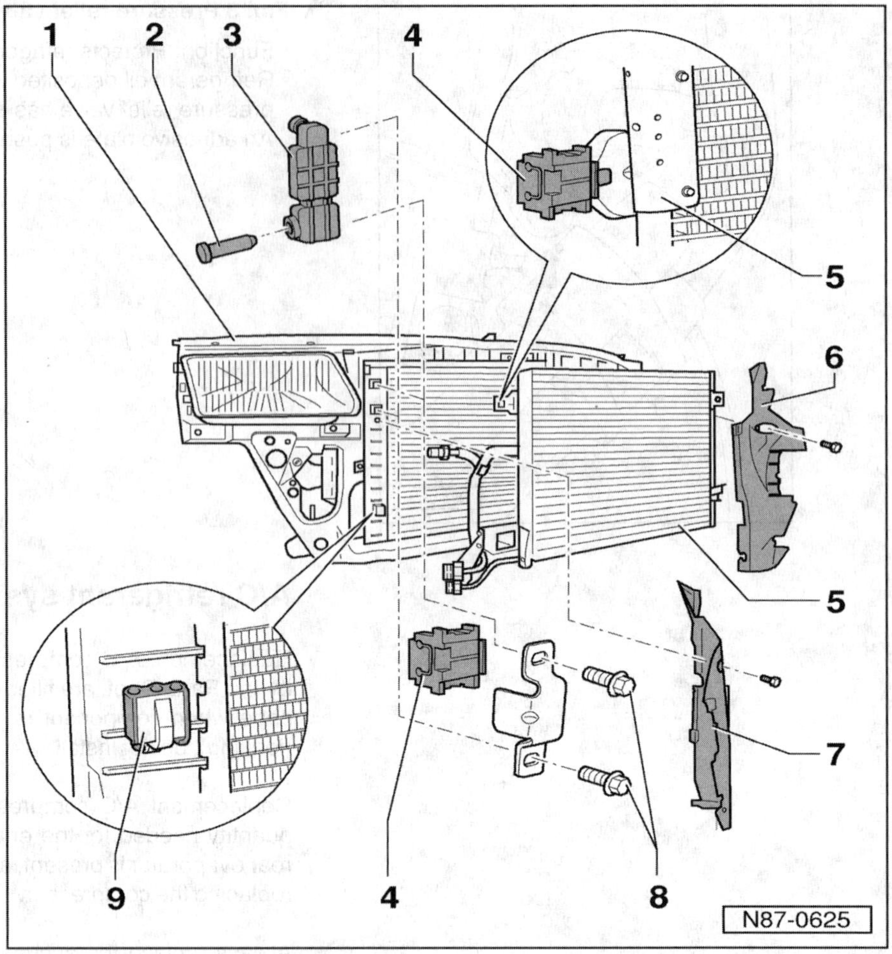

CAUTION—
Immediately plug open connections on A/C components and lines to prevent dirt and moisture contamination.

1. **Lock carrier**
2. **Clip**
   - Vehicles from 10/2000
3. **Bracket**
   - Vehicles from 10/2000
4. **Grommet, upper**
5. **Condensor**
6. **Left air guide**
7. **Right air guide**
8. **Bolt**
   - Vehicles from 10/2000
   - 8 Nm (71 in lb)
   - Qty.: 4
9. **Rubber grommet, lower**

Heating and A/C unit, removing and installing

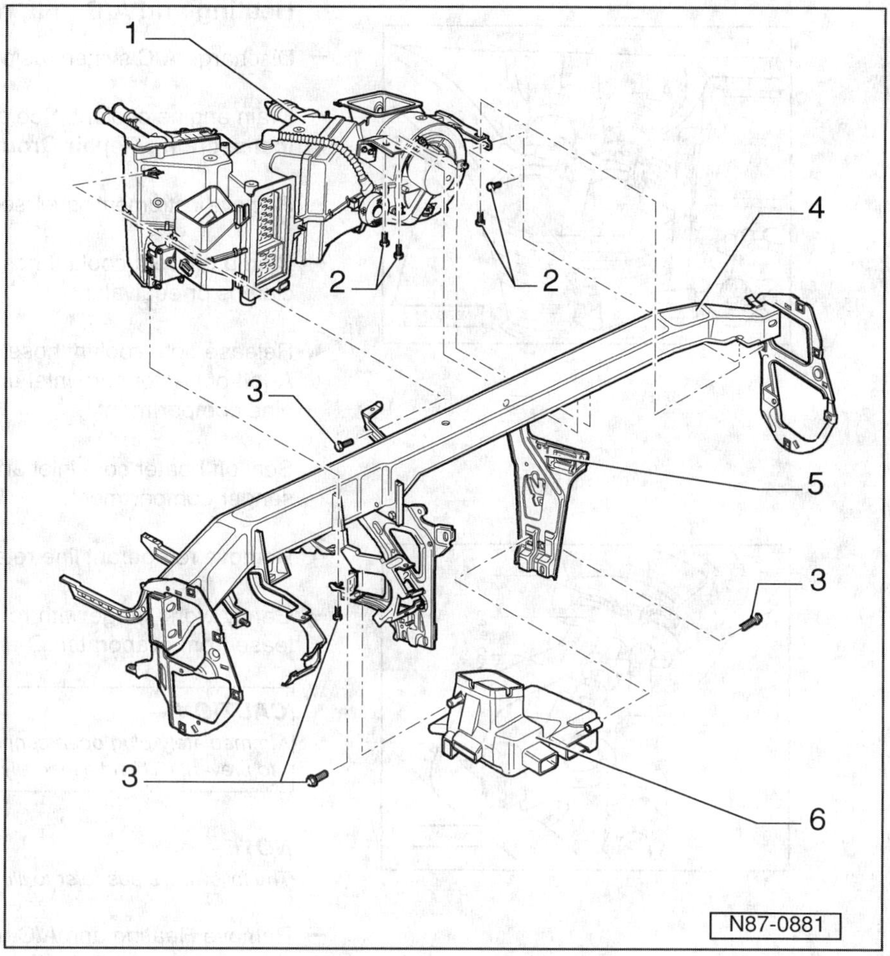

CAUTION—

Before beginning repairs on the electrical system:

- *Obtain the anti-theft radio security code.*
- *Switch off all electrical consumers.*
- *Switch the ignition off and remove ignition key*
- *Disconnect the battery Ground (GND) terminal.*
- *After reconnecting battery, re-code and check operation of anti-theft radio. Also check operation of clock and power windows according to Owner's Manual. Observe all applicable Notes and torque specifications.*
- *Before carrying out any work on the A/C refrigerant system, refer to all CAUTIONS and WARNINGS in A/C refrigerant system safety measures beginning on page V87-1*
- *Immediately plug open connections on A/C components and lines to prevent dirt and moisture contamination.*
- *During servicing, all couplings, fittings and related fasteners must be torqued to specifications.*

1. **Heating and A/C unit**
   - With heater core
   - With A/C evaporator
   - Removing, see **Heating and A/C unit, removing and installing**
2. **Mounting screws**
   - 12 Nm (106 in lb)
3. **Mounting screws**
   - 3.5 Nm (31 in lb)
4. **Instrument panel cross member**
5. **Support**
6. **Footwell vent assembly**

Heating and A/C unit, removing and installing

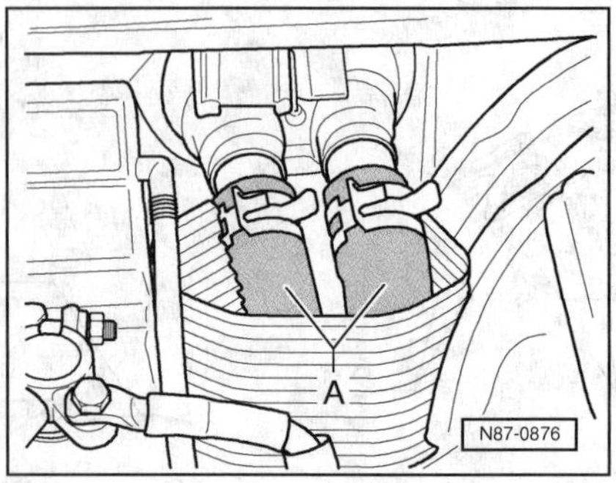

– Discharge A/C system using Kent More ACR 4 or equivalent.

– Drain engine coolant. See appropriate engine code in **Engine Mechanical section , Repair Group 19**.

– Remove instrument panel, see **Body-Interior section, Repair Group 70.**

– Clamp off both coolant hoses in engine compartment using 3094 hose clamps or equivalent.

◄ Release both coolant hose clamps using suitable pliers and pull hoses -A- off of heater core inlet and outlet noting installation orientation in engine compartment.

– Seal off heater core inlet and outlet to prevent coolant leakage into passenger compartment.

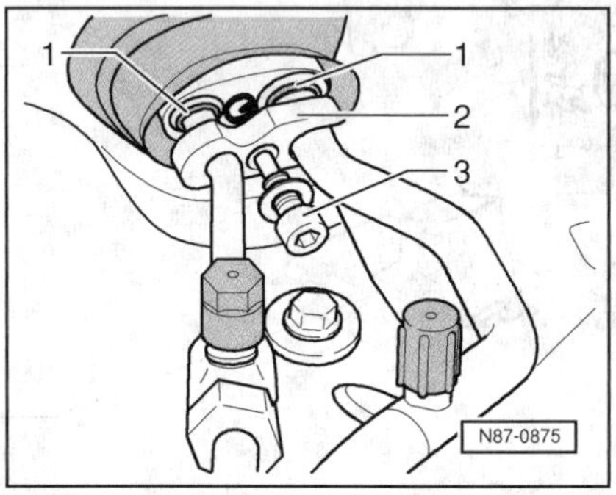

◄ Remove refrigerant line retaining screw -3- at evaporator.

– Gently work flange with refrigerant lines -2- up and down until they release from evaporator. O-rings -1- must always be replaced.

> ***CAUTION—***
> *Immediately plug open connections on all A/C components and lines to prevent dirt and especially mo4isture contamination.*

NOTE—
The following steps refer to illustration N87-0881 on the previous page.

– Remove Heating and A/C unit mounting screws -2- and -3-.

– Remove footwell vent assembly -6-.

– Remove support -5-.

– Disconnect all electrical connections and wiring harness tie straps at heating and A/C unit.

– Carefully pull out and remove complete housing - 1 - from vehicle.

– Loosen instrument panel cross member -4- if necessary

– Repair Heating and A/C unit as required, see **Heating and A/C unit, assembly.**

– Installion is the reverse of removal noting the following points:

- Inspect all heating and A/C unit housing seals for damage and replace as necessary before installation. These include evaporator inlet/outlet seal and evaporator seal. See **Heating and A/C unit, assembly.**
- Ensure all electrical harness tie straps are replaced and installed in their original locations.
- Replace evaporator inlet/outlet line O-rings.
- Ensure that refrigerant pipes are reinstalled stress free (without preload).
- Ensure that evaporator water drain and drain valve are correctly installed and function properly, see **Heating and A/C components in engine compartment, assembly.**

Heating and A/C unit, assembly

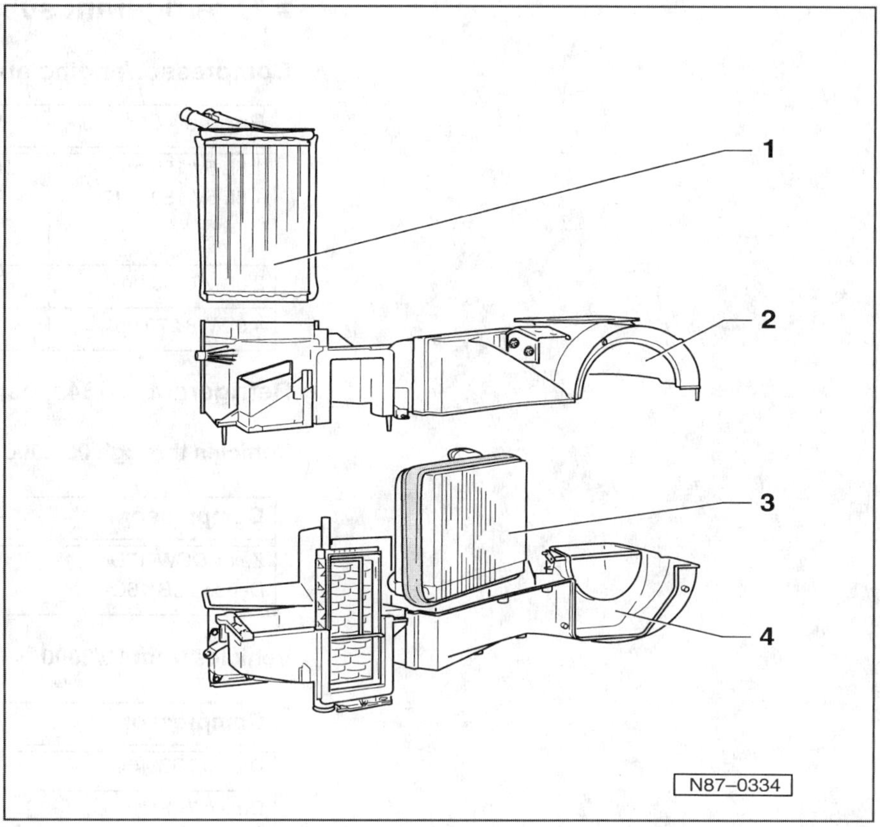

***CAUTION*—**
Before beginning repairs on the electrical system:

- *Obtain the anti-theft radio security code.*
- *Switch the ignition off.*
- *Disconnect the battery Ground (GND) strap.*
- *After reconnecting battery, re-code and check operation of anti-theft radio. Also check operation of clock and power windows according to Owner's Manual.*

Removing

– Discharge refrigerant system.

– Clamp off and disconnect coolant hoses to heater core see **Engine Mechanical section, Repair Group 19**.

– Disconnect refrigerant lines to evaporator.

***CAUTION*—**
Immediately plug open connections on A/C components and lines to prevent dirt and moisture contamination.

– Remove instrument panel, see **Body-Interior section, Repair Group 70**.

– Loosen instrument panel cross-member, see **Climatronic components in passenger compartment, assembly.**

– Remove heating and A/C unit.

Assembly

1. **Heater core**
   - Always use new coolant after replacement.
   - Removing and installing, see **Fig. 1**
2. **Housing, upper part**
3. **Evaporator**
   - Attaching inlet/outlet seal, see **Fig. 2**
   - Installing seal, see **Fig. 3**
4. **Housing, lower part**
   - With fresh/recirculating air flap
   - With central air flap

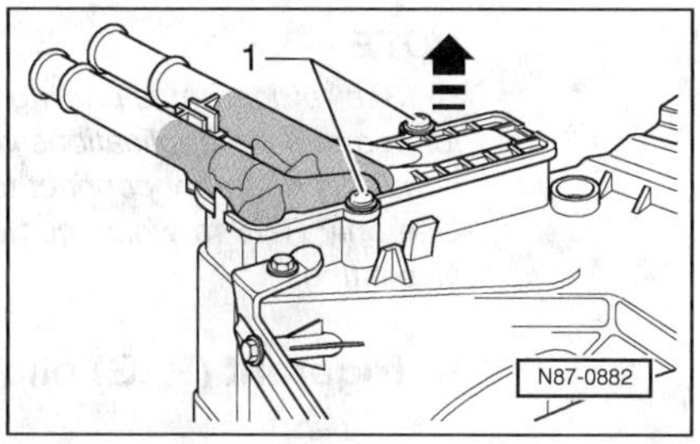

Fig. 1 Heater core, removing and installing

– Heating and A/C unit must be removed.

– Remove screws -1- and pull core out of housing in direction of arrow.

– Install in reverse order.

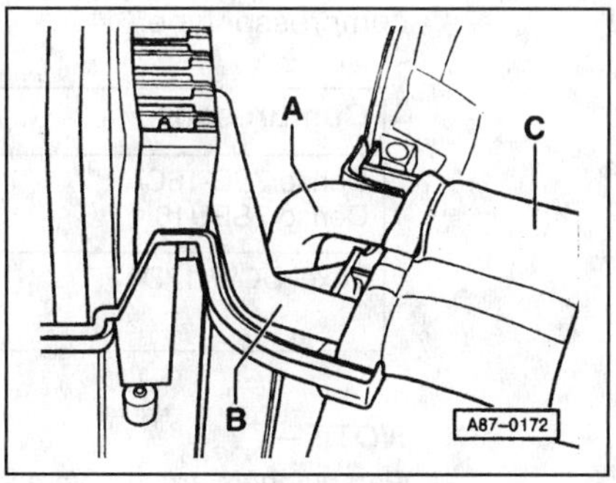

Fig. 2 Evaporator inlet/outlet seal, attaching

– Evaporator inlet -B- and outlet -A- pipes must be insulated between the halves of the housing with seal -C-.

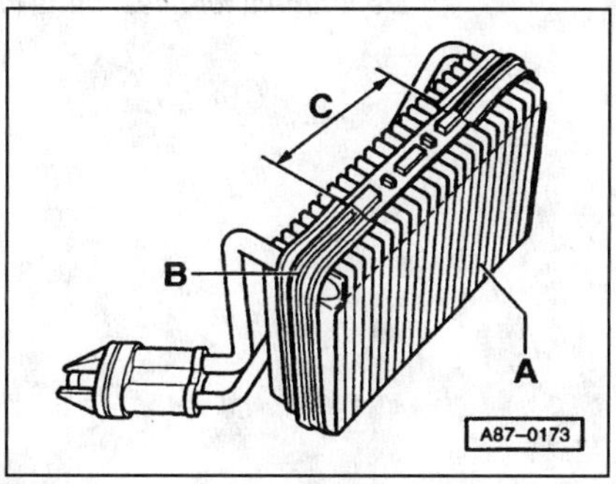

Fig. 3 Evaporator seal, installing

– Foam seal -B- must must be glued around circumference of evaporator -A-.

– To allow evaporator water to drain, do not attach foam seal on evaporator in area -C-.

A/C refrigerant system, capacities

Compressor/engine applications

Engine	Manufacturer	Engine code
1.8L 5V 150 hp 1.8L 5V 150 & 170 hp	Through 09/2000: Zexel DCW-17D From 10/2000: Denso 7SB-16C	AEB, ATW AUG, AWM
2.8L V6 190 hp	Denso 7SB-16C	AHA & ATQ
4.0L W8 270 hp	Denso 7SE-16	BDP

Refrigerant R-134a, capacity

Vehicles through 09/2000

Compressor	Capacity
Zexel DCW-17D Denso 7SB-16C	650 g + 50 g (23 oz + 1.8 oz)

Vehicles from 10/2000

Compressor	Capacity
Denso 7SB-16C	600 g ± 25 g (21 oz + .9 oz)
Denso 7SEU16	500 g + 25 g (18 oz + .9 oz)

NOTE—
R-134a refrigerant is packaged in different containers. Some are used only for commercial applications which are sold in cylinders using a 1/4" flare fitting. This does not connect to the vehicle fittings and servicing equipment. Use only R-134a which come in containers having the correct type of service fitting.

Refrigerant (PAG) oil R-134a

Use only the PAG oil(s) specified below for use with R-134a and compressor type.

Compressor	Refrigerant oil Part No.
Denso 7SB-16C/ Denso 7SEU16	G 052 300 A2
Zexel DCW-17D	G 052 154 A2 or G 052 200 A2

NOTE—
Part numbers are for reference only. Always check with for Parts Department for latest parts information.

PAG oil, capacity

Model	Total system capacity
Passat through 09/2000 Passat from 10/2000	250 cc (8.45 fl. oz.)
Passat W8	200 ± 10 cc (6.76 ±.33 fl. oz.)

PAG oil, distribution

Replacement A/C compressors supplied by the Parts Department are filled with 250 cc (fl. oz) of refrigerant oil. This is the total A/C system refrigerant oil capacity.

The total refrigerant oil quantity found in a new compressor is distributed in the refrigerant circuit the first time the compressor is engaged.

Before installation, all replacement refrigerant system components (except the compressor) should be filled with refrigerant (PAG) oil according to the system distribution (percentage) below:

Compressor: approx. 50%

Condensor: approx. 10%

Suction pipe: approx. 10%

Evaporator: approx. 20%

Accumulator: approx. 10%

Electrical Equipment

Section **W**

96 Lights, Switches - Interior, Anti-theft — Page

97 Wiring — Page

27 Battery, Starter, Generator, and Cruise Control

Repair Group

W 27

Battery

Battery type definitions

Two different types of batteries have been installed on Passat vehicles.For identification purposes, the following battery type definitions establish the technical differences between installed battery types:

- Low Maintenance Batteries: Lead-antimony coated electrodes (plates). With or without charge indicator. Electrolyte filling caps accessible.
- Maintenance Free Batteries: Lead-Calcium coated electrodes (plates). With charge indicator. Electrolyte filling caps not accessible.

Handling instructions

CAUTION—
- *Before beginning repairs on the electrical system:*
- *Obtain the anti-theft radio security code.*
- *Switch the ignition off.*
- *Disconnect the battery Ground (GND) strap.*
- *After reconnecting battery, re-code and check operation of anti-theft radio. Also check operation of clock and power windows according to Repair Manual and/or Owner's Manual.*

NOTE—
- *Battery posts and terminals must no longer be greased.*
- *In order to prevent damage to the battery housing, battery terminals must be placed on battery posts without using force (by hand only).*

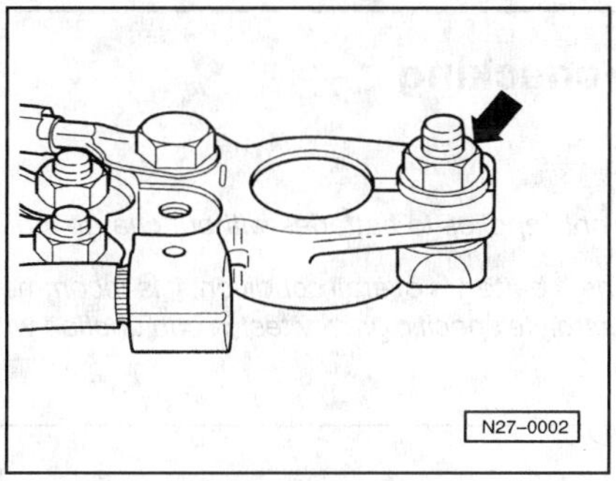

Battery post/terminal nut -arrow- tightening torque: 5 Nm (44 in lb)
- Additional battery post/terminals, tightening torque: 6 Nm (53 in lb).

WARNING—

1 Keep open flames and sparks away and DO NOT smoke near batteries!

• Avoid sparks when working with cables and electrical units.

• Always reinstall battery positive (B+) or negative (-) terminal covers if removed during servicing.

2 Always wear eye protection when working with electrolyte!

• If electrolyte gets into eyes, flush with large quantities of water. Seek medical attention immediately!

3 Keep children away from batteries!

4 Dispose of batteries properly!

• Waste batteries must only be disposed of in appropriate waste disposal sites. Refer to local regulations pertaining to battery disposal.

5 Never dispose of batteries in household waste!

6 Danger of explosion!

• Batteries produce explosive gasses while being charged.

• Keep open flames and sparks away and DO NOT smoke near batteries.

• The battery charger MUST be turned off when connecting or disconnecting the cables at the battery.

• Battery cell caps must NOT be removed while charging.

• Ensure that battery is charged in a well ventilated area.

• Avoid short circuits. Always reinstall battery positive (B+) or negative (-) terminal covers if removed during servicing.

7 Read and follow all instructions on battery, contained in Repair Manual, Electrical Equipment, and in Owner's Manual!

8 Battery acid can cause severe burns!

• When working with electrolyte always wear eye protection, rubber gloves and a suitable apron. If electrolyte is spilled on eyes, skin, clothing (or painted surfaces), flush at once with large quantities of water.

• Never tip a battery on edge. Electrolyte can spill from the ventilation openings.

Specific gravity, checking

NOTE—

• The following procedure only applies to batteries without charge indicator.

*• In order to better determine a battery's overall condition, it is recommended to perform always the electrolyte specific gravity test in conjunction with the load test, see **Load test**.*

WARNING—

Before working with electrolyte, refer to Warnings and Safety Measures

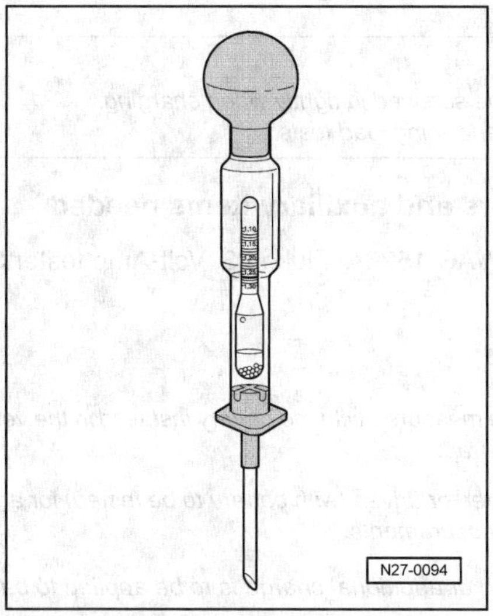
N27-0094

Special tools, testers and auxiliary items

- Commercially available hydrometer
- Shop rags

NOTE —
- *Minimum battery temperature of 10 °C (50 °F) is required when carrying out specific gravity tests.*
- *Specific gravity of all cells must be checked.*

– Switch ignition off

– Remove all battery cell caps.

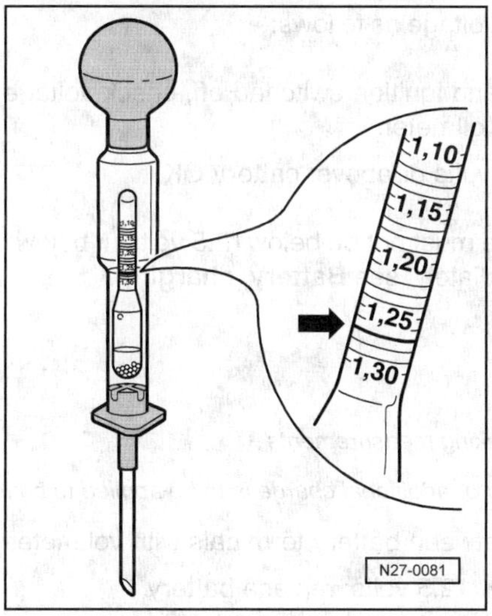

N27-0081

Dip end of hydrometer into cell and extract sufficient electrolyte to allow indicator float to swim freely (the greater the electrolyte density in kg/dm^3, the higher the indicator float position). Note each cell's specific gravity reading from float -arrow-. Repeat sequence until all cells are checked.

– Review noted values and compare with values in table below:

Charge condition (normal climate)	Specific gravity in kg/dm^3
Discharged	1.15
Half charged	1.22
Fully charged	1.28

- Electrolyte density must be at least 1.24kg/dm^3

– If specific density is too low, charge battery and recheck specific density afterwards.

- Electrolyte density must not vary more than 0.04kg/dm^3 between cells.

Examples of unacceptable variations in specific gravity:

Specific gravity variations per cell in kg/dm^3						
Cell Number:	1	2	3	4	5	6
Example 1:	1.24	1.25	1.25	1.10	1.24	1.25
Example 2:	1.26	1.26	1.25	1.14	1.18	1.24

Example 1: Specific gravity in cell 4 (1.10) is too low.

Example 2: Specific gravity in cells 4 (1.14) and 5 (1.18) are too low and variation between cells exceeds 0.04kg/dm^3.

- If specific gravity specifications are not met, replace battery.
- If specific gravity specifications are met, reinstall cell caps.

CAUTION —
Only the use of genuine VW battery cell caps ensures the different battery cover systems are sealed correctly. Only use genuine cell caps of the same construction if lost or damaged. All cell caps must be equipped with an O-ring seal.

No load voltage, checking

> ***CAUTION —***
> *Battery cell caps must be screwed in tightly when charging, measuring voltage or performing load tests.*

Special tools, testers and auxiliary items needed

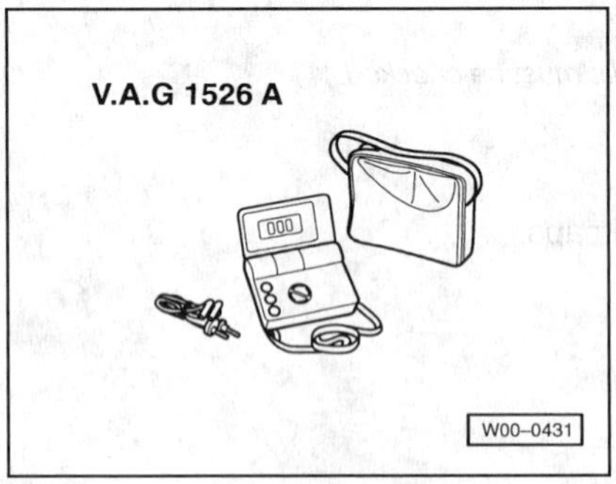

Multimeter VAG 1526, VAG 1526A, Fluke 83, Volt-Amp tester Sun VAT 40 or equivalent.

NOTE —
- *If no load voltage is to be measured with the battery installed in the vehicle, first disconnect ground strap.*
- *Vehicle must not be started or driven (with battery to be tested) for at least 2 hours before taking measurements.*
- *During this time, no load or additional charge is to be applied to battery.*

Check battery no load voltage as follows:

– With battery installed and ignition switched off, check voltage between battery terminals with voltmeter.
  - No load voltage 12.5 volts or above: battery OK.

  Battery no load voltage must not be below12.5 volts. If below12.5 volts, recharge battery immediately, see **Battery, charging**.

NOTE —
After charging battery:
- *Wait two hours before taking measurements*
- *During this time, no load or additional charge is to be applied to battery.*

– Again check voltage between battery terminals with voltmeter.
  - No load voltage below 12.5 volts: replace battery.

Load test

> ***WARNING —***
> - *Refer to Warnings and Safety Measures.*
> - *DO NOT load test batteries which are gassing. Otherwise there is a risk of explosion.*

NOTE —
- *In order to better determine a battery's overall condition, it is recommended to always perform the load test in conjunction with the electrolyte specific gravity test, see **Specific gravity, checking**.*
- *It is not necessary to remove battery from vehicle or disconnect battery erminals.*

Special tools, testers and auxiliary items needed

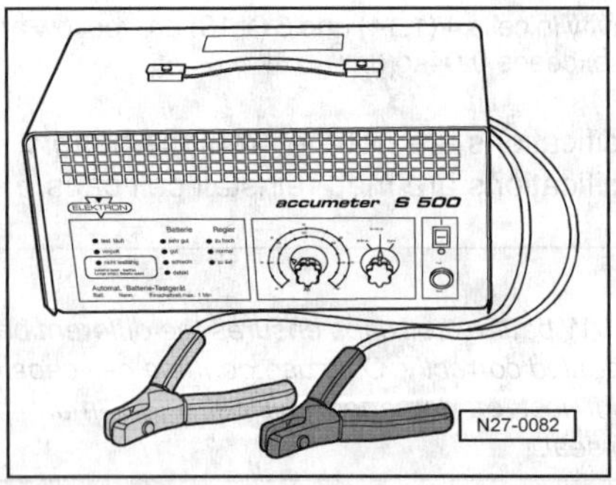

Battery tester VAS 1979, VAS 5033, Christie Automotive BT-9000 or equivalent

– Switch ignition off

– Connect tester leads to battery posts as described in the tester instruction manual. Tester leads must be securely fastened and make proper contact with the battery posts.

– Load current and minimum voltage specifications vary depending on battery capacity. Apply load as per tester instructions, or for 5 seconds at load specified below:

Capacity	Cold cranking amps	Load current	Minimum voltage
61Ah-80Ah	300A-380A	300A	9.0
81Ah-110Ah	380A-500A	300A	9.5

If the minimum voltage specification is not reached, replace battery.

Load test results

A battery is in good condition if the voltage measured during the load test drops gradually and does not fall below the specified value. The rate of decrease varies and depends on the battery capacity and cold cranking amp rating ("cold cranking amps" is the rated ability of the battery to perform at low temperatures and with high loads i.e.: high compression engines).

A battery is either faulty or in a low state of charge if the voltage measured during the load test drops quickly and falls below the specified voltage.

After performing the load test, a faulty battery will remain below the specified voltage and may only recover very slowly over a long period of time. The no load voltage (see **No load voltage, checking**) is never reached.

Such batteries no longer have sufficient reserve and load capacity and must be replaced.

Battery, charging

WARNING —

Danger of explosion!

- *Batteries produce explosive gasses while being charged.*
- *The battery charger MUST be turned off when connecting or disconnecting the cables at the battery.*
- *Battery cell caps must NOT be removed while charging.*
- *"Boosting" a sulfated battery a high charging rate can cause an explosion.*
- *Ensure that battery is charged in a well ventilated area.*
- *Precision tools must not be kept in the same room where batteries are being charged. Tools may corrode due to chemical reaction.*

CAUTION —

Before beginning repairs on the electrical system:

- *Obtain the anti-theft radio security code.*
- *Switch the ignition off.*
- *Disconnect the battery Ground (GND) strap.*
- *After reconnecting battery, re-code and check operation of anti-theft radio. Also check operation of clock and power windows according to Repair Manual and/or Owner's Manual.*

Special tools, testers and auxiliary items needed

- Battery charger: Christie Automotive 400 QDD or equivalent.

Battery charger, connecting

NOTE —

- *Before charging, battery must have a minimum temperature of 10 °C (50 °F)*
- *Always follow the battery charger operating instructions*
- *After charging, confirm battery no load voltage and electrolyte specific gravity before reuse in vehicle.*

– Switch off battery charger.

– Switch off ignition.

– Disconnect negative GND terminal from battery negative (-) pole

– Disconnect positive B+ terminal from battery positive (+) pole

– Connect battery charger to battery, positive to positive (+), negative to negative (-).

– Adjust charging rate according to battery capacity (refer to charger instructions).

- Switch on battery charger.

> ***WARNING —***
> *If battery begins gassing (boiling) violently when charging, REDUCE charging rate immediately.*

Battery, removing and installing

> ***CAUTION —***
> *Before beginning repairs on the electrical system:*
>
> *• Obtain the anti-theft radio security code.*
>
> *• Switch the ignition off.*
>
> *• Disconnect the battery Ground (GND) strap.*
>
> *• After reconnecting battery, re-code and check operation of anti-theft radio. Also check operation of clock and power windows according to Repair Manual and/or Owner's Manual.*

Special tools, testers and auxiliary items needed

• VAG 1331 Torque wrench (or equivalent 5 - 50 Nm)

Removing

- Open front hood and remove cowl panel.

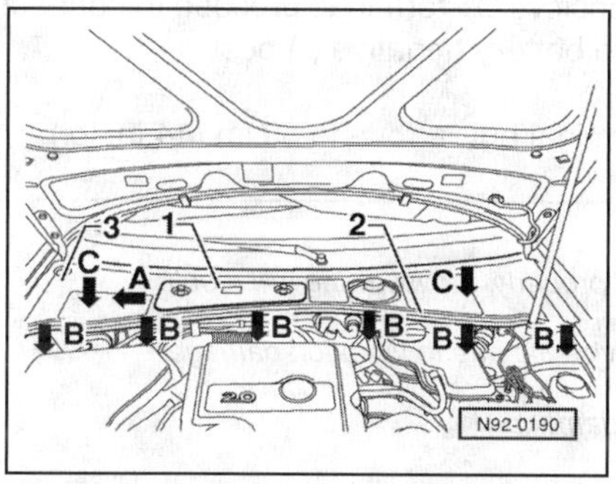

< Slide battery cover -1- in direction of -arrow A- and remove.

• Remove rubber seal -2- in direction of -arrow B-.
• Pull cover -3- forward and remove.

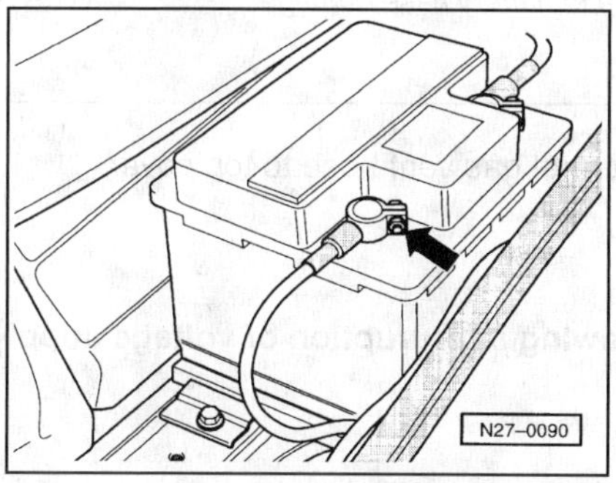

< Disconnect negative GND terminal -arrow- from battery negative (-) post.

- Then disconnect positive B+ terminal from battery positive (+) post.

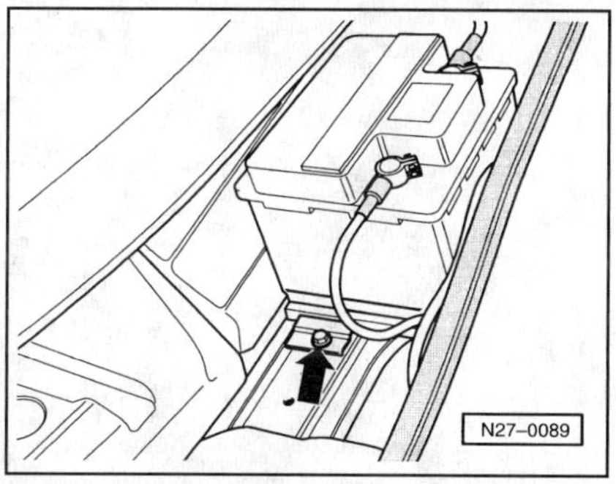

< Remove M8 x 25 bolt -arrow-.

- Remove battery hold-down bracket.
- If present, disconnect central gas vent hose from top cover.
- Carefully lift battery from plenum.

Installing

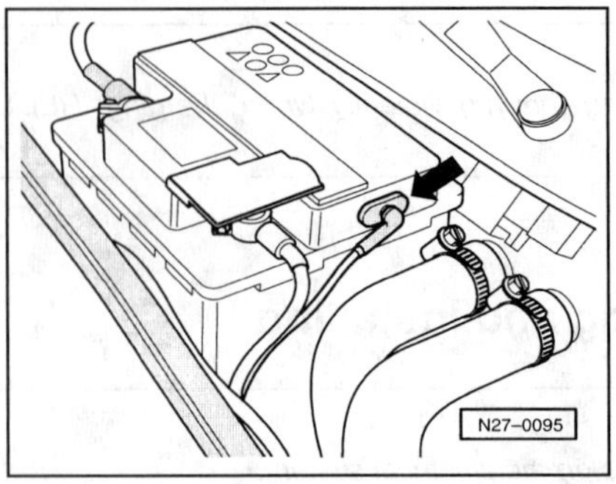

New generation batteries are equipped with a central gas venting system and integrated flame trap -arrow-. Function: Gases produced during charging escape through a central opening on the side of the upper cover. The integrated flame trap prevents flammable gases in the battery from igniting. The flame trap is comprised of a small round fiberglass mat with a diameter of 15 mm and a thickness of 2 mm. Working as a one-way-valve, it allows gases produced during charging to vent.

NOTE —
To ensure proper seating of installed battery, use only genuine Volkswagen approved replacement batteries, and adhere to the following procedures

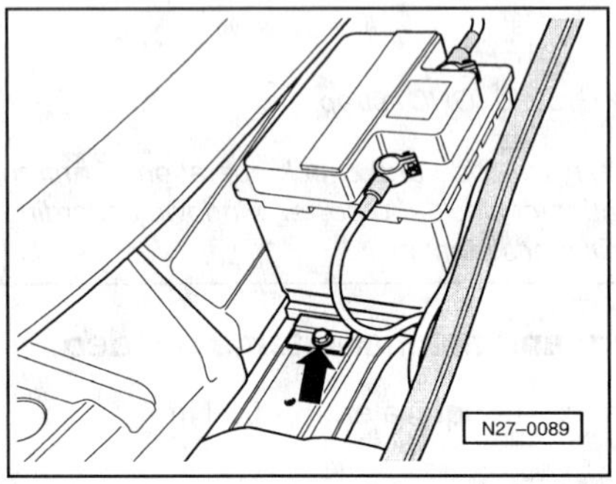

Install hold-down bracket and torque M8 x 25 bolt -arrow- to 22 Nm (16 ft.. lb.).

– Connect positive B+ terminal to battery positive (+) post.

NOTE —
- *In order to prevent damage to the battery housing, battery terminals must be placed on battery posts without using force (by hand only).*
- *Battery posts and terminals must no longer be greased.*

– Torque battery positive B+ post/terminal nut to 6 Nm (53 in. lb.).

– Only after connecting positive B+ terminal, proceed to connect negative GND terminal -arrow- to battery negative (-) post.

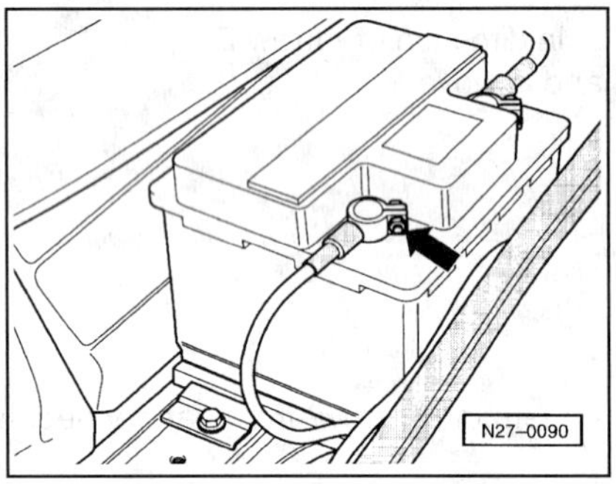

Torque battery post/terminal nut -arrow- to 5 Nm (44 in. lb.).

CAUTION —
If battery is not secured properly, the following will result:
- *Shortened battery service life due to vibration damage.*
- *Battery cell and plate damage.*
- *Electrolyte leakage due to battery case damage.*
- *Poor collision safety.*

– If present, reconnect central gas vent hose to top cover.

– Reinstall cowl panel.

– Follow **Procedure following interuruption of voltage supply**.

Procedure following the interruption of voltage supply

NOTE—

If the Voltage supply is interrupted, the Engine Control Module loses all the "learned" and/or "basic" values. As a result, severe driveability problems may occur depending on vehicle mileage.

Special tools, testers, measuring instruments and auxiliary items required

• VAG 1551/1552 Scan Tool with VAG 1551/3B adaptor cable

Work sequence

Perform the following operations after "Voltage supply open circuit" display appears

– Switch ON ignition for at least 10 seconds.

– Switch OFF ignition.

– Adapt the Engine Control Module to the Throttle Valve Control Module, see **Fuel Injection and Ignition section, Repair Group 24.**

– For vehicles equipped with automatic transmissions, the kick down shift point must be adapted, see **5-Speed Automatic Transmission section, Repair Group 37**.

– Test drive until any driveability complaints are eliminated.
During the road test the following operating conditions must be fulfilled:
• The Engine Coolant Temperature must exceed 80 °C.
• Once the required temperature is reached, the following operating conditions must be attained several times.
• Closed Throttle position
• Part Throttle position
• Wide Open Throttle position
• Overrun
• With Wide Open Throttle engine speed must exceed 3500 rpm.

Starter

Starter, removing and installing (4 & 6-cyl. engines)

Vehicles with 8-cyl. engines, see **Starter, removing and installing (8-cyl. engine)**

> ***CAUTION—***
> *Before beginning repairs on the electrical system:*
>
> - *Obtain the anti-theft radio security code.*
> - *Switch the ignition off.*
> - *Disconnect the battery Ground (GND) strap.*
> - *After reconnecting battery, re-code and check operation of anti-theft radio. Also check operation of clock and power windows according to Repair Manual and/or Owner's Manual.*

Special tools, testers and auxiliary items needed

- VAG 1331 Torque wrench (or equivalent 5 - 50 Nm)
- VAG 1332 Torque wrench (or equivalent 40 - 200 Nm)

Removing

– Disconnect battery Ground (GND) strap

NOTE—

- *On vehicles with restricted access to the engine compartment, it may become necessary to move the lock carrier/radiator support into service position, see **Body Exterior section, Repair Group 50***
- *Depending on engine type and additional equipment, it may be necessary to loosen the engine mounts and then lift the engine slightly with engine support bridge 10-222A.*
- *Depending on engine and transmission type, it may be necessary to loosen the a/c refrigerant hoses from their mounting locations, see **Heating and Air Conditioning section, Repair Group 87***
- *Depending on engine and transmission type, it may be necessary to loosen or remove the lines from the automatic transmission to the oil cooler, see **Repair Manual, Automatic Transmission, Repair Group 37***

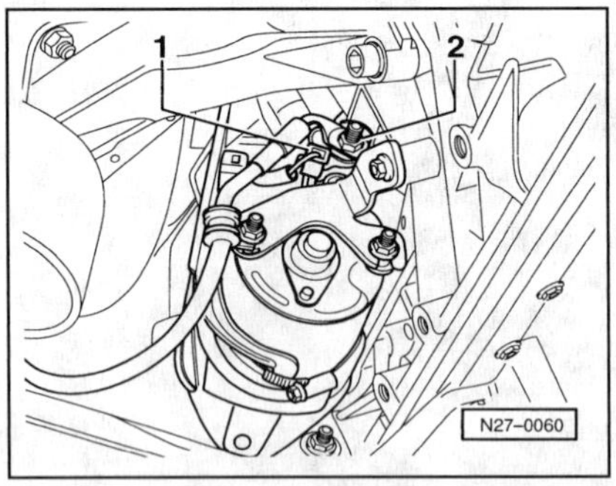

Unclip terminal 50 connector -1- and remove.

- Remove nut -2- at starter solenoid terminal 30 and remove connector eyelet from terminal. Tightening torque: 13 Nm (10 ft.. lb)

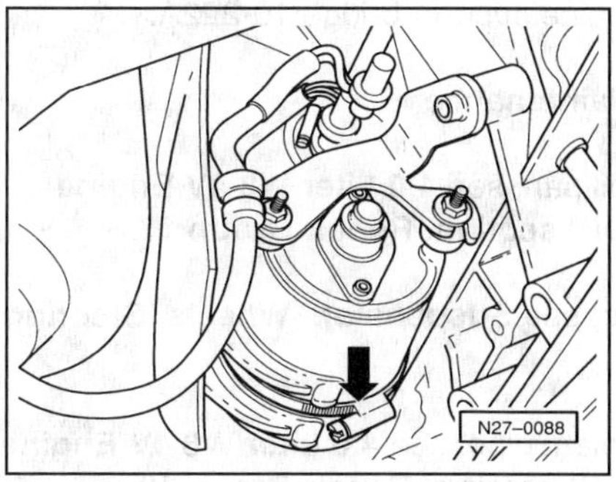

Loosen and remove clamp which secures heat shield to starter.

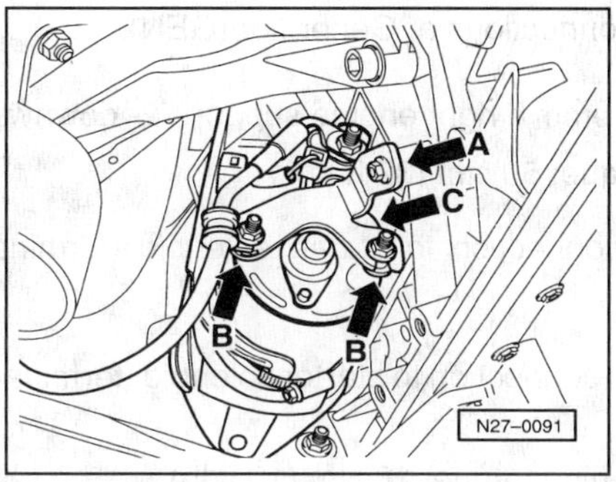

Mounting to engine block

NOTE —
Bracket -c- may not be installed on some engine/transmission types.

Where applicable, remove M8 nut -arrow A-.

- Tightening torque: 22 Nm (16 ft. lb.).

– Remove both M5 nuts -arrows B-, and remove retaining bracket.

- Tightening torque 4 Nm (35 in lb).

Mounting to transmission housing

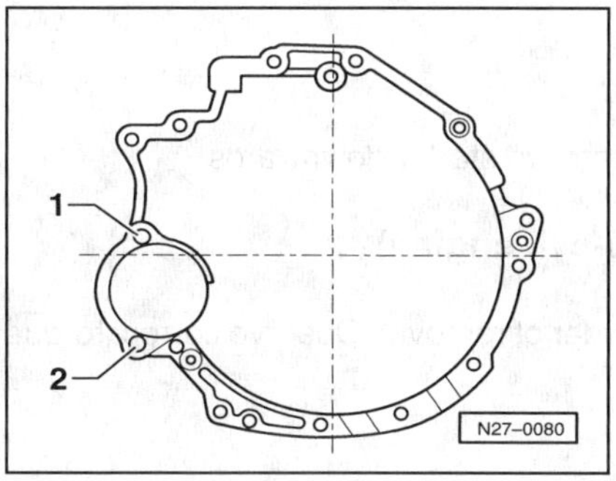

Remove upper M12x90 bolt -1- from transmission housing (end view of transmission housing illustrated).

- Tightening torque: 65 Nm (48 ft. lb).

– Remove lower M12x67 bolt -2- from from transmission housing.

- Tightening torque 65 Nm (48 ft. lb).

Starter, removing and installing (8-cyl. engine)

CAUTION —
Before beginning repairs on the electrical system:

- *Obtain the anti-theft radio security code.*
- *Switch the ignition off.*
- *Disconnect the battery Ground (GND) strap.*
- *After reconnecting battery, re-code and check operation of anti-theft radio. Also check operation of clock and power windows according to Repair Manual and/or Owner's Manual.*

Special tools, testers and auxiliary items needed

- VAG 1331 Torque wrench (or equivalent 5 - 50 Nm)
- VAG 1332 Torque wrench (or equivalent 40 - 200 Nm)

Removing

– Disconnect battery Ground (GND) strap.

– Drain engine coolant, see **4.0 Liter W8 4V Engine Mechanical (code: BDP) section, Repair Group 19.**

– Support engine with engine support bridge 10-222A.

– Drain engine oil, see **Maintenance**

– Remove oil pipe from oil pan, see **4.0 Liter W8 4V Engine Mechanical (code: BDP) section, Repair Group 17.**

– Remove front subframe, see **Suspension, Wheels, Steering section, Repair Group 40**

– Remove right lower coolant pipe, see **4.0 Liter W8 4V Engine Mechanical (code: BDP) section, Repair Group 19**

– Disconnect electrical connections at Generator (GEN).

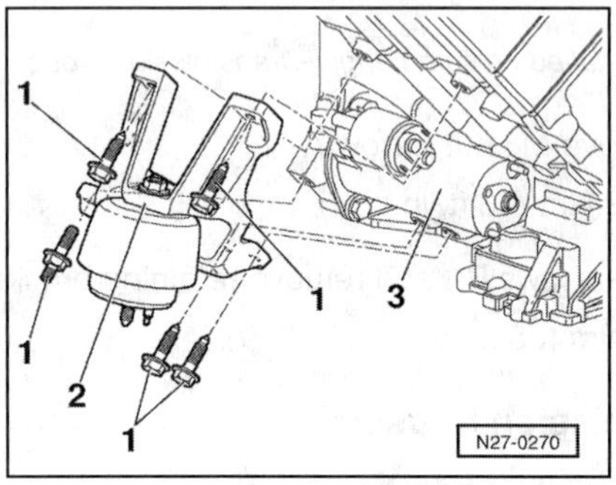

◄ Remove bolts -1- and remove right engine support -2- downwards.

- Bolt -1- tightening torque: 50 Nm (37 ft lb)

– Disconnect electrical connection to starter solenoid (terminal 50) on starter -3-.

– Remove main electrical connection to starter solenoid (terminal 30) from starter -3-

- Terminal 30 nut tightening torque: 13 Nm (10 ft lb)

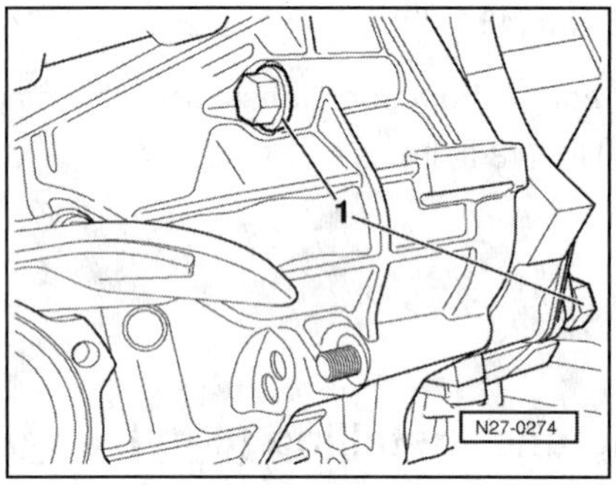

◄ Remove bolts -1- and remove starter downwards.

Starter, installing (8-cyl. engine)

Installation is reverse order of removal. Observe correct torque specifications.

Generator

> ***CAUTION*** —
> *Before beginning repairs on the electrical system:*
>
> - *Obtain the anti-theft radio security code.*
> - *Switch the ignition off.*
> - *Disconnect the battery Ground (GND) strap.*
> - *After reconnecting battery, re-code and check operation of anti-theft radio. Also check operation of clock and power windows according to Repair Manual and/or Owner's Manual.*

Special tools, testers and auxiliary items needed

- VAG 1331 Torque wrench (or equivalent 5 - 50 Nm)

Generator (GEN), removing and installing, 4-cylinder engine (through 08/1999)

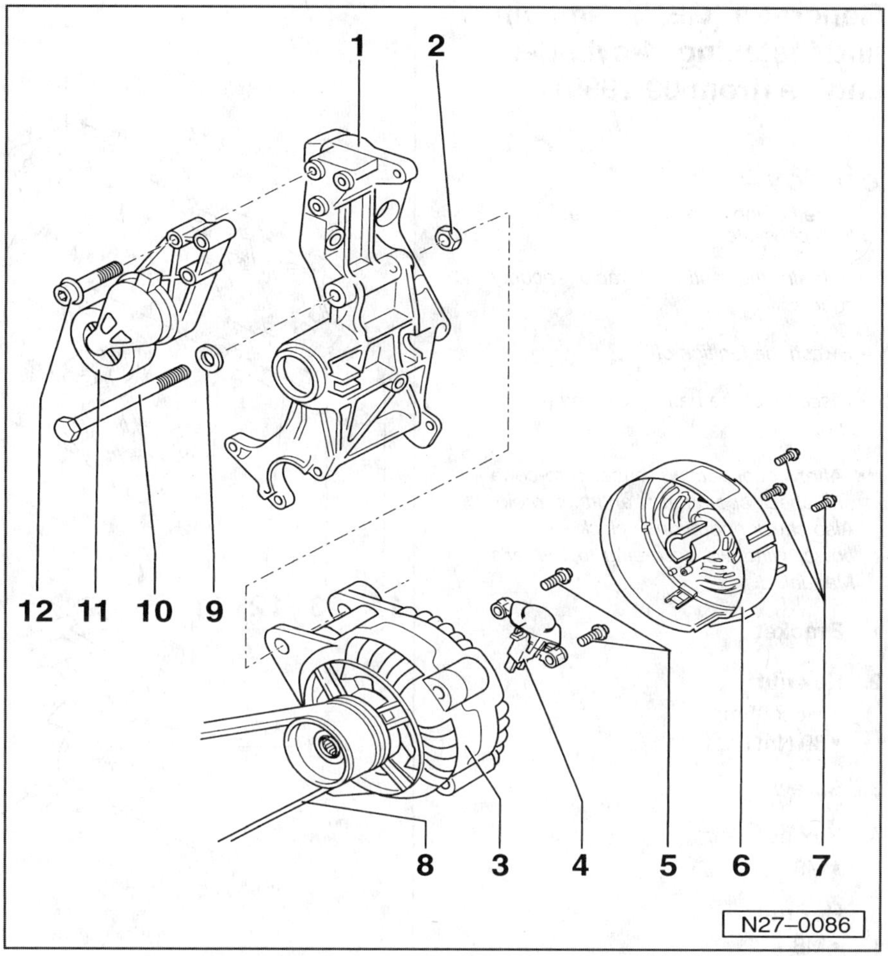

CAUTION —
Before beginning repairs on the electrical system:

- *Obtain the anti-theft radio security code.*
- *Switch the ignition off.*
- *Disconnect the battery Ground (GND) strap.*
- *After reconnecting battery, re-code and check operation of anti-theft radio. Also check operation of clock and power windows according to Owner's Manual.*

1. **Bracket**
2. **Nut**
   - M8 x 85mm
   - 30 Nm (22 ft. lb)
3. **Generator (GEN)**
   - Securing B+ wire to generator, see **Generator (GEN) B+ & D+ connections, securing (through 08/1999)**
4. **Voltage Regulator (VR)**
   - Removing and installing, see **Voltage Regulator (VR), removing and installing (through 08/1999)**
   - Checking carbon brushes, see **Voltage Regulator (VR) carbon brushes, checking (through 08/1999)**
5. **Screw**
   - M4 x 25mm
6. **Protective cover**
7. **Hex bolts**
   - M3 x 18mm
8. **Ribbed belt**
   - Routing, see **Ribbed belt, routing diagrams**
   - Removing and installing, see **Maintenance**
9. **Washer**
10. **Hex bolt**
    - M8 x 85mm
    - 25 Nm (18 ft. lb)
    - Remove viscous fan to access bolt, see **1.8 Liter 5V Turbo Engine Mechanical (code: AEB, ATW) section, Repair Group 19**.
11. **Tensioning roller**
12. **Hex socket bolt**
    - M8 x 85mm
    - 25 Nm (18 ft. lb)

Generator (GEN), removing and installing, 4-cylinder engine (from 09/1999)

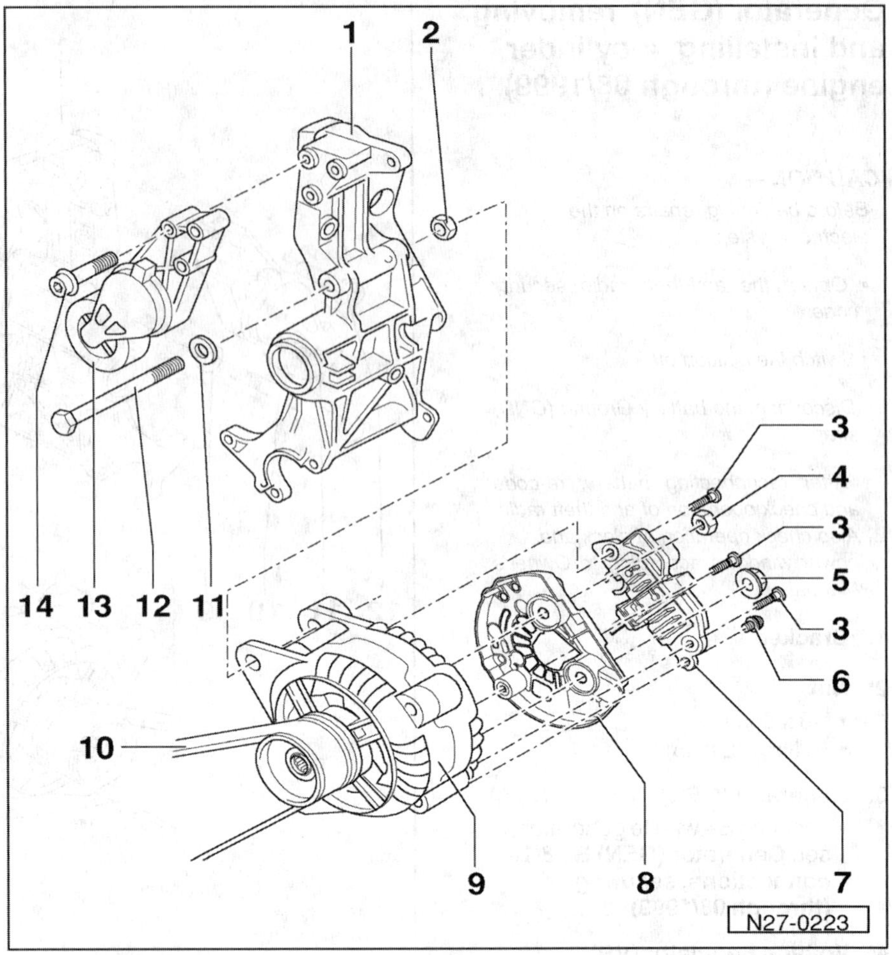

CAUTION—

Before beginning repairs on the electrical system:

- *Obtain the anti-theft radio security code.*
- *Switch the ignition off.*
- *Disconnect the battery Ground (GND) strap.*
- *After reconnecting battery, re-code and check operation of anti-theft radio. Also check operation of clock and power windows according to Owner's Manual.*

1. **Bracket**
2. **Hex nut**
   - M8 x 85mm
   - 30 Nm (22 ft. lb.)
3. **Screw**
4. **Hex nut**
   - M8
5. **Hex nut**
   - M8
6. **Screw**
7. **Voltage Regulator (VR)**
   - Carbon brushes, checking, see **Voltage Regulator (VR) carbon brushes, checking (from 09/1999)**
   - Removing and installing, see **Voltage Regulator (VR), removing and installing (from 09/1999)**
8. **Protective cap**
9. **Generator (GEN)**
   - Securing B1+ wire to generator, see **Generator (GEN) B+ connection, securing**
10. **Ribbed belt**
    - Routing, see **Ribbed belt, routing diagrams**
    - Removing and installing, see **Maintenance**
11. **Washer**
12. **Hex bolt**
    - M8 x 85mm
    - 25 Nm (18 ft. lb.)
    - Remove viscous fan to access bolt, see **1.8 Liter 5V Turbo Engine Mechanical section, Repair Group 19**.
13. **Tensioning roller**
14. **Hex socket bolt**
    - M8 x 85mm
    - 25 Nm (18 ft. lb.)

Generator (GEN), removing and installing, 6-cylinder engine (through 08/1999)

CAUTION —

Before beginning repairs on the electrical system:

- *Obtain the anti-theft radio security code.*
- *Switch the ignition off.*
- *Disconnect the battery Ground (GND) strap.*
- *After reconnecting battery, re-code and check operation of anti-theft radio. Also check operation of clock and power windows according to Owner's Manual.*

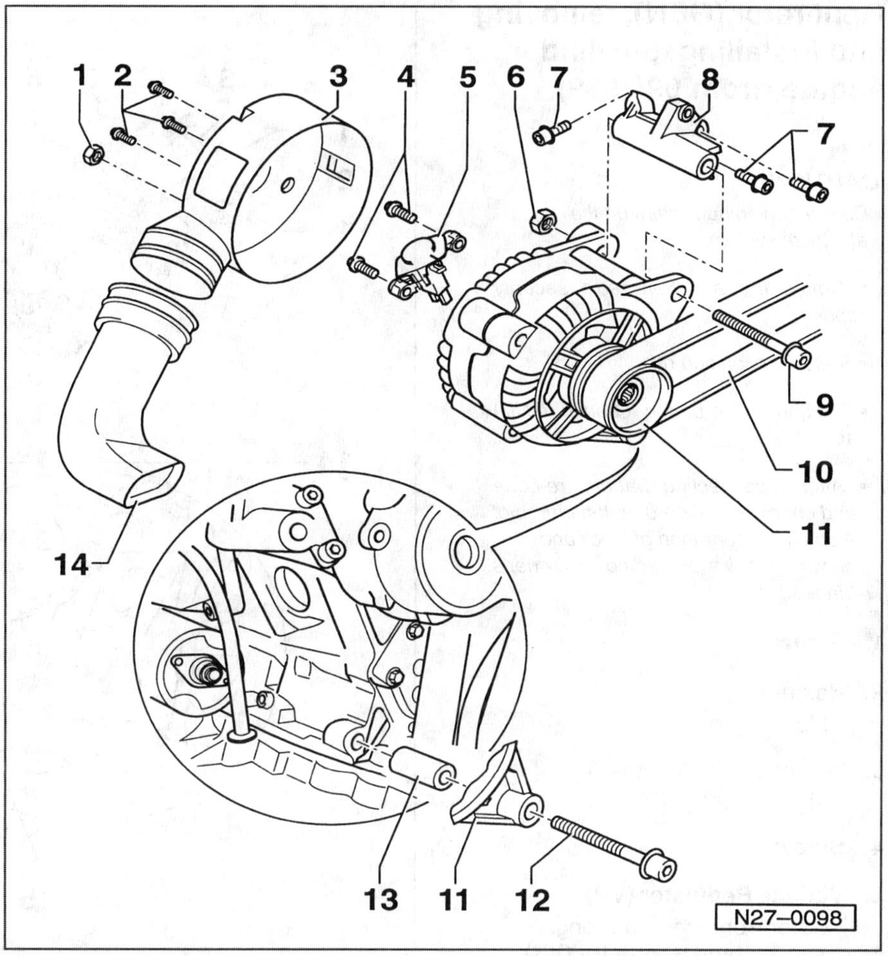

1. **Hex nut**
2. **Screw**
   - M3 x 18 mm
3. **Protective cover**
4. **Screw**
   - M4 x 25 mm
5. **Voltage Regulator (VR)**
   - Removing and installing, see **Voltage Regulator (VR), removing and installing (through 08/1999)**
   - Checking carbon brushes, see **Voltage Regulator (VR) carbon brushes, checking (through 08/1999)**
6. **Hex nut**
   - 45 Nm (33 ft. lb)
7. **Socket head bolt**
   - M8 x 20 mm
   - 25 Nm (18 ft. lb)
8. **Bracket**
9. **Socket head bolt**
   - M10 x 125 mm
10. **Ribbed belt**
    - Routing, see **Ribbed belt, routing diagrams**
    - Removing and installing, see **Maintenance**
11. **Generator (GEN)**
    - Tightening B+ and D+ wire to generator, see **Generator (GEN) B+ & D+ connections, securing (through 08/1999)**
12. **Socket head bolt**
    - M8 x 75
13. **Spacer sleeve**
14. **Air intake**

Generator (GEN), removing and installing, 6-cylinder engine (from 09/1999)

CAUTION—

Before beginning repairs on the electrical system:

- *Obtain the anti-theft radio security code.*
- *Switch the ignition off.*
- *Disconnect the battery Ground (GND) strap.*
- *After reconnecting battery, re-code and check operation of anti-theft radio. Also check operation of clock and power windows according to Owner's Manual.*

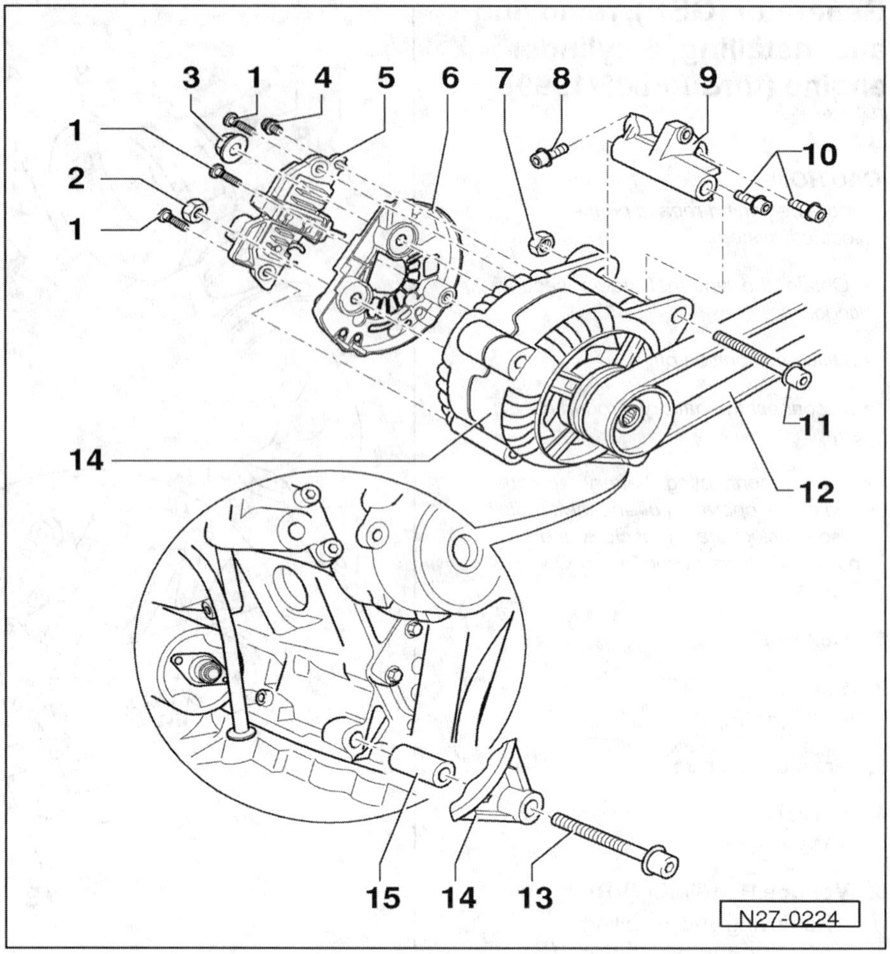

1. **Screw**
2. **Hex nut**
   - M8
3. **Hex nut**
   - M8
4. **Screw**
5. **Voltage Regulator (VR)**
   - Carbon brushes, checking, see **Voltage Regulator (VR) carbon brushes, checking (from 09/1999)**
   - Removing and installing, see **Voltage Regulator (VR), removing and installing (from 09/1999)**
6. **Protective cap**
7. **Hex nut**
   - 45 Nm
8. **Hex socket bolt**
   - M8 x 20mm
   - 25 Nm (18 ft lb)
9. **Bracket**
10. **Hex socket bolt**
    - M8 x 20mm
    - 25 Nm (18 ft lb)
11. **Hex socket bolt**
    - M10 x 125mm
12. **Ribbed belt**
    - Routing, see **Ribbed belt, routing diagrams, Fig. 2**.
    - Removing and installing, see **Maintenance**
13. **Hex socket bolt**
    - M8 x 75mm
14. **Generator (GEN)**
    - Securing B1+ wire to generator, see **Generator (GEN) B+ connection, securing**
15. **Spacer sleeve**

Generator (GEN), removing and installing, 8-cylinder engine

CAUTION *—*
Before beginning repairs on the electrical system:

- *Obtain the anti-theft radio security code.*
- *Switch the ignition off.*
- *Disconnect the battery Ground (GND) strap.*
- *After reconnecting battery, re-code and check operation of anti-theft radio. Also check operation of clock and power windows according to Owner's Manual.*

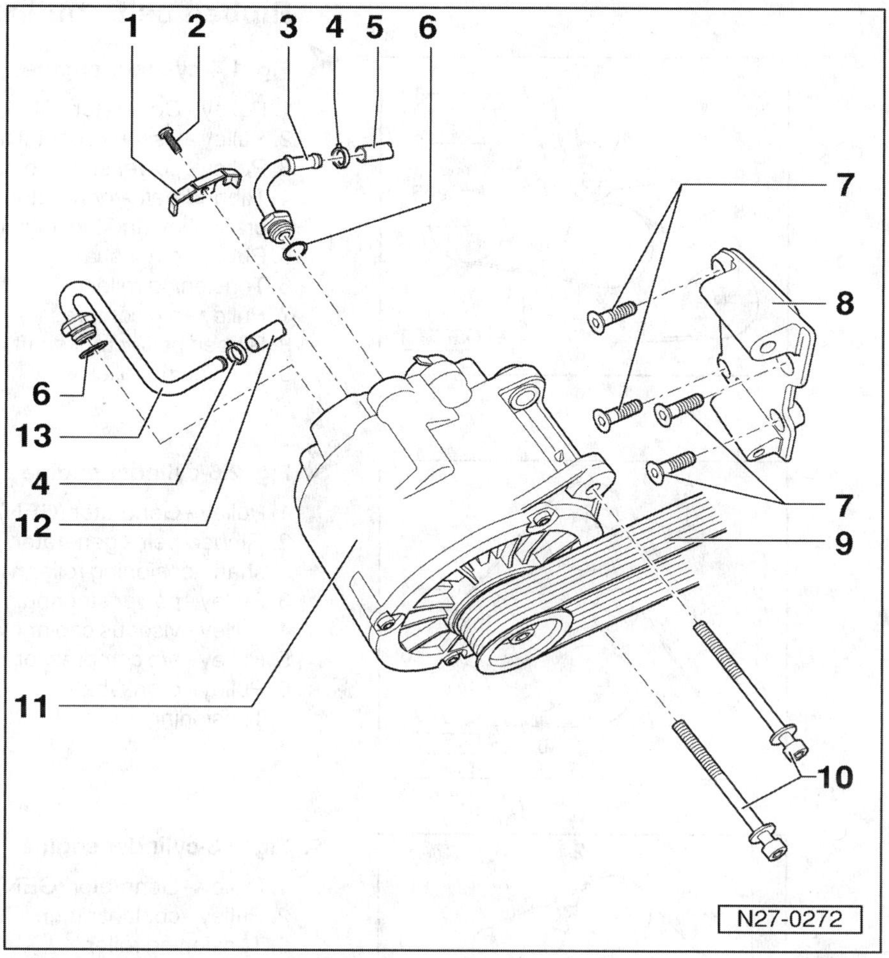

1. **Retainer**
2. **Torx head screw**
   - M6 x 20 mm
3. **Coolant pipe**

WARNING *—*
The cooling system is pressurized when the engine is warm. Wear gloves and other protection and carefully release system pressure if necessary, before performing repairs.

4. **Clamp**
5. **Coolant hose**
   - Coolant hose connection diagram, see **4.0 Liter W8 4V Engine Mechanical (code: BDP) section, Repair Group 19.**
6. **O ring**
   - Always replace
7. **Recessed hex socket bolts**
   - M8 x 32 mm
8. **Bracket**
9. **Ribbed belt**
   - Ribbed belt routing, see **Ribbed belt, routing diagrams**
   - Checking, see **Maintenance**
   - Removing and installing, see **Maintenance**
10. **Hex socket bolts**
    - M8 x 95 mm
11. **Generator (GEN)**
    - Securing B1+ wire to generator, see **Generator (GEN) B+ connection, securing**

NOTE *—*
Voltage regulator and/or brushes are not serviceable separately. When troubleshooting indicates a malfunction, replace generator complete.

12. **Coolant hose**
    - Coolant hose connection diagram, see **4.0 Liter W8 4V Engine Mechanical (code: BDP) section, Repair Group 19.**
13. **Coolant pipe**

WARNING *—*
The cooling system is pressurized when the engine is warm. Wear gloves and other protection and carefully release system pressure if necessary, before performing repairs.

Ribbed belt, routing diagrams

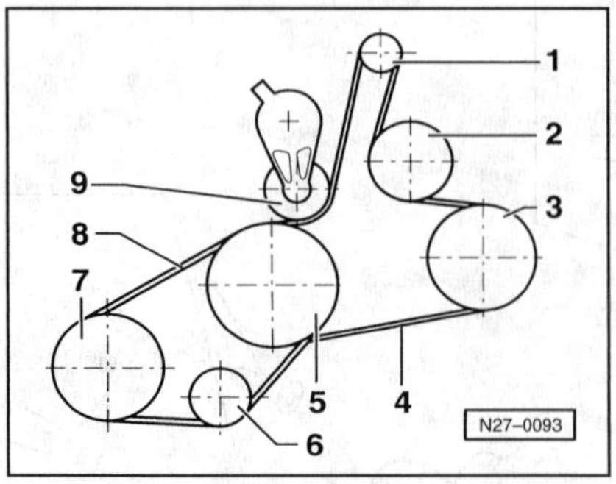

Fig. 1 4-cylinder engine

1. Pulley - Generator (GEN)
2. Pulley - viscous coolant fan
3. Pulley - power steering pump
4. Ribbed belt - generator, viscous coolant fan, power steering pump, crankshaft and tensioning roller
5. Pulley - crankshaft
6. Tensioning roller
7. Pulley - a/c compressor
8. Ribbed belt - crankshaft, tensioning roller and a/c compressor
9. Tensioning roller

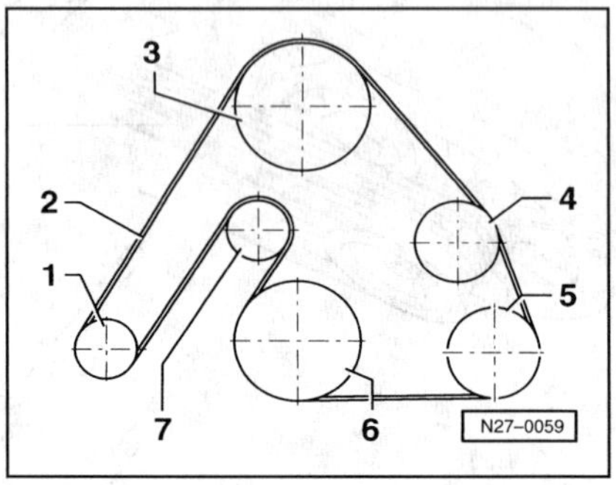

Fig. 2 6-cylinder engine

1. Pulley - Generator (GEN)
2. Ribbed belt - generator, power steering pump, coolant pump, crankshaft, tensioning roller, viscous fan and a/c compressor seetable
3. Pulley - power steering pump
4. Pulley - viscous coolant fan
5. Pulley - a/c compressor
6. Pulley - crankshaft
7. Tensioning roller

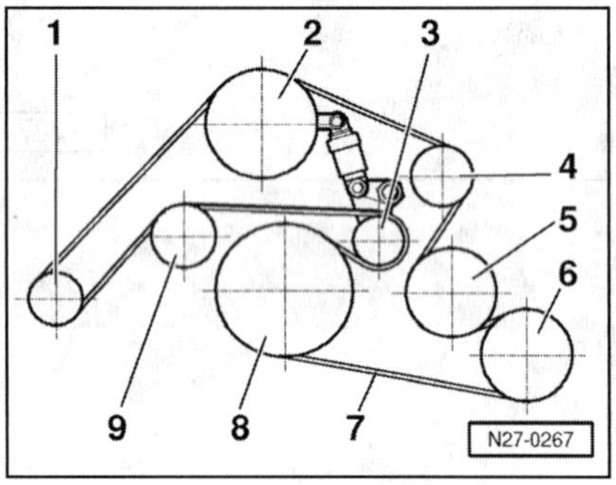

Fig. 3 8-cylinder engine

1. Pulley - Generator (GEN)
2. Pulley - coolant pump
3. Tensioning roller
4. Idler pulley
5. Pulley - a/c compressor
6. Pulley - power steering pump
7. Ribbed belt - generator, coolant pump, idler pulleys, power steering pump, a/c compressor, crankshaft and tensioning roller
8. Pulley - crankshaft
9. Idler pulley

Generator (GEN) B+ & D+ connections, securing (through 08/1999)

Various D+ terminal connector configurations exist on installed generators.

> ***CAUTION—***
> *If the B+ or D+ terminal nut is not torqued as specified, the following may result:*
>
> *• Battery will not be completely charged*
>
> *• Vehicle electrical system malfunctions*
>
> *• Danger of sparking*
>
> *• Damage to electronic components and control modules due to voltage spikes*

Special tools, testers and auxiliary items needed

- VAG 1331 Torque wrench (or equivalent 5 - 50 Nm)

Generator with D+ terminal plug connection

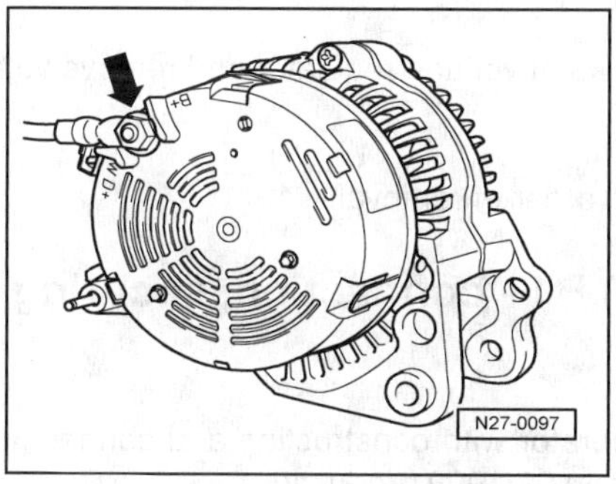

- Tightening torque for B+ terminal securing nut -arrow-: 5 Nm ± 1 Nm (11 ft lb ± 1).

Generator with D+ terminal nut connection

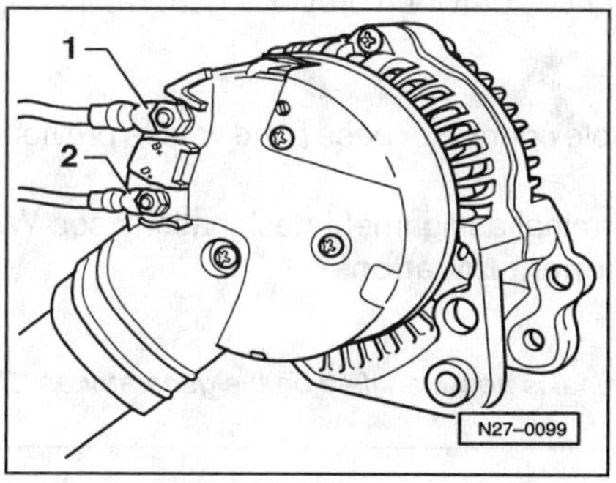

- Tightening torque for B+ terminal securing nut -1-: 15 Nm ± 1 Nm (11 ft lb ± 1).
- Tightening torque for D+ terminal securing nut -2-: 3.2 Nm ± 0.2 Nm (27 in lb).

Voltage Regulator (VR) carbon brushes, checking (through 08/1999)

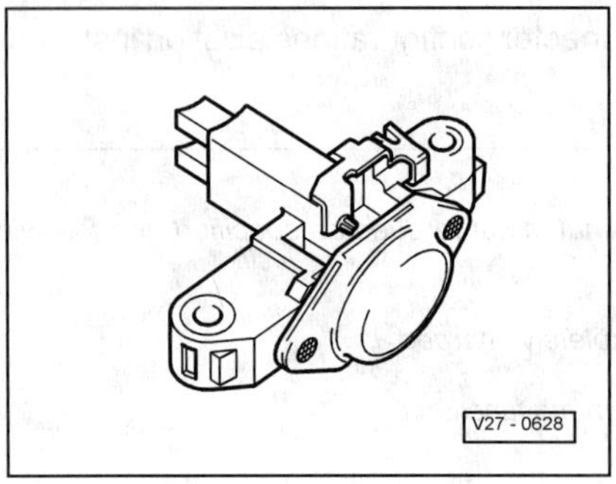

Length of new carbon brushes = 12mm (0.5")

- Wear limit = 5mm (0.19")
- Tolerance = ±1mm (0.03")

Voltage Regulator (VR), removing and installing (through 08/1999)

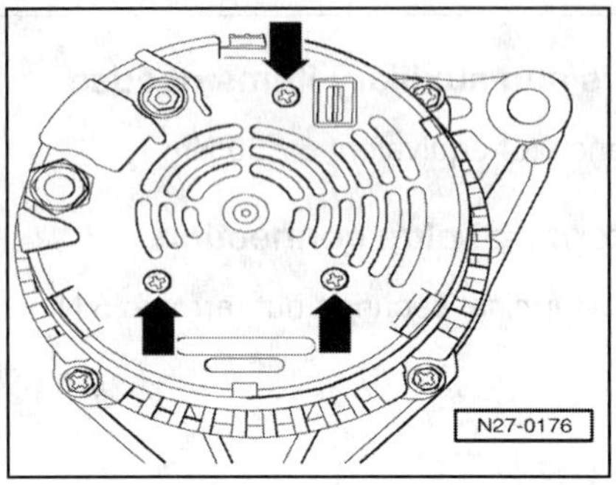

Remove screws -arrows- and protection cover.

NOTE —
On 6-cylinder engines, it may be necessary to also remove the generator air intake hose, see ***Generator (GEN), removing and installing, 6-cylinder engine (through 08/1999, item -14-****.*

– Remove mounting screws at voltage regulator and remove voltage regulator.

– Installation is in reverse order of removal.

Generator (GEN) B1+ connection, securing (from 09/1999)

A different version generator with construction and connector terminal changes are installed from 09/1999 production.

External changes/differences between previous generator versions are:

- Voltage regulator type
- Oval, radial sealed 2-pole connector base plate versus previous square base plate
- Different connector terminal assignment designations see **Wiring Diagrams and Component Locations**

NOTE —
The (battery) positive terminal is now identified on the generator as "B1+".

> ***CAUTION —***
> *If the B1+ terminal nut is not torqued as specified, the following may result:*
>
> *• Battery will not be completely charged*
>
> *• Vehicle electrical system malfunctions*
>
> *• Danger of sparking*
>
> *• Damage to electronic components and control modules due to voltage spikes*

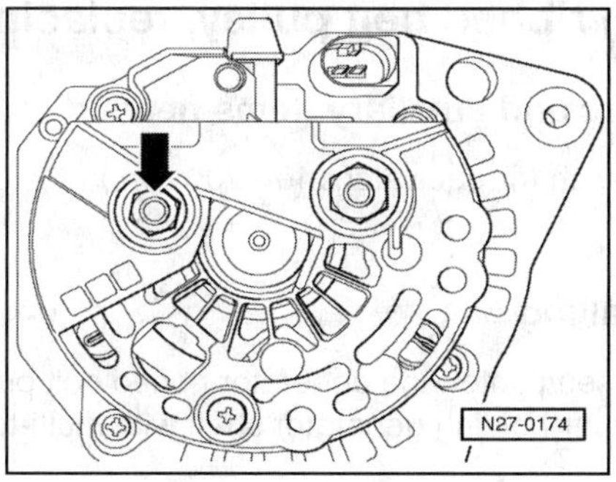

Generator with B1+ terminal nut connection (4 & 6-cyl. engines)

Tightening torque for B1+ terminal securing nut -arrow-: 15 Nm (11 ft lb).

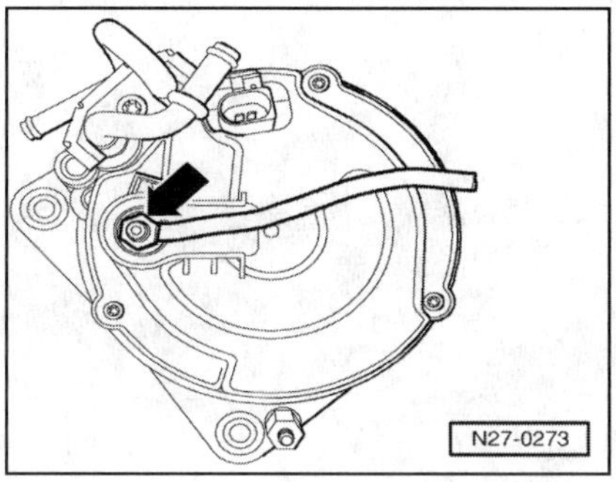

Generator with B1+ terminal nut connection (8-cyl. engine)

Tightening torque for B1+ terminal securing nut -arrow-: 15 Nm (11 ft lb).

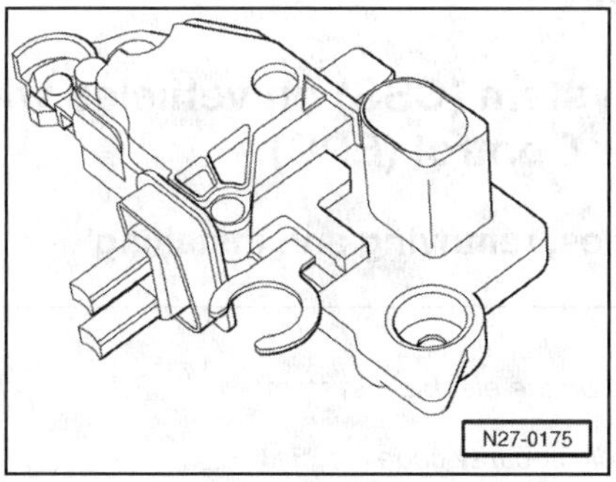

Voltage Regulator (VR) carbon brushes, checking (from 09/1999)

NOTE—

On 8-cyl. engines, the voltage regulator and/or brushes are not serviceable separately. When troubleshooting indicates a malfunction, replace generator complete.

Length of new carbon brushes = 12mm (0.5")

- Wear limit = 5mm (0.19")
- Tolerance = ±1mm (0.03")

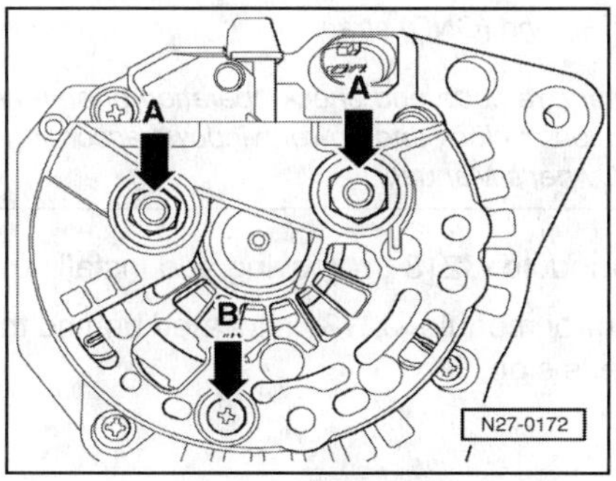

Voltage Regulator (VR), removing and installing (from 09/1999)

NOTE—

On 8-cyl. engines, the voltage regulator and/or brushes are not serviceable separately. When troubleshooting indicates a malfunction, replace generator complete.

Remove protective cover securing nuts -arrow A- and screw -arrow B- and remove protective cover.

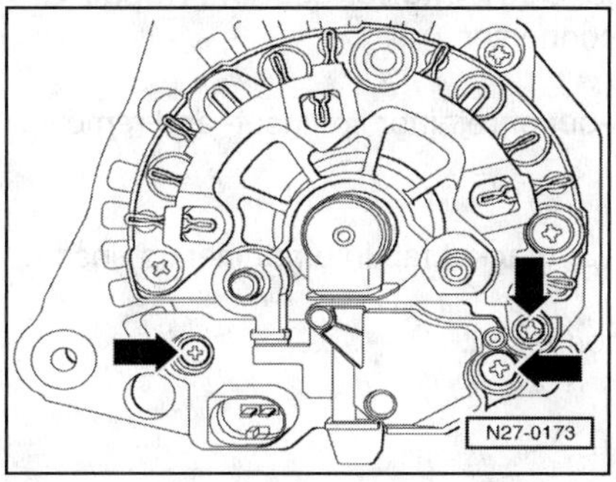

Remove mounting nuts -arrows- at voltage regulator and remove voltage regulator.

– Installation is in reverse order of removal.

Generator (GEN) ribbed belt pulley, replacing

Special tools, testers and auxiliary items needed

- VAG 1332 Torque wrench (or equivalent 40 - 200 Nm)
- Socket 3310

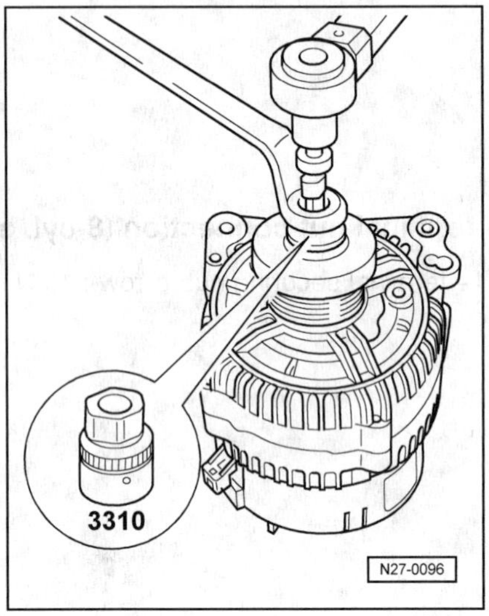

Removing and installing

Use socket 3310 to loosen/tighten the generator ribbed belt pulley securing nut while counter-holding generator shaft with multi-tooth socket.

- Pulley securing nut tightening torque: 65 Nm (48 ft. lb).

Cruise Control

Cruise control system (CSS) on vehicles without Electronic Power Control (EPC)

4 & 6 cylinder engines, removing and installing

> ***CAUTION—***
> *Before beginning repairs on the electrical system:*
>
> - *Obtain the anti-theft radio security code.*
> - *Switch the ignition off.*
> - *Disconnect the battery Ground (GND) strap.*
> - *After reconnecting battery, re-code and check operation of anti-theft radio. Also check operation of clock and power windows according to Repair Manual and/or Owner's Manual.*

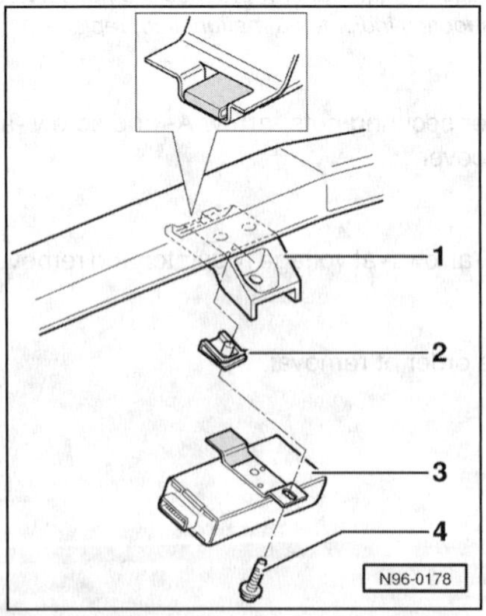

Cruise control, control module -J213-, removing and installing

- The cruise control, control module -J213- is located behind the airbag unit on front passenger's side.

– Removing:

- Remove glove box, see **Body-Interior section, Repair Group 70**
- Disconnect electrical connector
- Remove screw -4-.
- Pull control module -3-out of retainer lip -inset- and remove.

– Installing:

- Insert control module -3- bracket tab into retainer lip -inset-.
- Install screw -4-.
- Reconnect electrical connector.

Vacuum servo rod, adjusting

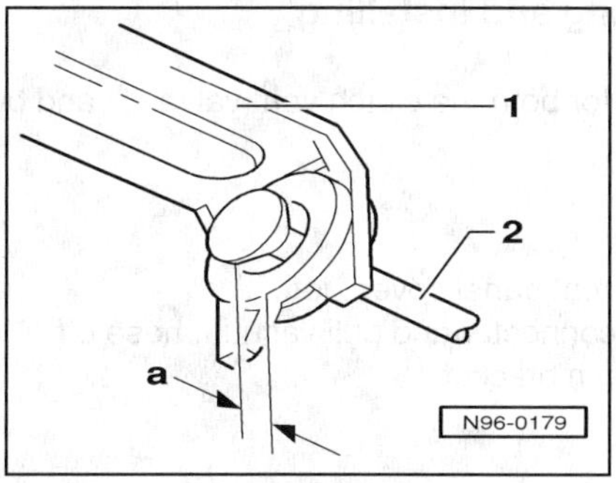

4-cylinder engine

- Start engine and run at idle (throttle housing in idling speed position)
- Turn rod -2- in onto stop and then turn back one turn (360°± 90°).
- Play -a- between rod -2- and vacuum servo linkage -1- must be between 0.5 mm and 1.0 mm (0.019" and 0.03")
- Secure servo linkage to keep from turning.

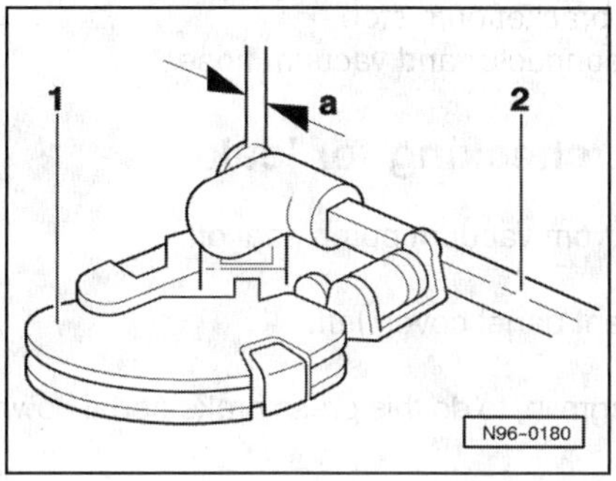

6-cylinder engine

- Adjust as per instructions above
- Play -a- between rod -2- and stop on throttle housing linkage -1- must be between 0.5 mm and 1.0 mm (0.019" and 0.03")
- Secure servo rod to keep from turning.

Vacuum servo, removing

Special tools, testers and auxiliary items needed

- VAG 1331 Torque wrench (or equivalent 5 - 50 Nm)

The actual installation position of the vacuum servo may vary from that illustrated.

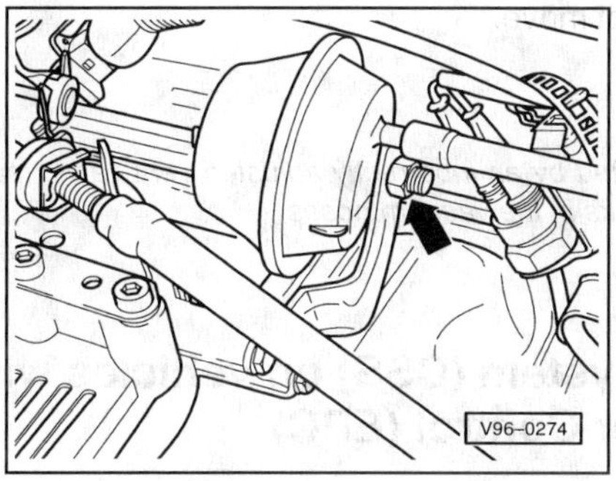

– Pull off vacuum pipe

Unscrew or unclip rod.

– Remove vacuum servo from bracket. Tightening torque: 25 Nm (18 Nm).

Vacuum pump, removing and installing

Special tools, testers and auxiliary items needed

- VAG 1331 Torque wrench (or equivalent 5 - 50 Nm)

The vacuum pump is located under the ABS assembly and can be removed outwards after removing the left wheel well liner.

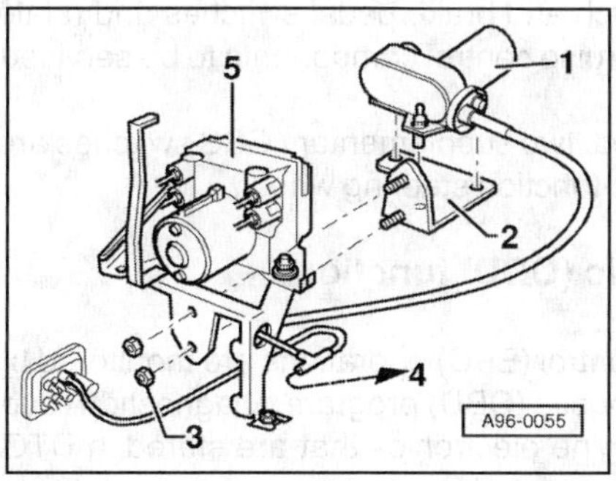

Remove securing nuts -3-, 5 Nm (44 in lb) and pull out pump from the wheel opening.

- Pull vacuum pump -1- upward from bracket -2- complete with rubber mountings.
- -4- hose routing to vacuum servo.

Vent valves, removing and installing

Procedure is the same for both the clutch vent valve -1- and brake vent valve -2-.

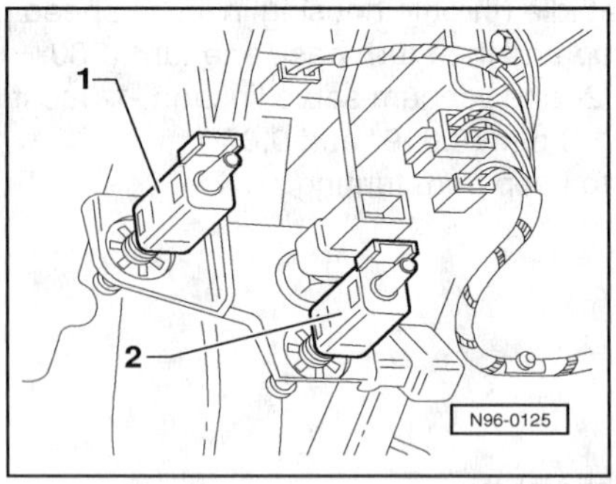

< Removing:

- Remove lower instrument panel cover , left.
- Disconnect electrical connector and pull vacuum hose off.
- Unscrew vent valve from bracket.

– Installing:

- Pedal in rest position.
- Screw vent valve into bracket onto stop.
- Reconnect electrical connector and vacuum hose.

Vacuum system, checking for leaks

– Remove vacuum hose from vacuum pump seal off.

– Remove lower instrument panel cover, left.

– Push servo unit diaphragm in, to do this press brake pedal down by hand until pressure point.

– Then release brake pedal.

The diaphragm must not move.

NOTE —
Malfunctions can be caused by an incorrectly adjusted vent valve, leaking servo unit or through cracks in the vacuum hoses.

Cruise control system (CSS) on vehicles with Electronic Power Control (EPC)

General information

All cruise control system functions are controlled by the applicable Engine Control Module (ECM). The processing of electronic engine controls and cruise control functions are integrated in the ECM. Other than the cruise control switches on the steering column or multifunction steering wheel, the clutch and brake pedal switches and related wiring, there are no separate cruise control components to be serviced.

On vehicles so equipped, two supplementary CSS switches are integrated into the multi-function steering wheel.

On Board Diagnostic (OBD) functions

As Electronic Power Control (EPC) operations are monitored by the engine On Board Diagnostic (OBD) program, Diagnostic Trouble Codes (DTC) pertaining to engine electronics that are stored in DTC memory may be relevant to cruise control function.

90 Instruments

Repair Group

W 90

Instrument cluster

General information

The instrument cluster contains an electronic speedometer, tachometer, liquid crystal (LCD) displays for odometer, trip odometer/clock, as well as analog coolant temperature and fuel level gauges. Control and warning lamps are situated within and between the speedometer and tachometer. Automatic transmission models also contain an LCD gear indicator display.

An LCD multi-function trip computer (MFI) with two operating modes is located in the tachometer display. The MFI includes selectable displays for trip time/mileage, average trip speed, average trip fuel consumption and outside temperature.

If available, a Navigation system display is integrated into the centrally located control and warning lamp display.

Notes on exchanging instrument cluster

- Do not disassemble instrument cluster.
- Instrument clusters with green illumination are equipped with replaceable control lamp bulbs. On these units, all other instrument cluster malfunctions require replacement of instrument cluster.
- Instrument clusters with blue illumination contain no field serviceable components. All malfunctions require replacement of instrument cluster.
- When the replacement of a malfunctioning instrument cluster is necessary, follow exchange part procedures.
- Complete report form and return together with inst. cluster.
- Use the original packaging from the new cluster when returning modules.
- The odometer reading must be matched to the new/exchange instrument cluster using Volkswagen factory scan tool. Consult an authorized Volkswagen service department for additional information.

On Board Diagnostic (OBD), function

The instrument cluster is controlled by an internal microprocessor with On Board Diagnostic (OBD) capability. If malfunctions occur in monitored sensors and components, Diagnostic Trouble Codes (DTC) will be stored in memory.

If OBD recognizes instrument cluster malfunctions related to the:

- odometer
- distance impulse number (K number)
- speedometer calibration

the trip odometer display will change to show "dEF.".

In this case, replace speedometer, see **Instrument cluster, removing and installing**.

Other malfunctions recognized by OBD will not cause the trip odometer display to change as described above.

Through 2000 m.y.

Before performing any troubleshooting or inspection, always begin by checking for DTCs using the On Board Diagnostic (OBD) program. DTCs stored in memory are retrieved/checked with either VAG 1551 or equivalent scan tool.

From 2001 m.y.

Before performing any troubleshooting or inspection, always begin by checking for DTCs using the On Board Diagnostic (OBD) program. DTCs stored in memory are retrieved/checked with VAS 5051 in mode "Self Diagnosis".

Additional information, instrument clusters from 08/1999

Instrument clusters from 08/1999 are integrated into the vehicle CAN Data Bus network (may also be known as "CAN-Bus" or "Data-Bus").

The Data Bus On Board Diagnostic Interface -J533- (which is integrated into the instrument cluster) enables data to be exchanged between the vehicle CAN Data-Bus network and the Data Link Connector (DLC) "K-wire".

The Data Bus On Board Diagnostic Interface -J533- has specific On Board Diagnostic (OBD) capabilities.

Notes on exchanging instrument cluster

- In addition to the normal service procedure of coding the instrument cluster and adapting the odometer reading, the Data Bus On Board Diagnostic Interface -J533- must also be separately coded according to vehicle market version and equipment level variables, see **Scan Tool section**.

Instrument cluster, removing and installing

> ***CAUTION —***
> *Before beginning repairs on the electrical system:*
>
> *• Obtain the anti-theft radio security code.*
>
> *• Switch the ignition off.*
>
> *• Disconnect the battery Ground (GND) strap.*
>
> *• After reconnecting battery, re-code and check operation of anti-theft radio. Also check operation of clock and power windows according to Repair Manual and/or Owner's Manual.*

> ***CAUTION —***
> *Special safety precautions apply to vehicles equipped with airbags. Refer to **Body-Interior section, Repair Group 69**.*

Special tools, testers and auxiliary items needed

- VAG 1332 Torque wrench (or equivalent 40 - 200 Nm)
- VAG 1551 Scan Tool (ST)
- VAG 1551/3A, or VAG 1551/3B Adapter cable
- Optional: VAG 1552 or VAS 5051 Diagnostic Testing and Information System

Removing

If replacing a malfunctioning instrument cluster with a new one, note odometer reading on existing cluster prior to removal.

– Remove driver's airbag, see **Body-Interior section, Repair Group 69**

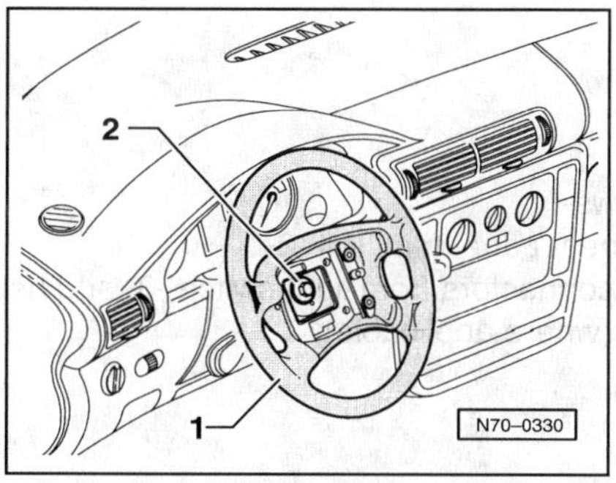

◄ Remove hex bolt -2- (always replace).

- Tightening torque: 60 Nm (44 ft lb).
- Position steering wheel -1- in central position (wheels straight ahead).
- Pull steering wheel -1- from steering column.

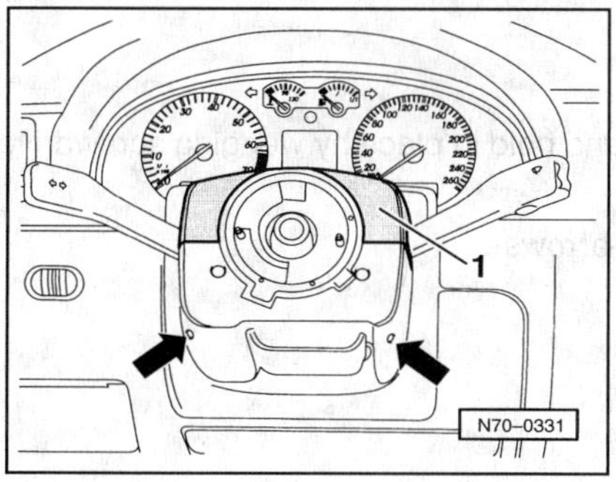

◄ Remove both (Phillips head) screws -arrows-.

- Remove upper steering column trim -1- upwards.

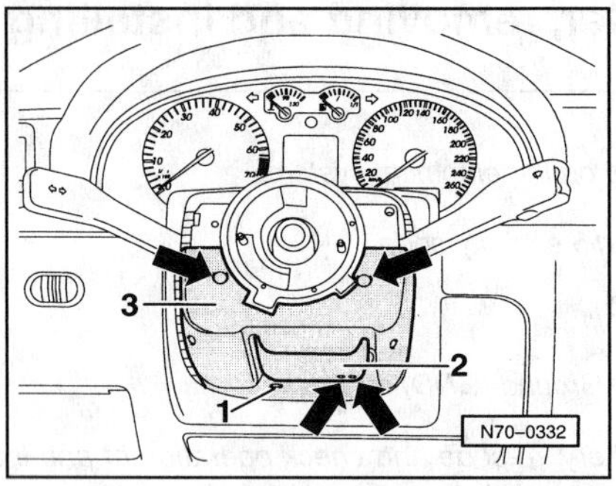

Remove four (Phillips head) screws -arrows-.

- Remove socket head bolt -1-.
- Release steering wheel height adjustment -2-.
- Remove lower steering column trim -3- downwards.

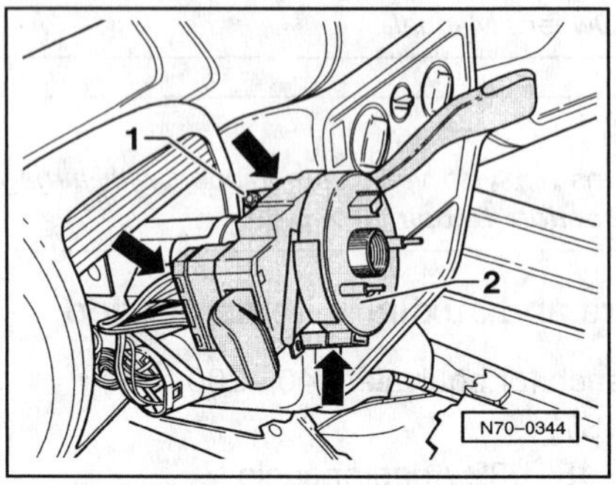

Loosen socket head bolt -1-.

– Disconnect electrical connectors -arrows- from steering column switches.

– Remove steering column switches

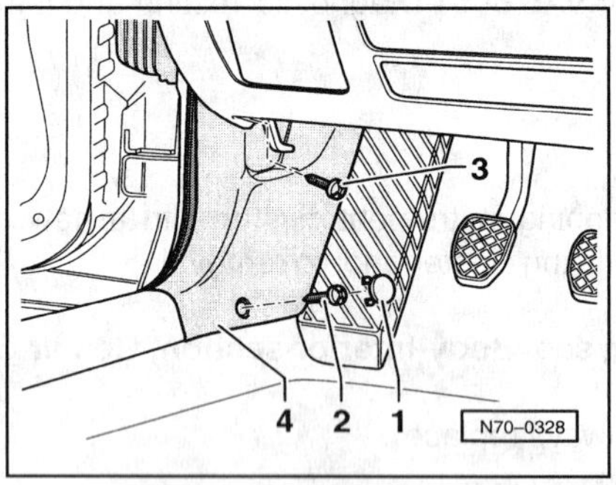

Pry off cap -1-.

- Remove screws -2- and -3-.
- Unclip kick panel trim -4- from driver's side sill trim strip and remove.

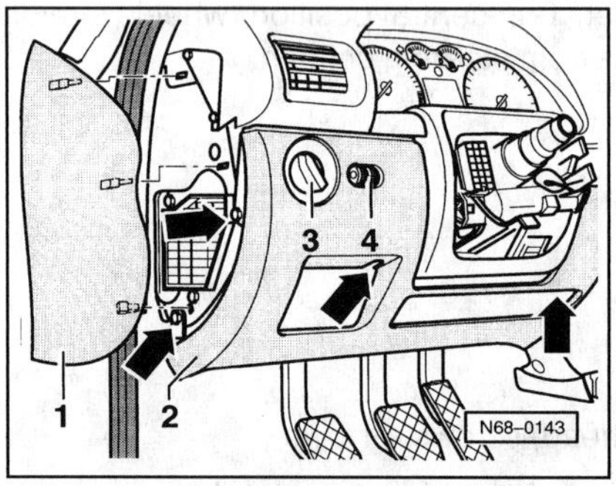

Unclip cover -1-.

- Remove screws -arrows-.
- Remove lower instrument panel trim, left -2-.
- Disconnect electrical connectors from light switch -3- and headlight range adjustment -4- (where applicable).

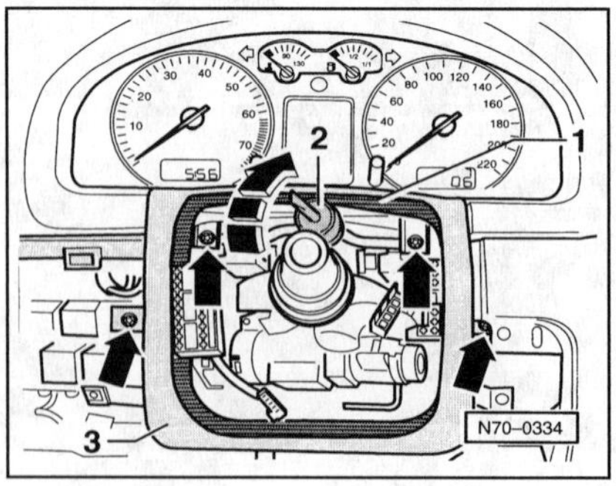

Slide trim -1- upwards and hold in place by wedging screwdriver handle -2-.

- Remove four screws -arrows-.
- Remove cover -3-.

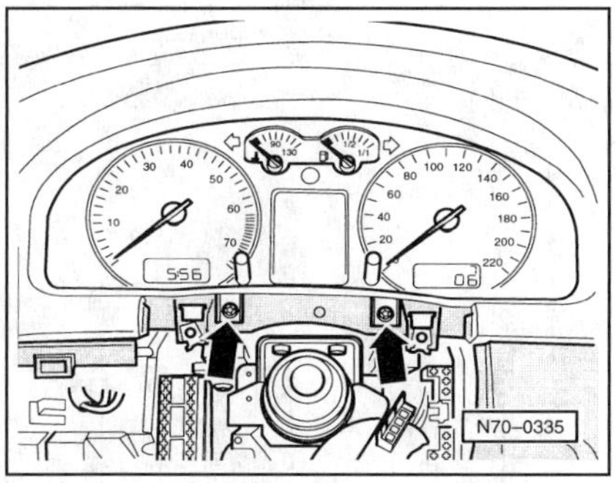

Remove two screws -arrows-.

– Pull out instrument cluster from opening.

– Disconnect electrical connections on rear of instrument cluster and remove cluster from instrument panel.

Instrument cluster, installing (through 2000 m.y.)

– Install in reverse order of removal.

– Perform functional check of instrument cluster.

- If replacing a malfunctioning instrument cluster, and a function check shows no further malfunctions, proceed as follows:

– Code instrument cluster using scan tool.

– On instrument clusters from 08/1999, the Data Bus On Board Diagnostic Interface -J533- must be also be coded according to vehicle market version and equipment level variables, see **Scan Tool section**.

– Adapt odometer reading using scan tool.

– Where applicable, adapt instrument cluster for Service Reminder Indicator (SRI) functions, see **Scan Tool section**.

– Where applicable, adapt instrument cluster for anti-theft immobilizer functions using scan tool.

Instrument cluster, installing (from 2001 m.y.)

– Install in reverse order of removal.

– Perform functional check of instrument cluster.

- If replacing a malfunctioning instrument cluster, and a function check shows no further malfunctions, proceed as follows:

– Connect VAS 5051 to vehicle Data Link Connector (DLC).

– Select mode "Vehicle Self-Diagnosis".

– Select "17- Dash panel insert".

Select and perform the following functions in sequence:

– Code instrument cluster

– On instrument clusters from 08/1999, the Data Bus On Board Diagnostic Interface -J533- must be also be coded according to vehicle market version and equipment level variables.

– Adapt odometer reading

– Where applicable, adapt instrument cluster for Service Reminder Indicator (SRI) functions

– Adapt instrument cluster for anti-theft immobilizer functions

Instrument cluster bulbs, replacing (green illumination only)

CAUTION—

Obtain the anti-theft radio security code.

- *Switch the ignition off.*
- *Disconnect the battery Ground (GND) strap.*
- *After reconnecting battery, re-code and check operation of anti-theft radio. Also check operation of clock and power windows according to Owner's Manual.*

NOTE—

- *Do not disassemble instrument cluster.*
- *Troubleshooting procedures and wiring diagrams, see* **Electrical Wiring Diagrams and Component Locations**

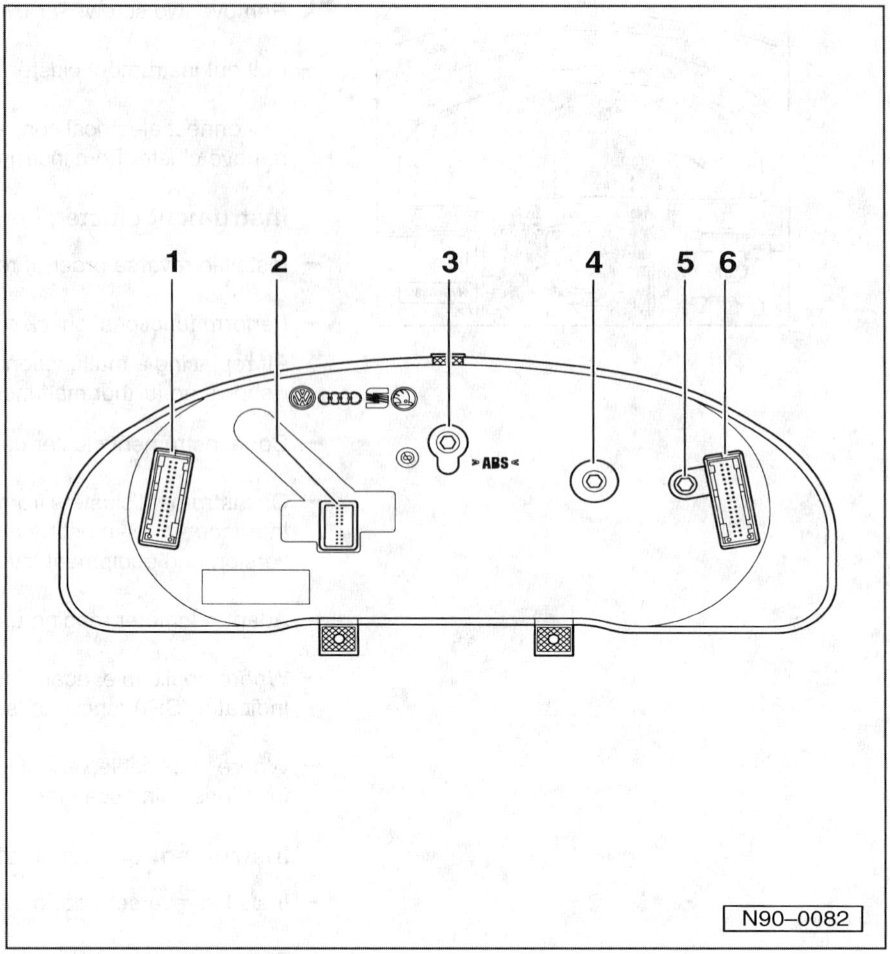

1. **32 pin multi-connector**
   - Green
   - Connector assignments through 08/1999, see **Instrument cluster multi-pin connector assignments (through 08/1999)**
   - Connector assignments from 08/1999 see **Instrument cluster multi-pin connector assignments (09/1999 to 05/2001)**
2. **20 pin multi-connector**
   - Where applicable
   - Red
   - Connector assignments through 08/1999, see **Instrument cluster multi-pin connector assignments (through 08/1999)**
   - Connector assignments from 08/1999 see **Instrument cluster multi-pin connector assignments (09/1999 to 05/2001)**
3. **Bulb with holder (headlight high beam indicator light -K1-)**
   - 12 V / 1.12 W
4. **Bulb with holder**
   - Malfunction Indicator Lamp (MIL)
   - 12 V / 1.12 W
5. **Bulb with holder (rear hatch/trunk lid ajar warning lamp -K102-)**
   - 12 V / 1.12 W
6. **32 pin multi-connector**
   - Blue
   - Connector assignments through 08/1999, see **Instrument cluster multi-pin connector assignments (through 08/1999)**
   - Connector assignments from 08/1999 see **Instrument cluster multi-pin connector assignments (09/1999 to 05/2001)**

Instrument cluster multi-pin connector assignments (through 08/1999)

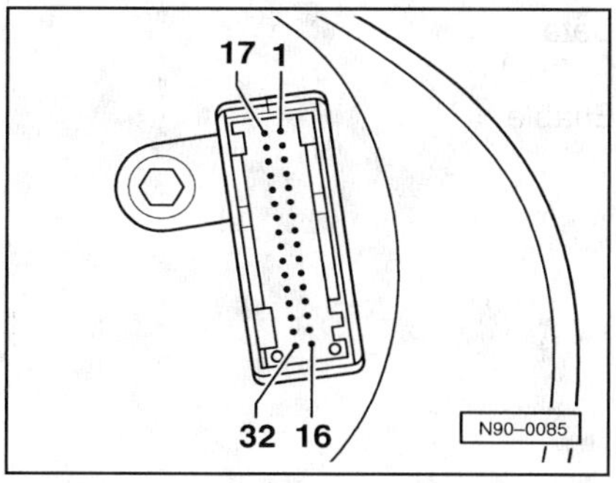

32 pin multi-connector -T32a-, blue

1. Terminal 15 - positive (ignition on)
2. Indicator lamp for turn signal, right
3. Output signal 1 from electronic speedometer
4. Open
5. Fuel gauge
6. Airbag Malfunction Indicator Lamp (MIL)
7. Terminal 31 - sensor/sender Ground (GND)
8. Engine Coolant Temperature (ECT) gauge
9. Terminal 31 - Ground (GND)
10. Oil pressure warning lamp
11. Engine Speed (RPM) signal
12. Terminal 61 - warning lamp for Generator (GEN),

13. - 15. Open

16. Warning lamp for rear hatch/trunk lid open
17. Terminal 56a - indicator lamp for high beam headlight
18. Indicator lamp for turn signal, left
19. Warning lamp for Anti-lock Brake System (ABS)
20. Terminal 58b - instrument cluster illumination
21. Driver's door open warning buzzer
22. Warning lamp for low Engine Coolant Level (ECL)
23. Terminal 30 - positive (B+)
24. Terminal 31 - Ground (GND)
25. Data Link Connector (DLC) K-wire
26. Parking lights, right (signal for "lights on" warning buzzer)
27. Parking lights, left, (signal for "lights on" warning buzzer)
28. Input signal - Speedometer Vehicle Speed Sensor (VSS)
29. Brake fluid level warning lamp
30. Terminal 86s - Ignition key warning buzzer
31. Seat belt warning lamp/buzzer
32. Malfunction Indicator Lamp (MIL)

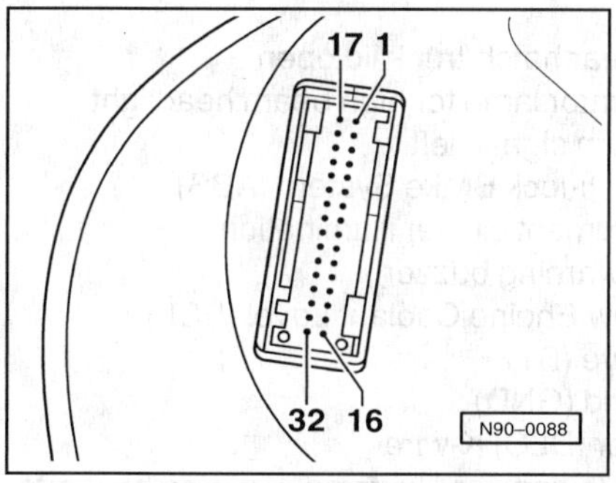

32 pin multi-connector -T32b-, green

1. - 4. Open

5. W-wire*
6. Indicator lamp for low windshield washer fluid*
7. Brake pad wear indicator*
8. External buzzer*
9. Open
10. Low fuel level warning*
11. Signal for vehicle stationary
12. A/C clutch cut-off - Engine Coolant Temperature (ECT) signal
13. Warning lamp for dual circuit brake system and handbrake
14. Indicator lamp for traction control

15. - 22. Open

23. Multi Function Trip Computer (MFI) - top row (display switches forwards)*) Where applicable
24. MFI - bottom row (display switches backwards)
25. MFI - memory switch: reset
26. MFI - ambient temperature display

27. - 29. Open

30. Output signal 2 from electronic speedometer
31. Automatic transmission selector lever display*
32. Fuel consumption signal

***NOTE*—**

**) Where applicable*

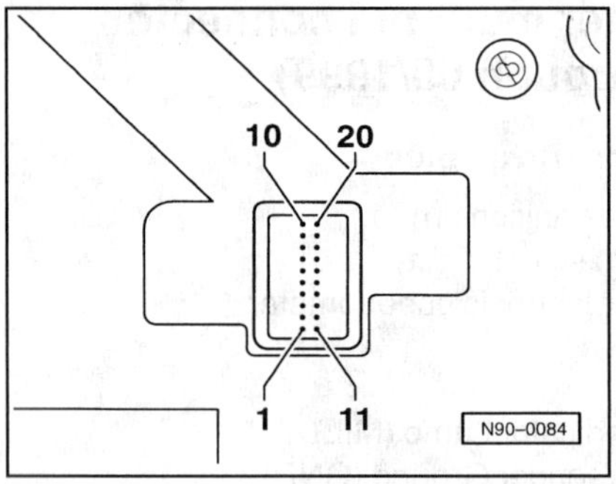

20 pin multi-connector -T20a-, red*

1. - 11. Open
12. Radio/Navigation - Clock
13. Radio/Navigation - Data
14. Open
15. Radio/Navigation - Enable
16. - 20. Open

NOTE —

**) Where applicable*

Instrument cluster multi-pin connector assignments (09/1999 to 05/2001)

NOTE —

Multi-pin connection allocation depends on installed equipment and engine variants.

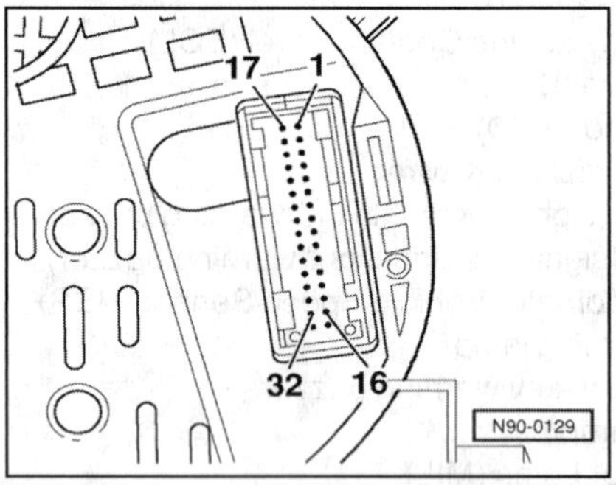

32 pin multi-connector -T32a-, blue

1. Terminal 15 - positive (ignition on)
2. Indicator lamp for turn signal, right
3. Output signal 1 from electronic speedometer
4. Open
5. Fuel gauge
6. Airbag Malfunction Indicator Lamp (MIL)*
7. Terminal 31 - sensor/sender Ground (GND)
8. Engine Coolant Temperature (ECT) gauge
9. Terminal 31 - Ground (GND)
10. Oil pressure warning lamp
11. Open
12. Terminal 61 - warning lamp for Generator (GEN),
13. - 15. Open
16. Warning lamp for rear hatch/trunk lid open
17. Terminal 56a - indicator lamp for high beam headlight
18. Indicator lamp for turn signal, left
19. Warning lamp for Anti-lock Brake System (ABS)
20. Terminal 58b - instrument cluster illumination
21. Driver's door open warning buzzer
22. Warning lamp for low Engine Coolant Level (ECL)
23. Terminal 30 - positive (B+)
24. Terminal 31 - Ground (GND)
25. Data Link Connector (DLC) K-wire
26. Parking lights, right (signal for "lights on" warning buzzer)
27. Parking lights, left, (signal for "lights on" warning buzzer)
28. Input signal - Speedometer Vehicle Speed Sensor (VSS)
29. Brake fluid level warning lamp
30. Terminal 86s - Ignition key warning buzzer
31. Seat belt warning lamp/buzzer
32. Open

*) Where applicable

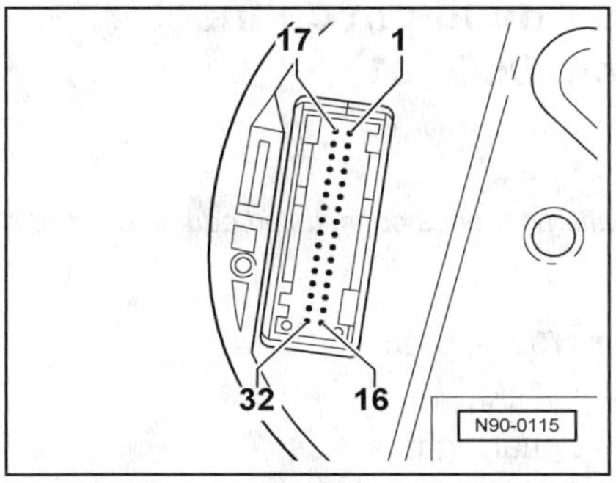

32 pin multi-connector -T32b-, green

1. Open
2. Reading coil for immobilizer 1*
3. Open
4. CAN Data Bus, shielded input*
5. W-wire*
6. Indicator lamp for low windshield washer fluid
7. Brake pad wear indicator
8. Open
9. Open
10. Low fuel level warning
11. Signal for vehicle stationary
12. A/C clutch cut-off - Engine Coolant Temperature (ECT) signal
13. Warning lamp for dual circuit brake system and handbrake
14. Indicator lamp for traction control/ESP*
15. - 16. Open
17. Reading coil for immobilizer 2*
18. Oil level thermal sensor*
19. Interface - CAN Data Hi-Bus
20. Interface - CAN Data Lo-Bus
21. Open
22. Hood alarm switch*
23. Multi Function Trip Computer (MFI) - top row (forwards)
24. MFI - bottom row (backwards)
25. MFI - memory switch: reset
26. MFI - ambient temperature display
27. On Board Diagnostic (OBD) Interface - CAN Data Hi-Bus
28. On Board Diagnostic (OBD) Interface - CAN Data Lo-Bus
29. - 32. Open

*) Where applicable

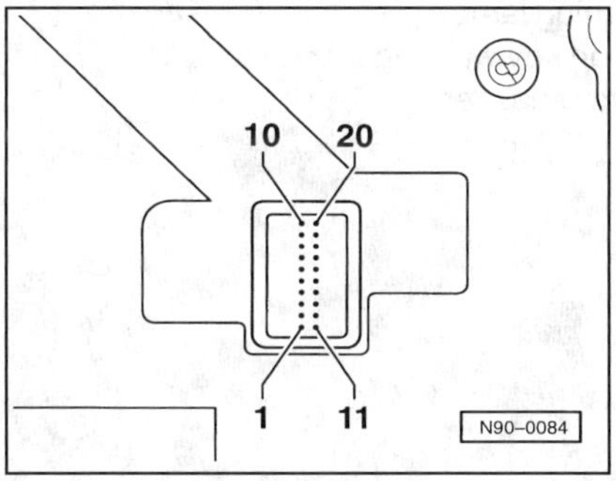

20 pin multi-connector -T20a-, red*

1. - 4. Open
5. Shielded Ground (GND) for Telematik*
6. -11. Open
12. Radio/Navigation - Clock
13. Radio/Navigation - Data
14. Open
15. Radio/Navigation - Enable
16. - 20. Open

*) Where applicable

Instrument cluster multi-pin connector assignments (from 06/2001)

NOTE —
Multi-pin connection allocation depends on installed equipment and engine variants.

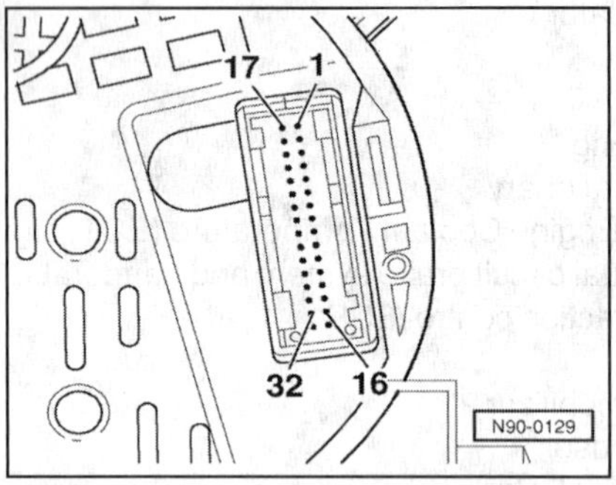

32 pin multi-connector -T32a-, blue

1. Terminal 15 - positive (ignition on)
2. Indicator lamp for turn signal, right
3. Output signal 1 from electronic speedometer
4. Open
5. Fuel gauge
6. Open
7. Terminal 31 - sensor/sender Ground (GND)
8. Engine Coolant Temperature (ECT) gauge
9. Terminal 31 - Ground (GND)
10. Oil pressure warning lamp
11. Open
12. Terminal 61 - warning lamp for Generator (GEN),

13. - 16. Open

17. Terminal 56a - indicator lamp for high beam headlight
18. Indicator lamp for turn signal, left
19. Warning lamp for Anti-lock Brake System (ABS)
20. Terminal 58d - instrument cluster illumination
21. Open
22. Warning lamp for low Engine Coolant Level (ECL)*
23. Terminal 30 - positive (B+)
24. Terminal 31 - Ground (GND)
25. Data Link Connector (DLC) K-wire
26. Parking lights, right (signal for "lights on" warning buzzer)
27. Parking lights, left, (signal for "lights on" warning buzzer)
28. Input signal - Speedometer Vehicle Speed Sensor (VSS)
29. Brake fluid level warning lamp
30. Terminal 86s - Ignition key warning buzzer
31. Seat belt warning lamp/buzzer
32. Open

*) Where applicable

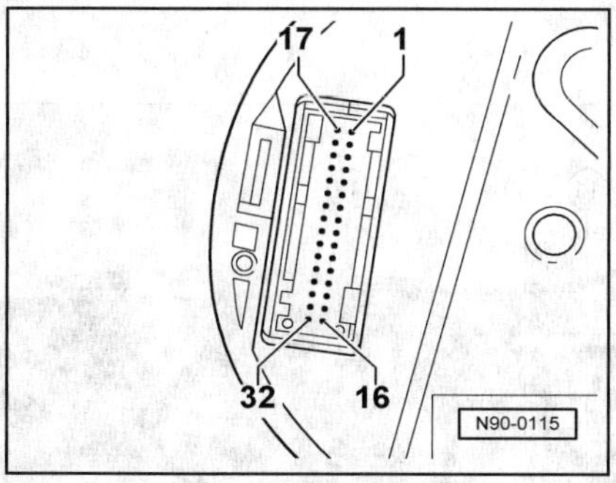

32 pin multi-connector -T32b-, green

1. Open
2. Reading coil for immobilizer 1
3. Open
4. CAN Data Bus, shielded input*
5. W-wire*
6. Indicator lamp for low windshield washer fluid
7. Brake pad wear indicator
8. CAN Data High Bus (Comfort System)*
9. CAN Data Low Bus (Comfort System)*
10. Open
11. Signal for vehicle stationary*
12. A/C clutch cut-off - Engine Coolant Temperature (ECT) signal
13. Warning lamp for dual circuit brake system and handbrake
14. Indicator lamp for traction control/ESP*
15. - 16. Open
17. Reading coil for immobilizer 2
18. Oil level thermal sensor*
19. Interface - CAN Data Hi-Bus
20. Interface - CAN Data Lo-Bus
21. Open
22. Hood alarm switch*
23. Multi Function Trip Computer (MFI) - top row (forwards)
24. MFI - bottom row (backwards)
25. MFI - memory switch: reset
26. MFI - ambient temperature display
27. On Board Diagnostic (OBD) Interface - CAN Data Hi-Bus
28. On Board Diagnostic (OBD) Interface - CAN Data Lo-Bus
29. - 32. Open

*) Where applicable

Instrument cluster layout

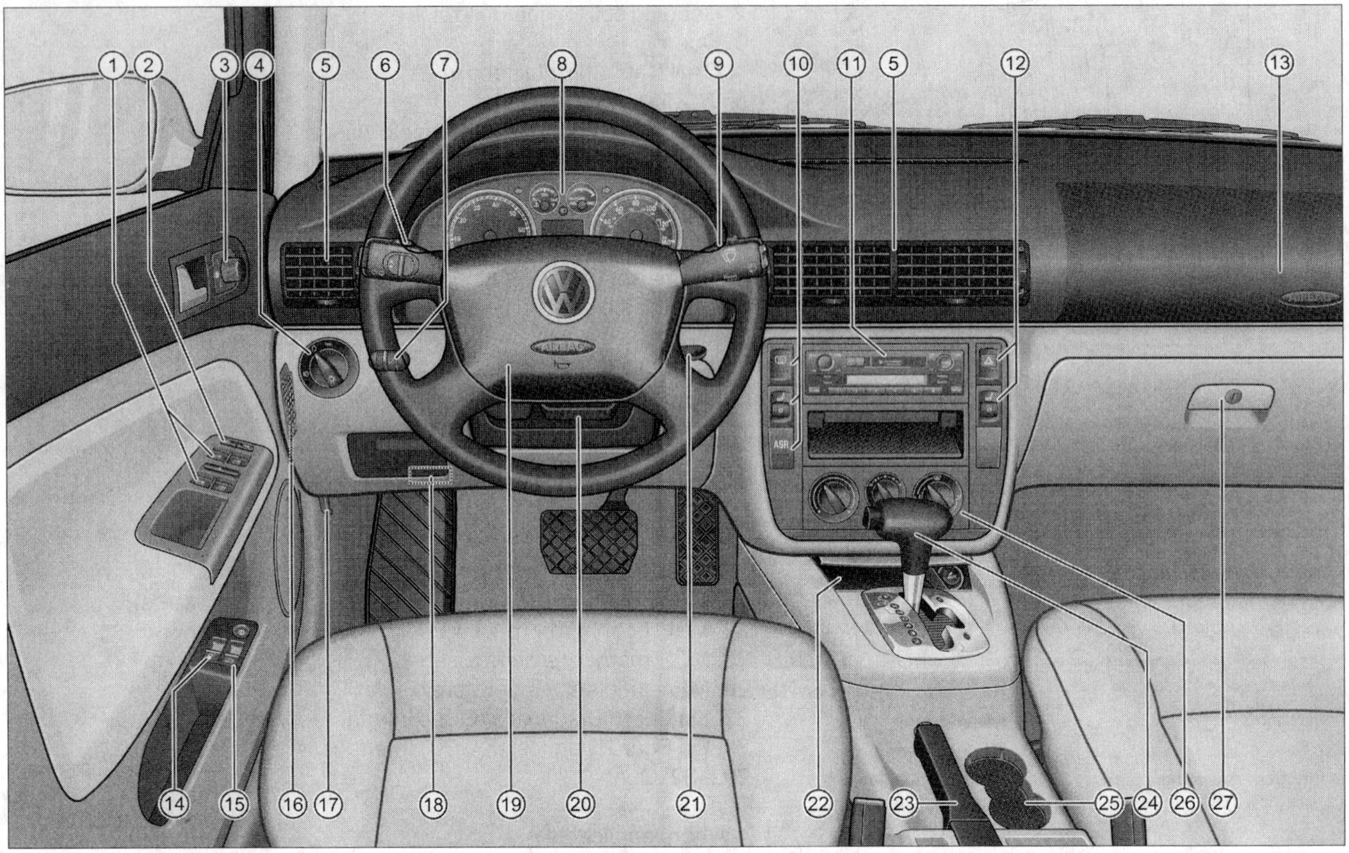

1. **Power window switches**
2. **Power locking switch**
3. **Outside mirrors adjuster**
4. **Light switch**
5. **Air vents**
6. **Turn signal / high beam lever / cruise control switches**
7. **Instrument cluster illumination**
8. **Instrument cluster**
9. **Windshield wiper / washer lever / multi-function indicator**
10. **Switches for:**
    - Rear window defogger
    - Heated driver's seat
    - Anti-slip regulation (optional)
11. **Radio**
12. **Switches for:**
    - Emergency flasher
    - Heated front passenger seat
13. **Front passenger airbag**
14. **Fuel filler flap release switch**
15. **Rear lid lock and switch**
16. **Fuse panel**
17. **Hood release lever**
18. **Data Link Connector (DLC) for On Board Diagnostics (OBD)**
19. **Horn / driver's airbag**
    - Horn only works when ignition is on
20. **Adjustable steering column lever**
21. **Ignition switch**
22. **Ashtray**
23. **Parking brake lever**
24. **Gear selector lever**
25. **Cup holder**
26. **Climate controls**
27. **Glove compartment**

NOTE—

Features and equipment may vary from the illustation shown. Variants are based on model and option packages. Switch and indicator light arrangement may also vary.

92 Windshield Wiper and Washer System

Repair Group

W 92

Windshield wiper system

Windshield wiper assembly, removing and installing

CAUTION—

Before beginning repairs on the electrical system:

- *Obtain the anti-theft radio security code.*
- *Switch the ignition off.*
- *Disconnect the battery Ground (GND) strap.*
- *After reconnecting battery, re-code and check operation of anti-theft radio. Also check operation of clock and power windows according to Repair Manual and/or Owner's Manual.*

Special tools, testers and auxiliary items needed

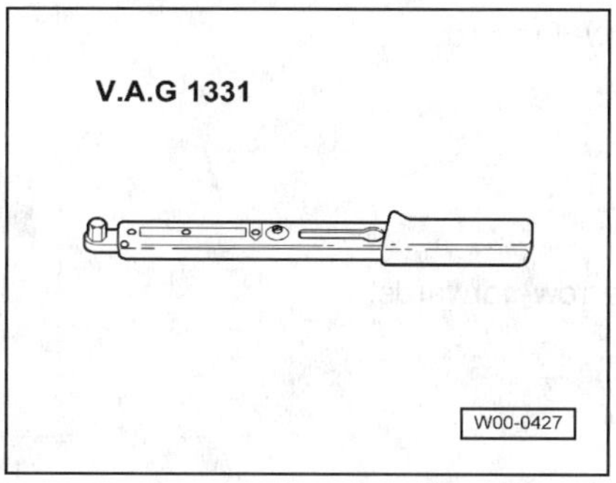

◄ VAG 1331 Torque wrench (or equivalent 5 - 50 Nm)

Wiper arms, removing

NOTE—

- *The wiper arms and windshield cowl panel must be removed in order to remove the wiper assembly (consisting of wiper motor, frame, wiper shafts and linkages).*
- *The housing for the control modules (in left side of plenum) must be also be released and pushed forwards.*
- *Before removing wiper arms, ensure that wiper motor is in park position. This ensures proper wiper arm/blade park position when reinstalling.*

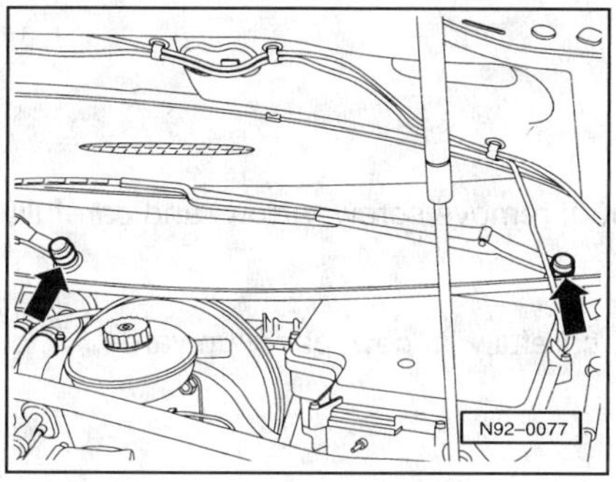

◄ Pry off wiper arm cap -arrow- with a screwdriver.

EDITOR'S NOTE—

For information on wiper blade park position, wiper blade contact angle, and adjustment of spray jets for wipers and headlights, see ***Maintenance****.*

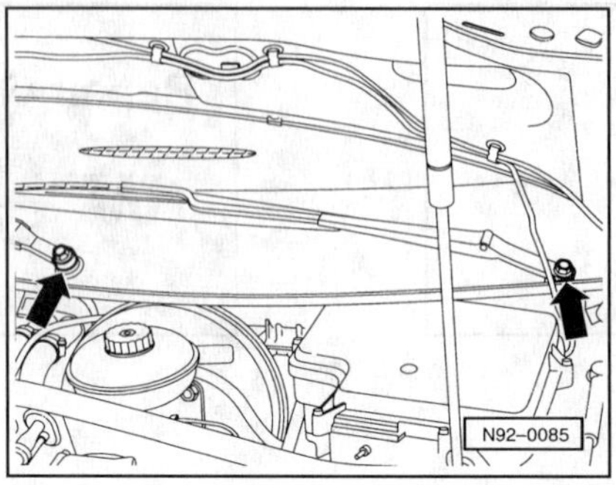

◄ Loosen M8 nuts -arrows-, but do not remove completely.

– Lift wiper arms lightly until they release from tapered shaft.

– Remove nut completely and remove wiper arms.

Air intake plenum cover, removing

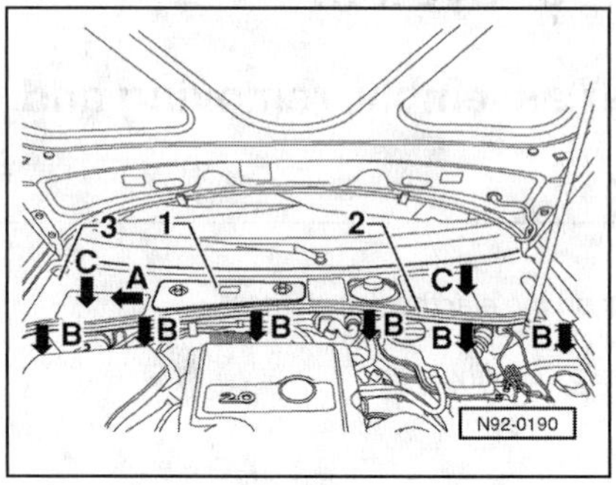

◄ Open front hood.

- Slide battery cover -1- in direction of -arrow A- and remove.
- Remove rubber seal -2- in direction of -arrow B-.
- Pull cover -3- forward and remove.

Windshield cowl panel, removing

NOTE —

On vehicles through 09/2000, the cowl panel is attached to the panel edge below the windshield and secured by two panel clips and a Phillips head screw.

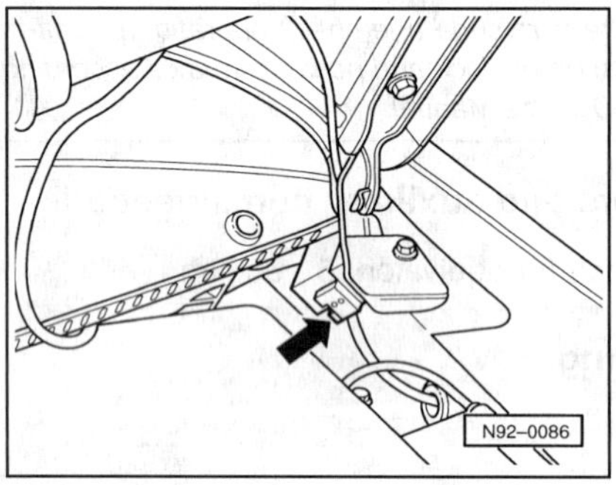

◄ Pull left panel clip -arrow- forwards.

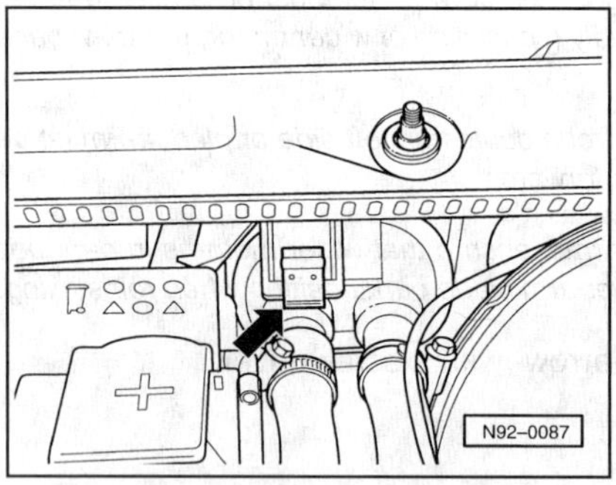

◄ Pull center panel clip -arrow- forwards.

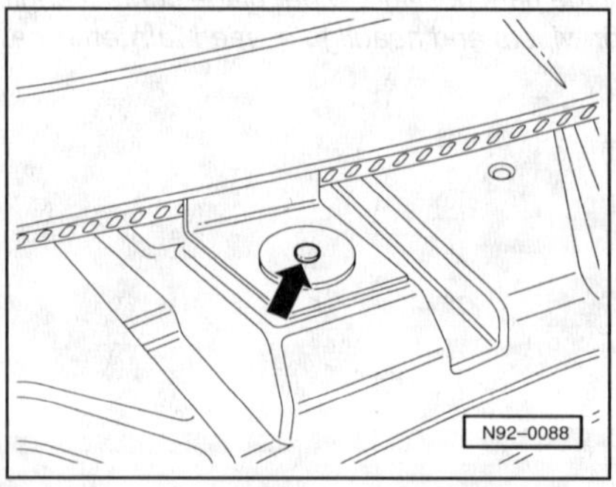

◄ Vehicles through 09/2000: remove screw -arrow- and carefully lift cowl panel upwards.

– Vehicles from 10/2000: carefully lift cowl panel upwards.

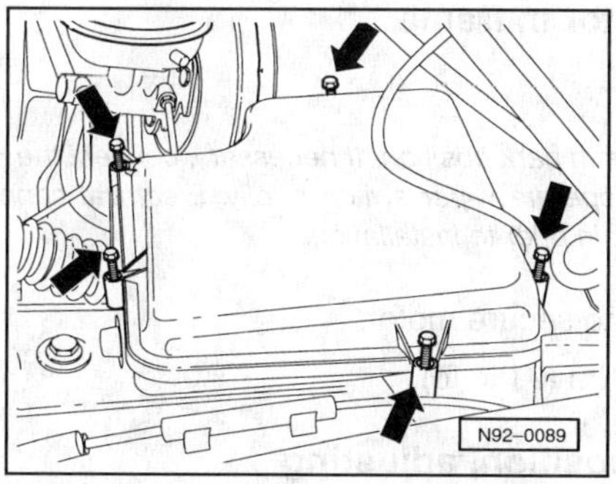

Remove screws -arrows- for ECM cover, lift cover and set aside.

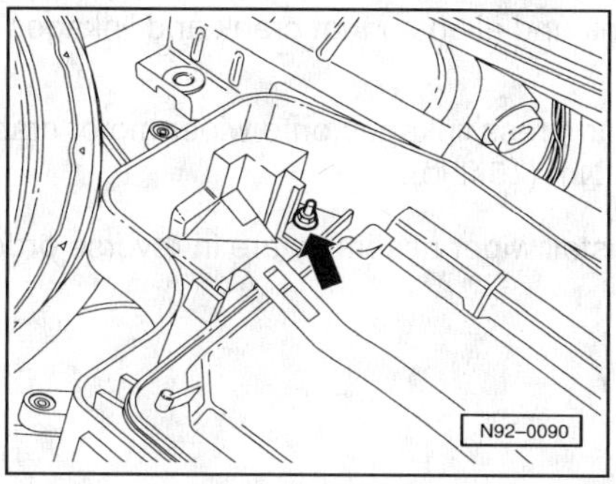

Remove hex nut -arrow- in housing.

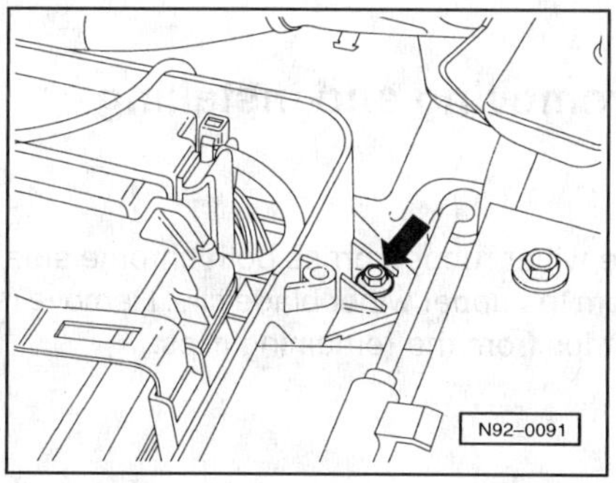

Remove hex nut -arrow- on housing.

- Pull housing from rubber mounting and push forwards.

Wiper assembly, removing

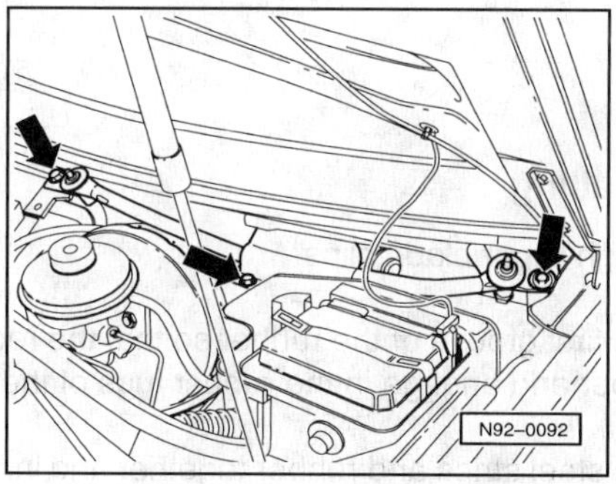

Disconnect electrical connector from wiper motor.

- Remove M6 hex bolts and washers -arrows-.
- Remove complete assembly.

Removing wiper motor from frame

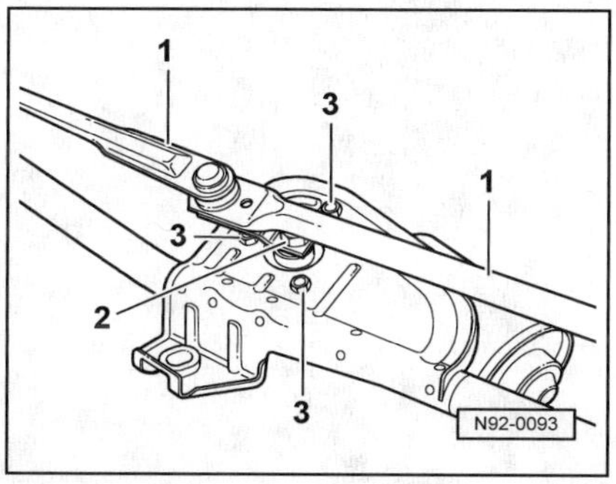

Pry linkages -1- from wiper motor crank using a large screwdriver.

- Remove M8 hex nut -2-.
- Remove wiper motor crank.
- Remove all M6 mounting bolts -3-.

Installing wiper motor in frame

NOTE —
Ensure that wiper motor is in park position. If necessary, connect the electrical connector and operate wiper switch briefly. Electrical connector must be disconnected again prior to installation.

– Install all M6 hex bolts to secure motor.

• Tightening torque: 8 Nm (71 in lb)

Wiper motor park position, adjusting

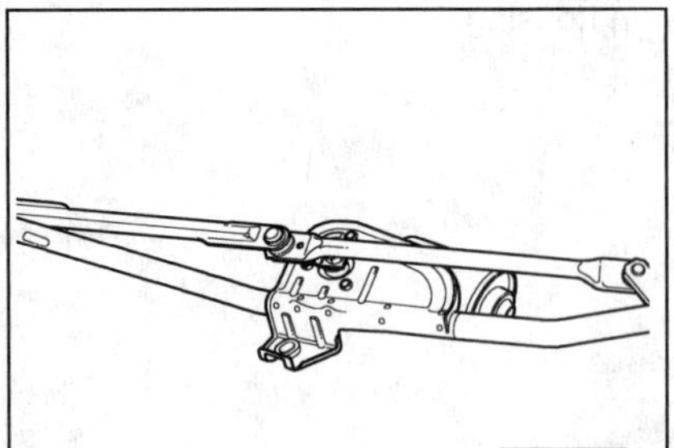
N92-0094

◄ Mount wiper motor crank and align so that crank and linkages line up, see illustration.

• Tighten M8 hex nut and press linkages onto wiper motor crank.
• Tightening torque: 20 Nm (15 ft lb).

• Remaining steps to install wiper assembly are in reverse order of removal.

Wiper rubbers, removing and installing

Removing

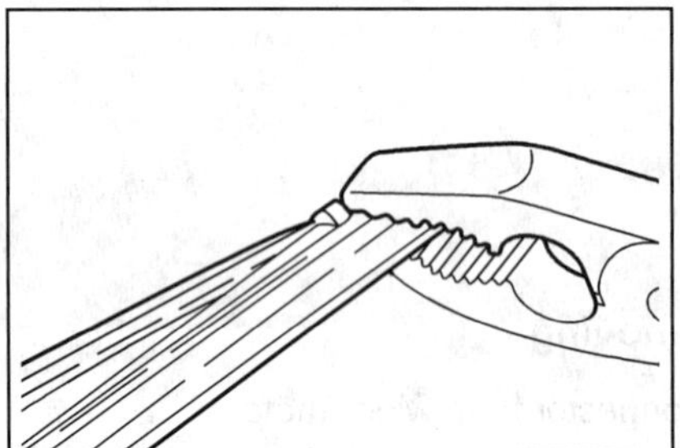
V92 - 0376

◄ At the closed end of the wiper rubber, press both chrome steel strips together and pull out from the upper wiper blade clip. Remove (slide out) rubber complete with strips from the remaining clips.

Wiper rubbers, installing

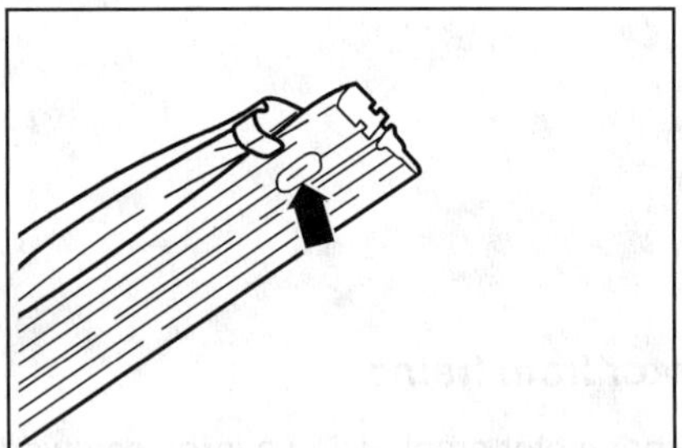
V92 - 0377

◄ Slide new rubber through wiper blade clips.

– Insert both strips in the first groove of the rubber so that the recesses in the strips face the rubber and engage in the rubber lugs of the groove.

– Using pliers press both steel strips and rubber together and insert in the clips so that the lugs of the clips on both sides engage in the retaining slots (arrow) of the wiper blade rubber.

Aero-wiper blades, removing and installing

An aero-wiper blade system is installed as a running production change during model year 2002.

> ***CAUTION —***
> - *To ensure proper wiper performance, only grab/lift Aero-wiper blades at the mounting pivot on wiper arm when removing and installing.*
> - *Driver and passenger wiper blades are different lengths. Damage will result if wiper blades are interchanged.*

– Bring wiper arm to park position.

– Switch off all electrical consumers, switch off ignition and remove ignition key.

– Lift aero-wiper blade/arm assembly from windshield, noting *CAUTION* above.

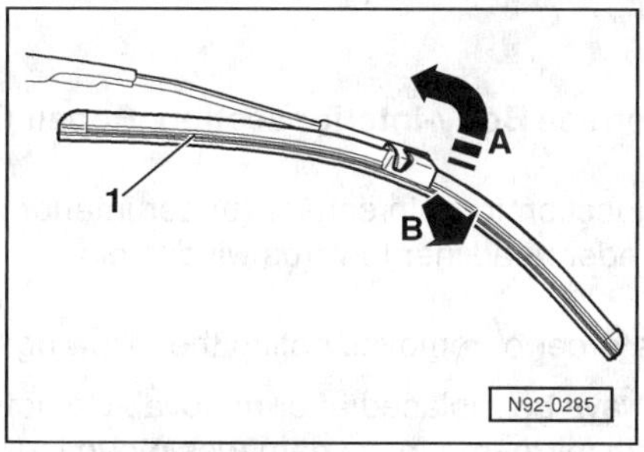

N92-0285

◄ Tilt wiper blade -1- on mounting pivot in direction of -arrow A- until it reaches stop.

– Detach wiper blade from mounting pivot in direction of -arrow B-.

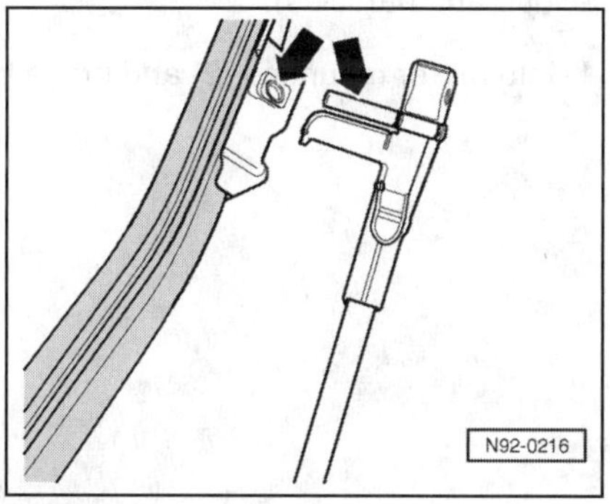
N92-0216

◄ To install, insert wiper blade onto wiper arm mounting pivot -arrows-.

– Engage blade by tilting on pivot until blade is parallel with arm.

– Return arm/blade assembly to windshield, noting *CAUTION* above

Rain sensor -G213-, removing and installing

– Remove interior rear-view mirror (with rain sensor), see **Body-Interior section, Repair Group 68**

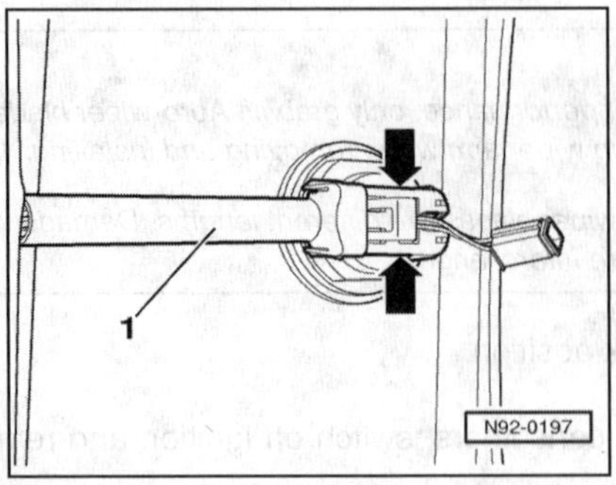

⮜ Unclip cable guide -arrows 1- and remove.

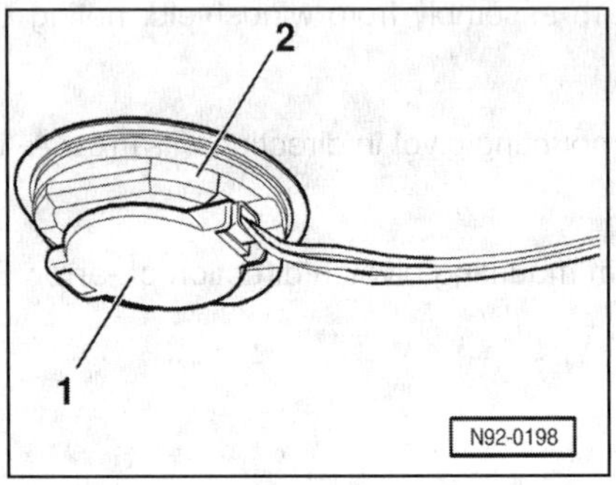

⮜ Separate rain sensor -1- from mirror mounting -2-.

– Remove interior light, see **Repair Group 96, Interior lights, removing and installing**.

– Remove center sun visor, see **Body-Interior section, Repair Group 68**

– Separate electrical connection for rain sensor (under interior light) and pull harness out from under headliner towards windshield.

– Installation is in reverse order of removal, noting the following:

- Rain sensors must always be replaced after removal. Do not reuse. Removal of sensor from mirror mounting damages mounting adhesive and subsequently affects proper function of sensor.
- Prior to installing sensor, thoroughly clean mounting surface must be free from all residue and particles).

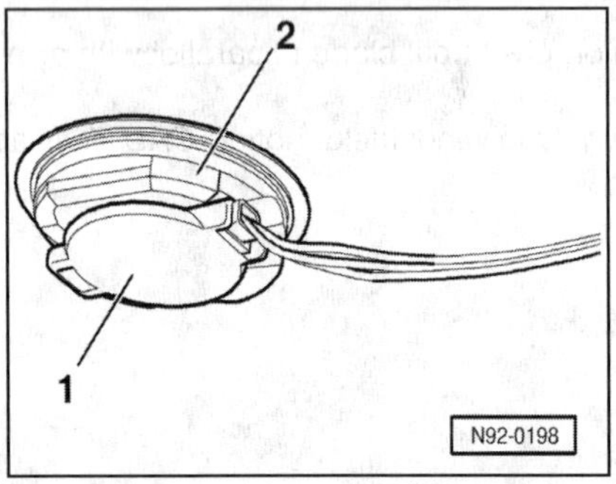

⮜ Insert new rain sensor -1- into mirror mounting -2- and press firmly against windshield.

Windshield and headlight washer system

> ***CAUTION—***
> *Before beginning repairs on the electrical system:*
>
> *• Obtain the anti-theft radio security code.*
>
> *• Switch the ignition off.*
>
> *• Disconnect the battery Ground (GND) strap.*
>
> *• After reconnecting battery, re-code and check operation of anti-theft radio. Also check operation of clock and power windows according to Owner's Manual.*

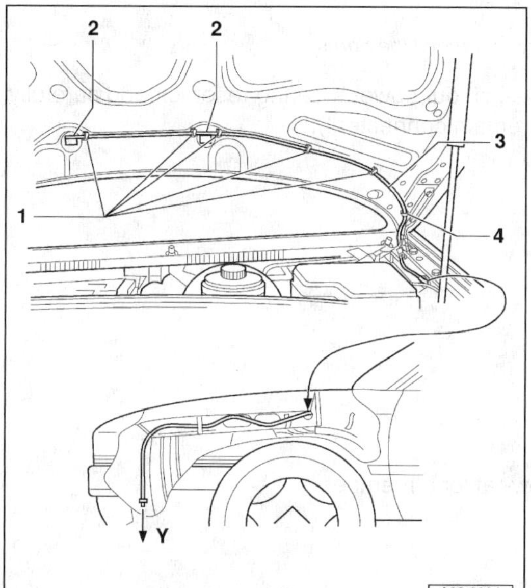

A92-0004

Windshield washer system, hose routing

1. **Hose clip**
   • Vehicles through 09/2000 only
2. **Jets**
   • Removing, see **Windshield washer jets, removing**
3. **Hose**
   • Vehicles from 10/2000, hose is routed inside front hood
4. **Hose clip**

Y. **Connection on windshield washer system pump**

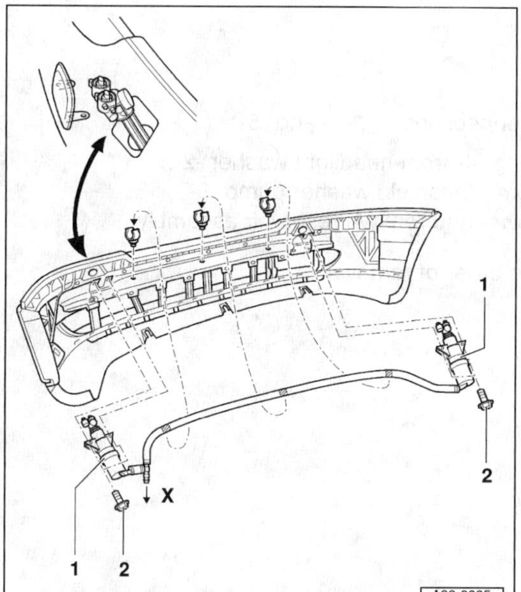

A92-0005

Headlight washer system, assembly (through 09/2000)

NOTE—
*The front bumper cover must be removed to reach the headlight washer system components illustrated, see **Body-Exterior section, Repair Group 63***

1. **Washer jet telescopic cylinder**
2. **Screws**
   • Tightening torque: 10 Nm (89 in lb)

X. **Hose to headlight washer system pump**

Windshield and headlight washer resevoir, removing and installing (from 10/2000)

Removing

– Move lock carrier into service position, see **Repair Manual, Body-Exterior section, Repair Group 50**

– Remove left front wheel housing liner, see **Body-Exterior section, Repair Group 66**

NOTE —
- *On vehicles equipped with 1.8L Turbo engines, it may also be necessary to remove the intercooler, see **1.8 Liter 5V Turbo Engine Mechanical section, Repair Group 21***
- *It may also be necessary to remove the horns.*

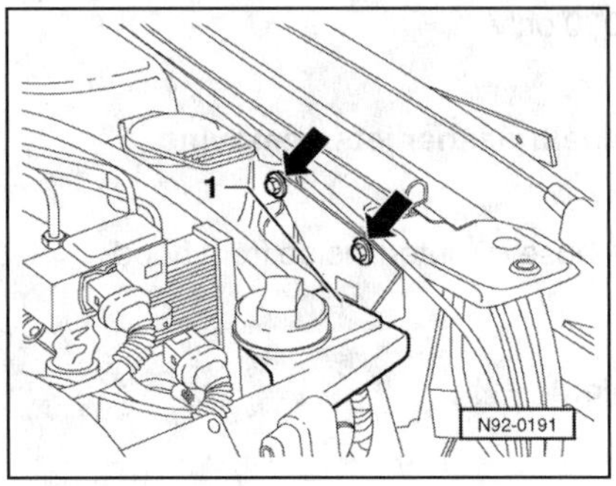

Remove screws -arrows-, lift out power steering reservoir and carefully lay to one side (hoses remain connected!).

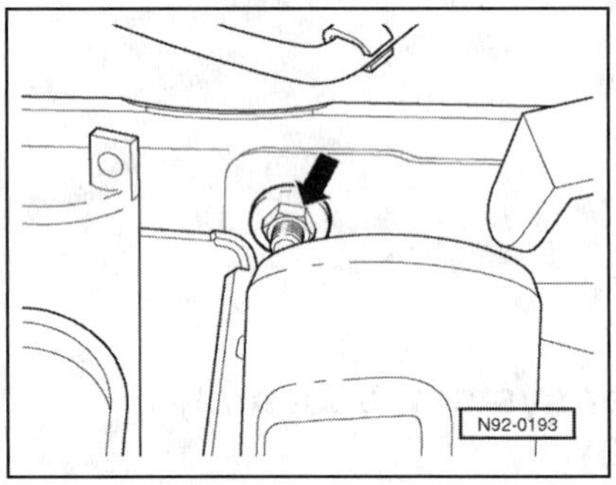

Remove fastener -arrow- at long member.

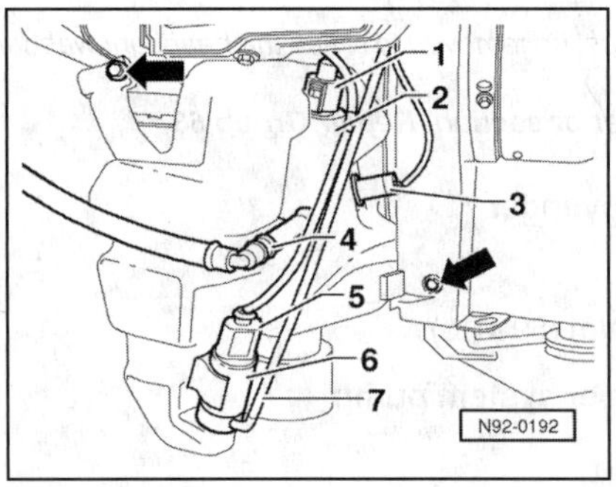

Disconnect electrical connections -1-, -3- and -5-.
- Separate hose coupling -4- from headlight washer -2-.
- Separate hose -7- from windshield washer pump -6- .
- Remove screws -arrows- and remove reservoir assembly.

– Installation is in reverse order of removal.

Headlight washer jet telescopic cylinder, removing and installing (from 10/2000)

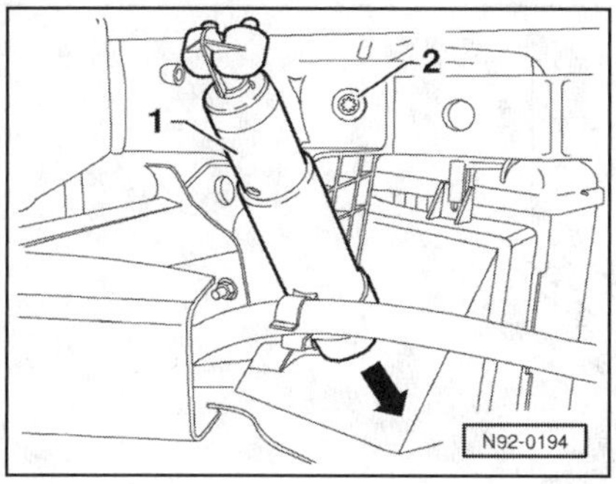

– Remove front bumper, see **Body-Exterior section, Repair Group 63**

< Remove mounting screw -2- at telescopic cylinder.

• Pull cylinder -1- down slightly -arrow-.

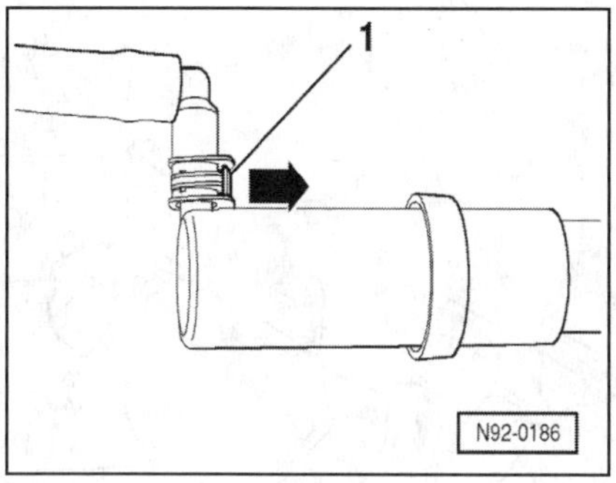

< Remove spring clip -1- at hose coupling -arrow-.

– Separate hose from cylinder.

– Remove cylinder downwards.

– Installation is in reverse order of removal, noting the following:

• Adjust headlight washer jets, see **Windshield washer jets, removing**

Headlight washer system, hose routing on front cross member (from 10/2000)

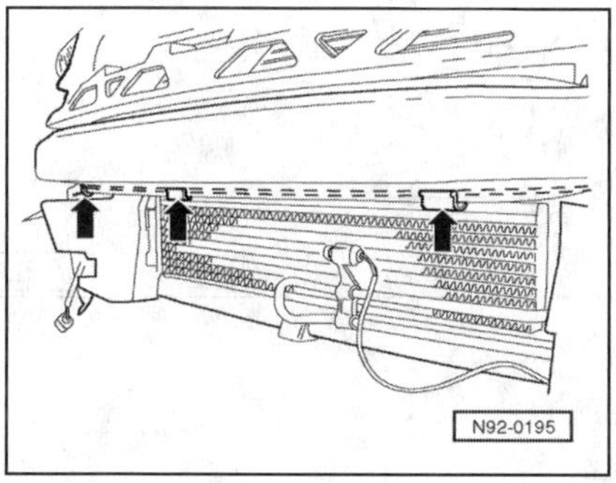

< Ensure washer hose is secured to cross member with clips -arrows-.

Cover for headlight washer jet telescopic cylinder, replacing (from 10/2000)

– Remove front bumper, see **Body-Exterior section, Repair Group 63**

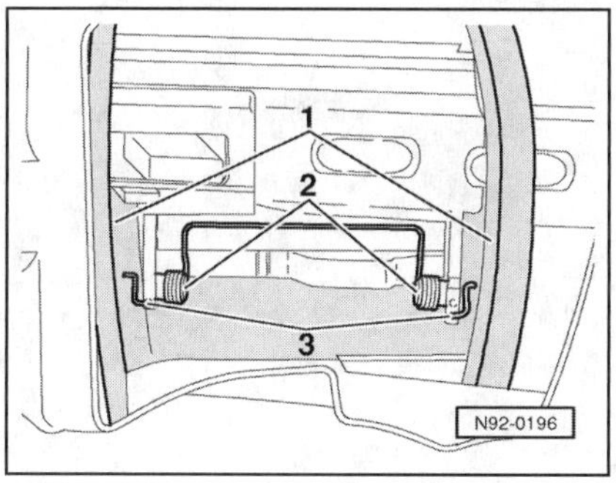

< Release spring from cover -3-.

• Remove spring from bolts -2-.
• Pull out cover -1- from bolts -2- and remove from bumper.

– Installation is in reverse order of removal.

Windshield and headlight washer system assembly (through 09/2000)

System with separate reservoirs for windshield washer and headlight washer systems

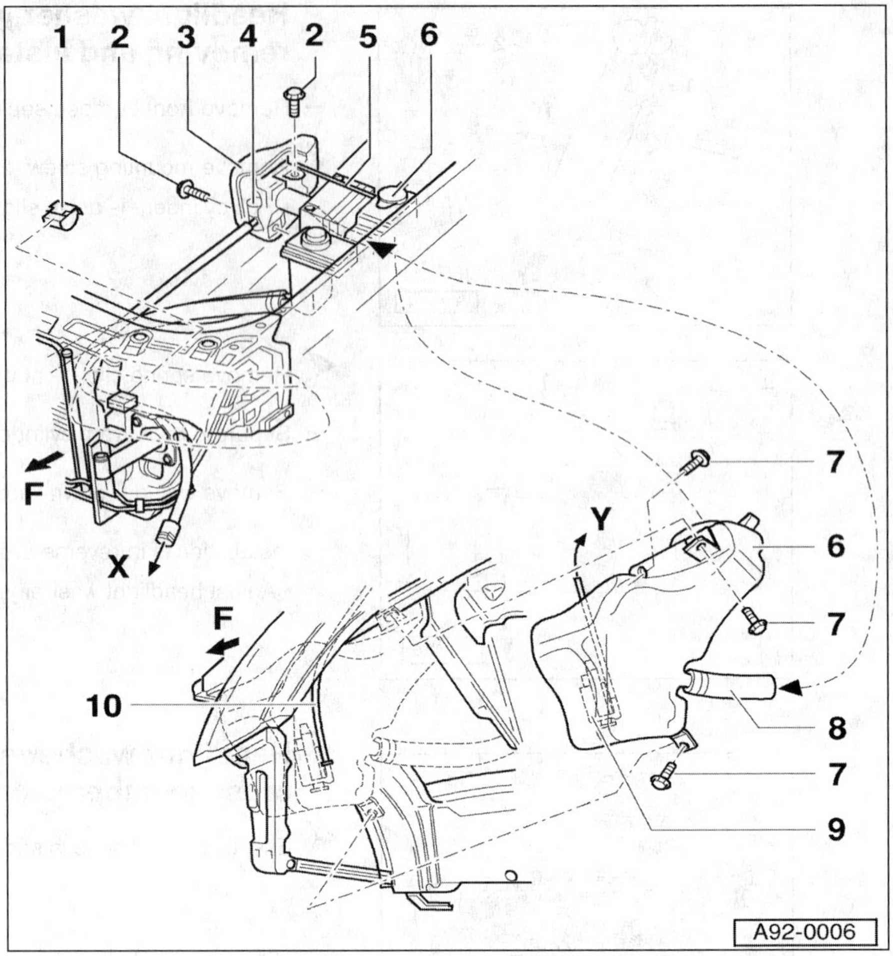

> *NOTE—*
> *Before removing reservoir, remove front left wheel housing liner.*

1. **Hose clip**
2. **Bolts**
   - Tightening torque: 7 Nm (62 in lb)
3. **Headlight washer system reservoir**
4. **Headlight washer system pump**
   - Pull pump upwards from rubber grommet
5. **Vent pipe**
6. **Windshield washer system reservoir**
7. **Bolts**
   - Tightening torque: 7 Nm (62 in lb)
8. **Fill hose for headlight washer system reservoir**
9. **Windshield washer system pump**
   - Pull pump and fluid level sender upwards from rubber grommet
10. **Hose to washer jets**
    - Ensure hose is clipped here and routed properly

System with single reservoir for windshield washer and headlight washer systems (not on vehicles with 1.8L Turbo engine)

NOTE—
Before removing reservoir, remove left front wheel housing liner.

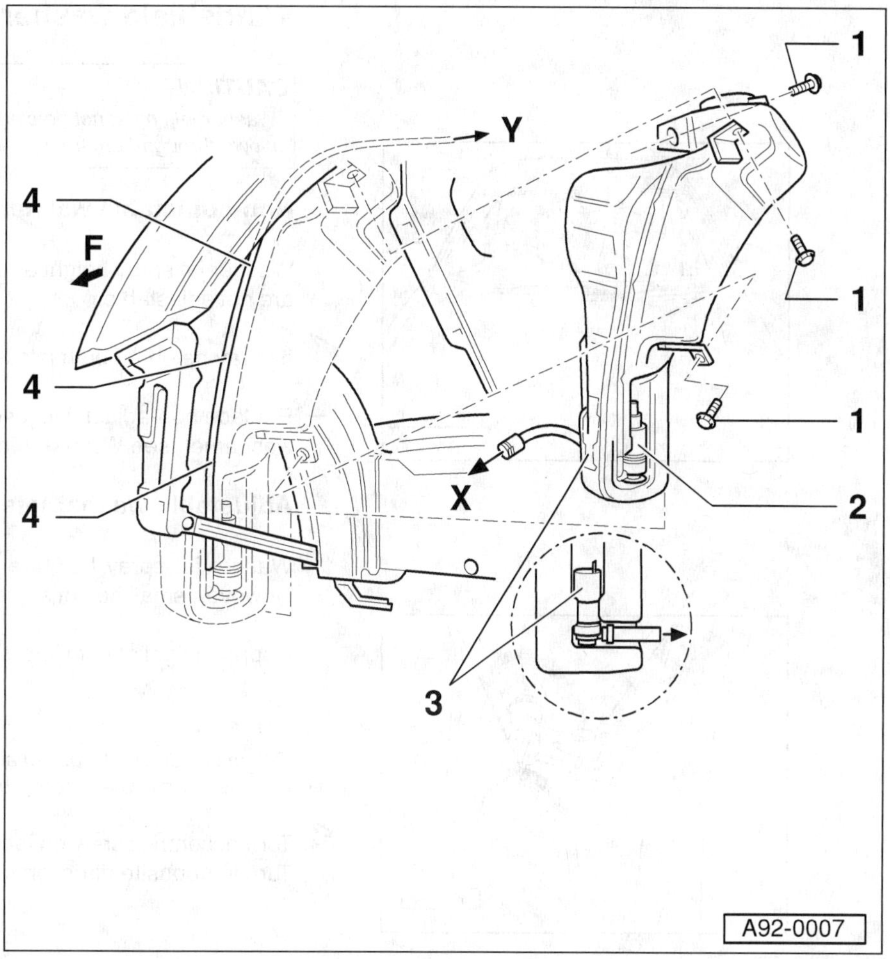

1. **Bolts**
   - Tightening torque 7 Nm (62 in lb)
2. **Windshield washer system pump**
   - Pull pump and fluid level sender upwards from rubber grommet
3. **Headlight washer system pump**
   - Pull pump upwards from rubber grommet
4. **Hose to washer jets**
   - Ensure hose is clipped here and routed properly

F. **Forwards in left wheel housing when viewed from below**

X. **Connection for headlight washer system jets**

Y. **Connection for windshield washer system jets**

Windshield washer jets, checking

> ***CAUTION —***
> *Washer jets must not be cleaned opposite to the spray direction, e.g.: blown through from front.*

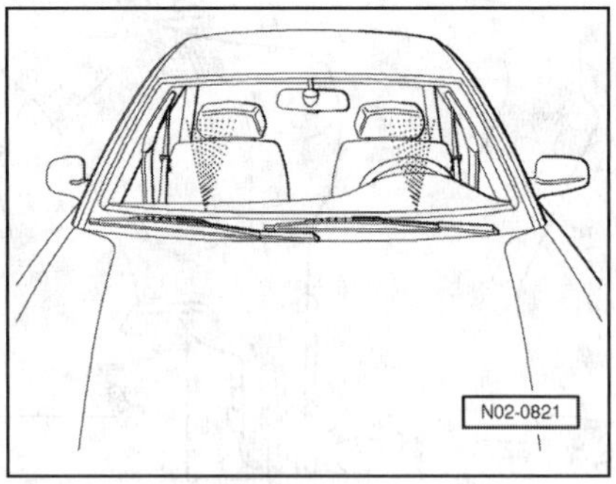

Non-adjustable washer jets

Washer jet spray height and lateral position are preset in production and are not adjustable.

◄ If spray height of both jets is not equal:

– Replace non-adjustable jet with lowest spray height with adjustable type washer jet, see **Windshield washer jets, checking**.

Adjustable washer jets

Washer jet spray height and lateral position are preset in production. However, small height adjustments are possible.

If spray height of both jets is not equal, adjust as follows:

NOTE —
One of two different types of adjustable washer jets may be installed. Use the appropriate procedure described below.

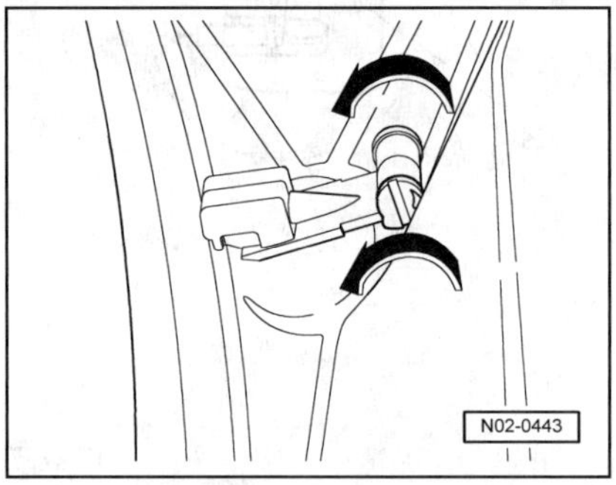

◄ Turn eccentric screw on jet in direction of -arrow- to raise spray pattern. Turn in opposite direction to lower pattern.

or

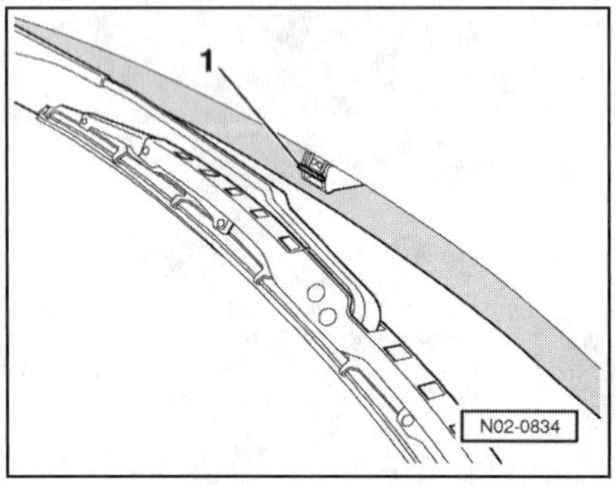

◄ Raise or lower adjustment tab -1- at jet by hand (or with small screwdriver) accordingly.

Windshield washer jets, removing

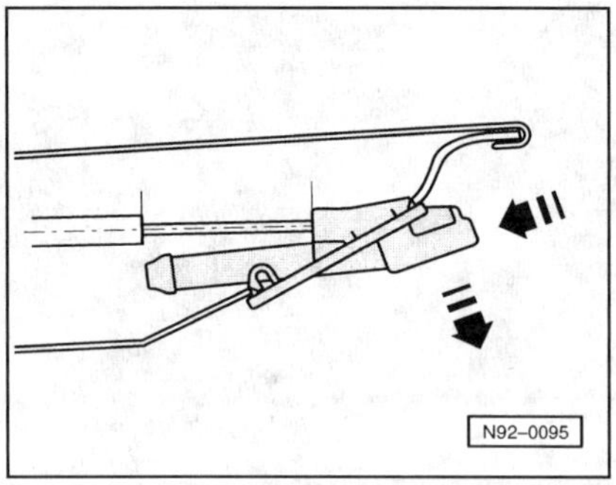

◄ Pull hose from jet and separate 2-pin electrical connector.

- Press jet forwards and remove downwards.

Rear window wiper/washer system

CAUTION—
Before beginning repairs on the electrical system:

- *Obtain the anti-theft radio security code.*
- *Switch the ignition off.*
- *Disconnect the battery Ground (GND) strap.*
- *After reconnecting battery, re-code and check operation of anti-theft radio. Also check operation of clock and power windows according to Owner's Manual.*

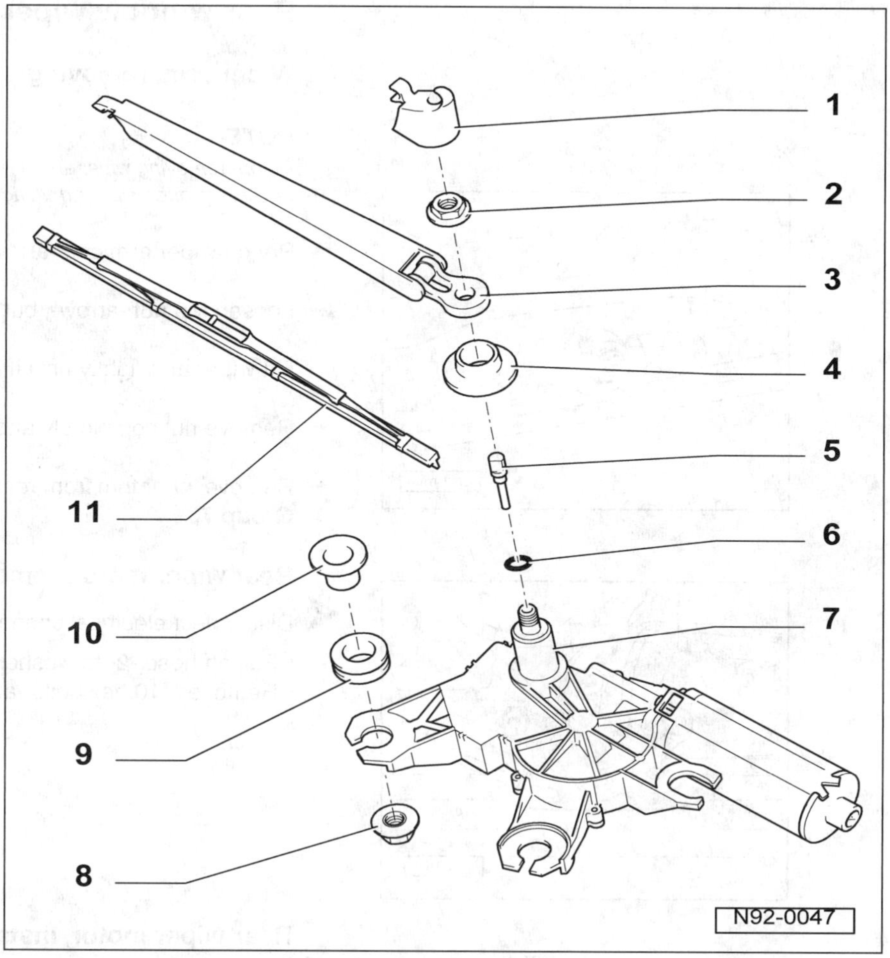

Rear window wiper, assembly

1. **Cap**
2. **Hex nut M13**
   - 15 Nm (11 ft lb)
3. **Wiper arm**
   - Adjusting park position, see **Rear window wiper park position, adjusting**
4. **Seal**
5. **Spray jet**
   - Adjusting, see **Spray jet, adjusting**
6. **Seal**
7. **Wiper motor**
   - See **Rear window wiper motor, removing**
8. **Hex nut M6**
   - 8 Nm (71 in lb)
9. **Grommet**
10. **Spacer**
11. **Wiper blade**
    - See **Wiper rubbers, removing and installing**

Rear window wiper motor, removing

Wiper arm, removing

NOTE —
Before removing wiper arm, ensure that wiper motor is in park position. This ensures proper wiper arm/blade park position when reinstalling.

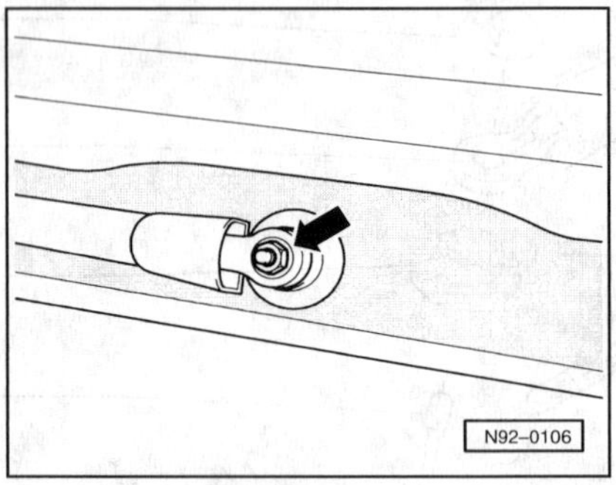

- Pry off wiper arm cap -arrow- with a screwdriver.
- Loosen M8 nut -arrow-, but do not remove completely.
- Lift wiper arm lightly until it releases from tapered shaft.
- Remove nut completely and remove wiper arm.
- Remove inner trim from rear hatch, see **Body-Interior section, Repair Group 70**

Rear wiper motor, removing

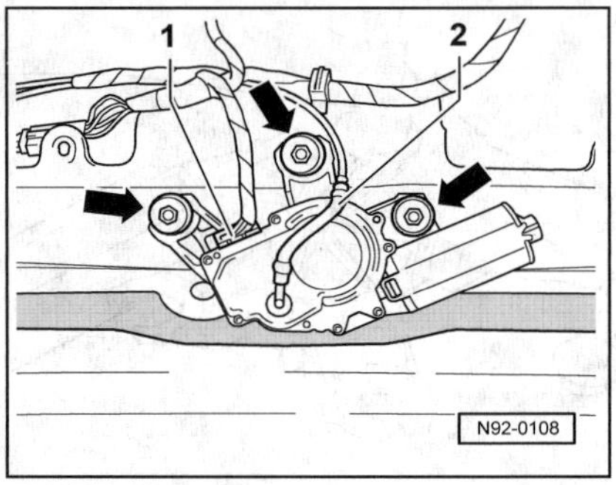

- Disconnect electrical connector -1- from wiper motor.
  - Pull off hose -2- to washer jets.
  - Remove M10 hex bolts -arrows- and remove wiper motor.

Rear wiper motor, installing

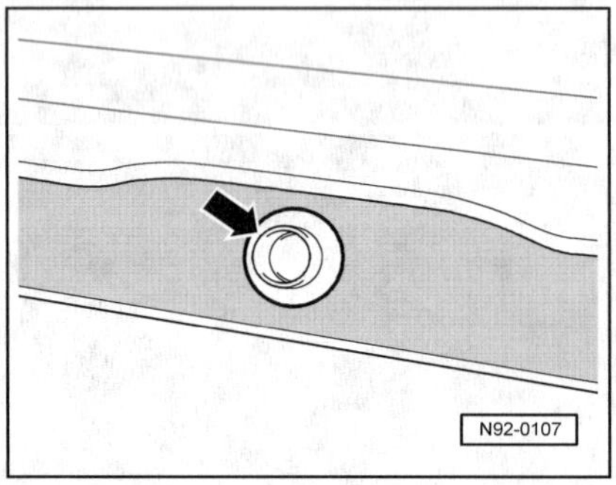

- When installing the wiper motor, ensure rear window seal seating as illustrated.
  - Installation is in reverse order of removal.

Rear window wiper park position, adjusting

Special tools, testers and auxiliary items needed

- VAG 1331 Torque wrench (or equivalent 5 - 50 Nm)

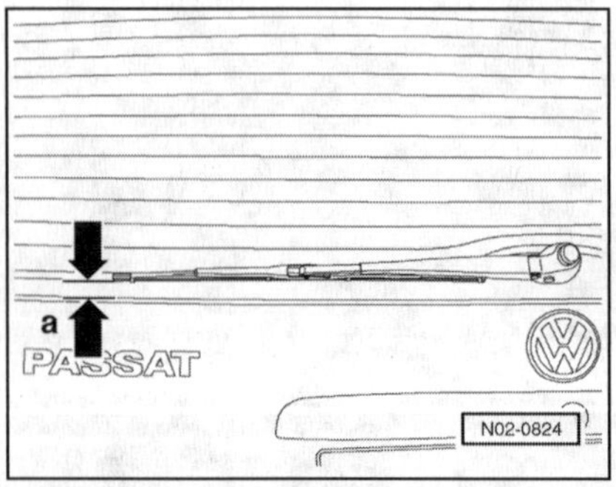

- Wiper blade must be 25 mm -arrows a- from lower edge of rear window.
  - If necessary, loosen wiper arm and adjust distance
  - Wiper arm nut tightening torque: 15 Nm (11 ft-lb)

94 Lights, Switches - Exterior

Repair Group

W 94

Headlights

> ***CAUTION—***
>
> *Before beginning repairs on the electrical system:*
>
> - *Obtain the anti-theft radio security code.*
> - *Switch the ignition off.*
> - *Disconnect the battery Ground (GND) strap.*
> - *After reconnecting battery, re-code and check operation of anti-theft radio. Also check operation of clock and power windows according to Repair Manual and/or Owner's Manual.*

EDITOR'S NOTE—
*For information on headlight and foglight adjustment, see **Maintenance**.*

Headlight bulbs, replacing (through 09/2000)

Low beam bulb, replacing

> ***CAUTION—***
>
> *Do not touch glass portion of bulb with bare hands. Even the smallest amount of moisture and/or contaminants from fingers that evaporates on the bulb during operation can cause the glass to cloud over.*

– Remove large cover cap on rear of headlight housing.

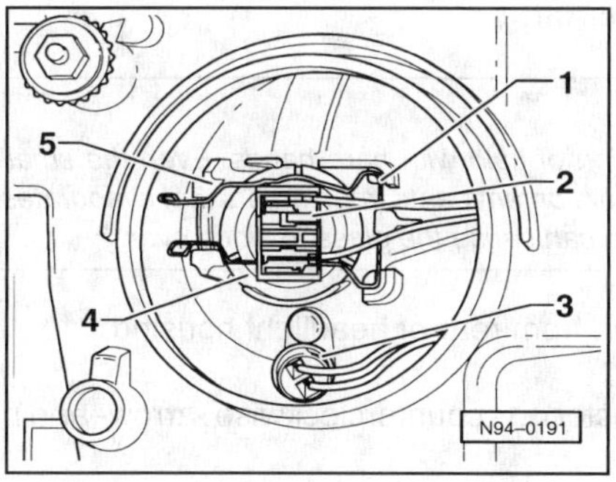

◄ Disconnect electrical connector -2- from bulb.

- Unclip retainer -1- from lugs -5- and flip over to side.
- Remove bulb -4- from reflector/housing.
- Insert new bulb (making sure that the bulb mounting aligns with the locating lugs in the housing) and re-clip retainer.
- Connect electrical connector and reinstall cover cap.
- Check headlight (bulb) function.

High beam bulb or dual filament bulb for high beam and fog light, replacing

> ***CAUTION***—
> *Do not touch glass portion of bulb with bare hands. Even the smallest amount of moisture and/or contaminants from fingers that evaporates on the bulb during operation can cause the glass to cloud over.*

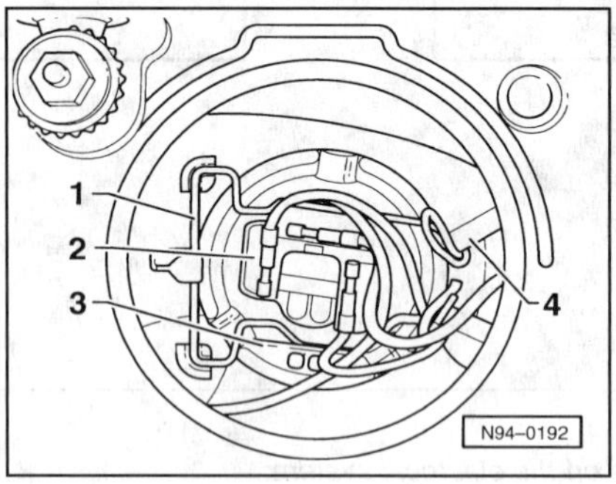

– Remove small cover cap on rear of headlight housing.

Disconnect electrical connector -2- from bulb.

- Unclip retainer -1- from lugs -4- flip over to side.
- Remove bulb -3- from reflector/housing.
- Insert new bulb (making sure that the bulb mounting aligns with the locating lugs in the housing) and re-clip retainer.
- Connect electrical connector and reinstall cover cap.
- Check headlight (bulb) function.

Headlight bulbs, replacing (from 10/2000)

Parking light bulb, replacing

> ***WARNING***—
> *Do not touch glass portion of bulb with bare hands. Even the smallest amount of moisture and/or contaminants from fingers that evaporates on the bulb during operation can cause the glass to cloud over.*

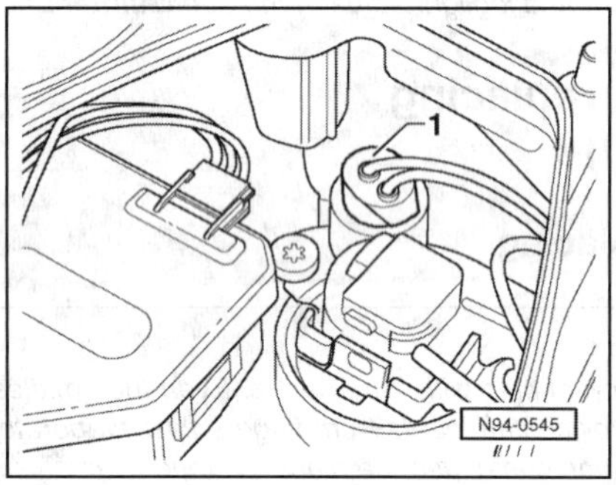

– Remove large cover cap on rear of headlight housing.

Pull out socket -1- (with bulb) from headlight housing.

– Pull out bulb from socket and replace bulb.

– Reinstall socket (with bulb). Push in all the way to stop.

– Reinstall cover cap.

– Check function of parking light.

Low beam bulb, replacing

> ***WARNING***—
> *Do not touch glass portion of bulb with bare hands. Even the smallest amount of moisture and/or contaminants from fingers that evaporates on the bulb during operation can cause the glass to cloud over.*

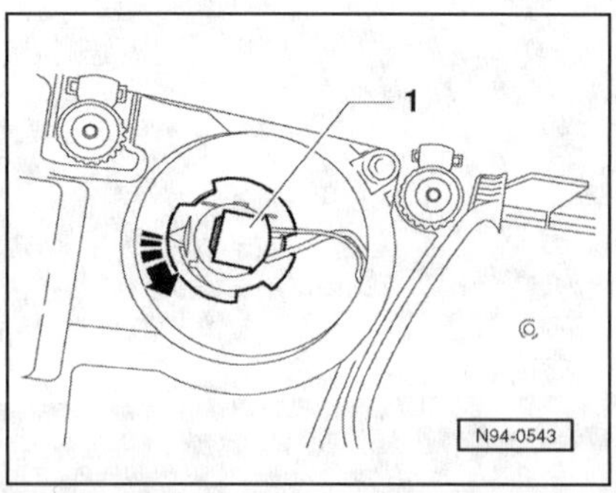

– Remove small cover cap from rear of headlight housing.

Turn socket (with connector) -1- counterclockwise -arrow- and pull out of headlight housing.

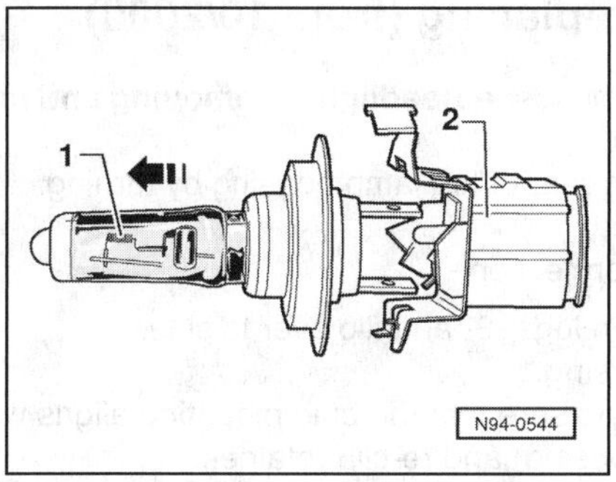

- Pull bulb -1- out of socket -2- in direction of -arrow-.
- Replace bulb.
- Reinstall socket (with connector) into headlight housing. Lock socket in place by turning clockwise.
- Reinstall cover cap.
- Check headlight (bulb) function.

High beam bulb, replacing

> ***WARNING —***
> *Do not touch glass portion of bulb with bare hands. Even the smallest amount of moisture and/or contaminants from fingers that evaporates on the bulb during operation can cause the glass to cloud over.*

- Remove large cover cap from rear of headlight housing.

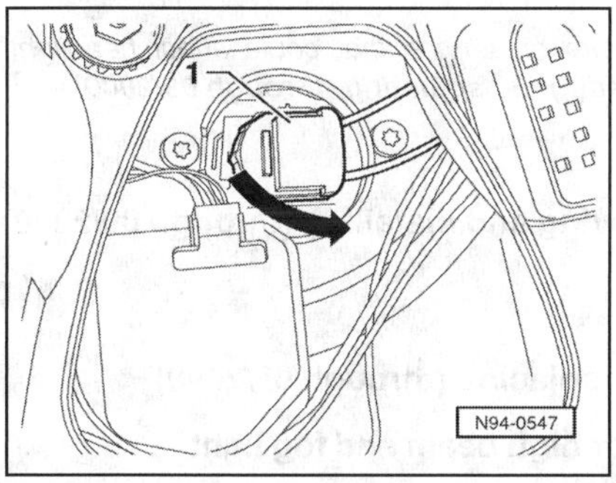

- Turn socket (with connector) -1- counterclockwise -arrow- and pull out of headlight housing.

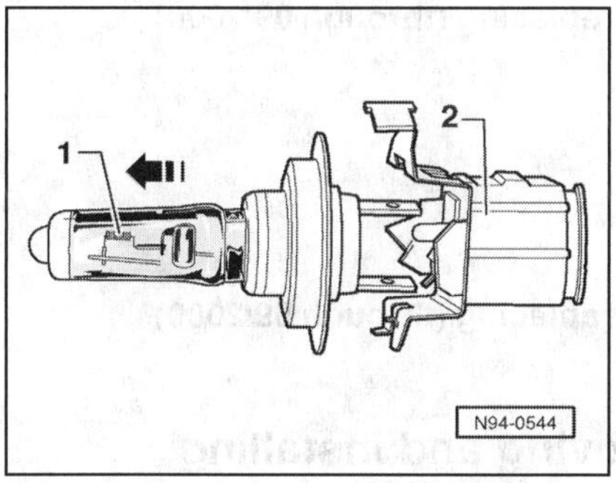

- Pull bulb -1- out of socket -2- in direction of -arrow-.
- Replace bulb.
- Reinstall socket (with connector) into headlight housing. Lock socket in place by turning clockwise.
- Reinstall cover cap.
- Check headlight (bulb) function.

Turn signal bulb, replacing

> ***WARNING —***
> *Do not touch glass portion of bulb with bare hands. Even the smallest amount of moisture and/or contaminants from fingers that evaporates on the bulb during operation can cause the glass to cloud over.*

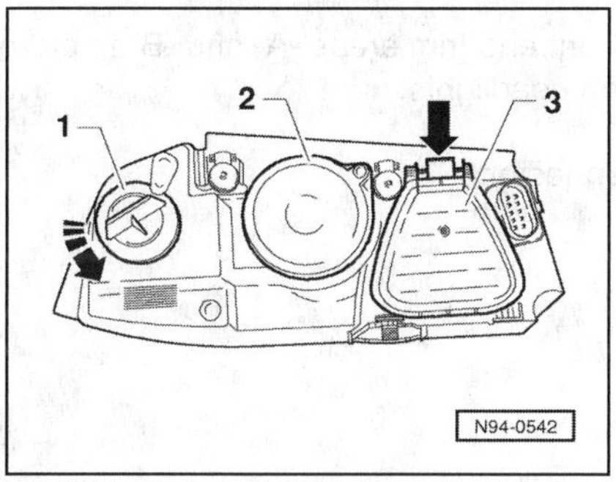

- Turn socket (with connector) -1- counterclockwise -arrow- and pull out of headlight housing.
  - Replace bulb.
  - Reinstall socket (with connector) into headlight housing. Lock socket in place by turning clockwise.
  - Check turn signal (bulb) function.

Fog lamp bulb, replacing (from 10/2000)

– Remove fog lamp assembly, see **Headlights, removing and installing**.

– Remove cover cap from rear of fog lamp housing by turning.

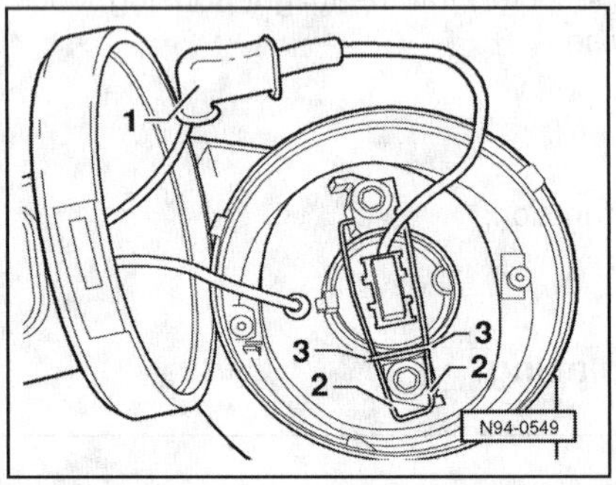

➤ Disconnect electrical connection -1-.

- Unclip retainer -2- from lugs -3- and flip over to side.
- Pull out bulb from housing.
- Insert new bulb (making sure that the bulb mounting aligns with the locating lugs in the housing) and re-clip retainer.

– Reinstall fog lamp housing, see **Headlights, removing and installing**.

– Check fog lamp function/aim, see **Headlights and fog lamps, adjusting (from 10/2000)**.

Headlight assembly (through 09/2000)

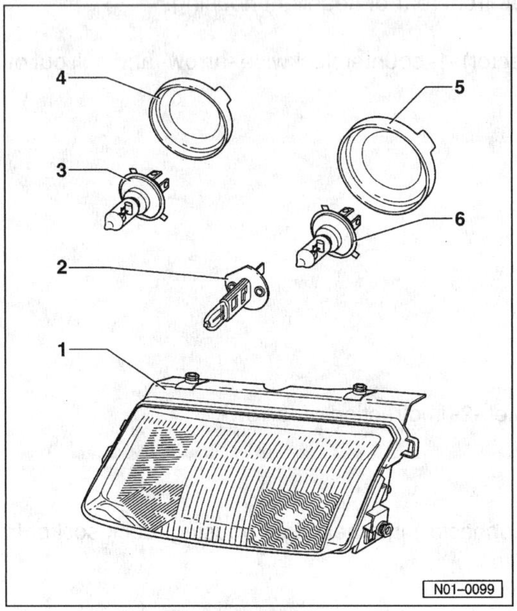

NOTE —
*After performing any repairs or service that could affect headlight aim, check/adjust aim, see **Headlights, adjusting (through 09/2000)**.*

1. **Headlight housing**
   - See **Headlights, removing and installing (through 09/2000)**
2. **High beam bulb**
   - Bulb H1 12 V, 60 W
   - See **Headlight bulbs, replacing (through 09/2000)**
3. **Double filament bulb for high beam and fog light**
   - Bulb H4 12 V, 60 W/55 W
   - See **Headlight bulbs, replacing (through 09/2000)**
4. **Small cover cap**
5. **Large cover cap**
6. **Low beam bulb**
   - Bulb H7 12 V, 55 W
   - See **Headlight bulbs, replacing (through 09/2000)**

Headlights, removing and installing (through 09/2000)

Removing

– Remove front turn signal housing, see **Front turn signal and side marker light housing, removing and installing (through 09/2000)**

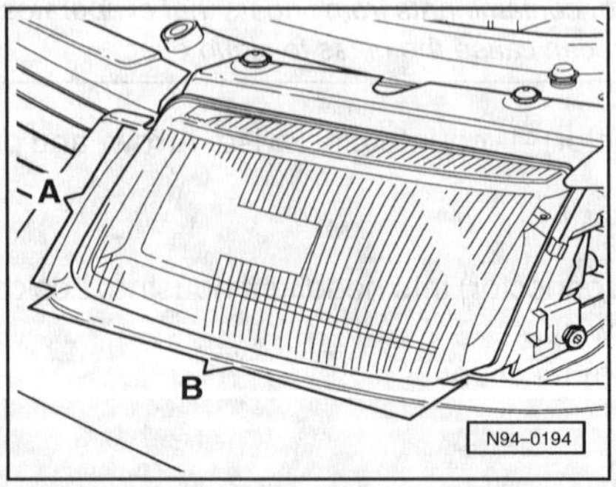

➤ Apply masking tape to front end trim areas -A- and -B- to protect paintwork while removing headlights.

– Disconnect electrical connector.

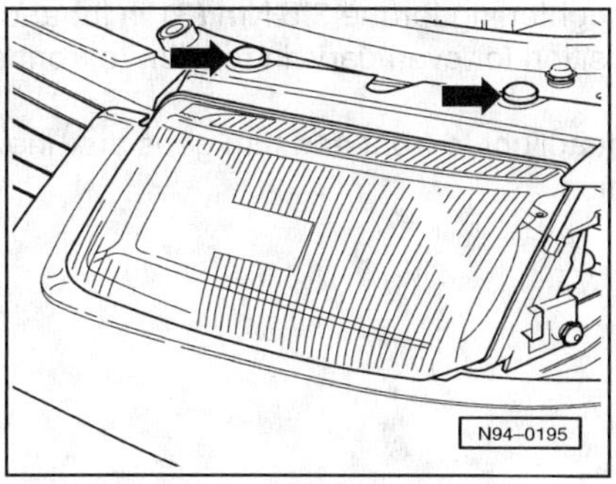

◀ Remove cap -arrows- from upper mounting screws.

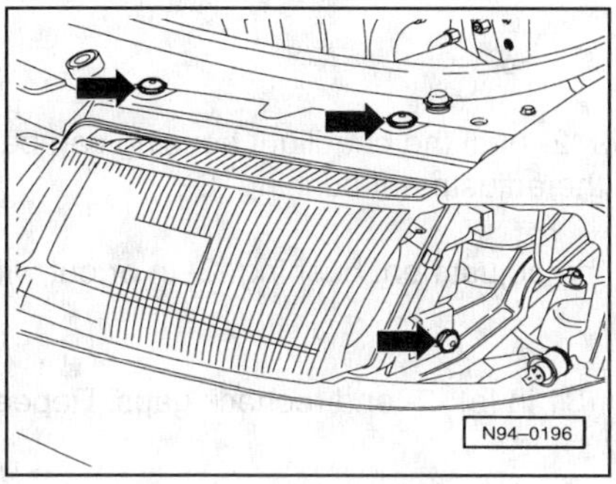

◀ Remove three mounting screws -arrows-.

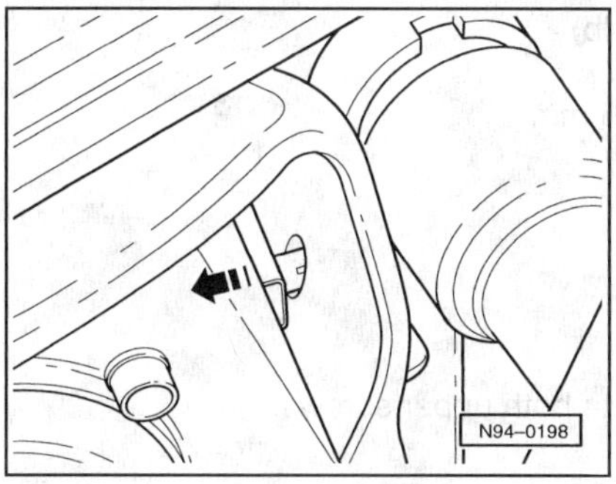

◀ Pull headlight housing sideways (outwards) and detach from retainer -arrow-.

NOTE —
When removing headlight, it may be necessary to press/move the front grille trim slightly to the side for clearance.

– Carefully remove headlight forwards, turning slightly if necessary.

Headlights, installing (through 09/2000)

– Apply masking tape to front end trim areas to protect paintwork while removing headlights.

– Carefully insert headlight into opening.

NOTE —
When installing headlight, it may be necessary to press/move the front grille trim slightly to the side for clearance.

◀ Push headlight sideways (inwards) onto retainer.

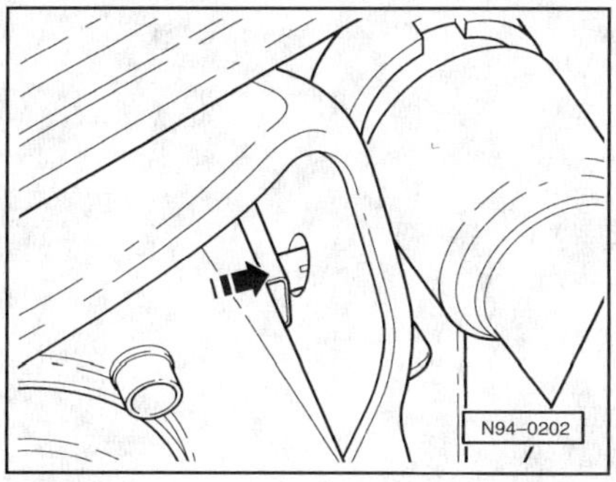

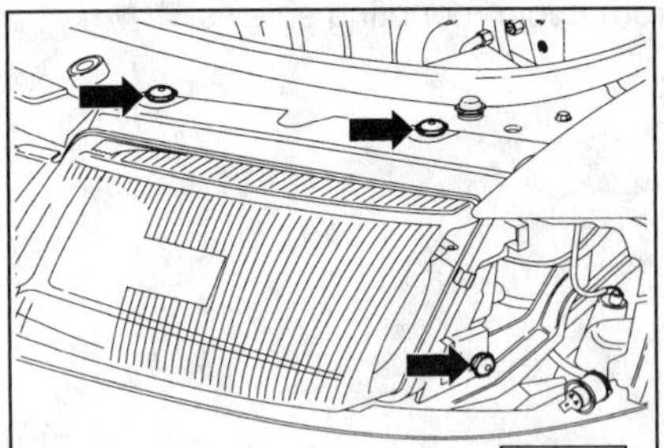

N94-0196

◄ Install screws -arrows-, tightening torque 3.5 Nm (31 in lb) and check headlight installation position for even gaps in relation to front end trim.

– If it is determined that headlight to front end trim gaps are uneven, correct installation position.

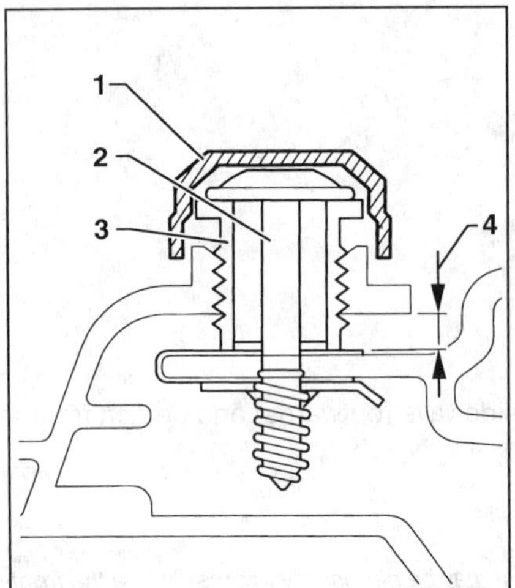

N94-0197

Proceed as follows:

◄ Loosen the three screws -2- until the headlight housing can be moved back and forth easily in the adjustment bushing -3-.

– Align gaps by screwing the adjustment bushing -3- in or out with a 17 mm wrench.

– Tighten screws, 3.5 Nm (31 in lb) -2- and recheck gaps. Repeat adjustment if necessary.

1 Cap
2 Screw, 3.5 Nm (31 in lb)
3 Adjustment bushing
4 Basic setting 3.5 ± 2.5 mm

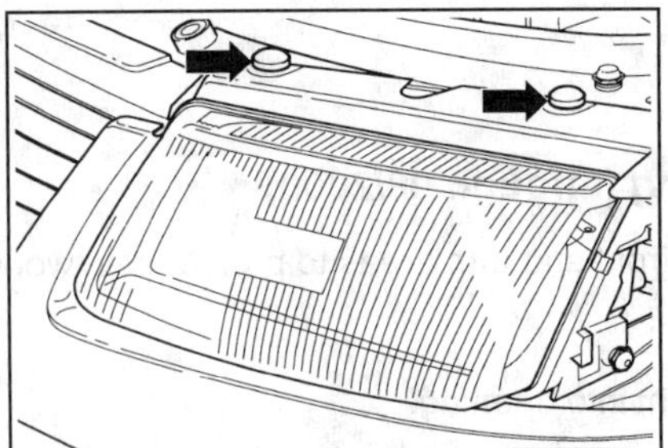

N94-0195

◄ Install caps -arrows- onto both upper screws.

– Connect electrical connector.

– Install front turn signal housing, see **Front turn signal and side marker light housing, removing and installing (through 09/2000).**

– Check headlight aim, see **Headlights, adjusting (through 09/2000).**

Headlight assembly (from 10/2000)

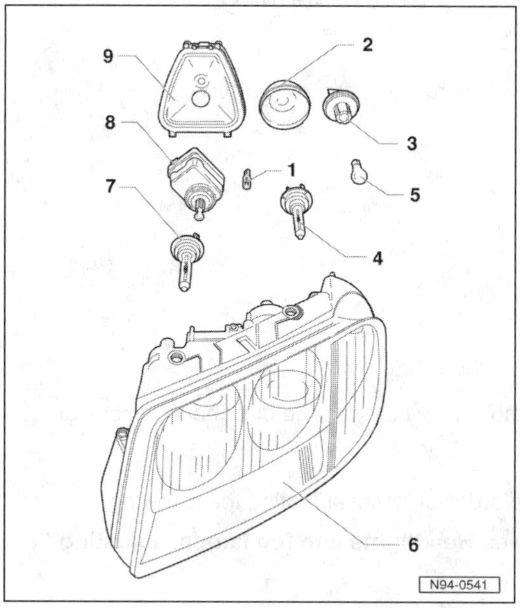

NOTE—
After performing any repairs or service that could affect headlight aim, check/adjust aim, see ***Maintenance.***

1. **Parking light bulb**
   - Bulb 12 V, 5 W
   - See **Headlight bulbs, replacing (from 10/2000)**
2. **Small cover cap**
3. **Turn signal bulb holder**
4. **Low beam bulb**
   - Bulb H7 12 V, 55 W
   - See **Headlight bulbs, replacing (from 10/2000)**
5. **Turn signal bulb**
   - Bulb 12 V, 21 W
   - See **Headlight bulbs, replacing (from 10/2000)**
6. **Headlight housing**
   - Replacing, see **Headlights, removing and installing (from 10/2000)**
   - Adjusting, see **Headlights and fog lamps, adjusting (from 10/2000)**
7. **High beam bulb**
   - Bulb H7 12 V, 55 W
   - See **Headlight bulbs, replacing (from 10/2000)**
8. **Headlight beam adjusting motor**
   - See **Headlight beam adjusting motor, removing and installing (from 10/2000)**
9. **Large cover cap**

Headlights, removing and installing (from 10/2000)

– Remove front bumper and plastic reinforcement behind bumper, see **Body-Exterior section, Repair Group 63**

– Disconnect electrical connections at headlight.

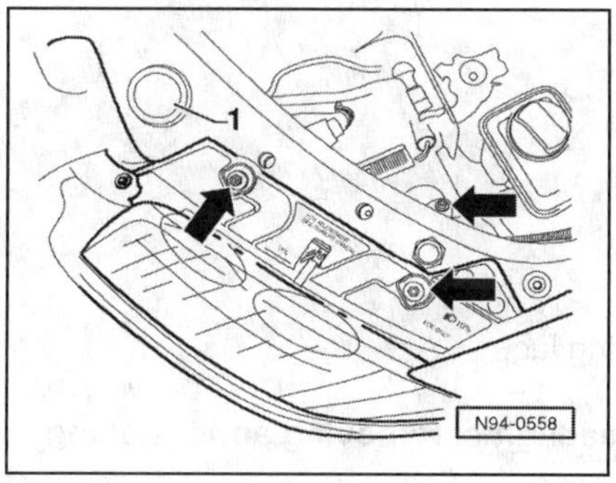

◄ Remove cap -1-.

– Remove screws -arrows- and additional screw located under removed cap -1-.

– Carefully remove headlight assembly forward from opening.

– Installation is in reverse order of removal, noting the following:

NOTE—
Check for even gaps between headlight assembly and front end trim. Adjust gaps as necessary by shifting housing slightly prior to tightening mounting screws.

- Install front bumper with plastic reinforcement, see **Body-Exterior section, Repair Group 63**
- Check headlight aim, see **Headlights and fog lamps, adjusting (from 10/2000)**.

Fog lamps, removing and installing (from 10/2000)

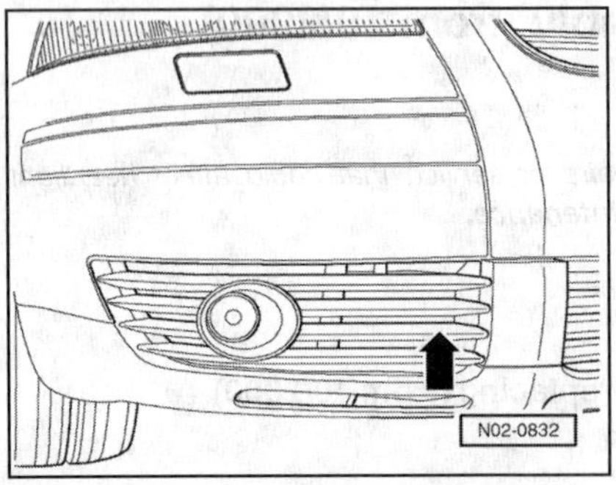

◄ Unclip trim from center outwards, and remove -arrow-.

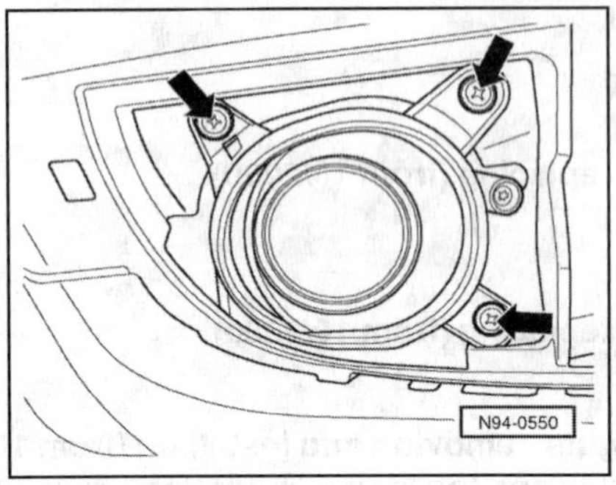

◄ Remove screws -arrow-.

– Pull out fog lamp assembly from bumper and disconnect electrical connection.

– Installation is in reverse order of removal, noting the following:

- Check fog lamp aim, see **Headlights and fog lamps, adjusting (from 10/2000)**.

Headlight securing lugs, servicing

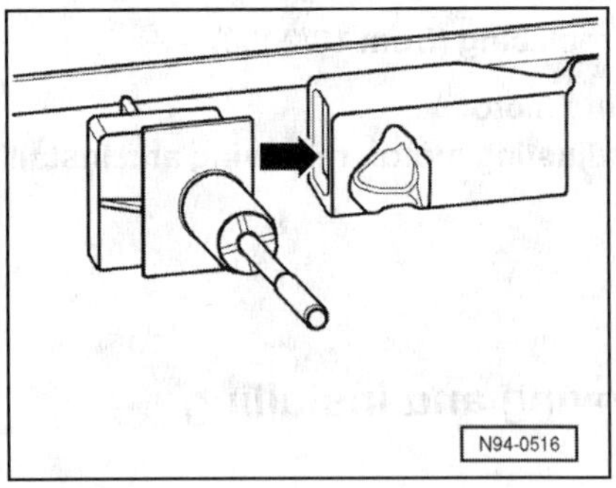

NOTE —
If headlight lower securing lugs break due to application of excessive force, it may still be possible to service headlight in a cost effective manner (though headlight housing must not be damaged).

– Remove headlight, see **Headlights, removing and installing**.

– Sand headlight tab socket breakage point smooth.

◄ Slide replacement part securing tab onto headlight tab -arrow-

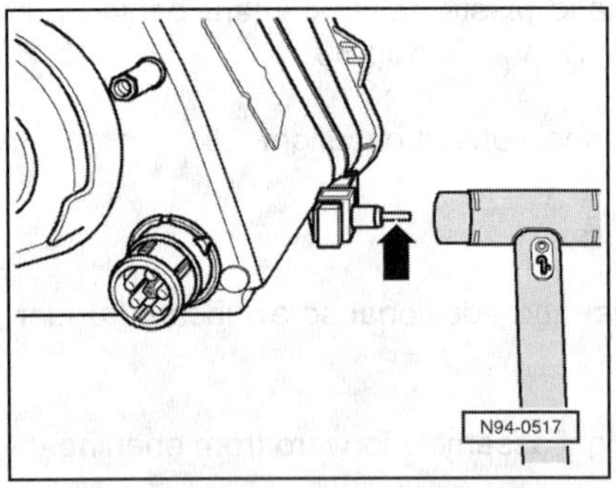

◄ Drive locking pin in carefully with small hammer.

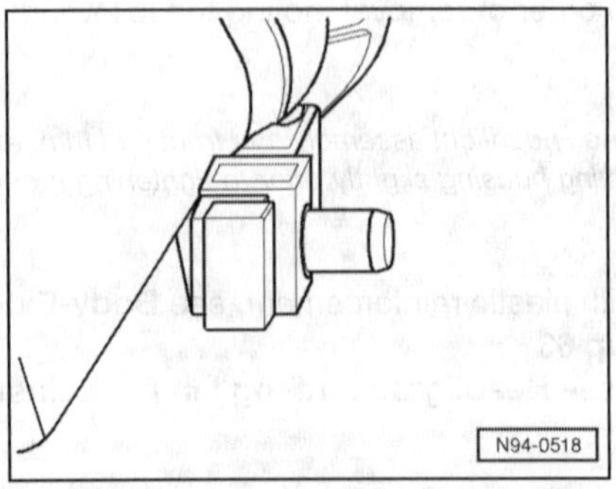

◄ View of repaired securing lug.

– Install headlight, see **Headlights, removing and installing**.

– Check headlight aim, see **Headlights and fog lamps, adjusting (from 10/2000)**.

Headlight driving direction setting, checking (from 10/2000)

Headlights may be equipped with a mechanism that allows the headlight beam pattern to be changed in accordance with market appropriate driving direction. Potential glare to oncoming traffic is minimized with the proper setting.

– Remove large cover cap from rear of headlight housing.

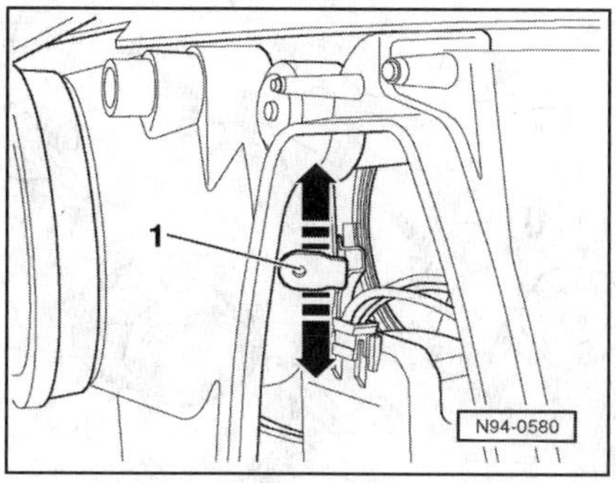

Confirm selector -1- to be in lower setting -down arrow-.

- Selector up: right-hand drive (e.g.: Great Britain - Australia etc.)
- Selector down: left-hand drive (e.g.: USA - Canada - Europe etc.)

NOTE—
High intensity gas discharge (HID) lamp driving direction setting, checking, see **High Intensity Gas Discharge (HID) driving direction setting, checking (from 10/2000).**

Headlights with high intensity gas discharge (HID) lamps (from 10/2000)

CAUTION—
Before beginning repairs on the electrical system:

- *Obtain the anti-theft radio security code.*
- *Switch the ignition off.*
- *Disconnect the battery Ground (GND) strap.*
- *After reconnecting battery, re-code and check operation of anti-theft radio. Also check operation of clock and power windows according to Repair Manual and/or Owner's Manual.*

WARNING—

- *HIGH VOLTAGE!*
- *Always disconnect battery Ground (GND) strap prior to servicing any parts of the HID lamp system marked with the yellow "HIGH VOLTAGE!"symbol.*
- *After disconnecting battery, briefly turn headlights on and off in order discharge any residual voltage.*

NOTE—

- *When removing and installing parking light and high beam bulbs, always remove complete headlight assembly from vehicle.*
- *HID lamp system has On Board Diagnostic (OBD) capabilities*

Headlights with high intensity gas discharge (HID) lamps, assembly (from 10/2000)

CAUTION—
Before beginning repairs on the electrical system:

- *Obtain the anti-theft radio security code.*
- *Switch the ignition off.*
- *Disconnect the battery Ground (GND) strap.*
- *After reconnecting battery, re-code and check operation of anti-theft radio. Also check operation of clock and power windows according to Owner's Manual.*

WARNING—
- *HIGH VOLTAGE!*
- *After disconnecting battery, briefly turn headlights on and off in order discharge any residual voltage.*

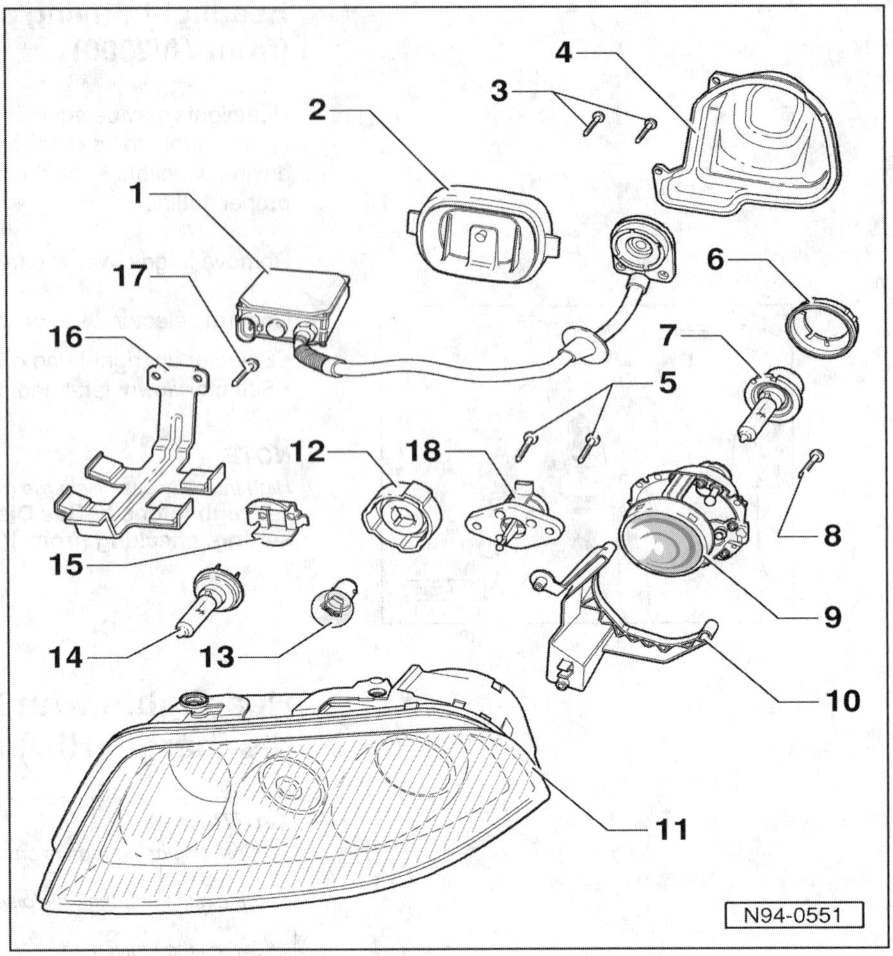

1. **Control Module for High Intensity Gas Discharge (HID) Lamp -J343- / -J344-**
   - See **Control module for High Intensity Gas Discharge (HID) lamp, removing and installing (from 10/2000)**
2. **Small cover cap**
3. **Mounting screws**
4. **Large cover cap**
5. **Mounting screws**
6. **Securing ring**
7. **HID low beam lamp**
   - 55 W
   - See **Headlights with high intensity gas discharge (HID) lamp, replacing bulbs (from 10/2000)**
8. **Mounting screw**
9. **HID lamp reflector**
10. **Mounting frame**
11. **Headlight housing**
    - Headlights, adjusting, see **Headlights and fog lamps, adjusting (from 10/2000)**
    - Removing and installing, see **Headlights, removing and installing (from 10/2000)**
12. **Bulb socket**
13. **Parking / turn signal light**
    - Bulb 12 V, 21W/5W
    - See **Headlights with high intensity gas discharge (HID) lamp, replacing bulbs (from 10/2000)**
14. **High beam bulb -M30- / -M32-**
    - Bulb H7 12 V, 55 W
    - See **Headlights with high intensity gas discharge (HID) lamp, replacing bulbs (from 10/2000)**
15. **Securing plate**
16. **Mounting frame**
    - For HID lamp control module
17. **Mounting screws**

Headlights with high intensity gas discharge (HID) lamp, replacing bulbs (from 10/2000)

HID lamp bulb, removing

WARNING —
- *HIGH VOLTAGE!*
- *Always disconnect battery Ground (GND) strap prior to servicing any parts of the HID lamp system marked with the yellow "HIGH VOLTAGE!"symbol.*
- *After disconnecting battery, briefly turn headlights on and off in order discharge any residual voltage.*
- *Internal pressure of HID glass bulb may exceed 10 bar.*
- *The physical and mechanical properties of HID bulbs are very sensitive. Always wear safety glasses when removing and installing HID bulb*

CAUTION —
Do not touch glass portion of bulb with bare hands. Even the smallest amount of moisture and/or contaminants from fingers that evaporates on the bulb during operation can cause the glass to cloud over.

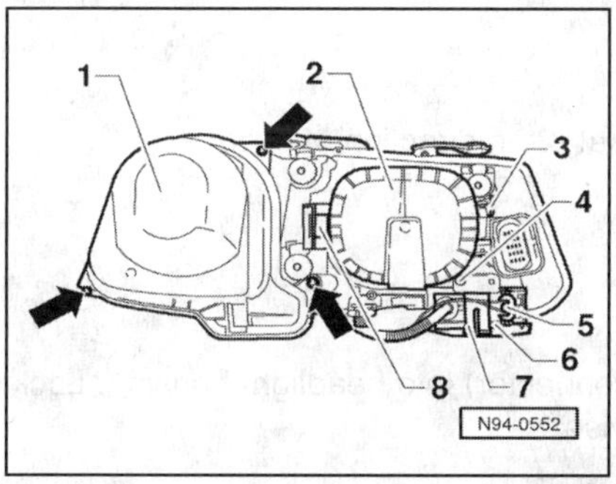

– Remove headlight housing, see **Headlights, removing and installing (from 10/2000).**

Remove three screws -arrow- and remove large cover cap -1-.

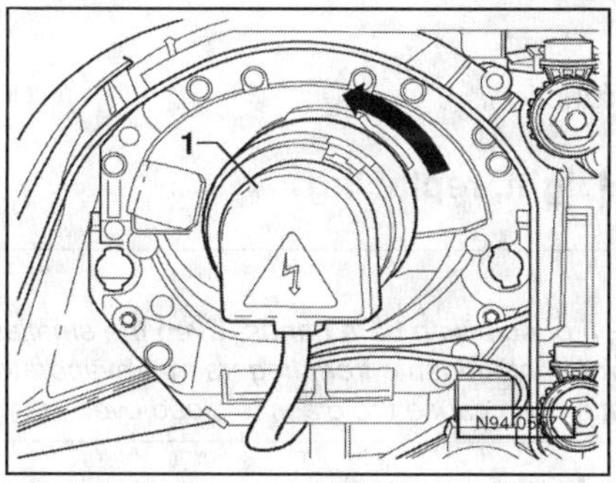

Turn connector coupling -1- counterclockwise 90°and pull off.

– Carefully remove HID bulb.

Installing

CAUTION —
The physical and mechanical properties of HID bulbs are very sensitive. Always use care when handling HID bulbs.

– Insert new bulb, making sure that the bulb mounting aligns with the locating lugs in the housing.

– Lock connector coupling into place by turning clockwise 90°.

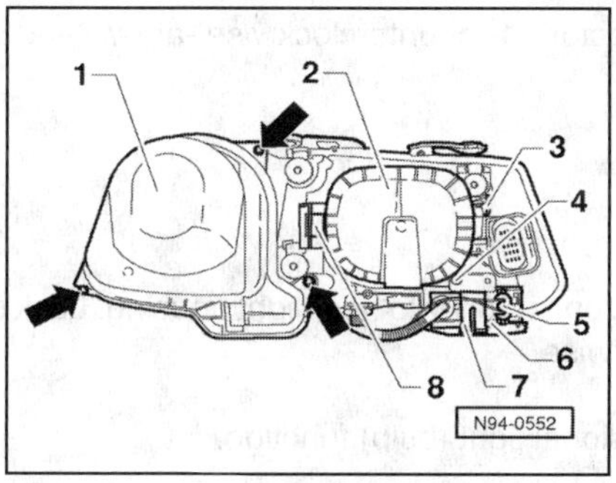

Reinstall cover cap -1- and secure with screws -arrows-.
- Reinstall headlight housing, see **Headlights, removing and installing (from 10/2000).**
- Check headlight (HID lamp) function.
- Check headlight aim, see **Headlights and fog lamps, adjusting (from 10/2000)**

High beam bulb, replacing

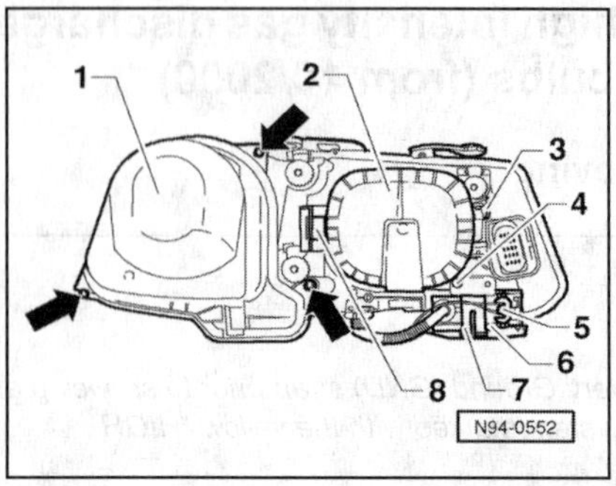

> ***CAUTION —***
> *Do not touch glass portion of bulb with bare hands. Even the smallest amount of moisture and/or contaminants from fingers that evaporates on the bulb during operation can cause the glass to cloud over.*

Removing

◄ Unclip retainer -3-, pull latch -8- from guide and remove cover cap from rear of headlight housing

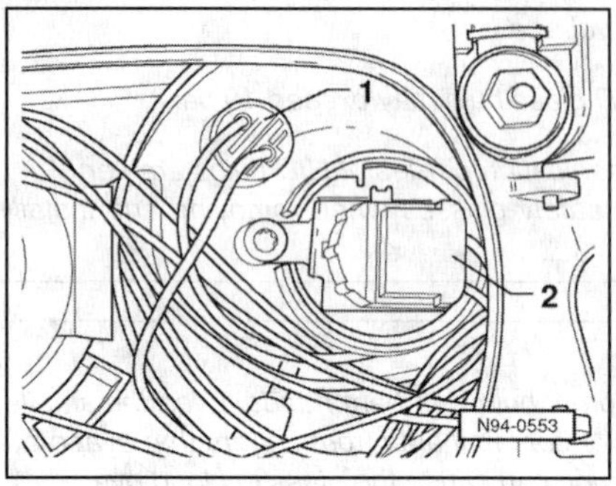

◄ Turn socket (with connector) -2- counterclockwise and pull out of headlight housing.

NOTE —
Bulb socket screws must not be loosened when replacing bulbs.

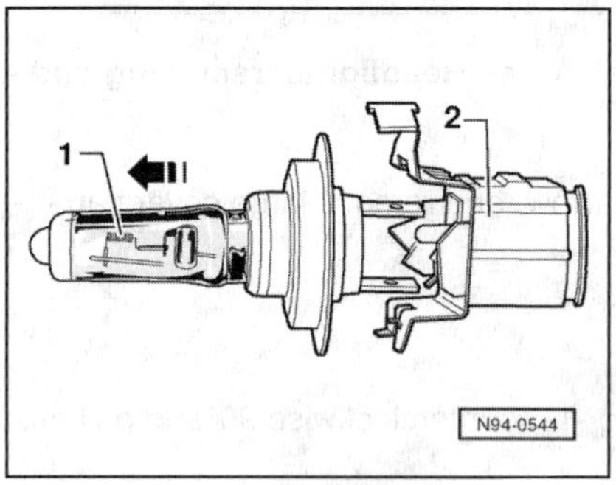

◄ Pull bulb -1- out of socket -2- in direction of -arrow-.

– Replace bulb.

Installing

– Reinstall socket (with connector) into headlight housing. Lock socket in place by turning clockwise.

– Reinstall cover cap.

– Check headlight (bulb) function.

Parking / turn signal light, replacing

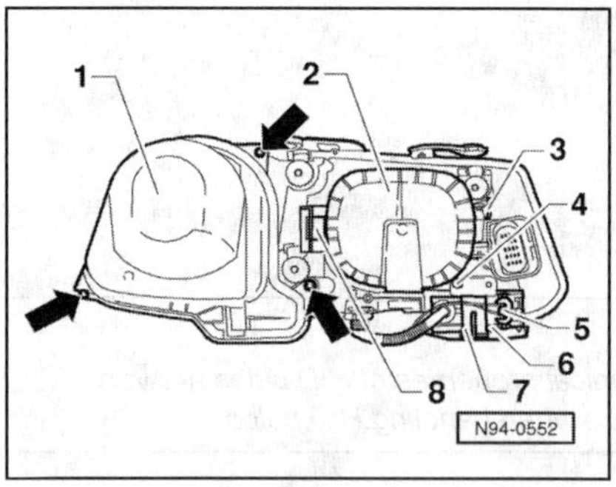

> ***CAUTION —***
> *Do not touch glass portion of bulb with bare hands. Even the smallest amount of moisture and/or contaminants from fingers that evaporates on the bulb during operation can cause the glass to cloud over.*

◄ Unclip retainer -3-, pull latch -8- from guide and remove cover cap from rear of headlight housing

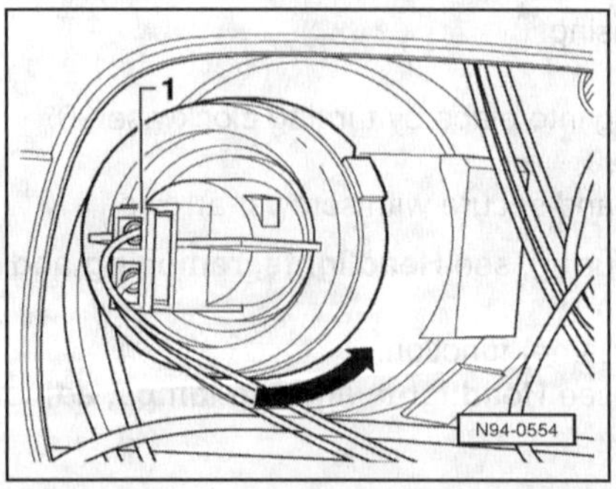

◄ Turn socket (with connector) -1- counterclockwise -arrow- and pull from housing.

– Remove bulb.

– Install bulb.

- Reinstall socket (with connector) into headlight housing. Lock socket in place by turning clockwise.
- Reinstall cover cap.
- Check parking / turn signal light (bulb) function.

Control module for high intensity gas discharge (HID) lamp, removing and installing (from 10/2000)

> ***WARNING —***
> - *HIGH VOLTAGE!*
> - *Always disconnect battery Ground (GND) strap prior to servicing any parts of the HID lamp system marked with the yellow "HIGH VOLTAGE!"symbol.*
> - *After disconnecting battery, briefly turn headlights on and off in order discharge any residual voltage.*

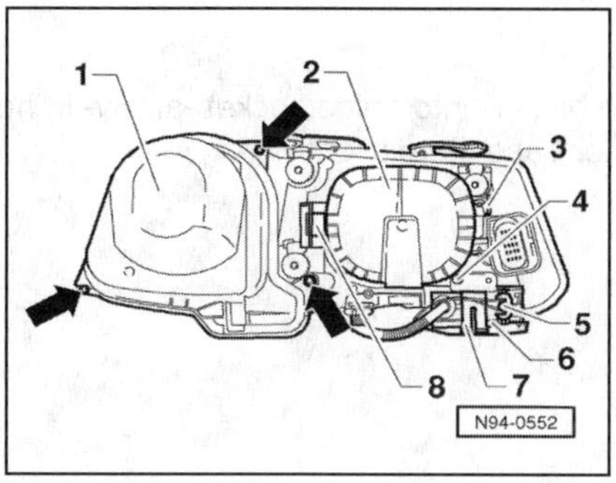

– Remove headlight assembly, see **Headlights, removing and installing (from 10/2000)**.

< Disconnect electrical connector -5-.

- Remove mounting screw -4-.
- Unclip retainer -7- and pull out control module -6-.
- Remove three screws -arrow- and remove large cover cap -1-.

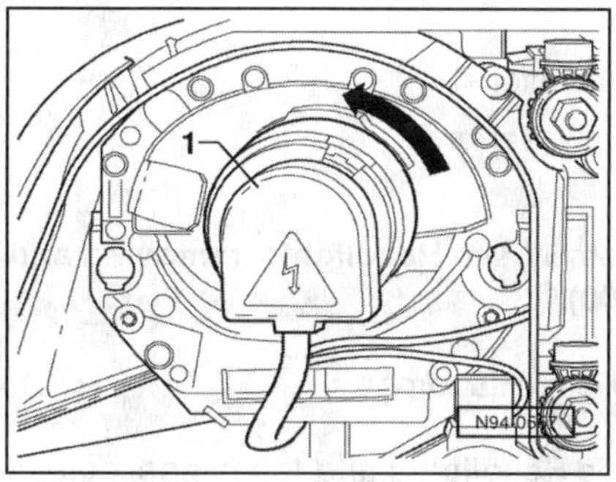

< Turn connector coupling -1- counterclockwise 90°and pull off.

– Pull electrical harness from control module to HID lamp out from headlight housing.

– Installation is in reverse order of removal, noting the following:

- Check function of headlight (HID lamp).
- Check headlight aim, see **Headlights and fog lamps, adjusting (from 10/2000)**.
- Aiming must be done in conjunction with the On Board Diagnostic (OBD) program "Basic Setting" (Scan Tool function 04), see **Scan Tool section.**

Headlight beam adjusting motor for headlights with high intensity discharge (HID) lamps, removing and installing (from 10/2000)

NOTE —
- *Removal and replacement is only possible with headlight housing removed from vehicle.*
- *After removal and installation of adjusting motor, headlight basic setting must always be reestablished using optical headlight aimer VAS 5107. The headlight inclination specification is embossed into the headlight housing.*

Removing

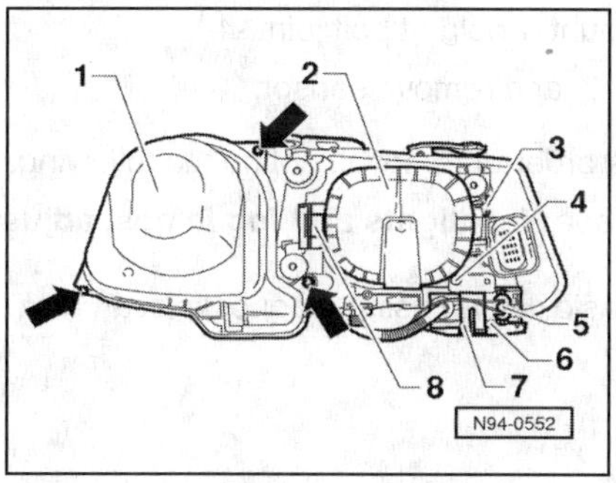

– Remove headlight assembly, see **Headlights, removing and installing (from 10/2000)**.

< Remove three mounting screws -arrows- and remove large cover cap -1-.

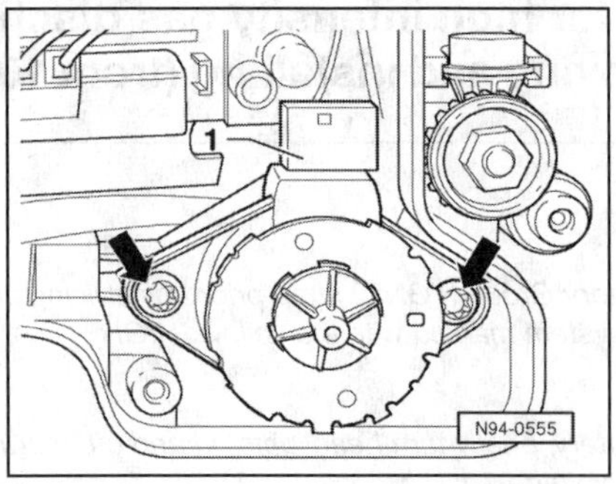

Disconnect electrical connector -1- at headlight beam adjusting motor.

- Remove mounting screws -arrow-.
- Remove positioner linkage ball from guide socket in headlight housing by lifting reflector base plate slightly.

Installing

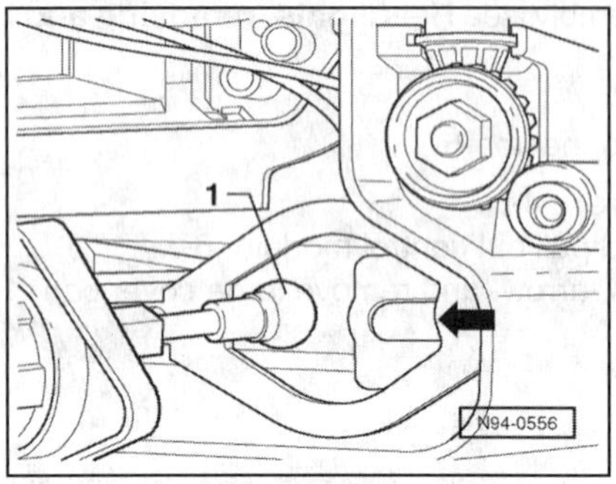

Insert positioner linkage ball -1- into guide socket -arrow- in headlight housing by lifting reflector base plate slightly.

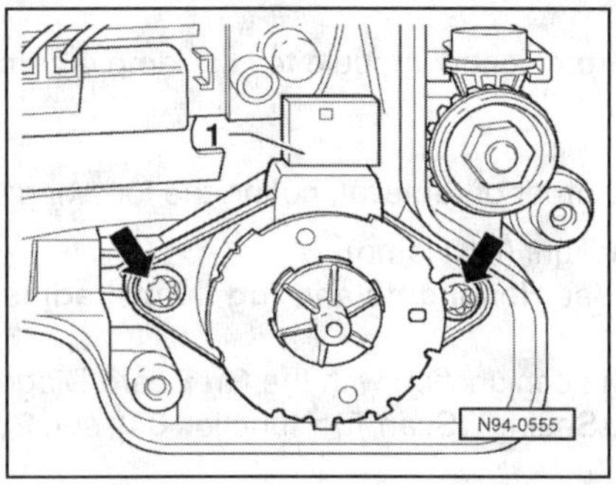

Reconnect electrical connector -1-.

- Reinstall mounting screws -arrows-.
- Reinstall cover cap.

– Reinstall headlight assembly, see **Headlights, removing and installing (from 10/2000)**.

– Check function of headlight (HID lamp).

– Check headlight aim, see **Headlights and fog lamps, adjusting (from 10/2000)**.

– Bring headlight into basic setting, see **Scan Tool section.**

Left front level control system sensor -G78-, removing and installing (from 10/2000)

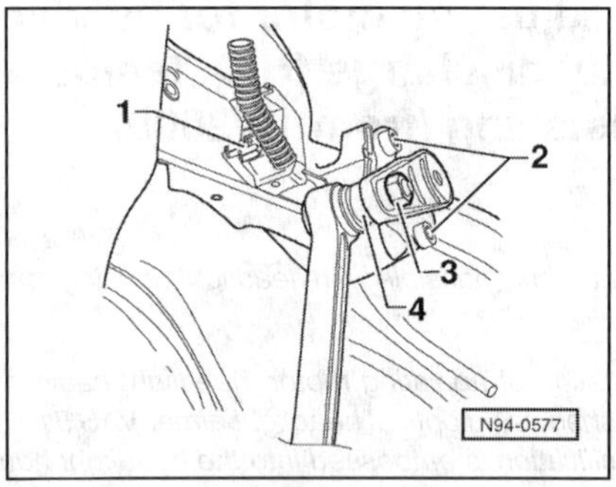

Sensor is mounted to stabilizer on left front suspension (in fender well) and is not externally adjustable.

Disconnect electrical connection -1- at sensor.

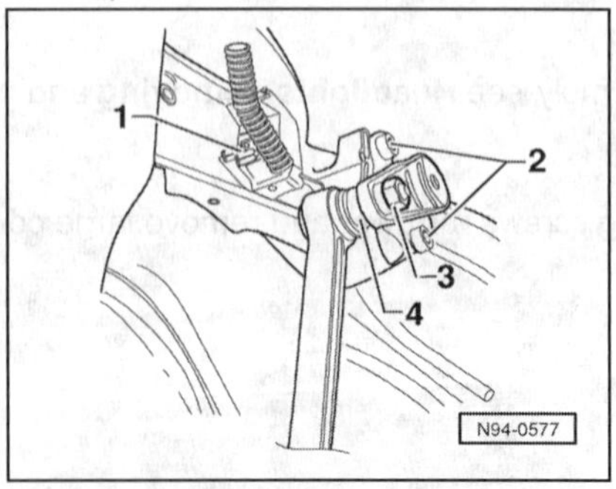

Remove hex bolt -3- (counter-hold at bolt/joint -4-).

- Remove Torx screws -2- and remove sensor.

– Installation is in reverse order of removal, noting the following:

- Check headlight aim, see **Headlights and fog lamps, adjusting (from 10/2000).**
- Bring headlight into basic setting using diagnostic scan tool.

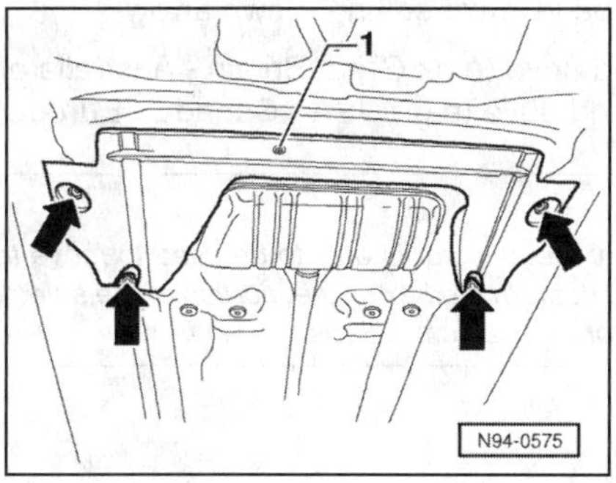

Left rear level control system sensor -G76-, removing and installing (from 10/2000)

Sensor is mounted to left rear suspension and is not externally adjustable.

Remove screw -1-, Torx screws -arrows- and remove rear suspension cover.

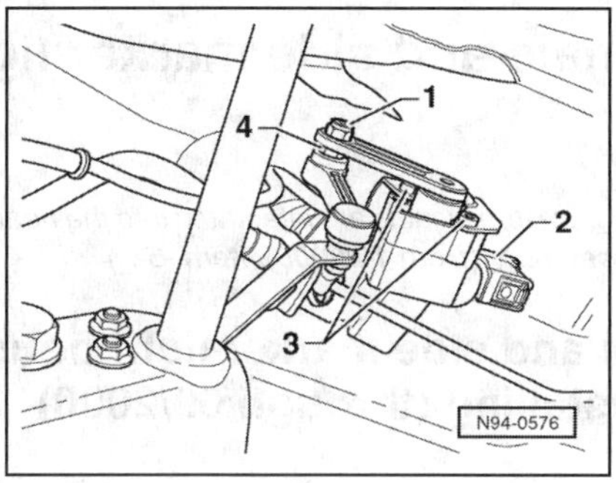

Disconnect electrical connection -2- at sensor.

- Remove hex bolt -1- (counter-hold at bolt/joint -4-).
- Remove Torx screws -3- and remove sensor.

– Installation is in reverse order of removal, noting the following:

- Check headlight aim, see **Headlights and fog lamps, adjusting (from 10/2000)**.
- Bring headlight into basic setting using diagnostic scan tool.

Control module for headlight beam adjusting -J431-, removing and installing (from 10/2000)

Control module is mounted in left luggage compartment behind left rear luggage compartment trim.

– Remove left rear luggage compartment trim, see **Body-Interior section, Repair Group 70.**

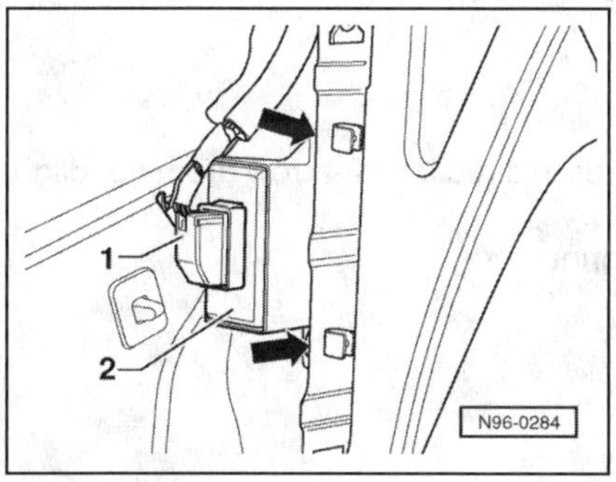

– If necessary, remove CD changer and changer mounting, see **Radio, section, Repair Group 91.**

Disconnect electrical connection -1-, remove screws -arrows- and remove control module -2-.

– Installation is in reverse order of removal, noting the following:

- After installation, check function of headlights.
- Code control module, see **Scan Tool section**.
- Bring headlight into basic settingusing diagnostic scan tool.

High intensity gas discharge (HID) lamp driving direction setting, checking (from 10/2000)

HID headlamps may be equipped with a mechanism that allows the headlight beam pattern to be changed in accordance with market appropriate driving direction. Potential glare to oncoming traffic is minimized with the proper setting.

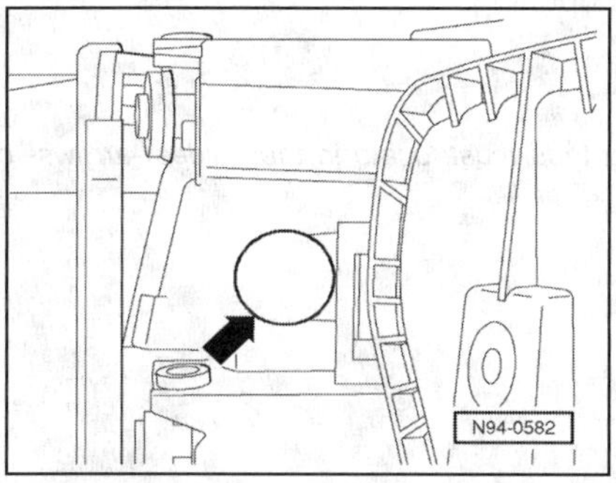

Confirm setting for the USA/CDN as follows:

Unclip cap -arrow- and remove.

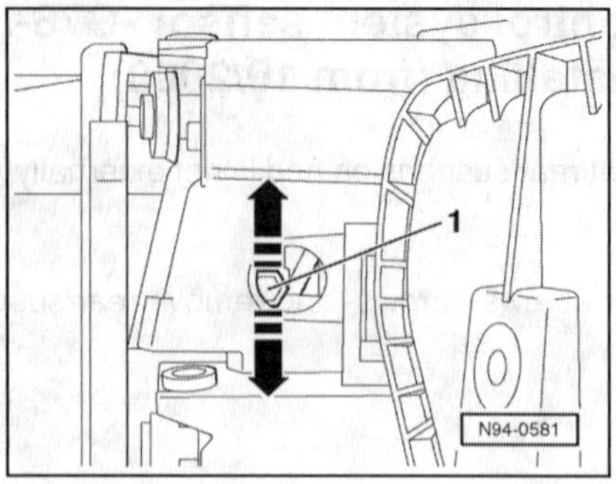

Confirm selector -1- to be in lower setting -down arrow-.

- Selector up: right-hand drive (e.g.: Great Britain - Australia etc.)
- Selector down: left-hand drive (e.g.: USA - Canada - Europe etc.)

CAUTION—
The hex socket portion of the selector is only to be used to move the selector into position. Do not turn the hex socket portion of the selector as damage to the selector mechanism will result.

Front turn signals and side marker lights

NOTE—
*On vehicles from 10/2000, front turn signals are integrated into the headlight housing, see **Headlight assembly (from 10/2000), item -5-**.*

Front turn signal and side marker light housing, removing and installing (through 09/2000)

Removing

NOTE—
To remove right-hand turn signal/side marker light housing, remove engine intake duct beforehand.

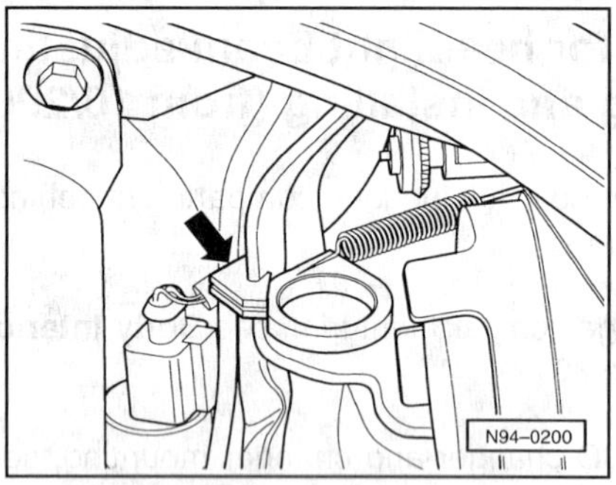

Unhook retainer spring -arrow- on inside of housing.

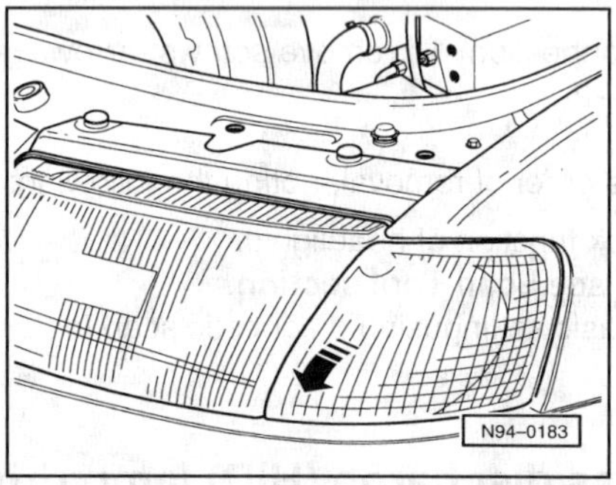

Pull light housing forwards in direction of -arrow- from headlight housing.

– Disconnect electrical connector.

Installing

– Reconnect electrical connector -1-.

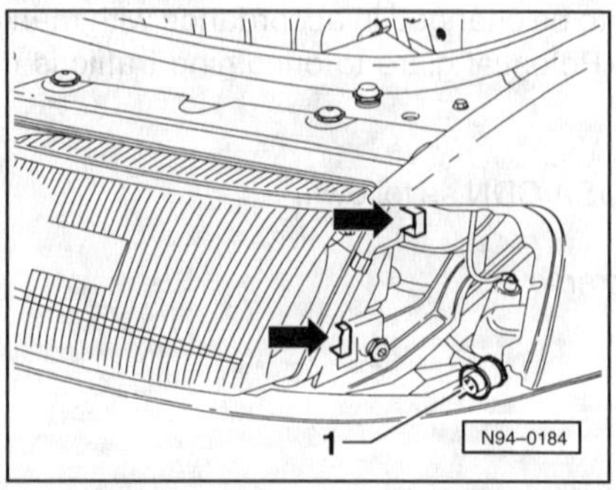

Insert light housing into opening.

NOTE—
The light housing locating tabs must locate in the guides -arrows- of the headlight housing.

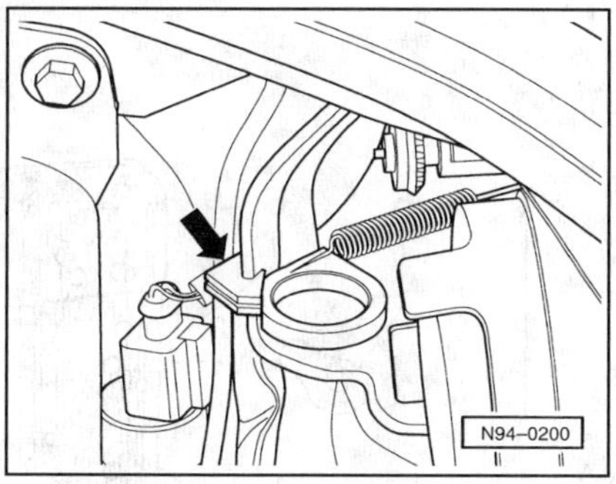

Attach retainer spring -arrow- on inside of light housing.

Turn signal bulb, replacing

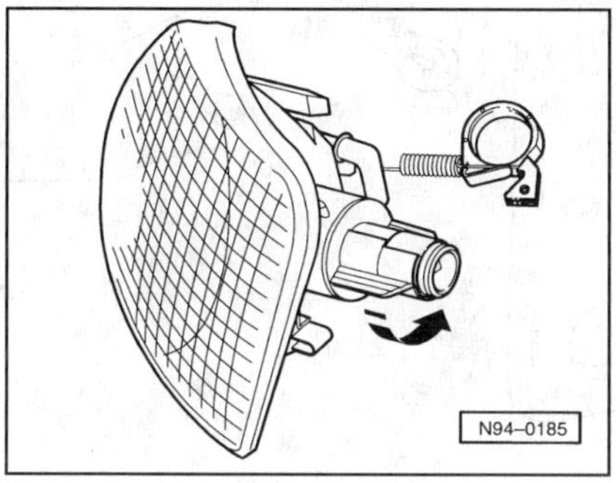

> ***CAUTION —***
> *Do not touch glass portion of bulb with bare hands. Even the smallest amount of moisture and/or contaminants from fingers that evaporates on the bulb during operation can cause the glass to cloud over.*

Turn bulb holder in direction of arrow and pull out from housing.

– Remove bulb from holder.

– When installing, ensure that the bulb holder engages in the housing by turning in opposite direction of arrow.

Side mounted turn signals

Side mounted turn signals, removing and installing

Removing

> ***CAUTION —***
> *Removal of turn signal lens/housing is only possible in one direction. The side of the housing equipped with retainer clip is not obvious in it's installed position. Use care when removing to prevent damage to retainer clip or housing.*

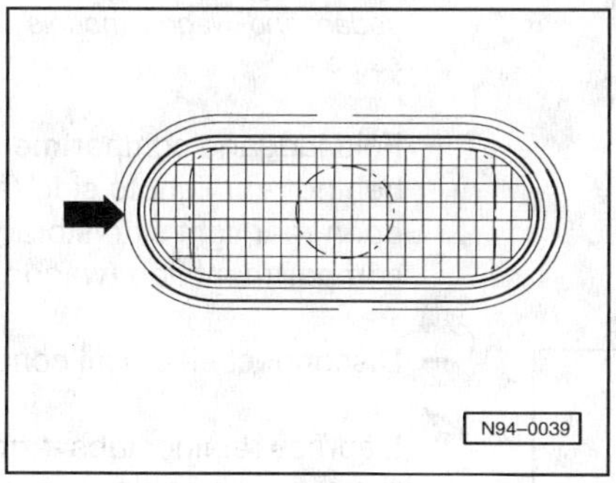

Carefully push turn signal lens from retainer side -1- towards retainer clip -arrow 2- and pull out from opening in fender (if necessary, apply masking tape to fender to prevent paint damage).

– Pull off rubber boot and disconnect electrical connector.

– Pull bulb (12V/5W) directly out of socket (do not turn).

Installing

– Replace bulb.

– Reconnect electrical connector and secure rubber boot.

– Insert (snap-in) turn signal into opening in fender.

– Check function of side turn signal (lamp).

Rear lights

Rear light assembly (through 09/2000)

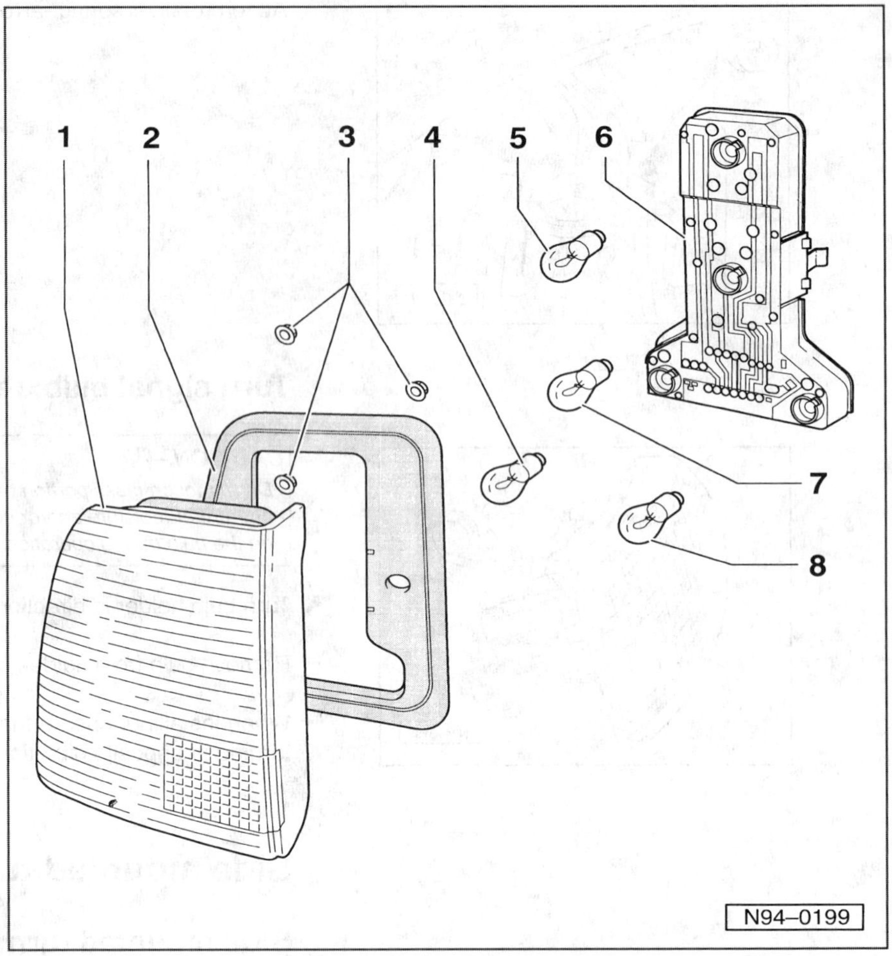

CAUTION—

Before beginning repairs on the electrical system:

- *Obtain the anti-theft radio security code.*
- *Switch the ignition off.*
- *Disconnect the battery Ground (GND) strap.*
- *After reconnecting battery, re-code and check operation of anti-theft radio. Also check operation of clock and power windows according to Owner's Manual.*

1. **Housing**
   - See **Rear light housing (sedan), removing and installing (from 10/2000)**
2. **Gasket**
   - Replace if damaged
3. **Hex nuts M5**
   - 3 Nm (27 in lb)
4. **Rear fog light bulb (driver's side only)**
   - 12V / 21W
5. **Dual filament bulb for brake and tail light**
   - 12V / 21W/5W
6. **Bulb holder**
   - See **Rear light bulb holder, removing and installing (through 09/2000)**
7. **Turn signal bulb**
   - 12V / 21W
8. **Back-up light bulb**
   - 12V / 21W

Rear light bulb holder, removing and installing (through 09/2000)

NOTE—

Apart from the different access required, removing and installing the rear light bulb holders on sedan and wagon models is the same.

– Fold luggage compartment trim behind rear light to side (sedan) or open rear light side storage compartment flap (wagon).

– Disconnect electrical connector.

◄ Depress retainer tabs -arrows- and remove bulb holder to rear.

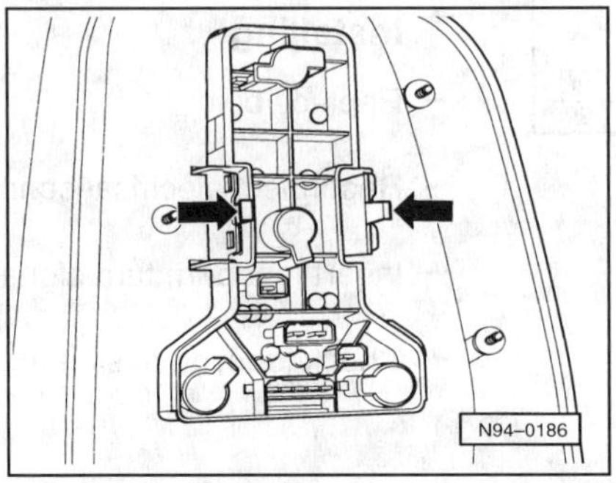

– Installation is reverse order of removal.

Rear light (sedan), assembly (from 10/2000)

CAUTION—

Before beginning repairs on the electrical system:

- *Obtain the anti-theft radio security code.*
- *Switch the ignition off.*
- *Disconnect the battery Ground (GND) strap.*
- *After reconnecting battery, re-code and check operation of anti-theft radio. Also check operation of clock and power windows according to Repair Manual and/or Owner's Manual.*

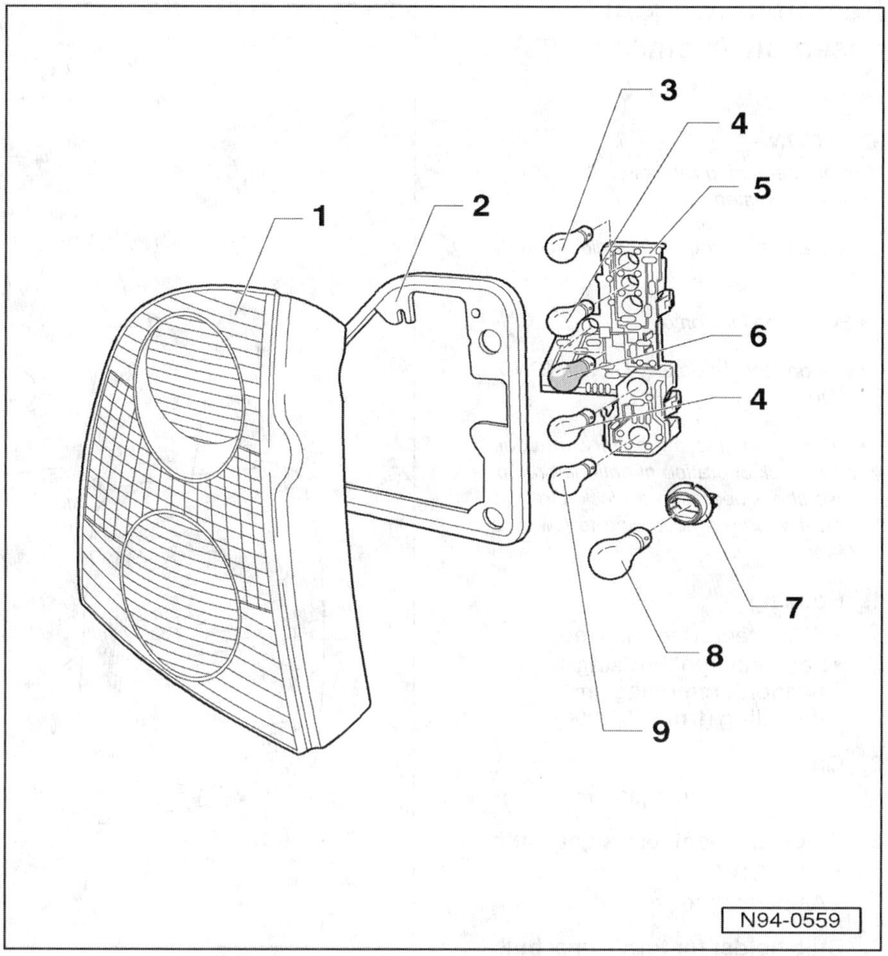

1. **Housing**
   - With integral lens and reflector
   - See **Rear light housing (sedan), removing and installing (from 10/2000)**
2. **Gasket**
   - Replace if damaged
3. **Dual filament bulb for brake and tail light**
   - 12V / 21W/5W
4. **Empty sockets (qty. 2)**
   - No bulb installed in these sockets for USA and Canadian vehicle
5. **Bulb holder**
   - See **Rear light bulb holder (sedan), removing and installing (from 10/2000)**
   - May have additional unused electrical connections depending on vehicle equipment
6. **Single filament turn signal bulb**
   - 12V / 21W
   - Amber-colored bulb
7. **Back-up light bulb socket**
   - Removing: rear light housing must be removed first.
8. **Single filament back-up light bulb**
   - 12V / 21W
9. **Dual filament bulb for brake and tail light**
   - 12V / 21W/5W

Rear light bulb holder (sedan), removing and installing (from 10/2000)

– Fold luggage compartment trim behind rear light to side.

– Disconnect electrical connector.

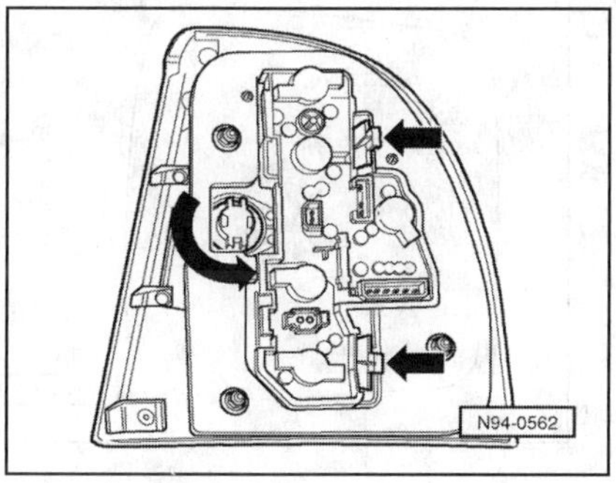

< Depress retainer tabs -arrows- and remove bulb holder to rear.

– To remove back-up light bulb, turn socket counterclockwise -arrow-.

– Installation is reverse order of removal.

NOTE—

*Removing back-up light bulb socket: remove rear light housing first, see **Rear light housing (sedan), removing and installing (from 10/2000).***

Rear light (wagon), assembly (from 10/2000)

CAUTION —
Before beginning repairs on the electrical system:

- *Obtain the anti-theft radio security code.*
- *Switch the ignition off.*
- *Disconnect the battery Ground (GND) strap.*
- *After reconnecting battery, re-code and check operation of anti-theft radio. Also check operation of clock and power windows according to Owner's Manual.*

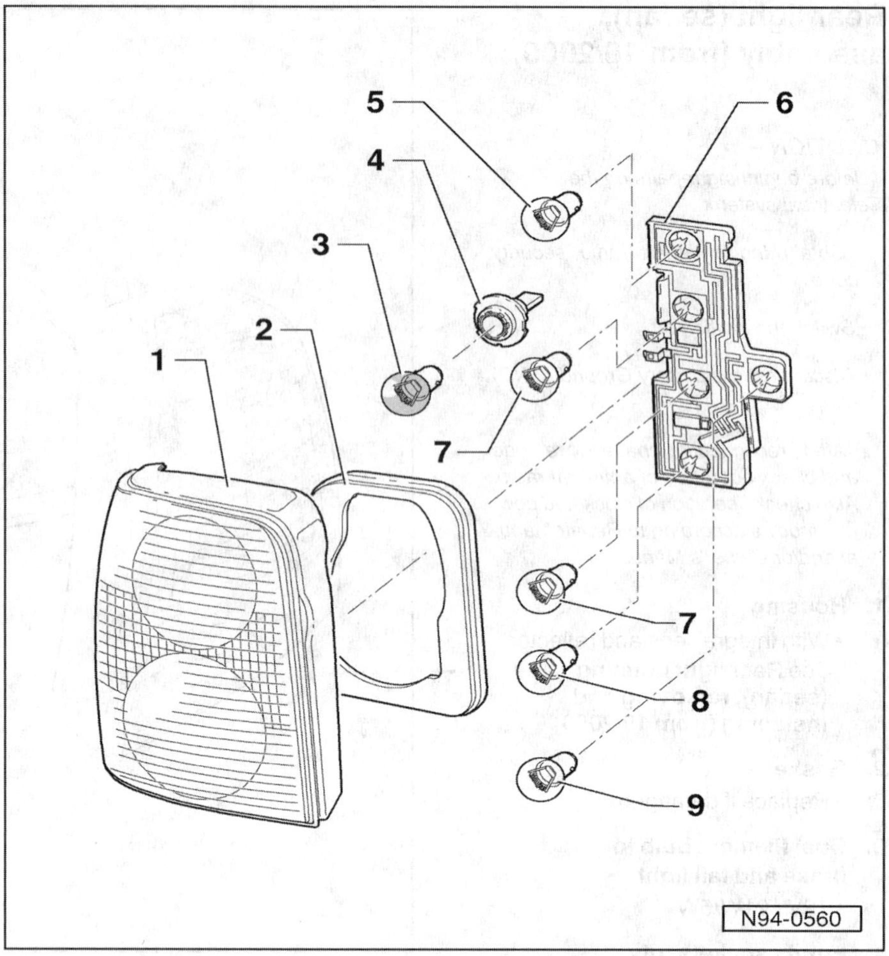

1. **Housing**
   - With integral lens and reflector
   - See **Rear light housing (wagon), removing and installing (from 10/2000)**
2. **Gasket**
   - Replace if damaged
3. **Single filament turn signal bulb**
   - 12V / 21W
   - Amber-colored bulb
4. **Bulb holder for turn signal bulb**
5. **Dual filament bulb for brake and tail light**
   - 12V / 21W/5W
6. **Bulb holder**
   - See **Rear light bulb holder (wagon), removing and installing (from 10/2000)**
   - May have additional unused electrical connections depending on vehicle equipment
7. **Empty sockets (qty. 2)**
   - No bulb installed in these sockets for USA and Canadian vehicle
8. **Dual filament bulb for brake and tail light**
   - 12V / 21W/5W
9. **Single filament back-up light bulb**
   - 12V / 21W

Rear light bulb holder (wagon), removing and installing (from 10/2000)

– Open rear light side storage flap.

– Disconnect electrical connector.

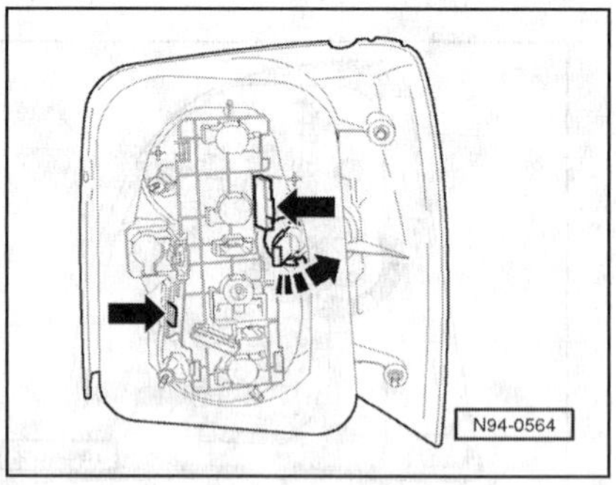

– Depress retainer tabs -arrows- and remove bulb holder to rear.

– To remove turn signal bulb, turn socket counterclockwise - arrow-.

– Installation is reverse order of removal.

Rear light housing, removing and installing (through 09/2000)

NOTE —

Apart from the different access required, removing and installing the rear light bulb holders on sedan and wagon models is the same.

Removing

– Fold luggage compartment trim behind rear light to side (sedan) or open rear light side storage compartment flap (wagon).

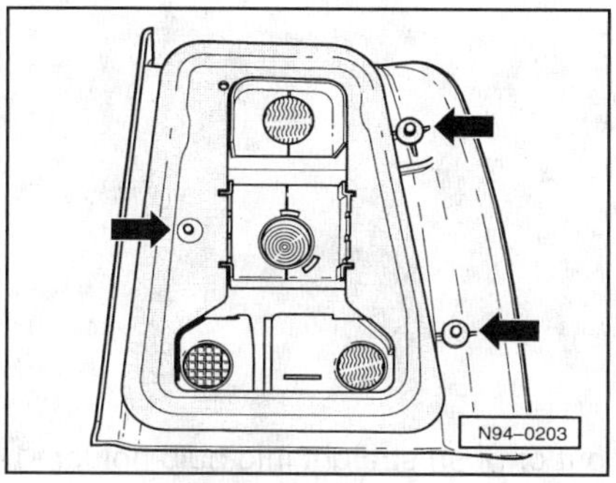

– Remove bulb holder, see **Rear light bulb holder, removing and installing (through 09/2000)**

◄ Remove three nuts -arrows- and remove housing.

Installing

– Insert housing into opening.

- Install and tighten securing nuts, 3 Nm (27 in lb).
- Install bulb holder and reconnect electrical connectors, see **Rear light bulb holder, removing and installing (through 09/2000)**.
- Refit luggage compartment trim (sedan) or close flap for storage compartment (wagon).

Rear light housing (sedan), removing and installing (from 10/2000)

Removing

– Fold luggage compartment trim behind rear light to side.

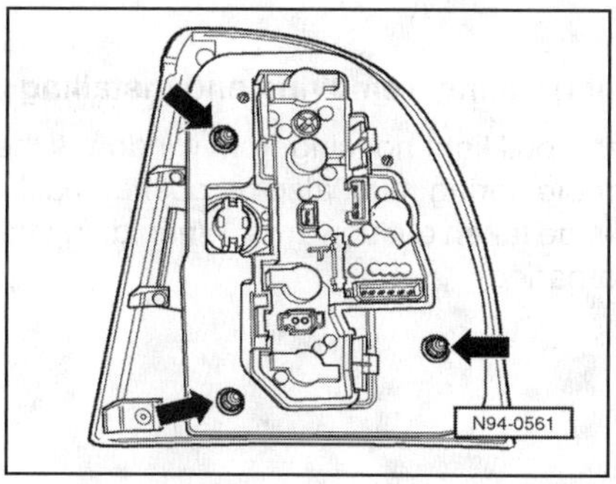

– Remove bulb holder, see **Rear light bulb holder (sedan), removing and installing (from 10/2000)**.

◄ Remove three screws -arrows- and remove housing.

Installing

– Insert housing into opening.

- Install and tighten securing nuts, 3 Nm (27 in lb).
- Install bulb holder and reconnect electrical connectors, see **Rear light bulb holder (sedan), removing and installing (from 10/2000)**.
- Refit luggage compartment trim.

Rear light housing (wagon), removing and installing (from 10/2000)

Removing

– Open rear light side storage flap.

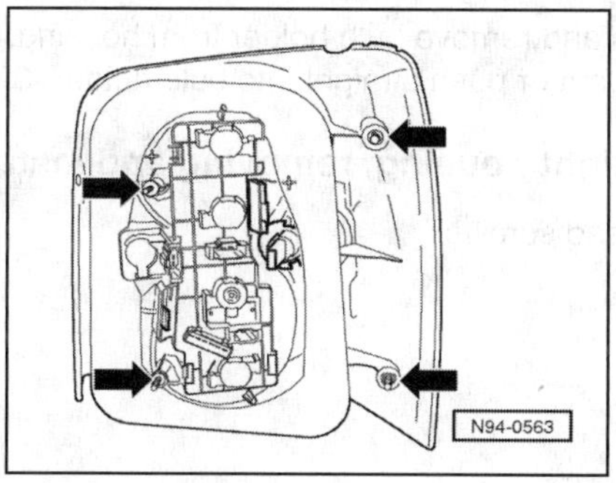

– Remove bulb holder, see **Rear light bulb holder (wagon), removing and installing (from 10/2000)**.

◄ Remove four screws -arrows- and remove housing.

Installing

– Insert housing into opening.

- Install and tighten securing nuts, 3 Nm (27 in lb).
- Install bulb holder and reconnect electrical connectors, see **Rear light bulb holder (wagon), removing and installing (from 10/2000)**.
- Refit luggage compartment trim.

High-mount brake light

High-mount brake light (sedan) (through 09/2000)

The high-mount brake light is installed in the rear of the rear shelf, directly against the rear window.

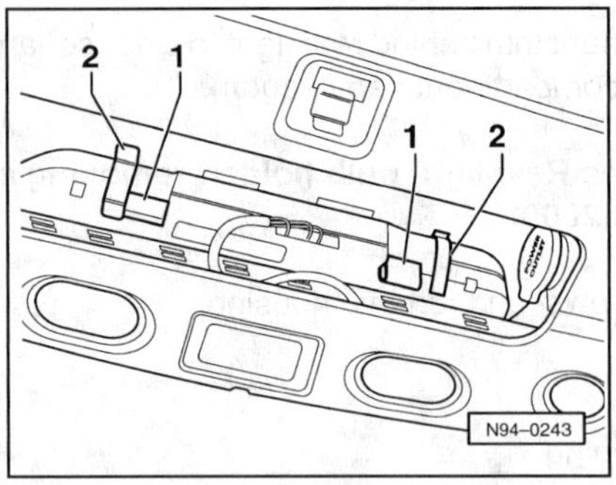

Bulb holder, removing and installing

- Disconnect electrical connector, press lock retaining spring -1- and remove bulb holder.

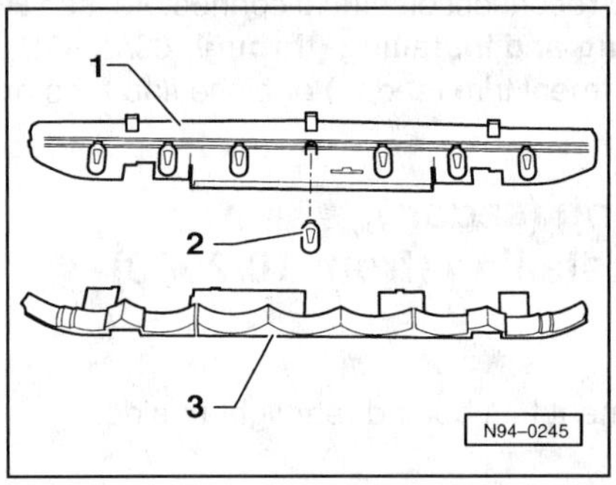

Bulbs, replacing

- Pull bulbs -2- straight out or push straight into bulb holder -1-.

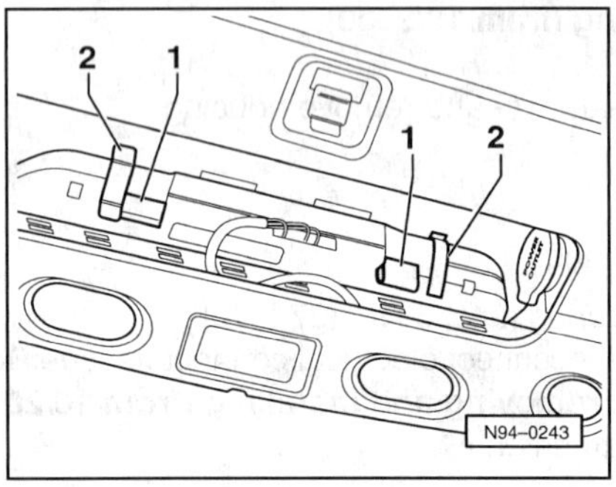

High-mount brake light housing, removing and installing

- Depress retainer -2- and pull light housing from window. When doing so, engaging device of the spring must also be pushed back.
- The brake light can now be taken out upwards. When doing this ensure the rear shelf is not damaged.

High-mount brake light (wagon) (through 09/2000)

– Remove inner trim on rear lid, see **Body-Interior section, Repair Group 70 (wagon)**

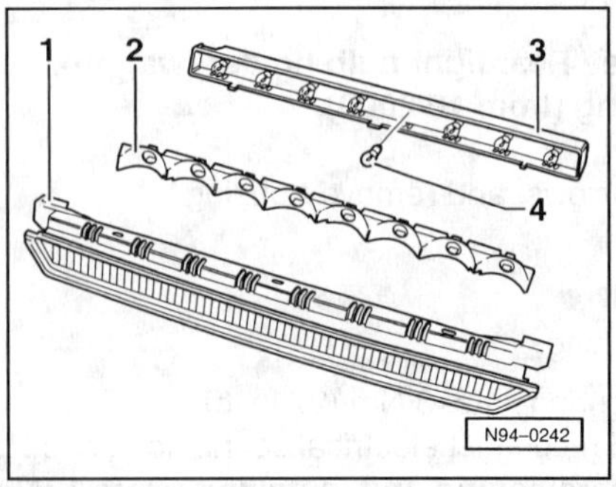

Bulbs, replacing

- Depress retainer tabs and remove bulb holder from housing.
- Pull bulbs -4- straight out or push straight into bulb holder -3-.

High-mount brake light housing, removing and installing

– Remove two Phillips head screws.

High mount brake light (sedan) (from 10/2000)

The high-mount brake light is installed in the rear of the rear shelf, directly against the rear window.

Bulb holder, removing and installing

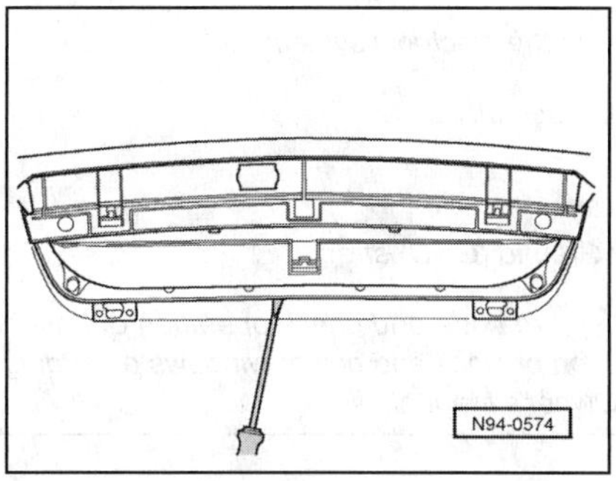

- Using screwdriver, unclip bulb holder from mounting.

NOTE—
The bulb holder base plate contains soldered-in light emitting diodes (LEDs) that are not replaceable. In the event of LED malfunction, replace complete bulb holder.

High mount brake light (wagon) (from 10/2000)

Bulb holder, removing and installing

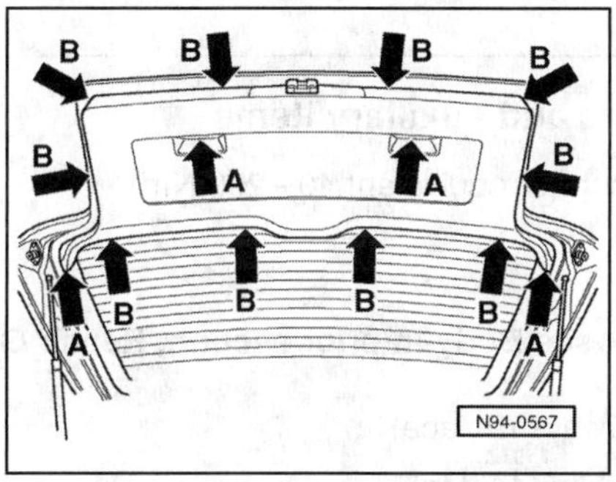

- Remove screws -arrows A-, unclip trim -arrows B- and remove.

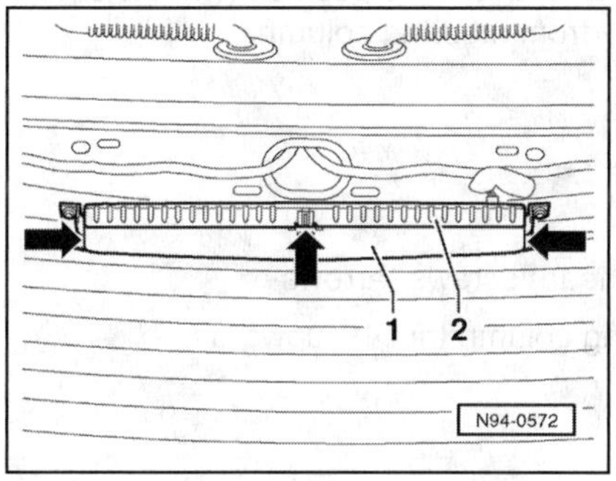

- Depress locking clips -arrows- and remove bulb holder - 1- from mounting -2-.

NOTE—
The bulb holder base plate contains soldered-in light emitting diodes (LEDs) that are not replaceable. In the event of LED malfunction, replace complete bulb holder.

License plate lights

License plate lights, removing and installing

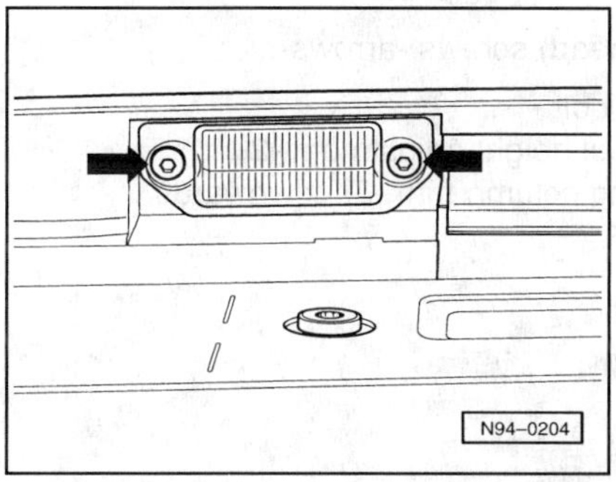

- Remove screws -arrows-.
- Disconnect electrical connector.
- Remove lens.
- Installation is reverse order of removal.

Steering column switches

Steering column switches, removing and installing

> ***CAUTION—***
> *Before beginning repairs on the electrical system:*
>
> *• Obtain the anti-theft radio security code.*
>
> *• Switch the ignition off.*
>
> *• Disconnect the battery Ground (GND) strap.*
>
> *• After reconnecting battery, re-code and check operation of anti-theft radio. Also check operation of clock and power windows according to Repair Manual and/or Owner's Manual.*

> ***CAUTION—***
> *Special safety precautions apply to vehicles equipped with airbags. Refer to Repair Manual, Body-Interior, Airbag: CAUTIONS and WARNINGS, Repair Group 69.*

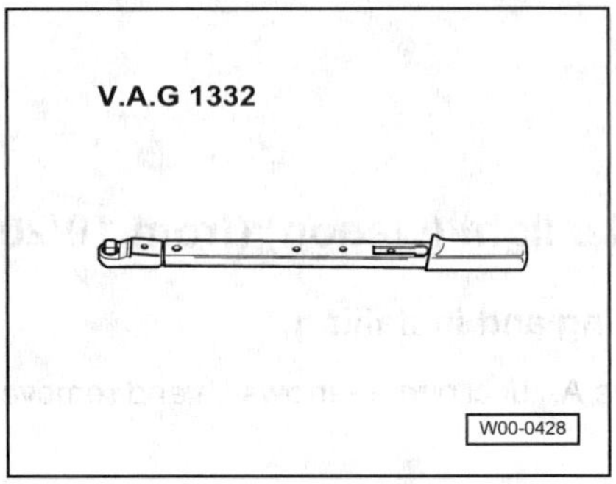

Special tools, testers and auxiliary items

VAG 1332 Torque wrench (or equivalent 40 - 200 Nm)

Removing

– Remove driver's airbag, see **Body-Interior section, Repair Group 69**

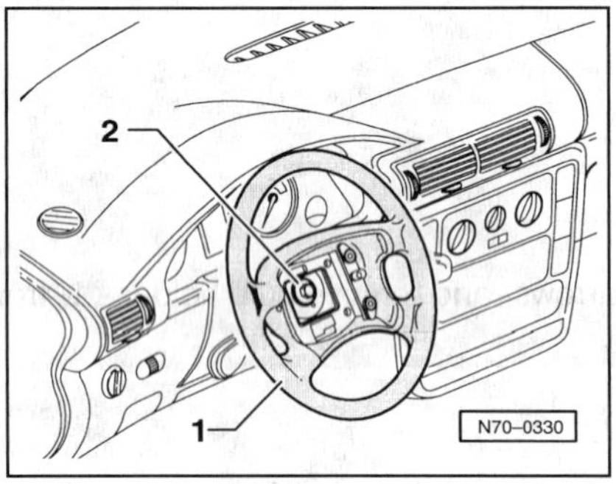

Remove hex bolt -2- (always replace).

• Tightening torque: 60 Nm (44 ft lb).
• Position steering wheel -1- in central position (wheels straight ahead).
• Pull steering wheel -1- from steering column.

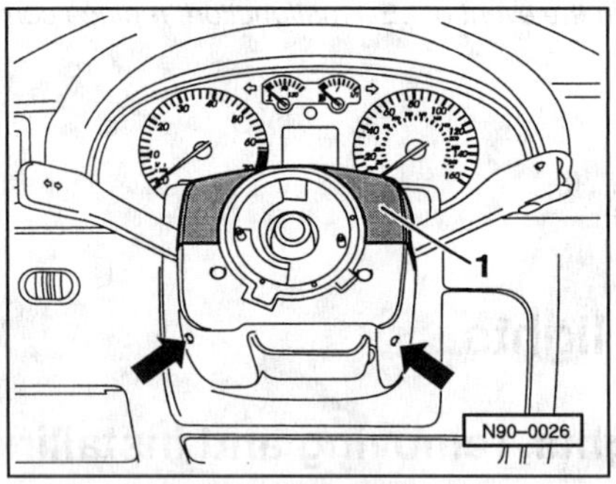

Remove both (Phillips head) screws -arrows-.

• Remove upper steering column trim -1- upwards.

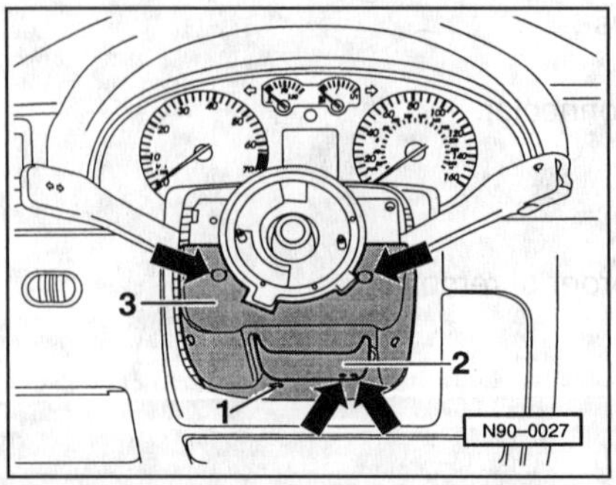

Remove four (Phillips head) screws -arrows-.

• Remove socket head bolt -1-.
• Release steering wheel height adjustment -2-.
• Remove lower steering column trim -3- downwards.

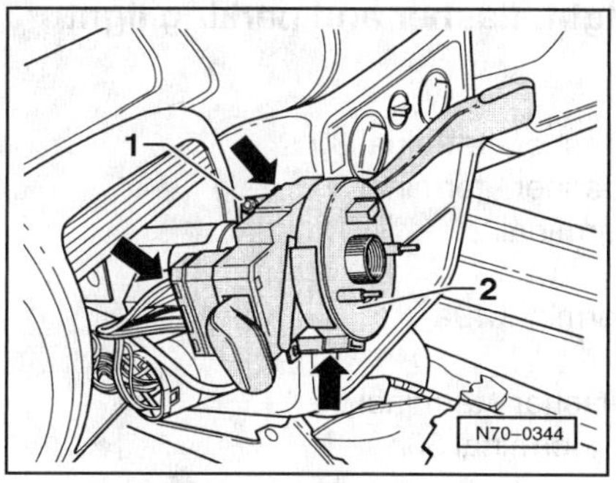

Loosen socket head bolt -1-.

- Disconnect electrical connectors -arrows- from steering column switches.

> ***CAUTION —***
> *The spiral spring remains on the steering column switch. Ensure spring is not turned from central position.*

– Remove steering column switches

Installing

– Install steering column switches loosely onto steering column only (do not tighten socket head bolt).

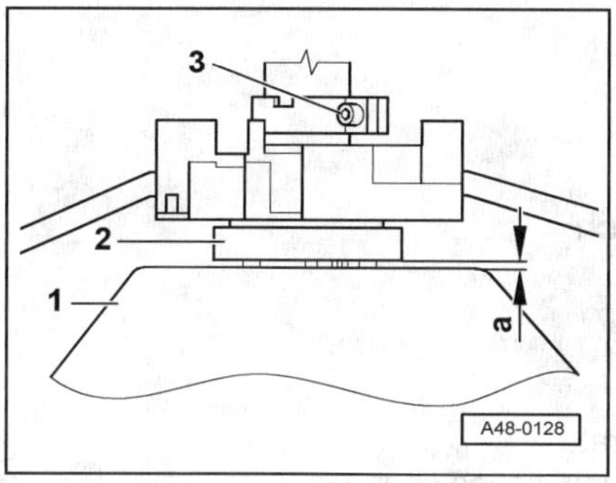

– Reconnect electrical connections.

– Slide steering wheel onto steering column shaft (do not install bolt).

Slide steering column switches on column so that gap -a- (3 mm) is maintained between switches -2- and steering wheel -1-.

- Tighten socket head bolt -3-.
- Remove steering wheel again.
- Install steering column trim
- Install steering wheel and hex bolt. Tightening torque: 60 Nm (44 ft lb).
- Install driver's airbag, see **Body-Interior section, Repair Group 69.**

Steering column switches, connector assignments

Windshield wiper/washer and multi-function indicator (MFI)

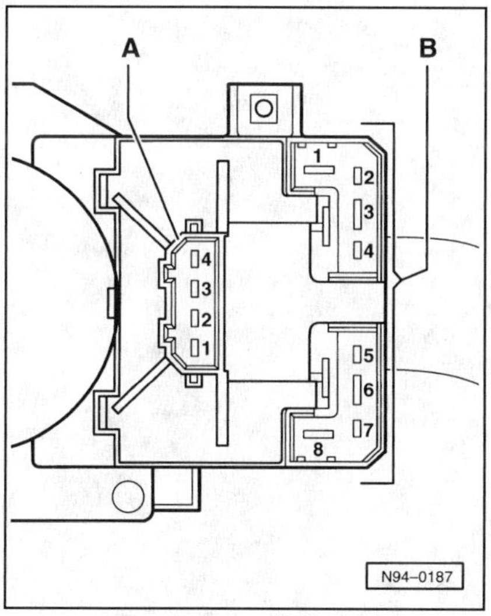

A - Connector, 4 pin

1. MFI call-up button - right
2. MFI call-up button - left
3. MFI call-up button - terminal 31 (ground)
4. MFI safe switch - Reset

- MFI = Multi-Function Indicator

B - Connector, 8 pin

1. Windshield wiper switch - terminal 53
2. Windshield wiper switch - terminal 31
3. Windshield wiper switch - terminal 53e
4. Windshield wiper switch - terminal 53c
5. Windshield wiper switch - rear window wiper
6. Windshield wiper switch - terminal 53b
7. Windshield wiper switch - intermittent operation
8. Windshield wiper switch - terminal 53a

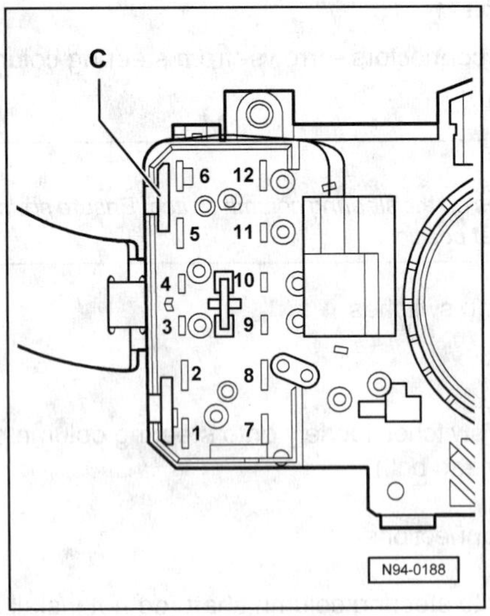

Turn signals, headlight, flasher and parking light

C - Connector, 12 pin

1. Switch for headlight/flasher - terminal 30
2. Switch for headlight/flasher - terminal 30
3. Turn signal switch - terminal L
4. Open
5. Turn signal switch - terminal 49a
6. Horn - terminal 71
7. Switch for headlight/flasher - terminal 56
8. Switch for dip/flasher - terminal 56b
9. Open
10. Open
11. Turn signal switch - terminal R
12. Switch for headlight/flasher - terminal 56a

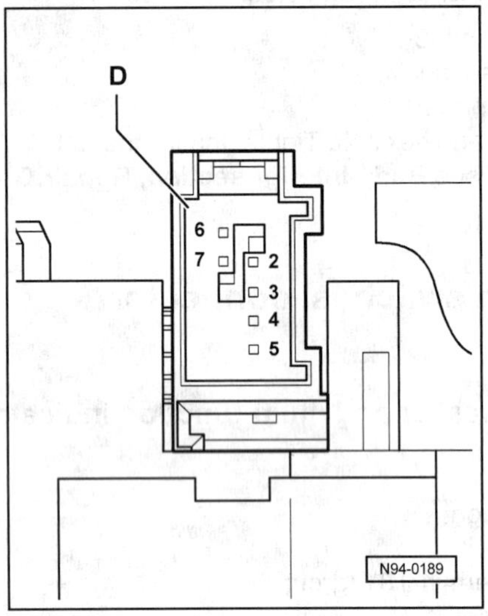

Cruise control switch

D - Connector, 10 pin

1. Open
2. Switch - On / Off
3. Switch - Set / accelerate
4. Switch - Reset / Resume saved speed
5. Switch - Reset / Resume saved speed
6. Switch - On / Off
7. Switch
8. Open
9. Open
10. Open

Ignition/starter switch and lock cylinder

Lock cylinder, removing and installing

> ***CAUTION—***
> *Before beginning repairs on the electrical system:*
>
> *• Obtain the anti-theft radio security code.*
>
> *• Switch the ignition off.*
>
> *• Disconnect the battery Ground (GND) strap.*
>
> *• After reconnecting battery, re-code and check operation of anti-theft radio. Also check operation of clock and power windows according to Repair Manual and/or Owner's Manual.*

Removing

– Remove upper and lower steering column switch trim, see **Steering column switches, removing and installing.**

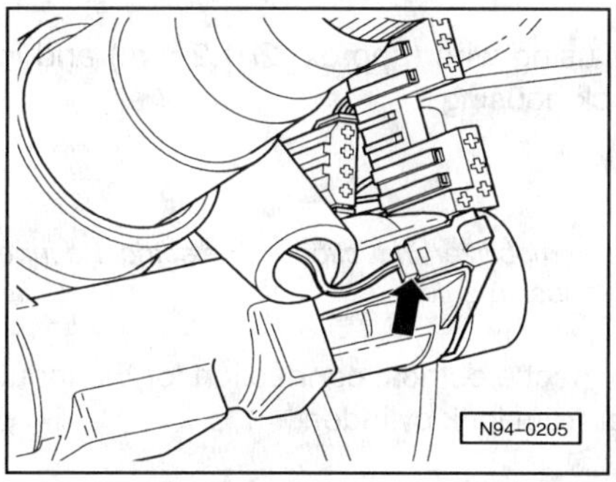

◄ Where applicable, carefully disconnect electrical connection for the induction coil for anti-theft immobilizer from lock cylinder.

NOTE—
If equipped, the induction coil for anti-theft immobilizer is integrated with the lock cylinder and cannot be serviced separately.

– Insert ignition key in lock cylinder and turn to "ON" position.

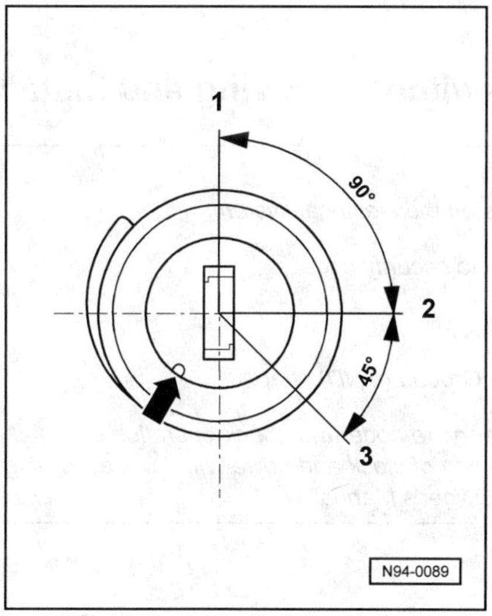

◄ Insert steel wire (approx. ∅ 1.2 mm) in drilling next to ignition key -arrow- .

NOTE—
Steering lock must be dismantled if the steel wire cannot be guided into the drilling correctly.

Position of key in lock cylinder:
1. 1 - Position "OFF"
2. 2 - Position "ON / DRIVE"
3. 3 - Position "START"

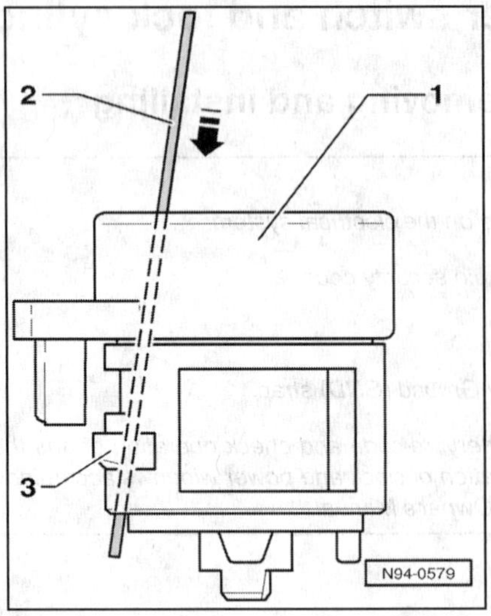

Release lock cylinder securing lever using wire -arrow- and remove lock cylinder from steering lock housing.

1. Lock cylinder
2. Wire (approx. ∅ 1.2 mm)
3. Securing lever

Installing

– Insert ignition key in lock cylinder and turn to position "ON".

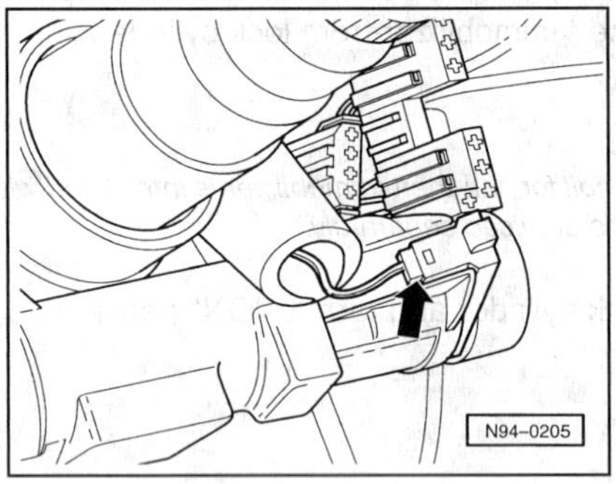

– Release securing lever using wire (approx. ∅ 1.2 mm) and insert lock cylinder into steering lock housing.

NOTE —
Where applicable, ensure immobilizer induction coil electrical connection orientation in lock cylinder housing guide.

Where applicable, reconnect electrical connection for the induction coil for anti-theft immobilizer from lock cylinder.

– Install steering column switches and upper/lower column trim pieces.

Ignition/starter switch, removing and installing

CAUTION —
Before beginning repairs on the electrical system:
- *Obtain the anti-theft radio security code.*
- *Switch the ignition off.*
- *Disconnect the battery Ground (GND) strap.*
- *After reconnecting battery, re-code and check operation of anti-theft radio. Also check operation of clock and power windows according to Repair Manual and/or Owner's Manual.*

Removing

– Remove upper and lower steering column switch trim. Removing steering column switches, see **Steering column switches, removing and installing**

– Disconnect electrical connector from ignition/starter switch.

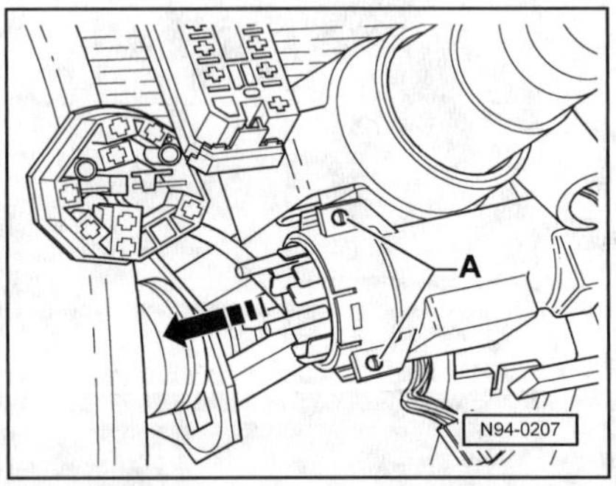

Remove locking paint from threaded holes of securing screws -A-.
- Loosen the two screws -A- slightly and pull ignition/starter switch out from steering lock housing in direction of arrow.

Installing

– Insert ignition/starter switch into steering lock housing.

NOTE —
When installing the ignition/starter lock switch ensure that the ignition/starter lock switch and the lock cylinder are in the same position, e.g. "ignition on".

– Tighten screws and seal securing holes with locking paint.

– Reconnect multi-pin connector to ignition/starter switch.

– Reinstall steering column switches, trim and steering wheel.

Ignition/starter switch, connector assignments

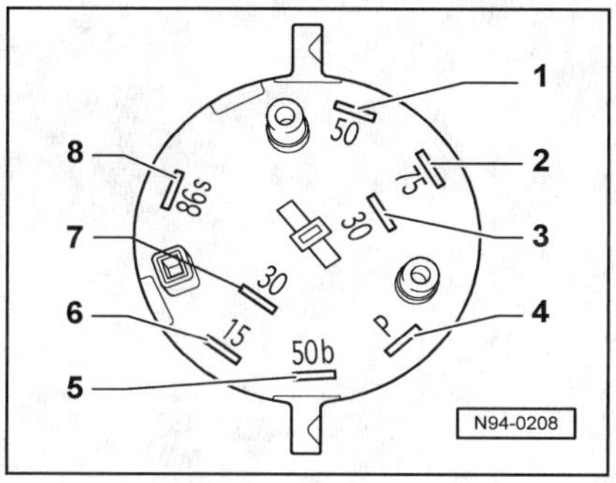

◀ **8 pin multi-connector -T8-**

1. Terminal 50
2. Terminal 75
3. Terminal 30
4. Terminal P
5. Terminal 50b
6. Terminal 15
7. Terminal 30
8. Terminal 86s

96 Lights, Switches - Interior, Anti-theft

Repair Group

W 96

Interior Lights and Switches

Light switch -E1-, removing and installing

Removing

– Turn rotary switch to "O" position.

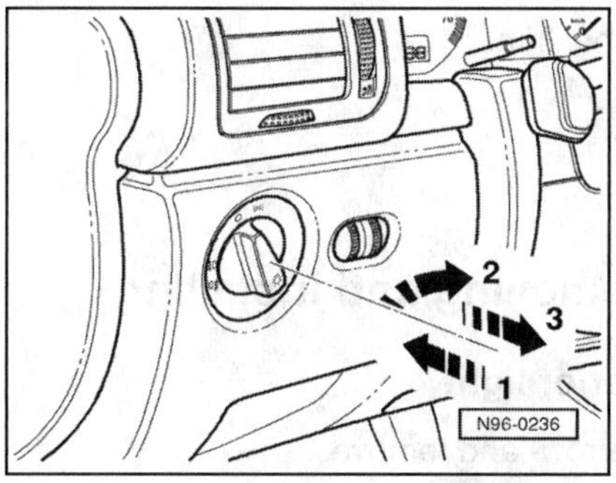

< Firmly push rotary switch inwards -arrow 1- and turn slightly to right -arrow 2-.

– Hold rotary switch in this position and pull entire switch from opening in instrument panel trim -arrow 3-

– Return rotary switch to "O" position.

– Disconnect electrical connector.

Installing

– Connect electrical connector to switch.

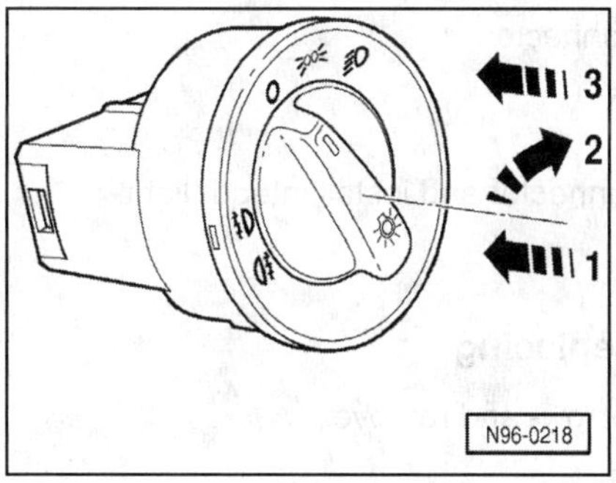

< Hold switch housing firmly, push rotary switch inwards -arrow 1- and turn slightly to right -arrow 2-.

– Hold rotary switch in this position and insert entire switch into opening in instrument panel trim -arrow 3-

– Return rotary switch to "O" position to engage.

Headlight adjuster -E102-* with instrument panel light dimmer switch -E20-, removing and installing

Removing

– Remove driver's side lower instrument panel trim.

– Disconnect electrical connection.

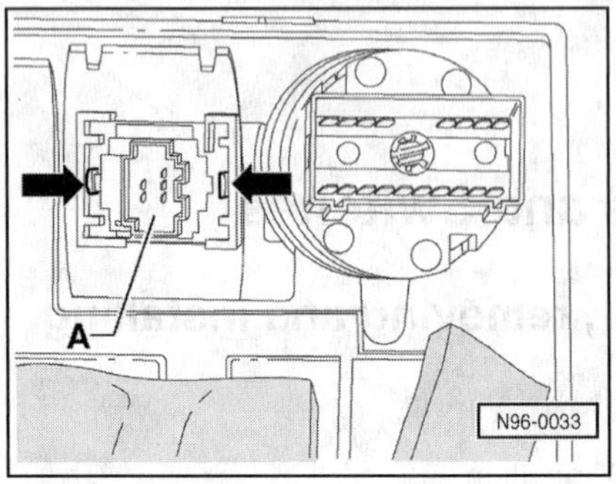

◄ Depress locking tabs -arrows- to release switch.

– Pull switch -A- through instrument panel frame from behind and remove.

Installing

– Install switch from behind instrument panel frame and line up with guide piece.

– Clip switch into place until locking tabs engage.

– Reconnect electrical connection.

– Reinstall instrument panel trim.

*) If equipped

Interior lights, removing and installing

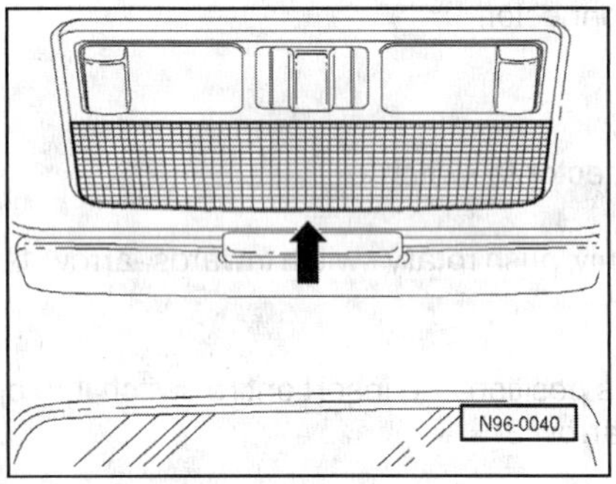

Removing front interior light

◄ Carefully pry off lens -arrow- and remove.

– Remove recessed mounting screws and pull down assembly.

– Disconnect electrical connector.

Installing

– Reconnect electrical connector and install interior light.

Interior light bulb, replacing

– Carefully pry off lens -arrow- and remove.

– Replace with 2V/10 W bulb.

– Insert lens again and locate properly.

Front reading/map light bulb, replacing

– Carefully pry off lens -arrow- and remove.

– Pull out bulb (12V/5W), do not turn.

– Insert lens again and locate properly

Sunroof regulator (switch) -E139-, removing and installing

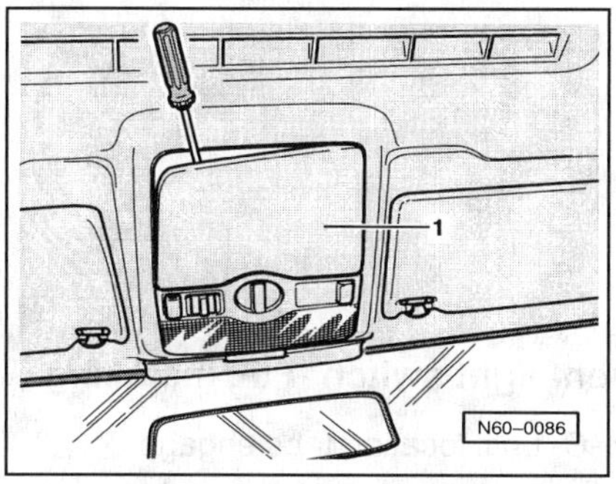

NOTE—

The sunroof regulator switch cannot be removed individually. The complete switch assembly and interior/reading light unit must be removed.

< Unclip cover for drive -1-.

– Carefully pry off lens and remove.

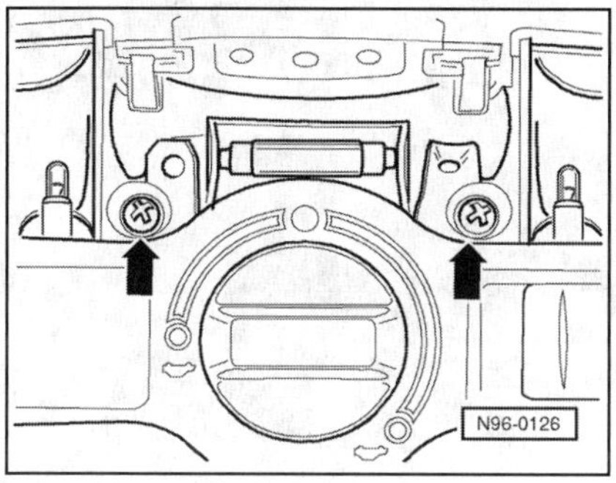

< Remove recessed mounting screws -arrows- and pull down assembly.

– Disconnect electrical connectors.

– Installation is reverse order of removal.

Luggage compartment light switch -F5-, removing and installing

NOTE—

On wagon models from 10/2000, light switch -F5- cannot be replaced separately.

The luggage compartment light switch -F5- is located in the rear lid lock assembly.

Removing

– Remove rear lid lock assembly, see **Body-Exterior section, Repair Group 55.**

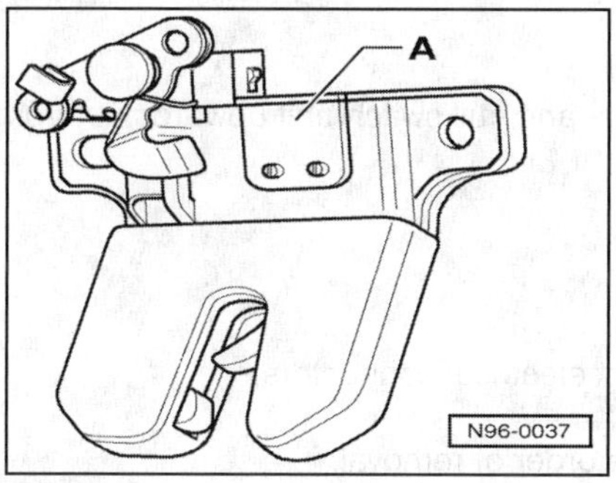

< Remove plastic protective cap -A-.

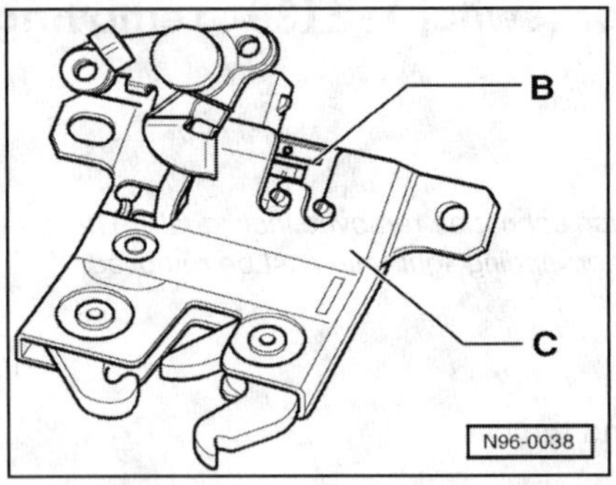

Release luggage compartment light switch -B- from locating tabs and pull upwards from rear lid lock -C-.

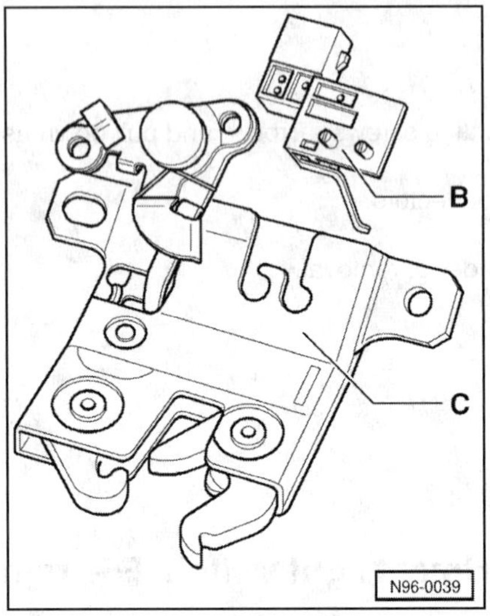

Luggage compartment light switch -F5-, installing

- Insert switch -B- in lock -C- until locating tabs engage.

Install protective cap.

Power window switches in driver's door, removing and installing (vehicles with green instrument cluster illumination)

NOTE —

On vehicles with green instrument cluster illumination, individual power window switches in the driver's door cannot be removed/replaced separately. The complete switch unit must be removed.

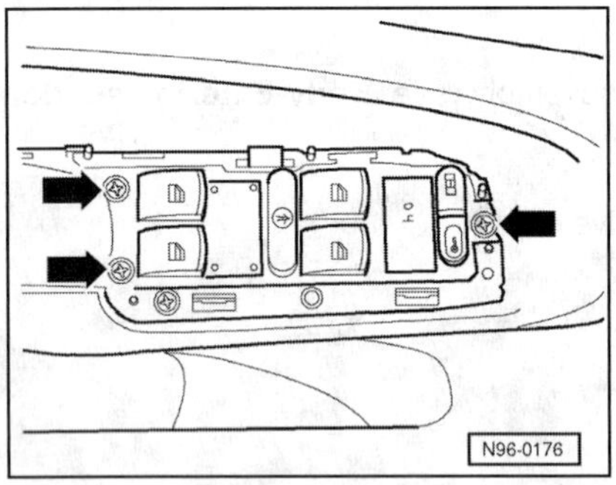

- Remove driver's door handle, see **Body-Interior section, Repair Group 70.**

Remove screws -arrows- and pull switch unit upwards out from door trim.

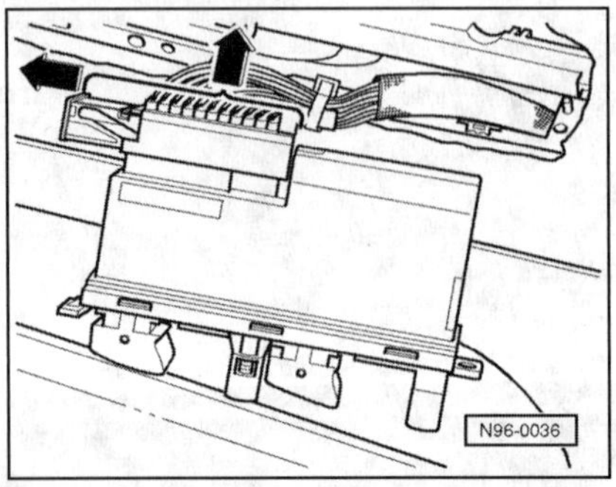

Release and disconnect electrical connectors.

- Installation is in reverse order of removal.

Power window switches in driver's door, removing and installing (vehicles with blue instrument cluster illumination)

NOTE —
On vehicles with blue (instrument cluster) illumination, individual power window switches in the driver's door can be removed/replaced separately.

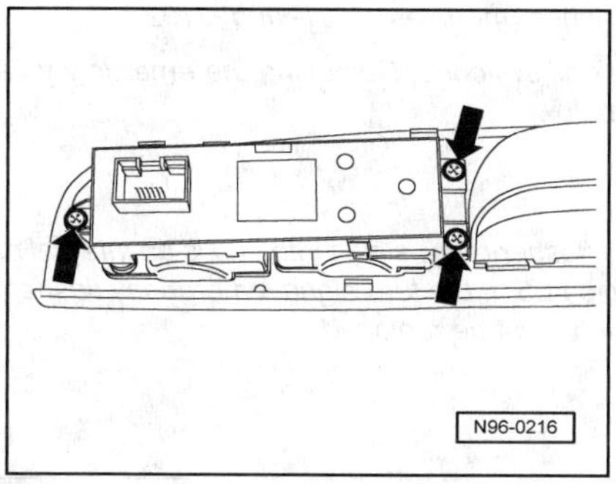

– Remove driver's door handle, see **Body-Interior section, Repair Group 70.**

– Lift release handle with controls upwards from door trim and carefully turn over.

– Disconnect electrical connections.

◄ Remove screws -arrows- and pull switch module from trim.

– Installation is in reverse order of removal.

Power window switches in passenger's doors, removing and installing

NOTE —
The procedure for removing and installing the

- *Right front window switch -E41-*
- *Left rear window switch -E52-*
- *Right rear window switch -E54-*

NOTE —
is essentially the same for all switches. Removing the right front window switch -E41- is illustrated below.

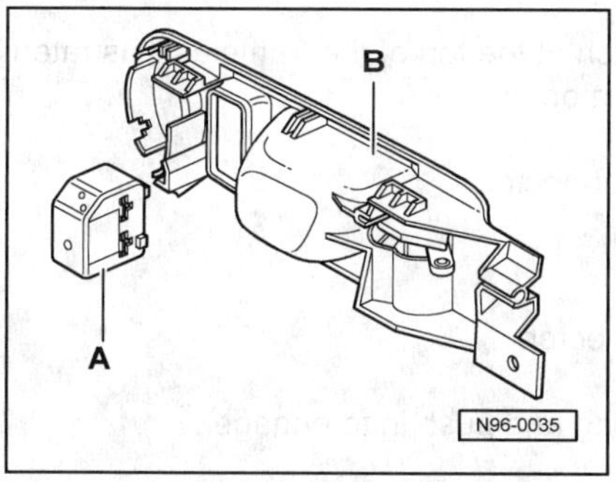

◄ Removing:

- Pull off connector
- Release switch housing -A- by pressing sideways out from locking device in mounting frame -B- and pull out.

– Installing:

- Insert switch -A- in the recess in the mounting frame -B- and engage.
- Push on connector

Mirror adjustment switch -E43- with mirror selector switch -E48-, removing and installing

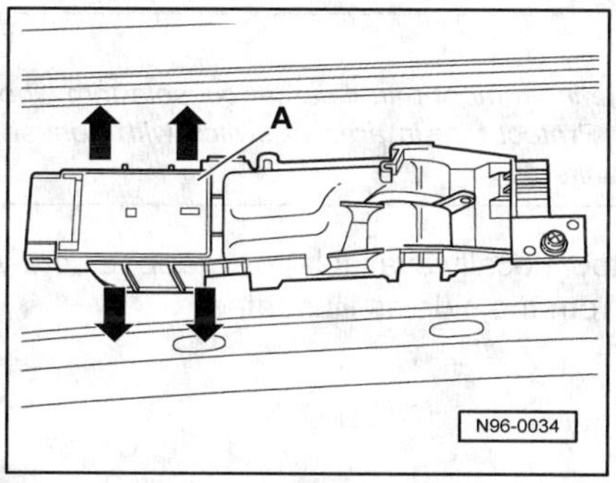

◄ Removing:

- Remove door panel, see **Body-Interior section, Repair Group 70.**
- Disconnect electrical connector.
- Bend out switch mounting frame on the long side -A- and push out switches from frame.

– Installing:

- Insert switch -A- in mounting frame and engage.
- Connect electrical connector.

Switches in instrument panel center console, removing and installing

NOTE —

The procedure for removing and installing the

- *Emergency flasher switch -E3-*
- *Rear window defogger switch -E15-*
- *Seat heater regulator switches (modules -J131- and -J132-)*

is essentially the same for all switches. Removing the emergency flasher switch -E3- is illustrated below.

EDITOR'S NOTE —

The turn signal/emergency flasher relay is built into and is an integral part of the Emergency flasher switch E3. If the turn signal/emergency flasher relay is defective, the entire switch must be replaced.

Removing

CAUTION —

- *Be careful not to damage instrument panel center console trim when removing switch. Protect trim in area of switch with adhesive tape before removing switch.*
- *Do Not pry out switches from the side. Only pull out from the top.*

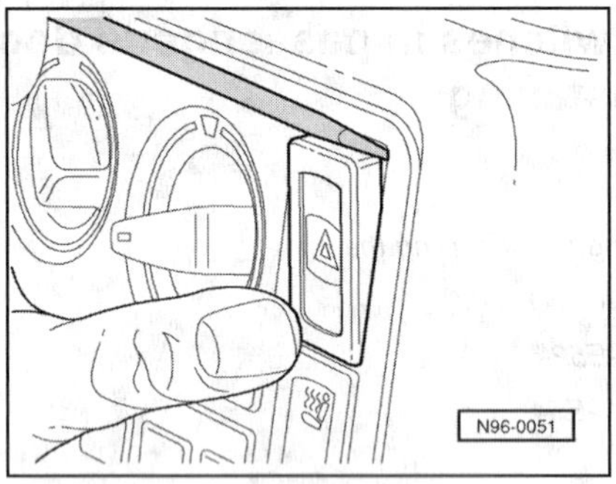

◄ Carefully loosen and remove switch faceplate using a plastic wedge tool (locally available) or wrapped screwdriver.

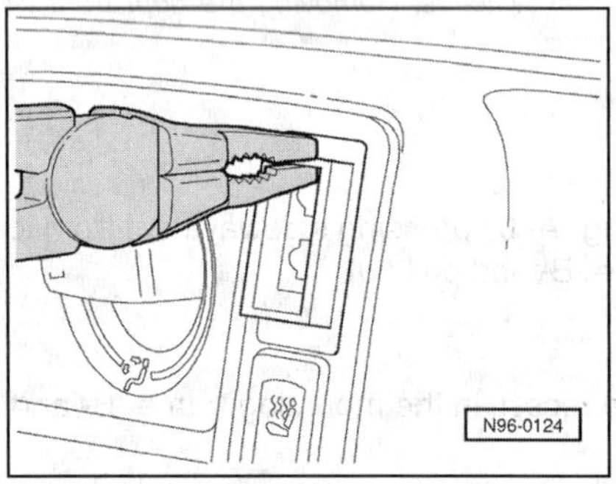

◄ Using pliers, grasp switch at the top of the frame as illustrated and carefully pull switch from opening.

– Disconnect electrical connector.

Installing

– Connect electrical connector.

– Insert switch into opening and push in to engage.

– Install switch faceplate.

Switch blanks, removing

CAUTION —

Be careful not to damage instrument panel center console trim when removing switch blanks. Protect trim in area of switch with adhesive tape before removing blanks.

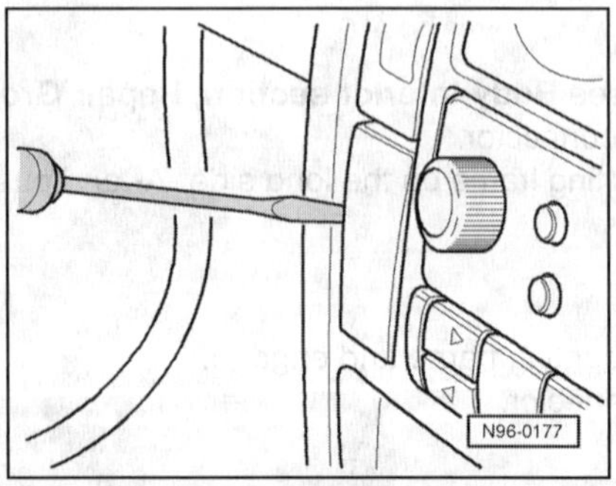

◄ Using a plastic wedge tool (locally available) or wrapped screwdriver, carefully pry out blank from the side as illustrated.

Door contact switch, removing and installing

Door contact switches are integrated into the door lock assembly and cannot be replaced individually.

Replace the complete door lock assembly if the door contact switch malfunctions, see **Body-Exterior section, Repair Group 57.**

Glove compartment light -W6-, removing and installing

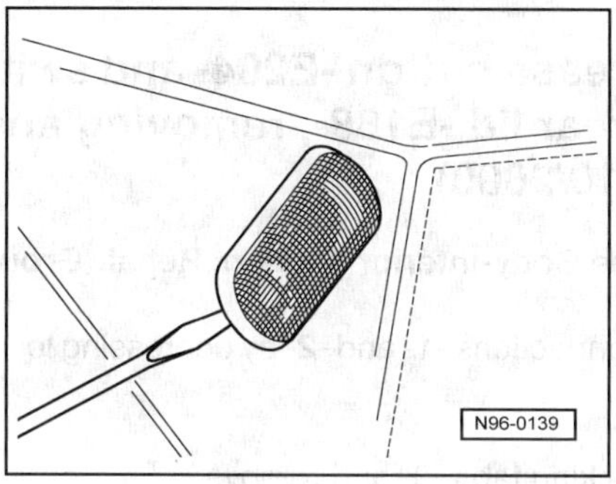
N96-0139

- Insert screwdriver behind lens and carefully pry out.
- Pull out lens with bulb holder and microswitch.
- Replace with 12V/3W bulb
- Insert lens with bulb holder and microswitch in glove box and locate properly.

NOTE —
Always perform a functional check of the switch. Light must not be on with lid closed!

Rear reading lights, removing and installing

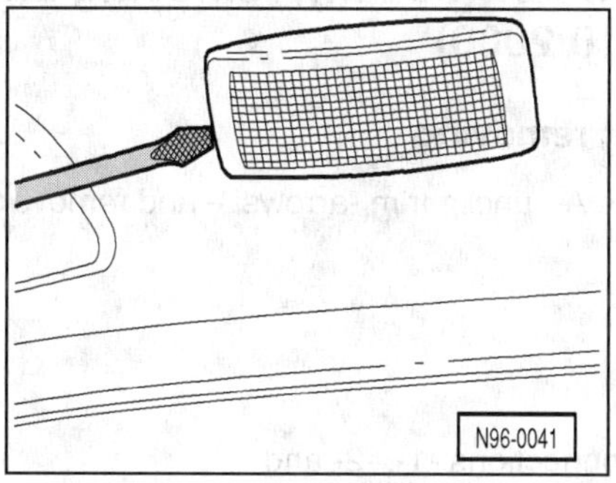
N96-0041

- Removing:
  - Insert small screwdriver into recess in lens and carefully pry out reading light from mounting frame.
  - Disconnect electrical connector.
- Installing:
  - Connect electrical connector.
  - Insert reading light into mounting frame and engage.

Luggage compartment light -W3-, removing and installing

- Removing:
  - Slide lens downwards and carefully disengage from opening.
  - Disconnect electrical connector.
  - Replace12V/3W bulb
- Installing:
  - Connect electrical connector
  - Insert light in mounting and engage.

Fuel tank flap release button -E204-, removing and installing (through 09/2000)

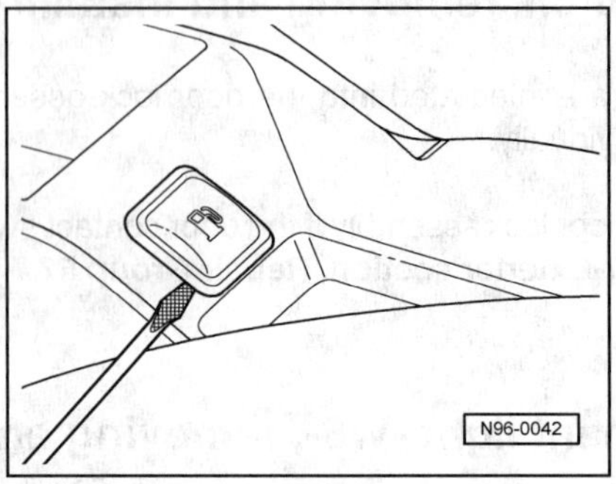

Removing:

- Insert small screwdriver behind switch housing and carefully pry out switch.
- Disconnect electrical connector.

– Installing:

- Connect electrical connector.
- Insert button into center console and engage.

Fuel tank flap release button -E204- and switch for remote unlock, rear lid -E188-, removing and installing (from 10/2000)

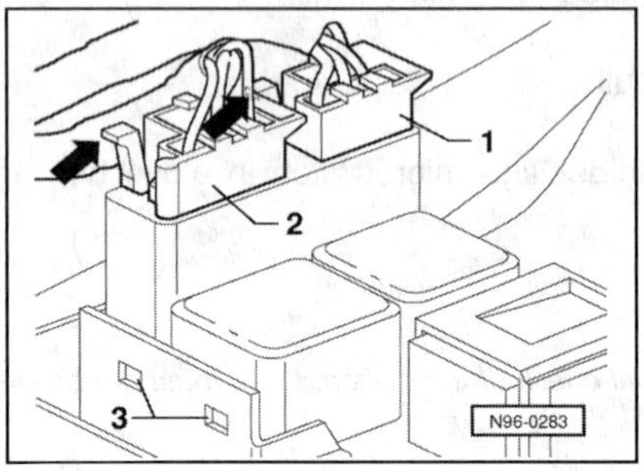

– Remove door panel, see **Body-Interior section, Repair Group 70.**

Disconnect electrical connections -1- and -2- by depressing locking tabs -arrow-.

– Unclip switches from locking tabs -3- and remove

– Installation is reverse order of removal.

Rear lid unlock motor (sedan), removing and installing (from 10/2000)

Rear lid lock carrier, removing

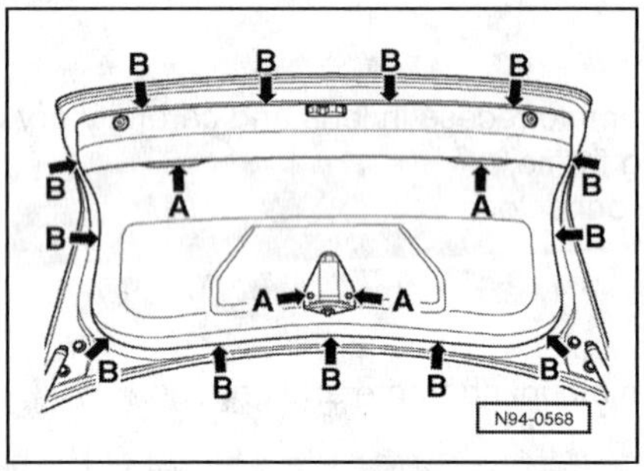

Remove screws -arrows A-, unclip trim -arrowsB- and remove trim.

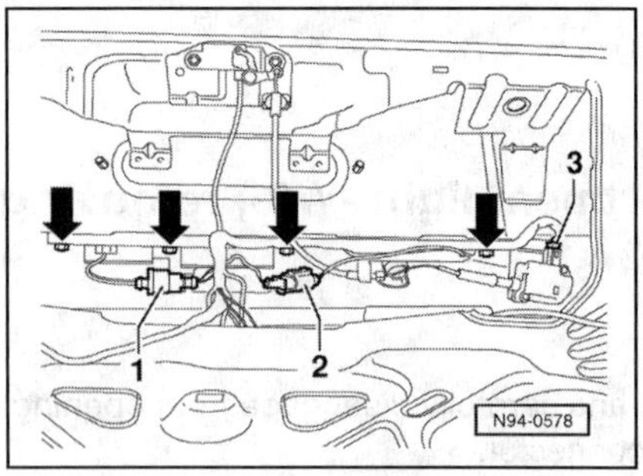

Disconnect electrical connections -1-, -2- and -3-.

– Unclip cable at lock assembly.

– Remove hex nuts -arrows- and remove lock carrier.

Rear lid unlock motor, removing

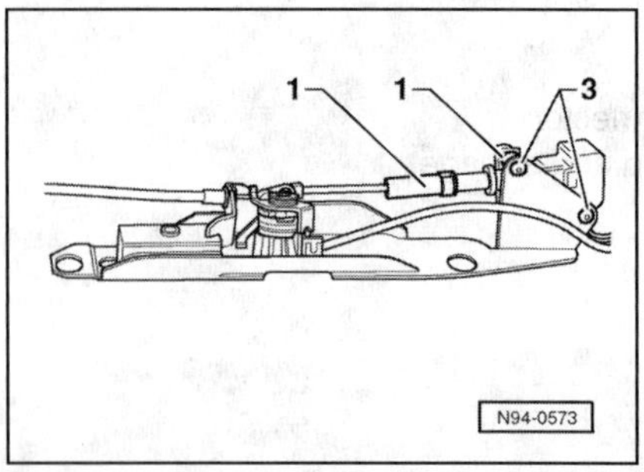

Remove mounting screws for motor -3-, remove motor and unclip cable -1-.

Installing

– Installation is reverse order of removal.

Rear lid unlock motor (wagon), removing and installing (from 10/2000)

Rear lid lock carrier, removing

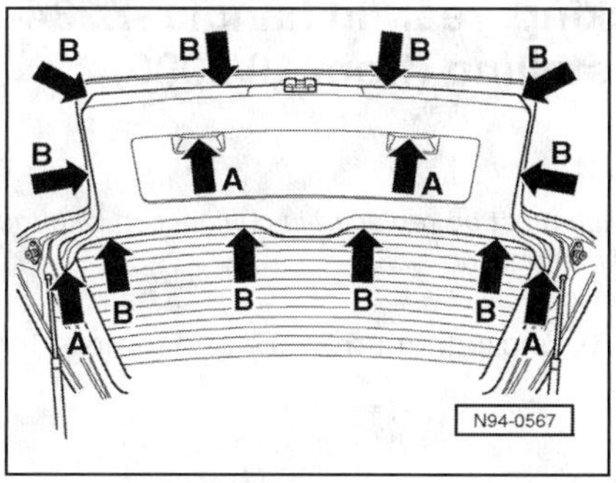

- Remove screws -arrows A-, unclip trim -arrows B- and remove trim.

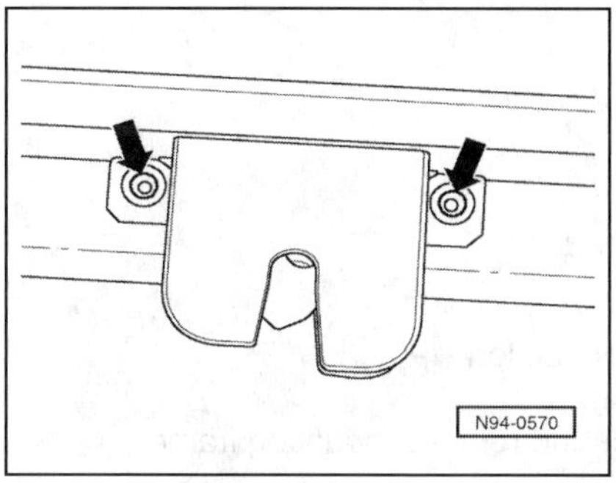

- Remove rear lock mounting screws -arrows- and partially pull out of rear lid.

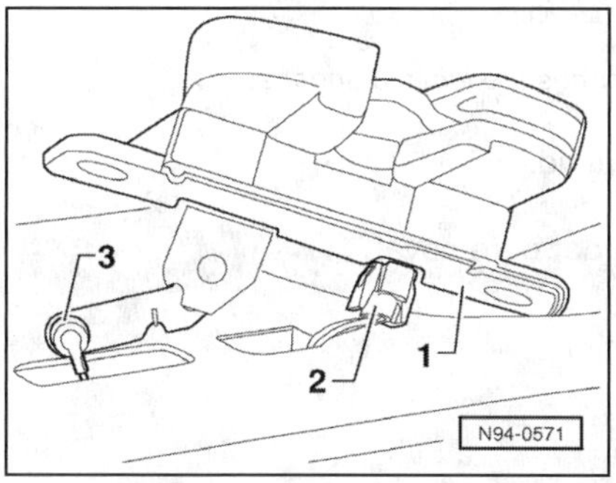

- Unclip cable -3-, disconnect electrical connection -2- and remove lock -1-.

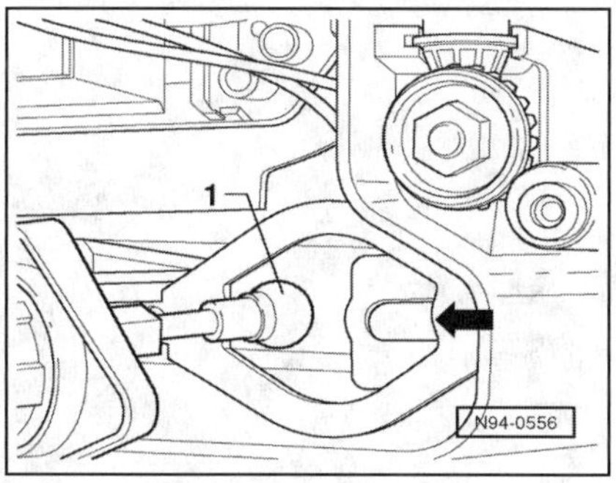

- Disconnect electrical connection -2-, remove hex nuts -arrows- and remove lock carrier -1-.

Rear lid unlock motor, removing

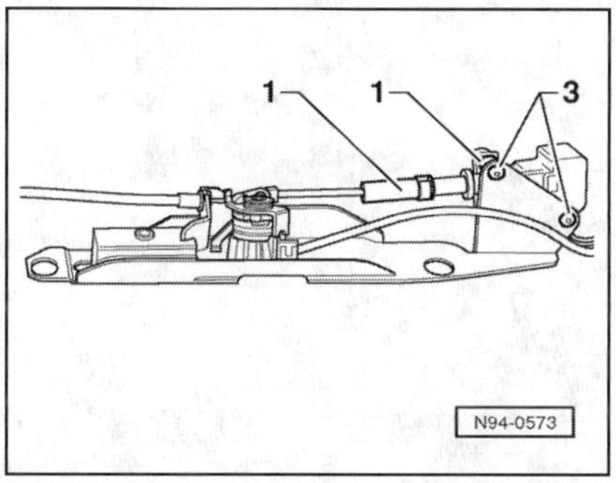

- Remove mounting screws for motor -3-, remove motor and unclip cable -1-.

Installing

– Installation is reverse order of removal.

Switch for unlocking, rear lid handle -E234- , removing and installing (from 10/2000)

NOTE —

On sedan models, the switch must be replaced in conjunction with the rear lock carrier in the event of malfunction.

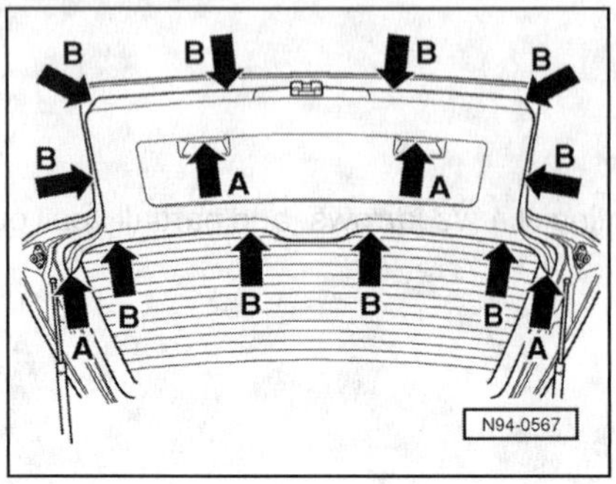

< Remove screws -arrowA-, unclip trim-arrowsB- and remove trim.

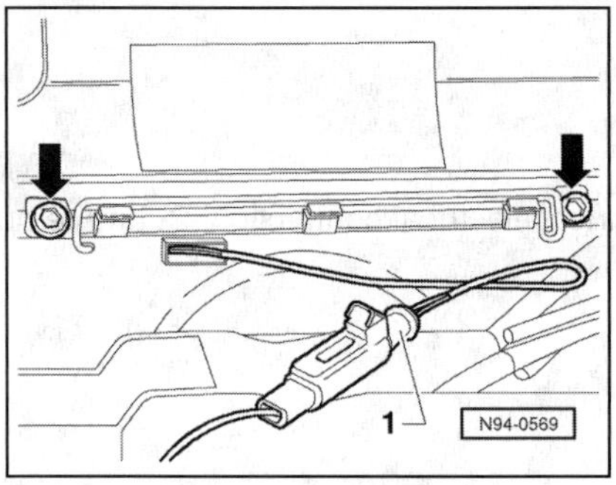

< Disconnect electrical connection -1-.

– Loosen screws -arrows- and remove mounting frame.

NOTE —

Do not remove screws. Springs are located under screws.

– Remove switch from rear lid.

– Installation is reverse order of removal.

97 Wiring

Repair Group

W 97

Fuse and relay panels

> ***CAUTION*** —
> *Before beginning repairs on the electrical system:*
>
> - *Obtain the anti-theft radio security code.*
> - *Switch the ignition off.*
> - *Disconnect the battery Ground (GND) strap.*
> - *After reconnecting battery, re-code and check operation of anti-theft radio. Also check operation of clock and power windows according to Repair Manual and/or Owner's Manual.*

Fuse panel, removing

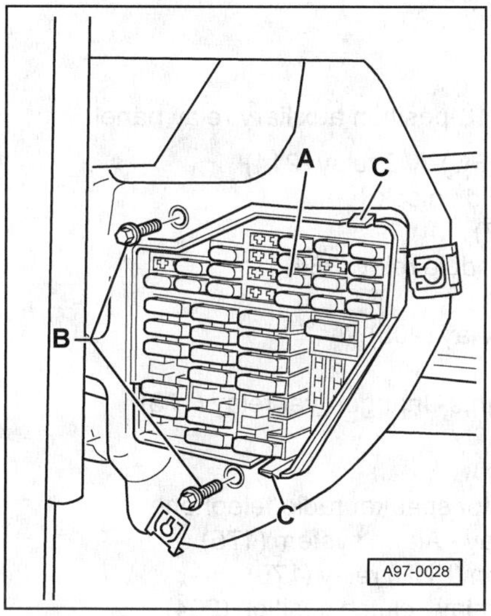

– Remove drivers' side lower instrument panel trim, see **Body-Interior section, Repair Group 70.**

– Carefully remove fuse panel cover at left side instrument panel.

◄ Remove screws -B-, 2 Nm (18 in lb), depress locking tabs -C- and remove fuse panel from the rear.

Relay panel (central electric) and 13-position auxiliary relay panel, removing

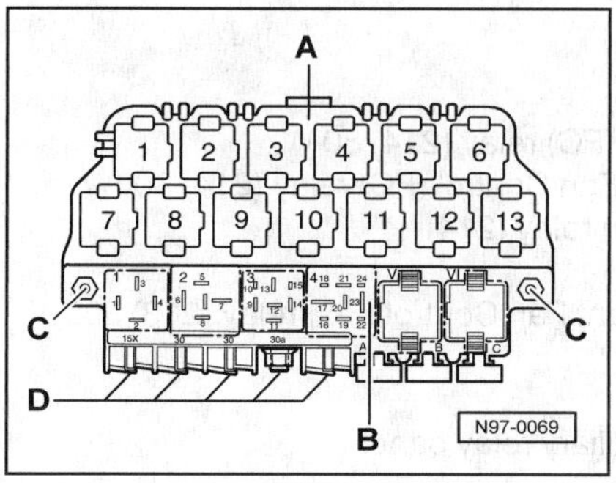

– Remove drivers' side lower instrument panel trim, see **Body-Interior section, Repair Group 70.**

◄ Remove screws -C-, 2 Nm (18 in lb) and unscrew/remove electrical connections -D-.

– Remove relay panel (central electric) -B- and auxiliary relay panel -A- from bottom.

Relay locations (central electric and 13-position auxiliary relay panel)

EDITOR'S NOTE —
Number in parenthesis is the control number which is uaually stamped on the relay housing.

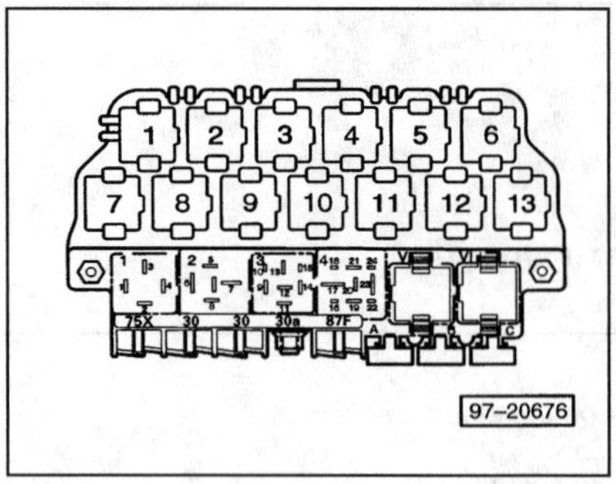

Relay arrangement on central electric panel

1. Dual Horn relay (204)
2. Load reduction Relay (213)
3. Open
4. Fuel Pump (FP) Relay (208)
V. Wiper/Washer Intermittent Relay (377)
VI. Wiper/Washer Intermittent Relay (377)

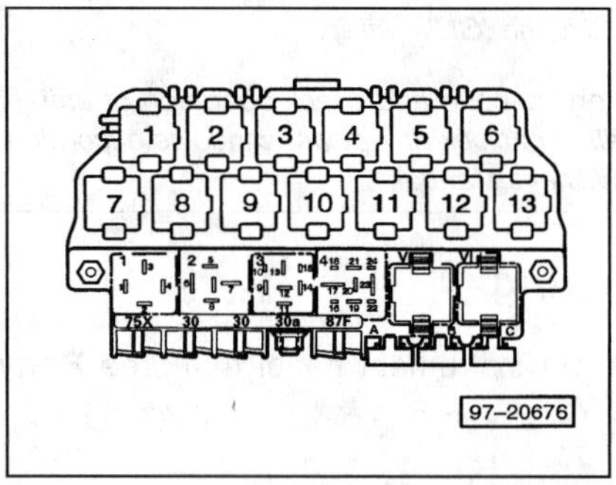

Relay arrangement on 13-position auxiliary relay panel

1. Coolant Fan control (FC)-A/C relay (214)
2. Relay for remote rar lid unlock
3. A/C Clutch relay (267)
4. A/C Clucth control module relay (384)
5. Open
6. Selector Lever light relay (205)
7. Fog Light relay (381)
8. Daytime Running Lights Change-over relay (173)
9. Open
10. Glow Plug relay (180)
11. Change Over relay for speaker radio/telephone
12. Starter Interlock relay - Alarm system (175)
13. Park/Neutral position (PNP) relay (175)
    or Starting Interlock relay - clutch position (204)

Relay and fuse locations (8-position auxiliary relay panel)

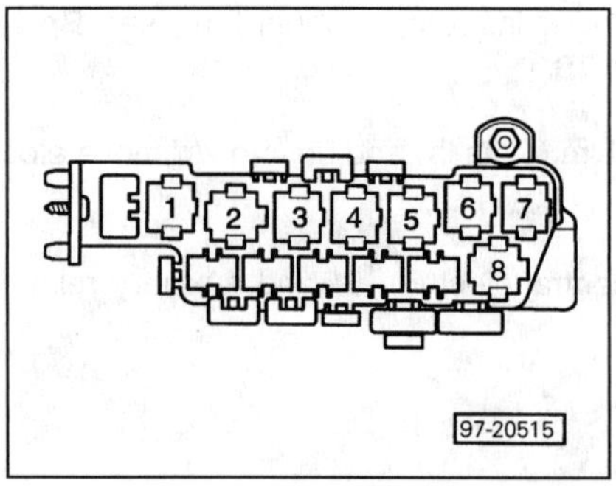

Relay arrangement on 8-position auxiliary relay panel (behind main relay panel)

1. Open
2. Open
3. Open
4. Open
5. Coolant Fan Control (FC) relay (214) 80 W
   First Speed Coolant Fan Control (FC) relay (214) 300W
6. Radiator Fan Afterrun relay (214)
7. Open
8. Second Speed Coolant Fan Control (FC) relay (213) 300W

Fuses at 8-position auxiliary relay panel

- Coolant Fan fuse 5A
- Coolant Fan fuse 30A
- ABS Hydraulic Pump fuse 50A

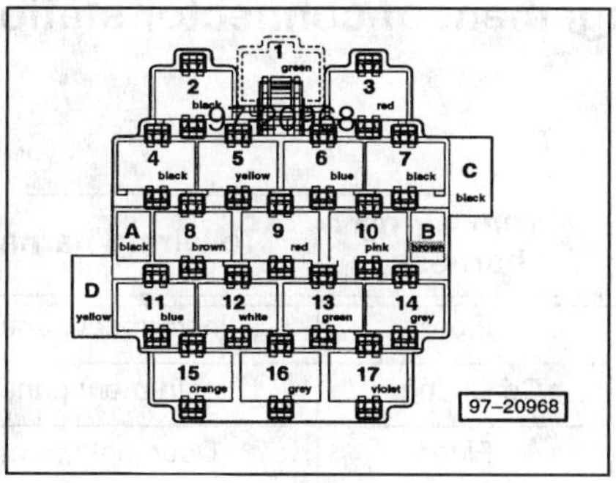

Connector arrangement of connector station A-plilar, left

Number	Color	From wiring harness	To wiring harness
1	green	Roof	Instrument panel
2	black	Floor	Door, left front
3	red	Floor	Roof
4	black	Cruise control	Instrument panel
5	yellow	Open	Open
6	blue	Floor	Instrument panel
7	black	Speaker	Instrument panel
8	brown	Floor	Instrument panel
9	red	Headlamp, left	Instrument panel
10	pink	headlamp, left	Instrument panel
11	blue	Floor	Instrument panel
12	white	Floor / Navigation	Instrument panel
13	green	Headlamp, left	Instrument panel
14	grey	Coolant fan	Instrument panel
15	orange	ABS	Instrument panel
16	grey	Floor	Door, left front
17	violet	Wiper motor	Instrument panel
A	black	Roof	Instrument panel
B	brown	Washer system	Instrument panel
C	black	Instrument panel	Outside Air Temp. display
D	yellow	Air bag	Instrument panel

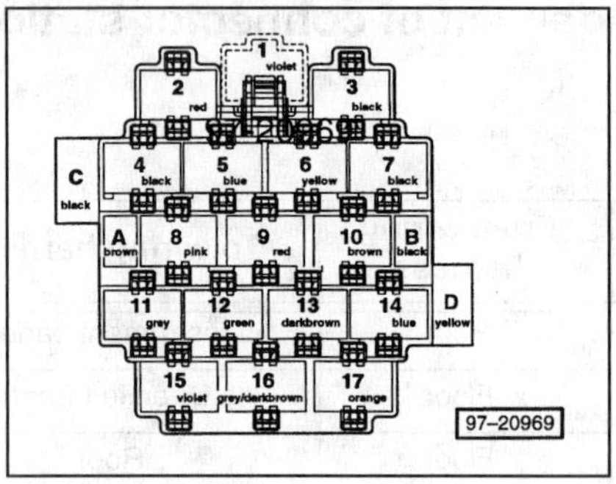

Connector arrangement of connector station A-plilar, right

Number	Color	From wiring harness	To wiring harness
1	vi0let	Floor	Instrument panel
2	red	Cd Changer	Instrument panel
3	black	Floor	Door, right front
4	black	Speaker	Instrument panel
5	blue	Open	Open
6	yellow	Open	Open
7	black	Floor / Rear Washer	Instrument panel
8	pink	Headlamp, right	Instrument panel
9	red	Headlamp, right	Instrument panel
10	brown	Automatic Transmission	Instrument panel
11	grey	ABS	Instrument panel
12	green	Air bag / heated seats	Instrument panel
13	white / dark brown	Floor	Instrument panel
14	blue	Automatic Transmission	Instrument panel
15	violet	Daytime Running lights	Instrument panel
16	grey / dark brown	Floor	Instrument panel
17	orange	Telephone	Instrument panel
A	brown	Air bag	Floor
B	black	Air bag	Instrument panel
C	black	Open	Open
D	yellow	Air bag	Instrument panel

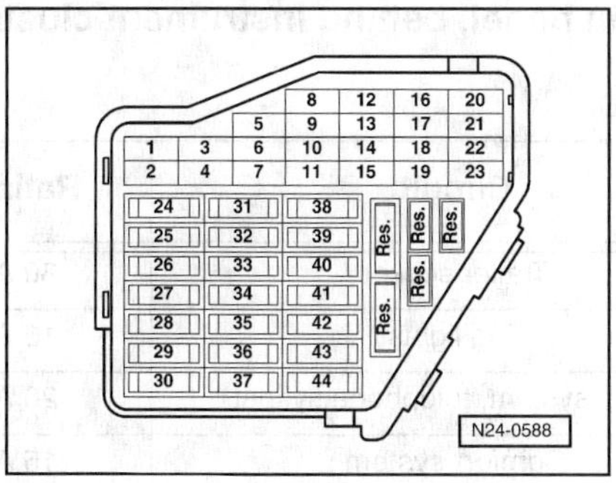

Fuse arrangement in panel, behind instrument cluster

Fues Position	Circuit	Rating
1	Heated washer nozzle	5 A
2	Turn signal system	10 A
3	Light for glove compartment, air conditioning, selector lever	5 A
4	License plate light	5 A
5	Instrument clusetr, heated seats, cruise control test plug, air conditioning	10 A
6	Control module comfort system	5 A
7	ABS	10 A
8	Automatic headlight beam adjusting, telephone system	5 A
9	Open	10 A
10	CD changer	5 A
11	Cruise control (automatic transmission)	5 A
12	B+ (battery vlotage positive) for on board diagnostic (OBD)	10 A
13	Brake lights	10 A
14	Comfort system	10 A
15	Instrument cluster, automatic transmission	10 A
16	Open	5 A
17	Navigation	10 A
18	Right headlight, high beam	10 A
19	Left headlight, high beam	10 A
20	Right headlight, low beam	15 A
21	Left headlight, low beam	15 A
22	Parking light, right	5 A
23	parking light, left	5 A
24	Wiper system	25 A
25	Fresh air blower, recirculating control, air conditioning	30 A
26	Rear window defogger	30 A
27	Rear window wiper system	15 A
28	Fuel pump	15 A
29	Engine control	20 A
30	Sunroof	20 A
31	Back up lights, cruise control (automatic transmission)	15 A
32	Engine control	20 A
33	Cigarette lighter	15 A
34	Engine control, injectors	15 A

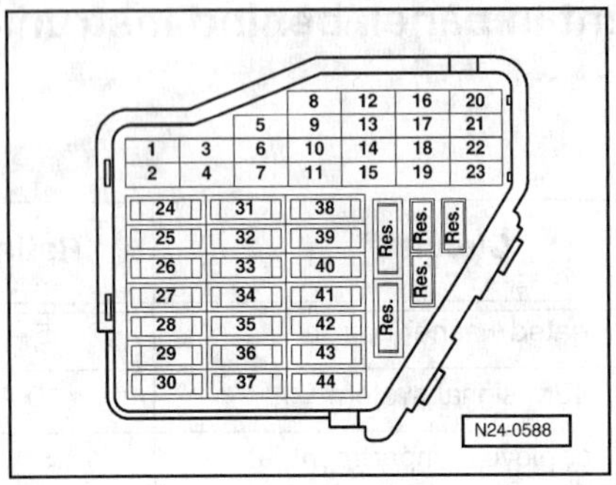

Fuse arrangement in panel, behind instrument cluster

Fues Position	Circuit	Rating
35	Trailer socket	30 A
36	Fog lights	15 A
37	Radio system, telephone syspet	20 A
38	Comfort system	15 A
39	Emergency flasher sysetm	15 A
40	Dual horn	25 A
41	Open	25 A
42	Open	25 A
43	Open	
44	Heated seats	30 A

Housing for control modules in plenum, removing and installing

– Remove cap in windshield cowl panel under the left wiper arm (move wiper arm if necessary).

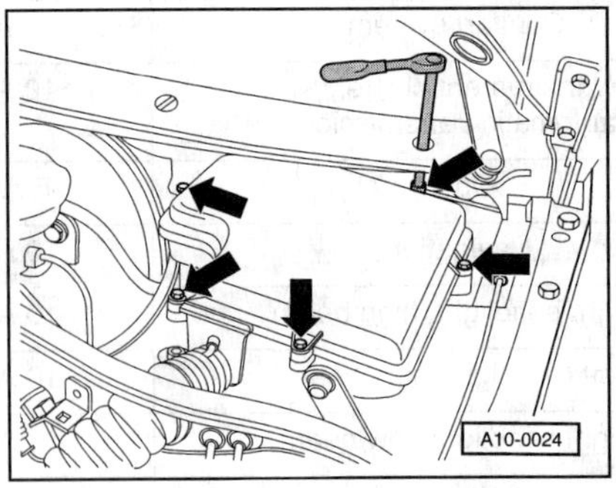

◄ Remove screw for housing cover underneath cowl panel as illustrated.

– Remove remaining screws -arrows- and take off housing cover.

– Remove Engine Control Module (ECM). (If necessary, remove eight position auxiliary relay panel and fuse panel, see **8-position auxiliary relay panel, removing**.

– Disconnect all electrical connectors at connector station.

– Slide rubber grommet and wiring harness for engine controls up and out of slot in housing.

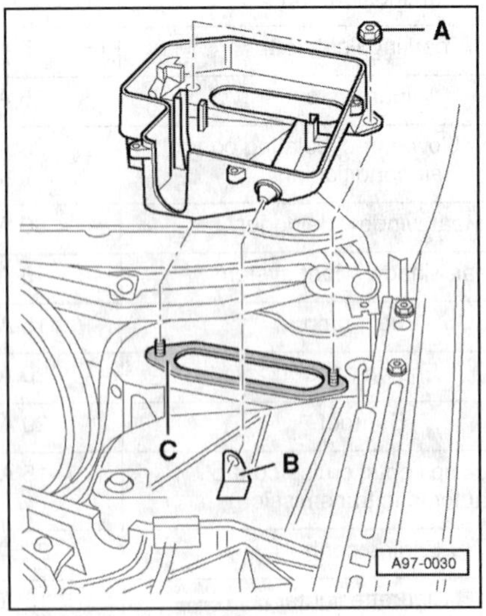

◄ Remove nuts -A-, 4 Nm (35 in lb).

– Push housing to rear (into plenum) to disengage from retainer bracket -B- and remove housing.

– Installation is in reverse order of removal.

NOTE—

- *Always replace gasket -C-. Do not allow gasket to block or interfere with opening in plenum for harnesses.*
- *When installing cover for housing, engage into tab and press firmly into place by hand first. Cover screws MUST be torqued to 4 Nm (35 in lb).*

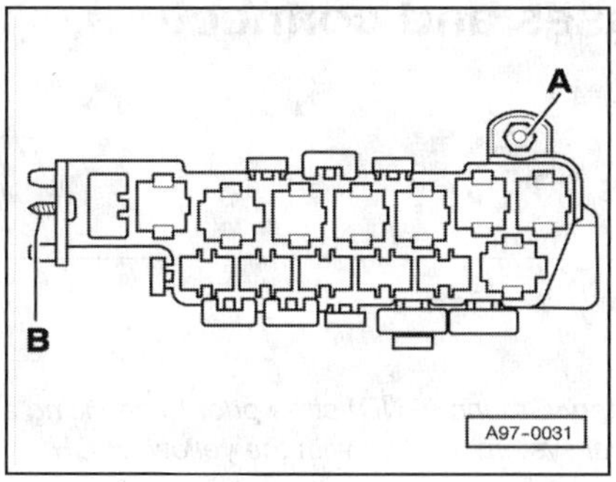

8-position auxiliary relay panel, removing

The 8-position auxiliary relay panel is located just behind the relay panel (central electric) and 13-position auxiliary relay panel in driver's footwell.

– Remove drivers' side lower instrument panel trim, see **Body-Interior section, Repair Group 70.**

< Remove nut -A-, 2 Nm (18 in lb) and screw -B-.

– Lift relay panel from holder and remove from bottom.

Engine Control Module (ECM), removing and installing

The engine control module regulates the fuel injection, throttle valve control module, oxygen sensor control, ignition, knock control, EVAP valve, engine speed limitation via injectors or fuel pump relay, camshaft adjustment and On Board Diagnostic (OBD).

Special tools and equipment

VAG1551 or VAG1552 Scan Tool (ST) with VAG1551/3 adapter cable

EDITOR'S NOTE —
Replacement of the ECM requires coding of the control module using VAG 1551, 1552, 5051, or an aftermarket scan tool. See ***Scan Tool section*** *for information on ECM coding.*

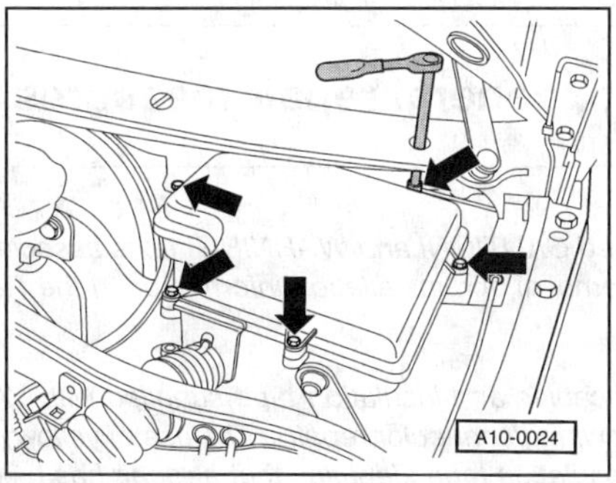

– Switch ignition off.

< Remove cover of protective housing for control modules.

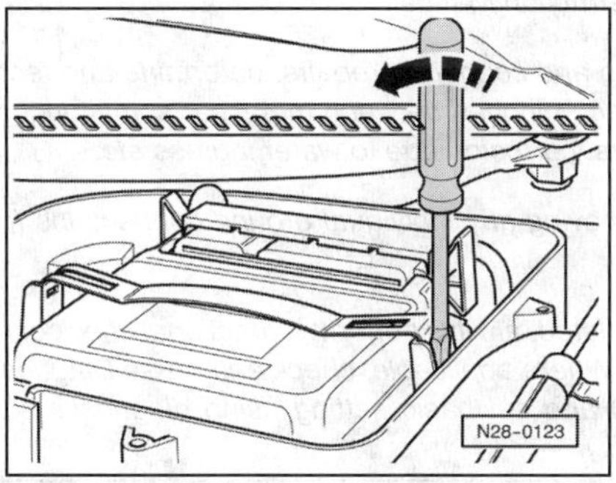

< Using screwdriver, carefully lift up mounting bracket (arrow).

– Disconnect harness connector clips and remove ECM harness connectors.

EDITOR'S NOTE —
Adaptation values are erased when connector is disconnected from the Engine Control Module (ECM), DTC memory content remains intact.

– Remove old control module and insert new one.

– Installation is reverse of removal, noting the following:

- Code Engine Control Module, see **Scan Tool section**.
- Carry out adaptation of throttle valve control module -J388- to ECM, see **Fuel Injection and Igniton section, Repair group 24, Throttle valve control module, adaptation to Engine Control Module (ECM).**
- Check DTC memory of new ECM and erase DTC memory if necessary.
- Adapt new control module module to immobilizer

EDITOR'S NOTE —
Coding the Engine Control Module to the immobilizer system requires VAG 5051 scan tool and factory security coding information. This procedure is beyond the scope of this manual, see a VW dealer or authorized service facility.

Wiring harnesses and connectors, repairing

Safety measures

> ***WARNING —***
> - *HIGH VOLTAGE!*
> - *Always disconnect battery Ground (GND) strap prior to servicing any parts of vehicle electrical system marked with the yellow "HIGH VOLTAGE!"symbol.*
> - *After disconnecting battery, briefly turn affected system on and off in order discharge any residual voltage seeRepair Manual of affected system.*

> ***CAUTION —***
> - *Airbag harness and connector repairs must only be performed using VAS 1978 Wiring Harness Repair Kit seeVAS 1978 Instruction Manual.*
> - *Airbag harness and connector repairs must only be performed using genuine Volkswagen replacement terminals, connectors and harnesses seeParts catalog*

Electrical system, general repair measures

> ***CAUTION —***
> - *Always refer to and heed CAUTION! and WARNING! texts associated with repair and/or removal and installation information in the Repair Manual.*
> - *Only the yellow repair cables and insulation tape supplied with VAS 1978 are to be used for wiring/connector repairs. The use of yellow repair wires and yellow insulation tape identifies that a repair has taken place seeVAS 1978 Instruction Manual.*
> - *Before beginning wiring and connector repairs, determine and rectify the cause of damage, eg.: pinched between metal parts, malfunctioning electrical consumers, corrosion due to water ingress etc.*
> - *Avoid loosening or removing any individual ground connections (potential for corrosion).*
> - *When repairs are completed, always perform a functional check of the component or system. Where applicable, check and erase Diagnostic Trouble Codes (DTCs) and set basic setting using diagnostic scan tool.*

Wiring harnesses and connectors, repair measures

> ***CAUTION*** *—*
> - *Soldering is not permitted!*
> - *Do not repair welded connections in the wiring harness. Fabricate and connect an appropriate overlay harness instead.*
> - *Wiring harness and connector repairs must only be performed using VAS 1978 Wiring Harness Repair Kit seeVAS 1978 Instruction Manual.*
> - *Only the yellow repair cables and insulation tape supplied with VAS 1978 are to be used for wiring/connector repairs. The use of yellow repair wires and yellow insulation tape identifies that a repair has taken place.*
> - *All shielded wiring (Eg.: knock sensors, antenna etc.) are Not to be repaired! Replace complete wiring harnesses as necessary.*

CAN-Bus wiring, repair measures

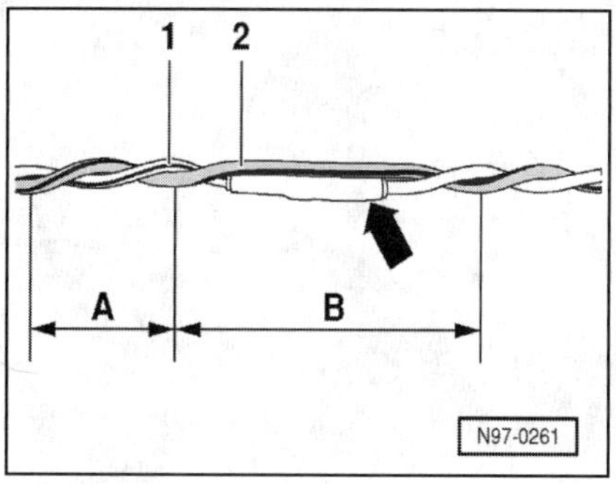

CAN Bus wiring consists of unshielded, two-cable wiring -1- and -2- with a cross section of 0.35 mm^2 or 0.5 mm^2.

- Color coding:

CAN high wire, powertrain	orange/black
CAN high wire, convenience	orange/green
CAN high wire, infotainment	orange/violet
CAN low wire, (all)	orange-brown

> ***CAUTION*** *—*
> - *Perform wiring harness repairs only with yellow wires from VAS 1978 Wiring Harness Repair Kit. Mark repair locations with yellow tape.*
> - *Both bus wires must be the same length. When both wires -1- and -2- are twisted, length -A- = 20 mm must be maintained for the twist.*
> - *No part of the bus wiring in the vicinity of heat-shrink sleeves -arrow- may be greater than B = 50 mm without the wires being twisted.*

Radio

Section

X

91 Radio

Repair Group

X 91

Radio systems

General information

> ***CAUTION—***
> *Before beginning repairs on the electrical system:*
>
> *• Obtain the anti-theft radio security code.*
>
> *• Switch the ignition off.*
>
> *• Disconnect the battery Ground (GND) strap.*
>
> *• After reconnecting battery, re-code and check operation of anti-theft radio. Also check operation of clock and power windows according to Repair Manual and/or Owner's Manual.*

The integrated "Premium" Radio/CD Player, optional CD Changer, "Monsoon" Amplifier, Diversity Antenna system, Multi-function steering wheel and Telematics systems are combined to form the Communication system.

Before troubleshooting or servicing, technicians must be familiar with the functions and operation specifics of the applicable Communication system. Always read the owner's manuals and review specific Communication system functions.

"Premium IV" & "Premium V" OBD program functions can be performed using VAG 1551/1552 Scan Tools (ST). Some Communication systems have extensive additional On Board Diagnostic (OBD) capabilities. In the event malfunctions occur in Communication systems with these additional OBD capabilities, troubleshooting can only be performed using the VAS 5051 Vehicle Diagnostic Testing and Information System in operating mode "Guided Fault Finding".

Various standard or optional radio systems are installed from 1998 model year:

- Radio systems "Premium III" & "Premium IV" (through 07/1999)
- Radio systems "Premium V" & "Monsoon" (from 08/1999)
- Radio systems "Premium VI" & "Monsoon" (from 2002 m.y.)

Before troubleshooting or servicing, technicians must be familiar with the functions and operation specifics of the applicable standard or optional radio system. Always read the radio owner's manual and review all radio, tape player and CD changer functions.

All radios are protected with anti-theft security coding (fixed code), see **Maintenance** for inputting anti-theft security code.

NOTE—
Premium VI and Premium VI - Monsoon radios are identified simply as Premium CD in the owner's literature.

The radio system consists of the receiver/tape unit and speakers located in the front and rear doors. A CD changer is available as optional equipment. CD changer wiring and mounting is generally pre-installed in production, see **CD changer, CD player**.

A 168 mm low-range/bass speaker is located in the lower portion of each front and rear door. High-range speakers (domed tweeters) are integrated into the forward window trim on both front doors. Tweeters are integrated into the rear door handle trim next to the power window switches.

The system is completed with an amplified roof mounted antenna, see **Roof mounted radio antenna, overview**.

Radio, "Premium III & IV" (through 08/1999)

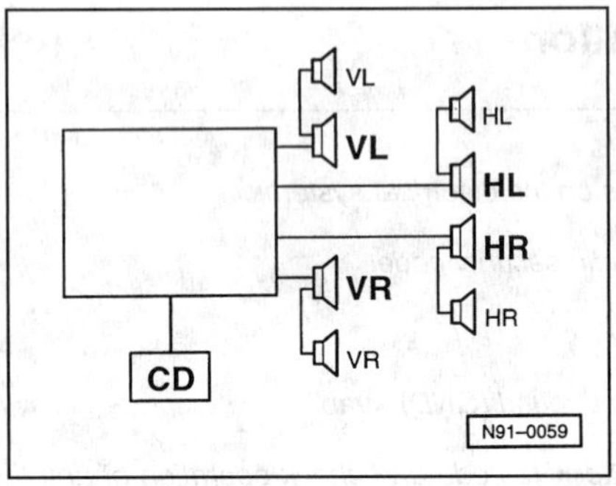

VL = Left front

HL = Left rear

VR = Right front

HR = Right rear

Radio system, "Premium III" is installed from start of production. Radio system, "Premium IV" is installed as a running change from 10/1997 through 08/1999 production.

Premium IV system has On Board Diagnostic (OBD) capabilities.

Radio, "Premium V" & "Premium V - Monsoon" (from 08/1999)

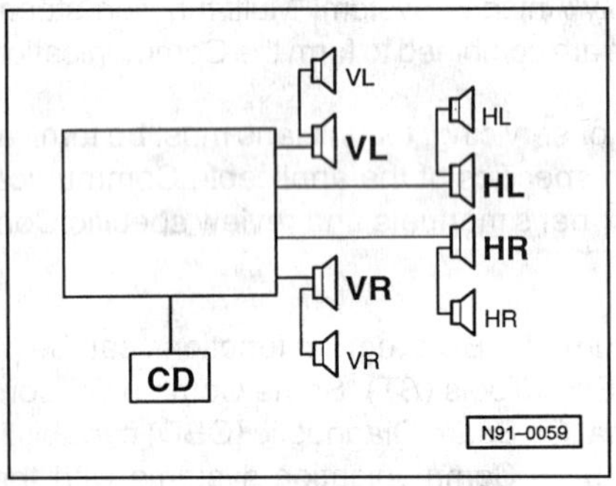

VL = Left front

HL = Left rear

VR = Right front

HR = Right rear

The standard radio system "Premium V" is installed from 08/1999 and consists of the receiver/tape unit with speakers located in the front and rear doors.

The optional "Monsoon" system from 08/1999 consists of the "Premium V" receiver/tape unit supplemented by a separate high power amplifier mounted in the luggage compartment. High power handling speakers specifically designed for the "Monsoon" system are located in the front and rear doors as with "Premium V".

The "Premium V" and "Premium V - Monsoon" radio systems have On Board Diagnostic (OBD) capabilities

CD Changer/player

A multiple disc CD changer and/or single disc CD player are available as optional equipment.

The luggage compartment mounted CD changer wiring and mounting is pre-installed in production, see **CD changer, CD player**.

The CD player is mounted in the instrument panel, see **CD changer, CD player**.

Radio, "Premium VI" & "Premium VI - Monsoon" (from 08/1999)

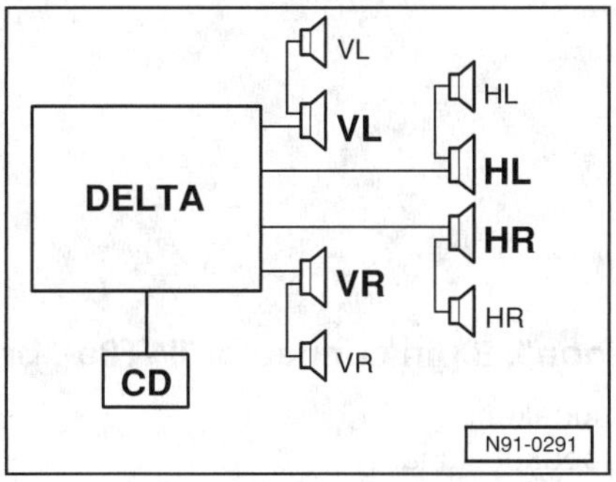

VL = Left front

HL = Left rear

VR = Right front

HR = Right rear

The radio system consists of an integrated receiver/CD/tape head unit and speakers located in the front and rear doors. The integrated head unit is in the "Double DIN" format. A 6-disc CD changer is available as an extra-cost option and is installed in the luggage compartment. The system uses an amplified roof mounted antenna or a diversity antenna system.

A 168 mm low-range/bass speaker is located in the lower portion of each front and rear door. High-range speakers (domed tweeters) are integrated into the forward window trim on both front doors. Tweeters are integrated into the rear door handle trim next to the power window switches.

The optional "Premium VI - Monsoon" system consists of the "Premium VI" receiver/CD/tape unit supplemented by a separate high power amplifier mounted in the luggage compartment.

High power handling speakers specifically designed for the "Monsoon" system are located in the front and rear doors as with "Premium VI".

The "Premium VI" and "Premium VI - Monsoon" radio systems have On Board Diagnostic (OBD) capabilities

Radio multi-pin connector assignments

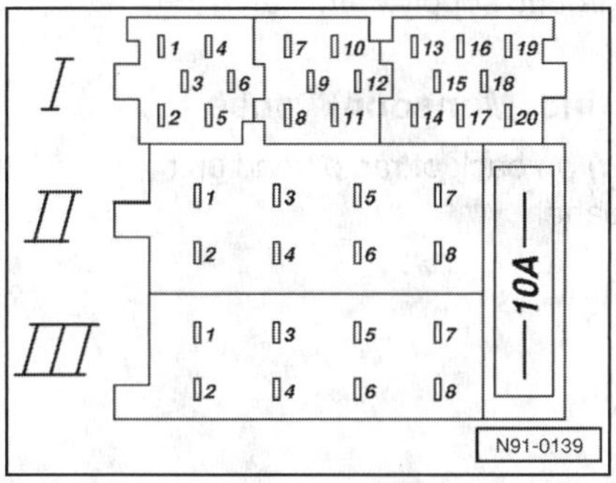

"Premium III, IV, V", Connector I (three separate connector sections):

20 pin multi-connector I -T20-, part 1, yellow*

1. Line out, left rear
2. Line out, right rear
3. Line out, Ground (GND)
4. Line out, left front
5. Line out, right front
6. Switched positive (B+) for amplifier

20 pin multi-connector I -T20-, part 2, green*

7. Open
8. Clock
9. DATA
10. ENA
11. Remote control (serial input)
12. Open

20 pin multi-connector I -T20-, part 3, blue*

13. CD changer - DATA IN
14. CD changer - DATA OUT
15. CD changer - CLOCK
16. CD changer - Power supply, positive (B+) supply
17. CD changer - Control signal
18. CD changer - Power supply, negative (-) ground supply
19. CD changer - LINE OUT, left
20. CD changer - LINE OUT, right

*) If equipped

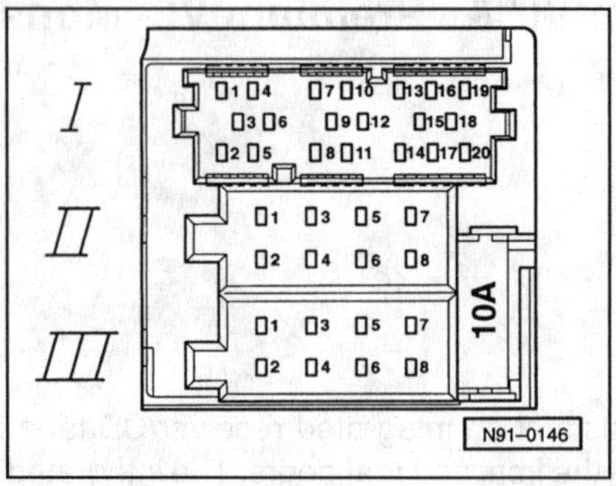

"Premium III, IV, V", 8 pin connector II -T8a-, brown

1. Speaker + right rear
2. Speaker - right rear
3. Speaker + right front
4. Speaker - right front
5. Speaker + left front
6. Speaker - left front
7. Speaker + left rear
8. Speaker - left rear

– "Premium V - Monsoon", 8 pin connector II -T8a-, brown

1. Amplifier + right rear speaker signal
2. Amplifier - right rear speaker signal
3. Amplifier + right front speaker signal
4. Amplifier - right front speaker signal
5. Amplifier + left front speaker signal
6. Amplifier - left front speaker signal
7. Amplifier + left rear speaker signal
8. Amplifier - left rear speaker signal

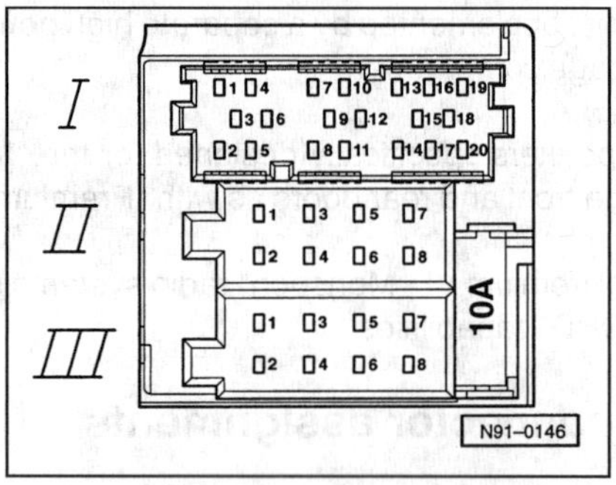

"Premium III, IV, V, (inc. Monsoon)", 8 pin connector III -T8-, black

1. Signal for anti-theft alarm system
2. Open
3. Data Link Connector (DLC) K-wire
4. Ignition switch - S contact
5. Battery, positive (B+) supply, terminal 30 (control input for "SAFE" display/anti-theft)
6. Illumination, terminal 58b
7. Battery, positive (B+) supply, terminal 30
8. Battery, negative (-) ground supply, terminal 31

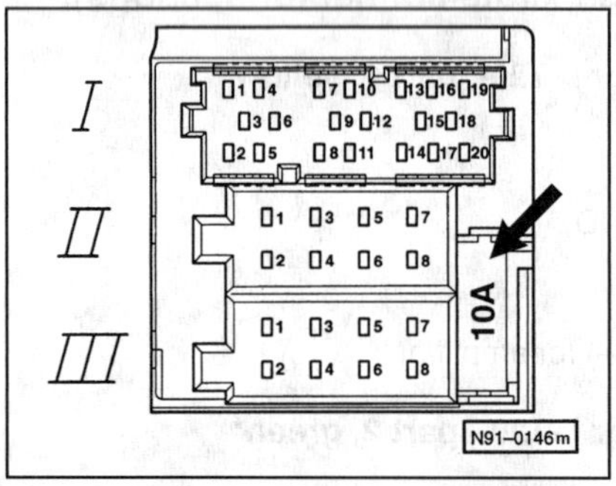

"Premium III, IV, V, (inc. Monsoon)", fuse

- Fuse - 10 Amp (**arrow**) on back of radio head unit
- Primary fuse in fuse panel

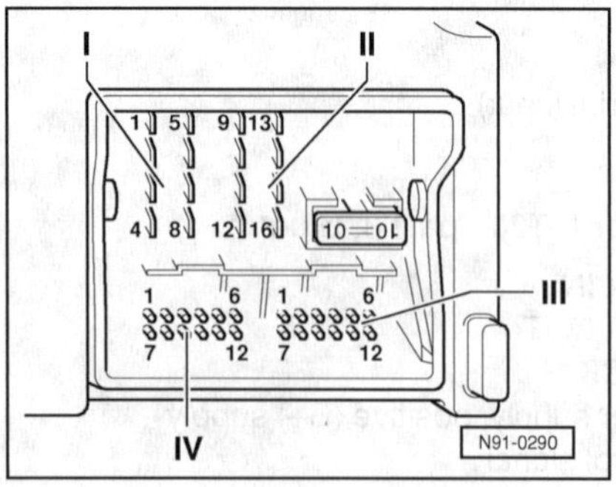

"Premium VI, (inc. Monsoon)", 16 pin, 2 part connector (I and II) -T16-, black

8 pin section I

1. Speaker (or line out*), (+) right rear
2. Speaker (or line out*), (+) right front
3. Speaker (or line out*), (+) left front
4. Speaker (or line out*), (+) left rear
5. Speaker (or line out*), (-) right rear
6. Speaker (or line out*), (-) right front
7. Speaker (or line out*), (-) left front
8. Speaker (or line out*), (-) left rear

*) If equipped with Monsoon amplifier

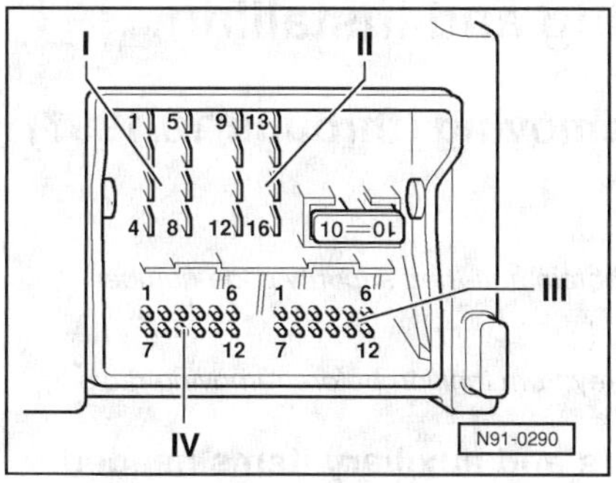

8 pin section II

9. CAN-BUS plus (+)
10. CAN-BUS plus (-)
11. Mute
12. Ground (GND)
13. Radio, control signal plus (+)
14. Contact, alarm system
15. Battery positive (30a)
16. Battery positive (30a)

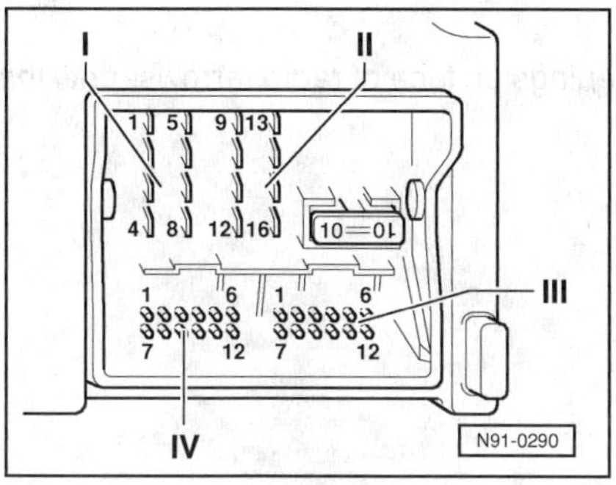

◀ "Premium VI, (inc. Monsoon)", 12 pin connector III -T12-

1. Open
2. Open
3. Open
4. Open
5. Open
6. Signal for telephone (-)*
7. Open
8. Open
9. Open
10. Open
11. Open
12. Signal for telephone (+)*

*) Where equipped

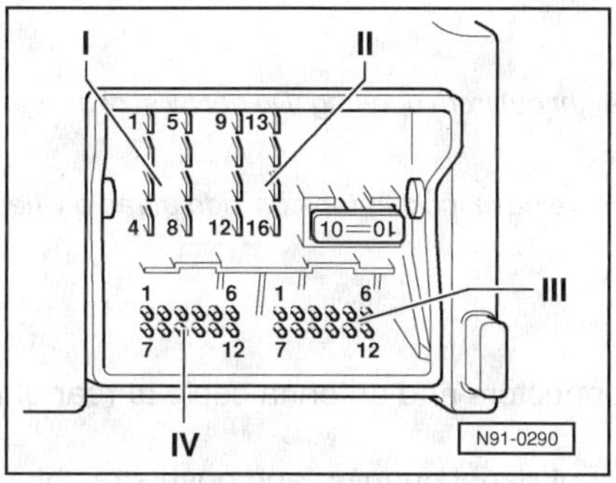

◀ "Premium VI, (inc. Monsoon)", 12 pin connector IV -T12-

1. Open
2. CD changer - left and right channel, Ground (GND)*
3. Open
4. CD changer - Positive (B+) supply*
5. Open
6. CD changer - DATA OUT*
7. Open
8. CD changer - left channel, CD/L*
9. CD changer - right channel, CD/L*
10. CD changer - Control signal*
11. CD changer - DATA in*
12. CD changer - CLOCK*

*) Where equipped

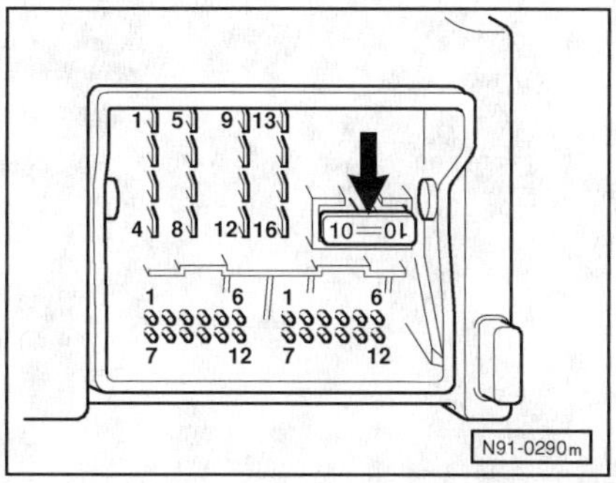

◀ "Premium VI, (inc. Monsoon)", fuse

- Fuse - 10 Amp (**arrow**) on back of radio head unit
- Primary fuse in fuse panel

Radio, removing and installing

"Premium III", removing (through 10/1997)

NOTE —
- *Before removing radio, obtain anti-theft security code number, see **Maintenance.***
- *Always remove ignition key from ignition before removing radio.*

Special tools, testers and auxiliary items needed.

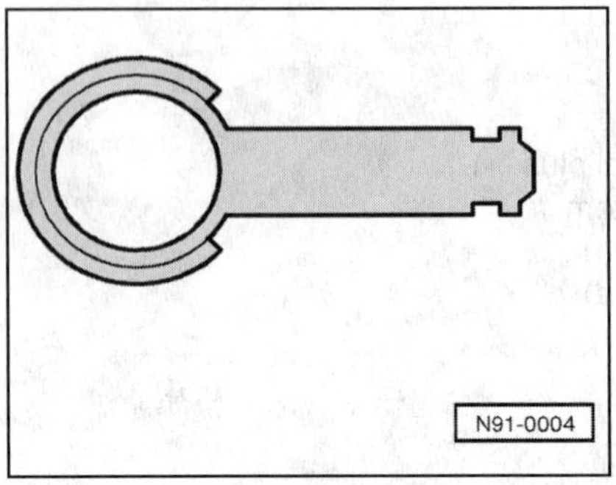

Radio removal tool 3316 (two identical pieces).

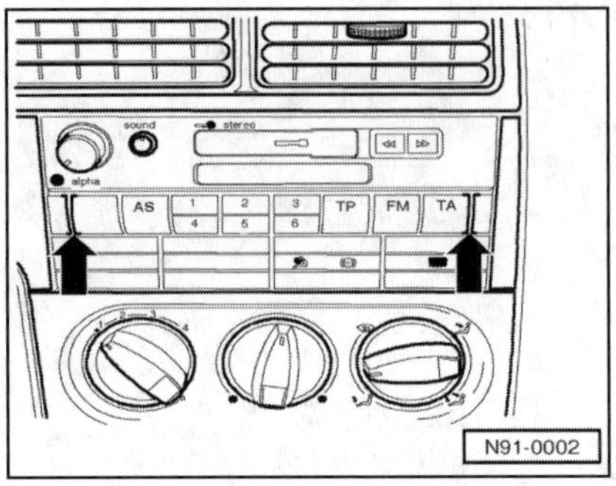

Insert tool 3316 into openings on face of radio -arrows- until they are felt to engage.

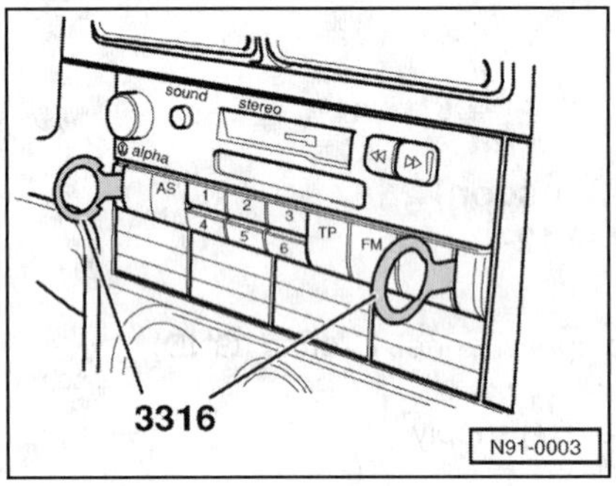

Pull radio from instrument panel opening using handles on tool 3316.

– Disconnect electrical connectors and antenna cable from rear of radio.

NOTE —
- *Radio must be pulled straight out from opening. Do not twist or tilt tool 3316 while removing.*
- *To remove tool from radio, release locking tabs on side of radio when removed.*

Installing

– Reconnect electrical connectors and antenna cable at rear of radio.

– Insert radio into instrument panel opening and push straight in until locking tabs engage.

– Input anti-theft security code, see **Maintenance**.

"Premium IV" (from 10/1997) & "Premium V" (inc. "Monsoon") (from 08/1999), removing

NOTE —
Always remove ignition key from ignition before removing radio.

Special tools, testers and auxiliary items needed.

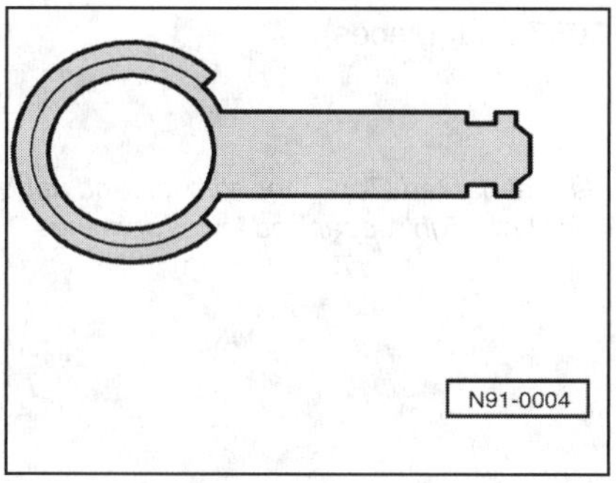

◀ Radio removal tool 3316 (two identical pieces).

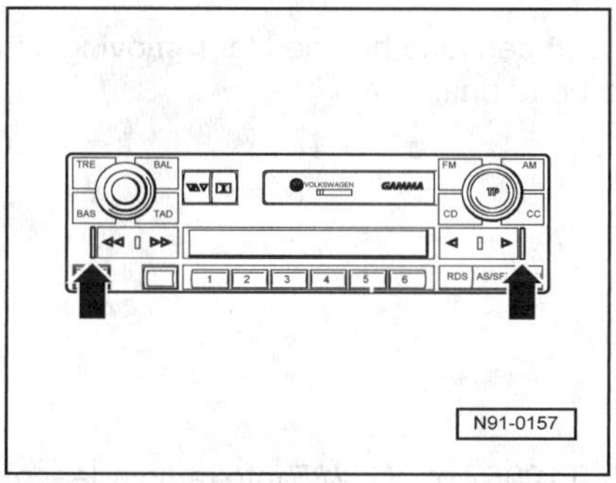

◀ Insert tool 3316 into openings on face of radio -arrows- until they are felt to engage.

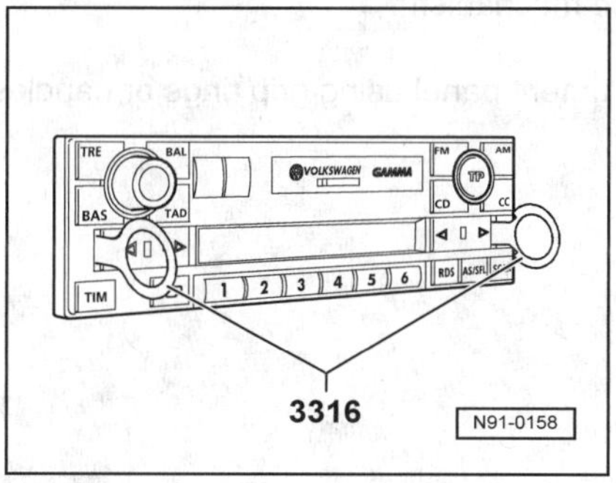

◀ Pull radio from instrument panel opening using handles on tool 3316.

– Disconnect electrical connectors and antenna cable from rear of radio.

NOTE —
- *Radio must be pulled straight out from opening. Do not twist or tilt tool 3316 while removing.*
- *To remove tool from radio, release locking tabs on side of radio when removed.*

Installing

– Reconnect electrical connectors and antenna cable at rear of radio.

– Insert radio into instrument panel opening and push straight in until locking tabs engage.

– Input anti-theft security code, see **Maintenance**.

– Code (radio) sound system functions using On Board Diagnostic program: "Code control module", function 07, using a diagnostic scan tool.

"Premium VI" (inc. "Monsoon"), removing and installing (from 2002 m.y.)

NOTE—
Always remove ignition key from ignition before removing radio.

Special tools, testers and auxiliary items needed.

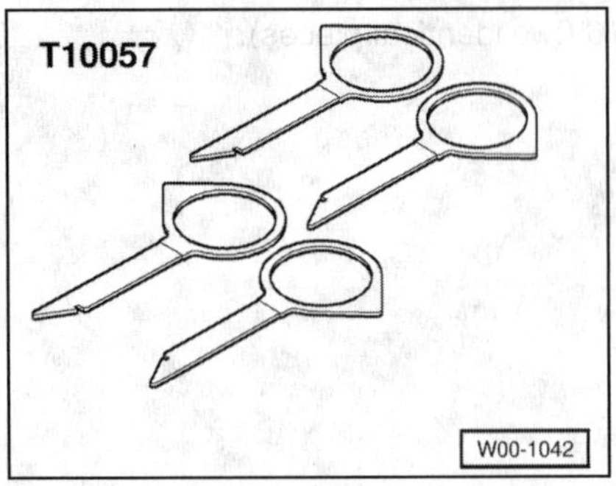

◄ Radio removal tools T10057 (four pieces).

NOTE—
Individual radio release tools are marked "Top L" for both left and upper and lower openings and "top R" for both right upper and lower openings.

◄ Radio removal tool JC4225 can also be used for removing all versions of the Premium VI radio head unit.

– Slide radio release tools T10057 or JC4225 into radio release slots until they engage the release mechanism.

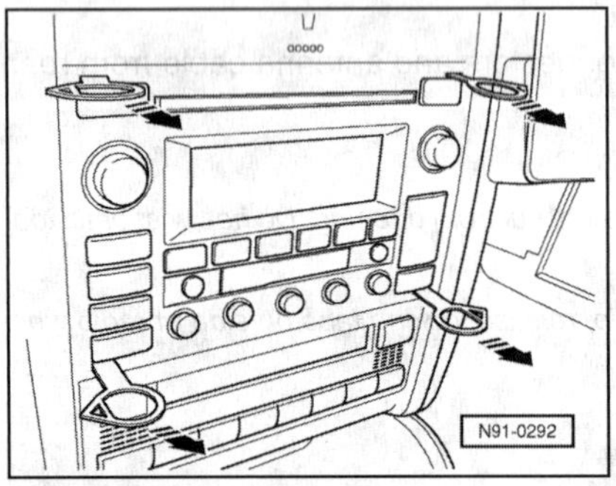

◄ Pull straight out of instrument panel using grip rings or handles of release tools.

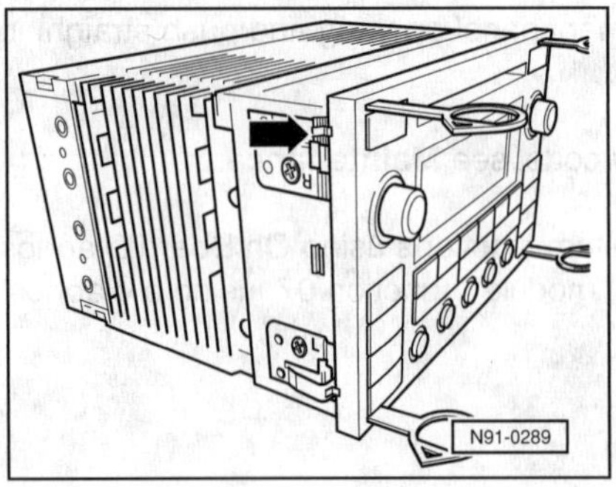

◄ Release removal tools from radio by pressing locking tabs on the side of radio once it is removed.

– To install radio:

- Radio must be installed without removal tools in place.
- Connect harness and antenna connectors to radio unit.
- Ensure proper operation of all radio functions prior to installation.
- Slide radio unit straight into instrument panel until it engages properly into assembly frame.
- Input anti-theft security code, see **Maintenance**.
- Confirm radio coding for sound system functions using diagnostic scan tool if necessary.

Amplifier (Monsoon), removing and installing

Sedan

The Monsoon system amplifier is located under the rear shelf, to the right rear. The amplifier is secured to a bracket on the rear shelf.

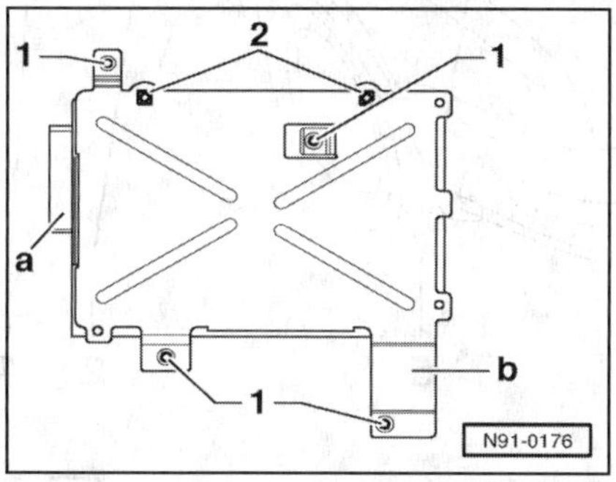

To remove the amplifier, perform the following steps:

- Open luggage compartment.
- Disconnect connectors.
- Remove amplifier securing screws -2- from within luggage compartment.
- Take amplifier out of bracket.

NOTE —

After amplifier replacement, code (radio) sound system functions using a diagnostic scan tool.

– If bracket -b- is to be removed, perform the following steps:

- Remove rear shelf cover.
- Remove nuts from threaded studs -1- using a suitable wrench. It may be necessary for a second person to counter-hold bracket -b- from the luggage compartment.

Wagon

The Monsoon system amplifier is located to the rear left of the luggage compartment. The amplifier is secured to a bracket above or below the CD changer.

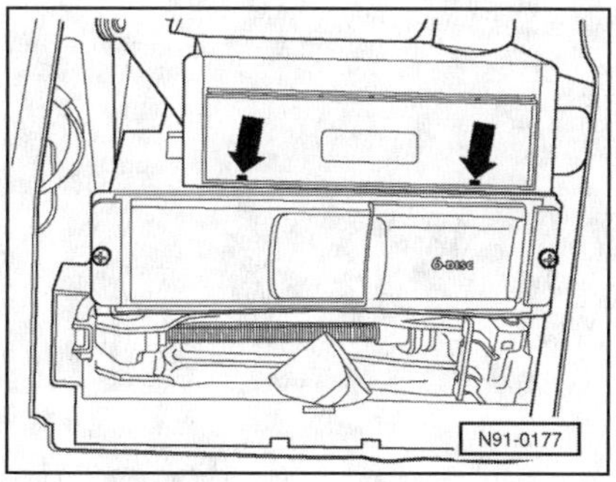

To remove the amplifier, perform the following steps:

- Open flap in left side luggage compartment trim panel.
- Disconnect connectors.
- Loosen securing screws -arrows-.
- Take amplifier out of bracket.

NOTE —

After amplifier replacement, code (radio) sound system functions using a diagnostic scan tool.

Speaker system, overview

NOTE—

- *Vehicles with radio system "Premium V - Monsoon" are equipped with high power handling speakers specifically designed for the "Monsoon" system.*
- *When replacing speakers, be aware of applicable radio system when referring to parts catalog.*

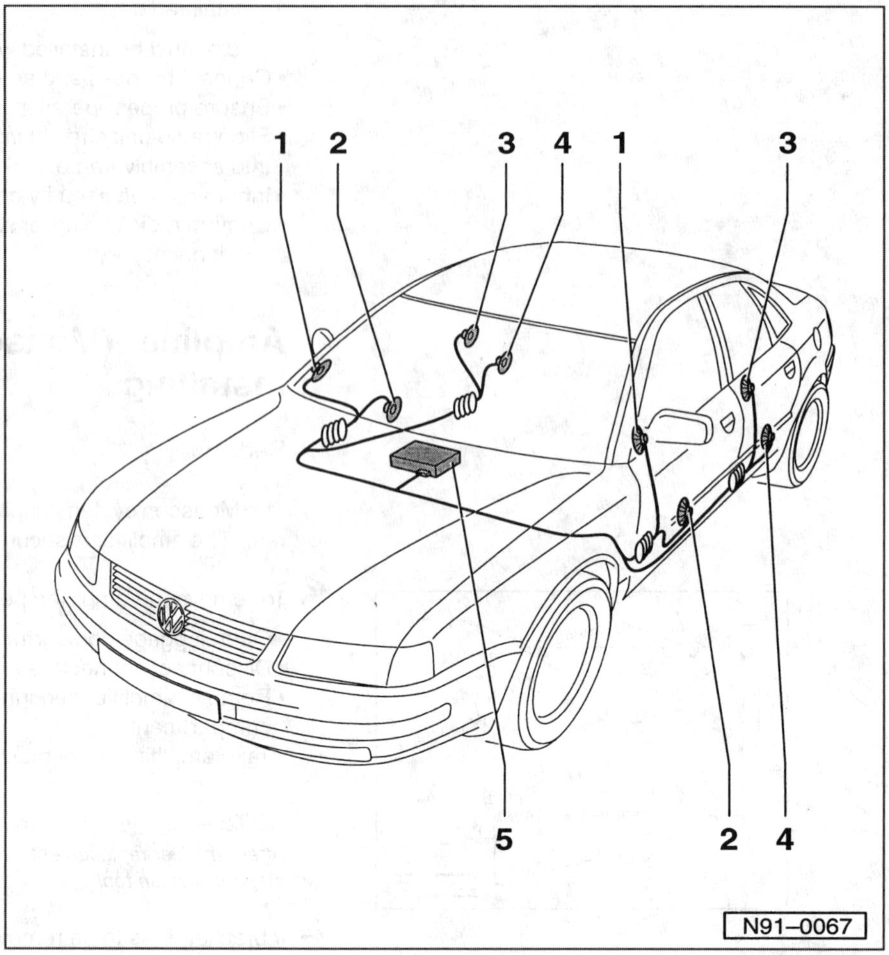

1. **High-range speaker (domed tweeter) in front doors**
   - Integrated with window trim
   - See **High-range speaker (domed tweeter) in front doors, removing**
2. **Mid-range/bass speaker in front doors**
   - See **Mid-range/bass speakers in front and rear doors, removing**
3. **High-range speaker (domed tweeter) in rear doors**
   - Integrated with inside door handle trim
   - See **High-range speaker (domed tweeter) in rear doors, removing**
4. **Mid-range/bass speaker in rear doors**
   - See **Mid-range/bass speakers in front and rear doors, removing**
5. **Radio**
   - In center of instrument panel
   - See **Radio, removing and installing**

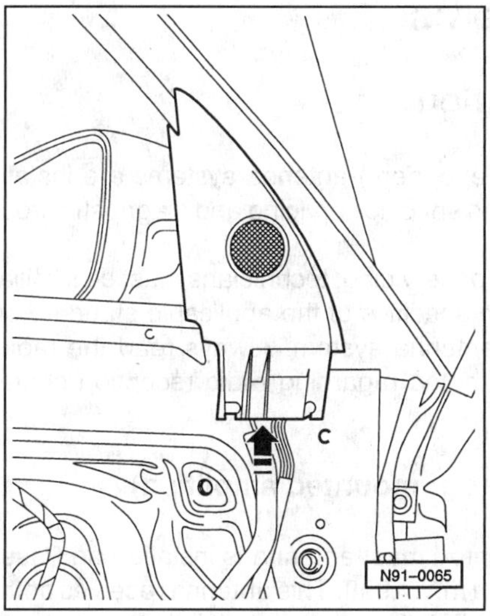

High-range speaker (domed tweeter) in front doors, removing

The high-range speaker is integrated into the forward window trim by the outside mirror mounting) on both front doors. The trim and speaker are constructed as a single unit and cannot be separated.

– Remove front door panel, see **Body-Interior section, Repair Group 70.**

– Disconnect electrical harness connection to speaker.

– Carefully push upward (arrow) and remove trim/speaker.

> ***CAUTION—***
> *To avoid damage, do not pull or pry out window trim. Only push upwards to disengage the inner locating tab.*

High-range speaker (domed tweeter) in rear doors, removing

The high-range speaker is integrated into the inside door handle trim on both rear doors. The trim and speaker are constructed as a single unit and cannot be separated.

– Remove rear door panel and inside door handle trim, see **Body-Interior section, Repair Group 70.**

– Disconnect electrical harness connections to speaker and power window switch.

– Unclip power window switch from door handle trim and remove trim/speaker.

Mid-range/bass speakers in front and rear doors, removing

– Remove door panel, see **Body-Interior section, Repair Group 70.**

– Disconnect electrical harness connection to speaker.

– Unclip power window switch from door handle trim and remove trim/speaker.

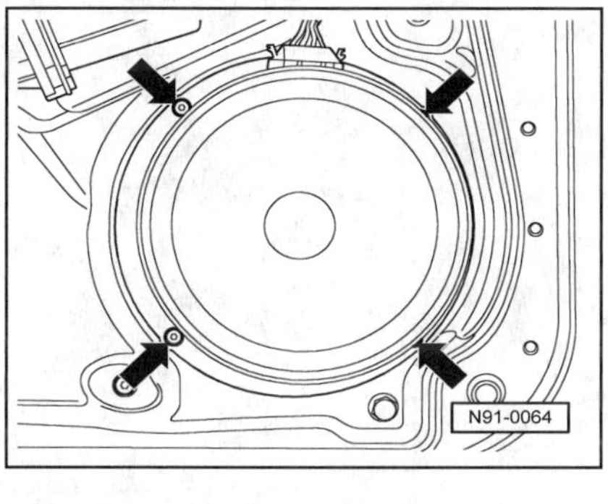

– Carefully drill out rivets at speaker frame -arrows- and remove speaker from opening.

Installing

– Mount speaker with locally available rivets of the proper dimension.

– Reconnect all electrical connections and reinstall all door panel trim, making sure the foam gasket between the speaker and door panel is in place.

Antenna systems

General information

Two different (model dependent) antenna systems are installed. Each antenna system requires specific servicing and diagnostic procedures.

Before troubleshooting or servicing, technicians must be familiar with the functions and operation specifics of the applicable standard or optional radio and associated antenna system. Always read the radio owner's manual and review all notes regarding radio reception characteristics and limitations.

Standard antenna (roof mounted amplified)

An amplified, roof mounted mast antenna is installed on Passat Sedan and Wagon models from m.y. 1998. This antenna receives both AM (amplitude modulation) and FM (frequency modulation) signals. Current to operate the amplifier is supplied by the radio up through the coaxial cable used to send the RF (radio frequency) signal back down to the radio head unit. Power supplied in this bi-directional manner is known as "phantom power supply" since there is no separate 12 volt power supply wire.

Diversity antenna (glass mounted amplified)

An amplified "diversity" antenna system integrated with the rear window(s) is installed on Passat W8 Sedan and Wagon models from m.y. 2002. This system consists of antenna wires embedded in the glass of the rear side and rear windows.

The diversity antenna system uses an amplifier for each glass embedded antenna wire. The amplifier output along is processed by an antenna selection control module (J515) before being sent to the radio head unit.

Current to operate the four individual amplifiers is supplied by the antenna selection control module up through the coaxial cable used to send the RF (radio frequency) signal back down to the antenna selection control module. Power supplied in this bi-directional manner is known as "phantom power supply" since there is no separate 12 volt power supply wire

Roof mounted radio antenna, overview

1. **Antenna mast**
2. **Antenna base**
   - Amplifier is integrated into antenna base
   - Power supplied by "phantom" method via coaxial cable
   - Removing and installing: Pull down headliner at rear, see **Body-Interior, section Repair Group 70.** Disconnect antenna cable -3- and remove nut -5-.
3. **Antenna cable**
   - Routed from antenna base to center console.
4. **Antenna cable**
   - Routed from center console to radio.
5. **Nut, special hex M14**
   - With integrated serrated washer.
   - Apply contact grease to roof panel when replacing.
6. **Gasket**
   - Ensure proper positioning to prevent water leakage

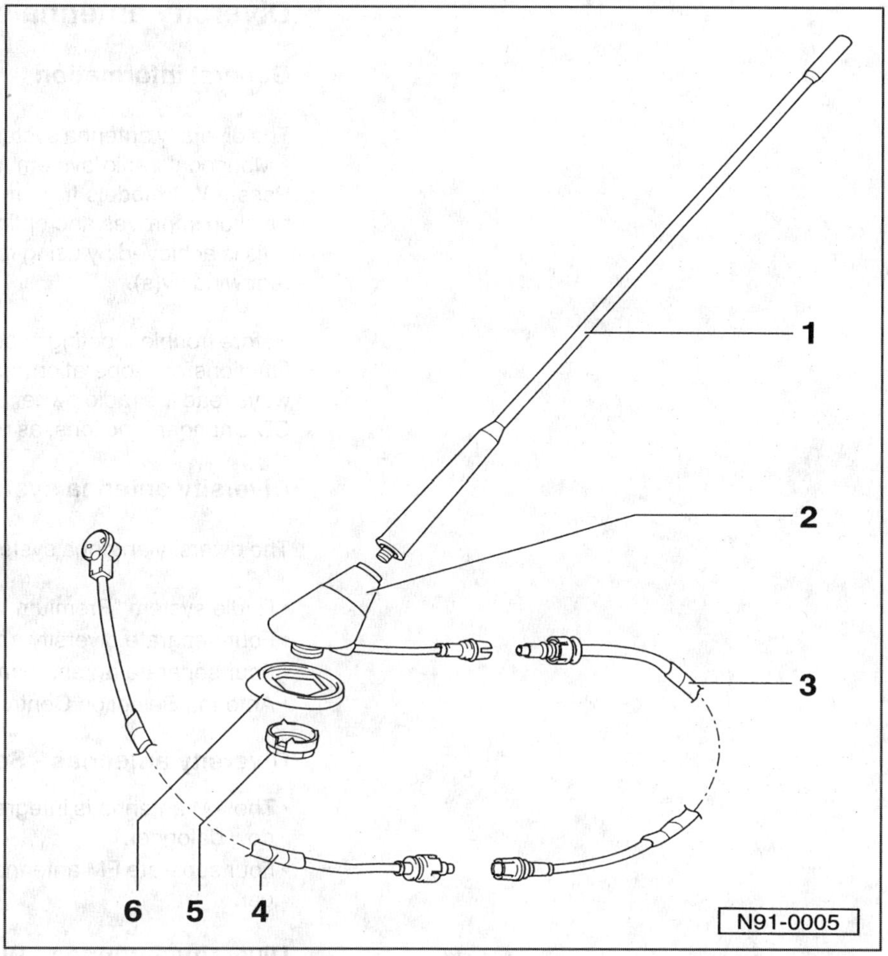

Reception interference, suppression measures

Most electrical consumers and features in the vehicle are developed to include radio interference measures. These include:

- Coolant fan -V7-
- Windshield wiper motor -V-
- Rear window wiper motor -V12-
- Braided ground strap from left fender mounting bolt to front hood.

Diversity antenna system (W8 only)

General information

The diversity antenna system is used in conjunction with the "Premium VI - Monsoon" radio system installed as standard equipment in US/CDN Passat W8 models from m.y. 2002. The antenna system with diversity function improves and optimizes the overall radio reception in a vehicle. This is achieved by using four separate FM antennas integrated into the rear window(s).

Before troubleshooting or servicing, technicians must be familiar with the functions and operation specifics of the diversity antenna system. Always read the radio owner's manual and review all radio, tape player and CD Changer functions, as required.

Diversity antenna system, components

The diversity antenna system consists of:

- Radio system "Premium VI - Monsoon"
- Four separate diversity antennas integrated into rear window(s)
- Four separate antenna amplifiers matched to the antennas
- Antenna Selection Control Module -J515-

Diversity antennas - Sedan, location

- The AM antenna is integrated into the rear window above the rear window defogger.
- Four separate FM antennas are integrated with the rear window defogger.

Diversity antennas - Wagon, location

- The antennas are integrated with the rear side windows.
- A combined AM/FM antenna and a separate FM antenna is located in the left rear window.
- The remaining FM antennas are located in the right rear window.

Function

The strength and quality of the amplified signals received by any of the four diversity system antennas is continuously monitored and evaluated by the radio unit. The evaluation of the antenna signal is output via an Intermediate Frequency (IF) to the Antenna Selection Control Module -J515- which routes the various antenna signals to the radio. The evaluation data is processed by -J515- and it actively selects the system antenna that provides the best reception under the existing conditions (diversity function).

The continuous, active switching between system antennas is not audible.

Diversity antenna system (W8 only), layout

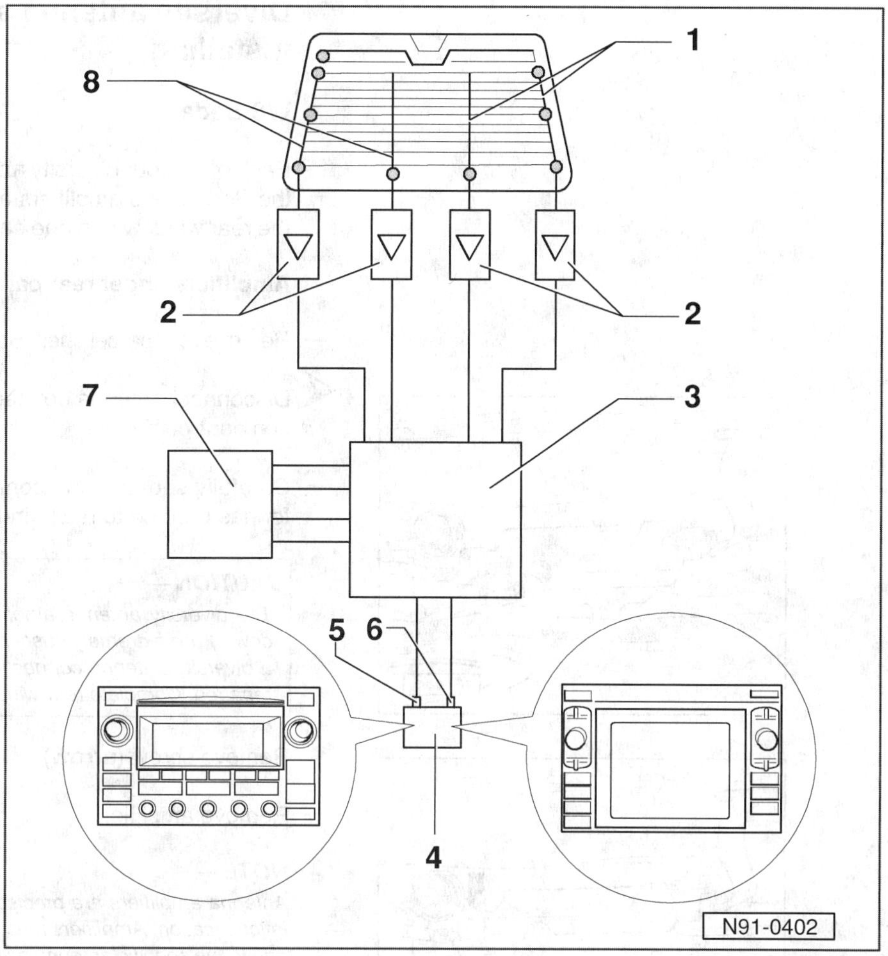

1. **Diversity antennas**
   - Quantity 4: AM/FM (R130), FM only (R131, R132, R133)
   - Sedan: integrated into rear window glass with AM/FM at left side, forward
   - Wagon: integrated into rear side window glass and rear window glass with AM/FM in the left side forward position and
   - Replacing: replace entire window glass as applicable
2. **Diversity antenna amplifiers**
   - Quantity 4: AM/FM (R24), FM only (R111, R112, R113)
   - Sedan: 2 located under rear parcel shelf with one each behind left and right rear D-pillar trim.
   - Wagon: 2 located behind each left and right luggage compartment side trim panels just below glass
   - Powered via phantom power supply
   - See: **Diversity antenna amplifiers, electrical connections**
   - See: **Diversity antenna amplifiers, removing and Installing**
3. **Antenna selection control module -J515-**
   - Sedan: under rear parcel shelf
   - Wagon: behind left luggage compartment side trim panel just below glass
   - See: **Antenna selection control module -J515-, electrical connections**
   - See: **Antenna selection control module -J515-, removing and Installing**
4. **Radio -R- "Premium VI - Monsoon"**
   - Radio/Navigation unit shown to the right is not applicable to the USA/Canadian market.
5. **Antenna input**
   - Connector code **B**
6. **Intermediate Frequency (IF) signal output**
   - Connector code **K**
7. **TV, navigation, communication antenna package**
   - Unit shown to the is not applicable to the USA/Canadian market.

* **Antenna cable system (not shown)**
   - Connects antenna selection control module, antenna amplifiers, and radio head unit
   - Repairing, see **Antenna cable system, assembly overview**

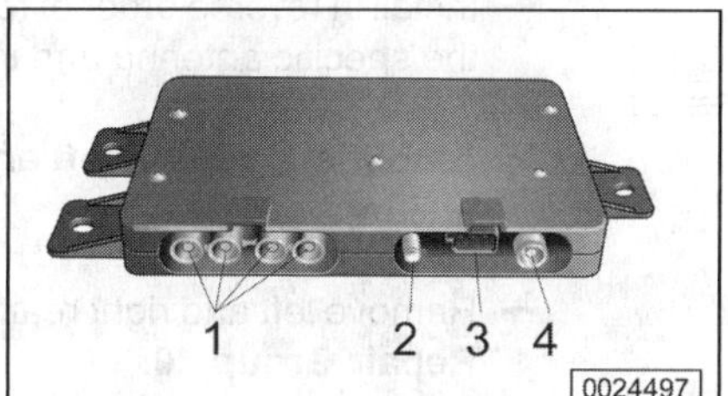

Antenna selection control module J515, electrical connections

1. Antenna inputs (from amplifiers)
2. Output to tuner (radio head unit)
3. Power supply
4. Input from tuner (radio head unit)

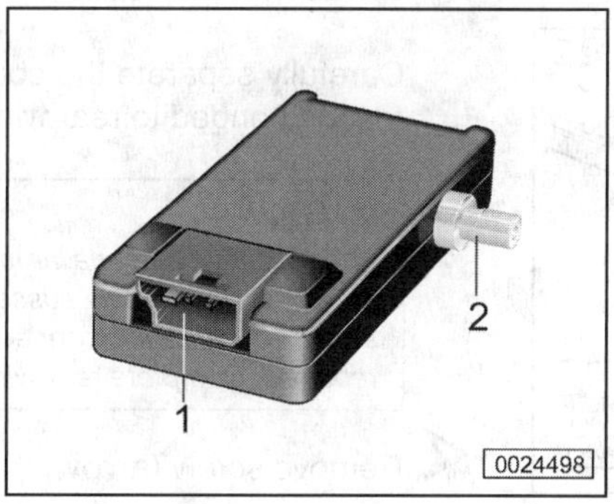

Diversity antenna amplifiers, electrical connections

1. Antenna input from embedded glass antenna
2. Antenna output to Antenna selection control module

Diversity antenna amplifiers, removing and installing

W8 Sedan

Each of the four diversity antennas is matched with it's own amplifier. On the Sedan, two amplifiers are located under the rear parcel shelf below the rear window and one each behind the left and right rear D-pillar trim.

Amplifiers under rear parcel shelf, removing and installing

– Remove rear parcel shelf. See **Body-Interior, section Repair Group 70.**

◄ Disconnect antenna connections between amplifier and antenna selection control module

– Carefully separate the connections between amplifier and diversity antennas bonded to rear window.

> ***CAUTION—***
> *The diversity antenna amplifier connections bonded to the rear window(s) are extremely susceptible to mechanical damage! In the event a diversity antenna connection is damaged or broken, it must not be repaired. Complete rear window replacement is required!*

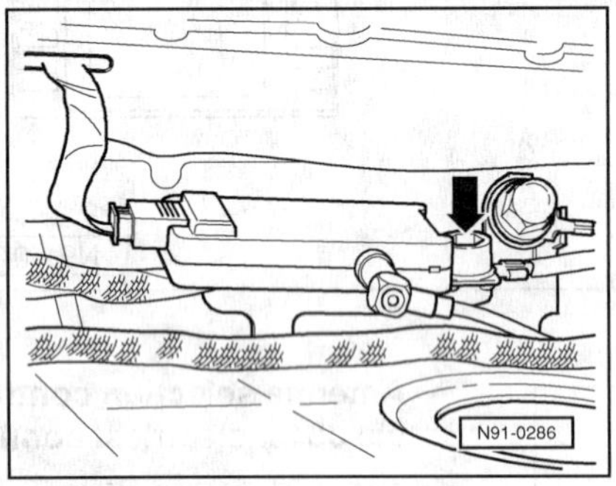

◄ Remove screw (**arrow**).

– Remove amplifier.

NOTE—
Antenna amplifiers are precisely matched to a specific antenna and installation location. Amplifiers must always be reinstalled in the same orientation and to the specific antenna as previous to removal, or reception will be poor.

– Install in reverse order of removal, making sure to match the amplifier with the specific antenna and installation orientation as previous to removal.

Amplifiers behind left and right rear D-pillar trim, removing and installing

– Remove left and right rear D-pillar side trim. See **Body-Interior, section Repair Group 70.**

– Disconnect antenna connections between amplifier and antenna selection control module

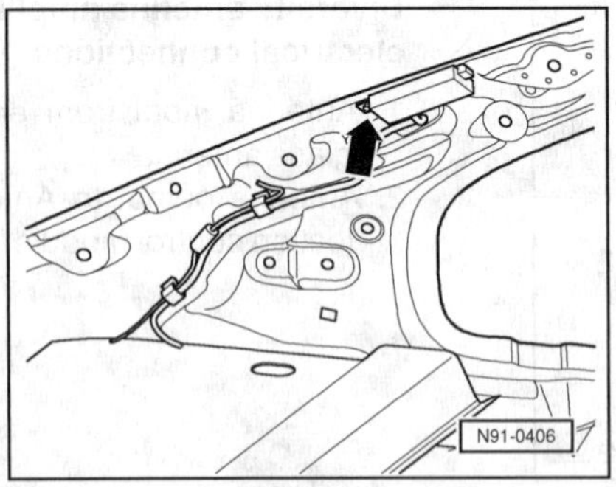

– Carefully separate the connections between amplifier and diversity antennas bonded to rear window.

> ***CAUTION—***
> *The diversity antenna amplifier connections bonded to the rear window(s) are extremely susceptible to mechanical damage! In the event a diversity antenna connection is damaged or broken, it must not be repaired. Complete rear window replacement is required!*

◄ Remove screw (**arrow**).

– Remove amplifier.

NOTE—
Antenna amplifiers are precisely matched to a specific antenna and installation location. Amplifiers must always be reinstalled in the same orientation and to the specific antenna as previous to removal, or reception will be poor.

– Install in reverse order of removal, making sure to match the amplifier with the specific antenna and installation orientation as previous to removal.

W8 Wagon.

Each of the four diversity antennas is matched with it's own amplifier. On the W8 Wagon, two amplifiers each are located behind the left and right rear D-pillar trim, below the side windows.

Amplifiers under rear side windows, removing and installing

– Remove left and right rear luggage compartment side trim. See **Body-Interior, section Repair Group 70.**

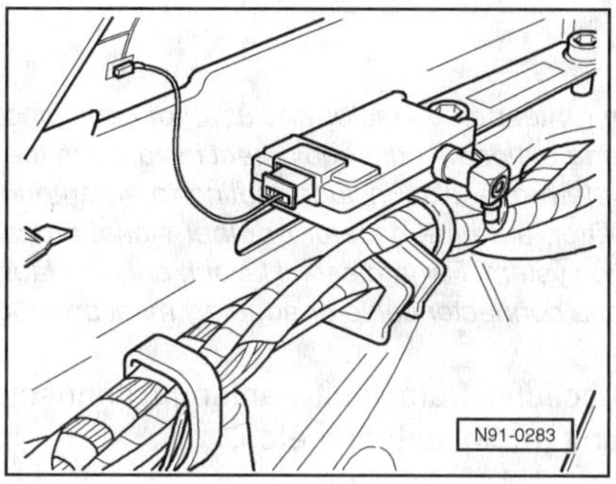

< Disconnect antenna connections between amplifier and antenna selection control module.

– Carefully separate the connections between amplifier and diversity antennas bonded to rear window.

> ***CAUTION—***
> *The diversity antenna amplifier connections bonded to the rear window(s) are extremely susceptible to mechanical damage! In the event a diversity antenna connection is damaged or broken, it must not be repaired. Complete rear window replacement is required!*

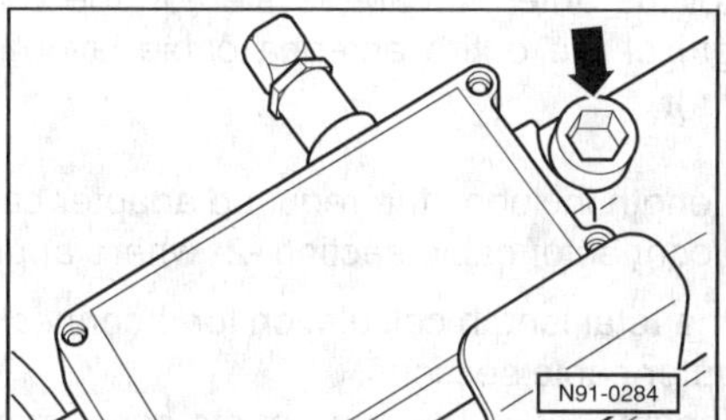

< Remove screw (**arrow**).

– Remove amplifier.

NOTE—
Antenna amplifiers are precisely matched to a specific antenna and installation location. Amplifiers must always be reinstalled in the same orientation and to the specific antenna as previous to removal, or reception will be poor.

– Install in reverse order of removal, making sure to match the amplifier with the specific antenna and installation orientation as previous to removal.

Antenna selection control module -J515-, removing and Installing

W8 Sedan, removing and installing

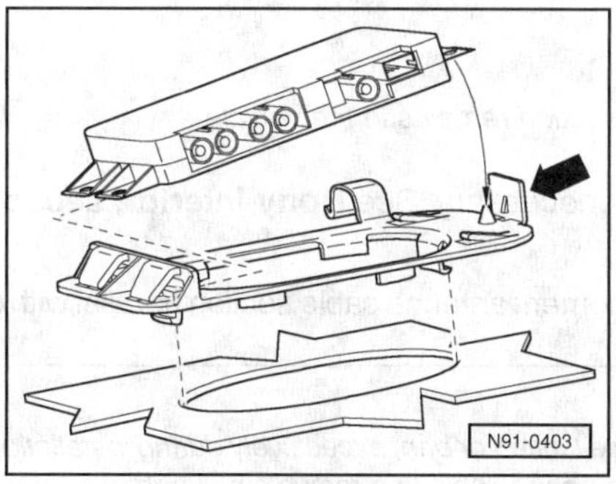

– Remove rear parcel shelf. See **Body-Interior, section Repair Group 70.**

< Depress clip on mounting (**arrow**) and lift-out control module.

– Disconnect electrical connections and antenna connections.

– Install in reverse order of removal.

W8 Wagon, removing and installing

– Remove left and right rear luggage compartment side trim. See **Body-Interior, section Repair Group 70.**

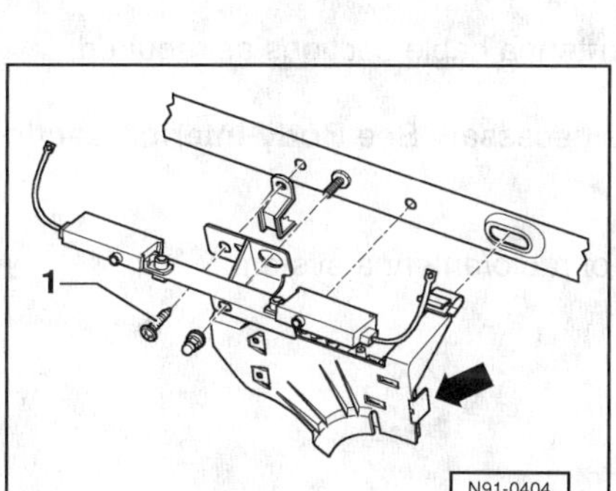

< Remove screw (**1**) at control module mounting plate.

– Disconnect electrical connections and antenna connections.

– Remove control module and mounting plate together.

– Depress clip on mounting plate (**arrow**) and separate control module from mounting plate.

– Install in reverse order of removal.

Antenna cable system, assembly overview

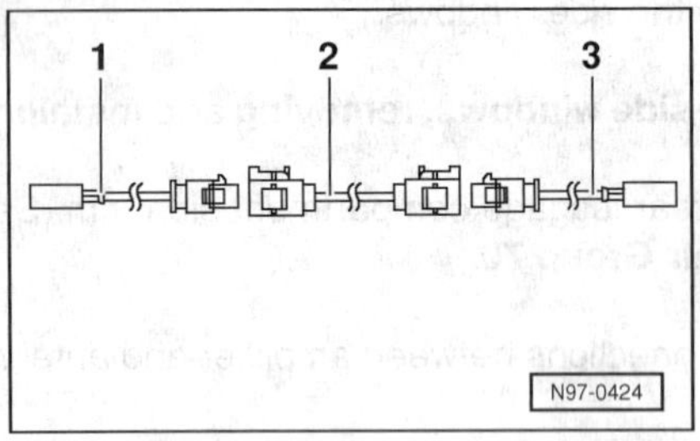

For servicing purposes, the antenna cable system consists of the following individual antenna cable sections:

1. Adapter cable to radio - Length approx. 30 cm.
2. Connector cable - Available in various lengths.
3. Adapter cable to antenna - Length approx. 30 cm.

Antenna connector cable, replacing

NOTE—
The antenna cable system (antenna connector and adaptor cable sections) depends on the available and optional audio equipment installed in the vehicle. The antenna cable system may also include routing to an antenna selection control module, splitter, antenna booster or other signal processing unit as required by the radio system. Regardless of layout, only the faulty antenna cable section (antenna connector cable or adapter) must be replaced.

– Disconnect cable connections from faulty antenna connector and/or adapter sections from unit (radio, antenna etc.).

– Determine the routing of the antenna cable system in the vehicle and calculate the total length of the entire antenna cable layout between units (radio, antenna etc.).

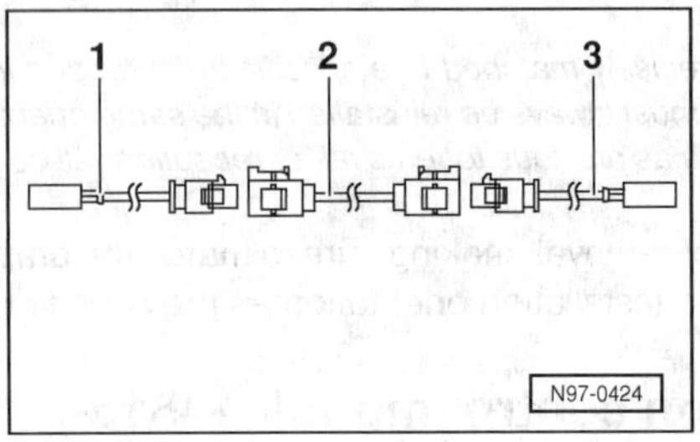

Calculation of the total length includes the required adapter cable sections -1 and/or -3-, and connector cable section -2- where applicable.

- Subtract 60 cm from the total length calculation for a connector cable to account for the adapter cable sections.
- Using the calculated length, obtain the appropriate connector cable (and adaptor cable if necessary) from the parts catalog.
- Cut the cable connection from the faulty antenna cable section.

NOTE—
The faulty antenna cable section remains in the vehicle.

– Remove interior trim as necessary. See **Body-Interior, section Repair Groups 68 and 70.**

– Overlay and install replacement antenna cable section parallel with old cable.

CAUTION—
Antenna cable must not be kinked or bent excessively during installation. Do not attempt to bend antenna cable in a radius smaller than 50mm.

– Connect replacement antenna cable sections as required.

– Reinstall interior trim as necessary See **Body-Interior, section Repair Groups 68 and 70.**

– Perform functional test of radio/antenna system.

CD changer, CD player

CD changer, removing and installing (through 10/2000)

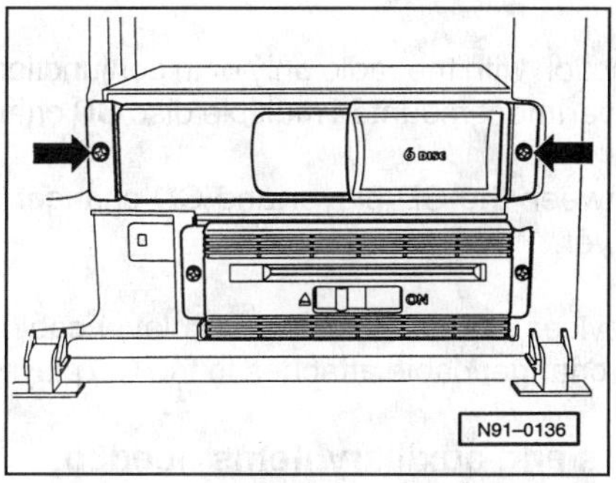

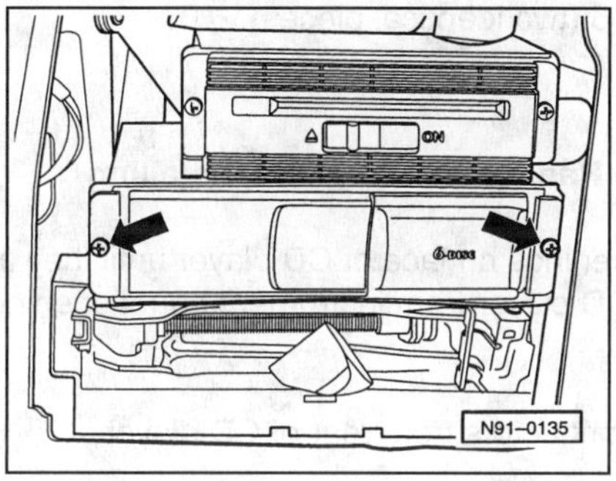

The optional CD changer unit is located in the left side of the luggage compartment behind the removable access panel.

– Remove screws -arrows- and slide CD changer unit from mounting frame.

– Disconnect electrical connection.

◄ Sedan

◄ Wagon

– Install in reverse order of removal, noting the following:

• Check and code (radio) sound system functions if necessary using On Board Diagnostic program: "Code control module", function 07 with a diagnostic scan tool.

CD changer, removing and installing (from 10/2000)

The optional CD changer unit is located in the left side of the luggage compartment behind the removable access panel.

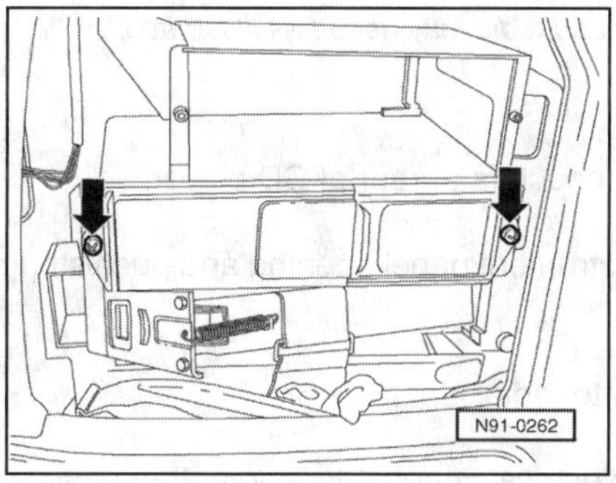

◄ Remove screws -arrows- and slide CD changer unit from mounting frame.

– Disconnect electrical connection.

– Install in reverse order of removal, noting the following:

• Check and code (radio) sound system functions if necessary using On Board Diagnostic program: "Code control module", function 07 with a diagnostic scan tool.

CD player, removing and installing (from 08/1999)

The optional single disc CD player (from 08/1999) is mounted in the instrument panel in place of the storage bin.

It can be used in conjunction with the radio only or in conjunction with the radio and luggage compartment mounted multiple disc CD changer.

Switching functions between the CD player and CD changer are done manually on the CD player.

In the event both CD devices are installed, the CD player cable attaches to the radio and the CD changer cable attaches to the CD player.

Special tools, testers and auxiliary items needed.

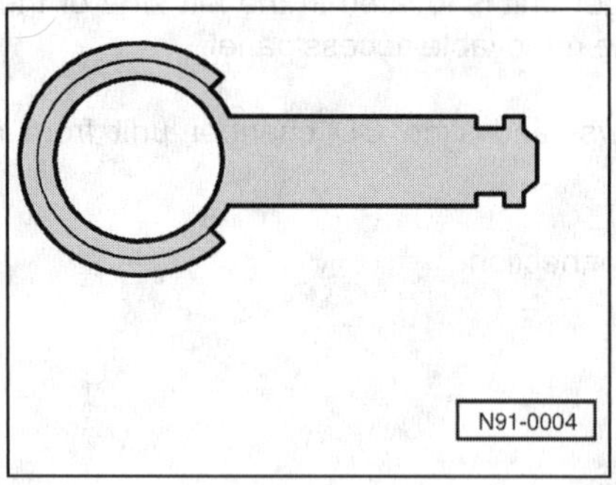

Radio removal tool 3316 (two identical pieces).

Removing

– First remove radio, see **Radio, removing and installing**

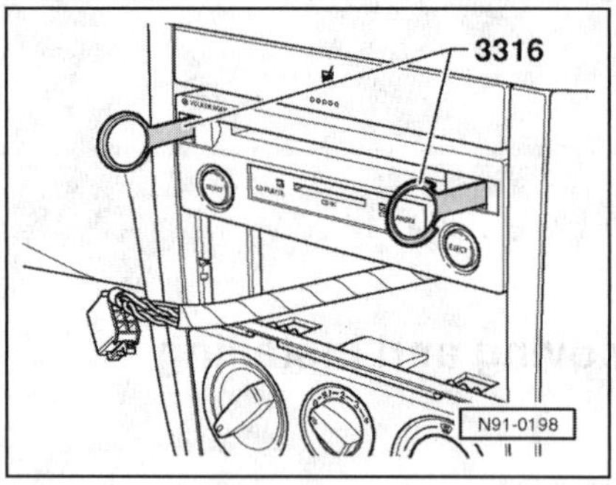

– Insert tool 3316 into openings on face of CD player until they are felt to engage and then pull CD player from instrument panel opening using handles on tool 3316.

– Disconnect electrical connectors from rear of CD player.

NOTE—

- *CD player must be pulled straight out from opening. Do not twist or tilt tool 3316 while removing.*
- *To remove tool from CD player, release locking tabs on side of radio when removed.*
- *Orientation of CD player and radio may not be as illustrated.*

Installing

– Reconnect electrical connectors at rear of CD player.

– Insert CD player into instrument panel opening and push straight in until locking tabs engage.

– Attach CD player cable to radio.

– Reconnect remaining electrical connectors and antenna cable at rear of radio.

– Insert radio into instrument panel opening and push straight in until locking tabs engage.

– Input radio anti-theft security code, see **Maintenance**.

– Check and code (radio) sound system functions if necessary using On Board Diagnostic program: "Code control module", function 07 with a diagnostic scan tool.

CD player multi-pin connector assignments

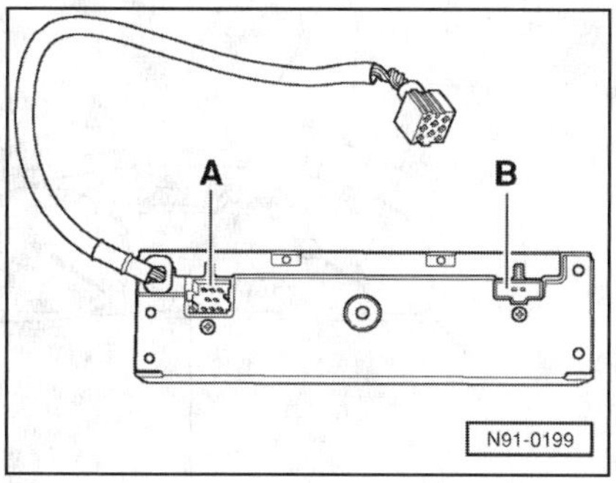

Multi-pin connector -A-

NOTE —
Multi-pin connector "A" is equivalent to 20 pin multi-connector I -T20-, part 3 (blue) on radio.

13. CD changer - DATA IN
14. CD changer - DATA OUT
15. CD changer - CLOCK
16. CD changer - Power supply, positive (B+) supply
17. CD changer - Control signal
18. CD changer - Power supply, negative (-) ground supply
19. CD changer - LINE OUT, left
20. CD changer - LINE OUT, right

Multi-pin connector -B-

1. CD player -Ground (GND), Terminal 31
2. CD player - illumination, Terminal 58d illumination

Multi-function steering wheel

CAUTION —
Before beginning repairs on the electrical system:

- *Obtain the anti-theft radio security code.*
- *Switch the ignition off.*
- *Disconnect the battery Ground (GND) strap.*
- *After reconnecting battery, re-code and check operation of anti-theft radio. Also check operation of clock and power windows according to Repair Manual and/or Owner's Manual.*

The multi-function steering allows functions of the radio and cruise control to be operated from the steering wheel.

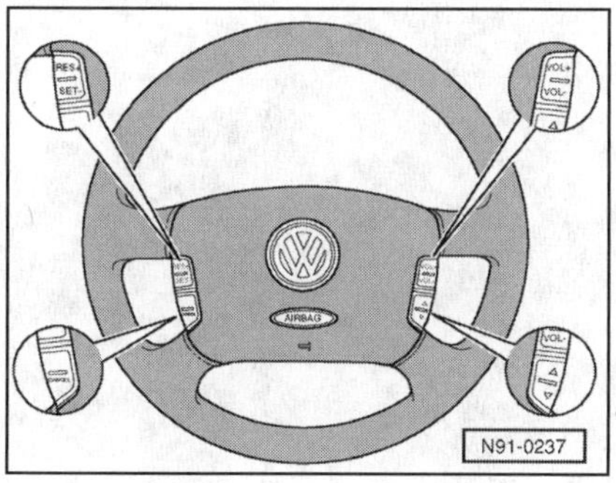

The multi-function steering wheel includes the following components:

- Operating unit in steering wheel with two key pads mounted left and the right on the steering wheel with integrated electronics.
- A control module for the multi-function steering wheel.
- In diesel vehicles, a damper weight is installed in the multi-function steering wheel in order to prevent transmission of vibrations.

NOTE —
It is necessary to know the function and operation of the multi-function steering wheel in order to service it.

Multi-function steering wheel, overview

1. **Multi-function steering wheel**
   - With integrated operating unit -E221-
   - See **Operating unit in steering wheel, removing and installing**
2. **Instrument cluster**
   - Control module with display unit in instrument cluster -J285-
   - Instrument cluster, removing and installing, see **Electrical Equipment section, Repair Group 90**
3. **Radio**
   - Installed in center console
4. **Control module for multi-function steering wheel -J453-**
   - Installed on relay panel
   - Removing and installing, see **Multi-function steering wheel control module, removing and installing**
   - Connector assignment, see **Connector assignment for multi-function steering wheel control module**
5. **Engine Control Module (ECM)**
   - Controls cruise control

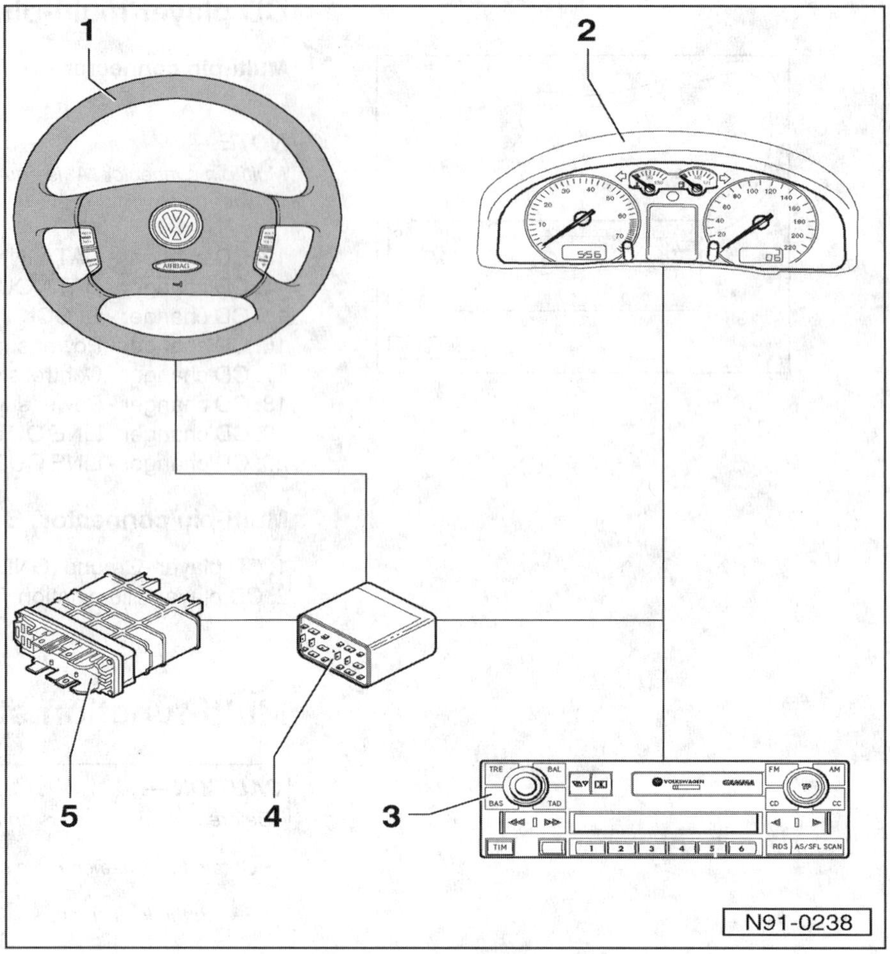

Operating unit in steering wheel, removing and installing

> ***CAUTION—***
> *Before beginning repairs on the electrical system:*
>
> - *Obtain the anti-theft radio security code.*
> - *Switch the ignition off.*
> - *Disconnect the battery Ground (GND) strap.*
> - *After reconnecting battery, re-code and check operation of anti-theft radio. Also check operation of clock and power windows according to Repair Manual and/or Owner's Manual.*

– Remove airbag unit on driver's side, see **Body-Interior section, Repair Group 69.**

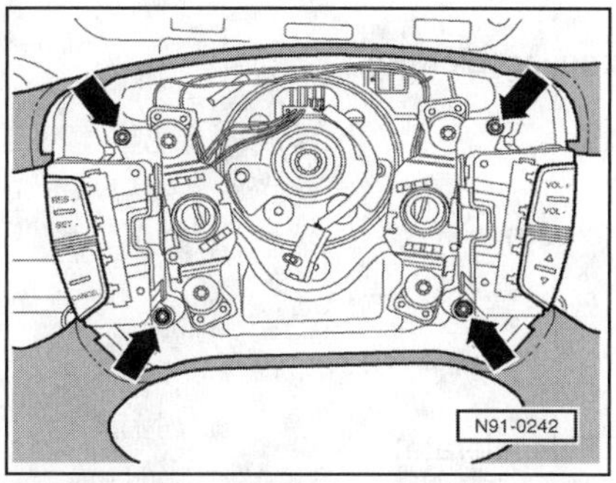

< Remove screws (arrows) from operating unit.

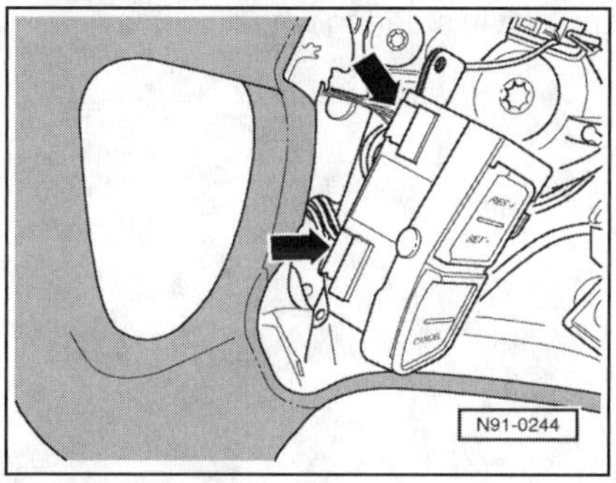

< Disconnect harness connectors (arrows) on operating unit.

– Remove operating unit.

– Installation is reverse order of removal.

Multi-function steering wheel control module, removing and installing

– Remove fuse box cover.

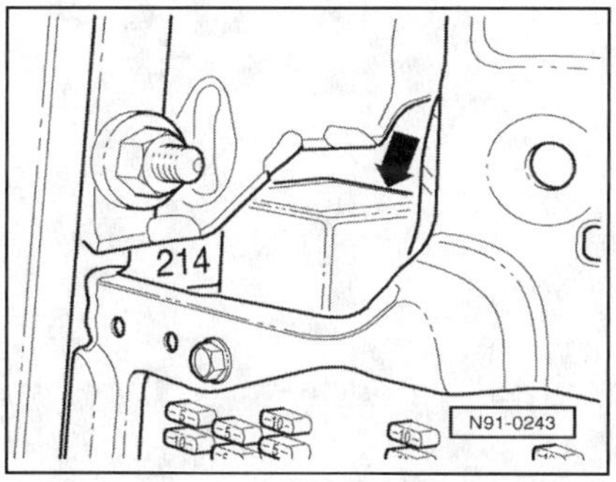

< Vehicles through 10/2000: Pull control module for multi-function steering wheel out from relay plate.

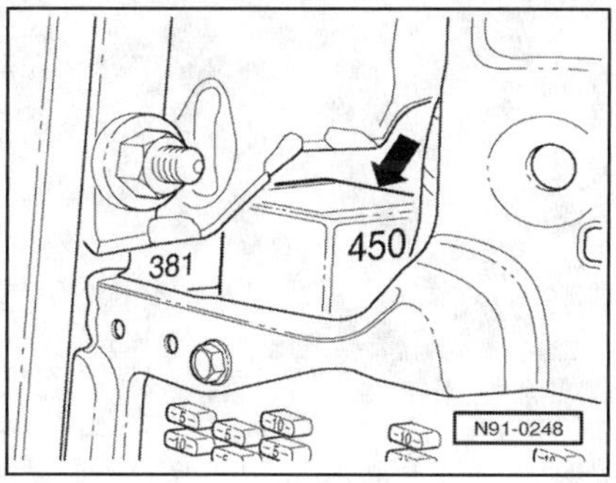

< Vehicles from 10/2000: Pull control module for multi-function steering wheel out from relay plate.

– Installation is reverse order of removal, noting the following:
- Code multi-function steering wheel control module.

Connector assignment for multi-function steering wheel control module

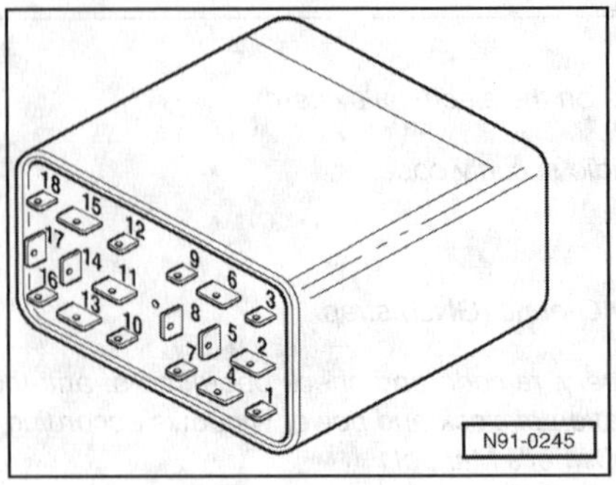

1. Comfort CAN Low
2. not assigned
3. Comfort CAN High
4. not assigned
5. Radio, serial data
6. K-wire, diagnosis
7. Cruise control - (resume, accelerate)
8. CAN ground
9. not assigned
10. Ground, terminal 31
11. Relay for double tone horn
12. Terminal 15
13. Terminal 30
14. not assigned
15. Terminal 58d
16. Cruise control + (SET/decelerate)
17. Input for Heating "ON"
18. CAN-Bus for operating unit in steering wheel

Wiring Diagrams and Component Locations

Section **Y**

WARNING!
Before proceeding with any work, read the warnings and cautions immediately following this index. Also read any warning and caution that accompanies a procedure in the manual.

Please read these WARNINGS and CAUTIONS before proceeding with maintenance and repair work.

WARNING—

• Some repairs may be beyond your capability. If you lack the skills, tools and equipment, or a suitable workplace for any procedure described in this manual, we suggest you leave such repairs to an authorized Audi dealer service department, or other qualified shop.

• Audi is constantly improving its cars. Sometimes these changes, both in parts and specifications, are made applicable to earlier models. Therefore, before starting any major jobs or repairs to components on which passenger safety may depend, consult your authorized Audi dealer about Technical Bulletins that may have been issued since the editorial closing of this manual.

• Do not re-use any fasteners that are worn or deformed in normal use. Some fasteners are designed to be used only once, and are unreliable and may fail if used a second time. This includes, but is not limited to, nuts, bolts, washers, circlips and cotter pins. Always replace these fasteners with new parts.

• Never work under a lifted car unless it is solidly supported on stands designed for the purpose. Do not support a car on cinder blocks, hollow tiles or other props that may crumble under continuous load. Never work under a car that is supported solely by a jack. Never work under the car while the engine is running.

• If you are going to work under a car on the ground, make sure that the ground is level. Block the wheels to keep the car from rolling. Disconnect the battery negative terminal (ground strap) to prevent others from starting the car while you are under it.

• Never run the engine unless the work area is well ventilated. Carbon monoxide (CO) kills.

• Do not attempt to work on your vehicle if you do not feel well. You increase the danger of injury to yourself and others if you are tired, upset or have taken medication or any other substance that may impair you or keep you from being fully alert.

• Disconnect the battery negative (–) terminal (ground strap) whenever you work on the fuel system or the electrical system. Do not smoke or work near heaters or other fire hazards. Keep an approved fire extinguisher handy.

• Batteries produce explosive hydrogen gas. Keep sparks, lighted matches and open flame away from the top of the battery. If hydrogen gas escaping from the cap vents is ignited, it will ignite gas trapped in the cells and cause the battery to explode.

• Connect and disconnect battery cables, jumper cables or a battery charger only with the ignition switched off, to prevent sparks. Do not quick-charge the battery (for boost starting) for longer than one minute and do not allow charging voltage to exceed 16.5 volts. Wait at least one minute before boosting the battery a second time.

• Illuminate your work area adequately but safely. Use a portable safety light for working inside or under the car. Make sure the bulb is enclosed by a wire cage. The hot filament of an accidentally broken bulb can ignite spilled fuel or oil.

• Tie long hair behind your head. Do not wear a necktie, a scarf, loose clothing, or a necklace when you work near machine tools or running engines. If your hair, clothing, or jewelry were to get caught in the machinery, severe injury could result.

• Finger rings, bracelets and other jewelry should be removed so that they cannot cause electrical shorts, get caught in running machinery, or be crushed by heavy parts.

• Catch draining fuel, oil, or brake fluid in suitable containers. Do not use food or beverage containers that might mislead someone into drinking from them. Store flammable fluids away from fire hazards. Wipe up spills at once, but do not store the oily rags, which can ignite and burn spontaneously.

• Always observe good workshop practices. Wear goggles when you operate machine tools or work with battery acid. Gloves or other protective clothing should be worn whenever the job requires working with harmful substances.

• The Air-Conditioning (A/C) system is filled with chemical refrigerant, which is hazardous. The A/C system should be serviced only by trained technicians using approved refrigerant recovery/recycling equipment, trained in related safety precautions, and familiar with regulations governing the discharging and disposal of automotive chemical refrigerants.

• Do not expose any part of the A/C system to high temperatures such as an open flame. Excessive heat will increase system pressure and may cause the system to burst.

• Some aerosol tire inflators are highly flammable. Be extremely cautious when repairing a tire that may have been inflated using an aerosol tire inflator. Keep sparks, open flame or other sources of ignition away from the tire repair area. Inflate and deflate the tire at least four times before breaking the bead from the rim. Completely remove the tire from the rim before attempting any repair.

• Friction materials such as brake or clutch discs may contain asbestos fibers. Do not create dust by grinding, sanding, or by cleaning with compressed air. Avoid breathing asbestos fibers and asbestos dust. Breathing asbestos can cause serious diseases such as asbestosis or cancer, and may result in death.

• Greases, lubricants and other automotive chemicals contain toxic substances, many of which are absorbed directly through the skin. Read manufacturer's instructions and warnings carefully. Use hand and eye protection. Avoid direct skin contact.

• An airbag is an explosive device. Handled improperly or without adequate safeguards, it can be accidently activated and cause serious personal injury. To prevent personal injury or airbag system failure, **only trained Audi dealer service technicians** should test, disassemble or service the airbag system.

WARNINGS and CAUTIONS continued on next page ▶

Please read these WARNINGS and CAUTIONS before proceeding with maintenance and repair work.

WARNING (continued) —

• ALWAYS disconnect the power supply before working on the airbag system, or when doing repairs that require removing airbag system components. Disconnect the battery negative (–) terminal and cover the battery.

• The airbag unit should not be removed, nor should any work be performed on the airbag system, without first reviewing the warnings and cautions that accompany the airbag system repair information in this manual.

• Never apply stickers or any other coverings on the steering wheel. Do not let chemical cleaners, oil or grease come into contact with vinyl covering of the airbag unit.

• Never open or otherwise attempt to repair airbag system parts. Always use new parts. Never leave airbag parts or the partially disassembled airbag system unattended.

• Never use a test light or volt meter to conduct electrical tests on the airbag system. The airbag system must only be tested by trained Audi Service technicians using special airbag tester equipment. The airbag system must only be tested while it is installed in the car.

• Do not expose an airbag unit to temperature above 194°F (90°C), even for a brief period, while handling it during the repair process. Keep the unit clear of heat sources such as hot plates, soldering irons, heat lamps, welding equipment, or the like.

• Always replace airbag system components that have been mechanically damaged (bubbles, cracks, etc.).

• Always set down a removed airbag unit so that the horn pad is facing upward.

• When driving or riding in an airbag-equipped vehicle, never hold test equipment in your hands or lap while the vehicle is in motion. Objects between you and the airbag can increase the risk of injury in an accident.

• The storage, transportation and disposal of airbag units are subject to the laws for flammable solids.

CAUTION—

• If you lack the skills, tools and equipment, or a suitable workshop for any procedure described in this manual, we suggest you leave such repairs to an authorized Audi dealer or other qualified shop. We especially urge you to consult an authorized Audi dealer before beginning repairs on any car that may still be covered wholly or in part by any of the extensive warranties issued by Audi of America, Inc.

• Audi is constantly improving its cars and sometimes these changes, both in parts and specifications, are made applicable to earlier models. Therefore part numbers in this manual are used for reference only. Always check with your authorized Audi dealer parts department for the latest information.

• Before starting a job, make certain that you have all the necessary tools and parts on hand. Read all the instructions thoroughly, do not attempt shortcuts. Use tools that are appropriate to the work and use only replacement parts meeting Audi specifications. Makeshift tools, parts and procedures will not make good repairs.

• Use pneumatic and electric tools only to loosen threaded parts and fasteners. Never use these tools to tighten fasteners, especially on light alloy parts. Always use a torque wrench to tighten fasteners to the tightening torque specification listed.

• Be mindful of the environment and ecology. Before you drain the crankcase, find out the proper way to dispose of the oil. Do not pour oil onto the ground, down a drain, or into a stream, pond or lake. Consult local ordinances that govern the disposal of wastes.

• When connecting cables or ground straps, all connections must be cleaned to bare metal. Be sure mounting bolts are tight.

• Do not replace fuse links with fuse links of other gauge or with standard sire. Always replace the entire length of the fuse link.

• Before doing any electrical welding on vehicles equipped with Anti-lock Brakes (ABS), disconnect the battery negative terminal (ground strap) and the ABS Control Module connector.

• On cars equipped with ABS, the ABS control unit cannot withstand temperatures from a paint-drying booth or a heat lamp in excess of 203°F (95°C) and should not be subjected to temperatures in excess of 185°F (85°C) for more than two hours.

• On cars equipped with anti-theft radios, make sure you know the correct radio activation code before disconnecting the battery or removing the radio. If the wrong code is entered into the radio when power is restored, that radio may lock up and be rendered inoperable, even if the correct code is then entered.

• When boost-charging the battery, first remove the fuses for the Engine Control Module (ECM), the Transmission Control Module (TCM), the ABS Control Module, and the On-Board computer. In cases where one or more of these components is not separately fused, disconnect the Control Module connector(s).

• Do not quick-charge the battery (for boost starting) for longer than one minute, and do not exceed 16.5 volts at the battery with the boosting cables attached. Wait at least one minute before boosting the battery a second time.

◀ **WARNINGS and CAUTIONS continued from previous page**

This electrical wiring diagram section has been organized based on the way the electrical wiring diagrams were originally distributed. Each wiring diagram page appears in its original and complete form, including original page numbers or technical bulletin numbers. In addition, every page contains a unique Repair Manual page number to facilitate indexing and easy access to individual circuits.

The electrical wiring diagram section consists of a general information group and comprehensive indexes designated by model year and identified by black page tabs. These indexes list the various components and electrical systems on a particular vehicle model. The actual wiring diagrams themselves follow the indexes.

The diagram below shows important wiring diagram information.

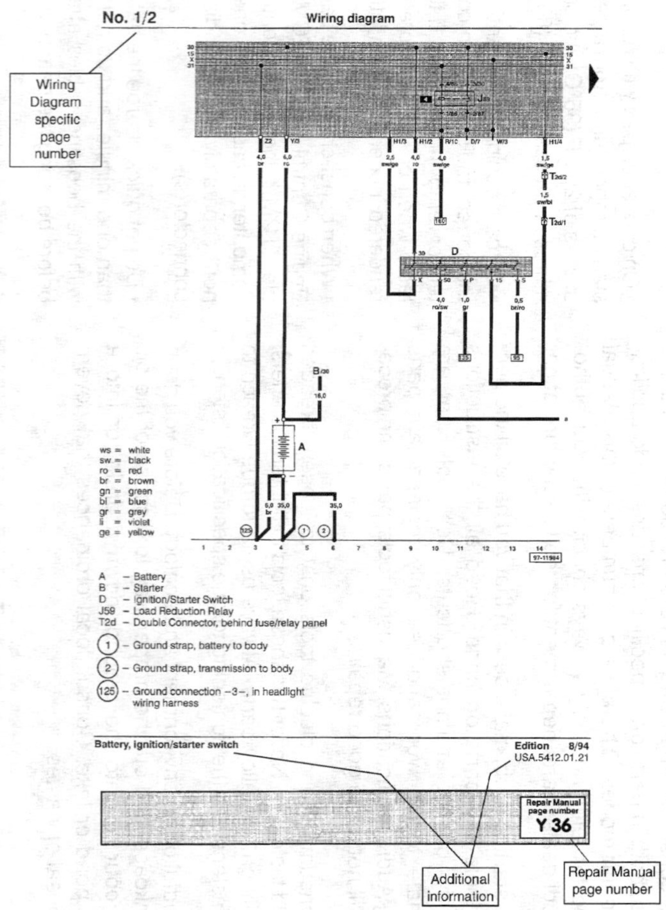

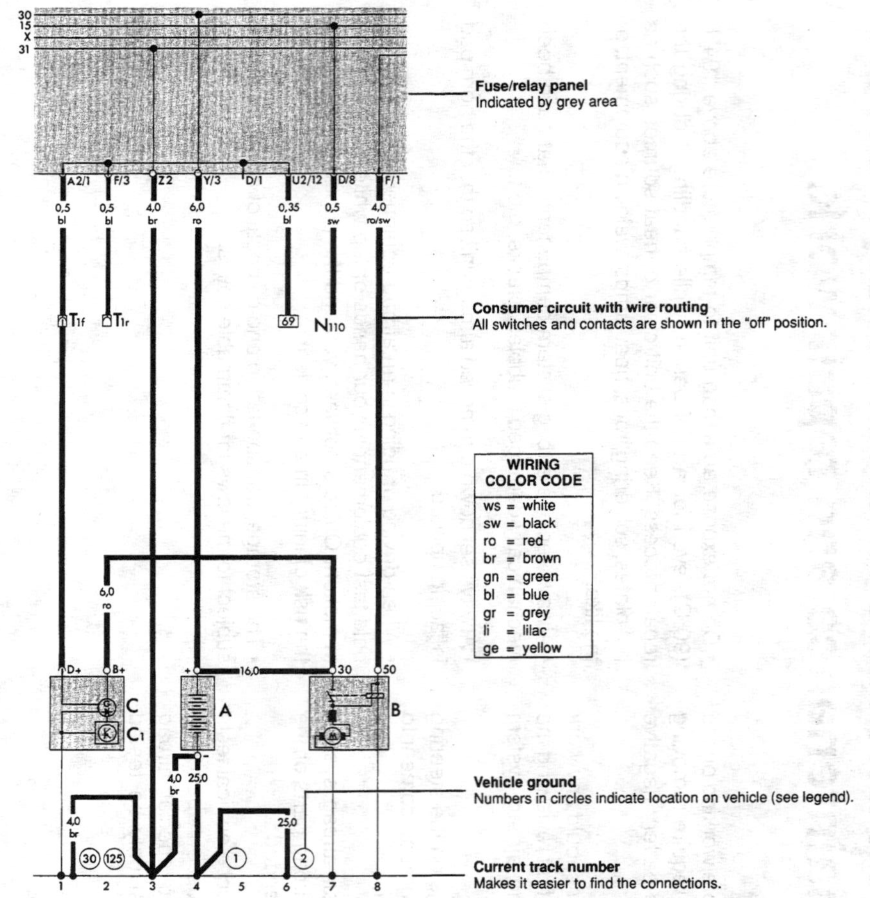

A – Battery
B – Starter
C – Generator (GEN)
C 1 – Voltage Regulator (VR)
N 110 – Shift Lock Solenoid
T 1f – Single connector, on transmission
T 1r – Single connector, behind fuse/relay panel
(1) – Ground strap, battery to body
(2) – Ground strap, transmission to body
(30) – Ground connection -1-, beside fuse/relay panel
(125) – Ground connection -3-, in headlight wiring harness

Legend
In all wiring diagrams the same component designation (code) is used for a particular component; for example, always **A** for battery.

Component identification

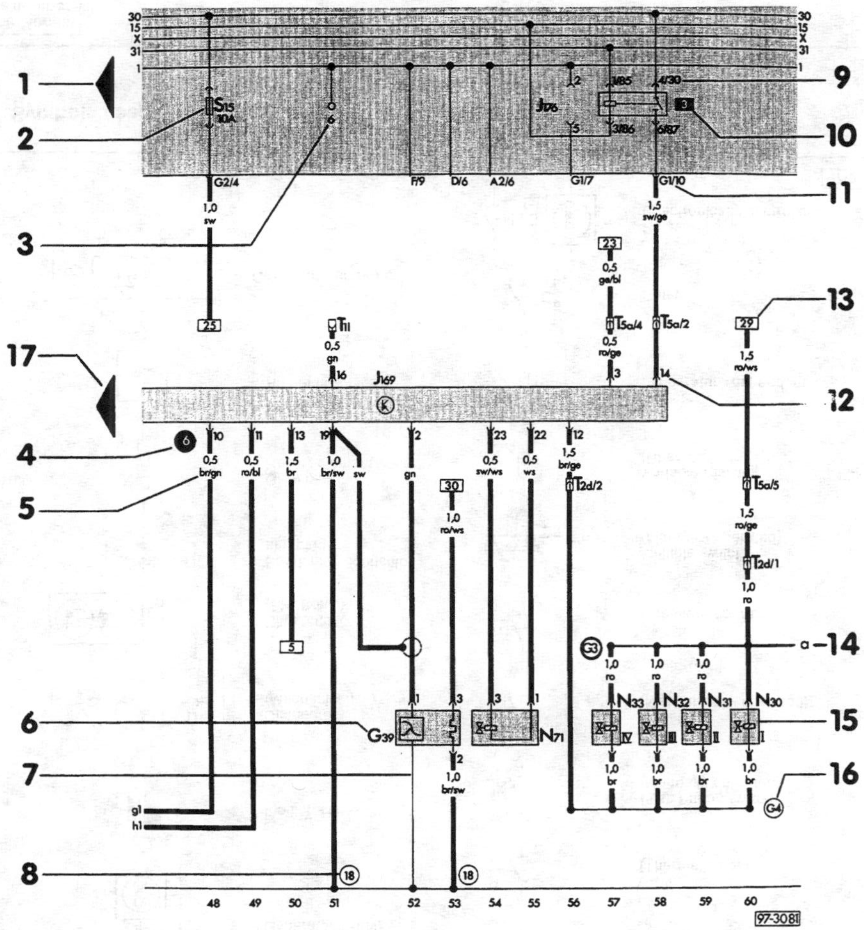

1 – **Arrow**
Indicates wiring circuit is continued on the previous and/or next page.

2 – **Fuse designation**
For example: **S15** = Fuse number **15**, 10 amps, in fuse/relay panel.

3 – **Test points on fuse/relay panel**

4 – **Test points used during troubleshooting**
Numbers in black circles may be found in a regular wiring diagram or in a wiring diagram included with troubleshooting.

5 – **Wire cross-section size (in mm²) and wire colors**
See chart at right for wire size conversion to American Wire Gauge (AWG). Wire color abbreviations are explained in color code chart on each page of the wiring diagram.

6 – **Component designation (code)**
Use legend at bottom of page to identify the component code.

7 – **Internal connections (thin lines)**
These connections are not always wires. Internal connections are current carrying and are listed to allow tracing of current flow inside components and wiring harnesses.

8 – **Wire harness ground connection**
Locations of ground connections are indicated in the legend.

9 – **Relay/control module connectors on fuse/relay panel**
Shows the individual terminals in a multi-point connector.
Example: 4/30
4 = contact **4** at location **3** on relay panel or relay carrier
30 = terminal **30 (B+)** on relay/control module

10 – **Relay location number**
Indicates relay location on fuse/relay panel.

11 – **Fuse/relay panel connectors**
Shows wiring of multi-point or single connectors on fuse/relay panel.
Example: G1/10 = multi-point connector **G1**, terminal **10**

12 – **Terminal designation**
Designation which appears on actual component and/or terminal number of a multi-point connector.

13 – **Wire continuation (current track number)**
Number in frame indicates current track where wire is continued.

14 – **Continuation of internal connections**
Letters indicate where connection continues on the previous and/or next page.

15 – **Component symbols (see page IV and V)**

16 – **Wiring harness wire connection**
Locations of wire connections are indicated in the legend.

17 – **Arrow**
Indicates component is continued on the previous and/or next page.

Wire Size

Wiring Diagrams identify wires by the metric wire size. Metric wire sizes indicate cross-sectional area in square millimeters (mm²). The chart below lists metric wire sizes and their approximate equivalents in American Wire Gauge (AWG) sizes.

Metric size Cross-section (mm²)	American Wire Gauge size (AWG)
0.35	22
0.50	20
0.75	18
1.00	16
1.50	14
2.50	12
4.00	10
6.00	8
16.00	4
25.00	2
35.00	2

Wiring diagram terminal (circuit) identification

Several wiring circuits in the vehicle's electrical system are identified with a number or letter designation. These circuits are identified the same in all wiring diagrams and are most commonly shown near the top of each page in the fuse/relay panel portion of the wiring diagram. The circuit designations may also be used to identify switch connector terminals (switch circuits). Following are the most common numbered/lettered circuits:

Terminal (circuit) 1- Ignition coil/ignition distributor low voltage (typically used as an Engine Speed (RPM) signal for the tachometer)
Terminal (circuit) 15- Switched Battery Positive Voltage (B+) from ignition/starter switch
Terminal (circuit) 30- Battery Positive Voltage (B+), hot at all times
Terminal (circuit) 31- Ground (GND)
Terminal (circuit) 50- Starter control; switched B+ from ignition/starter switch
Terminal (circuit) 56- Switched headlight B+ from light switch
Terminal (circuit) 58- Switched parking light, taillight, illumination B+ from light switch
Terminal (circuit) S (SU)- Key in ignition circuit; switched B+ from ignition/starter switch
Terminal (circuit) X- Load reduction circuit; switched B+ from load reduction relay

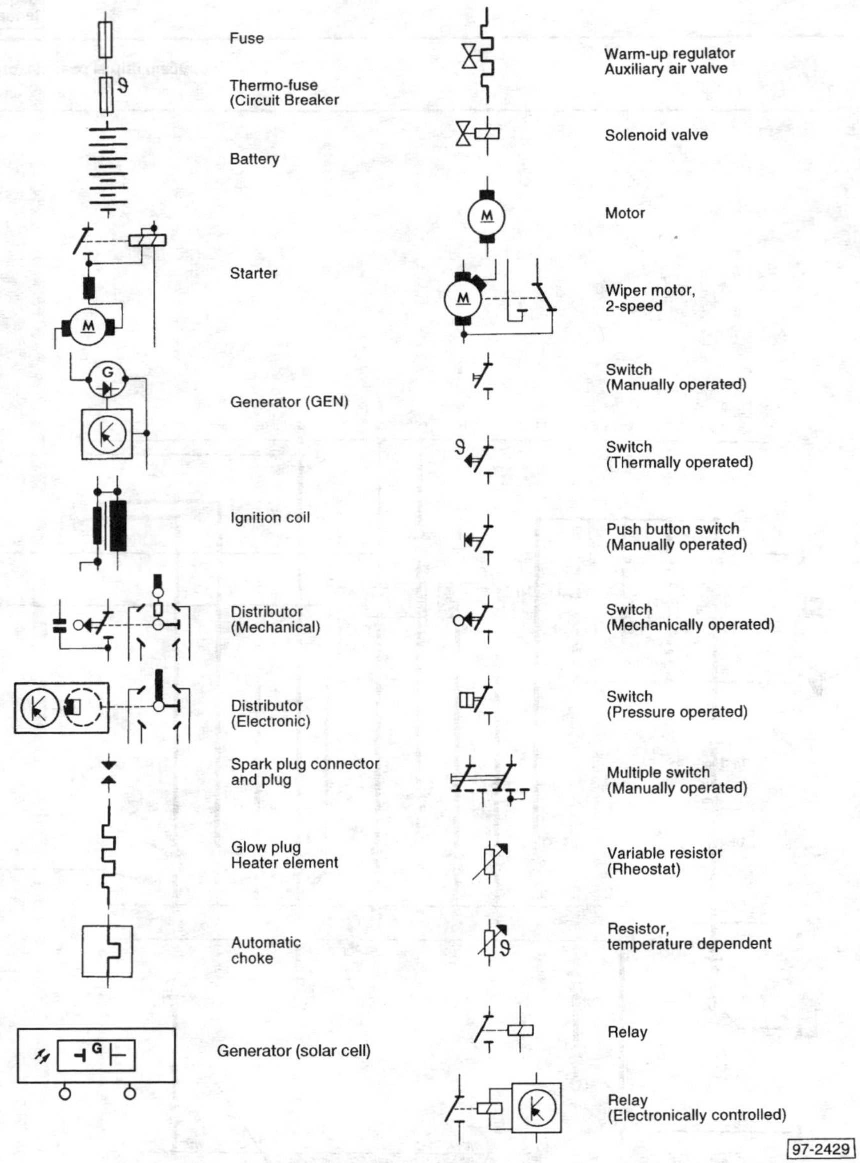

Symbols used in wiring diagrams

General Information

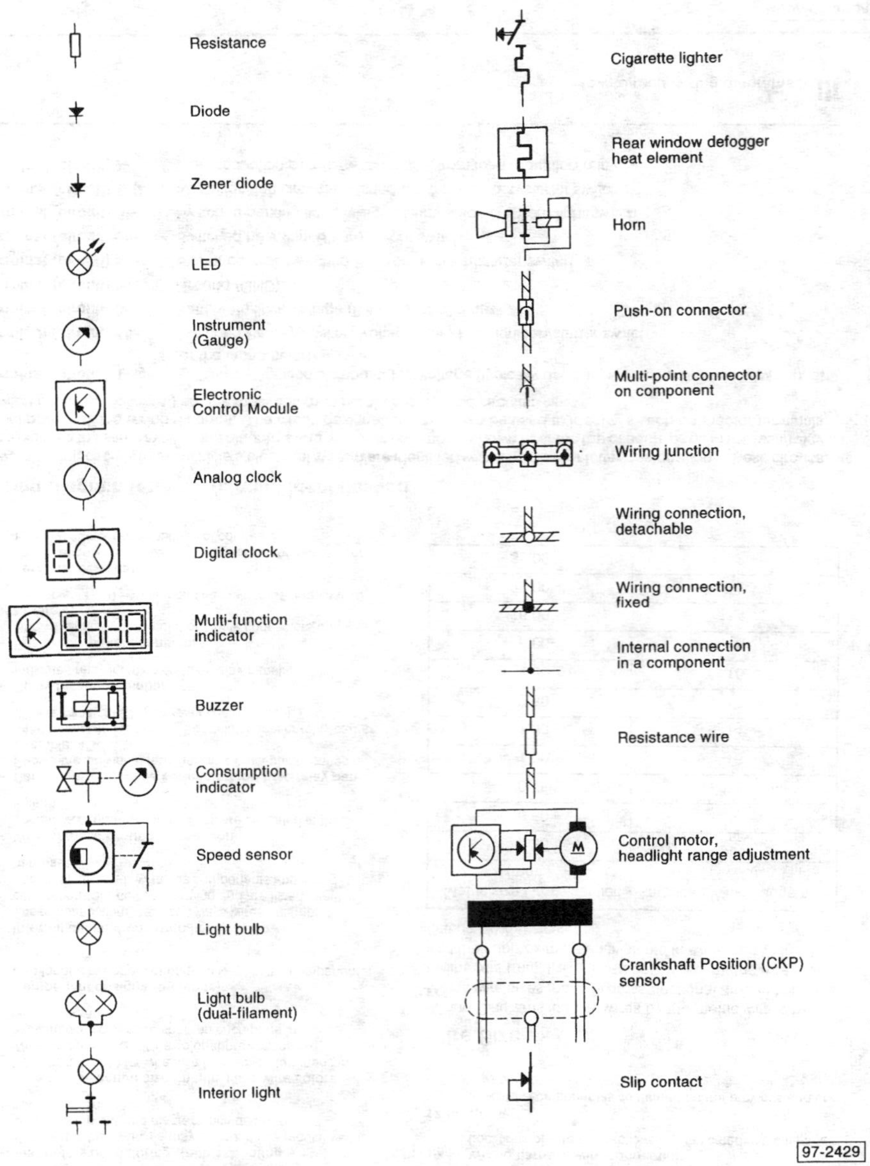

General Information

Symbols used in wiring diagrams

2. Test Equipment

Special Tools

The most basic electrical troubleshooting calls for specialized electrical test equipment. Dealer Service Department technicians do this work using Special Tools, many of which are designed specifically for working on electrical systems. Proper tools are essential to ensure accurate test results and to prevent damage to sensitive components.

The Special Tools shown in this manual are identified by the order numbers that can be used to special-order the tools through your authorized dealer Parts Department.

2.1 LED Test Light

A test light is an inexpensive tool used to perform many simple electrical tests that would otherwise require a multimeter. It lights to indicate when there is Voltage potential between any two test-points in a circuit.

With one probe contacting the battery negative (–) terminal or a known ground (GND) connection, the other probe can be used to check for Voltage at other points in the circuit. The test light will light when the test probe contacts a Voltage source, and current flows from the Voltage source to ground (GND).

With one probe contacting a known source of Voltage, the other probe can be used to check for continuity to ground (GND) at other points in the circuit. Current will flow and the test light will light when the test-point provides the known Voltage source with a path to ground (GND).

Caution

A common test light (with incandescent bulb) may no longer be used. The bulb's high current draw will damage sensitive electronic components.

LED Test Light (US 1115)

Used to check for Voltage reaching components or when searching for malfunctions in a circuit.

Voltage range: 3 to 48 VDC

CAUTION

A common test light (with incandescent bulb) may no longer be used. The bulb's high current draw will damage sensitive electronic components.

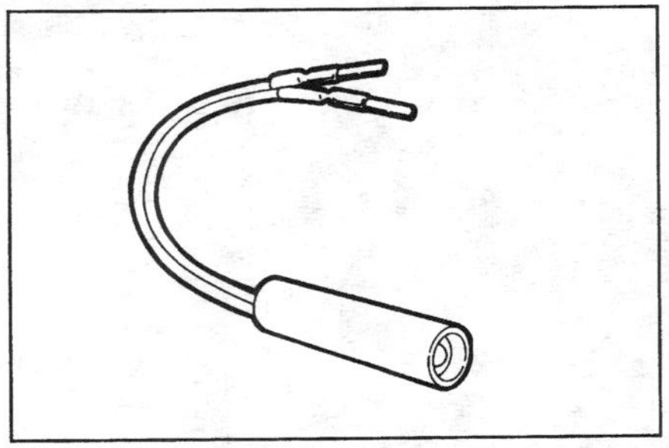

Note

This tool has been superseded by Special Tool VAG 1527B, LED Tester. (US 1115 is still considered an acceptable equivalent to VAG 1527B).

LED Tester (VAG 1527B)

Two-pole Voltage tester with two LEDs (light emitting diodes), suitable for measuring from 3 to 48 Volts DC or AC.

When used to test DC circuits, the positive (+) or negative (–) LED will light, depending on polarity. When used to test AC circuits, both LEDs will light.

Supplied with banana clip and detachable probe-can be used directly with VAG 1598 Test Box.

For troubleshooting all common electrical and electronic components in automobiles. Safe for testing electronic components and circuits because of extremely low current consumption (1.5 mA maximum).

Voltage range: 3 to 48 Volts (AC or DC)

CAUTION

A common test light (with incandescent bulb) may no longer be used. The bulb's high current draw will damage sensitive electronic components.

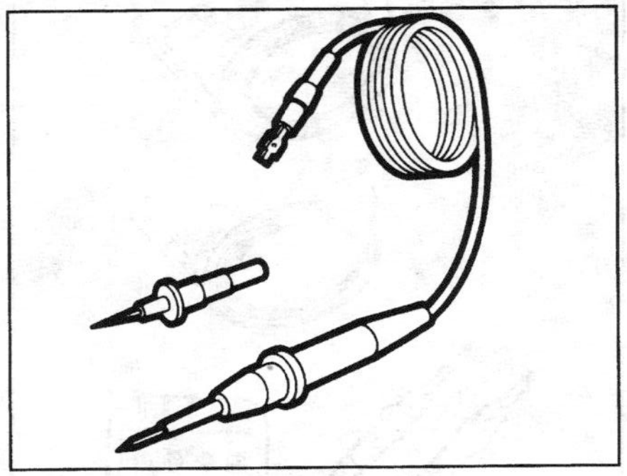

Note

This tool supersedes Special Tool US 1115 LED Test Light.

Making an LED Test Light

A lower-cost alternative to buying an LED Test Light such as US 1115 or VAG 1527B is to make one, using parts available from most any electronics supply outlet.

Assemble the components as shown. Use needle nose pliers to hold the parts and to act as a heat dam while soldering. Insulate the connections with heat-shrinkable tubing or electrical tape.

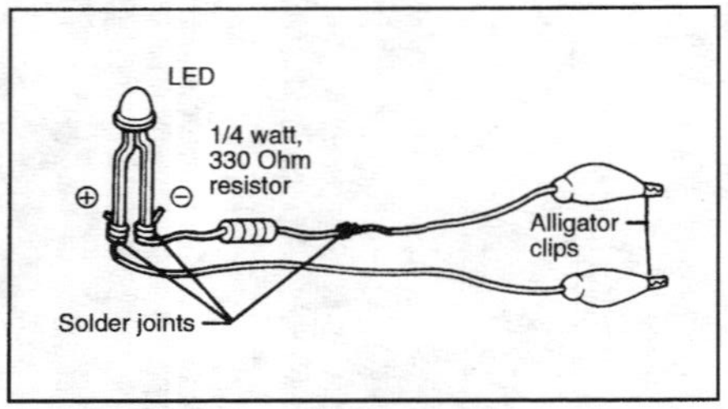

Do-it-yourself LED Test Light for safely testing sensitive solid-state circuits.

Parts:

1. LED (1)
2. 1/4 watt, 330 Ω resistor (1)
3. Alligator clips (2)
4. Wire, Solder and Soldering Iron
5. Heat-shrinkable tubing or electrical tape

2.2 Multimeter or Volt/Ohm Meter (VOM)

A multimeter or Volt-Ohm meter (VOM) is used to measure Voltage, resistance (Ohms) or current (amps or millamps). Two types of meters are in common use. The analog or swing-needle Volt-Ohm meter displays test values according to the position of a needle on the meter face. A digital multimeter displays test values as numbers. The meters shown here and recommended are digital multimeters.

Accuracy is an important consideration when choosing a meter. Analog or swing-needle Voltmeters are generally rated for accuracy as a percentage of full-scale on the meter face. A typical analog meter may be rated as accurate to ± 3% of full scale. Some analog meters offer 10-Volt and 50-Volt scales, with no range in between. For 12-Volt automotive electrical systems, this means testing on the 50-Volt scale, with a corresponding decrease in accuracy. On a 50-Volt scale, ± 3% accuracy is equal to ± 1.5 Volts. While still useful for testing circuits such as lights and horns, analog meters may not be accurate enough for use where precise measurements are required.

Digital multimeters are preferred for precise measurement and for electronics work because most digital multimeters are more accurate then most analog meters. Another advantage of digital multimeters is that they are less likely to be misread, since there is no needle position to be misinterpreted or distorted by reading at an angle.

Perhaps the most important consideration is input impedance-how much of a load the meter places on the circuit being tested. Meters with low input impedance place a greater load on the circuit because they allow more current flow through the meter. This can exceed design limits and be damaging to sensitive electronic components.

Most digital multimeters have very high input impedance-10 megohms (10,000,000 Ω) or higher-and make their measurements while allowing very little current flow through the meter. The meter itself is not a load in the circuit, and does not induce damaging current flow.

Multimeter Kit - digital (US 1119, VAG 1526)

Used for monitoring electronic engine controls (Example: setting/checking the idle speed control system) and for general electrical troubleshooting.

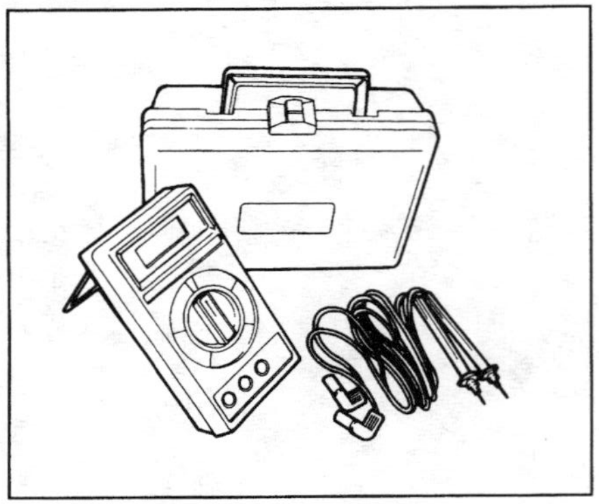

Note

This tool has been superseded by the Fluke 83® Digital Multimeter. (US 1119 or VAG 1526 are still considered acceptable equivalents).

Fluke 83® Digital Multimeter

Used for monitoring electronic engine controls (Example: setting/checking the idle speed control system) and for general electrical troubleshooting.

- 10 Megohm input impedance
- ± 0.3% DC Voltage accuracy
- fully overload-protected

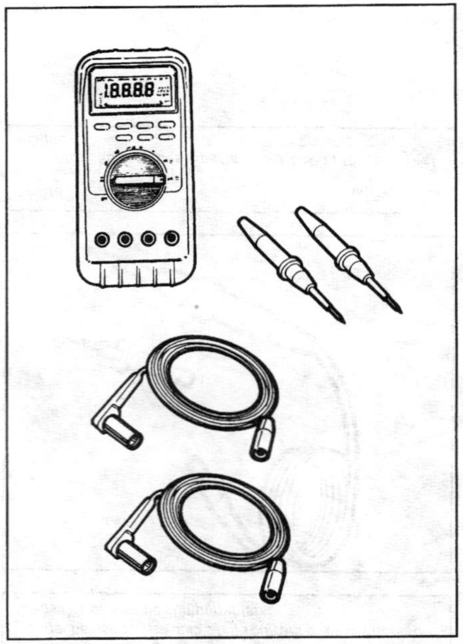

Note

This tool supersedes Special Tool US 1119 Digital Multimeter Kit.

2.3 Test Kits

Adaptor Kit (VW 1594)

Used to connect electrical measuring equipment to various harness connectors in the electrical system.

> ***CAUTION***
>
> *Test connections made carelessly or without proper tools can damage harness connectors, causing poor connections and future problems.*

Test Box (VAG 1598)

Used to perform static and dynamic tests of electrical and electronic systems without damaging miniature contacts, especially where limited access to connectors makes testing difficult.

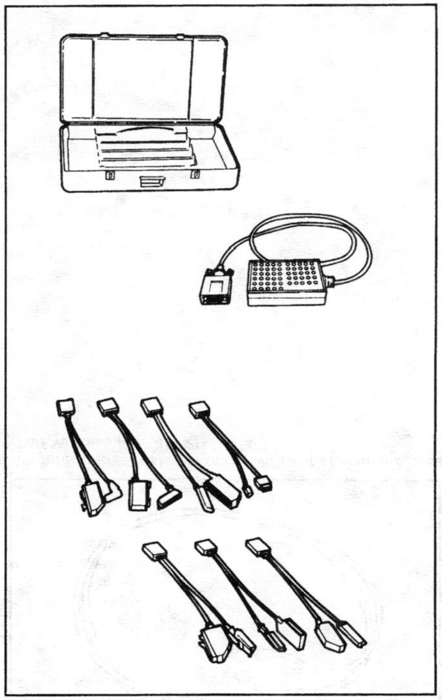

Tester (VAG 1466)

Used to systematically troubleshoot electrical circuits.

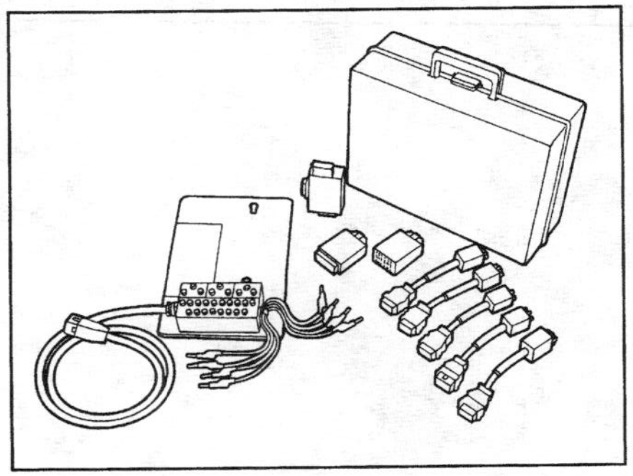

2.4 Memory Saver

VW DOES NOT recommend the use of any device sold or made for the purpose of retaining the memory of vehicle system computers or the radio when the battery is disconnected. These devices are commonly known as "memory savers" or "radio savers" and typically use a 9 volt battery plugged into the cigarette lighter. These devices often do not supply enough voltage to allow memory retention. They can, however, supply voltage high enough to cause accidental and unintended deployment of the airbag systems.

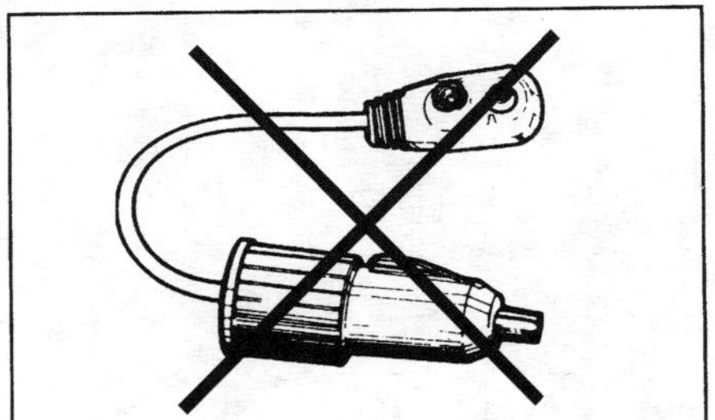

Computer Memory Saver

91-A051

3. Handling Components and Connectors

Harness Connectors

The harness connectors used throughout the vehicle are designed to positively lock into place. Press on wire lock as shown to release the connector.

CAUTION

To disconnect, pull only on the connector body. Never pull on the wires themselves.

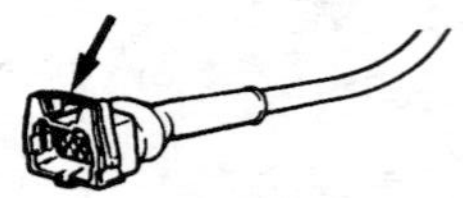

91-A051

Typical wiring harness connector. Push on wire lock (arrow) and gently pull on connector to release.

Making Test Connections

Many electrical troubleshooting tests will require hooking up to wiring harness connectors, or socket connectors on electrical components. Test connections made carelessly or without proper tools can easily damage the connectors, causing poor connections and future problems.

To avoid connector damage, test connections to wiring harness connectors must be made using small, flat-blade terminals which will mate properly with the connectors. The Special Tools recommended include special adapters which can be used to make test connections safely and prevent connector damage. See **2. Test Equipment.**

All test connections to harness connectors should be made using the proper adapter or correct size flat connectors that will not deform the connector cavities.

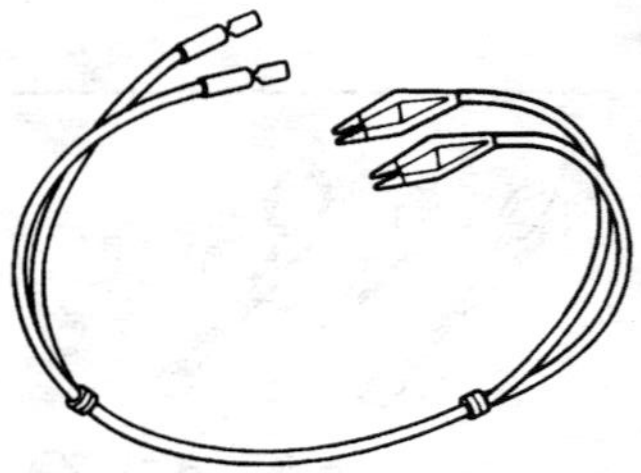

Alligator-clip test leads with flat connectors (Part Number: N 17 457 2)

Wiring Harness Repairs

If repairs must be made to Passat wiring harnesses, Volkswagen specifies the use of the VAS 1978 Wiring Harness Repair Kit. Use of this kit is described in the kit's enclosed operating instructions and in Section W, Electrical Equipment, Repair Group 97.

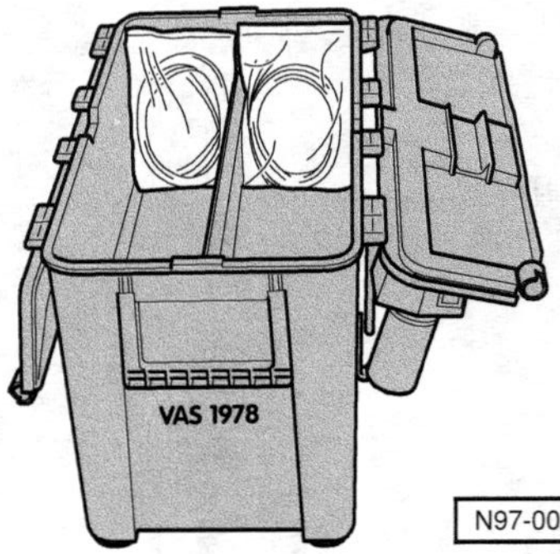

N97-0024

4. Troubleshooting

4.1 Basic Electricity

Electricity is defined by three basic elements: Voltage, Current and Resistance.

Voltage

Voltage is a measure of electromotive force, sometimes referred to as electrical "pressure". It can be described as the difference in potential (potential for the flow of electricity) between any two points in a circuit.

A typical automobile battery, for example, has a difference in potential of about 12 Volts between the positive (+) terminal and the negative (–) terminal.

The basic units of electrical potential are **Volts** (V). Very low Voltages are expressed as **millivolts** (mV).
1 V = 1000 mV; 1 mV = .001 V

Current

Current is the term describing the flow of electricity through a conductor. In a complete circuit, potential (Voltage) will cause current to flow from positive (+) to negative (–).

The basic units of current flow are amperes or **amps** (A). Small amounts of current flow are often measured in **millamps** (mA).
1 A = 1000 mA; 1 mA = .001 A

Resistance

Resistance resists or opposes the flow of electricity. Conductors are made from materials of low resistance that allow electricity to flow easily. Insulators are materials of very high resistance that inhibit the low of electricity.

The basic unit of resistance is the **Ohm** Ω. High resistance values are often expressed as Kilohms (KΩ).
1 KΩ = 1000Ω

Resistance vs. Current Flow

The basic rule of electricity (Ohm's Law) states that one unit of force (1 Volt) is required for one unit of current (1 am) to flow against one unit of resistance (1 Ohm). From Ohm's Law, we also know that:

Voltage = Current x Resistance

When Voltage is approximately constant, as in an automobile electrical system, current and resistance affect each other. As resistance increases, there will be less current flow. And lower resistance will permit higher current flow.

Higher resistance = lower current flow
Example: Corrosion on a headlight connector (higher resistance) causes the light to be dim (lower current flow).

Lower resistance = higher current flow
Example: A damaged wire shorted to ground (GND) (lower resistance) overloads circuit capacity (higher current flow) and blows a fuse.

Definition of a Circuit

Four things are required for current to flow in any electrical circuit, and for that circuit to function as intended:

- **Power Source** (Voltage)
- **Conductors** (wires, printed circuits, etc.)
- **Load or Consumer** (a user of electrical power)
- **Complete Circuit** (a connection to ground [GND])

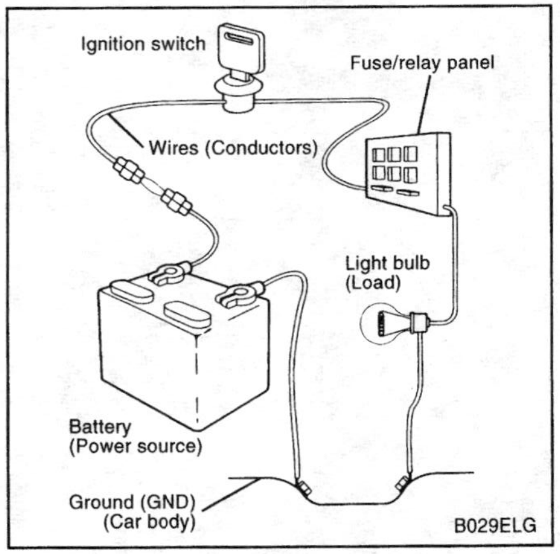

A complete circuit

Open Circuits

An open circuit is an incomplete circuit. An open circuit occurs when some kind of malfunction interrupts the circuit path and prevents current flow. Some common causes of open circuits are:

- broken wire
- loose or disconnected connector
- loose or damaged connector terminal
- corrosion
- malfunctioning fuse or component

Test for an open circuit by checking continuity using an Ohmmeter (multimeter), or by checking for Voltage at various points of the circuit using a test light or Voltmeter (multimeter). See **4.4 Checking Wiring and Components.**

Short Circuits

A short circuit is an unintended complete circuit. A short circuit occurs when some kind of malfunction causes current flow to follow the wrong path.

A short circuit to ground (GND) (grounded circuit) may prevent Voltage from reaching a component. If Voltage is shorted directly to ground (GND), bypassing any load, the unrestricted current flow will damage fuses wires or components. Some common causes of short circuits are:

- damaged wire or wiring harness
- malfunctioning insulation
- internally damaged component
- incorrect connection

Test for a short circuit to ground (GND) using a multimeter or a test light to indicate circuit malfunctions and abnormal current flow paths. See **4.5 Checking for Short Circuit to Ground (GND).**

Troubleshooting Procedure

Verify the complaint – Check the complaint. Try to understand the problem. If possible, let the driver show you what happens. Check all functions of the system and note the symptoms before starting any testing or disassembly.

Analyze the problem – Identify the part of the electrical system that is most likely to be causing the problem. Find the Circuit Section in the manual that applies to that part of the system. Find the wiring diagram that applies to the vehicle. By following the circuit from a ground (GND) back to the power source, get an understanding of how the circuit works.

Find the problem – You will find the problem if you follow a simple and logical step-by-step procedure. Test portions of the circuit one at a time, starting with he area or component most likely to be malfunctioning. Test first at points that you can reach most easily.

Repair the problem – When you find the cause of the problem, make the repair. Use appropriate tools and procedures.

Check the results – Be sure it works. Check the functions of all parts of the circuit that you worked on.

Working on the Electrical System

A test light or a multimeter can be very helpful for testing circuits. See **2. Test Equipment** for more information on test equipment.

Current flow is logical, always moving from the highest potential at the Voltage source (+) toward the lowest potential at ground (–). Using a wiring diagram to trace a circuit, you should start with the ground (GND) and then follow the wires back to the source of power.

To troubleshoot a circuit:

1. Inspect all connections, especially grounds (GND). Make sure they are clean, tight and corrosion-free.

2. Check the fuses.

> ***Note***
>
> *Repeated fuse failures are the sign of a malfunctioning wire, a failed component, or a short to ground (GND) somewhere in the circuit.*

3. Check for Voltage reaching particular components or points in the circuit.

4. Check continuity between points to look for breaks in the circuit (open circuit).

5. Check Voltage drop at connections, especially ground (GND) connections.

4.2 Checking for voltage

Checking for Voltage confirms that the circuit is uninterrupted between the Voltage source and the test point. The example illustrates troubleshooting the high-speed circuit for the radiator cooling fan.

Voltage: If the test light or multimeter indicates Voltage potential, then the circuit between the Voltage source and terminal 1 of the fan connector is OK.

NO Voltage: Power is not reaching the fan connector. The fan is probably OK. Look for a malfunction somewhere between the Voltage source and the fan connector. (Example: Check for Voltage reaching terminal E87La of the fuse/relay panel)

> ***CAUTION***
>
> *Direct contact with meter probes at the connector terminals can easily damage the small contacts, causing poor connections and risking future intermittent malfunctions.*
>
> *Special Tools include adapters for making test connections safely and preventing connector damage. See **2. Test Equipment.***

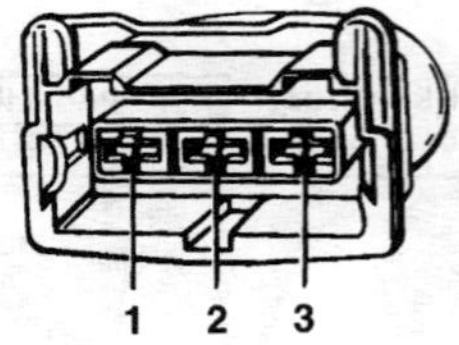

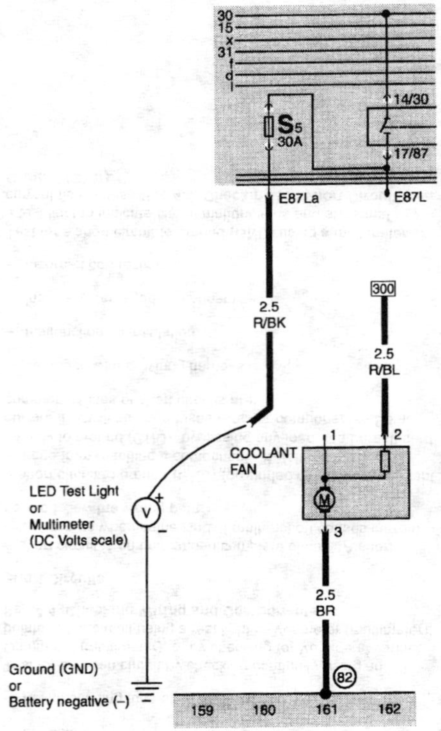

4.3 Checking Ground (GND) Connections

Checking ground (GND) connections as shown confirms that the circuit is complete - that the necessary path to ground (GND) is uninterrupted and current can flow in the circuit. The example illustrates troubleshooting the high-speed circuit for the radiator cooling fan.

Voltage: If the test light or multimeter indicates Voltage, then there is potential for current flow between the known Voltage source and ground (GND) at the test point. The ground (GND) side of the circuit, between terminal 3 of the fan connector and battery negative (–), is OK.

NO Voltage: The test point is not providing a path that completes the circuit to ground (GND). The fan is probably OK. Look for a malfunction somewhere in the wiring between the fan connector and chassis ground (GND). Also check the mechanical ground connection at the chassis (body).

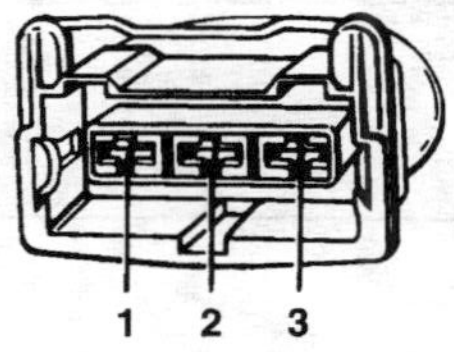

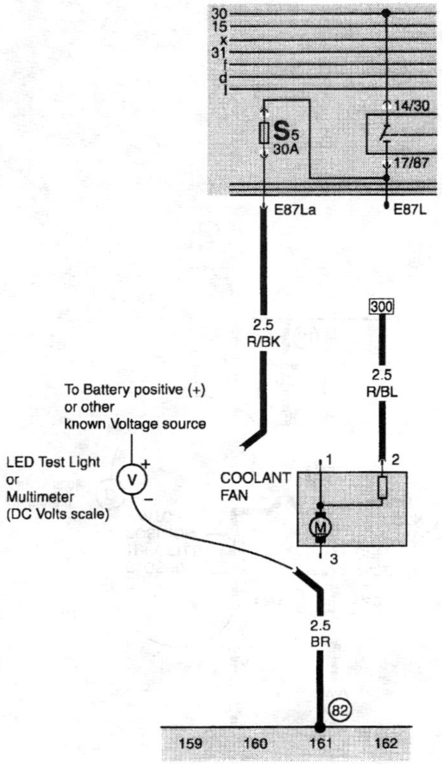

Continuity (approximately 0 Ω): Little or no resistance indicates that there is a continuous conductive path between the two test points - the circuit's ground (GND) path between terminal 3 and battery negative (–) is OK.

No Continuity: There is resistance to current flow in the ground (GND) side of the circuit. The fan is probably OK. Look for a malfunction somewhere in the wiring between the fan connector and chassis ground (GND). Also check the mechanical ground (GND) connection at the chassis (body).

> ***CAUTION***
>
> *direct contact with meter probes at the connector terminals can easily damage the small contacts, causing poor connections and risking future intermittent malfunctions.*
>
> *Special Tools include adapters for making test connections safely and preventing connector damage. See **2. Test Equipment.***

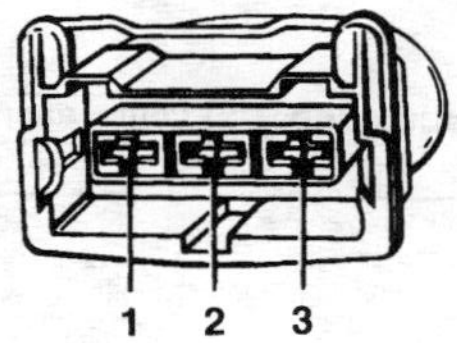

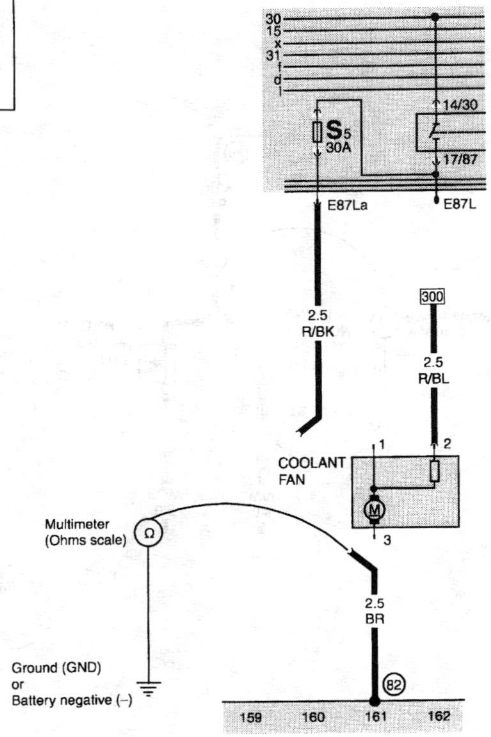

4.4 Checking Wiring and Components

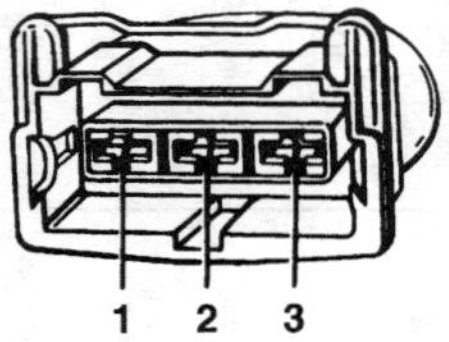

Resistance or Continuity

Checking a portion of the wiring harness or a component as shown indicates whether or not there is a continuous conductive path - whether current can flow between the two test points. The example illustrates troubleshooting the Close Throttle Position switch and the Wide Open Throttle Position switch in the fuel injection system.

> **CAUTION**
>
> *Resistance measurements and continuity checks must always be made with all power to the circuit or component switched OFF. When testing continuity in a circuit that is always powered (fuse/relay panel "30" circuit for example) disconnect the battery before testing.*
>
> *Always use a digital (low current) meter. An ohmmeter, or the Ohms scale of a multimeter, measures resistance by passing a small amount of current through the circuit or component being checked.*
>
> *Improper testing may damage sensitive electronic components.*

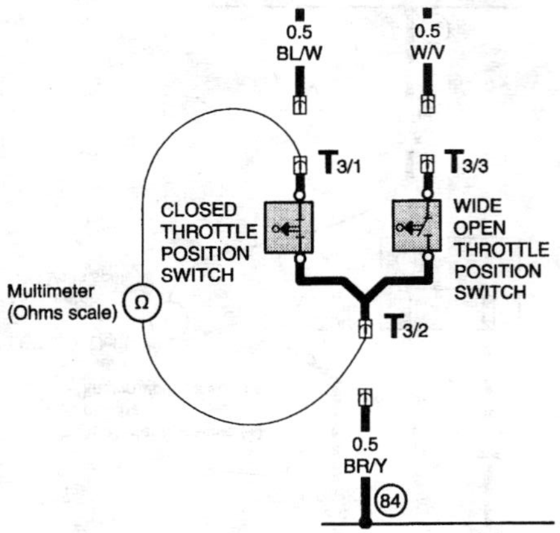

Continuity (approximately 0 Ω): Little or no resistance indicates that there is a continuous conductive path between the two test points. As shown, this is correct for the normally Closed Throttle Position switch in the Closed Throttle Position (switch not actuated). Also check that the switch opens (no continuity) when actuated by the throttle.

No Continuity: There is resistance to current flow through the switch. In the example, this indicates that the normally Closed Throttle Position switch or the wires between it and the connector (t3) are malfunctioning.

> **CAUTION**
>
> *Direct contact with meter probes at the connector terminals can easily damage the small contacts, causing poor connections and risking future intermittent malfunctions.*
>
> *Special Tools include adapters for making test connections safely and preventing connector damage. See **2. Test Equipment**.*

Continuity (approximately 0 Ω): Little or no resistance indicates that there is a continuous conductive path between the two test points. In the example, this indicates that the normally Wide Open Throttle Position switch is malfunctioning.

No Continuity: There is no connection-an open circuit. As shown, this is correct for the normally Wide Open Throttle Position switch. Also check that the switch closes, completing the circuit, when actuated by the Wide Open Throttle Postion switch.

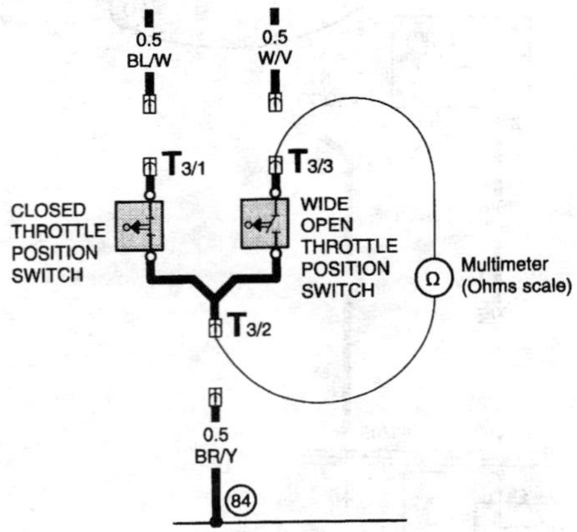

Voltage Drop

Checking Voltage drop across connections or components as shown will indicate whether there is abnormal resistance creating an additional load in the circuit - consuming power and dropping the Voltage available to other parts of the circuit.

> **Note**
>
> *Voltage drop measurements can only be made when the circuit is powered and there is normal current flow.*

The example illustrates troubleshooting the back-up light switch. The switch is in the circuit to switch power to the back-up lights On and Off. When Reverse gear is selected and the switch is closed, it should have very little resistance and not be a consumer.

If dirt or corrosion on the switch contacts creates resistance, some of the available battery Voltage goes into overcoming that resistance. Less Voltage is available to light the back-up lights, and they will not be as bright.

> **Note**
>
> *An ohmmeter can measure resistance (or check continuity) only when the circuit is not powered, i.e. when there is almost no load.*

> ***CAUTION***
>
> *Direct contact with meter probes at the connector terminals can easily damage the small contacts, causing poor connections and risking future intermittent malfunctions.*
>
> *Special Tools include adapters for making test connections safely and preventing connector damage. See **2. Test Equipment**.*

Troubleshooting | General Information

Low Voltage: A low Voltage reading across the two switch contacts indicates almost no difference in potential. Resistance across the switch is low - most of the battery Voltage is passing through the switch and is still available to power the lights.

High Voltage: Any significant Voltage reading indicates a difference in potential across the switch contacts. Excessive resistance is loading the circuit, causing a Voltage drop - Voltage is consumed overcoming the resistance of the switch, and less is left to power the lights.

Maximum allowable Voltage drops recommended by the Society of Automotive Engineers are:

- 0.0 Volt for small wire connections
- 0.1 Volt for high current connections (example: fuel pump, headlights)
- 0.1 Volt for ground (GND) connections
- 0.2 Volt for high-current cables (example: battery/starter cable)
- 0.3 Volt for switch or relay contacts

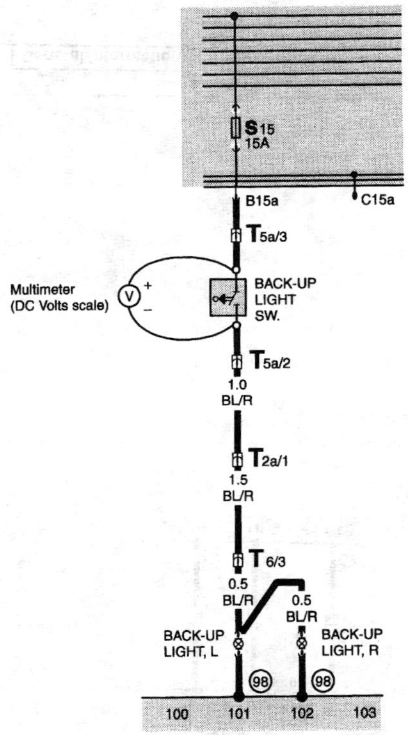

General Information | Troubleshooting

4.5 Checking for Short Circuit to Ground (GND)

Checking the circuit as shown will detect circuit malfunctions that are providing an unintended current flow path to ground (GND). The examples illustrate two methods of troubleshooting a short to ground (GND) that is causing a blown fuse in the circuit powering the license plate lights and the glove compartment light.

Using an Ohmmeter (Multimeter)

Step 1 - Remove the fuse

Step 2 - Disconnect the load (powered components) to eliminate the circuit's normal path to ground (GND)

> ***CAUTION***
>
> *Always use a digital (low current) meter. An ohmmeter, or the Ohms scale of a multimeter, measures resistance by passing a small amount of current through the circuit or component being checked. Improper testing may damage sensitive electronic components.*

Continuity (approximately 0 Ω): Little or no resistance indicates that there is a continuous conductive path between the isolated circuit and ground (GND), even though all the circuit's normal ground (GND) paths are eliminated. There is a short - an unintentional connection to ground (GND) - somewhere in the circuit.

No Continuity: The circuit's normal ground (GND) paths have been disconnected, and there is no other connection between the isolated circuit and ground (GND) - no short has been detected.

> ***CAUTION***
>
> *Direct contact with meter probes at the connector terminals can easily damage the small contacts, causing poor connections and risking future intermittent malfunctions.*
>
> *Special Tools include adapters for making test connections safely and preventing connector damage. See **2. Test Equipment.***

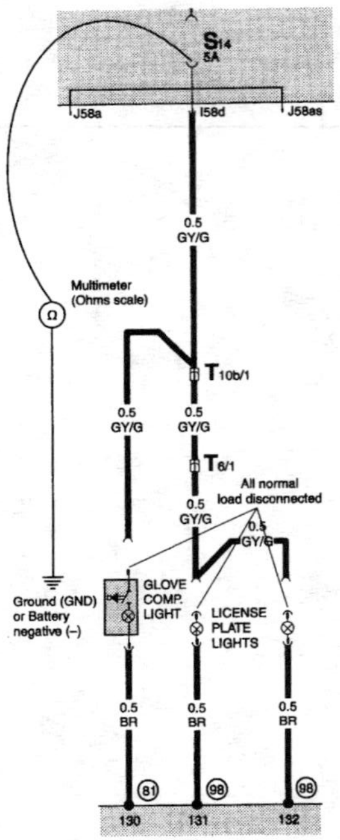

Using an LED Test Light or Voltmeter (Multimeter)

Step 1 - Remove the fuse

Step 2 - Disconnect the load (powered components) to eliminate the circuit's normal path to ground (GND)

Voltage: If the test light or multimeter indicates Voltage, then there is a complete circuit - a connection to ground (GND) even though all the circuit's Normal ground (GND) paths are eliminated. There is a short - an unintentional connection to ground (GND) - somewhere in the circuit.

NO Voltage: There is not a complete circuit. The circuit's normal ground (GND) paths have been disconnected, and there is no other connection between the isolated circuit and ground (GND) - no short has been detected.

> ***CAUTION***
>
> *Direct contact with meter probes at the connector terminals can easily damage the small contacts, causing poor connections and risking future intermittent malfunctions.*
>
> *Special Tools include adapters for making test connections safely and preventing connector damage. See **2. Test Equipment.***

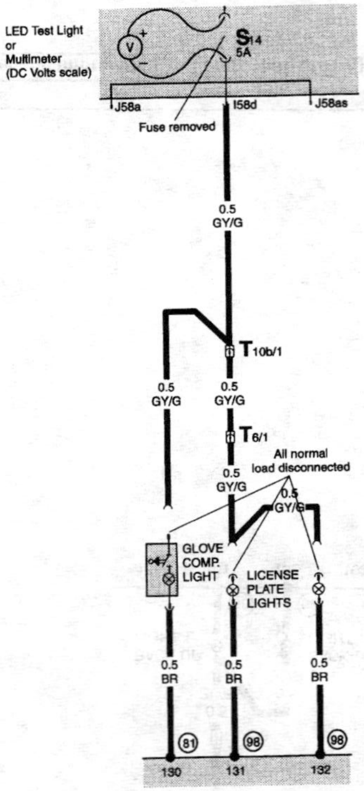

1.8 Liter engine (code AEB), 1998 m.y.

	Wiring Diagram page number	Repair Manual page number
● **Passat (1.8L-Engine - Motronic Multiport Fuel Injection (MFI)/110 kW), code AEB, from September 1997**	2/1	Y 75
- Battery, starter, generator (GEN)	2/2	Y 76
- Ignition system, motronic engine control module (ECM)	2/3	Y 77
- Motronic engine control module (ECM), barometric pressure (BARO) sensor, camshaft position (CMP) sensor, engine coolant temperature (ECT) sensor	2/4	Y 78
- Motronic engine control module (ECM), knock sensor (KS) 1, throttle valve control module, intake air temperature (IAT) sensor	2/5	Y 79
- Motronic engine control module (ECM), engine speed sensor, knock sensor (KS) 2	2/6	Y 80
- Motronic engine control module (ECM), fuel injectors, fuel pump (FP) relay	2/7	Y 81
- Motronic engine control module (ECM), mass air flow (MAF) sensor, heated oxygen sensor (HO2S), oxygen sensor (O2S)	2/8	Y 82
- Motronic engine control module (ECM), wastegate bypass regulator valve, leak detection pump (LDP)	2/9	Y 83
- Oil pressure switch, fuel pump (FP), sender for fuel gauge, speedometer vehicle speed sensor	2/10	Y 84
- Instrument cluster, engine coolant temperature (ECT) gauge, low fuel level warning light, oil pressure warning light, Malfunction Indicator Lamp (MIL)	2/11	Y 85
- Instrument cluster, multi-function indicator (MFI), multi-function indicator switches, spedometer, outside air temperature sensor	2/12	Y 86

2.8 Liter engine (code AHA), 1998 m.y.

	Wiring Diagram page number	Repair Manual page number
● **Passat (2.8L-Engine - Motronic Multiport Fuel Injection (MFI)/142 kW), code AHA, from September 1997**	5/1	Y 111
- Battery, starter, generator (GEN)	5/2	Y 112
- Starting interlock, battery	1/2	Y 56
- Ignition/starter switch	1/3	Y 57
- Ignition system, motronic engine control module (ECM), throttle valve control module	5/3	Y 113
- Motronic engine control module (ECM), engine speed sensor (RPM), knock sensor (KS)	5/4	Y 114
- Motronic engine control module (ECM), camshaft position (CMP) sensor, engine coolant temperature (ECT) sensor, intake air temperature (IAT) sensor	5/5	Y 115
- Motronic engine control module (ECM), mass air flow (MAF) sensor, heated oxygen sensors (HO2S)	5/6	Y 116
- Motronic engine control module (ECM), oxygen sensor (O2S), oxygen sensor (O2S) 2, valve for camshaft adjustment	5/7	Y 117
- Motronic engine control module (ECM), secondary air injection (AIR) system, Data Link Connector (DLC)	5/8	Y 118
- Motronic engine control module (ECM), fuel injectors, fuel pump (FP) relay	5/9	Y 119
- Leak detection pump (LDP), evaporative emission (EVAP) canister purge regulator valve, intake manifold change-over valve	5/10	Y 120
- Oil pressure switch, fuel pump (FP), sender for fuel gauge, speedometer vehicle speed sensor	5/11	Y 121
- Instrument cluster, engine coolant temperature (ECT) gauge, low fuel level warning light, oil pressure warning light, Malfunction Indicator Lamp (MIL)	5/12	Y 122
- Instrument cluster, multi-function indicator (MFI), multi-function indicator switches, speedometer, outside air temperature sensor	5/13	Y 123

Electrical Wiring Diagram Index

1998 Passat, standard equipment (green instrument illumination)

1998 Passat, standard equipment (blue instrument illumination)

1998 Passat, additional electrical wiring diagrams

THIS PAGE INTENTIONALLY LEFT

BLANK

Electrical Wiring Diagram Index

1.8 Liter engine (code AEB), 1999 m.y.

2.8 Liter engine (code AHA), 1999 m.y.

1999 Passat, standard equipment

	Wiring Diagram page number	Repair Manual page number
● **Passat Standard Equipment, only for cars with blue illuminated instrument cluster, from September 1997**	1/1	Y 55
- Starting interlock, battery	1/2	Y 56
- Ignition/starter switch	1/3	Y 57
- Instrument cluster, Data Link Connector (DLC), generator warning light	1/4	Y 58
- Instrument cluster, tachometer, speedometer vehicle speed sensor, Malfunction Indicator Lamp (MIL)	1/5	Y 59
- Instrument cluster, left and right turn signal indicator lights, parking brake warning light switch and brake fluid level warning light switch	1/6	Y 60
- Turn signal switch, headlight dimmer/flasher switch, daytime running lights change-over relay, series resistance wiring	1/7	Y 61
- Left front headlights, left front fog light, fog light relay	1/8	Y 62
- License plate light	1/9	Y 63
- Emergency flasher switch	1/10	Y 64
- Right front headlights, right front fog light	1/11	Y 65
- Left brake/tail light, left rear turn signal light, left back-up light, high-mount brake light	1/12	Y 66
- Brake light switch, back-up light switch, right rear turn signal light, right back-up light, right brake/tail light	1/13	Y 67
- Radio connection, cigarette lighter, panel vent illumination	1/14	Y 68
- Light switch	1/15	Y 69
- Instrument panel light dimmer switch, dual horns, dual horn relay	1/16	Y 70
- Fresh air blower switch, switch for fresh air blower and recirculated air, fresh air blower	1/17	Y 71
- Heated rear window, rear window defogger switch	1/18	Y 72
- Windshield wiper motor, windshield wiper intermittent regulator, rear window wiper motor	1/19	Y 73
- Wiper/washer intermittent relay, windshield wiper/washer switch, windshield and rear window washer pump, headlight washer	1/20	Y 74

1999 Passat, additional electrical wiring diagrams

	Wiring Diagram page number	Repair Manual page number
● **Airbag systems, only for vehicles with blue illuminated instrument cluster,** from September 1997	17/1	Y 193
● **Air conditioning, 1.8L-Engine - Motronic Multiport Fuel Injection (MFI)/110 kW, code AEB,** from September 1997	15/1	Y 179
● **Climatronic, 2.8L-Engine - Motronic Multiport Fuel Injection (MFI)/142 kW, code AHA,** from August 1998	33/1	Y 327
● **Air conditioning, 2.8L-Engine - Motronic Multiport Fuel Injection (MFI)/142 kW, code AHA,** from September 1997	15/1	Y 179
● **Air conditioning, 2.8L-Engine - Motronic Multiport Fuel Injection (MFI)/142 kW, code AHA,** from August 1998	43/1	Y 427
● **Anti-lock brake system (ABS) with electronic differential lock (EDL),** from September 1997	12/1	Y 157
● **Anti-lock brake system (ABS) with electronic differential lock (EDL) and anti-slip control (ASC)** from September 1997	11/1	Y 153
● **Automatic day/night interior mirror,** from August 1998	27/1	Y 265
● **Automatic transmission, 5 speed automatic with tiptronic, 1.8L-Engine - Motronic Multiport Fuel Injection (MFI)/110 kW, code AEB,** from August 1998	32/1	Y 320
● **Automatic transmission, 5 speed automatic with tiptronic, 2.8L-Engine - Motronic Multiport Fuel Injection (MFI)/142 kW, code AHA,** from August 1998	31/1	Y 312
● **Central locking system, see Comfort System**		
● **Comfort System, with anti-theft alarm system, only for vehicles with blue illuminated instrument cluster,** from September 1997	13/1	Y 161
● **Comfort System, with anti-theft alarm system,** from August 1998	29/1	Y 279
● **Coolant fan, 80 W,** from September 1997	6/1	Y 125
● **Coolant fan, two speeds, 300 W,** from September 1997	7/1	Y 127
● **Cruise control 1.8L/2,8L-Engine - Motronic Multiport Fuel Injection (MFI),** from September 1997	18/1	Y 197
● **Dual horns,** from September 1997	1/16	Y 70
● **Front interior light, see Comfort System**		
● **Fuel tank lid unlock system, see Comfort System**		
● **Heated seats,** from September 1997	20/1	Y 205
● **Headlight washer,** from September 1997	1/20	Y 74
● **Luggage compartment light, see Comfort System**		
● **Make-up mirror lights, see Comfort System**		
● **Power outlets (12V),** from September 1997	39/1	Y 410
● **Power seats, without Memory,** from November 1997	26/1	Y 261
● **Power seats and mirror adjustment with Memory,** from November 1997	25/1	Y 253
● **Power sunroof, see Comfort System**		
● **Radio System,** from August 1998	34/1	Y 335
● **Reading lights, see Comfort System**		
● **Rear lid unlock system, see Comfort System**		
● **Rear window wiper/washer system,** from September 1997	1/19	Y 73
● **Rainsensor,** from August 1998	27/1	Y 265
● **Warning light for rear lid unlocked, see Comfort System**		
● **Washer nozzle heaters,** from September 1997	14/1	Y 177

Electrical Wiring Diagram Index

1.8 Liter engine (code ATW), 2000 m.y.

2.8 Liter engine (code ATQ), 2000 m.y.

2000 Passat, Standard equipment

	Wiring Diagram page number	Repair Manual page number
● **Passat Standard Equipment, from August 1999**	35/1	Y 339
- Starting interlock, battery	35/2	Y 340
- Ignition/starter switch	35/3	Y 341
- Instrument cluster, Data Link Connector (DLC)	35/4	Y 342
- Instrument cluster, tachometer, speedometer vehicle speed sensor, Malfunction Indicator Lamp (MIL), seat belt warning system	35/5	Y 343
- Instrument cluster, generator warning light, outside air temperature sensor	35/6	Y 344
- Instrument cluster, left and right turn signal indicator lights, multi-function indicator switch	35/7	Y 345
- Windshield washer fluid level sensor, daytime running lights change-over relay, parking brake warning light switch and brake fluid level warning light switch	35/8	Y 346
- Turn signal switch, headlight dimmer/flasher switch	35/9	Y 347
- Left front headlights, left front fog light, fog light relay	35/10	Y 348
- License plate light	35/11	Y 349
- Emergency flasher switch	35/12	Y 350
- Right front headlights, right front fog light	35/13	Y 351
- Left brake/tail light, left rear turn signal light, left back-up light, high-mount brake light	35/14	Y 352
- Brake light switch, back-up light switch, right rear turn signal light, right back-up light, right brake/tail light	35/15	Y 353
- Radio connection, cigarette lighter, panel vent illumination	35/16	Y 354
- Light switch	35/17	Y 355
- Instrument panel light dimmer switch, dual horns, dual horn relay	35/18	Y 356
- Fresh air blower switch, switch for fresh air blower and recirculated air, fresh air blower	35/19	Y 357
- Heated rear window, rear window defogger switch	35/20	Y 358
- Windshield wiper motor, windshield wiper intermittent regulator, rear window wiper motor	35/21	Y 359
- Wiper/washer intermittent relay, windshield wiper/washer switch, windshield and rear window washer pump	35/22	Y 360

2000 Passat, additional electrical wiring diagrams

	Wiring Diagram page number	Repair Manual page number
● **Airbag systems,** from August 1999	42/1	Y 423
● **Air conditioning,** **2.8L-Engine - Motronic Multiport Fuel Injection (MFI)/142 kW,code AHA,** from August 1998 **1.8L-Engine - Motronic Multiport Fuel Injection (MFI)/110 kW,code ATW,** **2.8L-Engine - Motronic Multiport Fuel Injection (MFI)/142 kW,code ATQ,** from August 1999 **1.8L-Engine - Motronic Multiport Fuel Injection (MFI)/110 kW,code AUG,** from May 2000	43/1	Y 427
● **Climatronic, 2.8L-Engine - Motronic Multiport Fuel Injection (MFI)/142 kW, code AHA, ATQ,** from August 1998	33/1	Y 327
● **Anti-lock brake system (ABS) with electronic differential lock (EDL),** from September 1997	12/1	Y 157
● **Anti-lock brake system (ABS) with electronic differential lock (EDL) and anti-slip control (ASC)** from September 1997	11/1	Y 153
● **Automatic day/night interior mirror,** from August 1998	27/1	Y 265
● **Automatic transmission, 5 speed automatic with tiptronic,** **1.8L-Engine - Motronic Multiport Fuel Injection (MFI)/110 kW,code ATW,** **2.8L-Engine - Motronic Multiport Fuel Injection (MFI)/142 kW,code ATQ,** from August 1999 **1.8L-Engine - Motronic Multiport Fuel Injection (MFI)/110 kW,code AUG,** from May 2000	41/1	Y 415
● **Central locking system, see Comfort System**		
● **Comfort System, with alarm system,** from August 1999	38/1	Y 391
● **Coolant fan, 80 W,** from September 1997	6/1	Y 125
● **Coolant fan, two speeds, 300 W,** from September 1997	7/1	Y 127
● **Cruise control, 1.8L-Engine code ATW,** from August 1999	37/11	Y 388
● **Cruise control, 2.8L-Engine code ATQ,** from August 1999	36/6	Y 366
● **Daytime running Lights,** from August 1999	45/1	Y 439
● **Dual horns,** from August 1999	35/18	Y 356
● **Front interior light, see Comfort System**		
● **Fuel tank lid unlock system, see Comfort System**		
● **Heated seats,** from September 1997	20/1	Y 205
● **Headlight washer,** from August 1999	35/21	Y 359
● **Luggage compartment light, see Comfort System**		
● **Make-up mirror lights, see Comfort System**		
● **Monsoon-Soundsystem,** from August 1999	44/1	Y 433
● **Power outlets (12V),** from September 1997	39/1	Y 410
● **Power seats, without Memory,** from November 1997	26/1	Y 261
● **Power seats and mirror adjustment with Memory,** from November 1997	25/1	Y 253
● **Power sunroof, see Comfort System**		
● **Radio System,** from August 1999	40/1	Y 411
● **Reading lights, see Comfort System**		
● **Rear lid unlock system, see Comfort System**		
● **Rear window wiper,** from August 1999	35/21	Y 359
● **Rainsensor,** from August 1998	27/1	Y 265
● **Warning light for rear lid unlocked, see Comfort System**		
● **Washer nozzle heaters,** from September 1997	14/1	Y 177

Electrical Wiring Diagram Index

1.8 Liter engine (code AUG, from May 2000)

	Wiring Diagram page number	Repair Manual page number
● **Passat (1.8L-Engine - Motronic Multiport Fuel Injection (MFI)/110 kW), code AUG, from May 2000**	46/1	Y 443
- Battery, starter, generator (GEN)	46/2	Y 444
- Ignition system, motronic engine control module (ECM), Spark Plug Connectors, Spark Plug	46/3	Y 445
- Motronic engine control module (ECM), Secondary Air Injection (AIR) Pump	46/4	Y 446
- Motronic engine control module (ECM), Wastegate Bypass Regulator Valve, Evaporative Emission Canister Purge Regulator Valve, Secondary Air Injection Solenoid Valve, Recirculating valve for turbocharger, Heated Oxygen Sensor	46/5	Y 447
- Motronic engine control module (ECM), Oxygen Sensor, Mass Air Flow Sensor	46/6	Y 448
- Motronic engine control module (ECM), fuel injectors, Fuel Pump (FP) Relay, Fuel injectors	46/7	Y 449
- Motronic engine control module (ECM), Power Supply Relay, Brake booster	46/8	Y 450
- Motronic engine control module (ECM), Engine Coolant Temperature Sensor, Camshaft Position Sensor, Intake Air Temperature	46/9	Y 451
- Motronic engine control module (ECM), Knock Sensor1 and 2, Engine Speed Sensor	46/10	Y 452
- Motronic engine control module (ECM), Leak detection pump (LDP), Throttle Valve Control Module	46/11	Y 453
- Motronic engine control module (ECM), Cruise Control Switch, Brake Light Switch, Clutch Vacuum Vent Valve Switch	46/12	Y 454
- Motronic engine control module (ECM), Throttle Position Sensor, Sender -2- for accelerator pedal position, Data Link Connector (DLC)	46/13	Y 455
- Oil pressure switch, Hood Alarm Switch, fuel pump (FP), sender for fuel gauge, speedometer vehicle speed sensor, Engine Coolant Level Sensor, Control module with indicator unit	46/14	Y 456
- Control module with indicator unit in Instrument panel insert	46/15	Y 457

1.8 Liter engine (code AWM, from October 2000)

	Wiring Diagram page number	Repair Manual page number
● **Passat (1.8L-Engine - Motronic Multiport Fuel Injection (MFI)/125 kW), code AWM, from October 2000**	49/1	Y 488
- Battery, starter, generator (GEN)	49/2	Y 487
- Ignition system, motronic engine control module (ECM), Spark Plug Connectors, Spark Plug	49/3	Y 490
- Motronic engine control module (ECM), Secondary Air Injection (AIR) Pump Motor/Relay	49/4	Y 489
- Motronic engine control module (ECM), Wastegate Bypass Regulator Valve, Evaporative Emission Canister Purge Regulator Valve, Secondary Air Injection Solenoid Valve, Recirculating valve for turbocharger, Heated Oxygen Sensor	49/5	Y 492
- Motronic engine control module (ECM), Oxygen Sensor, Mass Air Flow Sensor	49/6	Y 491
- Motronic engine control module (ECM), Fuel Pump (FP) Relay, Fuel injectors	49/7	Y 494
- Motronic engine control module (ECM), Power Supply Relay, Brake booster	49/8	Y 493
- Motronic engine control module (ECM), Engine Coolant Temperature Sensor, Camshaft Position Sensor, Intake Air Temperature	49/9	Y 496
- Motronic engine control module (ECM), Knock Sensor1 and 2, Engine Speed Sensor	49/10	Y 495
- Motronic engine control module (ECM), Leak detection pump (LDP), Throttle Valve Control Module	49/11	Y 498
- Motronic engine control module (ECM), Cruise Control Switch, Brake Light Switch, Clutch Vacuum Vent Valve Switch	49/12	Y 497
- Motronic engine control module (ECM), Throttle Position Sensor, Sender -2- for accelerator pedal position, Data Link Connector (DLC)	49/13	Y 500
- Oil pressure switch, Hood Alarm Switch, fuel pump (FP), sender for fuel gauge, speedometer vehicle speed sensor, Engine Coolant Level Sensor, Control module with indicator unit	49/14	Y 499
- Control module with indicator unit in Instrument panel insert	49/15	Y 502

2.8 Liter engine (code ATQ, from August 1999)

	Wiring Diagram page number	Repair Manual page number
● **Passat (2.8L-Engine - Motronic Multiport Fuel Injection (MFI)/142 kW), code ATQ, from August 1999**	36/1	Y 361
- Battery, starter, generator (GEN)	36/2	Y 362
- Ignition system, motronic engine control module (ECM), throttle valve control module	36/3	Y 363
- Motronic engine control module (ECM), engine speed sensor (RPM), knock sensor (KS)	36/4	Y 364
- Motronic engine control module (ECM), camshaft position (CMP) sensor, engine coolant temperature (ECT) sensor, intake air temperature (IAT) sensor	36/5	Y 365
- Motronic engine control module (ECM), Button for cruise control, Cruise Control Switch, Brake Light Switch, Clutch Vacuum Vent Valve Switch	36/6	Y 366
- Motronic engine control module (ECM), valve for camshaft adjustment, Leak detection pump (LDP)	36/7	Y 367
- Motronic engine control module (ECM), mass air flow (MAF) sensor, Throttle Position Sensor Data Link Connector	36/8	Y 368
- Motronic engine control module (ECM), Fuel Injectors	36/9	Y 369
- Motronic engine control module (ECM), heated oxygen sensors	36/10	Y 370
- Motronic engine control module (ECM), oxygen sensors behind three way catalytic converter	36/11	Y 371
- Motronic engine control module (ECM), secondary air injection system	36/12	Y 372
- Speedometer vehicle speed sensor, engine coolant level (ECL) sensor, oil level thermal sensor hood alarm switch, oil pressure switch	36/13	Y 373
- Sender for fuel gauge, control module with indicator unit in instrument panel insert	36/14	Y 374
- Control module with indicator unit in instrument panel insert	36/15	Y 375

2.8 Liter engine (code ATQ, from October 2000)

	Wiring Diagram page number	Repair Manual page number
● **Passat (2.8L-Engine - Motronic Multiport Fuel Injection (MFI)/142 kW), code ATQ, from October 2000**	60/1	Y 582
- Battery, starter, generator (GEN)	60/2	Y 581
- Ignition system, motronic engine control module (ECM), throttle drive, spark plug connectors	60/3	Y 584
- Motronic engine control module (ECM), intake air temperature (IAT) sensor, knock sensor (KS)	60/4	Y 583
- Motronic engine control module (ECM), engine speed (RPM) sensor, lamp failture control module, brake system vacuum pump	60/5	Y 586
- Motronic engine control module (ECM), brake booster pressure sensor, camshaft position (CMP) sensor, engine coolant temperature (ECT) sensor	60/6	Y 585
- Motronic engine control module (ECM), Button for cruise control, Cruise Control Switch, Brake Light Switch, Clutch Vacuum Vent Valve Switch	60/7	Y 588
- Motronic engine control module (ECM), fuel injectors	60/8	Y 587
- Motronic engine control module (ECM), Throttle Position Sensor, Data Link Connector	60/9	Y 590
- Motronic engine control module (ECM), valve for camshaft adjustment, evaporative emmission (EVAP) canister purge regulator valve, mass air flow (MAF) sensor	60/10	Y 589
- Motronic engine control module (ECM), heated oxygen sensors	60/11	Y 592
- Motronic engine control module (ECM), oxygen sensors behind three way catalytic converter	60/12	Y 591
- Motronic engine control module (ECM), leak detection pump (LDP)	60/13	Y 594
- Motronic engine control module (ECM), secondary air injection system	60/14	Y 593
- Speedometer vehicle speed sensor (VSS), engine coolant level (ECL) sensor, oil level thermal sensor hood alarm switch, oil pressure switch	60/15	Y 596
- Sender for fuel gauge, control module with indicator unit in instrument panel insert, fuel pump (FP), fuel level sensor, fuel supply sensor	60/16	Y 595
- Control module with indicator unit in instrument panel insert, data bus on board diagnostic interface	60/17	Y 598

Electrical Wiring Diagram Index

2001 Passat, Standard equipment (from August 1999)

	Wiring Diagram page number	Repair Manual page number
● **Passat Standard Equipment, from August 1999**	35/1	Y 339
- Starting interlock, battery	35/2	Y 340
- Ignition/starter switch	35/3	Y 341
- Instrument cluster, Data Link Connector (DLC)	35/4	Y 342
- Instrument cluster, tachometer, speedometer vehicle speed sensor, Malfunction Indicator Lamp (MIL), seat belt warning system	35/5	Y 343
- Instrument cluster, generator warning light, outside air temperature sensor	35/6	Y 344
- Instrument cluster, left and right turn signal indicator lights, multi-function indicator switch	35/7	Y 345
- Windshield washer fluid level sensor, daytime running lights change-over relay, parking brake warning light switch and brake fluid level warning light switch	35/8	Y 346
- Turn signal switch, headlight dimmer/flasher switch	35/9	Y 347
- Left front headlights, left front fog light, fog light relay	35/10	Y 348
- License plate light	35/11	Y 349
- Emergency flasher switch	35/12	Y 350
- Right front headlights, right front fog light	35/13	Y 351
- Left brake/tail light, left rear turn signal light, left back-up light, high-mount brake light	35/14	Y 352
- Brake light switch, back-up light switch, right rear turn signal light, right back-up light, right brake/tail light	35/15	Y 353
- Radio connection, cigarette lighter, panel vent illumination	35/16	Y 354
- Light switch	35/17	Y 355
- Instrument panel light dimmer switch, dual horns, dual horn relay	35/18	Y 356
- Fresh air blower switch, switch for fresh air blower and recirculated air, fresh air blower	35/19	Y 357
- Heated rear window, rear window defogger switch	35/20	Y 358
- Windshield wiper motor, windshield wiper intermittent regulator, rear window wiper motor	35/21	Y 359
- Wiper/washer intermittent relay, windshield wiper/washer switch, windshield and rear window washer pump	35/22	Y 360

2001 Passat, Standard equipment (from October 2000)

	Wiring Diagram page number	Repair Manual page number
● **Passat Standard Equipment, from October 2000**	48/1	Y 464
- Locking relay for starter, battery, starter	48/2	Y 463
- Ignition/starter switch	48/3	Y 466
- Instrument cluster, Data Link Connector (DLC)	48/4	Y 465
- Instrument cluster, tachometer, speedometer vehicle speed sensor (VSS), Malfunction Indicator Lamp (MIL), seat belt warning system	48/5	Y 468
- Instrument cluster, generator warning light	48/6	Y 467
- Instrument coil of antitheft immonilizer, Instrument cluster, outside air temperature sensor	48/7	Y 470
- Instrument cluster, left and right turn signal indicator lights, multi-function indicator switch	48/8	Y 469
- Windshield washer fluid level sensor, daytime running lights change-over relay, parking brake warning light switch and brake fluid level warning light switch	48/9	Y 472
- Turn signal switch, headlight dimmer/flasher switch	48/10	Y 471
- Left front fog light, fog light relay	48/11	Y 474
- License plate light	48/12	Y 473
- Emergency flasher switch	48/13	Y 476
- Right front headlights, right front fog light	48/14	Y 475
- Brake Light Switch, Left brake/tail light, left rear turn signal light, left back-up light, high-mount brake light	48/15	Y 478
- Back-up light switch, right rear turn signal light, right back-up light, right brake/tail light	48/16	Y 477
- Radio connection, cigarette lighter, panel vent illumination	48/17	Y 480
- Light switch	48/18	Y 479
- Instrument panel light dimmer switch, dual horns, dual horn relay	48/19	Y 482
- Fresh air blower switch, switch for fresh air blower and recirculated air, fresh air blower	48/20	Y 481
- Heated rear window, rear window defogger switch	48/21	Y 484
- Windshield wiper motor, windshield wiper intermittent regulator, rear window wiper motor	48/22	Y 483
- Wiper/washer intermittent relay, windshield wiper/washer switch, windshield and rear window washer pump	48/23	Y 486

2001 Passat, additional electrical wiring diagrams

2001 Passat, additional electrical wiring diagrams

Electrical Wiring Diagram Index

1.8 Liter engine (code AWM), 2002-2005 m.y.

	Wiring Diagram page number	Repair Manual page number
● **Passat (1.8L-Engine - Motronic Multiport Fuel Injection (MFI)/125 kW), code AWM, from October 2000**	49/1	Y 488
- Battery, starter, generator (GEN)	49/2	Y 487
- Ignition system, motronic engine control module (ECM), Spark Plug Connectors, Spark Plug	49/3	Y 490
- Motronic engine control module (ECM), Secondary Air Injection (AIR) Pump Motor/Relay	49/4	Y 489
- Motronic engine control module (ECM), Wastegate Bypass Regulator Valve, Evaporative Emission Canister Purge Regulator Valve, Secondary Air Injection Solenoid Valve, Recirculating valve for turbocharger, Heated Oxygen Sensor	49/5	Y 492
- Motronic engine control module (ECM), Oxygen Sensor, Mass Air Flow Sensor	49/6	Y 491
- Motronic engine control module (ECM), Fuel Pump (FP) Relay, Fuel injectors	49/7	Y 494
- Motronic engine control module (ECM), Power Supply Relay, Brake booster	49/8	Y 493
- Motronic engine control module (ECM), Engine Coolant Temperature Sensor, Camshaft Position Sensor, Intake Air Temperature	49/9	Y 496
- Motronic engine control module (ECM), Knock Sensor1 and 2, Engine Speed Sensor	49/10	Y 495
- Motronic engine control module (ECM), Leak detection pump (LDP), Throttle Valve Control Module	49/11	Y 498
- Motronic engine control module (ECM), Cruise Control Switch, Brake Light Switch, Clutch Vacuum Vent Valve Switch	49/12	Y 497
- Motronic engine control module (ECM), Throttle Position Sensor, Sender -2- for accelerator pedal position, Data Link Connector (DLC)	49/13	Y 500
- Oil pressure switch, Hood Alarm Switch, fuel pump (FP), sender for fuel gauge, speedometer vehicle speed sensor, Engine Coolant Level Sensor, Control module with indicator unit	49/14	Y 499
- Control module with indicator unit in Instrument panel insert	49/15	Y 502

2.8 Liter engine (code ATQ), 2002-2005 m.y.

	Wiring Diagram page number	Repair Manual page number
● **Passat (2.8L-Engine - Motronic Multiport Fuel Injection (MFI)/142 kW), code ATQ, from June 2001**	69/1	Y 635
- Battery, starter, generator (GEN), Voltage Regulator (VR)	69/2	Y 636
- Motronic engine control module (ECM), throttle drive, spark plug connectors, Suppressor	69/3	Y 637
- Motronic engine control module (ECM), intake air temperature (IAT) sensor, knock sensor (KS)	69/4	Y 638
- Motronic engine control module (ECM), engine speed (RPM) sensor, brake booster relay, brake system vacuum pump	69/5	Y 639
- Motronic engine control module (ECM), brake booster pressure sensor, camshaft position (CMP) sensor, engine coolant temperature (ECT) sensor	69/6	Y 640
- Motronic engine control module (ECM), Button for cruise control, Cruise Control Switch, Brake Light Switch, Clutch Vacuum Vent Valve Switch	69/7	Y 641
- Motronic engine control module (ECM), cylinder fuel injectors	69/8	Y 642
- Motronic engine control module (ECM), Throttle Position Sensor, Data Link Connector	69/9	Y 643
- Motronic engine control module (ECM), valve for camshaft adjustment, evaporative emission (EVAP) canister purge regulator valve, mass air flow (MAF) sensor	69/10	Y 644
- Motronic engine control module (ECM), heated oxygen sensors	69/11	Y 645
- Motronic engine control module (ECM), oxygen sensors behind three way catalytic converter	69/12	Y 646
- Motronic engine control module (ECM), secondary air injection (AIR) pump, relay and motor secondary air injection (AIR), Solenoid Valve	69/13	Y 647
- Motronic engine control module (ECM), leak detection pump (LDP), oil pressure switch, speedometer Vehicle speed sensor, engine coolant level sensor	69/14	Y 648
- Sender for fuel gauge, control module with indicator unit, fuel pump, (FP), fuel level sensor, fuel supply sensor	69/15	Y 649
- Control module with indicator unit in instrument panel insert, data bus on board diagnostic interface	69/16	Y 650

4.0 Liter engine (code BDP), 2002-2005 m.y.

	Wiring Diagram page number	Repair Manual page number
● **Passat (4.0L-Engine - Motronic Multiport Fuel Injection (MFI)/202 kW), code BDP, from January 2002**	72/1	Y 657
- Battery, starter, generator (GEN), Voltage Regulator (VR)	72/2	Y 658
- Motronic engine control module (ECM), Ignition Coil 1-4, Spark plug connectors, Spark Plugs	72/3	Y 659
- Motronic engine control module (ECM), Motronic engine control module Power Supply Relay	72/4	Y 660
- Motronic engine control module (ECM), Ignition Coil 5-8, Spark Plug Connectors, Spark Plugs	72/5	Y 661
- Motronic engine control module (ECM), Cylinder 1-8 Fuel Injector	72/6	Y 662
- Motronic engine control module (ECM), Knock Sensor 1-4	72/7	Y 663
- Motronic engine control module (ECM), Engine Coolant, Temperature, Engine Speed Sensor, Camshaft Position Sensor 2-4	72/8	Y 664
- Motronic engine control module (ECM), heated oxygen sensors, Oxygen Sensor behind three way Catalytic Converter	72/9	Y 665
- Motronic engine control module (ECM), heated oxygen sensor 2, Oxygen Sensor 2 behind three way Catalytic Converter	72/10	Y 666
- Motronic engine control module (ECM), Cruise Control Switch, Button for Cruise Control, Brake Light and Brake Pedal Switch	72/11	Y 667
- Motronic engine control module (ECM), Auxiliary Engine Coolant Pump Relay, Left Electro-Hydraulic Mount Solenoid Valve, Coolant Pump	72/12	Y 668
- Motronic engine control module (ECM), Secondary Air Injection Pump Relay, Map Controlled Engine Cooling Thermostat, Valve 1 and 2	72/13	Y 669
- Motronic engine control module (ECM), Throttle Position Sensor, Sender 2 for accelerator pedal Position, Mass Air Flow Sensor	72/14	Y 670
- Motronic engine control module (ECM), Brake Booster Relay, Engine Coolant Temperature Sensor, Valve	72/15	Y 671
- Motronic engine control module (ECM), Fuel Pump Relay, Throttle drive, Angle sensor 1 and 2 for throttle drive	72/16	Y 672
- Motronic engine control module (ECM), Leak Detection Pump	72/17	Y 673
- Motronic engine control module (ECM), Control Module with indicator unit in instrument panel insert, Data Bus on Board Diagnostic Interface	72/18	Y 674
- Control Module with indicator unit in instrument panel insert, Speedometer, Vehicle Speed Sensor (VSS), Oil Pressure Switch	72/19	Y 675
- Control Module with indicator unit in instrument panel insert, Hood Alarm Switch	72/20	Y 676
- Fuel Pump Control Module, Sender for fuel gauge, Fuel Pump, Transfer Fuel Pump, Auxiliary Tank Fuel Level Empty Sensor	72/21	Y 677
- Fuel Pump Control Module, Fuel Level Sensor 2, Fuel Supply Sensor 3, Auxiliary Tank Fuel Level Sensor	72/22	Y 678

2002-2005 Passat, Standard equipment

	Wiring Diagram page number	Repair Manual page number
● **Passat Standard Equipment, from October 2000**	48/1	Y 464
- Locking relay for starter, battery, starter	48/2	Y 463
- Ignition/starter switch	48/3	Y 466
- Instrument cluster, Data Link Connector (DLC)	48/4	Y 465
- Instrument cluster, tachometer, speedometer vehicle speed sensor (VSS), Malfunction Indicator Lamp (MIL), seat belt warning system	48/5	Y 468
- Instrument cluster, generator warning light	48/6	Y 467
- Instrument coil of anti-theft immobilizer, Instrument cluster, outside air temperature sensor	48/7	Y 470
- Instrument cluster, left and right turn signal indicator lights, multi-function indicator switch	48/8	Y 469
- Windshield washer fluid level sensor, daytime running lights change-over relay, parking brake warning light switch and brake fluid level warning light switch	48/9	Y 472
- Turn signal switch, headlight dimmer/flasher switch	48/10	Y 471
- Left front fog light, fog light relay	48/11	Y 474
- License plate light	48/12	Y 473
- Emergency flasher switch	48/13	Y 476
- Right front headlights, right front fog light	48/14	Y 475
- Brake Light Switch, Left brake/tail light, left rear turn signal light, left back-up light, high-mount brake light	48/15	Y 478
- Back-up light switch, right rear turn signal light, right back-up light, right brake/tail light	48/16	Y 477
- Radio connection, cigarette lighter, panel vent illumination	48/17	Y 480
- Light switch	48/18	Y 479
- Instrument panel light dimmer switch, dual horns, dual horn relay	48/19	Y 482
- Fresh air blower switch, switch for fresh air blower and recirculated air, fresh air blower	48/20	Y 481
- Heated rear window, rear window defogger switch	48/21	Y 484
- Windshield wiper motor, windshield wiper intermittent regulator, rear window wiper motor	48/22	Y 483
- Wiper/washer intermittent relay, windshield wiper/washer switch, windshield and rear window washer pump	48/23	Y 486

Electrical Wiring Diagram Index

2002-2005 Passat, additional electrical wiring diagrams

2002-2005 Passat, additional electrical wiring diagrams

Passat - Standard Equipment, only for cars with blue illuminated instrument cluster

from September 1997

Deviate relay location and fuseplacements as well as the locations of multiple connectors see section "component locations".

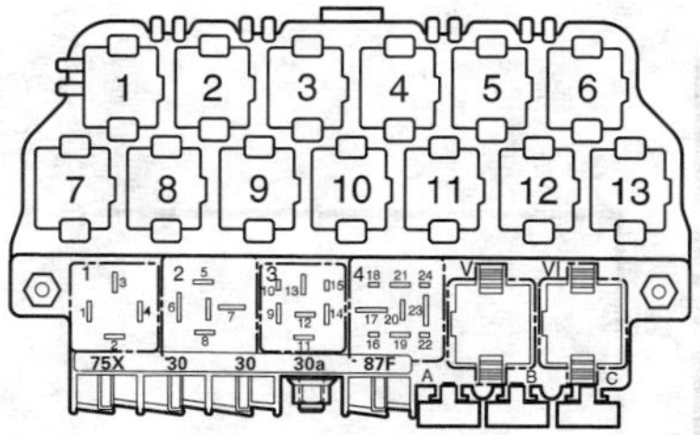

97-20676

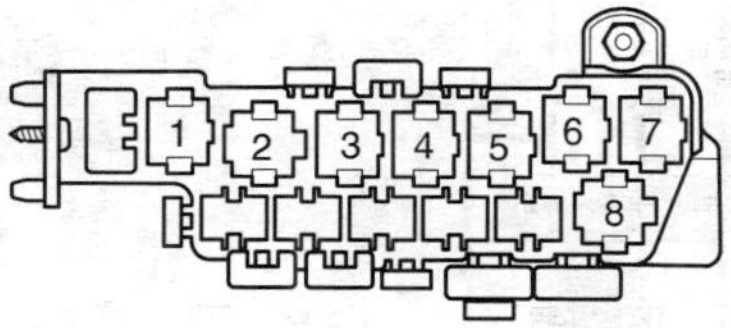

97-20515

Relay location on the thirteenfold auxiliary relay panel, above relay panel:

- **7** Fog Light Relay (381)
- **8** Daytime Running Lights Change-over Relay (173)
- **12** Starting Interlock Relay (alarm system) (186)
- **13** Starting Interlock Relay (clutch position) (204)

Relay panel:

- **1** Dual Horn Relay (204)
- **2** Load Reduction Relay (213)
- **V** Wiper/Washer Intermittent Relay (377)
- **VI** Wiper/Washer Intermittent Relay (377)

Relay location on the eightfold auxiliary relay panel, behind relay panel:

Note: Number in parentheses indicates production control number stamped on relay housing.

Fuse colors
30 A - green
25 A - white
20 A - yellow
15 A - blue
10 A - red
7,5 A - brown
5 A - beige

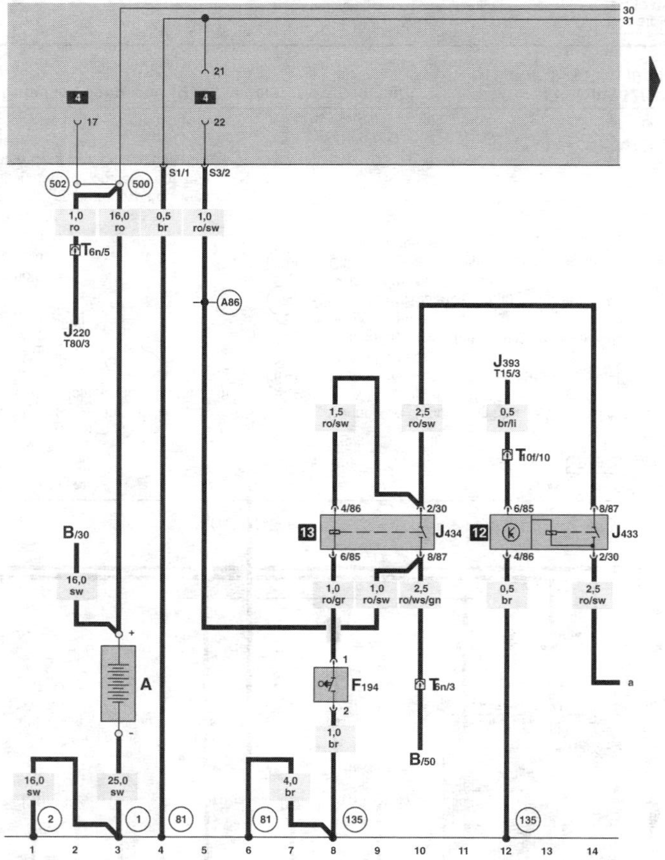

97-20970

ws = white
sw = black
ro = red
br = brown
gn = green
bl = blue
gr = grey
li = violet
ge = yellow

A - Battery
B - Starter
F194 - Clutch Pedal Position (CPP) Switch
J220 - Motronic Engine Control Module (ECM)
J393 - Central control module for comfort system
J433 - Starting Interlock Relay (alarm system), on auxiliary relay panel, above relay panel, production control number (186)
J434 - Starting Interlock Relay (clutch position), on auxiliary relay panel, above relay panel, production control number (204)
T6n - 6-Pin Connector, red, on protective housing for control module, in engine compartment, left
T10f - 10-Pin Connector, blue, connector station A-pillar, left
T15 - 15-Pin Connector, on central control module for comfort system
T80 - Connector, 80 point, on Engine Control Module (ECM)
(1) - Ground strap, battery to body
(2) - Ground strap, transmission to body
(81) - Ground connection -1-, in instrument panel wiring harnes
(135) - Ground connection -2-, in instrument panel wiring harness
(500) - Screw connection -1- (30), on relay panel
(502) - Screw connection (30a), on relay panel
(A86) - Connection (50a) in instrument panel wiring harness

Starting interlock, battery

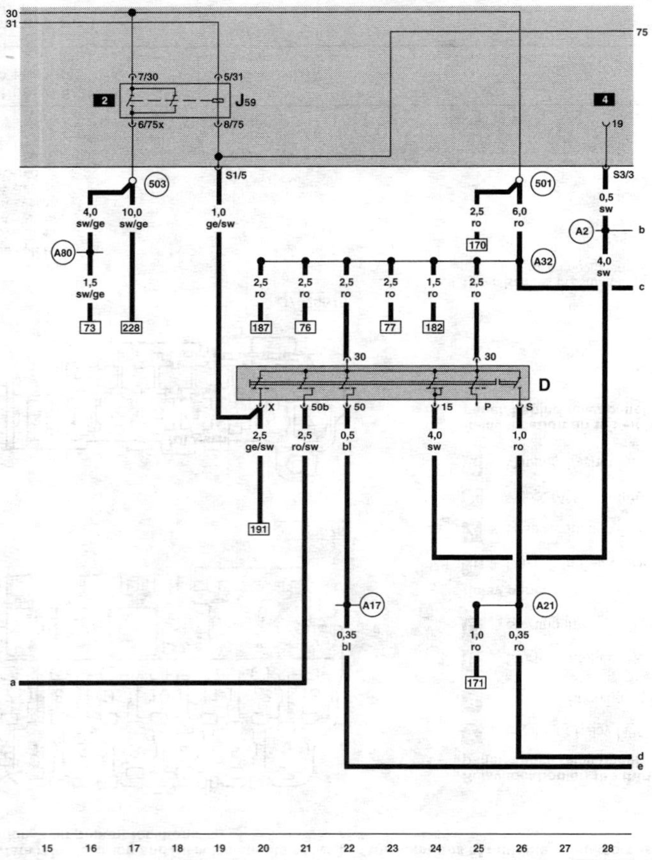

ws = white
sw = black
ro = red
br = brown
gn = green
bl = blue
gr = grey
li = violet
ge = yellow

97-20971

- D - Ignition/Starter Switch
- J59 - Load Reduction Relay, production control number (213)
- (501) - Screw connection -2- (30), on relay panel
- (503) - Screw connection (75x), on relay panel
- (A2) - plus connection (15), in instrument panel wiring harness
- (A17) - wire connection (61), in instrument panel wiring harness
- (A21) - wire connection (86s), in instrument panel wiring harness
- (A32) - plus connection (30), in instrument panel wiring harness
- (A80) - Connector -1- (X), in instrument panel wiring harness

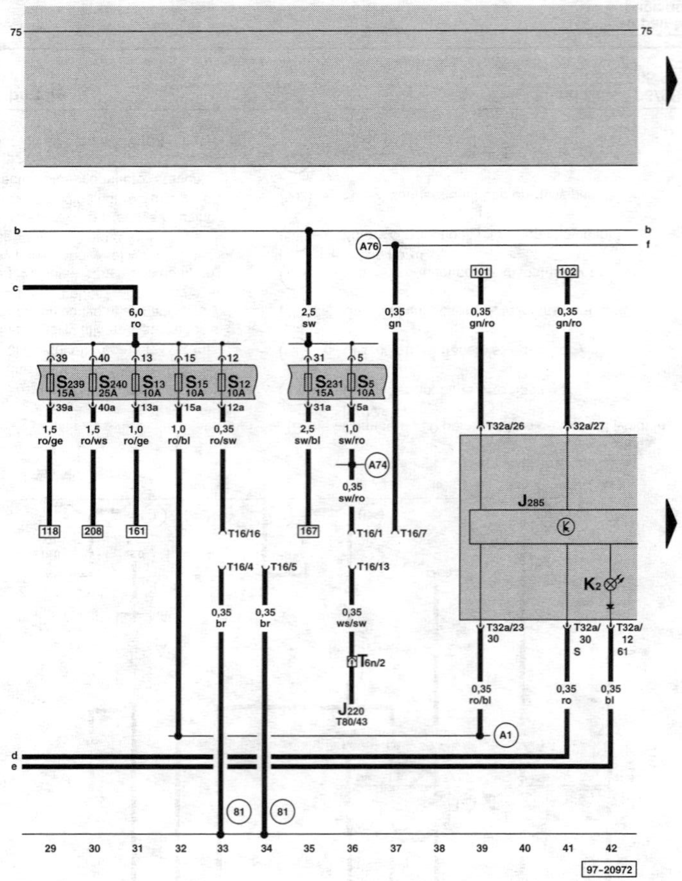

ws = white
sw = black
ro = red
br = brown
gn = green
bl = blue
gr = grey
li = violet
ge = yellow

97-20972

- J220 - Motronic Engine Control Module (ECM)
- J285 - Control module with indicator unit in instrument panel insert
- K2 - Generator (GEN) Warning Light
- S5 - Fuse 5 in fuse holder
- S12 - Fuse 12 in fuse holder
- S13 - Fuse 13 in fuse holder
- S15 - Fuse 15 in fuse holder
- S231 - Fuse 31 in fuse holder
- S239 - Fuse 39 in fuse holder
- S240 - Fuse 40 in fuse holder
- T6n - 6-Pin Connector, red, on protective housing for control module, in engine compartment, left
- T16 - 16-Pin Connector, Data Link Connector (DLC), below instrument panel, left
- T32a - 32-Pin Connector, blue, on instrument cluster
- T80 - Connector, 80 point, on Engine Control Module (ECM)
- (81) - Ground connection -1-, in instrument panel wiring harness
- (A1) - plus connection (30a), in instrument panel wiring harness
- (A74) - Connector (15a - fuse 5), in instrument panel wiring harness
- (A76) - Connector (K-diagnosis wire), in instrument panel wiring harness

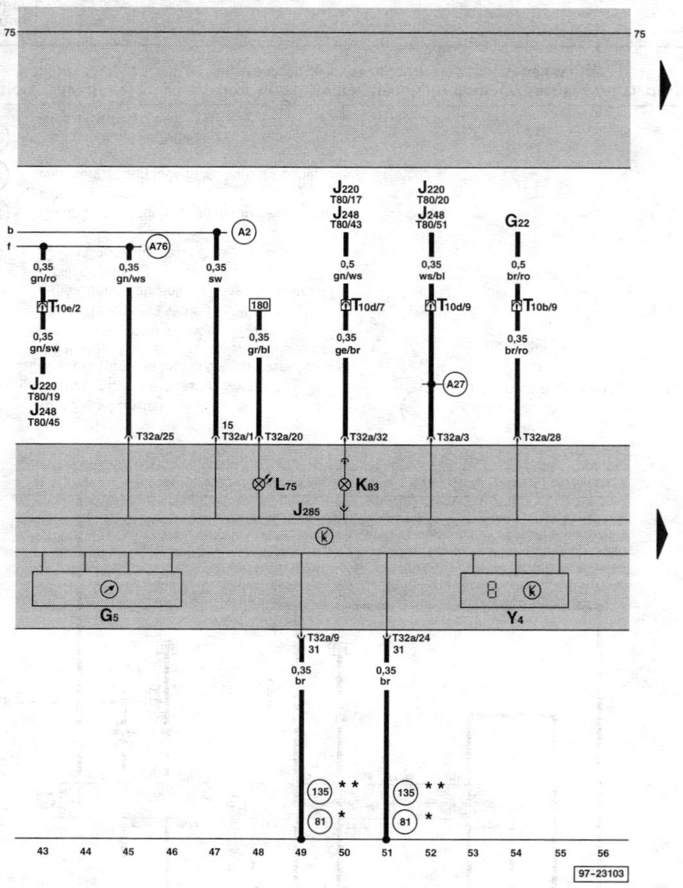

ws = white
sw = black
ro = red
br = brown
gn = green
bl = blue
gr = grey
li = violet
ge = yellow

G5 - Tachometer
J220 - Motronic Engine Control Module (ECM)
J248 - Diesel Direct Fuel Injection (DFI) Engine Control Module (ECM)
J285 - Control module with indicator unit in instrument panel insert
K83 - Malfunction Indicator Lamp (MIL)
L75 - Digital Display Light
T10b - 10-Pin Connector, black, on protective housing for control module, in engine compartment, left
T10d - 10-Pin Connector, brown, on protective housing for control module, in engine compartment, left
T10e - 10-Pin Connector, yellow, on protective housing for control module, in engine compartment, left
T32a - 32-Pin Connector, blue, on instrument cluster
T80 - Connector, 80 point, on Engine Control Module (ECM)

Y4 - Odometer Display

(81) - Ground connection -1-, in instrument panel wiring harness
(135) - Ground connection -2-, in instrument panel wiring harness
(A2) - plus connection (15), in instrument panel wiring harness
(A27) - Wire Connection (vehicle speed signal), in instrument panel wiring harness

* - to April 1998
** - from May 1998

Instrument cluster, tachometer, speedometer vehicle speed sensor, Malfunction Indicator Lamp (MIL)

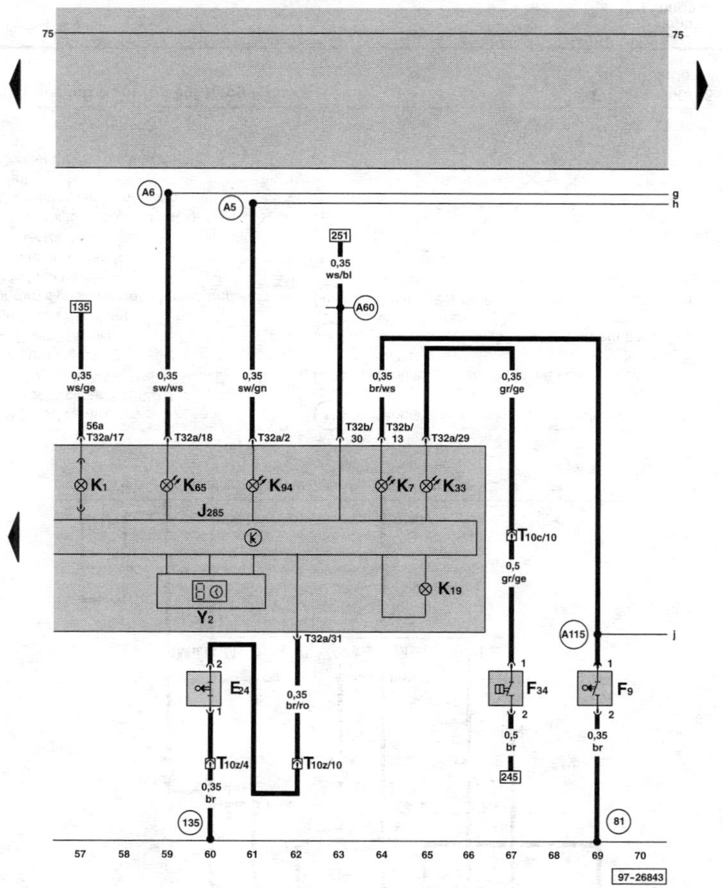

ws = white
sw = black
ro = red
br = brown
gn = green
bl = blue
gr = grey
li = violet
ge = yellow

E24 - Left Seat Belt Switch
F9 - Parking Brake Warning Light Switch
F34 - Brake Fluid Level Warning Switch
J285 - Control module with indicator unit in instrument panel insert
K1 - Headlight High Beam Indicator Light
K7 - Brake and Parking Brake Warning Light
K19 - Seat Belt Warning Light
K33 - Brake Fluid Level Warning Light
K65 - Left Turn Signal Indicator Light
K94 - Right Turn Signal Indicator Light
T10c - 10-Pin Connector, violet, connector station A-pillar, left
T10z - 10-Pin Connector, brown
T32a - 32-Pin Connector, blue, on instrument cluster
T32b - 32-Pin Connector, green, on instrument cluster

Y2 - Digital Clock

(81) - Ground connection -1-, in instrument panel wiring harness
(135) - Ground connection -2-, in instrument panel wiring harness
(A5) - plus connection (right turn signal), in instrument panel wiring harness
(A6) - plus connection (left turn signal), in instrument panel wiring harness
(A60) - Wire connection (Vehicle speed signal), in instrument cluster wiring harness
(A115) - Connection (parking brake warning light), in instrument panel wiring harness

Instrument cluster, left and right turn signal indicator lights, parking brake warning light switch and brake fluid level warning light switch

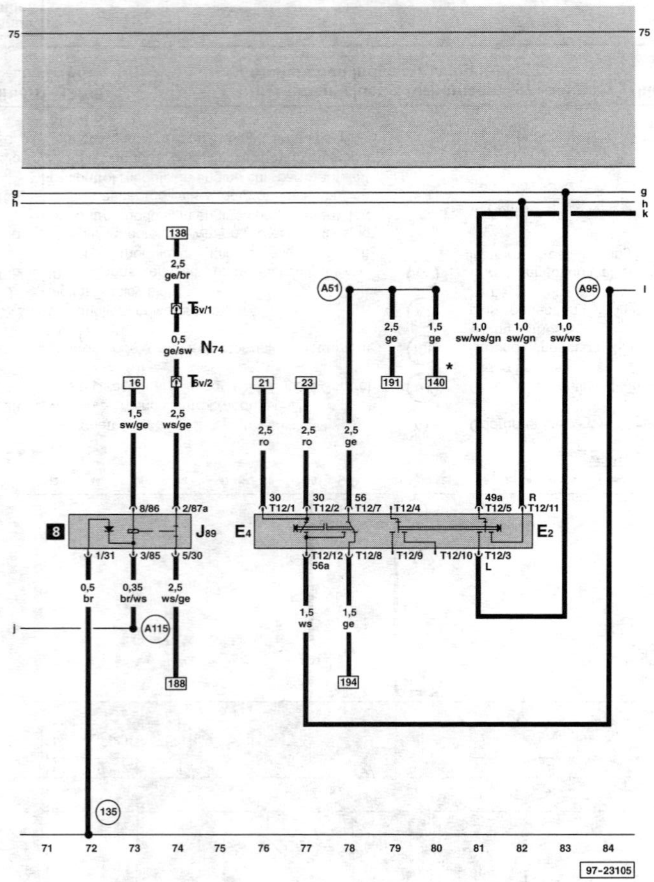

ws = white
sw = black
ro = red
br = brown
gn = green
bl = blue
gr = grey
li = violet
ge = yellow

E2 - Turn Signal Switch
E4 - Headlight Dimmer/Flasher Switch
J89 - Daytime Running Lights Change-over Relay on the thirteenfold relay panel, above relay panel, production control number (173)
N74 - Series Resistance Wiring
T6v - 6-Pin Connector, violet, connector station A-pillar, right
T12 - 12-Pin Connector, near steering column

(135) - Ground connection -2-, in instrument panel wiring harness
(A51) - wire connection (56), in instrument panel wiring harness
(A95) - wire connection -1- (56a), in instrument panel wiring harness

(A115) - Connection (parking brake warning light), in instrument panel wiring harness

* - to May 1998

Turn signal switch, headlight dimmer/flasher switch, daytime running lights change-over relay, series resistance wiring

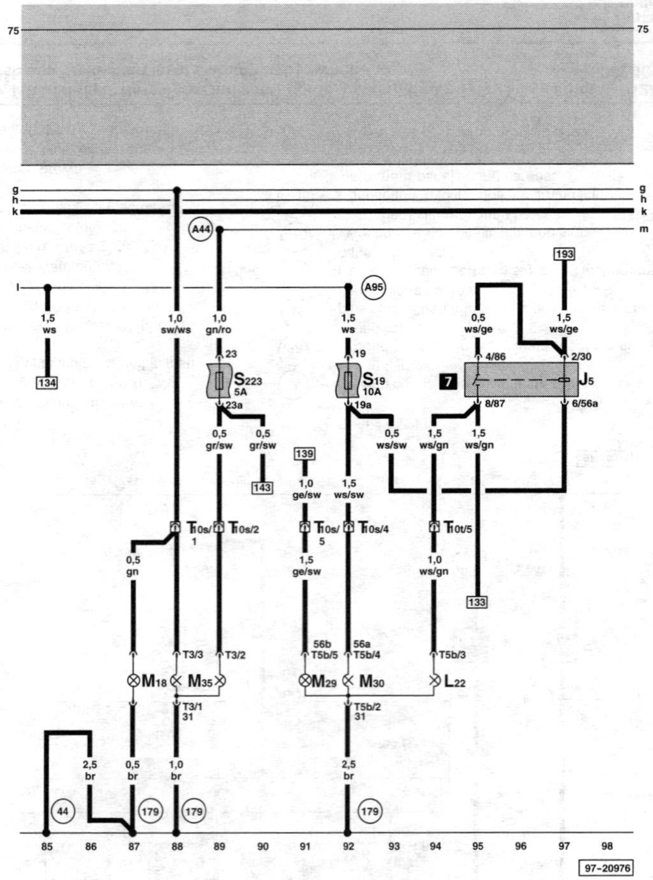

ws = white
sw = black
ro = red
br = brown
gn = green
bl = blue
gr = grey
li = violet
ge = yellow

J5 - Fog Light Relay, on auxiliary relay panel, production control number (381)
L22 - Left Front Fog Light
M18 - Left, Side Turn Signal Light
M29 - Left Low Beam Headlight
M30 - Left High Beam Headlight
M35 - Light for turn signal and side marker, front, left
S19 - Fuse 19 in fuse holder
S223 - Fuse 23 in fuse holder
T3 - 3-Pin Connector
T5b - 5-Pin Connector, on headlight left
T10s - 10-Pin Connector, pink, connector station A-pillar, left
T10t - 10-Pin Connector, red, connector station A-pillar, left

(44) - Ground connection, on left A-pillar, lower part
(179) - Ground connection, in left headlight wiring harness
(A44) - wire connection (57r), in instrument panel wiring harness
(A95) - Connection 1 (56a), in instrument panel wiring harness

Left front headlights, left front fog light, fog light relay

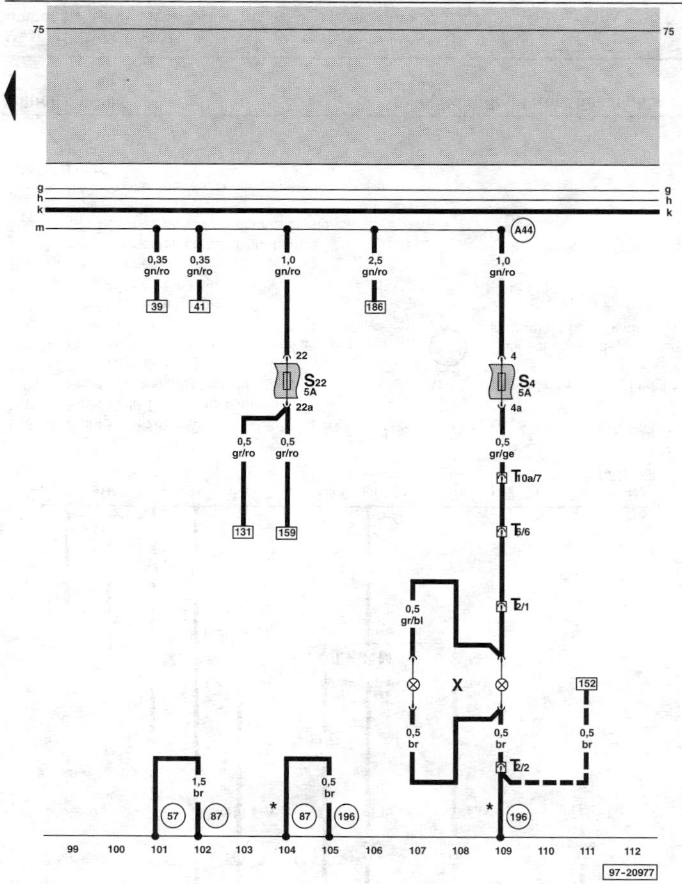

ws = white
sw = black
ro = red
br = brown
gn = green
bl = blue
gr = grey
li = violet
ge = yellow

S4 - Fuse 4 in fuse holder
S22 - Fuse 22 in fuse holder
T2 - Double Connector, in rear lid
T6 - 6-Pin Connector, black, in luggage compartment, left (wagon only)
T10a - 10-Pin Connector, brown, connector station A-pillar, left
X - License Plate Lights

(57) - Ground connection, on left rear pillar

(87) - Ground connection -2-, in rear wiring harness

(196) - Ground connection -3-, in wiring harness, rear

(A44) - wire connection (57r), in instrument panel wiring harness

- - - wagon only

* - sedan only

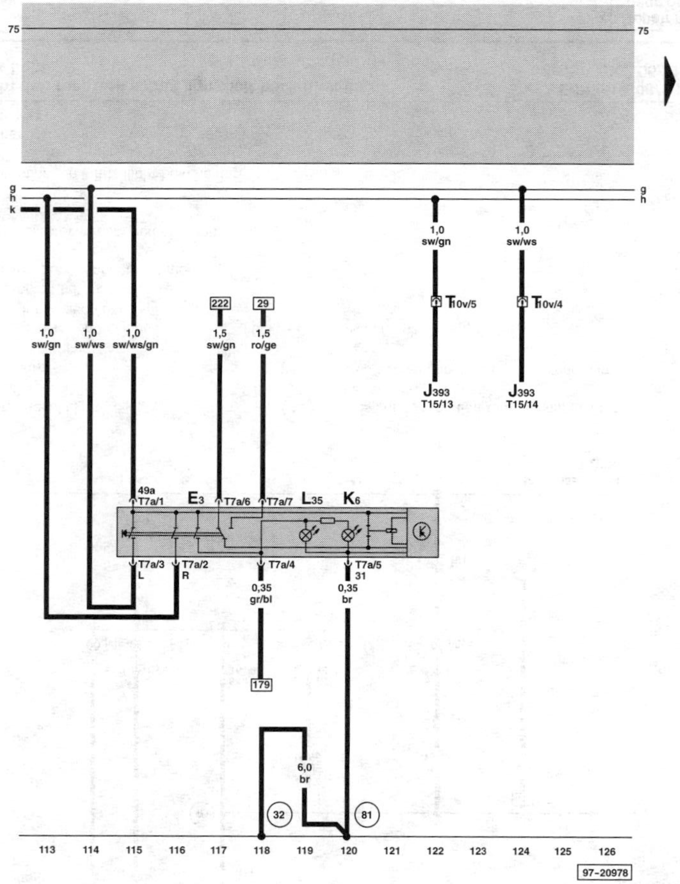

ws = white
sw = black
ro = red
br = brown
gn = green
bl = blue
gr = grey
li = violet
ge = yellow

E3 - Emergency Flasher Switch
J393 - Central control module for comfort system
K6 - Emergency Flasher Warning Light
L35 - Emergency Flasher Switch Light
T7a - 7-Pin Connector, on emergency flasher switch
T10v - 10-Pin Connector, violet, connector station A-pillar, right
T15 - 15-Pin Connector, on central control module for comfort system

(32) - Ground connection, behind instrument panel, left

(81) - Ground connection -1-, in instrument panel wiring harness

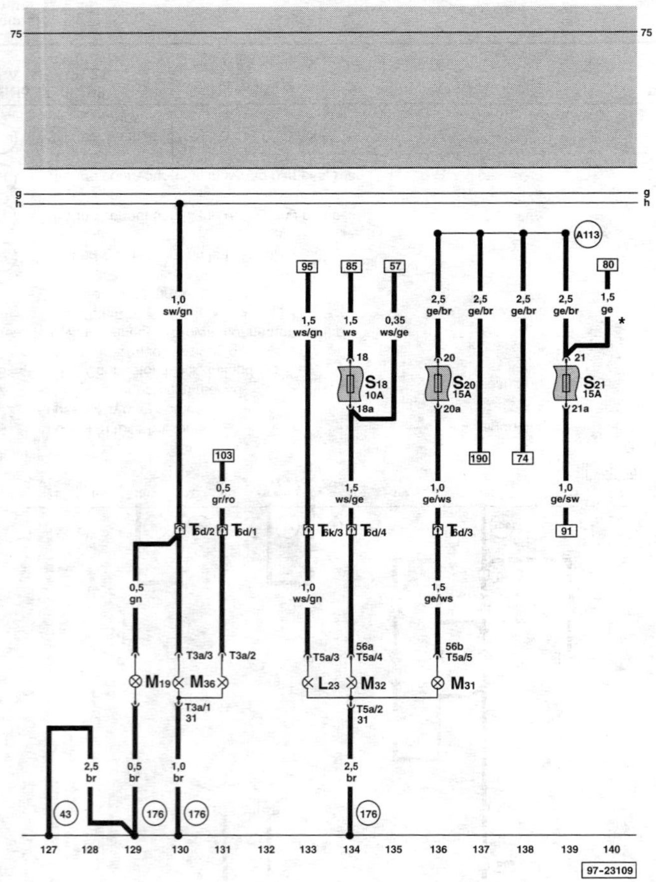

ws = white
sw = black
ro = red
br = brown
gn = green
bl = blue
gr = grey
li = violet
ge = yellow

L23 - Right Front Fog Light
M19 - Right, Side Turn Signal Light
M31 - Right Low Beam Headlight
M32 - Right High Beam Headlight
M36 - Light for turn signal and side marker, front, right
S18 - Fuse 18 in fuse holder
S20 - Fuse 20 in fuse holder
S21 - Fuse 21 in fuse holder
T3a - 3-Pin Connector
T5a - 5-Pin Connector, on headlight right
T6d - 6-Pin Connector, yellow, connector station A-pillar, right
T6k - 6-Pin Connector, red, connector station A-pillar, right

(43) - Ground connection, on right A-pillar, lower part

(176) - Ground connection, in right headlight wiring harness

(A113) - Connection -2- (daytime running lights), in instrument panel wiring harness

* - to May 1998

Right front headlights, right front foglight

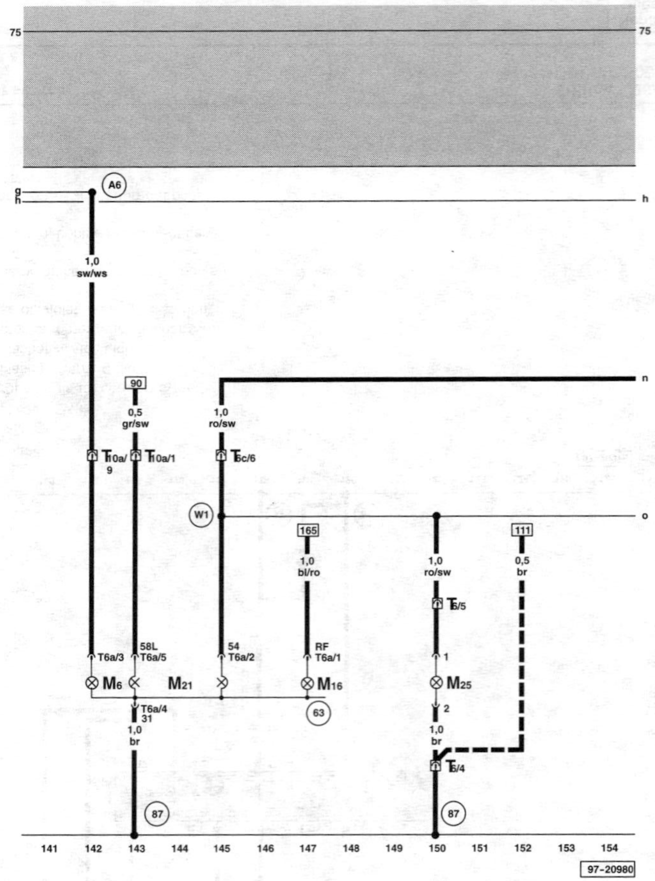

ws = white
sw = black
ro = red
br = brown
gn = green
bl = blue
gr = grey
li = violet
ge = yellow

M6 - Left Rear Turn Signal Light
M21 - Left Brake/Tail Light
M16 - Left Back-Up Light
M25 - High-mount Brake Light
T6 - 6-Pin Connector, black, in luggage compartment, left (wagon only)
T6a - 6-Pin Connector, on tail light, left
T6c - 6-Pin Connector, blue, connector station A-pillar, left
T10a - 10-Pin Connector, brown, connector station A-pillar, left

(63) - Ground connection, on left taillight bulb holder

(87) - Ground connection -2-, in rear wiring harness

(A6) - plus connection (left turn signal), in instrument panel wiring harness

(W1) - plus connection (54), in rear wiring harness

\- - - wagon only

Left brake/tail light, left rear turn signal light, left back-up light, high-mount brake light

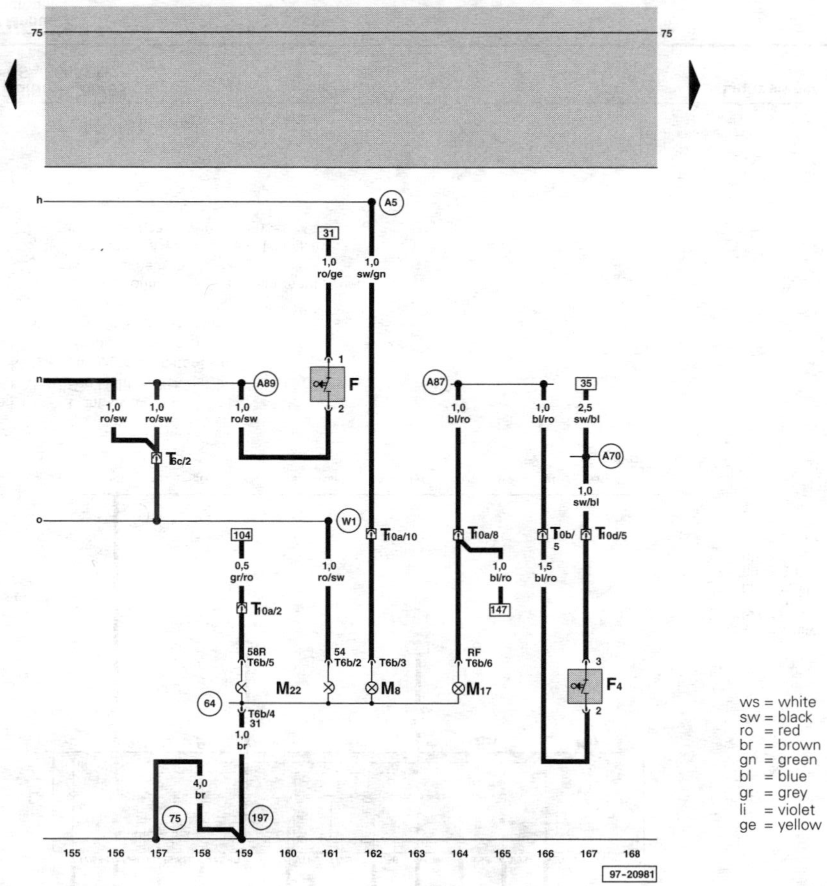

ws = white
sw = black
ro = red
br = brown
gn = green
bl = blue
gr = grey
li = violet
ge = yellow

- F - Brake Light Switch
- F4 - Back-Up Light Switch
- M8 - Right Rear Turn Signal Light
- M17 - Right Back-Up Light
- M22 - Right Brake/Tail Light
- T6b - 6-Pin Connector, on tail light, right
- T6c - 6-Pin Connector, blue, connector station A-pillar, left
- T10a - 10-Pin Connector, brown, connector station A-pillar, left
- T10b - 10-Pin Connector, black, on protective housing for control module, in engine compartment, left
- T10d - 10-Pin Connector, brown, on protective housing for control module, in engine compartment, left
- (64) - Ground connection, on right taillight bulb holder
- (75) - Ground connection, on right rear pillar
- (197) - Ground connection -4-, in wiring harness rear
- (A5) - plus connection (right turn signal), in instrument panel wiring harness
- (A70) - Connector (15a, fuse 231), in instrument panel wiring harness
- (A87) - Connector (reverse lamp), in instrument panel wiring harness
- (A89) - Connector (54), in instrument panel wiring harness
- (W1) - plus connection (54), in rear wiring harness

Brake light switch, back-up light switch, right rear turn signal light, right back-up light, right brake/tail light

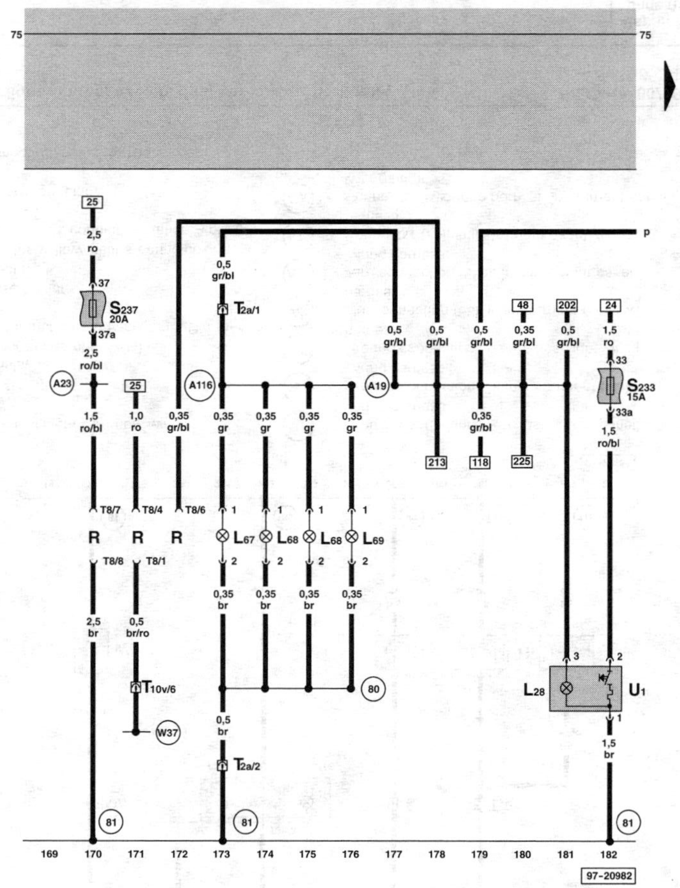

ws = white
sw = black
ro = red
br = brown
gn = green
bl = blue
gr = grey
li = violet
ge = yellow

- L28 - Cigarette Lighter Light
- L67 - Left Instrument Panel Vent Illumination
- L68 - Center Instrument Panel Vent Illumination
- L69 - Right Instrument Panel Vent Illumination
- R - Radio
- S233 - Fuse 33 in fuse holder
- S237 - Fuse 37 in fuse holder
- T2a - Double Connector
- T6n - 6-Pin Connector, red, on protective housing for control module, in engine compartment, left
- T8 - 8-Pin Connector, on radio
- T10v - 10-Pin Connector, violet, connector station A-pillar, right
- U1 - Cigarette Lighter
- (80) - Ground connection -1-, in instrument panel wiring harness
- (81) - Ground connection -1-, in instrument panel wiring harness
- (A19) - wire connection (58d), in instrument panel wiring harness
- (A23) - wire connection (30al), in instrument panel wiring harness
- (A116) - wire connection -2- (58d), in instrument panel wiring harness
- (W37) - Connector (alarm system), in wiring harness floor

Radio connection, cigarette lighter, panel vent illumination

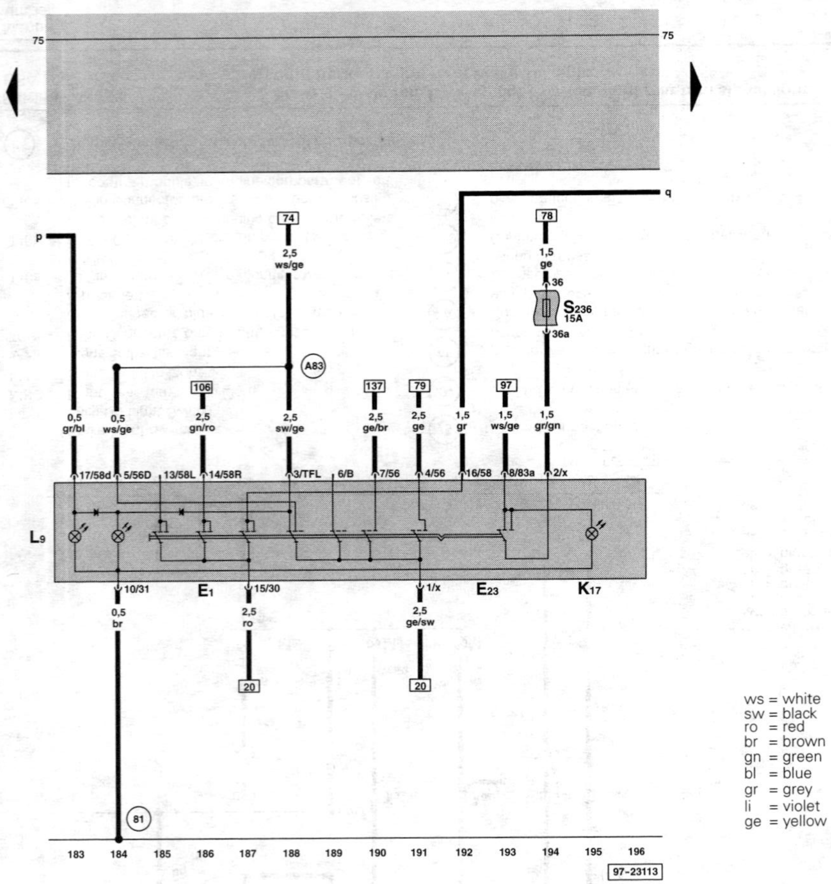

ws = white
sw = black
ro = red
br = brown
gn = green
bl = blue
gr = grey
li = violet
ge = yellow

E1 - Light switch
E23 - Fog Light Switch
K17 - Fog Light Indicator Light
L9 - Headlight Switch Licht
S236 - Fuse 36 in fuse holder

(81) - Ground connection -1-, in instrument panel wiring harness
(A83) - Connector (daytime running lights), in instrument panel wiring harness

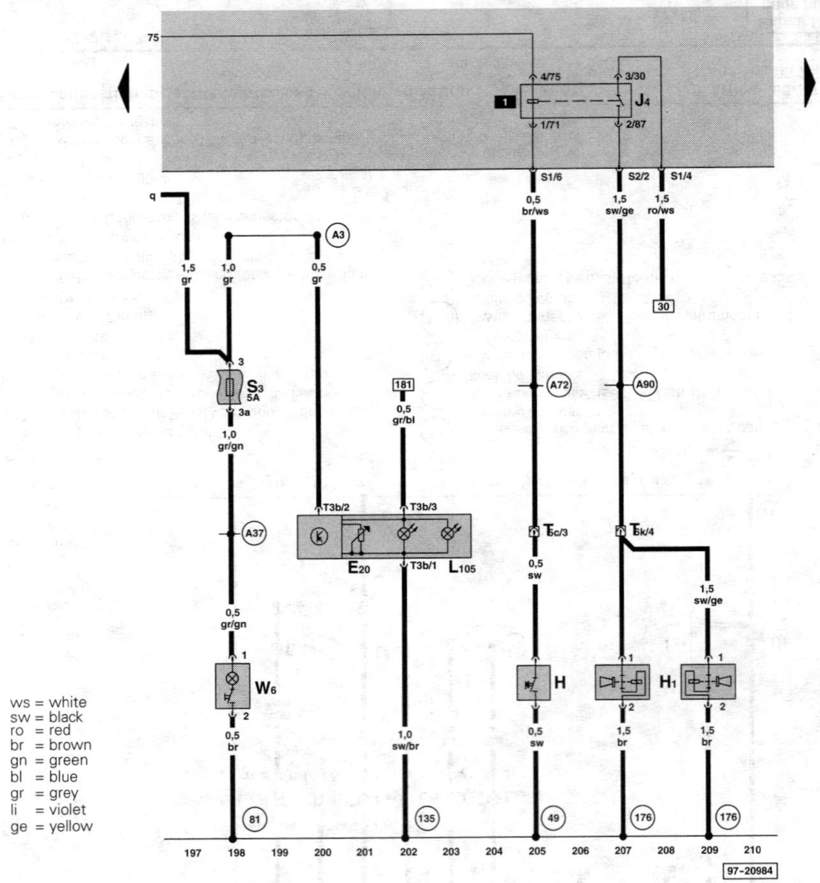

ws = white
sw = black
ro = red
br = brown
gn = green
bl = blue
gr = grey
li = violet
ge = yellow

E20 - Instrument Panel Light Dimmer Switch
H - Horn Button
H1 - Dual Horns
J4 - Dual Horn Relay, production control number (204)
L105 - Illumination for Instrument Panel Light Dimmer Switch
S3 - Fuse 3 in fuse holder
T3b - 3-Pin Connector
T5c - 5-Pin Connector, yellow, near steering column
T6k - 6-Pin Connector, red, connector station A-pillar, right
W6 - Glove Compartment Light

(49) - Ground connection, near steering wheel

(81) - Ground connection -1-, in instrument panel wiring harness
(135) - Ground connection -2-, in instrument panel wiring harness
(176) - Ground connection, in right headlight wiring harness
(A3) - plusconnection (58), in instrument panel wiring harness
(A37) - wire connection (58a), in instrument panel wiring harness
(A72) - Connector (71), in instrument panel wiring harness
(A90) - Connector (two-tone horn), in instrument panel wiring harness

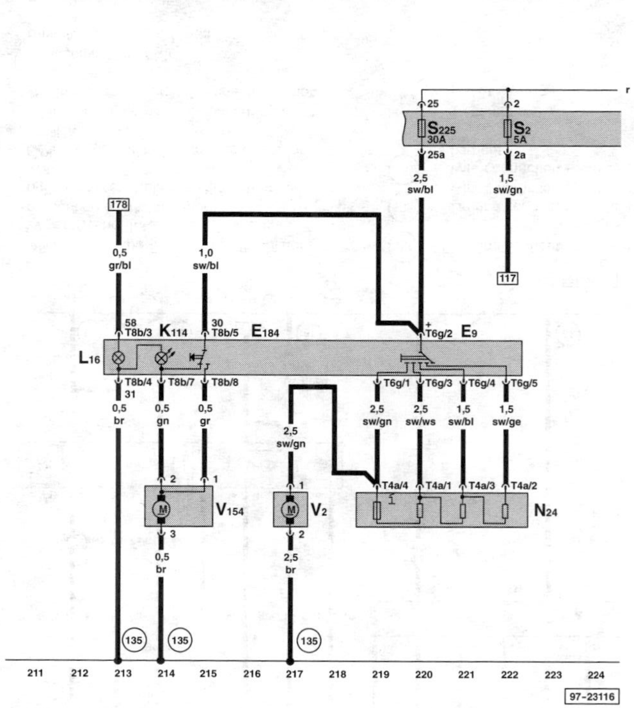

ws = white
sw = black
ro = red
br = brown
gn = green
bl = blue
gr = grey
li = violet
ge = yellow

- E9 - Fresh Air Blower Switch
- E184 - Switch for fresh air blower and recirculated air
- K114 - Fresh Air and Recirculating Air Mode Indicator Light
- L16 - Fresh Air Control Lever Light
- N24 - Fresh Air Blower Series Resistance With Fuse
- S2 - Fuse 2 in fuse holder
- S225 - Fuse 25 in fuse holder
- T4a - 4-Pin Connector
- T6g - 6-Pin Connector, on fresh air blower switch
- T8b - 8-Pin Connector, on switch for fresh air blower and recirculated air
- V2 - Fresh Air Blower
- V154 - Servo motor for fresh-/recirculating air door
- (135) - Ground connection -2-, in instrument panel wiring harness

Fresh air blower switch, switch for fresh air blower and recirculated air, fresh air blower

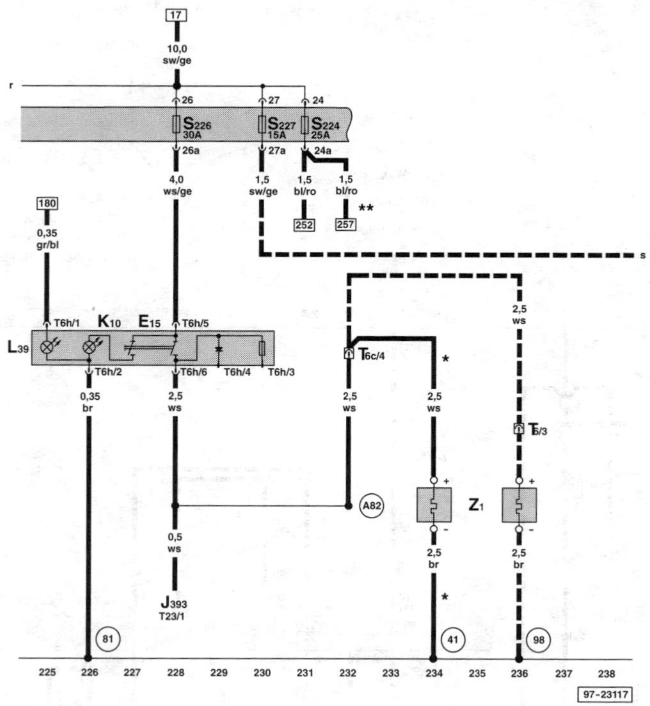

ws = white
sw = black
ro = red
br = brown
gn = green
bl = blue
gr = grey
li = violet
ge = yellow

- E15 - Rear window defogger switch
- J393 - Central control module for comfort system
- K10 - Rear Window Defogger Indicator Light
- L39 - Rear Window Defogger Switch Light
- S224 - Fuse 24 in fuse holder
- S226 - Fuse 26 in fuse holder
- S227 - Fuse 27 in fuse holder
- T6 - 6-Pin Connector, black, in luggage compartment, left (wagon only)
- T6c - 6-Pin Connector, blue, connector station A-pillar, left
- T6h - 6-Pin Connector, black, on rear window defogger switch
- T23 - 23-Pin Connector, on central control module for comfort system
- Z1 - Heated rear window

- (41) - Ground connection, below rear shelf
- (81) - Ground connection -1-, in instrument panel wiring harness
- (98) - Ground connection, in rear lid wiring harness
- (A82) - Connector (heated glass), in instrument panel wiring harness
- - - - wagon only
- * - sedan only
- ** - to May 1998

Heated rear window, rear window defogger switch

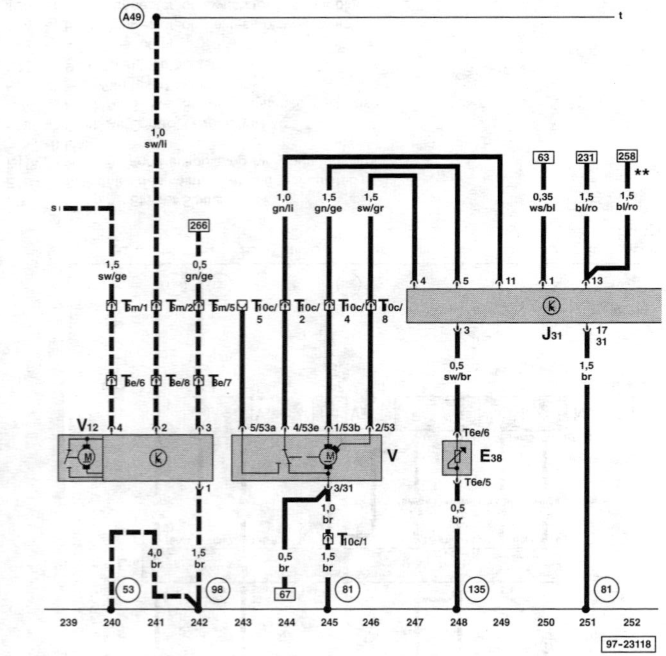

ws = white
sw = black
ro = red
br = brown
gn = green
bl = blue
gr = grey
li = violet
ge = yellow

E38 - Windshield Wiper Intermittent Regulator
J31 - Wiper/Washer Intermittent Relay, production control number (377)
T6e - 6-Pin Connector, black, behind steering column switch cover
T6m - 6-Pin Connector, black, connector station A-pillar, right
T8e - 8-Pin Connector, black, in rear lid, right
T10c - 10-Pin Connector, violet, connector station A-pillar, left
V - Windshield Wiper Motor
V12 - Rear Window Wiper Motor

(53) - Ground connection, in rear lid, right

(81) - Ground connection -1-, in instrument panel wiring harness

(98) - Ground connection, in rear lid wiring harness

(135) - Ground connection -2-, in instrument panel wiring harness

(A49) - wire connection -1-, in instrument panel wiring harness

\- - - wagon only
** from May 1998

Windshield wiper motor, windshield wiper intermittent regulator, rear window wiper motor

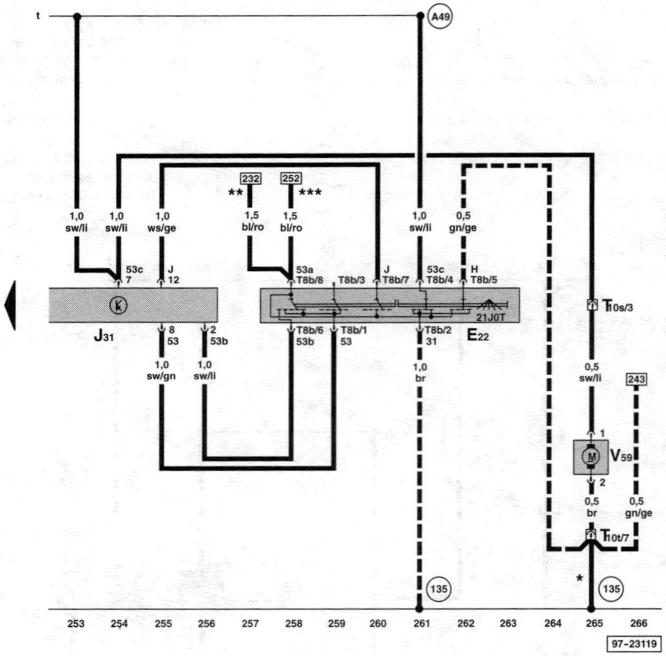

ws = white
sw = black
ro = red
br = brown
gn = green
bl = blue
gr = grey
li = violet
ge = yellow

E22 - Winshield Wiper/Washer Switch
J31 - Wiper/Washer Intermittent Relay, production control number (377)
T8b - 8-Pin Connector, on Windshield Wiper/Washer Switch
T10s - 10-Pin Connector, pink, connector station A-pillar, left
T10t - 10-Pin Connector, red, connector station A-pillar, left
V59 - Windshield and Rear Window Washer Pump

(135) - Ground connection -2-, in instrument panel wiring harness

(A49) - wire connection -1-, in instrument panel wiring harness

\- - - wagon only
* - sedan only
** - to May 1998
*** - from May 1998

Wiper/washer intermittent relay, windshield wiper/washer switch, windshield and rear window washer pump

1,8L- Engine – Motronic Multiport Fuel Injection (MFI)/110 kW, code AEB,

from September 1997

Deviate relay location and fuseplacements as well as the locations of multiple connectors see section "component locations".

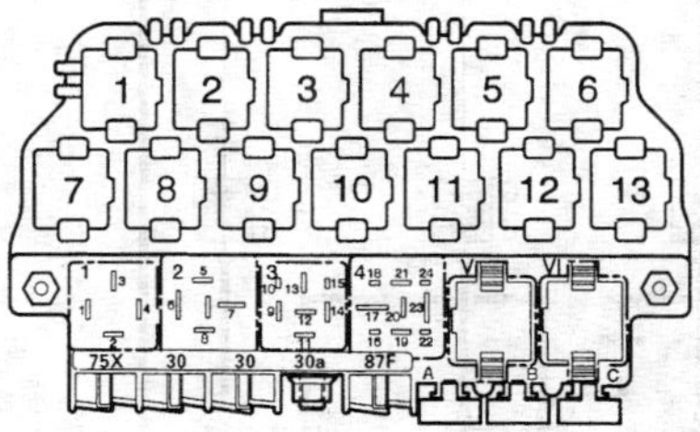

97–20676

Relay location on the thirteenfold auxiliary relay panel, above relay panel:

Relay panel:

4 Fuel Pump (FP) Relay (208)

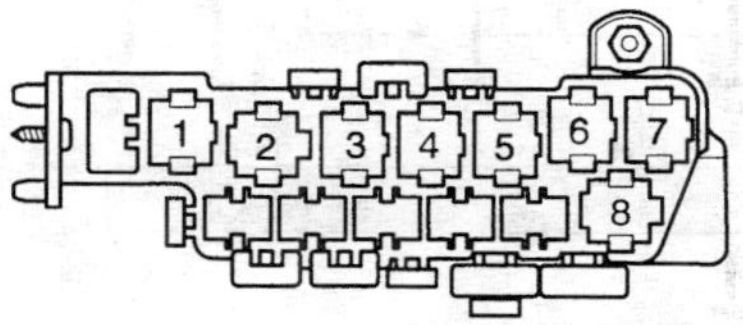

97–20515

Relay location on the eightfold auxiliary relay panel, behind relay panel:

Note: Number in parentheses indicates production control number stamped on relay housing.

Fuse colors

30 A – green
25 A – white
20 A – yellow
15 A – blue
10 A – red
7,5 A – brown
5 A – beige

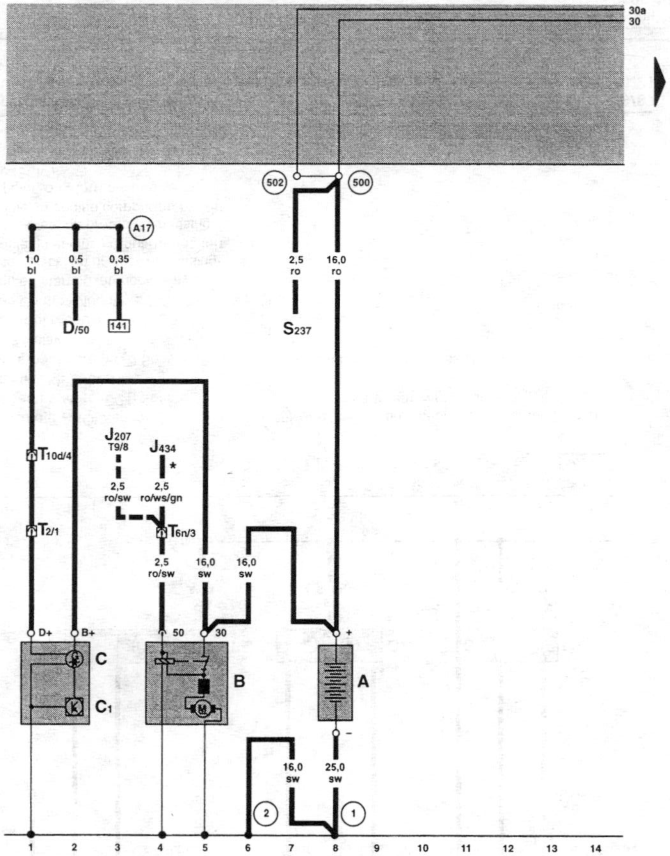

97–20688

ws = white
sw = black
ro = red
br = brown
gn = green
bl = blue
gr = grey
li = violet
ge = yellow

A – Battery
B – Starter
C – Generator (GEN)
C1 – Voltage Regulator (VR)
D – Ignition/Starter Switch
J207 – Starting Interlock Relay
J434 – Starting Interlock Realy (clutch position), manual transmission only
S237 – Fuse 37 in fuse holder
T2 – Double Connector, on engine right
T6n – 6-Pin Connector, red, on protective housing for control module, in engine compartment, left
T9 – 9-Pin Connector, on Starting Interlock Relay
T10d – 10-Pin Connector, brown, on protective housing for control module, in engine compartment, left

(1) – Ground strap, battery to body
(2) – Ground strap, transmission to body
(500) – Screw Connection -1- (30), on relay panel
(502) – Screw Connection (30a), on relay panel
(A17) – wire connection (61), in instrument panel wiring harness
* – manual transmission only
- - - automatic transmission only

Battery, starter, generator (GEN)

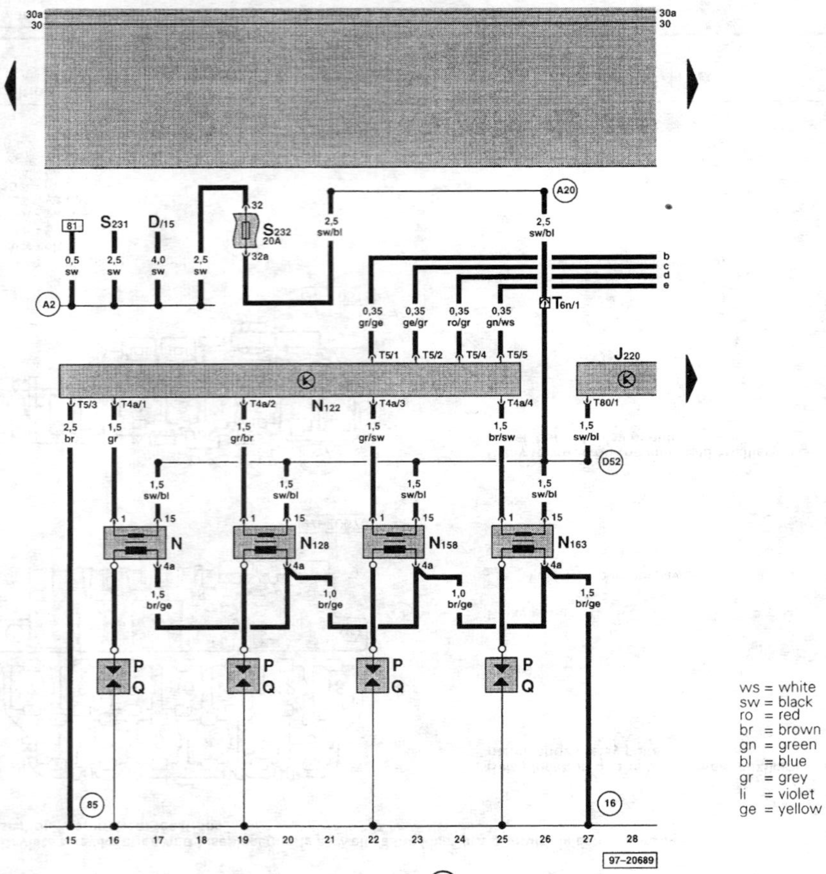

ws = white
sw = black
ro = red
br = brown
gn = green
bl = blue
gr = grey
li = violet
ge = yellow

D – Ignition/Starter Switch
J220 – Motronic Engine Control Module (ECM)
N – Ignition Coil
N122 – Power Output Stage
N128 – Ignition Coil 2
N158 – Ignition Coil 3
N163 – Ignition Coil 4
P – Spark Plug Connectors
Q – Spark Plugs
S231 – Fuse 31 in fuse holder
S232 – Fuse 32 in fuse holder
T4a – 4-Pin Connector, black, on power output stage
T5 – 5-Pin Connector, black, on power output stage
T6n – 6-Pin Connector, red, on protective housing for control module, in engine compartment, left
T80 – Connector, 80 point, on Motronic Engine Control Module (ECM)

(16) – Ground connection -1-, on valve cover

(85) – Ground connection -1-, in engine compartment wiring harness

(A2) – plus connection (15), in instrument panel wiring harness

(A20) – wire connection (15a), in instrument panel wiring harness

(D52) – plus connection (15a), in engine compartment wiring harness

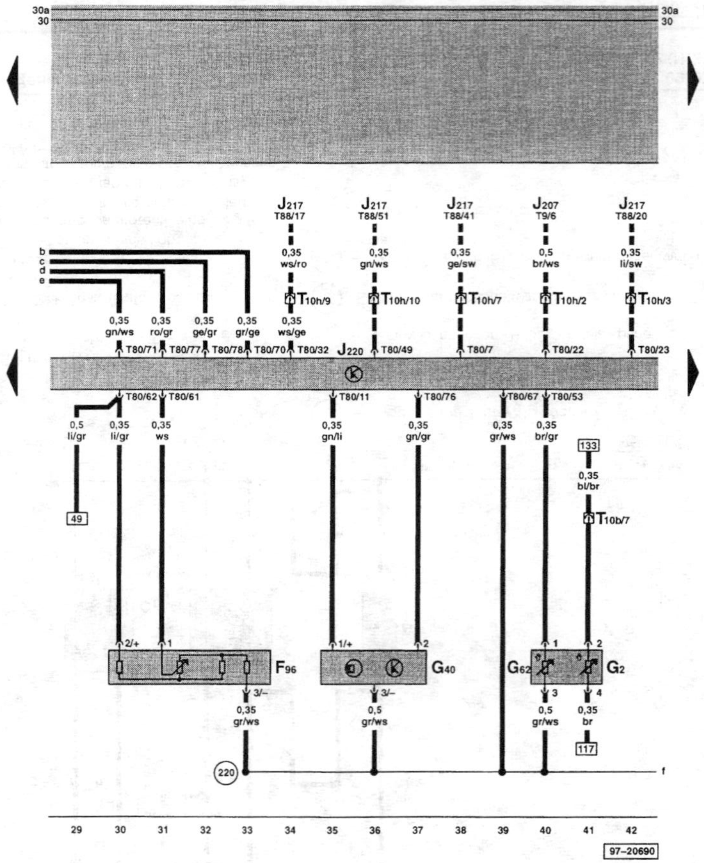

ws = white
sw = black
ro = red
br = brown
gn = green
bl = blue
gr = grey
li = violet
ge = yellow

F96 – Barometric Pressure (BARO) Sensor
G2 – Engine Coolant Temperature (ECT) Sensor
G40 – Camshaft Position (CMP) Sensor
G62 – Engine Coolant Temperature (ECT) Sensor
J207 – Starting Interlock Relay
J217 – Transmission Control Module (TCM)
J220 – Motronic Engine Control Module (ECM)
T9 – 9-Pin Connector, on starting interlock relay
T10b – 10-Pin Connector, black, on protective housing for control module, in engine compartment, left
T10h – 10-Pin Connector, blue, on protective housing for control module, in engine compartment, left
T80 – Connector, 80 point, on Motronic Engine Control Module (ECM)
T88 – Connector, 88 point, on Transmission Control Module (TCM)

(220) – Ground connection (sensor ground), in engine compartment wiring harness

- - - automatic transmission only

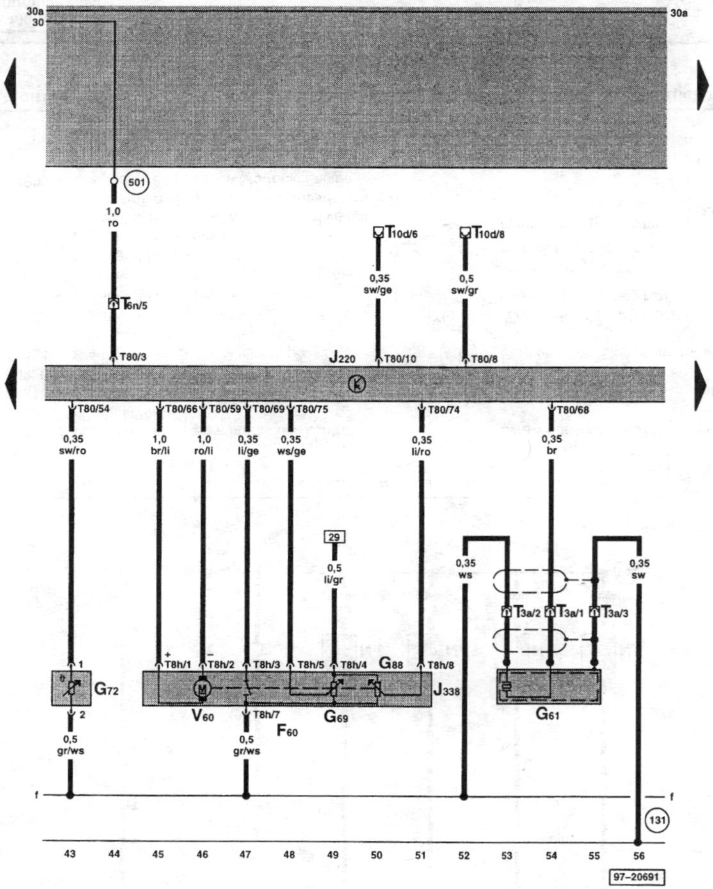

ws = white
sw = black
ro = red
br = brown
gn = green
bl = blue
gr = grey
li = violet
ge = yellow

F60 – Closed Throttle Position (CTP) Switch
G61 – Knock Sensor (KS) 1
G69 – Throttle Position (TP) Sensor
G72 – Intake Air Temperature (IAT) Sensor
G88 – Throttle Position (TP) Sensor
J220 – Motronic Engine Control Module (ECM)
J338 – Throttle Valve Control Module
T3a – 3-Pin Connector, green, on firewall, left
T6n – 6-Pin Connector, red, on protective housing for control module, in engine compartment, left
T8h – 8-Pin Connector, black, on throttle valve control module
T10d – 10-Pin Connector, brown, on protective housing for control module, in engine compartment, left
T80 – Connector, 80 point, on Motronic Engine Control Module (ECM)

V60 – Throttle Position (TP) Actuator

(131) – Ground connection -2-, in engine compartment wiring harness
(501) – Screw connection -2- (30), on relay panel

Motronic engine control module (ECM), knock sensor (KS) 1, throttle valve control module, intake air temperature (IAT) sensor

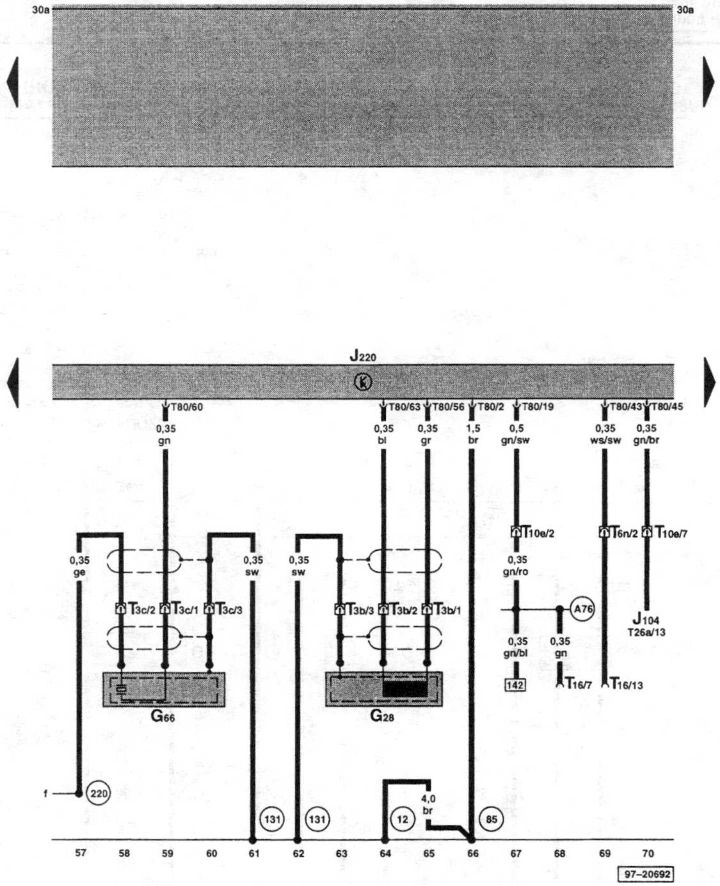

ws = white
sw = black
ro = red
br = brown
gn = green
bl = blue
gr = grey
li = violet
ge = yellow

G28 – Engine Speed (RPM) Sensor
G66 – Knock Sensor (KS) 2
J104 – ABS Control Module (w/EDL)
J220 – Motronic Engine Control Module (ECM)
T3b – 3-Pin Connector, grey, on firewall, left
T3c – 3-Pin Connector, blue, on firewall, left
T6n – 6-Pin Connector, red, on protective housing for control module, in engine compartment, left
T10e – 10-Pin Connector, yellow, on protective housing for control module, in engine compartment, left
T16 – 16-Pin Connector, On Board Diagnostic (OBD), below instrument panel, left
T26a – 26-Pin Connector, on ABS control module (with EDL)
T80 – Connector, 80 point, on Motronic Engine Control Module (ECM)

(12) – Ground connection, in engine compartment, left
(85) – Ground connection -1-, in engine compartment wiring harness
(131) – Ground connection -2-, in engine compartment wiring harness
(220) – Ground connection (sensor ground), in engine compartment wiring harness
(A76) – Connector (K-diagnosis wire), in instrument panel wiring harness

Motronic engine control module (ECM), engine speed sensor, knock sensor (KS) 2

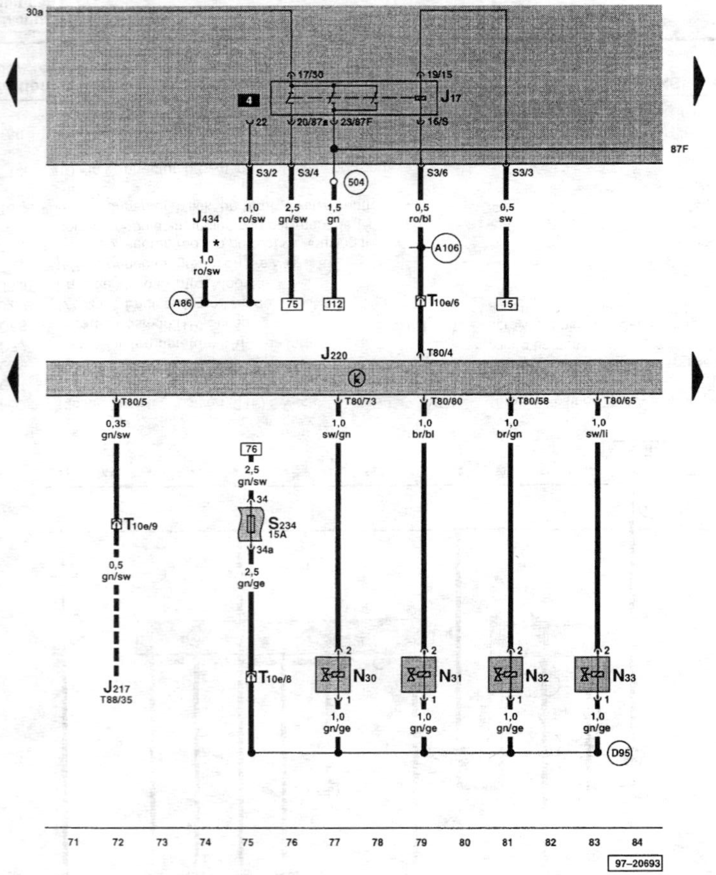

ws = white
sw = black
ro = red
br = brown
gn = green
bl = blue
gr = grey
li = violet
ge = yellow

J17 – Fuel Pump (FP) Relay, production control number (208)
J217 – Transmission Control Module (TCM)
J220 – Motronic Engine Control Module (ECM)
J434 – Starting Interlock Realy (clutch position), manual transmission only
N30 – Cylinder 1 Fuel Injector
N31 – Cylinder 2 Fuel Injector
N32 – Cylinder 3 Fuel Injector
N33 – Cylinder 4 Fuel Injector
S234 – Fuse 34 in fuse holder
T10e – 10-Pin Connector, yellow, on protective housing for control module, in engine compartment, left
T80 – Connector, 80 point, on Motronic Engine Control Module (ECM)
T88 – Connector, 88 point, on Transmission Control Module (TCM)

(504) – Screw connection (87F), on relay panel
(A86) – Connection (50a), in instrument panel wiring harness
(A106) – Connector -2- (86s), in instrument panel wiring harness
(D95) – wire connection (injectors), in engine compartment wiring harness

* – manual transmission only
- - - automatic transmission only

Motronic engine control module (ECM), fuel injectors, fuel pump (FP) relay

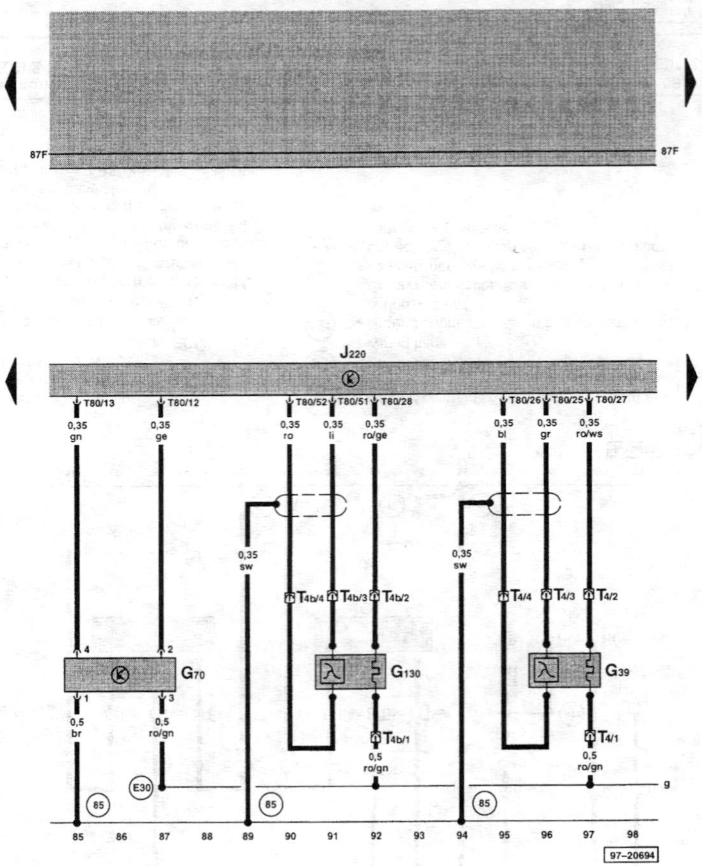

ws = white
sw = black
ro = red
br = brown
gn = green
bl = blue
gr = grey
li = violet
ge = yellow

G39 – Heated Oxygen Sensor (HO2S)
G70 – Mass Air Flow (MAF) Sensor
G130 – Oxygen Sensor (O2S) Behind Three Way Catalytic Converter (TWC)
J220 – Motronic Engine Control Module (ECM)
T4 – 4-Pin Connector
T4b – 4-Pin Connector
T80 – Connector, 80 point, on Motronic Engine Control Module (ECM)
(85) – Ground connection -1-, in engine compartment wiring harness
(E30) – Connector (87a), in wiring harness engine

Motronic engine control module (ECM), mass air flow (MAF) sensor, heated oxygen sensor (HO2S), oxygen sensor (O2S)

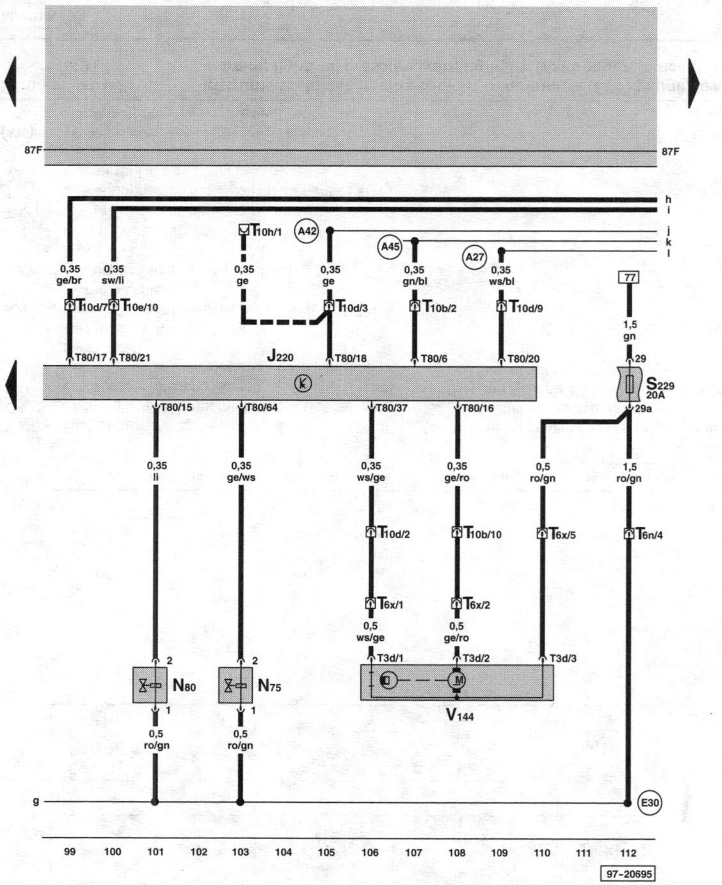

ws = white
sw = black
ro = red
br = brown
gn = green
bl = blue
gr = grey
li = violet
ge = yellow

J220 - Motronic Engine Control Module (ECM)
N75 - Wastegate Bypass Regulator Valve
N80 - Evaporative Emission (EVAP) Canister Purge Regulator Valve
S229 - Fuse 29 in fuse holder
T3d - 3-Pin Connector, black
T6n - 6-Pin Connector, red, on protective housing for control module, in engine compartment, left
T6x - 6-Pin Connector, brown, connector station A-pillar, right
T10b - 10-Pin Connector, black, on protective housing for control module, in engine compartment, left
T10d - 10-Pin Connector, brown, on protective housing for control module, in engine compartment, left
T10e - 10-Pin Connector, yellow, on protective housing for control module, in engine compartment, left
T10h - 10-Pin Connector, blue, on protective housing for control module, in engine compartment, left
T80 - Connector, 80 point, on Motronic Engine Control Module (ECM)
V144 - Leak Detection Pump (LDP)
(A27) - Wire Connection (vehicle speed signal), in instrument panel wiring harness
(A42) - plus connection (fuel gauge), in instrument panel wiring harness
(A45) - wire connection (RPM-signal), in instrument panel wiring harness
(E30) - Connector (87a), in wiring harness engine
- - - automatic transmission only

Motronic engine control module (ECM), wastegate bypass regulator valve, leak detection pump (LDP)

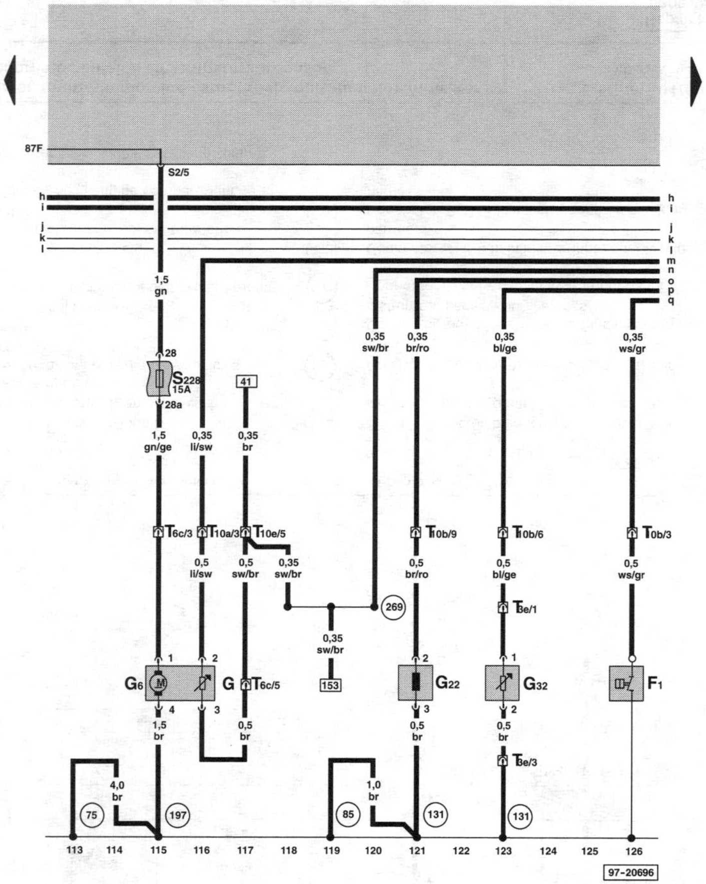

ws = white
sw = black
ro = red
br = brown
gn = green
bl = blue
gr = grey
li = violet
ge = yellow

F1 - Oil Pressure Switch
G - Sender for fuel gauge
G6 - Fuel Pump (FP)
G22 - Speedometer Vehicle Speed Sensor (VSS)
G32 - Engine Coolant Level (ECL) Sensor
S228 - Fuse 28 in fuse holder
T3e - 3-Pin Connector, in engine compartment, left front
T6c - 6-Pin Connector, blue, connector station A-pillar, left
T10a - 10-Pin Connector, brown, connector station A-pillar, left
T10b - 10-Pin Connector, black, on protective housing for control module, in engine compartment, left
T10e - 10-Pin Connector, yellow, on protective housing for control module, in engine compartment, left
(75) - Ground connection, on right rear pillar
(85) - Ground connection -1-, in engine compartment wiring harness
(131) - Ground connection -2-, in engine compartment wiring harness
(197) - Ground connection -4-, in wiring harness rear
(269) - Ground connection (sensor ground) -1-, in instrument panel wiring harness

Oil pressure switch, fuel pump (FP), sender for fuel gauge, speedometer vehicle speed sensor

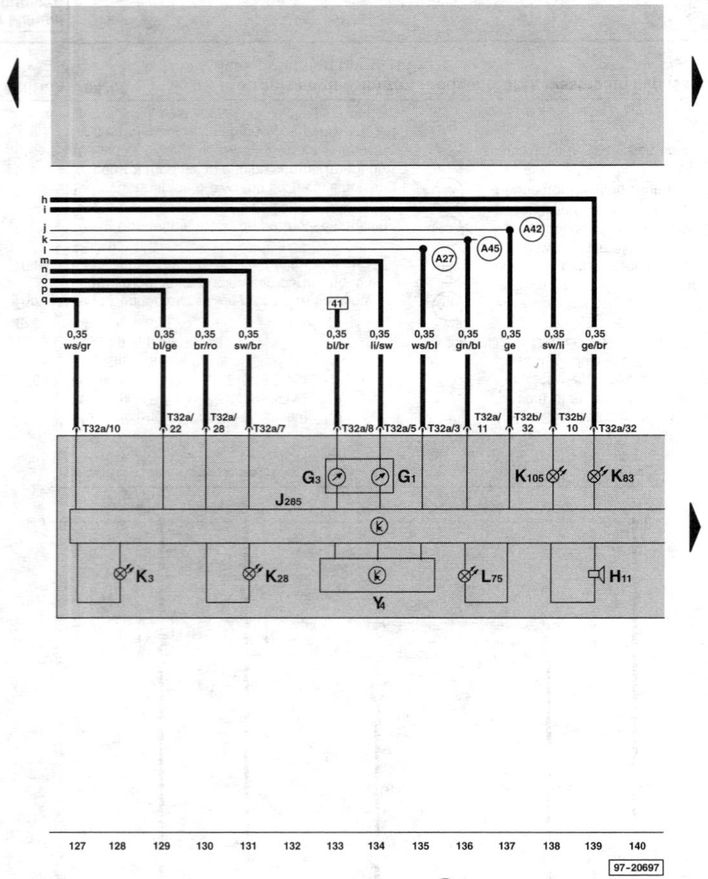

ws = white
sw = black
ro = red
br = brown
gn = green
bl = blue
gr = grey
li = violet
ge = yellow

G1 - Fuel gauge
G3 - Engine Coolant Temperature (ECT) Gauge
H11 - Oil Pressure Warning Buzzer
J285 - Control module with indicator unit in instrument panel insert
K3 - Oil Pressure Warning Light
K28 - Engine Coolant Level/Temperature (ECL/ECT) Warning Light
K83 - Malfunction Indicator Lamp (MIL)
K105 - Low Fuel Level Warning Light
L75 - Digital Display Light
T32a - 32-Pin Connector, blue, on instrument cluster
T32b - 32-Pin Connector, green, on instrument cluster
Y4 - Odometer Display

(A27) - Wire Connection (vehicle speed signal), in instrument panel wiring harness

(A42) - plus connection (fuel gauge), in instrument panel wiring harness
(A45) - wire connection (RPM-signal), in instrument panel wiring harness

Edition 04/00
USA.5062.11.21

Instrument cluster, engine coolant temperature (ECT) gauge, low fuel level warning light, oil pressure warning light, Malfunction Indicator Lamp (MIL)

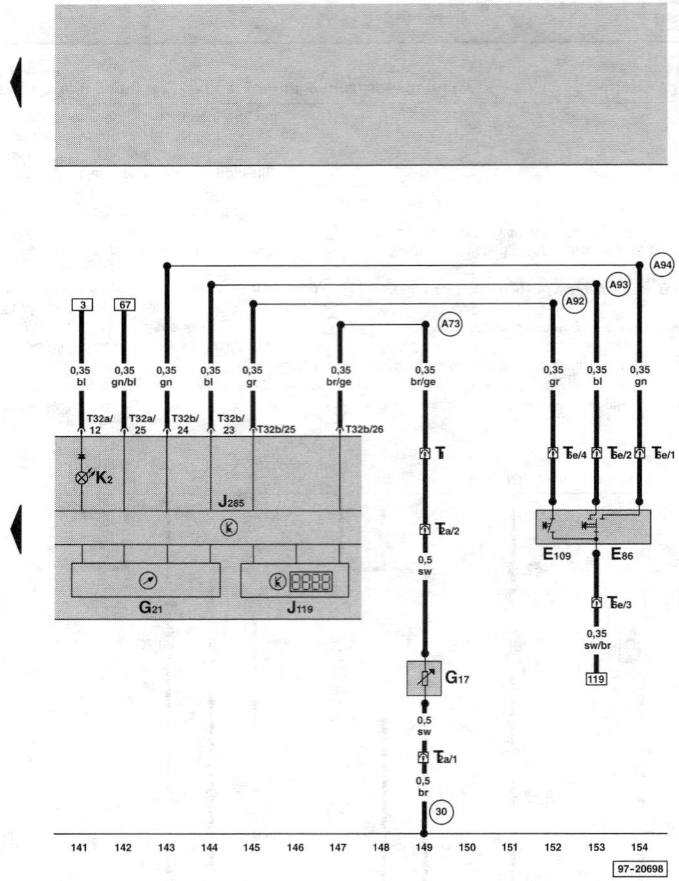

ws = white
sw = black
ro = red
br = brown
gn = green
bl = blue
gr = grey
li = violet
ge = yellow

E86 - Multi-Funktion Indicator Mode Select Switch
E109 - Multi-Funktion Indicator Memory Switch
G17 - Outside Air Temperature Sensor (use correspond wiring diagram by vehicles with air conditioning)
G21 - Speedometer
J119 - Multi-function Indicator (MFI)
J285 - Control module with indicator unit in instrument panel insert
K2 - Generator (GEN) Warning Light
T1 - Single Connector, brown, connector station A-pillar, left
T2a - Double Connector, in engine compartment, left front
T6e - 6-Pin Connector, black, behind steering column switch cover

T32a - 32-Pin Connector, blue, on instrument cluster
T32b - 32-Pin Connector, green, on instrument cluster

(30) - Ground connection, -1-, beside fuse/relay panel
(A73) - Connector (outside temperature indicator), in instrument panel wiring harness
(A92) - Connector (MFI-Reset), in instrument panel wiring harness
(A93) - Connector -1- (MFI), in instrument panel wiring harness
(A94) - Connector -2- (MFI), in instrument panel wiring harness

Instrument cluster, multi-function indicator (MFI), multi-function indicator switches, speedometer, outside air temperature sensor

Edition 04/00
USA.5062.11.21

1,9L-Engine – Turbo Diesel Fuel Injection (DFI)/66KW, code AHH, manual transmission only

from September 1997

Deviate relay location and fuseplacements as well as the locations of multiple connectors see section "component locations".

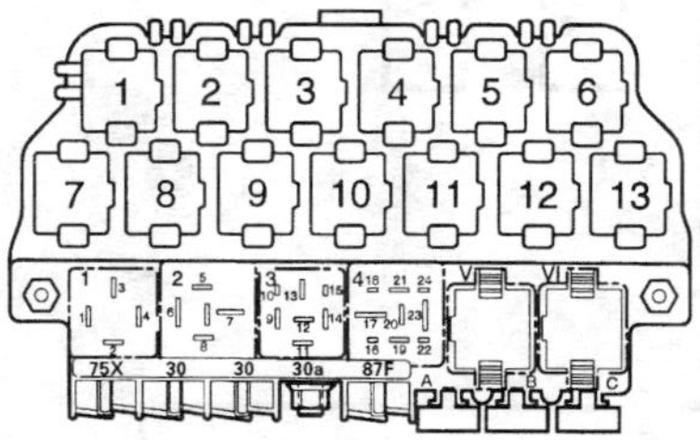

97-20676

Relay location on the thirteenfold auxiliary relay panel, above relay panel:

10 Glow Plug Relay (180)

Relay panel:

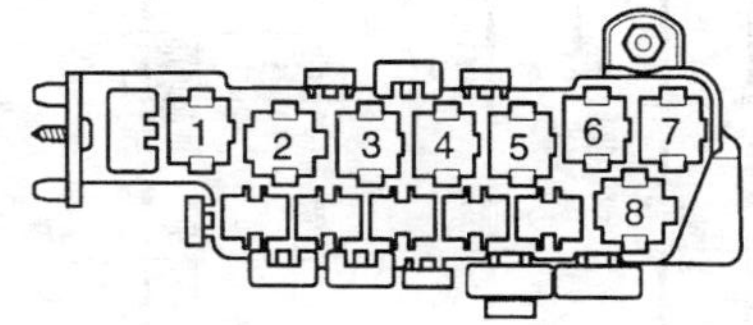

97-20515

Relay location on the eightfold auxiliary relay panel, behind relay panel:

Note: Number in parentheses indicates production control number stamped on relay housing.

Fuse colors

30 A – green
25 A – white
20 A – yellow
15 A – blue
10 A – red
7,5 A – brown
5 A – beige

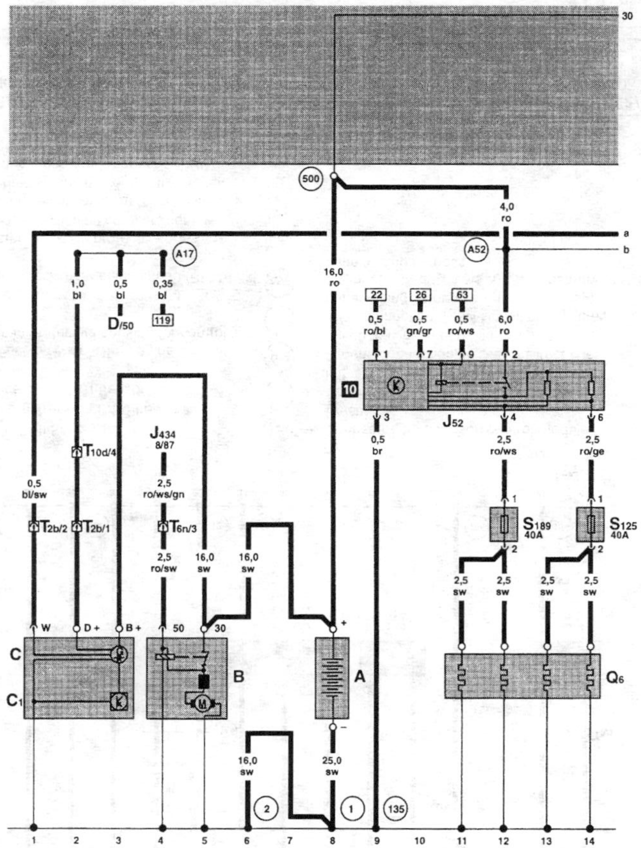

97-21030

ws = white
sw = black
ro = red
br = brown
gn = green
bl = blue
gr = grey
li = violet
ge = yellow

A – Battery
B – Starter
C – Generator (GEN)
C1 – Voltage Regulator (VR)
D – Ignition/Starter Switch
J52 – Glow Plug Relay, on the thirteenfold auxiliary relay panel, above relay panel, production control number (180)
J434 – Starting Interlock Relay (clutch position)
Q6 – Glow plugs (engine)
S125 – Fuse for glow plug, in protective housing for control module, in engine compartment, left
S189 – Fuse -2- for glow plug, in protective housing for control module, in engine compartment, left
T2b – Double Connector, on engine right
T6n – 6-Pin Connector, red, on protective housing for control module, in engine compartment, left
T10d – 10-Pin Connector, brown, on protective housing for control module, in engine compartment, left
(1) – Ground strap, battery to body
(2) – Ground strap, transmission to body
(135) – Ground connection -2-, in instrument panel wiring harness
(500) – Screw Connection -1- (30), on relay panel
(A17) – wire connection (61), in instrument panel wiring harness
(A52) – plus connection (30), in instrument panel wiring harness

Battery, starter, generator (GEN), glow plugs (engine)

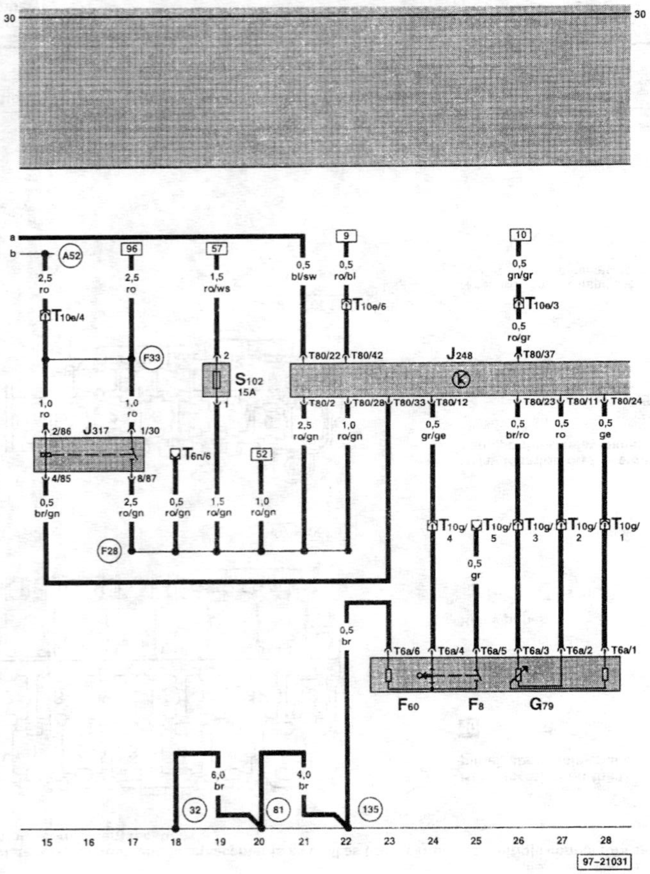

ws = white
sw = black
ro = red
br = brown
gn = green
bl = blue
gr = grey
li = violet
ge = yellow

F8 – Kick Down Switch
F60 – Closed Throttle Position (CTP) Switch
G79 – Throttle Position (TP) Sensor
J248 – Diesel Direct Fuel Injection (DFI) Engine Control Module (ECM), in protective housing for control module
J317 – Power Supply (Terminal 30, B+) Relay, in protective housing for control module, production control number (219)
S102 – Engine Control Module (ECM) Fuse, in protective housing for control module
T6a – 6-Pin Connector
T6n – 6-Pin Connector, red, on protective housing for control module, in engine compartment, left
T10e – 10-Pin Connector, yellow, on protective housing for control module, in engine compartment, left
T10g – 10-Pin Connector, white, on protective housing for control module, in engine compartment, left
T80 – Connector, 80 point, on control module for Diesel
(32) – Ground connection, behind instrument panel, left
(81) – Ground connection -1-, in instrument panel wiring harness
(135) – Ground connection -2-, in instrument panel wiring harness
(A52) – plus connection (30), in instrument panel wiring harness
(F28) – Wire connection -3-, in Diesel Direct Fuel Injection (DFI) system wiring harness
(F33) – Plus connector (30), in wiring harness Diesel direct injection System

Diesel direct fuel injection (DFI) engine control module (ECM), kick down switch, closed throttle position (CTP) switch, throttle position (TP) sensor

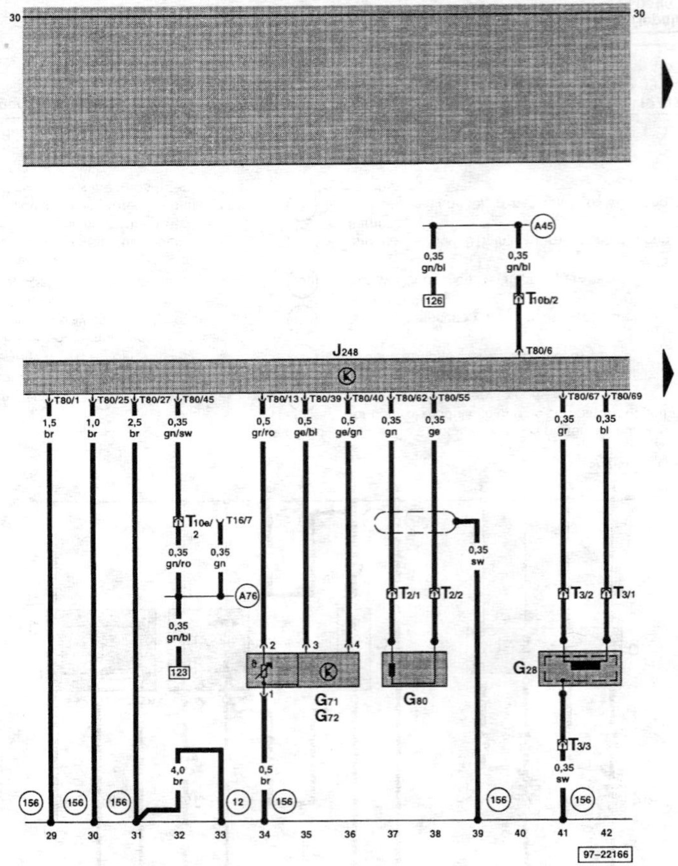

ws = white
sw = black
ro = red
br = brown
gn = green
bl = blue
gr = grey
li = violet
ge = yellow

G28 – Engine Speed (RPM) Sensor
G71 – Manifold Absolute Pressure (MAP) Sensor
G72 – Intake Air Temperature (IAT) Sensor
G80 – Needle Lift Sensor
J248 – Diesel Direct Fuel Injection (DFI) Engine Control Module (ECM), in protective housing for control module
T2 – Double Connector, brown
T3 – 3-Pin Connector, black, near egine speed (RPM) sensor
T10b – 10-Pin Connector, black, on protective housing for control module, in engine compartment, left
T10e – 10-Pin Connector, yellow, on protective housing for control module, in engine compartment, left
T16 – 16-Pin Connector, On Board Diagnostic (OBD), below instrument panel, left
T80 – Connector, 80 point, on control module for Diesel
(12) – Ground connection, in engine compartment, left
(156) – Ground connection, in Diesel Direct Fuel Injection (DFI) wiring harness
(A45) – wire connection (RPM-signal), in instrument panel wiring harness
(A76) – Connector (K-diagnosis wire), in instrument panel wiring harness

Engine speed (RPM) sensor, needle lift sensor, Manifold Absolute Pressure (MAP) Sensor, Intake Air Temperature (IAT) Sensor

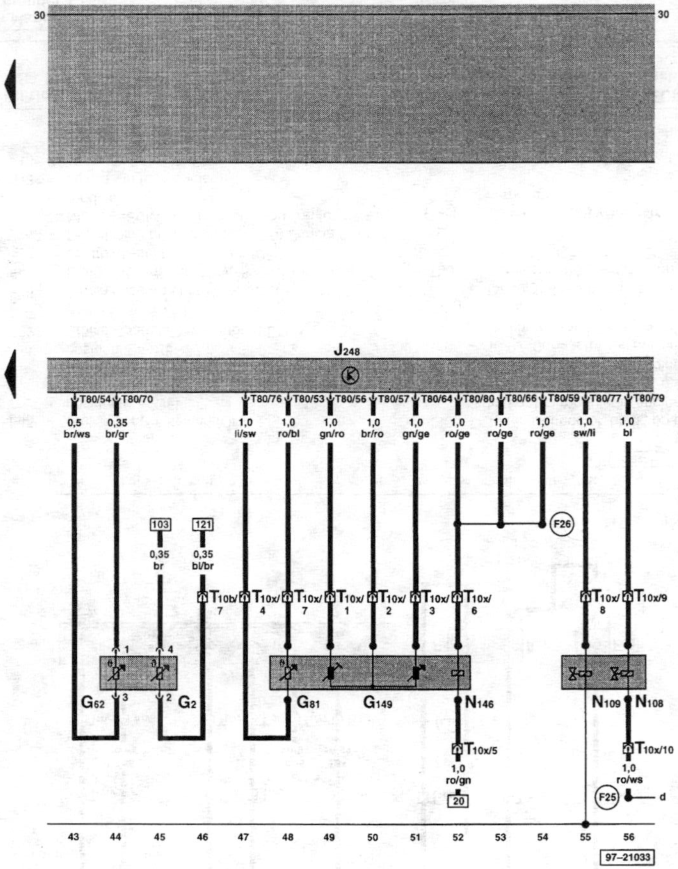

ws = white
sw = black
ro = red
br = brown
gn = green
bl = blue
gr = grey
li = violet
ge = yellow

G2 – Engine Coolant Temperature (ECT) Sensor
G62 – Engine Coolant Temperature (ECT) Sensor
G81 – Fuel Temperature Sensor
G149 – Modulating Piston Displacement Sensor
J248 – Diesel Direct Fuel Injection (DFI) Engine Control Module (ECM), in protective housing for control module
N108 – Cold Start Injector
N109 – Fuel Cut-off Valve
N146 – Quantity Adjuster
T10b – 10-Pin Connector, black, on protective housing for control module, in engine compartment, left
T10x – 10-Pin Connector, near fuel injection pump
T80 – Connector, 80 point, on control module for Diesel

(220) – Ground connection (sensor ground), in engine compartment wiring harness
(F25) – Wire connection -1-, in Diesel Direct Fuel Injection (DFI) system wiring harness
(F26) – Wire connection -2-, in Diesel Direct Fuel Injection (DFI) system wiring harness

Edition 09/97
USA.5062.01.21

Diesel direct fuel injection (DFI) engine control module (ECM), engine coolant temperature (ECT) sensor, quantity adjuster, cold start injector

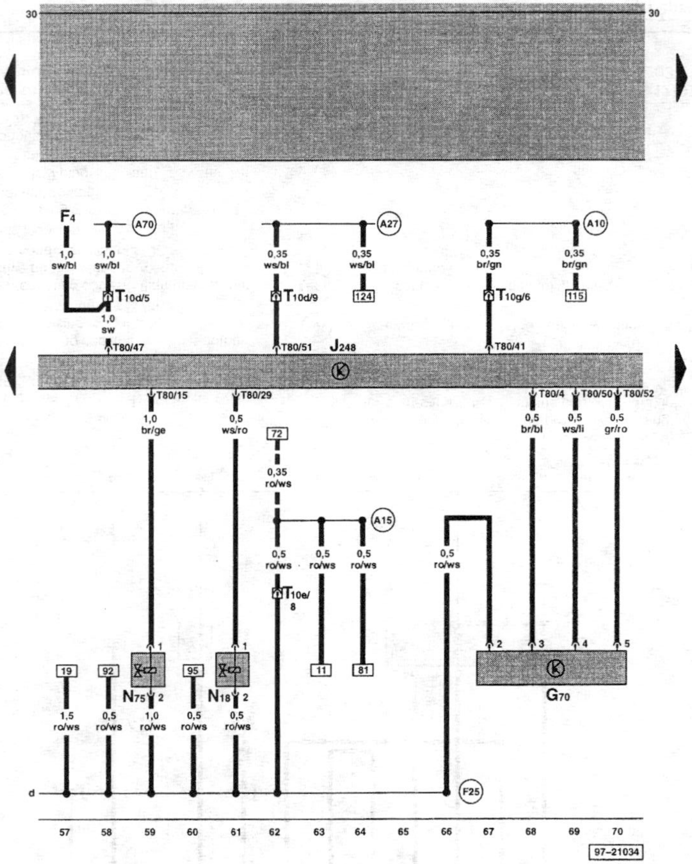

ws = white
sw = black
ro = red
br = brown
gn = green
bl = blue
gr = grey
li = violet
ge = yellow

F4 – Back-Up Light Switch
G70 – Mass Air Flow (MAF) Sensor
J248 – Diesel Direct Fuel Injection (DFI) Engine Control Module (ECM), in protective housing for control module
N18 – EGR Vacuum Regulator Solenoid Valve
N75 – Wastegate Bypass Regulator Valve
T10d – 10-Pin Connector, brown, on protective housing for control module, in engine compartment, left
T10e – 10-Pin Connector, yellow, on protective housing for control module, in engine compartment, left
T10g – 10-Pin Connector, white, on protective housing for control module, in engine compartment, left
T80 – Connector, 80 point, on control module for Diesel

(A10) – Wire connection (glow plug control) in instrument panel wiring harness
(A15) – plus connection (15), in instrument panel wiring harness
(A27) – Wire Connection (vehicle speed signal), in instrument panel wiring harness
(A70) – Connector (15a, fuse 231), in instrument panel wiring harness
(F25) – Wire connection -1-, in Diesel Direct Fuel Injection (DFI) system wiring harness

--- vehicles with cruise control only

Diesel direct fuel injection (DFI) engine control module (ECM), mass air flow (MAF) sensor, EGR vacuum regulator solenoid valve

Edition 09/97
USA.5062.01.21

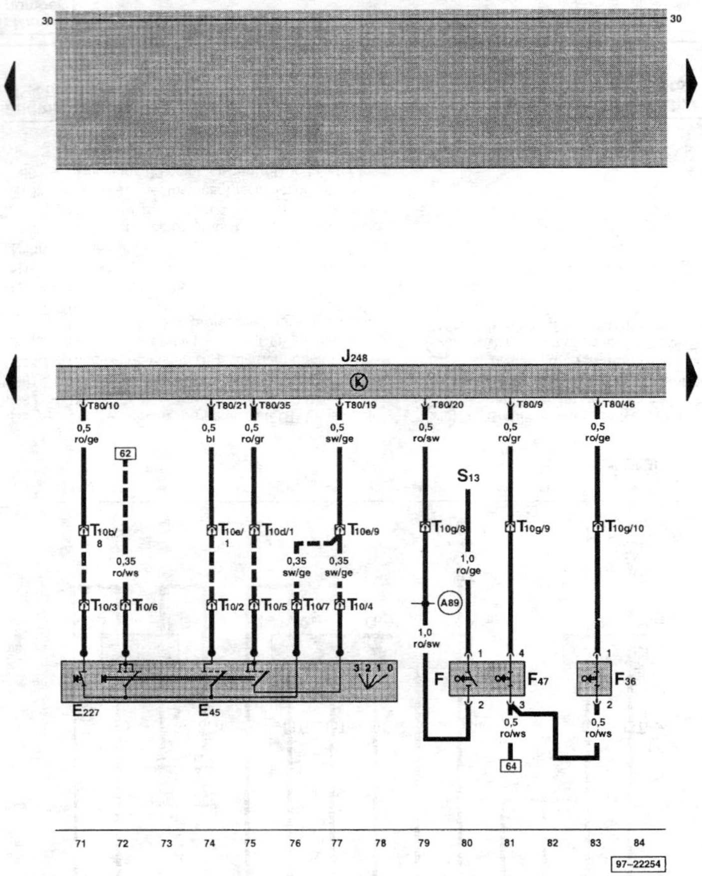

ws = white
sw = black
ro = red
br = brown
gn = green
bl = blue
gr = grey
li = violet
ge = yellow

E45 – Cruise Control Switch (ON/OFF)
position 0 = OFF
position 1 = OFF save
position 2 = ON
position 3 = RES
E227 – Cruise Control Push Button (SET)
F – Brake Light Switch
F36 – Clutch Vacuum Vent Valve Switch
F47 – Brake Vacuum Vent Valve Switch for cruise control/diesel fuel injection (DFI)
J248 – Diesel Direct Fuel Injection (DFI) Engine Control Module (ECM), in protective housing for control module
S13 – Fuse 13 in fuse holder
T10 – 10-Pin Connector, white, near steering column, right
T10b – 10-Pin Connector, black, on protective housing for control module, in engine compartment, left
T10d – 10-Pin Connector, brown, on protective housing for control module, in engine compartment, left
T10e – 10-Pin Connector, yellow, on protective housing for control module, in engine compartment, left
T10g – 10-Pin Connector, white, on protective housing for control module, in engine compartment, left
T80 – Connector, 80 point, on control module for Diesel
(A89) – Connector (54), in instrument panel wiring harness
- - - – vehicles with cruise control only

Diesel direct fuel injection (DFI) engine control module (ECM), cruise control switch, brake light switch, clutch vacuum vent valve switch

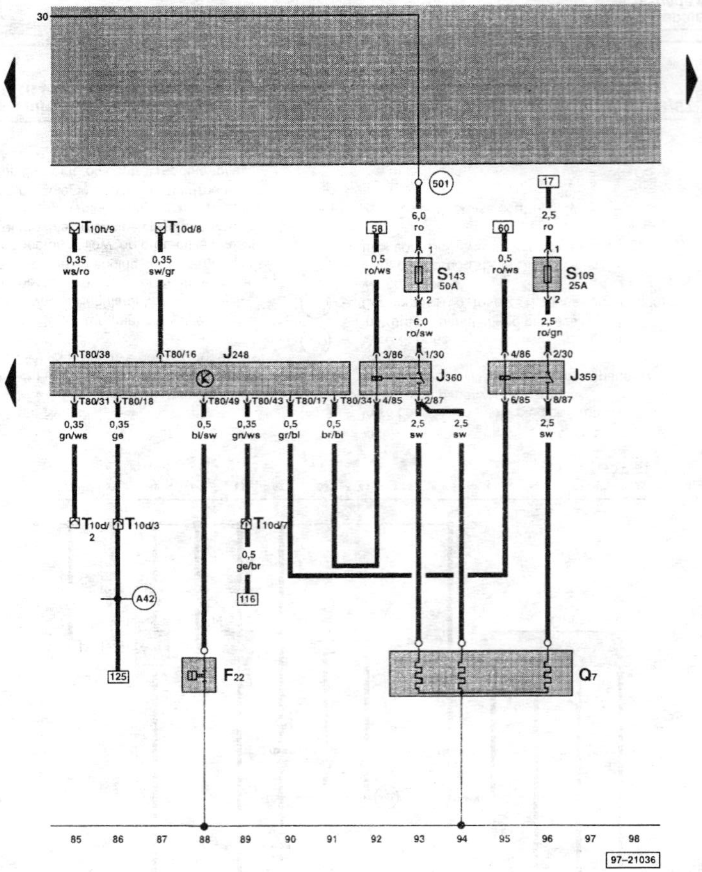

ws = white
sw = black
ro = red
br = brown
gn = green
bl = blue
gr = grey
li = violet
ge = yellow

F22 – 0,25bar Oil Pressure Switch
J248 – Diesel Direct Fuel Injection (DFI) Engine Control Module (ECM), in protective housing for control module
J359 – Relay for preheating coolant, low heat output, in protective housing for control module, production control number (373)
J360 – Relay for preheating coolant, high heat output, in protective housing for control module, production control number (370)
Q7 – Glow plugs (coolant)
S109 – Fuse for glow plug (coolant), in protective housing for control module
S143 – Fuse -2- for glow plugs (coolant), in protective housing for control module
T10d – 10-Pin Connector, brown, on protective housing for control module, in engine compartment, left
T10h – 10-Pin Connector, blue, on protective housing for control module, in engine compartment, left
T80 – Connector, 80 point, on control module for Diesel
(501) – Screw Connection -2- (30), on relay panel
(A42) – plus connection (fuel gauge), in instrument panel wiring harness

Diesel direct fuel injection (DFI) engine control module (ECM), coolant glow plug relay, glow plugs (coolant), 0,25bar oil pressure switch

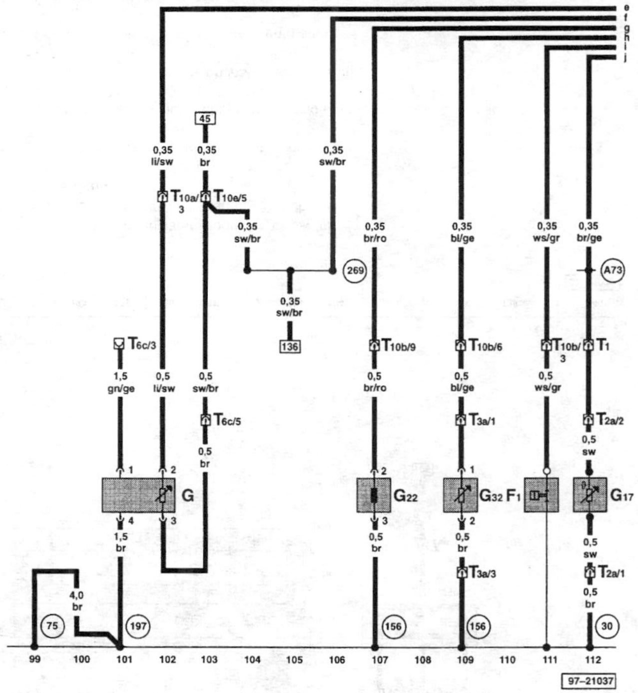

ws = white
sw = black
ro = red
br = brown
gn = green
bl = blue
gr = grey
li = violet
ge = yellow

F1 – Oil Pressure Switch
G – Sender for fuel gauge
G17 – Outside Air Temperature Sensor (use correspond wiring diagram by vehicles with air conditioning)
G22 – Speedometer Vehicle Speed Sensor (VSS)
G32 – Engine Coolant Level (ECL) Sensor
T1 – Single Connector, brown, connector station A-pillar, left
T2a – Double Connector, in engine compartment, left front
T3a – 3-Pin Connector, in engine compartment, left front
T6c – 6-Pin Connector, blue, connector station A-pillar, left
T10a – 10-Pin Connector, brown, connector station A-pillar, left
T10b – 10-Pin Connector, black, on protective housing for control module, in engine compartment, left
T10e – 10-Pin Connector, yellow, on protective housing for control module, in engine compartment, left
(30) – Ground connection, -1-, beside fuse/relay panel
(75) – Ground connection, on right rear pillar
(156) – Ground connection, in Diesel Direct Fuel Injection (DFI) wiring harness
(197) – Ground connection -4-, in wiring harness rear
(269) – Ground connection (sensor ground) -1-, in instrument panel wiring harness
(A73) – Connector (outside temperature indicator), in instrument panel wiring harness

Edition 09/97
USA.5062.01.21

Oil pressure switch, sender for fuel gauge, outside air temperature sensor, speedometer vehicle speed sensor (VSS), engine coolant level (ECL) sensor

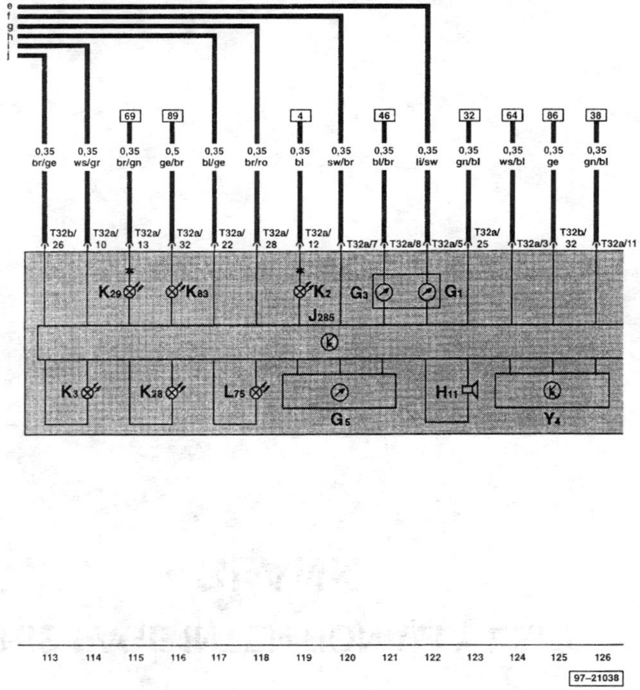

ws = white
sw = black
ro = red
br = brown
gn = green
bl = blue
gr = grey
li = violet
ge = yellow

G1 – Fuel gauge
G3 – Engine Coolant Temperature (ECT) Gauge
G5 – Tachometer
H11 – Oil Pressure Warning Buzzer
J285 – Control module with indicator unit in instrument panel insert
K2 – Generator (GEN) Warning Light
K3 – Oil Pressure Warning Light
K28 – Engine Coolant Level/Temperature (ECL/ECT) Warning Light
K29 – Glow Plug Indicator Light
K83 – Exhaust Warning Light
L75 – Digital Display Light
T32a – 32-Pin Connector, blue, on instrument cluster
T32b – 32-Pin Connector, green, on instrument cluster
Y4 – Odometer Display

Instrument cluster, engine coolant level/temperature (ECL/ECT) warning light, visual and acoustical oil pressure control, exhaust warning light

Edition 09/97
USA.5062.01.21

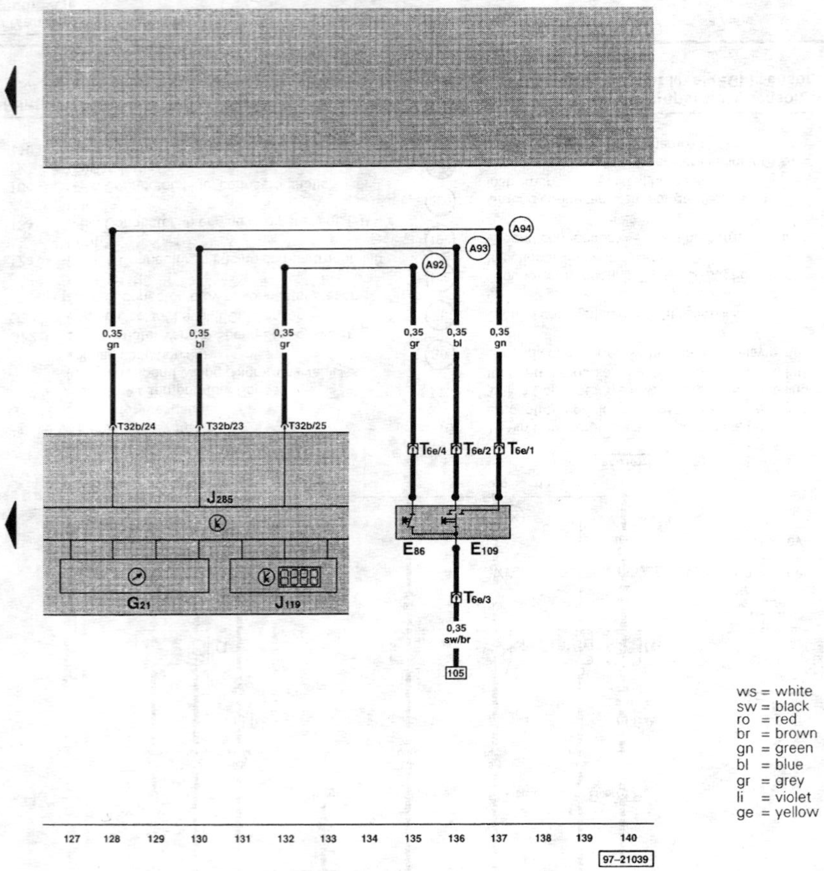

ws = white
sw = black
ro = red
br = brown
gn = green
bl = blue
gr = grey
li = violet
ge = yellow

E86 – Multi-Funktion Indicator Mode Select Switch
E109 – Multi-Funktion Indicator Memory Switch
G21 – Speedometer
J119 – Multi-function Indicator (MFI)
J285 – Control module with indicator unit in instrument panel insert
T6e – 6-Pin Connector, black, behind steering column switch cover
T32b – 32-Pin Connector, green, on instrument cluster

(A92) – Connector (MFI-Reset), in instrument panel wiring harness
(A93) – Connector -1- (MFI), in instrument panel wiring harness
(A94) – Connector -2- (MFI), in instrument panel wiring harness

THIS PAGE INTENTIONALLY LEFT
BLANK

1,9L-Engine – Turbo Diesel Fuel Injection (DFI)/66KW, code AHH, automatic transmission only

from September 1997

Deviate relay location and fuseplacements as well as the locations of multiple connectors see section "component locations".

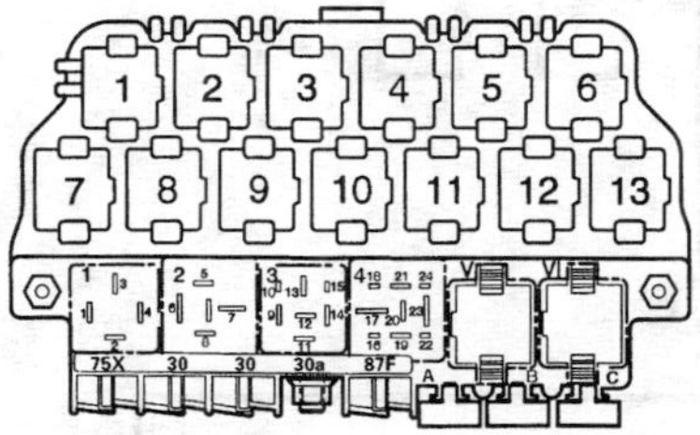

97–20676

Relay location on the thirteenfold auxiliary relay panel, above relay panel:

10 Glow Plug Relay (180)

Relay panel:

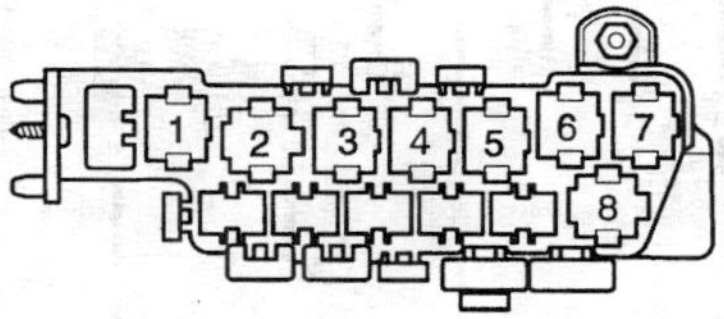

97–20515

Relay location on the eightfold auxiliary relay panel, behind relay panel:

Note: Number in parentheses indicates production control number stamped on relay housing.

Fuse colors
30 A – green
25 A – white
20 A – yellow
15 A – blue
10 A – red
7,5 A – brown
5 A – beige

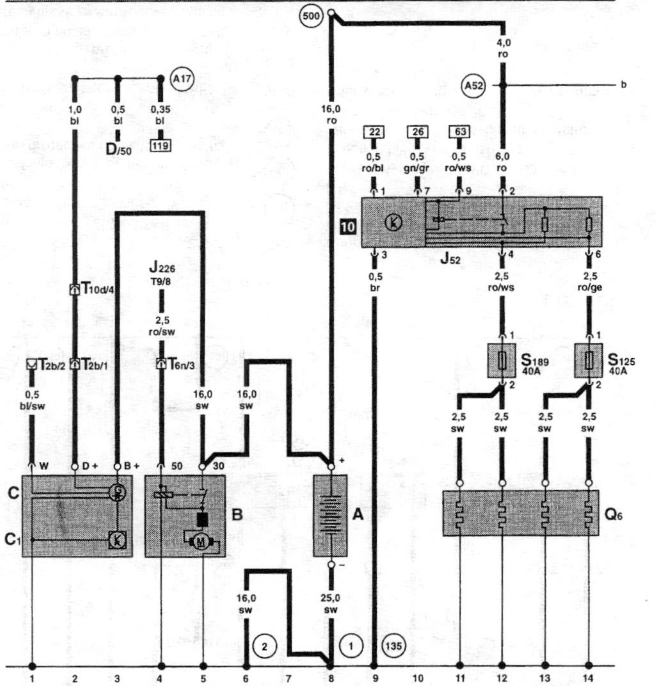

ws = white
sw = black
ro = red
br = brown
gn = green
bl = blue
gr = grey
li = violet
ge = yellow

A – Battery
B – Starter
C – Generator (GEN)
C1 – Voltage Regulator (VR)
D – Ignition/Starter Switch
J52 – Glow Plug Relay, on the thirteenfold auxiliary relay panel, above relay panel, production control number (180)
J226 – Park/Neutral Position (PNP) Relay
Q6 – Spark plugs (engine)
S125 – Fuse for glow plug, in protective housing for control module, in engine compartment, left
S189 – Fuse -2- for glow plug, in protective housing for control module, in engine compartment, left
T2b – Double Connector, on engine right
T6n – 6-Pin Connector, red, on protective housing for control module, in engine compartment, left
T9 – 9-Pin Connector, on Park/Neutral Position (PNP) Relay
T10d – 10-Pin Connector, brown, on protective housing for control module, in engine compartment, left
(1) – Ground strap, battery to body
(2) – Ground strap, transmission to body
(135) – Ground connection -2-, in instrument panel wiring harness
(500) – Screw Connection -1- (30), on relay panel
(A17) – wire connection (61), in instrument panel wiring harness
(A52) – plus connection (30), in instrument panel wiring harness

Battery, starter, generator (GEN), glow plug (engine)

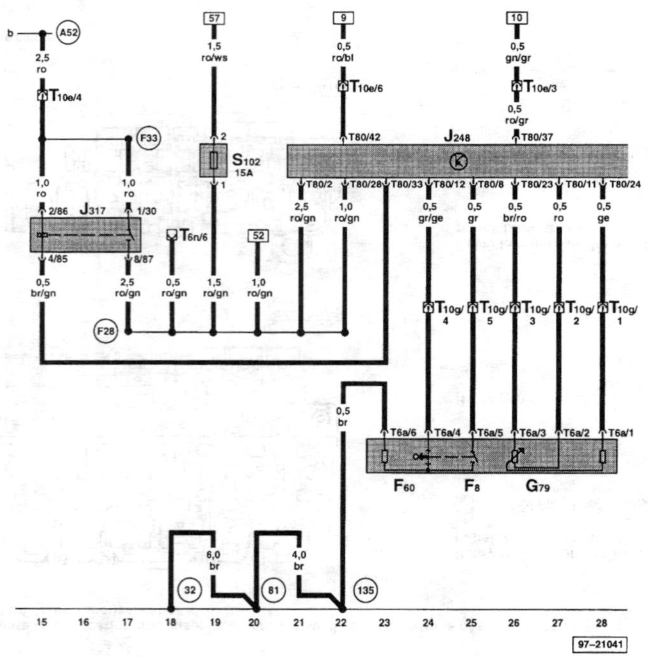

ws = white
sw = black
ro = red
br = brown
gn = green
bl = blue
gr = grey
li = violet
ge = yellow

F8 – Kick Down Switch
F60 – Closed Throttle Position (CTP) Switch
G79 – Throttle Position (TP) Sensor
J248 – Diesel Direct Fuel Injection (DFI) Engine Control Module (ECM), in protective housing for control module
J317 – Power Supply (Terminal 30, B+) Relay, in protective housing for control module, production control number (219)
S102 – Engine Control Module (ECM) Fuse, in protective housing for control module
T6a – 6-Pin Connector
T6n – 6-Pin Connector, red, on protective housing for control module, in engine compartment, left
T10e – 10-Pin Connector, yellow, on protective housing for control module, in engine compartment, left
T10g – 10-Pin Connector, white, on protective housing for control module, in engine compartment, left
T80 – Connector, 80 point, on control module for Diesel
(32) – Ground connection, behind instrument panel, left
(81) – Ground connection -1-, in instrument panel wiring harness
(135) – Ground connection -2-, in instrument panel wiring harness
(A52) – plus connection (30), in instrument panel wiring harness
(F28) – Wire connection -3-, in Diesel Direct Fuel Injection (DFI) system wiring harness
(F33) – Plus connector (30), in wiring harness Diesel direct injection System

Diesel direct fuel injection (DFI) engine control module (ECM), kick down switch, closed throttle position (CTP) switch, throttle position (TP) sensor

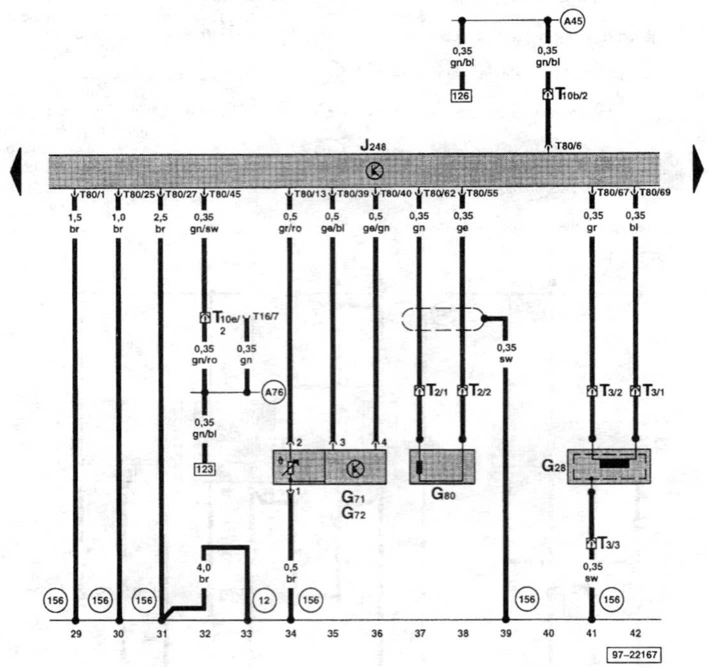

ws = white
sw = black
ro = red
br = brown
gn = green
bl = blue
gr = grey
li = violet
ge = yellow

G28 – Engine Speed (RPM) Sensor
G71 – Manifold Absolute Pressure (MAP) Sensor
G72 – Intake Air Temperature (IAT) Sensor
G80 – Needle Lift Sensor
J248 – Diesel Direct Fuel Injection (DFI) Engine Control Module (ECM), in protective housing for control module
T2 – Double Connector, brown
T3 – 3-Pin Connector, black, near egine speed (RPM) sensor
T10b – 10-Pin Connector, black, on protective housing for control module, in engine compartment, left
T10e – 10-Pin Connector, yellow, on protective housing for control module, in engine compartment, left
T16 – 16-Pin Connector, On Board Diagnostic (OBD), below instrument panel, left
T80 – Connector, 80 point, on control module for Diesel
(12) – Ground connection, in engine compartment, left
(156) – Ground connection, in Diesel Direct Fuel Injection (DFI) wiring harness
(A45) – wire connection (RPM-signal), in instrument panel wiring harness
(A76) – Connector (K-diagnosis wire), in instrument panel wiring harness

Engine speed (RPM) sensor, needle lift sensor, Manifold Absolute Pressure (MAP) Sensor, Intake Air Temperature (IAT) Sensor

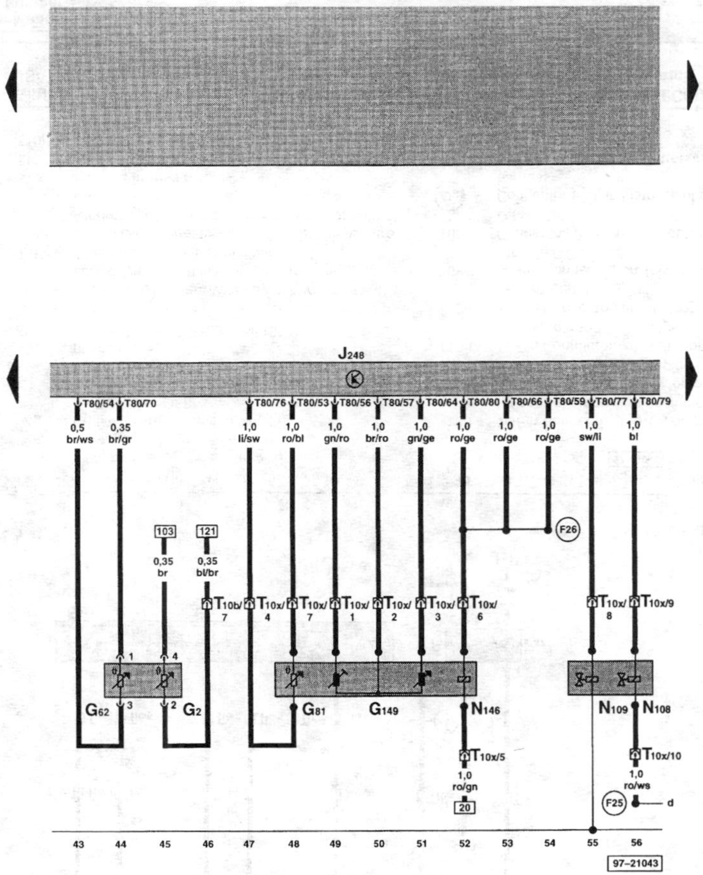

ws = white
sw = black
ro = red
br = brown
gn = green
bl = blue
gr = grey
li = violet
ge = yellow

G2 – Engine Coolant Temperature (ECT) Sensor
G62 – Engine Coolant Temperature (ECT) Sensor
G81 – Fuel Temperature Sensor
G149 – Modulating Piston Displacement Sensor
J248 – Diesel Direct Fuel Injection (DFI) Engine Control Module (ECM), in protective housing for control module
N108 – Cold Start Injector
N109 – Fuel Cut-off Valve
N146 – Quantity Adjuster
T10b – 10-Pin Connector, black, on protective housing for control module, in engine compartment, left
T10x – 10-Pin Connector, near fuel injection pump
T80 – Connector, 80 point,on control module for Diesel
(220) – Ground connection (sensor ground), in engine compartment wiring harness

(F25) – Wire connection -1-, in Diesel Direct Fuel Injection (DFI) system wiring harness
(F26) – Wire connection -2-, in Diesel Direct Fuel Injection (DFI) system wiring harness

Edition 09/97
USA.5062.01.21

Diesel direct fuel injection (DFI) engine control module (ECM), engine coolant temperature (ECT) sensor, quantity adjuster, cold start injector

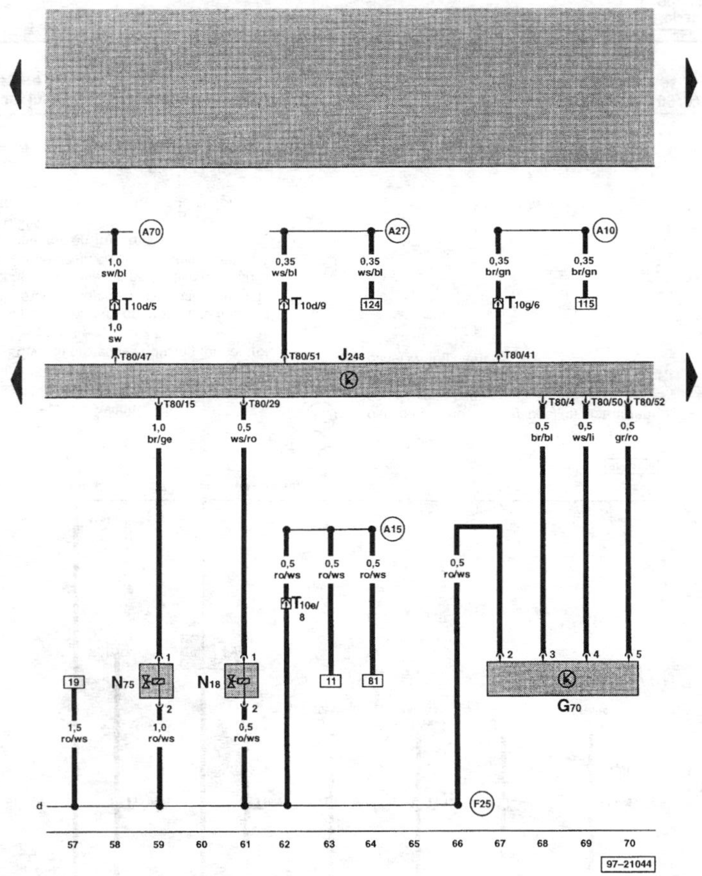

ws = white
sw = black
ro = red
br = brown
gn = green
bl = blue
gr = grey
li = violet
ge = yellow

G70 – Mass Air Flow (MAF) Sensor
J248 – Diesel Direct Fuel Injection (DFI) Engine Control Module (ECM), in protective housing for control module
N18 – EGR Vacuum Regulator Solenoid Valve
N75 – Wastegate Bypass Regulator Valve
T10d – 10-Pin Connector, brown, on protective housing for control module, in engine compartment, left
T10e – 10-Pin Connector, yellow, on protective housing for control module, in engine compartment, left
T10g – 10-Pin Connector, white, on protective housing for control module, in engine compartment, left
T80 – Connector, 80 point, on control module for Diesel
(A10) – Wire connection (glow plug control) in instrument panel wiring harness

(A15) – plus connection (15), in instrument panel wiring harness
(A27) – Wire Connection (vehicle speed signal), in instrument panel wiring harness
(A70) – Connector (15a, fuse 231), in instrument panel wiring harness
(F25) – Wire connection -1-, in Diesel Direct Fuel Injection (DFI) system wiring harness

Diesel direct fuel injection (DFI) engine control module (ECM), mass air flow (MAF) sensor, EGR vacuum regulator solenoid valve

Edition 09/97
USA.5062.01.21

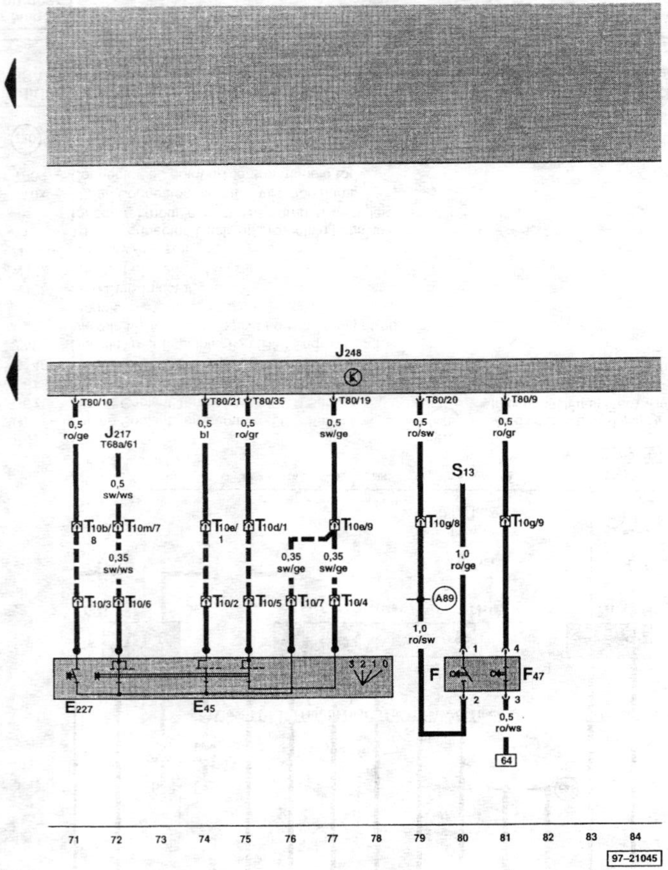

ws = white
sw = black
ro = red
br = brown
gn = green
bl = blue
gr = grey
li = violet
ge = yellow

E45 – Cruise Control Switch (ON/OFF)
position 0 = OFF
position 1 = OFF save
position 2 = ON
position 3 = RES
E227 – Cruise Control Push Button (SET)
F – Brake Light Switch
F47 – Brake Vacuum Vent Valve Switch for cruise control/diesel fuel injection (DFI)
J217 – Transmission Control Module (TCM)
J248 – Diesel Direct Fuel Injection (DFI) Engine Control Module (ECM), in protective housing for control module
S13 – Fuse 13 in fuse holder
T10 – 10-Pin Connector, white, near steering column
T10b – 10-Pin Connector, black, on protective housing for control module, in engine compartment, left
T10d – 10-Pin Connector, brown, on protective housing for control module, in engine compartment, left
T10e – 10-Pin Connector, yellow, on protective housing for control module, in engine compartment, left
T10g – 10-Pin Connector, white, on protective housing for control module, in engine compartment, left
T10m – 10-Pin Connector, brown, connector station A-pillar , right
T68a – 68-Pin Connector, on Transmission Control Module (TCM)
T80 – Connector, 80 point, on control module for Diesel
(A89) – Connector (54), in instrument panel wiring harness
– – – vehicles with cruise control only

Diesel direct fuel injection (DFI) engine control module (ECM), cruise control switch, brake light switch, clutch vacuum vent valve switch

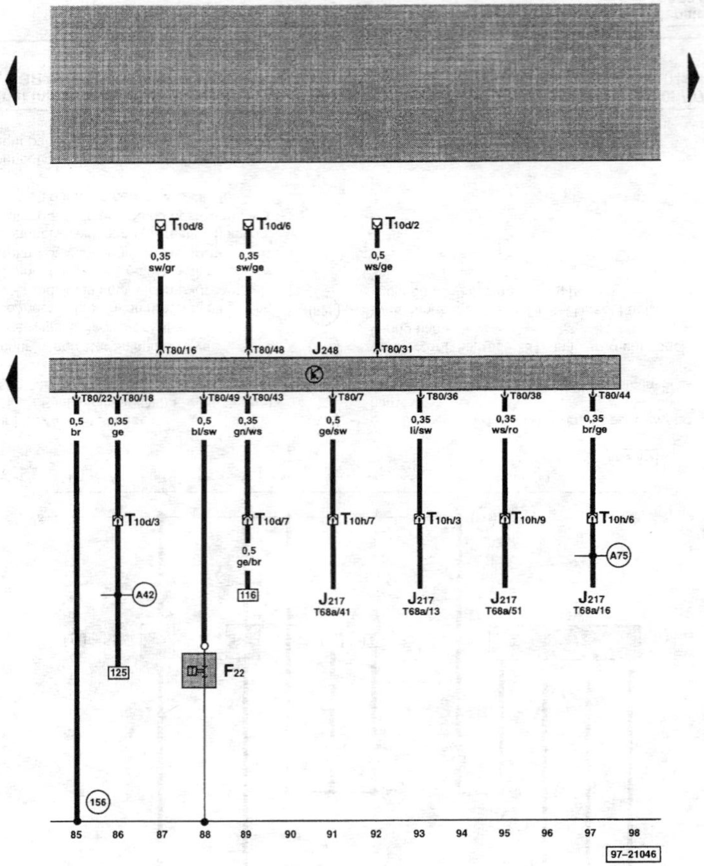

ws = white
sw = black
ro = red
br = brown
gn = green
bl = blue
gr = grey
li = violet
ge = yellow

F22 – 0,25bar Oil Pressure Switch
J217 – Transmission Control Module (TCM)
J248 – Diesel Direct Fuel Injection (DFI) Engine Control Module (ECM), in protective housing for control module
T10d – 10-Pin Connector, brown, on protective housing for control module, in engine compartment, left
T10h – 10-Pin Connector, blue, on protective housing for control module, in engine compartment, left
T68a – 68-Pin Connector, on Transmission Control Module (TCM)
T80 – Connector, 80 point, on control module for Diesel
(156) – Ground connection, in Diesel Direct Fuel Injection (DFI) wiring harness
(A42) – plus connection (fuel gauge), in instrument panel wiring harness
(A75) – Connector (kick-down-switch) in instrument panel wiring harness

Diesel direct fuel injection (DFI) engine control module (ECM), 0,25bar oilpressure switch

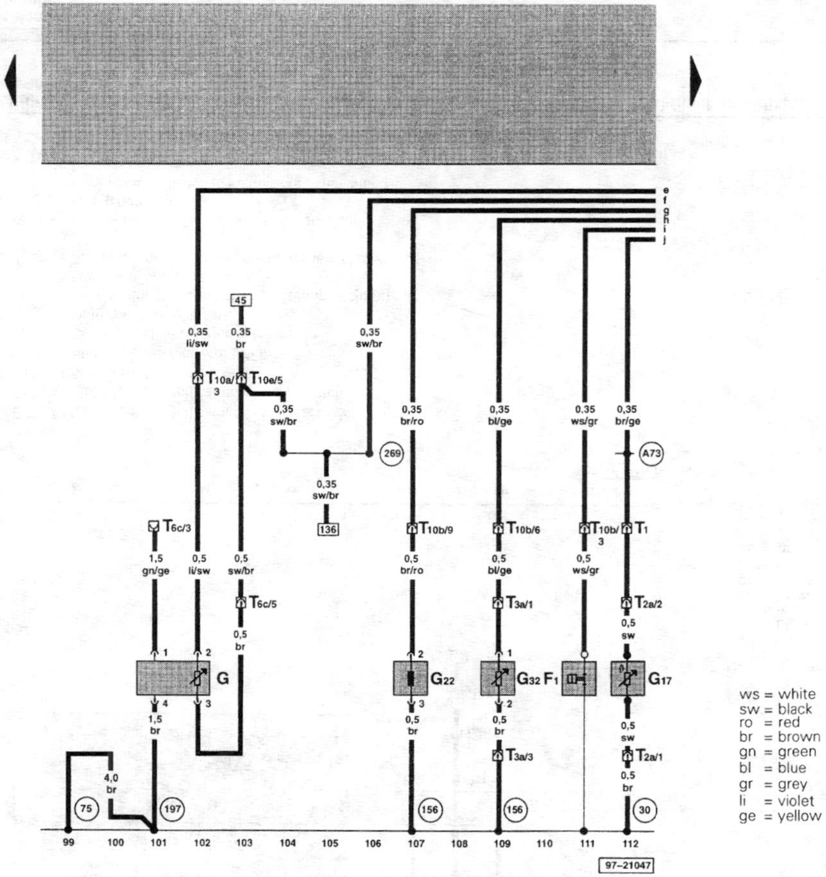

ws = white
sw = black
ro = red
br = brown
gn = green
bl = blue
gr = grey
li = violet
ge = yellow

F1 – Oil Pressure Switch
G – Sender for fuel gauge
G17 – Outside Air Temperature Sensor (use correspond wiring diagram by vehicles with air conditioning)
G22 – Speedometer Vehicle Speed Sensor (VSS)
G32 – Engine Coolant Level (ECL) Sensor
T1 – Single Connector, brown, connector station A-pillar, left
T2a – Double Connector, in engine compartment, left front
T3a – 3-Pin Connector, in engine compartment, left front
T6c – 6-Pin Connector, blue, connector station A-pillar, left
T10a – 10-Pin Connector, brown, connector station A-pillar, left
T10b – 10-Pin Connector, black, on protective housing for control module, in engine compartment, left
T10e – 10-Pin Connector, yellow, on protective housing for control module, in engine compartment, left
(30) – Ground connection, -1-, beside fuse/relay panel
(75) – Ground connection, on right rear pillar
(156) – Ground connection, in Diesel Direct Fuel Injection (DFI) wiring harness
(197) – Ground connection -4-, in wiring harness rear
(269) – Ground connection (sensor ground) -1-, in instrument panel wiring harness
(A73) – Connector (outside temperature indicator), in instrument panel wiring harness

Oil pressure switch, sender for fuel gauge, outside air temperature sensor, speedometer vehicle speed sensor (VSS), engine coolant level (ECL) sensor

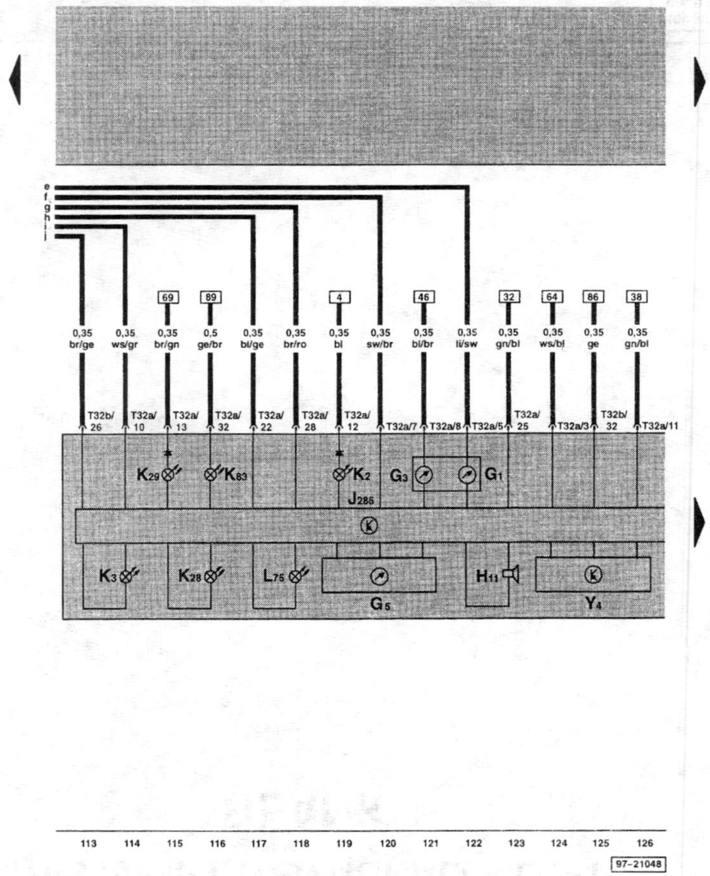

ws = white
sw = black
ro = red
br = brown
gn = green
bl = blue
gr = grey
li = violet
ge = yellow

G1 – Fuel gauge
G3 – Engine Coolant Temperature (ECT) Gauge
G5 – Tachometer
H11 – Oil Pressure Warning Buzzer
J285 – Control module with indicator unit in instrument panel insert
K2 – Generator (GEN) Warning Light .
K3 – Oil Pressure Warning Light
K28 – Engine Coolant Level/Temperature (ECL/ECT) Warning Light
K29 – Glow Plug Indicator Light
K83 – Exhaust Warning Light
L75 – Digital Display Light
T32a – 32-Pin Connector, blue, on instrument cluster
T32b – 32-Pin Connector, green, on instrument cluster
Y4 – Odometer Display

Instrument cluster, engine coolant level/temperature (ECL/ECT) warning light, visual and acoustical oil pressure control, exhaust warning light

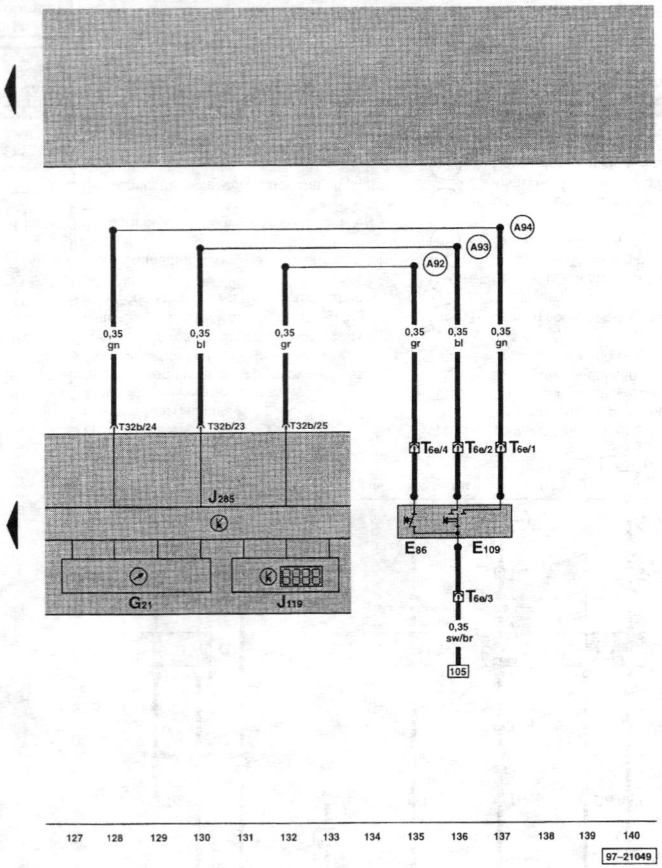

ws = white
sw = black
ro = red
br = brown
gn = green
bl = blue
gr = grey
li = violet
ge = yellow

E86 – Multi-Funktion Indicator Mode Select Switch
E109 – Multi-Funktion Indicator Memory Switch
G21 – Speedometer
J119 – Multi-function Indicator (MFI)
J285 – Control module with indicator unit in instrument panel insert
T6e – 6-Pin Connector, black, behind steering column switch cover
T32b – 32-Pin Connector, green, on instrument cluster

(A92) – Connector (MFI-Reset), in instrument panel wiring harness
(A93) – Connector -1- (MFI), in instrument panel wiring harness
(A94) – Connector -2- (MFI), in instrument panel wiring harness

Edition 09/97
USA.5062.01.21

Instrument cluster, multi-function indicator (MFI), multi-function indicator switches, speedometer

THIS PAGE INTENTIONALLY LEFT
BLANK

2,8L- Engine – Motronic Multiport Fuel Injection (MFI)/142 kW, code AHA,

from September 1997

Deviate relay location and fuseplacements as well as the locations of multiple connectors see section "component locations".

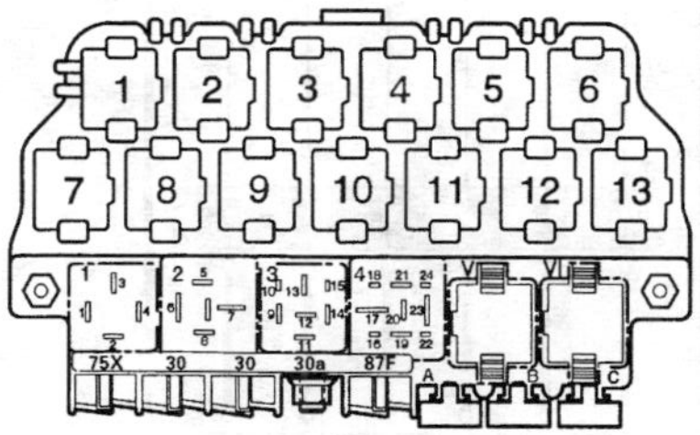

97–20676

Relay location on the thirteenfold auxiliary relay panel, above relay panel:

Relay panel:

4 Fuel Pump (FP) Relay (208)

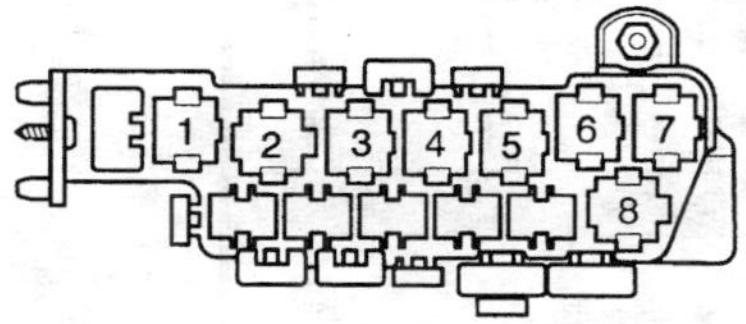

97–20515

Relay location on the eightfold auxiliary relay panel, behind relay panel:

Note: Number in parentheses indicates production control number stamped on relay housing.

Fuse colors
30 A – green
25 A – white
20 A – yellow
15 A – blue
10 A – red
7,5 A – brown
5 A – beige

Edition 09/97
USA.5062.01.21

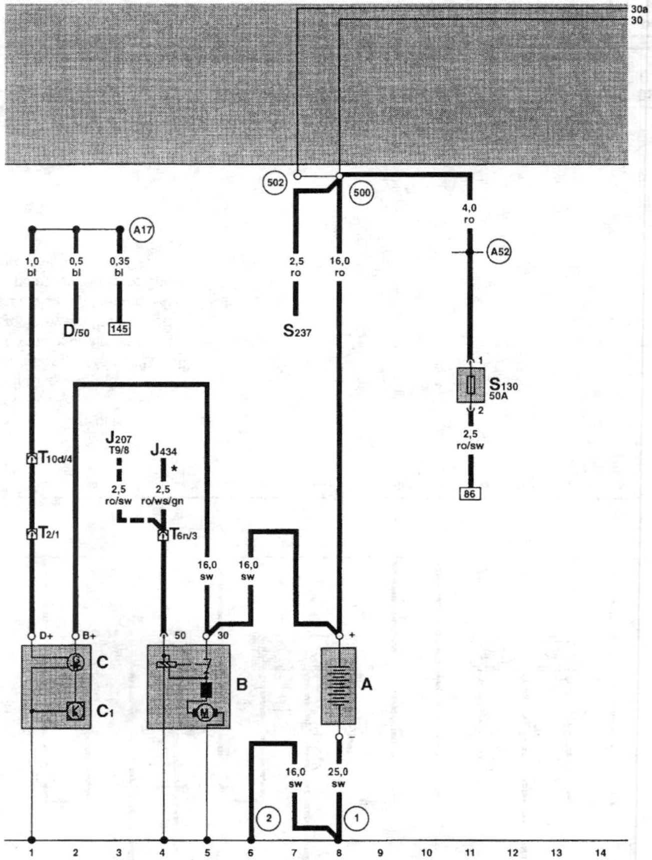

97–20808

ws = white
sw = black
ro = red
br = brown
gn = green
bl = blue
gr = grey
li = violet
ge = yellow

A – Battery
B – Starter
C – Generator (GEN)
C1 – Voltage Regulator (VR)
D – Ignition/Starter Switch
J207 – Starting Interlock Relay
J434 – Starting Interlock Relay (clutch position)
S130 – Fuse for secondary air pump, in fuse adaptor above relay panel
S237 – Fuse 37 in fuse holder
T2 – Double Connector, on engine right
T6n – 6-Pin Connector, red, on protective housing for control module, in engine compartment, left
T9 – 9-Pin Connector, on Starting Interlock Relay
T10d – 10-Pin Connector, brown, on protective housing for control module, in engine compartment, left

(1) – Ground strap, battery to body
(2) – Ground strap, transmission to body
(500) – Screw Connection -1- (30), on relay panel
(502) – Screw Connection (30a), on relay panel
(A17) – wire connection (61), in instrument panel wiring harness
(A52) – plus connection (30), in instrument panel wiring harness
* – manual transmission only
- - - automatic transmission only

Battery, starter, generator (GEN)

Edition 09/97
USA.5062.01.21

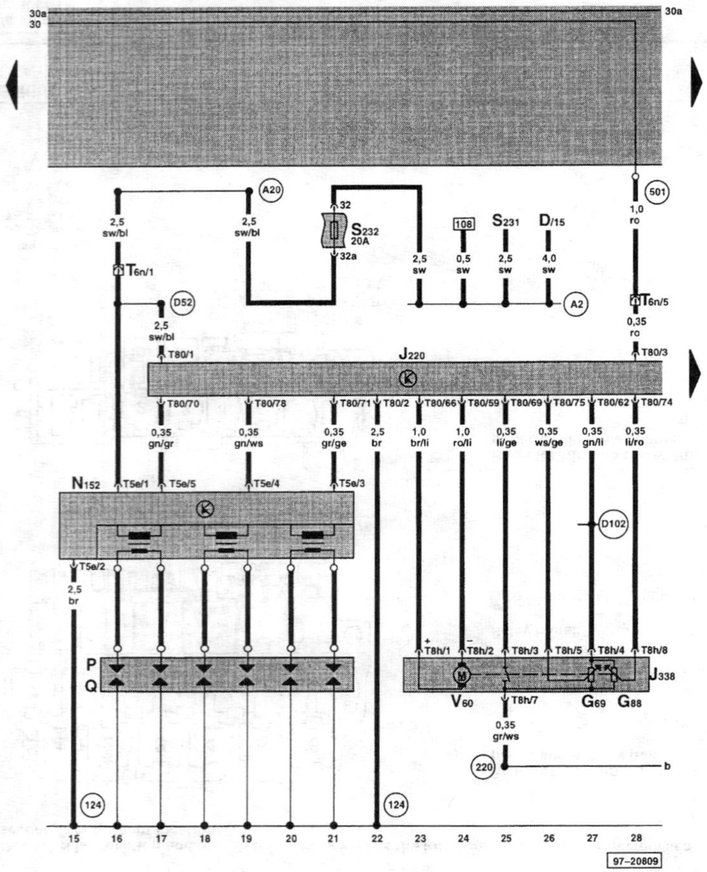

ws = white
sw = black
ro = red
br = brown
gn = green
bl = blue
gr = grey
li = violet
ge = yellow

D – Ignition/Starter Switch
G88 – Throttle Position (TP) Sensor
J220 – Motronic Engine Control Module (ECM)
J338 – Throttle Valve Control Module
N152 – Ignition Coil
P – Spark Plug Connectors
Q – Spark Plugs
S231 – Fuse 31 in fuse holder
S232 – Fuse 32 in fuse holder
T5e – 5-Pin Connector
T6n – 6-Pin Connector, red, on protective housing for control module, in engine compartment, left
T8h – 8-Pin Connector, on throttle valve control module
T80 – Connector, 80 point, on Motronic Engine Control Module (ECM)
V60 – Throttle Position (TP) Actuator

(124) – ground connection, in engine compartment right wiring harness
(220) – Ground connection (sensor ground), in engine compartment wiring harness
(501) – Screw Connection -2- (30), on relay panel
(A2) – plus connection (15), in instrument panel wiring harness
(A20) – wire connection (15a), in instrument panel wiring harness
(D52) – plus connection (15a), in engine compartment wiring harness
(D102) – Wire connection -2-, in engine compartment wiring harness

Ignition system, motronic engine control module (ECM), throttle valve control module

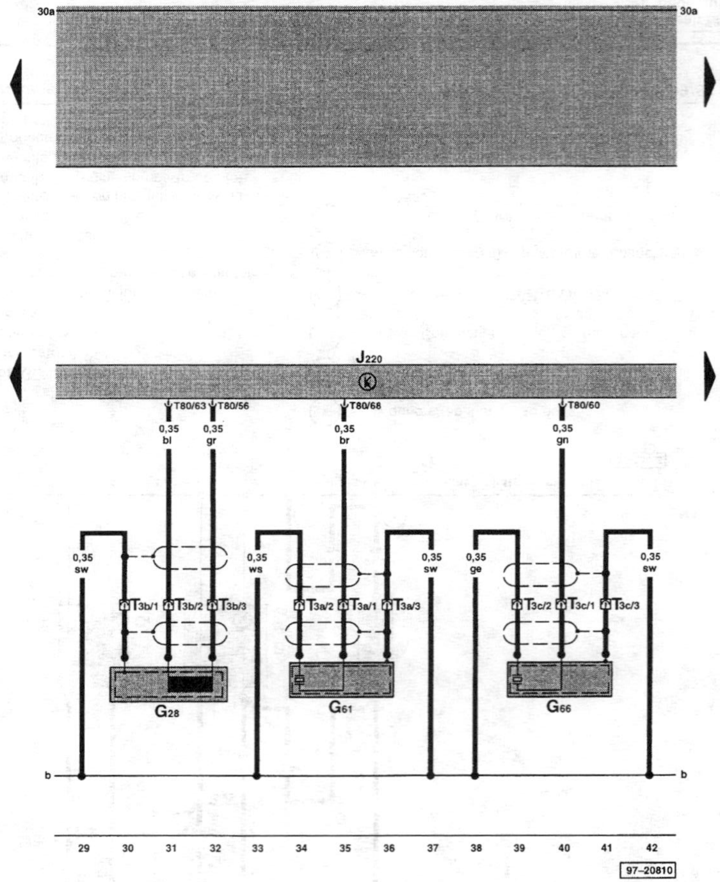

ws = white
sw = black
ro = red
br = brown
gn = green
bl = blue
gr = grey
li = violet
ge = yellow

G28 – Engine Speed (RPM) Sensor
G61 – Knock Sensor (KS) 1
G66 – Knock Sensor (KS) 2
J220 – Motronic Engine Control Module (ECM)
T3a – 3-Pin Connector, blue, on firewall, left
T3b – 3-Pin Connector, grey, on firewall, left
T3c – 3-Pin Connector, blue, on firewall, left
T80 – Connector, 80 point, on Motronic Engine Control Module (ECM)

Motronic engine control module (ECM), engine speed (RPM) sensor, knock sensor (KS)

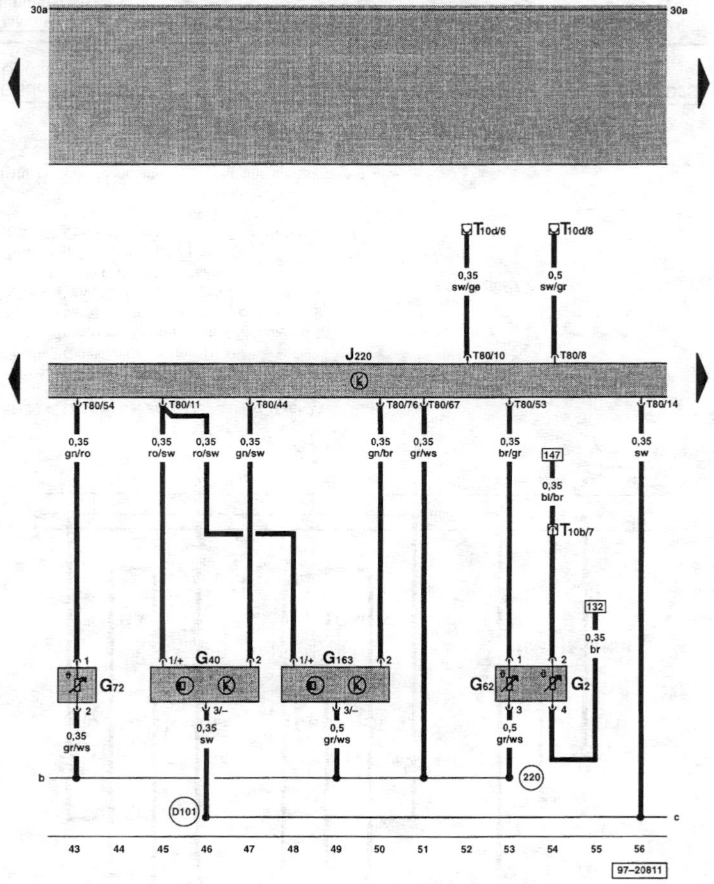

ws = white
sw = black
ro = red
br = brown
gn = green
bl = blue
gr = grey
li = violet
ge = yellow

G2 – Engine Coolant Temperature (ECT) Sensor
G40 – Camshaft Position (CMP) Sensor
G62 – Engine Coolant Temperature (ECT) Sensor
G72 – Intake Air Temperature (IAT) Sensor
G163 – Camshaft Position (CMP) Sensor 2
J220 – Motronic Engine Control Module (ECM)
T10b – 10-Pin Connector, black, on protective housing for control module, in engine compartment, left
T10d – 10-Pin Connector, brown, on protective housing for control module, in engine compartment, left
T80 – Connector, 80 point, on Motronic Engine Control Module (ECM)

(220) – Ground connection (sensor ground), in engine compartment wiring harness
(D101) – Wire connection -1-, in engine compartment wiring harness

Edition 09/97
USA.5062.01.21

Motronic engine control module (ECM), camshaft position (CMP) sensor, engine coolant temperature (ECT) sensor, intake air temperature (IAT) sensor

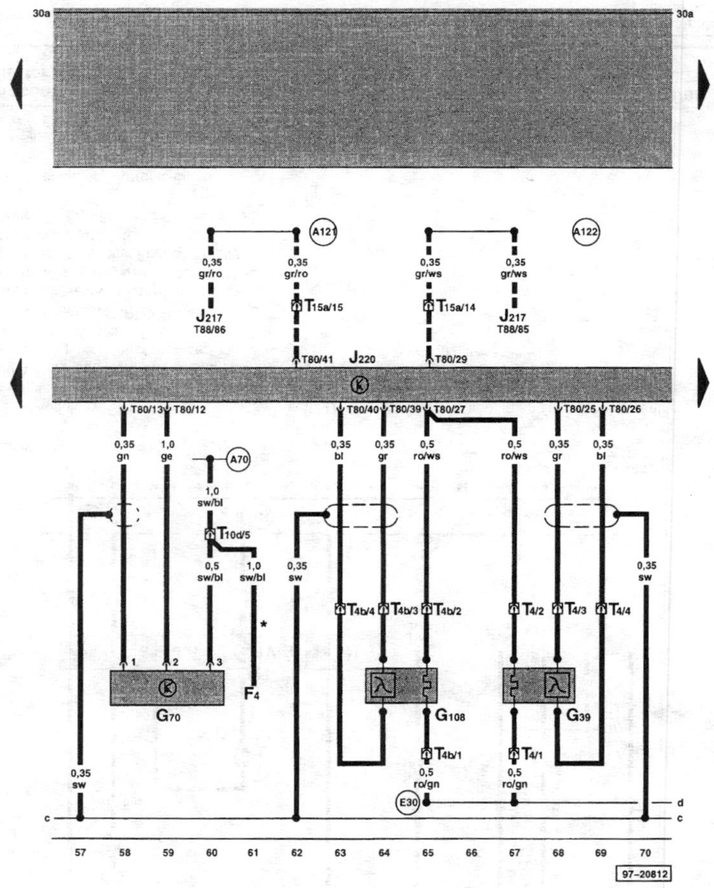

ws = white
sw = black
ro = red
br = brown
gn = green
bl = blue
gr = grey
li = violet
ge = yellow

F4 – Back-Up Light Switch
G39 – Heated Oxygen Sensor (HO2S)
G70 – Mass Air Flow (MAF) Sensor
G108 – Heated Oxygen Sensor (HO2S) 2
J217 – Transmission Control Module (TCM)
J220 – Motronic Engine Control Module (ECM)
T4 – 4-Pin Connector, black
T4b – 4-Pin Connector, black
T10d – 10-Pin Connector, brown, on protective housing for control module, in engine compartment, left
T15a – 15-Pin Connector, white, on protective housing for control module, in engine compartment,left
T80 – Connector, 80 point, on Motronic Engine Control Module (ECM)
T88 – Connector, 88 point, on Transmission Control Module (TCM)

(A70) – Connector (15a, fuse 231), in instrument panel wiring harness
(A121) – Connector (High-Bus), in instrument panel wiring harness
(A122) – Connector (Low-Bus), in instrument panel wiring harness
(E30) – Connector (87a), in wiring harness engine

- - - automatic transmission only
• – manual transmission only

Motronic engine control module (ECM), mass air flow (MAF) sensor, heated oxygen sensors (HO2S)

Edition 09/97
USA.5062.01.21

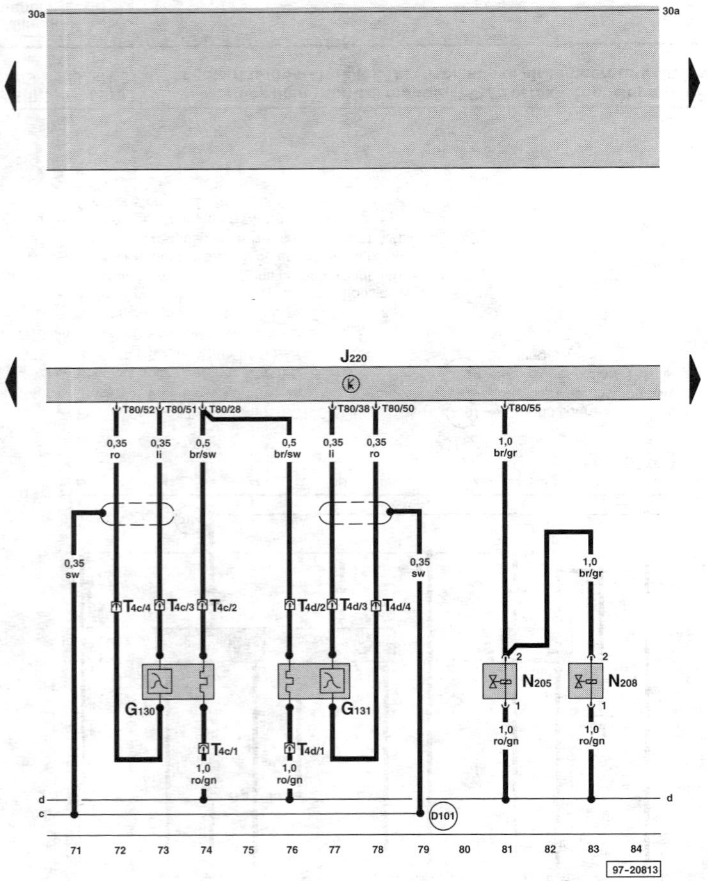

ws = white
sw = black
ro = red
br = brown
gn = green
bl = blue
gr = grey
li = violet
ge = yellow

G130 - Oxygen Sensor (O2S) Behind Three Way Catalytic Converter (TWC)
G131 - Oxygen Sensor (O2S) 2 Behind Three Way Catalytic Converter (TWC)
J220 - Motronic Engine Control Module (ECM)
N205 - Valve -1- for camshaft adjustment
N208 - Valve -2- for camshaft adjustment
T4c - 4-Pin Connector, green
T4d - 4-Pin Connector, brown
T80 - Connector, 80 point, on Motronic Engine Control Module (ECM)

(D101) - Wire connection -1-, in engine compartment wiring harness

Motronic engine control module (ECM), oxygen sensor (O2S), oxygen sensor (O2S) 2, valve for camshaft adjustment

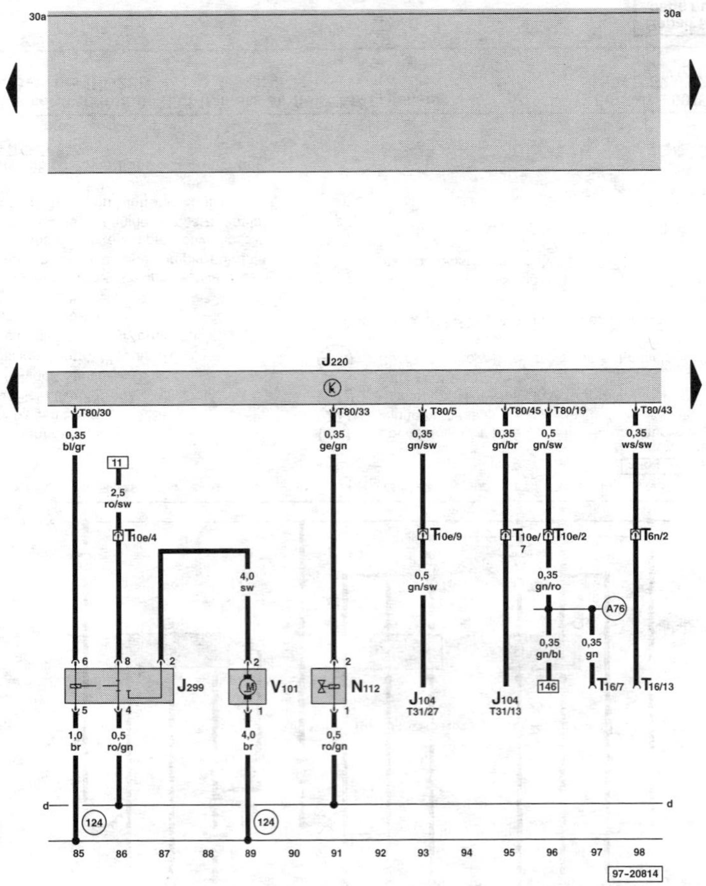

ws = white
sw = black
ro = red
br = brown
gn = green
bl = blue
gr = grey
li = violet
ge = yellow

J104 - ABS Control Module (w/EDL)
J220 - Motronic Engine Control Module (ECM)
J299 - Secondary Air Injection (AIR) Pump Relay, in protective housing for control module, production control number (203)
N112 - Secondary Air Injection (AIR) Solenoid Valve
T10e - 10-Pin Connector, yellow, on protective housing for control module, in engine compartment, left
T16 - 16-Pin Connector, Data Link Connector (DLC), under instrument panel, left
T31 - 31-Pin Connector, on ABS control module (with EDL)
T80 - Connector, 80 point, on Motronic Engine Control Module (ECM)
V101 - Secondary Air Injection (AIR) Pump Motor

(124) - ground connection, in engine compartment right wiring harness
(A76) Connector (K-diagnosis wire), in instrument panel wiring harness

Motronic engine control module (ECM), secondary air injection (AIR) system, Data Link Connector (DLC)

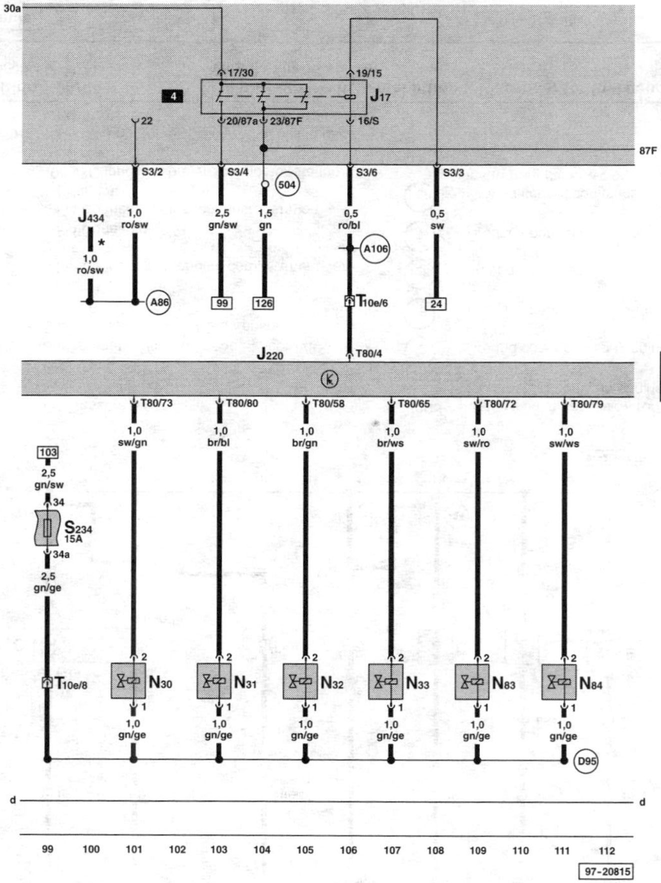

ws = white
sw = black
ro = red
br = brown
gn = green
bl = blue
gr = grey
li = violet
ge = yellow

J17 - Fuel Pump (FP) Relay, production control number (208)
J220 - Motronic Engine Control Module (ECM)
J434 - Starting Interlock Relay (clutch position)
N30 - Cylinder 1 Fuel Injector
N31 - Cylinder 2 Fuel Injector
N32 - Cylinder 3 Fuel Injector
N33 - Cylinder 4 Fuel Injector
N83 - Cylinder 5 Fuel Injector
N84 - Cylinder 6 Fuel Injector
S234 - Fuse 34 in fuse holder
T10e - 10-Pin Connector, yellow, on protective housing for control module, in engine compartment, left
T80 - Connector, 80 point, on Motronic Engine Control Module (ECM)

(504) - Screw Connection (87F), on relay panel
(A86) - Connection (50a) in instrument panel wiring harness
(A106) - Connector -2- (86s), in instrument panel wiring harness
(D95) - wire connection (injectors), in engine compartment wiring harness
* - manual transmission only

Edition 04/00
USA.5062.11.21

Motronic engine control module (ECM), fuel injectors, fuel pump (FP) relay

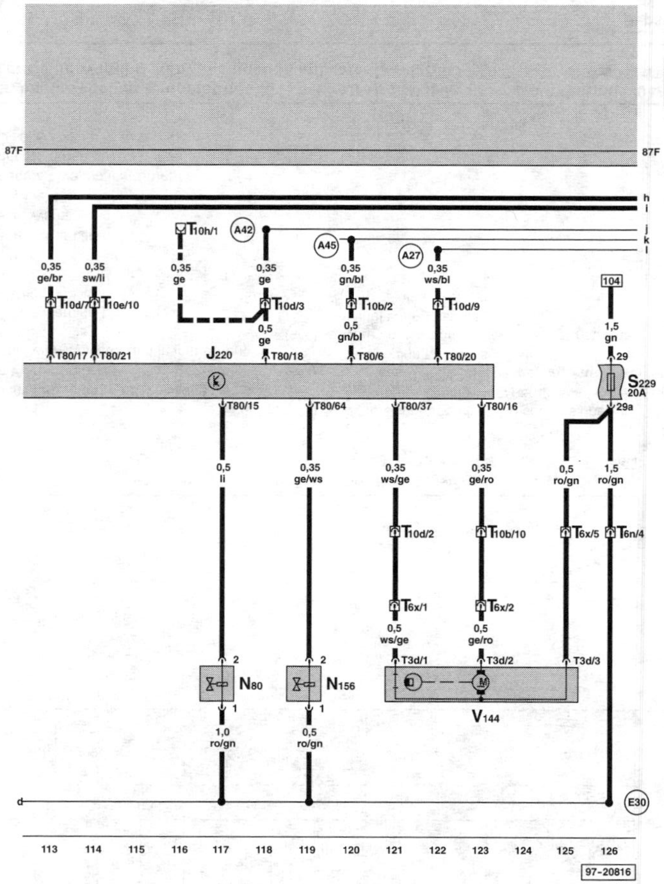

ws = white
sw = black
ro = red
br = brown
gn = green
bl = blue
gr = grey
li = violet
ge = yellow

J220 - Motronic Engine Control Module (ECM)
N80 - Evaporative Emission (EVAP) Canister Purge Regulator Valve
N156 - Intake Manifold Change-Over Valve
S229 - Fuse 29 in fuse holder
T3d - 3-Pin Connector
T6n - 6-Pin Connector, red, on protective housing for control module, in engine compartment, left
T6x - 6-Pin Connector, brown, connector station A-pillar, right
T10b - 10-Pin Connector, black, on protective housing for control module, in engine compartment, left
T10d - 10-Pin Connector, brown, on protective housing for control module, in engine compartment, left
T10e - 10-Pin Connector, yellow, on protective housing for control module, in engine compartment, left
T10h - 10-Pin Connector, blue, on protective housing for control module, in engine compartment, left
T80 - Connector, 80 point, on Motronic Engine Control Module (ECM)
V144 - Leak Detection Pump (LDP)

(A27) - Wire Connection (vehicle speed signal), in instrument panel wiring harness
(A42) - plus connection (fuel gauge), in instrument panel wiring harness
(A45) - wire connection (RPM-signal), in instrument panel wiring harness
(E30) - Connector (87a), in wiring harness engine
- - - automatic transmission only

Leak detection pump (LDP), evaporative emission (EVAP) canister purge regulator valve, intake manifold change-over valve

Edition 04/00
USA.5062.11.21

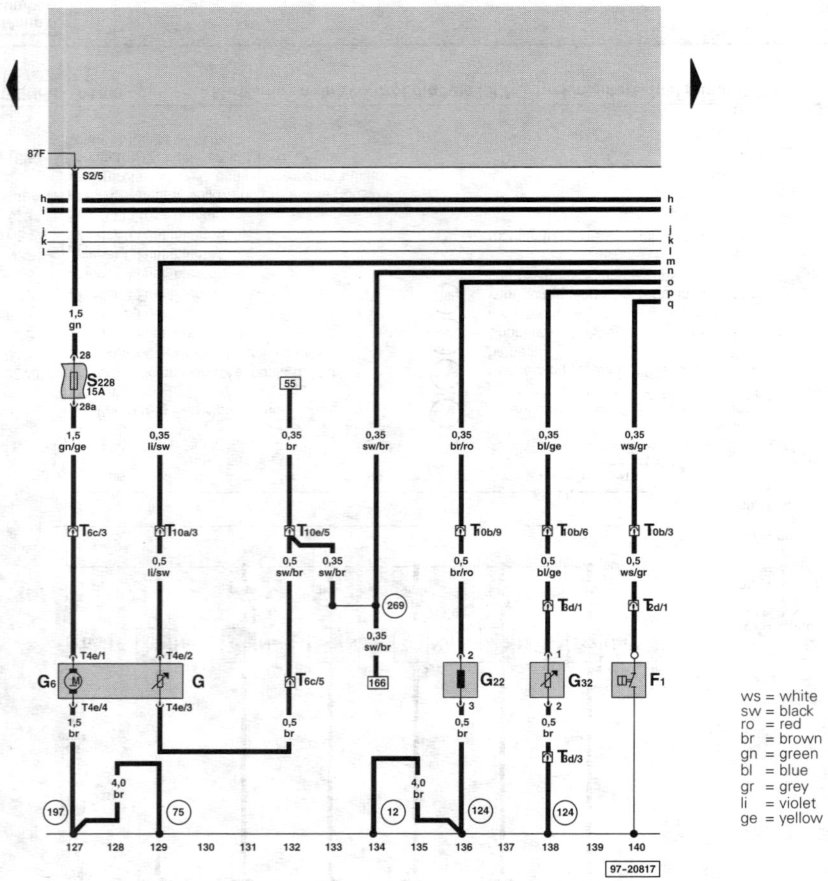

ws = white
sw = black
ro = red
br = brown
gn = green
bl = blue
gr = grey
li = violet
ge = yellow

F1 - Oil Pressure Switch
G - Sender for fuel gauge
G6 - Fuel Pump (FP)
G22 - Speedometer Vehicle Speed Sensor (VSS)
G32 - Engine Coolant Level (ECL) Sensor
S228 - Fuse 28 in fuse holder
T2d - Double Connector
T3d - 3-Pin Connector, in engine compartment, left front
T4e - 4-Pin Connector
T6c - 6-Pin Connector, blue, connector station A-pillar, left
T10a - 10-Pin Connector, brown, connector station A-pillar, left
T10b - 10-Pin Connector, black, on protective housing for control module, in engine compartment, left
T10e - 10-Pin Connector, yellow, on protective housing for control module, in engine compartment, left
(12) - Ground connection, in engine compartment, left
(75) - Ground connection, on right rear pillar
(124) - ground connection, in engine compartment right wiring harness
(197) - Ground connection -4-, in wiring harness rear
(269) - Ground connection (sensor ground) -1-, in instrument panel wiring harness

Oil pressure switch, fuel pump (FP), sender for fuel gauge, speedometer vehicle speed sensor

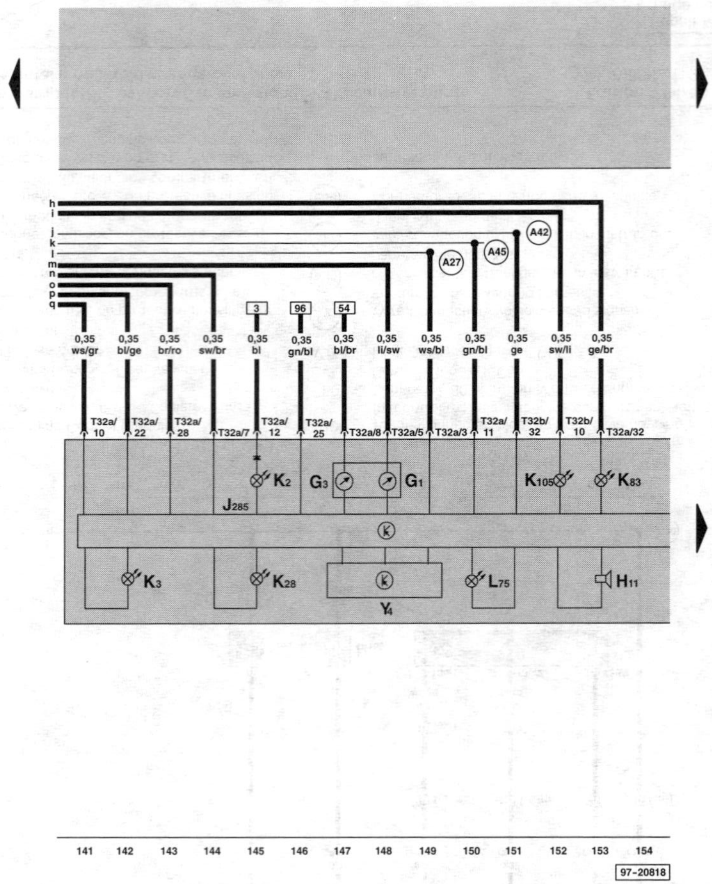

ws = white
sw = black
ro = red
br = brown
gn = green
bl = blue
gr = grey
li = violet
ge = yellow

G1 - Fuel gauge
G3 - Engine Coolant Temperature (ECT) Gauge
H11 - Oil Pressure Warning Buzzer
J285 - Control module with indicator unit in instrument panel insert
K2 - Generator (GEN) Warning Light
K3 - Oil Pressure Warning Light
K28 Engine Coolant Level/Temperature (ECL/ECT) Warning Light
K83 - Malfunction Indicator Lamp (MIL)
K105 - Low Fuel Level Warning Light
L75 - Digital Display Light
T32a - 32-Pin Connector, blue, on instrument cluster
T32b - 32-Pin Connector, green, on instrument cluster
Y4 - Odometer Display
(A27) - Wire Connection (vehicle speed signal), in instrument panel wiring harness
(A42) - plus connection (fuel gauge), in instrument panel wiring harness
(A45) - wire connection (RPM-signal), in instrument panel wiring harness

Instrument cluster, engine coolant temperature (ECT) gauge, low fuel level warning light, oil pressure warning light, Malfunction Indicator lamp (MIL)

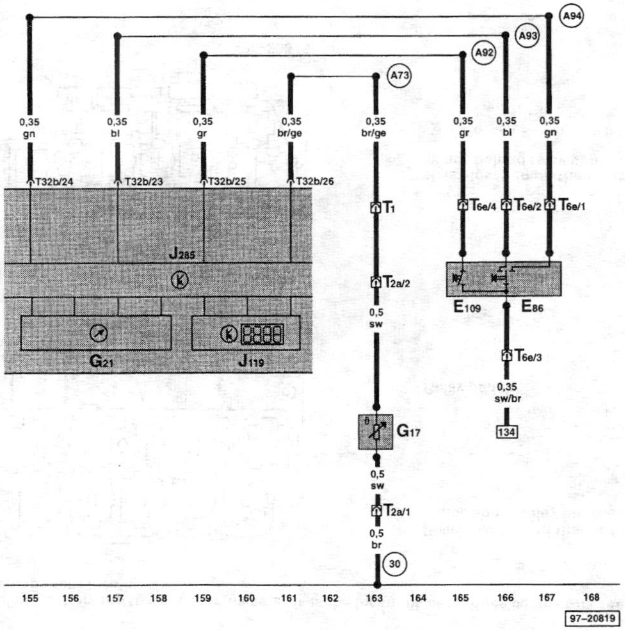

ws = white
sw = black
ro = red
br = brown
gn = green
bl = blue
gr = grey
li = violet
ge = yellow

E86 – Multi-Funktion Indicator Mode Select Switch
E109 – Multi-Funktion Indicator Memory Switch
G17 – Outside Air Temperature Sensor (use correspond wiring diagram by vehicles with air conditioning)
G21 – Speedometer
J119 – Multi-function Indicator (MFI)
J285 – Control module with indicator unit in instrument panel insert
T1 – Single Connector, brown, connector station A-pillar, left
T2a – Double Connector, in engine compartment left
T6e – 6-Pin Connector, black, behind steering column switch cover
T32b – 32-Pin Connector, green, on instrument cluster

(30) – Ground connection, -1-, beside fuse/relay panel
(A73) – Connector (outside temperature indicator), in instrument panel wiring harness
(A92) – Connector (MFI-Reset), in instrument panel wiring harness
(A93) – Connector -1- (MFI), in instrument panel wiring harness
(A94) – Connector -2- (MFI), in instrument panel wiring harness

Instrument cluster, multi-function indicator (MFI), multi-function indicator switches, speedometer, outside air temperature sensor

THIS PAGE INTENTIONALLY LEFT
BLANK

Coolant Fan, 80W,

from September 1997

Deviate relay location and fuseplacements as well as the locations of multiple connectors see section "component locations".

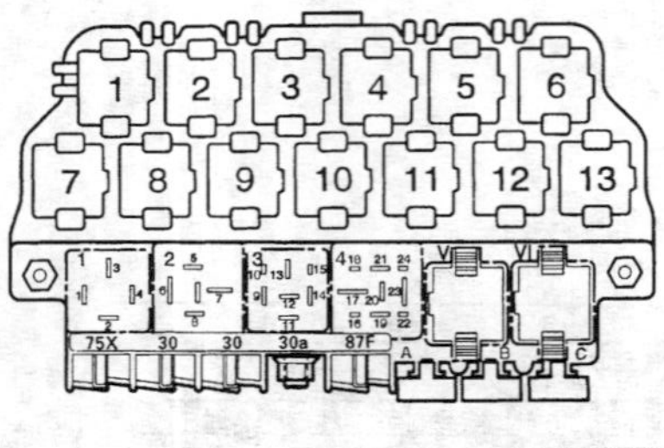

97-20676

Relay location on the thirteenfold auxiliary relay panel, above relay panel:

Relay panel:

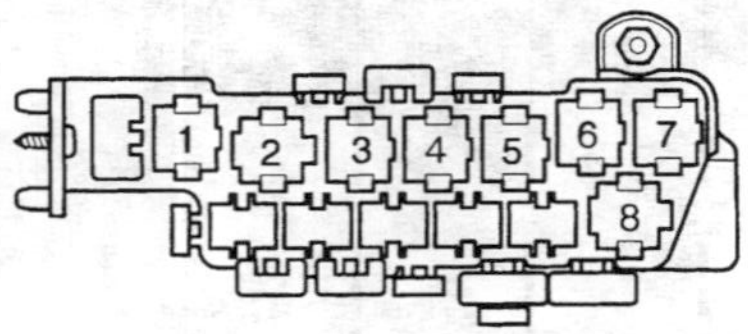

97-20515

Relay location on the eightfold auxiliary relay panel, behind relay panel:

5 Coolant Fan Control (FC) Relay (214)

Note: Number in parentheses indicates production control number stamped on relay housing.

Fuse colors
30 A – green
25 A – white
20 A – yellow
15 A – blue
10 A – red
7,5 A – brown
5 A – beige

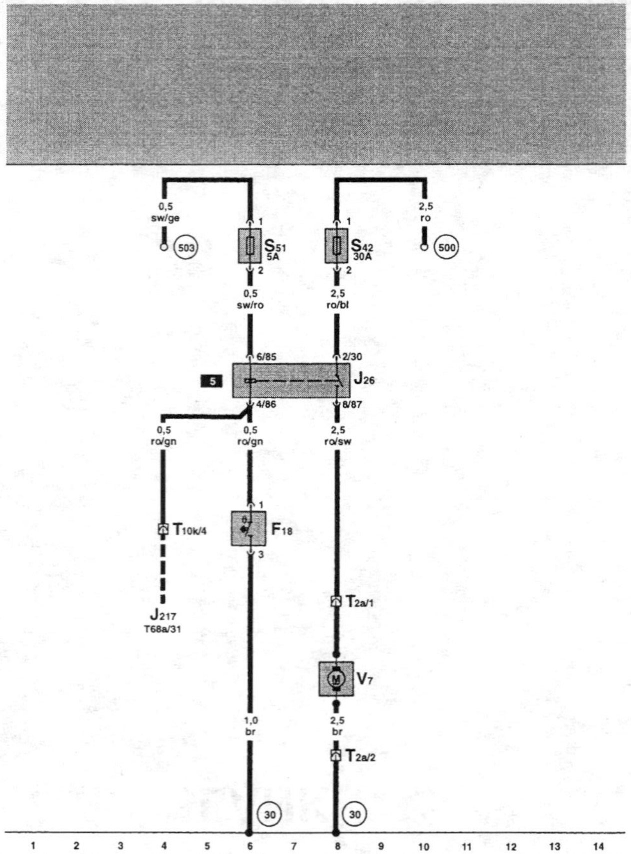

97-20847

ws = white
sw = black
ro = red
br = brown
gn = green
bl = blue
gr = grey
li = violet
ge = yellow

F18 – Coolant Fan Control (FC) Thermal Switch
J26 – Coolant Fan Control (FC) Relay, on auxiliary relay panel, behind fuse/relay panel, production control number (214)
J217 – Transmission Control Module (TCM)
S42 – Coolant Fan Fuse, on auxiliary relay panel, behind fuse/relay panel
S51 – Fuse for Coolant Fan control (FC) Relay, on auxiliary relay panel, behind fuse/relay panel
T2a – Double Connector, brown
T10k – 10-Pin Connector, grey, connector station A-pillar, left
T68a – 68-Pin Connector, on Transmission Control Module (TCM)
V7 – Coolant Fan

(30) – Ground connection, -1-, beside fuse/relay panel
(500) – Screw Connection -1- (30), on relay panel
(503) – Screw Connection (75x), on relay panel
--- code AHH with automatic transmission only

Coolant fan control (FC), 80W

Coolant Fan, 300W,

from September 1997

Deviate relay location and fuseplacements as well as the locations of multiple connectors see section "component locations".

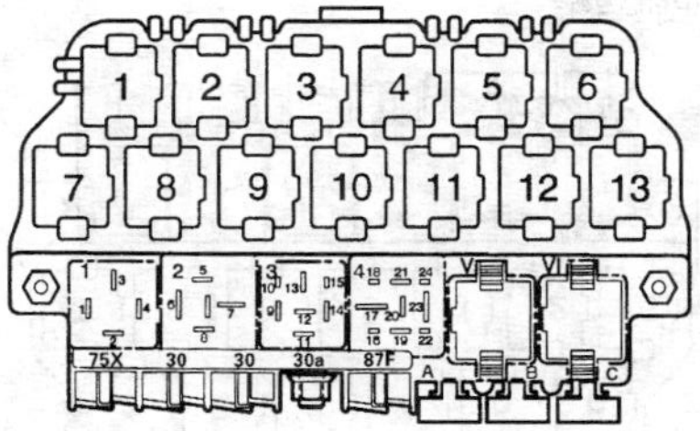

97–20676

Relay location on the thirteenfold auxiliary relay panel, above relay panel:

Relay panel:

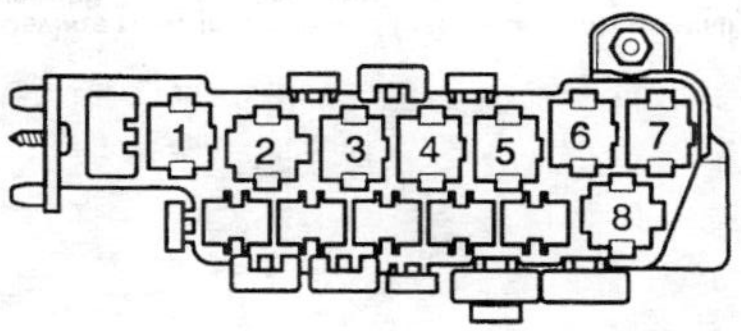

97–20515

Relay location on the eightfold auxiliary relay panel, behind relay panel:

- **5** First Speed Coolant Fan Control (FC) Relay (214)
- **8** Second Speed Coolant Fan Control (FC) Relay (213)

Note: Number in parentheses indicates production control number stamped on relay housing.

Fuse colors

30 A – green
25 A – white
20 A – yellow
15 A – blue
10 A – red
7,5 A – brown
5 A – beige

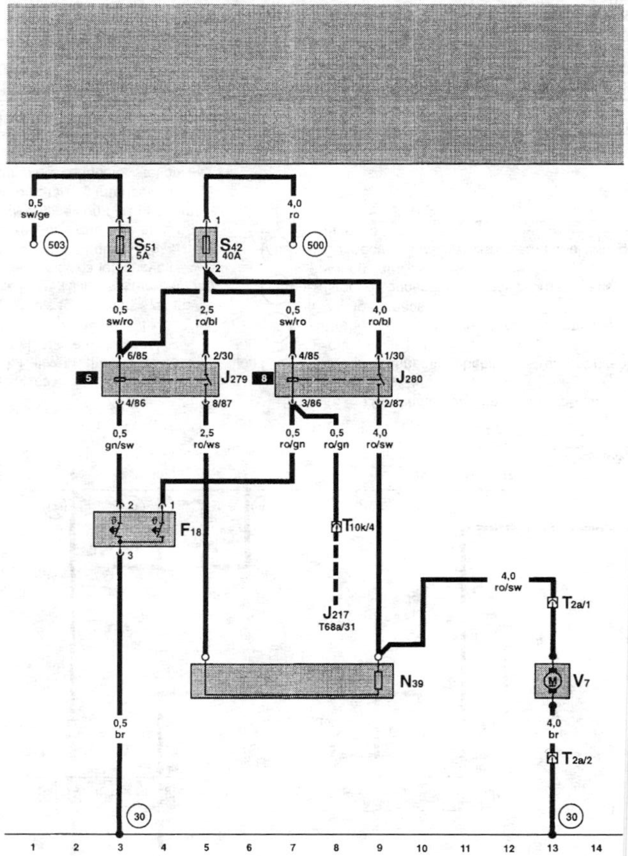

ws = white
sw = black
ro = red
br = brown
gn = green
bl = blue
gr = grey
li = violet
ge = yellow

F18 – Coolant Fan Control (FC) Thermal Switch
J279 – First Speed Coolant Fan Control (FC) Relay, on auxiliary relay panel, behind fuse/relay panel, production control number (214)
J280 – Second Speed Coolant Fan Control (FC) Relay, on auxiliary relay panel, behind fuse/relay panel, production control number (213)
J217 – Transmission Control Module (TCM)
N39 – Coolant Fan Control (FC) Series Resistance, unter dem linken Längsträger
S42 – Coolant Fan Fuse, on auxiliary relay panel, behind fuse/relay panel
S51 – Fuse for Coolant Fan control (FC) Relay, on auxiliary relay panel, behind fuse/relay panel
T2a – Double Connector, brown
T10k – 10-Pin Connector, grey, connector station A-pillar, left
T68a – 68-Pin Connector, on Transmission Control Module (TCM)
V7 – Coolant Fan
(30) – Ground connection, -1-, beside fuse/relay panel
(500) – Screw Connection -1- (30), on relay panel
(503) – Screw Connection (75x), on relay panel
- - - code AHH with automatic transmission only

Coolant fan control (FC), 300W

Automatic transmission, 5 speed automatic with tiptronic,

- **2,8L-Engine – Motronic Multiport Fuel Injection (MFI) / 142 kW, code AHA,**

from September 1997

Deviate relay location and fuseplacements as well as the locations of multiple connectors see section "component locations".

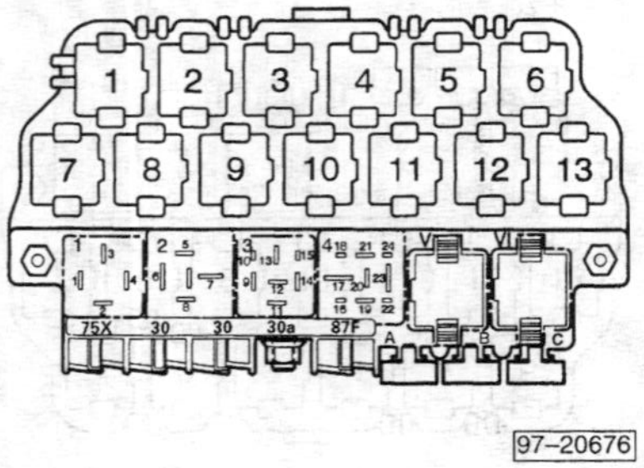

Relay location on the thirteenfold auxiliary relay panel, above relay panel:

- **6** Selector Lever Light Relay (205)
- **12** Starting Interlock Relay (alarm system) (186)
- **13** Starting Interlock Relay (204)

Relay panel:

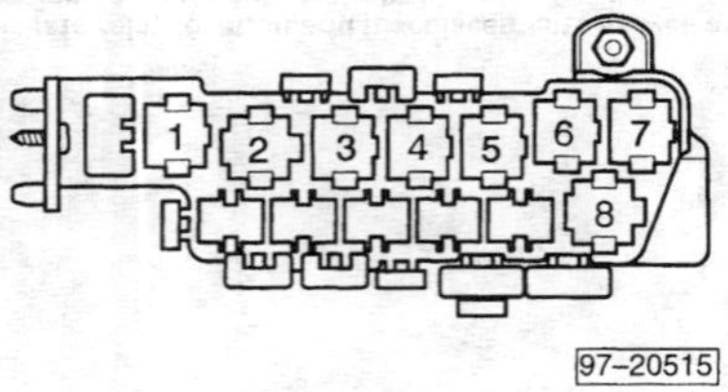

Relay location on the eightfold auxiliary relay panel, behind relay panel:

Note: Number in parentheses indicates production control number stamped on relay housing.

Fuse colors

30 A – green
25 A – white
20 A – yellow
15 A – blue
10 A – red
7,5 A – brown
5 A – beige

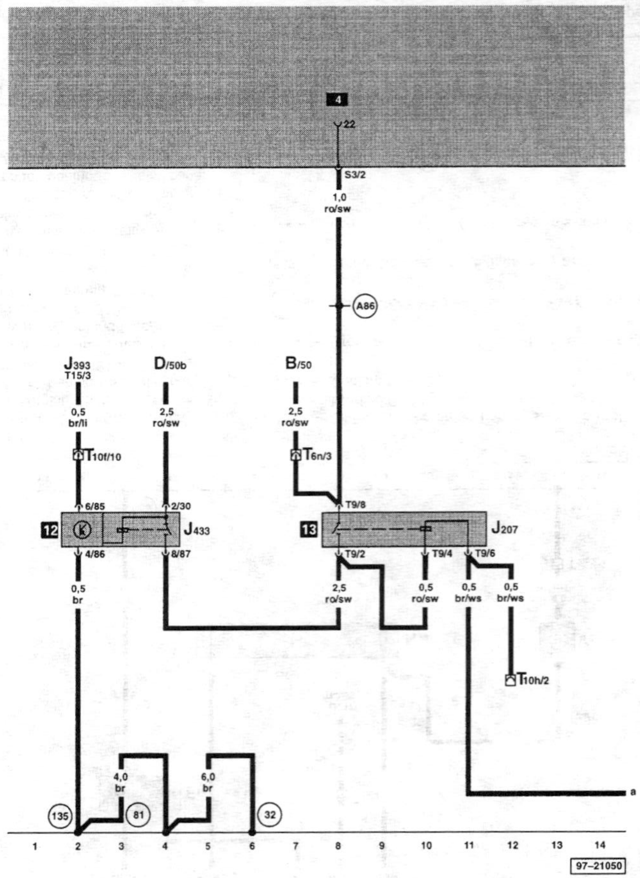

ws = white
sw = black
ro = red
br = brown
gn = green
bl = blue
gr = grey
li = violet
ge = yellow

- B – Starter
- D – Ignition/Starter Switch
- J207 – Starting Interlock Relay, on the thirteenfold auxiliary relay panel, above relay panel, production control number (204)
- J393 – Central control module for comfort system
- J433 – Starting Interlock Relay (alarm system), on the thirteenfold auxiliary relay panel, above relay panel, production control number (186)
- T6n – 6-Pin Connector, red, on protective housing for control module, in engine compartment, left
- T9 – 9-Pin Connector, on Starting Interlock Relay
- T10f – 10-Pin Connector, blue, connector station A-pillar, left
- T10h – 10-Pin Connector, blue, on protective housing for control module, in engine compartment, left
- T15 – 15-Pin Connector, on Central control module for comfort system
- (32) – Ground connection, behind instrument panel, left
- (81) – Ground connection -1-, in instrument panel wiring harness
- (135) – Ground connection -2-, in instrument panel wiring harness
- (A86) – Connection (50a), in instrument panel wiring harness

Starting interlock relay, Starting interlock relay (alarm system)

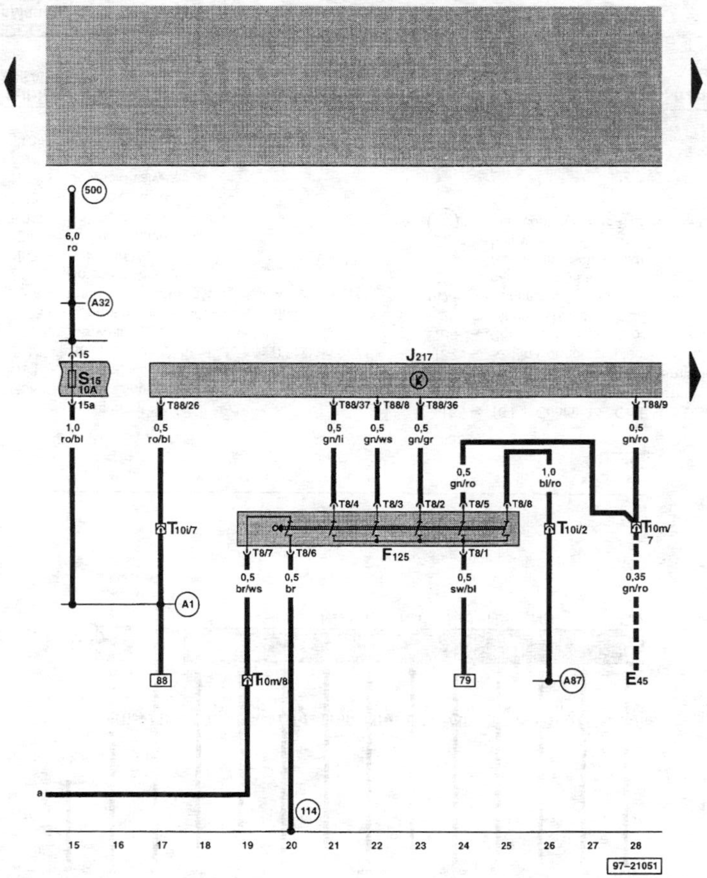

ws = white
sw = black
ro = red
br = brown
gn = green
bl = blue
gr = grey
li = violet
ge = yellow

E45 – Cruise Control Switch
F125 – Multi-Function Transmission Range (TR) Switch
J217 – Transmission Control Module (TCM), in front footwell, right
S15 – Fuse 15 in fuse holder
T8 – 8-Pin Connector, black
T10i – 10-Pin Connector, blue, connector station A-pillar, right
T10m – 10-Pin Connector, brown, connector station A-pillar, right
T88 – Connector, 88 point, on Transmission Control Module (TCM)
(114) – Ground connection, in automatic transmission wiring harness
(500) – Screw Connection -1- (30), on relay panel

(A1) – plus connection (30a), in instrument panel wiring harness
(A32) – plus connection (30), in instrument panel wiring harness
(A87) – Connector (reverse lamp), in instrument panel wiring harness

--- vehicles with cruise control only

Edition 09/97
USA.5062.01.21

Transmission control module (TCM), multi-function transmission range (TR) switch

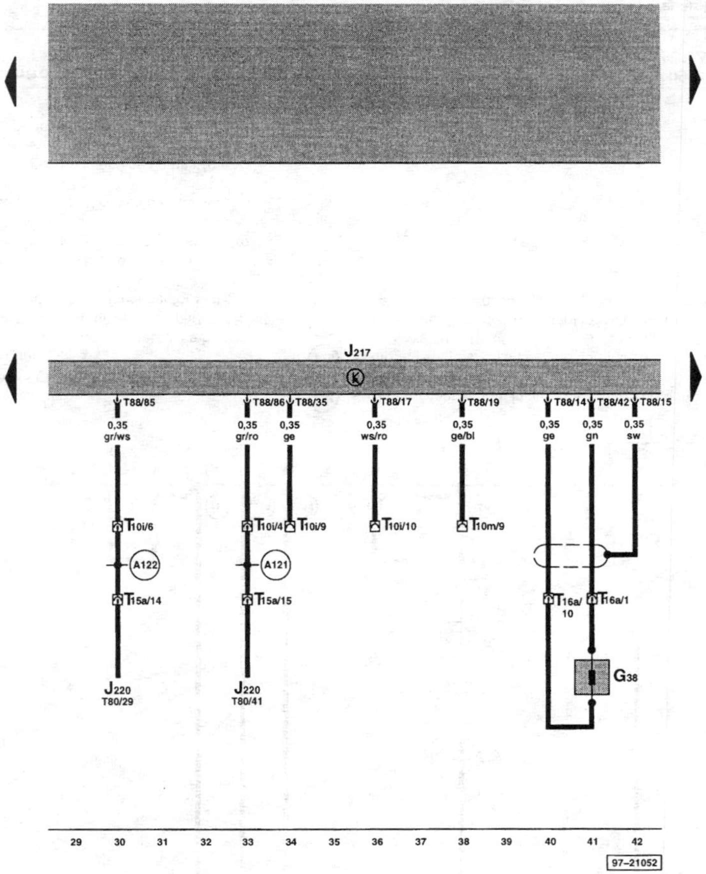

ws = white
sw = black
ro = red
br = brown
gn = green
bl = blue
gr = grey
li = violet
ge = yellow

G38 – Transmission Vehicle Speed Sensor (VSS)
J217 – Transmission Control Module (TCM), in front footwell, right
J220 – Motronic Engine Control Module (ECM)
T10i – 10-Pin Connector, blue, connector station A-pillar, right
T10m – 10-Pin Connector, brown, connector station A-pillar, right
T15a – 15-Pin Connector, white, on protective housing for control module, in engine compartment,left
T16a – 16-Pin Connector, on transmission
T80 – Connector, 80 point, on Motronic Engine Control Module (ECM)
T88 – Connector, 88 point, on Transmission Control Module (TCM)
(A121) – Connector (High-Bus), in instrument panel wiring harness

(A122) – Connector (Low-Bus), in instrument panel wiring harness

Transmission control module (TCM), transmission vehicle speed sensor (VSS)

Edition 09/97
USA.5062.01.21

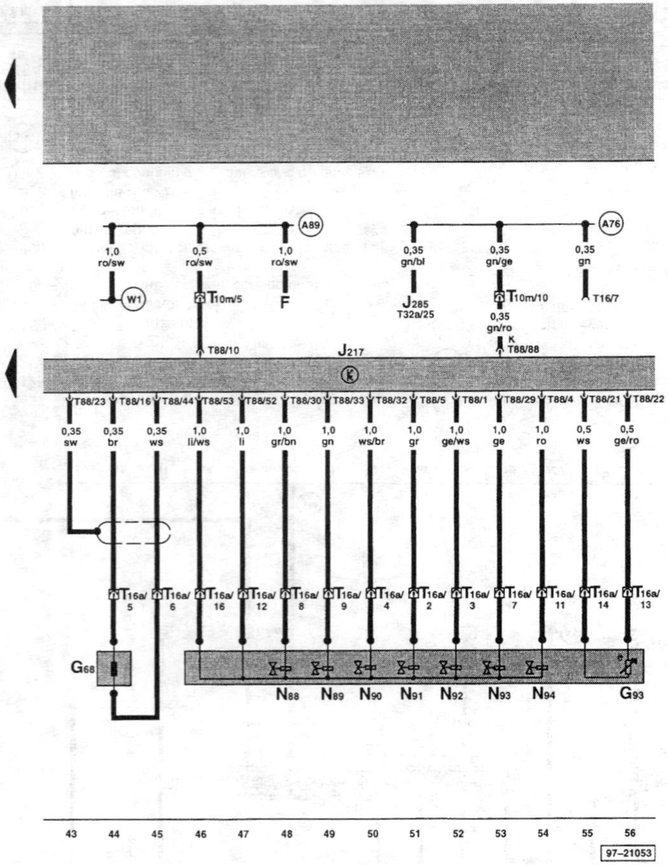

ws = white
sw = black
ro = red
br = brown
gn = green
bl = blue
gr = grey
li = violet
ge = yellow

F – Brake Light Switch
G68 – Vehicle Speed Sensor (VSS)
G93 – Transmission Fluid Temperature Sensor
J217 – Transmission Control Module (TCM), in front footwell, right
J285 – Control module with indicator unit in instrument panel insert
N88 – Solenoid Valve 1
N89 – Solenoid Valve 2
N90 – Solenoid Valve 3
N91 – Solenoid Valve 4
N92 – Solenoid Valve 5
N93 – Solenoid Valve 6
N94 – Solenoid Valve 7
T10m – 10-Pin Connector, brown, connector station A-pillar, right
T16 – 16-Pin Connector, On Board Diagnostic (OBD), below instrument panel, left
T16a – 16-Pin Connector, on transmission
T32a – 32-Pin Connector, blue, on instrument cluster
T88 – Connector, 88 point, on Transmission Control Module (TCM)
(A76) – Connector (K-diagnosis wire), in instrument panel wiring harness
(A89) – Connector (54), in instrument panel wiring harness
(W1) – plus connection (54), in rear wiring harness

Transmission control module (TCM), solenoid valves, transmission fluid temperature sensor, vehicle speed sensor (VSS)

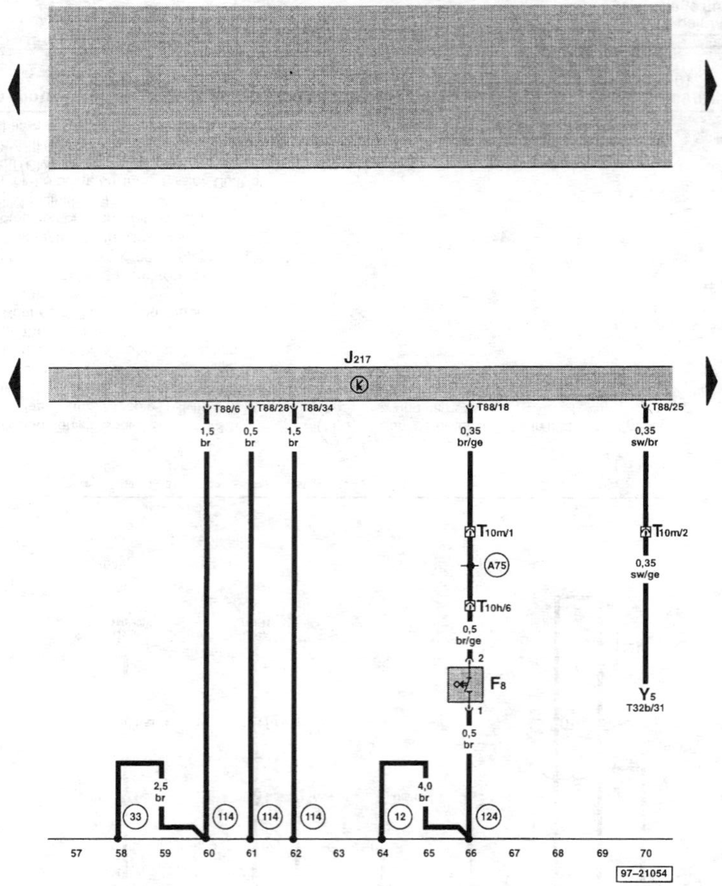

ws = white
sw = black
ro = red
br = brown
gn = green
bl = blue
gr = grey
li = violet
ge = yellow

F8 – Kick Down Switch
J217 – Transmission Control Module (TCM), in front footwell, right
T10h – 10-Pin Connector, blue, on protective housing for control module, in engine compartment, left
T10m – 10-Pin Connector, brown, connector station A-pillar, right
T32b – 32-Pin Connector, green, on instrument cluster
T88 – Connector, 88 point, on Transmission Control Module (TCM)
Y5 – Transmission Range (TR) Selector Lever Display
(12) – Ground connection, in engine compartment, left
(33) – Ground connection, behind instrument panel, right
(114) – Ground connection, in automatic transmission wiring harness
(124) – ground connection, in engine compartment right wiring harness
(A75) – Connector (kick-down-switch) in instrument panel wiring harness

Transmission control module (TCM), kick down switch

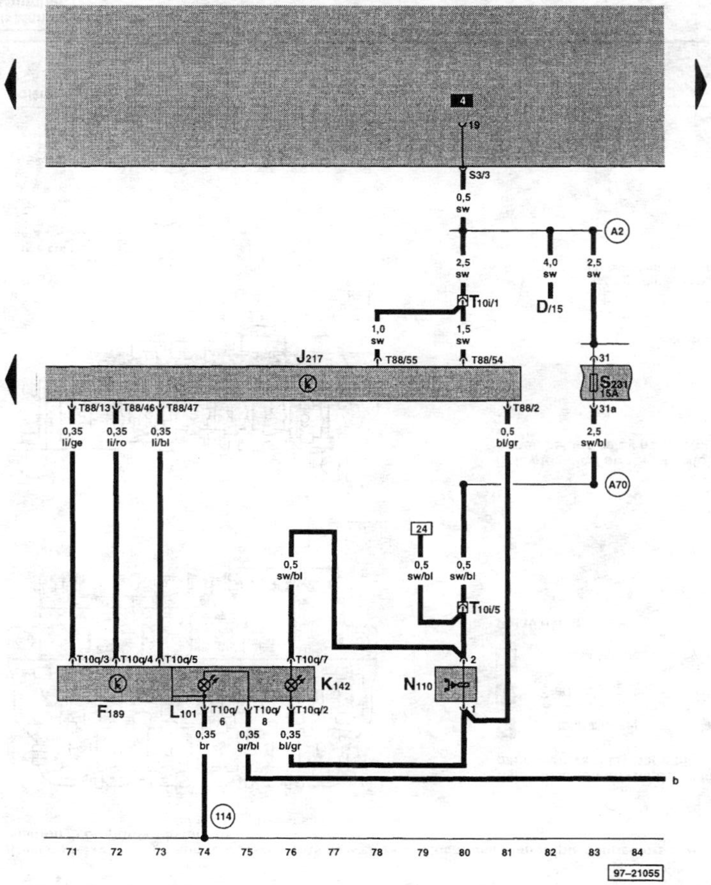

ws = white
sw = black
ro = red
br = brown
gn = green
bl = blue
gr = grey
li = violet
ge = yellow

D – Ignition/Starter Switch
F189 – Tiptronic Switch
K142 – Warning light for selector lever position P/N
L101 – Illumination for selector lever scale
J217 – Transmission Control Module (TCM), in front footwell, right
N110 – Shift Lock Solenoid
S231 – Fuse 31 in fuse holder
T10i – 10-Pin Connector, blue, connector station A-pillar, right
T10q – 10-Pin Connector
T88 – Connector, 88 point, on Transmission Control Module (TCM)
(114) – Ground connection, in automatic transmission wiring harness

(A2) – plus connection (15), in instrument panel wiring harness
(A70) – Connector (15a, fuse 231), in instrument panel wiring harness

Transmission control module (TCM), shift lock solenoid, tiptronic switch

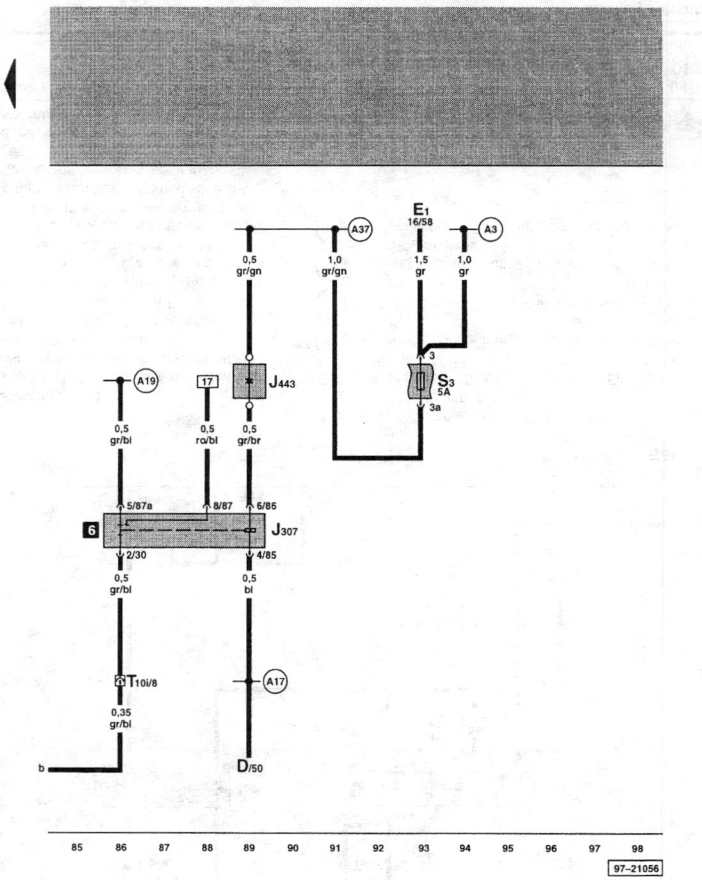

ws = white
sw = black
ro = red
br = brown
gn = green
bl = blue
gr = grey
li = violet
ge = yellow

D – Ignition/Starter Switch
E1 – Light switch
J307 – Selector Lever Light Relay, on the thirteenfold auxiliary relay panel, above relay panel, production control number (205)
J443 – Selector Lever Light Diode
S3 – Fuse 3 in fuse holder
T10i – 10-Pin Connector, blue, connector station A-pillar, right
(A3) – plus connection (58), in instrument panel wiring harness
(A17) – wire connection (61), in instrument panel wiring harness
(A19) – wire connection (58d), in instrument panel wiring harness
(A37) – wire connection (58a), in instrument panel wiring harness

Selector lever light

Automatic transmission, 5 speed automatic with tiptronic,

● 1,8L-Engine – Motronic Multiport Fuel Injection (MFI) / 110 kW, code AEB,

from September 1997

Deviate relay location and fuseplacements as well as the locations of multiple connectors see section "component locations".

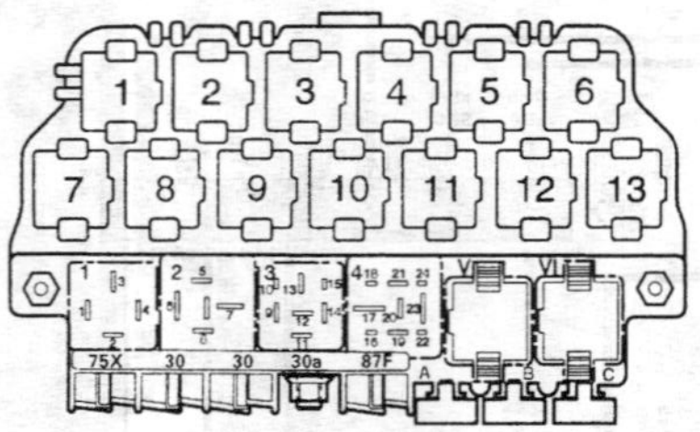

97–20676

Relay location on the thirteenfold auxiliary relay panel, above relay panel:

- **6** Selector Lever Light Relay (205)
- **12** Starting Interlock Relay (alarm system) (186)
- **13** Starting Interlock Relay (204)

Relay panel:

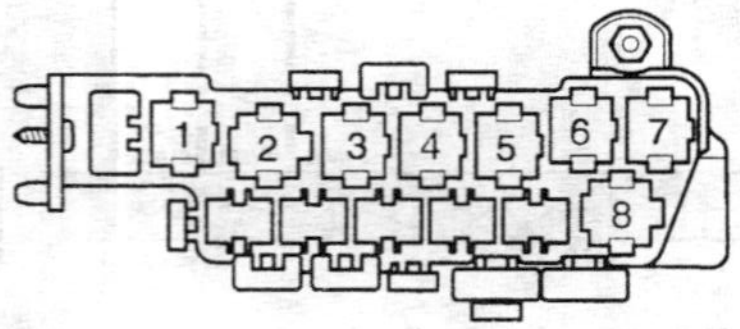

97–20515

Relay location on the eightfold auxiliary relay panel, behind relay panel:

Note: Number in parentheses indicates production control number stamped on relay housing.

Fuse colors
30 A – green
25 A – white
20 A – yellow
15 A – blue
10 A – red
7,5 A – brown
5 A – beige

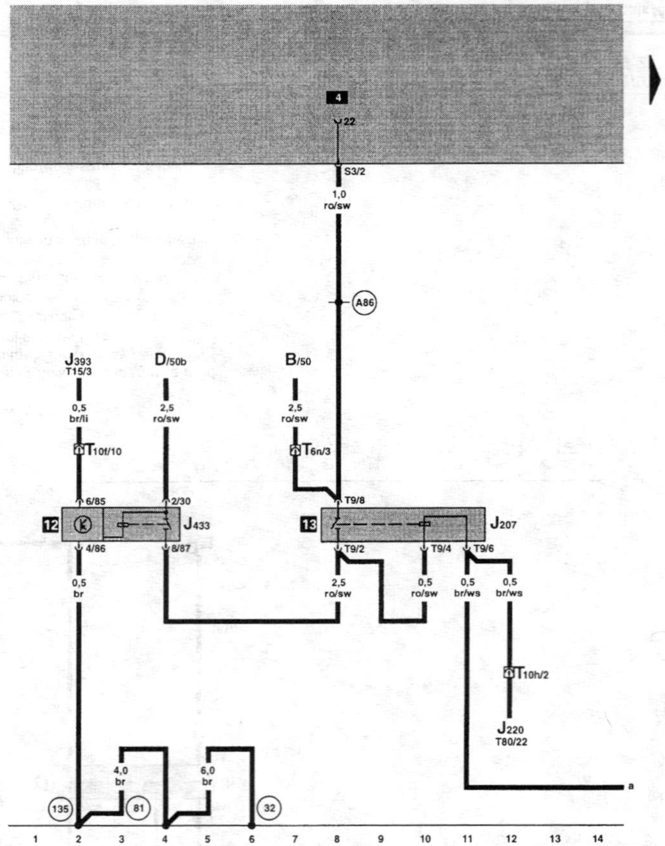

ws = white
sw = black
ro = red
br = brown
gn = green
bl = blue
gr = grey
li = violet
ge = yellow

97–21057

B – Starter
D – Ignition/Starter Switch
J207 – Starting Interlock Relay, on the thirteenfold auxiliary relay panel, above relay panel, production control number (204)
J220 – Motronic Engine Control Module (ECM)
J393 – Central control module for comfort system
J433 – Starting Interlock Relay (alarm system), on the thirteenfold auxiliary relay panel, above relay panel, production control number (186)
T6n – 6-Pin Connector, red, on protective housing for control module, in engine compartment, left
T9 – 9-Pin Connector, on Starting Interlock Relay
T10f – 10-Pin Connector, blue, connector station A-pillar, left
T10h – 10-Pin Connector, blue, on protective housing for control module, in engine compartment, left
T15 – 15-Pin Connector, on Central control module for comfort system
T80 – Connector, 80 point, on Motronic Engine Control Module (ECM)
(32) – Ground connection, behind instrument panel, left
(81) – Ground connection -1-, in instrument panel wiring harness
(135) – Ground connection -2-, in instrument panel wiring harness
(A86) – Connection (50a), in instrument panel wiring harness

Starting interlock relay, Starting interlock relay (alarm system)

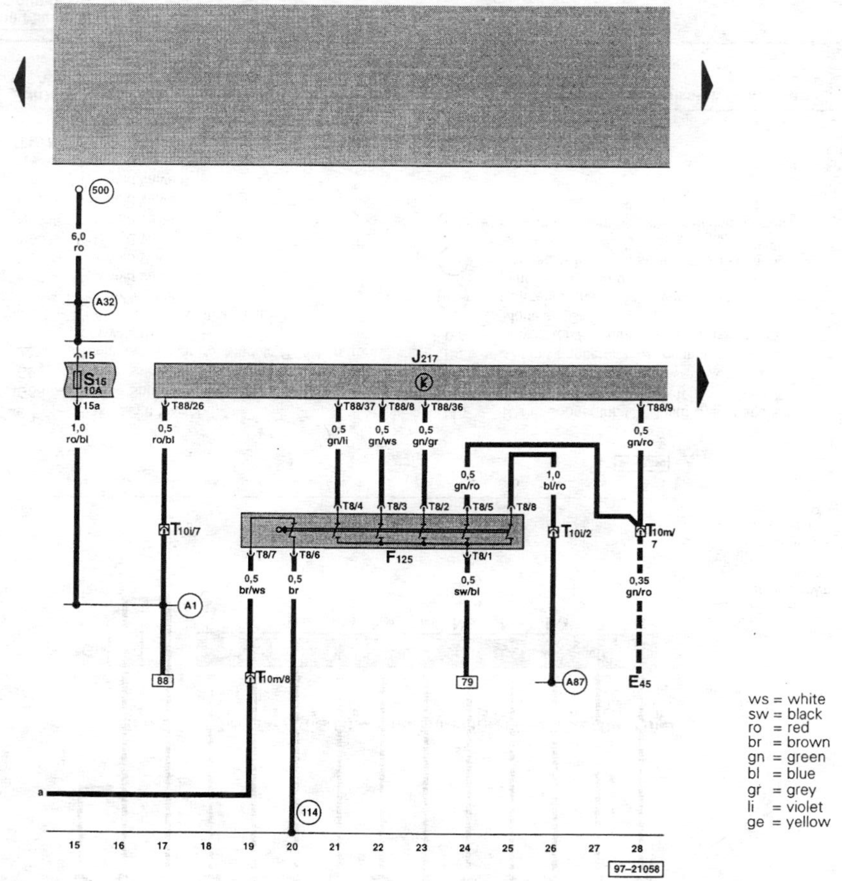

ws = white
sw = black
ro = red
br = brown
gn = green
bl = blue
gr = grey
li = violet
ge = yellow

E45 – Cruise Control Switch
F125 – Multi-Function Transmission Range (TR) Switch
J217 – Transmission Control Module (TCM), in front footwell, right
S15 – Fuse 15 in fuse holder
T8 – 8-Pin Connector, black
T10i – 10-Pin Connector, blue, connector station A-pillar, right
T10m – 10-Pin Connector, brown, connector station A-pillar, right
T88 – Connector 88-point, on Transmission Control Module (TCM)
(114) – Ground connection, in automatic transmission wiring harness
(500) – Screw Connection -1- (30), on relay panel
(A1) – plus connection (30a), in instrument panel wiring harness
(A32) – plus connection (30), in instrument panel wiring harness
(A87) – Connector (reverse lamp), in instrument panel wiring harness
- - - vehicles with cruise control only

Transmission control module (TCM), multi-function transmission range (TR) switch

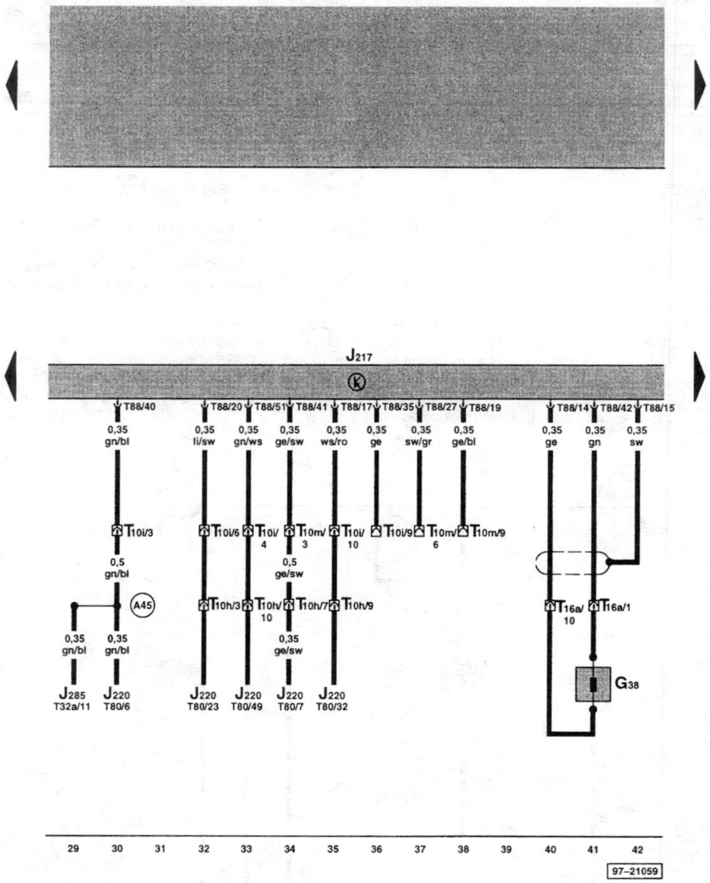

ws = white
sw = black
ro = red
br = brown
gn = green
bl = blue
gr = grey
li = violet
ge = yellow

G38 – Transmission Vehicle Speed Sensor (VSS)
J217 – Transmission Control Module (TCM), in front footwell, right
J220 – Motronic Engine Control Module (ECM)
J285 – Control module with indicator unit in instrument panel insert
T10i – 10-Pin Connector, blue, connector station A-pillar, right
T10h – 10-Pin Connector, blue, on protective housing for control module, in engine compartment, left
T10m – 10-Pin Connector, brown, connector station A-pillar, right
T16a – 16-Pin Connector, on transmission
T32a – 132-Pin Connector, blue, on instrument cluster
T80 – Connector, 80 point, on Motronic Engine Control Module (ECM)
T88 – Connector, 88 point, on Transmission Control Module (TCM)
(A45) – wire connection (RPM-signal), in instrument panel wiring harness

Transmission control module (TCM), transmission vehicle speed sensor (VSS)

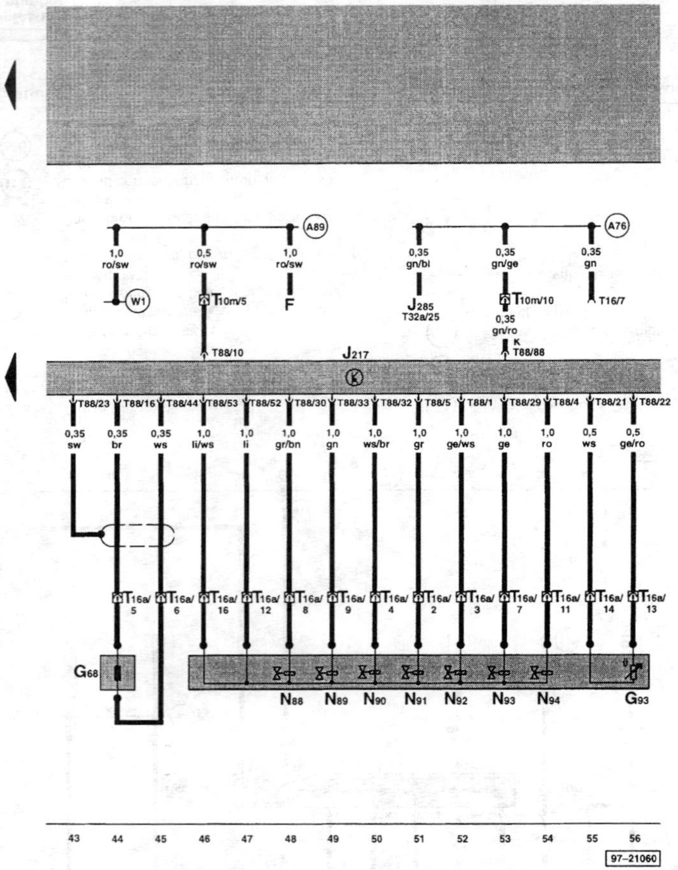

ws = white
sw = black
ro = red
br = brown
gn = green
bl = blue
gr = grey
li = violet
ge = yellow

- F – Brake Light Switch
- G68 – Vehicle Speed Sensor (VSS)
- G93 – Transmission Fluid Temperature Sensor
- J217 – Transmission Control Module (TCM), in front footwell, right
- J285 – Control module with indicator unit in instrument panel insert
- N88 – Solenoid Valve 1
- N89 – Solenoid Valve 2
- N90 – Solenoid Valve 3
- N91 – Solenoid Valve 4
- N92 – Solenoid Valve 5
- N93 – Solenoid Valve 6
- N94 – Solenoid Valve 7
- T10m – 10-Pin Connector, brown, connector station A-pillar, right
- T16 – 16-Pin Connector, On Board Diagnostic (OBD), below instument panel, left
- T16a – 16-Pin Connector, on transmission
- T32a – 32-Pin Connector, blue, on instrument cluster
- T88 – Connector, 88 point, on Transmission Control Module (TCM)
- (A76) – Connector (K-diagnosis wire), in instrument panel wiring harness
- (A89) – Connector (54), in instrument panel wiring harness
- (W1) – plus connection (54), in rear wiring harness

Transmission control module (TCM), solenoid valves, transmission fluid temperature sensor, vehicle speed sensor (VSS)

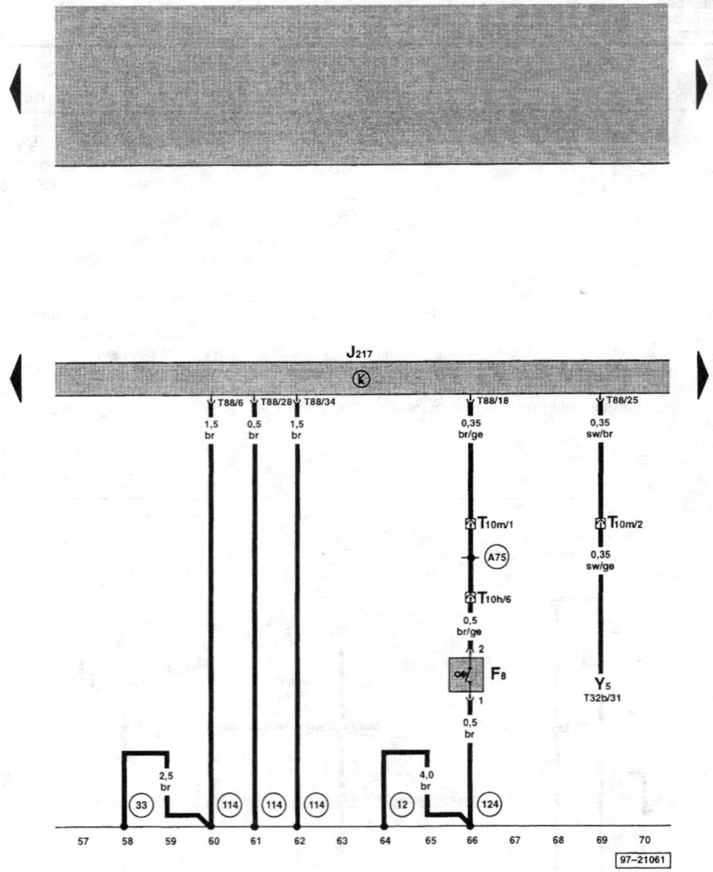

ws = white
sw = black
ro = red
br = brown
gn = green
bl = blue
gr = grey
li = violet
ge = yellow

- F8 – Kick Down Switch
- J217 – Transmission Control Module (TCM), in front footwell, right
- T10h – 10-Pin Connector, blue, on protective housing for control module, in engine compartment, left
- T10m – 10-Pin Connector, brown, connector station A-pillar, right
- T32b – 32-Pin Connector, green, on instrument cluster
- T88 – Connector, 88 point, on Transmission Control Module (TCM)
- Y5 – Transmission Range (TR) Selector Lever Display
- (12) – Ground connection, in engine compartment, left
- (33) – Ground connection, behind instrument panel, right
- (114) – Ground connection, in automatic transmission wiring harness
- (124) – ground connection, in engine compartment right wiring harness
- (A75) – Connector (kick-down-switch) in instrument panel wiring harness

Transmission control module (TCM), kick down switch

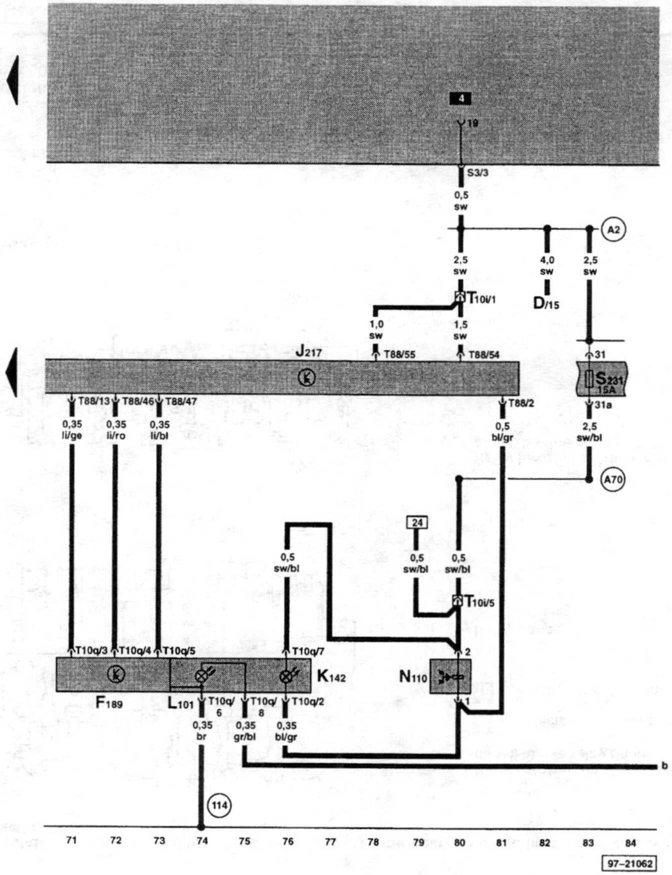

ws = white
sw = black
ro = red
br = brown
gn = green
bl = blue
gr = grey
li = violet
ge = yellow

D – Ignition/Starter Switch
F189 – Tiptronic Switch
K142 – Warning light for selector lever position P/N
L101 – Illumination for selector lever scale
J217 – Transmission Control Module (TCM), in front footwell, right
N110 – Shift Lock Solenoid
S231 – Fuse 31 in fuse holder
T10i – 10-Pin Connector, blue, connector station A-pillar, right
T10q – 10-Pin Connector
T88 – Connector, 88 point, on Transmission Control Module (TCM)

(114) – Ground connection, in automatic transmission wiring harness

(A2) – plus connection (15), in instrument panel wiring harness
(A70) – Connector (15a, fuse 231), in instrument panel wiring harness

Transmission control module (TCM), shift lock solenoid, tiptronic switch

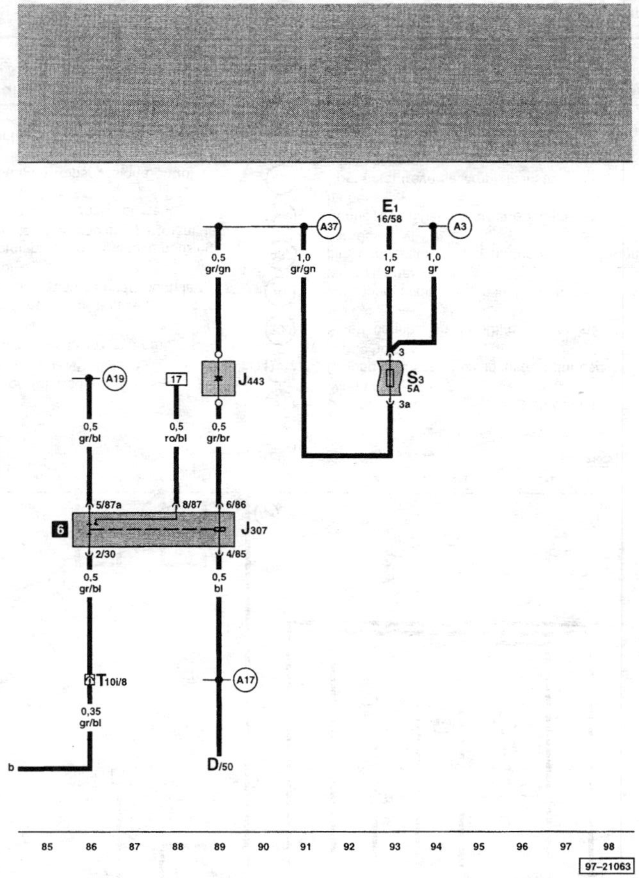

ws = white
sw = black
ro = red
br = brown
gn = green
bl = blue
gr = grey
li = violet
ge = yellow

D – Ignition/Starter Switch
E1 – Light switch
J307 – Selector Lever Light Relay, on the thirteenfold auxiliary relay panel, above relay panel, production control number (205)
J443 – Selector Lever Light Diode
S3 – Fuse 3 in fuse holder
T10i – 10-Pin Connector, blue, connector station A-pillar, right
(A3) – plus connection (58), in instrument panel wiring harness
(A17) – wire connection (61), in instrument panel wiring harness
(A19) – wire connection (58d), in instrument panel wiring harness
(A37) – wire connection (58a), in instrument panel wiring harness

Selector lever light

Automatic transmission, 4 speed automatic,

from September 1997

- **1,9l - Turbo Diesel Fuel Injection (DFI)/66kW, code AHH**

Deviate relay location and fuseplacements as well as the locations of multiple connectors see section "component locations".

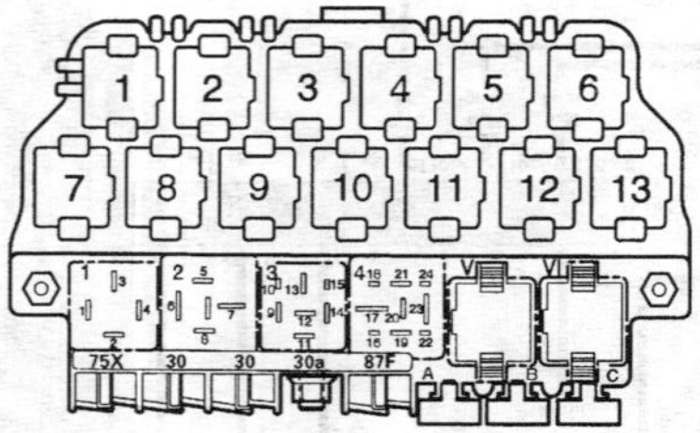

97–20676

Relay location on the thirteenfold auxiliary relay panel, above relay panel:

- **6** Selector Lever Light Relay (205)
- **13** Park/Neutral Position (PNP) Relay (175)

Relay panel:

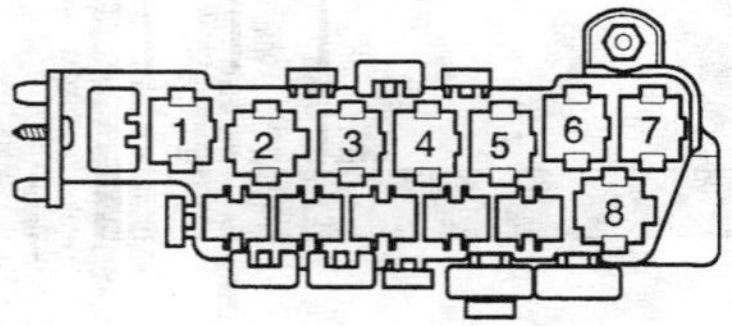

97–20515

Relay location on the eightfold auxiliary relay panel, behind relay panel:

Note: Number in parentheses indicates production control number stamped on relay housing.

Fuse colors

30 A – green
25 A – white
20 A – yellow
15 A – blue
10 A – red
7,5 A – brown
5 A – beige

Edition 09/97
USA.5062.01.21

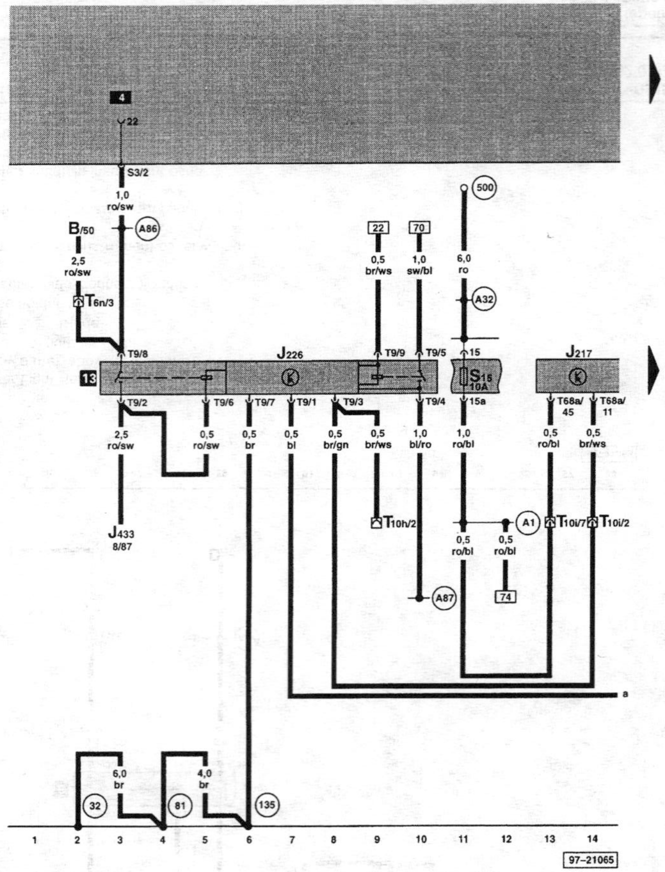

ws = white
sw = black
ro = red
br = brown
gn = green
bl = blue
gr = grey
li = violet
ge = yellow

B – Starter
J217 – Transmission Control Module (TCM)
J226 – Park/Neutral Position (PNP) Relay
J433 – Starting Interlock Relay (alarm system)
S15 – Fuse 15 in fuse holder
T6n – 6-Pin Connector, red, on protective housing for control module, in engine compartment, left
T9 – 9-Pin Connector
T10h – 10-Pin Connector, blue, on protective housing for control module, in engine compartment, left
T10i – 10-Pin Connector, blue, connector station A-pillar, right
T68a – 68-Pin Connector, on Transmission Control Module (TCM)
(32) – Ground connection, behind instrument panel, left
(81) – Ground connection -1-, in instrument panel wiring harness
(135) – Ground connection -2-, in instrument panel wiring harness
(500) – Screw connection -1- (30), on relay panel
(A1) – plus connection (30a), in instrument panel wiring harness
(A32) – plus connection (30), in instrument panel wiring harness
(A86) – Connection (50a) in instrument panel wiring harness
(A87) – Connector (reverse lamp), in instrument panel wiring harness

Transmission control module (TCM), Park/Neutral position (PNP) relay

Edition 09/97
USA.5062.01.21

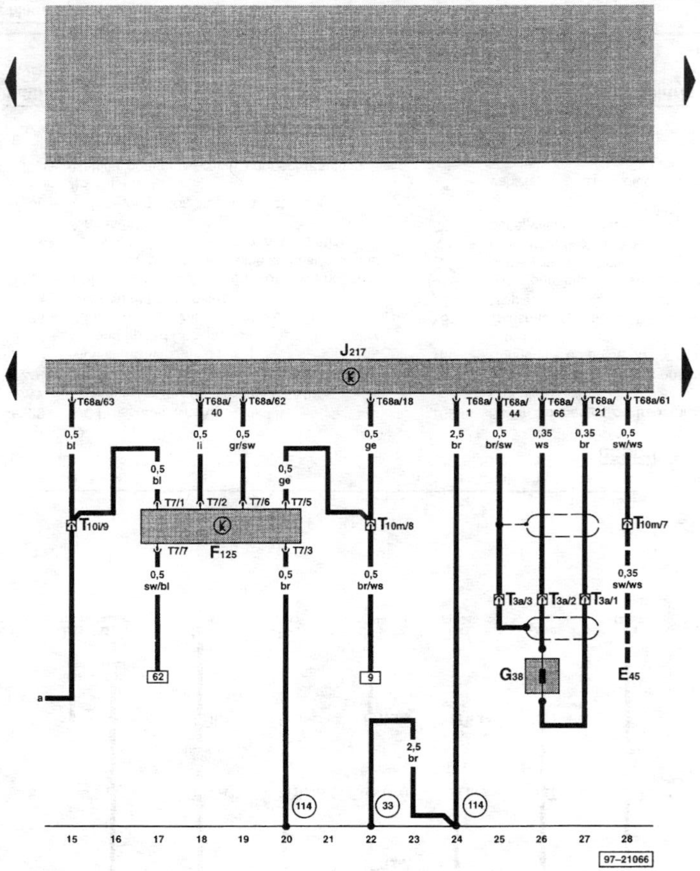

ws = white
sw = black
ro = red
br = brown
gn = green
bl = blue
gr = grey
li = violet
ge = yellow

- E45 – Cruise Control Switch
- F125 – Multi-Function Transmission Range (TR) Switch
- G38 – Transmission Vehicle Speed Sensor (VSS)
- J217 – Transmission Control Module (TCM)
- T3a – 3-Pin Connector, black
- T7 – 7-Pin Connector
- T10i – 10-Pin Connector, blue, connector station A-pillar, right
- T10m – 10-Pin Connector, brown, connector station A-pillar, right
- T68a – 68-Pin Connector, on Transmission Control Module (TCM)
- (33) – Ground connection, behind instrument panel, right
- (114) – Ground connection, in automatic transmission wiring harness

--- vehicles with cruise control only

Edition 09/97
USA.5062.01.21

Transmission control module (TCM), multi-function transmission range (TR) switch, transmission vehicle speed sensor (VSS)

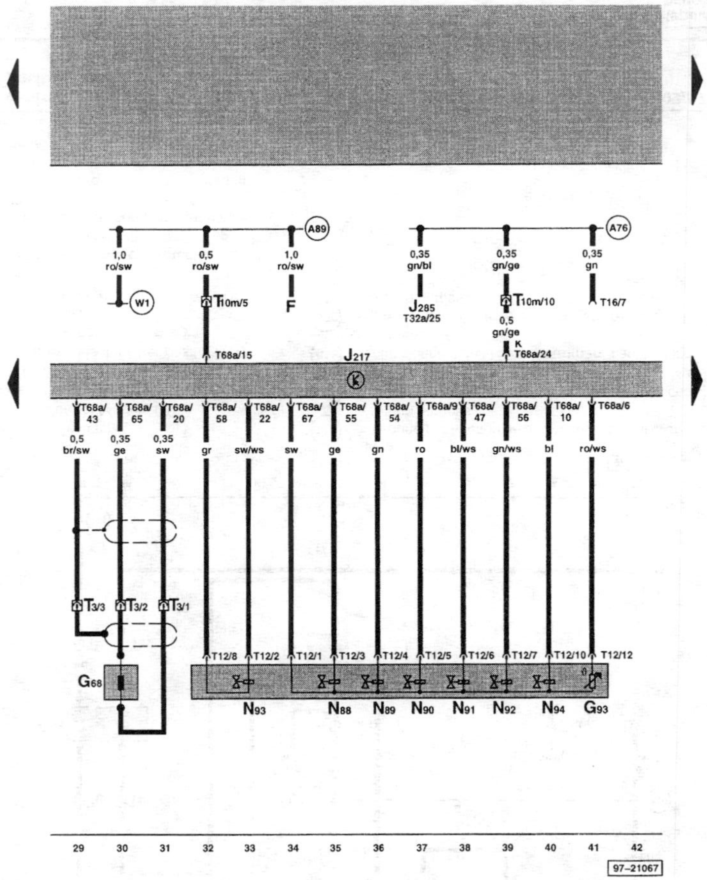

ws = white
sw = black
ro = red
br = brown
gn = green
bl = blue
gr = grey
li = violet
ge = yellow

- F – Brake Light Switch
- G68 – Vehicle Speed Sensor (VSS)
- G93 – Transmission Fluid Temperature Sensor
- J217 – Transmission Control Module (TCM)
- J285 – Control module with indicator unit in instrument panel insert
- N88 – Solenoid Valve 1
- N89 – Solenoid Valve 2
- N90 – Solenoid Valve 3
- N91 – Solenoid Valve 4
- N92 – Solenoid Valve 5
- N93 – Solenoid Valve 6
- N94 – Solenoid Valve 7
- T3 – 3-Pin Connector, brown
- T10m – 10-Pin Connector, brown, connector station A-pillar, right
- T12 – 12-Pin Connector, on transmission
- T16 – 16-Pin Connector, On Board Diagnostic (OBD), below instrument panel, left
- T32a – 32-Pin Connector, blue, on instrument cluster
- T68a – 68-Pin Connector, on Transmission Control Module (TCM)
- (A76) – Connector (K-diagnosis wire), in instrument panel wiring harness
- (A89) – Connector (54), in instrument panel wiring harness
- (W1) – plus connection (54), in rear wiring harness

Transmission control module (TCM), Transmission fluid temperature sensor, transmission

Edition 09/97
USA.5062.01.21

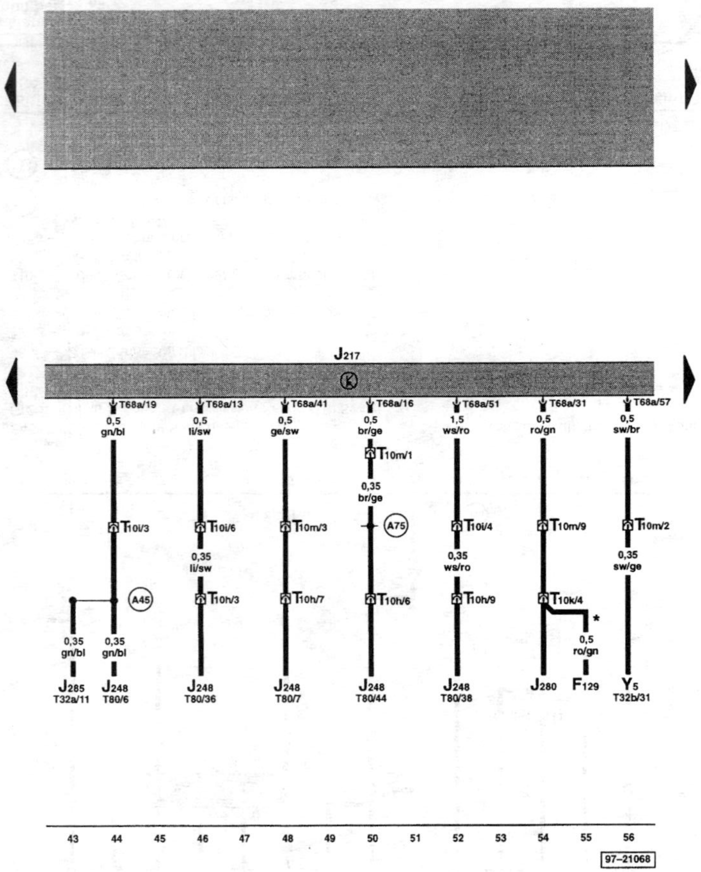

ws = white
sw = black
ro = red
br = brown
gn = green
bl = blue
gr = grey
li = violet
ge = yellow

F129 – A/C Pressure Switch
J217 – Transmission Control Module (TCM)
J248 – Diesel Direct Fuel Injection (DFI) Engine Control Module (ECM)
J280 – Second Speed Coolant Fan Control (FC) Relay
J285 – Control module with indicator unit in instrument panel insert
T10h – 10-Pin Connector, blue, on protective housing for control module, in engine compartment, left
T10k – 10-Pin Connector, grey, connector station A-pillar, left
T10i – 10-Pin Connector, blue, connector station A-pillar, right
T10m – 10-Pin Connector, brown, connector station A-pillar, right
T32a – 32-Pin Connector, blue, on instrument cluster
T68a – 68-Pin Connector, on Transmission Control Module (TCM)
T32b – 32-Pin Connector, green, on instrument cluster
T80 – Connector, 80 point, on control module for Diesel
Y5 – Transmission Range (TR) Selector Lever Display
(A45) – wire connection (RPM-signal), in instrument panel wiring harness
(A75) – Connector (kick-down-switch) in instrument panel wiring harness
* – vehicles with airconditioning only

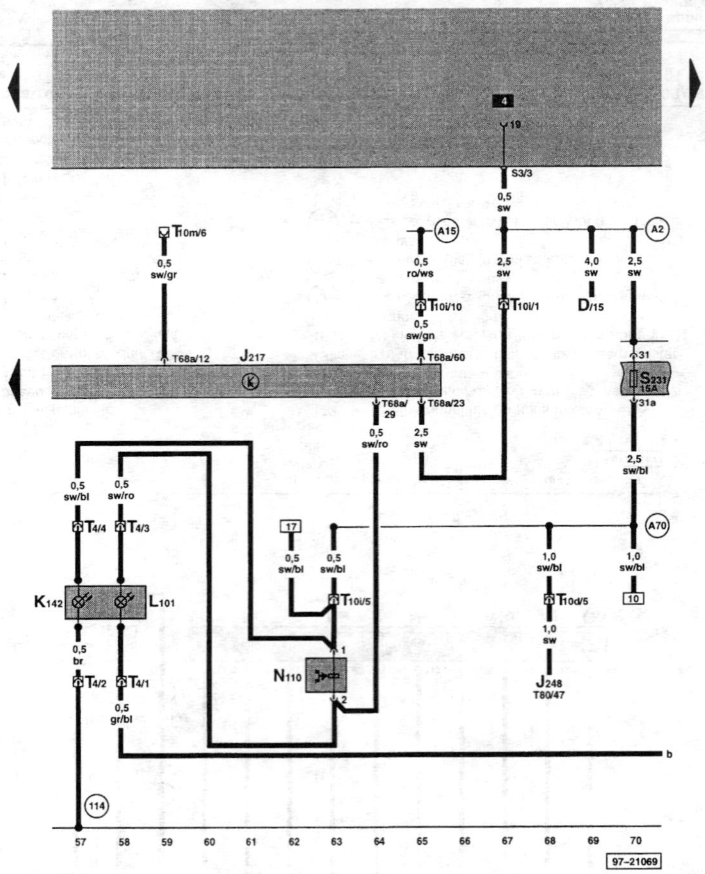

ws = white
sw = black
ro = red
br = brown
gn = green
bl = blue
gr = grey
li = violet
ge = yellow

D – Ignition/Starter Switch
J217 – Transmission Control Module (TCM)
J248 – Diesel Direct Fuel Injection (DFI) Engine Control Module (ECM)
K142 – Warning light for selector lever position P/N
L101 – Illumination for selector lever scale
N110 – Shift Lock Solenoid
S231 – Fuse 31 in fuse holder
T4 – 4-Pin Connector
T10d – 10-Pin Connector, brown, on protective housing for control module, in engine compartment, left
T10i – 10-Pin Connector, blue, connector station A-pillar, right
T10m – 10-Pin Connector, brown, connector station A-pillar, right
T80 – Connector, 80 point, on control module for Diesel
(114) – Ground connection, in automatic transmission wiring harness
(A2) – plus connection (15), in instrument panel wiring harness
(A15) – plus connection (15), in instrument panel wiring harness
(A70) – Connector (15a, fuse 231), in instrument panel wiring harness

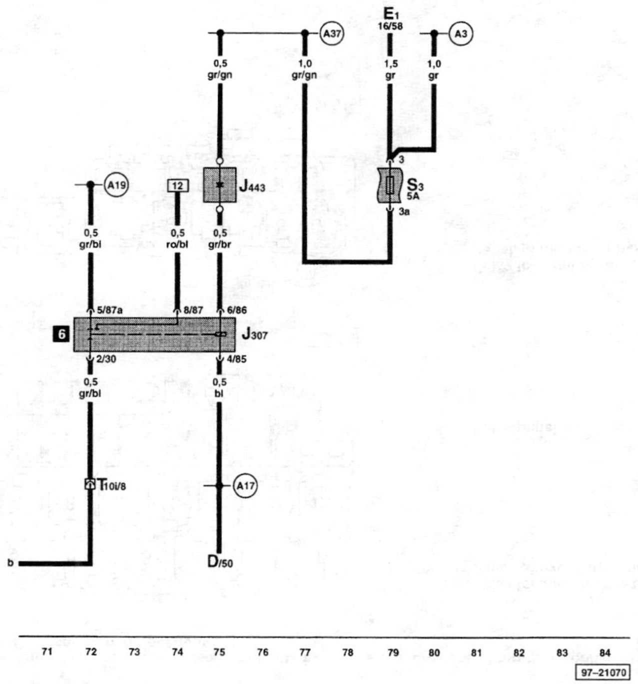

ws = white
sw = black
ro = red
br = brown
gn = green
bl = blue
gr = grey
li = violet
ge = yellow

- D – Ignition/Starter Switch
- E1 – Light switch
- J307 – Selector Lever Light Relay, on the thirteenfold auxiliary relay panel, above relay panel, production control number (205)
- J443 – Selektor Lever Light Diode
- S3 – Fuse 3 in fuse holder
- T10i – 10-Pin Connector, blue, connector station A-pillar, right
- (A3) – plus connection (58), in instrument panel wiring harness
- (A17) – wire connection (61), in instrument panel wiring harness
- (A19) – wire connection (58d), in instrument panel wiring harness
- (A37) – wire connection (58a), in instrument panel wiring harness

THIS PAGE INTENTIONALLY LEFT

BLANK

Anti-lock brake system (ABS) with electronic differential lock (EDL) and anti-slip regulation (ASR), front wheel drive only

from September 1997

Deviate relay location and fuseplacements as well as the locations of multiple connectors see section "component locations".

Relay location on the thirteenfold auxiliary relay panel, above relay panel:

Relay panel:

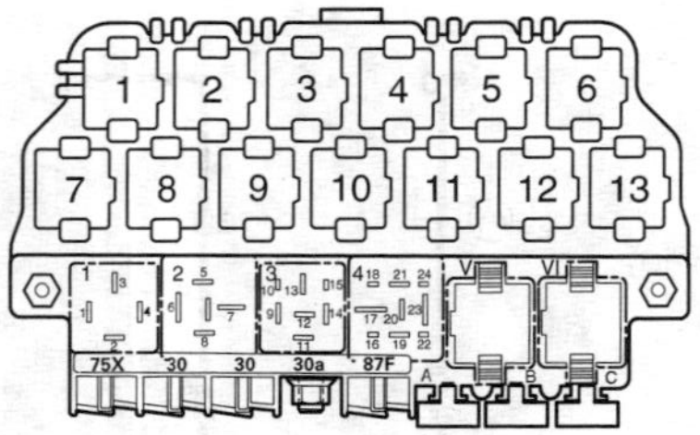

97-20676

Relay location on the eightfold auxiliary relay panel, behind relay panel:

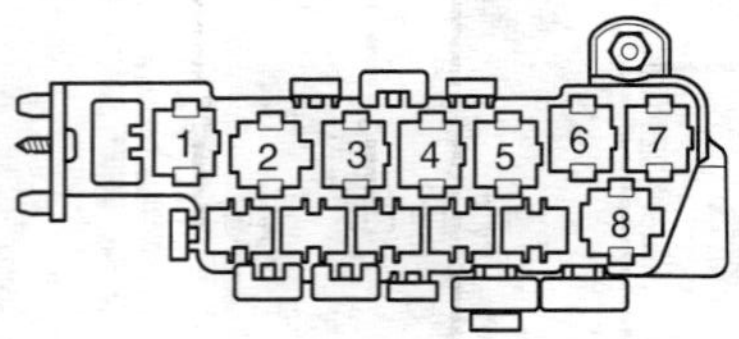

97-20515

Fuse colors
30 A - green
25 A - white
20 A - yellow
15 A - blue
10 A - red
7,5 A - brown
5 A - beige

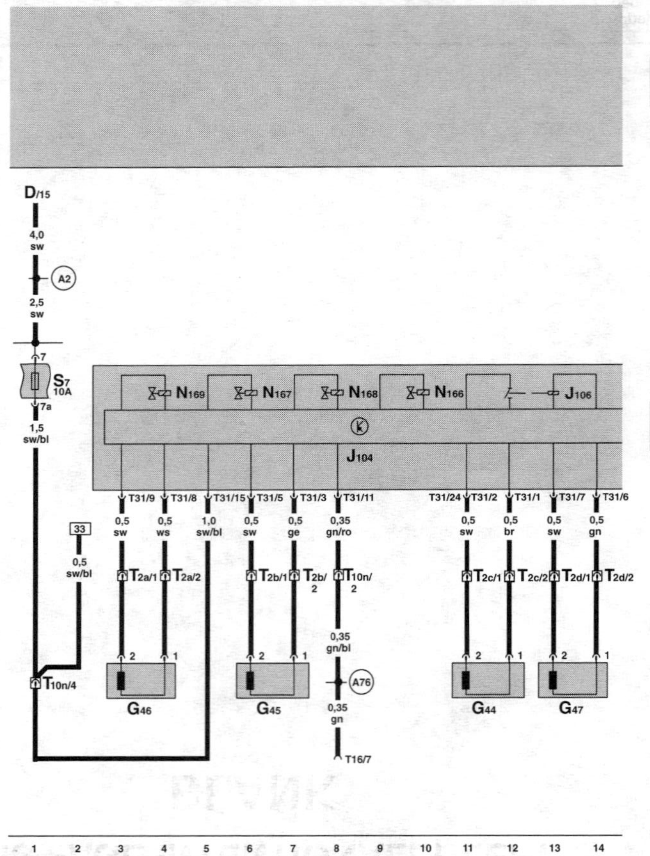

ws = white
sw = black
ro = red
br = brown
gn = green
bl = blue
gr = grey
li = violet
ge = yellow

97-25430

D - Ignition/Starter Switch
G44 - Right Rear ABS Wheel Speed Sensor
G45 - Right Front ABS Wheel Speed Sensor
G46 - Left Rear ABS Wheel Speed Sensor
G47 - Left Front ABS Wheel Speed Sensor
J104 - ABS Control Module (w/EDL)
J106 - ABS Solenoid Valve Relay
N166 - Right Front Traction Control Switch-Over Valve
N168 - Left Front Traction Control Switch-Over Valve
N169 - Left Front Traction Control Outlet Valve
S7 - Fuse 7 in fuse holder
T2a - 2-Pin Connector, (under rear seat)
T2b - 2-Pin Connector, (in right front wheel well)
T2c - 2-Pin Connector, (under rear seat)
T2d - 2-Pin Connector, (in left front wheel well)
T10n - 10-Pin Connector, orange, connector station A-pillar, left
T16 - 16-Pin Connector, Data Link Connector (DLC), below instrument panel, left
T31 - 31-Pin Connector

(A2) - plus connection (15), in instrument panel wiring harness
(A76) - Connector (K-diagnosis wire), in instrument panel wiring harness

ABS control module (with EDL), ABS wheel speed sensors

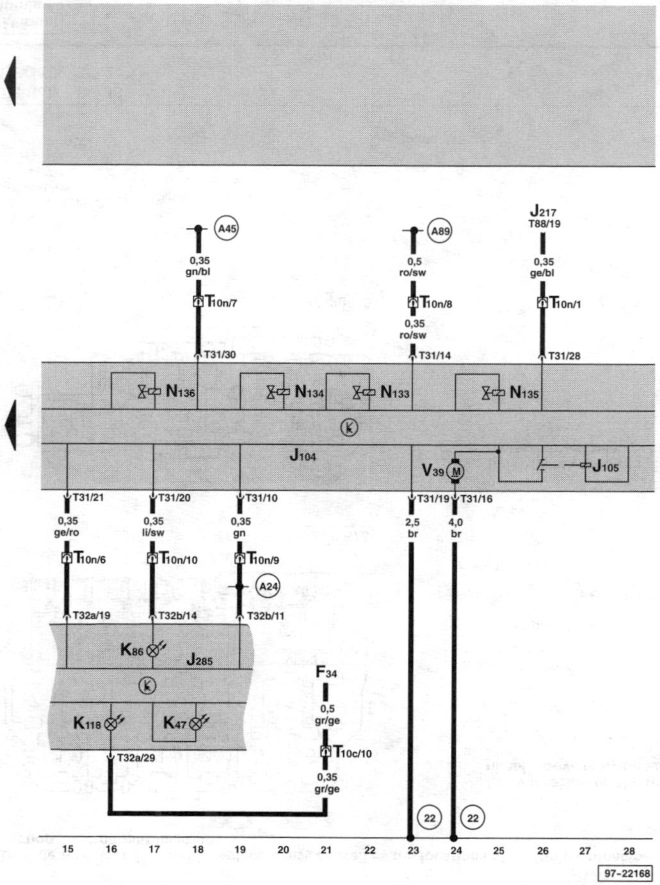

ws = white
sw = black
ro = red
br = brown
gn = green
bl = blue
gr = grey
li = violet
ge = yellow

F34 - Brake Fluid Level Warning Switch
J104 - ABS Control Module (w/EDL)
J105 - ABS Return Flow Pump Relay
J217 - Transmission Control Module (TCM)
J285 - Control module with indicator unit in instrument panel insert
K47 - ABS Warning Light
K86 - Traction Control Indicator Light
K118 - Warning light for brake system
N133 - Right Rear ABS Inlet Valve
N134 - Left Rear ABS Inlet Valve
N135 - Right Rear ABS Outlet Valve
N136 - Left Rear ABS Outlet Valve
T10c - 10-Pin Connector, violet, connector station A-pillar, left
T10n - 10-Pin Connector, orange, connector station A-pillar, left
T31 - 31-Pin Connector
T32a - 32-Pin Connector, blue, on instrument cluster
T32b - 32-Pin Connector, green, on instrument cluster
V39 - ABS Return Flow Pump
T88 - Connector, 88 point, on Transmission Control Module (TCM)
(22) - Ground connection, on hydraulic unit
(A24) - wire connection (brake system monitoring), in instrument panel wiring harness
(A45) - wire connection (RPM-signal), in instrument panel wiring harness
(A89) - Connector (54), in instrument panel wiring harness

Edition 12/97
USA.5062.03.21

ABS control module (with EDL), ABS return flow pump, ABS warning light, traction control indicator light, Warning light for brake system

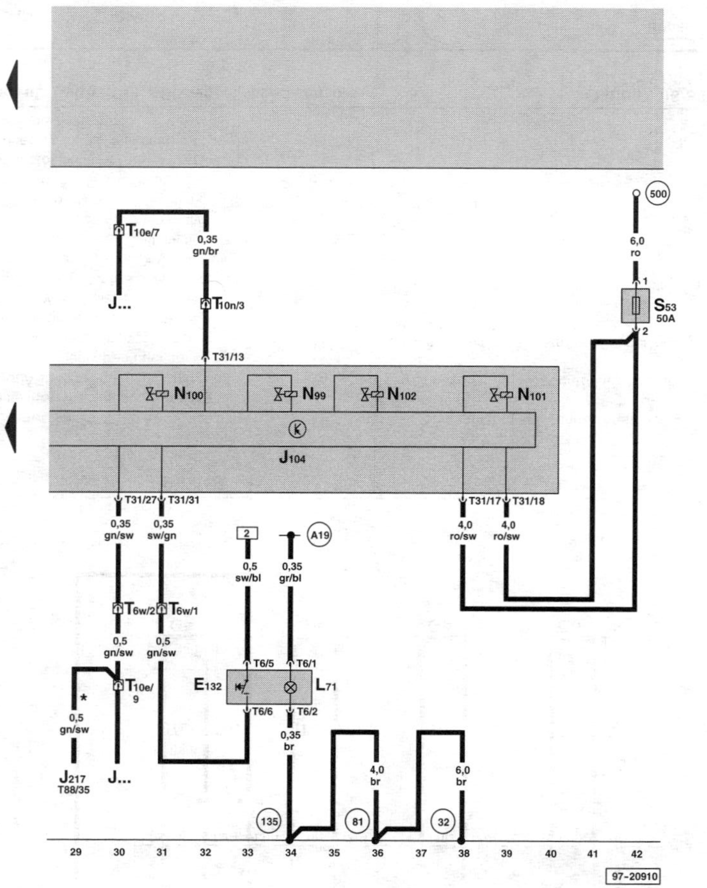

ws = white
sw = black
ro = red
br = brown
gn = green
bl = blue
gr = grey
li = violet
ge = yellow

E132 - Anti-Slip Control Switch
J... - Engine Control Module
J104 - ABS Control Module (w/EDL)
J217 - Transmission Control Module (TCM)
L71 - Anti-slip Control Switch Illumination
N99 - Right Front ABS Inlet Valve
N100 - Right Front ABS Outlet Valve
N101 - Left Front ABS Inlet Valve
N102 - Left Front ABS Outlet Valve
S53 - ABS Hydraulic Pump Fuse, on the thirteenfold auxiliray relay panel, above relay panel
T6 - 6-Pin Connector, black
T6w - 6-Pin Connector, grey, connector station A-pillar, right
T10e - 10-Pin Connector, yellow, on protective housing for control module, in engine compartment, left
T10n - 10-Pin Connector, orange, connector station A-pillar, left
T31 - 31-Pin Connector
T88 - Connector, 88 point, on Transmission Control Module (TCM)
(32) - Ground connection, behind instrument panel, left
(81) - Ground connection -1-, in instrument panel wiring harness
(135) - Ground connection -2-, in instrument panel wiring harness
(500) - Screw connection -1- (30), on relay panel
(A19) - wire connection (58d), in instrument panel wiring harness
* - automatic transmission only

ABS control module (with EDL), anti-slip control switch

Edition 12/97
USA.5062.03.21

Anti-lock brake system (ABS) with electronic differential lock (EDL), front and all wheel drive

from September 1997

Deviate relay location and fuseplacements as well as the locations of multiple connectors see section "component locations".

Relay location on the thirteenfold auxiliary relay panel, above relay panel:

Relay panel:

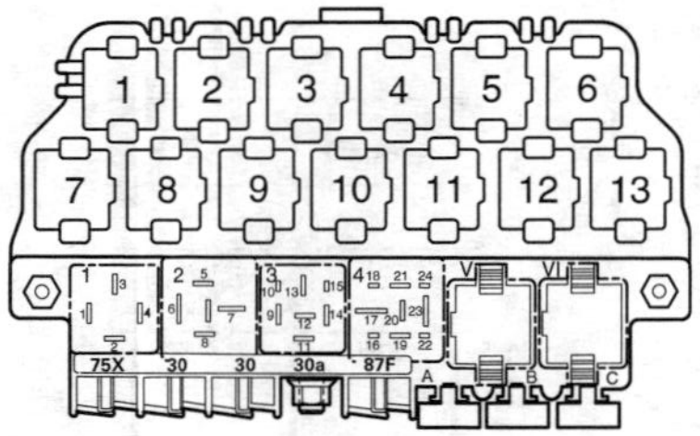

97-20676

Relay location on the eightfold auxiliary relay panel, behind relay panel:

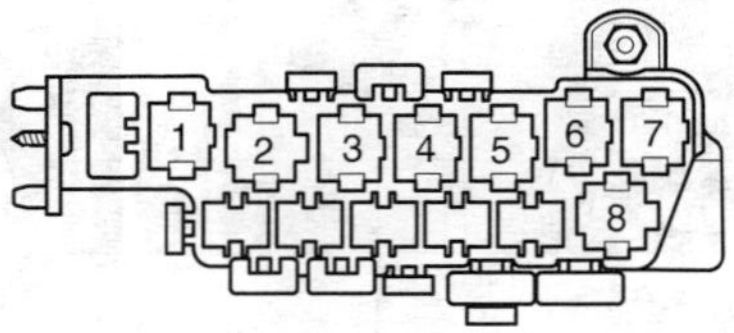

97-20515

Fuse colors

30 A - green
25 A - white
20 A - yellow
15 A - blue
10 A - red
7,5 A - brown
5 A - beige

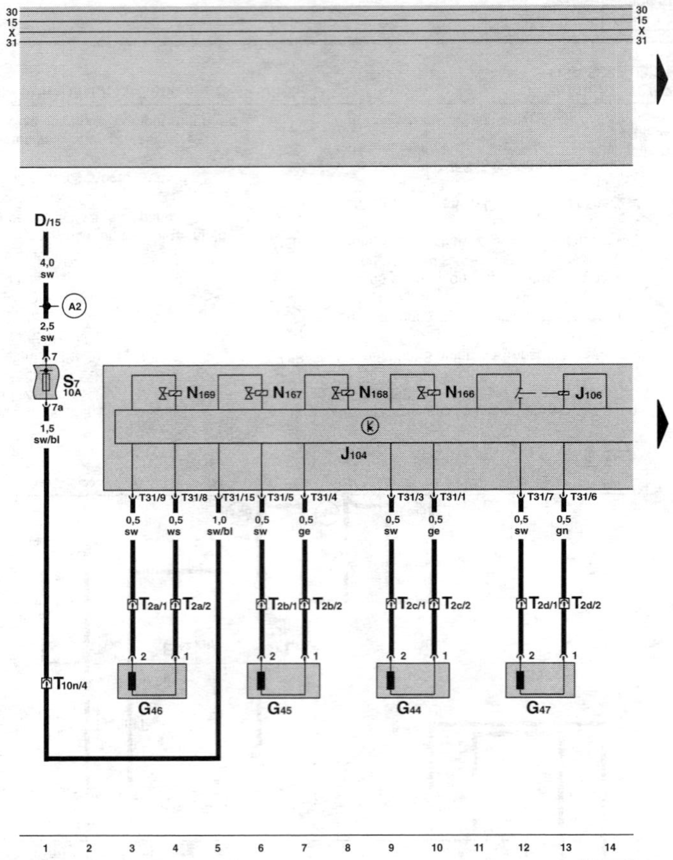

97-23252

ws = white
sw = black
ro = red
br = brown
gn = green
bl = blue
gr = grey
li = violet
ge = yellow

D - Ignition/Starter Switch
G44 - Right Rear ABS Wheel Speed Sensor
G45 - Right Front ABS Wheel Speed Sensor
G46 - Left Rear ABS Wheel Speed Sensor
G47 - Left Front ABS Wheel Speed Sensor
J104 - ABS Control Module (w/EDL)
J106 - ABS Solenoid Valve Relay
N166 - Right Front Traction Control Switch-Over Valve
N167 - Right Front Traction Control Outlet Valve
N168 - Left Front Traction Control Switch-Over Valve
N169 - Left Front Traction Control Outlet Valve
S7 - Fuse 7 in fuse holder
T2a - 2-Pin Connector (under rear seat)
T2b - 2-Pin Connector, (in right front wheel well)
T2c - 2-Pin Connector, (under rear seat)
T2d - 2-Pin Connector, (in left front wheel well)
T10n - 10-Pin Connector, orange, connector station A-pillar, left
T31 - 31-Pin Connector, on ABS Control Module (w/EDL)
(A2) - plus connection (15), in instrument panel wiring harness

ABS control module (with EDL), ABS wheel speed sensors

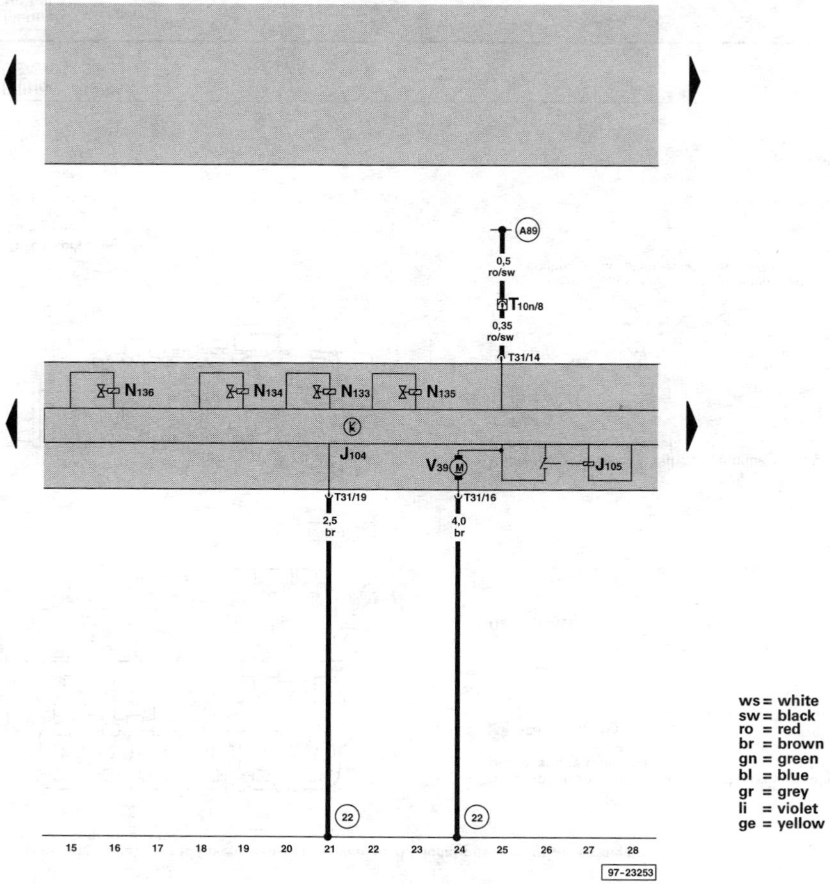

ws = white
sw = black
ro = red
br = brown
gn = green
bl = blue
gr = grey
li = violet
ge = yellow

J104 - ABS Control Module (w/EDL)
J105 - ABS Return Flow Pump Relay
N133 - Right Rear ABS Inlet Valve
N134 - Left Rear ABS Inlet Valve
N135 - Right Rear ABS Outlet Valve
N136 - Left Rear ABS Outlet Valve
T10n - 10-Pin Connector, orange, connector station A-pillar, left
T31 - 31-Pin Connector, on ABS Control Module (w/EDL)
V39 - ABS Return Flow Pump

(22) - Ground connection, on hydraulic unit

(A89) - Connector (54), in instrument panel wiring harness

ABS control module (with EDL), ABS return flow pump

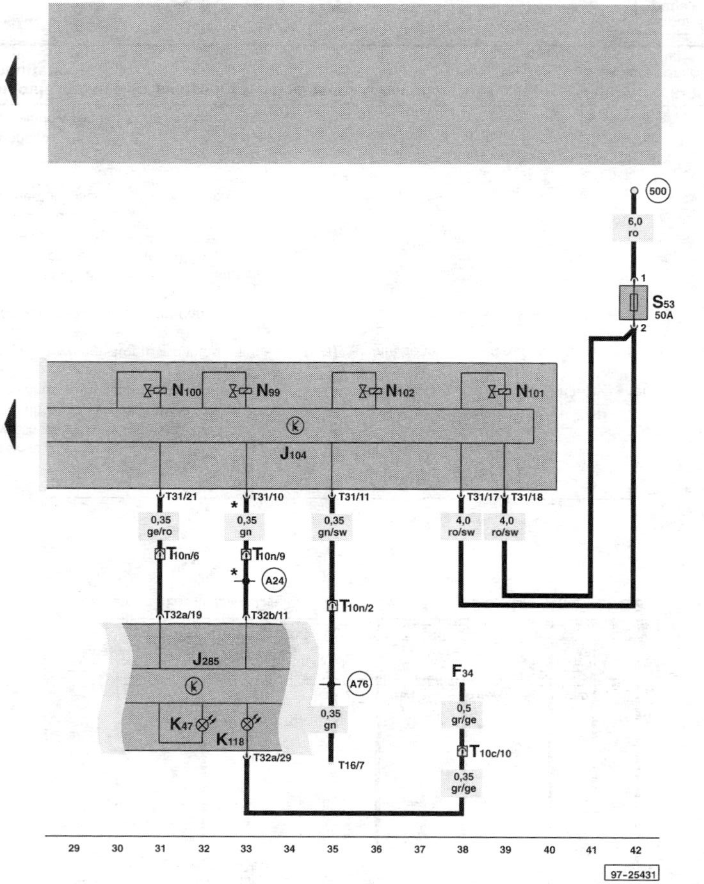

ws = white
sw = black
ro = red
br = brown
gn = green
bl = blue
gr = grey
li = violet
ge = yellow

F34 - Brake Fluid Level Warning Switch
J104 - ABS Control Module (w/EDL)
J285 - Control module with indicator unit in instrument panel insert
K47 - ABS Warning Light
K118 - Warning light for brake system
N99 - Right Front ABS Inlet Valve
N100 - Right Front ABS Outlet Valve
N101 - Left Front ABS Inlet Valve
N102 - Left Front ABS Outlet Valve
S53 - ABS Hydraulic Pump Fuse, on the thirteenfold auxiliary relay panel, above relay panel
T10c - 10-Pin Connector, violet, connector station A-pillar, left
T10n - 10-Pin Connector, orange, connector station A-pillar, left
T16 - 16-Pin Connector, Data Link Connector (DLC), below instrument panel, left
T31 - 31-Pin Connector, on ABS Control Module (w/EDL)
T32a - 32-Pin Connector, blue, on instrument cluster
T32b - 32-Pin Connector, green, on instrument cluster

(500) -Screw connection -1- (30), on relay panel

(A24) - wire connection (brake system monitoring), in instrument panel wiring harness

(A76) - Connector (K-diagnosis wire), in instrument panel wiring harness

* - with EDL only

ABS control module (with EDL), ABS warning light, Warning light for brake system

Comfort System, with Alarm System,

from September 1997

Deviate relay location and fuseplacements as well as the locations of multiple connectors see section "component locations".

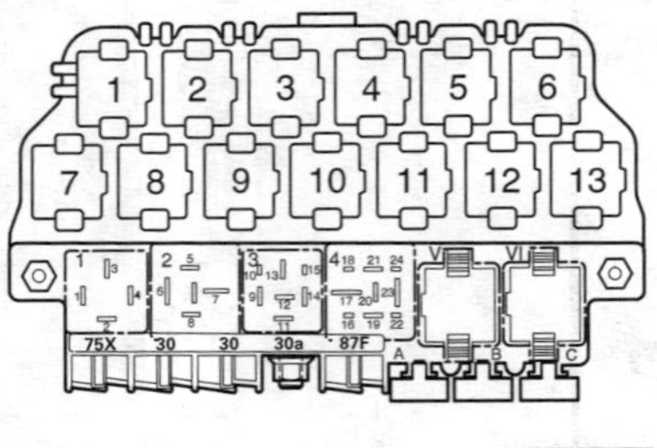

97-20676

Relay location on the thirteenfold auxiliary relay panel, above relay panel:

2 Relay for motor remote unlock rear lid (79)

Relay panel:

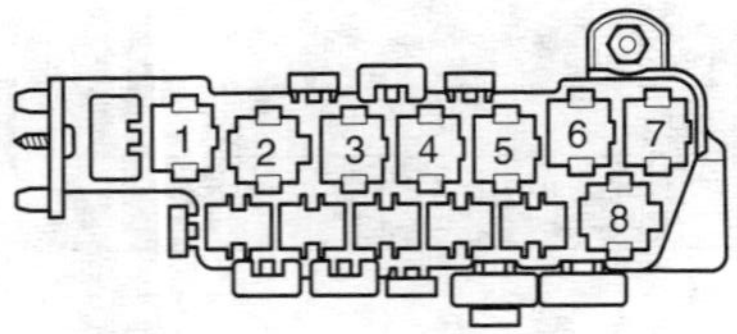

97-20515

Relay location on the eightfold auxiliary relay panel, behind relay panel:

Note: Number in parentheses indicates production control number stamped on relay housing.

Fuse colors

30 A - green
25 A - white
20 A - yellow
15 A - blue
10 A - red
7,5 A - brown
5 A - beige

Edition 06/98
USA.5062.05.21

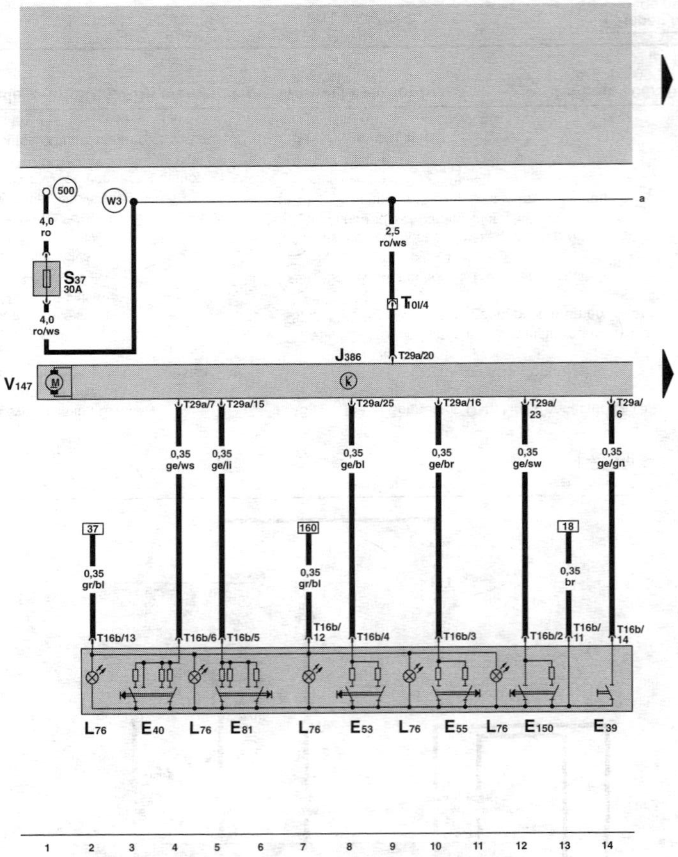

97-22851

ws = white
sw = black
ro = red
br = brown
gn = green
bl = blue
gr = grey
li = violet
ge = yellow

E39 - Safety switch for window regulator rear (00916)
E40 - Switch for window regulator front left (00912)
E53 - Switch for window regulator rear left, driver (00914)
E55 - Switch for window regulator rear right, driver (00915)
E81 - Switch for window regulator front right, driver (00913)
E150 - Switch for interior lock, driver side (01358)
J386 - Door control module, driver side (01331)
L76 - Push button light
S37 - Power window fuse
T10l - 10-pin connector, black, connector station A-pillar left
T16b - 16-pin connector
T29a - 29-pin connector
V147 - Motor for window regulator, driver side

(500) - Screw connection -1- (30), on relay panel
(W3) - wire connection, in rear wiring harness

Door control module, driver side, switch for window regulator in driver's door, switch for interior lock

Edition 06/98
USA.5062.05.21

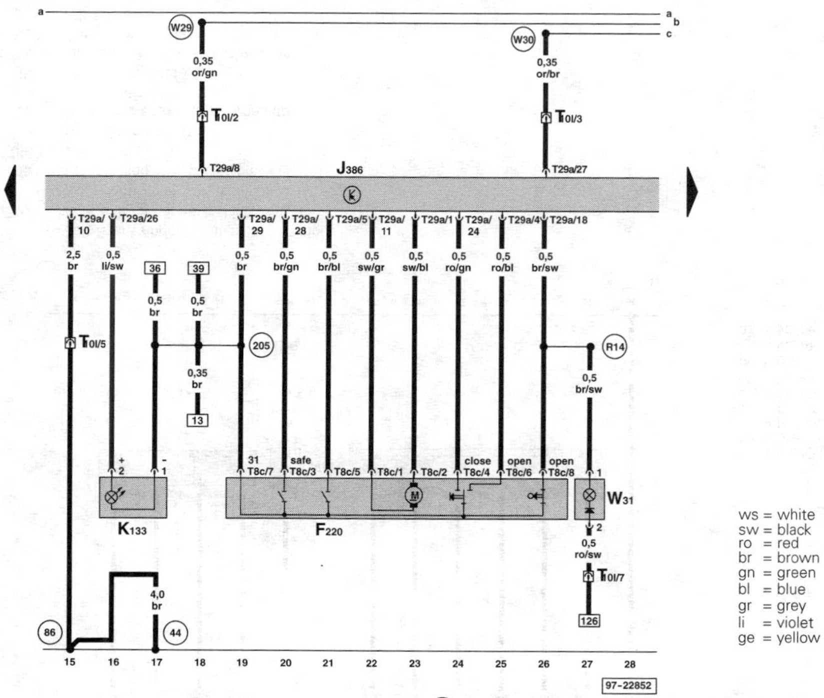

ws = white
sw = black
ro = red
br = brown
gn = green
bl = blue
gr = grey
li = violet
ge = yellow

F220 - Lock unit for central locking, driver side
J386 - Door control module, driver side
K133 - Warning light for central locking -SAFE-
T8c - 8-Pin Connector, black
T10l - 10-Pin Connector, black, connector station A-pillar, left
T29a - 29-Pin Connector
W31 - Entry light front, left

(44) - Ground connection, on left A-pillar, lower part

(86) - Ground connection -1-, in rear wiring harness

(205) - Ground connection, in driver's door wiring harness

(R14) - wire connection -1- (open), in driver's door wiring harness

(W29) - Plus connector -2- (A-Bus), in wiring harness, floor

(W30) - Plus connector -2- (B-Bus), in wiring harness, floor

Edition 06/98
USA.5062.05.21

Door control module, driver side, Central locking, driver's door, entry light front, left, Warning light for central locking -SAFE-

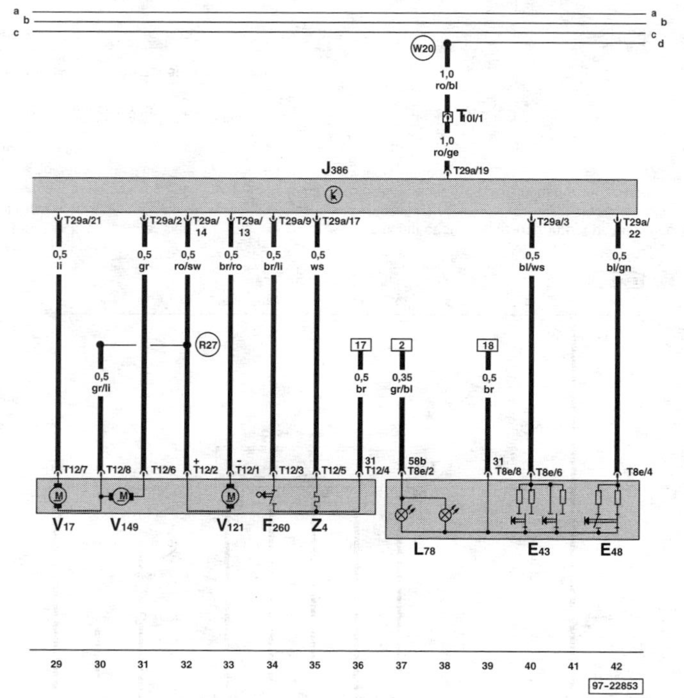

ws = white
sw = black
ro = red
br = brown
gn = green
bl = blue
gr = grey
li = violet
ge = yellow

E43 - Mirror Adjustment Switch
E48 - Mirror Selector Switch
F260 - Driver's Mirror Adjustment Contact Switch
J386 - Door control module, driver side
L78 - Mirror Adjusting Switch Light
T8e - 8-Pin Connector
T10l - 10-Pin Connector, black, connector station A-pillar, left
T12 - 12-Pin Connector
T29a - 29-Pin Connector
V17 - Driver's Side Mirror Adjustment Motor
V121 - Motor for fold-in mirror, driver side
V149 - Motor for mirror adjustment, driver side
Z4 - Heated outside mirror, driver side

(R27) - wire connection (power mirror), in driver's door wiring harness

(W20) - Plus connector (30a), in wiring harness, rear

Door control module, driver side, Mirror adjustment and fold-in mirror, driver side , heated outside mirror, driver side

Edition 06/98
USA.5062.05.21

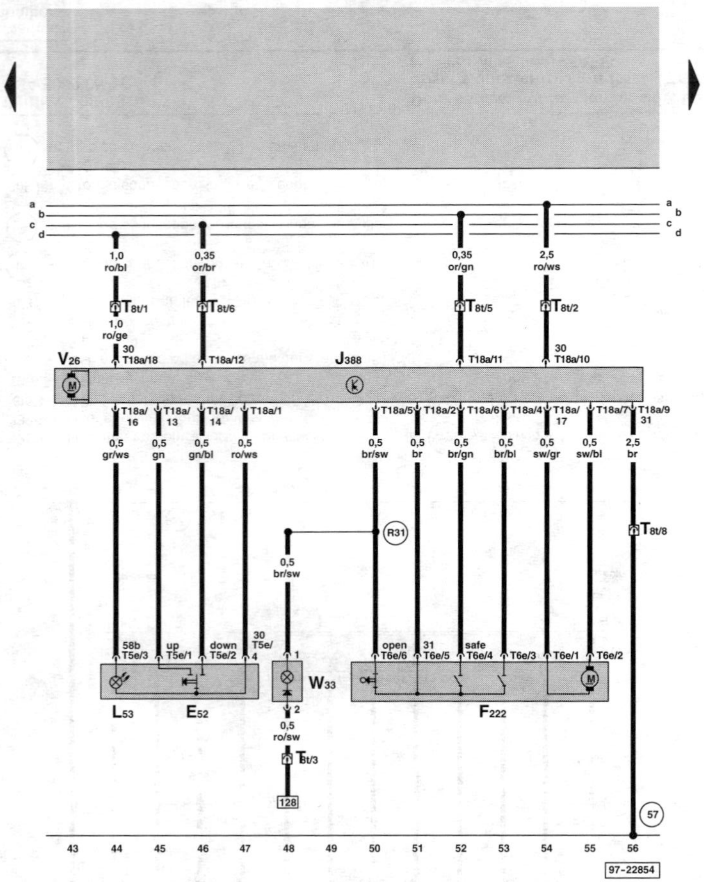

ws = white
sw = black
ro = red
br = brown
gn = green
bl = blue
gr = grey
li = violet
ge = yellow

- E52 - Left Rear Window Switch, (In LR Door)
- F222 - Lock unit for central locking, rear, left
- J388 - Door control module, rear, left
- L53 - Power Window Switch Light
- T5e - 5-Pin Connector
- T6e - 6-Pin Connector, brown
- T8t - 8-Pin Connector, black, connector station B-pillar, left
- T18a - 18-Pin Connector
- V26 - Motor for window regulator, rear, left
- W33 - Entry light rear, left

- (57) - Ground connection, on left rear pillar
- (R31) - Wire connection -1-, in left rear door wiring harness

Central locking rear left, power window rear left, entry light rear left

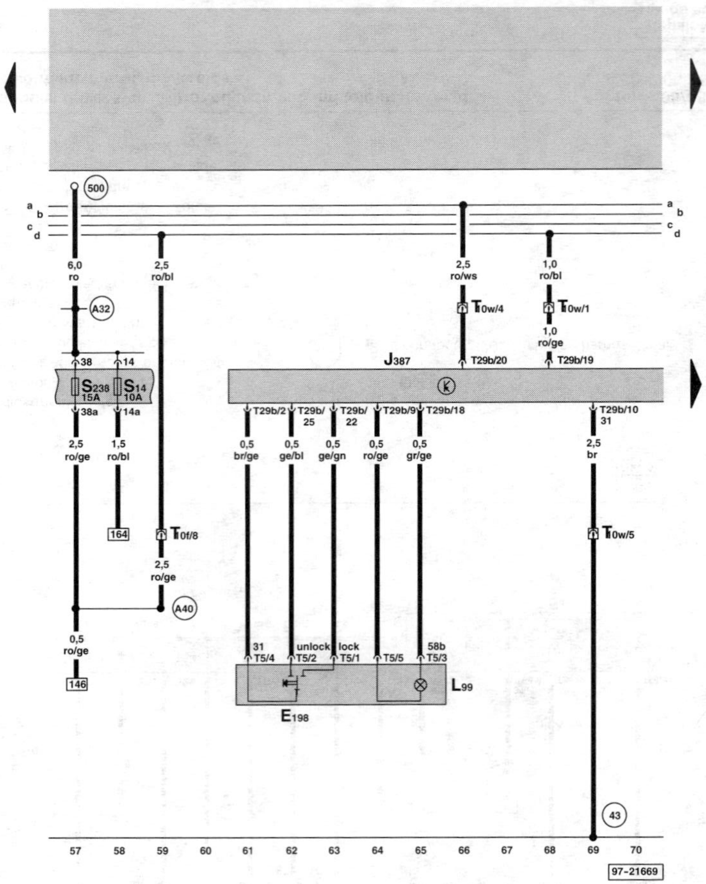

ws = white
sw = black
ro = red
br = brown
gn = green
bl = blue
gr = grey
li = violet
ge = yellow

- E198 - Switch for interior lock, passenger side
- J387 - Door control module, passenger side
- L99 - Lighting for switch interior lock
- S14 - Fuse 14 in fuse holder
- S238 - Fuse 38 in fuse holder
- T5 - 5-Pin Connector
- T10f - 10-Pin Connector, blue, connector station A-pillar, left
- T10w - 10-Pin Connector, black, connector station A-pillar, right
- T29b - 29-Pin Connector

- (43) - Ground connection, on right A-pillar, lower part
- (500) - Screw connection -1- (30), on relay panel
- (A32) - plus connection (30), in instrument panel wiring harness
- (A40) - plus connection -1- (30), in instrument panel wiring harness

Door control module, passenger side, switch for interior lock, passenger side

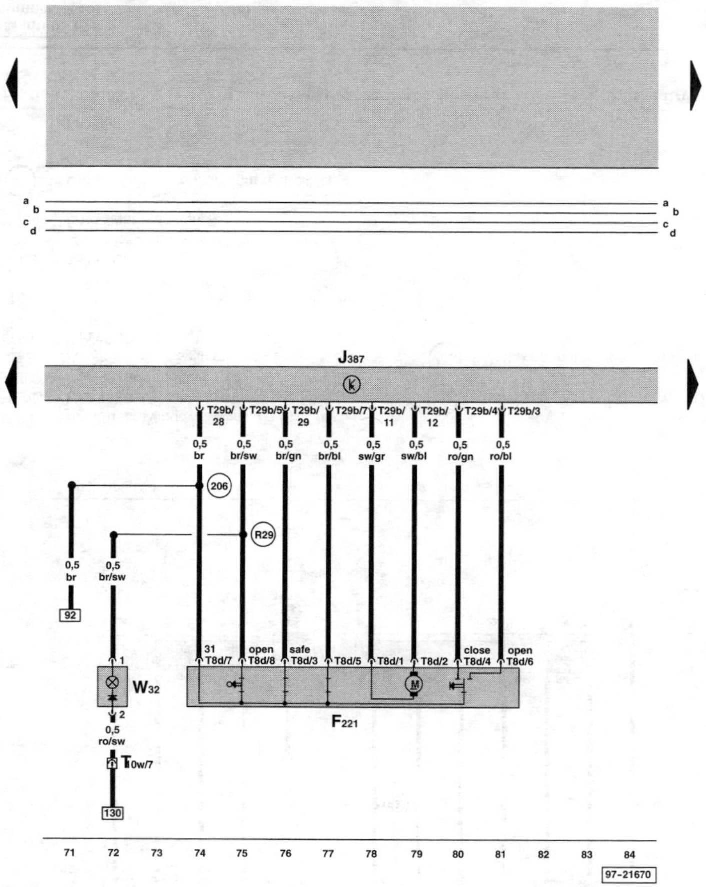

ws = white
sw = black
ro = red
br = brown
gn = green
bl = blue
gr = grey
li = violet
ge = yellow

F221 - Lock unit for central locking, passenger side
J387 - Door control module, passenger side
T8d - 8-Pin Connector, black
T10w - 10-Pin Connector, black, connector station A-pillar, right
T29b - 29-Pin Connector
W32 - Entry light front, right

(206) - Ground connection, in passenger's door wiring harness
(R29) - Wire connection -1- (open), in passenger's door wiring harness

Central locking passenger's door, entry light front, right

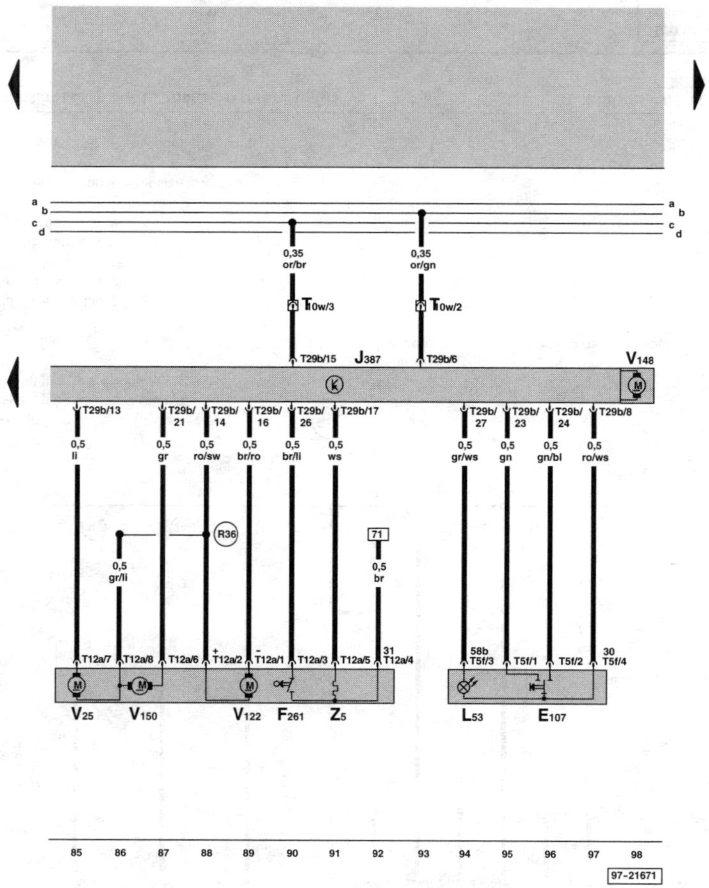

ws = white
sw = black
ro = red
br = brown
gn = green
bl = blue
gr = grey
li = violet
ge = yellow

E107 - Switch for window regulator, in passenger door (0936)
F222 - Switch for window regulator, in passenger door (0936)
F262 - Passenger's Mirror Tilt Contact Switch
J387 - Door control module, passenger side
L53 - Power Window Switch Light
T5f - 5-Pin Connector
T10w - 10-Pin Connector, black, connector station A-pillar, right
T12a - 12-Pin Connector, black
T29b - 29-Pin Connector
V25 - Passenger's Side Mirror Adjustment Motor
V122 - Motor for fold-in mirror, passenger side
V148 - Motor for window regulator, passenger side
V150 - Motor for mirror adjustment, passenger side
Z5 - Heated outside mirror, passenger side
(R36) - wire connection (power mirror), in passenger's door wiring harness

Mirror adjustment and fold-in mirror, passenger side , heated outside mirror, passenger side

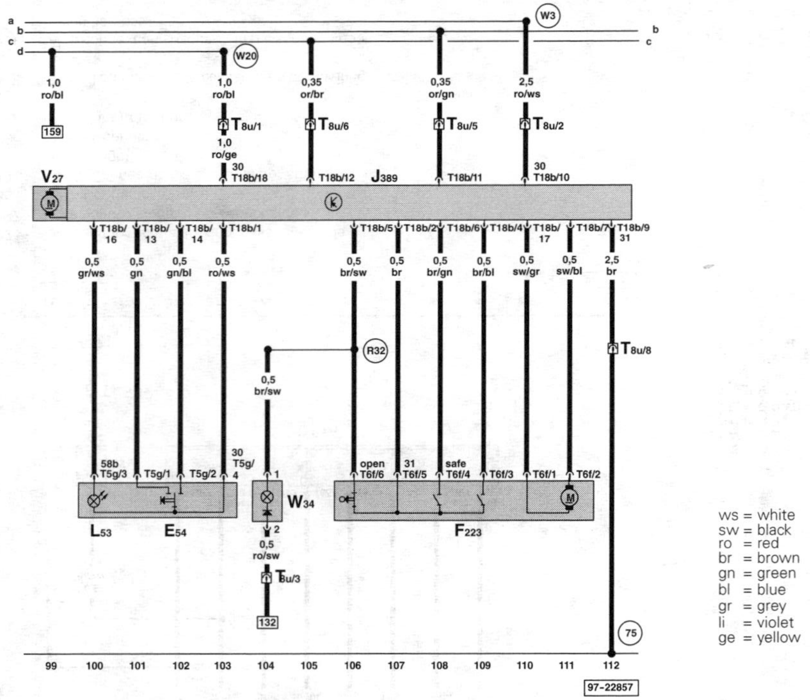

ws = white
sw = black
ro = red
br = brown
gn = green
bl = blue
gr = grey
li = violet
ge = yellow

E54 - Right Rear Window Switch, (In RR Door)
F223 - Lock unit for central locking, rear, right
J389 - Door control module, rear, right
L53 - Power Window Switch Light
T5g - 5-Pin Connector
T6f - 6-Pin Connector
T8u - 8-Pin Connector, brown, connector station B-pillar, right
T18b - 18-Pin Connector
V27 - Motor for window regulator, rear, right
W34 - Entry light rear, right

(75) - Ground connection, on right rear pillar

(R32) - Wire connection -2-, in right rear door wiring harness

(W3) - wire connection, in rear wiring harness

(W20) - Plus connector (30a), in wiring harness, rear

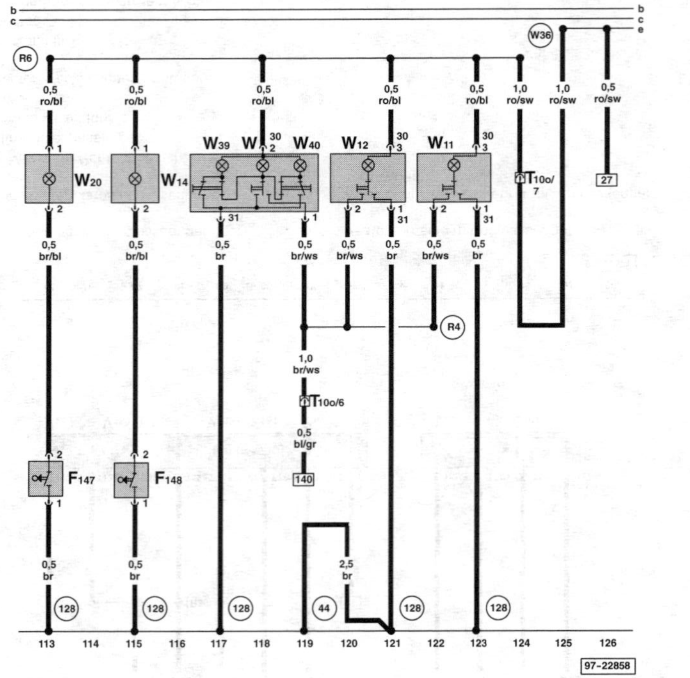

ws = white
sw = black
ro = red
br = brown
gn = green
bl = blue
gr = grey
li = violet
ge = yellow

F147 - Left Make-Up Mirror Light Switch
F148 - Right Make-Up Mirror Light Switch
T10o - 10-Pin Connector, red, connector station A-pillar, left
W - Front Interior Light
W11 - Left Rear Reading Light
W12 - Right Rear Reading Light
W14 - Right Make-up Mirror Light
W20 - Left Make-up Mirror Light
W39 - Reading light, center left
W40 - Reading light, center right

(44) - Ground connection, on left A-pillar, lower part

(128) - Ground connection -1-, in interrior light wiring harness

(R4) - wire connection -1-, in interior light/door contact switch wiring harness

(R6) - plus connection -1-, in interior light wiring harness

(W36) - Plus connector (30a), in wiring harness floor

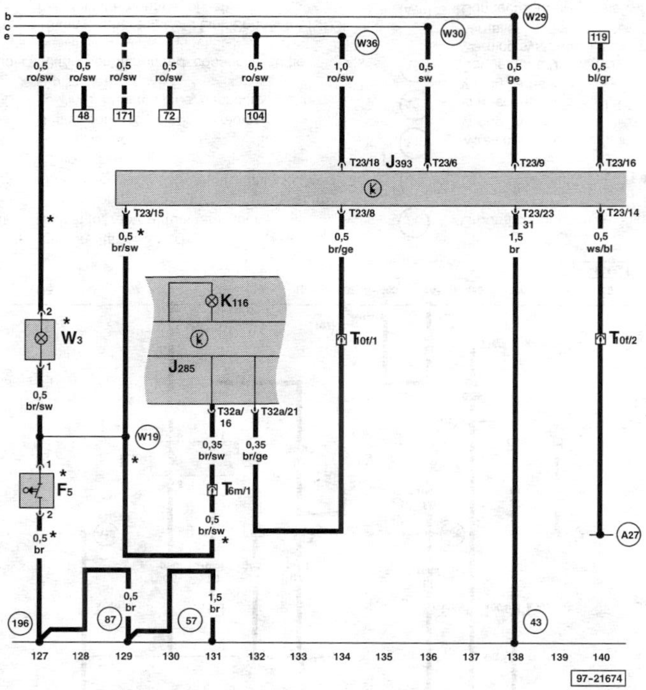

ws = white
sw = black
ro = red
br = brown
gn = green
bl = blue
gr = grey
li = violet
ge = yellow

F5 - Luggage Compartment Light Switch
J285 - Control module with indicator unit in instrument panel insert
J393 - Central control module for comfort system
K116 - Warning light for rear lid unlocked
T6m - 6-Pin Connector, black, connector station A-pillar, right
T10f - 10-Pin Connector, blue, connector station A-pillar, left
T23 - 23-Pin Connector
T32a - 32-Pin Connector, blue, on instrument cluster
W3 - Luggage compartment Light

(43) - Ground connection, on right A-pillar, lower part

(57) - Ground connection, on left rear pillar

(87) - Ground connection -2-, in rear wiring harness

(196) - Ground connection -3-, in wiring harness, rear

(A27) - Wire Connection (vehicle speed signal), in instrument panel wiring harness

(W19) - Connection -2- (luggage compartment light), in rear wiring harness

(W29) - Plus connector -2- (A-Bus), in wiring harness, floor

(W30) - Plus connector -2- (B-Bus), in wiring harness, floor

(W36) - Plus connector (30a), in wiring harness floor

* - sedan only

Central control module for comfort system, warning light for rear lid unlocked

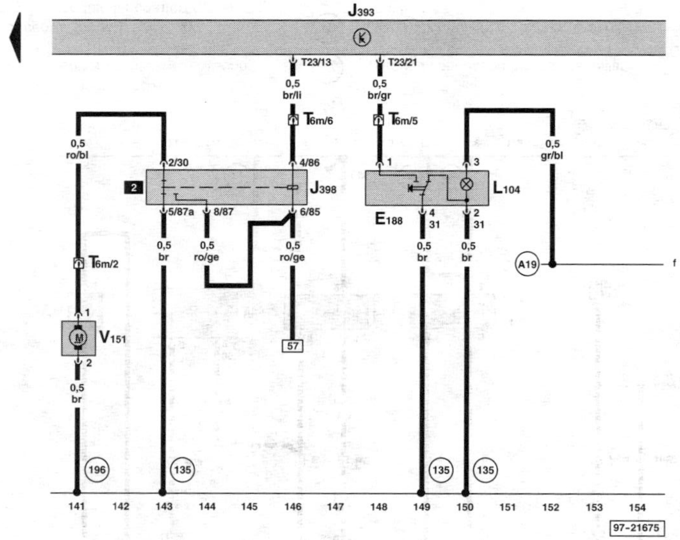

ws = white
sw = black
ro = red
br = brown
gn = green
bl = blue
gr = grey
li = violet
ge = yellow

E188 - Switch for remote unlock, rear lid
J393 - Central control module for comfort system
J398 - Relay for motor remote unlock rear lid, above relay panel, production control number (79)
L104 - Illumination for switch remote unlock
T6m - 6-Pin Connector, black, connector station A-pillar, right
T23 - 23-Pin Connector
V151 - Motor for remote unlock rear lid

(135) - Ground connection -2-, in instrument panel wiring harness

(196) - Ground connection -3-, in wiring harness, rear

(A19) - wire connection (58d), in instrument panel wiring harness

Central control module for comfort system, Relay for motor remote unlock rear lid, motor for remote unlock rear lid

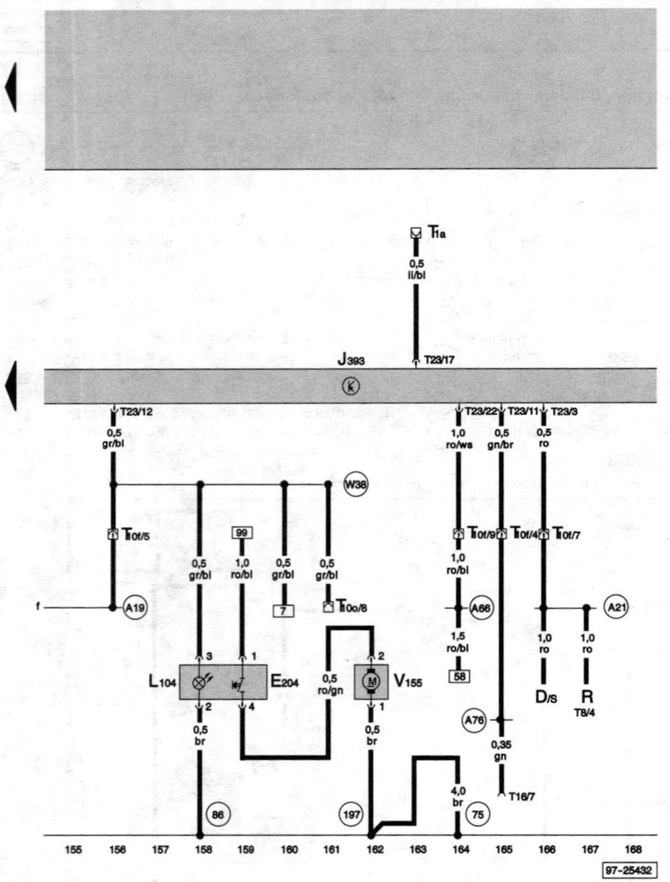

ws = white
sw = black
ro = red
br = brown
gn = green
bl = blue
gr = grey
li = violet
ge = yellow

D - Ignition/Starter Switch
E204 - Switch for remote/fuel tank door
J393 - Central control module for comfort system
L104 - Illumination for switch remote unlock
R - Radio
T1a - Single Connector
T8 - 8-Pin Connector, black, on radio
T10f - 10-Pin Connector, blue, connector station A-pillar, left
T10o - 10-Pin Connector, red, connector station A-pillar, left
T16 - 16-Pin Connector, Data Link Connector (DLC), below instrument panel, left
T23 - 23-Pin Connector
V155 - Motor for fuel tank lid unlock

(75) - Ground connection, on right rear pillar

(86) - Ground connection -1-, in rear wiring harness

(197) - Ground connection -4-, in wiring harness rear

(A19) - wire connection (58d), in instrument panel wiring harness

(A21) - wire connection (86s), in instrument panel wiring harness

(A66) - Connector (30a, central locking/anti-theft warning system/IR), in instrument panel wiring harness

(A76) - Connector (K-diagnosis wire), in instrument panel wiring harness

(W38) - Connector (58d), in wiring harness floor

Edition 06/99 **Central control module for comfort system, Switch for remote/fuel tank door**
USA.5062.09.21

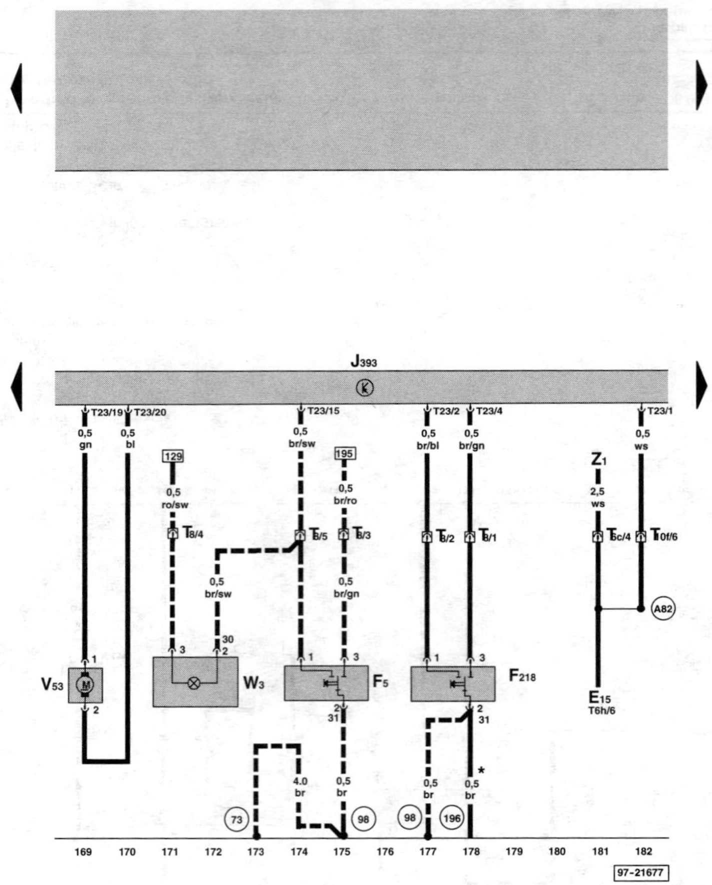

ws = white
sw = black
ro = red
br = brown
gn = green
bl = blue
gr = grey
li = violet
ge = yellow

E15 - Rear window defogger switch
F5 - Luggage Compartment light switch (wagon only)
F218 - Switch for central locking, trunk lid
J393 - Central control module for comfort system
T6c - 6-Pin Connector, blue, connector station A-pillar, left
T6h - 6-Pin Connector, black, am Schalter für beheizbare Heckscheibe
T8 - 8-Pin Connector, black, connector station D-pillar, right (wagon only)
T10f - 10-Pin Connector, blue, connector station A-pillar, left
T23 - 23-Pin Connector
V53 - Decklid Central Locking System Motor
W3 - Luggage compartment light (wagon only)
Z1 - Heated rear window

(57) - Ground connection, on left rear pillar

(73) - Ground connection, on roof bow, rear (wagon only)

(87) - Ground connection -2-, in rear wiring harness

(98) - Ground connection, in rear lid wiring harness

(196) - Ground connection -3-, in wiring harness, rear

(A82) - Connector (heated glass), in instrument panel wiring harness

* - sedan only

- - - wagon only

Central control module for comfort system, central locking for trunk lid **Edition 06/99**
USA.5062.09.21

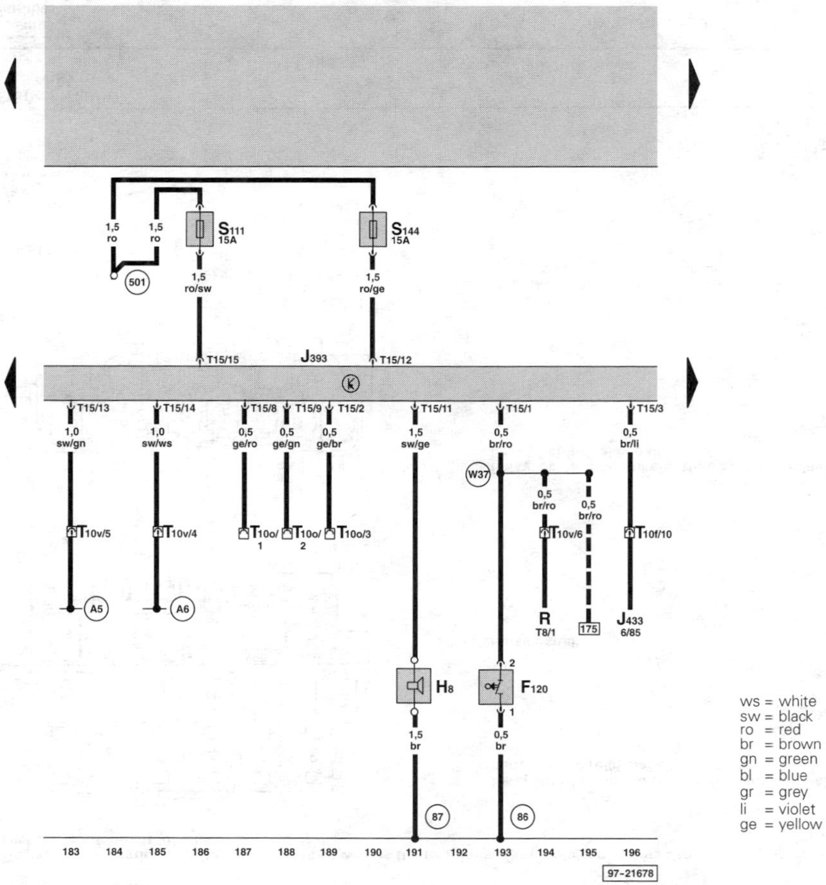

ws = white
sw = black
ro = red
br = brown
gn = green
bl = blue
gr = grey
li = violet
ge = yellow

- F120 - Hood Alarm Switch
- H8 - Alarm Horn
- J393 - Central control module for comfort system
- J433 - Starting Interlock Relay (alarm system), above relay panel
- R - Radio
- S111 - Fuse -1- for Alarm System, on the eightfold auxiliary relay panel, behind relay panel
- S144 - Fuse -2- for Alarm System, on the eightfold auxiliary relay panel, behind relay panel
- T8 - 8-Pin Connector, on radio
- T10f - 10-Pin Connector, blue, connector station A-pillar, left
- T10o - 10-Pin Connector, red, connector station A-pillar, left
- T10v - 10-Pin Connector, violet, connector station A-pillar, right
- T15 - 15-Pin Connector, on Central control module for comfort system
- (86) - Ground connection -1-, in rear wiring harness
- (87) - Ground connection -2-, in rear wiring harness
- (501) - Screw connection -2- (30), on relay panel
- (A5) - plus connection (right turn signal), in instrument panel wiring harness
- (A6) - plus connection (left turn signal), in instrument panel wiring harness
- (W37) - Connector (alarm system), in wiring harness floor
- - - - wagon only

Central control module for comfort system, alarm system

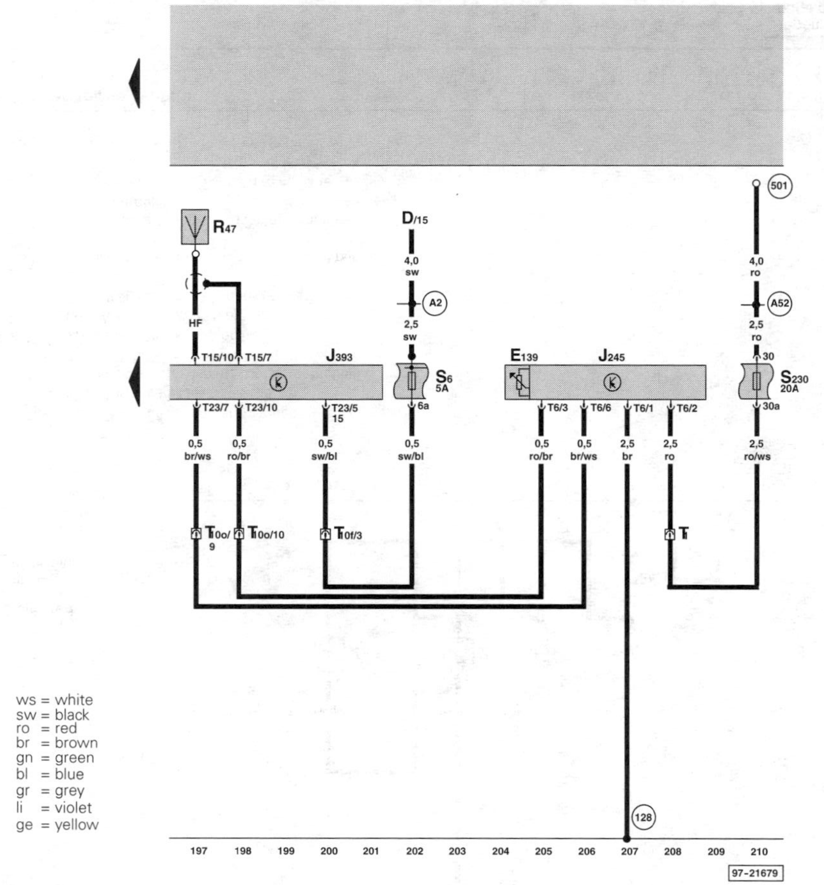

ws = white
sw = black
ro = red
br = brown
gn = green
bl = blue
gr = grey
li = violet
ge = yellow

- D - Ignition/Starter Switch
- E139 - Sunroof Regulator
- J245 - Power Sunroof Control Module
- J393 - Central control module for comfort system
- R47 - Antenna wire for central locking and alarm system, on B-pillar right
- S6 - Fuse 6 in fuse holder
- S230 - Fuse 30 in fuse holder
- T1 - Single Connector, black
- T6 - 6-Pin Connector
- T10f - 10-Pin Connector, blue, connector station A-pillar, left
- T10o - 10-Pin Connector, red, connector station A-pillar, left
- T15 - 15-Pin Connector, on Central control module for comfort system
- T23 - 23-Pin Connector
- (128) - Ground connection -1-, in interrior light wiring harness
- (501) - Screw connection -2- (30), on relay panel
- (A2) - plus connection (15), in instrument panel wiring harness
- (A52) - plus connection (30), in instrument panel wiring harness

Central control module for comfort system, power sunroof

Washer Nozzle Heaters,

from September 1997

Deviate relay location and fuseplacements as well as the locations of multiple connectors see section "component locations".

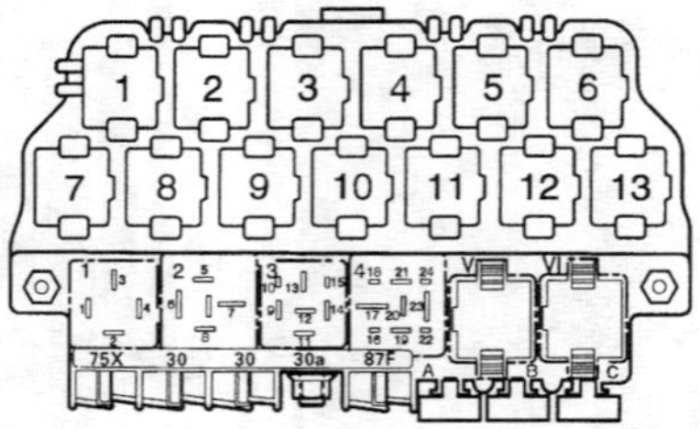

97–20676

Relay location on the thirteenfold auxiliary relay panel, above relay panel:

Relay panel:

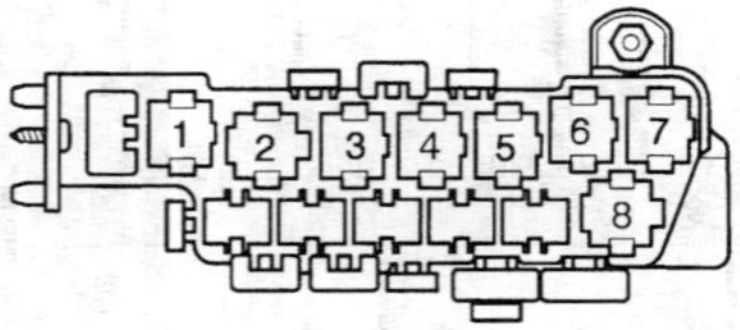

97–20515

Relay location on the eightfold auxiliary relay panel, behind relay panel:

Note: Number in parentheses indicates production control number stamped on relay housing.

Fuse colors
30 A – green
25 A – white
20 A – yellow
15 A – blue
10 A – red
7,5 A – brown
5 A – beige

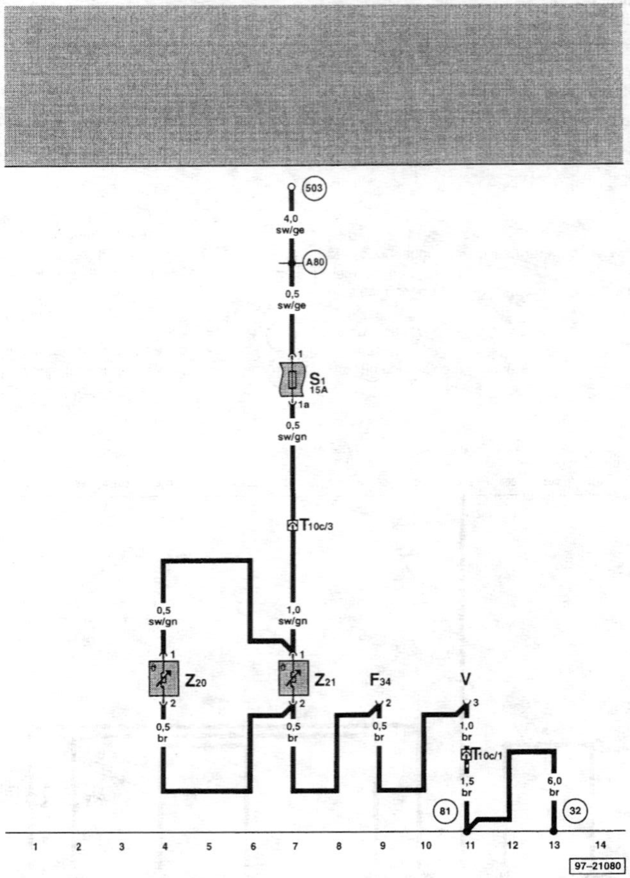

ws = white
sw = black
ro = red
br = brown
gn = green
bl = blue
gr = grey
li = violet
ge = yellow

F34 – Brake Fluid Level Warning Switch
S1 – Fuse 1 in fuse holder
T10c – 10-Pin Connector, violet, connector station A-pillar, left
V – Windshield Wiper Motor
Z20 – Left Washer Nozzle Heater
Z21 – Right Washer Nozzle Heater

(32) – Ground connection, behind instrument panel, left
(81) – Ground connection -1-, in instrument panel wiring harness
(503) – Screw Connection (75x), on relay panel
(A80) – Connector -1- (X), in instrument panel wiring harness

Air-Conditioning,

from September 1997

- **1,8L- Engine - Motronic Multiport Fuel Injection (MFI)/110 kW, code AEB,**
- **2,8l - Motronic Multiport Fuel Injection (MFI) / 142kW, code AHA,**

Deviate relay location and fuseplacements as well as the locations of multiple connectors see section "component locations".

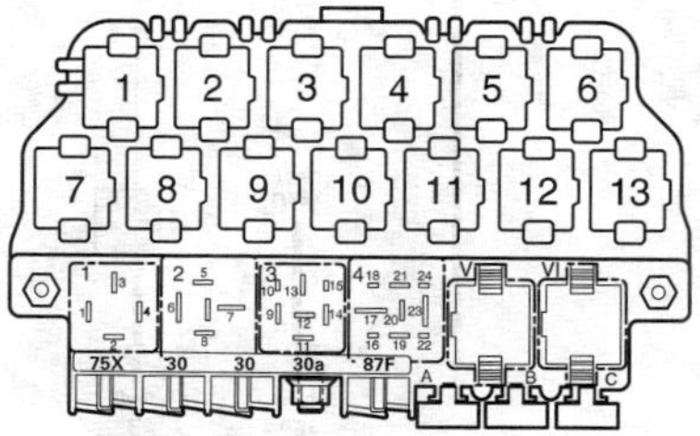

97-20676

Relay location on the thirteenfold auxiliary relay panel, above relay panel:

4 A/C Cut-Off Control Module (384)

Relay panel:

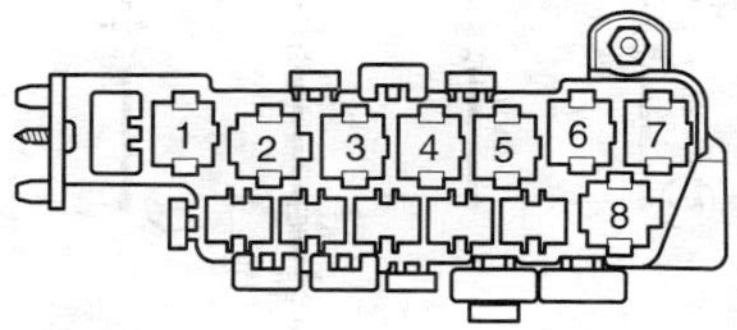

97-20515

Relay location on the eightfold auxiliary relay panel, behind relay panel:

5 First Speed Coolant Fan Control (FC) Relay (214)

8 Second Speed Coolant Fan Control (FC) Relay (213)

Note: Number in parentheses indicates production control number stamped on relay housing.

Fuse colors
30 A - green
25 A - white
20 A - yellow
15 A - blue
10 A - red
7,5 A - brown
5 A - beige

Edition 06/98
USA.5062.05.21

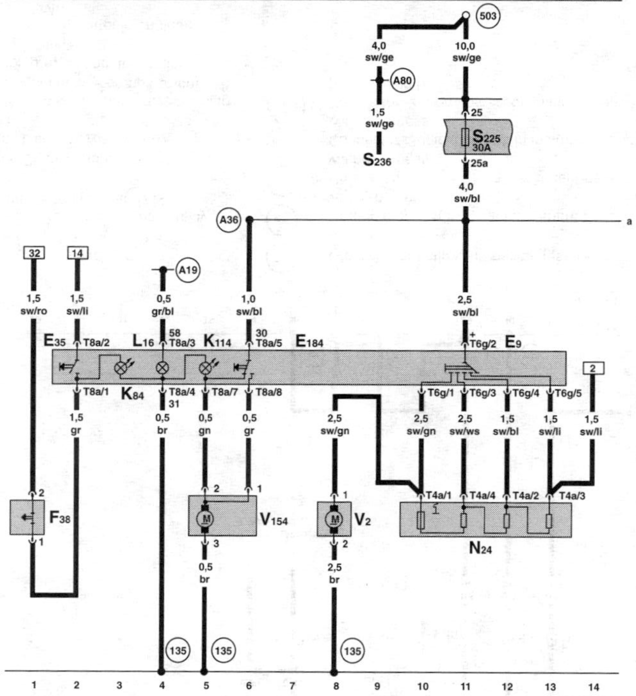

ws = white
sw = black
ro = red
br = brown
gn = green
bl = blue
gr = grey
li = violet
ge = yellow

97-21081

E9 - Fresh Air Blower Switch
E35 - A/C Switch
E184 - Switch for fresh air blower and recirculated air
F38 - Ambient Temperature Switch
K84 - A/C Indicator Light
K114 - Fresh Air and Recirculating Air Mode Indicator Light
L16 - Fresh Air Control Lever Light
N24 - Fresh Air Blower Series Resistance With Fuse
S225 - Fuse 25 in fuse holder
S236 - Fuse 36 in fuse holder
T4a - 4-Pin Connector
T6g - 6-Pin Connector
T8a - 8-Pin Connector
V2 - Fresh Air Blower
V154 - Servo motor for fresh-/recirculating air door

(135) - Ground connection -2-, in instrument panel wiring harness
(503) - Screw Connection (75x), on relay panel
(A19) - wire connection (58d), in instrument panel wiring harness
(A36) - wire connection (75a), in instrument panel wiring harness
(A80) - Connector -1- (X), in instrument panel wiring harness

Fresh air blower, A/C switch, ambient temperature switch, switch for fresh air recirculated air

Edition 06/98
USA.5062.05.21

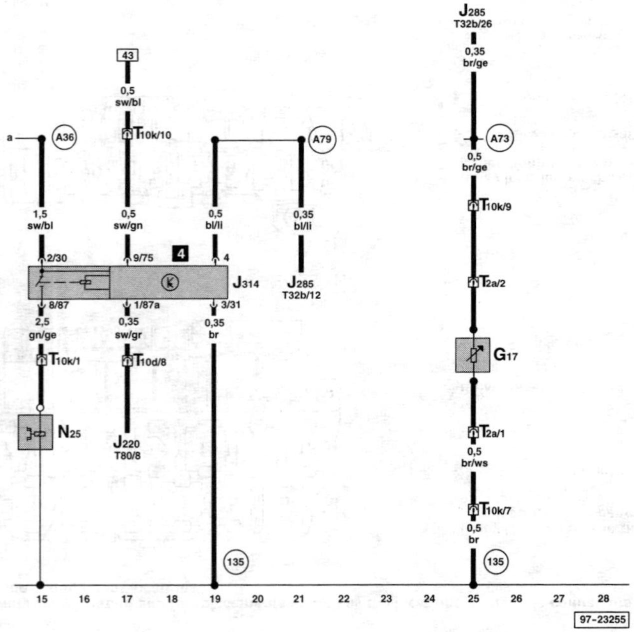

ws = white
sw = black
ro = red
br = brown
gn = green
bl = blue
gr = grey
li = violet
ge = yellow

G17 - Outside Air Temperature Sensor
J220 - Motronic Engine Control Module (ECM)
J285 - Control module with indicator unit in instrument panel insert
J314 - A/C Cut-Off Control Module, on thirteenfold auxiliary relay panel, above relay panel, production control number (384)
N25 - A/C Clutch
T2a - Double Connector
T10d - 10-Pin Connector, brown, on protective housing for control module, in engine compartment, left
T10k - 10-Pin Connector, grey, connector station A-pillar, left
T32b - 32-Pin Connector, green, on instrument cluster
T80 - Connector, 80 point, on Engine Control Module (ECM)

(135) - Ground connection -2-, in instrument panel wiring harness
(A36) - wire connection (75a), in instrument panel wiring harness
(A73) - Connector (outside temperature indicator), in instrument panel wiring harness
(A79) - Connector (engine overheat warning light), in instrument panel wiring harness

A/C cut-off control module, A/C clutch, outside air temperature sensor

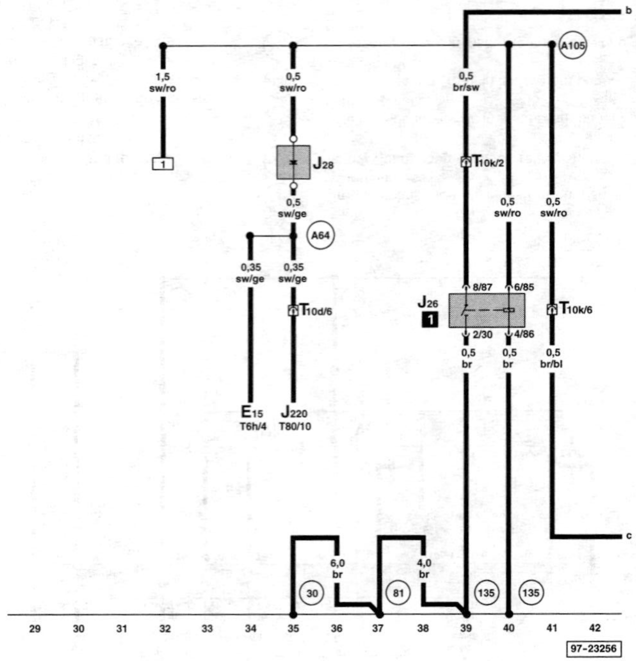

ws = white
sw = black
ro = red
br = brown
gn = green
bl = blue
gr = grey
li = violet
ge = yellow

E15 - Rear window defogger switch
J26 - Coolant Fan control (FC) Relay, on the thirteenfold auxiliary relay panel, above relay panel, production control number (214)
J28 - A/C Diode
J220 - Motronic Engine Control Module (ECM)
T6h - 6-Pin Connector, black, on rear window defogger switch
T10d - 10-Pin Connector, brown, on protective housing for control module, in engine compartment, left
T10k - 10-Pin Connector, grey, connector station A-pillar, left
T80 - Connector, 80 point, on Motronic Engine Control Module (ECM)

(30) - Ground connection, -1-, beside fuse/relay panel
(81) - Ground connection -1-, in instrument panel wiring harness
(135) - Ground connection -2-, in instrument panel wiring harness
(A64) - Connector (30, idle boost), in instrument panel wiring harness
(A105) - Connector -2- (rpm increase), in instrument panel wiring harness

Coolant fan control (FC) relay, A/C diode

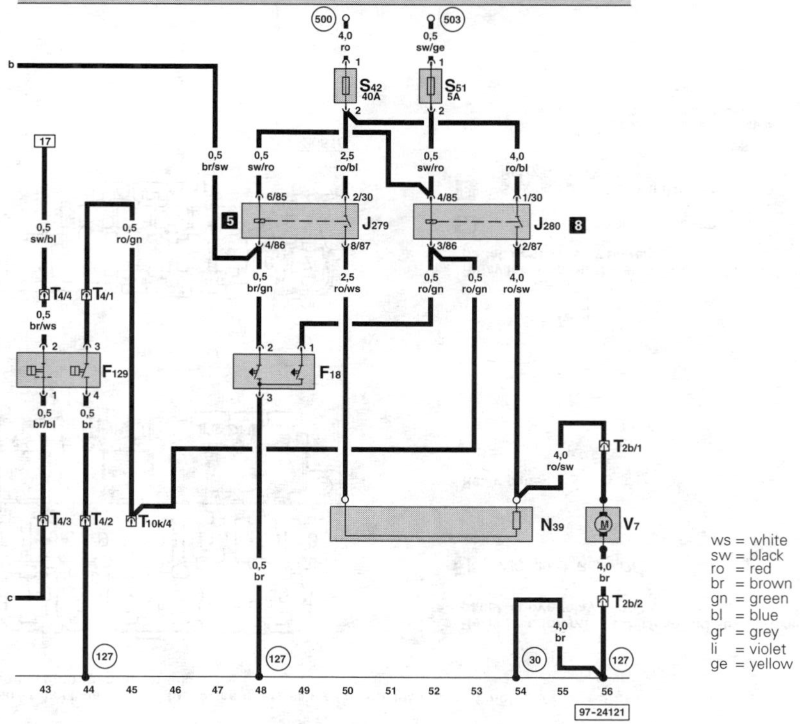

ws = white
sw = black
ro = red
br = brown
gn = green
bl = blue
gr = grey
li = violet
ge = yellow

F18 - Coolant Fan Control (FC) Thermal Switch
F129 - A/C Pressure Switch
J217 - Transmission Control Module (TCM)
J279 - First Speed Coolant Fan Control (FC) Relay, on eightfold auxiliary relay panel, production control number (214)
J280 - Second Speed Coolant Fan Control (FC) Relay, on eightfold auxiliary relay panel, production control number (213)
N39 - Coolant Fan Control (FC) Series Resistance, below side member, left
S42 - Coolant Fan Fuse, on eightfold auxiliary relay panel
S51 - Coolant Fan Fuse, on eightfold auxiliary relay panel
T2b - Double Connector, in engine compartment, left front
T4 - 4-Pin Connector, in engine compartment, left front
T10k - 10-Pin Connector, grey, connector station A-pillar, left
V7 - Coolant Fan
(30) - Ground connection, -1-, beside fuse/relay panel
(127) - Ground connection -1-, in A/C compressor wiring harness
(500) - Screw connection -1- (30), on relay panel
(503) - Screw Connection (75x), on relay panel

Edition 12/98
USA.5062.08.21

A/C pressure switch, coolant fan control (FC) thermal switch, first speed coolant fan control (FC) relay, second speed coolant fan control (FC) relay

THIS PAGE INTENTIONALLY LEFT BLANK

Climatronic,

from September 1997

- **2,8l - Motronic Multiport Fuel Injection (MFI) / 142kW, code AHA,**

Deviate relay location and fuseplacements as well as the locations of multiple connectors see section "component locations".

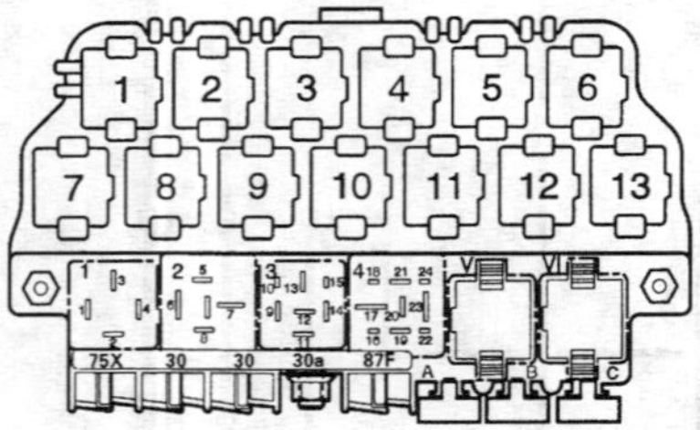

97–20676

Relay location on the thirteenfold auxiliary relay panel, above relay panel:

3 A/C Clutch Relay (267)

Relay panel:

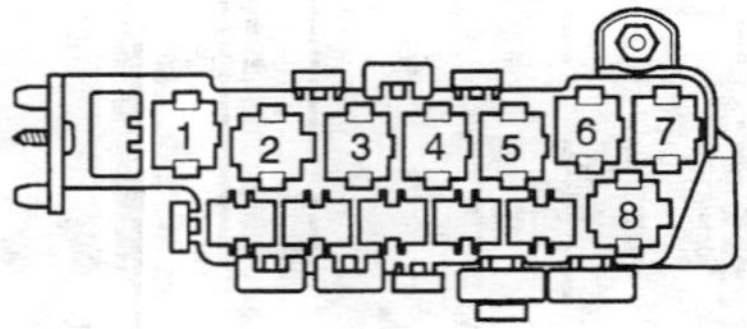

97–20515

Relay location on the eightfold auxiliary relay panel, behind relay panel:

5 First Speed Coolant Fan Control (FC) Relay (214)

8 Second Speed Coolant Fan Control (FC) Relay (213)

Note: Number in parentheses indicates production control number stamped on relay housing.

Fuse colors

30 A – green
25 A – white
20 A – yellow
15 A – blue
10 A – red
7,5 A – brown
5 A – beige

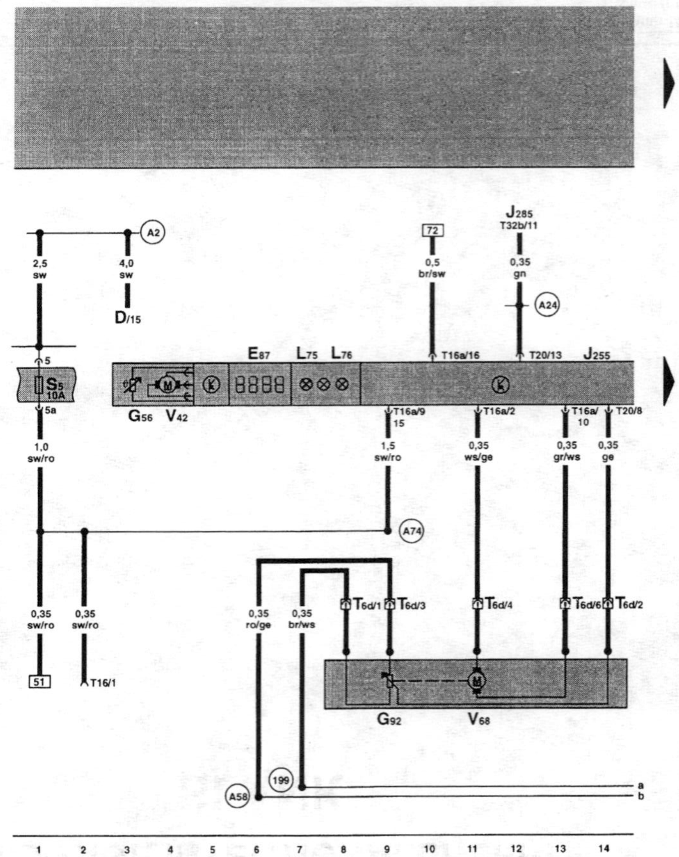

97–21085

ws = white
sw = black
ro = red
br = brown
gn = green
bl = blue
gr = grey
li = violet
ge = yellow

D – Ignition/Starter Switch
E87 – A/C Control Head
G56 – Instrument Panel Interior Temperature Sensor
G92 – Temperature Regulator Flap Motor Position Sensor
J255 – Climatronic Control Module
J285 – Control module with indicator unit in instrument panel insert
L75 – Digital Display Light
L76 – Push Button Light
S5 – Fuse 5 in fuse holder
T6d – 6-Pin Connector, red, behind instrument panel, center
T16 – 16-Pin Connector, On Board Diagnostic (OBD), below instrument panel, left
T16a – 16-Pin Connector
T20 – 20-Pin Connector
T32b – 32-Pin Connector, green, on instrument cluster
V42 – Interior Temperature Sensor Fan
V68 – Temperature Regulator Flap Motor

(199) – Ground connection -3-, in instrument panel wiring harness
(A2) – plus connection (15), in instrument panel wiring harness
(A24) – wire connection (brake system monitoring), in instrument panel wiring harness
(A58) – Wire connection (5Volt), in instrument cluster wiring harness
(A74) – Connector (15a - fuse 5), in instrument panel wiring harness

Climatronic control module, temperature regulator flap motor

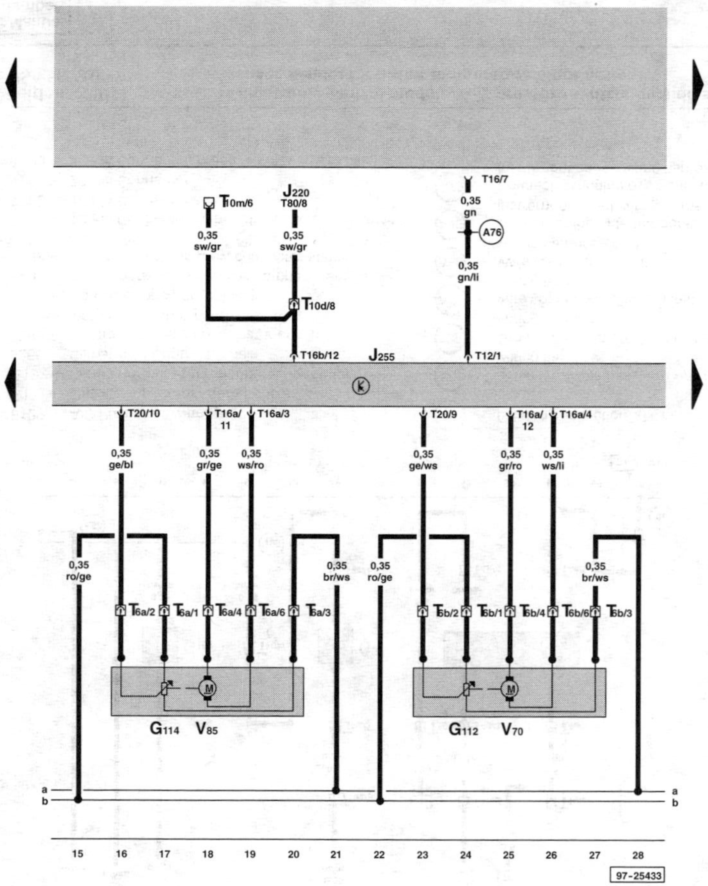

ws = white
sw = black
ro = red
br = brown
gn = green
bl = blue
gr = grey
li = violet
ge = yellow

G112 - Central Flap Motor Position Sensor
G114 - Footwell/Defrost Flap Motor Position Sensor
J220 - Motronic Engine Control Module (ECM)
J255 - Climatronic Control Module
T6a - 6-Pin Connector, behind instrument panel, center
T6b - 6-Pin Connector, behind instrument panel, center
T10d - 10-Pin Connector, brown, on protective housing for control module, in engine compartment, left
T10m - 10-Pin Connector, brown, connector station A-pillar, right
T12 - 12-Pin Connector
T16 - 16-Pin Connector, Data Link Connector (DLC), below instrument panel, left
T16a - 16-Pin Connector
T16b - 16-Pin Connector
T20 - 20-Pin Connector
T80 - Connector, 80 point, on Motronic Engine Control Module (ECM)
V70 - Central Air Flap Motor
V85 - Footwell/Defroster Flap Motor
(A76) - Connector (K-diagnosis wire), in instrument panel wiring harness

Climatronic control module, central air and footwell/defrost flap motors

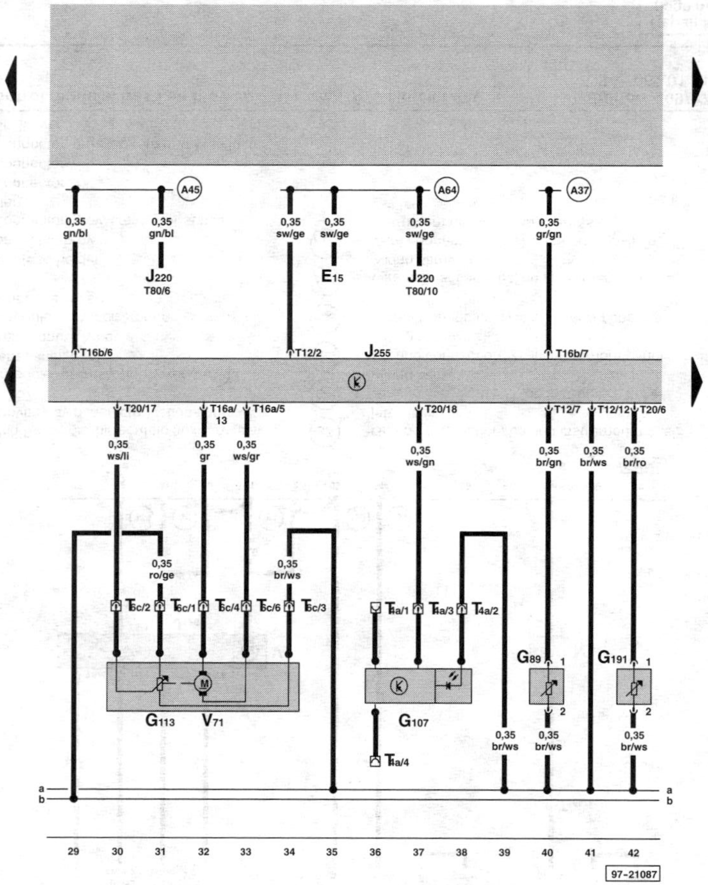

ws = white
sw = black
ro = red
br = brown
gn = green
bl = blue
gr = grey
li = violet
ge = yellow

E15 - Rear window defogger switch
G89 - Fresh Air Intake Duct Temperature Sensor
G107 - Sunlight Photo Sensor, on instrument panel, left
G113 - Back Pressure Flap Motor Position Sensor
G191 - Sender for outlet temperature, center
J220 - Motronic Engine Control Module (ECM)
J255 - Climatronic Control Module
T4a - 4-Pin Connector, black
T6c - 6-Pin Connector, black
T12 - 12-Pin Connector
T16a - 16-Pin Connector
T16b - 16-Pin Connector
T20 - 20-Pin Connector
T80 - Connector, 80 point, on Motronic Engine Control Module (ECM)
V71 - Air Flow Flap Motor
(A37) - wire connection (58a), in instrument panel wiring harness
(A45) - wire connection (RPM-signal), in instrument panel wiring harness
(A64) - Connector (30, idle boost), in instrument panel wiring harness

Climatronic control module, air flow flap motor, fresh air intake duct temperature sensor, sunlight photo sensor, sender for outlet temperature sensor, center

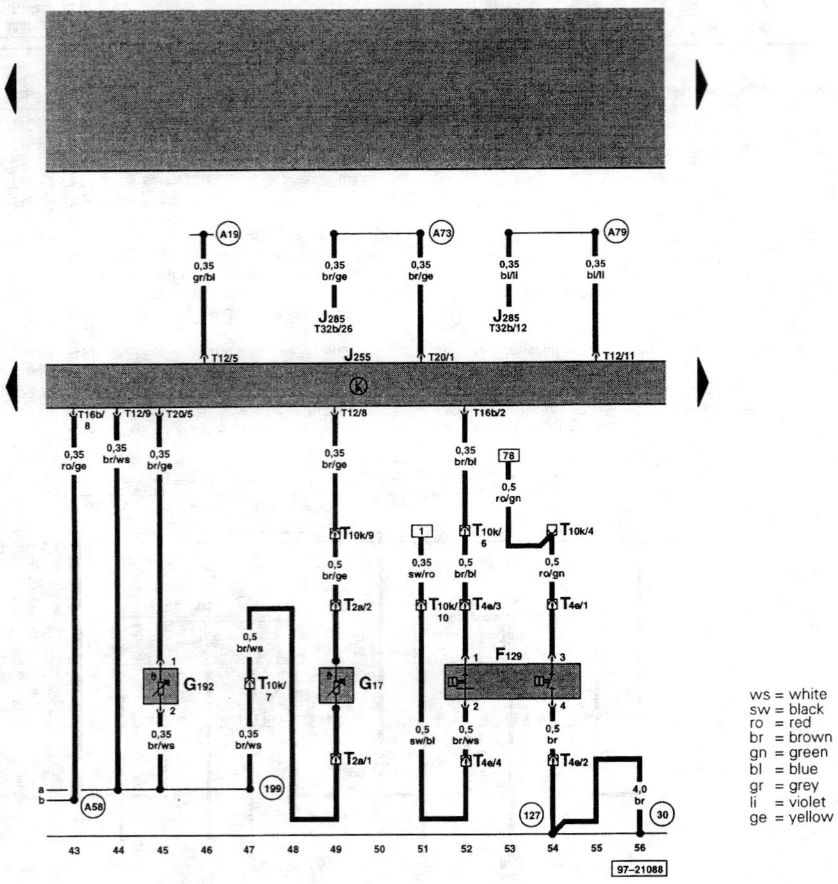

ws = white
sw = black
ro = red
br = brown
gn = green
bl = blue
gr = grey
li = violet
ge = yellow

F129 – A/C Pressure Switch
G17 – Outside Air Temperature Sensor
G192 – Sender for outlet temperature, floor outlet
J255 – Climatronic Control Module
J285 – Control module with indicator unit in instrument panel insert
T2a – Double Connector
T4e – 4-Pin Connector, in engine compartment, left
T10k – 10-Pin Connector, grey, connector station A-pillar, left
T12 – 12-Pin Connector, neben der Lenksäule rechts
T16b – 16-Pin Connector
T20 – 20-Pin Connector
T32b – 32-Pin Connector, green, on instrument cluster

(30) – Ground connection, -1-, beside fuse/relay panel
(127) – Ground connection -1-, in A/C compressor wiring harness
(199) – Ground connection -3-, in instrument panel wiring harness
(A19) – wire connection (58d), in instrument panel wiring harness
(A58) – Wire connection (5Volt), in instrument cluster wiring harness
(A73) – Connector (outside temperature indicator), in instrument panel wiring harness
(A79) – Connector (engine overheat warning light), in instrument panel wiring harness

Edition 09/97
USA.5062.01.21

Climatronic control module, A/C pressure switch, outside air temperature sensor, sender for outlet temperature, floor outlet

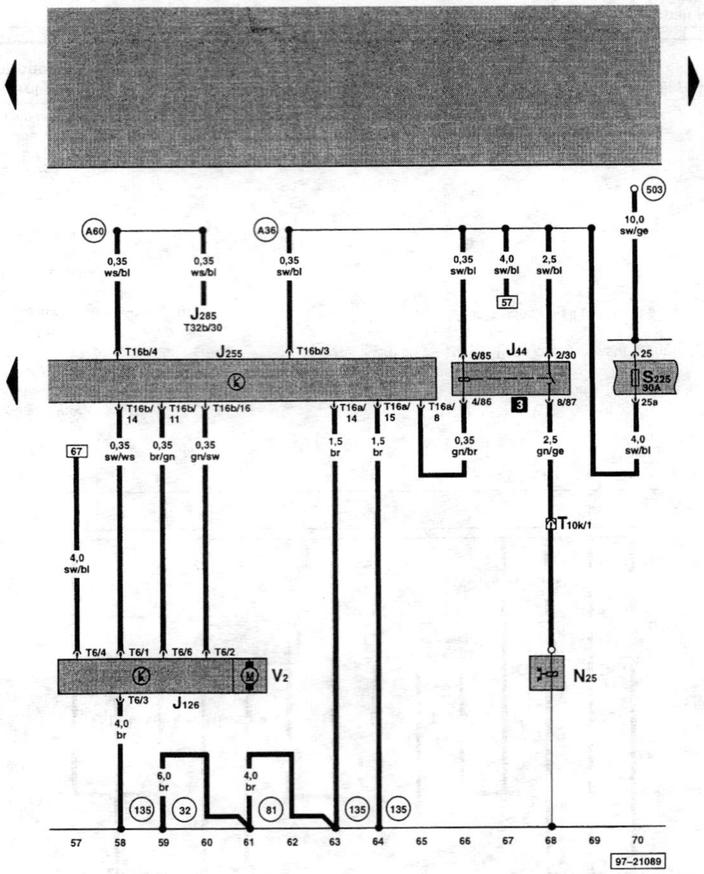

ws = white
sw = black
ro = red
br = brown
gn = green
bl = blue
gr = grey
li = violet
ge = yellow

J44 – A/C Clutch Relay, on thirteenfold auxiliary relay panel, above relay panel, production control number (267)
J126 – Control module for fresh air blower, behind instrument panel, right
J255 – Climatronic Control Module
J285 – Control module with indicator unit in instrument panel insert
N25 – A/C Clutch
S225 – Fuse 25 in fuse holder
T6 – 6-Pin Connector, brown
T10k – 10-Pin Connector, grey, connector station A-pillar, left
T16a – 16-Pin Connector
T16b – 16-Pin Connector
T32b – 32-Pin Connector, green, on instrument cluster
V2 – Fresh Air Blower

(32) – Ground connection, behind instrument panel, left
(81) – Ground connection -1-, in instrument panel wiring harness
(135) – Ground connection -2-, in instrument panel wiring harness
(503) – Screw Connection (75x), on relay panel
(A36) – wire connection (75a), in instrument panel wiring harness
(A60) – Wire connection (Vehicle speed signal), in instrument cluster wiring harnes

Climatronic control module, fresh air blower, A/C clutch, A/C clutch relay

Edition 09/97
USA.5062.01.21

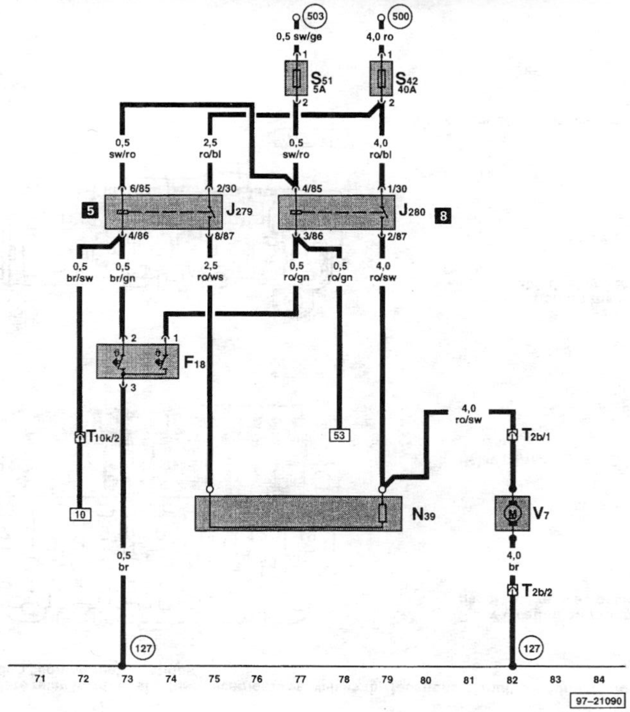

ws = white
sw = black
ro = red
br = brown
gn = green
bl = blue
gr = grey
li = violet
ge = yellow

F18 – Coolant Fan Control (FC) Thermal Switch
J279 – First Speed Coolant Fan Control (FC) Relay, on eightfold auxiliary relay panel, production control number (214)
J280 – Second Speed Coolant Fan Control (FC) Relay, on eightfold auxiliary relay panel, production control number (213)
N39 – Coolant Fan Control (FC) Series Resistance
S42 – Coolant Fan Fuse, on eightfold auxiliary relay panel
S51 – Coolant Fan Fuse, on eightfold auxiliary relay panel
T2b – Double Connector, in engine compartment, left
T10k – 10-Pin Connector, grey, connector station A-pillar, left
V7 – Coolant Fan

(127) – Ground connection -1-, in A/C compressor wiring harness
(500) – Screw Connection -1- (30), on relay panel
(503) – Screw Connection (75x), on relay panel

Edition 09/97
USA.5062.01.21

Coolant fan control (FC) thermal switch, first and second speed coolant fan control (FC) relay, coolant fan

THIS PAGE INTENTIONALLY LEFT
BLANK

Airbag Systems,

from September 1997

Deviate relay location and fuseplacements as well as the locations of multiple connectors see section "component locations".

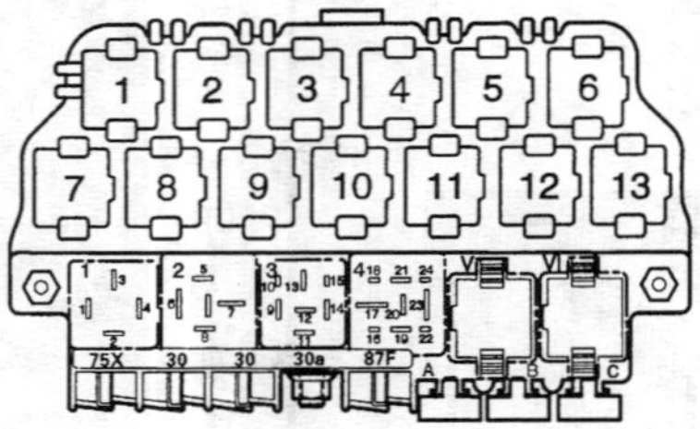

97-20676

Relay location on the thirteenfold auxiliary r panel, above relay panel:

Relay panel:

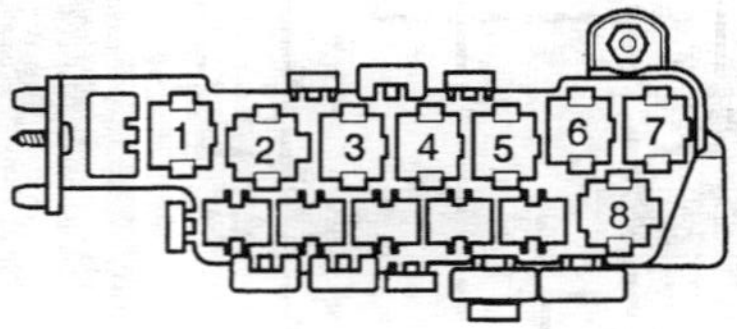

97-20515

Relay location on the eightfold auxiliary relay panel, behind relay panel:

Note: Number in parentheses indicates production control number stamped on relay housing.

Fuse colors

30 A – green
25 A – white
20 A – yellow
15 A – blue
10 A – red
7,5 A – brown
5 A – beige

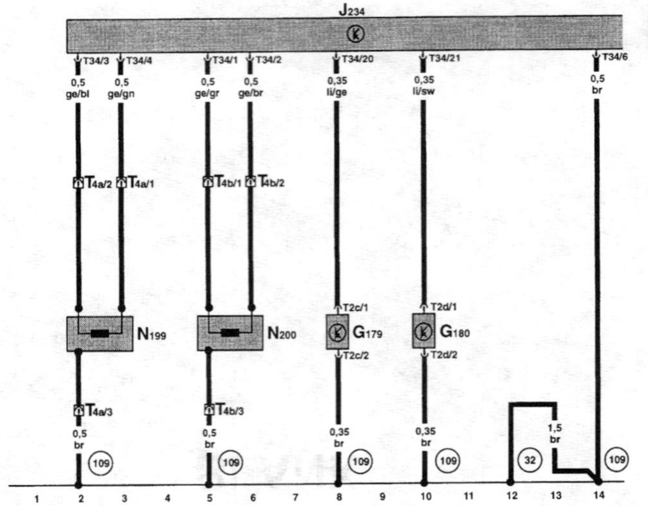

ws = white
sw = black
ro = red
br = brown
gn = green
bl = blue
gr = grey
li = violet
ge = yellow

G179 – Crash sensor for side airbag, driver's side
G180 – Crash sensor for side airbag, passenger side
J234 – Airbag Control Module, behind console
N199 – Igniter for side airbag, driver's side
N200 – Igniter for side airbag, passenger side
T2c – Double Connector
T2d – Double Connector
T4a – 4-Pin Connector, yellow, below driver's seat
T4b – 4-Pin Connector, yellow, below passenger's seat
T34 – 34-Pin Connector, on Airbag Control Module

(32) – Ground connection, behind instrument panel, left
(109) – Ground connection, in airbag wiring harness

Airbag control module, igniter for side airbag, crash sensor for side airbag

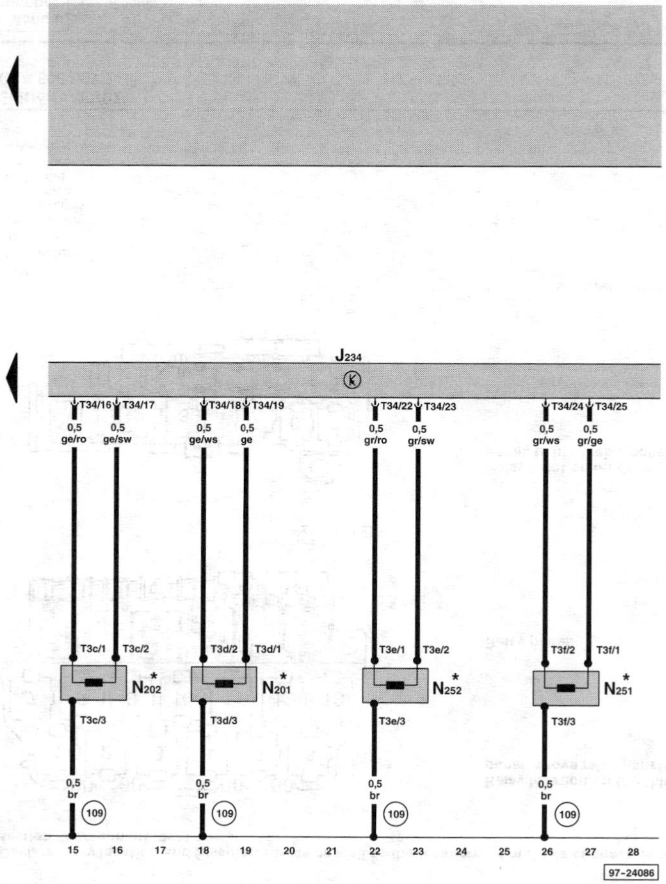

ws = white
sw = black
ro = red
br = brown
gn = green
bl = blue
gr = grey
li = violet
ge = yellow

J234 - Airbag Control Module, behind console
N201 - Igniter for side airbag, rear, driver's side
N202 - Igniter for side airbag, rear, passenger side
N251 - Igniter for head airbag, rear, driver side
N252 - Igniter for head airbag, rear, passenger side
T3c - 3-Pin Connector
T3d - 3-Pin Connector
T3e - 3-Pin Connector
T3f - 3-Pin Connector
T34 - 34-Pin Connector, on Airbag Control Module

(109) - Ground connection, in airbag wiring harness

* - from November 1998

Airbag control module, rear airbags, from November 1998

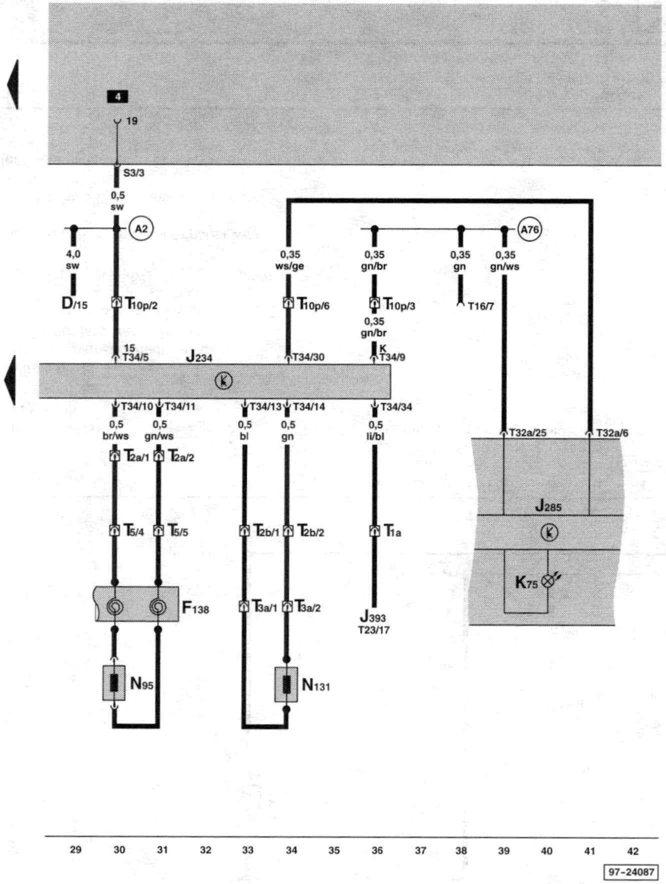

ws = white
sw = black
ro = red
br = brown
gn = green
bl = blue
gr = grey
li = violet
ge = yellow

D - Ignition/Starter Switch
F138 - Airbag Spiral Spring/Return Spring With Slip Ring
J234 - Airbag Control Module, behind console
J285 - Control module with indicator unit in instrument panel insert
J393 - Central control module for comfort system
K75 - Airbag Malfunction Indicator Lamp (MIL)
N95 - Driver's Side Airbag Igniter
N131 - Passenger's Side Airbag Igniter 1
T1a - Single Connector, black, connector station A-pillar, right
T2a - Double Connector, yellow, connector station A-pillar, left
T2b - Double Connector, yellow, connector station A-pillar, right
T3a - 3-Pin Connector, behind instrument panel, right
T5 - 5-Pin Connector, yellow, behind steering column switch cover
T10p - 10-Pin Connector, yellow, connector station A-pillar, right
T16 - 16-Pin Connector, Data Link Connector (DLC), below instrument panel, left
T23 - 23-Pin Connector, on central control module for comfort system
T32a - 32-Pin Connector, blue, on instrument cluster
T34 - 34-Pin Connector, on airbag control module

(A2) - plus connection (15), in instrument panel wiring harness

(A76) - Connector (K-diagnosis wire) in instrument panel wiring harness

Airbag control module, driver's side airbag igniter, passenger's side airbag igniter, airbag malfunction indicator lamp (MIL)

Cruise Control,

from September 1997

- **1,8L-Engine – Motronic Multiport Fuel Injection (MFI)/110 kW, code AEB,**
- **2,8L-Engine – Motronic Multiport Fuel Injection (MFI)/142 kW, code AHA,**

Deviate relay location and fuseplacements as well as the locations of multiple connectors see section "component locations".

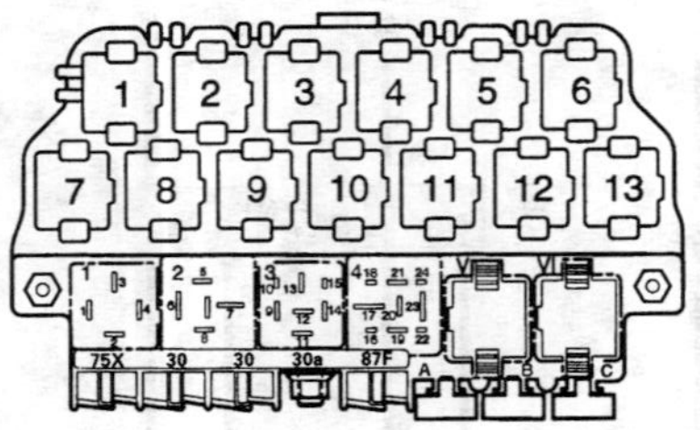

97–20676

Relay location on the thirteenfold auxiliary relay panel, above relay panel:

Relay panel:

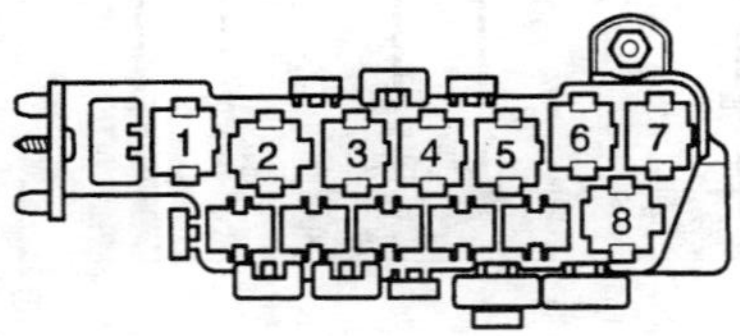

97–20515

Relay location on the eightfold auxiliary relay panel, behind relay panel:

Note: Number in parentheses indicates production control number stamped on relay housing.

Fuse colors
30 A – green
25 A – white
20 A – yellow
15 A – blue
10 A – red
7,5 A – brown
5 A – beige

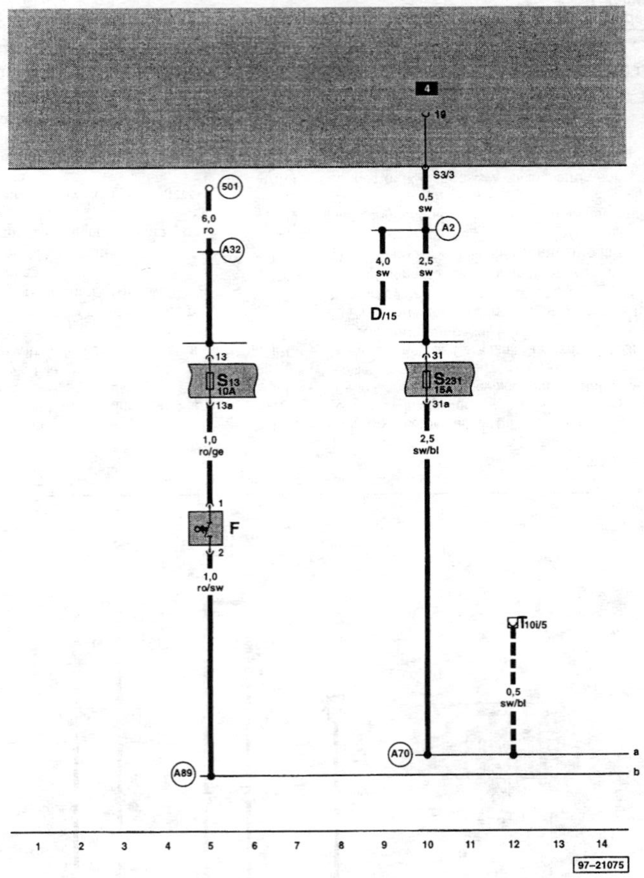

ws = white
sw = black
ro = red
br = brown
gn = green
bl = blue
gr = grey
li = violet
ge = yellow

D – Ignition/Starter Switch
F – Brake Light Switch
S13 – Fuse 13 in fuse holder
S231 – Fuse 31 in fuse holder
T10i – 10-Pin Connector, blue, connector station A-pillar, right, automatic transmission
(501) – Screw connection -2- (30), on relay panel
(A2) – plus connection (15), in instrument panel wiring harness
(A32) – plus connection (30), in instrument panel wiring harness
(A70) – Connector (15a, fuse 231), in instrument panel wiring harness
(A89) – Connector (54), in instrument panel wiring harness

--- automatic transmission only

Brake light switch

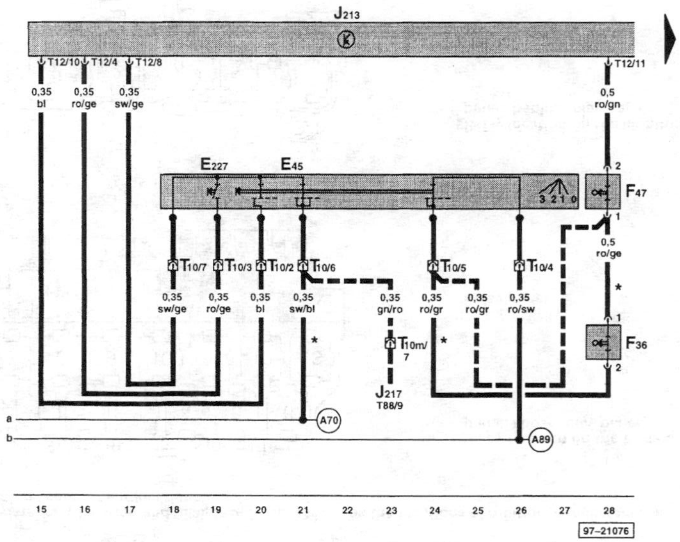

ws = white
sw = black
ro = red
br = brown
gn = green
bl = blue
gr = grey
li = violet
ge = yellow

E45 – Cruise Control Switch (ON/OFF)
position 0 = OFF
position 1 = OFF save
position 2 = ON
position 3 = RES
E227 – Cruise Control Push Button (SET)
F36 – Clutch Vacuum vent Valve Switch
F47 – Brake Vacuum Vent Valve Switch for cruise control
J213 – Cruise Control, Control Module, behind instrument panel, right, behind driver's side airbag
J217 – Transmission Control Module (TCM)
T10 – 10-Pin Connector, black, near steering column, right
T10m – 10-Pin Connector, brown, connector station A-pillar, right
T12 – 12-Pin Connector, on Cruise Control, Control Module
T88 – Connector, 88 point, on Transmission Control Module (TCM), 5 speed automatic transmission
(A70) – Connector (15a, fuse 231), in instrument panel wiring harness
(A89) – Connector (54), in instrument panel wiring harness
- - - – automatic transmission only
* – manual transmision only

Cruise control, control module, cruise control switch, clutch vacuum vent valve switch, brake vacuum vent valve switch

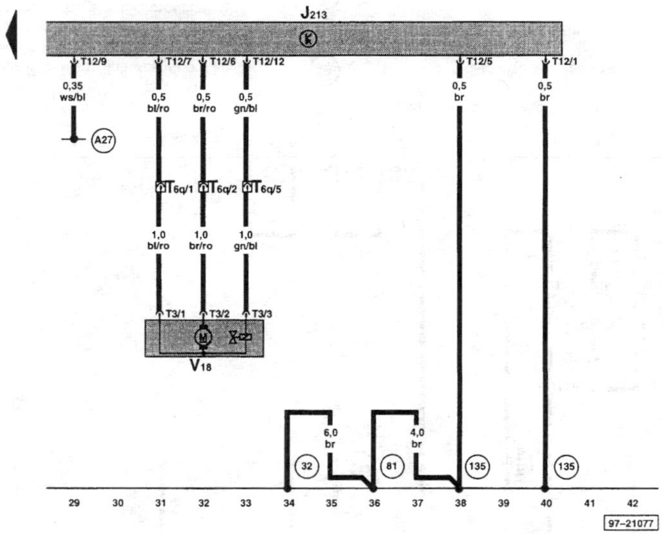

ws = white
sw = black
ro = red
br = brown
gn = green
bl = blue
gr = grey
li = violet
ge = yellow

J213 – Cruise Control, Control Module, behind instrument panel, right, behind driver's side airbag
T3 – 3-Pin Connector
T6q – 6-Pin Connector, black, connector station A-pillar, left
T12 – 12-Pin Connector, on Cruise Control, Control Module
V18 – Cruise Control Vacuum pump
(32) – Ground connection, behind instrument panel, left
(81) – Ground connection -1-, in instrument panel wiring harness
(135) – Ground connection -2-, in instrument panel wiring harness
(A27) – wire Connection (vehicle speed signal), in instrument panel wiring harness

Cruise control, control module, cruise control vacuum pump

Radio System,

from September 1997

Deviate relay location and fuseplacements as well as the locations of multiple connectors see section "component locations".

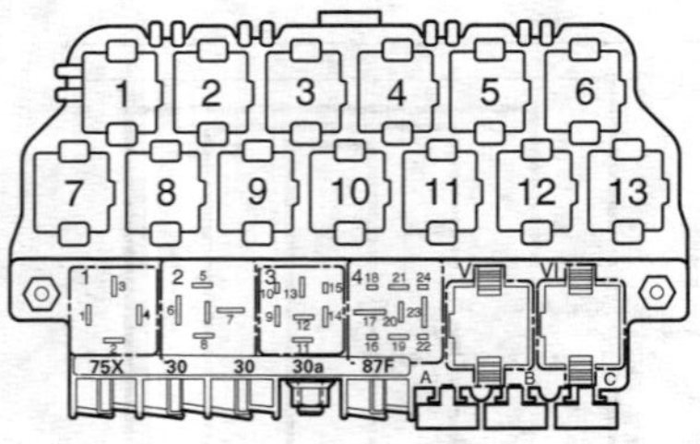

97-20676

Relay location on the thirteenfold auxiliary relay panel, above relay panel:

Relay panel:

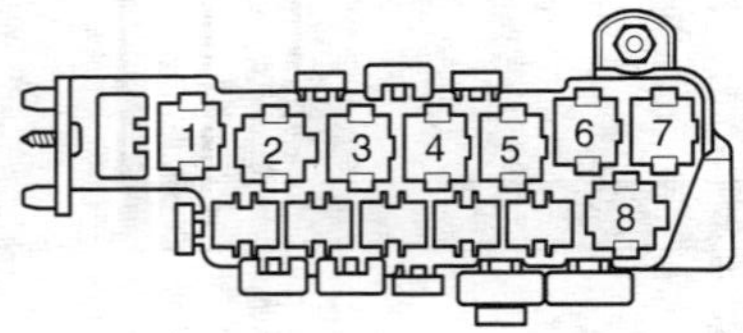

97-20515

Relay location on the eightfold auxiliary relay panel, behind relay panel:

Note: Number in parentheses indicates production control number stamped on relay housing.

Fuse colors

30 A - green
25 A - white
20 A - yellow
15 A - blue
10 A - red
7,5 A - brown
5 A - beige

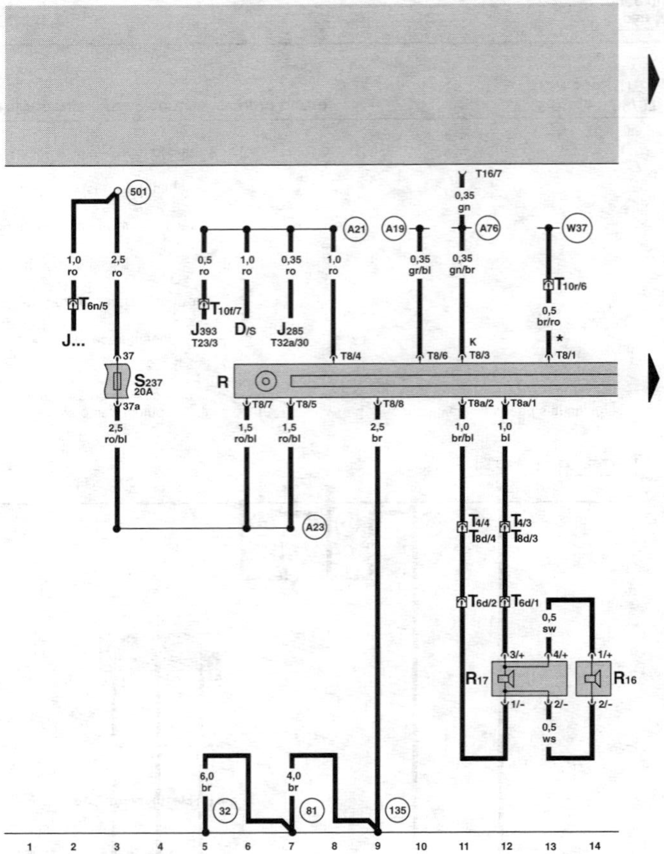

97-25434

ws = white
sw = black
ro = red
br = brown
gn = green
bl = blue
gr = grey
li = violet
ge = yellow

- J... - Engine Control Module (ECM)
- R - Radio
- R16 - Right Rear Tweeter, in right rear door
- R17 - Right Rear Woofer, in right rear door
- S237 - Fuse 37 in fuse holder
- T4 - 4-Pin Connector, black, connector station A-pillar, right
- T6d - 6-Pin Connector, red, on B-pillar, right
- T6n - 6-Pin Connector, red, on protective housing for control module
- T8 - 8-Pin Connector, black, on radio
- T8a - 8-Pin Connector, brown, on radio
- T8d - 8-Pin Connector, black, connector station A-pillar, right
- T10v - 10-Pin Connector, violett, connector station A-pillar, right
- T16 - 16-Pin Connector, Data Link Connector (DLC), below instrument panel, left

- (32) - Ground connection, behind instrument panel, left
- (A76) - Connector (K-diagnosis wire), in instrument panel wiring harness
- (81) - Ground connection -1-, in instrument panel wiring harness
- (135) - Ground connection -2-, in instrument panel wiring harness
- (501) - Screw Connection -2- (30), on relay panel
- (A19) - wire connection (58d), in instrument panel wiring harness
- (A23) - wire connection (30al), in instrument panel wiring harness
- (W37) - Connector (alarm system), in wiring harness floor

Radio, right rear speakers

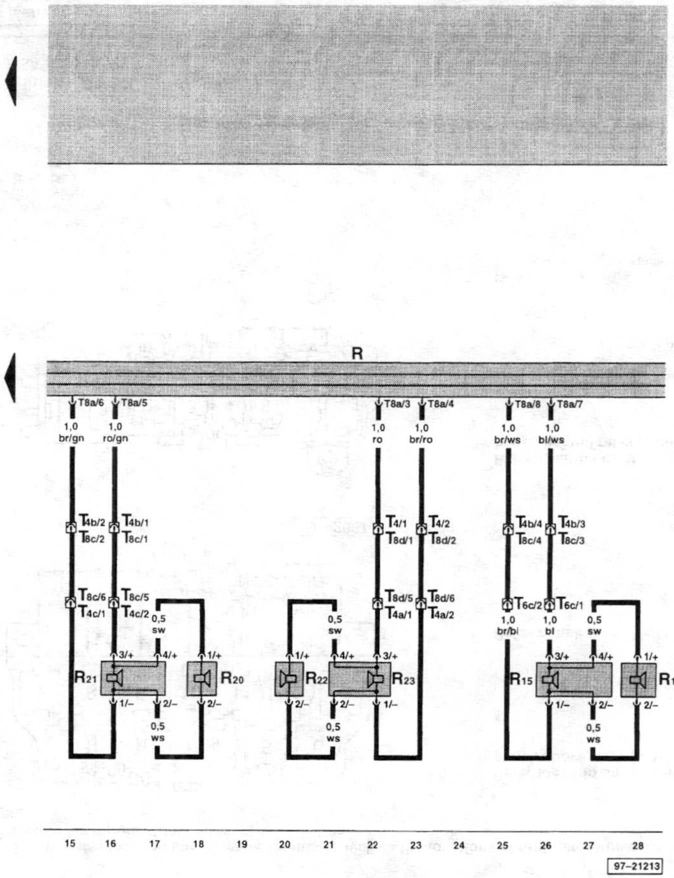

ws = white
sw = black
ro = red
br = brown
gn = green
bl = blue
gr = grey
li = violet
ge = yellow

R – Radio
R14 – Left Rear Tweeter, in left rear door
R15 – Left Rear Woofer, in left rear door
R20 – Left Front Tweeter, in driver's door
R21 – Left Front Woofer, in driver's door
R22 – Right Front Tweeter, in passenger's door
R23 – Right Front Tweeter, in passenger's door
T4 – 4-Pin Connector, black, connector station A-pillar, right
T4a – 4-Pin Connector, black, connector station A-pillar, right
T4b – 4-Pin Connector, black, connector station A-pillar, left
T4c – 4-Pin Connector, black, connector station A-pillar, left
T6c – 6-Pin Connector, red, on B-pillar, left
T8a – 8-Pin Connector, brown, on radio
T8c – 8-Pin Connector, black, connector station A-pillar, left
T8d – 8-Pin Connector, black, connector station A-pillar, right

Edition 09/97
USA.5062.01.21

Radio, front speakers, left rear speakers

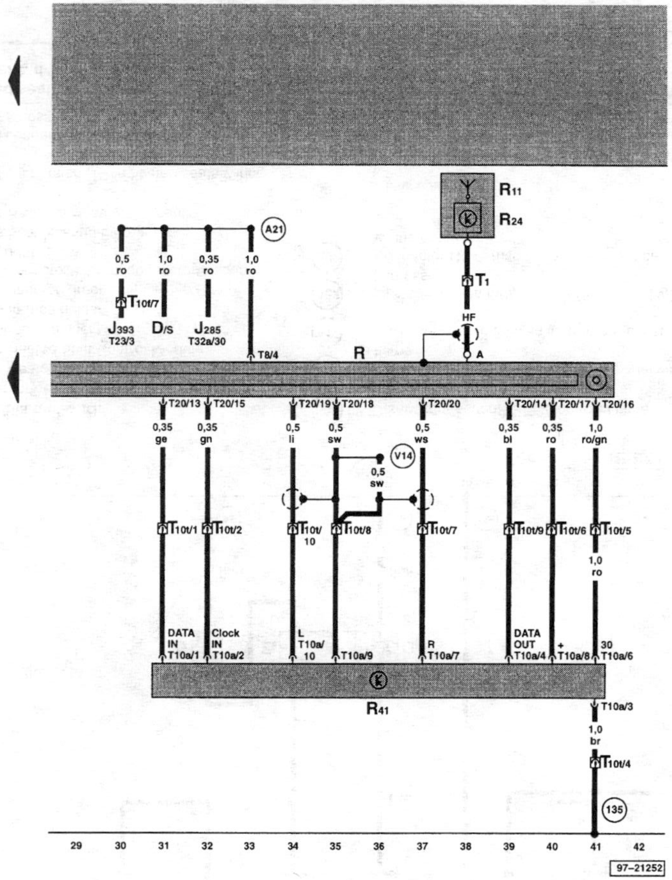

ws = white
sw = black
ro = red
br = brown
gn = green
bl = blue
gr = grey
li = violet
ge = yellow

D – Ignition/Starter Switch
J285 – Control module with indicator unit in instrument panel insert
J393 – Central control module for comfort system
R – Radio
R11 – Antenna
R24 – Antenna Amplifier
R41 – CD Changer Unit
T1 – 1-Pin Connector, black, below console
T8 – 8-Pin Connector, black, on radio
T10a – 10-Pin Connector, red, on CD Changer Unit
T10f – 10-Pin Connector, blue, connector station A-pillar, left
T10t – 10-Pin Connector, red, connector station A-pillar, right
T20 – 20-Pin Connector, on radio
T23 – 23-Pin Connector, on central control module for comfort system
T32a – 32-Pin Connector, blue, on instrument cluster
(135) – Ground connection -2-, in instrument panel wiring harness
(A21) – wire connection (86s), in instrument panel wiring harness
(V14) – wire connection (shielding), in CD-changer wiring harness

Radio, Antenna, Antenna Amplifier, CD Changer Unit

Edition 09/97
USA.5062.01.21

Heated seats,

from September 1997

Deviate relay location and fuseplacements as well as the locations of multiple connectors see section "component locations".

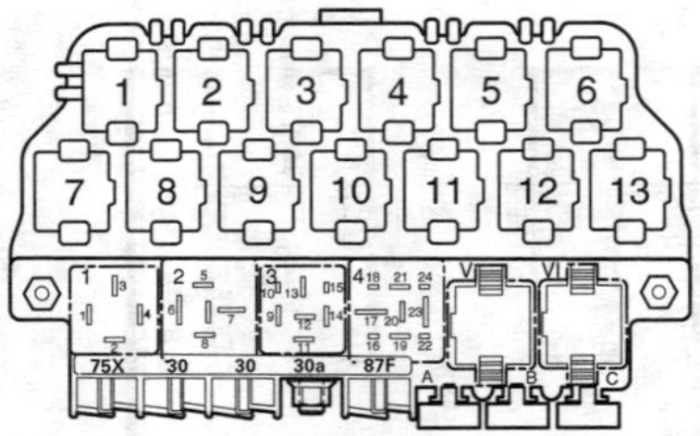

97-20676

Relay location on the thirteenfold auxiliary relay panel, above relay panel:

Relay panel:

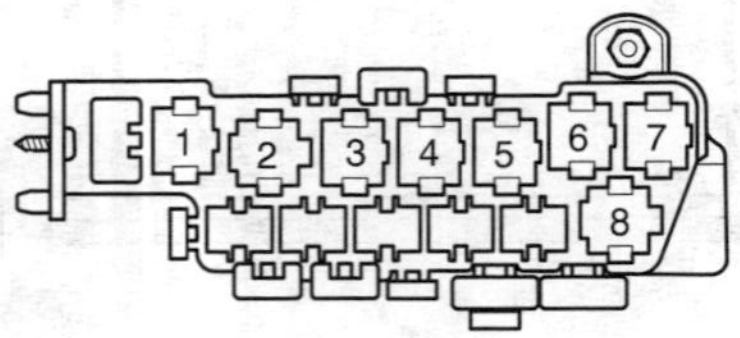

97-20515

Relay location on the eightfold auxiliary relay panel, behind relay panel:

Note: Number in parentheses indicates production control number stamped on relay housing.

Fuse colors
30 A - green
25 A - white
20 A - yellow
15 A - blue
10 A - red
7,5 A - brown
5 A - beige

Edition 08/00
USA.5062.12.21

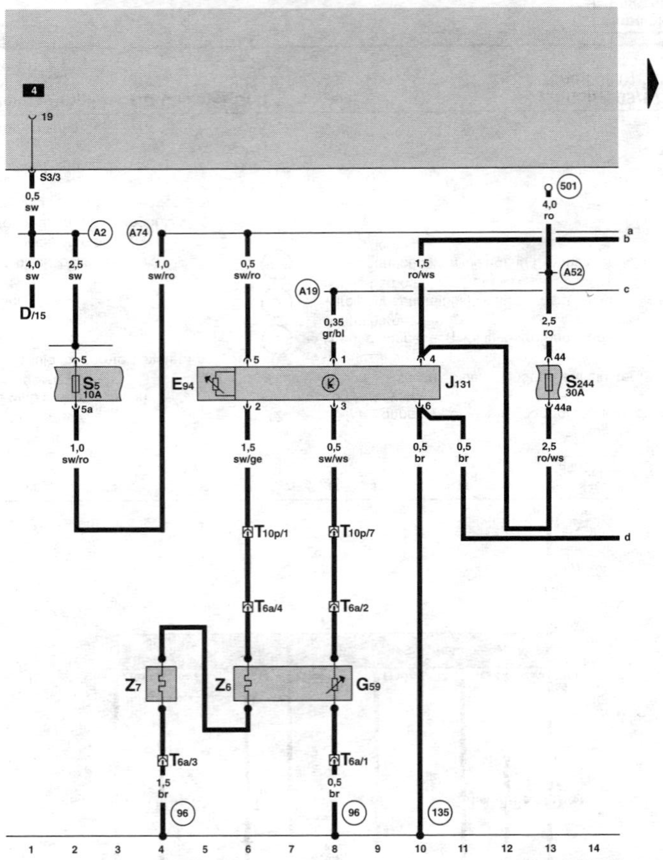

ws = white
sw = black
ro = red
br = brown
gn = green
bl = blue
gr = grey
li = violet
ge = yellow

97-21078

D - Ignition/Starter Switch
E94 - Adjuster for hested driver's seat
G59 - Driver's Heated Seat Temperature Sensor
J131 - Driver's Heated Seat Control Module
S5 - Fuse 5 in fuse holder
S244 - Fuse 44 in fuse holder
T6a - 6-Pin Connector, green
T10p - 10-Pin Connector, brown, connector station A-pillar, right
Z6 - Driver's Seat Heating Element
Z7 - Driver's Backrest Heating Element

(96) - Ground connection -1-, iin heated seats wiring harness
(135) - Ground connection -2-, in instrument panel wiring harness

(501) - Screw connection -2- (30), on relay panel
(A2) - plus connection (15), in instrument panel wiring harness
(A19) - wire connection (58d), in instrument panel wiring harness
(A52) - plus connection (30), in instrument panel wiring harness
(A74) - Connector (15a - fuse 5), in instrument panel wiring harness

Driver's heated seat control module, temperature sensor, driver's seat heating element, driver's backrest heating element

Edition 08/00
USA.5062.12.21

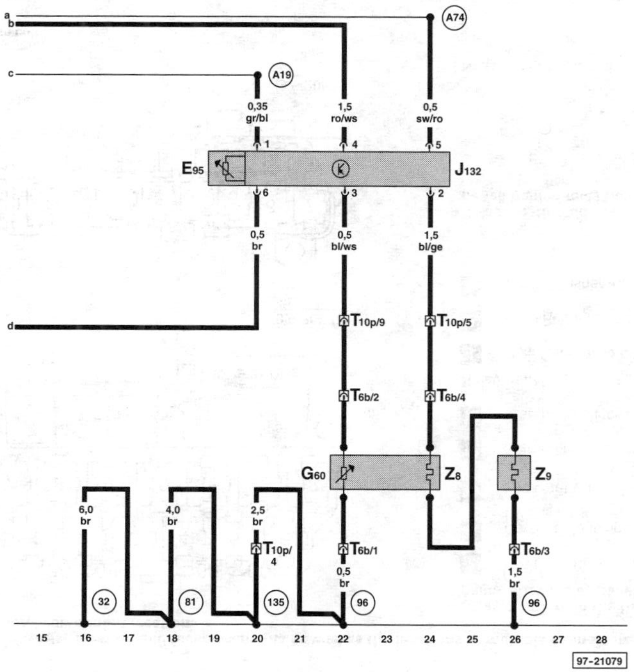

ws = white
sw = black
ro = red
br = brown
gn = green
bl = blue
gr = grey
li = violet
ge = yellow

E95 - Adjuster for hested passenger seat
G60 - Passenger's Heated Seat Temperature Sensor
J132 - Passenger's Heated Seat Control Module
T6b - 6-Pin Connector, green
T10p - 10-Pin Connector, brown, connector station A-pillar, right
Z8 - Passenger's Seat Heating Element
Z9 - Passenger's Backrest Heating Element

(32) - Ground connection, behind instrument panel, left
(81) - Ground connection -1-, in instrument panel wiring harness
(96) - Ground connection -1-, iin heated seats wiring harness
(135) - Ground connection -2-, in instrument panel wiring harness
(A19) - wire connection (58d), in instrument panel wiring harness
(A74) - Connector (15a - fuse 5), in instrument panel wiring harness

Edition 08/00
USA.5062.12.21

Passenger's heated seat control module, temperature sensor, passenger's seat heating element, passenger's backrest heating element

THIS PAGE INTENTIONALLY LEFT BLANK

Passat - Standard Equipment, only for cars with green illuminated instrument cluster

from September 1997

Deviate relay location and fuseplacements as well as the locations of multiple connectors see section "component locations".

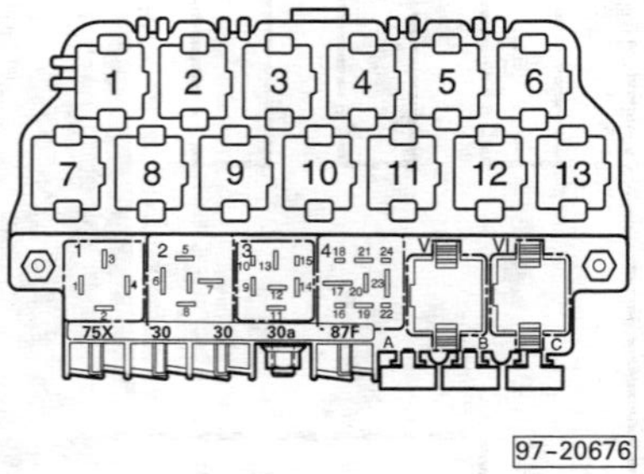

97-20676

Relay location on the thirteenfold auxiliary relay panel, above relay panel:

- 7 Fog Light Relay (381)
- 8 Daytime Running Lights Change-over Relay (173)
- 10 Fog Light Relay (381)
- 12 Starting Interlock Relay (alarm system) (186)
- 13 Starting Interlock Relay (clutch position) (204)

Relay panel:

- 1 Dual Horn Relay (204)
- 2 Load Reduction Relay (213)
- 3 Wiper/Washer Intermittent Relay (317)

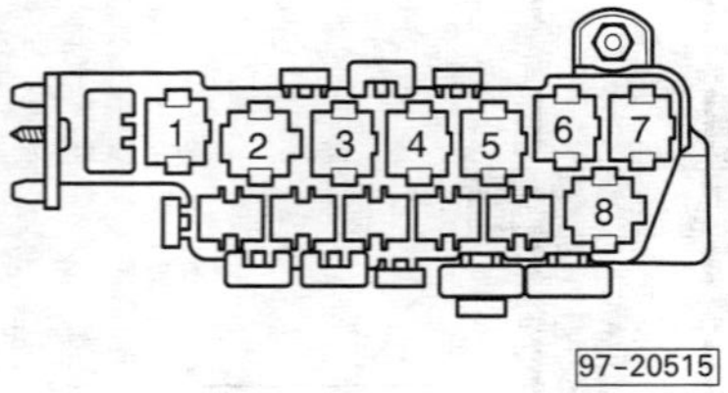

97-20515

Relay location on the eightfold auxiliary relay panel, behind relay panel:

Note: Number in parentheses indicates production control number stamped on relay housing.

Fuse colors
30 A - green
25 A - white
20 A - yellow
15 A - blue
10 A - red
7,5 A - brown
5 A - beige

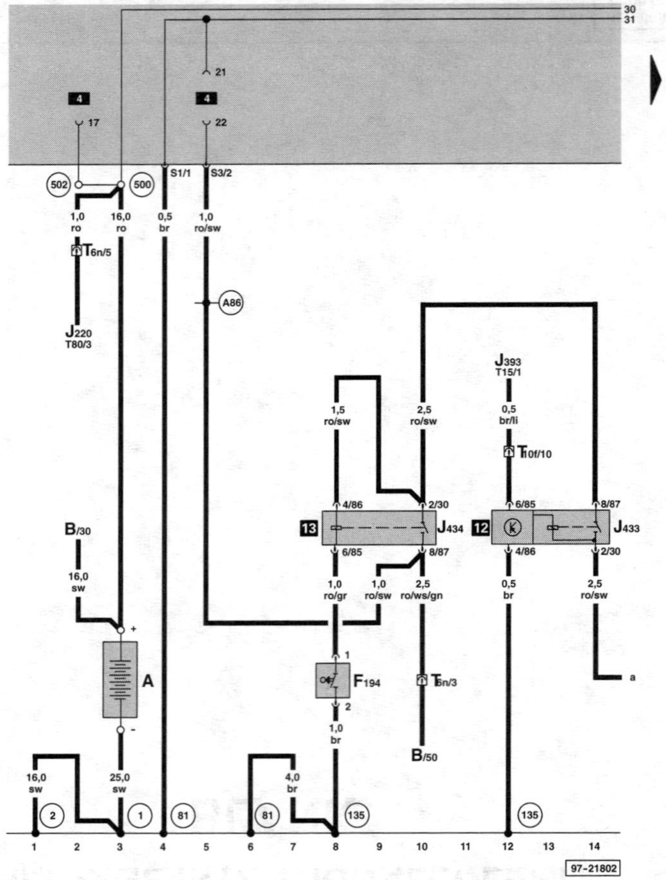

97-21802

ws = white
sw = black
ro = red
br = brown
gn = green
bl = blue
gr = grey
li = violet
ge = yellow

A - Battery
B - Starter
F194 - Clutch Pedal Position (CPP) Switch
J220 - Motronic Engine Control Module (ECM)
J393 - Central control module for comfort system
J433 - Lock relay for starter (theft warning system), on auxiliary relay panel, above relay panel, production control number (186)
J434 - Locking relay for starter (clutch pedal switch), on auxiliary relay panel, above relay panel, production control number (204)
T6n - 6-Pin Connector, red, on protective housing for control module, in engine compartment, left
T10f - 10-Pin Connector, blue, connector station A-pillar, left
T15 - 15-Pin Connector, on central control module for comfort system
T80 - Connector, 80 point, on Engine Control Module (ECM)
(1) - Ground strap, battery to body
(2) - Ground strap, transmission to body
(81) - Ground connection -1-, in instrument panel wiring harnes
(135) - Ground connection -2-, in instrument panel wiring harness
(500) - Screw connection -1- (30), on relay panel
(502) - Screw connection (30a), on relay panel
(A86) - Connection (50a) in instrument panel wiring harness

Starting interlock, battery

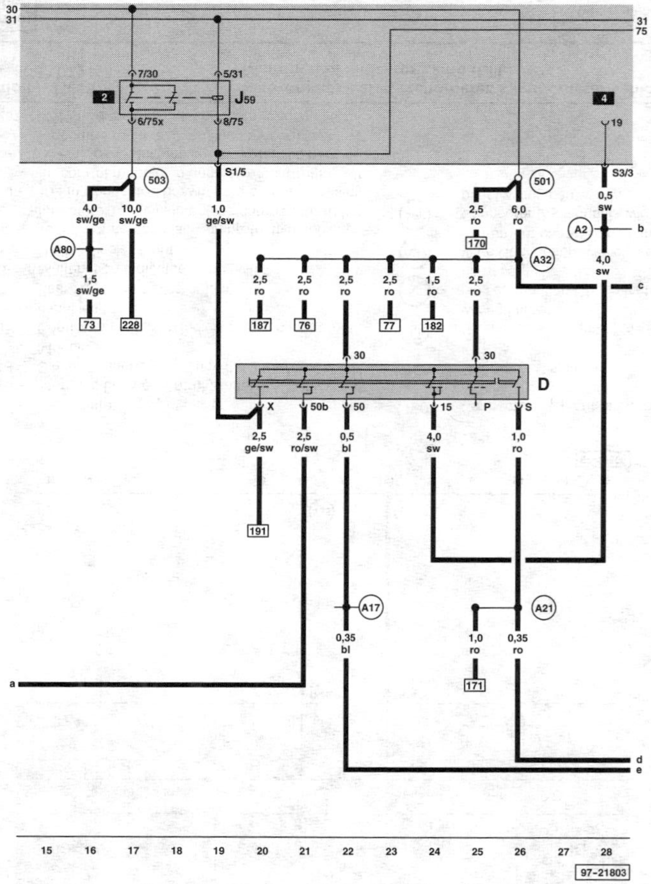

ws = white
sw = black
ro = red
br = brown
gn = green
bl = blue
gr = grey
li = violet
ge = yellow

D - Ignition/Starter Switch
J59 - Load Reduction Relay, production control number (213)
(501) - Screw connection -2- (30), on relay panel
(503) - Screw connection (75x), on relay panel
(A2) - plus connection (15), in instrument panel wiring harness
(A17) - wire connection (61), in instrument panel wiring harness
(A21) - wire connection (86s), in instrument panel wiring harness
(A32) - plus connection (30), in instrument panel wiring harness
(A80) - Connector -1- (X), in instrument panel wiring harness

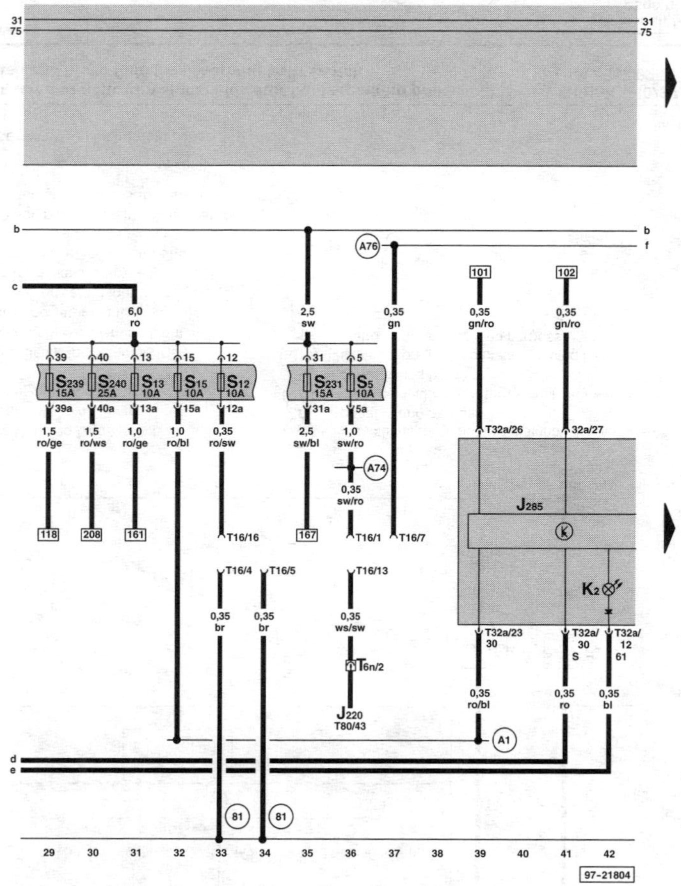

ws = white
sw = black
ro = red
br = brown
gn = green
bl = blue
gr = grey
li = violet
ge = yellow

J220 - Motronic Engine Control Module (ECM)
J285 - Control module with indicator unit in instrument panel insert
K2 - Generator (GEN) Warning Light
S5 - Fuse 5 in fuse holder
S12 - Fuse 12 in fuse holder
S13 - Fuse 13 in fuse holder
S15 - Fuse 15 in fuse holder
S231 - Fuse 31 in fuse holder
S239 - Fuse 39 in fuse holder
S240 - Fuse 40 in fuse holder
T6n - 6-Pin Connector, red, on protective housing for control module, in engine compartment, left
T16 - 16-Pin Connector, data link connector (DLC), below instrument panel, left
T32a - 32-Pin Connector, blue, on instrument cluster
T80 - Connector, 80 point, on Engine Control Module (ECM)
(81) - Ground connection -1-, in instrument panel wiring harness
(A1) - plus connection (30a), in instrument panel wiring harness
(A74) - Connector (15a - fuse 5), in instrument panel wiring harness
(A76) - Connector (K-diagnosis wire), in instrument panel wiring harness

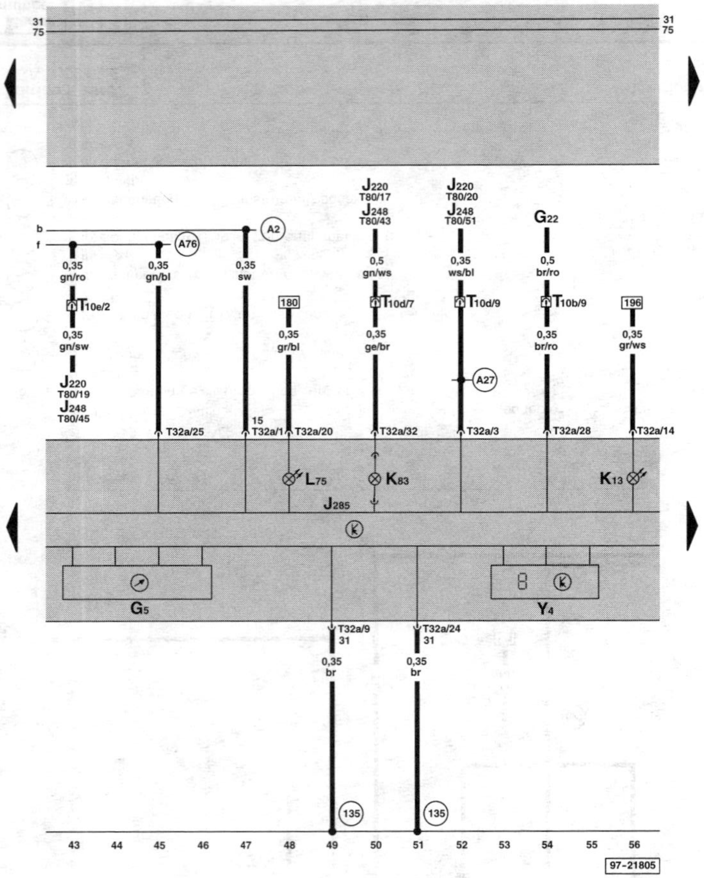

ws = white
sw = black
ro = red
br = brown
gn = green
bl = blue
gr = grey
li = violet
ge = yellow

G5 - Tachometer
J220 - Motronic Engine Control Module (ECM)
J248 - Diesel Direct Fuel Injection (DFI) Engine Control Module (ECM)
J285 - Control module with indicator unit in instrument panel insert
K13 - Rear Fog Light Indicator Light
K83 - Malfunction Indicator Lamp (MIL)
L75 - Digital Display Light
T10b - 10-Pin Connector, black, on protective housing for control module, in engine compartment, left
T10d - 10-Pin Connector, brown, on protective housing for control module, in engine compartment, left
T10e - 10-Pin Connector, yellow, on protective housing for control module, in engine compartment, left
T32a - 32-Pin Connector, blue, on instrument cluster
T80 - Connector, 80 point, on Engine Control Module (ECM)
Y4 - Odometer Display

(135) - Ground connection -2-, in instrument panel wiring harness
(A2) - plus connection (15), in instrument panel wiring harness
(A27) - Wire Connection (vehicle speed signal), in instrument panel wiring harness
(A76) - Connector (K-diagnosis wire), in instrument panel wiring harness

Edition 11/98
USA.5062.07.21

Instrument cluster, tachometer, speedometer vehicle speed sensor, Malfunction Indicator Lamp (MIL)

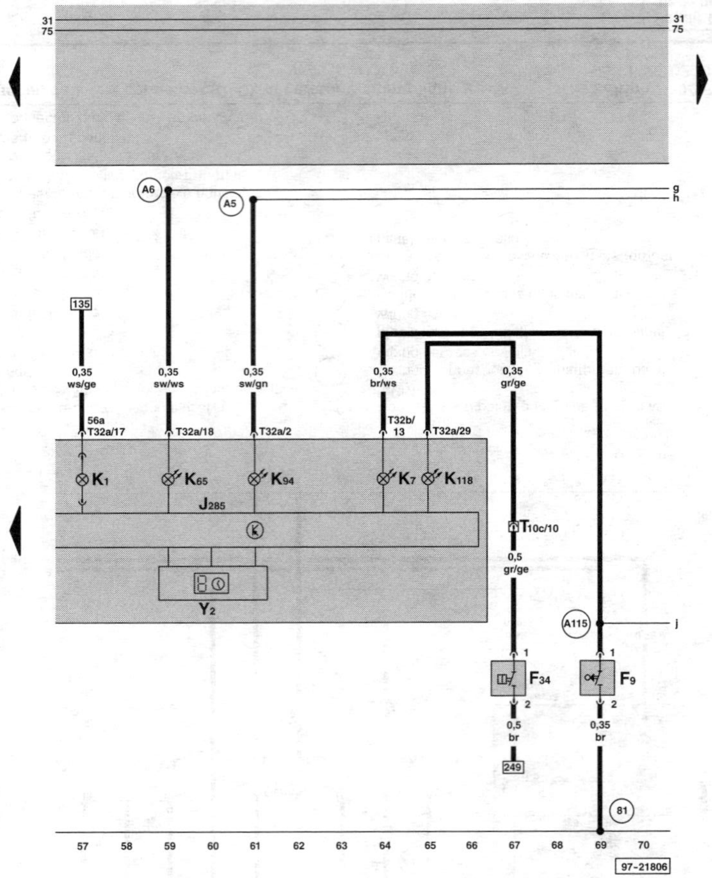

ws = white
sw = black
ro = red
br = brown
gn = green
bl = blue
gr = grey
li = violet
ge = yellow

F9 - Parking Brake Warning Light Switch
F34 - Brake Fluid Level Warning Switch
J285 - Control module with indicator unit in instrument panel insert
K1 - Headlight High Beam Indicator Light
K7 - Brake and Parking Brake Warning Light
K65 - Left Turn Signal Indicator Light
K94 - Right Turn Signal Indicator Light
K118 - Warning light for brake system
T10c - 10-Pin Connector, violet, connector station A-pillar, left
T32a - 32-Pin Connector, blue, on instrument cluster
T32b - 32-Pin Connector, green, on instrument cluster
Y2 - Digital Clock

(81) - Ground connection -1-, in instrument panel wiring harness

(A5) - plus connection (right turn signal), in instrument panel wiring harness
(A6) - plus connection (left turn signal), in instrument panel wiring harness
(A115) - Connection (parking brake warning light), in instrument panel wiring harness

Instrument cluster, left and right turn signal indicator lights, parking brake warning light switch and brake fluid level warning light switch

Edition 10/97
USA.5062.02.21

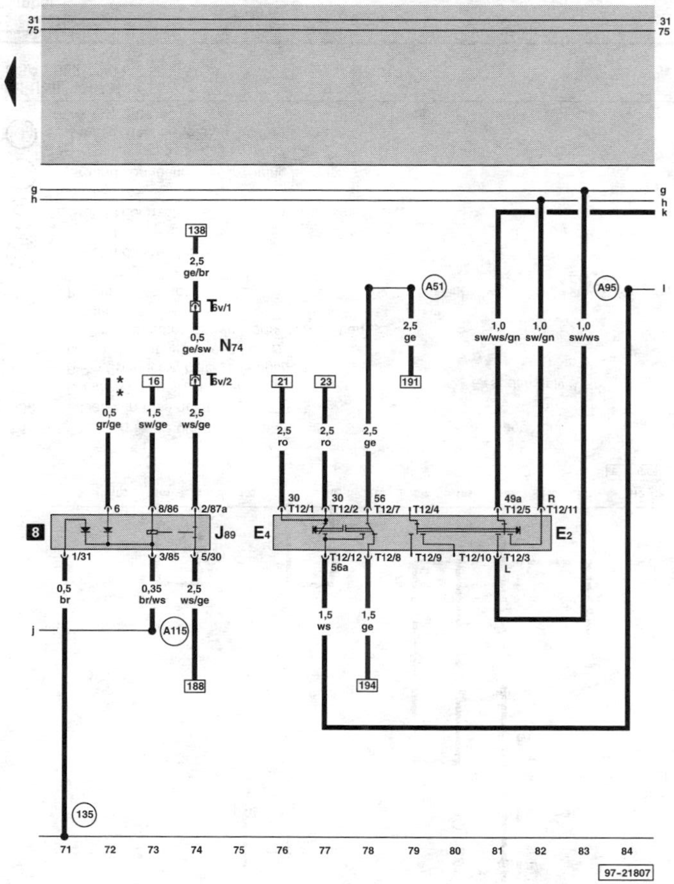

ws = white
sw = black
ro = red
br = brown
gn = green
bl = blue
gr = grey
li = violet
ge = yellow

E2 - Turn Signal Switch
E4 - Headlight Dimmer/Flasher Switch
J89 - Daytime Running Lights Change-over Relay on the thirteenfold relay panel, above relay panel, production control number (173)
N74 - Series Resistance Wiring
T6v - 6-Pin Connector, violet, connector station A-pillar, right
T12 - 12-Pin Connector, near steering column

(135) - Ground connection -2-, in instrument panel wiring harness
(A51) - wire connection (56), in instrument panel wiring harness
(A95) - wire connection -1- (56a), in instrument panel wiring harness

(A115) - Connection (parking brake warning light), in instrument panel wiring harness

** - wire is not connected, or terminal 6 of J89 is empty

Turn signal switch, headlight dimmer/flasher switch, daytime running lights change-over relay, series resistance wiring

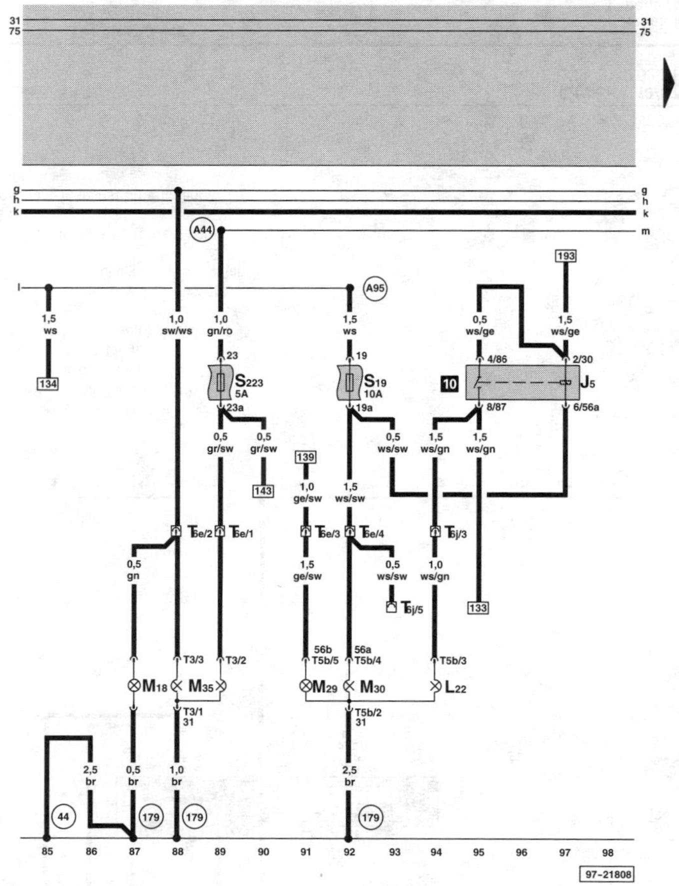

ws = white
sw = black
ro = red
br = brown
gn = green
bl = blue
gr = grey
li = violet
ge = yellow

J5 - Fog Light Relay, on auxiliary relay panel, production control number (381)
L22 - Left Front Fog Light
M18 - Left, Side Turn Signal Light
M29 - Left Low Beam Headlight
M30 - Left High Beam Headlight
M35 - Light for turn signal and side marker, front, left
S19 - Fuse 19 in fuse holder
S223 - Fuse 23 in fuse holder
T3 - 3-Pin Connector
T5b - 5-Pin Connector, on headlight left
T6e - 6-Pin Connector, yellow, connector station A-pillar, left
T6j - 6-Pin Connector, red, connector station A-pillar, left

(44) - Ground connection, on left A-pillar, lower part
(179) - Ground connection, in left headlight wiring harness
(A44) - wire connection (57r), in instrument panel wiring harness
(A95) - Connection 1 (56a), in instrument panel wiring harness

Left front headlights, left front fog light, fog light relay

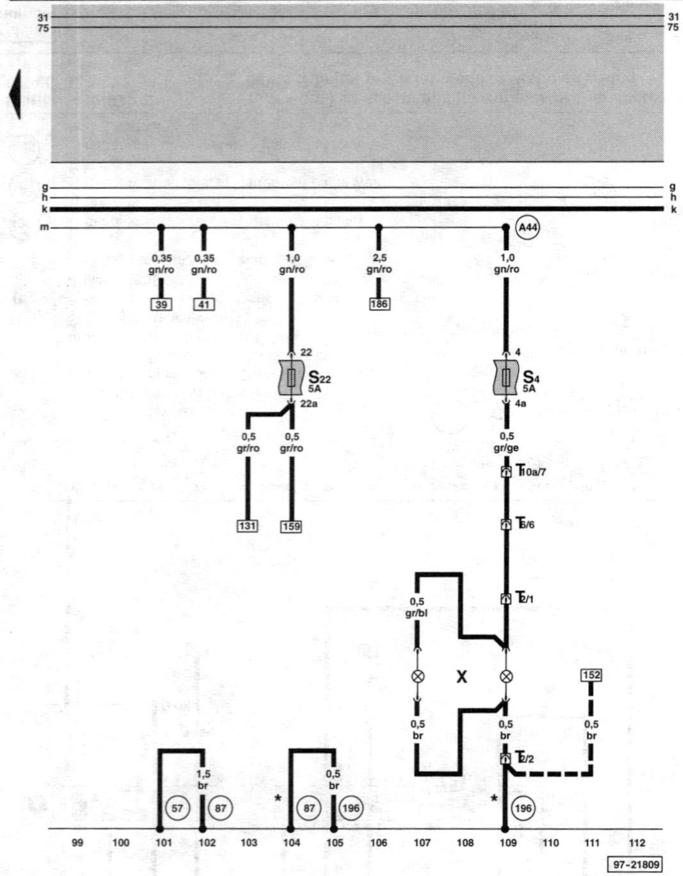

ws = white
sw = black
ro = red
br = brown
gn = green
bl = blue
gr = grey
li = violet
ge = yellow

S4 - Fuse 4 in fuse holder
S22 - Fuse 22 in fuse holder
T2 - Double Connector, in rear lid
T6 - 6-Pin Connector, black, in luggage compartment, left (wagon only)
T10a - 10-Pin Connector, brown, connector station A-pillar, left
X - License Plate Light
(57) - Ground connection, on left rear pillar
(87) - Ground connection -2-, in rear wiring harness
(196) - Ground connection -3-, in wiring harness, rear
(A44) - wire connection (57r), in instrument panel wiring harness

- - - - wagon only
* - sedan only

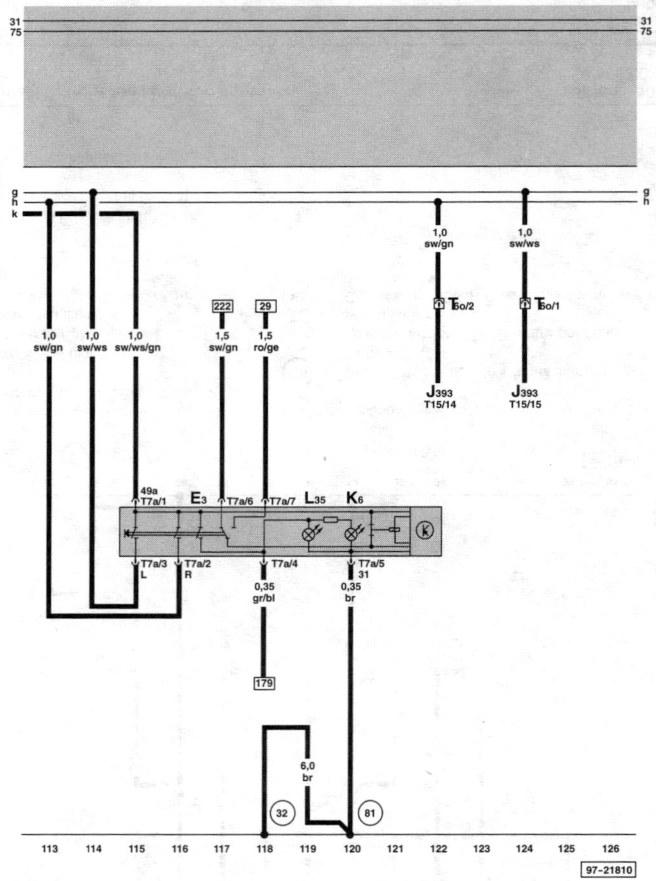

ws = white
sw = black
ro = red
br = brown
gn = green
bl = blue
gr = grey
li = violet
ge = yellow

E3 - Emergency Flasher Switch
J393 - Central control module for comfort system
K6 - Emergency Flasher Warning Light
L35 - Emergency Flasher Switch Light
T6o - 6-Pin Connector, white, connector station A-pillar, right
T7a - 7-Pin Connector, on emergency flasher switch
T15 - 15-Pin Connector, on central control module for comfort system
(32) - Ground connection, behind instrument panel, left
(81) - Ground connection -1-, in instrument panel wiring harness

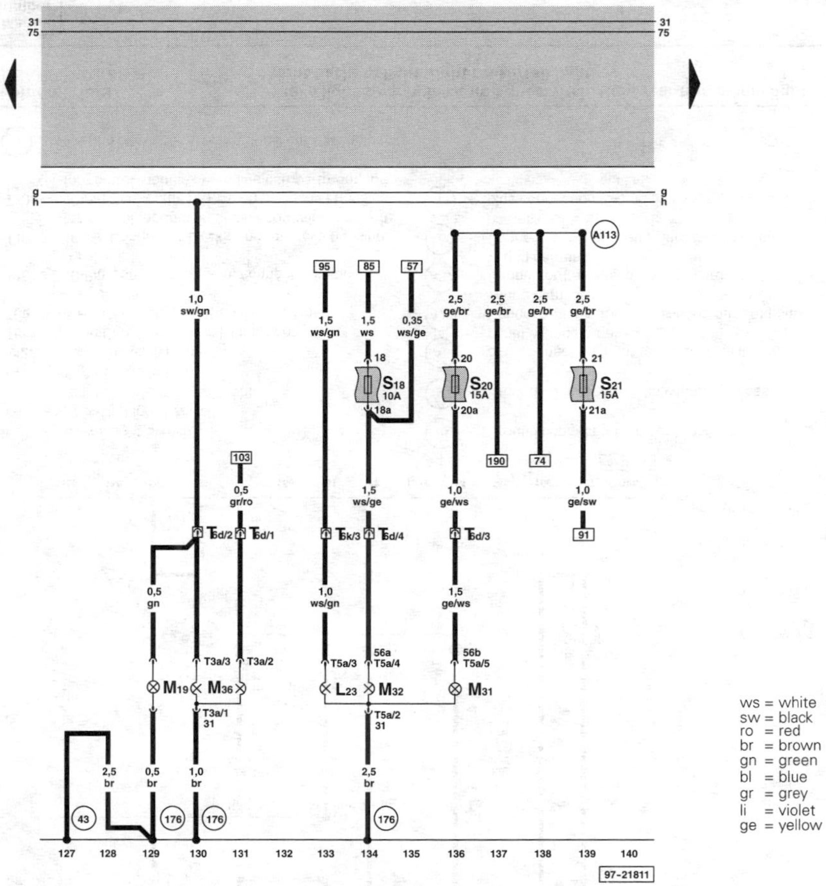

ws = white
sw = black
ro = red
br = brown
gn = green
bl = blue
gr = grey
li = violet
ge = yellow

L23 - Right Front Fog Light
M19 - Right, Side Turn Signal Light
M31 - Right Low Beam Headlight
M32 - Right High Beam Headlight
M36 - Light for turn signal and side marker, front, right
S18 - Fuse 18 in fuse holder
S20 - Fuse 20 in fuse holder
S21 - Fuse 21 in fuse holder
T3a - 3-Pin Connector
T5a - 5-Pin Connector, on headlight right
T6d - 6-Pin Connector, yellow, connector station A-pillar, right
T6k - 6-Pin Connector, red, connector station A-pillar, right

(43) - Ground connection, on right A-pillar, lower part

(176) - Ground connection, in right headlight wiring harness

(A113) - Connection -2- (daytime running lights), in instrument panel wiring harness

Right front headlights, right front foglight

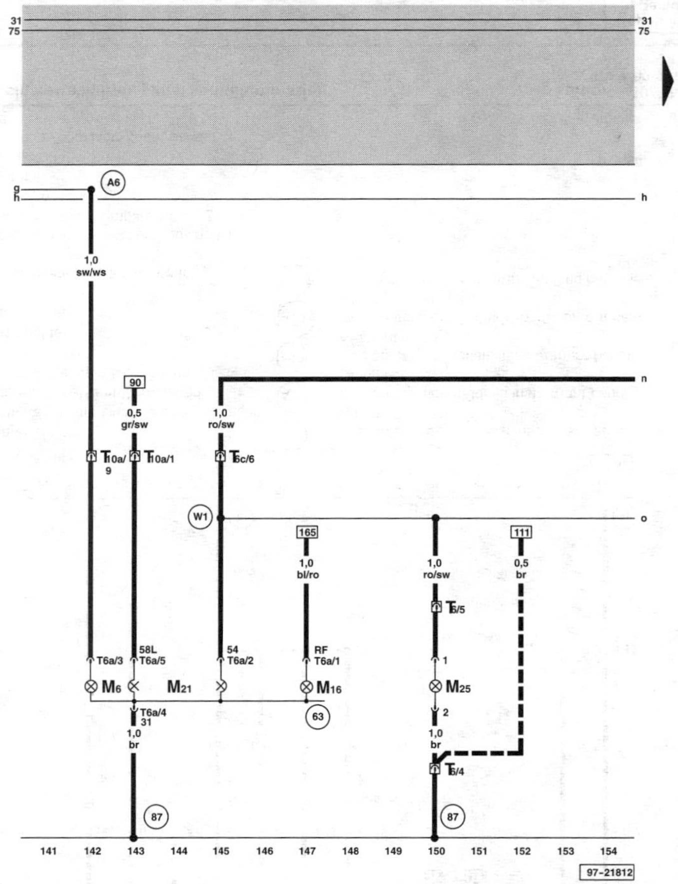

ws = white
sw = black
ro = red
br = brown
gn = green
bl = blue
gr = grey
li = violet
ge = yellow

M6 - Left Rear Turn Signal Light
M21 - Left Brake/Tail Light
M16 - Left Back-Up Light
M25 - High-mount Brake Light
T6 - 6-Pin Connector, black, in luggage compartment, left (wagon only)
T6a - 6-Pin Connector, on tail light, left
T6c - 6-Pin Connector, blue, connector station A-pillar, left
T10a - 10-Pin Connector, brown, connector station A-pillar, left

(63) - Ground connection, on left taillight bulb holder

(87) - Ground connection -2-, in rear wiring harness

(A6) - plus connection (left turn signal), in instrument panel wiring harness

(W1) - plus connection (54), in rear wiring harness

- - - wagon only

Left brake/tail light, left rear turn signal light, left back-up light, high-mount brake light

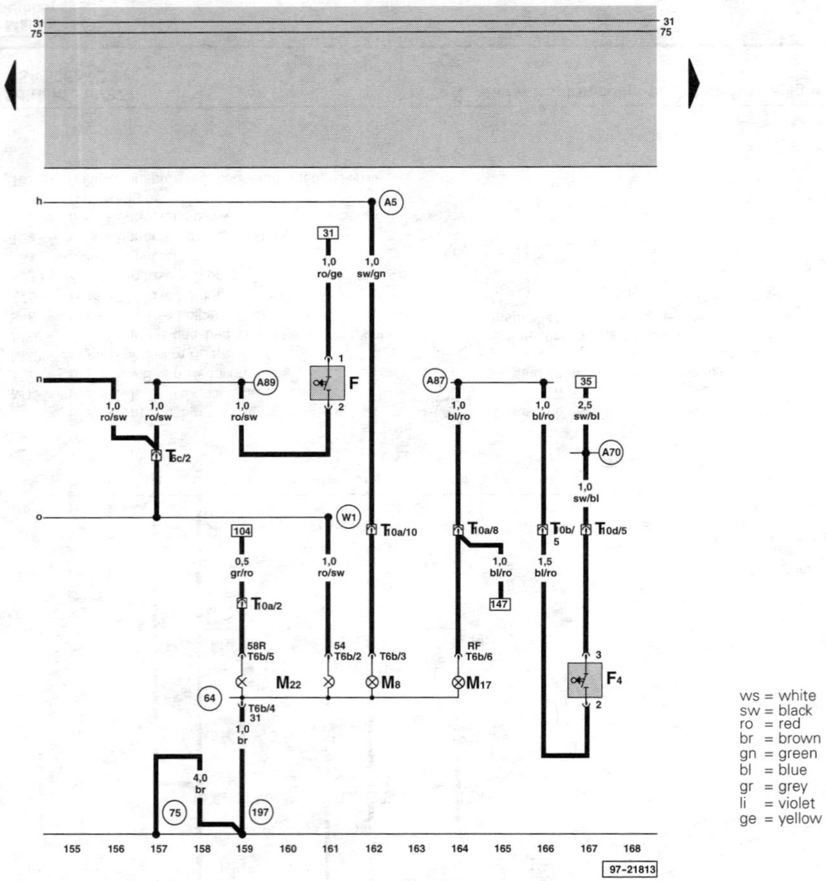

ws = white
sw = black
ro = red
br = brown
gn = green
bl = blue
gr = grey
li = violet
ge = yellow

F - Brake Light Switch
F4 - Back-Up Light Switch
M8 - Right Rear Turn Signal Light
M17 - Right Back-Up Light
M22 - Right Brake/Tail Light
T6b - 6-Pin Connector, on tail light, right
T6c - 6-Pin Connector, blue, connector station A-pillar, left
T10a - 10-Pin Connector, brown, connector station A-pillar, left
T10b - 10-Pin Connector, black, on protective housing for control module, in engine compartment, left
T10d - 10-Pin Connector, brown, on protective housing for control module, in engine compartment, left

(64) - Ground connection, on right taillight bulb holder

(75) - Ground connection, on right rear pillar

(197) - Ground connection -4-, in wiring harness rear

(A5) - plus connection (right turn signal), in instrument panel wiring harness

(A70) - Connector (15a, fuse 231), in instrument panel wiring harness

(A87) - Connector (reverse lamp), in instrument panel wiring harness

(A89) - Connector (54), in instrument panel wiring harness

(W1) - plus connection (54), in rear wiring harness

Brake light switch, back-up light switch, right rear turn signal light, right back-up light, right brake/tail light

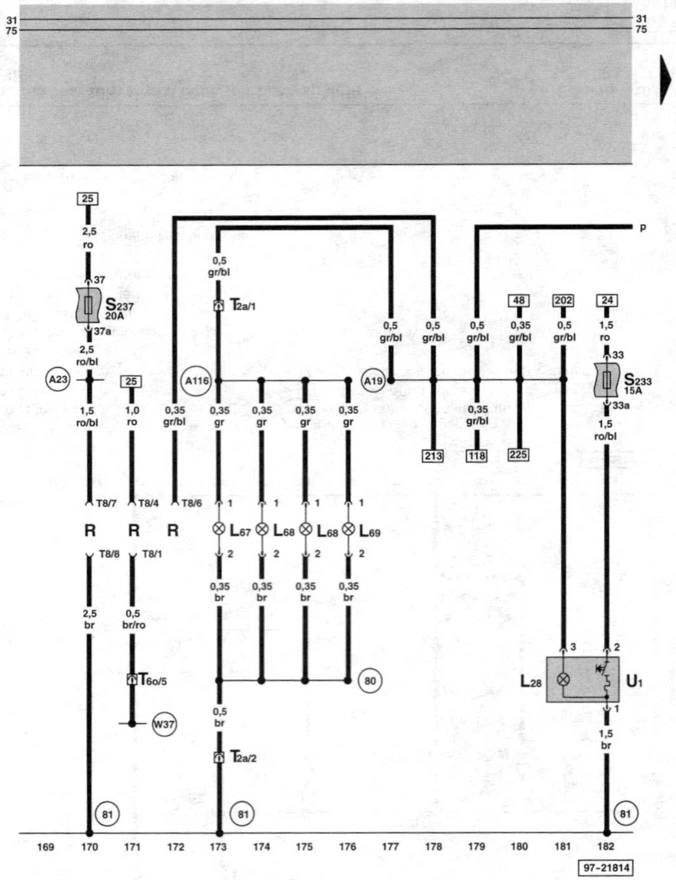

ws = white
sw = black
ro = red
br = brown
gn = green
bl = blue
gr = grey
li = violet
ge = yellow

L28 - Cigarette Lighter Light
L67 - Left Instrument Panel Vent Illumination
L68 - Center Instrument Panel Vent Illumination
L69 - Right Instrument Panel Vent Illumination
R - Radio
S233 - Fuse 33 in fuse holder
S237 - Fuse 37 in fuse holder
T2a - Double Connector
T6o - 6-Pin Connector, white, connector station A-pillar, right
T6n - 6-Pin Connector, red, on protective housing for control module, in engine compartment, left
T8 - 8-Pin Connector, on radio
U1 - Cigarette Lighter

(80) - Ground connection -1-, in instrument panel wiring harness

(81) - Ground connection -1-, in instrument panel wiring harness

(A19) - wire connection (58d), in instrument panel wiring harness

(A23) - wire connection (30al), in instrument panel wiring harness

(A116) - wire connection -2- (58d), in instrument panel wiring harness

(W37) - Connector (alarm system), in wiring harness floor

Radio connection, cigarette lighter, panel vent illumination

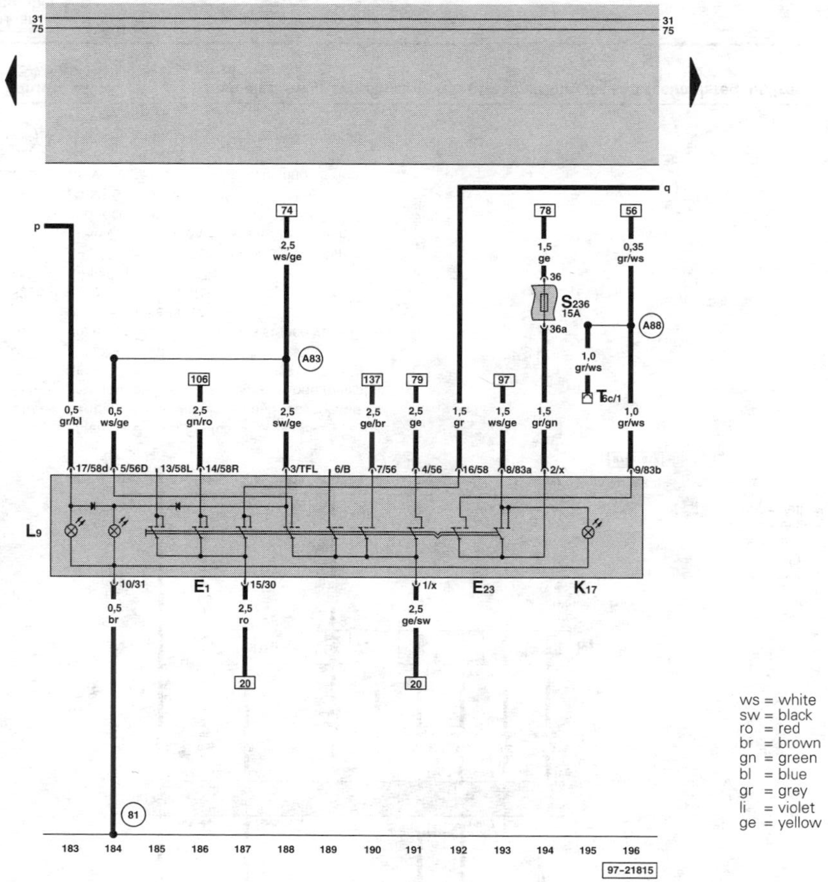

ws = white
sw = black
ro = red
br = brown
gn = green
bl = blue
gr = grey
li = violet
ge = yellow

E1 - Light switch
E23 - Fog Light Switch
K17 - Fog Light Indicator Light
L9 - Headlight Switch Licht
S236 - Fuse 36 in fuse holder
T6c - 6-pin Connector, blue connector station A-pillar, left

(81) - Ground connection -1-, in instrument panel wiring harness
(A83) - Connector (daytime running lights), in instrument panel wiring harness
(A88) - Connector fog light, in instrument panel wiring harness

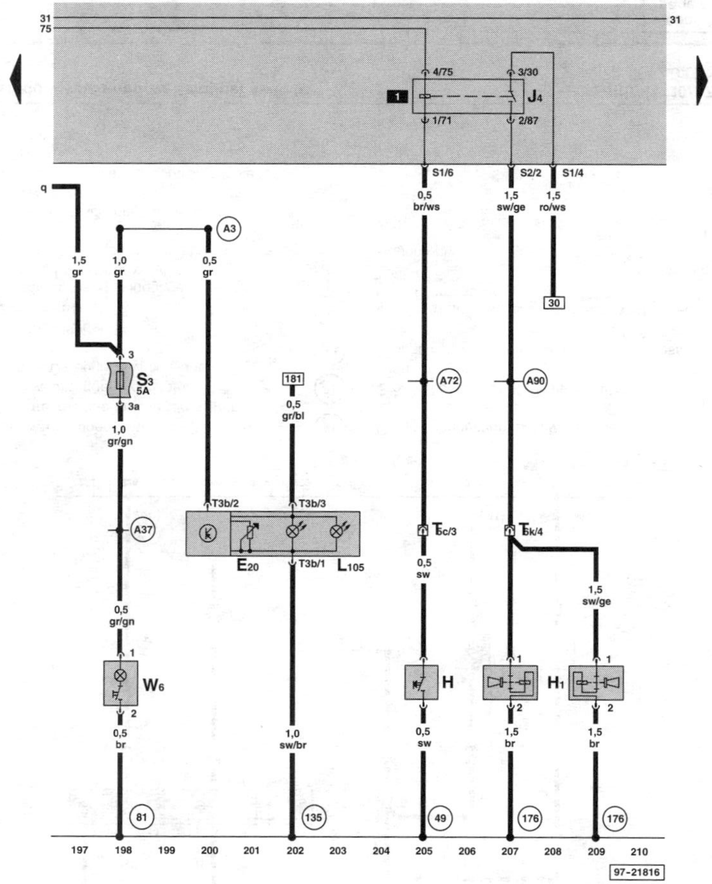

ws = white
sw = black
ro = red
br = brown
gn = green
bl = blue
gr = grey
li = violet
ge = yellow

E20 - Instrument Panel Light Dimmer Switch
H - Horn Button
H1 - Dual Horns
J4 - Dual Horn Relay, production control number (204)
L105 - Illumination for Instrument Panel Light Dimmer Switch
S3 - Fuse 3 in fuse holder
T3b - 3-Pin Connector
T5c - 5-Pin Connector, yellow, near steering column
T6k - 6-Pin Connector, red, connector station A-pillar, right
W6 - Glove Compartment Light

(49) - Ground connection, near steering wheel

(81) - Ground connection -1-, in instrument panel wiring harness
(135) - Ground connection -2-, in instrument panel wiring harness
(176) - Ground connection, in right headlight wiring harness
(A3) - plus connection (58), in instrument panel wiring harness
(A37) - wire connection (58a), in instrument panel wiring harness
(A72) - Connector (71), in instrument panel wiring harness
(A90) - Connector (two-tone horn), in instrument panel wiring harness

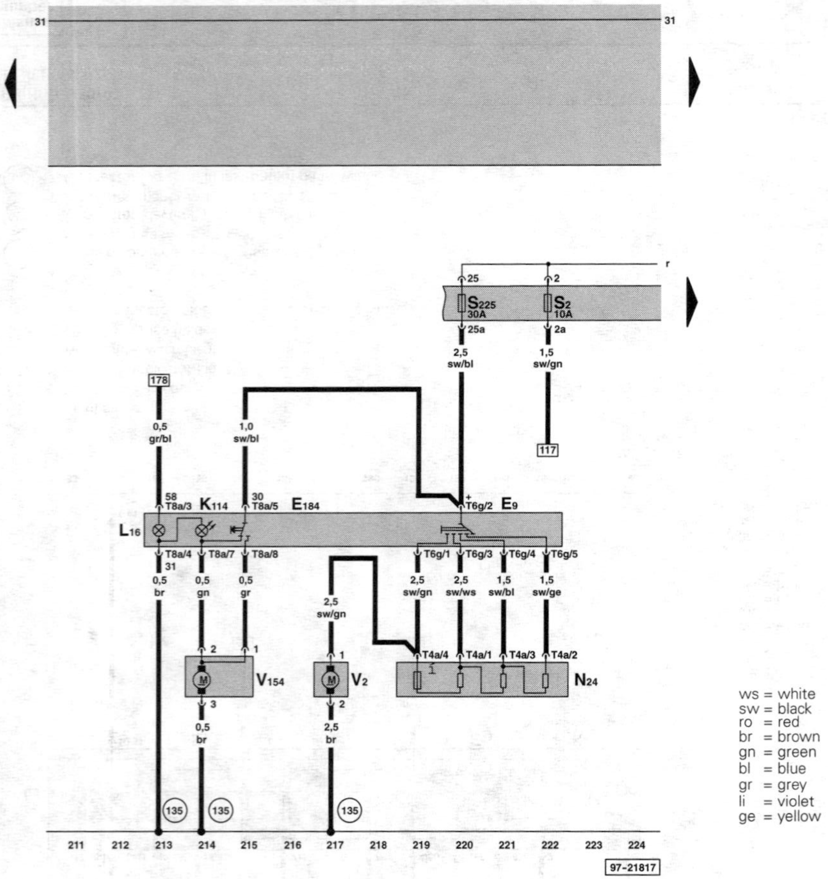

ws = white
sw = black
ro = red
br = brown
gn = green
bl = blue
gr = grey
li = violet
ge = yellow

E9 - Fresh Air Blower Switch
E184 - Switch for fresh air blower and recirculated air
K114 - Fresh Air and Recirculating Air Mode Indicator Light
L16 - Fresh Air Control Lever Light
N24 - Fresh Air Blower Series Resistance With Fuse
S2 - Fuse 2 in fuse holder
S225 - Fuse 25 in fuse holder
T4a - 4-Pin Connector
T6g - 6-Pin Connector, on fresh air blower switch
T8a - 8-Pin Connector, on switch for fresh air blower and recirculated air
V2 - Fresh Air Blower
V154 - Servo motor for fresh-/recirculating air door

(135) - Ground connection -2-, in instrument panel wiring harness

Fresh air blower switch, switch for fresh air blower and recirculated air, fresh air blower

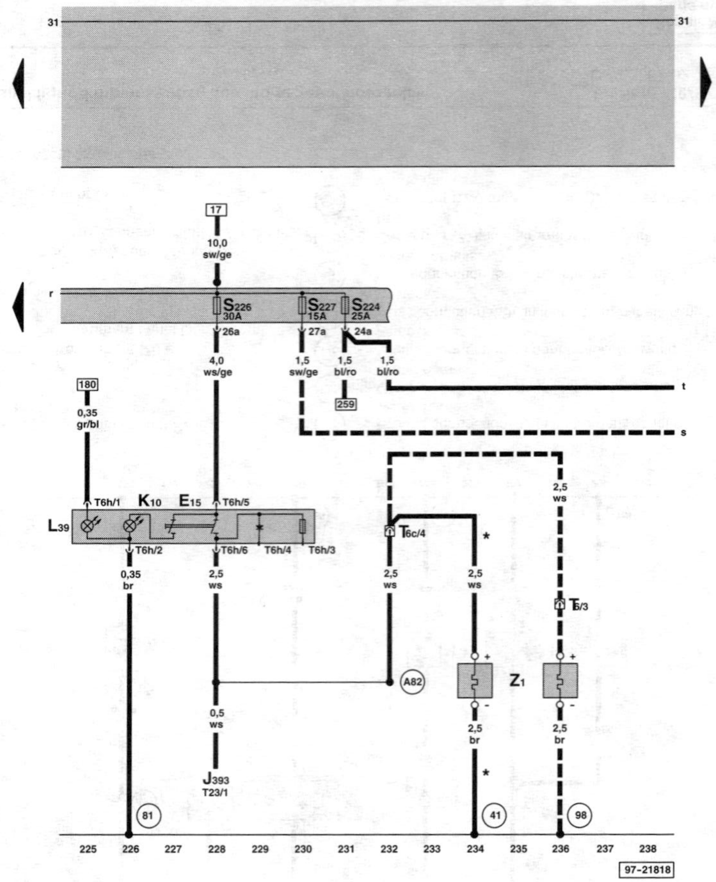

ws = white
sw = black
ro = red
br = brown
gn = green
bl = blue
gr = grey
li = violet
ge = yellow

E15 - Rear window defogger switch
J393 - Central control module for comfort system
K10 - Rear Window Defogger Indicator Light
L39 - Rear Window Defogger Switch Light
S224 - Fuse 24 in fuse holder
S226 - Fuse 26 in fuse holder
S227 - Fuse 27 in fuse holder
T6 - 6-Pin Connector, black, in luggage compartment, left (wagon only)
T6c - 6-Pin Connector, blue, connector station A-pillar, left
T6h - 6-Pin Connector, black, on rear window defogger switch
T23 - 23-Pin Connector, on central control module for comfort system
Z1 - Heated rear window

(41) - Ground connection, below rear shelf
(81) - Ground connection -1-, in instrument panel wiring harness
(98) - Ground connection, in rear lid wiring harness
(A82) - Connector (heated glass), in instrument panel wiring harness

- - - - wagon only
* - sedan only

Heated rear window, rear window defogger switch

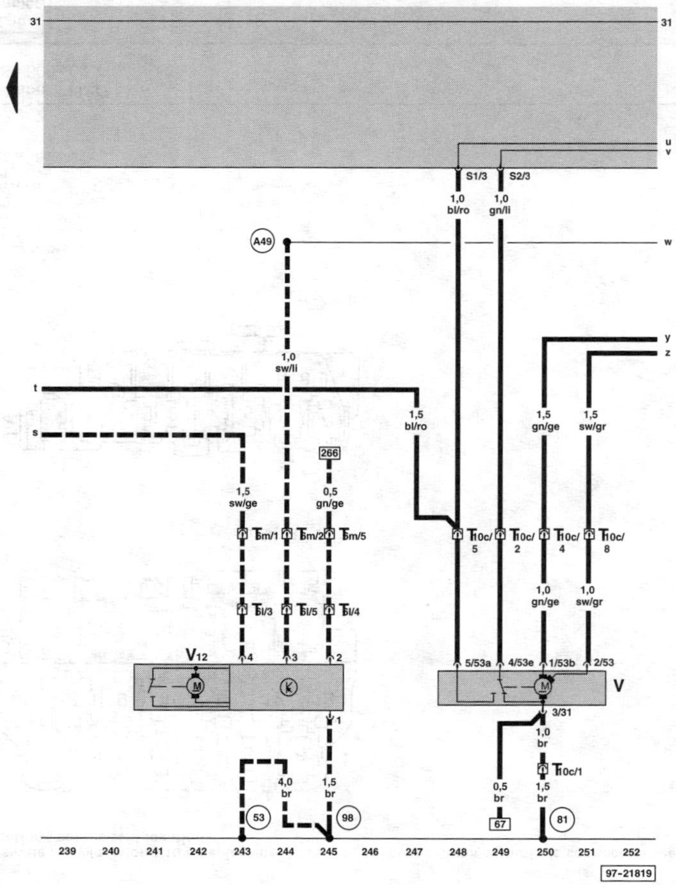

ws = white
sw = black
ro = red
br = brown
gn = green
bl = blue
gr = grey
li = violet
ge = yellow

T6l - 6-Pin Connector, black, in rear lid, right
T6m - 6-Pin Connector, black, connector station A-pillar, right
T10c - 10-Pin Connector, violet, connector station A-pillar, left
V - Windshield Wiper Motor
V12 - Rear Window Wiper Motor

(53) - Ground connection, in rear lid, right

(81) - Ground connection -1-, in instrument panel wiring harness

(98) - Ground connection, in rear lid wiring harness

(A49) - wire connection -1-, in instrument panel wiring harness

- - - wagon only

Edition 10/97
USA.5062.02.12

Windshield wiper motor, rear window wiper motor

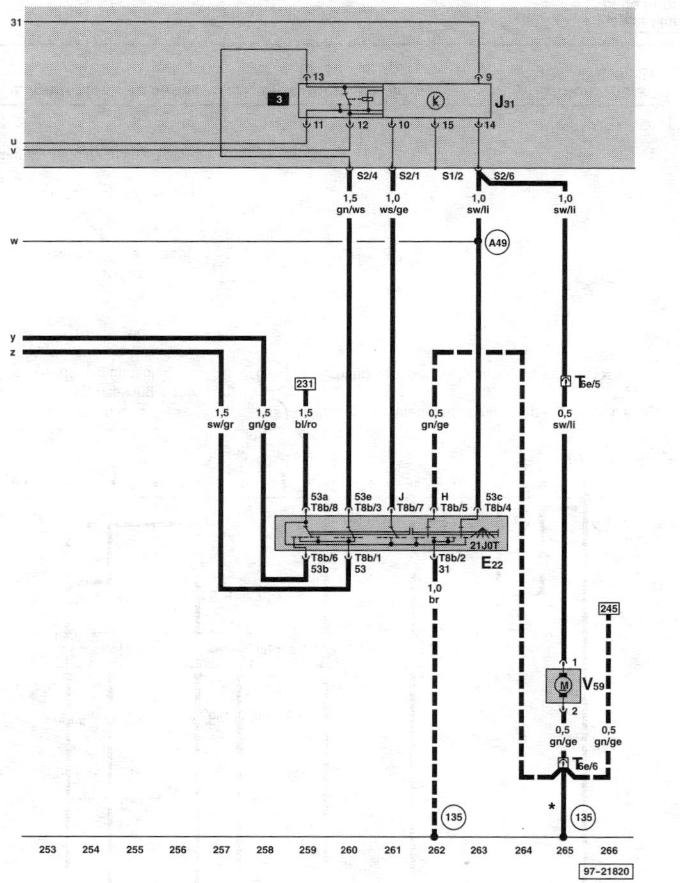

ws = white
sw = black
ro = red
br = brown
gn = green
bl = blue
gr = grey
li = violet
ge = yellow

E22 - Winshield Wiper/Washer Switch
J31 - Wiper/Washer Intermittent Relay, production control number (317)
T6e - 6-Pin Connector, yellow, connector station A-pillar, left
T8b - 8-Pin Connector, on Windshield Wiper/Washer Switch
V59 - Windshield and Rear Window Washer Pump

(135) - Ground connection -2-, in instrument panel wiring harness

(A49) - wire connection -1-, in instrument panel wiring harness

- - - - wagon only
* - sedan only

Windshield wiper/washer switch, wiper/washer intermittent relay, windshield and rear window washer pump

Edtion 10/97
USA.5062.02.21

Airbag Systems, only for cars with green illuminated instrument cluster

from September 1997

Deviate relay location and fuseplacements as well as the locations of multiple connectors see section "component locations".

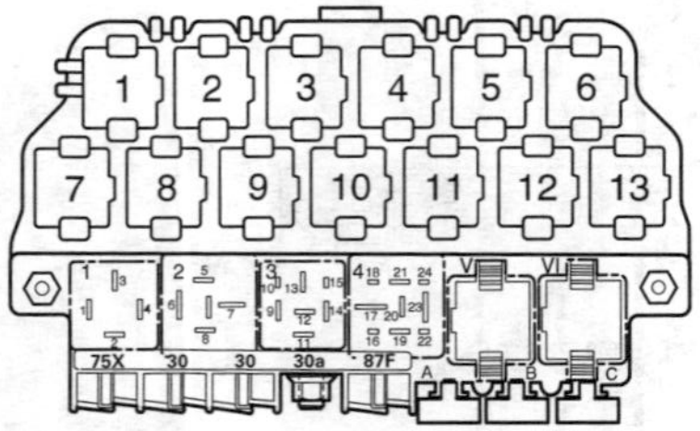

97-20676

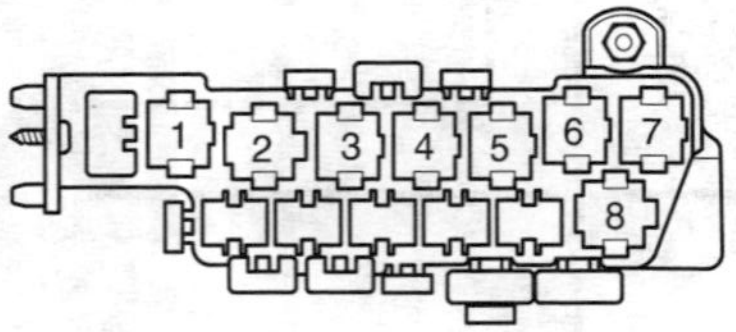

97-20515

Fuse colors
30 A - green
25 A - white
20 A - yellow
15 A - blue
10 A - red
7,5 A - brown
5 A - beige

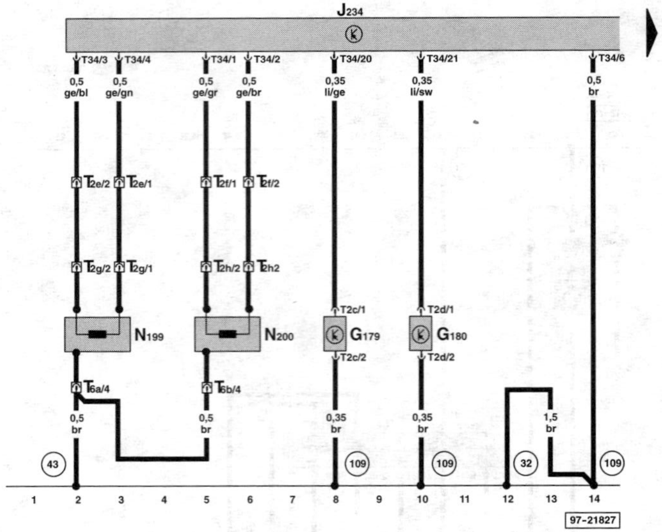

ws = white
sw = black
ro = red
br = brown
gn = green
bl = blue
gr = grey
li = violet
ge = yellow

G179 - Crash sensor for side airbag, driver's side
G180 - Crash sensor for side airbag, passenger side
J234 - Airbag Control Module, behind console
N199 - Igniter for side airbag, driver's side
N200 - Igniter for side airbag, passenger side
T2e - Double Connector
T2f - Double Connector
T2g - Double Connector, below driver's seat
T2h - Double Connector, below passenger's seat
T6a - 6-Pin Connector
T6b - 6-Pin Connector
T34 - 34-Pin Connector, on Airbag Control Module

(32) - Ground connection, behind instrument panel, left

(43) - Ground connection, on right A-pillar, lower part

(109) - Ground connection, in airbag wiring harness

Airbag control module, igniter for side airbag, crash sensor for side airbag

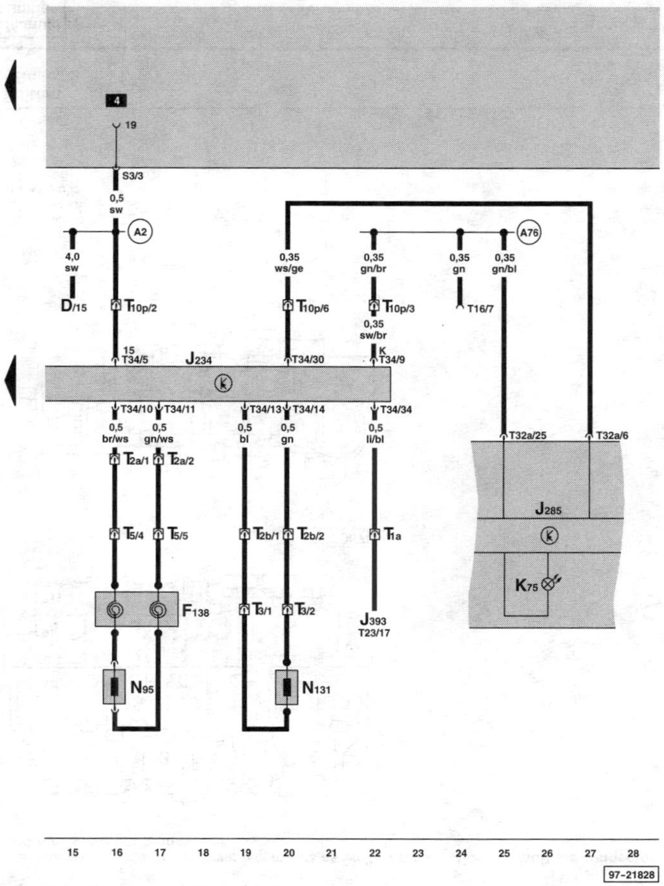

ws = white
sw = black
ro = red
br = brown
gn = green
bl = blue
gr = grey
li = violet
ge = yellow

- D - Ignition/Starter Switch
- F138 - Airbag Spiral Spring/Return Spring With Slip Ring
- J234 - Airbag Control Module, behind console
- J285 - Control module with indicator unit in instrument panel insert
- J393 - Central control module for comfort system
- K75 - Airbag Malfunction Indicator Lamp (MIL)
- N95 - Driver's Side Airbag Igniter
- N131 - Passenger's Side Airbag Igniter 1
- T1a - Single Connector, black, connector station A-pillar, right
- T2a - Double Connector, yellow, connector station A-pillar, left
- T2b - Double Connector, yellow, connector station A-pillar, right
- T3 - 3-Pin Connector, behind instrument panel, right
- T5 - 5-Pin Connector, black, behind steering column switch cover
- T10p - 10-Pin Connector, green, connector station A-pillar, right
- T16 - 16-Pin Connector, On Board Diagnostic (OBD), below instrument panel, left
- T23 - 23-Pin Connector, on central control module for comfort system
- T32a - 32-Pin Connector, blue, on instrument cluster
- T34 - 34-Pin Connector, on airbag control module
- (A2) - plus connection (15), in instrument panel wiring harness
- (A76) - Connector (K-diagnosis wire) in instrument panel wiring harness

Edition 10/97 USA.5062.02.21

Airbag control module, driver's side airbag igniter, passenger's side airbag igniter 1, airbag malfunction indicator lamp (MIL)

THIS PAGE INTENTIONALLY LEFT BLANK

Radio System, only for vehicles with green illuminated instrument cluster

from September 1997

Deviate relay location and fuseplacements as well as the locations of multiple connectors see section "component locations".

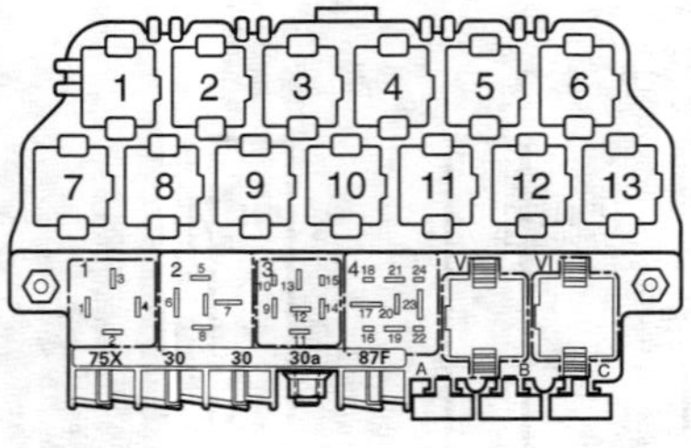

97-20676

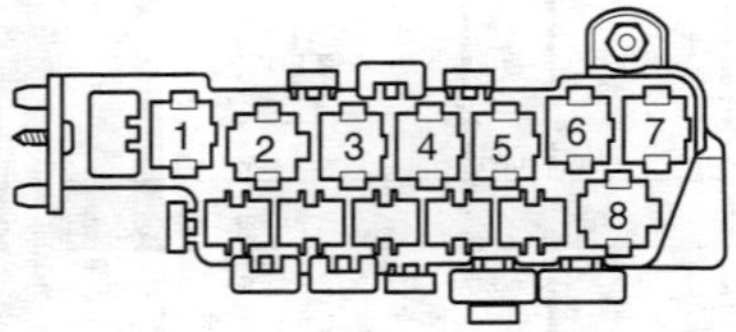

97-20515

Fuse colors
30 A - green
25 A - white
20 A - yellow
15 A - blue
10 A - red
7,5 A - brown
5 A - beige

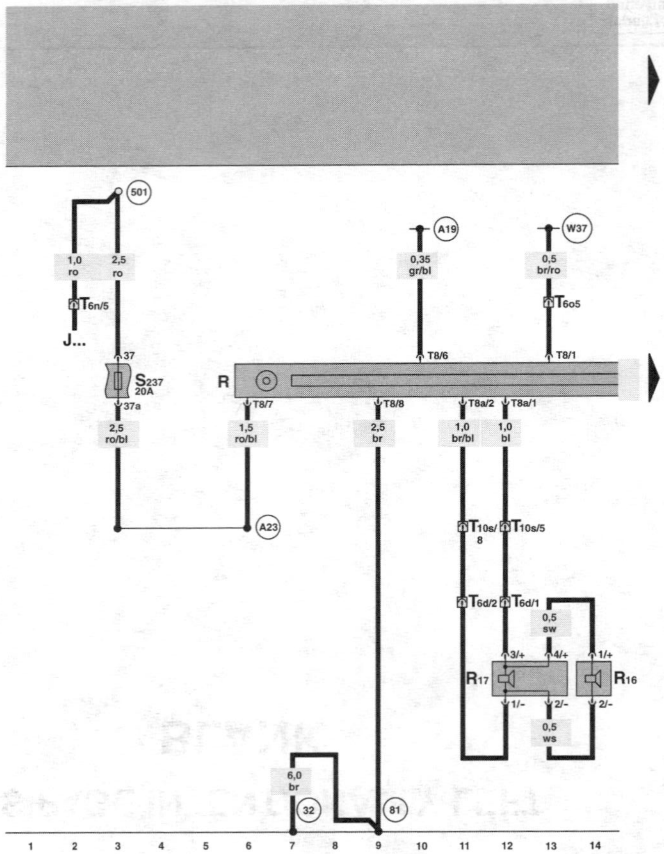

97-21842

ws = white
sw = black
ro = red
br = brown
gn = green
bl = blue
gr = grey
li = violet
ge = yellow

J... - Engine Control Module (ECM)
R - Radio
R16 - Treble speaker, right rear, in right rear door
R17 - Bass speaker, right rear, in right rear door
S237 - Fuse 37 in fuse holder
T6d - 6-Pin Connector, red, on B-pillar, right
T6n - 6-Pin Connector, red, on protective housing for control module
T6o - 6-Pin Connector, white, connector station A-pillar, right
T8 - 8-Pin Connector, black, on radio
T8a - 8-Pin Connector, brown, on radio
T10s - 10-Pin Connector, white, connector station A-pillar, right

(32) - Ground connection, behind instrument panel, left
(81) - Ground connection -1-, in instrument panel wiring harness
(501) - Threaded connection -2- (30) on the relay plate
(A19) - wire connection (58d), in instrument panel wiring harness
(A23) - wire connection (30al), in instrument panel wiring harness
(W37) - Connection (DWA), in floor wiring harness

Radio, right rear speakers

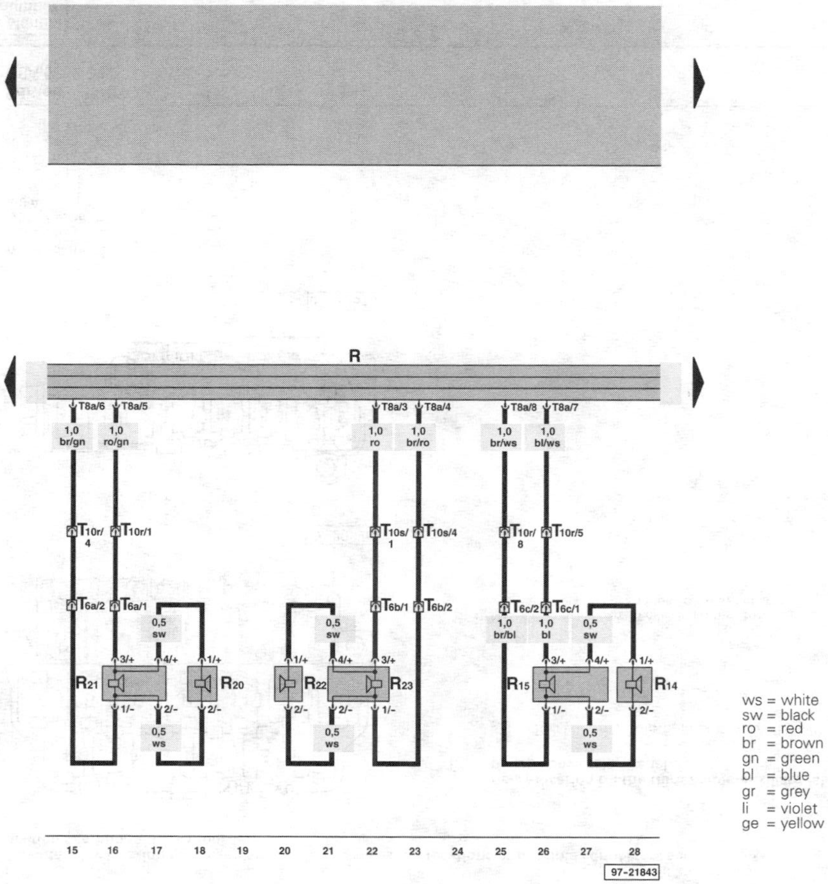

ws = white
sw = black
ro = red
br = brown
gn = green
bl = blue
gr = grey
li = violet
ge = yellow

R - Radio
R14 - Treble speaker, left rear, in left rear door
R15 - Bass speaker, left rear, in left rear door
R20 - Treble speaker, left front, in driver's door
R21 - Bass speaker, left front, in driver's door
R22 - Treble speaker, right front, in passenger's door
R23 - Bass speaker, right front, in passenger's door
T6a - 6-Pin Connector, yellow, connector station A-pillar, left
T6b - 6-Pin Connector, yellow, connector station A-pillar, right
T6c - 6-Pin Connector, red, on B-pillar, left
T8a - 8-Pin Connector, brown, on radio
T10r - 10-Pin Connector, white, connector station A-pillar, left
T10s - 10-Pin Connector, white, connector station A-pillar, right

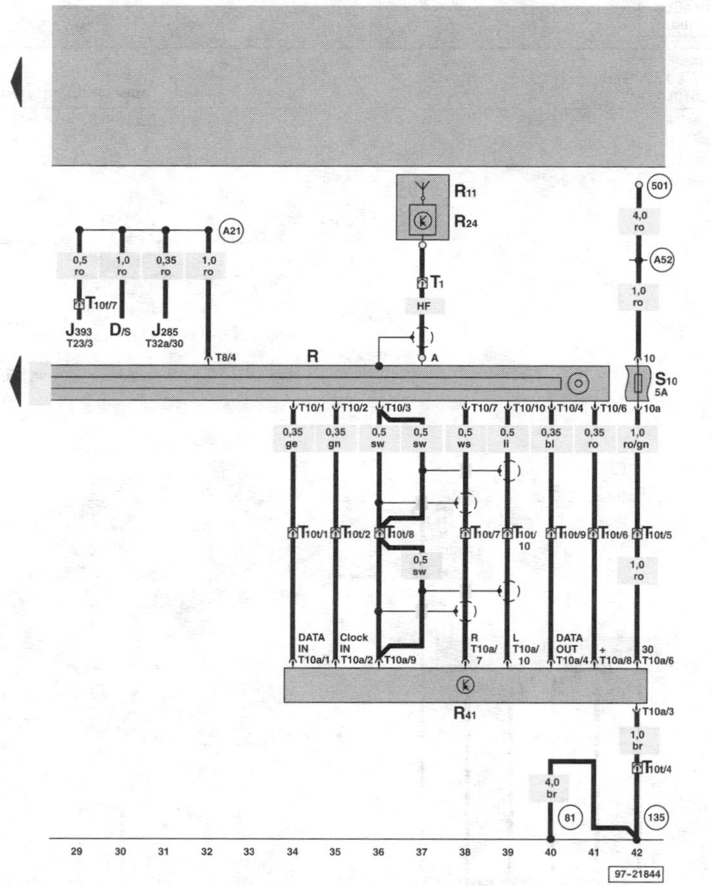

ws = white
sw = black
ro = red
br = brown
gn = green
bl = blue
gr = grey
li = violet
ge = yellow

D - Ignition/Starter Switch
J285 - Control module with indicator unit in instrument panel insert
J393 - Central control module for comfort system
R - Radio
R11 - Antenna
R24 - Antenna Amplifier
R41 - CD Changer Unit, in luggage compartment
S10 - Fuse 10 in fuse holder
T1 - 1-Pin Connector, black, below console
T8 - 8-Pin Connector, black, on radio
T10 - 10-Pin Connector, red, on radio
T10a - 10-Pin Connector, red, on CD Changer Unit
T10f - 10-Pin Connector, blue, connector station A-pillar, left
T10t - 10-Pin Connector, black, connector station A-pillar, right
T23 - 23-Pin Connector, on central control module for comfort system
T32a - 32-Pin Connector, blue, on instrument cluster
(81) - Ground connection -1-, in instrument panel wiring harness
(135) - Ground connection -2-, in instrument panel wiring harness
(501) - Threaded connection -2- (30) on the relay plate
(A21) - wire connection (86s), in instrument panel wiring harness
(A52) - plus connection (30), in instrument panel wiring harness

Comfort System, with Alarm System, only for vehicles with green illuminated instrument cluster

from September 1997

Deviate relay location and fuseplacements as well as the locations of multiple connectors see section "component locations".

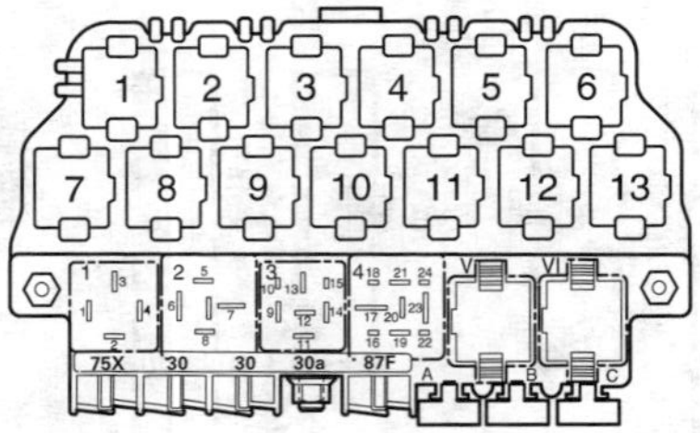

97-20676

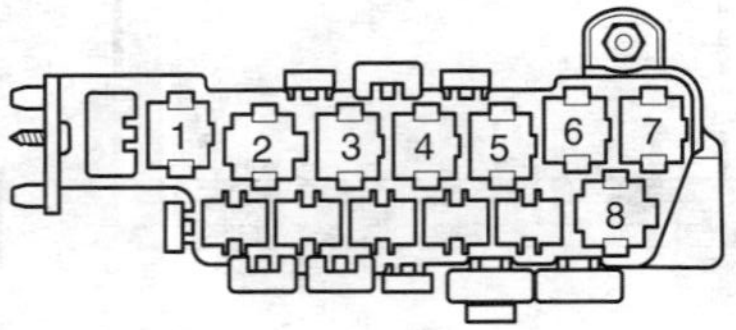

97-20515

Relay location on the thirteenfold auxiliary relay panel, above relay panel:

2 Relay for motor remote unlock rear lid (79)

Note: Number in parentheses indicates production control number stamped on relay housing.

Fuse colors

30 A - green
25 A - white
20 A - yellow
15 A - blue
10 A - red
7,5 A - brown
5 A - beige

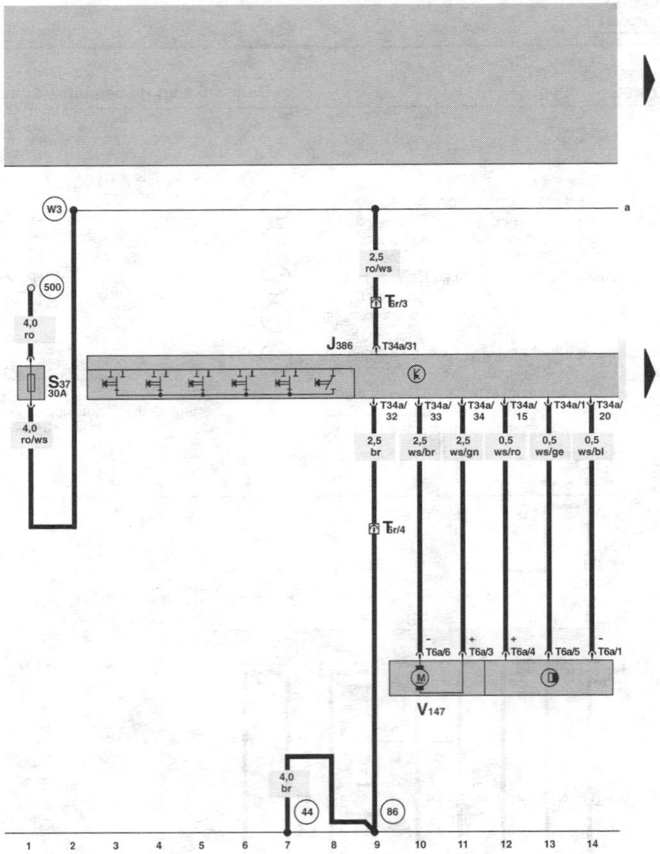

97-21870

ws = white
sw = black
ro = red
br = brown
gn = green
bl = blue
gr = grey
li = violet
ge = yellow

J386 - Door control module, driver side
S37 - Power Window Fuse, on eight position auxiliary relay panel, behind relay panel
T6a - 6-Pin Connector, black
T6r - 6-Pin Connector, brown, connector station A-pillar, left
T34a - 34-Pin Connector
V147 - Motor for window regulator, driver side

(44) - Ground connection, on left A-pillar, lower part

(86) - Ground connection -1-, in rear wiring harness

(500) - Threaded connection -1- (30) on the relay plate

(W3) - wire connection, in rear wiring harness

Door control module, driver side

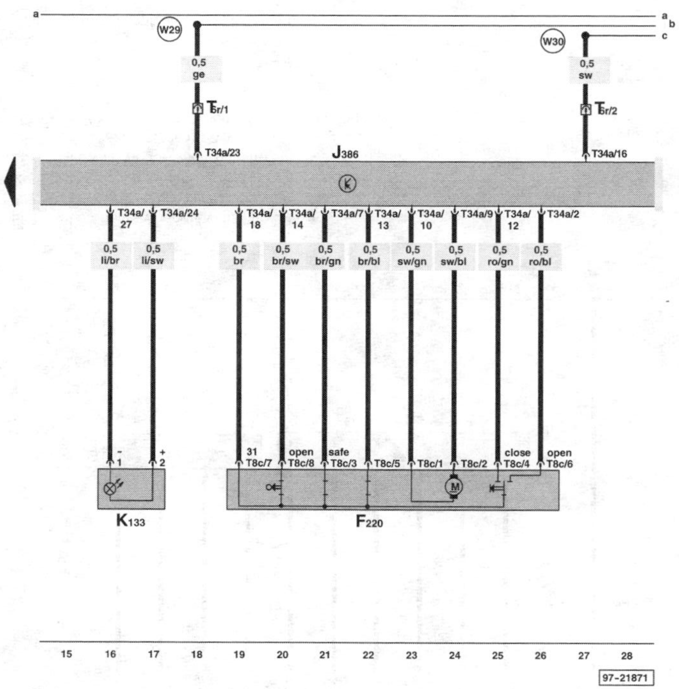

ws = white
sw = black
ro = red
br = brown
gn = green
bl = blue
gr = grey
li = violet
ge = yellow

F220 - Lock unit for central locking, driver side
J386 - Door control module, driver side
K133 - Warning light for central locking -SAFE-
T6r - 6-Pin Connector, brown, connector station A-pillar, left
T8c - 8-Pin Connector, black
T34a - 34-Pin Connector

(W29) - Connection (A-Bus), in floor wiring harness

(W30) - Connection (B-Bus), in floor wiring harness

Edition 10/97
USA.5062.02.21

Central locking, driver's door

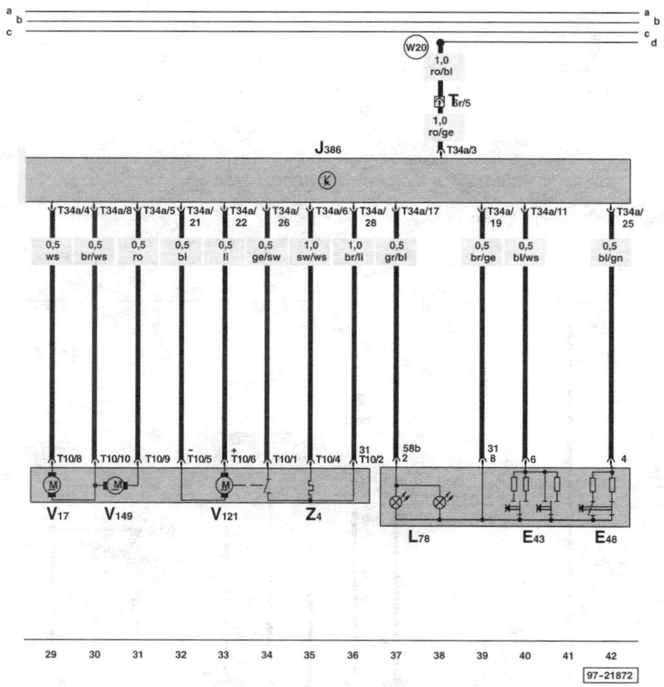

ws = white
sw = black
ro = red
br = brown
gn = green
bl = blue
gr = grey
li = violet
ge = yellow

E43 - Mirror Adjustment Switch
E48 - Mirror Selector Switch
J386 - Door control module, driver side
L78 - Mirror Adjusting Switch Light
T6r - 6-Pin Connector, brown, connector station A-pillar, left
T10 - 10-Pin Connector, black
T34a - 34-Pin Connector
V17 - Driver's Side Mirror Adjustment Motor
V121 - Motor for fold-in mirror, driver side
V149 - Motor for mirror adjustment, driver side
Z4 - Heated outside mirror, driver side

(W20) - Plus connector (30a), in wiring harness, rear

Mirror adjustment and fold-in mirror, driver side, heated outside mirror, driver side

Edition 10/97
USA.5062.02.21

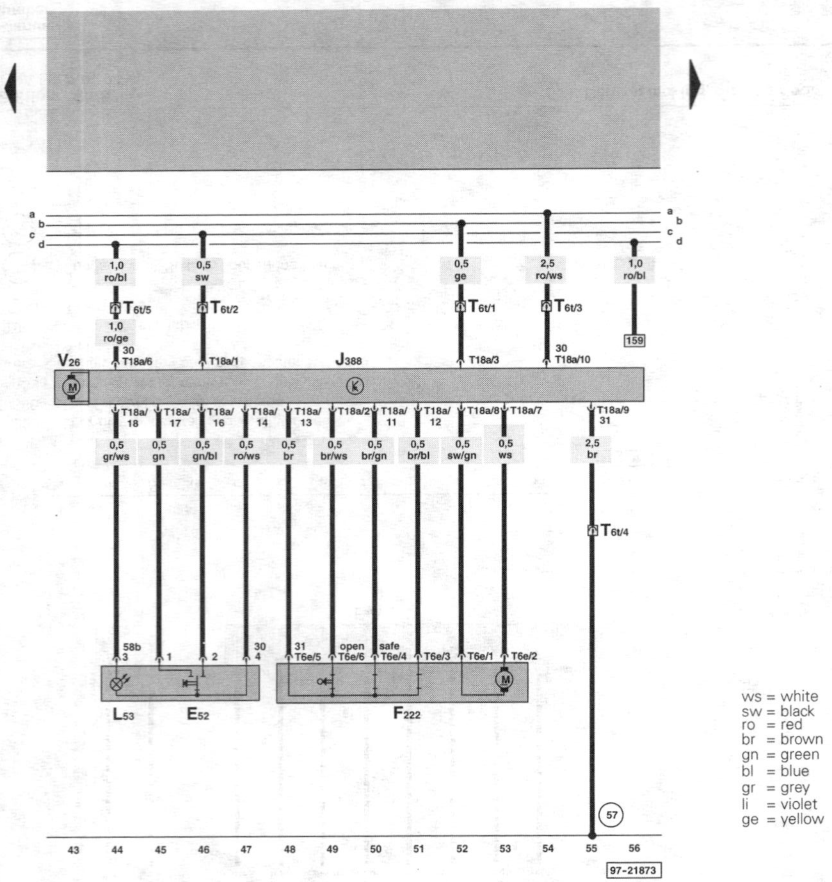

ws = white
sw = black
ro = red
br = brown
gn = green
bl = blue
gr = grey
li = violet
ge = yellow

E52 - Left Rear Window Switch, (In LR Door)
F222 - Lock unit for central locking, rear, left
J388 - Door control module, rear, left
L53 - Power Window Switch Light
T6e - 6-Pin Connector, brown
T6t - 6-Pin Connector, brown, connector station B-pillar, left
T18a - 18-Pin Connector
V26 - Motor for window regulator, rear, left

(57) - Ground connection, on left rear pillar

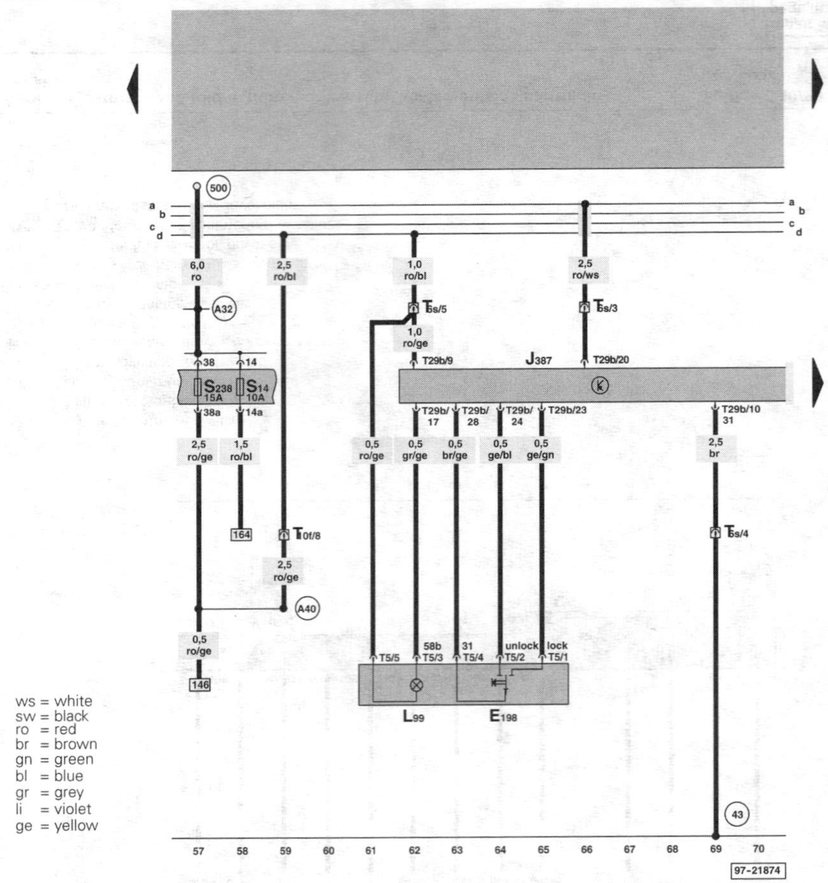

ws = white
sw = black
ro = red
br = brown
gn = green
bl = blue
gr = grey
li = violet
ge = yellow

E198 - Switch for interior lock, passenger side
J387 - Door control module, passenger side
L99 - Lighting for switch interior lock
S14 - Fuse 14 in fuse holder
S238 - Fuse 38 in fuse holder
T5 - 5-Pin Connector
T10f - 10-Pin Connector, blue, connector station A-pillar, left
T6s - 10-Pin Connector, brown, connector station A-pillar, right
T29b - 29-Pin Connector

(43) - Ground connection, on right A-pillar, lower part

(500) - Threaded connection -1- (30) on the relay plate

(A32) - plus connection (30), in instrument panel wiring harness

(A40) - plus connection -1- (30), in instrument panel wiring harness

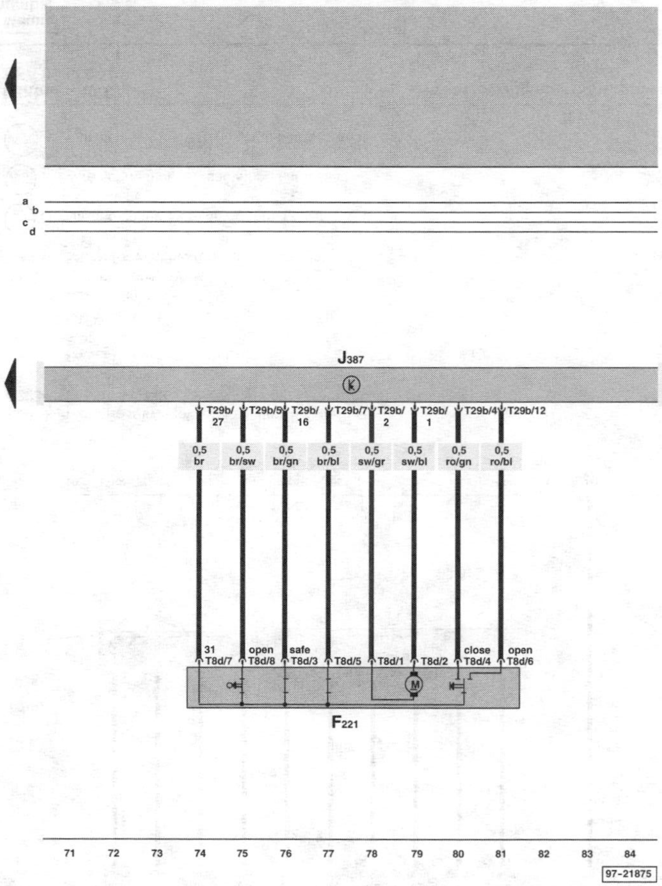

ws = white
sw = black
ro = red
br = brown
gn = green
bl = blue
gr = grey
li = violet
ge = yellow

F221 - Lock unit for central locking, passenger side
J387 - Door control module, passenger side
T8d - 8-Pin Connector, black
T29b - 29-Pin Connector

Edition 10/97
USA.5062.02.21

Central locking passenger's door

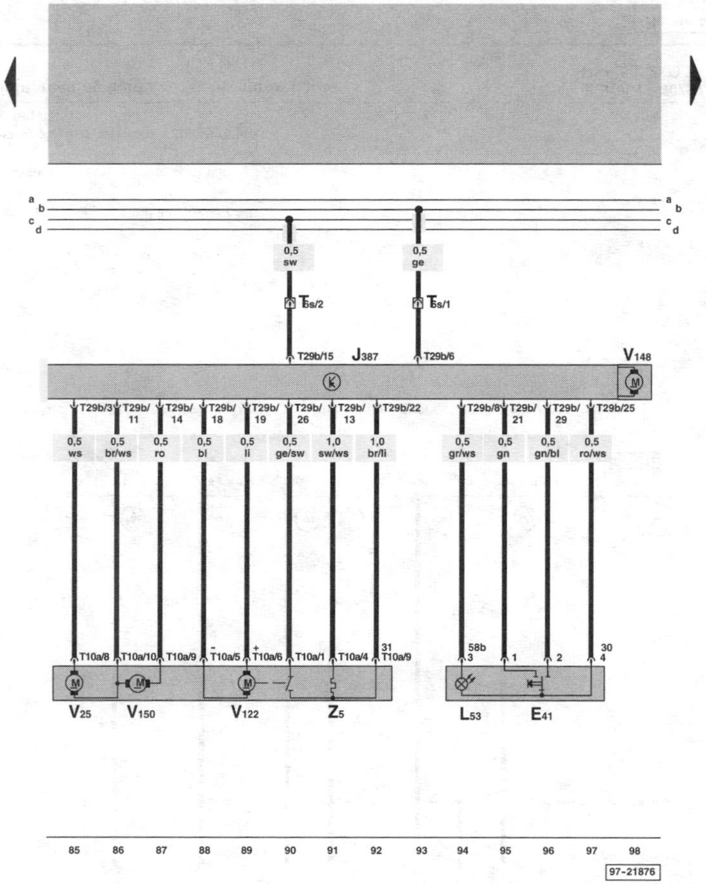

ws = white
sw = black
ro = red
br = brown
gn = green
bl = blue
gr = grey
li = violet
ge = yellow

E41 - Right Front Window Switch
J387 - Door control module, passenger side
L53 - Power Window Switch Light
T6s - 6-Pin Connector, brown, connector station A-pillar, right
T10a - 10-Pin Connector, black
T29b - 29-Pin Connector
V25 - Passenger's Side Mirror Adjustment Motor
V122 - Motor for fold-in mirror, passenger side
V148 - Motor for window regulator, passenger side
V150 - Motor for mirror adjustment, passenger side
Z5 - Heated outside mirror, passenger side

Mirror adjustment and fold-in mirror, passenger side , heated outside mirror, passenger side

Edition 10/97
USA.5062.02.21

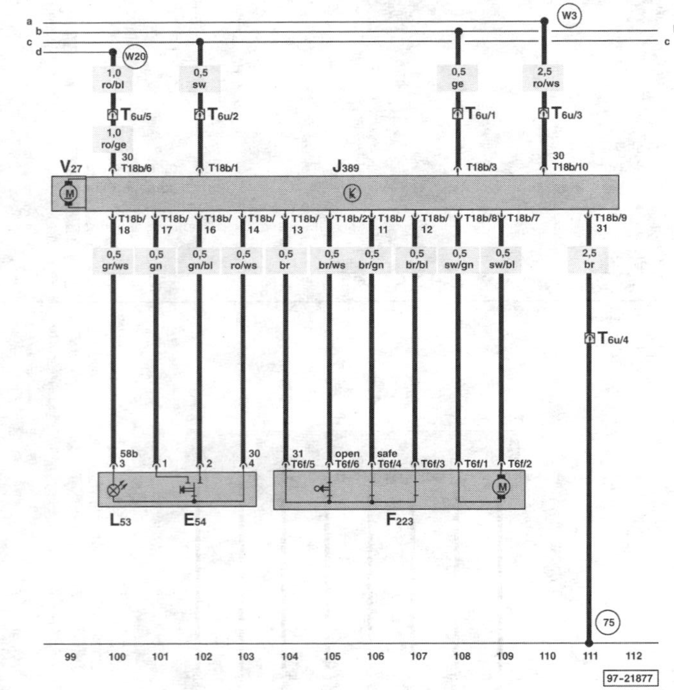

ws = white
sw = black
ro = red
br = brown
gn = green
bl = blue
gr = grey
li = violet
ge = yellow

E54 - Right Rear Window Switch, (In RR Door)
F223 - Lock unit for central locking, rear, right
J389 - Door control module, rear, right
L53 - Power Window Switch Light
T6f - 6-Pin Connector
T6u - 6-Pin Connector, brown, connector station B-pillar, right
T18b - 18-Pin Connector
V27 - Motor for window regulator, rear, right

(75) - Ground connection, on right rear pillar

(W3) - wire connection, in rear wiring harness

(W20) - Plus connector (30a), in wiring harness, rear

Central locking rear right, power window rear right

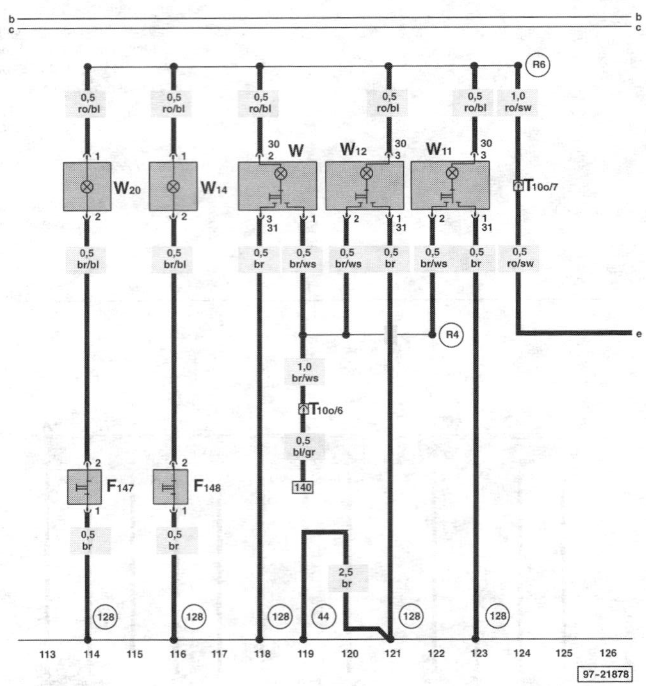

ws = white
sw = black
ro = red
br = brown
gn = green
bl = blue
gr = grey
li = violet
ge = yellow

F147 - Left Make-Up Mirror Light Switch
F148 - Right Make-Up Mirror Light Switch
T10o - 10-Pin Connector, red, connector station A-pillar, left
W - Front Interior Light
W11 - Left Rear Reading Light
W12 - Right Rear Reading Light
W14 - Right Make-up Mirror Light
W20 - Left Make-up Mirror Light

(44) - Ground connection, on left A-pillar, lower part

(128) - Ground connection -1-, in interior light wiring harness

(R4) - wire connection -1-, in interior light/door contact switch wiring harness

(R6) - plus connection -1-, in interior light wiring harness

Front interior light, reading lights, make-up mirror lights

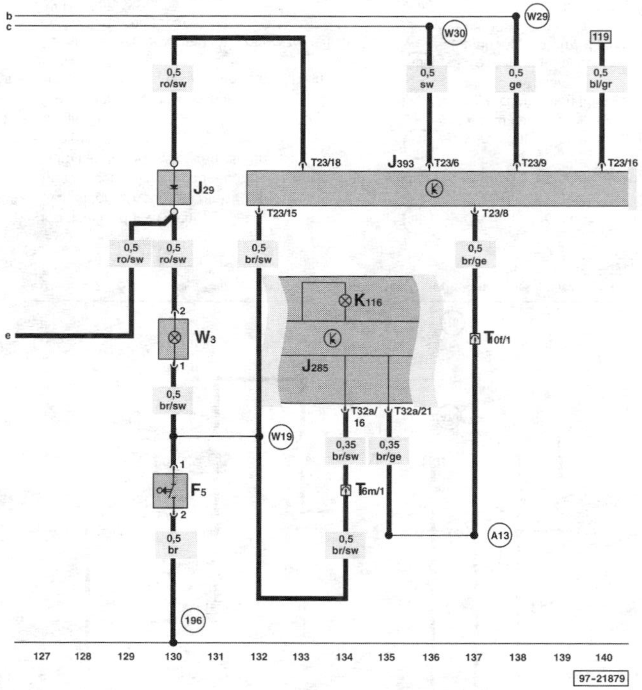

ws = white
sw = black
ro = red
br = brown
gn = green
bl = blue
gr = grey
li = violet
ge = yellow

F5 - Luggage Compartment Light Switch
J29 - Blocking Diode, in wiring harness floor, on C-pillar, left, lower part
J285 - Control module with indicator unit in instrument panel insert
J393 - Central control module for comfort system
K116 - Warning light for rear lid unlocked
T6m - 6-Pin Connector, black, connector station A-pillar, right
T10f - 10-Pin Connector, blue, connector station A-pillar, left
T23 - 23-Pin Connector
T32a - 32-Pin Connector, blue, on instrument cluster
W3 - Luggage compartment Light

(196) - Ground connection -3-, in wiring harness, rear
(A13) - wire connection (door contact switch), in instrument panel wiring harness
(W19) - Connection -2- (luggage compartment light), in rear wiring harness
(W29) - Connection (A-Bus), in floor wiring harness
(W30) - Connection (B-Bus), in floor wiring harness

Edition 10/97
USA.5062.02.21

Central control module for comfort system, warning light for rear lid unlocked, luggage compartment light

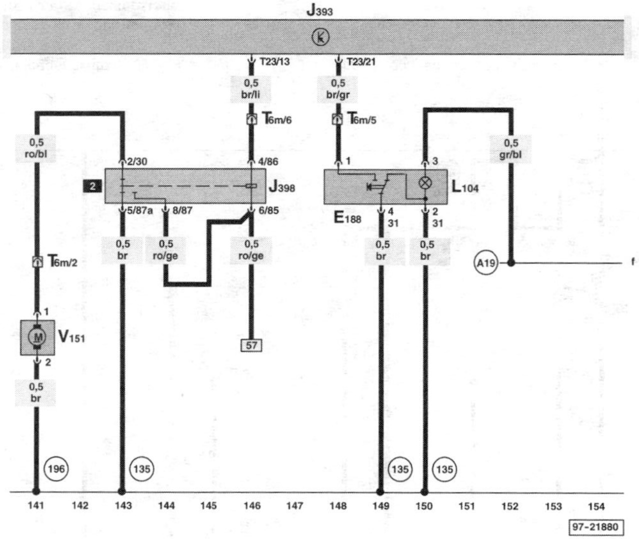

ws = white
sw = black
ro = red
br = brown
gn = green
bl = blue
gr = grey
li = violet
ge = yellow

E188 - Switch for remote unlock, rear lid
J393 - Central control module for comfort system
J398 - Relay for motor remote unlock rear lid, above relay panel, production control number (79)
L104 - Illumination for switch remote unlock
T6m - 6-Pin Connector, black, connector station A-pillar, right
T23 - 23-Pin Connector
V151 - Motor for remote unlock rear lid

(135) - Ground connection -2-, in instrument panel wiring harness
(196) - Ground connection -3-, in wiring harness, rear
(A19) - wire connection (58d), in instrument panel wiring harness

Central control module for comfort system, relay for motor remote unlock rear lid, motor for remote unlock rear lid

Edition 10/97
USA.5062.02.21

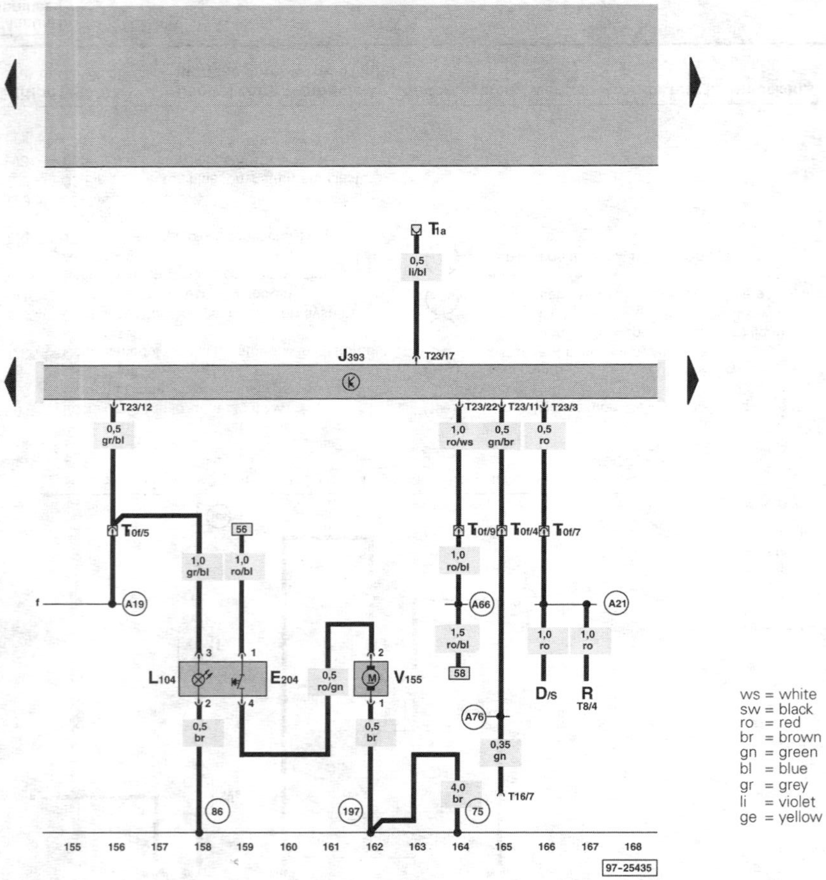

ws = white
sw = black
ro = red
br = brown
gn = green
bl = blue
gr = grey
li = violet
ge = yellow

D - Ignition/Starter Switch
E204 - Switch for remote/fuel tank door
J393 - Central control module for comfort system
L104 - Illumination for switch remote unlock
R - Radio
T1a - Single Connector
T8 - 8-Pin Connector, black, on radio
T10f - 10-Pin Connector, blue, connector station A-pillar, left
T16 - 16-Pin Connector, Data Link Connector (DLC), below instrument panel, left
T23 - 23-Pin Connector
V155 - Motor for fuel tank lid unlock

(75) - Ground connection, on right rear pillar

(86) - Ground connection -1-, in rear wiring harness

(197) - Ground connection -4-, in wiring harness rear

(A19) - wire connection (58d), in instrument panel wiring harness

(A21) - wire connection (86s), in instrument panel wiring harness

(A66) - Connector (30a, central locking/anti-theft warning system/IR), in instrument panel wiring harness

(A76) - Connector (K-diagnosis wire), in instrument panel wiring harness

Central control module for comfort system, Switch for remote/fuel tank door

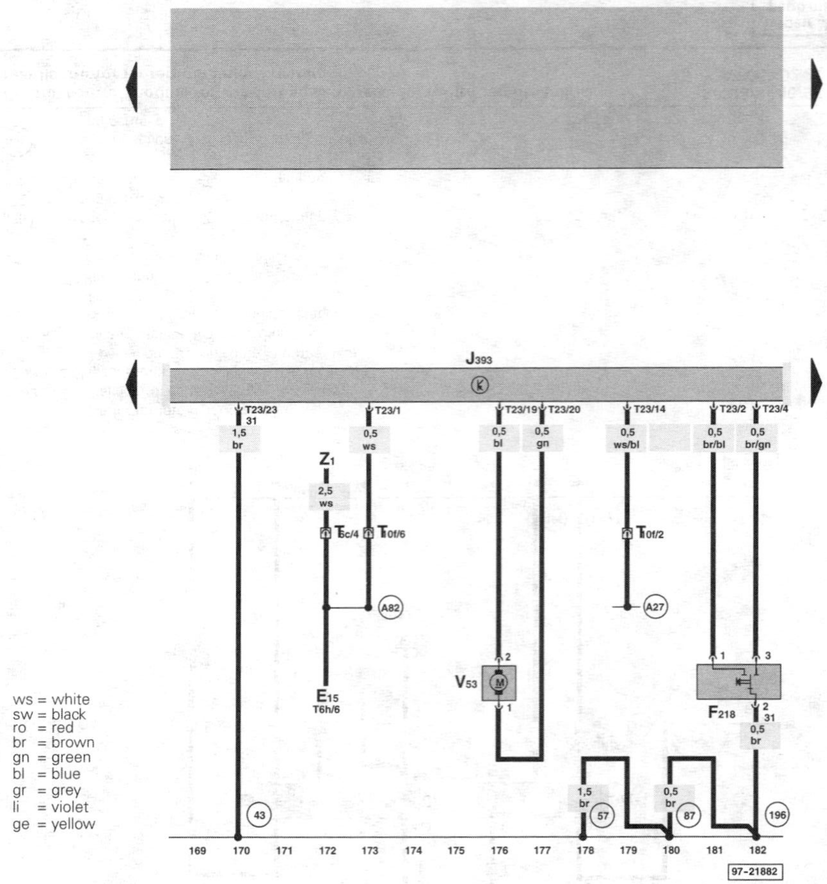

ws = white
sw = black
ro = red
br = brown
gn = green
bl = blue
gr = grey
li = violet
ge = yellow

E15 - Rear window defogger switch
F218 - Switch for central locking, trunk lid
J393 - Central control module for comfort system
T6c - 6-Pin Connector, blue, connector station A-pillar, left
T6h - 6-Pin Connector, black, on rear window defogger switch -E15-
T10f - 10-Pin Connector, blue, connector station A-pillar, left
T23 - 23-Pin Connector
V53 - Decklid Central Locking System Motor
Z1 - Heated rear window

(43) - Ground connection, on right A-pillar, lower part

(57) - Ground connection, on left rear pillar

(87) - Ground connection -2-, in rear wiring harness

(196) - Ground connection -3-, in wiring harness, rear

(A27) - wire Connection (vehicle speed signal), in instrument panel wiring harness

(A82) - Connector (heated glass), in instrument panel wiring harness

Central control module for comfort system, central locking for trunk lid

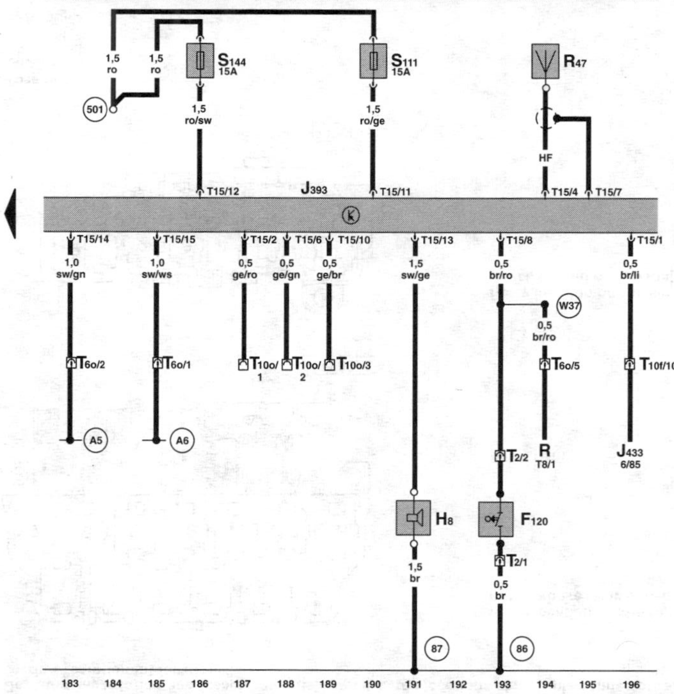

ws = white
sw = black
ro = red
br = brown
gn = green
bl = blue
gr = grey
li = violet
ge = yellow

97-30005

F120 - Hood Alarm Switch
H8 - Alarm Horn
J393 - Central control module for comfort system
J433 - Lock relay for starter (theft warning system), above relay panel
R - Radio
R47 - Antenna wire for central locking and anti-theft warning system, on A-pillar right
S111 - Fuse -1- for Alarm System, on eight position auxiliary relay panel, behind relay panel
S144 - Fuse -2- for Alarm System, on eight position auxiliary relay panel, behind relay panel
T6o - 6-Pin Connector, white, connector station A-pillar, right
T2 - Double Connector
T8 - 8-Pin Connector, on radio
T10f - 10-Pin Connector, blue, connector station A-pillar, left
T10o - 10-Pin Connector, red, connector station A-pillar, left
T15 - 15-Pin Connector, on Central control module for comfort system
(86) - Ground connection -1-, in rear wiring harness
(87) - Ground connection -2-, in rear wiring harness
(501) - Threaded connection -2- (30) on the relay plate
(A5) - plus connection (right turn signal), in instrument panel wiring harness
(A6) - plus connection (left turn signal), in instrument panel wiring harness
(W37) - Connection (alarm), in floor wiring harness

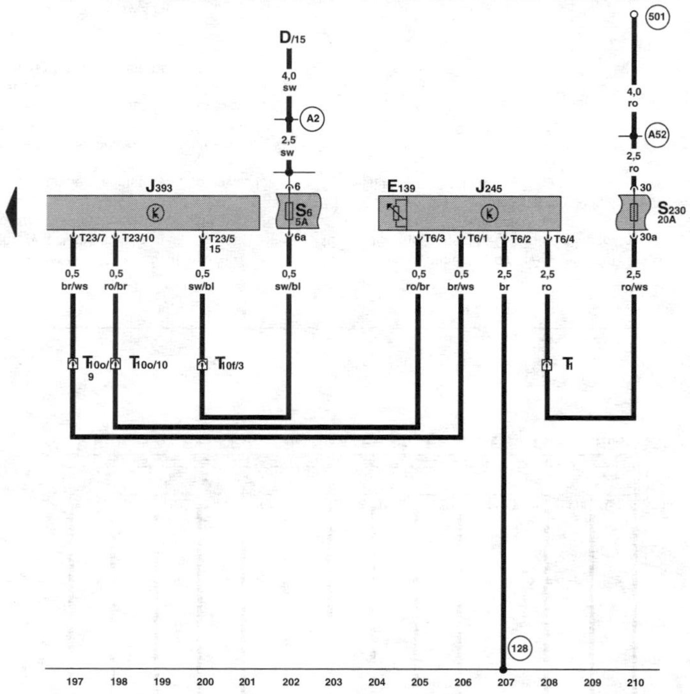

ws = white
sw = black
ro = red
br = brown
gn = green
bl = blue
gr = grey
li = violet
ge = yellow

97-21884

D - Ignition/Starter Switch
E139 - Sunroof Regulator
J245 - Power Sunroof Control Module
J393 - Central control module for comfort system
S6 - Fuse 6 in fuse holder
S230 - Fuse 30 in fuse holder
T1 - Single Connector, black
T4k - 4-Pin Connector
T6 - 6-Pin Connector
T10f - 10-Pin Connector, blue, connector station A-pillar, left
T10o - 10-Pin Connector, red, connector station A-pillar, left
T23 - 23-Pin Connector
(128) - Ground connection -1-, in interior light wiring harness
(501) - Threaded connection -2- (30) on the relay plate
(A2) - plus connection (15), in instrument panel wiring harness
(A52) - plus connection (30), in instrument panel wiring harness

Power seats and mirror adjustment with Memory

from November 1997

Deviate relay location and fuseplacements as well as the locations of multiple connectors see section "component locations".

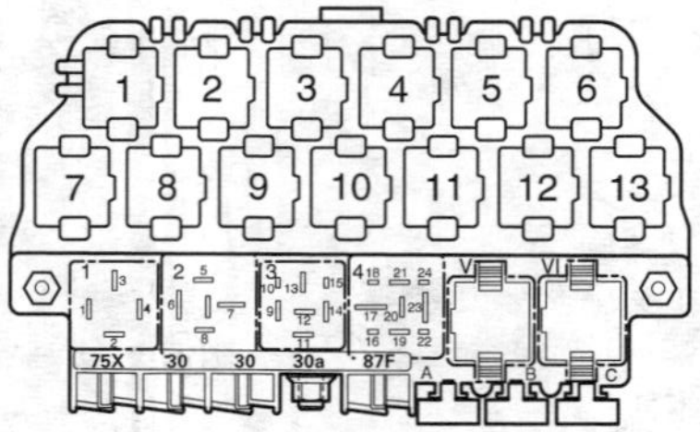

97-20676

Relay location on the thirteenfold auxiliary relay panel, above relay panel:

Relay panel:

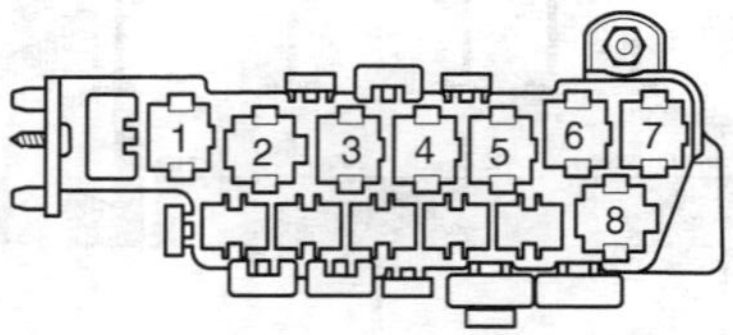

97-20515

Relay location on the eightfold auxiliary relay panel, behind relay panel:

Fuse colors

30 A - green
25 A - white
20 A - yellow
15 A - blue
10 A - red
7,5 A - brown
5 A - beige

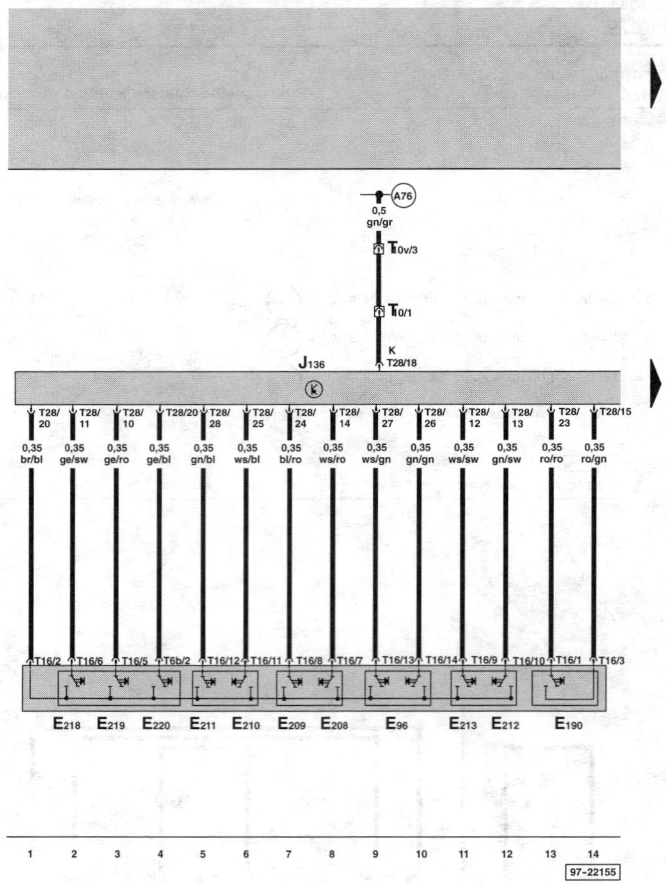

ws = white
sw = black
ro = red
br = brown
gn = green
bl = blue
gr = grey
li = violet
ge = yellow

E96 - Driver's Backrest Adjustment Switch
E190 - Seat memory lock-out switch
E208 - Button for front height adjustment, upwards, driver seat
E209 - Button for front height adjustment, downwards, driver seat
E210 - Button for rear height adjustment, upwards, driver seat
E211 - Button for rear height adjustment, downwards, driver seat
E213 - Button for length adjustment, backward, driver seat
E218 - Button for driver 1, memory seat
E219 - Button for driver 2, memory seat
E220 - Button for driver 3, memory seat
J136 - Memory Seat Control Module
T10 - 10-Pin Connector, brown, near driver's seat
T10v - 10-Pin Connector, pink, A-pillar, right
T16 - 16-Pin Connector, On Board Diagnostic (OBD), below instrument panel, left
(A76) - Connector (K-diagnosis wire), in instrument panel wiring harness

Memory seat control module, buttons for front height adjustment, buttons for front rear adjustment, buttons for length adjustment

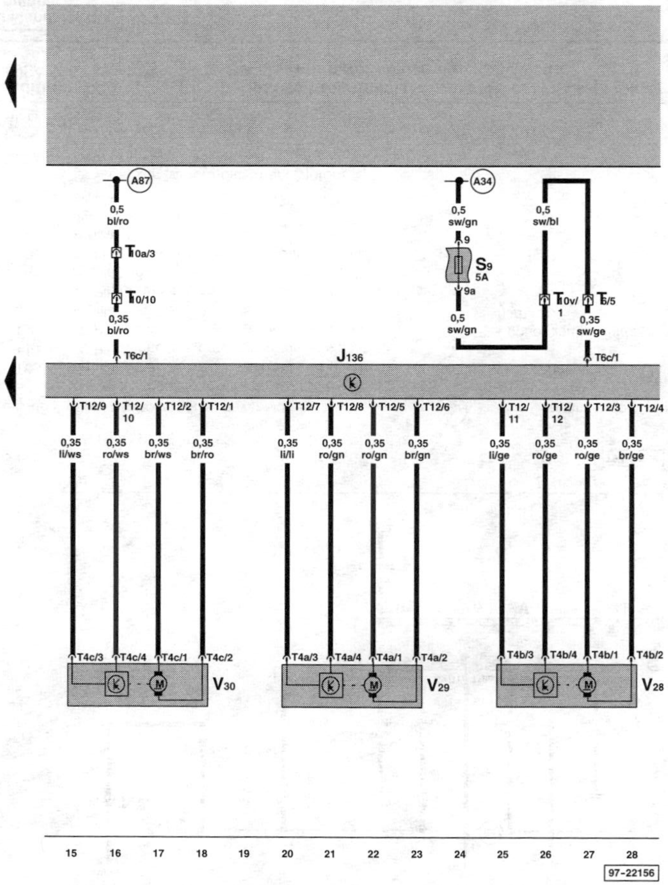

ws = white
sw = black
ro = red
br = brown
gn = green
bl = blue
gr = grey
li = violet
ge = yellow

- J136 - Memory Seat Control Module
- S9 - Fuse 9 in fuse holder
- T4a - 4-Pin Connector
- T4b - 4-Pin Connector
- T4c - 4-Pin Connector
- T6 - 6-Pin Connector, red, near driver's seat
- T6c - 6-Pin Connector
- T10 - 10-Pin Connector, brown, near driver's seat
- T10a - 10-Pin Connector, grey, connector station A-pillar, left
- T10v - 10-Pin Connector, pink, on A-pillar, right
- T12 - 12-Pin Connector
- V28 - Driver's Seat Fore/Aft Adjusting Motor
- V29 - Driver's Seat Front Height Adjusting Motor
- V30 - Driver's Seat Rear Height Adjusting Motor
- (A34) - wire connection (75x), in instrument panel wiring harness
- (A87) - Connector (reverse lamp), in instrument panel wiring harness

Memory seat control module, driver's seat adjusting motors

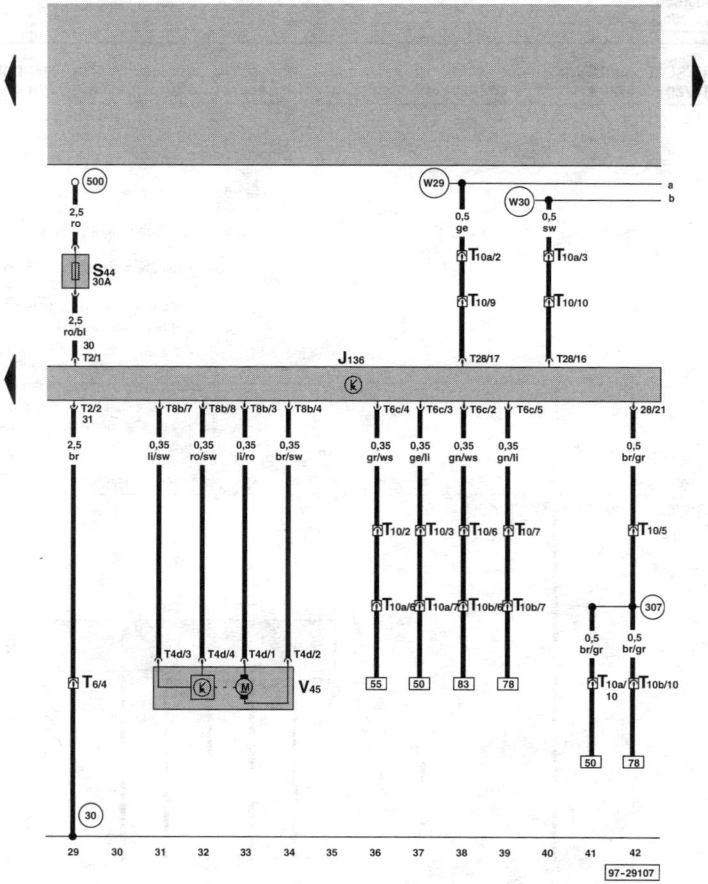

ws = white
sw = black
ro = red
br = brown
gn = green
bl = blue
gr = grey
li = violet
ge = yellow

- J136 - Memory Seat Control Module
- S44 - Power (Memory) Seat Circuit Breaker
- T2 - Double Connector
- T4d - 4-Pin Connector, black
- T6 - 6-Pin Connector
- T6c - 6-Pin Connector
- T8b - 8-Pin Connector
- T10 - 10-Pin Connector, brown
- T10a - 10-Pin Connector, grey, connector station A-pillar, left
- T10b - 10-Pin Connector, green, connector station A-pillar, left
- V45 - Driver's Backrest Adjusting Motor
- (30) - Ground connection, -1-, beside fuse/relay panel
- (307) - Ground connection for sensor (floor wiring harness)
- (500) - Threaded connection -1- (30) on the relay plate
- (W29) - Plus connector -2- (A-Bus), in wiring harness, floor
- (W30) - Plus connector -2- (B-Bus), in wiring harness, floor

Memory seat control module, driver's backrest adjusting motor

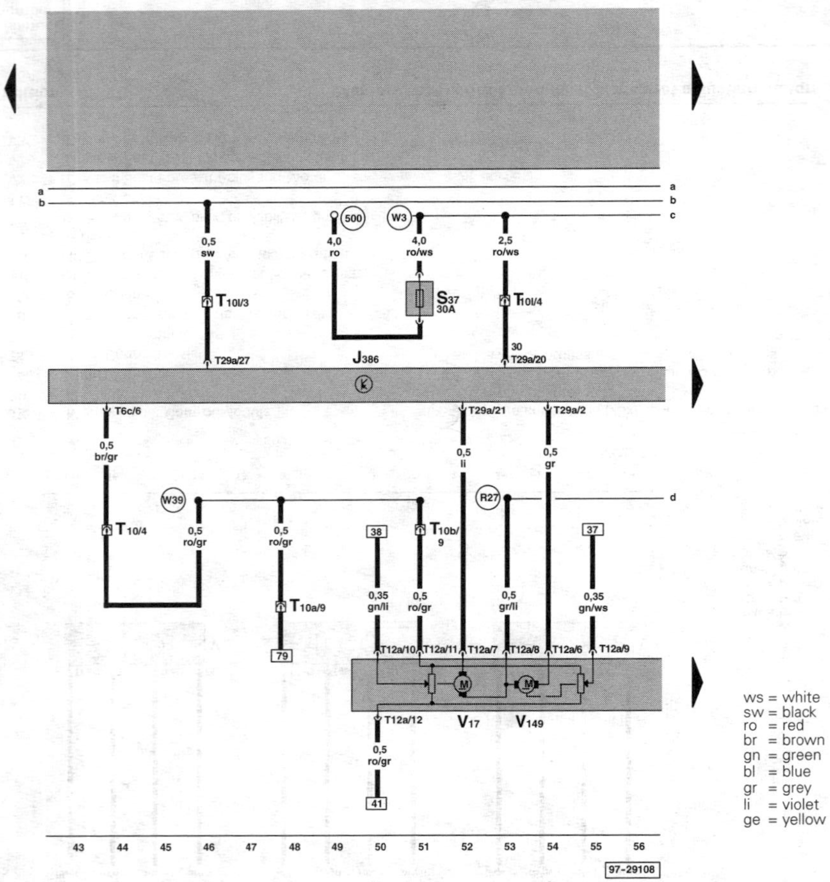

ws = white
sw = black
ro = red
br = brown
gn = green
bl = blue
gr = grey
li = violet
ge = yellow

J386 - Door control module, driver side
S37 - Power Window Fuse
T6c - 6-Pin Connector
T10 - 10-Pin Connector, brown, near driver's seat
T10a - 10-Pin Connector, grey, connector station A-pillar, left
T10b - 10-Pin Connector, grey, connector station A-pillar, right
T10l - 10-Pin Connector, brown, connector station A-pillar, left
T12a - 12-Pin Connector
T29a - 29-Pin Connector
V17 - Driver's Side Mirror Adjustment Motor
V149 - Motor for mirror adjustment, driver side

(500) - Threaded connection -1- (30) on the relay plate

(R27) - wire Connection (power mirror), in driver's door wiring harness
(W3) - wire connection, in rear wiring harness
(W39) - Connection for mirror memory (floor wiring harness)

Door control module, driver side, driver's side mirror adjustment motor, motor for mirror adjustment, driver side

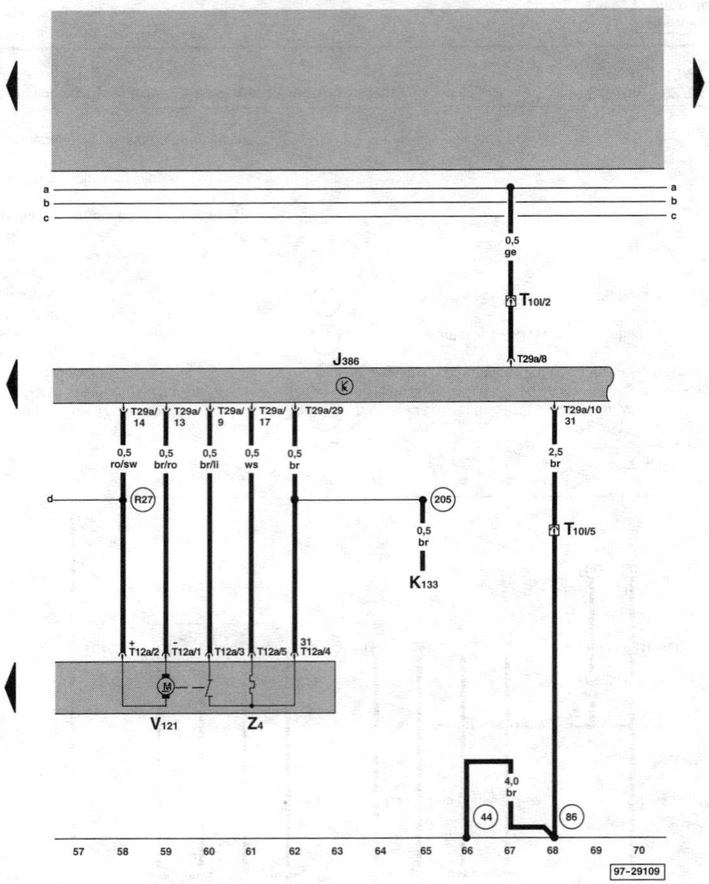

ws = white
sw = black
ro = red
br = brown
gn = green
bl = blue
gr = grey
li = violet
ge = yellow

J386 - Door control module, driver side
K133 - Warning light for central locking -SAFE-
T10l - 10-Pin Connector, brown, A-pillar, left
T12a - 12-Pin Connector
T29a - 29-Pin Connector
V121 - Motor for fold-in mirror, driver side
Z4 - Heated outside mirror, driver side

(44) - Ground connection, on left A-pillar, lower part

(86) - Ground connection -1-, in rear wiring harness

(205) - Ground connection, in drivers door wiring harness

(R27) - wire Connection (power mirror), in driver's door wiring harness

Door control module, driver side, motor for fold-in mirror, driver side, heated outside mirror, driver side

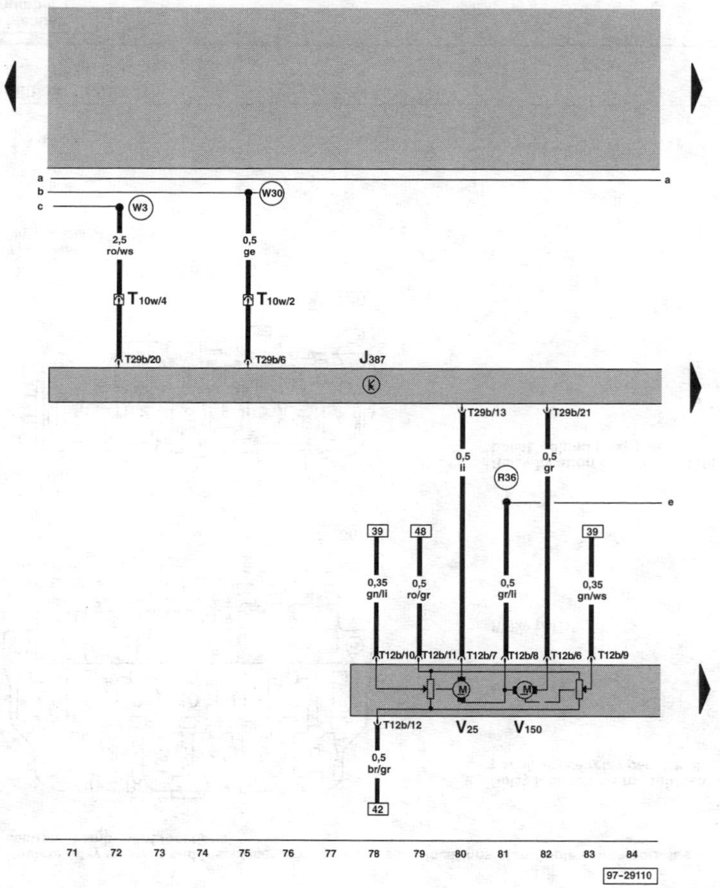

ws = white
sw = black
ro = red
br = brown
gn = green
bl = blue
gr = grey
li = violet
ge = yellow

J387 - Door control module, passenger side
T10w - 10-Pin Connector, black, connector station A-pillar, right
T12b - 12-Pin Connector
T29b - 29-Pin Connector
V25 - Passenger's Side Mirror Adjustment Motor
V150 - Motor for mirror adjustment, passenger side

(R36) - connection (mirror adjustment), in door wiring harness, passenger side
(W3) - wire connection, in rear wiring harness
(W30) - Plus connector -2- (B-Bus), in wiring harness, floor

Edition 02/01
USA.5062.13.21

Door control module, passenger side, passenger's side mirror adjustment motor, motor for mirror adjustment, passenger side

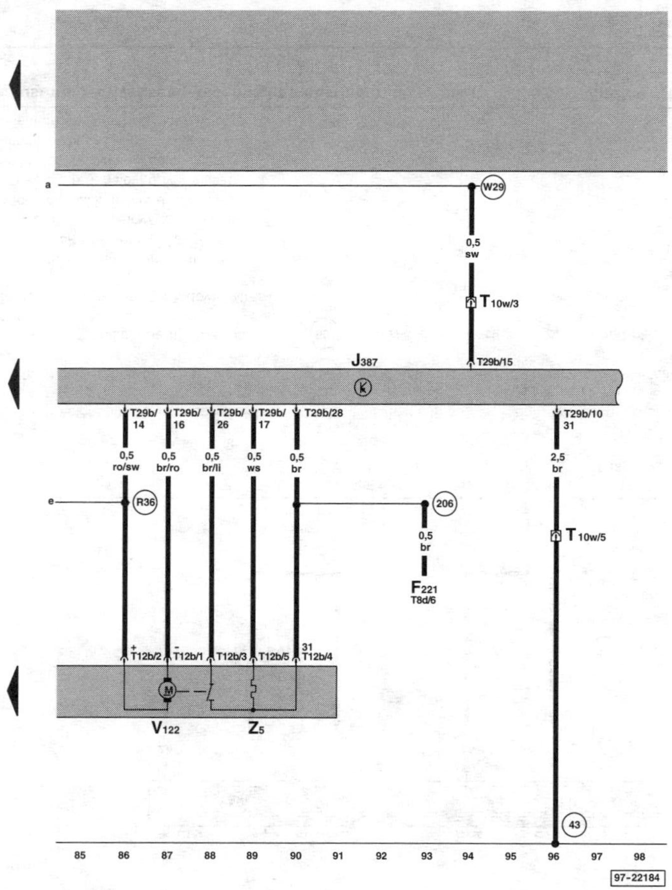

ws = white
sw = black
ro = red
br = brown
gn = green
bl = blue
gr = grey
li = violet
ge = yellow

F221 - Lock unit for central locking, passenger side
J387 - Door control module, passenger side
T8d - 8-Pin Connector, black, on lock unit for central locking, passenger side
T10w - 10-Pin Connector, black, connector station A-pillar, right
T12b - 12-Pin Connector
T29b - 29-Pin Connector
V122 - Motor for fold-in mirror, passenger side
Z5 - Heated outside mirror, passenger side

(43) - Ground connection, on right A-pillar, lower part
(206) - Ground connection, in passenger's door wiring harness
(R36) - connection (mirror adjustment), in door wiring harness, passenger side

(W29) - connection (A-Bus), in floor wiring harness

Door control module, motor for fold-in mirror, passenger side, heated outside mirror, passenger side

Edition 02/01
USA.5062.13.21

Power seats without Memory

from November 1997

Deviate relay location and fuseplacements as well as the locations of multiple connectors see section "component locations".

Relay location on the thirteenfold auxiliary relay panel, above relay panel:

Relay panel:

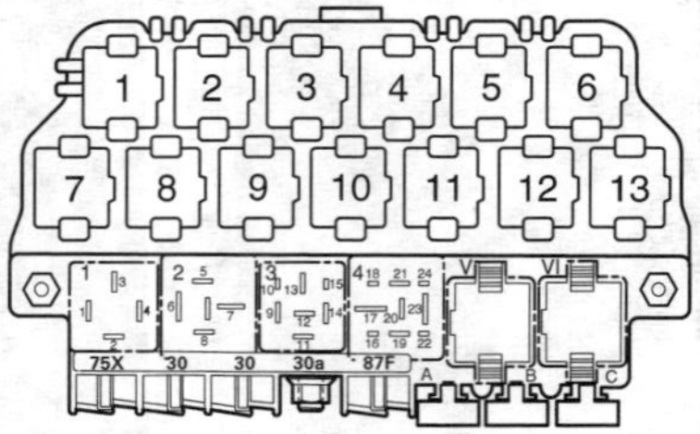

97-20676

Relay location on the eightfold auxiliary relay panel, behind relay panel:

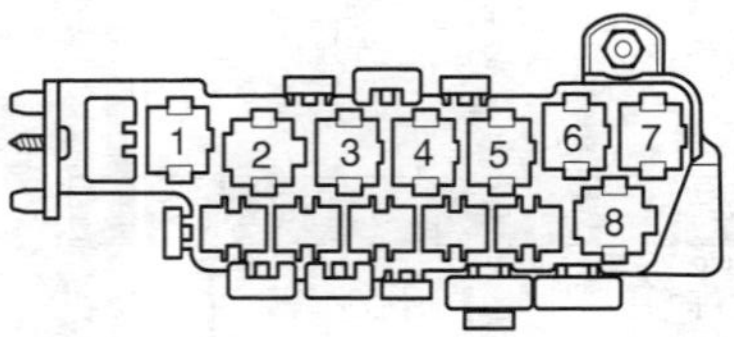

97-20515

Fuse colors

30 A - green
25 A - white
20 A - yellow
15 A - blue
10 A - red
7,5 A - brown
5 A - beige

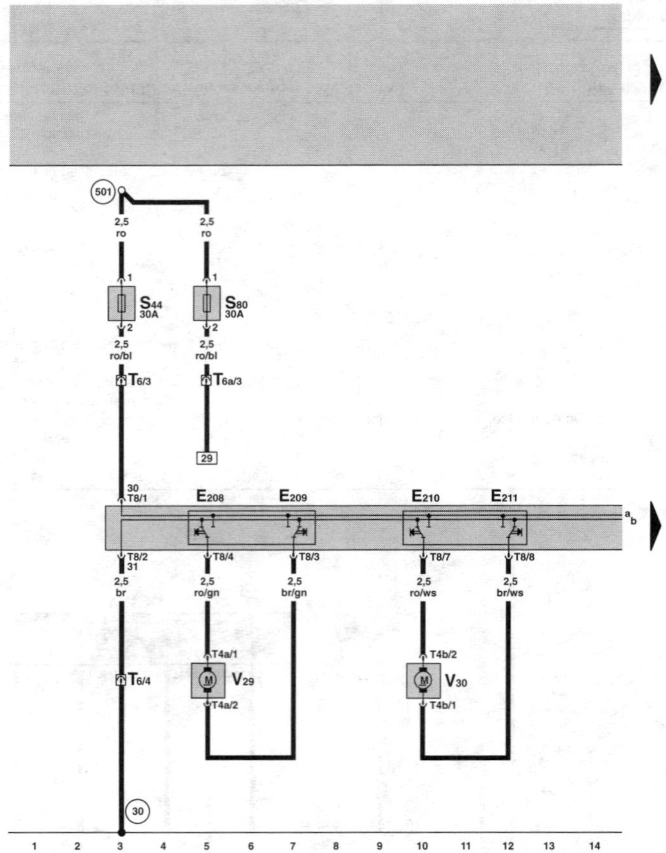

97-22240

ws = white
sw = black
ro = red
br = brown
gn = green
bl = blue
gr = grey
li = violet
ge = yellow

E208 - Button for front height adjustment, upwards, driver seat
E209 - Button for front height adjustment, downwards, driver seat
E210 - Button for rear height adjustment, upwards, driver seat
E211 - Button for rear height adjustment, downwards, driver seat
S44 - Power Seat Circuit Breaker (driver's)
S80 - Power Seat Circuit Breaker (passenger's)
T4a - 4-Pin Connector, black, below driver's seat
T4b - 4-Pin Connector, black, below driver's seat
T6 - 6-Pin Connector, red, below driver's seat
T6a - 6-Pin Connector, red, below driver's seat
T8 - 8-Pin Connector, black
V29 - Driver's Seat Front Height Adjusting Motor
V30 - Driver's Seat Rear Height Adjusting Motor
(30) - Ground connection -1-, beside fuse/relay panel
(501) - Threaded connection -2-, on the relay plate

Front height adjustment, driver seat, rear height adjustment, driver seat

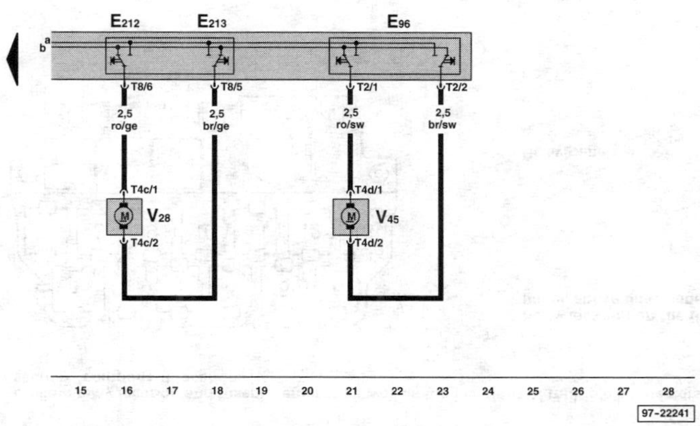

ws = white
sw = black
ro = red
br = brown
gn = green
bl = blue
gr = grey
li = violet
ge = yellow

E96 - Driver's Backrest Adjustment Switch
E212 - Button for length adjustment, forward, driver seat
E213 - Button for length adjustment, backward, driver seat
T2 - Double Connector, black
T4c - 4-Pin Connector, black, below driver's seat
T4d - 4-Pin Connector, black, below driver's seat
T8 - 8-Pin Connector, black
V28 - Driver's Seat Fore/Aft Adjusting Motor
V45 - Driver's Backrest Adjusting Motor

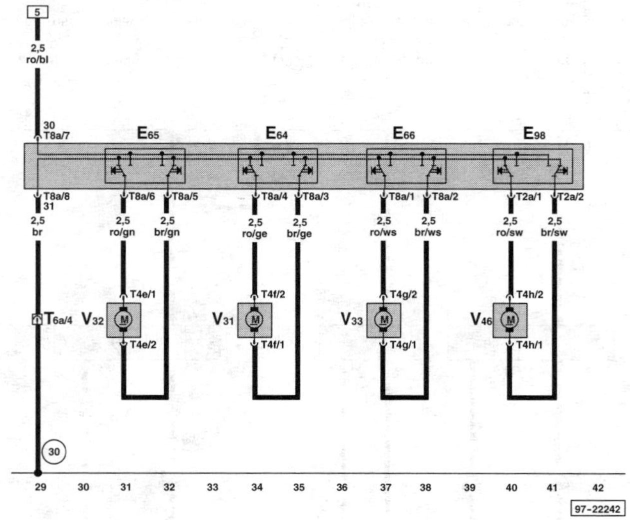

ws = white
sw = black
ro = red
br = brown
gn = green
bl = blue
gr = grey
li = violet
ge = yellow

E64 - Passenger's Seat Fore/Aft Adjusting Switch
E65 - Passenger's Seat Front Height Adjusting Switch
E66 - Passenger's Seat Rear Height Adjusting Switch
E98 - Passenger's Backrest Adjustment Switch
T2a - Double Connector, black
T4e - 4-Pin Connector, black, below driver's seat
T4f - 4-Pin Connector, black, below driver's seat
T4g - 4-Pin Connector, black, below driver's seat
T4h - 4-Pin Connector, black, below driver's seat
T6a - 6-Pin Connector, red, below driver's seat
T8a - 8-Pin Connector, black
V31 - Passenger's Seat Fore/Aft Adjusting Motor
V32 - Passenger's Seat Front Height Adjusting Motor
V33 - Passenger's Seat Rear Height Adjusting Motor
V46 - Passenger's Backrest Adjusting Motor

(30) - Ground connection, -1-, beside fuse/relay panel

Automatic day/night interior mirror,

Rainsensor

from August 1998

Deviate relay location and fuseplacements as well as the locations of multiple connectors see section "component locations".

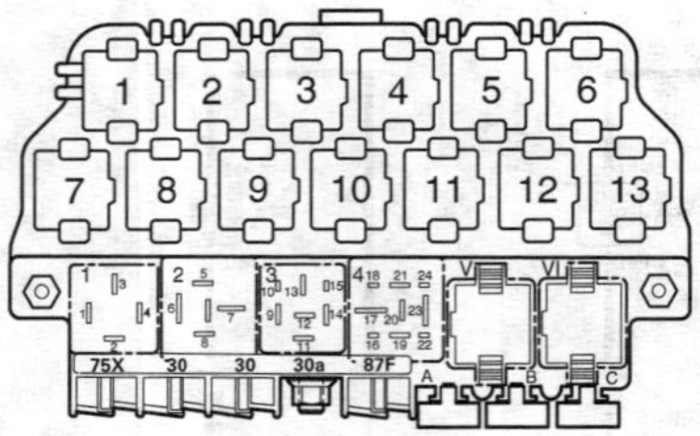
97-20676

Relay location on the thirteenfold auxiliary relay panel, above relay panel:

Relay panel:

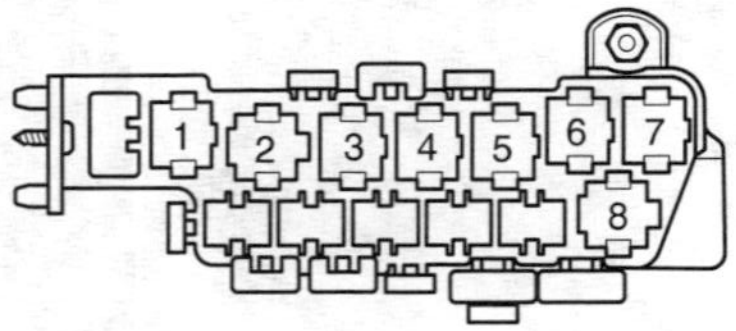
97-20515

Relay location on the eightfold auxiliary relay panel, behind relay panel:

Fuse colors
30 A - green
25 A - white
20 A - yellow
15 A - blue
10 A - red
7,5 A - brown
5 A - beige

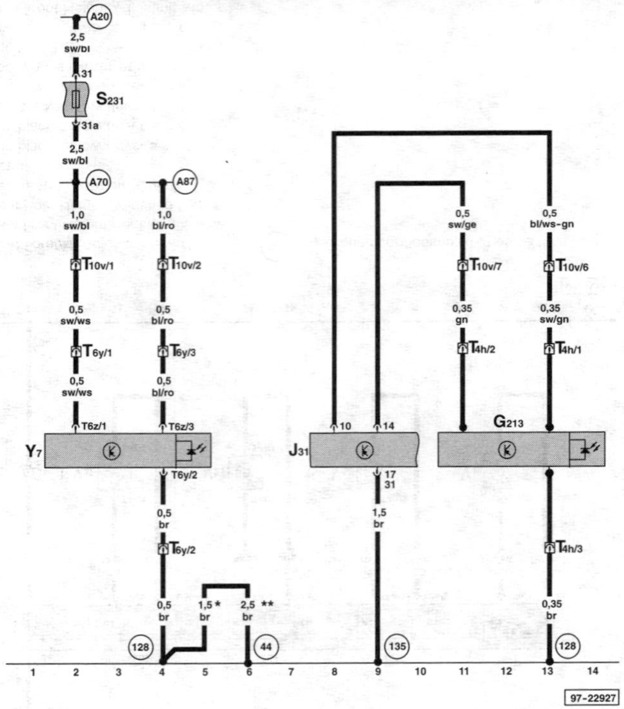

ws = white
sw = black
ro = red
br = brown
gn = green
bl = blue
gr = grey
li = violet
ge = yellow

G213 - Rainsensor (on automatic day/night interior mirror)
J31 - Wiper/Washer Intermittent Relay
S231 - Fuse 31 in fuse holder
T4h - 4-Pin Connector, black, near mirror under the roof
T6y - 6-Pin Connector, black, near mirror under the roof
T6z - 6-Pin Connector
T10v - 6-Pin Connector, green, connector station A-pillar, left
Y7 - Automatic day/night interior mirror
(44) - Ground connection, on left A-pillar, lower part
(128) - Ground connection -1-, in interior light wiring harness
(135) - Ground connection -2-, in instrument panel wiring harness
(A20) - wire conection (15a), in instrument panel wiring harness
(A70) - Connector (15a, fuse 231) , in instrument panel wiring harness
(A87) - Connector (reverse lamp), im instrument panel wiring harness

* - valid for vehicles without sunroof
** - valid for vehicles equiped with sunroof

1,8L- Engine - Motronic Multiport Fuel Injection (MFI)/110 kW, code AEB,

from August 1998

Deviate relay location and fuseplacements as well as the locations of multiple connectors see section "component locations".

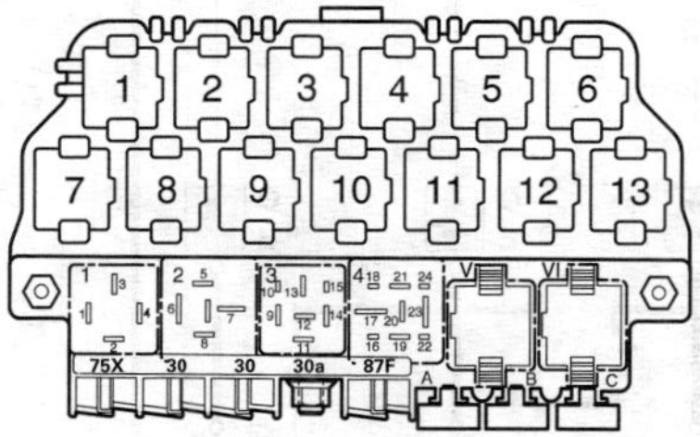

97-20676

Relay location on the thirteenfold auxiliary relay panel, above relay panel:

Relay panel:

4 Fuel Pump (FP) Relay (208)

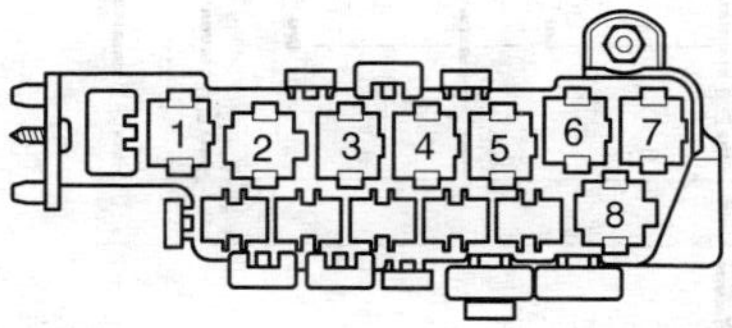

97-20515

Relay location on the eightfold auxiliary relay panel, behind relay panel:

Note: Number in parentheses indicates production control number stamped on relay housing.

Fuse colors
30 A - green
25 A - white
20 A - yellow
15 A - blue
10 A - red
7,5 A - brown
5 A - beige

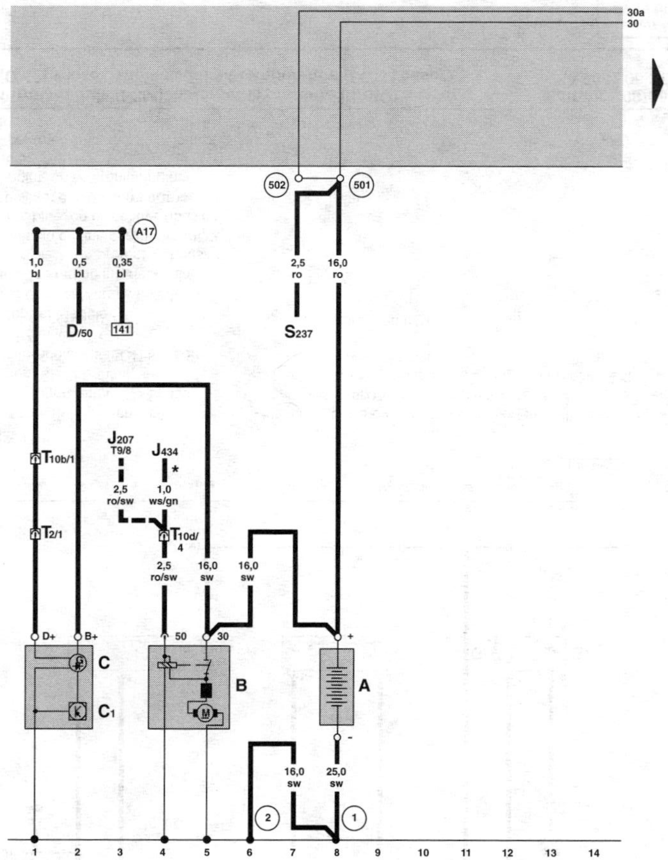

97-23452

ws = white
sw = black
ro = red
br = brown
gn = green
bl = blue
gr = grey
li = violet
ge = yellow

A - Battery
B - Starter
C - Generator (GEN)
C1 - Voltage Regulator (VR)
D - Ignition/Starter Switch
J207 - Starting Interlock Relay
J434 - Starting Interlock Realy (clutch position), manual transmission only
S237 - Fuse 37 in fuse holder
T2 - Double Connector, on engine right
T9 - 9-Pin Connector, on Starting Interlock Relay
T10b - 10-Pin Connector, black, on protective housing for control module, in engine compartment, left
T10d - 10-Pin Connector, brown, on protective housing for control module, in engine compartment, left

(1) - Ground strap, battery to body
(2) - Ground strap, transmission to body
(501) - Screw Connection -2- (30), on relay panel
(502) - Screw Connection (30a), on relay panel
(A17) - wire connection (61), in instrument panel wiring harness
* - manual transmission only
- - - automatic transmission only

Battery, starter, generator (GEN)

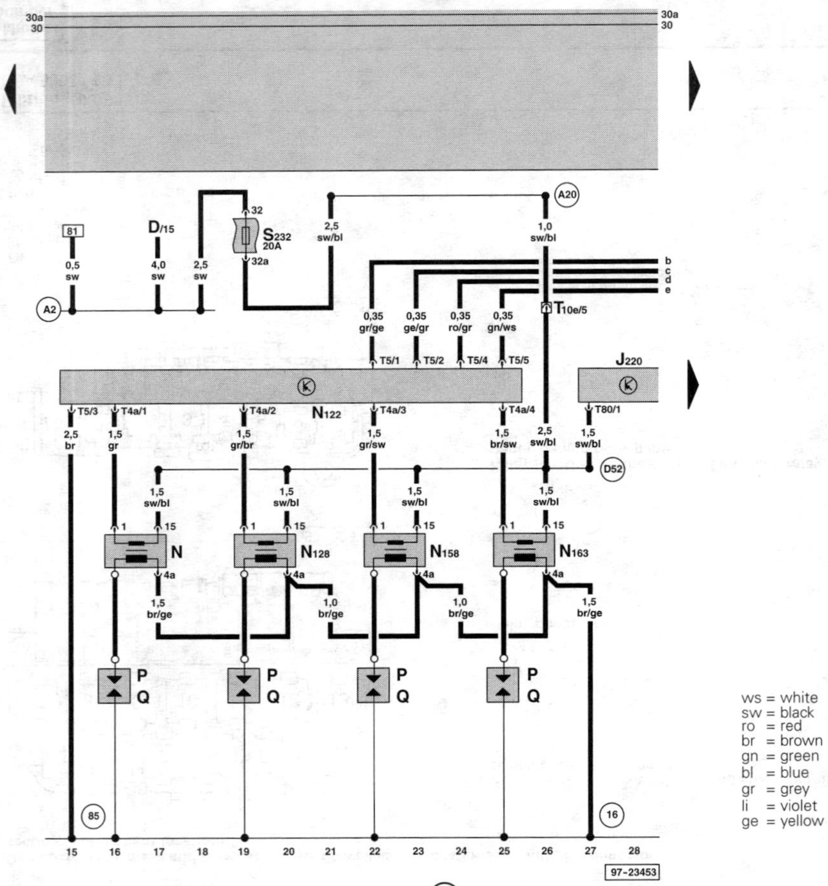

ws = white
sw = black
ro = red
br = brown
gn = green
bl = blue
gr = grey
li = violet
ge = yellow

- D - Ignition/Starter Switch
- J220 - Motronic Engine Control Module (ECM)
- N - Ignition Coil
- N122 - Power Output Stage
- N128 - Ignition Coil 2
- N158 - Ignition Coil 3
- N163 - Ignition Coil 4
- P - Spark Plug Connectors
- Q - Spark Plugs
- S231 - Fuse 31 in fuse holder
- S232 - Fuse 32 in fuse holder
- T4a - 4-Pin Connector, black, on power output stage
- T5 - 5-Pin Connector, black, on power output stage
- T10e - 10-Pin Connector, black, on protective housing for control module, in engine compartment, left
- T80 - Connector, 80 point, on Motronic Engine Control Module (ECM)

- (16) - Ground connection -1-, on valve cover
- (85) - Ground connection -1-, in engine compartment wiring harness
- (A2) - plus connection (15), in instrument panel wiring harness
- (A20) - wire connection (15a), in instrument panel wiring harness
- (D52) - plus connection (15a), in engine compartment wiring harness

Ignition-system, Motronic engine control module (ECM)

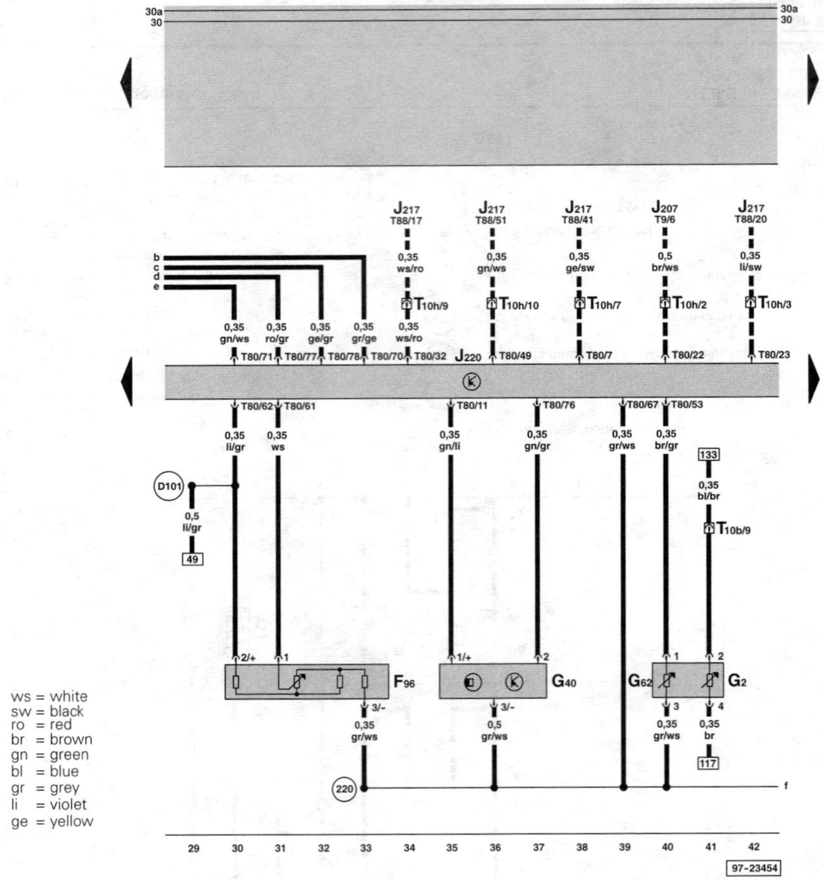

ws = white
sw = black
ro = red
br = brown
gn = green
bl = blue
gr = grey
li = violet
ge = yellow

- F96 - Barometric Pressure (BARO) Sensor
- G2 - Engine Coolant Temperature (ECT) Sensor
- G40 - Camshaft Position (CMP) Sensor
- G62 - Engine Coolant Temperature (ECT) Sensor
- J207 - Starting Interlock Relay
- J217 - Transmission Control Module (TCM)
- J220 - Motronic Engine Control Module (ECM)
- T9 - 9-Pin Connector, on starting interlock relay
- T10b - 10-Pin Connector, black, on protective housing for control module, in engine compartment, left
- T10h - 10-Pin Connector, blue, on protective housing for control module, in engine compartment, left
- T80 - Connector, 80 point, on Motronic Engine Control Module (ECM)
- T88 - Connector, 88 point, on Transmission Control Module (TCM)

- (220) - Ground connection (sensor ground), in engine compartment wiring harness
- (D101) - Wire connection -1-, in engine compartment wiring harness
- - - - automatic transmission only

Motronic engine control module (ECM), barometric pressure (BARO) sensor, camshaft position (CMP) sensor, engine coolant temperature (ECT) sensor

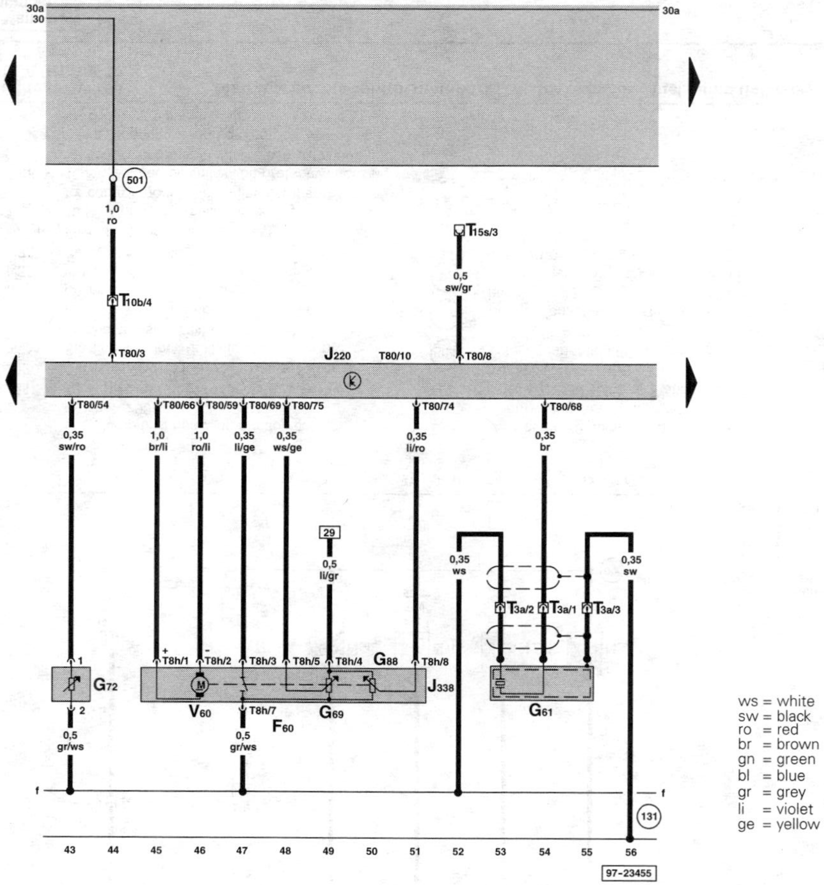

ws = white
sw = black
ro = red
br = brown
gn = green
bl = blue
gr = grey
li = violet
ge = yellow

F60 - Closed Throttle Position (CTP) Switch
G61 - Knock Sensor (KS) 1
G69 - Throttle Position (TP) Sensor
G72 - Intake Air Temperature (IAT) Sensor
G88 - Throttle Position (TP) Sensor
J220 - Motronic Engine Control Module (ECM)
J338 - Throttle Valve Control Module
T3a - 3-Pin Connector, green, on firewall, left
T8h - 8-Pin Connector, black, on throttle valve control module
T10b - 10-Pin Connector, black, on protective housing for control module, in engine compartment, left
T15s - 15-Pin Connector, red, on protective housing for control module, in engine compartment, left
T80 - Connector, 80 point, on Motronic Engine Control Module (ECM)
V60 - Throttle Position (TP) Actuator
(131) - Ground connection -2-, in engine compartment wiring harness
(501) - Screw connection -2- (30), on relay panel

Edition 04/00
USA.5062.11.21

Motronic engine control module (ECM), knock sensor (KS) 1, throttle valve control module, intake air temperature (IAT) sensor

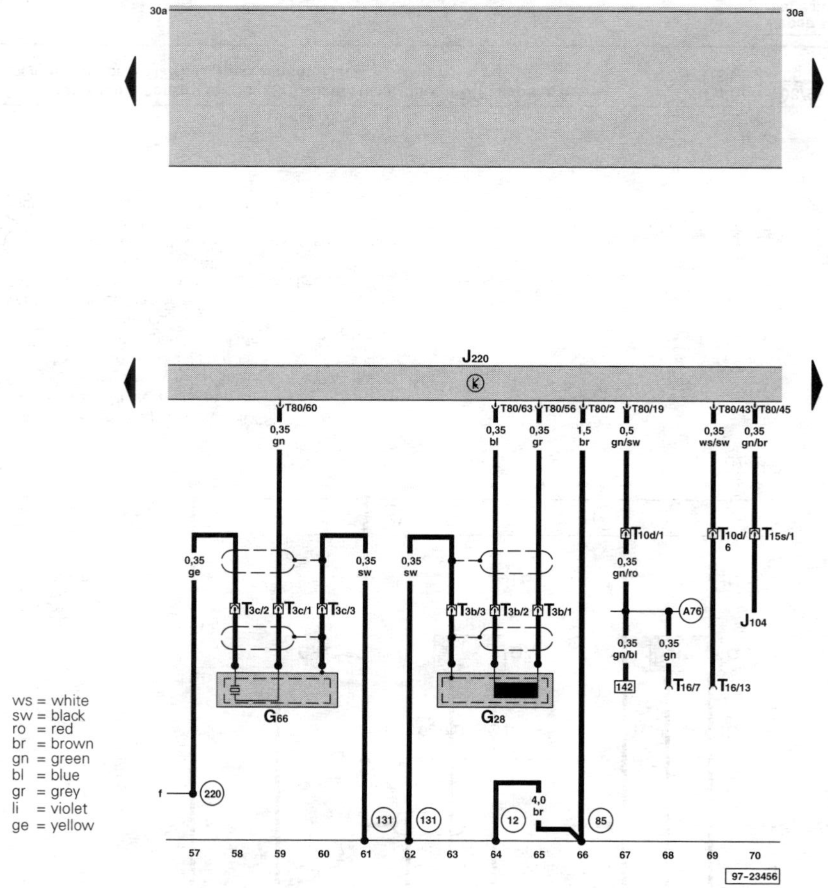

ws = white
sw = black
ro = red
br = brown
gn = green
bl = blue
gr = grey
li = violet
ge = yellow

G28 - Engine Speed (RPM) Sensor
G66 - Knock Sensor (KS) 2
J104 - ABS Control Module (w/EDL)
J220 - Motronic Engine Control Module (ECM)
T3b - 3-Pin Connector, grey, on firewall, left
T3c - 3-Pin Connector, blue, on firewall, left
T6n - 6-Pin Connector, red, on protective housing for control module, in engine compartment, left
T10d - 10-Pin Connector, brown, on protective housing for control module, in engine compartment, left
T10e - 10-Pin Connector, yellow, on protective housing for control module, in engine compartment, left
T15s - 15-Pin Connector, red, on protective housing for control module, in engine compartment, left
T16 - 16-Pin Connector, Data Link Connector (DLC), below instrument panel, left
T80 - Connector, 80 point, on Motronic Engine Control Module (ECM)
(12) - Ground connection, in engine compartment, left
(85) - Ground connection -1-, in engine compartment wiring harness
(131) - Ground connection -2-, in engine compartment wiring harness
(220) - Ground connection (sensor ground), in engine compartment wiring harness
(A76) - Connector (K-diagnosis wire), in instrument panel wiring harness

Motronic engine control module (ECM), engine speed sensor, knock sensor (KS) 2

Edition 04/00
USA.5062.11.21

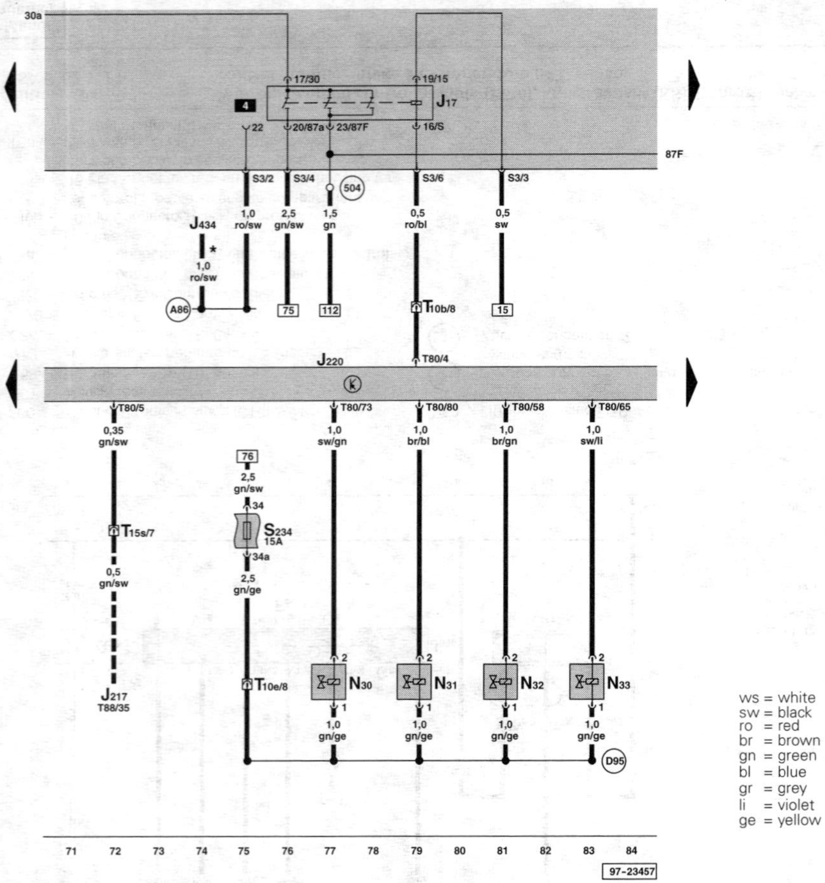

ws = white
sw = black
ro = red
br = brown
gn = green
bl = blue
gr = grey
li = violet
ge = yellow

J17 - Fuel Pump (FP) Relay, production control number (208)
J217 - Transmission Control Module (TCM)
J220 - Motronic Engine Control Module (ECM)
J434 - Starting Interlock Realy (clutch position), manual transmission only
N30 - Cylinder 1 Fuel Injector
N31 - Cylinder 2 Fuel Injector
N32 - Cylinder 3 Fuel Injector
N33 - Cylinder 4 Fuel Injector
S234 - Fuse 34 in fuse holder
T10b - 10-Pin Connector, black, on protective housing for control module, in engine compartment, left
T10e - 10-Pin Connector, yellow, on protective housing for control module, in engine compartment, left
T15s - 10-Pin Connector, red, on protective housing for control module, in engine compartment, left
T80 - Connector, 80 point, on Motronic Engine Control Module (ECM)
T88 - Connector, 88 point, on Transmission Control Module (TCM)
(504) - Screw connection (87F), on relay panel
(A86) - Connection (50a), in instrument panel wiring harness
(D95) - wire connection (injectors), in engine compartment wiring harness
* - manual transmission only
- - - automatic transmission only

Motronic engine control module (ECM), fuel injectors, fuel pump (FP) relay

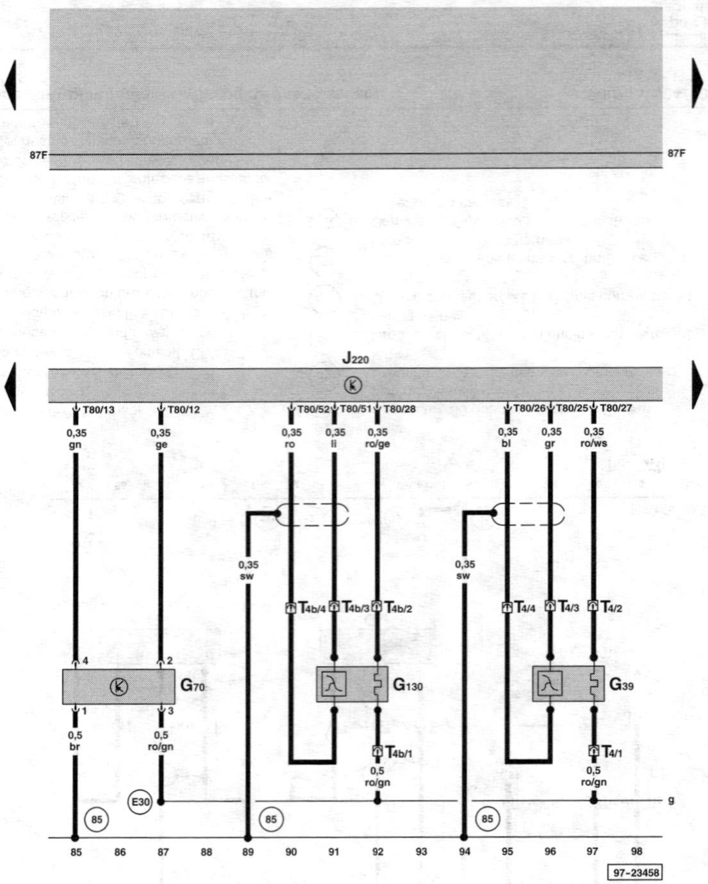

ws = white
sw = black
ro = red
br = brown
gn = green
bl = blue
gr = grey
li = violet
ge = yellow

G39 - Heated Oxygen Sensor (HO2S)
G70 - Mass Air Flow (MAF) Sensor
G130 - Oxygen Sensor (O2S) Behind Three Way Catalytic Converter (TWC)
J220 - Motronic Engine Control Module (ECM)
T4 - 4-Pin Connector
T4b - 4-Pin Connector
T80 - Connector, 80 point, on Motronic Engine Control Module (ECM)
(85) - Ground connection -1-, in engine compartment wiring harness
(E30) - Connector (87a), in wiring harness engine

Motronic engine control module (ECM), mass air flow (MAF) sensor, heated oxygen sensor (HO2S), oxygen sensor (O2S)

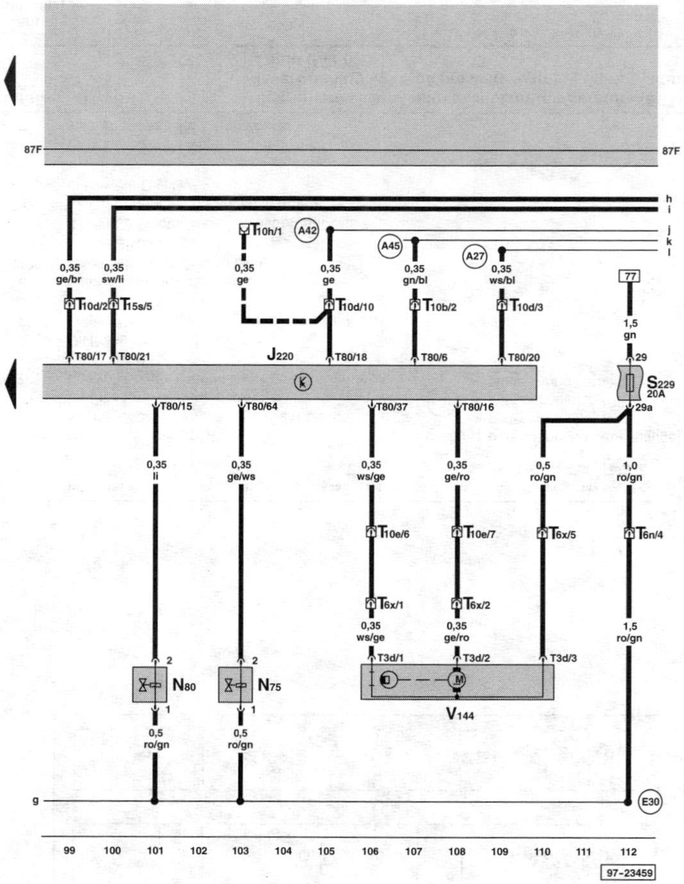

ws = white
sw = black
ro = red
br = brown
gn = green
bl = blue
gr = grey
li = violet
ge = yellow

J220 - Motronic Engine Control Module (ECM)
N75 - Wastegate Bypass Regulator Valve
N80 - Evaporative Emission (EVAP) Canister Purge Regulator Valve
S229 - Fuse 29 in fuse holder
T3d - 3-Pin Connector, black
T6n - 6-Pin Connector, red, on protective housing for control module, in engine compartment, left
T6x - 6-Pin Connector, brown, connector station A-pillar, right
T10b - 10-Pin Connector, black, on protective housing for control module, in engine compartment, left
T10d - 10-Pin Connector, brown, on protective housing for control module, in engine compartment, left
T10e - 10-Pin Connector, orange, on protective housing for control module, in engine compartment, left
T15s - 15-Pin Connector, red, on protective housing for control module, in engine compartment, left
T80 - Connector, 80 point, on Motronic Engine Control Module (ECM)
V144 - Leak Detection Pump (LDP)
(A27) - Wire Connection (vehicle speed signal), in instrument panel wiring harness
(A42) - plus connection (fuel gauge), in instrument panel wiring harness
(A45) - wire connection (RPM-signal), in instrument panel wiring harness
(E30) - Connector (87a), in wiring harness engine
- - - automatic transmission only

Motronic engine control module (ECM), wastegate bypass regulator valve, Leak Detection Pump (LDP)

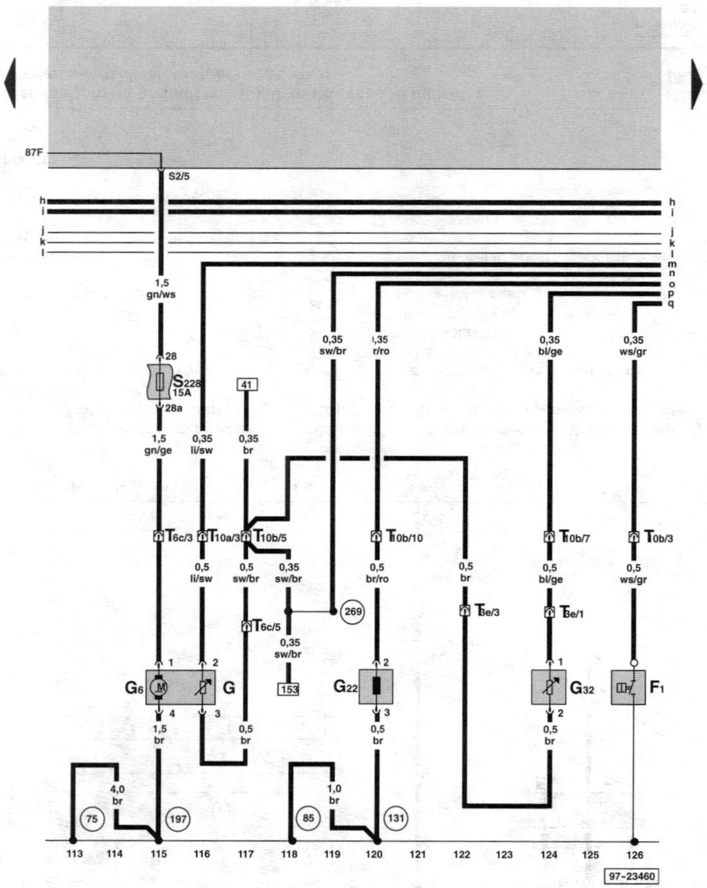

ws = white
sw = black
ro = red
br = brown
gn = green
bl = blue
gr = grey
li = violet
ge = yellow

F1 - Oil Pressure Switch
G - Sender for fuel gauge
G6 - Fuel Pump (FP)
G22 - Speedometer Vehicle Speed Sensor (VSS)
G32 - Engine Coolant Level (ECL) Sensor
S228 - Fuse 28 in fuse holder
T3e - 3-Pin Connector, in engine compartment, left front
T6c - 6-Pin Connector, blue, connector station A-pillar, left
T10a - 10-Pin Connector, brown, connector station A-pillar, left
T10b - 10-Pin Connector, black, on protective housing for control module, in engine compartment, left
T10e - 10-Pin Connector, yellow, on protective housing for control module, in engine compartment, left
(75) - Ground connection, on right rear pillar
(85) - Ground connection -1-, in engine compartment wiring harness
(131) - Ground connection -2-, in engine compartment wiring harness
(197) - Ground connection -4-, in wiring harness rear
(269) - Ground connection (sensor ground) -1-, in instrument panel wiring harness

Oil pressure switch, fuel pump (FP), sender for fuel gauge, speedometer vehicle speed sensor

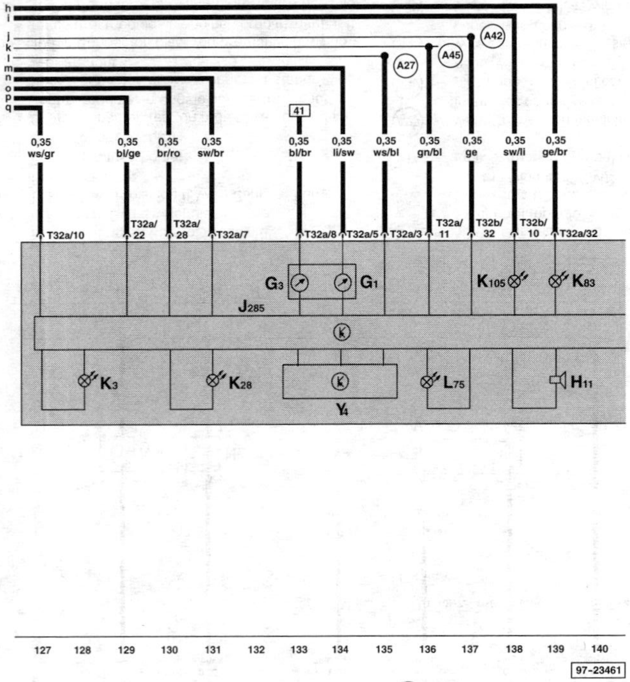

ws = white
sw = black
ro = red
br = brown
gn = green
bl = blue
gr = grey
li = violet
ge = yellow

G1 - Fuel gauge
G3 - Engine Coolant Temperature (ECT) Gauge
H11 - Oil Pressure Warning Buzzer
J285 - Control module with indicator unit in instrument panel insert
K3 - Oil Pressure Warning Light
K28 - Engine Coolant Level/Temperature (ECL/ECT) Warning Light
K83 - Malfunction Indicator Lamp (MIL)
K105 - Low Fuel Level Warning Light
L75 - Digital Display Light
T32a - 32-Pin Connector, blue, on instrument cluster
T32b - 32-Pin Connector, green, on instrument cluster
Y4 - Odometer Display

(A27) - Wire Connection (vehicle speed signal), in instrument panel wiring harness

(A42) - plus connection (fuel gauge), in instrument panel wiring harness
(A45) - wire connection (RPM-signal), in instrument panel wiring harness

Instrument cluster, engine coolant temperature (ECT) gauge, low fuel level warning light, oil pressure warning light, Malfunction Indicator Lamp (MIL)

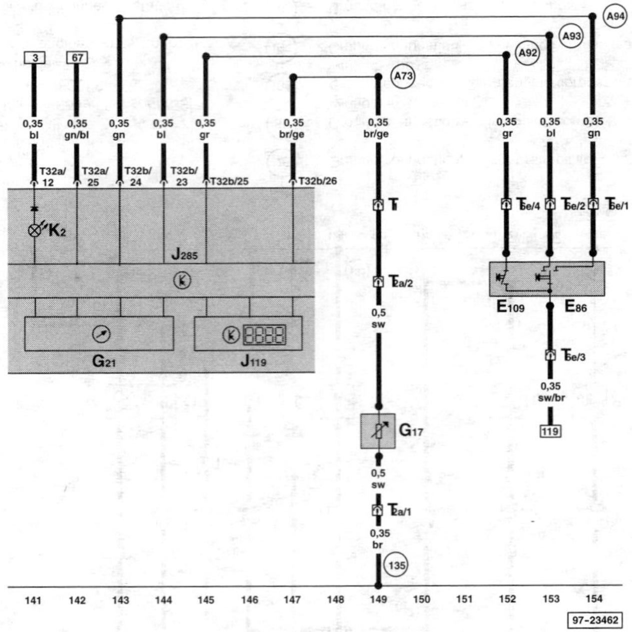

ws = white
sw = black
ro = red
br = brown
gn = green
bl = blue
gr = grey
li = violet
ge = yellow

E86 - Multi-Function Indicator Mode Select Switch
E109 - Multi-Function Indicator Memory Switch
G17 - Outside Air Temperature Sensor (use correspond wiring diagram by vehicles with air conditioning)
G21 - Speedometer
J119 - Multi-function Indicator (MFI)
J285 - Control module with indicator unit in instrument panel insert
K2 - Generator (GEN) Warning Light
T1 - Single Connector, brown, connector station A-pillar, left
T2a - Double Connector, in engine compartment, left front
T6e - 6-Pin Connector, black, behind steering column switch cover
T32a - 32-Pin Connector, blue, on instrument cluster
T32b - 32-Pin Connector, green, on instrument cluster

(135) - Ground connection, -2-,in instrument panel wiring harness
(A73) - Connector (outside temperature indicator), in instrument panel wiring harness
(A92) - Connector (MFI-Reset), in instrument panel wiring harness
(A93) - Connector -1- (MFI), in instrument panel wiring harness
(A94) - Connector -2- (MFI), in instrument panel wiring harness

Instrument cluster, multi-function indicator (MFI), multi-function indicator switches, speedometer, outside air temperature sensor

Comfort System, with Alarm System,

from September 1998

Deviate relay location and fuseplacements as well as the locations of multiple connectors see section "component locations".

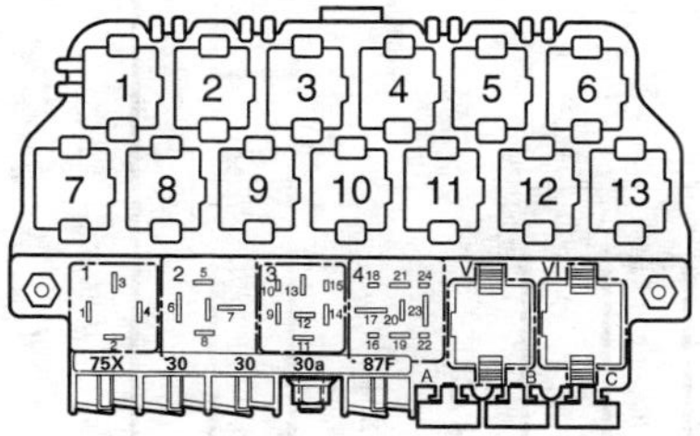

97-20676

Relay location on the thirteenfold auxiliary relay panel, above relay panel:

2 Relay for motor remote unlock rear lid (79)

Relay panel:

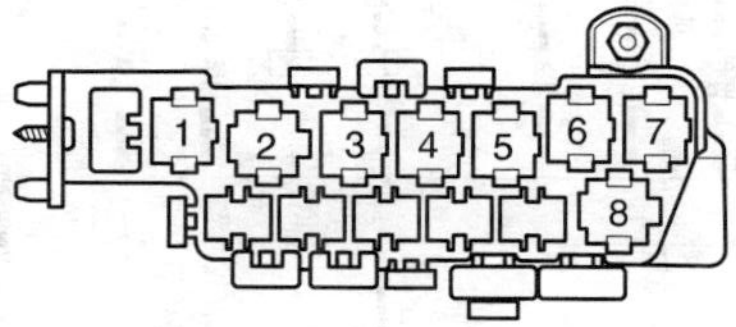

97-20515

Relay location on the eightfold auxiliary relay panel, behind relay panel:

Note: Number in parentheses indicates production control number stamped on relay housing.

Fuse colors

30 A - green
25 A - white
20 A - yellow
15 A - blue
10 A - red
7,5 A - brown
5 A - beige

Edition 08/98
USA.5062.06.21

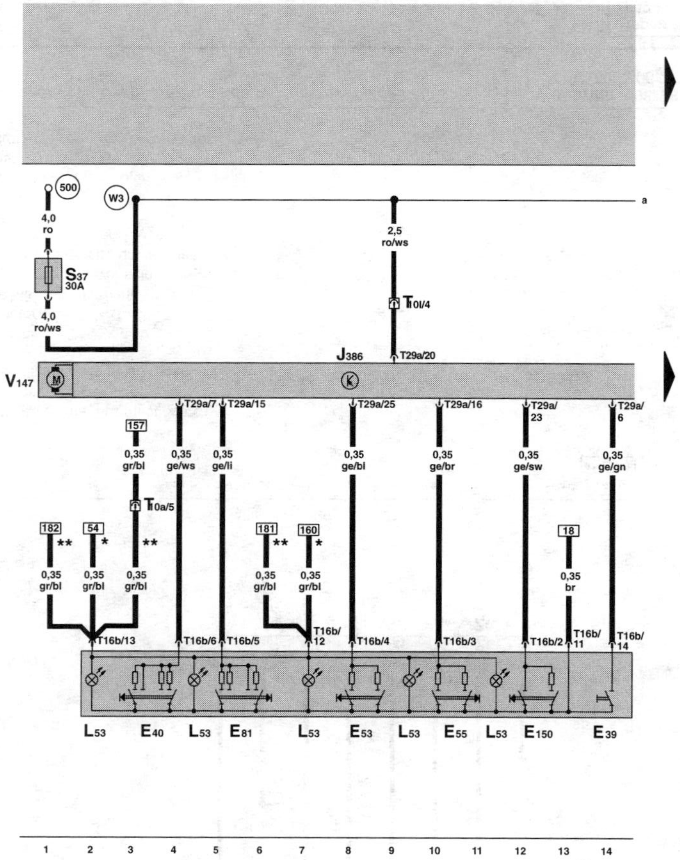

ws = white
sw = black
ro = red
br = brown
gn = green
bl = blue
gr = grey
li = violet
ge = yellow

E39 - Safety switch for window regulator rear
E40 - Switch for window regulator front left
E53 - Switch for window regulator rear left, driver
E55 - Switch for window regulator rear right, driver
E81 - Switch for window regulator front right, driver
E150 - Switch for interior lock, driver side
J386 - Door control module, driver side
L53 - Power Window Switch Light
S37 - Power window fuse
T10a - 10-pin connector, grey, connector station A-pillar left
T10l - 10-pin connector, black, connector station A-pillar left
T16b - 16-pin connector
T29a - 29-pin connector
V147 - Motor for window regulator, driver side

(500) - Screw connection -1- (30), on relay panel
(W3) - wire connection, in rear wiring harness

* - only vehicles without switches for remote unlock rear lid and remote fuel tank door placed in driver's door
** - only vehicles with switches for remote unlock rear lid and remote fuel tank door placed in driver's door

Door control module, driver side, switch for window regulator in driver's door, switch for interior lock

Edition 08/98
USA.5062.06.21

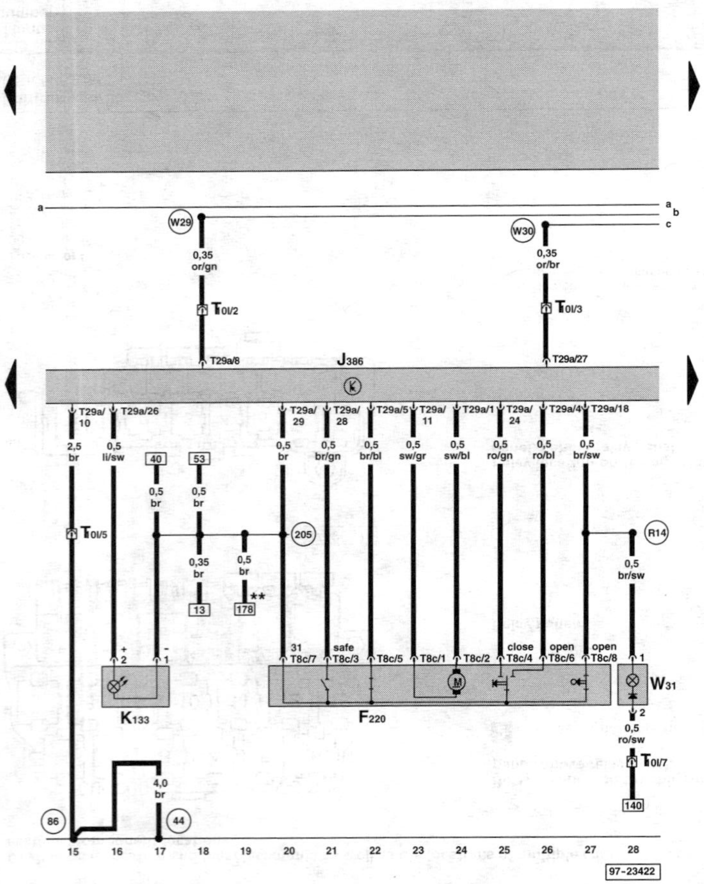

ws = white
sw = black
ro = red
br = brown
gn = green
bl = blue
gr = grey
li = violet
ge = yellow

- F220 - Lock unit for central locking, driver side, switches are shown with doors in closed position
- J386 - Door control module, driver side
- K133 - Warning light for central locking -SAFE-
- T8c - 8-Pin Connector, black
- T10l - 10-Pin Connector, black, connector station A-pillar, left
- T29a - 29-Pin Connector
- W31 - Entry light front, left
- (44) - Ground connection, on left A-pillar, lower part
- (86) - Ground connection -1-, in rear wiring harness
- (205) - Ground connection, in driver's door wiring harness
- (R14) - wire connection -1- (open), in driver's door wiring harness
- (W29) - Plus connector -2- (A-Bus), in wiring harness, floor
- (W30) - Plus connector -2- (B-Bus), in wiring harness, floor
- ** - only vehicles with switches for remote unlock rear lid and remote fuel tank door placed in driver's door

Door control module, driver side, Central locking, driver's door, entry light front, left, Warning light for central locking -SAFE-

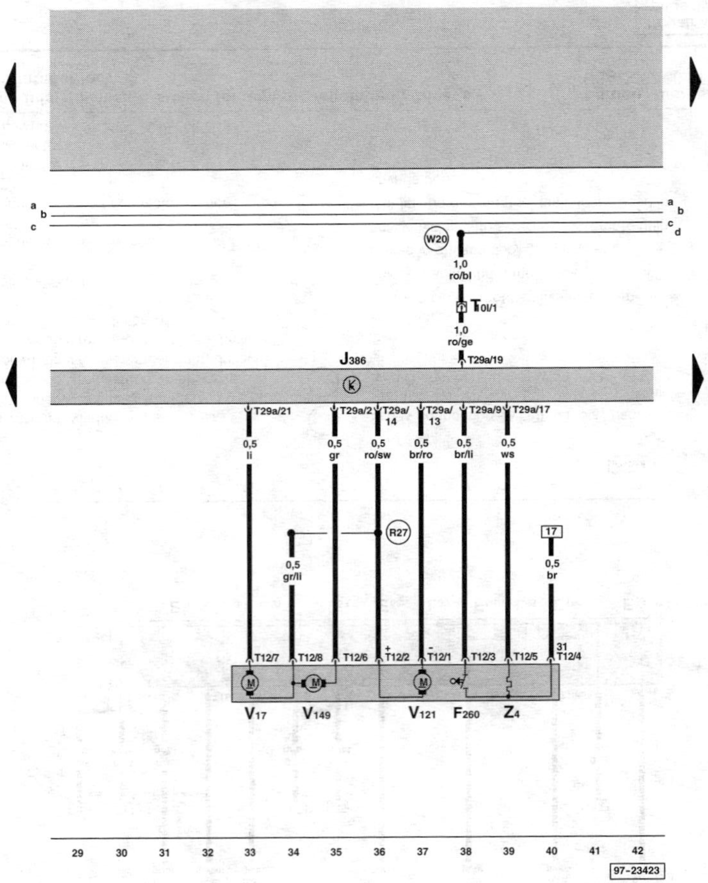

ws = white
sw = black
ro = red
br = brown
gn = green
bl = blue
gr = grey
li = violet
ge = yellow

- F260 - Driver's Mirror Adjustment Contact Switch
- J386 - Door control module, driver side
- T10l - 10-Pin Connector, black, connector station A-pillar, left
- T12 - 12-Pin Connector
- T29a - 29-Pin Connector
- V17 - Driver's Side Mirror Adjustment Motor
- V121 - Motor for fold-in mirror, driver side
- V149 - Motor for mirror adjustment, driver side
- Z4 - Heated outside mirror, driver side
- (R27) - wire connection (power mirror), in driver's door wiring harness
- (W20) - Plus connector (30a), in wiring harness, rear

Door control module, driver side, heated outside mirror, driver side.

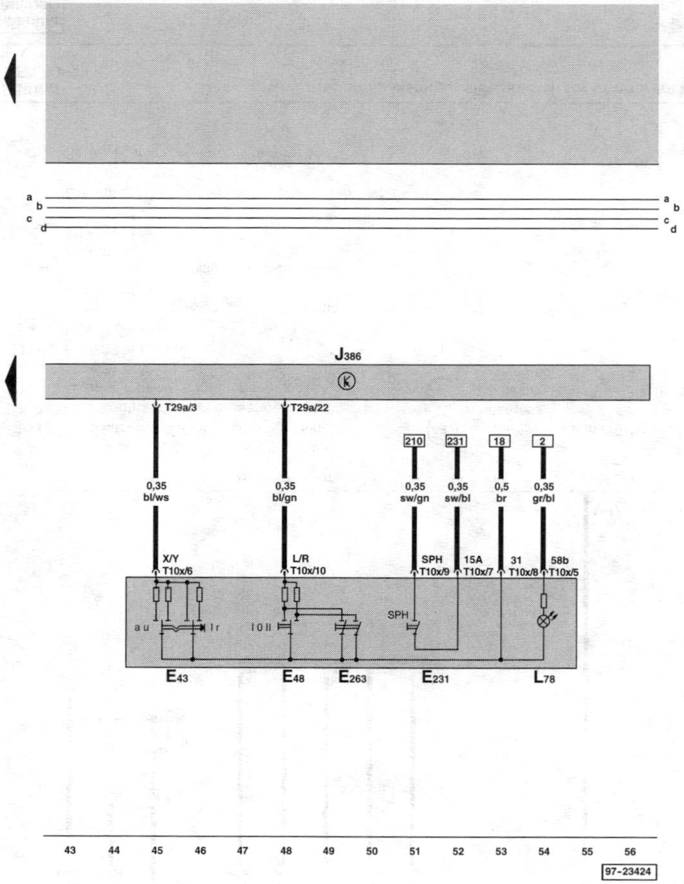

ws = white
sw = black
ro = red
br = brown
gn = green
bl = blue
gr = grey
li = violet
ge = yellow

E43 - Mirror Adjustment Switch
E48 - Mirror Selector Switch
E231 - Switch for outside mirror heating
E263 - Mirror Folding Switch
J386 - Door control module, driver side
L78 - Mirror Adjusting Switch Light
T10x - 10-Pin Connector
T29a - 29-Pin Connector

Door control module, driver side, Mirror adjustment and fold-in mirror, driver side

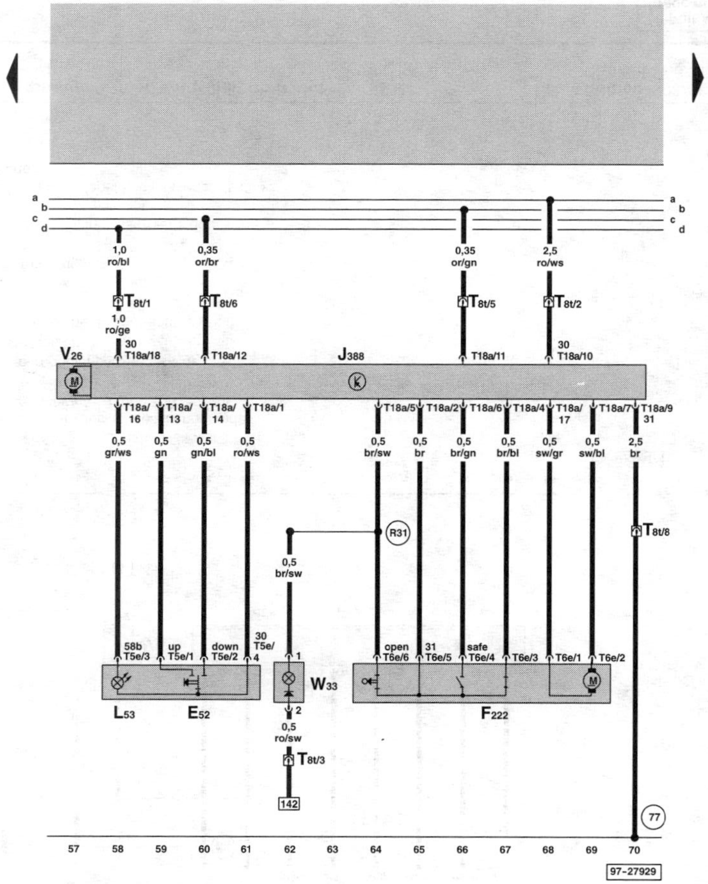

ws = white
sw = black
ro = red
br = brown
gn = green
bl = blue
gr = grey
li = violet
ge = yellow

E52 - Left Rear Window Switch, (In LR Door)
F222 - Lock unit for central locking, rear, left, switches are shown with doors in closed position
J388 - Door control module, rear, left
L53 - Power Window Switch Light
T5e - 5-Pin Connector
T6e - 6-Pin Connector, brown
T8t - 8-Pin Connector, black, connector station B-pillar, left
T18a - 18-Pin Connector
V26 - Motor for window regulator, rear, left
W33 - Entry light rear, left

(77) - Ground connection, (lower left B-pillar)

(R31) - Wire connection -1-, in left rear door wiring harness

Central locking rear left, power window rear left, entry light rear left

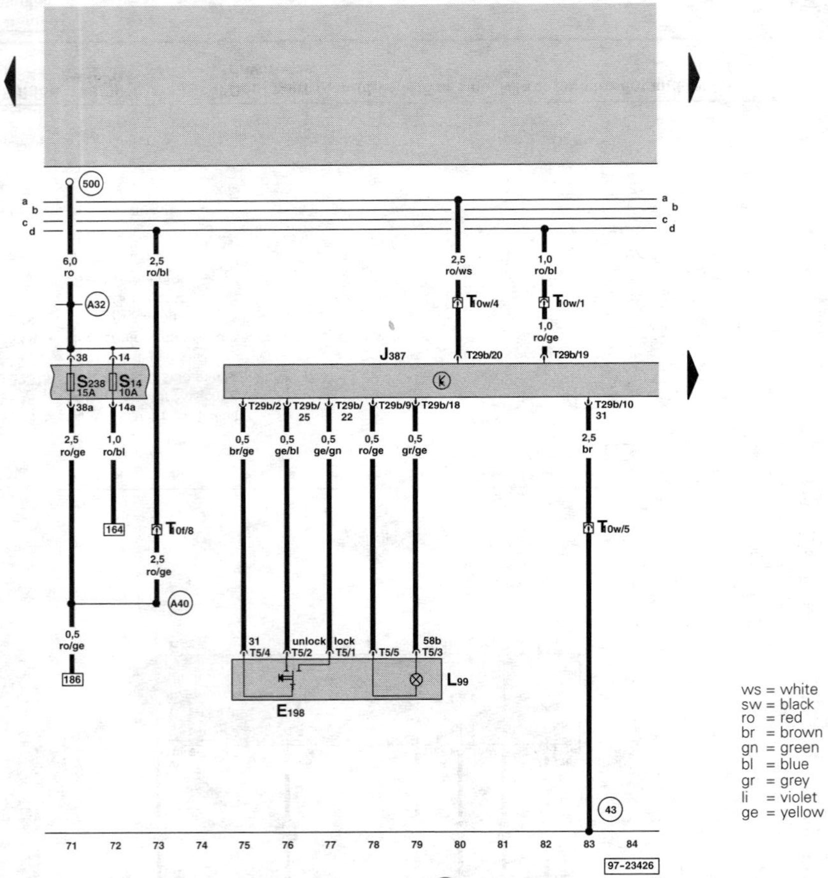

ws = white
sw = black
ro = red
br = brown
gn = green
bl = blue
gr = grey
li = violet
ge = yellow

E198 - Switch for interior lock, passenger side
J387 - Door control module, passenger side
L99 - Lighting for switch interior lock
S14 - Fuse 14 in fuse holder
S238 - Fuse 38 in fuse holder
T5 - 5-Pin Connector
T10f - 10-Pin Connector, blue, connector station A-pillar, left
T10w - 10-Pin Connector, black, connector station A-pillar, right
T29b - 29-Pin Connector

(43) - Ground connection, on right A-pillar, lower part

(500) - Screw connection -1- (30), on relay panel

(A32) - plus connection (30), in instrument panel wiring harness
(A40) - plus connection -1- (30), in instrument panel wiring harness

Door control module, passenger side, switch for interior lock, passenger side

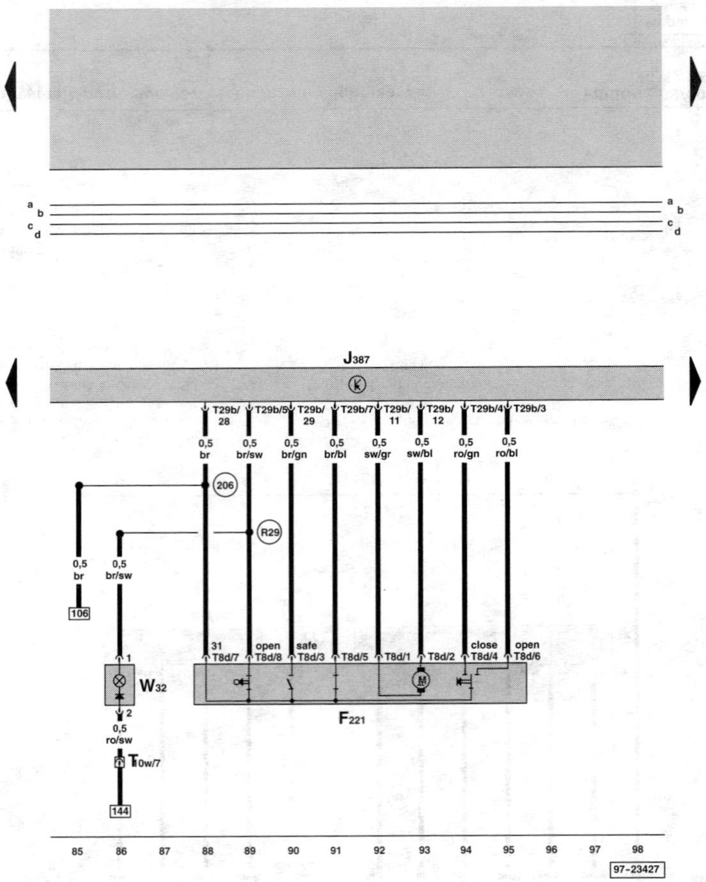

ws = white
sw = black
ro = red
br = brown
gn = green
bl = blue
gr = grey
li = violet
ge = yellow

F221 - Lock unit for central locking, passenger side, switches are shown with doors in closed position
J387 - Door control module, passenger side
T8d - 8-Pin Connector, black
T10w - 10-Pin Connector, black, connector station A-pillar, right
T29b - 29-Pin Connector
W32 - Entry light front, right

(206) - Ground connection, in passenger's door wiring harness
(R29) - Wire connection -1- (open), in passenger's door wiring harness

Central locking passenger's door, entry light front, right

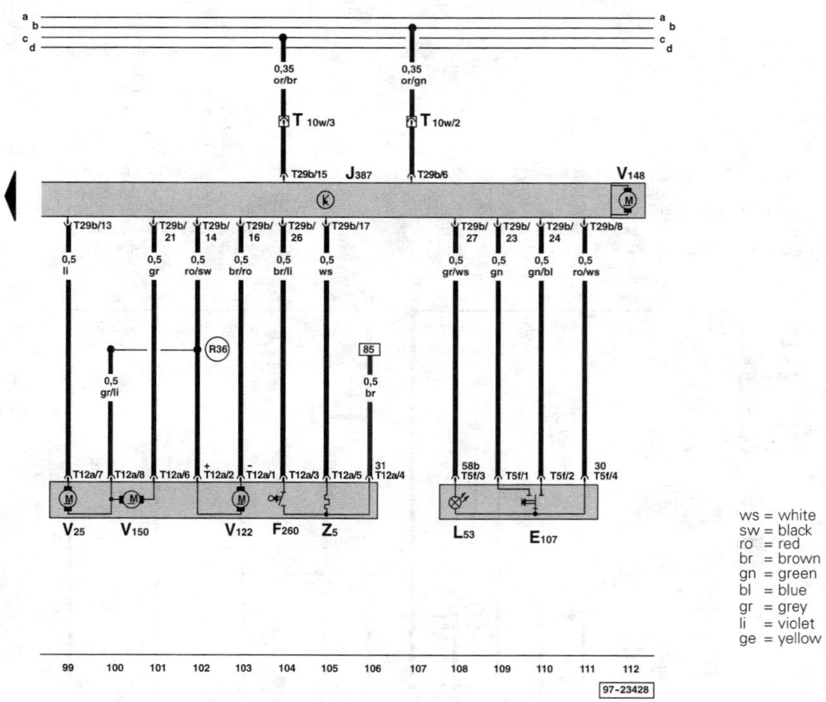

ws = white
sw = black
ro = red
br = brown
gn = green
bl = blue
gr = grey
li = violet
ge = yellow

E107 - Switch for window regulator, in passenger door
F261 - Passenger's Mirror Tilt Contact Switch
J387 - Door control module, passenger side
L53 - Power Window Switch Light
T5f - 5-Pin Connector
T10w - 10-Pin Connector, black, connector station A-pillar, right
T12a - 12-Pin Connector, black
T29b - 29-Pin Connector
V25 - Passenger's Side Mirror Adjustment Motor
V122 - Motor for fold-in mirror, passenger side
V148 - Motor for window regulator, passenger side
V150 - Motor for mirror adjustment, passenger side
Z5 - Heated outside mirror, passenger side

(R36) - wire connection (power mirror), in passenger's door wiring harness

Edition 08/00
USA.5062.12.21

Mirror adjustment and fold-in mirror, passenger side , heated outside mirror, passenger side

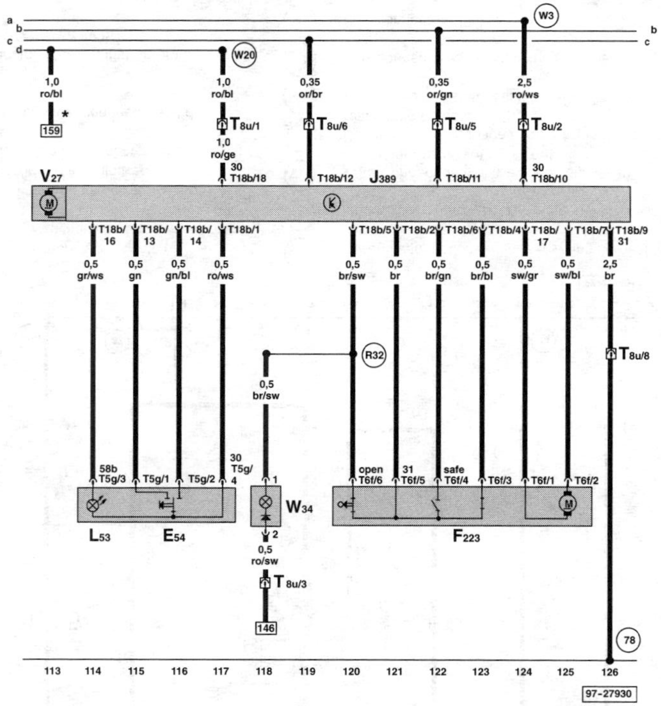

ws = white
sw = black
ro = red
br = brown
gn = green
bl = blue
gr = grey
li = violet
ge = yellow

E54 - Right Rear Window Switch, (In RR Door), switches are shown with doors in closed position
F223 - Lock unit for central locking, rear, right
J389 - Door control module, rear, right
L53 - Power Window Switch Light
T5g - 5-Pin Connector
T6f - 6-Pin Connector
T8u - 8-Pin Connector, brown, connector station B-pillar, right
T18b - 18-Pin Connector
V27 - Motor for window regulator, rear, right
W34 - Entry light rear, right

(78) - Ground connection, (lower right B-pillar)

(R32) - Wire connection -2-, in right rear door wiring harness
(W3) - wire connection, in rear wiring harness
(W20) - Plus connector (30a), in wiring harness, rear
* - only vehicles with switches for remote unlock rear lid and remote fuel tank door placed in driver's door

Central locking rear right, power window rear right, entry light rear right

Edition 08/00
USA.5062.12.21

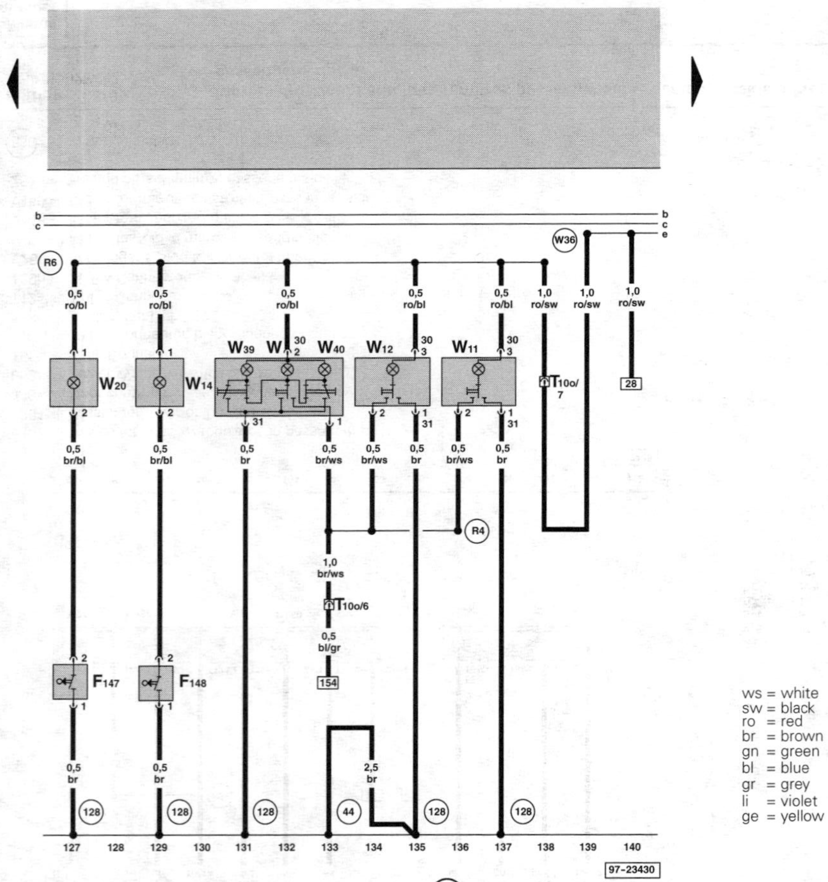

ws = white
sw = black
ro = red
br = brown
gn = green
bl = blue
gr = grey
li = violet
ge = yellow

F147 - Left Make-Up Mirror Light Switch
F148 - Right Make-Up Mirror Light Switch
T10o - 10-Pin Connector, red, connector station A-pillar, left
W - Front Interior Light
W11 - Left Rear Reading Light
W12 - Right Rear Reading Light
W14 - Right Make-up Mirror Light
W20 - Left Make-up Mirror Light
W39 - Reading light, center left
W40 - Reading light, center right

(44) - Ground connection, on left A-pillar, lower part

(128) - Ground connection -1-, in interrior light wiring harness

(R4) - wire connection -1-, in interior light/door contact switch wiring harness

(R6) - plus connection -1-, in interior light wiring harness

(W36) - Plus connector (30a), in wiring harness floor

Front interior light, reading lights, make-up mirror lights

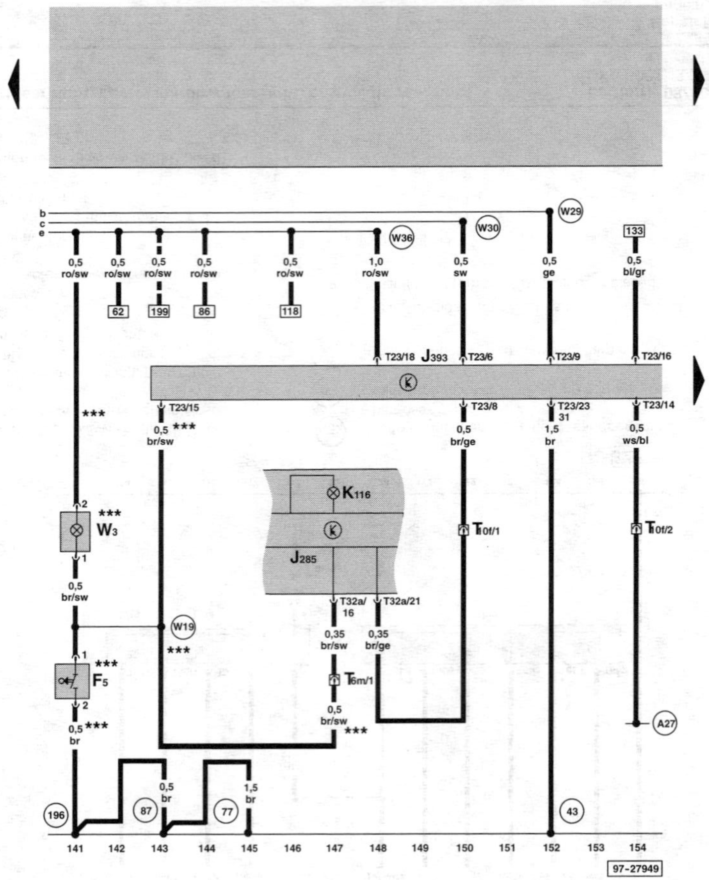

ws = white
sw = black
ro = red
br = brown
gn = green
bl = blue
gr = grey
li = violet
ge = yellow

F5 - Luggage Compartment Light Switch
J285 - Control module with indicator unit in instrument panel insert
J393 - Central control module for comfort system
K116 - Warning light for rear lid unlocked
T6m - 6-Pin Connector, black, connector station A-pillar, right
T10f - 10-Pin Connector, blue, connector station A-pillar, left
T23 - 23-Pin Connector
T32a - 32-Pin Connector, blue, on instrument cluster
W3 - Luggage compartment Light

(43) - Ground connection, on right A-pillar, lower part

(77) - Ground connection, (lower left B-pillar)

(87) - Ground connection -2-, in rear wiring harness

(196) - Ground connection -3-, in wiring harness, rear

(A27) - Wire Connection (vehicle speed signal), in instrument panel wiring harness

(W19) - Connection -2- (luggage compartment light), in rear wiring harness

(W29) - Plus connector -2- (A-Bus), in wiring harness, floor

(W30) - Plus connector -2- (B-Bus), in wiring harness, floor

(W36) - Plus connector (30a), in wiring harness floor

*** - sedan only

Central control module for comfort system, warning light for rear lid unlocked

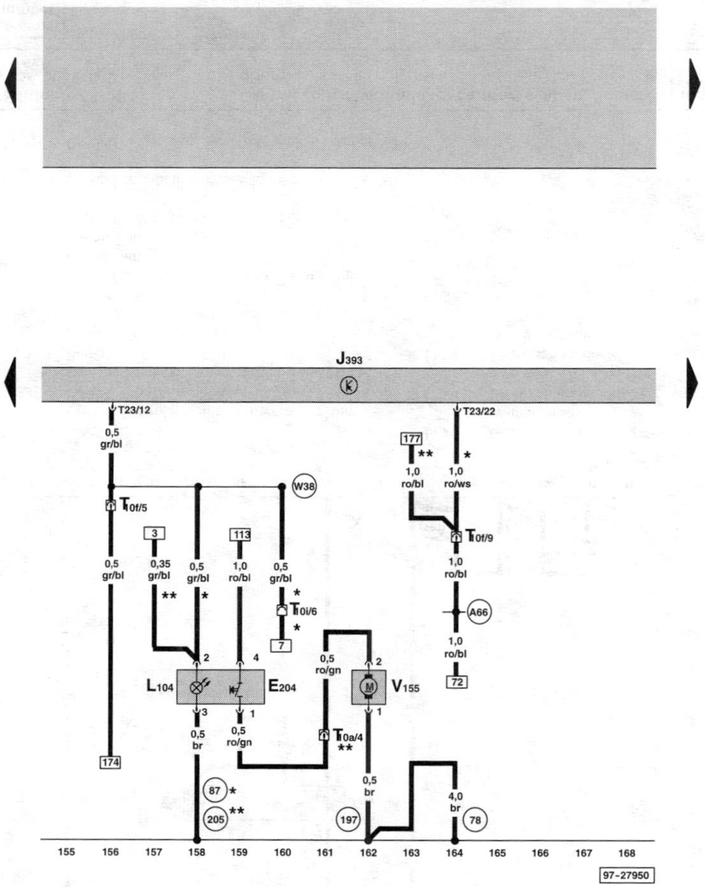

ws = white
sw = black
ro = red
br = brown
gn = green
bl = blue
gr = grey
li = violet
ge = yellow

E204 - Switch for remote/fuel tank door
J393 - Central control module for comfort system
L104 - Illumination for switch remote unlock
T10a - 10-Pin Connector, grey, connector station A-pillar, left, from November 1998
T10f - 10-Pin Connector, blue, connector station A-pillar, left
T10i - 10-Pin Connector, black, connector station driver's door
T23 - 23-Pin Connector
V155 - Motor for fuel tank lid unlock
(78) - Ground connection, (lower right B-pillar)
(87) - Ground connection -2-, in rear wiring harness
(197) - Ground connection -4-, in wiring harness rear
(205) - Ground connection, in driver's door wiring harness rear
(A66) - Connector (30a, central locking/anti-theft warning system/IR), in instrument panel wiring harness
(W38) - Connector (58d), in wiring harness floor
* - only vehicles without switches for remote unlock rear lid and remote fuel tank door placed in driver's door
** - only vehicles with switches for remote unlock rear lid and remote fuel tank door placed in driver's door

Central control module for comfort system, Relay for motor remote unlock rear lid, motor for remote unlock rear lid

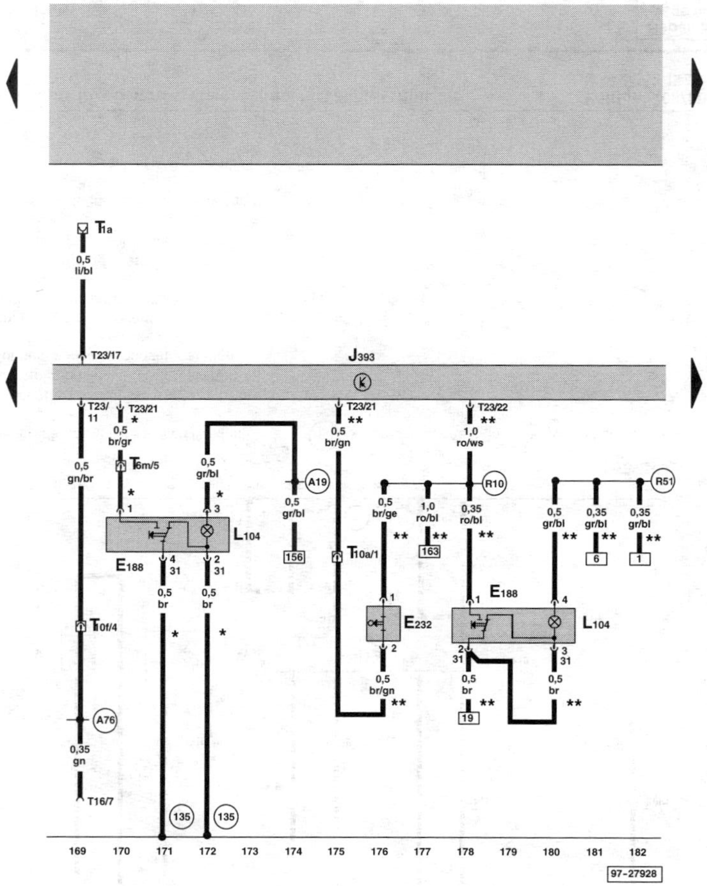

ws = white
sw = black
ro = red
br = brown
gn = green
bl = blue
gr = grey
li = violet
ge = yellow

E188 - Switch for remote unlock, rear lid
E232 - Rear lid Remote Lock Key Switch
J393 - Central control module for comfort system
L104 - Illumination for switch remote unlock
T1a - Single Connector
T6m - 6-Pin Connector, black, connector station A-pillar, right
T10a - 10-Pin Connector, grey, connector station A-pillar, left
T16 - 16-Pin Connector, Data Link Connector (DLC), below instrument panel, left
T23 - 23-Pin Connector
(135) - Ground connection -2-, in instrument panel wiring harness
(R51) - Connection (58b) (driver's side door wiring harness)
(A19) - wire connection (58d), in instrument panel wiring harness
(A76) - Connector (K-diagnosis wire), in instrument panel wiring harness
(R10) - plus connection -1- (30), in driver's door wiring harness
* - only vehicles without switches for remote unlock rear lid and remote fuel tank door placed in driver's door
** - only vehicles with switches for remote unlock rear lid and remote fuel tank door placed in driver's door

Central control module for comfort system, Switch for remote/fuel tank door

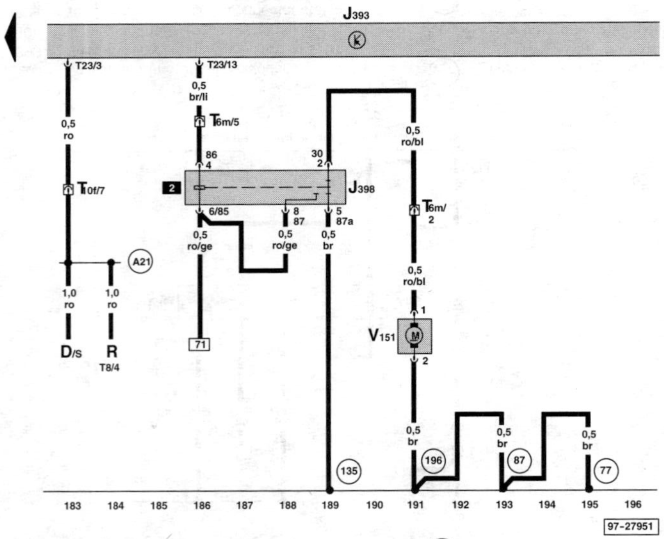

ws = white
sw = black
ro = red
br = brown
gn = green
bl = blue
gr = grey
li = violet
ge = yellow

- D - Ignition/Starter Switch
- J393 - Central control module for comfort system
- J398 - Relay for motor remote unlock rear lid, above relay panel, production control number (79)
- R - Radio
- T8 - 8-Pin Connector, on radio
- T6m - 6-Pin Connector, black, connector station A-pillar, right
- T10f - 10-Pin Connector, blue, connector station A-pillar, left
- T23 - 23-Pin Connector
- V151 - Motor for remote unlock rear lid
- (77) - Ground connection, (lower left B-pillar)
- (87) - Ground connection -2-, in rear wiring harness
- (135) - Ground connection -2-, in instrument panel wiring harness
- (196) - Ground connection -3-, in wiring harness, rear
- (A21) - wire connection (86s), in instrument panel wiring harness

Central control module for comfort system, Switch for remote/fuel tank door

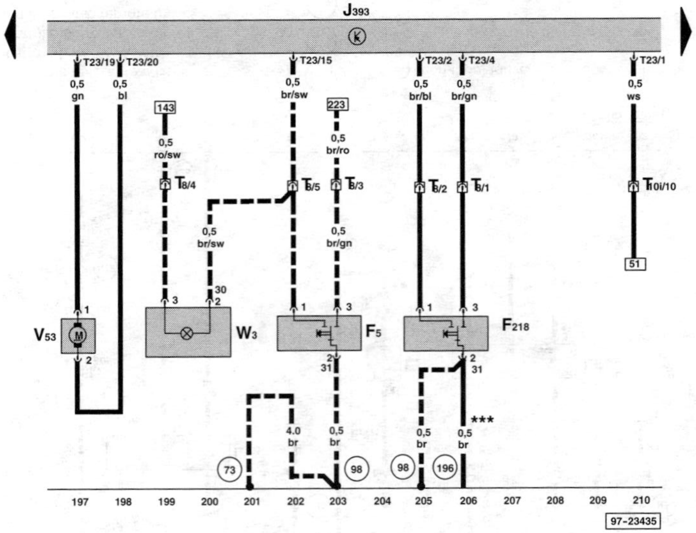

ws = white
sw = black
ro = red
br = brown
gn = green
bl = blue
gr = grey
li = violet
ge = yellow

- F5 - Luggage Compartment light switch (wagon only)
- F218 - Switch for central locking, trunk lid
- J393 - Central control module for comfort system
- T10i - 10-Pin Connector, black, connector station driver's door
- T8 - 8-Pin Connector, black, connector station D-pillar, right (wagon only)
- T23 - 23-Pin Connector
- V53 - Decklid Central Locking System Motor
- W3 - Luggage compartment light (wagon only)
- (73) - Ground connection, on roof bow, rear (wagon only)
- (98) - Ground connection, in rear lid wiring harness
- (196) - Ground connection -3-, in wiring harness, rear
- *** - sedan only
- - - - wagon only

Central control module for comfort system, central locking for trunk lid

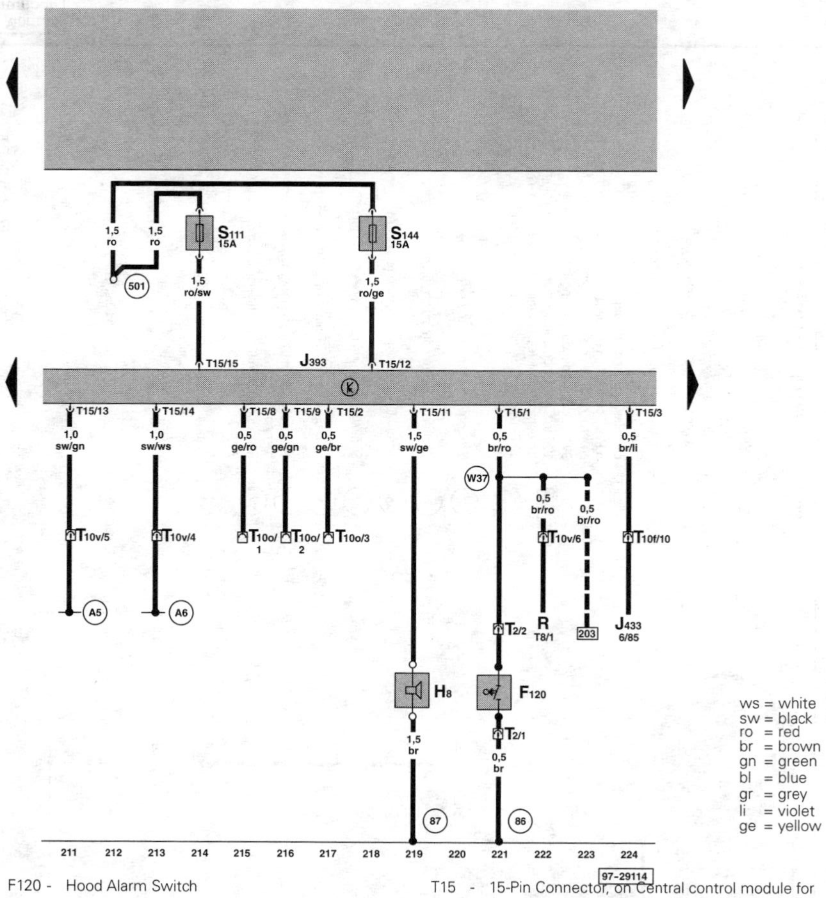

ws = white
sw = black
ro = red
br = brown
gn = green
bl = blue
gr = grey
li = violet
ge = yellow

F120 - Hood Alarm Switch
H8 - Alarm Horn
J393 - Central control module for comfort system
J433 - Starting Interlock Relay (alarm system), above relay panel
R - Radio
S111 - Fuse -1- for Alarm System, on the eightfold auxiliary relay panel, behind relay panel
S144 - Fuse -2- for Alarm System, on the eightfold auxiliary relay panel, behind relay panel
T8 - Double Connector
T8 - 8-Pin Connector, on radio
T10f - 10-Pin Connector, blue, connector station A-pillar, left
T10o - 10-Pin Connector, red, connector station A-pillar, left
T10v - 10-Pin Connector, violet, connector station A-pillar, right
T15 - 15-Pin Connector, on Central control module for comfort system
(86) - Ground connection -1-, in rear wiring harness
(87) - Ground connection -2-, in rear wiring harness
(501) - Screw connection -2- (30), on relay panel
(A5) - plus connection (right turn signal), in instrument panel wiring harness
(A6) - plus connection (left turn signal), in instrument panel wiring harness
(W37) - Connector (alarm system), in wiring harness floor
- - - wagon only

Central control module for comfort system, alarm system

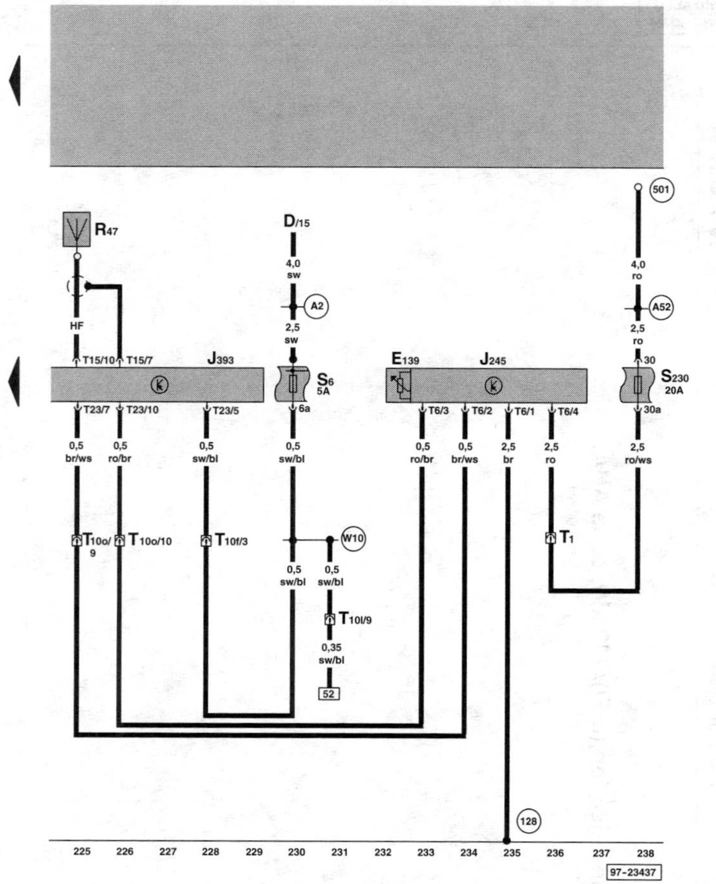

ws = white
sw = black
ro = red
br = brown
gn = green
bl = blue
gr = grey
li = violet
ge = yellow

D - Ignition/Starter Switch
E139 - Sunroof Regulator
J245 - Power Sunroof Control Module
J393 - Central control module for comfort system
R47 - Antenna wire for central locking and alarm system, on B-pillar right
S6 - Fuse 6 in fuse holder
S230 - Fuse 30 in fuse holder
T1 - Single Connector, black
T6 - 6-Pin Connector
T10f - 10-Pin Connector, blue, connector station A-pillar, left
T10l - 10-Pin Connector, black, connector station A-pillar, left
T10o - 10-Pin Connector, red, connector station A-pillar, left
T15 - 15-Pin Connector, on Central control module for comfort system
T23 - 23-Pin Connector
(128) - Ground connection -1-, in interior light wiring harness
(501) - Screw connection -2- (30), on relay panel
(A2) - plus connection (15), in instrument panel wiring harness
(A52) - plus connection (30), in instrument panel wiring harness
(W10) - plus connection (15), in rear wiring harness from October 1998

Central control module for comfort system, power sunroof

2.8L - Engine - Motronic Multiport Fuel Injection (MFI)/142 kW, code AHA,

from August 1998

Notes:

Information for

- *Relay and fuse arrangements*
- *Connector arrangements*
- *Control Modules and Relays*
- *Ground connections*

⇒ Component Locations!

Information for

- *Troubleshooting Procedures*

⇒ Guided Fault Finding using VAS 5051/5052

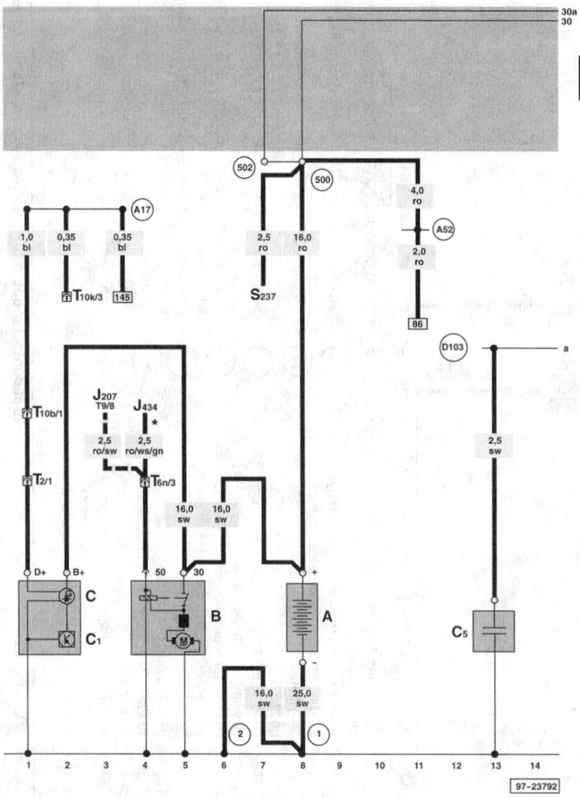

ws = white
sw = black
ro = red
br = brown
gn = green
bl = blue
gr = grey
li = lilac
ge = yellow
or = orange

Battery, Starter, Generator (GEN)

A - Battery
B - Starter
C - Generator (GEN)
C1 - Voltage Regulator (VR)
C5 - Suppressor (Terminal 15), on engine
J207 - Starting Interlock Relay
J434 - Starting Interlock Relay (clutch position)
S237- Fuse 37 in fuse holder
T2 - Double Connector, on engine right
T6n - 6-Pin Connector, red, on protective housing for control module, in engine compartment, left
T9 - 9-Pin Connector, on Starting Interlock Relay
T10b - 10-Pin Connector, black, on protective housing for control module, in engine compartment, left
T10k - 10-Pin Connector, grey, coupling statiom, A-pillar, left

(1) - Ground strap, battery to body

(2) - Ground strap, transmission to body

(500) - Screw Connection -1- (30), on relay panel

(502) - Screw Connection (30a), on relay panel

(A17) - wire connection (61), in instrument panel wiring harness

(A52) - plus connection (30), in instrument panel wiring harness

(D103) - plus connection -3-, in engine compartment wiring harness

* - manual transmission only
- - - automatic transmission only

Passat — Wiring diagram — No. 30/3

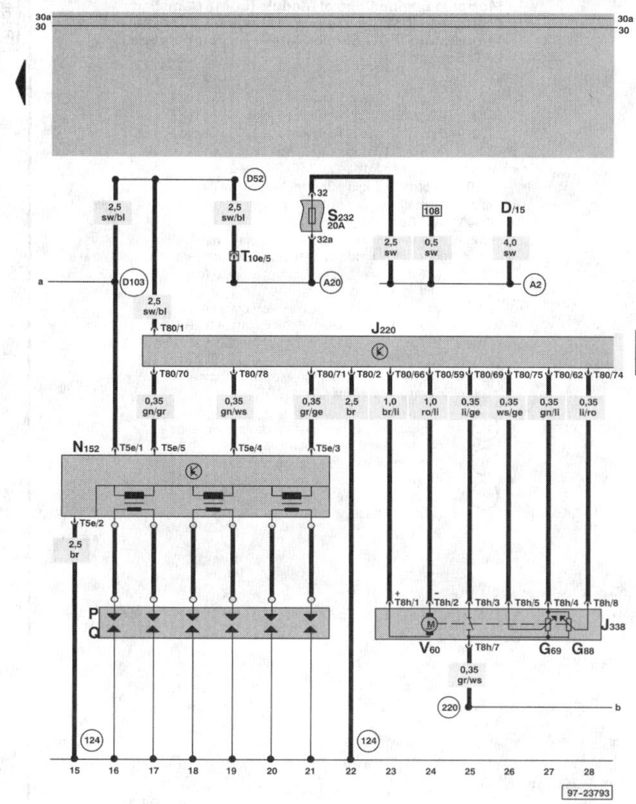

Ignition System, Motronic Engine Control Module (ECM), Throttle Valve Control Module

D - Ignition/Starter Switch
G69 - Throttle Position (TP) Sensor
G88 - Throttle Position (TP) Sensor
J220 - Motronic Engine Control Module (ECM)
J338 - Throttle Valve Control Module
N152 - Ignition Coil
P - Spark Plug Connectors
Q - Spark Plugs
S232 - Fuse 32 in fuse holder
T5e - 5-Pin Connector
T8h - 8-Pin Connector, on throttle valve control module
T10e - 6-Pin Connector, orange, on protective housing for control module, in engine compartment, left
T80 - Connector, 80 point, on Motronic Engine Control Module (ECM)
V60 - Throttle Position (TP) Actuator

(124) - ground connection, in engine compartment right wiring harness
(220) - Ground connection (sensor ground), in engine compartment wiring harness
(A2) - plus connection (15), in instrument panel wiring harness
(A20) - wire connection (15a), in instrument panel wiring harness
(D52) - plus connection (15a), in engine compartment wiring harness
(D103) - plus connection -3-, in engine compartment wiring harness

ws = white
sw = black
ro = red
br = brown
gn = green
bl = blue
gr = grey
li = lilac
ge = yellow
or = orange

Edition 04.04

Passat — Wiring diagram — No. 30/4

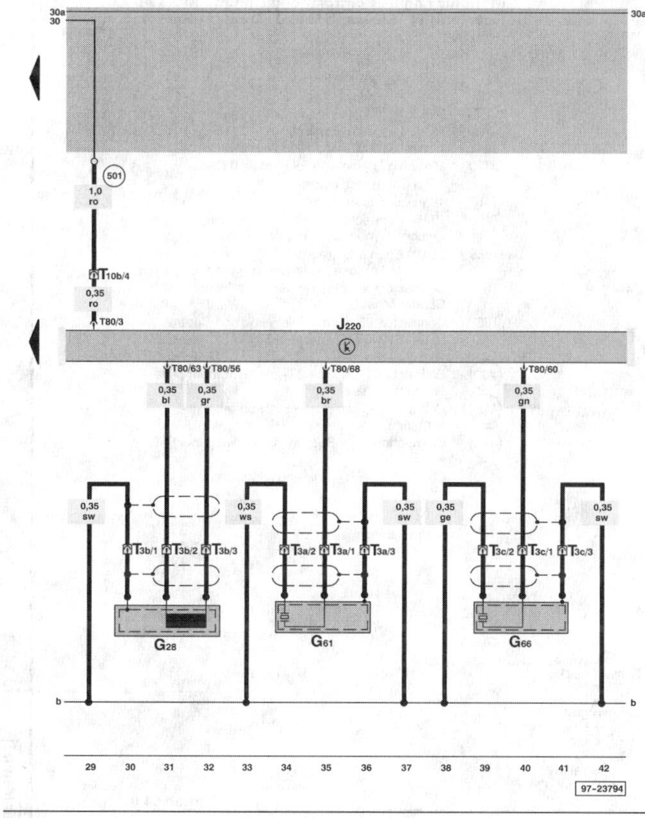

Motronic Engine Control Module (ECM), Engine Speed (RPM) Sensor, Knock Sensor (KS)

G28 - Engine Speed (RPM) Sensor
G61 - Knock Sensor (KS) 1
G66 - Knock Sensor (KS) 2
J220 - Motronic Engine Control Module (ECM)
T3a - 3-Pin Connector, blue, on firewall, left
T3b - 3-Pin Connector, grey, on firewall, left
T3c - 3-Pin Connector, blue, on firewall, left
T10b - 6-Pin Connector, black, on protective housing for control module, in engine compartment, left
T80 - Connector, 80 point, on Motronic Engine Control Module (ECM)

(501) - Screw Connection -2- (30), on relay panel

ws = white
sw = black
ro = red
br = brown
gn = green
bl = blue
gr = grey
li = lilac
ge = yellow
or = orange

Edition 04.04

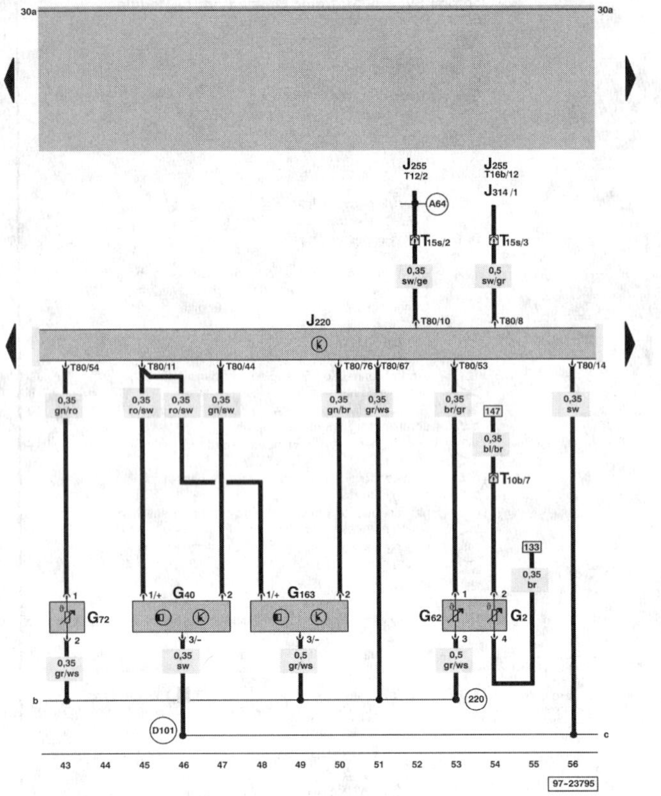

Motronic Engine Control Module (ECM), Camshaft Position (CMP) Sensor, Engine Coolant Temperature (ECT) Sensor, Intake Air Temperature (IAT) Sensor

G2 - Engine Coolant Temperature (ECT) Sensor
G40 - Camshaft Position (CMP) Sensor
G62 - Engine Coolant Temperature (ECT) Sensor
G72 - Intake Air Temperature (IAT) Sensor
G163 - Camshaft Position (CMP) Sensor 2
J220 - Motronic Engine Control Module (ECM)
J255 - Climatronic Control Module
J314 - A/C Cut-Off Control Module
T10b - 10-Pin Connector, black, on protective housing for control module, in engine compartment, left
T12 - 12-Pin Connector, on Control Module for Climatronic
T15s - 10-Pin Connector, red, on protective housing for control module, in engine compartment, left
T16b - 16-Pin Connector, on Control Module for Climatronic
T80 - Connector, 80 point, on Motronic Engine Control Module (ECM)

(220) - Ground connection (sensor ground), in engine compartment wiring harness
(A64) - Connection (30, idle boost), in instrument panel wiring harness
(D101) - Wire connection -1-, in engine compartment wiring harness

ws = white
sw = black
ro = red
br = brown
gn = green
bl = blue
gr = grey
li = lilac
ge = yellow
or = orange

Edition 04.04

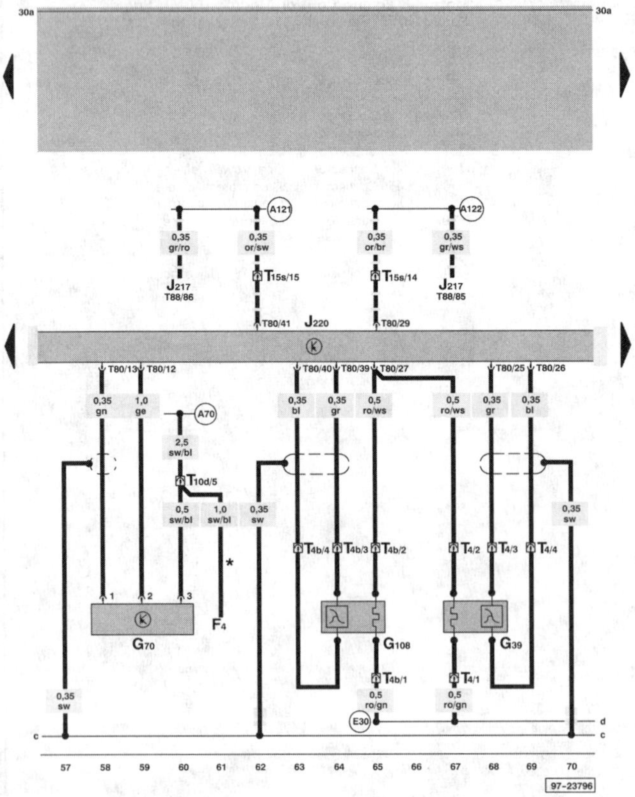

Motronic Engine Control Module (ECM), Mass Air Flow (MAF) Sensor, Heated Oxygen Sensors (HO2S)

F4 - Back-Up Light Switch
G39 - Heated Oxygen Sensor (HO2S)
G70 - Mass Air Flow (MAF) Sensor
G108 - Heated Oxygen Sensor (HO2S) 2
J217 - Transmission Control Module (TCM)
J220 - Motronic Engine Control Module (ECM)
T4 - 4-Pin Connector, black
T4b - 4-Pin Connector, black
T10d - 10-Pin Connector, brown, on protective housing for control module, in engine compartment, left
T15s - 15-Pin Connector, red, on protective housing for control module, in engine compartment,left
T80 - Connector, 80 point, on Motronic Engine Control Module (ECM)
T88 - Connector, 88 point, on Transmission Control Module (TCM)

(A70) - Connection (15a, fuse 231), in instrument panel wiring harness
(A121) - Connection (High-Bus), in instrument panel wiring harness
(A122) - Connection (Low-Bus), in instrument panel wiring harness
(E30) - Connection (87a), in wiring harness engine

- - - automatic transmission only
* - manual transmission only

ws = white
sw = black
ro = red
br = brown
gn = green
bl = blue
gr = grey
li = lilac
ge = yellow
or = orange

Edition 04.04

Motronic Engine Control Module (ECM), Oxygen Sensor (O2S) Behind Three Way Catalytic Converter (TWC), Valve for Camshaft Adjustment

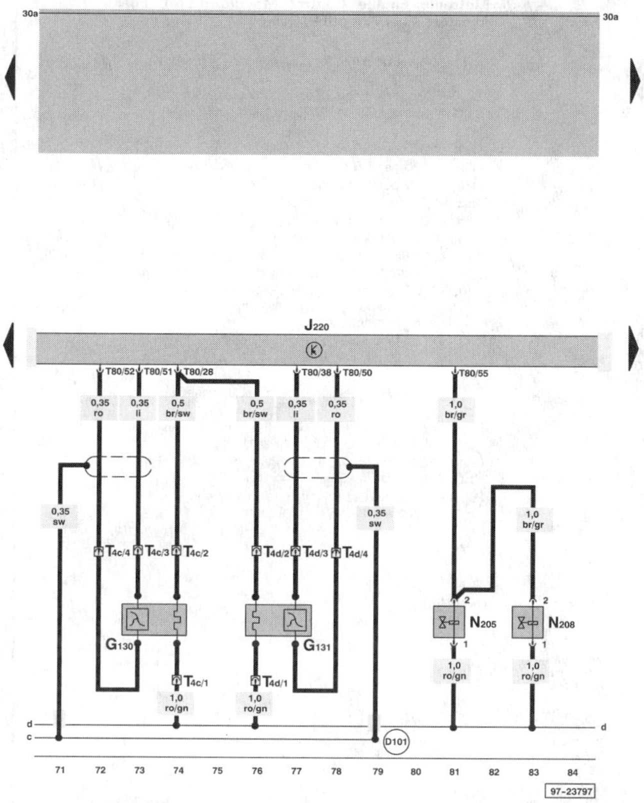

- G130 - Oxygen Sensor (O2S) Behind Three Way Catalytic Converter (TWC)
- G131 - Oxygen Sensor (O2S) 2 Behind Three Way Catalytic Converter (TWC)
- J220 - Motronic Engine Control Module (ECM)
- N205 - Valve -1- for camshaft adjustment
- N208 - Valve -2- for camshaft adjustment
- T4c - 4-Pin Connector, green
- T4d - 4-Pin Connector, brown
- T80 - Connector, 80 point, on Motronic Engine Control Module (ECM)
- (D101) - Wire connection -1-, in engine compartment wiring harness

ws = white
sw = black
ro = red
br = brown
gn = green
bl = blue
gr = grey
li = lilac
ge = yellow
or = orange

Motronic Engine Control Module (ECM), Secondary Air Injection (AIR) System, Data Link Connector (DLC)

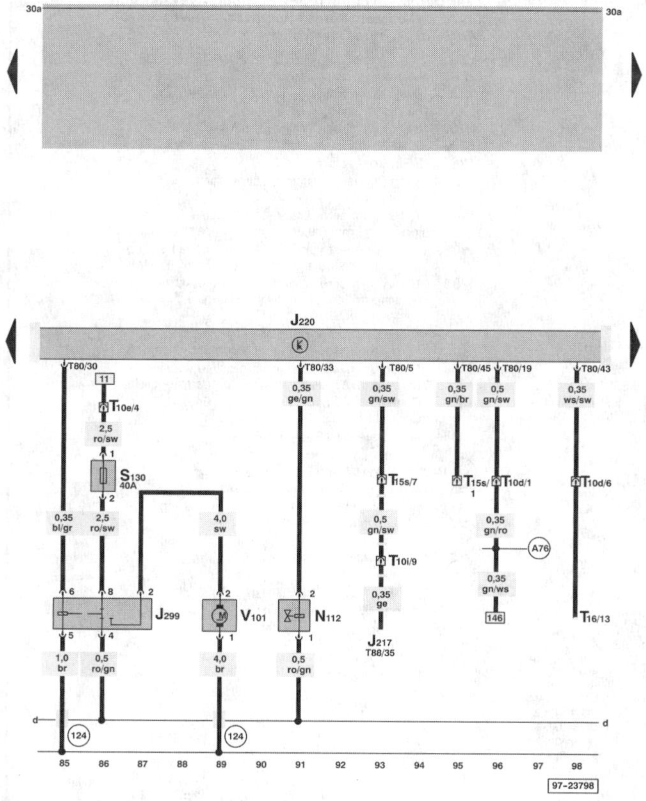

- J217 - Transmission Control Module (TCM)
- J220 - Motronic Engine Control Module (ECM)
- J299 - Secondary Air Injection (AIR) Pump Relay (203), in protective housing for control module
- N112 - Secondary Air Injection (AIR) Solenoid Valve, green
- S130 - Fuse for secondary air pump, in fuse adaptor above relay panel
- T10d - 10-Pin Connector, brown, on protective housing for control module, in engine compartment, left
- T10e - 10-Pin Connector, orange, on protective housing for control module, in engine compartment, left
- T10i - 10-Pin Connector, blue, on coupling station A-pillar, right
- T15s - 15-Pin Connector, red, on protective housing for control module, in engine compartment, left
- T16 - 16-Pin Connector, On Board Diagnostic (OBD), below instrument panel, left
- T80 - Connector, 80 point, on Motronic Engine Control Module (ECM)
- T88 - Connector, 88 point, on Transmission Control Module (TCM)
- V101 - Secondary Air Injection (AIR) Pump Motor
- (124) - ground connection, in engine compartment right wiring harness
- (A76) Connector (K-diagnosis wire), in instrument panel wiring harness
- --- - automatic transmission only

ws = white
sw = black
ro = red
br = brown
gn = green
bl = blue
gr = grey
li = lilac
ge = yellow
or = orange

Passat — Wiring diagram — No. 30/9

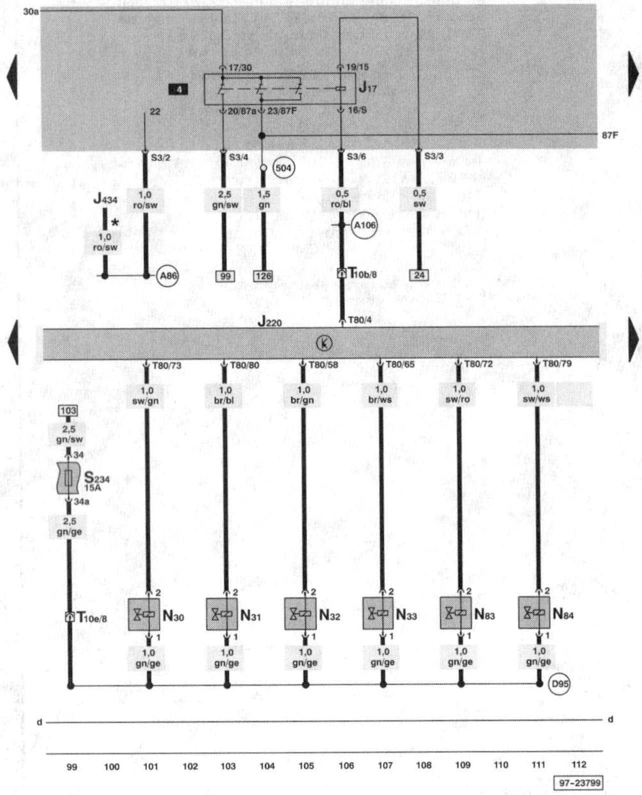

Motronic Engine Control Module (ECM), Fuel Injectors, Fuel Pump (FP) Relay

J17 - Fuel Pump (FP) Relay, production control number (208)
J220 - Motronic Engine Control Module (ECM)
J434 - Starting Interlock Relay (clutch position)
N30 - Cylinder 1 Fuel Incjector
N31 - Cylinder 2 Fuel Injector
N32 - Cylinder 3 Fuel Injector
N33 - Cylinder 4 Fuel Injector
N83 - Cylinder 5 Fuel Injector
N84 - Cylinder 6 Fuel Injector
S234 - Fuse 34 in fuse holder
T10b - 10-Pin Connector, black, on protective housing for control module, in engine compartment, left
T10e - 10-Pin Connector, orange, on protective housing for control module, in engine compartment, left
T80 - Connector, 80 point, on Motronic Engine Control Module (ECM)

(504) - Screw Connection (87F), on relay panel

(A86) - Connection (50a) in instrument panel wiring harness
(A106) - Connection -2- (86s), in instrument panel wiring harness
(D95) - wire connection (injectors), in engine compartment wiring harness

* - manual transmission only

ws = white
sw = black
ro = red
br = brown
gn = green
bl = blue
gr = grey
li = lilac
ge = yellow
or = orange

97-23799

Edition 04.04

Passat — Wiring diagram — No. 30/10

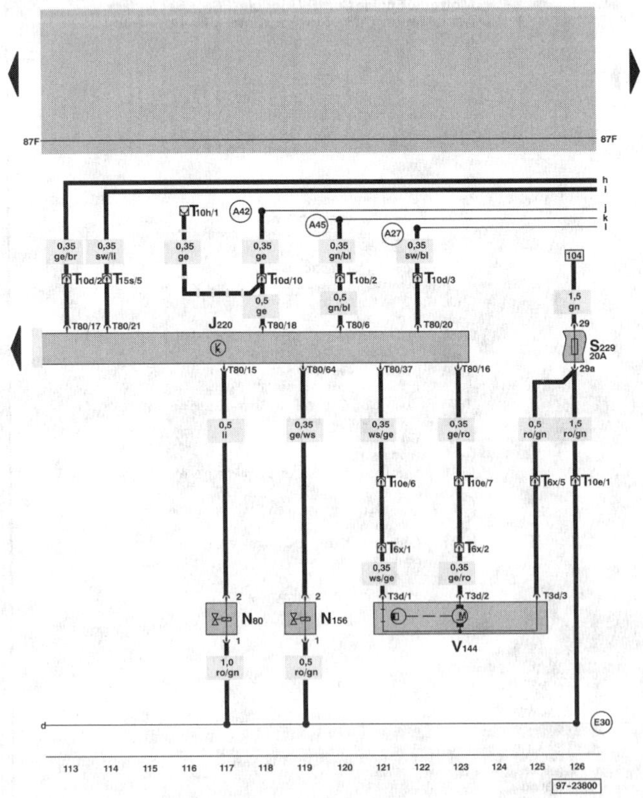

Leak Detection Pump (LDP), Eevaporative Emission (EVAP) Canister Purge Regulator Valve, Intake Manifold Change-Over Valve

J220 - Motronic Engine Control Module (ECM)
N80 - Evaporative Emission (EVAP) Canister Purge Regulator Valve
N156 - Intake Manifold Change-Over Valve
S229 - Fuse 29 in fuse holder
T3d - 3-Pin Connector
T6x - 6-Pin Connector, brown, connector station A-pillar, right
T10b - 10-Pin Connector, black, on protective housing for control module, in engine compartment, left
T10d - 10-Pin Connector, brown, on protective housing for control module, in engine compartment, left
T10e - 10-Pin Connector, orange, on protective housing for control module, in engine compartment, left
T10h - 10-Pin Connector, blue, on protective housing for control module, in engine compartment, left
T15s - 15-Pin Connector, red, on protective housing for control module, in engine compartment,left
T80 - Connector, 80 point, on Motronic Engine Control Module (ECM)
V144 - Leak Detection Pump (LDP)

(A27) - Wire Connection (vehicle speed signal), in instrument panel wiring harness
(A42) - plus connection (fuel gauge), in instrument panel wiring harness
(A45) - wire connection (RPM-signal), in instrument panel wiring harness
(E30) - Connection (87a), in wiring harness engine

- - - - automatic transmission only

ws = white
sw = black
ro = red
br = brown
gn = green
bl = blue
gr = grey
li = lilac
ge = yellow
or = orange

97-23800

Edition 04.04

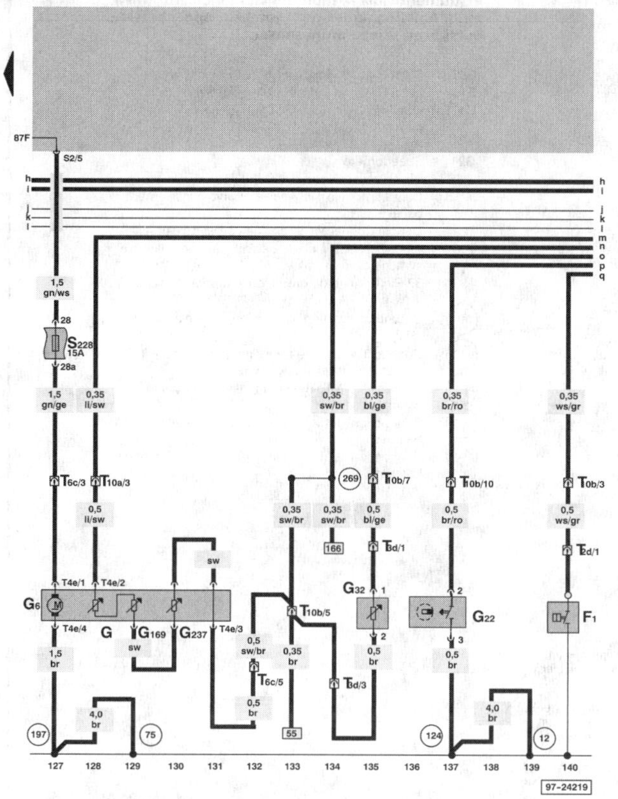

Oil pressure switch, Fuel Pump (FP), sender for fuel gauge, sender -2- and -3- for fuel supply, speedometer vehicle speed sensor

F1 - Oil Preassure Switch
G - Sender for fuel gauge
G6 - Fuel Pump (FP)
G22 - Speedometer Vehicle Speed Sensor (VSS), hall sender on transmission, left
G32 - Engine Coolant Level (ECL) Sensor
G169 - Sender -2- for fuel supply, only Syncro
G237 - Sender -3- for fuel supply, only Syncro
S228 - Fuse 28 in fuse holder
T2d - Double Connector
T3d - 3-Pin Connector, in engine compartment, left front
T4e - 4-Pin Connector
T6c - 6-Pin Connector, blue, connector station A-pillar, left
T10a 10-Pin Connector, brown, connector station A-pillar, left
T10b - 10-Pin Connector, black, on protective housing for control module, in engine compartment, left

(12) - Ground connection, in engine compartment, left
(75) - Ground connection, on right rear pillar
(124) - ground connection, in engine compartment right wiring harness
(197) - Ground connection -4-, in wiring harness rear
(269) - Ground connection (sensor ground) -1-, in instrument panel wiring harness

ws = white
sw = black
ro = red
br = brown
gn = green
bl = blue
gr = grey
li = lilac
ge = yellow
or = orange

Edition 04.04

Repair Manual page number Y 308

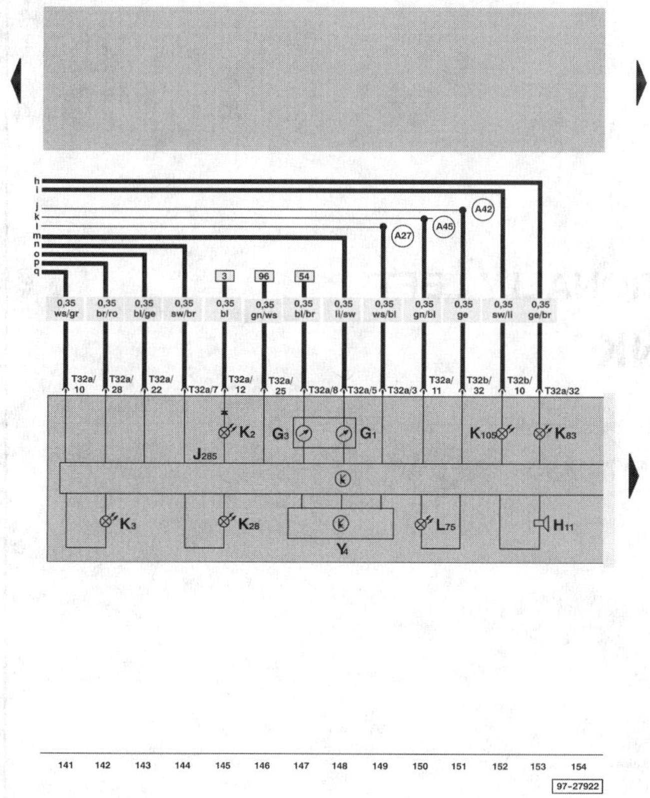

Instrument cluster, engine coolant temperature (ECT) gauge, low fuel level warning light, oil pressure warning light, Malfunction Indicator Lamp (MIL)

G1 - Fuel Gauge
G3 - Engine Coolant Temperature (ECT) Gauge
H11 - Oil Pressure Warning Buzzer
J285 - Control module with indicator unit in instrument panel insert
K2 - Generator (GEN) Warning Light
K3 - Oil Pressure Warning Light
K28 Engine Coolant Level/Temperature (ECL/ECT) Warning Light
K83 - Malfunction Indicator Lamp (MIL)
K105 - Low Fuel Level Warning Light
L75 - Digital Display Light
T32a - 32-Pin Connector, blue, on instrument cluster
T32b - 32-Pin Connector, green, on instrument cluster
Y4 - Odometer Display

(A27) - Wire Connection (vehicle speed signal), in instrument panel wiring harness
(A42) - plus connection (fuel gauge), in instrument panel wiring harness
(A45) - wire connection (RPM-signal), in instrument panel wiring harness

ws = white
sw = black
ro = red
br = brown
gn = green
bl = blue
gr = grey
li = lilac
ge = yellow
or = orange

Edition 04.04

Repair Manual page number Y 307

Instrument cluster, multi-function indicator (MFI), multi-function indicator switches, speedometer, outside air temperature sensor

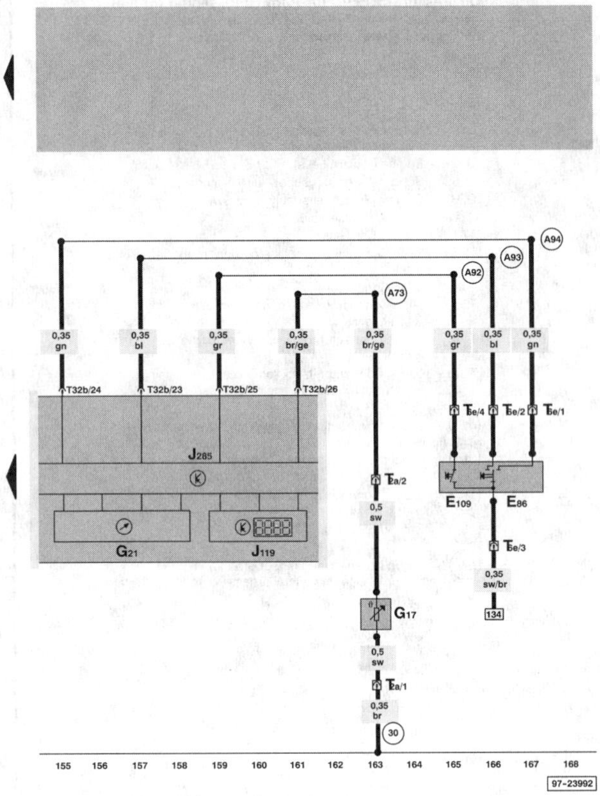

- E86 - Multi-Function Indicator Mode Select Switch
- E109 - Multi-Function Indicator Memory Switch
- G17 - Outside Air Temperature Sensor (use corresponding wiring diagram on vehicles with A/C)
- G21 - Speedometer
- J119 - Multi-function Indicator (MFI)
- J285 - Control module with indicator unit in instrument panel insert
- T2a - Double Connector, in engine compartment left
- T6e - 6-Pin Connector, black, behind steering column switch cover
- T32b - 32-Pin Connector, green, on instrument cluster
- (30) - Ground connection, -1-, beside fuse/relay panel
- (A73) - Connection (outside temperature indicator), in instrument panel wiring harness
- (A92) - Connection (MFI-Reset), in instrument panel wiring harness
- (A93) - Connection -1- (MFI), in instrument panel wiring harness
- (A94) - Connection -2- (MFI), in instrument panel wiring harness

ws = white
sw = black
ro = red
br = brown
gn = green
bl = blue
gr = grey
li = lilac
ge = yellow
or = orange

Edition 04.04

THIS PAGE INTENTIONALLY LEFT BLANK

Automatic transmission, 5 speed automatic with tiptronic

- **2,8L-Engine - Motronic Multiport Fuel Injection (MFI) / 142 kW, code AHA,**

from August 1998

Deviate relay location and fuseplacements as well as the locations of multiple connectors see section "component locations"

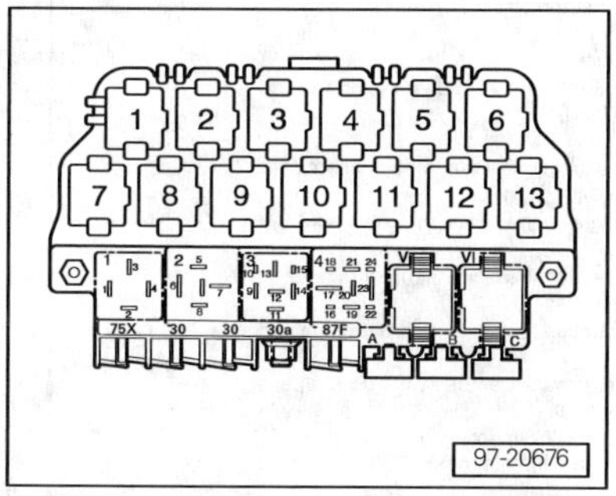

Relay location on the thirteen position auxiliary relay panel, above relay panel:

6 - Selector Lever Light Relay (205)
12 - Starting Interlock Relay (alarm system) (186)
13 - Starting Interlock Relay (204)

Relay panel:

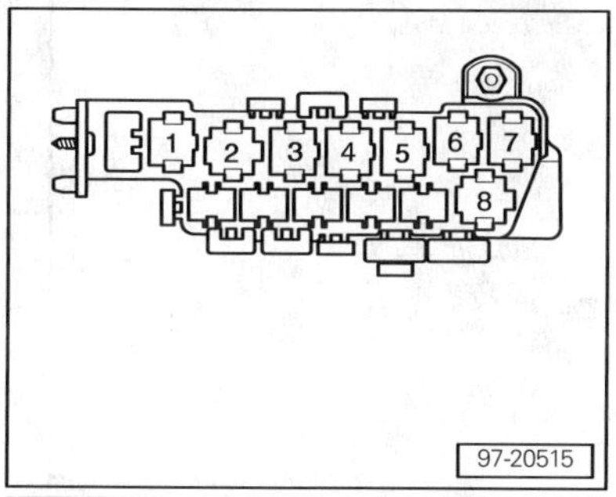

Relay location on the eight position auxiliary relay panel, above relay panel:

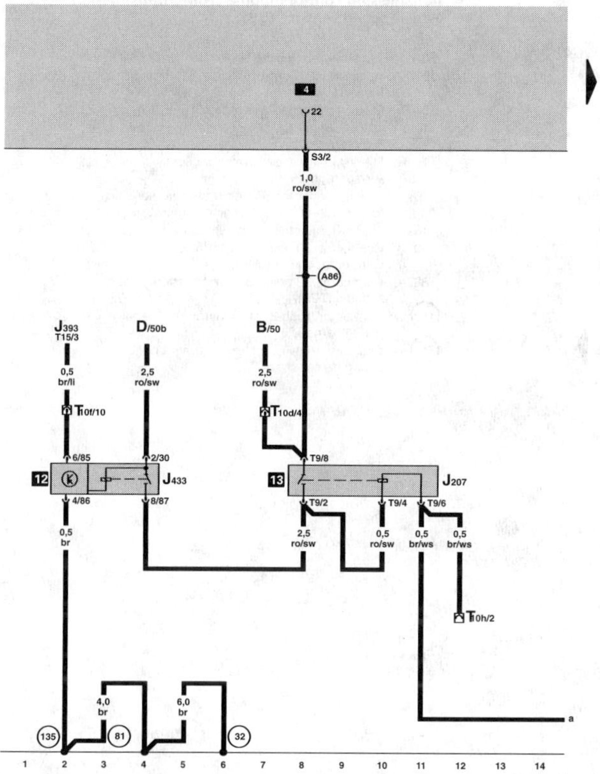

ws = white
sw = black
ro = red
br = brown
gn = green
bl = blue
gr = grey
li = lilac
ge = yellow
or = orange

Starting interlock relay, Starting interlock relay (alarm system)

B - Starter
D - Ignition/Starter Switch
J207 - Starting Interlock Relay, on the thirteenfold auxiliary relay panel, above relay panel, production control number (204)
J393 - Central control module for comfort system
J433 - Starting Interlock Relay (alarm system), on the thirteenfold auxiliary relay panel, above relay panel, production control number (186)
T9 - 9-Pin Connector, on Starting Interlock Relay
T10d - 10-Pin Connector, brown, on protective housing for control module, in engine compartment, left
T10f - 10-Pin Connector, blue, connector station A-pillar, left
T10h - 10-Pin Connector, blue, on protective housing for control module, in engine compartment, left
T15 - 15-Pin Connector, on Central control module for comfort system
(32) - Ground connection, behind instrument panel, left
(81) - Ground connection -1-, in instrument panel wiring harness
(135) - Ground connection -2-, in instrument panel wiring harness
(A86) - Connection (50a), in instrument panel wiring harness

Transmission control module (TCM), multi-function transmission range (TR) switch

E45 - Cruise Control Switch
F125 - Multi-Function Transmission Range (TR) Switch
J217 - Transmission Control Module (TCM), in front footwell, right
S15 - Fuse 15 in fuse holder
T3m - 3-Pin Connector, brown, connector station A-pillar, right
T8 - 8-Pin Connector, black
T15m - 15-Pin Connector, brown, connector station A-pillar, right
T88 - Connector, 88 point, on Transmission Control Module (TCM)

(114) - Ground connection, in automatic transmission wiring harness
(500) - Screw Connection -1- (30), on relay panel

(A1) - plus connection (30a), in instrument panel wiring harness
(A32) - plus connection (30), in instrument panel wiring harness
(A87) - Connector (reverse lamp), in instrument panel wiring harness
(U10) - Cruise Control Connection (in automatic transmission wiring harness)

- - - - vehicles with cruise control only

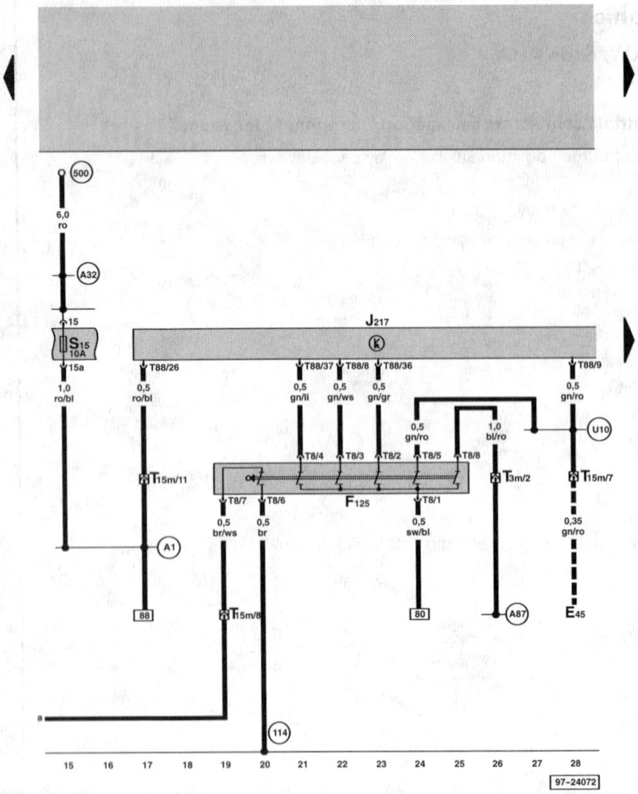

ws = white
sw = black
ro = red
br = brown
gn = green
bl = blue
gr = grey
li = lilac
ge = yellow
or = orange

Edition 10.02

Transmission control module (TCM), transmission vehicle speed sensor (VSS)

G38 - Transmission Vehicle Speed Sensor (VSS)
J217 - Transmission Control Module (TCM), in front footwell, right
J220 - Motronic Engine Control Module (ECM)
T10i - 10-Pin Connector, blue, connector station A-pillar, right
T15m - 15-Pin Connector, brown, connector station A-pillar, right
T15s - 15-Pin Connector, red, on protective housing for control module, in engine compartment, left
T16a - 16-Pin Connector, on transmission
T80 - Connector, 80 point, on Motronic Engine Control Module (ECM)
T88 - Connector, 88 point, on Transmission Control Module (TCM)
(A121) - Connector (High-Bus), in instrument panel wiring harness
(A122) - Connector (Low-Bus), in instrument panel wiring harness

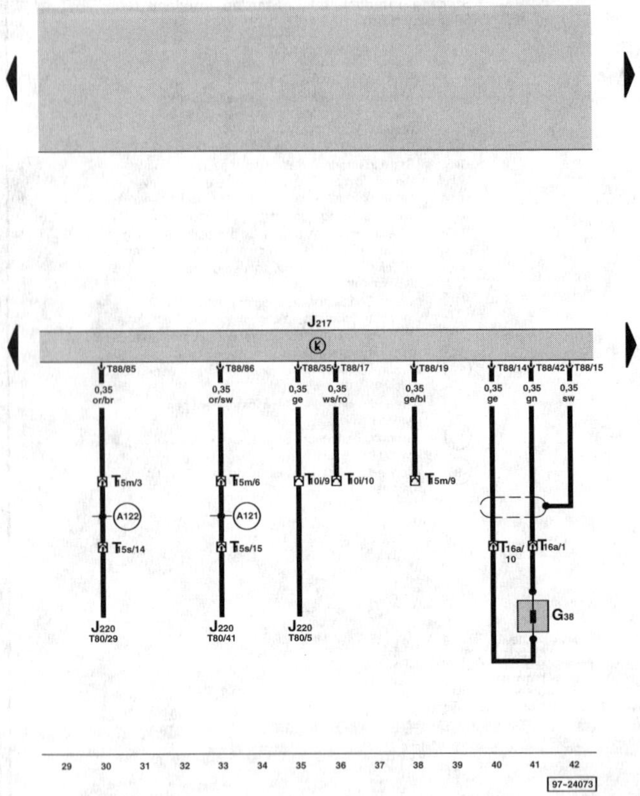

ws = white
sw = black
ro = red
br = brown
gn = green
bl = blue
gr = grey
li = lilac
ge = yellow
or = orange

Edition 10.02

Repair Manual page number Y 314

Repair Manual page number Y 313

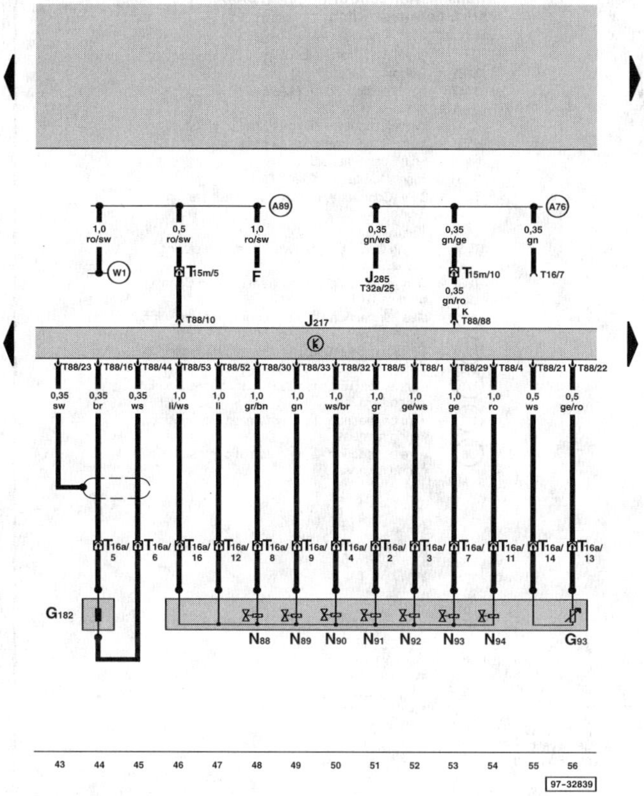

Transmission control module (TCM), solenoid valves, transmission fluid temperature sensor, Sensor for Transmission

F - Brake Light Switch
G93 - Transmission Fluid Temperature Sensor
G182 - Sensor for Transmission (RPM)
J217 - Transmission Control Module (TCM), in front footwell, right
J285 - Control module with indicator unit in instrument panel insert
N88 - Solenoid Valve 1
N89 - Solenoid Valve 2
N90 - Solenoid Valve 3
N91 - Solenoid Valve 4
N92 - Solenoid Valve 5
N93 - Solenoid Valve 6
N94 - Solenoid Valve 7
T15m - 15-Pin Connector, brown, connector station A-pillar, right
T16 - 16-Pin Connector, On Board Diagnostic (OBD), below instrument panel, left
T16a - 16-Pin Connector, on transmission
T32a - 32-Pin Connector, blue, on instrument cluster
T88 - Connector, 88 point, on Transmission Control Module (TCM)

(A76) - Connector (K-diagnosis wire), in instrument panel wiring harness
(A89) - Connector (54), in instrument panel wiring harness
(W1) - plus connection (54), in rear wiring harness

ws = white
sw = black
ro = red
br = brown
gn = green
bl = blue
gr = grey
li = lilac
ge = yellow
or = orange

Edition 10.02

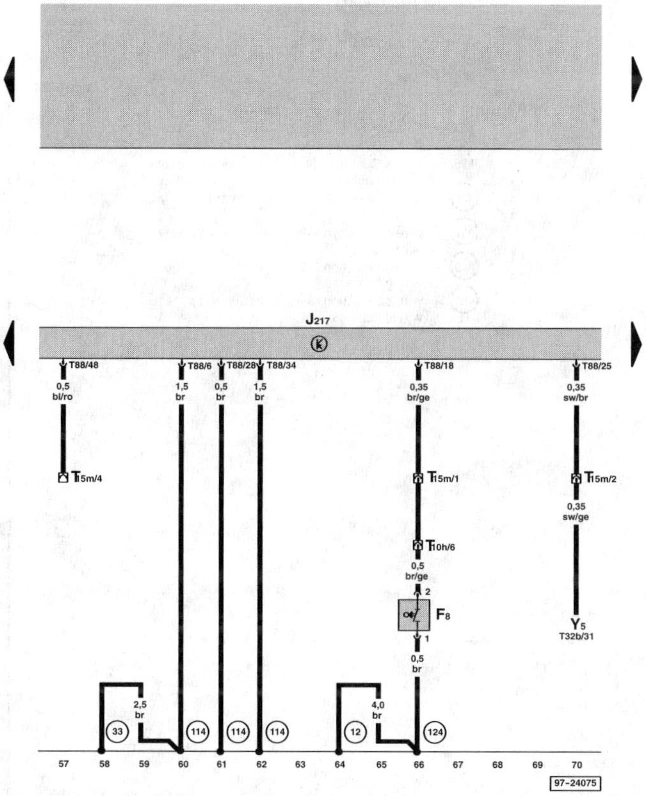

Transmission control module (TCM), kick down switch

F8 - Kick Down Switch
J217 - Transmission Control Module (TCM), in front footwell, right
T10h - 10-Pin Connector, blue, on protective housing for control module, in engine compartment, left
T15m - 15-Pin Connector, brown, connector station A-pillar, right
T32b - 32-Pin Connector, green, on instrument cluster
T88 - Connector, 88 point, on Transmission Control Module (TCM)
Y5 - Transmission Range (TR) Selector Lever Display

(12) - Ground connection, in engine compartment, left
(33) - Ground connection, behind instrument panel, right
(114) - Ground connection, in automatic transmission wiring harness
(124) - ground connection, in engine compartment right wiring harness

ws = white
sw = black
ro = red
br = brown
gn = green
bl = blue
gr = grey
li = lilac
ge = yellow
or = orange

Edition 10.02

Repair Manual page number Y 316
Repair Manual page number Y 315

Transmission control module (TCM), shift lock solenoid, tiptronic switch

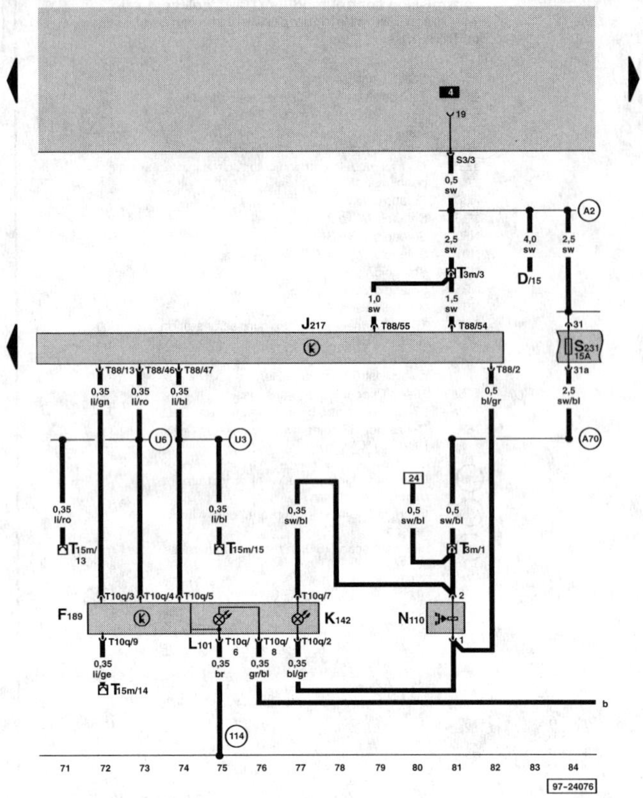

- D - Ignition/Starter Switch
- F189 - Tiptronic Switch
- J217 - Transmission Control Module (TCM), in front footwell, right
- K142 - Warning light for selector lever position P/N
- L101 - Illumination for selector lever scale
- N110 - Shift Lock Solenoid
- S231 - Fuse 31 in fuse holder
- T3m - 3-Pin Connector, brown, connector station A-pillar, right
- T10q - 10-Pin Connector
- T15m - 15-Pin Connector, brown, connector station A-pillar, right
- T88 - Connector, 88 point, on Transmission Control Module (TCM)
- (A2) - plus connection (15), in instrument panel wiring harness
- (114) - Ground connection, in automatic transmission wiring harness
- (A70) - Connector (15a, fuse 231), in instrument panel wiring harness
- (U3) - wire connection -1-, in automatic transmission wiring harness
- (U6) - wire connection -2-, in automatic transmission wiring harness

ws = white
sw = black
ro = red
br = brown
gn = green
bl = blue
gr = grey
li = lilac
ge = yellow
or = orange

Edition 10.02

Selector lever light

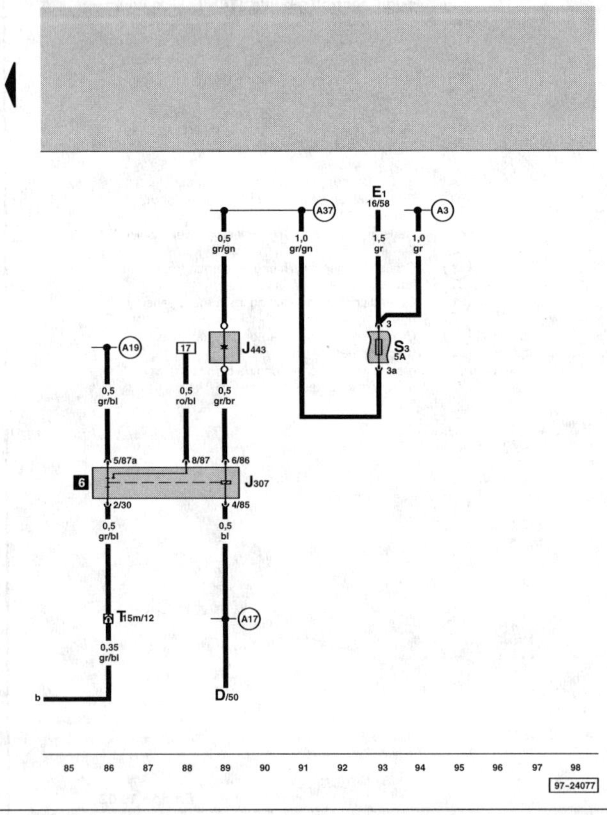

- D - Ignition/Starter Switch
- E1 - Light switch
- J307 - Selector Lever Light Relay, on the thirteenfold auxiliary relay panel, above relay panel, production control number (205)
- J443 - Selector Lever Light Diode
- S3 - Fuse 3 in fuse holder
- T15m - 15-Pin Connector, brown, connector station A-pillar, right
- (A3) - plus connection (58), in instrument panel wiring harness
- (A17) - wire connection (61), in instrument panel wiring harness
- (A19) - wire connection (58d), in instrument panel wiring harness
- (A37) - wire connection (58a), in instrument panel wiring harness

ws = white
sw = black
ro = red
br = brown
gn = green
bl = blue
gr = grey
li = lilac
ge = yellow
or = orange

Edition 10.02

Automatic transmission, 5 speed automatic with tiptronic

- **1,8L-Engine - Motronic Multiport Fuel Injection (MFI) / 110 kW, code AEB,**

from August 1998

Deviate relay location and fuseplacements as well as the locations of multiple connectors see section "component locations"

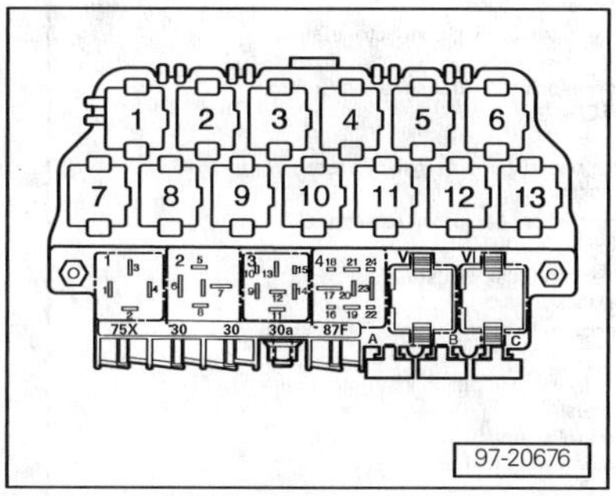

Relay location on the thirteen position auxiliary relay panel, above relay panel:

6 - Selector Lever Light Relay (205)
12 - Starting Interlock Relay (alarm system) (186)
13 - Starting Interlock Relay (204)

Relay panel:

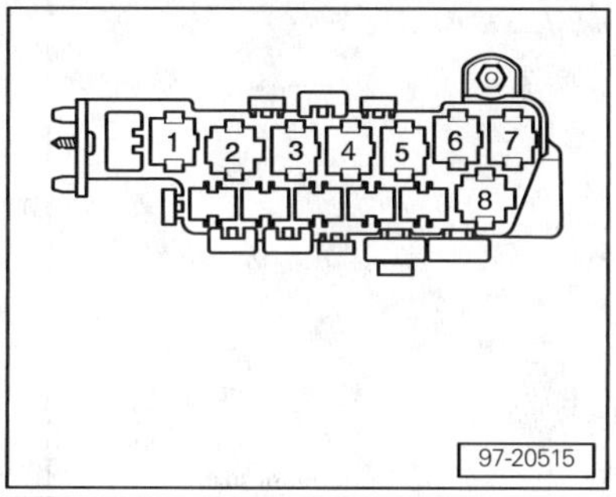

Relay location on the eight position auxiliary relay panel, above relay panel:

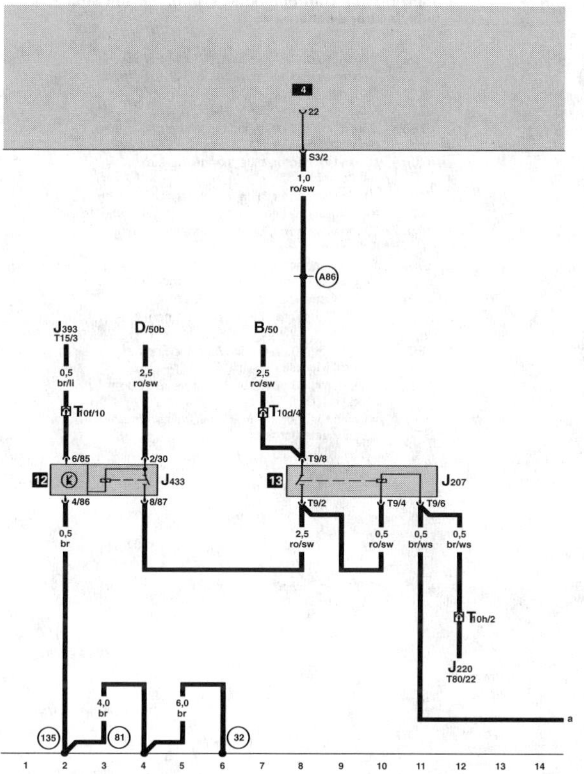

Starting interlock relay, Starting interlock relay (alarm system)

- B - Starter
- D - Ignition/Starter Switch
- J207 - Starting Interlock Relay, on the thirteenfold auxiliary relay panel, above relay panel, production control number (204)
- J220 - Motronic Engine Control Module (ECM)
- J393 - Central control module for comfort system
- J433 - Starting Interlock Relay (alarm system), on the thirteenfold auxiliary relay panel, above relay panel, production control number (186)
- T9 - 9-Pin Connector, on Starting Interlock Relay
- T10d - 10-Pin Connector, brown, on protective housing for control module, in engine compartment, left
- T10f - 10-Pin Connector, blue, connector station A-pillar, left
- T10h - 10-Pin Connector, blue, on protective housing for control module, in engine compartment, left
- T15 - 15-Pin Connector, on Central control module for comfort system
- T80 - Connector, 80 point, on Motronic Engine Control Module (ECM)
- (32) - Ground connection, behind instrument panel, left
- (81) - Ground connection -1-, in instrument panel wiring harness
- (135) - Ground connection -2-, in instrument panel wiring harness
- (A86) - Connection (50a), in instrument panel wiring harness

ws = white
sw = black
ro = red
br = brown
gn = green
bl = blue
gr = grey
li = lilac
ge = yellow
or = orange

Transmission control module (TCM), multi-function transmission range (TR) switch

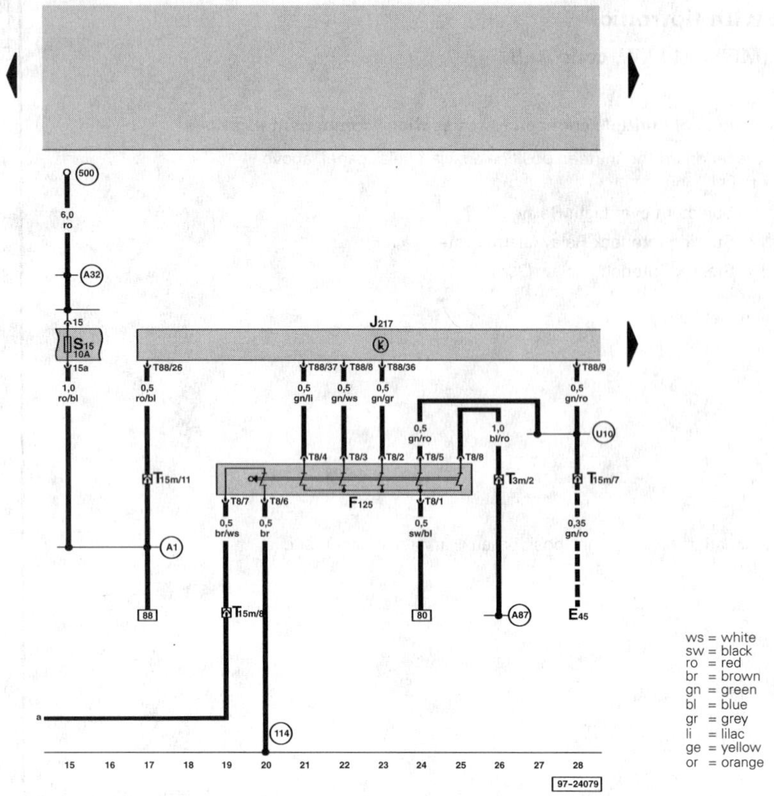

- E45 - Cruise Control Switch
- F125 - Multi-Function Transmission Range (TR) Switch
- J217 - Transmission Control Module (TCM), in front footwell, right
- S15 - Fuse 15 in fuse holder
- T3m - 3-Pin Connector, brown, connector station A-pillar, right
- T8 - 8-Pin Connector, black
- T15m - 15-Pin Connector, brown, connector station A-pillar, right
- T88 - Connector 88-point, on Transmission Control Module (TCM)
- (114) - Ground connection, in automatic transmission wiring harness
- (500) - Screw Connection -1- (30), on relay panel
- (A1) - plus connection (30a), in instrument panel wiring harness
- (A32) - plus connection (30), in instrument panel wiring harness
- (A87) - Connector (reverse lamp), in instrument panel wiring harness
- (U10) - Cruise Control Connection (in automatic transmission wiring harness)
- - - - vehicles with cruise control only

Transmission control module (TCM), transmission vehicle speed sensor (VSS)

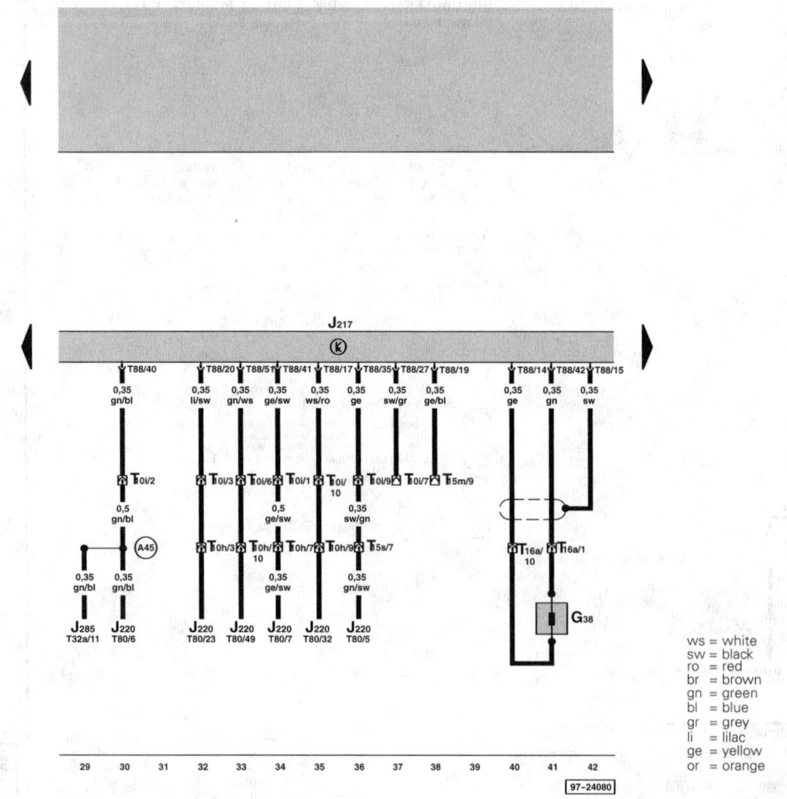

- G38 - Transmission Vehicle Speed Sensor (VSS)
- J217 - Transmission Control Module (TCM), in front footwell, right
- J220 - Motronic Engine Control Module (ECM)
- J285 - Control module with indicator unit in instrument panel insert
- T10i - 10-Pin Connector, blue, connector station A-pillar, right
- T10h - 10-Pin Connector, blue, on protective housing for control module, in engine compartment, left
- T15m - 15-Pin Connector, brown, connector station A-pillar, right
- T15s - 15-Pin Connector, red, on protective housing for control module, in engine compartment,left
- T16a - 16-Pin Connector, on transmission
- T32a - 32-Pin Connector, blue, on instrument cluster
- T80 - Connector, 80 point, on Motronic Engine Control Module (ECM)
- T88 - Connector, 88 point, on Transmission Control Module (TCM)
- (A45) - wire connection (RPM-signal), in instrument panel wiring harness

Passat

Wiring diagram

No. 32/5

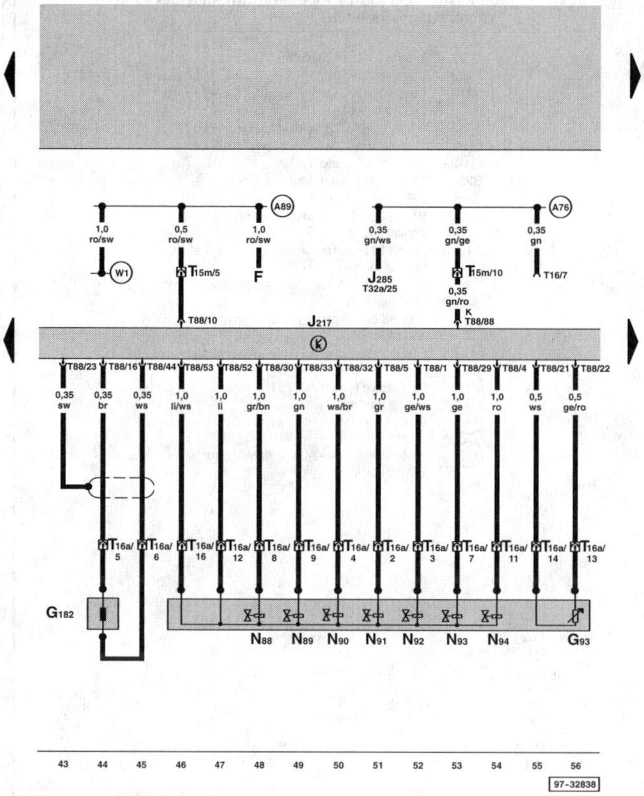

Transmission control module (TCM), solenoid valves, transmission fluid temperature sensor, Sensor for transmission

- F - Brake Light Switch
- G93 - Transmission Fluid Temperature Sensor
- G182 - Sensor for transmission RPM
- J217 - Transmission Control Module (TCM), in front footwell, right
- J285 - Control module with indicator unit in instrument panel insert
- N89 - Solenoid Valve 2
- N89 - Solenoid Valve 2
- N90 - Solenoid Valve 3
- N91 - Solenoid Valve 4
- N92 - Solenoid Valve 5
- N93 - Solenoid Valve 6
- N94 - Solenoid Valve 7
- T15m - 15-Pin Connector, brown, connector station A-pillar, right
- T16 - 16-Pin Connector, On Board Diagnostic (OBD), below instrument panel, left
- T16a - 16-Pin Connector, on transmission
- T32a - 32-Pin Connector, blue, on instrument cluster
- T88 - Connector, 88 point, on Transmission Control Module (TCM)
- (A76) - Connector (K-diagnosis wire), in instrument panel wiring harness
- (A89) - Connector (54), in instrument panel wiring harness
- (W1) - plus connection (54), in rear wiring harness

ws = white
sw = black
ro = red
br = brown
gn = green
bl = blue
gr = grey
li = lilac
ge = yellow
or = orange

Edition 10.02

Passat

Wiring diagram

No. 32/6

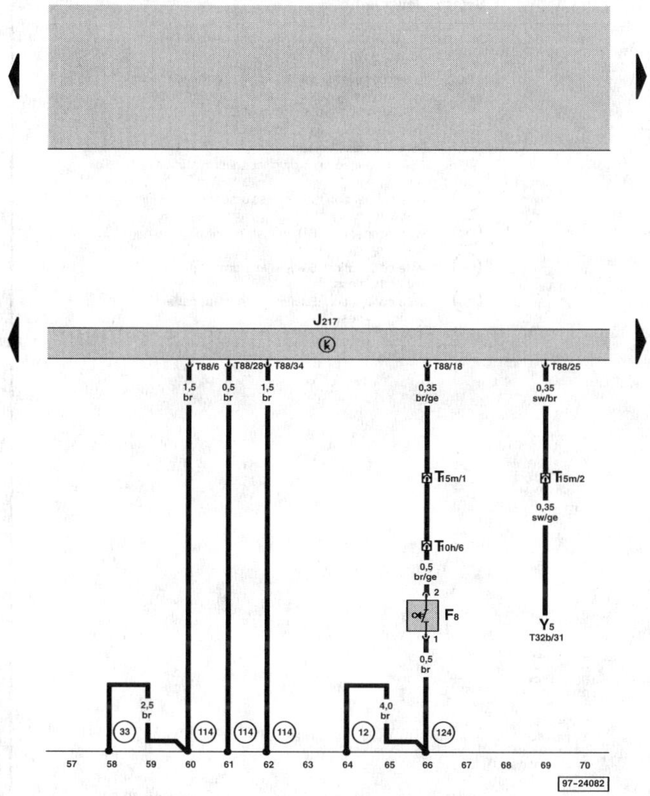

Transmission control module (TCM), kick down switch

- F8 - Kick Down Switch
- J217 - Transmission Control Module (TCM), in front footwell, right
- T10h - 10-Pin Connector, blue, on protective housing for control module, in engine compartment, left
- T15m - 15-Pin Connector, brown, connector station A-pillar, right
- T32b - 32-Pin Connector, green, on instrument cluster
- T88 - Connector, 88 point, on Transmission Control Module (TCM)
- Y5 - Transmission Range (TR) Selector Lever Display
- (12) - Ground connection, in engine compartment, left
- (33) - Ground connection, behind instrument panel, right
- (114) - Ground connection, in automatic transmission wiring harness
- (124) - ground connection, in engine compartment right wiring harness

ws = white
sw = black
ro = red
br = brown
gn = green
bl = blue
gr = grey
li = lilac
ge = yellow
or = orange

Edition 10.02

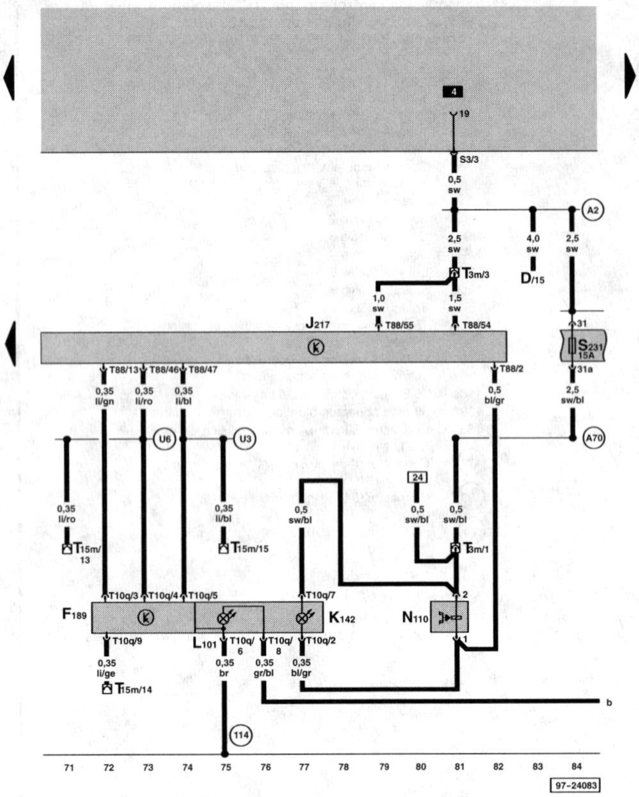

Transmission control module (TCM), shift lock solenoid, tiptronic switch

- D - Ignition/Starter Switch
- F189 - Tiptronic Switch
- J217 - Transmission Control Module (TCM), in front footwell, right
- K142 - Warning light for selector lever position P/N
- L101 - Illumination for selector lever scale
- N110 - Shift Lock Solenoid
- S231 - Fuse 31 in fuse holder
- T3m - 3-Pin Connector
- T10q - 10-Pin Connector, brown, connector station A-pillar, right
- T15m - 15-Pin Connector, brown, connector station A-pillar, right
- T88 - Connector, 88 point, on Transmission Control Module (TCM)
- (114) - Ground connection, in automatic transmission wiring harness
- (A2) - plus connection (15), in instrument panel wiring harness
- (A70) - Connector (15a, fuse 231), in instrument panel wiring harness
- (U3) - wire connection -1-, in automatic transmission wiring harness
- (U6) - wire connection -2-, in automatic transmission wiring harness

ws = white
sw = black
ro = red
br = brown
gn = green
bl = blue
gr = grey
li = lilac
ge = yellow
or = orange

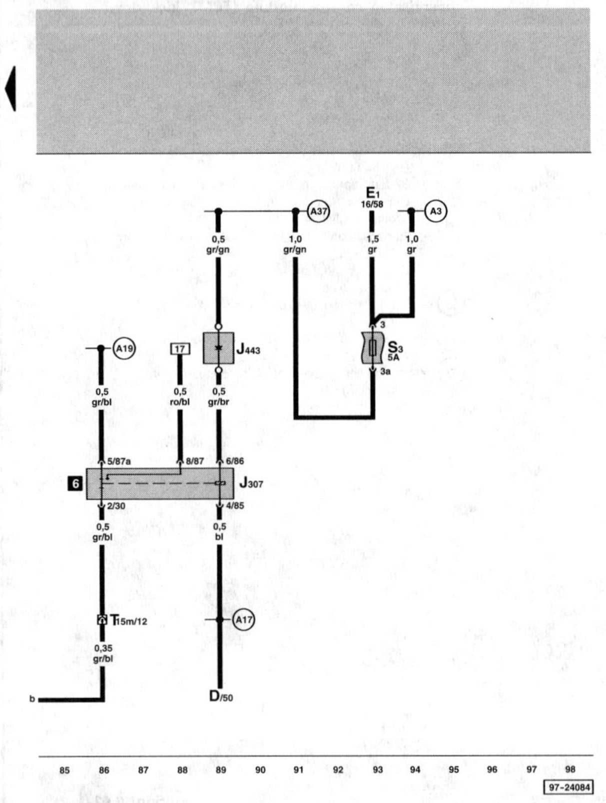

Selector lever light

- D - Ignition/Starter Switch
- E1 - Light switch
- J307 - Selector Lever Light Relay, on the thirteenfold auxiliary relay panel, above relay panel, production control number (205)
- J443 - Selector Lever Light Diode
- S3 - Fuse 3 in fuse holder
- T15m - 15-Pin Connector, brown, connector station A-pillar, right
- (A3) - plus connection (58), in instrument panel wiring harness
- (A17) - wire connection (61), in instrument panel wiring harness
- (A19) - wire connection (58d), in instrument panel wiring harness
- (A37) - wire connection (58a), in instrument panel wiring harness

ws = white
sw = black
ro = red
br = brown
gn = green
bl = blue
gr = grey
li = lilac
ge = yellow
or = orange

Climatronic,

- **2,8L - Engine - Motronic Multiport Fuel Injection (MFI) / 142kW, codes AHA**

from August 1998

- **2,8L - Engine - Motronic Multiport Fuel Injection (MFI) / 142kW, codes ATQ**

from August 1999

Deviate relay location and fuseplacements as well as the locations of multiple connectors see section "component locations".

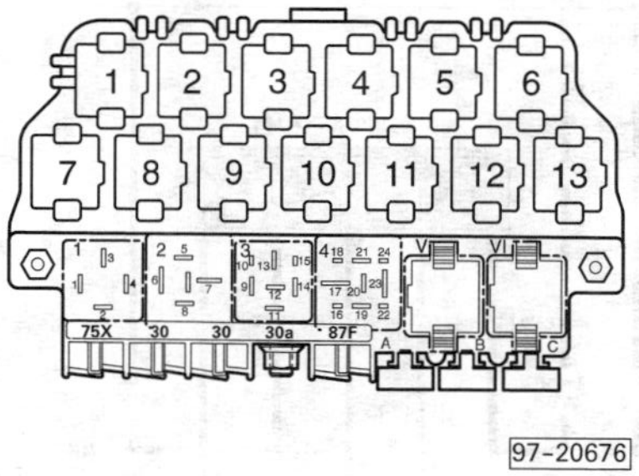

Relay location on the thirteenfold auxiliary relay panel, above relay panel:

3 A/C Clutch Relay (267)

Relay panel:

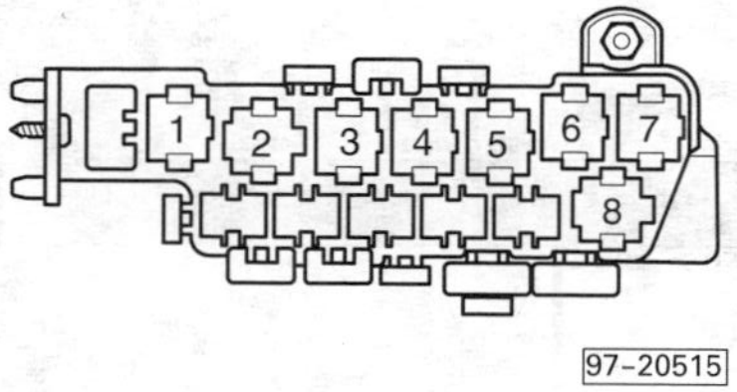

Relay location on the eightfold auxiliary relay panel, behind relay panel:

5 First Speed Coolant Fan Control (FC) Relay (214)

8 Second Speed Coolant Fan Control (FC) Relay (213)

Note: Number in parentheses indicates production control number stamped on relay housing.

Fuse colors

30 A - green
25 A - white
20 A - yellow
15 A - blue
10 A - red
7,5 A - brown
5 A - beige

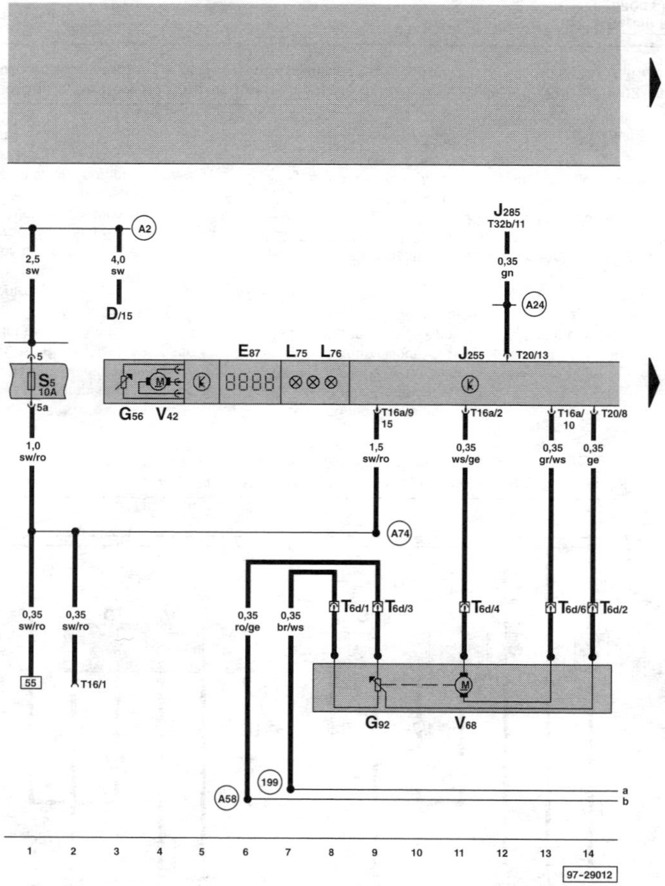

ws = white
sw = black
ro = red
br = brown
gn = green
bl = blue
gr = grey
li = violet
ge = yellow

- D - Ignition/Starter Switch
- E87 - A/C Control Head
- G56 - Instrument Panel Interior Temperature Sensor
- G92 - Temperature Regulator Flap Motor Position Sensor
- J255 - Climatronic Control Module
- J285 - Control module with indicator unit in instrument panel insert
- L75 - Digital Display Light
- L76 - Push Button Light
- S5 - Fuse 5 in fuse holder
- T6d - 6-Pin Connector, red, behind instrument panel, center
- T16 - 16-Pin Connector, On Board Diagnostic (OBD), below instrument panel, left
- T16a - 16-Pin Connector
- T20 - 20-Pin Connector
- T32b - 32-Pin Connector, green, on instrument cluster
- V42 - Interior Temperature Sensor Fan
- V68 - Temperature Regulator Flap Motor
- (199) - Ground connection -3-, in instrument panel wiring harness
- (A2) - plus connection (15), in instrument panel wiring harness
- (A24) - wire connection (brake system monitoring), in instrument panel wiring harness
- (A58) - Wire connection (5Volt), in instrument cluster wiring harness
- (A74) - Connector (15a - fuse 5), in instrument panel wiring harness

Climatronic control module, temperature regulator flap motor

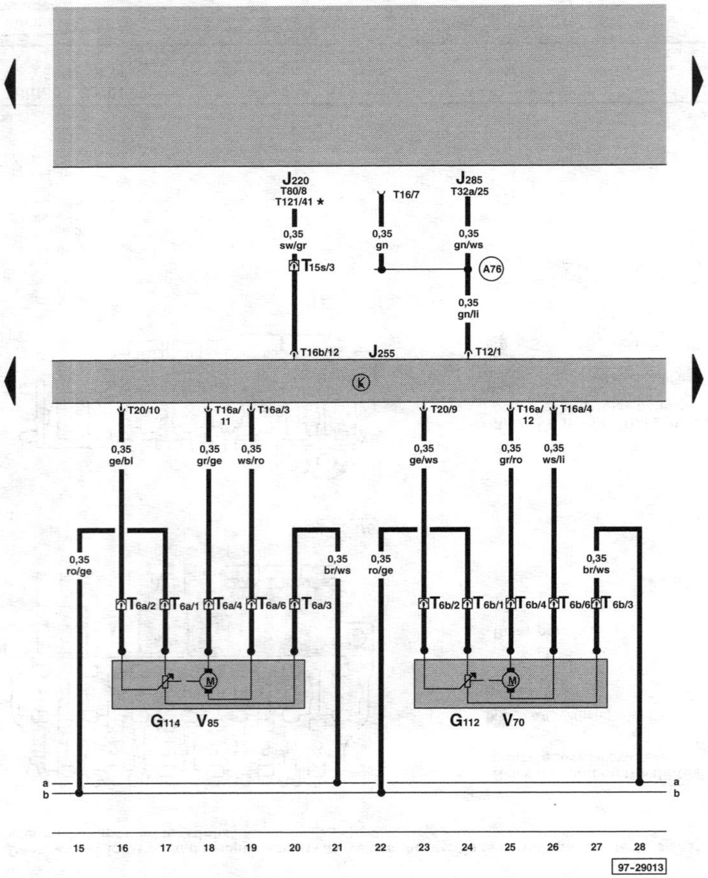

ws = white
sw = black
ro = red
br = brown
gn = green
bl = blue
gr = grey
li = violet
ge = yellow

G112 - Central Flap Motor Position Sensor
G114 - Footwell/Defrost Flap Motor Position Sensor
J220 - Motronic Engine Control Module (ECM)
J255 - Climatronic Control Module
J285 - Control module with indicator unit in instrument panel insert
T6a - 6-Pin Connector, behind instrument panel, center
T6b - 6-Pin Connector, behind instrument panel, center
T12 - 12-Pin Connector
T15s - 15-Pin Connector, red, on protective housing for control module, in engine compartment, left
T16 - 16-Pin Connector, On Board Diagnostic (OBD), below instrument panel, left
T16a - 16-Pin Connector
T16b - 16-Pin Connector
T20 - 20-Pin Connector
T32a - 32-Pin Connector, blue, on Control Module with indicator unit in instrument panel insert
T80 - Connector, 80 point, on Motronic Engine Control Module (ECM)
T121 - Connector, 121 point, on Motronic Engine Control Module (ECM)
V70 - Central Air Flap Motor
V85 - Footwell/Defroster Flap Motor
(A76) - Connector (K-diagnosis wire), in instrument panel wiring harness
* - from August 1999

Climatronic control module, central air and footwell/defrost flap motors

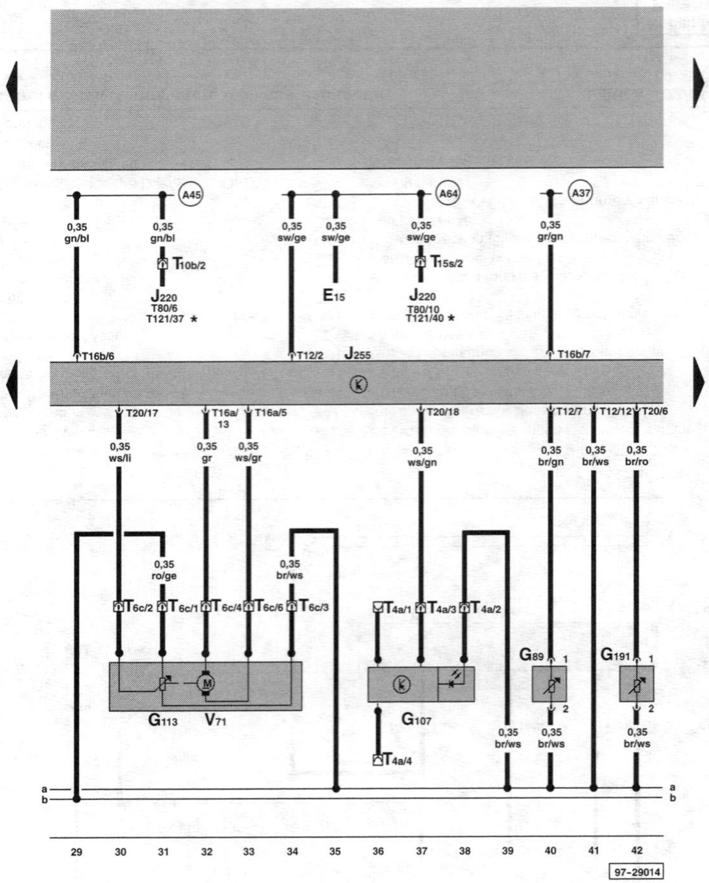

ws = white
sw = black
ro = red
br = brown
gn = green
bl = blue
gr = grey
li = violet
ge = yellow

E15 - Rear window defogger switch
G89 - Fresh Air Intake Duct Temperature Sensor
G107 - Sunlight Photo Sensor, on instrument panel, left
G113 - Back Pressure Flap Motor Position Sensor
G191 - Sender for outlet temperature, center
J220 - Motronic Engine Control Module (ECM)
J255 - Climatronic Control Module
T4a - 4-Pin Connector, black
T6c - 6-Pin Connector, black
T10b - 10-Pin Connector, black, on protective housing for control module, in engine compartment, left
T15s - 15-Pin Connector, red, on protective housing for control module, in engine compartment, left
T12 - 12-Pin Connector
T16a - 16-Pin Connector
T16b - 16-Pin Connector
T20 - 20-Pin Connector
T80 - Connector, 80 point, on Motronic Engine Control Module (ECM)
T121 - Connector, 121 point, on Motronic Engine Control Module (ECM)
V71 - Air Flow Flap Motor
(A37) - wire connection (58a), in instrument panel wiring harness
(A45) - wire connection (RPM-signal), in instrument panel wiring harness
(A64) - Connector (30, idle boost), in instrument panel wiring harness
* - from August 1999

Climatronic control module, air flow flap motor, fresh air intake duct temperature sensor, sunlight photo sensor, sender for outlet temperature sensor, center

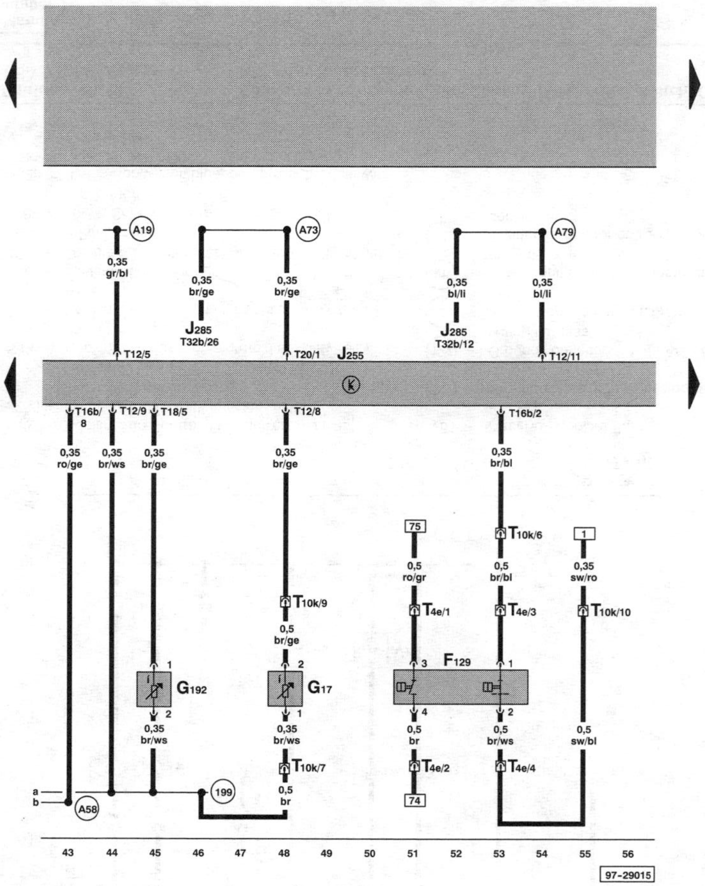

ws = white
sw = black
ro = red
br = brown
gn = green
bl = blue
gr = grey
li = violet
ge = yellow

F129 - A/C Pressure Switch
G17 - Outside Air Temperature Sensor
G192 - Sender for outlet temperature, floor outlet
J255 - Climatronic Control Module
J285 - Control module with indicator unit in instrument panel insert
T4e - 4-Pin Connector, in engine compartment, left
T10k - 10-Pin Connector, grey, connector station A-pillar, left
T12 - 12-Pin Connector, neben der Lenksäule rechts
T16b - 16-Pin Connector
T20 - 20-Pin Connector
T32b - 32-Pin Connector, green, on instrument cluster

(199) - Ground connection -3-, in instrument panel wiring harness

(A19) - wire connection (58d), in instrument panel wiring harness
(A58) - Wire connection (5Volt), in instrument cluster wiring harness
(A73) - Connector (outside temperature indicator), in instrument panel wiring harness
(A79) - Connector (engine overheat warning light), in instrument panel wiring harness

Edition 02/01
USA.5062.13.21

Climatronic control module, A/C pressure switch, outside air temperature sensor, sender for outlet temperature, floor outlet

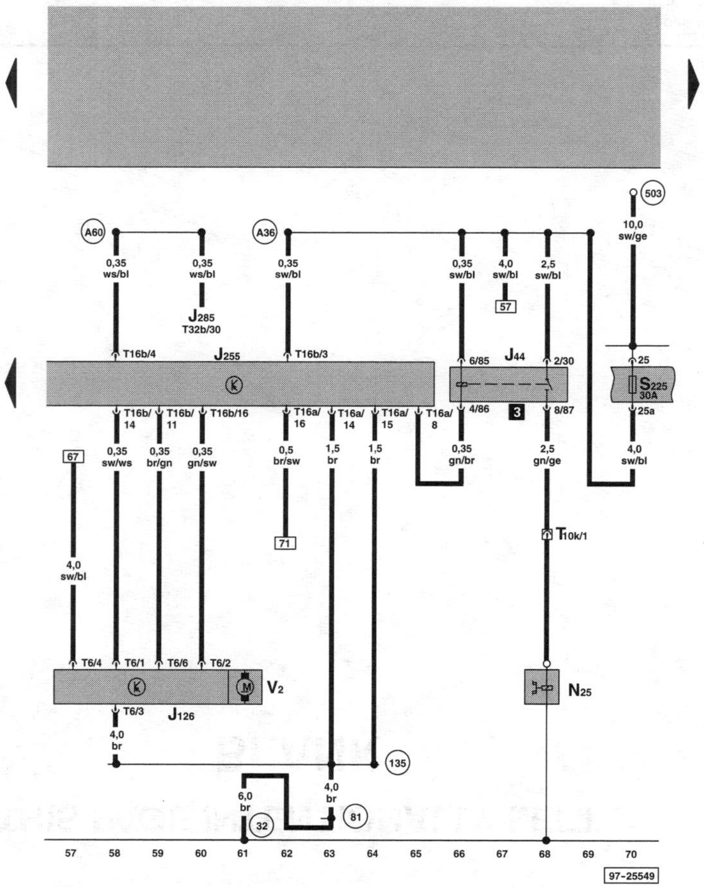

ws = white
sw = black
ro = red
br = brown
gn = green
bl = blue
gr = grey
li = violet
ge = yellow

J44 - A/C Clutch Relay, on thirteenfold auxiliary relay panel, above relay panel, production control number (267)
J126 - Control module for fresh air blower, behind instrument panel, right
J255 - Climatronic Control Module
J285 - Control module with indicator unit in instrument panel insert
N25 - A/C Clutch
S225 - Fuse 25 in fuse holder
T6 - 6-Pin Connector, brown
T10k - 10-Pin Connector, grey, connector station A-pillar, left
T16a - 16-Pin Connector
T16b - 16-Pin Connector
T32b - 32-Pin Connector, green, on instrument cluster
V2 - Fresh Air Blower

(32) - Ground connection, behind instrument panel, left
(81) - Ground connection -1-, in instrument panel wiring harness
(135) - Ground connection -2-, in instrument panel wiring harness
(503) - Screw Connection (75x), on relay panel
(A36) - wire connection (75a), in instrument panel wiring harness
(A60) - Wire connection (Vehicle speed signal), in instrument cluster wiring harnes

Climatronic control module, fresh air blower, A/C clutch, A/C clutch relay

Edition 02/01
USA.5062.13.21

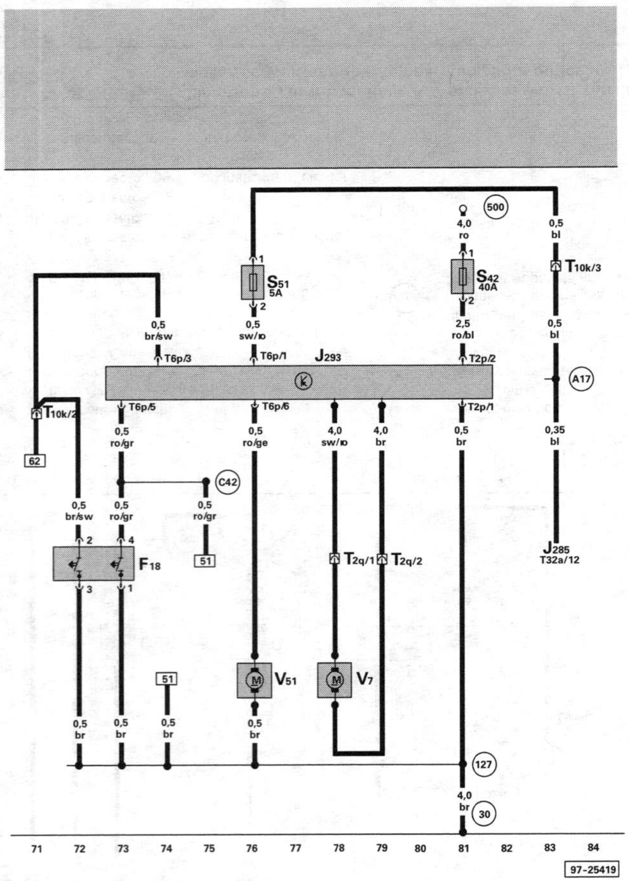

ws = white
sw = black
ro = red
br = brown
gn = green
bl = blue
gr = grey
li = violet
ge = yellow

F18 - Coolant Fan Control (FC) Thermal Switch
J285 - Control Module with indicator unit in instrument panel insert
J293 - Coolant FC (Fan Control) Control Module
S42 - Coolant Fan Fuse, on eightfold auxiliary relay panel
S51 - Coolant Fan Fuse, on eightfold auxiliary relay panel
T2p - Double Connector
T2q - Double Connector, in engine cpmpartment, left
T6p - 6-Pin Connector
T10k - 10-Pin Connector, grey, connector station A-pillar, left
T32a - 32-Pin Connector, blue, on Control Module with indicator unit in instrument panel insert
V7 - Coolant Fan
V51 - After-Run Coolant Pump

(30) - Ground connection -1-,beside fuse/relay panel
(127) - Ground connection -1-, in A/C compressor wiring harness
(500) - Screw Connection -1- (30), on relay panel
(A17) - wire connection (61), in instrument panel wiring harness
(C42) - Connector (fan for coolant) ,in wiring harnes coolant

THIS PAGE INTENTIONALLY LEFT
BLANK

Radio System,

from August 1998

Deviate relay location and fuseplacements as well as the locations of multiple connectors see section "component locations".

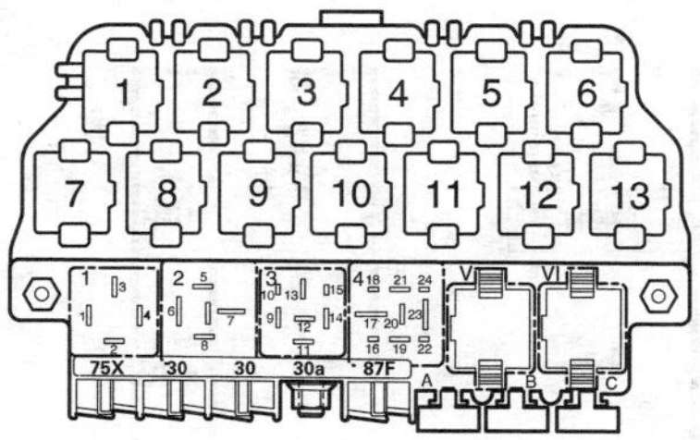

97-20676

Relay location on the thirteenfold auxiliary relay panel, above relay panel:

Relay panel:

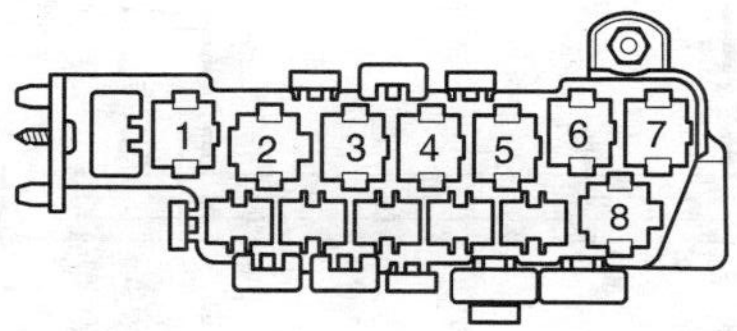

97-20515

Relay location on the eightfold auxiliary relay panel, behind relay panel:

Note: Number in parentheses indicates production control number stamped on relay housing.

Fuse colors
30 A - green
25 A - white
20 A - yellow
15 A - blue
10 A - red
7,5 A - brown
5 A - beige

Edition 06/99
USA.5062.09.21

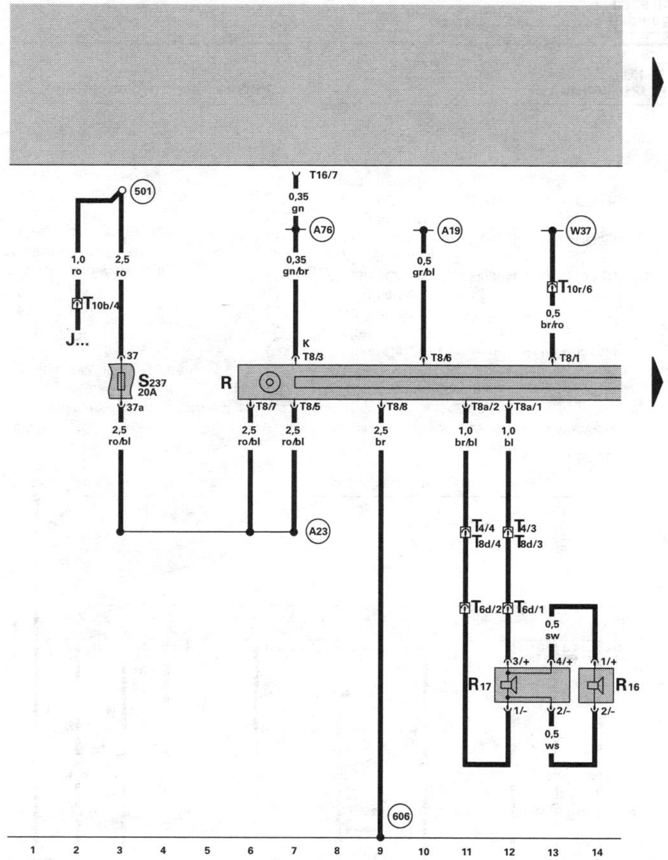

97-25437

ws = white
sw = black
ro = red
br = brown
gn = green
bl = blue
gr = grey
li = violet
ge = yellow

J... - Engine Control Module (ECM)
R - Radio
R16 - Right Rear Tweeter, in right rear door
R17 - Right Rear Woofer, in right rear door
S237 - Fuse 37 in fuse holder
T4 - 4-Pin Connector, black, connector station A-pillar, right
T6d - 6-Pin Connector, red, on B-pillar, right
T8 - 8-Pin Connector, black, on radio
T8a - 8-Pin Connector, brown, on radio
T8d - 8-Pin Connector, black, connector station A-pillar, right
T10b - 10-Pin Connector, black, on protective housing for control module, in engine compartment, left
T10r - 10-Pin Connector, violett, connector station A-pillar, right

T16 - 16-Pin Connector, Data Link Connector (DLC), below instrument panel, left
(501) - Screw Connection -2- (30), on relay panel
(606) - Ground Connection (under center console at shifter)
(A19) - wire connection (58d), in instrument panel wiring harness
(A23) - wire connection (30al), in instrument panel wiring harness
(A76) - Connector (K-diagnosis wire), in instrument panel wiring harness
(W37) - Connector (alarm system), in wiring harness floor

Radio, right rear speakers

Edition 06/99
USA.5062.09.21

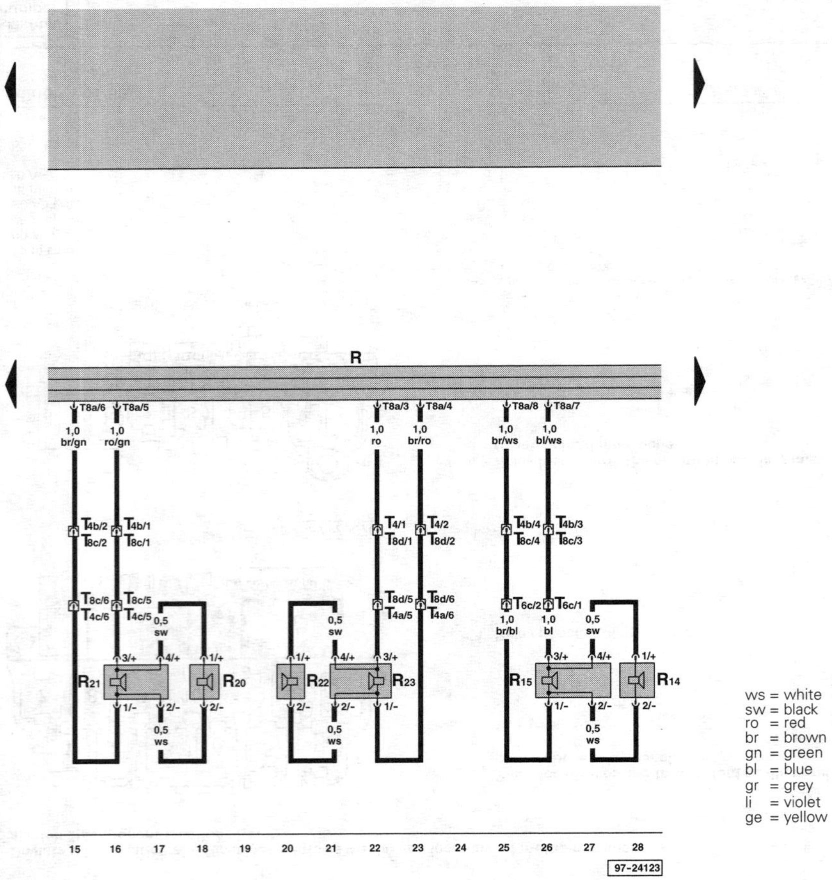

ws = white
sw = black
ro = red
br = brown
gn = green
bl = blue
gr = grey
li = violet
ge = yellow

R - Radio
R14 - Left Rear Tweeter, in left rear door
R15 - Left Rear Woofer, in left rear door
R20 - Left Front Tweeter, in driver's door
R21 - Left Front Woofer, in driver's door
R22 - Right Front Tweeter, in passenger's door
R23 - Right Front Tweeter, in passenger's door
T4 - 4-Pin Connector, black, connector station A-pillar, right
T4a - 4-Pin Connector, black, connector station A-pillar, right
T4b - 4-Pin Connector, black, connector station A-pillar, left
T4c - 4-Pin Connector, black, connector station A-pillar, left
T6c - 6-Pin Connector, red, on B-pillar, left
T8a - 8-Pin Connector, brown, on radio
T8c - 8-Pin Connector, black, connector station A-pillar, left
T8d - 8-Pin Connector, black, connector station A-pillar, right

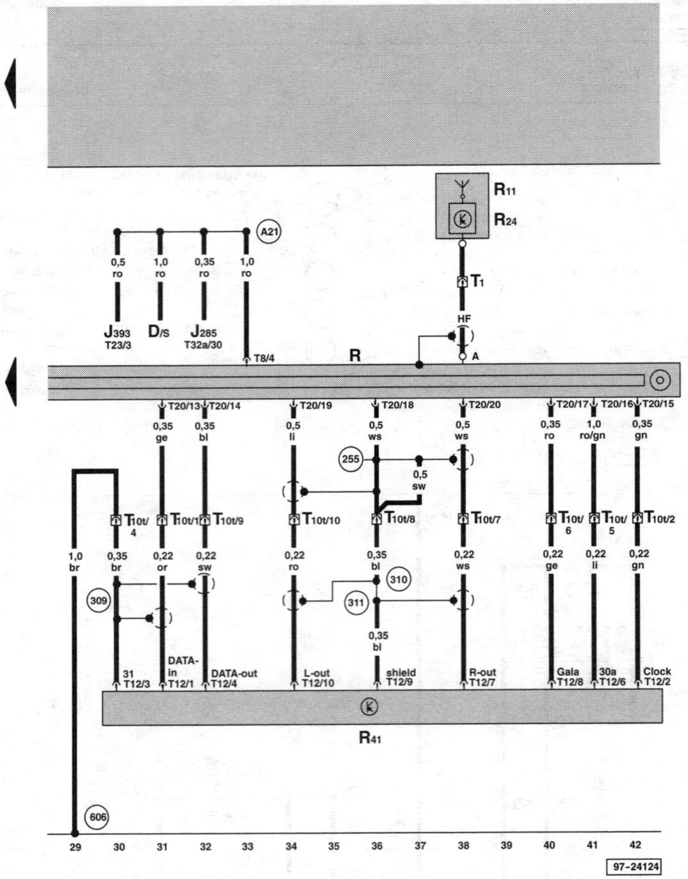

ws = white
sw = black
ro = red
br = brown
gn = green
bl = blue
gr = grey
li = violet
ge = yellow

D - Ignition/Starter Switch
J285 - Control module with indicator unit in instrument panel insert
J393 - Central control module for comfort system
R - Radio
R11 - Antenna
R24 - Antenna Amplifier
R41 - CD Changer Unit
T1 - 1-Pin Connector, black, below console
T8 - 8-Pin Connector, black, on radio
T10t - 10-Pin Connector, red, connector station A-pillar, right
T12 - 12-Pin Connector, red, on CD Changer Unit
T20 - 20-Pin Connector, on radio
T23 - 23-Pin Connector, on central control module for comfort system
T32a - 32-Pin Connector, blue, on instrument cluster
(255) - Ground connection, in wiring harness Radio
(606) - Ground Connection (under center console at shifter)
(A21) - wire connection (86s), in instrument panel wiring harness
(309) - Ground Connection 1 (shielding) (in CD changer wiring harness)
(310) - Ground Connection 2 (shielding) (in CD changer wiring harness)
(311) - Ground Connection 3 (shielding) (in CD changer wiring harness)

Passat - Standard Equipment,

from August 1999

Deviate relay location and fuseplacements as well as the locations of multiple connectors see section "component locations".

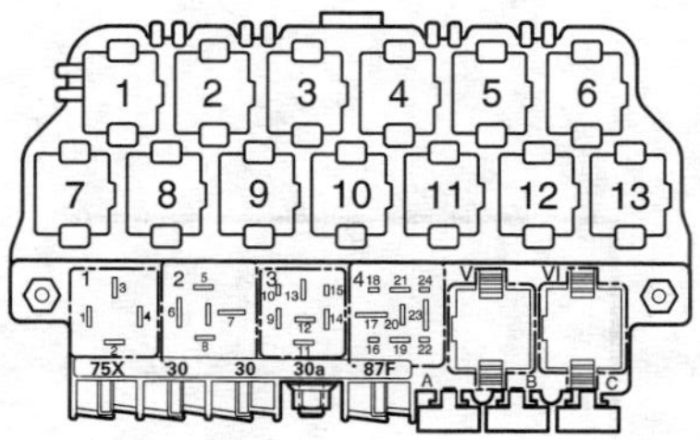

97-20676

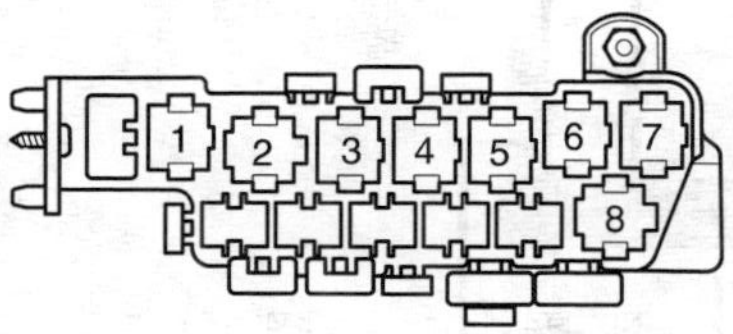

97-20515

Fuse colors

30 A - green
25 A - white
20 A - yellow
15 A - blue
10 A - red
7,5 A - brown
5 A - beige

Relay location on the thirteenfold auxiliary relay panel, above relay panel:

- **7** Fog Light Relay (381)
- **8** Daytime Running Lights Change-over Relay (173)
- **12** Starting Interlock Relay (alarm system) (186)
- **13** Starting Interlock Relay (clutch position) (204)

Relay panel:

- **1** Dual Horn Relay (204)
- **2** Load Reduction Relay (213)
- **V** Wiper/Washer Intermittent Relay (377)
- **VI** Wiper/Washer Intermittent Relay (377)

Relay location on the eightfold auxiliary relay panel, behind relay panel:

Note: Number in parentheses indicates production control number stamped on relay housing.

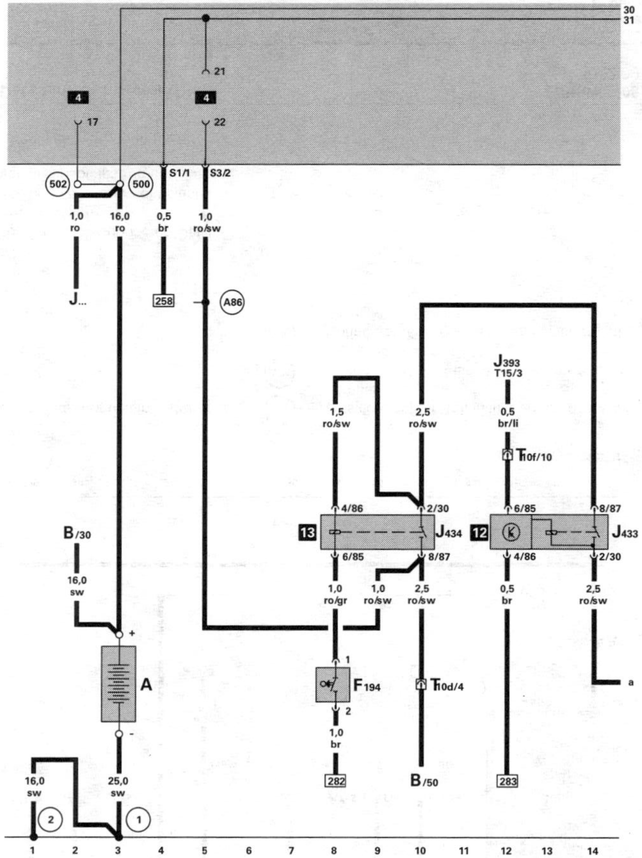

97-25347

ws = white
sw = black
ro = red
br = brown
gn = green
bl = blue
gr = grey
li = violet
ge = yellow

A - Battery
B - Starter
F194 - Clutch Pedal Position (CPP) Switch
J... - Engine control module (ECM)
J393 - Central control module for comfort system
J433 - Starting Interlock Relay (alarm system), on auxiliary relay panel, above relay panel, production control number (186)
J434 - Starting Interlock Relay (clutch position), on auxiliary relay panel, above relay panel, production control number (204)
T10d - 10-Pin Connector, brown, on protective housing for control module, in engine compartment, left
T10f - 10-Pin Connector, blue, connector station A-pillar, left
T15 - 15-Pin Connector, on central control module for comfort system

(1) - Ground strap, battery to body
(2) - Ground strap, transmission to body
(500) - Screw connection -1- (30), on relay panel
(502) - Screw connection (30a), on relay panel
(A86) - Connection (50a) in instrument panel wiring harness

Starting interlock, battery

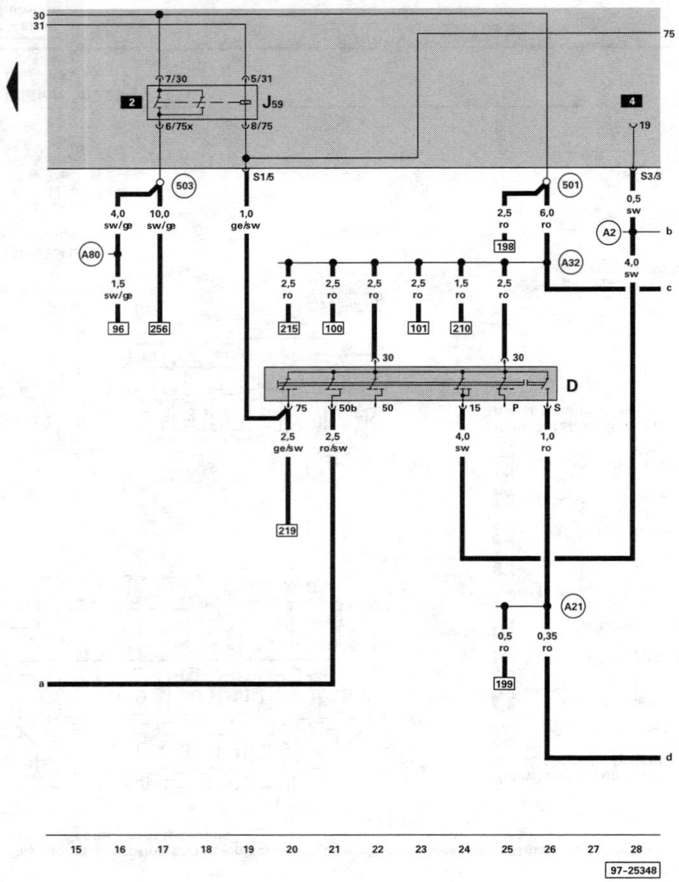

ws = white
sw = black
ro = red
br = brown
gn = green
bl = blue
gr = grey
li = violet
ge = yellow

D - Ignition/Starter Switch
J59 - Load Reduction Relay, production control number (213)

(501) - Screw connection -2- (30), on relay panel

(503) - Screw connection (75x), on relay panel

(A2) - plus connection (15), in instrument panel wiring harness
(A21) - wire connection (86s), in instrument panel wiring harness
(A32) - plus connection (30), in instrument panel wiring harness
(A80) - Connector -1- (X), in instrument panel wiring harness

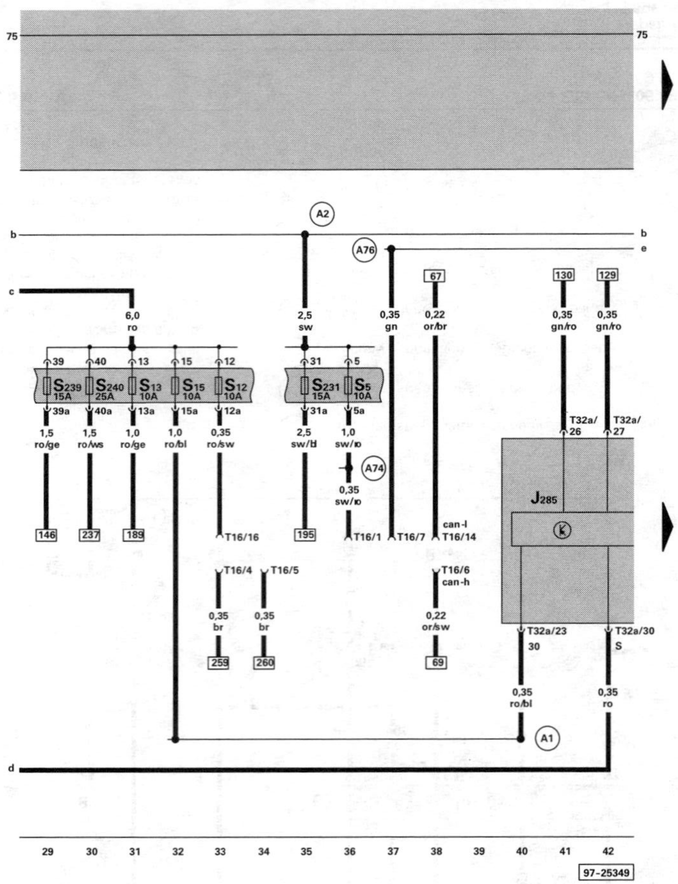

ws = white
sw = black
ro = red
br = brown
gn = green
bl = blue
gr = grey
li = violet
ge = yellow
or = orange

J285 - Control module with indicator unit in instrument panel insert
S5 - Fuse 5 in fuse holder
S12 - Fuse 12 in fuse holder
S13 - Fuse 13 in fuse holder
S15 - Fuse 15 in fuse holder
S231 - Fuse 31 in fuse holder
S239 - Fuse 39 in fuse holder
S240 - Fuse 40 in fuse holder
T16 - 16-Pin Connector, Data Link Connector (DLC), below instrument panel, left
T32a - 32-Pin Connector, blue, on instrument cluster
T32b - 32-Pin Connector, green, on instrument cluster

(A1) - plus connection (30a), in instrument panel wiring harness

(A2) - plus connection (15), in instrument panel wiring harness
(A74) - Connector (15a - fuse 5), in instrument panel wiring harness
(A76) - Connector (K-diagnosis wire), in instrument panel wiring harness

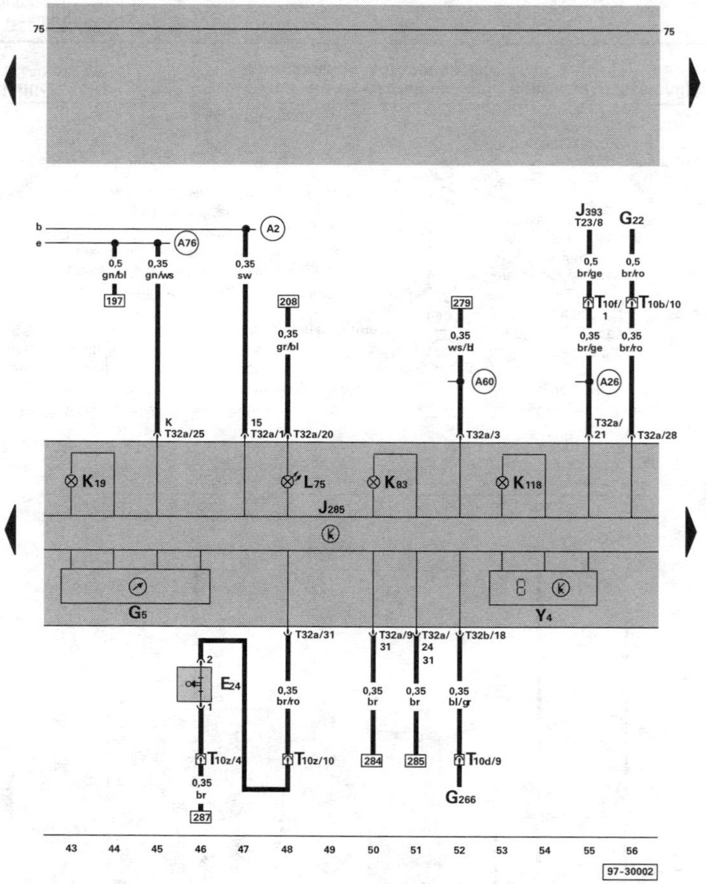

ws = white
sw = black
ro = red
br = brown
gn = green
bl = blue
gr = grey
li = violet
ge = yellow

E24 - Left Seat Belt Switch
G5 - Tachometer
G22 - Speedometer Vehicle Speed Sensor (VSS)
G266 - Oil Level Thermal Sensor (SRI)
J285 - Control module with indicator unit in instrument panel insert
J393 - Central control module for comfort system
K19 - Seat Belt Warning Light
K83 - Malfunction Indicator Lamp (MIL)
K118 - Warning Light for brake system
L75 - Digital Display Light
T10b - 10-Pin Connector, black, on protective housing for control module, in engine compartment, left
T10d - 10-Pin Connector, brown, on protective housing for control module, in engine compartment, left
T10f - 10-Pin Connector, blue, connector station A-pillar, left
T10z - 10-Pin Connector, brown
T23 - 23-Pin Connector, on central control module for comfort system
T32a - 32-Pin Connector, blue, on instrument cluster
T32b - 32-Pin Connector, green, on instrument cluster
Y4 - Odometer Display
(A2) - plus connection (15), in instrument panel wiring harness
(A26) - wire connection (driver's door contact switch), in instrument panel wiring harness
(A60) - plus connection (Vehicle speed signal), in instrument cluster wiring harness
(A76) - Connector (K-diagnosis wire), in instrument panel wiring harness

Edition 07.01
USA.5062.15.21

Instrument cluster, Tachometer, Speedometer vehicle speed sensor (VSS), Malfunction Indicator Lamp (MIL), Seat belt warning system

Repair Manual page number
Y 343

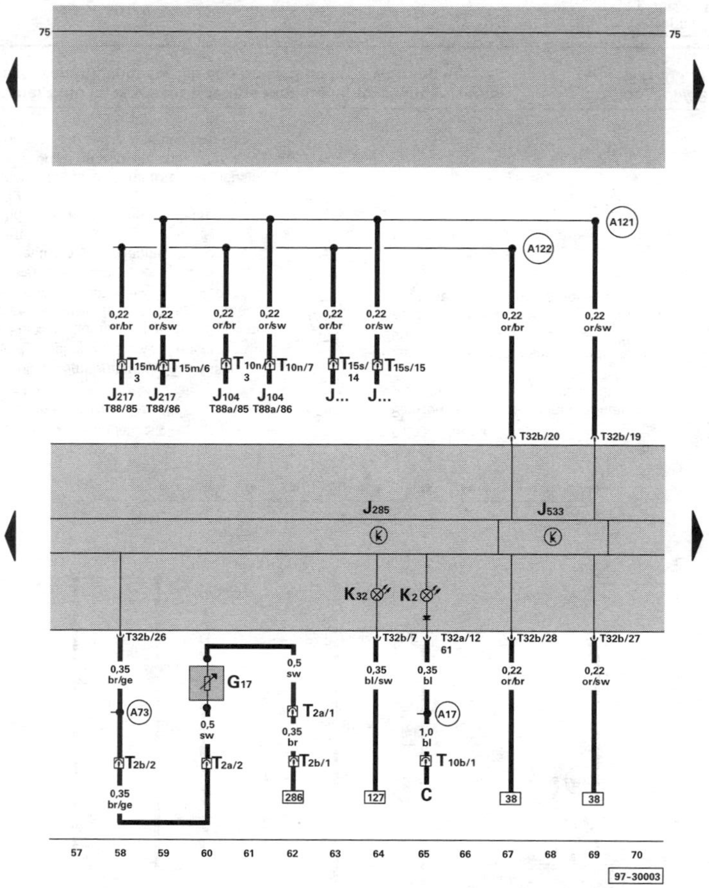

ws = white
sw = black
ro = red
br = brown
gn = green
bl = blue
gr = grey
li = violet
ge = yellow
or = orange

C - Generator (GEN)
G17 - Outside Air Temperature Sensor
J... - Engine Control Module (ECM)
J104 - ABS Control Module (w/EDL)
J217 - Transmission Control Module (TCM)
J285 - Control module with indicator unit in instrument panel insert
J533 - Interface for Can Bus (Diagnosis)
K2 - Generator (GEN) Warning Light
K32 - Brake Pad Wear Indicator Light
T2a - 2-Pin Connector,
T2b - 2-Pin Connector,
T10b - 10-Pin Connector, black, on protective housing for control module, in engine compartment, left
T10n - 10-Pin Connector,
T15m - 15-Pin Connector,
T15s - 15-Pin Connector, red, on protective housing for control module, in engine compartment, left
T32a - 32-Pin Connector, blue, on instrument cluster
T32b - 32-Pin Connector, green, on instrument cluster
T88 - Connector, 88 point, on Transmission Control Module (TCM)
T88a - Connector, (outside temperature indicator), in instrument panel wiring harness
(A17) - wire connection (61), in instrument panel wiring harness
(A73) - Connector (outside temperature indicator), in instrument panel wiring harness
(A121) Connector (High-Bus), in instrument panel wiring harness
(A122) Connector (Low-Bus), in instrument panel wiring harness

Instrument cluster, generator warning light, outside air temperature sensor

Edition 07.01
USA.5062.15.21

Repair Manual page number
Y 344

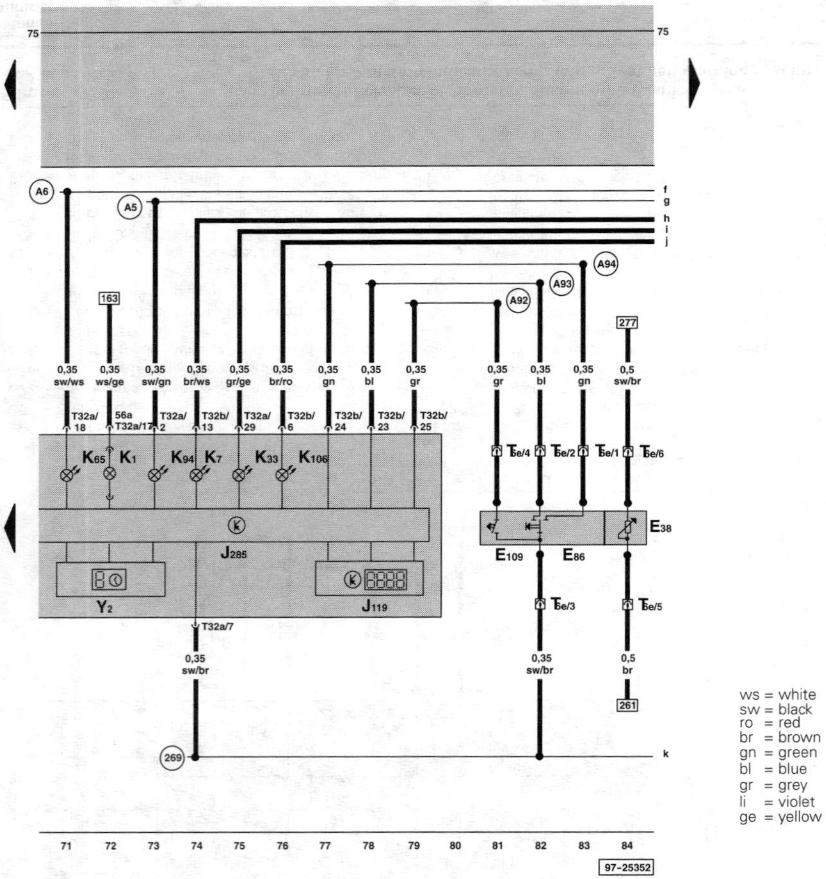

ws = white
sw = black
ro = red
br = brown
gn = green
bl = blue
gr = grey
li = violet
ge = yellow

E38 - Windshield Wiper Intermittent Regulator
E86 - Multi-Function Indicator Mode Select Switch
E109 - Multi-Function Indicator Memory Switch
J119 - Multi-Function Indicator (MFA)
J285 - Control module with indicator unit in instrument panel insert
K1 - Headlight High Beam Indicator Light
K7 - Brake and Parking Brake Warning Light
K33 - Brake Fluid Level Warning Light
K65 - Left Turn Signal Indicator Light
K94 - Right Turn Signal Indicator Light
K106 - Low Windshield Washer Fluid Level Indicator Light
T6e - 6-Pin Connector, black, near steering column, right
T32a - 32-Pin Connector, blue, on instrument cluster
T32b - 32-Pin Connector, green, on instrument cluster
Y2 - Digital Clock

(A5) - plus connection (right turn signal), in instrument panel wiring harness
(A6) - plus connection (left turn signal), in instrument panel wiring harness
(A92) - Connector (MFI-Reset), in instrument panel wiring harness
(A93) - Connector -1- (MFI), in instrument panel wiring harness
(A94) - Connector -1- (MFI), in instrument panel wiring harness
(269) - Ground connector (sensor ground) -1-, in instrument panel wiring harness

Instrument cluster, left and right turn signal indicator lights, multi-function indicator switch

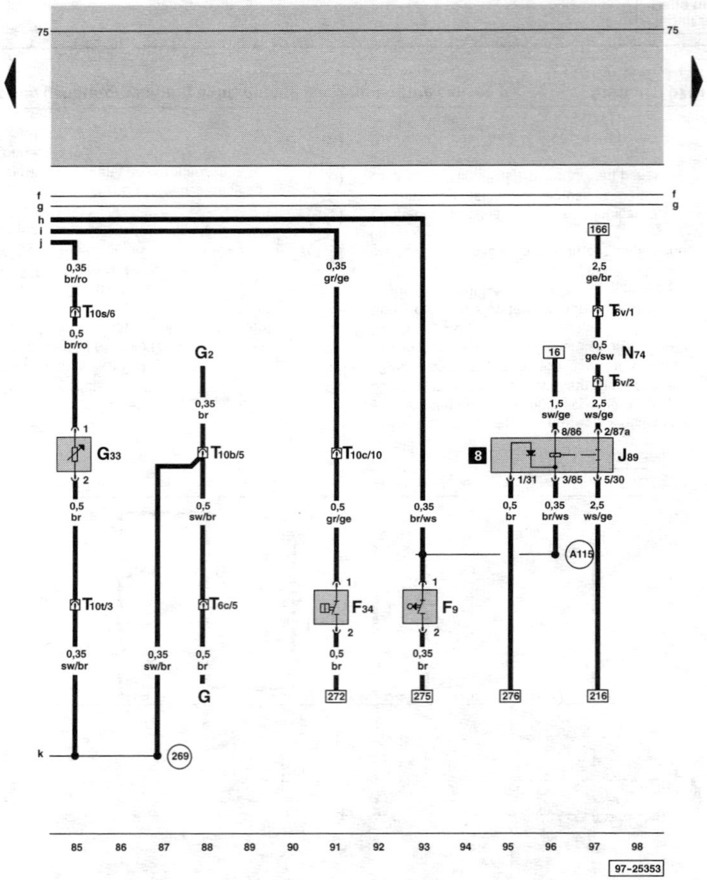

ws = white
sw = black
ro = red
br = brown
gn = green
bl = blue
gr = grey
li = violet
ge = yellow

F9 - Parking Brake Warning Light Switch
F34 - Brake Fluid Level Warning Switch
G - Sender for fuel gauge
G2 - Engine Coolant Temperature (ECT) Sensor
G33 - Windshield Washer Fluid Level Sensor
J89 - Daytime Running Lights Change-over Relay on the thirteenfold relay panel, above relay panel, production control number (173)
N74 - Series Resistance Wiring
T6c - 6-Pin Connector, blue, connector station A-pillar, left
T6v - 6-Pin Connector, violet, connector station A-pillar, right
T10b - 10-Pin Connector, black, on protective housing for control module, in engine compartment, left
T10c - 10-Pin Connector, violet, connector station A-pillar, left
T10s - 10-Pin Connector,
T10t - 10-Pin Connector, red, connector station A-pillar, left

(269) - Ground connector (sensor ground) -1-, in instrument panel wiring harness
(A115) - Connection (parking brake warning light), in instrument panel wiring harness

Windshield washer fluid level sensor, daytime running lights change-over relay, parking brake warning light switch and brake fluid level warning light switch

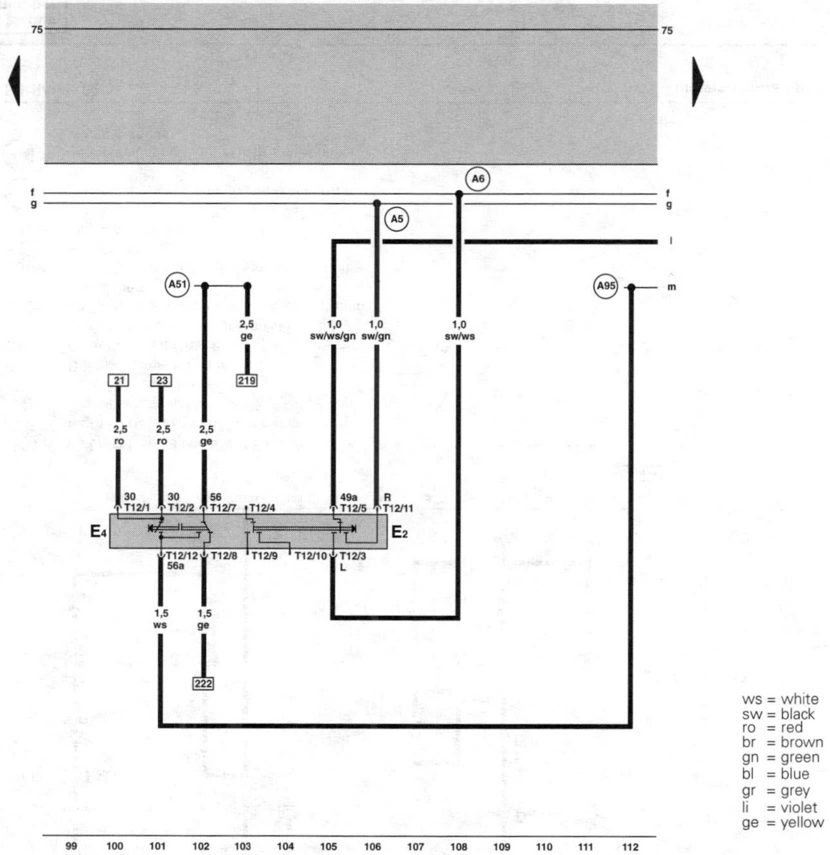

ws = white
sw = black
ro = red
br = brown
gn = green
bl = blue
gr = grey
li = violet
ge = yellow

97-25354

E2 - Turn Signal Switch
E4 - Headlight Dimmer/Flasher Switch
T12 - 12-Pin Connector, near steering column

(A5) - plus connection (right turn signal), in instrument panel wiring harness
(A6) - plus connection (left turn signal), in instrument panel wiring harness
(A51) - wire connection (56), in instrument panel wiring harness
(A95) - wire connection -1- (56a), in instrument panel wiring harness

Turn signal switch, headlight dimmer/flasher switch

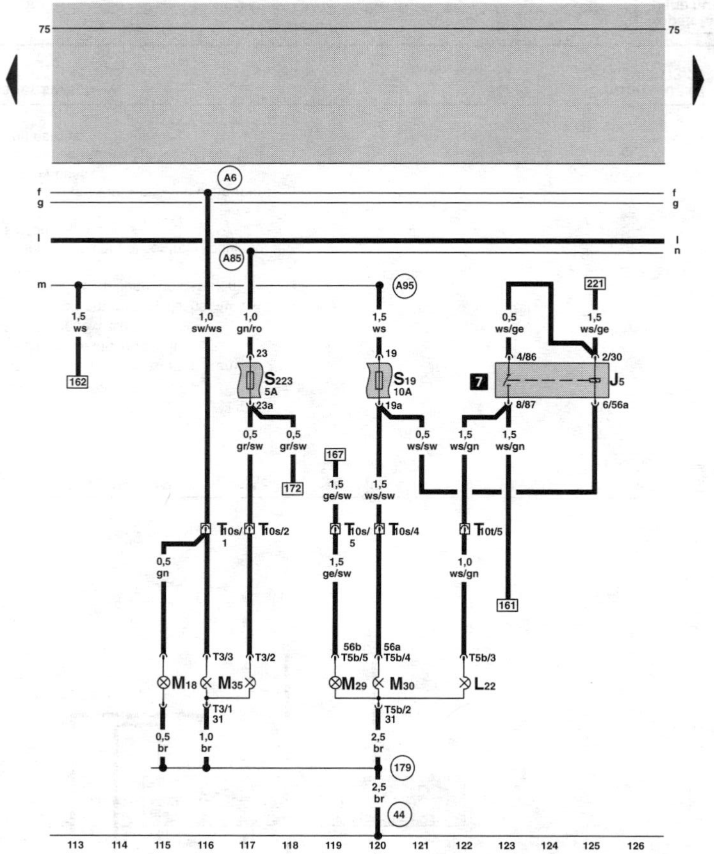

ws = white
sw = black
ro = red
br = brown
gn = green
bl = blue
gr = grey
li = violet
ge = yellow

97-25355

J5 - Fog Light Relay, on auxiliary relay panel, production control number (381)
L22 - Left Front Fog Light
M18 - Left, Side Turn Signal Light
M29 - Left Low Beam Headlight
M30 - Left High Beam Headlight
M35 - Light for turn signal and side marker, front, left
S19 - Fuse 19 in fuse holder
S223 - Fuse 23 in fuse holder
T3 - 3-Pin Connector
T5b - 5-Pin Connector, on headlight left
T10s - 10-Pin Connector, pink, connector station A-pillar, left
T10t - 10-Pin Connector, red, connector station A-pillar, left

(44) - Ground connection, on left A-pillar, lower part
(179) - Ground connection, in left headlight wiring harness
(A6) - plus connection (left turn signal), in instrument panel wiring harness
(A85) - wire connection (58r), in instrument panel wiring harness
(A95) - Connection 1 (56a), in instrument panel wiring harness

Left front headlights, left front fog light, fog light relay

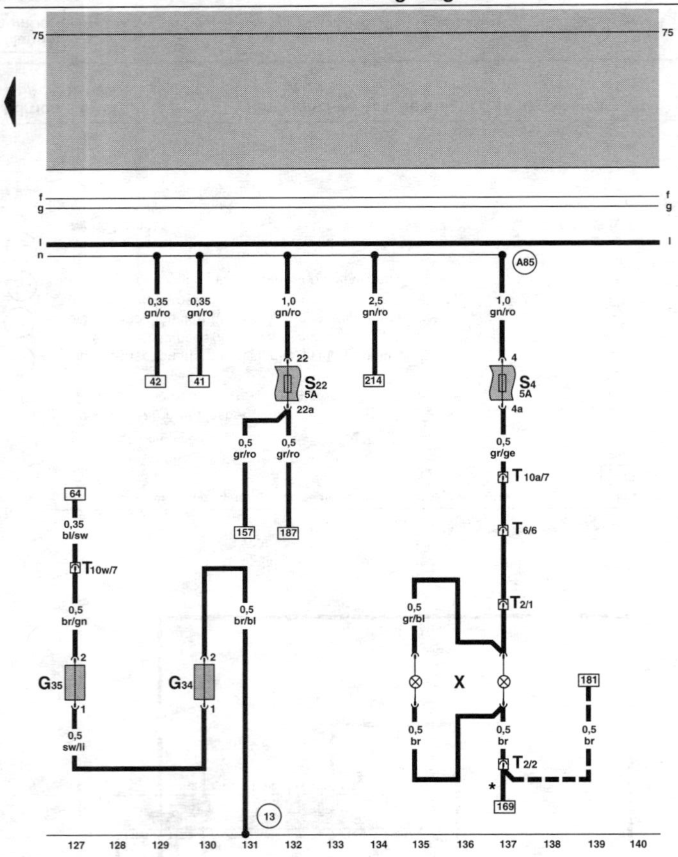

ws = white
sw = black
ro = red
br = brown
gn = green
bl = blue
gr = grey
li = violet
ge = yellow

- G34 - Left Front Brake Pad Wear Sensor
- G35 - Right Front Brake Pad Wear Sensor
- S4 - Fuse 4 in fuse holder
- S22 - Fuse 22 in fuse holder
- T2 - Double Connector, in rear lid
- T6 - 6-Pin Connector, black, in luggage compartment, left (wagon only)
- T10a - 10-Pin Connector, brown, connector station A-pillar, left
- T10w - 10-Pin Connector, grey, connector station A-pillar, right
- X - License Plate Light
- (13) - Ground connection, in engine compartment, right
- (A85) - wire connection (58r), in instrument panel wiring harness

--- wagon only

* - sedan only

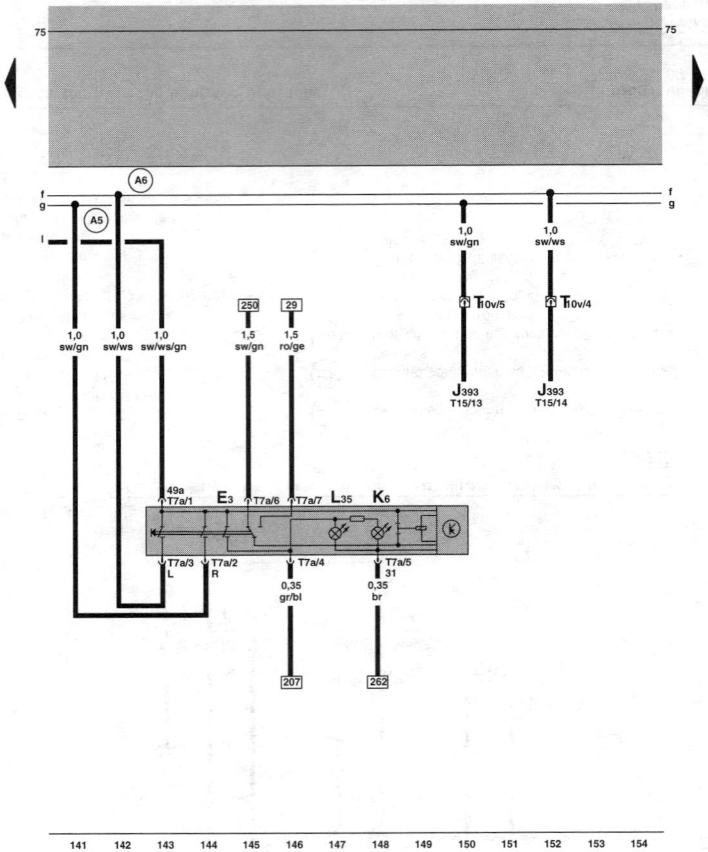

ws = white
sw = black
ro = red
br = brown
gn = green
bl = blue
gr = grey
li = violet
ge = yellow

- E3 - Emergency Flasher Switch
- J393 - Central control module for comfort system
- K6 - Emergency Flasher Warning Light
- L35 - Emergency Flasher Switch Light
- T7a - 7-Pin Connector, on emergency flasher switch
- T10v - 10-Pin Connector, violet, connector station A-pillar, right
- T15 - 15-Pin Connector, on central control module for comfort system
- (A5) - plus connection (right turn signal), in instrument panel wiring harness
- (A6) - plus connection (left turn signal), in instrument panel wiring harness

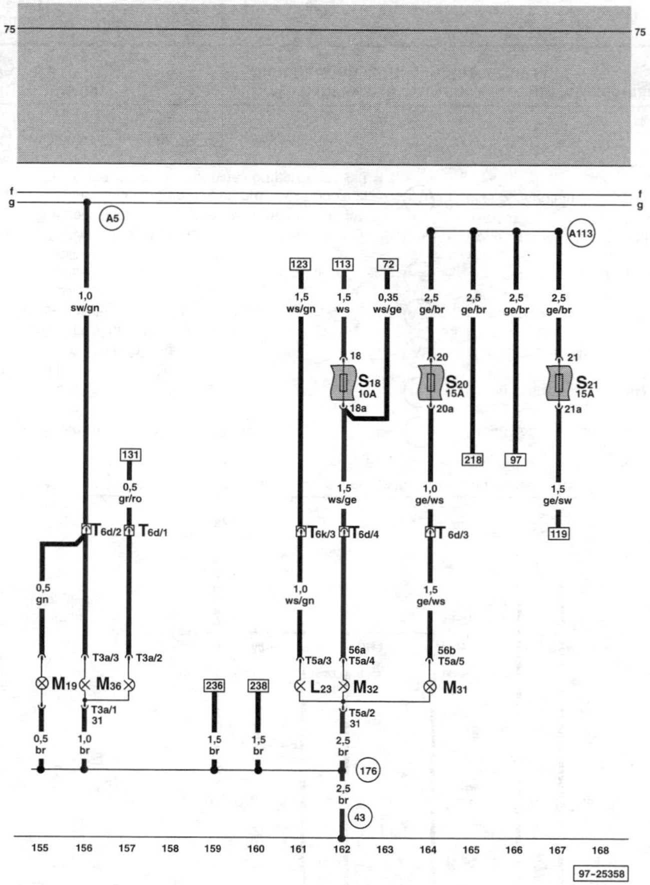

ws = white
sw = black
ro = red
br = brown
gn = green
bl = blue
gr = grey
li = violet
ge = yellow

- L23 - Right Front Fog Light
- M19 - Right, Side Turn Signal Light
- M31 - Right Low Beam Headlight
- M32 - Right High Beam Headlight
- M36 - Light for turn signal and side marker, front, right
- S18 - Fuse 18 in fuse holder
- S20 - Fuse 20 in fuse holder
- S21 - Fuse 21 in fuse holder
- T3a - 3-Pin Connector
- T5a - 5-Pin Connector, on headlight right
- T6d - 6-Pin Connector, pink, connector station A-pillar, right
- T6k - 6-Pin Connector, red, connector station A-pillar, right

- (43) - Ground connection, on right A-pillar, lower part
- (176) - Ground connection, in right headlight wiring harness
- (A5) - plus connection (right turn signal), in instrument panel wiring harness
- (A113) - Connection -2- (daytime running lights), in instrument panel wiring harness

Edition 02/01
USA.5062.13.21

Right front headlights, right front foglight

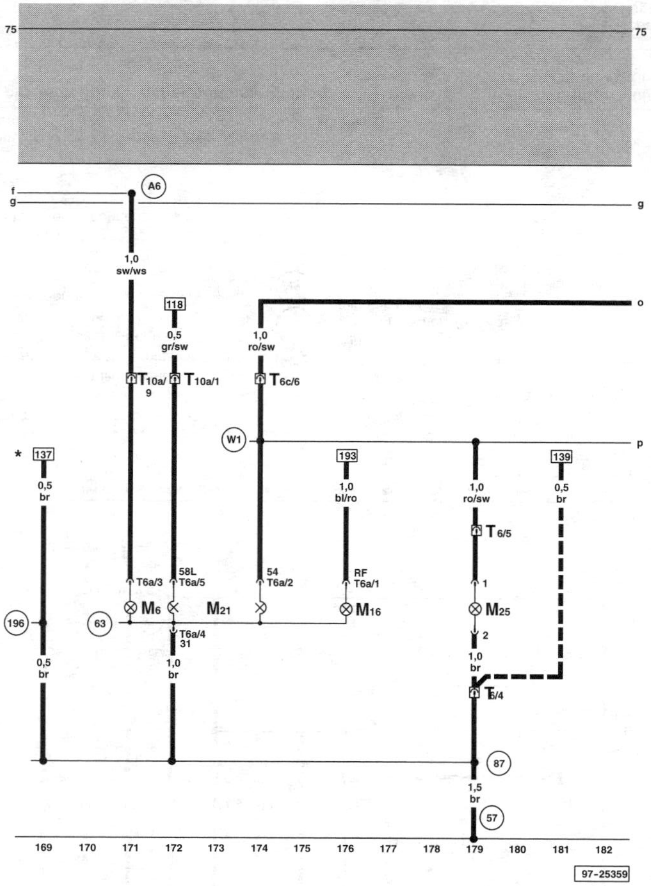

ws = white
sw = black
ro = red
br = brown
gn = green
bl = blue
gr = grey
li = violet
ge = yellow

- M6 - Left Rear Turn Signal Light
- M16 - Left Back-Up Light
- M21 - Left Brake/Tail Light
- M25 - High-mount Brake Light
- T6 - 6-Pin Connector, black, in luggage compartment, left (wagon only)
- T6a - 6-Pin Connector, on tail light, left
- T6c - 6-Pin Connector, blue, connector station A-pillar, left
- T10a - 10-Pin Connector, brown, connector station A-pillar, left

- (63) - Ground connection, on left taillight bulb holder
- (57) - Ground connection, on left rear window wiper

- (87) - Ground connection -2-, in rear wiring harness
- (196) - Ground connection -3-, in wiring harness, rear
- (A6) - plus connection (left turn signal), in instrument panel wiring harness
- (W1) - plus connection (54), in rear wiring harness
- - - - wagon only
- * - sedan only

Left brake/tail light, left rear turn signal light, left back-up light, high-mount brake light

Edition 02.01
USA.5062.13.21

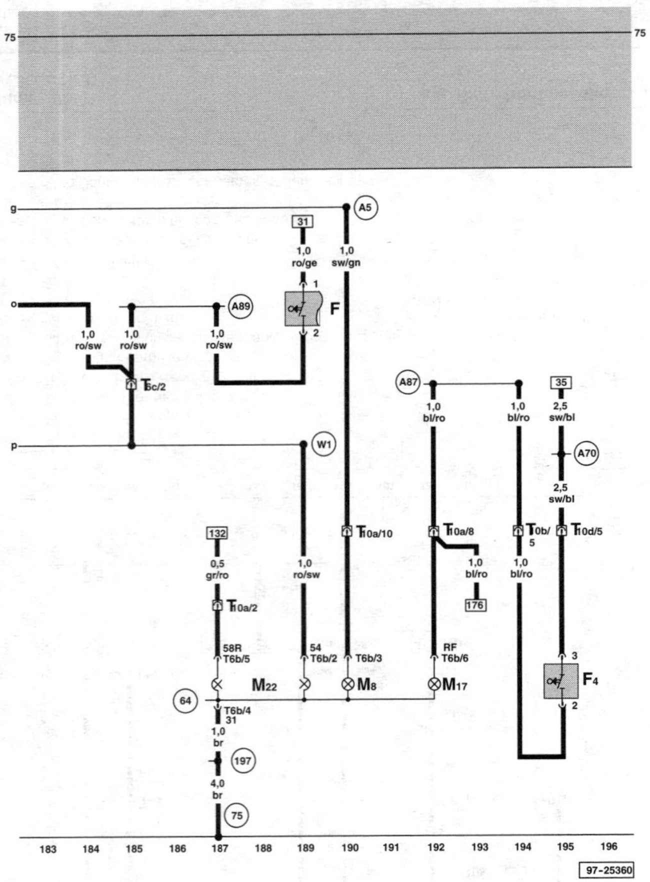

ws = white
sw = black
ro = red
br = brown
gn = green
bl = blue
gr = grey
li = violet
ge = yellow

F - Brake Light Switch
F4 - Back-Up Light Switch
M8 - Right Rear Turn Signal Light
M17 - Right Back-Up Light
M22 - Right Brake/Tail Light
T6b - 6-Pin Connector, on tail light, right
T6c - 6-Pin Connector, blue, connector station A-pillar, left
T10a - 10-Pin Connector, brown, connector station A-pillar, left
T10b - 10-Pin Connector, black, on protective housing for control module, in engine compartment, left
T10d - 10-Pin Connector, brown, on protective housing for control module, in engine compartment, left

(64) - Ground connection, on right taillight bulb holder

(75) - Ground connection, on right rear pillar

(197) - Ground connection -4-, in wiring harness rear

(A5) - plus connection (right turn signal), in instrument panel wiring harness

(A70) - Connector (15a, fuse 231), in instrument panel wiring harness

(A87) - Connector (reverse lamp), in instrument panel wiring harness

(A89) - Connector (54), in instrument panel wiring harness

(W1) - plus connection (54), in rear wiring harness

Brake light switch, back-up light switch, right rear turn signal light, right back-up light, right brake/tail light

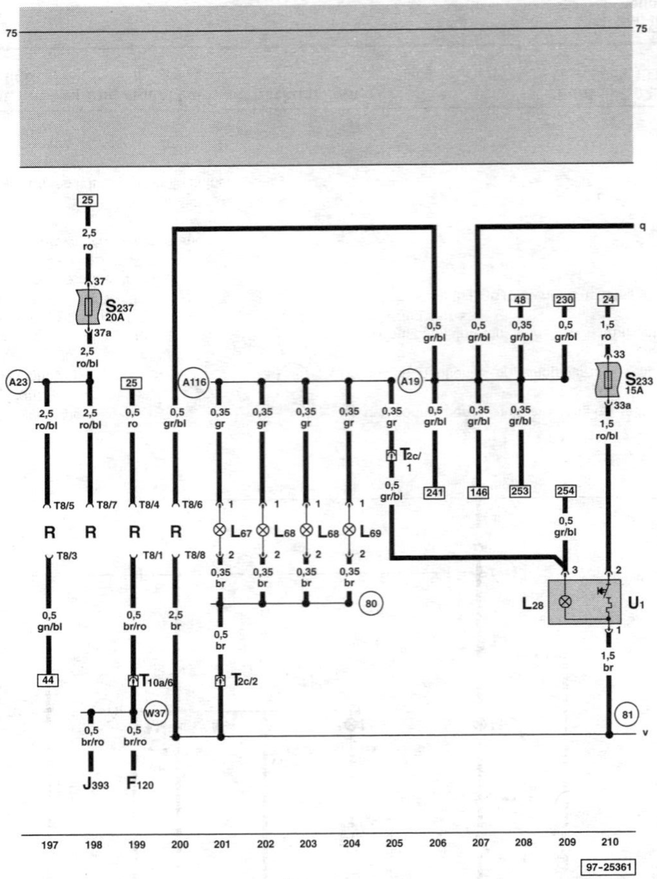

ws = white
sw = black
ro = red
br = brown
gn = green
bl = blue
gr = grey
li = violet
ge = yellow

F120 - Hood Alarm Switch
J393 - Central control module for comfort system
L28 - Cigarette Lighter Light
L67 - Left Instrument Panel Vent Illumination
L68 - Center Instrument Panel Vent Illumination
L69 - Right Instrument Panel Vent Illumination
R - Radio
S233 - Fuse 33 in fuse holder
S237 - Fuse 37 in fuse holder
T2c - Double Connector
T8 - 8-Pin Connector, on radio
T10a - 10-Pin Connector, brown, connector station A-pillar, left
U1 - Cigarette Lighter

(80) - Ground connection -1-, in instrument panel wiring harness

(81) - Ground connection -1-, in instrument panel wiring harness

(A19) - wire connection (58d), in instrument panel wiring harness

(A23) - wire connection (30al), in instrument panel wiring harness

(A116) - wire connection -2- (58d), in instrument panel wiring harness

(W37) - Connector (alarm system), in wiring harness floor

Radio connection, cigarette lighter, panel vent illumination

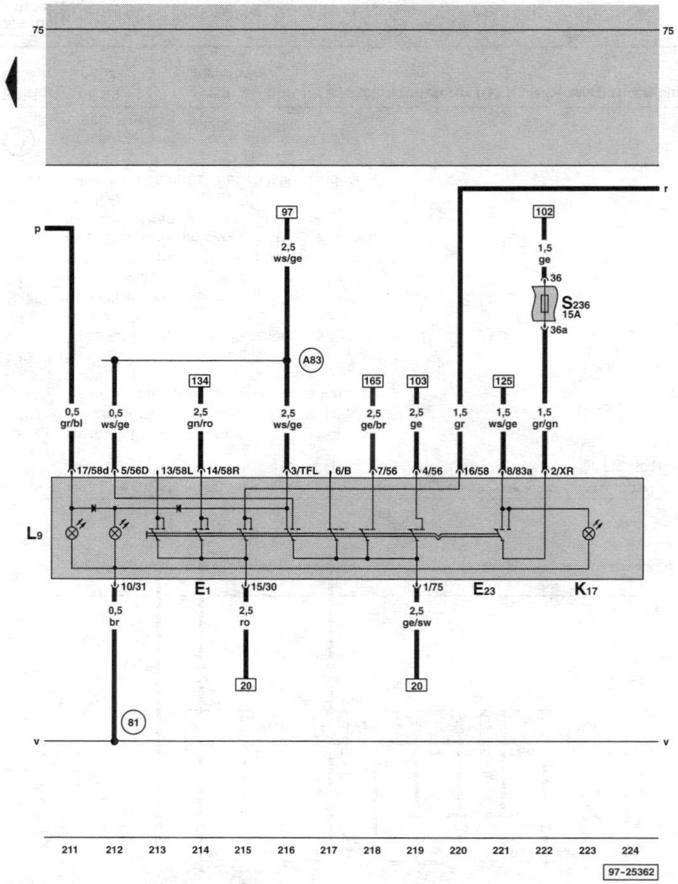

ws = white
sw = black
ro = red
br = brown
gn = green
bl = blue
gr = grey
li = violet
ge = yellow

E1 - Light switch
E23 - Fog Light Switch
K17 - Fog Light Indicator Light
L9 - Headlight Switch Licht
S236 - Fuse 36 in fuse holder

(81) - Ground connection -1-, in instrument panel wiring harness
(A83) - Connector (daytime running lights), in instrument panel wiring harness

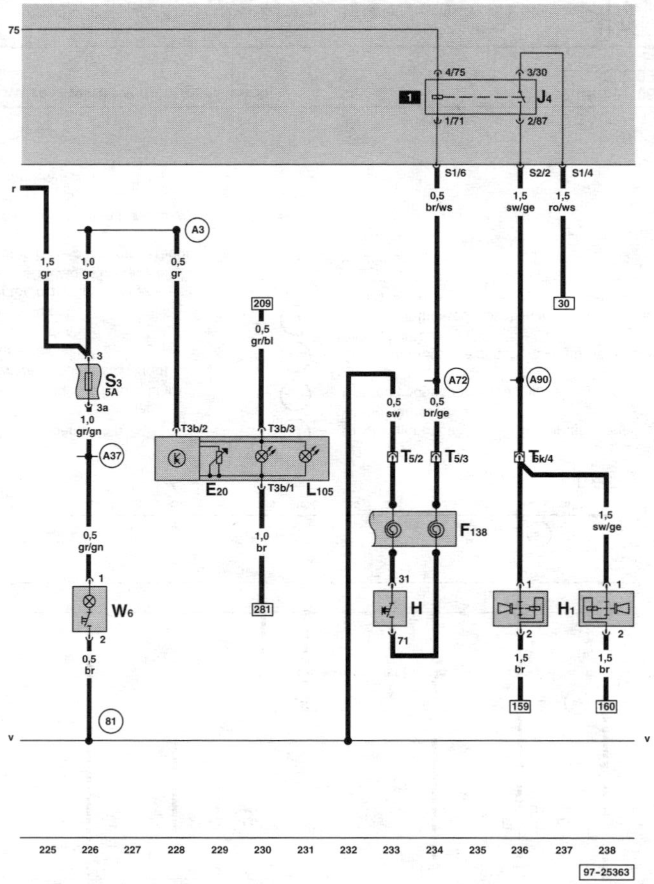

ws = white
sw = black
ro = red
br = brown
gn = green
bl = blue
gr = grey
li = violet
ge = yellow

E20 - Instrument Panel Light Dimmer Switch
F138 - Airbag Spiral Spring/Return Spring With Slip Ring
H - Horn Button
H1 - Dual Horns
J4 - Dual Horn Relay, production control number (204)
L105 - Illumination for Instrument Panel Light Dimmer Switch
S3 - Fuse 3 in fuse holder
T3b - 3-Pin Connector
T5 - 5-Pin Connector, yellow, near steering column
T6k - 6-Pin Connector, red, connector station A-pillar, right
W6 - Glove Compartment Light

(81) - Ground connection -1-, in instrument panel wiring harness
(A3) - plusconnection (58), in instrument panel wiring harness
(A37) - wire connection (58a), in instrument panel wiring harness
(A72) - Connector (71), in instrument panel wiring harness
(A90) - Connector (two-tone horn), in instrument panel wiring harness

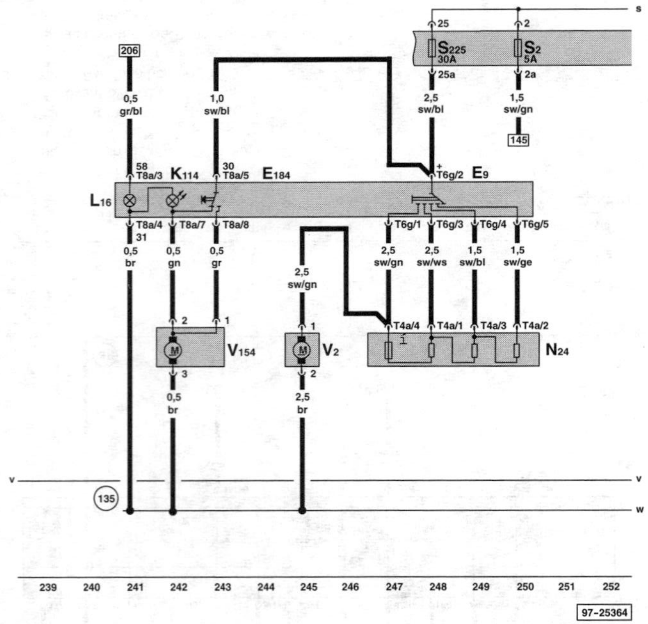

ws = white
sw = black
ro = red
br = brown
gn = green
bl = blue
gr = grey
li = violet
ge = yellow

- E9 - Fresh Air Blower Switch
- E184 - Switch for fresh air blower and recirculated air
- K114 - Fresh Air and Recirculating Air Mode Indicator Light
- L16 - Fresh Air Control Lever Light
- N24 - Fresh Air Blower Series Resistance With Fuse
- S2 - Fuse 2 in fuse holder
- S225 - Fuse 25 in fuse holder
- T4a - 4-Pin Connector
- T6g - 6-Pin Connector, on fresh air blower switch
- T8a - 8-Pin Connector, on switch for fresh air blower and recirculated air
- V2 - Fresh Air Blower
- V154 - Servo motor for fresh-/recirculating air door
- (135) - Ground connection -2-, in instrument panel wiring harness

Edition 06.99
USA.5062.09.21

Fresh air blower switch, switch for fresh air blower and recirculated air, fresh air blower

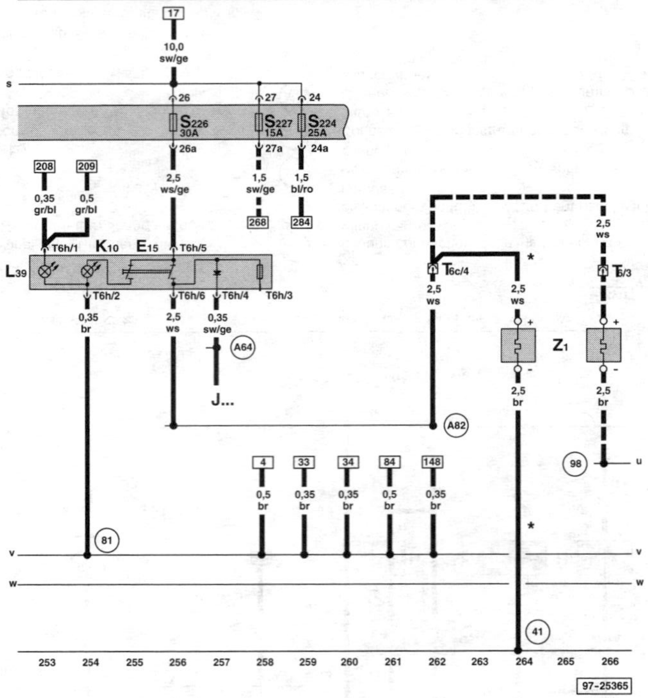

ws = white
sw = black
ro = red
br = brown
gn = green
bl = blue
gr = grey
li = violet
ge = yellow

- E15 - Rear window defogger switch
- J... -
- K10 - Rear Window Defogger Indicator Light
- L39 - Rear Window Defogger Switch Light
- S224 - Fuse 24 in fuse holder
- S226 - Fuse 26 in fuse holder
- S227 - Fuse 27 in fuse holder
- T6 - 6-Pin Connector, black, in luggage compartment, left (wagon only)
- T6c - 6-Pin Connector, blue, connector station A-pillar, left
- T6h - 6-Pin Connector, black, on rear window defogger switch
- Z1 - Heated rear window
- (41) - Ground connection, below rear shelf
- (81) - Ground connection -1-, in instrument panel wiring harness
- (98) - Ground connection, in rear lid wiring harness
- (A64) - Connector (30, idle boost), in instrument panel wiring harness
- (A82) - Connector (heated glass), in instrument panel wiring harness
- - - - wagon only
- * - sedan only

Heated rear window, rear window defogger switch

Edition 06.99
USA.5062.09.21

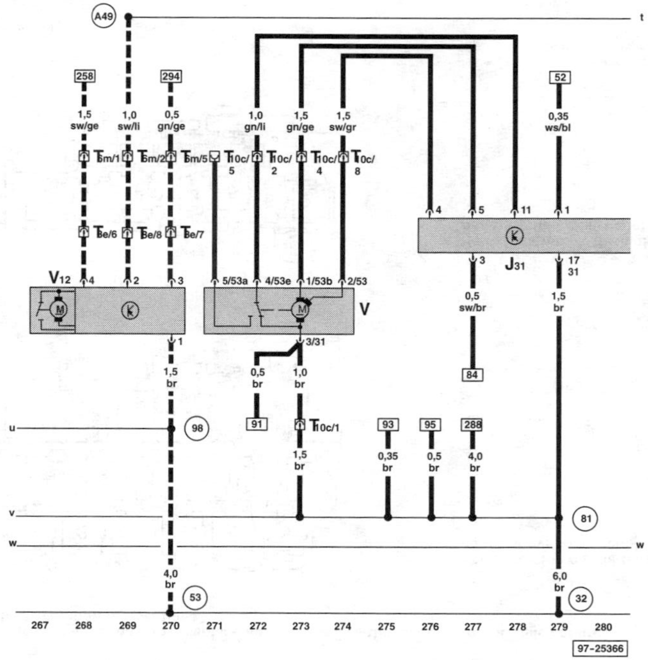

ws = white
sw = black
ro = red
br = brown
gn = green
bl = blue
gr = grey
li = violet
ge = yellow

J31 - Wiper/Washer Intermittent Relay, production control number (377)
T6m - 6-Pin Connector, black, connector station A-pillar, right
T8e - 8-Pin Connector, black, in rear lid, right
T10c - 10-Pin Connector, violet, connector station A-pillar, left
V - Windshield Wiper Motor
V12 - Rear Window Wiper Motor

(32) - Ground connection, behind instrument panel, left
(53) - Ground connection, in rear lid, right
(81) - Ground connection -1-, in instrument panel wiring harness
(98) - Ground connection, in rear lid wiring harness
(A49) - wire connection -1-, in instrument panel wiring harness

- - - wagon only

Windshield wiper motor, windshield wiper intermittent regulator, rear window wiper motor

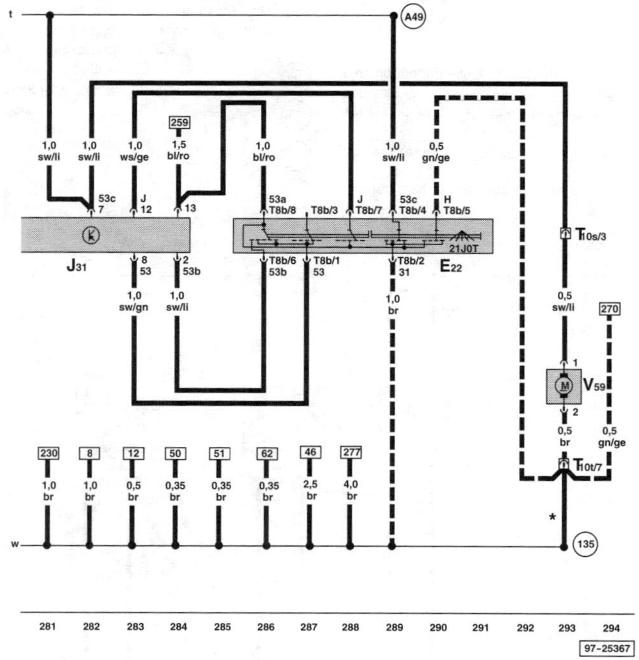

ws = white
sw = black
ro = red
br = brown
gn = green
bl = blue
gr = grey
li = violet
ge = yellow

E22 - Winshield Wiper/Washer Switch
J31 - Wiper/Washer Intermittent Relay, production control number (377)
T8b - 8-Pin Connector, on Windshield Wiper/Washer Switch
T10s - 10-Pin Connector, pink, connector station A-pillar, left
T10t - 10-Pin Connector, red, connector station A-pillar, left
V59 - Windshield and Rear Window Washer Pump

(135) - Ground connection -2-, in instrument panel wiring harness
(A49) - wire connection -1-, in instrument panel wiring harness

- - - wagon only
* - sedan only

Wiper/washer intermittent relay, windshield wiper/washer switch, windshield and rear window washer pump

2,8L- Engine - Motronic Multiport Fuel Injection (MFI)/142 kW, code ATQ

from August 1999

Deviate relay location and fuseplacements as well as the locations of multiple connectors see section "component locations".

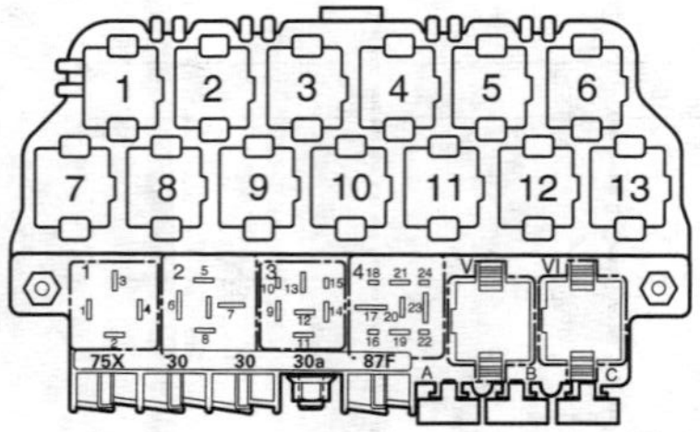

97-20676

Relay location on the thirteenfold auxiliary relay panel, above relay panel:

Relay panel:

4 Fuel Pump (FP) Relay (372)

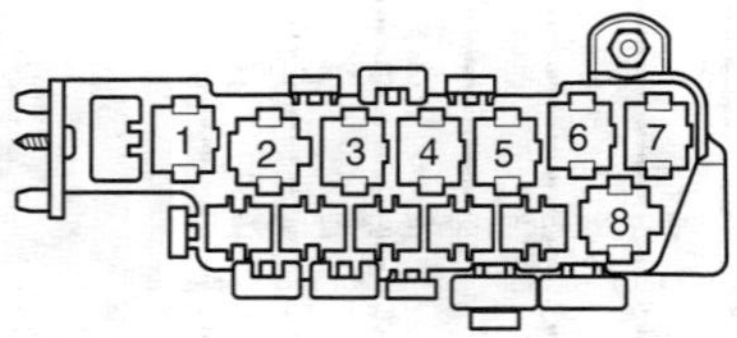

97-20515

Relay location on the eightfold auxiliary relay panel, behind relay panel:

Note: Number in parentheses indicates production control number stamped on relay housing.

Fuse colors
30 A - green
25 A - white
20 A - yellow
15 A - blue
10 A - red
7,5 A - brown
5 A - beige

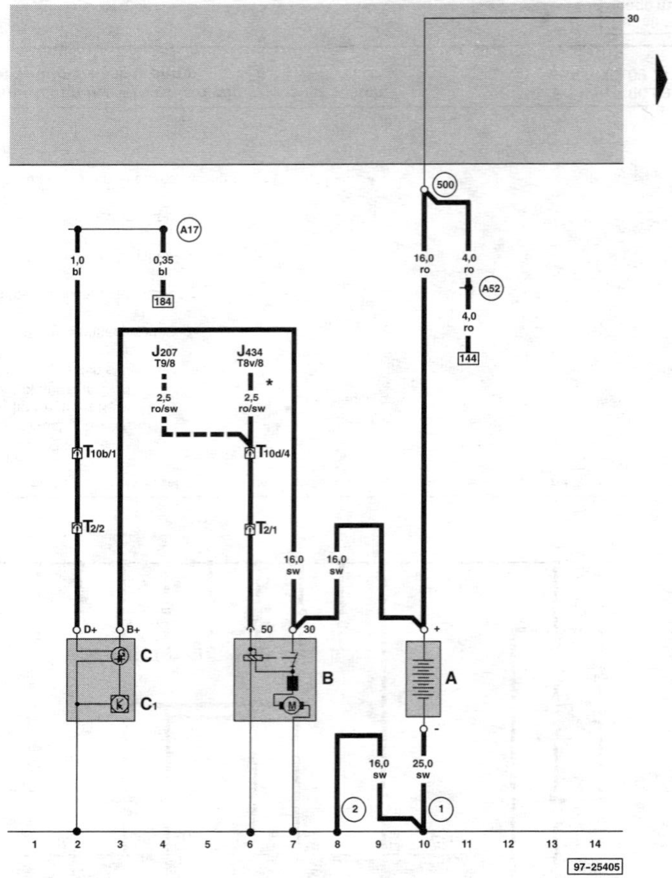

ws = white
sw = black
ro = red
br = brown
gn = green
bl = blue
gr = grey
li = violet
ge = yellow

- A - Battery
- B - Starter
- C - Generator (GEN)
- C1 - Voltage Regulator (VR)
- J207 - Starting Interlock Relay
- J434 - Starting Interlock Relay (clutch position)
- T2 - Double Connector, on engine right
- T8v - 8-Pin Connector, on Starting Interlock Relay
- T9 - 9-Pin Connector, on Starting Interlock Relay
- T10b - 10-Pin Connector, black, on protective housing for control module, in engine compartment, left
- T10d - 10-Pin Connector, brown, on protective housing for control module, in engine compartment, left
- (1) - Ground strap, battery to body
- (2) - Ground strap, transmission to body
- (500) - Screw Connection -1- (30), on relay panel
- (A17) - wire connection (61), in instrument panel wiring harness
- (A52) - plus connection (30), in instrument panel wiring harness
- --- - automatic transmission only
- * - manual transmission only

Battery, starter, generator (GEN)

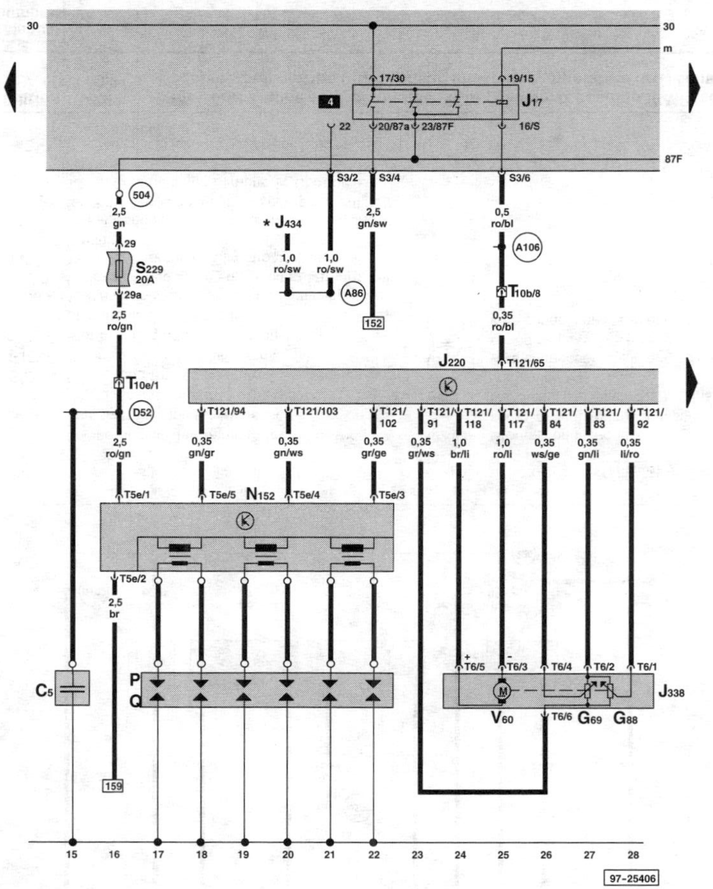

ws = white
sw = black
ro = red
br = brown
gn = green
bl = blue
gr = grey
li = violet
ge = yellow

C5 - Suppressor (Terminal 15)
G69 - Throttle Position (TP) Sensor
G88 - Throttle Position (TP) Sensor
J17 - Fuel Pump (FP) Relay, production control number (208)
J220 - Motronic Engine Control Module (ECM)
J338 - Throttle Valve Control Module
J434 - Starting Interlock Relay (clutch position)
N152 - Ignition Coil
P - Spark Plug Connectors
Q - Spark Plugs
S229 - Fuse in fuse holder
T5e - 5-Pin Connector, black, on Ignition Coil
T6 - 6-Pin Connector, black, on Throttel Valve Control Module
T10b - 10-Pin Connector, black, on protective housing for control module, in engine compartment, left
T10e - 10-Pin Connector, orange, on protective housing for control module, in engine compartment, left
T121 - Connector, 121 point, on Motronic Engine Control Module (ECM)
V60 - Throttle Position (TP) Actuator

(504) - Screw Connection (87F), on relay panel
(A86) - Connection (50a) in instrument panel wiring harness
(A106) - Connector -2- (86s), in instrument panel wiring harness
(D52) - plus connection (15a), in engine compartment wiring harness
* - manual transmission only

Edition 06.99
USA.5062.09.21

Ignition system, Motronic engine control module (ECM), Throttle valve control module

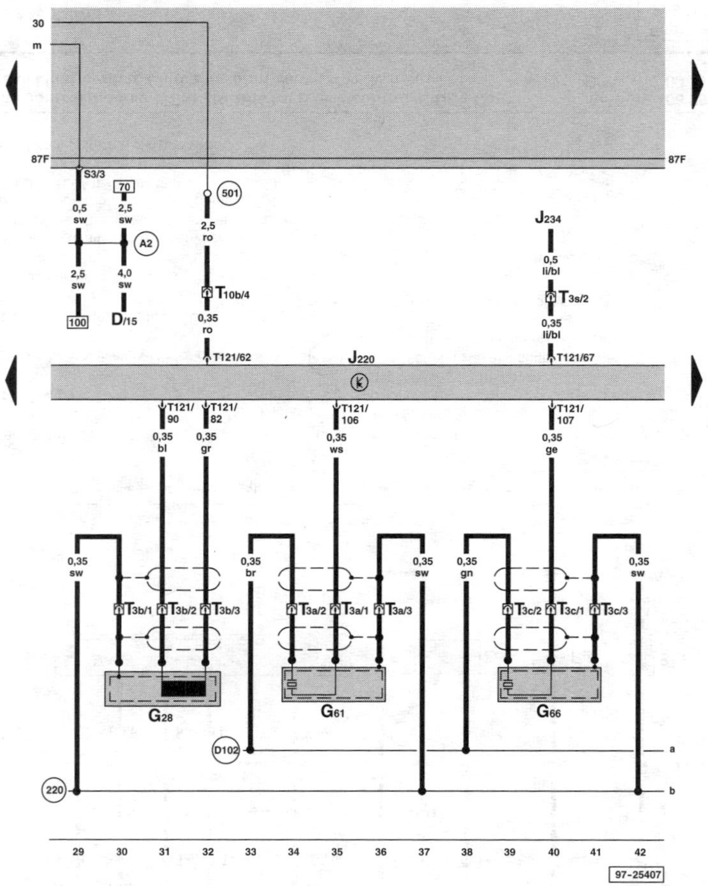

ws = white
sw = black
ro = red
br = brown
gn = green
bl = blue
gr = grey
li = violet
ge = yellow

D - Ignition/Starter Switch
G28 - Engine Speed (RPM) Sensor
G61 - Knock Sensor (KS) 1
G66 - Knock Sensor (KS) 2
J220 - Motronic Engine Control Module (ECM)
J234 - Airbag Control Module
T3a - 3-Pin Connector, blue, on firewall, left
T3b - 3-Pin Connector, grey, on firewall, left
T3c - 3-Pin Connector, blue, on firewall, left
T3s - 3-Pin Connector, red, on protective housing for control module, in engine compartment, left
T121 - Connector, 121 point, on Motronic Engine Control Module (ECM)

(220) - Ground connection (sensor ground), in engine compartment wiring harness

(501) - Screw Connection -2- (30), on relay panel
(A2) - plus connection (15), in instrument panel wiring harness
(D102) - Wire connection -2-, in engine compartment wiring harness

Motronic engine control module (ECM), engine speed (RPM) sensor, knock sensor (KS)

Edition 06.99
USA.5062.09.21

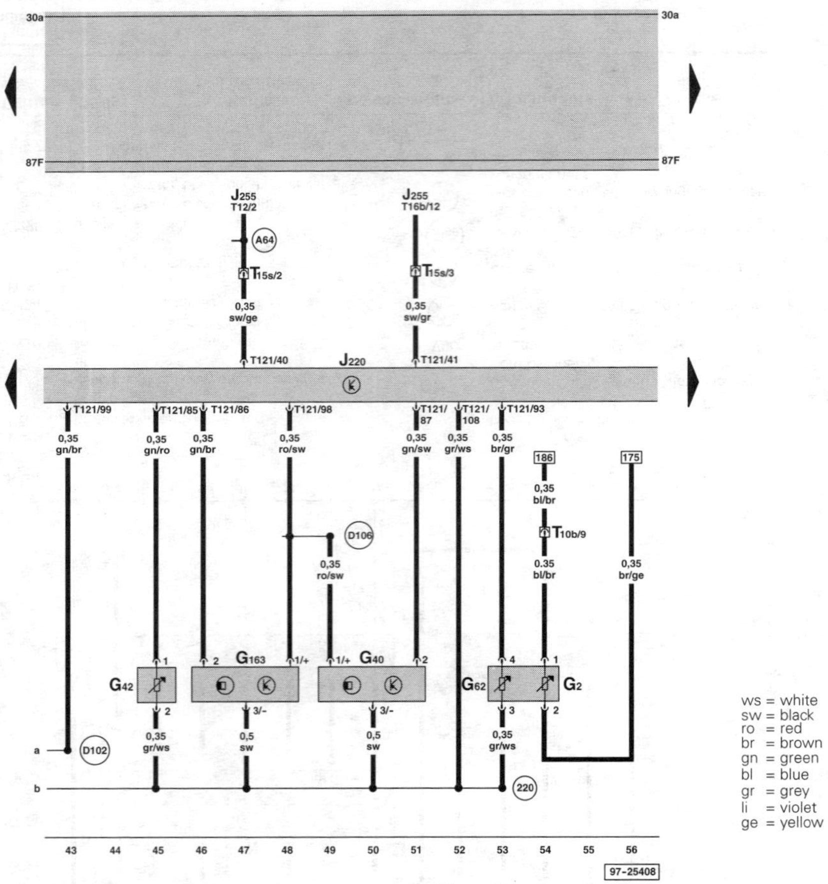

ws = white
sw = black
ro = red
br = brown
gn = green
bl = blue
gr = grey
li = violet
ge = yellow

G2 - Engine Coolant Temperature (ECT) Sensor
G40 - Camshaft Position (CMP) Sensor
G42 - Intake Air Temperature (IAT) Sensor
G62 - Engine Coolant Temperature (ECT) Sensor
G163 - Camshaft Position (CMP) Sensor 2
J220 - Motronic Engine Control Module (ECM)
J255 - Climatronic Control Module
T10b - 10-Pin Connector, black, on protective housing for control module, in engine compartment, left
T12 - 12-Pin Connector, on Climatronic Control Module
T15s - 15-Pin Connector, red, on protective housing for control module, in engine compartment, left
T16b - 16-Pin Connector, on Climatronic Control Module
T121 - Connector, 121 point, on Motronic Engine Control Module (ECM)

(220) - Ground connection (sensor ground), in engine compartment wiring harness
(A64) - Connector (30, idle boost), in instrument panel wiring harness
(D102) - wire connection -2-, in engine compartment wiring harness
(D106) - wire connection -4-, in engine compartment wiring harness

Edition 06.99
USA.5062.09.21

Motronic engine control module (ECM), camshaft position (CMP) sensor, engine coolant temperature (ECT) sensor, intake air temperature (IAT) sensor

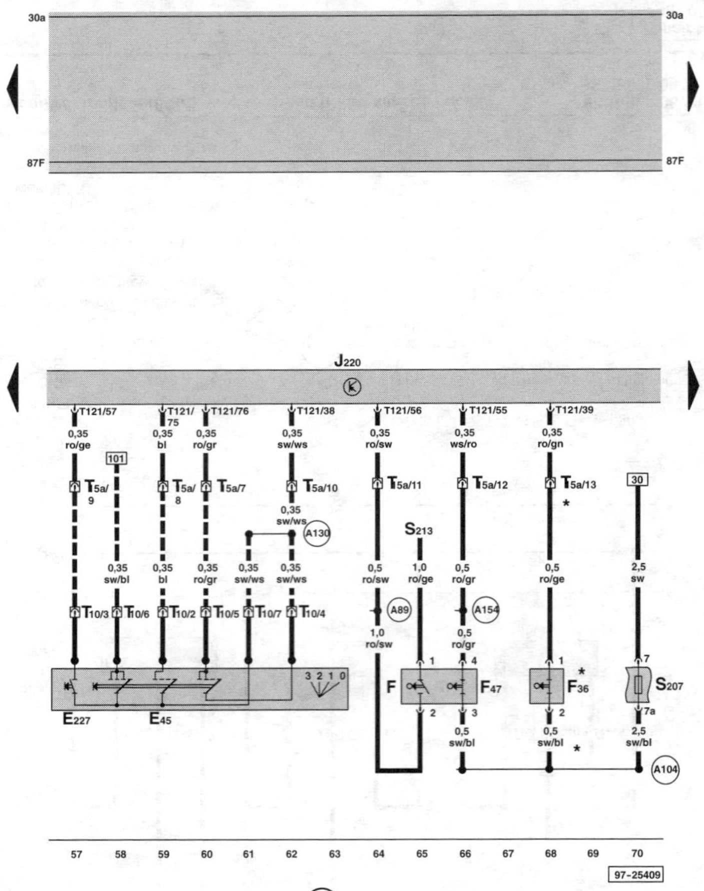

ws = white
sw = black
ro = red
br = brown
gn = green
bl = blue
gr = grey
li = violet
ge = yellow

E45 - Cruise Control Switch
E227 - Button for cruise control (set)
F - Brake Light Switch
F36 - Clutch Vacuum Vent Valve Switch
F47 - Brake Vacuum Vent Valve Switch for cruise control/diesel
J220 - Motronic Engine Control Module (ECM)
S207 - Fuse 7 in fuse holer
S213 - Fuse 13 in fuse holder
T10 - 10-Pin Connector
T15a - 15-Pin Connector, white, on protective housing for control module, in engine compartment,left
T121 - Connector, 121 point, on Motronic Engine Control Module (ECM)

(A89) - Connector (54) in instrument panel wiring harness

(A104) - Plus connector -2- (15), in instrument panel wiring harness
(A130) - Connection for Cruise Control (instrument panel wiring harness)
(A154) - Connection (Brake Vacuum Vent Valve Switch) in instrument panel wiring harness

- - - automatic transmission only

* - manual transmission only

Motronic engine control module (ECM), Button for cruise control, Cruise Control Switch, Brake Light Switch, Clutch Vacuum Vent Valve Switch

Edition 06.99
USA.5062.09.21

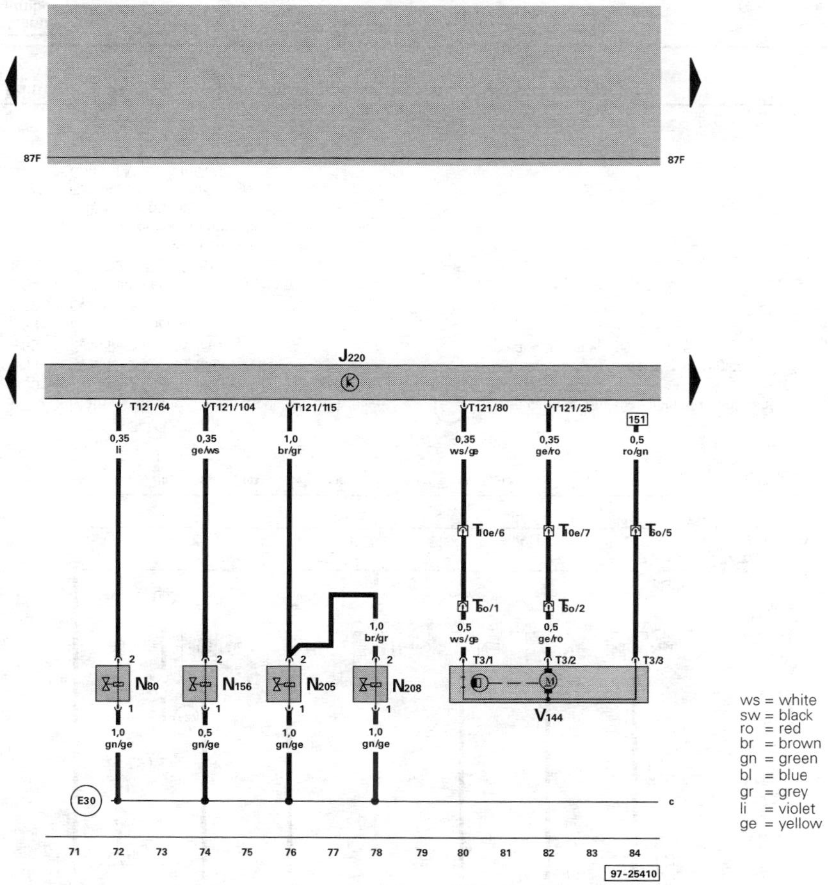

ws = white
sw = black
ro = red
br = brown
gn = green
bl = blue
gr = grey
li = violet
ge = yellow

J220 - Motronic Engine Control Module (ECM)
N80 - Evaporative Emission (EVAP) Canister Purge Regulator Valve
N156 - Intake Manifold Change-Over Valve
N205 - Valve -1- for camshaft adjustment
N208 - Valve -2- for camshaft adjustment
T6o - 6-Pin Connector, brown, connector station A-pillar, right
T10e - 10-Pin Connector, orange, on protective housing for control module, in engine compartment, left
T121 - Connector, 121 point, on Motronic Engine Control Module (ECM)
V144 - Leak Detection Pump (LDP)

(E30) - Connector (87a), in wiring harness engine

Motronic engine control module (ECM), valve for camshaft adjustment, Leak detection pump (LDP)

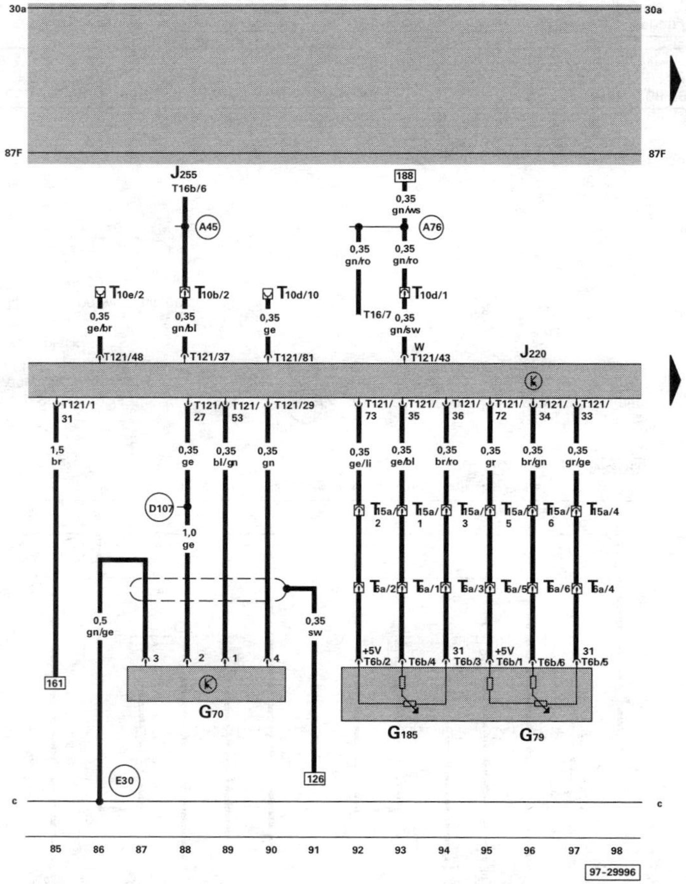

ws = white
sw = black
ro = red
br = brown
gn = green
bl = blue
gr = grey
li = violet
ge = yellow

G70 - Mass Air Flow (MAF) Sensor
G79 - Throttle Position (TP) Sensor
G185 - Sender -2- for accelerator pedal position
J220 - Motronic Engine Control Module (ECM)
J255 - Climatronic Control Module
T6a - 6-Pin Connector
T10b - 10-Pin Connector, black, on protective housing for control module, in engine compartment, left
T10d - 10-Pin Connector, brown, on protective housing for control module, in engine compartment, left
T10e - 10-Pin Connector, orange, on protective housing for control module, in engine compartment, left
T15a - 15-Pin Connector, white, on protective housing for control module, in engine compartment, left
T16 - 16-Pin Connector, Data Link Connector (DLC), below instrument panel, left
T16b - 16-Pin Connector, on Climatronic Control Module
T121 - Connector, 121 point, on Motronic Engine Control Module (ECM)

(A45) - wire connection (RPM-signal), in instrument panel wiring harness
(A76) - Connector (K-diagnosis wire), in instrument panel wiring harness
(D107) - wire connection -5-, in engine compartment wiring harness
(E30) - Connector (87a), in wiring harness engine

Motronic engine control module (ECM), Mass Air Flow Sensor, Throttle Position Sensor, Data Link Connector (DLC)

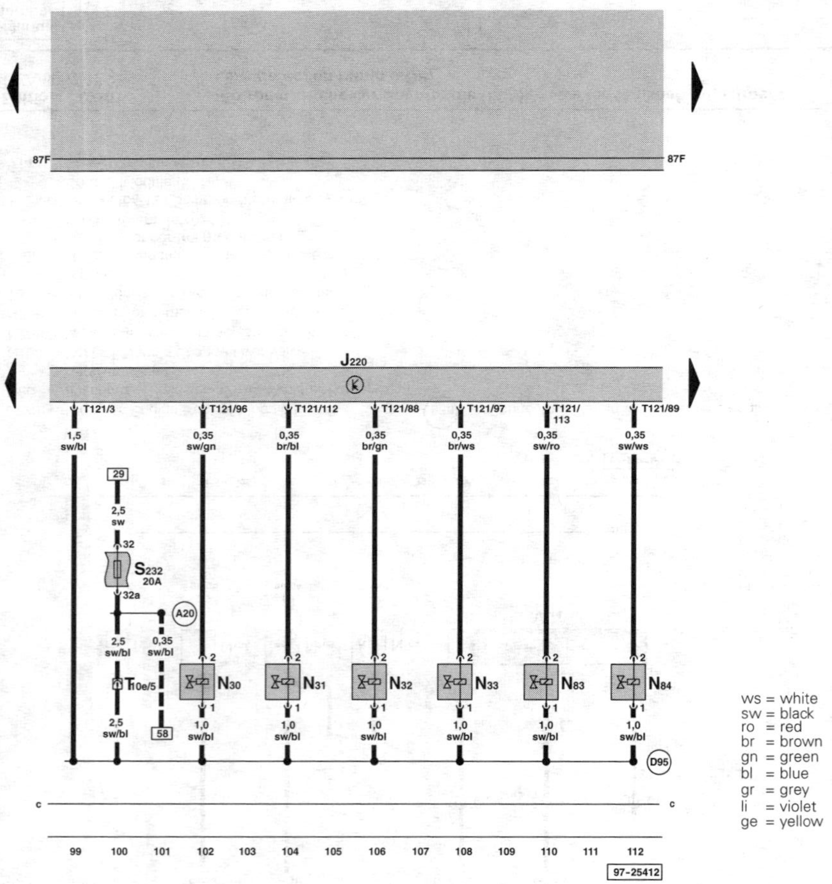

ws = white
sw = black
ro = red
br = brown
gn = green
bl = blue
gr = grey
li = violet
ge = yellow

J220 - Motronic Engine Control Module (ECM)
N30 - Cylinder 1 Fuel Injector
N31 - Cylinder 2 Fuel Injector
N32 - Cylinder 3 Fuel Injector
N33 - Cylinder 4 Fuel Injector
N83 - Cylinder 5 Fuel Injector
N84 - Cylinder 6 Fuel Injector
S232 - Fuse 32 in fuse holder
T10e - 10-Pin Connector, orange, on protective housing for control module, in engine compartment, left
T121 - Connector, 121 point, on Motronic Engine Control Module (ECM)

(A20) - wire connection (15a), in instrument panel wiring harness
(D95) - wire connection (injectors), in engine compartment wiring harness

- - - cruise control only

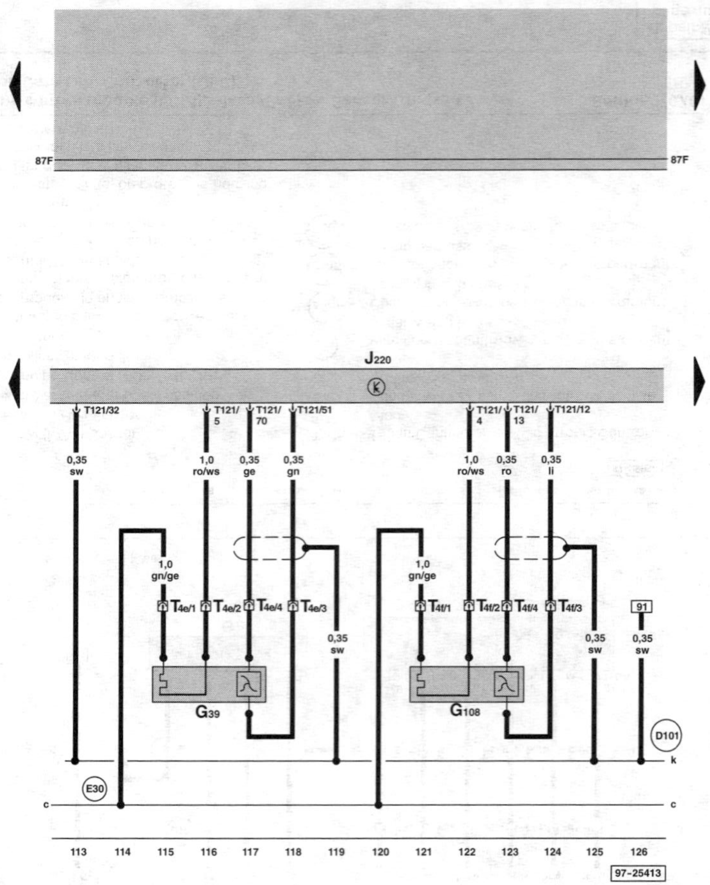

ws = white
sw = black
ro = red
br = brown
gn = green
bl = blue
gr = grey
li = violet
ge = yellow

G39 - Heated Oxygen Sensor (HO2S)
G108 - Heated Oxygen Sensor (HO2S) 2
J220 - Motronic Engine Control Module (ECM)
T4e - 4-Pin Connector, **black**
T4f - 4-Pin Connector, black
T121 - Connector, 121 point, on Motronic Engine Control Module (ECM)

(D101) - wire connection -1-, in engine compartment wiring harness
(E30) - Connector (87a), in wiring harness engine

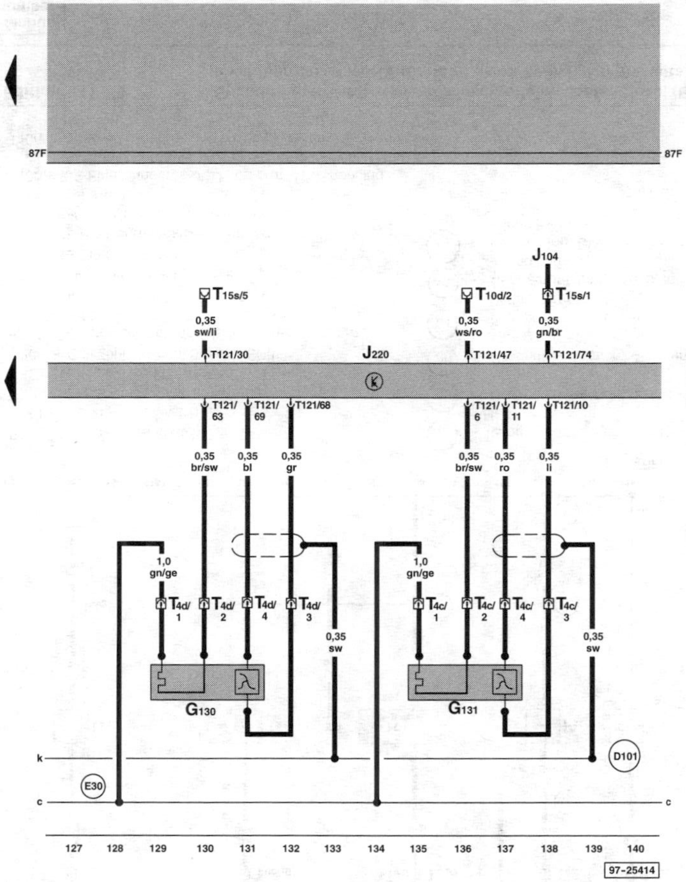

ws = white
sw = black
ro = red
br = brown
gn = green
bl = blue
gr = grey
li = violet
ge = yellow

G130 - Oxygen Sensor (O2S) Behind Three Way Catalytic Converter (TWC)
G131 - Oxygen Sensor (O2S) 2 Behind Three Way Catalytic Converter (TWC)
J104 - ABS Control Module (w/EDL)
J220 - Motronic Engine Control Module (ECM)
T4c - 4-Pin Connector, green
T4d - 4-Pin Connector, brown
T10d - 10-Pin Connector, brown, on protective housing for control module, in engine compartment, left
T15s - 15-Pin Connector, red, on protective housing for control module, in engine compartment, left
T121 - Connector, 121 point, on Motronic Engine Control Module (ECM)

(D101) - wire connection -1-, in engine compartment wiring harness
(E30) - Connector (87a), in wiring harness engine

Edition 04/01
USA.5062.14.21

Motronic engine control module (ECM), oxygen sensors behind three way catalytic converter

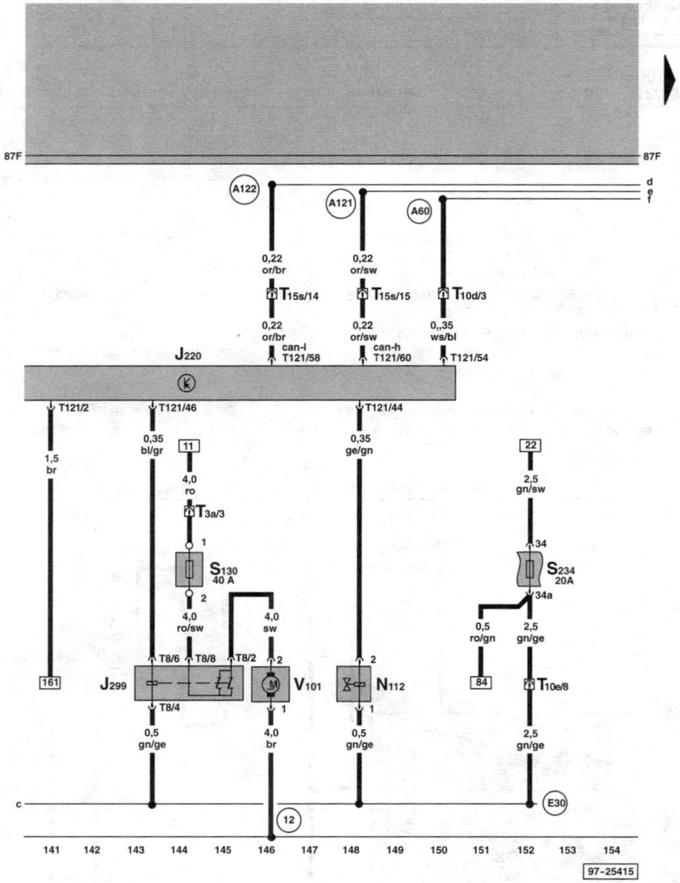

ws = white
sw = black
ro = red
br = brown
gn = green
bl = blue
gr = grey
li = violet
ge = yellow
or = orange

J220 - Motronic Engine Control Module (ECM)
J299 - Secondary Air Injection (AIR) Pump Relay
N112 - Secondary Air Injection (AIR) Solenoid Valve
S130 - Fuse for secondary air pump, in fuse adaptor above relay panel
S234 - Fuse 34 in fuse holder
T3a - 3-Pin Connector,
T8 - 8-Pin Connector, black on Radio
T10d - 10-Pin Connector, brown, on protective housing for control module, in engine compartment, left
T10e - 10-Pin Connector, yellow, on protective housing for control module, in engine compartment, left
T15s - 15-Pin Connector, red, , on protective housing for control module, in engine compartment, left
T121 - Connector, 121 point, on Motronic Engine Control Module (ECM)
V101 - Secondary Air Injection (AIR) Pump Motor

(12) - Ground connection, in engine compartment, left
(A60) - Wire connection (Vehicle speed signal), in instrument cluster wiring harness
(A121) - Connection (high bus) in instrument panel wiring harness
(A122) - Connection (low bus) in instrument panel wiring harness
(E30) - Connector (87a), in wiring harness engine

Motronic engine control module (ECM), secondary air injection system

Edition 04/01
USA.5062.14.21

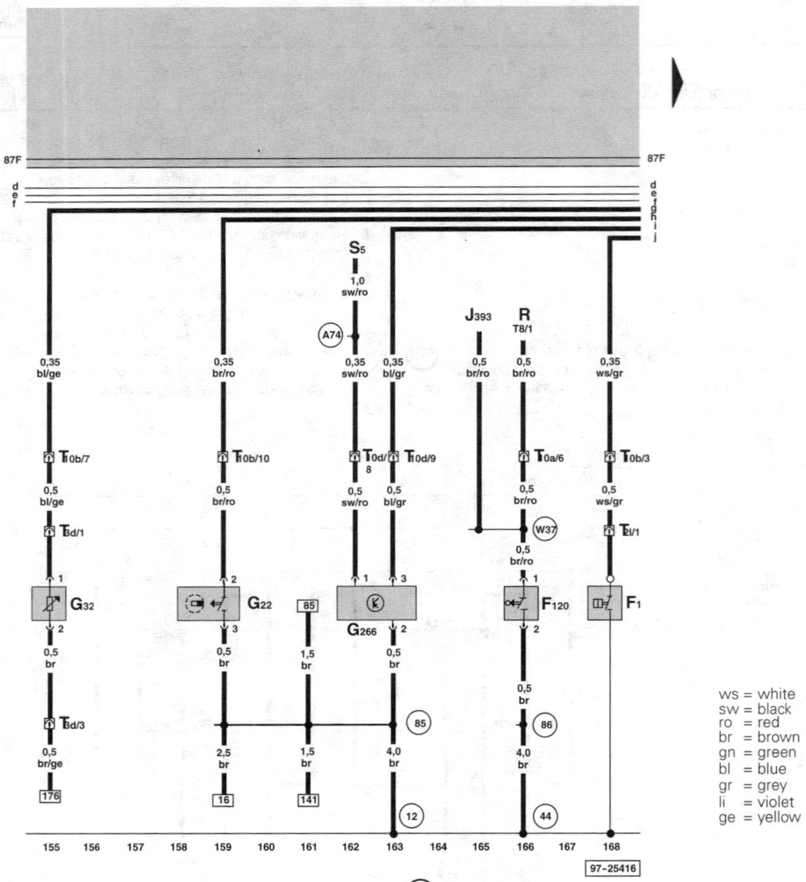

ws = white
sw = black
ro = red
br = brown
gn = green
bl = blue
gr = grey
li = violet
ge = yellow

F1 - Oil Pressure Switch
F120 - Hood Alarm Switch
G22 - Speedometer Vehicle Speed Sensor (VSS)
G32 - Engine Coolant Level (ECL) Sensor
G266 - Oil Level Thermal Sensor (SRI)
J393 - Central control module for comfort system
R - Radio
S5 - Fuse
T2l - Double Connector, black
T3d - 3-Pin Connector, in engine compartment, left
T8 - 8-Pin Connector, black, on radio
T10a - 10-Pin Connector,brown, connector station A-pillar, left
T10b - 10-Pin Connector, black, on protective housing for control module, in engine compartment, left
T10d - 10-Pin Connector, brown, on protective housing for control module, in engine compartment, left

(12) - Ground connection, in engine compartment, left
(44) - Ground connection, on left A-pillar, lower part
(85) - Ground connection, -1-, in engine compartment wiring harness
(86) - Ground connection, -1-, in rear compartment wiring harness
(A74) - Connector (15a - fuse 5), in instrument panel wiring harness
(W37) - Connection (DWA - alarm system) in floor wiring harness

Edition 12/99
USA.5062.10.21

Speedometer vehicle speed sensor, Engine coolant level (ECL) sensor, Oil level thermal sensor(SRI), Hood alarm switch, Oil pressure switch

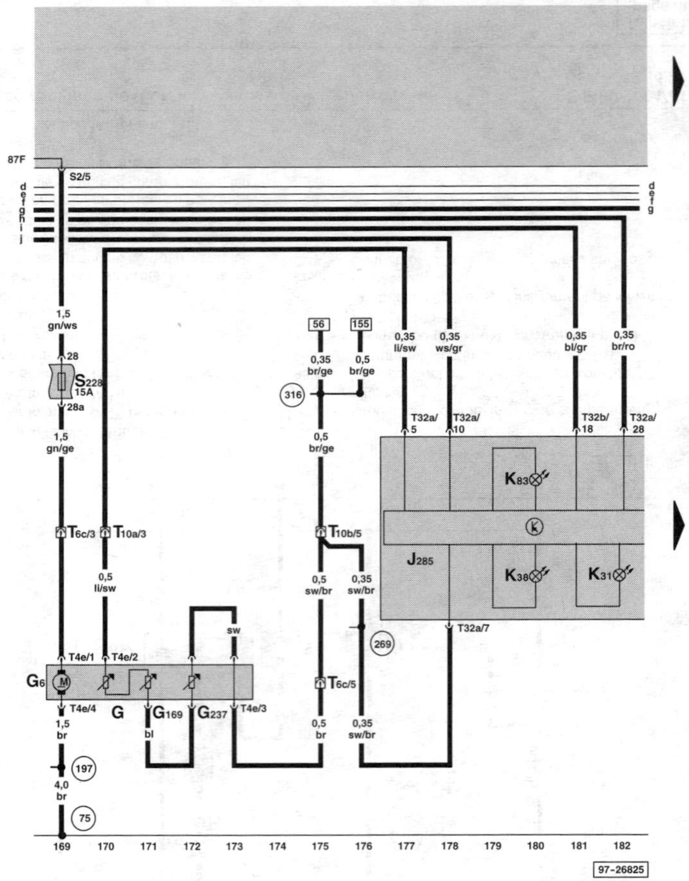

ws = white
sw = black
ro = red
br = brown
gn = green
bl = blue
gr = grey
li = violet
ge = yellow

G - Sender for fuel gauge
G6 - Fuel Pump (FP)
G169 - Sender -2- for fuel supply
G237 - Sensor -3- for fuel supply
J285 - Control module with indicator unit in instrument panel insert
K31 - Cruise Control Indicator Light
K38 - Engine Oil Level Indicator Light
K83 - Malfunction Indicator Lamp (MIL)
S228 – Fuse 28 in fuse holder
T4e - 4-Pin Connector, on Sender for fuel gauge
T6c - 6-Pin Connector, blue, connector station A-pillar, left
T10a 10-Pin Connector, brown, connector station A-pillar, left
T10b - 10-Pin Connector, black, on protective housing for control module, in engine compartment, left
T32a - 32-Pin Connector, blue, on instrument cluster
T32b - 32-Pin Connector, green, on instrument cluster

(75) - Ground connection, on right rear pillar
(197) - Ground connection -4-, in wiring harness rear
(269) - Ground connection (sensor ground) -1-, in instrument panel wiring harness
(316) - Ground connection (sensor ground -2-), in engine compartment wiring harness

Sender for fuel gauge, control module with indicator unit in instrument panel insert

Edition 12/99
USA.5062.10.21

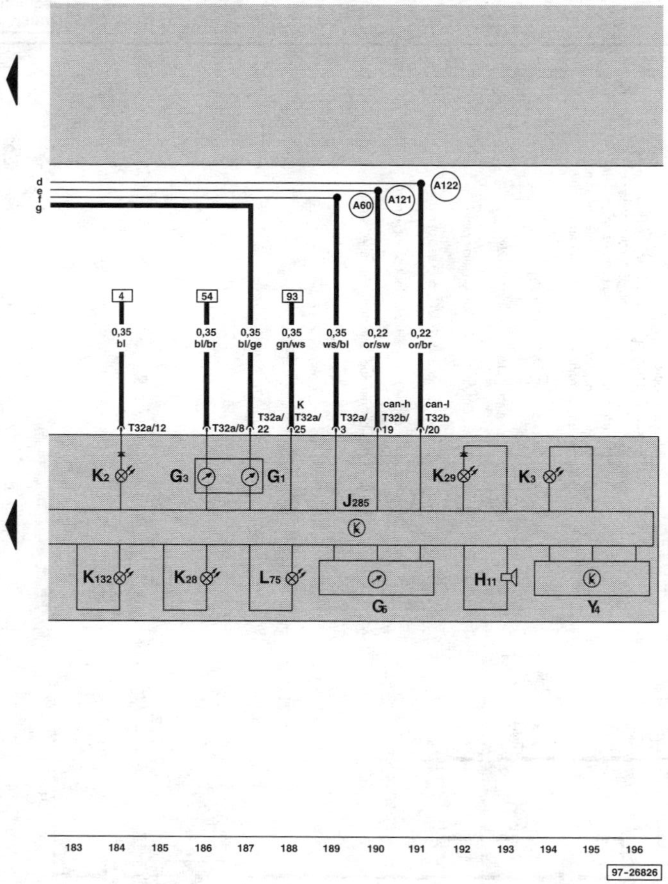

ws = white
sw = black
ro = red
br = brown
gn = green
bl = blue
gr = grey
li = violet
ge = yellow
or = orange

G1 - Fuel gauge
G3 - Engine Coolant Temperature (ECT) Gauge
H11 - Oil Pressure Warning Buzzer
J285 - Control module with indicator unit in instrument panel insert
K2 - Generator (GEN) Warning Light
K3 - Oil Pressure Warning Light
K28 - Engine Coolant Level/Temperature (ECL/ECT) Warning Light
K29 - Glow Plug Indicator Light
K132 - Electronic Power Control (EPC) Warning Lamp
L75 - Digital Display Light
T32a - 32-Pin Connector, blue, on instrument cluster
T32b - 32-Pin Connector, green, on instrument cluster
Y4 - Odometer Display

(A60) - Wire connection (Vehicle speed signal), in instrument cluster wiring harness
(A121) - Connection (high bus) in instrument panel wiring harness
(A122) - Connection (low bus) in instrument panel wiring harness

Edition 12/99
USA.5062.10.21

Control module with indicator unit in instrument panel insert

THIS PAGE INTENTIONALLY LEFT BLANK

1,8L-Engine - Motronic Multipoint Fuel Injection (MFI)/110 kW, code ATW

from August 1999

Deviate relay location and fuseplacements as well as the locations of multiple connectors see section "component locations"

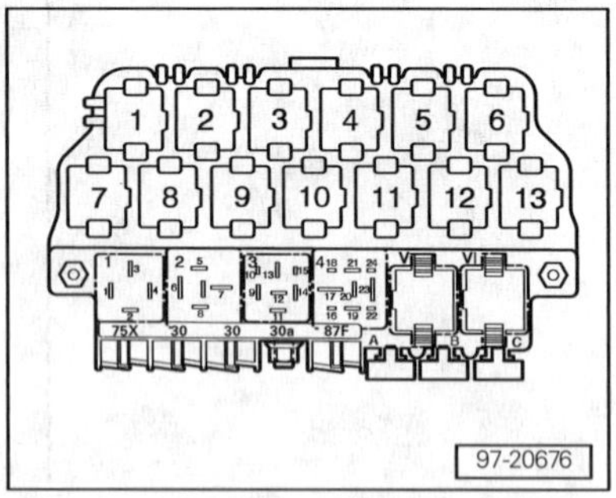

Relay location on the thirteen position auxiliary relay panel, above relay panel:

Relay panel:

4 - Fuel Pump (FP) Relay (372)

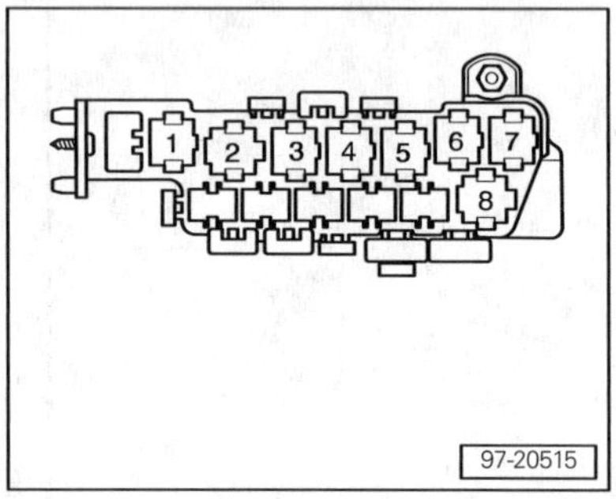

Relay location on the eight position auxiliary relay panel, above relay panel:

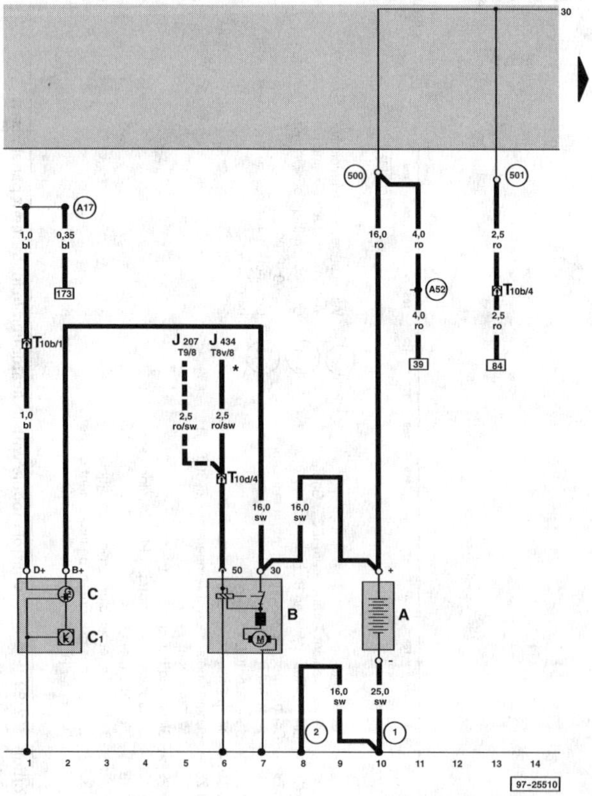

Battery, starter, generator (GEN)

A - Battery
B - Starter
C - Generator (GEN)
C1 - Voltage Regulator (VR)
J207 - Starting Interlock Realay
J434 - Locking relay for starter
T8v - 8-Pin Connector
T9 - 9-Pin Connector, on Starting Interlock Relay
T10b - 10-Pin Connector, black, on protective housing for control module, in engine compartment,left
T10d - 10-Pin Connector, brown, on protective housing for control module, in engine compartment,left

(1) - Ground strap, battery to body
(2) - Ground strap, transmission to body
(500) - Screw Connection -1- (30), on relay panel
(501) - Screw Connection -2- (30), on relay panel
(A17) - wire connection, in instrument panel wiring harness
(A52) - plus connection, in instrument panel wiring harness

- - - - automatic transmission only
* - manual transmission only

ws = white
sw = black
ro = red
br = brown
gn = green
bl = blue
gr = grey
li = lilac
ge = yellow
or = orange

Passat — Wiring diagram — No. 37/3

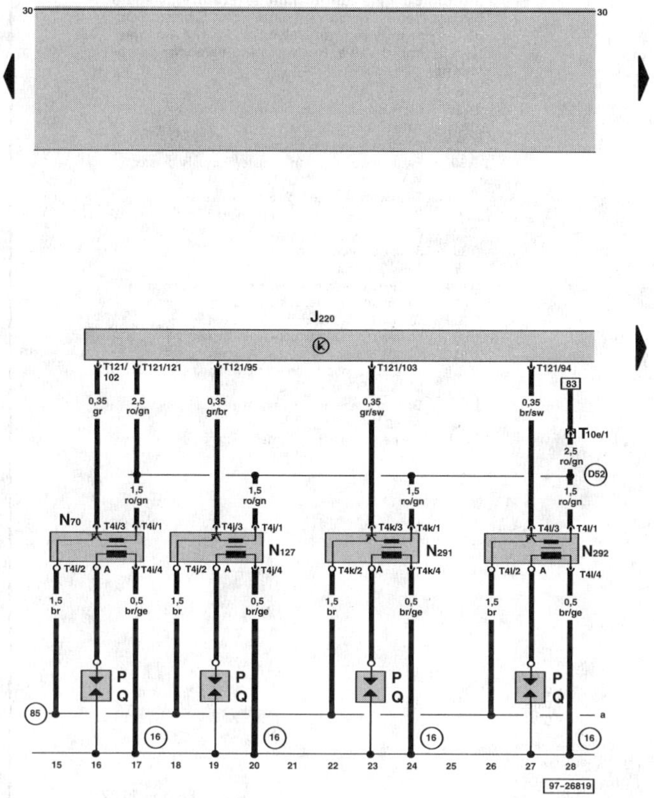

Ignition system, Motronic engine control module (ECM), Spark plug connectors, Spark plugs

J220 - Motronic Engine Control Module (ECM)
N70 - Ignition Coil 1 with Power Output Stage
N127 - Ignition Coil 2 with Power Output Stage
N291 - Ignition Coil 3 with Power Output Stage
N292 - Ignition Coil 4 with Power Output Stage
P - Spark Plug Connectors
Q - Spark Plugs
T4i - 4-Pin Connector
T4j - 4-Pin Connector
T4k - 4-Pin Connector
T4l - 4-Pin Connector
T10e - 10-Pin Connector, orange, on protective housing for control module, in engine compartment, left
T121 - 121-Pin Connector, on Motronic Engine Control Module (ECM)

(16) - Ground connection -1-, on valve cover

(85) - Ground connection -1-, in engine compartment wiring harness

(D52) - plus connection (15a), in engine compartment wiring harness

ws = white
sw = black
ro = red
br = brown
gn = green
bl = blue
gr = grey
li = lilac
ge = yellow
or = orange

Edition 10.02

Passat — Wiring diagram — No. 37/4

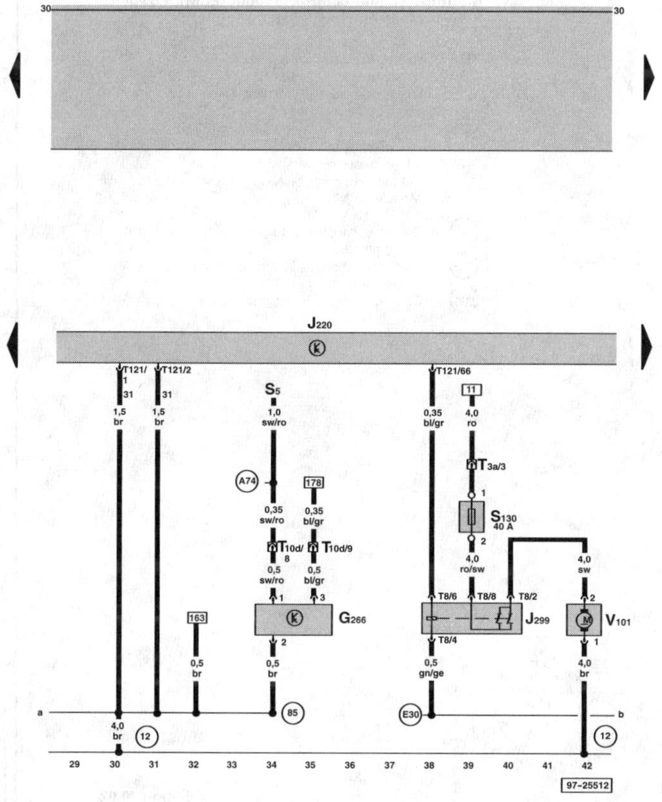

Motronic engine control module (ECM), Secondary Air Injection (AIR) Pump Motor/Relay

G266 - Oil Level Thermal Sensor
J220 - Motronic Engine Control Module (ECM)
J299 - Secondary Air Injection (AIR) Pump Relay
S5 - Fuse
S130 - Fuse for secondary air pump
T3a - 3-Pin Connector, white
T8 - 8-Pin Connector, brown
T10d - 10-Pin Connector, brown, on protective housing for control module, in engine compartment, left
T121 - 121-Pin Connector, on Motronic Engine Control Module (ECM)
V101 - Secondary Air Injection (AIR) Pump Motor

(12) - Ground connection, in engine compartment, left

(85) - Ground connection -1-, in engine compartment wiring harness

(A74) - Connector (kick-down-switch) in instrument panel wiring harness

(E30) - Connector, in wiring harness engine

ws = white
sw = black
ro = red
br = brown
gn = green
bl = blue
gr = grey
li = lilac
ge = yellow
or = orange

Edition 10.02

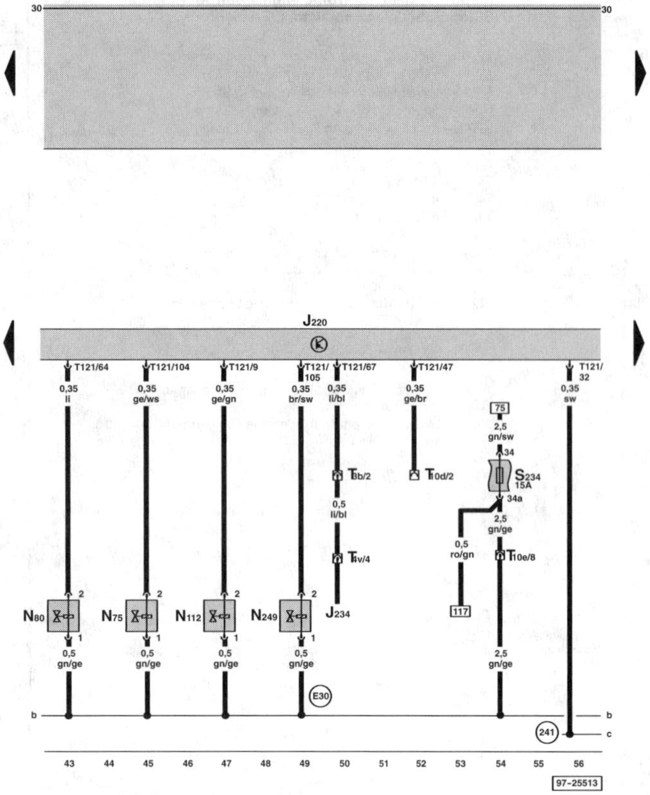

Motronic engine control module (ECM), Wastegate Bypass Regulator Valve, Ecaporative Emission Canister Purge Regulator Valve,Secondary Air Injection Solenoid Valve, Recirculating valve for turbocharger

J220 - Motronic Engine Control Module (ECM)
J234 - Airbag Control Module
N75 - Wastegate Bypass Regulator Valve
N80 - Evaporative Emission Canister Purge Regulator Valve
N112 - Secondary Air Injection Solenoid Valve
N249 - Recirculating valve for turbocharger
S234 - Fuse in fuse holder
T3b - 3-Pin Connector, red
T4v - 4-Pin Connector
T10d - 10-Pin Connector, brown, on protective housing for control module, in engine compartment, left
T10e - 10-Pin Connector, orange, on protective housing for control module, in engine compartment, left
T121 - 121-Pin Connector, on Motronic Engine Control Module (ECM)
(241) - Ground connection 2, engine ground in engine wiring harness
(E30) - Connector, in wiring harness engine

ws = white
sw = black
ro = red
br = brown
gn = green
bl = blue
gr = grey
li = lilac
ge = yellow
or = orange

Edition 10.02

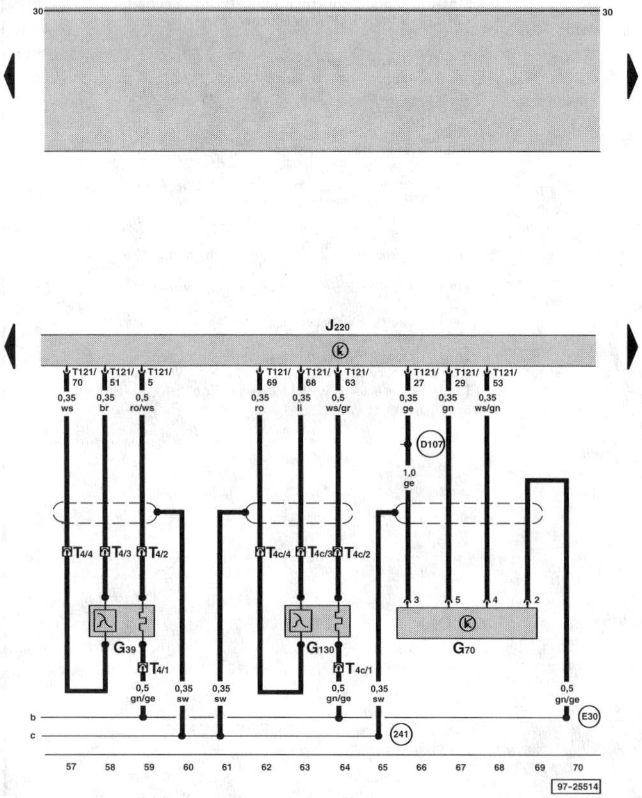

Motronic engine control module (ECM), Oxygen Sensor, Mass Air Flow Sensor

G39 - Heated Oxygen Sensor (HO2S)
G70 - Mass Air Flow (MAF) Sensor
G130 - Oxygen Sensor (O2S) Behind Three Way Catalytic Converter (TWC)
J220 - Motronic Engine Control Module (ECM)
T4 - 4-Pin Connector, black
T4c - 4-Pin Connector, black
T121 - 121-Pin Connector, on Motronic Engine Control Module (ECM)

(241) - Ground connection 2, engine ground in engine wiring harness
(D107) - wire connection -5-, in engine compartment wiring harness
(E30) - Connector, in wiring harness engine

ws = white
sw = black
ro = red
br = brown
gn = green
bl = blue
gr = grey
li = lilac
ge = yellow
or = orange

Edition 10.02

Motronic engine control module (ECM), fuel pump relay, power supply relay, fuel injectors

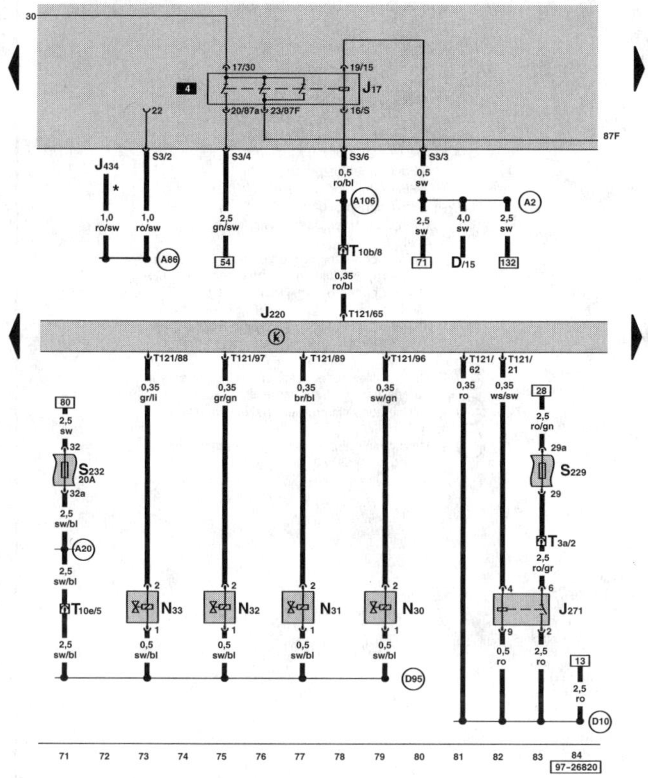

D - Ignition/Starter Switch
J17 - Fuel Pump (FP) Relay (372)
J220 - Motronic Engine Control Module (ECM)
J271 - Motronic Engine Control (ECM) Power Supply Relay
J434 - Locking relay for starter
N30 - Cylinder 1 Fuel Injector
N31 - Cylinder 2 Fuel Injector
N32 - Cylinder 3 Fuel Injector
N33 - Cylinder 4 Fuel Injector
S229 - Fuse in fuse holder
S232 - Fuse in fuse holder
T3a - 3-Pin Connector, white
T10b - 10-Pin Connector, black, on protective housing for control module, in engine compartment,left
T10e - 10-Pin Connector, orange, on protective housing for control module, in engine compartment, left
T121 - 121-Pin Connector, on Motronic Engine Control Module (ECM)
A2 - plus connection (15), in instrument panel wiring harness
A20 - wire connection (15a), in instrument panel wiring harness
A86 - Connection (50a), in instrument panel wiring harness
A106 - Connector -2- (86s), in instrument panel wiring harness
D10 - plus connection (30), in right front wiring harness
D95 - wire connection (injectors), in engine compartment wiring harness
* - manual transmission only

ws = white
sw = black
ro = red
br = brown
gn = green
bl = blue
gr = grey
li = lilac
ge = yellow
or = orange

Edition 10.02

Motronic engine control module (ECM), Engine Coolant Temperature Sensor, Camshaft Position Sensor, Intake Air Temperature

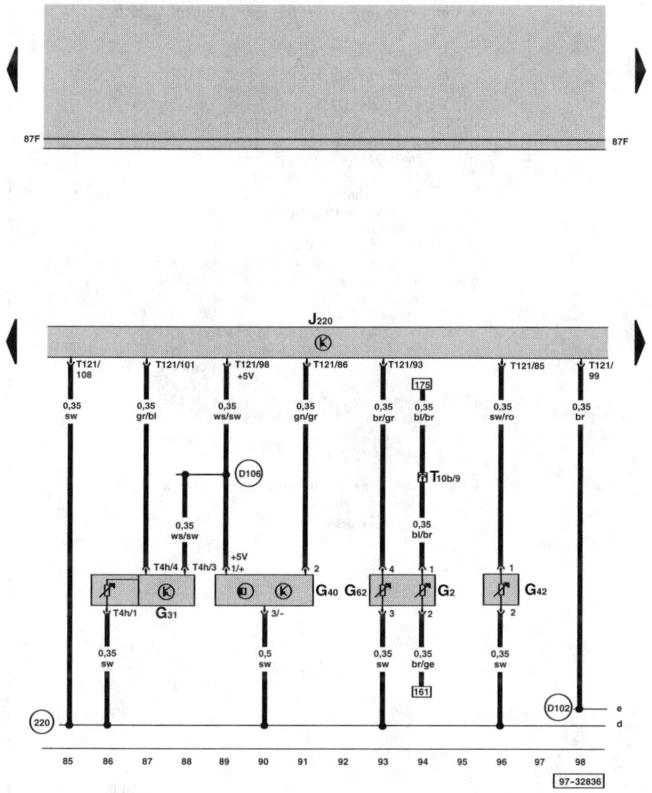

G2 - Engine Coolant Temperature (ECT) Sensor
G40 - Camshaft Position (CMP) Sensor
G42 - Intake Air Temperature
G62 - Engine Coolant Temperature (ECT) Sensor
G31 - Charge Air Pressure Sensor
J220 - Motronic Engine Control Module (ECM)
T4h - 4-Pin Connector, black
T10b - 10-Pin Connector, black, on protective housing for control module, in engine compartment,left
T121 - 121-Pin Connector, on Motronic Engine Control Module (ECM)

(220) - Ground connection (sensor ground), in engine compartment wiring harness
(D102) - Wire connection -2-, in engine compartment wiring harness
(D106) - Wire connection -4-, in engine compartment wiring harness

ws = white
sw = black
ro = red
br = brown
gn = green
bl = blue
gr = grey
li = lilac
ge = yellow
or = orange

Edition 10.02

Motronic engine control module (ECM), Knock Sensor 1 and 2, Engine Speed Sensor

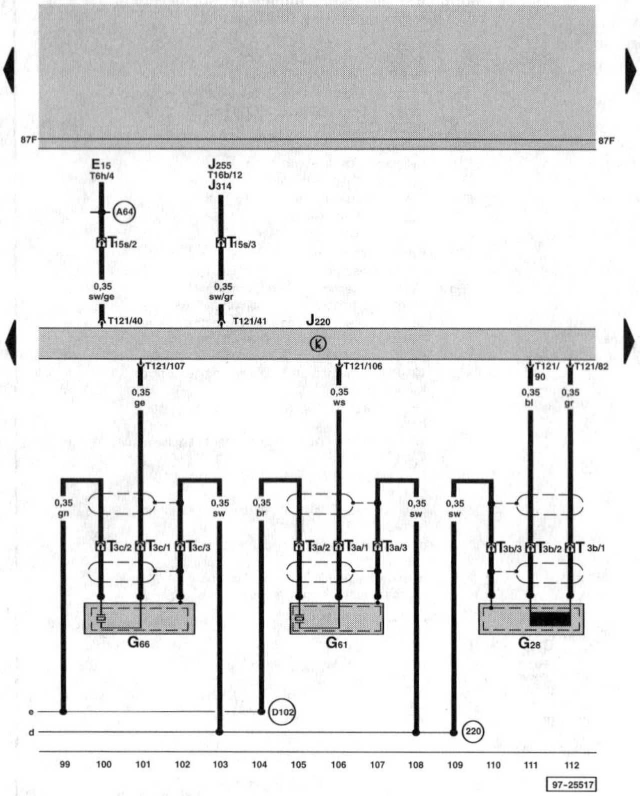

E 15 - Rear window defogger switch
G28 - Engine Speed Sensor
G61 - Knock Sensor (KS) 1
G66 - Knock Sensor (KS) 2
J220 - Motronic Engine Control Module (ECM)
J255 - Climatronic Control Module
J314 - A/C Cut-Off Control Module
T3a - 3-Pin Connector, green
T3b - 3-Pin Connector, grey
T3c - 3-Pin Connector,blue
T6h - 6-Pin Connector, on rear window defogger switch
T15s - 15-Pin Connector, red, on protective housing for control module, in engine compartment,left
T16b - 16-Pin Connector, on Climatronic Control Module
T121 - 121-Pin Connector, on Motronic Engine Control Module (ECM)

(220) - Ground connection (sensor ground), in engine compartment wiring harness
(A64) - Connector, in instrument panel wiring harness
(D102) - Wire connection -2-, in engine compartment wiring harness

ws = white
sw = black
ro = red
br = brown
gn = green
bl = blue
gr = grey
li = lilac
ge = yellow
or = orange

Motronic Engine Control Module (ECM), Leak Detection Pump (LDP), Throttle Valve Control Module

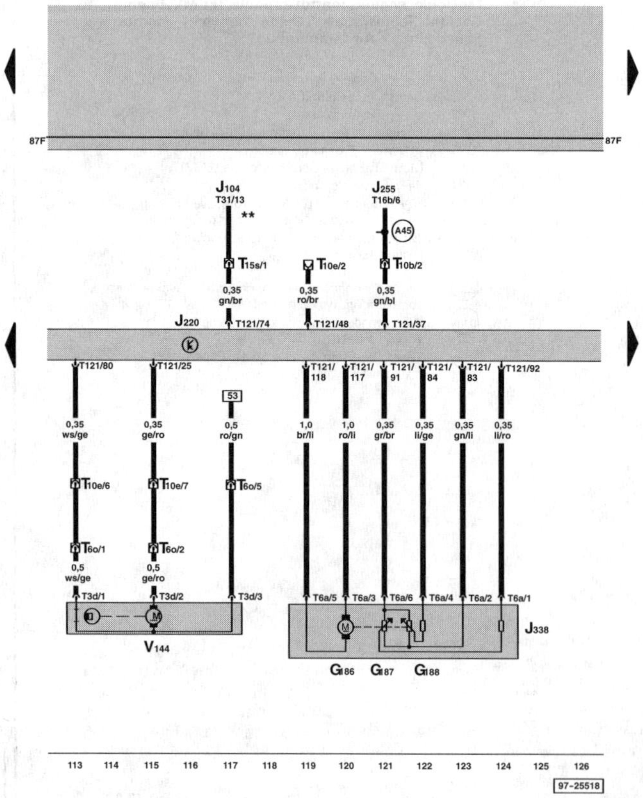

G186 - Throttle senor
G187 - Angle senor -1- for throttle drive
G188 - Angle senor -2- for throttle drive
J104 - ABS Control Module
J220 - Motronic Engine Control Module (ECM)
J255 - Climatronic Control Module
J338 - Throttle Valve Control Module
T3d - 3-Pin Connector, black
T6a - 6-Pin Connector, black
T6o - 6-Pin Connector, brown, connector station A-pillar, right
T10b - 10-Pin Connector, black, on protective housing for control module, in engine compartment, left
T10e - 10-Pin Connector, orange, on protective housing for control module, in engine compartment, left
T15s - 15-Pin Connector, red, on protective housing for control module, in engine compartment,left
T16b - 16-Pin Connector, on Climatronic Control Module
T31 - 31-Pin Connector, on ABS Control Module
T121 - 121-Pin Connector, on Motronic Engine Control Module (ECM)
V144 - Leak Detection Pump (LDP)

(A45) - wire connection (RPM-signal), in instrument panel wiring harness
** - only on vehicles with ASR

ws = white
sw = black
ro = red
br = brown
gn = green
bl = blue
gr = grey
li = lilac
ge = yellow
or = orange

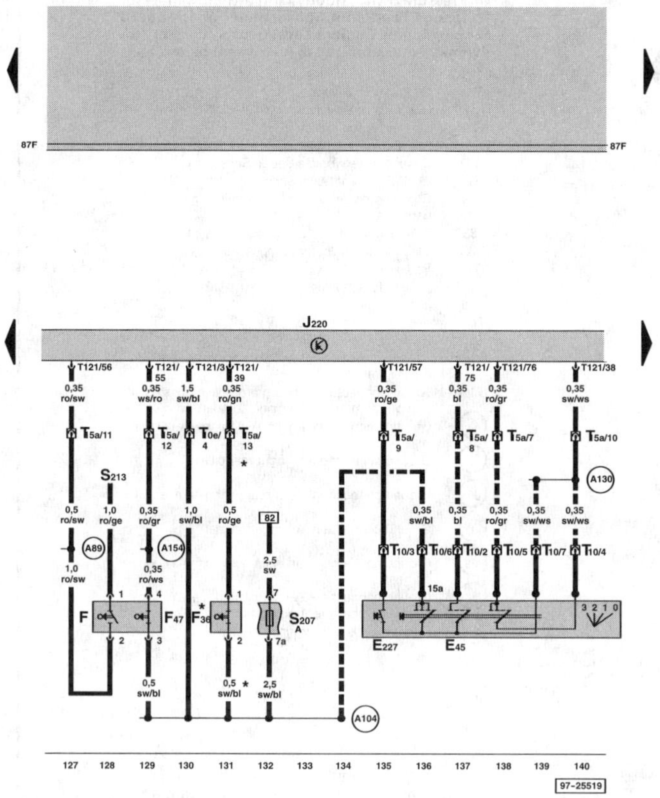

Motronic engine control module (ECM), Cruise control switch, Brake light switch, Clutch vacuum vent valve switch

E45 - Cruise Control Switch
E227 - Button for cruise control
F - Brake Light Switch
F36 - Clutch Vacuum Vent Valve Switch
F47 - Brake Vacuum Vent Valve Switch for cruise control/diesel
J220 - Motronic Engine Control Module (ECM)
S207 - Fuse 7 in Fuseholder
S213 - Fuse 13 in Fuseholder
T10 - 10-Pin Connector, black
T10e - 10-Pin Connector, orange, on protective housing for control module, in engine compartment, left
T15a - 15-Pin Connector, white
T121 - 121-Pin Connector, on Motronic Engine Control Module (ECM)

(A89) - Connector, in instrument panel wiring harness
(A104) - Plus Connector -2-, in instrument panel wiring harness
(A130) - Connection for Cruise Control (instrument panel wiring harness)
(A154) - Connection (Brake Vacuum Vent Valve Switch) in instrument panel wiring harness

* - manual transmission only
- - - - cruise control only

ws = white
sw = black
ro = red
br = brown
gn = green
bl = blue
gr = grey
li = lilac
ge = yellow
or = orange

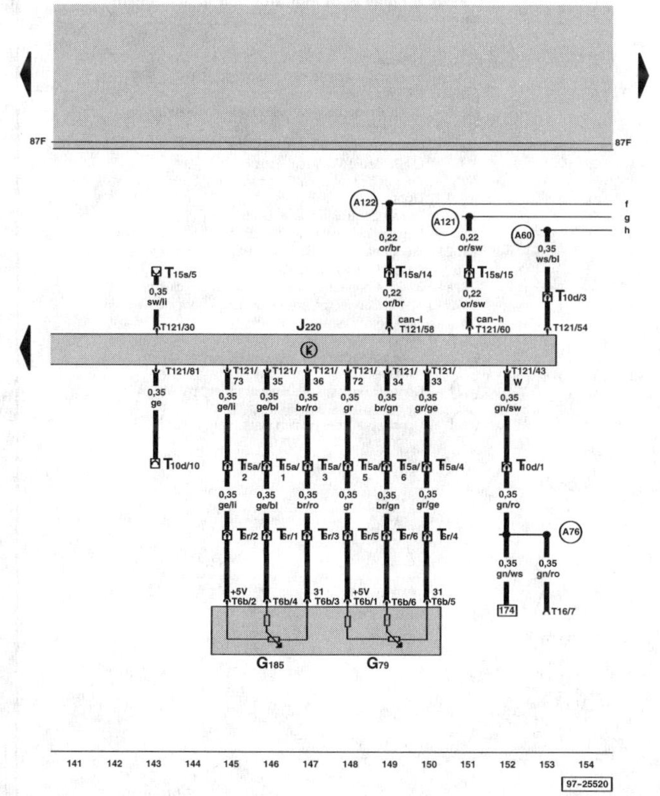

Motronic engine control module (ECM), Throttle position sensor, Sender -2- for accelerator pedal position, Data Link Connector (DLC)

G79 - Throttle Position (TP) Sensor
G185 - Sender -2- for accelerator pedal position
J220 - Motronic Engine Control Module (ECM)
T6b - 6-Pin Connector, on Throttle Position Sensor
T6r - 6-Pin Connector, black
T10d - 10-Pin Connector, brown, on protective housing for control module, in engine compartment, left
T15a - 15-Pin Connector,white,on protective housing for control module, in engine compartment,left
T15s - 15-Pin Connector, red, on protective housing for control module, in engine compartment,left
T16 - 16-Pin Connector, Data Link Connector (DLC), below instrument panel, left
T121 - 121-Pin Connector, on Motronic Engine Control Module (ECM)

(A60) - Wire connection, in instrument cluster wiring harness
(A76) - Connector in instrument panel wiring harness
(A121) - Connection (high bus) in instrument panel wiring harness
(A122) - Connection (low bus) in instrument panel wiring harness

ws = white
sw = black
ro = red
br = brown
gn = green
bl = blue
gr = grey
li = lilac
ge = yellow
or = orange

Oil Pressure Swich, Hood Alarm Switch, Sender for fuel gauge, Fuel Pump, Speedometer Vehicle Speed Sensor, Engine Coolant Level Sensor, Control module with indicator unit in instrument panel insert

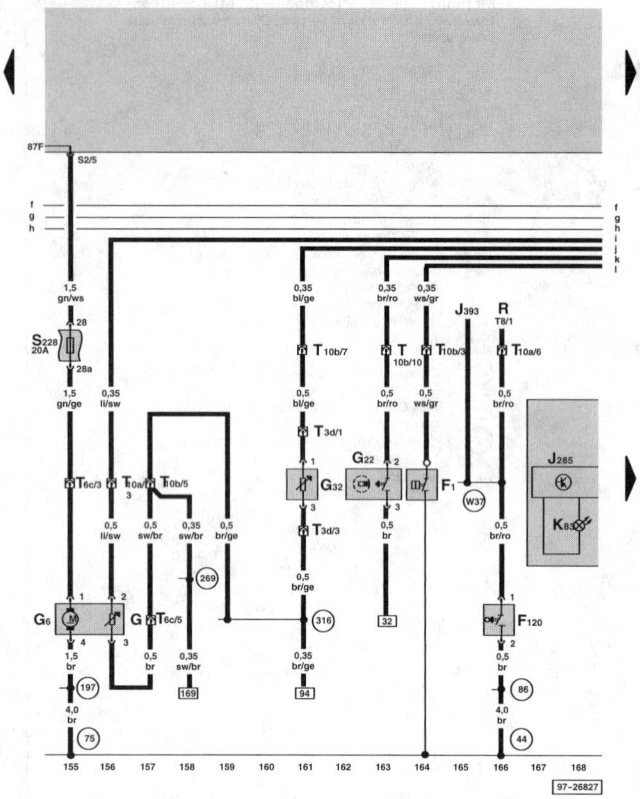

- F1 - Oil Pressure Switch
- F120 - Hood Alarm Switch
- G - Sender for fuel gauge
- G6 - Fuel Pump
- G22 - Speedometer Vehicle Speed Sensor
- G32 - Engine Coolant Level Sensor
- J285 - Control module with indicator unit in instrument panel insert
- J393 - Central control module for comfort system
- K83 - Malfunction Indicator Lamp (MIL)
- R - Radio
- S228 - Fuse in fuse holder
- T3d - 3-Pin Connector
- T6c - 6-Pin Connector, blue
- T8 - 8-Pin Connector, black
- T10a - 10-Pin Connector, brown, connector station A-pillar, left
- T10b - 10-Pin Connector, black, on protective housing for control module, in engine compartment, left
- (44) - Ground connection, on left A-pillar, lower part
- (75) - Ground connection, on right rear pillar
- (86) - Ground connection -1-, in rear wiring harness
- (197) - Ground connection -4-, in wiring rear
- (269) - Ground connetor -1-, in instrument panel wiring harness
- (316) - Ground connection (sensor ground -2-), in engine compartment wiring harness
- (W37) - Connection in floor wiring harness

ws = white
sw = black
ro = red
br = brown
gn = green
bl = blue
gr = grey
li = lilac
ge = yellow
or = orange

97-26827

Edition 10.02

Control module with indicator unit in instrument panel insert

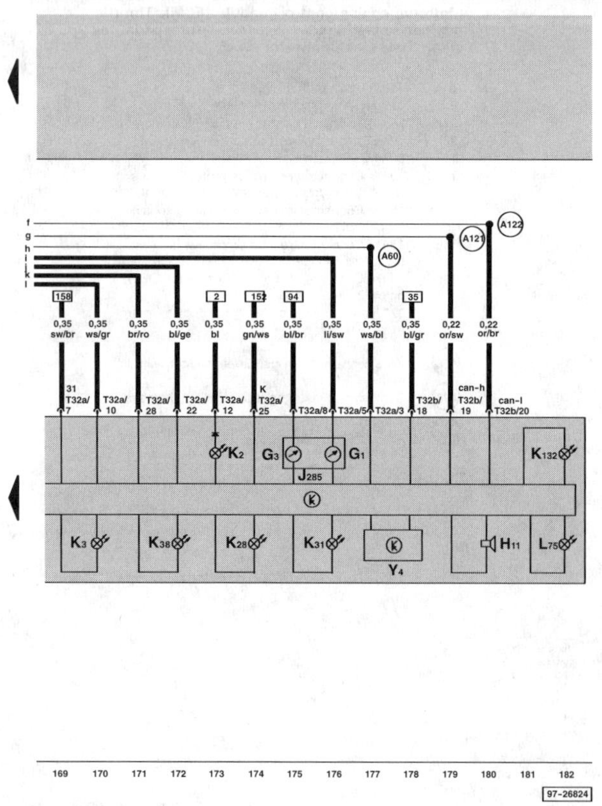

- G1 - Fuel Gauge
- G3 - Engine Coolant Temperature Gauge
- H11 - Oil Pressure Warning Buzzer
- J285 - Control module with indicator unit in instrument panel insert
- K2 - Generator (GEN) Warning Light
- K3 - Oil Pressure Warning Light
- K28 - Engine Collant Level/Temperature Warning Light
- K31 - Cruise Control Indicator Light
- K38 - Engine Oil Level Indicator Light
- K132 - Electronic Power Control (EPC) Warning Lamp
- L75 - Digital Display Light
- T32a - 32-Pin Connector, blue, on instrument cluster
- T32b - 32-Pin Connector, green, on instrument cluster
- Y4 - Odometer Display
- (A60) - Wire connection, in instrument cluster wiring harness
- (A121) - Connection (high bus) in instrument panel wiring harness
- (A122) - Connection (low bus) in instrument panel wiring harness

ws = white
sw = black
ro = red
br = brown
gn = green
bl = blue
gr = grey
li = lilac
ge = yellow
or = orange

97-26824

Edition 10.02

Comfort System, with Alarm System

from August 1999

Deviate relay location and fuseplacements as well as the locations of multiple connectors see section "component locations".

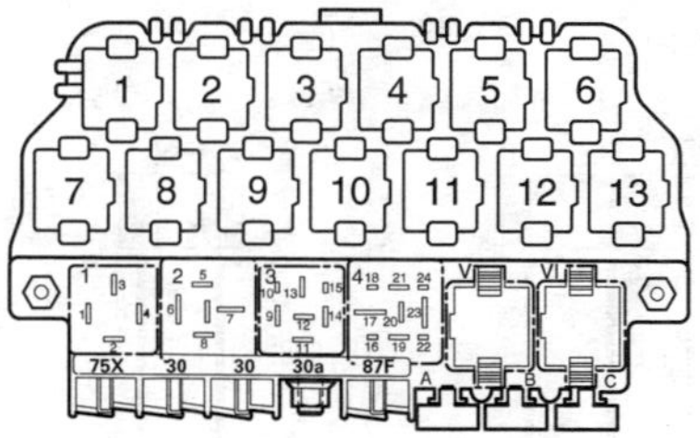

97-20676

Relay location on the thirteenfold auxiliary relay panel, above relay panel:

2 Relay for motor remote unlock rear lid (79)

Relay panel:

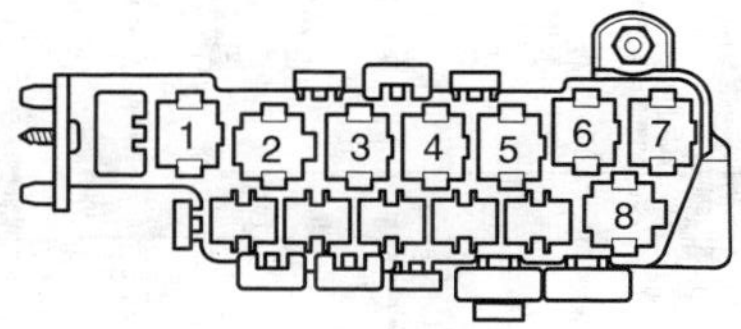

97-20515

Relay location on the eightfold auxiliary relay panel, behind relay panel:

Note: Number in parentheses indicates production control number stamped on relay housing.

Fuse colors
30 A - green
25 A - white
20 A - yellow
15 A - blue
10 A - red
7,5 A - brown
5 A - beige

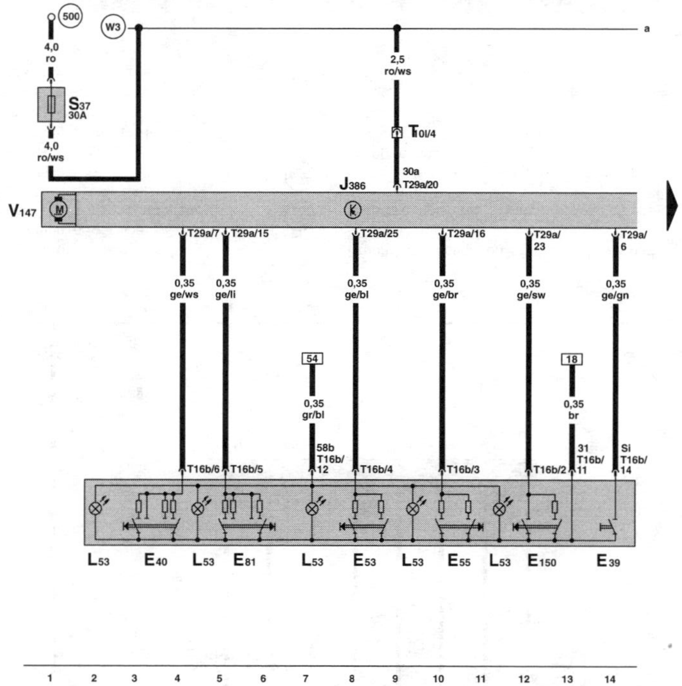

ws = white
sw = black
ro = red
br = brown
gn = green
bl = blue
gr = grey
li = violet
ge = yellow

97-25470

- E39 - Safety switch for window regulator rear
- E40 - Switch for window regulator front left
- E53 - Switch for window regulator rear left, driver
- E55 - Switch for window regulator rear right, driver
- E81 - Switch for window regulator front right, driver
- E150 - Switch for interior lock, driver side
- J386 - Door control module, driver side
- L53 - Power Window Switch Light
- S37 - Power window fuse
- T10l - 10-pin connector, black, connector station A-pillar left
- T16b - 16-pin connector, brown
- T29a - 29-pin connector
- V147 - Motor for window regulator, driver side

- (500) - Screw connection -1- (30), on relay panel
- (W3) - wire connection, in rear wiring harness

Door control module, driver side, switch for window regulator in driver's door, switch for interior lock

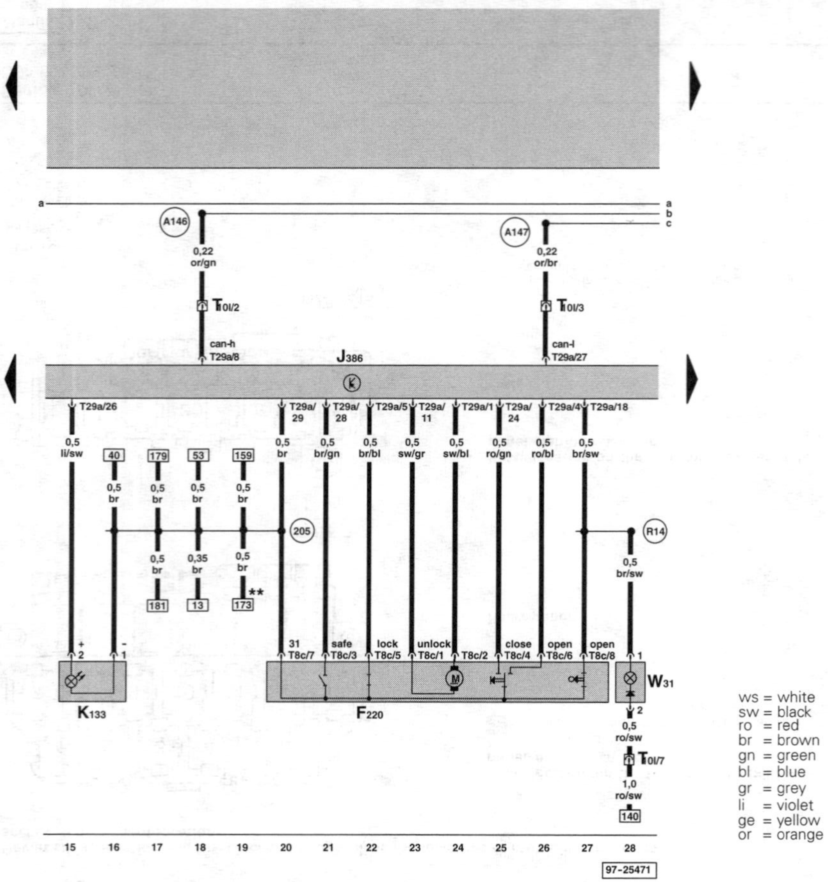

ws = white
sw = black
ro = red
br = brown
gn = green
bl = blue
gr = grey
li = violet
ge = yellow
or = orange

F220 - Lock unit for central locking, driver side, switches are shown with doors in closed position
J386 - Door control module, driver side
K133 - Warning light for central locking -SAFE-
T8c - 8-Pin Connector, black, on lock unit for central locking, driver side
T10l - 10-Pin Connector, black, connector station A-pillar, left
T29a - 29-Pin Connector
W31 - Entry light front, left

(205) - Ground connection, in driver's door wiring harness
(R14) - wire connection -1- (open), in driver's door wiring harness
(A146) - Connection (CAN-Bus Comfort, High), in instrument panel wiring harness
(A147) - Connection (CAN-Bus Comfort, Low), in instrument panel wiring harness

** - only vehicles with switches for remote unlock rear lid and remote fuel tank door placed in driver's door

Door control module, driver side, Central locking, driver's door, entry light front, left, Warning light for central locking -SAFE-

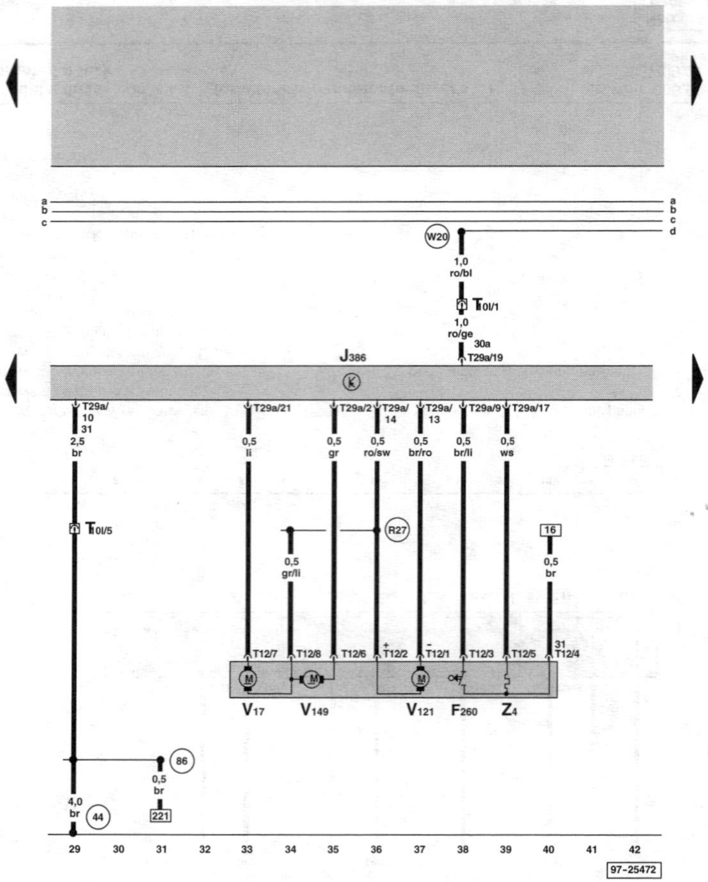

ws = white
sw = black
ro = red
br = brown
gn = green
bl = blue
gr = grey
li = violet
ge = yellow

F260 - Driver's Folding Mirror Contact Switch
J386 - Door control module, driver side
T10l - 10-Pin Connector, black, connector station A-pillar, left
T12 - 12-Pin Connector,black
T29a - 29-Pin Connector
V17 - Driver's Side Mirror Adjustment Motor
V121 - Motor for fold-in mirror, driver side
V149 - Motor for mirror adjustment, driver side
Z4 - Heated outside mirror, driver side

(44) - Ground connection, on left A-pillar, lower part
(86) - Ground connection -1-, in rear wiring harness
(R27) - wire connection (power mirror), in drivr's door wiring harness
(W20) - Plus connector (30a), in wiring harness, rear

Door control module, driver side, heated outside mirror, driver side.

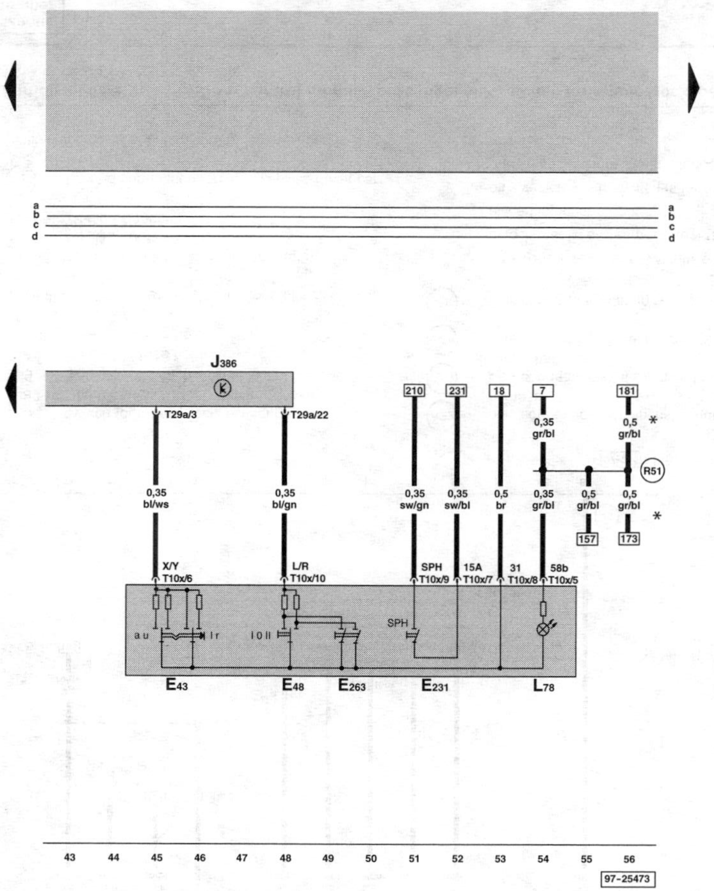

ws = white
sw = black
ro = red
br = brown
gn = green
bl = blue
gr = grey
li = violet
ge = yellow

- E43 - Mirror Adjustment Switch
- E48 - Mirror Selector Switch
- E231 - Switch for outside mirror heating
- E263 - Mirror Folding Switch
- J386 - Door control module, driver side
- L78 - Mirror Adjusting Switch Light
- T10x - 10-Pin Connector, black
- T29a - 29-Pin Connector

(R51) - Connection (58b) (driver's side door wiring harness)

* - only vehicles with switches for remote unlock rear lid and remote fuel tank door placed in driver's door

Edition 08.00
USA.5062.12.21

Door control module, driver side, Mirror adjustment and selector switch

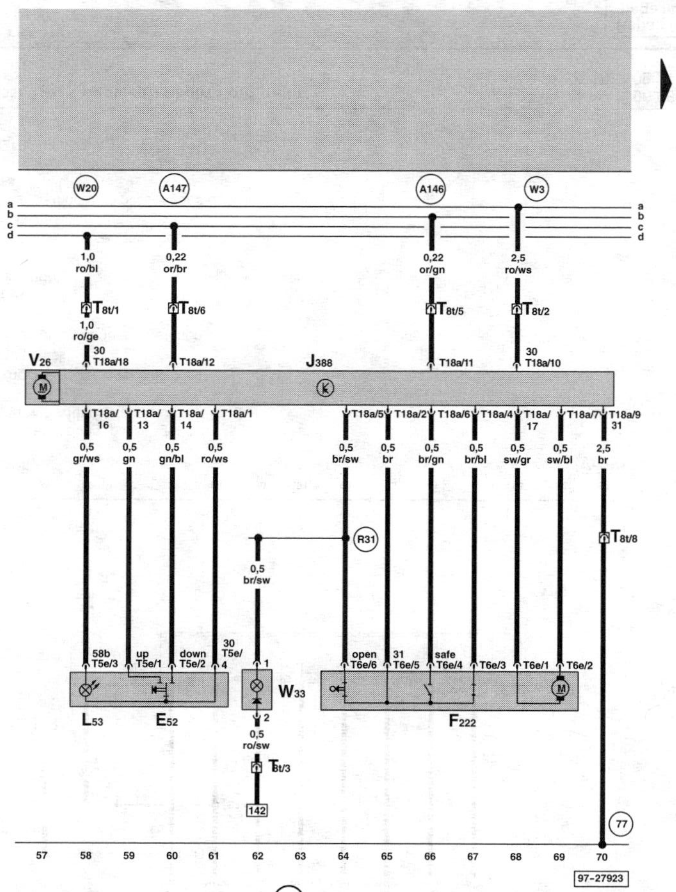

ws = white
sw = black
ro = red
br = brown
gn = green
bl = blue
gr = grey
li = violet
ge = yellow
or = orange

- E52 - Left Rear Window Switch, (In LR Door)
- F222 - Lock unit for central locking, rear, left, switches are shown with doors in closed position
- J388 - Door control module, rear, left
- L53 - Power Window Switch Light
- T5e - 5-Pin Connector, black
- T6e - 6-Pin Connector, black
- T8t - 8-Pin Connector, black, connector station B-pillar, left
- T18a - 18-Pin Connector
- V26 - Motor for window regulator, rear, left
- W33 - Entry light rear, left

(77) - Ground connection, (lower left B-pillar)

(R31) - Wire connection -1-, in left rear door wiring harness

(W3) - wire connection, in rear wiring harness

(W20) - Plus connector (30a), in wiring harness, rear

(A146) - Connection (CAN-Bus Comfort, High), in instrument panel wiring harness

(A147) - Connection (CAN-Bus Comfort, Low), in instrument panel wiring harness

Central locking rear left, power window rear left, entry light rear left, Lock unit for central locking, rear, left

Edition 08.00
USA.5062.12.21

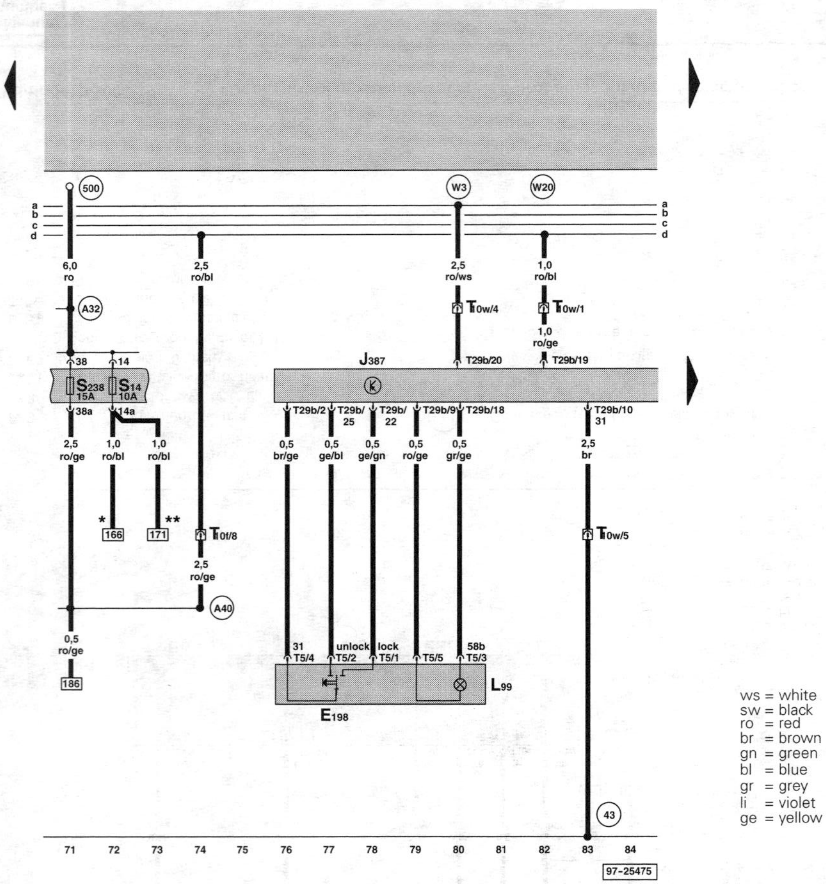

ws = white
sw = black
ro = red
br = brown
gn = green
bl = blue
gr = grey
li = violet
ge = yellow

E198 - Switch for interior lock, passenger side
J387 - Door control module, passenger side
L99 - Lighting for switch interior lock
S14 - Fuse 14 in fuse holder
S238 - Fuse 38 in fuse holder
T5 - 5-Pin Connector
T10f - 10-Pin Connector, blue, connector station A-pillar, left
T10w - 10-Pin Connector, black, connector station A-pillar, right
T29b - 29-Pin Connector

(43) - Ground connection, on right A-pillar, lower part

(500) - Screw connection -1- (30), on relay panel

(A32) - plus connection (30), in instrument panel wiring harness

(A40) - plus connection -1- (30), in instrument panel wiring harness

(W3) - wire connection, in rear wiring harness

(W20) - Plus connector (30a), in wiring harness, rear

* - only vehicles without switches for remote unlock rear lid and remote fuel tank door placed in driver's door

** - only vehicles with switches for remote unlock rear lid and remote fuel tank door placed in driver's door

Door control module, passenger side, switch for interior lock, passenger side

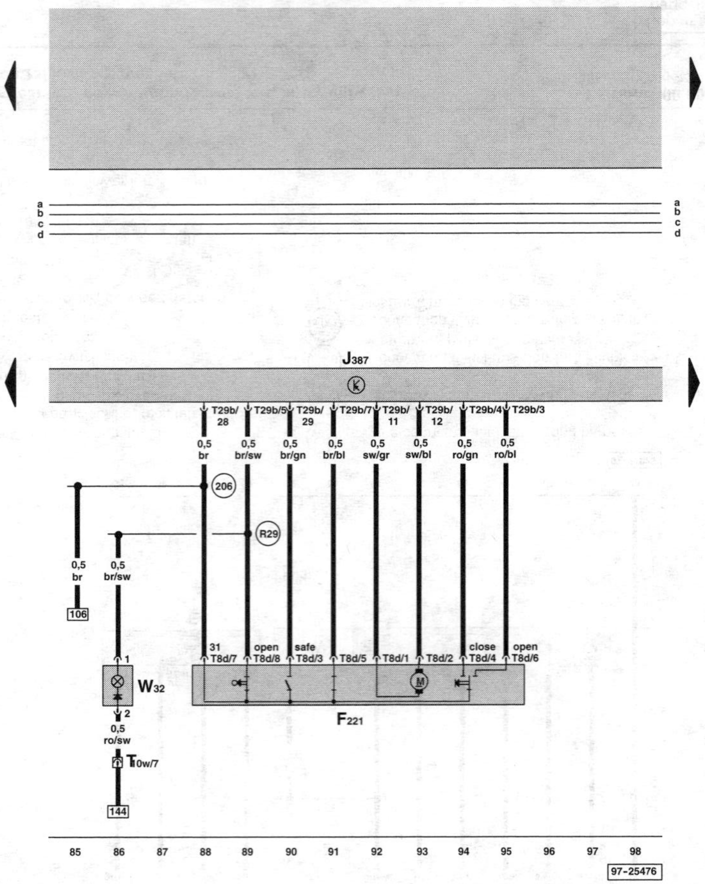

ws = white
sw = black
ro = red
br = brown
gn = green
bl = blue
gr = grey
li = violet
ge = yellow

F221 - Lock unit for central locking, passenger side, switches are shown with doors in closed position
J387 - Door control module, passenger side
T8d - 8-Pin Connector, black
T10w - 10-Pin Connector, black, connector station A-pillar, right
T29b - 29-Pin Connector
W32 - Entry light front, right

(206) - Ground connection, in passenger's door wiring harness

(R29) - Wire connection -1- (open), in passenger's door wiring harness

Central locking passenger's door, entry light front, right

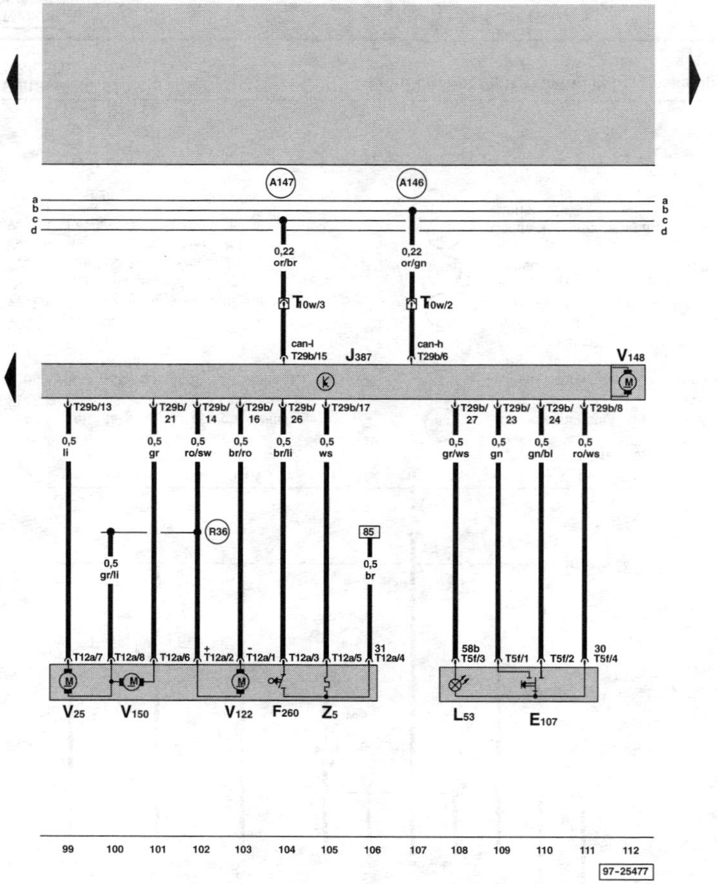

ws = white
sw = black
ro = red
br = brown
gn = green
bl = blue
gr = grey
li = violet
ge = yellow
or = orange

E107 - Switch for window regulator, in passenger door
F260 - Driver's Folding Mirror Contact Switch
J387 - Door control module, passenger side
L53 - Power Window Switch Light
T5f - 5-Pin Connector
T10w - 10-Pin Connector, black, connector station A-pillar, right
T12a - 12-Pin Connector, black
T29b - 29-Pin Connector
V25 - Passenger's Side Mirror Adjustment Motor
V122 - Motor for fold-in mirror, passenger side
V148 - Motor for window regulator, passenger side
V150 - Motor for mirror adjustment, passenger side
Z5 - Heated outside mirror, passenger side

(A146) - Connection (CAN-Bus Comfort, High), in instrument panel wiring harness
(A147) - Connection (CAN-Bus Comfort, Low), in instrument panel wiring harness
(R36) - wire connection (power mirror), in passenger's door wiring harness

Mirror adjustment and fold-in mirror, passenger side, heated outside mirror, passenger side, switch for window regulator

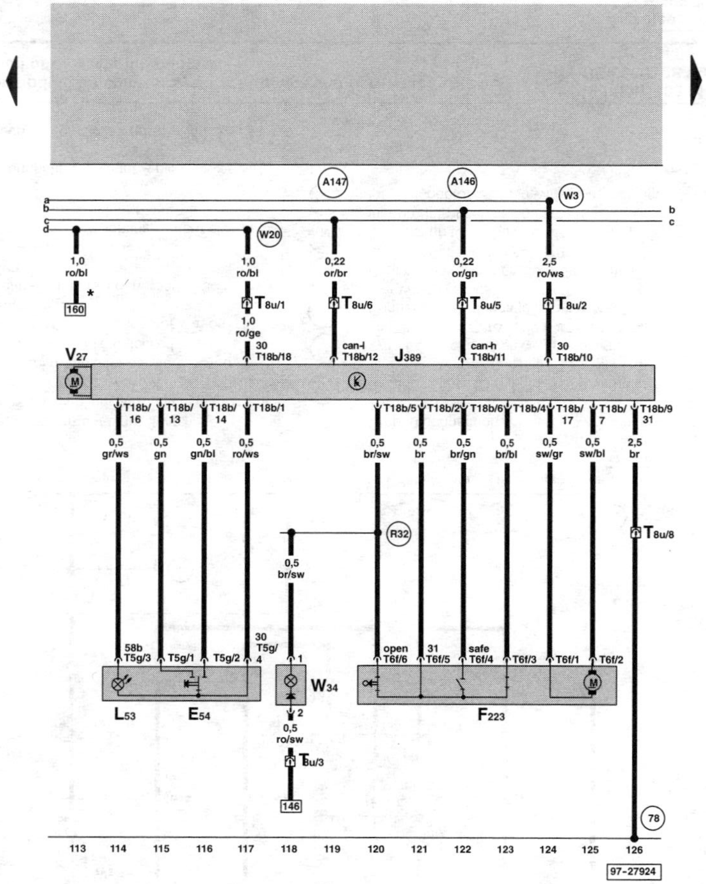

ws = white
sw = black
ro = red
br = brown
gn = green
bl = blue
gr = grey
li = violet
ge = yellow
or = orange

E54 - Right Rear Window Switch, (In RR Door), switches are shown with doors in closed position
F223 - Lock unit for central locking, rear, right
J389 - Door control module, rear, right
L53 - Power Window Switch Light
T5g - 5-Pin Connector
T6f - 6-Pin Connector
T8u - 8-Pin Connector, brown, connector station B-pillar, right
T18b - 18-Pin Connector
V27 - Motor for window regulator, rear, right
W34 - Entry light rear, right

(78) - Ground connection, (lower right B-pillar)
(A146) - Connection (CAN-Bus Comfort, High), in instrument panel wiring harness
(A147) - Connection (CAN-Bus Comfort, Low), in instrument panel wiring harness
(R32) - Wire connection -2-, in right rear door wiring harness
(W3) - wire connection, in rear wiring harness
(W20) - Plus connector (30a), in wiring harness, rear
* - only vehicles with switches for remote unlock rear lid and remote fuel tank door placed in driver's door

Central locking rear right, power window rear right, entry light rear right

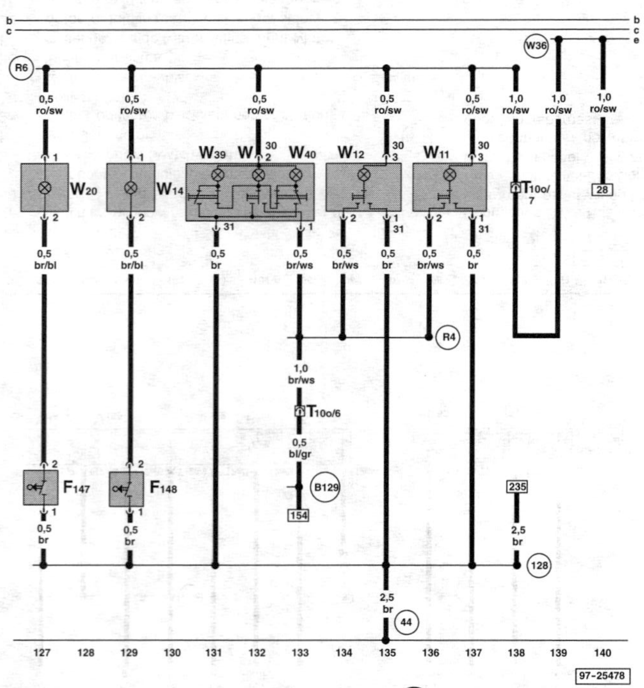

ws = white
sw = black
ro = red
br = brown
gn = green
bl = blue
gr = grey
li = violet
ge = yellow

F147 - Left Make-Up Mirror Light Switch
F148 - Right Make-Up Mirror Light Switch
T10o - 10-Pin Connector, green, connector station A-pillar, left
W - Front Interior Light
W11 - Left Rear Reading Light
W12 - Right Rear Reading Light
W14 - Right Make-up Mirror Light
W20 - Left Make-up Mirror Light
W39 - Reading light, center left
W40 - Reading light, center right

(44) - Ground connection, on left A-pillar, lower part

(128) - Ground connection -1-, in interior light wiring harness

(R4) - wire connection -1-, in interior light/door contact switch wiring harness

(R6) - plus connection -1-, in interior light wiring harness

(W36) - Plus connector (30a), in wiring harness floor

(B129) - Connector (interior light 3/L) in wiring harness interior

Edition 02.01
USA.5062.13.21

Front interior light, reading lights, make-up mirror lights

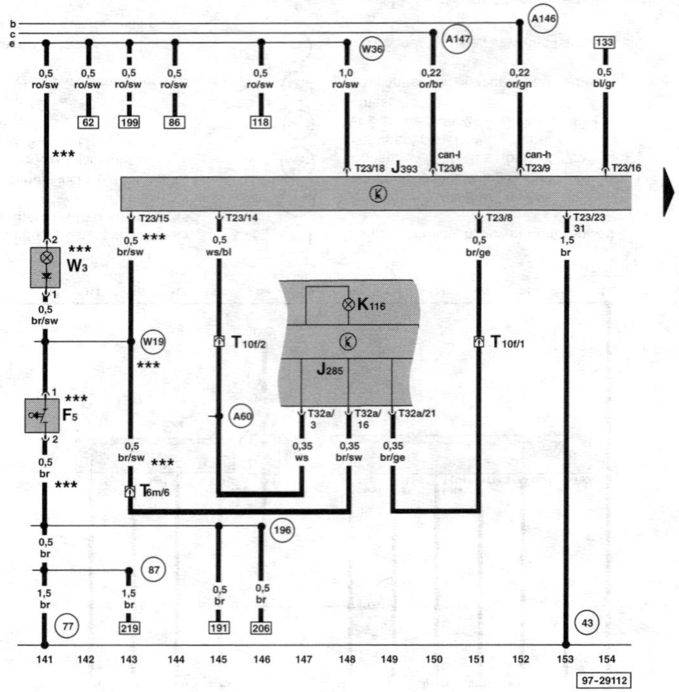

ws = white
sw = black
ro = red
br = brown
gn = green
bl = blue
gr = grey
li = violet
ge = yellow
or = orange

F5 - Luggage Compartment Light Switch
J285 - Control module with indicator unit in instrument panel insert
J393 - Central control module for comfort system
K116 - Warning light for rear lid unlocked
T6m - 6-Pin Connector, black, connector station A-pillar, right
T10f - 10-Pin Connector, blue, connector station A-pillar, left
T23 - 23-Pin Connector
T32a - 32-Pin Connector, blue, on instrument cluster
W3 - Luggage compartment Light

(43) - Ground connection, on right A-pillar, lower part

(77) - Ground connection, (lower left B-pillar)

(87) - Ground connection -2-, in rear wiring harness

(196) - Ground connection -3-, in wiring harness, rear

(A60) - Wire Connection (Vehicle speed signal), in instrument cluster wiring harness

(W19) - Connection -2- (luggage compartment light), in rear wiring harness

(A146) - Connection (CAN-Bus Comfort, High), in instrument panel wiring harness

(A147) - Connection (CAN-Bus Comfort, Low), in instrument panel wiring harness

(W36) - Plus connector (30a), in wiring harness floor

*** - sedan only

- - - wagon only

Central control module for comfort system, Luggage compartment light switch, warning light for rear lid unlocked

Edition 02.01
USA.5062.13.21

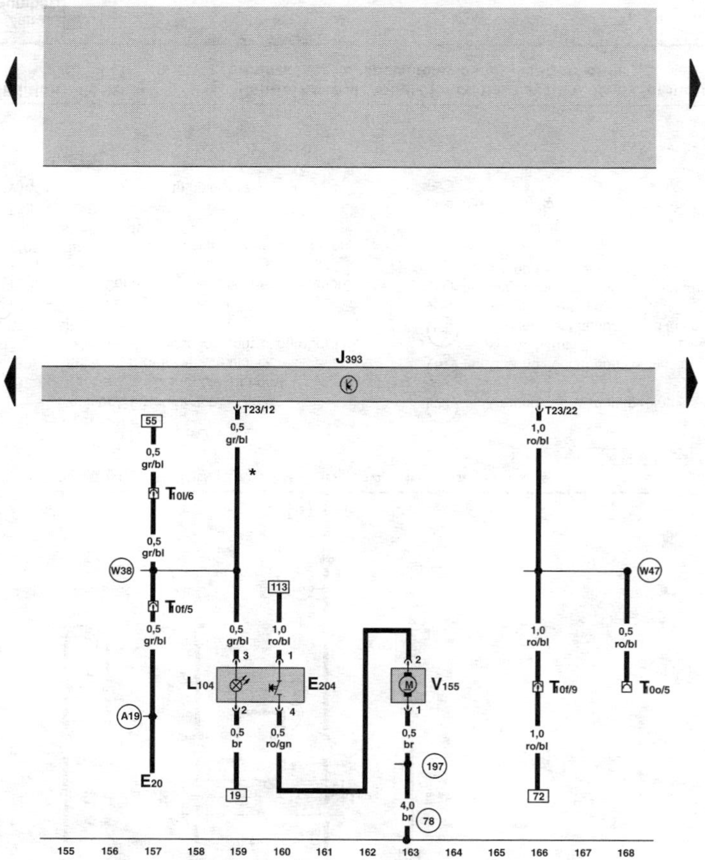

97-27926

ws = white
sw = black
ro = red
br = brown
gn = green
bl = blue
gr = grey
li = violet
ge = yellow

E20 - Instrument Panel Light Dimmer Switch
E204 - Switch for remote/fuel tank door
J393 - Central control module for comfort system
L104 - Illumination for switch remote unlock
T10a - 10-Pin Connector, grey, connector station A-pillar, left
T10f - 10-Pin Connector, blue, connector station A-pillar, left
T10l - 10-pin connector, black, connector station A-pillar left
T10o - 10-Pin Connector, green, connector station A-pillar, left
T23 - 23-Pin Connector
V155 - Motor for fuel tank lid unlock

(78) - Ground connection, (lower right B-pillar)

(197) - Ground connection -4-, in wiring harness rear

(A19) - wire connection (58d), in instrument panel wiring harness

(W38) - Connector (58d), in wiring harness floor

(W47) - Plus-Connection -2- (30a), in wiring harness floor

* - wagon only

Central control module for comfort system, Remote fuel tank door - wagon only

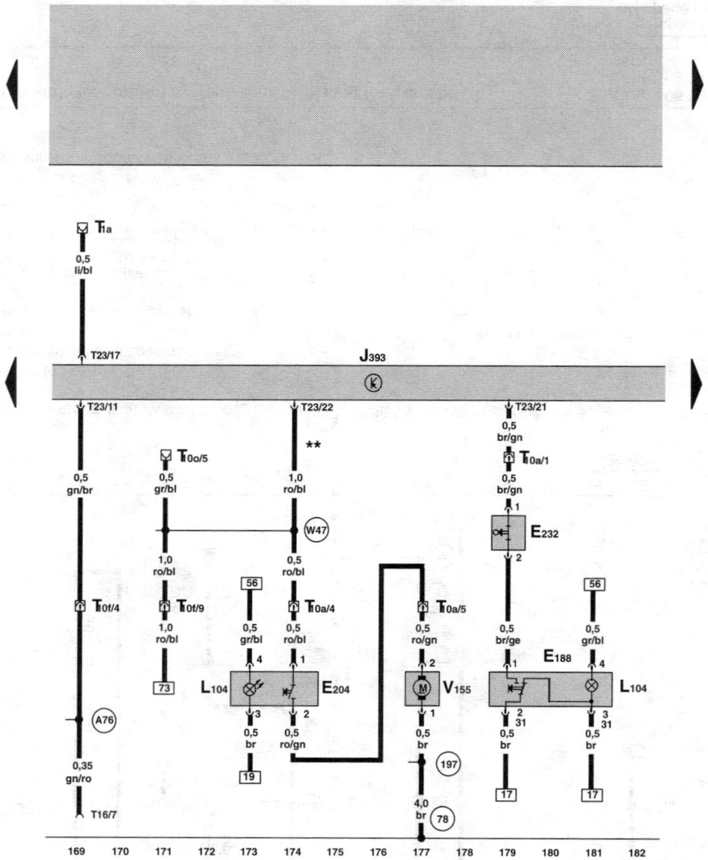

97-27927

ws = white
sw = black
ro = red
br = brown
gn = green
bl = blue
gr = grey
li = violet
ge = yellow

E188 - Switch for remote unlock, rear lid
E204 - Switch for remote/fuel tank door
E232 - Rear lid Remote Lock Key Switch
J393 - Central control module for comfort system
L104 - Illumination for switch remote unlock
T1a - Single Connector, brown
T10a - 10-Pin Connector, grey, connector station A-pillar, left
T10f - 10-Pin Connector, blue, connector station A-pillar, left
T10o - 10-Pin Connector, green, connector station A-pillar, left
T16 - 16-Pin Connector, Data Link Connector (DLC), below instrument panel, left
T23 - 23-Pin Connector
V155 - Motor for fuel tank lid unlock

(78) - Ground connection, (lower right B-pillar)

(197) - Ground connection -4-, in wiring harness rear

(A76) - Connector (K-diagnosis wire), in instrument panel wiring harness

(W47) - Plus-Connection -2- (30a), in wiring harness floor

** - sedan only

Central control module for comfort system, Remote unlock rear lid and remote fuel tank door - sedan only, Data Link Connector (DLC)

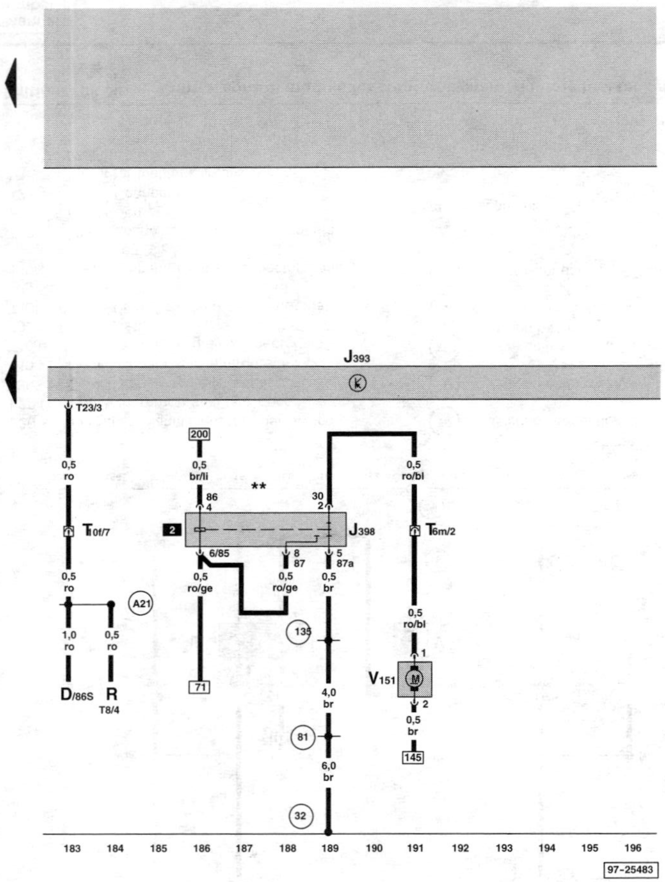

ws = white
sw = black
ro = red
br = brown
gn = green
bl = blue
gr = grey
li = violet
ge = yellow

D - Ignition/Starter Switch
J393 - Central control module for comfort system
J398 - Relay for motor remote unlock rear lid, above relay panel, production control number (79)
R - Radio
T8 - 8-Pin Connector, on radio
T6m - 6-Pin Connector, black, connector station A-pillar, right
T10f - 10-Pin Connector, blue, connector station A-pillar, left
T23 - 23-Pin Connector
V151 - Motor for remote unlock rear lid

(32) - Ground connection, behind instrument panel, left
(81) - Ground connection -1-, in instrument panel wiring harness
(135) - Ground connection -2-, in instrument panel wiring harness
(A21) - wire connection (86s), in instrument panel wiring harness

** - wagon only

Edition 06.99
USA.5062.09.21

Central control module for comfort system, Relay for motor remote unlock rear lid, above relay panel - wagon only

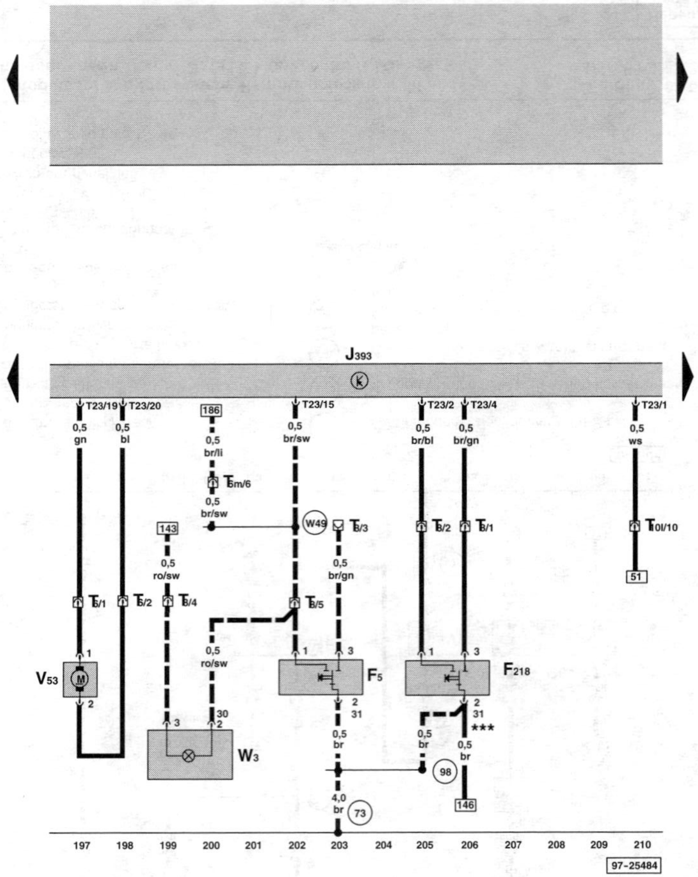

ws = white
sw = black
ro = red
br = brown
gn = green
bl = blue
gr = grey
li = violet
ge = yellow

F5 - Luggage Compartment light switch (wagon only)
F218 - Switch for central locking, trunk lid
J393 - Central control module for comfort system
T6 - 6-Pin Connector, black
T6m - 6-Pin Connector, black, connector station A-pillar, right
T8 - 8-Pin Connector, black, connector station D-pillar, right (wagon only)
T10l - 10-pin connector, black, connector station A-pillar left
T23 - 23-Pin Connector
V53 - Decklid Central Locking System Motor
W3 - Luggage compartment light (wagon only)

(73) - Ground connection, on roof bow, rear (wagon only)
(98) - Ground connection, in rear lid wiring harness
(W49) - Connection -1-, in wiring harness floor

*** - sedan only
- - - wagon only

Central control module for comfort system, central locking for trunk lid

Edition 06.99
USA.5062.09.21

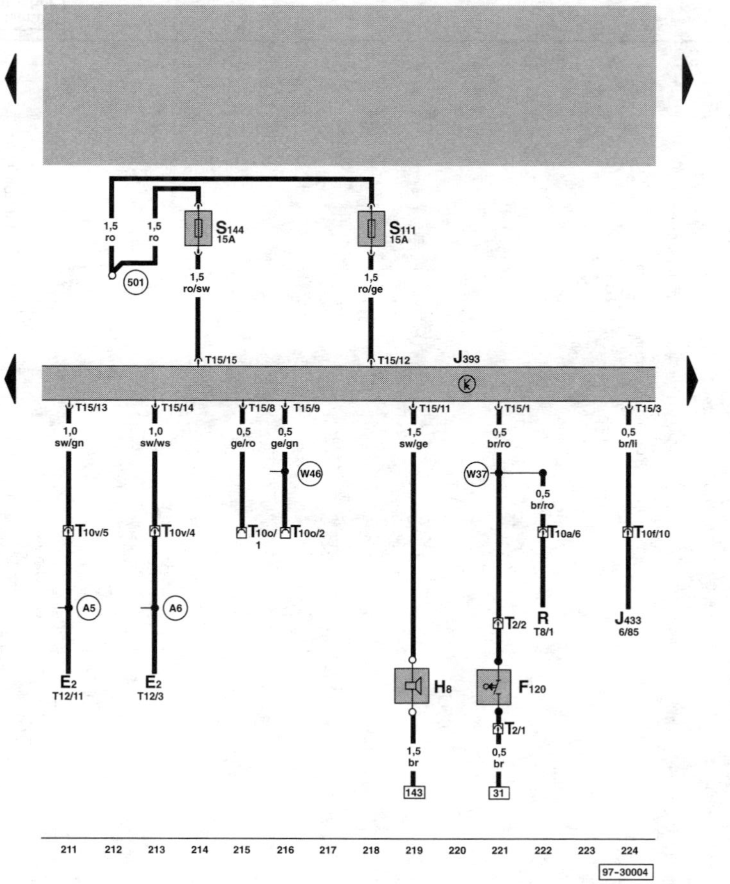

ws = white
sw = black
ro = red
br = brown
gn = green
bl = blue
gr = grey
li = violet
ge = yellow

E2 - Turn signal switch
F120 - Hood Alarm Switch
H8 - Alarm Horn
J393 - Central control module for comfort system
J433 - Starting Interlock Relay (alarm system), above relay panel
R - Radio
S111 - Fuse -1- for Alarm System, on the eight position auxiliary relay panel, behind relay panel
S144 - Fuse -2- for Alarm System, on the eight position auxiliary relay panel, behind relay panel
T2 - Double Connector
T8 - 8-Pin Connector, on radio
T10a - 10-Pin Connector, grey, connector station A-pillar, left
T10o - 10-Pin Connector, red, connector station A-pillar, left
T10v - 10-Pin Connector, violet, connector station A-pillar, right
T12 - 12-Pin Connector, near steering column
T15 - 15-Pin Connector, on Central control module for comfort system
(501) - Screw connection -2- (30), on relay panel
(A5) - plus connection (right turn signal), in instrument panel wiring harness
(A6) - plus connection (left turn signal), in instrument panel wiring harness
(W37) - Connector (alarm system), in wiring harness floor
(W46) - Connection -1- (Interior Monitoring), in wiring harness floor

Central control module for comfort system, alarm system

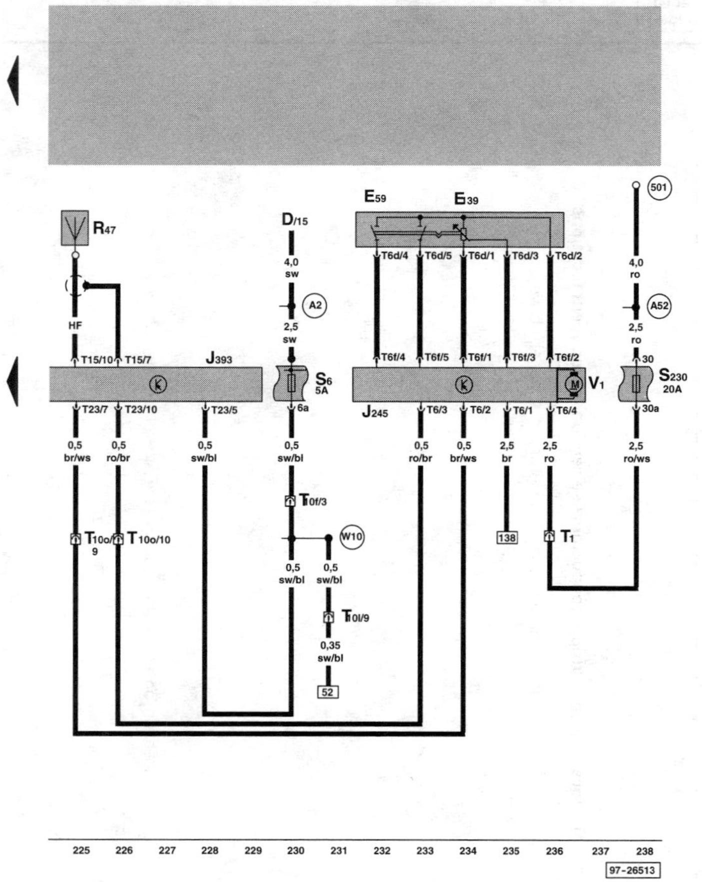

ws = white
sw = black
ro = red
br = brown
gn = green
bl = blue
gr = grey
li = violet
ge = yellow

D - Ignition/Starter Switch
E59 - Raise/Lower Sunroof Switch
E139 - Sunroof Regulator
J245 - Power Sunroof Control Module
J393 - Central control module for comfort system
R47 - Antenna wire for central locking and alarm system, on B-pillar right
S6 - Fuse 6 in fuse holder
S230 - Fuse 30 in fuse holder
T1 - Single Connector, black
T6 - 6-Pin Connector
T6d - 6-Pin Connector
T6f - 6-Pin Connector
T10f - 10-Pin Connector, blue, connector station A-pillar, left
T10l - 10-Pin Connector, black, connector station A-pillar, left
T10o - 10-Pin Connector, red, connector station A-pillar, left
T15 - 15-Pin Connector, on Central control module for comfort system
T23 - 23-Pin Connector
V1 - Sunroof Motor
(501) - Screw connection -2- (30), on relay panel
(A2) - plus connection (15), in instrument panel wiring harness
(A52) - plus connection (30), in instrument panel wiring harness
(W10) - plus connection (15), in rear wiring harness from October 1998

Central control module for comfort system, power sunroof

Power outlets (12V)

from September 1997

Deviate relay location and fuseplacements as well as the locations of multiple connectors see section "component locations"

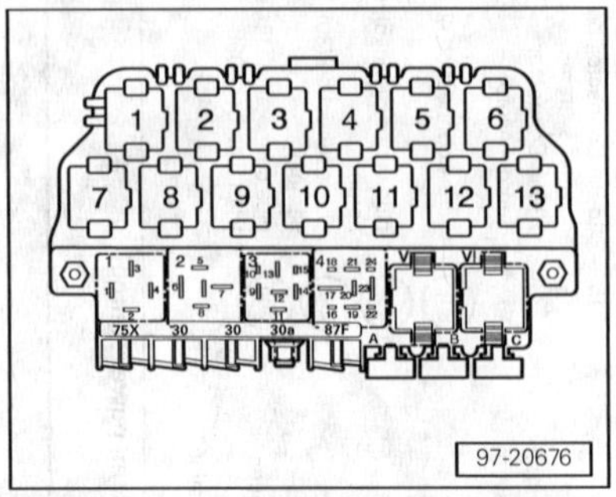

Relay location on the thirteen position auxiliary relay panel, above relay panel:

Relay panel:

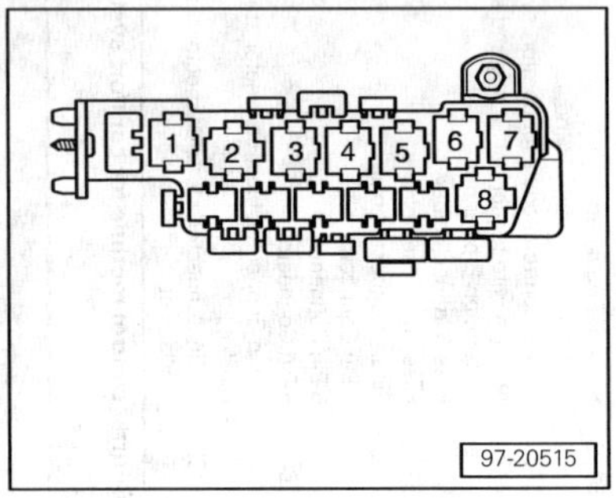

Relay location on the eight position auxiliary relay panel, above relay panel:

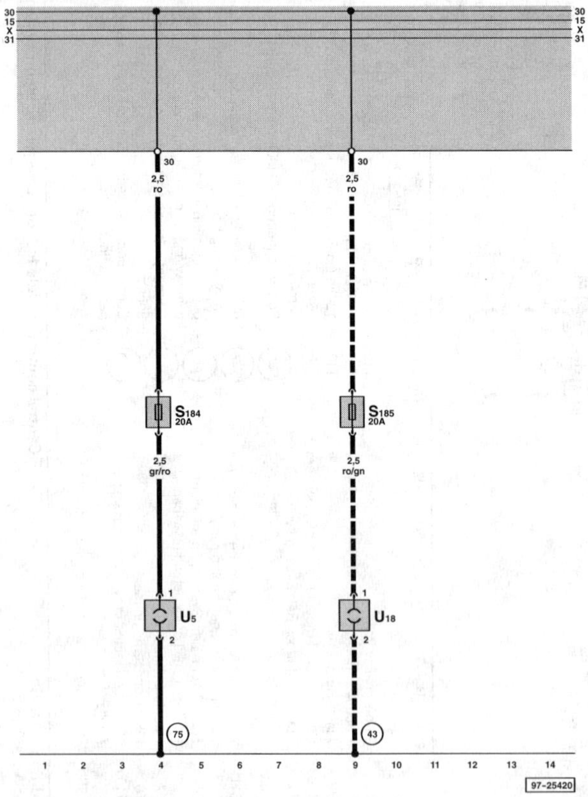

Power outlets (12V) - in luggage compartment

S184 - Fuse for outlet I
S185 - Fuse for outlet II
U5 - 12V Socket
U18 - 12V outlet -2-

(43) - Ground connection, on right B pillar, lower part

(75) - Ground connection, on right rear pillar

- - - - Sedan only

ws = white
sw = black
ro = red
br = brown
gn = green
bl = blue
gr = grey
li = lilac
ge = yellow
or = orange

Radio System

from August 1999

Deviate relay location and fuseplacements as well as the locations of multiple connectors see section "component locations".

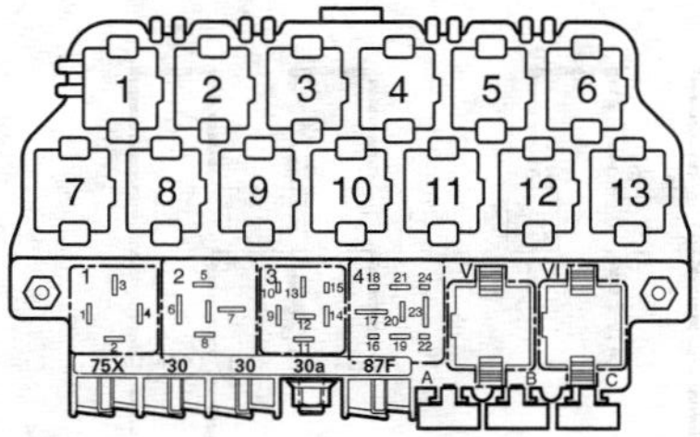

97-20676

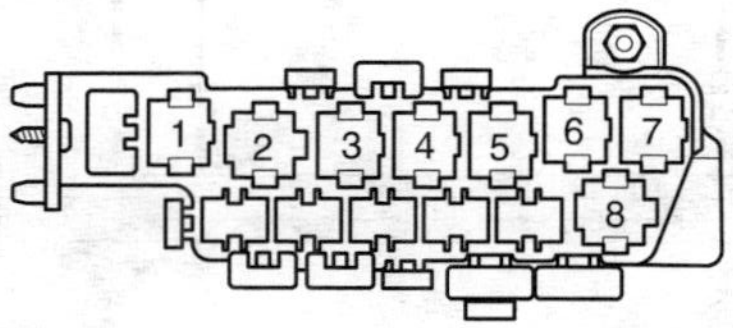

97-20515

Fuse colors
30 A - green
25 A - white
20 A - yellow
15 A - blue
10 A - red
7,5 A - brown
5 A - beige

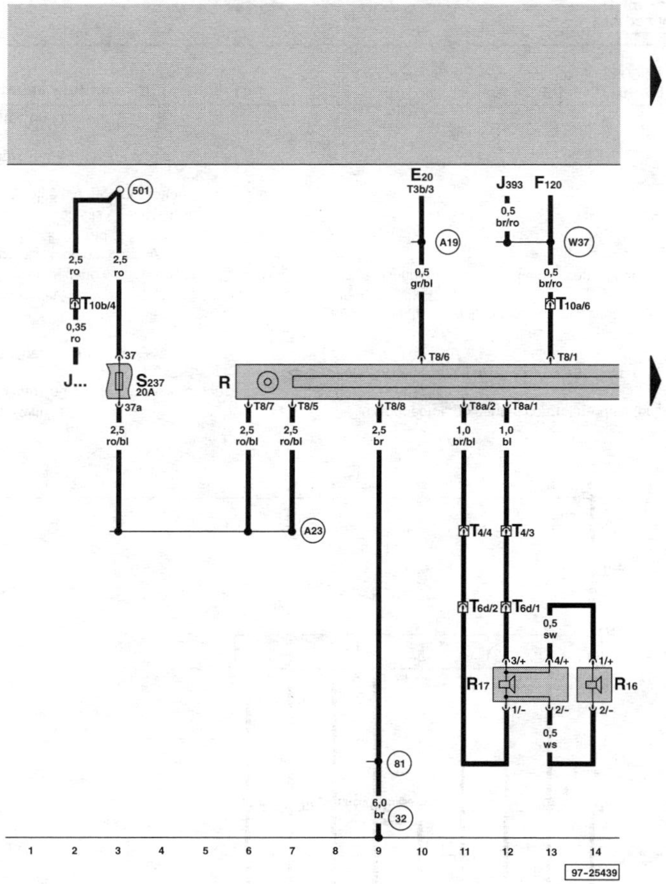

ws = white
sw = black
ro = red
br = brown
gn = green
bl = blue
gr = grey
li = violet
ge = yellow

E20 - Instrument Panel Light Dimmer Switch
F120 - Hood Alarm Switch
J... - Engine Control Module (ECM)
J393 - Central control module for comfort system
R - Radio
R16 - Treble speaker, right rear, in right rear door
R17 - Bass speaker, right rear, in right rear door
S237 - Fuse 37 in fuse holder
T3b - 3-Pin Connector
T4 - 10-Pin Connector, black, connector station A-pillar, right
T6d - 6-Pin Connector, red, on B-pillar, right
T8 - 8-Pin Connector, black, on radio
T8a - 8-Pin Connector, brown, on radio
T10a - 10-Pin Connector, brown, connector station A-pillar, left
T10b - 10-Pin Connector, black, on protective housing for control module, in engine compartment, left

(32) - Ground connection, behind instrument panel, left
(81) - Ground connection -1-, in instrument panel wiring harness
(501) - Threaded connection -2- (30) on the relay plate
(A19) - wire connection (58d), in instrument panel wiring harness
(A23) - wire connection (30al), in instrument panel wiring harness
(W37) - Connection (DWA - alarm system), in floor wiring harness

Radio, right rear speakers

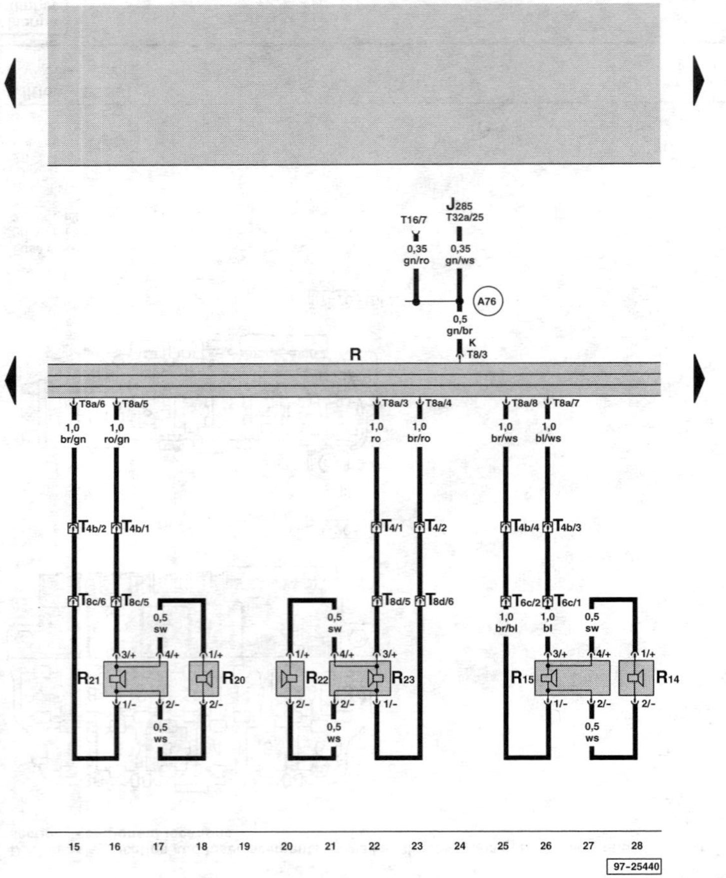

ws = white
sw = black
ro = red
br = brown
gn = green
bl = blue
gr = grey
li = violet
ge = yellow

J285 - Control module with indicator unit in instrument panel insert
R - Radio
R14 - Treble speaker, left rear, in left rear door
R15 - Bass speaker, left rear, in left rear door
R20 - Treble speaker, left front, in driver's door
R21 - Bass speaker, left front, in driver's door
R22 - Treble speaker, right front, in passenger's door
R23 - Bass speaker, right front, in passenger's door
T4 - 10-Pin Connector, black, connector station A-pillar, right
T4b - 10-Pin Connector, black, connector station A-pillar, left
T6c - 6-Pin Connector, red, on B-pillar, left
T8 - 8-Pin Connector, black, on radio
T8a - 8-Pin Connector, brown, on radio
T8c - 8-Pin Connector, black, connector station A-pillar, left
T8d - 8-Pin Connector, black, connector station A-pillar, right
T16 - 16-Pin Connector, Data Link Connector (DLC), below instrument panel, left
T32a - 32-Pin Connector, blue, on instrument cluster

(A76) - Connector (K-diagnosis wire), in instrument panel wiring harness

Radio, front speakers, left rear speakers, Data Link Connector (DLC)

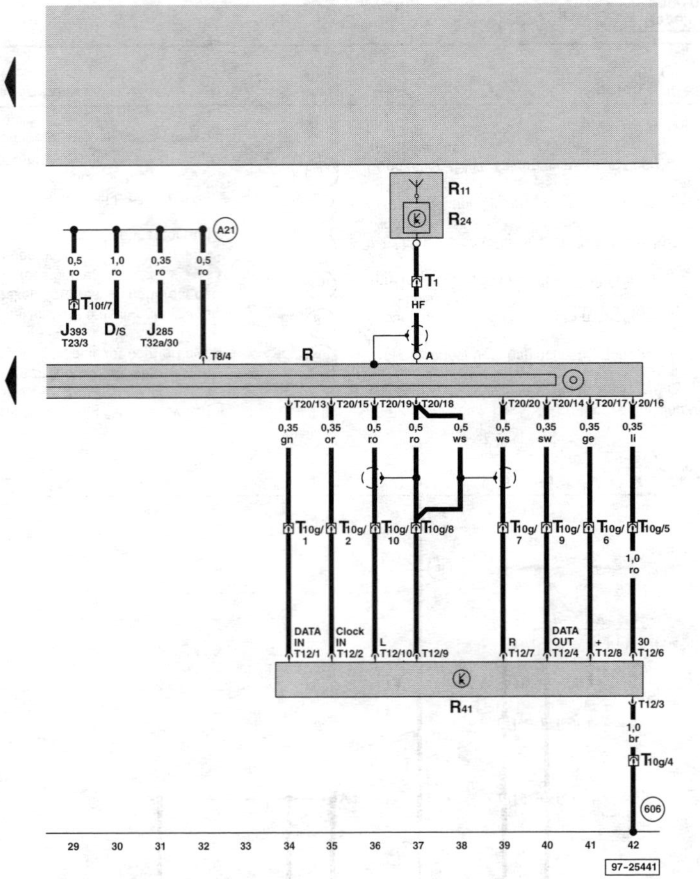

ws = white
sw = black
ro = red
br = brown
gn = green
bl = blue
gr = grey
li = violet
ge = yellow
or = orange

D - Ignition/Starter Switch
J285 - Control module with indicator unit in instrument panel insert
J393 - Central control module for comfort system
R - Radio
R11 - Antenna
R24 - Antenna Amplifier
R41 - CD Changer Unit, in luggage compartment
T1 - 1-Pin Connector, black, below console
T8 - 8-Pin Connector, black, on radio
T12 - 12-Pin Connector, red, on CD Changer Unit
T10f - 10-Pin Connector, blue, connector station A-pillar, left
T10g - 10-Pin Connector, red, connector station A-pillar, left
T20 - 20-Pin Connector, on radio
T23 - 23-Pin Connector, on central control module for comfort system
T32a - 32-Pin Connector, blue, on instrument cluster

(A21) - wire connection (86s), in instrument panel wiring harness
(606) - Ground Connection, (under center console at shifter)

Radio, Antenna, Antenna Amplifier, CD Changer Unit

Automatic transmission, 5 speed automatic with tiptronic,

- 1,8L- Engine - Motronic Multiport Fuel Injection (MFI)/110 kW, code ATW
- 2,8L - Engine - Motronic Multiport Fuel Injection (MFI)/142 kW, code ATQ

from August 1999

- 1,8L- Engine - Motronic Multiport Fuel Injection (MFI)/110 kW, code AUG

from May 2000

Deviate relay location and fuseplacements as well as the locations of multiple connectors see section "component locations".

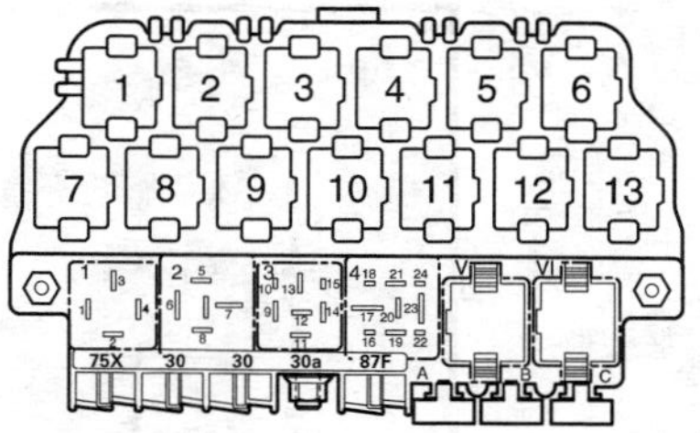

97-20676

Relay location on the thirteenfold auxiliary relay panel, above relay panel:

6 Selector Lever Light Relay (205)

12 Starting Interlock Relay (alarm system) (186)

13 Starting Interlock Relay (204)

Relay panel:

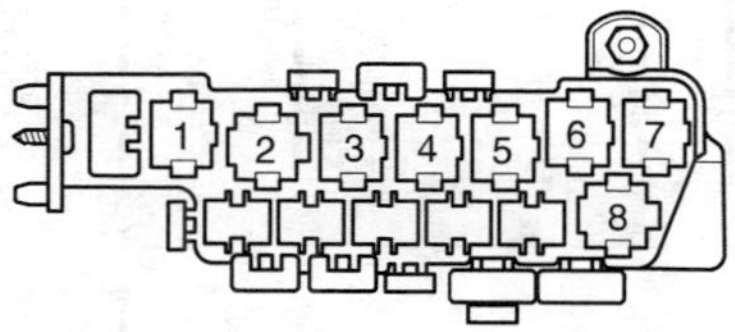

97-20515

Relay location on the eightfold auxiliary relay panel, behind relay panel:

Note: Number in parentheses indicates production control number stamped on relay housing.

Fuse colors
30 A - green
25 A - white
20 A - yellow
15 A - blue
10 A - red
7,5 A - brown
5 A - beige

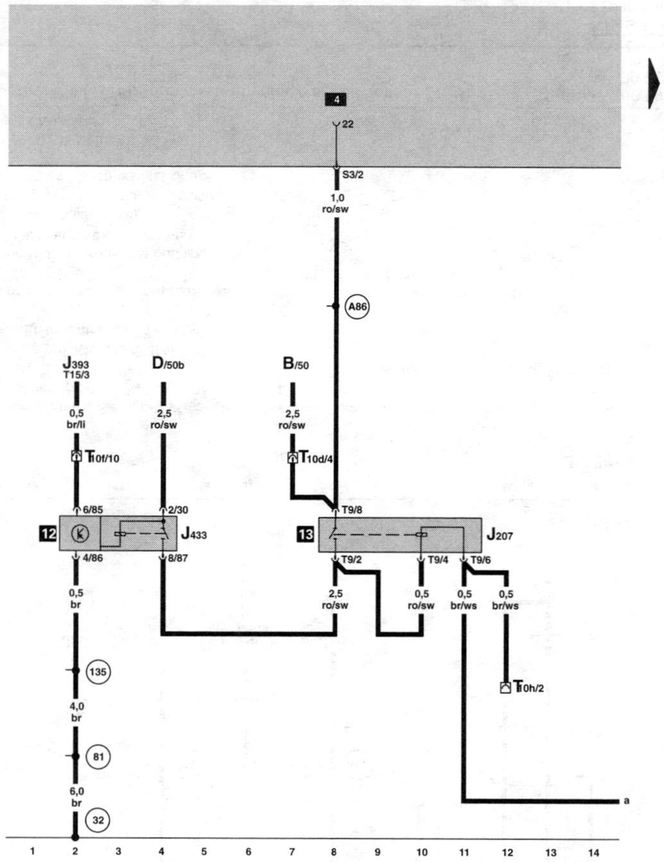

ws = white
sw = black
ro = red
br = brown
gn = green
bl = blue
gr = grey
li = violet
ge = yellow

B - Starter
D - Ignition/Starter Switch
J207 - Starting Interlock Relay, on the thirteenfold auxiliary relay panel, above relay panel, production control number (204)
J393 - Central control module for comfort system
J433 - Starting Interlock Relay (alarm system), on the thirteenfold auxiliary relay panel, above relay panel, production control number (186)
T9 - 9-Pin Connector, on Starting Interlock Relay
T10d - 10-Pin Connector, brown, on protective housing for control module, in engine compartment, left
T10f - 10-Pin Connector, blue, connector station A-pillar, left
T10h - 10-Pin Connector, blue, on protective housing for control module, in engine compartment, left
T15 - 15-Pin Connector, on Central control module for comfort system
(32) - Ground connection, behind instrument panel, left
(81) - Ground connection -1-, in instrument panel wiring harness
(135) - Ground connection -2-, in instrument panel wiring harness
(A86) - Connection (50a), in instrument panel wiring harness

Starting interlock relay, Starting interlock relay (alarm system)

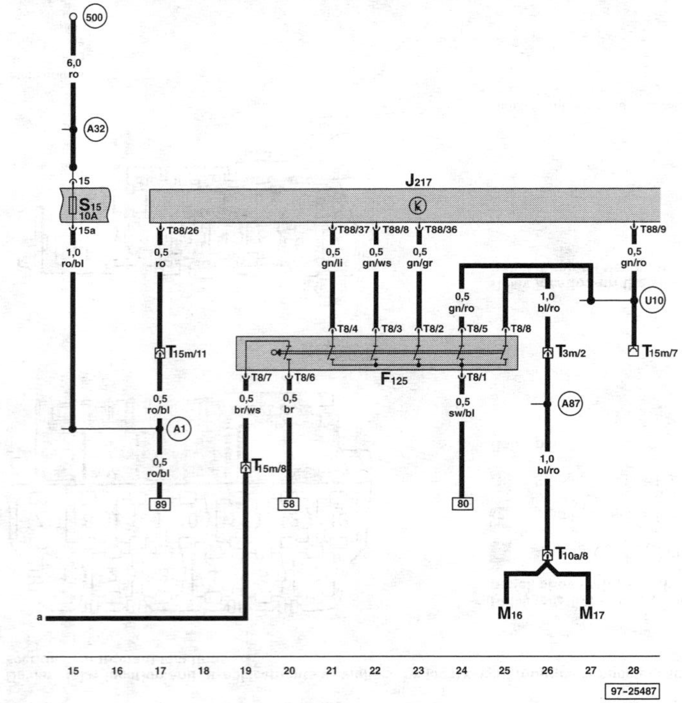

ws = white
sw = black
ro = red
br = brown
gn = green
bl = blue
gr = grey
li = violet
ge = yellow

F125 - Multi-Function Transmission Range (TR) Switch
J217 - Transmission Control Module (TCM), in front footwell, right
M16 - Left Back-Up Light
M17 - Right Back-Up Light
S15 - Fuse 15 in fuse holder
T3m - 3-Pin Connector, brown, connector station A-pillar, right
T8 - 8-Pin Connector, black
T10a - 10-Pin Connector, brown, connector station A-pillar, left
T15m- 15-Pin Connector, brown, connector station A-pillar, right
T88 - Connector, 88 point, on Transmission Control Module (TCM)

(500) - Screw Connection -1- (30), on relay panel
(A1) - plus connection (30a), in instrument panel wiring harness
(A32) - plus connection (30), in instrument panel wiring harness
(A87) - Connector (reverse lamp), in instrument panel wiring harness
(U10) - Cruise Control Connection (in automatic transmission wiring harness)

Transmission control module (TCM), multi-function transmission range (TR) switch

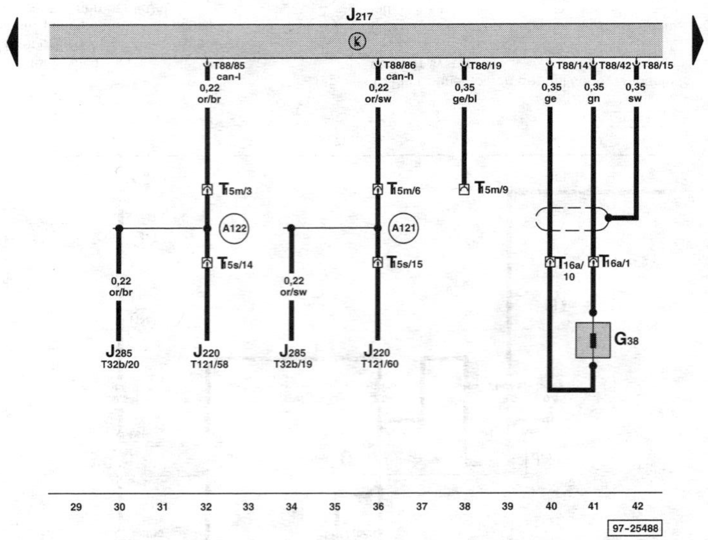

ws = white
sw = black
ro = red
br = brown
gn = green
bl = blue
gr = grey
li = violet
ge = yellow
or = orange

G38 - Transmission Vehicle Speed Sensor (VSS)
J217 - Transmission Control Module (TCM), in front footwell, right
J220 - Motronic Engine Control Module (ECM)
J285 - Control module with indicator unit in instrument panel insert
T15m- 15-Pin Connector, brown, connector station A-pillar, right
T15s - 15-Pin Connector, red, on protective housing for control module, in engine compartment,left
T16a - 16-Pin Connector, on transmission
T32b - 32-Pin Connector, green, on instrument cluster
T88 - Connector, 88 point, on Transmission Control Module (TCM)
T121 - Connector, 121 point, on Motronic Engine Control Module (ECM)

(A121) - Connector (High-Bus), in instrument panel wiring harness
(A122) - Connector (Low-Bus), in instrument panel wiring harness

Transmission control module (TCM), transmission vehicle speed sensor (VSS)

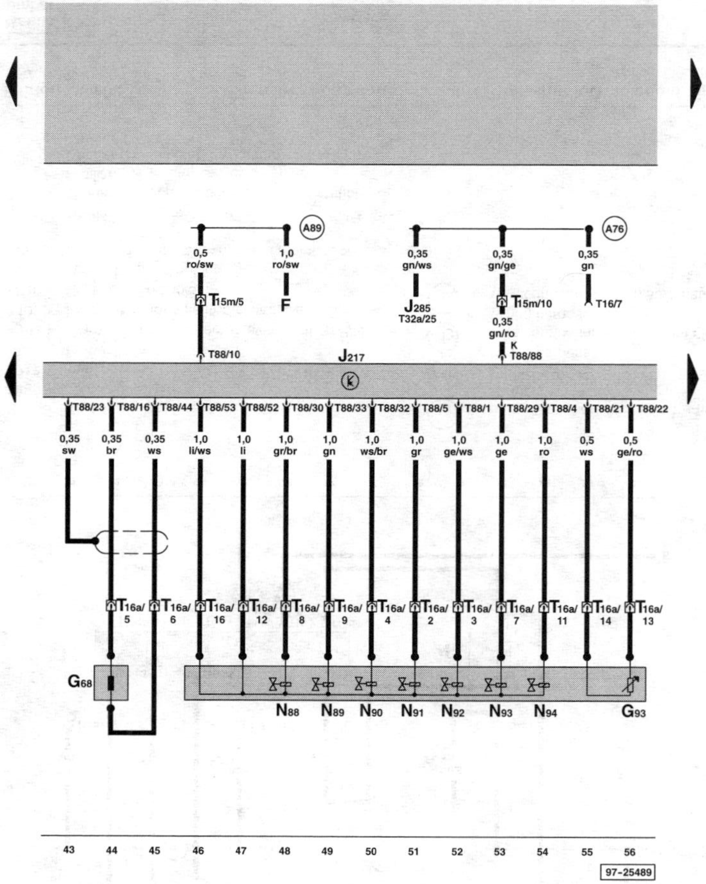

ws = white
sw = black
ro = red
br = brown
gn = green
bl = blue
gr = grey
li = violet
ge = yellow

- F - Brake Light Switch
- G68 - Vehicle Speed Sensor (VSS)
- G93 - Transmission Fluid Temperature Sensor
- J217 - Transmission Control Module (TCM), in front footwell, right
- J285 - Control module with indicator unit in instrument panel insert
- N88 - Solenoid Valve 1
- N89 - Solenoid Valve 2
- N90 - Solenoid Valve 3
- N91 - Solenoid Valve 4
- N92 - Solenoid Valve 5
- N93 - Solenoid Valve 6
- N94 - Solenoid Valve 7
- T15m - 15-Pin Connector, brown, connector station A-pillar, right
- T16 - 16-Pin Connector, Data Link Connector (DLC), below instrument panel, left
- T16a - 16-Pin Connector, on transmission
- T32a - 32-Pin Connector, blue, on instrument cluster
- T88 - Connector, 88 point, on Transmission Control Module (TCM)
- (A76) - Connector (K-diagnosis wire), in instrument panel wiring harness
- (A89) - Connector (54), in instrument panel wiring harness

Transmission control module (TCM), solenoid valves, transmission fluid temperature sensor, vehicle speed sensor (VSS)

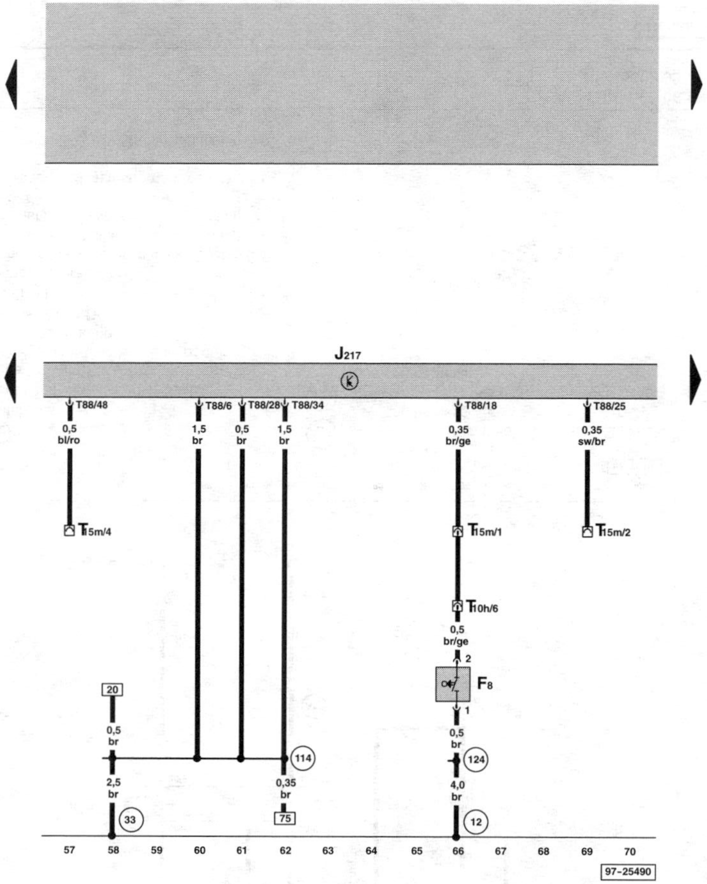

ws = white
sw = black
ro = red
br = brown
gn = green
bl = blue
gr = grey
li = violet
ge = yellow

- F8 - Kick Down Switch
- J217 - Transmission Control Module (TCM), in front footwell, right
- T10h - 10-Pin Connector, blue, on protective housing for control module, in engine compartment, left
- T15m - 15-Pin Connector, brown, connector station A-pillar, right
- T88 - Connector, 88 point, on Transmission Control Module (TCM)
- (12) - Ground connection, in engine compartment, left
- (33) - Ground connection, behind instrument panel, right
- (114) - Ground connection, in automatic transmission wiring harness
- (124) - ground connection, in engine compartment right wiring harness

Transmission control module (TCM), kick down switch

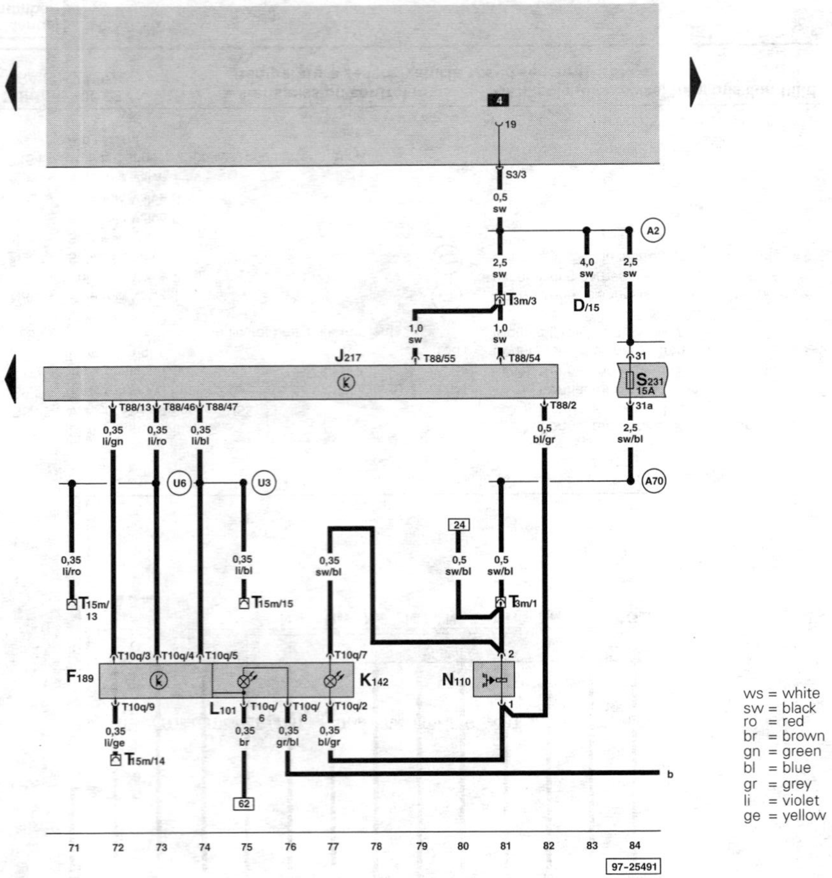

ws = white
sw = black
ro = red
br = brown
gn = green
bl = blue
gr = grey
li = violet
ge = yellow

- D - Ignition/Starter Switch
- F189 - Tiptronic Switch
- J217 - Transmission Control Module (TCM), in front footwell, right
- K142 - Warning light for selector lever position P/N
- L101 - Illumination for selector lever scale
- N110 - Shift Lock Solenoid
- S231 - Fuse 31 in fuse holder
- T3m - 3-Pin Connector, brown, connector station A-pillar, right
- T10q - 10-Pin Connector
- T15m - 15-Pin Connector, brown, connector station A-pillar, right
- T88 - Connector, 88 point, on Transmission Control Module (TCM)

- (A2) - plus connection (15), in instrument panel wiring harness
- (A70) - Connector (15a, fuse 231), in instrument panel wiring harness
- (U3) - wire connection -1-, in automatic transmission wiring harness
- (U6) - wire connection -2-, in automatic transmission wiring harness

Transmission control module (TCM), shift lock solenoid, tiptronic switch

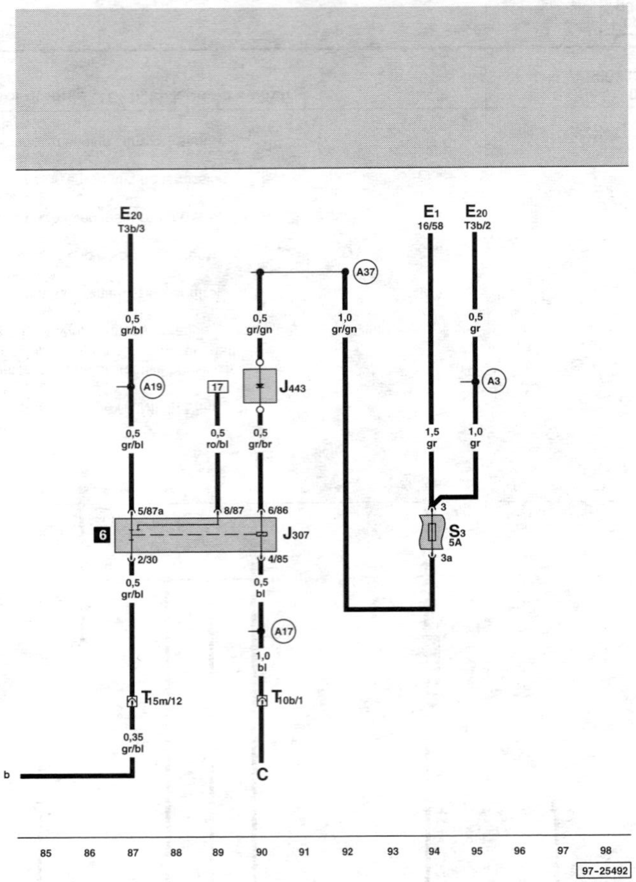

ws = white
sw = black
ro = red
br = brown
gn = green
bl = blue
gr = grey
li = violet
ge = yellow

- C - Generator (GEN)
- E1 - Light switch
- E20 - Instrument Panel Light Dimmer Switch
- J307 - Selector Lever Light Relay, on the thirteenfold auxiliary relay panel, above relay panel, production control number (205)
- J443 - Selector Lever Light Diode
- S3 - Fuse 3 in fuse holder
- T3b - 3-Pin Connector, grey
- T10b - 10-Pin Connector, black, on protective housing for control module, in engine compartment, left
- T15m - 15-Pin Connector, brown, connector station A-pillar, right
- (A3) - plus connection (58), in instrument panel wiring harness
- (A17) - wire connection (61), in instrument panel wiring harness
- (A19) - wire connection (58d), in instrument panel wiring harness
- (A37) - wire connection (58a), in instrument panel wiring harness

Selector lever light

Airbag Systems

from August 1999

Deviate relay location and fuseplacements as well as the locations of multiple connectors see section "component locations".

Relay location on the thirteenfold auxiliary relay panel, above relay panel:

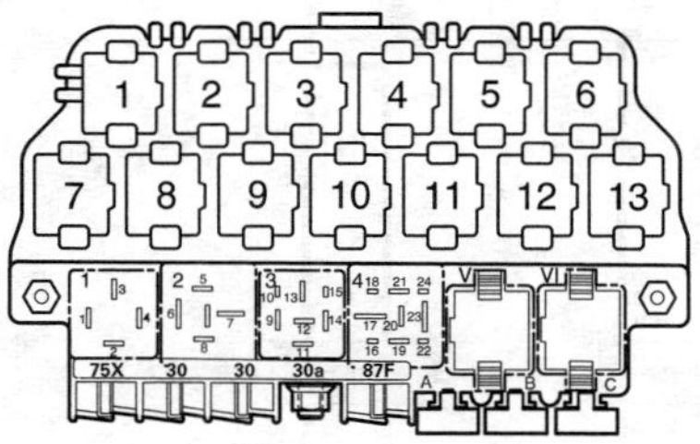

97-20676

Relay panel:

Relay location on the eightfold auxiliary relay panel, behind relay panel:

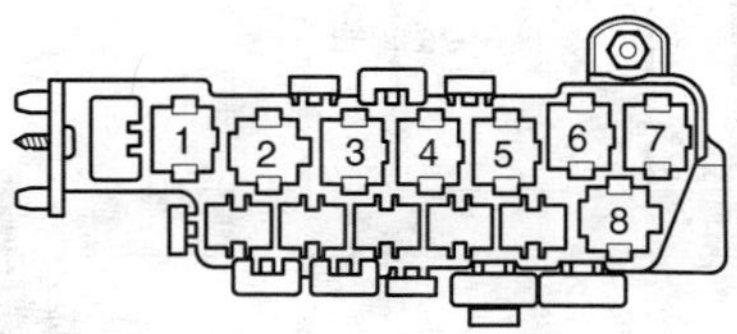

97-20515

Note: Number in parentheses indicates production control number stamped on relay housing.

Fuse colors

30 A - green
25 A - white
20 A - yellow
15 A - blue
10 A - red
7,5 A - brown
5 A - beige

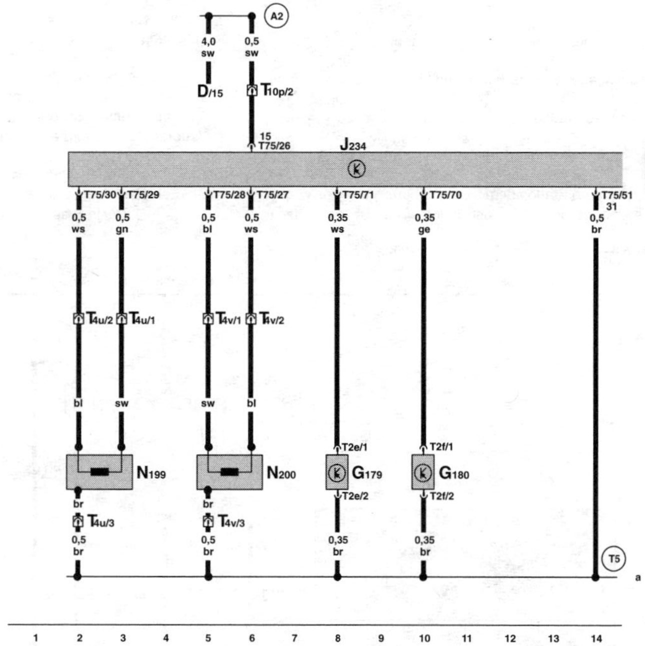

ws = white
sw = black
ro = red
br = brown
gn = green
bl = blue
gr = grey
li = violet
ge = yellow

D	-	Ignition/Starter Switch
G179	-	Crash sensor for side airbag, driver's side
G180	-	Crash sensor for side airbag, passenger side
J234	-	Airbag Control Module, behind console
N199	-	Igniter for side airbag, driver's side
N200	-	Igniter for side airbag, passenger side
T2e	-	Double Connector
T2f	-	Double Connector
T4u	-	4-Pin Connector, yellow, below driver's seat
T4v	-	4-Pin Connector, yellow, below passenger's seat
T10p	-	10-Pin Connector, yellow, connector station A-pillar, right
T75	-	75-Pin Connector, on Airbag Control Module

(A2)	-	plus connection (15), in instrument panel wiring harness
(T5)	-	wire connection -2-, in airbag wiring harness

Airbag control module, igniter for side airbag, crash sensor for side airbag

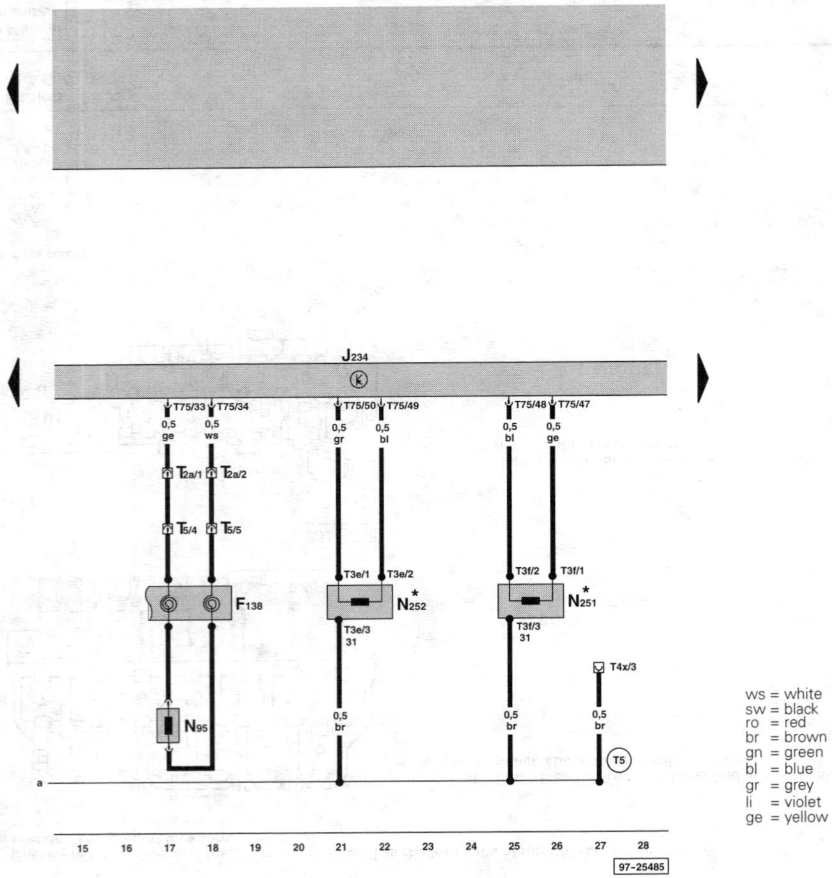

ws = white
sw = black
ro = red
br = brown
gn = green
bl = blue
gr = grey
li = violet
ge = yellow

F138 - Airbag Spiral Spring/Return Spring With Slip Ring
J234 - Airbag Control Module, behind console
N95 - Driver's Side Airbag Igniter
N251 - Igniter for head airbag, rear, driver side
N252 - Igniter for head airbag, rear, passenger side
T2a - Double Connector, yellow, connector station A-pillar, left
T3e - 3-Pin Connector
T3f - 3-Pin Connector
T4x - 4-Pin Connector, yellow, connector station A-pillar, right
T5 - 5-Pin Connector, yellow, behind steering column switch cover
T75 - 75-Pin Connector, on Airbag Control Module

(T5) - wire connection -2-, in airbag wiring harness

* - vehicles with head airbag only

Airbag control module, driver's side airbag igniter, igniter for head airbag, airbag spiral spring/return spring with slip ring

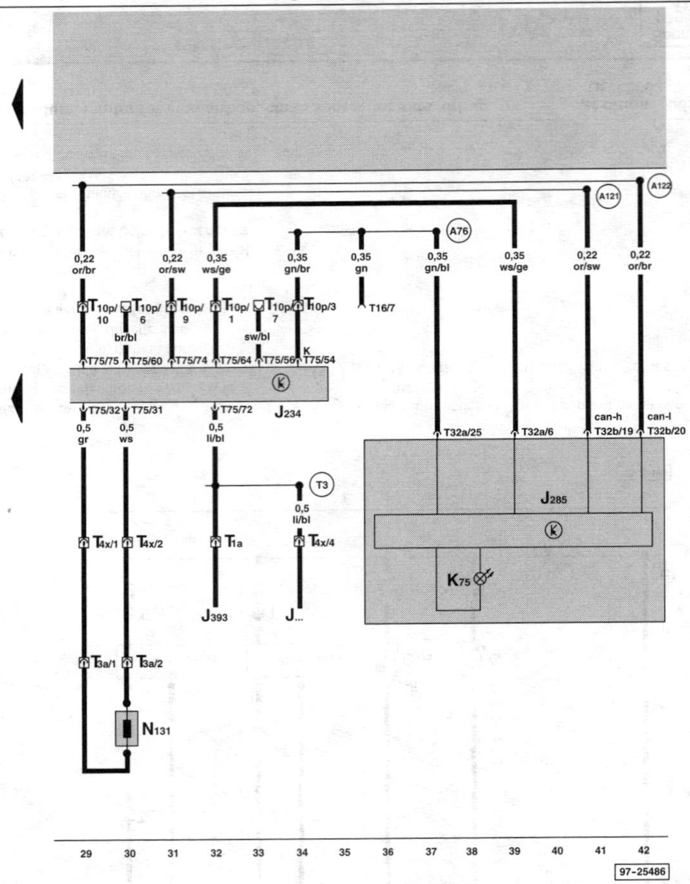

ws = white
sw = black
ro = red
br = brown
gn = green
bl = blue
gr = grey
li = violet
ge = yellow
or = orange

J... - Engine Control Modules (ECM)
J234 - Airbag Control Module, behind console
J285 - Control module with indicator unit in instrument panel insert
J393 - Central control module for comfort system
K75 - Airbag Malfunction Indicator Lamp (MIL)
N131 - Passenger's Side Airbag Igniter 1
T1a - Single Connector, black, connector station A-pillar, right
T3a - 3-Pin Connector, behind instrument panel, right
T4x - 4-Pin Connector, yellow, connector station A-pillar, right
T10p - 10-Pin Connector, yellow, connector station A-pillar, right
T16 - 16-Pin Connector, Data Link Connector (DLC), below instrument panel, left
T32a - 32-Pin Connector, blue, on instrument cluster
T32b - 32-Pin Connector, green, on instrument cluster
T75 - 75-Pin Connector, on airbag control module

(A76) - Connector (K-diagnosis wire) in instrument panel wiring harness
(A121) - Connector (high bus) in instrument panel wiring harness
(A122) - Connector (low bus) in instrument panel wiring harness
(T3) - wire connection, in airbag wiring harness

Airbag control module, driver's side airbag igniter, passenger's side airbag igniter, airbag malfunction indicator lamp (MIL)

Air-Conditioning

- **2,8L - Engine - Motronic Multiport Fuel Injection (MFI)/142 kW, code AHA**

from August 1998

- **1,8L- Engine - Motronic Multiport Fuel Injection (MFI)/110 kW, code ATW**
- **2,8L - Engine - Motronic Multiport Fuel Injection (MFI)/142 kW, code ATQ**

from August 1999

- **1,8L- Engine - Motronic Multiport Fuel Injection (MFI)/110 kW, code AUG**

from May 2000

Deviate relay location and fuseplacements as well as the locations of multiple connectors see section "component locations".

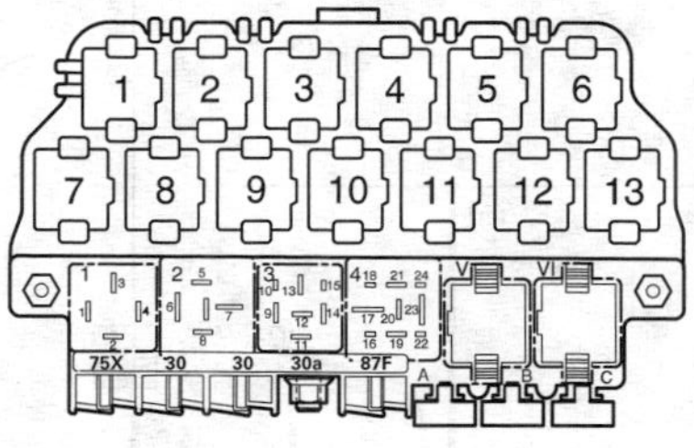

97-20676

Relay location on the thirteenfold auxiliary relay panel, above relay panel:

4 A/C Cut-Off Control Module (384)

Relay panel:

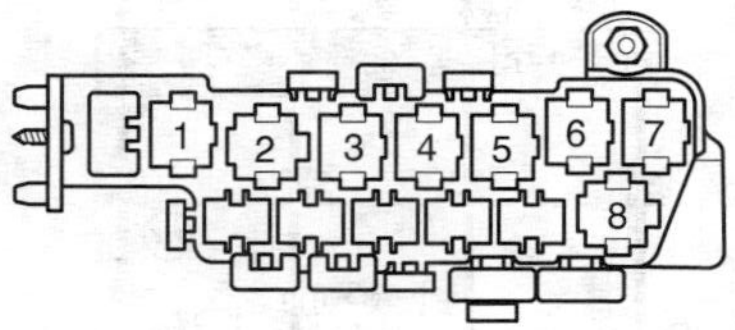

97-20515

Relay location on the eightfold auxiliary relay panel, behind relay panel:

5 First Speed Coolant Fan Control (FC) Relay (214)

8 Second Speed Coolant Fan Control (FC) Relay (213)

Note: Number in parentheses indicates production control number stamped on relay housing.

Fuse colors
30 A - green
25 A - white
20 A - yellow
15 A - blue
10 A - red
7,5 A - brown
5 A - beige

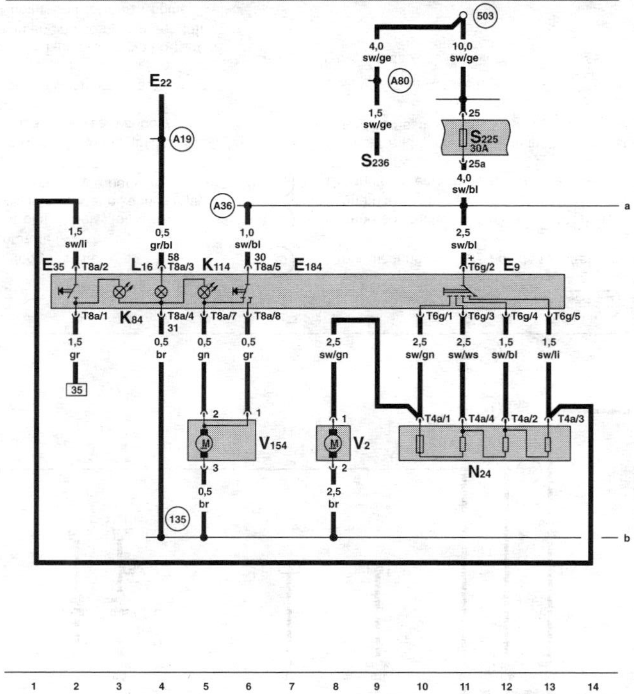

ws = white
sw = black
ro = red
br = brown
gn = green
bl = blue
gr = grey
li = violet
ge = yellow

E9 - Fresh Air Blower Switch
E22 - Windshield Wiper/Washer Switch
E35 - A/C Switch
E184 - Switch for fresh air blower and recirculated air
K84 - A/C Indicator Light
K114 - Fresh Air and Recirculating Air Mode Indicator Light
L16 - Fresh Air Control Lever Light
N24 - Fresh Air Blower Series Resistance With Fuse
S225 - Fuse 25 in fuse holder
S236 - Fuse 36 in fuse holder
T4a - 4-Pin Connector
T6g - 6-Pin Connector
T8a - 8-Pin Connector
V2 - Fresh Air Blower
V154 - Servo motor for fresh-/recirculating air door

(135) - Ground connection -2-, in instrument panel wiring harness
(503) - Screw Connection (75x), on relay panel
(A19) - wire connection (58d), in instrument panel wiring harness
(A36) - wire connection (75a), in instrument panel wiring harness
(A80) - Connector -1- (X), in instrument panel wiring harness

Fresh air blower, A/C switch, ambient temperature switch, switch for fresh air recirculated air

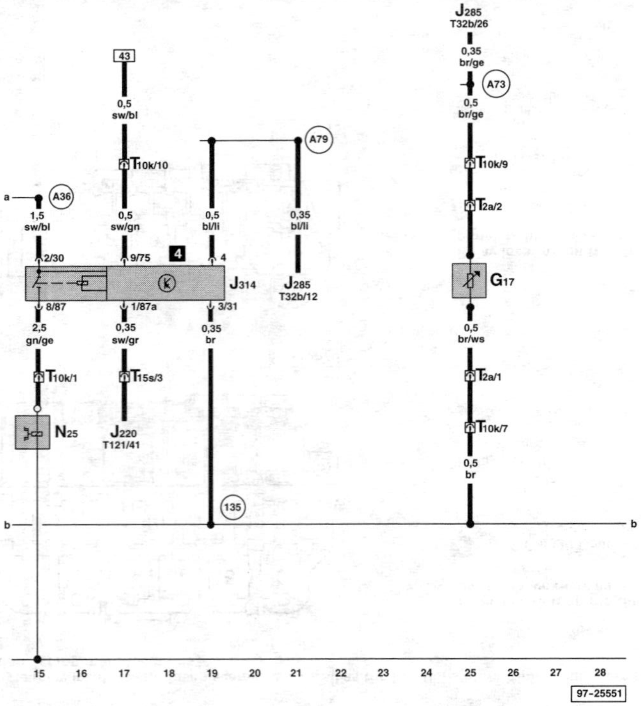

ws = white
sw = black
ro = red
br = brown
gn = green
bl = blue
gr = grey
li = violet
ge = yellow

G17 - Outside Air Temperature Sensor
J220 - Motronic Engine Control Module (ECM)
J285 - Control module with indicator unit in instrument panel insert
J314 - A/C Cut-Off Control Module, on thirteenfold auxiliary relay panel, above relay panel, production control number (384)
N25 - A/C Clutch
T2a - Double Connector
T10k - 10-Pin Connector, grey, connector station A-pillar, left
T15s - 15-Pin Connector, red, on protective housing for control module, in engine compartment, left
T32b - 32-Pin Connector, green, on instrument cluster
T121 - Connector, 121 point, on Engine Control Module (ECM)

(135) - Ground connection -2-, in instrument panel wiring harness
(A36) - wire connection (75a), in instrument panel wiring harness
(A73) - Connector (outside temperature indicator), in instrument panel wiring harness
(A79) - Connector (engine overheat warning light), in instrument panel wiring harness

Edition 08.00
USA.5062.12.21

A/C cut-off control module, A/C clutch, outside air temperature sensor

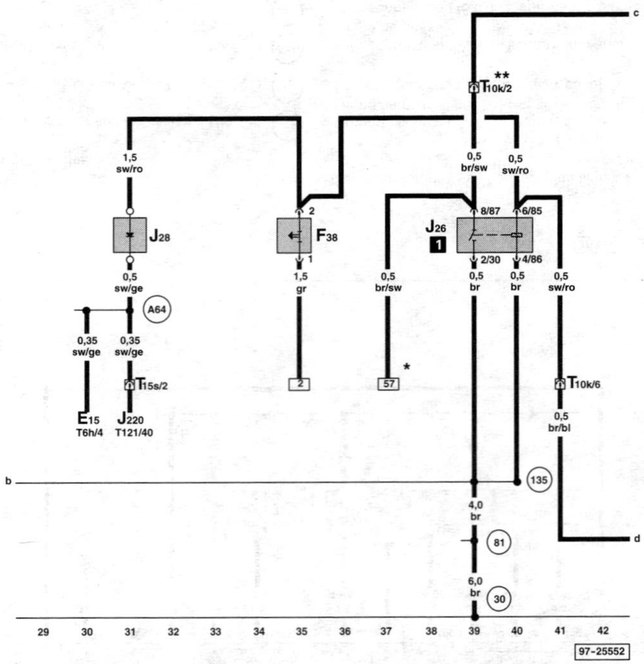

ws = white
sw = black
ro = red
br = brown
gn = green
bl = blue
gr = grey
li = violet
ge = yellow

E15 - Rear window defogger switch
F38 - Ambient Temperature Switch
J26 - Coolant Fan control (FC) Relay, on the thirteenfold auxiliary relay panel, above relay panel, production control number (214)
J28 - A/C Diode
J220 - Motronic Engine Control Module (ECM)
T6h - 6-Pin Connector, black, on rear window defogger switch
T10k - 10-Pin Connector, grey, connector station A-pillar, left
T15s - 15-Pin Connector, red, on protective housing for control module, in engine compartment, left
T121 - Connector, 121 point, on Motronic Engine Control Module (ECM)

(30) - Ground connection, -1-, beside fuse/relay panel
(81) - Ground connection -1-, in instrument panel wiring harness
(135) - Ground connection -2-, in instrument panel wiring harness
(A64) - Connector (30, idle boost), in instrument panel wiring harness

* - 2.8L V6, code ATQ only
** - 1.8L T, code ATW, AUG only

Coolant fan control (FC) relay, A/C diode

Edition 08/00
USA.5062.12.21

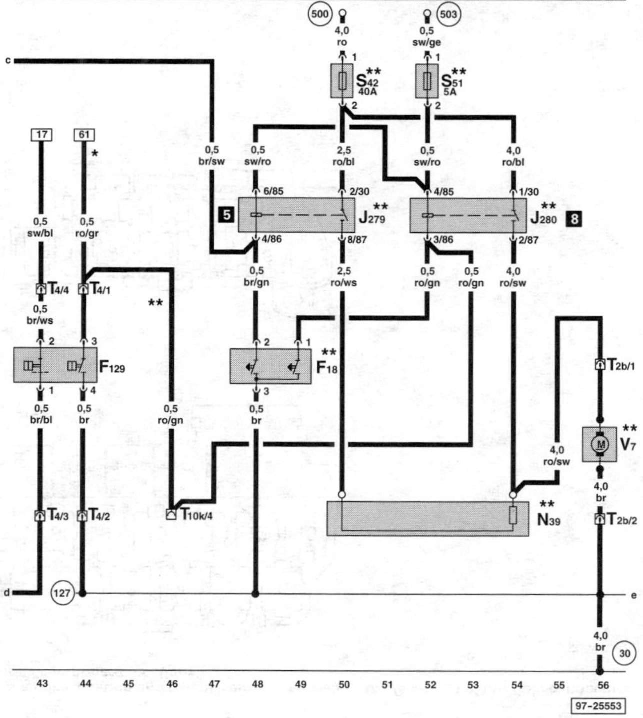

ws = white
sw = black
ro = red
br = brown
gn = green
bl = blue
gr = grey
li = violet
ge = yellow

F18 - Coolant Fan Control (FC) Thermal Switch
F129 - A/C Pressure Switch
J279 - First Speed Coolant Fan Control (FC) Relay, on eightfold auxiliary relay panel, production control number (214)
J280 - Second Speed Coolant Fan Control (FC) Relay, on eightfold auxiliary relay panel, production control number (213)
N39 - Coolant Fan Control (FC) Series Resistance, below side member, left
S42 - Coolant Fan Fuse, on eightfold auxiliary relay panel
S51 - Coolant Fan Fuse, on eightfold auxiliary relay panel
T2b - Double Connector, in engine compartment, left
T4 - 4-Pin Connector, in engine compartment, left front
T10k - 10-Pin Connector, grey, connector station A-pillar, left
V7 - Coolant Fan
(30) - Ground connection, -1-, beside fuse/relay panel
(127) - Ground connection -1-, in A/C compressor wiring harness
(500) - Screw connection -1- (30), on relay panel
(503) - Screw Connection (75x), on relay panel
* - 2.8L V6, code ATQ only
** - 1.8L T, code ATW, AUG only

Edition 08.00
USA.5062.12.21

A/C pressure switch, coolant fan control (FC) thermal switch, first speed coolant fan control (FC) relay, second speed coolant fan control (FC) relay

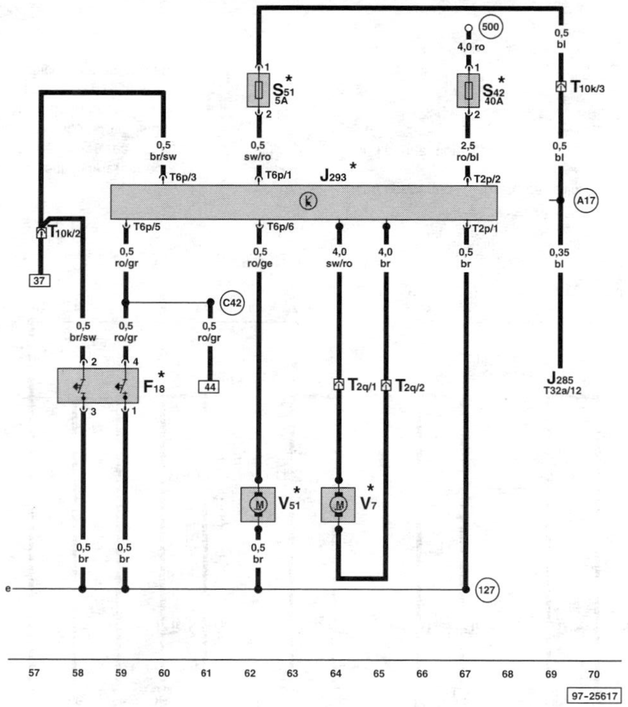

ws = white
sw = black
ro = red
br = brown
gn = green
bl = blue
gr = grey
li = violet
ge = yellow

F18 - Coolant Fan Control (FC) Thermal Switch
J285 - Control Module with indicator unit in instrument panel insert
J293 - Coolant FC (Fan Control) Control Module
S42 - Coolant Fan Fuse, on eightfold auxiliary relay panel
S51 - Coolant Fan Fuse, on eightfold auxiliary relay panel
T2p - Double Connector
T2q - Double Connector, in engine cpmpartment, left
T6p - 6-Pin Connector
T10k - 10-Pin Connector, grey, connector station A-pillar, left
T32a - 32-Pin Connector, blue, on Control Module with indicator unit in instrument panel insert
V7 - Coolant Fan
V51 - After-Run Coolant Pump
(127) - Ground connection -1-, in A/C compressor wiring harness
(500) - Screw Connection -1- (30), on relay panel
(A17) - wire connection (61), in instrument panel wiring harness
(C42) - Connector (fan for coolant) ,in wiring harnes coolant
* - 2.8L V6, code ATQ only

Coolant fan control module, Coolant fan thermal switch, coolant fan, After-run coolant pump

Edition 08.00
USA.5062.12.21

Monsoon-Soundsystem

from August 1999

Deviate relay location and fuseplacements as well as the locations of multiple connectors see section "component locations".

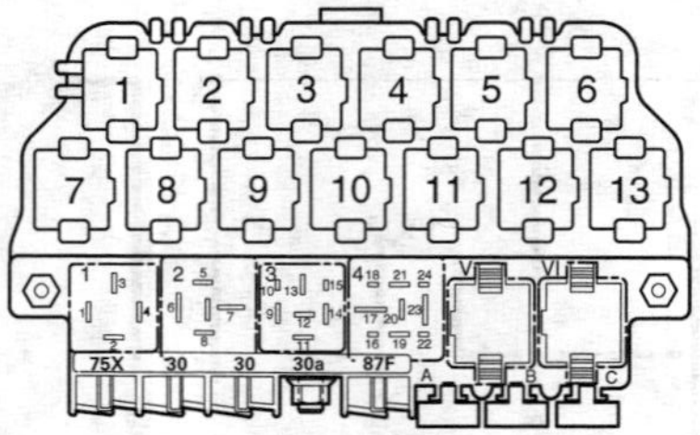

97-20676

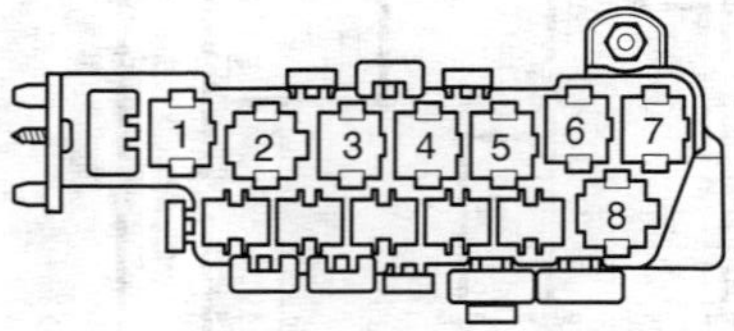

97-20515

Fuse colors
30 A - green
25 A - white
20 A - yellow
15 A - blue
10 A - red
7,5 A - brown
5 A - beige

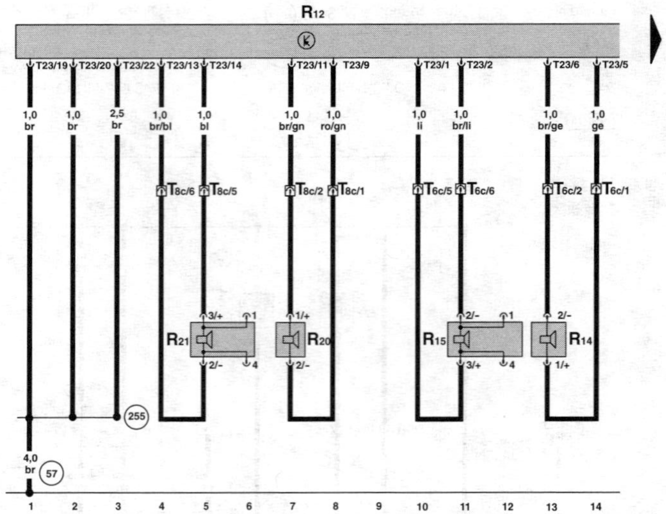

97-26062

ws = white
sw = black
ro = red
br = brown
gn = green
bl = blue
gr = grey
li = violet
ge = yellow

R12 - Amplifier
R14 - Treble speaker, left rear, in left rear door
R15 - Bass speaker, left rear, in left rear door
R20 - Treble speaker, left front, in driver's door
R21 - Bass speaker, left front, in driver's door
T6c - 6-Pin Connector, red, on B-pillar, left
T8c - 8-Pin Connector, black, connector station A-pillar, left
T23 - 23-Pin Connector, on Amplifier

(57) - Ground connection , on left rear pillar

(255) - Ground connection -1-, in wiring harness Radio

Amplifier, left front and rear speakers

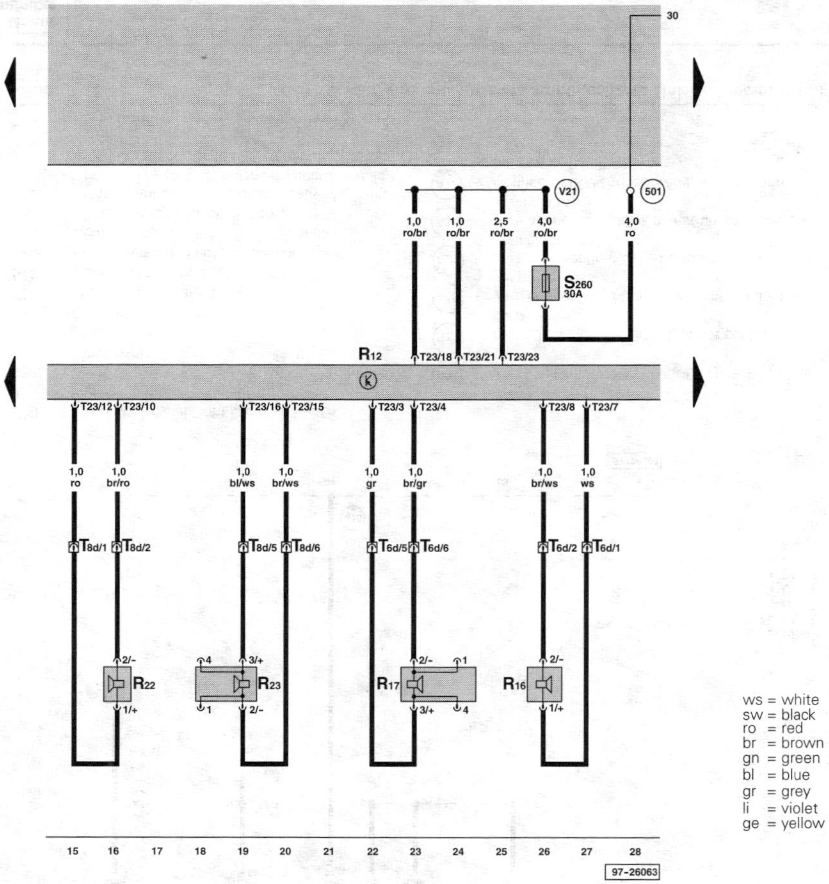

ws = white
sw = black
ro = red
br = brown
gn = green
bl = blue
gr = grey
li = violet
ge = yellow

R12 - Amplifier
R16 - Treble speaker, right rear, in right rear door
R17 - Bass speaker, right rear, in right rear door
R22 - Treble speaker, right front, in passenger's door
R23 - Bass speaker, right front, in passenger's door
S260 - Amplifier/DSP Fuse
T6d - 6-Pin Connector, red, on B-pillar, right
T8d - 8-Pin Connector, black, connector station A-pillar, right
T23 - 23-Pin Connector, on Amplifier

(501) - Threaded connection -2- (30) on the relay plate
(V21) - Plus connector (30), in wiring harness Radio -Amplifier

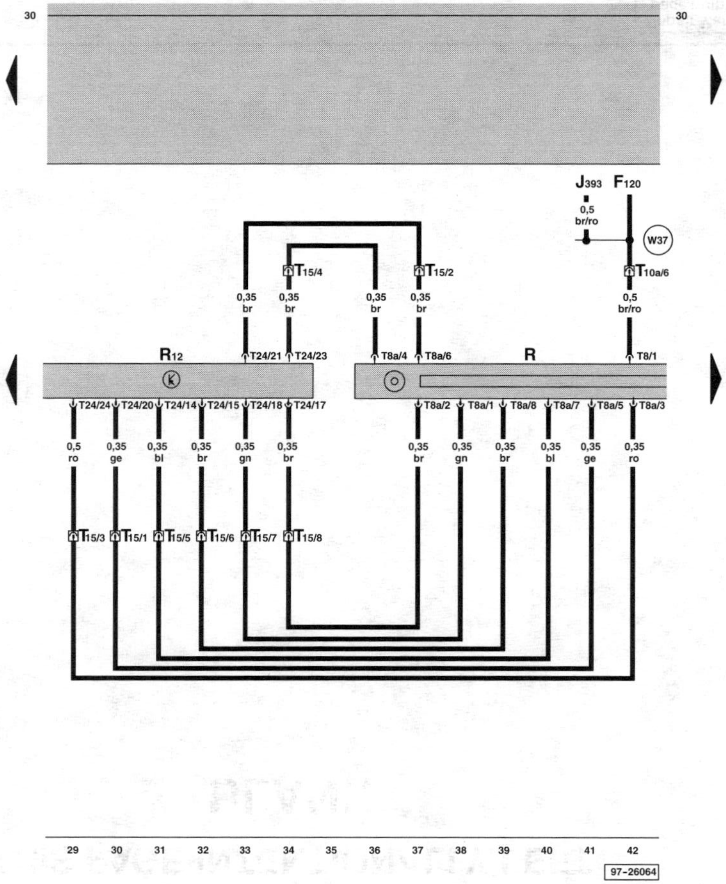

ws = white
sw = black
ro = red
br = brown
gn = green
bl = blue
gr = grey
li = violet
ge = yellow
or = orange

F120 - Hood Alarm Switch
J393 - Central control module for comfort system
R - Radio
R12 - Amplifier
T8 - 8-Pin Connector, black, on radio
T8a - 8-Pin Connector, brown, on radio
T10a - 10-Pin Connector, brown, connector station A-pillar, left
T15 - 15-Pin Connector, white, connector station A-pillar, left
T24 - 24-Pin Connector, on Amplifier

(W37) - Connection (alarm), in floor wiring harness

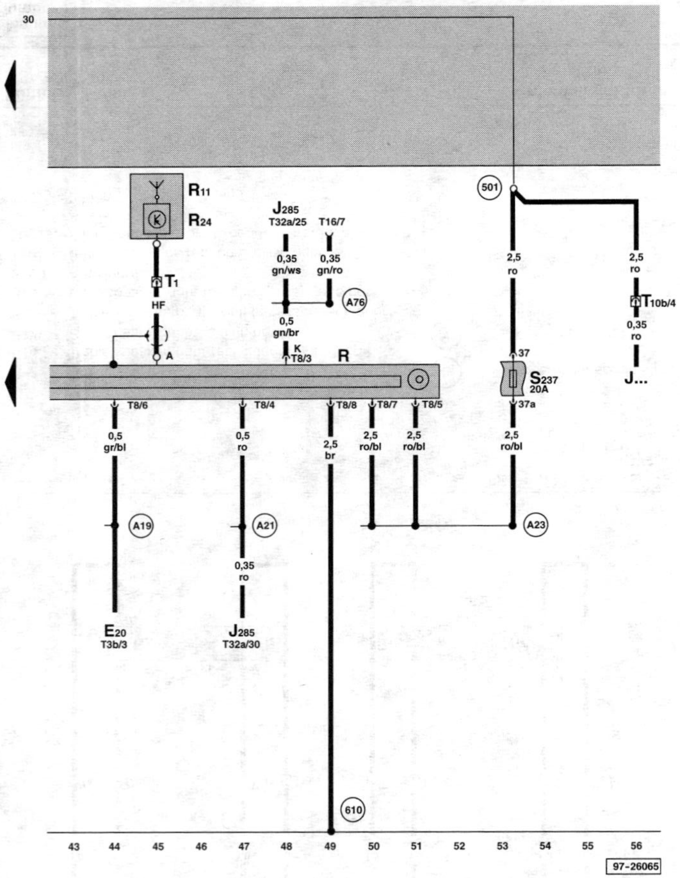

ws = white
sw = black
ro = red
br = brown
gn = green
bl = blue
gr = grey
li = violet
ge = yellow

E20 - Instrument Panel Light Dimmer Switch
J... - Engine Control Module (ECM)
J285 - Control module with indicator unit in instrument panel insert
R - Radio
R11 - Antenna
R24 - Antenna Amplifier
S237 - Fuse 37 in fuse holder
T1 - 1-Pin Connector, black, below console
T3b - 3-Pin Connector
T8 - 8-Pin Connector, black, on radio
T10b - 10-Pin Connector, black, on protective housing for control module, in engine compartment, left
T16 - 16-Pin Connector, Data Link Connector (DLC), below instrument panel, left
T32a - 32-Pin Connector, blue, on instrument cluster

(501) - Threaded connection -2- (30) on the relay plate
(610) - ground connection (audio), under center conside
(A19) - wire connection (58d), in instrument panel wiring harness
(A21) - wire connection (86s), in instrument panel wiring harness
(A23) - wire connection (30al), in instrument panel wiring harness
(A76) - Connector (K-diagnosis wire), in instrument panel wiring harness

THIS PAGE INTENTIONALLY LEFT BLANK

Daytime Running Lights (DRL),

from August 1999

Notes:

Information for

- *Relay and fuse arrangements*
- *Connector arrangements*
- *Control Modules and Relays*
- *Ground connections*

⇒ Component Locations!

Information for

- *Troubleshooting Procedures*

⇒ Guided Fault Finding using VAS 5051/5052

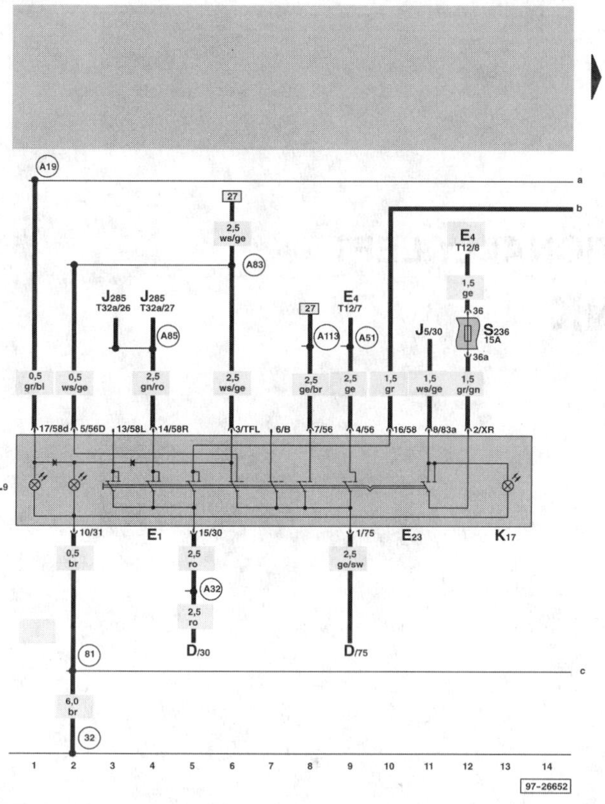

ws = white
sw = black
ro = red
br = brown
gn = green
bl = blue
gr = grey
li = lilac
ge = yellow
or = orange

Light Switch

D - Ignition/Starter Switch
E1 - Light switch
E4 - Headlight Dimmer/Flasher Switch
E23 - Fog Light Switch
J5 - Fog Light Relay, on auxiliary relay panel, production control number (381)
K17 - Fog Light Indicator Light
L9 - Headlight Switch Light
S236 - Fuse 36 in fuse holder
T12 - 12-Pin Connector
T32a - 32-Pin Connector, blue, on instrument cluster

(32) - Ground connection, befind instrument panel, left
(81) - Ground connection -1-, in instrument panel wiring harness
(A19) - wire connection (58d), in instrument panel wiring harness
(A32) - plus connection (30), in instrument panel wiring harness
(A51) - wire connection (56) in instrument panel wiring harness
(A83) - Connection (daytime running lights), in instrument panel wiring harness
(A85) - Connection (58R) in instriment panel wiring harness
(A113) - Connection 2 (daytime running lights), in instrument panel wiring harness

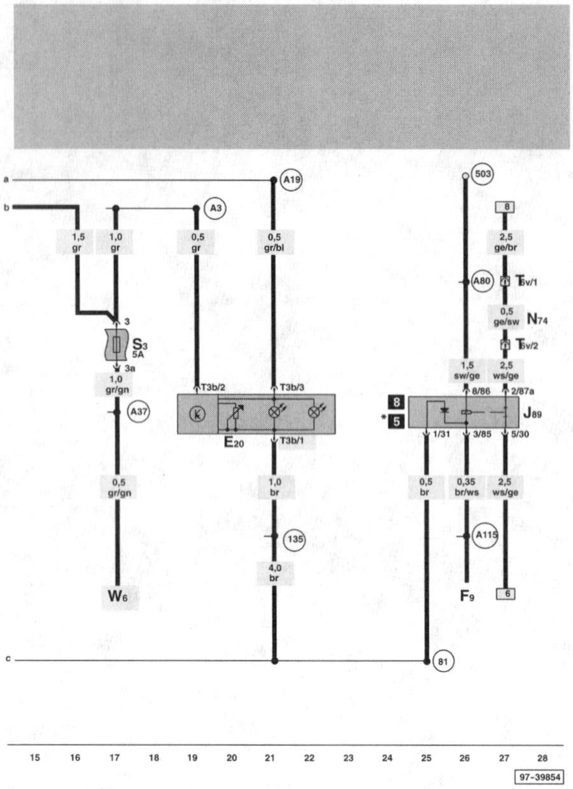

Instrument panel light dimmer switch, daytime running lights change-over relay

E20 - Instrument Panel Light Dimmer Switch
F9 - Parking Brake Warning Light Switch
J89 - Daytime Running Lights Change-over Relay (173) on the thirteen position relay panel, above relay panel
N74 - Series Resistance Wiring
S3 - Fuse 3 in fuse holder
T3b - 3-Pin Connector
T6v - 6-Pin Connector, violet, connector station A-pillar, left
W6 - Glove Compartment Light

(81) - Ground connection -1-,in instrument panel wiring harness
(135) - Ground connection -2-,in instrument panel wiring harness
(503) - Screw connection (75x), on relay panel

(A3) - plus connection (58), in instrument panel wiring harness
(A19) - wire connection (58d), in instrument panel wiring harness
(A37) - wire connection (58a), in instrument panel wiring harness
(A80) - Connection -1- (X), in instrument panel wiring harness
(A115) - Connection (parking brake warning light), in instrument panel wiring harness

* - when equipped with Multi-Function Steering Wheel

ws = white
sw = black
ro = red
br = brown
gn = green
bl = blue
gr = grey
li = lilac
ge = yellow
or = orange

Edition 04.04

Repair Manual page number Y 442

THIS PAGE INTENTIONALLY LEFT

BLANK

Repair Manual page number Y 441

1,8L- Engine - Motronic Multiport Fuel Injection (MFI)/110 kW, code AUG

from May 2000

Deviate relay location and fuseplacements as well as the locations of multiple connectors see section "component locations".

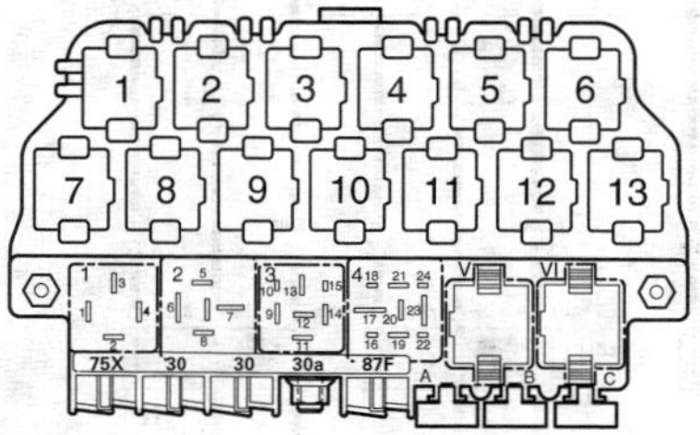

97-20676

Relay location on the thirteenfold auxiliary relay panel, above relay panel:

Relay panel:

- **4** Fuel Pump (FP) Relay (372)
- **10** Brake Booster Relay (569)

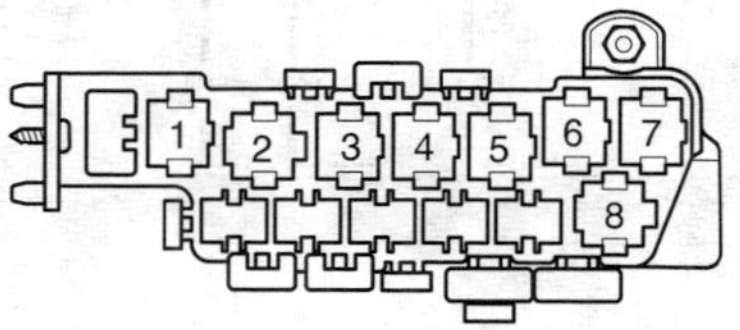

97-20515

Relay location on the eightfold auxiliary relay panel, behind relay panel:

Note: Number in parentheses indicates production control number stamped on relay housing.

Fuse colors

30 A - green
25 A - white
20 A - yellow
15 A - blue
10 A - red
7,5 A - brown
5 A - beige

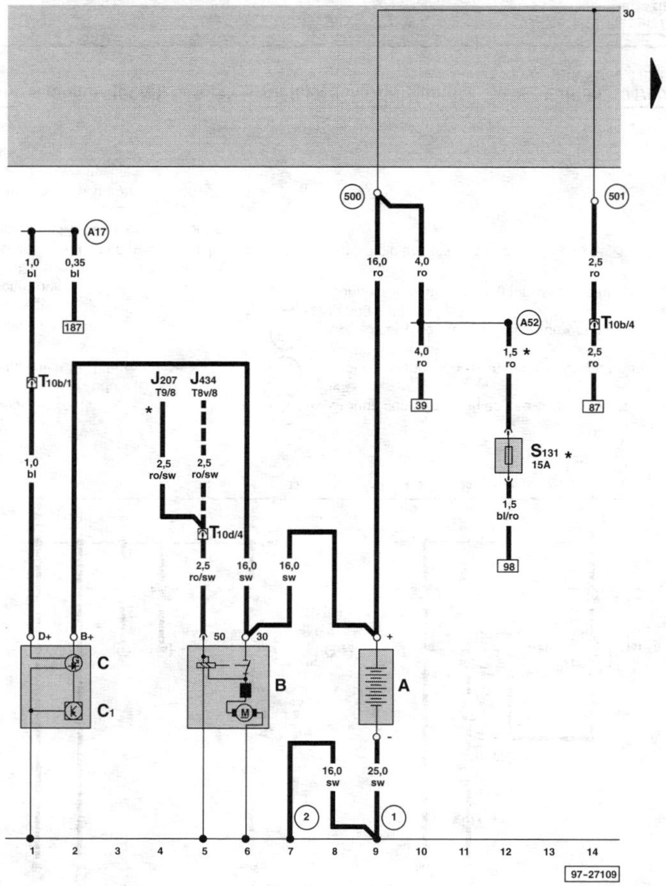

ws = white
sw = black
ro = red
br = brown
gn = green
bl = blue
gr = grey
li = violet
ge = yellow

A - Battery
B - Starter
C - Generator (GEN)
C1 - Voltage Regulator (VR)
J207 - Starting Interlock Relay
J434 - Locking relay for starter (clutch pedal switch)
S131 - Safety fuse 1
T8v - 8-Pin Connector
T9 - 9-Pin Connector, on Starting Interlock Relay
T10b - 10-Pin Connector, black, on protective housing for control module, in engine compartment, left
T10d - 10-Pin Connector, brown, on protective housing for control module, in engine compartment, left

(1) - Ground strap, battery to body
(2) - Ground strap, transmission to body
(500) - Threaded connection -1- (30), on relay plate
(501) - Threaded connection -2- (30), on relay plate
(A17) - wire connection (61), in instrument panel wiring harness
(A52) - plus connection 2 (30), in instrument panel wiring harness

* automatic transmission only
--- manual transmission only

Battery, starter, generator (GEN)

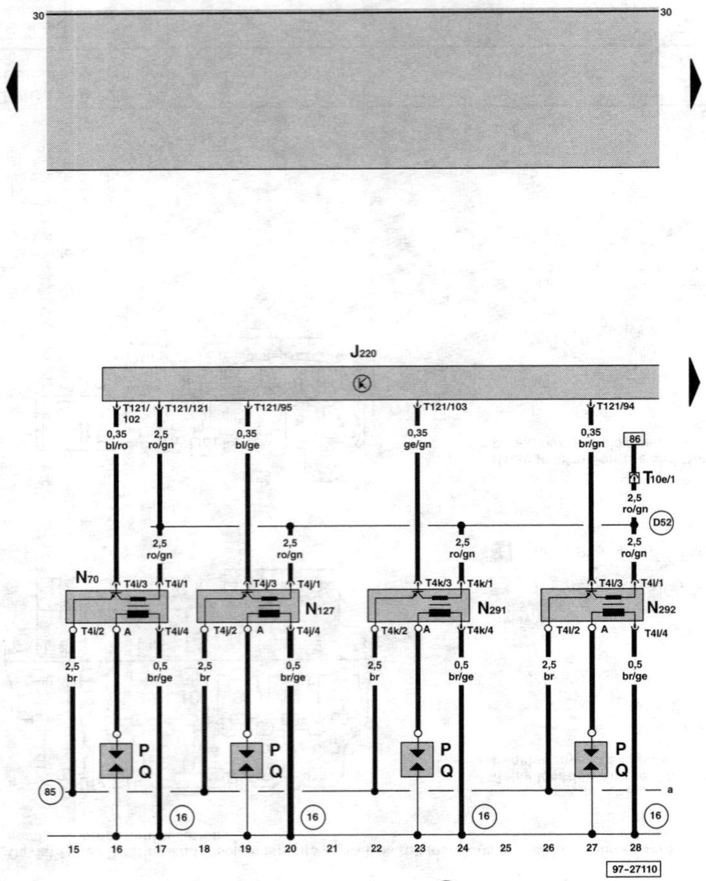

ws = white
sw = black
ro = red
br = brown
gn = green
bl = blue
gr = grey
li = violet
ge = yellow

J220 - Motronic Engine Control Module (ECM)
N70 - Ignition Coil 1 with Power Output Stage
N127 - Ignition Coil 2 with Power Output Stage
N291 - Ignition Coil 3 with Power Output Stage
N292 - Ignition Coil 4 with Power Output Stage
P - Spark Plug Connectors
Q - Spark Plugs
T4i - 4-Pin Connector, black
T4j - 4-Pin Connector, black
T4k - 4-Pin Connector, black
T4l - 4-Pin Connector, black
T10e - 10-Pin Connector, orange, on protective housing for control module, in engine compartment, left
T121 - 121-Pin Connector, on Motronic Engine Control Module (ECM)

(16) - Ground connection -1-, on valve cover
(85) - Ground connection -1-, in engine compartment wiring harness
(D52) - plus connection (15a), in engine compartment wiring harness

Ignition system, Motronic engine control module (ECM), Spark plug connectors, Spark plugs

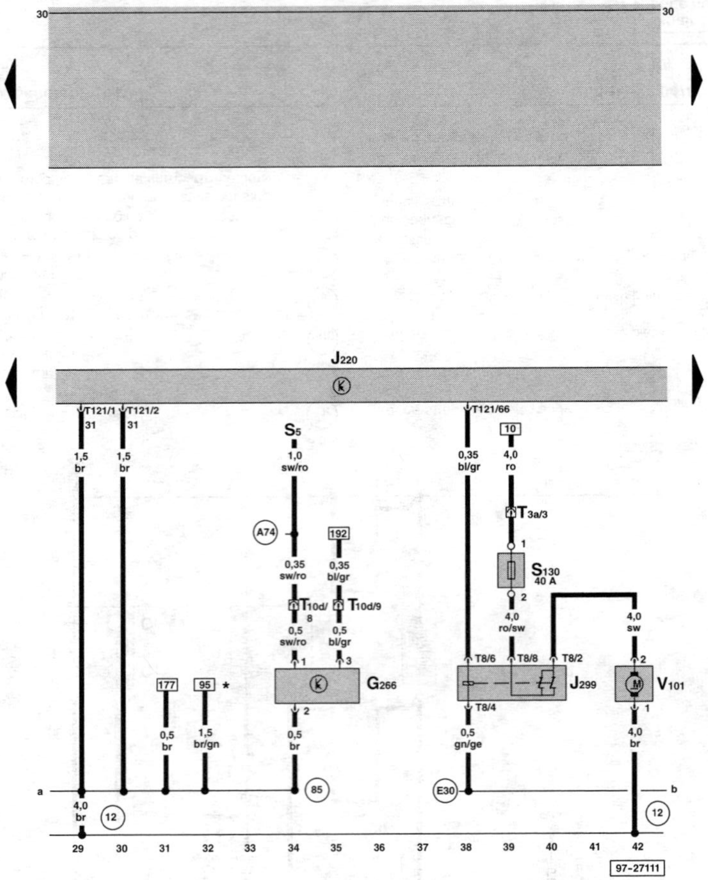

ws = white
sw = black
ro = red
br = brown
gn = green
bl = blue
gr = grey
li = violet
ge = yellow

G266 - Oil Level Thermal Sensor
J220 - Motronic Engine Control Module (ECM)
J299 - Secondary Air Injection (AIR) Pump Relay
S5 - Fuse
S130 - Fuse for secondary air pump
T3a - 3-Pin Connector, white
T8 - 8-Pin Connector, brown
T10d - 10-Pin Connector, brown, on protective housing for control module, in engine compartment, left
T121 - 121-Pin Connector, on Motronic Engine Control Module (ECM)
V101 - Secondary Air Injection (AIR) Pump Motor

(12) - Ground connection, in engine compartment, left
(85) - Ground connection -1-, in engine compartment wiring harness
(A74) - Connector (15a - fuse 5) in instrument panel wiring harness
(E30) - Connector (87a), in wiring harness engine
* - automatic transmission only

Motronic engine control module (ECM), Secondary Air Injection (AIR) Pump Motor/Relay

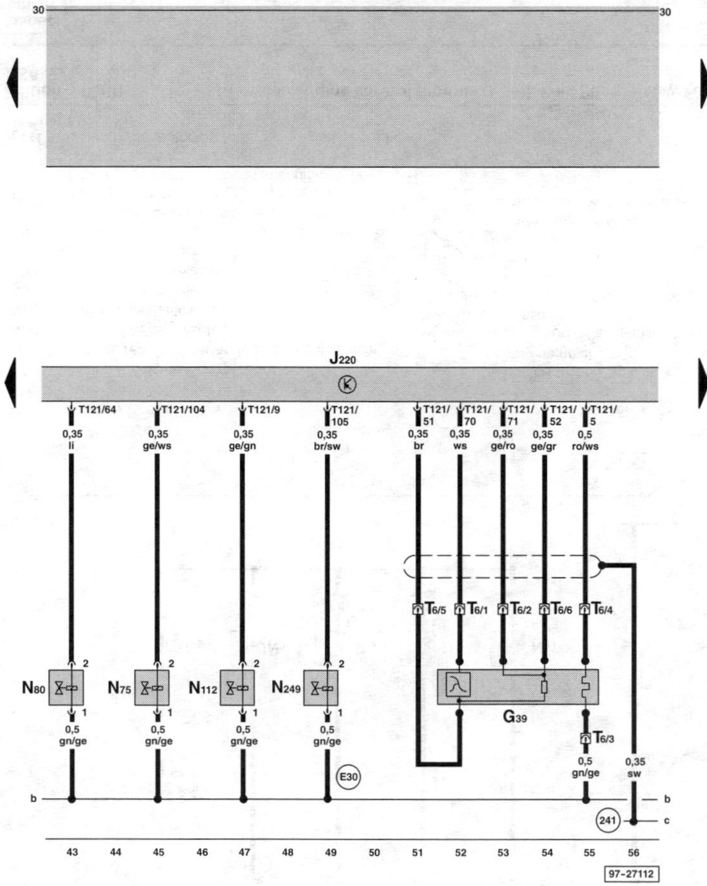

ws = white
sw = black
ro = red
br = brown
gn = green
bl = blue
gr = grey
li = violet
ge = yellow

G39 - Heated Oxygen Sensor (HO2S)
J220 - Motronic Engine Control Module (ECM)
N75 - Wastegate Bypass Regulator Valve
N80 - Evaporative Emission (EVAP) Canister Purge Regulator Valve
N112 - Secondary Air Infection (AIR) Solenoid Valve
N249 - Recirculating valve for turbocharger
T6 - 6-Pin Connector, black
T121 - 121-Pin Connector, on Motronic Engine Control Module (ECM)

(241) - Ground connection 2, engine ground in engine wiring harness
(E30) - Connector (87a), in wiring harness engine

Edition 04/00
USA.5062.11.21

Wastegate Bypass Reg. Valve, EVAP Canister Purge Reg. Valve, Secondary Air Injection Solenoid Valve, Recirc.valve for turbocharger, HO2S

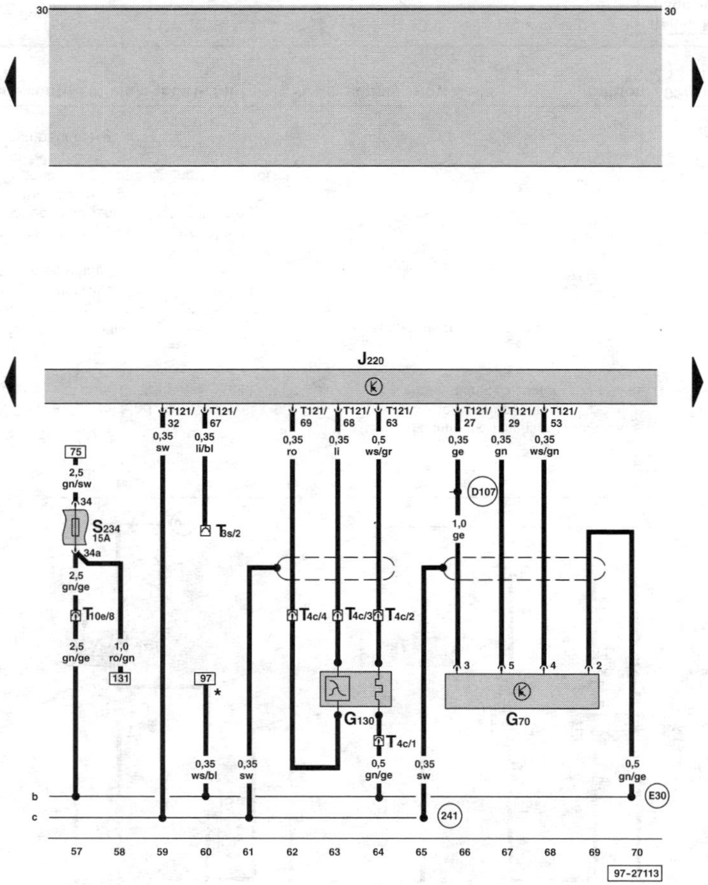

ws = white
sw = black
ro = red
br = brown
gn = green
bl = blue
gr = grey
li = violet
ge = yellow

G70 - Mass Air Flow (MAF) Sensor
G130 - Oxygen Sensor (O2S) Behind Three Way Catalytic Converter (TWC)
J220 - Motronic Engine Control Module (ECM)
S234 - Fuse in fuse holder
T3s - 3-Pin Connector, red, on protective housing for control module, in engine compartment,left
T4c - 4-Pin Connector, black
T10e - 10-Pin Connector, orange, on protective housing for control module, in engine compartment, left
T121 - 121-Pin Connector, on Motronic Engine Control Module (ECM)

(241) - Ground connection 2, engine ground in engine wiring harness
(D107) - wire connection -5-, in engine compartment wiring harness
(E30) - Connector (87a), in wiring harness engine
* automatic transmission only

Motronic engine control module (ECM), Oxygen Sensor, Mass Air Flow Sensor

Edition 04/00
USA.5062.11.21

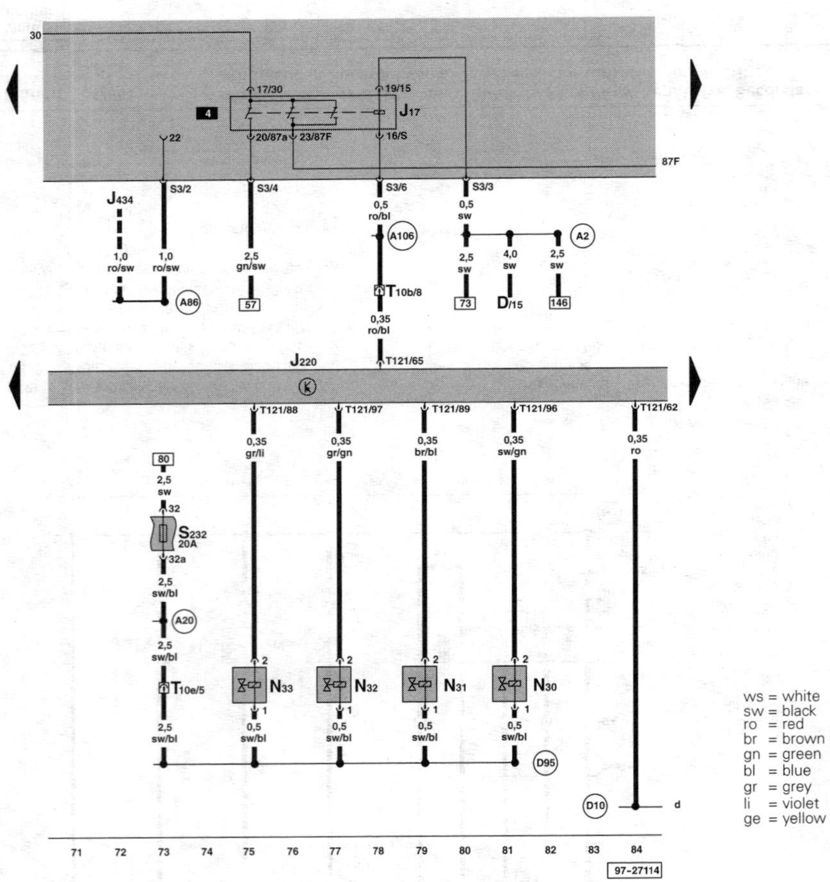

ws = white
sw = black
ro = red
br = brown
gn = green
bl = blue
gr = grey
li = violet
ge = yellow

D - Ignition/Starter Switch
J17 - Fuel Pump (FP) Relay (372)
J220 - Motronic Engine Control Module (ECM)
J434 - Locking relay for starter (clutch pedal switch)
N30 - Cylinder 1 Fuel Injector
N31 - Cylinder 2 Fuel Injector
N32 - Cylinder 3 Fuel Injector
N33 - Cylinder 4 Fuel Injector
S232 - Fuse in fuse holder
T10b - 10-Pin Connector, black, on protective housing for control module, in engine compartment,left
T10e - 10-Pin Connector, orange, on protective housing for control module, in engine compartment, left
T121 - 121-Pin Connector, on Motronic Engine Control Module (ECM)

(A2) - plus connection (15), in instrument panel wiring harness
(A20) - wire connection (15a), in instrument panel wiring harness
(A86) - Connection (50a), in instrument panel wiring harness
(A106) - Connector -2- (86s), in instrument panel wiring harness
(D10) - plus connection (30), in right front wiring harness
(D95) - wire connection (injectors), in engine compartment wiring harness
- - - manual transmission only

Motronic engine control module (ECM), Fuel Pump Relay, Fuel injectors

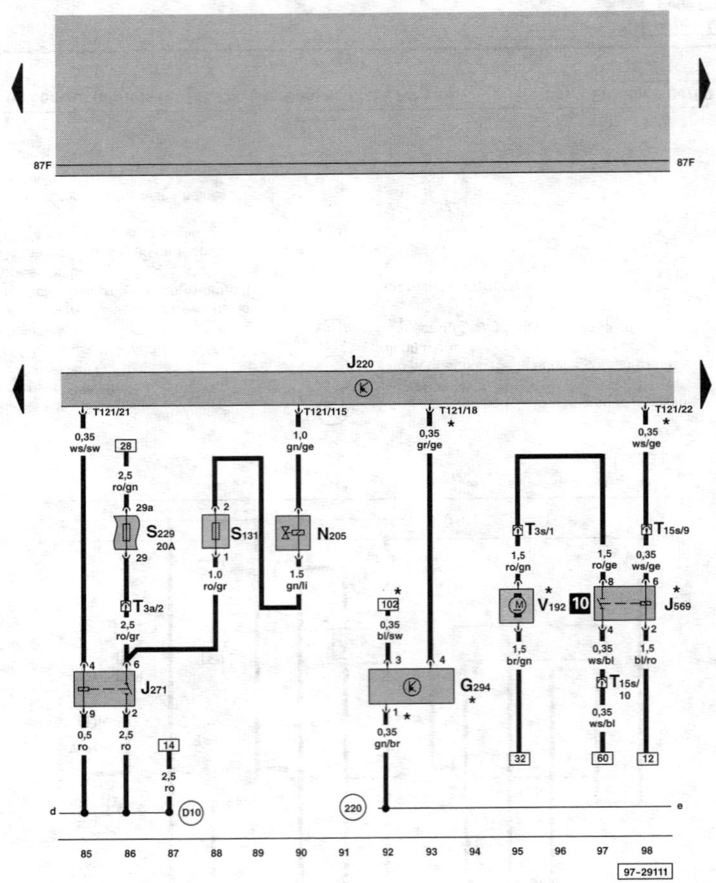

ws = white
sw = black
ro = red
br = brown
gn = green
bl = blue
gr = grey
li = violet
ge = yellow

G294 - Sensor for brake booster
J220 - Motronic Engine Control Module (ECM)
J271 - Motronic Engine Control (ECM) Power Supply Relay
J569 - Relay for brake booster
N205 - Valve -1- for camshaft adjustment
S131 - Safety fuse 1
S229 - Fuse in fuse holder
T3a - 3-Pin Connector, white
T3s - 3-Pin Connector, red, on protective housing for control module, in engine compartment,left
T15s - 15-Pin Connector, red, on protective housing for control module, in engine compartment,left
T121 - 121-Pin Connector, on Motronic Engine Control Module (ECM)
V192 - Vacuum pump for brake

(220) - Ground connection (sensor ground), in engine compartment wiring harness
(D10) - plus connection (30), in right front wiring harness
* - automatic transmission only

Motronic engine control module (ECM), Motronic Power supply relay, Brake booster sensor

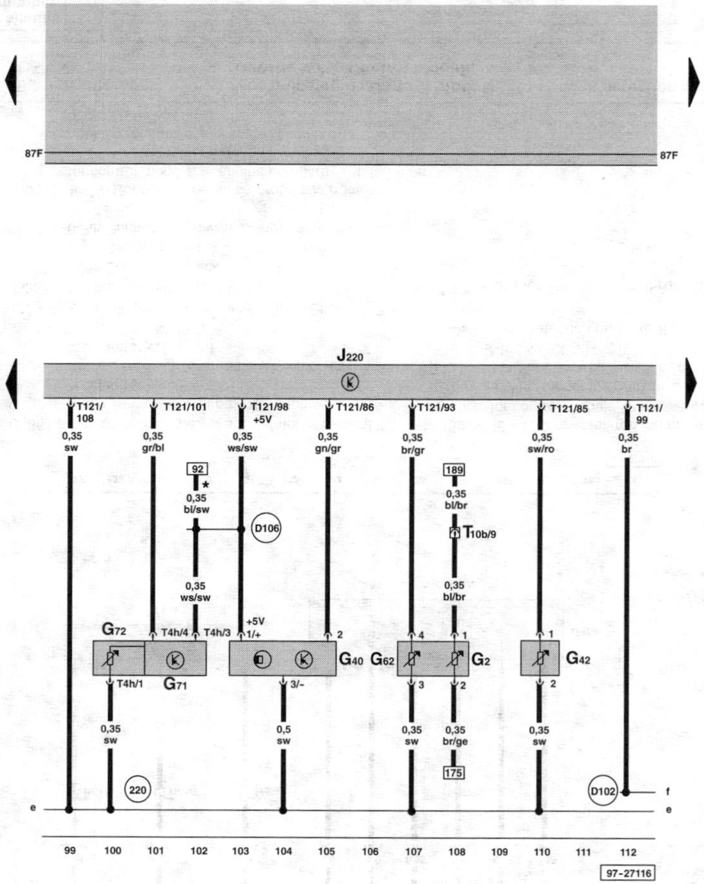

ws = white
sw = black
ro = red
br = brown
gn = green
bl = blue
gr = grey
li = violet
ge = yellow

G2 - Engine Coolant Temperature (ECT) Sensor
G40 - Camshaft Position (CMP) Sensor
G42 - Intake Air Temperature (IAT) Sensor
G62 - Engine Coolant Temperature (ECT) Sensor
G71 - Manifold Absolute Pressure (MAP) Sensor
G72 - Intake Air Temperature (IAT) Sensor
J220 - Motronic Engine Control Module (ECM)
T4h - 4-Pin Connector, black
T10b - 10-Pin Connector, black, on protective housing for control module, in engine compartment,left
T121 - 121-Pin Connector, on Motronic Engine Control Module (ECM)

(220) - Ground connection (sensor ground), in engine compartment wiring harness
(D102) - Wire connection -2-, in engine compartment wiring harness
(D106) - Wire connection -4-, in engine compartment wiring harness
* - automatic transmission only

Edition 04/00
USA.5062.11.21

Motronic engine control module (ECM), Engine Coolant Temperature Sensor, Camshaft Position Sensor, Intake Air Temperature

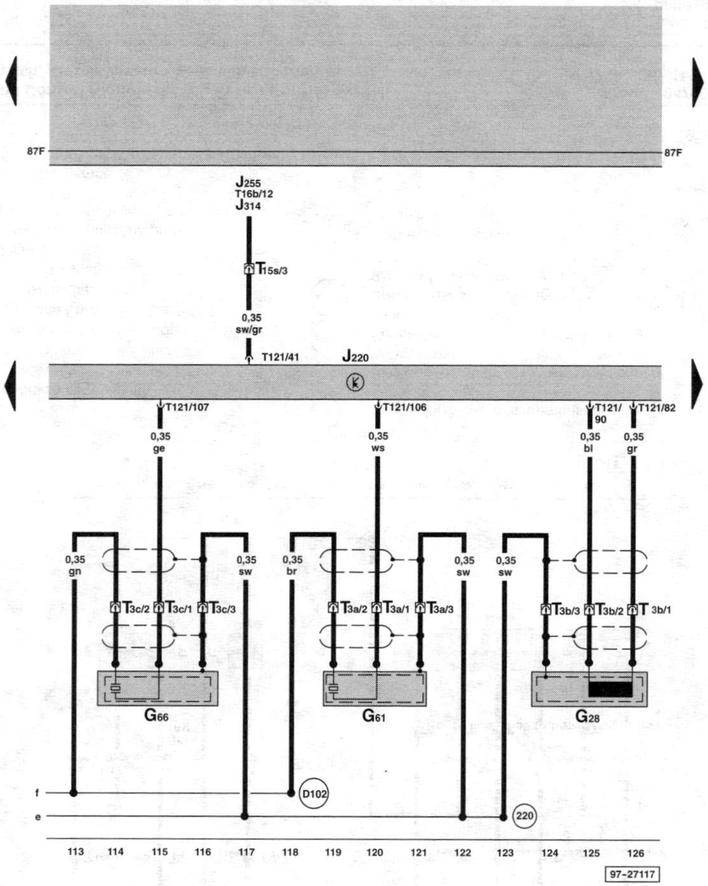

ws = white
sw = black
ro = red
br = brown
gn = green
bl = blue
gr = grey
li = violet
ge = yellow

G28 - Engine Speed (RPM) Sensor
G61 - Knock Sensor (KS) 1
G66 - Knock Sensor (KS) 2
J220 - Motronic Engine Control Module (ECM)
J255 - Climatronic Control Module
J314 - A/C Cut-Off Control Module
T3a - 3-Pin Connector, green
T3b - 3-Pin Connector, grey
T3c - 3-Pin Connector,blue
T15s - 15-Pin Connector, red, on protective housing for control module, in engine compartment,left
T16b - 16-Pin Connector, on Climatronic Control Module
T121 - 121-Pin Connector, on Motronic Engine Control Module (ECM)

(220) - Ground connection (sensor ground), in engine compartment wiring harness
(D102) - Wire connection -2-, in engine compartment wiring harness

Motronic engine control module (ECM), Knock Sensor 1 and 2, Engine Speed Sensor

Edtion 04/00
USA.5062.11.21

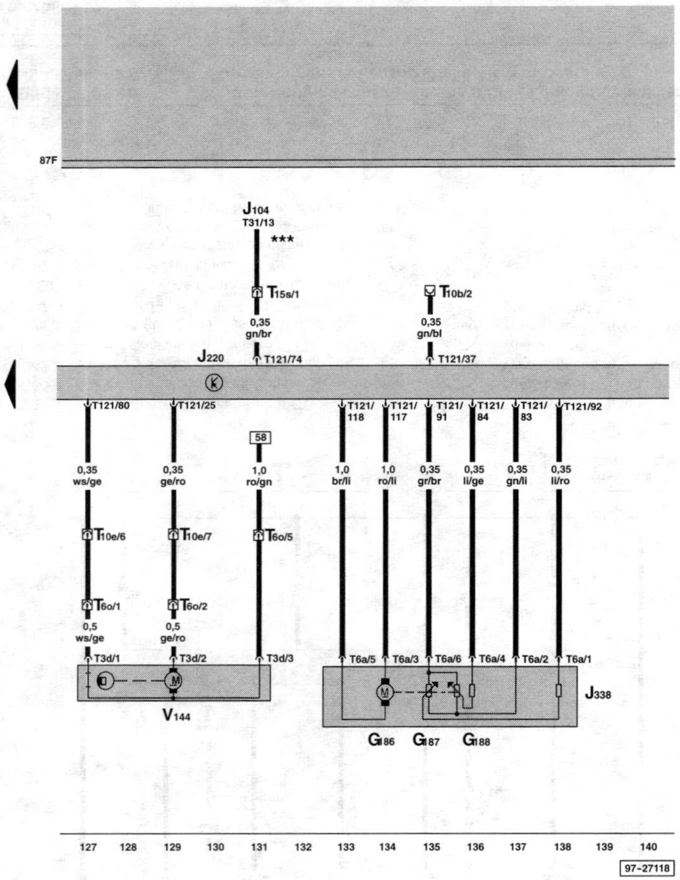

ws = white
sw = black
ro = red
br = brown
gn = green
bl = blue
gr = grey
li = violet
ge = yellow

G186 - Throttle sensor (power accelerator actuation)
G187 - Angle senor -1- for throttle drive (power accelerator actuation)
G188 - Angle sensor -2- for throttle drive (power accelerator actuation)
J104 - ABS Control Module
J220 - Motronic Engine Control Module (ECM)
J338 - Throttle Valve Control Module
T3d - 3-Pin Connector, black
T6a - 6-Pin Connector, black
T6o - 6-Pin Connector, brown, connector station A-pillar, right
T10b - 10-Pin Connector, black, on protective housing for control module, in engine compartment, left
T10e - 10-Pin Connector, orange, on protective housing for control module, in engine compartment, left
T15s - 15-Pin Connector, red, on protective housing for control module, in engine compartment,left
T31 - 31-Pin Connector, on ABS Control Module
T121 - 121-Pin Connector, on Motronic Engine Control Module (ECM)
V144 - Leak detection pump (LDP)

*** - only for vevicles with ASR

Motronic Engine Control Module (ECM), Leak detection pump (LDP), Throttle Valve Control Module

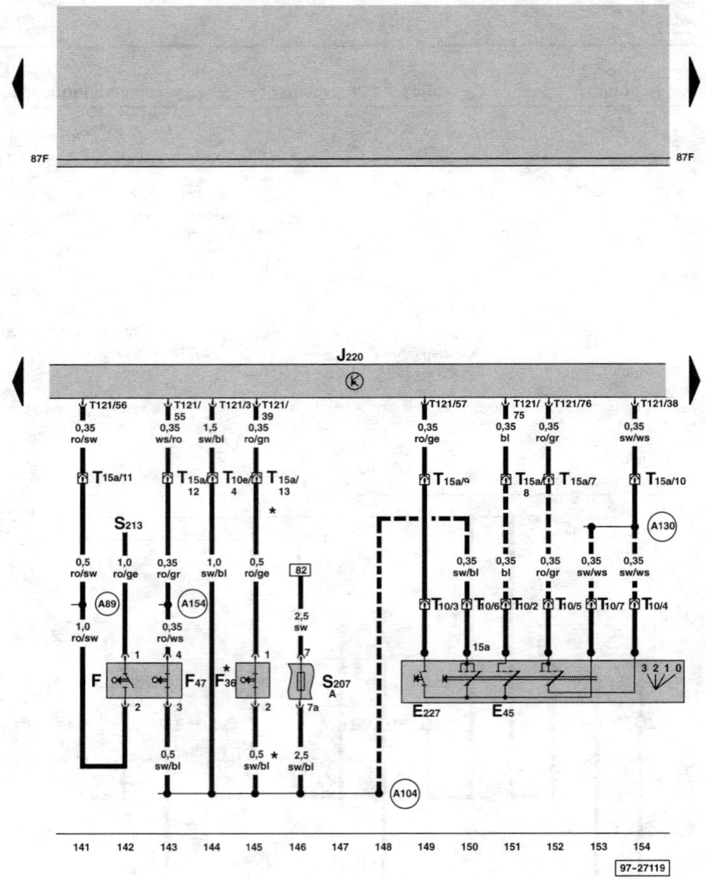

ws = white
sw = black
ro = red
br = brown
gn = green
bl = blue
gr = grey
li = violet
ge = yellow

E45 - Cruise Control Switch
E227 - Button for cruise control (set)
F - Brake Light Switch
F36 - Clutch Vacuum Vent Valve Switch
F47 - Brake Light Switch for Cruise Control Indicator
J220 - Motronic Engine Control Module (ECM)
S207 - Fuse 7 in Fuseholder
S213 - Fuse 13 in Fuseholder
T10 - 10-Pin Connector, black
T10e - 10-Pin Connector, orange, on protective housing for control module, in engine compartment, left
T15a - 15-Pin Connector, white
T121 - 121-Pin Connector, on Motronic Engine Control Module (ECM)

(A89) - Connector 2 (54) in instrument panel wiring harness
(A104) - Plus Connector -2- 15 in instrument panel wiring harness
(A130) - Connection for Cruise Control (instrument panel wiring harness)
(A154) - Brake Pedal Switch connection (in instrument panel wiring harness

* - manual transmission only
- - - - cruise control only

Motronic Engine Control Module (ECM), Cruise Control Switch, Brake Light Switch, Clutch Vacuum Vent Valve Switch

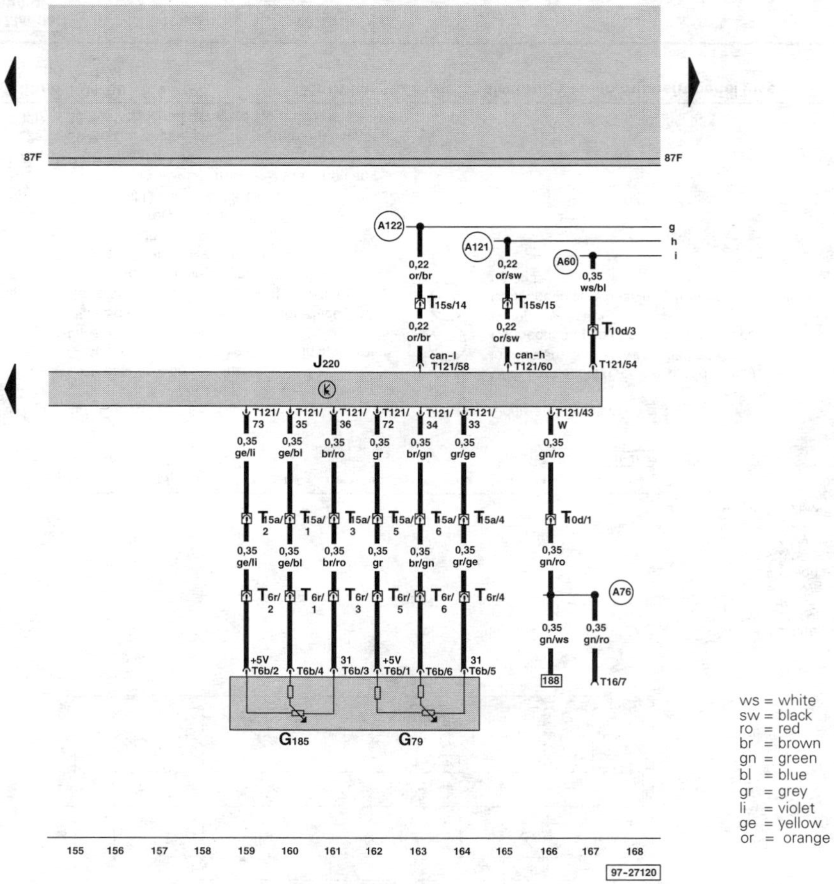

ws = white
sw = black
ro = red
br = brown
gn = green
bl = blue
gr = grey
li = violet
ge = yellow
or = orange

G79 - Throttle Position (TP) Sensor
G185 - Sender -2- for accelerator pedal position
J220 - Motronic Engine Control Module (ECM)
T6b - 6-Pin Connector, on Throttle Position Sensor
T6r - 6-Pin Connector, black
T10d - 10-Pin Connector, brown, on protective housing for control module, in engine compartment, left
T15a - 15-Pin Connector,white,on protective housing for control module, in engine compartment,left
T15s - 15-Pin Connector, red, on protective housing for control module, in engine compartment,left
T16 - 16-Pin Connector, Data Link Connector (DLC), below instrument panel, left
T121 - 121-Pin Connector, on Motronic Engine Control Module (ECM)

(A60) - Wire connection (Vehicle speed signal) , in instrument cluster wiring harness
(A76) - Connector (K-diagnosis wire) in instrument panel wiring harness
(A121) - Connection (high bus) in instrument panel wiring harness
(A122) - Connection (low bus) in instrument panel wiring harness

Motronic Engine Control Module (ECM), Throttle Position Sensor, Sender -2- for accelerator pedal position, Data Link Connector (DLC)

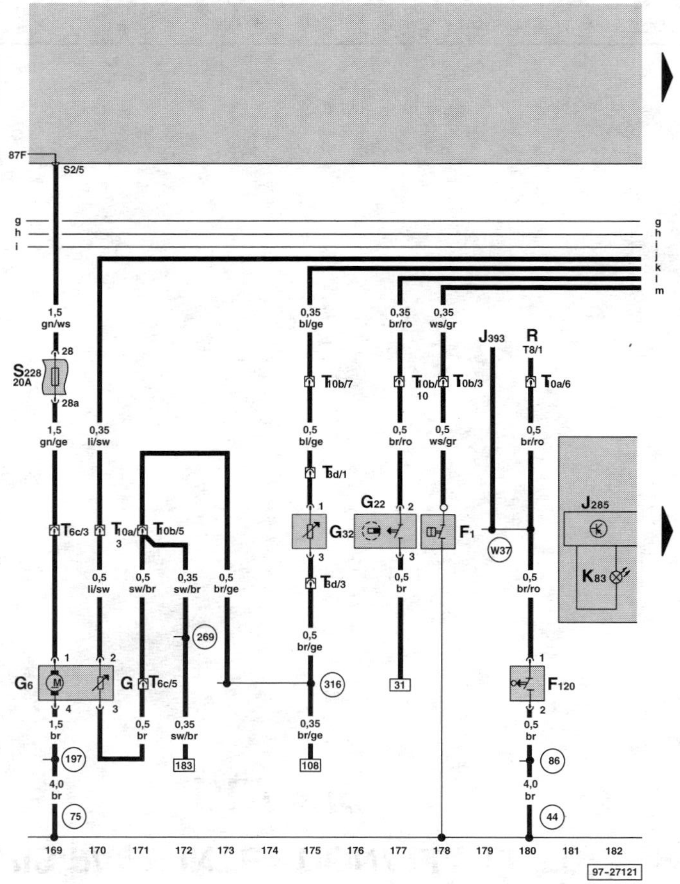

ws = white
sw = black
ro = red
br = brown
gn = green
bl = blue
gr = grey
li = violet
ge = yellow
or = orange

F1 - Oil Pressure Switch
F120 - Hood Alarm Switch
G - Sender for fuel gauge
G6 - Fuel Pump (FP)
G22 - Speedometer Vehicle Speed Sensor (VSS)
G32 - Engine Coolant Level (ECL) Sensor
J285 - Control module with indicator unit in instrument panel insert
J393 - Central control module for comfort system
K83 - Malfunction Indicator Lamp (MIL)
R - Radio
S228 - Fuse in fuse holder
T3d - 3-Pin Connector
T6c - 6-Pin Connector, blue
T8 - 8-Pin Connector, black
T10a - 10-Pin Connector, brown, connector station A-pillar, left
T10b - 10-Pin Connector, black, on protective housing for control module, in engine compartment, left

(44) - Ground connection, on left A-pillar, lower part
(75) - Ground connection, on right rear pillar
(86) - Ground connection -1-, in rear wiring harness
(197) - Ground connection -4-, in wiring harness rear
(269) - Ground connetor (sensor ground) -1-, in instrument panel wiring harness
(316) - Sensor ground connection 2 (in engine compartment wiring harness)
(W37) - Connection (DWA) in floor wiring harness

Oil Pressure Swich, Hood Alarm Switch, Sender for fuel gauge, Fuel Pump, Speedometer Vehicle Speed Sensor, Engine Coolant Level

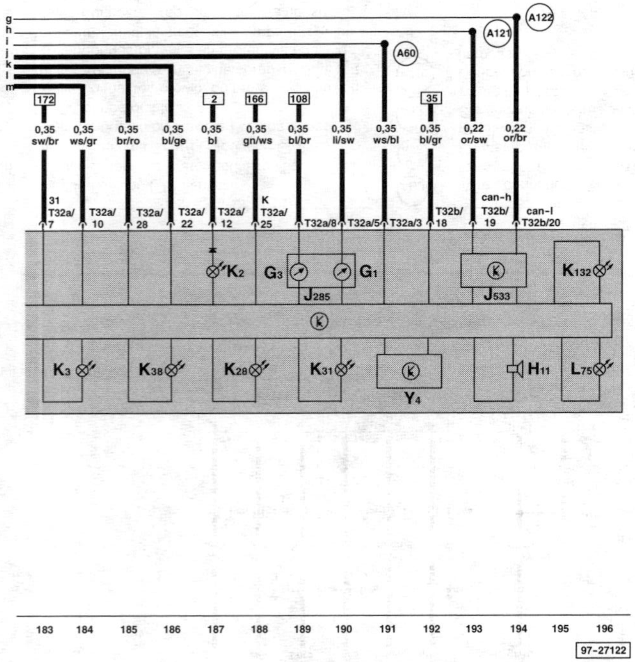

ws = white
sw = black
ro = red
br = brown
gn = green
bl = blue
gr = grey
li = violet
ge = yellow
or = orange

G1 - Fuel Gauge
G3 - Engine Coolant Temperature (ECT) Gauge
H11 - Oil Pressure Warning Buzzer
J285 - Control module with indicator unit in instrument panel insert
J533 - Data Bus On Board Diagnostic Interface
K2 - Generator (GEN) Warning Light
K3 - Oil Pressure Warning Light
K28 - Engine Coolant Level/Temperature (ECL/ECT) Warning Light
K31 - Cruise Control Indicator Light
K38 - Engine Oil Level Indicator Light
K132 - Electronic Power Control (EPC) Warning Lamp
L75 - Digital Display Light
T32a - 32-Pin Connector, blue, on instrument cluster
T32b - 32-Pin Connector, green, on instrument cluster

Y4 - Odometer Display

(A60) - Wire connection (Vehicle speed signal), in instrument cluster wiring harness
(A121) - Connection (high bus) in instrument panel wiring harness
(A122) - Connection (low bus) in instrument panel wiring harness

Control module with indicator unit in instrument panel insert

THIS PAGE INTENTIONALLY LEFT
BLANK

Multi-Function Steering Wheel

from May 2000

Deviate relay location and fuseplacements as well as the locations of multiple connectors see section "component locations"

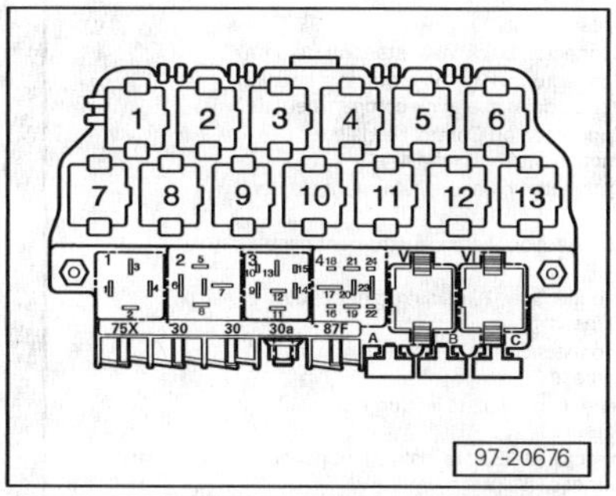

Relay location on the thirteen position auxiliary relay panel, above relay panel:

8 - Control Unit Multifunction Steering Wheel

9 - Control Unit Multifunction Steering Wheel

Relay panel:

1 - Dual Horn Relay (204)

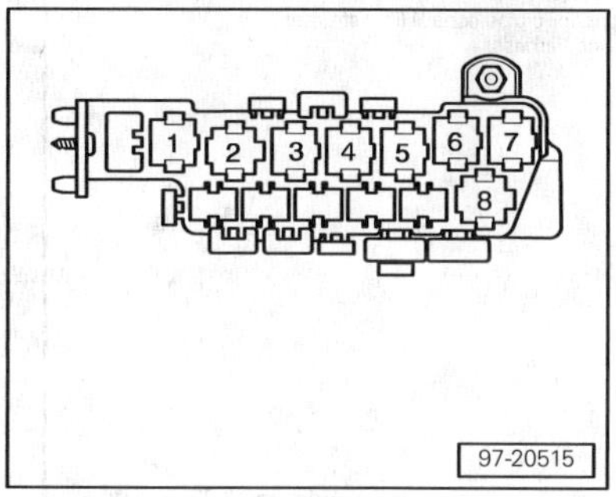

Relay location on the eight position auxiliary relay panel, above relay panel:

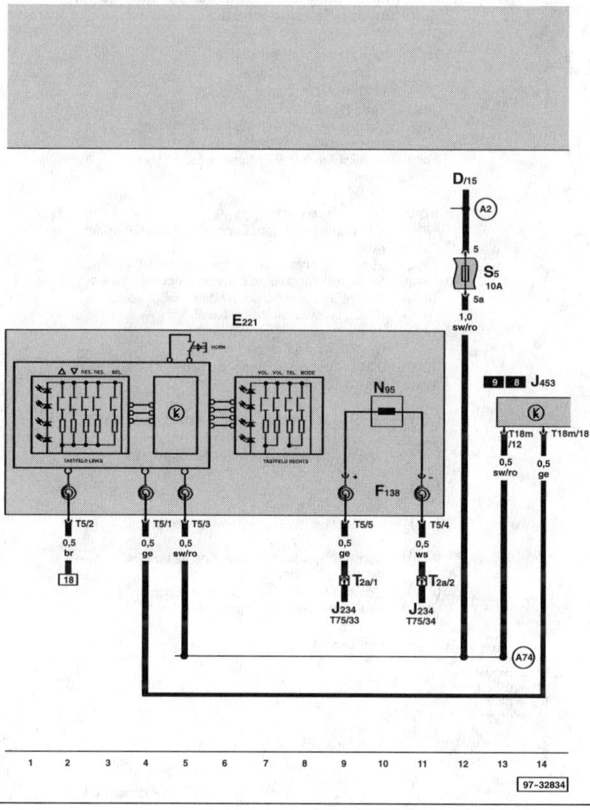

Control Unit in Steering Wheel, Control Module for Multi-Function Steering Wheel

- D - Ignition/Starter Switch
- E221 - Control unit in steering wheel
- F138 - Airbag Spiral Spring/Return Spring with Slip Ring
- J234 - Airbag Control Module
- J453 - Control Module for Multi-Function Steering wheel (450)
- N95 - Driver's Side Airbag Igniter
- S5 - Fuse
- T2a - Double Connector, yellow, Connector station A-pillar, left
- T5 - 5-Pin Connector, black, behind steering colum switch cover
- T18m- 18-Pin Connector, on Control Module for Multi-Function Steering wheel
- T75 - 75-Pin Connector, on Airbag Control Module
- (A2) - Plus Connection (15), in instrument panel wiring harness
- (A74) - Connection (15a, Fuse 5), in instrument panel wiring harness

ws = white
sw = black
ro = red
br = brown
gn = green
bl = blue
gr = grey
li = lilac
ge = yellow
or = orange

Control Module for Multi-Function Steering Wheel, Cruise Control Switch

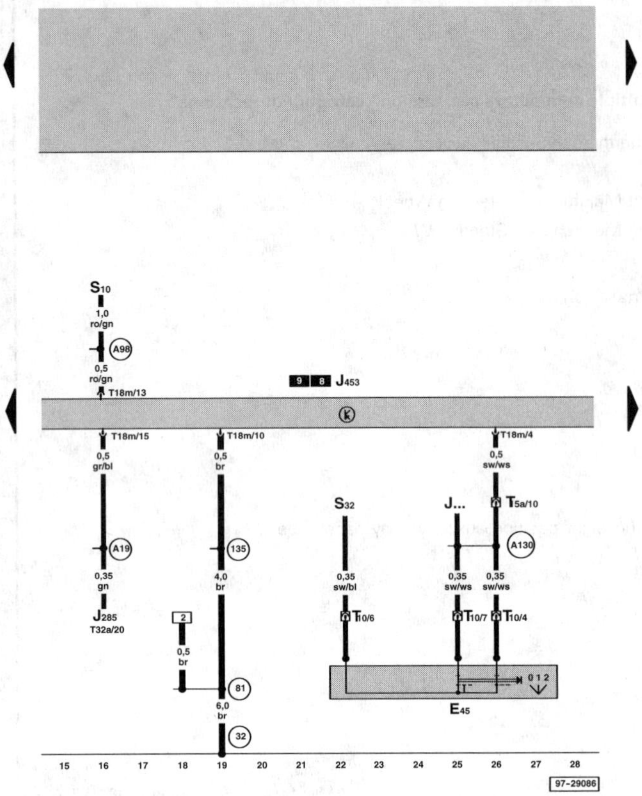

E45 - Cruise Control Switch
J... - Engine Control Module
J285 - Control Module with Indicator unit in instrument panel insert
J453 - Control Module for Multi-Function Steering wheel (450)
S10 - Fuse
S32 - Power Antenne Fuse
T10 - 10-Pin Connector, black, near steering Column
T15a - 15-Pin Connector, white, on protective housing for control module, in engine compartment, left
T18m- 18-Pin Connector, on Control Module for Multi-Function Steering wheel
T32a - 32-Pin Connector, blue, on instrument cluster

(32) - Ground connection, behind instrument panel

(81) - Ground connection -1-, in instrument panel wiring harness

(135) - Ground connection -2-, in instrument panel wiring harness

(A19) - Wire connection (58d), in instrument panel wiring harness

(A98) - Plus connector -4- (30), in instrument panel wiring harness

(A130) - Connection for cruise control (in instrument panel wiring harness)

ws = white
sw = black
ro = red
br = brown
gn = green
bl = blue
gr = grey
li = lilac
ge = yellow
or = orange

Edition 10.02

Signal horn/ dual-tone horn, Control Module for Multi-function Steering wheel

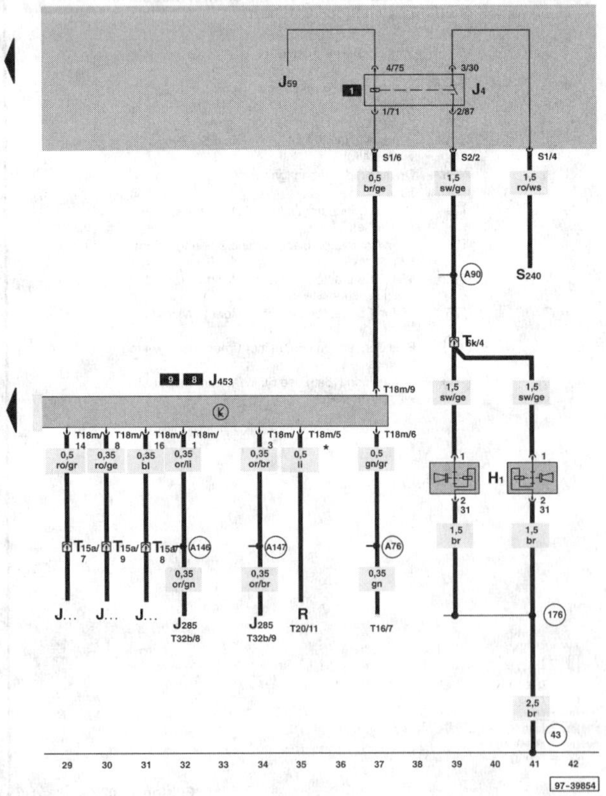

H1 - Signal Horn/dual Tone Horn
J... - Engine Control Module
J4 - Horn Relay
J59 - Load Reduction Relay
J285 - Control module with indicator unit in instrument panel insert
J453 - Control Module for Multi-Function Steering wheel
R - Radio
S240 - Fuse in fuse holder
T6k - 6-Pin Connector, red, Connector Station A-pillar, right
T15a - 15-Pin Connector, white, on protective housing for control module, in engine compartment, left
T16 - 16-Pin Connector, black, Data Link Connector (DLC), below instrument panel, left
T18m- 18-Pin Connector, on Control Module for Multi-Function Steering wheel
T20 - 20-Pin Connector, green, on Radio
T32b - 32-Pin Connector, green

(43) - Ground connection, on right A-pillar, lower part

(176) - Ground connection, in right headlight wiring harness

(A76) - Connection (K-diagnosis wire), in instrument panel wiring harness

(A90) - Connection (two-tone horn), in instrument panel wiring harness

(A146) - Comfort System High-bus Connection (in instrument panel wiring harness

(A147) - Comfort System Low-bus Connection (in instrument panel wiring harness

* - not with CAN-Bus

ws = white
sw = black
ro = red
br = brown
gn = green
bl = blue
gr = grey
li = lilac
ge = yellow
or = orange

Edition 04.04

Passat - Standard Equipment

from October 2000

Deviate relay location and fuseplacements as well as the locations of multiple connectors see section "component locations"

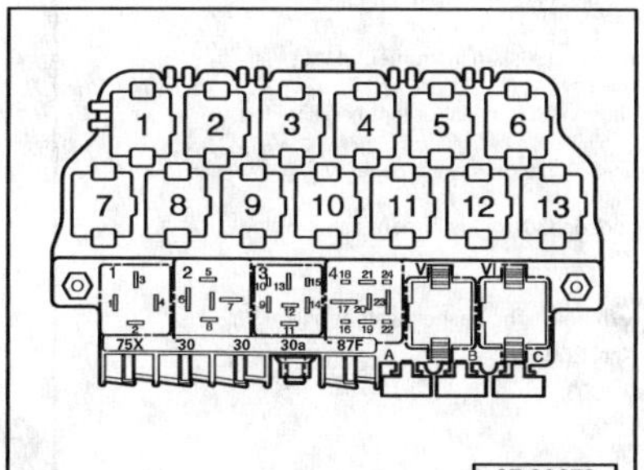

97-20676

Relay location on the thirteen position auxiliary relay panel, above relay panel:

- 5 - Daytime Running Lights Change-over Relay (173)
- 7 - Fog Light Relay (381)
- 12 - Starting Interlock Relay (alarm system) (185)
- 13 - Starting Interlock Relay (clutch system) (53)

Relay panel:

- 1 - Dual Horn Relay (53)
- 2 - Load Reduction Relay (213)
- 4 - Fuel Pump (FP) Relay (372)
- V - Wiper/Washer Intermittent Relay (377)
- VI - Wiper/Washer Intermittent Relay (377)

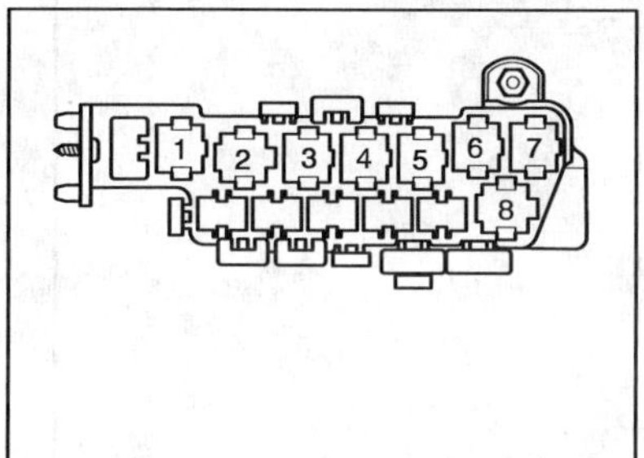

97-20515

Relay location on the eight position auxiliary relay panel, above relay panel:

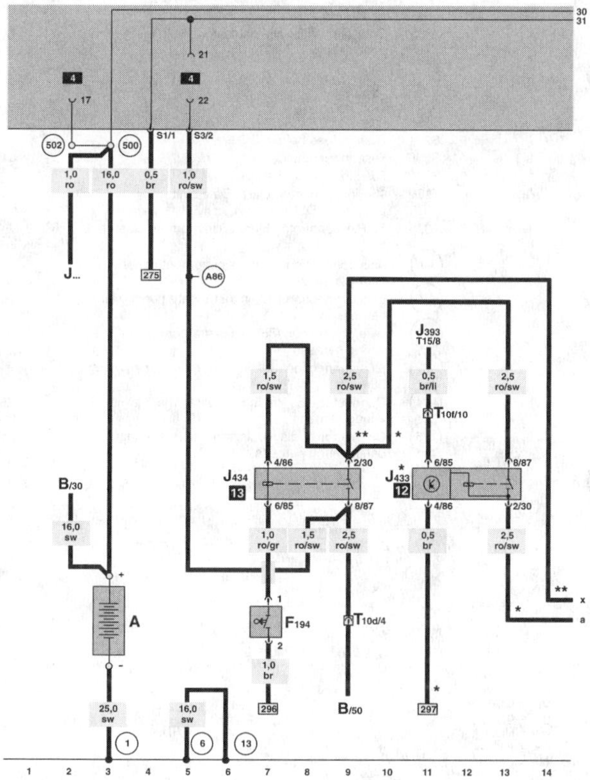

Locking relay for starter, Battery, Starter

- A - Battery
- B - Starter
- F194 - Clutch Pedal Position (CPP) Switch
- J... - Engine control module (ECM)
- J393 - Central control module for comfort system
- J433 - Locking relay for starter (anti - theft immobilizer)
- J434 - Locking relay for starter (clutch pedal switch)
- T10d - 10-Pin Connector, brown, on protective housing for control module, in engine compartment, left
- T10f - 10-Pin Connector, blue, connector station A-pillar, left
- T15 - 15-Pin Connector, on central control module for comfort system

- (1) - Ground strap, battery to body
- (6) - Ground strap, engine to generator
- (13) - Ground connection, in engine compartment, right
- (500) - Threaded connection -1- (30), on relay plate
- (502) - Threaded connection (30a), on relay plate
- (A86) - Connection (50a) in instrument panel wiring harness

- * - through April 2003
- ** - from May 2003

ws = white
sw = black
ro = red
br = brown
gn = green
bl = blue
gr = grey
li = lilac
ge = yellow
or = orange

Passat

Wiring diagram

No. 48/3

Ignition/Starter Switch

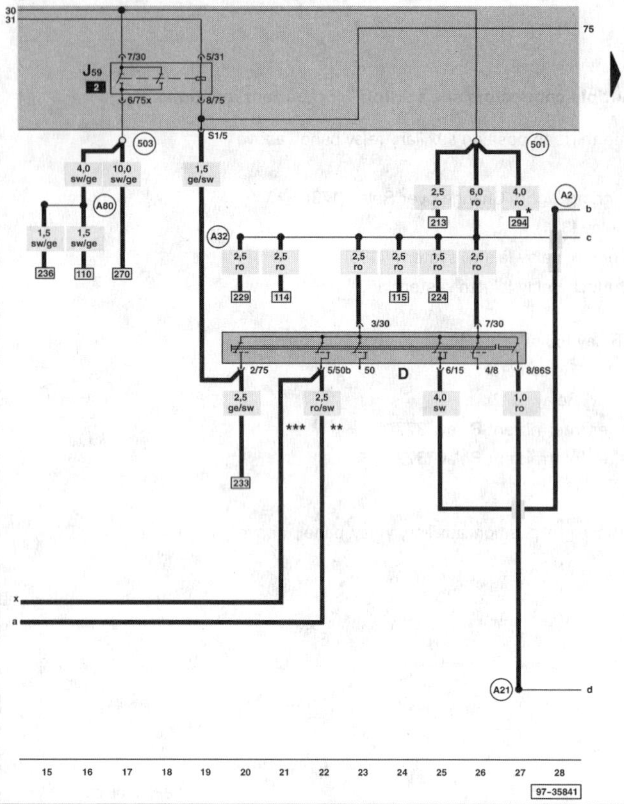

D - Ignition/Starter Switch
J59 - Load Reduction Relay

(501) - Threaded connection -2- (30), on relay plate
(503) - Bolted connection (75x) (at relay panel)
(A2) - plus connection (15), in instrument panel wiring harness
(A21) - wire connection (86s), in instrument panel wiring harness
(A32) - plus connection (30), in instrument panel wiring harness
(A80) - Connector -1- (X) in instrument panel wiring harness
(A32) - plus connection (30), in instrument panel wiring harness

* - models with headlight washer system only
** - through April 2003
*** - from May 2003

ws = white
sw = black
ro = red
br = brown
gn = green
bl = blue
gr = grey
li = lilac
ge = yellow
or = orange

Edition 06.03

Passat

Wiring diagram

No. 48/4

Instrument cluster, Data Link Connector (DLC)

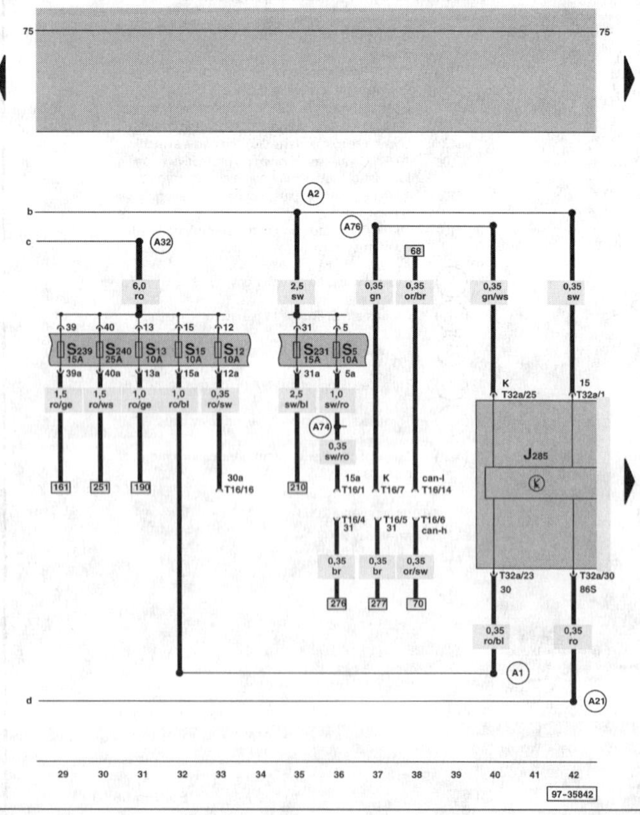

J285 - Control module with indicator unit in instrument panel insert
S5 - Fuse
S12 - Fuse
S13 - Fuse
S15 - Fuse
S231 - Fuse in fuse holder
S239 - Fuse in fuse holder
S240 - Fuse in fuse holder
T16 - 16-Pin Connector, black, Data Link Connector(DLC)
T32a - 32-Pin Connector, blue, on instrument cluster

(A1) - plus connection (30a), in instrument panel wiring harness
(A2) - plus connection (15), in instrument panel wiring harness
(A21) - wire connection (86s), in instrument panel wiring harness
(A32) - plus connection (30), in instrument panel wiring harness
(A74) - Connector (15a - fuse 5), in instrument panel wiring harness
(A76) - Connector (K-diagnosis wire), in instrument panel wiring harness

ws = white
sw = black
ro = red
br = brown
gn = green
bl = blue
gr = grey
li = lilac
ge = yellow
or = orange

Edition 06.03

Passat — Wiring diagram — No. 48/5

Instrument cluster, Tachometer, Speedometer Vehicle Speed Sensor (VSS), Malfunction Indicator Lamp (MIL), Seat Belt Warning System

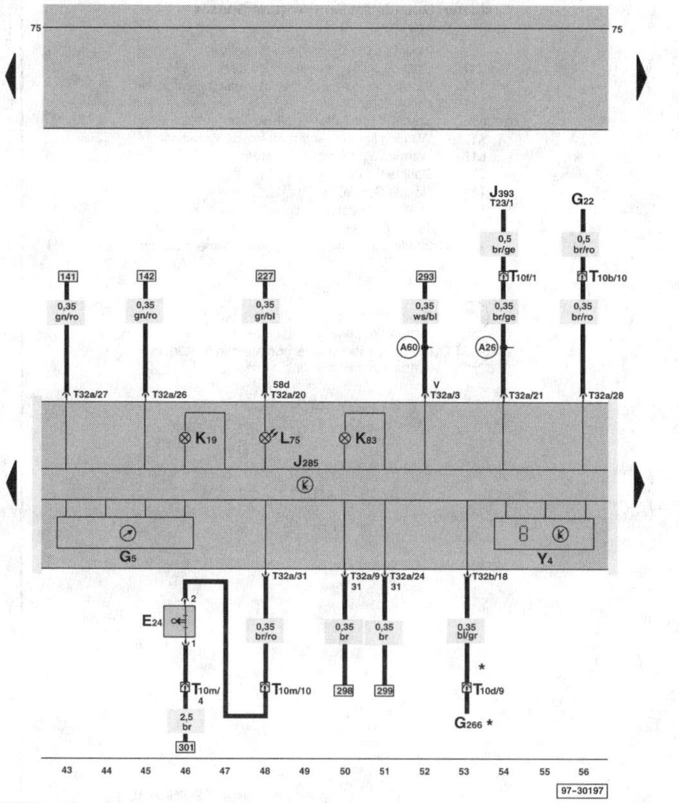

- E24 - Left Seat Belt Switch
- G5 - Tachometer
- G22 - Speedometer Vehicle Speed Sensor (VSS)
- G266 - Oil Level Thermal Sensor
- J285 - Control module with indicator unit in instrument panel insert
- J393 - Central control module for comfort system
- K19 - Seat Belt Warning Light
- K83 - Malfunction Indicator Lamp (MIL)
- L75 - Digital Display Light
- T10b - 10-Pin Connector, black, on protective housing for control module, in engine compartment, left
- T10d - 10-Pin Connector, brown, on protective housing for control module, in engine compartment, left
- T10f - 10-Pin Connector, blue, connector station A-pillar, left
- T10m - 10-Pin Connector, brown
- T23 - 23-Pin Connector, on central control module for comfort system
- T32a - 32-Pin Connector, blue, on instrument cluster
- T32b - 32-Pin Connector, green, on instrument cluster
- Y4 - Odometer Display
- (A26) - wire connection (driver's door contact switch), in instrument panel wiring harness
- (A60) - wire connection (Vehicle speed signal), in instrument cluster wiring harness
- * - discontinued from June 2001

ws = white
sw = black
ro = red
br = brown
gn = green
bl = blue
gr = grey
li = lilac
ge = yellow
or = orange

Edition 06.03

Passat — Wiring diagram — No. 48/6

Instrument cluster, Generator Warning Light

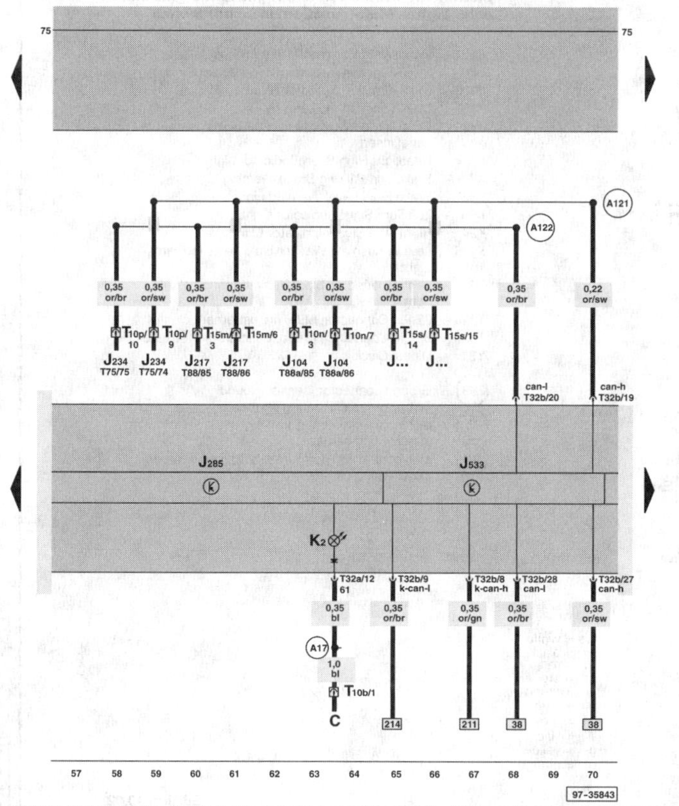

- C - Generator (GEN)
- J... - Engine Control Module (ECM)
- J104 - ABS Control Module (w/EDL)
- J217 - Transmission Control Module (TCM)
- J234 - Airbag Control Module
- J285 - Control module with indicator unit in instrument panel insert
- J533 - Data Bus On Board Diagnostic Interface
- K2 - Generator (GEN) Warning Light
- T10b - 10-Pin Connector, black, on protective housing for control module, in engine compartment, left
- T10n - 10-Pin Connector,
- T10p - 10-Pin Connector, yellow
- T15m - 15-Pin Connector,
- T15s - 15-Pin Connector, red, on protective housing for control module, in engine compartment,left
- T32a - 32-Pin Connector, blue, on instrument cluster
- T32b - 32-Pin Connector, green, on instrument cluster
- T75 - 75-Pin Connector
- T88 - Connector, 88 point, on Transmission Control Module (TCM)
- T88a - Connector, (outside temperature indicator) , in instrument panel wiring harness
- (A17) - wire connection (61), in instrument panel wiring harness
- (A121) - Connector (high-Bus), in instrument panel wiring harness
- (A122) - Connector (low-Bus), in instrument panel wiring harness

ws = white
sw = black
ro = red
br = brown
gn = green
bl = blue
gr = grey
li = lilac
ge = yellow
or = orange

Edition 06.03

Induction coil of antitheft immonilizer, Instrument cluster, Outside Air Temperature Sensor

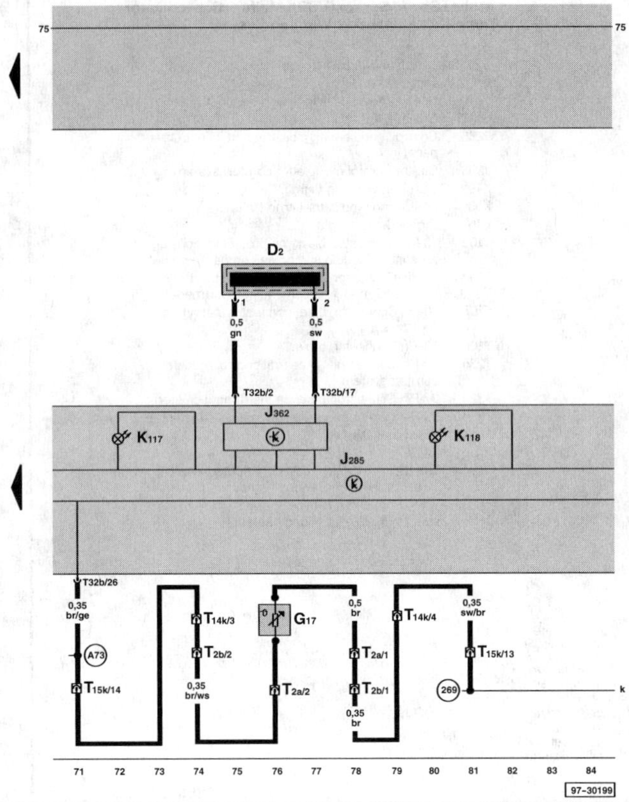

D2 - Induction coil of antiheft immonilizer
G17 - Outside Air Temperature Sensor
J285 - Control module with indicator unit in instrument panel insert
J362 - Control Module for Anti-theft Immobilizer
K117 - Warning light for anti-theft imobilizer sensor
K118 - Warning light for brake system
T2a - Double Connector
T2b - Double Connector
T14k - 14-Pin Connector, black
T15k - 15-Pin Connector, grey
T32b - 32-Pin Connector, green, on instrument cluster

(269) - Ground connector (sensor ground) -1-, in instrument panel wiring harness
(A73) - Connector (outside temperature indicator), in instrument panel wiring harness

ws = white
sw = black
ro = red
br = brown
gn = green
bl = blue
gr = grey
li = lilac
ge = yellow
or = orange

Edition 10.02

Instrument cluster, Left and Right Turn Signal Indicator Lights, Multi-Function, Indicator Switch

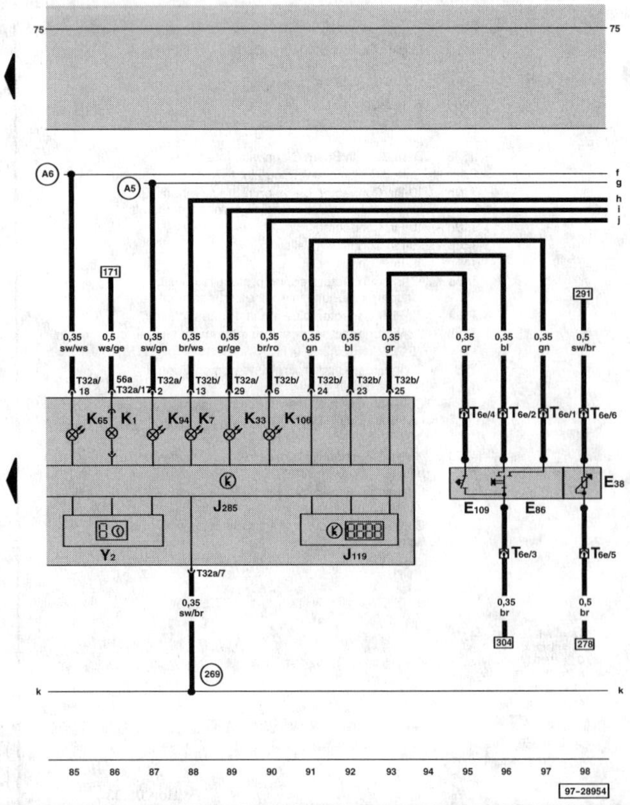

E38 - Windshield Wiper Intermittent Regulator
E86 - Multi-Function Indicator Mode Select Switch
E109 - Multi-Function Indicator Memory Switch
J119 - Multi-Function Indicator (MFA)
J285 - Control module with indicator unit in instrument panel insert
K1 - Headlight High Beam Indicator Light
K7 - Brake and Parking Brake Warning Light
K33 - Brake Fluid Level Warning Light
K65 - Left Turn Signal Indicator Light
K94 - Right Turn Signal Indicator Light
K106 - Low Windshield Washer Fluid Level Indicator Light
T6e - 6-Pin Connector, black, near steering column, right
T32a - 32-Pin Connector, blue, on instrument cluster
T32b - 32-Pin Connector, green, on instrument cluster
Y2 - Digital Clock

(269) - Ground connector (sensor ground) -1-, in instrument panel wiring harness
(A5) - plus connection (right turn signal), in instrument panel wiring harness
(A6) - plus connection (left turn signal), in instrument panel wiring harness

ws = white
sw = black
ro = red
br = brown
gn = green
bl = blue
gr = grey
li = lilac
ge = yellow
or = orange

Edition 10.02

Windshield Washer Fluid Level Sensor, Daytime Running Lights Change-over Relay, Parking Brake Warning Light Switch and Brake Fluid Level Warning Light Switch

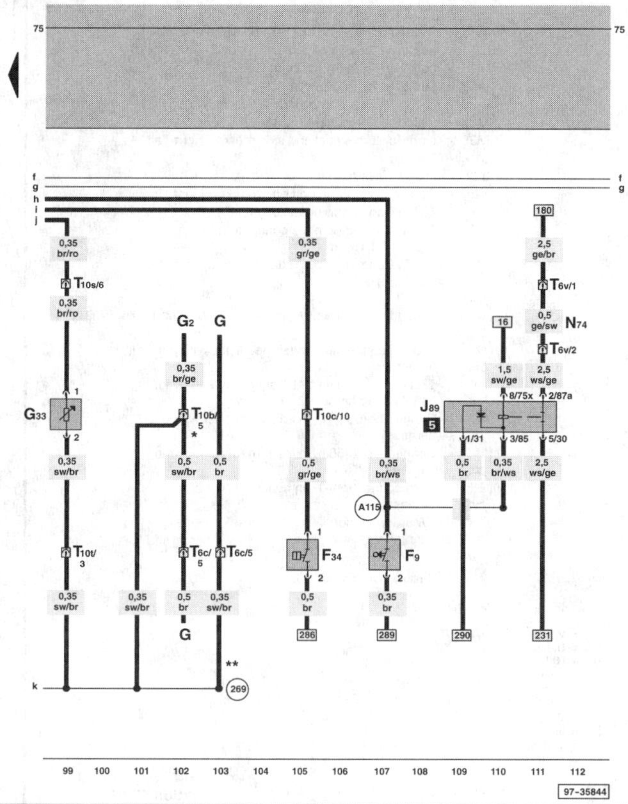

- F9 - Parking Brake Warning Light Switch
- F34 - Brake Fluid Level Warning Switch
- G - Sender for fuel gauge
- G2 - Engine Coolant Temperature (ECT) Sensor
- G33 - Windshield Washer Fluid Level Sensor
- J89 - Daytime Running Lights Change-over Relay
- N74 - Series Resistance Wiring
- T6c - 6-Pin Connector, blue, connector station A-pillar, left
- T6v - 6-Pin Connector, violet, connector station A-pillar, right
- T10b - 10-Pin Connector, black, on protective housing for control module, in engine compartment, left
- T10c - 10-Pin Connector, violet, connector station A-pillar, left
- T10s - 10-Pin Connector,rose, connector station A-pillar, left
- T10t - 10-Pin Connector, red, connector station A-pillar, left

- (269) - Ground connector (sensor ground) -1-, in instrument panel wiring harness
- (A115) - Connection (parking brake control), in instrument panel wiring harness

- * - through April 2003
- ** - from May 2003

ws = white
sw = black
ro = red
br = brown
gn = green
bl = blue
gr = grey
li = lilac
ge = yellow
or = orange

Turn Signal Switch, Headlight Dimmer/Flasher Switch

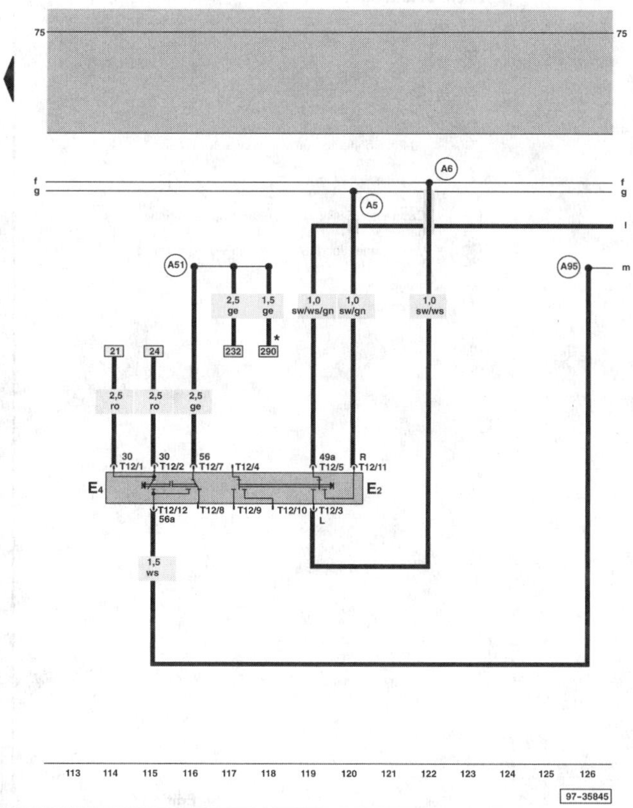

- E2 - Turn Signal Switch
- E4 - Headlight Dimmer/Flasher Switch
- T12 - 12-Pin Connector, near steering column

- (A5) - plus connection (right turn signal), in instrument panel wiring harness
- (A6) - plus connection (left turn signal), in instrument panel wiring harness
- (A51) - wire connection (56), in instrument panel wiring harness
- (A95) - Connection 1 (56a), in instrument panel wiring harness

- * - models with headlight washer system only

ws = white
sw = black
ro = red
br = brown
gn = green
bl = blue
gr = grey
li = lilac
ge = yellow
or = orange

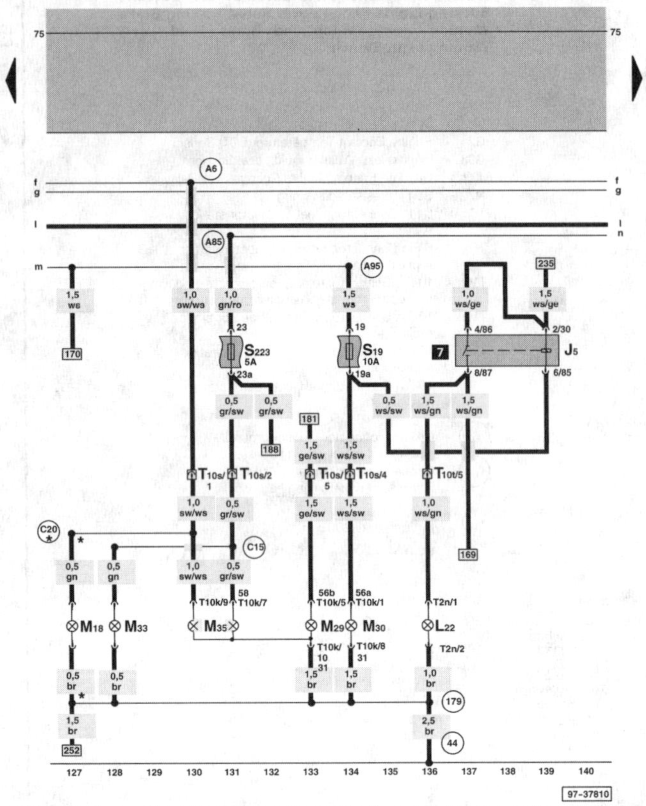

Left Front Fog Light, Fog Light Relay

J5 - Fog Light Relay, on auxiliary relay panel
L22 - Left Front Fog Light
M18 - Left, Side Turn Signal Light
M29 - Left Low Beam Headlight
M30 - Left High Beam Headlight
M33 - Light for side marker front left
M35 - Light for turn signal and side marker, front, left
S19 - Fuse
S223 - Fuse in fuse holder
T2n - Double Connector, on Left,Front Fog Light
T10k - 10-Pin Connector, on headlight left
T10s - 10-Pin Connector, pink, connector station A-pillar, left
T10t - 10-Pin Connector, red, connector station A-pillar, left

(44) - Ground connection, on left A-pillar, lower part
(179) - Ground connection, in left headlight wiring harness
(A6) - plus connection (left turn signal), in instrument panel wiring harness
(A85) - Connector (58R), in Instrument panel wiring harness
(A95) - Connection 1 (56a), in instrument panel wiring harness
(C15) - Wire connection -1-, in headlight wiring harness
(C20) - Wire Connection (left turn signal), in headlight wiring harness

* - through October 2003

ws = white
sw = black
ro = red
br = brown
gn = green
bl = blue
gr = grey
li = lilac
ge = yellow
or = orange

Edition 12.03

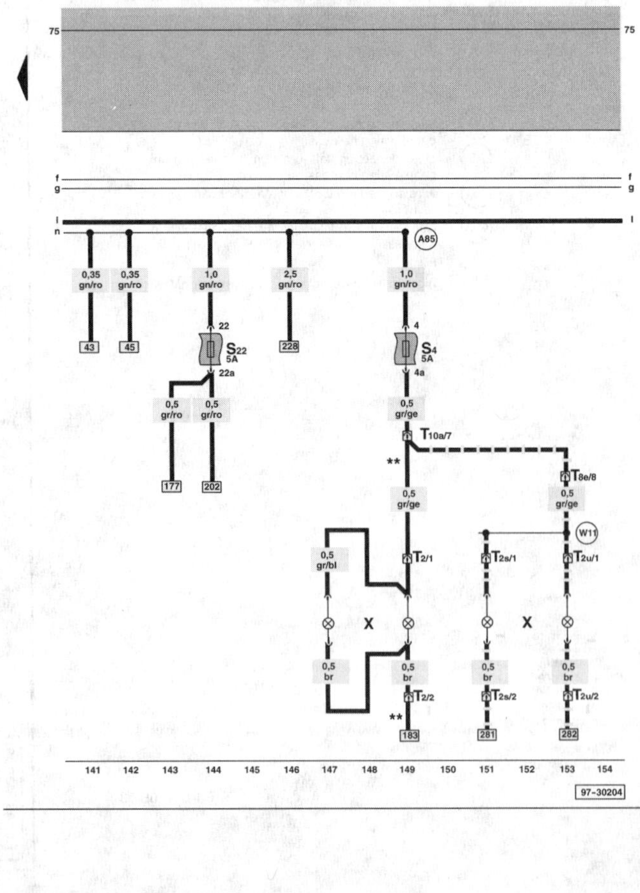

License Plate Light

S4 - Fuse
S22 - Fuse
T2 - Double Connector, in rear lid
T2s - Double Connector, black
T2u - Double Connector, black
T8e - 8-Pin Connector, black, in rear lid, right
T10a - 10-Pin Connector, brown, connector station A-pillar, left
X - License Plate Light

(A85) - Connector (58R), in Instrument panel wiring harness
(W11) - wire connection (58), in rear lid wiring harness

- - - - wagon only
** - sedan only

ws = white
sw = black
ro = red
br = brown
gn = green
bl = blue
gr = grey
li = lilac
ge = yellow
or = orange

Edition 06.03

Passat — Wiring diagram — No. 48/13

Emergency Flasher Switch

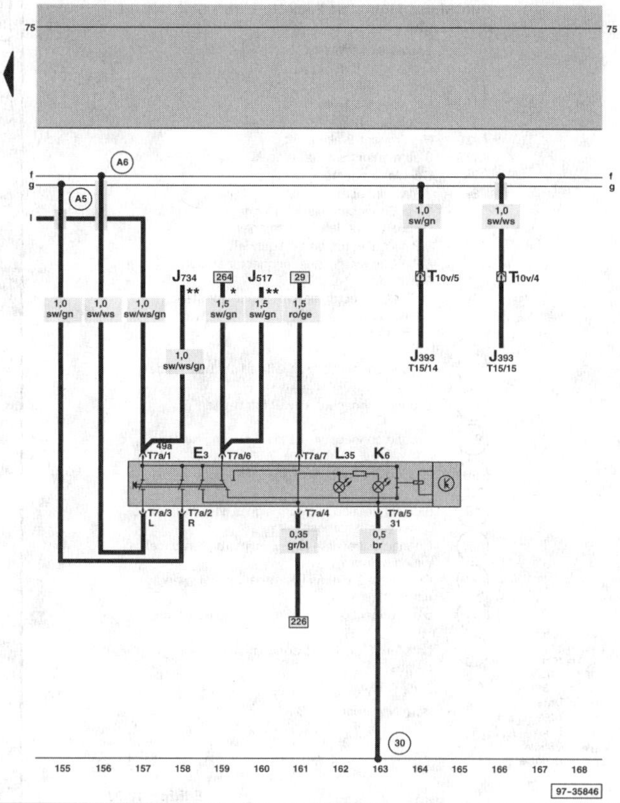

- E3 - Emergency Flasher Switch
- J393 - Central control module for comfort system
- J517 - Auxiliary Emergency Flasher Relay **
- J734 - Warning Lamp Auxiliary Relay **
- K6 - Indicator light for emergency flasher system
- L35 - Emergency Flasher Switch Light
- T7a - 7-Pin Connector, on emergency flasher switch
- T10v - 10-Pin Connector, violet, connector station A-pillar, right
- T15 - 15-Pin Connector, on central control module for comfort system

- (30) - Ground connection, -1-, beside fuse/relay panel
- (A5) - plus connection (right turn signal), in instrument panel wiring harness
- (A6) - plus connection (left turn signal), in instrument panel wiring harness

- * - through April 2003
- ** - from May 2003

ws = white
sw = black
ro = red
br = brown
gn = green
bl = blue
gr = grey
li = lilac
ge = yellow
or = orange

97-35846

Edition 06.03

Passat — Wiring diagram — No. 48/14

Right Front Headlights, Right Front Fog Light

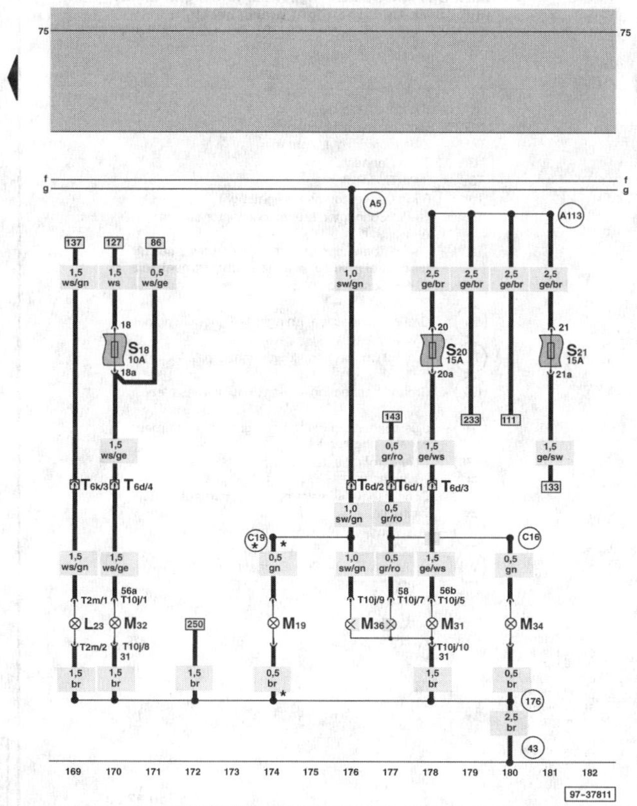

- L23 - Right Front Fog Light
- M19 - Right, Side Turn Signal Light
- M31 - Right Low Beam Headlight
- M32 - Right High Beam Headlight
- M34 - Light for side marker front left
- M36 - Light for turn signal and side marker, front, right
- S18 - Fuse
- S20 - Fuse
- S21 - Fuse
- T2m - Double Connector
- T6d - 6-Pin Connector, pink, connector station A-pillar, right
- T6k - 6-Pin Connector, red, connector station A-pillar, right
- T10j - 10-Pin Connector

- (43) - Ground connection, on right A-pillar, lower part
- (176) - Ground connection, in right headlight wiring harness
- (A5) - plus connection (right turn signal), in instrument panel wiring harness
- (A113) - Connection 2 (daytime running lights), in instrument panel wiring harness
- (C16) - Wire connection -2-, in headlight wiring harness
- (C19) - Connection -1- (right turn signal), in headlight wiring harness

- * - through October 2003

ws = white
sw = black
ro = red
br = brown
gn = green
bl = blue
gr = grey
li = lilac
ge = yellow
or = orange

97-37811

Edition 12.03

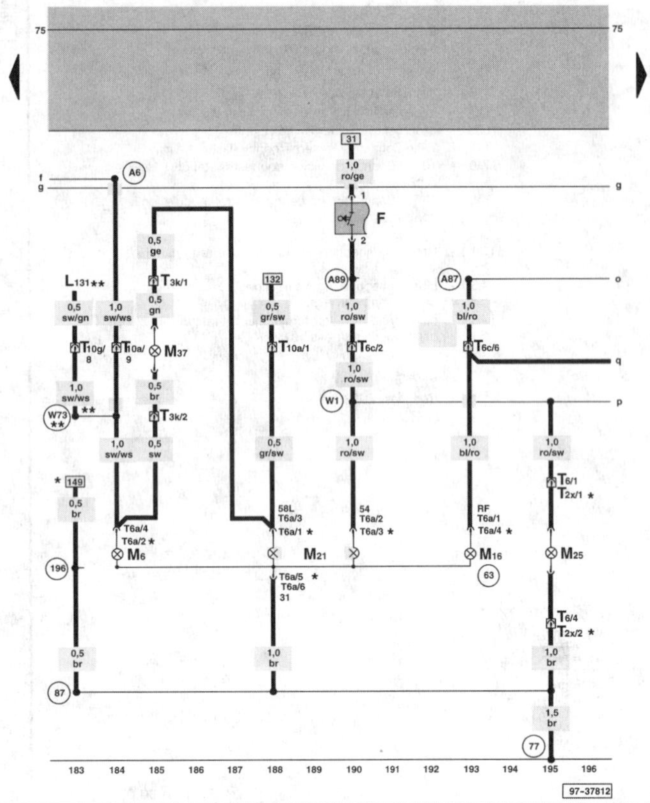

Brake Light Switch, Left Brake/Tail Light, Left Rear Turn Signal Light, left Back-Up Light, High-mount Brake Light

- F - Brake Light Switch
- L131 - Left Mirror Turn Signal Lamp
- M6 - Left Rear Turn Signal Light
- M16 - Left Back-Up Light
- M21 - Left Brake/Tail Light
- M25 - High-mount Brake Light
- M37 - Side marker light, left rear
- T2x - Double Connector
- T3k - 3-Pin Connector
- T6 - 6-Pin Connector, black, in luggage compartment, left (wagon only)
- T6a - 6-Pin Connector, on tail light, left
- T6c - 6-Pin Connector, blue, connector station A-pillar, left
- T10a - 10-Pin Connector, brown, connector station A-pillar, left
- T10g - 10-Pin Connector
- (63) - Ground connection, on left taillight bulb holder
- (77) - Ground connection (lower left B-pillar)
- (87) - Ground connection -2-, in rear wiring harness
- (196) - Ground connection -3-, in wiring harness, rear
- (A6) - plus connection (left turn signal), in instrument panel wiring harness
- (A87) - Connector (reverse lamp), in instrument panel wiring harness
- (A89) - Connector 2 (54) (in instrument panel wiring harness)
- (W1) - plus connection (54), in rear wiring harness
- (W73) - Left Turn Signal Plus Connection (in floor wiring harness)
- * - sedan only
- ** - from November 2003

ws = white
sw = black
ro = red
br = brown
gn = green
bl = blue
gr = grey
li = lilac
ge = yellow
or = orange

97-37812

Edition 12.03

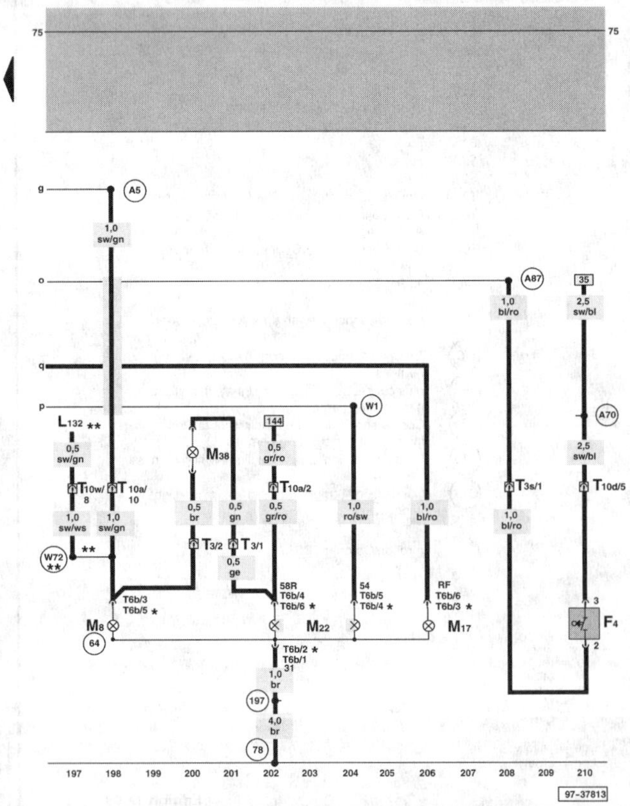

Back-Up Light Switch, Right Rear Turn Signal Light, Right Back-Up Light, Right Brake/Tail Light

- F4 - Back-Up Light Switch
- L132 - Right Mirror Turn Signal Lamp
- M8 - Right Rear Turn Signal Light
- M17 - Right Back-Up Light
- M22 - Right Brake/Tail Light
- M38 - Side marker light, right rear
- T3 - 3-Pin Connector
- T3s - 3-Pin Connector
- T6b - 6-Pin Connector, on tail light, right
- T10a - 10-Pin Connector, brown, connector station A-pillar, left
- T10d - 10-Pin Connector, brown, on protective housing for control module, in engine compartment, left
- T10w - 10-Pin Connector, black
- (64) - Ground connection, on right taillight bulb holder
- (78) - Ground Connection, (lower right B-pillar)
- (197) - Ground connection -4-, in wiring harness rear
- (A5) - plus connection (right turn signal), in instrument panel wiring harness
- (A70) - Connector (15a, fuse 231), in instrument panel wiring harness
- (A87) - Connector (reverse lamp), in instrument panel wiring harness
- (W1) - plus connection 1 (54), in rear wiring harness
- (W72) - Right Turn Signal Plus Connection (in floor wiring harness)
- * - sedan only
- ** - from November 2003

ws = white
sw = black
ro = red
br = brown
gn = green
bl = blue
gr = grey
li = lilac
ge = yellow
or = orange

97-37813

Edition 12.03

Passat

Wiring diagram

No. 48/17

Radio connection, Cigarette Lighter, Panel Vent Illumination

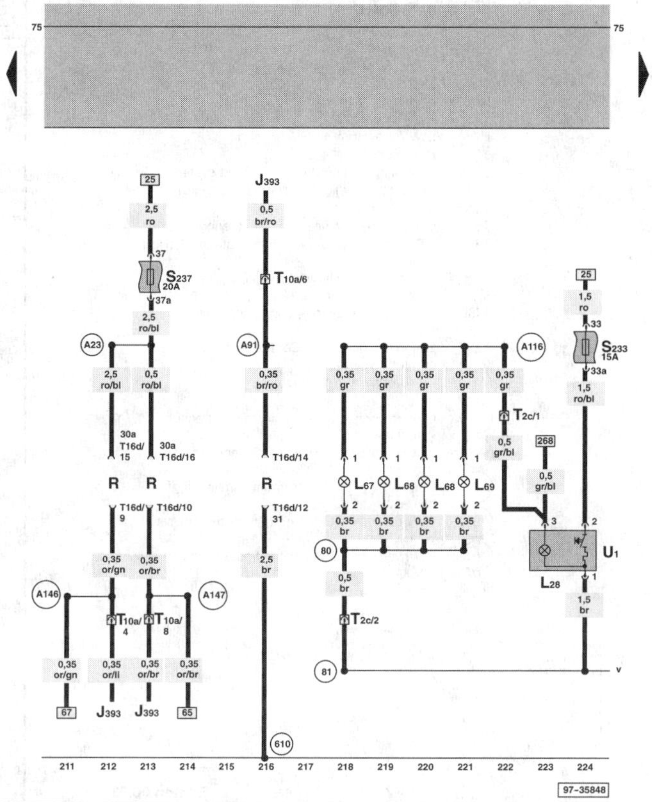

J393 - Central control module for comfort system
L28 - Cigarette Lighter Light
L67 - Left Instrument Panel Vent Illumination
L68 - Center Instrument Panel Vent Illumination
L69 - Right Instrument Panel Vent Illumination
R - Radio
S233 - Fuse in fuse holder
S237 - Fuse in fuse holder
T2c - Double Connector (behind glove box, right side)
T10a - 10-Pin Connector, brown, connector station A-pillar, left
T16d - 16-Pin Connector, black, on radio
U1 - Cigarette Lighter

(80) - Ground connection -1-, in instrument panel wiring harness
(81) - Ground connection -1-, in instrument panel wiring harness
(610) - Ground connection (Audio) (under center console, front)
(A23) - wire connection (30al), in instrument panel wiring harness
(A91) - Connector (signal for instrument panel insert), in instrument panal wiring harness
(A116) - Connection -2- (58d), in instrument panel wiring harness
(A146) - Comfort System High-bus Connection (in instrument panal wiring harness)
(A147) - Comfort System Low-bus Connection (in instrument panal wiring harness)

ws = white
sw = black
ro = red
br = brown
gn = green
bl = blue
gr = grey
li = lilac
ge = yellow
or = orange

Edition 06.03

Passat

Wiring diagram

No. 48/18

Light switch

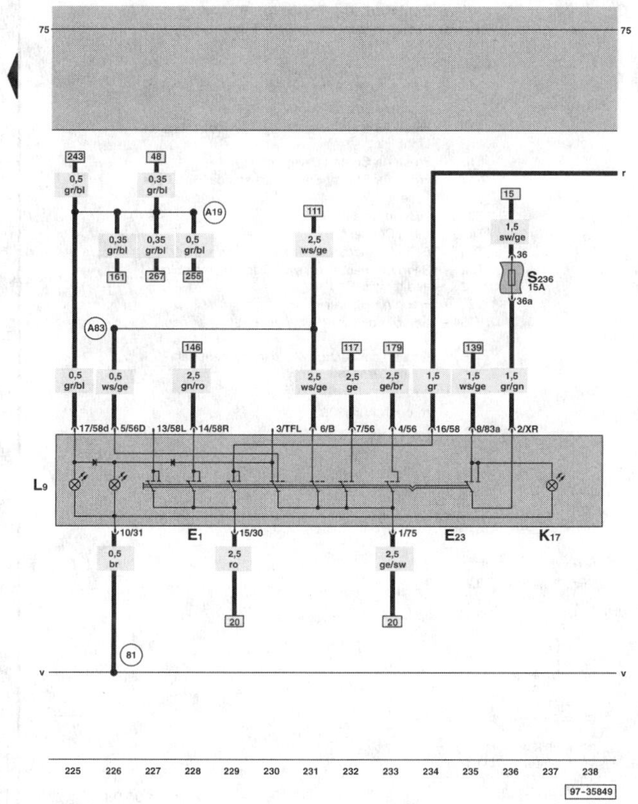

E1 - Light switch
E23 - Fog Light Switch
K17 - Fog Light Indicator Light
L9 - Headlight Switch Licht
S236 - Fuse in fuse holder

(81) - Ground connection -1-, in instrument panel wiring harness
(A19) - wire connection (58d), in instrument panal wiring harness
(A83) - Connector (daytime running lights), in instrument panel wiring harness

ws = white
sw = black
ro = red
br = brown
gn = green
bl = blue
gr = grey
li = lilac
ge = yellow
or = orange

Edition 06.03

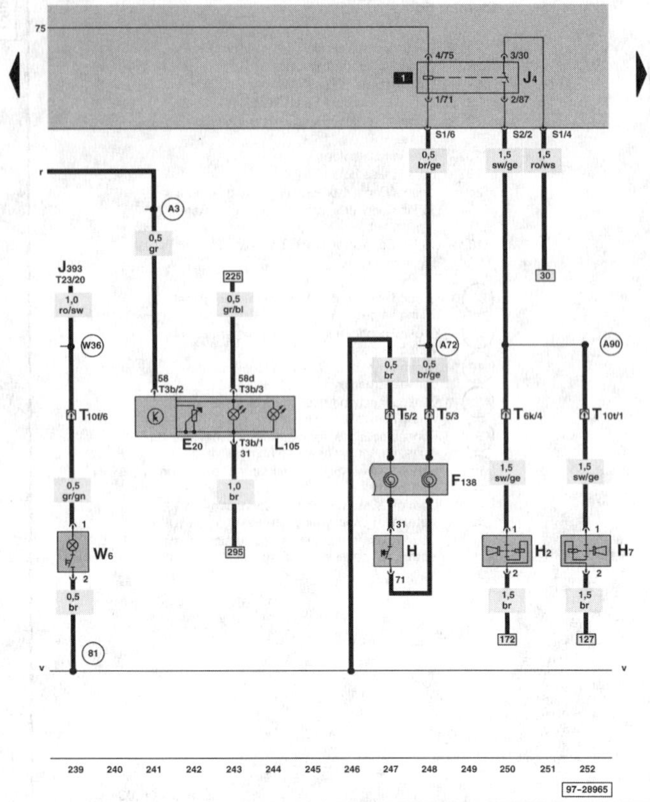

Instrument Panel Light Dimmer Switch, Dual Horns, Dual Horn Relay

- E20 - Instrument Panel Light Dimmer Switch
- F138 - Airbag Spiral Spring/Return Spring With Slip Ring
- H - Signal horn activation
- H2 - High Tone Horn
- H7 - Low Tone Horn
- J4 - Horn Relay
- J393 - Central control module for comfort system
- L105 - Illumination for lighting controller
- T3b - 3-Pin Connector
- T5 - 5-Pin Connector, yellow, near steering column
- T6k - 6-Pin Connector, red, connector station A-pillar, right
- T10f - 10-Pin Connector, blue, connector station A-pillar, left
- T10t - 10-Pin Connector, red, connector station A-pillar, left
- T23 - 23-Pin Connector, on Central control module for comfort system
- W6 - Glove Compartment Light

- (81) - Ground connection -1-, in instrument panel wiring harness
- (A3) - plus connection (58), in instrument panel wiring harness
- (A72) - Connector (71), in instrument panel wiring harness
- (A90) - Connector (two-tone horn), in instrument panel wiring harness
- (W36) - Positive connection (30a) in floor wiring harness

ws = white
sw = black
ro = red
br = brown
gn = green
bl = blue
gr = grey
li = lilac
ge = yellow
or = orange

Edition 06.03

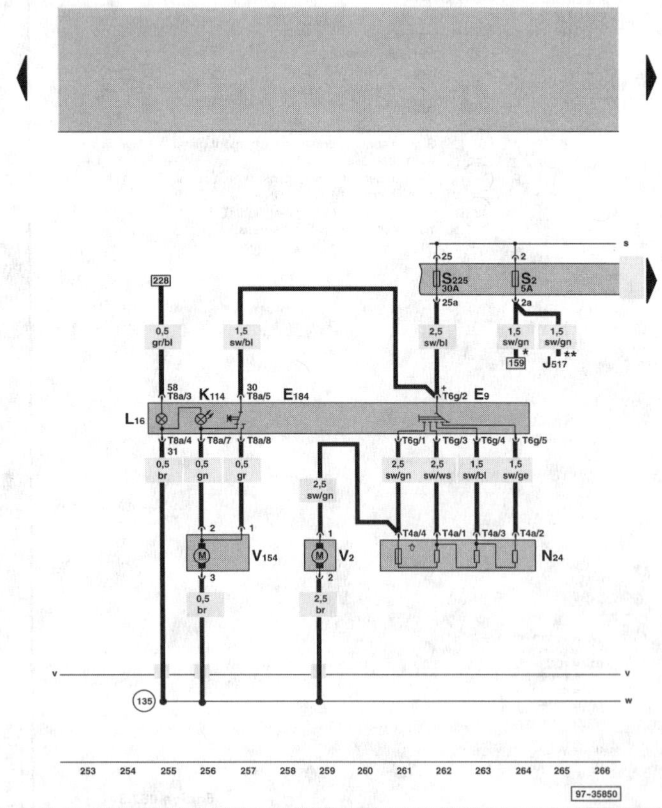

Fresh Air Blower Switch, Switch for fresh air blower and recirculated air, Fresh Air Blower, Fresh Air Blower Series Resistance with Fuse

- E9 - Fresh Air Blower Switch
- E184 - Switch for fresh air blower and recirculated air
- J517 - Auxiliary Emergency Flasher Relay **
- K114 - Fresh Air and Recirculating Air Mode Indicator Light
- L16 - Fresh Air Control Lever Light
- N24 - Fresh Air Blower Series Resistance With Fuse
- S2 - Fuse
- S225 - Fuse in fuse holder
- T4a - 4-Pin Connector
- T6g - 6-Pin Connector, on fresh air blower switch
- T8a - 8-Pin Connector, on switch for fresh air blower and recirculated air
- V2 - Fresh Air Blower
- V154 - Servo motor for fresh-/recirculating air door

- (135) - Ground connection -2-, in instrument panel wiring harness

- * - through April 2003
- ** - from May 2003

ws = white
sw = black
ro = red
br = brown
gn = green
bl = blue
gr = grey
li = lilac
ge = yellow
or = orange

Edition 06.03

Heated rear window, Rear window defogger switch

Repair Manual page number Y 484

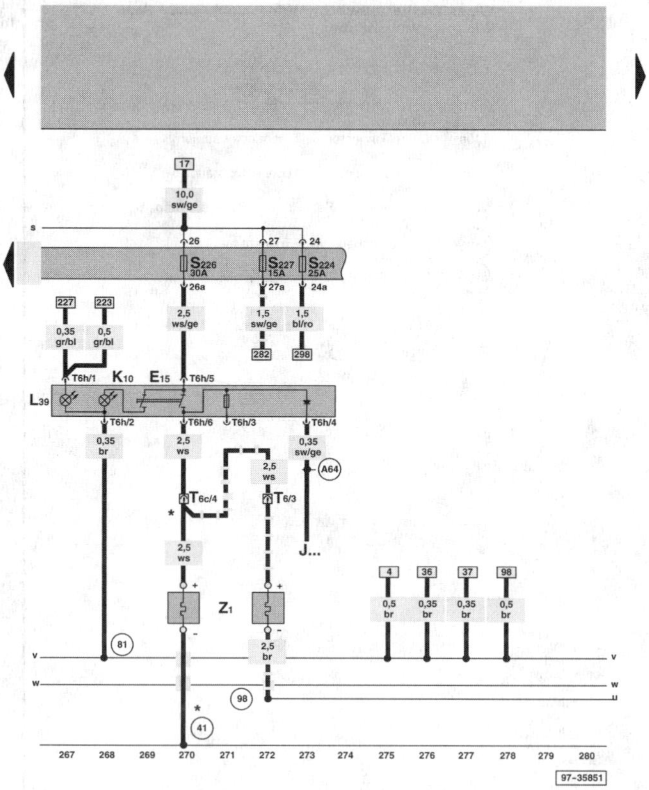

- E15 - Rear window defogger switch
- J... - Engine control module (ECM)
- K10 - Indicator light for heated rear windshield
- L39 - Rear Window Defogger Switch Light
- S224 - Fuse 24 in fuse holder
- S226 - Fuse 26 in fuse holder
- S227 - Fuse 27 in fuse holder
- T6 - 6-Pin Connector, black, in luggage compartment, left (wagon only)
- T6c - 6-Pin Connector, blue, connector station A-pillar, left
- T6h - 6-Pin Connector, black, on rear window defogger switch
- Z1 - Heated rear window
- (41) - Ground connection, below rear shelf
- (81) - Ground connection -1-, in instrument panel wiring harness
- (98) - Ground connection, in rear lid wiring harness
- (A64) - Connector (30, idle boost), in instrument panel wiring harness
- - - - wagon only
- * - sedan only

ws = white
sw = black
ro = red
br = brown
gn = green
bl = blue
gr = grey
li = lilac
ge = yellow
or = orange

97-35851

Edition 06.03

Windshield wiper motor, Wiper/Washer Intermittent Relay, Motor for rear windshield wiper

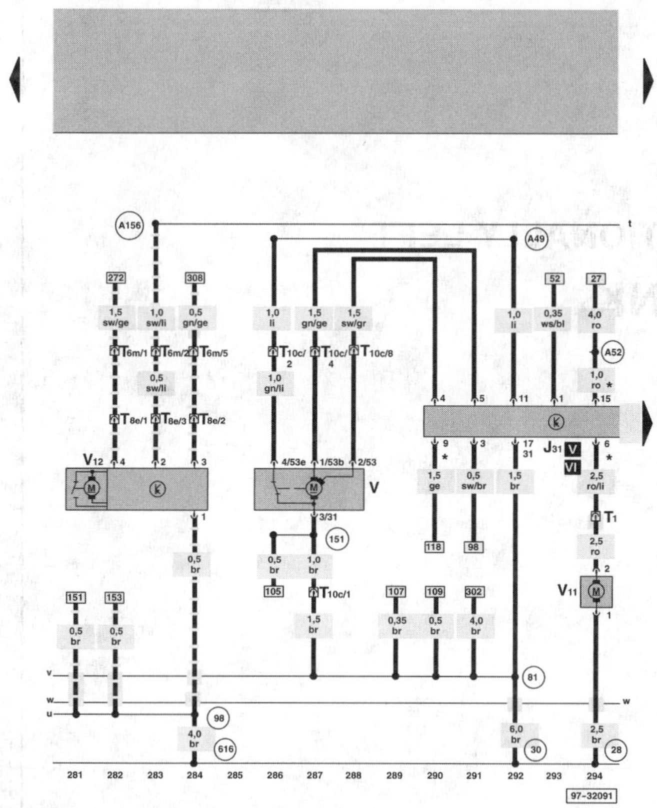

- J31 - Wiper/Washer Intermittent Relay
- T6m - 6-Pin Connector, black, connector station A-pillar, right
- T1 - 1-Pin Connector, brown
- T8e - 8-Pin Connector, black, in rear lid, right
- T10c - 10-Pin Connector, violet, connector station A-pillar, left
- V - Windshield Wiper Motor
- V11 - Headlight Washer Pump
- V12 - Motor for rear windshield wiper
- (28) - Ground connection, on firewall
- (30) - Ground connection, -1-, beside fuse/relay panel
- (81) - Ground connection -1-, in instrument panel wiring harness
- (98) - Ground connection, in rear lid wiring harness
- (151) - Ground connection -1-, in front wiring harness
- (616) - Ground connection (on luggage compartment light)
- (A49) - wire connection -1-, in instrument panel wiring harness
- (A52) - plus connection 2 (30) (in instrument panel wiring harness)
- (A156) - Sensor Ground Connection (in engine wiring harness
- - - - wagon only
- * - models with headlight washer system only

ws = white
sw = black
ro = red
br = brown
gn = green
bl = blue
gr = grey
li = lilac
ge = yellow
or = orange

97-32091

Edition 06.03

Repair Manual page number Y 483

Wiper/washer Intermittent Relay, Windshield Wiper/Washer Switch, Windshield and Rear Window Washer Pump

E22 - Windshield Wiper/Washer Switch
J31 - Wiper/Washer Intermittent Relay
T8b - 8-Pin Connector, on Windshield Wiper/Washer Switch
T10s - 10-Pin Connector, pink, connector station A-pillar, left
T10t - 10-Pin Connector, red, connector station A-pillar, left
V59 - Windshield and Rear Window Washer Pump

(135) - Ground connection -2-, in instrument panel wiring harness
(A156) - Sensor Ground Connection (in engine wiring harness

--- wagon only
* sedan only

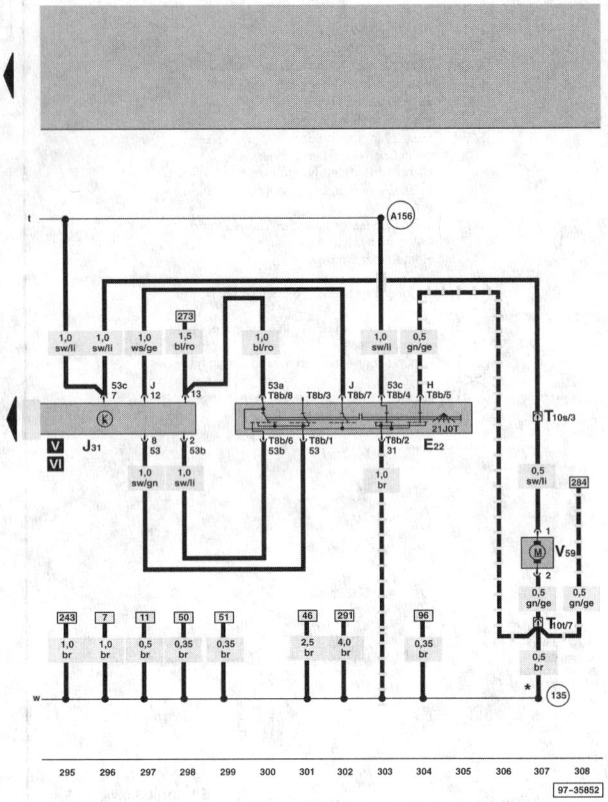

ws = white
sw = black
ro = red
br = brown
gn = green
bl = blue
gr = grey
li = lilac
ge = yellow
or = orange

Edition 06.03

THIS PAGE INTENTIONALLY LEFT

BLANK

1,8L-Engine - Motronic Multiport Fuel Injection (MFI)/125 kW, code AWM

from October 2000

Deviate relay location and fuseplacements as well as the locations of multiple connectors see section "component locations"

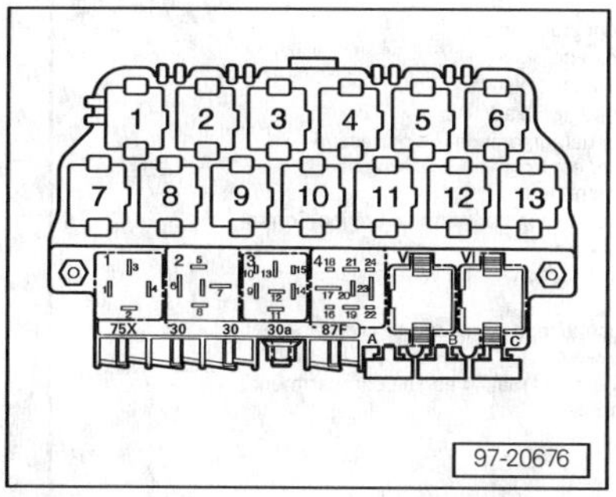

Relay location on the thirteen position auxiliary relay panel, above relay panel:

10 - Brake Booster Relay (373)

Relay panel:

4 - Fuel Pump (FP) Relay (372)

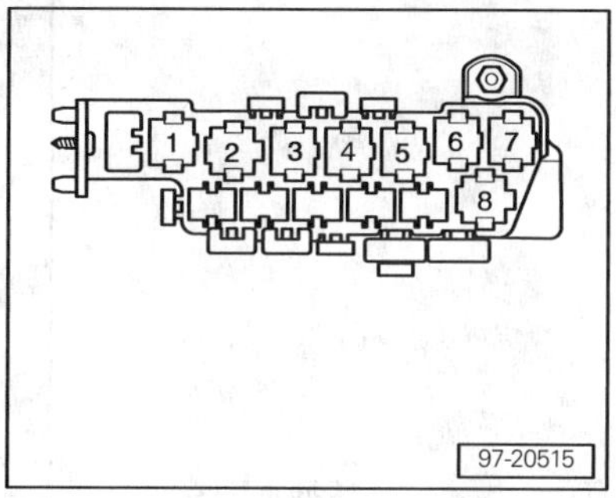

Relay location on the eight position auxiliary relay panel, above relay panel:

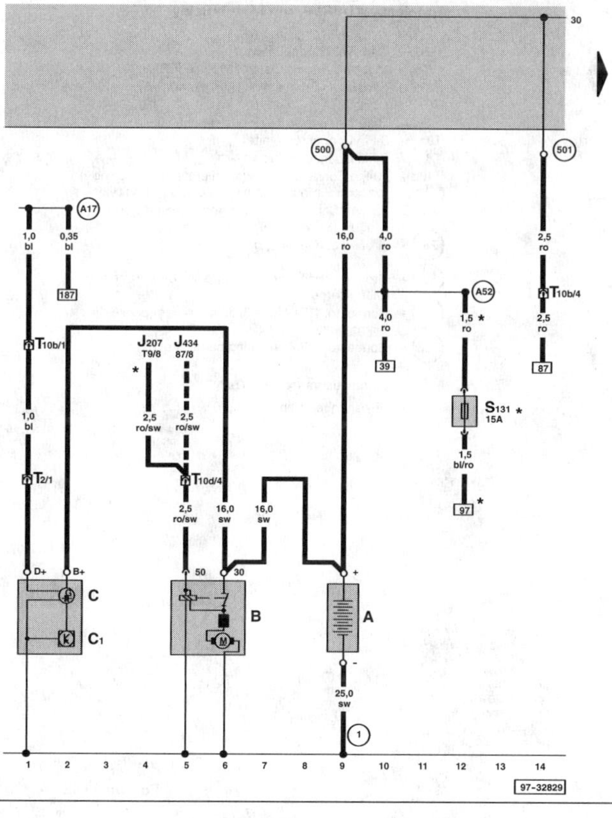

Battery, Starter, Generator (GEN)

A	-	Battery
B	-	Starter
C	-	Generator (GEN)
C1	-	Voltage Regulator (VR)
J207	-	Starting Interlock Relay
J434	-	Locking relay for starter (clutch pedal switch)
S131	-	Safety fuse 1
T2	-	Double Connector, black
T9	-	9-Pin Connector, on Starting Interlock Relay
T10b	-	10-Pin Connector, black, on protective housing for control module, in engine compartment,left
T10d	-	10-Pin Connector, brown, on protective housing for control module, in engine compartment,left
(1)	-	Ground strap, battery to body
(500)	-	Threaded connection -1- (30), on the relay plate
(501)	-	Threaded connection -2- (30), on the relay plate
(A17)	-	wire connection (61), in instrument panel wiring harness
(A52)	-	Plus Connection 2 (30) (in instrument panel wiring harness)
*	-	automatic transmission only
- - -	-	manual transmission only

ws = white
sw = black
ro = red
br = brown
gn = green
bl = blue
gr = grey
li = lilac
ge = yellow
or = orange

Ignition system, Motronic Engine Control Module (ECM), Spark Plug Connectors, Spark Plugs

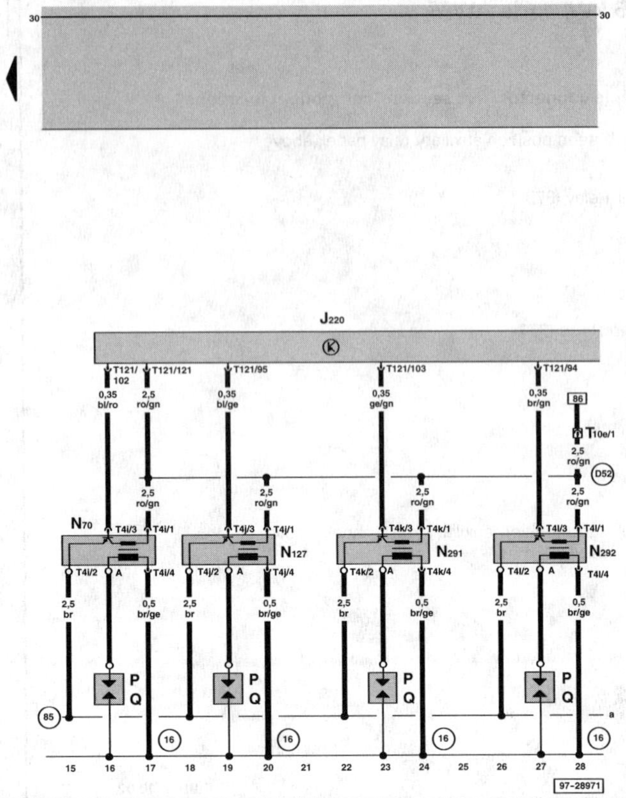

J220 - Motronic Engine Control Module (ECM)
N70 - Ignition Coil 1 with Power Output Stage
N127 - Ignition Coil 2 with Power Output Stage
N291 - Ignition Coil 3 with Power Output Stage
N292 - Ignition Coil 4 with Power Output Stage
P - Spark Plug Connectors
Q - Spark Plugs
T4i - 4-Pin Connector, black
T4j - 4-Pin Connector, black
T4k - 4-Pin Connector, black
T4l - 4-Pin Connector, black
T10e - 10-Pin Connector, orange, on protective housing for control module, in engine compartment, left
T121 - 121-Pin Connector, on Motronic Engine Control Module (ECM)
(16) - Ground connection -1-, on valve cover
(85) - Ground connection -1-, in engine compartment wiring harness
(D52) - plus connection (15a), in engine compartment wiring harness

ws = white
sw = black
ro = red
br = brown
gn = green
bl = blue
gr = grey
li = lilac
ge = yellow
or = orange

Edition 10.02

Motronic Engine Control Module (ECM), Secondary Air Injection (AIR) Pump Motor/Relay

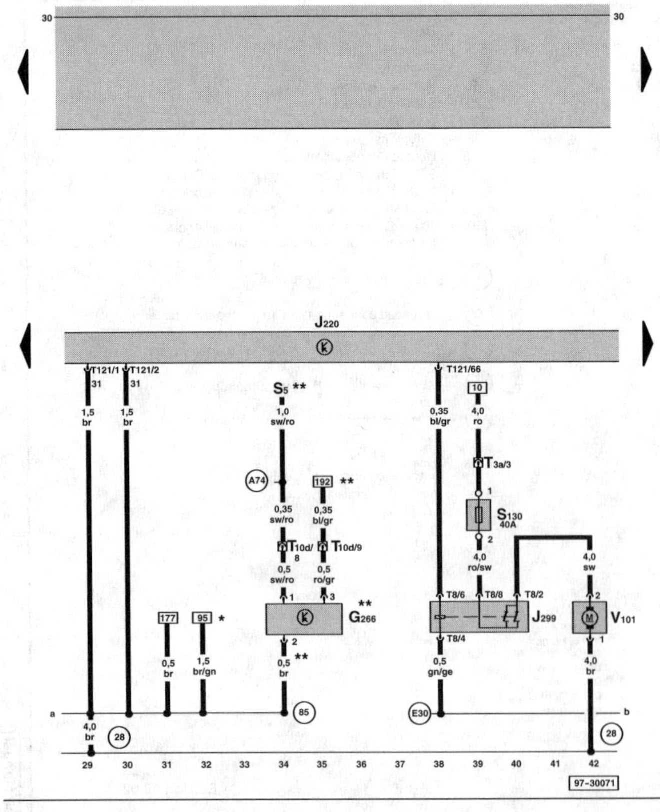

G266 - Oil Level Thermal Sensor
J220 - Motronic Engine Control Module (ECM)
J299 - Secondary Air Injection (AIR) Pump Relay
S5 - Fuse
S130 - Fuse for secondary air pump
T3a - 3-Pin Connector, white
T8 - 8-Pin Connector, brown
T10d - 10-Pin Connector, brown, on protective housing for control module, in engine compartment, left
T121 - 121-Pin Connector, on Motronic Engine Control Module (ECM)
(28) - Ground connection, on firewall
(85) - Ground connection -1-, in engine compartment wiring harness
(A74) - Connector (15a - fuse 5) in instrument panel wiring harness
(E30) - Connector (87a), in wiring harness engine

* - automatic transmission only
** - discontinued from June 2001

ws = white
sw = black
ro = red
br = brown
gn = green
bl = blue
gr = grey
li = lilac
ge = yellow
or = orange

Edition 10.02

Passat — Wiring diagram — No. 49/5

Motronic ECM, Wastegate Bypass Reg. Valve, EVAP Canister Purge Reg. Valve, Secondary Air Injection Solenoid Valve, Recirc.valve for turbocharger, HO2S sensor

G39 - Heated Oxygen Sensor (HO2S)
J220 - Motronic Engine Control Module (ECM)
N75 - Wastegate Bypass Regulator Valve
N80 - Evaporative Emission (EVAP) Canister Purge Regulator Valve
N112 - Secondary Air Infection (AIR) Solenoid Valve
N249 - Recirculating valve for turbocharger
T6 - 6-Pin Connector, black
T121 - 121-Pin Connector, on Motronic Engine Control Module (ECM)

(241) - Ground connection 2, engine ground in engine wiring harness
(E30) - Connector (87a), in wiring harness engine

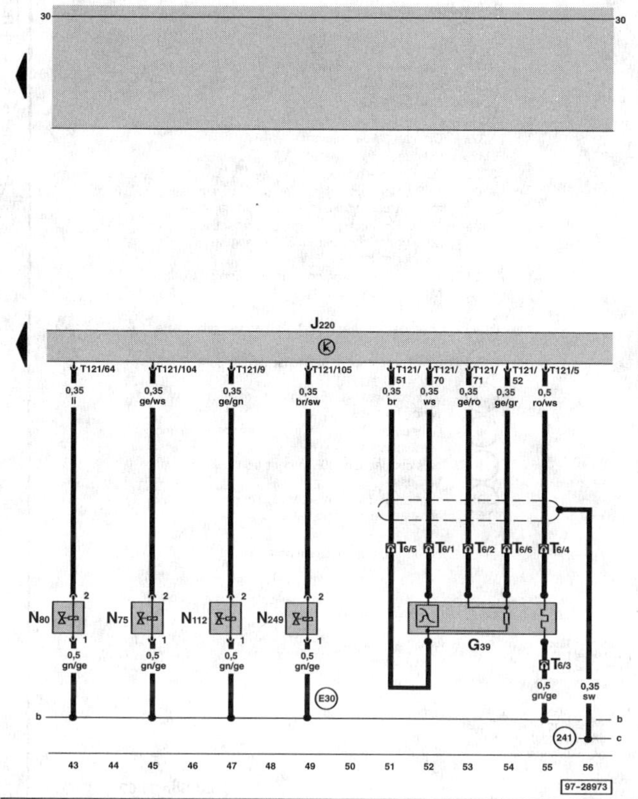

ws = white
sw = black
ro = red
br = brown
gn = green
bl = blue
gr = grey
li = lilac
ge = yellow
or = orange

Edition 10.02

Passat — Wiring diagram — No. 49/6

Motronic Engine Control Module (ECM), Oxygen Sensor, Mass Air Flow Sensor

G70 - Mass Air Flow (MAF) Sensor
G130 - Oxygen Sensor (O2S) Behind Three Way Catalytic Converter (TWC)
J220 - Motronic Engine Control Module (ECM)
S234 - Fuse in fuse holder
T3s - 3-Pin Connector, red, on protective housing for control module, in engine compartment,left
T4c - 4-Pin Connector, black
T10e - 10-Pin Connector, orange, on protective housing for control module, in engine compartment, left
T121 - 121-Pin Connector, on Motronic Engine Control Module (ECM)

(241) - Ground connection 2, engine ground in engine wiring harness
(D107) - wire connection -5-, in engine compartment wiring harness
(E30) - Connector (87a), in wiring harness engine

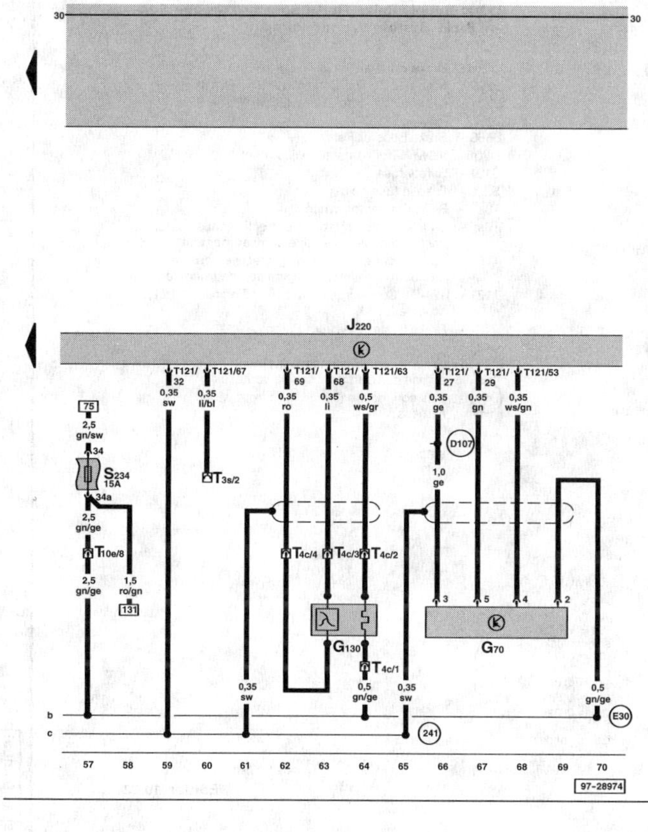

ws = white
sw = black
ro = red
br = brown
gn = green
bl = blue
gr = grey
li = lilac
ge = yellow
or = orange

Edition 10.02

Motronic Engine Control Module (ECM), Fuel Pump Relay, Fuel Injectors

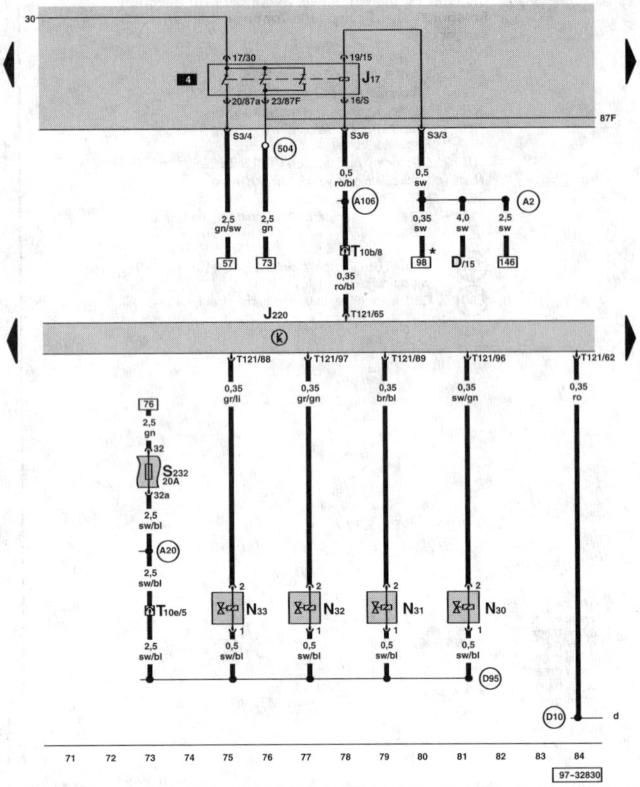

D - Ignition/Starter Switch
J17 - Fuel Pump (FP) Relay (372)
J220 - Motronic Engine Control Module (ECM)
N30 - Cylinder 1 Fuel Injector
N31 - Cylinder 2 Fuel Injector
N32 - Cylinder 3 Fuel Injector
N33 - Cylinder 4 Fuel Injector
S232 - Fuse in fuse holder
T10b - 10-Pin Connector, black, on protective housing for control module, in engine compartment, left
T10e - 10-Pin Connector, orange, on protective housing for control module, in engine compartment, left
T121 - 121-Pin Connector, on Motronic Engine Control Module (ECM)

(504) - Threaded connection -1- (87) on the relay plate
(A2) - plus connection (15), in instrument panel wiring harness
(A20) - wire connection (15a), in instrument panel wiring harness
(A106) - Connector -2- (86s), in instrument panel wiring harness
(D10) - plus connection (30), in right front wiring harness
(D95) - wire connection (injectors), in engine compartment wiring harness

* - automatic transmission only

ws = white
sw = black
ro = red
br = brown
gn = green
bl = blue
gr = grey
li = lilac
ge = yellow
or = orange

97-32830

Motronic Engine Control Module (ECM), Motronic Power Supply Relay, Brake Booster Pressure Sensor, Brake System Vaccum Pump

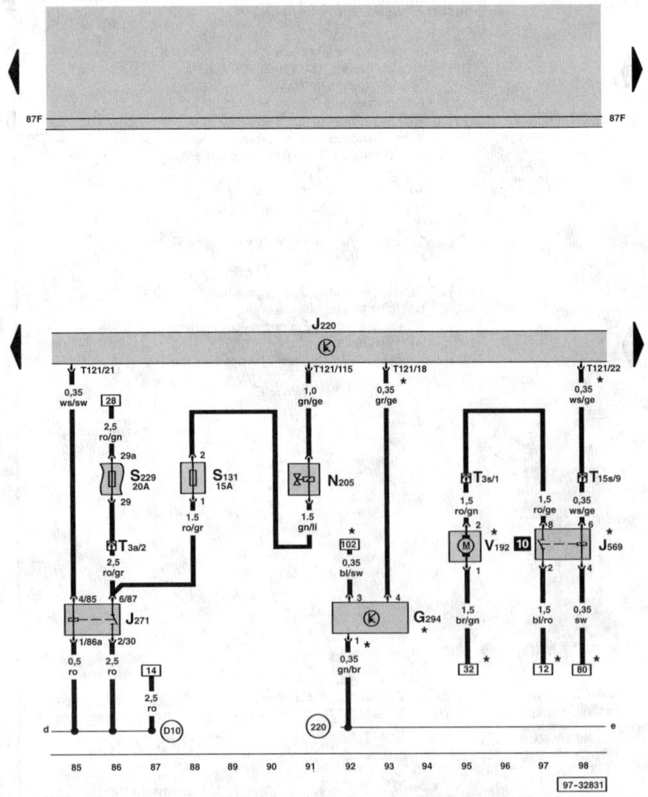

G294 - Brake Booster Pressure Sensor
J220 - Motronic Engine Control Module (ECM)
J271 - Motronic Engine Control (ECM) Power Supply Relay
J569 - Brake Booster Relay
N205 - Valve -1- for camshaft adjustment
S131 - Safety fuse 1
S229 - Fuse in fuse holder
T3a - 3-Pin Connector, white
T3s - 3-Pin Connector, red, on protective housing for control module, in engine compartment, left
T15s - 15-Pin Connector, red, on protective housing for control module, in engine compartment, left
T121 - 121-Pin Connector, on Motronic Engine Control Module (ECM)
V192 - Brake System Vacuum Pump

(220) - Ground connection (sensor ground), in engine compartment wiring harness
(D10) - plus connection (30), in right front wiring harness

* - automatic transmission only

ws = white
sw = black
ro = red
br = brown
gn = green
bl = blue
gr = grey
li = lilac
ge = yellow
or = orange

97-32831

Passat

Wiring diagram

No. 49/9

Motronic Engine Control Module (ECM), Engine Coolant Temperature Sensor, Camshaft Position Sensor, Intake Air Temperature

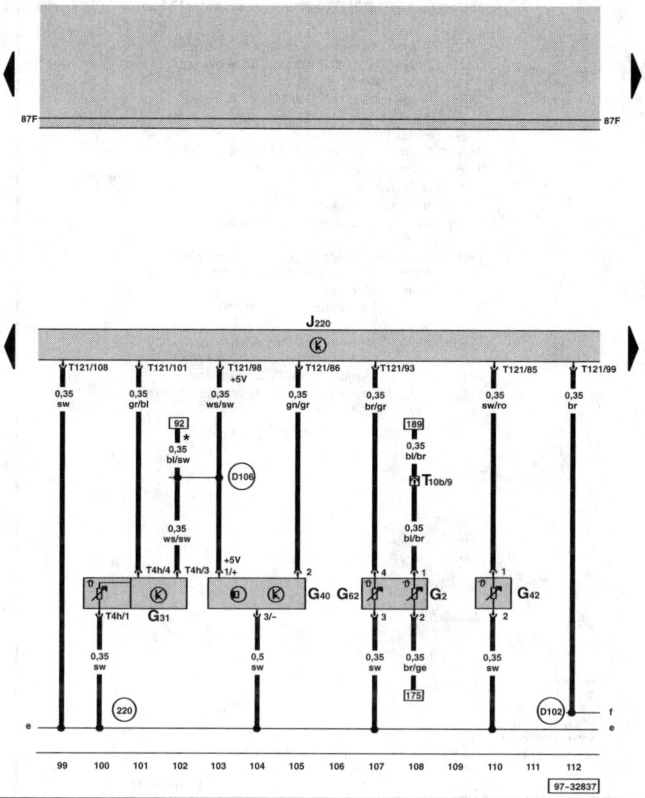

- G2 - Engine Coolant Temperature (ECT) Sensor
- G40 - Camshaft Position (CMP) Sensor
- G42 - Intake Air Temperature (IAT) Sensor
- G62 - Engine Coolant Temperature (ECT) Sensor
- G31 - Charge Air Pressure Sensor
- J220 - Motronic Engine Control Module (ECM)
- T4h - 4-Pin Connector, black
- T10b - 10-Pin Connector, black, on protective housing for control module, in engine compartment,left
- T121 - 121-Pin Connector, on Motronic Engine Control Module (ECM)
- (220) - Ground connection (sensor ground), in engine compartment wiring harness
- (D102) - Wire connection -2-, in engine compartment wiring harness
- (D106) - Wire connection -4-, in engine compartment wiring harness
- * - automatic transmission only

ws = white
sw = black
ro = red
br = brown
gn = green
bl = blue
gr = grey
li = lilac
ge = yellow
or = orange

Edition 10.02

Passat

Wiring diagram

No. 49/10

Motronic Engine Control Module (ECM), Knock Sensor 1 and 2, Engine Speed Sensor

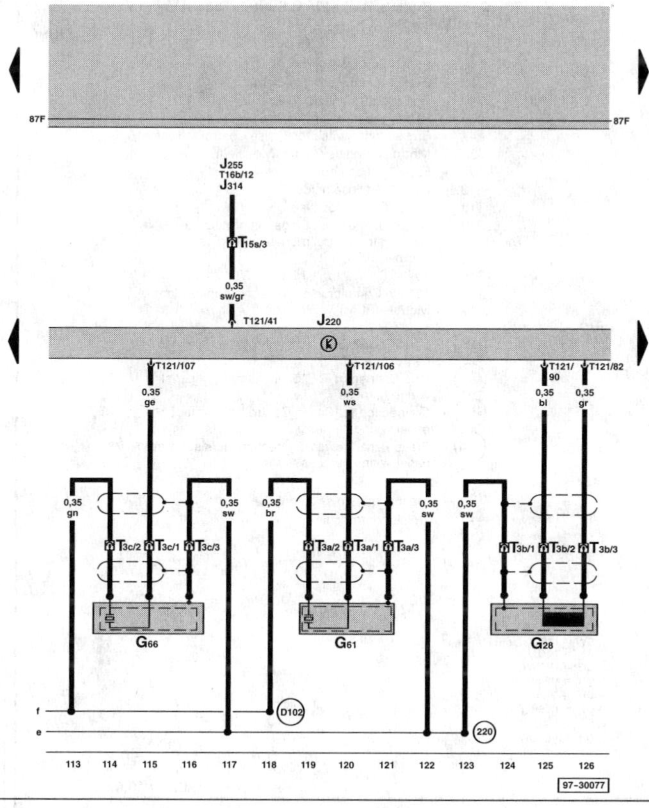

- G28 - Engine Speed (RPM) Sensor
- G61 - Knock Sensor (KS) 1
- G66 - Knock Sensor (KS) 2
- J220 - Motronic Engine Control Module (ECM)
- J255 - Climatronic Control Module
- J314 - A/C Cut-Off Control Module
- T3a - 3-Pin Connector, green
- T3b - 3-Pin Connector, grey
- T3c - 3-Pin Connector,blue
- T15s - 15-Pin Connector, red, on protective housing for control module, in engine compartment,left
- T16b - 16-Pin Connector, on Climatronic Control Module
- T121 - 121-Pin Connector, on Motronic Engine Control Module (ECM)
- (220) - Ground connection (sensor ground), in engine compartment wiring harness
- (D102) - Wire connection -2-, in engine compartment wiring harness

ws = white
sw = black
ro = red
br = brown
gn = green
bl = blue
gr = grey
li = lilac
ge = yellow
or = orange

Edition 10.02

Motronic Engine Control Module (ECM), Leak detection pump (LDP), Throttle Valve Control Module

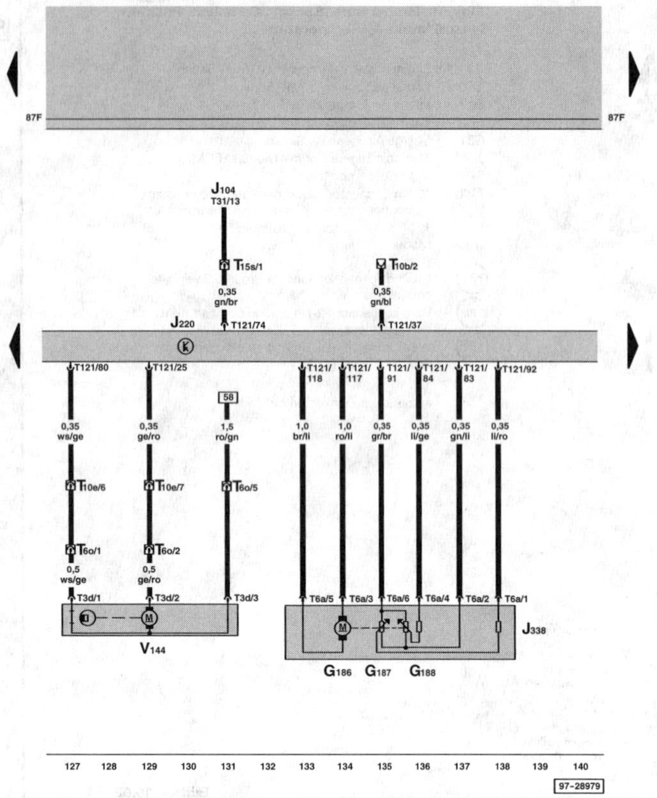

G186 - Throttle drive (power accelerator actuation)
G187 - Angle sensor -1- for throttle drive (power accelerator actuation)
G188 - Angle sensor -2- for throttle drive (power accelerator actuation)
J104 - ABS Control Module
J220 - Motronic Engine Control Module (ECM)
J338 - Throttle Valve Control Module
T3d - 3-Pin Connector, black
T6a - 6-Pin Connector, black
T6o - 6-Pin Connector, brown, connector station A-pillar, right
T10b - 10-Pin Connector, black, on protective housing for control module, in engine compartment, left
T10e - 10-Pin Connector, orange, on protective housing for control module, in engine compartment, left
T15s - 15-Pin Connector, red, on protective housing for control module, in engine compartment, left
T31 - 31-Pin Connector, on ABS Control Module
T121 - 121-Pin Connector, on Motronic Engine Control Module (ECM)
V144 - Leak detection pump (LDP)

ws = white
sw = black
ro = red
br = brown
gn = green
bl = blue
gr = grey
li = lilac
ge = yellow
or = orange

Motronic Engine Control Module (ECM), Cruise Control Switch, Brake Light Switch, Clutch Vacuum Vent Valve Switch

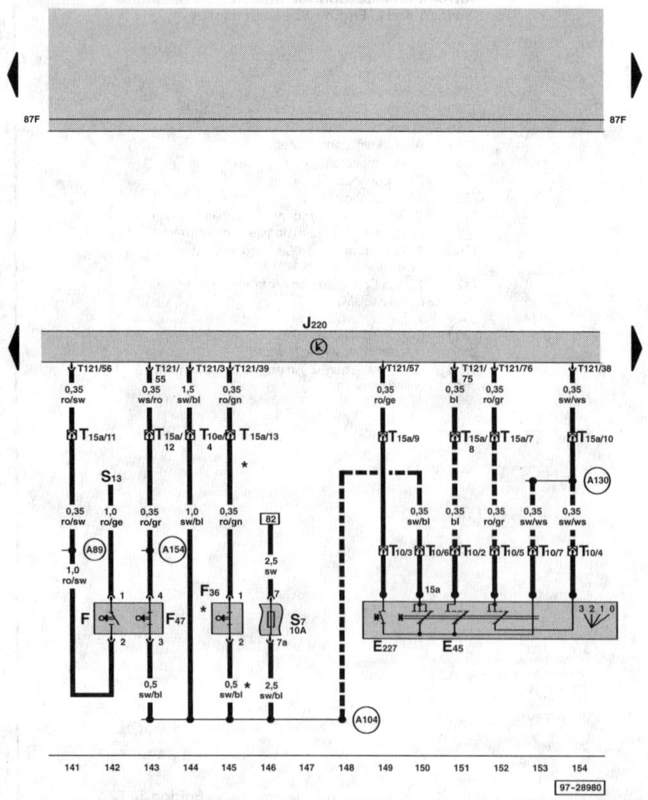

E45 - Cruise Control Switch
E227 - Button for cruise control (set)
F - Brake Light Switch
F36 - Clutch Vacuum Vent Valve Switch
F47 - Brake Light Switch for Cruise Control Indicator
J220 - Motronic Engine Control Module (ECM)
S7 - Fuse 7 in Fuse holder
S13 - Fuse 13 in Fuse holder
T10 - 10-Pin Connector, black
T10e - 10-Pin Connector, orange, on protective housing for control module, in engine compartment, left
T15a - 15-Pin Connector, white
T121 - 121-Pin Connector, on Motronic Engine Control Module (ECM)

(A89) - Connector 2 (54) (in instrument panel wiring harness)
(A104) - Plus Connector -2- (15), in instrument panel wiring harness
(A130) - Connection for Cruise Control (instrument panel wiring harness)
(A154) - Brake Pedal Switch connection (in instrument panel wiring harness

* - manual transmission only
- - - - cruise control only

ws = white
sw = black
ro = red
br = brown
gn = green
bl = blue
gr = grey
li = lilac
ge = yellow
or = orange

Motronic Engine Control Module (ECM), Throttle Position Sensor, Sender -2- for accelerator pedal position, Data Link Connector (DLC)

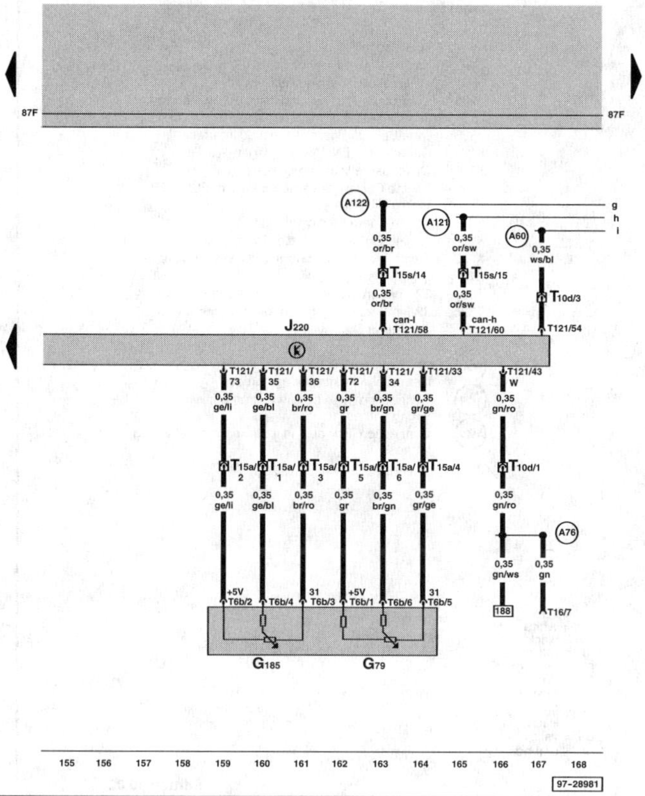

G79 - Throttle Position (TP) Sensor
G185 - Sender -2- for accelerator pedal position
J220 - Motronic Engine Control Module (ECM)
T6b - 6-Pin Connector, on Throttle Position Sensor
T10d - 10-Pin Connector, brown, on protective housing for control module, in engine compartment, left
T15a - 15-Pin Connector, white, on protective housing for control module, in engine compartment, left
T15s - 15-Pin Connector, red, on protective housing for control module, in engine compartment, left
T16 - 16-Pin Connector, Data Link Connector (DLC), below instrument panel, left
T121 - 121-Pin Connector, on Motronic Engine Control Module (ECM)

(A60) - Wire connection (Vehicle speed signal), in instrument cluster wiring harness
(A76) - Connector (K-diagnosis wire) in instrument panel wiring harness
(A121) - Connection (high bus) in instrument panel wiring harness
(A122) - Connection (low bus) in instrument panel wiring harness

ws = white
sw = black
ro = red
br = brown
gn = green
bl = blue
gr = grey
li = lilac
ge = yellow
or = orange

Oil Pressure Swich, Hood Alarm Switch, Sender for fuel gauge, Fuel Pump, Speedometer VSS, ECL Sensor, Control module with indicator unit in instrument panel insert

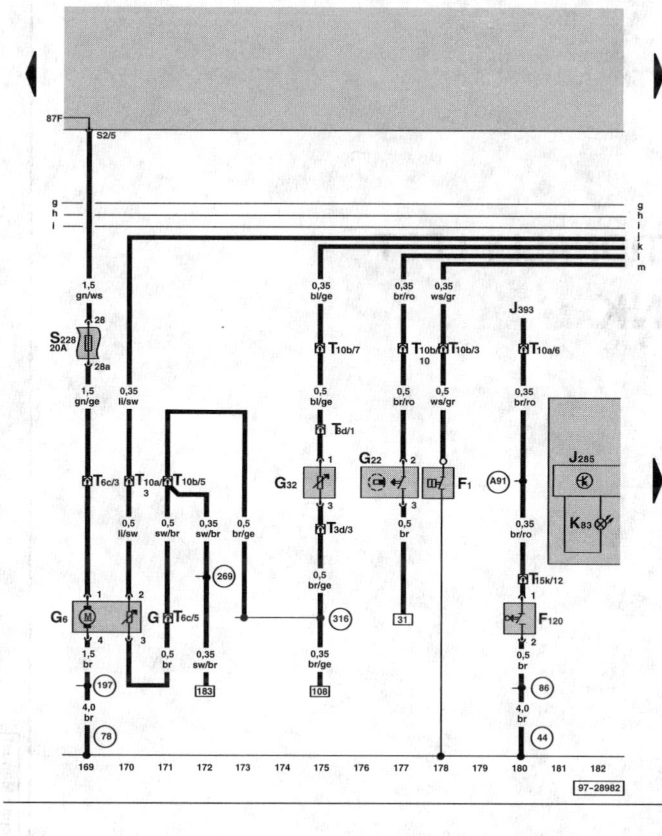

F1 - Oil Pressure Switch
F120 - Hood Alarm Switch
G - Sender for fuel gauge
G6 - Fuel Pump (FP)
G22 - Speedometer Vehicle Speed Sensor (VSS)
G32 - Engine Coolant Level (ECL) Sensor
J285 - Control module with indicator unit in instrument panel insert
J393 - Central control module for comfort system
K83 - Malfunction Indicator Lamp (MIL)
S228 - Fuse in fuse holder
T3d - 3-Pin Connector
T6c - 6-Pin Connector, blue
T10a - 10-Pin Connector, brown, connector station A-pillar, left
T10b - 10-Pin Connector, black, on protective housing for control module, in engine compartment, left
T15k - 15-Pin Connector, green

(44) - Ground connection, on left A-pillar, lower part
(78) - Ground connection, (lower right B-pillar)
(86) - Ground connection -1-, in rear wiring harness
(197) - Ground connection -4-, in wiring harness rear
(269) - Ground connetor (sensor ground) -1-, in instrument panel wiring harness
(316) - Sensor Ground Connection 2 (in engine compartment wiring harness)
(A91) - Connector (signal for instrument panel insert), in instrument panel wiring harness

ws = white
sw = black
ro = red
br = brown
gn = green
bl = blue
gr = grey
li = lilac
ge = yellow
or = orange

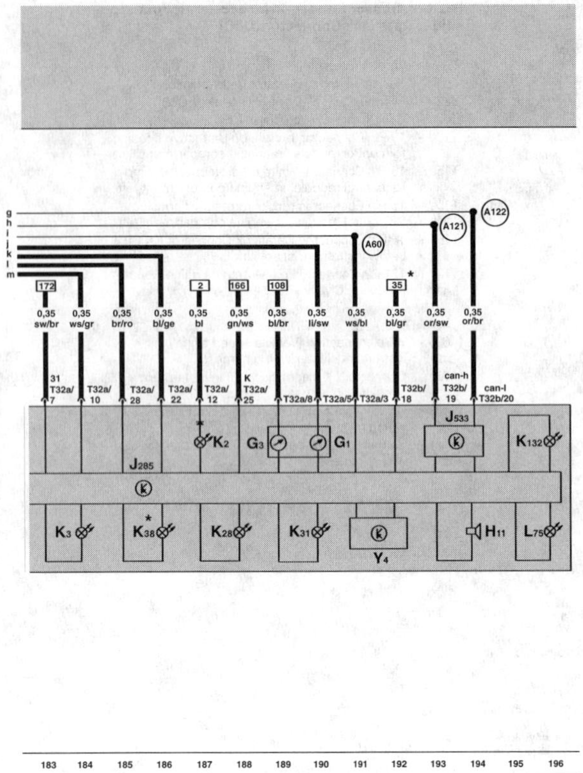

Control module with indicator unit in instrument panel insert

G1 - Fuel Gauge
G3 - Engine Coolant Temperature (ECT) Gauge
H11 - Oil Pressure Warning Buzzer
J285 - Control module with indicator unit in instrument panel insert
J533 - Data Bus On Board Diagnostic Interface
K2 - Generator (GEN) Warning Light
K3 - Oil Pressure Warning Light
K28 - Engine Coolant Level/Temperature (ECL/ECT) Warning Light
K31 - Cruise Control Indicator Light
K38 - Engine Oil Level Indicator Light *
K132 - Electronic Power Control (EPC) Warning Lamp
L75 - Digital Display Light
T32a - 32-Pin Connector, blue, on instrument cluster
T32b - 32-Pin Connector, green, on instrument cluster
Y4 - Odometer Display

(A60) - Wire connection (Vehicle speed signal), in instrument cluster wiring harness
(A121) - Connection (high bus) in instrument panel wiring harness
(A122) - Connection (low bus) in instrument panel wiring harness
* - discontinued from June 2001

ws = white
sw = black
ro = red
br = brown
gn = green
bl = blue
gr = grey
li = lilac
ge = yellow
or = orange

97-30082

THIS PAGE INTENTIONALLY LEFT BLANK

Automatic transmission, 5 speed automatic with tiptronic,

- **1.8L - Engine - Motronic Multiport Fuel Injection (MFI)/125 kW, code AWM**

from October 2000

- **2.0L - Engine - Turbo Diesel PD Fuel Injection (DFI)/100kW, code BHW**

from November 2003

Notes:

Information for

- *Relay and fuse arrangements*
- *Connector arrangements*
- *Control Modules and Relays*
- *Ground connections*

⇒ *Component Locations!*

Information for

- *Troubleshooting Procedures*

⇒ *Guided Fault Finding using VAS 5051/5052*

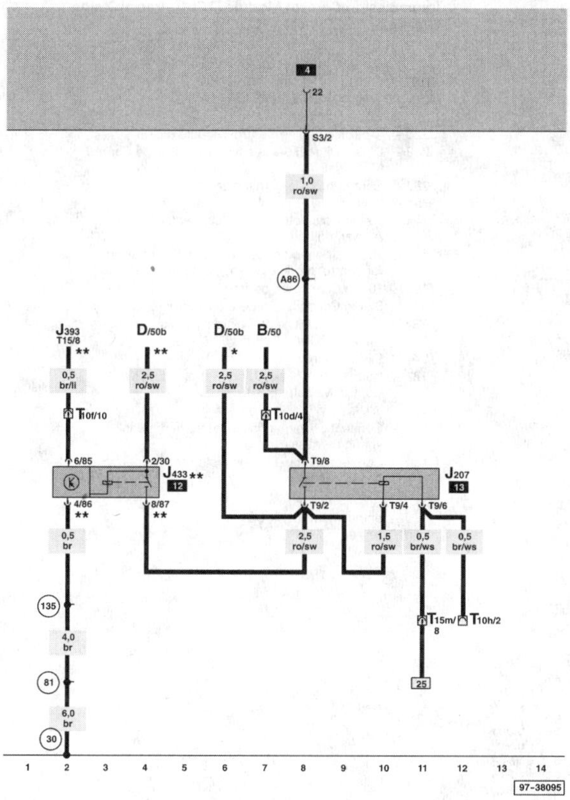

Starting Interlock Relay, Locking relay for starter (anti-theft immobilizer)

- B - Starter
- D - Ignition/Starter Switch
- J207 - Starting Interlock Relay
- J393 - Central control module for comfort system
- J433 - Locking relay for starter (anti-theft immobilizer)
- T9 - 9-Pin Connector, on Starting Interlock Relay
- T10d - 10-Pin Connector, brown, on protective housing for control module, in engine compartment, left
- T10f - 10-Pin Connector, blue, connector station A-pillar, left
- T10h - 10-Pin Connector, blue, on protective housing for control module, in engine compartment, left
- T15 - 15-Pin Connector, on Central control module for comfort system
- T15m - 15-Pin Connector, brown, connector station A-pillar, right

- (30) - Ground connection, -1-, beside fuse/relay panel
- (81) - Ground connection -1-, in instrument panel wiring harness
- (135) - Ground connection -2-, in instrument panel wiring harness
- (A86) - Connection (50a), in instrument panel wiring harness

- * - from May 2003 where applicable
- ** - through April 2003 where applicable

ws = white
sw = black
ro = red
br = brown
gn = green
bl = blue
gr = grey
li = lilac
ge = yellow
or = orange

Transmission Control Module (TCM), Multi-Function Transmission Range (TR) Switch

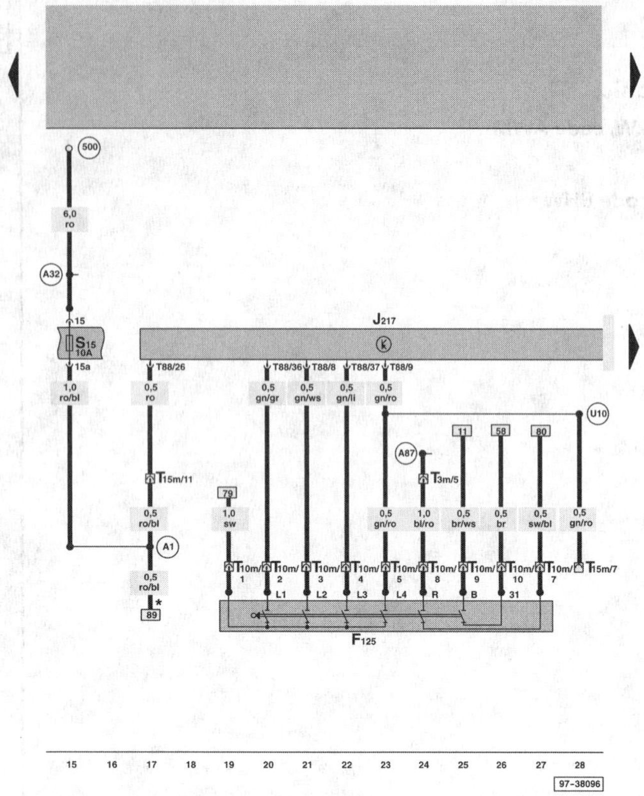

F125 - Multi-Function Transmission Range (TR) Switch
J217 - Transmission Control Module (TCM), in front footwell, right
S15 - Fuse
T3m - 3-Pin Connector, brown, connector station A-pillar, right
T10m - 10-Pin Connector
T15m - 15-Pin Connector, brown, connector station A-pillar, right
T88 - Connector, 88 point, on Transmission Control Module (TCM)

(500) - Threaded Connection -1- (30), on relay plate
(A1) - plus connection (30a), in instrument panel wiring harness
(A32) - plus connection (30), in instrument panel wiring harness
(A87) - Connector (reverse lamp), in instrument panel wiring harness
(U10) - Cruise Control Connection (in automatic transmission wiring harness)

* - through April 2003

ws = white
sw = black
ro = red
br = brown
gn = green
bl = blue
gr = grey
li = lilac
ge = yellow
or = orange

Edition 12.03

Transmission Control Module (TCM), Transmission Vehicle Speed Sensor (VSS)

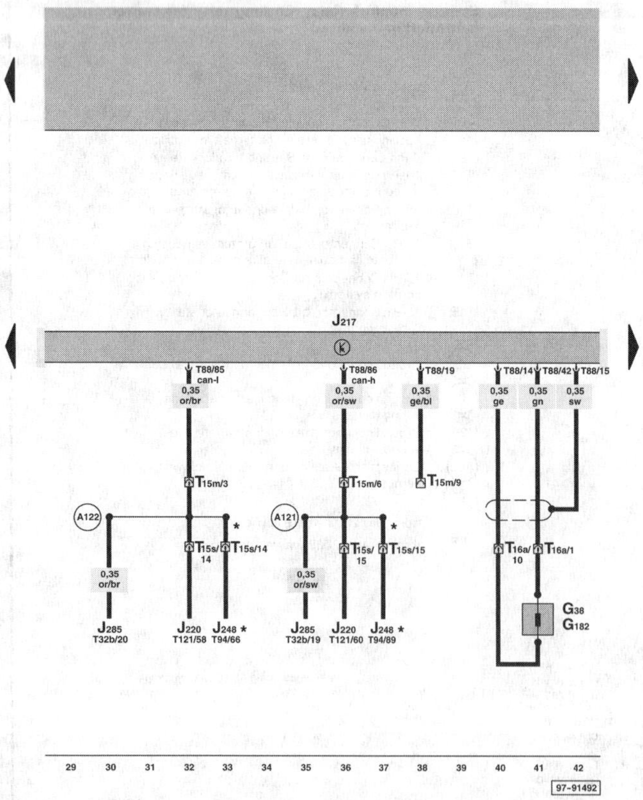

G38 - Transmission Vehicle Speed Sensor (VSS)
G182 - Sensor for transmission RPM
J217 - Transmission Control Module (TCM), in front footwell, right
J220 - Motronic Engine Control Module (ECM)
J248 - Diesel Direct Fuel Injection (DFI) Engine Control Module (ECM)
J285 - Control module with indicator unit in instrument panel insert
T15m - 15-Pin Connector, brown, connector station A-pillar, right
T15s - 15-Pin Connector, red, on protective housing for control module, in engine compartment,left
T16a - 16-Pin Connector, on transmission
T32b - 32-Pin Connector, green, on instrument cluster
T88 - Connector, 88 point, on Transmission Control Module (TCM)
T94 - 94-Pin connector
T121 - Connector, 121 point, on Motronic Engine Control Module (ECM)

(A121) - Connection (high-bus), in instrument panel wiring harness
(A122) - Connection (low-bus), in instrument panel wiring harness

* - engine code BHW only

ws = white
sw = black
ro = red
br = brown
gn = green
bl = blue
gr = grey
li = lilac
ge = yellow
or = orange

Edition 06.04

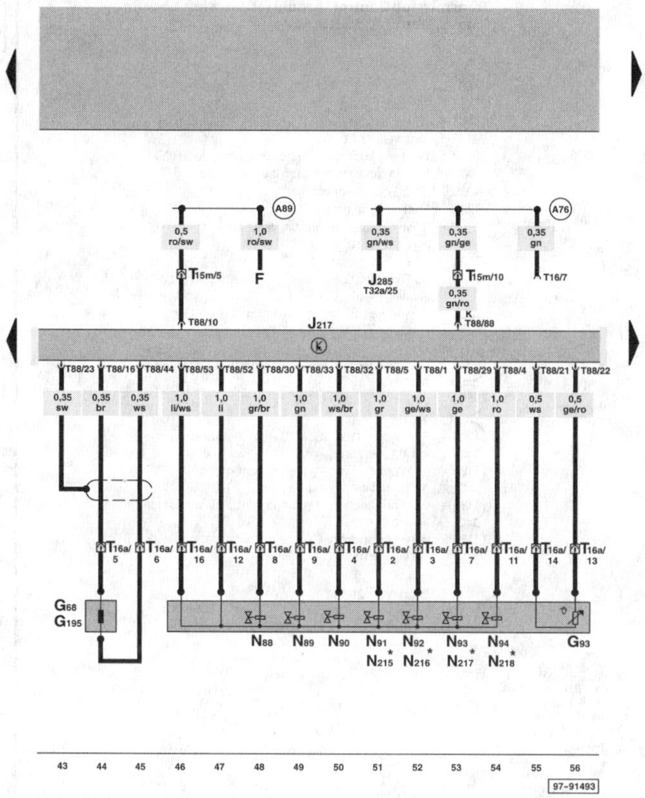

Transmission Control Module (TCM), Solenoid Valves, Transmission Fluid Temperature Sensor, Vehicle Speed Sensor (VSS)

- F - Brake Light Switch
- G68 - Vehicle Speed Sensor (VSS)
- G93 - Transmission Fluid Temperature Sensor
- G195 - Sensor for Transmission output RPM
- J217 - Transmission Control Module (TCM), in front footwell, right
- J285 - Control module with indicator unit in instrument panel insert
- N88 - Solenoid Valve 1
- N89 - Solenoid Valve 2
- N90 - Solenoid Valve 3
- N91 - Solenoid Valve 4
- N92 - Solenoid Valve 5
- N93 - Solenoid Valve 6
- N94 - Solenoid Valve 7
- N215 - Pressure Control Valve 1 for Auto. transmission
- N216 - Pressure Control Valve 2 for Auto. transmission
- N217 - Pressure Control Valve 3 for Auto. transmission
- N218 - Pressure Control Valve 4 for Auto. transmission
- T15m - 15-Pin Connector, brown, connector station A-pillar, right
- T16 - 16-Pin Connector, black, Data Link Connector (DLC) below instrument panel, left
- T16a - 16-Pin Connector, on transmission
- T32a - 32-Pin Connector, blue, on instrument cluster
- T88 - Connector, 88 point, on Transmission Control Module (TCM)
- (A76) - Connector (K-diagnosis wire), in instrument panel wiring harness
- (A89) - Connector (54) (in instrument panel wiring harness)
- * - from October 2000

ws = white
sw = black
ro = red
br = brown
gn = green
bl = blue
gr = grey
li = lilac
ge = yellow
or = orange

Edition 06.04

Repair Manual page number Y 508

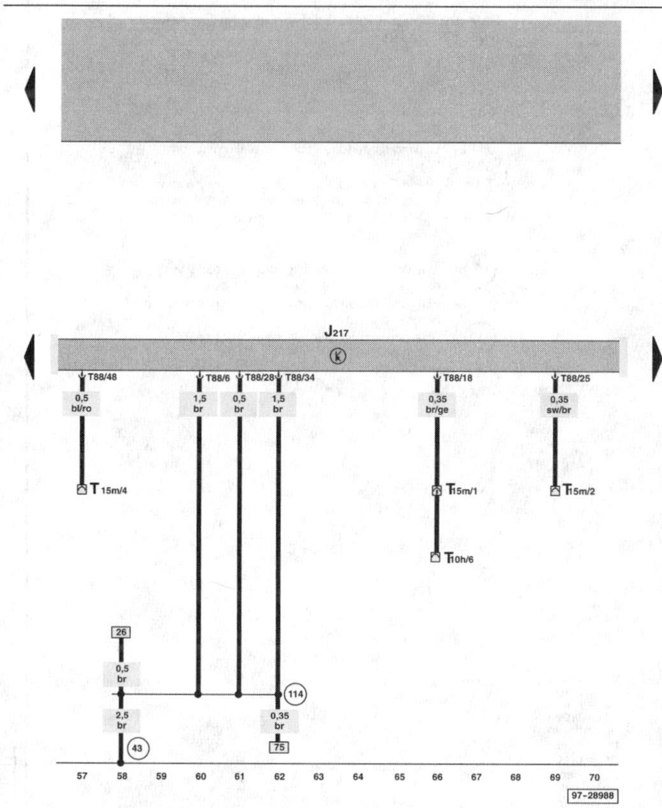

Transmission Control Module (TCM)

- J217 - Transmission Control Module (TCM), in front footwell, right
- T10h - 10-Pin Connector, blue, on protective housing for control module, in engine compartment, left
- T15m - 15-Pin Connector, brown, connector station A-pillar, right
- T88 - Connector, 88 point, on Transmission Control Module (TCM)
- (43) - Ground connection, on right A-pillar, lower part
- (114) - Ground connection, in automatic transmission wiring harness

ws = white
sw = black
ro = red
br = brown
gn = green
bl = blue
gr = grey
li = lilac
ge = yellow
or = orange

Edition 12.03

Repair Manual page number Y 507

Transmission Control Module (TCM), Shift Lock Solenoid, Tiptronic Switch

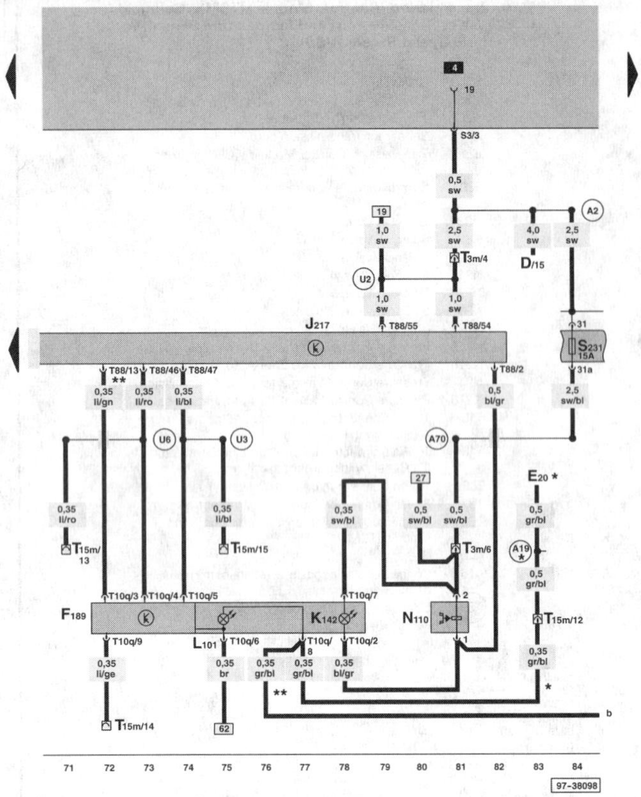

D - Ignition/Starter Switch
E20 - Instrument Panel Light Dimmer Switch *
F189 - Tiptronic Switch
J217 - Transmission Control Module (TCM), in front footwell, right
K142 - Warning light for selector lever position P/N
L101 - Illumination for selector lever scale
N110 - Shift Lock Solenoid
S231 - Fuse in fuse holder
T3m - 3-Pin Connector, brown, connector station A-pillar, right
T10q - 10-Pin Connector
T15m - 15-Pin Connector, brown, connector station A-pillar, right
T88 - Connector, 88 point, on Transmission Control Module (TCM)

(A2) - plus connection (15), in instrument panel wiring harness
(A19) - wire connection (58d), in instrument panel wiring harness *
(A70) - Connector (15a, fuse 231), in instrument panel wiring harness
(U2) - wire connection -1- (15), in automatic transmission wiring harness
(U3) - wire connection -1-, in automatic transmission wiring harness
(U6) - wire connection -2-, in automatic transmission wiring harness

* - from May 2003
** - through April 2003

ws = white
sw = black
ro = red
br = brown
gn = green
bl = blue
gr = grey
li = lilac
ge = yellow
or = orange

Edition 12.03

Selector Lever Light, Locking diode for selector lever illumination

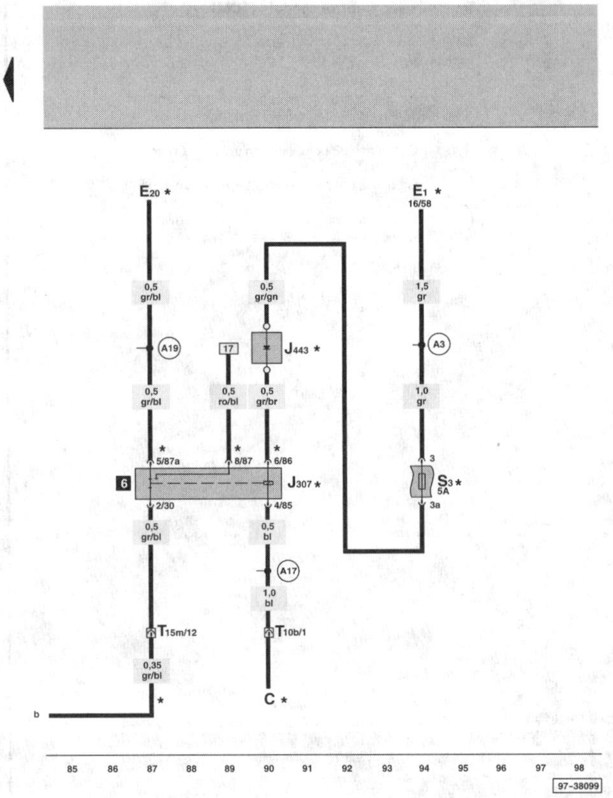

C - Generator (GEN)
E1 - Light switch
E20 - Instrument Panel Light Dimmer Switch
J307 - Selector Lever Light Relay
J443 - Locking diode for selector lever illumination
S3 - Fuse
T10b - 10-Pin Connector, black, on protective housing for control module, in engine compartment,left
T15m - 15-Pin Connector, brown, connector station A-pillar, right

(A3) - plus connection (58), in instrument panel wiring harness
(A17) - wire connection (61), in instrument panel wiring harness
(A19) - wire connection (58d), in instrument panel wiring harness

* - through April 2003

ws = white
sw = black
ro = red
br = brown
gn = green
bl = blue
gr = grey
li = lilac
ge = yellow
or = orange

Edition 12.03

Comfort System, with Alarm System,

from October 2000

Notes:

Information for

- *Relay and fuse arrangements*
- *Connector arrangements*
- *Control Modules and Relays*
- *Ground connections*

⇒ Component Locations!

Information for

- *Troubleshooting Procedures*

⇒ Guided Fault Finding using VAS 5051/5052

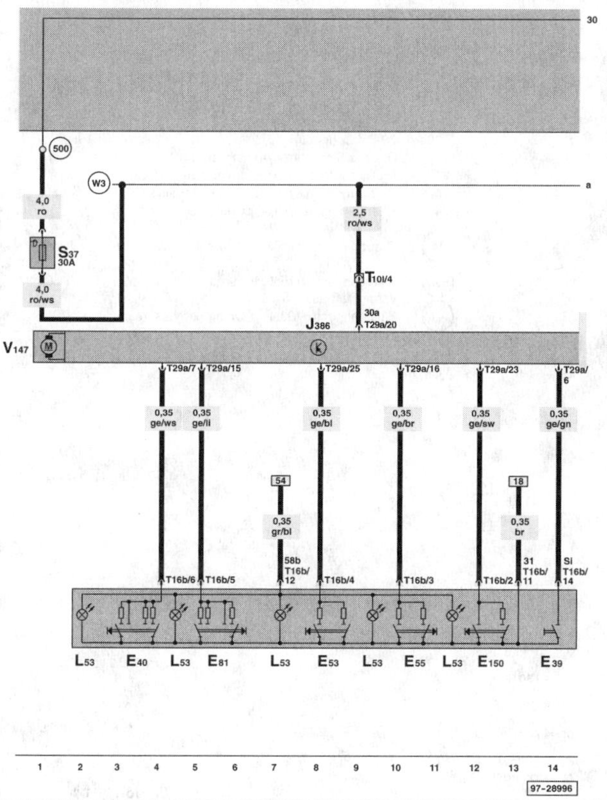

Door control module, driver side, Switch for window regulator in driver's door, Switch for interior lock

E39 - Safety switch for window regulator rear
E40 - Switch for window regulator front left
E53 - Switch for window regulator rear left, driver
E55 - Switch for window regulator rear right, driver
E81 - Switch for window regulator front right, driver
E150 - Switch for interior lock, driver side
J386 - Door control module, driver side
L53 - Power Window Switch Light
S37 - Power Window Fuse
T10l - 10-pin connector, black, connector station A-pillar left
T16b - 16-pin connector, brown
T29a - 29-pin connector
V147 - Motor for window regulator, driver side

(500) - Threaded connection -1- (30), on relay plate

(W3) - wire connection, in rear wiring harness

ws = white
sw = black
ro = red
br = brown
gn = green
bl = blue
gr = grey
li = lilac
ge = yellow
or = orange

Passat — Wiring diagram — No. 51/3

Lock unit for central locking, Door control module, driver side, Warning light for central locking -SAFE-

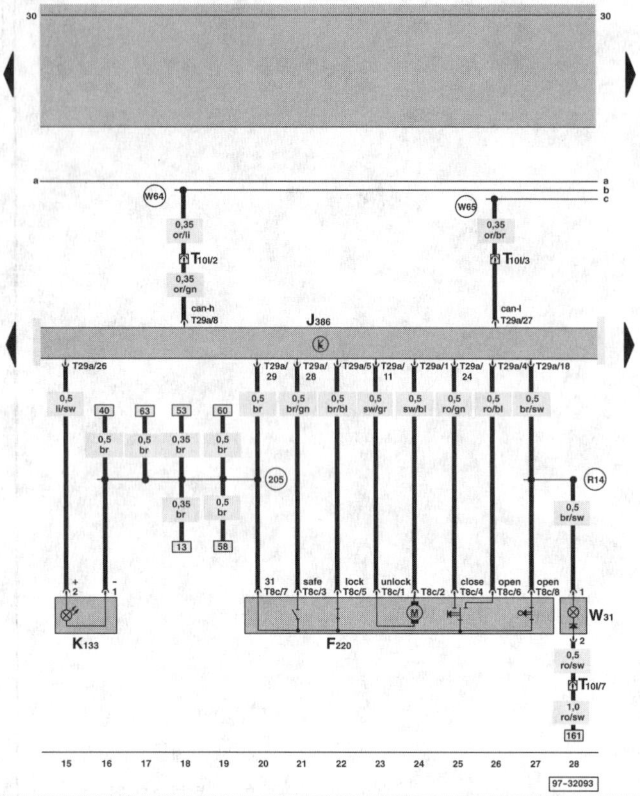

- F220 - Lock unit for central locking, driver side
- J386 - Door control module, driver side
- K133 - Warning light for central locking -SAFE-
- T8c - 8-Pin Connector, black, on lock unit for centrallocking, driver side
- T10l - 10-Pin Connector, black, connector station A-pillar, left
- T29a - 29-Pin Connector
- W31 - Entry light front, left
- (205) - Ground connection, in driver's door wiring harness
- (R14) - wire connection -1- (open), in driver's door wiring harness
- (W64) - CAN-Bus High Connection - Convenience (in floor wiring harness)
- (W65) - CAN-Bus Low Connection - Convenience (in floor wiring harness)

ws = white
sw = black
ro = red
br = brown
gn = green
bl = blue
gr = grey
li = lilac
ge = yellow
or = orange

Edition 06.04

Passat — Wiring diagram — No. 51/4

Door control module, driver side, Heated outside mirror, driver side

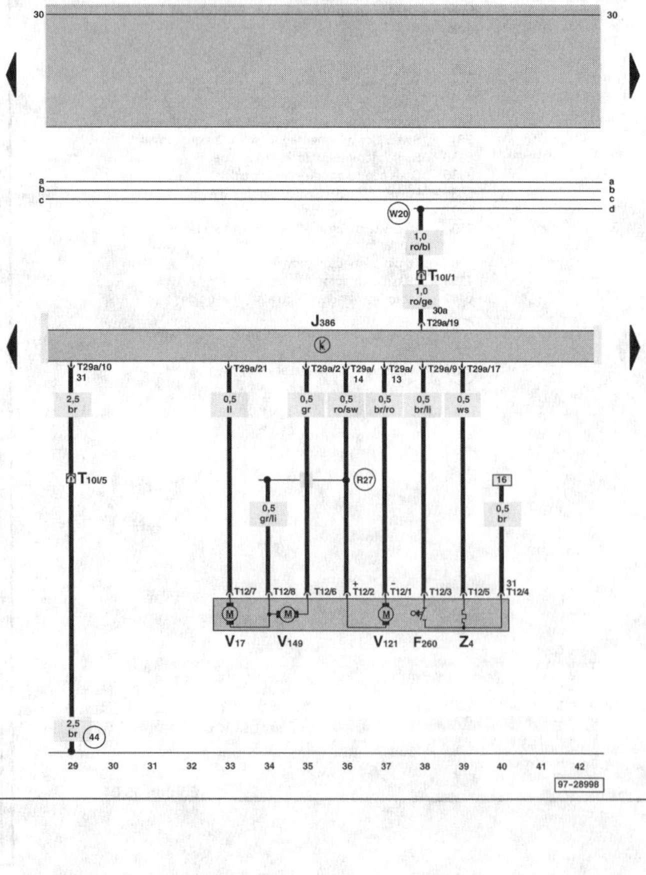

- F260 - Driver's Folding Mirror Contact Switch
- J386 - Door control module, driver side
- T10l - 10-Pin Connector, black, connector station A-pillar, left
- T12 - 12-Pin Connector,black
- T29a - 29-Pin Connector
- V17 - Driver's Side Mirror Adjustment Motor
- V121 - Motor for fold-in mirror, driver side
- V149 - Motor for mirror adjustment, driver side
- Z4 - Heated outside mirror, driver side
- (44) - Ground connection, on left A-pillar, lower part
- (R27) - wire connection (power mirror), in drivr's door wiring harness
- (W20) - Plus connector (30a), in wiring harness, rear

ws = white
sw = black
ro = red
br = brown
gn = green
bl = blue
gr = grey
li = lilac
ge = yellow
or = orange

Edition 06.04

Passat — Wiring diagram — No. 51/5

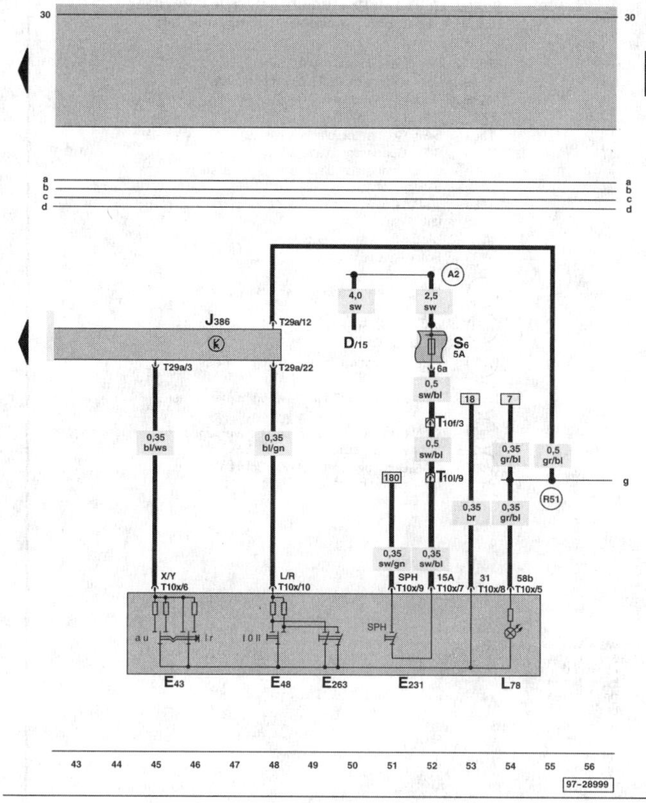

Door control module, driver side, Mirror Adjustment and Selector Switch

- D - Ignition/Starter Switch
- E43 - Mirror Adjustment Switch
- E48 - Mirror Selector Switch
- E231 - Switch for outside mirror heating
- E263 - Mirror Folding Switch
- J386 - Door control module, driver side
- L78 - Mirror Adjusting Switch Light
- S6 - Fuse
- T10f - 10-Pin Connector, blue, connector station A-pillar, left
- T10l - 10-Pin Connector, black, connector station A-pillar, left
- T10x - 10-Pin Connector, black
- T29a - 29-Pin Connector

- (A2) - plus connection (15), in instrument panel wiring harness
- (R51) - Connection (58b) (driver's side door wiring harness)

ws = white
sw = black
ro = red
br = brown
gn = green
bl = blue
gr = grey
li = lilac
ge = yellow
or = orange

Edition 06.04

Passat — Wiring diagram — No. 51/6

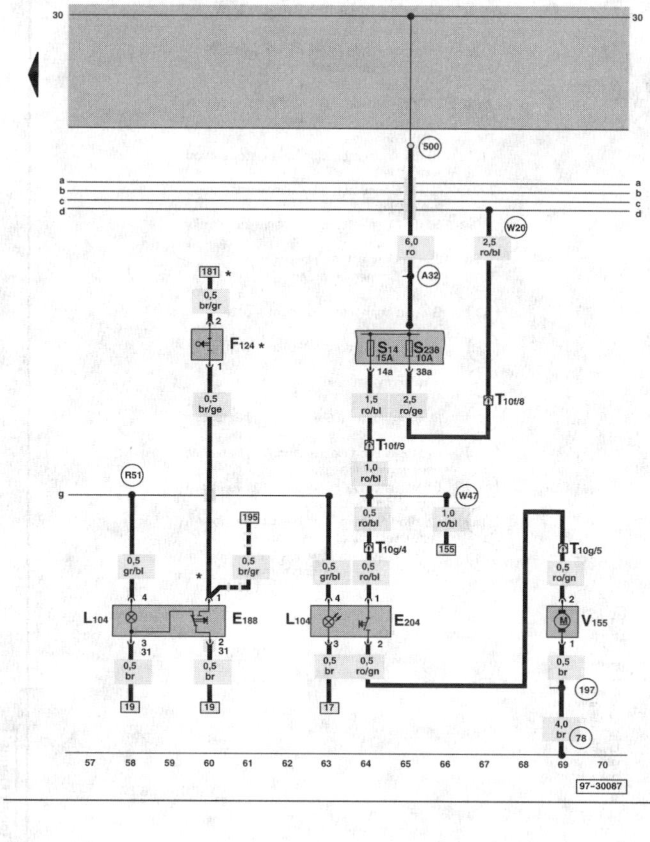

Switch for remote unlock and fuel tank door, Trunk Lock Alarm/Central Locking Switch, Illumination for switch remote unlock, Motor for fuel tank lid unlock

- E188 - Switch for remote unlock, rear lid
- E204 - Switch for remote/fuel tank door
- F124 - Trunk Lock Alarm/Central Locking Switch
- L104 - Illumination for switch remote unlock
- T10f - 10-Pin Connector, blue, connector station A-pillar, left
- T10g - 10-Pin Connector, grey
- V155 - Motor for fuel tank lid unlock

- (78) - Ground connection (lower right B-pillar)
- (197) - Ground connection -4-, in wiring harness rear
- (500) - Threaded connection -1- (30) on the relay plate
- (A32) - plus connection (30), in instrument panel wiring harness
- (R51) - Connection (58d) (driver's side door wiring harness)
- (W20) - Plus-Connector (30a) in wiring harness, rear
- (W47) - Plus-Connection -2- (30a), in wiring harness floor
- * - sedan only
- --- - wagon only

ws = white
sw = black
ro = red
br = brown
gn = green
bl = blue
gr = grey
li = lilac
ge = yellow
or = orange

Edition 06.04

Repair Manual page number Y 516

Repair Manual page number Y 515

Passat — Wiring diagram — No. 51/7

Door control module,rear, left, Power Window Switch Light, Left Rear Window Switch, Entry light rear, left, Lock unit for central locking, rear, left

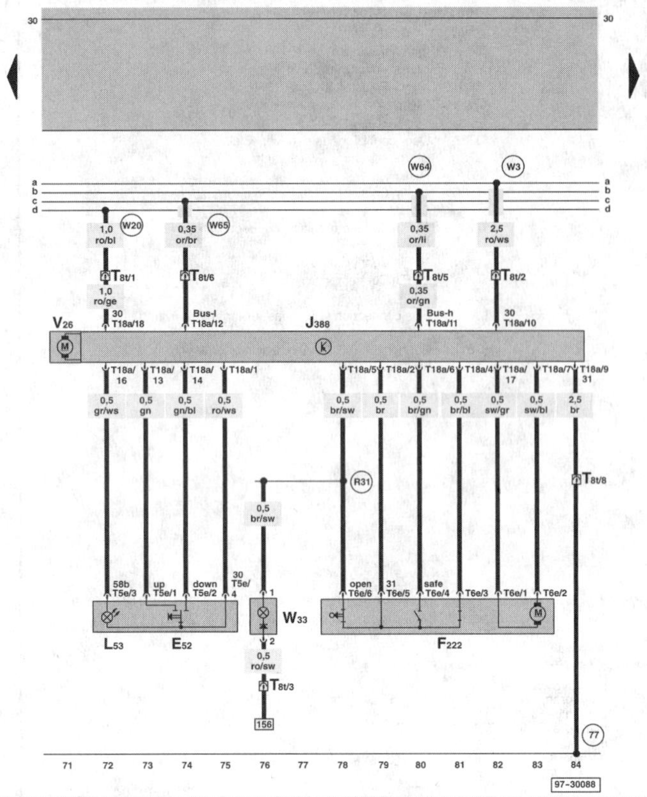

- E52 - Left Rear Window Switch, (In LR Door)
- F222 - Lock unit for central locking, rear, left
- J388 - Door control module, rear, left
- L53 - Power Window Switch Light
- T5e - 5-Pin Connector, black
- T6e - 6-Pin Connector, black
- T8t - 8-Pin Connector, black, connector station B-pillar, left
- T18a - 18-Pin Connector, on Door control module, rear, left
- V26 - Motor for window regulator, rear, left
- W33 - Entry light rear, left
- (77) - Ground connection (lower left B-pillar)
- (R31) - Wire connection -1-, in left rear door wiring harness
- (W3) - wire connection, in rear wiring harness
- (W20) - Plus connector (30a), in wiring harness, rear
- (W64) - CAN-Bus High Connection - Convenience (in floor wiring harness)
- (W65) - CAN-Bus Low Connection - Convenience (in floor wiring harness)

ws = white
sw = black
ro = red
br = brown
gn = green
bl = blue
gr = grey
li = lilac
ge = yellow
or = orange

Edition 06.04

Passat — Wiring diagram — No. 51/8

Door control module, Passenger Side Interior Lock Switch, Lighting forswitch interior lock

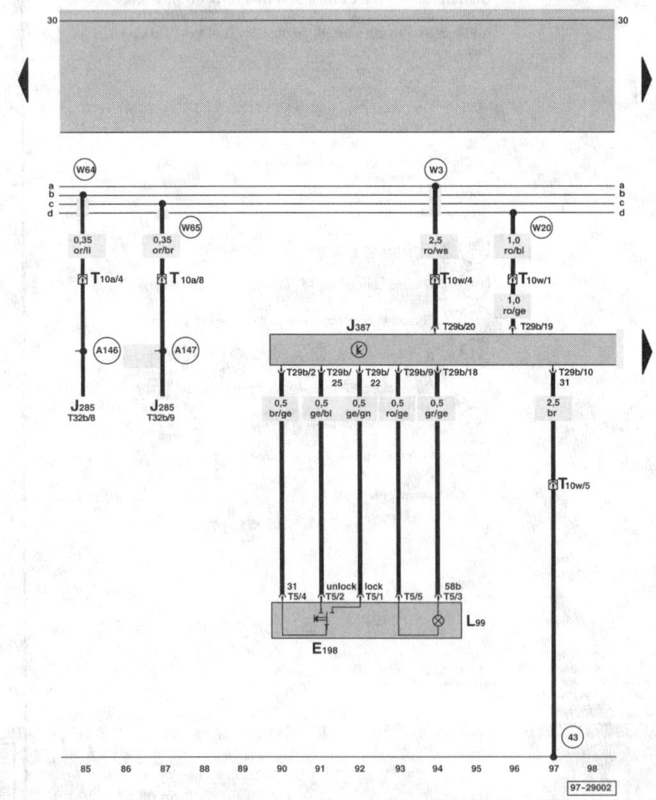

- E198 - Passenger Side Interior Lock Switch
- J285 - Control module with indicator unit in instrument panelinsert
- J387 - Door control module, passenger side
- L99 - Lighting for switch interior lock
- T5 - 5-Pin Connector
- T10f - 10-Pin Connector, blue, connector station A-pillar, left
- T10a - 10-Pin Connector, grey, connector station A-pillar, left
- T10w - 10-Pin Connector, black, connector station A-pillar, right
- T29b - 29-Pin Connector
- T32b - 32-Pin Connector, green
- (43) - Ground connection, on right A-pillar, lower part
- (500) - Screw connection -1- (30), on relay panel
- (W3) - wire connection, in rear wiring harness
- (W20) - Plus connector (30a), in wiring harness, rear
- (W64) - CAN-Bus High Connection - Convenience (in floor wiring harness)
- (W65) - CAN-Bus Low Connection - Convenience (in floor wiring harness)
- (A146) - Comfort System High-bus Connection (in instrument panel wiring harness)
- (A147) - Comfort System Low-bus Connection (in instrument panel wiring harness)

ws = white
sw = black
ro = red
br = brown
gn = green
bl = blue
gr = grey
li = lilac
ge = yellow
or = orange

Edition 06.04

Passat — Wiring diagram — No. 51/9

Central locking passenger's door, Entry light front, right

F221 - Lock unit for central locking, passenger side
J387 - Door control module, passenger side
T8d - 8-Pin Connector, black
T10w - 10-Pin Connector, black, connector station A-pillar, right
T29b - 29-Pin Connector
W32 - Entry light front, right

(206) - Ground connection, in passenger's door wiring harness
(R29) - Wire connection -1- (open), in passenger's door wiring harness

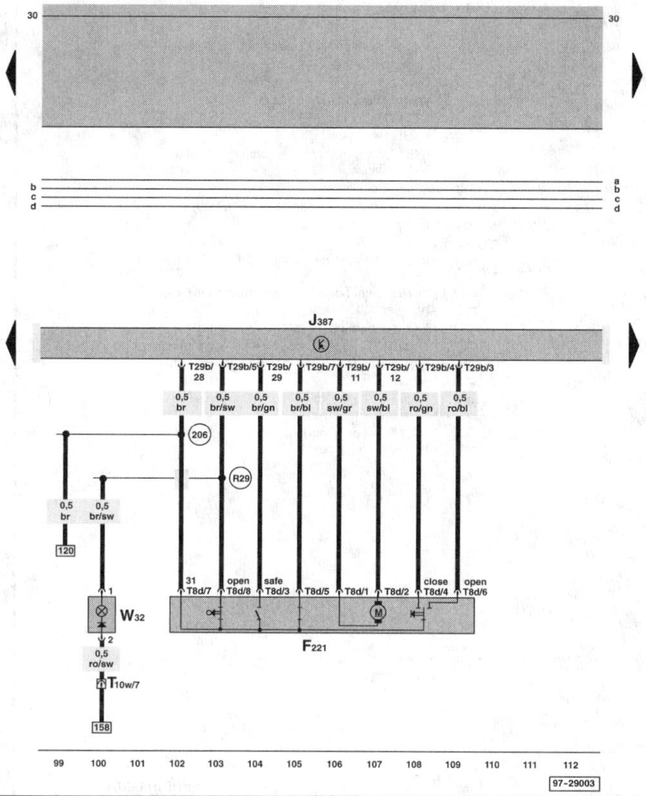

ws = white
sw = black
ro = red
br = brown
gn = green
bl = blue
gr = grey
li = lilac
ge = yellow
or = orange

Edition 06.04

Passat — Wiring diagram — No. 51/10

Door control module, Mirror adjustment and fold-in mirror, Heated outside mirror, passenger side, Switch for window regulator

E107 - Switch for window regulator, in passenger door
F261 - Passenger's Folding Mirror Contact Switch
J387 - Door control module, passenger side
L53 - Power Window Switch Light
T5f - 5-Pin Connector
T10w - 10-Pin Connector, black, connector station A-pillar, right
T12a - 12-Pin Connector, black
T29b - 29-Pin Connector, on Door control module, passenger side
V25 - Passenger's Side Mirror Adjustment Motor
V122 - Motor for fold-in mirror, passenger side
V148 - Motor for window regulator, passenger side
V150 - Motor for mirror adjustment, passenger side
Z5 - Heated outside mirror, passenger side

(R36) - Connection (mirror adjustment) in door wiring harness, passenger side
(W64) - CAN-Bus High Connection - Convenience (in floor wiring harness)
(W65) - CAN-Bus Low Connection - Convenience (in floor wiring harness)

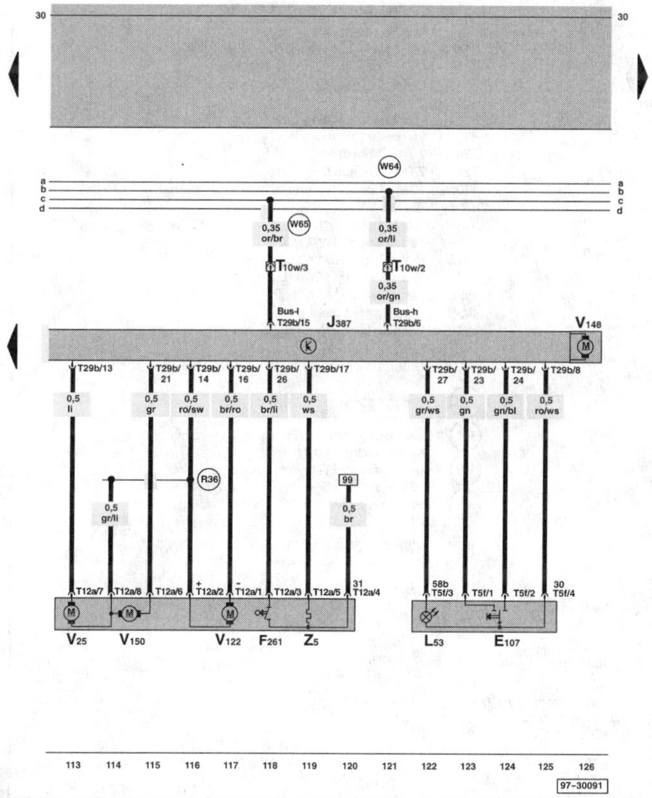

ws = white
sw = black
ro = red
br = brown
gn = green
bl = blue
gr = grey
li = lilac
ge = yellow
or = orange

Edition 06.04

Central locking rear right, Power window rear right, Entry light rear, right

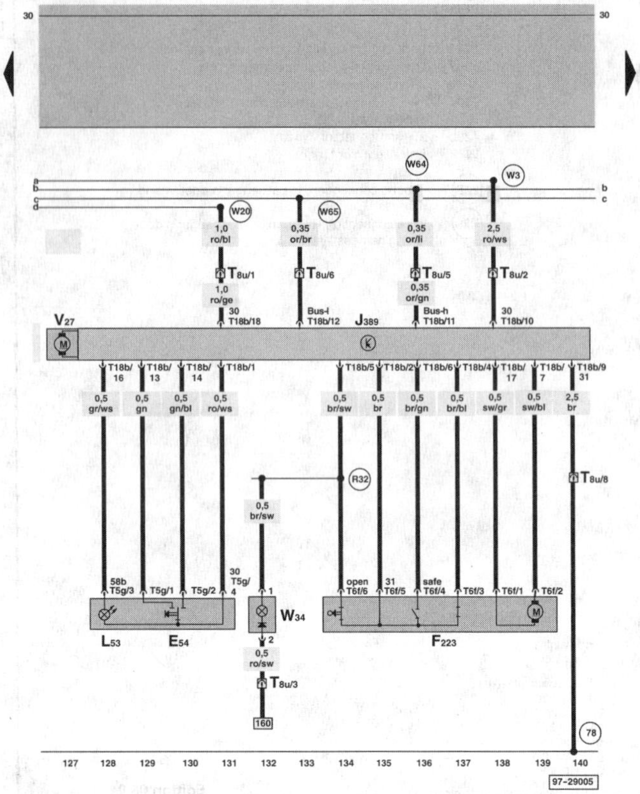

E54 - Right Rear Window Switch, (In RR Door)
F223 - Lock unit for central locking, rear, right
J389 - Door control module, rear, right
L53 - Power Window Switch Light
T5g - 5-Pin Connector
T6f - 6-Pin Connector
T8u - 8-Pin Connector, brown, connector station B-pillar, right
T18b - 18-Pin Connector
V27 - Motor for window regulator, rear, right
W34 - Entry light rear, right

(78) - Ground connection (lower right B-pillar)
(R32) - Wire connection -2-, in right rear door wiring harness
(W3) - wire connection, in rear wiring harness
(W20) - Plus connector (30a), in wiring harness, rear
(W64) - CAN-Bus High Connection - Convenience (in floor wiring harness)
(W65) - CAN-Bus Low Connection - Convenience (in floor wiring harness)

ws = white
sw = black
ro = red
br = brown
gn = green
bl = blue
gr = grey
li = lilac
ge = yellow
or = orange

Front Interior Light, Reading lights, Make-Up Mirror Lights, Sliding Cover Contact Switch

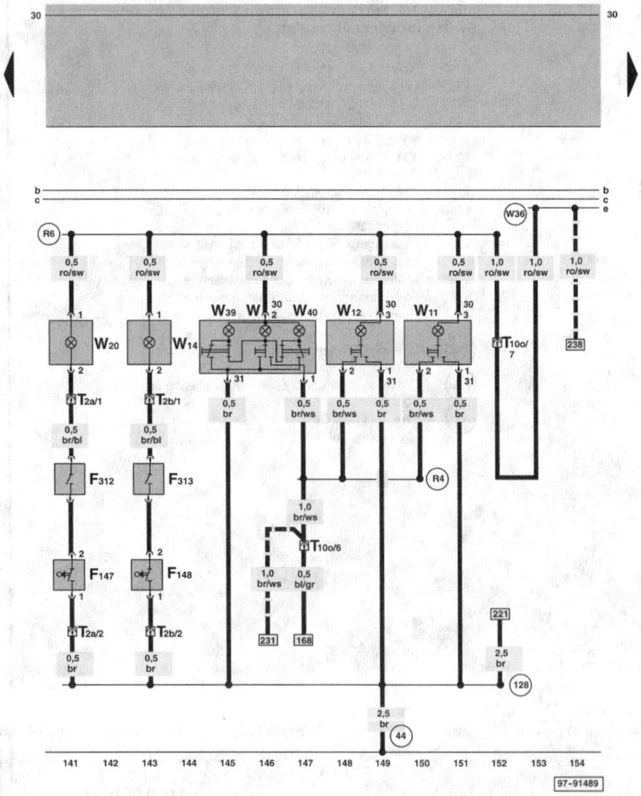

F147 - Left Make-Up Mirror Light Switch
F148 - Right Make-Up Mirror Light Switch
F312 - Driver's Sliding Cover Contact Switch (make-up mirror)
F313 - Passenger's Sliding Cover Contact Switch (make-up mirror)
T2a - 2-Pin Connector
T2b - 2-Pin Connector
T10o - 10-Pin Connector, green, connector station A-pillar, left
W - Front Interior Light
W11 - Left Rear Reading Light
W12 - Right Rear Reading Light
W14 - Right Make-up Mirror Light
W20 - Left Make-up Mirror Light
W39 - Reading light, center left
W40 - Reading light, center right

(44) - Ground connection, on left A-pillar, lower part
(128) - Ground connection -1-, in interrior light wiring harness
(R4) - wire connection -1-, in interior light/door contact switch wiring harness
(R6) - plus connection -1-, in interior light wiring harness
(W36) - Positive connection (30a) in floor wiring harness

- - - - engine code BDN only

ws = white
sw = black
ro = red
br = brown
gn = green
bl = blue
gr = grey
li = lilac
ge = yellow
or = orange

Repair Manual page number **Y 524**

Central control module for comfort system, Warning light for rear lid unlocked, Data Link Connector (DLC)

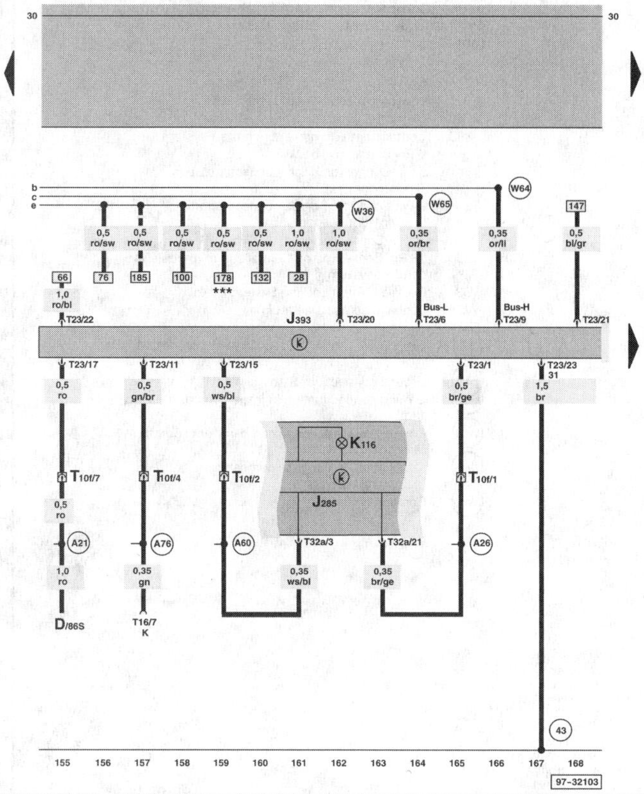

D - Ignition/Starter Switch
J285 - Control module with indicator unit in instrument panel insert
J393 - Central control module for comfort system
K116 - Warning light for rear lid unlocked
T10f - 10-Pin Connector, blue, connector station A-pillar, left
T16 - 16-Pin Connector, Data Link Connector (DLC), below instrument panel, left
T23 - 23-Pin Connector
T32a - 32-Pin Connector, blue, on instrument cluster

(43) - Ground connection, on right A-pillar, lower part

(A21) - wire connection (86s), in instrument panel wiring harness
(A26) - wire connection (driver's door contact switch), in instrument panel wiring harness
(A60) - Wire Connection (Vehicle speed signal), in instrument cluster wiring harness
(A76) - Connector (K-diagnosis wire) in instrument panel wiring harness
(W36) - Positive connection (30a) in floor wiring harness

(W64) - CAN-Bus High Connection - Convenience (in floor wiring harness)
(W65) - CAN-Bus Low Connection - Convenience (in floor wiring harness)

*** - sedan only
- - - - wagon only

ws = white
sw = black
ro = red
br = brown
gn = green
bl = blue
gr = grey
li = lilac
ge = yellow
or = orange

Edition 06.04

Central control module for comfort system, Luggage Compartment Light Switch, Decklid Central Locking System Motor

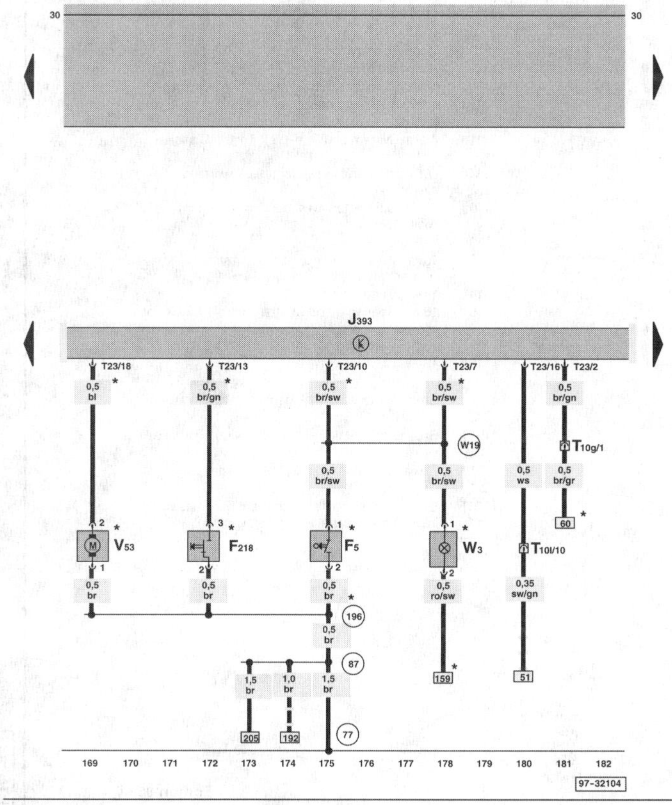

F5 - Luggage Compartment Light Switch
F218 - Switch for Central locking, trunk lid
J393 - Central control module for comfort system
T10g - 10-Pin Connector, grey
T10l - 10-Pin Connector, black, connector station A-pillar left
T23 - 23-Pin Connector, on Central control module for comfort system
V53 - Decklid Central Locking System Motor
W3 - Luggage compartment Light

(77) - Ground connection (lower left B-pillar)

(87) - Ground connection -2-, in rear wiring harness

(196) - Ground connection -3-, in wiring harness, rear

(W19) - Connection -2- (luggage compartment light), in rear wiring harness

* - sedan only
- - - - wagon only

ws = white
sw = black
ro = red
br = brown
gn = green
bl = blue
gr = grey
li = lilac
ge = yellow
or = orange

Edition 06.04

Repair Manual page number **Y 523**

Passat

Wiring diagram

No. 51/15

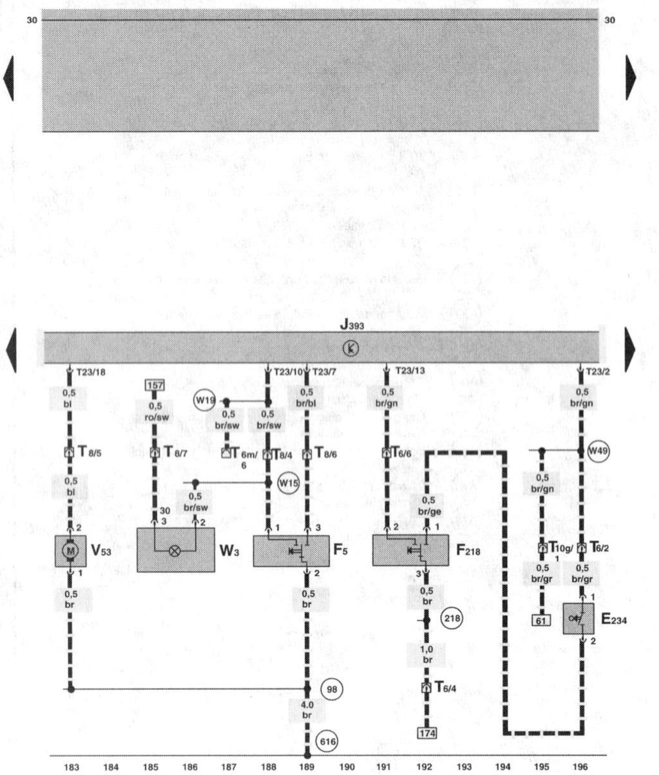

Central control module for comfort system, Luggage Compartment Light Switch, Luggage compartment light, Decklid Central Locking System Motor

E234 - Switch for unlocking, rear lid handle
F5 - Luggage Compartment Light Switch
F218 - Switch for central locking, trunk lid
J393 - Central control module for comfort system
T6 - 6-Pin Connector, black
T6m - 6-Pin Connector, black, connector station A-pillar, right
T8 - 8-Pin Connector, black, connector station D-pillar, right (wagon only)
T10g - 10-Pin Connector, grey
T23 - 23-Pin Connector, on Central control module for comfort system
V53 - Decklid Central Locking System Motor
W3 - Luggage compartment Light

(98) - Ground connection, in rear lid wiring harness

(218) - Ground connection -1-, in rear lid wiring harness
(616) - Ground connection (on luggage compartment light)
(W15) - Connection -1- (luggage compartment light), in rear wiring harness
(W19) - Connection -2- (luggage compartment light), in rear wiring harness
(W49) - Connection (in floor wiring harness

- - - - wagon only

ws = white
sw = black
ro = red
br = brown
gn = green
bl = blue
gr = grey
li = lilac
ge = yellow
or = orange

97-32105

Edition 06.04

Passat

Wiring diagram

No. 51/16

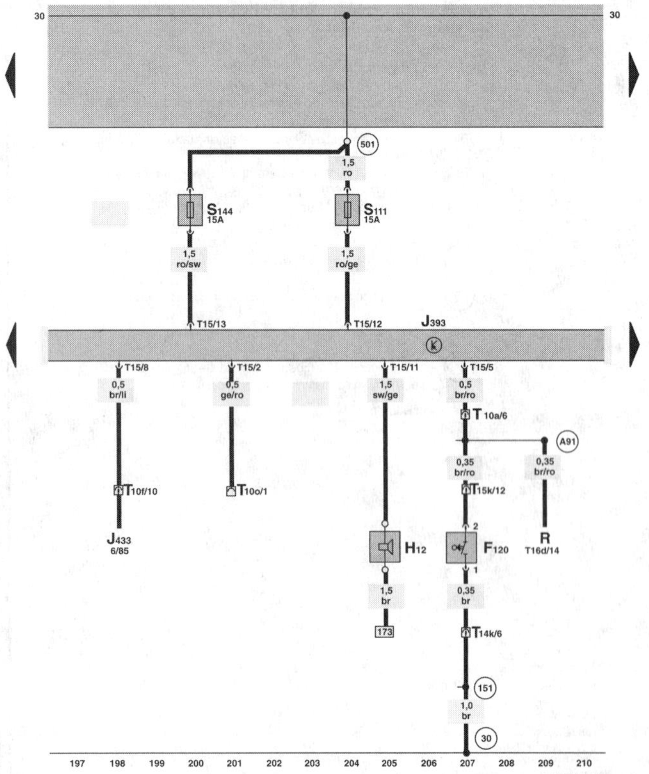

Central control module for comfort system, Alarm Switch, Hood Alarm Switch

F120 - Hood Alarm Switch
H12 - Alarm Switch
J393 - Central control module for comfort system
J433 - Locking relay for starter (anti-theft immobilizer)
R - Radio
S111 - Alarm System and Anti-Theft Engine Disable Fuse
S144 - Fuse for central locking/anti-theft warning system
T10a - 10-Pin Connector, grey, connector station A-pillar, left
T10f - 10-Pin Connector
T10o - 10-Pin Connector, red, connector station A-pillar, left
T14k - 14-Pin Connector
T15 - 15-Pin Connector, on Central control module for comfort system
T15k - 15-Pin Connector
T16d - 16-Pin Connector

(30) - Ground connection -1-, beside fuse/relay panel

(151) - Ground connection -1-, in front wiring harness

(501) - Threaded connection -2- (30), on relay plate

(A91) - Connector (signal for inatrument panel insert), in instrument panel wiring harness

ws = white
sw = black
ro = red
br = brown
gn = green
bl = blue
gr = grey
li = lilac
ge = yellow
or = orange

97-30097

Edition 06.04

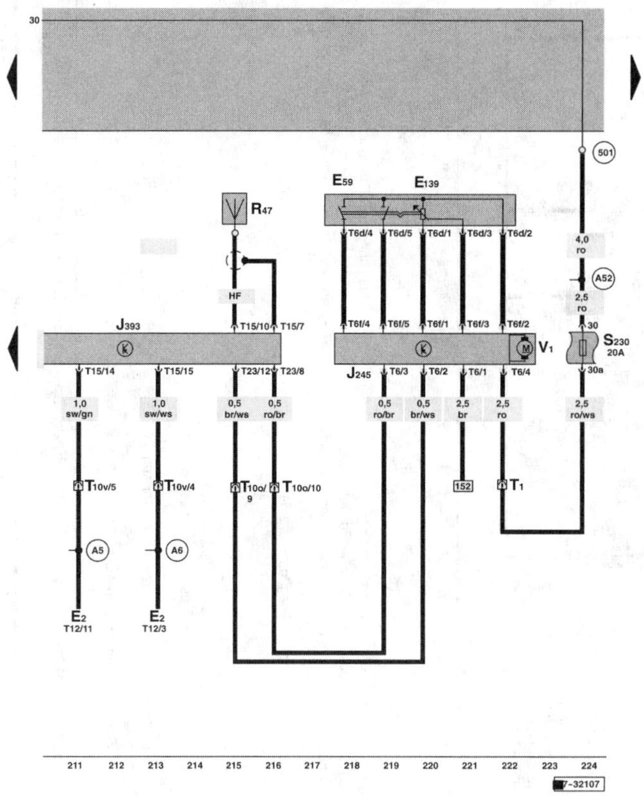

Central control module for comfort system, Power Sunroof

E2 - Turn signal switch
E59 - Raise/Lower Sunroof Switch
E139 - Sunroof Regulator
J245 - Power Sunroof Control Module
J393 - Central control module for comfort system
R47 - Central Locking and Anti-theft Alarm System Antenna
S230 - Fuse in fuse holder
T1 - Single Connector, black
T6 - 6-Pin Connector
T6d - 6-Pin Connector
T6f - 6-Pin Connector
T10v - 10-Pin Connector, violet, connector station A-pillar, right
T10o - 10-Pin Connector, red, connector station A-pillar, left
T12 - 12-Pin Connector, near steering column
T15 - 15-Pin Connector, on Central control module for comfort system
T23 - 23-Pin Connector
V1 - Sunroof Motor

(501) - Threaded connection -2- (30), on relay plate
(A52) - Plus Connection 2 (30) (in instrument panel wiring harness)
(A5) - plus connection (right turn signal), in instrument panel wiring harness
(A6) - plus connection (left turn signal), in instrument panel wiring harness

ws = white
sw = black
ro = red
br = brown
gn = green
bl = blue
gr = grey
li = lilac
ge = yellow
or = orange

Edition 06.04

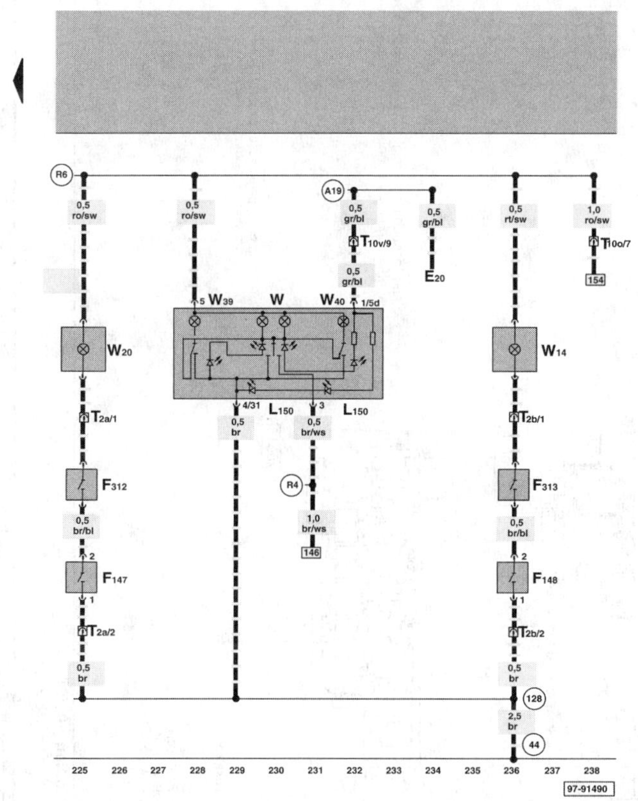

Front Interior Light, Make-up Mirror Light, Reading light

E20 - Instrument Panel Light Dimmer Switch
F147 - Left Make-Up Mirror Light Switch
F148 - Right Make-Up Mirror Light Switch
F312 - Driver's Sliding Cover Contact Switch (make-up mirror)
F313 - Passenger's Sliding Cover Contact Switch (make-up mirror)
L150 - Center Console Illumination
T2a - 2-Pin Connector
T2b - 2-Pin Connector
T10o - 10-Pin Connector, green, connector station A-pillar, left
T10v - 10-Pin Connector
W - Front Interior Light
W14 - Right Make-up Mirror Light
W20 - Left Make-up Mirror Light
W39 - Reading light, center left
W40 - Reading light, center right

(44) - Ground connection, on left A-pillar, lower part
(128) - Ground connection -1-, in interior light wiring harness
(A19) - wire connection (58d), in instrument panel wiring harness
(R4) - wire connection -1-, in interior light/door contact switch wiring harness
(R6) - plus connection -1-, in interior light wiring harness

- - - - engine code BDN only

ws = white
sw = black
ro = red
br = brown
gn = green
bl = blue
gr = grey
li = lilac
ge = yellow
or = orange

Edition 06.04

Airbag Systems

from October 2000

Deviate relay location and fuseplacements as well as the locations of multiple connectors see section "component locations".

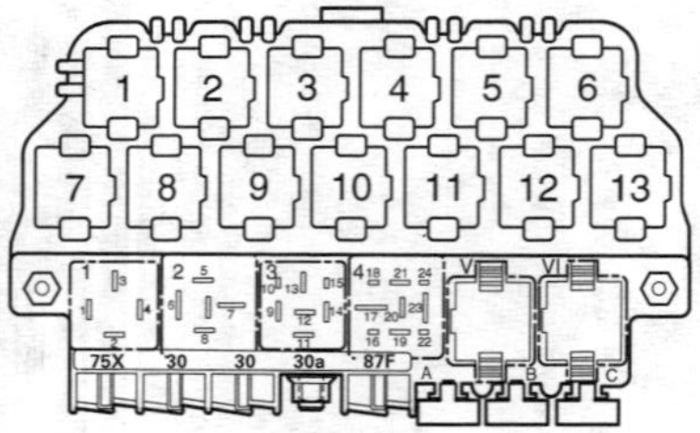

97-20676

Relay location on the thirteen position auxiliary relay panel, above relay panel:

Relay panel:

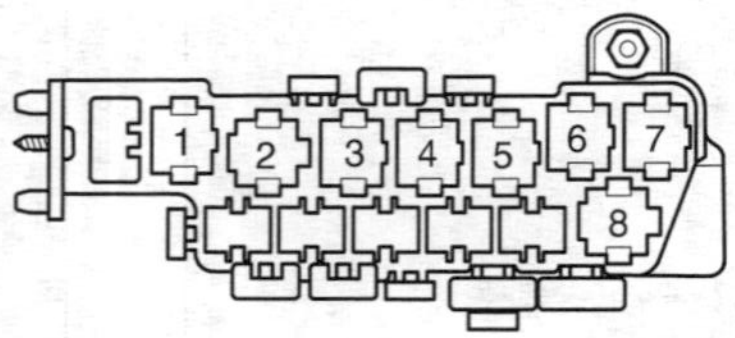

97-20515

Relay location on the eight position auxiliary relay panel, behind relay panel:

Note: Number in parentheses indicates production control number stamped on relay housing.

Fuse colors

30 A - green
25 A - white
20 A - yellow
15 A - blue
10 A - red
7,5 A - brown
5 A - beige

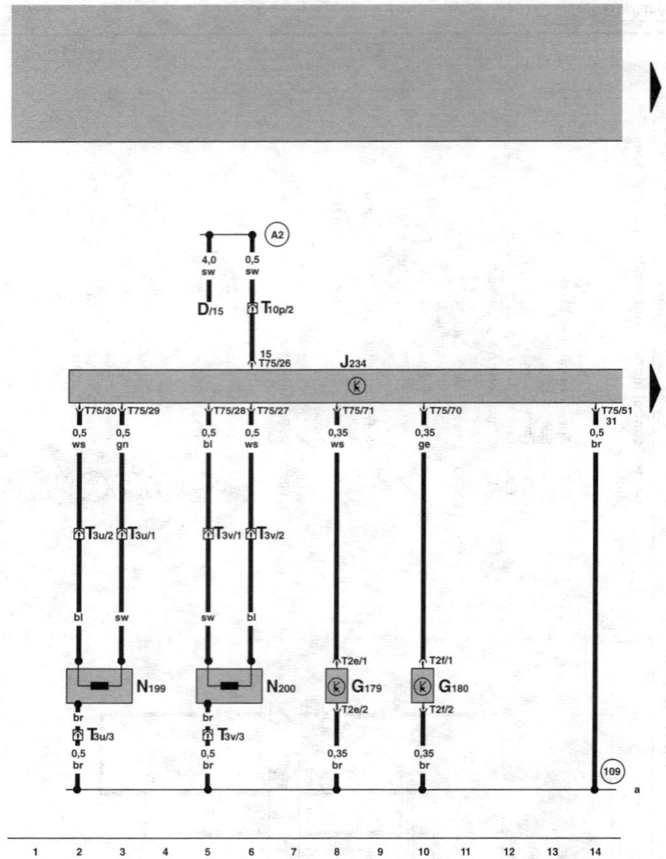

ws = white
sw = black
ro = red
br = brown
gn = green
bl = blue
gr = grey
li = lilac
ge = yellow
or = orange

D - Ignition/Starter Switch
G179 - Crash sensor for side airbag, driver's side
G180 - Crash sensor for side airbag, passenger side
J234 - Airbag Control Module
N199 - Igniter for side airbag, driver's side
N200 - Igniter for side airbag, passenger side
T2e - Double Connector
T2f - Double Connector
T3u - 3-Pin Connector, yellow, below driver's seat
T3v - 3-Pin Connector, yellow, below passenger's seat
T10p - 10-Pin Connector, yellow, connector station A-pillar, right
T75 - 75-Pin Connector, on Airbag Control Module

(A2) - plus connection (15), in instrument panel wiring harness
(109) - Ground connection, in airbag wiring harness

Airbag Control Module, Igniter for side airbag, Crash sensor for side airbag

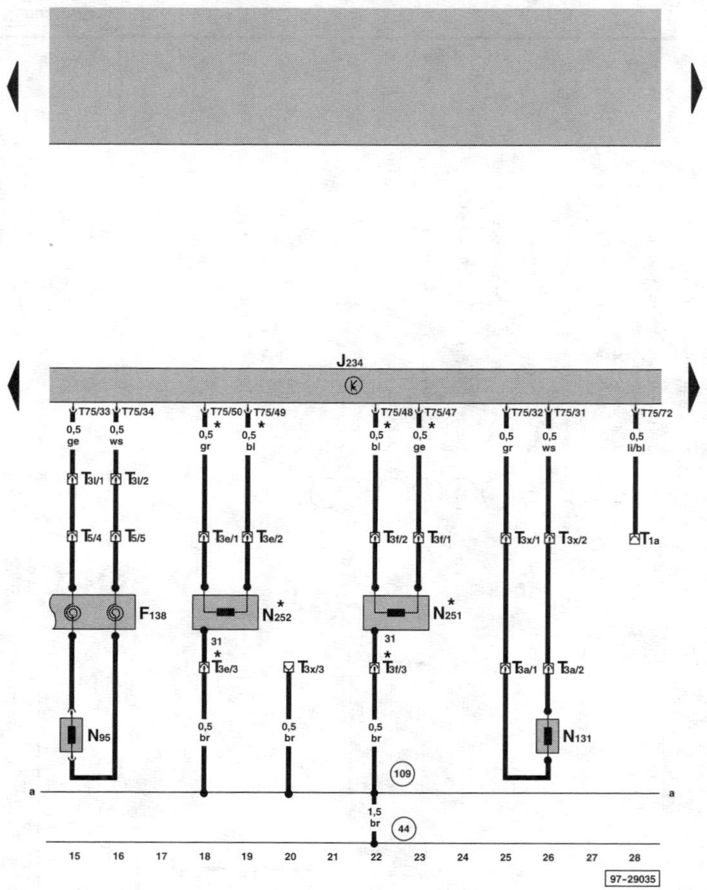

ws = white
sw = black
ro = red
br = brown
gn = green
bl = blue
gr = grey
li = lilac
ge = yellow
or = orange

- F138 - Airbag Spiral Spring/Return Spring With Slip Ring
- J234 - Airbag Control Module, behind console
- N95 - Driver's Side Airbag Igniter
- N131 - Passenger's Side Airbag Igniter 1
- N251 - Igniter for head airbag, rear, driver side
- N252 - Front Passenger Head Airbag Igniter
- T1a - Single Connector, black, connector station A-pillar, right
- T3a - 3-Pin Connector, behind instrument panel, right
- T3e - 3-Pin Connector
- T3f - 3-Pin Connector
- T3l - 3-Pin Connector, yellow, connector station A-pillar, left
- T3x - 3-Pin Connector, yellow, connector station A-pillar, right
- T5 - 5-Pin Connector, yellow, behind steering column switch cover
- T75 - 75-Pin Connector, on Airbag Control Module
- (44) - Ground connection, on left A-pillar, lower part
- (109) - Ground connection, in airbag wiring harness
- * - vehicles with head airbag only

Airbag Control Module, Driver's Side Airbag, Igniter for head airbag, Front Passenger's Side Airbag Igniter

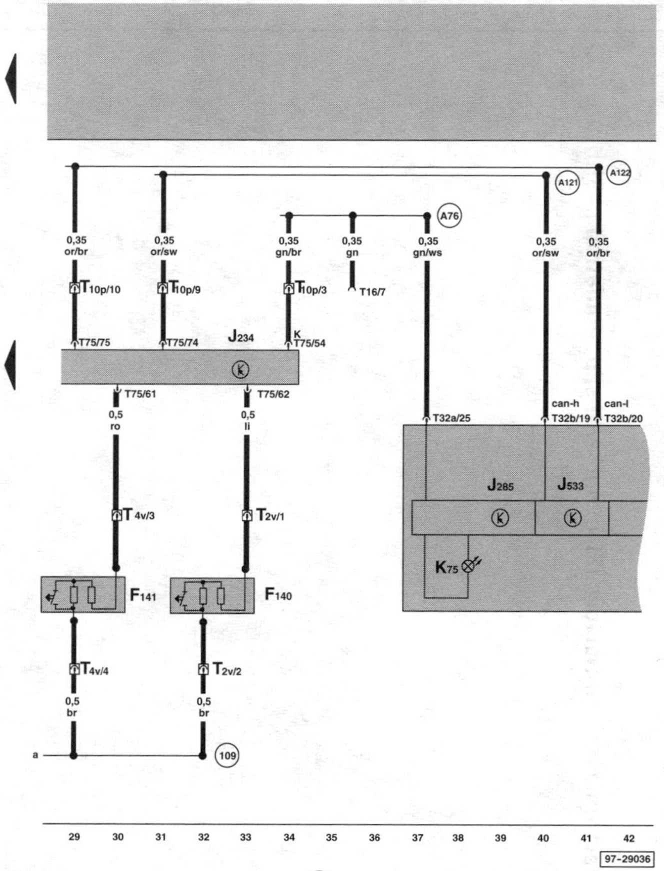

ws = white
sw = black
ro = red
br = brown
gn = green
bl = blue
gr = grey
li = lilac
ge = yellow
or = orange

- F140 - Left Front Seat Belt Microswitch*
- F141 - Right Front Seat Belt Microswitch*
- J234 - Airbag Control Module, behind console
- J285 - Control module with indicator unit in instrument panel insert
- J533 - Data Bus On Board Diagnostic Interface
- K75 - Airbag Malfunction Indicator Lamp (MIL)
- T10p - 10-Pin Connector, yellow, connector station A-pillar, right
- T2v - Double Connector
- T4v - 4-Pin Connector
- T16 - 16-Pin Connector, Data Link Connector (DLC) below instrument panel, left
- T32a - 32-Pin Connector, blue, on instrument cluster
- T32b - 32-Pin Connector, green, on instrument cluster
- T75 - 75-Pin Connector, on Airbag Control Module
- (109) - Ground connection, in airbag wiring harness
- (A76) - Connector (K-diagnosis wire) in instrument panel wiring harness
- (A121) - Connection (high bus) in instrument panel wiring harness
- (A122) - Connection (low bus) in instrument panel wiring harness
- * - Scan Tool display text for these component codes may be different than indicated

Airbag Control Module, Control module with indicator unit in instrument panel insert, Data Link Connector (DLC) , Front seat belt microswitches

Anti-lock brake system (ABS) with electronic differential lock (EDL) and anti-slip regulation (ASR), front wheel drive only,

from October 2000

Notes:

Information for

- *Relay and fuse arrangements*
- *Connector arrangements*
- *Control Modules and Relays*
- *Ground connections*

⇒ Component Locations!

Information for

- *Troubleshooting Procedures*

⇒ Guided Fault Finding using VAS 5051/5052

ABS Control Module, Anti-Slip Control Switch

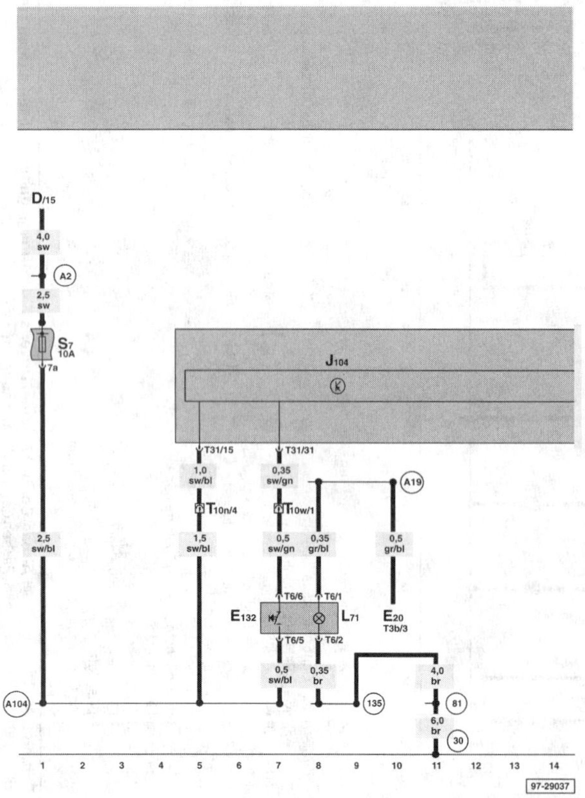

D - Ignition/Starter Switch
E20 - Instrument Panel Light Dimmer Switch
E132 - Anti-Slip Control Switch
J104 - ABS Control Module
L71 - Anti-slip Control Switch Illumination
S7 - Fuse
T3b - 3-Pin Connector, black on Instrument Panel Light Dimmer Switch
T6 - 6-Pin Connector, black
T10n - 10-Pin Connector, orange, connector station A-pillar, left
T10w - 10-Pin Connector, grey, connector station A-pillar, right
T31 - 31-Pin Connector

(30) - Ground connection, -1-, beside fuse/relay panel
(81) - Ground connection -1-, in instrument panel wiring harness
(135) - Ground connection -2-, in instrument panel wiring harness
(A2) - plus connection (15), in instrument panel wiring harness
(A19) - wire connection (58d), in instrument panel wiring harness
(A104) - Plus connection -2- (15), in instrument panel wiring harness

ws = white
sw = black
ro = red
br = brown
gn = green
bl = blue
gr = grey
li = lilac
ge = yellow
or = orange

ABS Control Module, ABS Wheel Speed Sensor

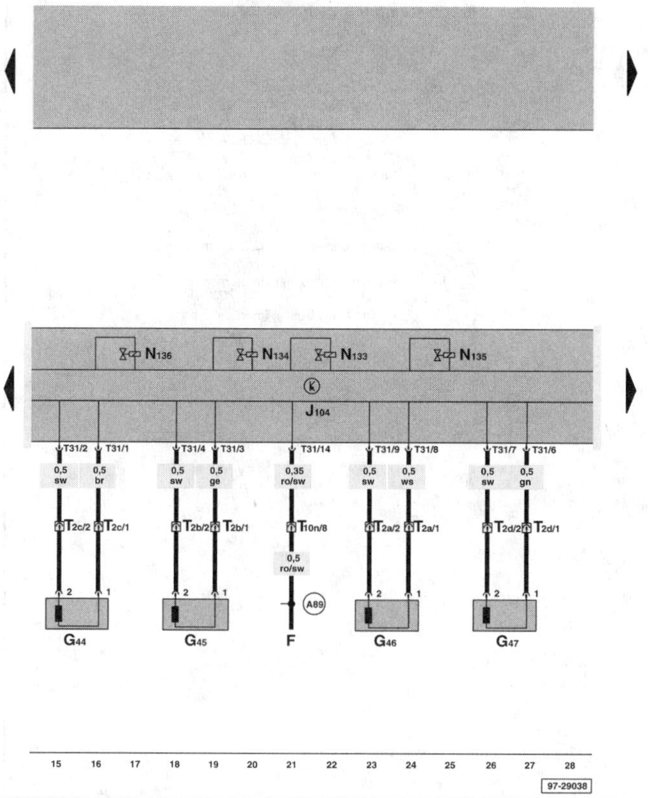

- F - Brake Light Switch
- G44 - Right Rear ABS Wheel Speed Sensor
- G45 - Right Front ABS Wheel Speed Sensor
- G46 - Left Rear ABS Wheel Speed Sensor
- G47 - Left Front ABS Wheel Speed Sensor
- J104 - ABS Control Module
- N133 - Right Rear ABS Inlet Valve
- N134 - Left Rear ABS Inlet Valve
- N135 - Right Rear ABS Outlet Valve
- N136 - Left Rear ABS Outlet Valve
- T2a - Double Connector (under rear seat)
- T2b - Double Connector (in right front wheel well)
- T2c - Double Connector (under rear seat)
- T2d - Double Connector (in left front wheel well)
- T10n - 10-Pin Connector, orange, connector station A-pillar, left
- T31 - 31-Pin Connector
- (A89) - Connection 2 (54) (in instrument panel wiring harness)

ws = white
sw = black
ro = red
br = brown
gn = green
bl = blue
gr = grey
li = lilac
ge = yellow
or = orange

ABS Control Module, Data Link Connector (DLC)

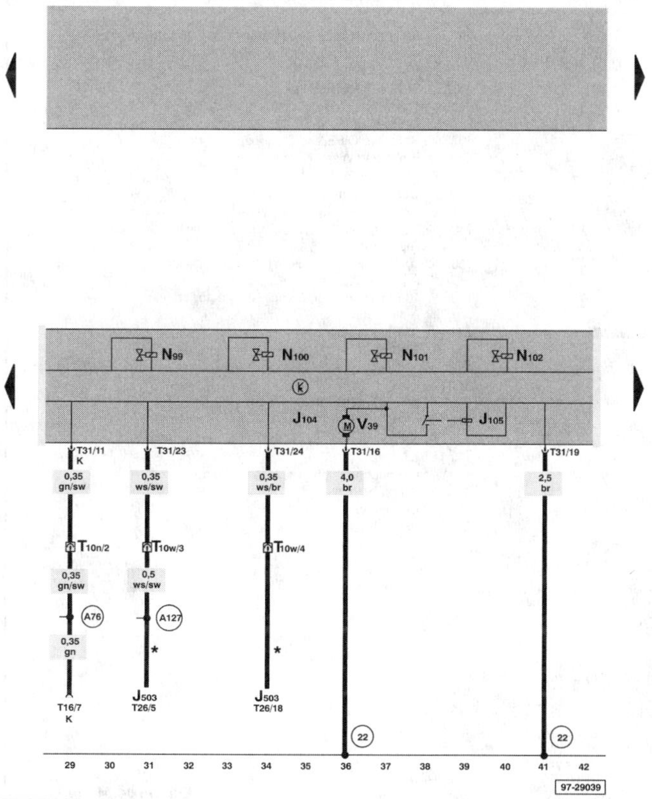

- J104 - ABS Control Module
- J105 - ABS Return Flow Pump Relay
- J503 - Radio/Navigation Display Control Module
- N99 - Right Front ABS Inlet Valve
- N100 - Right Front ABS Outlet Valve
- N101 - Left Front ABS Inlet Valve
- N102 - Left Front ABS Outlet Valve
- T10n - 10-Pin Connector, orange, connector station A-pillar, left
- T10w - 10-Pin Connector, grey, connector station A-pillar, right
- T16 - 16-Pin Connector, black, Data Link Connector (DLC), below instrument panel, left
- T26 - 26-Pin Connector, on Radio/Navigation Display Control Module
- T31 - 31-Pin Connector
- V39 - ABS Return Flow Pump
- (22) - Ground connection, on hydraulic unit
- (A76) - Connection (K-diagnosis wire) in instrument panel wiring harness
- (A127) - Connection (left rear speed sensor) in instrument panel wiring harness
- * - Navigation only

ws = white
sw = black
ro = red
br = brown
gn = green
bl = blue
gr = grey
li = lilac
ge = yellow
or = orange

ABS Control Module, ABS Solenoid Valve Relay, Traction Control Switch-over Valves

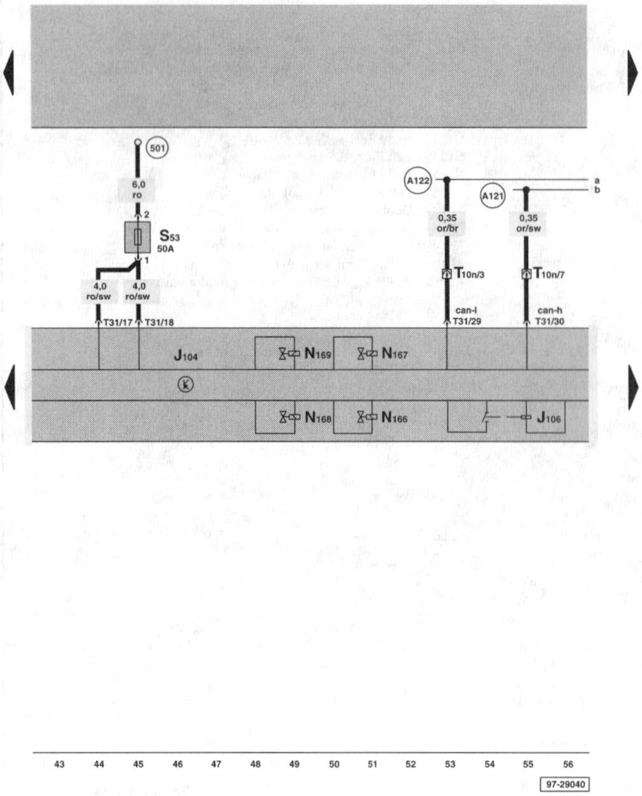

J104 - ABS Control Module
J106 - ABS Solenoid Valve Relay
N166 - Right Front Traction Control Switch-Over Valve
N167 - Right Front Traction Control Outlet Valve
N168 - Left Front Traction Control Switch-Over Valve
N169 - Left Front Traction Control Outlet Valve
S53 - ABS Hydraulic Pump Fuse
T10n - 10-Pin Connector, orange, connector station A-pillar, left
T31 - 31-Pin Connector

(501) - Threaded connection -2- (30) on the relay plate

(A121) - Connection (high bus) in instrument panel wiring harness

(A122) - Connection (low bus) in instrument panel wiring harness

ws = white
sw = black
ro = red
br = brown
gn = green
bl = blue
gr = grey
li = lilac
ge = yellow
or = orange

ABS Control Module, Control module with indicator unit

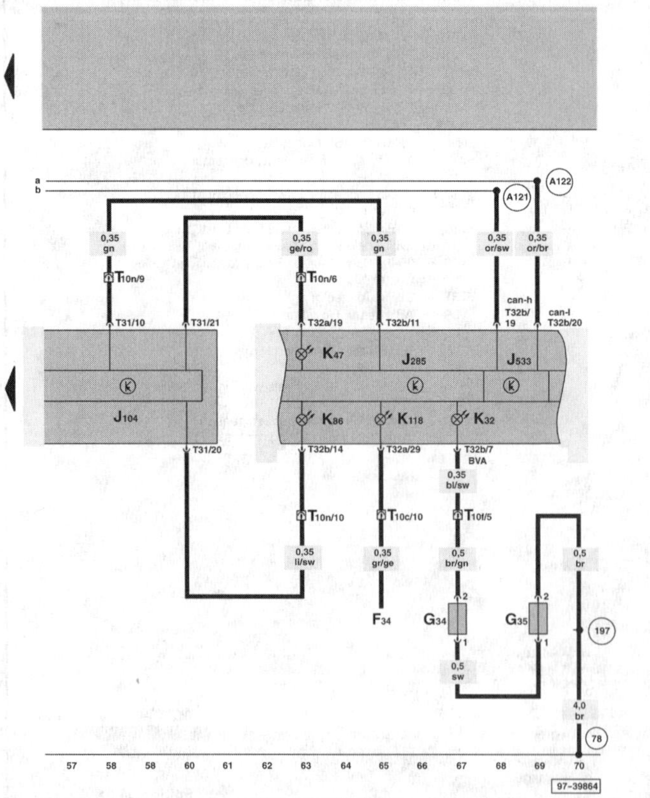

F34 - Brake Fluid Level Warning Switch
G34 - Left Front Brake Pad Wear Sensor
G35 - Right Front Brake Pad Wear Sensor
J104 - ABS Control Module (w/EDL)
J285 - Control module with indicator unit in instrument panel insert
J533 - Data Bus On Board Diagnostic Interface
K32 - Brake Pad Wear Indicator Light
K47 - ABS Warning Light
K86 - Traction Control Indicator Light
K118 - Warning light for brake system
T10c - 10-Pin Connector, lilac, connector station A-pillar, left
T10f - 10-Pin Connector, blue, connector station A-pillar, left
T10n - 10-Pin Connector, orange, connector station A-pillar, left
T31 - 31-Pin Connector
T32a - 32-Pin Connector, blue, on instrument cluster
T32b - 32-Pin Connector, green, on instrument cluster

(78) - Ground connection (lower right B-pillar)

(197) - Ground connection 4, in wiring harness rear

(A121) - Connection (high bus) in instrument panel wiring harness

(A122) - Connection (low bus) in instrument panel wiring harness

ws = white
sw = black
ro = red
br = brown
gn = green
bl = blue
gr = grey
li = lilac
ge = yellow
or = orange

Anti-lock brake system (ABS) with electronic differential lock (EDL),

from October 2000

Notes:

Information for

- *Relay and fuse arrangements*
- *Connector arrangements*
- *Control Modules and Relays*
- *Ground connections*

⇒ Component Locations!

Information for

- *Troubleshooting Procedures*

⇒ Guided Fault Finding using VAS 5051/5052

ABS Control Module, ABS Wheel Speed Sensors

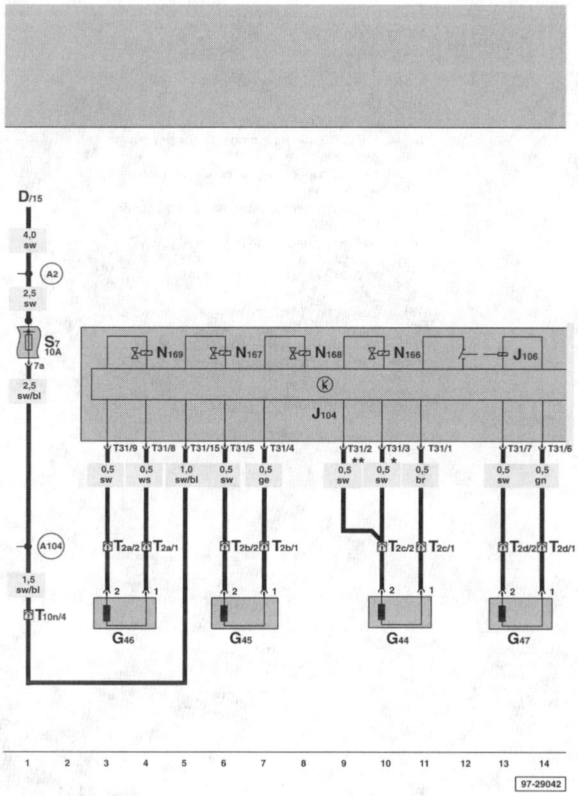

D - Ignition/Starter Switch
G44 - Right Rear ABS Wheel Speed Sensor
G45 - Right Front ABS Wheel Speed Sensor
G46 - Left Rear ABS Wheel Speed Sensor
G47 - Left Front ABS Wheel Speed Sensor
J104 - ABS Control Module
J106 - ABS Solenoid Valve Relay
N166 - Right Front Traction Control Switch-Over Valve
N167 - Right Front Traction Control Outlet Valve
N168 - Left Front Traction Control Switch-Over Valve
N169 - Left Front Traction Control Outlet Valve
S7 - Fuse
T2a - Double Connector (under rear seat)
T2b - Double Connector (in right front wheel well)
T2c - Double Connector (under rear seat)
T2d - Double Connector (in left front wheel well)
T10n - 10-Pin Connector, orange, connector station A-pillar, left
T31 - 31-Pin Connector

(A2) - plus connection (15), in instrument panel wiring harness
(A104) - Plus connection -2- (15), in instrument panel wiring harness

* - front wheel drive only
** - 4-Motion only

ws = white
sw = black
ro = red
br = brown
gn = green
bl = blue
gr = grey
li = lilac
ge = yellow
or = orange

ABS Control Module, ABS Return Flow Pump

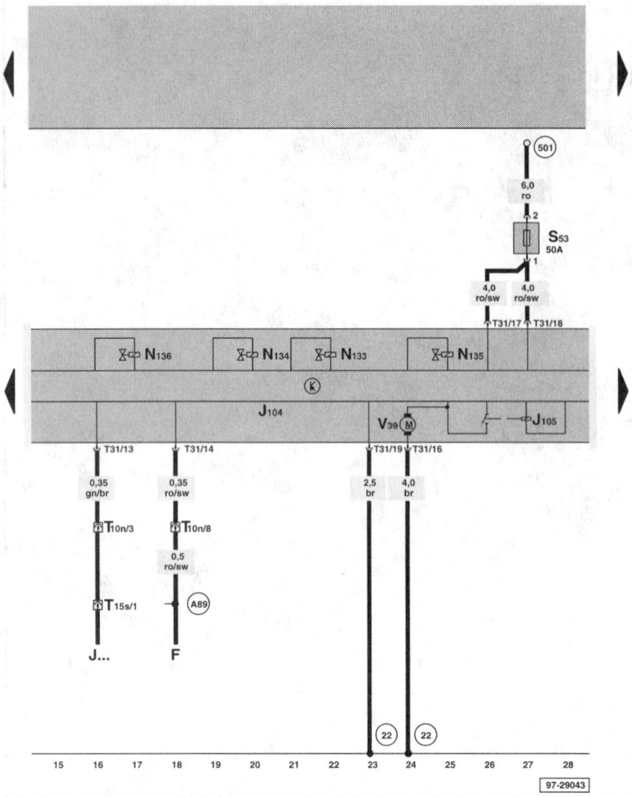

- J... - Engine Control Module (ECM)
- F - Brake Light Switch
- J104 - ABS Control Module
- J105 - ABS Return Flow Pump Relay
- N133 - Right Rear ABS Inlet Valve
- N134 - Left Rear ABS Inlet Valve
- N135 - Right Rear ABS Outlet Valve
- N136 - Left Rear ABS Outlet Valve
- S53 - ABS Hydraulic Pump Fuse
- T10n - 10-Pin Connector, orange, connector station A-pillar, left
- T15s - 15-Pin Connector, red
- T31 - 31-Pin Connector
- V39 - ABS Return Flow Pump

- (22) - Ground connection, on hydraulic unit
- (501) - Threaded connection -2- (30) on the relay plate
- (A89) - Connection 2 (54) (in instrument panel wiring harness)

ws = white
sw = black
ro = red
br = brown
gn = green
bl = blue
gr = grey
li = lilac
ge = yellow
or = orange

Edition 04.04

ABS Control Module, Data Link Connector (DLC)

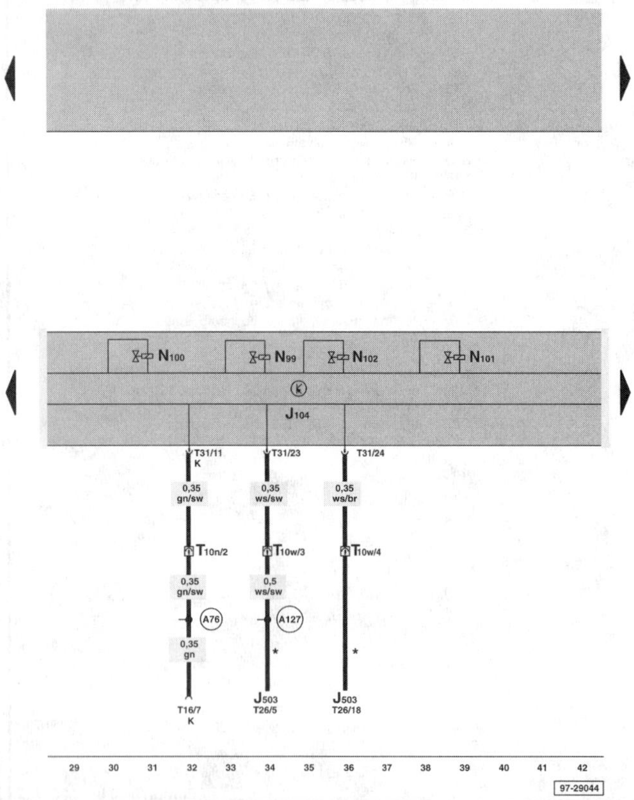

- J104 - ABS Control Module
- J503 - Radio/Navigation Display Control Module
- N99 - Right Front ABS Inlet Valve
- N100 - Right Front ABS Outlet Valve
- N101 - Left Front ABS Inlet Valve
- N102 - Left Front ABS Outlet Valve
- T10n - 10-Pin Connector, orange, connector station A-pillar, left
- T10w - 10-Pin Connector, grey, connector station A-pillar, right
- T16 - 16-Pin Connector, black, Data Link Connector (DLC), below instrument panel, left
- T26 - 26-Pin Connector, on Radio/Navigation Display Control Module
- T31 - 31-Pin Connector

- (A76) - Connection (K-diagnosis wire) in instrument panel wiring harness
- (A127) - Connection (left rear speed sensor) in instrument panel wiring harness

- * - Navigation only

ws = white
sw = black
ro = red
br = brown
gn = green
bl = blue
gr = grey
li = lilac
ge = yellow
or = orange

Edition 04.04

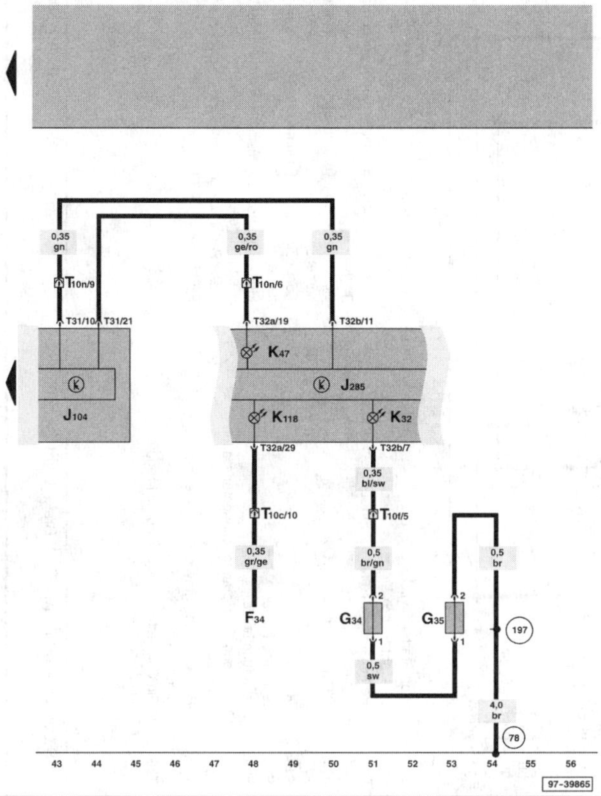

ABS Control Module, Control module with indicator unit

F34 - Brake Fluid Level Warning Switch
G34 - Left Front Brake Pad Wear Sensor
G35 - Right Front Brake Pad Wear Sensor
J104 - ABS Control Module (w/EDL)
J285 - Control module with indicator unit in instrument panel insert
K32 - Brake Pad Wear Indicator Light
K47 - ABS Warning Light
K118 - Warning light for brake system
T10c - 10-Pin Connector, lilac, connector station A-pillar, left
T10f - 10-Pin Connector, blue, connector station A-pillar, left
T10n - 10-Pin Connector, orange, connector station A-pillar, left
T31 - 31-Pin Connector
T32a - 32-Pin Connector, blue, on instrument cluster
T32b - 32-Pin Connector, green, on instrument cluster

(78) - Ground connection (lower right B-pillar)

(197) - Ground connection 4, in wiring harness rear

ws = white
sw = black
ro = red
br = brown
gn = green
bl = blue
gr = grey
li = lilac
ge = yellow
or = orange

Edition 04.04

Repair Manual page number Y 544

THIS PAGE INTENTIONALLY LEFT BLANK

Repair Manual page number Y 543

Anti-lock brake system (ABS) with electronic differential lock (EDL), anti-slip regulation (ASR) and electronic stabilization program (ESP)

from October 2000

Deviate relay location and fuseplacements as well as the locations of multiple connectors see section component locations.

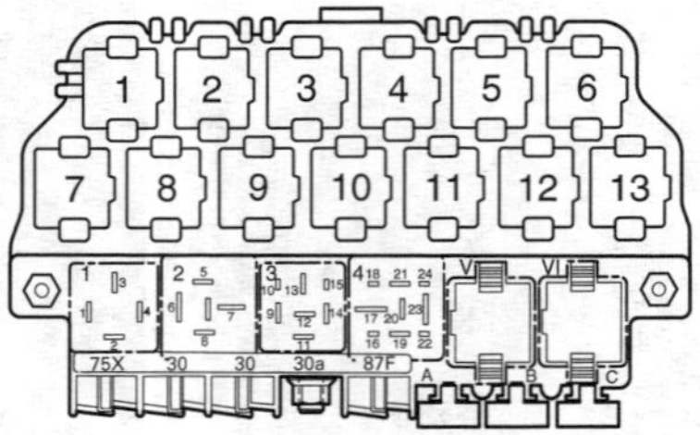

97-20676

Relay location on the thirteen position auxiliary relay panel, above relay panel:

Relay panel:

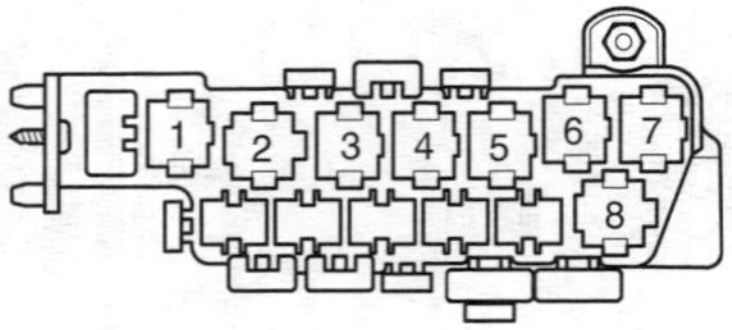

97-20515

Relay location on the eight position auxiliary relay panel, behind relay panel:

4 ABS Solenoid Valve Relay (373)

7 ABS Return Flow Pump Relay (373)

Note: Number in parentheses indicates production control number stamped on relay housing.

Fuse colors
30 A - green
25 A - white
20 A - yellow
15 A - blue
10 A - red
7,5 A - brown
5 A - beige

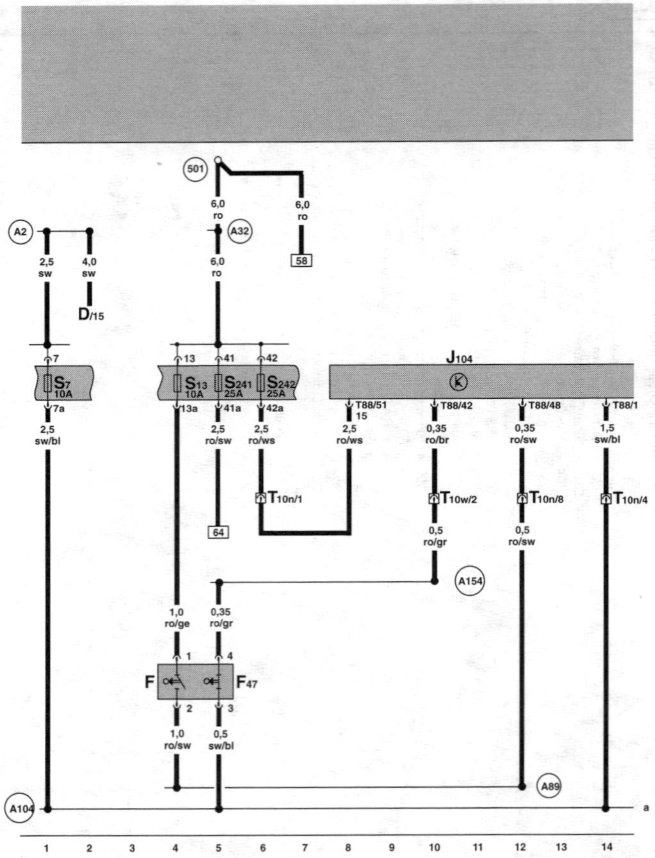

ws = white
sw = black
ro = red
br = brown
gn = green
bl = blue
gr = grey
li = lilac
ge = yellow
or = orange

D - Ignition/Starter Switch
F - Brake Light Switch
F47 - Brake Pedal Switch (cruise control/Diesel Direct Fuel Injection
J104 - ABS Control Module (w/EDL)
S7 - Fuse
S13 - Fuse
S241 - Fuse in fuse holder
S242 - Fuse in fuse holder
T10n - 10-Pin Connector, orange, connector station A-pillar, left
T10w - 10-Pin Connector, grey, connector station A-pillar, right
T88 - 88-Pin Connector, on ABS Control Module (w/EDL)

(501) - Threaded connection -2- (30) on the relay plate
(A2) - plus connection (15), in instrument panel wiring harness
(A32) - plus connection (30), in instrument panel wiring harness
(A89) - Connection 2 (54) (in instrument panel wiring harness)
(A104) - Plus connector -2- (15), in instrument panel wiring harness
(A154) - Brake Pedal Switch Connection (in instrument panel wiring harness)

ABS Control Module (w/EDL), Brake Light Switch

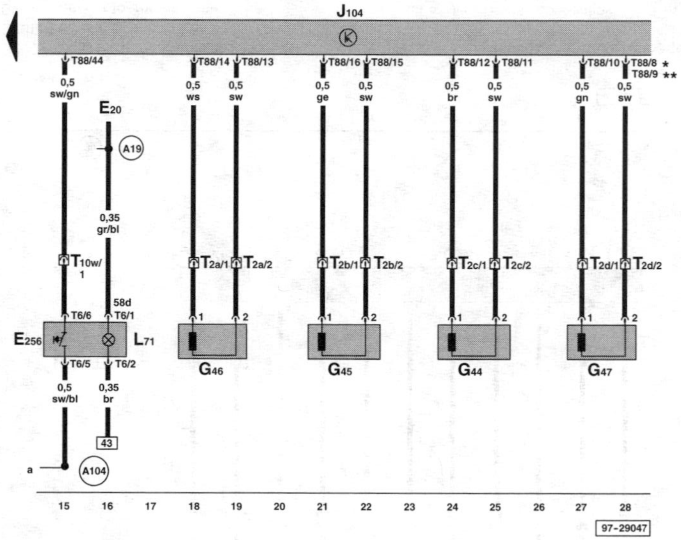

ws = white
sw = black
ro = red
br = brown
gn = green
bl = blue
gr = grey
li = lilac
ge = yellow
or = orange

E20 - Instrument Panel Light Dimmer Switch
E256 - ASR/ESP Button
G44 - Right Rear ABS Wheel Speed Sensor
G45 - Right Front ABS Wheel Speed Sensor
G46 - Left Rear ABS Wheel Speed Sensor
G47 - Left Front ABS Wheel Speed Sensor
J104 - ABS Control Module (w/EDL)
L71 - Anti-slip Control Switch Illumination
T2a - Doubel Connector (under rear seat)
T2b - Doubel Connector (in right front wheel well)
T2c - Doubel Connector (under rear seat)
T6 - 6-Pin Connector, red on ASR/ESP Button
T10w - 10-Pin Connector, grey, connector station A-pillar, right
T88 - 88-Pin Connector, on ABS Control Module (w/EDL)

(A19) - wire connection (58d), in instrument panel wiring harness
(A104) - Plus connector -2- (15), in instrument panel wiring harness
* all wheel drive only
** 4-Motion only

ABS Control Module (w/EDL), Wheel Speed Sensor, ASR/ESP Button

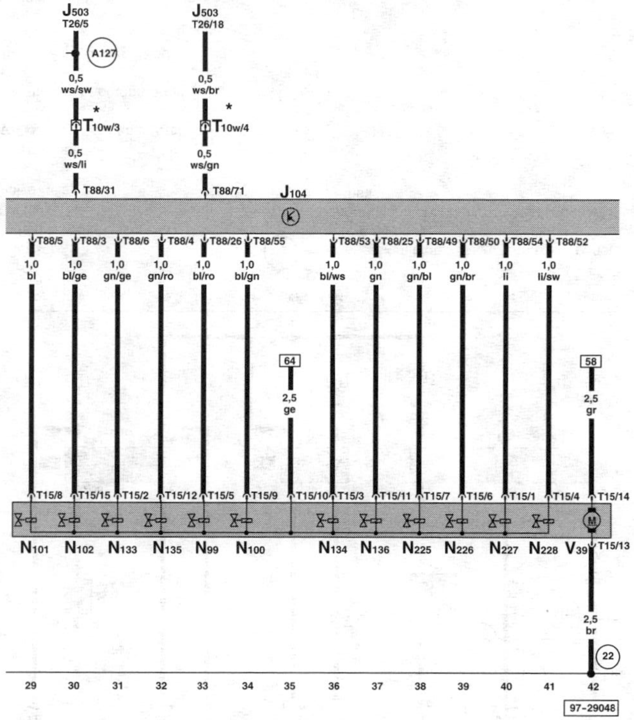

ws = white
sw = black
ro = red
br = brown
gn = green
bl = blue
gr = grey
li = lilac
ge = yellow
or = orange

J104 - ABS Control Module (w/EDL)
J503 - Radio/Navigation Display Control Module
N99 - Right Front ABS Inlet Valve
N100 - Right Front ABS Outlet Valve
N101 - Left Front ABS Inlet Valve
N102 - Left Front ABS Outlet Valve
N133 - Right Rear ABS Inlet Valve
N134 - Left Rear ABS Inlet Valve
N135 - Right Rear ABS Outlet Valve
N136 - Left Rear ABS Outlet Valve
N225 - Pilot valve -1- traction control
N226 - Pilot valve -2- traction control
N227 - High pressure switch valve -1- traction control
N228 - High pressure switch valve -2- traction control
T10w - 10-Pin Connector, grey, connector station A-pillar, right

T15 - 15-Pin Connector, black
T26 - 26-Pin Connector, on Radio/Navigation Display Control Module
T88 - 88-Pin Connector, on ABS Control Module (w/EDL)
V39 - ABS Return Flow Pump

(22) - Ground connection, on hydraulic unit
(A127) - Connector (left rear speed sensor) in instrument panel wiring harness
* Navigation only

ABS Control Module (w/EDL), ABS Inlet Valve, ABS Outlet Valve

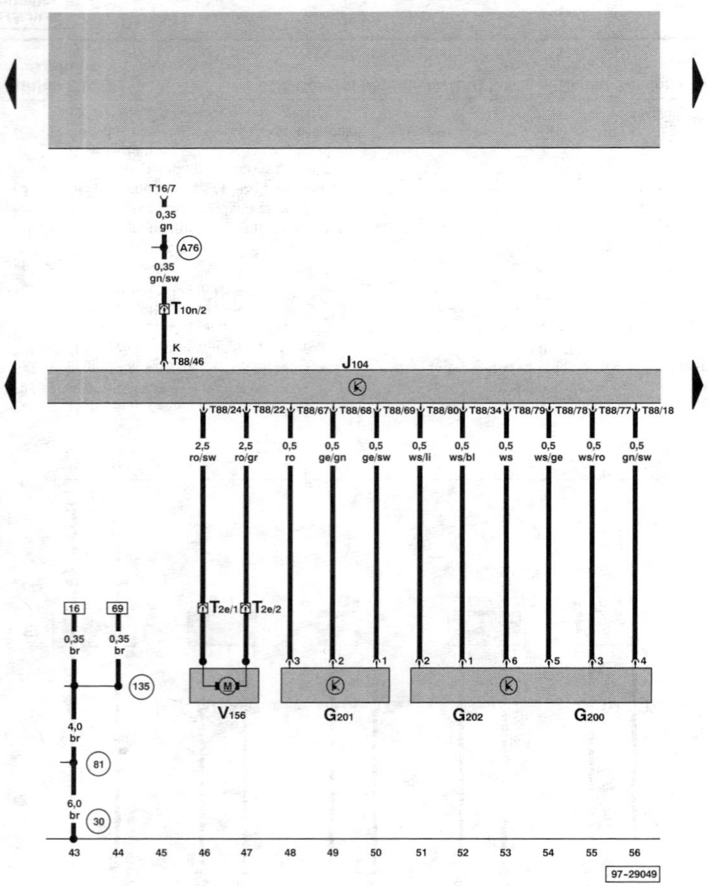

ws = white
sw = black
ro = red
br = brown
gn = green
bl = blue
gr = grey
li = lilac
ge = yellow
or = orange

G200 - Sensor for transverse acceleration
G201 - Sender 1 for brake booster
G202 - Sender for rotation rate
J104 - ABS Control Module (w/EDL)
T2e - Double Connector, black
T10n - 10-Pin Connector, orange, connector station A-pillar, left
T16 - 16-Pin Connector, Data Link Connector(DLC), below instrument panel, left
T88 - 88-Pin Connector, on ABS Control Module (w/EDL)
V156 - Hydraulic pump for traction control

(30) - Ground connection, -1-, beside fuse/relay panel
(81) - Ground connection -1-, in instrument panel wiring harness
(135) - Ground connection -2-, in instrument panel wiring harness
(A76) - Connector (K-diagnosis wire) in instrument panel wiring harness

ABS Control Module (w/EDL), Sensor for transverse acceleration, Sender 1 for brake booster, Sender for rotation rate, Data Link Connector (DLC)

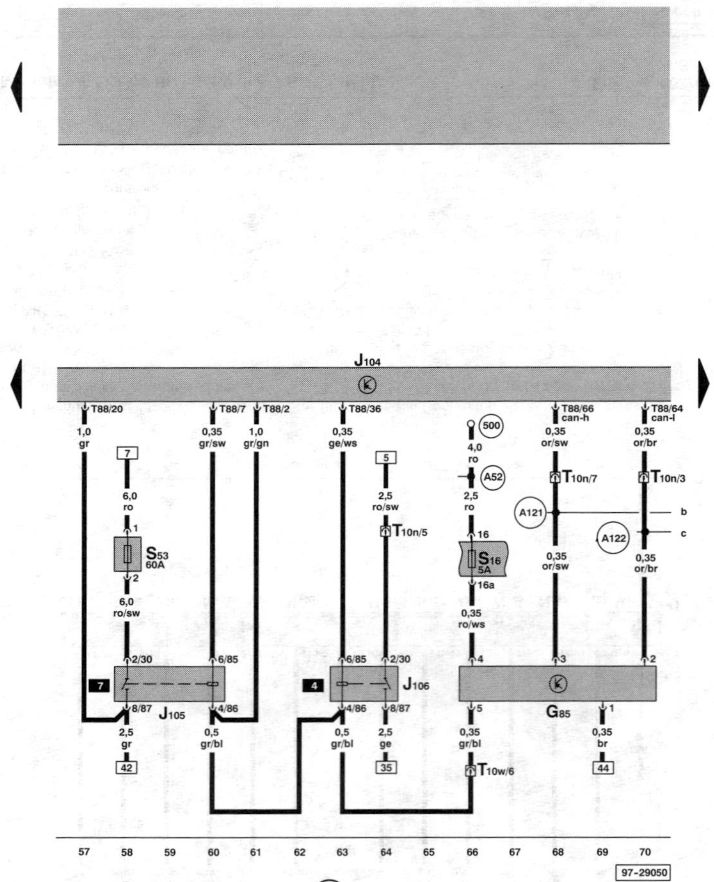

ws = white
sw = black
ro = red
br = brown
gn = green
bl = blue
gr = grey
li = lilac
ge = yellow
or = orange

G85 - Steering Angle Sensor
J104 - ABS Control Module (w/EDL)
J105 - - ABS Return Flow Pump Relay
J106 - ABS Solenoid Valve Relay
S16 - Fuse
S53 - ABS Hydraulic Pump Fuse
T10n - 10-Pin Connector, orange, connector station A-pillar, left
T10w - 10-Pin Connector, grey, connector station A-pillar, right
T88 - 88-Pin Connector, on ABS Control Module (w/EDL)

(500) - Threaded connection -1- (30) on the relay plate
(A52) - Plus Connection 2 (30) (in instrument panel wiring harness)
(A121) - Connection (high bus) in instrument panel wiring harness
(A122) - Connection (low bus) in instrument panel wiring harness

ABS Control Module (w/EDL), ABS Return Flow Pump Relay, ABS Solenoid Valve Relay, Steering Angle Sensor

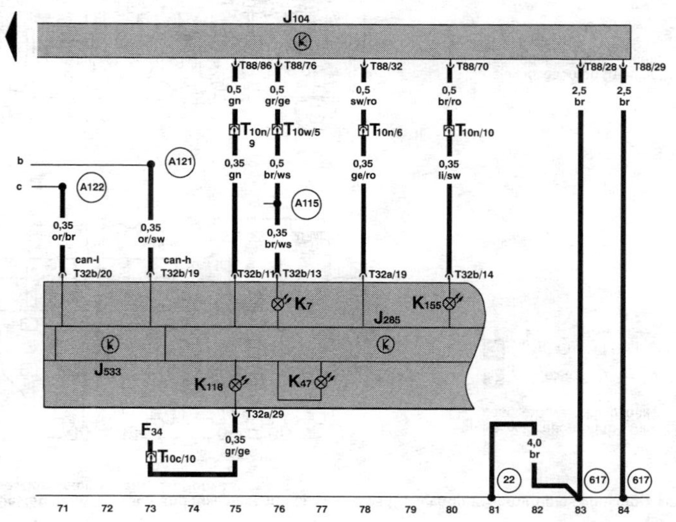

97-29051

F34 - Brake Fluid Level Warning Switch
J104 - ABS Control Module (w/EDL)
J285 - Control module with indicator unit in instrument panel insert
J533 - Data Bus On Board Diagnostic Interface
K47 - ABS Warning Light
K118 - Warning light for brake system
T10c - 10-Pin Connector, lilac, connector station A-pillar, left
T10n - 10-Pin Connector, orange, connector station A-pillar, left
T10w - 10-Pin Connector, grey, connector station A-pillar, right
T32a - 32-Pin Connector, blue, on instrument cluster
T32b - 32-Pin Connector, green, on instrument cluster
T88 - 88-Pin Connector, on ABS Control Module (w/EDL)
(22) - Ground connection, on hydraulic unit
(617) - Ground Connection 2 (on right lower A-pillar)
(A115) - Connection (parking brake control) in instrument panel wiring harness
(A121) - Connection (high bus) in instrument panel wiring harness
(A122) - Connection (low bus) in instrument panel wiring harness

ws = white
sw = black
ro = red
br = brown
gn = green
bl = blue
gr = grey
li = lilac
ge = yellow
or = orange

THIS PAGE INTENTIONALLY LEFT

BLANK

Air-Conditioning

2,8L - Engine - Motronic Multiport Fuel Injection (MFI)/142 kW, code ATQ

from October 2000

Deviate relay location and fuseplacements as well as the locations of multiple connectors see section "component locations".

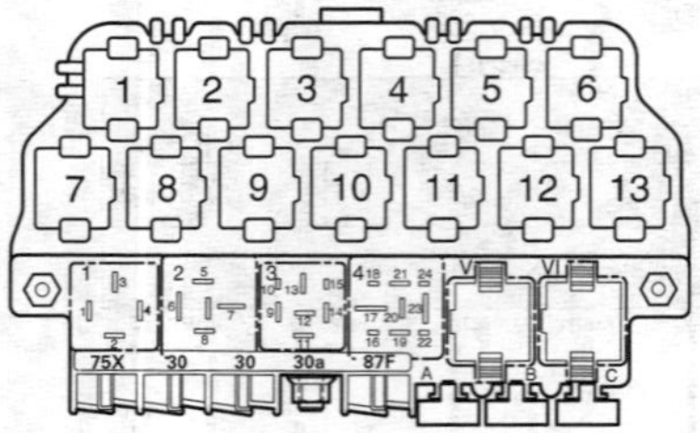

97-20676

Relay location on the thirteen position auxiliary relay panel, above relay panel:

- **1** Coolant Fan Control (FC) Relay (214)
- **3** A/C Cut-Off Control Module (384)

Relay panel:

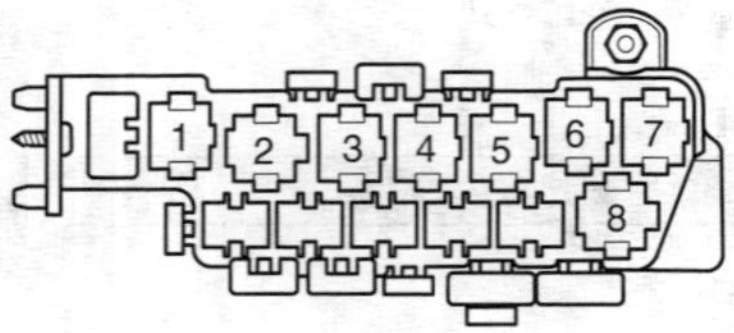

97-20515

Relay location on the eight position auxiliary relay panel, behind relay panel:

Note: Number in parentheses indicates production control number stamped on relay housing.

Fuse colors
30 A - green
25 A - white
20 A - yellow
15 A - blue
10 A - red
7,5 A - brown
5 A - beige

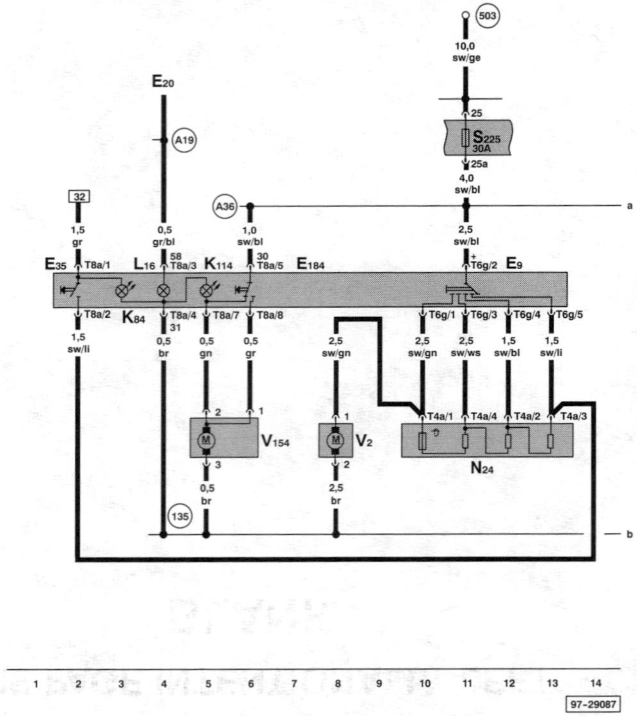

ws = white
sw = black
ro = red
br = brown
gn = green
bl = blue
gr = grey
li = lilac
ge = yellow
or = orange

E9 - Fresh Air Blower Switch
E20 - Instrument Panel Light Dimmer Switch
E35 - A/C Switch
E184 - Switch for fresh air blower and recirculated air
K84 - A/C Indicator Light
K114 - Fresh Air and Recirculating Air Mode Indicator Light
L16 - Fresh Air Control Lever Light
N24 - Fresh Air Blower Series Resistance With Fuse
S225 - Fuse in fuse holder
T4a - 4-Pin Connector
T6g - 6-Pin Connector
T8a - 8-Pin Connector
V2 - Fresh Air Blower
V154 - Servo motor for fresh-/recirculating air door

(135) - Ground connection -2-, in instrument panel wiring harness
(503) - Bolted Connection (75x) (at relay panel)
(A19) - wire connection (58d), in instrument panel wiring harness
(A36) - wire connection (75a), in instrument panel wiring harness

Fresh Air Blower, A/C Switch, ambient temperature switch, Switch for fresh air blower and recirculated air, Servomotor for fresh-/recirculated air door

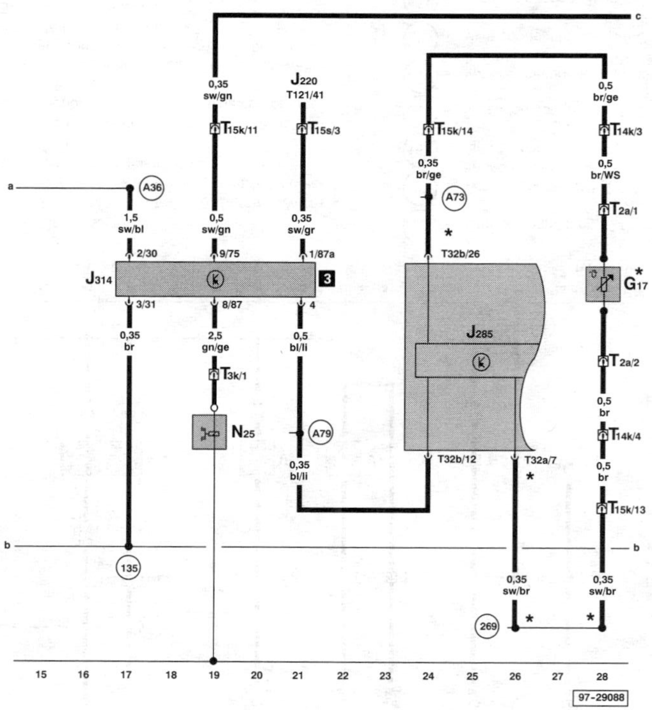

ws = white
sw = black
ro = red
br = brown
gn = green
bl = blue
gr = grey
li = lilac
ge = yellow
or = orange

G17 - Outside Air Temperature Sensor
J220 - Motronic Engine Control Module (ECM)
J285 - Control module with indicator unit in instrument panel insert
J314 - A/C Cut-Off Control Module
N25 - A/C Clutch
T2a - Double Connector
T3k - 3-Pin Connector, grey
T14k - 14-Pin Connector, black
T15k - 15-Pin Connector
T15s - 15-Pin Connector, red, on protective housing for control module, in engine compartment, left
T32a - 32-Pin Connector, bue, on instrument cluster
T32b - 32-Pin Connector, green, on instrument cluster
T121 - Connector 121 point, on Engine Control Module (ECM)

(135) - Ground connection -2-, in instrument panel wiring harness
(269) - Ground connector (sensor ground) -1-, in instrument panel wiring harness
(A36) - wire connection (75a), in instrument panel wiring harness
(A73) - Connector (outside temperature indicator), in instrument panel wiring harness
(A79) - Connector (engine overheat warning light) instrument wiring harness
* multi-function display only

Edition 02/01
USA.5062.13.21

Control module with indicator unit, A/C Cut Off Control Module, Outside Air Temperature Sensor, A/C Clutch

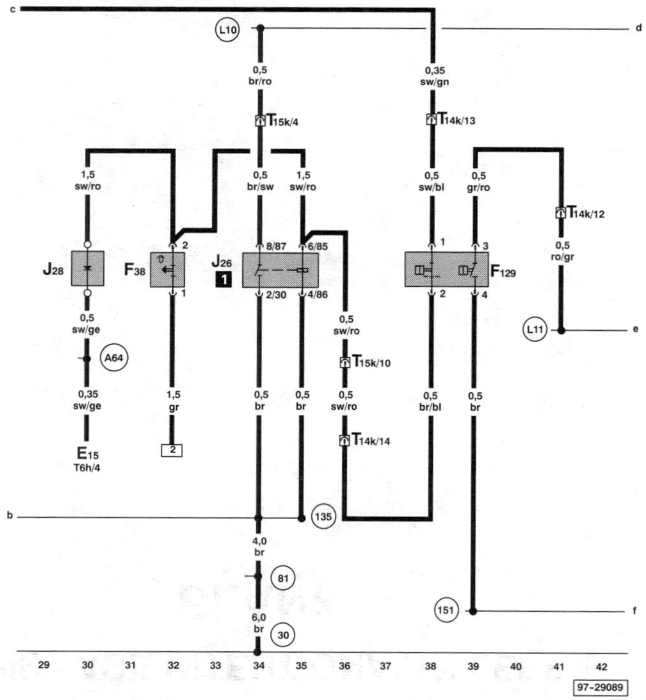

ws = white
sw = black
ro = red
br = brown
gn = green
bl = blue
gr = grey
li = lilac
ge = yellow
or = orange

E15 - Rear window defogger switch
F38 - Ambient Temperature Switch
F129 - A/C Pressure Switch
J26 - Coolant Fan control (FC) Relay
J28 - A/C Diode
T6h - 6-Pin Connector, black, on rear window defogger switch
T14k - 14-Pin Connector, black
T15k - 15-Pin Connector

(30) - Ground connection, -1-, beside fuse/relay panel
(81) - Ground connection -1-, in instrument panel wiring harness
(135) - Ground connection -2-, in instrument panel wiring harness
(151) - Ground connection -1-, in front wiring harness
(A64) - Connector (30, idle boost), in instrument panel wiring harness
(L10) - wire connection -2-, in A/C wiring harness
(L11) - wire connection -3-, in A/C wiring harness

Coolant Fan control (FC) Relay, A/C Diode, Ambient Temperature Switch, A/C Pressure Switch

Edition 02/01
USA.5062.13.21

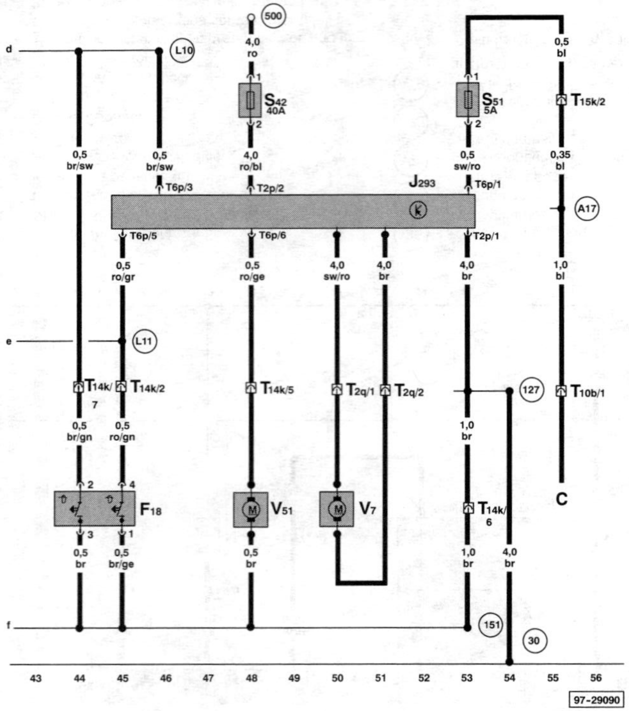

ws = white
sw = black
ro = red
br = brown
gn = green
bl = blue
gr = grey
li = lilac
ge = yellow
or = orange

C - Generator
F18 - Coolant Fan Control (FC) Thermal Switch
J293 - Coolant FC (Fan Control) Control Module
S42 - Coolant Fan Fuse
S51 - Fuse
T2p - Double Connector
T2q - Double Connector, in engine cpmpartment, left
T6p - 6-Pin Connector
T10b - 10-Pin Connector, black, on protective housing for control module, in engine compartment, left
T14k - 14-Pin Connector, black
T15k - 15-Pin Connector
V7 - Coolant Fan
V51 - After-Run Coolant Pump

(30) - Ground connection, -1-, beside fuse/relay panel
(127) - Ground connection -1-, in A/C compressor wiring harness
(151) - Ground connection -1-, in front wiring harness
(500) - Threaded connection -1- (30) on the relay plate
(A17) - wire connection (61), in instrument panel wiring harness
(L10) - wire connection -2-, in A/C wiring harness
(L11) - wire connection -3-, in A/C wiring harness

Edition 02/01
USA.5062.13.21

Coolant FC (Fan Control) Control Module, Coolant Fan, After -Run Coolant Pump, Coolant Fan Control(FC) Thermal Switch

THIS PAGE INTENTIONALLY LEFT BLANK

Air-Conditioning

● 1,8L-Engine - Motronic Multiport Fuel Injection (MFI)/125 kW, code AWM

from October 2000

Deviate relay location and fuseplacements as well as the locations of multiple connectors see section "component locations"

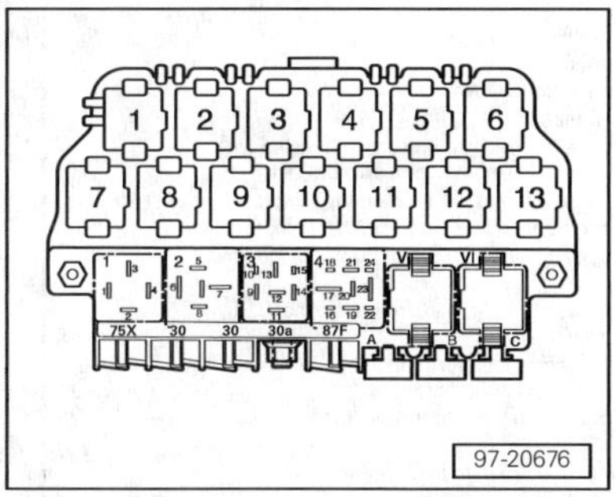

97-20676

Relay location on the thirteen position auxiliary relay panel, above relay panel:

1 - Collant Fan Control (FC) Relay (214)

3 - A/C Cut-Off Control Module Relay (384)

Relay panel:

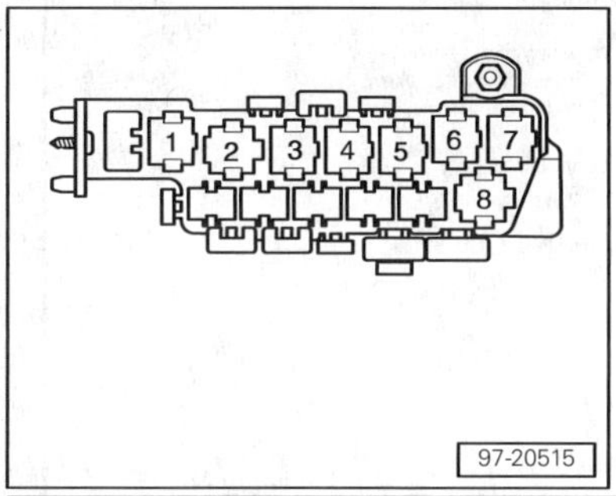

97-20515

Relay location on the eight position auxiliary relay panel, above relay panel:

2 - Second Speed Coolant Fan Control (FC) Relay (V7)

5 - First Speed Engine Coolant (EC) Fan Control (FC) Relay (V7, V35)

Fresh Air Blower, A/C Switch, ambient temperature switch, Fresh Air/Recirculating Flap Switch, Servomotor for fresh-/recirculated air door

- E9 - Fresh Air Blower Switch
- E20 - Instrument Panel Light Dimmer Switch
- E35 - A/C Switch
- E159 - Fresh Air/Recirculating Flap Switch
- K84 - A/C Indicator Light
- K114 - Fresh Air and Recirculating Air Mode Indicator Light
- L16 - Fresh Air Control Lever Light
- N24 - Fresh Air Blower Series Resistance With Fuse
- T4a - 4-Pin Connector
- T6g - 6-Pin Connector
- T8a - 8-Pin Connector
- V2 - Fresh Air Blower
- V154 - Servo motor for fresh-/recirculating air door

- (135) - Ground connection -2-, in instrument panel wiring harness
- (A19) - wire connection (58d), in instrument panel wiring harness
- (A36) - wire connection (75a), in instrument panel wiring harness

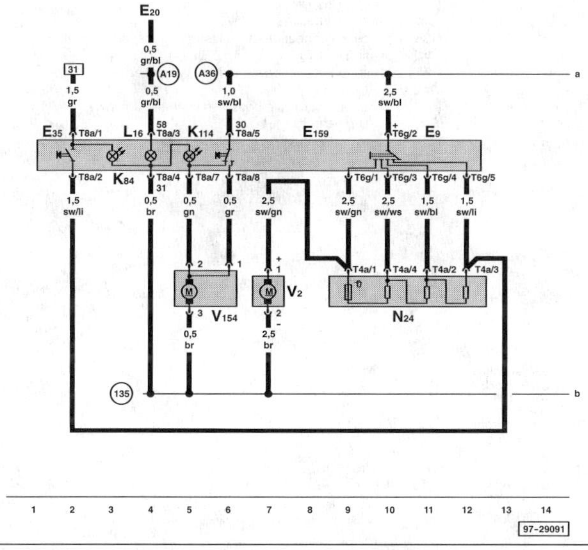

97-29091

ws = white
sw = black
ro = red
br = brown
gn = green
bl = blue
gr = grey
li = lilac
ge = yellow
or = orange

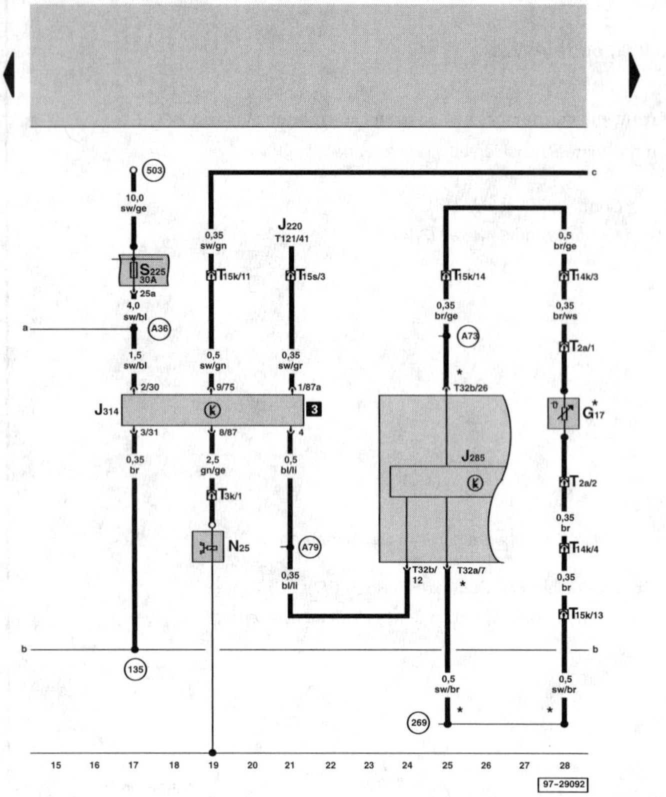

Control module with indicator unit, A/C Cut Off Control Module, Outside Air Temperature Sensor, A/C Clutch

G17 - Outside Air Temperature Sensor
J220 - Motronic Engine Control Module (ECM)
J285 - Control module with indicator unit in instrument panel insert
J314 - A/C Cut-Off Control Module
N25 - A/C Clutch
S225 - Fuse in fuse holder
T2a - Double Connector
T3k - 3-Pin Connector, grey
T14k - 14-Pin Connector, black
T15k - 15-Pin Connector
T15s - 15-Pin Connector, red, on protective housing for control module, in engine compartment, left
T32a - 32-Pin Connector, bue, on instrument cluster
T32b - 32-Pin Connector, green, on instrument cluster
T121 - Connector 121 point, on Engine Control Module (ECM)
(135) - Ground connection -2-, in instrument panel wiring harness
(269) - Ground connector (sensor ground) -1-, in instrument panel wiring harness
(503) - Bolted Connection (75x) (at relay panel)
(A36) - wire connection (75a), in instrument panel wiring harness
(A73) - Connector (outside temperature indicator), in instrument panel wiring harness
(A79) - Connector (engine overheat warning light) instrument wiring harness

* - multi-function display only

ws = white
sw = black
ro = red
br = brown
gn = green
bl = blue
gr = grey
li = lilac
ge = yellow
or = orange

97-29092

Edition 10.02

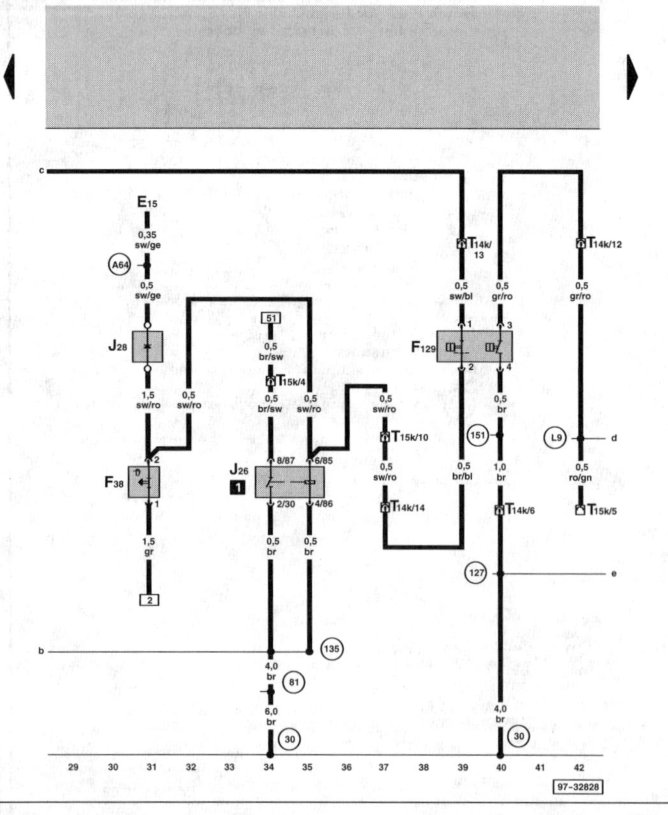

Coolant Fan control (FC) Relay, A/C Diode, Ambient Temperature Switch,A/C Pressure Switch

E15 - Rear window defogger switch
F38 - Ambient Temperature Switch
F129 - A/C Pressure Switch
J26 - Coolant Fan control (FC) Relay
J28 - A/C Diode
T14k - 14-Pin Connector, black
T15k - 15-Pin Connector
(30) - Ground connection, -1-, beside fuse/relay panel
(81) - Ground connection -1-, in instrument panel wiring harness
(127) - Ground connection -1-, in A/C compressor wiring harness
(135) - Ground connection -2-, in instrument panel wiring harness
(151) - Ground connection -1-, in front wiring harness
(A64) - Connector (30, idle boost), in instrument panel wiring harness
(L9) - wire connection -1-, in A/C wiring harness

ws = white
sw = black
ro = red
br = brown
gn = green
bl = blue
gr = grey
li = lilac
ge = yellow
or = orange

97-32828

Edition 10.02

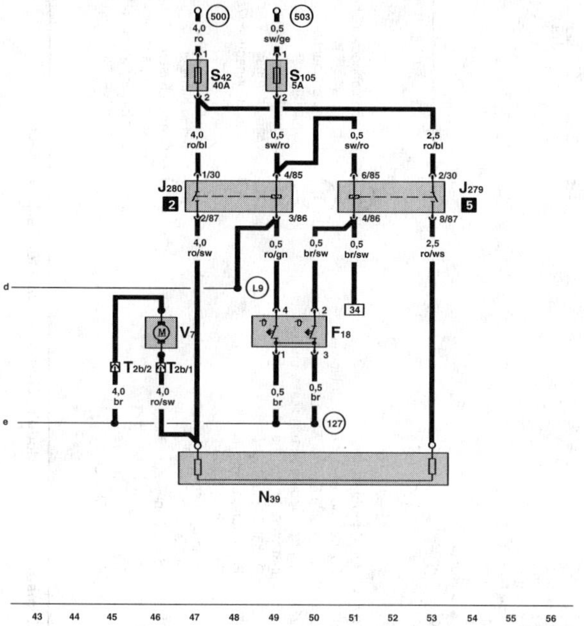

Coolant Fan Control (FC) Thermal Switch, First Speed Engine Coolant (EC) Fan Control Relay, Second Speed Coolant Fan Control Relay

- F18 - Coolant Fan Control (FC) Thermal Switch
- J279 - First Speed Engine Coolant (EC) Fan Control (FC) Relay (V7, V35)
- J280 - Second Speed Coolant Fan Control (FC) Relay (V7)
- N39 - Coolant Fan Control (FC) Series Resistance
- S42 - Coolant Fan Fuse
- S105 - First Speed Coolant Fan Fuse
- T2b - Double Connector
- V7 - Coolant Fan
- (127) - Ground connection -1-, in A/C compressor wiring harness
- (500) - Threaded connection -1- (30) on the relay plate
- (503) - Bolted Connection (75x) (at relay panel)
- (L9) - wire connection -1-, in A/C wiring harness

ws = white
sw = black
ro = red
br = brown
gn = green
bl = blue
gr = grey
li = lilac
ge = yellow
or = orange

THIS PAGE INTENTIONALLY LEFT BLANK

Climatronic

2,8L - Engine - Motronic Multiport Fuel Injection (MFI) / 142kW, code ATQ

from October 2000

Deviate relay location and fuseplacements as well as the locations of multiple connectors see section "component locations".

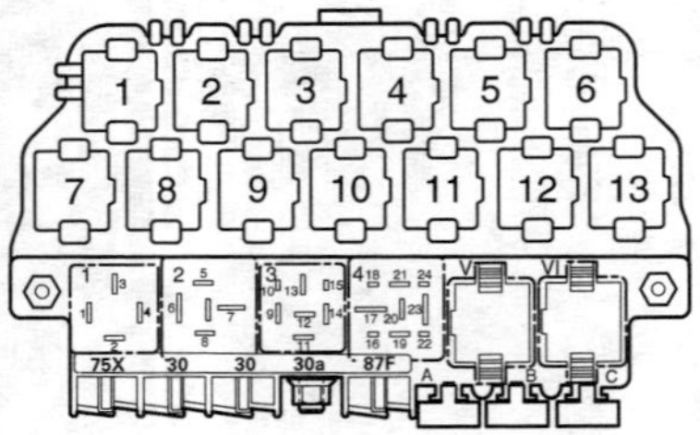

97-20676

Relay location on the thirteen position auxiliary relay panel, above relay panel:

3 A/C Clutch Relay (267)

Relay panel:

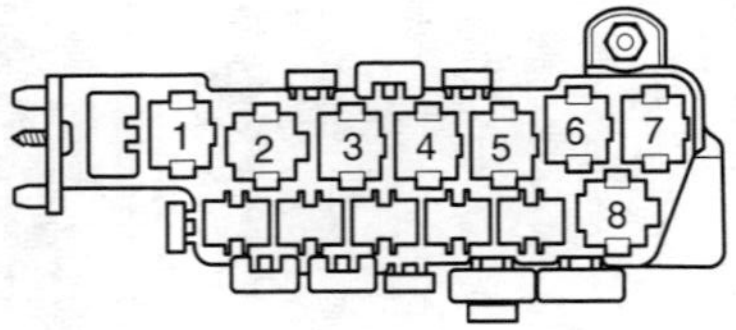

97-20515

Relay location on the eight position auxiliary relay panel, behind relay panel:

Note: Number in parentheses indicates production control number stamped on relay housing.

Fuse colors
30 A - green
25 A - white
20 A - yellow
15 A - blue
10 A - red
7,5 A - brown
5 A - beige

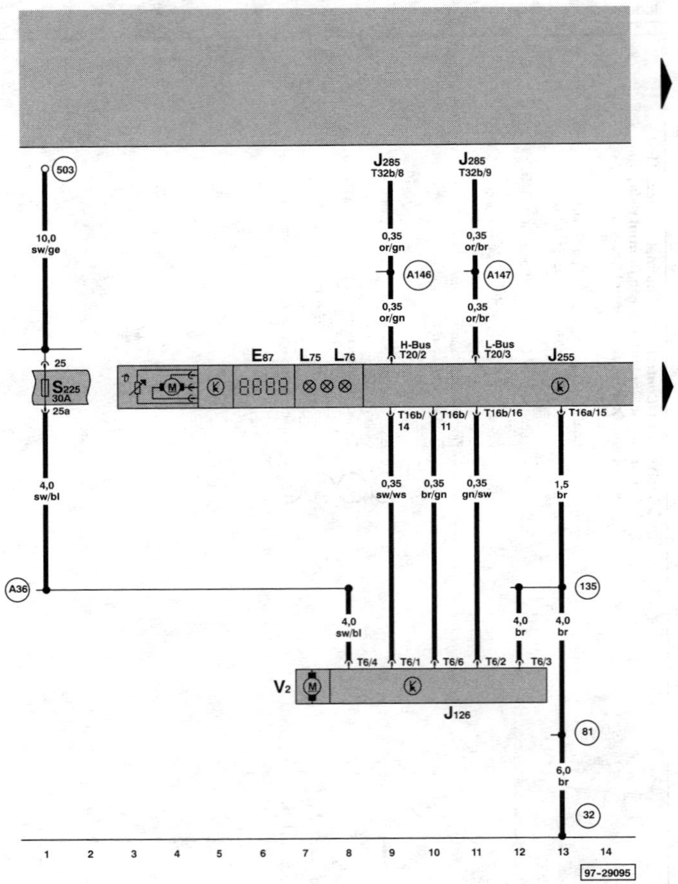

ws = white
sw = black
ro = red
br = brown
gn = green
bl = blue
gr = grey
li = lilac
ge = yellow
or = orange

E87 - A/C Control Head
J126 - Control module for fresh air blower
J255 - Climatronic Control Module
J285 - Control module with indicator unit in instrument panel insert
L75 - Digital Display Light
L76 - Push Button Light
S225 - Fuse in fuse holder
T6 - 6-Pin Connector, on Control module for fresh air blower
T16a - 16-Pin Connector, black
T16b - 16-Pin Connector, brown
T20 - 20-Pin Connector, red
V2 - Fresh Air Blower

(32) - Ground connection, behind instrument panel, left
(81) - Ground connection -1-, in instrument panel wiring harness
(135) - Ground connection -2-, in instrument panel wiring harness
(503) - Bolted Connection (75x) (at relay panel)
(A36) - wire connection (75a), in instrument panel wiring harness
(A146) - Comfort System High-bus Connection (in instrument panel wiring harness)
(A147) - Comfort System Low-bus Connection (in instrument panel wiring harness)

Climatronic Control Module, Control module for fresh air blower

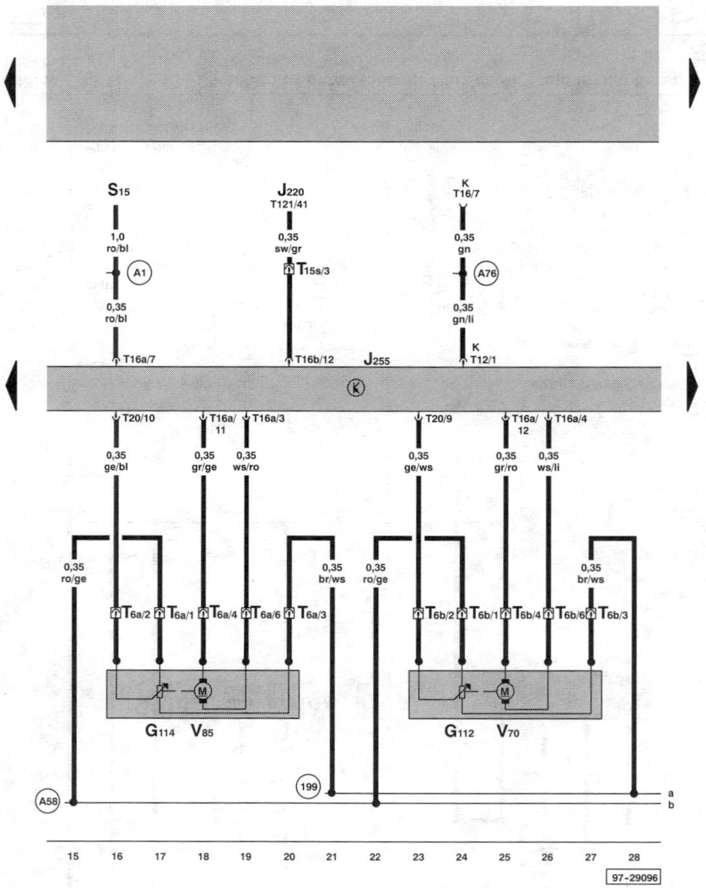

ws = white
sw = black
ro = red
br = brown
gn = green
bl = blue
gr = grey
li = lilac
ge = yellow
or = orange

G112 - Central Flap Motor Position Sensor
G114 - Footwell/Defrost Flap Motor Position Sensor
J220 - Motronic Engine Control Module (ECM)
J255 - Climatronic Control Module
S15 - Fuse
T6a - 6-Pin Connector, behind instrument panel, center
T6b - 6-Pin Connector, behind instrument panel, center
T12 - 12-Pin Connector
T15s - 15-Pin Connector, red, on protective housing for control module, in engine compartment, left
T16 - 16-Pin Connector, Data Link Connector (DLC), below instrument panel, left
T16a - 16-Pin Connector, black
T16b - 16-Pin Connector, brown
T20 - 20-Pin Connector, red
T121 - Connector, 121 point, on Motronic Engine Control Module (ECM)
V70 - Central Air Flap Motor
V85 - Footwell/Defroster Flap Motor

(199) - Ground connection -3-, in instrument panel wiring harness
(A1) - plus connection (30a), in instrument panel wiring harness
(A58) - Wire connection (5 V), in instrument cluster wiring harness
(A76) - Connector (K-diagnosis wire), in instrument panel wiring harness

Edition 02/01 **Climatronic Control Module, Central Air and Footwell/Defrost Flap Motors**
USA.5062.13.21

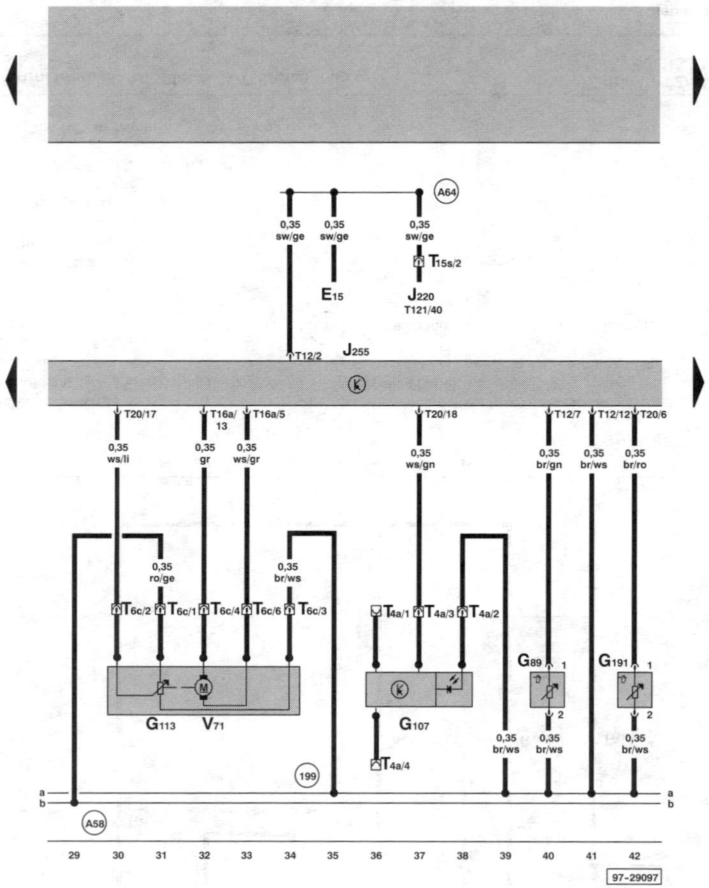

ws = white
sw = black
ro = red
br = brown
gn = green
bl = blue
gr = grey
li = lilac
ge = yellow
or = orange

E15 - Rear window defogger switch
G89 - Fresh Air Intake Duct Temperature Sensor
G107 - Sunlight Photo Sensor
G113 - Back Pressure Flap Motor Position Sensor
G191 - Sender for outlet temperature, center
J220 - Motronic Engine Control Module (ECM)
J255 - Climatronic Control Module
T4a - 4-Pin Connector, black
T6c - 6-Pin Connector, black
T15s - 15-Pin Connector, red, on protective housing for control module, in engine compartment, left
T12 - 12-Pin Connector, black
T16a - 16-Pin Connector, black
T20 - 20-Pin Connector, red
T121 - Connector, 121 point, on Motronic Engine Control Module (ECM)
V71 - Air Flow Flap Motor

(199) - Ground connection -3-, in instrument panel wiring harness
(A58) - Wire connection (5 V), in instrument cluster wiring harness
(A64) - Connector (30, idle boost), in instrument panel wiring harness

Climatronic Control Module, Air Flow Flap Motor, Fresh Air intake Duct Temperature Sensor, Sunlight Photo Sensor, sender for outlet temperature sensor, center **Edition 02/01**
USA.5062.13.21

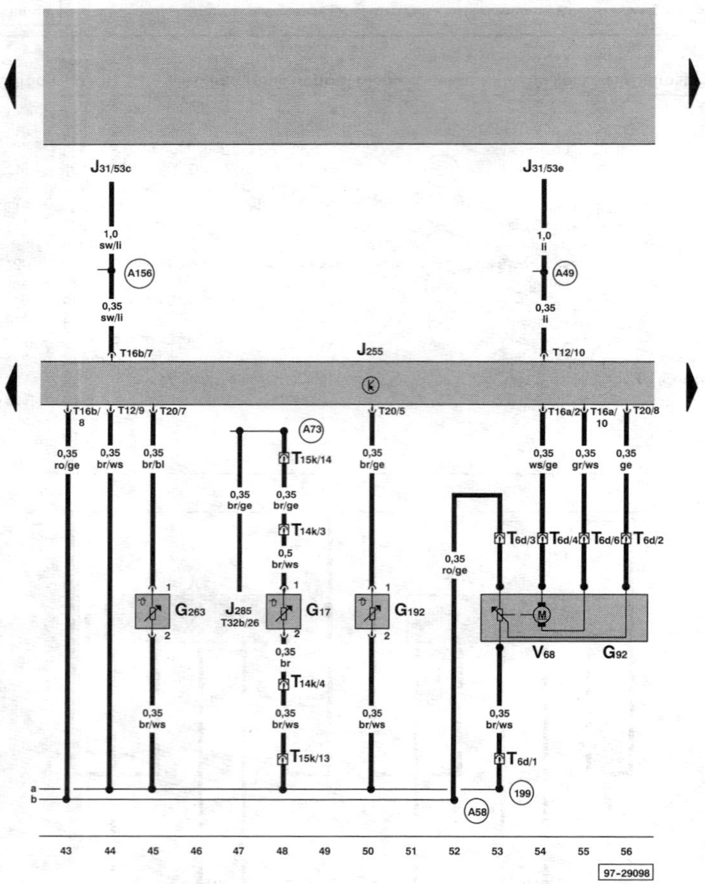

ws = white
sw = black
ro = red
br = brown
gn = green
bl = blue
gr = grey
li = lilac
ge = yellow
or = orange

G17 - Outside Air Temperature Sensor
G92 - Temperature Regulator Flap Motor Position Sensor
G192 - Sender for outlet temperature, floor outlet
G263 - Evaporator Vent Temperature Sensor
J31 - Wiper/Washer Intermittent Relay
J255 - Climatronic Control Module
J285 - Control module with indicator unit in instrument panel insert
T6d - 6-Pin Connector
T12 - 12-Pin Connector, black
T14k - 14-Pin Connector, black
T15k - 15-Pin Connector
T16a - 16-Pin Connector, black
T16b - 16-Pin Connector, brown
T20 - 20-Pin Connector, red
T32b - 32-Pin Connector, green, on instrument cluster
V68 - Temperature Regulator Flap Motor

(199) - Ground connection -3-, in instrument panel wiring harness
(A49) - wire connection -1-, in instrument panel wiring harness
(A58) - Wire connection (5 V), in instrument cluster wiring harness
(A73) - Connector (outside temperature indicator), in instrument panel wiring harness
(A156) - Sensor Ground Connection (in engine wiring harness)

Climatronic Control Module, Outside Air Temperature Sensor, Sender for outlet temperature, floor outlet

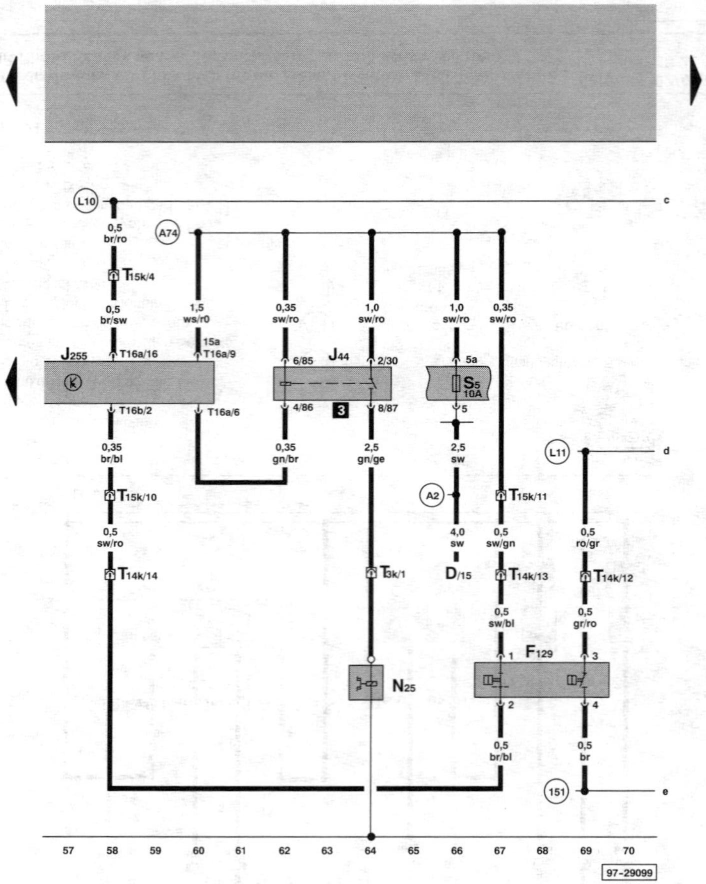

ws = white
sw = black
ro = red
br = brown
gn = green
bl = blue
gr = grey
li = lilac
ge = yellow
or = orange

D - Ignition/Starter Switch
F129 - A/C Pressure Switch
J44 - A/C Clutch Relay
J255 - Climatronic Control Module
N25 - A/C Clutch
S5 - Fuse
T3k - 3-Pin Connector
T14k - 14-Pin Connector, black
T15k - 15-Pin Connector
T16a - 16-Pin Connector, black
T16b - 16-Pin Connector, brown

(151) - Ground connection -1-, in front wiring harness
(A2) - plus connection (15), in instrument panel wiring harness
(A74) - Connector (15a - fuse 5) in instrument panel wiring harness
(L10) - wire connection -2-, in A/C wiring harness
(L11) - wire connection -3-, in A/C wiring harness

Climatronic Control Module, A/C Clutch, A/C Clutch Relay

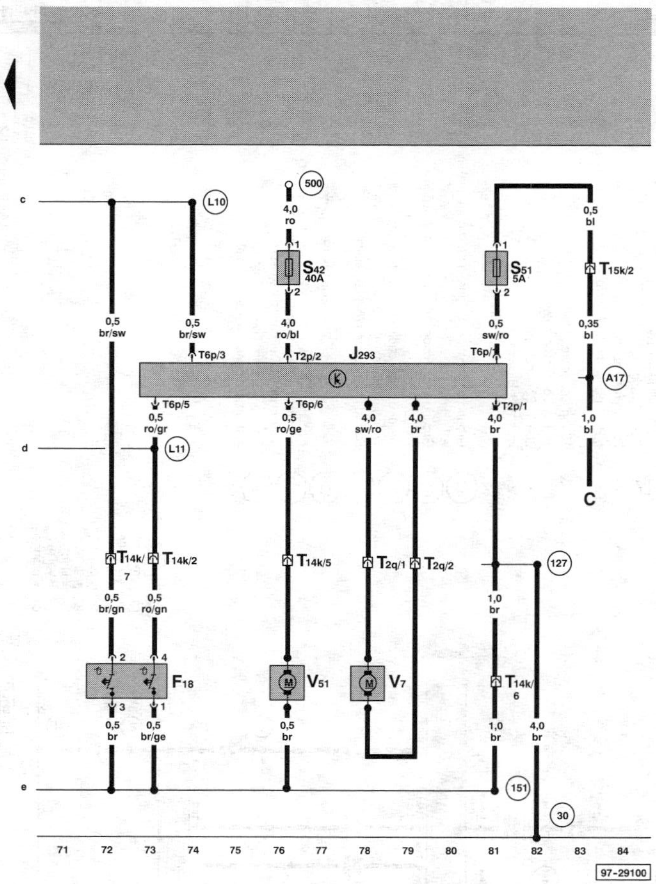

ws = white
sw = black
ro = red
br = brown
gn = green
bl = blue
gr = grey
li = lilac
ge = yellow
or = orange

C - Generator (GEN)
F18 - Coolant Fan Control (FC) Thermal Switch
J293 - Coolant FC (Fan Control) Control Module
S42 - Coolant Fan Fuse, on eightfold auxiliary relay panel
S51 - Coolant Fan Fuse
T2p - Double Connector
T2q - Double Connector, in engine compartment, left
T6p - 6-Pin Connector
T14k - 14-Pin Connector, black
T15k - 15-Pin Connector
V7 - Coolant Fan
V51 - After-Run Coolant Pump

(30) - Ground connection, -1-, beside fuse/relay panel

(127) - Ground connection -1-, in A/C compressor wiring harness
(151) - Ground connection -1-, in front wiring harness
(500) - Threaded connection -1- (30) on the relay plate
(A17) - wire connection (61), in instrument panel wiring harness
(L10) - wire connection -2-, in A/C wiring harness
(L11) - wire connection -3-, in A/C wiring harness

Edition 02/01
USA.5062.13.21

Coolant FC (Fan Control) Control Module, Coolant Fan, After Run Coolant Pump, Coolant Fan Control (FC) Thermal Switch

THIS PAGE INTENTIONALLY LEFT
BLANK

Climatronic

- **1,8L - Engine - Motronic Multiport Fuel Injection (MFI) / 125kW, code AWM**

from October 2000

- **2,0L - Engine - Turbo Diesel PD Fuel Injection (DFI) / 100kW, code BHW**

from October 2000

Additional Fuse, Relay, Control Module and Connector locations ⇒ Component Locations
Troubleshooting Procedures ⇒ Guided Fault Finding using VAS 5051/5052

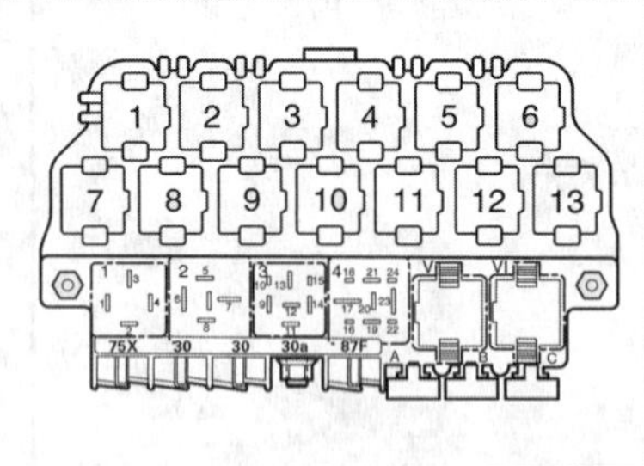

97-20676

Relay location on the thirteen position auxiliary relay panel, above relay panel:

3 - A/C Cluch Relay (267)

Relay panel:

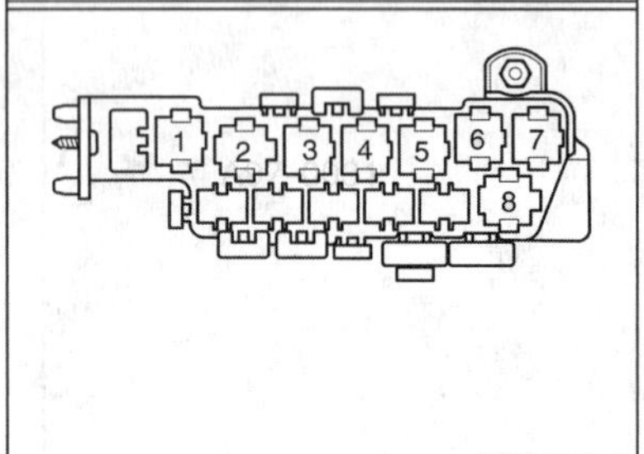

97-20515

Relay location on the eight position auxiliary relay panel, above relay panel:

2 - Second Speed Coolant Fan Control (FC) Relay (V7)

5 - First Speed Engine Coolant (EC) Fan Control (FC) Relay (V7, V35)

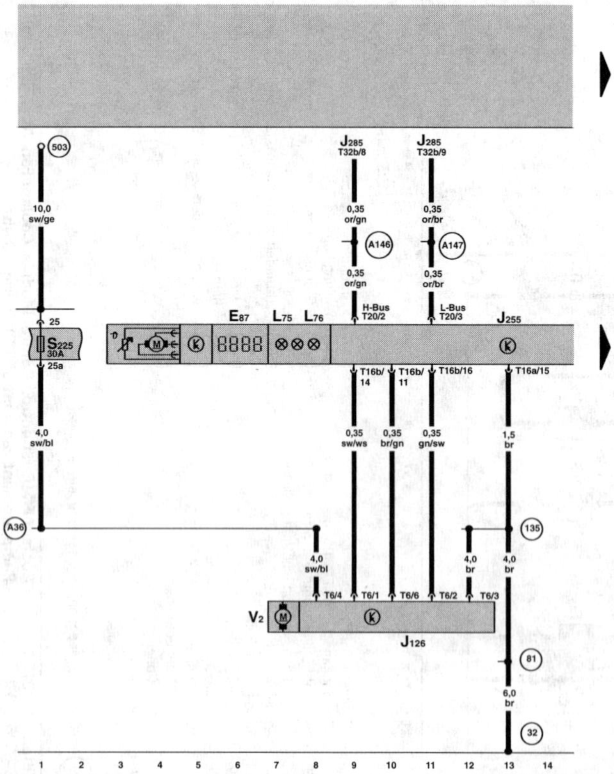

Climatronic Control Module, Control module for fresh air blower

E87 - A/C Control Head
J126 - Control module for fresh air blower
J255 - Climatronic Control Module
J285 - Control module with indicator unit in instrument panel insert
L75 - Digital Display Light
L76 - Push Button Light
S225 - Fuse in fuse holder
T6 - 6-Pin Connector, on Control module for fresh air blower
T16a - 16-Pin Connector, black
T16b - 16-Pin Connector, brown
T20 - 20-Pin Connector, red
V2 - Fresh Air Blower

(32) - Ground connection, behind instrument panel, left
(81) - Ground connection -1-, in instrument panel wiring harness
(135) - Ground connection -2-, in instrument panel wiring harness
(503) - Bolted Connection (75x) (at relay panel)
(A36) - wire connection (75a), in instrument panel wiring harness
(A146) - Comfort System High-bus Connection (in instrument panel wiring harness)
(A147) - Comfort System Low-bus Connection (in instrument panel wiring harness)

ws = white
sw = black
ro = red
br = brown
gn = green
bl = blue
gr = grey
li = lilac
ge = yellow
or = orange

Passat — Wiring diagram — No. 59/3

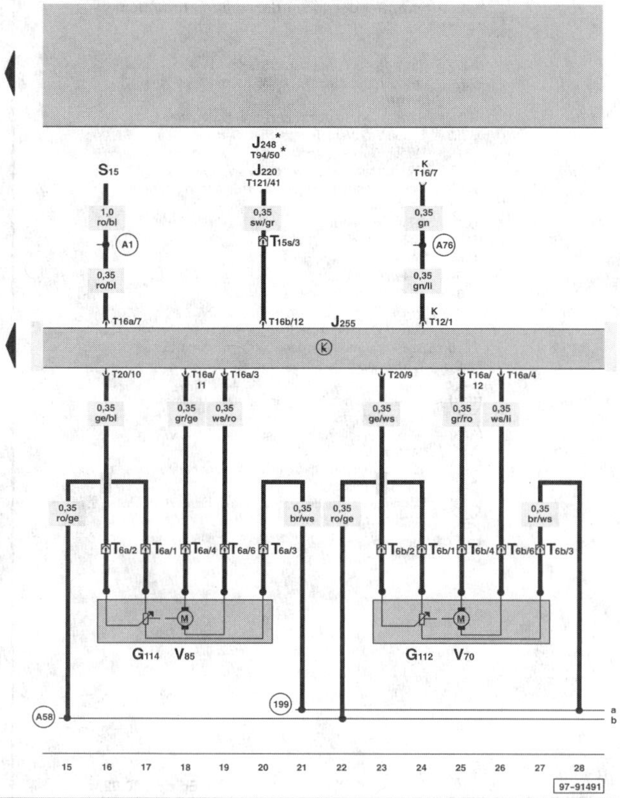

Climatronic Control Module, Central Air and Footwell/Defrost Flap Motors

G112 - Central Flap Motor Position Sensor
G114 - Footwell/Defrost Flap Motor Position Sensor
J220 - Motronic Engine Control Module (ECM)
J248 - Diesel Direct Fuel Injection (DFI) Engine Control Module (ECM)
J255 - Climatronic Control Module
S15 - Fuse
T6a - 6-Pin Connector, behind instrument panel, center
T6b - 6-Pin Connector, behind instrument panel, center
T12 - 12-Pin Connector
T15s - 15-Pin Connector, red, on protective housing for control module, in engine compartment, left
T16 - 16-Pin Connector, Data Link Connector (DLC), below instrument panel, left
T16a - 16-Pin Connector, black
T16b - 16-Pin Connector, brown
T20 - 20-Pin Connector, red
T94 - 94-Pin Connector
T121 - Connector, 121 point, on Motronic Engine Control Module (ECM)
V70 - Central Air Flap Motor
V85 - Footwell/Defroster Flap Motor

(199) - Ground connection -3-, in instrument panel wiring harness
(A1) - plus connection (30a), in instrument panel wiring harness
(A58) - Wire connection (5 V), in instrument cluster wiring harness
(A76) - Connector (K-diagnosis wire), in instrument panel wiring harness

* - engine code BHW only

ws = white
sw = black
ro = red
br = brown
gn = green
bl = blue
gr = grey
li = lilac
ge = yellow
or = orange

97-91491

Edition 06.04

Passat — Wiring diagram — No. 59/4

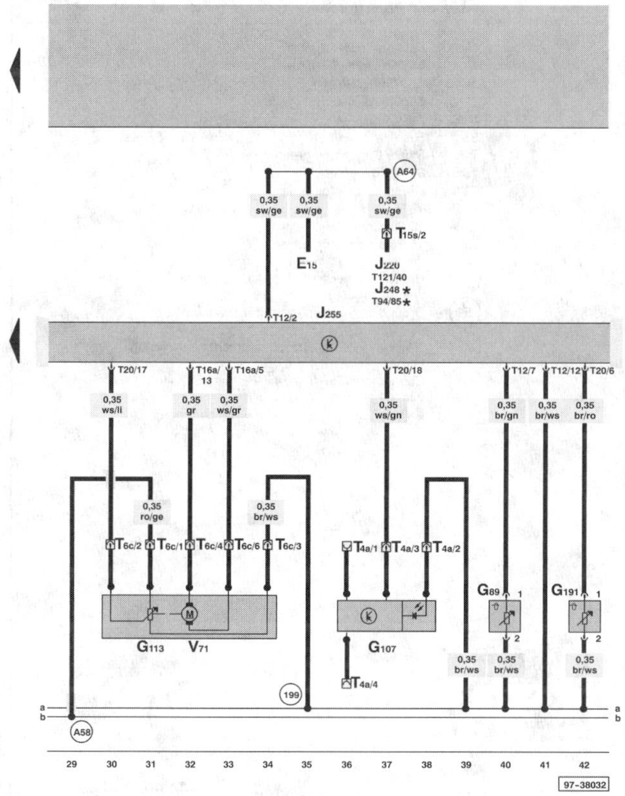

Climatronic Control Module, Air Flow Flap Motor, Fresh Air intake Duct Temperature Sensor, Sunlight Photo Sensor, sender for outlet temperature sensor, center

E15 - Rear window defogger switch
G89 - Fresh Air Intake Duct Temperature Sensor
G107 - Sunlight Photo Sensor
G113 - Back Pressure Flap Motor Position Sensor
G191 - Sender for outlet temperature, center
J220 - Motronic Engine Control Module (ECM)
J248 - Diesel Direct Fuel Injection (DFI) Engine Control Module (ECM)
J255 - Climatronic Control Module
T4a - 4-Pin Connector, black
T6c - 6-Pin Connector, black
T15s - 15-Pin Connector, red, on protective housing for control module, in engine compartment, left
T12 - 12-Pin Connector, black
T16a - 16-Pin Connector, black
T20 - 20-Pin Connector, red
T94 - 94-Pin Connector
T121 - Connector, 121 point, on Motronic Engine Control Module (ECM)
V71 - Air Flow Flap Motor

(199) - Ground connection -3-, in instrument panel wiring harness
(A58) - Wire connection (5 V), in instrument cluster wiring harness
(A64) - Connector (30, idle boost), in instrument panel wiring harness

* - engine code BHW only

ws = white
sw = black
ro = red
br = brown
gn = green
bl = blue
gr = grey
li = lilac
ge = yellow
or = orange

97-38032

Edition 12.03

Passat

Wiring diagram

No. 59/5

Climatronic Control Module, Outside Air Temperature Sensor, Sender for outlet temperature, floor outlet

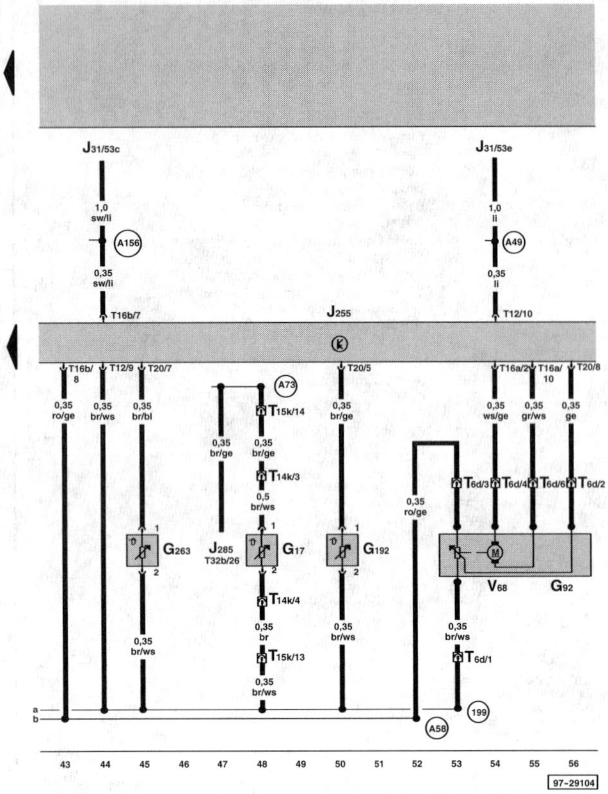

- G17 - Outside Air Temperature Sensor
- G92 - Temperature Regulator Flap Motor Position Sensor
- G192 - Sender for outlet temperature, floor outlet
- G263 - Evaporator Vent Temperature Sensor
- J31 - Wiper/Washer Intermittent Relay
- J255 - Climatronic Control Module
- J285 - Control module with indicator unit in instrument panel insert
- T6d - 6-Pin Connector
- T12 - 12-Pin Connector, black
- T14k - 14-Pin Connector, black
- T15k - 15-Pin Connector
- T16a - 16-Pin Connector, black
- T16b - 16-Pin Connector, brown
- T20 - 20-Pin Connector, red
- T32b - 32-Pin Connector, green, on instrument cluster
- V68 - Temperature Regulator Flap Motor

- (199) - Ground connection -3-, in instrument panel wiring harness
- (A49) - wire connection -1-, in instrument panel wiring harness
- (A58) - Wire connection (5 V), in instrument cluster wiring harness
- (A73) - Connector (outside temperature indicator), in instrument panel wiring harness
- (A156) - Sensor Ground Connection (in engine wiring harness)

ws = white
sw = black
ro = red
br = brown
gn = green
bl = blue
gr = grey
li = lilac
ge = yellow
or = orange

Edition 10.02

Passat

Wiring diagram

No. 59/6

Climatronic Control Module, A/C Clutch, A/C Clutch Relay, A/C Pressure Switch

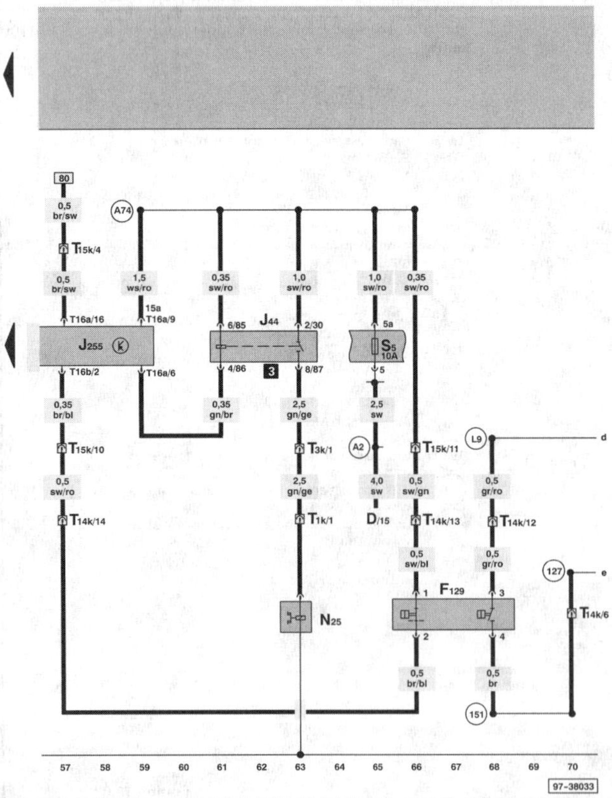

- D - Ignition/Starter Switch
- F129 - A/C Pressure Switch
- J44 - A/C Clutch Relay
- J255 - Climatronic Control Module
- N25 - A/C Clutch
- S5 - Fuse
- T1k - Single Connector, green
- T3k - 3-Pin Connector
- T14k - 14-Pin Connector, black
- T15k - 15-Pin Connector
- T16a - 16-Pin Connector, black
- T16b - 16-Pin Connector, brown

- (127) - Ground connection -1-, in A/C compressor wiring harness
- (151) - Ground connection -1-, in front wiring harness
- (A2) - plus connection (15), in instrument panel wiring harness
- (A74) - Connector (15a - fuse 5) in instrument panel wiring harness
- (L9) - wire connection -1-, in A/C wiring harness

ws = white
sw = black
ro = red
br = brown
gn = green
bl = blue
gr = grey
li = lilac
ge = yellow
or = orange

Edition 12.03

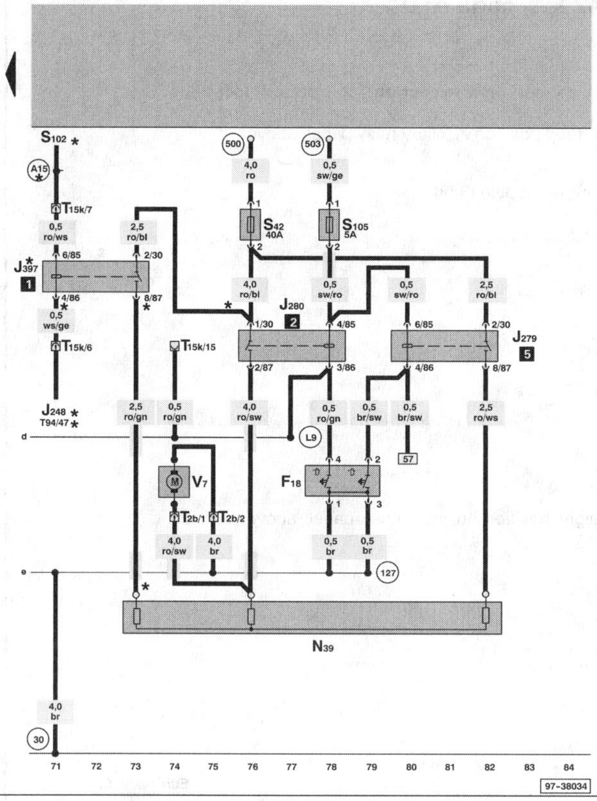

Coolant Fan Control (FC) Thermal Switch, First Speed Engine Coolant (EC) Fan Control Relay, Second Speed Coolant Fan Control Relay

F18 - Coolant Fan Control (FC) Thermal Switch
J248 - Diesel Direct Fuel Injection (DFI) Engine Control Module (ECM)
J279 - First Speed Engine Coolant (EC) Fan Control (FC) Relay (V7, V35)
J280 - Second Speed Coolant Fan Control (FC) Relay (V7)
J397 - Relay for Radiator Fan Afterrun
N39 - Coolant Fan Control (FC) Series Resistance
S42 - Coolant Fan Fuse
S102 - Engine Control Module (ECM) Fuse
S105 - First Speed Coolant Fan Fuse
T2b - Double Connector
T15k - 15-Pin Connector
T94 - 94-Pin Connector
V7 - Coolant Fan

(30) - Ground connection, -1-, beside fuse/relay panel
(127) - Ground connection -1-, in A/C compressor wiring harness
(500) - Threaded connection -1- (30) on the relay plate
(503) - Bolted Connection (75x) (at relay panel)
(A15) - plus connection (15) in instrument panel wiring harness
(L9) - wire connection -1-, in A/C wiring harness

* - engine code BHW only

ws = white
sw = black
ro = red
br = brown
gn = green
bl = blue
gr = grey
li = lilac
ge = yellow
or = orange

Edition 12.03

THIS PAGE INTENTIONALLY LEFT

BLANK

2,8L - Engine - Motronic Multiport Fuel Injection (MFI) / 142 kW, code ATQ

from October 2000

Deviate relay location and fuseplacements as well as the locations of multiple connectors see section "component locations"

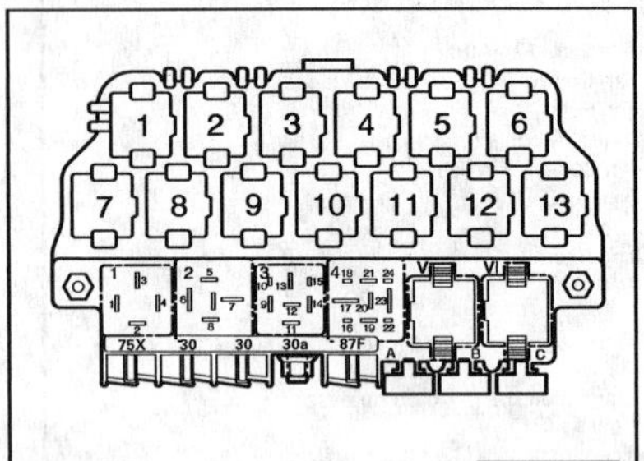

97-20676

Relay location on the thirteen position auxiliary relay panel, above relay panel:

10 - Lamp Failure Control Module (373)

Relay panel:

4 - Fuel Pump (FP) Relay (372)

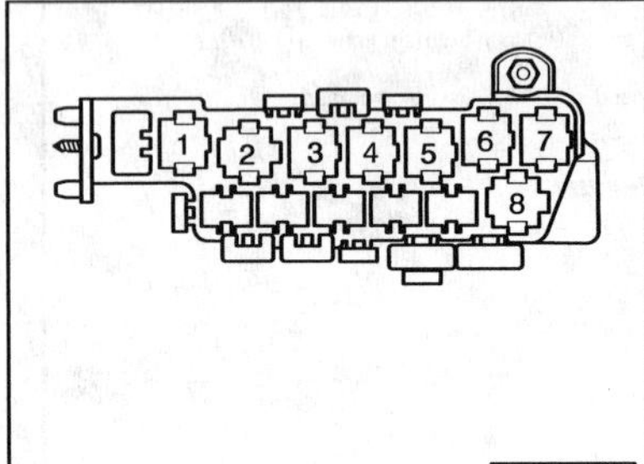

97-20515

Relay location on the eight position auxiliary relay panel, above relay panel:

Battery, Starter, Generator (GEN)

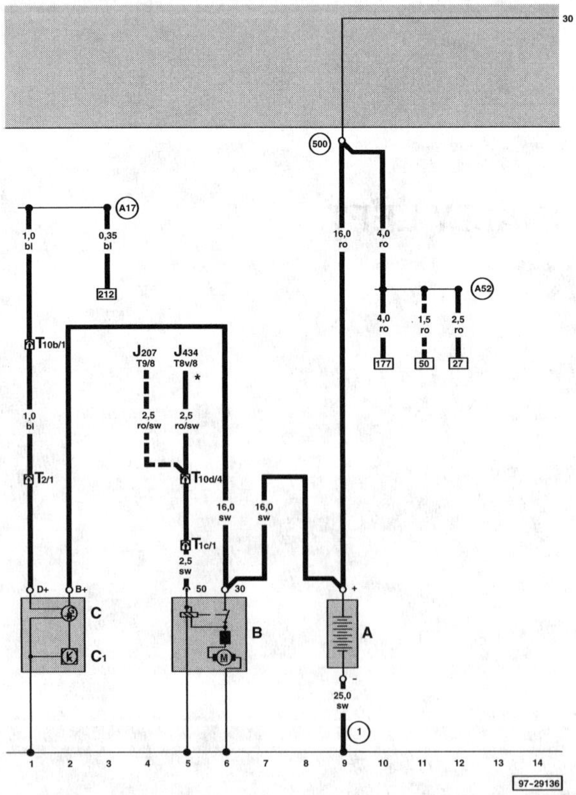

97-29136

- A - Battery
- B - Starter
- C - Generator (GEN)
- C1 - Voltage Regulator (VR)
- J207 - Starting Interlock Relay
- J434 - Locking relay for starter (clutch pedal switch)
- T1c - Single Connector, black
- T2 - Double Connector, black
- T8v - 8-Pin Connector, on Locking relay for starte
- T9 - 9-Pin Connector, on Starting Interlock Relay
- T10b - 10-Pin Connector, black, on protective housing for control module, in engine compartment, left
- T10d - 10-Pin Connector, brown, on protective housing for control module, in engine compartment, left

- (1) - Ground strap, battery to body
- (500) - Threaded connection -1- (30) on the relay plate
- (A17) - wire connection (61), in instrument panel wiring harness
- (A52) - Plus Connection 2 (30) (in instrument panel wiring harness)

- * - manual transmission only
- - - - - automatic transmission only

ws = white
sw = black
ro = red
br = brown
gn = green
bl = blue
gr = grey
li = lilac
ge = yellow
or = orange

Motronic Engine Control Module (ECM), Throttle drive, Spark Plug Connectors

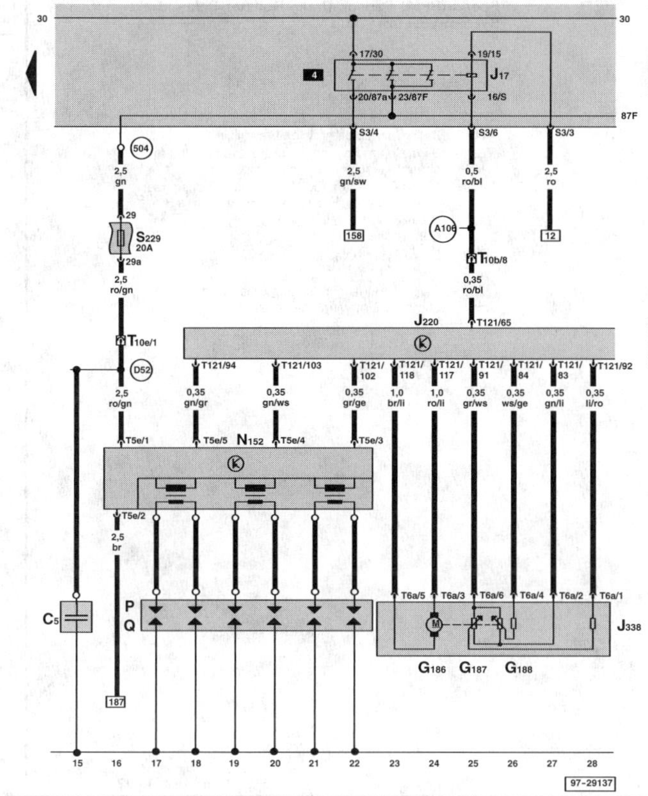

- C5 - Suppressor (Terminal 15)
- G186 - Throttle drive (power accelerator actuation)
- G187 - Angle sensor -1- for throttle drive (power accelerator actuation)
- G188 - Angle sensor -2- for throttle drive (power accelerator actuation)
- J17 - Fuel Pump (FP) Relay
- J220 - Motronic Engine Control Module (ECM)
- J338 - Throttle Valve Control Module
- N152 - Ignition Coil
- P - Spark Plug Connectors
- Q - Spark Plugs
- S229 - Fuse in fuse holder
- T5e - 5-Pin Connector, black, on Ignition Coil
- T6a - 6-Pin Connector, black, on Throttle Valve Control Module
- T10b - 10-Pin Connector, black, on protective housing for control module, in engine compartment, left
- T10e - 10-Pin Connector, orange, on protective housing for control module, in engine compartment, left
- T121 - Connector, 121 point, on Motronic Engine Control Module (ECM)
- (504) - Threaded connection -1- (87) on the relay plate
- (A106) - Connector -2- (86s), in instrument panel wiring harness
- (D52) - plus connection (15a), in engine compartment wiring harness

ws = white
sw = black
ro = red
br = brown
gn = green
bl = blue
gr = grey
li = lilac
ge = yellow
or = orange

Edition 10.02

Repair Manual page number Y 584

Motronic Control Module (ECM), Intake Air Temperature (IAT) Sensor, Knock Sensor

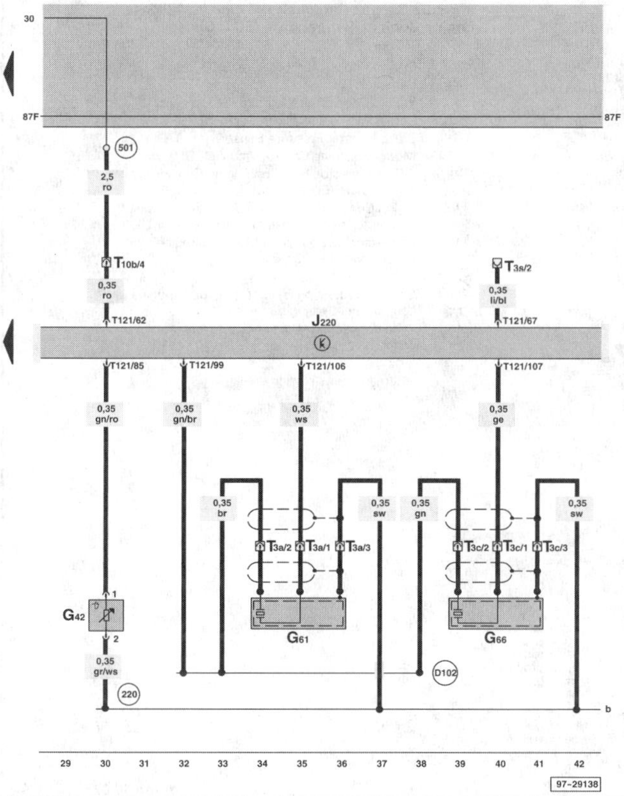

- G42 - Intake Air Temperature (IAT) Sensor
- G61 - Knock Sensor (KS) 1
- G66 - Knock Sensor (KS) 2
- J220 - Motronic Engine Control Module (ECM)
- T3a - 3-Pin Connector, brown, on firewall, left
- T3c - 3-Pin Connector, brown, on firewall, left
- T3s - 3-Pin Connector, red, on protective housing for control module, in engine compartment, left
- T10b - 10-Pin Connector, black, on protective housing for control module, in engine compartment, left
- T121 - Connector, 121 point, on Motronic Engine Control Module (ECM)
- (220) - Ground connection (sensor ground), in engine compartment wiring harness
- (501) - Threaded connection -2- (30) on the relay plate
- (D102) - Wire connection -2-, in engine compartment wiring harness

ws = white
sw = black
ro = red
br = brown
gn = green
bl = blue
gr = grey
li = lilac
ge = yellow
or = orange

Edition 04.04

Repair Manual page number Y 583

Motronic Engine Control Module (ECM), Engine Speed (RPM) Sensor, Brake Booster Relay, Brake System Vacuum Pump

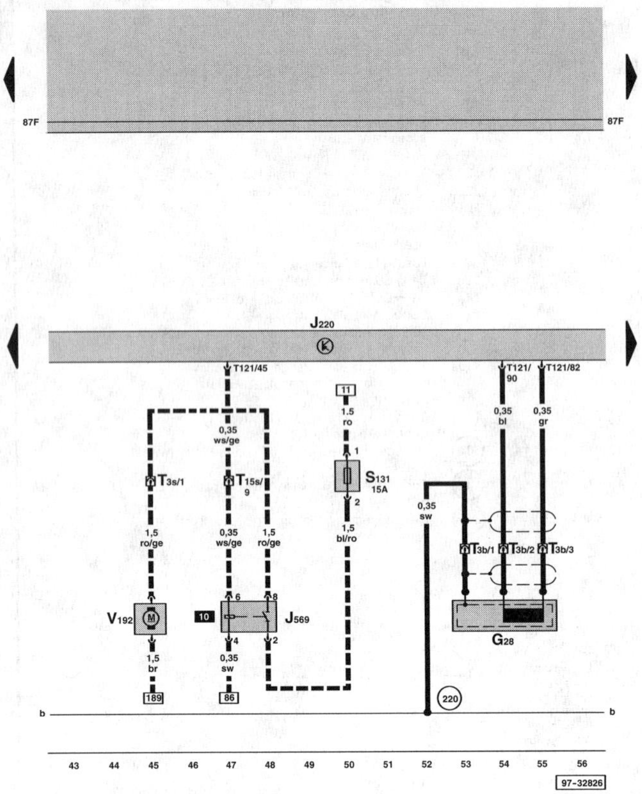

G28 - Engine Speed (RPM) Sensor
J220 - Motronic Engine Control Module (ECM)
J569 - Brake Booster Relay
S131 - Safety fuse 1
T3b - 3-Pin Connector, grey, on firewall, left
T3s - 3-Pin Connector, red, on protective housing for control module, in engine compartment, left
T15s - 15-Pin Connector, red, on protective housing for control module, in engine compartment, left
T121 - Connector, 121 point, on Motronic Engine Control Module (ECM)
V192 - Brake System Vacuum Pump

(220) - Ground connection (sensor ground), in engine compartment wiring harness

- - - - automatic transmission only

ws = white
sw = black
ro = red
br = brown
gn = green
bl = blue
gr = grey
li = lilac
ge = yellow
or = orange

Edition 10.02

Motronic Control Module (ECM), Brake Booster Pressure Sensor, Camshaft Position (CMP) Sensor, Engine Coolant Temperature (ECT) Sensor

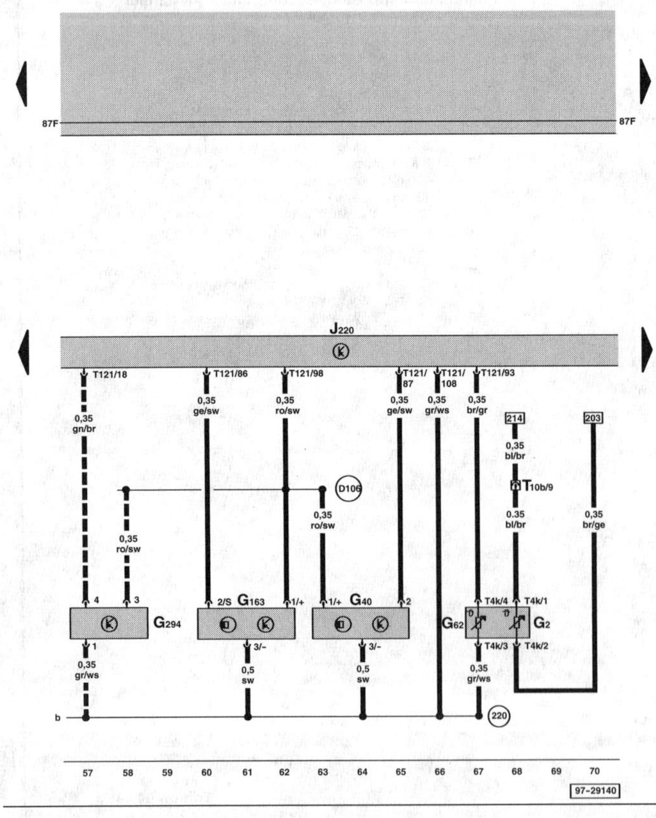

G2 - Engine Coolant Temperature (ECT) Sensor
G40 - Camshaft Position (CMP) Sensor
G62 - Engine Coolant Temperature (ECT) Sensor
G163 - Camshaft Position (CMP) Sensor 2
G294 - Brake Booster Pressure Sensor
J220 - Motronic Engine Control Module (ECM)
T4k - 4-Pin Connector, black, on Engine Coolant Temperature (ECT) Sensor
T10b - 10-Pin Connector, black, on protective housing for control module, in engine compartment, left
T121 - Connector, 121 point, on Motronic Engine Control Module (ECM)

(220) - Ground connection (sensor ground), in engine compartment wiring harness
(D106) - wire connection -4-, in engine compartment wiring harness

- - - - automatic transmission only

ws = white
sw = black
ro = red
br = brown
gn = green
bl = blue
gr = grey
li = lilac
ge = yellow
or = orange

Edition 10.02

Passat

Wiring diagram

No. 60/7

Motronic Engine Control Module (ECM), Button for cruise control, Cruise Control Switch, Brake Light Switch, Clutch Pedal Switch, Brake Pedal Switch

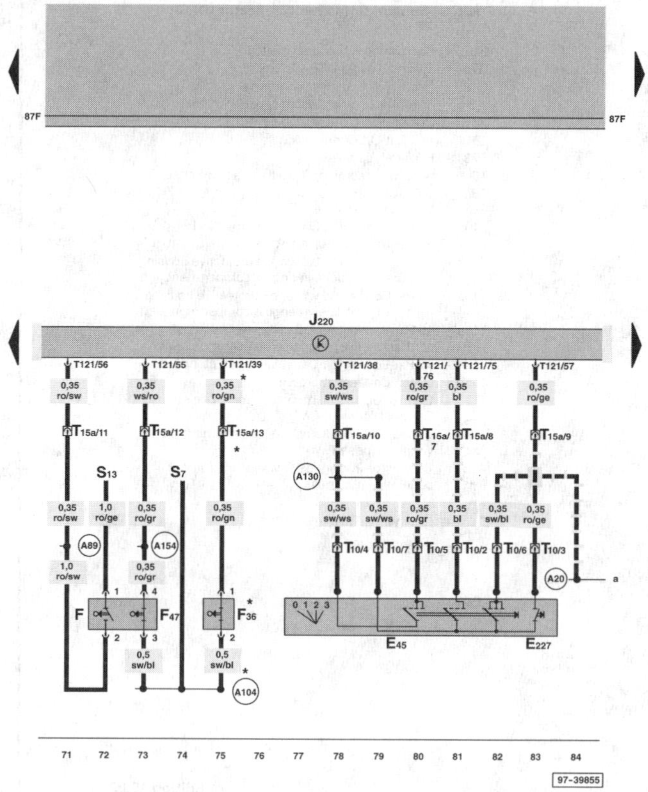

E45 - Cruise Control Switch
Position 0=OFF looked
Position 1=OFF pushed
Position 2=ON
Position 3=RES
E227 - Button for cruise control (set)
F - Brake Light Switch
F36 - Clutch Pedal Switch
F47 - Brake Pedal Switch
J220 - Motronic Engine Control Module (ECM)
S7 - Fuse
S13 - Fuse
T10 - 10-Pin Connector, black
T15a - 15-Pin Connector, white, on protective housing for control module, in engine compartment, left
T121 - Connector, 121 point, on Motronic Engine Control Module (ECM)

(A20) - wire connection (15a), in instrument panel wiring harness
(A89) - Connection 2 (54) (in instrument panel wiring harness)
(A104) - Plus connection -2- (15), in instrument panel wiring harness
(A130) - Connection for Cruise Control (instrument panel wiring harness)
(A154) - Brake Pedal Switch Connection (in instrument panel wiring harness)

- - - - cruise control only
* - manual transmission only

ws = white
sw = black
ro = red
br = brown
gn = green
bl = blue
gr = grey
li = lilac
ge = yellow
or = orange

Edition 04.04

Passat

Wiring diagram

No. 60/8

Motronic Engine Control Module (ECM), Fuel injectors

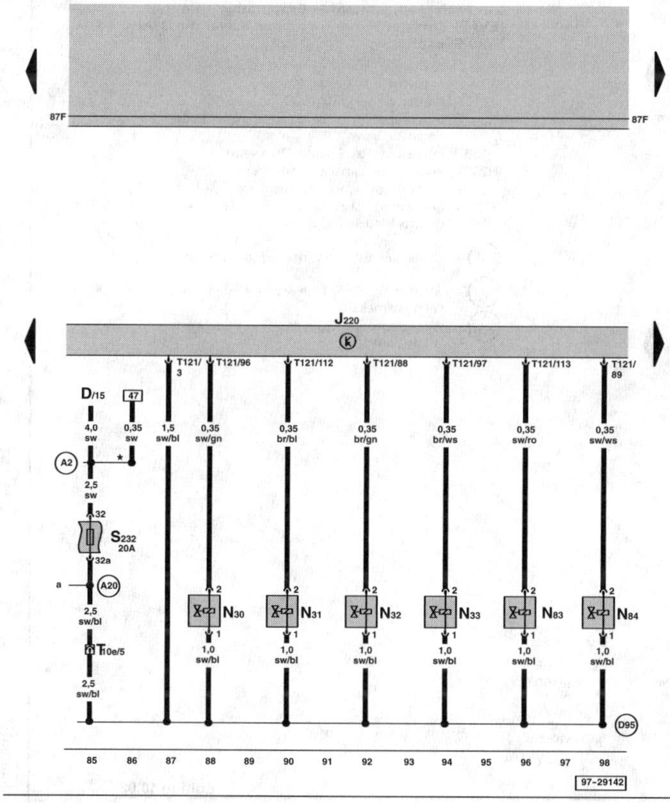

D - Ignition/Starter Switch
J220 - Motronic Engine Control Module (ECM)
N30 - Cylinder 1 Fuel Injector
N31 - Cylinder 2 Fuel Injector
N32 - Cylinder 3 Fuel Injector
N33 - Cylinder 4 Fuel Injector
N83 - Cylinder 5 Fuel Injector
N84 - Cylinder 6 Fuel Injector
S232 - Fuse in fuse holder
T10e - 10-Pin Connector, orange, on protective housing for control module, in engine compartment, left
T121 - Connector, 121 point, on Motronic Engine Control Module (ECM)

(A2) - plus connection (15), in instrument panel wiring harness
(A20) - wire connection (15a), in instrument panel wiring harness
(D95) - wire connection (injectors), in engine compartment wiring harness

* - automatic transmission only

ws = white
sw = black
ro = red
br = brown
gn = green
bl = blue
gr = grey
li = lilac
ge = yellow
or = orange

Edition 10.02

Passat

Wiring diagram

No. 60/9

Motronic Engine Control Module (ECM), Throttle Position Sensor, Data Link Connector (DLC)

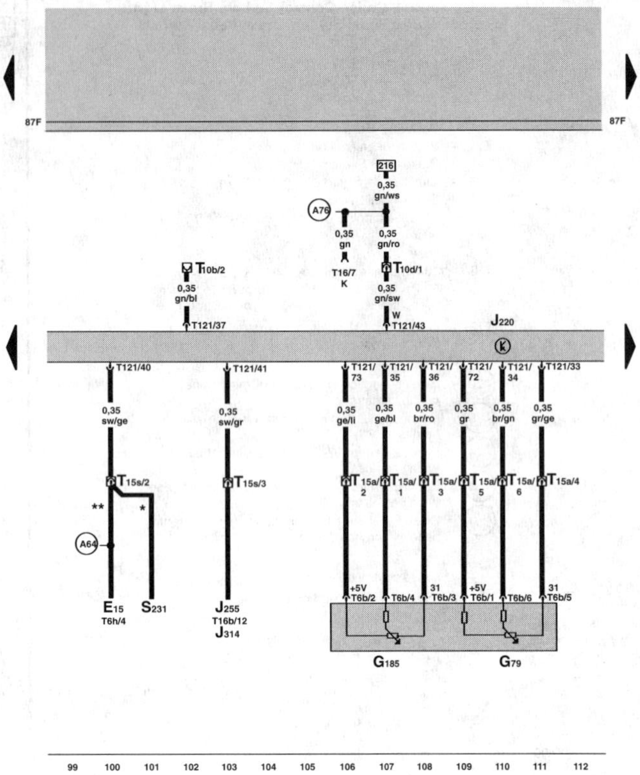

- E15 - Rear window defogger switch
- G79 - Throttle Position (TP) Sensor
- G185 - Sender -2- for accelerator pedal position
- J220 - Motronic Engine Control Module (ECM)
- J255 - Climatronic Control Module
- J314 - A/C Cut-Off Control Module
- S231 - Fuse in fuse holder
- T6b - 6-Pin Connector, on Throttle Position Sensor
- T6h - 6-Pin Connector, black, on Rear window defogger switch
- T10b - 10-Pin Connector, black, on protective housing for control module, in engine compartment, left
- T10d - 10-Pin Connector, brown, on protective housing for control module, in engine compartment, left
- T15a - 15-Pin Connector, while, on protective housing for control module, in engine compartment, left
- T15s - 15-Pin Connector, red, on protective housing for control module, in engine compartment, left
- T16 - 16-Pin Connector, Data Link Connector (DLC)
- T16b - 16-Pin Connector, brown
- T121 - Connector, 121 point, on Motronic Engine Control Module (ECM)
- (A64) - Connector (30, idle boost), in instrument panel wiring harness
- (A76) - Connector (K-diagnosis wire) in instrument panel wiring harness

- * - front wheel drive only
- ** - 4-Motion only

ws = white
sw = black
ro = red
br = brown
gn = green
bl = blue
gr = grey
li = lilac
ge = yellow
or = orange

97-29143

Edition 10.02

Passat

Wiring diagram

No. 60/10

Motronic Engine Control Module (ECM), Valve for camshaft adjustment, Evaporative Emission (EVAP) Canister Purge Regulator Valve, Mass Air Flow Sensor

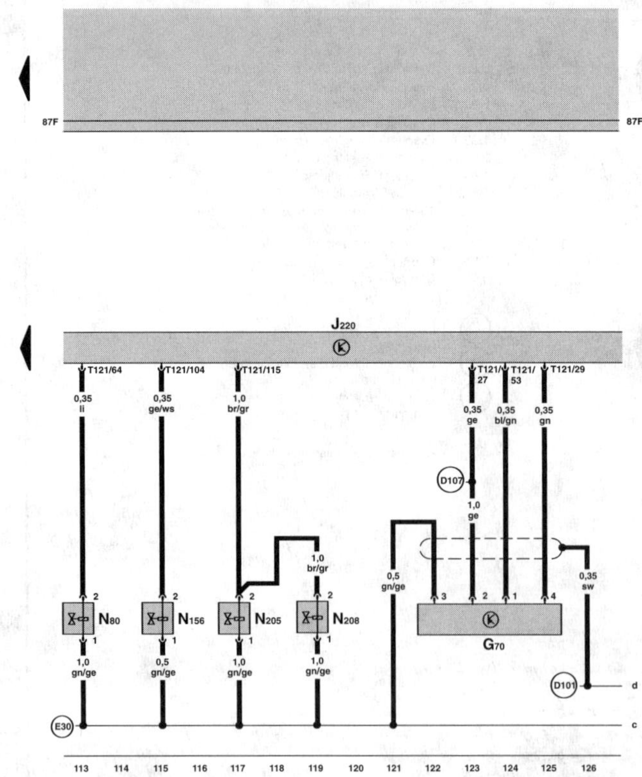

- G70 - Mass Air Flow (MAF) Sensor
- J220 - Motronic Engine Control Module (ECM)
- N80 - Evaporative Emission (EVAP) Canister Purge Regulator Valve
- N156 - Intake Manifold Change-Over Valve
- N205 - Valve -1- for camshaft adjustment
- N208 - Valve -2- for camshaft adjustment
- T121 - Connector, 121 point, on Motronic Engine Control Module (ECM)

- (E30) - Connector (87a), in wiring harness engine
- (D101) - Wire connection -1-, in engine compartment wiring harness
- (D107) - wire connection -5-, in engine compartment wiring harness

ws = white
sw = black
ro = red
br = brown
gn = green
bl = blue
gr = grey
li = lilac
ge = yellow
or = orange

97-29144

Edition 10.02

Passat — Wiring diagram — No. 60/11

Motronic Engine Control Module (ECM), Heated Oxygen Sensors

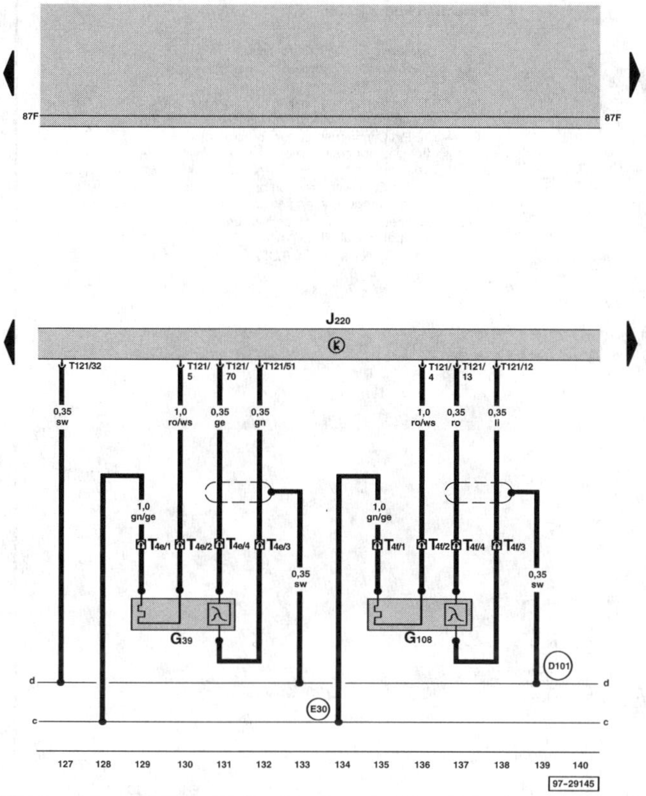

G39 - Heated Oxygen Sensor (HO2S)
G108 - Heated Oxygen Sensor (HO2S) 2
J220 - Motronic Engine Control Module (ECM)
T4e - 4-Pin Connector, black
T4f - 4-Pin Connector, black
T121 - Connector, 121 point, on Motronic Engine Control Module (ECM)

(E30) - Connector (87a), in wiring harness engine

(D101) - Wire connection -1-, in engine compartment wiring harness

ws = white
sw = black
ro = red
br = brown
gn = green
bl = blue
gr = grey
li = lilac
ge = yellow
or = orange

Edition 10.02

Passat — Wiring diagram — No. 60/12

Motronic Engine Control Module (ECM), Oxygen Sensors behind Three Way Catalytic Converter

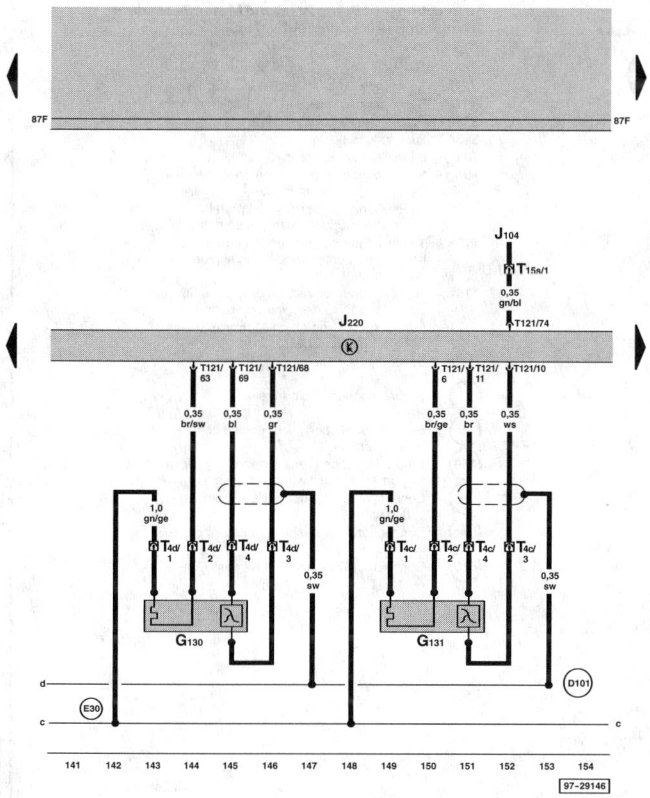

G130 - Oxygen Sensor (O2S) Behind Three Way Catalytic Converter (TWC)
G131 - Oxygen Sensor (O2S) 2 Behind Three Way Catalytic Converter (TWC)
J104 - ABS Control Module (w/EDL)
J220 - Motronic Engine Control Module (ECM)
T4c - 4-Pin Connector, brown
T4d - 4-Pin Connector, green
T15s - 15-Pin Connector, red, on protective housing for control module, in engine compartment, left
T121 - Connector, 121 point, on Motronic Engine Control Module (ECM)

(E30) - Connector (87a), in wiring harness engine

(D101) - Wire connection -1-, in engine compartment wiring harness

ws = white
sw = black
ro = red
br = brown
gn = green
bl = blue
gr = grey
li = lilac
ge = yellow
or = orange

Edition 10.02

Motronic Engine Control Module (ECM), Leak Detection Pump (LDP)

- J220 - Motronic Engine Control Module (ECM)
- T6o - 6-Pin Connector, brown, connector station A-pillar, right
- T10d - 10-Pin Connector, brown, on protective housing for control module, in engine compartment, left
- T10e - 10-Pin Connector, orange, on protective housing for control module, in engine compartment, left
- T15s - 15-Pin Connector, red, on protective housing for control module, in engine compartment, left
- T121 - Connector, 121 point, on Motronic Engine Control Module (ECM)
- V144 - Leak Detection Pump (LDP)

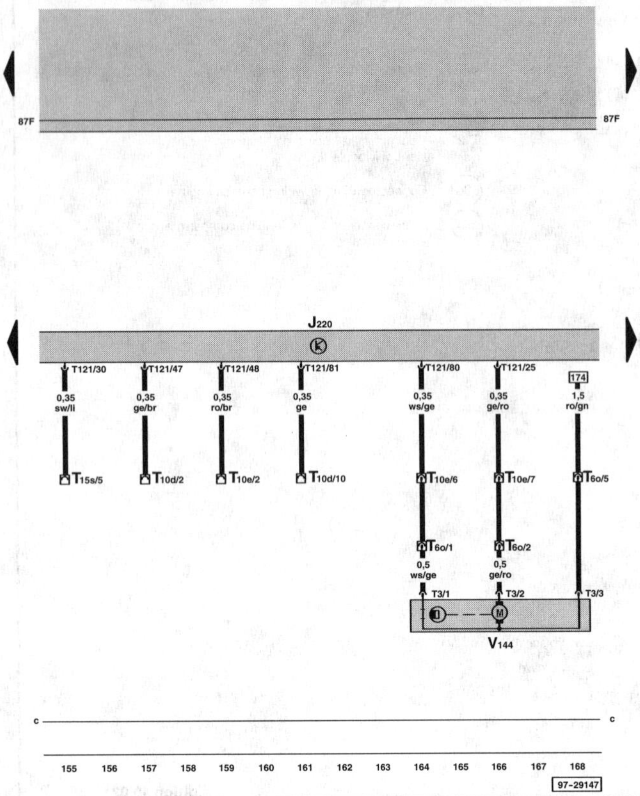

ws = white
sw = black
ro = red
br = brown
gn = green
bl = blue
gr = grey
li = lilac
ge = yellow
or = orange

Motronic Engine Control Module (ECM), Secondary Air Injection (Air) Pump Relayand Motor, Secondary Air Injection (AIR) Solenoid Valve

- J220 - Motronic Engine Control Module (ECM)
- J299 - Secondary Air Injection (AIR) Pump Relay
- N112 - Secondary Air Injection (AIR) Solenoid Valve
- S130 - Fuse for secondary air pump
- S234 - Fuse in fuse holder
- T3a - 3-Pin Connector, green
- T8 - 8-Pin Connector, brown, on Secondary Air Injection (AIR) Pump Relay
- T10d - 10-Pin Connector, brown, on protective housing for control module, in engine compartment, left
- T10e - 10-Pin Connector, orange, on protective housing for control module, in engine compartment, left
- T15s - 15-Pin Connector, red, on protective housing for control module, in engine compartment, left
- T121 - Connector, 121 point, on Motronic Engine Control Module (ECM)
- V101 - Secondary Air Injection (AIR) Pump Motor
- (28) - Ground connection, on firewall
- (85) - Ground connection -1-, in engine compartment wiring harness
- (A60) - Wire connection (Vehicle speed signal), in instrument cluster wiring harness
- (A121) - Connection (high bus) in instrument panel wiring harness
- (A122) - Connection (low bus) in instrument panel wiring harness
- (E30) - Connector (87a), in wiring harness engine

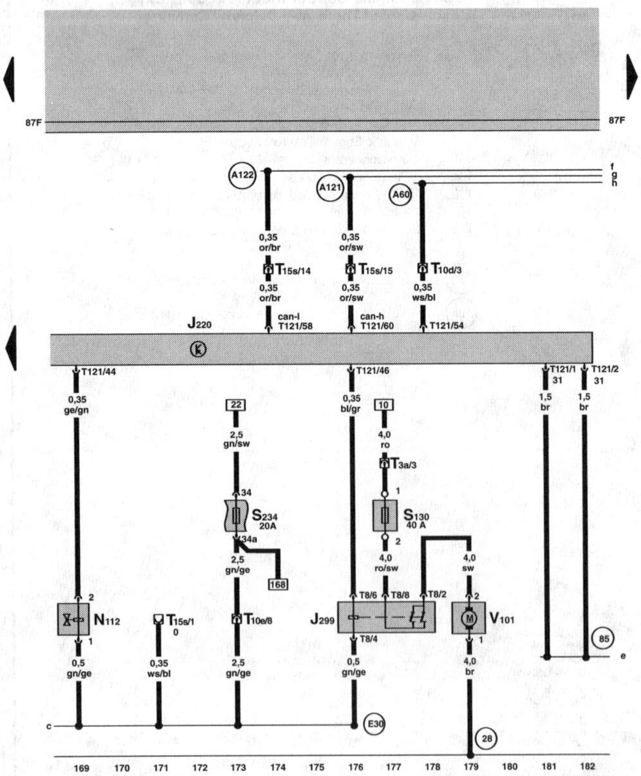

ws = white
sw = black
ro = red
br = brown
gn = green
bl = blue
gr = grey
li = lilac
ge = yellow
or = orange

Oil Pressure Switch, Hood Alarm Switch, Speedometer Vehicle Speed Sensor (VSS), Engine Coolant Level (ECL) Sensor, Oil Level Thermal Sensor

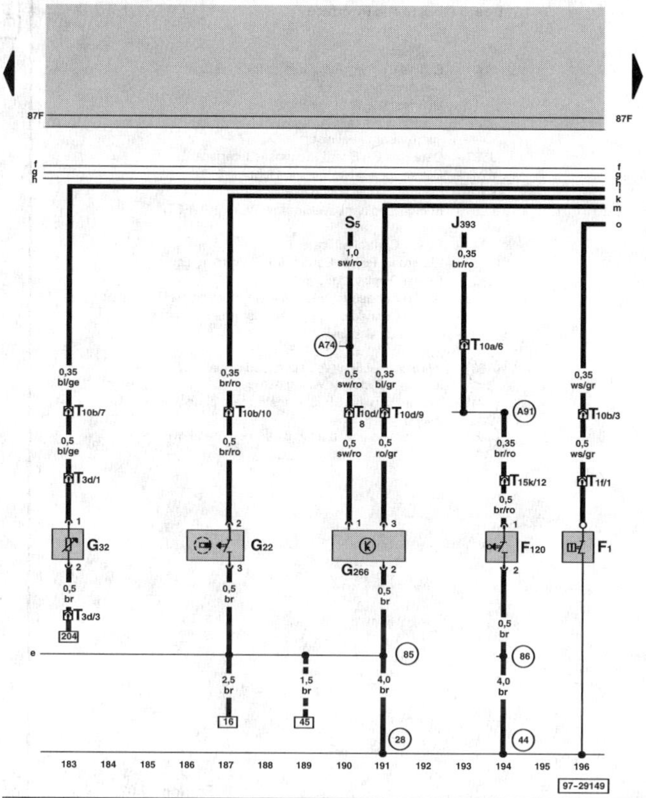

F1 - Oil Pressure Switch
F120 - Hood Alarm Switch
G22 - Speedometer Vehicle Speed Sensor (VSS)
G32 - Engine Coolant Level (ECL) Sensor
G266 - Oil Level Thermal Sensor
J393 - Central control module for comfort system
S5 - Fuse
T1f - Single Connector, black
T3d - 3-Pin Connector,
T10a - 10-Pin Connector, brown
T10b - 10-Pin Connector, black, on protective housing for control module, in engine compartment, left
T10d - 10-Pin Connector, brown, on protective housing for control module, in engine compartment, left
T15k - 15-Pin Connector, grey

(28) - Ground connection, on firewall
(44) - Ground connection, on left A-pillar, lower part
(85) - Ground connection -1-, in engine compartment wiring harness
(86) - Ground connection -1-, in rear wiring harness
(A74) - Connector (15a - fuse 5) in instrument panel wiring harness
(A91) - Connector (signal for instrument panel insert), in instrument panel wiring harness

--- - automatic transmission only

ws = white
sw = black
ro = red
br = brown
gn = green
bl = blue
gr = grey
li = lilac
ge = yellow
or = orange

97-29149

Control module with indicator unit, Fuel Pump (FP), Fuel Level Sensor, Fuel Supply Sensor, Sender for fuel gauge

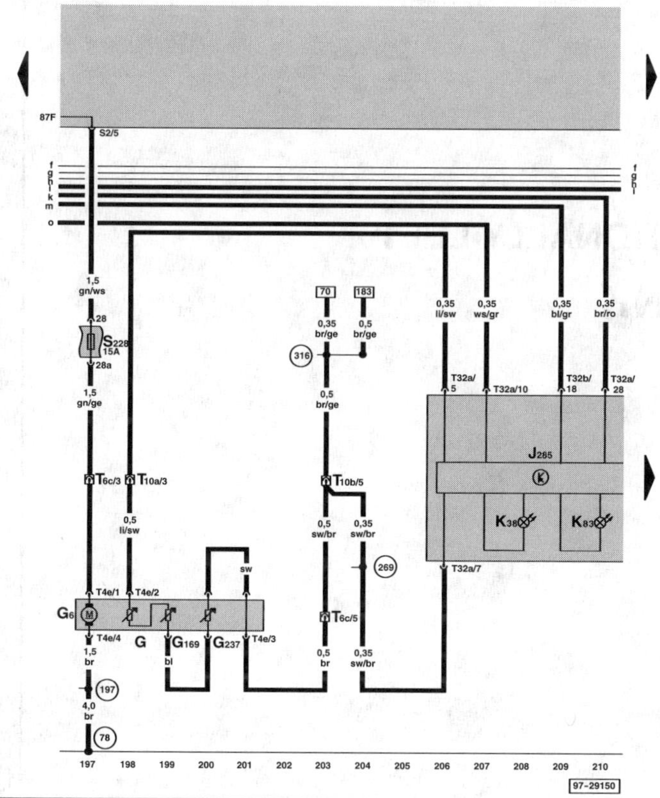

G - Sender for fuel gauge
G6 - Fuel Pump (FP)
G169 - Fuel Level Sensor 2
G237 - Fuel Supply Sensor 3
S228 - Fuse in fuse holder
J285 - Control module with indicator unit in instrument panel insert
K38 - Engine Oil Level Indicator Light
K83 - Malfunction Indicator Lamp (MIL)
T4e - 4-Pin Connector, near Fuel Pump (FP)
T6c - 6-Pin Connector, blue
T10a - 10-Pin connector, black
T10b - 10-Pin Connector, black, on protective housing for control module, in engine compartment, left
T32a - 32-Pin Connector, bue, on instrument cluster
T32b - 32-Pin Connector, green, on instrument cluster

(78) - Ground Connection (lower right B-pillar)
(197) - Ground connection -4-, in wiring harness rear
(269) - Ground connector (sensor ground) -1-, in instrument panel wiring harness
(316) - Sensor Ground Connection 2 (in engine wiring harness)

ws = white
sw = black
ro = red
br = brown
gn = green
bl = blue
gr = grey
li = lilac
ge = yellow
or = orange

97-29150

Control module with indicator unit, Data Bus On Board Diagnostic Interface

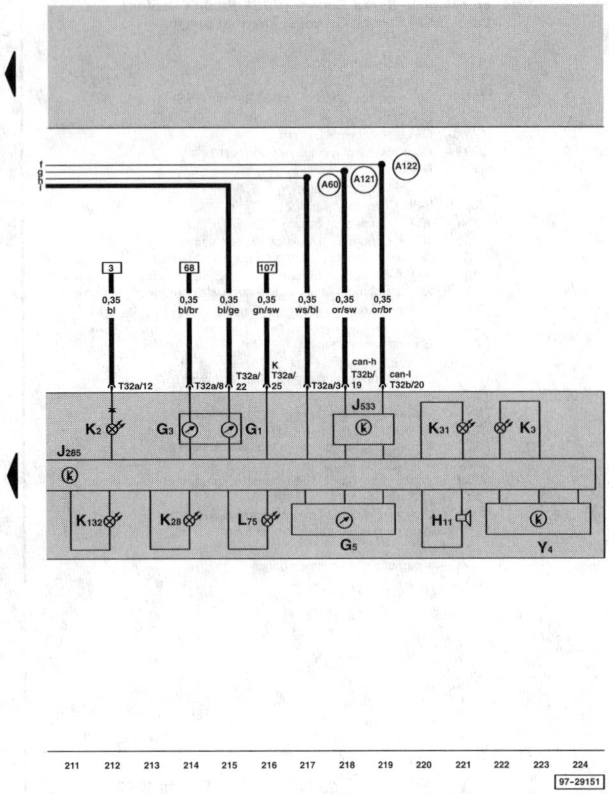

G1 - Fuel Gauge
G3 - Engine Coolant Temperature (ECT) Gauge
G5 - Tachometer
H11 - Oil Pressure Warning Buzzer
J285 - Control module with indicator unit in instrument panel insert
J533 - Data Bus On Board Diagnostic Interface
K2 - Generator (GEN) Warning Light
K3 - Oil Pressure Warning Light
K28 - Engine Coolant Level/Temperature (ECL/ECT) Warning Light
K31 - Cruise Control Indicator Light
K132 - Electronic Power Control (EPC) Warning Lamp
L75 - Digital Display Light
T32a - 32-Pin Connector, bue, on instrument cluster
T32b - 32-Pin Connector, green, on instrument cluster
Y4 - Odometer Display

(A60) - Wire connection (Vehicle speed signal), in instrument cluster wiring harness
(A121) - Connection (high bus) in instrument panel wiring harness
(A122) - Connection (low bus) in instrument panel wiring harness

ws = white
sw = black
ro = red
br = brown
gn = green
bl = blue
gr = grey
li = lilac
ge = yellow
or = orange

Edition 10.02

Repair Manual page number Y 598

THIS PAGE INTENTIONALLY LEFT

BLANK

Repair Manual page number Y 597

Radio System

from October 2000

Deviate relay location and fuseplacements as well as the locations of multiple connectors see section "component locations".

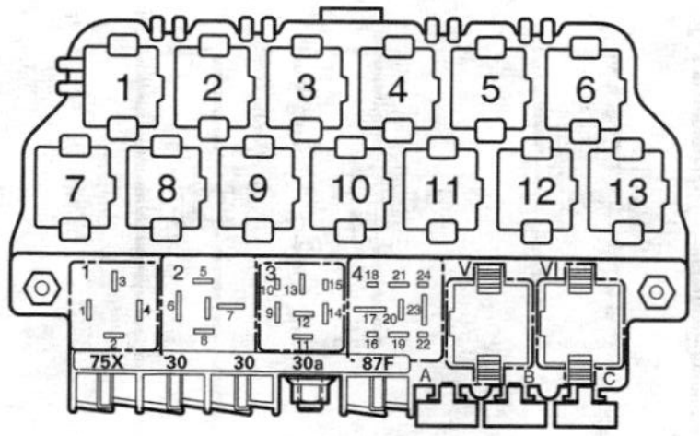

97-20676

Relay location on the thirteen position auxiliary relay panel, above relay panel:

Relay panel:

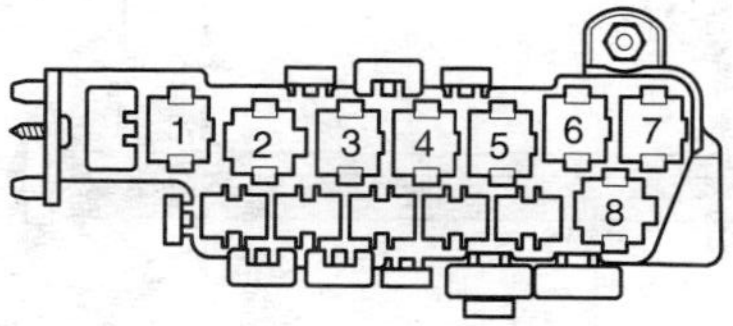

97-20515

Relay location on the eight position auxiliary relay panel, behind relay panel:

Note: Number in parentheses indicates production control number stamped on relay housing.

Fuse colors

30 A - green
25 A - white
20 A - yellow
15 A - blue
10 A - red
7,5 A - brown
5 A - beige

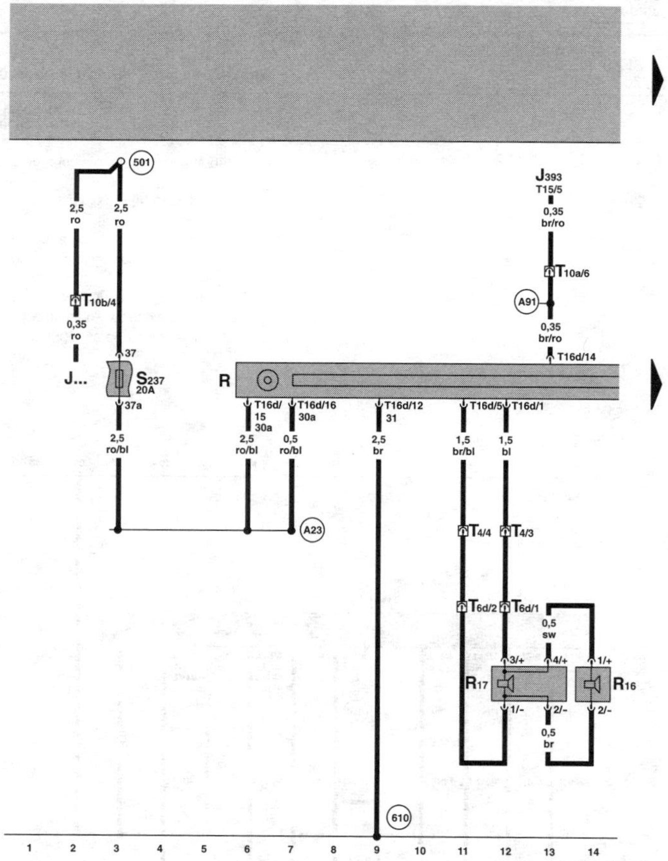

97-29152

ws = white
sw = black
ro = red
br = brown
gn = green
bl = blue
gr = grey
li = lilac
ge = yellow
or = orange

J... - Engine Control Module (ECM)
J393 - Central control module for comfort system
R - Radio
R16 - Treble speaker, right rear, in right rear door
R17 - Bass speaker, right rear, in right rear door
S237 - Fuse in fuse holder
T4 - 4-Pin Connector, black, connector station A-pillar, right
T6d - 6-Pin Connector, red, on B-pillar, right
T10a - 10-Pin Connector, brown, connector station A-pillar, left
T10b - 10-Pin Connector, black, on protective housing for control module, in engine compartment, left
T15 - 15-Pin Connector, on Central control module for comfort system
T16d - 16-Pin Connector, black, on radio

(501) - Threaded connection -2- (30) on the relay plate
(610) - Ground Connection (Audio) (under center console, front)
(A23) - wire connection (30al), in instrument panel wiring harness
(A91) - Connector (signal for instrument panel insert), in instrument panel wiring harness

Radio, right rear speakers

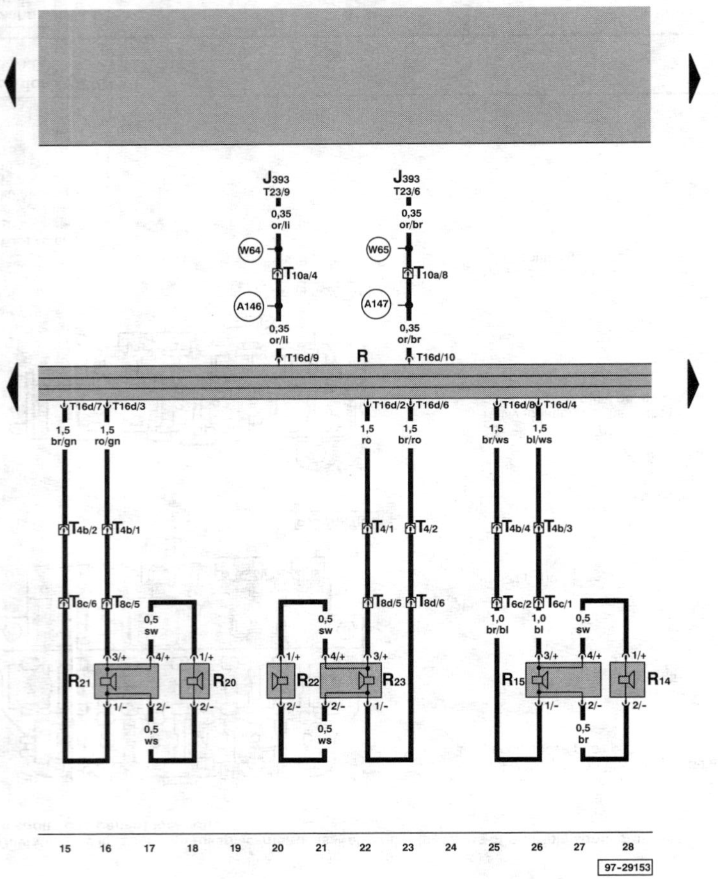

ws = white
sw = black
ro = red
br = brown
gn = green
bl = blue
gr = grey
li = lilac
ge = yellow
or = orange

J393 - Central control module for comfort system
R - Radio
R14 - Treble speaker, left rear, in left rear door
R15 - Bass speaker, left rear, in left rear door
R20 - Treble speaker, left front, in driver's door
R21 - Bass speaker, left front, in driver's door
R22 - Treble speaker, right front, in passenger's door
R23 - Bass speaker, right front, in passenger's door
T4 - 4-Pin Connector, black, connector station A-pillar, right
T4b - 4-Pin Connector, black, connector station A-pillar, left
T6c - 6-Pin Connector, red, on B-pillar, left
T8c - 8-Pin Connector, black, connector station A-pillar, left
T8d - 8-Pin Connector, black, connector station A-pillar, right
T10a - 10-Pin Connector, brown, connector station A-pillar, left
T16d - 16-Pin Connector, black, on radio
T23 - 23-Pin Connector, on Central control module for comfort system

(A146) - Comfort System High-bus Connection (in instrument panel wiring harness)
(A147) - Comfort System Low-bus Connection (in instrument panel wiring harness)
(W64) - CAN-Bus High Connection - Convenience (in floor wiring harness)
(W65) - CAN-Bus Low Connection - Convenience (in floor wiring harness)

Radio, front speakers, left rear speakers

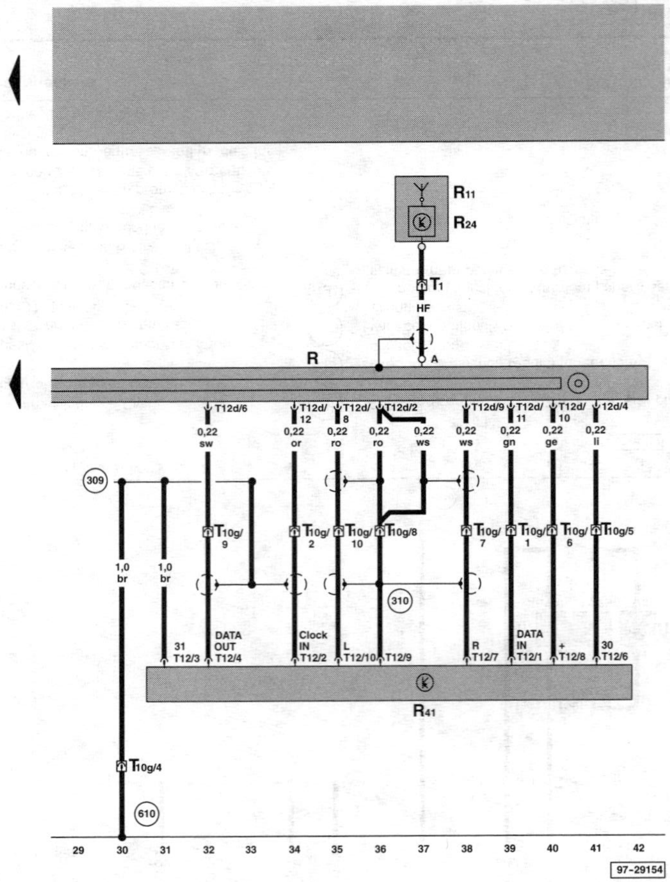

ws = white
sw = black
ro = red
br = brown
gn = green
bl = blue
gr = grey
li = lilac
ge = yellow
or = orange

R - Radio
R11 - Antenna
R24 - Antenna Amplifier
R41 - CD Changer Unit, in luggage compartment
T1 - 1-Pin Connector, black, below console
T10g - 10-Pin Connector, brown, connector station A-pillar, left
T12 - 12-Pin Connector, red, on CD Changer Unit
T12d - 12-Pin Connector, on radio

(309) - Ground Connection 1 (shielding) (in CD changer wiring harness)
(310) - Ground Connection 2 (shielding) (in CD changer wiring harness)
(610) - Ground Connection (Audio) (under center console, front)

Radio, Antenna, Antenna Amplifier, CD Changer Unit

Parking aid

from February 2001

Deviate relay location and fuseplacements as well as the locations of multiple connectors see section component locations.

Relay location on the thirteen position auxiliary relay panel, above relay panel:

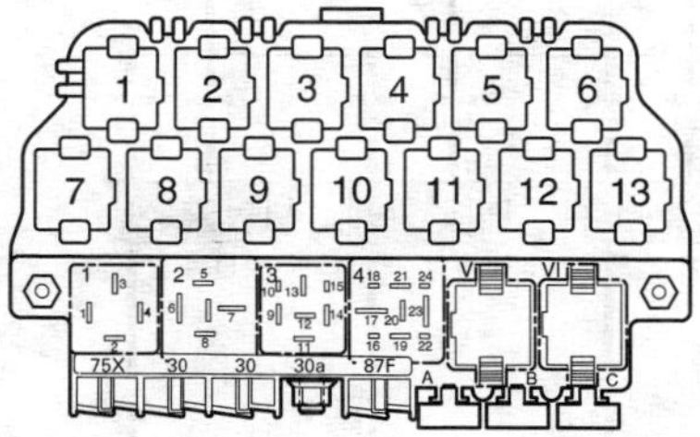

97-20676

Relay panel:

Relay location on the eight position auxiliary relay panel, behind relay panel:

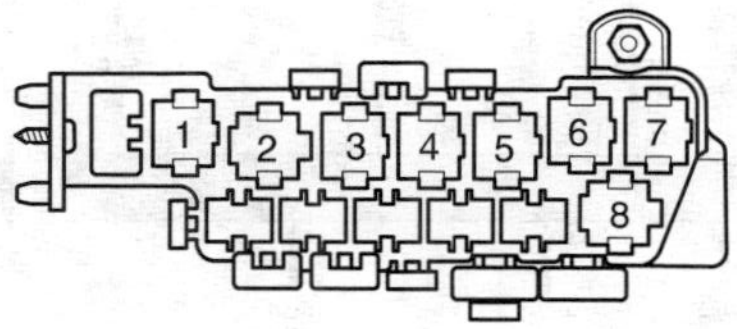

97-20515

Note: Number in parentheses indicates production control number stamped on relay housing.

Fuse colors
30 A - green
25 A - white
20 A - yellow
15 A - blue
10 A - red
7,5 A - brown
5 A - beige

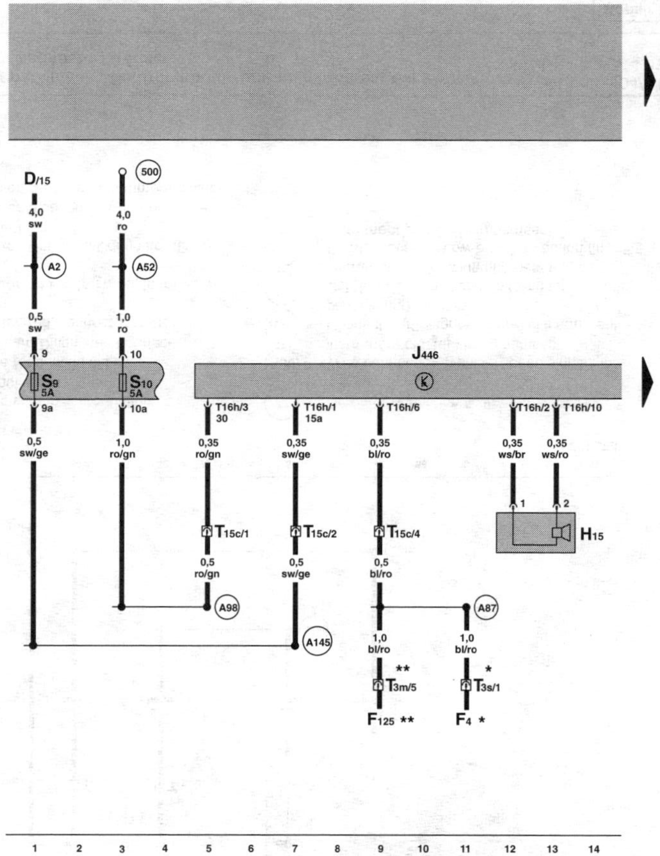

ws = white
sw = black
ro = red
br = brown
gn = green
bl = blue
gr = grey
li = lilac
ge = yellow
or = orange

- D - Ignition/Starter Switch
- F4 - Back-Up Light Switch
- F125 - Multi-Function Transmission Range (TR) Switch
- H15 - Warning buzzer for parking
- J446 - Control module for parking aid
- S9 - Fuse
- S10 - Fuse
- T3m - 3-Pin Connector, brown, connector station A-pillar, right
- T3s - 3-Pin Connector, red, on protective housing for control module, in engine compartment,left
- T15c - 15-Pin Connector, red, connector station A-pillar, right
- T16h - 16-Pin Connector, on Control module for parking aid

- (500) - Threaded connection -1- (30) on the relay plate
- (A2) - plus connection (15), in instrument panel wiring harness
- (A52) - Plus Connection 2 (30) (in instrument panel wiring harness)
- (A87) - Connector (reverse lamp), in instrument panel wiring harness
- (A98) - Plus connector -4- (30), in instrument panel wiring harness
- (A145) - Plus Connection 4 (15) (in instrument panel wiring harness)

* manual transmission only

** automatic transmission only

Control module for parking aid, Warning buzzer for parking

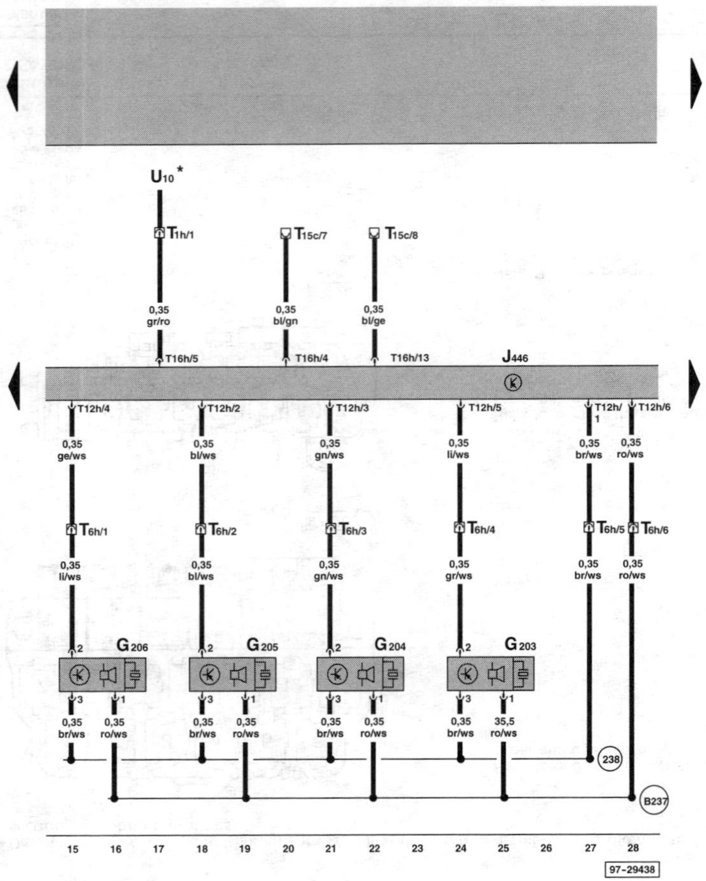

ws = white
sw = black
ro = red
br = brown
gn = green
bl = blue
gr = grey
li = lilac
ge = yellow
or = orange

G203 - Sensor for parking, rear, left
G204 - Sensor for parking, rear, left, center
G205 - Sensor for parking, rear, right, center
G206 - Sensor for parking, rear, right
J446 - Control module for parking aid
T1h - Single Connector
T6h - 6-Pin Connector
T12h - 12-Pin Connector, on Control module for parking aid
T15c - 15-Pin Connector, red, connector station A-pillar, right
T16h - 16-Pin Connector, on Control module for parking aid
U10 - Trailer Socket

(238) - Ground connection -1-, in wiring harness interior
(B237) - Parking Aid Plus Connection (in interior wiring harness)

* Trailer Socket only

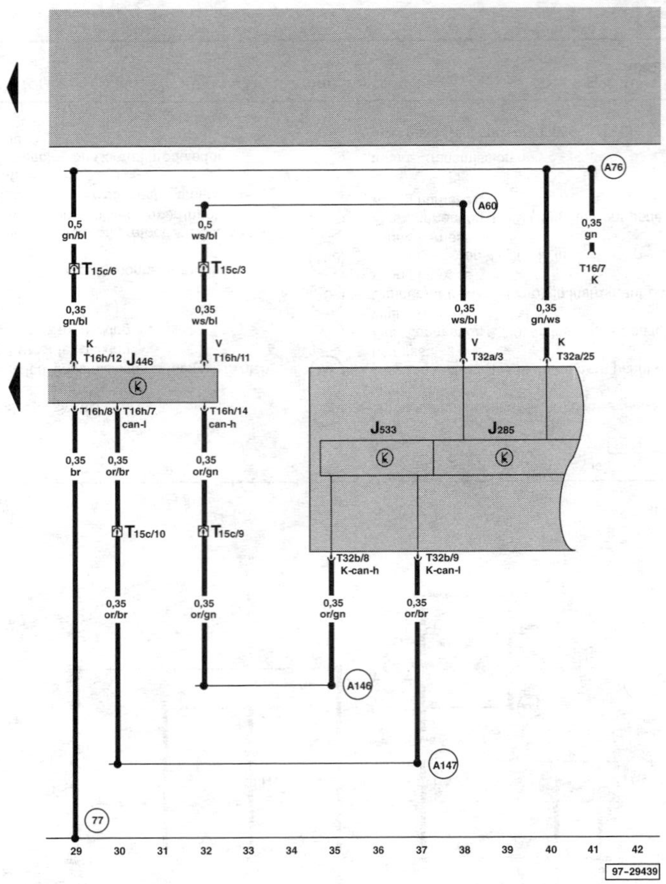

ws = white
sw = black
ro = red
br = brown
gn = green
bl = blue
gr = grey
li = lilac
ge = yellow
or = orange

J285 - Control module with indicator unit in instrument panel insert
J446 - Control module for parking aid
J533 - Data Bus On Board Diagnostic Interface
T15c - 15-Pin Connector, red, connector station A-pillar, right
T16h - 16-Pin Connector, on Control module for parking aid
T16 - 16-Pin Connector, Data Link Connector (DLC), below instrument panel, left
T32a - 32-Pin Connector, blue, on instrument cluster
T32b - 32-Pin Connector, green, on instrument cluster

(77) - Ground Connection (lower left B-pillar)
(A60) - Wire connection (Vehicle speed signal), in instrument cluster wiring harness
(A76) - Connector (K-diagnosis wire) in instrument panel wiring harness
(A146) - Comfort System High-bus Connection (in instrument panel wiring harness)
(A147) - Comfort System Low-bus Connection (in instrument panel wiring harness)

Power seats without Memory

from October 2000

Deviate relay location and fuseplacements as well as the locations of multiple connectors see section component locations.

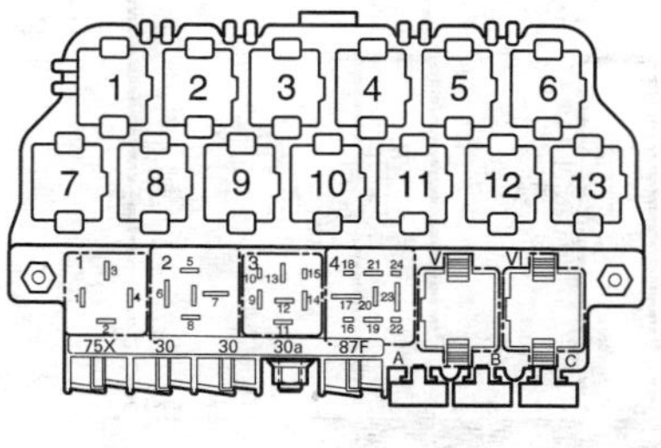

97-20676

Relay location on the thirteen position auxiliary relay panel, above relay panel:

Relay panel:

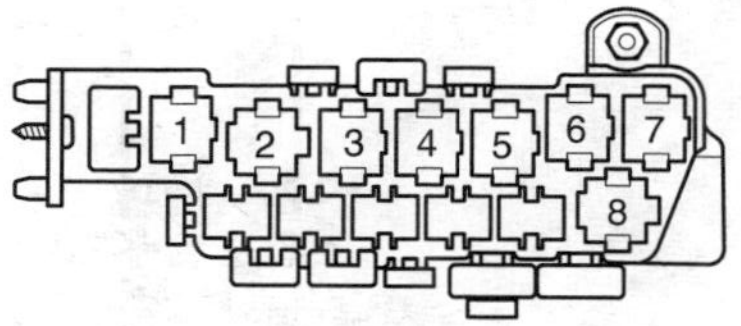

97-20515

Relay location on the eight position auxiliary relay panel, behind relay panel:

Note: Number in parentheses indicates production control number stamped on relay housing.

Fuse colors
30 A - green
25 A - white
20 A - yellow
15 A - blue
10 A - red
7,5 A - brown
5 A - beige

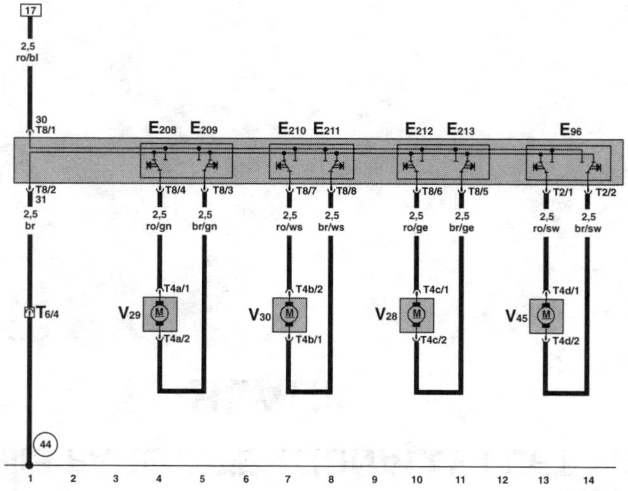

97-29440

ws = white
sw = black
ro = red
br = brown
gn = green
bl = blue
gr = grey
li = lilac
ge = yellow
or = orange

- E96 - Driver s Backrest Adjustment Switch
- E208 - Button for front height adjustment, upwards, driver seat
- E209 - Button for front height adjustment, downwards, driver seat
- E210 - Button for rear height adjustment, upwards, driver seat
- E211 - Button for rear height adjustment, downwards, driver seat
- E212 - Button for length adjustment, forward, driver seat
- E213 - Button for length adjustment, backward, driver seat
- T2 - Double Connector, black
- T4a - 4-Pin Connector, black, below driver s seat
- T4b - 4-Pin Connector, black, below driver s seat
- T4c - 4-Pin Connector, black, below driver s seat
- T4d - 4-Pin Connector, black, below driver s seat
- T6 - 6-Pin Connector, red, below driver s seat
- T8 - 8-Pin Connector, black
- V28 - Driver s seat fore/aft adjusting motor
- V29 - Motor for front height adjustment, driver seat
- V30 - Motor for rear height adjustment, driver seat
- V45 - Motor for seatback adjustment, driver seat
- (44) - Ground connection, on left A-pillar, lower part

Driver s Backrest Adjustment Switch, Button for front and rear height adjustment, Motor for front, rear and seatback adjustment

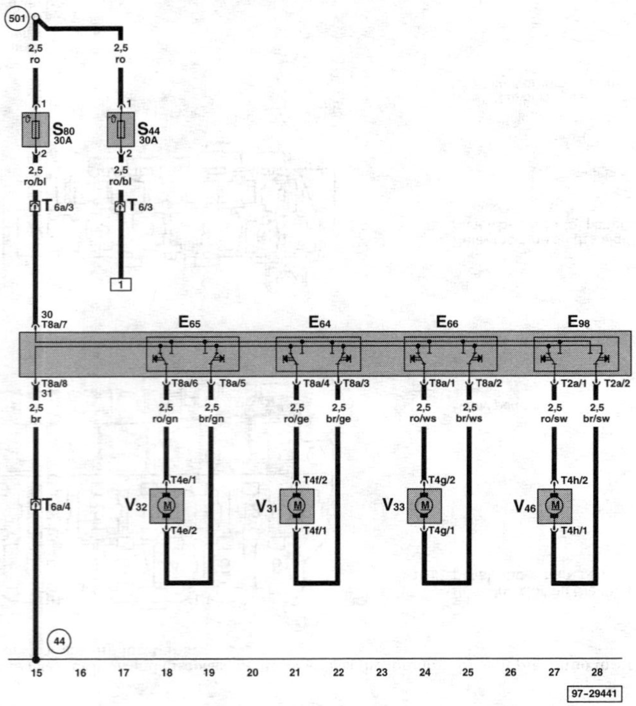

ws = white
sw = black
ro = red
br = brown
gn = green
bl = blue
gr = grey
li = lilac
ge = yellow
or = orange

- E64 - Passenger s Seat Fore/Aft Adjusting Switch
- E65 - Passenger s Seat Front Height Adjusting Switch
- E66 - Passenger s Seat Rear Height Adjusting Switch
- E98 - Passenger Backrest Adjustment Switch
- S44 - Power (Memory) Seat Circuit Breaker
- S80 - Power Seat Circuit Breaker
- T2a - Double Connector, black
- T4e - 4-Pin Connector, black, below driver s seat
- T4f - 4-Pin Connector, black, below driver s seat
- T4g - 4-Pin Connector, black, below driver s seat
- T4h - 4-Pin Connector, black, below driver s seat
- T6 - 6-Pin Connector, red, below driver s seat
- T6a - 6-Pin Connector, red, below driver s seat
- T8a - 8-Pin Connector, black
- V31 - Passenger s Seat Fore/Aft Adjusting Motor
- V32 - Passenger s Seat Front Height Adjusting Motor
- V33 - Passenger s Seat Rear Height Adjusting Motor
- V46 - Passenger s Backrest Adjusting Motor
- (44) - Ground connection, on left A-pillar, lower part
- (501) - Threaded connection -2- (30) on the relay plate

THIS PAGE INTENTIONALLY LEFT

BLANK

Power seats and mirror adjustment with Memory,

from October 2000

Notes:

Information for

- *Relay and fuse arrangements*
- *Connector arrangements*
- *Control Modules and Relays*
- *Ground connections*

⇒ Component Locations!

Information for

- *Troubleshooting Procedures*

⇒ Guided Fault Finding using VAS 5051/5052

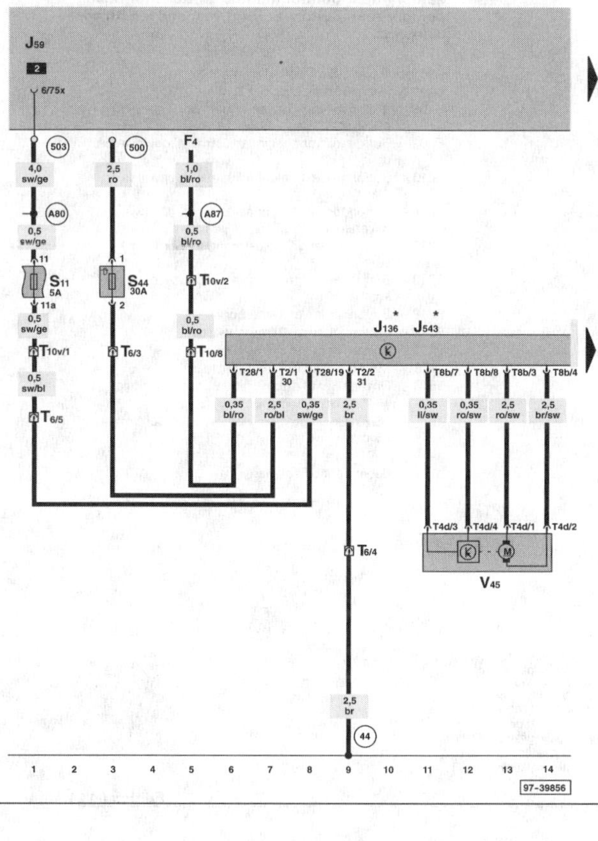

Memory Seat Control Module, Motor for seatback adjustment, driver seat

- F4 - Back-Up Light Switch
- J59 - Load Reduction Relay
- J136 - Memory Seat Control Module
- J543 - Driver's Seat/Mirror Position Control Module
- S11 - Fuse
- S44 - Power (Memory) Seat Circuit Breaker
- T2 - Double Connector
- T4d - 4-Pin Connector, black, on Motor for seatback adjustment, driver seat
- T6 - 6-Pin Connector, red
- T8b - 8-Pin Connector, on Memory Seat Control Module
- T10 - 10-Pin Connector, brown
- T10v - 10-Pin Connector, pink, on A-pillar, right
- T28 - 28-Pin Connector, on Memory Seat Control Module
- V45 - Motor for seatback adjustment, driver seat
- (44) - Ground connection, on left A-pillar, lower part
- (500) - Threaded connection -1- (30) on the relay plate
- (503) - Bolted Connection (75x) (at relay panel)
- (A80) - Connection -1- (X) in instrument panel wiring harness
- (A87) - Connection (reverse lamp), in instrument panel wiring harness
- * - dependent on software

ws = white
sw = black
ro = red
br = brown
gn = green
bl = blue
gr = grey
li = lilac
ge = yellow
or = orange

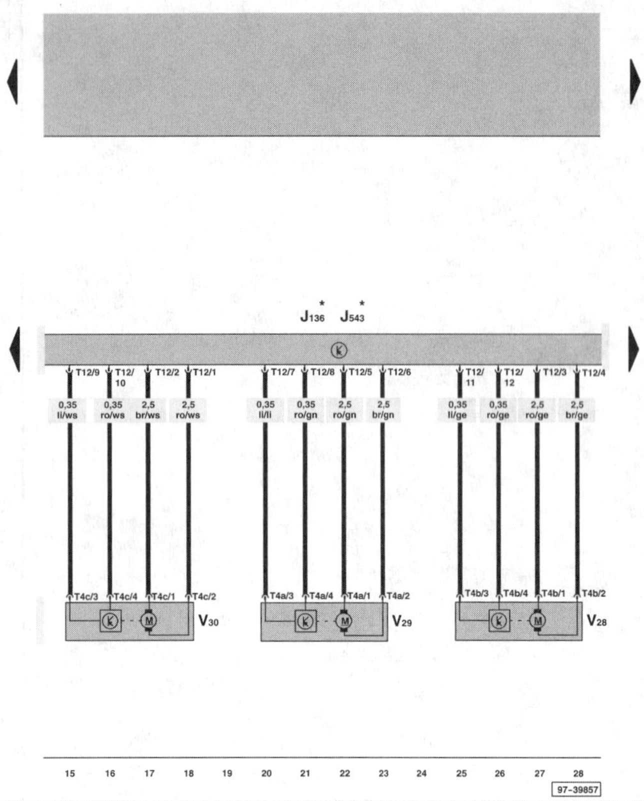

Memory Seat Control Module, Motor for front and rear height adjustment, driver seat, Driver's seat fore/aft adjusting motor

- J136 - Memory Seat Control Module
- J543 - Driver's Seat/Mirror Position Control Module
- T4a - 4-Pin Connector, black, on Motor for front height adjustment, driver seat
- T4b - 4-Pin Connector, black, on Driver's seat fore/aft adjusting motor
- T4c - 4-Pin Connector, black, on Motor for rear height adjustment, driver seat
- T12 - 12-Pin Connector, black, on Memory Seat Control Module
- V28 - Driver's seat fore/aft adjusting motor
- V29 - Motor for front height adjustment, driver seat
- V30 - Motor for rear height adjustment, driver seat

* - dependent on software

ws = white
sw = black
ro = red
br = brown
gn = green
bl = blue
gr = grey
li = lilac
ge = yellow
or = orange

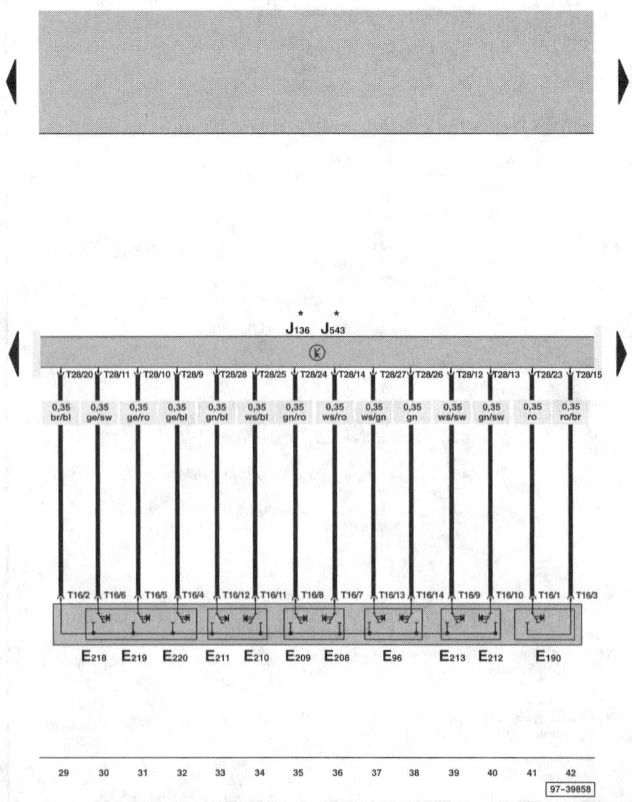

Memory Seat Control Module, Buttons for driver, memory seat, Button for front, rear and length adjustment

- E96 - Driver's Backrest Adjustment Switch
- E190 - Seat Memory Lock-out Button
- E208 - Button for front height adjustment, upwards, driver seat
- E209 - Button for front height adjustment, downwards, driver seat
- E210 - Button for rear height adjustment, upwards, driver seat
- E211 - Button for rear height adjustment, downwards, driver seat
- E212 - Button for length adjustment, forward, driver seat
- E213 - Button for length adjustment, backward, driver seat
- E218 - Button for driver 1, memory seat
- E219 - Button for driver 2, memory seat
- E220 - Button for driver 3, memory seat
- J136 - Memory Seat Control Module
- J543 - Driver's Seat/Mirror Position Control Module
- T16 - 16-Pin Connector, black
- T28 - 28-Pin Connector, on Memory Seat Control Module

* - dependent on software

ws = white
sw = black
ro = red
br = brown
gn = green
bl = blue
gr = grey
li = lilac
ge = yellow
or = orange

Memory Seat Control Module, Door control module, driver side

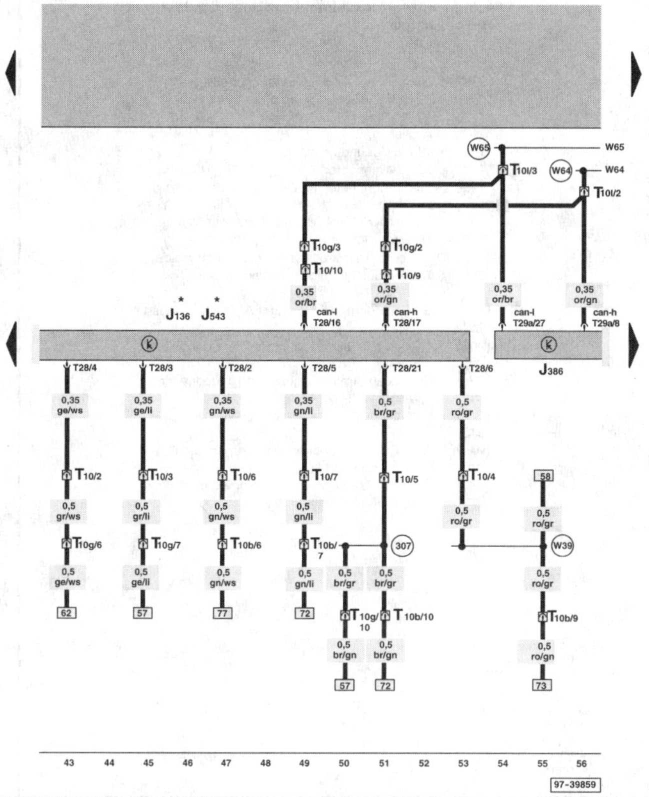

- J136 - Memory Seat Control Module
- J543 - Driver's Seat/Mirror Position Control Module
- J386 - Door control module, driver side
- T10 - 1–Pin Connector, brown, under driver's seat
- T10b - 10-Pin Connector, green, connector station A-pillar, right
- T10g - 10-Pin Connector, grey, connector station A-pillar, left
- T10l - 10-Pin Connector, black
- T28 - 28-Pin Connector, on Memory Seat Control Module
- T29a - 29-Pin Connector, on Door control module, driver side

- (307) - Ground Connection for Sensor (floor wiring harness)
- (W39) - Connection for Mirror Memory (floor wiring harness)
- (W64) - CAN-Bus High Connection - Convenience (in floor wiring harness)
- (W65) - CAN-Bus Low Connection - Convenience (in floor wiring harness

- * - dependent on software

ws = white
sw = black
ro = red
br = brown
gn = green
bl = blue
gr = grey
li = lilac
ge = yellow
or = orange

97-39859

Edition 04.04

Door Control module, driver side, Motor for fold-in mirror, Motor for mirror, Heated outside mirror, driver side

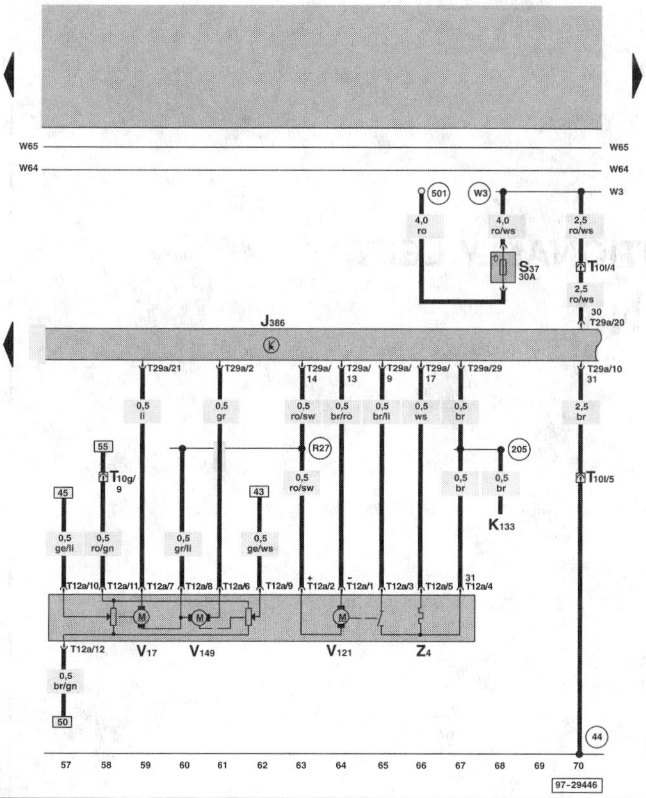

- J386 - Door control module, driver side
- K133 - Warning light for central locking -SAFE-
- S37 - Power Window Fuse
- T10g - 10-Pin Connector, grey, connector station A-pillar, left
- T10l - 10-Pin Connector, black
- T12a - 12-Pin Connector, black
- T29a - 29-Pin Connector, on Door control module, driver side
- V17 - Driver's Side Mirror Adjustment Motor
- V121 - Motor for fold-in mirror, driver side
- V149 - Motor for mirror adjustment, driver side
- Z4 - Heated outside mirror, driver side

- (44) - Ground connection, on left A-pillar, lower part
- (205) - Ground connection, in driver's door wiring harness
- (501) - Threaded connection -2- (30) on the relay plate
- (R27) - wire connection (power mirror), in driver's door wiring harness
- (W3) - wire connection, in rear wiring harness

ws = white
sw = black
ro = red
br = brown
gn = green
bl = blue
gr = grey
li = lilac
ge = yellow
or = orange

97-29446

Edition 04.04

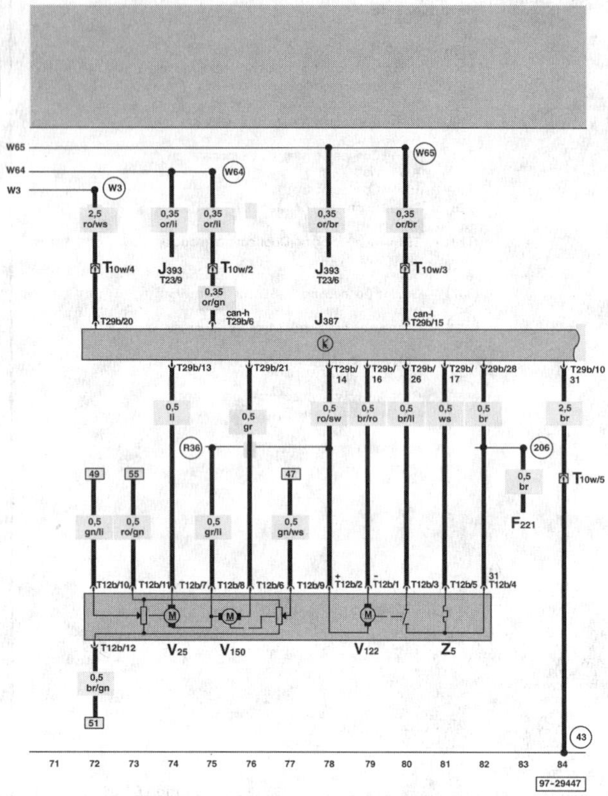

ws = white
sw = black
ro = red
br = brown
gn = green
bl = blue
gr = grey
li = lilac
ge = yellow
or = orange

Door control module, passenger side, Motor for fold-in mirror , Motor for mirror, Heated outside mirror, passenger side

F221 - Lock unit for central locking, passenger side
J387 - Door control module, passenger side
J393 - Central control module for comfort system
T10w - 10-Pin Connector, black, connector station A-pillar, right
T12b - 12-Pin Connector, black
T23 - 23-Pin Connector
T29b - 29-Pin Connector, on Door control module, passenger side
V25 - Passenger's Side Mirror Adjustment Motor
V122 - Motor for fold-in mirror, passenger side
V150 - Motor for mirror adjustment, passenger side
Z5 - Heated outside mirror, passenger side

(43) - Ground connection, on right A-pillar, lower part
(206) - Ground connection, in passenger's door wiring harness
(R36) - Connection (mirror adjustment) in door wiring harness, passenger side
(W3) - wire connection, in rear wiring harness
(W64) - CAN-Bus High Connection - Convenience (in floor wiring harness)
(W65) - CAN-Bus Low Connection - Convenience (in floor wiring harness

Edition 04.04

THIS PAGE INTENTIONALLY LEFT

BLANK

Heated seats

from October 2000

Deviate relay location and fuseplacements as well as the locations of multiple connectors see section component locations.

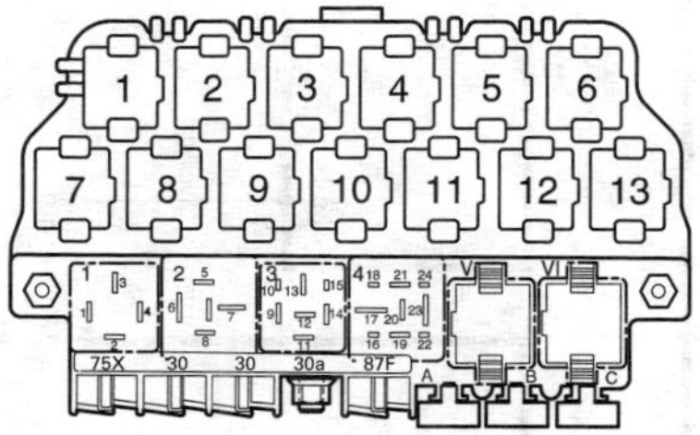

97-20676

Relay location on the thirteen position auxiliary relay panel, above relay panel:

Relay panel:

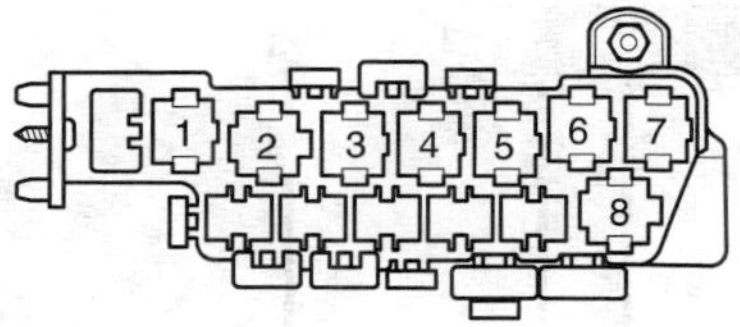

97-20515

Relay location on the eight position auxiliary relay panel, behind relay panel:

Note: Number in parentheses indicates production control number stamped on relay housing.

Fuse colors
30 A - green
25 A - white
20 A - yellow
15 A - blue
10 A - red
7,5 A - brown
5 A - beige

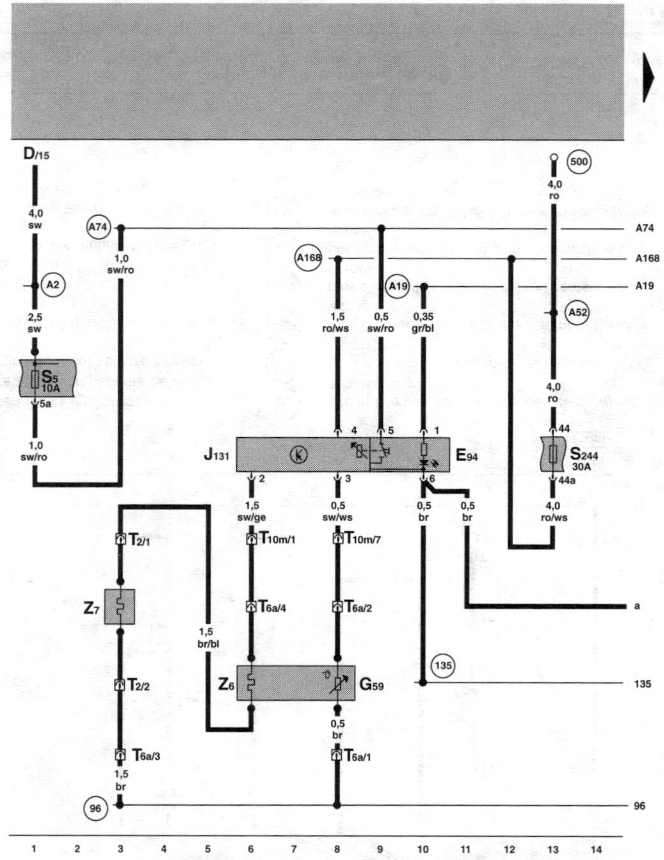

ws = white
sw = black
ro = red
br = brown
gn = green
bl = blue
gr = grey
li = lilac
ge = yellow
or = orange

97-29448

D - Ignition/Starter Switch
E94 - Adjuster for heated driver s seat
G59 - Driver s Heated Seat Temperature Sensor
J131 - Driver s Heated Seat Control Module
S5 - Fuse
S244 - Fuse in fuse holder
T2 - Double Connector, under Driver s Heated Seat
T6a - 6-Pin Connector, green
T10m - 10-Pin Connector, brown, connector station, A-pillar, right
Z6 - Driver s Seat Heating Element
Z7 - Driver s Backrest Heating Element

(96) - Ground connection -1-, in heated seats wiring harness
(135) - Ground connection -2-, in instrument panel wiring harness
(500) - Threaded connection -1-(30) on the relay plate
(A2) - plus connection (15), in instrument panel wiring harness
(A19) - wire connection (58d), in instrument panel wiring harness
(A52) - Plus Connection 2 (30) (in instrument panel wiring harness)
(A74) - Connector (15a - fuse 5) in instrument panel wiring harness
(A168) - Plus Connection 4 (30a) (in instrument panel wiring harness)

Driver s Heated Seat Control Module, Driver s Heated Seat Temperature Sensor, Driver s Seat/Backrest Heating Element

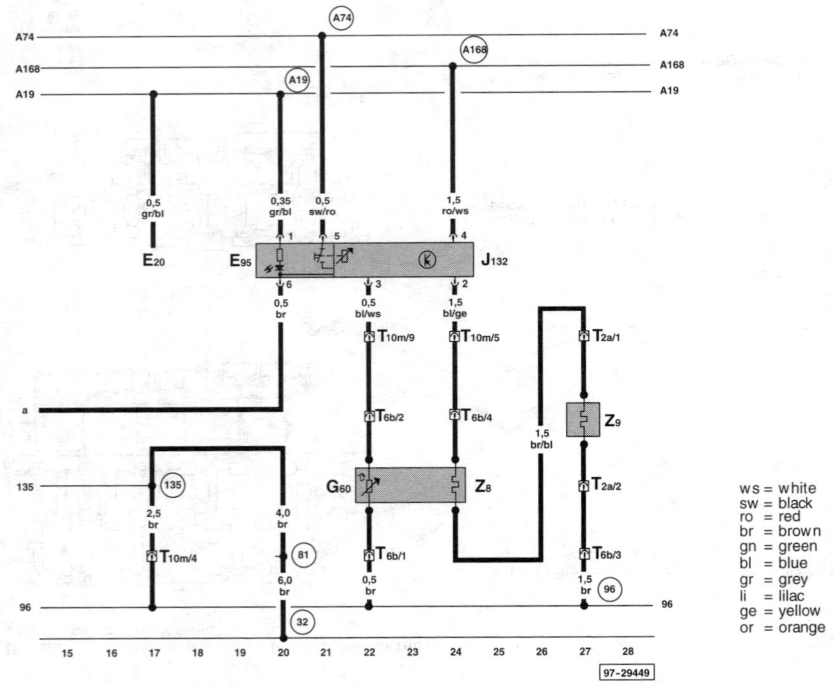

ws = white
sw = black
ro = red
br = brown
gn = green
bl = blue
gr = grey
li = lilac
ge = yellow
or = orange

E20 - Instrument Panel Light Dimmer Switch
E95 - Adjuster for heated passenger seat
G60 - Passenger s Heated Seat Temperature Sensor
J132 - Passenger s Heated Seat Control Module
T2a - Double Connector, under Passenger s Heated Seat
T6b - 6-Pin Connector, green
T10m - 10-Pin Connector, brown, connector station, A-pillar, right
Z8 - Passenger s Seat Heating Element
Z9 - Passenger s Backrest Heating Element

(32) - Ground connection, behind instrument panel, left
(81) - Ground connection -1-, in instrument panel wiring harness
(96) - Ground connection -1-, in heated seats wiring harness
(135) - Ground connection -2-, in instrument panel wiring harness
(A19) - wire connection (58d), in instrument panel wiring harness
(A74) - Connector (15a - fuse 5) in instrument panel wiring harness
(A168) - Plus Connection 4 (30a) (in instrument panel wiring harness)

Passenger s Heated Seat Control Module and Temperature Sensor, Adjuster for heated passenger seat, Passenger s Seat/Backrest Heating Element

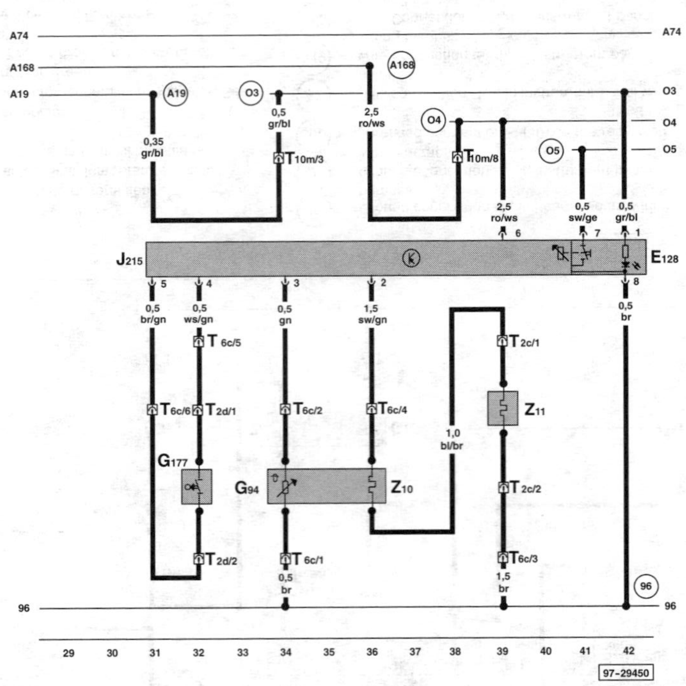

ws = white
sw = black
ro = red
br = brown
gn = green
bl = blue
gr = grey
li = lilac
ge = yellow
or = orange

E128 - Left Rear Heated Seat Regulating Switch
G94 - Left Rear Heated Seat Temperature Sensor
G177 - Sensor for occupied sensor seat, rear driver side
J215 - Left Rear Heated Seat Control Module
T2c - Double Connector
T2d - Double Connector
T6c - 6-Pin connector, green
T10m - 10-Pin Connector, brown, connector station, A-pillar, right
Z10 - Left Rear Seat Heating Element
Z11 - Left Rear Backrest Heating Element

(96) - Ground connection -1-, in heated seats wiring harness
(A19) - wire connection (58d), in instrument panel wiring harness
(A168) - Plus Connection 4 (30a) (in instrument panel wiring harness)
(O3) - plus connection (58b), in heated seats wiring harness
(O4) - plus connection (30 as), in heated seats wiring harness
(O5) - wire connection -1- (15a), in heated seats wiring harness

Left Rear Heated Seat Control Module and Regulating/Temperature Switch, Left Rear Seat/Backrest Heating Element, Sensor for occupied sensor sear

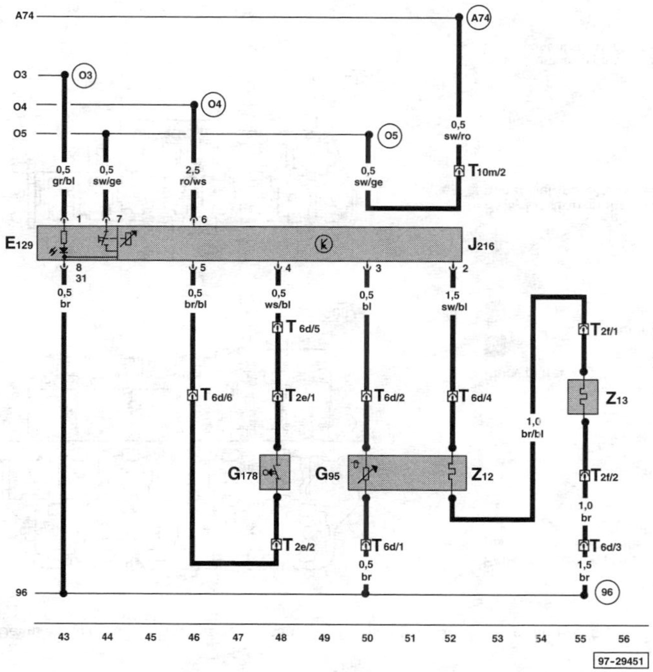

ws = white
sw = black
ro = red
br = brown
gn = green
bl = blue
gr = grey
li = lilac
ge = yellow
or = orange

E129 - Right Rear Heated Seat Regulating Switch
G95 - Left Rear Heated Seat Temperature Sensor
G178 - Sensor for occupied seat, rear, passenger side
J216 - Right Rear Heated Seat Control Module
T2e - Double Connector
T2f - Double Connector
T6d - 6-Pin Connector, green
T10m - 10-Pin Connector, brown, connector station, A-pillar, right
Z12 - Right Rear Seat Heating Element
Z13 - Right Rear Backrest Heating Element

(96) - Ground connection -1-, in heated seats wiring harness
(A74) - Connector (15a - fuse 5) in instrument panel wiring harness
(O3) - plus connection (58b), in heated seats wiring harness
(O4) - plus connection (30 as), in heated seats wiring harness
(O5) - wire connection -1- (15a), in heated seats wiring harness

Edition 04/01
USA.5062.14.21

Right Rear Heated Seat Control and Regulating/Temperature Switch, Sensor for occupied seat, Left Rear Heated Seat Temperature Sensor

THIS PAGE INTENTIONALLY LEFT

BLANK

Automatic day/night interior mirror, Rainsensor

from October 2000

Deviate relay location and fuseplacements as well as the locations of multiple connectors see section component locations.

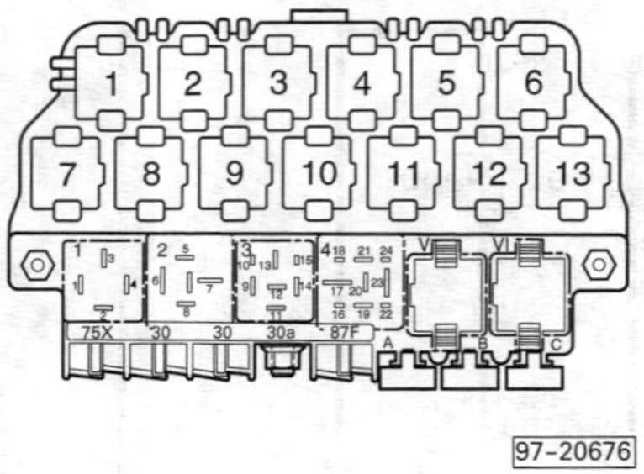

Relay location on the thirteen position auxiliary relay panel, above relay panel:

Relay panel:

V Wiper/Washer Intermittent Relay(377)

VI

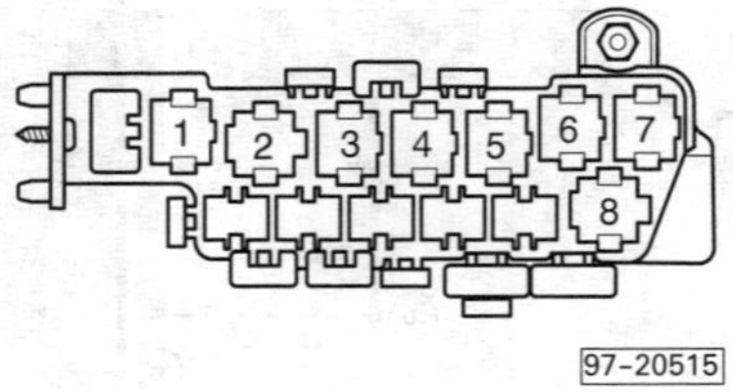

Relay location on the eight position auxiliary relay panel, behind relay panel:

Note: Number in parentheses indicates production control number stamped on relay housing.

Fuse colors
30 A - green
25 A - white
20 A - yellow
15 A - blue
10 A - red
7,5 A - brown
5 A - beige

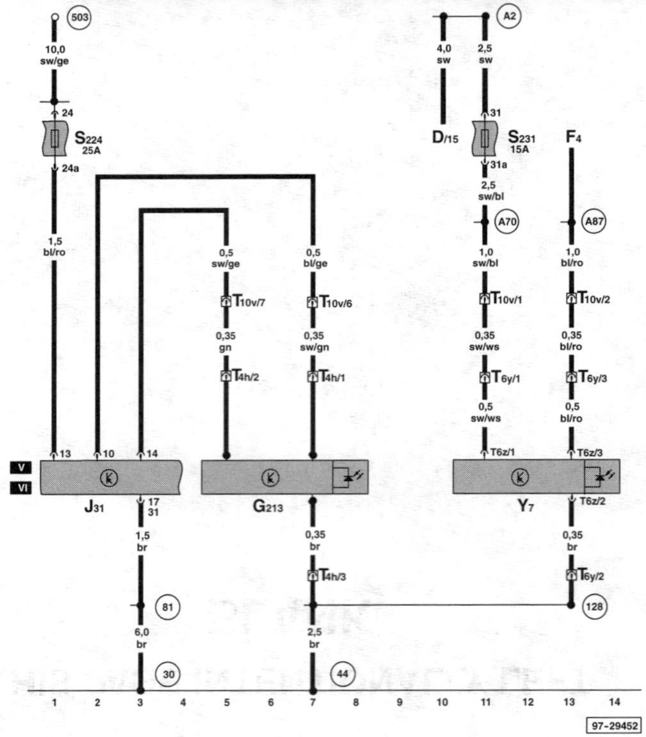

ws = white
sw = black
ro = red
br = brown
gn = green
bl = blue
gr = grey
li = lilac
ge = yellow
or = orange

D - Ignition/Starter Switch
F4 - Back-Up Light Switch
G213 - Rain Sensor
J31 - Wiper/Washer Intermittent Relay
S224 - Fuse in fuse holder
S231 - Fuse in fuse holder
T4h - 4-Pin Connector, black, near mirror under the roof
T6y - 6-Pin Connector, black, near mirror under the roof
T6z - 6-Pin Connector
T10v - 10-Pin Connector, green, connector station A-pillar, left
Y7 - Automatic Day/Night Interior Mirror

(30) - Ground connection, -1-, beside fuse/relay panel
(44) - Ground connection, on left A-pillar, lower part
(81) - Ground connection -1-, in instrument panel wiring harness
(128) - Ground connection -1-, in interior light wiring harness
(503) - Bolted Connection (75x) (at relay panel)
(A2) - plus connection (15), in instrument panel wiring harness
(A70) - Connector (15a, fuse 231), in instrument panel wiring harness
(A87) - Connector (reverse lamp), in instrument panel wiring harness

Power Outlet (12 V), Rear Console

from October 2000

Deviate relay location and fuseplacements as well as the locations of multiple connectors see section "component locations"

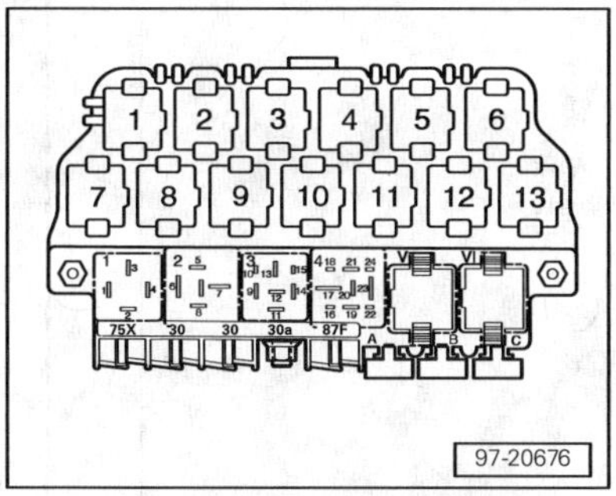

Relay location on the thirteen position auxiliary relay panel, above relay panel:

Relay panel:

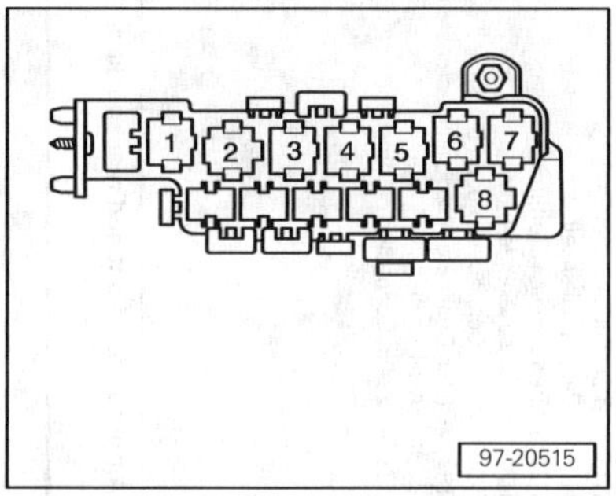

Relay location on the eight position auxiliary relay panel, above relay panel:

12 V outlet, Soket Light Bulb

E20 - Instrument Panel Light Dimmer Switch
L42 - Socket Light Bulb
S281 - Power Outlet Fuse 3, in fuse holder, behind left front footwell kick-panel
U19 - 12V outlet -3-

(30) - Ground connection, -1-, beside fuse/relay panel

(81) - Ground connection -1-, in instrument panel wiring harness

(135) - Ground connection -2-, in instrument panel wiring harness

(500) - Threaded connection -1- (30) on the relay plate

(A19) - wire connection (58d), in instrument panel wiring harness

(A52) - Plus Connection 2 (30) (in instrument panel wiring harness)

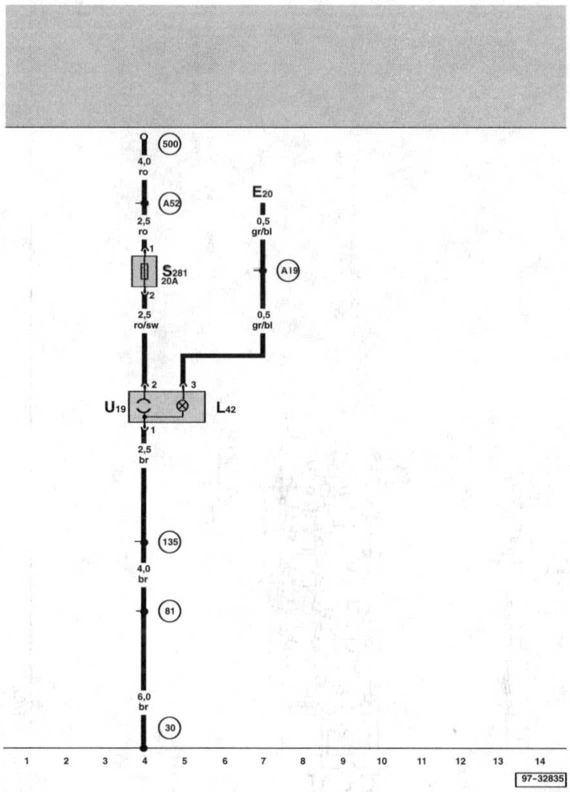

ws = white
sw = black
ro = red
br = brown
gn = green
bl = blue
gr = grey
li = lilac
ge = yellow
or = orange

Radio System - Monsoon

from October 2000

Deviate relay location and fuseplacements as well as the locations of multiple connectors see section "component locations".

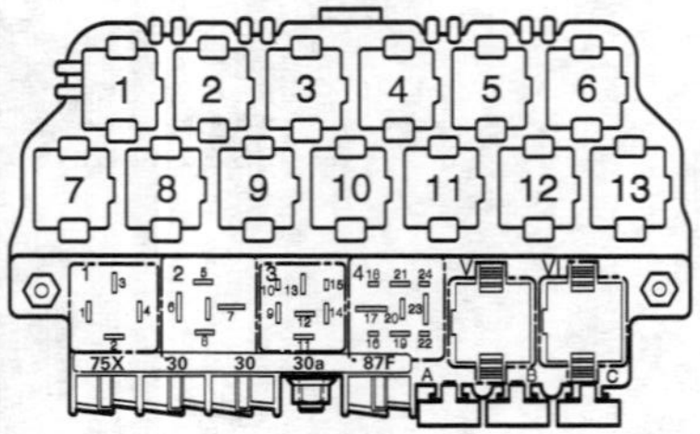

97-20676

Relay location on the thirteen position auxiliary relay panel, above relay panel:

Relay panel:

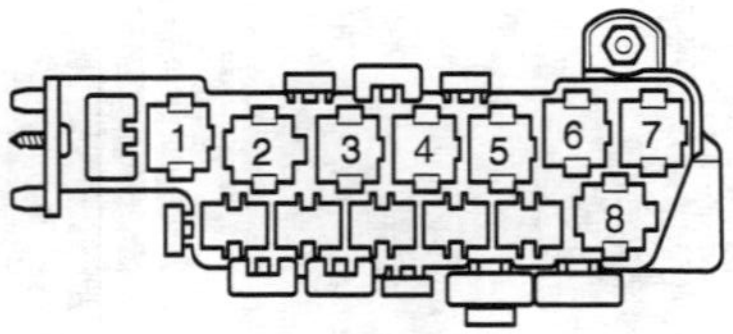

97-20515

Relay location on the eight position auxiliary relay panel, behind relay panel:

Note: Number in parentheses indicates production control number stamped on relay housing.

Fuse colors

30 A - green
25 A - white
20 A - yellow
15 A - blue
10 A - red
7,5 A - brown
5 A - beige

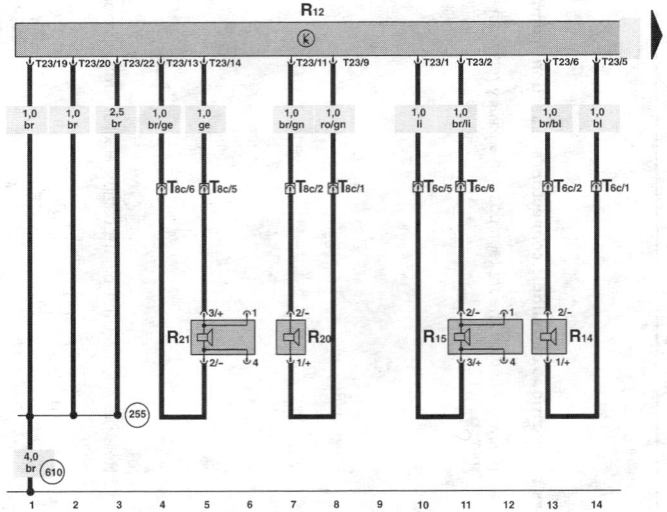

97-29551

ws = white
sw = black
ro = red
br = brown
gn = green
bl = blue
gr = grey
li = lilac
ge = yellow
or = orange

R12 - Amplifier
R14 - Treble speaker, left rear
R15 - Bass speaker, left rear
R20 - Treble speaker, left front
R21 - Bass speaker, left front
T6c - 6-Pin Connector, red, on B-pillar, left
T8c - 8-Pin Connector, black, connector station A-pillar, left
T23 - 23-Pin Connector, on Amplifier

(255) - Ground connection, -1-, in wiring harness Radio

(610) - Ground Connection (Audio) (under center console, front)

Amplifier, left front and rear speakers

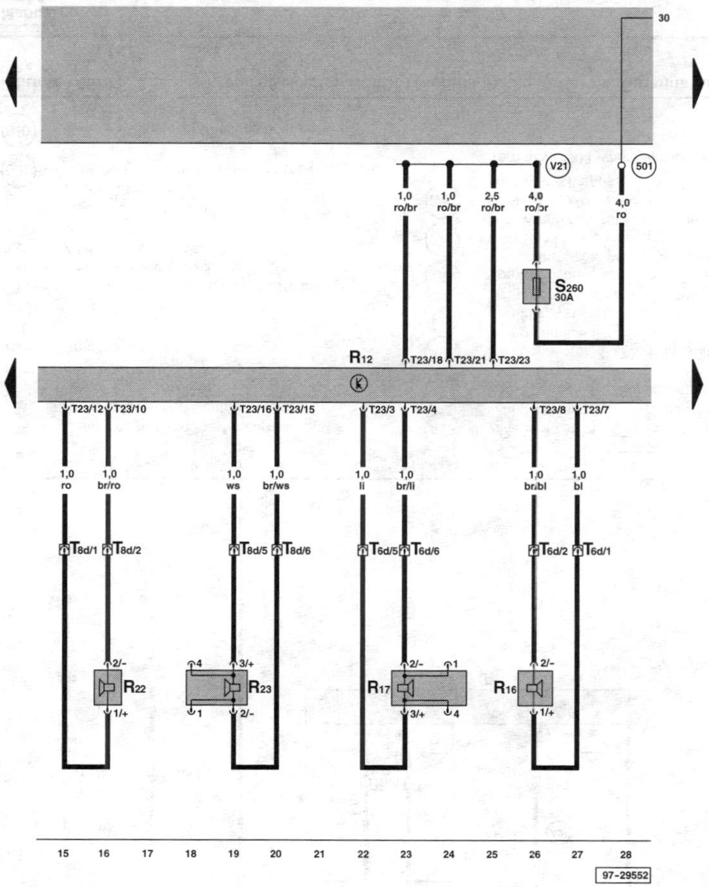

ws = white
sw = black
ro = red
br = brown
gn = green
bl = blue
gr = grey
li = lilac
ge = yellow
or = orange

R12 - Amplifier
R16 - Treble speaker, right rear
R17 - Bass speaker, right rear
R22 - Treble speaker, right front
R23 - Bass speaker, right front
S260 - Amplifier/DSP Fuse
T6d - 6-Pin Connector, red, on B-pillar, right
T8d - 8-Pin Connector, black, connector station A-pillar, right
T23 - 23-Pin Connector, on Amplifier

(501) - Threaded connection -2- (30) on the relay plate

(V21) - Plus connector (30), in wiring harness Radio -Amplifier

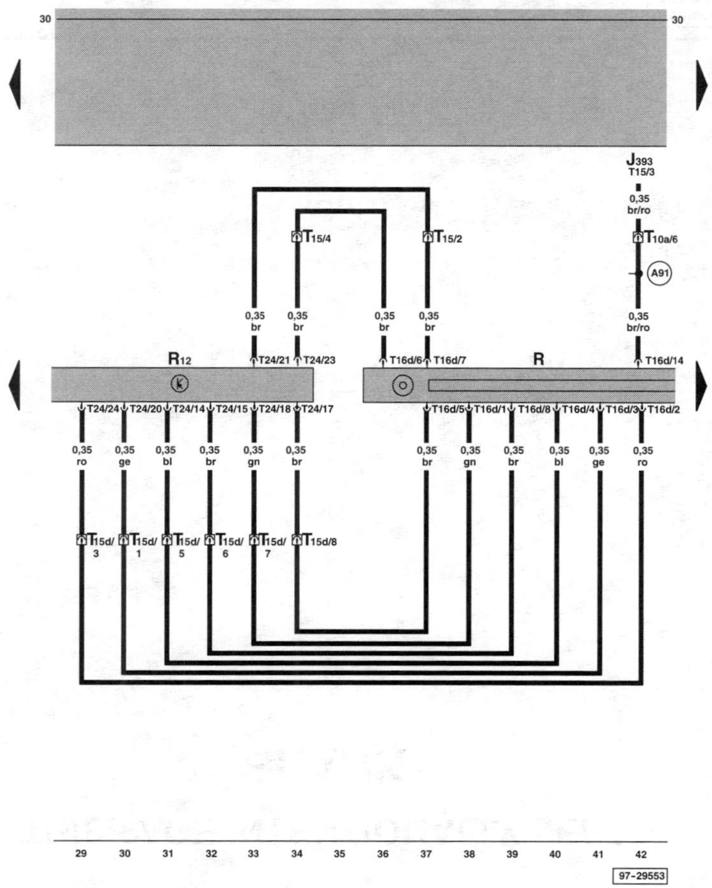

ws = white
sw = black
ro = red
br = brown
gn = green
bl = blue
gr = grey
li = lilac
ge = yellow
or = orange

J393 - Central control module for comfort system
R - Radio
R12 - Amplifier
T10a - 10-Pin Connector, brown, connector station A-pillar, left
T15 - 15-Pin Connector, on Central control module for comfort system
T15d - 15-Pin Connector, white, connector station A-pillar, left
T16d - 16-Pin Connector, black, on radio
T24 - 24-Pin Connector, on Amplifier

(A91) - Connector (signal for instrument panel insert), in instrument panel wiring harness

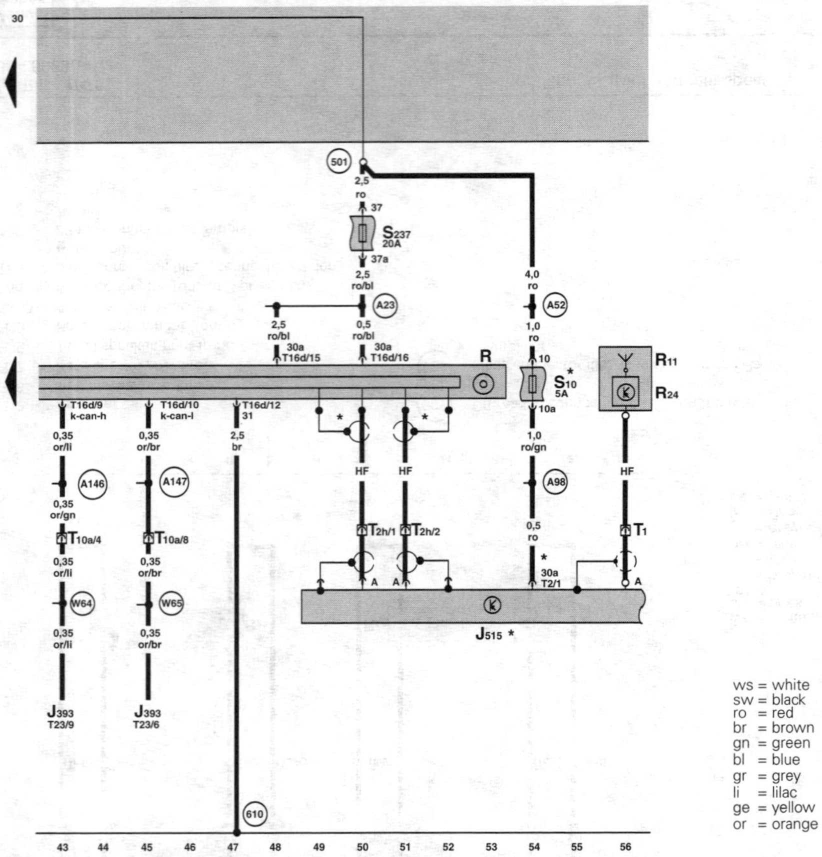

ws = white
sw = black
ro = red
br = brown
gn = green
bl = blue
gr = grey
li = lilac
ge = yellow
or = orange

J393 - Central control module for comfort system
J515 - Antenna Selection Control Module
R - Radio
R11 - Antenna
R24 - Antenna Amplifier
S10 - Fuse
S237 - Fuse 37 in fuse holder
T1 - 1-Pin Connector, black, below console
T2 - 2-Pin Connector, black
T2h - 2-Pin Connector, black
T10a - 10-Pin Connector, brown
T16d - 16-Pin Connector, black, on radio
T23 - 23-Pin Connector
(501) - Threaded connection -2- (30) on the relay plate
(610) - Ground Connection (Audio) (under center console, front)
(A23) - wire connection (30al), in instrument panel wiring harness
(A52) - Plus Connection 2 (30) (in instrument panel wiring harness)
(A98) - Plus connector -4- (30), in instrument panel wiring harness
(A146) - Comfort System High-bus Connection (in instrument panel wiring harness)
(A147) - Comfort System Low-bus Connection (in instrument panel wiring harness)
(W64) - CAN-Bus High Connection - Convenience (in floor wiring harness)
(W65) - CAN-Bus Low Connection - Convenience (in floor wiring harness)
* - from m.y. 2003 where applicable

Edition 05/02 **Antenna Selection Control Module, Radio, Antenna, Antenna Amplifier**
USA.5062.17.21

THIS PAGE INTENTIONALLY LEFT BLANK

2.8L - Engine - Motronic Multiport Fuel Injection (MFI) / 142kW, code ATQ,

from June 2001

Notes:

Information for

- *Relay and fuse arrangements*
- *Connector arrangements*
- *Control Modules and Relays*
- *Ground connections*

⇒ Component Locations!

Information for

- *Troubleshooting Procedures*

⇒ Guided Fault Finding using VAS 5051/5052

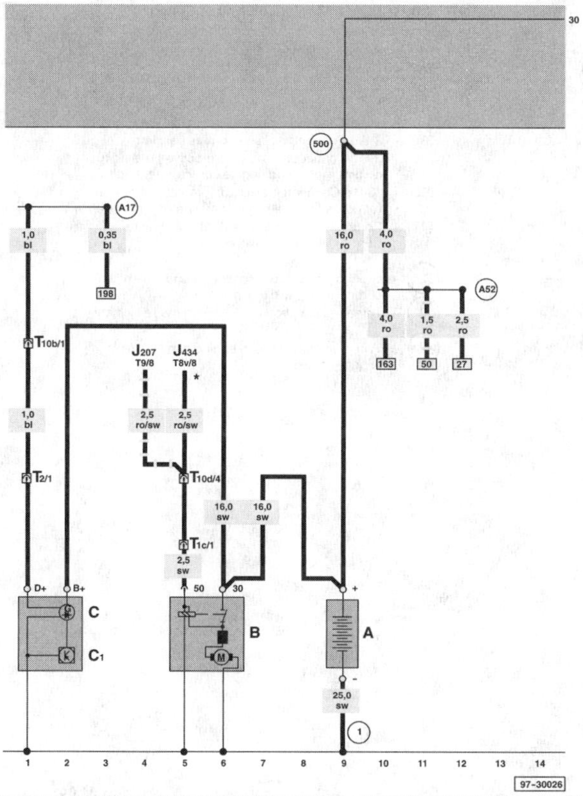

ws = white
sw = black
ro = red
br = brown
gn = green
bl = blue
gr = grey
li = lilac
ge = yellow
or = orange

Battery, Starter, Generator (GEN), Voltage Regulator (VR)

- A - Battery
- B - Starter
- C - Generator (GEN)
- C1 - Voltage Regulator (VR)
- J207 - Starting Interlock Relay
- J434 - Locking relay for starter (clutch pedal switch)
- T1c - Single Connector, black
- T2 - Double Connector, black
- T8v - 8-Pin Connector, on Locking relay for starter
- T9 - 9-Pin Connector, on Starting Interlock Relay
- T10b - 10-Pin Connector, black, on protective housing for control module, in engine compartment, left
- T10d - 10-Pin Connector, brown, on protective housing for control module, in engine compartment, left
- (1) - Ground strap, battery to body
- (500) - Threaded connection -1- (30) on the relay plate
- (A17) - wire connection (61), in instrument panel wiring harness
- (A52) - Plus Connection 2 (30) (in instrument panel wiring harness)
- * - manual transmission only
- --- - automatic transmission only

Motronic Engine Control Module (ECM), Throttle Valve Control Module, Ignition Coil, Throttle drive, Spark Plug Connectors, Suppressor

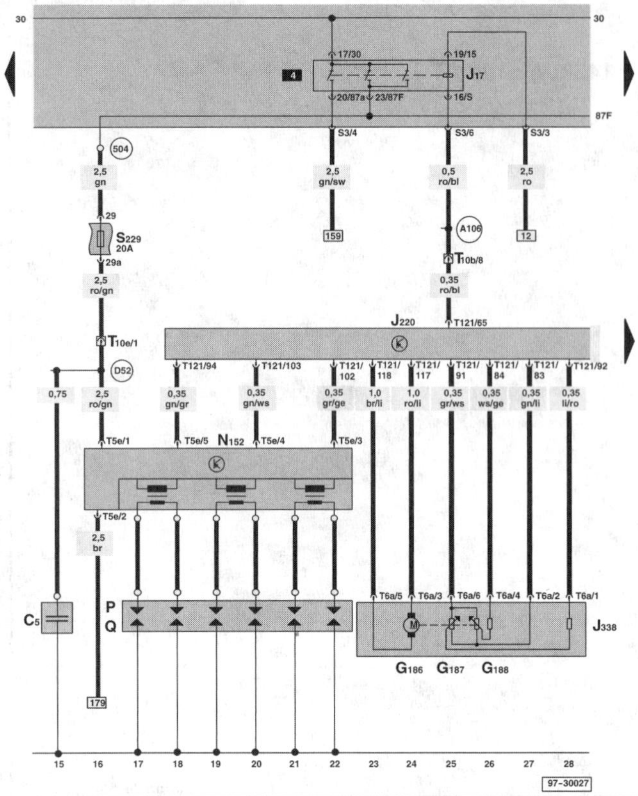

- C5 - Suppressor (Terminal 15)
- G186 - Throttle drive (power accelerator actuation)
- G187 - Angle sensor -1- for throttle drive (power accelerator actuation)
- G188 - Angle sensor -2- for throttle drive (power accelerator actuation)
- J17 - Fuel Pump (FP) Relay
- J220 - Motronic Engine Control Module (ECM)
- J338 - Throttle Valve Control Module
- N152 - Ignition Coil
- P - Spark Plug Connectors
- Q - Spark Plugs
- S229 - Fuse in fuse holder
- T5e - 5-Pin Connector, black, on Ignition Coil
- T6a - 6-Pin Connector, black, on Throttle Valve Control Module
- T10b - 10-Pin Connector, black, on protective housing for control module, in engine compartment, left
- T10e - 10-Pin Connector, orange, on protective housing for control module, in engine compartment, left
- T121 - Connector, 121 point, on Motronic Engine Control Module (ECM)

- (504) - Threaded connection -1- (87) on the relay plate
- (A106) - Connection -2- (86s), in instrument panel wiring harness
- (D52) - plus connection (15a), in engine compartment wiring harness

ws = white
sw = black
ro = red
br = brown
gn = green
bl = blue
gr = grey
li = lilac
ge = yellow
or = orange

Motronic Engine Control Module (ECM), Intake Air Temperature (IAT) Sensor, Knock Sensor

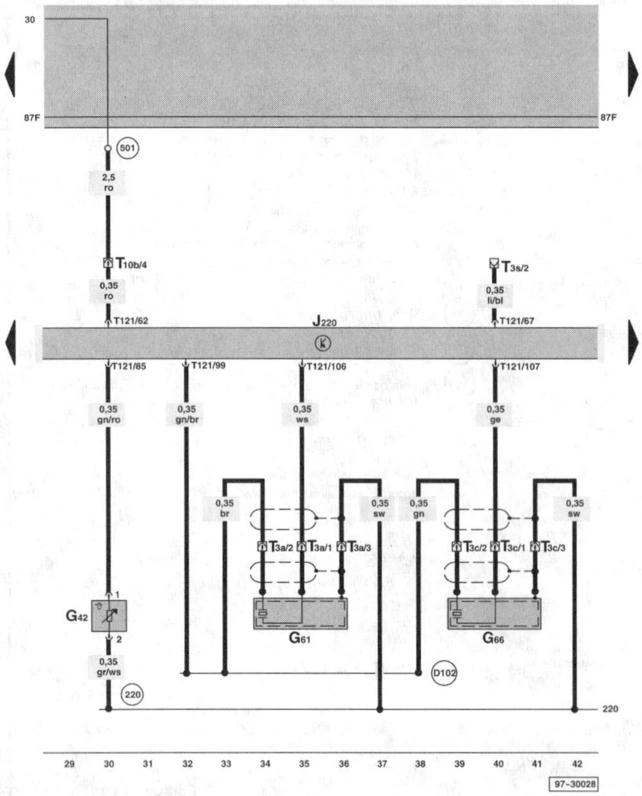

- G42 - Intake Air Temperature (IAT) Sensor
- G61 - Knock Sensor (KS) 1
- G66 - Knock Sensor (KS) 2
- J220 - Motronic Engine Control Module (ECM)
- T3a - 3-Pin Connector, brown, on firewall, left
- T3c - 3-Pin Connector, brown, on firewall, left
- T3s - 3-Pin Connector, red, on protective housing for control module, in engine compartment, left
- T10b - 10-Pin Connector, black, on protective housing for control module, in engine compartment, left
- T121 - Connector, 121 point, on Motronic Engine Control Module (ECM)

- (220) - Ground connection (sensor ground), in engine compartment wiring harness
- (501) - Threaded connection -2- (30) on the relay plate
- (D102) - Wire connection -2-, in engine compartment wiring harness

ws = white
sw = black
ro = red
br = brown
gn = green
bl = blue
gr = grey
li = lilac
ge = yellow
or = orange

Motronic Engine Control Module (ECM), Engine Speed (RPM) Sensor, Brake Booster Relay, Brake System Vacuum Pump

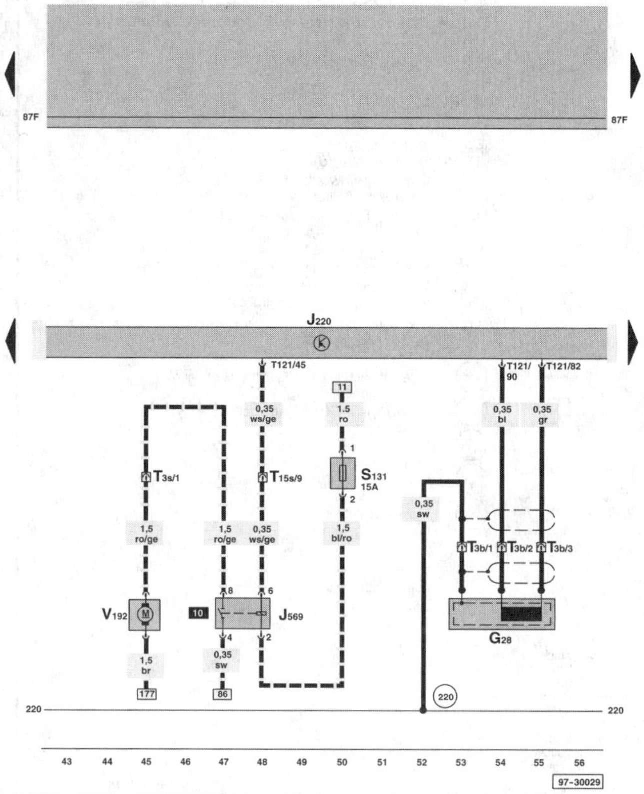

- G28 - Engine Speed (RPM) Sensor
- J220 - Motronic Engine Control Module (ECM)
- J569 - Brake Booster Relay
- S131 - Safety fuse 1
- T3b - 3-Pin Connector, grey, on firewall, left
- T3s - 3-Pin Connector, red, on protective housing for control module, in engine compartment, left
- T15s - 15-Pin Connector, red, on protective housing for control module, in engine compartment, left
- T121 - Connector, 121 point, on Motronic Engine Control Module (ECM)
- V192 - Brake System Vacuum Pump
- (220) - Ground connection (sensor ground), in engine compartment wiring harness
- - - - automatic transmission only

ws = white
sw = black
ro = red
br = brown
gn = green
bl = blue
gr = grey
li = lilac
ge = yellow
or = orange

97-30029

Edition 04.04

Motronic Engine Control Module (ECM), Brake Booster Pressure Sensor, Camshaft Position (CMP) Sensor, Engine Coolant Temperature (ECT) Sensor

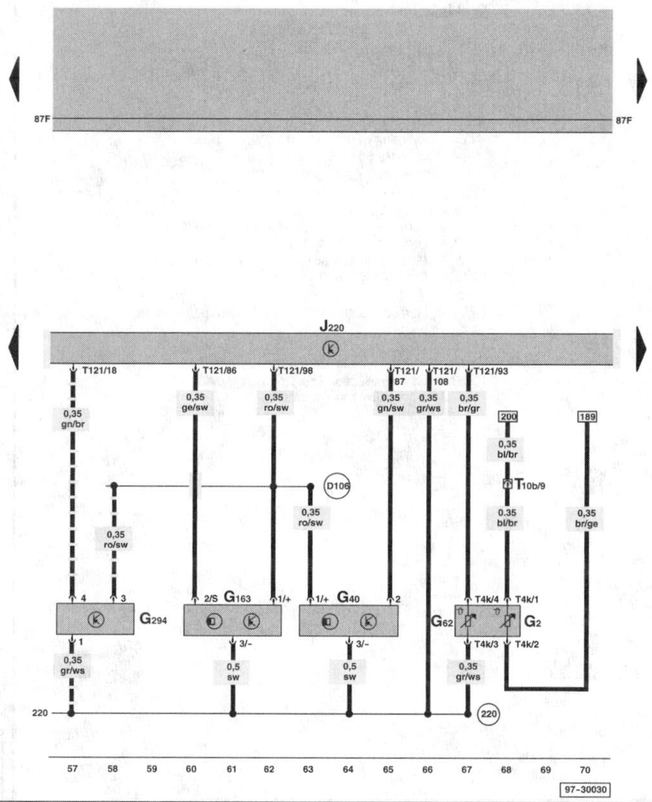

- G2 - Engine Coolant Temperature (ECT) Sensor
- G40 - Camshaft Position (CMP) Sensor
- G62 - Engine Coolant Temperature (ECT) Sensor
- G163 - Camshaft Position (CMP) Sensor 2
- G294 - Brake Booster Pressure Sensor
- J220 - Motronic Engine Control Module (ECM)
- T4k - 4-Pin Connector, black, on Engine Coolant Temperature (ECT) Sensor
- T10b - 10-Pin Connector, black, on protective housing for control module, in engine compartment, left
- T121 - Connector, 121 point, on Motronic Engine Control Module (ECM)
- (220) - Ground connection (sensor ground), in engine compartment wiring harness
- (D106) - wire connection -4-, in engine compartment wiring harness
- - - - automatic transmission only

ws = white
sw = black
ro = red
br = brown
gn = green
bl = blue
gr = grey
li = lilac
ge = yellow
or = orange

97-30030

Edition 04.04

Motronic Engine Control Module (ECM), Button for cruise control, Cruise Control Switch, Brake Light Switch, Clutch Pedal Switch, Brake Pedal Switch

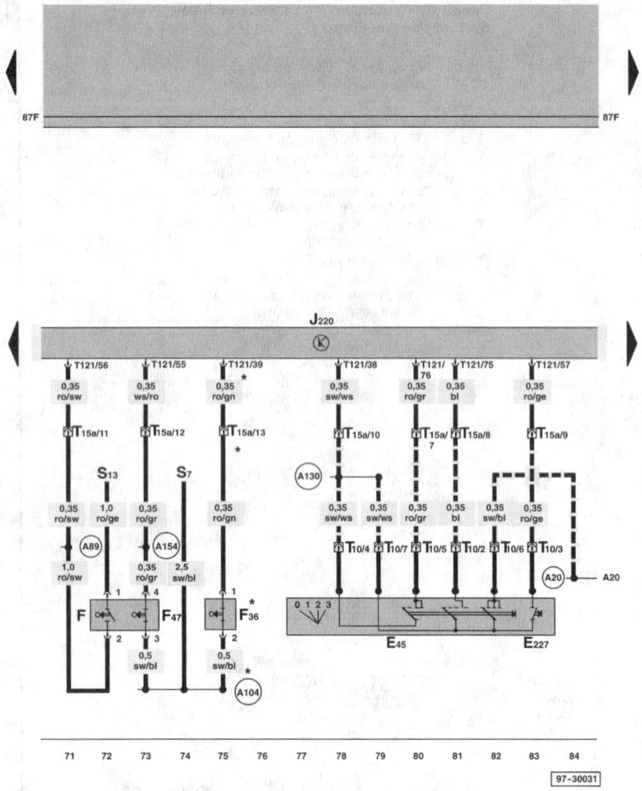

E45 - Cruise Control Switch
Position 0=OFF locked
Position 1=OFF pushed
Position 2=ON
Position 3=RES
E227 - Button for cruise control (set)
F - Brake Light Switch
F36 - Clutch Pedal Switch
F47 - Brake Pedal Switch
J220 - Motronic Engine Control Module (ECM)
S7 - Fuse
S13 - Fuse
T10 - 10-Pin Connector, black
T15a - 15-Pin Connector, white, on protective housing for control module, in engine compartment, left
T121 - Connector, 121 point, on Motronic Engine Control Module (ECM)

(A20) - wire connection (15a), in instrument panel wiring harness
(A89) - Connection 2 (54) (in instrument panel wiring harness)
(A104) - Plus connection -2- (15), in instrument panel wiring harness
(A130) - Connection for Cruise Control (instrument panel wiring harness)
(A154) - Brake Pedal Switch Connection (in instrument panel wiring harness)

--- - cruise control only
* - manual transmission only

ws = white
sw = black
ro = red
br = brown
gn = green
bl = blue
gr = grey
li = lilac
ge = yellow
or = orange

Edition 04.04

Motronic Engine Control Module (ECM), Cylinder Fuel injectors

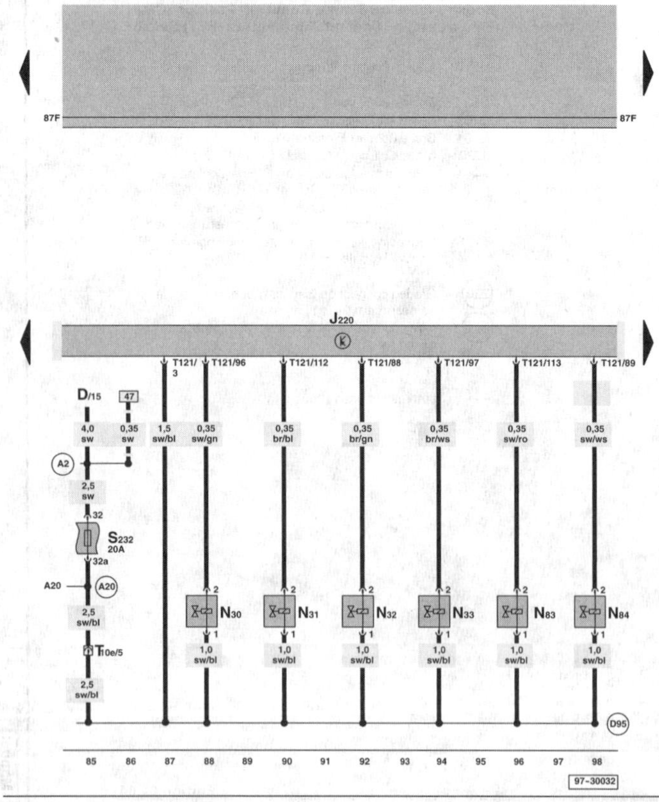

D - Ignition/Starter Switch
J220 - Motronic Engine Control Module (ECM)
N30 - Cylinder 1 Fuel Injector
N31 - Cylinder 2 Fuel Injector
N32 - Cylinder 3 Fuel Injector
N33 - Cylinder 4 Fuel Injector
N83 - Cylinder 5 Fuel Injector
N84 - Cylinder 6 Fuel Injector
S232 - Fuse in fuse holder
T10e - 10-Pin Connector, orange, on protective housing for control module, in engine compartment, left
T121 - Connector, 121 point, on Motronic Engine Control Module (ECM)

(A2) - plus connection (15), in instrument panel wiring harness
(A20) - wire connection (15a), in instrument panel wiring harness
(D95) - wire connection (injectors), in engine compartment wiring harness

---- automatic transmission only

ws = white
sw = black
ro = red
br = brown
gn = green
bl = blue
gr = grey
li = lilac
ge = yellow
or = orange

Edition 04.04

Motronic Engine Control Module (ECM), Throttle Position (TP) Sensor, Sender -2- for accelerator pedal position, Data Link Connector (DLC)

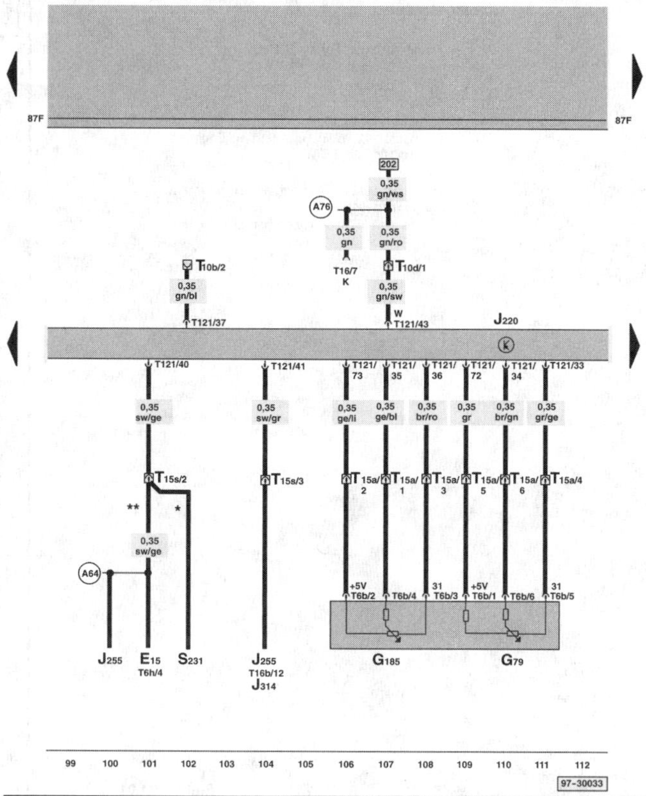

- E15 - Rear window defogger switch
- G79 - Throttle Position (TP) Sensor
- G185 - Sender -2- for accelerator pedal position
- J220 - Motronic Engine Control Module (ECM)
- J255 - Climatronic Control Module
- J314 - A/C Cut-Off Control Module
- S231 - Fuse in fuse holder
- T6b - 6-Pin Connector, on Throttle Position Sensor
- T6h - 6-Pin Connector, black, on Rear window defogger switch
- T10b - 10-Pin Connector, black, on protective housing for control module, in engine compartment, left
- T10d - 10-Pin Connector, brown, on protective housing for control module, in engine compartment, left
- T15a - 15-Pin Connector, while, on protective housing for control module, in engine compartment, left
- T15s - 15-Pin Connector, red, on protective housing for control module, in engine compartment, left
- T16 - 16-Pin Connector, black, Data Link Connector (DLC), below instrument panel, left
- T16b - 16-Pin Connector, brown
- T121 - Connector, 121 point, on Motronic Engine Control Module (ECM)

- (A64) - Connection (30, idle boost), in instrument panel wiring harness
- (A76) - Connection (K-diagnosis wire) in instrument panel wiring harness

- * - front wheel drive only
- ** - 4-Motion only

ws = white
sw = black
ro = red
br = brown
gn = green
bl = blue
gr = grey
li = lilac
ge = yellow
or = orange

Edition 04.04

Motronic Engine Control Module (ECM), Valve for camshaft adjustment, Evaporative Emission (EVAP) Canister Purge Regulator Valve, Mass Air Flow (MAF) Sensor

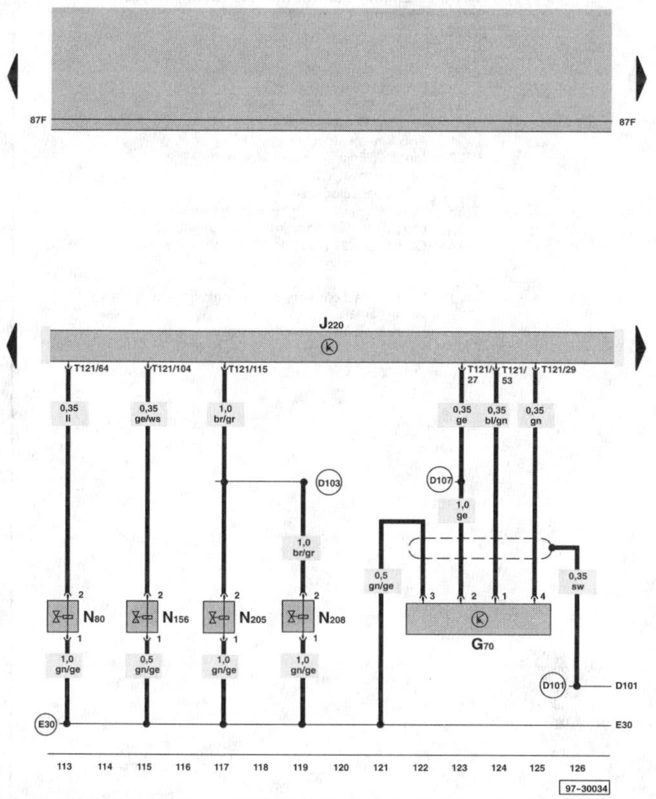

- G70 - Mass Air Flow (MAF) Sensor
- J220 - Motronic Engine Control Module (ECM)
- N80 - Evaporative Emission (EVAP) Canister Purge Regulator Valve
- N156 - Intake Manifold Change-Over Valve
- N205 - Valve -1- for camshaft adjustment
- N208 - Valve -2- for camshaft adjustment
- T121 - Connector, 121 point, on Motronic Engine Control Module (ECM)

- (D101) - Wire connection -1-, in engine compartment wiring harness
- (D103) - wire connection -3-, in engine compartment wiring harness
- (D107) - wire connection -5-, in engine compartment wiring harness
- (E30) - Connection (87a), in wiring harness engine

ws = white
sw = black
ro = red
br = brown
gn = green
bl = blue
gr = grey
li = lilac
ge = yellow
or = orange

Edition 04.04

Motronic Engine Control Module (ECM), Heated Oxygen Sensors

G39 - Heated Oxygen Sensor (HO2S)
G108 - Heated Oxygen Sensor (HO2S) 2
J220 - Motronic Engine Control Module (ECM)
T4e - 4-Pin Connector, black
T4f - 4-Pin Connector, black
T121 - Connector, 121 point, on Motronic Engine Control Module (ECM)

(D101) - Wire connection -1-, in engine compartment wiring harness
(E30) - Connection (87a), in wiring harness engine

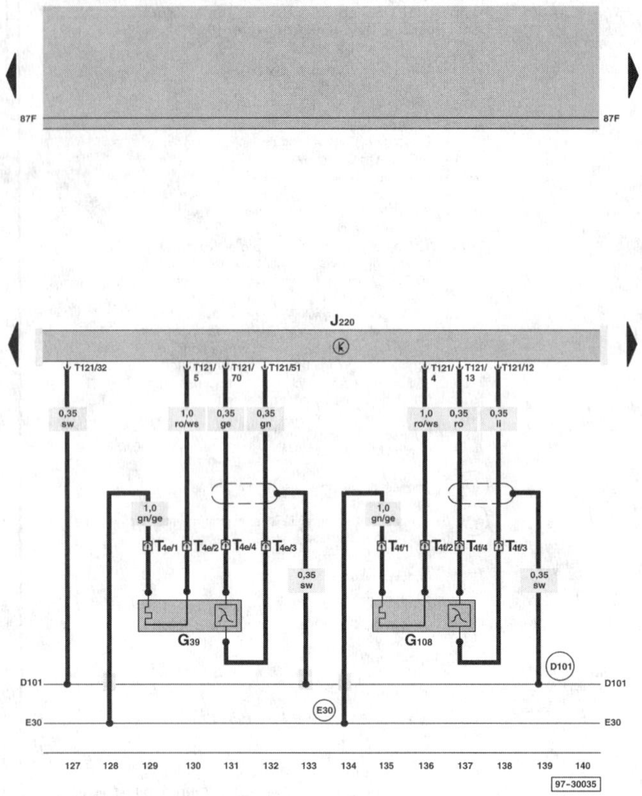

ws = white
sw = black
ro = red
br = brown
gn = green
bl = blue
gr = grey
li = lilac
ge = yellow
or = orange

Edition 04.04

Motronic Engine Control Module (ECM), Oxygen Sensors Behind Three Way Catalytic Converter (TWC)

G130 - Oxygen Sensor (O2S) Behind Three Way Catalytic Converter (TWC)
G131 - Oxygen Sensor (O2S) 2 Behind Three Way Catalytic Converter (TWC)
J104 - ABS Control Module (w/EDL)
J220 - Motronic Engine Control Module (ECM)
T4c - 4-Pin Connector, brown
T4d - 4-Pin Connector, green
T15s - 15-Pin Connector, red, on protective housing for control module, in engine compartment, left
T121 - Connector, 121 point, on Motronic Engine Control Module (ECM)

(D101) - Wire connection -1-, in engine compartment wiring harness
(E30) - Connection (87a), in wiring harness engine

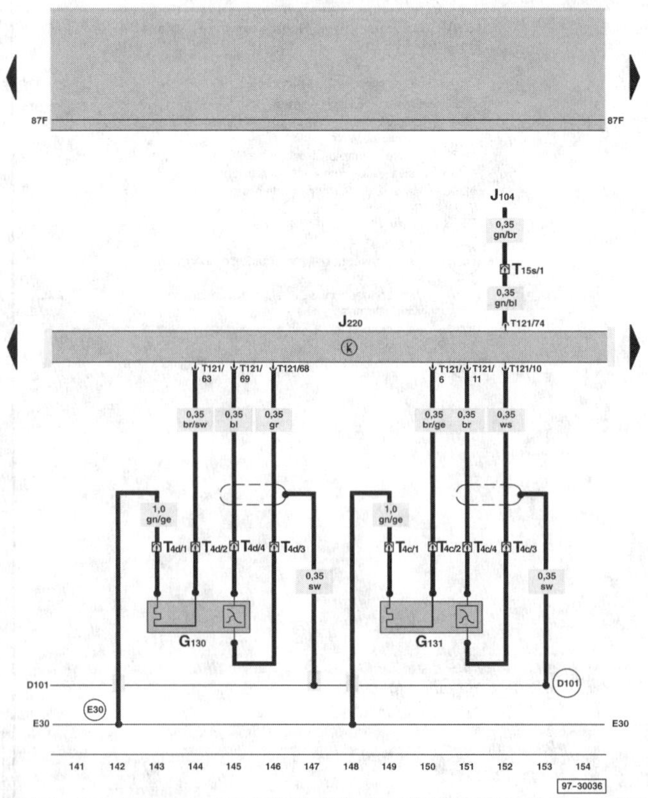

ws = white
sw = black
ro = red
br = brown
gn = green
bl = blue
gr = grey
li = lilac
ge = yellow
or = orange

Edition 04.04

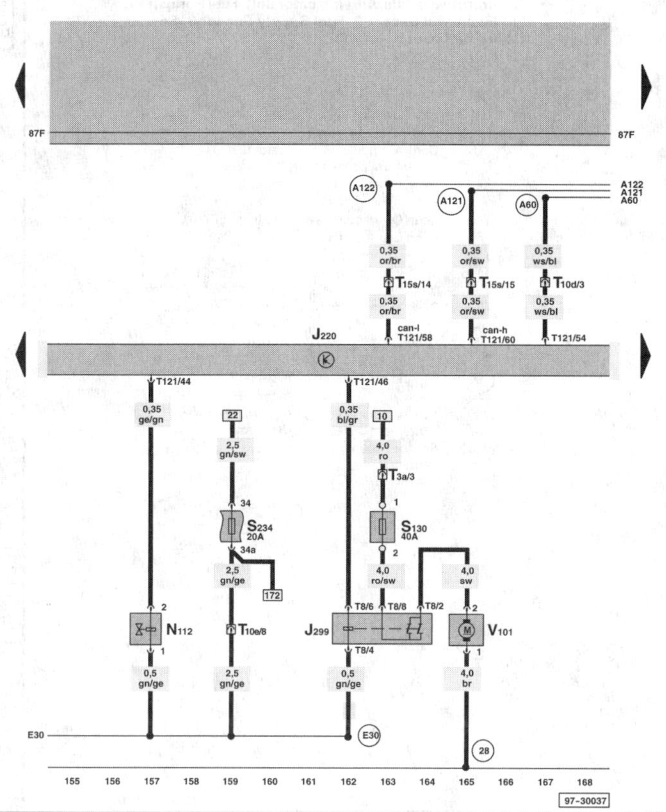

ws = white
sw = black
ro = red
br = brown
gn = green
bl = blue
gr = grey
li = lilac
ge = yellow
or = orange

Motronic Engine Control Module (ECM), Secondary Air Injection (AIR) Pump Relay and Motor, Secondary Air Injection (AIR) Solenoid Valve

- J220 - Motronic Engine Control Module (ECM)
- J299 - Secondary Air Injection (AIR) Pump Relay
- N112 - Secondary Air Injection (AIR) Solenoid Valve
- S130 - Fuse for secondary air pump
- S234 - Fuse in fuse holder
- T3a - 3-Pin Connector, green
- T8 - 8-Pin Connector, brown, on Secondary Air Injection (AIR) Pump Relay
- T10d - 10-Pin Connector, brown, on protective housing for control module, in engine compartment, left
- T10e - 10-Pin Connector, orange, on protective housing for control module, in engine compartment, left
- T15s - 15-Pin Connector, red, on protective housing for control module, in engine compartment, left
- T121 - Connector, 121 point, on Motronic Engine Control Module (ECM)
- V101 - Secondary Air Injection (AIR) Pump Motor
- (28) - Ground connection, on firewall
- (A60) - Wire connection (Vehicle speed signal), in instrument cluster wiring harness
- (A121) - Connection (high bus) in instrument panel wiring harness
- (A122) - Connection (low bus) in instrument panel wiring harness
- (E30) - Connection (87a), in wiring harness engine

Edition 04.04

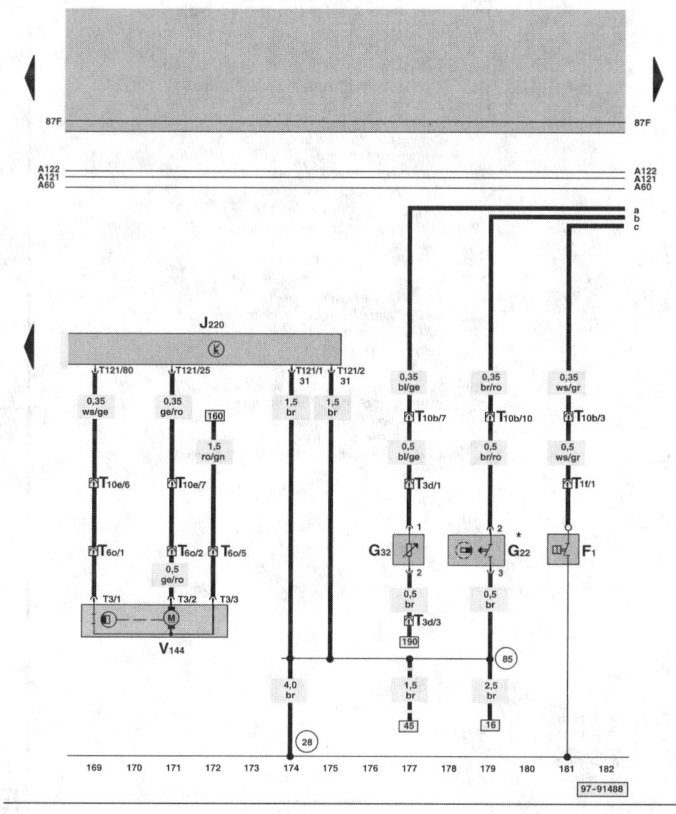

ws = white
sw = black
ro = red
br = brown
gn = green
bl = blue
gr = grey
li = lilac
ge = yellow
or = orange

Motronic Engine Control Module (ECM), Leak Detection Pump (LDP), Oil Pressure Switch, Speedometer Vehicle Speed Sensor, Engine Coolant Level Sensor

- F1 - Oil Pressure Switch
- G22 - Speedometer Vehicle Speed Sensor (VSS)
- G32 - Engine Coolant Level (ECL) Sensor
- J220 - Motronic Engine Control Module (ECM)
- T1f - Single Connector, black
- T3 - 3-Pin Connector, on Leak Detection Pump
- T3d - 3-Pin Connector
- T6o - 6-Pin Connector, brown, connector station A-pillar, right
- T10b - 10-Pin Connector, black, on protective housing for control module, in engine compartment, left
- T10e - 10-Pin Connector, orange, on protective housing for control module, in engine compartment, left
- T121 - Connector, 121 point, on Motronic Engine Control Module (ECM)
- V144 - Leak Detection Pump (LDP)
- (28) - Ground connection, on firewall
- (85) - Ground connection -1-, in engine compartment wiring harness
- * - manual transmission only
- --- - automatic transmission only

Edition 06.04

Control module with indicator unit, Fuel Pump (FP), Fuel Level Sensor 2, Fuel Supply Sensor 3, Sender for fuel gauge

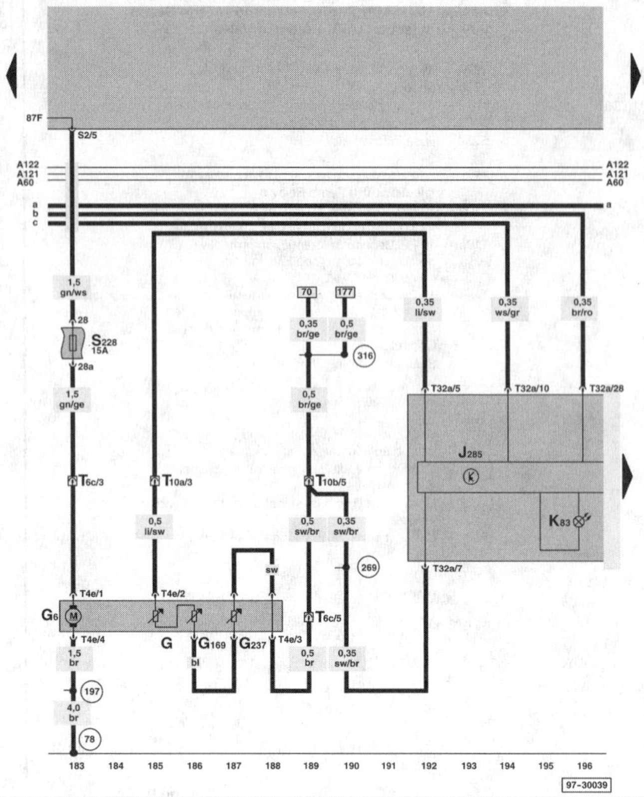

- G - Sender for fuel gauge
- G6 - Fuel Pump (FP)
- G169 - Fuel Level Sensor 2
- G237 - Fuel Supply Sensor 3
- J285 - Control module with indicator unit in instrument panel insert
- K83 - Malfunction Indicator Lamp (MIL)
- S228 - Fuse in fuse holder
- T4e - 4-Pin Connector, near Fuel Pump (FP)
- T6c - 6-Pin Connector, blue
- T10a - 10-Pin connector, black
- T10b - 10-Pin Connector, black, on protective housing for control module, in engine compartment, left
- T32a - 32-Pin Connector, bue, on instrument cluster

- (78) - Ground Connection (lower right B-pillar)
- (197) - Ground connection -4-, in wiring harness rear
- (269) - Ground connection (sensor ground) -1-, in instrument panel wiring harness
- (316) - Sensor Ground Connection 2 (in engine wiring harness)

ws = white
sw = black
ro = red
br = brown
gn = green
bl = blue
gr = grey
li = lilac
ge = yellow
or = orange

Control module with indicator unit in instrument panel insert, Data Bus On Board Diagnostic Interface

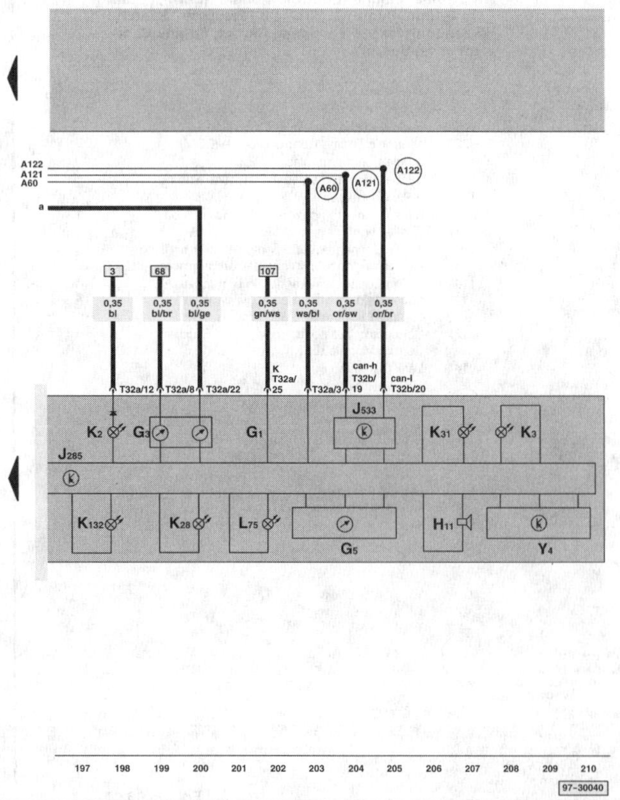

- G1 - Fuel Gauge
- G3 - Engine Coolant Temperature (ECT) Gauge
- G5 - Tachometer
- H11 - Oil Pressure Warning Buzzer
- J285 - Control module with indicator unit in instrument panel insert
- J533 - Data Bus On Board Diagnostic Interface
- K2 - Generator (GEN) Warning Light
- K3 - Oil Pressure Warning Light
- K28 - Engine Coolant Level/Temperature (ECL/ECT) Warning Light
- K31 - Cruise Control Indicator Light
- K132 - Electronic Power Control (EPC) Warning Lamp
- L75 - Digital Display Light
- T32a - 32-Pin Connector, bue, on instrument cluster
- T32b - 32-Pin Connector, green, on instrument cluster
- Y4 - Odometer Display

- (A60) - Wire connection (Vehicle speed signal), in instrument cluster wiring harness
- (A121) - Connection (high bus) in instrument panel wiring harness
- (A122) - Connection (low bus) in instrument panel wiring harness

ws = white
sw = black
ro = red
br = brown
gn = green
bl = blue
gr = grey
li = lilac
ge = yellow
or = orange

Garage door opener - HomeLink

from October 2000

Deviate relay location and fuseplacements as well as the locations of multiple connectors see section component locations.

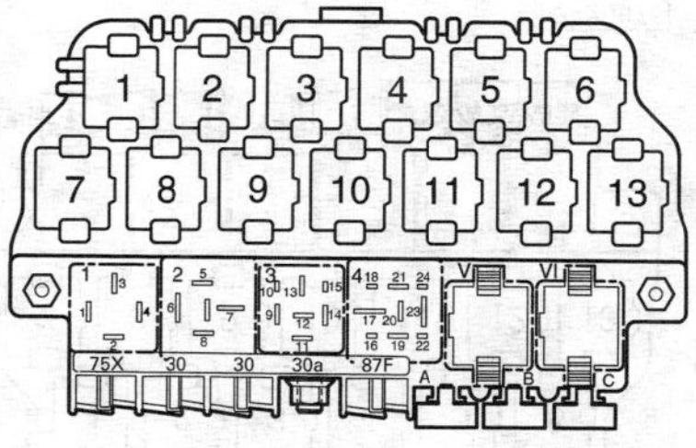

97-20676

Relay location on the thirteen position auxiliary relay panel, above relay panel:

Relay panel:

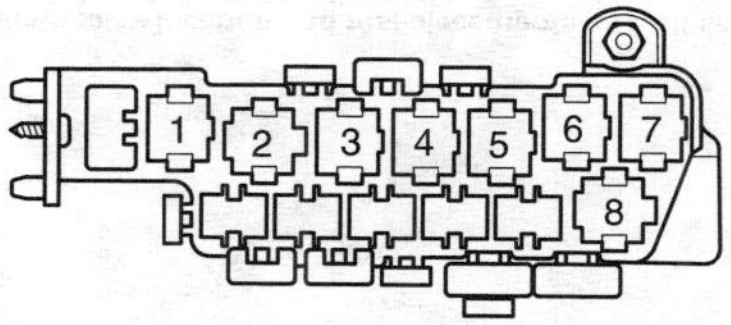

97-20515

Relay location on the eight position auxiliary relay panel, behind relay panel:

Note: Number in parentheses indicates production control number stamped on relay housing.

Fuse colors
30 A - green
25 A - white
20 A - yellow
15 A - blue
10 A - red
7,5 A - brown
5 A - beige

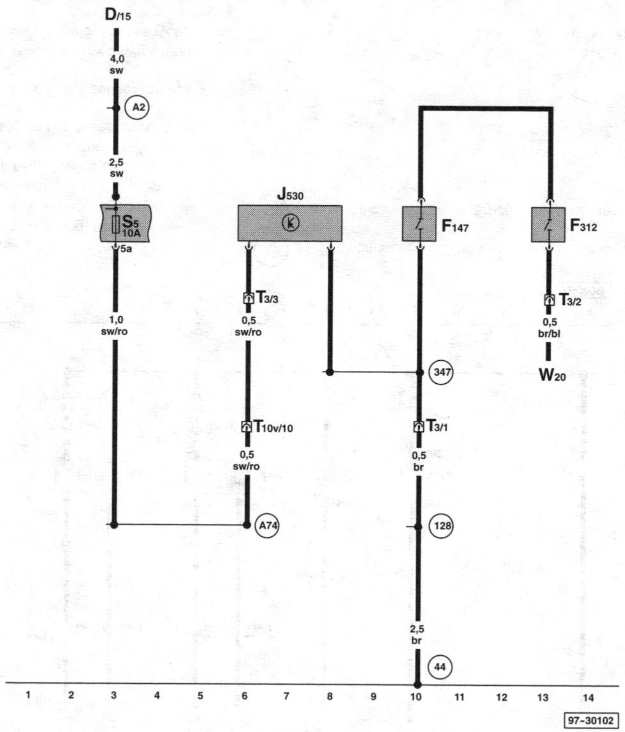

ws = white
sw = black
ro = red
br = brown
gn = green
bl = blue
gr = grey
li = lilac
ge = yellow
or = orange

D - Ignition/Starter Switch
J530 - Garage Door Opener Control Module
S5 - Fuse
T3 - 3-Pin Connector
T10v - 10-Pin Connector, green, connector station A-pillar, left
W20 - Left Make-Up Mirror Light
F147 - Left Make-Up Mirror Light Switch
F312 - Mirror sliding cover contact Switch, drivers side

(44) - Ground connection, on left A-pillar, lower part
(128) - Ground connection -1-, in interior light wiring harness
(347) - Ground Connection (in roof wiring harness)
(A2) - plus connection (15), in instrument panel wiring harness
(A74) - Connector (15a - fuse 5) in instrument panel wiring harness

Garage Door Opener Control Module, mirror sliding cover contact switch, drivers side, Left Make - Up Mirror Light/Switch

Airbag Systems

from June 2001

Deviate relay location and fuseplacements as well as the locations of multiple connectors see section "component locations".

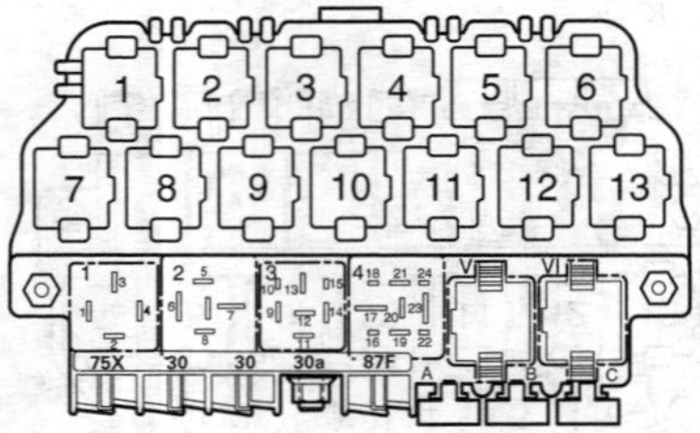

Relay location on the thirteen position auxiliary relay panel, above relay panel:

Relay panel:

97-20676

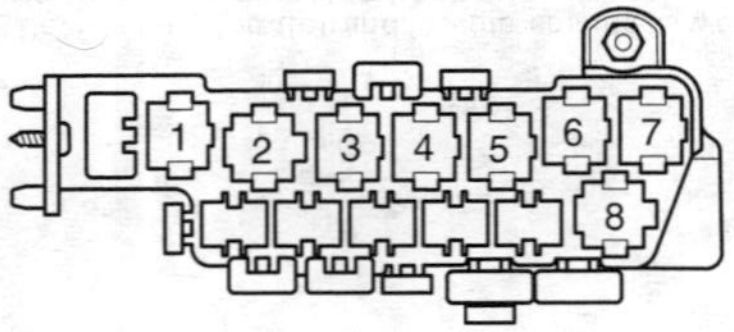

Relay location on the eight position auxiliary relay panel, behind relay panel:

97-20515

Note: Number in parentheses indicates production control number stamped on relay housing.

Fuse colors
30 A - green
25 A - white
20 A - yellow
15 A - blue
10 A - red
7,5 A - brown
5 A - beige

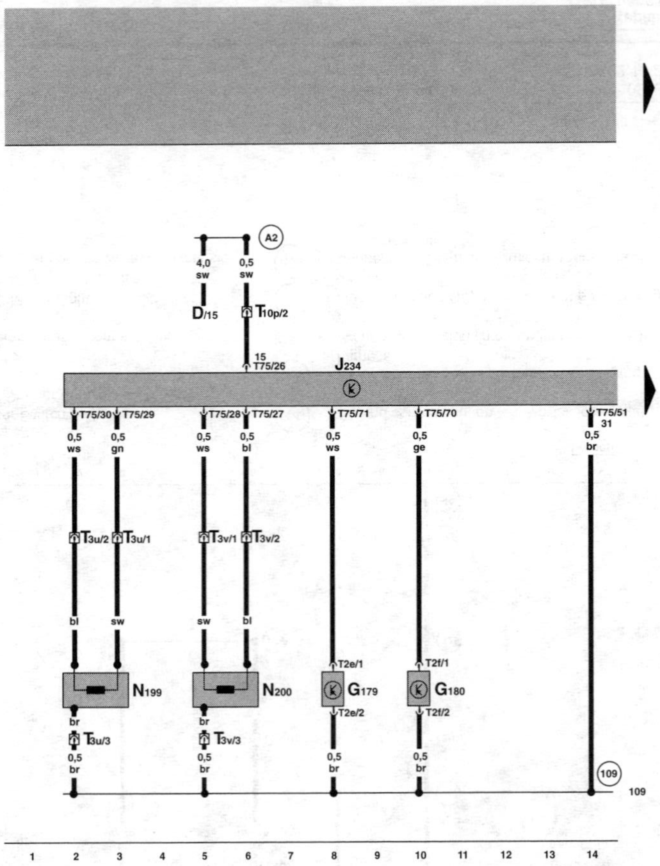

97-30099

ws = white
sw = black
ro = red
br = brown
gn = green
bl = blue
gr = grey
li = lilac
ge = yellow
or = orange

D - Ignition/Starter Switch
G179 - Crash sensor for side airbag, driver's side
G180 - Crash sensor for side airbag, passenger side
J234 - Airbag Control Module
N199 - Igniter for side airbag, driver's side
N200 - Igniter for side airbag, passenger side
T2e - Double Connector, on Crash sensor for side airbag, driver's side
T2f - Double Connector, on Crash sensor for side airbag, passenger side
T3u - 3-Pin Connector, yellow, below driver's seat
T3v - 3-Pin Connector, yellow, below passenger's seat
T10p - 10-Pin Connector, yellow, connector station A-pillar, right
T75 - 75-Pin Connector, on Airbag Control Module

(A2) - plus connection (15), in instrument panel wiring harness
(109) - Ground connection, in airbag wiring harness

Airbag Control Module, Igniter for side airbag, Crash sensor for side airbag

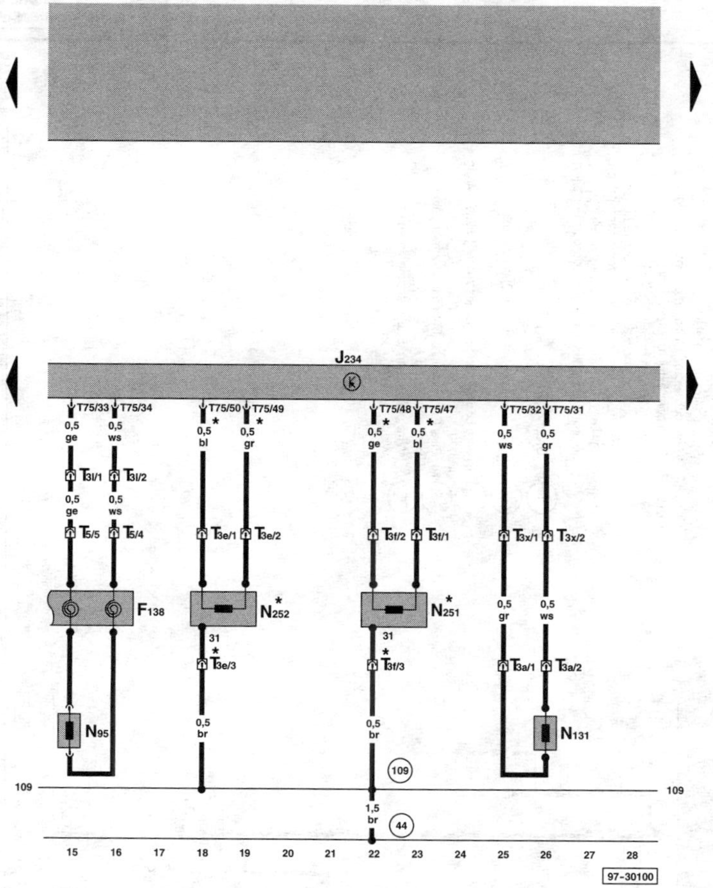

ws = white
sw = black
ro = red
br = brown
gn = green
bl = blue
gr = grey
li = lilac
ge = yellow
or = orange

F138 - Airbag Spiral Spring/Return Spring With Slip Ring
J234 - Airbag Control Module, behind console
N95 - Driver's Side Airbag Igniter
N131 - Passenger's Side Airbag Igniter 1
N251 - Igniter for head airbag, rear, driver side
N252 - Front Passenger Head Airbag Igniter
T3a - 3-Pin Connector, behind instrument panel, right
T3e - 3-Pin Connector
T3f - 3-Pin Connector
T3l - 3-Pin Connector, yellow, connector station A-pillar, left
T3x - 3-Pin Connector, yellow, connector station A-pillar, right
T5 - 5-Pin Connector, yellow, behind steering column switch cover

T75 - 75-Pin Connector, on Airbag Control Module
(44) - Ground connection, on left A-pillar, lower part
(109) - Ground connection, in airbag wiring harness
* - vehicles with head airbag only

Airbag Control Module, Driver's Side Airbag, Igniter for head airbag, Front Passenger's Side Airbag Igniter

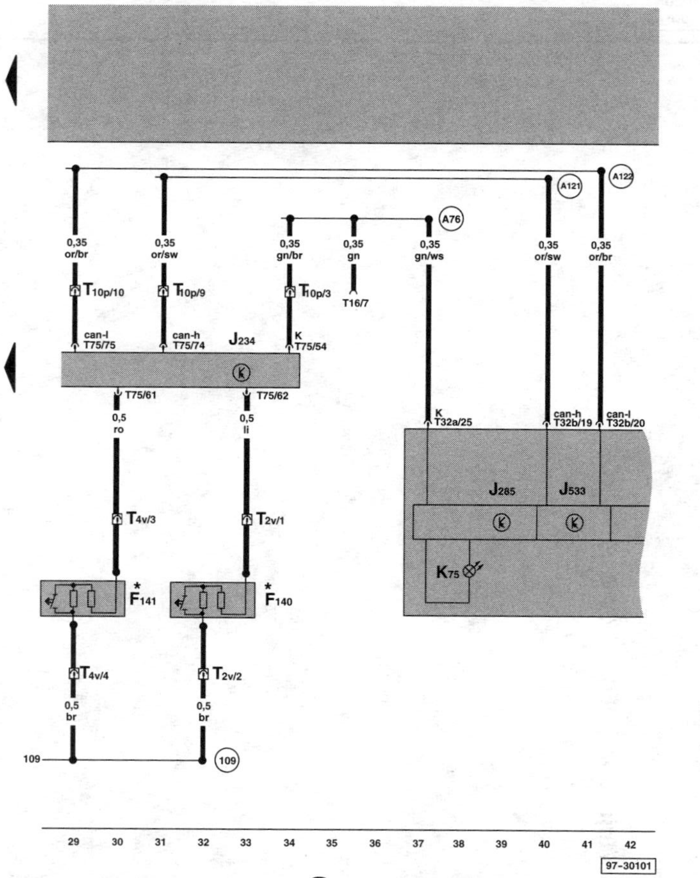

ws = white
sw = black
ro = red
br = brown
gn = green
bl = blue
gr = grey
li = lilac
ge = yellow
or = orange

F140 - Left Front Seat Belt Microswitch*
F141 - Right Front Seat Belt Microswitch*
J234 - Airbag Control Module, behind console
J285 - Control module with indicator unit in instrument panel insert
J533 - Data Bus On Board Diagnostic Interface
K75 - Airbag Malfunction Indicator Lamp (MIL)
T2v - Double Connector
T4v - 4-Pin Connector
T10p - 10-Pin Connector, yellow, connector station A-pillar, right
T16 - 16-Pin Connector, Data Link Connector (DLC) below instrument panel, left
T32a - 32-Pin Connector, blue, on instrument cluster
T32b - 32-Pin Connector, green, on instrument cluster
T75 - 75-Pin Connector, on Airbag Control Module

(109) - Ground connection, in airbag wiring harness
(A76) - Connector (K-diagnosis wire) in instrument panel wiring harness
(A121) - Connection (high bus) in instrument panel wiring harness
(A122) - Connection (low bus) in instrument panel wiring harness
* - Scan Tool display text for these component codes may be different than indicated

Airbag Control Module, Control module with indicator unit in instrument panel insert, Data Link Connector (DLC) , Front seat belt microswitches

4.0L - Engine - Motronic Multiport Fuel Injection (MFI) /202kW, code BDP,

from January 2002

Notes:

Information for

- *Relay and fuse arrangements*
- *Connector arrangements*
- *Control Modules and Relays*
- *Ground connections*

⇒ Component Locations!

Information for

- *Troubleshooting Procedures*

⇒ Guided Fault Finding using VAS 5051/5052

Battery, Starter, Generator, Voltage Regulator

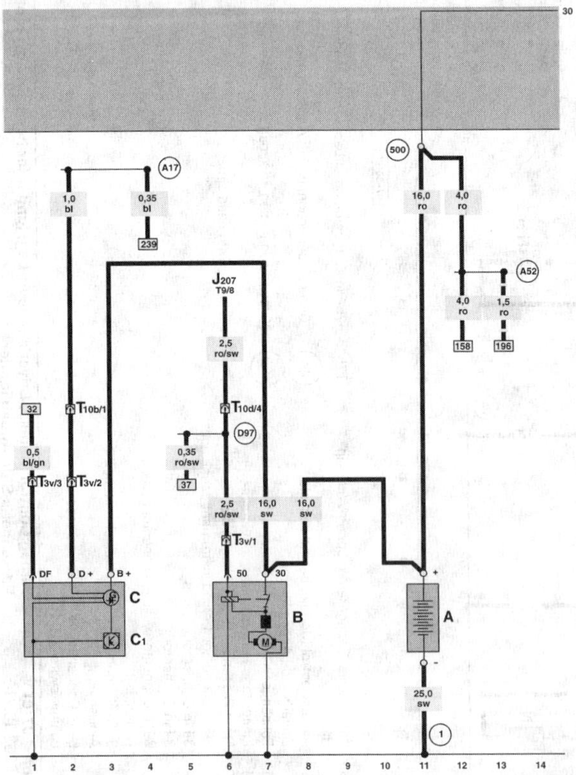

A - Battery
B - Starter
C - Generator (GEN)
C1 - Voltage Regulator (VR)
J207 - Starting Interlock Relay
T3v - 3-Pin connector, black, on engine block right
T9 - 9-Pin connector, black, on starting interlock relay
T10b - 10-Pin connector, black, on protective housing for control module, in engine compartment, left
T10d - 10-Pin connector, brown, on protective housing for control module, in engine compartment, left

(1) - Ground strap, battery to body
(500) - Threaded connection -1- (30) on the relay plate
(A17) - wire connection (61), in instrument panel wiring harness
(A52) - Plus Connection 2 (30) (in instrument panel wiring harness)
(D97) - Wire connection (50), in engine compartment wiring harness, right

ws = white
sw = black
ro = red
br = brown
gn = green
bl = blue
gr = grey
li = lilac
ge = yellow
or = orange

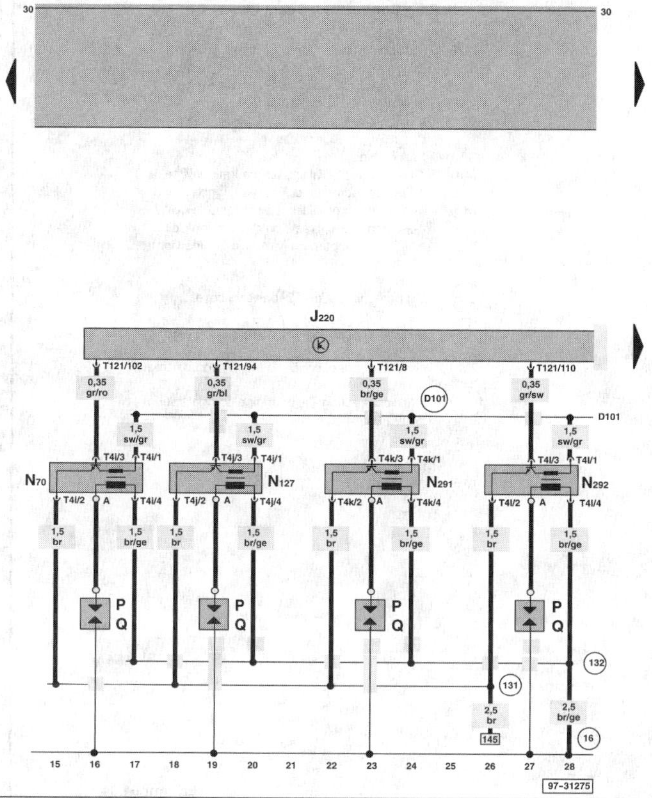

Motronic Engine Control Module, Ignition Coil 1-4, Spark Plug Connectors, Spark Plugs

J220 - Motronic Engine Control Module (ECM)
N70 - Ignition Coil 1 with Power Output Stage
N127 - Ignition Coil 2 with Power Output Stage
N291 - Ignition Coil 3 with Power Output Stage
N292 - Ignition Coil 4 with Power Output Stage
P - Spark Plug Connectors
Q - Spark Plugs
T4i - 4-Pin connector, black, on Ignition Coil 1
T4j - 4-Pin connector, black, on Ignition Coil 2
T4k - 4-Pin connector, black, on Ignition Coil 3
T4l - 4-Pin connector, black, on Ignition Coil 4
T121 - 121-Pin connector, on Motronic Engine Control Module

(16) - Ground connection, -1-, on valve cover
(131) - Ground connection -2-, in engine compartment wiring harness
(132) - Ground connection -3-, in engine compartment wiring harness
(D101) - Wire connection -1-, in engine compartment wiring harness

ws = white
sw = black
ro = red
br = brown
gn = green
bl = blue
gr = grey
li = lilac
ge = yellow
or = orange

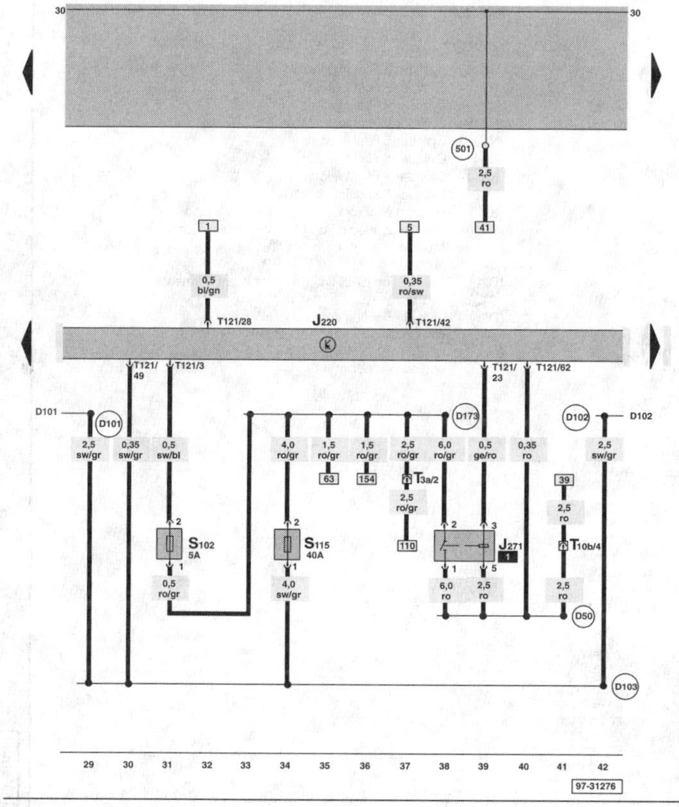

Motronic Engine Control Module, Motronic Engine Control Module Power Supply Relay

J220 - Motronic Engine Control Module (ECM)
J271 - Motronic Engine Control Module (ECM) Power Supply Relay
S102 - Engine Control Module (ECM) Fuse
S115 - Fuse for ignition coil termial 15
T3a - 3-Pin connector, white, on protective housing for control module, in engine compartment, left
T10b - 10-Pin connector, black, on protective housing for control module, in engine compartment, left
T121 - 121-Pin connector, on Motronic Engine Control Module

(501) - Threaded connection -2-(30) on the relay plate
(D50) - plus connection (30), in engine compartment wiring harness
(D101) - Wire connection -1-, in engine compartment wiring harness
(D102) - Wire connection -2-, in engine compartment wiring harness
(D103) - wire connection -3-, in engine compartment wiring harness
(D173) - Connection (87)(in engine wiring harness)

ws = white
sw = black
ro = red
br = brown
gn = green
bl = blue
gr = grey
li = lilac
ge = yellow
or = orange

Motronic Engine Control Module, Ignition Coil 5-8, Spark Plug Connectors, Spark Plugs

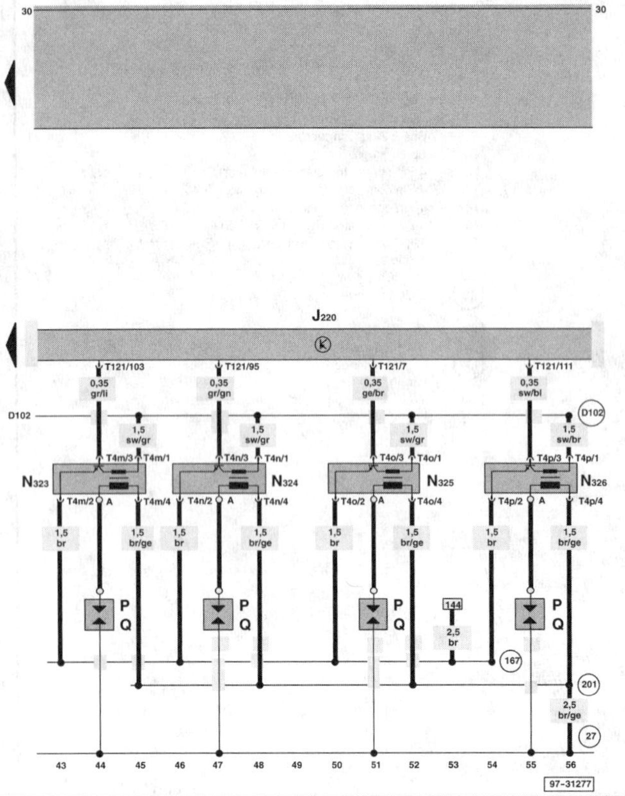

J220 - Motronic Engine Control Module (ECM)
N323 - Ignition Coil 5 with Power Output Stage
N324 - Ignition Coil 6 with Power Output Stage
N325 - Ignition Coil 7 with Power Output Stage
N326 - Ignition Coil 8 with Power Output Stage
P - Spark Plug Connectors
Q - Spark Plugs
T4m - 4-Pin connector, black, on the Ignition Coil 5
T4n - 4-Pin connector, black, on the Ignition Coil 6
T4o - 4-Pin connector, black, on the Ignition Coil 7
T4p - 4-Pin connector, black, on the Ignition Coil 8
T121 - 121-Pin connector, on Motronic Engine Control Module

(27) - Ground connection, -2- on valve cover
(167) - Ground connection -4-, in engine compartment wiring harness
(201) - Ground connection -5-, in engine compartment wiring harness
(D102) - Wire connection -2-, in engine compartment wiring harness

ws = white
sw = black
ro = red
br = brown
gn = green
bl = blue
gr = grey
li = lilac
ge = yellow
or = orange

Edition 04.04

Motronic Engine Control Module, Cylinder 1-8 Fuel Injector

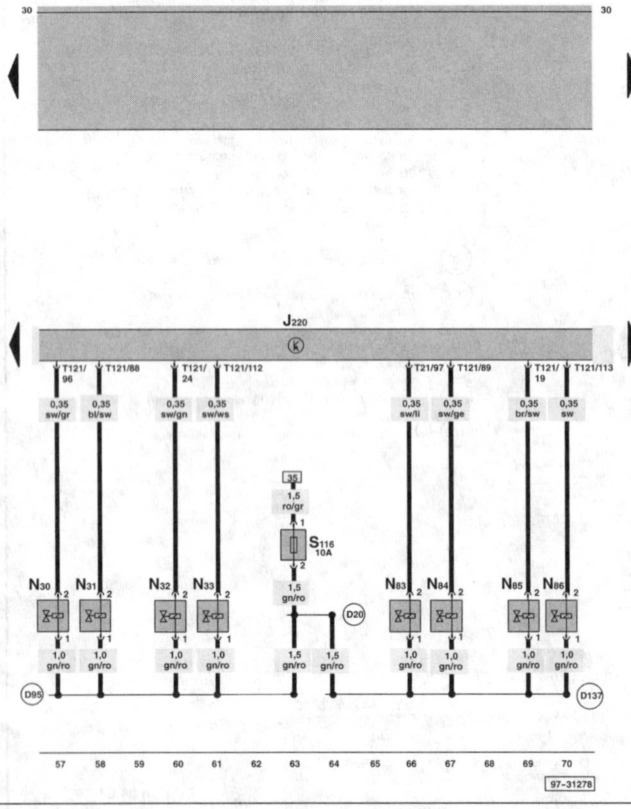

J220 - Motronic Engine Control Module (ECM)
N30 - Cylinder 1 Fuel Injector
N31 - Cylinder 2 Fuel Injector
N32 - Cylinder 3 Fuel Injector
N33 - Cylinder 4 Fuel Injector
N83 - Cylinder 5 Fuel Injector
N84 - Cylinder 6 Fuel Injector
N85 - Cylinder 7 Fuel Injector
N86 - Cylinder 8 Fuel Injector
S116 - Fuse for injectors
T121 - 121-Pin connector, on Motronic Engine Control Module

(D20) - Connection (87a) (in right front wiring harness)
(D95) - wire connection (injectors), in engine compartment wiring harness
(D137) - Connection -2- (injectors), wiring harness, engine compartment

ws = white
sw = black
ro = red
br = brown
gn = green
bl = blue
gr = grey
li = lilac
ge = yellow
or = orange

Edition 04.04

Motronic Engine Control Module, Knock Sensor 1-4

G61 - Knock Sensor (KS) 1
G66 - Knock Sensor (KS) 2
G198 - Knock sensor 3
G199 - Knock sensor 4
J220 - Motronic Engine Control Module (ECM)
T3b - 3-Pin connector, black, on Knock Sensor 1
T3c - 3-Pin connector, brown, on Knock Sensor 2
T3d - 3-Pin connector, orange, on Knock Sensor 3
T3e - 3-Pin connector, grey, on Knock Sensor 4
T121 - 121-Pin connector, on Motronic Engine Control Module

(D106) - wire connection -4-, in engine compartment wiring harness
(D107) - wire connection -5-, in engine compartment wiring harness

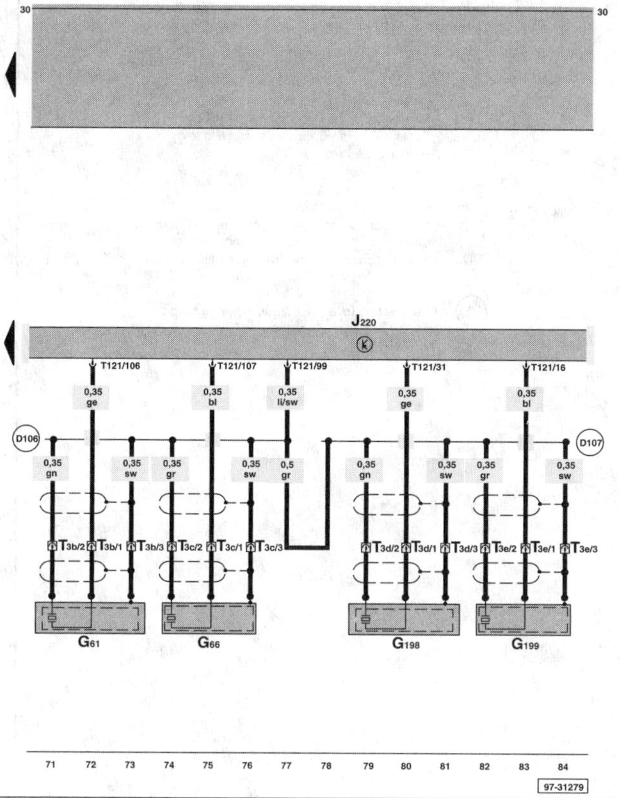

ws = white
sw = black
ro = red
br = brown
gn = green
bl = blue
gr = grey
li = lilac
ge = yellow
or = orange

Edition 04.04

Motronic Engine Control Module, Engine Coolant Temperature, Engine Speed Sensor, Camshaft Position Sensor 2-4

G28 - Engine Speed (RPM) Sensor
G40 - Camshaft Position (CMP) Sensor
G62 - Engine Coolant Temperature (ECT) Sensor
G163 - Camshaft Position (CMP) Sensor 2
G300 - Camshaft Position (CMP) Sensor 3
G301 - Camshaft Position (CMP) Sensor 4
J220 - Motronic Engine Control Module (ECM)
T121 - 121-Pin connector, on Motronic Engine Control Module

(220) - Ground connection (sensor ground), in engine compartment wiring harness
(D141) - Connection (5V), in wiring harness, engine pre-wiring

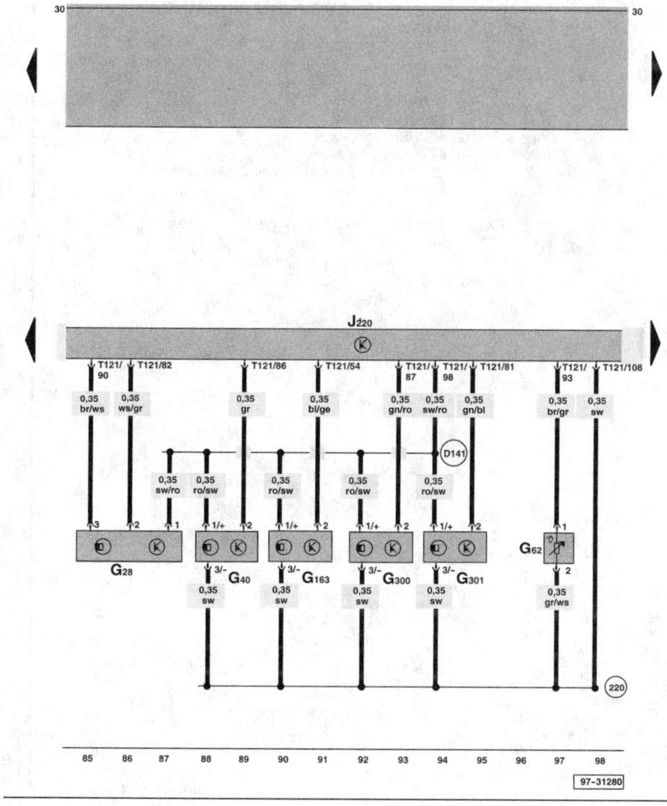

ws = white
sw = black
ro = red
br = brown
gn = green
bl = blue
gr = grey
li = lilac
ge = yellow
or = orange

Edition 04.04

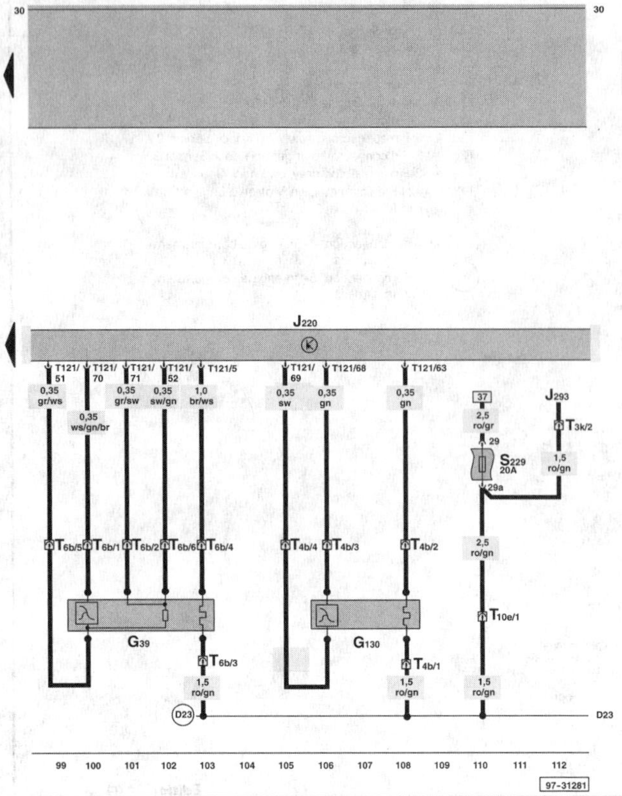

Motronic Engine Control Module, Heated Oxygen Sensor, Oxygen Sensor Behind Three Way Catalytic Converter

G39 - Heated Oxygen Sensor (HO2S)
G130 - Oxygen Sensor (O2S) Behind Three Way Catalytic Converter (TWC)
J220 - Motronic Engine Control Module (ECM)
J293 - Coolant FC (Fan Control) Control Module
S229 - Fuse in fuse holder
T3k - 3-Pin connector, grey
T4b - 4-Pin connector, black
T6b - 6-Pin connector, black
T10e - 10-Pin connector, orange, on protective housing for control module, in engine compartment, left
T121 - 121-Pin connector, on Motronic Engine Control Module

(D23) - Connection (over fuse 229), in wiring harness front, right

ws = white
sw = black
ro = red
br = brown
gn = green
bl = blue
gr = grey
li = lilac
ge = yellow
or = orange

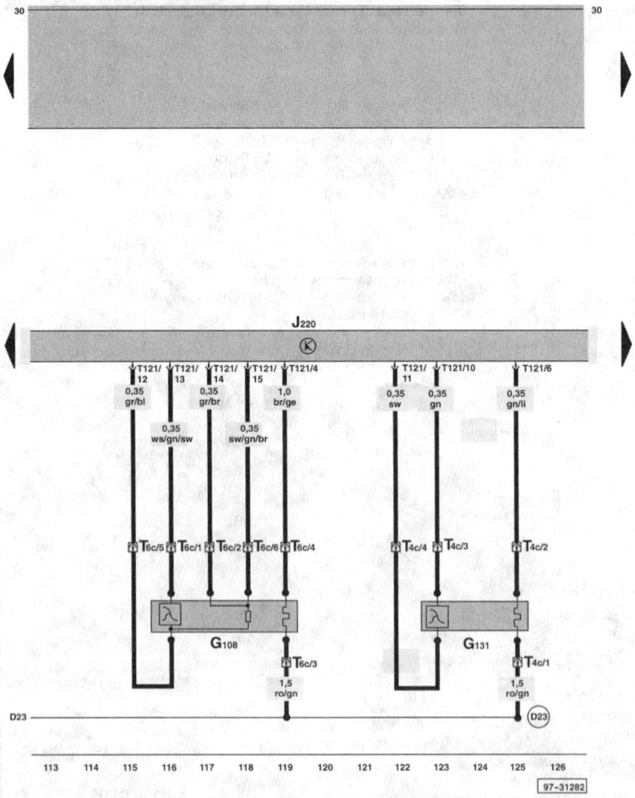

Motronic Engine Control Module, Heated Oxygen Sensor 2, Oxygen Sensor 2 Behind Three Way Catalytic Converter

G108 - Heated Oxygen Sensor (HO2S) 2
G131 - Oxygen Sensor (O2S) 2 Behind Three Way Catalytic Converter (TWC)
J220 - Motronic Engine Control Module (ECM)
T4c - 4-Pin connector, brown
T6c - 6-Pin connector, brown
T121 - 121-Pin connector, on Motronic Engine Control Module

(D23) - Connection (over fuse 229), in wiring harness front, right

ws = white
sw = black
ro = red
br = brown
gn = green
bl = blue
gr = grey
li = lilac
ge = yellow
or = orange

Repair Manual page number **Y 668**

Motronic Engine Control Module, Cruise Control Switch, Button for cruise control, Brake Light Switch, Clutch Pedal Switch, Brake Pedal Switch

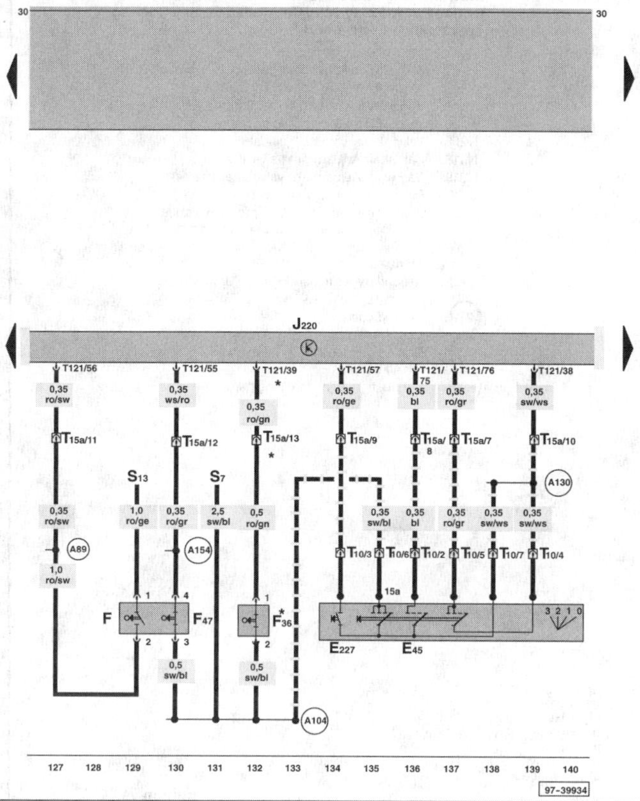

E45 - Cruise Control Switch
E227 - Button for cruise control (set)
F - Brake Light Switch
F36 - Clutch Pedal Switch
F47 - Brake Pedal Switch
J220 - Motronic Engine Control Module (ECM)
S7 - Fuse
S13 - Fuse
T10 - 10-Pin connector, black, near steering column
T15a - 15-Pin connector, white, on protective housing for control module, in engine compartment, left
T121 - 121-Pin connector, on Motronic Engine Control Module

(A89) - Connection 2 (54) (in instrument panel wiring harness)
(A104) - Plus connection -2- (15), in instrument panel wiring harness
(A130) - Connection for Cruise Control (instrument panel wiring harness)
(A154) - Brake Pedal Switch Connection (in instrument panel wiring harness)

- - - - cruise control only
* - manual transmission only

ws = white
sw = black
ro = red
br = brown
gn = green
bl = blue
gr = grey
li = lilac
ge = yellow
or = orange

Edition 04.04

Motronic Engine Control Module, Auxilary Engine Coolant Pump Relay, Left Electro-Hydraulic Mount Solenoid Valve, Coolant Pump

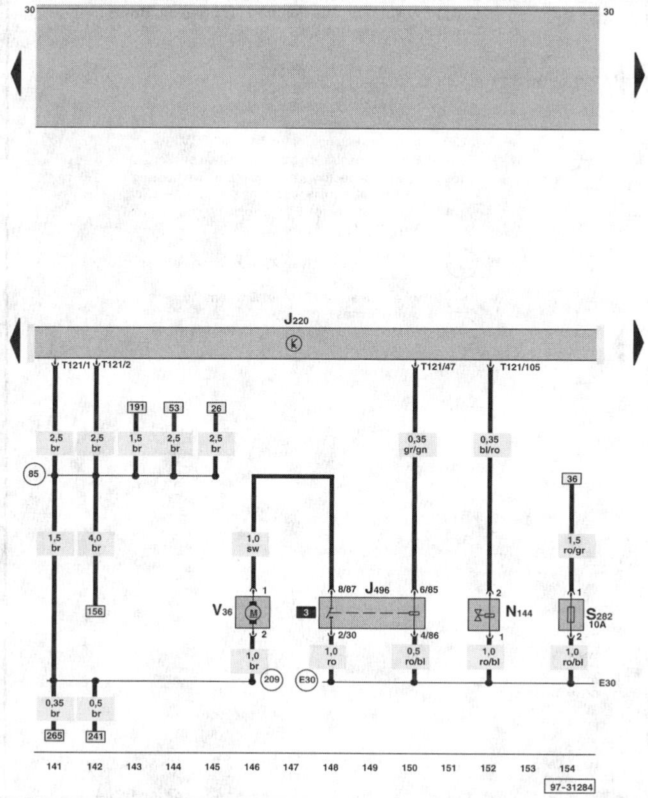

J220 - Motronic Engine Control Module (ECM)
J496 - Auxiliary Engine Coolant (EC) Pump Relay
N144 - Left Electro-Hydraulic Engine Mount Solenoid Valve
S282 - Engine Electronics Fuse
T121 - 121-Pin connector, on Motronic Engine Control Module
V36 - Coolant Pump

(85) - Ground connection -1-, in engine compartment wiring harness
(209) - Ground connection -6-, in engine compartment wiring harness
(E30) - Connection (87a), in wiring harness engine

ws = white
sw = black
ro = red
br = brown
gn = green
bl = blue
gr = grey
li = lilac
ge = yellow
or = orange

Edition 04.04

Repair Manual page number **Y 667**

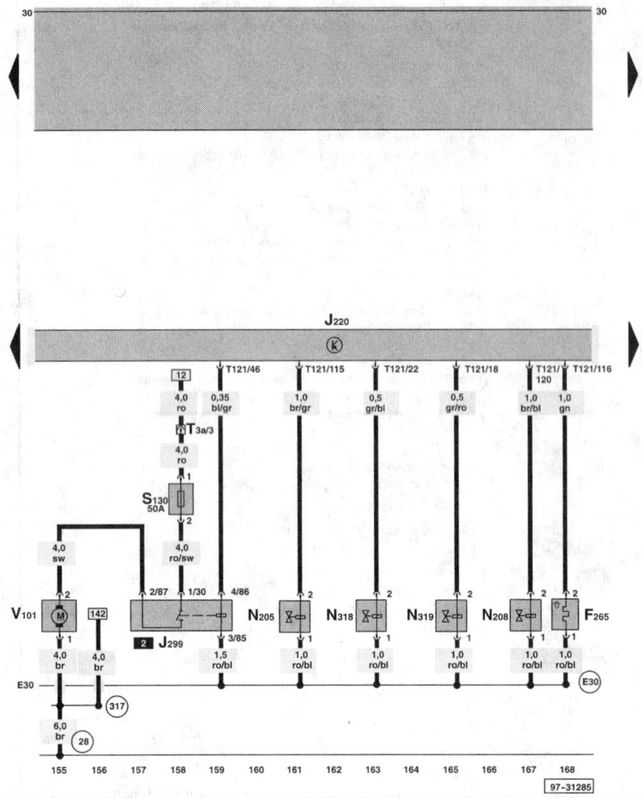

Motronic Engine Control Module, Secondary Air Injection Pump Relay, Map Controlled Engine Cooling Thermostat, Valve 1 and 2

F265 - Map Controlled Engine Cooling Thermostat
J220 - Motronic Engine Control Module (ECM)
J299 - Secondary Air Injection (AIR) Pump Relay
N205 - Valve -1- for camshaft adjustment
N208 - Valve -2- for camshaft adjustment
N318 - Camshaft Adjustment Valve 1 (exhaust)
N319 - Camshaft Adjustment Valve 2 (exhaust)
S130 - Fuse for secondary air pump
T3a - 3-Pin connector, white, on protective housing for control module, in engine compartment, left
T121 - 121-Pin connector, on Motronic Engine Control Module
V101 - Secondary Air Injection (AIR) Pump Motor

(28) - Ground connection, on firewall
(317) - Ground connection -7-, in engine compartment wiring harness
(E30) - Connection (87a), in wiring harness engine

ws = white
sw = black
ro = red
br = brown
gn = green
bl = blue
gr = grey
li = lilac
ge = yellow
or = orange

97-31285

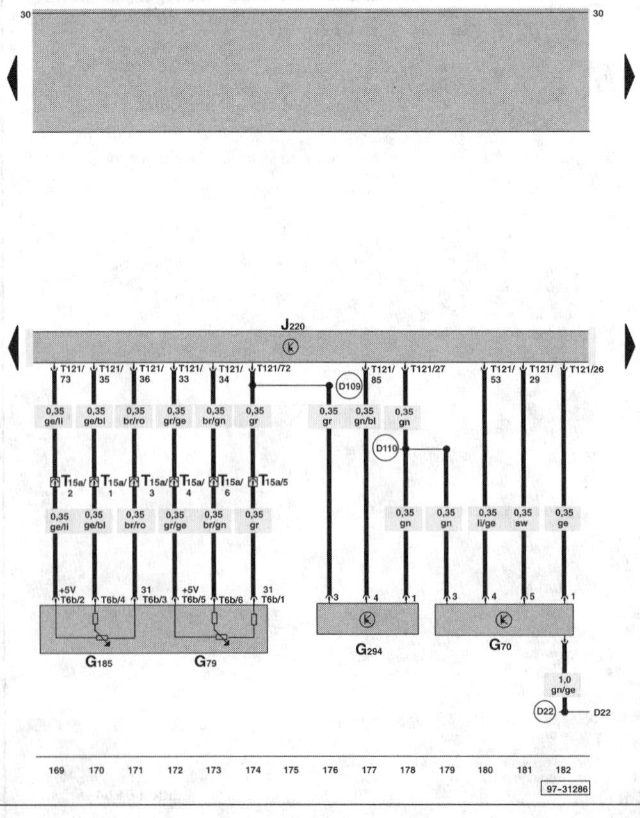

Motronic Engine Control Module, Mass Air Flow (MAF) Sensor, Throttle Position (TP) Sensor, Brake Booster Pressure Sensor

G70 - Mass Air Flow (MAF) Sensor
G79 - Throttle Position (TP) Sensor
G185 - Sender -2- for accelerator pedal position
G294 - Brake Booster Pressure Sensor
J220 - Motronic Engine Control Module (ECM)
T6b - 6-Pin connector, on throttle position sensor
T15a - 15-Pin connector, white, on protective housing for control module, in engine compartment, left
T121 - 121-Pin connector, on Motronic Engine Control Module

(D22) - Connection (over fuse 234), in wiring harness front, right
(D109) - wire connection -7-, in engine compartment wiring harness
(D110) - wire connection -8-, in engine compartment wiring harness

ws = white
sw = black
ro = red
br = brown
gn = green
bl = blue
gr = grey
li = lilac
ge = yellow
or = orange

97-31286

Passat Wiring diagram No. 72/15

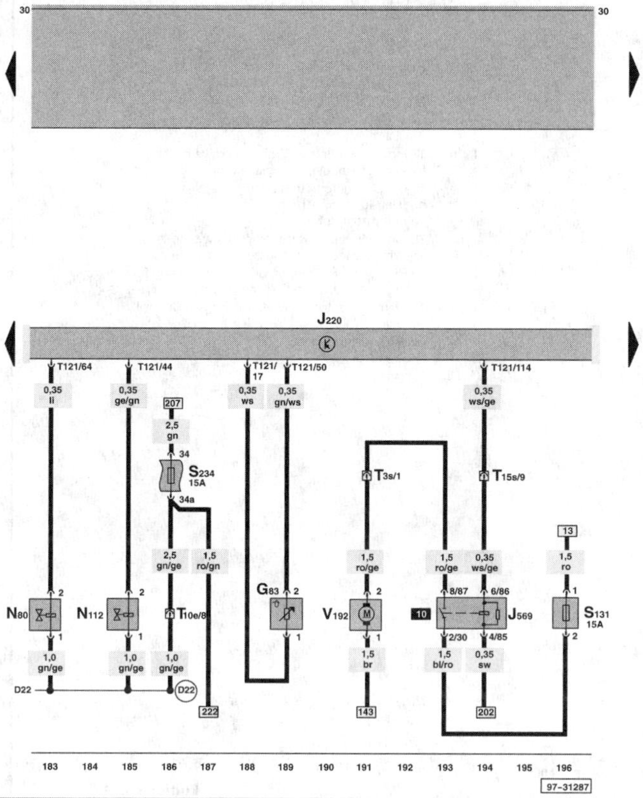

Motronic Engine Control Module, Brake Booster Relay, Engine Coolant Temperature Sensor, Valve

- G83 - Engine Coolant Temperature (ECT) Sensor (on Radiator)
- J220 - Motronic Engine Control Module (ECM)
- J569 - Brake Booster Relay
- N80 - Evaporative Emission (EVAP) Canister Purge Regulator Valve
- N112 - Secondary Air Injection (AIR) Solenoid Valve
- S131 - Safety fuse 1
- S234 - Fuse in fuse holder
- T3s - 3-Pin connector, red, on protective housing for control module, in engine compartment, left
- T10e - 10-Pin connector, orange, on protective housing for control module, in engine compartment, left
- T15s - 15-Pin connector, red, on protective housing for control module, in engine compartment, left
- T121 - 121-Pin connector, on Motronic Engine Control Module
- V192 - Brake System Vacuum Pump
- (D22) - Connection (over fuse 234), in wiring harness front, right

ws = white
sw = black
ro = red
br = brown
gn = green
bl = blue
gr = grey
li = lilac
ge = yellow
or = orange

Edition 04.04

Passat Wiring diagram No. 72/16

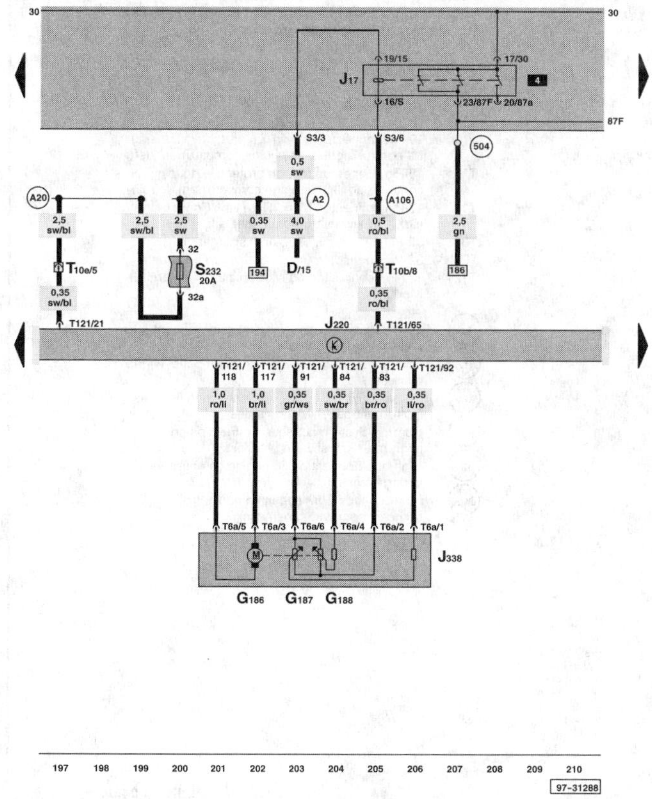

Motronic Engine Control Module, Fuel Pump Relay, Throttle drive, Angle sensor 1 and 2 for throttle drive

- D - Ignition/Starter Switch
- G186 - Throttle drive (power accelerator actuation)
- G187 - Angle sensor -1- for throttle drive (power accelerator actuation)
- G188 - Angle sensor -2- for throttle drive (power accelerator actuation)
- J17 - Fuel Pump (FP) Relay
- J220 - Motronic Engine Control Module (ECM)
- J338 - Throttle Valve Control Module
- S232 - Fuse in fuse holder
- T6a - 6-Pin connector, black, connector station A-pillar
- T10b - 10-Pin connector, black, on protective housing for control module, in engine compartment, left
- T10e - 10-Pin connector, orange, on protective housing for control module, in engine compartment, left
- T121 - 121-Pin connector, on Motronic Engine Control Module
- (504) - Threaded connection -1- (87) on the relay plate
- (A2) - Plus connection (15), in instrument panel wiring harness
- (A20) - Plus Connection (15a) (in instrument panel wiring harness)
- (A106) - Connection -2- (86s), in instrument panel wiring harness

ws = white
sw = black
ro = red
br = brown
gn = green
bl = blue
gr = grey
li = lilac
ge = yellow
or = orange

Edition 04.04

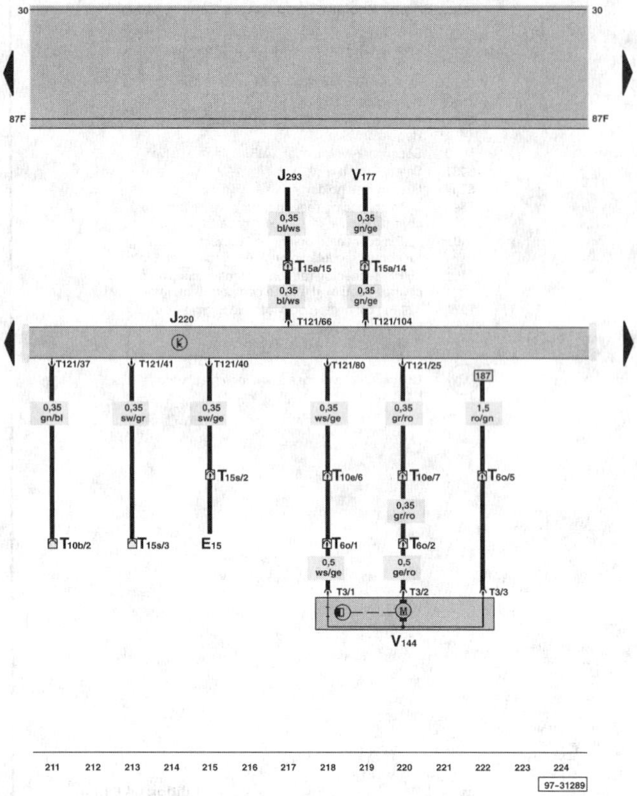

Motronic Engine Control Module, Leak Detection Pump

- E15 - Rear window defogger switch
- J220 - Motronic Engine Control Module (ECM)
- J293 - Coolant FC (Fan Control) Control Module
- T6o - 6-Pin Connector, brown, connector station A-pillar, right
- T10b - 10-Pin connector, black, on protective housing for control module, in engine compartment, left
- T10e - 10-Pin Connector, orange, on protective housing for control module, in engine compartment, left
- T15a - 15-Pin connector, white, on protective housing for control module, in engine compartment, left
- T15s - 15-Pin connector, red, on protective housing for control module, in engine compartment, left
- T121 - 121-Pin connector, on Motronic Engine Control Module
- V144 - Leak Detection Pump (LDP)
- V177 - Coolant Fan -2-

ws = white
sw = black
ro = red
br = brown
gn = green
bl = blue
gr = grey
li = lilac
ge = yellow
or = orange

Edition 04.04

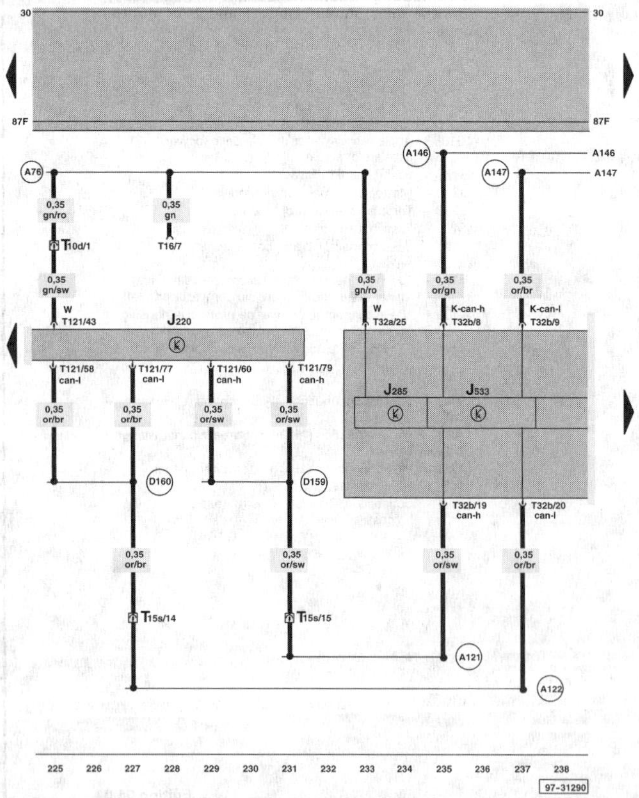

Motronic Engine Control Module, Control Module with indicator unit in instrument panel insert, Data Bus On Board Diagnostic Interface

- J220 - Motronic Engine Control Module (ECM)
- J285 - Control module with indicator unit in instrument panel insert
- J533 - Data Bus On Board Diagnostic Interface
- T10d - 10-Pin connector, brown, on protective housing for control module, in engine compartment, left
- T15s - 15-Pin connector, red, on protective housing for control module, in engine compartment, left
- T16 - 16-Pin connector, black, Data Link Connector (DLC), below instrument panel, left
- T32a - 32-Pin connector, blue, on instrument cluster
- T32b - 32-Pin connector, green, on instrument cluster
- T121 - 121-Pin connector, on Motronic Engine Control Module
- (A76) - Connection (K-diagnosis wire) in instrument panel wiring harness
- (A122) - Connection (low bus) in instrument panel wiring harness
- (A146) - Comfort System High-bus Connection (in instrument panel wiring harness)
- (A147) - Comfort System Low-bus Connection (in instrument panel wiring harness)
- (D159) - High-bus Connection (in engine compartment wiring harness)
- (D160) - Low-bus Plus Connection (in engine compartment wiring harness)

ws = white
sw = black
ro = red
br = brown
gn = green
bl = blue
gr = grey
li = lilac
ge = yellow
or = orange

Edition 04.04

Passat — Wiring diagram — No. 72/19

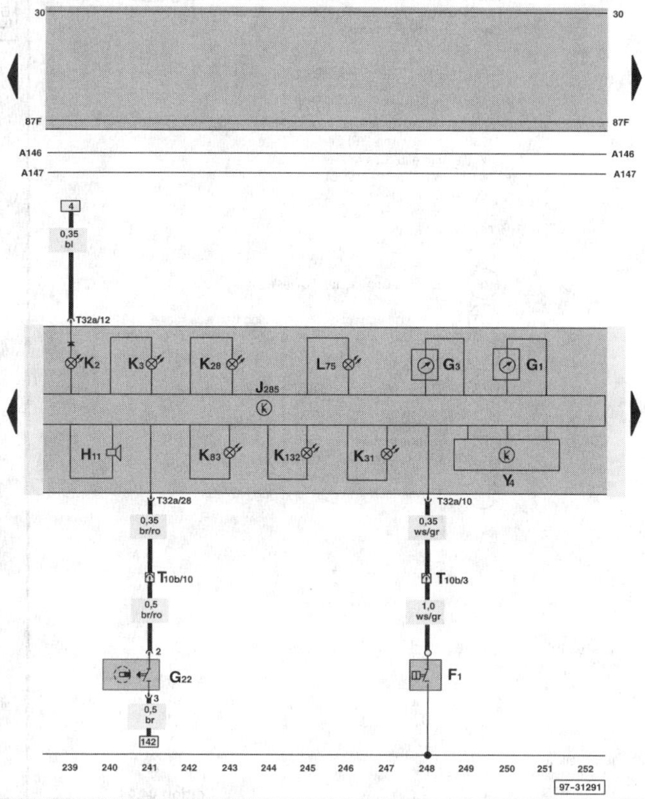

ws = white
sw = black
ro = red
br = brown
gn = green
bl = blue
gr = grey
li = lilac
ge = yellow
or = orange

Control Module with indicator unit in instrument panel insert, Speedometer Vehicle Speed Sensor (VSS), Oil Pressure Switch

- F1 - Oil Pressure Switch
- G1 - Fuel Gauge
- G3 - Engine Coolant Temperature (ECT) Gauge
- G22 - Speedometer Vehicle Speed Sensor (VSS)
- H11 - Oil Pressure Warning Buzzer
- J285 - Control module with indicator unit in instrument panel insert
- K2 - Generator (GEN) Warning Light
- K3 - Oil Pressure Warning Light
- K28 - Engine Coolant Level/Temperature (ECL/ECT) Warning Light
- K31 - Cruise Control Indicator Light
- K83 - Malfunction Indicator Lamp (MIL)
- K132 - Electronic Power Control (EPC) Warning Lamp
- L75 - Digital Display Light
- T10b - 10-Pin connector, black, on protective housing for control module, in engine compartment, left
- T32a - 32-Pin connector, blue, on instrument cluster
- Y4 - Odometer Display

Edition 04.04

Passat — Wiring diagram — No. 72/20

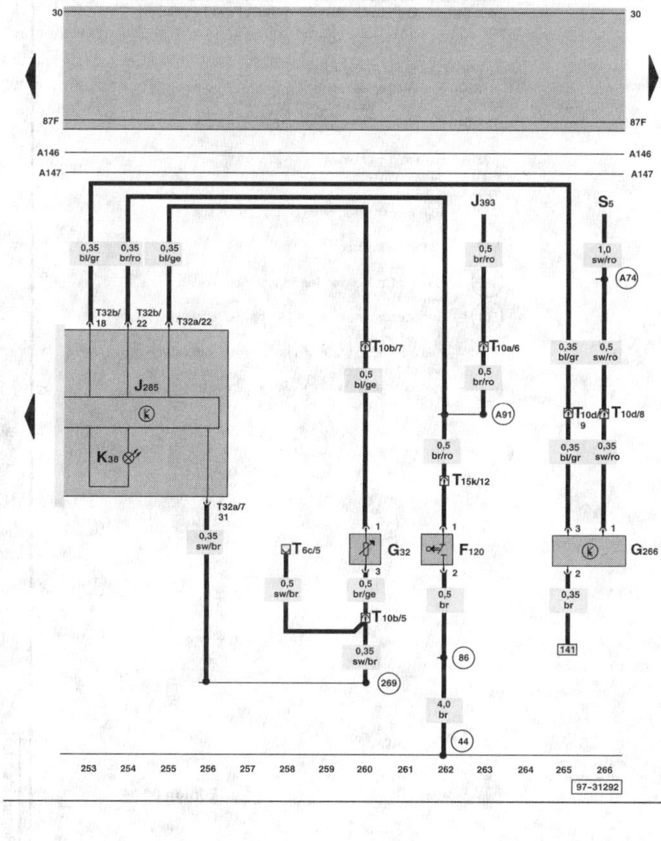

ws = white
sw = black
ro = red
br = brown
gn = green
bl = blue
gr = grey
li = lilac
ge = yellow
or = orange

Control Module with indicator unit in instrument panel insert, Hood Alarm Switch

- F120 - Hood Alarm Switch
- G32 - Engine Coolant Level (ECL) Sensor
- G266 - Oil Level Thermal Sensor
- J285 - Control module with indicator unit in instrument panel insert
- J293 - Coolant FC (Fan Control) Control Module
- K38 - Engine Oil Level Indicator Light
- S5 - Fuse
- T6c - 6-Pin connector, blue, connector station A-pillar, left
- T10a - 10-Pin connector, brown, connector station A-pillar, left
- T10b - 10-Pin connector, black, on protective housing for control module, in engine compartment, left
- T10d - 10-Pin connector, brown, on protective housing for control module, in engine compartment, left
- T15k - 15-Pin connector, grey, connector station A-pillar, left
- T32a - 32-Pin connector, blue, on instrument cluster
- T32b - 32-Pin connector, green, on instrument cluster
- (44) - Ground connection, on left A-pillar, lower part
- (86) - Ground connection -1-, in rear wiring harness
- (269) - Ground connection (sensor ground) -1-, in instrument panel wiring harness
- (A74) - Connection (15a - fuse 5) in instrument panel wiring harness
- (A91) - Connection (signal for instrument panel insert), in instrument panel wiring harness

Edition 04.04

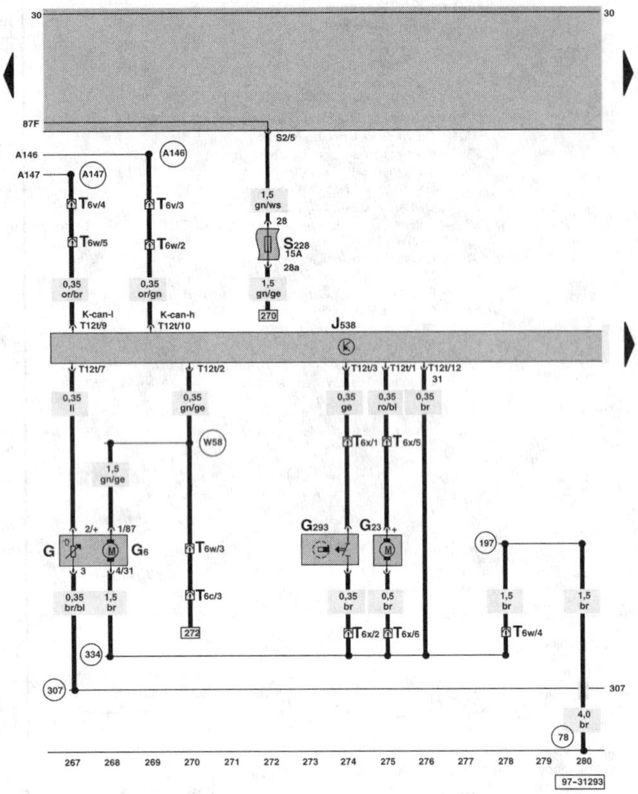

Fuel Pump Control Module, Sender for fuel gauge, Fuel Pump, Transfer Fuel Pump, Auxiliary Tank Fuel Level Empty Sensor

- G - Sender for fuel gauge
- G6 - Fuel Pump (FP)
- G23 - Transfer Fuel Pump (FP)
- G293 - Auxiliary Tank Fuel Level Empty Sensor
- J538 - Fuel Pump (FP) Control Module
- S228 - Fuse in fuse holder
- T6c - 6-Pin connector, blue
- T6v - 6-Pin connector, lilac
- T6w - 6-Pin connector
- T6x - 6-Pin connector
- T12t - 12-Pin connector

- (78) - Ground Connection (lower right B-pillar)
- (197) - Ground connection -4-, in wiring harness rear
- (307) - Ground Connection for Sensor (floor wiring harness)
- (334) - Ground Connection 6 (in rear wiring harness)
- (A146) - Comfort System High-bus Connection (in instrument panel wiring harness)
- (A147) - Comfort System Low-bus Connection (in instrument panel wiring harness)
- (W58) - Connection (87) (in rear wiring harness)

ws = white
sw = black
ro = red
br = brown
gn = green
bl = blue
gr = grey
li = lilac
ge = yellow
or = orange

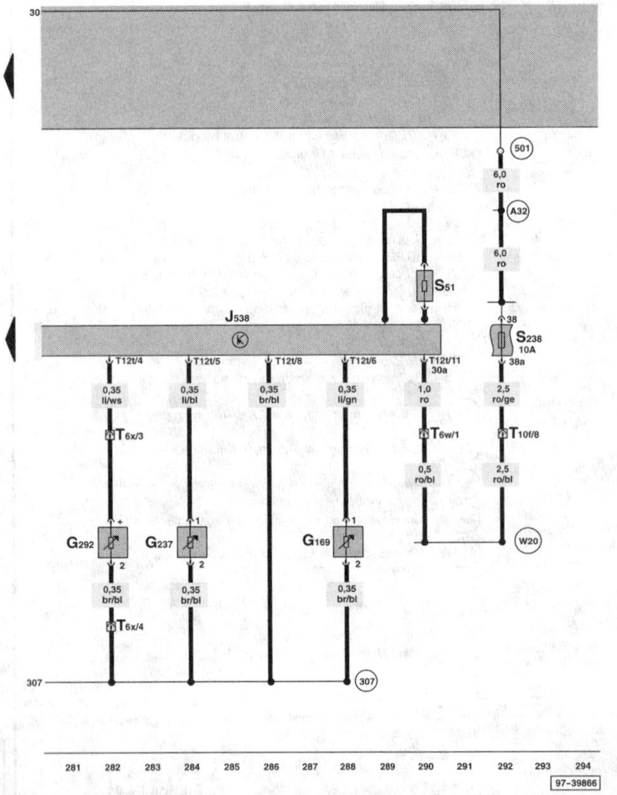

Fuel Pump Control Module, Fuel Level Sensor 2, Fuel Supply Sensor 3, Auxiliary Tank Fuel Level Sensor

- G169 - Fuel Level Sensor 2
- G237 - Fuel Supply Sensor 3
- G292 - Auxiliary Tank Fuel Level Sensor
- J538 - Fuel Pump (FP) Control Module
- S51 - Fuse
- S238 - Fuse in fuse holder
- T6w - 6-Pin connector
- T6x - 6-Pin connector
- T10f - 10-Pin connector, blue, connector station A-pillar, left
- T12t - 12-Pin connector

- (307) - Ground Connection for Sensor (floor wiring harness)
- (501) - Threaded connection -2- (30) on the relay plate
- (A32) - Plus connection (30), in instrument panel wiring harness
- (W20) - Plus connection (30a) in wiring harness, rear

ws = white
sw = black
ro = red
br = brown
gn = green
bl = blue
gr = grey
li = lilac
ge = yellow
or = orange

Automatic transmission, 5-speed automatic with tiptronic,

● 4.0L - Engine - Motronic Multiport Fuel Injection (MFI)/202kW, code BDP

from January 2002

Notes:

Information for

- ♦ *Relay and fuse arrangements*
- ♦ *Connector arrangements*
- ♦ *Control Modules and Relays*
- ♦ *Ground connections*
- ⇒ *Component Locations!*

Information for

- ♦ *Troubleshooting Procedures*
- ⇒ *Guided Fault Finding using VAS 5051/5052*

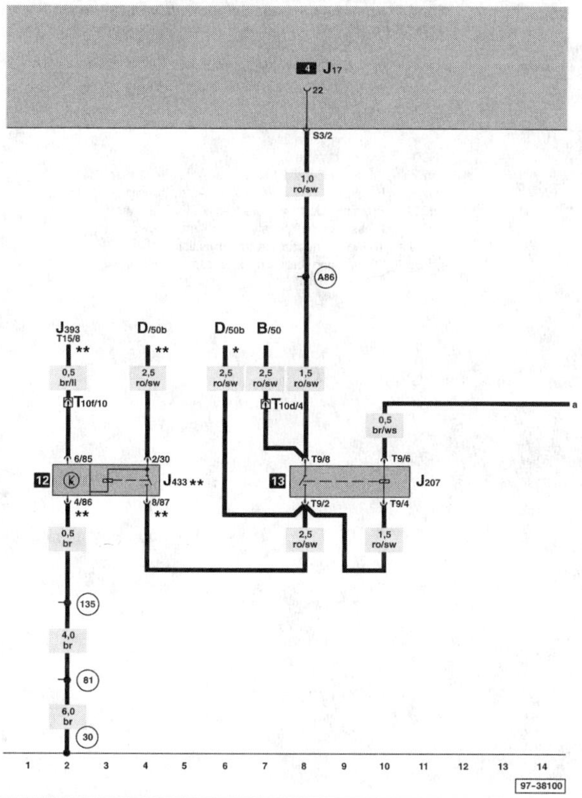

Starting Interlock Relay, Locking relay for starter (anti-theft immobilizer)

- B - Starter
- D - Ignition/Starter Switch
- J17 - Fuel Pump (FP) Relay
- J207 - Starting Interlock Relay
- J393 - Central control module for comfort system
- J433 - Locking relay for starter (anti-theft immobilizer) **
- T9 - 9-Pin Connector, on Starting Interlock Relay
- T10d - 10-Pin Connector, brown, on protective housing for control module, in engine compartment, left
- T10f - 10-Pin Connector, blue, connector station A-pillar, left
- T15 - 15-Pin Connector, on Central control module for comfort system
- (30) - Ground connection, -1-, beside fuse/relay panel
- (81) - Ground connection -1-, in instrument panel wiring harness
- (135) - Ground connection -2-, in instrument panel wiring harness
- (A86) - Connection (50a), in instrument panel wiring harness
- * - from May 2003
- ** - through April 2003

ws = white
sw = black
ro = red
br = brown
gn = green
bl = blue
gr = grey
li = lilac
ge = yellow
or = orange

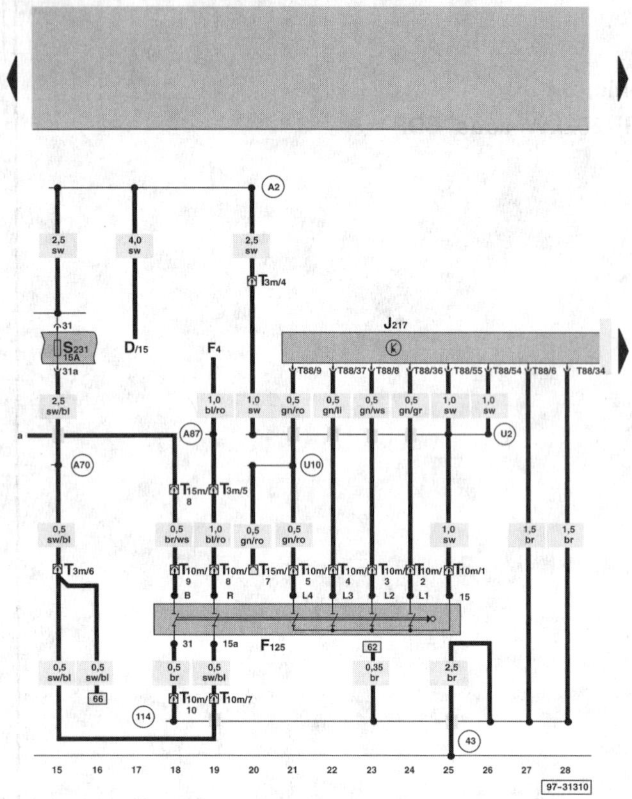

Multi-Function Transmission Range (TR) Switch, Transmission Control Module (TCM)

- D - Ignition/Starter Switch
- F4 - Back-Up Light Switch
- F125 - Multi-Function Transmission Range (TR) Switch
- J217 - Transmission Control Module (TCM)
- S231 - Fuse in fuse holder
- T3m - 3-Pin connector, brown, connector station A-pillar, right
- T10m - 10-Pin connector
- T15m - 15-Pin connector, brown, connector station A-pillar, right
- T88 - 88-Pin connector, on transmission control Module (TCM)
- (43) - Ground connection, on right A-pillar, lower part
- (114) - Ground connection, in automatic transmission wiring harness
- (A2) - plus connection (15), in instrument panel wiring harness
- (A70) - Connector (15a, fuse 231), in instrument panel wiring harness
- (A87) - Connector (reverse lamp), in instrument panel wiring harness
- (U2) - wire connection -1- (15), in automatic transmission wiring harness
- (U10) - Cruise Control Connection (in automatic transmission wiring harness)

ws = white
sw = black
ro = red
br = brown
gn = green
bl = blue
gr = grey
li = lilac
ge = yellow
or = orange

Edition 12.03

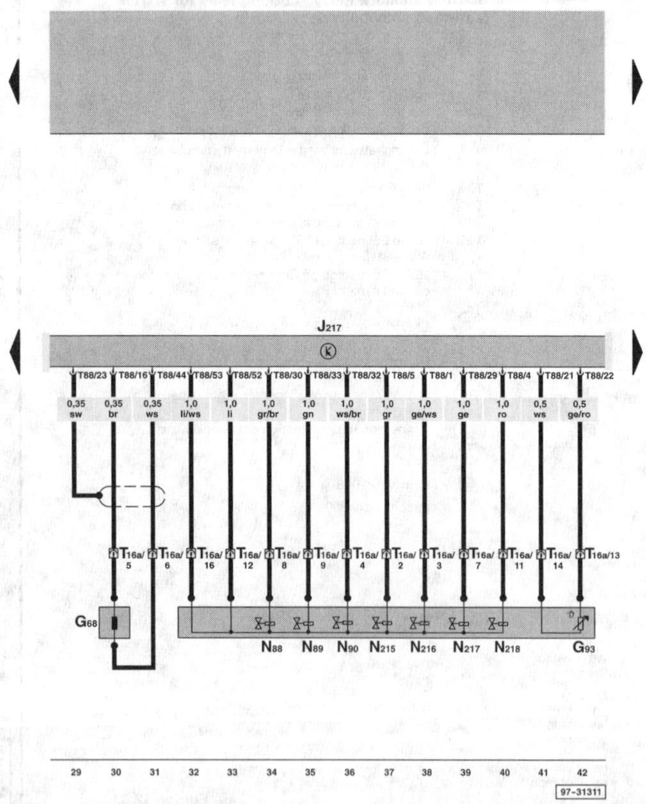

Transmission Control Module (TCM), Transmission Fluid Temperature Sensor, Solenoid Valve 1-3, Pressure Control Valve 1-4, Vehicle Speed Sensor (VSS)

- G68 - Vehicle Speed Sensor (VSS)
- G93 - Transmission Fluid Temperature Sensor
- J217 - Transmission Control Module (TCM)
- N88 - Solenoid Valve 1
- N89 - Solenoid Valve 2
- N90 - Solenoid Valve 3
- N215 - Pressure Control Valve 1 for Auto. Transmission
- N216 - Pressure Control Valve 2 for Auto. Transmission
- N217 - Pressure Control Valve 3 for Auto. Transmission
- N218 - Pressure Control Valve 4 for Auto. Transmission
- T16a - 16-Pin connector, on transmission
- T88 - 88-Pin connector, on transmission control Module (TCM)

ws = white
sw = black
ro = red
br = brown
gn = green
bl = blue
gr = grey
li = lilac
ge = yellow
or = orange

Edition 12.03

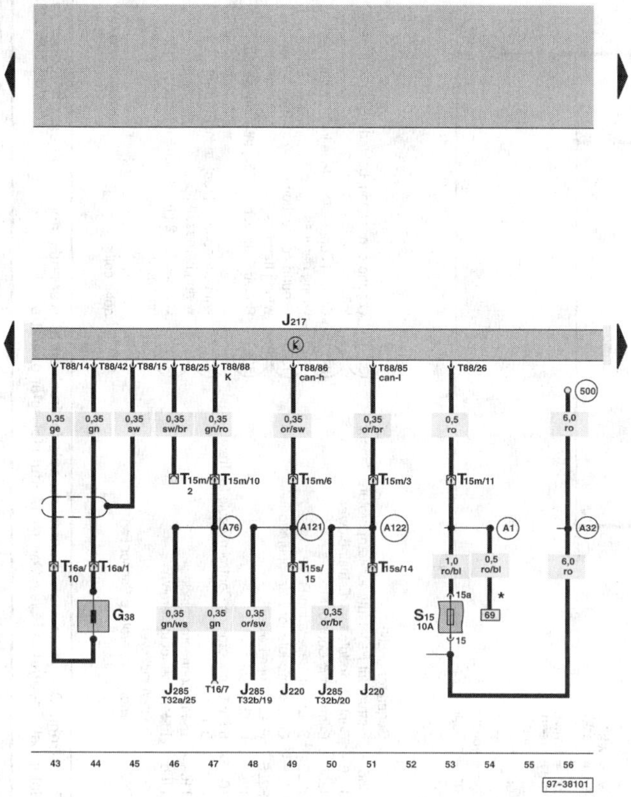

Transmission Control Module (TCM), Transmission Vehicle Speed Sensor (VSS), below instrument panel

G38 - Transmission Vehicle Speed Sensor (VSS)
J217 - Transmission Control Module (TCM)
J220 - Motronic Engine Control Module (ECM)
J285 - Control module with indicator unit in instrument panel insert
S15 - Fuse
T15m- 15-Pin connector, brown, connector station A-pillar, right
T15s - 15-Pin connector
T16 - 16-Pin connector, black, Data Link Connector (DLC), below instrument panel, left
T16a - 16-Pin connector, on transmission
T32a - 32-Pin connector, blue, on instrument cluster
T32b - 32-Pin connector, green, on instrument cluster
T88 - 88-Pin connector, on transmission control Module (TCM)

(500) - Threaded connection -1- (30) on the relay plate

(A1) - plus connection (30a), in instrument panel wiring harness
(A32) - plus connection (30), in instrument panel wiring harness
(A76) - Connector (K-diagnosis wire) in instrument panel wiring harness
(A121) - Connection (high bus) in instrument panel wiring harness
(A122) - Connection (low bus) in instrument panel wiring harness

* - through April 2003

ws = white
sw = black
ro = red
br = brown
gn = green
bl = blue
gr = grey
li = lilac
ge = yellow
or = orange

Edition 12.03

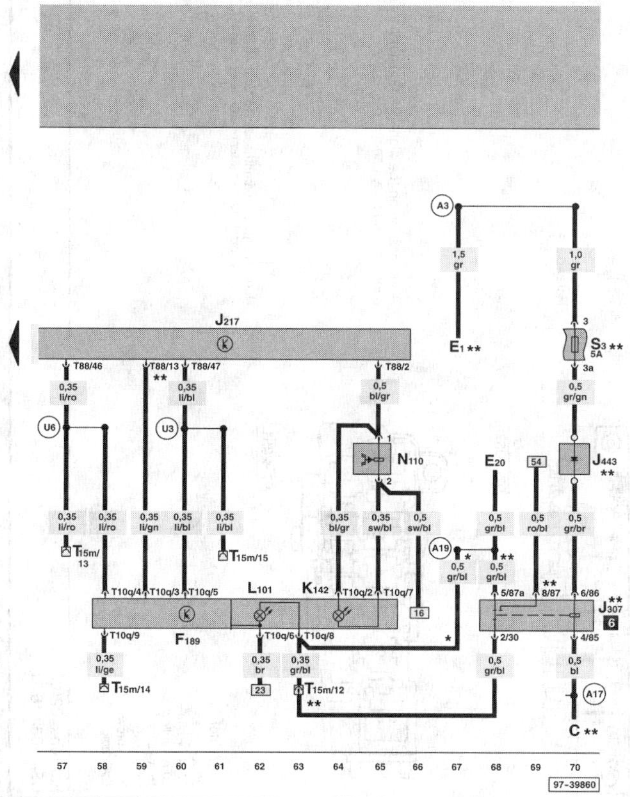

Transmission Control Module (TCM), Selector Lever Light Relay, Tiptronic Switch, Warning light for selector lever position P/N

C - Generator (GEN)
E1 - Light switch
E20 - Instrument Panel Light Dimmer Switch
F189 - Tiptronic Switch
J217 - Transmission Control Module (TCM)
J307 - Selector Lever Light Relay
J443 - Locking diode for selector lever illumination
K142 - Warning light for selector lever position P/N
L101 - Illumination for selector lever scale
N110 - Shift Lock Solenoid
S3 - Fuse
T10q - 10-Pin connector
T15m- 15-Pin connector, brown, connector station A-pillar, right
T88 - 88-Pin connector, on transmission control Module (TCM)

(A3) - plus connection (58), in instrument panel wiring harness
(A17) - wire connection (61), in instrument panel wiring harness
(A19) - wire connection (58d), in instrument panel wiring harness
(U3) - wire connection -1-, in automatic transmission wiring harness
(U6) - Wire connection -2-, in automatic transmission wiring harness

* - from May 2003
** - through April 2003

ws = white
sw = black
ro = red
br = brown
gn = green
bl = blue
gr = grey
li = lilac
ge = yellow
or = orange

Edition 04.04

Climatronic

● 4,0 l - Engine - Motronic Multiport Fuel Injection (MFI)/202kW, code BDP

from January 2002

Deviate relay location and fuseplacements as well as the locations of multiple connectors see section component locations .

Relay location on the thirteen position auxiliary relay panel, above relay panel:

2 Solar Cell Separation Relay

Relay panel:

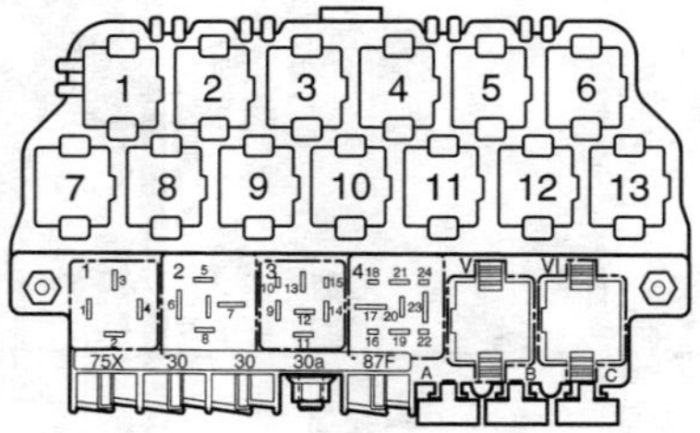

97-20676

Relay location on the eight position auxiliary relay panel, behind relay panel:

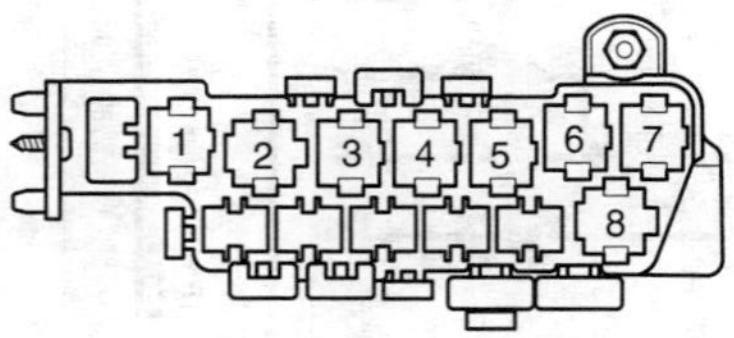

97-20515

Note: Number in parentheses indicates production control number stamped on relay housing.

Fuse colors
30 A - green
25 A - white
20 A - yellow
15 A - blue
10 A - red
7,5 A - brown
5 A - beige

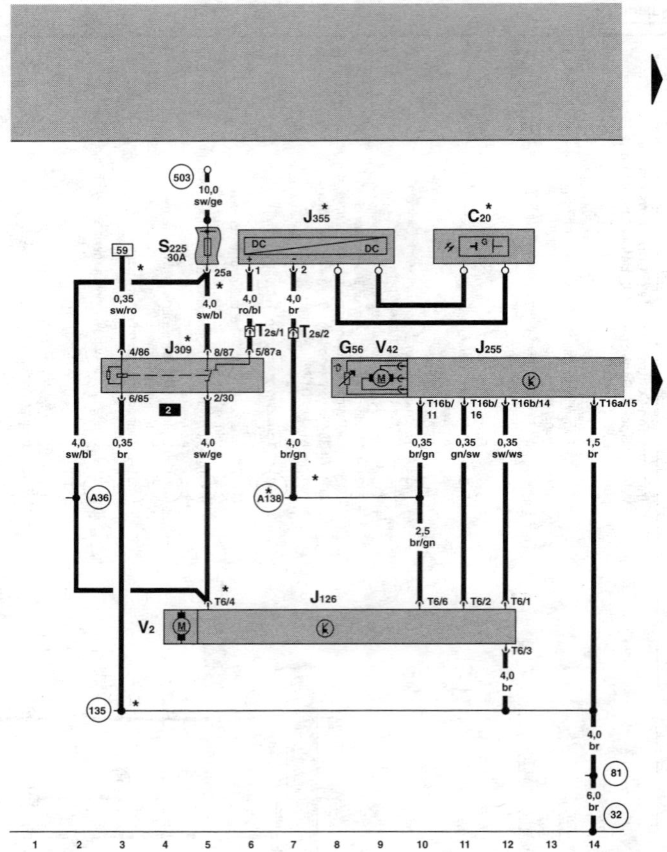

97-31319

ws = white
sw = black
ro = red
br = brown
gn = green
bl = blue
gr = grey
li = lilac
ge = yellow
or = orange

C20 - Generator (GEN)
G56 - Instrument Panel Interior Temperature Sensor
J126 - Control module for fresh air blower
J255 - Climatronic Control Module
J309 - Solar Cell Separation Relay
J355 - Control module for solar operation
S225 - Fuse in fuse holder
T2s - Double Connector
T6 - 6-Pin Connector, on Control module for fresh air blower
T16a - 16-Pin Connector, black
T16b - 16-Pin Connector, brown
V2 - Fresh Air Blower
V42 - Interior Temperature Sensor Fan

(32) - Ground connection, behind instrument panel, left
(81) - Ground connection -1-, in instrument panel wiring harness
(135) - Ground connection -2-, in instrument panel wiring harness
(503) - Bolted Connection (75x) (at relay panel)
(A36) - wire connection (75a), in instrument panel wiring harness
(A138) - Connection 2 (in instrument panel wiring harness)

* - solar cell

Climatronic Control Module, Control module for fresh air blower and solor operation, Solar Cell Separation Relay, Control Module for solar operation

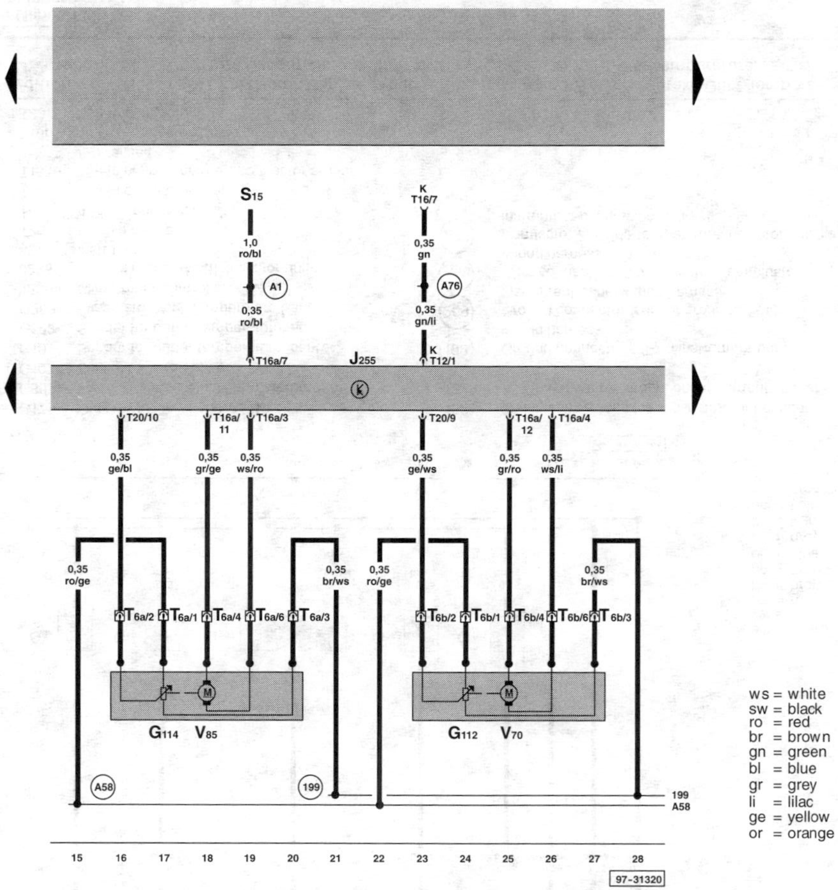

ws = white
sw = black
ro = red
br = brown
gn = green
bl = blue
gr = grey
li = lilac
ge = yellow
or = orange

G112 - Central Flap Motor Position Sensor
G114 - Footwell/Defrost Flap Motor Position Sensor
J255 - Climatronic Control Module
S15 - Fuse
T6a - 6-Pin Connector, black, behind instrument panel, center
T6b - 6-Pin Connector, blue, behind instrument panel, center
T12 - 12-Pin Connectorblack
T16 - 16-Pin Connector, Data Link Connector (DLC), below instrument panel
T16a - 16-Pin Connector, black
T20 - 20-Pin Connector, red
V70 - Central Air Flap Motor
V85 - Footwell/Defroster Flap Motor

(199) - Ground connection -3-, in instrument panel wiring harness
(A1) - plus connection (30a), in instrument panel wiring harness
(A58) - Wire connection (5 V), in instrument cluster wiring harness
(A76) - Connector (K-diagnosis wire) in instrument panel wiring harness

Edition 02/02
USA.5062.16.21

Climatronic Control Module, Central Flap Motor Position Sensor, Footwell/ Defrost Flap Motor Position Sensor

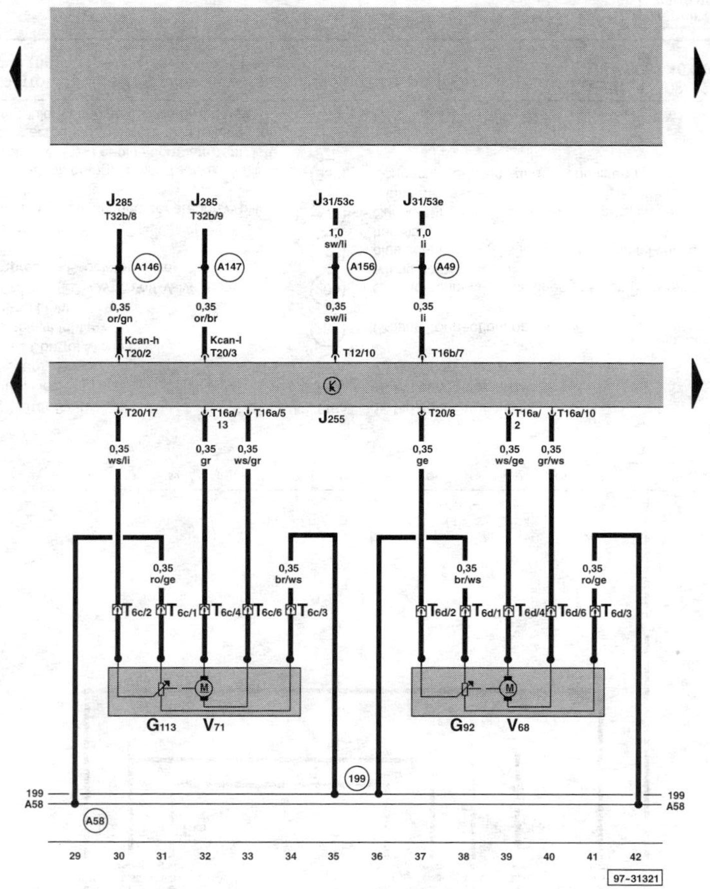

ws = white
sw = black
ro = red
br = brown
gn = green
bl = blue
gr = grey
li = lilac
ge = yellow
or = orange

G92 - Temperature Regulator Flap Motor Position Sensor
G113 - Back Pressure Flap Motor Position Sensor
J31 - Wiper/Washer Intermittent Relay
J255 - Climatronic Control Module
J285 - Control module with indicator unit in instrument panel insert
T6c - 6-Pin connector, black
T6d - 6-Pin connector, red
T12 - 12-Pin connector, black
T16a - 16-Pin connector, black
T16b - 16-Pin connector, brown
T20 - 20-Pin connector, red
T32b - 32-Pin connector, on instrument cluster
V68 - Temperature Regulator Flap Motor
V71 - Air Flow Flap Motor

(199) - Ground connection -3-, in instrument panel wiring harness
(A49) - wire connection -1-, in instrument panel wiring harness
(A58) - Wire connection (5 V), in instrument cluster wiring harness
(A146) - Comfort System High-bus Connection (in instrument panel wiring harness)
(A147) - Comfort System Low-bus Connection (in instrument panel wiring harness)
(A156) - Sensor Ground Connection (in engine wiring harness)

Climatronic Control Module, Temperature Regulator and Back Pressure Flap Motor Position Sensor, Air Flow Flap Motor

Edition 02/02
USA.5062.16.21

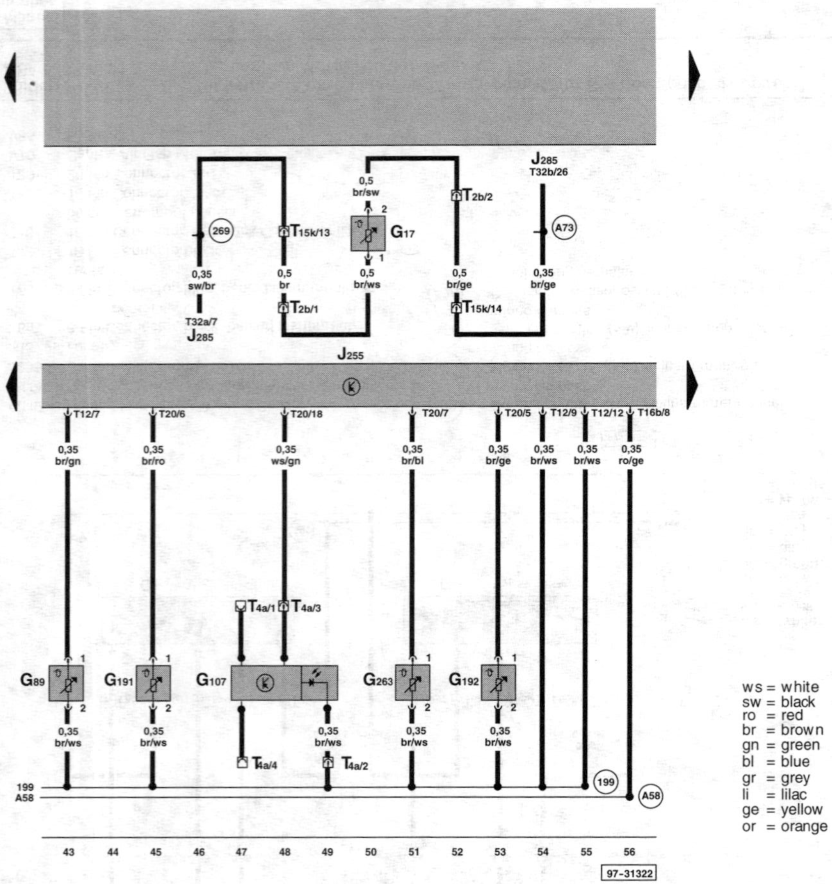

ws = white
sw = black
ro = red
br = brown
gn = green
bl = blue
gr = grey
li = lilac
ge = yellow
or = orange

G17 - Outside Air Temperature Sensor
G89 - Fresh Air Intake Duct Temperature Sensor
G107- Sunlight Photo Sensor
G191- Sender for outlet temperature, center
G192- Sender for outlet temperature, floor outlet
G263- Evaporator Vent Temperature Sensor
J255 - Climatronic Control Module
J285 - Control module with indicator unit in instrument panel insert
T2b - Double connector
T4a - 4-Pin connector, black
T12 - 12-Pin connector, black
T15k - 15-Pin connector, grey, connector station A-pillar, left
T16b - 16-Pin connector, brown
T20 - 20-Pin connector, red

T32a - 32-Pin connector, blue, on instrument cluster
T32b - 32-Pin connector, green, on instrument cluster
(199) - Ground connection -3-, in instrument panel wiring harness
(269) - Ground connector (sensor ground) -1-, in instrument panel wiring harness
(A58) - Wire connection (5 V), in instrument cluster wiring harness
(A73) - Connector (outside temperature indicator), in instrument panel wiring harness

Climatronic Control Module, Outside Air and Fresh Air Intake Duct Temperature Sensor, Sunglight Photo Sensor, Sender for outlet temperature

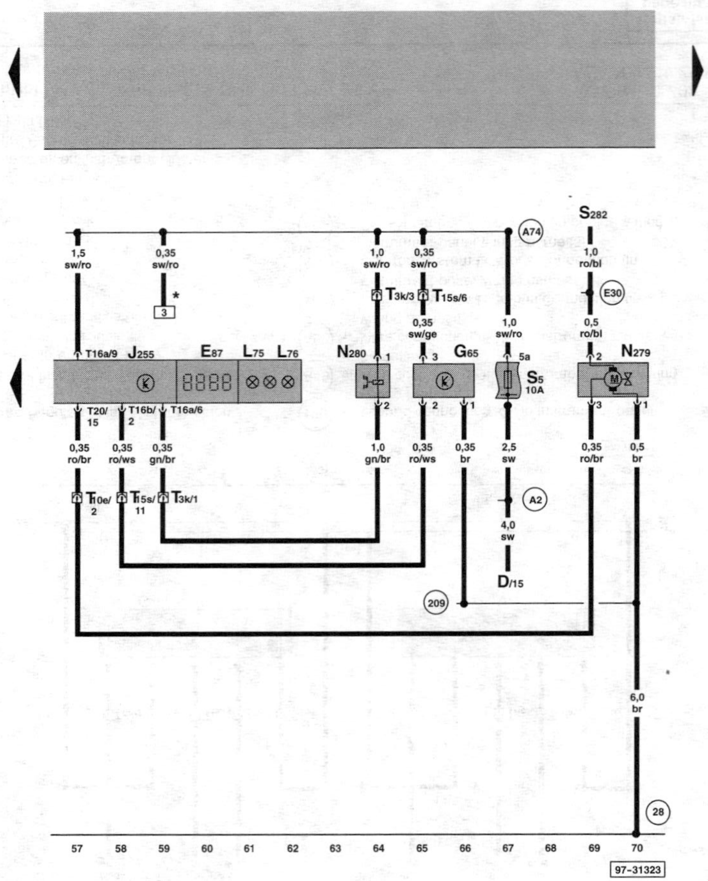

ws = white
sw = black
ro = red
br = brown
gn = green
bl = blue
gr = grey
li = lilac
ge = yellow
or = orange

D - Ignition/Starter Switch
E87 - A/C Control Head
G65 - High Pressure Sensor
J255 - Climatronic Control Module
L75 - Digital Display Light
L76 - Push Button Light
N279- Engine Coolant (EC) Switch-off Valve (heater)
N280- A/C Compressor Regulator Valve
S5 - Fuse
S282 - Engine Electronics Fuse
T3k - 3-Pin connector, grey, connector station A-pillar, left
T10e - 10-Pin connector, orange, on protective housing for control module, in engine compartment, left
T15s - 15-Pin connector, red, on protective housing for control module, in engine compartment, left

T16a - 16-Pin connector, black
T16b - 16-Pin connector, brown
T20 - 20-Pin Connector, red
(28) - Ground connection, on firewall
(209) - Ground connection -6-, in engine compartment wiring harness
(A2) - plus connection (15), in instrument panel wiring harness
(A74) - Connector (15a - fuse 5) in instrument panel wiring harness
(E30) - Connector (87a), in wiring harness engine
* - solar cell

Climatronic Control Module, High Pressure Sensor, Engine Coolant (EC) Switch-off Valve (heater), A/C Compressor Regulator Valve

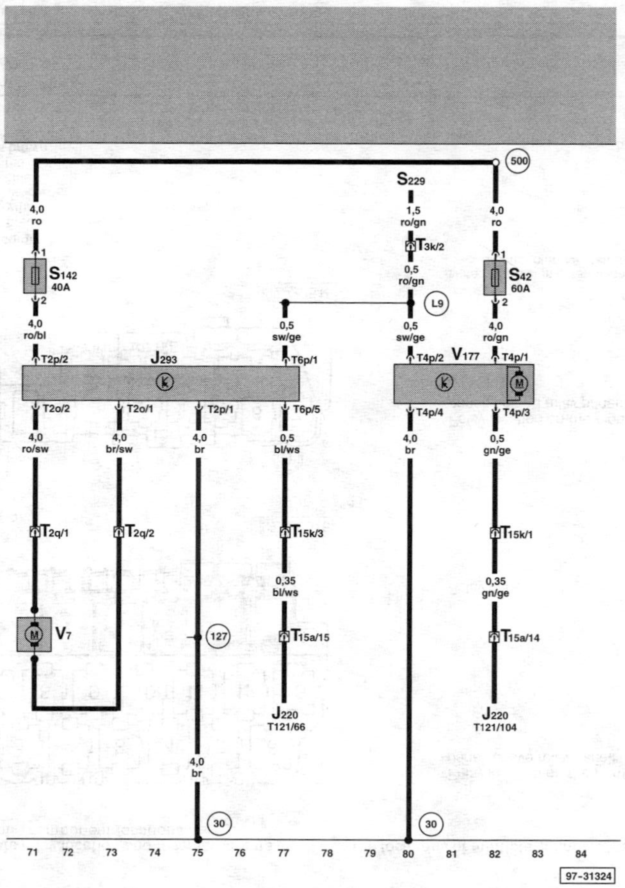

ws = white
sw = black
ro = red
br = brown
gn = green
bl = blue
gr = grey
li = lilac
ge = yellow
or = orange

J220 - Motronic Engine Control Module (ECM)
J293 - Coolant FC (Fan Control) Control Module
S42 - Coolant Fan Fuse
S142 - Control module fuse for coolant fan
S229 - Fuse in fuse holder
T2o - Double connector
T2p - Double connector
T2q - Double connector, in engine compartment, front left
T3k - 3-Pin connector, grey, connector station A-pillar, left
T4p - 4-Pin connector
T6p - 6-Pin connector
T15a - 15-Pin connector, white
T15k - 15-Pin connector, grey, connector station A-pillar, left
T121 - Connector, 121 point, on Motronic Engine Control Module (ECM)
V7 - Coolant Fan
V177 - Coolant Fan -2-
(30) - Ground connection, -1-, beside fuse/relay panel
(127) - Ground connection, -1-, in A/C compressor wiring harness
(500) - Threaded connection -1- (30) on the relay plate
(L9) - wire connection -1-, in A/C wiring harness

Edition 02/02
USA.5062.16.21

Coolant FC (Fan Control) Control Module, Coolant Fan, Coolant Fan 2

THIS PAGE INTENTIONALLY LEFT

BLANK

Anti-lock brake system (ABS) with electronic differential lock (EDL), anti-slip regulation (ASR) and electronic stabilization program (ESP)

from January 2002

Deviate relay location and fuseplacements as well as the locations of multiple connectors see section component locations.

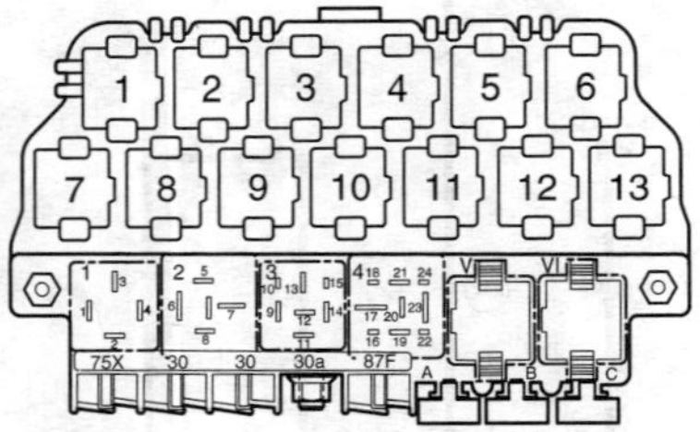

97-20676

Relay location on the thirteen position auxiliary relay panel, above relay panel:

Relay panel:

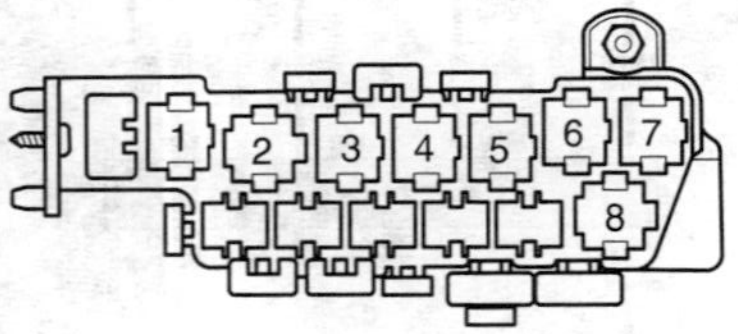

97-20515

Relay location on the eight position auxiliary relay panel, behind relay panel:

Note: Number in parentheses indicates production control number stamped on relay housing.

Fuse colors
30 A - green
25 A - white
20 A - yellow
15 A - blue
10 A - red
7,5 A - brown
5 A - beige

Edition 02/02
USA.5062.16.21

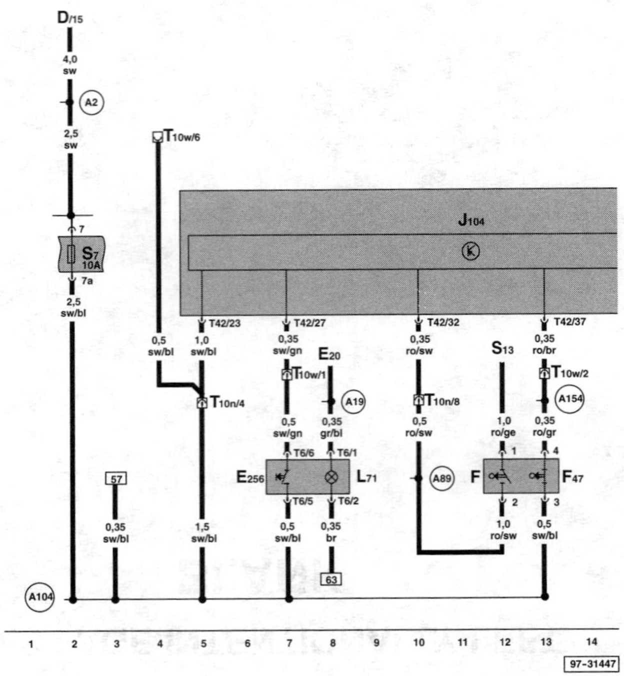

ws = white
sw = black
ro = red
br = brown
gn = green
bl = blue
gr = grey
li = lilac
ge = yellow
or = orange

- D - Ignition/Starter Switch
- E20 - Instrument Panel Light Dimmer Switch
- E256- ASR/ESP Button
- F - Brake Light Switch
- F47 - Brake Pedal Switch (cruise control/Diesel Direct Fuel Injection
- J104 - ABS Control Module (w/EDL)
- L71 - Anti-slip Control Switch Illumination
- S7 - Fuse
- S13 - Fuse
- T6 - 6-Pin Connector, red, on ASR/ESP Button
- T10n - 10-Pin Connector, orange, connector station A-pillar, left
- T10w - 10-Pin Connector, grey, connector station A-pillar, right
- T42 - 42-Pin Connector, on ABS Control Module

- (A19) - wire connection (58d), in instrument panel wiring harness
- (A89) - Connection 2 (54) (in instrument panel wiring harness)
- (A104) - Plus connector -2- (15), in instrument panel wiring harness
- (A154) - Brake Pedal Switch Connection (in instrument panel wiring harness)

ABS Control Module, Brake Light Switch, ASR/ESP Button

Edition 02/02
USA.5062.16.21

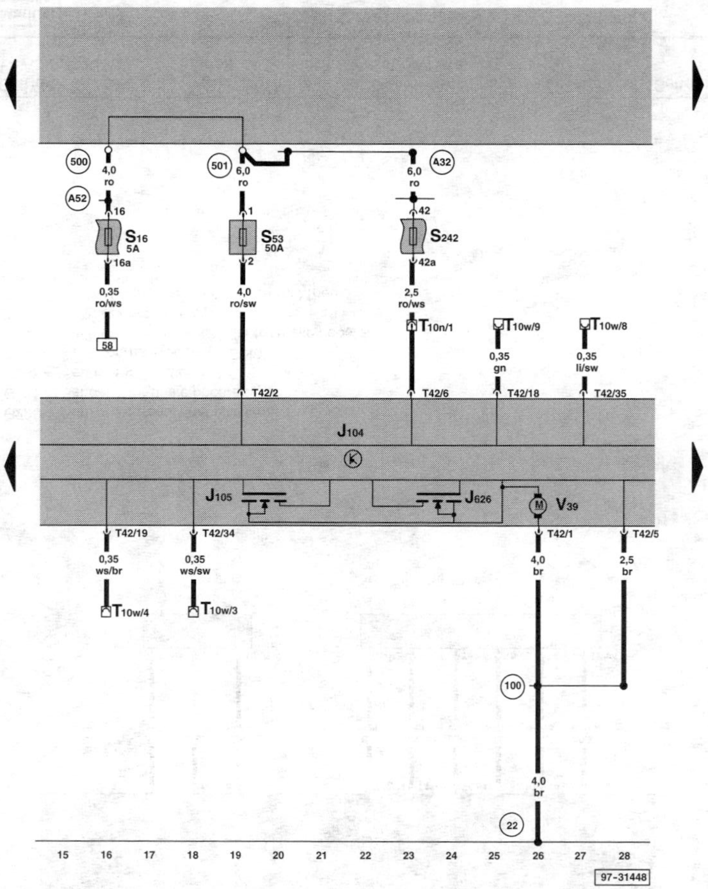

ws = white
sw = black
ro = red
br = brown
gn = green
bl = blue
gr = grey
li = lilac
ge = yellow
or = orange

J104 - ABS Control Module (w/EDL)
J105 - ABS Return Flow Pump Relay
J626 - ABS Return Flow Pump Relay 2
S16 - Fuse
S53 - ABS Hydraulic Pump Fuse
S242 - Fuse in fuse holder
T10n - 10-Pin Connector, orange, connector station A-pillar, left
T10w - 10-Pin Connector, grey, connector station A-pillar, right
T42 - 42-Pin Connector, on ABS Control Module
V39 - ABS Return Flow Pump

(22) - Ground connection, on hydraulic unit
(100) - Ground connection -1-, in ABS wiring harness
(500) - Threaded connection -1- (30) on the relay plate
(501) - Threaded connection -2- (30) on the relay plate
(A32) - plus connection (30), in instrument panel wiring harness
(A52) - Plus Connection 2 (30) (in instrument panel wiring harness)

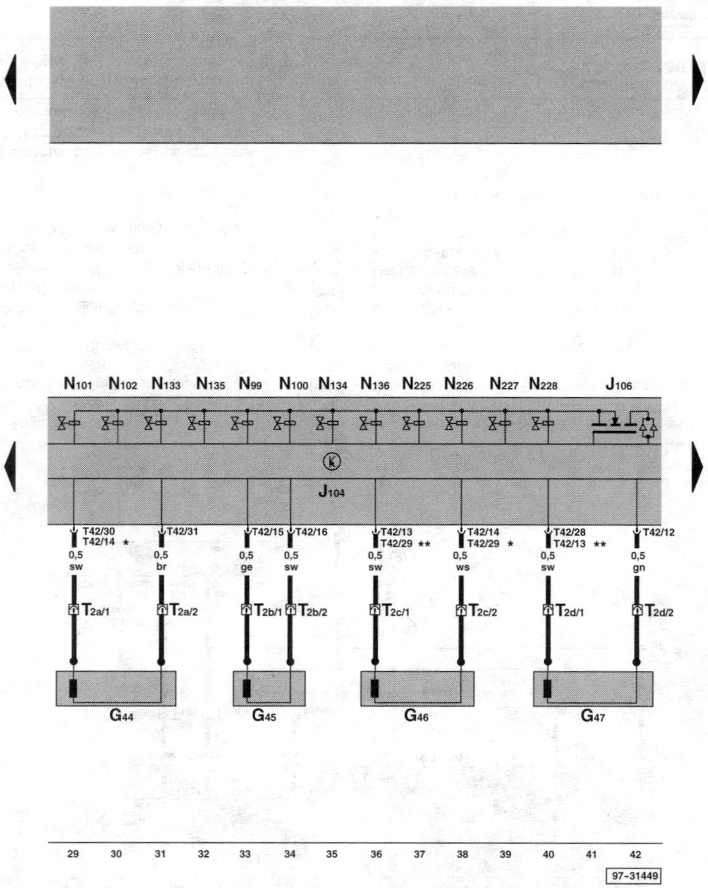

ws = white
sw = black
ro = red
br = brown
gn = green
bl = blue
gr = grey
li = lilac
ge = yellow
or = orange

G44 - Right Rear ABS Wheel Speed Sensor
G45 - Right Front ABS Wheel Speed Sensor
G46 - Left Rear ABS Wheel Speed Sensor
G47 - Left Front ABS Wheel Speed Sensor
J104 - ABS Control Module (w/EDL)
J106 - ABS Solenoid Valve Relay
N99 - Right Front ABS Inlet Valve
N100 - Right Front ABS Outlet Valve
N101 - Left Front ABS Inlet Valve
N102 - Left Front ABS Outlet Valve
N133 - Right Rear ABS Inlet Valve
N134 - Left Rear ABS Inlet Valve
N135 - Right Rear ABS Outlet Valve
N136 - Left Rear ABS Outlet Valve
N225 - Pilot valve -1- traction control
N226 - Pilot valve -2- traction control
N227 - High pressure switch valve -1- traction control
N228 - High pressure switch valve -2- traction control
T2a - 2-Pin Connector
T2b - 2-Pin Connector
T2c - 2-Pin Connector
T2d - 2-Pin connector
T42 - 42-Pin Connector, on ABS Control Module

* - 4 -motion only
** - code ATQ only

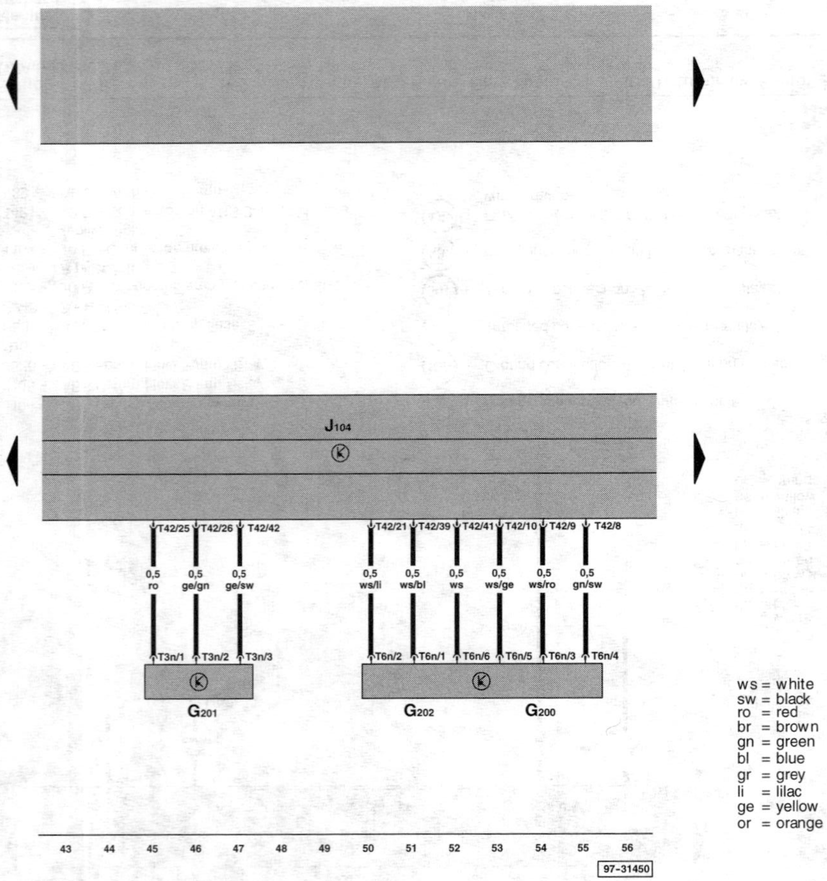

ws = white
sw = black
ro = red
br = brown
gn = green
bl = blue
gr = grey
li = lilac
ge = yellow
or = orange

G200 - Sensor for transverse acceleration
G201 - Sender 1 for brake booster
G202 - Sender for Rotation Rate
J104 - ABS Control Module (w/EDL)
T3n - 3-Pin Connector, on Sender for brake booster
T6n - 6-Pin Connector, black
T42 - 42-Pin Connector, on ABS Control Module

ABS Control Module, Sensor for transverse acceleretion, Sender 1 for brake booster, Sender for Rotation Rate

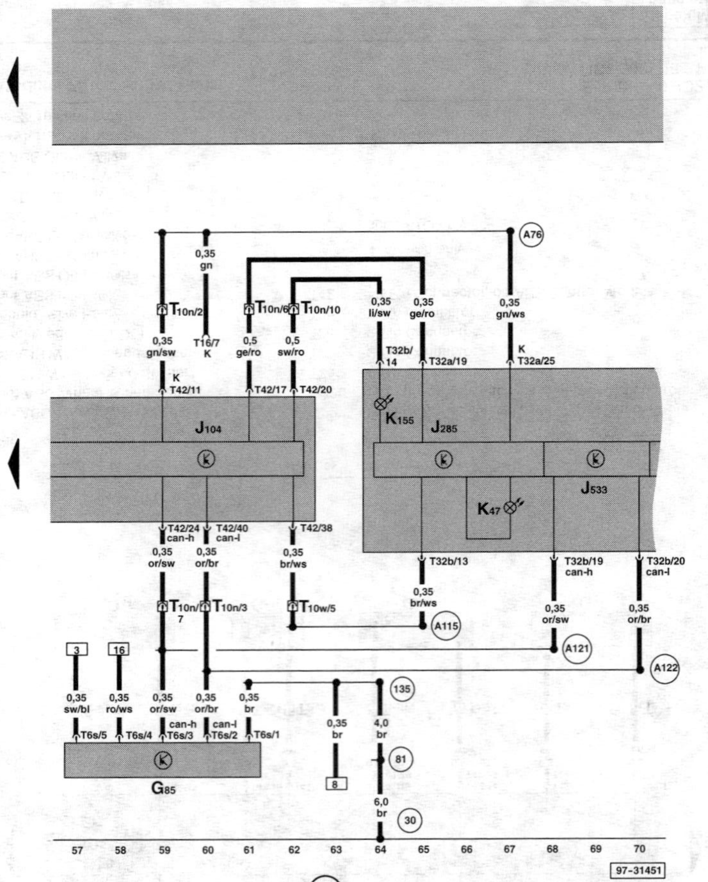

ws = white
sw = black
ro = red
br = brown
gn = green
bl = blue
gr = grey
li = lilac
ge = yellow
or = orange

G85 - Steering Angle Sensor
J104 - ABS Control Module (w/EDL)
J285 - Control module with indicator unit in instrument panel insert
J533 - Data Bus On Board Diagnostic Interface
K47 - ABS Warning Light
K155 - ASR/ESP Control Lamp
T6s - 6-Pin connector, black
T16 - 16-Pin connector, Data Link Connector (DLC), below instrument panel, left
T10n - 10-Pin Connector, orange, connector station A-pillar, left
T10w - 10-Pin Connector, grey, connector station A-pillar, right
T32a - 32-Pin Connector, blue, on instrument cluster
T32b - 32-Pin Connector, green, on instrument cluster
T42 - 42-Pin Connector, on ABS Control Module

(30) - Ground connection, -1-, beside fuse/relay panel
(81) - Ground connection -1-, in instrument panel wiring harness
(135) - Ground connection -2-, in instrument panel wiring harness
(A76) - Connector (K-diagnosis wire) in instrument panel wiring harness
(A115) - Connection (parking brake control) in instrument panel wiring harness
(A121) - Connection (high bus) in instrument panel wiring harness
(A122) - Connection (low bus) in instrument panel wiring harness

ABS Control Module, Control Module with indicator unit in instrument panel insert, Steering Angle Sensor, Data Link Connector (DLC)

High Intensity Gas Discharge (HID) Headlights with Automatic Vertical Headlight Aim Control

from January 2002

Deviate relay location and fuseplacements as well as the locations of multiple connectors see section "component locations".

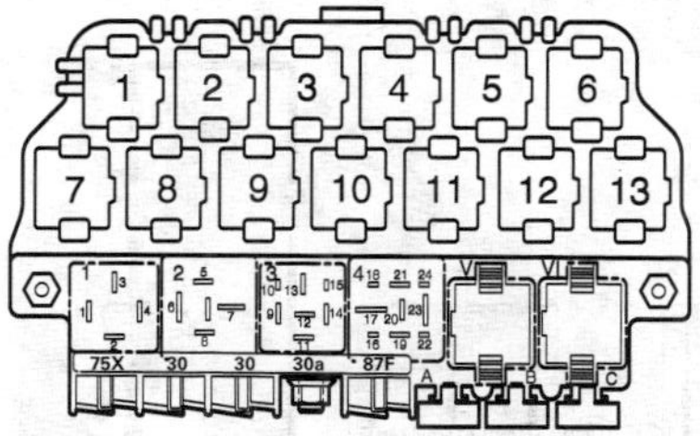

97-20676

Relay location on the thirteen position auxiliary relay panel, above relay panel:

Relay panel:

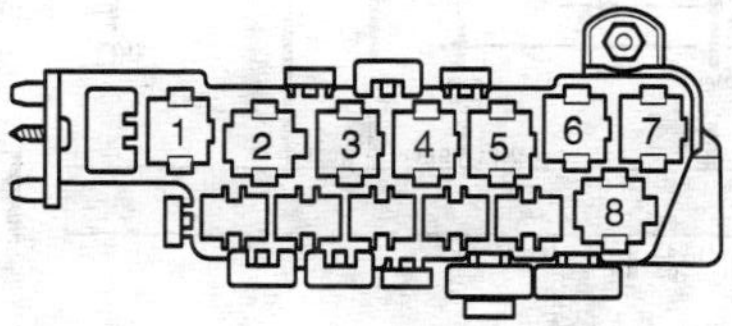

97-20515

Relay location on the eight position auxiliary relay panel, behind relay panel:

Note: Number in parentheses indicates production control number stamped on relay housing.

Fuse colors

30 A - green
25 A - white
20 A - yellow
15 A - blue
10 A - red
7,5 A - brown
5 A - beige

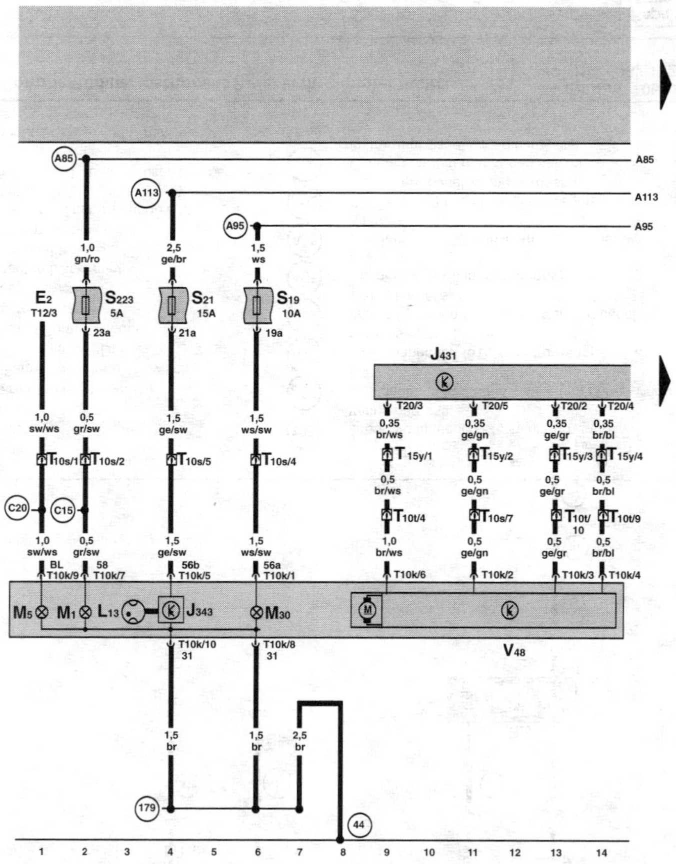

97-32085

ws = white
sw = black
ro = red
br = brown
gn = green
bl = blue
gr = grey
li = lilac
ge = yellow
or = orange

E2 - Turn signal switch
J343 - Left High-intensity Gas Discharge Lamp Control Module
J431 - Headlight Range Control Module
L13 - High intensity gas discharge lamp
M1 - Left Parking Light
M5 - Left Front Turn Signal Light
M30 - Left High Beam Headlight
S19 - Fuse
S21 - Fuse
S223 - Fuse in fuse holder
T10k - 10-Pin Connector, on headlight left
T10s - 10-Pin Connector, rose
T10t - 10-Pin Connector, red
T12 - 12-Pin Connector, black
T15y - 15-Pin Connector, black
T20 - 20-Pin Connector
V48 - Left Headlight Beam Adjusting Motor
(44) - Ground connection, on left A-pillar, lower part
(179) - Ground connection, in left headlight wiring harness
(A85) - Connector (58R) in Instrument panel wiring harness
(A95) - Connection 1 (56a), in instrument panel wiring harness
(A113) - Connection 2 (daytime running lights), in instrument panel wiring harness
(C15) - Wire connection -1-, in headlight wiring harness
(C20) - Wire connection (left turn signal), in headlight wiring harness

Headlight Range Control Module, Left High - intensity Gas Discharge Lamp Control

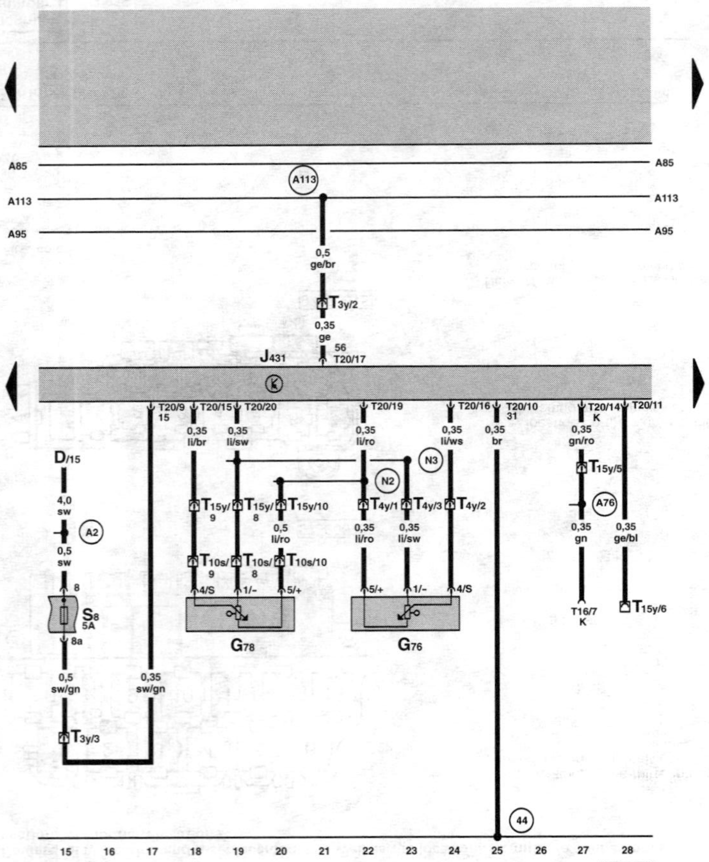

ws = white
sw = black
ro = red
br = brown
gn = green
bl = blue
gr = grey
li = lilac
ge = yellow
or = orange

D - Ignition/Starter Switch
G76 - Left Rear Level Control System Sensor
G78 - Left Front Level Control System Sensor
J431 - Headlight Range Control Module
S8 - Fuse
T3y - 3-Pin Connector, black
T4y - 4-Pin Connector
T10s - 10-Pin Connector, rose
T15y - 15-Pin Connector, black
T16 - 16-Pin Connector, Data Link Connector (DLC), below instrument panel, left
T20 - 20-Pin Connector
(44) - Ground connection, on left A-pillar, lower part
(A2) - plus connection (15), in instrument panel wiring harness
(A76) - Connector (K-diagnosis wire) in instrument panel wiring harness
(A113) - Connection 2 (daytime running lights), in instrument panel wiring harness
(N2) - Connection 1 (in headlight range control wiring harness)
(N3) - Connection 2 (in headlight range control wiring harness)

Headlight Range Control Module, Left Rear Level Control System Sensor, Left Front Level Control System Sensor

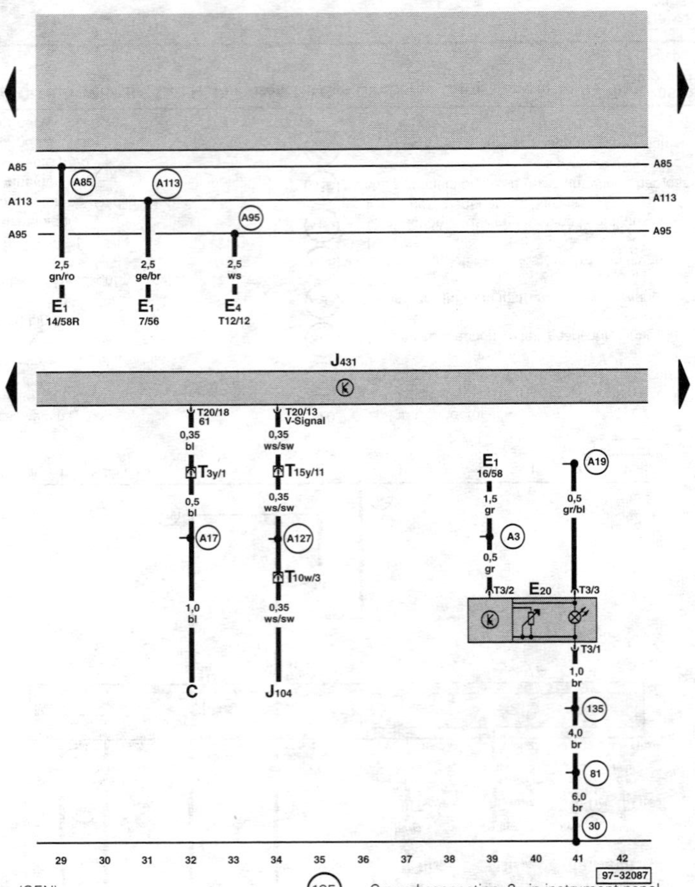

ws = white
sw = black
ro = red
br = brown
gn = green
bl = blue
gr = grey
li = lilac
ge = yellow
or = orange

C - Generator (GEN)
E1 - Light switch
E4 - Headlight Dimmer/Flasher Switch
E20 - Instrument Panel Light Dimmer Switch
J104 - ABS Control Module (w/EDL)
J431 - Headlight Range Control Module
T3 - 3-Pin Connector, black
T3y - 3-Pin Connector, black
T10w - 10-Pin Connector, black, grey
T12 - 12-Pin Connector, black
T15y - 15-Pin Connector, black
T20 - 20-Pin Connector, on control module for Headlight Range
(30) - Ground connection, -1-, beside fuse/relay panel
(81) - Ground connection -1-, in instrument panel wiring harness
(135) - Ground connection -2-, in instrument panel wiring harness
(A3) - plus connection (58), in instrument panel wiring harness
(A17) - wire connection (61), in instrument panel wiring harness
(A19) - wire connection (58d), in instrument panel wiring harness
(A85) - Connector (58R) in Instrument panel wiring harness
(A95) - Connection 1 (56a), in instrument panel wiring harness
(A113) - Connection 2 (daytime running lights), in instrument panel wiring harness
(A127) - Connection (left rear speed sensor) in instrument panel wiring harness

Headlight Range Control Module, Instrument Panel Light Dimmer Switch

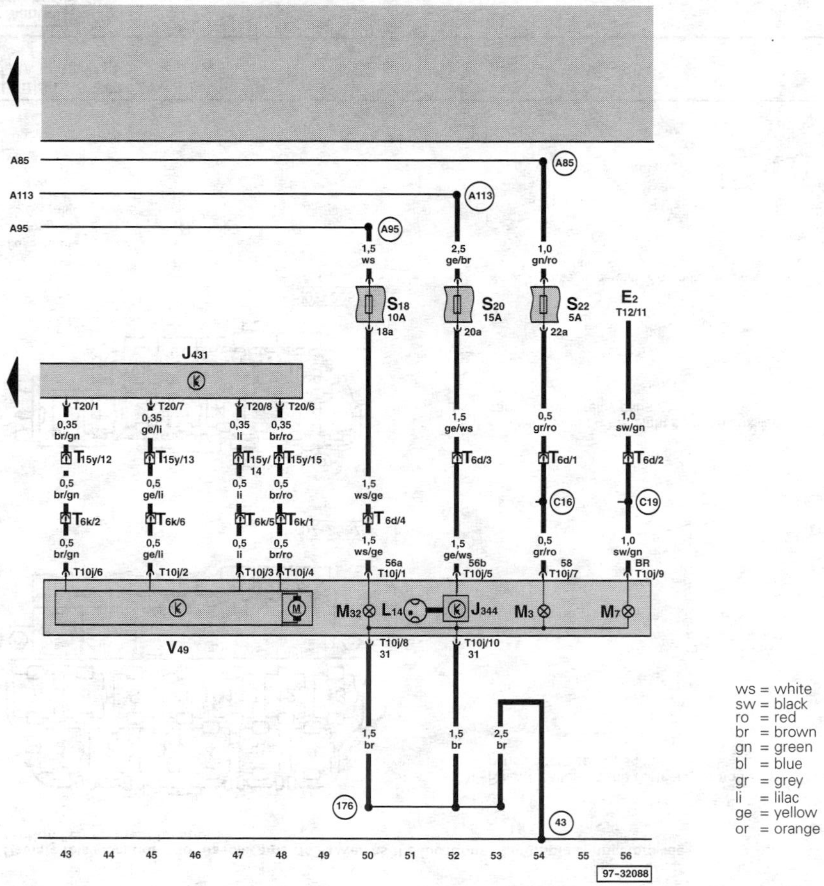

ws = white
sw = black
ro = red
br = brown
gn = green
bl = blue
gr = grey
li = lilac
ge = yellow
or = orange

E2 - Turn signal switch
J344 - Right High-intensity Gas Discharge Lamp Control Module
J431 - Headlight Range Control Module
L14 - High intensity gas discharge lamp
M3 - Right Parking Light
M7 - Right Front Turn Signal Light
M32 - Right High Beam Headlight
S18 - Fuse
S20 - Fuse
S22 - Fuse
T6d - 6-Pin Connector
T6k - 6-Pin Connector, red
T10j - 10-Pin Connector, on headlight left
T12 - 12-Pin Connector, black
T15y - 15-Pin Connector, black
T20 - 20-Pin Connector
V49 - Right Headlight Beam Adjusting Motor
(43) - Ground connection, on right A-pillar, lower part
(176) - Ground connection, in right headlight wiring harness
(A85) - Connector (58R) in Instrument panel wiring harness
(A95) - Connection 1 (56a), in instrument panel wiring harness
(A113) - Connection 2 (daytime running lights), in instrument panel wiring harness
(C16) - Wire connection -2-, in headlight wiring harness
(C19) - Connector -1- (right turn signal), in wiring harness headlamp

Edition 05/02 **Headlight Range Control Module, Right High - intensity Gas Discharge Lamp**
USA.5062.17.21

THIS PAGE INTENTIONALLY LEFT
BLANK

Diversity Antenna System

from January 2002

Deviate relay location and fuseplacements as well as the locations of multiple connectors see section "component locations".

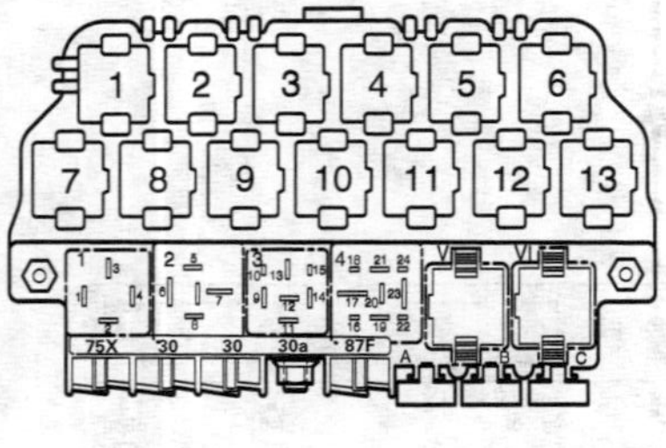

97-20676

Relay location on the thirteen position auxiliary relay panel, above relay panel:

Relay panel:

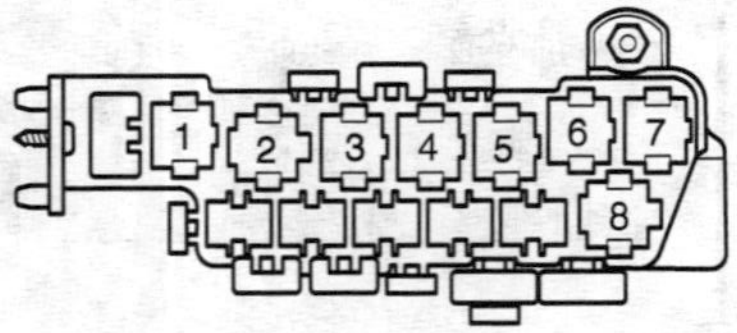

97-20515

Relay location on the eight position auxiliary relay panel, behind relay panel:

Note: Number in parentheses indicates production control number stamped on relay housing.

Fuse colors
30 A - green
25 A - white
20 A - yellow
15 A - blue
10 A - red
7,5 A - brown
5 A - beige

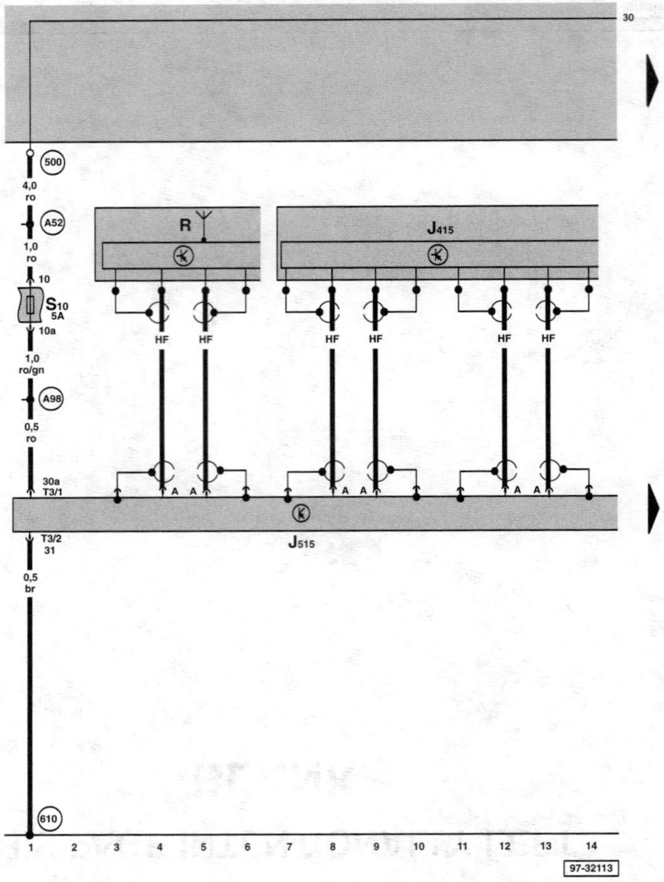

ws = white
sw = black
ro = red
br = brown
gn = green
bl = blue
gr = grey
li = lilac
ge = yellow
or = orange

J415 - Tuner for navigation/TV*
J515 - Antenna Selection Control Module
R - Radio
S10 - Fuse
T3 - 3-Pin Connector

* - Where applicable

(500) - Threaded connection -1- (30) on the relay plate

(610) - Ground Connection (Audio) (under center console, front)

(A52) - Plus Connection 2 (30) (in instrument panel wiring harness)

(A98) - Plus connector -4- (30), in instrument panel wiring harness

Antenna Selection Control Module, Tuner for navigation/TV, Radio

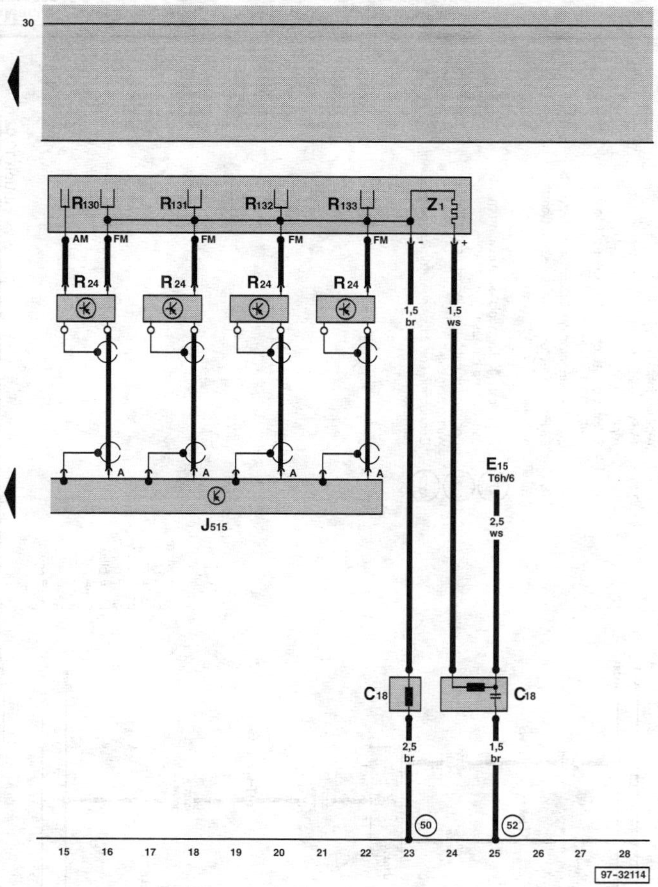

ws = white
sw = black
ro = red
br = brown
gn = green
bl = blue
gr = grey
li = lilac
ge = yellow
or = orange

C18 - Windshield Antenna Suppression Filter
E15 - Rear window defogger switch
J515 - Antenna Selection Control Module
R24 - Antenna Amplifier
R130 - Rear Window Antenna 1
R131 - Rear Window Antenna 2
R132 - Rear Window Antenna 3
R133 - Rear Window Antenna 4
T6h - 6-Pin Connector
Z1 - Heated rear window

(50) - Ground connection, in luggage compartment, left
(52) - Ground connection, in rear lid, left

Edition 05/02
USA.5062.17.21

Antenna Selection Control Module, Windshield Antenna Suppression Filter, Rear Window Antenna 1-4

THIS PAGE INTENTIONALLY LEFT BLANK

CAN-Bus - Powertrain, Comfort System and Data Link Connector (DLC)

from October 2000

Deviate relay location and fuseplacements as well as the locations of multiple connectors see section "component locations"

Relay location on the thirteen position auxiliary relay panel, above relay panel:

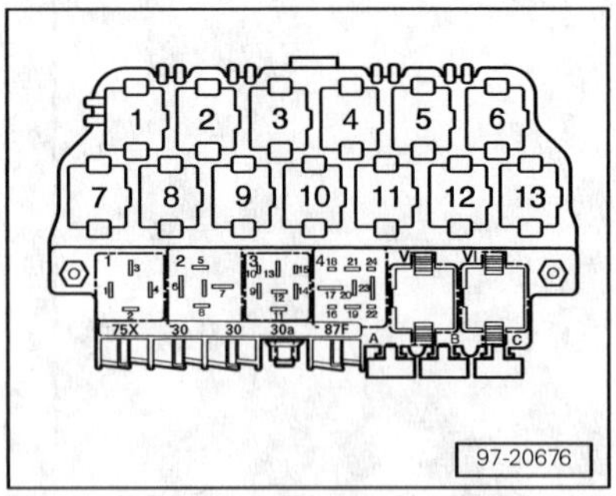

Relay panel:

Relay location on the eight position auxiliary relay panel, above relay panel:

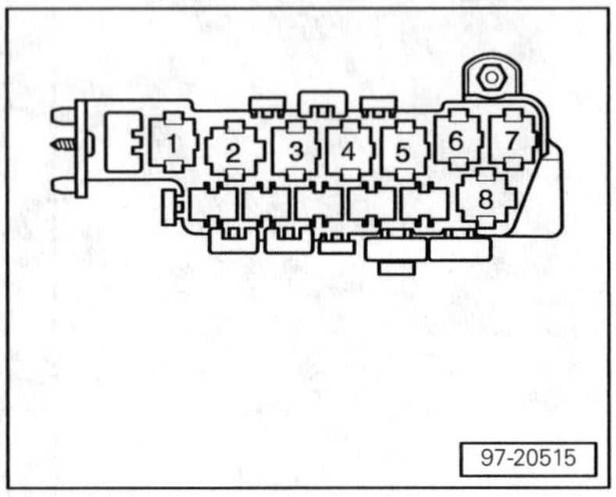

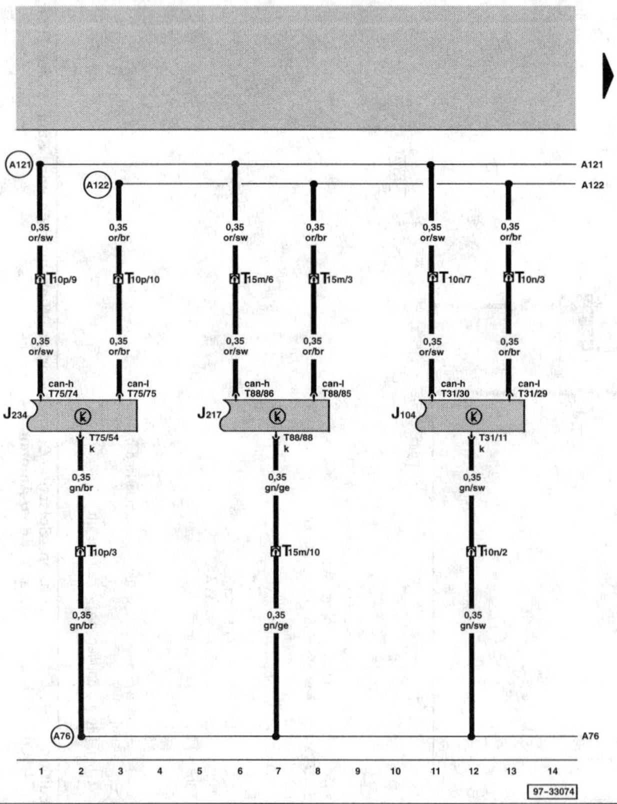

ABS Control Module, Transmission Control Module, Airbag Control Module

- J104 - ABS Control Module (w/EDL)
- J217 - Transmission Control Module (TCM)
- J234 - Airbag Control Module
- T10n - 10-Pin Connector, orange, connector station A-pillar, left
- T10p - 10-Pin Connector, yellow, connector station A-pillar, right
- T15m- 15-Pin Connector, brown, connector station A-pillar, right
- T31 - 31-Pin Connector, on ABS Control Module (w/EDL)
- T75 - 75-Pin Connector, on Airbag Control Module
- T88 - 88-Pin Connector, on Transmission Control Module (TCM)
- (A76) - Connector (K-diagnosis wire) in instrument panel wiring harness
- (A121) - Connection (high bus) in instrument panel wiring harness
- (A122) - Connection (low bus) in instrument panel wiring harness

ws = white
sw = black
ro = red
br = brown
gn = green
bl = blue
gr = grey
li = lilac
ge = yellow
or = orange

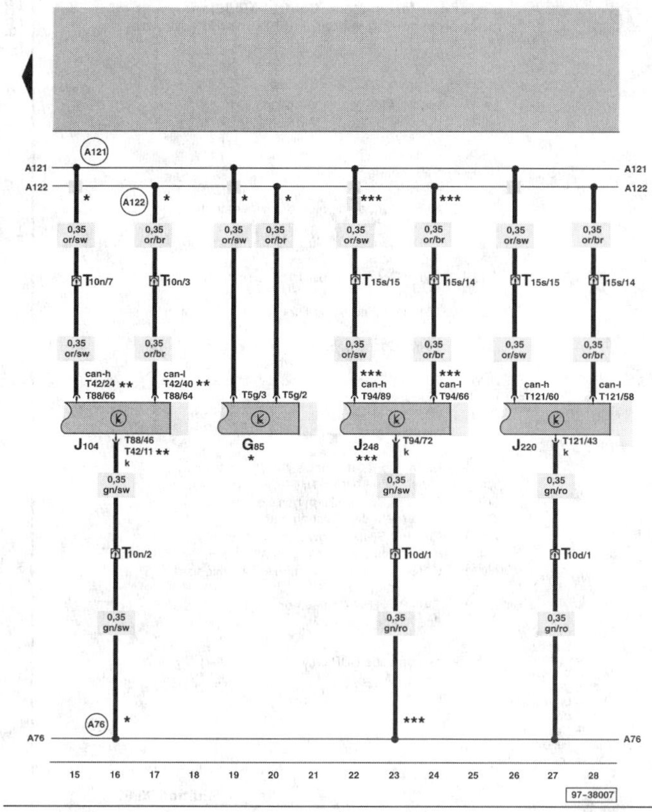

ABS Control Module, Motronic Engine Control Module, Steering Angle Sensor, Diesel Direct Fuel Injection Engine Control Module

- G85 - Steering Angle Sensor
- J104 - ABS Control Module (w/EDL) *
- J220 - Motronic Engine Control Module (ECM), on protective housing for control module, in engine compartment, left
- J248 - Diesel Direct Fuel Injection (DFI) Engine Control Module (ECM)
- T5g - 5-Pin Connector
- T10d - 10-Pin Connector, brown, on protective housing for control module, in engine compartment, left
- T10n - 10-Pin Connector, orange, connector station A-pillar, left
- T15s - 15-Pin Connector, red, on protective housing for control module, in engine compartment,left
- T42 - 42-Pin connector
- T88 - 88-Pin Connector, on Transmission Control Module (TCM)
- T94 - 94-Pin connector
- T121 - 121-Pin Connector, on Motronic Engine Control Module (ECM)

- (A76) - Connector (K-diagnosis wire) in instrument panel wiring harness
- (A121) - Connection (high bus) in instrument panel wiring harness
- (A122) - Connection (low bus) in instrument panel wiring harness

- * - ESP only
- ** - from January 2002
- *** - engine code BHW only

ws = white
sw = black
ro = red
br = brown
gn = green
bl = blue
gr = grey
li = lilac
ge = yellow
or = orange

Edition 12.03

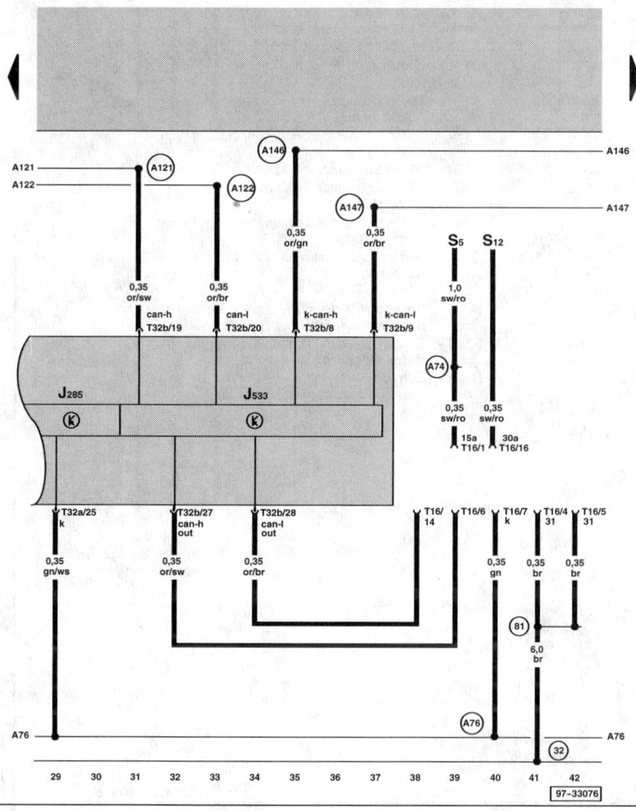

Control Module with indicator unit in instrument panel insert, Data Bus On Board Diagnostic Interface, Data Link Connector (DLC)

- J285 - Control module with indicator unit in instrument panel insert
- J533 - Data Bus On Board Diagnostic Interface
- S5 - Fuse
- S12 - Fuse
- T16 - 16-Pin Connector, Data Link Connector (DLC)
- T32a - 32-Pin Connector, blue, on instrument cluster
- T32b - 32-Pin Connector, green, on instrument cluster

- (32) - Ground connection, behind instrument panel, left
- (81) - Ground connection -1-, in instrument panel wiring harness
- (A74) - Connector (15a - fuse 5) in instrument panel wiring harness
- (A76) - Connector (K-diagnosis wire) in instrument panel wiring harness
- (A121) - Connection (high bus) in instrument panel wiring harness
- (A122) - Connection (low bus) in instrument panel wiring harness
- (A146) - Comfort System High-bus Connection (in instrument panel wiring harness)
- (A147) - Comfort System Low-bus Connection (in instrument panel wiring harness)

ws = white
sw = black
ro = red
br = brown
gn = green
bl = blue
gr = grey
li = lilac
ge = yellow
or = orange

Edition 10.02

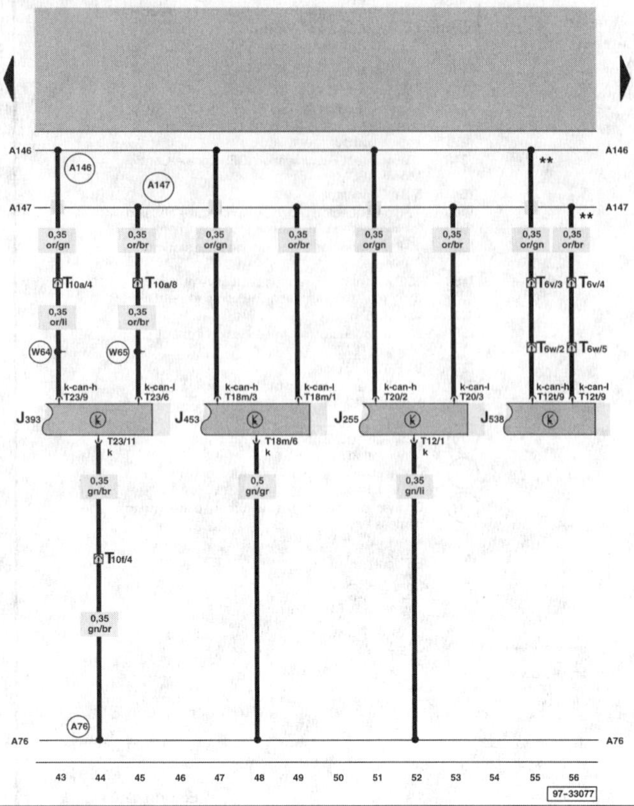

Climatronic Control Module, Central Control Module for comfort system, Control Module for multi-function streering wheel, Fuel Pump (FP) Control Module

- J255 - Climatronic Control Module
- J393 - Central control module for comfort system
- J453 - Control module for multi-function steering wheel
- J538 - Fuel Pump (FP) Control Module
- T6v - 6-Pin connector, lilac
- T6w - 6-Pin connector
- T10a - 10-Pin Connector, brown, connector station A-pillar, left
- T10f - 10-Pin Connector, blue, connector station A-pillar, left
- T12 - 12-Pin Connector, black, on Climatronic Control Module
- T12t - 12-Pin connector, on Fuel Pump (FP) Control Module
- T18m- 18-Pin Connector, black, on Control module for multi-function steering wheel
- T20 - 20-Pin Connector, red, on Climatronic Control Module
- T23 - 23-Pin Connector, black, on Central control module for comfort system

- (A76) - Connector (K-diagnosis wire) in instrument panel wiring harness
- (A146) - Comfort System High-bus Connection (in instrument panel wiring harness)
- (A147) - Comfort System Low-bus Connection (in instrument panel wiring harness)
- (W64) - CAN-Bus High Connection - Convenience (in floor wiring harness)
- (W65) - CAN-Bus Low Connection - Convenience (in floor wiring harness)

** - engine code BDP only

ws = white
sw = black
ro = red
br = brown
gn = green
bl = blue
gr = grey
li = lilac
ge = yellow
or = orange

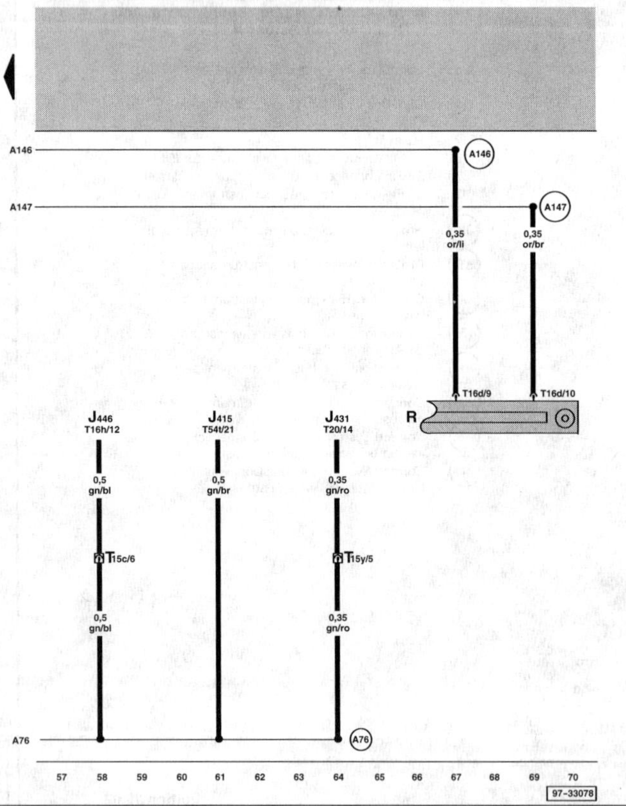

Radio

- J415 - Tuner for navigation/TV
- J431 - Headlight Range Control Module
- J446 - Control module for parking aid
- R - Radio
- T15c - 15-Pin Connector, red, connector station A-pillar, right
- T15y - 15-Pin Connector, black, connector station A-pillar, right
- T16d - 16-Pin Connector, black, on Radio
- T16h - 16-Pin Connector, on Control module for parking aid
- T20 - 20-Pin Connector, red, on Headlight Range Control Module
- T54t - 54-Pin Connector, black, on Tuner for navigation/TV

- (A76) - Connector (K-diagnosis wire) in instrument panel wiring harness
- (A146) - Comfort System High-bus Connection (in instrument panel wiring harness)
- (A147) - Comfort System Low-bus Connection (in instrument panel wiring harness)

ws = white
sw = black
ro = red
br = brown
gn = green
bl = blue
gr = grey
li = lilac
ge = yellow
or = orange

High Intensity Gas Discharge (HID) Headlights, only engine code BDP

from January 2002

Deviate relay location and fuseplacements as well as the locations of multiple connectors see section "component locations"

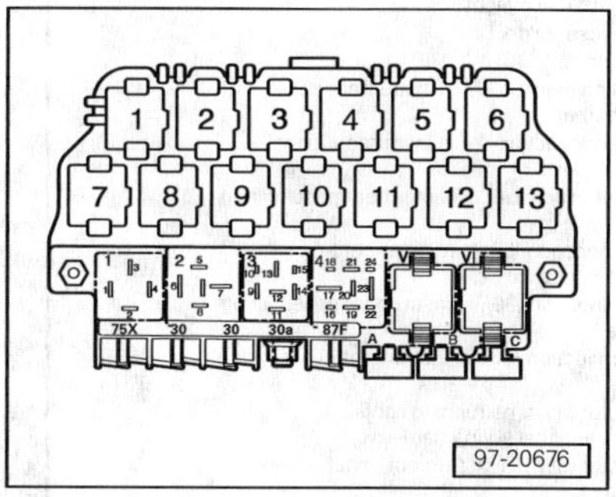

97-20676

Relay location on the thirteen position auxiliary relay panel, above relay panel:

5 - J89 - Daytime running Lights Change-over Relay (173)

7 - J9 - Rear Window Defogger Relay (381)

Relay panel:

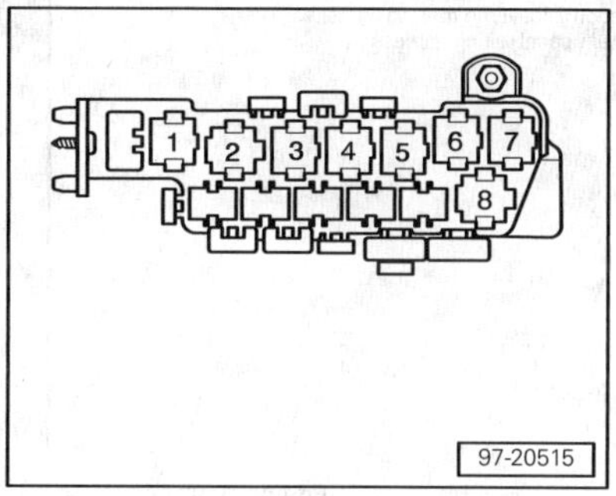

97-20515

Relay location on the eight position auxiliary relay panel, above relay panel:

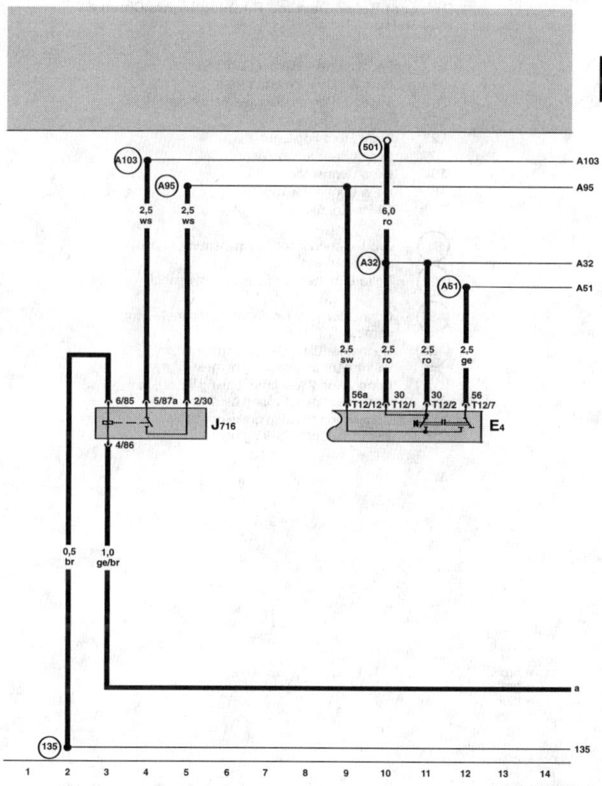

97-33079

Headlight Dimmer/Flasher Switch, Additional High Beam Switch - off Relay

E4 - Headlight Dimmer/Flasher Switch
J716 - Additional High Beam Switch - off Relay
T12 - 12-Pin Connector, black

(135) - Ground connection -2-, in instrument panel wiring harness
(501) - Threaded connection -2- (30) on the relay plate
(A32) - plus connection (30), in instrument panel wiring harness
(A51) - wire connection (56), in instrument panel wiring harness
(A95) - Connection 1 (56a), in instrument panel wiring harness
(A103) - Connector -2- (56), in instrument panel wiring harness

ws = white
sw = black
ro = red
br = brown
gn = green
bl = blue
gr = grey
li = lilac
ge = yellow
or = orange

Passat — Wiring diagram — No. 79/3

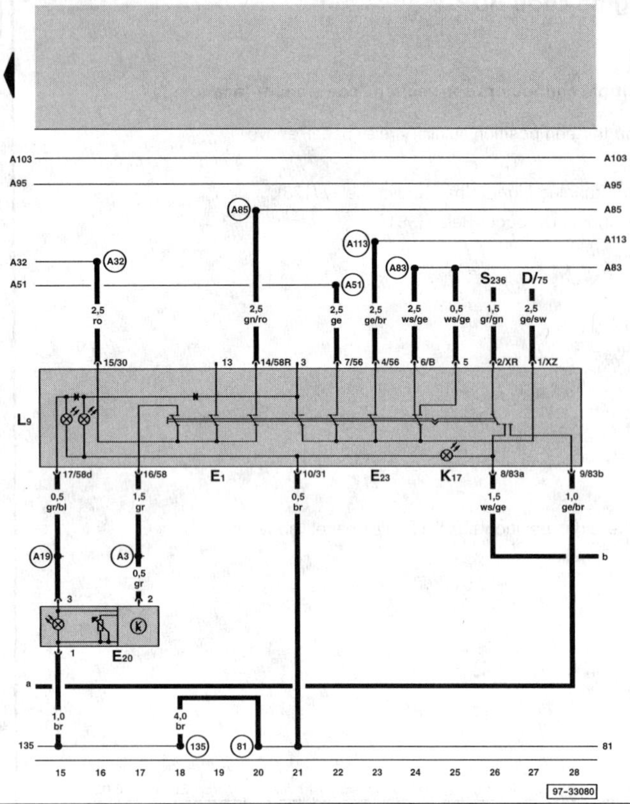

Light switch, Instrument Panel Light Dimmer Switch, Fog Light Switch, Fog Light Indicator Light, Headlight Switch Light

D - Ignition/Starter Switch
E1 - Light switch
E20 - Instrument Panel Light Dimmer Switch
E23 - Fog Light Switch
K17 - Fog Light Indicator Light
L9 - Headlight Switch Light
S236 - Fuse in fuse holder

(81) - Ground connection -1-, in instrument panel wiring harness
(135) - Ground connection -2-, in instrument panel wiring harness
(A3) - plus connection (58), in instrument panel wiring harness
(A19) - wire connection (58d), in instrument panel wiring harness
(A32) - plus connection (30), in instrument panel wiring harness
(A51) - wire connection (56), in instrument panel wiring harness
(A83) - Connector (daytime running lights), in instrument panel wiring harness
(A85) - Connector (58R) in Instrument panel wiring harness
(A113) - Connection 2 (daytime running lights), in instrument panel wiring harness

ws = white
sw = black
ro = red
br = brown
gn = green
bl = blue
gr = grey
li = lilac
ge = yellow
or = orange

Edition 10.02

Passat — Wiring diagram — No. 79/4

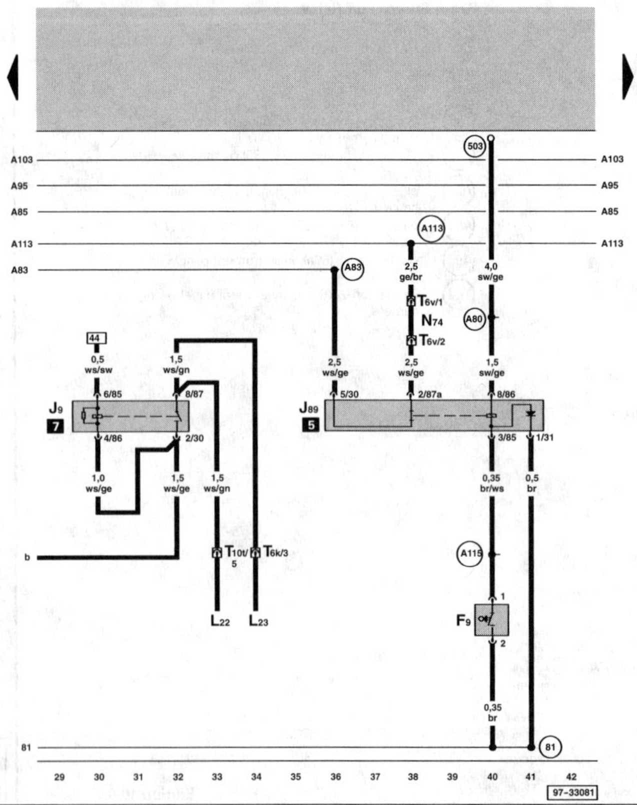

Parking Brake Warning Light Switch, Rear Window Defogger Relay, Daytime Running Lights Change - over Relay

F9 - Parking Brake Warning Light Switch
J9 - Rear Window Defogger Relay
J89 - Daytime Running Lights Change-over Relay
L22 - Left Front Fog Light
L23 - Right Front Fog Light
N74 - Series Resistance Wiring
T6v - 6-Pin Connector, lilac
T6k - 6-Pin Connector, red
T10t - 10-Pin Connector, red

(81) - Ground connection -1-, in instrument panel wiring harness
(503) - Bolted Connection (75x) (at relay panel)
(A80) - Connector -1- (X) in instrument panel wiring harness
(A83) - Connector (daytime running lights), in instrument panel wiring harness
(A113) - Connection 2 (daytime running lights), in instrument panel wiring harness
(A115) - Connection (parking brake control) in instrument panel wiring harness

ws = white
sw = black
ro = red
br = brown
gn = green
bl = blue
gr = grey
li = lilac
ge = yellow
or = orange

Edition 10.02

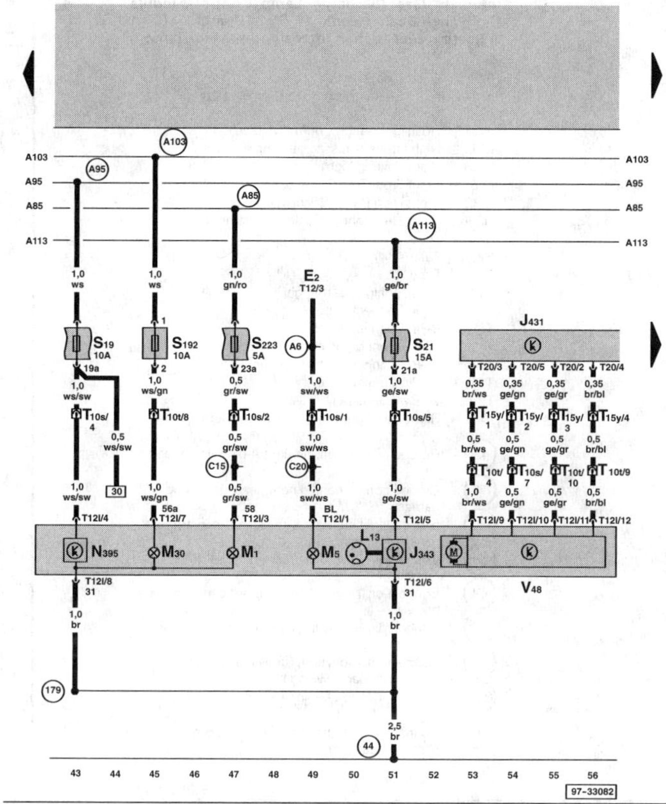

ws = white
sw = black
ro = red
br = brown
gn = green
bl = blue
gr = grey
li = lilac
ge = yellow
or = orange

Headlight Range Control Module, Left High - intensity Gas Discharge Lamp Control Module, Left High Beam Headlight, Left Headlight Beam Adjusting Motor, High intensity gas discharge lamp

- E2 - Turn signal switch
- J343 - Left High-intensity Gas Discharge Lamp Control Module
- J431 - Headlight Range Control Module
- L13 - High intensity gas discharge lamp
- M1 - Left Parking Light
- M5 - Left Front Turn Signal Light
- M30 - Left High Beam Headlight
- N395 - Left Headlight Reflector Adjustment Solenoid
- S19 - Fuse
- S21 - Fuse
- S192 - Fuse for high beam, left
- S223 - Fuse in fuse holder
- T10s - 10-Pin Connector, rose
- T10t - 10-Pin Connector, red
- T12 - 12-Pin Connector, black
- T12l - 12-Pin Connector, on Left Headlight
- T15y - 15-Pin Connector, black
- T20 - 20-Pin Connector
- V48 - Left Headlight Beam Adjusting Motor

- (44) - Ground connection, on left A-pillar, lower part
- (179) - Ground connection, in right headlight wiring harness
- (A6) - plus connection (left turn signal), in instrument panel wiring harness
- (A85) - Connector (58R) in Instrument panel wiring harness
- (A95) - Connection 1 (56a), in instrument panel wiring harness
- (A103) - Connector -2- (56), in instrument panel wiring harness
- (A113) - Connection 2 (daytime running lights), in instrument panel wiring harness
- (C15) - Wire connection -1-, in headlight wiring harness
- (C20) - Wire connection (left turn signal), in headlight wiring harness

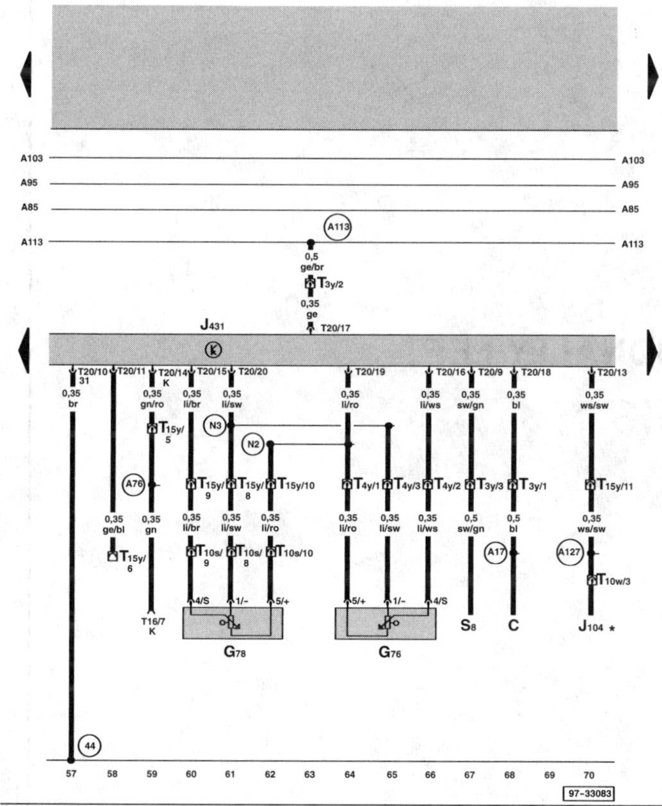

ws = white
sw = black
ro = red
br = brown
gn = green
bl = blue
gr = grey
li = lilac
ge = yellow
or = orange

Headlight Range Control Module, Left Rear Level Control System Sensor, Left Front Level Control System Sensor, Data Link Connector (DLC)

- C - Generator (GEN)
- G76 - Left Rear Level Control System Sensor
- G78 - Left Front Level Control System Sensor
- J104 - ABS Control Module (w/EDL)
- J431 - Headlight Range Control Module
- S8 - Fuse
- T3y - 3-Pin Connector, black
- T4y - 4-Pin Connector
- T10s - 10-Pin Connector, rose
- T10w - 10-Pin Connector
- T15y - 15-Pin Connector, black
- T16 - 16-Pin Connector, Data Link Connector (DLC), below instrument panel, left
- T20 - 20-Pin Connector

- (44) - Ground connection, on left A-pillar, lower part
- (A17) - wire connection (61), in instrument panel wiring harness
- (A76) - Connector (K-diagnosis wire) in instrument panel wiring harness
- (A113) - Connection 2 (daytime running lights), in instrument panel wiring harness
- (A127) - Connection (left rear speed sensor) in instrument panel wiring harness
- (N2) - Connection 1 (in headlight range control wiring harness)
- (N3) - Connection 2 (in headlight range control wiring harness)

* - Navigation only

Headlight Range Control Module, Right High - intensity Gas Discharge Lamp Control Module, Right High Beam Headlight, Right Headlight Beam Adjusting Motor, High intensity dischaege lamp

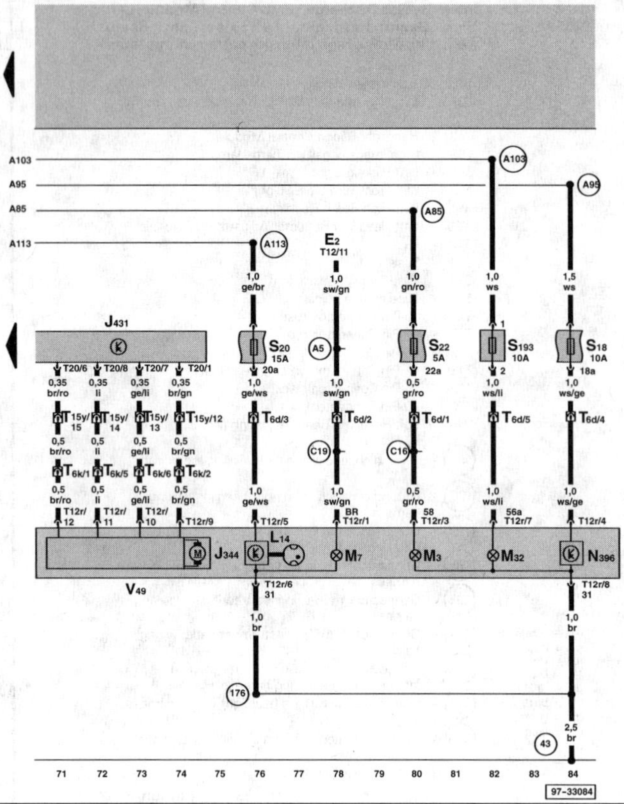

E2 - Turn signal switch
J344 - Right High-intensity Gas Discharge Lamp Control Module
J431 - Headlight Range Control Module
L14 - High intensity discharge lamp
M3 - Right Parking Light
M7 - Right Front Turn Signal Light
M32 - Right High Beam Headlight
N396 - Right Headlight Reflector Adjustment Solenoid
S18 - Fuse
S20 - Fuse
S22 - Fuse
S193 - Fuse for high beam, right
T6d - 6-Pin Connector
T6k - 6-Pin Connector, red
T12 - 12-Pin Connector, black
T12r - 12-Pin Connector, on headlight left
T15y - 15-Pin Connector, black
T20 - 20-Pin Connector
V49 - Right Headlight Beam Adjusting Motor

(43) - Ground connection, on right A-pillar, lower part
(176) - Ground connection, in right headlight wiring harness
(A5) - plus connection (right turn signal), in instrument panel wiring harness
(A85) - Connector (58R) in Instrument panel wiring harness
(A95) - Connection 1 (56a), in instrument panel wiring harness
(A103) - Connector -2- (56), in instrument panel wiring harness
(A113) - Connection 2 (daytime running lights), in instrument panel wiring harness
(C16) - Wire connection -2-, in headlight wiring harness
(C19) - Connector -1- (right turn signal), in wiring harness headlamp

ws = white
sw = black
ro = red
br = brown
gn = green
bl = blue
gr = grey
li = lilac
ge = yellow
or = orange

Edition 10.02

THIS PAGE INTENTIONALLY LEFT BLANK

Heated Leather Seats

from October 2000

Deviate relay location and fuseplacements as well as the locations of multiple connectors see section "component locations"

Relay location on the thirteen position auxiliary relay panel, above relay panel:

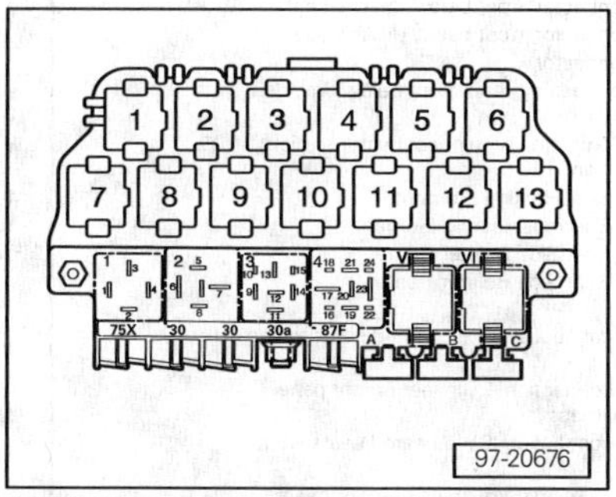

Relay panel:

Relay location on the eight position auxiliary relay panel, above relay panel:

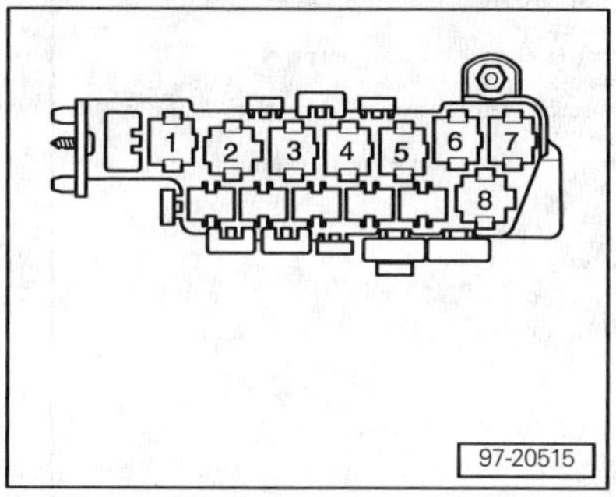

Instrument Panel Light Dimmer Switch

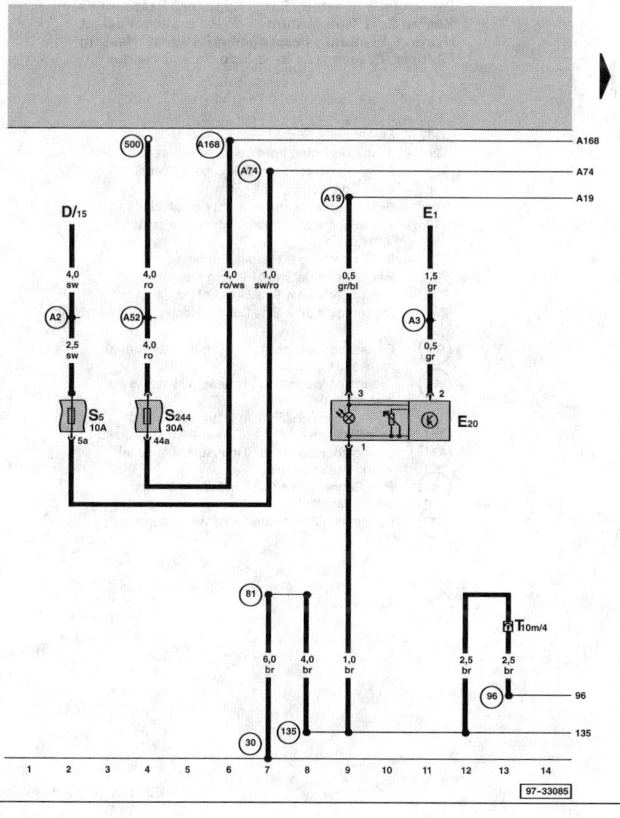

E1 - Light switch
E20 - Instrument Panel Light Dimmer Switch
S5 - Fuse
S244 - Fuse in fuse holder
T10m- 10-Pin Connector, brown, connector station A-pillar, right

(30) - Ground connection, -1-, beside fuse/relay panel
(81) - Ground connection -1-, in instrument panel wiring harness
(96) - Ground connection -1-, in heated seats wiring harness
(135) - Ground connection -2-, in instrument panel wiring harness
(500) - Threaded connection -1- (30) on the relay plate
(A2) - plus connection (15), in instrument panel wiring harness
(A3) - plus connection (58), in instrument panel wiring harness
(A19) - wire connection (58d), in instrument panel wiring harness
(A52) - Plus Connection 2 (30) (in instrument panel wiring harness)
(A74) - Connector (15a - fuse 5) in instrument panel wiring harness
(A168) - Plus Connection 4 (30a) (in instrument panel wiring harness)

ws = white
sw = black
ro = red
br = brown
gn = green
bl = blue
gr = grey
li = lilac
ge = yellow
or = orange

Driver's Heated Seat Control Module, Driver's Heated Seat Adjuster, Driver's Heated Seat Temperature Sensor, Driver's Seat Heating Element, Driver's Backrest Heating Element, Driver's Seat Side Bolster Heater

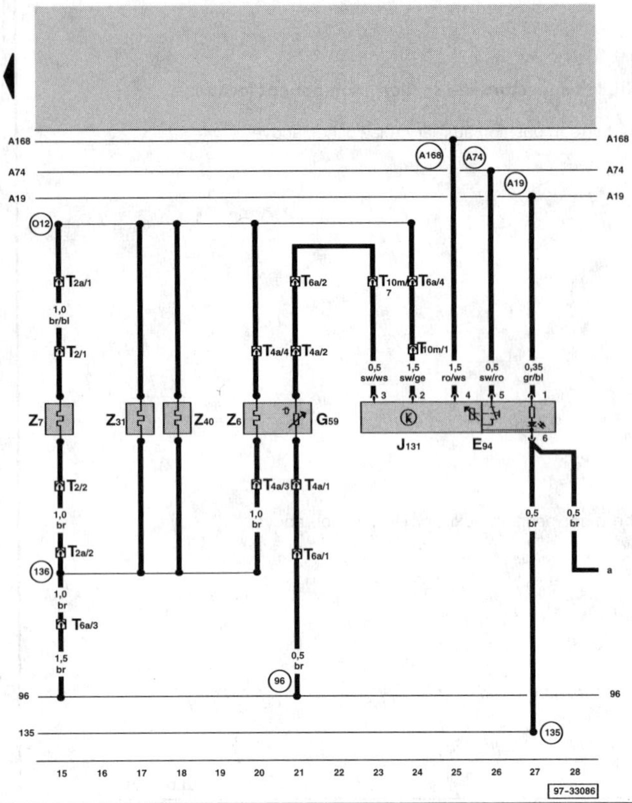

E94 - Driver's Heated Seat Adjuster
G59 - Driver's Heated Seat Temperature Sensor
J131 - Driver's Heated Seat Control Module
T2 - 2-Pin Connector, front below driver's seat
T2a - 2-Pin Connector, front below driver's seat
T4a - 4-Pin Connector
T6a - 6-Pin Connector, green, front below driver's seat
T10m - 10-Pin Connector, brown, connector station A-pillar, right
Z6 - Driver's Seat Heating Element
Z7 - Driver's Backrest Heating Element
Z31 - Driver's Seat Side Bolster Heater
Z40 - Driver's Seat Side Bolster Heater 2

(96) - Ground connection -1-, in heated seats wiring harness
(135) - Ground connection -2-, in instrument panel wiring harness
(136) - Ground connection -2-, in heated seats wiring harness
(A19) - wire connection (58d), in instrument panel wiring harness
(A74) - Connector (15a - fuse 5) in instrument panel wiring harness
(A168) - Plus Connection 4 (30a) (in instrument panel wiring harness)
(O12) - Connection 1 (in heated seat wiring harness)

ws = white
sw = black
ro = red
br = brown
gn = green
bl = blue
gr = grey
li = lilac
ge = yellow
or = orange

Edition 10.02

Passenger's Heated Seat Control Module, Passenger's Heated Seat Adjuster, Passenger's Heated Seat Temperature Sensor, Passenger's Seat Heating Element, Passenger's Backrest Heating Element, Passenger's Seat Side Bolster Heater

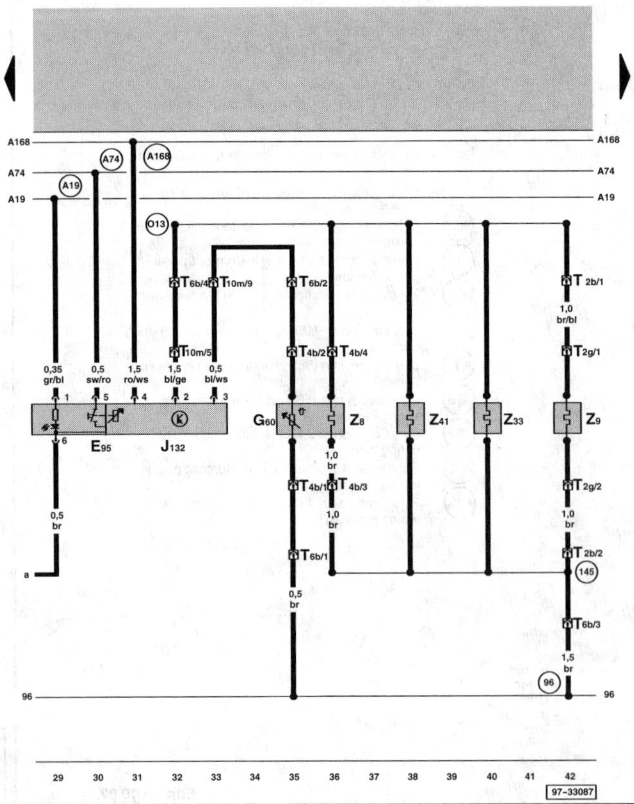

E95 - Passenger's Heated Seat Adjuster
G60 - Passenger's Heated Seat Temperature Sensor
J132 - Passenger's Heated Seat Control Module
T2b - 2-Pin Connector, front below rider's seat
T2g - 2-Pin Connector, front below rider's seat
T4b - 4-Pin Connector
T6b - 6-Pin Connector, green, front below rider's seat
T10m - 10-Pin Connector, brown, connector station A-pillar, right
Z8 - Passenger's Seat Heating Element
Z9 - Passenger's Backrest Heating Element
Z33 - Passenger's Seat Side Bolster Heater
Z41 - Passenger's Seat Side Bolster Heater 2

(96) - Ground connection -1-, in heated seats wiring harness
(145) - Ground connection -3-, in heated seats wiring harness
(A19) - wire connection (58d), in instrument panel wiring harness
(A74) - Connector (15a - fuse 5) in instrument panel wiring harness
(A168) - Plus Connection 4 (30a) (in instrument panel wiring harness)
(O13) - Connection 2 (in heated seat wiring harness)

ws = white
sw = black
ro = red
br = brown
gn = green
bl = blue
gr = grey
li = lilac
ge = yellow
or = orange

Edition 10.02

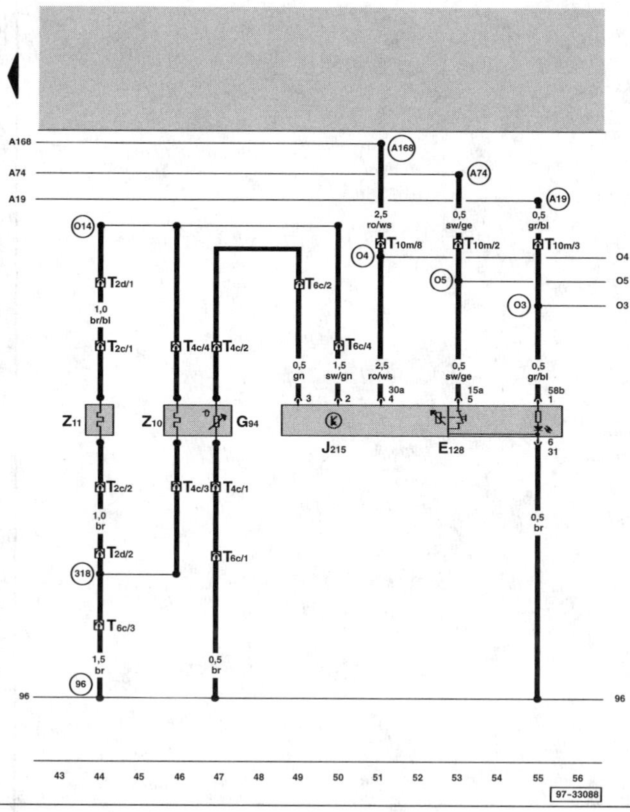

Left Rear Heated Seat Control Module, Left Rear Heated Seat Regulating Switch, Left Rear Heated Seat Temperature Sensor, Left Rear Seat Heating Element, Left Rear Backrest Heating Element

- E128 - Left Rear Heated Seat Regulating Switch
- G94 - Left Rear Heated Seat Temperature Sensor
- J215 - Left Rear Heated Seat Control Module
- T2c - 2-Pin Connector
- T2d - 2-Pin Connector
- T4c - 4-Pin Connector
- T6c - 6-Pin Connector, green
- T10m - 10-Pin Connector, brown, connector station A-pillar, right
- Z10 - Left Rear Seat Heating Element
- Z11 - Left Rear Backrest Heating Element

- (96) - Ground connection -1-, in heated seats wiring harness
- (318) - Ground Connection 4 (in heated seats wiring harness)
- (A19) - wire connection (58d), in instrument panel wiring harness
- (A74) - Connector (15a - fuse 5) in instrument panel wiring harness
- (A168) - Plus Connection 4 (30a) (in instrument panel wiring harness)
- (O3) - plus connection (58b), in heated seats wiring harness
- (O4) - plus connection (30 as), in heated seats wiring harness
- (O5) - wire connection -1- (15a), in heated seats wiring harness
- (O14) - Connection 3 (in heated seat wiring harness)

ws = white
sw = black
ro = red
br = brown
gn = green
bl = blue
gr = grey
li = lilac
ge = yellow
or = orange

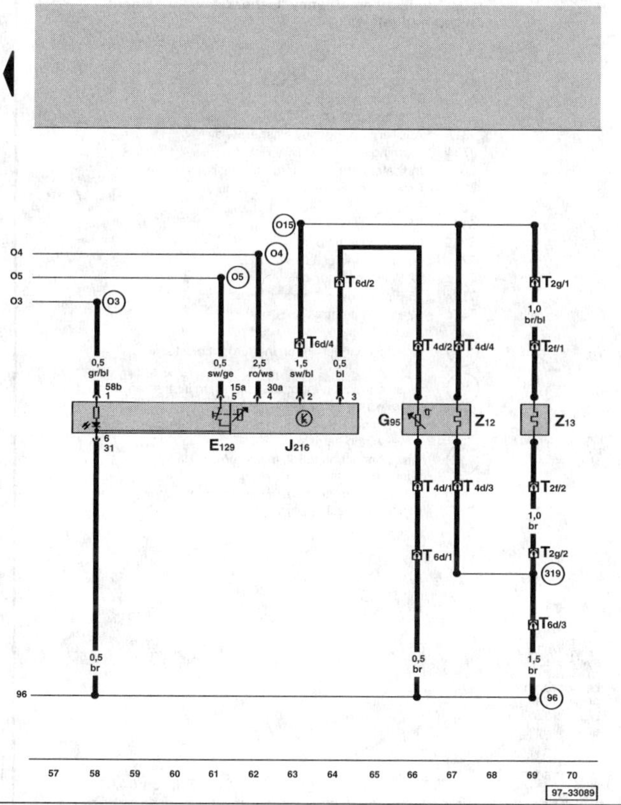

Right Rear Heated Seat Control Module, Right Rear Heated Seat Regulating Switch, Right Rear Heated Seat Temperature Sensor, Right Rear Seat Heating Element, Right Rear Backrest Heating Element

- E129 - Right Rear Heated Seat Regulating Switch
- G95 - Right Rear Heated Seat Temperature Sensor
- J216 - Right Rear Heated Seat Control Module
- T2f - 2-Pin Connector
- T2g - 2-Pin Connector
- T4d - 4-Pin Connector
- T6d - 6-Pin Connector, green
- Z12 - Right Rear Seat Heating Element
- Z13 - Right Rear Backrest Heating Element

- (96) - Ground connection -1-, in heated seats wiring harness
- (319) - Ground Connection 5 (in heated seats wiring harness)
- (O3) - plus connection (58b), in heated seats wiring harness
- (O4) - plus connection (30 as), in heated seats wiring harness
- (O5) - wire connection -1- (15a), in heated seats wiring harness
- (O15) - Connection 4 (in seat heater wiring harness)

ws = white
sw = black
ro = red
br = brown
gn = green
bl = blue
gr = grey
li = lilac
ge = yellow
or = orange

Telematics,

from May 2003

Note:

Information for

- *Relay and fuse arrangements*
- *Connector arrangements*
- *Control Modules and Relays*
- *Ground connections*

⇒ Component Locations!

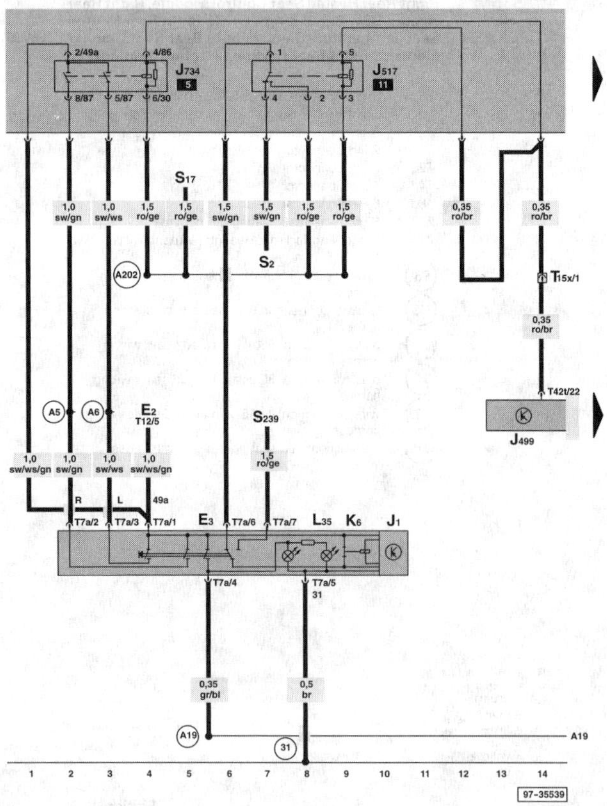

Telematics Control Module, warning lamp auxiliary relay, auxiliary emergency flasher relay, Turn signal switch and relay

- E2 - Turn signal switch
- E3 - Emergency Flasher Switch
- J1 - Turn Signal Relay
- J499 - Telematics Control Module
- J517 - Auxiliary Emergency Flasher Relay
- J734 - Warning Lamp Auxiliary Relay
- K6 - Indicator light for emergency flasher system
- L35 - Emergency Flasher Switch Light
- S2 - Fuse
- S17 - Fuse
- S239 - Fuse in fuse holder
- T7a - 7-Pin Connector, black
- T12 - 12-Pin Connector
- T15x - 15-Pin Connector, red
- T42t - 42-Pin Connector
- (31) - Ground connection, on instrument cluster
- (A5) - Plus connection (right turn signal), in instrument panel wiring harness
- (A6) - Plus connection (left turn signal), in instrument panel wiring harness
- (A19) - wire connection (58d), in instrument panel wiring harness
- (A202) - Telematic Connection 1 (in instrument panel wiring harness)

ws = white
sw = black
ro = red
br = brown
gn = green
bl = blue
gr = grey
li = lilac
ge = yellow
or = orange

Telematics Control Module, Telematics Control Head, Roadside Assistance Button, Emergency Call Button, End Conversation Button (Telematics USA), Telematics Indicator Lamp

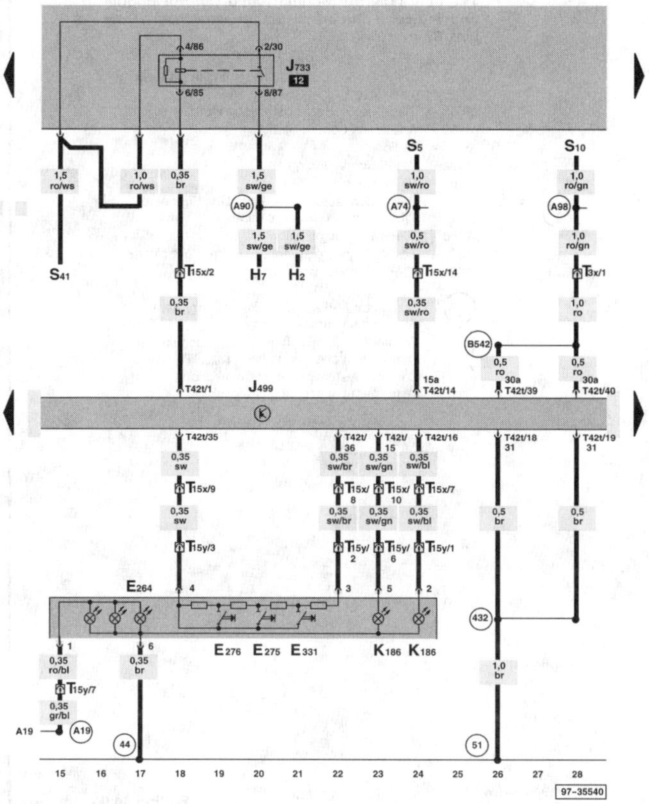

E264 - Telematics Control Head
E275 - Roadside Assistance Button
E276 - Emergency Call Button
E331 - End Conversation Button (Telematics USA)
H2 - High Tone Horn
H7 - Low Tone Horn
J499 - Telematics Control Module
J733 - Dual Horn Auxiliary Relay
K186 - Telematics Indicator Lamp
S5 - Fuse
S10 - Fuse
S41 - Individual fuse for heated rear glass
T3x - 3-Pin Connector, red
T15x - 15-Pin Connector, red
T15y - 15-Pin Connector, green
T42t - 42-Pin Connector

(44) - Ground connection, on left A-pillar, lower part
(51) - Ground connection, in luggage compartment, right
(432) - Ground Connection (in telematic wiring harness)
(A19) - wire connection (58d), in instrument panel wiring harness
(A74) - Connector (15a - fuse 5) in instrument panel wiring harness
(A90) - Connector (two-tone horn), in instrument panel wiring harness
(A98) - Plus connector -4- (30), in instrument panel wiring harness
(B542) - Plus Connection (in Telematic wiring harness)

ws = white
sw = black
ro = red
br = brown
gn = green
bl = blue
gr = grey
li = lilac
ge = yellow
or = orange

97-35540

Telematics Control Module, Antenna for Navigation System (GPS), Telephone Microphone

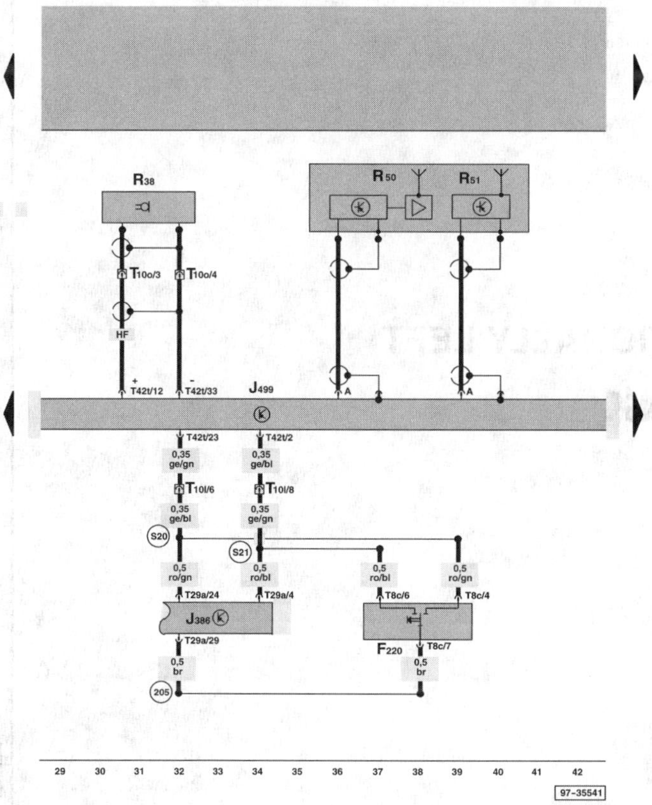

F220 - Driver's Door Central Locking Lock Unit
J386 - Door Control Module, driver side
J499 - Telematics Control Module
R38 - Telephone Microphone
R50 - Antenna for Navigation System (GPS)
R51 - Radio/Telephone Antenna
T8c - 8-Pin Connector
T10o - 10-Pin Connector
T10l - 10-Pin Connector, black, connector station A-pillar, left
T29a - 29-Pin Connector
T42t - 42-Pin Connector

(205) - Ground connection, in driver's door wiring harness
(S20) - Connector -4- in wiring harness central locking
(S21) - Connector -5- in wiring harness central locking

ws = white
sw = black
ro = red
br = brown
gn = green
bl = blue
gr = grey
li = lilac
ge = yellow
or = orange

97-35541

Telematics Control Module, Radio, Control module with indicator unit in instrument panel insert, Data Link Connector (DLC)

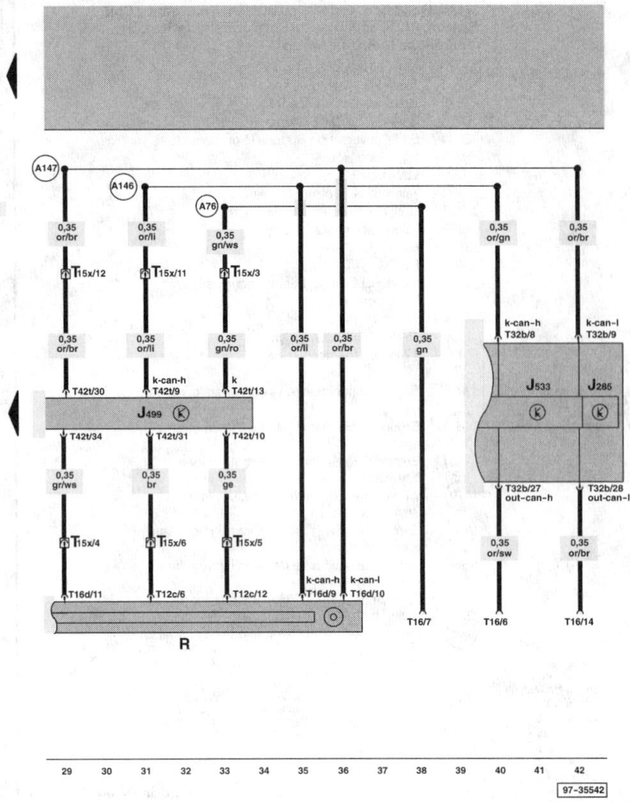

J285 - Control module with indicator unit in instrument panel insert
J533 - Data Bus On Board Diagnostic Interface
J499 - Telematics Control Module
R - Radio
T12c - 12-Pin Connector, green
T15x - 15-Pin Connector, red
T16 - 16-Pin Connector, black, Data Link Connector (DLC), under instrument panel. left
T16d - 16-Pin Connector, black
T32b - 32-Pin Connector, green, on instrument cluster
T42t - 42-Pin Connector

(A76) - Connector (K-diagnosis wire) in instrument panel wiring harness
(A146) - Comfort System High-bus Connection (in instrument panel wiring harness)
(A147) - Comfort System Low-bus Connection (in instrument panel wiring harness)

ws = white
sw = black
ro = red
br = brown
gn = green
bl = blue
gr = grey
li = lilac
ge = yellow
or = orange

Edition 04.04

THIS PAGE INTENTIONALLY LEFT
BLANK

Comfort System, with Alarm System,

from May 2003

Note:

Informations for

- *Relay and fuse arrangement*
- *Connector arrangement*
- *Control units and relays*
- *Ground connections*

⇒ Component Locations!

Lock unit for central locking, Door control module, driver side, Telematic Control Module, Warning light for central locking -SAFE-

- F220 - Lock unit for central locking, driver side
- J386 - Door control module, driver side
- J499 - Telematic Control Module, below spare tire
- K133 - Warning light for central locking -SAFE-
- T8c - 8-Pin Connector, black, on lock unit for centrallocking, driver side
- T10l - 10-Pin Connector, black, connector station A-pillar, left
- T29a - 29-Pin Connector
- T42t - 42-Pin Connector
- V147 - Motor for window regulator, driver side
- W31 - Entry light front, left
- (205) - Ground connection, in driver's door wiring harness
- (R14) - wire connection -1- (open), in driver's door wiring harness
- (S20) - Connector -4- in wiring harness central locking
- (S21) - Connector -5- in wiring harness central locking

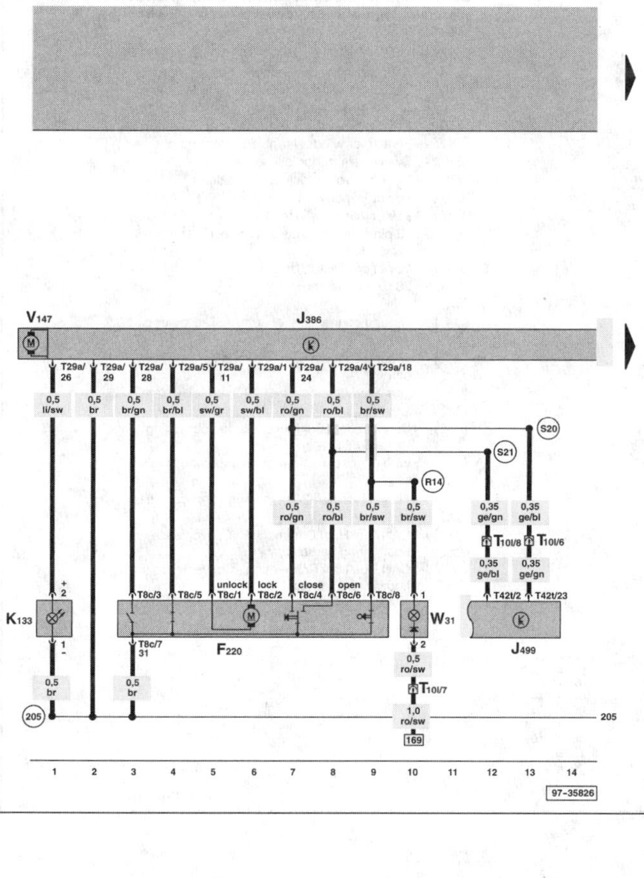

ws = white
sw = black
ro = red
br = brown
gn = green
bl = blue
gr = grey
li = lilac
ge = yellow
or = orange

Passat — Wiring diagram — No. 82/3

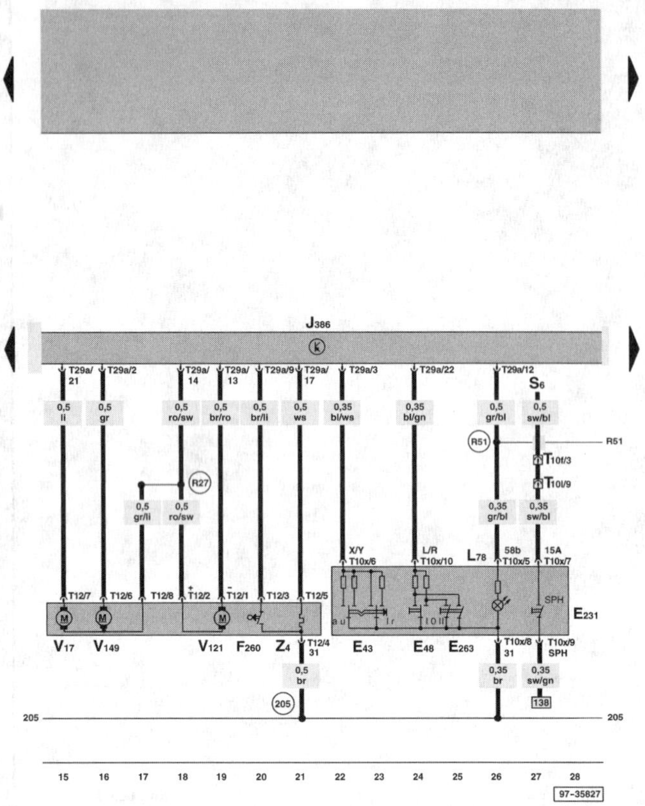

Door control module, driver side, Heated outside mirror, driver side, Mirror Adjustment and Selector Switch

E43 - Mirror Adjustment Switch
E48 - Mirror Selector Switch
E231 - Switch for outside mirror heating
E263 - Mirror Folding Switch
F260 - Driver's Folding Mirror Contact Switch
J386 - Door control module, driver side
L78 - Mirror Adjusting Switch Light
S6 - Fuse
T10f - 10-Pin Connector, blue, connector station A-pillar, left
T10l - 10-Pin Connector, black, connector station A-pillar, left
T10x - 10-Pin Connector, black
T12 - 12-Pin Connector,black
T29a - 29-Pin Connector
V17 - Driver's Side Mirror Adjustment Motor
V121 - Motor for fold-in mirror, driver side
V149 - Motor for mirror adjustment, driver side
Z4 - Heated outside mirror, driver side

(205) - Ground connection, in driver's door wiring harness
(R27) - wire connection (power mirror), in drivr's door wiring harness
(R51) - Connection (58b) (driver's side door wiring harness)

ws = white
sw = black
ro = red
br = brown
gn = green
bl = blue
gr = grey
li = lilac
ge = yellow
or = orange

Edition 06.03

Passat — Wiring diagram — No. 82/4

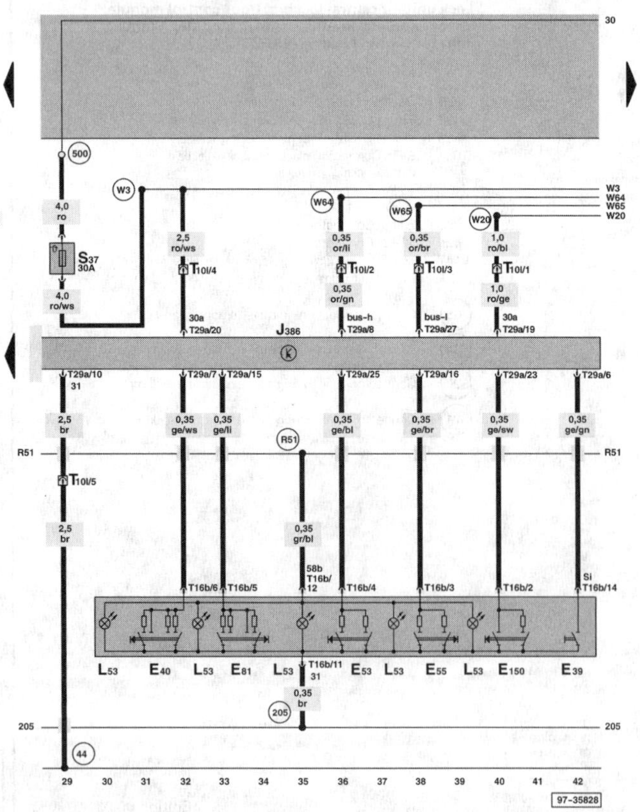

Door control module, driver side, Switch for window regulator in driver's door, Switch for interior lock

E39 - Safety switch for window regulator rear
E40 - Switch for window regulator front left
E53 - Switch for window regulator rear left, driver
E55 - Switch for window regulator rear right, driver
E81 - Switch for window regulator front right, driver
E150 - Switch for interior lock, driver side
J386 - Door control module, driver side
L53 - Power Window Switch Light
S37 - Power Window Fuse
T10l - 10-pin connector, black, connector station A-pillar left
T16b - 16-pin connector, brown
T29a - 29-pin connector

(44) - Ground connection, on left A-pillar, lower part
(205) - Ground connection, in driver's door wiring harness
(500) - Threaded connection -1- (30), on relay plate
(R51) - Connection (58d) (driver's side door wiring harness)
(W3) - wire connection, in rear wiring harness
(W20) - Plus connector (30a), in wiring harness, rear
(W64) - CAN-Bus High Connection - Convenience (in floor wiring harness)
(W65) - CAN-Bus Low Connection - Convenience (in floor wiring harness)

ws = white
sw = black
ro = red
br = brown
gn = green
bl = blue
gr = grey
li = lilac
ge = yellow
or = orange

Edition 06.03

Passat — Wiring diagram — No. 82/5

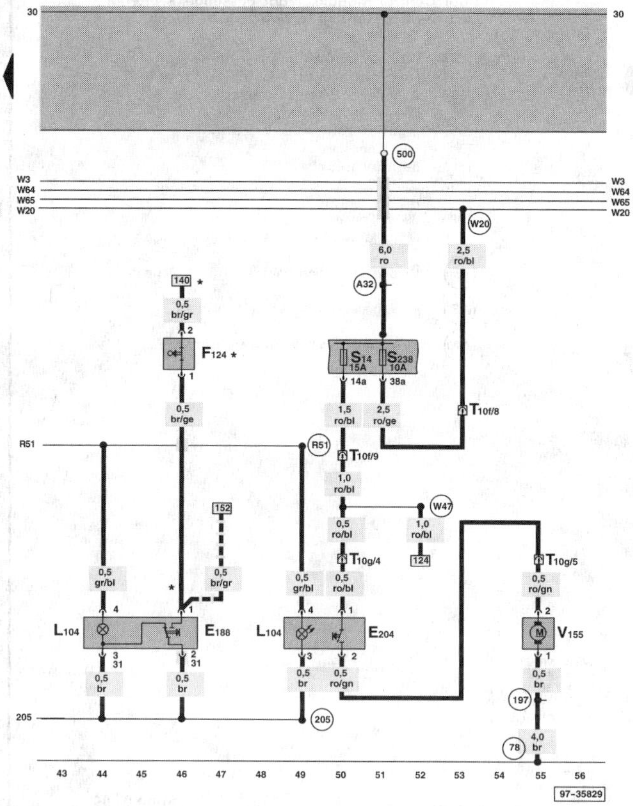

Switch for remote unlock and fuel tank door, Trunk Lock Alarm/Central Locking Switch, Illumination for switch remote unlock, Motor for fuel tank lid unlock

E188 - Switch for remote unlock, rear lid
E204 - Switch for remote/fuel tank door
F124 - Trunk Lock/Alarm/Central Locking Switch
L104 - Remote Unlock Switch Illumination
S14 - Fuse
S238 - Fuse in fuse holder
T10f - 10-Pin Connector, blue, connector station A-pillar, left
T10g - 10-Pin Connector, grey
V155 - Motor for fuel tank lid unlock

(78) - Ground Connection (lower right B-pillar)
(197) - Ground connection -4-, in wiring harness rear
(205) - Ground connection, in driver's door wiring harness
(500) - Threaded connection -1- (30) on the relay plate
(A32) - plus connection (30), in instrument panel wiring harness
(R51) - Connection (58d) (driver's side door wiring harness)
(W20) - Plus-Connector (30a) in wiring harness, rear
(W47) - Plus-Connection -2- (30a), in wiring harness floor

* - sedan only
--- - wagon only

ws = white
sw = black
ro = red
br = brown
gn = green
bl = blue
gr = grey
li = lilac
ge = yellow
or = orange

Edition 06.03

Passat — Wiring diagram — No. 82/6

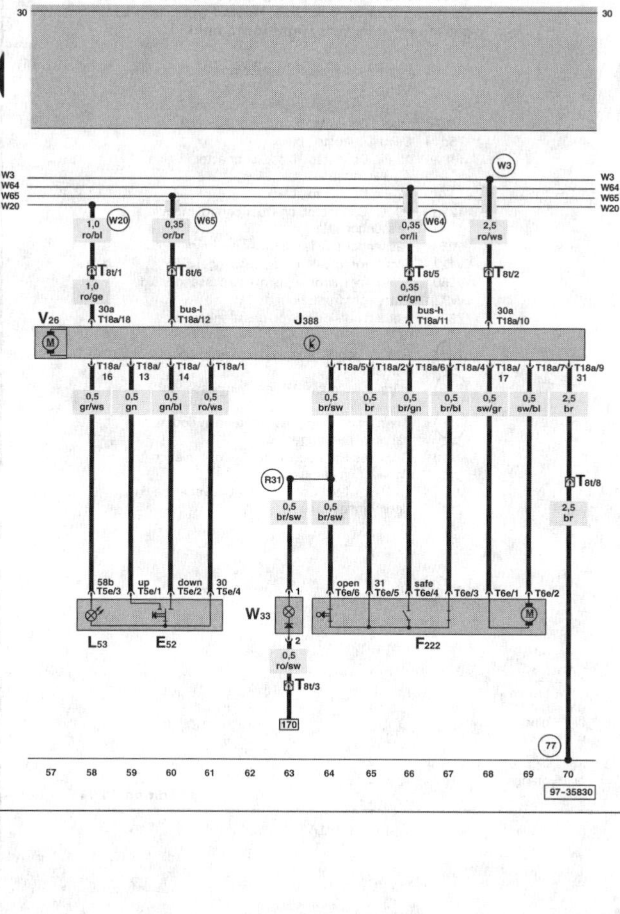

Door control module, rear, left, Power Window Switch Light, Left Rear Window Switch, Entry light rear, left, Lock unit for central locking, rear, left

E52 - Left Rear Window Switch, (In LR Door)
F222 - Lock unit for central locking, rear, left
J388 - Door control module, rear, left
L53 - Power Window Switch Light
T5e - 5-Pin Connector, black
T6e - 6-Pin Connector, black
T8t - 8-Pin Connector, black, connector station B-pillar, left
T18a - 18-Pin Connector, on Door control module, rear, left
V26 - Motor for window regulator, rear, left
W33 - Entry light rear, left

(77) - Ground connection (lower left B-pillar)
(R31) - Wire connection -1-, in left rear door wiring harness
(W3) - wire connection, in rear wiring harness
(W20) - Plus connector (30a), in wiring harness, rear
(W64) - CAN-Bus High Connection - Convenience (in floor wiring harness)
(W65) - CAN-Bus Low Connection - Convenience (in floor wiring harness)

ws = white
sw = black
ro = red
br = brown
gn = green
bl = blue
gr = grey
li = lilac
ge = yellow
or = orange

Edition 06.03

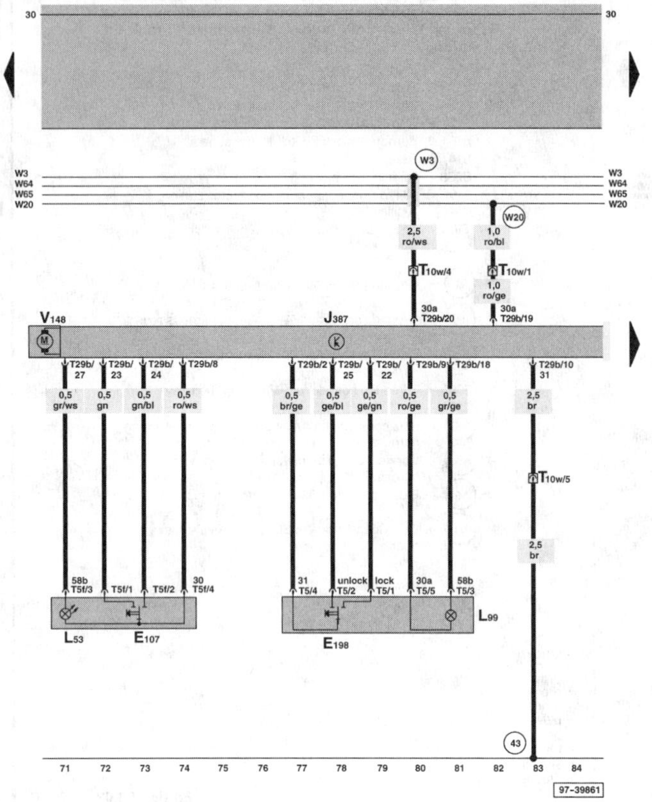

Door Control Module, Front Passenger's Interior Lock Switch, Lighting forswitch interior lock,

E107 - Window Regulator Switch (in RF door)
E198 - Front Passenger's Interior Lock Switch
J387 - Door Control Module, passenger's side
L53 - Power Window Switch Light
L99 - Lighting for switch interior lock
T5 - 5-Pin Connector
T5f - 5-Pin Connector, black
T10w - 10-Pin Connector, black, connector station A-pillar, right
T29b - 29-Pin Connector
V148 - Motor for window regulator, passenger side

(43) - Ground connection, on right A-pillar, lower part
(W3) - wire connection, in rear wiring harness
(W20) - Plus connection (30a), in wiring harness, rear

ws = white
sw = black
ro = red
br = brown
gn = green
bl = blue
gr = grey
li = lilac
ge = yellow
or = orange

Edition 04.04

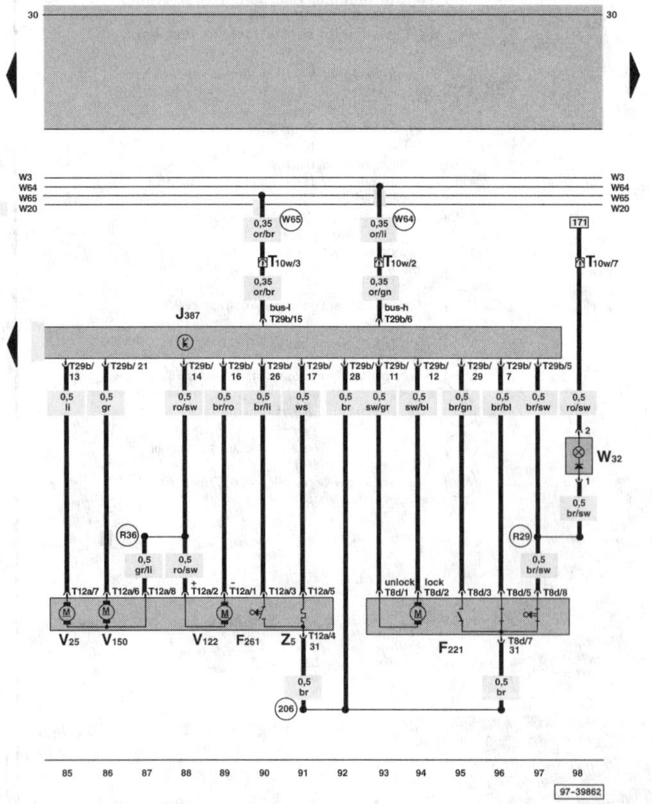

Door control module, Front Passenger's Door Central Locking Lock Unit, Heated outside mirror, passenger side, Entry light front, right

F221 - Front Passenger's Door Central Locking Lock Unit
F261 - Passenger's Folding Mirror Contact Switch
J387 - Door Control Module, passenger side
T8d - 8-Pin Connector, black
T10w - 10-Pin Connector, black, connector station A-pillar, right
T12a - 12-Pin Connector, black
T29b - 29-Pin Connector, on Door control module, passenger side
V25 - Passenger's Side Mirror Adjustment Motor
V122 - Motor for fold-in mirror, passenger side
V150 - Motor for mirror adjustment, passenger side
W32 - Entry light front, right
Z5 - Heated outside mirror, passenger side

(206) - Ground connection, in passenger's door wiring harness
(R29) - Wire connection -1- (open), in passenger's door wiring harness
(R36) - Connection (mirror adjustment) in door wiring harness, passenger side
(W64) - CAN-Bus High Connection - Convenience (in floor wiring harness)
(W65) - CAN-Bus Low Connection - Convenience (in floor wiring harness)

ws = white
sw = black
ro = red
br = brown
gn = green
bl = blue
gr = grey
li = lilac
ge = yellow
or = orange

Edition 04.04

No. 82/9

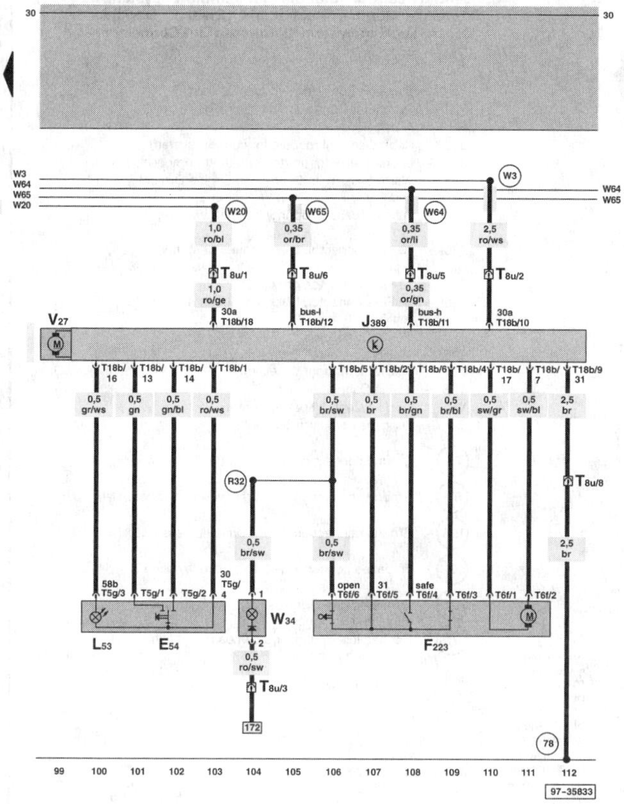

Door control module, Lock unit for central locking rear right, Power window Switch Light, Entry light rear, right

- E54 - Right Rear Window Switch, (In RR Door)
- F223 - Lock unit for central locking, rear, right
- J389 - Door control module, rear, right
- L53 - Power Window Switch Light
- T5g - 5-Pin Connector
- T6f - 6-Pin Connector
- T8u - 8-Pin Connector, brown, connector station B-pillar, right
- T18b - 18-Pin Connector
- V27 - Motor for window regulator, rear, right
- W34 - Entry light rear, right
- (78) - Ground connection (lower right B-pillar)
- (R32) - Wire connection -2-, in right rear door wiring harness
- (W3) - wire connection, in rear wiring harness
- (W20) - Plus connector (30a), in wiring harness, rear
- (W64) - CAN-Bus High Connection - Convenience (in floor wiring harness)
- (W65) - CAN-Bus Low Connection - Convenience (in floor wiring harness)

ws = white
sw = black
ro = red
br = brown
gn = green
bl = blue
gr = grey
li = lilac
ge = yellow
or = orange

Edition 06.03

No. 82/10

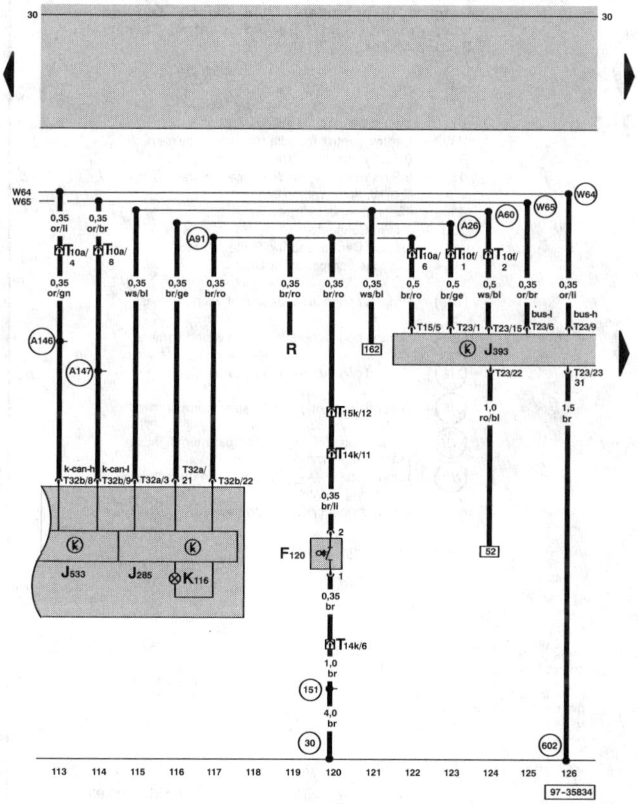

Central control module for comfort system, Warning light for rear lid unlocked

- F120 - Hood Alarm Switch
- J285 - Control module with indicator unit in instrument panel insert
- J393 - Central control module for comfort system
- J533 - Data Bus On Board Diagnostic Interface
- K116 - Warning light for rear lid unlocked
- R - Radio
- T10a - 10-Pin Connector, brown, connector station A-pillar, left
- T10f - 10-Pin Connector, blue, connector station A-pillar, left
- T14k - 14-Pin Connector
- T15 - 15-Pin Connector
- T15k - 15-Pin Connector, grey
- T23 - 23-Pin Connector
- T32a - 32-Pin Connector, blue, on instrument cluster
- T32b - 32-Pin Connector, green, on instrument cluster
- (30) - Ground connection, -1-, beside fuse/relay panel
- (151) - Ground connection -1-, in front wiring harness
- (602) - Ground Point in Floor out let front, left
- (A26) - wire connection (driver's door contact switch), in instrument panel wiring harness
- (A60) - Wire Connection (Vehicle speed signal), in instrument cluster wiring harness
- (A91) - Connector (signal for instrument panel insert), in instrument panel wiring harness
- (A146) - Comfort System High-bus Connection (in instrument panel wiring harness)
- (A147) - Comfort System Low-bus Connection (in instrument panel wiring harness)
- (W64) - CAN-Bus High Connection - Convenience (in floor wiring harness)
- (W65) - CAN-Bus Low Connection - Convenience (in floor wiring harness)

ws = white
sw = black
ro = red
br = brown
gn = green
bl = blue
gr = grey
li = lilac
ge = yellow
or = orange

Edition 06.03

Central control module for comfort system, Luggage Compartment Light Switch, Decklid Central Locking System Motor, Data Link Connector (DLC)

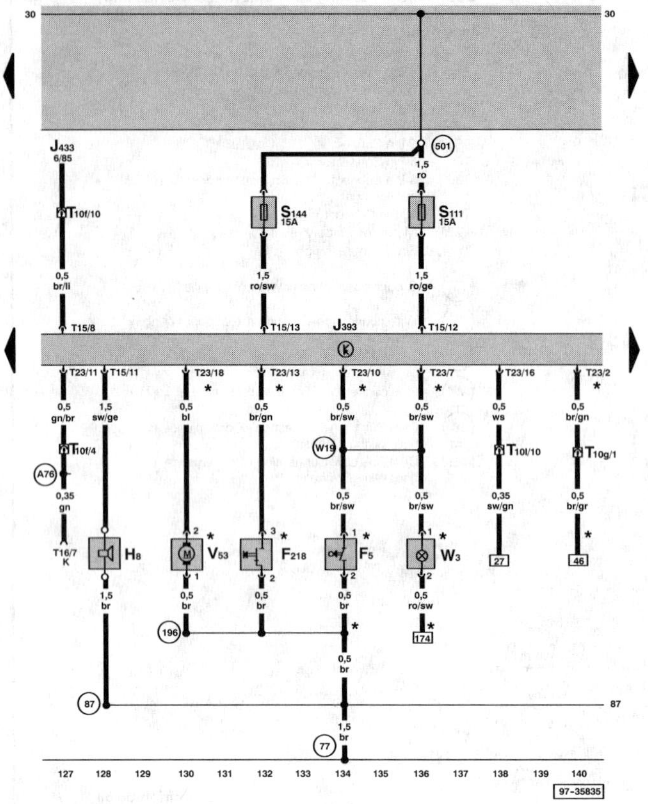

- F5 - Luggage Compartment Light Switch
- F218 - Switch for Central locking, trunk lid
- H8 - Alarm Horn
- J393 - Central control module for comfort system
- J433 - Locking relay for starter (anti-theft immobilizer)
- S111 - Alarm System and Anti-Theft Engine Disable Fuse
- S144 - Fuse for central locking/anti-theft warning system
- T10f - 10-Pin Connector, blue, connector station A-pillar, left
- T10g - 10-Pin Connector, grey
- T10l - 10-Pin Connector, black, connector station A-pillar left
- T15 - 15-Pin Connector
- T16 - 16-Pin Connector, Data Link Connector (DLC), below instrument panel, left
- T23 - 23-Pin Connector
- V53 - Decklid Central Locking System Motor
- W3 - Luggage compartment Light
- (77) - Ground connection (lower left B-pillar)
- (87) - Ground connection -2-, in rear wiring harness
- (196) - Ground connection -3-, in wiring harness, rear
- (501) - Threaded connection -2- (30), on relay plate
- (A76) - Connector (K-diagnosis wire) in instrument panel wiring harness
- (W19) - Connection -2- (luggage compartment light), in rear wiring harness
- * - sedan only

ws = white
sw = black
ro = red
br = brown
gn = green
bl = blue
gr = grey
li = lilac
ge = yellow
or = orange

Edition 06.03

Central control module for comfort system, Luggage Compartment Light Switch, Luggage compartment light, Decklid Central Locking System Motor

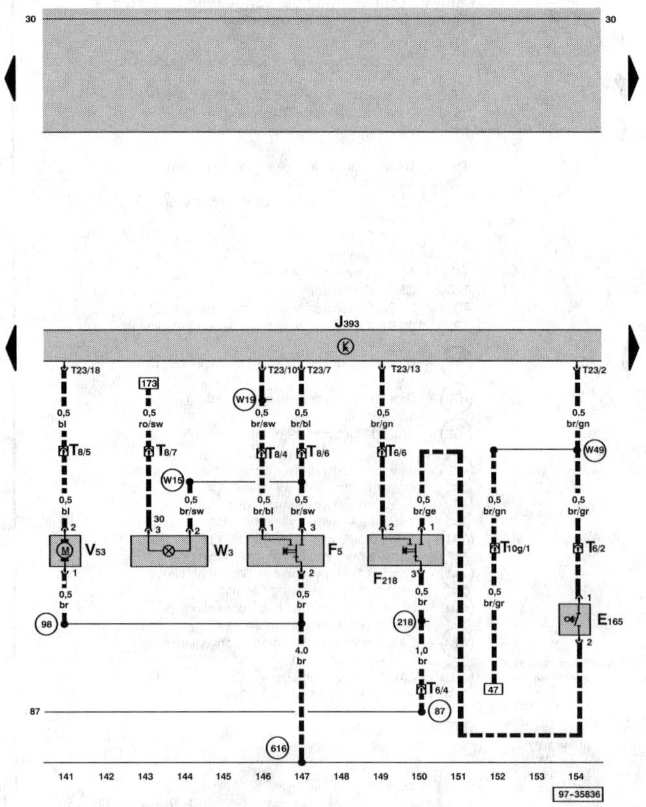

- E165 - Trunk Lid Release Switch
- F5 - Luggage Compartment Light Switch
- F218 - Switch for central locking, trunk lid
- J393 - Central control module for comfort system
- T6 - 6-Pin Connector, black
- T8 - 8-Pin Connector, black, connector station D-pillar, right (wagon only)
- T10g - 10-Pin Connector, grey
- T23 - 23-Pin Connector
- V53 - Decklid Central Locking System Motor
- W3 - Luggage compartment Light
- (87) - Ground connection -2-, in rear wiring harness
- (98) - Ground connection, in rear lid wiring harness
- (218) - Ground connection -1-, in rear lid wiring harness
- (616) - Ground connection (on luggage compartment light)
- (W15) - Connection -1- (luggage compartment light), in rear wiring harness
- (W19) - Connection -2- (luggage compartment light), in rear wiring harness
- (W49) - Connection (in floor wiring harness
- - - - - wagon only

ws = white
sw = black
ro = red
br = brown
gn = green
bl = blue
gr = grey
li = lilac
ge = yellow
or = orange

Edition 06.03

Passat — Wiring diagram — No. 82/13

Central control module for comfort system, Power Sunroof Control Module, Sunroof Regulator, Power Sunroof

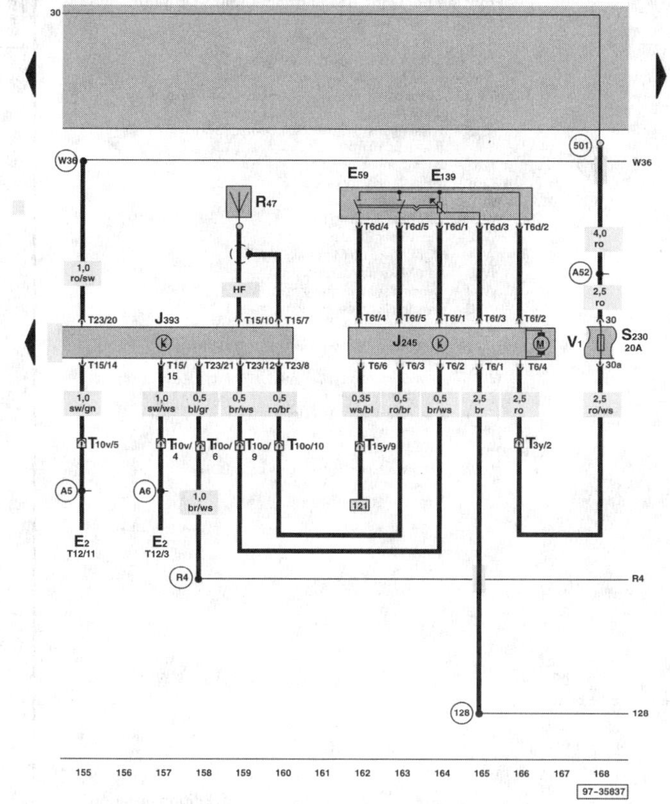

- E2 - Turn signal switch
- E59 - Raise/Lower Sunroof Switch
- E139 - Sunroof Regulator
- J245 - Power Sunroof Control Module
- J393 - Central control module for comfort system
- R47 - Central Locking and Anti-theft Alarm System Antenna, A-pillar, left
- S230 - Fuse in fuse holder
- T3y - 3-Pin Connector, green
- T6 - 6-Pin Connector, blue
- T6d - 6-Pin Connector
- T6f - 6-Pin Connector
- T10v - 10-Pin Connector, violet, connector station A-pillar, right
- T10o - 10-Pin Connector, green, connector station A-pillar, left
- T12 - 12-Pin Connector, near steering column
- T15 - 15-Pin Connector
- T15y - 15-Pin Connector, green
- T23 - 23-Pin Connector
- V1 - Sunroof Motor

- (128) - Ground connection -1-, in interior light wiring harness
- (501) - Threaded connection -2- (30), on relay plate

- (A5) - plus connection (right turn signal), in instrument panel wiring harness
- (A6) - plus connection (left turn signal), in instrument panel wiring harness
- (A52) - Plus Connection 2 (30) (in instrument panel wiring harness)
- (R4) - wire connection -1-, in interior light/door contact switch wiring harness
- (W36) - Positive connection (30a) in floor wiring harness

ws = white
sw = black
ro = red
br = brown
gn = green
bl = blue
gr = grey
li = lilac
ge = yellow
or = orange

97-35837

Edition 04.04

Passat — Wiring diagram — No. 82/14

Front Interior Light, Reading light, center left and right, Center Console Illumination

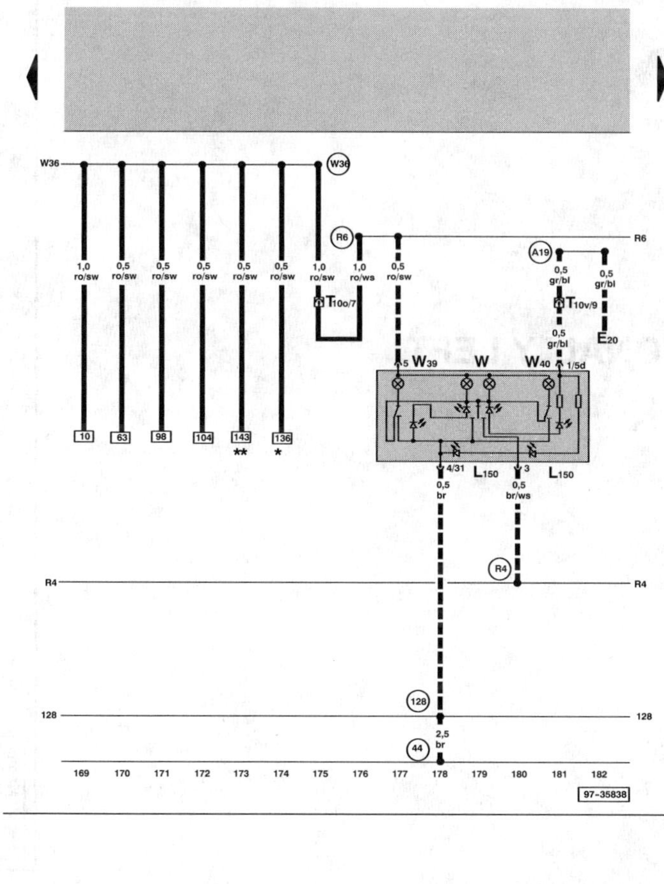

- E20 - Instrument Panel Light Dimmer Switch
- L150 - Center Console Illumination
- T10o - 10-Pin Connector, green, connector station A-pillar, left
- T10v - 10-Pin Connector
- W - Front Interior Light
- W39 - Reading light, center left
- W40 - Reading light, center right

- (44) - Ground connection, on left A-pillar, lower part
- (128) - Ground connection -1-, in interior light wiring harness
- (A19) - wire connection (58d), in instrument panel wiring harness
- (R4) - wire connection -1-, in interior light/door contact switch wiring harness
- (R6) - plus connection -1-, in interior light wiring harness
- (W36) - Positive connection (30a) in floor wiring harness

- * - sedan only
- ** - wagon only
- - - - - engine code BDP only

ws = white
sw = black
ro = red
br = brown
gn = green
bl = blue
gr = grey
li = lilac
ge = yellow
or = orange

97-35838

Edition 06.03

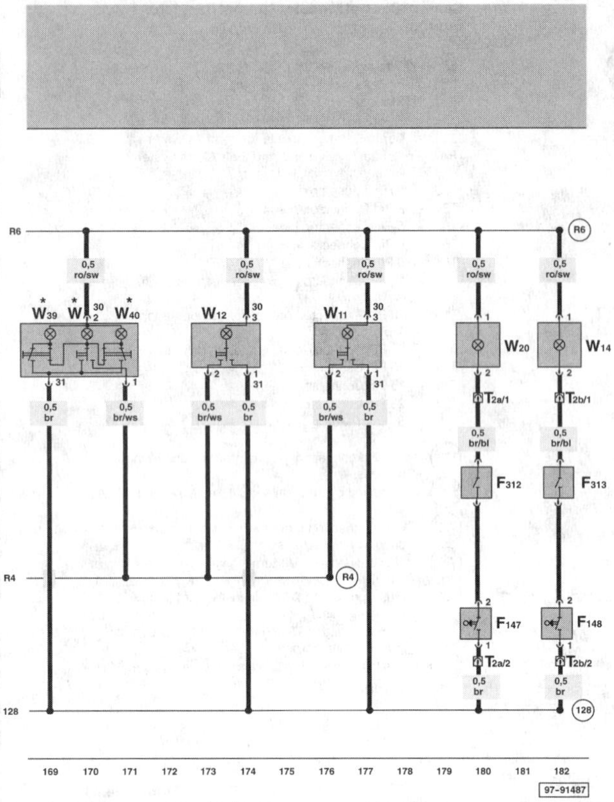

Front Interior Light, Reading lights, Make-Up Mirror Lights, Sliding Cover Contact Switch

F147 - Left Make-Up Mirror Light Switch
F148 - Right Make-Up Mirror Light Switch
F312 - Driver's Sliding Cover Contact Switch (make-up mirror)
F313 - Passenger's Sliding Cover Contact Switch (make-up mirror)
T2a - 2-Pin Connector
T2b - 2-Pin Connector
W - Front Interior Light
W11 - Left Rear Reading Light
W12 - Right Rear Reading Light
W14 - Right Make-up Mirror Light
W20 - Left Make-up Mirror Light
W39 - Reading light, center left
W40 - Reading light, center right

(128) - Ground connection -1-, in interrior light wiring harness
(R4) - wire connection -1-, in interior light/door contact switch wiring harness
(R6) - plus connection -1-, in interior light wiring harness

* - not for vehicles with engine code BDN

ws = white
sw = black
ro = red
br = brown
gn = green
bl = blue
gr = grey
li = lilac
ge = yellow
or = orange

Edition 06.04

THIS PAGE INTENTIONALLY LEFT BLANK

2.0L - Engine - Turbo Diesel PD Fuel Injection (DFI)/100 kW, code BHW,

from November 2003

Notes:

Information for

- *Relay and fuse arrangements*
- *Connector arrangements*
- *Control Modules and Relays*
- *Ground connections*

⇒ Component Locations!

Information for

- *Troubleshooting Procedures*

⇒ Guided Fault Finding using VAS 5051/5052

Battery, Starter, Generator, Voltage Regulator

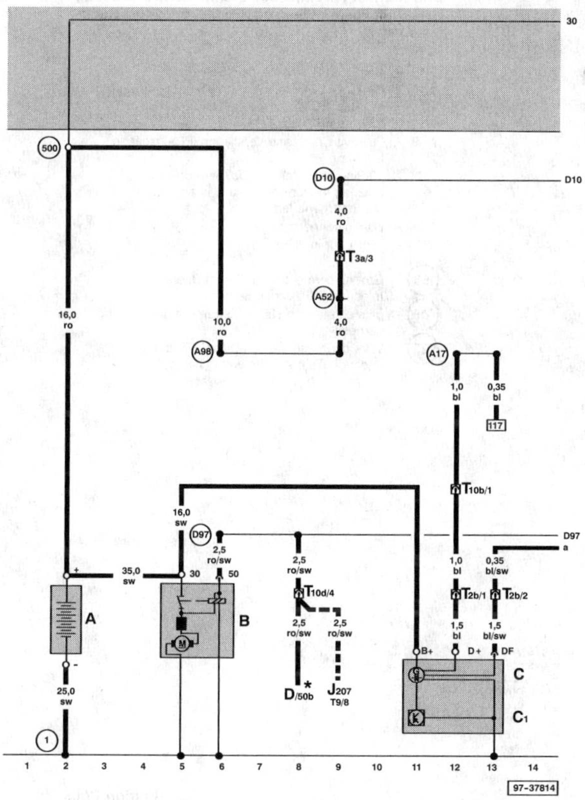

ws = white
sw = black
ro = red
br = brown
gn = green
bl = blue
gr = grey
li = lilac
ge = yellow
or = orange

A - Battery
B - Starter
C - Generator (GEN)
C1 - Voltage Regulator (VR)
D - Ignition/Starter Switch
J207 - Starting Interlock Relay
T2b - Double connector, black, on engine block right
T3a - 3-Pin connector, white, on protectiive housing for control module, in engine compartment, left
T9 - 9-Pin connector, black,
T10b - 10-Pin connector, black, on protectiive housing for control module, in engine compartment, left
T10d - 10-Pin connector, brown, on protectiive housing for control module, in engine compartment, left

(1) - Ground strap, battery to body

(500) - Threaded connection -1- (30) on the relay plate

(A17) - wire connection (61), in instrument panel wiring harness
(A52) - Plus Connection 2 (30) (in instrument panel wiring harness)
(A98) - Plus connector -4- (30), in instrument panel wiring harness
(D10) - plus connection (30), in right front wiring harness
(D97) - Wire connection (50), in engine compartment wiring harness, right

* - manual transmission only
--- - automatic transmission only

Automatic Glow Time Control Module, Diesel Direct Fuel Injection Engine Control Module, Fuel Pump Relay, Glow Plug

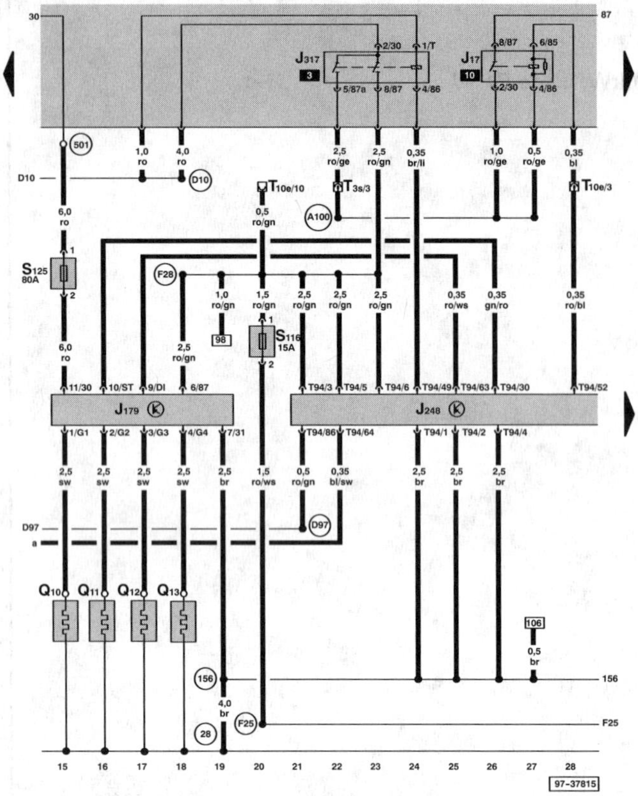

J17 - Fuel Pump (FP) Relay
J179 - Automatic Glow Time Control Module
J248 - Diesel Direct Fuel Injection (DFI) Engine Control Module (ECM)
J317 - Power Supply (Terminal 30, B+) Relay
S116 - Fuse for injectors
S125 - Fuse for glow plug
T3s - 3-Pin connector
T10e - 10-Pin connector, orange, on protectiive housing for control module, in engine compartment, left
T94 - 94-Pin connector
Q10 - Glow Plug 1
Q11 - Glow Plug 2
Q12 - Glow Plug 3
Q13 - Glow Plug 4

(28) - Ground connection, on firewall

(156) - Ground connection, in Diesel Direct Fuel Injection (DFI) wiring harness

(501) - Threaded connection -2- (30) on the relay plate

(A100) - Connector -2- (87), in instrument panel wiring harness

(D10) - plus connection (30), in right front wiring harness

(D97) - Wire connection (50), in engine compartment wiring harness, right

(F25) - Wire connection -1-, in Diesel Direct Fuel Injection (DFI) system wiring harness

(F28) - Wire connection -3-, in Diesel Direct Fuel Injection (DFI) system wiring harness

ws = white
sw = black
ro = red
br = brown
gn = green
bl = blue
gr = grey
li = lilac
ge = yellow
or = orange

Diesel Direct Fuel Injection Engine Control Module, Camshaft Position Sensor, Wastegate Bypass Regulator Valve

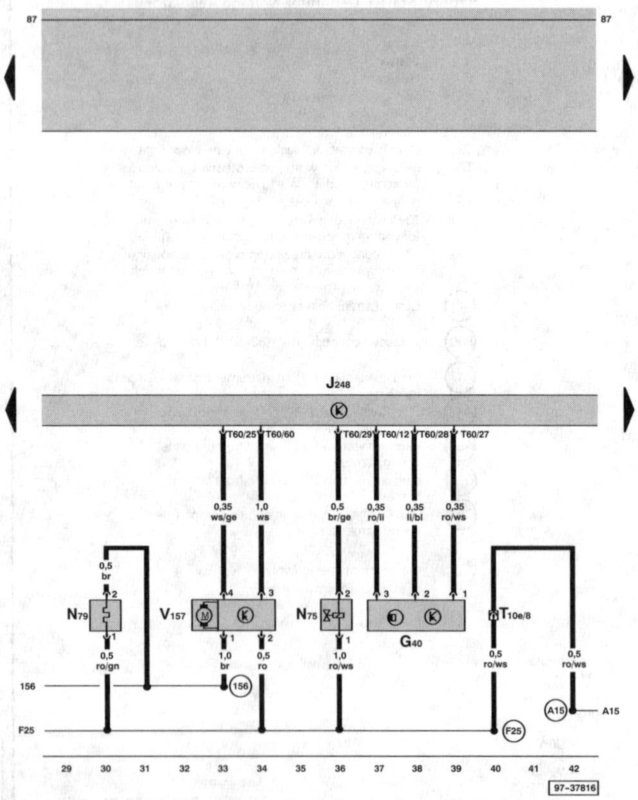

G40 - Camshaft Position (CMP) Sensor
J248 - Diesel Direct Fuel Injection (DFI) Engine Control Module (ECM)
N75 - Wastegate Bypass Regulator Valve
N79 - Positive Crankcase Ventilation (PCV) Heating Element
T10e - 10-Pin connector, orange, on protectiive housing for control module, in engine compartment, left
T60 - 60-Pin connector
V157 - Motor for intake flap

(156) - Ground connection, in Diesel Direct Fuel Injection (DFI) wiring harness

(A15) - plus connection (15) in instrument panel wiring harness

(F25) - Wire connection -1-, in Diesel Direct Fuel Injection (DFI) system wiring harness

ws = white
sw = black
ro = red
br = brown
gn = green
bl = blue
gr = grey
li = lilac
ge = yellow
or = orange

Diesel Direct Fuel Injection Engine Control Module, Cruise Control Switch, Brake Light Switch, Clutch Vacuum Venr Valve Switch

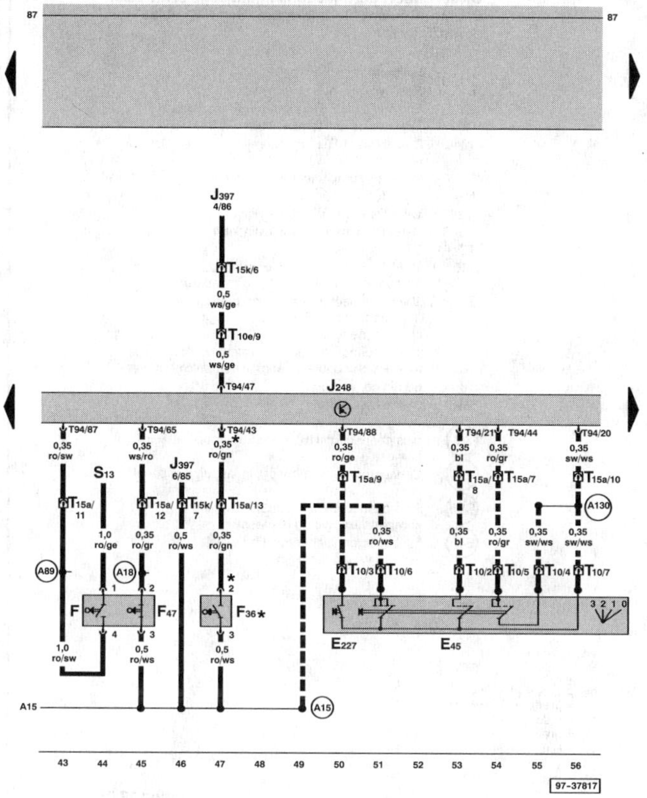

- E45 - Cruise Control Switch
- E227 - Button for cruise control (set)
- F - Brake Light Switch
- F36 - Clutch Vacuum Vent Valve Switch
- F47 - Brake Pedal Switch (cruise control/Diesel Direct Fuel Injection)
- J248 - Diesel Direct Fuel Injection (DFI) Engine Control Module (ECM)
- J397 - Relay for Radiator Fan Afterrun
- S13 - Fuse
- T10 - 10-Pin connector, black
- T10e - 10-Pin connector, orange, on protectiive housing for control module, in engine compartment, left
- T15a - 15-Pin connector, white, on protectiive housing for control module, in engine compartment, left
- T15k - 15-Pin connector, grey
- T94 - 94-Pin connector

- (A15) - plus connection (15) in instrument panel wiring harness
- (A18) - wire connection (54), in instrument panel wiring harness
- (A89) - Connection 2 (54) (in instrument panel wiring harness)
- (A130) - Connection for Cruise Control (instrument panel wiring harness)

- * - manual transmission only
- - - - - cruise control only

ws = white
sw = black
ro = red
br = brown
gn = green
bl = blue
gr = grey
li = lilac
ge = yellow
or = orange

Edition 12.03

Diesel Direct Fuel Injection Engine Control Module, Kick Down Switch, Closed Throttle Position Switch, Mass Air Flow Sensor, Engine Speed Sensor

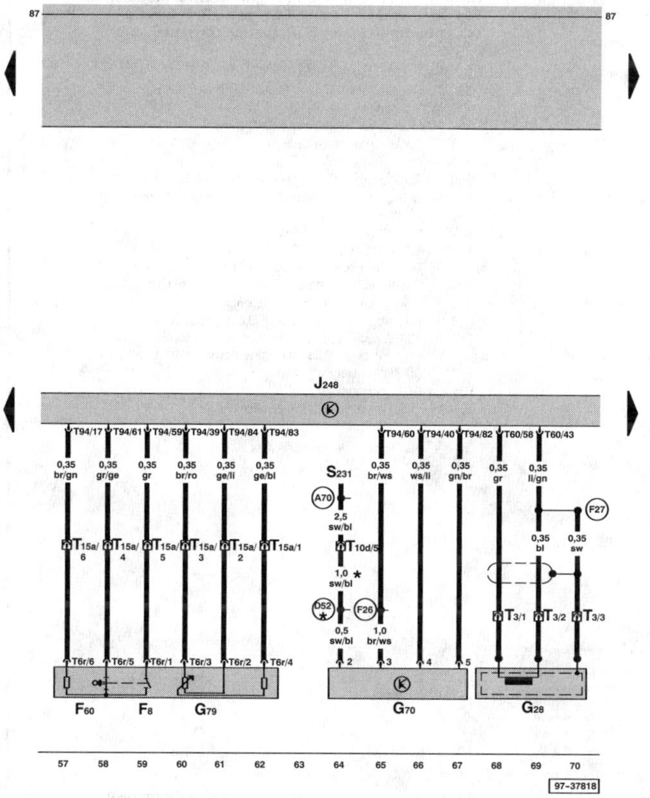

- F8 - Kick Down Switch
- F60 - Closed Throttle Position (CTP) Switch
- G28 - Engine Speed (RPM) Sensor
- G70 - Mass Air Flow (MAF) Sensor
- G79 - Throttle Position (TP) Sensor
- J248 - Diesel Direct Fuel Injection (DFI) Engine Control Module (ECM)
- S231 - Fuse in fuse holder
- T3 - 3-Pin connector
- T6r - 6-Pin connector, black
- T10d - 10-Pin connector, brown, on protectiive housing for control module, in engine compartment, left
- T15a - 15-Pin connector, white, on protectiive housing for control module, in engine compartment, left
- T60 - 60-Pin connector
- T94 - 94-Pin connector

- (A70) - Connector (15a, fuse 231), in instrument panel wiring harness
- (D52) - plus connection (15a), in engine compartment wiring harness
- (F26) - Wire connection -2-, in Diesel Direct Fuel Injection (DFI) system wiring harness
- (F27) - Wire connection (screening), in Diesel Direct Fuel Injection (DFI) system wiring harness

- * - manual transmission only

ws = white
sw = black
ro = red
br = brown
gn = green
bl = blue
gr = grey
li = lilac
ge = yellow
or = orange

Edition 12.03

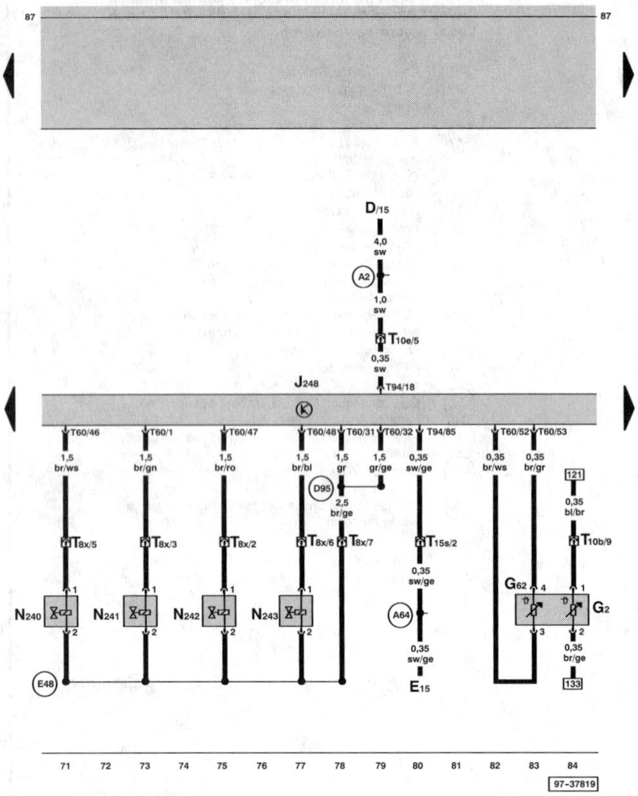

Diesel Direct Fuel Injection Engine Control Module, Engine Coolant Temperature Sensor, Valve for pump/injector

- D - Ignition/Starter Switch
- E15 - Rear window defogger switch
- G2 - Engine Coolant Temperature (ECT) Sensor
- G62 - Engine Coolant Temperature (ECT) Sensor
- J248 - Diesel Direct Fuel Injection (DFI) Engine Control Module (ECM)
- N240 - Valve for pump/injector, cylinder 1
- N241 - Valve for pump/injector, cylinder 2
- N242 - Valve for pump/injector, cylinder 3
- N243 - Valve for pump/injector, cylinder 4
- T8x - 8-Pin connector
- T10b - 10-Pin connector, black, on protectiive housing for control module, in engine compartment, left
- T10e - 10-Pin connector, orange, on protectiive housing for control module, in engine compartment, left
- T15s - 15-Pin connector, red, on protectiive housing for control module, in engine compartment, left
- T60 - 60-Pin connector
- T94 - 94-Pin connector
- (A2) - plus connection (15), in instrument panel wiring harness
- (A64) - Connector (30, idle boost), in instrument panel wiring harness
- (D95) - wire connection (injectors), in engine compartment wiring harness
- (E48) - Fuel Injector Connection (in fuel injection system wiring harness)

ws = white
sw = black
ro = red
br = brown
gn = green
bl = blue
gr = grey
li = lilac
ge = yellow
or = orange

97-37819

Edition 12.03

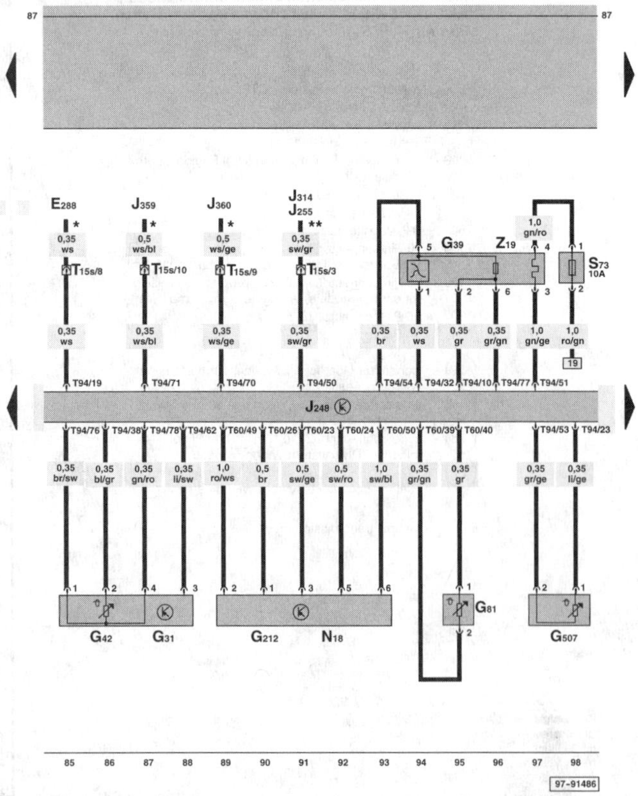

Diesel Direct Fuel Injection Engine Control Module, Fuel Temperature Sensor, Charge Air Pressure Sensor, Potentiometer for EGR, Heated Oxygen Sensor

- E288 - Additional Heater Button (ECON)
- G31 - Charge Air Pressure Sensor
- G39 - Heated Oxygen Sensor (HO2S)
- G42 - Intake Air Temperature (IAT) Sensor
- G81 - Fuel Temperature Sensor
- G212 - Potentiometer for EGR
- G507 - Turbocharger (TC) Intake Temperatur Sensor
- J248 - Diesel Direct Fuel Injection (DFI) Engine Control Module (ECM)
- J255 - Climatronic Control Module
- J314 - A/C Cut-Off Control Module
- J359 - Relay for Preheating Coolant, Low Heat Output
- J360 - Relay for Preheating Coolant, High Heat Output
- N18 - EGR Vacuum Regulator Solenoid Valve
- S73 - Heated Oxygen Sensor (HO2S) Circuit Breaker
- T15s - 15-Pin connector, red, on protectiive housing for control module, in engine compartment, left
- T60 - 60-Pin connector
- T94 - 94-Pin connector
- Z19 - Oxygen Sensor (O2S) Heater

* - Not applicable to USA/CDN
** - air conditioning only

ws = white
sw = black
ro = red
br = brown
gn = green
bl = blue
gr = grey
li = lilac
ge = yellow
or = orange

97-91486

Edition 06.04

Diesel Direct Fuel Injection Engine Control Module, Control module with indicator unit in instrument panel insert, Speedometer Vehicle Speed Sensor

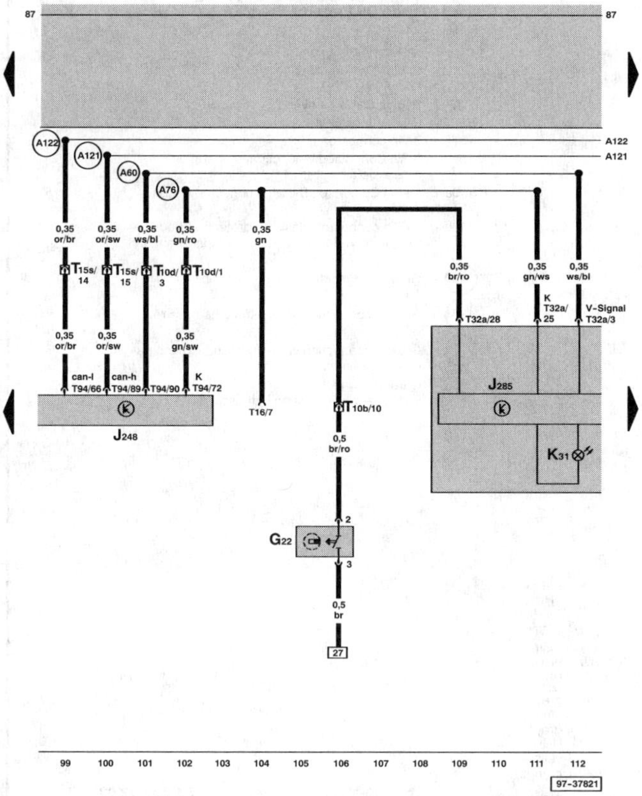

- G22 - Speedometer Vehicle Speed Sensor (VSS)
- J248 - Diesel Direct Fuel Injection (DFI) Engine Control Module (ECM)
- J285 - Control module with indicator unit in instrument panel insert
- K31 - Cruise Control Indicator Light
- T10b - 10-Pin connector, black, on protectiive housing for control module, in engine compartment, left
- T10d - 10-Pin connector, brown, on protectiive housing for control module, in engine compartment, left
- T15s - 15-Pin connector, red, on protectiive housing for control module, in engine compartment, left
- T16 - 16-Pin Connector, data link connector (DLC),
- T32a - 32-Pin connector, blue, on instrument cluster
- T94 - 94-Pin connector

- A60 - Wire connection (Vehicle speed signal), in instrument cluster wiring harness
- A76 - Connector (K-diagnosis wire) in instrument panel wiring harness
- A121 - Connection (high bus) in instrument panel wiring harness
- A122 - Connection (low bus) in instrument panel wiring harness

ws = white
sw = black
ro = red
br = brown
gn = green
bl = blue
gr = grey
li = lilac
ge = yellow
or = orange

Control module with indicator unit in instrument panel insert, Fuel Gauge, Tachometer, Odometer Display

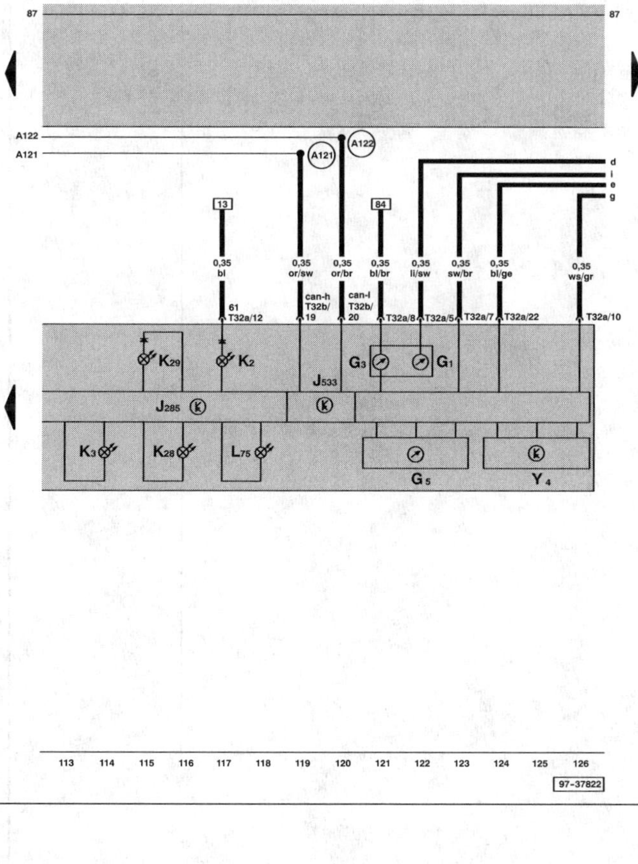

- G1 - Fuel Gauge
- G3 - Engine Coolant Temperature (ECT) Gauge
- G5 - Tachometer
- J285 - Control module with indicator unit in instrument panel insert
- J533 - Data Bus On Board Diagnostic Interface
- K2 - Generator (GEN) Warning Light
- K3 - Oil Pressure Warning Light
- K28 - Engine Coolant Level/Temperature (ECL/ECT) Warning Light
- K29 - Glow Plug Indicator Light
- L75 - Digital Display Light
- T32a - 32-Pin connector, blue, on instrument cluster
- T32b - 32-Pin connector, green, on instrument cluster
- Y4 - Odometer Display

- A121 - Connection (high bus) in instrument panel wiring harness
- A122 - Connection (low bus) in instrument panel wiring harness

ws = white
sw = black
ro = red
br = brown
gn = green
bl = blue
gr = grey
li = lilac
ge = yellow
or = orange

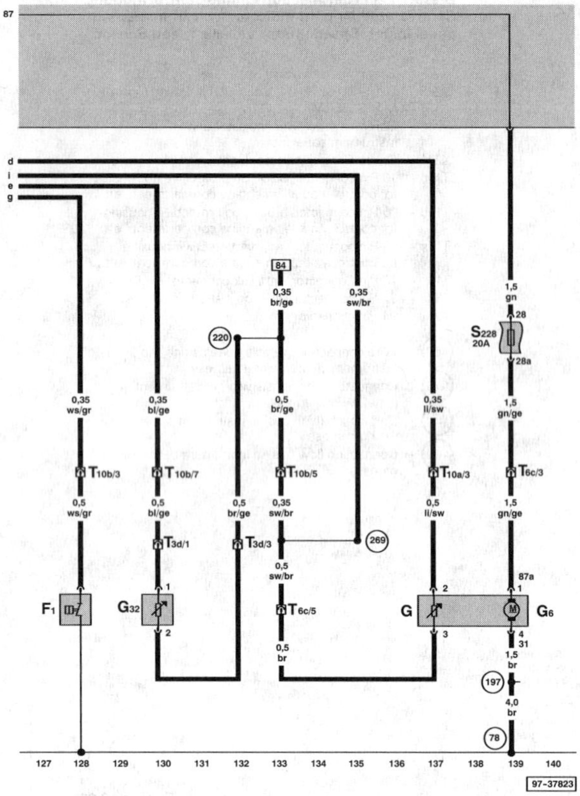

Sender for fuel gauge, Fuel Pump, Engine Coolant Level Sensor, Oil Pressure Switch

F1 - Oil Pressure Switch
G - Sender for fuel gauge
G6 - Fuel Pump (FP)
G32 - Engine Coolant Level (ECL) Sensor
S228 - Fuse in fuse holder
T3d - 3-Pin connector
T6c - 6-Pin Connector, blue, connector station A-pillar, left
T10a - 10-Pin Connector, brown, connector station A-pillar, left
T10b - 10-Pin connector, black, on protectiive housing for control module, in engine compartment, left

(78) - Ground Connection (lower right B-pillar)

(197) - Ground connection -4-, in wiring harness rear

(220) - Ground connection (sensor ground), in engine compartment wiring harness

(269) - Ground connector (sensor ground) -1-, in instrument panel wiring harness

ws = white
sw = black
ro = red
br = brown
gn = green
bl = blue
gr = grey
li = lilac
ge = yellow
or = orange

Edition 12.03

Service Information

Repair Manual Technical Bulletin

Passat 1998 ▸

Bulletin-No.: 22

Electrical Wiring Diagrams, Troubleshooting and Component Locations

Follow these instructions when processing this Technical Bulletin:

- Replace Wiring Diagrams No. 50/4, 50/5,
- Replace complete Wiring Diagram No. 51.
- Replace Wiring Diagrams No. 59/1, 59/3, 69/14, 72/10, 82/15 & 83/8.

Edition 06.04

Component Location and Identification

Component	Code	OBD	Application	Location and additional information
ABS control module (w/EDL)	J104		all	on left side of engine compartment near coolant reservoir, on bottom of hydraulic modulator unit
ABS hydraulic unit			all	in engine compartment, driver's side, forward of upper suspension mount
ABS inlet valve, left front	N101		all	in engine compartment, driver's side, component of ABS hydraulic unit
ABS inlet valve, right front	N99		all	in engine compartment, driver's side, component of ABS hydraulic unit
ABS inlet valve, right rear	N133		all	in engine compartment, driver's side, component of ABS hydraulic unit
ABS inlet valve, right rear	N135		all	in engine compartment, driver's side, component of ABS hydraulic unit
ABS outlet valve, left front	N102		all	in engine compartment, driver's side, component of ABS hydraulic unit
ABS outlet valve, left rear	N134		all	in engine compartment, driver's side, component of ABS hydraulic unit
ABS outlet valve, left rear	N136		all	in engine compartment, driver's side, component of ABS hydraulic unit
ABS outlet valve, right front	N100		all	in engine compartment, driver's side, component of ABS hydraulic unit
ABS return flow pump	V39		all	in engine compartment, driver's side, forward of upper suspension mount, component of J104
ABS return flow pump	V39		all	in engine compartment, driver's side, component of ABS hydraulic unit
ABS return flow pump relay	J105		as applicable	in auxiliary relay panel
ABS return flow pump relay	J105		as applicable	in engine compartment, driver's side, forward of upper suspension mount, component of J104
ABS return flow pump relay 2	J626		as applicable	in engine compartment, driver's side, forward of upper suspension mount, component of J104
ABS solenoid valve relay	J106		all	in auxiliary relay panel
ABS wheel speed sensor left front	G47		all	in passenger's side front wheel bearing housing
ABS wheel speed sensor, left rear	G46		all	in driver's side rear wheel bearing housing
ABS wheel speed sensor, right front	G45		all	in driver's side front wheel bearing housing
ABS wheel speed sensors, right rear	G44		all	in passenger's side rear wheel bearing housing
A/C clutch relay	J44		all	in auxiliary relay panel
A/C Control head	E87		Climatronic	in center console, below radio
A/C cut-off control module	J314		manual A/C	in auxiliary relay panel
A/C diode	J28		manual A/C	in wiring harness near F38
A/C pressure switch (F129)	F129		all	in A/C line at condensor, below passenger's side headlight
Accelerator pedal position sensor	G185		all with Motronic ME7	behind instrument panel, driver's side, above accelerator pedal mount integral with G79
Air flow flap motor	V71		all	above blower motor, right side, integral with G113
Airbag control module	J234		all	on center tunnel, below center console, behind footwell heater duct
Airbag igniter, driver side	N95		all	on airbag assembly, in steering wheel
Airbag igniter, driver side head-level	N251		as applicable	on left rear D pillar, behind trim
Airbag igniter, passenger side 1	N131		all	on airbag assembly, instrument panel, above glovebox
Airbag igniter, passenger side head-level	N252		as applicable	on right rear D pillar, behind trim
Airbag spiral spring	F138		all	on steering column behind steering wheel, with slip ring for horn contact
Alarm horn	H8, H12		all	at rear seat/luggage compartment bulkhead, left rear, behind trim
Ambient temperature switch	F38		manual A/C	below dust/pollen filter, in lower air intake housing
Auxiliary Engine Coolant Pump (EC) Relay	J496	EC relay	4.0L	
Auxiliary Tank Fuel Level Empty Sensor	G293		4.0L	
Auxiliary Tank Fuel Level Sensor	G292		4.0L	
Back pressure flap motor position sensor	G113		all	above blower motor, right side, integral with V71
Back-up light switch	F4		all manual	on left side of transmission, above drive flange, above and behind G22

Component	Code	OBD	Application	Location and additional information
Barometric pressure sensor	F96		1.8L	in e-box in fresh air plenum, driver's side
Brake booster pressure sensor	G294		automatic trans. as applicable	in brake booster vacuum line
Brake booster relay	J569		automatic trans. as applicable	
Brake fluid level warning switch	F34		all	in engine compartment, integral with brake fluid reservoir
Brake light switch	F		➤1999	behind driver's side kneebar on pedal cluster, above brake pedal
Brake light switch	F		2000➤	behind driver's side kneebar on pedal cluster, integral with F47
Brake system vacuum pump	V192		automatic trans. as applicable	in engine compartment, driver's side, forward of ABS hydraulic unit
Brake vacuum vent valve switch	F47		➤1999	behind driver's side kneebar on pedal cluster, below F
Brake vacuum vent valve switch	F47		2000➤	behind driver's side kneebar on pedal cluster, integral with F
Camshaft position sensor	G40	CMP	1.8L	on cylinder head, front, behind upper timing cover and sensor housing
Camshaft position sensor, bank 1	G163	CMP	2.8L	on passenger's side cylinder, front
Camshaft position sensor, bank 2	G40	CMP	2.8L	on driver's side cylinder head, rear
Camshaft position sensor, bank 2	G40	CMP	4.0L	
Camshaft position sensor, bank 2	G62	CMP	4.0L	
Camshaft position sensor, bank 2	G300	CMP	4.0L	
Camshaft position sensor, bank 2	G301	CMP	4.0L	
Central air flap motor	V70		all	on heater box bottom, forward of footwell outlet duct, integral with G112
Central air flap motor position sensor	G112		all	on heater box bottom, forward of footwell outlet duct, integral with V70
Central control module, comfort system	J393		all	in plastic box, under carpet on floorpan, under driver's seat
Change-over valve for intake manifold	N156	IMT	2.8L	at rear of intake manifold, on mounting plate for intake air duct
Climatronic control module	J255		as applicable	In central console, below radio, includes switches and digital display
Closed throttle position switch	F60	CTP	1.8L	on intake manifold, integral with J338
Closed throttle position switch	F60	CTP	2.8L	on intake manifold, rear, integral with J338
Clutch pedal position switch	F194		all manual	behind driver's side kneebar on pedal cluster
Clutch vacuum vent valve switch	F36		all manual	behind driver's side kneebar on pedal cluster
Cold start injector	N108		1.9L	integral with Diesel injection pump
Connector, 3-pin blue	T3c		2.8L	in engine compartment, on bulkhead, right side, for G66
Connector, 3-pin green, blue, or brown	T3a		2.8L	in engine compartment, on bulkhead, right side, for G61
Connector, 4-pin black	T4		2.8L	in engine compartment, on bulkhead, right side, for G39 and Z19
Connector, 4-pin black	T4b		2.8L	in engine compartment, on bulkhead, left side, behind coolant reservoir, for G108 and Z28
Connector, 4-pin brown	T4d		2.8L	in engine compartment, on bulkhead, right side, for G131 and Z30
Connector, 4-pin green	T4c		2.8L	in engine compartment, on bulkhead, right side, for G130 and Z29
Coolant after-run pump	V51		as applicable	in engine compartment, lower right side, to the right of engine
Coolant Fan Control module	J293	FC module	Climatronic	on left front frame rail bottom
Coolant Fan Control relay	J26	FC relay	manual A/C	in auxiliary relay panel
Coolant Fan Control relay, first speed	J279	FC relay	manual A/C	in auxiliary relay panel
Coolant Fan Control relay, second speed	J280	FC relay	manual A/C	in auxiliary relay panel
Coolant Fan Control series resistance	N39	FC	all	on left front frame rail bottom
Coolant Fan Control thermal switch	F18	FC	all	on radiator, lower left side
Coolant glow plug relay, high heat output	J360		1.9L	in e-box in fresh air plenum, driver's side
Coolant glow plug relay, low heat output	J359		1.9L	in e-box in fresh air plenum, driver's side
Coolant glow plugs	Q7		1.9L	in coolant outlet flange, at rear of cylinder head
Crash sensor for side airbag, driver's side	G179		all	under driver's seat and carpet, mounted on back side of floorpan crossmember

Component	Code	OBD	Application	Location and additional information
Crash sensor for side airbag, passenger's side	G180		all	under passenger's seat and carpet, mounted on back side of floorpan crossmember
Cruise control ECM	J213	CSS	all without Motronic ME7	under instrument panel, passenger's side, forward of glove box
Cruise control vacuum pump	V18	CSS	all without Motronic ME7	in engine compartment, left front, under the ABS hydraulic unit and behind wheel well liner
Data bus, On Board Diagnostic Interface	J533		as applicable	integral with instrument panel
Daytime running light relay	J89		as applicable	in auxiliary relay panel
Diesel Direct Fuel Injection, Engine Control Module	J248	DFI ECM	1.9L	in e-box in fresh air plenum, driver's side
Diesel Direct Fuel Injection, Engine Control Module power supply relay	J317	DFI ECM relay	1.9L	in e-box in fresh air plenum, driver's side
Diode, relay protection	J443		manual A/C	in wiring harness near J307
Door control module, L/F	J386		early (with green illumination)	in driver's door, behind trim panel, integral with switch assembly
Door control module, L/F	J386		later (with blue illumination)	in driver's door, behind trim panel, integral with motor assembly
Door control module, L/R	J388		all	in L/R door, behind trim panel, integral with motor assembly
Door control module, R/F	J387		all	in passenger's front door, behind trim panel, integral with motor assembly
Door control module, R/R	J389		all	in R/R door, behind trim panel, integral with motor assembly
Engine Control Module	J220	ECM	all	in e-box in fresh air plenum, driver's side
Engine Coolant Level Sensor	G32	ECL	all	In bottom of coolant reservoir
Engine Coolant switch-off valve (heater valve)	N279	EC valve	Climatronic	in heater/evaporator housing
Engine coolant temperature sensor	G62	ECT	2.8L	in coolant pipe, behind right cylinder head, integral with G2
Engine coolant temperature sensor	G83	ECT	4.0L	on radiator
Engine coolant temperature sensor	G2	ECT	2.8L	in coolant pipe, behind right cylinder head, integral with G62
Engine coolant temperature sensor	G62	ECT	1.8L	in coolant flange on rear of cylinder head, integral with G2
Engine Coolant Temperature sensor	G2	ECT	1.8L	in coolant flange on rear of cylinder head, integral with G62
Engine oil temperature sensor	G8		2.8	on left front of engine, threaded into oil pump housing
Engine oil temperature sensor	G8		1.8L	on driver's side of cylinder block, in oil filter housing
Engine speed sensor	G28	RPM	1.8L	on drivers's side front of engine block, lower left, near transmission flange
Engine speed sensor	G28	RPM	4.0L	
Engine speed sensor	G28	RPM	2.8L	in transmission bellhousing, front, drivers's side
Evaporative emission canister purge regulator valve	N80	EVAP	1.8L	in engine compartment, passengers's side, on top of air cleaner housing
Evaporative emission canister purge regulator valve (N80)	N80	EVAP	2.8L	in engine compartment, passenger's side, on top of air cleaner housing
Exhaust Gas Recirculation vacuum regulator solenoid valve	N18	EGR	1.9L	attached to connecting pipe, charge air cooler to intake manifold
Fog light relay	J5		as applicable	in auxiliary relay panel
Footwell /defrost flap motor	V85		as applicable	on left side of heater box, above footwell outlet, integral with G114
Footwell /defrost flap motor position sensor	G114		as applicable	on left side of heater box, above footwell outlet, integral with flap motor V85
Fresh air blower	V2		all	in fresh air inlet housing, behind (in front of) glove box, integral with J126
Fresh air blower fan control module	J126		Climatronic	in fresh air inlet housing, behind (in front of) glove box, integral with V2
Fresh air blower series resistance	N24		manual A/C	in fresh air inlet housing, behind (in front of) glove box, integral with V2
Fresh air intake duct temperature sensor	G89		Climatronic	in fresh air plenum, right side, under dust and pollen filter, in duct

Component	Code	OBD	Application	Location and additional information
Fresh/recirculating air door flap servo motor	V154		manual A/C	in intake air duct, right side, under dust and pollen filter
Fuel gauge sending unit (level sensor 2)	G169		as applicable	in fuel tank, with access through luggage compartment, integral with G6
Fuel gauge sending unit (level sensor 3)	G237		as applicable	in fuel tank, with access through luggage compartment, integral with G6
Fuel gauge sending unit (level sensor)	G		all	in fuel tank, with access through luggage compartment, integral with G6
Fuel injectors (1 - 4)	N30, N32, N32, N33		1.8L	on driver's side of cylinder head, on fuel rail, cylinder #1 injector closest to front (belt end) of engine
Fuel injectors (1 - 6)	N30, N32, N32, N33, N83, N84		2.8L	between intake manifold and cylinder head, on fuel rail, cylinder #1 injector closest to front on passenger's side of engine
Fuel injectors (1 - 6)	N30, N32, N32, N33, N83, N84, N85, N86		4.0L	
Fuel pump	G6	FP	all	in fuel tank, with access through luggage compartment, integral with G
Fuel Pump Control Module	J538	FP Control Module	4.0L	
Fuel pump relay	J17	FP relay	all	in relay panel
Fuel pump, transfer	G23		4.0L	
Fuel tank lid unlock motor	V155		all	in luggage compartment, right side
Fuel temperature sensor	G81		1.9L	integral with Diesel injection pump
Fuel temperature sensor	G81		1.9L	integral with Diesel injection pump
Generator (alternator)	C	GEN	1.8L	in engine compartment, belt-driven, front, to the left of the cylinder head, integral with C1
Generator (alternator)	C	GEN	2.8L, 4.0L	in engine compartment, belt-driven, lower right front, integral with C1
Generator, (solar cell)	C20	GEN	4.0L	
Glow plug relay	J59		1.9L	in auxiliary relay panel
Glow plug(s)	Q6		1.9L	in cylinder head, near fuel injector for individual cylinder
Heated seat control module, L/F	J131		as applicable	In central console, to the left of radio, integral with E94
Heated seat control module, L/R	J215		as applicable	In center console, at the L/R, integral with E128
Heated seat control module, R/F	J132		as applicable	In central console, to the right of radio, integral with E95
Heated seat control module, R/R	J216		as applicable	In center console, at the R/R, integral with E129
Heated seat seat adjuster, L/F	E94		as applicable	In central console, to the left of radio, integral with J131
Heated seat seat adjuster, L/R	E128		as applicable	In center console, at the L/R, integral with J215
Heated seat seat adjuster, R/F	E95		as applicable	In central console, to the right of radio, integral with J132
Heated seat seat adjuster, R/R	E129		as applicable	In center console, at the R/R, integral with J216
Heated seat temperature sensor, L/F	G59		as applicable	in heating element pad for L/F seat cushion, at the rear, integral with Z6
Heated seat temperature sensor, L/R	G94		as applicable	in heating element pad for L/R seat cushion, at the rear, integral with Z10
Heated seat temperature sensor, R/F	G60		as applicable	in heating element pad for R/F seat cushion, at the rear, integral with Z8
Heated seat temperature sensor, R/R	G95		as applicable	in heating element pad for R/R seat cushion, at the rear, integral with Z12
Heated seat, seat occupied sensor, L/R	G177		as applicable	between upholstery and bottom cushion, L/R seat
Heated seat, seat occupied sensor, R/R	G178		as applicable	between upholstery and bottom cushion, R/R seat
Heating element, L/F seat backrest	Z7		as applicable	between upholstery and backrest cushion, L/F seat,

Component	Code	OBD	Application	Location and additional information
Heating element, L/F seat cushion	Z6		as applicable	between upholstery and bottom cushion, L/F seat, integral with G59
Heating element, L/R seat backrest	Z11		as applicable	between upholstery and backrest cushion, L/R seat,
Heating element, L/R seat cushion	Z10		as applicable	between upholstery and bottom cushion, L/R seat, integral with G94
Heating element, R/F seat backrest	Z9		as applicable	between upholstery and backrest cushion, R/F seat,
Heating element, R/F seat cushion	Z8		as applicable	between upholstery and bottom cushion, R/F seat, integral with G60
Heating element, R/R seat backrest	Z13		as applicable	between upholstery and backrest cushion, R/R seat,
Heating element, R/R seat cushion	Z12		as applicable	between upholstery and bottom cushion, R/R seat, integral with G95
Home link - garage door opener control module	J530		as applicable	in driver's sun visor
Hood alarm switch	F120		as applicable	on hood lock carrier, integral with hood latch assembly
Horn button	H		all	in steering wheel
Horn, high tone	H2		as applicable	right front, below headlight
Horn, low tone	H7		as applicable	right front, below headlight
Horns, dual	H1		as applicable	right front, below headlight
Hydraulic pump for traction control	V156		as applicable	on left side of engine compartment near coolant reservoir, on bottom of hydraulic modulator unit
Ignition coil (1-6, 2-4, 3-5)	N152		2.8L	in front of engine, center, in front of intake manifold, integral with paired Power Output Stages
Ignition coils (1, 2, 3, 4)	N, N128, N158, N163		1.8L ➤1999	on top of cylinder head, above cylinder number 1, farthest forward remaining coil on top of cylinder head, above individual cylinders
Ignition coils (1, 2, 3, 4)	N70, N127, N291, N292		1.8L 2000➤	on top of cylinder head, above cylinder number 1, farthest forward remaining coil on top of cylinder head, above individual cylinders, integral with individual Power Output Stages
Ignition coils (1, 2, 3, 4, 5, 6, 7, 8)	N70, N127, N291, N292, N323, N324, N325, N326		4.0L W8	on top of cylinder head, above individual cylinders, integral with individual Power Output Stages
Ignition suppressor, terminal #15	C5		2.8L	in ignition coil wiring
Induction coil for immobilizer	D2		all	in housing with ignition switch
Instrument panel indicator unit	J285		as applicable	integral with instrument panel
Instrument panel interior temperature sensor	G56		as applicable	In Climatronic control head, integral with E87, V42, and J255
Intake air temperature sensor	G72	IAT	1.8L	in intake manifold, behind, throttle valve housing
Intake air temperature sensor	G72	IAT	1.9L	in intake duct, just before intake manifold, integral with G71
Intake air temperature sensor	G42	IAT	2.8L	in intake air duct, at rear of intake manifold
Interior temperature sensor fan	V42		as applicable	In Climatronic control head, integral with E87, G56, and J255
Intermittent wiper washer relay	J31		all	in relay panel
Kick down switch for automatic transmission	F8		all automatic transmission	in engine compartment, rear, on bulkhead, integral with accelerator cable
Knock sensor 1	G61	KS1	1.8L	on driver's side of cylinder block, near coolant outlet housing
Knock sensor 1	G61	KS1	2.8L	below intake manifold, in valley area on right side (bank 1) of cylinder block
Knock sensor 1	G61	KS1	4.0L	
Knock sensor 2	G61	KS2	4.0L	
Knock sensor 2	G66	KS2	1.8L	on driver's side of cylinder block, rear
Knock sensor 2	G66	KS2	2.8L	below intake manifold, in valley area on left side (bank 2) of cylinder block
Knock sensor 3	G198	KS3	4.0L	
Knock sensor 4	G199	KS4	4.0L	

Component	Code	OBD	Application	Location and additional information
Leak detection pump	V144	LDP	as applicable	inside left-rear wheelhousing, behind wheelhousing liner
Left Electro-Hydraulic Engine Mount Solenoid Valve,	N144		4.0L	
Load reduction relay	J59		all	in relay panel
Lock unit for central locking, L/F	F220		all	in L/F door, behind trim panel
Lock unit for central locking, L/R	F222		all	in L/R door, behind trim panel
Lock unit for central locking, L/R	F223		all	in R/R door, behind trim panel
Lock unit for central locking, R/F	F221		all	in R/F door, behind trim panel
Manifold Absolute Pressure sensor	G71	MAP	1.9L	in intake duct, just before intake manifold, integral with G72
Map Controlled Engine Cooling Thermostat	F265		4.0L	
Mass air flow sensor	G70	MAF	1.8L	in engine compartment, passenger's side, inside air cleaner housing
Mass air flow sensor	G70	MAF	2.8L	in engine compartment, passenger's side, attached to air cleaner housing
Memory seat adjustment switches	E218, E219, E220, E211, E210, E209, E208, E96, E213, E212, E190		with memory seats	on L/F seat frame bottom, in common housing
Memory seat control module	J136		with memory seats	on L/F seat frame bottom
Modulating Piston Displacement Sensor	G149	MPDS	1.9L	integral with Diesel injection pump
Motronic engine control module power supply relay	J271	ECM relay	all	in e-box in fresh air plenum, driver's side
Multi-function steering wheel control module	J453		as applicable	in auxiliary relay panel
Multi-function transmission range switch	F125	TR switch	all automatics	on transmission case, driver's side, behind mounting bracket
Needle lift sensor	G80		1.9L	integral with cylinder 3 fuel injector
Oil level thermal sensor	G266		as applicable	on bottom of oil pan
Oil pressure switch	F22		1,9L	on rear of cylinder head, for DFI ECM
Oil pressure switch	F1		1.8L, 1,9L	on driver's side of engine block, center, in oil filter housing
Oil pressure switch	F1		4.0L	
Oil pressure switch	F1		2.8L	on lower driver's side on oil pan, near oil filter
Outside air temperature sensor	G17		all	below front bumper, center, behind lower grille
Oxygen sensor heater, post-cat B1S2	Z29	O2S	2.8L	in exhaust system, right side, in pipe at outlet of catalytic converter, integral with G130
Oxygen sensor heater, post-cat B2S2	Z30	O2S	2.8L	in exhaust system, left side, in pipe at outlet of catalytic converter, integral with G131
Oxygen sensor heater, post-cat, B1S2	Z29	O2S	1.8L	in exhaust system, in pipe at outlet of catalytic converter, integral with G130
Oxygen sensor heater, pre-cat B2S1	Z28	O2S	2.8L	in exhaust manifold, left side, bank 2, ahead of catalytic converter, integral with G108
Oxygen sensor heater, pre-cat, B1S1	Z19	O2S	1.8L	in exhaust system, in pipe at inlet of catalytic converter, integral with G39
Oxygen sensor heater, pre-cat, B1S1	Z19	O2S	2.8L	in exhaust manifold, right side, bank 1, ahead of catalytic converter, integral with G39
Oxygen sensor post-cat B1S2, heated	G130	HO2S	1.8L	in exhaust system, in pipe at outlet of catalytic converter, integral with Z29
Oxygen sensor post-cat B1S2, heated	G130	HO2S	2.8L	in exhaust system, right side, in pipe at outlet of catalytic converter, integral with Z29
Oxygen sensor post-cat B1S2, heated	G130	HO2S	4.0L	integral with Z29

Component	Code	OBD	Application	Location and additional information
Oxygen sensor post-cat B2S2, heated	G131	HO2S	2.8L	in exhaust system, left side, in pipe at outlet of catalytic converter, integral with Z30
Oxygen sensor post-cat B2S2, heated	G131	HO2S	4.0L	integral with Z30
Oxygen sensor, pre-cat, B1S1, heated	G39	HO2S	2.8L	in exhaust manifold, right side, bank 1, ahead of catalytic converter, integral with Z19
Oxygen sensor, pre-cat, B1S1, heated	G39	HO2S	4.0L	integral with Z19
Oxygen sensor, pre-cat, B1S1, heated	G39	HO2S	1.8L	in exhaust system, in pipe at inlet of catalytic converter, integral with Z19
Oxygen sensor, pre-cat, B2S1, heated	G108	HO2S	2.8L	in exhaust manifold, left side, bank 2, ahead of catalytic converter, integral with Z28
Oxygen sensor, pre-cat, B2S1, heated	G108	HO2S	4.0L	integral with Z28
Park/Neutral Position relay	J226	PNP relay	all	in auxiliary relay panel
Parking aid control module	J446		as applicable	in luggage compartment, behind trim panel, left rear
Parking aid sensors	G203, G204, G205, G206		as applicable	in rear bumper
Parking aid warning buzzer	H15		as applicable	in instrument cluster
Parking brake warning switch	F9		all	under center console, central tunnel, below parking brake lever
Power output stage	N122		1.8L ➤1999	in engine compartment, passenger's side, on top of air cleaner housing
Quantity adjuster	G146		1.9L	integral with Diesel injection pump
Rain sensor	G213		as applicable	integral with interior rear-view mirror housing
Rear lid central locking system motor	V53		all	in rear lid, attached to rear lock assembly
Rear lid remote unlock motor	V151		all	in rear lid, attached to rear lock assembly
Rear lid remote unlock motor relay	J398		as applicable	on auxiliary relay panel
Rear window wiper motor	V12		Wagon	in rear hatch, center, behind trim panel
Recirculating valve for turbocharger	N249		1.8L	left side of engine, under intake manifold, on bracket
Seat belt microswitch, L/F	F140		as applicable	in L/F seat belt buckle
Seat belt microswitch, R/F	F141		as applicable	in R/F seat belt buckle
Secondary air injection motor	V101	AIR	all	in engine compartment, right front, accessed from below
Secondary air injection pump relay	J299	AIR	all	in e-box in fresh air plenum, driver's side
Secondary air injection solenoid valve	N112	AIR	2.8L	at rear of intake manifold, on mounting plate for air intake duct
Secondary air injection solenoid valve	N112	AIR	1.8L	left side of engine, under intake manifold, on bracket
Sender 1 for brake booster	G201		as applicable	in engine compartment, driver's side, component of ABS hydraulic unit
Sensor for rotational rate (yaw)	G202		with ESP	under rear seat, integral with G200
Sensor for transverse acceleration (lateral)	G200		with ESP	under rear seat, integral with G202
Shift lock solenoid	N110		all automatics	on central tunnel, integral with gear selector mechanism
Side airbag igniter, driver side	N199		all	on left side of driver seat backrest bolster, beneath seat cover
Side airbag igniter, passenger side	N200		all	on right side of passenger seat backrest bolster, beneath seat cover
Solar Cell Separation Relay	J309		4.0L	
Solar Operation Control Module	J355		4.0L	
Starter interlock relay (anti-theft, immobilizer)	J433		all	on auxiliary relay panel
Starter interlock relay (clutch pedal)	J434		all manual trans.	on auxiliary relay panel
Starter interlock relay (park / neutral)	J207		all automatics	on auxiliary relay panel
Steering angle sensor	G85		with ESP	in steering column, under slip ring and spiral spring
Sunlight photo sensor	G107		as applicable	in center of instrument panel, inside defroster grille
Sunroof control module	J245		as applicable	on main roof panel, front center, forward of sunroof, integral with sunroof motor V1

Component	Code	OBD	Application	Location and additional information
Temperature flap motor position sensor	G92		as applicable	on heater box bottom, forward of footwell outlet duct, integral with V68
Temperature regulator flap motor	V68		as applicable	on heater box bottom, forward of footwell outlet duct, integral with G92
Temperature sensor, center air outlet	G191		as applicable	on bottom of heater box, center, above center tunnel
Temperature sensor, evaporator vent	G263		Climatronic	in the air duct on the outlet side of the evaporator
Temperature sensor, floor air outlet	G192		as applicable	in footwell duct, bottom, driver's side
Throttle drive (EPC actuator)	G186		1.8L with Motronic ME7	in engine compartment, on intake manifold, integral with J338
Throttle drive (EPC actuator) angle sensor 1	G187		1.8 with Motronic ME7	in engine compartment, on intake manifold, integral with J338
Throttle drive (EPC actuator) angle sensor 2	G188		1.8L with Motronic ME7	in engine compartment, on intake manifold, integral with J338
Throttle position actuator	V60		1.8L, 2.8L with Motronic M5.9.2	on intake manifold, component of throttle valve control module J338
Throttle position sensor	G69		1.8L, 2.8L with Motronic M5.9.2	on intake manifold, component of throttle valve control module J339
Throttle position sensor	G88		1.8L, 2.8L with Motronic M5.9.2	on intake manifold, component of throttle valve control module J338
Throttle position sensor	G79		all with Motronic ME7	behind instrument panel, driver's side, above accelerator pedal mount integral with G185
Throttle valve control module	J338		all with Motronic ME7	on intake manifold, integral with throttle valve housing and G186, G187, G188
Throttle valve control module	J338		all with Motronic M5.9.2	on intake manifold, integral with throttle valve housing and V60, G69, F60
Tiptronic switch	F189		all automatics	on central tunnel, integral with gear selector mechanism
Traction control outlet valve, L/F	N169		as applicable	in engine compartment, driver's side, forward of upper suspension mount, component of J104
Traction control outlet valve, R/F	N167		as applicable	in engine compartment, driver's side, forward of upper suspension mount, component of J104
Traction control switch-over valve, L/F	N168		as applicable	in engine compartment, driver's side, forward of upper suspension mount, component of J104
Traction control switch-over valve, R/F	N166		as applicable	in engine compartment, driver's side, forward of upper suspension mount, component of J104
Transmission control module	J217	TCM	all automatics	in passenger's compartment, just forward of front passenger's seat, on floorpan under carpet, in black plastic protective case
Transmission Fluid Temperature Sensor	G93		4.0L	
Transmission selector lever light relay	J307		automatic transmission as applicable	on auxiliary relay panel
Trunk lock alarm switch	F218		all	in rear lid, mounted on locking mechanism
Turn signal, emergency flasher relay			all	integral with E3, emergency flasher switch, in central dash console
Valve -1- for camshaft adjustment	N205		1.8L as applicable	at rear of cylinder head
Valve -1- for camshaft adjustment	N205		2.8L	at rear of cylinder head, passenger's side
Valve -1- for camshaft adjustment	N205		4.0L	
Valve -1- for exhaust camshaft adjustment	N318		4.0L	
Valve -2- for camshaft adjustment	N206		4.0L	
Valve -2- for camshaft adjustment	N206		2.8L	at front of cylinder head, driver's side
Valve -2- for exhaust camshaft adjustment	N319		4.0L	
Vehicle speed sensor, speedometer	G22	VSS	all	on left side of transmission housing, at transmission output flange

Component	Code	OBD	Application	Location and additional information
Vehicle speed sensor, transmission	G68	VSS	automatic transmission	in transmission main housing, on top of valve body
Vehicle speed sensor, transmission	G38	VSS	automatic transmission	on transmission main housing, left side
Voltage Regulator	C1	VR	all	on back of generator, integral with C
Wastegate bypass regulator valve	N75		1.8L	in engine compartment, passenger's side, attached to intake air duct
Wastegate bypass regulator valve	N75		1.9L	in engine compartment, passenger's side
Window regulator motor, L/F	V147		all	in driver's door, behind trim panel, integral with J386 in vehicles with red instrument illumination, separate in vehicles with green instrument illumination
Window regulator motor, L/R	V26		all	in rear driver's side door, behind trim panel, integral with J388
Window regulator motor, R/F	V148		all	in front passenger's side door, behind trim panel, integral with J387
Window regulator motor, R/R	V27		all	in rear passenger's side door, behind trim panel, integral with J389
Windshield washer fluid level sensor	G33		all	in engine compartment, on windshield washer fluid reservoir
Windshield washer pump	V59		all	in engine compartment, on windshield washer fluid reservoir, left front

Notes

Scan Tool

Section

Z

Page

General

All vehicles covered by this manual are equipped with the Government required diagnostic system known as On-Board Diagnostics II (OBD II). This system monitors operation and function of all engine management system activity and automatic transmission operation to insure compliance with specified emission levels.

Vehicle emission levels are constantly monitored by the OBD II system and malfunctions are recognized and recorded. A Malfunction Indicator Light (MIL) in the instrument cluster alerts the driver to the fault and the need to have the system checked for fault codes. These codes follow a standard format and are known as Diagnostic Trouble Codes (DTCs).

DTCs are assigned two codes. The first code is a numerical code assigned by the factory. The second code is referred to as a P-code and follows a structure required by law and defined by the Society of Automotive Engineers (SAE). This standard uses a letter to designate the system and four numbers to further identify and detail the malfunction as listed below.

First digit structure is as follows:

- Pxxxx for powertrain
- Bxxxx for body
- Cxxxx for chassis
- Uxxxx for future systems

Second digit structure is:

- P0xxx Government required codes
- P1xxx Manufacturer codes for additional emission system function; not required but reported to the government

Third digit structure is:

- Px1xx measurement of air and fuel
- Px2xx measurement of air and fuel
- Px3xx ignition system
- Px4xx additional emission control
- Px5xx speed and idle regulation
- Px6xx computer and output signals
- Px7xx transmission
- Px8xx transmission
- Px9xx control modules, input and output signals

The fourth and fifth digits designate the individual component and system.

DTCs and data can be retrieved with Volkswagen/Audi Factory Scan Tools such as the VAG 1551/1552, the new diagnostic computers VAS 5051/5052, or aftermarket scan tools, through a Data Link Connector (DLC). For location of the data link connector, see **Maintenance** section. Some aftermarket scan tools and computer programs are capable of emulating factory scan tools and diagnostic modes. These factory modes allow the scan tool to be used for other system diagnostic functions such as component coding and information retrieval.

Some aftermarket scan tools only retrieve and erase codes in a generic mode. The generic mode is not capable of factory testing and control module coding, but allows scan tools to be used simply to read DTCs. Generic scan tool mode displays DTCs as SAE P-codes.

Diagnostic Trouble Code Table

NOTE —

The following table contains a list of available powertrain scan tool codes for Volkswagen and Audi vehicles available at the time of publication. Not all of the codes apply to the vehicles covered by this manual.

Diagnostic Trouble Codes (Powertrain)

V.A.G. code	SAE P-code	Description
16394	P0010	-A- Camshaft Pos. Actuator Circ. Bank 1 Malfunction
16395	P0020	-A- Camshaft Pos. Actuator Circ. Bank 2 Malfunction
16449	P0065	Air Assisted Injector Control Range/Performance
16450	P0066	Air Assisted Injector Control Low Input/Short to ground
16451	P0067	Air Assisted Injector Control Input/Short to B+
16485	P0101	Mass or Volume Air Flow Circ Range/Performance
16486	P0102	Mass or Volume Air Flow Circ Low Input
16487	P0103	Mass or Volume Air Flow Circ High Input
16489	P0105	Manifold Abs.Pressure or Bar.Pressure Voltage supply
16490	P0106	Manifold Abs.Pressure or Bar.Pressure Range/Performance
16491	P0107	Manifold Abs.Pressure or Bar.Pressure Low Input
16492	P0108	Manifold Abs.Pressure or Bar.Pressure High Input
16496	P0112	Intake Air Temp.Circ Low Input
16497	P0113	Intake Air Temp.Circ High Input
16500	P0116	Engine Coolant Temp.Circ Range/Performance
16501	P0117	Engine Coolant Temp.Circ Low Input
16502	P0118	Engine Coolant Temp.Circ High Input
16504	P0120	Throttle/Pedal Pos.Sensor A Circ Malfunction
16505	P0121	Throttle/Pedal Pos.Sensor A Circ Range/Performance
16506	P0122	Throttle/Pedal Pos.Sensor A Circ Low Input
16507	P0123	Throttle/Pedal Pos.Sensor A Circ High Input
16509	P0125	Insufficient Coolant Temp.for Closed Loop Fuel Control
16512	P0128	Coolant Thermostat/Valve Temperature below control range
16514	P0130	O2 Sensor Circ.,Bank1-Sensor1 Malfunction
16515	P0131	O2 Sensor Circ.,Bank1-Sensor1 Low Voltage
16516	P0132	O2 Sensor Circ.,Bank1-Sensor1 High Voltage
16517	P0133	O2 Sensor Circ.,Bank1-Sensor1 Slow Response
16518	P0134	O2 Sensor Circ.,Bank1-Sensor1 No Activity Detected
16519	P0135	O2 Sensor Heater Circ.,Bank1-Sensor1 Malfunction
16520	P0136	O2 Sensor Circ.,Bank1-Sensor2 Malfunction
16521	P0137	O2 Sensor Circ.,Bank1-Sensor2 Low Voltage
16522	P0138	O2 Sensor Circ.,Bank1-Sensor2 High Voltage

Diagnostic Trouble Codes (Powertrain)

V.A.G. code	SAE P-code	Description
16523	P0139	O2 Sensor Circ.,Bank1-Sensor2 Slow Response
16524	P0140	O2 Sensor Circ.,Bank1-Sensor2 No Activity Detected
16525	P0141	O2 Sensor Heater Circ.,Bank1-Sensor2 Malfunction
16534	P0150	O2 Sensor Circ.,Bank2-Sensor1 Malfunction
16535	P0151	O2 Sensor Circ.,Bank2-Sensor1 Low Voltage
16536	P0152	O2 Sensor Circ.,Bank2-Sensor1 High Voltage
16537	P0153	O2 Sensor Circ.,Bank2-Sensor1 Slow Response
16538	P0154	O2 Sensor Circ.,Bank2-Sensor1 No Activity Detected
16539	P0155	O2 Sensor Heater Circ.,Bank2-Sensor1 Malfunction
16540	P0156	O2 Sensor Circ.,Bank2-Sensor2 Malfunction
16541	P0157	O2 Sensor Circ.,Bank2-Sensor2 Low Voltage
16542	P0158	O2 Sensor Circ.,Bank2-Sensor2 High Voltage
16543	P0159	O2 Sensor Circ.,Bank2-Sensor2 Slow Response
16544	P0160	O2 Sensor Circ.,Bank2-Sensor2 No Activity Detected
16545	P0161	O2 Sensor Heater Circ.,Bank2-Sensor2 Malfunction
16554	P0170	Fuel Trim,Bank1 Malfunction
16555	P0171	Fuel Trim,Bank1 System too Lean
16556	P0172	Fuel Trim,Bank1 System too Rich
16557	P0173	Fuel Trim,Bank2 Malfunction
16558	P0174	Fuel Trim,Bank2 System too Lean
16559	P0175	Fuel Trim,Bank2 System too Rich
16566	P0182	Fuel temperature sender-G81 Short to ground
16567	P0183	Fuel temperature sender-G81 Interruption/Short to B+
16581	P0197	Engine Oil Temperature Circuit Low Input
16582	P0198	Engine Oil Temperature Circuit High Input
16585	P0201	Cyl.1, Injector Circuit Fault in electrical circuit
16586	P0202	Cyl.2, Injector Circuit Fault in electrical circuit
16587	P0203	Cyl.3, Injector Circuit Fault in electrical circuit
16588	P0204	Cyl.4, Injector Circuit Fault in electrical circuit
16589	P0205	Cyl.5 Injector Circuit Fault in electrical circuit
16590	P0206	Cyl.6 Injector Circuit Fault in electrical circuit
16591	P0207	Cyl.7 Injector Circuit Fault in electrical circuit
16592	P0208	Cyl.8 Injector Circuit Fault in electrical circuit
16599	P0215	Engine Shut-Off Solenoid Malfunction
16600	P0216	Injector/Injection Timing Control Malfunction
16603	P0219	Engine Overspeed Condition

Diagnostic Trouble Codes (Powertrain)

V.A.G. code	SAE P-code	Description
16605	P0221	Throttle Pos. Sensor -B- Circuit Range/Performance
16606	P0222	Throttle Pos. Sensor -B- Circuit Low Input
16607	P0223	Throttle Pos. Sensor -B- Circuit High Input
16609	P0225	Throttle Pos. Sensor -C- Circuit Voltage supply
16610	P0226	Throttle Pos. Sensor -C- Circuit Range/Performance
16611	P0227	Throttle Pos. Sensor -C- Circuit Low Input
16612	P0228	Throttle Pos. Sensor -C- Circuit High Input
16614	P0230	Fuel Pump Primary Circuit Fault in electrical circuit
16618	P0234	Turbocharger Overboost Condition Control limit exceeded
16619	P0235	Turbocharger Boost Sensor (A) Circ Control limit not reached
16620	P0236	Turbocharger Boost Sensor (A) Circ Range/Performance
16621	P0237	Turbocharger Boost Sensor (A) Circ Low Input
16622	P0238	Turbocharger Boost Sensor (A) Circ High Input
16627	P0243	Turbocharger Wastegate Solenoid (A) Open/Short Circuit to Ground
16629	P0245	Turbocharger Wastegate Solenoid (A) Low Input/Short to ground
16630	P0246	Turbocharger Wastegate Solenoid (A) High Input/Short to B+
16636	P0252	Injection Pump Metering Control (A) Range/Performance
16645	P0261	Cyl.1 Injector Circuit Low Input/Short to ground
16646	P0262	Cyl.1 Injector Circuit High Input/Short to B+
16648	P0264	Cyl.2 Injector Circuit Low Input/Short to ground
16649	P0265	Cyl.2 Injector Circuit High Input/Short to B+
16651	P0267	Cyl.3 Injector Circuit Low Input/Short to ground
16652	P0268	Cyl.3 Injector Circuit High Input/Short to B+
16654	P0270	Cyl.4 Injector Circuit Low Input/Short to ground
16655	P0271	Cyl.4 Injector Circuit High Input/Short to B+
16657	P0273	Cyl.5 Injector Circuit Low Input/Short to ground
16658	P0274	Cyl.5 Injector Circuit High Input/Short to B+
16660	P0276	Cyl.6 Injector Circuit Low Input/Short to ground
16661	P0277	Cyl.6 Injector Circuit High Input/Short to B+
16663	P0279	Cyl.7 Injector Circuit Low Input/Short to ground
16664	P0280	Cyl.7 Injector Circuit High Input/Short to B+
16666	P0282	Cyl.8 Injector Circuit Low Input/Short to ground
16667	P0283	Cyl.8 Injector Circuit High Input/Short to B+
16684	P0300	Random/Multiple Cylinder Misfire Detected
16685	P0301	Cyl.1 Misfire Detected
16686	P0302	Cyl.2 Misfire Detected

Diagnostic Trouble Codes (Powertrain)

V.A.G. code	SAE P-code	Description
16687	P0303	Cyl.3 Misfire Detected
16688	P0304	Cyl.4 Misfire Detected
16689	P0305	Cyl.5 Misfire Detected
16690	P0306	Cyl.6 Misfire Detected
16691	P0307	Cyl.7 Misfire Detected
16692	P0308	Cyl.8 Misfire Detected
16697	P0313	Misfire Detected Low Fuel Level
16698	P0314	Single Cylinder Misfire
16705	P0321	Ign./Distributor Eng.Speed Inp.Circ Range/Performance
16706	P0322	Ign./Distributor Eng.Speed Inp.Circ No Signal
16709	P0325	Knock Sensor 1 Circuit Electrical Fault in Circuit
16710	P0326	Knock Sensor 1 Circuit Range/Performance
16711	P0327	Knock Sensor 1 Circ Low Input
16712	P0328	Knock Sensor 1 Circ High Input
16716	P0332	Knock Sensor 2 Circ Low Input
16717	P0333	Knock Sensor 2 Circ High Input
16719	P0335	Crankshaft Pos. Sensor (A) Circ Malfunction
16720	P0336	Crankshaft Pos. Sensor (A) Circ Range/Performance/Missing tooth
16721	P0337	Crankshaft Pos.Sensor (A) Circ Low Input
16724	P0340	Camshaft Pos. Sensor (A) Circ Incorrect allocation
16725	P0341	Camshaft Pos.Sensor Circ Range/Performance
16726	P0342	Camshaft Pos.Sensor Circ Low Input
16727	P0343	Camshaft Pos.Sensor Circ High Input
16735	P0351	Ignition Coil (A) Cyl.1 Prim./Sec. Circ Malfunction
16736	P0352	Ignition Coil (B) Cyl.2 Prim./Sec. Circ Malfunction
16737	P0353	Ignition Coil (C) Cyl.3 Prim./Sec. Circ Malfunction
16738	P0354	Ignition Coil (D) Cyl.4 Prim./Sec. Circ Malfunction
16739	P0355	Ignition Coil (E) Cyl.5 Prim./Sec. Circ Malfunction
16740	P0356	Ignition Coil (F) Cyl.6 Prim./Sec. Circ Malfunction
16741	P0357	Ignition Coil (G) Cyl.7 Prim./Sec. Circ Malfunction
16742	P0358	Ignition Coil (H) Cyl.8 Prim./Sec. Circ Malfunction
16764	P0380	Glow Plug/Heater Circuit (A) Electrical Fault in Circuit
16784	P0400	Exhaust Gas Recirc.Flow Malfunction
16785	P0401	Exhaust Gas Recirc.Flow Insufficient Detected
16786	P0402	Exhaust Gas Recirc.Flow Excessive Detected
16787	P0403	Exhaust Gas Recirc. Contr. Circ Malfunction

Diagnostic Trouble Codes (Powertrain)

V.A.G. code	SAE P-code	Description
16788	P0404	Exhaust Gas Recirc. Contr. Circ Range/Performance
16789	P0405	Exhaust Gas Recirc. Sensor (A) Circ Low Input
16790	P0406	Exhaust Gas Recirc. Sensor (A) Circ High Input
16791	P0407	Exhaust Gas Recirc. Sensor (B) Circ Low Input
16792	P0408	Exhaust Gas Recirc. Sensor (B) Circ High Input
16794	P0410	Sec.Air Inj.Sys Malfunction
16795	P0411	Sec.Air Inj.Sys. Incorrect Flow Detected
16796	P0412	Sec.Air Inj.Sys.Switching Valve A Circ Malfunction
16802	P0418	Sec. Air Inj. Sys. Relay (A) Contr. Circ Malfunction
16804	P0420	Catalyst System,Bank1 Efficiency Below Threshold
16806	P0422	Main Catalyst,Bank1 Below Threshold
16811	P0427	Catalyst Temperature Sensor, Bank 1 Low Input/Short to ground
16812	P0428	Catalyst Temperature Sensor, Bank 1 High Input/Open/Short Circuit to B+
16816	P0432	Main Catalyst,Bank2 Efficiency Below Threshold
16820	P0436	Catalyst Temperature Sensor, Bank 2 Range/Performance
16821	P0437	Catalyst Temperature Sensor, Bank 2 Low Input/Short to ground
16822	P0438	Catalyst Temperature Sensor, Bank 2 High Input/Open/Short Circuit to B+
16824	P0440	EVAP Emission Contr.Sys. Malfunction
16825	P0441	EVAP Emission Contr.Sys.Incorrect Purge Flow
16826	P0442	EVAP Emission Contr.Sys.(Small Leak) Leak Detected
16827	P0443	EVAP Emiss. Contr. Sys. Purge Valve Circ Electrical Fault in Circuit
16836	P0452	EVAP Emission Contr.Sys.Press.Sensor Low Input
16837	P0453	EVAP Emission Contr.Sys.Press.Sensor High Input
16839	P0455	EVAP Emission Contr.Sys.(Gross Leak) Leak Detected
16845	P0461	Fuel Level Sensor Circ Range/Performance
16846	P0462	Fuel Level Sensor Circuit Low Input
16847	P0463	Fuel Level Sensor Circuit High Input
16885	P0501	Vehicle Speed Sensor Range/Performance
16887	P0503	Vehicle Speed Sensor Intermittent/Erratic/High Input
16889	P0505	Idle Control System Malfunction
16890	P0506	Idle Control System RPM Lower than Expected
16891	P0507	Idle Control System Higher than Expected
16894	P0510	Closed Throttle Pos.Switch Malfunction
16915	P0531	A/C Refrigerant Pressure Sensor Circuit Range/Performance
16916	P0532	A/C Refrigerant Pressure Sensor Circuit Low Input
16917	P0533	A/C Refrigerant Pressure Sensor Circuit High Input

Diagnostic Trouble Codes (Powertrain)

V.A.G. code	SAE P-code	Description
16935	P0551	Power Steering Pressure Sensor Circuit Range/Performance
16944	P0560	System Voltage Malfunction
16946	P0562	System Voltage Low Voltage
16947	P0563	System Voltage High Voltage
16952	P0568	Cruise Control Set Signal Incorrect Signal
16955	P0571	Cruise/Brake Switch (A) Circ Malfunction
16984	P0600	Serial Comm. Link (Data Bus) Message Missing
16985	P0601	Internal Contr.Module Memory Check Sum Error
16986	P0602	Control Module Programming Error/Malfunction
16987	P0603	Internal Contr.Module (KAM) Error
16988	P0604	Internal Contr.Module Random Access Memory (RAM) Error
16989	P0605	Internal Contr.Module ROM Test Error
16990	P0606	ECM/PCM Processor
17026	P0642	Knock Control Module Malfunction
17029	P0645	A/C Clutch Relay Control Circuit
17034	P0650	MIL Control Circuit Electrical Fault in Circuit
17038	P0654	Engine RPM Output Circuit Electrical Fault in Circuit
17040	P0656	Fuel Level Output Circuit Electrical Fault in Circuit
17084	P0700	Transm.Contr.System Malfunction
17086	P0702	Transm.Contr.System Electrical
17087	P0703	Torque Converter/Brake Switch B Circ Malfunction
17089	P0705	Transm.Range Sensor Circ.(PRNDL Inop.) Malfunction
17090	P0706	Transm.Range Sensor Circ Range/Performance
17091	P0707	Transm.Range Sensor Circ Low Input
17092	P0708	Transm.Range Sensor Circ High Input
17094	P0710	Transm.Fluid Temp.Sensor Circ. Malfunction
17095	P0711	Transm.Fluid Temp.Sensor Circ. Range/Performance
17096	P0712	Transm.Fluid Temp.Sensor Circ. Low Input
17097	P0713	Transm.Fluid Temp.Sensor Circ. High Input
17099	P0715	Input Turbine/Speed Sensor Circ. Malfunction
17100	P0716	Input Turbine/Speed Sensor Circ. Range/Performance
17101	P0717	Input Turbine/Speed Sensor Circ. No Signal
17105	P0721	Output Speed Sensor Circ Range/Performance
17106	P0722	Output Speed Sensor Circ No Signal
17109	P0725	Engine Speed Inp.Circ. Malfunction
17110	P0726	Engine Speed Inp.Circ. Range/Performance

Diagnostic Trouble Codes (Powertrain)

V.A.G. code	SAE P-code	Description
17111	P0727	Engine Speed Inp.Circ. No Signal
17114	P0730	Gear Incorrect Ratio
17115	P0731	Gear 1 Incorrect Ratio
17116	P0732	Gear 2 Incorrect Ratio
17117	P0733	Gear 3 Incorrect Ratio
17118	P0734	Gear 4 Incorrect Ratio
17119	P0735	Gear 5 Incorrect Ratio
17124	P0740	Torque Converter Clutch Circ Malfunction
17125	P0741	Torque Converter Clutch Circ Performance or Stuck Off
17132	P0748	Pressure Contr.Solenoid Electrical
17134	P0750	Shift Solenoid A malfunction
17135	P0751	Shift Solenoid A Performance or Stuck Off
17136	P0752	Shift Solenoid A Stuck On
17137	P0753	Shift Solenoid A Electrical
17140	P0756	Shift Solenoid B Performance or Stuck Off
17141	P0757	Shift Solenoid B Stuck On
17142	P0758	Shift Solenoid B Electrical
17145	P0761	Shift Solenoid C Performance or Stuck Off
17146	P0762	Shift Solenoid C Stuck On
17147	P0763	Shift Solenoid C Electrical
17152	P0768	Shift Solenoid D Electrical
17157	P0773	Shift Solenoid E Electrical
17174	P0790	Normal/Performance Switch Circ Malfunction
17509	P1101	O2 Sensor Circ.,Bank1-Sensor1Voltage too Low/Air Leak
17510	P1102	O2 Sensor Heating Circ.,Bank1-Sensor1 Short to B+
17511	P1103	O2 Sensor Heating Circ.,Bank1-Sensor1 Output too Low
17512	P1104	Bank1-Sensor2 Voltage too Low/Air Leak
17513	P1105	O2 Sensor Heating Circ.,Bank1-Sensor2 Short to B+
17514	P1106	O2 Sensor Circ.,Bank2-Sensor1 Voltage too Low/Air Leak
17515	P1107	O2 Sensor Heating Circ.,Bank2-Sensor1 Short to B+
17516	P1108	O2 Sensor Heating Circ.,Bank2-Sensor1 Output too Low
17517	P1109	O2 Sensor Circ.,Bank2-Sensor2 Voltage too Low/Air Leak
17518	P1110	O2 Sensor Heating Circ.,Bank2-Sensor2 Short to B+
17519	P1111	O2 Control (Bank 1) System too lean
17520	P1112	O2 Control (Bank 1) System too rich
17521	P1113	Bank1-Sensor1 Internal Resistance too High

Diagnostic Trouble Codes (Powertrain)

V.A.G. code	SAE P-code	Description
17522	P1114	Bank1-Sensor2 Internal Resistant too High
17523	P1115	O2 Sensor Heater Circ.,Bank1-Sensor1 Short to Ground
17524	P1116	O2 Sensor Heater Circ.,Bank1-Sensor1 Open
17525	P1117	O2 Sensor Heater Circ.,Bank1-Sensor2 Short to Ground
17526	P1118	O2 Sensor Heater Circ.,Bank1-Sensor2 Open
17527	P1119	O2 Sensor Heater Circ.,Bank2-Sensor1 Short to Ground
17528	P1120	O2 Sensor Heater Circ.,Bank2-Sensor1 Open
17529	P1121	O2 Sensor Heater Circ.,Bank2-Sensor2 Short to Ground
17530	P1122	O2 Sensor Heater Circ.,Bank2-Sensor2 Open
17531	P1123	Long Term Fuel Trim Add.Air.,Bank1 System too Rich
17532	P1124	Long Term Fuel Trim Add.Air.,Bank1 System too Lean
17533	P1125	Long Term Fuel Trim Add.Air.,Bank2 System too Rich
17534	P1126	Long Term Fuel Trim Add.Air.,Bank2 System too Lean
17535	P1127	Long Term Fuel Trim mult.,Bank1 System too Rich
17536	P1128	Long Term Fuel Trim mult.,Bank1 System too Lean
17537	P1129	Long Term Fuel Trim mult.,Bank2 System too Rich
17538	P1130	Long Term Fuel Trim mult.,Bank2 System too Lean
17539	P1131	Bank2-Sensor1 Internal Resistance too High
17540	P1132	O2 Sensor Heating Circ.,Bank1+2-Sensor1 Short to B+
17541	P1133	O2 Sensor Heating Circ.,Bank1+2-Sensor1 Electrical Malfunction
17542	P1134	O2 Sensor Heating Circ.,Bank1+2-Sensor2 Short to B+
17543	P1135	O2 Sensor Heating Circ.,Bank1+2-Sensor2 Electrical Malfunction
17544	P1136	Long Term Fuel Trim Add.Fuel,Bank1 System too Lean
17545	P1137	Long Term Fuel Trim Add.Fuel,Bank1 System too Rich
17546	P1138	Long Term Fuel Trim Add.Fuel,Bank2 System too Lean
17547	P1139	Long Term Fuel Trim Add.Fuel,Bank2 System too Rich
17548	P1140	Bank2-Sensor2 Internal Resistance too High
17549	P1141	Load Calculation Cross Check Range/Performance
17550	P1142	Load Calculation Cross Check Lower Limit Exceeded
17551	P1143	Load Calculation Cross Check Upper Limit Exceeded
17552	P1144	Mass or Volume Air Flow Circ Open/Short to Ground
17553	P1145	Mass or Volume Air Flow Circ Short to B+
17554	P1146	Mass or Volume Air Flow Circ Supply Malfunction
17555	P1147	O2 Control (Bank 2) System too lean
17556	P1148	O2 Control (Bank 2) System too rich
17557	P1149	O2 Control (Bank 1) Out of range

Diagnostic Trouble Codes (Powertrain)

V.A.G. code	SAE P-code	Description
17558	P1150	O2 Control (Bank 2) Out of range
17559	P1151	Bank1, Long Term Fuel Trim, Range 1 Leanness Lower Limit Exceeded
17560	P1152	Bank1, Long Term Fuel Trim, Range 2 Leanness Lower Limit Exceeded
17562	P1154	Manifold Switch Over Malfunction
17563	P1155	Manifold Abs.Pressure Sensor Circ. Short to B+
17564	P1156	Manifold Abs.Pressure Sensor Circ. Open/Short to Ground
17565	P1157	Manifold Abs.Pressure Sensor Circ. Power Supply Malfunction
17566	P1158	Manifold Abs.Pressure Sensor Circ. Range/Performance
17568	P1160	Manifold Temp.Sensor Circ. Short to Ground
17569	P1161	Manifold Temp.Sensor Circ. Open/Short to B+
17570	P1162	Fuel Temp.Sensor Circ. Short to Ground
17571	P1163	Fuel Temp.Sensor Circ. Open/Short to B+
17572	P1164	Fuel Temperature Sensor Range/Performance/Incorrect Signal
17573	P1165	Bank1, Long Term Fuel Trim, Range 1 Rich Limit Exceeded
17574	P1166	Bank1, Long Term Fuel Trim, Range 2 Rich Limit Exceeded
17579	P1171	Throttle Actuation Potentiometer Sign.2 Range/Performance
17580	P1172	Throttle Actuation Potentiometer Sign.2 Signal too Low
17581	P1173	Throttle Actuation Potentiometer Sign.2 Signal too High
17582	P1174	Fuel Trim, Bank 1 Different injection times
17584	P1176	O2 Correction Behind Catalyst,B1 Limit Attained
17585	P1177	O2 Correction Behind Catalyst,B2 Limit Attained
17586	P1178	Linear 02 Sensor / Pump Current Open Circuit
17587	P1179	Linear 02 Sensor / Pump Current Short to ground
17588	P1180	Linear 02 Sensor / Pump Current Short to B+
17589	P1181	Linear 02 Sensor / Reference Voltage Open Circuit
17590	P1182	Linear 02 Sensor / Reference Voltage Short to ground
17591	P1183	Linear 02 Sensor / Reference Voltage Short to B+
17592	P1184	Linear 02 Sensor / Common Ground Wire Open Circuit
17593	P1185	Linear 02 Sensor / Common Ground Wire Short to ground
17594	P1186	Linear 02 Sensor / Common Ground Wire Short to B+
17595	P1187	Linear 02 Sensor / Compens. Resistor Open Circuit
17596	P1188	Linear 02 Sensor / Compens. Resistor Short to ground
17597	P1189	Linear 02 Sensor / Compens. Resistor Short to B+
17598	P1190	Linear 02 Sensor / Reference Voltage Incorrect Signal
17604	P1196	O2 Sensor Heater Circ.,Bank1-Sensor1 Electrical Malfunction
17605	P1197	O2 Sensor Heater Circ.,Bank2-Sensor1 Electrical Malfunction

Diagnostic Trouble Codes (Powertrain)

V.A.G. code	SAE P-code	Description
17606	P1198	O2 Sensor Heater Circ.,Bank1-Sensor2 Electrical Malfunction
17607	P1199	O2 Sensor Heater Circ.,Bank2-Sensor2 Electrical Malfunction
17609	P1201	Cyl.1-Fuel Inj.Circ. Electrical Malfunction
17610	P1202	Cyl.2-Fuel Inj.Circ. Electrical Malfunction
17611	P1203	Cyl.3-Fuel Inj.Circ. Electrical Malfunction
17612	P1204	Cyl.4-Fuel Inj.Circ. Electrical Malfunction
17613	P1205	Cyl.5-Fuel Inj.Circ. Electrical Malfunction
17614	P1206	Cyl.6-Fuel Inj.Circ. Electrical Malfunction
17615	P1207	Cyl.7-Fuel Inj.Circ. Electrical Malfunction
17616	P1208	Cyl.8-Fuel Inj.Circ. Electrical Malfunction
17617	P1209	Intake valves for cylinder shut-off Short circuit to ground
17618	P1210	Intake valves for cylinder shut-off Short to B+
17619	P1211	Intake valves for cylinder shut-off Open circuit
17621	P1213	Cyl.1-Fuel Inj.Circ. Short to B+
17622	P1214	Cyl.2-Fuel Inj.Circ. Short to B+
17623	P1215	Cyl.3-Fuel Inj.Circ. Short to B+
17624	P1216	Cyl.4-Fuel Inj.Circ. Short to B+
17625	P1217	Cyl.5-Fuel Inj.Circ. Short to B+
17626	P1218	Cyl.6-Fuel Inj.Circ. Short to B+
17627	P1219	Cyl.7-Fuel Inj.Circ. Short to B+
17628	P1220	Cyl.8-Fuel Inj.Circ. Short to B+
17629	P1221	Cylinder shut-off exhaust valves Short circuit to ground
17630	P1222	Cylinder shut-off exhaust valves Short to B+
17631	P1223	Cylinder shut-off exhaust valves Open circuit
17633	P1225	Cyl.1-Fuel Inj.Circ. Short to Ground
17634	P1226	Cyl.2-Fuel Inj.Circ. Short to Ground
17635	P1227	Cyl.3-Fuel Inj.Circ. Short to Ground
17636	P1228	Cyl.4-Fuel Inj.Circ. Short to Ground
17637	P1229	Cyl.5-Fuel Inj.Circ. Short to Ground
17638	P1230	Cyl.6-Fuel Inj.Circ. Short to Ground
17639	P1231	Cyl.7-Fuel Inj.Circ. Short to Ground
17640	P1232	Cyl.8-Fuel Inj.Circ. Short to Ground
17645	P1237	Cyl.1-Fuel Inj.Circ. Open Circ.
17646	P1238	Cyl.2-Fuel Inj.Circ. Open Circ.
17647	P1239	Cyl.3-Fuel Inj.Circ. Open Circ.
17648	P1240	Cyl.4-Fuel Inj.Circ. Open Circ.

Diagnostic Trouble Codes (Powertrain)

V.A.G. code	SAE P-code	Description
17649	P1241	Cyl.5-Fuel Inj.Circ. Open Circ.
17650	P1242	Cyl.6-Fuel Inj.Circ. Open Circ.
17651	P1243	Cyl.7-Fuel Inj.Circ. Open Circ.
17652	P1244	Cyl.8-Fuel Inj.Circ. Open Circ.
17653	P1245	Needle Lift Sensor Circ. Short to Ground
17654	P1246	Needle Lift Sensor Circ. Range/Performance
17655	P1247	Needle Lift Sensor Circ. Open/Short to B+
17656	P1248	Injection Start Control Deviation
17657	P1249	Fuel consumption signal Electrical Fault in Circuit
17658	P1250	Fuel Level Too Low
17659	P1251	Start of Injection Solenoid Circ Short to B+
17660	P1252	Start of Injection Solenoid Circ Open/Short to Ground
17661	P1253	Fuel consumption signal Short to ground
17662	P1254	Fuel consumption signal Short to B+
17663	P1255	Engine Coolant Temp.Circ Short to Ground
17664	P1256	Engine Coolant Temp.Circ Open/Short to B+
17665	P1257	Engine Coolant System Valve Open
17666	P1258	Engine Coolant System Valve Short to B+
17667	P1259	Engine Coolant System Valve Short to Ground
17688	P1280	Fuel Inj.Air Contr.Valve Circ. Flow too Low
17691	P1283	Fuel Inj.Air Contr.Valve Circ. Electrical Malfunction
17692	P1284	Fuel Inj.Air Contr.Valve Circ. Open
17693	P1285	Fuel Inj.Air Contr.Valve Circ. Short to Ground
17694	P1286	Fuel Inj.Air Contr.Valve Circ. Short to B+
17695	P1287	Turbocharger bypass valve open
17696	P1288	Turbocharger bypass valve short to B+
17697	P1289	Turbocharger bypass valve short to ground
17704	P1296	Cooling system malfunction
17705	P1297	Connection turbocharger - throttle valve pressure hose
17708	P1300	Misfire detected Reason: Fuel level too low
17721	P1319	Knock Sensor 1 Circ. Short to Ground
17728	P1320	Knock Sensor 2 Circ. Short to Ground
17729	P1321	Knock Sensor 3 Circ. Low Input
17730	P1322	Knock Sensor 3 Circ. High Input
17731	P1323	Knock Sensor 4 Circ. Low Input
17732	P1324	Knock Sensor 4 Circ. High Input

Diagnostic Trouble Codes (Powertrain)

V.A.G. code	SAE P-code	Description
17733	P1325	Cyl.1-Knock Contr. Limit Attained
17734	P1326	Cyl.2-Knock Contr. Limit Attained
17735	P1327	Cyl.3-Knock Contr. Limit Attained
17736	P1328	Cyl.4-Knock Contr. Limit Attained
17737	P1329	Cyl.5-Knock Contr. Limit Attained
17738	P1330	Cyl.6-Knock Contr. Limit Attained
17739	P1331	Cyl.7-Knock Contr. Limit Attained
17740	P1332	Cyl.8-Knock Contr. Limit Attained
17743	P1335	Engine Torque Monitoring 2 Control Limit Exceeded
17744	P1336	Engine Torque Monitoring Adaptation at limit
17745	P1337	Camshaft Pos.Sensor,Bank1 Short to Ground
17746	P1338	Camshaft Pos.Sensor,Bank1 Open Circ./Short to B+
17747	P1339	Crankshaft Pos./Engine Speed Sensor Cross Connected
17748	P1340	Crankshaft-/Camshaft Pos.Sens.Signals Out of Sequence
17749	P1341	Ignition Coil Power Output Stage 1 Short to Ground
17750	P1342	Ignition Coil Power Output Stage 1 Short to B+
17751	P1343	Ignition Coil Power Output Stage 2 Short to Ground
17752	P1344	Ignition Coil Power Output Stage 2 Short to B+
17753	P1345	Ignition Coil Power Output Stage 3 Short to Ground
17754	P1346	Ignition Coil Power Output Stage 3 Short to B+
17755	P1347	Bank2,Crankshaft-/Camshaft os.Sens.Sign. Out of Sequence
17756	P1348	Ignition Coil Power Output Stage 1 Open Circuit
17757	P1349	Ignition Coil Power Output Stage 2 Open Circuit
17758	P1350	Ignition Coil Power Output Stage 3 Open Circuit
17762	P1354	Modulation Piston Displ.Sensor Circ. Malfunction
17763	P1355	Cyl. 1, ignition circuit Open Circuit
17764	P1356	Cyl. 1, ignition circuit Short to B+
17765	P1357	Cyl. 1, ignition circuit Short to ground
17766	P1358	Cyl. 2, ignition circuit Open Circuit
17767	P1359	Cyl. 2, ignition circuit Short Circuit to B+
17768	P1360	Cyl. 2, ignition circuit Short Circuit to Ground
17769	P1361	Cyl. 3, ignition circuit Open Circuit
17770	P1362	Cyl. 3, ignition circuit Short Circuit to B+
17771	P1363	Cyl. 3, ignition circuit Short Circuit to ground
17772	P1364	Cyl. 4 ignition circuit Open Circuit
17773	P1365	Cyl. 4 ignition circuit Short circuit to B+

Diagnostic Trouble Codes (Powertrain)

V.A.G. code	SAE P-code	Description
17774	P1366	Cyl. 4 ignition circuit Short circuit to ground
17775	P1367	Cyl. 5, ignition circuit Open Circuit
17776	P1368	Cyl. 5, ignition circuit Short Circuit to B+
17777	P1369	Cyl. 5, ignition circuit short to ground
17778	P1370	Cyl. 6, ignition circuit Open Circuit
17779	P1371	Cyl. 6, ignition circuit Short Circuit to B+
17780	P1372	Cyl. 6, ignition circuit short to ground
17781	P1373	Cyl. 7, ignition circuit Open Circuit
17782	P1374	Cyl. 7, ignition circuit Short Circuit to B+
17783	P1375	Cyl. 7, ignition circuit short to ground
17784	P1376	Cyl. 8, ignition circuit Open Circuit
17785	P1377	Cyl. 8, ignition circuit Short Circuit to B+
17786	P1378	Cyl. 8, ignition circuit short to ground
17794	P1386	Internal Control Module Knock Control Circ.Error
17795	P1387	Internal Contr. Module altitude sensor error
17796	P1388	Internal Contr. Module drive by wire error
17799	P1391	Camshaft Pos.Sensor,Bank2 Short to Ground
17800	P1392	Camshaft Pos.Sensor,Bank2 Open Circ./Short to B+
17801	P1393	Ignition Coil Power Output Stage 1 Electrical Malfunction
17802	P1394	Ignition Coil Power Output Stage 2 Electrical Malfunction
17803	P1395	Ignition Coil Power Output Stage 3 Electrical Malfunction
17804	P1396	Engine Speed Sensor Missing Tooth
17805	P1397	Engine speed wheel Adaptation limit reached
17806	P1398	Engine RPM signal, TD Short to ground
17807	P1399	Engine RPM signal, TD Short Circuit to B+
17808	P1400	EGR Valve Circ Electrical Malfunction
17809	P1401	EGR Valve Circ Short to Ground
17810	P1402	EGR Valve Circ Short to B+
17811	P1403	EGR Flow Deviation
17812	P1404	EGR Flow Basic Setting not carried out
17814	P1406	EGR Temp.Sensor Range/Performance
17815	P1407	EGR Temp.Sensor Signal too Low
17816	P1408	EGR Temp.Sensor Signal too High
17817	P1409	Tank Ventilation Valve Circ. Electrical Malfunction
17818	P1410	Tank Ventilation Valve Circ. Short to B+
17819	P1411	Sec.Air Inj.Sys.,Bank2 Flow too Flow

Diagnostic Trouble Codes (Powertrain)

V.A.G. code	SAE P-code	Description
17820	P1412	EGR Different.Pressure Sensor Signal too Low
17821	P1413	EGR Different.Pressure Sensor Signal too High
17822	P1414	Sec.Air Inj.Sys.,Bank2 Leak Detected
17825	P1417	Fuel Level Sensor Circ Signal too Low
17826	P1418	Fuel Level Sensor Circ Signal too High
17828	P1420	Sec.Air Inj.Valve Circ Electrical Malfunction
17829	P1421	Sec.Air Inj.Valve Circ Short to Ground
17830	P1422	Sec.Air Inj.Sys.Contr.Valve Circ Short to B+
17831	P1423	Sec.Air Inj.Sys.,Bank1 Flow too Low
17832	P1424	Sec.Air Inj.Sys.,Bank1 Leak Detected
17833	P1425	Tank Vent.Valve Short to Ground
17834	P1426	Tank Vent.Valve Open
17840	P1432	Sec.Air Inj.Valve Open
17841	P1433	Sec.Air Inj.Sys.Pump Relay Circ. open
17842	P1434	Sec.Air Inj.Sys.Pump Relay Circ. Short to B+
17843	P1435	Sec.Air Inj.Sys.Pump Relay Circ. Short to ground
17844	P1436	Sec.Air Inj.Sys.Pump Relay Circ. Electrical Malfunction
17847	P1439	EGR Potentiometer Error in Basic Setting
17848	P1440	EGR Valve Power Stage Open
17849	P1441	EGR Valve Circ Open/Short to Ground
17850	P1442	EGR Valve Position Sensor Signal too high
17851	P1443	EGR Valve Position Sensor Signal too low
17852	P1444	EGR Valve Position Sensor range/performance
17853	P1445	Catalyst Temp.Sensor 2 Circ. Range/Performance
17854	P1446	Catalyst Temp.Circ Short to Ground
17855	P1447	Catalyst Temp.Circ Open/Short to B+
17856	P1448	Catalyst Temp.Sensor 2 Circ. Short to Ground
17857	P1449	Catalyst Temp.Sensor 2 Circ. Open/Short to B+
17858	P1450	Sec.Air Inj.Sys.Circ Short to B+
17859	P1451	Sec.Air Inj.Sys.Circ Short to Ground
17860	P1452	Sec.Air Inj.Sys. Open Circ.
17861	P1453	Exhaust gas temperature sensor 1 open/short to B+
17862	P1454	Exhaust gas temperature sensor short 1 to ground
17863	P1455	Exhaust gas temperature sensor 1 range/performance
17864	P1456	Exhaust gas temperature control bank 1 limit attained
17865	P1457	Exhaust gas temperature sensor 2 open/short to B+

Diagnostic Trouble Codes (Powertrain)

V.A.G. code	SAE P-code	Description
17866	P1458	Exhaust gas temperature sensor 2 short to ground
17867	P1459	Exhaust gas temperature sensor 2 range/performance
17868	P1460	Exhaust gas temperature control bank 2 limit attained
17869	P1461	Exhaust gas temperature control bank 1 Range/Performance
17870	P1462	Exhaust gas temperature control bank 2 Range/Performance
17873	P1465	Additive Pump Short Circuit to B+
17874	P1466	Additive Pump Open/Short to Ground
17875	P1467	EVAP Canister Purge Solenoid Valve Short Circuit to B+
17876	P1468	EVAP Canister Purge Solenoid Valve Short Circuit to Ground
17877	P1469	EVAP Canister Purge Solenoid Valve Open Circuit
17878	P1470	EVAP Emission Contr.LDP Circ Electrical Malfunction
17879	P1471	EVAP Emission Contr.LDP Circ Short to B+
17880	P1472	EVAP Emission Contr.LDP Circ Short to Ground
17881	P1473	EVAP Emission Contr.LDP Circ Open Circ.
17882	P1474	EVAP Canister Purge Solenoid Valve electrical malfunction
17883	P1475	EVAP Emission Contr.LDP Circ Malfunction/Signal Circ.Open
17884	P1476	EVAP Emission Contr.LDP Circ Malfunction/Insufficient Vacuum
17885	P1477	EVAP Emission Contr.LDP Circ Malfunction
17886	P1478	EVAP Emission Contr.LDP Circ Clamped Tube Detected
17908	P1500	Fuel Pump Relay Circ. Electrical Malfunction
17909	P1501	Fuel Pump Relay Circ. Short to Ground
17910	P1502	Fuel Pump Relay Circ. Short to B+
17911	P1503	Load signal from Alternator Term. DF Range/performance/Incorrect Signal
17912	P1504	Intake Air Sys.Bypass Leak Detected
17913	P1505	Closed Throttle Pos. Does Not Close/Open Circ
17914	P1506	Closed Throttle Pos.Switch Does Not Open/Short to Ground
17915	P1507	Idle Sys.Learned Value Lower Limit Attained
17916	P1508	Idle Sys.Learned Value Upper Limit Attained
17917	P1509	Idle Air Control Circ. Electrical Malfunction
17918	P1510	Idle Air Control Circ. Short to B+
17919	P1511	Intake Manifold Changeover Valve circuit electrical malfunction
17920	P1512	Intake Manifold Changeover Valve circuit Short to B+
17921	P1513	Intake Manifold Changeover Valve2 circuit Short to B+
17922	P1514	Intake Manifold Changeover Valve2 circuit Short to ground
17923	P1515	Intake Manifold Changeover Valve circuit Short to Ground
17924	P1516	Intake Manifold Changeover Valve circuit Open

Diagnostic Trouble Codes (Powertrain)

V.A.G. code	SAE P-code	Description
17925	P1517	Main Relay Circ. Electrical Malfunction
17926	P1518	Main Relay Circ. Short to B+
17927	P1519	Intake Camshaft Contr.,Bank1 Malfunction
17928	P1520	Intake Manifold Changeover Valve2 circuit Open
17929	P1521	Intake Manifold Changeover Valve2 circuit electrical malfunction
17930	P1522	Intake Camshaft Contr.,Bank2 Malfunction
17931	P1523	Crash Signal from Airbag Control Unit range/performance
17933	P1525	Intake Camshaft Contr.Circ.,Bank1 Electrical Malfunction
17934	P1526	Intake Camshaft Contr.Circ.,Bank1 Short to B+
17935	P1527	Intake Camshaft Contr.Circ.,Bank1 Short to Ground
17936	P1528	Intake Camshaft Contr.Circ.,Bank1 Open
17937	P1529	Camshaft Control Circuit Short to B+
17938	P1530	Camshaft Control Circuit Short to ground
17939	P1531	Camshaft Control Circuit open
17941	P1533	Intake Camshaft Contr.Circ.,Bank2 Electrical Malfunction
17942	P1534	Intake Camshaft Contr.Circ.,Bank2 Short to B+
17943	P1535	Intake Camshaft Contr.Circ.,Bank2 Short to Ground
17944	P1536	Intake Camshaft Contr.Circ.,Bank2 Open
17945	P1537	Engine Shutoff Solenoid Malfunction
17946	P1538	Engine Shutoff Solenoid Open/Short to Ground
17947	P1539	Clutch Vacuum Vent Valve Switch Incorrect signal
17948	P1540	Vehicle Speed Sensor High Input
17949	P1541	Fuel Pump Relay Circ Open
17950	P1542	Throttle Actuation Potentiometer Range/Performance
17951	P1543	Throttle Actuation Potentiometer Signal too Low
17952	P1544	Throttle Actuation Potentiometer Signal too High
17953	P1545	Throttle Pos.Contr Malfunction
17954	P1546	Boost Pressure Contr.Valve Short to B+
17955	P1547	Boost Pressure Contr.Valve Short to Ground
17956	P1548	Boost Pressure Contr.Valve Open
17957	P1549	Boost Pressure Contr.Valve Short to Ground
17958	P1550	Charge Pressure Deviation
17959	P1551	Barometric Pressure Sensor Circ. Short to B+
17960	P1552	Barometric Pressure Sensor Circ. Open/Short to Ground
17961	P1553	Barometric/manifold pressure signal ratio out of range
17962	P1554	Idle Speed Contr.Throttle Pos. Basic Setting Conditions not met

Diagnostic Trouble Codes (Powertrain)

V.A.G. code	SAE P-code	Description
17963	P1555	Charge Pressure Upper Limit exceeded
17964	P1556	Charge Pressure Contr. Negative Deviation
17965	P1557	Charge Pressure Contr. Positive Deviation
17966	P1558	Throttle Actuator Electrical Malfunction
17967	P1559	Idle Speed Contr.Throttle Pos. Adaptation Malfunction
17968	P1560	Maximum Engine Speed Exceeded
17969	P1561	Quantity Adjuster Deviation
17970	P1562	Quantity Adjuster Upper Limit Attained
17971	P1563	Quantity Adjuster Lower Limit Attained
17972	P1564	Idle Speed Contr.Throttle Pos. Low Voltage During Adaptation
17973	P1565	Idle Speed Control Throttle Position lower limit not attained
17974	P1566	Load signal from A/C compressor range/performance
17975	P1567	Load signal from A/C compressor no signal
17976	P1568	Idle Speed Contr.Throttle Pos. mechanical Malfunction
17977	P1569	Cruise control switch Incorrect signal
17978	P1570	Contr.Module Locked
17979	P1571	Left Eng. Mount Solenoid Valve Short to B+
17980	P1572	Left Eng. Mount Solenoid Valve Short to ground
17981	P1573	Left Eng. Mount Solenoid Valve Open circuit
17982	P1574	Left Eng. Mount Solenoid Valve Electrical fault in circuit
17983	P1575	Right Eng. Mount Solenoid Valve Short to B+
17984	P1576	Right Eng. Mount Solenoid Valve Short to ground
17985	P1577	Right Eng. Mount Solenoid Valve Open circuit
17986	P1578	Right Eng. Mount Solenoid Valve Electrical fault in circuit
17987	P1579	Idle Speed Contr.Throttle Pos. Adaptation not started
17988	P1580	Throttle Actuator B1 Malfunction
17989	P1581	Idle Speed Contr.Throttle Pos. Basic Setting Not Carried Out
17990	P1582	Idle Adaptation at Limit
17991	P1583	Transmission mount valves Short to B+
17992	P1584	Transmission mount valves Short to ground
17993	P1585	Transmission mount valves Open circuit
17994	P1586	Engine mount solenoid valves Short to B+
17995	P1587	Engine mount solenoid valves Short to ground
17996	P1588	Engine mount solenoid valves Open circuit
18008	P1600	Power Supply (B+) Terminal 15 Low Voltage
18010	P1602	Power Supply (B+) Terminal 30 Low Voltage

Diagnostic Trouble Codes (Powertrain)

V.A.G. code	SAE P-code	Description
18011	P1603	Internal Control Module Malfunction
18012	P1604	Internal Control Module Driver Error
18013	P1605	Rough Road/Acceleration Sensor Electrical Malfunction
18014	P1606	Rough Road Spec Engine Torque ABS-ECU Electrical Malfunction
18015	P1607	Vehicle speed signal Error message from instrument cluster
18016	P1608	Steering angle signal Error message from steering angle sensor
18017	P1609	Crash shut-down activated
18019	P1611	MIL Call-up Circ./Transm.Contr.Module Short to Ground
18020	P1612	Electronic Control Module Incorrect Coding
18021	P1613	MIL Call-up Circ Open/Short to B+
18022	P1614	MIL Call-up Circ./Transm.Contr.Module Range/Performance
18023	P1615	Engine Oil Temperature Sensor Circuit range/performance
18024	P1616	Glow Plug/Heater Indicator Circ. Short to B+
18025	P1617	Glow Plug/Heater Indicator Circ. Open/Short to Ground
18026	P1618	Glow Plug/Heater Relay Circ. Short to B+
18027	P1619	Glow Plug/Heater Relay Circ. Open/Short to Ground
18028	P1620	Engine coolant temperature signal open/short to B+
18029	P1621	Engine coolant temperature signal short to ground
18030	P1622	Engine coolant temperature signal range/performance
18031	P1623	Data Bus Powertrain No Communication
18032	P1624	MIL Request Sign.active
18033	P1625	Data-Bus Powertrain Implausible Message from Transm.Contr.
18034	P1626	Data-Bus Powertrain Missing Message from Transm.Contr.
18035	P1627	Data-Bus Powertrain missing message from fuel injection pump
18036	P1628	Data-Bus Powertrain missing message from steering sensor
18037	P1629	Data-Bus Powertrain missing message from distance control
18038	P1630	Accelera.Pedal Pos.Sensor 1 Signal too Low
18039	P1631	Accelera.Pedal Pos.Sensor 1 Signal too High
18040	P1632	Accelera.Pedal Pos.Sensor 1 Power Supply Malfunction
18041	P1633	Accelera.Pedal Pos.Sensor 2 Signal too Low
18042	P1634	Accelera.Pedal Pos.Sensor 2 Signal too High
18043	P1635	Data Bus Powertrain missing message f.air condition control
18044	P1636	Data Bus Powertrain missing message from Airbag control
18045	P1637	Data Bus Powertrain missing message f.central electr.control
18046	P1638	Data Bus Powertrain missing message from clutch control
18047	P1639	Accelera.Pedal Pos.Sensor 1+2 Range/Performance

Diagnostic Trouble Codes (Powertrain)

V.A.G. code	SAE P-code	Description
18048	P1640	Internal Contr.Module (EEPROM) Error
18049	P1641	Please check DTC Memory of Air Condition ECU
18050	P1642	Please check DTC Memory of Airbag ECU
18051	P1643	Please check DTC Memory of central electric ECU
18052	P1644	Please check DTC Memory of clutch ECU
18053	P1645	Data Bus Powertrain missing message f.all wheel drive contr.
18054	P1646	Please Check DTC Memory of all wheel drive ECU
18055	P1647	Please check coding of ECUs in Data Bus Powertrain
18056	P1648	Data Bus Powertrain Malfunction
18057	P1649	Data Bus Powertrain Missing message from ABS Control Module
18058	P1650	Data Bus Powertrain Missing message fr.instrument panel ECU
18059	P1651	Data Bus Powertrain missing messages
18060	P1652	Please check DTC Memory of transmission ECU
18061	P1653	Please check DTC Memory of ABS Control Module
18062	P1654	Please check DTC Memory of control panel ECU
18063	P1655	Please check DTC Memory of ADR Control Module
18064	P1656	A/C clutch relay circuit short to ground
18065	P1657	A/C clutch relay circuit short to B+
18066	P1658	Data Bus Powertrain Incorrect signal from ADR Control Module
18084	P1676	Drive by Wire-MIL Circ. Electrical Malfunction
18085	P1677	Drive by Wire-MIL Circ. Short to B+
18086	P1678	Drive by Wire-MIL Circ. Short to Ground
18087	P1679	Drive by Wire-MIL Circ. Open
18089	P1681	Contr.Unit Programming, Programming not Finished
18092	P1684	Contr.Unit Programming Communication Error
18094	P1686	Contr.Unit Error Programming Error
18098	P1690	Malfunction Indication Light Malfunction
18099	P1691	Malfunction Indication Light Open
18100	P1692	Malfunction Indication Light Short to Ground
18101	P1693	Malfunction Indication Light Short to B+
18102	P1694	Malfunction Indication Light Open/Short to Ground
18112	P1704	Kick Down Switch Malfunction
18113	P1705	Gear/Ratio Monitoring Adaptation limit reached
18119	P1711	Wheel Speed Signal 1 Range/Performance
18124	P1716	Wheel Speed Signal 2 Range/Performance
18129	P1721	Wheel Speed Signal 3 Range/Performance

Diagnostic Trouble Codes (Powertrain)

V.A.G. code	SAE P-code	Description
18131	P1723	Starter Interlock Circ. Open
18132	P1724	Starter Interlock Circ. Short to Ground
18134	P1726	Wheel Speed Signal 4 Range/Performance
18136	P1728	Different Wheel Speed Signals Range/Performance
18137	P1729	Starter Interlock Circ. Short to B+
18141	P1733	Tiptronic Switch Down Circ. Short to Ground
18147	P1739	Tiptronic Switch up Circ. Short to Ground
18148	P1740	Clutch temperature control
18149	P1741	Clutch pressure adaptation at limit
18150	P1742	Clutch torque adaptation at limit
18151	P1743	Clutch slip control signal too high
18152	P1744	Tiptronic Switch Recognition Circ. Short to Ground
18153	P1745	Transm.Contr.Unit Relay Short to B+
18154	P1746	Transm.Contr.Unit Relay Malfunction
18155	P1747	Transm.Contr.Unit Relay Open/Short to Ground
18156	P1748	Transm.Contr.Unit Self-Check
18157	P1749	Transm.Contr.Unit Incorrect Coded
18158	P1750	Power Supply Voltage Low Voltage
18159	P1751	Power Supply Voltage High Voltage
18160	P1752	Power Supply Malfunction
18168	P1760	Shift Lock Malfunction
18169	P1761	Shift Lock Short to Ground
18170	P1762	Shift Lock Short to B+
18171	P1763	Shift Lock Open
18172	P1764	Transmission temperature control
18173	P1765	Hydraulic Pressure Sensor 2 adaptation at limit
18174	P1766	Throttle Angle Signal Stuck Off
18175	P1767	Throttle Angle Signal Stuck On
18176	P1768	Hydraulic Pressure Sensor 2 Too High
18177	P1769	Hydraulic Pressure Sensor 2 Too Low
18178	P1770	Load Signal Range/Performance
18179	P1771	Load Signal Stuck Off
18180	P1772	Load Signal Stuck On
18181	P1773	Hydraulic Pressure Sensor 1 Too High
18182	P1774	Hydraulic Pressure Sensor 1 Too Low
18183	P1775	Hydraulic Pressure Sensor 1 adaptation at limit

Diagnostic Trouble Codes (Powertrain)

V.A.G. code	SAE P-code	Description
18184	P1776	Hydraulic Pressure Sensor 1 range/performance
18185	P1777	Hydraulic Pressure Sensor 2 range/performance
18186	P1778	Solenoid EV7 Electrical Malfunction
18189	P1781	Engine Torque Reduction Open/Short to Ground
18190	P1782	Engine Torque Reduction Short to B+
18192	P1784	Shift up/down Wire Open/Short to Ground
18193	P1785	Shift up/down Wire Short to B+
18194	P1786	Reversing Light Circ. Open
18195	P1787	Reversing Light Circ. Short to Ground
18196	P1788	Reversing Light Circ. Short to B+
18197	P1789	Idle Speed Intervention Circ. Error Message from Engine Contr.
18198	P1790	Transmission Range Display Circ. Open
18199	P1791	Transmission Range Display Circ. Short to Ground
18200	P1792	Transmission Range Display Circ. Short to B+
18201	P1793	Output Speed Sensor 2 Circ. No Signal
18203	P1795	Vehicle Speed Signal Circ. Open
18204	P1796	Vehicle Speed Signal Circ. Short to Ground
18205	P1797	Vehicle Speed Signal Circ. Short to B+
18206	P1798	Output Speed Sensor 2 Circ. Range/Performance
18207	P1799	Output Speed Sensor 2 Circ. RPM too High
18221	P1813	Pressure Contr.Solenoid 1 Electrical
18222	P1814	Pressure Contr.Solenoid 1 Open/Short to Ground
18223	P1815	Pressure Contr.Solenoid 1 Short to B+
18226	P1818	Pressure Contr.Solenoid 2 Electrical
18227	P1819	Pressure Contr.Solenoid 2 Open/Short to Ground
18228	P1820	Pressure Contr.Solenoid 2 Short to B+
18231	P1823	Pressure Contr.Solenoid 3 Electrical
18232	P1824	Pressure Contr.Solenoid 3 Open/Short to Ground
18233	P1825	Pressure Contr.Solenoid 3 Short to B+
18236	P1828	Pressure Contr.Solenoid 4 Electrical
18237	P1829	Pressure Contr.Solenoid 4 Open/Short to Ground
18238	P1830	Pressure Contr.Solenoid 4 Short to B+
18242	P1834	Pressure Contr.Solenoid 5 Open/Short to Ground
18243	P1835	Pressure Contr.Solenoid 5 Short to B+
18249	P1841	Engine/Transmission Control Modules Versions do not match
18250	P1842	Please check DTC Memory of instrument panel ECU

Diagnostic Trouble Codes (Powertrain)

V.A.G. code	SAE P-code	Description
18251	P1843	Please check DTC Memory of ADR Control Module
18252	P1844	Please check DTC Memory of central electric control ECU
18255	P1847	Please check DTC Memory of brake system ECU
18256	P1848	Please check DTC Memory of engine ECU
18257	P1849	Please check DTC Memory of transmission ECU
18258	P1850	Data-Bus Powertrain Missing Message from Engine Contr.
18259	P1851	Data-Bus Powertrain Missing Message from Brake Contr.
18260	P1852	Data-Bus Powertrain Implausible Message from Engine Contr.
18261	P1853	Data-Bus Powertrain Implausible Message from Brake Contr.
18262	P1854	Data-Bus Powertrain Hardware Defective
18263	P1855	Data-Bus Powertrain Software version Contr.
18264	P1856	Throttle/Pedal Pos.Sensor A Circ. Error Message from Engine Contr.
18265	P1857	Load Signal Error Message from Engine Contr.
18266	P1858	Engine Speed Input Circ. Error Message from Engine Contr.
18267	P1859	Brake Switch Circ. Error Message from Engine Contr.
18268	P1860	Kick Down Switch Error Message from Engine Contr.
18269	P1861	Throttle Position (TP) sensor Error Message from ECM
18270	P1862	Data Bus Powertrain Missing message from instr. panel ECU
18271	P1863	Data Bus Powertrain Missing Message from St. Angle Sensor
18272	P1864	Data Bus Powertrain Missing message from ADR control module
18273	P1865	Data Bus Powertrain Missing message from central electronics
18274	P1866	Data Bus Powertrain Missing messages

Diagnostic scan tool suppliers

NOTE—

The following suppliers offer diagnostic scan tools and/or scan tool programs that can be used to retrieve diagnostic trouble codes as listed in the table above.

Assenmacher Specialty Tools, Inc.
6440 Odell Place
Boulder, CO 80301
(800)525-2943
www.asttool.com

Baum Tools Unlimited, Inc.
PO Box 5867
Sarasota, FL 34277
(800)848-6657
www.baumtools.com

Equipment Solutions
PO Box 1450
Kenosha, WI 53141-1450
800-892-9650

Metalnerd
5113 Crestview Drive, Suite B
Greensberg, PA 15601
(412)601-42705
www.metalnerd.com

Ross-Tech
920 South Broad St.
Lansdale, PA 19446
215-361-8942
www.ross-tech.com

Samstag Sales
115 Main St. N., Suite 216
Carthage, TN 37030
(615)735-3388
www.samstagsales.com

Shade Tree Software
4186 Culebra Ct.
Boulder, CO 80301
(303)449-1664
(303)940-2468
www.shadetreesoftware.com

ZDMAK Tools
P.O. Box 5100
Sarasota, FL 34277
(877)938-6657
www.zdmak.com

Zelenda Automotive Inc.
66-02 Austin Street
Forest Hills, NY 11374
888-892-8348
www.zelenda.com

Control Module Coding

Control modules are produced in a few major variations and electronically matched to the exact system needs through a process known as coding. Coding is automatically displayed when the control module responds to the scan tool inquiry and can also be called up through Function 01 of the Function Menu of the VAG 1551/1552. The coding process requires a properly set-up scan tool or scan tool computer program and is accessed through Function 07.

Most codes are 5 digits and require assembly from the tables in this section. Careful attention must be paid to assemble accurate codes. Failure to code, or insertion of an incorrect code can cause any or all of the following conditions:

- Driveability problems
- Increased fuel consumption
- Increased emissions
- Storage of non-existent or erroneous DTCs resulting in Malfunction Indicator Light (MIL) illumination.
- Necessary functions not being performed (oxygen sensor regulation, activation of the EVAP canister system, non-functioning of anti-slip regulation (ASR) etc.).
- Necessary functions being performed incorrectly (mileage recorded instead of kilometers, warning light failure, etc.).
- Reduction in component service life.

NOTE —

- *New or replacement control units will usually require coding of the control module before they will function properly.*
- *In some instances, a control module will require input of a 7 digit code, but the table lists only a 5 digit code. In these instances, enter 00 before the correct code to satisfy the 7 digit requirement.*
- *After a new code is entered it will usually be necessary to switch the ignition off and back on for the new code to be recorded.*

WARNING —

- *DO NOT attempt to connect a VAG 1551/1552 scan tool, VAS 5051/5052 diagnostic tool, aftermarket scan tool or scan tool program to any vehicle in which an incorrect radio has been installed.*
- *Installation of aftermarket radios or Volkswagen radios not produced for use in model year 1999 and newer vehicles will*
  - *Affect various vehicle systems connected to the CAN-BUS.*
  - *Cause severe damage to VAG 1551 and VAG 1552 scan tools.*
  - *Cause severe damage to VAS 5051 and VAS 5052 diagnostic tools.*
  - *Cause severe damage to aftermarket scan tools.*
  - *Cause severe damage to computers running scan tool programs.*
- *Inspect radio; if radio is not correct for the vehicle or if there is any doubt, remove and disconnect wiring before connecting scan tools.*

Engine Control Module Coding Variations, 1.8L Turbo (engine code: AEB)			
Country or Emission Level	**Drive or Additional Functions**	**Transmission**	**Vehicle Type**
00 =	**0** = Front Wheel Drive without traction control	**0** = 5-Speed Manual	**0** =
01 =	**1** = Front Wheel Drive with Traction Control	**1** =	**1** = B Class (B5 Passat)
02 =	**2** =	**2** =	**2** =
03 =	**3** =	**3** =	**3** =
04 =	**4** =	**4** =	**4** =
05 =	**5** =	**5** = 5-Speed Automatic 01V (Tiptronic)	**5** =
06 = Emission Standard TLEV	**6** =	**6** =	**6** =
Notes: Disregard empty table cells. ECM will only recognize the values listed above as valid. Access through Address Word 01.			

Example, engine code AEB				
Passat				1
5-Speed Manual Transmission			0	
Front Wheel Drive without Traction Control		0		
TLEV Emission Standard	06			
The correct code number for the above vehicle is:	**06001**			

Engine Control Module Coding Variations, 1.8L Turbo (engine codes: ATW, AUG, AWM)			
Country or Emission Level	**Drive or Additional Functions**	**Transmission**	**Vehicle Type**
00 =	**0** = Front Wheel Drive without Anti-Slip Regulation (ASR)	**0** = 5-Speed Manual	**0** =
01 =	**1** =	**1** = 6-Speed Manual	**1** = B Class (B5 Passat)
02 =	**2** = All Wheel Drive (4Motion) without Anti-Slip Regulation (ASR)	**2** =	**2** =
03 =	**3** =	**3** =	**3** =
04 =	**4** =	**4** =	**4** =
05 =	**5** = Front Wheel Drive with Anti-Slip Regulation (ASR)	**5** = 5-Speed Automatic 01V (Tiptronic)	**5** =
06 =	**6** =	**6** =	**6** =
07 = Emission Standard LEV	**7** = All Wheel Drive (4Motion) with Anti-Slip Regulation (ASR)	**7** =	**7** =
16 = Emission Standard ULEV	**8** =	**8** =	**8** =
Note: Some model years display a 7 digit code instead of a 5 digit code. If this is the case with your vehicle, enter 00 and then the correct 5 digit code. Disregard empty table cells. ECM will only recognize the values listed above as valid. Access through Address Word 01.			

Example, engine codes ATW/AUG/AWM				
Passat Wagon				1
5-Speed Automatic Transmission			5	
Front Wheel Drive without Traction Control		0		
ULEV Emission Standard	16			
The correct code number for the above vehicle is:	**16051**			

Engine Control Module Coding Variations, 2.8L 5V V6 (engine code: AHA)			
Country or Emission Level	**Drive or Additional Functions**	**Transmission**	**Vehicle Type**
00 =	**0** = Front Wheel Drive without Anti-Slip Regulation (ASR)	**0** = 5-Speed Manual	**0** =
01 =	**1** = Front Wheel Drive with Anti-Slip Regulation (ASR)	**1** =	**1** = B Class (B5 Passat)
02 =	**2** = All Wheel Drive (4Motion) without Anti-Slip Regulation (ASR)	**2** =	**2** =
03 =	**3** =	**3** =	**3** =
04 =	**4** =	**4** =	**4** =
05 =	**5** =	**5** = 5-Speed Automatic 01V (Tiptronic)	**5** =
06 = USA OBD II, with Evaporative Emissions System (EVAP) and Leak Detection Pump (LDP)	**6** =	**6** =	**6** =
07 =	**7** =	**7** =	**7** =
08 =	**8** =	**8** =	**8** =
Notes: Disregard empty table cells. ECM will only recognize the values listed above as valid. Access through Address Word 01.			

Example, engine code AHA				
Passat Sedan				1
5-Speed Automatic Transmission			5	
Front Wheel Drive without ASR		0		
With EVAP and LDP systems	06			
The correct code number for the above vehicle is:	**06051**			

Engine Control Module Coding Variations, 2.8L 5V V6 (engine code: ATQ)			
Country or Emission Level	**Drive or Additional Functions**	**Transmission**	**Vehicle Type**
00 =	**0** = Front Wheel Drive without Anti-Slip Regulation (ASR)	**0** = 5-Speed Manual	**0** =
01 =	**1** =	**1** =	**1** = B Class (B5 Passat)
02 =	**2** = All Wheel Drive (4Motion) without Anti-Slip Regulation (ASR)	**2** =	**2** =
03 =	**3** =	**3** =	**3** =
04 =	**4** =	**4** =	**4** =
05 =	**5** = Front Wheel Drive with Anti-Slip Regulation (ASR)	**5** = 5-Speed Automatic 01V (Tiptronic)	**5** =
06 =	**6** =	**6** =	**6** =
07 = Emission Standard LEV	**7** = All Wheel Drive (4Motion) with Electronic Stability Program (ESP)	**7** =	**7** =
Note: Some model years display a 7 digit code instead of a 5 digit code. If this is the case with your vehicle, enter 00 and then the correct 5 digit code. Disregard empty table cells. ECM will only recognize the values listed above as valid. Access through Address Word 01.			

Example, engine code ATQ				
Passat Sedan				1
5-Speed Automatic Transmission			5	
Front Wheel Drive with ASR		5		
LEV Emission Standard	07			
The correct code number for the above vehicle is:	**07551**			

TCM Coding

Transmission Control Module Coding Variations[3]

Electronic Power Control, EPC (drive-by-wire)	Vehicle Equipment/Version	Market	Tiptronic Strategy	TCM Code[4, 5]
Without EPC	All engine codes without EPC all transmission codes (Also 2.5L V6 TDI with EPC)	Front Wheel Drive (all) All Wheel Drive 4Motion (all)	Old Strategy[1]	00000
Without EPC	Not USA or Canada	Front Wheel Drive (all) All Wheel Drive 4Motion (all)	n/a	00010
With EPC	Not a valid code for EPC vehicles	n/a	n/a	00000
With EPC	All engine codes with EPC all transmission codes	USA All Wheel Drive 4Motion (all)	New Strategy[2]	00102
With EPC	All engine codes with EPC all transmission codes	USA Front Wheel Drive (all)	New Strategy[2]	00104

Notes:

1. Tiptronic strategy "Old" means: when the selector lever is shifted into the Tiptronic gate, the transmission shifts up and down when the selector lever is shifted forward (+) or rearward (-). If the vehicle is braked to a standstill, first gear is automatically engaged.
2. Tiptronic strategy "New" means: when the selector lever is shifted into the Tiptronic gate, the transmission automatically shifts up or down before the RPM limit is reached. It also downshifts automatically when the accelerator pedal is depressed to kick down.
3. Access through Address Word 02.
4. TCM will only recognize the values listed above as valid.
5. Some model years display a 7 digit code instead of a 5 digit code. If this is the case with your vehicle, enter 00 and then the correct 5 digit code.

ABS/EDL/ASR Coding Table

ABS Control Module Coding Variations[3], Bosch ABS/EDL/ASR 5.3		
Engine Code	**Equipment Level[1]**	**Code[2]**
V6	Manual transmission, ASR	00000
V6	Automatic transmission, ASR	00003
1.8L	Manual transmission, ASR, CAN-BUS	00022
1.8L	Automatic transmission, ASR, CAN-BUS	00021
V6	Manual transmission, ASR, CAN-BUS	00023
V6	Automatic transmission, ASR, CAN-BUS	00031

Notes:

1. Anti-Slip Regulation is not installed on vehicles with 4Motion all wheel drive. Coding of the ABS control module is not possible in these vehicles.
2. Only the codes listed above are recognized as valid.
3. Access through Address Word 03.

Convenience System Coding Table

Convenience System Control Module Coding Variations[3], ➤10.00		
Vehicle Equipment	**Central Locking Mode[1]**	**Code Number[2]**
Central Locking, Electric Windows	Selective unlocking (1 door)	00256
Central Locking, Electric Windows	Simultaneous unlocking (all)	00257
Central Locking, Electric Windows, Seat Memory	Selective unlocking (1 door)	00258
Central Locking, Electric Windows, Seat Memory	Simultaneous unlocking (all)	00259

Notes:

1. Selective unlocking: Turn key or activate remote one time to unlock just driver's door, turn key or active remote twice to unlock all doors. Simultaneous unlocking: Turn key or activate remote one time to unlock all doors.
2. Only the codes listed above are recognized as valid.
3. Access through Address Word 46. Up to 5 Control Modules will respond, only first unit requires coding.

Airbag System Coding Table

Vehicle	Airbag Control Module Part Number	Color Code/Index	Coding	Front Airbags	Front Seat Belt Tensioners	Side Airbags	Side Curtain Airbags	Seat Belt Switches	Crash Sensors
Passat 1998-2001	6Q0909601 A	03	12339	YES	NO	NO	NO	NO	N/A
Passat 1998-2001	6Q0909605 B	03	12339	YES	NO	YES	NO	NO	N
Passat 1998-2001	6Q0909605 C	03	12339	YES	NO	YES	YES	NO	N
Passat 1998-2001	6Q0909601 A	0M	12365	YES	NO	NO	NO	NO	N/A
Passat 1998-2001	6Q0909605 B	0C	12355	YES	NO	YES	NO	NO	N
Passat 1998-2001	1C0909601A	0M	12365	YES	NO	NO	NO	NO	N/A
Passat 1998-2001	1C0909605B	0C	12355	YES	NO	YES	NO	NO	N
Passat 1998-2001	1C0909605C	03	12339	YES	NO	YES	YES	NO	N
Passat 2001> (Facelift)	6Q0909605 B	0B	12354	YES	NO	YES	NO	YES	04
Passat 2001> (Facelift)	6Q0909605 C	09	12345	YES	NO	YES	YES	YES	04
Passat 2001> (Facelift)	1C0909605B	0B	12354	YES	NO	YES	NO	YES	04
Passat 2001> (Facelift)	1C0909605C	09	12345	YES	NO	YES	YES	YES	04

Note: Only the codes listed above are recognized as valid. Access through Address Word 15. Passat Facelift vehicles are also unofficially known as Passat B5½.

Climatronic Coding Table

Climatronic Electronic Control Unit Coding		
Vehicle Type	**Equipment Level**	**Code**
Passat, USA/Canada, ➤11.98	Automatic Climate Control (Climatronic)	02000
Passat, USA/Canada, 11.98 ➤10.00	Automatic Climate Control (Climatronic)	05000
Passat, USA/Canada, ➤10.00	Automatic Climate Control (Climatronic)	17200
Notes: Coding is determined by production date. Only the codes listed above are recognized as valid. Access through Address Word 08.		

Instrument Cluster Control Module Coding Variations[5], through production ➤ 08.99				
Designations	**Code Assembly Position**[4]			
	1 + 2[1]	**3**[2]	**4**[3]	**5**
Available Equipment				
Brake pad wear warning	01			
Seat belt warning	02			
Washer fluid level warning	04			
Navigation system	16			
Market Version				
USA		2		
Canada		3		
Engine Cylinders				
4-cylinder			4	
5-cylinder			5	
6-cylinder			6	
Engine Versions				
Diesel				0
Gasoline				2

Notes:

1. Values in this column are added together where applicable to determine value. For example, if vehicle in question has seat belt warning (02) and washer fluid warning (04), values are added together (02 + 04 = 06) to get a value of 06 for positions 1 and 2 of the instrument cluster code. If vehicle has all of the equipment listed, add all of the values together.
2. Only select from the listed codes.
3. Only select from the listed codes.
4. Code will be 5 digits. For example, a Passat with brake pad wear warning, seat belt warning, and washer fluid level warning, for the USA market, with a V6 engine, and a speedometer impulse number of 3538, will have a code of 07262.
5. Access through Address Word 17.

Instrument Cluster Control Module Coding Variations[6], production 08.99 ➤ 09.00				
Designations	**Code Assembly Position[5]**			
	1 + 2[1]	**3[2]**	**4[3]**	**5[4]**
Available Equipment				
None	00			
Brake pad wear warning	01			
Seat belt warning	02			
Washer fluid level warning	04			
Navigation system	16			
Market Version				
USA		2		
Canada		3		
Engine Cylinders				
4-cylinder			4	
6-cylinder			6	
Distance Impulse Number (K number)				
4182				5

Notes:

1. Values in this column are added together where applicable to determine value. For example, if vehicle in question has seat belt warning (02), washer fluid warning (04), and navigation system, values are added together (02 + 04 + 16 = 22) to get a value of 22 for positions 1 and 2 of the instrument cluster code. If vehicle has all of the equipment listed, add all of the values together.
2. Only select from the listed codes.
3. Only select from the listed codes.
4. Distance impulse number is a constant used by the speedometer to calculate speed indication and distance travelled. K number for all vehicles covered by this table is always 5. Distance impulse number can only be read/confirmed using Scan Tool function 10, “Adaptation”.
5. Code will be 5 digits. For example, a Passat with brake pad wear warning, seat belt warning, washer fluid level warning, and navigation system, for the USA market, with a V6 engine, and a speedometer impulse number of 4182, will have a code of 23265.
6. Access through Address Word 17.

Instrument Cluster Control Module Coding Variations[6], production 10.00 ➤ 11.01

Designations	Code Assembly Position[5]			
	1 + 2[1]	3[2]	4[3]	5[4]
Available Equipment				
None	00			
Brake pad wear warning	01			
Seat belt warning	02			
Washer fluid level warning	04			
Market Version				
USA		2		
Canada		3		
Service Interval Display				
USA and Canadian vehicles with fixed maintenance intervals, **WITHOUT** oil level thermal sensor (G266) installed in oil pan.			0	
USA and Canadian vehicles with fixed maintenance intervals, **WITH** oil level thermal sensor (G266) installed in oil pan.			2	
Distance Impulse Number (K number)				
4182				5

Notes:

1. Values in this column are added together where applicable to determine value. For example, if vehicle in question has seat belt warning (02) and washer fluid warning (04), values are added together (02 + 04 = 06) to get a value of 06 for positions 1 and 2 of the instrument cluster code. If vehicle has all of the equipment listed, add all of the values together.
2. Only select from the listed codes.
3. Only select from the listed codes.
4. Distance impulse number is a constant used by the speedometer to calculate speed indication and distance travelled. K number for all vehicles covered by this table is always 5. Distance impulse number can only be read/confirmed using Scan Tool function 10, "Adaptation"
5. Code will be 5 digits. For example, a Passat with brake pad wear warning, seat belt warning, and washer fluid level warning, for the USA market, with an oil level thermal sensor, and a speedometer impulse number of 4182, will have a code of 07225.
6. Access through Address Word 17.

Instrument Cluster Control Module Coding, from production 12.01 ➤

Note: The OBD program text and/or data generated by control modules installed on vehicles from 12.01 production may not be recognized by VAG 1551/1552 Scan Tools with the latest program card or by scan tool programs. For example, the scan tool display may read "text 799", "01529 /references", or similar. Only the VAS 5051/5052 Vehicle Diagnostic Testing and Information System is capable of processing all display text/data on these vehicles. OBD program functions on vehicles from 12.01 production must be performed using the VAS 5051 Vehicle Diagnostic Testing and Information System in mode "Guided Fault Finding".

Gateway On Board Diagnostic Data Bus Interface Coding Tables

Gateway Data Bus Interface Coding[3], through production ➤ 09.00

Vehicle Type[1]	Control Modules on "Powertrain" CAN-BUS[4]	Code[2]
Passat Sedan and Wagon	Automatic Transmission (TCM)	00001
	Airbag	00004

Notes:

1. Only the codes listed above are recognized as valid.
2. Values in this column are added together where applicable to determine value. For example, if vehicle in question has an automatic transmission (00001) and airbag (00004), values are added together (00001 + 00004 = 00005) to get a value of 00005 for the code.
3. Access through Address Word 19.
4. Through January 2000 production, Gateway coding software did not account for the presence of the ABS control module as part of the CAN-BUS network. As a result, the DTC "01044 Control Module incorrectly coded" is stored in memory and will always be displayed. This DTC can be ignored for vehicles through January 2000 production as long as coding is otherwise correct.
   For vehicles through January 2000, DO NOT re-code the ABS control module or attempt to use Gateway code values other than those given in the table above! This software glitch was corrected from February 2000 production. DO NOT re-code the ABS control module for these vehicle in an attempt to work around the this DTC.

Gateway Data Bus Interface Coding[3], production 10.00 ➤ 11.01

Vehicle Type[1]	Control Modules on "Powertrain" CAN-BUS[2]	Code[3]
Passat Sedan and Wagon	Anti-Lock Brakes System (ABS)	00002
	Airbag	00004

Notes:

1. Only the codes listed above are recognized as valid.
2. The presence of an Automatic Transmission Control Module (TCM) is automatically detected by the On Board Diagnostic Data Bus Interface and therefore does not need to be coded separately.
3. Values in this column are added together where applicable to determine value. For example, if vehicle in question has ABS (00002) and airbag (00004), values are added together (00002 + 00004 = 00006) to get a value of 00006 for the code. If vehicle has all of the equipment listed, add all of them together.
4. Access through Address Word 19.

Gateway Data Bus Interface Coding, from production 12.01 ➤

Note: The OBD program text and/or data generated by control modules installed on vehicles from 12.01 production may not be recognized by VAG 1551/1552 Scan Tools with the latest program card or by scan tool programs. For example, the scan tool display may read "text 799", "01529 /references", or similar. Only the VAS 5051/5052 Vehicle Diagnostic Testing and Information System is capable of processing all display text/data on these vehicles. OBD program functions on vehicles from 12.01 production must be performed using the VAS 5051 Vehicle Diagnostic Testing and Information System in mode "Guided Fault Finding".

Radio and Sound System Coding Table

Radio Head Unit Coding Variations[6], Premium IV, Premium V and Monsoon				
Designations	**Code Assembly Position**[4,5]			
	1 + 2	**3**	**4**	**5**
Sound Field Tuning[1]				
Premium IV and V	00			
Monsoon	04			
Speaker System				
Amplified (Monsoon only)		0		
Passive (Premium IV and V)		4		
Sound System Correction[2]				
Premium IV and V			0	
Monsoon			4	
Radio System Equipment Configuration[3]				
Without CD changer/player				1
With CD changer/player				3
Without CD changer/player, WITH radio display in instrument cluster				5
With CD changer/player, WITH radio display in instrument cluster				7

Notes:
1. Sound field tuning provides various frequency response characteristics that are matched to the vehicle interior volume.
2. Sound system correction provides various frequency response characteristics that are matched to the vehicle interior shape.
3. Sound system can have in-dash player or rear-mounted changer or both.
4. Code will be 5 digits. For example, a Passat with "Premium V, CD changer, and instrument cluster display is coded 00407.
5. Only select from the listed codes.
6. Access through Address Word 56.

Multi-function Steering Wheel (MSW) Coding Table

Multi-function Steering Wheel Coding[3]		
Control Module Identification[1]	**Equipment**	**Code**[2]
1J0 907 XXX (no suffix letter)	Radio and Cruise Control System (CSS)	00008
1J0 907 XXX A	Radio and Cruise Control System (CSS)	00118
1J0 907 XXX B	Radio and Cruise Control System (CSS)	00008

Notes:
1. Control module part number is displayed on scan tool from function 01. Part number prefix (1J0) may vary.
2. Only select from the listed codes.
3. Access through Address Word 16.

Readiness Codes

The inspection and maintenance (I/M) legislation requiring On Board Diagnostics (OBD II) includes provisions to prevent tampering or manipulating an Inspection/Maintenance (I/M) test. The readiness code will be stored after complete diagnostic monitoring of all components and systems has been carried out. This function was put into place to prevent manipulating an I/M emission test procedure by clearing faults codes or disconnecting the ECM or battery.

An additional benefit of this legislation is the ability of the engine management system to check for proper operation of up to 8 functions or systems. For the Motronic M5.9.2 and ME7.X engine management systems used on Passats, the exact number will depend on the equipment level of each particular vehicle.

Most of the time, the readiness code will set itself after the appropriate conditions have been met. These may include a cold or a hot start, operation at a certain load or speed for a certain period of time, or operation at different temperatures. After repairs, it is usually advantageous to verify the proper operation of a component or system. Sometimes this can be accomplished by driving the vehicle and observing the Malfunction Indicator Light (MIL), but a more efficient way to confirm a repair is by setting the readiness codes.

Setting the readiness code requires Volkswagen scan tools VAG 1551, VAG 1552, or the new diagnostic computers VAS 5051 and VAS 5052, or an equivalent aftermarket scan tool or scan tool computer program. A thorough understanding of your equipment operation and it's capabilities is also required. Readiness codes are then generated according to instructions provided by the factory and the scan tool manufacturer.

Significance of the 8 digit readiness code,

Readiness code values[2]								
Digit position[1]								Diagnostic Function
1	2	3	4	5	6	7	8	
							0	Three Way Catalyst Efficiency
						0		Catalyst Heating (currently always 0)
					0			Evaporative Emissions System (Fuel tank venting system)
				0				Secondary Air Injection System
			0					Air Conditioning (R12, no current diagnostic function, always 0)
		0						Oxygen Sensors
	0							Oxygen Sensor Heaters
0								Exhaust Gas Recirculation (EGR)

Notes:

1. Value can only be a 1 or a 0.
   A value of 1 indicates that test has not yet run or has run and failed.
   A value of 0 indicates that test has run and passed or function is not applicable.
2. Readiness code is properly set when all 8 positions are zero, i.e. "00000000".

The readiness code can be displayed in any of several areas within the ECM's diagnostic memory depending on the individual scan tool. It will always have a format consisting of 8 digits. If the 8 digit code contains characters other than the number 0, such as 10100110, than the readiness code is NOT set. When the code displays all zeros, i.e. 00000000, the readiness code IS set and no further action is required. While the format is required to be the same for all engines, certain equipment may not be used on each engine. For example, most Passat engines do not use EGR and as such, digit position number 8 will always show 0. Other Volkswagen engines that use EGR will show either a 1 or a 0 as appropriate.

Readiness Code Quick Reference Tables

The following tables outline the procedures for setting the readiness code for each Passat engine type. The code can be read from function 08 in the appropriate display group. Most ECM versions will also display the readiness code separately through function 15.

Quick Reference:
Readiness code setting, engine code: AEB

			Procedure VAG 1551 Settings after successful Diagnosis				
Step	Diagnostic Status	Engine Condition	Function/ Display-Group	Channel 1	Channel 2	Channel 3	Channel 4
0	Clear DTC Memory: 02 -05	Engine OFF Ignition ON	-	-	-	-	-
1	Check Readiness Code Setting	Engine OFF Ignition ON	08/086	11100101 / 01100101	11111101	00011111	00000000
2	Adaptation Throttle Valve Control Module	Engine OFF Ignition ON	04/060	0 - 25 <°	0 - 25 <°	idle	ADP.OK
3	Pre and Post Heated Oxygen Sensor, Readiness	Engine ON 1800…..2200/min	08/030	111	111	-	-
4	Response Time Front Heated Oxygen Sensor	Engine ON 1800…..2200/min	04/034	1800 - 2200 RPM*	0.8 - 2.0 ms*	0.0 - 3.3 s	B1-S1 OK
5	Three Way Catalytic Converter	Engine ON 1800…..2200/min	04/046	0.0 - 1.0	0 - 2	58.6 s	CATB1 OK
6	EVAP Canister Purge Regulator Valve Check	Engine ON Idle	04/070	-	-5.4 - +9.4 %	0.0 - 0.33 g/s	EVAP OK
7	Fuel Tank System, Leakage Check**	Engine ON Idle	04/071	Reed cl.	-	Check END	Syst. OK
8	Check Readiness Code Setting	Engine ON Idle	08/086	00000000	00000000	00000000 / 00000100	00000000

*** only during Diagnostic Performance**
**** could take up to 1 min. to "Test ON"**

Confirm Readiness Code generating with a VAG 1551 Print-Out !

Quick Reference:
Readiness code setting, engine code: ATW

Step	Diagnostic Status	Engine Condition	Procedure VAG 1551 Settings after successful Diagnosis				
			Function/ Display-Group	Channel 1	Channel 2	Channel 3	Channel 4
1	Check DTC Memory	Engine OFF	02	-	-	-	-
2	Clear DTC Memory	Ignition ON	05				
3	Adaptation Throttle Valve Control Module	Engine OFF Ignition ON	04/060	8% - 60%	60 - 94%	0 - 8	ADP.OK
4	Adaption Kick Down Only AT-vehicles	Engine OFF Ignition ON	04/063	12% - 97%	4% - 94%	Kick Down	ADP.OK
5	Response Time Front Heated Oxygen Sensor	Engine ON 1800.....2200/min**	04/034	1800 - 2200 RPM**	larger than 350°C	0 – 1.0s	B1-S1 OK
6	Operational Readiness of Oxygen Sen. behind Cat. Conv.	Engine ON Idle	04/036	0.5V - 0.95V	B1-S2 OK	-	-
7	Oxygen Sensor Control System	Engine ON Idle	04/037	15% - 30%	0.1V–0.9 V	-150ms – +150ms	Syst. OK
8	Oxygen Sensor aging behind Cat. Conv.	Engine ON 1800.....2200/min**	04/043	1800 - 2200 RPM**	more than 350°C	0.1 – 0.9V	B1-S2 OK
9	Three Way Catalytic Converter	Engine ON 1800.....2200/min**	04/046	1800 - 2200 RPM**	more than 350°C	0.0 - 0.4	CATB1 OK
10	EVAP Canister Purge Regulator Valve Check	Engine ON Idle	04/070	0% - 100%	-7% - +7%	-	EVAP OK
11	Fuel System Leakage Check	Engine ON Idle	04/071	-	-	Check END	Syst. OK
12	Secondary Air System	Engine ON Idle	04/077	740 – 920 RPM	60 – 100 °C	4 – 10 g/s	Syst. OK
13	Check Readiness Code Setting	Engine ON Idle	08/086	00000000	00000000	00000000	00000000

Requirement: Coolant Temperature more than 80°C (08/001/2)
*** to warm up engine if necessary ** only during Diagnostic Performance**
<u>Confirm Readiness Code generating with a VAG 1551 Print-Out !</u>

Quick Reference:
Readiness code setting, engine codes: AUG, AWM (m.y. 2001)

			Procedure VAG 1551 Settings after successful Diagnosis				
Step	**Diagnostic Status**	**Engine Condition**	**Function/ Display-Group**	**Channel 1**	**Channel 2**	**Channel 3**	**Channel 4**
1 2	Check DTC Memory Clear DTC Memory	Engine OFF Ignition ON	02 05	-	-	-	-
3	Adaptation Throttle Valve Control Module	Engine OFF Ignition ON	04/060	8% - 60%	60% - 94%	0 - 8	ADP.OK
4	Adaptation Kick Down (only for Automatic Transmission)	Engine OFF Ignition ON	04/63	79% - 94%	79% - 94%	-	ADP.OK
5	EVAP Canister Purge Regulator Valve Check	Engine ON Idle	04/070	0% - 100%	-	-	EVAP OK
6	Response Time Front Heated Oxygen Sensor	Engine ON 2300.....2800/min**	04/034	2300 - 2800 RPM**	more than 350°C	0.5 – 2.5	B1-S1 OK
7	Operational Readiness of Oxygen Sen. behind Cat. Conv.	Engine ON Idle	04/036	0.1V–0.9V	B1-S2 OK	-	-
8	Oxygen Sensor Control System	Engine ON Idle	04/037	13% - 45%	0.1V–0.9 V	less than 0.02	Syst. OK
9	Oxygen Sensor aging behind Cat. Conv.	Engine ON 2300.....2800/min*	04/043	2300 - 2800 RPM*	more than 350°C	0.1 – 0.9V	B1-S2 OK
10	Fuel System Leakage Check	Engine ON Idle	04/071	-	-	Check END	Syst. OK
11	Secondary Air System	Engine ON Idle	04/77	740 – 920 RPM	2 – 5 g/s	more than - 20%	Syst. OK
12	Three Way Catalytic Converter	Engine ON 2300.....2500/min**	04/046	2300 - 2500 RPM**	550°C – 700°C	more than 2.5	CATB1 OK
13	Lambda Control	Engine ON Idle	04/107	800 – 900 RPM	-8% - +8%	-	Syst. OK
14	Diagnosis Camshaft Timing Check	Engine ON more than 1080/min**	04/094	more than 1080RPM	CS-ctr. ON	Syst. OK	-
15	Misfire Activity Check	Engine ON Idle	08/014	740 – 920 RPM	13%-45%	0	activated
16	Heated Oxygen Sensor	Engine ON Idle	08/041	-	Htg.bC ON	less than 2.0 kOhm	Htg.aC ON
17	Check Readiness Code Setting	Engine ON Idle	08/086	00000000	00000000	00000000	00000000

Requirement: Coolant Temperature more than 80°C (08/001/2)

*** to warm up engine if necessary   ** only during Diagnostic Performance**

<u>Confirm Readiness Code generating with a VAG 1551 Print-Out !</u>

Quick Reference:
Readiness code setting, engine code AWM (from m.y. 2002)

				Procedure VAS 5051 - Settings after successful Diagnosis			
Step	**Diagnostic Status**	**Engine Condition**	**Function/ Display-Group**	**Channel 1**	**Channel 2**	**Channel 3**	**Channel 4**
1	Check DTC Memory	Engine OFF	02	-	-	-	-
2	Clear DTC Memory	Ignition ON	05				
3	Adaptation Throttle Valve Control Module	Engine OFF Ignition ON	04/060 "Activate" to start test	8% - 60%	60% - 94%	0 - 8	ADP.OK
4	Adaptation Kick Down (only for Automatic Transmission)	Engine OFF Ignition ON	04/063 "Activate" to start test	79% - 94%	79% - 94%	-	ADP.OK
5	Check Measurement Block (Basic Settings)	Engine ON !!!!warm up engine!!!!	04/001	700 – 6600 RPM	80-110°C	-10 - +10	11111111
6	Lambda Control	Engine ON Idle	04/107 "Activate" to start test	700 – 860 RPM	-8%- +8%	-	Syst. OK ? then rev. engine over 3000 rpm once
7	EVAP Canister Purge Regulator Valve Check	Engine ON Idle	04/070 "Activate" to start test	0% - 100%	-7% or 7%		TBV OK
8	Knock Sensor Test	Engine ON *press throttle and brake pedals simultaneously	04/028 "Activate" to start test	preset RPM	> 13%	> 80°C	Syst. OK
9	Response Time Front Heated Oxygen Sensor	Engine ON *press throttle and brake pedals simultaneously	04/034 "Activate" to start test	preset RPM	more than 350°C	0.5 – 2.5	B1- P1 OK
10	Operational Readiness of Oxygen Sensor Behind Cat. Conv.	Engine ON Idle	04/036 "Activate" to start test	0.1V – 0.9V	B1- P2 OK	-	-
11	Oxygen Sensor Control System	Engine ON Idle	04/037 "Activate" to start test	13% - 45%	0.1V – 0.9V	-0.02 - +0.02	Syst. OK
12	Oxygen Sensor aging behind Cat. Conv.	Engine ON *press throttle and brake pedals simultaneously	04/043 "Activate" to start test	preset RPM	more than 350°C	0.1V – 0.9V	B1- P2 OK
13	Three Way Catalytic Converter	Engine ON *press throttle and brake pedals simultaneously	04/046 "Activate" to start test	preset RPM	550°C - 700°C	> 2.0	CAT B1 OK
14	Diagnosis Camshaft Timing Check	Engine ON *press throttle and brake pedals simultaneously	04/094 "Activate" to start test	preset RPM	CS – ctr. ON	Syst. OK	-
15	Fuel System Leakage Check	Engine ON Idle	04/071 "Activate" to start test	-	-	Check. End	Syst. OK
16	Secondary Air System	Engine ON Idle	04/077 "Activate" to start test	700 – 860 RPM	1 – 10 g/s	-50% - +30%	Syst. OK
17	Misfire Activity Check	Engine ON Idle	04/ 014	700 – 860 RPM	13% - 45%	< 5	activated
18	Heated Oxygen Sensor	Engine ON Idle	04/041		Htg. before Cat. ON /OFF (fluctuates)	< 0.02	Htg. after Cat. ON
19	Check Readiness Code Settings	Engine ON - Idle	04/086	00000000			

Requirement: Emergency brake engaged, AC off, Manual transmission in Neutral, Automatic transmission in "P" / "N", Coolant Temperature more than 80°C (08/001/2)
*only during Diagnostic Performance

Quick Reference:
Readiness code setting, engine code AHA

			Procedure VAG 1551 Settings after successful Diagnosis				
Step	Diagnostic Status	Engine Condition	Function/ Display-Group	Channel 1	Channel 2	Channel 3	Channel 4
0	Clear DTC Memory: 02 -05	Engine OFF Ignition ON	-	-	-	-	-
1	Check Readiness Code Setting	Engine OFF Ignition ON	08/086	01101101	11111111	00111111	00011111
2	Adaptation Throttle Valve Control Module	Engine OFF Ignition ON	04/098	3.255 - 4.745 V	0.902 - 4.863 V	Idle	ADP.OK
3	Pre and Post Heated Oxygen Sensor, Readiness	Engine ON 2000…2500/min	08/030	111	111	111	111
4	Response Time Front Heated Oxygen Sensor (B1-S1)	Engine ON 1800…2200/min	04/034	1800 - 2200 /min*	0.8 - 2.5 ms*	0.00 - 2.1 s	B1-S1 OK
5	Response Time Front Heated Oxygen Sensor (B2-S1)	Engine ON 1800…2200/min	04/035	1800 - 2200 /min*	0.8 - 2.5 ms*	0.00 - 2.1 s	B2-S1 OK
6	Three Way Catalytic Converter (CATB1)	Engine ON 1800…2200/min	04/046	0.00 - 0.28	0 - 2	58.6s	CATB1 OK
7	Three Way Catalytic Converter (CATB2)	Engine ON 1800…2200/min	04/047	0.00 - 0.28	0 - 2	58.6s	CATB2 OK
8	EVAP Canister Purge Regulator Valve Check	Engine ON Idle	04/070	- %	-5.5 - +6.3 %	max. 0.2 g/s	EVAP OK
9	Fuel Tank System, Leakage Check**	Engine ON Idle	04/071	Reed cl.	-	Check. END	Syst. OK
10	Secondary Air Injection	Engine ON Idle	04/77	- %	- %	-°C	Syst. OK
11	Check Readiness Code Setting	Engine ON Idle	08/086	00000000	00000000	00000000 / 00000100	00000000 / 00000100

* only during Diagnostic Performance
** could take up to 1 min. to "Test ON"

Confirm Readiness Code generating with a VAG 1551 Print-Out !

Quick Reference:
Readiness code setting, engine code ATQ (through m.y. 2000)

Step	Diagnostic Status	Engine Condition	Procedure VAG 1551 Function/ Display-Group	Settings after successful Diagnosis Channel 1	Channel 2	Channel 3	Channel 4
1 2	Check DTC Memory Clear DTC Memory	Engine OFF Ignition ON	02 05	-	-	-	-
3	Adaptation Throttle Valve Control Module	Engine OFF Ignition ON	04/060	8% - 18%	80 - 90%	0 - 8	ADP.OK
4	Adaptation Kick Down	Engine OFF Ignition ON	04/063	12% - 97%	4 - 94%	Kick Down	ADP.OK
5	Fuel System Leakage Check	Engine ON Idle	04/071	-	-	Check END	Syst. OK
6	EVAP Canister Purge Regulator Valve Check	Engine ON Idle	04/070	0% - 100 %	-	-	EVAP OK
7	Lambda Control	Engine ON Idle	04/107	630 – 820 RPM	-10-+10%	-10-+10%	Syst. OK
8 9	Heated Oxygen Sensor	Engine ON 1800....2200/min*	04/041 04/042	0–0.9 kOhm	Htg.bC ON	0–0.9 kOhm	Htg.aC ON
10	Operational Readiness of Oxygen Sen. behind Cat. Conv.	Engine ON Idle	04/036	0–1V	B1-S2 OK	0–1V	B2-S2 OK
11 12	Response Time Front Heated Oxygen Sensor	Engine ON 1000.....2400/min**	04/034 04/035	1000 - 2400 RPM**	more than 380°C	0.1 – 1.8s	B1-S1 OK B2-S1 OK
13 14	Oxygen Sensor Control System	Engine ON Idle	04/037 04/038	12 % - 26 %	0 V - 1 V	-800ms – +800ms	Syst. OK
15 16	Oxygen Sensor aging behind Cat. Conv.	Engine ON Idle	04/043 04/044	630 – 820 RPM	more than 380°C	0 – 1 V	B1-S2 OK B2-S2 OK
17 18	Three Way Catalytic Converter	Engine ON 1800.....2400/min**	04/046 04/047	1800 - 2400 RPM**	more than 420°C	0 - 0.50*** 0 - 0.55****	CAT B1 OK CAT B2 OK
19 20	Secondary Air System	Engine ON Idle	04/077 04/078	630 – 820 RPM	0 – 5 g/s	-80 - +20 %	Syst. OK
21	Check Readiness Code Setting	Engine ON Idle	08/086	00000000	00000000	00000000	00000000

Requirement: Coolant Temperature more than 80°C (08/001/2)
*** to warm up engine if necessary, test is complete when values are present in channel 1 and 3 ** only during Diagnostic Performance**
***** Automatic Transmission **** Manual Transmission**

Quick Reference:
Readiness code setting, engine code ATQ (from m.y. 2001)

			Procedure VAG 1551	Settings after successful Diagnosis			
Step	Diagnostic Status	Engine Condition	Function/ Display-Group	Channel 1	Channel 2	Channel 3	Channel 4
1 2	Check DTC Memory Clear DTC Memory	Engine OFF Ignition ON	02 05	-	-	-	-
3	Adaptation Throttle Valve Control Module	Engine OFF Ignition ON	04/060	8% - 18%	80 - 90%	0 - 8	ADP.OK
4	Adaptation Kick Down (only with Automatic Transmission)	Engine OFF Ignition ON	04/063	12% - 97%	4 - 94%	Kick Down	ADP.OK
5	Fuel System Leakage Check	Engine ON Idle	04/071	-	-	Check END	Syst. OK
6	EVAP Canister Purge Regulator Valve Check	Engine ON Idle	04/070	0% - 100 %	-	-	EVAP OK
7	Lambda Control	Engine ON Idle	04/107	630 – 820 RPM	-10-+10%	-10-+10%	Syst. OK
8 9	Heated Oxygen Sensor	Engine ON 1800….2200/min*	04/041 04/042	0–0.9 kOhm	Htg.bC ON	0–0.9 kOhm	Htg.aC ON
10	Operational Readiness of Oxygen Sen. behind Cat. Conv.	Engine ON Idle	04/036	0–1V	B1-S2 OK	0–1V	B2-S2 OK
11 12	Response Time Front Heated Oxygen Sensor	Engine ON 1000…..2400/min**	04/034 04/035	1000 - 2400 RPM**	more than 380°C	0.1 – 1.8s	B1-S1 OK B2-S1 OK
13 14	Oxygen Sensor Control System	Engine ON Idle	04/037 04/038	12 % - 26 %	0 V – 1 V	-800ms – +800ms	Syst. OK
15 16	Oxygen Sensor aging behind Cat. Conv.	Engine ON Idle	04/043 04/044	630 – 820 RPM	more than 380°C	0 – 1 V	B1-S2 OK B2-S2 OK
17 18	Three Way Catalytic Converter	Engine ON 1800…..2400/min**	04/046 04/047	1800 - 2400 RPM**	more than 420°C	0 - 0.58*** 0 - 0.62****	CAT B1 OK CAT B2 OK
19 20	Secondary Air System	Engine ON Idle	04/077 04/078	630 – 820 RPM	0 – 5 g/s	-80 - +20 %	Syst. OK
21	Check Readiness Code Setting	Engine ON Idle	08/086	00000000	00000000	00000000	00000000

Requirement: Coolant Temperature more than 80°C (08/001/2)
*** to warm up engine if necessary, test is complete when values are present in channel 1 and 3 ** only during Diagnostic Performance**
***** Automatic Transmission **** Manual Transmission**

Quick Reference: Readiness code setting, engine code: BDP

Step	Diagnostic Status	Engine Condition	Function/ Display-Group	Procedure VAS 5051 - Settings after successful Diagnosis			
				Channel 1	Channel 2	Channel 3	Channel 4
1	Check DTC Memory	Engine OFF		-	-	-	-
2	Clear DTC Memory	Ignition ON					
3	Adaptation Throttle Valve Control Module	Engine OFF Ignition ON	04/060	3% - 16%	82% - 97%	0 - 8	ADP.OK
4	Adaptation Kick Down (only for Automatic Transmission)	Engine OFF Ignition ON	04/063	xx%	xx%	Kick Down	ADP.OK
5	Fuel System Leakage Check *	Engine ON at Idle	04/202			Text	xxxxx100
6	O2 Sensor b. C. Heating Bank 1	Depress Brake Pedal and Hold	04/205				xxxxx100
7	O2 Sensor b. C. Heating Bank 2	Press Accelerator to Full Throttle	04/206				xxxxx100
8	O2 Sensor b. C. Aging Bank 1	Depress Brake Pedal and Hold	04/215			>1	xxxxx100
9	O2 Sensor b. C. Aging Bank 2	Press Accelerator to Full Throttle	04/216			>1	xxxxx100
10	Fuel Mixture Diag. Bank 1	Depress Brake Pedal and Hold	04/219				xxxxx100
11	Partial Load Bank 2	Press Accelerator to Full Throttle	04/220				xxxxx100
12	Air Mass Filling Detection Diag. *	Depress Brake Pedal and Hold Press Accelerator to Full Throttle	04/221				xxxxx100
13	O2 Sensor a. C. Heating Bank 1	Depress Brake Pedal and Hold	04/222			Between 200°-650°C	xxxxx100
14	O2 Sensor a. C. Heating Bank 2	Press Accelerator to Full Throttle	04/223				xxxxx100
15	Diag. Location of O2 sensor a. C.	Depress Brake Pedal and Hold Press Accelerator to Full Throttle	04/226				xxxxx100
16	O2 Sensor a. C. Aging Bank 1	Depress Brake Pedal and Hold	04/227				xxxxx100
17	O2 Sensor a. C. Aging Bank 2	Press Accelerator to Full Throttle	04/228				xxxxx100
18	Catalytic Converter Bank 1	Depress Brake Pedal and Hold	04/229				xxxxx100
19	Catalytic Converter Bank 2	Press Accelerator to Full Throttle	04/230				xxxxx100
20	Fuel Mixture Diag. Bank 1	Engine ON	04/219				xxxx1100
21	at Idle Bank 2	at Idle	04/220				xxxx1100
22	EVAP System	Engine ON at Idle	04/231				xxxxx100
23	Secondary Air Bank 1	Engine ON	04/232				xxxxx100
24	Secondary Air Bank 2	at Idle	04/233				xxxxx100
25	Check Readiness Code	Engine ON at Idle	15	00000000			

Automatic transmission selector lever in "P" or "N" position. All electrical consumers shut off.

Requirement: Coolant temp must be at least 80°C, intake temp less than 60°C. (See display group 04 field 3 and 4)

***These tests make take several minutes**

V.A.G. 1551/1552 OPERATING OVERVIEW FLOW CHART

MODE MENU

1 - Rapid Data =>
2 - Blink Code
3 - Self Test
4 - Dealership Code
5 - LT - Diagnosis

ADDRESS WORDS MENU (V9)

01 - Engine Electronics =>
11 - Engine Electronics II
41 - Diesel Pump Electronics
51 - Electric Drive
61 - Battery Control
71 - Battery Charger
02 - Transmission Electronics
12 - Clutch Electronics
22 - 4 Wheel Drive Electronics
03 - Brake Electronics
13 - Distance Control
33 - OBD II
14 - Suspension Electronics
24 - Anti - Slip Control
34 - Level Control
44 - Steering Assist
54 - Rear Spoiler
15 - Airbag
25 - Anti-Theft Immobilizer Sensor
35 - Central Locking
45 - Interior Monitor
55 - Headlamp Vert. Aim Control
65 - Tire Pressure Monitoring
75 - Emergency Call Module
16 - Steering Wheel Electronics
26 - Automatic Roof Control
36 - Seat Adjustment Driver's Side
46 - Comfort System (AC CM)
56 - Radio
66 - Seat & Mirror Adjustment
76 - Parking Aid
17 - Instrument Cluster
37 - Navigation System
47 - Sound System
57 - Airbag (Cabriolet only)
67 - Language Control
08 - A/C & Heating Electronics
18 - Auxiliary Heater
09 - Electronic Control Module
19 - Diagnostic Interface for Data Bus
29 - Left Light Control
39 - Right Light Control
49 - Auto Light Switch
00 - Automatic Test Sequence

FUNCTIONS MENU

01 - Check Control Module Versions
02 - Check DTC Memory
03 - Output Diagnostic Test Mode
04 - Basic Setting
    Display Group xxx =>
05 - Erase DTC Memory
06 - End Output
07 - Code Control Module
08 - Read Measuring Value Block
    Display Group xxx =>
09 - Read Individual Measuring Values
    Channel xx
10 - Adaptation
    Channel xx
11 - Log-In Procedure
15 - Readiness Code

DISPLAY GROUP FORMAT

Display Group 000 (only)
xx xx xx xx xx xx xx xx xx xx

Display Group 001 (and higher)
xxxxx xxxxx xxxxx xxxxx

WARNING

Your common sense, good judgement, and general alertness are crucial to safe and successful service work. Before attempting any work on your Volkswagen, read the warnings and cautions on page vi and the copyright page at the front of the manual. Review these warnings and cautions each time you prepare to work on your Volkswagen. Please also read any warnings and cautions that accompany the procedures in the manual.

A

B

C

WARNING

Your common sense, good judgement, and general alertness are crucial to safe and successful service work. Before attempting any work on your Volkswagen, read the warnings and cautions on page vi and the copyright page at the front of the manual. Review these warnings and cautions each time you prepare to work on your Volkswagen. Please also read any warnings and cautions that accompany the procedures in the manual.

> **WARNING**
>
> *Your common sense, good judgement, and general alertness are crucial to safe and successful service work. Before attempting any work on your Volkswagen, read the warnings and cautions on page vi and the copyright page at the front of the manual. Review these warnings and cautions each time you prepare to work on your Volkswagen. Please also read any warnings and cautions that accompany the procedures in the manual.*

V

W

Notes

Notes

Notes

Notes

Notes

Notes

Notes

Selected Books and Repair Information From Bentley Publishers

Engineering

Bosch Fuel Injection and Engine Management *Charles O. Probst, SAE*
ISBN 0-8376-0300-5

Maximum Boost: Designing, Testing, and Installing Turbocharger Systems *Corky Bell* ISBN 0-8376-0160-6

Race Car Aerodynamics *Joseph Katz*
ISBN 0-8376-0142-8

Scientific Design of Exhaust and Intake Systems *Phillip H. Smith & John C. Morrison* ISBN 0-8376-0309-9

Audi Repair Manuals (paper)

Audi A4 Service Manual: 1996-2001, 1.8L Turbo, 2.8L, including Avant and quattro *Bentley Publishers*
ISBN 0-8376-0371-4

Audi A6 Electrical Wiring Manual: 1998-2000 *Bentley Publishers*
ISBN 0-8376-0166-5

Audi 80, 90 Coupe Quattro Repair Manual: 1988-1992 *Audi of America*
ISBN 0-8376-0368-4

Audi Repair Manuals (electronic)

Audi A3: 2006 Repair Manual on CD-ROM *Audi of America*
ISBN 0-8376-1363-9

Audi A6: 2005-2006 Repair Manual on CD-ROM *Audi of America*
ISBN 0-8376-1362-0

Audi A4/S4: 1996-2001 Manual on CD-ROM *Audi of America*
ISBN 0-8376-1255-1

Audi TT Coupe: 2000-2005, Roadster: 2001-2005 Repair Manual on CD-ROM *Audi of America*
ISBN 0-8376-1261-6

BMW Repair Manuals (paper only)

BMW 3 Series (E46) Service Manual: 1999-2005 *Bentley Publishers*
ISBN 0-8376-1277-2

BMW Z3 (E36/7)Service Manual: 1996-2002 *Bentley Publishers*
ISBN 0-8376-1250-0

BMW 5 Series (E39) Service Manual: 1997-2002 *Bentley Publishers*
ISBN 0-8376-0317-X

BMW 3 Series (E36) Service Manual: 1992-1998 *Bentley Publishers*
ISBN 0-8376-0326-9

BMW 5 Series (E34) Service Manual: 1989-1995 *Bentley Publishers*
ISBN 0-8376-0319-6

BMW 7 Series (E32) Service Manual: 1988-1994 *Bentley Publishers*
ISBN 0-8376-0328-5

BMW 3 Series (E30) Service Manual: 1984-1990 *Bentley Publishers*
ISBN 0-8376-0325-0

Porsche Repair Manuals (paper only)

Porsche Boxster Service Manual: 1997-2004 *Bentley Publishers*
ISBN 0-8376-1333-7

Porsche 911 Carrera Service Manual: 1984-1989 *Bentley Publishers*
ISBN 0-8376-0291-2

Porsche 911 SC Service Manual: 1987-1983 *Bentley Publishers*
ISBN 0-8376-0290-4

Volkswagen Repair Manuals (paper)

Volkswagen Jetta Service Manual: 2005-2006 *Bentley Publishers*
ISBN 0-8376-1335-3

Volkswagen Jetta, Golf, GTI Service Manual: 1999-2005 *Bentley Publishers*
ISBN 0-8376-1251-9

Volkswagen Passat Service Manual: 1998-2004 *Bentley Publishers*
ISBN 0-8376-0369-2

Volkswagen Jetta, Golf, GTI: 1993-1999 Cabrio: 1995-2002 Service Manual *Bentley Publishers*
ISBN 0-8376-0366-8

Volkswagen GTI, Golf, Jetta Service Manual: 1985-1992 *Bentley Publishers*
ISBN 0-8376-0342-0

Volkswagen Vanagon Repair Manual: 1980-1991 *Volkswagen of America*
ISBN 0-8376-0336-6

Volkswagen Corrado Repair Manual: 1990-1994 *Volkswagen of America*
ISBN 0-8376-0387-0

Volkswagen Rabbit, Scirocco, Jetta Service Manual: 1980-1984 Gasoline Models *Bentley Publishers*
ISBN 0-8376-0183-5

Volkswagen Super Beetle, Beetle and Karmann Ghia Service Manual Type 1: 1970-1979 *Volkswagen of America*
ISBN 0-8376-0096-0

Volkswagen 1200 Workshop Manual: 1961-1965, Type 11, 14 and 15, Beetle, Beetle Convertible, Karmann Ghia Coupe, and Karmann Ghia Convertible *Volkswagen of America*
ISBN 0-8376-0390-0

Volkswagen Repair Manuals (electronic)

Volkswagen Jetta: 2005-2006 Repair Manual on CD-ROM *Volkswagen of America* ISBN 0-8376-1360-4

Volkswagen Golf, GTI, Jetta: 1999-2005, Jetta Wagon 2001-2005 Repair Manual on CD-ROM *Volkswagen of America* ISBN 0-8376-1264-0

Volkswagen Passat, Passat Wagon: 1998-2005 Repair Manual on CD-ROM *Volkswagen of America*
ISBN 0-8376-1267-5

Volkswagen Passat, Passat Wagon: 1995-1997 Repair Manual on CD-ROM *Volkswagen of America*
ISBN 0-8376-1266-7

Volkswagen New Beetle: 1998-2005, New Beetle Convertible 2003-2005 Repair Manual on CD-ROM *Volkswagen of America*
ISBN 0-8376-1265-9

BentleyPublishers.com

Automotive Reference

Bentley Publishers has published service manuals and automobile books since 1950. For more information, please contact Bentley Publishers at 1734 Massachusetts Avenue, Cambridge, MA 02138, visit our web site at **www.BentleyPublishers.com** or call 1-800-423-4595 for a free copy of our complete catalog.

eBahn.com

eBahn® Desktop – Automotive Repair Information for your PC

eBahn® lets users access a wide variety of information about their car, including repair procedures, wiring diagrams, diagnostic trouble codes and much more. Available on CD-ROM, DVD-ROM, download and through online subscriptions.

eBahn® Version 3.0 Features

- Factory repair information with accurate and detailed repair procedures for vehicle systems
- Comprehensive wiring diagrams to help with troubleshooting and repair procedures
- Electrical component locations – Factory drawings and photos, including fuses and relays, to help you efficiently locate hard-to-find electrical components
- DTC Xpress™ – The handy tool developed by Bentley Publishers to quickly decipher Diagnostic Trouble Codes (DTCs) and match them to the relevant service procedures
- VIN Decoding – Simply enter your VIN to filter and select repair information specific to your vehicle.
- Electronic editions of original factory training materials
- Official Factory Technical Bulletins - Read or print out the Technical Service Bulletins (TSBs) that the manufacturer issues to quickly disseminate information to dealer technicians
- Check For Updates – Easy online updates of repair information and software

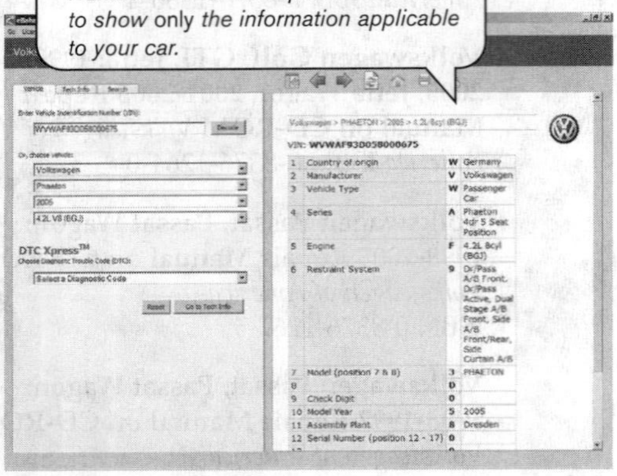

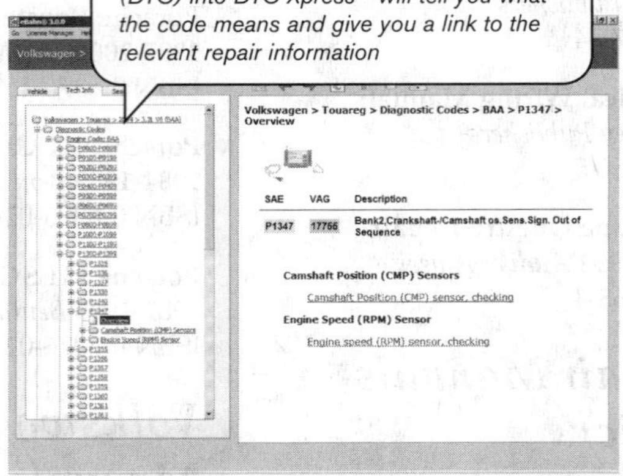

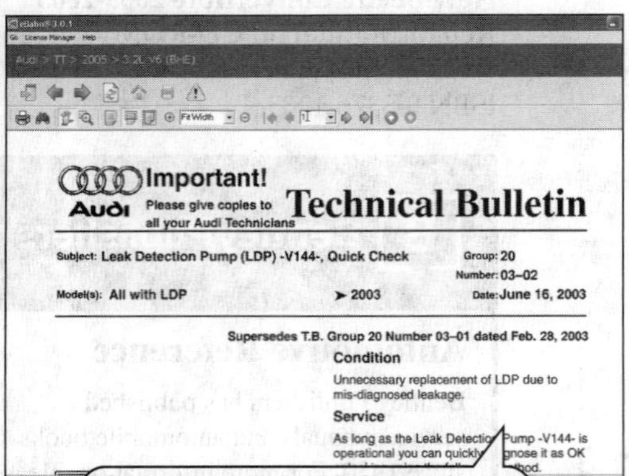

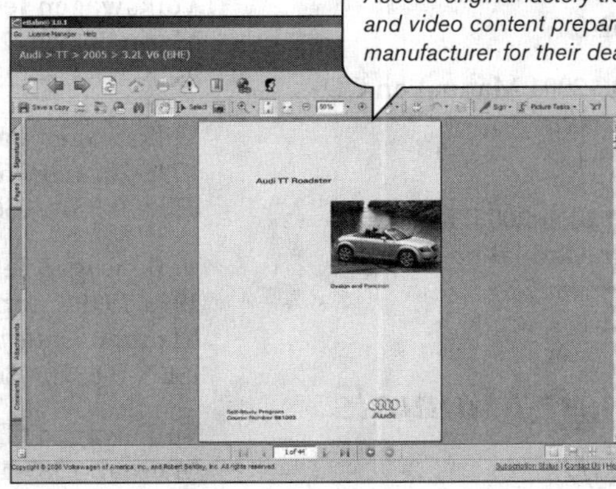

Visit **www.BentleyPublishers.com/ebahn** for much more information